CRC Handbook of Chemistry and Physics

82nd Edition

EDITORIAL ADVISORY BOARD

CRC Handbook of Chemistry and Physics

A Ready-Reference Book of Chemical and Physical Data

HANDBOOK OF CHEMISTRY AND PHYSICS * CRC PRESS LLC *

2001-2002

82nd

EDITION

Editor-in-Chief

David R. Lide, Ph.D.
Former Director, Standard Reference Data
National Institute of Standards and Technology

CRC Press
Boca Raton London New York Washington, D.C.

Direct all inquiries to CRC Press LLC, 2000 N.W. Corporate Blvd., Boca Raton, Florida 33431.

International Standard Book Number 0-8493-0482-2
Library of Congress Card Number 13-11056
Printed in the United States of America 1 2 3 4 5 6 7 8 9 0
Printed on acid-free paper

PREFACE

This is the first edition of the *CRC Handbook of Chemistry and Physics* for the 21st Century (as "century" is officially defined). Few would dispute that the 20th Century was the century of science; major paradigm shifts took place first in physics and chemistry and, in the second half of the century, in biology. The "*Rubber Handbook*", so-called after the original name of its publisher, the Chemical Rubber Company, was a fixture for almost all of that eventful century. When the first edition appeared in 1913, the electron had been known for only 17 years, there were 81 elements, and the Bohr theory of the hydrogen atom was still in press. The *Handbook* was a significant innovation. While systematic compilation of data from the chemistry and physics literature had begun earlier, especially in Germany, the massive tomes that were published by Beilstein, Gmelin, and Landolt-Börnstein were strictly for libraries. The *Rubber Handbook* appears to have been the first compact, low-price volume of reference data suitable for students and individual researchers to keep on their desk or laboratory bench. It quickly became a standard and grew from its original 116 pages to the present size of over 2500 pages.

Generations of students have relied on the *CRC Handbook* as a resource in their studies, but the impact has been much broader. Senior research scientists, engineers, and workers in other fields have used the book extensively. Linus Pauling, perhaps the most influential chemist of the 20th century, made the following comments a few years before his death:

> People who have interviewed me have commented on the extensive knowledge that I have about the properties of substances, especially inorganic compounds, including minerals. I attribute this knowledge in part to the fact that I possessed the *Rubber Handbook.* I remember clearly the five summers, beginning in 1919, when I worked as a paving-plant inspector, supervising the laying of bituminous pavement in the mountainous region of southern Oregon. For much of the time I was free to read, just keeping an eye on the operation of the paving plant. I remember the book that I read over and over was the *Rubber Handbook.* I puzzled over the tables of properties — hardness, color, melting and boiling point, density, magnetic properties, and others — trying to think of reasonable explanations of the empirical data. Only in the 1920s and 1930s did I have some success in this effort.

Pauling's "success" was the first step in our ability to relate the physical and chemical behavior of bulk materials to their molecular structure in a quantitative manner.

One factor in the success of the *Handbook* has been its annual revisions. This practice, followed throughout the century except for a few wartime years, permitted the replacement of old data with new and more accurate values, as well as the introduction of new topics that became important as science moved forward. This policy has helped the *Handbook* meet the needs of the scientific community in a timely fashion.

The 82nd Edition continues the tradition of updates and improvements. The major change is a revised and expanded table of Physical Constants of Inorganic Compounds. The number of compounds has been increased by 12%, the format improved, and the constants updated. In addition, quantitative data on solubility in water are now included in the table, and a formula index has been added. Other tables that have been expanded and updated include:

- Critical Constants
- Aqueous Solubility and Henry's Law Constants of Organic Compounds
- Chemical Carcinogens
- Threshold Limits for Airborne Contaminants
- Nomenclature for Organic Polymers
- Standard Atomic Weights
- Atomic Masses and Abundances
- Table of the Isotopes

New topics covered in this edition include:

- Surface Tension of Aqueous Mixtures
- Viscosity of Carbon Dioxide along the Saturation Line
- Gibbs Energy of Formation for Important Biological Species
- Optical Properties of Inorganic and Organic Solids

- Interstellar Molecules
- Allocation of Frequencies in the Radio Spectrum
- Units for Magnetic Properties

The Editor will appreciate suggestions on new topics for the *Handbook* and notification of any errors. Address all comments to Editor, *Handbook of Chemistry and Physics*, CRC Press LLC, 2000 N.W. Corporate Blvd., Boca Raton, FL 33431. Comments may also be sent by electronic mail to drlide@post.harvard.edu.

The *Handbook of Chemistry and Physics* is dependent on the efforts of many contributors throughout the world. The list of current contributors follows this Preface. I appreciate the valuable suggestions that have come from the Editorial Advisory Board and from many users. I should also like to thank Susan Fox and the rest of the production team at CRC Press for their excellent support.

David R. Lide
January 1, 2001

This Edition is Dedicated to the Memory of
David Reynolds Lide (1901-1976) and Kate Simmons Lide (1896-1991)

Note on the Ordering of Chemical Compounds: The decision on the order in which to list chemical compounds in a table is always difficult. An alphabetical list by name has the disadvantage that several different synonyms are often in common use, with the result that a reader may conclude incorrectly that a compound is not present if he looks it up under the wrong name. Listing by common formula is satisfactory for simple inorganic compounds, but is cumbersome for organics. A listing by molecular formula is attractive because clear rules can be given for locating a compound, but the user may have to go to some effort to determine the molecular formula. In this book the choice is made on pragmatic grounds. The long tables, Physical Constants of Organic Compounds and Physical Constants of Inorganic Compounds, are ordered by systematic name, but indexes to synonyms, formulas, and CAS Registry Numbers are provided. If the table is very short and includes only familiar substances, the listing is usually alphabetical by common formula or name. The remaining tables are ordered by molecular formula using a modification of the Hill convention. In this convention the molecular formula is written with C first, H second, and then all other elements in alphabetical order of their chemical symbols. For tables with organic compounds only, the sequence of entries is determined by the alphabetical order of elements in the molecular formula and the number of atoms of each element, in ascending order, e.g., C_3H_7Cl, C_3H_7N, C_3H_7NO, $C_3H_7NO_2$, etc. (For organic compounds, a quick way to determine the molecular formula is to use the Physical Constants of Organic Compounds table, which starts on Page 3-1, and its synonym index on Page 3-586.) In tables containing non-carbon compounds, those are listed first, followed by a separate listing of compounds that do contain carbon. This is in contrast to the strict Hill convention as followed by Chemical Abstracts Service, where the molecular formulas beginning with A and B precede the formulas for carbon-containing compounds, while those beginning with D... Z follow. For tabular displays, as opposed to an index, it appears more convenient to the user if the non-carbon compounds are listed as a block, rather than being split by the longer list of carbon compounds.

CURRENT CONTRIBUTORS

Lev I. Berger
California Institute of Electronics and Materials Science
2115 Flame Tree Way
Hemet, California 92545

A. K. Covington
Department of Chemistry
University of Newcastle
Newcastle upon Tyne NE1 7RU
England

Robert B. Fox
6115 Wiscassett Rd.
Bethesda, Maryland 20816

H. P. R. Frederikse
9625 Dewmar Lane
Kensington, Maryland 20895

J.R. Fuhr
Atomic Physics Division
National Institute of Standards and Technology
Gaithersburg, Maryland 20899

Karl A. Gschneidner
Ames Laboratory
Energy and Mineral Resources Research Institute
Iowa State University
Ames, Iowa 50011

C. R. Hammond
17 Greystone Rd.
West Hartford, Connecticut 06107

Robert F. Hampson
Chemical Kinetics Division
National Institute of Standards and Technology
Gaithersburg, Maryland 20899

Norman E. Holden
National Nuclear Data Center
Brookhaven National Laboratory
Upton, New York 11973

H. Donald Brooke Jenkins
Department of Chemistry
University of Warwick
Coventry CV4 7AL England

Henry V. Kehiaian
ITODYS
1 rue Guy de la Brosse
75005 Paris
France

J. Alistair Kerr
School of Chemistry
University of Birmingham
Birmingham B15 2TT England

Sharon G. Lias
National Institute of Standards and Technology
Gaithersburg, Maryland 20899

Frank J. Lovas
8616 Melwood Rd.
Bethesda, Maryland 20817

William C. Martin
Atomic Physics Division
National Institute of Standards and Technology
Gaithersburg, Maryland 20899

Joel S. Miller
Department of Chemistry
University of Utah
Salt Lake City, Utah 84112

Thomas M. Miller
Air Force Research Laboratory/VSBP
29 Randolph Rd.
Hanscom AFB, Massachusetts 01731-3010

Joseph Reader
Atomic Physics Division
National Institute of Standards and Technology
Gaithersburg, Maryland 20899

H. K. Roobottom
Department of Chemistry
University of Warwick
Coventry CV4 7AL England

B. L. Sharma
E-1/15 Model Town
Delhi — 110009 India

Lewis E. Snyder
Astronomy Department
University of Illinois
Urbana, Illinois 61801

David W. Stocker
School of Chemistry
University of Leeds
Leeds LS2 9JT England

B. N. Taylor
Physics Laboratory
National Institute of Standards and Technology
Gaithersburg, Maryland 20899

Thomas G. Trippe
Particle Data Group
Lawrence Berkeley Laboratory
1 Cyclotron Road
Berkeley, California 94720

Petr Vanýsek
Department of Chemistry
Northern Illinois University
DeKalb, Illinois 60115

Wolfgang L. Wiese
Atomic Physics Division
National Institute of Standards and Technology
Gaithersburg, Maryland 20899

Edward S. Wilks
E.I. du Pont de Nemours and Company Inc.
Barley Mills Plaza 14/1290
Wilmington, Delaware 19880-0014

Christian Wohlfarth
Institut für Physikalische Chemie
Martin Luther University
D-06217 Merseburg
Germany

TABLE OF CONTENTS

SECTION 1: BASIC CONSTANTS, UNITS, AND CONVERSION FACTORS

Fundamental Physical Constants 1-1
Standard Atomic Weights (1995) 1-12
Atomic Masses and Abundances 1-15
Electron Configuration of Neutral Atoms in the Ground State 1-19
International Temperature Scale of 1990 (ITS-90) 1-21
Conversion of Temperatures from the 1948 and 1968 Scales to ITS-90 1-23
International System of Units (SI) 1-25
Units for Magnetic Properties 1-33
Conversion Factors 1-34
Conversion of Temperatures 1-46
Conversion Factors for Energy Units 1-47
Conversion Factors for Pressure Units 1-48
Conversion Factors for Thermal Conductivity Units 1-49
Conversion Factors for Electrical Resistivity Units 1-50
Conversion Factors for Chemical Kinetics 1-51
Conversion Factors for Ionizing Radiation 1-52
Values of the Gas Constant in Different Unit Systems 1-54

SECTION 2: SYMBOLS, TERMINOLOGY, AND NOMENCLATURE

Symbols and Terminology for Physical and Chemical Quantities 2-1
Nomenclature of Chemical Compounds 2-22
Nomenclature for Inorganic Ions and Ligands 2-23
Organic Substituent Groups and Ring Systems 2-25
Scientific Abbreviations and Symbols 2-29
Greek, Russian, and Hebrew Alphabets 2-37
Definitions of Scientific Terms 2-38

SECTION 3: PHYSICAL CONSTANTS OF ORGANIC COMPOUNDS

Physical Constants of Organic Compounds 3-1
Structure Diagrams for Table of Physical Constants 3-331
Synonym Index of Organic Compounds 3-586
Molecular Formula Index of Organic Compounds 3-631
CAS Registry Number Index of Organic Compounds 3-709
Melting Point Index of Organic Compounds 3-737
Boiling Point Index of Organic Compounds 3-751
Diamagnetic Susceptibility of Selected Organic Compounds 3-764

SECTION 4: PROPERTIES OF THE ELEMENTS AND INORGANIC COMPOUNDS

The Elements 4-1
Physical Constants of Inorganic Compounds 4-37
Formula Index of Inorganic Compounds 4-97
CAS Registry Number Index of Inorganic Compounds 4-112
Physical Properties of the Rare Earth Metals 4-122
Melting, Boiling, and Critical Temperatures of the Elements 4-132
Heat Capacity of the Elements at 25°C 4-133
Vapor Pressure of the Metallic Elements 4-134
Density of Molten Elements and Representative Salts 4-137
Magnetic Susceptibility of the Elements and Inorganic Compounds 4-141
Index of Refraction of Inorganic Liquids 4-147
Physical and Optical Properties of Minerals 4-149
Crystallographic Data on Minerals 4-156

SECTION 5: THERMOCHEMISTRY, ELECTROCHEMISTRY, AND KINETICS

CODATA Key Values for Thermodynamics 5-1
Standard Thermodynamic Properties of Chemical Substances 5-4
Thermodynamic Properties as a Function of Temperature 5-61
Thermodynamic Properties of Aqueous Systems 5-85
Heat of Combustion 5-89
Electrical Conductivity of Water 5-90
Standard KCl Solutions for Calibrating Conductivity Cells 5-91
Molar Conductivity of Aqueous HF, HCl, HBr, and HI 5-92
Equivalent Conductivity of Electrolytes in Aqueous Solution 5-94
Ionic Conductivity and Diffusion at Infinite Dilution 5-95
Activity Coefficients of Acids, Bases, and Salts 5-97
Mean Activity Coefficients of Electrolytes as a Function of Concentration 5-99
Enthalpy of Dilution of Acids 5-104
Enthalpy of Solution of Electrolytes 5-105
Chemical Kinetic Data for Stratospheric Modeling 5-106

SECTION 6: FLUID PROPERTIES

Thermodynamic Properties of Air 6-1
Properties of Water in the Range 0-100°C 6-3
Enthalpy of Vaporization of Water 6-3
Fixed Point Properties of H_2O and D_2O 6-4
Thermal Conductivity of Saturated H_2O and D_2O 6-4
Standard Density of Water 6-5
Properties of Ice and Supercooled Water 6-6
Volumetric Properties of Aqueous Sodium Chloride Solutions 6-8
Density of D_2O 6-9
Vapor Pressure of Ice 6-9
Vapor Pressure of Water from 0 to 370°C 6-10
Boiling Point of Water at Various Pressures 6-12
Melting Point of Ice as a Function of Pressure 6-12
Properties of Water and Steam as a Function of Temperature and Pressure 6-13
Permittivity (Dielectric Constant) of Water at Various Frequencies 6-15
Thermophysical Properties of Fluids 6-16
Virial Coefficients of Selected Gases 6-25
Van der Waals Constants for Gases 6-45
Mean Free Path and Related Properties of Gases 6-46
Influence of Pressure on Freezing Points 6-47
Critical Constants 6-48
Sublimation Pressure of Solids 6-62
Vapor Pressure 6-65
Vapor Pressure of Fluids at Temperatures below 300 K 6-93
Vapor Pressure of Saturated Salt Solutions 6-104
IUPAC Recommended Data for Vapor Pressure Calibration 6-105
Enthalpy of Vaporization 6-106
Enthalpy of Fusion 6-122
Pressure and Temperature Dependence of Liquid Density 6-137
Properties of Cryogenic Fluids 6-139
Properties of Liquid Helium 6-140
Halocarbon Refrigerants 6-141
Density and Specific Volume of Mercury 6-144
Thermal Properties of Mercury 6-145
Surface Tension of Common Liquids 6-146
Surface Tension of Aqueous Mixtures 6-150

Permittivity (Dielectric Constant) of Liquids 6-151
Permittivity (Dielectric Constant) of Gases 6-174
Azeotropic Data for Binary Mixtures 6-176
Viscosity of Gases 6-180
Viscosity of Liquids 6-182
Viscosity of Carbon Dioxide along the Saturation Line 6-187
Thermal Conductivity of Gases 6-188
Thermal Conductivity of Liquids 6-190
Diffusion in Gases 6-192
Diffusion of Gases in Water 6-194
Diffusion Coefficients in Liquids at Infinite Dilution 6-195

SECTION 7: BIOCHEMISTRY

Properties of Amino Acids 7-1
Structures of Common Amino Acids 7-3
Properties of Purine and Pyrimidine Bases 7-4
The Genetic Code 7-5
Properties of Fatty Acids 7-6
Carbohydrate Names and Symbols 7-7
Standard Transformed Gibbs Energy of Formation for Important Biochemical Species 7-9
Biological Buffers 7-12
Typical pH Values of Biological Materials and Foods 7-13
Chemical Composition of the Human Body 7-14

SECTION 8: ANALYTICAL CHEMISTRY

Preparation of Special Analytical Reagents 8-1
Standard Solutions of Acids, Bases, and Salts 8-5
Standard Solutions of Oxidation and Reduction Reagents 8-7
Organic Analytical Reagents for the Determination of Inorganic Substances 8-8
Acid-Base Indicators 8-16
Fluorescent Indicators 8-19
Conversion Formulas for Concentration of Solutions 8-20
Electrochemical Series 8-21
Reduction and Oxidation Potentials for Certain Ion Radicals 8-32
pH Scale for Aqueous Solutions 8-35
Practical pH Measurements on Natural Waters 8-41
Buffer Solutions Giving Round Values of pH at 25°C 8-43
Dissociation Constants of Inorganic Acids and Bases 8-44
Dissociation Constants of Organic Acids and Bases 8-46
Concentrative Properties of Aqueous Solutions: Density, ... 8-57
Refractive Index, Freezing Point Depression, and Viscosity Ion Product of Water Substance 8-84
Ionization Constant of Normal and Heavy Water 8-85
Solubility of Selected Gases in Water 8-86
Solubility of Carbon Dioxide in Water at Various Temperatures and Pressures 8-90
Aqueous Solubility and Henry's Law Constants of Organic Compounds 8-91
Aqueous Solubility of Inorganic Compounds at Various Temperatures 8-108
Solubility Product Constants 8-117
Solubility of Common Salts at Ambient Temperatures 8-121
Solubility Chart 8-122
Reduction of Weighings in Air to Vacuo 8-125
Volume of One Gram of Water 8-125
Properties of Carrier Gases for Gas Chromatography 8-126
Solvents for Ultraviolet Spectrophotometry 8-127
^{13}C Chemical Shifts of Useful NMR Solvents 8-128
Mass Spectral Peaks of Common Organic Solvents 8-129

SECTION 9: MOLECULAR STRUCTURE AND SPECTROSCOPY

Bond Lengths in Crystalline Organic Compounds 9-1
Bond Lengths in Organometallic Compounds 9-15
Bond Lengths and Angles in Gas-Phase Molecules 9-17
Dipole Moments 9-44
Strengths of Chemical Bonds 9-51
Electronegativity 9-75
Force Constants for Bond Stretching 9-76
Fundamental Vibrational Frequencies of Small Molecules 9-77
Spectroscopic Constants of Diatomic Molecules 9-81
Infrared Correlation Charts 9-87
Nuclear Spins, Moments, and Other Data Related to NMR Spectroscopy 9-92
Proton NMR Chemical Shifts for Characteristic Organic Structures 9-95
^{13}C NMR Absorptions of Major Functional Groups 9-96

SECTION 10: ATOMIC, MOLECULAR, AND OPTICAL PHYSICS

Line Spectra of the Elements 10-1
NIST Atomic Transition Probability Tables 10-88
Electron Affinities 10-147
Atomic and Molecular Polarizabilities 10-160
Ionization Potentials of Atoms and Atomic Ions 10-175
Ionization Energies of Gas-Phase Molecules 10-178
X-Ray Atomic Energy Levels 10-196
Electron Binding Energies of the Elements 10-200
Natural Width of X-Ray Lines 10-206
Photon Attenuation Coefficients 10-207
Classification of Electromagnetic Radiation 10-212
Black Body Radiation 10-214
Characteristics of Infrared Detectors 10-216
Refractive Index and Transmittance of Representative Glasses 10-217
Index of Refraction of Water 10-218
Index of Refraction of Liquids for Calibration Purposes 10-219
Index of Refraction of Air 10-220
Characteristics of Laser Sources 10-221
Infrared Laser Frequencies 10-228
Infrared and Far-Infrared Absorption Frequency Standards 10-236

SECTION 11: NUCLEAR AND PARTICLE PHYSICS

Summary Tables of Particle Properties 11-1
Table of the Isotopes 11-50
Neutron Scattering and Absorption Properties 11-199
Cosmic Radiation 11-216

SECTION 12: PROPERTIES OF SOLIDS

Techniques for Materials Characterization 12-1
Symmetry of Crystals 12-7
Ionic Radii in Crystals 12-14
Polarizability of Atoms and Ions in Solids 12-17
Crystal Structures and Lattice Parameters of Allotropes of the Elements 12-19
Lattice Energies 12-22
The Madelung Constant and Crystal Lattice Energy 12-36
Elastic Constants of Single Crystals 12-37
Electrical Resistivity of Pure Metals 12-45
Electrical Resistivity of Selected Alloys 12-48

Permittivity (Dielectric Constant) of Inorganic Solids 12-52
Curie Temperature of Selected Ferroelectric Crystals 12-61
Properties of Antiferroelectric Crystals 12-62
Dielectric Constants of Glasses 12-63
Properties of Superconductors 12-64
High Temperature Superconductors 12-91
Organic Superconductors 12-94
Properties of Semiconductors 12-97
Diffusion Data for Semiconductors 12-108
Properties of Magnetic Materials 12-117
Organic Magnets 12-126
Electron Work Function of the Elements 12-130
Secondary Electron Emission 12-131
Optical Properties of Selected Elements 12-133
Optical Properties of Selected Inorganic and Organic Solids 12-157
Elasto-optic, Electro-optic, and Magneto-optic Constants 12-179
Nonlinear Optical Constants 12-196
Phase Diagrams 12-200
Heat Capacity of Selected Solids 12-218
Thermal and Physical Properties of Pure Metals 12-219
Thermal Conductivity of Metals and Semiconductors as a Function of Temperature 12-221
Thermal Conductivity of Alloys as a Function of Temperature 12-223
Thermal Conductivity of Crystalline Dielectrics 12-224
Thermal Conductivity of Ceramics and Other Insulating Materials 12-226
Thermal Conductivity of Glasses 12-228
Commercial Metals and Alloys 12-232
Hardness of Minerals and Ceramics 12-233

SECTION 13: POLYMER PROPERTIES

Nomenclature for Organic Polymers 13-1
Solvents for Common Polymers 13-6
Glass Transition Temperature for Selected Polymers 13-7
Dielectric Constant of Selected Polymers 13-15
Pressure-Volume-Temperature Relationship for Polymer Melts 13-16

SECTION 14: GEOPHYSICS, ASTRONOMY, AND ACOUSTICS

Astronomical Constants 14-1
Properties of the Solar System 14-2
Satellites of the Planets 14-4
Interstellar Molecules 14-6
Mass, Dimensions, and other Parameters of the Earth 14-9
Geological Time Scale 14-11
Acceleration Due to Gravity 14-12
Density, Pressure, and Gravity as a Function of Depth within the Earth 14-13
Ocean Pressure as a Function of Depth and Latitude 14-14
Properties of Seawater 14-15
Abundance of Elements in the Earth's Crust and in the Sea 14-17
Solar Spectral Irradiance 14-18
U.S. Standard Atmosphere (1976) 14-19
Geographical and Seasonal Variation in Solar Radiation 14-26
Infrared Absorption by the Earth's Atmosphere 14-27
Atmospheric Concentration of Carbon Dioxide, 1958-1996 14-28
Mean Temperatures in the United States, 1900-1992 14-30
Global Temperature Trend, 1866-1996 14-32

Atmospheric Electricity ... 14-33
Speed of Sound in Various Media ... 14-41
Attenuation and Speed of Sound in Air as a Function of Humidity and Frequency ... 14-43
Speed of Sound in Dry Air ... 14-44
Musical Scales ... 14-45
Characteristics of Human Hearing ... 14-47

SECTION 15: PRACTICAL LABORATORY DATA
Standard ITS-90 Thermocouple Tables ... 15-1
Properties of Common Laboratory Solvents ... 15-14
Dependence of Boiling Point on Pressure ... 15-19
Ebullioscopic Constants for Calculation of Boiling Point Elevation ... 15-20
Cryoscopic Constants for Calculation of Freezing Point Depression ... 15-21
Freezing Point Lowering by Electrolytes in Aqueous Solution ... 15-22
Determination of Relative Humidity from Dew Point ... 15-23
Determination of Relative Humidity from Wet and Dry Bulb Temperatures ... 15-24
Constant Humidity Solutions ... 15-25
Standard Salt Solutions for Humidity Calibration ... 15-26
Low Temperature Baths for Maintaining Constant Temperature ... 15-26
Wire Tables ... 15-27
Characteristics of Particles and Particle Dispersoids ... 15-28
Density of Various Solids ... 15-29
Dielectric Strength of Insulating Materials ... 15-30
Allocation of Frequencies in the Radio Spectrum ... 15-36

SECTION 16: HEALTH AND SAFETY INFORMATION
Handling and Disposal of Chemicals in Laboratories ... 16-1
Flammability of Chemical Substances ... 16-16
Threshold Limits for Airborne Contaminants ... 16-32
Octanol-Water Partition Coefficients ... 16-43
Protection Against Ionizing Radiation ... 16-48
Annual Limits on Intakes of Radionuclides ... 16-49
Chemical Carcinogens ... 16-53

APPENDIX A: MATHEMATICAL TABLES
Miscellaneous Mathematical Constants ... A-1
Exponential and Hyperbolic Functions and their Common Logarithms ... A-2
Natural Trigonometric Functions to Four Places ... A-7
Relation of Angular Functions in Terms of One Another ... A-11
Derivatives ... A-12
Integration ... A-15
Integrals ... A-20
Differential Equations ... A-65
Fourier Series ... A-75
The Fourier Transforms ... A-80
Series Expansion ... A-84
Vector Analysis ... A-87
Moment of Inertial for Various Bodies of Mass ... A-96
Bessel Functions ... A-96
The Gamma Function ... A-99
The Beta Function ... A-101
The Error Function ... A-101
Orthogonal Polynomials ... A-102

Normal Probability Function **A**-104
Percentage Points, Student's *t*-Distribution **A**-105
Percentage Points, Chi-Square Distribution **A**-105
Percentage Points, F-Distribution **A**-106

INDEX **I**-1

Section 1
Basic Constants, Units, and Conversion Factors

Fundamental Physical Constants 1-1
Standard Atomic Weights (1995) 1-12
Atomic Masses and Abundances 1-15
Electron Configuration of Neutral Atoms in the Ground State 1-19
International Temperature Scale of 1990 (ITS-90) 1-21
Conversion of Temperatures from the 1948 and 1968 Scales to ITS-90 1-23
International System of Units (SI) 1-25
Units for Magnetic Properties 1-33
Conversion Factors 1-34
Conversion of Temperatures 1-46
Conversion Factors for Energy Units 1-47
Conversion Factors for Pressure Units 1-48
Conversion Factors for Thermal Conductivity Units 1-49
Conversion Factors for Electrical Resistivity Units 1-50
Conversion Factors for Chemical Kinetics 1-51
Conversion Factors for Ionizing Radiation 1-52
Values of the Gas Constant in Different Unit Systems 1-54

FUNDAMENTAL PHYSICAL CONSTANTS

Peter J. Mohr and Barry N. Taylor

These tables give the 1998 self-consistent set of values of the basic constants and conversion factors of physics and chemistry recommended by the Committee on Data for Science and Technology (CODATA) for international use. The 1998 set replaces the previous set of constants recommended by CODATA in 1986; assigned uncertainties have been reduced by a factor of 1/5 to 1/12 (and sometimes even greater) relative to the 1986 uncertainties. The recommended set is based on a least-squares adjustment involving all of the relevant experimental and theoretical data available through December 31, 1998. Full details of the input data and the adjustment procedure are given in Reference 1.

The 1998 adjustment was carried out by P. J. Mohr and B. N. Taylor of the National Institute of Standards and Technology (NIST) under the auspices of the CODATA Task Group on Fundamental Constants. The Task Group was established in 1969 with the aim of periodically providing the scientific and technological communities with a self-consistent set of internationally recommended values of the fundamental physical constants based on all applicable information available at a given point in time. The first set was published in 1973 and was followed by a revised set first published in 1986; the current 1998 set first appeared in 1999. In the future, the CODATA Task Group plans to take advantage of the high level of automation developed for the current set in order to issue a new set of recommended values at least every four years.

At the time of completion of the 1998 adjustment, the membership of the Task Group was as follows:

F. Cabiati, Istituto Elettrotecnico Nazionale "Galileo Ferraris," Italy
E. R. Cohen, Science Center, Rockwell International (retired), United States of America
T. Endo, Electrotechnical Laboratory, Japan
R. Liu, National Institute of Metrology, China (People's Republic of)
B. A. Mamyrin, A. F. Ioffe Physical-Technical Institute, Russian Federation
P. J. Mohr, National Institute of Standards and Technology, United States of America
F. Nez, Laboratoire Kastler-Brossel, France
B. W. Petley, National Physical Laboratory, United Kingdom
T. J. Quinn, Bureau International des Poids et Mesures
B. N. Taylor, National Institute of Standards and Technology, United States of America
V. S. Tuninsky, D. I. Mendeleyev All-Russian Research Institute for Metrology, Russian Federation
W. Wöger, Physikalisch-Technische Bundesanstalt, Germany
B. M. Wood, National Research Council, Canada

REFERENCE

1. Mohr, Peter J., and Taylor, Barry N., *J. Phys Chem. Ref. Data* **28**, No. 6, 1999; *Rev. Mod. Phys.* **72**, No. 2, 2000. The 1998 set of recommended values is also available at the Web site of the Fundamental Constants Data Center of the NIST Physics Laboratory: http://physics.nist.gov/constants.

Fundamental Physical Constants

Quantity	Symbol	Value	Unit	Relative std. uncert. u_r
UNIVERSAL				
speed of light in vacuum	c, c_0	299 792 458	$m\ s^{-1}$	(exact)
magnetic constant	μ_0	$4\pi \times 10^{-7}$	$N\ A^{-2}$	
		$= 12.566\,370\,614... \times 10^{-7}$	$N\ A^{-2}$	(exact)
electric constant $1/\mu_0 c^2$	ε_0	$8.854\,187\,817... \times 10^{-12}$	$F\ m^{-1}$	(exact)
characteristic impedance of vacuum $\sqrt{\mu_0/\epsilon_0} = \mu_0 c$	Z_0	376.730 313 461...	Ω	(exact)
Newtonian constant of gravitation	G	$6.673(10) \times 10^{-11}$	$m^3\ kg^{-1}\ s^{-2}$	1.5×10^{-3}
	$G/\hbar c$	$6.707(10) \times 10^{-39}$	$(GeV/c^2)^{-2}$	1.5×10^{-3}
Planck constant	h	$6.626\,068\,76(52) \times 10^{-34}$	J s	7.8×10^{-8}
in eV s		$4.135\,667\,27(16) \times 10^{-15}$	eV s	3.9×10^{-8}
$h/2\pi$	$\hbar$	$1.054\,571\,596(82) \times 10^{-34}$	J s	7.8×10^{-8}
in eV s		$6.582\,118\,89(26) \times 10^{-16}$	eV s	3.9×10^{-8}
Planck mass $(\hbar c/G)^{1/2}$	m_P	$2.1767(16) \times 10^{-8}$	kg	7.5×10^{-4}
Planck length $\hbar/m_P c = (\hbar G/c^3)^{1/2}$	l_P	$1.6160(12) \times 10^{-35}$	m	7.5×10^{-4}
Planck time $l_P/c = (\hbar G/c^5)^{1/2}$	t_P	$5.3906(40) \times 10^{-44}$	s	7.5×10^{-4}
ELECTROMAGNETIC				
elementary charge	e	$1.602\,176\,462(63) \times 10^{-19}$	C	3.9×10^{-8}
	e/h	$2.417\,989\,491(95) \times 10^{14}$	$A\ J^{-1}$	3.9×10^{-8}
magnetic flux quantum $h/2e$	Φ_0	$2.067\,833\,636(81) \times 10^{-15}$	Wb	3.9×10^{-8}
conductance quantum $2e^2/h$	G_0	$7.748\,091\,696(28) \times 10^{-5}$	S	3.7×10^{-9}
inverse of conductance quantum	G_0^{-1}	12 906.403 786(47)	Ω	3.7×10^{-9}
Josephson constant[a] $2e/h$	K_J	$483\,597.898(19) \times 10^9$	$Hz\ V^{-1}$	3.9×10^{-8}
von Klitzing constant[b] $h/e^2 = \mu_0 c/2\alpha$	R_K	25 812.807 572(95)	Ω	3.7×10^{-9}
Bohr magneton $e\hbar/2m_e$	μ_B	$927.400\,899(37) \times 10^{-26}$	$J\ T^{-1}$	4.0×10^{-8}
in eV T^{-1}		$5.788\,381\,749(43) \times 10^{-5}$	$eV\ T^{-1}$	7.3×10^{-9}
	μ_B/h	$13.996\,246\,24(56) \times 10^9$	$Hz\ T^{-1}$	4.0×10^{-8}
	μ_B/hc	46.686 4521(19)	$m^{-1}\ T^{-1}$	4.0×10^{-8}
	μ_B/k	0.671 7131(12)	$K\ T^{-1}$	1.7×10^{-6}
nuclear magneton $e\hbar/2m_p$	μ_N	$5.050\,783\,17(20) \times 10^{-27}$	$J\ T^{-1}$	4.0×10^{-8}
in eV T^{-1}		$3.152\,451\,238(24) \times 10^{-8}$	$eV\ T^{-1}$	7.6×10^{-9}
	μ_N/h	7.622 593 96(31)	$MHz\ T^{-1}$	4.0×10^{-8}
	μ_N/hc	$2.542\,623\,66(10) \times 10^{-2}$	$m^{-1}\ T^{-1}$	4.0×10^{-8}
	μ_N/k	$3.658\,2638(64) \times 10^{-4}$	$K\ T^{-1}$	1.7×10^{-6}
ATOMIC AND NUCLEAR				
General				
fine-structure constant $e^2/4\pi\epsilon_0\hbar c$	α	$7.297\,352\,533(27) \times 10^{-3}$		3.7×10^{-9}
inverse fine-structure constant	α^{-1}	137.035 999 76(50)		3.7×10^{-9}

Fundamental Physical Constants

Quantity	Symbol	Value	Unit	Relative std. uncert. u_r
Rydberg constant $\alpha^2 m_e c/2h$	R_∞	10 973 731.568 549(83)	m^{-1}	7.6×10^{-12}
	$R_\infty c$	$3.289\,841\,960\,368(25) \times 10^{15}$	Hz	7.6×10^{-12}
	$R_\infty hc$	$2.179\,871\,90(17) \times 10^{-18}$	J	7.8×10^{-8}
$R_\infty hc$ in eV		13.605 691 72(53)	eV	3.9×10^{-8}
Bohr radius $\alpha/4\pi R_\infty = 4\pi\epsilon_0\hbar^2/m_e e^2$	a_0	$0.529\,177\,2083(19) \times 10^{-10}$	m	3.7×10^{-9}
Hartree energy $e^2/4\pi\epsilon_0 a_0 = 2R_\infty hc$ $= \alpha^2 m_e c^2$	E_h	$4.359\,743\,81(34) \times 10^{-18}$	J	7.8×10^{-8}
in eV		27.211 3834(11)	eV	3.9×10^{-8}
quantum of circulation	$h/2m_e$	$3.636\,947\,516(27) \times 10^{-4}$	$m^2\ s^{-1}$	7.3×10^{-9}
	h/m_e	$7.273\,895\,032(53) \times 10^{-4}$	$m^2\ s^{-1}$	7.3×10^{-9}
		Electroweak		
Fermi coupling constant[c]	$G_F/(\hbar c)^3$	$1.166\,39(1) \times 10^{-5}$	GeV^{-2}	8.6×10^{-6}
weak mixing angle[d] θ_W (on-shell scheme) $\sin^2\theta_W = s_W^2 \equiv 1 - (m_W/m_Z)^2$	$\sin^2\theta_W$	0.2224(19)		8.7×10^{-3}
		Electron, e^-		
electron mass	m_e	$9.109\,381\,88(72) \times 10^{-31}$	kg	7.9×10^{-8}
in u, $m_e = A_r(e)$ u (electron relative atomic mass times u)		$5.485\,799\,110(12) \times 10^{-4}$	u	2.1×10^{-9}
energy equivalent	$m_e c^2$	$8.187\,104\,14(64) \times 10^{-14}$	J	7.9×10^{-8}
in MeV		0.510 998 902(21)	MeV	4.0×10^{-8}
electron-muon mass ratio	m_e/m_μ	$4.836\,332\,10(15) \times 10^{-3}$		3.0×10^{-8}
electron-tau mass ratio	m_e/m_τ	$2.875\,55(47) \times 10^{-4}$		1.6×10^{-4}
electron-proton mass ratio	m_e/m_p	$5.446\,170\,232(12) \times 10^{-4}$		2.1×10^{-9}
electron-neutron mass ratio	m_e/m_n	$5.438\,673\,462(12) \times 10^{-4}$		2.2×10^{-9}
electron-deuteron mass ratio	m_e/m_d	$2.724\,437\,1170(58) \times 10^{-4}$		2.1×10^{-9}
electron to alpha particle mass ratio	m_e/m_α	$1.370\,933\,5611(29) \times 10^{-4}$		2.1×10^{-9}
electron charge to mass quotient	$-e/m_e$	$-1.758\,820\,174(71) \times 10^{11}$	$C\ kg^{-1}$	4.0×10^{-8}
electron molar mass $N_A m_e$	$M(e)$, M_e	$5.485\,799\,110(12) \times 10^{-7}$	$kg\ mol^{-1}$	2.1×10^{-9}
Compton wavelength $h/m_e c$	λ_C	$2.426\,310\,215(18) \times 10^{-12}$	m	7.3×10^{-9}
$\lambda_C/2\pi = \alpha a_0 = \alpha^2/4\pi R_\infty$	$\bar{\lambda}_C$	$386.159\,2642(28) \times 10^{-15}$	m	7.3×10^{-9}
classical electron radius $\alpha^2 a_0$	r_e	$2.817\,940\,285(31) \times 10^{-15}$	m	1.1×10^{-8}
Thomson cross section $(8\pi/3)r_e^2$	σ_e	$0.665\,245\,854(15) \times 10^{-28}$	m^2	2.2×10^{-8}
electron magnetic moment	μ_e	$-928.476\,362(37) \times 10^{-26}$	$J\ T^{-1}$	4.0×10^{-8}
to Bohr magneton ratio	μ_e/μ_B	$-1.001\,159\,652\,1869(41)$		4.1×10^{-12}
to nuclear magneton ratio	μ_e/μ_N	$-1\,838.281\,9660(39)$		2.1×10^{-9}
electron magnetic moment anomaly $\lvert\mu_e\rvert/\mu_B - 1$	a_e	$1.159\,652\,1869(41) \times 10^{-3}$		3.5×10^{-9}
electron g-factor $-2(1 + a_e)$	g_e	$-2.002\,319\,304\,3737(82)$		4.1×10^{-12}
electron-muon magnetic moment ratio	μ_e/μ_μ	206.766 9720(63)		3.0×10^{-8}

Fundamental Physical Constants

Quantity	Symbol	Value	Unit	Relative std. uncert. u_r
electron-proton magnetic moment ratio	μ_e/μ_p	$-658.2106875(66)$		1.0×10^{-8}
electron to shielded proton magnetic moment ratio (H_2O, sphere, 25 °C)	μ_e/μ'_p	$-658.2275954(71)$		1.1×10^{-8}
electron-neutron magnetic moment ratio	μ_e/μ_n	$960.92050(23)$		2.4×10^{-7}
electron-deuteron magnetic moment ratio	μ_e/μ_d	$-2143.923498(23)$		1.1×10^{-8}
electron to shielded helion[e] magnetic moment ratio (gas, sphere, 25 °C)	μ_e/μ'_h	$864.058255(10)$		1.2×10^{-8}
electron gyromagnetic ratio $2\lvert\mu_e\rvert/\hbar$	γ_e	$1.760859794(71) \times 10^{11}$	$s^{-1}\ T^{-1}$	4.0×10^{-8}
	$\gamma_e/2\pi$	$28024.9540(11)$	$MHz\ T^{-1}$	4.0×10^{-8}
		Muon, μ^-		
muon mass	m_μ	$1.88353109(16) \times 10^{-28}$	kg	8.4×10^{-8}
in u, $m_\mu = A_r(\mu)$ u (muon relative atomic mass times u)		$0.1134289168(34)$	u	3.0×10^{-8}
energy equivalent	$m_\mu c^2$	$1.69283332(14) \times 10^{-11}$	J	8.4×10^{-8}
in MeV		$105.6583568(52)$	MeV	4.9×10^{-8}
muon-electron mass ratio	m_μ/m_e	$206.7682657(63)$		3.0×10^{-8}
muon-tau mass ratio	m_μ/m_τ	$5.94572(97) \times 10^{-2}$		1.6×10^{-4}
muon-proton mass ratio	m_μ/m_p	$0.1126095173(34)$		3.0×10^{-8}
muon-neutron mass ratio	m_μ/m_n	$0.1124545079(34)$		3.0×10^{-8}
muon molar mass $N_A m_\mu$	$M(\mu)$, M_μ	$0.1134289168(34) \times 10^{-3}$	$kg\ mol^{-1}$	3.0×10^{-8}
muon Compton wavelength $h/m_\mu c$	$\lambda_{C,\mu}$	$11.73444197(35) \times 10^{-15}$	m	2.9×10^{-8}
$\lambda_{C,\mu}/2\pi$	$\bar{\lambda}_{C,\mu}$	$1.867594444(55) \times 10^{-15}$	m	2.9×10^{-8}
muon magnetic moment	μ_μ	$-4.49044813(22) \times 10^{-26}$	$J\ T^{-1}$	4.9×10^{-8}
to Bohr magneton ratio	μ_μ/μ_B	$-4.84197085(15) \times 10^{-3}$		3.0×10^{-8}
to nuclear magneton ratio	μ_μ/μ_N	$-8.89059770(27)$		3.0×10^{-8}
muon magnetic moment anomaly $\lvert\mu_\mu\rvert/(e\hbar/2m_\mu) - 1$	a_μ	$1.16591602(64) \times 10^{-3}$		5.5×10^{-7}
muon g-factor $-2(1 + a_\mu)$	g_μ	$-2.0023318320(13)$		6.4×10^{-10}
muon-proton magnetic moment ratio	μ_μ/μ_p	$-3.18334539(10)$		3.2×10^{-8}
		Tau, τ^-		
tau mass[f]	m_τ	$3.16788(52) \times 10^{-27}$	kg	1.6×10^{-4}
in u, $m_\tau = A_r(\tau)$ u (tau relative atomic mass times u)		$1.90774(31)$	u	1.6×10^{-4}
energy equivalent	$m_\tau c^2$	$2.84715(46) \times 10^{-10}$	J	1.6×10^{-4}
in MeV		$1777.05(29)$	MeV	1.6×10^{-4}

Fundamental Physical Constants

Quantity	Symbol	Value	Unit	Relative std. uncert. u_r
tau-electron mass ratio	m_τ/m_e	3 477.60(57)		1.6×10^{-4}
tau-muon mass ratio	m_τ/m_μ	16.8188(27)		1.6×10^{-4}
tau-proton mass ratio	m_τ/m_p	1.893 96(31)		1.6×10^{-4}
tau-neutron mass ratio	m_τ/m_n	1.891 35(31)		1.6×10^{-4}
tau molar mass $N_A m_\tau$	$M(\tau)$, M_τ	$1.907\,74(31) \times 10^{-3}$	kg mol^{-1}	1.6×10^{-4}
tau Compton wavelength $h/m_\tau c$	$\lambda_{C,\tau}$	$0.697\,70(11) \times 10^{-15}$	m	1.6×10^{-4}
$\lambda_{C,\tau}/2\pi$	$ƛ_{C,\tau}$	$0.111\,042(18) \times 10^{-15}$	m	1.6×10^{-4}
		Proton, p		
proton mass	m_p	$1.672\,621\,58(13) \times 10^{-27}$	kg	7.9×10^{-8}
in u, $m_p = A_r(p)$ u (proton relative atomic mass times u)		1.007 276 466 88(13)	u	1.3×10^{-10}
energy equivalent	m_pc^2	$1.503\,277\,31(12) \times 10^{-10}$	J	7.9×10^{-8}
in MeV		938.271 998(38)	MeV	4.0×10^{-8}
proton-electron mass ratio	m_p/m_e	1 836.152 6675(39)		2.1×10^{-9}
proton-muon mass ratio	m_p/m_μ	8.880 244 08(27)		3.0×10^{-8}
proton-tau mass ratio	m_p/m_τ	0.527 994(86)		1.6×10^{-4}
proton-neutron mass ratio	m_p/m_n	0.998 623 478 55(58)		5.8×10^{-10}
proton charge to mass quotient	e/m_p	$9.578\,834\,08(38) \times 10^{7}$	C kg^{-1}	4.0×10^{-8}
proton molar mass $N_A m_p$	M(p), M_p	$1.007\,276\,466\,88(13) \times 10^{-3}$	kg mol^{-1}	1.3×10^{-10}
proton Compton wavelength h/m_pc	$\lambda_{C,p}$	$1.321\,409\,847(10) \times 10^{-15}$	m	7.6×10^{-9}
$\lambda_{C,p}/2\pi$	$ƛ_{C,p}$	$0.210\,308\,9089(16) \times 10^{-15}$	m	7.6×10^{-9}
proton magnetic moment	μ_p	$1.410\,606\,633(58) \times 10^{-26}$	J T^{-1}	4.1×10^{-8}
to Bohr magneton ratio	μ_p/μ_B	$1.521\,032\,203(15) \times 10^{-3}$		1.0×10^{-8}
to nuclear magneton ratio	μ_p/μ_N	2.792 847 337(29)		1.0×10^{-8}
proton g-factor $2\mu_p/\mu_N$	g_p	5.585 694 675(57)		1.0×10^{-8}
proton-neutron magnetic moment ratio	μ_p/μ_n	−1.459 898 05(34)		2.4×10^{-7}
shielded proton magnetic moment (H_2O, sphere, 25 °C)	μ'_p	$1.410\,570\,399(59) \times 10^{-26}$	J T^{-1}	4.2×10^{-8}
to Bohr magneton ratio	μ'_p/μ_B	$1.520\,993\,132(16) \times 10^{-3}$		1.1×10^{-8}
to nuclear magneton ratio	μ'_p/μ_N	2.792 775 597(31)		1.1×10^{-8}
proton magnetic shielding correction $1 - \mu'_p/\mu_p$ (H_2O, sphere, 25 °C)	σ'_p	$25.687(15) \times 10^{-6}$		5.7×10^{-4}
proton gyromagnetic ratio $2\mu_p/\hbar$	γ_p	$2.675\,222\,12(11) \times 10^{8}$	s^{-1} T^{-1}	4.1×10^{-8}
	$\gamma_p/2\pi$	42.577 4825(18)	MHz T^{-1}	4.1×10^{-8}
shielded proton gyromagnetic ratio $2\mu'_p/\hbar$ (H_2O, sphere, 25 °C)	γ'_p	$2.675\,153\,41(11) \times 10^{8}$	s^{-1} T^{-1}	4.2×10^{-8}
	$\gamma'_p/2\pi$	42.576 3888(18)	MHz T^{-1}	4.2×10^{-8}
		Neutron, n		

Fundamental Physical Constants

Quantity	Symbol	Value	Unit	Relative std. uncert. u_r
neutron mass	m_n	$1.674\,927\,16(13) \times 10^{-27}$	kg	7.9×10^{-8}
in u, $m_n = A_r(n)$ u (neutron relative atomic mass times u)		$1.008\,664\,915\,78(55)$	u	5.4×10^{-10}
energy equivalent	m_nc^2	$1.505\,349\,46(12) \times 10^{-10}$	J	7.9×10^{-8}
in MeV		$939.565\,330(38)$	MeV	4.0×10^{-8}
neutron-electron mass ratio	m_n/m_e	$1\,838.683\,6550(40)$		2.2×10^{-9}
neutron-muon mass ratio	m_n/m_μ	$8.892\,484\,78(27)$		3.0×10^{-8}
neutron-tau mass ratio	m_n/m_τ	$0.528\,722(86)$		1.6×10^{-4}
neutron-proton mass ratio	m_n/m_p	$1.001\,378\,418\,87(58)$		5.8×10^{-10}
neutron molar mass N_Am_n	$M(n)$, M_n	$1.008\,664\,915\,78(55) \times 10^{-3}$	kg mol^{-1}	5.4×10^{-10}
neutron Compton wavelength h/m_nc	$\lambda_{C,n}$	$1.319\,590\,898(10) \times 10^{-15}$	m	7.6×10^{-9}
$\lambda_{C,n}/2\pi$	$\bar{\lambda}_{C,n}$	$0.210\,019\,4142(16) \times 10^{-15}$	m	7.6×10^{-9}
neutron magnetic moment	μ_n	$-0.966\,236\,40(23) \times 10^{-26}$	J T^{-1}	2.4×10^{-7}
to Bohr magneton ratio	μ_n/μ_B	$-1.041\,875\,63(25) \times 10^{-3}$		2.4×10^{-7}
to nuclear magneton ratio	μ_n/μ_N	$-1.913\,042\,72(45)$		2.4×10^{-7}
neutron g-factor $2\mu_n/\mu_N$	g_n	$-3.826\,085\,45(90)$		2.4×10^{-7}
neutron-electron magnetic moment ratio	μ_n/μ_e	$1.040\,668\,82(25) \times 10^{-3}$		2.4×10^{-7}
neutron-proton magnetic moment ratio	μ_n/μ_p	$-0.684\,979\,34(16)$		2.4×10^{-7}
neutron to shielded proton magnetic moment ratio (H_2O, sphere, 25 °C)	μ_n/μ'_p	$-0.684\,996\,94(16)$		2.4×10^{-7}
neutron gyromagnetic ratio $2\lvert\mu_n\rvert/\hbar$	γ_n	$1.832\,471\,88(44) \times 10^{8}$	s^{-1} T^{-1}	2.4×10^{-7}
	$\gamma_n/2\pi$	$29.164\,6958(70)$	MHz T^{-1}	2.4×10^{-7}
		Deuteron, d		
deuteron mass	m_d	$3.343\,583\,09(26) \times 10^{-27}$	kg	7.9×10^{-8}
in u, $m_d = A_r(d)$ u (deuteron relative atomic mass times u)		$2.013\,553\,212\,71(35)$	u	1.7×10^{-10}
energy equivalent	m_dc^2	$3.005\,062\,62(24) \times 10^{-10}$	J	7.9×10^{-8}
in MeV		$1\,875.612\,762(75)$	MeV	4.0×10^{-8}
deuteron-electron mass ratio	m_d/m_e	$3\,670.482\,9550(78)$		2.1×10^{-9}
deuteron-proton mass ratio	m_d/m_p	$1.999\,007\,500\,83(41)$		2.0×10^{-10}
deuteron molar mass N_Am_d	$M(d)$, M_d	$2.013\,553\,212\,71(35) \times 10^{-3}$	kg mol^{-1}	1.7×10^{-10}
deuteron magnetic moment	μ_d	$0.433\,073\,457(18) \times 10^{-26}$	J T^{-1}	4.2×10^{-8}
to Bohr magneton ratio	μ_d/μ_B	$0.466\,975\,4556(50) \times 10^{-3}$		1.1×10^{-8}
to nuclear magneton ratio	μ_d/μ_N	$0.857\,438\,2284(94)$		1.1×10^{-8}
deuteron-electron magnetic moment ratio	μ_d/μ_e	$-4.664\,345\,537(50) \times 10^{-4}$		1.1×10^{-8}
deuteron-proton magnetic moment ratio	μ_d/μ_p	$0.307\,012\,2083(45)$		1.5×10^{-8}

Fundamental Physical Constants

Quantity	Symbol	Value	Unit	Relative std. uncert. u_r
deuteron-neutron magnetic moment ratio	μ_d/μ_n	−0.448 206 52(11)		2.4×10^{-7}
		Helion, h		
helion mass[e]	m_h	$5.006\,411\,74(39) \times 10^{-27}$	kg	7.9×10^{-8}
in u, $m_h = A_r(h)$ u (helion relative atomic mass times u)		3.014 932 234 69(86)	u	2.8×10^{-10}
energy equivalent	m_hc^2	$4.499\,538\,48(35) \times 10^{-10}$	J	7.9×10^{-8}
in MeV		2 808.391 32(11)	MeV	4.0×10^{-8}
helion-electron mass ratio	m_h/m_e	5 495.885 238(12)		2.1×10^{-9}
helion-proton mass ratio	m_h/m_p	2.993 152 658 50(93)		3.1×10^{-10}
helion molar mass $N_A m_h$	$M(h)$, M_h	$3.014\,932\,234\,69(86) \times 10^{-3}$	kg mol^{-1}	2.8×10^{-10}
shielded helion magnetic moment (gas, sphere, 25 °C)	μ'_h	$-1.074\,552\,967(45) \times 10^{-26}$	J T^{-1}	4.2×10^{-8}
to Bohr magneton ratio	μ'_h/μ_B	$-1.158\,671\,474(14) \times 10^{-3}$		1.2×10^{-8}
to nuclear magneton ratio	μ'_h/μ_N	−2.127 497 718(25)		1.2×10^{-8}
shielded helion to proton magnetic moment ratio (gas, sphere, 25 °C)	μ'_h/μ_p	−0.761 766 563(12)		1.5×10^{-8}
shielded helion to shielded proton magnetic moment ratio (gas/H_2O, spheres, 25 °C)	μ'_h/μ'_p	−0.761 786 1313(33)		4.3×10^{-9}
shielded helion gyromagnetic ratio $2\lvert\mu'_h\rvert/\hbar$ (gas, sphere, 25 °C)	γ'_h	$2.037\,894\,764(85) \times 10^{8}$	s^{-1} T^{-1}	4.2×10^{-8}
	$\gamma'_h/2\pi$	32.434 1025(14)	MHz T^{-1}	4.2×10^{-8}
		Alpha particle, α		
alpha particle mass	m_α	$6.644\,655\,98(52) \times 10^{-27}$	kg	7.9×10^{-8}
in u, $m_\alpha = A_r(\alpha)$ u (alpha particle relative atomic mass times u)		4.001 506 1747(10)	u	2.5×10^{-10}
energy equivalent	$m_\alpha c^2$	$5.971\,918\,97(47) \times 10^{-10}$	J	7.9×10^{-8}
in MeV		3 727.379 04(15)	MeV	4.0×10^{-8}
alpha particle to electron mass ratio	m_α/m_e	7 294.299 508(16)		2.1×10^{-9}
alpha particle to proton mass ratio	m_α/m_p	3.972 599 6846(11)		2.8×10^{-10}
alpha particle molar mass $N_A m_\alpha$	$M(\alpha)$, M_α	$4.001\,506\,1747(10) \times 10^{-3}$	kg mol^{-1}	2.5×10^{-10}
		PHYSICO-CHEMICAL		
Avogadro constant	N_A, L	$6.022\,141\,99(47) \times 10^{23}$	mol^{-1}	7.9×10^{-8}
atomic mass constant $m_u = \frac{1}{12}m(^{12}C) = 1$ u $= 10^{-3}$ kg mol$^{-1}/N_A$	m_u	$1.660\,538\,73(13) \times 10^{-27}$	kg	7.9×10^{-8}
energy equivalent	m_uc^2	$1.492\,417\,78(12) \times 10^{-10}$	J	7.9×10^{-8}
in MeV		931.494 013(37)	MeV	4.0×10^{-8}
Faraday constant[g] $N_A e$	F	96 485.3415(39)	C mol^{-1}	4.0×10^{-8}

Fundamental Physical Constants

Quantity	Symbol	Value	Unit	Relative std. uncert. u_r
molar Planck constant	$N_A h$	$3.990\,312\,689(30) \times 10^{-10}$	J s mol^{-1}	7.6×10^{-9}
	$N_A hc$	$0.119\,626\,564\,92(91)$	J m mol^{-1}	7.6×10^{-9}
molar gas constant	R	$8.314\,472(15)$	J mol^{-1} K^{-1}	1.7×10^{-6}
Boltzmann constant R/N_A	k	$1.380\,6503(24) \times 10^{-23}$	J K^{-1}	1.7×10^{-6}
in eV K^{-1}		$8.617\,342(15) \times 10^{-5}$	eV K^{-1}	1.7×10^{-6}
	k/h	$2.083\,6644(36) \times 10^{10}$	Hz K^{-1}	1.7×10^{-6}
	k/hc	$69.503\,56(12)$	m^{-1} K^{-1}	1.7×10^{-6}
molar volume of ideal gas RT/p				
$T = 273.15$ K, $p = 101.325$ kPa	V_m	$22.413\,996(39) \times 10^{-3}$	m^3 mol^{-1}	1.7×10^{-6}
Loschmidt constant N_A/V_m	n_0	$2.686\,7775(47) \times 10^{25}$	m^{-3}	1.7×10^{-6}
$T = 273.15$ K, $p = 100$ kPa	V_m	$22.710\,981(40) \times 10^{-3}$	m^3 mol^{-1}	1.7×10^{-6}
Sackur-Tetrode constant (absolute entropy constant)[h]				
$\frac{5}{2} + \ln[(2\pi m_u kT_1/h^2)^{3/2} kT_1/p_0]$				
$T_1 = 1$ K, $p_0 = 100$ kPa	S_0/R	$-1.151\,7048(44)$		3.8×10^{-6}
$T_1 = 1$ K, $p_0 = 101.325$ kPa		$-1.164\,8678(44)$		3.7×10^{-6}
Stefan-Boltzmann constant				
$(\pi^2/60)k^4/\hbar^3 c^2$	σ	$5.670\,400(40) \times 10^{-8}$	W m^{-2} K^{-4}	7.0×10^{-6}
first radiation constant $2\pi hc^2$	c_1	$3.741\,771\,07(29) \times 10^{-16}$	W m^2	7.8×10^{-8}
first radiation constant for spectral radiance $2hc^2$	c_{1L}	$1.191\,042\,722(93) \times 10^{-16}$	W m^2 sr^{-1}	7.8×10^{-8}
second radiation constant hc/k	c_2	$1.438\,7752(25) \times 10^{-2}$	m K	1.7×10^{-6}
Wien displacement law constant				
$b = \lambda_{max} T = c_2/4.965\,114\,231...$	b	$2.897\,7686(51) \times 10^{-3}$	m K	1.7×10^{-6}

[a] See the "Adopted values" table for the conventional value adopted internationally for realizing representations of the volt using the Josephson effect.

[b] See the "Adopted values" table for the conventional value adopted internationally for realizing representations of the ohm using the quantum Hall effect.

[c] Value recommended by the Particle Data Group, Caso et al., Eur. Phys. J. C 3(1-4), 1-794 (1998).

[d] Based on the ratio of the masses of the W and Z bosons m_W/m_Z recommended by the Particle Data Group (Caso et al., 1998). The value for $\sin^2\theta_W$ they recommend, which is based on a particular variant of the modified minimal subtraction ($\overline{MS}$) scheme, is $\sin^2\hat{\theta}_W(M_Z) = 0.231\,24(24)$.

[e] The helion, symbol h, is the nucleus of the ^{3}He atom.

[f] This and all other values involving m_τ are based on the value of $m_\tau c^2$ in MeV recommended by the Particle Data Group, Caso et al., Eur. Phys. J. C 3(1-4), 1-794 (1998), but with a standard uncertainty of 0.29 MeV rather than the quoted uncertainty of −0.26 MeV, +0.29 MeV.

[g] The numerical value of F to be used in coulometric chemical measurements is 96 485.3432(76) [7.9×10^{-8}] when the relevant current is measured in terms of representations of the volt and ohm based on the Josephson and quantum Hall effects and the internationally adopted conventional values of the Josephson and von Klitzing constants K_{J-90} and R_{K-90} given in the "Adopted values" table.

[h] The entropy of an ideal monoatomic gas of relative atomic mass A_r is given by $S = S_0 + \frac{3}{2}R \ln A_r - R \ln(p/p_0) + \frac{5}{2}R \ln(T/\mathrm{K})$.

Fundamental Physical Constants — Adopted values

Quantity	Symbol	Value	Unit	Relative std. uncert. u_r
molar mass of ^{12}C	$M(^{12}\mathrm{C})$	12×10^{-3}	kg mol^{-1}	(exact)
molar mass constant[a] $M(^{12}\mathrm{C})/12$	M_u	1×10^{-3}	kg mol^{-1}	(exact)
conventional value of Josephson constant[b]	K_{J-90}	483 597.9	GHz V^{-1}	(exact)
conventional value of von Klitzing constant[c]	R_{K-90}	25 812.807	Ω	(exact)
standard atmosphere		101 325	Pa	(exact)
standard acceleration of gravity	g_n	9.806 65	m s^{-2}	(exact)

[a] The relative atomic mass $A_r(X)$ of particle X with mass $m(X)$ is defined by $A_r(X) = m(X)/m_u$, where $m_u = m(^{12}\mathrm{C})/12 = M_u/N_A = 1$ u is the atomic mass constant, N_A is the Avogadro constant, and u is the atomic mass unit. Thus the mass of particle X in u is $m(X) = A_r(X)$ u and the molar mass of X is $M(X) = A_r(X)M_u$.

[b] This is the value adopted internationally for realizing representations of the volt using the Josephson effect.

[c] This is the value adopted internationally for realizing representations of the ohm using the quantum Hall effect.

Energy Equivalents

	J	kg	m^{-1}	Hz
1 J	$(1\ \mathrm{J}) =$ 1 J	$(1\ \mathrm{J})/c^2 =$ $1.112\,650\,056 \times 10^{-17}$ kg	$(1\ \mathrm{J})/hc =$ $5.034\,117\,62(39) \times 10^{24}\ \mathrm{m^{-1}}$	$(1\ \mathrm{J})/h =$ $1.509\,190\,50(12) \times 10^{33}$ Hz
1 kg	$(1\ \mathrm{kg})c^2 =$ $8.987\,551\,787 \times 10^{16}$ J	$(1\ \mathrm{kg}) =$ 1 kg	$(1\ \mathrm{kg})c/h =$ $4.524\,439\,29(35) \times 10^{41}\ \mathrm{m^{-1}}$	$(1\ \mathrm{kg})c^2/h =$ $1.356\,392\,77(11) \times 10^{50}$ Hz
$1\ \mathrm{m^{-1}}$	$(1\ \mathrm{m^{-1}})hc =$ $1.986\,445\,44(16) \times 10^{-25}$ J	$(1\ \mathrm{m^{-1}})h/c =$ $2.210\,218\,63(17) \times 10^{-42}$ kg	$(1\ \mathrm{m^{-1}}) =$ $1\ \mathrm{m^{-1}}$	$(1\ \mathrm{m^{-1}})c =$ 299 792 458 Hz
1 Hz	$(1\ \mathrm{Hz})h =$ $6.626\,068\,76(52) \times 10^{-34}$ J	$(1\ \mathrm{Hz})h/c^2 =$ $7.372\,495\,78(58) \times 10^{-51}$ kg	$(1\ \mathrm{Hz})/c =$ $3.335\,640\,952 \times 10^{-9}\ \mathrm{m^{-1}}$	$(1\ \mathrm{Hz}) =$ 1 Hz
1 K	$(1\ \mathrm{K})k =$ $1.380\,6503(24) \times 10^{-23}$ J	$(1\ \mathrm{K})k/c^2 =$ $1.536\,1807(27) \times 10^{-40}$ kg	$(1\ \mathrm{K})k/hc =$ $69.503\,56(12)\ \mathrm{m^{-1}}$	$(1\ \mathrm{K})k/h =$ $2.083\,6644(36) \times 10^{10}$ Hz
1 eV	$(1\ \mathrm{eV}) =$ $1.602\,176\,462(63) \times 10^{-19}$ J	$(1\ \mathrm{eV})/c^2 =$ $1.782\,661\,731(70) \times 10^{-36}$ kg	$(1\ \mathrm{eV})/hc =$ $8.065\,544\,77(32) \times 10^{5}\ \mathrm{m^{-1}}$	$(1\ \mathrm{eV})/h =$ $2.417\,989\,491(95) \times 10^{14}$ Hz
1 u	$(1\ \mathrm{u})c^2 =$ $1.492\,417\,78(12) \times 10^{-10}$ J	$(1\ \mathrm{u}) =$ $1.660\,538\,73(13) \times 10^{-27}$ kg	$(1\ \mathrm{u})c/h =$ $7.513\,006\,658(57) \times 10^{14}\ \mathrm{m^{-1}}$	$(1\ \mathrm{u})c^2/h =$ $2.252\,342\,733(17) \times 10^{23}$ Hz
$1\ E_\mathrm{h}$	$(1\ E_\mathrm{h}) =$ $4.359\,743\,81(34) \times 10^{-18}$ J	$(1\ E_\mathrm{h})/c^2 =$ $4.850\,869\,19(38) \times 10^{-35}$ kg	$(1\ E_\mathrm{h})/hc =$ $2.194\,746\,313\,710(17) \times 10^{7}\ \mathrm{m^{-1}}$	$(1\ E_\mathrm{h})/h =$ $6.579\,683\,920\,735(50) \times 10^{15}$ Hz

Derived from the relations $E = mc^2 = hc/\lambda = h\nu = kT$, and based on the 1998 CODATA adjustment of the values of the constants; $1\ \mathrm{eV} = (e/\mathrm{C})\ \mathrm{J}$, $1\ \mathrm{u} = m_\mathrm{u} = \frac{1}{12}m(^{12}\mathrm{C}) = 10^{-3}\ \mathrm{kg\ mol^{-1}}/N_\mathrm{A}$, and $E_\mathrm{h} = 2R_\infty hc = \alpha^2 m_\mathrm{e}c^2$ is the Hartree energy (hartree).

Energy Equivalents

	K	eV	u	E_h
1 J	$(1\ \text{J})/k =$ 7.242 964(13) $\times 10^{22}$ K	$(1\ \text{J}) =$ 6.241 509 74(24) $\times 10^{18}$ eV	$(1\ \text{J})/c^2 =$ 6.700 536 62(53) $\times 10^{9}$ u	$(1\ \text{J}) =$ 2.293 712 76(18) $\times 10^{17}$ E_h
1 kg	$(1\ \text{kg})c^2/k =$ 6.509 651(11) $\times 10^{39}$ K	$(1\ \text{kg})c^2 =$ 5.609 589 21(22) $\times 10^{35}$ eV	$(1\ \text{kg}) =$ 6.022 141 99(47) $\times 10^{26}$ u	$(1\ \text{kg})c^2 =$ 2.061 486 22(16) $\times 10^{34}$ E_h
1 m^{-1}	$(1\ \text{m}^{-1})hc/k =$ 1.438 7752(25) $\times 10^{-2}$ K	$(1\ \text{m}^{-1})hc =$ 1.239 841 857(49) $\times 10^{-6}$ eV	$(1\ \text{m}^{-1})h/c =$ 1.331 025 042(10) $\times 10^{-15}$ u	$(1\ \text{m}^{-1})hc =$ 4.556 335 252 750(35) $\times 10^{-8}$ E_h
1 Hz	$(1\ \text{Hz})h/k =$ 4.799 2374(84) $\times 10^{-11}$ K	$(1\ \text{Hz})h =$ 4.135 667 27(16) $\times 10^{-15}$ eV	$(1\ \text{Hz})h/c^2 =$ 4.439 821 637(34) $\times 10^{-24}$ u	$(1\ \text{Hz})h =$ 1.519 829 846 003(12) $\times 10^{-16}$ E_h
1 K	$(1\ \text{K}) =$ 1 K	$(1\ \text{K})k =$ 8.617 342(15) $\times 10^{-5}$ eV	$(1\ \text{K})k/c^2 =$ 9.251 098(16) $\times 10^{-14}$ u	$(1\ \text{K})k =$ 3.166 8153(55) $\times 10^{-6}$ E_h
1 eV	$(1\ \text{eV})/k =$ 1.160 4506(20) $\times 10^{4}$ K	$(1\ \text{eV}) =$ 1 eV	$(1\ \text{eV})/c^2 =$ 1.073 544 206(43) $\times 10^{-9}$ u	$(1\ \text{eV}) =$ 3.674 932 60(14) $\times 10^{-2}$ E_h
1 u	$(1\ \text{u})c^2/k =$ 1.080 9528(19) $\times 10^{13}$ K	$(1\ \text{u})c^2 =$ 931.494 013(37) $\times 10^{6}$ eV	$(1\ \text{u}) =$ 1 u	$(1\ \text{u})c^2 =$ 3.423 177 709(26) $\times 10^{7}$ E_h
1 E_h	$(1\ E_h)/k =$ 3.157 7465(55) $\times 10^{5}$ K	$(1\ E_h) =$ 27.211 3834(11) eV	$(1\ E_h)/c^2 =$ 2.921 262 304(22) $\times 10^{-8}$ u	$(1\ E_h) =$ 1 E_h

Derived from the relations $E = mc^2 = hc/\lambda = h\nu = kT$, and based on the 1998 CODATA adjustment of the values of the constants; 1 eV = (e/C) J, 1 u = $m_u = \frac{1}{12}m(^{12}\text{C}) = 10^{-3}$ kg mol$^{-1}/N_A$, and $E_h = 2R_\infty hc = \alpha^2 m_e c^2$ is the Hartree energy (hartree).

STANDARD ATOMIC WEIGHTS (1997)

This table of atomic weights is reprinted from the 1997 report of the IUPAC Commission on Atomic Weights and Isotopic Abundances. The Standard Atomic Weights apply to the elements as they exist naturally on Earth, and the uncertainties take into account the isotopic variation found in most laboratory samples. Further comments on the variability are given in the footnotes.

The number in parentheses following the atomic weight value gives the uncertainty in the last digit. An entry in brackets indicates the mass number of the longest-lived isotope of an element that has no stable isotopes and for which a Standard Atomic Weight cannot be defined because of wide variability in isotopic composition (or complete absence) in nature.

REFERENCE

Vocke, R. D. (for IUPAC Commission on Atomic Weights and Isotopic Abundances), Atomic Weights of the Elements 1997, *Pure Appl. Chem.*, 71, 1593, 1999.

Name	Symbol	At. no.	Atomic Weight	Footnotes		
Actinium	Ac	89	[227]			
Aluminum	Al	13	26.981538(2)			
Americium	Am	95	[243]			
Antimony	Sb	51	121.760(1)	g		
Argon	Ar	18	39.948(1)	g		r
Arsenic	As	33	74.92160(2)			
Astatine	At	85	[210]			
Barium	Ba	56	137.327(7)			
Berkelium	Bk	97	[247]			
Beryllium	Be	4	9.012182(3)			
Bismuth	Bi	83	208.98038(2)			
Bohrium	Bh	107	[264]			
Boron	B	5	10.811(7)	g	m	r
Bromine	Br	35	79.904(1)			
Cadmium	Cd	48	112.411(8)	g		
Calcium	Ca	20	40.078(4)	g		
Californium	Cf	98	[251]			
Carbon	C	6	12.0107(8)	g		r
Cerium	Ce	58	140.116(1)	g		
Cesium	Cs	55	132.90545(2)			
Chlorine	Cl	17	35.4527(9)		m	
Chromium	Cr	24	51.9961(6)			
Cobalt	Co	27	58.933200(9)			
Copper	Cu	29	63.546(3)			r
Curium	Cm	96	[247]			
Dubnium	Db	105	[262]			
Dysprosium	Dy	66	162.50(3)	g		
Einsteinium	Es	99	[252]			
Erbium	Er	68	167.26(3)	g		
Europium	Eu	63	151.964(1)	g		
Fermium	Fm	100	[257]			
Fluorine	F	9	18.9984032(5)			
Francium	Fr	87	[223]			
Gadolinium	Gd	64	157.25(3)	g		
Gallium	Ga	31	69.723(1)			
Germanium	Ge	32	72.61(2)			
Gold	Au	79	196.96655(2)			
Hafnium	Hf	72	178.49(2)			
Hassium	Hs	108	[269]			
Helium	He	2	4.002602(2)	g		r
Holmium	Ho	67	164.93032(2)			
Hydrogen	H	1	1.00794(7)	g	m	r
Indium	In	49	114.818(3)			
Iodine	I	53	126.90447(3)			

STANDARD ATOMIC WEIGHTS (1997) (continued)

Name	Symbol	At. no.	Atomic Weight	Footnotes		
Iridium	Ir	77	192.217(3)			
Iron	Fe	26	55.845(2)			
Krypton	Kr	36	83.80(1)	g	m	
Lanthanum	La	57	138.9055(2)	g		
Lawrencium	Lr	103	[262]			
Lead	Pb	82	207.2(1)	g		r
Lithium	Li	3	6.941(2)*	g	m	r
Lutetium	Lu	71	174.967(1)	g		
Magnesium	Mg	12	24.3050(6)			
Manganese	Mn	25	54.938049(9)			
Meitnerium	Mt	109	[268]			
Mendelevium	Md	101	[258]			
Mercury	Hg	80	200.59(2)			
Molybdenum	Mo	42	95.94(1)	g		
Neodymium	Nd	60	144.24(3)	g		
Neon	Ne	10	20.1797(6)	g	m	
Neptunium	Np	93	[237]			
Nickel	Ni	28	58.6934(2)			
Niobium	Nb	41	92.90638(2)			
Nitrogen	N	7	14.00674(7)	g		r
Nobelium	No	102	[259]			
Osmium	Os	76	190.23(3)	g		
Oxygen	O	8	15.9994(3)	g		r
Palladium	Pd	46	106.42(1)	g		
Phosphorus	P	15	30.973761(2)			
Platinum	Pt	78	195.078(2)			
Plutonium	Pu	94	[244]			
Polonium	Po	84	[209]			
Potassium	K	19	39.0983(1)	g		
Praseodymium	Pr	59	140.90765(2)			
Promethium	Pm	61	[145]			
Protactinium	Pa	91	231.03588(2)			
Radium	Ra	88	[226]			
Radon	Rn	86	[222]			
Rhenium	Re	75	186.207(1)			
Rhodium	Rh	45	102.90550(2)			
Rubidium	Rb	37	85.4678(3)	g		
Ruthenium	Ru	44	101.07(2)	g		
Rutherfordium	Rf	104	[261]			
Samarium	Sm	62	150.36(3)	g		
Scandium	Sc	21	44.955910(8)			
Seaborgium	Sg	106	[266]			
Selenium	Sc	34	78.96(3)			
Silicon	Si	14	28.0855(3)			r
Silver	Ag	47	107.8682(2)	g		
Sodium	Na	11	22.989770(2)			
Strontium	Sr	38	87.62(1)	g		r
Sulfur	S	16	32.066(6)	g		r
Tantalum	Ta	73	180.9479(1)			
Technetium	Tc	43	[98]			
Tellurium	Te	52	127.60(3)	g		
Terbium	Tb	65	158.92534(2)			
Thallium	Tl	81	204.3833(2)			
Thorium	Th	90	232.0381(1)	g		
Thulium	Tm	69	168.93421(2)			
Tin	Sn	50	118.710(7)	g		
Titanium	Ti	22	47.867(1)			
Tungsten	W	74	183.84(1)			

STANDARD ATOMIC WEIGHTS (1997) (continued)

Name	Symbol	At. no.	Atomic Weight	Footnotes		
Uranium	U	92	238.0289(1)	g		m
Vanadium	V	23	50.9415(1)			
Xenon	Xe	54	131.29(2)	g		m
Ytterbium	Yb	70	173.04(3)	g		
Yttrium	Y	39	88.90585(2)			
Zinc	Zn	30	65.39(2)			
Zirconium	Zr	40	91.224(2)	g		

* Commercially available Li materials have atomic weights that are known to range between 6.939 and 6.996; if a more accurate value is required, it must be determined for the specific material.

g geological specimens are known in which the element has an isotopic composition outside the limits for normal material. The difference between the atomic weight of the element in such specimens and that given in the table may exceed the stated uncertainty.

m modified isotopic compositions may be found in commercially available material because it has been subjected to an undisclosed or inadvertent isotopic fractionation. Substantial deviations in atomic weight of the element from that given the table can occur.

r range in isotopic composition of normal terrestrial material prevents a more precise atomic weight being given; the tabulated atomic weight value should be applicable to any normal material.

ATOMIC MASSES AND ABUNDANCES

This table lists the mass (in atomic mass units, symbol u) and the natural abundance (in percent) of the stable nuclides and a few important radioactive nuclides. A complete table of all nuclides may be found in Section 11 ("Table of the Isotopes").

The atomic masses are based on the 1995 evaluation of Audi and Wapstra (Reference 2). The number in parentheses following the mass value is the uncertainty in the last digit(s) given.

Natural abundance values are also followed by uncertainties in the last digit(s) of the stated values. This uncertainty includes both the estimated measurement uncertainty and the reported range of variation in different terrestrial sources of the element (see Reference 3 and 4 for more details). The absence of an entry in the Abundance column indicates a radioactive nuclide not present in nature or an element whose isotopic composition varies so widely that a meaningful natural abundance cannot be defined.

An electronic version of these data is available on the Web site of the NIST Physics Laboratory (Reference 5).

REFERENCES

1. Holden, N. E., "Table of the Isotopes", in Lide, D. R., Ed., *CRC Handbook of Chemistry and Physics, 82nd Ed.*, CRC Press, Boca Raton FL, 2001.
2. Audi, G., and Wapstra, A. H., *Nucl. Phys.*, A595, 409, 1995.
3. Rosman, K. J. R., and Taylor, P. D. P., *J. Phys. Chem. Ref. Data,* 27, 1275, 1998.
4. R. D. Vocke (for IUPAC Commission on Atomic Weights and Isotopic Abundances), *Pure Appl. Chem.*, 71, 1593, 1999.
5. Coursey, J. S., and Dragoset, R. A., *Atomic Weights and Isotopic Compositions* (version 2.1). Available: http://physics.nist.gov/Compositions/ National Institute of Standards and Technology, Gaithersburg, MD.

Z	Isotope	Mass in u	Abundance in %
1	^{1}H	1.0078250321(4)	99.9850(70)
	^{2}D	2.0141017780(4)	0.0115(70)
	^{3}T	3.0160492675(11)	
2	^{3}He	3.0160293097(9)	0.000137(3)
	^{4}He	4.0026032497(10)	99.999863(3)
3	^{6}Li	6.0151223(5)	7.59(4)
	^{7}Li	7.0160040(5)	92.41(4)
4	^{9}Be	9.0121821(4)	100
5	^{10}B	10.0129370(4)	19.9(7)
	^{11}B	11.0093055(5)	80.1(7)
6	^{12}C	12.0000000(0)	98.93(8)
	^{13}C	13.0033548378(10)	1.07(8)
7	^{14}N	14.0030740052(9)	99.632(7)
	^{15}N	15.0001088984(9)	0.368(7)
8	^{16}O	15.9949146221(15)	99.757(16)
	^{17}O	16.99913150(22)	0.038(1)
	^{18}O	17.9991604(9)	0.205(14)
9	^{19}F	18.99840320(7)	100
10	^{20}Ne	19.9924401759(20)	90.48(3)
	^{21}Ne	20.99384674(4)	0.27(1)
	^{22}Ne	21.99138551(23)	9.25(3)
11	^{23}Na	22.98976967(23)	100
12	^{24}Mg	23.98504190(20)	78.99(4)
	^{25}Mg	24.98583702(20)	10.00(1)
	^{26}Mg	25.98259304(21)	11.01(3)
13	^{27}Al	26.98153844(14)	100
14	^{28}Si	27.9769265327(20)	92.2297(7)
	^{29}Si	28.97649472(3)	4.6832(5)
	^{30}Si	29.97377022(5)	3.0872(5)
15	^{31}P	30.97376151(20)	100
16	^{32}S	31.97207069(12)	94.93(31)
	^{33}S	32.97145850(12)	0.76(2)
	^{34}S	33.96786683(11)	4.29(28)
	^{36}S	35.96708088(25)	0.02(1)
17	^{35}Cl	34.96885271(4)	75.78(4)
	^{37}Cl	36.96590260(5)	24.22(4)
18	^{36}Ar	35.96754628(27)	0.3365(30)
	^{38}Ar	37.9627322(5)	0.0632(5)
	^{40}Ar	39.962383123(3)	99.6003(30)
19	^{39}K	38.9637069(3)	93.2581(44)
	^{40}K	39.96399867(29)	0.0117(1)
	^{41}K	40.96182597(28)	6.7302(44)
20	^{40}Ca	39.9625912(3)	96.941(156)
	^{42}Ca	41.9586183(4)	0.647(23)
	^{43}Ca	42.9587668(5)	0.135(10)
	^{44}Ca	43.9554811(9)	2.086(110)
	^{46}Ca	45.9536928(25)	0.004(3)
	^{48}Ca	47.952534(4)	0.187(21)
21	^{45}Sc	44.9559102(12)	100
22	^{46}Ti	45.9526295(12)	8.25(3)
	^{47}Ti	46.9517638(10)	7.44(2)
	^{48}Ti	47.9479471(10)	73.72(3)
	^{49}Ti	48.9478708(10)	5.41(2)
	^{50}Ti	49.9447921(11)	5.18(2)
23	^{50}V	49.9471628(14)	0.250(4)
	^{51}V	50.9439637(14)	99.750(4)
24	^{50}Cr	49.9460496(14)	4.345(13)
	^{52}Cr	51.9405119(15)	83.789(18)
	^{53}Cr	52.9406538(15)	9.501(17)
	^{54}Cr	53.9388849(15)	2.365(7)
25	^{55}Mn	54.9380496(14)	100
26	^{54}Fe	53.9396148(14)	5.845(35)
	^{56}Fe	55.9349421(15)	91.754(36)
	^{57}Fe	56.9353987(15)	2.119(10)
	^{58}Fe	57.9332805(15)	0.282(4)
27	^{59}Co	58.9332002(15)	100
28	^{58}Ni	57.9353479(15)	68.0769(89)
	^{60}Ni	59.9307906(15)	26.2231(77)
	^{61}Ni	60.9310604(15)	1.1399(6)
	^{62}Ni	61.9283488(15)	3.6345(17)
	^{64}Ni	63.9279696(16)	0.9256(9)
29	^{63}Cu	62.9296011(15)	69.17(3)
	^{65}Cu	64.9277937(19)	30.83(3)
30	^{64}Zn	63.9291466(18)	48.63(60)
	^{66}Zn	65.9260368(16)	27.90(27)
	^{67}Zn	66.9271309(17)	4.10(13)

ATOMIC MASSES AND ABUNDANCES (continued)

Z	Isotope	Mass in u	Abundance in %
	^{68}Zn	67.9248476(17)	18.75(51)
	^{70}Zn	69.925325(4)	0.62(3)
31	^{69}Ga	68.925581(3)	60.108(9)
	^{71}Ga	70.9247050(19)	39.892(9)
32	^{70}Ge	69.9242504(19)	20.84(87)
	^{72}Ge	71.9220762(16)	27.54(34)
	^{73}Ge	72.9234594(16)	7.73(5)
	^{74}Ge	73.9211782(16)	36.28(73)
	^{76}Ge	75.9214027(16)	7.61(38)
33	^{75}As	74.9215964(18)	100
34	^{74}Se	73.9224766(16)	0.89(4)
	^{76}Se	75.9192141(16)	9.37(29)
	^{77}Se	76.9199146(16)	7.63(16)
	^{78}Se	77.9173095(16)	23.77(28)
	^{80}Se	79.9165218(20)	49.61(41)
	^{82}Se	81.9167000(22)	8.73(22)
35	^{79}Br	78.9183376(20)	50.69(7)
	^{81}Br	80.916291(3)	49.31(7)
36	^{78}Kr	77.920386(7)	0.35(1)
	^{80}Kr	79.916378(4)	2.28(6)
	^{82}Kr	81.9134846(28)	11.58(14)
	^{83}Kr	82.914136(3)	11.49(6)
	^{84}Kr	83.911507(3)	57.00(4)
	^{86}Kr	85.9106103(12)	17.30(22)
37	^{85}Rb	84.9117893(25)	72.17(2)
	^{87}Rb	86.9091835(27)	27.83(2)
38	^{84}Sr	83.913425(4)	0.56(1)
	^{86}Sr	85.9092624(24)	9.86(1)
	^{87}Sr	86.9088793(24)	7.00(1)
	^{88}Sr	87.9056143(24)	82.58(1)
39	^{89}Y	88.9058479(25)	100
40	^{90}Zr	89.9047037(23)	51.45(40)
	^{91}Zr	90.9056450(23)	11.22(5)
	^{92}Zr	91.9050401(23)	17.15(8)
	^{94}Zr	93.9063158(25)	17.38(28)
	^{96}Zr	95.908276(3)	2.80(9)
41	^{93}Nb	92.9063775(24)	100
42	^{92}Mo	91.906810(4)	14.84(35)
	^{94}Mo	93.9050876(20)	9.25(12)
	^{95}Mo	94.9058415(20)	15.92(13)
	^{96}Mo	95.9046789(20)	16.68(2)
	^{97}Mo	96.9060210(20)	9.55(8)
	^{98}Mo	97.9054078(20)	24.13(31)
	^{100}Mo	99.907477(6)	9.63(23)
43	^{97}Tc	96.906365(5)	
	^{98}Tc	97.907216(4)	
	^{99}Tc	98.9062546(21)	
44	^{96}Ru	95.907598(8)	5.54(14)
	^{98}Ru	97.905287(7)	1.87(3)
	^{99}Ru	98.9059393(21)	12.76(14)
	^{100}Ru	99.9042197(22)	12.60(7)
	^{101}Ru	100.9055822(22)	17.06(2)
	^{102}Ru	101.9043495(22)	31.55(14)
	^{104}Ru	103.905430(4)	18.62(27)
45	^{103}Rh	102.905504(3)	100
46	^{102}Pd	101.905608(3)	1.02(1)
	^{104}Pd	103.904035(5)	11.14(8)
	^{105}Pd	104.905084(5)	22.33(8)
	^{106}Pd	105.903483(5)	27.33(3)
	^{108}Pd	107.903894(4)	26.46(9)
	^{110}Pd	109.905152(12)	11.72(9)
47	^{107}Ag	106.905093(6)	51.839(8)
	^{109}Ag	108.904756(3)	48.161(8)
48	^{106}Cd	105.906458(6)	1.25(6)
	^{108}Cd	107.904183(6)	0.89(3)
	^{110}Cd	109.903006(3)	12.49(18)
	^{111}Cd	110.904182(3)	12.80(12)
	^{112}Cd	111.9027572(30)	24.13(21)
	^{113}Cd	112.9044009(30)	12.22(12)
	^{114}Cd	113.9033581(30)	28.73(42)
	^{116}Cd	115.904755(3)	7.49(18)
49	^{113}In	112.904061(4)	4.29(5)
	^{115}In	114.903878(5)	95.71(5)
50	^{112}Sn	111.904821(5)	0.97(1)
	^{114}Sn	113.902782(3)	0.66(1)
	^{115}Sn	114.903346(3)	0.34(1)
	^{116}Sn	115.901744(3)	14.54(9)
	^{117}Sn	116.902954(3)	7.68(7)
	^{118}Sn	117.901606(3)	24.22(9)
	^{119}Sn	118.903309(3)	8.59(4)
	^{120}Sn	119.9021966(27)	32.58(9)
	^{122}Sn	121.9034401(29)	4.63(3)
	^{124}Sn	123.9052746(15)	5.79(5)
51	^{121}Sb	120.9038180(24)	57.21(5)
	^{123}Sb	122.9042157(22)	42.79(5)
52	^{120}Te	119.904020(11)	0.09(1)
	^{122}Te	121.9030471(20)	2.55(12)
	^{123}Te	122.9042730(19)	0.89(3)
	^{124}Te	123.9028195(16)	4.74(14)
	^{125}Te	124.9044247(20)	7.07(15)
	^{126}Te	125.9033055(20)	18.84(25)
	^{128}Te	127.9044614(19)	31.74(8)
	^{130}Te	129.9062228(21)	34.08(62)
53	^{127}I	126.904468(4)	100
54	^{124}Xe	123.9058958(21)	0.09(1)
	^{126}Xe	125.904269(7)	0.09(1)
	^{128}Xe	127.9035304(15)	1.92(3)
	^{129}Xe	128.9047795(9)	26.44(24)
	^{130}Xe	129.9035079(10)	4.08(2)
	^{131}Xe	130.9050819(10)	21.18(3)
	^{132}Xe	131.9041545(12)	26.89(6)
	^{134}Xe	133.9053945(9)	10.44(10)
	^{136}Xe	135.907220(8)	8.87(16)
55	^{133}Cs	132.905447(3)	100
56	^{130}Ba	129.906310(7)	0.106(1)
	^{132}Ba	131.905056(3)	0.101(1)
	^{134}Ba	133.904503(3)	2.417(18)
	^{135}Ba	134.905683(3)	6.592(12)
	^{136}Ba	135.904570(3)	7.854(24)
	^{137}Ba	136.905821(3)	11.232(24)
	^{138}Ba	137.905241(3)	71.698(42)
57	^{138}La	137.907107(4)	0.090(1)
	^{139}La	138.906348(3)	99.910(1)
58	^{136}Ce	135.907140(50)	0.185(2)
	^{138}Ce	137.905986(11)	0.251(2)
	^{140}Ce	139.905434(3)	88.450(51)

Z	Isotope	Mass in u	Abundance in %
	^{142}Ce	141.909240(4)	11.114(51)
59	^{141}Pr	140.907648(3)	100
60	^{142}Nd	141.907719(3)	27.2(5)
	^{143}Nd	142.909810(3)	12.2(2)
	^{144}Nd	143.910083(3)	23.8(3)
	^{145}Nd	144.912569(3)	8.3(1)
	^{146}Nd	145.913112(3)	17.2(3)
	^{148}Nd	147.916889(3)	5.7(1)
	^{150}Nd	149.920887(4)	5.6(2)
61	^{145}Pm	144.912744(4)	
	^{147}Pm	146.915134(3)	
62	^{144}Sm	143.911995(4)	3.07(7)
	^{147}Sm	146.914893(3)	14.99(18)
	^{148}Sm	147.914818(3)	11.24(10)
	^{149}Sm	148.917180(3)	13.82(7)
	^{150}Sm	149.917271(3)	7.38(1)
	^{152}Sm	151.919728(3)	26.75(16)
	^{154}Sm	153.922205(3)	22.75(29)
63	^{151}Eu	150.919846(3)	47.81(3)
	^{153}Eu	152.921226(3)	52.19(3)
64	^{152}Gd	151.919788(3)	0.20(1)
	^{154}Gd	153.920862(3)	2.18(3)
	^{155}Gd	154.922619(3)	14.80(12)
	^{156}Gd	155.922120(3)	20.47(9)
	^{157}Gd	156.923957(3)	15.65(2)
	^{158}Gd	157.924101(3)	24.84(7)
	^{160}Gd	159.927051(3)	21.86(19)
65	^{159}Tb	158.925343(3)	100
66	^{156}Dy	155.924278(7)	0.06(1)
	^{158}Dy	157.924405(4)	0.10(1)
	^{160}Dy	159.925194(3)	2.34(8)
	^{161}Dy	160.926930(3)	18.91(24)
	^{162}Dy	161.926795(3)	25.51(26)
	^{163}Dy	162.928728(3)	24.90(16)
	^{164}Dy	163.929171(3)	28.18(37)
67	^{165}Ho	164.930319(3)	100
68	^{162}Er	161.928775(4)	0.14(1)
	^{164}Er	163.929197(4)	1.61(3)
	^{166}Er	165.930290(3)	33.61(35)
	^{167}Er	166.932045(3)	22.93(17)
	^{168}Er	167.932368(3)	26.78(26)
	^{170}Er	169.935460(3)	14.93(27)
69	^{169}Tm	168.934211(3)	100
70	^{168}Yb	167.933894(5)	0.13(1)
	^{170}Yb	169.934759(3)	3.04(15)
	^{171}Yb	170.936322(3)	14.28(57)
	^{172}Yb	171.9363777(30)	21.83(67)
	^{173}Yb	172.9382068(30)	16.13(27)
	^{174}Yb	173.9388581(30)	31.83(92)
	^{176}Yb	175.942568(3)	12.76(41)
71	^{175}Lu	174.9407679(28)	97.41(2)
	^{176}Lu	175.9426824(28)	2.59(2)
72	^{174}Hf	173.940040(3)	0.16(1)
	^{176}Hf	175.9414018(29)	5.26(7)
	^{177}Hf	176.9432200(27)	18.60(9)
	^{178}Hf	177.9436977(27)	27.28(7)
	^{179}Hf	178.9458151(27)	13.62(2)
	^{180}Hf	179.9465488(27)	35.08(16)

Z	Isotope	Mass in u	Abundance in %
73	^{180}Ta	179.947466(3)	0.012(2)
	^{181}Ta	180.947996(3)	99.988(2)
74	^{180}W	179.946706(5)	0.12(1)
	^{182}W	181.948206(3)	26.50(16)
	^{183}W	182.9502245(29)	14.31(4)
	^{184}W	183.9509326(29)	30.64(2)
	^{186}W	185.954362(3)	28.43(19)
75	^{185}Re	184.9529557(30)	37.40(2)
	^{187}Re	186.9557508(30)	62.60(2)
76	^{184}Os	183.952491(3)	0.02(1)
	^{186}Os	185.953838(3)	1.59(3)
	^{187}Os	186.9557479(30)	1.96(2)
	^{188}Os	187.9558360(30)	13.24(8)
	^{189}Os	188.9581449(30)	16.15(5)
	^{190}Os	189.958445(3)	26.26(2)
	^{192}Os	191.961479(4)	40.78(19)
77	^{191}Ir	190.960591(3)	37.3(2)
	^{193}Ir	192.962924(3)	62.7(2)
78	^{190}Pt	189.959930(7)	0.014(1)
	^{192}Pt	191.961035(4)	0.782(7)
	^{194}Pt	193.962664(3)	32.967(99)
	^{195}Pt	194.964774(3)	33.832(10)
	^{196}Pt	195.964935(3)	25.242(41)
	^{198}Pt	197.967876(4)	7.163(55)
79	^{197}Au	196.966552(3)	100
80	^{196}Hg	195.965815(4)	0.15(1)
	^{198}Hg	197.966752(3)	9.97(20)
	^{199}Hg	198.968262(3)	16.87(22)
	^{200}Hg	199.968309(3)	23.10(19)
	^{201}Hg	200.970285(3)	13.18(9)
	^{202}Hg	201.970626(3)	29.86(26)
	^{204}Hg	203.973476(3)	6.87(15)
81	^{203}Tl	202.972329(3)	29.524(14)
	^{205}Tl	204.974412(3)	70.476(14)
82	^{204}Pb	203.973029(3)	1.4(1)
	^{206}Pb	205.974449(3)	24.1(1)
	^{207}Pb	206.975881(3)	22.1(1)
	^{208}Pb	207.976636(3)	52.4(1)
83	^{209}Bi	208.980383(3)	100
84	^{209}Po	208.982416(3)	
	^{210}Po	209.982857(3)	
85	^{210}At	209.987131(9)	
	^{211}At	210.987481(4)	
86	^{211}Rn	210.990585(8)	
	^{220}Rn	220.0113841(29)	
	^{222}Rn	222.0175705(27)	
87	^{223}Fr	223.0197307(29)	
88	^{223}Ra	223.018497(3)	
	^{224}Ra	224.0202020(29)	
	^{226}Ra	226.0254026(27)	
	^{228}Ra	228.0310641(27)	
89	^{227}Ac	227.0277470(29)	
90	^{230}Th	230.0331266(22)	
	^{232}Th	232.0380504(22)	100
91	^{231}Pa	231.0358789(28)	100
92	^{233}U	233.039628(3)	
	^{234}U	234.0409456(21)	0.0055(2)
	^{235}U	235.0439231(21)	0.7200(51)

Z	Isotope	Mass in u	Abundance in %
	^{236}U	236.0455619(21)	
	^{238}U	238.0507826(21)	99.2745(106)
93	^{237}Np	237.0481673(21)	
	^{239}Np	239.0529314(23)	
94	^{238}Pu	238.0495534(21)	
	^{239}Pu	239.0521565(21)	
	^{240}Pu	240.0538075(21)	
	^{241}Pu	241.0568453(21)	
	^{242}Pu	242.0587368(21)	
	^{244}Pu	244.064198(5)	
95	^{241}Am	241.0568229(21)	
	^{243}Am	243.0613727(23)	
96	^{243}Cm	243.0613822(24)	
	^{244}Cm	244.0627463(21)	
	^{245}Cm	245.0654856(29)	
	^{246}Cm	246.0672176(24)	
	^{247}Cm	247.070347(5)	
	^{248}Cm	248.072342(5)	
97	^{247}Bk	247.070299(6)	

Z	Isotope	Mass in u	Abundance in %
	^{249}Bk	249.074980(3)	
98	^{249}Cf	249.074847(3)	
	^{250}Cf	250.0764000(24)	
	^{251}Cf	251.079580(5)	
	^{252}Cf	252.081620(5)	
99	^{252}Es	252.082970(50)	
100	^{257}Fm	257.095099(7)	
101	^{256}Md	256.094050(60)	
	^{258}Md	258.098425(5)	
102	^{259}No	259.101020(110)*	
103	^{262}Lr	262.109690(320)*	
104	^{261}Rf	261.108750(110)*	
105	^{262}Db	262.114150(200)*	
106	^{263}Sg	263.118310(130)*	
107	^{264}Bh	264.124730(300)*	
108	^{265}Hs	265.130000(320)*	
109	^{268}Mt	268.138820(340)*	
110	^{269}Uun	269.145140(310)*	
111	^{272}Uuu	272.153480(360)*	

*Mass values derived not purely from experimental data, but at least partly from systematic trends.

ELECTRON CONFIGURATION OF NEUTRAL ATOMS IN THE GROUND STATE

		K	L		M			N				O				P			Q
Atomic	*n* =	1	2		3			4				5				6			7
no.	Element	s	s	p	s	p	d	s	p	d	f	s	p	d	f	s	p	d	s
1	H	1																	
2	He	2																	
3	Li	2	1																
4	Be	2	2																
5	B	2	2	1															
6	C	2	2	2															
7	N	2	2	3															
8	O	2	2	4															
9	F	2	2	5															
10	Ne	2	2	6															
11	Na	2	2	6	1														
12	Mg	2	2	6	2														
13	Al	2	2	6	2	1													
14	Si	2	2	6	2	2													
15	P	2	2	6	2	3													
16	S	2	2	6	2	4													
17	Cl	2	2	6	2	5													
18	Ar	2	2	6	2	6													
19	K	2	2	6	2	6		1											
20	Ca	2	2	6	2	6		2											
21	Sc	2	2	6	2	6	1	2											
22	Ti	2	2	6	2	6	2	2											
23	V	2	2	6	2	6	3	2											
24	Cr	2	2	6	2	6	5*	1											
25	Mn	2	2	6	2	6	5	2											
26	Fe	2	2	6	2	6	6	2											
27	Co	2	2	6	2	6	7	2											
28	Ni	2	2	6	2	6	8	2											
29	Cu	2	2	6	2	6	10*	1											
30	Zn	2	2	6	2	6	10	2											
31	Ga	2	2	6	2	6	10	2	1										
32	Ge	2	2	6	2	6	10	2	2										
33	As	2	2	6	2	6	10	2	3										
34	Se	2	2	6	2	6	10	2	4										
35	Br	2	2	6	2	6	10	2	5										
36	Kr	2	2	6	2	6	10	2	6										
37	Rb	2	2	6	2	6	10	2	6			1							
38	Sr	2	2	6	2	6	10	2	6			2							
39	Y	2	2	6	2	6	10	2	6	1		2							
40	Zr	2	2	6	2	6	10	2	6	2		2							
41	Nb	2	2	6	2	6	10	2	6	4*		1							
42	Mo	2	2	6	2	6	10	2	6	5		1							
43	Tc	2	2	6	2	6	10	2	6	5		2							
44	Ru	2	2	6	2	6	10	2	6	7		1							
45	Rh	2	2	6	2	6	10	2	6	8		1							
46	Pd	2	2	6	2	6	10	2	6	10*									
47	Ag	2	2	6	2	6	10	2	6	10		1							
48	Cd	2	2	6	2	6	10	2	6	10		2							
49	In	2	2	6	2	6	10	2	6	10		2	1						
50	Sn	2	2	6	2	6	10	2	6	10		2	2						
51	Sb	2	2	6	2	6	10	2	6	10		2	3						
52	Te	2	2	6	2	6	10	2	6	10		2	4						
53	I	2	2	6	2	6	10	2	6	10		2	5						
54	Xe	2	2	6	2	6	10	2	6	10		2	6						
55	Cs	2	2	6	2	6	10	2	6	10		2	6			1			
56	Ba	2	2	6	2	6	10	2	6	10		2	6			2			

ELECTRON CONFIGURATION OF NEUTRAL ATOMS IN THE GROUND STATE (continued)

Atomic	n =	K 1	L 2		M 3			N 4				O 5				P 6			Q 7
no.	Element	s	s	p	s	p	d	s	p	d	f	s	p	d	f	s	p	d	s
57	La	2	2	6	2	6	10	2	6	10		2	6	1		2			
58	Ce	2	2	6	2	6	10	2	6	10	1*	2	6	1		2			
59	Pr	2	2	6	2	6	10	2	6	10	3	2	6			2			
60	Nd	2	2	6	2	6	10	2	6	10	4	2	6			2			
61	Pm	2	2	6	2	6	10	2	6	10	5	2	6			2			
62	Sm	2	2	6	2	6	10	2	6	10	6	2	6			2			
63	Eu	2	2	6	2	6	10	2	6	10	7	2	6			2			
64	Gd	2	2	6	2	6	10	2	6	10	7	2	6	1		2			
65	Tb	2	2	6	2	6	10	2	6	10	9*	2	6			2			
66	Dy	2	2	6	2	6	10	2	6	10	10	2	6			2			
67	Ho	2	2	6	2	6	10	2	6	10	11	2	6			2			
68	Er	2	2	6	2	6	10	2	6	10	12	2	6			2			
69	Tm	2	2	6	2	6	10	2	6	10	13	2	6			2			
70	Yb	2	2	6	2	6	10	2	6	10	14	2	6			2			
71	Lu	2	2	6	2	6	10	2	6	10	14	2	6	1		2			
72	Hf	2	2	6	2	6	10	2	6	10	14	2	6	2		2			
73	Ta	2	2	6	2	6	10	2	6	10	14	2	6	3		2			
74	W	2	2	6	2	6	10	2	6	10	14	2	6	4		2			
75	Re	2	2	6	2	6	10	2	6	10	14	2	6	5		2			
76	Os	2	2	6	2	6	10	2	6	10	14	2	6	6		2			
77	Ir	2	2	6	2	6	10	2	6	10	14	2	6	7		2			
78	Pt	2	2	6	2	6	10	2	6	10	14	2	6	9		1			
79	Au	2	2	6	2	6	10	2	6	10	14	2	6	10		1			
80	Hg	2	2	6	2	6	10	2	6	10	14	2	6	10		2			
81	Tl	2	2	6	2	6	10	2	6	10	14	2	6	10		2	1		
82	Pb	2	2	6	2	6	10	2	6	10	14	2	6	10		2	2		
83	Bi	2	2	6	2	6	10	2	6	10	14	2	6	10		2	3		
84	Po	2	2	6	2	6	10	2	6	10	14	2	6	10		2	4		
85	At	2	2	6	2	6	10	2	6	10	14	2	6	10		2	5		
86	Rn	2	2	6	2	6	10	2	6	10	14	2	6	10		2	6		
87	Fr	2	2	6	2	6	10	2	6	10	14	2	6	10		2	6		1
88	Ra	2	2	6	2	6	10	2	6	10	14	2	6	10		2	6		2
89	Ac	2	2	6	2	6	10	2	6	10	14	2	6	10		2	6	1	2
90	Th	2	2	6	2	6	10	2	6	10	14	2	6	10		2	6	2	2
91	Pa	2	2	6	2	6	10	2	6	10	14	2	6	10	2*	2	6	1	2
92	U	2	2	6	2	6	10	2	6	10	14	2	6	10	3	2	6	1	2
93	Np	2	2	6	2	6	10	2	6	10	14	2	6	10	4	2	6	1	2
94	Pu	2	2	6	2	6	10	2	6	10	14	2	6	10	6*	2	6		2
95	Am	2	2	6	2	6	10	2	6	10	14	2	6	10	7	2	6		2
96	Cm	2	2	6	2	6	10	2	6	10	14	2	6	10	7*	2	6	1	2
97	Bk	2	2	6	2	6	10	2	6	10	14	2	6	10	9	2	6		2
98	Cf	2	2	6	2	6	10	2	6	10	14	2	6	10	10	2	6		2
99	Es	2	2	6	2	6	10	2	6	10	14	2	6	10	11	2	6		2
100	Fm	2	2	6	2	6	10	2	6	10	14	2	6	10	12	2	6		2
101	Md	2	2	6	2	6	10	2	6	10	14	2	6	10	13	2	6		2
102	No	2	2	6	2	6	10	2	6	10	14	2	6	10	14	2	6		2
103	Lr	2	2	6	2	6	10	2	6	10	14	2	6	10	14	2	6	1	2
104	Rf	2	2	6	2	6	10	2	6	10	14	2	6	10	14	2	6	2	2

* Note irregularity.

REFERENCE

W. L. Wiese and G. A. Martin, in *A Physicist's Desk Reference*, American Institute of Physics, New York, 1989, 94.

INTERNATIONAL TEMPERATURE SCALE OF 1990 (ITS-90)

B. W. Mangum

A new temperature scale, the International Temperature Scale of 1990 (ITS-90), was officially adopted by the Comité International des Poids et Mesures (CIPM), meeting 26—28 September 1989 at the Bureau International des Poids et Mesures (BIPM). The ITS-90 was recommended to the CIPM for its adoption following the completion of the final details of the new scale by the Comité Consultatif de Thermométrie (CCT), meeting 12—14 September 1989 at the BIPM in its 17th Session. The ITS-90 became the official international temperature scale on 1 January 1990. The ITS-90 supersedes the present scales, the International Practical Temperature Scale of 1968 (IPTS-68) and the 1976 Provisional 0.5 to 30 K Temperature Scale (EPT-76).

The ITS-90 extends upward from 0.65 K, and temperatures on this scale are in much better agreement with thermodynamic values that are those on the IPTS-68 and the EPT-76. The new scale has subranges and alternative definitions in certain ranges that greatly facilitate its use. Furthermore, its continuity, precision, and reproducibility throughout its ranges are much improved over that of the present scales. The replacement of the thermocouple with the platinum resistance thermometer at temperatures below 961.78°C resulted in the biggest improvement in reproducibility.

The ITS-90 is divided into four primary ranges:

1. Between 0.65 and 3.2 K, the ITS-90 is defined by the vapor pressure-temperature relation of ^{3}He, and between 1.25 and 2.1768 K (the λ point) and between 2.1768 and 5.0 K by the vapor pressure-temperature relations of ^{4}He. T_{90} is defined by the vapor pressure equations of the form:

$$T_{90}/\mathrm{K} = A_0 + \sum_{i=1}^{9} A_i \left[\left(\ln(p/\mathrm{Pa}) - B \right) / C \right]^i$$

   The values of the coefficients A_i, and of the constants A_o, B, and C of the equations are given below.
2. Between 3.0 and 24.5561 K, the ITS-90 is defined in terms of a ^{3}He or ^{4}He constant volume gas thermometer (CVGT). The thermometer is calibrated at three temperatures — at the triple point of neon (24.5561 K), at the triple point of equilibrium hydrogen (13.8033 K), and at a temperature between 3.0 and 5.0 K, the value of which is determined by using either ^{3}He or ^{4}He vapor pressure thermometry.
3. Between 13.8033 K (–259.3467°C) and 1234.93 K (961.78°C), the ITS-90 is defined in terms of the specified fixed points given below, by resistance ratios of platinum resistance thermometers obtained by calibration at specified sets of the fixed points, and by reference functions and deviation functions of resistance ratios which relate to T_{90} between the fixed points.
4. Above 1234.93 K, the ITS-90 is defined in terms of Planck's radiation law, using the freezing-point temperature of either silver, gold, or copper as the reference temperature.

Full details of the calibration procedures and reference functions for various subranges are given in:

The International Temperature Scale of 1990, *Metrologia*, 27, 3, 1990; errata in *Metrologia*, 27, 107, 1990.

Defining Fixed Points of the ITS-90

Material[a]	Equilibrium state[b]	Temperature	
		T_{90} (K)	t_{90} (°C)
He	VP	3 to 5	–270.15 to –268.15
e-H_2	TP	13.8033	–259.3467
e-H_2 (or He)	VP (or CVGT)	≈17	≈ –256.15
e-H_2 (or He)	VP (or CVGT)	≈20.3	≈ –252.85
Ne[c]	TP	24.5561	–248.5939
O_2	TP	54.3584	–218.7916
Ar	TP	83.8058	–189.3442
Hg[c]	TP	234.3156	–38.8344
H_2O	TP	273.16	0.01
Ga[c]	MP	302.9146	29.7646
In[c]	FP	429.7485	156.5985
Sn	FP	505.078	231.928
Zn	FP	692.677	419.527
Al[c]	FP	933.473	660.323
Ag	FP	1234.93	961.78
Au	FP	1337.33	1064.18
Cu[c]	FP	1357.77	1084.62

Defining Fixed Points of the ITS-90 (continued)

[a] e-H_2 indicates equilibrium hydrogen, that is, hydrogen with the equilibrium distribution of its ortho and para states. Normal hydrogen at room temperature contains 25% para hydrogen and 75% ortho hydrogen.

[b] VP indicates vapor pressure point; CVGT indicates constant volume gas thermometer point; TP indicates triple point (equilibrium temperature at which the solid, liquid, and vapor phases coexist); FP indicates freezing point, and MP indicates melting point (the equilibrium temperatures at which the solid and liquid phases coexist under a pressure of 101 325 Pa, one standard atmosphere). The isotopic composition is that naturally occurring.

[c] Previously, these were secondary fixed points.

Values of Coefficients in the Vapor Pressure Equations for Helium

Coef.or constant	^{3}He 0.65—3.2 K	^{4}He 1.25—2.1768 K	^{4}He 2.1768—5.0 K
A_0	1.053 447	1.392 408	3.146 631
A_1	0.980 106	0.527 153	1.357 655
A_2	0.676 380	0.166 756	0.413 923
A_3	0.372 692	0.050 988	0.091 159
A_4	0.151 656	0.026 514	0.016 349
A_5	−0.002 263	0.001 975	0.001 826
A_6	0.006 596	−0.017 976	−0.004 325
A_7	0.088 966	0.005 409	−0.004 973
A_8	−0.004 770	0.013 259	0
A_9	−0.054 943	0	0
B	7.3	5.6	10.3
C	4.3	2.9	1.9

CONVERSION OF TEMPERATURES FROM THE 1948 AND 1968 SCALES TO ITS-90

This table gives temperature corrections from older scales to the current International Temperature Scale of 1990 (see the preceding table for details on ITS-90). The first part of the table may be used for converting Celsius temperatures in the range -180 to 4000°C from IPTS-68 or IPTS-48 to ITS-90. Within the accuracy of the corrections, the temperature in the first column may be identified with either t_{68}, t_{48}, or t_{90}. The second part of the table is designed for use at lower temperatures to convert values expressed in kelvins from EPT-76 or IPTS-68 to ITS-90.

The references give analytical equations for expressing these relations. Note that Reference 1 supersedes Reference 2 with respect to corrections in the 630 to 1064°C range.

REFERENCES

1. Burns, G. W. et al., in *Temperature: Its Measurement and Control in Science and Industry,* Vol. 6, Schooley, J. F., Ed., American Institute of Physics, New York, 1993.
2. Goldberg, R. N. and Weir, R. D., *Pure and Appl. Chem.*, 1545, 1992.

t/°C	$t_{90}-t_{68}$	$t_{90}-t_{48}$	t/°C	$t_{90}-t_{68}$	$t_{90}-t_{48}$	t/°C	$t_{90}-t_{68}$	$t_{90}-t_{48}$
-180	0.008	0.020	270	-0.039	0.028	720	0.00	0.45
-170	0.010	0.017	280	-0.039	0.030	730	0.02	0.49
-160	0.012	0.007	290	-0.039	0.032	740	0.03	0.53
-150	0.013	0.000	300	-0.039	0.034	750	0.03	0.56
-140	0.014	0.001	310	-0.039	0.035	760	0.04	0.60
-130	0.014	0.008	320	-0.039	0.036	770	0.05	0.63
-120	0.014	0.017	330	-0.040	0.036	780	0.05	0.66
-110	0.013	0.026	340	-0.040	0.037	790	0.05	0.69
-100	0.013	0.035	350	-0.041	0.036	800	0.05	0.72
-90	0.012	0.041	360	-0.042	0.035	810	0.05	0.75
-80	0.012	0.045	370	-0.043	0.034	820	0.04	0.76
-70	0.011	0.045	380	-0.045	0.032	830	0.04	0.79
-60	0.010	0.042	390	-0.046	0.030	840	0.03	0.81
-50	0.009	0.038	400	-0.048	0.028	850	0.02	0.83
-40	0.008	0.032	410	-0.051	0.024	860	0.01	0.85
-30	0.006	0.024	420	-0.053	0.022	870	0.00	0.87
-20	0.004	0.016	430	-0.056	0.019	880	-0.02	0.87
-10	0.002	0.008	440	-0.059	0.015	890	-0.03	0.89
0	0.000	0.000	450	-0.062	0.012	900	-0.05	0.90
10	-0.002	-0.006	460	-0.065	0.009	910	-0.06	0.92
20	-0.005	-0.012	470	-0.068	0.007	920	-0.08	0.93
30	-0.007	-0.016	480	-0.072	0.004	930	-0.10	0.94
40	-0.010	-0.020	490	-0.075	0.002	940	-0.11	0.96
50	-0.013	-0.023	500	-0.079	0.000	950	-0.13	0.97
60	-0.016	-0.026	510	-0.083	-0.001	960	-0.15	0.97
70	-0.018	-0.026	520	-0.087	-0.002	970	-0.16	0.99
80	-0.021	-0.027	530	-0.090	-0.001	980	-0.18	1.00
90	-0.024	-0.027	540	-0.094	0.000	990	-0.19	1.02
100	-0.026	-0.026	550	-0.098	0.002	1000	-0.20	1.04
110	-0.028	-0.024	560	-0.101	0.007	1010	-0.22	1.05
120	-0.030	-0.023	570	-0.105	0.011	1020	-0.23	1.07
130	-0.032	-0.020	580	-0.108	0.018	1030	-0.23	1.10
140	-0.034	-0.018	590	-0.112	0.025	1040	-0.24	1.12
150	-0.036	-0.016	600	-0.115	0.035	1050	-0.25	1.14
160	-0.037	-0.012	610	-0.118	0.047	1060	-0.25	1.17
170	-0.038	-0.009	620	-0.122	0.060	1070	-0.25	1.19
180	-0.039	-0.005	630	-0.125	0.075	1080	-0.26	1.20
190	-0.039	-0.001	640	-0.11	0.12	1090	-0.26	1.20
200	-0.040	0.003	650	-0.10	0.15	1100	-0.26	1.2
210	-0.040	0.007	660	-0.09	0.19	1200	-0.30	1.4
220	-0.040	0.011	670	-0.07	0.24	1300	-0.35	1.5
230	-0.040	0.014	680	-0.05	0.29	1400	-0.39	1.6
240	-0.040	0.018	690	-0.04	0.32	1500	-0.44	1.8
250	-0.040	0.021	700	-0.02	0.37	1600	-0.49	1.9
260	-0.040	0.024	710	-0.01	0.41	1700	-0.54	2.1

t/°C	t_{90}-t_{68}	t_{90}-t_{48}
1800	-0.60	2.2
1900	-0.66	2.3
2000	-0.72	2.5
2100	-0.79	2.7
2200	-0.85	2.9
2300	-0.93	3.1
2400	-1.00	3.2
2500	-1.07	3.4
2600	-1.15	3.7
2700	-1.24	3.8
2800	-1.32	4.0
2900	-1.41	4.2
3000	-1.50	4.4
3100	-1.59	4.6
3200	-1.69	4.8
3300	-1.78	5.1
3400	-1.89	5.3
3500	-1.99	5.5
3600	-2.10	5.8
3700	-2.21	6.0
3800	-2.32	6.3
3900	-2.43	6.6
4000	-2.55	6.8

T/K	T_{90}-T_{76}	T_{90}-T_{68}
5	-0.0001	
6	-0.0002	
7	-0.0003	
8	-0.0004	
9	-0.0005	
10	-0.0006	
11	-0.0007	
12	-0.0008	
13	-0.0010	
14	-0.0011	-0.006
15	-0.0013	-0.003
16	-0.0014	-0.004
17	-0.0016	-0.006
18	-0.0018	-0.008
19	-0.0020	-0.009
20	-0.0022	-0.009
21	-0.0025	-0.008
22	-0.0027	-0.007
23	-0.0030	-0.007
24	-0.0032	-0.006
25	-0.0035	-0.005
26	-0.0038	-0.004
27	-0.0041	-0.004

T/K	T_{90}-T_{76}	T_{90}-T_{68}
28		-0.005
29		-0.006
30		-0.006
31		-0.007
32		-0.008
33		-0.008
34		-0.008
35		-0.007
36		-0.007
37		-0.007
38		-0.006
39		-0.006
40		-0.006
41		-0.006
42		-0.006
43		-0.006
44		-0.006
45		-0.007
46		-0.007
47		-0.007
48		-0.006
49		-0.006
50		-0.006
51		-0.005
52		-0.005
53		-0.004
54		-0.003
55		-0.002
56		-0.001
57		0.000
58		0.001
59		0.002
60		0.003
61		0.003
62		0.004
63		0.004
64		0.005
65		0.005
66		0.006
67		0.006
68		0.007
69		0.007
70		0.007
71		0.007
72		0.007
73		0.007
74		0.007
75		0.008
76		0.008

T/K	T_{90}-T_{76}	T_{90}-T_{68}
77		0.008
78		0.008
79		0.008
80		0.008
81		0.008
82		0.008
83		0.008
84		0.008
85		0.008
86		0.008
87		0.008
88		0.008
89		0.008
90		0.008
91		0.008
92		0.008
93		0.008
94		0.008
95		0.008
96		0.008
97		0.009
98		0.009
99		0.009
100		0.009
110		0.011
120		0.013
130		0.014
140		0.014
150		0.014
160		0.014
170		0.013
180		0.012
190		0.012
200		0.011
210		0.010
220		0.009
230		0.008
240		0.007
250		0.005
260		0.003
270		0.001
273.16		0.000
300		-0.006
400		-0.031
500		-0.040
600		-0.040
700		-0.055
800		-0.089
900		-0.124

INTERNATIONAL SYSTEM OF UNITS (SI)

1 SI base units

Table 1 gives the seven base quantities, assumed to be mutually independent, on which the SI is founded; and the names and symbols of their respective units, called ``SI base units.'' Definitions of the SI base units are given in Appendix A. The kelvin and its symbol K are also used to express the value of a temperature interval or a temperature difference.

Table 1. SI base units

Base quantity	SI base unit: Name	SI base unit: Symbol
length	meter	m
mass	kilogram	kg
time	second	s
electric current	ampere	A
thermodynamic temperature	kelvin	K
amount of substance	mole	mol
luminous intensity	candela	cd

2 SI derived units

Derived units are expressed algebraically in terms of base units or other derived units (including the radian and steradian which are the two supplementary units – see Sec. 3). The symbols for derived units are obtained by means of the mathematical operations of multiplication and division. For example, the derived unit for the derived quantity molar mass (mass divided by amount of substance) is the kilogram per mole, symbol kg/mol. Additional examples of derived units expressed in terms of SI base units are given in Table 2.

Table 2. Examples of SI derived units expressed in terms of SI base units

Derived quantity	SI derived unit: Name	SI derived unit: Symbol
area	square meter	m^2
volume	cubic meter	m^3
speed, velocity	meter per second	m/s
acceleration	meter per second squared	m/s^2
wave number	reciprocal meter	m^{-1}
mass density (density)	kilogram per cubic meter	kg/m^3
specific volume	cubic meter per kilogram	m^3/kg
current density	ampere per square meter	A/m^2
magnetic field strength	ampere per meter	A/m
amount-of-substance concentration (concentration)	mole per cubic meter	mol/m^3
luminance	candela per square meter	cd/m^2

2.1 SI derived units with special names and symbols

Certain SI derived units have special names and symbols; these are given in Tables 3a and 3b. As discussed in Sec. 3, the radian and steradian, which are the two supplementary units, are included in Table 3a.

Table 3a. SI derived units with special names and symbols, including the radian and steradian

	SI derived unit			
Derived quantity	Special name	Special symbol	Expression in terms of other SI units	Expression in terms of SI base units
plane angle	radian	rad		$m \cdot m^{-1} = 1$
solid angle	steradian	sr		$m^2 \cdot m^{-2} = 1$
frequency	hertz	Hz		s^{-1}
force	newton	N		$m \cdot kg \cdot s^{-2}$
pressure, stress	pascal	Pa	N/m^2	$m^{-1} \cdot kg \cdot s^{-2}$
energy, work, quantity of heat	joule	J	$N \cdot m$	$m^2 \cdot kg \cdot s^{-2}$
power, radiant flux	watt	W	J/s	$m^2 \cdot kg \cdot s^{-3}$
electric charge, quantity of electricity	coulomb	C		$s \cdot A$
electric potential, potential difference, electromotive force	volt	V	W/A	$m^2 \cdot kg \cdot s^{-3} \cdot A^{-1}$
capacitance	farad	F	C/V	$m^{-2} \cdot kg^{-1} \cdot s^4 \cdot A^2$
electric resistance	ohm	Ω	V/A	$m^2 \cdot kg \cdot s^{-3} \cdot A^{-2}$
electric conductance	siemens	S	A/V	$m^{-2} \cdot kg^{-1} \cdot s^3 \cdot A^2$
magnetic flux	weber	Wb	$V \cdot s$	$m^2 \cdot kg \cdot s^{-2} \cdot A^{-1}$
magnetic flux density	tesla	T	Wb/m^2	$kg \cdot s^{-2} \cdot A^{-1}$
inductance	henry	H	Wb/A	$m^2 \cdot kg \cdot s^{-2} \cdot A^{-2}$
Celsius temperature[(a)]	degree Celsius	°C		K
luminous flux	lumen	lm	$cd \cdot sr$	$cd \cdot sr$[(b)]
illuminance	lux	lx	lm/m^2	$m^{-2} \cdot cd \cdot sr$[(b)]

[(a)] See Sec. 2.1.1.
[(b)] The steradian (sr) is not an SI base unit. However, in photometry the steradian (sr) is maintained in expressions for units (see Sec. 3).

Table 3b. SI derived units with special names and symbols admitted for reasons of safeguarding human health[(a)]

	SI derived unit			
Derived quantity	Special name	Special symbol	Expression in terms of other SI units	Expression in terms of SI base units
activity (of a radionuclide)	becquerel	Bq		s^{-1}
absorbed dose, specific energy (imparted), kerma	gray	Gy	J/kg	$m^2 \cdot s^{-2}$
dose equivalent, ambient dose equivalent, directional dose equivalent, personal dose equivalent, equivalent dose	sievert	Sv	J/kg	$m^2 \cdot s^{-2}$

[(a)] The derived quantities to be expressed in the gray and the sievert have been revised in accordance with the recommendations of the International Commission on Radiation Units and Measurements (ICRU).

2.1.1 Degree Celsius In addition to the quantity thermodynamic temperature (symbol T), expressed in the unit kelvin, use is also made of the quantity Celsius temperature (symbol t) defined by the equation

$$t = T - T_0 ,$$

where $T_0 = 273.15$ K by definition. To express Celsius temperature, the unit degree Celsius, symbol °C, which is equal in magnitude to the unit kelvin, is used; in this case, ``degree Celsius'' is a special name used in place of ``kelvin.'' An interval or difference of Celsius temperature can, however, be expressed in the unit kelvin as well as in the unit degree Celsius. (Note that the thermodynamic temperature T_0 is exactly 0.01 K below the thermodynamic temperature of the triple point of water.)

2.2 Use of SI derived units with special names and symbols

Examples of SI derived units that can be expressed with the aid of SI derived units having special names and symbols (including the radian and steradian) are given in Table 4.

Table 4. Examples of SI derived units expressed with the aid of SI derived units having special names and symbols

Derived quantity	SI derived unit: Name	Symbol	Expression in terms of SI base units
angular velocity	radian per second	rad/s	$m \cdot m^{-1} \cdot s^{-1} = s^{-1}$
angular acceleration	radian per second squared	rad/s^2	$m \cdot m^{-1} \cdot s^{-2} = s^{-2}$
dynamic viscosity	pascal second	Pa · s	$m^{-1} \cdot kg \cdot s^{-1}$
moment of force	newton meter	N · m	$m^2 \cdot kg \cdot s^{-2}$
surface tension	newton per meter	N/m	$kg \cdot s^{-2}$
heat flux density, irradiance	watt per square meter	W/m^2	$kg \cdot s^{-3}$
radiant intensity	watt per steradian	W/sr	$m^2 \cdot kg \cdot s^{-3} \cdot sr^{-1}$ [a]
radiance	watt per square meter steradian	$W/(m^2 \cdot sr)$	$kg \cdot s^{-3} \cdot sr^{-1}$ [a]
heat capacity, entropy	joule per kelvin	J/K	$m^2 \cdot kg \cdot s^{-2} \cdot K^{-1}$
specific heat capacity, specific entropy	joule per kilogram kelvin	J/(kg · K)	$m^2 \cdot s^{-2} \cdot K^{-1}$
specific energy	joule per kilogram	J/kg	$m^2 \cdot s^{-2}$
thermal conductivity	watt per meter kelvin	W/(m · K)	$m \cdot kg \cdot s^{-3} \cdot K^{-1}$
energy density	joule per cubic meter	J/m^3	$m^{-1} \cdot kg \cdot s^{-2}$
electric field strength	volt per meter	V/m	$m \cdot kg \cdot s^{-3} \cdot A^{-1}$
electric charge density	coulomb per cubic meter	C/m^3	$m^{-3} \cdot s \cdot A$
electric flux density	coulomb per square meter	C/m^2	$m^{-2} \cdot s \cdot A$
permittivity	farad per meter	F/m	$m^{-3} \cdot kg^{-1} \cdot s^4 \cdot A^2$
permeability	henry per meter	H/m	$m \cdot kg \cdot s^{-2} \cdot A^{-2}$
molar energy	joule per mole	J/mol	$m^2 \cdot kg \cdot s^{-2} \cdot mol^{-1}$
molar entropy, molar heat capacity	joule per mole kelvin	J/(mol · K)	$m^2 \cdot kg \cdot s^{-2} \cdot K^{-1} \cdot mol^{-1}$
exposure (x and γ rays)	coulomb per kilogram	C/kg	$kg^{-1} \cdot s \cdot A$
absorbed dose rate	gray per second	Gy/s	$m^2 \cdot s^{-3}$

[a] The steradian (sr) is not an SI base unit. However, in radiometry the steradian (sr) is maintained in expressions for units (see Sec. 3).

The advantages of using the special names and symbols of SI derived units are apparent in Table 4. Consider, for example, the quantity molar entropy: the unit J/(mol · K) is obviously more easily understood than its SI base-unit equivalent, $m^2 \cdot kg \cdot s^{-2} \cdot K^{-1} \cdot mol^{-1}$. Nevertheless, it should always be recognized that the special names and symbols exist for convenience; either the form in which special names or symbols are used for certain combinations of units or the form in which they are not used is correct. For example, because of the descriptive value implicit in the compound-unit form, communication is sometimes facilitated if magnetic flux (see Table 3a) is expressed in terms of the volt second (V · s) instead of the weber (Wb).

Tables 3a, 3b, and 4 also show that the values of several different quantities are expressed in the same SI unit. For example, the joule per kelvin (J/K) is the SI unit for heat capacity as well as for entropy. Thus the name of the unit is not sufficient to define the quantity measured.

A derived unit can often be expressed in several different ways through the use of base units and derived units with special names. In practice, with certain quantities, preference is given to using certain units with special names, or combinations of units, to facilitate the distinction between quantities whose values have identical expressions in terms of SI base units. For example, the SI unit of frequency is specified as the hertz (Hz) rather than the reciprocal second (s^{-1}), and the SI unit of moment of force is specified as the newton meter (N · m) rather than the joule (J).

Similarly, in the field of ionizing radiation, the SI unit of activity is designated as the becquerel (Bq) rather than the reciprocal second (s^{-1}), and the SI units of absorbed dose and dose equivalent are designated as the gray (Gy) and the sievert (Sv), respectively, rather than the joule per kilogram (J/kg).

3 SI supplementary units

As previously stated, there are two units in this class: the radian, symbol rad, the SI unit of the quantity plane angle; and the steradian, symbol sr, the SI unit of the quantity solid angle. Definitions of these units are given in Appendix A.

The SI supplementary units are now interpreted as so-called dimensionless derived units for which the CGPM allows the freedom of using or not using them in expressions for SI derived units.[3] Thus the radian and steradian are not given in a separate table but have been included in Table 3a together with other derived units with special names and symbols (seeSec.2.1). This interpretation of the supplementary units implies that plane angle and solid angle are considered derived quantities of dimension one (so-called dimensionless quantities), each of which has the which has the unit one, symbol 1, as its coherent SI unit. However, in practice, when one expresses the values of derived quantities involving plane angle or solid angle, it often aids understanding if the special names (or symbols) ``radian'' (rad) or ``steradian'' (sr) are used in place of the number 1. For example, although values of the derived quantity angular velocity (plane angle divided by time) may be expressed in the unit s^{-1}, such values are usually expressed in the unit rad/s.

Because the radian and steradian are now viewed as so-called dimensionless derived units, the Consultative Committee for Units (CCU, *Comité Consultatif des Unités*) of the CIPM as result of a 1993 request it received from ISO/TC12, recommended to the CIPM that it request the CGPM to abolish the class of supplementary units as a separate class in the SI. The CIPM accepted the CCU recommendation, and if the abolishment is approved by the CGPM as is likely (the question will be on the agenda of the 20th CGPM, October 1995), the SI will consist of only two classes of units: base units and derived units, with the radian and steradian subsumed into the class of derived units of the SI. (The option of using or not using them in expressions for SI derived units, as is convenient, would remain unchanged.)

4 Decimal multiples and submultiples of SI units: SI prefixes

Table 5 gives the SI prefixes that are used to form decimal multiples and submultiples of SI units. They allow very large or very small numerical values to be avoided. A prefix attaches directly to the name of a unit, and a prefix symbol attaches directly to the symbol for a unit. For example, one kilometer, symbol 1 km, is equal to one thousand meters, symbol 1000 m or 10^3 m. When prefixes are attached to SI units, the units so formed are called ``multiples and submultiples of SI units'' in order to distinguish them from the coherent system of SI units.

Note: Alternative definitions of the SI prefixes and their symbols are not permitted. For example, it is unacceptable to use kilo (k) to represent $2^{10} = 1024$, mega (M) to represent $2^{20} = 1\,048\,576$, or giga (G) to represent $2^{30} = 1\,073\,741\,824$.

[3] This interpretation was given in 1980 by the CIPM . It was deemed necessary because Resolution 12 of the 11th CGPM, which established the SI in 1960, did not specify the nature of the supplementary units. The interpretation is based on two principal considerations: that plane angle is generally expressed as the ratio of two lengths and solid angle as the ratio of an area and the square of a length, and are thus quantities of dimension one (so-called dimensionless quantities); and that treating the radian and steradian as SI base units – a possibility not disallowed by Resolution 12 – could compromise the internal coherence of the SI based on only seven base units. (See ISO 31-0 for a discussion of the concept of dimension.)

Table 5. SI prefixes

Factor	Prefix	Symbol	Factor	Prefix	Symbol
$10^{24} = (10^3)^8$	yotta	Y	10^{-1}	deci	d
$10^{21} = (10^3)^7$	zetta	Z	10^{-2}	centi	c
$10^{18} = (10^3)^6$	exa	E	$10^{-3} = (10^3)^{-1}$	milli	m
$10^{15} = (10^3)^5$	peta	P	$10^{-6} = (10^3)^{-2}$	micro	μ
$10^{12} = (10^3)^4$	tera	T	$10^{-9} = (10^3)^{-3}$	nano	n
$10^{9} = (10^3)^3$	giga	G	$10^{-12} = (10^3)^{-4}$	pico	p
$10^{6} = (10^3)^2$	mega	M	$10^{-15} = (10^3)^{-5}$	femto	f
$10^{3} = (10^3)^1$	kilo	k	$10^{-18} = (10^3)^{-6}$	atto	a
10^{2}	hecto	h	$10^{-21} = (10^3)^{-7}$	zepto	z
10^{1}	deka	da	$10^{-24} = (10^3)^{-8}$	yocto	y

5 Units Outside the SI

Units that are outside the SI may be divided into three categories:

- those units that are accepted for use with the SI;
- those units that are temporarily accepted for use with the SI; and
- those units that are not accepted for use with the SI and thus must strictly be avoided.

5.1 Units accepted for use with the SI

The following sections discuss in detail the units that are acceptable for use with the SI.

5.1.1 Hour, degree, liter, and the like

Certain units that are not part of the SI are essential and used so widely that they are accepted by the CIPM for use with the SI. These units are given in Table 6. The combination of units of this table with SI units to form derived units should be restricted to special cases in order not to lose the advantages of the coherence of SI units.

Additionally, it is recognized that it may be necessary on occasion to use time-related units other than those given in Table 6; in particular, circumstances may require that intervals of time be expressed in weeks, months, or years. In such cases, if a standardized symbol for the unit is not available, the name of the unit should be written out in full.

Table 6. Units accepted for use with the SI

Name		Symbol	Value in SI units
minute	time	min	1 min = 60 s
hour	time	h	1 h = 60 min = 3600 s
day	time	d	1 d = 24 h = 86 400 s
degree	plane angle	°	1° = (π/180) rad
minute	plane angle	'	1' = (1/60)° = (π/10 800) rad
second	plane angle	"	1" = (1/60)' = (π/648 000) rad
liter		l, L[(b)]	1 L = 1 dm³ = 10^{-3} m³
metric ton[(c)]		t	1 t = 10^3 kg

[(b)] The alternative symbol for the liter, L, was adopted by the CGPM in order to avoid the risk of confusion between the letter l and the number 1 . Thus, although both l and L are internationally accepted symbols for the liter, to avoid this risk the symbol to be used in the United States is L . The script letter ℓ is not an approved symbol for the liter.

[(c)] This is the name to be used for this unit in the United States; it is also used in some other English-speaking countries. However, ``tonne'' is used in many countries.

5.1.2 Neper, bel, shannon, and the like

There are a few highly specialized units not listed in Table 6 that are given by the International Organization for Standardization (ISO) or the International Electrotechnical Commission (IEC) and which are also acceptable for use with the SI. They include the neper (Np), bel (B), octave, phon, and sone, and units used in information technology, including the baud (Bd), bit (bit), erlang (E), hartley (Hart), and shannon (Sh).[4] It is the position of NIST that the only such additional units that may be used with the SI are those given in either the International Standards on quantities and units of ISO or of IEC .

5.1.3 Electronvolt and unified atomic mass unit

The CIPM also finds it necessary to accept for use with the SI the two units given in Table 7. These units are used in specialized fields; their values in SI units must be obtained from experiment and, therefore, are not known exactly.

Note: In some fields the unified atomic mass unit is called the dalton, symbol Da; however, this name and symbol are not accepted by the CGPM, CIPM, ISO, or IEC for use with the SI. Similarly, AMU is not an acceptable unit symbol for the unified atomic mass unit. The only allowed name is ``unified atomic mass unit'' and the only allowed symbol is u.

Table 7. Units accepted for use with the SI whose values in SI units are obtained experimentally

Name	Symbol	Definition
electronvolt	eV	(*a*)
unified atomic mass unit	u	(*b*)

(*a*) The electronvolt is the kinetic energy acquired by an electron in passing through a potential difference of 1 V in vacuum; 1 eV = $1.602\,177\,33 \times 10^{-19}$ J with a combined standard uncertainty of $0.000\,000\,49 \times 10^{-19}$ J .

(*b*) The unified atomic mass unit is equal to 1/12 of the mass of an atom of the nuclide ^{12}C; 1 u = $1.660\,540\,2 \times 10^{-27}$ kg with a combined standard uncertainty of $0.000\,001\,0 \times 10^{-27}$ kg .

5.1.4 Natural and atomic units

In some cases, particularly in basic science, the values of quantities are expressed in terms of fundamental constants of nature or so-called natural units.The use of these units with the SI is permissible when it is necessary for the most effective communication of information. In such cases, the specific natural units that are used must be identified. This requirement applies even to the system of units customarily called ``atomicunits'' used in theoretical atomic physics and chemistry, inasmuch as there are several different systems that have the appellation ``atomic units." Examples of physical quantities used as natural units are given in Table 8.

NIST also takes the position that while theoretical results intended primarily for other theorists may be left in natural units, if they are also intended for experimentalists, they must also be given in acceptable units.

[4] The symbol in parentheses following the name of the unit is its internationally accepted unit symbol, but the octave, phon, and sone have no such unit symbols. For additional information on the neper and bel, see Sec. 0.5 of ISO 31-2. The question of the byte (B) is under international consideration.

Table 8. Examples of physical quantities sometimes used as natural units

Kind of quantity	Physical quantity used as a unit	Symbol
action	Planck constant divided by 2π	$\hbar$
electric charge	elementary charge	e
energy	Hartree energy	E_h
length	Bohr radius	a_0
length	Compton wavelength (electron)	λ_C
magnetic flux	magnetic flux quantum	Φ_0
magnetic moment	Bohr magneton	μ_B
magnetic moment	nuclear magneton	μ_N
mass	electron rest mass	m_e
mass	proton rest mass	m_p
speed	speed of electromagnetic waves in vacuum	c

5.2 Units temporarily accepted for use with the SI

Because of existing practice in certain fields or countries, in 1978 the CIPM considered that it was permissible for the units given in Table 9 to continue to be used with the SI until the CIPM considers that their use is no longer necessary. However, these units must not be introduced where they are not presently used. Further, NIST strongly discourages the continued use of these units except for the nautical mile, knot, are, and hectare; and except for the curie, roentgen, rad, and rem until the year 2000 (the cessation date suggested by the Committee for Ineragency Radiation Research and Policy Coordination or CIRRPC, a United States Government interagency group).[5]

Table 9. Units temporarily accepted for use with the SI[(a)]

Name	Symbol	Value in SI units
nautical mile		1 nautical mile = 1852 m
knot		1 nautical mile per hour = (1852/3600) m/s
ångström	Å	1 Å = 0.1 nm = 10^{-10}m
are[(b)]	a	1 a = 1 dam^2 = 10^2 m^2
hectare[(b)]	ha	1 ha = 1 hm^2 = 10^4 m^2
barn	b	1 b = 100 fm^2 = 10^{-28} m^2
bar	bar	1 bar=0.1 MPa=100 kPa=1000 hPa=10^5 Pa
gal	Gal	1 Gal = 1 cm/s^2 = 10^{-2} m/s^2
curie	Ci	1 Ci = 3.7×10^{10} Bq
roentgen	R	1 R = 2.58×10^{-4} C/kg
rad	rad[(c)]	1 rad = 1 cGy = 10^{-2} Gy
rem	rem	1 rem = 1 cSv = 10^{-2} Sv

[(a)] See Sec. 5.2 regarding the continued use of these units.
[(b)] This unit and its symbol are used to express agrarian areas.
[(c)] When there is risk of confusion with the symbol for the radian, rd may be used as the symbol for rad.

[5] In 1993 the CCU (see Sec. 3) was requested by ISO/TC 12 to consider asking the CIPM to deprecate the use of the units of Table 9 except for the nautical mile and knot, and possibly the are and hectare. The CCU discussed this request at its February 1995 meeting.

Appendix A. Definitions of the SI Base Units and the Radian and Steradian

A.1 Introduction

The following definitions of the SI base units are taken from NIST SP 330; the definitions of the SI supplementary units, the radian and steradian, which are now interpreted as SI derived units (see Sec. 3), are those generally accepted and are the same as those given in ANSI/IEEE Std 268-1992. SI derived units are uniquely defined only in terms of SI base units; for example, $1\ \text{V} = 1\ \text{m}^2 \cdot \text{kg} \cdot \text{s}^{-3} \cdot \text{A}^{-1}$.

A.2 Meter (17th CGPM, 1983)

The meter is the length of the path travelled by light in vacuum during a time interval of 1/299 792 458 *of a second.*

A.3 Kilogram (3d CGPM, 1901)

The kilogram is the unit of mass; it is equal to the mass of the international prototype of the kilogram.

A.4 Second (13th CGPM, 1967)

The second is the duration of 9 192 631 770 *periods of the radiation corresponding to the transition between the two hyperfine levels of the ground state of the cesium-* 133 *atom.*

A.5 Ampere (9th CGPM, 1948)

The ampere is that constant current which, if maintained in two straight parallel conductors of infinite length, of negligible circular cross section, and placed 1 *meter apart in vacuum, would produce between these conductors a force equal to* 2×10^{-7} *newton per meter of length.*

A.6 Kelvin (13th CGPM, 1967)

The kelvin, unit of thermodynamic temperature, is the fraction 1/273.16 *of the thermodynamic temperature of the triple point of water.*

A.7 Mole (14th CGPM, 1971)

1. *The mole is the amount of substance of a system which contains as many elementary entities as there are atoms in* 0.012 *kilogram of carbon* 12.

2. *When the mole is used, the elementary entities must be specified and may be atoms, molecules, ions, electrons, other particles, or specified groups of such particles.*

In the definition of the mole, it is understood that unbound atoms of carbon 12, at rest and in their ground state, are referred to.

Note that this definition specifies at the same time the nature of the quantity whose unit is the mole.

A.8 Candela (16th CGPM, 1979)

The candela is the luminous intensity, in a given direction, of a source that emits monochromatic radiation of frequency 540×10^{12} *hertz and that has a radiant intensity in that direction of* (1/683) *watt per steradian.*

A.9 Radian

The radian is the plane angle between two radii of a circle that cut off on the circumference an arc equal in length to the radius.

A.10 Steradian

The steradian is the solid angle that, having its vertex in the center of a sphere, cuts off an area of the surface of the sphere equal to that of a square with sides of length equal to the radius of the sphere.

UNITS FOR MAGNETIC PROPERTIES

Quantity	Symbol	Gaussian & cgs emu [a]	Conversion factor, C [b]	SI & rationalized mks [c]
Magnetic flux density, magnetic induction	B	gauss (G) [d]	10^{-4}	tesla (T), Wb/m^2
Magnetic flux	Φ	maxwell (Mx), $G\cdot cm^2$	10^{-8}	weber (Wb), volt second (V·s)
Magnetic potential difference, magnetomotive force	U, F	gilbert (Gb)	$10/4\pi$	ampere (A)
Magnetic field strength, magnetizing force	H	oersted (Oe),[e] Gb/cm	$10^3/4\pi$	A/m [f]
(Volume) magnetization [g]	M	emu/cm^3 [h]	10^3	A/m
(Volume) magnetization	$4\pi M$	G	$10^3/4\pi$	A/m
Magnetic polarization, intensity of magnetization	J, I	emu/cm^3	$4\pi \times 10^{-4}$	T, Wb/m^2 [i]
(Mass) magnetization	σ, M	emu/g	1 $4\pi \times 10^{-7}$	$A\cdot m^2/kg$ Wb·m/kg
Magnetic moment	m	emu, erg/G	10^{-3}	$A\cdot m^2$, joule per tesla (J/T)
Magnetic dipole moment	j	emu, erg/G	$4\pi \times 10^{-10}$	Wb·m [i]
(Volume) susceptibility	χ, κ	dimensionless, emu/cm^3	4π $(4\pi)^2 \times 10^{-7}$	dimensionless henry per meter (H/m), Wb/(A·m)
(Mass) susceptibility	χ_ρ, κ_ρ	cm^3/g, emu/g	$4\pi \times 10^{-3}$ $(4\pi)^2 \times 10^{-10}$	m^3/kg $H\cdot m^2/kg$
(Molar) susceptibility	χ_{mol}, κ_{mol}	cm^3/mol, emu/mol	$4\pi \times 10^{-6}$ $(4\pi)^2 \times 10^{-13}$	m^3/mol $H\cdot m^2/mol$
Permeability	μ	dimensionless	$4\pi \times 10^{-7}$	H/m, Wb/(A·m)
Relative permeability [j]	μ_r	not defined		dimensionless
(Volume) energy density, energy product [k]	W	erg/cm^3	10^{-1}	J/m^3
Demagnetization factor	D, N	dimensionless	$1/4\pi$	dimensionless

a. Gaussian units and cgs emu are the same for magnetic properties. The defining relation is $B=H+4\pi M$.
b. Multiply a number in Gaussian units by C to convert it to SI (e.g., $1\ G \times 10^{-4}\ T/G = 10^{-4}\ T$).
c. SI (*Système International d'Unités*) has been adopted by the National Bureau of Standards. Where two conversion factors are given, the upper one is recognized under, or consistent with, SI and is based on the definition $B=\mu_0(H+M)$, where $\mu_0 = 4\pi \times 10^{-7}$ H/m. The lower one is not recognized under SI and is based on the definition $B=\mu_0 H+J$, where the symbol I is often used in place of J.
d. 1 gauss $= 10^5$ gamma (γ).
e. Both oersted and gauss are expressed as $cm^{-1/2}\cdot g^{1/2}\cdot s^{-1}$ in terms of base units.
f. A/m was often expressed as "ampere turn per meter" when used for magnetic field strength.
g. Magnetic moment per unit volume.
h. The designation "emu" is not a unit.
i. Recognized under SI, even though based on the definition $B=\mu_0 H+J$. See footnote *c*.
j. $\mu_r = \mu/\mu_0 = 1+\chi$, all in SI. μ_r is equal to Gaussian μ.
k. $B\cdot H$ and $\mu_0 M\cdot H$ have SI units J/m^3; $M\cdot H$ and $B\cdot H/4\pi$ have Gaussian units erg/cm^3.

R. B. Goldfarb and F. R. Fickett, U.S. Department of Commerce, National Bureau of Standards, Boulder, Colorado 80303, March 1985
NBS Special Publication 696 For sale by the Superintendent of Documents, U.S. Government Printing Office, Washington, DC 20402

CONVERSION FACTORS

The following table gives conversion factors from various units of measure to SI units. It is reproduced from NIST Special Publication 811, *Guide for the Use of the International System of Units (SI)*. The table gives the factor by which a quantity expressed in a non-SI unit should be multiplied in order to calculate its value in the SI. The SI values are expressed in terms of the base, supplementary, and derived units of SI in order to provide a coherent presentation of the conversion factors and facilitate computations (see the table "International System of Units" in this Section). If desired, powers of ten can be avoided by using SI Prefixes and shifting the decimal point if necessary.

Conversion from a non-SI unit to a different non-SI unit may be carried out by using this table in two stages, e.g.,

$$1 \text{ cal}_{th} = 4.184 \text{ J}$$

$$1 \text{ Btu}_{IT} = 1.055056 \text{ E+03 J}$$

Thus,

$$1 \text{ Btu}_{IT} = (1.055056 \text{ E+03} \div 4.184) \text{ cal}_{th} = 252.164 \text{ cal}_{th}$$

Conversion factors are presented for ready adaptation to computer readout and electronic data transmission. The factors are written as a number equal to or greater than one and less than ten with six or fewer decimal places. This number is followed by the letter E (for exponent), a plus or a minus sign, and two digits which indicate the power of 10 by which the number must be multiplied to obtain the correct value. For example:

$$3.523\,907 \text{ E-02 is } 3.523\,907 \times 10^{-2}$$

or

$$0.035\,239\,07$$

Similarly:

$$3.386\,389 \text{ E+03 is } 3.386\,389 \times 10^{3}$$

or

$$3\,386.389$$

A factor in boldface is exact; i.e., all subsequent digits are zero. All other conversion factors have been rounded to the figures given in accordance with accepted practice. Where less than six digits after the decimal point are shown, more precision is not warranted.

It is often desirable to round a number obtained from a conversion of units in order to retain information on the precision of the value. The following rounding rules may be followed:

(1) If the digits to be discarded begin with a digit less than 5, the digit preceding the first discarded digit is not changed.

Example: 6.974 951 5 rounded to 3 digits is 6.97

(2) If the digits to be discarded begin with a digit greater than 5, the digit preceding the first discarded digit is increased by one.

Example: 6.974 951 5 rounded to 4 digits is 6.975

(3) If the digits to be discarded begin with a 5 and at least one of the following digits is greater than 0, the digit preceding the 5 is increased by 1.

Example: 6.974 851 rounded to 5 digits is 6.974 9

(4) If the digits to be discarded begin with a 5 and all of the following digits are 0, the digit preceding the 5 is unchanged if it is even and increased by one if it is odd. (Note that this means that the final digit is always even.)

Examples: 6.974 951 5 rounded to 7 digits is 6.974 952
6.974 950 5 rounded to 7 digits is 6.974 950

REFERENCE

Taylor, B. N., *Guide for the Use of the International System of Units (SI)*, NIST Special Publication 811, 1995 Edition, Superintendent of Documents, U.S. Government Printing Office, Washington, DC 20402, 1995.

Factors in **boldface** are exact

To convert from	to	Multiply by	
abampere	ampere (A)	**1.0**	**E+01**
abcoulomb	coulomb (C)	**1.0**	**E+01**
abfarad	farad (F)	**1.0**	**E+09**
abhenry	henry (H)	**1.0**	**E−09**
abmho	siemens (S)	**1.0**	**E+09**
abohm	ohm (Ω)	**1.0**	**E−09**
abvolt	volt (V)	**1.0**	**E−08**
acceleration of free fall, standard (g_n)	meter per second squared (m/s^2)	**9.806 65**	**E+00**
acre (based on U.S. survey foot)[9]	square meter (m^2)	4.046 873	E+03
acre foot (based on U.S. survey foot)[9]	cubic meter (m^3)	1.233 489	E+03
ampere hour (A · h)	coulomb (C)	**3.6**	**E+03**
ångström (Å)	meter (m)	**1.0**	**E−10**
ångström (Å)	nanometer (nm)	**1.0**	**E−01**
are (a)	square meter (m^2)	**1.0**	**E+02**
astronomical unit (AU)	meter (m)	1.495 979	E+11
atmosphere, standard (atm)	pascal (Pa)	**1.013 25**	**E+05**
atmosphere, standard (atm)	kilopascal (kPa)	**1.013 25**	**E+02**
atmosphere, technical (at)[10]	pascal (Pa)	**9.806 65**	**E+04**
atmosphere, technical (at)[10]	kilopascal (kPa)	**9.806 65**	**E+01**
bar (bar)	pascal (Pa)	**1.0**	**E+05**
bar (bar)	kilopascal (kPa)	**1.0**	**E+02**
barn (b)	square meter (m^2)	**1.0**	**E−28**
barrel [for petroleum, 42 gallons (U.S.)](bbl)	cubic meter (m^3)	1.589 873	E−01
barrel [for petroleum, 42 gallons (U.S.)](bbl)	liter (L)	1.589 873	E+02
biot (Bi)	ampere (A)	**1.0**	**E+01**
British thermal unit$_{IT}$ (Btu$_{IT}$)[11]	joule (J)	1.055 056	E+03
British thermal unit$_{th}$ (Btu$_{th}$)[11]	joule (J)	1.054 350	E+03
British thermal unit (mean) (Btu)	joule (J)	1.055 87	E+03
British thermal unit (39 °F) (Btu)	joule (J)	1.059 67	E+03
British thermal unit (59 °F) (Btu)	joule (J)	1.054 80	E+03
British thermal unit (60 °F) (Btu)	joule (J)	1.054 68	E+03
British thermal unit$_{IT}$ foot per hour square foot degree Fahrenheit [Btu$_{IT}$ · ft/(h · ft^2 · °F)]	watt per meter kelvin [W/(m · K)]	1.730 735	E+00
British thermal unit$_{th}$ foot per hour square foot degree Fahrenheit [Btu$_{th}$ · ft/(h · ft^2 · °F)]	watt per meter kelvin [W/(m · K)]	1.729 577	E+00
British thermal unit$_{IT}$ inch per hour square foot degree Fahrenheit [Btu$_{IT}$ · in/(h · ft^2 · °F)]	watt per meter kelvin [W/(m · K)]	1.442 279	E−01
British thermal unit$_{th}$ inch per hour square foot degree Fahrenheit [Btu$_{th}$ · in/(h · ft^2 · °F)]	watt per meter kelvin [W/(m · K)]	1.441 314	E−01
British thermal unit$_{IT}$ inch per second square foot degree Fahrenheit [Btu$_{IT}$ · in/(s · ft^2 · °F)]	watt per meter kelvin [W/(m · K)]	5.192 204	E+02

[9] The U.S. survey foot equals (1200/3937) m. 1 international foot = 0.999998 survey foot.

[10] One technical atmosphere equals one kilogram-force per square centimeter (1 at = 1 kgf/cm^2).

[11] The Fifth International Conference on the Properties of Steam (London, July 1956) defined the International Table calorie as 4.1868 J. Therefore the exact conversion factor for the International Table Btu is 1.055 055 852 62 kJ. Note that the notation for International Table used in this listing is subscript "IT". Similarily, the notation for thermochemical is subscript "th." Further, the thermochemical Btu, Btu$_{th}$, is based on the thermochemical calorie, cal$_{th}$, where cal$_{th}$ = 4.184 J exactly.

To convert from	to	Multiply by	
British thermal unit$_{th}$ inch per second square foot degree Fahrenheit [$Btu_{th} \cdot in/(s \cdot ft^2 \cdot °F)$]	watt per meter kelvin [$W/(m \cdot K)$]	5.188 732	E+02
British thermal unit$_{IT}$ per cubic foot (Btu_{IT}/ft^3)	joule per cubic meter (J/m^3)	3.725 895	E+04
British thermal unit$_{th}$ per cubic foot (Btu_{th}/ft^3)	joule per cubic meter (J/m^3)	3.723 403	E+04
British thermal unit$_{IT}$ per degree Fahrenheit ($Btu_{IT}/°F$)	joule per kelvin (J/k)	1.899 101	E+03
British thermal unit$_{th}$ per degree Fahrenheit ($Btu_{th}/°F$)	joule per kelvin (J/k)	1.897 830	E+03
British thermal unit$_{IT}$ per degree Rankine ($Btu_{IT}/°R$)	joule per kelvin (J/k)	1.899 101	E+03
British thermal unit$_{th}$ per degree Rankine ($Btu_{th}/°R$)	joule per kelvin (J/k)	1.897 830	E+03
British thermal unit$_{IT}$ per hour (Btu_{IT}/h)	watt (W)	2.930 711	E−01
British thermal unit$_{th}$ per hour (Btu_{th}/h)	watt (W)	2.928 751	E−01
British thermal unit$_{IT}$ per hour square foot degree Fahrenheit [$Btu_{IT}/(h \cdot ft^2 \cdot °F)$]	watt per square meter kelvin [$W/(m^2 \cdot K)$]	5.678 263	E+00
British thermal unit$_{th}$ per hour square foot degree Fahrenheit [$Btu_{th}/(h \cdot ft^2 \cdot °F)$]	watt per square meter kelvin [$W/(m^2 \cdot K)$]	5.674 466	E+00
British thermal unit$_{th}$ per minute (Btu_{th}/min)	watt (W)	1.757 250	E+01
British thermal unit$_{IT}$ per pound (Btu_{IT}/lb)	joule per kilogram (J/kg)	**2.326**	**E+03**
British thermal unit$_{th}$ per pound (Btu_{th}/lb)	joule per kilogram (J/kg)	2.324 444	E+03
British thermal unit$_{IT}$ per pound degree Fahrenheit [$Btu_{IT}/(lb \cdot °F)$]	joule per kilogram kelvin ($J/(kg \cdot K)$]	**4.1868**	**E+03**
British thermal unit$_{th}$ per pound degree Fahrenheit [$Btu_{th}/(lb \cdot °F)$]	joule per kilogram kelvin [$J/(kg \cdot K)$]	**4.184**	**E+03**
British thermal unit$_{IT}$ per pound degree Rankine [$Btu_{IT}/(lb \cdot °R)$]	joule per kilogram kelvin [$J/(kg \cdot K)$]	**4.1868**	**E+03**
British thermal unit$_{th}$ per pound degree Rankine [$Btu_{th}/(lb \cdot °R)$]	joule per kilogram kelvin [$J/(kg \cdot K)$]	**4.184**	**E+03**
British thermal unit$_{IT}$ per second (Btu_{IT}/s)	watt (W)	1.055 056	E+03
British thermal unit$_{th}$ per second (Btu_{th}/s)	watt (W)	1.054 350	E+03
British thermal unit$_{IT}$ per second square foot degree Fahrenheit [$Btu_{IT}/(s \cdot ft^2 \cdot °F)$]	watt per square meter kelvin [$W/(m^2 \cdot K)$]	2.044 175	E+04
British thermal unit$_{th}$ per second square foot degree Fahrenheit [$Btu_{th}/(s \cdot ft^2 \cdot °F)$]	watt per square meter kelvin [$W/(m^2 \cdot K)$]	2.042 808	E+04
British thermal unit$_{IT}$ per square foot (Btu_{IT}/ft^2)	joule per square meter (J/m^2)	1.135 653	E+04
British thermal unit$_{th}$ per square foot (Btu_{th}/ft^2)	joule per square meter (J/m^2)	1.134 893	E+04
British thermal unit$_{IT}$ per square foot hour [($Btu_{IT}/(ft^2 \cdot h)$]	watt per square meter (W/m^2)	3.154 591	E+00
British thermal unit$_{th}$ per square foot hour [$Btu_{th}/(ft^2 \cdot h)$]	watt per square meter (W/m^2)	3.152 481	E+00
British thermal unit$_{th}$ per square foot minute [$Btu_{th}/(ft^2 \cdot min)$]	watt per square meter (W/m^2)	1.891 489	E+02
British thermal unit$_{IT}$ per square foot second [($Btu_{IT}/(ft^2 \cdot s)$]	watt per square meter (W/m^2)	1.135 653	E+04
British thermal unit$_{th}$ per square foot second [$Btu_{th}/(ft^2 \cdot s)$]	watt per square meter (W/m^2)	1.134 893	E+04
British thermal unit$_{th}$ per square inch second [$Btu_{th}/(in^2 \cdot s)$]	watt per square meter (W/m^2)	1.634 246	E+06

To convert from	to	Multiply by	
bushel (U.S.) (bu)	cubic meter (m^3)	3.523 907	E−02
bushel (U.S.) (bu)	liter (L)	3.523 907	E+01
calorie$_{IT}$ (cal_{IT})[11]	joule (J)	**4.1868**	**E+00**
calorie$_{th}$ (cal_{th})[11]	joule (J)	**4.184**	**E+00**
calorie (cal) (mean)	joule (J)	4.190 02	E+00
calorie (15 °C) (cal_{15})	joule (J)	4.185 80	E+00
calorie (20 °C) (cal_{20})	joule (J)	4.181 90	E+00
calorie$_{IT}$, kilogram (nutrition)[12]	joule (J)	**4.1868**	**E+03**
calorie$_{th}$, kilogram (nutrition)[12]	joule (J)	**4.184**	**E+03**
calorie (mean), kilogram (nutrition)[12]	joule (J)	4.190 02	E+03
calorie$_{th}$ per centimeter second degree Celsius [$cal_{th}/(cm \cdot s \cdot °C)$]	watt per meter kelvin [$W/(m \cdot K)$]	**4.184**	**E+02**
calorie$_{IT}$ per gram (cal_{IT}/g)	joule per kilogram (J/kg)	**4.1868**	**E+03**
calorie$_{th}$ per gram (cal_{th}/g)	joule per kilogram (J/kg)	**4.184**	**E+03**
calorie$_{IT}$ per gram degree Celsius [$cal_{IT}/(g \cdot °C)$]	joule per kilogram kelvin [$J/(kg \cdot K)$]	**4.1868**	**E+03**
calorie$_{th}$ per gram degree Celsius [$cal_{th}/(g \cdot °C)$]	joule per kilogram kelvin [$J/(kg \cdot K)$]	**4.184**	**E+03**
calorie$_{IT}$ per gram kelvin [$cal_{IT}/(g \cdot K)$]	joule per kilogram kelvin [$J/(kg \cdot K)$]	**4.1868**	**E+03**
calorie$_{th}$ per gram kelvin [$cal_{th}/(g \cdot K)$]	joule per kilogram kelvin [$J/(kg \cdot K)$]	**4.184**	**E+03**
calorie$_{th}$ per minute (cal_{th}/min)	watt (W)	6.973 333	E−02
calorie$_{th}$ per second (cal_{th}/s)	watt (W)	**4.184**	**E+00**
calorie$_{th}$ per square centimeter (cal_{th}/cm^2)	joule per square meter (J/m^2)	**4.184**	**E+04**
calorie$_{th}$ per square centimeter minute [$cal_{th}/(cm^2 \cdot min)$]	watt per square meter (W/m^2)	6.973 333	E+02
calorie$_{th}$ per square centimeter second [$cal_{th}/(cm^2 \cdot s)$]	watt per square meter (W/m^2)	**4.184**	**E+04**
candela per square inch (cd/in^2)	candela per square meter (cd/m^2)	1.550 003	E+03
carat, metric	kilogram (kg)	**2.0**	**E−04**
carat, metric	gram (g)	**2.0**	**E−01**
centimeter of mercury (0 °C)[13]	pascal (Pa)	1.333 22	E+03
centimeter of mercury (0 °C)[13]	kilopascal (kPa)	1.333 22	E+00
centimeter of mercury, conventional (cmHg)[13]	pascal (Pa)	1.333 224	E+03
centimeter of mercury, conventional (cmHg)[13]	kilopascal (kPa)	1.333 224	E+00
centimeter of water (4 °C)[13]	pascal (Pa)	9.806 38	E+01
centimeter of water, conventional (cmH_2O)[13]	pascal (Pa)	**9.806 65**	**E+01**
centipoise (cP)	pascal second ($Pa \cdot s$)	**1.0**	**E−03**
centistokes (cSt)	meter squared per second (m^2/s)	**1.0**	**E−06**
chain (based on U.S. survey foot) (ch)[9]	meter (m)	2.011 684	E+01
circular mil	square meter (m^2)	5.067 075	E−10
circular mil	square millimeter (mm^2)	5.067 075	E−04
clo	square meter kelvin per watt ($m^2 \cdot K/W$)	1.55	E−01
cord (128 ft^3)	cubic meter (m^3)	3.624 556	E+00
cubic foot (ft^3)	cubic meter (m^3)	2.831 685	E−02
cubic foot per minute (ft^3/min)	cubic meter per second (m^3/s)	4.719 474	E−04
cubic foot per minute (ft^3/min)	liter per second (L/s)	4.719 474	E−01
cubic foot per second (ft^3/s)	cubic meter per second (m^3/s)	2.831 685	E−02

[12] The kilogram calorie or "large calorie" is an obsolete term used for the kilocalorie, which is the calorie used to express the energy content of foods. However, in practice, the prefix "kilo" is usually omitted.

[13] Conversion factors for mercury manometer pressure units are calculated using the standard value for the acceleration of gravity and the density of mercury at the stated temperature. Additional digits are not justified because the definitions of the units do not take into account the compressibility of mercury or the change in density caused by the revised practical temperature scale, ITS-90. Similar comments also apply to water manometer pressure units. Conversion factors for conventional mercury and water manometer pressure units are based on ISO 31-3.

To convert from	to	Multiply by	
cubic inch (in^3)[14]	cubic meter (m^3)	1.638 706	E−05
cubic inch per minute (in^3/min)	cubic meter per second (m^3/s)	2.731 177	E−07
cubic mile (mi^3)	cubic meter (m^3)	4.168 182	E+09
cubic yard (yd^3)	cubic meter (m^3)	7.645 549	E−01
cubic yard per minute (yd^3/min)	cubic meter per second (m^3/s)	1.274 258	E−02
cup (U.S.)	cubic meter (m^3)	2.365 882	E−04
cup (U.S.)	liter (L)	2.365 882	E−01
cup (U.S.)	milliliter (mL)	2.365 882	E+02
curie (Ci)	becquerel (Bq)	**3.7**	**E+10**
darcy[15]	meter squared (m^2)	9.869 233	E−13
day (d)	second (s)	**8.64**	**E+04**
day (sidereal)	second (s)	8.616 409	E+04
debye (D)	coulomb meter (C · m)	3.335 641	E−30
degree (angle) (°)	radian (rad)	1.745 329	E−02
degree Celsius (temperature) (°C)	kelvin (K)	T/K = t/°C+**273.15**	
degree Celsius (temperature interval) (°C)	kelvin (K)	**1.0**	**E+00**
degree centigrade (temperature)[16]	degree Celsius (°C)	t/°C ≈ t/deg. cent.	
degree centigrade (temperature interval)[16]	degree Celsius (°C)	1.0	E+00
degree Fahrenheit (temperature) (°F)	degree Celsius (°C)	**t/°C = (t/°F − 32)/1.8**	
degree Fahrenheit (temperature) (°F)	kelvin (K)	**T/K = (t/°F + 459.67)/1.8**	
degree Fahrenheit (temperature interval)(°F)	degree Celsius (°C)	5.555 556	E−01
degree Fahrenheit (temperature interval)(°F)	kelvin (K)	5.555 556	E−01
degree Fahrenheit hour per British thermal unit$_{IT}$ (°F · h/Btu_{IT})	kelvin per watt (K/W)	1.895 634	E+00
degree Fahrenheit hour per British thermal unit$_{th}$ (°F · h/Btu_{th})	kelvin per watt (K/W)	1.896 903	E+00
degree Fahrenheit hour square foot per British thermal unit$_{IT}$ (°F · h · ft^2/Btu_{IT})	square meter kelvin per watt (m^2 · K/W)	1.761 102	E−01
degree Fahrenheit hour square foot per British thermal unit$_{th}$ (°F · h · ft^2/Btu_{th})	square meter kelvin per watt (m^2 · K/W)	1.762 280	E−01
degree Fahrenheit hour square foot per British thermal unit$_{IT}$ inch [°F · h · ft^2/(Btu_{IT} · in)]	meter kelvin per watt (m · K/W)	6.933 472	E+00
degree Fahrenheit hour square foot per British thermal unit$_{th}$ inch [°F · h · ft^2/(Btu_{th} · in)]	meter kelvin per watt (m · K/W)	6.938 112	E+00
degree Fahrenheit second per British thermal unit$_{IT}$ (°F · s/Btu_{IT})	kelvin per watt (K/W)	5.265 651	E−04
degree Fahrenheit second per British thermal unit$_{th}$ (°F · s/Btu_{th})	kelvin per watt (K/W)	5.269 175	E−04
degree Rankine (°R)	kelvin (K)	T/K = (T/°R)/1.8	
degree Rankine (temperature interval) (°R)	kelvin (K)	5.555 556	E−01
denier	kilogram per meter (kg/m)	1.111 111	E−07
denier	gram per meter (g/m)	1.111 111	E−04
dyne (dyn)	newton (N)	**1.0**	**E−05**
dyne centimeter (dyn · cm)	newton meter (N · m)	**1.0**	**E−07**
dyne per square centimeter (dyn/cm^2)	pascal (Pa)	**1.0**	**E−01**
electronvolt (eV)	joule (J)	1.602 177	E−19
EMU of capacitance (abfarad)	farad (F)	**1.0**	**E+09**
EMU of current (abampere)	ampere (A)	**1.0**	**E+01**
EMU of electric potential (abvolt)	volt (V)	**1.0**	**E−08**
EMU of inductance (abhenry)	henry (H)	**1.0**	**E−09**

[14] The exact conversion factor is 1.638 706 4 E−05.

[15] The darcy is a unit for expressing the permeability of porous solids, not area.

[16] The centigrade temperature scale is obsolete; the degree centigrade is only approximately equal to the degree Celsius.

To convert from	to	Multiply by	
EMU of resistance (abohm)	ohm (Ω)	**1.0**	**E−09**
erg (erg)	joule (J)	**1.0**	**E−07**
erg per second (erg/s)	watt (W)	**1.0**	**E−07**
erg per square centimeter second |1obrkt§1rul/(cm^2 · s)]	watt per square meter (W/m^2)	**1.0**	**E−03**
ESU of capacitance (statfarad)	farad (F)	1.112 650	E−12
ESU of current (statampere)	ampere (A)	3.335 641	E−10
ESU of electric potential (statvolt)	volt (V)	2.997 925	E+02
ESU of inductance (stathenry)	henry (H)	8.987 552	E+11
ESU of resistance (statohm)	ohm (Ω)	8.987 552	E+11
faraday (based on carbon 12)	coulomb (C)	9.648 531	E+04
fathom (based on U.S. survey foot)[9]	meter (m)	1.828 804	E+00
fermi	meter (m)	**1.0**	**E−15**
fermi	femtometer (fm)	**1.0**	**E+00**
fluid ounce (U.S.) (fl oz)	cubic meter (m^3)	2.957 353	E−05
fluid ounce (U.S.) (fl oz)	milliliter (mL)	2.957 353	E+01
foot (ft)	meter (m)	**3.048**	**E−01**
foot (U.S. survey) (ft)[9]	meter (m)	3.048 006	E−01
footcandle	lux (lx)	1.076 391	E+01
footlambert	candela per square meter (cd/m^2)	3.426 259	E+00
foot of mercury, conventional (ftHg)[13]	pascal (Pa)	4.063 666	E+04
foot of mercury, conventional (ftHg)[13]	kilopascal (kPa)	4.063 666	E+01
foot of water (39.2 °F)[13]	pascal (Pa)	2.988 98	E+03
foot of water (39.2 °F)[13]	kilopascal (kPa)	2.988 98	E+00
foot of water, conventional (ftH_2O)[13]	pascal (Pa)	2.989 067	E+03
foot of water, conventional (ftH_2O)[13]	kilopascal (kPa)	2.989 067	E+00
foot per hour (ft/h)	meter per second (m/s)	8.466 667	E−05
foot per minute (ft/min)	meter per second (m/s)	**5.08**	**E−03**
foot per second (ft/s)	meter per second (m/s)	**3.048**	**E−01**
foot per second squared (ft/s^2)	meter per second squared (m/s^2)	**3.048**	**E−01**
foot poundal	joule (J)	4.214 011	E−02
foot pound-force (ft · lbf)	joule (J)	1.355 818	E+00
foot pound-force per hour (ft · lbf/h)	watt (W)	3.766 161	E−04
foot pound-force per minute (ft · lbf/min)	watt (W)	2.259 697	E−02
foot pound-force per second (ft · lbf/s)	watt (W)	1.355 818	E+00
foot to the fourth power (ft^4)[17]	meter to the fourth power (m^4)	8.630 975	E−03
franklin (Fr)	coulomb (C)	3.335 641	E−10
gal (Gal)	meter per second squared (m/s^2)	**1.0**	**E−02**
gallon [Canadian and U.K. (Imperial)] (gal)	cubic meter (m^3)	**4.546 09**	**E−03**
gallon [Canadian and U.K. (Imperial)] (gal)	liter (L)	**4.546 09**	**E+00**
gallon (U.S.) (gal)	cubic meter (m^3)	3.785 412	E−03
gallon (U.S.) (gal)	liter (L)	3.785 412	E+00
gallon (U.S.) per day (gal/d)	cubic meter per second (m^3/s)	4.381 264	E−08
gallon (U.S.) per day (gal/d)	liter per second (L/s)	4.381 264	E−05
gallon (U.S.) per horsepower hour [gal/(hp · h)]	cubic meter per joule (m^3/J)	1.410 089	E−09
gallon (U.S.) per horsepower hour [gal/(hp · h)]	liter per joule (L/J)	1.410 089	E−06
gallon (U.S.) per minute (gpm)(gal/min)	cubic meter per second (m^3/s)	6.309 020	E−05
gallon (U.S.) per minute (gpm)(gal/min)	liter per second (L/s)	6.309 020	E−02

[17] This is a unit for the quantity second moment of area, which is sometimes called the ''moment of section'' or ''area moment of inertia'' of a plane section about a specified axis.

To convert from	to	Multiply by	
gamma (γ)	tesla (T)	**1.0**	**E−09**
gauss (Gs, G)	tesla (T)	**1.0**	**E−04**
gilbert (Gi)	ampere (A)	7.957 747	E−01
gill [Canadian and U.K. (Imperial)] (gi)	cubic meter (m^3)	1.420 653	E−04
gill [Canadian and U.K. (Imperial)] (gi)	liter (L)	1.420 653	E−01
gill (U.S.) (gi)	cubic meter (m^3)	1.182 941	E−04
gill (U.S.) (gi)	liter (L)	1.182 941	E−01
gon (also called grade) (gon)	radian (rad)	1.570 796	E−02
gon (also called grade) (gon)	degree (angle) (°)	**9.0**	**E−01**
grain (gr)	kilogram (kg)	**6.479 891**	**E−05**
grain (gr)	milligram (mg)	**6.479 891**	**E+01**
grain per gallon (U.S.) (gr/gal)	kilogram per cubic meter (kg/m^3)	1.711 806	E−02
grain per gallon (U.S.) (gr/gal)	milligram per liter (mg/L)	1.711 806	E+01
gram-force per square centimeter (gf/cm^2)	pascal (Pa)	**9.806 65**	**E+01**
gram per cubic centimeter (g/cm^3)	kilogram per cubic meter (kg/m^3)	**1.0**	**E+03**
hectare (ha)	square meter (m^2)	**1.0**	**E+04**
horsepower (550 ft · lbf/s) (hp)	watt (W)	7.456 999	E+02
horsepower (boiler)	watt (W)	9.809 50	E+03
horsepower (electric)	watt (W)	**7.46**	**E+02**
horsepower (metric)	watt (W)	7.354 988	E+02
horsepower (U.K.)	watt (W)	7.4570	E+02
horsepower (water)	watt (W)	7.460 43	E+02
hour (h)	second (s)	**3.6**	**E+03**
hour (sidereal)	second (s)	3.590 170	E+03
hundredweight (long, 112 lb)	kilogram (kg)	5.080 235	E+01
hundredweight (short, 100 lb)	kilogram (kg)	4.535 924	E+01
inch (in)	meter (m)	**2.54**	**E−02**
inch (in)	centimeter (cm)	**2.54**	**E+00**
inch of mercury (32 °F)[13]	pascal (Pa)	3.386 38	E+03
inch of mercury (32 °F)[13]	kilopascal (kPa)	3.386 38	E+00
inch of mercury (60 °F)[13]	pascal (Pa)	3.376 85	E+03
inch of mercury (60 °F)[13]	kilopascal (kPa)	3.376 85	E+00
inch of mercury, conventional (inHg)[13]	pascal (Pa)	3.386 389	E+03
inch of mercury, conventional (inHg)[13]	kilopascal (kPa)	3.386 389	E+00
inch of water (39.2 °F)[13]	pascal (Pa)	2.490 82	E+02
inch of water (60 °F)[13]	pascal (Pa)	2.4884	E+02
inch of water, conventional (inH_2O)[13]	pascal (Pa)	2.490 889	E+02
inch per second (in/s)	meter per second (m/s)	**2.54**	**E−02**
inch per second squared (in/s^2)	meter per second squared (m/s^2)	**2.54**	**E−02**
inch to the fourth power (in^4)[17]	meter to the fourth power (m^4)	4.162 314	E−07
kayser (K)	reciprocal meter (m^{-1})	**1.0**	**E+02**
kelvin (K)	degree Celsius (°C)	$t/°C = T/K - $ **273.15**	
$kilocalorie_{IT}$ ($kcal_{IT}$)	joule (J)	**4.1868**	**E+03**
$kilocalorie_{th}$ ($kcal_{th}$)	joule (J)	**4.184**	**E+03**
kilocalorie (mean) (kcal)	joule (J)	4.190 02	E+03
$kilocalorie_{th}$ per minute ($kcal_{th}/min$)	watt (W)	6.973 333	E+01
$kilocalorie_{th}$ per second ($kcal_{th}/s$)	watt (W)	**4.184**	**E+03**
kilogram-force (kgf)	newton (N)	**9.806 65**	**E+00**
kilogram-force meter (kgf · m)	newton meter (N · m)	**9.806 65**	**E+00**

To convert from	to	Multiply by	
kilogram-force per square centimeter (kgf/cm^2)	pascal (Pa)	**9.806 65**	**E+04**
kilogram-force per square centimeter (kgf/cm^2)	kilopascal (kPa)	**9.806 65**	**E+01**
kilogram-force per square meter (kgf/m^2)	pascal (Pa)	**9.806 65**	**E+00**
kilogram-force per square millimeter (kgf/mm^2)	pascal (Pa)	**9.806 65**	**E+06**
kilogram-force per square millimeter (kgf/mm^2)	megapascal (MPa)	**9.806 65**	**E+00**
kilogram-force second squared per meter (kgf · s^2/m)	kilogram (kg)	**9.806 65**	**E+00**
kilometer per hour (km/h)	meter per second (m/s)	2.777 778	E−01
kilopond (kilogram-force) (kp)	newton (N)	**9.806 65**	**E+00**
kilowatt hour (kW · h)	joule (J)	**3.6**	**E+06**
kilowatt hour (kW · h)	megajoule (MJ)	**3.6**	**E+00**
kip (1 kip=1000 lbf)	newton (N)	4.448 222	E+03
kip (1 kip=1000 lbf)	kilonewton (kN)	4.448 222	E+00
kip per square inch (ksi) (kip/in^2)	pascal (Pa)	6.894 757	E+06
kip per square inch (ksi) (kip/in^2)	kilopascal (kPa)	6.894 757	E+03
knot (nautical mile per hour)	meter per second (m/s)	5.144 444	E−01
lambert[18]	candela per square meter (cd/m^2)	3.183 099	E+03
langley (cal$_{th}$/cm^2)	joule per square meter (J/m^2)	**4.184**	**E+04**
light year (l.y.)[19]	meter (m)	9.460 73	E+15
liter (L)[20]	cubic meter (m^3)	**1.0**	**E−03**
lumen per square foot (lm/ft^2)	lux (lx)	1.076 391	E+01
maxwell (Mx)	weber (Wb)	**1.0**	**E−08**
mho	siemens (S)	**1.0**	**E+00**
microinch	meter (m)	**2.54**	**E−08**
microinch	micrometer (μm)	**2.54**	**E−02**
micron (μ)	meter (m)	**1.0**	**E−06**
micron (μ)	micrometer (μm)	**1.0**	**E+00**
mil (0.001 in)	meter (m)	**2.54**	**E−05**
mil (0.001 in)	millimeter (mm)	**2.54**	**E−02**
mil (angle)	radian (rad)	9.817 477	E−04
mil (angle)	degree (°)	**5.625**	**E−02**
mile (mi)	meter (m)	**1.609 344**	**E+03**
mile (mi)	kilometer (km)	**1.609 344**	**E+00**
mile (based on U.S. survey foot) (mi)[9]	meter (m)	1.609 347	E+03
mile (based on U.S. survey foot) (mi)[9]	kilometer (km)	1.609 347	E+00
mile, nautical[21]	meter (m)	**1.852**	**E+03**
mile per gallon (U.S.) (mpg) (mi/gal)	meter per cubic meter (m/m^3)	4.251 437	E+05
mile per gallon (U.S.) (mpg) (mi/gal)	kilometer per liter (km/L)	4.251 437	E−01
mile per gallon (U.S.) (mpg) (mi/gal)[22]	liter per 100 kilometer (L/100 km)	divide 235.215 by number of miles per gallon	
mile per hour (mi/h)	meter per second (m/s)	**4.4704**	**E−01**
mile per hour (mi/h)	kilometer per hour (km/h)	**1.609 344**	**E+00**

[18] The exact conversion factor is $10^4/\pi$.

[19] This conversion factor is based on 1 d = 86 400 s; and 1 Julian century = 36 525 d. (See *The Astronomical Almanac for the Year 1995*, page K6, U.S. Government Printing Office, Washington, DC, 1994).

[20] In 1964 the General Conference on Weights and Measures reestablished the name "liter" as a special name for the cubic decimeter. Between 1901 and 1964 the liter was slightly larger (1.000 028 dm^3); when one uses high-accuracy volume data of that time, this fact must be kept in mind.

[21] The value of this unit, 1 nautical mile = 1852 m, was adopted by the First International Extraordinary Hydrographic Conference, Monaco, 1929, under the name "International nautical mile."

[22] For converting fuel economy, as used in the U.S., to fuel consumption.

To convert from	to	Multiply by	
mile per minute (mi/min)	meter per second (m/s)	**2.682 24**	**E+01**
mile per second (mi/s)	meter per second (m/s)	**1.609 344**	**E+03**
millibar (mbar)	pascal (Pa)	**1.0**	**E+02**
millibar (mbar)	kilopascal (kPa)	**1.0**	**E−01**
millimeter of mercury, conventional (mmHg)[13]	pascal (Pa)	1.333 224	E+02
millimeter of water, conventional (mmH_2O)[13]	pascal (Pa)	**9.806 65**	**E+00**
minute (angle) (′)	radian (rad)	2.908 882	E−04
minute (min)	second (s)	**6.0**	**E+01**
minute (sidereal)	second (s)	5.983 617	E+01
oersted (Oe)	ampere per meter (A/m)	7.957 747	E+01
ohm centimeter (Ω · cm)	ohm meter (Ω · m)	**1.0**	**E−02**
ohm circular-mil per foot	ohm meter (Ω · m)	1.662 426	E−09
ohm circular-mil per foot	ohm square millimeter per meter ($\Omega \cdot mm^2/m$)	1.662 426	E−03
ounce (avoirdupois) (oz)	kilogram (kg)	2.834 952	E−02
ounce (avoirdupois) (oz)	gram (g)	2.834 952	E+01
ounce (troy or apothecary) (oz)	kilogram (kg)	3.110 348	E−02
ounce (troy or apothecary) (oz)	gram (g)	3.110 348	E+01
ounce [Canadian and U.K. fluid (Imperial)] (fl oz)	cubic meter (m^3)	2.841 306	E−05
ounce [Canadian and U.K. fluid (Imperial)] (fl oz)	milliliter (mL)	2.841 306	E+01
ounce (U.S. fluid) (fl oz)	cubic meter (m^3)	2.957 353	E−05
ounce (U.S. fluid) (fl oz)	millimeter (mL)	2.957 353	E+01
ounce (avoirdupois)-force (ozf)	newton (N)	2.780 139	E−01
ounce (avoirdupois)-force inch (ozf · in)	newton meter (N · m)	7.061 552	E−03
ounce (avoirdupois)-force inch (ozf · in)	millinewton meter (mN · m)	7.061 552	E+00
ounce (avoirdupois) per cubic inch (oz/in^3)	kilogram per cubic meter (kg/m^3)	1.729 994	E+03
ounce (avoirdupois) per gallon [Canadian and U.K. (Imperial)] (oz/gal)	kilogram per cubic meter (kg/m^3)	6.236 023	E+00
ounce (avoirdupois) per gallon [Canadian and U.K. (Imperial)] (oz/gal)	gram per liter (g/L)	6.236 023	E+00
ounce (avoirdupois) per gallon (U.S.) (oz/gal)	kilogram per cubic meter (kg/m^3)	7.489 152	E+00
ounce (avoirdupois) per gallon (U.S.) (oz/gal)	gram per liter (g/L)	7.489 152	E+00
ounce (avoirdupois) per square foot (oz/ft^2)	kilogram per square meter (kg/m^2)	3.051 517	E−01
ounce (avoirdupois) per square inch (oz/in^2)	kilogram per square meter (kg/m^2)	4.394 185	E+01
ounce (avoirdupois) per square yard (oz/yd^2)	kilogram per square meter (kg/m^2)	3.390 575	E−02
parsec (pc)	meter (m)	3.085 678	E+16
peck (U.S.) (pk)	cubic meter (m^3)	8.809 768	E−03
peck (U.S.) (pk)	liter (L)	8.809 768	E+00
pennyweight (dwt)	kilogram (kg)	1.555 174	E−03
pennyweight (dwt)	gram (g)	1.555 174	E+00
perm (0 °C)	kilogram per pascal second square meter [$kg/(Pa \cdot s \cdot m^2)$]	5.721 35	E−11
perm (23 °C)	kilogram per pascal second square meter [$kg/(Pa \cdot s \cdot m^2)$]	5.745 25	E−11
perm inch (0 °C)	kilogram per pascal second meter [kg/(Pa · s · m)]	1.453 22	E−12
perm inch (23 °C)	kilogram per pascal second meter [kg/(Pa · s · m)]	1.459 29	E−12

To convert from	to	Multiply by	
phot (ph)	lux (lx)	**1.0**	**E+04**
pica (computer) (1/6 in)	meter (m)	4.233 333	E−03
pica (computer) (1/6 in)	millimeter (mm)	4.233 333	E+00
pica (printer's)	meter (m)	4.217 518	E−03
pica (printer's)	millimeter (mm)	4.217 518	E+00
pint (U.S. dry) (dry pt)	cubic meter (m^3)	5.506 105	E−04
pint (U.S. dry) (dry pt)	liter (L)	5.506 105	E−01
pint (U.S. liquid) (liq pt)	cubic meter (m^3)	4.731 765	E−04
pint (U.S. liquid) (liq pt)	liter (L)	4.731 765	E−01
point (computer) (1/72 in)	meter (m)	3.527 778	E−04
point (computer) (1/72 in)	millimeter (mm)	3.527 778	E−01
point (printer's)	meter (m)	3.514 598	E−04
point (printer's)	millimeter (mm)	3.514 598	E−01
poise (P)	pascal second (Pa · s)	**1.0**	**E−01**
pound (avoirdupois) (lb)[23]	kilogram (kg)	4.535 924	E−01
pound (troy or apothecary) (lb)	kilogram (kg)	3.732 417	E−01
poundal	newton (N)	1.382 550	E−01
poundal per square foot	pascal (Pa)	1.488 164	E+00
poundal second per square foot	pascal second (Pa · s)	1.488 164	E+00
pound foot squared (lb · ft^2)	kilogram meter squared (kg · m^2)	4.214 011	E−02
pound-force (lbf)[24]	newton (N)	4.448 222	E+00
pound-force foot (lbf · ft)	newton meter (N · m)	1.355 818	E+00
pound-force foot per inch (lbf · ft/in)	newton meter per meter (N · m/m)	5.337 866	E+01
pound-force inch (lbf · in)	newton meter (N · m)	1.129 848	E−01
pound-force inch per inch (lbf · in/in)	newton meter per meter (N · m/m)	4.448 222	E+00
pound-force per foot (lbf/ft)	newton per meter (N/m)	1.459 390	E+01
pound-force per inch (lbf/in)	newton per meter (N/m)	1.751 268	E+02
pound-force per pound (lbf/lb) (thrust to mass ratio)	newton per kilogram (N/kg)	**9.806 65**	**E+00**
pound-force per square foot (lbf/ft^2)	pascal (Pa)	4.788 026	E+01
pound-force per square inch (psi) (lbf/in^2)	pascal (Pa)	6.894 757	E+03
pound-force per square inch (psi) (lbf/in^2)	kilopascal (kPa)	6.894 757	E+00
pound-force second per square foot (lbf · s/ft^2)	pascal second (Pa · s)	4.788 026	E+01
pound-force second per square inch (lbf · s/in^2)	pascal second (Pa · s)	6.894 757	E+03
pound inch squared (lb · in^2)	kilogram meter squared (kg · m^2)	2.926 397	E−04
pound per cubic foot (lb/ft^3)	kilogram per cubic meter (kg/m^3)	1.601 846	E+01
pound per cubic inch (lb/in^3)	kilogram per cubic meter (kg/m^3)	2.767 990	E+04
pound per cubic yard (lb/yd^3)	kilogram per cubic meter (kg/m^3)	5.932 764	E−01
pound per foot (lb/ft)	kilogram per meter (kg/m)	1.488 164	E+00
pound per foot hour [lb/(ft · h)]	pascal second (Pa · s)	4.133 789	E−04
pound per foot second [lb/(ft · s)]	pascal second (Pa · s)	1.488 164	E+00
pound per gallon [Canadian and U.K. (Imperial)] (lb/gal)	kilogram per cubic meter (kg/m^3)	9.977 637	E+01
pound per gallon [Canadian and U.K. (Imperial)] (lb/gal)	kilogram per liter (kg/L)	9.977 637	E−02
pound per gallon (U.S.) (lb/gal)	kilogram per cubic meter (kg/m^3)	1.198 264	E+02
pound per gallon (U.S.) (lb/gal)	kilogram per liter (kg/L)	1.198 264	E−01
pound per horsepower hour [lb/(hp · h)]	kilogram per joule (kg/J)	1.689 659	E−07
pound per hour (lb/h)	kilogram per second (kg/s)	1.259 979	E−04

[23] The exact conversion factor is 4.535 923 7 E−01. All units that contain the pound refer to the avoirdupois pound.

[24] If the local value of the acceleration of free fall is taken as g_n=9.806 65 m/s^2 (the standard value), the exact conversion factor is 4.448 221 615 260 5 E+00.

To convert from	to	Multiply by	
pound per inch (lb/in)	kilogram per meter (kg/m)	1.785 797	E+01
pound per minute (lb/min)	kilogram per second (kg/s)	7.559 873	E−03
pound per second (lb/s)	kilogram per second (kg/s)	4.535 924	E−01
pound per square foot (lb/ft^2)	kilogram per square meter (kg/m^2)	4.882 428	E+00
pound per square inch (*not* pound-force) (lb/in^2)	kilogram per square meter (kg/m^2)	7.030 696	E+02
pound per yard (lb/yd)	kilogram per meter (kg/m)	4.960 546	E−01
psi (pound-force per square inch) (lbf/in^2)	pascal (Pa)	6.894 757	E+03
psi (pound-force per square inch) (lbf/in^2)	kilopascal (kPa)	6.894 757	E+00
quad (10^{15} Btu_{IT})[11]	joule (J)	1.055 056	E+18
quart (U.S. dry) (dry qt)	cubic meter (m^3)	1.101 221	E−03
quart (U.S. dry) (dry qt)	liter (L)	1.101 221	E+00
quart (U.S. liquid) (liq qt)	cubic meter (m^3)	9.463 529	E−04
quart (U.S. liquid) (liq qt)	liter (L)	9.463 529	E−01
rad (absorbed dose) (rad)	gray (Gy)	**1.0**	**E−02**
rem (rem)	sievert (Sv)	**1.0**	**E−02**
revolution (r)	radian (rad)	6.283 185	E+00
revolution per minute (rpm) (r/min)	radian per second (rad/s)	1.047 198	E−01
rhe	reciprocal pascal second [(Pa · s)$^{-1}$]	**1.0**	**E+01**
rod (based on U.S. survey foot) (rd)[9]	meter (m)	5.029 210	E+00
roentgen (R)	coulomb per kilogram (C/kg)	**2.58**	**E−04**
rpm (revolution per minute) (r/min)	radian per second (rad/s)	1.047 198	E−01
second (angle) (")	radian (rad)	4.848 137	E−06
second (sidereal)	second (s)	9.972 696	E−01
shake	second (s)	**1.0**	**E−08**
shake	nanosecond (ns)	**1.0**	**E+01**
slug (slug)	kilogram (kg)	1.459 390	E+01
slug per cubic foot (slug/ft^3)	kilogram per cubic meter (kg/m^3)	5.153 788	E+02
slug per foot second [slug/(ft · s)]	pascal second (Pa · s)	4.788 026	E+01
square foot (ft^2)	square meter (m^2)	**9.290 304**	**E−02**
square foot per hour (ft^2/h)	square meter per second (m^2/s)	**2.580 64**	**E−05**
square foot per second (ft^2/s)	square meter per second (m^2/s)	**9.290 304**	**E−02**
square inch (in^2)	square meter (m^2)	**6.4516**	**E−04**
square inch (in^2)	square centimeter (cm^2)	**6.4516**	**E+00**
square mile (mi^2)	square meter (m^2)	2.589 988	E+06
square mile (mi^2)	square kilometer (km^2)	2.589 988	E+00
square mile (based on U.S. survey foot) (mi^2)[9]	square meter (m^2)	2.589 998	E+06
square mile (based on U.S. survey foot) (mi^2)[9]	square kilometer (km^2)	2.589 998	E+00
square yard (yd^2)	square meter (m^2)	8.361 274	E−01
statampere	ampere (A)	3.335 641	E−10
statcoulomb	coulomb (C)	3.335 641	E−10
statfarad	farad (F)	1.112 650	E−12
stathenry	henry (H)	8.987 552	E+11
statmho	siemens (S)	1.112 650	E−12
statohm	ohm (Ω)	8.987 552	E+11
statvolt	volt (V)	2.997 925	E+02
stere (st)	cubic meter (m^3)	**1.0**	**E+00**
stilb (sb)	candela per square meter (cd/m^2)	**1.0**	**E+04**
stokes (St)	meter squared per second (m^2/s)	**1.0**	**E−04**

To convert from	to	Multiply by	
tablespoon	cubic meter (m^3)	1.478 676	E−05
tablespoon	milliliter (mL)	1.478 676	E+01
teaspoon	cubic meter (m^3)	4.928 922	E−06
teaspoon	milliliter (mL)	4.928 922	E+00
tex	kilogram per meter (kg/m)	**1.0**	**E−06**
therm (EC)[25]	joule (J)	**1.055 06**	**E+08**
therm (U.S.)[25]	joule (J)	**1.054 804**	**E+08**
ton, assay (AT)	kilogram (kg)	2.916 667	E−02
ton, assay (AT)	gram (g)	2.916 667	E+01
ton-force (2000 lbf)	newton (N)	8.896 443	E+03
ton-force (2000 lbf)	kilonewton (kN)	8.896 443	E+00
ton, long (2240 lb)	kilogram (kg)	1.016 047	E+03
ton, long, per cubic yard	kilogram per cubic meter (kg/m^3)	1.328 939	E+03
ton, metric (t)	kilogram (kg)	**1.0**	**E+03**
tonne (called "metric ton" in U.S.) (t)	kilogram (kg)	**1.0**	**E+03**
ton of refrigeration (12 000 Btu_{IT}/h)	watt (W)	3.516 853	E+03
ton of TNT (energy equivalent)[26]	joule (J)	**4.184**	**E+09**
ton, register	cubic meter (m^3)	2.831 685	E+00
ton, short (2000 lb)	kilogram (kg)	9.071 847	E+02
ton, short, per cubic yard	kilogram per cubic meter (kg/m^3)	1.186 553	E+03
ton, short, per hour	kilogram per second (kg/s)	2.519 958	E−01
torr (Torr)	pascal (Pa)	1.333 224	E+02
unit pole	weber (Wb)	1.256 637	E−07
watt hour (W · h)	joule (J)	**3.6**	**E+03**
watt per square centimeter (W/cm^2)	watt per square meter (W/m^2)	**1.0**	**E+04**
watt per square inch (W/in^2)	watt per square meter (W/m^2)	1.550 003	E+03
watt second (W · s)	joule (J)	**1.0**	**E+00**
yard (yd)	meter (m)	**9.144**	**E−01**
year (365 days)	second (s)	**3.1536**	**E+07**
year (sidereal)	second (s)	3.155 815	E+07
year (tropical)	second (s)	3.155 693	E+07

[25] The therm (EC) is legally defined in the Council Directive of 20 December 1979, Council of the European Communities (now the European Union, EU). The therm (U.S.) is legally defined in the Federal Register of July 27, 1968. Although the therm (EC), which is based on the International Table Btu, is frequently used by engineers in the United States, the therm (U.S.) is the legal unit used by the U.S. natural gas industry.

[26] Defined (not measured) value.

CONVERSION OF TEMPERATURES

From	To	
Celsius	Fahrenheit	$t_F/°F = (9/5)\ t/°C + 32$
	Kelvin	$T/K = t/°C + 273.15$
	Rankine	$T/°R = (9/5)\ (t/°C + 273.15)$
Fahrenheit	Celsius	$t/°C = (5/9)\ [(t_F/°F) - 32]$
	Kelvin	$T/K = (5/9)\ [(t_F/°F) - 32] + 273.15$
	Rankine	$T/°R = t_F/°F + 459.67$
Kelvin	Celsius	$t/°C = T/K - 273.15$
	Rankine	$T/°R = (9/5)\ T/K$
Rankine	Fahrenheit	$t_F/°F = T/°R - 459.67$
	Kelvin	$T/K = (5/9)\ T/°R$

Definition of symbols:

T = thermodynamic (absolute) temperature
t = Celsius temperature (the symbol q is also used for Celsius temperature)
t_F = Fahrenheit temperature

DESIGNATION OF LARGE NUMBERS

	U.S.A.	Other Countries
10^6	million	million
10^9	billion	milliard
10^{12}	trillion	billion
10^{15}	quadrillion	billiard
10^{18}	quintillion	trillion

CONVERSION FACTORS FOR ENERGY UNITS

If greater accuracy is required, use the *Energy Equivalents* section of the *Fundamental Physical Constants* table.

	Wavenumber $\bar{\nu}$ cm^{-1}	Frequency ν MHz	Energy E aJ	Energy E eV	Energy E E_h	Molar energy E_m kJ/mol	Molar energy E_m kcal/mol	Temperature T K
$\bar{\nu}$: 1 cm^{-1}	≑ 1	2.997925×10^{4}	1.986447×10^{-5}	1.239842×10^{-4}	4.556335×10^{-6}	11.96266×10^{-3}	2.85914×10^{-3}	1.438769
ν: 1 MHz	≑ 3.33564×10^{-5}	1	6.626076×10^{-10}	4.135669×10^{-9}	1.519830×10^{-10}	3.990313×10^{-7}	9.53708×10^{-8}	4.79922×10^{-5}
1 aJ	≑ 50341.1	1.509189×10^{9}	1	6.241506	0.2293710	602.2137	143.9325	7.24292×10^{4}
E: 1 eV	≑ 8065.54	2.417988×10^{8}	0.1602177	1	3.674931×10^{-2}	96.4853	23.0605	1.16045×10^{4}
E_h	≑ 219474.63	6.579684×10^{9}	4.359748	27.2114	1	2625.500	627.510	3.15773×10^{5}
E_m: 1 kJ/mol	≑ 83.5935	2.506069×10^{6}	1.660540×10^{-3}	1.036427×10^{-2}	3.808798×10^{-4}	1	0.239006	120.272
1 kcal/mol	≑ 349.755	1.048539×10^{7}	6.947700×10^{-3}	4.336411×10^{-2}	1.593601×10^{-3}	4.184	1	503.217
T: 1 K	≑ 0.695039	2.08367×10^{4}	1.380658×10^{-5}	8.61738×10^{-5}	3.16683×10^{-6}	8.31451×10^{-3}	1.98722×10^{-3}	1

Examples of the use of this table: 1 aJ ≑ 50341 cm^{-1}
1 eV ≑ 96.4853 kJ mol^{-1}

The symbol ≑ should be read as meaning "corresponds to" or "is equivalent to".
$E = h\nu = hc\bar{\nu} = kT$; $E_m = N_A E$; E_h is the Hartree energy

CONVERSION FACTORS FOR PRESSURE UNITS

	Pa	kPa	MPa	bar	atmos	Torr	μmHg	psi
Pa	1	0.001	0.000001	0.00001	9.8692×10^{-6}	0.0075006	7.5006	0.0001450377
kPa	1000	1	0.001	0.01	0.0098692	7.5006	7500.6	0.1450377
MPa	1000000	1000	1	10	9.8692	7500.6	7500600	145.0377
bar	100000	100	0.1	1	0.98692	750.06	750060	14.50377
atmos	101325	101.325	0.101325	1.01325	1	760	760000	14.69594
Torr	133.322	0.133322	0.000133322	0.00133322	0.00131579	1	1000	0.01933672
μmHg	0.133322	0.000133322	1.33322×10^{-7}	1.33322×10^{-6}	1.31579×10^{-6}	0.001	1	1.933672×10^{-5}
psi	6894.757	6.894757	0.006894757	0.06894757	0.068046	51.7151	51715.1	1

To convert a pressure value from a unit in the left hand column to a new unit, multiply the value by the factor appearing in the column for the new unit. For example:

1 kPa = 9.8692×10^{-3} atmos
1 Torr = 1.33322×10^{-4} MPa

Notes: μmHg is often referred to as "micron"
Torr is essentially identical to mmHg
psi is an abbreviation for the unit pound–force per square inch
psia (as a term for a physical quantity) implies the true (absolute) pressure
psig implies the true pressure minus the local atmospheric pressure

CONVERSION FACTORS FOR THERMAL CONDUCTIVITY UNITS

MULTIPLY ↓ by appropriate factor to OBTAIN→	Btu$_{IT}$ h^{-1} ft^{-1} °F^{-1}	Btu$_{IT}$ in. h^{-1} ft^{-2} °F^{-1}	Btu$_{th}$ h^{-1} ft^{-1} °F^{-1}	Btu$_{th}$ in. h^{-1} ft^{-2} °F^{-1}	cal$_{IT}$ s^{-1} cm^{-1} °C^{-1}	cal$_{th}$ s^{-1} cm^{-1} °C^{-1}	kcal$_{th}$ h^{-1} m^{-1} °C^{-1}	J s^{-1} cm^{-1} K^{-1}	W cm^{-1} K^{-1}	W m^{-1} K^{-1}	mW cm^{-1} K^{-1}
Btu$_{IT}$ h^{-1} ft^{-1} °F^{-1}	1	12	1.00067	12.0080	4.13379×10^{-3}	4.13656×10^{-3}	1.48916	1.73073×10^{-2}	1.73073×10^{-2}	1.73073	17.3073
Btu$_{IT}$ in. h^{-1} ft^{-2} °F^{-1}	8.33333×10^{-2}	1	8.33891×10^{-2}	1.00067	3.44482×10^{-4}	3.44713×10^{-4}	0.124097	1.44228×10^{-3}	1.44228×10^{-3}	0.144228	1.44228
Btu$_{th}$ h^{-1} ft^{-1} °F^{-1}	0.999331	11.9920	1	12	4.13102×10^{-3}	4.13379×10^{-3}	1.48816	1.72958×10^{-2}	1.72958×10^{-2}	1.72958	17.2958
Btu$_{th}$ in. h^{-1} ft^{-2} °F^{-1}	8.32776×10^{-2}	0.999331	8.33333×10^{-2}	1	3.44252×10^{-4}	3.44482×10^{-4}	0.124014	1.44131×10^{-3}	1.44131×10^{-3}	0.144131	1.44131
cal$_{IT}$ s^{-1} cm^{-1} °C^{-1}	2.41909×10^{2}	2.90291×10^{3}	2.42071×10^{2}	2.90485×10^{3}	1	1.00067	3.60241×10^{2}	4.1868	4.1868	4.1868×10^{2}	4.1868×10^{3}
cal$_{th}$ s^{-1} cm^{-1} °C^{-1}	2.41747×10^{2}	2.90096×10^{3}	2.41909×10^{2}	2.90291×10^{3}	0.999331	1	3.6×10^{2}	4.184	4.184	4.184×10^{2}	4.184×10^{3}
kcal$_{th}$ h^{-1} m^{-1} °C^{-1}	0.671520	8.05824	0.671969	8.06363	2.77592×10^{-3}	2.77778×10^{-3}	1	1.16222×10^{-2}	1.16222×10^{-2}	1.16222	11.6222
J s^{-1} cm^{-1} K^{-1}	57.7789	6.93347×10^{2}	57.8176	6.93811×10^{2}	0.238846	0.239006	86.0421	1	1	1×10^{2}	1×10^{3}
W cm^{-1} K^{-1}	57.7789	6.93347×10^{2}	57.8176	6.93811×10^{2}	0.238846	0.239006	86.0421	1	1	1×10^{2}	1×10^{3}
W m^{-1} K^{-1}	0.577789	6.93347	0.578176	6.93811	2.38846×10^{-3}	2.39006×10^{-3}	0.860421	1×10^{-2}	1×10^{-2}	1	10
mW cm^{-1} K^{-1}	5.77789×10^{-2}	0.693347	5.78176×10^{-2}	0.693811	2.38846×10^{-4}	2.39006×10^{-4}	8.60421×10^{-2}	1×10^{-3}	1×10^{-3}	0.1	1

CONVERSION FACTORS FOR ELECTRICAL RESISTIVITY UNITS

To convert from ↓ multiply by appropriate factor to Obtain →	abΩ cm	μΩ cm	Ω cm	StatΩ cm	Ω m	Ω cir. mil ft^{-1}	Ω in.	Ω ft
abohm centimeter	1	1×10^{-3}	10^{-9}	1.113×10^{-21}	10^{-11}	6.015×10^{-3}	3.937×10^{-10}	3.281×10^{-11}
microohm centimeter	10^{3}	1	10^{-6}	1.113×10^{-18}	10^{-6}	6.015	3.937×10^{-7}	3.281×10^{-6}
ohm centimeter	10^{8}	10^{6}	1	1.113×10^{-12}	1×10^{-2}	6.015×10^{6}	3.937×10^{-1}	3.281×10^{-2}
statohm centimeter (esu)	8.987×10^{20}	8.987×10^{17}	8.987×10^{11}	1	8.987×10^{9}	5.406×10^{18}	3.538×10^{11}	2.949×10^{10}
ohm meter	10^{11}	10^{8}	10^{2}	1.113×10^{-10}	1	6.015×10^{8}	3.937×10^{1}	3.281
ohm circular mil per foot	1.662×10^{2}	1.662×10^{-1}	1.662×10^{-7}	1.850×10^{-19}	1.662×10^{-9}	1	6.54×10^{-6}	5.45×10^{-9}
ohm inch	2.54×10^{9}	2.54×10^{6}	2.54	2.827×10^{-12}	2.54×10^{-2}	1.528×10^{7}	1	8.3×10^{-2}
ohm foot	3.048×10^{10}	3.048×10^{7}	3.048×10^{-1}	3.3924×10^{-11}	3.048×10^{-1}	1.833×10^{8}	12	1

CONVERSION FACTORS FOR CHEMICAL KINETICS

Equivalent second order rate constants

A \ B	$cm^3\ mol^{-1}\ s^{-1}$	$dm^3\ mol^{-1}\ s^{-1}$	$m^3\ mol^{-1}\ s^{-1}$	$cm^3\ molecule^{-1}\ s^{-1}$	$(mm\ Hg)^{-1}\ s^{-1}$	$atm^{-1}\ s^{-1}$	$ppm^{-1}\ min^{-1}$	$m^2\ kN^{-1}\ s^{-1}$
$1\ cm^3\ mol^{-1}\ s^{-1} =$	1	10^{-3}	10^{-6}	1.66×10^{-24}	$1.604 \times 10^{-5} T^{-1}$	$1.219 \times 10^{-2} T^{-1}$	2.453×10^{-9}	$1.203 \times 10^{-4} T^{-1}$
$1\ dm^3\ mol^{-1}\ s^{-1} =$	10^3	1	10^{-3}	1.66×10^{-21}	$1.604 \times 10^{-2} T^{-1}$	$12.19\ T^{-1}$	2.453×10^{-6}	$1.203 \times 10^{-1} T^{-1}$
$1\ m^3\ mol^{-1}\ s^{-1} =$	10^6	10^3	1	1.66×10^{-18}	$16.04\ T^{-1}$	$1.219 \times 10^4 T^{-1}$	2.453×10^{-3}	$120.3\ T^{-1}$
$1\ cm^3\ molecule^{-1}\ s^{-1} =$	6.023×10^{23}	6.023×10^{20}	6.023×10^{17}	1	$9.658 \times 10^{18} T^{-1}$	$7.34 \times 10^{21} T^{-1}$	1.478×10^{15}	$7.244 \times 10^{19} T^{-1}$
$1\ (mm\ Hg)^{-1}\ s^{-1} =$	$6.236 \times 10^4 T$	$62.36\ T$	$6.236 \times 10^{-2} T$	$1.035 \times 10^{-19} T$	1	760	4.56×10^{-2}	7.500
$1\ atm^{-1}\ s^{-1}$	$82.06\ T$	$8.206 \times 10^{-2} T$	$8.206 \times 10^{-5} T$	$1.362 \times 10^{-22} T$	1.316×10^{-3}	1	6×10^{-5}	9.869×10^{-3}
$1\ ppm^{-1}\ min^{-1} =$ at 298 K, 1 atm total pressure	4.077×10^8	4.077×10^5	407.7	6.76×10^{-16}	21.93	1.667×10^4	1	164.5
$1\ m^2\ kN^{-1}\ s^{-1} =$	$8314\ T$	$8.314\ T$	$8.314 \times 10^{-3} T$	$1.38 \times 10^{-20} T$	0.1333	101.325	6.079×10^{-3}	1

To convert a rate constant from one set of units A to a new set B find the conversion factor for the row A under column B and multiply the old value by it, e.g. to convert $cm^3\ molecule^{-1}\ s^{-1}$ to $m^3\ mol^{-1}\ s^{-1}$ multiply by 6.023×10^{17}.

Table adapted from High Temperature Reaction Rate Data No. 5, The University, Leeds (1970).

Equivalent third order rate constants

A \ B	$cm^6\ mol^{-2}\ s^{-1}$	$dm^6\ mol^{-2}\ s^{-1}$	$m^6\ mol^{-2}\ s^{-1}$	$cm^6\ molecule^{-2}\ s^{-1}$	$(mm\ Hg)^{-2}\ s^{-1}$	$atm^{-2}\ s^{-1}$	$ppm^{-2}\ min^{-1}$	$m^4\ kN^{-2}\ s^{-1}$
$1\ cm^6\ mol^{-2}\ s^{-1} =$	1	10^{-6}	10^{-12}	2.76×10^{-48}	$2.57 \times 10^{-10} T^{-2}$	$1.48 \times 10^{-4} T^{-2}$	1.003×10^{-19}	$1.447 \times 10^{-8} T^{-2}$
$1\ dm^6\ mol^{-2}\ s^{-1} =$	10^6	1	10^{-6}	2.76×10^{-42}	$2.57 \times 10^{-4} T^{-2}$	$148\ T^{-2}$	1.003×10^{-13}	$1.447 \times 10^{-2} T^{-2}$
$1\ m^6\ mol^{-2}\ s^{-1} =$	10^{12}	10^6	1	2.76×10^{-36}	$257\ T^{-2}$	$1.48 \times 10^8 T^{-2}$	1.003×10^{-7}	$1.447 \times 10^4 T^{-2}$
$1\ cm^6 molecule^{-2}\ s^{-1} =$	3.628×10^{47}	3.628×10^{41}	3.628×10^{35}	1	$9.328 \times 10^{37} T^{-2}$	$5.388 \times 10^{43} T^{-2}$	3.64×10^{28}	$5.248 \times 10^{39} T^{-2}$
$1\ (mm\ Hg)^{-2}\ s^{-1} =$	$3.89 \times 10^9 T^2$	$3.89 \times 10^3 T^2$	$3.89 \times 10^{-3} T^2$	$1.07 \times 10^{-38} T^2$	1	5.776×10^5	3.46×10^{-5}	56.25
$1\ atm^{-2}\ s^{-1} =$	$6.733 \times 10^3 T^2$	$6.733 \times 10^{-3} T^2$	$6.733 \times 10^{-9} T^2$	$1.86 \times 10^{-44} T^2$	1.73×10^{-6}	1	6×10^{-11}	9.74×10^{-5}
$1\ ppm^{-2}\ min^{-1} =$ at 298 K, 1 atm total pressure	9.97×10^{18}	9.97×10^{12}	9.97×10^6	2.75×10^{-29}	2.89×10^4	1.667×10^{10}	1	1.623×10^6
$1\ m^4 kN^{-2}\ s^{-1} =$	$6.91 \times 10^7 T^2$	$6.91\ T^2$	$69.1 \times 10^{-5} T^2$	$1.904 \times 10^{-40} T^2$	0.0178	1.027×10^4	6.16×10^{-7}	1

From *J. Phys. Chem. Ref. Data*, 9, 470, 1980, by permission of the authors and the copyright owner, the American Institute of Physics.

CONVERSION FACTORS FOR IONIZING RADIATION

CONVERSION BETWEEN SI AND OTHER UNITS

Quantity	Symbol for quantity	Expression in SI units	Expression in symbols for SI units	Special name for SI units	Symbols using special names	Conventional units	Symbol for conventional unit	Value of conventional unit in SI units
Activity	A	1 per second	s^{-1}	becquerel	Bq	curie	Ci	3.7×10^{10} Bq
Absorbed dose	D	joule per kilogram	J kg^{-1}	gray	Gy	rad	rad	0.01 Gy
Absorbed dose rate	$\dot{D}$	joule per kilogram second	J kg^{-1} s^{-1}		Gy s^{-1}	rad	rad s^{-1}	0.01 Gy s^{-1}
Average energy per ion pair	W	joule	J			electronvolt	eV	1.602×10^{-19} J
Dose equivalent	H	joule per kilogram	J kg^{-1}	sievert	Sv	rem	rem	0.01 Sv
Dose equivalent rate	$\dot{H}$	joule per kilogram second	J kg^{-1} s^{-1}		Sv s^{-1}	rem per second	rem s^{-1}	0.01 Sv s^{-1}
Electric current	I	ampere	A			ampere	A	1.0 A
Electric potential difference	U, V	watt per ampere	W A^{-1}	volt	V	volt	V	1.0 A
Exposure	X	coulomb per kilogram	C kg^{-1}			roentgen	R	2.58×10^{-4} C kg^{-1}
Exposure rate	$\dot{X}$	coulomb per kilogram second	C kg^{-1} s^{-1}			roentgen	R s^{-1}	2.58×10^{-4} C kg^{-1} s^{-1}
Fluence	ϕ	1 per meter squared	m^{-2}			1 per centimeter squared	cm^{-2}	1.0×10^{4} m^{-2}
Fluence rate	Φ	1 per meter squared second	m^{-2} s^{-1}			1 per centimeter squared second	cm^{-2} s^{-1}	1.0×10^{4} m^{-2} s^{-1}
Kerma	K	joule per kilogram	J kg^{-1}	gray	Gy	rad	rad	0.01 Gy
Kerma rate	$\dot{K}$	joule per kilogram second	J kg^{-1} s^{-1}		Gy s^{-1}	rad per second	rad s^{-1}	0.01 Gy s^{-1}
Lineal energy	y	joule per meter	J m^{-1}			kiloelectron volt per micrometer	keV μm^{-1}	1.602×10^{-10} J m^{-1}
Linear energy transfer	L	joule per meter	J m^{-1}			kiloelectron volt per micrometer	keV μm^{-1}	1.602×10^{-10} J m^{-1}
Mass attenuation coefficient	μ/ρ	meter squared per kilogram	m^2 kg^{-1}			centimeter squared per gram	cm^2 g^{-1}	0.1 m^2 kg^{-1}
Mass energy transfer coefficient	μ_{tr}/ρ	meter squared per kilogram	m^2 kg^{-1}			centimeter squared per gram	cm^2 g^{-1}	0.1 m^2 kg^{-1}
Mass energy absorption coefficient	μ_{en}/ρ	meter squared per kilogram	m^2 kg^{-1}			centimeter squared per gram	cm^2 g^{-1}	0.1 m^2 kg^{-1}
Mass stopping power	S/ρ	joule meter squared per kilogram	J m^2 kg^{-1}			MeV centimeter squared per gram	MeV cm^2 g^{-1}	1.602×10^{-14} J m^2 kg^{-1}
Power	P	joule per second	J s^{-1}	watt	W	watt	W	1.0 W
Pressure	p	newton per meter squared	N m^{-2}	pascal	Pa	torr	torr	(101325/760)Pa
Radiation chemical yield	G	mole per joule	mol J^{-1}			molecules per 100 electron volts	molecules $(100\ eV)^{-1}$	1.04×10^{-7} mol J^{-1}
Specific energy	z	joule per kilogram	J kg^{-1}	gray	Gy	rad	rad	0.01 Gy

CONVERSION FACTORS FOR IONIZING RADIATION (continued)

CONVERSION OF RADIOACTIVITY UNITS FROM MBq TO mCi AND μCi

MBq	mCi	MBq	mCi	MBq	μCi	MBq	μCi
7000	189.	500	13.5	30	810	1	27
6000	162.	400	10.8	20	540	0.9	24
5000	135.	300	8.1	10	270	0.8	21.6
4000	108.	200	5.4	9	240	0.7	18.9
3000	81.	100	2.7	8	220	0.6	16.2
2000	54.	90	2.4	7	189	0.5	13.5
1000	27.	80	2.16	6	162	0.4	10.8
900	24.	70	1.89	5	135	0.3	8.1
800	21.6	60	1.62	4	108	0.2	5.4
700	18.9	50	1.35	3	81	0.1	2.7
600	16.2	40	1.08	2	54		

CONVERSION OF RADIOACTIVITY UNITS FROM mCi AND μCi TO MBq

mCi	MBq	mCi	MBq	μCi	MBq	μCi	MBq
200	7400	10	370	1000	37.0	80	2.96
150	5550	9	333	900	33.3	70	2.59
100	3700	8	296	800	29.6	60	2.22
90	3330	7	259	700	25.9	50	1.85
80	2960	6	222	600	22.2	40	1.48
70	2590	5	185	500	18.5	30	1.11
60	2220	4	148	400	14.8	20	0.74
50	1850	3	111	300	11.1	10	0.37
40	1480	2	74.0	200	7.4	5	0.185
30	1110	1	37.0	100	3.7	2	0.074
20	740			90	3.33	1	0.037

CONVERSION OF RADIOACTIVITY UNITS

100 TBq (10^{14} Bq)	=	2.7 kCi (2.7×10^3 Ci)
10 TBq (10^{13} Bq)	=	270 Ci (2.7×10^2 Ci)
1 TBq (10^{12} Bq)	=	27 Ci (2.7×10^1 Ci)
100 GBq (10^{11} Bq)	=	2.7 Ci (2.7×10^0 Ci)
10 GBq (10^{10} Bq)	=	270 mCi (2.7×10^{-1} Ci)
1 GBq (10^9 Bq)	=	27 mCi (2.7×10^{-2} Ci)
100 MBq (10^8 Bq)	=	2.7 mCi (2.7×10^{-3} Ci)
10 MBq (10^7 Bq)	=	270 μCi (2.7×10^{-4} Ci)
1 MBq (10^6 Bq)	=	27 μCi (2.7×10^{-5} Ci)
100 kBq (10^5 Bq)	=	2.7 μCi (2.7×10^{-6} Ci)
10 kBq (10^4 Bq)	=	270 nCi (2.7×10^{-7} Ci)
1 kBq (10^3 Bq)	=	27 nCi (2.7×10^{-8} Ci)
100 Bq (10^2 Bq)	=	2.7 nCi (2.7×10^{-9} Ci)
10 Bq (10^1 Bq)	=	270 pCi (2.7×10^{-10} Ci)
1 Bq (10^0 Bq)	=	27 pCi (2.7×10^{-11} Ci)
100 mBq (10^{-1} Bq)	=	2.7 pCi (2.7×10^{-12} Ci)
10 mBq (10^{-2} Bq)	=	270 fCi (2.7×10^{-13} Ci)
1 mBq (10^{-3} Bq)	=	27 fCi (2.7×10^{-14} Ci)

CONVERSION OF ABSORBED DOSE UNITS

SI Units		Conventional
100 Gy (10^2 Gy)	=	10,000 rad (10^4 rad)
10 Gy (10^1 Gy)	=	1,000 rad (10^3 rad)
1 Gy (10^0 Gy)	=	100 rad (10^2 rad)
100 mGy (10^{-1} Gy)	=	10 rad (10^1 rad)
10 mGy (10^{-2} Gy)	=	1 rad (10^0 rad)
1 mGy (10^{-3} Gy)	=	100 mrad (10^{-1} rad)
100 μGy (10^{-4} Gy)	=	10 mrad (10^{-2} rad)
10 μGy (10^{-5} Gy)	=	1 mrad (10^{-3} rad)
1 μGy (10^{-6} Gy)	=	100 μrad (10^{-4} rad)
100 nGy (10^{-7} Gy)	=	10 μrad (10^{-5} rad)
10 nGy (10^{-8} Gy)	=	1 μrad (10^{-6} rad)
1 nGy (10^{-9} Gy)	=	100 nrad (10^{-7} rad)

CONVERSION OF DOSE EQUIVALENT UNITS

100 Sv (10^2 Sv)	=	10,000 rem (10^4 rem)
10 Sv (10^1 Sv)	=	1,000 rem (10^3 rem)
1 Sv (10^0 Sv)	=	100 rem (10^2 rem)
100 mSv (10^{-1} Sv)	=	10 rem (10^1 rem)
10 mSv (10^{-2} Sv)	=	1 rem (10^0 rem)
1 mSv (10^{-3} Sv)	=	100 mrem (10^{-1} rem)
100 μSv (10^{-4} Sv)	=	10 mrem (10^{-2} rem)
10 μSv (10^{-5} Sv)	=	1 mrem (10^{-3} rem)
1 μSv (10^{-6} Sv)	=	100 μrem (10^{-4} rem)
100 nSv (10^{-7} Sv)	=	10 μrem (10^{-5} rem)
10 nSv (10^{-8} Sv)	=	1 μrem (10^{-6} rem)
1 nSv (10^{-9} Sv)	=	100 nrem (10^{-7} rem)

VALUES OF THE GAS CONSTANT IN DIFFERENT UNIT SYSTEMS

In SI units the value of the gas constant, R, is:

R = 8.314510 Pa m³ K⁻¹ mol⁻¹

= 8314.510 Pa L K⁻¹ mol⁻¹

= 0.08314510 bar L K⁻¹ mol⁻¹

This table gives the appropriate value of R for use in the ideal gas equation, $PV = nRT$, when the variables are expressed in other units. The following conversion factors for pressure units were used in generating the table:

1 atm = 101325 Pa

1 psi = 6894.757 Pa

1 torr (mmHg) = 133.322 Pa [at 0°C]

1 in Hg = 3386.38 Pa [at 0°C]

1 in H_2O = 249.082 Pa [at 4°C]

1 ft H_2O = 2988.98 Pa [at 4°C]

The advice of Prabir K. Chandra is appreciated.

Units of V, T, n			Units of P						
V	**T**	**n**	**kPa**	**atm**	**psi**	**mmHg**	**in Hg**	**in H_2O**	**ft H_2O**
ft³	K	mol	0.2936241	0.00289785	0.0425866	2.20237	0.0867074	1.17882	0.0982355
		lb·mol	133.1857	1.31444	19.3169	998.978	39.3298	534.706	44.5589
	°R	mol	0.1631245	0.00160991	0.0236592	1.22354	0.0481708	0.654903	0.0545753
		lb·mol	73.99204	0.730245	10.7316	554.987	21.8499	297.059	24.7549
cm³	K	mol	8314.510	82.0578	1205.92	62364.1	2455.28	33380.6	2781.72
		lb·mol	3771398	37220.8	546995	28287900	1113700	15141200	1261770
	°R	mol	4619.172	45.5877	669.954	34646.7	1364.04	18544.8	1545.40
		lb·mol	2095221	20678.2	303886	15715500	618720	8411770	700982
L	K	mol	8.314510	0.0820578	1.20592	62.3641	2.45528	33.3806	2.78172
		lb·mol	3771.398	37.2208	546.995	28287.9	1113.70	15141.2	1261.77
	°R	mol	4.619172	0.0455877	0.669954	34.6467	1.36404	18.5448	1.54540
		lb·mol	2095.221	20.6782	303.886	15715.5	618.720	8411.77	700.982
m³	K	mol	0.008314510	0.0000820578	0.00120592	0.0623641	0.00245528	0.0333806	0.00278172
		lb·mol	3.771398	0.0372208	0.546995	28.2879	1.11370	15.1412	1.26177
	°R	mol	0.004619172	0.0000455877	0.000669954	0.0346467	0.00136404	0.0185448	0.00154540
		lb·mol	2.095221	0.0206782	0.303886	15.7155	0.618720	8.41177	0.700982

Section 2
Symbols, Terminology, and Nomenclature

Symbols and Terminology for Physical and Chemical Quantities 2-1
Nomenclature of Chemical Compounds 2-22
Nomenclature for Inorganic Ions and Ligands 2-23
Organic Substituent Groups and Ring Systems 2-25
Scientific Abbreviations and Symbols 2-29
Greek, Russian, and Hebrew Alphabets 2-37
Definitions of Scientific Terms 2-38

SYMBOLS AND TERMINOLOGY FOR PHYSICAL AND CHEMICAL QUANTITIES

The International Organization for Standardization (ISO), International Union of Pure and Applied Chemistry (IUPAC), and the International Union of Pure and Applied Physics (IUPAP) have jointly developed a set of recommended symbols for physical and chemical quantities. Consistent use of these recommended symbols helps assure unambiguous scientific communication. The list below is reprinted from Reference 1 with permission from IUPAC. Full details may be found in the following references:

1. Ian Mills, Ed., *Quantities, Units, and Symbols in Physical Chemistry*, Blackwell Scientific Publications, Oxford, 1988.
2. E. R. Cohen and P. Giacomo, *Symbols, Units, Nomenclature, and Fundamental Constants in Physics*, Document IUPAP-25, 1987; also published in *Physica*, 146A, 1—68, 1987.
3. *ISO Standards Handbook 2: Units of Measurement*, International Organization of Standardization, Geneva, 1982.

GENERAL RULES

The value of a physical quantity is expressed as the product of a numerical value and a unit, e.g.:

$T = 300\ \mathrm{K}$
$V = 26.2\ \mathrm{cm^3}$
$C_p = 45.3\ \mathrm{J\ mol^{-1}\ K^{-1}}$

The symbol for a physical quantity is always given in italic (sloping) type, while symbols for units are given in roman type. Column headings in tables and axis labels on graphs may conveniently be written as the physical quantity symbol divided by the unit symbol, e.g.:

T/K
$V/\mathrm{cm^3}$
$C_p/\mathrm{J\ mol^{-1}\ K^{-1}}$

The values in the table or graph axis are then pure numbers.

Subscripts to symbols for physical quantities should be italic if the subscript refers to another physical quantity or to a number, e.g.:

C_p — heat capacity at constant pressure
B_n — nth virial coefficient

Subscripts which have other meanings should be in roman type:

m_p — mass of the proton
E_k — kinetic energy

The following tables give the recommended symbols for the major classes of physical and chemical quantities. The expression in the Definition column is given as an aid in identifying the quantity but is not necessarily the complete or unique definition. The SI Unit gives one (not necessarily unique) expression for the coherent SI unit for the quantity. Other equivalent unit expressions, including those which involve SI prefixes, may be used.

Name	*Symbol*	*Definition*	*SI unit*
SPACE AND TIME			
cartesian space coordinates	x, y, z		m
spherical polar coordinates	r, θ, ϕ		m, 1, 1
generalized coordinate	q, q_i		(varies)
position vector	$\boldsymbol{r}$	$\boldsymbol{r} = x\boldsymbol{i} + y\boldsymbol{j} + z\boldsymbol{k}$	m
length	l		m
special symbols:			
height	h		
breadth	b		
thickness	d, δ		
distance	d		

Name	*Symbol*	*Definition*	*SI unit*
radius	r		
diameter	d		
path length	s		
length of arc	s		
area	A, A_s, S		m^2
volume	$V, (v)$		m^3
plane angle	$\alpha, \beta, \gamma, \theta, \phi \ldots$	$\alpha = s/r$	rad, 1
solid angle	ω, Ω	$\omega = A/r^2$	sr, 1
time	t		s
period	T	$T = t/N$	s
frequency	ν, f	$\nu = 1/T$	Hz
circular frequency, angular frequency	ω	$\omega = 2\pi\nu$	rad s^{-1}, s^{-1}
characteristic time interval, relaxation time, time constant	τ, T	$\tau = \lvert dt/d\ln x \rvert$	s
angular velocity	ω	$\omega = d\phi/dt$	rad s^{-1}, s^{-1}
velocity	$\boldsymbol{v}, \boldsymbol{u}, \boldsymbol{w}, \boldsymbol{c}, \dot{\boldsymbol{r}}$	$\boldsymbol{v} = d\boldsymbol{r}/dt$	m s^{-1}
speed	v, u, w, c	$v = \lvert \boldsymbol{v} \rvert$	m s^{-1}
acceleration	$\boldsymbol{a}, (\boldsymbol{g})$	$\boldsymbol{a} = d\boldsymbol{v}/dt$	m s^{-2}

CLASSICAL MECHANICS

Name	*Symbol*	*Definition*	*SI unit*
mass	m		kg
reduced mass	μ	$\mu = m_1 m_2/(m_1 + m_2)$	kg
density, mass density	ρ	$\rho = m/V$	kg m^{-3}
relative density	d	$d = \rho/\rho^{\circ}$	1
surface density	ρ_A, ρ_S	$\rho_A = m/A$	kg m^{-2}
specific volume	v	$v = V/m = 1/\rho$	m^3 kg^{-1}
momentum	$\boldsymbol{p}$	$\boldsymbol{p} = m\boldsymbol{v}$	kg m s^{-1}
angular momentum, action	$\boldsymbol{L}$	$\boldsymbol{L} = \boldsymbol{r} \times \boldsymbol{p}$	J s
moment of inertia	I, J	$I = \sum m_i r_i^2$	kg m^2
force	$\boldsymbol{F}$	$\boldsymbol{F} = d\boldsymbol{p}/dt = m\boldsymbol{a}$	N
torque, moment of a force	$\boldsymbol{T}, (\boldsymbol{M})$	$\boldsymbol{T} = \boldsymbol{r} \times \boldsymbol{F}$	N m
energy	E		J
potential energy	E_p, V, Φ	$E_p = -\int \boldsymbol{F} \cdot d\boldsymbol{s}$	J
kinetic energy	E_k, T, K	$E_k = \frac{1}{2}mv^2$	J
work	W, w	$W = \int \boldsymbol{F} \cdot d\boldsymbol{s}$	J
Hamilton function	H	$H(q, p) = T(q, p) + V(q)$	J
Lagrange function	L	$L(q, \dot{q}) = T(q, \dot{q}) - V(q)$	J

Name	*Symbol*	*Definition*	*SI unit*
pressure	p, P	$p = F/A$	Pa, N m^{-2}
surface tension	γ, σ	$\gamma = dW/dA$	N m^{-1}, J m^{-2}
weight	$G, (W, P)$	$G = mg$	N
gravitational constant	G	$F = Gm_1 m_2/r^2$	N m^2 kg^{-2}
normal stress	σ	$\sigma = F/A$	Pa
shear stress	τ	$\tau = F/A$	Pa
linear strain, relative elongation	ε, e	$\varepsilon = \Delta l/l$	1
modulus of elasticity, Young's modulus	E	$E = \sigma/\varepsilon$	Pa
shear strain	γ	$\gamma = \Delta x/d$	1
shear modulus	G	$G = \tau/\gamma$	Pa
volume strain, bulk strain	θ	$\theta = \Delta V/V_0$	1
bulk modulus, compression modulus	K	$K = -V_0(dp/dV)$	Pa
viscosity, dynamic viscosity	η, μ	$\tau_{x,z} = \eta(dv_x/dz)$	Pa s
fluidity	ϕ	$\phi = 1/\eta$	m kg^{-1} s
kinematic viscosity	ν	$\nu = \eta/\rho$	m^2 s^{-1}
friction coefficient	$\mu, (f)$	$F_{frict} = \mu F_{norm}$	1
power	P	$P = dW/dt$	W
sound energy flux	P, P_a	$P = dE/dt$	W
acoustic factors,			
reflection factor	ρ	$\rho = P_r/P_0$	1
acoustic absorption factor	$\alpha_a, (\alpha)$	$\alpha_a = 1 - \rho$	1
transmission factor	τ	$\tau = P_{tr}/P_0$	1
dissipation factor	δ	$\delta = \alpha_a - \tau$	1

ELECTRICITY AND MAGNETISM

Name	*Symbol*	*Definition*	*SI unit*
quantity of electricity, electric charge	Q		C
charge density	ρ	$\rho = Q/V$	C m^{-3}
surface charge density	σ	$\sigma = Q/A$	C m^{-2}
electric potential	V, ϕ	$V = dW/dQ$	V, J C^{-1}
electric potential difference	$U, \Delta V, \Delta\phi$	$U = V_2 - V_1$	V
electromotive force	E	$E = \int(F/Q)\cdot ds$	V
electric field strength	E	$E = F/Q = -\text{grad } V$	V m^{-1}
electric flux	Ψ	$\Psi = \int D \cdot dA$	C
electric displacement	D	$D = \varepsilon E$	C m^{-2}
capacitance	C	$C = Q/U$	F, C V^{-1}
permittivity	ε	$D = \varepsilon E$	F m^{-1}

Name	*Symbol*	*Definition*	*SI unit*
permittivity of vacuum	ε_0	$\varepsilon_0 = \mu_0^{-1} c_0^{-2}$	$\mathrm{F\,m^{-1}}$
relative permittivity	ε_r	$\varepsilon_r = \varepsilon/\varepsilon_0$	1
dielectric polarization (dipole moment per volume)	P	$P = D - \varepsilon_0 E$	$\mathrm{C\,m^{-2}}$
electric susceptibility	χ_e	$\chi_e = \varepsilon_r - 1$	1
electric dipole moment	p, μ	$p = Qr$	C m
electric current	I	$I = \mathrm{d}Q/\mathrm{d}t$	A
electric current density	j, J	$I = \int j \cdot \mathrm{d}A$	$\mathrm{A\,m^{-2}}$
magnetic flux density, magnetic induction	B	$F = Qv \times B$	T
magnetic flux	Φ	$\Phi = \int B \cdot \mathrm{d}A$	Wb
magnetic field strength	H	$B = \mu H$	$\mathrm{A\,m^{-1}}$
permeability	μ	$B = \mu H$	$\mathrm{N\,A^{-2}}$, $\mathrm{H\,m^{-1}}$
permeability of vacuum	μ_0		$\mathrm{H\,m^{-1}}$
relative permeability	μ_r	$\mu_r = \mu/\mu_0$	1
magnetization (magnetic dipole moment per volume)	M	$M = B/\mu_0 - H$	$\mathrm{A\,m^{-1}}$
magnetic susceptibility	$\chi, \kappa, (\chi_m)$	$\chi = \mu_r - 1$	1
molar magnetic susceptibility	χ_m	$\chi_m = V_m \chi$	$\mathrm{m^3\,mol^{-1}}$
magnetic dipole moment	m, μ	$E_p = -m \cdot B$	$\mathrm{A\,m^2}$, $\mathrm{J\,T^{-1}}$
electrical resistance	R	$R = U/I$	Ω
conductance	G	$G = 1/R$	S
loss angle	δ	$\delta = (\pi/2) + \phi_I - \phi_U$	1, rad
reactance	X	$X = (U/I)\sin\delta$	Ω
impedance (complex impedance)	Z	$Z = R + \mathrm{i}X$	Ω
admittance (complex admittance)	Y	$Y = 1/Z$	S
susceptance	B	$Y = G + \mathrm{i}B$	S
resistivity	ρ	$\rho = E/j$	Ω m
conductivity	κ, γ, σ	$\kappa = 1/\rho$	$\mathrm{S\,m^{-1}}$
self-inductance	L	$E = -L(\mathrm{d}I/\mathrm{d}t)$	H
mutual inductance	M, L_{12}	$E_1 = L_{12}(\mathrm{d}I_2/\mathrm{d}t)$	H
magnetic vector potential	A	$B = \nabla \times A$	$\mathrm{Wb\,m^{-1}}$
Poynting vector	S	$S = E \times H$	$\mathrm{W\,m^{-2}}$

QUANTUM MECHANICS

momentum operator	$\hat{p}$	$\hat{p} = -\mathrm{i}\hbar\nabla$	$\mathrm{m^{-1}\,J\,s}$
kinetic energy operator	$\hat{T}$	$\hat{T} = -(\hbar^2/2m)\nabla^2$	J

Name	*Symbol*	*Definition*	*SI unit*
hamiltonian operator	$\hat{H}$	$\hat{H}=\hat{T}+V$	J
wavefunction, state function	Ψ, ψ, ϕ	$\hat{H}\psi=E\psi$	($m^{-3/2}$)
probability density	P	$P=\psi^*\psi$	(m^{-3})
charge density of electrons	ρ	$\rho=-eP$	($C\ m^{-3}$)
probability current density	S	$S=-i\hbar(\psi^*\nabla\psi-\psi\nabla\psi^*)/2m_e$	($m^{-2}\ s^{-1}$)
electric current density of electrons	j	$j=-eS$	($A\ m^{-2}$)
matrix element of operator $\hat{A}$	$A_{ij}, \langle i\vert\hat{A}\vert j\rangle$	$A_{ij}=\int\psi_i^*\hat{A}\psi_j\,d\tau$	(varies)
expectation value of operator $\hat{A}$	$\langle A\rangle, \bar{A}$	$\langle A\rangle=\int\psi^*\hat{A}\psi\,d\tau$	(varies)
hermitian conjugate of $\hat{A}$	$\hat{A}^\dagger$	$(\hat{A}^\dagger)_{ij}=(A_{ji})^*$	(varies)
commutator of $\hat{A}$ and $\hat{B}$	$[\hat{A},\hat{B}], [\hat{A},\hat{B}]_-$	$[\hat{A},\hat{B}]=\hat{A}\hat{B}-\hat{B}\hat{A}$	(varies)
anticommutator	$[\hat{A},\hat{B}]_+$	$[\hat{A},\hat{B}]_+=\hat{A}\hat{B}+\hat{B}\hat{A}$	(varies)
spin wavefunction	$\alpha; \beta$		1
coulomb integral	H_{AA}	$H_{AA}=\int\psi_A{}^*\hat{H}\psi_A\,d\tau$	J
resonance integral	H_{AB}	$H_{AB}=\int\psi_A{}^*\hat{H}\psi_B\,d\tau$	J
overlap integral	S_{AB}	$S_{AB}=\int\psi_A{}^*\psi_B\,d\tau$	1

ATOMS AND MOLECULES

Name	*Symbol*	*Definition*	*SI unit*
nucleon number, mass number	A		1
proton number, atomic number	Z		1
neutron number	N	$N=A-Z$	1
electron rest mass	m_e		kg
mass of atom, atomic mass	m_a, m		kg
atomic mass constant	m_u	$m_u=m_a(^{12}C)/12$	kg
mass excess	Δ	$\Delta=m_a-Am_u$	kg
elementary charge, proton charge	e		C
Planck constant	h		J s
Planck constant/2π	$\hbar$	$\hbar=h/2\pi$	J s
Bohr radius	a_0	$a_0=4\pi\varepsilon_0\hbar^2/m_e e^2$	m
Hartree energy	E_h	$E_h=\hbar^2/m_e a_0{}^2$	J
Rydberg constant	R_∞	$R_\infty=E_h/2hc$	m^{-1}
fine structure constant	α	$\alpha=e^2/4\pi\varepsilon_0\hbar c$	1
ionization energy	E_i		J

Name	*Symbol*	*Definition*	*SI unit*
electron affinity	E_{ea}		J
dissociation energy	E_d, D		J
from the ground state	D_0		J
from the potential minimum	D_e		J
principal quantum number (H atom)	n	$E=-hcR/n^2$	1
angular momentum quantum numbers	see under Spectroscopy		
magnetic dipole moment of a molecule	m, μ	$E_p=-m\cdot B$	$J\,T^{-1}$
magnetizability of a molecule	ξ	$m=\xi B$	$J\,T^{-2}$
Bohr magneton	μ_B	$\mu_B=e\hbar/2m_e$	$J\,T^{-1}$
nuclear magneton	μ_N	$\mu_N=(m_e/m_p)\mu_B$	$J\,T^{-1}$
magnetogyric ratio (gyromagnetic ratio)	γ	$\gamma=\mu/L$	$C\,kg^{-1}$
g factor	g		1
Larmor circular frequency	ω_L	$\omega_L=(e/2m)B$	s^{-1}
Larmor frequency	ν_L	$\nu_L=\omega_L/2\pi$	Hz
longitudinal relaxation time	T_1		s
transverse relaxation time	T_2		s
electric dipole moment of a molecule	p, μ	$E_p=-p\cdot E$	C m
quadrupole moment of a molecule	Q; Θ	$E_p=\frac{1}{2}Q:V''=\frac{1}{3}\Theta:V''$	$C\,m^2$
quadrupole moment of a nucleus	eQ	$eQ=2\langle\Theta_{zz}\rangle$	$C\,m^2$
electric field gradient tensor	q	$q_{\alpha\beta}=-\partial^2V/\partial\alpha\partial\beta$	$V\,m^{-2}$
quadrupole interaction energy tensor	χ	$\chi_{\alpha\beta}=eQq_{\alpha\beta}$	J
electric polarizability of a molecule	α	p (induced)$=\alpha E$	$C\,m^2\,V^{-1}$
activity (of a radioactive substance)	A	$A=-dN_B/dt$	Bq
decay (rate) constant, disintegration (rate) constant	λ	$A=\lambda N_B$	s^{-1}
half life	$t_{\frac{1}{2}}$, $T_{\frac{1}{2}}$		s
mean life	τ		s
level width	Γ	$\Gamma=\hbar/\tau$	J

Name	*Symbol*	*Definition*	*SI unit*
disintegration energy	Q		J
cross section (of a nuclear reaction)	σ		m^2

SPECTROSCOPY

Name	*Symbol*	*Definition*	*SI unit*
total term	T	$T = E_{tot}/hc$	m^{-1}
transition wavenumber	$\tilde{\nu}$, (ν)	$\tilde{\nu} = T' - T''$	m^{-1}
transition frequency	ν	$\nu = (E' - E'')/h$	Hz
electronic term	T_e	$T_e = E_e/hc$	m^{-1}
vibrational term	G	$G = E_{vib}/hc$	m^{-1}
rotational term	F	$F = E_{rot}/hc$	m^{-1}
spin orbit coupling constant	A	$T_{s.o.} = A\langle \hat{L} \cdot \hat{S} \rangle$	m^{-1}
principal moments of inertia	$I_A; I_B; I_C$	$I_A \leqslant I_B \leqslant I_C$	$kg\ m^2$
rotational constants,			
in wavenumber	$\tilde{A}; \tilde{B}; \tilde{C}$	$\tilde{A} = h/8\pi^2 c I_A$	m^{-1}
in frequency	$A; B; C$	$A = h/8\pi^2 I_A$	Hz
inertial defect	Δ	$\Delta = I_C - I_A - I_B$	$kg\ m^2$
asymmetry parameter	κ	$\kappa = \dfrac{(2B - A - C)}{(A - C)}$	1
centrifugal distortion constants,			
S reduction	$D_J; D_{JK}; D_K; d_1; d_2$		m^{-1}
A reduction	$\Delta_J; \Delta_{JK}; \Delta_K; \delta_J; \delta_K$		m^{-1}
harmonic vibration wavenumber	$\omega_e; \omega_r$		m^{-1}
vibrational anharmonicity constant	$\omega_e x_e; x_{rs}; g_{tt'}$		m^{-1}
vibrational quantum numbers	$v_r; l_t$		1
Coriolis zeta constant	ζ_{rs}^{α}		1
angular momentum quantum numbers	see additional information below		
degeneracy, statistical weight	g, d, β		1
electric dipole moment of a molecule	p, μ	$E_p = -p \cdot E$	C m
transition dipole moment of a molecule	M, R	$M = \int \psi' p \psi'' d\tau$	C m
molecular geometry, interatomic distances,			
equilibrium distance	r_e		m
zero-point average distance	r_z		m
ground state distance	r_0		m

Name	*Symbol*	*Definition*	*SI unit*
substitution structure distance	r_s		m
vibrational coordinates,			
internal coordinates	R_i, r_i, θ_j, etc.		(varies)
symmetry coordinates	S_i		(varies)
normal coordinates			
mass adjusted	Q_r		$kg^{\frac{1}{2}}$ m
dimensionless	q_r		1
vibrational force constants,			
diatomic	f, (k)	$f=\partial^2 V/\partial r^2$	$J\ m^{-2}$
polyatomic,			
internal coordinates	f_{ij}	$f_{ij}=\partial^2 V/\partial r_i \partial r_j$	(varies)
symmetry coordinates	F_{ij}	$F_{ij}=\partial^2 V/\partial S_i \partial S_j$	(varies)
dimensionless normal coordinates	$\phi_{rst\ldots}$, $k_{rst\ldots}$		m^{-1}
nuclear magnetic resonance (NMR),			
magnetogyric ratio	γ	$\gamma=\mu/I\hbar$	$C\ kg^{-1}$
shielding constant	σ_A	$B_A=(1-\sigma_A)B$	1
chemical shift, δ scale	δ	$\delta=10^6(\nu-\nu_0)/\nu_0$	1
(indirect) spin–spin coupling constant	J_{AB}	$\hat{H}/h=J_{AB}\hat{I}_A\cdot\hat{I}_B$	Hz
direct (dipolar) coupling constant	D_{AB}		Hz
longitudinal relaxation time	T_1		s
transverse relaxation time	T_2		s
electron spin resonance, electron paramagnetic resonance (ESR, EPR),			
magnetogyric ratio	γ	$\gamma=\mu/s\hbar$	$C\ kg^{-1}$
g factor	g	$h\nu=g\mu_B B$	1
hyperfine coupling constant,			
in liquids	a, A	$\hat{H}_{hfs}/h=a\hat{S}\cdot\hat{I}$	Hz
in solids	T	$\hat{H}_{hfs}/h=\hat{S}\cdot T\cdot\hat{I}$	Hz

Angular momentum	*Operator symbol*	*Quantum number symbol* *Total*	*Z-axis*	*z-axis*
electron orbital	$\hat{L}$	L	M_L	Λ
one electron only	$\hat{l}$	l	m_l	λ

Angular momentum	*Operator symbol*	*Quantum number symbol Total*	*Z-axis*	*z-axis*
electron spin	$\hat{S}$	S	M_S	Σ
one electron only	$\hat{s}$	s	m_s	σ
electron orbital + spin	$\hat{L}+\hat{S}$			$\Omega=\Lambda+\Sigma$
nuclear orbital (rotational)	$\hat{R}$	R		K_R, k_R
nuclear spin	$\hat{I}$	I	M_I	
internal vibrational				
spherical top	$\hat{l}$	$l(l\zeta)$		K_l
other	$\hat{j}$, $\hat{\pi}$			$l(l\zeta)$
sum of $R+L(+j)$	$\hat{N}$	N		K, k
sum of $N+S$	$\hat{J}$	J	M_J	K, k
sum of $J+I$	$\hat{F}$	F	M_F	

ELECTROMAGNETIC RADIATION

Name	*Symbol*	*Definition*	*SI unit*
wavelength	λ		m
speed of light			
in vacuum	c_0		$m\ s^{-1}$
in a medium	c	$c=c_0/n$	$m\ s^{-1}$
wavenumber in vacuum	$\tilde{\nu}$	$\tilde{\nu}=\nu/c_0=1/n\lambda$	m^{-1}
wavenumber (in a medium)	σ	$\sigma=1/\lambda$	m^{-1}
frequency	ν	$\nu=c/\lambda$	Hz
circular frequency, pulsatance	ω	$\omega=2\pi\nu$	s^{-1}, rad s^{-1}
refractive index	n	$n=c_0/c$	1
Planck constant	h		J s
Planck constant/2π	$\hbar$	$\hbar=h/2\pi$	J s
radiant energy	Q, W		J
radiant energy density	ρ, w	$\rho=Q/V$	$J\ m^{-3}$
spectral radiant energy density			
in terms of frequency	ρ_ν, w_ν	$\rho_\nu=d\rho/d\nu$	$J\ m^{-3}\ Hz^{-1}$
in terms of wavenumber	$\rho_{\tilde{\nu}}$, $w_{\tilde{\nu}}$	$\rho_{\tilde{\nu}}=d\rho/d\tilde{\nu}$	$J\ m^{-2}$
in terms of wavelenglth	ρ_λ, w_λ	$\rho_\lambda=d\rho/d\lambda$	$J\ m^{-4}$
Einstein transition probabilities			
spontaneous emission	A_{nm}	$dN_n/dt=-A_{nm}N_n$	s^{-1}
stimulated emission	B_{nm}	$dN_n/dt=-\rho_{\tilde{\nu}}(\tilde{\nu}_{nm})\times B_{nm}N_n$	$s\ kg^{-1}$
stimulated absorption	B_{mn}	$dN_n/dt=\rho_{\tilde{\nu}}(\tilde{\nu}_{nm})B_{mn}N_m$	$s\ kg^{-1}$
radiant power, radiant energy per time	Φ, P	$\Phi=dQ/dt$	W

Name	*Symbol*	*Definition*	*SI unit*
radiant intensity	I	$I = d\Phi/d\Omega$	$W\ sr^{-1}$
radiant exitance, (emitted radiant flux)	M	$M = d\Phi/dA_{source}$	$W\ m^{-2}$
irradiance, (radiant flux received)	E, (I)	$E = d\Phi/dA$	$W\ m^{-2}$
emittance	ε	$\varepsilon = M/M_{bb}$	1
Stefan–Boltzmann constant	σ	$M_{bb} = \sigma T^4$	$W\ m^{-2}\ K^{-4}$
first radiation constant	c_1	$c_1 = 2\pi h c_0{}^2$	$W\ m^2$
second radiation constant	c_2	$c_2 = hc_0/k$	K m
transmittance, transmission factor	τ, T	$\tau = \Phi_{tr}/\Phi_0$	1
absorptance, absorption factor	α	$\alpha = \Phi_{abs}/\Phi_0$	1
reflectance, reflection factor	ρ	$\rho = \Phi_{refl}/\Phi_0$	1
(decadic) absorbance	A	$A = -\lg(1-\alpha_i)$	1
napierian absorbance	B	$B = -\ln(1-\alpha_i)$	1
absorption coefficient			
(linear) decadic	a, K	$a = A/l$	m^{-1}
(linear) napierian	α	$\alpha = B/l$	m^{-1}
molar (decadic)	ε	$\varepsilon = a/c = A/cl$	$m^2\ mol^{-1}$
molar napierian	κ	$\kappa = \alpha/c = B/cl$	$m^2\ mol^{-1}$
absorption index	k	$k = \alpha/4\pi\tilde{\nu}$	1
complex refractive index	$\hat{n}$	$\hat{n} = n + ik$	1
molar refraction	R, R_m	$R = \frac{(n^2-1)}{(n^2+2)} V_m$	$m^3\ mol^{-1}$
angle of optical rotation	α		1, rad

SOLID STATE

Name	*Symbol*	*Definition*	*SI unit*
lattice vector	R, R_0		m
fundamental translation vectors for the crystal lattice	a_1; a_2; a_3, a; b; c	$R = n_1 a_1 + n_2 a_2 + n_3 a_3$	m
(circular) reciprocal lattice vector	G	$G \cdot R = 2\pi m$	m^{-1}
(circular) fundamental translation vectors for the reciprocal lattice	b_1; b_2; b_3, a^*; b^*; c^*	$a_i \cdot b_k = 2\pi\delta_{ik}$	m^{-1}
lattice plane spacing	d		m
Bragg angle	θ	$n\lambda = 2d \sin\theta$	1, rad

Name	*Symbol*	*Definition*	*SI unit*
order of reflection	n		1
order parameters			
short range	σ		1
long range	s		1
Burgers vector	$\boldsymbol{b}$		m
particle position vector	$\boldsymbol{r}, \boldsymbol{R}_j$		m
equilibrium position vector of an ion	$\boldsymbol{R}_0$		m
displacement vector of an ion	$\boldsymbol{u}$	$\boldsymbol{u}=\boldsymbol{R}-\boldsymbol{R}_0$	m
Debye–Waller factor	B, D		1
Debye circular wavenumber	q_D		m^{-1}
Debye circular frequency	ω_D		s^{-1}
Grüneisen parameter	γ, Γ	$\gamma=\alpha V/\kappa C_V$	1
Madelung constant	$\alpha, \mathcal{M}$	$E_{coul}=\frac{\alpha N_A z_+ z_- e^2}{4\pi\varepsilon_0 R_0}$	1
density of states	N_E	$N_E=dN(E)/dE$	$J^{-1}\,m^{-3}$
(spectral) density of vibrational modes	N_ω, g	$N_\omega=dN(\omega)/d\omega$	$s\,m^{-3}$
resistivity tensor	ρ_{ik}	$\boldsymbol{E}=\boldsymbol{\rho}\cdot\boldsymbol{j}$	$\Omega\,m$
conductivity tensor	σ_{ik}	$\boldsymbol{\sigma}=\boldsymbol{\rho}^{-1}$	$S\,m^{-1}$
thermal conductivity tensor	λ_{ik}	$\boldsymbol{J}_q=-\boldsymbol{\lambda}\cdot\text{grad}\,T$	$W\,m^{-1}\,K^{-1}$
residual resistivity	ρ_R		$\Omega\,m$
relaxation time	τ	$\tau=l/v_F$	s
Lorenz coefficient	L	$L=\lambda/\sigma T$	$V^2\,K^{-2}$
Hall coefficient	A_H, R_H	$\boldsymbol{E}=\boldsymbol{\rho}\cdot\boldsymbol{j}+R_H(\boldsymbol{B}\times\boldsymbol{j})$	$m^3\,C^{-1}$
thermoelectric force	E		V
Peltier coefficient	Π		V
Thomson coefficient	$\mu, (\tau)$		$V\,K^{-1}$
work function	Φ	$\Phi=E_\infty-E_F$	J
number density, number concentration	$n, (p)$		m^{-3}
gap energy	E_g		J
donor ionization energy	E_d		J
acceptor ionization energy	E_a		J
Fermi energy	E_F, ε_F		J
circular wave vector, propagation vector	$\boldsymbol{k}, \boldsymbol{q}$	$k=2\pi/\lambda$	m^{-1}
Bloch function	$u_k(\boldsymbol{r})$	$\psi(\boldsymbol{r})=u_k(\boldsymbol{r})\exp(i\boldsymbol{k}\cdot\boldsymbol{r})$	$m^{-3/2}$
charge density of electrons	ρ	$\rho(\boldsymbol{r})=-e\psi^*(\boldsymbol{r})\psi(\boldsymbol{r})$	$C\,m^{-3}$

Name	*Symbol*	*Definition*	*SI unit*
effective mass	m^*		kg
mobility	μ	$\mu = v_{\mathrm{drift}}/E$	$\mathrm{m^2\,V^{-1}\,s^{-1}}$
mobility ratio	b	$b = \mu_n/\mu_p$	1
diffusion coefficient	D	$\mathrm{d}N/\mathrm{d}t = -DA(\mathrm{d}n/\mathrm{d}x)$	$\mathrm{m^2\,s^{-1}}$
diffusion length	L	$L = \sqrt{D\tau}$	m
characteristic (Weiss) temperature	θ, θ_W		K
Curie temperature	T_C		K
Néel temperature	T_N		K

STATISTICAL THERMODYNAMICS

Name	*Symbol*	*Definition*	*SI unit*
number of entities	N		1
number density of entities, number concentration	n, C	$n = N/V$	$\mathrm{m^{-3}}$
Avogadro constant	L, N_A		$\mathrm{mol^{-1}}$
Boltzmann constant	k, k_B		$\mathrm{J\,K^{-1}}$
gas constant (molar)	R	$R = Lk$	$\mathrm{J\,K^{-1}\,mol^{-1}}$
molecular position vector	$\boldsymbol{r}\,(x, y, z)$		m
molecular velocity vector	$\boldsymbol{c}(c_x, c_y, c_z)$, $\boldsymbol{u}(u_x, u_y, u_z)$	$\boldsymbol{c} = \mathrm{d}\boldsymbol{r}/\mathrm{d}t$	$\mathrm{m\,s^{-1}}$
molecular momentum vector	$\boldsymbol{p}(p_x, p_y, p_z)$	$\boldsymbol{p} = m\boldsymbol{c}$	$\mathrm{kg\,m\,s^{-1}}$
velocity distribution function (Maxwell)	$f(c_x)$	$f(c_x) = (m/2\pi kT)^{\frac{1}{2}} \times \exp(-mc_x^2/2kT)$	$\mathrm{m^{-1}\,s}$
speed distribution function (Maxwell–Boltzmann)	$F(c)$	$F(c) = (m/2\pi kT)^{3/2} \times 4\pi c^2 \exp(-mc^2/2kT)$	$\mathrm{m^{-1}\,s}$
average speed	$\bar{c}$, $\bar{u}$, $\langle c \rangle$, $\langle u \rangle$	$\bar{c} = \int cF(c)\mathrm{d}c$	$\mathrm{m\,s^{-1}}$
generalized coordinate	q		(m)
generalized momentum	p	$p = \partial L/\partial \dot{q}$	$(\mathrm{kg\,m\,s^{-1}})$
volume in phase space	Ω	$\Omega = (1/h)\int p\,\mathrm{d}q$	1
probability	P		1
statistical weight, degeneracy	g, d, W, ω, β		1
density of states	$\rho(E)$	$\rho(E) = \mathrm{d}N/\mathrm{d}E$	$\mathrm{J^{-1}}$
partition function, sum over states,			
for a single molecule	q, z	$q = \sum_i g_i \exp(-\varepsilon_i/kT)$	1
for a canonical ensemble (system, or assembly)	Q, Z		1

Name	*Symbol*	*Definition*	*SI unit*
microcanonical ensemble	Ω		1
grand (canonical ensemble)	Ξ		1
symmetry number	σ, s		1
reciprocal temperature parameter	β	$\beta = 1/kT$	J^{-1}
characteristic temperature	Θ		K

GENERAL CHEMISTRY

Name	*Symbol*	*Definition*	*SI unit*
number of entities (e.g. molecules, atoms, ions, formula units)	N		1
amount (of substance)	n	$n_B = N_B/L$	mol
Avogadro constant	L, N_A		mol^{-1}
mass of atom, atomic mass	m_a, m		kg
mass of entity (molecule, or formula unit)	m_f, m		kg
atomic mass constant	m_u	$m_u = m_a(^{12}C)/12$	kg
molar mass	M	$M_B = m/n_B$	$kg\ mol^{-1}$
relative molecular mass (relative molar mass, molecular weight)	M_r	$M_{r,B} = m_B/m_u$	1
molar volume	V_m	$V_{m,B} = V/n_B$	$m^3\ mol^{-1}$
mass fraction	w	$w_B = m_B/\Sigma m_i$	1
volume fraction	ϕ	$\phi_B = V_B/\Sigma V_i$	1
mole fraction, amount fraction, number fraction	x, y	$x_B = n_B/\Sigma n_i$	1
(total) pressure	p, P		Pa
partial pressure	p_B	$p_B = y_B p$	Pa
mass concentration (mass density)	γ, ρ	$\gamma_B = m_B/V$	$kg\ m^{-3}$
number concentration, number density of entities	C, n	$C_B = N_B/V$	m^{-3}
amount concentration, concentration	c	$c_B = n_B/V$	$mol\ m^{-3}$
solubility	s	$s_B = c_B$ (saturated solution)	$mol\ m^{-3}$
molality (of a solute)	m, (b)	$m_B = n_B/m_A$	$mol\ kg^{-1}$

Name	*Symbol*	*Definition*	*SI unit*
surface concentration	Γ	$\Gamma_B = n_B/A$	mol m^{-2}
stoichiometric number	ν		1
extent of reaction, advancement	ξ	$\Delta\xi = \Delta n_B/\nu_B$	mol
degree of dissociation	α		1

CHEMICAL THERMODYNAMICS

Name	*Symbol*	*Definition*	*SI unit*
heat	q, Q		J
work	w, W		J
internal energy	U	$\Delta U = q + w$	J
enthalpy	H	$H = U + pV$	J
thermodynamic temperature	T		K
Celsius temperature	θ, t	$\theta/°C = T/K - 273.15$	°C
entropy	S	$dS \geqslant dq/T$	J K^{-1}
Helmholtz energy, (Helmholtz function)	A	$A = U - TS$	J
Gibbs energy, (Gibbs function)	G	$G = H - TS$	J
Massieu function	J	$J = -A/T$	J K^{-1}
Planck function	Y	$Y = -G/T$	J K^{-1}
surface tension	γ, σ	$\gamma = (\partial G/\partial A_s)_{T,p}$	J m^{-2}, N m^{-1}
molar quantity X	X_m	$X_m = X/n$	(varies)
specific quantity X	x	$x = X/m$	(varies)
pressure coefficient	β	$\beta = (\partial p/\partial T)_V$	Pa K^{-1}
relative pressure coefficient	α_p	$\alpha_p = (1/p)(\partial p/\partial T)_V$	K^{-1}
compressibility,			
isothermal	κ_T	$\kappa_T = -(1/V)(\partial V/\partial p)_T$	Pa^{-1}
isentropic	κ_S	$\kappa_S = -(1/V)(\partial V/\partial p)_S$	Pa^{-1}
linear expansion coefficient	α_l	$\alpha_l = (1/l)(\partial l/\partial T)$	K^{-1}
cubic expansion coefficient	α, α_V, γ	$\alpha = (1/V)(\partial V/\partial T)_p$	K^{-1}
heat capacity,			
at constant pressure	C_p	$C_p = (\partial H/\partial T)_p$	J K^{-1}
at constant volume	C_V	$C_V = (\partial U/\partial T)_V$	J K^{-1}
ratio of heat capacities	γ, (κ)	$\gamma = C_p/C_V$	1
Joule–Thomson coefficient	μ, μ_{JT}	$\mu = (\partial T/\partial p)_H$	K Pa^{-1}
second virial coefficient	B	$pV_m = RT(1 + B/V_m + \cdots)$	m^3 mol^{-1}
compression factor (compressibility factor)	Z	$Z = pV_m/RT$	1

Name	*Symbol*	*Definition*	*SI unit*
partial molar quantity X	X_B, (X'_B)	$X_B = (\partial X/\partial n_B)_{T,p,n_{j \neq B}}$	(varies)
chemical potential (partial molar Gibbs energy)	μ	$\mu_B = (\partial G/\partial n_B)_{T,p,n_{j \neq B}}$	$J\ mol^{-1}$
absolute activity	λ	$\lambda_B = \exp(\mu_B/RT)$	1
standard chemical potential	$\mu^\ominus$, μ°		$J\ mol^{-1}$
standard partial molar enthalpy	$H_B^\ominus$	$H_B^\ominus = \mu_B^\ominus + TS_B^\ominus$	$J\ mol^{-1}$
standard partial molar entropy	$S_B^\ominus$	$S_B^\ominus = -(\partial \mu_B^\ominus/\partial T)_p$	$J\ mol^{-1}\ K^{-1}$
standard reaction Gibbs energy (function)	$\Delta_r G^\ominus$	$\Delta_r G^\ominus = \sum_B \nu_B \mu_B^\ominus$	$J\ mol^{-1}$
affinity of reaction	A, ($\mathscr{A}$)	$A = -(\partial G/\partial \xi)_{p,T}$ $= -\sum_B \nu_B \mu_B$	$J\ mol^{-1}$
standard reaction enthalpy	$\Delta_r H^\ominus$	$\Delta_r H^\ominus = \sum_B \nu_B H_B^\ominus$	$J\ mol^{-1}$
standard reaction entropy	$\Delta_r S^\ominus$	$\Delta_r S^\ominus = \sum_B \nu_B S_B^\ominus$	$J\ mol^{-1}\ K^{-1}$
equilibrium constant	$K^\ominus$, K	$K^\ominus = \exp(-\Delta_r G^\ominus/RT)$	1
equilibrium constant,			
pressure basis	K_p	$K_p = \prod_B p_B^{\nu_B}$	$Pa^{\Sigma\nu}$
concentration basis	K_c	$K_c = \prod_B c_B^{\nu_B}$	$(mol\ m^{-3})^{\Sigma\nu}$
molality basis	K_m	$K_m = \prod_B m_B^{\nu_B}$	$(mol\ kg^{-1})^{\Sigma\nu}$
fugacity	f, $\tilde{p}$	$f_B = \lambda_B \lim_{p \to 0} (p_B/\lambda_B)_T$	Pa
fugacity coefficient	ϕ	$\phi_B = f_B/p_B$	1
activity and activity coefficient referenced to Raoult's law,			
(relative) activity	a	$a_B = \exp\left[\frac{\mu_B - \mu_B^*}{RT}\right]$	1
activity coefficient	f	$f_B = a_B/x_B$	1
activities and activity coefficients referenced to Henry's law, (relative) activity,			
molality basis	a_m	$a_{m,B} = \exp\left[\frac{\mu_B - \mu_B^\ominus}{RT}\right]$	1
concentration basis	a_c	$a_{c,B} = \exp\left[\frac{\mu_B - \mu_B^\ominus}{RT}\right]$	1

Name	*Symbol*	*Definition*	*SI Unit*
mole fraction basis	a_x	$a_{x,\mathrm{B}}=\exp\left[\dfrac{\mu_\mathrm{B}-\mu_\mathrm{B}^*}{RT}\right]$	1
activity coefficient,			
molality basis	γ_m	$a_{m,\mathrm{B}}=\gamma_{m,\mathrm{B}}m_\mathrm{B}/m^\circ$	1
concentration basis	γ_c	$a_{c,\mathrm{B}}=\gamma_{c,\mathrm{B}}c_\mathrm{B}/c^\circ$	1
mole fraction basis	γ_x	$a_{x,\mathrm{B}}=\gamma_{x,\mathrm{B}}x_\mathrm{B}$	1
ionic strength,			
molality basis	I_m, I	$I_m=\frac{1}{2}\Sigma m_\mathrm{B}z_\mathrm{B}^2$	mol kg^{-1}
concentration basis	I_c, I	$I_c=\frac{1}{2}\Sigma c_\mathrm{B}z_\mathrm{B}^2$	mol m^{-3}
osmotic coefficient,			
molality basis	ϕ_m	$\phi_m=(\mu_\mathrm{A}^*-\mu_\mathrm{A})/(RTM_\mathrm{A}\Sigma m_\mathrm{B})$	1
mole fraction basis	ϕ_x	$\phi_x=(\mu_\mathrm{A}-\mu_\mathrm{A}^*)/(RT\ln x_\mathrm{A})$	1
osmotic pressure	Π	$\Pi=c_\mathrm{B}RT$ (ideal dilute solution)	Pa

(i) *Symbols used as subscripts to denote a chemical process or reaction*

These symbols should be printed in roman (upright) type, without a full stop (period).

vaporization, evaporation (liquid→gas)	vap
sublimation (solid→gas)	sub
melting, fusion (solid→liquid)	fus
transition (between two phases)	trs
mixing of fluids	mix
solution (of solute in solvent)	sol
dilution (of a solution)	dil
adsorption	ads
displacement	dpl
immersion	imm
reaction in general	r
atomization	at
combustion reaction	c
formation reaction	f

(ii) *Recommended superscripts*

standard	$\ominus$, o
pure substance	*
infinite dilution	∞
ideal	id
activated complex, transition state	‡
excess quantity	E

Name	*Symbol*	*Definition*	*SI unit*
		CHEMICAL KINETICS	
rate of change of quantity X	$\dot{X}$	$\dot{X}=\mathrm{d}X/\mathrm{d}t$	(varies)
rate of conversion	$\dot{\xi}$	$\dot{\xi}=\mathrm{d}\xi/\mathrm{d}t$	$\mathrm{mol\ s^{-1}}$
rate of concentration change (due to chemical reaction)	r_B, v_B	$r_B=\mathrm{d}c_B/\mathrm{d}t$	$\mathrm{mol\ m^{-3}\ s^{-1}}$
rate of reaction (based on amount concentration)	v	$v=\dot{\xi}/V$ $=\nu_B^{-1}\mathrm{d}c_B/\mathrm{d}t$	$\mathrm{mol\ m^{-3}\ s^{-1}}$
partial order of reaction	n_B	$v=k\Pi c_B^{n_B}$	1
overall order of reaction	n	$n=\Sigma n_B$	1
rate constant, rate coefficient	k	$v=k\Pi c_B^{n_B}$	$\mathrm{(mol^{-1}\ m^3)^{n-1}s^{-1}}$
Boltzmann constant	k, k_B		$\mathrm{J\ K^{-1}}$
half life	$t_{\frac{1}{2}}$	$c(t_{\frac{1}{2}})=c_0/2$	s
relaxation time	τ	$\tau=1/(k_1+k_{-1})$	s
energy of activation, activation energy	E_a, E	$E_a=RT^2\ \mathrm{d}\ln k/\mathrm{d}T$	$\mathrm{J\ mol^{-1}}$
pre-exponential factor	A	$k=A\exp(-E_a/RT)$	$\mathrm{(mol^{-1}\ m^3)^{n-1}s^{-1}}$
volume of activation	$\Delta^{\ddagger}V$	$\Delta^{\ddagger}V=-RT\times$ $(\partial\ln k/\partial p)_T$	$\mathrm{m^3\ mol^{-1}}$
collision diameter	d	$d_{AB}=r_A+r_B$	m
collision cross-section	σ	$\sigma_{AB}=\pi d_{AB}^2$	$\mathrm{m^2}$
collision frequency	Z_A		$\mathrm{s^{-1}}$
collision number	Z_{AB}, Z_{AA}		$\mathrm{m^{-3}s^{-1}}$
collision frequency factor	z_{AB}, z_{AA}	$z_{AB}=Z_{AB}/Lc_Ac_B$	$\mathrm{m^3\ mol^{-1}\ s^{-1}}$
standard enthalpy of activation	$\Delta^{\ddagger}H^{\circ}, \Delta H^{\ddagger}$		$\mathrm{J\ mol^{-1}}$
standard entropy of activation	$\Delta^{\ddagger}S^{\circ}, \Delta S^{\ddagger}$		$\mathrm{J\ mol^{-1}\ K^{-1}}$
standard Gibbs energy of activation	$\Delta^{\ddagger}G^{\circ}, \Delta G^{\ddagger}$		$\mathrm{J\ mol^{-1}}$
quantum yield, photochemical yield	ϕ		1
		ELECTROCHEMISTRY	
elementary charge (proton charge)	e		C
Faraday constant	F	$F=eL$	$\mathrm{C\ mol^{-1}}$

Name	*Symbol*	*Definition*	*SI unit*
charge number of an ion	z	$z_B = Q_B/e$	1
ionic strength	I_c, I	$I_c = \frac{1}{2}\sum c_i z_i^2$	$mol\ m^{-3}$
mean ionic activity	$a_\pm$	$a_\pm = m_\pm \gamma_\pm / m^\circ$	1
mean ionic molality	$m_\pm$	$m_\pm^{(\nu_+ + \nu_-)} = m_+^{\nu_+} m_-^{\nu_-}$	$mol\ kg^{-1}$
mean ionic activity coefficient	$\gamma_\pm$	$\gamma_\pm^{(\nu_+ + \nu_-)} = \gamma_+^{\nu_+} \gamma_-^{\nu_-}$	1
charge number of electrochemical cell reaction	n, (z)		1
electric potential difference (of a galvanic cell)	ΔV, E, U	$\Delta V = V_R - V_L$	V
emf, electromotive force	E	$E = \lim_{I \to 0} \Delta V$	V
standard emf, standard potential of the electrochemical cell reaction	E°	$E^\circ = -\Delta_r G^\circ / nF$ $= (RT/nF) \ln K^\circ$	V
standard electrode potential	E°		V
emf of the cell, potential of the electrochemical cell reaction	E	$E = E^\circ - (RT/nF) \times \sum \nu_i \ln a_i$	V
pH	pH	$pH \approx -\lg\left[\frac{c(H^+)}{mol\ dm^{-3}}\right]$	1
inner electric potential	ϕ	$\nabla\phi = -E$	V
outer electric potential	ψ	$\psi = Q/4\pi\varepsilon_0 r$	V
surface electric potential	χ	$\chi = \phi - \psi$	V
Galvani potential difference	$\Delta\phi$	$\Delta_\alpha^\beta \phi = \phi^\beta - \phi^\alpha$	V
volta potential difference	$\Delta\psi$	$\Delta_\alpha^\beta \psi = \psi^\beta - \psi^\alpha$	V
electrochemical potential	$\tilde{\mu}$	$\tilde{\mu}_B{}^\alpha = (\partial G / \partial n_B{}^\alpha)$	$J\ mol^{-1}$
electric current	I	$I = dQ/dt$	A
(electric) current density	j	$j = I/A$	$A\ m^{-2}$
(surface) charge density	σ	$\sigma = Q/A$	$C\ m^{-2}$
electrode reaction rate constant	k	$k_{ox} = I_a / (nFA \prod_i c_i^{n_i})$	(varies)
mass transfer coefficient, diffusion rate constant	k_d	$k_{d,B} = \lvert\nu_B\rvert I_{l,B} / nFcA$	$m\ s^{-1}$
thickness of diffusion layer	δ	$\delta_B = D_B / k_{d,B}$	m

Name	*Symbol*	*Definition*	*SI unit*
transfer coefficient (electrochemical)	α	$\alpha_c = \frac{-\lvert\nu\rvert RT}{nF}\frac{\partial \ln \lvert I_c\rvert}{\partial E}$	1
overpotential	η	$\eta = E_I - E_{I=0} - IR_u$	V
electrokinetic potential (zeta potential)	ζ		V
conductivity	$\kappa, (\sigma)$	$\kappa = j/E$	$S\,m^{-1}$
conductivity cell constant	K_{cell}	$K_{cell} = \kappa R$	m^{-1}
molar conductivity (of an electrolyte)	Λ	$\Lambda_B = \kappa/c_B$	$S\,m^2\,mol^{-1}$
ionic conductivity, molar conductivity of an ion	λ	$\lambda_B = \lvert z_B\rvert F u_B$	$S\,m^2\,mol^{-1}$
electric mobility	$u, (\mu)$	$u_B = v_B/E$	$m^2\,V^{-1}\,s^{-1}$
transport number	t	$t_B = j_B/\Sigma j_i$	1
reciprocal radius of ionic atmosphere	κ	$\kappa = (2F^2 I/\varepsilon RT)^{\frac{1}{2}}$	m^{-1}

COLLOID AND SURFACE CHEMISTRY

Name	*Symbol*	*Definition*	*SI unit*
specific surface area	a, a_s, s	$a = A/m$	$m^2\,kg^{-1}$
surface amount of B, adsorbed amount of B	n_B^s, n_B^a		mol
surface excess of B	n_B^σ		mol
surface excess concentration of B	$\Gamma_B, (\Gamma_B^\sigma)$	$\Gamma_B = n_B^\sigma/A$	$mol\,m^{-2}$
total surface excess concentration	$\Gamma, (\Gamma^\sigma)$	$\Gamma = \sum_i \Gamma_i$	$mol\,m^{-2}$
area per molecule	a, σ	$a_B = A/N_B^\sigma$	m^2
area per molecule in a filled monolayer	a_m, σ_m	$a_{m,B} = A/N_{m,B}$	m^2
surface coverage	θ	$\theta = N_B^\sigma/N_{m,B}$	1
contact angle	θ		1, rad
film thickness	t, h, δ		m
thickness of (surface or interfacial) layer	τ, δ, t		m
surface tension, interfacial tension	γ, σ	$\gamma = (\partial G/\partial A_s)_{T,p}$	$N\,m^{-1}$, $J\,m^{-2}$
film tension	Σ_f	$\Sigma_f = 2\gamma_f$	$N\,m^{-1}$
reciprocal thickness of the double layer	κ	$\kappa = [2F^2 I_c/\varepsilon RT]^{\frac{1}{2}}$	m^{-1}
average molar masses			
number-average	M_n	$M_n = \Sigma n_i M_i/\Sigma n_i$	$kg\,mol^{-1}$
mass-average	M_m	$M_m = \Sigma n_i M_i^2/\Sigma n_i M_i$	$kg\,mol^{-1}$

Name	*Symbol*	*Definition*	*SI unit*
Z-average	M_Z	$M_Z = \Sigma n_i M_i^3 / \Sigma n_i M_i^2$	kg mol^{-1}
sedimentation coefficient	s	$s = v/a$	s
van der Waals constant	λ		J
retarded van der Waals constant	β, B		J
van der Waals–Hamaker constant	A_H		J
surface pressure	π^s, π	$\pi^s = \gamma^0 - \gamma$	N m^{-1}

TRANSPORT PROPERTIES

flux (of a quantity X)	J_X, J	$J_X = A^{-1}\, dX/dt$	(varies)
volume flow rate	q_V, $\dot{V}$	$q_v = dV/dt$	m^3 s^{-1}
mass flow rate	q_m, $\dot{m}$	$q_m = dm/dt$	kg s^{-1}
mass transfer coefficient	k_d		m s^{-1}
heat flow rate	ϕ	$\phi = dq/dt$	W
heat flux	J_q	$J_q = \phi/A$	W m^{-2}
thermal conductance	G	$G = \phi/\Delta T$	W K^{-1}
thermal resistance	R	$R = 1/G$	K W^{-1}
thermal conductivity	λ, k	$\lambda = J_q/(dT/dl)$	W m^{-1} K^{-1}
coefficient of heat transfer	h, (k, K, α)	$h = J_q/\Delta T$	W m^{-2} K^{-1}
thermal diffusivity	a	$a = \lambda/\rho c_p$	m^2 s^{-1}
diffusion coefficient	D	$D = J_n/(dc/dl)$	m^2 s^{-1}

The following symbols are used in the definitions of the dimensionless quantities: mass (m), time (t), volume (V), area (A), density (ρ), speed (v), length (l), viscosity (η), pressure (p), acceleration of free fall (g), cubic expansion coefficient (α), temperature (T), surface tension (γ), speed of sound (c), mean free path (λ), frequency (f), thermal diffusivity (a), coefficient of heat transfer (h), thermal conductivity (k), specific heat capacity at constant pressure (c_p), diffusion coefficient (D), mole fraction (x), mass transfer coefficient (k_d), permeability (μ), electric conductivity (κ), and magnetic flux density (B).

Name	*Symbol*	*Definition*	*SI unit*
Reynolds number	Re	$Re = \rho v l/\eta$	1
Euler number	Eu	$Eu = \Delta p/\rho v^2$	1
Froude number	Fr	$Fr = v/(lg)^{\frac{1}{2}}$	1
Grashof number	Gr	$Gr = l^3 g\alpha\Delta T\rho^2/\eta^2$	1
Weber number	We	$We = \rho v^2 l/\gamma$	1
Mach number	Ma	$Ma = v/c$	1
Knudsen number	Kn	$Kn = \lambda/l$	1
Strouhal number	Sr	$Sr = lf/v$	1
Fourier number	Fo	$Fo = at/l^2$	1
Péclet number	Pe	$Pe = vl/a$	1
Rayleigh number	Ra	$Ra = l^3 g\alpha\Delta T\,\rho/\eta a$	1
Nusselt number	Nu	$Nu = hl/k$	1
Stanton number	St	$St = h/\rho v c_p$	1

Name	*Symbol*	*Definition*	*SI units*
Fourier number for mass transfer	Fo^*	$Fo^* = Dt/l^2$	1
Péclet number for mass transfer	Pe^*	$Pe^* = vl/D$	1
Grashof number for mass transfer	Gr^*	$Gr^* = l^3 g \left(\frac{\partial \rho}{\partial x}\right)_{T,p} \left(\frac{\Delta x \rho}{\eta}\right)$	1
Nusselt number for mass transfer	Nu^*	$Nu^* = k_d l/D$	1
Stanton number for mass transfer	St^*	$St^* = k_d/v$	1
Prandtl number	Pr	$Pr = \eta/\rho a$	1
Schmidt number	Sc	$Sc = \eta/\rho D$	1
Lewis number	Le	$Le = a/D$	1
magnetic Reynolds number	Rm, Re_m	$Rm = v\mu\kappa l$	1
Alfvén number	Al	$Al = v(\rho\mu)^{\frac{1}{2}}/B$	1
Hartmann number	Ha	$Ha = Bl(\kappa/\eta)^{\frac{1}{2}}$	1
Cowling number	Co	$Co = B^2/\mu\rho v^2$	1

NOMENCLATURE OF CHEMICAL COMPOUNDS

The International Union of Pure and Applied Chemistry (IUPAC) maintains several commissions that deal with the naming of chemical substances. In general, the approach of IUPAC is to present rules for arriving at names in a systematic manner, rather than recommending a unique name for each compound. Thus there are often several alternative "IUPAC names", depending on which nomenclature system is used, each of which may have advantages in specific applications. However, each of these names will be unambiguous.

Organizations such as the Chemical Abstacts Service and the Beilstein Institute that prepare indexes to the chemical literature must adopt a system for selecting unique names in order to avoid excessive cross referencing. Chemical Abstracts Service uses a system which groups together compounds derived from a single parent compound. Thus most index names are inverted (e.g., Benzene, bromo rather than bromobenzene; Acetic acid, sodium salt rather than sodium acetate). In this *Handbook* the CAS Index Names are used only in the table "Physical Constants of Organic Compounds". Other tables use more familiar names which, with a few possible exceptions, conform to one of the IUPAC naming systems.

Recommended names for the most common substituent groups, ligands, ions, and organic rings are given in the two following tables, "Nomenclature for Inorganic Ions and Ligands" and "Organic Substituent Groups and Ring Systems". For the basics of macromolecular nomenclature, see "Naming Organic Polymers" in Section 13.

Some of the most useful recent guides to chemical nomenclature, prepared by IUPAC and other organizations such as the International Union of Biochemistry and Molecular Biology (IUBMB) and the American Chemical Society are listed below . These books contain citations to the more detailed nomenclature documents in each area.

Inorganic Chemistry

International Union of Pure and Applied Chemistry, *Nomenclature of Inorganic Chemistry, Recommendations 1990,* edited by Leigh, G.J., Blackwell Scientific Publications, Oxford, 1990.

Block, B.P., Powell, W.H., and Fernelius, W.C., *Inorganic Chemical Nomenclature, Principles and Practice*, American Chemical Society, Washington, 1990.

Organic Chemistry

International Union of Pure and Applied Chemistry, *A Guide to IUPAC Nomenclature of Organic Compounds, Recommendations 1993,* edited by Panico, R., Powell, W.H., and Richer, J.-C., Blackwell Scientific Publications, Oxford, 1993.

International Union of Pure and Applied Chemistry, *Glossary of Class Names of Organic Compounds and Reactive Intermediates Based on Structure*, edited by Moss, G.P., Smith, P.A.S., and Tavernier, D., *Pure & Appl. Chem*, 67, 1307, 1995.

Rhodes, P.H., *The Organic Chemist's Desk Reference*, Chapman & Hall, London, 1995.

International Union of Pure and Applied Chemistry, *Basic Terminology of Stereochemistry*, edited by Moss, G.P., *Pure & Applied Chemistry*, 68, 2193, 1966.

Macromolecular Chemistry

International Union of Pure and Applied Chemistry, *Compendium of Macromolecular Nomenclature*, edited by, Metanomski, W.V., Blackwell Scientific Publications, Oxford, 1991.

International Union of Pure and Applied Chemistry, *Glossary of Basic Terms in Polymer Science*, edited by Jenkins, A.D., Kratochvil, P., Stepto, R.F.T., and Suter, U.W., *Pure & Appl. Chem*, 68, in press.

Biochemistry

International Union of Biochemistry and Molecular Biology, *Biochemical Nomenclature and Related Documents, 2nd Edition, 1992*, Portland Press, London, 1993; includes recommendations of the IUPAC-IUBMB Joint Commission on Biochemical Nomenclature.

International Union of Biochemistry and Molecular Biology, *Enzyme Nomenclature, 1992*, Academic Press, Orlando, FL, 1992.

IUPAC-IUBMB Joint Commission on Biochemical Nomenclature, *Nomenclature of Carbohydrates, Recommendations 1996*, edited by McNaught, A.D., *Pure & Appl. Chem.*, 68, 1919, 1996.

General

Chemical Abstracts Service, *Naming and Indexing Chemical Substances for Chemical Abstracts, Appendix IV, Chemical Abstracts 1994 Index Guide.*

NOMENCLATURE FOR INORGANIC IONS AND LIGANDS

See the table *Nomenclature of Chemical Compounds* for references. The assistance of Warren H. Powell in preparing this list is gratefully acknowledged.

Group	As cation	As anion	As ligand	As prefix in organic compounds
H	hydrogen	hydride	hydrido	
F	fluorine	fluoride	fluoro	fluoro
Cl	chlorine	chloride	chloro	chloro
Br	bromine	bromide	bromo	bromo
I	iodine	iodide	iodo	iodo
ClO	chlorosyl	hypochlorite	hypochlorito	chlorosyl
ClO_2	chloryl	chlorite	chlorito	chloryl
ClO_3	perchloryl	chlorate	chlorato	perchloryl
ClO_4		perchlorate		
IO	iodosyl	hypoiodite		iodoso
IO_2	iodyl			iodyl; iodoxy
O		oxide	oxo	oxo
O_2		peroxide (O_2^{2-}) hyperoxide (O_2^-)	peroxo	peroxy
HO		hydroxide	hydroxo	hydroxy
HO_2		hydrogen peroxide	hydrogen peroxo	hydroperoxy
S		sulfide	thio; sulfido	thio; thioxo
HS		hydrogen sulfide	thiolo	mercapto
S_2		disulfide	disulfido	
SO	sulfinyl; thionyl			sulfinyl
SO_2	sulfonyl; sulfuryl	sulfoxylate		sulfonyl
SO_3		sulfite	sulfito	
HSO_3		hydrogen sulfite	hydrogen sulfito	
S_2O_3		thiosulfate	thiosulfato	
SO_4		sulfate	sulfato	
Se		selenide	seleno	seleno; selenoxo
SeO	seleninyl			seleninyl
SeO_2	selenonyl			selenonyl
SeO_3		selenite	selenito	
SeO_4		selenate	selenato	
Te		telluride	telluro	telluro; telluroxo
CrO_2	chromyl			
UO_2	uranyl			
NpO_2	neptunyl			
PuO_2	plutoryl			
AmO_2	americyl			
N		nitride	nitrido	
N_3		azide	azido	
NH		imide	imido	imino
NH_2		amide	amido	amino
NHOH		hydroxylamide	hydroxylamido	hydroxyamino
N_2H_3		hydrazide	hydrazido	hydrazino; diazanyl
NO	nitrosyl		nitrosyl	nitroso
NO_2	nitryl		nitro	nitro
ONO		nitrite	nitrito	
NS	thionitrosyl			
NO_3		nitrate	nitrato	
N_2O_3		hyponitrite	hyponitrito	
P		phosphide	phosphido	phosphinidyne
PO	phosphoryl			phosphoroso; phosphinylidyne
PO_2	phospho			
PS	thiophosphoryl			phosphinothioylidyne; thiophosphorozo
PH_2O_3		hypophosphite	hypophosphito	
PHO_3		phosphite	phosphito	

NOMENCLATURE FOR INORGANIC IONS AND LIGANDS (continued)

Group	As cation	As anion	As ligand	As prefix in organic compounds
PO_4		phosphate	phosphato	
AsO_4		arsenate	arsenato	
VO	vanadyl			
CO	carbonyl		carbonyl	carbonyl
CS	thiocarbonyl			thiocarbonyl
CH_3O		methanolate	methoxo	methoxy
C_2H_5O		ethanolate	ethoxo	ethoxy
CH_3S		methanethiolate	methanethiolato	methylthio
C_2H_5S		ethanethiolate	ethanethiolato	ethylthio
CN	cyanogen	cyanide	cyano	cyano
OCN		cyanate	cyanato	cyanato
SCN		thiocyanate	thiocyanato	thiocyanato
SeCN		selenocyanate	selenocyanato	selenocyanato
TeCN		tellurocyanate	tellurocyanato	tellurocyanato
CO_3		carbonate	carbonato	
HCO_3		hydrogen carbonate	hydrogen carbonato	carboxycarbonyl
C_2O_4		oxalate	oxalato	

ORGANIC SUBSTITUENT GROUPS AND RING SYSTEMS

The first part of this table lists substituent groups and their line formulas. A substituent group is defined by IUPAC as a group that replaces one or more hydrogen atoms attached to a parent structure. Such groups are sometimes called radicals, but IUPAC now reserves the term radical for a free molecular species with unpaired electrons. IUPAC does not recommend some of these names, which are marked here with asterisks (e.g., amyl*), but they are included in this list because they are often encountered in the older literature. Substituent group names which are formed by systematic rules (e.g., methyl from methane, ethyl from ethane, etc.) are included here only for the first few members of a homologous series.

In the second part of the table a number of common organic ring compounds are shown, with the conventional numbering of the ring positions indicated.

The help of Warren H. Powell in preparing this table is greatly appreciated. Pertinent references may be found in the table *Nomenclature of Chemical Compounds*.

SUBSTITUENT GROUPS

acetamido (acetylamino)	CH_3CONH-
acetoacetyl	CH_3COCH_2CO-
acetonyl	CH_3COCH_2-
acetyl	CH_3CO-
acryloyl* (1-oxo-2-propenyl)	CH_2=CHCO-
alanyl (from alanine)	$CH_3CH(NH_2)CO$-
β-alanyl	$H_2N(CH_2)_2CO$-
allyl (2-propenyl)	CH_2=$CHCH_2$-
allylidene (2-propenylidene)	CH_2=CHCH=
amidino (aminoiminomethyl)	H_2NC(=NH)-
amino	H_2N-
amyl* (pentyl)	$CH_3(CH_2)_4$-
anilino (phenylamino)	C_6H_5NH-
anisidino	$CH_3OC_6H_4NH$-
anthranoyl (2-aminobenzoyl)	2-$H_2NC_6H_4CO$-
arsino	AsH_2-
azelaoyl (from azelaic acid)	-$OC(CH_2)_7CO$-
azido	N_3-
azino	=N-N=
azo	-N=N-
azoxy	-N(O)=N-
benzal* (benzylidene)	C_6H_5CH=
benzamido (benzoylamino)	C_6H_5CONH-
benzhydryl (diphenylmethyl)	$(C_6H_5)_2CH$-
benzoxy* (benzoyloxy)	C_6H_5COO-
benzoyl	C_6H_5CO-
benzyl	$C_6H_5CH_2$-
benzylidene	C_6H_5CH=
benzylidyne	C_6H_5C≡
biphenylyl	$C_6H_5C_6H_5$-
biphenylene	-C_6H_4-C_6H_4-
butoxy	C_4H_9O-
sec-butoxy (1-methylpropoxy)	$C_2H_5CH(CH_3)O$-
tert-butoxy (1,1-dimethylethoxy)	$(CH_3)_3CO$-
butyl	$CH_3(CH_2)_3$-
sec-butyl (1-methylpropyl)	$CH_3CH_2CH(CH_3)$-
tert-butyl (1,1-dimethylethyl)	$(CH_3)_3C$-
butyryl (1-oxobutyl)	$CH_3(CH_2)_2CO$-
caproyl* (hexanoyl)	$CH_3(CH_2)_4CO$-
capryl* (decanoyl)	$CH_3(CH_2)_8CO$-
capryloyl* (octanoyl)	$CH_3(CH_2)_6CO$-
carbamido (carbamoylamino)	H_2NCONH-
carbamoyl (aminocarbonyl)	H_2NCO-
carbamyl (aminocarbonyl)	H_2NCO
carbazoyl (hydrazinocarbonyl)	H_2NNHCO-
carbethoxy (ethoxycarbonyl)	C_2H_5OCO-
carbonyl	=C=O
carboxy	HOOC-
cetyl* (hexadecyl)	$CH_3(CH_2)_{15}$-
chloroformyl (chlorcarbonyl)	ClCO-
cinnamoyl	C_6H_5CH=CHCO-
cinnamyl (3-phenyl-2-propenyl)	C_6H_5CH=$CHCH_2$-
cinnamylidene	C_6H_5CH=CHCH=
cresyl* (hydroxymethylphenyl)	$HO(CH_3)C_6H_4$-
crotonoyl	CH_3CH=CHCO-
crotyl (2-butenyl)	CH_3CH=$CHCH_2$-
cyanamido (cyanoamino)	NCNH-
cyanato	NCO-
cyano	NC-
decanedioyl	-$OC(CH_2)_8CO$-
decanoyl	$CH_3(CH_2)_8CO$-
diazo	N_2=
diazoamino	-NHN=N-
disilanyl	H_3SiSiH_2-
disiloxanyloxy	$H_3SiOSiH_2O$-
disulfinyl	-S(O)S(O)-
dithio	-SS-
enanthoyl* (heptanoyl)	$CH_3(CH_2)_5CO$-
epoxy	-O-
ethenyl (vinyl)	CH_2=CH-
ethynyl	HC≡C-
ethoxy	C_2H_5O-
ethyl	CH_3CH_2-
ethylene	-CH_2CH_2-
ethylidene	CH_3CH=
ethylthio	C_2H_5S-
formamido (formylamino)	HCONH-
formyl	HCO-
furmaroyl (from fumaric acid)	-OCCH=CHCO-
furfuryl (2-furanylmethyl)	$OC_4H_3CH_2$-
furfurylidene (2-furanylmethylene)	OC_4H_3CH=
glutamoyl (from glutamic acid)	-$OC(CH_2)_2CH(NH_2)CO$-
glutaryl (from glutaric acid)	-$OC(CH_2)_3CO$-
glycylamino	H_2NCH_2CONH-
glycoloyl; glycolyl (hydroxyacetyl)	$HOCH_2CO$-
glycyl (aminoacetyl)	H_2NCH_2CO-
glyoxyloyl; glyoxylyl (oxoacetyl)	HCOCO-
guanidino	H_2NC(=NH)NH-
guanyl (aminoiminomethyl)	H_2NC(=NH)-
heptadecanoyl	$CH_3(CH_2)_{15}CO$-
heptanamido	$CH_3(CH_2)_5CONH$-
heptanedioyl	-$OC(CH_2)_5CO$-
heptanoyl	$CH_3(CH_2)_5CO$-
hexadecanoyl	$CH_3(CH_2)_{14}CO$-
hexamethylene (1,6-hexanediyl)	-$(CH_2)_6$-
hexanedioyl	-$OC(CH_2)_4CO$-
hippuryl (N-benzoylglycyl)	$C_6H_5CONHCH_2CO$-
hydrazino	H_2NNH-
hydrazo	-HNNH-
hydrocinnamoyl	$C_6H_5(CH_2)_2CO$-
hydroperoxy	HOO-
hydroxyamino	HONH-
hydroxy	HO-
imino	HN=
iodoso* (iodosyl)	OI-
iodyl	O_2I-
isoamyl* (isopentyl; 3-methylbutyl)	$(CH_3)_2CH(CH_2)_2$-
isobutenyl (2-methyl-1-propenyl)	$(CH_3)_2$C=CH-
isobutoxy (2-methylpropoxy)	$(CH_3)_2CHCH_2O$-
isobutyl (2-methylpropyl)	$(CH_3)_2CHCH_2$-
isobutylidene (3-methylpropylidene)	$(CH_3)_2$CHCH=
isobutyryl (2-methyl-1-oxopropyl)	$(CH_3)_2CHCO$-

isocyanato	OCN-
isocyano	CN-
isohexyl (4-methylpentyl)	$(CH_3)_2CH(CH_2)_3$-
isoleucyl (from isoleucine)	$C_2H_5CH(CH_3)CH(NH_2)CO$-
isonitroso* (hydroxyamino)	HON=
isopentyl (3-methylbutyl)	$(CH_3)_2CH(CH_2)_2$-
isopentylidene (3-methylbutylidene)	$(CH_3)_2CHCH_2CH$=
isopropenyl (1-methylethenyl)	CH_2=$C(CH_3)$-
isopropoxy (1-methylethoxy)	$(CH_3)_2CHO$-
isopropyl (1-methylethyl)	$(CH_3)_2CH$-
isopropylidene (1-methylethylidene)	$(CH_3)_2C$=
isothiocyanato (isothiocyano)	SCN-
isovaleryl* (3-methyl-1-oxobutyl)	$(CH_3)_2CHCH_2CO$-
lactoyl (from lactic acid)	$CH_3CH(OH)CO$-
lauroyl (from lauric acid)	$CH_3(CH_2)_{10}CO$-
lauryl (dodecyl)	$CH_3(CH_2)_{11}$-
leucyl (from leucine)	$(CH_3)_2CHCH_2CH(NH_2)CO$-
levulinoyl (from levulinic acid)	$CH_3CO(CH_2)_2CO$-
malonyl (from malonic acid)	-$OCCH_2CO$-
mandeloyl (from mandelic acid)	$C_6H_5CH(OH)CO$-
mercapto	HS-
mesityl	2,4,6-$(CH_3)_3C_6H_2$-
methacryloyl (from methacrylic acid)	CH_2=$C(CH_3)CO$-
methallyl (2-methyl-2-propenyl)	CH_2=$C(CH_3)CH_2$-
methionyl (from methionine)	$CH_3SCH_2CH_2CH(NH_2)CO$-
methoxy	CH_3O-
methyl	H_3C-
methylene	H_2C=
methylthio	CH_3S-
myristoyl (from myristic acid)	$CH_3(CH_2)_{12}CO$-
myristyl (tetradecyl)	$CH_3(CH_2)_{13}$-
naphthyl	$(C_{10}H_7)$-
naphthylene	-$(C_{10}H_6)$-
neopentyl (2,2-dimethylpropyl)	$(CH_3)_3CCH_2$-
nitramino (nitroamino)	O_2NNH-
nitro	O_2N-
nitrosamino (nitrosoamino)	ONNH-
nitrosimino (nitrosoimino)	ONN=
nitroso	ON-
nonanoyl (from nonanoic acid)	$CH_3(CH_2)_7CO$-
oleoyl (from oleic acid)	$CH_3(CH_2)_7CH$=$CH(CH_2)_7CO$-
oxalyl (from oxalic acid)	-OCCO-
oxo	O=
palmitoyl (from palmitic acid)	$CH_3(CH_2)_{14}CO$-
pentamethylene (1,5-pentanediyl)	-$(CH_2)_5$-
pentyl	$CH_3(CH_2)_4$-
tert-pentyl	$CH_3CH_2C(CH_3)_2$-
phenacyl	$C_6H_5COCH_2$-
phenacylidene	C_6H_5COCH=
phenethyl (2-phenylethyl)	$C_6H_5CH_2CH_2$-
phenoxy	C_6H_5O-
phenyl	C_6H_5-
phenylene (benzenediyl)	-C_6H_4-
phosphino* (phosphanyl)	H_2P-
phosphinyl* (phosphinoyl)	$H_2P(O)$-
phospho	O_2P-
phosphono	$(HO)_2P(O)$-
phthaloyl (from phthalic acid)	1,2-$C_6H_4(CO\text{-})_2$
picryl (2,4,6-trinitrophenyl)	2,4,6-$(NO_2)_3C_6H_2$-
pimeloyl (from pimelic acid)	-$OC(CH_2)_5CO$-
piperidino (1-piperidinyl)	$C_5H_{10}N$-
pivaloyl (from pivalic acid)	$(CH_3)_3CCO$-
prenyl (3-methyl-2-butenyl)	$(CH_3)_2C$=$CHCH_2$-
propargyl (2-propynyl)	$HC{\equiv}CCH_2$-
1-propenyl	-CH=$CHCH_2$
2-propenyl (allyl)	CH_2=$CHCH_2$-
propionyl* (propanyl)	CH_3CH_2CO-
propoxy	$CH_3CH_2CH_2O$-
propyl	$CH_3CH_2CH_2$-
propylidene	CH_3CH_2CH=
pyrryl (pyrrolyl)	C_3H_4N-
salicyloyl (2-hydroxybenzoyl)	2-HOC_6H_4CO-
selenyl* (selanyl; hydroseleno)	HSe-
seryl (from serine)	$HOCH_2CH(NH_2)CO$-
siloxy	H_3SiO-
silyl	H_3Si-
silylene	H_2Si=
sorboyl (from sorbic acid)	CH_3CH=CHCH=CHCO-
stearoyl (from stearic acid)	$CH_3(CH_2)_{14}CO$-
stearyl (octadecyl)	$CH_3(CH_2)_{17}$-
styryl (2-phenylethenyl)	C_6H_5CH=CH-
suberoyl (from suberic acid)	-$OC(CH_2)_6CO$-
succinyl (from succinic acid)	-$OCCH_2CH_2CO$-
sulfamino (sulfoamino)	$HOSO_2NH$-
sulfamoyl (sulfamyl)	H_2NSO_2-
sulfanilyl [(4-aminophenyl)sulfonyl]	4-$H_2NC_6H_4SO_2$-
sulfeno	HOS-
sulfhydryl (mercapto)	HS-
sulfinyl	OS=
sulfo	HO_3S-
sulfonyl (sulfuryl)	-SO_2-
terephthaloyl	1,4-$C_6H_4(CO\text{-})_2$
tetramethylene	-$(CH_2)_4$-
thienyl (from thiophene)	(C_4H_3S)-
thiocarbonyl (carbothionyl)	=CS
thiocarboxy	HOSC-
thiocyanato (thiocyano)	NCS-
thionyl* (sulfinyl)	-SO-
threonyl (from threonine)	$CH_3CH(OH)CH(NH_2)CO$-
toluidino [(methylphenyl)amino]	$CH_3C_6H_4NH$-
toluoyl (methylbenzoyl)	$CH_3C_6H_4CO$-
tolyl (methylphenyl)	$CH_3C_6H_4$-
α-tolyl (benzyl)	$C_6H_5CH_2$-
tolylene (methylphenylene)	-$(CH_3C_6H_3)$-
tosyl [(4-methylphenyl) sulfonyl)]	4-$CH_3C_6H_4SO_2$-
triazano	H_2NNHNH-
trimethylene (1,3-propanediyl)	-$(CH_2)_3$-
trityl (triphenylmethyl)	$(C_6H_5)_3C$-
valeryl* (pentanoyl)	$CH_3(CH_2)_3CO$-
valyl (from valine)	$(CH_3)_2CHCH(NH_2)CO$-
vinyl (ethenyl)	CH_2=CH-
vinylidene (ethenylidene)	CH_2=C=
xylidino [(dimethylphenyl)amino]	$(CH_3)_2C_6H_3NH$-
xylyl (dimethylphenyl)	$(CH_3)_2C_6H_3$-
xylylene [phenelenebis(methylene)]	-$CH_2C_6H_4CH_2$-

ORGANIC SUBSTITUENT GROUPS AND RING SYSTEMS (continued)

ORGANIC RING COMPOUNDS

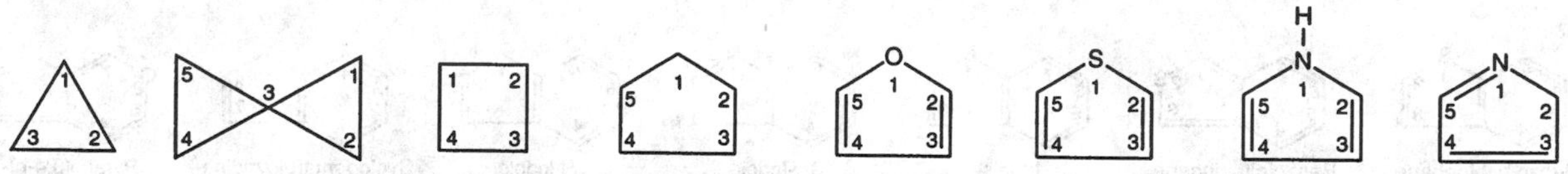

Cyclopropane | Spiropentane | Cyclobutane | Cyclopentane | Furan | Thiophene | Pyrrole (Azole) | 2*H*-Pyrrole (2*H*-Azole)

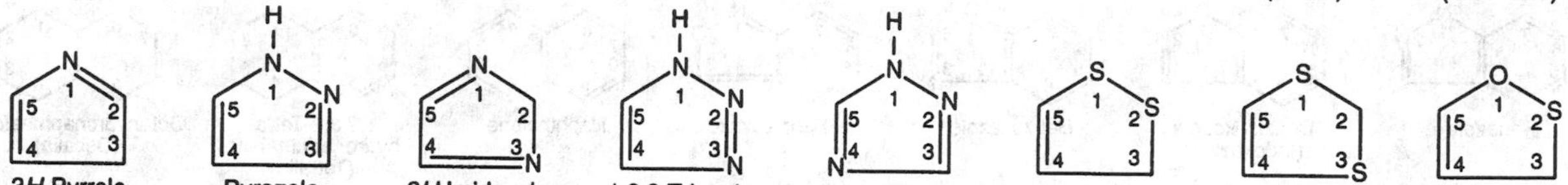

3*H*-Pyrrole (3*H*-Azole) | Pyrazole (1,2-Diazole) | 2*H*-Imidazole (1,3-Diazole) | 1,2,3-Triazole | 1,2,4-Triazole | 1,2-Dithiole | 1,3-Dithiole | 3*H*-1,2-Oxathiole

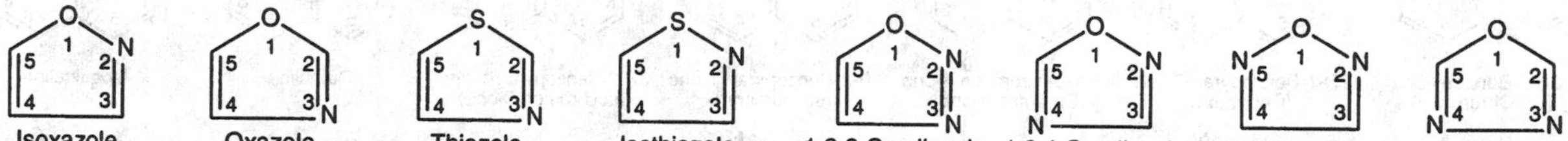

Isoxazole (1,2-Oxazole) | Oxazole (1,3-Oxazole) | Thiazole (1,3-Thiazole) | Isothiazole (1,2-Thiazole) | 1,2,3-Oxadiazole | 1,2,4-Oxadiazole | 1,2,5-Oxadiazole (Furazan) | 1,3,4-Oxadiazole

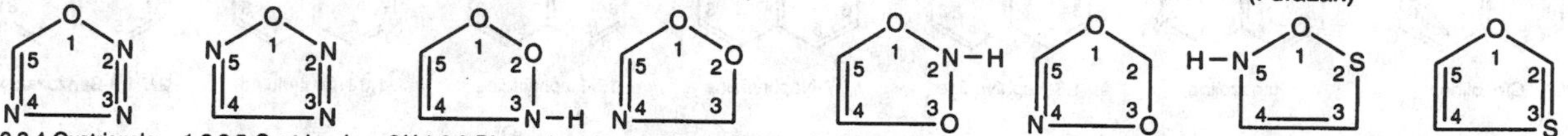

1,2,3,4-Oxatriazole | 1,2,3,5-Oxatriazole | 3*H*-1,2,3-Dioxazole | 1,2,4-Dioxazole | 1,3,2-Dioxazole | 1,3,4-Dioxazole | 5*H*-1,2,5-Oxathiazole | 1,3-Oxathiole

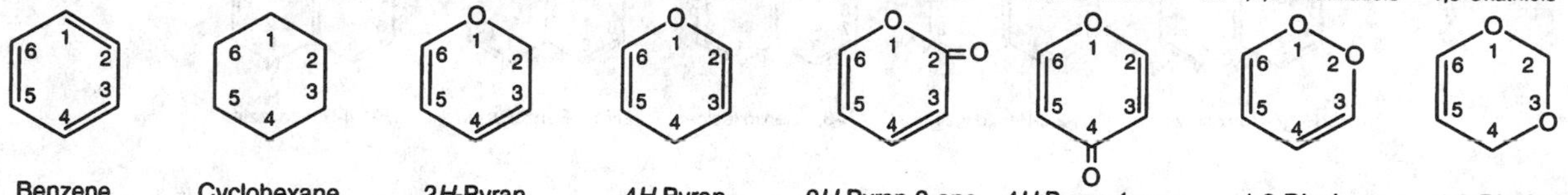

Benzene | Cyclohexane | 2*H*-Pyran | 4*H*-Pyran | 2*H*-Pyran-2-one (2-Pyrone) | 4*H*-Pyran-4-one (4-Pyrone) | 1,2-Dioxin | 1,3-Dioxin

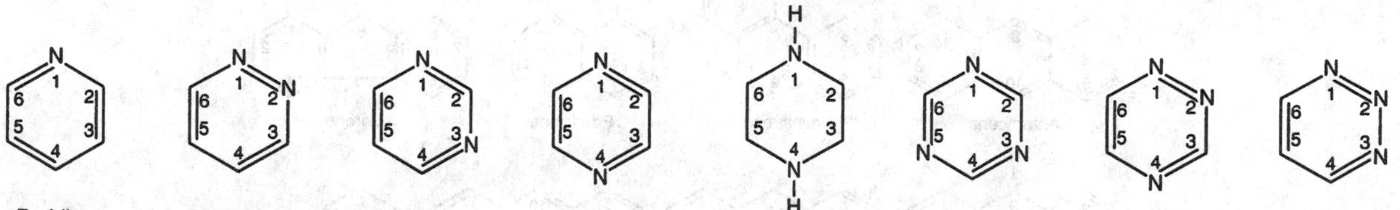

Pyridine | Pyridazine | Pyrimidine | Pyrazine | Piperazine | 1,3,5-Triazine (*s*-Triazine) | 1,2,4-Triazine (*as*-Triazine) | 1,2,3-Triazine (*v*-Triazine)

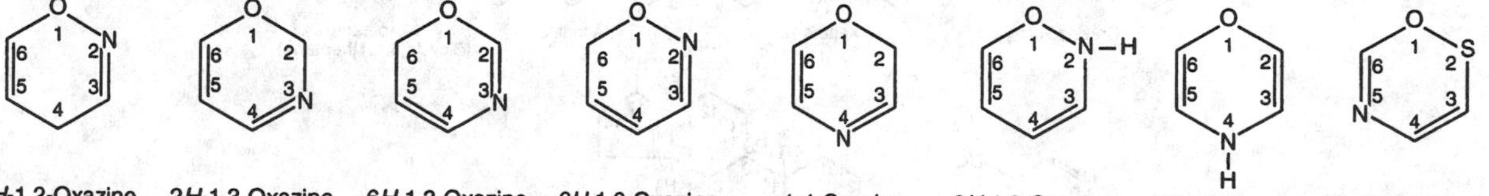

4*H*-1,2-Oxazine | 2*H*-1,3-Oxazine | 6*H*-1,3-Oxazine | 6*H*-1,2-Oxazine | 1,4-Oxazine | 2*H*-1,2-Oxazine | 4*H*-1,4-Oxazine | 1,2,5-Oxathiazine

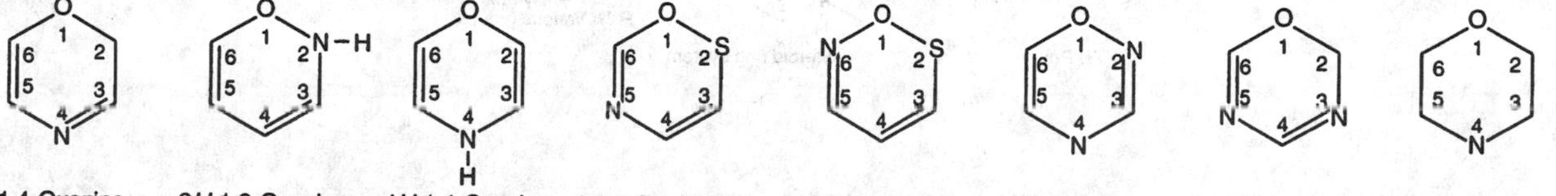

1,4-Oxazine | 2*H*-1,2-Oxazine | 4*H*-1,4-Oxazine | 1,2,5-Oxathiazine | 1,2,6-Oxathiazine | 1,2,4-Oxadiazine | 1,3,5-Oxadiazine | Morpholine

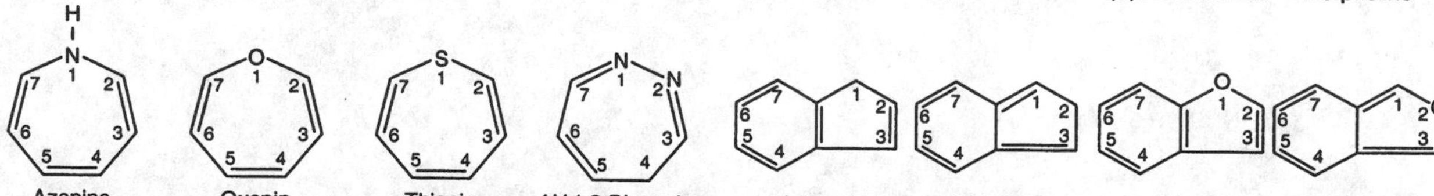

Azepine | Oxepin | Thiepin | 4*H*-1,2-Diazepine | Indene | 2*H*-Indene (Isoindene) | Benzofuran | Isobenzofuran

ORGANIC RING COMPOUNDS (continued)

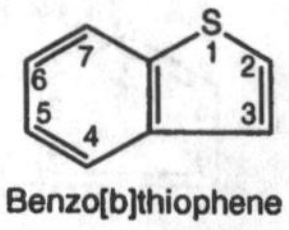
Benzo[b]thiophene

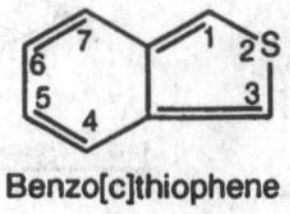
Benzo[c]thiophene

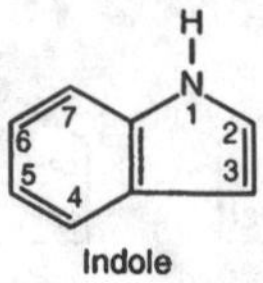
Indole

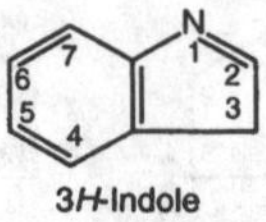
3*H*-Indole

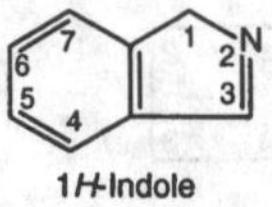
1*H*-Indole

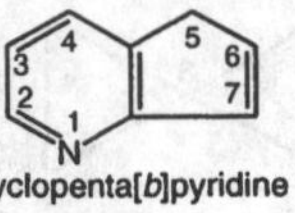
Cyclopenta[*b*]pyridine

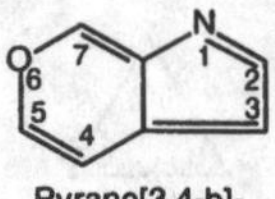
Pyrano[3,4-b]-pyrrole

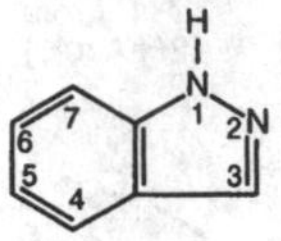
Indazole

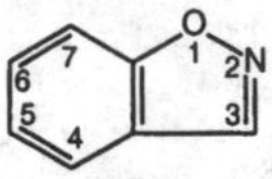
Benzisoxazole (Indoxazine)

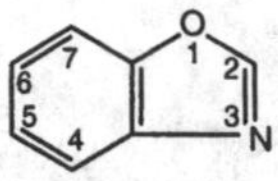
Benzoxazole

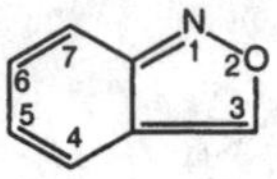
2,1-Benzisoxazole

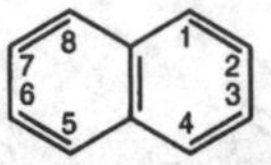
Naphthalene

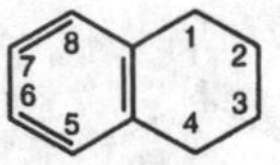
1,2,3,4-Tetra-hydronaphthalene (Tetralin)

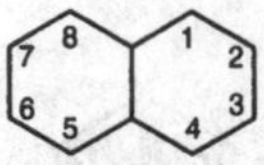
Octahydronaphthalene (Decalin)

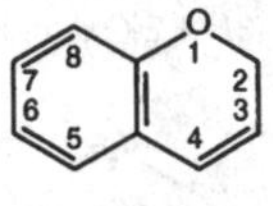
2*H*-1-Benzopyran (2*H*-Chromene)

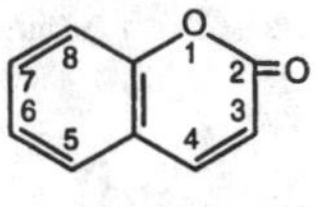
2*H*-1-Benzopyran-2-one (Coumarin)

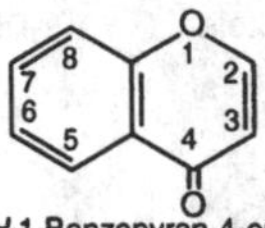
4*H*-1-Benzopyran-4-one (Chromen-4-one)

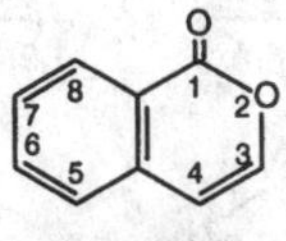
1*H*-2-Benzopyran-1-one (Isocoumarin)

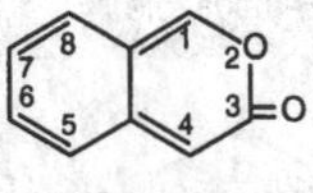
3*H*-2-Benzopyran-1-one (Isochromen-3-one)

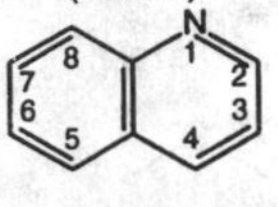
Quinoline

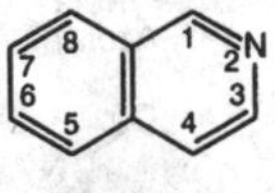
Isoquinoline

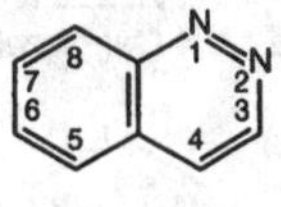
Cinnoline

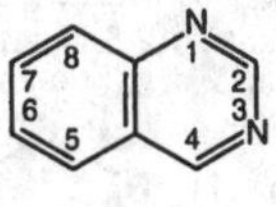
Quinazoline

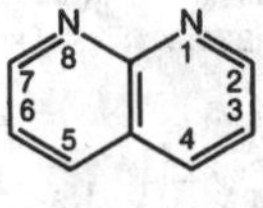
1,8-Naphtyridine

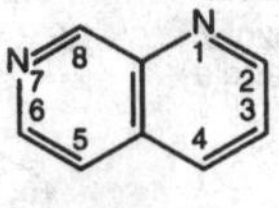
1,7-Naphtyridine

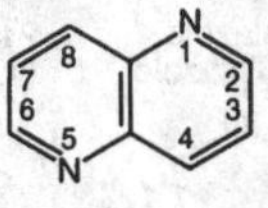
1,5-Naphtyridine

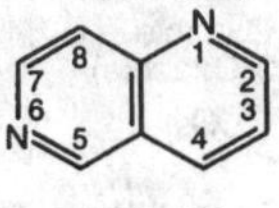
1,6-Naphtyridine

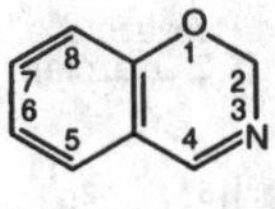
2*H*-1,3-Benzoxazine

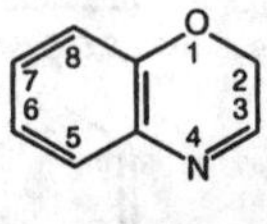
2*H*-1,4-Benzoxazine

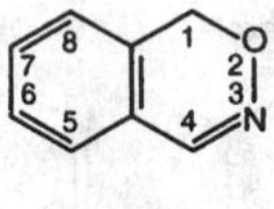
1*H*-2,3-Benzoxazine

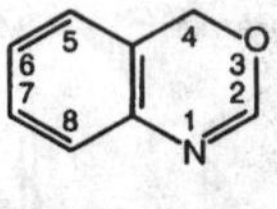
4*H*-3,1-Benzoxazine

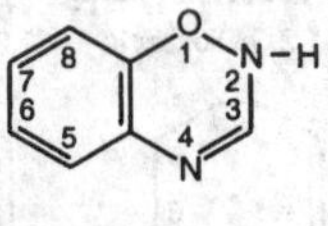
2*H*-1,2-Benzoxazine

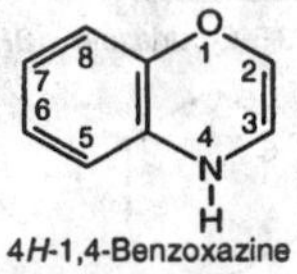
4*H*-1,4-Benzoxazine

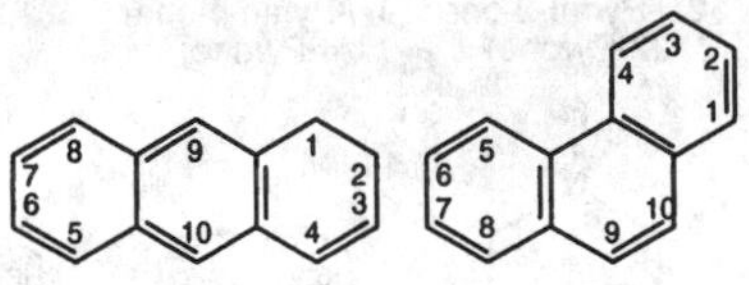
Anthracene

Phenanthrene

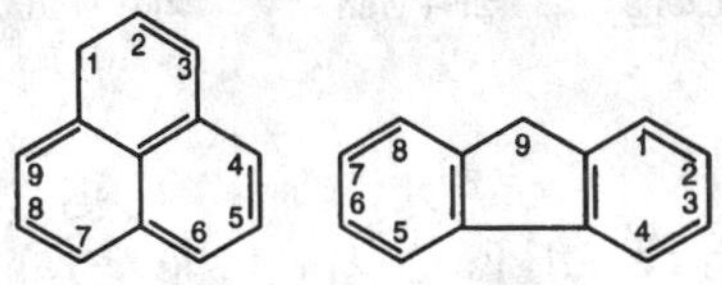
Phenalene

Fluorene

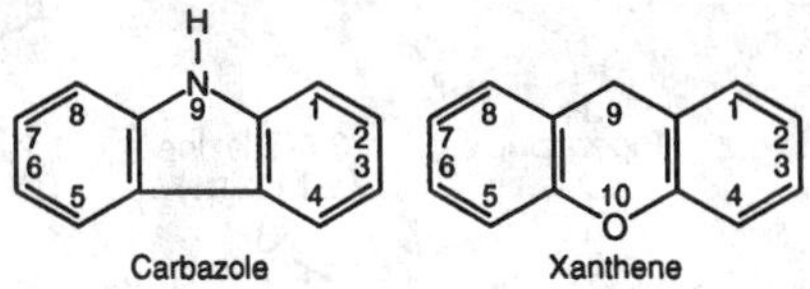
Carbazole

Xanthene

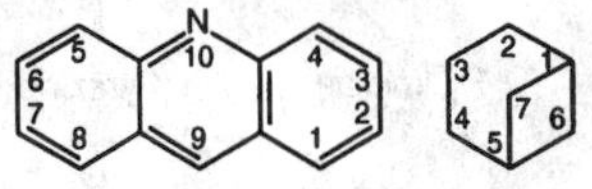
Acridine

Norpinane (Bicyclo[3.1.1]heptane)

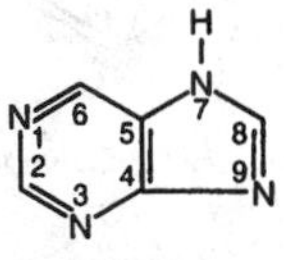
7*H*-Purine

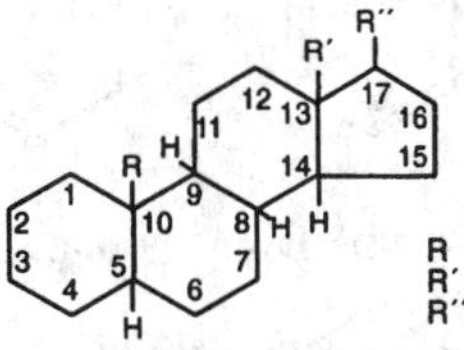
R = Nearly always methyl
R′ = Usually methyl
R′′ = Various groups

Steroid ring system

SCIENTIFIC ABBREVIATIONS AND SYMBOLS

This table lists some symbols, abbreviations, and acronyms encountered in the physical sciences. Most entries in italic type are symbols for physical quantities; for more details on these, see the table "Symbols and Terminology for Physical and Chemical Quantities" in this section. Additional information on units may be found in the table "International System of Units" in Section 1. Many of the terms to which these abbreviations refer are included in the tables "Definitions of Scientific Terms" in Section 2 and "Techniques for Materials Characterization" in Section 12.

Publication practices vary with regard to the use of capital or lower case letters for many abbreviations. An effort has been made to follow the most common practices in this table, but much variation is found in the literature. Likewise, policies on the use of periods in an abbreviation vary considerably. Periods are generally omitted in this table unless they are necessary for clarity. Periods should never appear in SI units. The SI prefixes (m, k ,M, etc.) are not listed here, since they should never be used alone, but selected combinations with SI units (e.g., mg, kV, MW) are included.

Abbreviations are listed in alphabetical order without regard to case. Entries beginning with Greek letters fall at the end of the table.

a	absorption coefficient, acceleration, activity
a_0	Bohr radius
A	ampere, adenine (in genetic code)
Å	ångstrom
A	absorbance, area, Helmholtz energy, mass number
A_H	Hall coefficient
A_r	atomic weight (relative atomic mass)
AAS	atomic absorption spectroscopy
Abe	abequose
abs	absolute
ac	alternating current
Ac	acetyl
AcOH	acetic acid
ACT	activated complex theory
ACTH	adrenocorticotropic hormone
Ade	adenine
ADP	adenosine diphosphate
ads	adsorption
ae	eon (10^9 years)
AES	atomic emission spectroscopy, Auger electron spectroscopy
AF	audio frequency
AFM	atomic force microscopy
AI	artificial intelligence
AIM	atoms in molecules
Al	Alfen number
Ala	alanine
alc	alcohol
aliph.	aliphatic
alk.	alkaline
All	allose
Alt	altrose
am	amorphous solid
Am	amyl
AM	amplitude modulation
AMP	adenosine 5'-monophosphate
amu	atomic mass unit (recommended symbol is u)
anh, anhyd	anhydrous
antilog	antilogarithm
AO	atomic orbital
AOM	angular overlap model
Api	apiose
APS	appearance potential spectroscopy
APW	augmented plane wave
aq	aqueous
Ar	aryl
Ara	arabinose
Ara-ol	arabinitol
Arg	arginine
as, asym	asymmetrical (as chemical descriptor)
ASCII	American National Standard Code for Information Interchange
ASE	aromatic stabilization energy
Asn	asparagine
Asp	aspartic acid
at	atomization
atm	standard atmosphere
ATP	adenosine 5'-triphosphate
ATR	attenuated total internal reflection
at.wt.	atomic weight
AU	astronomical unit
av	average
avdp	avoirdupois
b	barn
B	magnetic flux density, second virial coefficient, susceptance
bar	bar (pressure unit)
bbl	barrel
bcc	body centered cubic
BCS	Bardeen-Cooper-Schrieffer (theory)
BDE	bond dissociation energy
Bé	Baumé
BET	Brunauer-Emmett-Teller (method)
BeV	billion electronvolt
Bhn	Brinell hardness number
Bi	biot
BN	bond number
BNS	nuclear backscattering spectroscopy
BO	bond order, Born-Oppenheimer (approximation)
BOD	biochemical oxygen demand
bp	boiling point
bpy	2,2'-bipyridine
Bq	becquerel
BRE	bond resonance energy
BSSE	basis set superposition error
Btu	British thermal unit
bu	bushel
Bu	butyl
Bz	benzoyl
Bzl	benzyl
c	combustion reaction
c	amount concentration, specific heat, velocity
c_0	speed of light in vacuum
C	coulomb, cytosine (in genetic code)
°C	degree Celsius
C	capacitance, heat capacity, number concentration
ca.	approximately
cal	calorie

calc	calculated
CARS	coherent anti-Stokes Raman spectroscopy
CAS RN	Chemical Abstracts Service Registry Number
CAT	clear-air turbulence, computerized axial tomography
CBS	complete basis set
cc	cubic centimeter
CCD	charge-coupled device
cd	candela, condensed phase
c.d.	current density
CD	circular dichroism
CDP	cytidine 5'-diphosphate
CEPA	couplet electron pair approximation
cf.	compare
cfm	cubic feet per minute
cgs	centimeter-gram-second system
CHF	coupled Hartree-Fock (method)
Ci	curie
CI	configuration interaction, chemical ionization
CIDEP	chemically induced dynamic electron polarization
CIDNP	chemically induced dynamic nuclear polarization
cir	circular
CKFF	Cotton-Kraihanzel force field
CL	cathode luminescence
cm	centimeter
c.m.	center of mass
c.m.c.	critical micelle concentration
CMO	canonical molecular orbital
CMP	cytidine 5'-monophosphate
CN	coordination number
CNDO	complete neglect of differential overlap
Co	Cowling number
COD	chemical oxygen demand
conc	concentrated, concentration
const	constant
cos	cosine
cosh	hyperbolic cosine
COSY	correlation spectroscopy (in NMR)
cot	cotangent
coth	hyperbolic cotangent
cp	candle power
cP	centipoise
Cp	cyclopentadienyl
CP	chemically pure
CPA	coherent potential approximation
cpd	contact potential difference
cps	cycles per second
CPT	charge conjugation-space inversion-time reversal (theorem)
CPU	central processing unit
cr, cryst	crystalline (phase)
CRU	constitutional repeating unit
csc	cosecant
ct	carat
CT	charge transfer
CTEM	conventional transmission electron microscopy
CTP	cytidine 5'-triphosphate
CTR	controlled thermonuclear reaction
cu	cubic
CV	cyclic voltammetry
CVD	chemical vapor deposition
cw	continuous wave

cwt	hundredweight (112 pounds)
Cy	cyclohexyl
cyl	cylinder
Cys	cysteine
d	day, deuteron
d	distance, density, dextrorotatory
D	debye unit
D	diffusion coefficient, dissociation energy, electric displacement
Da	dalton
DA	donor-acceptor (complex)
dB	decibel
dc	direct current
DE	delocalization energy
dec	decomposes
deg	degree
den	density
det	determinant
dev	deviation
DFT	density functional theory
diam	diameter
dil	dilute, dilution
DIM	diatomics in molecules
dm	decimeter
dmf, DMF	*N,N*-dimethylformamide
dmso, DMSO	dimethylsulfoxide
DNA	deoxyribonucleic acid
DNase	deoxyribonuclease
DNMR	dynamic nuclear magnetic resonance
DOS	density of states
doz	dozen
d.p.	degree of polymerization
dpl	displacement
dpm	disintegrations per minute
dps	disintegrations per second
dr	dram
dRib	2-deoxyribose
DRIFT	diffuse reflectance infrared Fourier transform
DRS	diffuse reflectance spectroscopy
DSC	differential scanning calorimetry
DTA	differential thermal analysis
dyn	dyne
e	electron, base of natural logarithms
e	elementary charge, linear strain
E	electric field strength, electromotive force, energy, modulus of elasticity, entgegen (*trans* configuration)
E_h	Hartree energy
EA	electron affinity
EAN	effective atomic number
ECP	effective core potential
ECR	electron cyclotron resonance
ED	electron diffraction, effective dose
EDS	energy dispersive X-ray spectroscopy
EDTA	ethylenediaminetetraacetic acid
EELS	electron energy loss spectroscopy
EFFF	energy factored force field
EHMO	extended Hückel molecular orbital
EHT	extended Hückel theory
emf	electromotive force
emu	electromagnetic unit system
en	ethylenediamine

ENDOR	electron-nuclear double resonance
EOS	equation of state
EPMA	electron probe microanalysis
EPR	electron paramagnetic (spin) resonance
eq, eqn	equation
eqQ	quadrupole coupling constant
erf	error function
erg	erg
ESCA	electron spectroscopy for chemical analysis
e.s.d.	estimated standard deviation
ESD	electron stimulated desorption
ESR	electron spin resonance
est	estimate, estimated
esu	electrostatic unit system
Et	ethyl
ET	electron transfer, ephemeris time
Et_2O	diethyl ether
e.u.	entropy unit
Eu	Euler number
eV	electronvolt
EWG	electron withdrawing group
EXAFS	extended x-ray absorption fine structure
EXELFS	extended energy loss fine structure
exp	exponential function
expt	experimental
ext	external
f	formation reaction
f	activity coefficient, aperture ratio, focal length, force constant, frequency, fugacity
F	farad
°F	degree Fahrenheit
F	Faraday constant, force, angular momentum
FAD	flavin adenine dinucleotide
fcc	face centered cubic
FEL	free electron laser
FEM	field emission microscopy
FEMO	free electron molecular orbital
FET	field effect transistor
fid	free induction decay
FIM	field ion microscopy
FIR	far infrared
fl	fluid (phase)
FM	frequency modulation
Fo	Fourier number
fp	freezing point
fpm	feet per minute
fps	feet per second, foot-pound-second system
Fr	franklin
Fr	Froude number
Fru	fructose
FSGO	floating spherical Gaussian orbital
ft	foot
ft-lb	foot pound
FT	Fourier transform
FTIR	Fourier transform infrared spectroscopy
Fuc	fucose
Fuc-ol	fucitol
fus	fusion (melting)
g	gram, gas
g	acceleration due to gravity, degeneracy, statistical weight, Landé g-factor
G	gauss, guanine (in genetic code)
G	electrical conductance, Gibbs energy, gravitational constant, sheer modulus
gal	gallon
Gal	gal, galileo, galactose
GalN	galactosamine
GC	gas chromatography
GC-MS	gas chromatography-mass spectrometry
GDMS	glow discharge mass spectroscopy
gem	geminal (on the same carbon atom)
GeV	gigaelectronvolt
GIAO	gauge invariant atomic orbital
gl	glacial
GLC	gas-liquid chromatography
Glc	glucose
GlcN	glucosamine
Glc-ol	glucitol
Gln	glutamine
Glu	glutamic acid
Gly	glycine
GMP	guanosine 5'-triphosphate
GMT	Greenwich mean time
gpm	gallons per minute
gps	gallons per second
gr	grain
Gr	Grashof number
GTO	gaussian type atomic orbital
Gua	guanine
Gul	gulose
GUT	grand unified theory
GVB	generalized valence bond
GWS	Glashow-Weinberg-Salam (theory)
Gy	gray, gigayear
h	helion, hour
h	Planck constant
H	henry
H	enthalpy, Hamiltonian function, magnetic field
H_0	Hubble constant
ha	hectare
Ha	Hartmann number
Hacac	acetylacetone
HAM	hydrogenic atoms in molecules
hav	haversine
Hb	hemoglobin
hcp	hexagonal closed packed
Hea	ethanolamine
HEIS	high energy ion scattering
HEP	high energy physics
HF	high frequency
hfs	hyperfine structure
Him	imidazole
His	histidine
HMO	Hückel molecular orbital
HOMO	highest occupied molecular orbital
hp	horsepower
HPLC	high-performance liquid chromatography
Hpz	pyrazole
hr	hour
HREELS	high resolution electron energy loss spectroscopy
HREM	high resolution electron microscopy
HSAB	hard-soft acid-base (theory)

HSE	homodesmotic stabilization energy
Hz	hertz
i	square root of minus one
I	electric current, ionic strength, moment of inertia, nuclear spin angular momentum, radiant intensity
IAT	international atomic time
i-Bu	isobutyl
IC	integrated circuit
ICP	inductively coupled plasma
ICR	ion cyclotron resonance
id	ideal (solution)
ID	inside diameter
Ido	idose
IDP	inosine 5'-diphosphate
IE	ionization energy
i.e.p.	isoelectric point
IEPA	independent electron pair approximation
IF	intermediate frequency
IGLO	individual gauge for localized orbitals
Ile	isoleucine
Im	imaginary part
imm	immersion
IMPATT	impact ionization avalanche transit time
in.	inch
INDO	intermediate neglect of differential overlap
INS	inelastic neutron scattering, ion neutralization spectroscopy
int	internal
I/O	input/output
IP	ionization potential
IPN	interpenetrating polymer network
i-Pr	isopropyl
IPR	isotopic perturbation of resonance
IPTS	International Practical Temperature Scale
IR	infrared
IRAS	reflection-absorption infrared spectroscopy
IRC	intrinsic reaction coordinate
isc	intersystem crossing
ISE	isodesmic stabilization energy
ISS	ion scattering spectroscopy
ITP	inosine 5'-triphosphate
ITS	International Temperature Scale (1990)
IU	international unit
j	angular momentum, electric current density
J	joule
J	angular momentum, electric current density, flux, Massieu function
k	absorption index, Boltzmann constant, rate constant, thermal conductivity, wave vector
K	kelvin
K	absorption coefficient, bulk modulus, equilibrium constant, kinetic energy
kb	kilobar, kilobase (DNA or RNA)
kcal	kilocalorie
KE	kinetic energy
keV	kiloelectronvolt
kg	kilogram
kgf	kilogram force
kJ	kilojoule
km	kilometer
Kn	Knudsen number
kPa	kilopascal
kt	karat
kV	kilovolt
kva	kilovolt ampere
kW	kilowatt
kwh	kilowatt hour
l	liquid, liter
l	angular momentum, length, levorotatory
L	liter, lambert
L	Avogadro constant, inductance, Lagrange function
lat.	latitude
lb	pound
lbf	pound force
lc	liquid crystal
LC	liquid chromatography
LCAO	linear combination of atomic orbitals
LD	lethal dose
Le	Lewis function
LED	light emitting diode
LEED	low-energy electron diffraction
LEIS	low energy ion scattering
Leu	leucine
LFER	linear free energy relationship
lim	limit
LIMS	laser ionization mass spectroscopy, laboratory information management system
liq	liquid
lm	lumen
ln	logarithm (natural)
LNDO	local neglect of differential overlap
log	logarithm (common)
long.	longitude
LST	local sidereal time
LT	local time
LTE	local thermodynamic equilibrium
LUMO	lowest unoccupied molecular orbital
lut	lutidine
lx	lux
ly	langley
l.y.	light year
Lys	lysine
Lyx	lyxose
m	meter, molal (as in 0.1 m solution), metastable (isotope)
m	magnetic dipole moment, mass, molality, angular momentum component, *meta* (as chemical descriptor)
M	molar (as in 0.1 M solution), metal (in chemical formulas)
M	magnetization, molar mass, mutual inductance, torque, angular momentum component
M_r	molecular weight (relative molar mass)
Ma	Mach number
Man	mannose
MASNMR	magic angle spinning nuclear magnetic resonance
max	maximum
MBE	molecular beam epitaxy
MBPT	many body perturbation theory
MC	Monte Carlo (method)
MCD	magnetic circular dichroism
MCPF	modified couple pair functional
MCSCF	multi-configurational self-consistent field

SCIENTIFIC ABBREVIATIONS AND SYMBOLS (continued)

MD	molecular dynamics
Me	methyl
MEP	molecular electrostatic potential
MERP	minimum energy reaction path
Mes	mesityl
MESFET	metal-semiconductor field-effect transistor
Met	methionine
meV	millielectronvolt
MeV	megaelectronvolt
MF	molecular formula
mg	milligram
MHD	magnetohydrodynamics
mi	mile
MIM	molecules-in-molecules
min	minimum, minute
MINDO	modified intermediate neglect of differential overlap
MIR	mid-infrared
misc	miscible
MKS	meter-kilogram-second system
MKSA	meter-kilogram-second-ampere system
mL, ml	milliliter
mm	millimeter
MM	molecular mechanics
mmf	magnetomotive force
mmHg	millimeter of mercury
MO	molecular orbital
mol	mole
mol.wt.	molecular weight
mon	monomeric form
MOS	metal-oxide semiconductor
MOSFET	metal-oxide semiconductor field-effect transistor
mp	melting point
MPa	megapascal
MPA	Mulliken population analysis
Mpc	megaparsec
MRI	magnetic resonance imaging
mRNA	messenger RNA
ms	millisecond
MS	mass spectroscopy
MSL	mean sea level
Mur	muramic acid
mV	millivolt
mW	milliwatt
MW	megawatt, microwave, molecular weight
Mx	maxwell
n	neutron
n	amount of substance, number density, principal quantum number, refractive index, normal (in chemical formulas)
N	newton
N	angular momentum, neutron number
N_A	Avogadro constant
N_E	density of states
NAA	neutron activation analysis
NAD	nicotinamide adenine dinucleotide
NADH	reduced NAD
NADP	nicotinamide adenine dinucleotide phosphate
NAO	natural atomic orbital
NBO	natural bond order
nbp	normal boiling point
Neu	neuraminic acid
NEXAFS	near-edge x-ray absorption fine structure
ng	nanogram
NIR	near infrared
nm	nanometer
NMR	nuclear magnetic resonance
NNDO	neglect of nonbonded differential overlap
NO	natural orbital
NOE	nuclear Overhauser effect
NPA	natural population analysis
NQR	nuclear quadrupole resonance
NRA	prompt nuclear reaction analysis
ns	nanosecond
NTP	normal temperature and pressure
Nu	nucleophile
Nu	Nusselt number
o	*ortho* (as chemical descriptor)
obs, obsd	observed
OD	optical density, outside diameter
Oe	oersted
ORD	optical rotatory dispersion
oz	ounce
p	proton
p	dielectric polarization, electric dipole moment, momentum, pressure, *para* (as chemical descriptor)
P	poise
P	power, pressure, probability, sound energy flux
Pa	pascal
PA	proton affinity
PAS	photoacoustic spectroscopy
pc	parsec
PCR	polymerase chain reaction
PD	potential difference
pdl	poundal
pe	probable error
Pe	Péclet number
PES	photoelectron spectroscopy
PET	positron emission tomography
peth	petroleum ether
pf	power factor
pg	picogram
pH	negative log of hydrogen ion concentration
Ph	phenyl
Phe	phenylalanine
pI	isoelectric point
pip	piperidine
p*K*	negative log of ionization constant
pm	picometer
PMO	perturbational molecular orbital
PNDO	partial neglect of differential overlap
PNRA	prompt nuclear reaction analysis
pol	polymeric form
ppb	parts per billion
ppm	parts per million
PPP	Pariser-Parr-Pople (method)
ppt	parts per thousand, precipitate
Pr	propyl
Pr	Prandtl number
PRDDO	partial retention of diatomic differential overlap
Pro	proline
ps	picosecond
PS	photoelectron spectroscopy

SCIENTIFIC ABBREVIATIONS AND SYMBOLS (continued)

PSD	photon stimulated desorption
Psi	psicose
psi	pounds per square inch
psia	pounds per square inch absolute
psig	pounds per square inch gage
pt	pint
PVT	pressure-volume-temperature
py	pyridine
q	electric field gradient, flow rate, heat, wave vector (phonons)
Q	electric charge, heat, partition function, quadrupole moment, radiant energy, vibrational normal coordinate
QCD	quantum chromodynamics
QED	quantum electrodynamics
Q.E.D.	quod erat demonstrandum (which was to be proved)
QSAR	quantitative structure-activity relationship
QSO	quasi-stellar object (quasar)
qt	quart
quad	quadrillion Btu (= 1.05510^{18} J)
Qui	quinovose
q.v.	quod vide (which you should see)
r	reaction
r	position vector, radius
R	roentgen, alkyl radical (in chemical formulas)
°R	degree Rankine
R	electrical resistance, gas constant, molar refraction, Rydberg constant
RA	right ascension
rad	radian
RAIRS	reflection-absorption infrared spectroscopy
RAM	random access memory
RBS	Rutherford backscattering spectroscopy
RE	resonance energy
Re	real part
RED	radial electron distribution
REM	reflection electron microscopy
rem	roentgen equivalent man
RF	radiofrequency
Rha	rhamnose
RHEED	reflection high-energy electron diffraction
RHF	restricted Hartree-Fock (theory)
RIA	radioimmunoassay
Rib	ribose
Ribulo	ribulose
rms	root mean square
RNA	ribonucleic acid
RNase	ribonuclease
rRNA	ribosomal RNA
ROHF	restricted open shell Hartree-Fock
ROM	read only memory
RPA	random phase approximation
rpm	revolutions per minute
rps	revolutions per second
RRK	Rice-Ramsperger-Kassel (theory)
RRKM	Rice-Ramsperger-Kassel-Marcus (theory)
RRS	resonance Raman spectroscopy
RS	Raman spectroscopy
Ry	rydberg
s	second, solid
s	path length, solubility, spin angular momentum, symmetry number, symmetrical (as stereochemical descriptor)
S	siemens
S	area, entropy, probability current density, Poynting vector, symmetry coordinate, spin angular momentum
SALC	symmetry adapted linear combinations
SALI	surface analysis by laser ionization
SAM	scanning Auger microscopy
SANS	small angle neutron scattering
Sar	sarcosine
sat, satd	saturated
SAXS	small angle x-ray scattering
s-Bu	*sec*-butyl
Sc	Schmidt number
SCE	saturated calomel electrode
SCF	self-consistent field
SCR	silicon-controlled rectifier
sd	standard deviation
sec	secant, second
sec	secondary (in chemical name)
SEELFS	surface sensitive energy loss fine structure
SEM	scanning electron microscope
sepn	separation
Ser	serine
SERS	surface-enhanced Raman spectroscopy
SET	single electron transfer
SEXAFS	surface extended x-ray absorption fine structure
Sh	Sherwood number
SI	International System of Units
SIMS	secondary ion mass spectroscopy
sin	sine
SINDO	symmetrically orthogonalized INDO method
sinh	hyperbolic sine
SIPN	semi-interpenetrating polymer network
SLAM	scanning laser acoustic microscopy
sln	solution
SMO	semiempirical molecular orbital
SMOW	Standard Mean Ocean Water
SNMS	sputtered neutral mass spectroscopy
SNU	solar neutrino unit
SO	spin orbital
sol	soluble, solution
soln	solution
SOMO	singly occupied molecular orbital
Sor	sorbose
sp gr	specific gravity
SPM	scanned probe microscopy
sq	square
sr	steradian
Sr	Strouhal number
SSMS	spark source mass spectroscopy
St	stoke
St	Stanton number
std, stnd	standard (state)
STEM	scanning transmission electron microscope
STM	scanning tunneling microscopy
STO	Slater type orbital
STP	standard temperature and pressure
sub	sublimation, sublimes
Sv	sievert
t	metric tonne, triton
t	Celsius temperature, thickness, time, transport number
T	tesla

T	kinetic energy, period, term value, temperature (thermodynamic), torque, transmittance
Tag	tagatose
Tal	talose
tan	tangent
tanh	hyperbolic tangent
t-Bu	*tert*-butyl
TCA	trichloroacetic acid
TCE	trichloroethylene
tcne	tetracyanoethylene
TCSCF	two configuration self-consistent field
TE	transverse electric
TED	transmission electron diffraction, transferred electron device
TEM	transmission electron microscopy, transverse electromagnetic
temp	temperature
tert	tertiary (in chemical name)
TFD	Thomas-Fermi-Dirac (method)
TGA	thermo-gravimetric analysis
theor	theoretical
thf, THF	tetrahydrofuran
Thr	threonine
Thy	thymine
TL	thermoluminescence
TLC	thin-layer chromatography
TM	transverse magnetic
Tol	tolyl
Torr	torr
tRNA	transfer RNA
Trp	tryptophan
trs	transition
TS	transition state
tsp	teaspoon
Tyr	tyrosine
u	unified atomic mass unit
u	Bloch function, electric mobility, velocity
U	uracil (in genetic code)
U	electric potential difference, internal energy
UDP	uridine 5'-diphosphate
UHF	ultrahigh frequency, unrestricted Hartree-Fock (theory)
UMP	uridine 5'-monophosphate
uns, unsym	unsymmetrical (as chemical descriptor)
UPES	ultraviolet photoelectron spectroscopy
UPS	ultraviolet photoelectron spectroscopy
ur	urea
Ura	uracil
USP	United States Pharmacopeia
UT	universal time
UTP	uridine 5'-triphosphate
UV	ultraviolet
v	reaction rate, specific volume, velocity, vibrational quantum number, vicinal (as chemical descriptor)
V	volt
V	electric potential, potential energy, volume
Val	valine
vap	vaporization
VB	valence band, valence bond
VCD	vibrational circular dichroism
VHF	very high frequency
vic	vicinal (on adjacent carbon atoms)
VIS	visible region of the spectrum
vit	vitreous
VSEPR	valence shell electron pair repulsion
VSLI	very large scale integrated (circuit)
VUV	vacuum ultraviolet
v/v	volume per volume (volume of solute divided by volume of solution, expressed as percent)
w	energy density, mass fraction, velocity, work
W	watt
W	radiant energy, statistical weight, work
WAXS	wide angle x-ray scattering
Wb	weber
We	Weber number
WKB	Wentzel-Kramers-Brillouin (method)
wt	weight
w/v	weight per volume (mass of solute divided by volume of solution, generally expressed as g/100 mL)
w/w	weight per weight (mass of solute divided by mass of solution, expressed as percent)
x	mole fraction
X	X unit, halogen (in chemical formula)
X	reactance
XAFS	x-ray absorption fine structure
XANES	x-ray absorption near-edge structure
XPES	x-ray photoelectron spectroscopy
XPS	x-ray photoelectron spectroscopy
XRD	x-ray diffraction
XRF	x-ray fluorescence
XRS	x-ray spectroscopy
Xyl	xylose
y, yr	year
Y	admittance, Planck function, Young's modulus
yd	yard
z	charge number (of an ion), collision frequency factor
Z	atomic number, compression factor, collision number, impedance, partition function, zusammen (*cis* configuration)
ZDO	zero differential overlap
ZPE, ZPVE	zero point vibrational energy
ZULU	Greenwich mean time
α	alpha particle
α	absorption coefficient, degree of dissociation, electric polarizability, expansion coefficient, fine structure constant
β	beta particle
γ	photon
γ	activity coefficient, conductivity, magnetogyric ratio, mass concentration, ratio of heat capacities, surface tension
Γ	Gruneisin parameter, level width, surface concentration
δ	chemical shift, Dirac delta function, Kronecker delta, loss angle
Δ	inertia defect, mass defect
ε	emittance, Levi-Civita symbol, linear strain, molar absorption coefficient, permittivity
η	overpotential, viscosity
θ	Bragg angle, temperature, scattering angle, surface coverage
Θ	quadrupole moment

κ	compressibility, conductivity, magnetic susceptibility, molar absorption coefficient, transmission coefficient
λ	absolute activity, radioactive decay constant, thermal conductivity, wavelength
Λ	angular momentum, ionic conductivity
μ	muon
μ	chemical potential, electric dipole moment, electric mobility, friction coefficient, Joule-Thompson coefficient, magnetic dipole moment, mobility, permeability
μF	microfarad
μg	microgram
μm	micrometer
μs	microsecond
ν	frequency, kinematic velocity, stoichiometric number, wavenumber
ν_e	neutrino
π	pion
Π	osmotic pressure, Peltier coefficient
ρ	density, reflectance, resistivity
σ	electrical conductivity, cross section, normal stress, shielding constant (NMR), Stefan-Boltzmann constant, surface tension
τ	transmittance, chemical shift, shear stress, relaxation time
ϕ	electrical potential, fugacity coefficient, osmotic coefficient, quantum yield, volume fraction, wavefunction
Φ	magnetic flux, potential energy, radiant power, work function
χ	magnetic susceptibility, electronegativity
χ_e	electric susceptibility
ψ	wavefunction
ω	circular frequency, angular velocity, harmonic vibration wavenumber, statistical weight
Ω	ohm
Ω	axial angular momentum, solid angle

GREEK, RUSSIAN, AND HEBREW ALPHABETS

The following table presents the Hebrew, Greek, and Russian alphabets, their letters, the names of the letters, and the English equivalents.

HEBREW[1,3]

א	aleph	’ [2]
ב	beth	b, bh
ג	gimel	g, gh
ד	daleth	d, dh
ה	he	h
ו	waw	w
ז	zayin	z
ח	heth	ḥ
ט	teth	ṭ
י	yodh	y
כ ך	kaph	k, kh
ל	lamedh	l
מ ם	mem	m
נ ן	nun	n
ס	samekh	s
ע	ayin	ʿ
פ ף	pe	p, ph
צ ץ	sadhe	ṣ
ק	qoph	q
ר	resh	r
שׂ	sin	ś
שׁ	shin	sh
ת	taw	t, th

GREEK[4]

Α α	alpha	a
Β β	beta	b
Γ γ	gamma	g, n
Δ δ	delta	d
Ε ϵ	epsilon	e
Ζ ζ	zeta	z
Η η	eta	ē
Θ θ	theta	th
Ι ι	iota	i
Κ κ	kappa	k
Λ λ	lambda	l
Μ μ	mu	m
Ν ν	nu	n
Ξ ξ	xi	x
Ο ο	omicron	o
Π π	pi	p
Ρ ρ	rho	r, rh
Σ σ ς	sigma	s
Τ τ	tau	t
Υ υ	upsilon	y, u
Φ φ	phi	ph
Χ χ	chi	ch
Ψ ψ	psi	ps
Ω ω	omega	ō

RUSSIAN

А а	a
Б б	b
В в	v
Г г	g
Д д	d
Е е	e
Ж ж	zh
З з	z
И и Й й	i, ĭ
К к	k
Л л	l
М м	m
Н н	n
О о	o
П п	p
Р р	r
С с	s
Т т	t
У у	u
Ф ф	f
Х х	kh
Ц ц	ts
Ч ч	ch
Ш ш	sh
Щ щ	shch
Ъ ъ [5]	”
Ы ы	y
Ь ь [6]	’
Э э	e
Ю ю	yu
Я я	ya

1 Where two forms of a letter are given, the second one is the form used at the end of a word.
2 Not represented in transliteration when initial.
3 The Hebrew letters are primarily consonants; a few of them are also used secondarily to represent certain vowels, when provided at all, is by means of a system of dots or strokes adjacent to the consonated characters.
4 The letter gamma is transliterated "n" only before velars; the letter upsilon is transliterated "u" only as the final element in diphthongs.
5 This sign indicates that the immediately preceding consonant is not palatized even though immediately followed by a palatized vowel.
6 This sign indicates that the immediately preceding consonant is palatized even though not immediately followed by a palatized vowel.

DEFINITIONS OF SCIENTIFIC TERMS

Brief definitions of selected terms of importance in chemistry, physics, and related fields of science are given in this section. The selection process emphasizes the following types of terms:

- Physical quantities
- Units of measure
- Classes of chemical compounds and materials
- Important theories, laws, and basic concepts.

Individual chemical compounds are not included.

Definitions have taken wherever possible from the recommendations of international or national bodies, especially the International Union of Pure and Applied Chemistry (IUPAC) and International Organization for Standardization (ISO). For physical quantities and units, the recommended symbol is also given. The source of such definitions is indicated by the reference number in brackets following the definition. In many cases these official definitions have been edited in the interest of stylistic consistency and economy of space. The user is referred to the original source for further details.

* An asterisk following a term indicates that further information can be found by consulting the index of this handbook under the entry for that term.

REFERENCES

1. *ISO Standards Handbook 2, Units of Measurement*, International Organization for Standardization, Geneva, 1992.
2. *Quantities, Units, and Symbols in Physical Chemistry, Second Edition*, International Union of Pure and Applied Chemistry, Blackwell Scientific Publications, Oxford, 1993.
3. *Compendium of Chemical Terminology*, International Union of Pure and Applied Chemistry, Blackwell Scientific Publications, Oxford, 1987.
4. *A Guide to IUPAC Nomenclature of Organic Compounds*, International Union of Pure and Applied Chemistry, Blackwell Scientific Publications, Oxford, 1993.
5. *Glossary of Class Names of Organic Compounds and Reactive Intermediates Based on Structure, Pure and Applied Chemistry*, 67, 1307, 1995.
6. *Compendium of Analytical Nomenclature*, International Union of Pure and Applied Chemistry, Blackwell Scientific Publications, Oxford, 1987.
7. *Nomenclature of Inorganic Chemistry*, International Union of Pure and Applied Chemistry, Blackwell Scientific Publications, Oxford, 1990.
8. *Glossary of Basic Terms in Polymer Science, Pure and Applied Chemistry*, 68, 2287, 1996.
9. *The International Temperature Scale of 1990, Metrologia*, 27, 107, 1990.
10. *Compilation of ASTM Standard Definitions*, American Society of Testing and Materials, Philadelphia, 1990.
11. *ASM Metals Reference Book*, American Society for Metals, Metals Park, OH, 1983.

***Ab initio* method** - An approach to quantum-mechanical calculations on molecules which starts with the Schrödinger equation and carries out a complete integration, without introducing empirical factors derived from experimental measurement.

Absorbance (*A*) - Defined as $-\log(1-\alpha) = \log(1/\tau)$, where α is the absorptance and τ the transmittance of a medium through which a light beam passes. [2]

Absorbed dose (*D*) - For any ionizing radiation, the mean energy imparted to an element of irradiated matter divided by the mass of that element. [1]

Absorptance (α) - Ratio of the radiant or luminous flux in a given spectral interval absorbed in a medium to that of the incident radiation. Also called absorption factor. [1]

Absorption coefficient (*a*) - The relative decrease in the intensity of a collimated beam of electromagnetic radiation, as a result of absorption by a medium, during traversal of an infinitesimal layer of the medium, divided by the length traversed. [1]

Absorption coefficient, molar (ε) - Absorption coefficient divided by amount-of-substance concentration of the absorbing material in the sample solution ($\varepsilon = a/c$). The SI unit is m^2/mol. Also called extinction coefficient, but usually in units of $mol^{-1}dm^3cm^{-1}$. [2]

Acceleration - Rate of change of velocity with respect to time.

Acceleration due to gravity (*g*)* - The standard value ($9.80665\ m/s^2$) of the acceleration experienced by a body in the earth's gravitational field. [1]

Acenes - Polycyclic aromatic hydrocarbons consisting of fused benzene rings in a rectilinear arrangement. [5]

Acid - Historically, a substance that yields an H^+ ion when it dissociates in solution, resulting in a pH<7. In the Brönsted definition, an acid is a substance that donates a proton in any type of reaction. The most general definition, due to G.N. Lewis, classifies any chemical species capable of accepting an electron pair as an acid.

Acid dissociation constant (K_a)* - The equilibrium constant for the dissociation of an acid HA through the reaction $HA + H_2O \rightleftharpoons A^- + H_3O^+$. The quantity $pK_a = -\log K_a$ is often used to express the acid dissociation constant.

Actinides - The elements of atomic number 89 through 103, e.g., Ac, Th, Pa, U, Np, Pu, Am, Cm, Bk, Cf, Es, Fm, Md, No, Lr. [7]

Activation energy* - In general, the energy that must be added to a system in order for a process to occur, even though the process may already be thermodynamically possible. In chemical kinetics, the activation energy is the height of the potential barrier separating the products and reactants. It determines the temperature dependence of the reaction rate.

Activity - For a mixture of substances, the absolute activity λ of substance B is defined as $\lambda_B = \exp(\mu_B/RT)$, where μ_B is the chemical potential of substance B, *R* the gas constant, and *T* the thermodynamic temperature. The relative activity *a* is defined as $a_B = \exp[(\mu_B - \mu_B{}^\circ)/RT]$, where $\mu_B{}^\circ$ designates the chemical potential in the standard state. [2]

Activity coefficient (γ)* - Ratio of the activity a_B of component B of a mixture to the concentration of that component. The value of γ depends on the method of stating the composition. For mole fraction x_B, the relation is $a_B = \gamma_B x_B$; for molarity c_B, it is $a_B = \gamma_B c_B/c^\circ$, where c° is the standard state composition (typically chosen as 1 mol/L); for molality

m_B, it is $a_B = \gamma_B m_B/m°$, where $m°$ is the standard state molality (typically 1 mol/kg). [2]

Activity, of radioactive substance (A) - The average number of spontaneous nuclear transitions from a particular energy state occurring in an amount of a radionuclide in a small time interval divided by that interval. [1]

Acyl groups - Groups formed by removing the hydroxy groups from oxoacids that have the general structure RC(=O)(OH) and replacement analogues of such acyl groups. [5]

Adiabatic process - A thermodynamic process in which no heat enters or leaves the system.

Admittance (Y) - Reciprocal of impedance. $Y = G + iB$, where G is conductance and B is susceptance. [1]

Adsorption - A process in which molecules of gas, of dissolved substances in liquids, or of liquids adhere in an extremely thin layer to surfaces of solid bodies with which they are in contact. [10]

Albedo* - The ratio of the light reflected or scattered from a surface to the intensity of incident light. The term is often used in reference to specific types of terrain or to entire planets.

Alcohols - Compounds in which a hydroxy group, -OH, is attached to a saturated carbon atom. [5]

Aldehydes - Compounds RC(=O)H, in which a carbonyl group is bonded to one hydrogen atom and to one R group. [5]

Aldoses - Aldehydic parent sugars (polyhydroxyaldehydes $H[CH(OH)]_nC(=O)H$, n>1) and their intramolecular hemiacetals. [5]

Aldoximes - Oximes of aldehydes: RCH=NOH. [5]

Alfvén number (Al) - A dimensionless quantity used in plasma physics, defined by $Al = v(\rho\mu)^{1/2}/B$, where ρ is density, v is velocity, μ is permeability, and B is magnetic flux density. [2]

Alfven waves - Very low frequency waves which can exist in a plasma in the presence of a uniform magnetic field. Also called magnetohydrodynamic waves.

Alicyclic compounds - Aliphatic compounds having a carbocyclic ring structure which may be saturated or unsaturated, but may not be a benzenoid or other aromatic system. [5]

Aliphatic compounds - Acyclic or cyclic, saturated or unsaturated carbon compounds, excluding aromatic compounds. [5]

Alkali metals - The elements lithium, sodium, potassium, rubidium, cesium, and francium.

Alkaline earth metals - The elements calcium, strontium, barium, and radium. [7]

Alkaloids - Basic nitrogen compounds (mostly heterocyclic) occurring mostly in the plant kingdom (but not excluding those of animal origin). Amino acids, peptides, proteins, nucleotides, nucleic acids, and amino sugars are not normally regarded as alkaloids. [5]

Alkanes - Acyclic branched or unbranched hydrocarbons having the general formula C_nH_{2n+2}, and therefore consisting entirely of hydrogen atoms and saturated carbon atoms. [5]

Alkenes - Acyclic branched or unbranched hydrocarbons having one carbon-carbon double bond and the general formula C_nH_{2n}. Acyclic branched or unbranched hydrocarbons having more than one double bond are alkadienes, alkatrienes, etc. [5]

Alkoxides - Compounds, ROM, derivatives of alcohols, ROH, in which R is saturated at the site of its attachment to oxygen and M is a metal or other cationic species. [5]

Alkyl groups - Univalent groups derived from alkanes by removal of a hydrogen atom from any carbon atom: C_nH_{2n+1}-. The groups derived by removal of a hydrogen atom from a terminal carbon atom of unbranched alkanes form a subclass of normal alkyl (n-alkyl) groups. The groups RCH_2-, R_2CH-, and R_3C- (R not equal to H) are primary, secondary, and tertiary alkyl groups, respectively. [5]

Alkynes - Acyclic branched or unbranched hydrocarbons having a carbon-carbon triple bond and the general formula C_nH_{2n-2}, RC≡CR′. Acyclic branched or unbranched hydrocarbons having more than one triple bond are known as alkadiynes, alkatriynes, etc. [5]

Allotropy - The occurrence of an element in two or more crystalline forms.

Allylic groups - The group CH_2=$CHCH_2$- (allyl) and derivatives formed by substitution. The term 'allylic position' or 'allylic site' refers to the saturated carbon atom. A group, such as -OH, attached at an allylic site is sometimes described as "allylic". [5]

Amagat volume unit - A non-SI unit previously used in high pressure science. It is defined as the molar volume of a real gas at one atmosphere pressure and 273.15 K. The approximate value is 22.4 L/mol.

Amides - Derivatives of oxoacids R(C=O)(OH) in which the hydroxy group has been replaced by an amino or substituted amino group. [5]

Amine oxides - Compounds derived from tertiary amines by the attachment of one oxygen atom to the nitrogen atom: R_3N^+-O^-. By extension the term includes the analogous derivatives of primary and secondary amines. [5]

Amines - Compounds formally derived from ammonia by replacing one, two, or three hydrogen atoms by hydrocarbyl groups, and having the general structures RNH_2 (primary amines), R_2NH (secondary amines), R_3N (tertiary amines). [5]

Amino acids* - Compounds containing both a carboxylic acid group (-COOH) and an amino group ($-NH_2$). The most important are the α-amino acids, in which the $-NH_2$ group in attached to the C atom adjacent to the -COOH group. In the ß-amino acids, there is an intervening carbon atom. [4]

Ampere (A)* - The SI base unit of electric current. [1]

Ampere's law - The defining equation for the magnetic induction $\boldsymbol{B}$, viz., $dF = Idl \times \boldsymbol{B}$, where dF is the force produced by a current I flowing in an element of the conductor dl pointing in the direction of the current.

Ångström (Å) - A unit of length used in spectroscopy, crystallography, and molecular structure, equal to 10^{-10} m.

Angular momentum (L) - The angular momentum of a particle about a point is the vector product of the radius vector from this point to the particle and the momentum of the particle; i.e., $\boldsymbol{L} = \boldsymbol{r} \times \boldsymbol{p}$. [1]

Angular velocity (ω) - The angle through which a body rotates per unit time.

Anilides - Compounds derived from oxoacids R(C=O)(OH) by replacing the -OH group by the -NHPh group or derivative formed by ring substitution. Also used for salts formed by replacement of a nitrogen-bound hydrogen of aniline by a metal. [5]

Anion - A negatively charged atomic or molecular particle.

Antiferroelectricity* - An effect analogous to antiferromagnetism in which electric dipoles in a crystal are ordered in two sublattices that are polarized in opposite directions, leading to zero net polarization. The effect vanishes above a critical temperature.

Antiferromagnetism* - A type of magnetism in which the magnetic moments of atoms in a solid are ordered into two antiparallel aligned sublattices. Antiferromagnets are characterized by a zero or small positive magnetic susceptibility. The susceptibility increases with temperature up to a critical value, the Neel temperature, above which the material becomes paramagnetic.

Antiparticle - A particle having the same mass as a given elementary particle and a charge equal in magnitude but opposite in sign.

Appearance potential* - The lowest energy which must be imparted to the parent molecule to cause it to produce a particular specified parent ion. This energy, usually stated in eV, may be imparted by electron impact, photon impact, or in other ways. More properly called appearance energy. [3]

Appearance potential spectroscopy (APS) - See Techniques for Materials Characterization, page **12-1**.

Are (a) - A unit of area equal to 100 m^2. [1]

Arenes - Monocyclic and polycyclic aromatic hydrocarbons. See aromatic compounds. [5]

Aromatic compounds - Compounds whose structure includes a cyclic delocalized π-electron system. Historical use of the term implies a ring containing only carbon (e.g., benzene, naphthalene), but it is often generalized to include heterocyclic structures such as pyridine and thiophene. [5]

Arrhenius equation - A key equation in chemical kinetics which expresses the rate constant k as $k = A\exp(-E_a/RT)$, where E_a is the activation energy, R the molar gas constant, and T the temperature. A is called the preexponential factor and, for simple gas phase reactions, may be identified with the collision frequency.

Arsines - AsH_3 and compounds derived from it by substituting one, two or three hydrogen atoms by hydrocarbyl groups. $RAsH_2$, R_2AsH, R_3As (R not equal to H) are called primary, secondary and tertiary arsines, respectively. [5]

Aryl groups - Groups derived from arenes by removal of a hydrogen atom from a ring carbon atom. Groups similarly derived from heteroarenes are sometimes subsumed in this definition. [5]

Astronomical unit (AU)* - The mean distance of the earth from the sun, equal to $1.49597870 \times 10^{11}$ m.

Atomic absorption spectroscopy (AAS) - See Techniques for Materials Characterization, page **12**-1.

Atomic emission spectroscopy (AES) - See Techniques for Materials Characterization, page **12**-1.

Atomic force microscopy (AFM) - See Techniques for Materials Characterization, page **12**-1.

Atomic mass* - The mass of a nuclide, normally expressed in unified atomic mass units (u).

Atomic mass unit (u)* - A unit of mass used in atomic, molecular, and nuclear science, defined as the mass of one atom of ^{12}C divided by 12. Its approximate value is 1.66054×10^{-27} kg. Also called the unified atomic mass unit. [1]

Atomic number (Z) - A characteristic property of an element, equal to the number of protons in the nucleus.

Atomic weight (A_r)* - The ratio of the average mass per atom of an element to 1/12 of the mass of nuclide ^{12}C. An atomic weight can be defined for a sample of any given isotopic composition. The standard atomic weight refers to a sample of normal terrestrial isotopic composition. The term relative atomic mass is synonymous with atomic weight. [2]

Attenuated total reflection (ATR) - See Techniques for Materials Characterization, page **12**-1.

Auger effect - An atomic process in which an electron from a higher energy level fills a vacancy in an inner shell, transferring the released energy to another electron which is ejected.

Aurora - An atmospheric phenomenon in which streamers of light are produced when electrons from the sun are guided into the thermosphere by the earth's magnetic field. It occurs in the polar regions at altitudes of 95—300 km.

Avogadro constant (N_A)* - The number of elementary entities in one mole of a substance.

Azeotrope - A liquid mixture in a state where the variation of vapor pressure with composition at constant temperature (or, alternatively, the variation of normal boiling point with composition) shows either a maximum or a minimum. Thus when an azeotrope boils the vapor has the same composition as the liquid.

Azides - Compounds bearing the group $-N_3$, viz. $-N=N^+=N^-$; usually attached to carbon, e.g. PhN_3, phenyl azide or azidobenzene. Also used for salts of hydrazoic acid, HN_3, e.g. NaN_3, sodium azide. [5]

Azines - Condensation products, $R_2C=NN=CR_2$, of two moles of a carbonyl compound with one mole of hydrazine. [5]

Azo compounds - Derivatives of diazene (diimide), HN=NH, wherein both hydrogens are substituted by hydrocarbyl groups, e.g., PhN=NPh, azobenzene or diphenyldiazene. [5]

Balmer series - The series of lines in the spectrum of the hydrogen atom which corresponds to transitions between the state with principal quantum number $n = 2$ and successive higher states. The wavelengths are given by $1/\lambda = R_H(1/4 - 1/n^2)$, where $n = 3,4,...$ and R_H is the Rydberg constant for hydrogen. The first member of the series ($n = 2 \rightleftharpoons 3$), which is often called the H_α line, falls at a wavelength of 6563 Å.

Bar (bar) - A unit of pressure equal to 10^5 Pa.

Bardeen-Cooper-Schrieffer (BCS) theory - A theory of superconductivity which is based upon the formation of electron pairs as a result of an electron-lattice interaction. The theory relates the superconducting transition temperature to the density of states and the Debye temperature.

Barn (b) - A unit used for expressing cross sections of nuclear processes, equal to 10^{-28} m^2.

Barrel - A unit of volume equal to 158.9873 L.

Baryon - Any elementary particle built up from three quarks. Examples are the proton, neutron, and various short-lived hyperons. Baryons have odd half-integer spins.

Base - Historically, a substance that yields an OH^- ion when it dissociates in solution, resulting in a pH>7. In the Brönsted definition, a base is a substance capable of accepting a proton in any type of reaction. The more general definition, due to G.N. Lewis, classifies any chemical species capable of donating an electron pair as a base.

Becquerel (Bq)* - The SI unit of radioactivity (disintegrations per unit time), equal to s^{-1}. [1]

Beer's law - An approximate expression for the change in intensity of a light beam that passes through an absorbing medium, viz., $\log(I/I_0) = -\varepsilon cl$, where I_0 is the incident intensity, I is the final intensity, ε is the molar (decadic) absorption coefficient, c is the molar concentration of the absorbing substance, and l is the path length. Also called the Beer-Lambert law

Binding energy* - A generic term for the energy required to decompose a system into two or more of its constituent parts. In nuclear physics, the binding energy is the energy difference between a nucleus and the separated nucleons of which it is composed (the energy equivalent of the mass defect). In atomic physics, it is the energy required to remove an electron from an atom.

Biot (Bi) - A name sometimes used for the unit of current in the emu system.

Birefringence - A property of certain crystals in which two refracted rays result from a single incident light ray. One, the ordinary ray, follows the normal laws of refraction, while the other, the extraordinary ray, exhibits a variable refractive index which depends on the direction in the crystal.

Black body radiation* - The radiation emitted by a perfect black body, i.e., a body which absorbs all radiation incident on it and reflects none. The wavelength dependence of the radiated energy density ρ (energy per unit volume per unit wavelength range) is given by the Planck formula

$$\rho = \frac{8\pi hc}{\lambda^5\left(e^{hc/\lambda kT} - 1\right)}$$

where λ is the wavelength, h is Planck's constant, c is the speed of light, k is the Boltzmann constant, and T is the temperature.

Black hole - A very dense object, formed in a supernova explosion, whose gravitational field is so large that no matter or radiation can escape from the object.

Bloch wave function - A solution of the Schrödinger equation for an electron moving in a spatially periodic potential; used in the band theory of solids.

Bohr magneton (μ_B)* - The atomic unit of magnetic moment, defined as

$eh/4\pi m_e$, where h is Planck's constant, m_e the electron mass, and e the elementary charge. It is the moment associated with a single electron spin.

Bohr, bohr radius (a_0)* - The radius of the lowest orbit in the Bohr model of the hydrogen atom, defined as $\varepsilon_0 h^2/\pi m_e e^2$, where ε_0 is the permittivity of a vacuum, h is Planck's constant, m_e the electron mass, and e the elementary charge. It is customarily taken as the unit of length when using atomic units.

Boiling point - The temperature at which the liquid and gas phases of a substance are in equilibrium at a specified pressure. The normal boiling point is the boiling point at normal atmospheric pressure (101.325 kPa).

Boltzmann constant (k)* - The molar gas constant R divided by Avogadro's constant.

Boltzmann distribution - An expression for the equilibrium distribution of molecules as a function of their energy, in which the number of molecules in a state of energy E is proportional to $\exp(-E/kT)$, where k is the Boltzmann constant and T is the temperature.

Bond strength - See Dissociation energy.

Born-Haber cycle* - A thermodynamic cycle in which a crystalline solid is converted to gaseous ions and then reconverted to the solid. The cycle permits calculation of the lattice energy of the crystal.

Bose-Einstein distribution - A modification of the Boltzmann distribution which applies to a system of particles that are bosons. The number of particles of energy E is proportional to $[e^{(E-\mu)/kT}-1]^{-1}$, where μ is a normalization constant, k is the Boltzmann constant, and T is the temperature.

Boson - A particle that obeys Bose-Einstein Statistics; specifically, any particle with spin equal to zero or an integer. This includes the photon, pion, deuteron, and all nuclei of even mass number.

Boyle's law - The empirical law, exact only for an ideal gas, which states that the volume of a gas is inversely proportional to its pressure at constant temperature.

Bragg angle (θ) - Defined by the equation $n\lambda = 2d\sin\theta$, which relates the angle θ between a crystal plane and the diffracted x-ray beam, the wavelength λ of the x-rays, the crystal plane spacing d, and the diffraction order n (any integer).

Bravais lattices* - The 14 distinct crystal lattices that can exist in three dimensions. They include three in the cubic crystal system, two in the tetragonal, four in the orthorhombic, two in the monoclinic, and one each in the triclinic, hexagonal, and trigonal systems.

Breakdown voltage - The potential difference at which an insulating substance undergoes a physical or chemical change that causes it to become a conductor, thus allowing current to flow through the sample.

Bremsstrahlung - Electromagnetic radiation generated when the velocity of a charged particle is reduced (literally, "braking radiation"). An example is the x-ray continuum resulting from collisions of electrons with the target in an x-ray tube.

Brewster angle - The angle of incidence for which the maximum degree of plane polarization occurs when a beam of unpolarized light is incident on the surface of a medium of refractive index n. At this angle, the angle between the reflected and refracted beams is 90°. The value of the Brewster angle is $\tan^{-1}n$.

Brillouin scattering - The scattering of light by acoustic phonons in a solid or liquid.

Brillouin zone - A region of allowed wave vectors and energy levels in a crystalline solid, which plays a part in the propagation of waves through the lattice.

British thermal unit (Btu) - A non-SI unit of energy, equal to approximately 1055 J. Several values of the Btu, defined in slightly different ways, have been used.

Brownian motion - The random movements of small particles suspended in a fluid, which arise from collisions with the fluid molecules.

Brunauer-Emmett-Teller method (BET) - See Techniques for Materials Characterization, page 12-1.

Buffer* - A solution designed to maintain a constant pH when small amounts of a strong acid or base are added. Buffers usually consist of a fairly weak acid and its salt with a strong base. Suitable concentrations are chosen so that the pH of the solution remains close to the pK_a of the weak acid.

Calorie (cal) - A non-SI unit of energy, originally defined as the heat required to raise the temperature of 1 g of water by 1°C. Several calories of slightly different values have been used. The thermochemical calorie is now defined as 4.184 J.

Candela (cd)* - The SI base unit of luminous intensity. [1]

Capacitance (C) - Ratio of the charge acquired by a body to the change in potential. [1]

Carbamates - Salts or esters of carbamic acid, $H_2NC(=O)OH$, or of N-substituted carbamic acids: $R_2NC(=O)OR'$, (R' = hydrocarbyl or a cation). The esters are often called urethanes or urethans, a usage that is strictly correct only for the ethyl esters. [5]

Carbenes - The electrically neutral species H_2C: and its derivatives, in which the carbon is covalently bonded to two univalent groups of any kind or a divalent group and bears two nonbonding electrons, which may be spin-paired (singlet state) or spin-non-paired (triplet state). [5]

Carbinols - An obsolete term for substituted methanols, in which the name carbinol is synonymous with methanol. [5]

Carbohydrates - Originally, compounds such as aldoses and ketoses, having the stoichiometric formula $C_n(H_2O)_n$ (hence "hydrates of carbon"). The generic term carbohydrate now includes mono-, oligo-, and polysaccharides, as well as their reaction products and derivatives. [5]

Carboranes - A contraction of carbaboranes. Compounds in which a boron atom in a polyboron hydride is replaced by a carbon atom with maintenance of the skeletal structure. [5]

Carboxylic acids - Oxoacids having the structure RC(=O)OH. The term is used as a suffix in systematic name formation to denote the -C(=O)OH group including its carbon atom. [5]

Carnot cycle - A sequence of reversible changes in a heat engine using a perfect gas as the working substance, which is used to demonstrate that entropy is a state function. The Carnot cycle also provides a means to calculate the efficiency of a heat engine.

Catalyst - A substance that participates in a particular chemical reaction and thereby increases its rate but without a net change in the amount of that substance in the system. [3]

Catenanes, catena compounds - Hydrocarbons having two or more rings connected in the manner of links of a chain, without a covalent bond. More generally, the class catena compounds embraces functional derivatives and hetero analogues. [5]

Cation - A positively charged atomic or molecular particle.

Centipoise (cP) - A common non-SI unit of viscosity, equal to mPa s.

Centrifugal distortion - An effect in molecular spectroscopy in which rotational levels are lowered in energy, relative to the values of a rigid rotor, as the rotational angular momentum increases. The effect may be understood classically as a stretching of the bonds in the molecule as it rotates faster, thus increasing the moment of inertia.

Ceramic - A nonmetallic material of very high melting point.

Cerenkov radiation - Light emitted when a beam of charged particles travels through a medium at a speed greater than the speed of light in the medium. It is typically blue in color.

Cgs system of units - A system of units based upon the centimeter, gram, and second. The cgs system has been supplanted by the International System (SI).

Chalcogens - The Group VIA elements (oxygen, sulfur, selenium, tellurium, and polonium). Compounds of these elements are called chalcogenides. [7]

Chaotic system - A complex system whose behavior is governed by deterministic laws but whose evolution can vary drastically when small changes are made in the initial conditions.

Charge - See Electric charge.

Charles' law - The empirical law, exact only for an ideal gas, which states that the volume of a gas is directly proportional to its temperature at constant pressure.

Charm - A quantum number introduced in particle physics to account for certain properties of elementary particles and their reactions.

Chelate - A compound characterized by the presence of bonds from two or more bonding sites within the same ligand to a central metal atom. [3]

Chemical potential - For a mixture of substances, the chemical potential of constituent B is defined as the partial derivative of the Gibbs energy *G* with respect to the amount (number of moles) of B, with temperature, pressure, and amounts of all other constituents held constant. Also called partial molar Gibbs energy. [2]

Chemical shift* - A small change in the energy levels (and hence in the spectra associated with these levels) resulting from the effects of chemical binding in a molecule. The term is used in fields such as NMR, Mössbauer, and photoelectron spectroscopy, where the energy levels are determined primarily by nuclear or atomic effects.

Chiral molecule - A molecule which cannot be superimposed on its mirror image. A common example is an organic molecule containing a carbon atom to which four different atoms or groups are attached. Such molecules exhibit optical activity, i.e., they rotate the plane of a polarized light beam.

Chlorocarbons - Compounds consisting solely of chlorine and carbon. [5]

Chromatography* - A method for separation of the components of a sample in which the components are distributed between two phases, one of which is stationary while the other moves. In gas chromatography the gas moves over a liquid or solid stationary phase. In liquid chromatography the liquid mixture moves through another liquid, a solid, or a gel. The mechanism of separation of components may be adsorption, differential solubility, ion-exchange, permeation, or other mechanisms. [6]

Clapeyron equation - A relation between pressure and temperature of two phases of a pure substance that are in equilibrium, viz., $dp/dT = \Delta_{trs}S/\Delta_{trs}V$, where $\Delta_{trs}S$ is the difference in entropy between the phases and $\Delta_{trs}V$ the corresponding difference in volume.

Clathrates - Inclusion compounds in which the guest molecule is in a cage formed by the host molecule or by a lattice of host molecules. [5]

Clausius (Cl) - A non-SI unit of entropy or heat capacity defined as cal/K = 4.184 J/K. [2]

Clausius-Clapeyron equation - An approximation to the Clapeyron equation applicable to liquid-gas and solid-gas equilibrium, in which one assumes an ideal gas with volume much greater than the condensed phase volume. For the liquid-gas case, it takes the form $d(\ln p)/dT = \Delta_{vap}H/RT^2$, where R is the molar gas constant and $\Delta_{vap}H$ is the molar enthalpy of vaporization. For the solid-gas case, $\Delta_{vap}H$ is replaced by the molar enthalpy of sublimation, $\Delta_{sub}H$.

Clausius-Mosotti equation - A relation between the dielectric constant ε_r at optical frequencies and the polarizability α:

$$\frac{\varepsilon_r - 1}{\varepsilon_r + 2} = \frac{\rho N_A \alpha}{3M\varepsilon_0}$$

where ρ is density, N_A is Avogadro's number, M is molar mass, and ε_0 is the permittivity of a vacuum.

Clebsch-Gordon coefficients - A set of coefficients used to describe the vector coupling of angular momenta in atomic and nuclear physics.

Codon - A set of three bases, chosen from the four primary bases found in the DNA molecule (uracil, cytosine, adenine, and guanine), which specifies the production of a particular amino acid or carries some other genetic instruction. For example, the codon UCA specifies the amino acid serine, CAG specifies glutamine, etc. There are a total of 64 codons.

Coercive force - The magnetizing force at which the magnetic flux density is equal to zero. [10]

Coercivity* - The maximum value of coercive force that can be attained when a magnetic material is symmetrically magnetized to saturation induction. [10]

Coherent anti-Stokes Raman spectroscopy (CARS) - See Techniques for Materials Characterization, page **12**-1.

Colloid - Molecules or polymolecular particles dispersed in a medium that have, at least in one direction, a dimension roughly between 1 nm and 1 μm. [3]

Color center - A defect in a crystal that gives rise to optical absorption, thus changing the color of the material. A common type is the F-center, which results when an electron occupies the site of a negative ion.

Compressibility (κ)* - The fractional change of volume as pressure is increased, viz., $\kappa = -(1/V)(dV/dp)$. [1]

Compton wavelength (λ_C)* - In the scattering of electromagnetic radiation by a free particle (e.g., electron, proton), $\lambda_C = h/mc$ is the increase in wavelength, at a 90° scattering angle, corresponding to the transfer of energy from radiation to particle. Here h is Planck's constant, c the speed of light, and m the mass of the particle.

Conductance (G)* - For direct current, the reciprocal of resistance. More generally, the real part of admittance. [1]

Conductivity, electrical (σ)* - The reciprocal of the resistivity. [1]

Conductivity, thermal - See Thermal conductivity.

Congruent transformation - A phase transition (melting, vaporization, etc.) in which the substance preserves its exact chemical composition.

Constitutional repeating unit (CRU) - In polymer science, the smallest constitutional unit, the repetition of which constitutes a regular macromolecule, i.e., a macromolecule with all units connected identically with respect to directional sense. [8]

Copolymer - A polymer derived from more than one species of monomer. [8]

Coriolis effect - The deviation from simple trajectories when a mechanical system is described in a rotating coordinate system. It affects the motion of projectiles on the earth and in molecular spectroscopy leads to an important interaction between the rotational and vibrational motions. The effect may be described by an additional term in the equations of motion, called the Coriolis force.

Cosmic rays* - High energy nuclear particles, electrons, and photons, originating mostly outside the solar system, which continually bombard the earth's atmosphere.

Coulomb (C)* - The SI unit of electric charge, equal to A s. [1]

Coulomb's law - The statement that the force F between two electrical charges q_1 and q_2 separated by a distance r is $F = (4\pi\varepsilon_0)^{-1}q_1q_2/r^2$, where ε_0 is the permittivity of a vacuum.

Covalent bond - A chemical bond between two atoms whose stability results from the sharing of two electrons, one from each atom.

Cowling number (*Co*) - A dimensionless quantity used in plasma physics, defined by $Co = B^2/\mu\rho v^2$, where ρ is density, v is velocity, μ is permeability, and B is magnetic flux density. [2]

CPT theorem - A theorem in particle physics which states that any local Lagrangian theory that is invariant under proper Lorentz transformations is also invariant under the combined operations of charge conjugation, C, space inversion, P, and time reversal, T, taken in any order.

Critical point* - In general, the point on the phase diagram of a two-phase system at which the two coexisting phases have identical properties and therefore represent a single phase. At the liquid-gas critical point of a pure substance, the distinction between liquid and gas vanishes, and the vapor pressure curve ends. The coordinates of this point are called the critical temperature and critical pressure. Above the critical temperature, it is not possible to liquefy the substance.

Cross section (σ)* - A measure of the probability of collision (or other interaction) between a beam of particles and a target which it encounters. In rough terms it is the effective area the target particles present to the incident ones; however, the precise definition depends on the nature of the interaction. A general definition of σ is the number of encounters per unit time divided by nv, where n is the concentration of incident particles and v their velocity.

Crosslink - In polymer science, a small region in a macromolecule from which at least four chains emanate, and formed by reactions involving sites or groups on existing macromolecules or by interactions between existing macromolecules. [8]

Crown compounds - Macrocyclic polydentate compounds, usually uncharged, in which three or more coordinating ring atoms (usually oxygen or nitrogen) are or may become suitably close for easy formation of chelate complexes with metal ions or other cationic species. [5]

Crust* - The outer layer of the solid earth, above the Mohorovicic discontinuity. Its thickness averages about 35 km on the continents and about 7 km below the ocean floor.

Cryoscopic constant (E_f)* - The constant that expresses the amount by which the freezing point T_f of a solvent is lowered by a non-dissociating solute, through the relation $\Delta T_f = E_f\, m$, where m is the molality of the solute.

Curie (Ci) - A non-SI unit of radioactivity (disintegrations per unit time), equal to 3.7×10^{10} s^{-1}.

Curie temperature (T_C)* - For a ferromagnetic material, the critical temperature above which the material becomes paramagnetic. Also applied to the temperature at which the spontaneous polarization disappears in a ferroelectric solid. [1]

Cyanohydrins - Alcohols substituted by a cyano group, most commonly, but not limited to, examples having a CN and an OH group attached to the same carbon atom. They are formally derived from aldehydes or ketones by the addition of hydrogen cyanide. [5]

Cycloalkanes - Saturated monocyclic hydrocarbons (with or without side chains). See alicyclic compounds. Unsaturated monocyclic hydrocarbons having one endocyclic double or one triple bond are called cycloalkenes and cycloalkynes, respectively. [5]

Cyclotron resonance - The resonant absorption of energy from a system in which electrons or ions that are orbiting in a uniform magnetic field are subjected to radiofrequency or microwave radiation. The resonance frequency is given by $\nu = eH/2\pi m^*c$, where e is the elementary charge, H is the magnetic field strength, m^* is the effective mass of the charged particle, and c is the speed of light. The effect occurs in both solids (involving electrons or holes) and in low pressure gasses (involving ions)

Dalton (Da) - A name sometimes used in biochemistry for the unified atomic mass unit (u).

De Broglie wavelength - The wavelength associated with the wave representation of a moving particle, given by h/mv, where h is Planck's constant, m the particle mass, and v the velocity.

De Haas-Van Alphen effect - An effect observed in certain metals and semiconductors at low temperatures and high magnetic fields, characterized by a periodic variation of magnetic susceptibility with field strength.

Debye equation* - The relation between the relative permittivity (dielectric constant) ε_r, polarizability α, and permanent dipole moment μ in a dielectric material whose molecules are free to rotate. It takes the form

$$\frac{\varepsilon_r - 1}{\varepsilon_r + 2} = \frac{\rho N_A}{3M\varepsilon_0}\left(\alpha + \frac{\mu^2}{3kT}\right)$$

where ρ is density, N_A is Avogadro's number, M is molar mass, and ε_0 is the permittivity of a vacuum.

Debye length - In the Debye-Hückel theory of ionic solutions, the effective thickness of the cloud of ions of opposite charge which surrounds each given ion and shields the Coulomb potential produced by that ion.

Debye temperature (θ_D)* - In the Debye model of the heat capacity of a crystalline solid, $\theta_D = h\nu_D/k$, where h is Planck's constant, k is the Boltzmann constant, and ν_D is the maximum vibrational frequency the crystal can support. For $T \ll \theta_D$, the heat capacity is proportional to T^3.

Debye unit (D) - A non-SI unit of electric dipole moment used in molecular physics, equal to 3.335641×10^{-30} C m.

Debye-Waller factor (*D*) - The factor by which the intensity of a diffraction line is reduced because of lattice vibrations. [1]

Defect - Any departure from the regular structure of a crystal lattice. A Frenkel defect results when an atom or ion moves to an interstitial position and leaves behind a vacancy. A Schottky defect involves either a vacancy where the atom has moved to the surface or a structure where a surface atom has moved to an interstitial position.

Degree of polymerization - The number of monomeric units in a macromolecule or an oligomer molecule. [8]

Dendrite - A tree-like crystalline pattern often observed, for example, in ice crystals and alloys in which the crystal growth branches repeatedly.

Density (ρ)* - In the most common usage, mass density or mass per unit volume. More generally, the amount of some quantity (mass, charge, energy, etc.) divided by a length, area, or volume.

Density of states (N_E, ρ) - The number of one-electron states in an infinitesimal interval of energy, divided by the range of that interval and by volume. [1]

Dew point* - The temperature at which liquid begins to condense as the temperature of a gas mixture is lowered. In meteorology, it is the temperature at which moisture begins to condense on a surface in contact with the air.

Diamagnetism - A type of magnetism characterized by a negative magnetic susceptibility, so that the material, when placed in an external magnetic field, becomes weakly magnetized in the direction opposite to the field. This magnetization is independent of temperature.

Diazo compounds - Compounds having the divalent diazo group, $=N^+=N^-$, attached to a carbon atom, e.g., $CH_2=N_2$ diazomethane. [5]

Dielectric constant (ε)* - Ratio of the electric displacement in a medium to the electric field strength. Also called permittivity. [1]

Dienes - Compounds that contain two fixed double bonds (usually assumed to be between carbon atoms). Dienes in which the two double-bond units are linked by one single bond are termed conjugated. [5]

Differential scanning calorimetry (DSC) - See Techniques for Materials Characterization, page **12-1**.

Differential thermal analysis (DTA) - See Techniques for Materials Characterization, page **12-1**.

Diffusion* - The migration of atoms, molecules, ions, or other particles as a result of some type of gradient (concentration, temperature, etc.).

Diopter - A unit used in optics, formally equal to m^{-1}. It is used in expressing dioptic power, which is the reciprocal of the focal length of a lens.

Dipole moment, electric (p,μ)* - For a distribution of equal positive and negative charge, the magnitude of the dipole moment vector is the positive charge multiplied by the distance between the centers of positive and negative charge distribution. The direction is given by the line from the center of negative charge to the center of positive charge.

Dipole moment, magnetic (m,μ) - Formally defined in electromagnetic theory as a vector quantity whose vector product with the magnetic flux density equals the torque. The magnetic dipole generated by a current I flowing in a small loop of area A has a magnetic moment of magnitude IA. In atomic and nuclear physics, a magnetic moment is associated with the angular momentum of a particle; e.g., an electron with orbital angular momentum l exhibits a magnetic moment of $-el/2m_e$ where e is the elementary charge and m_e the mass of the electron. [1]

Disaccharides - Compounds in which two monosaccharides are joined by a glycosidic bond. [5]

Dislocation - An extended displacement of a crystal from a regular lattice. An edge dislocation results when one portion of the crystal has partially slipped with respect to the other, resulting in an extra plane of atoms extending through part of the crystal. A screw dislocation transforms successive atomic planes into the surface of a helix.

Dispersion - Splitting of a beam of light (or other electromagnetic radiation) of mixed wavelengths into the constituent wavelengths as a result of the variation of refractive index of the medium with wavelength.

Dissociation constant* - The equilibrium constant for a chemical reaction in which a compound dissociates into its constituent parts.

Dissociation energy (D_e)* - For a diatomic molecule, the difference between the energies of the free atoms at rest and the minimum in the potential energy curve. The term bond dissociation energy (D_0), which can be applied to polyatomic molecules as well, is used for the difference between the energies of the fragments resulting when a bond is broken and the energy of the original molecule in its lowest energy state. The term bond strength implies differences in enthalpy rather than energy.

Domain - A small region of a solid in which the magnetic or electric moments of the individual units (atoms, molecules, or ions) are aligned in the same direction.

Domain wall - The transition region between adjacent ferromagnetic domains, generally a layer with a thickness of a few hundred ångström units. Also called Bloch wall.

Doppler effect - The change in the apparent frequency of a wave (sound, light, or other) when the source of the wave is moving relative to the observer.

Dose equivalent (H) - The product of the absorbed dose of radiation at a point of interest in tissue and various modifying factors which depend on the type of tissue and radiation. [1]

Drift velocity - The velocity of charge carriers (electrons, ions, etc.) moving under the influence of an electric field in a medium which subjects the carriers to some frictional force.

Dyne (dyn) - A non-SI (cgs) unit of force, equal to 10^{-5} N.

Ebullioscopic constant (E_b)* - The constant that expresses the amount by which the boiling point T_b of a solvent is raised by a non-dissociating solute, through the relation $\Delta T_b = E_b\, m$, where m is the molality of the solute.

Eddy currents - Circulating currents set up in conducting bulk materials or sheets by varying magnetic fields.

Effinghausen effect - The appearance of a temperature gradient in a current carrying conductor that is placed in a transverse magnetic field. The direction of the gradient is perpendicular to the current and the field.

Eigenvalue - An allowed value of the constant a in the equation $Au = au$, where A is an operator acting on a function u (which is called an eigenfunction). In quantum mechanics, the outcome of any observation is an eigenvalue of the corresponding operator. Also called characteristic value.

Einstein - A non-SI unit used in photochemistry, equal to one mole of photons.

Einstein temperature (θ_V) - In the Einstein theory of the heat capacity of a crystalline solid, $\theta_V = h\nu/k$, where h is Planck's constant, k is the Boltzmann constant, and ν is the vibrational frequency of the crystal.

Einstein transition probability - A constant in the Einstein relation $A_{ij} + B_{ij}\rho$ for the probability of a transition between two energy levels i and j in a radiation field of energy density ρ. The A_{ij} coefficient describes the probability of spontaneous emission, while B_{ij} and B_{ji} govern the probability of stimulated emission and absorption, respectively ($B_{ij} = B_{ji}$).

Elastic limit - The greatest stress which a material is capable of sustaining without any permanent strain remaining after complete release of the stress. [10]

Elastic modulus - See Young's modulus.

Electric charge (Q) - The quantity of electricity; i.e., the property that controls interactions between bodies through electrical forces.

Electric current (I) - The charge passing through a circuit per unit time. [1]

Electric displacement (D) - A vector quantity whose magnitude equals the electric field strength multiplied by the permittivity of the medium and whose direction is the same as that of the field strength.

Electric field strength (E) - The force exerted by an electric field on a point charge divided by the electric charge. [1]

Electric potential (V) - A scalar quantity whose gradient is equal to the negative of the electric field strength.

Electrical conductance - See Conductance

Electrical resistance - See Resistance

Electrical resistivity - See Resistivity.

Electrochemical series* - An arrangement of reactions which produce or consume electrons in an order based on standard electrode potentials. A common arrangement places metals in decreasing order of their tendency to give up electrons.

Electrode potential* - The electromotive force of a cell in which the electrode on the left is the standard hydrogen electrode and that on the right is the electrode in question. [2]

Electrolysis - The decomposition of a substance as a result of passing an electric current between two electrodes immersed in the sample.

Electromotive force (emf) - The energy supplied by a source divided by the charge transported through the source. [1]

Electron* - An elementary particle in the family of leptons, with negative charge and spin of 1/2.

Electron affinity* - The energy difference between the ground state of a gas-phase atom or molecule and the lowest state of the corresponding negative ion.

Electron cyclotron resonance (ECR) - See Techniques for Materials Characterization, page **12**-1.

Electron energy loss spectroscopy (EELS) - See Techniques for Materials Characterization, page **12**-1.

Electron nuclear double resonance (ENDOR) - See Techniques for Materials Characterization, page **12**-1.

Electron paramagnetic resonance (EPR) - See Techniques for Materials Characterization, page **12**-1.

Electron probe microanalysis (EPMA) - See Techniques for Materials Characterization, page **12**-1.

Electron spectroscopy for chemical analysis (ESCA) - See Techniques for Materials Characterization, page **12**-1.

Electron spin (s) - The quantum number, equal to 1/2, that specifies the intrinsic angular momentum of the electron.

Electron stimulated desorption (ESD) - See Techniques for Materials Characterization, page **12-1**.

Electron volt (eV)* - A non-SI unit of energy used in atomic and nuclear physics, equal to approximately 1.602177×10^{-19} J. The electron volt is defined as the kinetic energy acquired by an electron upon acceleration through a potential difference of 1 V. [1]

Electronegativity* - A parameter originally introduced by Pauling which describes, on a relative basis, the power of an atom or group of atoms to attract electrons from the same molecular entity. [3]

Electrophoresis - The motion of macromolecules or colloidal particles in an electric field. [3]

Emissivity (ε)* - Ratio of the radiant flux emitted per unit area to that of an ideal black body at the same temperature. Also called emittance. [1]

Emu - The electromagnetic system of units, based upon the cm, g, and s plus the emu of current (sometimes called the abampere).

Enantiomers - A chiral molecule and its non-superposable mirror image. The two forms rotate the plane of polarized light by equal amounts in opposite directions. Also called optical isomers.

Energy (*E,U*)* - The characteristic of a system that enables it to do work.

Energy gap* - In the theory of solids, the region between two energy bands, in which no bound states can occur.

Enols, alkenols - The term refers specifically to vinylic alcohols, which have the structure $HOCR'{=}CR_2$. Enols are tautomeric with aldehydes (R′ = H) or ketones (R′ not equal to H). [5]

Enthalpy (*H*)* - A thermodynamic function, especially useful when dealing with constant-pressure processes, defined by $H = E + PV$, where E is energy, P pressure, and V volume. [1]

Enthalpy of combustion* - The enthalpy change in a combustion reaction. Its negative is the heat released in combustion.

Enthalpy of formation, standard* - The enthalpy change for the reaction in which a substance is formed from its constituent elements, each in its standard reference state (normally refers to 1 mol, sometimes to 1 g, of the substance).

Enthalpy of fusion* - The enthalpy change in the transition from solid to liquid state.

Enthalpy of sublimation - The enthalpy change in the transition from solid to gas state.

Enthalpy of vaporization* - The enthalpy change in the transition from liquid to gas state.

Entropy (*S*)* - A thermodynamic function defined such that when a small quantity of heat dQ is received by a system at temperature T, the entropy of the system is increased by dQ/T, provided that no irreversible change takes place in the system. [1]

Entropy unit (e.u.) - A non-SI unit of entropy, equal to 4.184 J/K mol.

Ephemeris time - Time measured in tropical years from January 1, 1900.

Epoxy compounds - Compounds in which an oxygen atom is directly attached to two adjacent or non-adjacent carbon atoms of a carbon chain or ring system; thus cyclic ethers. [5]

Equation of continuity - Any of a class of equations that express the fact that some quantity (mass, charge, energy, etc.) cannot be created or destroyed. Such equations typically specify that the rate of increase of the quantity in a given region of space equals the net current of the quantity flowing into the region.

Equation of state* - An equation relating the pressure, volume, and temperature of a substance or system.

Equilibrium constant (*K*)* - For a chemical reaction $a\mathrm{A} + b\mathrm{B} \rightleftharpoons c\mathrm{C} + d\mathrm{D}$, the equilibrium constant is defined by:

$$K = \frac{a_\mathrm{C}{}^c \cdot a_\mathrm{D}{}^d}{a_\mathrm{A}{}^a \cdot a_\mathrm{B}{}^b}$$

where a_i is the activity of component i. To a certain approximation, the activities can be replaced by concentrations. The equilibrium constant is related to $\Delta_r G°$, the standard Gibbs energy change in the reaction, by $RT \ln K = -\Delta_r G°$.

Equivalent conductance - See Conductivity, electrical

Erg (erg) - A non-SI (cgs) unit of energy, equal to 10^{-7} J.

Esters - Compounds formally derived from an oxoacid RC(=O)(OH) and an alcohol, phenol, heteroarenol, or enol by linking, with formal loss of water from an acidic hydroxy group of the former and a hydroxy group of the latter. [5]

Esu - The electrostatic system of units, based upon the cm, g, and s plus the esu of charge (sometimes called the statcoulomb or franklin).

Ethers - Compounds with formula ROR, where R is not equal to H. [5]

Euler number (*Eu*) - A dimensionless quantity used in fluid mechanics, defined by $Eu = \Delta p/\rho v^2$, where p is pressure, ρ is density, and v is velocity. [2]

Eutectic - The point on a two-component solid-liquid phase diagram which represents the lowest melting point of any possible mixture. A liquid having the eutectic composition will freeze at a single temperature without change of composition.

Excitance (*M*) - Radiant energy flux leaving an element of a surface divided by the area of that element. [1]

Exciton - A localized excited state consisting of a bound electron-hole pair in a molecular or ionic crystal. The exciton can propagate through the crystal.

Exosphere - The outermost part of the earth's atmosphere, beginning at about 500 to 1000 km above the surface. It is characterized by densities so low that air molecules can escape into outer space.

Expansion coefficient - See thermal expansion coefficient.

Extended electron energy loss fine structure (EXELFS) - See Techniques for Materials Characterization, page **12-1**.

Extended x-ray absorption fine structure (EXAFS) - See Techniques for Materials Characterization, page **12-1**.

Extinction coefficient - See Absorption coefficient, molar

F-Center - See Color center

Fahrenheit temperature (°F) - The temperature scale based on the assignment of 32°F = 0°C and a temperature interval of °F =(5/9)°C; i.e., t/°F = (9/5)t/°C + 32.

Farad (F)* - The SI unit of electric capacitance, equal to C/V. [1]

Faraday constant (*F*)* - The electric charge of 1 mol of singly charged positive ions; i.e., $F = N_\mathrm{A} e$, where N_A is Avogadro's constant and e is the elementary charge. [1]

Faraday effect* - The rotation of the plane of plane-polarized light by a medium placed in a magnetic field parallel to the direction of the light beam. The effect can be observed in solids, liquids, and gasses.

Fatty acids - Aliphatic monocarboxylic acids derived from or contained in esterified form in an animal or vegetable fat, oil, or wax. Natural fatty acids commonly have a chain of 4 to 28 carbons (usually unbranched and even-numbered), which may be saturated or unsaturated. By extension, the term is sometimes used to embrace all acyclic aliphatic carboxylic acids. [5]

Fermat's principle - The law that a ray of light traversing one or more media will follow a path which minimizes the time required to pass between two given points.

Fermi (f) - Name sometimes used in nuclear physics for the femtometer.

Fermi level - The highest energy of occupied states in a solid at zero temperature. Sometimes called Fermi energy. The Fermi surface is the surface in momentum space formed by electrons occupying the Fermi level.

Fermi resonance - An effect observed in vibrational spectroscopy when an overtone of one fundamental vibration closely coincides in energy with another fundamental of the same symmetry species. It leads to a splitting of vibrational bands.

Fermi-Dirac distribution - A modification of the Boltzmann distribution which takes into account the Pauli exclusion principle. The number of particles of energy E is proportional to $[e^{(E-\mu)/kT}+1]^{-1}$, where μ is a normalization constant, k the Boltzmann constant, and T the temperature. The distribution is applicable to a system of fermions.

Fermion - A particle that obeys Fermi-Dirac statistics. Specifically, any particle with spin equal to an odd multiple of 1/2. Examples are the electron, proton, neutron, muon, etc.

Ferrimagnetism* - A type of magnetism in which the magnetic moments of atoms in a solid are ordered into two nonequivalent sublattices with unequal magnetic moments, leading to a nonzero magnetic susceptibility.

Ferrite - A ferrimagnetic material of nominal formula MFe_2O_4, where M is a divalent metal; widely used in microwave switches and other solid state devices.

Ferroelectricity* - The retention of electric polarization by certain materials after the external field that produced the polarization has been removed.

Ferromagnetism* - A type of magnetism in which the magnetic moments of atoms in a solid are aligned within domains which can in turn be aligned with each other by a weak magnetic field. Some ferromagnetic materials can retain their magnetization when the external field is removed, as long as the temperature is below a critical value, the Curie temperature. They are characterized by a large positive magnetic susceptibility.

Fick's law - The statement that the flux J of a diffusing substance is proportional to the concentration gradient, i.e., $J = -D(\mathrm{d}c/\mathrm{d}x)$, where D is called the diffusion coefficient.

Field - A mathematical construct which describes the interaction between particles resulting from gravity, electromagnetism, or other physical phenomena. In classical physics a field is described by equations. Quantum field theory introduces operators to represent the physical observables.

Field emission microscopy (FEM) - See Techniques for Materials Characterization, page **12-1**.

Field ion microscopy (FIM) - See Techniques for Materials Characterization, page **12-1**.

Fine structure - The splitting in spectral lines that results from interactions of the electron spin with the orbital angular momentum.

Fine structure constant (α)* - Defined as $e^2/2hc\varepsilon_0$, where e is the elementary charge, h Planck's constant, c the speed of light, and ε_0 the permittivity of a vacuum. It is a measure of the strength of the electromagnetic interaction between particles.

First radiation constant (c_1)* - Constant ($=2\pi hc^2$) in the equation for the radiant excitance M_λ of a black body:

$$M_\lambda = \frac{c_1 \lambda^{-5} \Delta\lambda}{e^{c_2/\lambda T} - 1}$$

where λ is the wavelength, T is the temperature, and $c_2 = hc/k$ is the second radiation constant.

Flash point - The lowest temperature at which vapors above a volatile combustible substance will ignite in air when exposed to a flame. [10]

Fluence (F) - Term used in photochemistry to specify the energy per unit area delivered in a given time interval, for example by a laser pulse. [2]

Fluorocarbons - Compounds consisting solely of fluorine and carbon. [5]

Fluxoid - The quantum of magnetic flux in superconductivity theory, equal to $hc/2e$, where h is Planck's constant, c the velocity of light, and e the elementary charge.

Force (F) - The rate of change of momentum with time. [1]

Force constants (f, k)* - In molecular vibrations, the coefficients in the expression of the potential energy in terms of atom displacements from their equilibrium positions. In a diatomic molecule, $f = \mathrm{d}^2V/\mathrm{d}r^2$, where $V(r)$ is the potential energy and r is the interatomic distance. [2]

Fourier number (Fo) - A dimensionless quantity used in fluid mechanics, defined by $Fo = at/l^2$, where a is thermal diffusivity, t is time, and l is length. [2]

Fourier transform infrared spectroscopy (FTIR) - A technique for obtaining an infrared spectrum by use of an interferometer in which the path length of one of the beams is varied. A Fourier transformation of the resulting interferogram yields the actual spectrum. The technique is also used for NMR and other types of spectroscopy.

Fractals - Geometrical objects that are self-similar under a change of scale; i.e., they appear similar at all levels of magnification. They can be considered to have fractional dimensionality. Examples occur in diverse fields such as geography (rivers and shorelines), biology (trees), and solid state physics (amorphous materials).

Franck-Condon principle - An important principle in molecular spectroscopy which states that the nuclei in a molecule remain essentially stationary while an electronic transition is taking place. The physical interpretation rests on the fact that the electrons move much more rapidly than the nuclei because of their much smaller mass.

Franklin (Fr) - Name sometimes given to the unit of charge in the esu system.

Fraunhofer diffraction - Diffraction of light in situations where the source and observation point are so far removed that the wave surfaces may be considered planar.

Fraunhofer lines - Sharp absorption lines in the spectrum of sunlight, caused by absorption of the solar blackbody radiation by atoms near the sun's surface.

Free radical - See Radicals. The term "free radical" is often used more broadly for molecules that have a paramagnetic ground state (e.g., O_2) and sometimes for any transient or highly reactive molecular species.

Freezing point - See Melting point

Frequency (ν)* - Number of cycles of a periodic phenomenon divided by time. [1]

Fresnel diffraction - Diffraction of light in a situation where the source and observation point are sufficiently close together that the curvature of the wave surfaces must be taken into account.

Froude number (Fr) - A dimensionless quantity used in fluid mechanics, defined by $Fr = v/(lg)^{1/2}$, where v is velocity, l is length, and g is acceleration due to gravity. [2]

Fugacity (f_B) - For a gas mixture, the fugacity of component B is defined as the absolute activity λ_B times the limit, as the pressure p approaches zero at constant temperature, of p_B/λ_B. [2]

Fullerenes - Compounds composed solely of an even number of carbon atoms, which form a cage-like fused-ring polycyclic system with twelve five-membered rings and the rest six-membered rings. The archetypal example is [60]fullerene, where the atoms and bonds delineate a truncated icosahedron. The term has been broadened to include any closed cage structure consisting entirely of three-coordinate carbon atoms. [5]

Fulvalenes - The hydrocarbon fulvalene and its derivatives formed by substitution (and by extension, analogues formed by replacement of one or more carbon atoms of the fulvalene skeleton by a heteroatom). [5]

Fulvenes - The hydrocarbon fulvene and its derivatives formed by substitution (and by extension, analogues formed by replacement of one or more carbon atoms of the fulvene skeleton by a heteroatom). [5]

Fundamental vibrational frequencies* - In molecular spectroscopy, the characteristic vibrational frequencies obtained when the vibrational energy is expressed in normal coordinates. They determine the primary features of the infrared and Raman spectra of the molecule.

γ - Name sometimes used for microgram.

γ-rays* - Electromagnetic radiation (photons) with energy greater than about 0.1 MeV (wavelength less than about 1 pm).

g*-Factor of the electron - The proportionality factor in the equation relating the magnetic moment μ of an electron to its total angular momentum quantum number J, i.e., $\mu = -g\mu_B J$, where μ_B is the Bohr magneton. Also called Landé factor.

Gal - A non-SI unit of acceleration, equal to 0.01 m/s. Also called galileo.

Gallon (US) - A unit of volume equal to 3.785412 L.

Gallon (UK, Imperial) - A unit of volume equal to 4.546090 L.

Gauss (G) - A non-SI unit of magnetic flux density (B) equal to 10^{-4} T.

Gaussian system of units - A hybrid system used in electromagnetic theory, which combines features of both the esu and emu systems.

Gel - A colloidal system with a finite, but usually rather small, yield stress (the sheer stress at which yielding starts abruptly). [3]

Genetic code* - The set of relations between each of the 64 codons of DNA and a specific amino acid (or other genetic instruction).

Gibbs energy (*G*)* - An important function in chemical thermodynamics, defined by $G = H\text{-}TS$, where H is the enthalpy, S the entropy, and T the thermodynamic temperature. Sometimes called Gibbs free energy and, in older literature, simply "free energy". [2]

Gibbs phase rule - The relation $F = C - P + 2$, where C is the number of components in a mixture, P is the number of phases, and F is the degrees of freedom, i.e., the number of intensive variables that can be changed independently without affecting the number of phases.

Glass transition temperature* - The temperature at which an amorphous polymer is transformed, in a reversible way, from a viscous or rubbery condition to a hard and relatively brittle one. [10]

Glow discharge mass spectroscopy (GDMS) - See Techniques for Materials Characterization, page **12**-1.

Gluon - A hypothetical particle postulated to take part in the binding of quarks, in analogy to the role of the photon in electromagnetic interactions.

Glycerides - Esters of glycerol (propane-1,2,3-triol) with fatty acids, widely distributed in nature. They are by long-established custom subdivided into triglycerides, 1,2- or 1,3-diglycerides, and 1- or 2-monoglycerides, according to the number and positions of acyl groups. [5]

Glycols - Dihydric alcohols in which two hydroxy groups are on different carbon atoms, usually but not necessarily adjacent. Also called diols. [5]

Grain (gr) - A non-SI unit of mass, equal to 64.79891 mg.

Grain boundary - The interface between two regions of different crystal orientation.

Grashof number (*Gr*) - A dimensionless quantity used in fluid mechanics, defined by $Gr = l^3 g\alpha\Delta T\rho^2/\eta^2$, where T is temperature, ρ is density, l is length, η is viscosity, α is cubic expansion coefficient, and g is acceleration of gravity. [2]

Gravitational constant (*G*)* - The universal constant in the equation for the gravitational force between two particles, $F = Gm_1m_2/r^2$, where r is the distance between the particles and m_1 and m_2 are their masses. [1]

Gray (Gy)* - The SI unit of absorbed dose of radiation, equal to J/kg. [1]

Gregorian calendar - The modification of the Julian calendar introduced in 1582 by Pope Gregory XII which specified that a year divisible by 100 is a leap year only if divisible by 400.

Grignard reagents - Organomagnesium halides, RMgX, having a carbon-magnesium bond (or their equilibrium mixtures in solution with $R_2Mg + MgX_2$). [5]

Gruneisen parameter (γ) - Defined by $\gamma = \alpha_V/\kappa\, c_V\, \rho$, where α_V is the cubic thermal expansion coefficient, κ is the isothermal compressibility, c_V is the specific heat capacity at constant volume, and ρ is the mass density. γ is independent of temperature for most crystalline solids. [1]

Gyromagnetic ratio (γ) - Ratio of the magnetic moment of a particle to its angular momentum. Also called magnetogyric ratio.

Hadron - Any elementary particle that can take part in the strong interaction. Hadrons are subdivided into baryons, with odd half integer spins, and mesons, which have zero or integral spin.

Hall effect* - The development of a transverse potential difference V in a conducting material when subjected to a magnetic field H perpendicular to the direction of the current. The potential difference is given by $V = R_H BJt$, where B is the magnetic induction, J the current density, t the thickness of the specimen in the direction of the potential difference, and R_H is called the Hall coefficient.

Halocarbon - A compound containing no elements other than carbon, hydrogen, and one or more halogens. In common practice, the term is used mainly for compounds of no more than four or five carbon atoms.

Halogens - The elements F, Cl, Br, I, and At. Compounds of these elements are called halogenides or halides. [7]

Hamiltonian (*H*) - An expression for the total energy of a mechanical system in terms of the momenta and positions of constituent particles. In quantum mechanics, the Hamiltonian operator appears in the eigenvalue equation $H\psi = E\psi$, where E is an energy eigenvalue and ψ the corresponding eigenfunction.

Hardness* - The resistance of a material to deformation, indentation, or scratching. Hardness is measured on various scales, such as Mohs, Brinell, Knoop, Rockwell, and Vickers. [10]

Hartmann number (*Ha*) - A dimensionless quantity used in plasma physics, defined by $Ha = Bl(\kappa/\eta)^{1/2}$, where B is magnetic flux density, l is length, κ is electric conductivity, and η is viscosity. [2]

Hartree (E_h)* - An energy unit used in atomic and molecular science, equal to approximately $4.3597482 \times 10^{-18}$ J.

Hartree-Fock method - A iterative procedure for solving the Schrödinger equation for an atom or molecule in which the equation is solved for each electron in an initial assumed potential from all the other electrons. The new potential that results is used to repeat the calculation and the procedure continued until convergence is reached. Also called self-consistent field (SCF) method.

Heat capacity* - Defined in general as dQ/dT, where dQ is the amount of heat that must be added to a system to increase its temperature by a small amount dT. The heat capacity at constant pressure is $C_p = (\partial H/\partial T)_p$; that at constant volume is $C_V = (\partial E/\partial T)_V$, where H is enthalpy, E is internal energy, p is pressure, V is volume, and T is temperature. An upper case C normally indicates the molar heat capacity, while a lower case c is used for the specific (per unit mass) heat capacity. [1]

Heat of formation, vaporization, etc. - See corresponding terms under Enthalpy.

Hectare (ha) - A unit of area equal to 10^4 m^2. [1]

Heisenberg uncertainty principle - The statement that two observable properties of a system that are complementary, in the sense that their quantum-mechanical operators do not commute, cannot be specified simultaneously with absolute precision. An example is the position and momentum of a particle; according to this principle, the uncertainties in position Δq and momentum Δp must satisfy the relation $\Delta p\Delta q \geq h/4\pi$, where h is Planck's constant.

Heitler-London model - An early quantum-mechanical model of the hydrogen atom which introduced the concept of the exchange interaction between electrons as the primary reason for stability of the chemical bond.

Helicon - A low-frequency wave generated when a metal at low temperature is exposed to a uniform magnetic field and a circularly polarized electric field.

Helmholz energy (*A*) - A thermodynamic function defined by $A = E\text{-}TS$, where E is the energy, S the entropy, and T the thermodynamic temperature. [2]

Hemiacetals - Compounds having the general formula $R_2C(OH)OR'$ (R′ not equal to H). [5]

Henry (H)* - The SI unit of inductance, equal to Wb/A. [1]

Henry's law * - An expression which applies to an ideal dilute solution in which one or more gasses are dissolved, viz., $p_i = H_i x_i$, where p_i is the partial pressure of component i above the solution, x_i is its mole fraction in the solution, and H_i is the Henry's law constant (a characteristic of the given gas and solvent, as well as the temperature).

Hermitian operator - An operator A that satisfies the relation $\int u_m{}^*Au_n \mathrm{d}x = (\int u_n{}^*Au_m \mathrm{d}x)^*$, where * indicates the complex conjugate. The eigenvalues of Hermitian operators are real, and eigenfunctions belonging to different eigenvalues are orthogonal.

Hertz (Hz) - The SI unit of frequency, equal to s^{-1}. [1]

Heterocyclic compounds - Cyclic compounds having as ring members atoms of at least two different elements, e.g., quinoline, 1,2-thiazole, bicyclo[3.3.1]tetrasiloxane. [5]

Heusler alloys - Alloys of manganese, copper, aluminum, nickel, and sometimes other metals which find important uses as permanent magnets.

Holography - A technique for creating a three-dimensional image of a object by recording the interference pattern between a light beam diffracted from the object and a reference beam. The image can be reconstructed from this pattern by a suitable optical system.

Homopolymer - A polymer derived from one species of (real, implicit, or hypothetical) monomer. [8]

Hooke's law - The statement that the ratio of stress to strain is a constant in a totally elastic medium.

Horse power - A non-SI unit of energy, equal to approximately 746 W.

Hubble constant - The ratio of the recessional velocity of an extragalactic object to the distance of that object. Its value is about 2×10^{-18} s^{-1}.

Huckel theory - A simple approximation for calculating the energy of conjugated molecules in which only the resonance integrals between neighboring bonds are considered. Also called CNDO method (complete neglect of differential overlap).

Hume-Rothery rules - A set of empirical rules for predicting the occurrence of solid solutions in metallic systems. The rules involve size, crystal structure, and electronegativity.

Hund's rules - A series of rules for predicting the sequence of energy states in atoms and molecules. One of the important results is that when two electrons exist in different orbitals, the state with their spins parallel (triplet state) lies at lower energy than the state with antiparallel spins (singlet).

Hydrazines - Hydrazine (diazane), H_2NNH_2, and its hydrocarbyl derivatives. When one or more substituents are acyl groups, the compound is a hydrazide. [5]

Hydrocarbon - A compound containing only carbon and hydrogen. [5]

Hydrolysis - A reaction occurring in water in which a chemical bond is cleaved and a new bond formed with the oxygen atom of water.

Hyperfine structure - Splitting of energy levels and spectral lines into several closely spaced components as a result of interaction of nuclear spin angular momentum with other angular momenta in the atom or molecule.

Hysteresis* - An irreversible response of a system (parameter A) as a function of an external force (parameter F), usually symmetric with respect to the origin of the A vs. F graph after the initial application of the force. A common example is magnetic induction vs. magnetic field strength in a ferromagnet.

Ideal gas law - The equation of state $pV = RT$, which defines an ideal gas, where p is pressure, V molar volume, T temperature, and R the molar gas constant.

Ideal solution - A solution in which solvent-solvent and solvent-solute interactions are identical, so that properties such as volume and enthalpy are exactly additive. Ideal solutions follow Raoult's law, which states that the vapor pressure p_i of component i is $p_i = x_i p_i^*$, where x_i is the mole fraction of component i and p_i^* the vapor pressure of the pure substance i.

Ignition temperature* - The lowest temperature at which combustion of a material will occur spontaneously under specified conditions. Sometimes called autoignition temperature, kindling point. [10]

Imides - Diacyl derivatives of ammonia or primary amines, especially those cyclic compounds derived from diacids. Also used for salts having the anion RN_2^-. [5]

Impedence (Z) - The complex representation of potential difference divided by the complex representation of current. In terms of reactance X and resistance R, the impedance is given by $Z = R + \mathrm{i}X$. [1]

Index of refraction (n)* - For a non-absorbing medium, the ratio of the velocity of electromagnetic radiation *in vacuo* to the phase velocity of radiation of a specified frequency in the medium. [1]

Inductance - The ratio of the electromagnetic force induced in a coil by a current to the rate of change of the current.

Inductive coupled plasma mass spectroscopy (ICPMS) - See Techniques for Materials Characterization, page **12-1**.

Inertial defect - In molecular spectroscopy, the quantity I_c-I_a-I_b for a molecule whose equilibrium configuration is planar, where I_a, I_b, and I_c are the effective principal moments of inertia. The inertial defect for a rigid planar molecule would be zero, but vibration-rotation interactions in a real molecule lead to a positive inertial defect.

Insulator - A material in which the highest occupied energy band (valence band) is completely filled with electrons, while the next higher band (conduction band) is empty. Solids with an energy gap of 5 eV or more are generally considered as insulators at room temperature. Their conductivity is less than 10^{-6} S/m and increases with temperature.

Intercalation compounds - Compounds resulting from reversible inclusion, without covalent bonding, of one kind of molecule in a solid matrix of another compound, which has a laminar structure. The host compound, a solid, may be macromolecular, crystalline, or amorphous. [5]

International System of Units (SI)* - The unit system adopted by the General Conference on Weights and Measures in 1960. It consists of seven base units (meter, kilogram, second, ampere, kelvin, mole, candela), plus derived units and prefixes. [1]

International Temperature Scale (ITS-90)* - The official international temperature scale adopted in 1990. It consists of a set of fixed points and equations which enable the thermodynamic temperature to be determined from operational measurements. [9]

Ion - An atomic or molecular particle having a net electric charge. [3]

Ion exchange - A process involving the adsorption of one or several ionic species accompanied by the simultaneous desorption (displacement) of one or more other ionic species. [3]

Ion neutralization spectroscopy (INS) - See Techniques for Materials Characterization, page **12-1**.

Ionic strength (I) - A measure of the total concentration of ions in a solution, defined by $I = 1/2\Sigma_i z_i^2 m_i$, where z_i is the charge of ionic species i and m_i is its molality. For a 1-1 electrolyte at molality m, $I = m$.

Ionization constant* - The equilibrium constant for a reaction in which a substance in solution dissociates into ions.

Ionization potential* - The minimum energy required to remove an electron from an isolated atom or molecule (in its vibrational ground state) in the gaseous phase. More properly called ionization energy. [3]

Irradiance (E) - The radiant energy flux incident on an element of a surface, divided by the area of that element. [1]

Isentropic process - A thermodynamic process in which the entropy of the system does not change.

Ising model - A model describing the coupling between two atoms in a ferromagnetic lattice, in which the interaction energy is proportional to the negative of the product of the spin components along a specified axis.

Isobar - A line connecting points of equal pressure on a graphical representation of a physical system.

Isochore - A line or surface of constant volume on a graphical representation of a physical system.

Isoelectric point* - The pH of a solution or dispersion at which the net charge on the macromolecules or colloidal particles is zero. In electrophoresis there is no motion of the particles in an electric field at the isoelectric point.

Isomers - In chemistry, compounds that have identical molecular formulas but differ in the nature or sequence of bonding of their atoms or in the arrangement of their atoms in space. In physics, nuclei of the same atomic number *Z* and mass number *A* but in different energy states. [3]

Isomorphs - Substances of different chemical nature but having the same crystal structure.

Isotactic macromolecule - A tactic macromolecule, essentially comprising only one species of repeating unit which has chiral or prochiral atoms in the main chain in a unique arrangement with respect to its adjacent constitutional units. [8]

Isotherm - A line connecting points of equal temperature on a graphical representation of a physical system.

Isothermal process - A thermodynamic process in which the temperature of the system does not change.

Isotones - Nuclides having the same neutron number *N* but different atomic number Z. [3]

Isotopes - Two or more nuclides with the same atomic number *Z* but different mass number *A*. The term is sometimes used synonymously with nuclide, but it is preferable to reserve the word nuclide for a species of specific *Z* and *A*. [3]

Jahn-Teller effect - An interaction of vibrational and electronic motions in a nonlinear molecule which removes the degeneracy of certain electronic energy levels. It can influence the spectrum, crystal structure, and magnetic properties of the substance.

Johnson noise - Electrical noise generated by random thermal motion of electrons in a conductor or semiconductor. Also called thermal noise.

Josephson effect - The tunneling of electron pairs through a thin insulating layer which separates two superconductors. When a potential difference is applied to the superconductors, an alternating current is generated whose frequency is precisely proportional to the potential difference. This effect has important applications in metrology and determination of fundamental physical constants.

Joule (J)* - The SI unit of energy, equal to N m. [1]

Joule-Thomson coefficient (μ) - A parameter which describes the temperature change when a gas expands adiabatically through a nozzle from a high pressure to a low pressure region. It is defined by $\mu = (\partial T/\partial p)_H$, where *H* is enthalpy.

Julian calendar - The calendar introduced by Julius Caeser in 46 B.C. which divided the year into 365 days with a leap year of 366 days every fourth year.

Julian date (JD) - The number of days elapsed since noon Greenwich Mean Time on January 1, 4713 B.C. Thus January 1, 2000, 0h (midnight) will be JD 2,451,543.5. This dating system was introduced by Joseph Scaliger in 1582.

Kaon - One of the elementary particles in the family of mesons. Kaons have a spin of zero and may be neutral or charged.

Kelvin (K)* - The SI base unit of thermodynamic temperature. [1]

Kepler's laws - The three laws of planetary motion, which established the elliptical shape of planetary orbits and the relation between orbital dimensions and the period of rotation.

Kerr effect* - An electrooptical effect in which birefringence is induced in a liquid or gas when a strong electric field is applied perpendicular to the direction of an incident light beam. The Kerr constant *k* is given by $n_1\text{-}n_2 = k\lambda E^2$, where λ is the wavelength, *E* is the electric field strength, and n_1 and n_2 are the indices of refraction of the ordinary and extraordinary rays, respectively.

Ketenes - Compounds in which a carbonyl group is connected by a double bond to an alkylidene group: $R_2C{=}C{=}O$. [5]

Ketones - Compounds in which a carbonyl group is bonded to two carbon atoms: $R_1R_2C{=}O$ (neither R may be H). [5]

Kilogram (kg)* - The SI base unit of mass. [1]

Kinetic energy (E_k, *T*) - The energy associated with the motion of a system of particles in a specified reference frame. For a single particle of mass *m* moving at velocity *v*, $E_k = 1/2mv^2$.

Kirchhoff's laws - Basic rules for electric circuits, which state (a) the algebraic sum of the currents at a network node is zero and (b) the algebraic sum of the voltage drops around a closed path is zero.

Klein-Gordon equation - A relativistic extension of the Schrödinger equation.

Klein-Nishima formula - An expression for the scattering cross section of a photon by an unbound electron, based upon the Dirac electron theory.

Knight shift - The change in magnetic resonance frequency of a nucleus in a metal relative to the same nucleus in a diamagnetic solid. The effect is due to the polarization of the conduction electrons in the metal.

Knudsen number (*Kn*) - A dimensionless quantity used in fluid mechanics, defined by $Kn = \lambda/l$, where λ is mean free path and *l* is length. [2]

Kondo effect - A large increase in electrical resistance observed at low temperatures in certain dilute alloys of a magnetic metal in a nonmagnetic material.

Kramers-Kronig relation - A set of equations relating the real and imaginary parts of the index of refraction of a medium

Lactams - Cyclic amides of amino carboxylic acids, having a 1-azacycloalkan-2-one structure, or analogues having unsaturation or heteroatoms replacing one or more carbon atoms of the ring. [5]

Lactones - Cyclic esters of hydroxy carboxylic acids, containing a 1-oxacycloalkan-2-one structure, or analogues having unsaturation or heteroatoms replacing one or more carbon atoms of the ring. [5]

Lagrangian function (*L*) - A function used in classical mechanics, defined as the kinetic energy minus the potential energy for a system of particles.

Lamb shift - The small energy difference between the $^2S_{1/2}$ and $^2P_{1/2}$ levels in the hydrogen atom, which results from interactions between the electron and the radiation field.

Laminar flow - Smooth, uniform, non-turbulent flow of a gas or liquid in parallel layers, with little mixing between layers. It is characterized by small values of the Reynolds number.

Landé *g*-factor - See *g*-Factor of the electron

Langevin function - The mathematical function $L(x) = (e^x+e^{-x})/(e^x\text{-}e^{-x})\text{-}1/x$, which occurs in the expression for the average dipole moment of a group of rotating polar molecules in an electric field: $\mu_{av} = \mu L(\mu E/kT)$, where μ is the electric dipole moment of a single molecule, *E* is the electric field strength, *k* is the Boltzmann constant, and *T* is the temperature.

Lanthanides - The elements of atomic number 57 through 71, which share common chemical properties: La, Ce, Pr, Nd, Pm, Sm, Eu, Gd, Tb, Dy, Ho, Er, Tm, Yb, Lu. [7]

Larmor frequency (ν_L) - The precession frequency of a magnetic dipole in an applied magnetic field. In particular, a nucleus in a magnetic field of strength *B* has a Larmor frequency of $\gamma B/2\pi$, where γ is the magnetogyric ratio of the nucleus.

Laser* - A device in which an optical cavity is filled with a medium where a population inversion can be produced by some means. When the resonant frequency of the cavity bears the proper relation to the separation of the inverted energy levels, stimulated emission occurs, producing a highly monochromatic, coherent beam of light.

Laser ionization mass spectroscopy (LIMS) - See Techniques for Materials Characterization, page **12-1**.

Lattice constants* - Parameters specifying the dimensions of a unit cell in a crystal lattice, specifically the lengths of the cell edges and the angles between them.

Lattice energy* - The energy per ion pair required to separate completely the ions in a crystal lattice at a temperature of absolute zero.

Laue diagram - A diffraction pattern produced when an x-ray beam passes through a thin slice of a crystal and impinges on a detector behind the crystal.

Lenz's law - The statement that the current induced in a circuit by a change in magnetic flux is so directed as to oppose the change in flux

Leonard-Jones potential - A simple but useful function for approximating the interaction between two neutral atoms or molecules separated by a distance r by writing the potential energy as $U(r) = 4\varepsilon\{(r_0/r)^{12} - (r_0/r)^6\}$, where ε and r_0 are adjustable parameters. In this form the depth of the potential well is ε and the minimum occurs at $2^{1/6}r_0$. The $(1/r)^{12}$ term is often replaced by other powers of $1/r$.

Lepton - One of the class of elementary particles that do not take part in the strong interaction. Included are the electron, muon, and neutrino. All leptons have a spin of 1/2.

Lewis number (*Le*) - A dimensionless quantity used in fluid mechanics, defined by $Le = a/D$, where a is thermal diffusivity and D is diffusion coefficient. [2]

Ligand field theory - A description of the structure of crystals containing a transition metal ion surrounded by nonmetallic ions (ligands). It is based on construction of molecular orbitals involving the *d*-orbitals of the central metal ion and combinations of atomic orbitals of the ligands.

Light year (l.y.) - A unit of distance used in astronomy, defined as the distance light travels in one year in a vacuum. Its approximate value is 9.46073×10^{15} m.

Lignins - Macromolecular constituents of wood related to lignans, composed of phenolic propylbenzene skeletal units, linked at various sites and apparently randomly. [5]

Ligroin - The petroleum fraction consisting mostly of C_7 and C_8 hydrocarbons and boiling in the range 90-140°C; commonly used as a laboratory solvent.

Lipids - A loosely defined term for substances of biological origin that are soluble in nonpolar solvents. They consist of saponifiable lipids, such as glycerides (fats and oils) and phospholipids, as well as nonsaponifiable lipids, principally steroids. [5]

Lipoproteins - Clathrate complexes consisting of a lipid enwrapped in a protein host without covalent binding, in such a way that the complex has a hydrophilic outer surface consisting of all the protein and the polar ends of any phospholipids. [5]

Liter (L)* - A synonym for cubic decimeter. [1]

Lithosphere* - The outer layer of the solid earth, extending from the base of the mantle to the surface of the crust.

Lorentz contraction - The reduction in length of a moving body in the direction of motion, given by the factor $(1-v^2/c^2)^{1/2}$, where v is the velocity of the body and c the velocity of light. Also known as the FitzGerald-Lorentz contraction.

Lorentz force - The force exerted on a point charge Q moving at velocity $\boldsymbol{v}$ in the presence of external fields $\boldsymbol{E}$ and $\boldsymbol{B}$. It is given (in SI units) by $\boldsymbol{F} = Q(\boldsymbol{E} + \boldsymbol{v} \times \boldsymbol{B})$.

Loss angle (δ) - For a dielectric material in an alternating electromagnetic field, δ is the phase difference between the current and the potential difference. The function $\tan \delta$ is a measure of the ratio of the power dissipated in the dielectric to the power stored.

Low energy electron diffraction (LEED) - See Techniques for Materials Characterization, page **12-1**.

Lumen (lm)* - The SI unit of luminous flux, equal to cd sr. [1]

Luminous flux (Φ) - The intensity of light from a source multiplied by the solid angle. The SI unit is lumen. [1]

Lux (lx)* - The SI unit of illuminance, equal to cd sr m^{-2}. [1]

Lyddane-Sachs-Teller relation - A relation between the phonon frequencies and dielectric constants of an ionic crystal which states that $(\omega_T/\omega_L)^2 = \varepsilon(\infty)/\varepsilon(0)$, where ω_T is the angular frequency of transverse optical phonons, ω_L that of longitudinal optical phonons, $\varepsilon(0)$ is the static dielectric constant, and $\varepsilon(\infty)$ the dielectric constant at optical frequencies.

Lyman series - The series of lines in the spectrum of the hydrogen atom which corresponds to transitions between the ground state (principal quantum number n = 1) and successive excited states. The wavelengths are given by $1/\lambda = R_H(1-1/n^2)$, where $n = 2,3,4,...$ and R_H is the Rydberg constant for hydrogen. The first member of the series ($n = 1 \leftrightarrow 2$), which is often called the Lyman-α line, falls at a wavelength of 1216 Å, and the series converges at 912 Å, the ionization limit of hydrogen.

Mach number (*Ma*) - A dimensionless quantity used in fluid mechanics, defined by $Ma = v/c$, where v is velocity and c is the speed of sound. [2]

Macromolecule - A molecule of high relative molecular mass (molecular weight), the structure of which essentially comprises the multiple repetition of units derived, actually or conceptually, from molecules of low relative molecular mass [8]

Madelung constant* - A constant characteristic of a particular crystalline material which gives a measure of the electrostatic energy binding the ions in the crystal.

Magnetic field strength (*H*) - An axial vector quantity, the curl of which is equal to the current density, including the displacement current. [1]

Magnetic induction (*B*) - An axial vector quantity such that the force exerted on an element of current is equal to the vector product of this element and the magnetic induction. [1]

Magnetic moment - See Dipole moment, magnetic.

Magnetic susceptibility (χ_m, κ)* - Defined by $\chi_m = (\mu-\mu_0)/\mu_0$, where μ is the permeability of the medium and μ_0 the permeability of a vacuum. [1]

Magnetization (*M*) - Defined by $M = (B/\mu_0)-H$, where B is magnetic induction, H magnetic field strength, and μ_0 the permeability of a vacuum. [1]

Magnetogyric ratio (γ) - Ratio of the magnetic moment of a particle to its angular momentum. Also called gyromagnetic ratio.

Magneton - See Bohr magneton, Nuclear magneton.

Magnetostriction* - The change in dimensions of a solid sample when it is placed in a magnetic field.

Magnon - A quantum of magnetic energy associated with a spin wave in a ferromagnetic or antiferromagnetic crystal.

Mantle - The layer of the earth between the crust and the liquid outer core, which begins about 2900 km below the earth's surface.

Maser - A device in which a microwave cavity is filled with a medium where a population inversion can be produced by some means. When the resonant frequency of the cavity bears the proper relation to the separation of the inverted energy levels, the device can serve as an amplifier or oscillator at that frequency.

Mass (*m*)* - Quantity of matter. Mass can also be defined as "resistance to acceleration".

Mass defect (*B*) - Defined by $B = Zm(^1\text{H}) + Nm_n - m_a$, where Z is the atomic number, $m(^1\text{H})$ is the mass of the hydrogen atom, N is the neutron number, m_n is the rest mass of the neutron, and m_a is the mass of the atom in question. Thus Bc^2 can be equated to the binding energy of the nucleus if the binding energy of atomic electrons is neglected. [1]

Mass excess (Δ) - Defined by $\Delta = m_a - Am_u$, where m_a is the mass of the atom, A the number of nucleons, and m_u the unified atomic mass constant (m_u = 1 u). [1]

Mass fraction (w_B) - The ratio of the mass of substance B to the total mass of a mixture. [1]

Mass number (A) - A characteristic property of a specific isotope of an element, equal to the sum of the number of protons and neutrons in the nucleus.

Mass spectrometry - An analytical technique in which ions are separated according to the mass/charge ratio and detected by a suitable detector. The ions may be produced by electron impact on a gas, a chemical reaction, energetic vaporization of a solid, etc. [6]

Massieu function - A thermodynamic function defined by $J = -A/T$, where A is the Helmholz energy and T the thermodynamic temperature. [2]

Matthiessen's rule - The statement that the electrical resistivity ρ of a metal can be written as $\rho = \rho_L + \rho_i$, where ρ_L is due to scattering of conduction electrons by lattice vibrations and ρ_i to scattering by impurities and imperfections. If the impurity concentration is small, ρ_i is temperature independent.

Maxwell (Mx)* - A non-SI unit of magnetic field strength (H) equal to 10^{-8} Wb. [1]

Maxwell's equations - The fundamental equations of electromagnetism. In a form appropriate to SI units, they are:

curl $\boldsymbol{H} = \partial \boldsymbol{D}/\partial t + \boldsymbol{j}$

div $\boldsymbol{B} = 0$

curl $\boldsymbol{E} = -\partial \boldsymbol{B}/\partial t$

div $\boldsymbol{D} = \rho$

where $\boldsymbol{H}$ is the magnetic field strength, $\boldsymbol{B}$ the magnetic induction, $\boldsymbol{E}$ the electric field strength, $\boldsymbol{D}$ the electric displacement, $\boldsymbol{j}$ the current density, ρ the charge density, and t is time.

Maxwell-Boltzmann distribution - An expression for the fraction of molecules $f(v)$ in a gas that have velocity v within a specified interval. It takes the form

$$f(v) = 4\pi(M/2\pi RT)^{3/2} v^2 e^{-Mv^2/2RT}$$

where M is the molar mass, R the molar gas constant, and T the temperature.

Mean free path* - The average distance a gas molecule travels between collisions.

Meissner effect - The complete exclusion of magnetic induction from the interior of a superconductor.

Melting point* - The temperature at which the solid and liquid phases of a substance are in equilibrium at a specified pressure (normally taken to be atmospheric unless stated otherwise).

Mercaptans - A traditional term abandoned by IUPAC, synonymous with thiols. This term is still widely used. [5]

Meson - Any elementary particle that has zero or integral spin. Mesons are responsible for the forces between protons and neutrons in the nucleus.

Mesosphere - The part of the earth's atmosphere extending from the top of the stratosphere (about 50 km above the surface) to 80-90 km. It is characterized by a decrease in temperature with increasing altitude.

Metal - A material in which the highest occupied energy band (conduction band) is only partially filled with electrons. The electrical conductivity of metals generally decreases with temperature.

Metallocenes - Organometallic coordination compounds in which one atom of a transition metal such as iron, ruthenium or osmium is bonded to and only to the face of two cyclopentadienyl ligands which lie in parallel planes. [5]

Meter (m)* - The SI base unit of length. [1]

Methine group - In organic compounds, the -C= group. [5]

Mho - An archaic name for the SI unit siemens (reciprocal ohm).

Micelle - A particle formed by the aggregation of surfactant molecules (typically, 10 to 100 molecules) in solution. For aqueous solutions, the hydrophilic end of the molecule is on the surface of the micelle, while the hydrophobic end (often a hydrocarbon chain) points toward the center. At the critical micelle concentration (cmc) the previously dissolved molecules aggregate into a micelle.

Micron (μ) - An obsolete name for micrometer.

Mie scattering - The scattering of light by spherical dielectric particles whose diameter is comparable to the wavelength of the light.

Milky way - The band of light in the night sky resulting from the stars in the galactic plane. The term is also used to denote the galaxy in which the sun is located.

Miller indices (*hkl*) - A set of indices used to label planes in a crystal lattice. [2]

Millimeter of mercury (mmHg) - A non-SI unit of pressure, equal to 133.322 Pa. The name is generally considered interchangeable with torr.

Mobility (μ)* - In solid state physics, the drift velocity of electrons or holes in a solid divided by the applied electric field strength. The term is used in a similar sense in other fields.

Molality (m) - A measure of concentration of a solution in which one states the amount of substance (i.e., number of moles) of solute per kilogram of solvent. Thus a 0.1 molal solution (often written as 0.1 m) has m = 0.1 mol/kg.

Molar mass - The mass of one mole of a substance. It is normally expressed in units of g/mol, in which case its numerical value is identical with the molecular weight (relative molecular mass). [1]

Molar quantity - It is often convenient to express an extensive quantity (e.g., volume, enthalpy, heat capacity, etc.) as the actual value divided by amount of substance (number of moles). The resulting quantity is called molar volume, molar enthalpy, etc

Molar refraction (R) - A property of a dielectric defined by the equation $R = V_m[(n^2-1)/(n^2+2)]$, where n is the index of refraction of the medium (at optical wavelengths) and V_m the molar volume. It is related to the polarizability α of the molecules that make up the medium by the Lorenz-Lorentz equation, $R = N_A\alpha/3\varepsilon_0$, where N_A is Avogadro's constant and ε_0 is the permittivity of a vacuum.

Molarity (c) - A measure of concentration of a solution in which one states the amount of substance (i.e., number of moles) of solute per liter of solution. Thus a 0.1 molar solution (often referred to as 0.1 M) has a concentration c = 0.1 mol/L.

Mole (mol)* - The SI base unit of amount of substance. [1]

Mole fraction (x_B) - The ratio of the amount of substance (number of moles) of substance B to the total amount of substance in a mixture. [1]

Molecular orbital - See Orbital.

Molecular weight (M_r)* - The ratio of the average mass per molecule or specified entity of a substance to 1/12 of the mass of nuclide ^{12}C. Also called relative molar (or molecular) mass. [1]

Moment of inertia (I) - The moment of inertia of a body about an axis is the sum (or integral) of the products of its elements of mass and the squares of their distances from the axis. [1]

Momentum (p) - The product of mass and velocity. [1]

Monomer - A substance consisting of molecules which can undergo polymerization, thereby contributing constitutional units to the essential structure of a macromolecule. [8]

Monosaccharides - A term which includes aldoses, ketoses, and a wide variety of derivatives. [5]

Mössbauer effect - The recoilless emission of γ-rays from nuclei bound in a crystal under conditions where the recoil energy associated with the γ emission is taken up by the crystal as a whole. This results in a very narrow line width, which can be exploited in various types of precise measurements.

Muon* - An unstable elementary particle of spin 1/2 and mass about 200 times that of the electron.

Naphtha - The petroleum fraction consisting mostly of C_6 to C_8 hydrocarbons and boiling in the range 80-120°C. Solvents derived from this fraction include ligroin and petroleum ether.

Nautical mile - A non-SI unit of length, equal to exactly 1852 m.

Navier-Stokes equations - A set of complex equations for the motion of a viscous fluid subject to external forces.

Néel temperature (T_N)* - The critical temperature above which an antiferromagnetic substance becomes paramagnetic. [1]

Nernst effect - The production of an electric field in a conductor subject to an applied magnetic field and containing a transverse temperature gradient. The electric field is perpendicular to the magnetic field and the temperature gradient.

Network - In polymer science, a highly ramified macromolecule in which essentially each constitutional unit is connected to each other constitutional unit and to the macroscopic phase boundary by many permanent paths through the macromolecule, the number of such paths increasing with the number of intervening bonds. The paths must on the average be coextensive with the macromolecule. [8]

Neutrino - A stable elementary particle in the lepton family. Neutrinos have zero (or at least near-zero) rest mass and spin 1/2.

Neutron* - An elementary particle on spin 1/2 and zero charge. The free neutron has a mean lifetime of 887 seconds. Neutrons and protons, which are collectively called nucleons, are the constituents of the nucleus.

Neutron activation analysis (NAA) - See Techniques for Materials Characterization, page **12**-1.

Neutron number (*N*) - A characteristic property of a specific isotope of an element, equal to the number of neutrons in the nucleus.

Newton (N)* - The SI unit of force, equal to m kg s^{-2}. [1]

Nitriles - Compounds having the structure RC≡N; thus C-substituted derivatives of hydrocyanic acid, HC≡N. [5]

Nitrosamines - N-Nitroso amines: compounds of the structure R_2NNO. Compounds RNHNO are not ordinarily isolatable, but they, too, are nitrosamines. The name is a contraction of N-nitrosoamine and, as such, does not require the N locant. [5]

Nuclear magnetic resonance (NMR)* - A widely used technique in which the resonant absorption of radiofrequency radiation by magnetic nuclei in a magnetic field is measured. The results give important information on the local environment of each nucleus.

Nuclear magneton (μ_N)* - The unit of nuclear magnetic moment, defined as $eh/4\pi m_p$, where h is Planck's constant, m_p the proton mass, and e the elementary charge.

Nuclear quadrupole resonance (NQR) - See Techniques for Materials Characterization, page **12**-1.

Nuclear reaction analysis (NRA) - See Techniques for Materials Characterization, page **12**-1.

Nuclear spin (*I*) - The quantum number that specifies the intrinsic angular momentum of a particular nucleus. The magnitude of the angular momentum is given by $[I(I+1)]^{1/2}\, h/2\pi$, where h is Planck's constant.

Nucleic acids* - Macromolecules, the major organic matter of the nuclei of biological cells, made up of nucleotide units, and hydrolyzable into certain pyrimidine or purine bases (usually adenine, cytosine, guanine, thymine, uracil), D-ribose or 2-deoxy-D-ribose. [5]

Nucleon - A collective term for the proton and neutron.

Nucleosides - Ribosyl or deoxyribosyl derivatives (rarely, other glycosyl derivatives) of certain pyrimidine or purine bases. They are thus glycosylamines or N-glycosides related to nucleotides by the lack of phosphorylation. [5]

Nucleotides - Compounds formally obtained by esterification of the 3′ or 5′ hydroxy group of nucleosides with phosphoric acid. They are the monomers of nucleic acids and are formed from them by hydrolytic cleavage. [5]

Nuclide - A species of atoms in which each atom has identical atomic number *Z* and identical mass number *A*. [3]

Nusselt number (*Nu*) - A dimensionless quantity used in fluid mechanics, defined by $Nu = hl/k$, where h is coefficient of heat transfer, l is length, and k is thermal conductivity. [2]

Nyquist theorem - An expression for the mean square thermal noise voltage across a resistor, given by $4RkT\Delta f$ where R is the resistance, k the Boltzmann constant, T the temperature, and Δf the frequency band within which the voltage is measured.

Octanol-water partition coefficient (*P*)* - A measure of the way in which a compound will partition itself between the octanol and water phases in the two-phase octanol-water system, and thus an indicator of certain types of biological activity. Specifically, *P* is the ratio of the concentration (in moles per liter) of the compound in the octanol phase to that in the water phase at infinite dilution. The quantity normally reported is log *P*.

Oersted (Oe) - A non-SI unit of magnetic field (*H*), equal to 79.57747 A/m.

Ohm (Ω)* - The SI unit of electric resistance, equal to V/A. [1]

Ohm's law - A relation among electric current *I*, potential difference *V*, and resistance *R*, viz., $I = V/R$. The resistance is constant at constant temperature to high precision for many materials.

Olefins - Acyclic and cyclic hydrocarbons having one or more carbon-carbon double bonds, apart from the formal ones in aromatic compounds. The class olefins subsumes alkenes and cycloalkenes and the corresponding polyenes. [5]

Oligomer - A substance consisting of molecules of intermediate relative molecular mass (molecular weight), the structure of which essentially comprises the multiple repetition of units derived, actually or conceptually, from molecules of low relative molecular mass. In contrast to a polymer, the properties of an oligomer can vary significantly with the removal of one or a few of its units. [8]

Oligopeptides - Peptides containing from three to nine amino groups. [5]

Onsager relations - An important set of equations in the thermodynamics of irreversible processes. They express the symmetry between the transport coefficients describing reciprocal processes in systems with a linear dependence of flux on driving forces.

Optical rotary power - Angle by which the plane of polarization of a light beam is rotated by an optically active medium, divided by path length and by concentration of the active constituent. Depending on whether mass or molar concentration is used, the modifier "specific" or "molar" is attached. [2]

Orbital - A one-electron wavefunction. Atomic orbitals are classified as *s*-, *p*-, *d*,- or *f*-orbitals according to whether the angular momentum quantum number $l = 0, 1, 2,$ or 3. Molecular orbitals, which are usually constructed as linear combinations of atomic orbitals, describe the distribution of electrons over the entire molecule.

Oscillator strength (*f*) - A measure of the intensity of a spectroscopic transition, defined by

$$f = \frac{8\pi^2 M e \nu}{3he^2}\left|\mu_{ij}\right|^2$$

where ν is the frequency, μ_{ij} the transition dipole moment, m_e the mass of the electron, e the elementary charge, and h Planck's constant.

Osmosis - The flow of a solvent in a system in which two solutions of different concentration are separated by a semipermeable membrane which cannot pass solute molecules. The solvent will flow from the side of lower concentration to that of higher concentration, thus tending to equalize the concentrations. The pressure that must be applied to the more concentrated side to stop the flow is called the osmotic pressure.

Osmotic coefficient (ϕ) - Defined by $\phi = \ln a_A/(M_A \Sigma m_B)$, where M_A is the molar mass of substance A (normally the solvent), a_A is its activity, and the m_B are molalities of the solutes. [1]

Osmotic pressure (Π) - The excess pressure necessary to maintain osmotic equilibrium between a solution and the pure solvent separated by a membrane permeable only to the solvent. In an ideal dilute solution $\Pi = c_B RT$, where c_B is the amount-of-substance concentration of the solute, R is the molar gas constant, and T the temperature. [1,2]

Ostwald dilution law - A relation for the concentration dependence of the molar conductivity Λ of an electrolyte solution, viz.,

$$\frac{1}{\Lambda} = \frac{1}{\Lambda^\circ} + \frac{\Lambda c}{K(\Lambda^\circ)^2}$$

where c is the solute concentration, K is the equilibrium constant for dissociation of the solute, and Λ° is the conductivity at $c\Lambda = 0$.

Ounce (oz) - A non-SI unit of mass. The avoirdupois once equals 28.34952 g, while the troy ounce equals 31.10348 g.

Overpotential (η) - In an electrochemical cell, the difference between the potential of an electrode and its zero-current value.

Oximes - Compounds of structure $R_2C{=}NOH$ derived from condensation of aldehydes or ketones with hydroxylamine. Oximes from aldehydes may be called aldoximes; those from ketones may be called ketoximes. [5]

Oxo compounds - Compounds containing an oxygen atom, =O, doubly bonded to carbon or another element. The term thus embraces aldehydes, carboxylic acids, ketones, sulfonic acids, amides and esters. [5]

Ozonides - The 1,2,4-trioxolanes formed by the reaction of ozone at a carbon-carbon double bond, or the analogous compounds derived from acetylenic compounds. [5]

Pair production - A process in which a photon is converted into a particle and its antiparticle (e.g., an electron and positron) in the electromagnetic field of a nucleus.

Paraffins - Obsolescent term for saturated hydrocarbons, commonly but not necessarily acyclic. Still widely used in the petrochemical industry, where the term designates acyclic saturated hydrocarbons, and stands in contradistinction to naphthenes. [5]

Paramagnetism* - A type of magnetism characterized by a positive magnetic susceptibility, so that the material becomes weakly magnetized in the direction of an external field. The magnetization disappears when the field in removed. In the simplest approximation (Curie's law) the susceptibility is inversely proportional to temperature.

Parity - The property of a quantum-mechanical wave function that describes its behavior under the symmetry operation of coordinate inversion. A parity of +1 (or even) is assigned if the wave function does not change sign when the signs of all the coordinates are changed; the parity is -1 (or odd) if the wave function changes sign under this operation.

Parsec (pc) - A unit of distance defined as the distance at which 1 astronomical unit (AU) subtends an angle of 1 second of arc. It is equal to 206264.806 AU or 3.085678×10^{16} m.

Particle induced x-ray emission (PIXE) - See Techniques for Materials Characterization, page **12-1**.

Partition function (q,z) - For a single molecule, $q = \Sigma_i g_i \exp(\varepsilon_i/kT)$, where ε_i is an energy level of degeneracy g_i, k the Boltzmann constant, and T the absolute temperature; the summation extends over all energy states. For a system of N non-interacting molecules which are indistinguishable, as in an ideal gas, the canonical partition function $Q = qN/N!$.

Pascal (Pa)* - The SI unit of pressure, equal to N/m^2. [1]

Paschen series - The series of lines in the spectrum of the hydrogen atom which corresponds to transitions between the state with principal quantum number $n = 3$ and successive higher states. The wavelengths are given by $1/\lambda = R_H(1/9 - 1/n^2)$, where $n = 4,5,6,...$ and R_H is the Rydberg constant. The first member of the series ($n = 3 \leftrightarrow 4$), which is often called the P_α line, falls in the infrared at a wavelength of 1.875 µm.

Paschen-Back effect - In atomic spectroscopy, the decoupling of electron spin from orbital angular momentum as the strength of an external magnetic field is increased.

Pauli exclusion principle - The statement that two electrons in an atom cannot have identical quantum numbers; thus if there are two electrons in the same orbital, their spin quantum numbers must be of opposite sign.

Pearson symbol - A code for designating crystallographic information, including the crystal system, the lattice type, and the number of atoms per unit cell.

Péclet number (*Pe*) - A dimensionless quantity used in fluid mechanics, defined by $Pe = vl/a$, where v is velocity, l is length, and a is thermal diffussivity. [2]

Peltier effect - The absorption or generation of heat (depending on the current direction) which occurs when an electric current is passed through a junction between two materials.

Peptides - Amides derived from two or more amino carboxylic acid molecules (the same or different) by formation of a covalent bond from the carbonyl carbon of one to the nitrogen atom of another with formal loss of water. [5]

Permeability (μ) - Magnetic induction divided by magnetic field strength; i.e. $\mu = B/H$. The relative permeability $\mu_r = \mu/\mu_0$, where μ_0 is the permeability of a vacuum. [1]

Permittivity (ε) - Ratio of the electric displacement in a medium to the electric field strength. Also called dielectric constant. [1]

Peroxides - Compounds of structure ROOR in which R may be any organic group. In inorganic chemistry, salts of the anion O_2^{-2} [5]

Peroxy acids - Acids in which an acidic -OH group has been replaced by an -OOH group; e.g. $CH_3C({=}O)OOH$ peroxyacetic acid, $PhS({=}O)_2OOH$ benzeneperoxysulfonic acid. [5]

Petroleum ether - The petroleum fraction consisting of C_5 and C_6 hydrocarbons and boiling in the range 35-60°C; commonly used as a laboratory solvent.

pH* - A convenient measure of the acid-base character of a solution, usually defined by pH = -log [$c(H^+)$/mol L^{-1})], where $c(H^+)$ is the concentration of hydrogen ions. The more precise definition is in terms af activity rather than concentration. [2]

Phenols - Compounds having one or more hydroxy groups attached to a benzene or other arene ring. [5]

Phonon - A quantum of energy associated with a vibrational mode of a crystal lattice.

Phosphines - PH_3 and compounds derived from it by substituting one, two or three hydrogen atoms by hydrocarbyl groups. RPH_2, R_2PH and R_3P (R not equal to H) are called primary, secondary and tertiary phosphines, respectively. [5]

Phosphonium compounds - Salts (and hydroxides) $[R_4P]^+X^-$ containing tetracoordinate phosphonium ion and the associated anion. [5]

Phosphonium ylides - Compounds having the structure $R_3P^+{-}C^-R_2 \rightleftharpoons R_3P{=}CR_2$. Also known as Wittig reagents. [5]

Phosphoresence - The process by which a molecule is excited by light to a higher electronic state and then undergoes a radiationless transition to a state of different multiplicity from which it decays, after some delay, to the ground state. The emitted light is normally of longer wavelength than the exciting light because vibrational energy has been dissipated.

Photoelectric effect - The complete absorption of a photon by a solid with the emission of an electron.

Photon - An elementary particle of zero mass and spin 1/2. The photon is involved in electromagnetic interactions and is the quantum of electromagnetic radiation.

Photon stimulated desorption (PSD) - See Techniques for Materials Characterization, page **12-1**.

Pinacols - Tetra(hydrocarbyl)ethane-1,2-diols, $R_2C(OH)C(OH)R_2$, of which the tetramethyl example is the simplest one and is itself commonly known as pinacol. [5]

Pion - An elementary particle in the family of mesons. Pions have zero spin and may be neutral or charged. They participate in the strong interaction which holds the nucleus together.

p*K** - The negative logarithm (base 10) of an equilibrium constant K. For pK_a, see Acid dissociation constant.

Planck constant (*h*)* - The elementary quantum of action, which relates energy to frequency through the equation $E = h\nu$.

Planck distribution - See Black body radiation

Planck function - A thermodynamic function defined by $Y = -G/T$, where G is Gibbs energy and T thermodynamic temperature. [2]

Plasma - A highly ionized gas in which the charge of the electrons is balanced by the charge of the positive ions, so that the system as a whole is electrically neutral.

Plasmon - A quantum associated with a plasma oscillation in the electron gas of a solid.

Point group* - A group of symmetry operations (rotations, reflections, etc.) that leave a molecule invariant. Every molecular conformation can be assigned to a specific point group, which plays a major role in determining the spectrum of the molecule.

Poise (P) - A non-SI unit of viscosity, equal to 0.1 Pa s.

Poiseuille's equation - A formula for the rate of flow of a viscous fluid through a tube:

$$\frac{dV}{dt} = \frac{\left(p_1^2 - p_2^2\right)\pi r^4}{16 l \eta p_0}$$

where V is the volume as measured at pressure p_0; p_1 and p_2 are the pressures at each end of the tube; r is the radius and l the length of the tube; and η is the viscosity.

Poisson ratio (μ) - The absolute value of the ratio of the transverse strain to the corresponding axial strain resulting from uniformly distributed axial stress below the proportional limit (i.e., where Hooke's law is valid). [10]

Polariton - A quantum associated with the coupled modes of photons and optical phonons in an ionic crystal.

Polarizability (α)* - The change in dipole moment of a molecule produced by an external electric field; specifically, $\alpha_{ab} = \partial p_a/\partial E_b$, where p_a is the dipole moment component on the a axis and E_b is the component of the electric field strength along the b axis. [2]

Polymer - A substance composed of molecules of high relative molecular mass (molecular weight), the structure of which essentially comprises the multiple repetition of units derived, actually or conceptually, from molecules of low relative molecular mass. A single molecule of a polymer is called a macromolecule. [8]

Polypeptides - Peptides containing 10 or more amino acid residues. See also Peptides. [5]

Polysaccharides - Compounds consisting of a large number of monosaccharides linked glycosidically. This term is commonly used only for those containing more than ten monosaccharide residues. Also called glycans. [5]

Porphyrins - Natural pigments containing a fundamental skeleton of four pyrrole nuclei united through the α-positions by four methine groups to form a macrocyclic structure (porphyrin is designated porphine in Chemical Abstracts indexes). [5]

Positron - The antiparticle of the electron. It has the same mass and spin as an electron, and an equal but opposite charge.

Positronium - The hydrogen-like "atom" formed from a positron nucleus and an electron. Its lifetime is very short because of annihilation of the positron and electron.

Potential - See Electric potential

Potential energy (E_p, *V*, *U*) - The portion of the energy of a system that is associated with its position in a force field.

Pound (lb) - A non-SI unit of mass, equal to 0.4535924 kg.

Power (*P*) - Rate of energy transfer. For electrical circuits, this is equal to the product of current and potential difference, $P = IV$. [1]

Poynting vector (*S*) - For electromagnetic radiation, the vector product of the electric field strength and the magnetic field strength. [1]

Prandtl number (*Pr*) - A dimensionless quantity used in fluid mechanics, defined by $Pr = \eta/\rho a$, where η is viscosity, ρ is density, and a is thermal diffusivity. [2]

Pressure* - Force divided by area. [1]

Proteins - Naturally occurring and synthetic polypeptides having molecular weights greater than about 10,000 (the limit is not precise). See also Peptides. [5]

Proton* - A stable elementary particle of unit positive charge and spin 1/2. Protons and neutrons, which are collectively called nucleons, are the constituents of the nucleus.

Pulsar - A neutron star which rotates rapidly and emits electromagnetic radiation in regular pulses at a frequency related to the rotation period.

Purine bases* - Purine and its substitution derivatives, especially naturally occurring examples. [5]

Pyrimidine bases* - Pyrimidine and its substitution derivatives, especially naturally occurring examples. [5]

Q-switching - A technique for obtaining very high power from a laser by keeping the Q factor of the laser cavity low while the population inversion builds up, then suddenly increasing the Q to initiate the stimulated emission.

Quad - A unit of energy defined as 10^{15} Btu, equal to approximately 1.055056×10^{18} J.

Quadrupole moment - A coefficient of the third term (after monopole and dipole) in the power series expansion of the electric potential of an array of charges. A nucleus of spin greater than 1/2 has a non-vanishing nuclear quadrupole moment which can interact with the electric field gradient of the surrounding electrons. Molecular quadrupole moments have an influence on intermolecular forces.

Quality factor (*Q*) - The ratio of the absolute value of the reactance of an electrical system to the resistance; thus a measure of the energy stored per cycle relative to the energy dissipated.

Quantum yield - In photochemistry, the number of moles transformed in a specific process, either physically (e.g., by emission of photons) or chemically, per mole of photons absorbed by the system. [3]

Quark - An elementary entity which has not been directly observed but is considered a constituent of protons, neutrons, and other hadrons.

Quasar - An extragalactic object emitting electromagnetic radiation at a very high power level and showing a very large red shift, thus indicating that the object is receding at a speed approaching the speed of light.

Quasicrystal - A solid having conventional crystalline properties but whose lattice does not display translational periodicity.

Quaternary ammonium compounds - Derivatives of ammonium compounds, $NH_4^+Y^-$, in which all four of the hydrogens bonded to nitrogen have been replaced with hydrocarbyl groups. Compounds having a carbon-nitrogen double bond (i.e. $R_2C{=}N^+R_2Y^-$) are more accurately called iminium compounds. [5]

Quinones - Compounds having a fully conjugated cyclic dione structure, such as that of benzoquinones, derived from aromatic compounds by conversion of an even number of -CH= groups into -C(=O)- groups with any necessary rearrangement of double bonds. [5]

Racemic mixture - A mixture of equal amounts of a pair of enantiomers (optical isomers); such a mixture is not optically active.

Rad - A non-SI unit of absorbed dose of radiation, equal to 0.01 Gy.

Radiance (*L*) - The radiant intensity in a given direction from an element of a surface, divided by the area of the orthogonal projection of this element on a plane perpendicular to the given direction. [1]

Radiant intensity (*I*) - The radiant energy flux leaving an element of a source within an element of solid angle, divided by that element of solid angle. [1]

Radicals - Molecular entities possessing an unpaired electron, such as $\cdot CH_3$, $\cdot SnH_3$, $\cdot Cl$. (In these formulas the dot, symbolizing the unpaired electron, should be placed so as to indicate the atom of highest spin density, if this is possible). [5]

Raman effect - The inelastic scattering of light by a molecule, in which the incident photon either gives up to, or receives energy from, one of the internal vibrational modes of the molecule. The scattered light thus has either a lower frequency (Stokes radiation) or higher frequency (anti-Stokes radiation) than the incident light. These shifts provide a measure of the normal vibrational frequencies of the molecule.

Rankine cycle - A thermodynamic cycle which can be used to calculate the ideal performance of a heat engine that uses a condensable vapor as the working fluid (e.g., a steam engine or a heat pump).

Rankine temperature - A thermodynamic temperature scale based on a temperature interval °R = (5/9) K; i.e., $T/°R = (9/5)T/K = t/°F + 459.67$.

Raoult's law - The expression for the vapor pressure p_i of component i in an ideal solution, viz., $p_i = x_i\, p_{i0}$, where x_i is the mole fraction of component i and p_{i0} the vapor pressure of the pure substance i.

Rare earth elements - The elements Sc, Y, and the lanthanides (La, Ce, Pr, Nd, Pm, Sm, Eu, Gd, Tb, Dy, Ho, Er, Tm, Yb, Lu). [7]

Rayleigh number (*Ra*) - A dimensionless quantity used in fluid mechanics, defined by $Ra = l^3 g\alpha\Delta T\rho/\eta a$, where l is length, g is acceleration of gravity, α is cubic expansion coefficient, T is temperature, ρ is density, η is viscosity, and a is thermal diffusivity. [2]

Rayleigh scattering - The scattering of light by particles which are much smaller than the wavelength of the light. It is characterized by a scattered intensity which varies as the inverse fourth power of the wavelength.

Rayleigh wave - A guided elastic wave along the surface of a solid; also called surface acoustic wave.

Reactance (*X*) - The imaginary part of impedance. For an inductive reactance L and a capacitive reactance C in series, the reactance is $X = L\omega - 1/(C\omega)$, where ω is 2π times the frequency of the current. [1]

Red shift - A displacement of a spectral line toward longer wavelengths. This can occur through the Doppler effect (e.g., in the light from receding galaxies) or, in the general theory of relativity, from the effects of a star's gravitational field.

Reflectance (ρ) - Ratio of the radiant or luminous flux at a given wavelength that is reflected to that of the incident radiation. Also called reflection factor. [1]

Reflection high energy electron diffraction (RHEED) - See Techniques for Materials Characterization, page **12-1**.

Relative humidity* - The ratio of the partial pressure of water vapor in air to the saturation vapor pressure of water at the same temperature, expressed as a percentage. [10]

Relative molar mass - See Molecular weight.

Rem - A non-SI unit of dose equivalent, equal to 0.01 Sv.

Resistance (*R*) - Electric potential difference divided by current when there is no electromotive force in the conductor. This definition applies to direct current. More generally, resistance is defined as the real part of impedance. [1]

Resistivity (ρ) - Electric field strength divided by current density when there is no electromotive force in the conductor. Resistivity is an intrinsic property of a material. For a conductor of uniform cross section with area A and length L, and whose resistance is R, the resistivity is given by $\rho = RA/L$. [1]

Reynolds number (*Re*) - A dimensionless quantity used in fluid mechanics, defined by $Re = \rho v l/\eta$, where ρ is density, v is velocity, l is length, and η is viscosity. [2]

Rheology - The study of the flow of liquids and deformation of solids. Rheology addresses such phenomena as creep, stress relaxation, anelasticity, nonlinear stress deformation, and viscosity.

Ribonucleic acids (RNA) - Naturally occurring polyribonucleotides. See also nucleic acids, nucleosides, nucleotides, ribonucleotides. [5]

Ribonucleotides - Nucleotides in which the glycosyl group is a ribosyl group. See also nucleotides. [5]

Roentgen (R) - A unit used for expressing the charge (positive or negative) liberated by x-ray or γ radiation in air, divided by the mass of air. A roentgen is defined as 2.58×10^{-4} C/kg.

Rotational constants - In molecular spectroscopy, the constants appearing in the expression for the rotational energy levels as a function of the angular momentum quantum numbers. These constants are proportional to the reciprocals of the principal moments of inertia, averaged over the vibrational motion.

Rutherford back scattering (RBS) - See Techniques for Materials Characterization, page **12-1**.

Rydberg constant (R_∞)* - The fundamental constant which appears in the equation for the energy levels of hydrogen-like atoms; i.e., $E_n = hcR_\infty Z^2\mu/n^2$, where h is Planck's constant, c the speed of light, Z the atomic number, μ the reduced mass of nucleus and electron, and n the principal quantum number ($n = 1, 2, \ldots$).

Rydberg series - A regular series of lines in the spectrum of an atom or molecule, with the spacing between successive lines becoming smaller as the frequency increases (wavelength decreases). The series eventually converges to a limit which usually corresponds to the complete removal of an electron from the atom or molecule.

Sackur-Tetrode equation* - An equation for the molar entropy S_m of an ideal monatomic gas: $S_m = R\ln(e^{5/2}\, V/N_A\Lambda^3)$, where R is the molar gas constant, V is the volume, and N_A is Avogadro's number. The constant Λ is given by $\Lambda = h/(2\pi mkT)^{1/2}$, where h is Planck's constant, m the atomic mass, k the Boltzmann constant, and T the temperature.

Salinity (*S*)* - A parameter used in oceanography to describe the concentration of dissolved salts in seawater. It is defined in terms of electrical conductivity relative to a standard solution of KCl. When expressed in units of parts per thousand, S may be roughly equated to the concentration of dissolved material in grams per kilogram of seawater.

Salt - An ionic compound formed by the reaction of an acid and a base.

Scanned probe microscopy (SPM) - See Techniques for Materials Characterization, page **12-1**.

Scanning electron microscopy (SEM) - See Techniques for Materials Characterization, page **12-1**.

Scanning laser acoustic microscopy (SLAM) - See Techniques for Materials Characterization, page **12-1**.

Scanning transmission electron microscopy (STEM) - See Techniques for Materials Characterization, page **12-1**.

Scanning tunneling microscopy (STM) - See Techniques for Materials Characterization, page **12-1**.

Schiff bases - Imines bearing a hydrocarbyl group on the nitrogen atom: $R_2C{=}NR'$ (R′ not equal to H). Considered by many to be synonymous with azomethines. [5]

Schmidt number (Sc) - A dimensionless quantity used in fluid mechanics, defined by $Sc = \eta/\rho D$, where η is viscosity, ρ is density, and D is diffusion coefficient. [2]

Schottky barrier - A potential barrier associated with a metal-semiconductor contact. It forms the basis for the rectifying device known as the Schottly diode.

Schrödinger equation - The basic equation of wave mechanics which, for systems not dependent on time, takes the form:

$$-(\hbar/2m)\nabla^2\psi + V\psi = E\psi$$

where ψ is the wavefunction, V is the potential energy expressed as a function of the spatial coordinates, E is an energy eigenvalue, ∇^2 is the Laplacian operator, $\hbar$ is Planck's constant divided by 2π, and m is the mass.

Second (s)* - The SI base unit of time. [1]

Second radiation constant (c_2)* - See First radiation constant.

Secondary ion mass spectroscopy (SIMS) - See Techniques for Materials Characterization, page **12**-1.

Seebeck effect - The development of a potential difference in a circuit where two different metals or semiconductors are joined and their junctions maintained at different temperatures. It is the basis of the thermocouple.

Selenides - Compounds having the structure RSeR (R not equal to H). They are thus selenium analogues of ethers. Also used for metal salts of H_2Se. [5]

Semicarbazones - Compounds having the structure $R_2C{=}NNHC({=}O)NH_2$, formally derived by condensation of aldehydes or ketones with semicarbazide [$NH_2NHC({=}O)NH_2$]. [5]

Semiconductor - A material in which the highest occupied energy band (valence band) is completely filled with electrons at $T = 0$ K, and the energy gap to the next highest band (conduction band) ranges from 0 to 4 or 5 eV. With increasing temperature electrons are excited into the conduction band, leading to an increase in the electrical conductivity.

Semiquinones - Radical anions having the structure -O-Z-O· where Z is an ortho- or para-arylene group or analogous heteroarylene group; they are formally generated by the addition of an electron to a quinone. [5]

SI units* - The International System of Units adopted in 1960 and recommended for use in all scientific and technical fields. [1]

Siemens (S)* - The SI unit of electric conductance, equal to Ω^{-1}. [1]

Sievert (Sv)* - The SI unit of dose equivalent (of radiation), equal to J/kg. [1]

Silanes - Saturated silicon hydrides, analogues of the alkanes; i.e. compounds of the general formula Si_nH_{2n+2}. Silanes may be subdivided into silane, oligosilanes, and polysilanes. Hydrocarbyl derivatives are often referred to loosely as silanes. [5]

Silicones - Polymeric or oligomeric siloxanes, usually considered unbranched, of general formula $[\text{-OSiR}_2\text{-}]_n$ (R not equal to H). [5]

Siloxanes - Saturated silicon-oxygen hydrides with unbranched or branched chains of alternating silicon and oxygen atoms (each silicon atom is separated from its nearest silicon neighbors by single oxygen atoms). [5]

Skin effect - The concentration of high frequency alternating currents near the surface of a conductor.

Slater orbital - A particular mathematical expression for the radial part of the wave function of a single electron, which is used in quantum-mechanical calculations of the energy and other properties of atoms and molecules.

Small angle neutron scattering (SANS) - See Techniques for Materials Characterization, page **12**-1.

Snell's law - The relation between the angle of incidence i and the angle of refraction r of a light beam which passes from a medium of refractive index n_0 to a medium of index n_1, viz., $\sin i/\sin r = n_1/n_0$.

Solar constant* - The mean radiant energy flux from the sun on a unit surface normal to the direction of the rays at the mean distance of the earth from the sun. The value is approximately 1373 W/m^2.

Solar wind - The stream of high velocity hydrogen and helium ions emitted by the sun which flows through the solar system and beyond.

Soliton - A spatially localized wave in a solid or liquid that can interact strongly with other solitons but will afterwards regain its original form.

Solubility* - A quantity expressing the maximum concentration of some material (the solute) that can exist in another liquid or solid material (the solvent) at thermodynamic equilibrium at specified temperature and pressure. Common measures of solubility include the mass of solute per unit mass of solution (mass fraction), mole fraction of solute, molality, molarity, and others.

Solubility product constant (K_{sp})* - The equilibrium constant for the dissolution of a sparsely soluble salt into its constituent ions.

Space group* - A group of symmetry operations (reflections, rotations, etc.) that leave a crystal invariant. A total of 230 space groups have been identified.

Spark source mass spectroscopy (SSMS) - See Techniques for Materials Characterization, page **12**-1.

Specific gravity - Ratio of the mass density of a material to that of water. Since one must specify the temperature of both the sample and the water to have a precisely defined quantity, the use of this term is now discouraged.

Specific heat - Heat capacity divided by mass. See Heat capacity.

Specific quantity - It is often convenient to express an extensive quantity (e.g., volume, enthalpy, heat capacity, etc.) as the actual value divided by mass. The resulting quantity is called specific volume, specific enthalpy, etc.

Specific rotation $[\alpha]^\theta_\lambda$ - For an optically active substance, defined by $[\alpha]^\theta_\lambda = \alpha/\gamma l$, where α is the angle through which plane polarized light is rotated by a solution of mass concentration γ and path length l. Here θ is the Celsius temperature and λ the wavelength of the light at which the measurement is carried out. Also called specific optical rotatory power. [2]

Spin (s, I)* - A measure of the intrinsic angular momentum of a particle, which it possesses independent of its orbital motion. The symbol s is used for the spin quantum number of an electron, while I is generally used for nuclear spin.

Spiro compounds - Compounds having one atom (usually a quaternary carbon) as the only common member of two rings. [5]

Stacking fault - An error in the normal sequence of layer growth in a crystal.

Standard mean ocean water (SMOW) - A standard sample of pure water of accurately known isotopic composition which is maintained by the International Atomic Energy Agency. It is used for precise calibration of density and isotopic composition measurements.

Standard reduction potential (E°) - The zero-current potential of a cell in which the specified reduction reaction occurs at the right-hand electrode and the left-hand electrode is the standard hydrogen electrode. Also called Standard electrode potential.

Standard state - A defined state (specified temperature, pressure, concentration, etc.) for tabulating thermodynamic functions and carrying out thermodynamic calculations. The standard state pressure is usually taken as 100,000 Pa (1 bar), but various standard state temperatures are used. [2]

Stanton number (St) - A dimensionless quantity used in fluid mechanics, defined by $St = h/\rho v c_p$, where h is coefficient of heat transfer, ρ is density, v is velocity, and c_p is specific heat capacity at constant pressure. [2]

Stark effect - The splitting of an energy level of an atom or molecule, and hence a splitting of spectral lines arising from that level, as a result of the application of an external electric field.

Statistical weight (*g*) - The number of distinct states corresponding to the same energy level. Also called degeneracy.

Stefan-Boltzmann constant (σ)* - Constant in the equation for the radiant exitance M (radiant energy flux per unit area) from a black body at thermodynamic temperature T, viz. $M = \sigma T^4$. [1]

Stibines - SbH_3 and compounds derived from it by substituting one, two or three hydrogen atoms by hydrocarbyl groups: R_3Sb. $RSbH_2$, R_2SbH, and R_3Sb (R not equal to H) are called primary, secondary and tertiary stibines, respectively. [5]

Stochastic process - A process which involves random variables and whose outcome can thus be described only in terms of probabilities.

Stoichiometric number (ν) - The number appearing before the symbol for each compound in the equation for a chemical reaction. By convention, it is negative for reactants and positive for products. [2]

Stokes (St) - A non-SI unit of kinematic viscosity, equal to 10^{-4} m^2/s.

Stokes' law - The statement, valid under certain conditions, that the viscous force F experienced by a sphere of radius a moving at velocity v in a medium of viscosity η is given by $F = -6\pi\eta av$.

Strain - The deformation of a body that results from an applied stress.

Stratosphere - The part of the earth's atmosphere extending from the top of the troposphere (typically 10 to 15 km above the surface) to about 50 km. It is characterized by an increase in temperature with increasing altitude.

Stress - Force per unit area (pressure) applied to a body. Tensile stress tends to stretch or compress the body in the direction of the applied force. Sheer stress results from a tangential force which tends to twist the body.

Strong interaction - The short range (order of 1 fm) attractive forces between protons, neutrons, and other hadrons which are responsible for the stability of the nucleus.

Strouhal number (*Sr*) - A dimensionless quantity used in fluid mechanics, defined by $Sr = lf/v$, where l is length, f is frequency, and v is velocity. [2]

Structure factor - In x-ray crystallography, the sum of the scattering factors of all the atoms in a unit cell, weighted by an appropriate phase factor. The intensity of a given reflection is proportional to the square of the structure factor.

Sublimation pressure - The pressure of a gas in equilibrium with a solid at a specified temperature.

Sulfides - Compounds having the structure RSR (R not equal to H). Such compounds were once called thioethers. In an inorganic sense, salts or other derivatives of hydrogen sulfide. [5]

Sulfones - Compounds having the structure, $RS(=O)_2R$ (R not equal to H), e.g. $C_2H_5S(=O)_2CH_3$, ethyl methyl sulfone. [5]

Sulfonic acids - $HS(=O)_2OH$, sulfonic acid, and its S-hydrocarbyl derivatives. [5]

Sulfoxides - Compounds having the structure $R_2S{=}O$ (R not equal to H), e.g. $Ph_2S{=}O$, diphenyl sulfoxide. [5]

Superconductor - A material that experiences a nearly total loss of electrical resistivity below a critical temperature T_c. The effect can occur in pure metals, alloys, semiconductors, organic compounds, and certain inorganic solids.

Superfluid - A fluid with near-zero viscosity and extremely high thermal conductivity. Liquid helium exhibits these properties below 2.186 K (the λ point).

Supernova - A star in the process of exploding because of instabilities which follow the exhaustion of its nuclear fuel.

Surface analysis by laser ionization (SALI) - See Techniques for Materials Characterization, page **12**-1.

Surface tension (γ,σ)* - The force per unit length in the plane of the interface between a liquid and a gas, which resists an increase in the area of that surface. It can also be equated to the surface Gibbs energy per unit area.

Surfactant - A substance which lowers the surface tension of the medium in which it is dissolved, and/or the interfacial tension with other phases, and accordingly is positively adsorbed at the liquid-vapor or other interfaces. [3]

Susceptance (*B*) - Imaginary part of admittance. [1]

Svedberg - A non-SI unit of time, used to express sedimentation coefficients, equal to 10^{-13} s.

Syndiotactic macromolecule - A tactic macromolecule, essentially comprising alternating enantiomeric configurational base units which have chiral or prochiral atoms in the main chain in a unique arrangement with respect to their adjacent constitutional units. In this case the repeating unit consists of two configurational base units that are enantiomeric. [8]

Tacticity - The orderliness of the succession of configurational repeating units of a macromolecule or oligomer molecule. In a tactic macromolecule essentially all the configurational repeating units are identical with respect to directional sense. See Configurational repeating unit, Isotactic, Syndiotactic. [8]

Tautomerism - Isomerism of the general form G-X-Y=Z $\rightleftharpoons$ X=Y-Z-G, where the isomers (called tautomers) are readily interconvertible; the atoms connecting the groups X, Y, Z are typically any of C, H, O, or S, and G is a group which becomes an electrofuge (i.e., a group that does not carry away the bonding electron pair when it leaves its position in the molecule) or nucleofuge (a group that does carry away the bonding electrons when leaving) during isomerization. The commonest case, when the electrofuge is H^+, is also known as prototropy. A common example, written so as to illustrate the general pattern given above, is keto-enol tautomerism, such as

$$H\text{-}O\text{-}C(CH_3)=CH\text{-}CO_2Et \text{ (enol)} \rightleftharpoons (CH_3)C(=O)\text{-}CH_2\text{-}CO_2Et \text{ (keto)}$$

In some cases the interconversion rate between tautomers is slow enough to permit isolation of the separate keto and enol forms. [5]

Tensile strength* - In tensile testing, the ratio of maximum load a body can bear before breaking to original cross-sectional area. Also called ultimate strength. [11]

Terpenes - Hydrocarbons of biological origin having carbon skeletons formally derived from isoprene $[CH_2=C(CH_3)CH=CH_2]$. [5]

Terpenoids - Natural products and related compounds formally derived from isoprene units. They contain oxygen in various functional groups. The skeleton of terpenoids may differ from strict additivity of isoprene units by the loss or shift of a methyl (or other) group. [5]

Tesla (T)* - The SI unit of magnetic flux density (*B*), equal to V s/m^2. [1]

Thermal conductivity* - Rate of heat flow divided by area and by temperature gradient. [1]

Thermal diffusivity - Thermal conductivity divided by density and by specific heat capacity at constant pressure. [1]

Thermal expansion coefficient (α)* - The linear expansion coefficient is defined by $\alpha_l = (1/l)(dl/dT)$; the volume expansion coefficient by $\alpha_V = (1/V)(dV/dT)$. [1]

Thermionic emission - The emission of electrons from a solid as a result of heat. The effect requires a high enough temperature to impart sufficient kinetic energy to the electrons to exceed the work function of the solid.

Thermodynamic laws - The foundation of the science of thermodynamics:

First law: The internal energy of an isolated system is constant; if energy is supplied to the system in the form of heat dq and work dw, then the change in energy dU = dq + dw.

Second law: No process is possible in which the only result is the transfer of heat from a reservoir and its complete conversion to work.

Third law: The entropy of a perfect crystal approaches zero as the thermodynamic temperature approaches zero.

Thermoelectric power - For a bar of a pure material whose ends are at different temperatures, the potential difference divided by the difference in temperature of the ends. See also Seeback effect.

Thermogravimetric analysis (TGA) - See Techniques for Materials Characterization, page **12-1**.

Thermosphere - The layer of the earth's atmosphere extending from the top of the mesosphere (typically 80-90 km above the surface) to about 500 km. It is characterized by a rapid increase in temperature with increasing altitude up to about 200 km, followed by a leveling off in the 300-500 km region.

Thiols - Compounds having the structure RSH (R not equal to H). Also known by the term mercaptans (abandoned by IUPAC); e.g. CH_3CH_2SH, ethanethiol. [5]

Thomson coefficient (μ, τ) - The heat power developed in the Thomson effect (whereby heat is evolved in a conductor when a current is flowing in the presence of a temperature gradient), divided by the current and the temperature difference. [1]

Tonne (t) - An alternative name for megagram (1000 kg). [1]

Torque (T) - For a force $\boldsymbol{F}$ that produces a torsional motion, $T = \boldsymbol{r} \times \boldsymbol{F}$, where $\boldsymbol{r}$ is a vector from some reference point to the point of application of the force.

Torr - A non-SI unit of pressure, equal to 133.322 Pa. The name is generally considered interchangeable with millimeter of mercury.

Townsend coefficient - In a radiation counter, the number of ionizing collisions by an electron per unit path length in the direction of an applied electric field.

Transducer - Any device that converts a signal from acoustical, optical, or some other form of energy into an electrical signal (or vice versa) while preserving the information content of the original signal.

Transistor - A voltage amplifier using controlled electron currents inside a semiconductor.

Transition metals - Elements characterized by a partially filled d subshell. The First Transition Series comprises Sc, Ti, V, Cr, Mn, Fe, Co, Ni, Cu. The Second and Third Transition Series include the lanthanides and actinides, respectively. [7]

Transition probability* - See Einstein transition probability.

Transmittance (τ) - Ratio of the radiant or luminous flux at a given wavelength that is transmitted to that of the incident radiation. Also called transmission factor. [1]

Tribology - The study of frictional forces between solid surfaces.

Triple point* - The point in p,T space where the solid, liquid, and gas phases of a substance are in thermodynamic equilibrium. The corresponding temperature and pressure are called the triple point temperature and triple point pressure.

Troposphere - The lowest part of the earth's atmosphere, extending to 10-15 km above the surface. It is characterized by a decrease in temperature with increasing altitude. The exact height varies with latitude and season.

Tunnel diode - A device involving a p-n junction in which both sides are so heavily doped that the Fermi level on the p-side lies in the valence band and on the n-side in the conduction band. This leads to a current-voltage curve with a maximum, so that the device exhibits a negative resistance in some regions.

Ultraviolet photoelectron spectroscopy (UPS) - See Techniques for Materials Characterization, page **12-1**.

Umklapp process - A process involving the interaction of three or more waves (lattice or electron) in a solid in which the sum of the wave vectors does not equal zero.

Unified atomic mass unit (u)* - A unit of mass used in atomic, molecular, and nuclear science, defined as the mass of one atom of ^{12}C divided by 12. Its approximate value is 1.66054×10^{-27} kg. [1]

Universal time (t_U, **UT**) - Mean solar time counted from midnight at the Greenwich meridian. Also called Greenwich mean time (GMT). The interval of mean solar time is based on the average, over one year, of the time between successive transits of the sun across the observer's meridian.

Vacancy - A missing atom or ion in a crystal lattice.

Van Allen belts - Two toroidal regions above the earth's atmosphere containing protons and electrons. The outer belt at about 25,000 km above the surface is probably of solar origin. The inner belt at about 3000 km contains more energetic particles from outside the solar system.

Van der Waals' equation* - An equation of state for fluids which takes the form:

$$pV_m = RT\left(\frac{1}{V_m - b} - \frac{a}{V_m{}^2}\right)$$

where p is pressure, V_m is molar volume, T is temperature, R is the molar gas constant, and a and b are characteristic parameters of the substance which describe the effect of attractive and repulsive intermolecular forces, respectively.

Van der Waals' force - The weak attractive force between two molecules which arises from electric dipole interactions. It can lead to the formation of stable but weakly bound dimer molecules or clusters.

Van't Hoff equation - The equation expressing the temperature dependence of the equilibrium constant K of a chemical reaction:

$$\frac{\mathrm{d}\ln k}{\mathrm{d}T} = \frac{\Delta_r H^\circ}{RT^2}$$

where $\Delta_r H^\circ$ is the standard enthalpy of reaction, R the molar gas constant, and T the temperature. Also called van't Hoff isochore.

Vapor pressure* - The pressure of a gas in equilibrium with a liquid (or, in some usage, a solid) at a specified temperature.

Varistor - A device that utilizes the properties of certain metal oxides with small amounts of impurities, which show abrupt nonlinearities at specific voltages where the material changes from a semiconductor to an insulator.

Velocity (v) - Rate of change of distance with time.

Verdet constants (V)* - Angle of rotation of a plane polarized light beam passing through a medium in a magnetic field, divided by the field strength and by the path length.

Virial equation of state* - An equation relating the pressure p, molar volume V_m, and temperature T of a real gas in the form of an expansion in powers of the molar volume, viz., $pV_m = RT(1+BV_m^{-1}+CV_m^{-2}+\ldots)$, where R is the molar gas constant. B is called the second virial coefficient, C the third virial coefficient, etc. The virial coefficients are functions of temperature.

Viscosity (η)* - The proportionality factor between sheer rate and sheer stress, defined through the equation $F = \eta A(\mathrm{d}v/\mathrm{d}x)$, where F is the tangential force required to move a planar surface of area A at velocity v relative to a parallel surface separated from the first by a distance x. Sometimes called dynamic or absolute viscosity. The term kinematic viscosity (symbol ν) is defined as η divided by the mass density.

Volt (V)* - The SI unit of electric potential, equal to W/A. [1]

Volume fraction (ϕ_j) - Defined as $V_j/\Sigma_i V_i$, where V_j is the volume of the specified component and the V_i are the volumes of all the components of a mixture prior to mixing. [2]

Watt (W)* - The SI unit of power, equal to J/s. [1]

Wave function - A function of the coordinates of all the particles in a quantum mechanical system (and, in general, of time) which fully describes the state of the system. The product of the wave function and its complex conjugate is proportional to the probability of finding a particle at a particular point in space.

Weak interaction - The weak forces (order of 10^{-12} of the strong interaction) between elementary particles which are responsible for beta decay and other nuclear effects.

Weber (Wb)* - The SI unit of magnetic flux, equal to V s. [1]

Weber number (*We*) - A dimensionless quantity used in fluid mechanics, defined by $We = \rho v^2 l/\gamma$, where ρ is density, v is velocity, l is length, and γ is surface tension. [2]

Weight - That force which, when applied to a body, would give it an acceleration equal to the local acceleration of gravity. [1]

Wiedeman-Franz law - The law stating that the thermal conductivity k and electrical conductivity σ of a pure metal are related by $k = L\sigma T$, where T is the temperature and L (called the Lorenz ratio) has the approximate value 2.45×10^{-8} V^2/K^2.

Wien displacement law - The relation, which can be derived from the Planck formula for black body radiation, that $\lambda_{max}T = 0.0028978$ m K, where λ_{max} is the wavelength of maximum radiance at temperature T.

Wigner-Seitz method - A method of calculating electron energy levels in a solid using a model in which each electron is subject to a spherically symmetric potential.

Wittig reagents - See phosphonium ylides.

Work (*W*) - Force multiplied by the displacement in the direction of the force. [1]

Work function (Φ)* - The energy difference between an electron at rest at infinity and an electron at the Fermi level in the interior of a substance. It is thus the minimum energy required to remove an electron from the interior of a solid to a point just outside the surface. [1]

X unit (X) - A unit of length used in x-ray crystallography, equal to approximately 1.002×10^{-13} m.

X-ray photoelectron spectroscopy (XPS) - See Techniques for Materials Characterization, page **12-1**.

Yield strength - The stress at which a material exhibits a specified deviation (often chosen as 0.2% for metals) from proportionality of stress and strain. [11]

Young's modulus (*E*) - In tension or compression of a body below its elastic limit, the ratio of stress to corresponding strain. Since strain is normally expressed on a fractional basis, Young's modulus has dimensions of pressure. Also called elastic modulus. [11]

Zeeman effect - The splitting of an energy level of an atom or molecule, and hence a splitting of spectral lines arising from that level, as a result of the application of an external magnetic field.

Zener diode - A control device utilizing a p-n junction with a well defined reverse-bias avalanche breakdown voltage.

Zeotrope - A liquid mixture that shows no maximum or minimum when vapor pressure is plotted against composition at constant temperature. See Azeotrope.

Zero-point energy - The energy possessed by a quantum mechanical system as a result of the uncertainty principle even when it is in its lowest energy state; e.g., the difference between the lowest energy level of a harmonic oscillator and the minimum in the potential well.

Zeta potential (ζ) - The electric potential at the surface of a colloidal particle relative to the potential in the bulk medium at a long distance. Also called electrokinetic potential.

Zwitterions - Neutral compounds having formal unit electrical charges of opposite sign. Some chemists restrict the term to compounds with the charges on non-adjacent atoms. Sometimes referred to as inner salts, dipolar ions (a misnomer). [5]

Section 3
Physical Constants of Organic Compounds

Physical Constants of Organic Compounds 3-1
Structure Diagrams for Table of Physical Constants 3-331
Synonym Index of Organic Compounds 3-586
Molecular Formula Index of Organic Compounds 3-631
CAS Registry Number Index of Organic Compounds 3-709
Melting Point Index of Organic Compounds 3-737
Boiling Point Index of Organic Compounds 3-751
Diamagnetic Susceptibility of Selected Organic Compounds 3-764

PHYSICAL CONSTANTS OF ORGANIC COMPOUNDS

The basic physical constants for over 12,000 organic compounds are presented in this table, along with structures and references to other sources of information. An effort has been made to include the compounds most frequently encountered in the laboratory, the workplace, and the environment. The selection was based mainly on the appearance of the compounds in various specialized tables in this *Handbook* and in other widely used reference sources, such as the *Merck Index* and the *DIPPR Database of Pure Compound Properties.* The occurrence of a compound on regulatory lists of hazardous chemicals was also taken into consideration, as was the availability of reliable physical constant data. Clearly, criteria of this type are somewhat subjective, and compounds considered important by some users have undoubtedly been omitted. Suggestions for additional compounds or other improvements are welcomed.

The data in the table have been taken from many sources, including both compilations and the primary literature. Where conflicts were found, the value deemed most reliable was chosen. Some of the useful compilations of physical property data are listed at the end of this introduction.

The table is arranged alphabetically by the primary name, which is generally the Index Name from the 8th or 9th Collective Index of Chemical Abstracts Service (CAS). In a few cases, especially pesticides and pharmaceuticals, the common name is used rather than the more complex systematic name. By convention, CAS Index Names are written in inverted order, e.g., chloromethane is listed as methane, chloro and ethyl acetate as acetic acid, ethyl ester. Furthermore, certain important compounds are listed under Index Names which differ from the names by which they are commonly known (e.g. aniline appears as benzenamine and acetone as 2-propanone). In order to facilitate the location of compounds in the table, three indexes are provided:

- **Synonym Index:** Includes common synonyms, but not the primary name by which the table is arranged.
- **Molecular Formula Index:** Lists compounds by molecular formula in the Hill order (see Preface to this *Handbook*).
- **CAS Registry Number Index:** Lists compounds by Chemical Abstracts Service Registry Number.

Two lines of data appear for each compound. The explanation of the data fields follows.

Top Line:

- **No.:** An identification number used in the indexes and to identify the structure diagrams.
- **Name:** Primary name, generally the CAS Index Name.
- **Mol. Form.:** The molecular formula written in the Hill convention.
- **CAS RN:** The Chemical Abstracts Service Registry Number assigned by CAS as a unique identifier for the compound.
- **Merck No:** Monograph Number in *The Merck Index, Eleventh Edition.* It should be noted that this is not a unique identifier for a single compound, since several derivatives or isomers of a compound may be included in the same Monograph.
- **Beil. Ref:** Citation to the *Beilstein Handbook of Organic Chemistry.* An entry of 5-18-11-01234, for example, indicates that the compound may be found in the 5th Series, Volume 18, Subvolume 11, page 1234.
- **Solubility:** Solubility in common solvents on a relative scale: 1 = insoluble; 2 = slightly soluble; 3 = soluble; 4 = very soluble; 5 = miscible; 6 = decomposes. See List of Abbreviations for the solvent abbreviations.

Bottom line:

- **Synonym:** A synonym in common use. When the primary name is non-systematic, the systematic name appears here.
- **Mol. Wt.:** Molecular weight (relative molar mass) as calculated with the 1991 IUPAC Standard Atomic Weights.
- **mp/°C:** Normal melting point in °C. Although some values are quoted to 0.1°C, uncertainties are typically several degrees Celsius. A value is sometimes followed by "dec", indicating decomposition is observed at the stated temperature (so that it is probably not a true melting point). See the List of Abbreviations for other abbreviations.
- **bp/°C:** Boiling point in °C. When available, the normal boiling point is given first, without a superscript. This is the temperature at which the liquid phase is in equilibrium with the vapor at a pressure of 760 mmHg (101.325 kPa). Boiling point values at reduced pressure are also given in many cases; here the superscript indicates the pressure in mmHg. A "dec" or "exp" following the value indicates decomposition or explosion has been observed at the boiling point. A simple entry of "exp" (sometimes followed by a temperature) indicates explosion may occur on heating, even below the boiling point. An entry of "sub" indicates that no boiling point is available, but measurable vapor (sublimation) pressure has been observed upon heating the solid. A temperature may be given, but no precise meaning can be attached because the pressure is not specified.
- **den/g cm^{-3}:** density (mass per unit volume) in g/cm^3. The superscript indicates the temperature in °C. Values are given only for the liquid and solid phases, and all values are true densities, not specific gravities. The number of decimal places gives a rough estimate of the accuracy of the value.
- **n_D:** Refractive index, at the temperature indicated by the superscript. Unless otherwise indicated, all values refer to a wavelength of 589 nm (sodium D line). Values are given only for liquids and solids.

Structures are given, when available, in the section following the main table, using the **No.** in the first column as the linking identifier.

LIST OF ABBREVIATIONS

Ac	Acetyl	EtOH	Ethanol
ace	Acetone	exp	Explodes
acid	Acids	gl	Glacial
AcOEt	Ethyl acetate	HOAc	Acetic acid
alk	Alkalis	hp	Heptane
anh	Anhydrous	hx	Hexane
aq	Aqueous	lig	Ligroin
bz	Benzene	MeOH	Methanol
chl	Chloroform	os	Organic solvents
con	Concentrated	peth	Petroleum ether
ctc	Carbon tetrachloride	$PhNO_2$	Nitrobenzene
cyhex	Cyclohexane	PrOH	1-Propanol
dec	Decomposes	py	Pyridine
dil	Dilute	sub	Sublimes
diox	Dioxane	sulf	Sulfuric acid
DMF	Dimethyl formamide	temp.	Temperature
DMSO	Dimethyl sulfoxide	tfa	Trifluoroacetic acid
DMSO-d_6	Deuterated dimethyl sulfoxide	tol	Toluene
eth	Ethyl ether	xyl	Xylene

REFERENCES

1. Lide, D. R., and Milne, G. W. A., Editors, *Handbook of Data on Organic Compounds, Third Edition*, CRC Press, Boca Raton, FL, 1993.
2. Luckenbach, R., Editor, *Beilstein Handbook of Organic Chemistry*, Springer-Verlag, Heidelberg. Data can also be accessed through the on-line Beilstein Database.
3. Budavari, S., Editor, *The Merck Index, Eleventh Edition*, Merck & Co., Rahway, NJ, 1989.
4. Daubert, T. E., Danner, R. P., Sibul, H. M., and Stebbins, C. C., *Physical and Thermodynamic Properties of Pure Compounds: Data Compilation*, extant 1994 (core with 4 supplements), Taylor & Francis, Bristol, PA (also available as database).
5. *TRC Thermodynamic Tables*, Thermodynamic Research Center, Texas A & M University, College Station, TX.
6. Riddick, J. A., Bunger, W. B., and Sakano, T. K., *Organic Solvents, Fourth Edition*, John Wiley & Sons, New York, 1986.
7. *Physical Constants of Hydrocarbon and Non-Hydrocarbon Compounds,* ASTM Data Series DS 4B, ASTM, Philadelphia, 1988.
8. Stevenson, R. M., and Malanowski, S., *Handbook of the Thermodynamics of Organic Compounds*, Elsevier, New York, 1987.
9. Buckingham, J., Editor, *Dictionary of Organic Compounds, Fifth Edition* (and Supplements), Chapman and Hall, New York, 1982-1993.
10. Lide, D. R., and Kehiaian, H. V., *Handbook of Thermophysical and Thermochemical Data*, CRC Press, Boca Raton, FL, 1994.

PHYSICAL CONSTANTS OF ORGANIC COMPOUNDS (continued)

No.	Name Synonym	Mol. Form. Mol. Wt.	CAS RN mp/°C	Merck No. bp/°C	Beil. Ref. den/g cm^{-3}	Solubility n_D
1	Abate Phosphorothioic acid, *O,O'*-(thiodi-4,1-phenylene) *O,O,O',O'*-tetramethyl ester	$C_{16}H_{20}O_6P_2S_3$ 466.48	3383-96-8 30	9075	 1.32	
2	1-Acenaphthylenamine, 1,2-dihydro- 1-Acenaphthenamine	$C_{12}H_{11}N$ 169.23	40745-44-6 135	 sub	2-12-00-00764	bz 4; EtOH 4
3	Acenaphthylene Acenaphthalene	$C_{12}H_8$ 152.20	208-96-8 92.5	 280; 150^{28}	4-05-00-02138 0.8987^{16}	H_2O 1; EtOH 4; eth 4; bz 4
4	Acenaphthylene, 5-bromo-1,2-dihydro- 5-Bromoacenaphthene	$C_{12}H_9Br$ 233.11	2051-98-1 52	 335	4-05-00-01839 1.4392^{52}	H_2O 1; EtOH 2; chl 2 1.6565^{54}
5	Acenaphthylene, 5-chloro-1,2-dihydro- 5-Chloroacenaphthene	$C_{12}H_9Cl$ 188.66	5209-33-6 70.5	 319; 163^{13}	4-05-00-01837 1.1954^{70}	 1.6169^{100}
6	Acenaphthylene, 1,2-dihydro- Acenaphthene	$C_{12}H_{10}$ 154.21	83-32-9 93.4	23 279	4-05-00-01834 1.222^{20}	H_2O 1; EtOH 2; bz 4; chl 2 1.6048^{95}
7	Acenaphthylene, 1,2-dihydro-5-iodo- 5-Iodoacenaphthene	$C_{12}H_9I$ 280.11	6861-64-9 64.3		2-05-00-00497 1.5000^{20}	bz 4; EtOH 4 1.6909^{65}
8	1,2-Acenaphthylenedione	$C_{12}H_6O_2$ 182.18	82-86-0 261	 sub	4-07-00-02498 1.4800^{20}	H_2O 1; EtOH 2; bz 2; lig 3
9	Acenaphthylene, 1,2,2a,3,4,5-hexahydro-	$C_{12}H_{14}$ 158.24	480-72-8 12.8	 249	4-05-00-01566 1.0290^{20}	chl 3
10	Acephate Phosphoramidothioic acid, acetyl-, *O,S*-dimethyl ester	$C_4H_{10}NO_3PS$ 183.17	30560-19-1 88		 1.35^{20}	
11	Acetaldehyde Ethanal	C_2H_4O 44.05	75-07-0 -123	32 20.1	4-01-00-03094 0.7834^{18}	H_2O 5; EtOH 5; eth 5; bz 5 1.3316^{20}
12	Acetaldehyde, chloro- Chloroacetaldehyde	C_2H_3ClO 78.50	107-20-0 -16.3	2108 85.5	4-01-00-03134	eth 3
13	Acetaldehyde, chloro-, trimer Chloroacetaldehyde, trimer	$C_6H_9Cl_3O_3$ 235.49	65438-35-9 87	 142^{10}	4-19-00-04718	eth 4
14	Acetaldehyde, ethoxy- α-Ethoxyacetaldehyde	$C_4H_8O_2$ 88.11	22056-82-2		3-01-00-03181 0.942^{20}	H_2O 4; ace 4; EtOH 4 1.3956^{20}
15	Acetaldehyde, hydroxy- Glycolaldehyde	$C_2H_4O_2$ 60.05	141-46-8 97		4-01-00-03955 1.366^{100}	chl 3 1.4772^{19}
16	Acetaldehyde, methoxy- Methoxyacetaldehyde	$C_3H_6O_2$ 74.08	10312-83-1	 92	4-01-00-03955 1.005^{25}	H_2O 4; ace 4; eth 4; EtOH 4 1.3950^{20}
17	Acetaldehyde, oxime Acetaldoxime	C_2H_5NO 59.07	107-29-9 45	35 115	4-01-00-03121 0.9656^{20}	H_2O 3; EtOH 5; eth 5; chl 3 1.4264^{20}
18	Acetaldehyde, phenylhydrazone	$C_8H_{10}N_2$ 134.18	935-07-9 99.5	 150^{40}; 135^{21}	3-15-00-00079	EtOH 4
19	Acetaldehyde, tetramer Metaldehyde II (tetramer)	$C_4H_{12}O_4$ 124.14	66056-06-2 47	 110	4-19-00-05643	ace 4; bz 4; eth 4; EtOH 4
20	Acetaldehyde, tribromo- Bromal	C_2HBr_3O 280.74	115-17-3	1372 174	4-01-00-03155 2.6649^{25}	ace 4; eth 4; EtOH 4 1.5939^{20}
21	Acetaldehyde, trichloro- Chloral	C_2HCl_3O 147.39	75-87-6 -57.5	9538 97.8	4-01-00-03142 1.512^{20}	H_2O 4; EtOH 3; eth 3 1.4580^{20}
22	Acetamide Ethanamide	C_2H_5NO 59.07	60-35-5 81	36 222.0	4-02-00-00399 0.9986^{85}	H_2O 4; EtOH 4 1.4278
23	Acetamide, *N*-acetyl-	$C_4H_7NO_2$ 101.11	625-77-4 79	 223.5	4-02-00-00416	H_2O 3; EtOH 3; eth 3; chl 3
24	Acetamide, *N*-acetyl-*N*-ethyl-	$C_6H_{11}NO_2$ 129.16	1563-83-3	 197	4-04-00-00352 1.0092^{20}	EtOH 4 1.4513^{20}
25	Acetamide, *N*-acetyl-*N*-methyl-	$C_5H_9NO_2$ 115.13	1113-68-4 -25	 195; 114.5^{61}	4-04-00-00183 1.0663^{25}	H_2O 5; eth 1 1.4502^{25}
26	Acetamide, *N*-acetyl-*N*-phenyl-	$C_{10}H_{11}NO_2$ 177.20	1563-87-7 37.5	 200^{100}; 142^{11}	4-12-00-00385	H_2O 2; EtOH 3; bz 3; tol 3
27	Acetamide, *N*-(aminocarbonyl)-	$C_3H_6N_2O_2$ 102.09	591-07-1 218	 sub 180	3-03-00-00119	H_2O 2; EtOH 3; eth 2
28	Acetamide, *N*-(2-aminoethyl)-	$C_4H_{10}N_2O$ 102.14	1001-53-2 51	 136^{27}	4-04-00-01193	H_2O 3; EtOH 3; eth 1; bz 3
29	Acetamide, *N*-(4-aminophenyl)- *p*-Aminoacetanilide	$C_8H_{10}N_2O$ 150.18	122-80-5 166.5	422 267	4-13-00-00137	H_2O 3; EtOH 4; eth 4
30	Acetamide, *N*-[(4-aminophenyl)sulfonyl]- Sulfacetamide	$C_8H_{10}N_2O_3S$ 214.25	144-80-9 183	8870	4-14-00-02662	H_2O 2; EtOH 3; eth 1; ace 4
31	Acetamide, *N*-[4-(aminosulfonyl)phenyl]- Acetylsulfanilamide	$C_8H_{10}N_2O_3S$ 214.25	121-61-9 219.5	97	4-14-00-02704	H_2O 3; EtOH 3; ace 3
32	Acetamide, *N*-[5-(aminosulfonyl)-1,3,4-thiadiazol-2-yl]- Acetazolamide	$C_4H_6N_4O_3S_2$ 222.25	59-66-5 260.5	45	4-27-00-08219	H_2O 2
33	Acetamide, *N*-(aminothioxomethyl)- Acetylthiourea	$C_3H_6N_2OS$ 118.16	591-08-2 165		4-03-00-00354	H_2O 2; EtOH 3; eth 2
34	Acetamide, *N*-[4-[[(aminothioxomethyl)hydrazono]methyl]phenyl]- Thiacetazone	$C_{10}H_{12}N_4OS$ 236.30	104-06-3 225 dec	9218	4-14-00-00075	H_2O 1; os 1; CS_2 1
35	Acetamide, *N*-[1,1'-biphenyl]-2-yl-	$C_{14}H_{13}NO$ 211.26	2113-47-5 121	 355	4-12-00-03224	H_2O 1; EtOH 4; eth 4
36	Acetamide, *N*-bromo- *N*-Bromoacetamide	C_2H_4BrNO 137.96	79-15-2 103.5	1386	4-02-00-00417	eth 4

PHYSICAL CONSTANTS OF ORGANIC COMPOUNDS (continued)

No.	Name	Synonym	Mol. Form.	Mol. Wt.	CAS RN	mp/°C	Merck No.	bp/°C	Beil. Ref.	den/g cm^{-3}	Solubility	n_D
37	Acetamide, *N*-(4-bromophenyl)-	*p*-Bromoacetanilide	C_8H_8BrNO	214.06	103-88-8	168	1387		4-12-00-01504	1.717^{25}	H_2O 1; EtOH 3; eth 2; bz 2	
38	Acetamide, *N*-butyl-		$C_6H_{13}NO$	115.18	1119-49-9			229	4-04-00-00565	0.8960^{25}		1.4388^{25}
39	Acetamide, *N*-butyl-*N*-phenyl-		$C_{12}H_{17}NO$	191.27	91-49-6	24.5		281	4-12-00-00379	0.9912^{20}	chl 2	1.5146^{20}
40	Acetamide, *N,N'*-carbonylbis-		$C_5H_8N_2O_3$	144.13	638-20-0	154.5		sub	3-03-00-00121		H_2O 2; EtOH 2; eth 3; ace 3	
41	Acetamide, 2-chloro-	Chloroacetamide	C_2H_4ClNO	93.51	79-07-2	121	2109	225	4-02-00-00490		H_2O 3; EtOH 4; eth 2	
42	Acetamide, 2-chloro-*N,N*-bis(1-methylethyl)-		$C_8H_{16}ClNO$	177.67	7403-66-9	48.5		163^{55}; $86^{2.7}$	4-04-00-00516			1.4619^{25}
43	Acetamide, 2-chloro-*N*-3-chloroallyl-	Chloroacetamide, *N*-3-chloroallyl	$C_5H_7Cl_2NO$	168.02	100130-25-4	52.5		$112^{0.5}$	4-04-00-01087		peth 4	
44	Acetamide, 2-chloro-*N,N*-di-2-propenyl-	Allidochlor	$C_8H_{12}ClNO$	173.64	93-71-0		250	116^{1}; $92^{0.7}$	4-04-00-01064		H_2O 2; EtOH 3	1.4932^{25}
45	Acetamide, 2-chloro-*N*-hexyl-	α-Chloro-*N*-hexylacetamide	$C_8H_{16}ClNO$	177.67	5326-81-8	108.5		$100^{0.2}$	4-04-00-00713		peth 4	
46	Acetamide, 2-chloro-*N*-(3-methoxypropyl)-	α-Chloro-*N*-(3-methoxypropyl)acetamide	$C_6H_{12}ClNO_2$	165.62	1709-03-1	30		$88^{0.5}$	4-04-00-01645			1.4712^{25}
47	Acetamide, 2-chloro-*N*-(1-methylpropyl)-		$C_6H_{12}ClNO$	149.62	32322-73-9	45.5		$68^{0.7}$	4-04-00-00621		peth 4	
48	Acetamide, *N*-(4-chlorophenyl)-		C_8H_8ClNO	169.61	539-03-7	179		333	4-12-00-01178	1.385^{22}	H_2O 1; EtOH 3; eth 4; ctc 2	
49	Acetamide, 2-cyano-	Cyanoacetamide	$C_3H_4N_2O$	84.08	107-91-5	121.5	2695		4-02-00-01891		H_2O 4	
50	Acetamide, *N,N*-diacetyl-	Triacetamide	$C_6H_9NO_3$	143.14	641-06-5	79			4-02-00-00416		eth 4	
51	Acetamide, 2,2-dichloro-		$C_2H_3Cl_2NO$	127.96	683-72-7	99.4		234	4-02-00-00505		H_2O 3; EtOH 3; eth 3; ace 2	
52	Acetamide, 2,2-dichloro-*N*-[2-hydroxy-1-(hydroxymethyl)-2-(4-nitrophenyl)ethyl]-, [*R*-(*R**,*R**)]-	Chloramphenicol	$C_{11}H_{12}Cl_2N_2O_5$	323.13	56-75-7	150.5	2068	sub	3-13-00-02268		ace 4; EtOH 4; chl 4	
53	Acetamide, *N,N*-diethyl-		$C_6H_{13}NO$	115.18	685-91-6			185.5	4-04-00-00349	0.9130^{17}	H_2O 3; EtOH 3; eth 5; ace 5	1.4374^{17}
54	Acetamide, 2-(diethylamino)-*N*-(2,6-dimethylphenyl)-	Lidocaine	$C_{14}H_{22}N_2O$	234.34	137-58-6	68.5	5359	$180\text{-}2^{4}$	4-12-00-02538		bz 4; eth 4; EtOH 4; chl 4	
55	Acetamide, *N,N*-dimethyl-	*N,N*-Dimethylethanamide	C_4H_9NO	87.12	127-19-5	-18.59	3216	165	4-04-00-00180	0.9366^{25}	H_2O 5; EtOH 5; eth 5; ace 5	1.4380^{20}
56	Acetamide, *N*-(2,4-dimethylphenyl)-		$C_{10}H_{13}NO$	163.22	2050-43-3	129.3		170^{10}	4-12-00-02548		EtOH 4; chl 4	
57	Acetamide, *N,N*-diphenyl-	Diphenylacetamide	$C_{14}H_{13}NO$	211.26	519-87-9	103	3315	sub	4-12-00-00381		H_2O 2; EtOH 3; eth 2; chl 2	
58	Acetamide, *N,N*-dipropyl-		$C_8H_{17}NO$	143.23	1116-24-1			209.5	4-04-00-00476	0.8992^{17}	EtOH 4	1.4419^{17}
59	Acetamide, *N*-(2-ethoxyphenyl)-		$C_{10}H_{13}NO_2$	179.22	581-08-8	79		>240	3-13-00-00779		H_2O 1; EtOH 3; eth 3; chl 3	
60	Acetamide, *N*-(4-ethoxyphenyl)-	Phenacetin	$C_{10}H_{13}NO_2$	179.22	62-44-2	137.5	7155		4-13-00-01092		H_2O 2; EtOH 3; eth 2; ace 3	1.571
61	Acetamide, *N*-ethyl-	*N*-Ethylacetamide	C_4H_9NO	87.12	625-50-3			205; 104^{18}	4-04-00-00347	0.942^{4}	H_2O 5; EtOH 5; chl 3; HOAc 3	1.4338^{20}
62	Acetamide, *N*-ethyl-*N*-phenyl-		$C_{10}H_{13}NO$	163.22	529-65-7	55		260	4-12-00-00379	0.9938^{60}	H_2O 3; eth 3; ctc 3	
63	Acetamide, 2-fluoro-	Fluoroacetic acid amide	C_2H_4FNO	77.06	640-19-7	108	4095	sub	4-02-00-00454		H_2O 3; ace 3; chl 2	
64	Acetamide, *N*-9H-fluoren-2-yl-	2-(Acetylamino)fluorene	$C_{15}H_{13}NO$	223.27	53-96-3	193	4083		4-12-00-03373		H_2O 1; EtOH 3; eth 3; HOAc 3	
65	Acetamide, *N*-(2-hydroxyethyl)-		$C_4H_9NO_2$	103.12	142-26-7	63.5		$166\text{-}7^{8}$	4-04-00-01535	1.1079^{25}	H_2O 5; ace 3; bz 2; lig 2	1.4674^{20}
66	Acetamide, *N*-(2-hydroxy-1-naphthalenyl)-		$C_{12}H_{11}NO_2$	201.22	117-93-1	235 dec		sub	4-13-00-02095		ace 4; bz 4; eth 4; EtOH 4	
67	Acetamide, *N*-(4-hydroxyphenyl)-	Acetaminophen	$C_8H_9NO_2$	151.17	103-90-2	170	40		4-13-00-01091	1.293^{21}	H_2O 1; EtOH 4	
68	Acetamide, 2-mercapto-*N*-2-naphthalenyl-	Thionalide	$C_{12}H_{11}NOS$	217.29	93-42-5	111.5	9273		3-12-00-03052		H_2O 1; EtOH 4; os 4	
69	Acetamide, *N*-[2-(5-methoxy-1H-indol-3-yl)ethyl]-	Melatonin	$C_{13}H_{16}N_2O_2$	232.28	73-31-4	117	5695		5-22-12-00042			
70	Acetamide, *N*-(2-methoxyphenyl)-		$C_9H_{11}NO_2$	165.19	93-26-5	87.5		304	4-13-00-00833		H_2O 4; EtOH 4; eth 3; ace 3	
71	Acetamide, *N*-(4-methoxyphenyl)-	*p*-Acetanisidine	$C_9H_{11}NO_2$	165.19	51-66-1	131	43		4-13-00-01092		ace 4; EtOH 4; chl 4	
72	Acetamide, *N*-methyl-		C_3H_7NO	73.09	79-16-3	28		205	4-04-00-00176	0.9371^{25}	ace 4; bz 4; eth 4; EtOH 4	1.4301^{20}
73	Acetamide, *N*-[(methylamino)carbonyl]-		$C_4H_8N_2O_2$	116.12	623-59-6	180.5		dec	4-04-00-00207		H_2O 3; EtOH 2; eth 2; chl 3	

No.	Name Synonym	Mol. Form. Mol. Wt.	CAS RN mp/°C	Merck No. bp/°C	Beil. Ref. den/g cm^{-3}	Solubility n_D
74	Acetamide, *N*-(1-methylethyl)-*N*-phenyl- *N*-Isopropylacetanilide	$C_{11}H_{15}NO$ 177.25	5461-51-8 40.5	 264	3-12-00-00466	lig 4
75	Acetamide, *N*-methyl-*N*-(2-methylphenyl)-	$C_{10}H_{13}NO$ 163.22	573-26-2 55.5	 260	3-12-00-01854	EtOH 3; chl 3
76	Acetamide, *N*-(2-methylphenyl)-	$C_9H_{11}NO$ 149.19	120-66-1 110	 296	4-12-00-01755 1.168^{15}	H_2O 2; EtOH 3; eth 3; ace 3
77	Acetamide, *N*-(3-methylphenyl)-	$C_9H_{11}NO$ 149.19	537-92-8 65.5	 303	4-12-00-01823 1.141^{15}	H_2O 2; EtOH 4; eth 4; chl 3
78	Acetamide, *N*-(4-methylphenyl)-	$C_9H_{11}NO$ 149.19	103-89-9 152	 307	4-12-00-01902 1.2120^{15}	eth 4; EtOH 4
79	Acetamide, *N*-methyl-*N*-phenyl- *N*-Methylacetanilide	$C_9H_{11}NO$ 149.19	579-10-2 103	5931 256	4-12-00-00378 1.0036^{105}	H_2O 3; EtOH 3; eth 3; chl 3 1.576
80	Acetamide, 2-nitro-*N*-phenyl- α-Nitroacetanilide	$C_8H_8N_2O_3$ 180.16	10151-95-8 94	 $100^{0.1}$	1-12-00-00193 1.419^{15}	bz 4; EtOH 4; lig 4
81	Acetamide, *N*-(2-nitrophenyl)-	$C_8H_8N_2O_3$ 180.16	552-32-9 94	 $100^{0.1}$	4-12-00-01574 1.419^{15}	H_2O 3; EtOH 3; eth 4; bz 3
82	Acetamide, *N*-(3-nitrophenyl)-	$C_8H_8N_2O_3$ 180.16	122-28-1 155	 $100^{0.0074}$	4-12-00-01595	H_2O 3; EtOH 3; eth 1; chl 3
83	Acetamide, *N*-(4-nitrophenyl)-	$C_8H_8N_2O_3$ 180.16	104-04-1 216	 $100^{0.008}$	4-12-00-01632	H_2O 2; EtOH 3; eth 2; chl 2
84	Acetamide, *N*-(5-nitro-2-propoxyphenyl)- 5'-Nitro-2'-propoxyacetanilide	$C_{11}H_{14}N_2O_4$ 238.24	553-20-8 102.5	6550	3-13-00-00881	
85	Acetamide, *N*-(5-nitro-2-thiazolyl)- Aminitrozole	$C_5H_5N_3O_3S$ 187.18	140-40-9 264.5	421	4-27-00-04676	alk 3
86	Acetamide, *N*-phenyl- Acetanilide	C_8H_9NO 135.17	103-84-4 114.3	42 304	4-12-00-00373 1.2190^{15}	H_2O 2; EtOH 4; eth 3; ace 4
87	Acetamide, *N*-(phenylmethyl)-	$C_9H_{11}NO$ 149.19	588-46-5 61	 157^2	4-12-00-02230	EtOH 4; eth 4
88	Acetamide, *N*-phenyl-*N*-propyl- *N*-Propylacetanilide	$C_{11}H_{15}NO$ 177.25	2437-98-1 49	 268	4-12-00-00379	eth 4; EtOH 4; lig 4
89	Acetamide, *N*-3-pyridinyl-	$C_7H_8N_2O$ 136.15	5867-45-8 133	 326.5	5-22-09-00017	H_2O 3; EtOH 3; eth 2; bz 2
90	Acetamide, *N*-4-quinolinyl-	$C_{11}H_{10}N_2O$ 186.21	32433-28-6 176	 sub	5-22-10-00250	EtOH 4
91	Acetamide, *N,N*'-(sulfonyldi-4,1-phenylene)bis- Acedapsone	$C_{16}H_{16}N_2O_4S$ 332.38	77-46-3 290	16	4-13-00-01361	EtOH 4
92	Acetamide, *N*-(5,6,7,9-tetrahydro-10-hydroxy-1,2,3-trimethoxy-9-oxobenzo[a]heptalen-7-yl)-, (*S*)- Colchiceine	$C_{21}H_{23}NO_6$ 385.42	477-27-0 178.5	2469	4-14-00-00943 1.24^{25}	H_2O 2; EtOH 4; eth 1; bz 1
93	Acetamide, *N*-(5,6,7,9-tetrahydro-1,2,3,10-tetramethoxy-9-oxobenzo[a]heptalen-7-yl)-, (*S*)- Colchicine	$C_{22}H_{25}NO_6$ 399.44	64-86-8 156	2470	4-14-00-00946	H_2O 4; EtOH 4
94	Acetamide, 2,2,2-tribromo- Tribromoacetamide	$C_2H_2Br_3NO$ 295.76	594-47-8 121.5	 sub	3-02-00-00485	H_2O 4; eth 4; EtOH 4
95	Acetamide, 2,2,2-trichloro-	$C_2H_2Cl_3NO$ 162.40	594-65-0 142	 240	4-02-00-00520	H_2O 2; EtOH 4; eth 4
96	Acetamide, 2,2,2-trichloro-*N,N*-diethyl-	$C_6H_{10}Cl_3NO$ 218.51	2430-00-4 27	 112^{12}	4-04-00-00351	 1.4900^{24}
97	Acetamide, 2,2,2-trichloro-*N,N*-dimethyl-	$C_4H_6Cl_3NO$ 190.46	7291-33-0 12	 231 dec	4-04-00-00182 1.390^{20}	bz 4; chl 4 1.5017^{25}
98	Acetamide, 2,2,2-trichloro-*N*-phenyl-	$C_8H_6Cl_3NO$ 238.50	2563-97-5 96	 169^{19}	4-12-00-00377	EtOH 4
99	Acetamide, 2,2,2-trifluoro-*N,N*-dimethyl-	$C_4H_6F_3NO$ 141.09	1547-87-1	 133	4-04-00-00182 1.232^{21}	 1.3611^{25}
100	Acetic acid Ethanoic acid	$C_2H_4O_2$ 60.05	64-19-7 16.6	47 117.9	4-02-00-00094 1.0446^{25}	H_2O 5; EtOH 5; eth 5; ace 5 1.3720^{20}
101	Acetic acid, (acetyloxy)-	$C_4H_6O_4$ 118.09	13831-30-6 67.5	 145^{12}	4-03-00-00576	H_2O 4; EtOH 4; eth 4; ace 4
102	Acetic acid, (acetyloxy)-, ethyl ester	$C_6H_{10}O_4$ 146.14	623-86-9	 179	4-03-00-00584 1.0880^{20}	eth 4; EtOH 4; HOAc 4 1.4112^{20}
103	Acetic acid, [2-(aminocarbonyl)phenoxy]- Salicylamide *O*-acetic acid	$C_9H_9NO_4$ 195.17	25395-22-6	8298	4-10-00-00184	
104	Acetic acid, aminooxo- Oxamic acid	$C_2H_3NO_3$ 89.05	471-47-6 210 dec	6870	4-02-00-01857	H_2O 2; EtOH 1; eth 1
105	Acetic acid, aminooxo-, hydrazide Semioxamazide	$C_2H_5N_3O_2$ 103.08	515-96-8 221 dec	8397	4-02-00-01868	H_2O 2; EtOH 1; eth 1; alk 4
106	Acetic acid, ammonium salt Ammonium, acetate	$C_2H_7NO_2$ 77.08	631-61-8 114	514 dec	 1.17^{20}	H_2O 4; EtOH 3; ace 2
107	Acetic acid, anhydride Acetic anhydride	$C_4H_6O_3$ 102.09	108-24-7 -73	48 139.5	4-02-00-00386 1.082^{20}	H_2O 4; EtOH 3; eth 5; bz 3 1.3901^{20}
108	Acetic acid, anhydride with nitric acid Acetyl nitrate	$C_2H_3NO_4$ 105.05	591-09-3	92 22^{70}; exp 60	4-02-00-00435 1.24^{15}	
109	Acetic acid, azido-	$C_2H_3N_3O_2$ 101.06	18523-48-3 16	 116^{12}; 92^3	4-02-00-00541 1.354^{33}	
110	Acetic acid, barium salt	$C_4H_6BaO_4$ 255.42	543-80-6	975	4-02-00-00094 2.468^{25}	H_2O 3

PHYSICAL CONSTANTS OF ORGANIC COMPOUNDS (continued)

No.	Name	Synonym	Mol. Form.	Mol. Wt.	CAS RN	mp/°C	Merck No.	bp/°C	Beil. Ref.	den/g cm^{-3}	Solubility	n_D
111	Acetic acid, bromo-	Bromoacetic acid	$C_2H_3BrO_2$	138.95	79-08-3	50	1388	208	4-02-00-00526	1.9335^{50}	H_2O 5; EtOH 5; eth 5; ace 3	1.4804^{50}
112	Acetic acid, bromochloro-	Bromochloroacetic acid	$C_2H_2BrClO_2$	173.39	5589-96-8	31.5		215; 104^{11}	4-02-00-00532	1.9848^{31}		
113	Acetic acid, bromochloro-, ethyl ester	Ethyl bromochloroacetate	$C_4H_6BrClO_2$	201.45	22524-32-9			174 dec	4-02-00-00532	1.5890^{22}	eth 4; EtOH 4	1.4639^{24}
114	Acetic acid, bromo-, 1,1-dimethylethyl ester		$C_6H_{11}BrO_2$	195.06	5292-43-3			$73\text{-}4^{25}$	3-02-00-00482		eth 4; EtOH 4	1.4430^{20}
115	Acetic acid, bromo-, ethyl ester	Ethyl bromacetate	$C_4H_7BrO_2$	167.00	105-36-2			168.5	4-02-00-00527	1.5032^{20}	H_2O 1; EtOH 5; eth 5; ace 3	1.4489^{20}
116	Acetic acid, bromofluoro-	Bromofluoroacetic acid	$C_2H_2BrFO_2$	156.94	359-25-1	50		183	4-02-00-00531		H_2O 4; EtOH 4; chl 4	
117	Acetic acid, bromo-, methyl ester	Methyl 2-bromoacetate	$C_3H_5BrO_2$	152.98	96-32-2			132	4-02-00-00527	1.6350^{20}	H_2O 1; EtOH 3; eth 3; ace 3	1.4520^{20}
118	Acetic acid, bromo-, 2-methylpropyl ester		$C_6H_{11}BrO_2$	195.06	59956-48-8			188; 74.5^{10}	1-02-00-00096	1.3269^{20}	ace 4; eth 4; EtOH 4	
119	Acetic acid, bromo-, phenyl ester		$C_8H_7BrO_2$	215.05	620-72-4	32		140^{20}	3-06-00-00599		eth 4; EtOH 4	
120	Acetic acid, bromo-, propyl ester	Propyl bromoacetate	$C_5H_9BrO_2$	181.03	35223-80-4			176	4-02-00-00528	1.4099^{20}	ace 4; eth 4; EtOH 4	1.4518^{20}
121	Acetic acid, butyl ester	Butyl acetate	$C_6H_{12}O_2$	116.16	123-86-4	-78	1535	126.1	4-02-00-00143	0.8825^{20}	H_2O 2; EtOH 5; eth 5; ace 3	1.3941^{20}
122	Acetic acid, calcium salt		$C_4H_6CaO_4$	158.17	62-54-4	dec	1643		4-02-00-00094		H_2O 3; EtOH 2	1.55
123	Acetic acid, chloro-	Chloroacetic acid (β)	$C_2H_3ClO_2$	94.50	79-11-8	63	2111	189.3	4-02-00-00474	1.4043^{40}	H_2O 4; EtOH 3; eth 3; bz 3	1.4351^{55}
124	Acetic acid, chloro-, anhydride	Chloroacetic anhydride	$C_4H_4Cl_2O_3$	170.98	541-88-8	46	2112	203	4-02-00-00487	1.5497^{20}		
125	Acetic acid, chloro-, butyl ester		$C_6H_{11}ClO_2$	150.60	590-02-3			183	4-02-00-00482	1.0704^{20}	eth 4; EtOH 4	1.4297^{20}
126	Acetic acid, chloro-, 2-chloroethyl ester		$C_4H_6Cl_2O_2$	157.00	3848-12-2			203; 102^{20}	4-02-00-00481	1.3600^{25}	eth 4	1.4619^{25}
127	Acetic acid, chlorodifluoro-	Chlorodifluoroacetic acid	$C_2HClF_2O_2$	130.48	76-04-0	25		122	4-02-00-00497		chl 3	1.3559^{20}
128	Acetic acid, chloro-, 1,1-dimethylethyl ester	*tert*-Butyl chloroacetate	$C_6H_{11}ClO_2$	150.60	107-59-5		1563	150; 50^{10}	3-02-00-00444		H_2O 6	1.4260^{20}
129	Acetic acid, chloro-, 1,2-ethanediyl ester		$C_6H_8Cl_2O_4$	215.03	6941-69-1	45.5		157^{8}	3-02-00-00448		H_2O 1; eth 4	
130	Acetic acid, chloro-, ethyl ester	Ethyl chloroacetate	$C_4H_7ClO_2$	122.55	105-39-5	-21	3741	144.3	4-02-00-00481	1.1585^{20}	H_2O 1; EtOH 5; eth 5; ace 5	1.4215^{20}
131	Acetic acid, chlorofluoro-	Chlorofluoroacetic acid	$C_2H_2ClFO_2$	112.49	471-44-3			162	4-02-00-00493	1.532^{25}		1.4085^{25}
132	Acetic acid, chlorofluoro-, butyl ester	Chlorofluoroacetic acid, butyl ester	$C_6H_{10}ClFO_2$	168.60	368-34-3			165.5	3-02-00-00453	1.124^{25}		1.4067^{25}
133	Acetic acid, chlorofluoro-, ethyl ester		$C_4H_6ClFO_2$	140.54	401-56-9			129	4-02-00-00493	1.225^{20}		1.3927^{20}
134	Acetic acid, chlorofluoro-, methyl ester	Methyl chlorofluoroacetate	$C_3H_4ClFO_2$	126.51	433-52-3			116	4-02-00-00493	1.323^{25}		1.3903^{25}
135	Acetic acid, chlorofluoro-, propyl ester	Chlorofluoroacetic acid, propyl ester	$C_5H_8ClFO_2$	154.57	348-70-9			147	3-02-00-00453	1.170^{25}		1.3994^{25}
136	Acetic acid, chloro-, 2-hydroxyethyl ester		$C_4H_7ClO_3$	138.55	35280-53-6			240 dec; $86^{1.5}$	3-02-00-00447	1.3300^{20}	H_2O 4; EtOH 4	1.4609^{20}
137	Acetic acid, chloro-, 2-methoxyethyl ester		$C_5H_9ClO_3$	152.58	13361-36-9			$86\text{-}6^{9}$	3-02-00-00447	1.2015^{20}	eth 4	1.4382^{20}
138	Acetic acid, chloro-, methyl ester	Chloroacetic acid, methyl ester	$C_3H_5ClO_2$	108.52	96-34-4	-32.1	5965	129.5	4-02-00-00480	1.236^{20}	ace 4; bz 4; eth 4; EtOH 4	1.4218^{20}
139	Acetic acid, chloro-, 1-methylethyl ester	Isopropyl chloroacetate	$C_5H_9ClO_2$	136.58	105-48-6			150.5	4-02-00-00482	1.0888^{20}	eth 4	1.4382^{20}
140	Acetic acid, chloro-, 4-methylphenyl ester		$C_9H_9ClO_2$	184.62	15150-39-7	32		254; 162^{41}	4-06-00-02112	1.1840^{35}	eth 4; EtOH 4	
141	Acetic acid, (4-chloro-2-methylphenoxy)-	MCPA	$C_9H_9ClO_3$	200.62	94-74-6	120	5645		4-06-00-01991		H_2O 2; EtOH 4; eth 4; bz 3	
142	Acetic acid, chloro-, 1-methylpropyl ester		$C_6H_{11}ClO_2$	150.60	17696-64-9			163.5	3-02-00-00443	1.060^{20}	eth 4; EtOH 4	1.4251^{19}
143	Acetic acid, chloro-, 2-methylpropyl ester		$C_6H_{11}ClO_2$	150.60	13361-35-8			170	4-02-00-00483	1.0612^{20}	ace 4; eth 4	1.4255^{20}
144	Acetic acid, chlorooxo-, ethyl ester		$C_4H_5ClO_3$	136.53	4755-77-5			137	4-02-00-01853	1.2226^{20}	bz 4; eth 4	
145	Acetic acid, chlorooxo-, methyl ester		$C_3H_3ClO_3$	122.51	5781-53-3			119	3-02-00-01583	1.3316^{20}		1.4189^{20}
146	Acetic acid, (4-chlorophenoxy)-	(*p*-Chlorophenoxy)acetic acid	$C_8H_7ClO_3$	186.59	122-88-3	156.5			4-06-00-00845		H_2O 4; chl 2	
147	Acetic acid, chloro-, phenyl ester	Phenyl chloroacetate	$C_8H_7ClO_2$	170.60	620-73-5	44.5		232.5	3-06-00-00598	1.2202^{44}	H_2O 1; EtOH 4; eth 4	1.5146^{44}
148	Acetic acid, chloro-, phenylmethyl ester		$C_9H_9ClO_2$	184.62	140-18-1			147.5^{9}; $85^{0.4}$	4-06-00-02264	1.1223^{4}	eth 4; EtOH 4	1.5426^{18}
149	Acetic acid, chloro-, propyl ester	Propyl chloroacetate	$C_5H_9ClO_2$	136.58	5396-24-7			161	4-02-00-00482	1.1022^{20}	eth 4	1.4261^{20}

PHYSICAL CONSTANTS OF ORGANIC COMPOUNDS (continued)

No.	Name Synonym	Mol. Form. Mol. Wt.	CAS RN mp/°C	Merck No. bp/°C	Beil. Ref. den/g cm^{-3}	Solubility n_D
150	Acetic acid, cyano- Cyanoacetic acid	$C_3H_3NO_2$ 85.06	372-09-8 66	2696 108$^{0.15}$; 160 dec	4-02-00-01888	H_2O 3; EtOH 3; eth 3; chl 2
151	Acetic acid, cyano-, butyl ester	$C_7H_{11}NO_2$ 141.17	5459-58-5	 231; 115^{15}	4-02-00-01891 1.0010^{20}	 1.4200^{20}
152	Acetic acid, cyano-, ethyl ester Ethyl cyanoacetate	$C_5H_7NO_2$ 113.12	105-56-6 -22.5	3744 205	4-02-00-01889 1.0654^{20}	eth 4; EtOH 4 1.4175^{20}
153	Acetic acid, cyano-, hydrazide Cyacetacide	$C_3H_5N_3O$ 99.09	140-87-4 114.5	2688	4-02-00-01895	H_2O 4; EtOH 4
154	Acetic acid, cyano-, methyl ester Methyl cyanoacetate	$C_4H_5NO_2$ 99.09	105-34-0 -22.5	 200.5	4-02-00-01889 1.1225^{25}	eth 4; EtOH 4 1.4176^{20}
155	Acetic acid, cyclohexyl ester Cyclohexyl acetate	$C_8H_{14}O_2$ 142.20	622-45-7	 173; 96^{75}	4-06-00-00036 0.968^{20}	eth 4; EtOH 4 1.442^{20}
156	Acetic acid, cyclohexylidene-, methyl ester Methyl cyclohexylideneacetate	$C_9H_{14}O_2$ 154.21	40203-74-5	 99^{20}	4-09-00-00124 1.0021^{19}	 1.4838^{19}
157	Acetic acid, decyl ester Decyl acetate	$C_{12}H_{24}O_2$ 200.32	112-17-4 -15	 244	4-02-00-00168 0.8671^{20}	H_2O 1; EtOH 3; eth 3; bz 3 1.4273^{20}
158	Acetic acid, diazo-, ethyl ester Diazoacetic ester	$C_4H_6N_2O_2$ 114.10	623-73-4 -22	2980 140 dec	4-03-00-01495 1.0852^{18}	H_2O 2; EtOH 5; eth 5; bz 5 1.4605^{20}
159	Acetic acid, dibromo- Dibromoacetic acid	$C_2H_2Br_2O_2$ 217.84	631-64-1 49	195^{250}; 130^{16}	4-02-00-00533	H_2O 4; EtOH 4; eth 3
160	Acetic acid, dibromo-, ethyl ester Ethyl dibromoacetate	$C_4H_6Br_2O_2$ 245.90	617-33-4	 194	4-02-00-00533 1.8991^{20}	H_2O 1; EtOH 5; eth 5 1.5017^{13}
161	Acetic acid, dibromofluoro-, ethyl ester Ethyl dibromofluoroacetate	$C_4H_5Br_2FO_2$ 263.89	565-53-7	 83^{33}	4-02-00-00534 1.894^{25}	 1.4633^{25}
162	Acetic acid, dichloro- Dichloroacetic acid	$C_2H_2Cl_2O_2$ 128.94	79-43-6 13.5	3037 194; 102^{20}	4-02-00-00498 1.5634^{20}	H_2O 5; EtOH 5; eth 5; ace 3 1.4658^{20}
163	Acetic acid, dichloro-, butyl ester Butyl dichloroacetate	$C_6H_{10}Cl_2O_2$ 185.05	29003-73-4	 193.5	4-02-00-00502 1.1820^{20}	eth 4; EtOH 4 1.4420^{20}
164	Acetic acid, dichloro-, ethyl ester Ethyl dichloroacetate	$C_4H_6Cl_2O_2$ 157.00	535-15-9	 155; 56^{10}	4-02-00-00501 1.2827^{20}	H_2O 2; EtOH 5; eth 5; ace 3 1.4386^{20}
165	Acetic acid, dichloro-, 2-hydroxyethyl ester	$C_4H_6Cl_2O_3$ 173.00	17704-30-2	 81-2$^{0.5}$	3-02-00-00460 1.438^{20}	EtOH 4 1.4735^{20}
166	Acetic acid, dichloro-, methyl ester Methyl dichloroacetate	$C_3H_4Cl_2O_2$ 142.97	116-54-1 -51.9	 142.9	4-02-00-00501 1.3774^{20}	H_2O 1; EtOH 3; ctc 3 1.4429^{20}
167	Acetic acid, dichloro-, 1-methylethyl ester	$C_5H_8Cl_2O_2$ 171.02	25006-60-4	 163.5	4-02-00-00502 1.2053^{20}	eth 4; EtOH 4 1.4328^{20}
168	Acetic acid, [2,3-dichloro-4-(2-methylene-1-oxobutyl)phenoxy]- Ethacrynic acid	$C_{13}H_{12}Cl_2O_4$ 303.14	58-54-8 122.5	3669		
169	Acetic acid, (2,4-dichlorophenoxy)- 2,4-D	$C_8H_6Cl_2O_3$ 221.04	94-75-7 140.5	2802 160$^{0.4}$	4-06-00-00908	H_2O 1; EtOH 3; bz 2
170	Acetic acid, (2,4-dichlorophenoxy)-, 2-butoxyethyl ester 2,4-D 2-Butoxyethyl ester	$C_{14}H_{18}Cl_2O_4$ 321.20	1929-73-3	 156-62^{1}	4-06-00-00911 1.232^{20}	
171	Acetic acid, (2,4-dichlorophenoxy)-, butyl ester 2,4-D Butyl ester	$C_{12}H_{14}Cl_2O_3$ 277.15	94-80-4	2802	3-06-00-00706	
172	Acetic acid, (2,4-dichlorophenoxy)-, methyl ester 2,4-D Methyl ester	$C_9H_8Cl_2O_3$ 235.07	1928-38-7 119	 141^{18}	4-06-00-00909	
173	Acetic acid, (2,4-dichlorophenoxy)-, 1-methylethyl ester 2,4-D Isopropyl ester	$C_{11}H_{12}Cl_2O_3$ 263.12	94-11-1 5	 140^{1}	4-06-00-00910 1.26^{25}	 1.5209^{25}
174	Acetic acid, diethoxy-, ethyl ester	$C_8H_{16}O_4$ 176.21	6065-82-3	 199	4-03-00-01494 0.985^{25}	 1.4100^{20}
175	Acetic acid, difluoro- Difluoroacetic acid	$C_2H_2F_2O_2$ 96.03	381-73-7 -1	 133	4-02-00-00455 1.526^{25}	 1.3470^{20}
176	Acetic acid, difluoro-, ethyl ester Ethyl difluoroacetate	$C_4H_6F_2O_2$ 124.09	454-31-9	 100	4-02-00-00455 1.1893^{9}	H_2O 1
177	Acetic acid, 1,1-dimethylethyl ester *tert*-Butyl acetate	$C_6H_{12}O_2$ 116.16	540-88-5	1537 95.1	4-02-00-00151 0.8665^{20}	EtOH 3; eth 3; chl 3; HOAc 3 1.3855^{20}
178	Acetic acid, dodecyl ester	$C_{14}H_{28}O_2$ 220.00	112-66-3 1.3	 265; 150^{40}	4-02-00-00170 0.8652^{22}	 1.4400^{20}
179	Acetic acid ethenyl ester Vinyl acetate	$C_4H_6O_2$ 86.09	108-05-4 -93.2	9896 72.5	4-02-00-00176 0.9317^{20}	H_2O 1; EtOH 5; eth 3; ace 3 1.3959^{20}
180	Acetic acid, ethoxy- 2-Ethoxyacetic acid	$C_4H_8O_3$ 104.11	627-03-2	 206.5	4-03-00-00574 1.1021^{20}	H_2O 4; EtOH 4; eth 4; chl 3 1.4194^{20}
181	Acetic acid, ethoxy-, ethyl ester	$C_6H_{12}O_3$ 132.16	817-95-8	 158	4-03-00-00581 0.9702^{20}	EtOH 3; eth 3; ace 3 1.4039^{20}
182	Acetic acid, ethoxy-, methyl ester Methyl ethoxyacetate	$C_5H_{10}O_3$ 118.13	17640-26-5	 147	4-03-00-00578 1.0112^{15}	ace 4; eth 4; EtOH 4
183	Acetic acid, ethoxy-, 5-methyl-2-(1-methylethyl)cyclohexyl ester, (1α,2β,5α)-	$C_{14}H_{26}O_3$ 242.36	579-94-2	 155^{20}	2-06-00-00047 0.9545^{20}	eth 4; EtOH 4; chl 4
184	Acetic acid, (ethoxysulfonyl)-, ethyl ester Acetic acid, sulfo, diethyl ester	$C_6H_{12}O_5S$ 196.22	59376-54-4	 116$^{0.5}$	3-04-00-00058 1.232^{16}	 1.436^{21}
185	Acetic acid, 2-ethylbutyl ester 2-Ethylbutyl acetate	$C_8H_{16}O_2$ 144.21	10031-87-5 <-100	 162.5	4-02-00-00161 0.8790^{20}	H_2O 1; EtOH 3; eth 3; ctc 3 1.4109^{20}
186	Acetic acid ethyl ester Ethyl acetate	$C_4H_8O_2$ 88.11	141-78-6 -83.6	3713 77.11	4-02-00-00127 0.9003^{20}	H_2O 3; EtOH 5; eth 5; ace 4 1.3723^{20}

PHYSICAL CONSTANTS OF ORGANIC COMPOUNDS (continued)

No.	Name Synonym	Mol. Form. Mol. Wt.	CAS RN mp/°C	Merck No. bp/°C	Beil. Ref. den/g cm^{-3}	Solubility n_D
187	Acetic acid, 2-ethylhexyl ester 2-Ethylhexyl acetate	$C_{10}H_{20}O_2$ 172.27	103-09-3 -80	6683 199	4-02-00-00166 0.8718^{20}	H_2O 1; EtOH 3; eth 3 1.4204^{20}
188	Acetic acid, (ethylthio)-	$C_4H_8O_2S$ 120.17	627-04-3 -8.5	164^{83}; 109^{5}	4-03-00-00603 1.1497^{20}	H_2O 4; EtOH 4; eth 4
189	Acetic acid, fluoro- Fluoroethanoic acid	$C_2H_3FO_2$ 78.04	144-49-0 35.2	4096 168	4-02-00-00446 1.3693^{36}	H_2O 3; EtOH 3
190	Acetic acid, fluoro-, 1,1-dimethylethyl ester	$C_6H_{11}FO_2$ 134.15	406-74-6	132	4-02-00-00450 0.9904^{20}	1.386^{20}
191	Acetic acid, fluoro-, ethyl ester	$C_4H_7FO_2$ 106.10	459-72-3	120	4-02-00-00447 1.098^{25}	H_2O 4 1.3755^{20}
192	Acetic acid, fluoro-, methyl ester Methyl fluoroacetate	$C_3H_5FO_2$ 92.07	453-18-9 -35	104.5	4-02-00-00447 1.1613^{15}	
193	Acetic acid, heptyl ester Heptyl acetate	$C_9H_{18}O_2$ 158.24	112-06-1 -50.2	193	4-02-00-00162 0.8750^{15}	H_2O 1; EtOH 3; eth 3; ctc 3 1.4150^{20}
194	Acetic acid, hexyl ester	$C_8H_{16}O_2$ 144.21	142-92-7 -80.9	171.5	4-02-00-00159 0.8779^{15}	eth 4; EtOH 4 1.4092^{20}
195	Acetic acid, hydrazide	$C_2H_6N_2O$ 74.08	1068-57-1 67	137^{25}	4-02-00-00435	H_2O 3; EtOH 3; eth 2
196	Acetic acid, hydroxy- Glycolic acid	$C_2H_4O_3$ 76.05	79-14-1 79.5	4394 100	4-03-00-00571	H_2O 3; EtOH 3; eth 3
197	Acetic acid, hydroxy-, ethyl ester	$C_4H_8O_3$ 104.11	623-50-7	160	4-03-00-00580 1.0826^{23}	eth 4; EtOH 4 1.4180^{20}
198	Acetic acid, hydroxy-, methyl ester	$C_3H_6O_3$ 90.08	96-35-5	149; 52^{17}	4-03-00-00578 1.1677^{18}	H_2O 3; EtOH 5; eth 5
199	Acetic acid, hydroxy-, propyl ester Hydroxyacetic acid, propyl ester	$C_5H_{10}O_3$ 118.13	90357-58-7	170.5	4-03-00-00588 1.0631^{18}	1.4231^{18}
200	Acetic acid, iodo- Iodoacetic acid	$C_2H_3IO_2$ 185.95	64-69-7 82.5	4918 dec	4-02-00-00534	H_2O 3; EtOH 3; eth 2; chl 2
201	Acetic acid, iodo-, ethyl ester	$C_4H_7IO_2$ 214.00	623-48-3	179	4-02-00-00535 1.8173^{13}	EtOH 3; eth 3 1.5079^{13}
202	Acetic acid, mercapto- Thioglycolic acid	$C_2H_4O_2S$ 92.12	68-11-1 -16.5	9265 120^{20}	4-03-00-00600 1.3253^{20}	H_2O 5; EtOH 5; eth 5; chl 2 1.5080^{20}
203	Acetic acid, mercapto-, ethyl ester	$C_4H_8O_2S$ 120.17	623-51-8	157	4-03-00-00616 1.0964^{15}	EtOH 3; eth 3; ctc 2 1.4582^{20}
204	Acetic acid, mercapto-, methyl ester	$C_3H_6O_2S$ 106.15	2365-48-2	42-3^{10}	4-03-00-00614	eth 4; EtOH 4 1.4657^{20}
205	Acetic acid, methoxy-	$C_3H_6O_3$ 90.08	625-45-6	203.5	4-03-00-00574 1.1768^{20}	H_2O 3; EtOH 3; eth 3 1.4168^{20}
206	Acetic acid, methoxy-, ethyl ester Ethyl methoxyacetate	$C_5H_{10}O_3$ 118.13	3938-96-3	142	3-03-00-00381 1.0118^{15}	H_2O 2; EtOH 4; eth 4; ctc 3 1.4050^{20}
207	Acetic acid, methoxy-, methyl ester	$C_4H_8O_3$ 104.11	6290-49-9	131	4-03-00-00578 1.0511^{20}	H_2O 2; EtOH 4; eth 4; ace 3 1.3962^{20}
208	Acetic acid, methyl ester Methyl acetate	$C_3H_6O_2$ 74.08	79-20-9 -98	5932 56.87	4-02-00-00122 0.9342^{20}	H_2O 4; eth 4; EtOH 4 1.3614^{20}
209	Acetic acid, 1-methylethyl ester Isopropyl acetate	$C_5H_{10}O_2$ 102.13	108-21-4 -73.4	5093 88.6	4-02-00-00141 0.8718^{20}	H_2O 3; EtOH 3; eth 5; ace 3 1.3773^{20}
210	Acetic acid, [[5-methyl-2-(1-methylethyl)cyclohexyl]oxy]-, [1*R*-(1α,2β,5α)]- *l*-Methoxyacetic acid	$C_{12}H_{22}O_3$ 214.30	40248-63-3 54	171^{11}	4-06-00-00157	H_2O 2; EtOH 4; eth 3; ace 4
211	Acetic acid, 2-methylphenyl ester *o*-Cresyl acetate	$C_9H_{10}O_2$ 150.18	533-18-6	2588 208	4-06-00-01960 1.0533^{15}	eth 4; EtOH 4 1.5002^{20}
212	Acetic acid, 3-methylphenyl ester *m*-Cresyl acetate	$C_9H_{10}O_2$ 150.18	122-46-3 12	2587 212	4-06-00-02047 1.043^{20}	bz 4; eth 4; EtOH 4 1.4978^{20}
213	Acetic acid, 4-methylphenyl ester	$C_9H_{10}O_2$ 150.18	140-39-6	212.5	4-06-00-02112 1.0512^{17}	H_2O 2; EtOH 3; eth 3; ctc 2 1.5163^{22}
214	Acetic acid, 1-methylpropyl ester *sec*-Butyl acetate	$C_6H_{12}O_2$ 116.16	105-46-4 -98.9	1536 112	4-02-00-00148 0.8748^{20}	H_2O 1; EtOH 3; eth 3; ctc 2 1.3888^{20}
215	Acetic acid, 2-methylpropyl ester Isobutyl acetate	$C_6H_{12}O_2$ 116.16	110-19-0 -98.8	5014 116.5	4-02-00-00149 0.8712^{20}	H_2O 2; EtOH 5; eth 5; ace 3 1.3902^{20}
216	Acetic acid, (methylthio)-	$C_3H_6O_2S$ 106.15	2444-37-3	130-1^{27}	4-03-00-00603 1.221^{20}	1.495^{20}
217	Acetic acid, (2-naphthalenyloxy)- 2-Naphthoxyacetic acid	$C_{12}H_{10}O_3$ 202.21	120-23-0 156	6317	4-06-00-04274	H_2O 3; EtOH 3; eth 3; HOAc 3
218	Acetic acid, nitro-, ethyl ester Ethyl nitroacetate	$C_4H_7NO_4$ 133.10	626-35-7	106^{25}; 83^{6}	4-02-00-00537 1.1953^{20}	H_2O 2; EtOH 5; eth 4; dil alk 3 1.4250^{20}
219	Acetic acid, 2-nitrophenyl ester	$C_8H_7NO_4$ 181.15	610-69-5 40.5	253 dec; 141^{11}	4-06-00-01256	H_2O 3; EtOH 4; eth 4; ace 4
220	Acetic acid, nonyl ester Nonyl acetate	$C_{11}H_{22}O_2$ 186.29	143-13-5 -26	6597 210	4-02-00-00167 0.8785^{15}	1.426^{20}
221	Acetic acid, octadecyl ester	$C_{20}H_{40}O_2$ 312.54	822-23-1 34.5	208^{9}	4-02-00-00171 0.8510^{30}	EtOH 4
222	Acetic acid, octyl ester	$C_{10}H_{20}O_2$ 172.27	112-14-1 -38.5	210	4-02-00-00165 0.8705^{20}	H_2O 1; EtOH 3; eth 3; ctc 2 1.4150^{20}
223	Acetic acid, oxo- Glyoxylic acid	$C_2H_2O_3$ 74.04	298-12-4 98	4407	4-03-00-01489	H_2O 4; EtOH 2; eth 2; bz 2
224	Acetic acid, oxo(phenylamino)-, ethyl ester	$C_{10}H_{11}NO_3$ 193.20	1457-85-8 65.5	280	3-12-00-00550	ace 4; bz 4; eth 4; EtOH 4

PHYSICAL CONSTANTS OF ORGANIC COMPOUNDS (continued)

No.	Name	Synonym	Mol. Form.	Mol. Wt.	CAS RN	mp/°C	Merck No.	bp/°C	Beil. Ref.	den/g cm^{-3}	Solubility	n_D
225	Acetic acid, [(4-oxo-2-phenyl-4H-1-benzopyran-7-yl)oxy]-, ethyl ester	Efloxate	$C_{19}H_{16}O_5$	324.33	119-41-5	123.7	3490		4-18-00-00695		chl 3	
226	Acetic acid, 2,2'-oxybis-		$C_4H_6O_5$	134.09	110-99-6	148		dec	4-03-00-00577		H_2O 4; eth 4; EtOH 4	
227	Acetic acid, pentyl ester	Pentyl acetate	$C_7H_{14}O_2$	130.19	628-63-7	-70.8		149.2	4-02-00-00152	0.8756^{20}	H_2O 2; EtOH 5; eth 5; ctc 3	1.4023^{20}
228	Acetic acid, phenoxy-	Phenoxyacetic acid	$C_8H_8O_3$	152.15	122-59-8	98.5	7223	285 dec	4-06-00-00634		H_2O 3; EtOH 4; eth 4; bz 4	
229	Acetic acid, phenoxy-, ethyl ester	Ethyl phenoxyacetate	$C_{10}H_{12}O_3$	180.20	2555-49-9			$109-12^{3}$	4-06-00-00635	1.1082^{20}		1.5080^{20}
230	Acetic acid, phenoxy-, methyl ester	Methyl phenoxyacetate	$C_9H_{10}O_3$	166.18	2065-23-8			245	4-06-00-00635	1.1493^{20}	eth 4; EtOH 4	1.5155^{20}
231	Acetic acid, phenyl ester	Phenyl acetate	$C_8H_8O_2$	136.15	122-79-2		7238	196; 75^{8}	4-06-00-00611	1.0780^{20}	H_2O 2; EtOH 5; eth 5; ctc 3	1.5035^{20}
232	Acetic acid, 2-phenylethyl ester		$C_{10}H_{12}O_2$	164.20	103-45-7	-31.1		232.6	4-06-00-03073	1.0883^{20}	eth 4; EtOH 4	1.5171^{20}
233	Acetic acid, phenylmethyl ester	Benzyl acetate	$C_9H_{10}O_2$	150.18	140-11-4	-51.3	1137	213	4-06-00-02262	1.0550^{20}	H_2O 2; EtOH 5; eth 3; ace 3	1.5232^{20}
234	Acetic acid, potassium salt	Potassium acetate	$C_2H_3KO_2$	98.14	127-08-2	292	7580		4-02-00-00094	1.57^{25}	H_2O 4; EtOH 3; eth 1; ace 1	
235	Acetic acid, 2-propenyl ester	Allyl acetate	$C_5H_8O_2$	100.12	591-87-7			103.5	4-02-00-00180	0.9275^{20}	H_2O 2; EtOH 5; eth 5; ace 3	1.4049^{20}
236	Acetic acid, propyl ester	Propyl acetate	$C_5H_{10}O_2$	102.13	109-60-4	-93	7853	101.54	4-02-00-00138	0.8878^{20}	H_2O 2; EtOH 5; eth 5; ctc 3	1.3842^{20}
237	Acetic acid, sodium salt	Sodium acetate	$C_2H_3NaO_2$	82.03	127-09-3	324	8513		4-02-00-00094	1.528^{25}	H_2O 4; EtOH 2	1.464
238	Acetic acid, sulfo-	Sulfoacetic acid	$C_2H_4O_5S$	140.12	123-43-3	85	8932	245 dec	4-04-00-00102		H_2O 4; ace 4; EtOH 4	
239	Acetic acid, 2,2'-sulfonylbis-	Sulfonyldiacetic acid	$C_4H_6O_6S$	182.15	123-45-5	187	8940		3-03-00-00424		H_2O 4; EtOH 4; eth 3; sulf 3	
240	Acetic acid, 2,2'-thiobis-	Thiodiglycolic acid	$C_4H_6O_4S$	150.16	123-93-3	129	9260		4-03-00-00612		H_2O 2; EtOH 4; bz 3	
241	Acetic acid, tribromo-	Tribromoacetic acid	$C_2HBr_3O_2$	296.74	75-96-7	132	9521	245 dec	4-02-00-00534		H_2O 3; EtOH 3; eth 3	
242	Acetic acid, tribromo-, ethyl ester	Ethyl tribromoacetate	$C_4H_5Br_3O_2$	324.79	599-99-5			225	3-02-00-00485	2.2260^{20}	EtOH 3; eth 3	1.5438^{13}
243	Acetic acid, trichloro-	Trichloroacetic acid	$C_2HCl_3O_2$	163.39	76-03-9	57.5	9539	196.5	4-02-00-00508	1.6126^{64}	H_2O 4; EtOH 3; eth 3; ctc 2	1.4603^{61}
244	Acetic acid, trichloro-, anhydride	Trichloroacetic anhydride	$C_4Cl_6O_3$	308.76	4124-31-6			223 dec	4-02-00-00518	1.6908^{20}	eth 4; HOAc 4	
245	Acetic acid, trichloro-, butyl ester	Butyl trichloroacetate	$C_6H_9Cl_3O_2$	219.49	3657-07-6			204	4-02-00-00515	1.2778^{20}	ctc 3	1.4525^{25}
246	Acetic acid, trichloro-, 2-chloroethyl ester		$C_4H_4Cl_4O_2$	225.89	4974-21-4			217	4-02-00-00515	1.5357^{20}	eth 4	1.4813^{20}
247	Acetic acid, trichloro-, 1,1-dimethylethyl ester		$C_6H_9Cl_3O_2$	219.49	1860-21-5	25.5		$54-5^{7}$	4-02-00-00516	1.2363^{25}	eth 4; EtOH 4	1.4398^{25}
248	Acetic acid, trichloro-, ethyl ester	Ethyl trichloroacetate	$C_4H_5Cl_3O_2$	191.44	515-84-4		9539	167.5	4-02-00-00514	1.3836^{20}	H_2O 1; EtOH 3; eth 3; bz 3	1.4505^{20}
249	Acetic acid, trichloro-, 2-hydroxyethyl ester		$C_4H_5Cl_3O_3$	207.44	33560-17-7			$130-4^{12}$	3-02-00-00474	1.532^{20}	EtOH 4	1.4775^{20}
250	Acetic acid, trichloro-, 2-methoxyethyl ester	2-Methoxyethyl trichloroacetate	$C_5H_7Cl_3O_3$	221.47	35449-34-4	14.5		$98-9^{17}$	3-02-00-00474	1.3826^{20}	bz 4; eth 4; EtOH 4	1.4563^{20}
251	Acetic acid, trichloro-, 1-methylpropyl ester		$C_6H_9Cl_3O_2$	219.49	4484-80-4			$93-4^{24}$	3-02-00-00472	1.2636^{20}	bz 4; eth 4; EtOH 4	1.4483^{20}
252	Acetic acid, trichloro-, 2-methylpropyl ester	Isobutyl trichloroacetate	$C_6H_9Cl_3O_2$	219.49	33560-15-5			188	4-02-00-00515	1.2636^{20}	bz 4; eth 4; EtOH 4	1.4483^{20}
253	Acetic acid, trichloro-, methyl ester	Methyl trichloroacetate	$C_3H_3Cl_3O_2$	177.41	598-99-2	-17.5		153.8	4-02-00-00513	1.4874^{20}	H_2O 1; EtOH 4; eth 4; ctc 3	1.4572^{20}
254	Acetic acid, trichloro-, 1-methylethyl ester	Isopropyl trichloroacetate	$C_5H_7Cl_3O_2$	205.47	3974-99-0			175; 66^{15}	4-02-00-00515	1.3034^{20}	bz 4; eth 4; EtOH 4	1.4428^{20}
255	Acetic acid, trichloro-, 3-methylbutyl ester		$C_7H_{11}Cl_3O_2$	233.52	57392-55-9			217	4-02-00-00516	1.2314^{20}	eth 4; EtOH 4	1.4521^{20}
256	Acetic acid, (2,4,5-trichlorophenoxy)-	2,4,5-T	$C_8H_5Cl_3O_3$	255.48	93-76-5	153	8999		4-06-00-00973		H_2O 1; EtOH 3; bz 4	
257	Acetic acid, (2,4,5-trichlorophenoxy)-, 2-butoxyethyl ester	2,4,5-T Butoxyethyl ester	$C_{14}H_{17}Cl_3O_4$	355.64	2545-59-7			$163-6^{1}$	4-06-00-00976	1.280^{20}	ctc 3	
258	Acetic acid, (2,4,5-trichlorophenoxy)-, butyl ester	2,4,5-T Butyl ester	$C_{12}H_{13}Cl_3O_3$	311.59	93-79-8	28.5			4-06-00-00975			
259	Acetic acid, (2,4,5-trichlorophenoxy)-, 1-methylethyl ester	2,4,5-T Isopropyl ester	$C_{11}H_{11}Cl_3O_3$	297.56	93-78-7	45			4-06-00-00974		chl 4	
260	Acetic acid, (2,4,5-trichlorophenoxy)-, 1-methylpropyl ester	2,4,5-T *sec*-Butyl ester	$C_{12}H_{13}Cl_3O_3$	311.59	61792-07-2	39			4-06-00-00975			
261	Acetic acid, trichloro-, propyl ester		$C_5H_7Cl_3O_2$	205.47	13313-91-2			187	4-02-00-00515	1.3221^{20}	eth 4; EtOH 4	1.4501^{20}
262	Acetic acid, trichloro-, trichloromethyl ester		$C_3Cl_6O_2$	280.75	6135-29-1	34		191.5	2-03-00-00015	1.6733^{35}	bz 4; eth 4; chl 4; lig 4	

PHYSICAL CONSTANTS OF ORGANIC COMPOUNDS (continued)

No.	Name Synonym	Mol. Form. Mol. Wt.	CAS RN mp/°C	Merck No. bp/°C	Beil. Ref. den/g cm^{-3}	Solubility n_D
263	Acetic acid, trifluoro- Trifluoroacetic acid	$C_2HF_3O_2$ 114.02	76-05-1 -15.2	9595 73	4-02-00-00458 1.5351^{25}	H_2O 3; EtOH 3; eth 3; ace 3
264	Acetic acid, trifluoro-, anhydride	$C_4F_6O_3$ 210.03	407-25-0 -65	 39.5	4-02-00-00469 1.490^{25}	 1.269^{25}
265	Acetic acid, trifluoro-, butyl ester Butyl trifluoroacetate	$C_6H_9F_3O_2$ 170.13	367-64-6	 102	4-02-00-00464 1.0268^{22}	chl 3 1.353^{22}
266	Acetic acid, trifluoro-, 1,1-dimethylethyl ester	$C_6H_9F_3O_2$ 170.13	400-52-2	 83	4-02-00-00465	chl 3 1.3300^{25}
267	Acetic acid, trifluoro-, ethyl ester	$C_4H_5F_3O_2$ 142.08	383-63-1	 61	4-02-00-00463 1.194^{20}	 1.308^{20}
268	Acetic acid, trifluoro-, 2-methylpentyl ester 2-Methylpentyl trifluoroacetate	$C_8H_{13}F_3O_2$ 198.19	10042-29-2	 140	4-02-00-00466 1.0504^{20}	chl 2
269	Acetic acid, trifluoro-, propyl ester Propyl trifluoroacetate	$C_5H_7F_3O_2$ 156.10	383-66-4	 82.5	4-02-00-00464 1.1285^{25}	chl 3 1.3233^{22}
270	Acetochlor Acetamide, 2-chloro-*N*-(ethoxymethyl)-*N*-(2-ethyl-6-methylphenyl)-	$C_{14}H_{20}ClNO_2$ 269.77	34256-82-1			
271	Acetonitrile Methyl cyanide	C_2H_3N 41.05	75-05-8 -43.8	62 81.6	4-02-00-00419 0.7857^{20}	H_2O 5; EtOH 5; eth 5; ace 5 1.3442^{30}
272	Acetonitrile, amino- Aminoacetonitrile	$C_2H_4N_2$ 56.07	540-61-4	423 58^{15}	4-04-00-02363	EtOH 4
273	Acetonitrile, (butylamino)-	$C_6H_{12}N_2$ 112.17	3010-04-6	 85^{9}	4-04-00-02382 0.891^{25}	 1.4337^{20}
274	Acetonitrile, chloro- Chloromethyl cyanide	C_2H_2ClN 75.50	107-14-2	 126.5	4-02-00-00492 1.1930^{20}	eth 4; EtOH 4 1.4202^{25}
275	Acetonitrile, (cyclohexylamino)-	$C_8H_{14}N_2$ 138.21	1074-58-4 18	 76^{1}	4-12-00-00073 0.9657^{25}	
276	Acetonitrile, cyclohexylidene- Cyclohexylideneacetonitrile	$C_8H_{11}N$ 121.18	4435-18-1	 $107\text{-}8^{22}$	4-09-00-00124 0.9483^{15}	eth 4; EtOH 4 1.4382^{25}
277	Acetonitrile, dibromo- Dibromoacetonitrile	C_2HBr_2N 198.84	3252-43-5	 $67\text{-}9^{24}$	4-02-00-00533 2.296^{25}	 1.5393^{20}
278	Acetonitrile, dichloro- Dichloroacetonitrile	C_2HCl_2N 109.94	3018-12-0	 112.5	4-02-00-00506 1.369^{20}	MeOH 3 1.4391^{25}
279	Acetonitrile, dichlorofluoro- Dichlorofluoroacetonitrile	C_2Cl_2FN 127.93	353-82-2 -110	 33	4-02-00-00508 1.3909^{25}	 1.3682^{20}
280	Acetonitrile, (diethylamino)-	$C_6H_{12}N_2$ 112.17	3010-02-4	 170	4-04-00-02378 0.8660^{20}	H_2O 3 1.4260^{20}
281	Acetonitrile, (dimethylamino)- Glycinonitrile, *N,N*-dimethyl-	$C_4H_8N_2$ 84.12	926-64-7	 137.5	4-04-00-02368 0.8649^{20}	H_2O 4; EtOH 4 1.4095^{20}
282	Acetonitrile, hydroxy- Glyconitrile	C_2H_3NO 57.05	107-16-4 <-72	 183 dec; 119^{24}	4-03-00-00598	H_2O 4; EtOH 4; eth 4; bz 1 1.4117^{19}
283	Acetonitrile, iodo- Iodoacetonitrile	C_2H_2IN 166.95	624-75-9	 185	2-02-00-00206 2.307^{25}	 1.5744^{20}
284	Acetonitrile, methoxy-	C_3H_5NO 71.08	1738-36-9	 119	3-03-00-00399 0.9492^{20}	H_2O 2; EtOH 3; eth 3; ace 3 1.3831^{20}
285	Acetonitrile, (methyleneamino)- Methyleneaminoacetonitrile	$C_3H_4N_2$ 68.08	109-82-0 129	5976		
286	Acetonitrile, (methylthio)- Cyanomethyl methyl sulfide	C_3H_5NS 87.15	35120-10-6	 $61\text{-}3^{15}$	4-03-00-00630 1.039^{25}	 1.4826^{20}
287	Acetonitrile, phenoxy- Phenoxyacetonitrile	C_8H_7NO 133.15	3598-14-9	 239.5	4-06-00-00640 1.0991^{20}	eth 4; EtOH 4 1.5246^{20}
288	Acetonitrile, trichloro- Trichloroacetonitrile	C_2Cl_3N 144.39	545-06-2 -42	9540 85.7	4-02-00-00524 1.4403^{25}	H_2O 1 1.4409^{20}
289	Acetonitrile, trifluoro- Trifluoroacetonitrile	C_2F_3N 95.02	353-85-5	 -68.8	4-02-00-00472	
290	Acetonitrile, trinitro- Trinitroacetonitrile	$C_2N_4O_6$ 176.05	630-72-8 41.5	 exp 220	0-02-00-00229	eth 4
291	Acetyl bromide Acetic acid, bromide	C_2H_3BrO 122.95	506-96-7 -96	76 76	4-02-00-00398 1.6625^{16}	eth 5; ace 3; bz 5; chl 5 1.4486^{20}
292	Acetyl bromide, bromo- Bromoacetyl bromide	$C_2H_2Br_2O$ 201.85	598-21-0	 148.5	4-02-00-00530 2.312^{22}	ace 3; ctc 3 1.5449^{20}
293	Acetyl bromide, trichloro- Trichloroacetyl bromide	C_2BrCl_3O 226.28	34069-94-8	 143	3-02-00-00476 1.898^{15}	bz 4; eth 4
294	Acetyl chloride Ethanoyl chloride	C_2H_3ClO 78.50	75-36-5 -112.8	79 50.7	4-02-00-00395 1.1051^{20}	eth 5; ace 5; bz 5; ctc 3 1.3886^{20}
295	Acetyl chloride, chloro- Chloroacetyl chloride	$C_2H_2Cl_2O$ 112.94	79-04-9 -22	2054 106	4-02-00-00488 1.4202^{20}	eth 5; ace 3; ctc 3 1.4530^{20}
296	Acetyl chloride, chlorofluoro- Chlorofluoroacetyl chloride	C_2HCl_2FO 130.93	359-32-0	 69.5	3-02-00-00453 1.468^{25}	 1.3992^{25}
297	Acetyl chloride, dichloro- Dichloroacetyl chloride	C_2HCl_3O 147.39	79-36-7	3040 108	4-02-00-00504 1.5315^{16}	H_2O 6; EtOH 6; eth 5 1.4591^{20}
298	Acetyl chloride, ethoxy-	$C_4H_7ClO_2$ 122.55	14077-58-8	 125	3-03-00-00396 1.1170^{25}	ace 4; eth 4 1.4204^{20}
299	Acetyl chloride, methoxy-	$C_3H_5ClO_2$ 108.52	38870-89-2	 112.5	3-03-00-00396 1.1871^{20}	eth 3; ace 3; ctc 3; chl 4 1.4199^{20}
300	Acetyl chloride, trichloro- Trichloroacetyl chloride	C_2Cl_4O 181.83	76-02-8	 117.9	4-02-00-00519 1.6202^{20}	eth 5 1.4695^{20}
301	Acetyl fluoride Acetic acid, fluoride	C_2H_3FO 62.04	557-99-3 -84	 20.8	4-02-00-00393 1.032^{25}	EtOH 5; eth 5; bz 3; chl 3

No.	Name Synonym	Mol. Form. Mol. Wt.	CAS RN mp/°C	Merck No. bp/°C	Beil. Ref. den/g cm^{-3}	Solubility n_D
302	Acetyl iodide	C_2H_3IO 169.95	507-02-8	88 108	4-02-00-00399 2.0673^{20}	eth 4 1.5491^{20}
303	Acetyl isothiocyanate Isothiocyanic acid, acetyl	C_3H_3NOS 101.13	13250-46-9	 132.5	3-03-00-00278 1.1523^{13}	eth 3; CS_2 3 1.5231^{18}
304	Acifluorfen Benzoic acid, 5-[2-chloro-4-(trifluoromethyl)phenoxy]-2-nitro-	$C_{14}H_7ClF_3NO_5$ 361.66	50594-66-6 150	105		
305	Aconitane-3,8,13,14,15-pentol, 20-ethyl-1,6,16-trimethoxy-4-(methoxymethyl)-, Aconine	$C_{25}H_{41}NO_9$ 499.60	509-20-6 132	110	5-21-06-00310	H_2O 3; EtOH 3; eth 2; chl 3
306	Aconitane-3,8,13,14,15-pentol, 20-ethyl-1,6,16-trimethoxy-4-(methoxymethyl)-, 8-acetate 14-benzoate, Aconitine	$C_{34}H_{47}NO_{11}$ 645.75	302-27-2 204	113	5-21-06-00310	bz 4; EtOH 4; chl 4
307	Aconitane-3,8,13,14-tetrol, 20-ethyl-1,6,16-trimethoxy-4-(methoxymethyl)-, 8-acetate 14-benzoate, (1α,3alp Indaconitine	$C_{34}H_{47}NO_{10}$ 629.75	4491-19-4 202-3 dec	4842	5-21-06-00308	eth 4; EtOH 4; chl 4
308	Aconitane-3,8,13,14-tetrol, 20-ethyl-1,6,16-trimethoxy-4-(methoxymethyl)-, 8-acetate 14-(3,4-dimethoxybenzoate),(1α,3α,6α,14α,16β)- Pseudoaconitine	$C_{36}H_{51}NO_{12}$ 689.80	127-29-7 214	7924	5-21-06-00309	eth 4; EtOH 4
309	Aconitane-4,8,9-triol, 20-ethyl-1,14,16-trimethoxy-, 4-[2-(acetylamino)benzoate], (1α,14α,16β)- Lappaconitine	$C_{32}H_{44}N_2O_8$ 584.71	32854-75-4 217.5	5237	5-21-06-00291	H_2O 1; EtOH 2; eth 2; bz 3
310	Aconitane-8,13,14-triol, 20-ethyl-1,6,16-trimethoxy-4-(methoxymethyl)-, 8-acetate 14-(3,4-dimethoxybenzoate), Bikhaconitine	$C_{36}H_{51}NO_{11}$ 673.80	6078-26-8 164	1234	5-21-06-00303	eth 4; EtOH 4; chl 4
311	Aconitane-8,13,14-triol, 1,6,16-trimethoxy-4-(methoxymethyl)-20-methyl-, 8-acetate 14-benzoate, Delphinine	$C_{33}H_{45}NO_9$ 599.72	561-07-9 199	2867	5-21-06-00302	H_2O 1; chl 3; ace 3; eth 3
312	4-Acridinamine	$C_{13}H_{10}N_2$ 194.24	578-07-4 108	 183.5	5-22-11-00008	ace 4; bz 4; eth 4; EtOH 4
313	9-Acridinamine Aminacrine	$C_{13}H_{10}N_2$ 194.24	90-45-9 241	418	5-22-11-00008	EtOH 3; ace 3; dil HCl 4
314	Acridine Dibenzo[b,e]pyridine	$C_{13}H_9N$ 179.22	260-94-6 106(form a); 110(form b)	117 344.86	5-20-08-00199 1.005^{20}	H_2O 2; EtOH 4; eth 4; bz 4
315	Acridine, 9-chloro- 9-Chloroacridine	$C_{13}H_8ClN$ 213.67	1207-69-8 121	 sub	5-20-08-00210	H_2O 4; EtOH 4
316	3,6-Acridinediamine Proflavine	$C_{13}H_{11}N_3$ 209.25	92-62-6 285	7780	5-22-11-00322	H_2O 3; EtOH 4; eth 2; bz 2
317	3,9-Acridinediamine, 7-ethoxy- Ethacridine	$C_{15}H_{15}N_3O$ 253.30	442-16-0 226	3668	5-22-12-00243	
318	Acridine, 9-phenyl- 9-Phenylacridine	$C_{19}H_{13}N$ 255.32	602-56-2 184	 404	5-20-08-00564	H_2O 1; EtOH 2; eth 3; bz 4
319	Adenosine β-*D*-Ribofuranoside, adenine-9	$C_{10}H_{13}N_5O_4$ 267.24	58-61-7 235.5	143	4-26-00-03598	H_2O 2; EtOH 1
320	5'-Adenylic acid Adenosine 5'-monophosphate	$C_{10}H_{14}N_5O_7P$ 347.22	61-19-8 195 dec	148	4-26-00-03615	H_2O 4; EtOH 3
321	Ajmalan-17,21-diol, (17*R*,21α)- Ajmaline	$C_{20}H_{26}N_2O_2$ 326.44	4360-12-7 206	184	4-23-00-03212	H_2O 1; EtOH 3; eth 2; bz 2
322	Alachlor Acetamide, 2-chloro-*N*-(2,6-diethylphenyl)-*N*-(methoxymethyl)-	$C_{14}H_{20}ClNO_2$ 269.77	15972-60-8 40	193 $100^{0.02}$	 1.133^{25}	
323	β-Alanine 3-Aminopropanoic acid	$C_3H_7NO_2$ 89.09	107-95-9 200 dec	196	4-04-00-02526 1.437^{19}	H_2O 3; EtOH 2; eth 1; ace 1
324	*DL*-Alanine *DL*-2-Aminopropanoic acid	$C_3H_7NO_2$ 89.09	302-72-7 300 dec	195 sub 250	4-04-00-02481 1.424^{25}	H_2O 3; EtOH 4
325	*L*-Alanine (*S*)-2-Aminopropanoic acid	$C_3H_7NO_2$ 89.09	56-41-7 297 dec	195 sub 250	4-04-00-02480 1.432^{22}	H_2O 3; EtOH 2; eth 1; ace 1
326	Alanine, 3-amino- 2,3-Diaminopropionic acid	$C_3H_8N_2O_2$ 104.11	515-94-6 110	2962	4-04-00-02581	H_2O 4
327	*DL*-Alanine, *N*-benzoyl-	$C_{10}H_{11}NO_3$ 193.20	1205-02-3 165.5	 dec	4-09-00-00794	H_2O 3; EtOH 3; eth 2; DMSO 2
328	β-Alanine, *N*,*N*-diethyl-, ethyl ester Ethyl 3-(diethylamino)propionate	$C_9H_{19}NO_2$ 173.26	5515-83-3	 $83\text{-}4^{12}$	4-04-00-02538 0.881^{25}	 1.4253^{20}
329	β-Alanine, *N*-(2,4-dihydroxy-3,3-dimethyl-1-oxobutyl)-, (*R*)- Pantothenic acid	$C_9H_{17}NO_5$ 219.24	79-83-4	6964	4-04-00-02569	H_2O 4; bz 4; eth 4
330	β-Alanine, *N*-(3-ethoxy-3-oxopropyl)-*N*-methyl-, ethyl ester Propionic acid, 3,3'-iminodi-*N*-methyl, diethyl ester	$C_{11}H_{21}NO_4$ 231.29	6315-60-2	 $136\text{-}8^4$	3-04-00-01278 1.0172^{20}	eth 4; EtOH 4 1.4421^{20}
331	Alanine, 2-methyl- α-Aminoisobutyric acid	$C_4H_9NO_2$ 103.12	62-57-7 335	457 sub 280	4-04-00-02616	H_2O 4; EtOH 2; eth 1

PHYSICAL CONSTANTS OF ORGANIC COMPOUNDS (continued)

No.	Name Synonym	Mol. Form. Mol. Wt.	CAS RN mp/°C	Merck No. bp/°C	Beil. Ref. den/g cm^{-3}	Solubility n_D
332	β-Alanine, *N*-methyl-, ethyl ester	$C_6H_{13}NO_2$ 131.17	2213-08-3	80^{21}	3-04-00-01264 1.0064^{20}	eth 4; EtOH 4 1.4443^{20}
333	*L*-Alanine, 3-(2-propenylsulfinyl)-, (*S*)- Alliin	$C_6H_{11}NO_3S$ 177.22	556-27-4 165	251	4-04-00-03147	H_2O 4
334	Aldicarb Propanal, 2-methyl-2-(methylthio)-, *O*-[(methylamino)carbonyl]oxime	$C_7H_{14}N_2O_2S$ 190.27	116-06-3 99	216	1.195^{25}	
335	Aldoxycarb Propanal, 2-methyl-2-(methylsulfonyl)-, *O*-[(methylamino)carbonyl]oxime	$C_7H_{14}N_2O_4S$ 222.27	1646-88-4			
336	Allethrin 2-Methyl-4-oxo-3-(2-propenyl)-2-cyclopenten-1-yl 2,2-dimethyl-3-(2-methyl-1-propenyl) cyclopropane carboxylate	$C_{19}H_{26}O_3$ 302.41	584-79-2		1.010^{20}	
337	*D*-Allose	$C_6H_{12}O_6$ 180.16	2595-97-3 128	279	4-01-00-04299	H_2O 4
338	*D*-Altrose	$C_6H_{12}O_6$ 180.16	1990-29-0 103.5	319	4-01-00-04300	H_2O 4
339	Aluminum, trimethyl Trimethyl aluminum	C_3H_9Al 72.09	75-24-1 15.4	130; 20^{8}	0.752^{20}	
340	Aluminum, tris(2-hydroxypropanoato-*O*1,*O*2)- Aluminum lactate	$C_9H_{15}AlO_9$ 294.19	18917-91-4	350		H_2O 3
341	Ametryn *N*-Ethyl-*N'*-(1-methylethyl)-6-(methylthio)1,3,5-triazine-2,4-diamine	$C_9H_{17}N_5S$ 227.33	834-12-8 88	402		
342	Amitraz *N*-Methylbis(2,4-xylyliminomethyl)amine	$C_{19}H_{23}N_3$ 293.41	33089-61-1 86	503	1.128^{20}	
343	Androsta-1,4-dien-3-one, 17-hydroxy-, (17β)- Boldenone	$C_{19}H_{26}O_2$ 286.41	846-48-0 165	1327		
344	Androsta-1,4-dien-3-one, 17-hydroxy-17-methyl-, (17β)- Methandrostenolone	$C_{20}H_{28}O_2$ 300.44	72-63-9 166	5862	4-08-00-01119	
345	Androstane	$C_{19}H_{32}$ 260.46	24887-75-0 50	668 $60^{0.003}$	4-05-00-01211	ace 4; eth 4; EtOH 4; peth 4
346	Androstane-17-carboxylic acid, (5β,17β)- Etiocholanic acid	$C_{20}H_{32}O_2$ 304.47	438-08-4 228.5	3824 sub 160	3-09-00-02644	
347	Androstan-17-one, 3-hydroxy-, (3β,5α)- Epiandrosterone	$C_{19}H_{30}O_2$ 290.45	481-29-8 178	3562	4-08-00-00641	
348	Androstan-17-one, 3-hydroxy-, (3α,5α)- Androsterone	$C_{19}H_{30}O_2$ 290.45	53-41-8 185	673	4-08-00-00642	H_2O 2; EtOH 3; eth 3; ace 3
349	Androstan-3-one, 17-hydroxy-, (5α,17β)- Stanolone	$C_{19}H_{30}O_2$ 290.45	521-18-6 181	8753 sub 135	4-08-00-00634	
350	Androstan-3-one, 17-hydroxy-17-methyl-, (5α,17β)- Mestanolone	$C_{20}H_{32}O_2$ 304.47	521-11-9 192.5	5816	4-08-00-00656	AcOEt 2
351	Androstan-3-one, 2-methyl-17-(1-oxopropoxy)-, (2α,5α,17β)- Dromostanolone propionate	$C_{23}H_{36}O_3$ 360.54	521-12-0 128	3436	4-08-00-00654	
352	Androst-4-ene-3,17-dione 4-Androstene-3,17-dione	$C_{19}H_{26}O_2$ 286.41	63-05-8 143(form a); 173(form b)	671	4-07-00-02381	
353	Androst-4-ene-3,11,17-trione Adrenosterone	$C_{19}H_{24}O_3$ 300.40	382-45-6 222	165 sub	4-07-00-02796	H_2O 2; EtOH 3; eth 3; ace 3
354	Androst-4-en-3-one, 4-chloro-17-hydroxy-, (17β)- Clostebol	$C_{19}H_{27}ClO_2$ 322.87	1093-58-9 189	2409	4-08-00-00989	
355	Androst-4-en-3-one, 9-fluoro-11,17-dihydroxy-17-methyl-, (11β,17β)- Fluoxymesterone	$C_{20}H_{29}FO_3$ 336.45	76-43-7 270	4113	4-08-00-02057	
356	Androst-4-en-3-one, 17-hydroxy-, (17β)- Testosterone	$C_{19}H_{28}O_2$ 288.43	58-22-0 155	9109	4-08-00-00974	H_2O 1; EtOH 3; eth 3; ace 3
357	Androst-4-en-3-one, 17-hydroxy-7,17-dimethyl-, (7β,17β)- Calusterone	$C_{28}H_{48}O$ 400.69	17021-26-0 157.5	1730	4-08-00-01029	
358	Androst-4-en-3-one, 17-hydroxy-17-methyl-, (17β)- 17-Methyltestosterone	$C_{20}H_{30}O_2$ 302.46	58-18-4 163.5	6044	4-08-00-01010	eth 4; EtOH 4
359	Androst-4-en-3-one, 17-(1-oxopropoxy)-, (17β)- Testosterone-17-propionate	$C_{22}H_{32}O_3$ 344.49	57-85-2 120	9115	3-08-00-00897	eth 4; py 4; EtOH 4
360	Anilazine 2,4-Dichloro-6-(o-chloroanilino)-s-triazine	$C_9H_5Cl_3N_4$ 275.52	101-05-3 160	685	1.8^{20}	
361	Aniline, 3,3'-azobis[*N*,*N*-dimethyl- Azobenzene, 3,3'-bis(dimethylamino)	$C_{16}H_{20}N_4$ 268.36	21232-53-1 118	sub	0-16-00-00305	bz 4; EtOH 4; chl 4
362	Aniline, *p*-isopentyl-	$C_{11}H_{17}N$ 163.26	104177-72-2	257	4-12-00-02858 0.928^{15}	1.5305^{20}
363	Anisole, 3-isopropyl-2-methyl- *o*-Cresol, 3-isopropyl, methyl ether	$C_{11}H_{16}O$ 164.25	31202-12-7 -0.5	67^{3}	4-06-00-03328 0.9540^{25}	1.5148^{20}
364	Anisole, 2,3,5-trinitro- 2,3,5-Trinitroanisole	$C_7H_5N_3O_7$ 243.13	7539-25-5 107.4		2-06-00-00253 1.618^{15}	ace 4; bz 4; EtOH 4

No.	Name Synonym	Mol. Form. Mol. Wt.	CAS RN mp/°C	Merck No. bp/°C	Beil. Ref. den/g cm^{-3}	Solubility n_D
365	2-Anthracenamine	$C_{14}H_{11}N$ 193.25	613-13-8 238.8	 sub	4-12-00-03435	H_2O 1; EtOH 3; os 2; con sulf 1
366	Anthracene	$C_{14}H_{10}$ 178.23	120-12-7 215.0	712 339.9	4-05-00-02281 1.28^{25}	H_2O 1; EtOH 2; eth 2; ace 2
367	Anthracene, 9,10-bis(phenylmethyl)- 9,10-Dibenzylanthracene	$C_{28}H_{22}$ 358.48	3613-42-1 246.5		3-05-00-02640 1.1787^{16}	H_2O 1; eth 2; bz 2; lig 1
368	1-Anthracenecarboxylic acid 1-Anthroic acid	$C_{15}H_{10}O_2$ 222.24	607-42-1 251.5	 sub	4-09-00-02670	H_2O 1; EtOH 3; eth 3; bz 2
369	2-Anthracenecarboxylic acid 2-Anthroic acid	$C_{15}H_{10}O_2$ 222.24	613-08-1 281	 sub	4-09-00-02671	HOAc 4
370	9-Anthracenecarboxylic acid 9-Anthroic acid	$C_{15}H_{10}O_2$ 222.24	723-62-6 217 dec	 sub	4-09-00-02671	H_2O 1; EtOH 3
371	2-Anthracenecarboxylic acid, 9,10-dihydro-4,5-dihydroxy-9,10-dioxo- Rhein	$C_{15}H_8O_6$ 284.22	478-43-3 321	8175 sub	4-10-00-04088	H_2O 2; EtOH 2; eth 2; ace 2
372	2-Anthracenecarboxylic acid, 9,10-dihydro-9,10-dioxo-	$C_{15}H_8O_4$ 252.23	117-78-2 291	 sub	4-10-00-03200	EtOH 2; eth 1; ace 3; bz 1
373	2-Anthracenecarboxylic acid, 7-β-*D*-glucopyranosyl-9,10-dihydro-3,5,6,8-tetrahydroxy-1-methyl-9,10-dioxo- Carminic acid	$C_{22}H_{20}O_{13}$ 492.39	1260-17-9 136 dec	1850	3-10-00-04874	H_2O 3; EtOH 3; eth 2; bz 1
374	Anthracene, 1-chloro- 1-Chloroanthracene	$C_{14}H_9Cl$ 212.68	4985-70-0 83.5		4-05-00-02292 1.1707^{100}	H_2O 1; EtOH 3; eth 3; bz 3 1.6959^{100}
375	Anthracene, 9,10-dibromo- 9,10-Dibromoanthracene	$C_{14}H_8Br_2$ 336.03	523-27-3 226	3000 sub	4-05-00-02295	H_2O 1; EtOH 2; eth 2; bz 2
376	Anthracene, 9,10-dihydro- 9,10-Dihydroanthracene	$C_{14}H_{12}$ 180.25	613-31-0 111	 305	4-05-00-02182 1.215^{20}	H_2O 1; EtOH 3; eth 3; bz 3
377	Anthracene, 1,3-dimethyl- 1,3-Dimethylanthracene	$C_{16}H_{14}$ 206.29	610-46-8 83	 143^{2}	4-05-00-02327	eth 4; EtOH 4
378	Anthracene, 2,10-dimethyl- Anthracene, 9,3-dimethyl	$C_{16}H_{14}$ 206.29	27532-76-9 84	 $140\text{-}5^{2}$	4-05-00-02328	H_2O 1; EtOH 4; eth 4
379	9,10-Anthracenedione Anthraquinone	$C_{14}H_8O_2$ 208.22	84-65-1 286	717 377	4-07-00-02556 1.438^{20}	H_2O 1; EtOH 2; eth 2; bz 2
380	9,10-Anthracenedione, 1-amino- 1-Aminoanthraquinone	$C_{14}H_9NO_2$ 223.23	82-45-1 253.5	428 sub	4-14-00-00429	ace 4; bz 4; EtOH 4; chl 4
381	9,10-Anthracenedione, 2-amino- 2-Aminoanthraquinone	$C_{14}H_9NO_2$ 223.23	117-79-3 304.5	 sub	4-14-00-00447	H_2O 1; EtOH 2; eth 1; ace 3
382	9,10-Anthracenedione, 3-amino-1,2-dihydroxy-	$C_{14}H_9NO_4$ 255.23	3963-78-8 >300	 sub	2-14-00-00185	aq NH_3 4
383	9,10-Anthracenedione, 2-amino-1-hydroxy- Anthraquinone, 2-amino-1-hydroxy	$C_{14}H_9NO_3$ 239.23	568-99-0 258	 sub	4-14-00-00891	bz 4; eth 4; EtOH 4
384	9,10-Anthracenedione, 1-amino-2-methyl- C.I. Solvent Orange 35	$C_{15}H_{11}NO_2$ 237.26	82-28-0 205.5		4-14-00-00496	H_2O 1; EtOH 3; eth 2; bz 3
385	9,10-Anthracenedione, 1-bromo-	$C_{14}H_7BrO_2$ 287.11	632-83-7 186.5	 sub	3-07-00-04076	bz 4; EtOH 4
386	9,10-Anthracenedione, 2-bromo-	$C_{14}H_7BrO_2$ 287.11	572-83-8 205.8	 sub	3-07-00-04076	bz 4
387	9,10-Anthracenedione, 1-chloro-	$C_{14}H_7ClO_2$ 242.66	82-44-0 163	 sub	4-07-00-02559	H_2O 1; EtOH 2; eth 5; bz 3
388	9,10-Anthracenedione, 2-chloro-	$C_{14}H_7ClO_2$ 242.66	131-09-9 211	 sub	4-07-00-02559	H_2O 1; EtOH 2; eth 1; bz 2
389	9,10-Anthracenedione, 1,5-diamino-	$C_{14}H_{10}N_2O_2$ 238.25	129-44-2 319	 sub	4-14-00-00479	H_2O 1; EtOH 2; eth 2; ace 2
390	9,10-Anthracenedione, 2,7-dibromo- 2,7-Dibromoanthraquinone	$C_{14}H_6Br_2O_2$ 366.01	605-42-5 249	 sub	2-07-00-00718	bz 4; HOAc 4
391	9,10-Anthracenedione, 1,2-dihydroxy- Alizarin	$C_{14}H_8O_4$ 240.22	72-48-0 289.5	237	4-08-00-03256	H_2O 2; EtOH 3; eth 3; ace 3
392	9,10-Anthracenedione, 1,4-dihydroxy- Quinizarin	$C_{14}H_8O_4$ 240.22	81-64-1 200	8094	4-08-00-03260	H_2O 3; EtOH 3; eth 3; bz 3
393	9,10-Anthracenedione, 1,5-dihydroxy- Anthrarufin	$C_{14}H_8O_4$ 240.22	117-12-4 280	719 sub	4-08-00-03268	H_2O 1; EtOH 2; eth 2; ace 2
394	9,10-Anthracenedione, 1,7-dihydroxy- 9,10-Anthraquinone, 1,7-dihydroxy	$C_{14}H_8O_4$ 240.22	569-08-4 293.3	 sub	4-08-00-03270	H_2O 1; EtOH 3; eth 3; bz 3
395	9,10-Anthracenedione, 1,8-dihydroxy- Danthron	$C_{14}H_8O_4$ 240.22	117-10-2 193	2813 sub	4-08-00-03271	H_2O 1; EtOH 3; eth 3; ace 3
396	9,10-Anthracenedione, 2,7-dihydroxy-	$C_{14}H_8O_4$ 240.22	572-93-0 353.8	 sub	4-08-00-03273	H_2O 1; EtOH 3; eth 2; bz 2
397	9,10-Anthracenedione, 1,8-dihydroxy-3-(hydroxymethyl)- Aloe-emodol	$C_{15}H_{10}O_5$ 270.24	481-72-1 223.5	303 sub	4-08-00-03578	bz 4; eth 4; EtOH 4
398	9,10-Anthracenedione, 1,8-dihydroxy-3-methyl- Chrysophanic acid	$C_{15}H_{10}O_4$ 254.24	481-74-3 196	2263 sub	4-08-00-03277 0.92^{25}	bz 4; HOAc 4
399	9,10-Anthracenedione, 1,2-dihydroxy-3-nitro- Alizarine Orange	$C_{14}H_7NO_6$ 285.21	568-93-4 244 dec	240 sub	4-08-00-03259	H_2O 2; EtOH 3; bz 3; chl 3
400	9,10-Anthracenedione, 1,2-dihydroxy-4-nitro-	$C_{14}H_7NO_6$ 285.21	2243-71-2 289 dec	 sub	4-08-00-03259	bz 4; EtOH 4; chl 4; HOAc 4

PHYSICAL CONSTANTS OF ORGANIC COMPOUNDS (continued)

No.	Name Synonym	Mol. Form. Mol. Wt.	CAS RN mp/°C	Merck No. bp/°C	Beil. Ref. den/g cm^{-3}	Solubility n_D
401	9,10-Anthracenedione, 1,8-dihydroxy-2,4,5,7-tetranitro- Chrysamminic acid	$C_{14}H_4N_4O_{12}$ 420.21	517-92-0 exp	2253 dec	1-08-00-00723	eth 4; EtOH 4
402	9,10-Anthracenedione, 1,4-dimethyl-	$C_{16}H_{12}O_2$ 236.27	1519-36-4 140.5	sub	4-07-00-02583	H_2O 1; EtOH 2; bz 3; xyl 3
403	9,10-Anthracenedione, 2,3-dimethyl-	$C_{16}H_{12}O_2$ 236.27	6531-35-7 210.8	sub	3-07-00-04131	EtOH 3; bz 3; chl 2; xyl 3
404	9,10-Anthracenedione, 2,6-dimethyl-	$C_{16}H_{12}O_2$ 236.27	3837-38-5 242	sub	3-07-00-04132	EtOH 4; HOAc 4; tol 4
405	9,10-Anthracenedione, 1,5-dinitro-	$C_{14}H_6N_2O_6$ 298.21	82-35-9 385	sub	4-07-00-02562	H_2O 1; EtOH 2; eth 2; bz 2
406	9,10-Anthracenedione, 1,2,3,5,6,7-hexahydroxy- Rufigallol	$C_{14}H_8O_8$ 304.21	82-12-2	8267 sub	4-08-00-03750	H_2O 1; EtOH 2; eth 2; ace 3
407	9,10-Anthracenedione, 1-hydroxy-	$C_{14}H_8O_3$ 224.22	129-43-1 193.8	sub	4-08-00-02586	H_2O 1; EtOH 3; eth 3; bz 3
408	9,10-Anthracenedione, 2-hydroxy-	$C_{14}H_8O_3$ 224.22	605-32-3 306	sub	4-08-00-02598	H_2O 1; EtOH 3; eth 3; KOH 3
409	9,10-Anthracenedione, 2-methyl- 2-Methylanthraquinone	$C_{15}H_{10}O_2$ 222.24	84-54-8 177	5943 sub	4-07-00-02574	bz 4; EtOH 4; HOAc 4
410	9,10-Anthracenedione, 1-nitro-	$C_{14}H_7NO_4$ 253.21	82-34-8 231.5	270^{7}	4-07-00-02561	H_2O 1; EtOH 2; eth 2; ace 3
411	9,10-Anthracenedione, 2-nitro-	$C_{14}H_7NO_4$ 253.21	605-27-6 184.5	sub	4-07-00-02561	H_2O 1; EtOH 2; eth 2; ace 3
412	9,10-Anthracenedione, 1,2,5,6-tetrahydroxy-	$C_{14}H_8O_6$ 272.21	632-77-9 340	sub	3-08-00-04288	py 4; EtOH 4; HOAc 4
413	9,10-Anthracenedione, 1,2,5,8-tetrahydroxy- Quinalizarin	$C_{14}H_8O_6$ 272.21	81-61-8 >275	8059	4-08-00-03683	H_2O 2; EtOH 2; eth 2; ace 2
414	9,10-Anthracenedione, 1,3,5,7-tetrahydroxy-	$C_{14}H_8O_6$ 272.21	632-82-6 360	sub	4-08-00-03683	ace 4; bz 4; EtOH 4; HOAc 4
415	9,10-Anthracenedione, 1,2,3-trihydroxy- Anthragallol	$C_{14}H_8O_5$ 256.21	602-64-2 313	713 sub 290	4-08-00-03567	H_2O 2; EtOH 3; eth 3; CS_2 3
416	9,10-Anthracenedione, 1,2,4-trihydroxy- Purpurin	$C_{14}H_8O_5$ 256.21	81-54-9 259	7962 sub	4-08-00-03568	H_2O 2; EtOH 4; eth 3; bz 4
417	9,10-Anthracenedione, 1,2,6-trihydroxy-	$C_{14}H_8O_5$ 256.21	82-29-1 330	459	4-08-00-03568	H_2O 3; EtOH 3; eth 2; bz 3
418	9,10-Anthracenedione, 1,2,7-trihydroxy-	$C_{14}H_8O_5$ 256.21	602-65-3 374	462	4-08-00-03569	bz 4; EtOH 4; HOAc 4
419	9,10-Anthracenedione, 1,3,8-trihydroxy-6-methyl- Emodin	$C_{15}H_{10}O_5$ 270.24	518-82-1 257	3518 sub	4-08-00-03575	eth 4; EtOH 4
420	Anthracene, 9-ethyl- 9-Ethylanthracene	$C_{16}H_{14}$ 206.29	605-83-4 60		4-05-00-02326 1.0413^{99}	H_2O 1; EtOH 3; eth 3 1.6767^{99}
421	Anthracene, 9-ethyl-9,10-dihydro- 9-Ethyl-9,10-dihydroanthracene	$C_{16}H_{16}$ 208.30	605-82-3	321.5	0-05-00-00649 1.048^{18}	bz 4; eth 4; EtOH 4
422	Anthracene, 1,2,3,4,5,6-hexahydro- 1,2,3,4,5,6-Hexahydroanthracene	$C_{14}H_{16}$ 184.28	6109-22-4 68.5	160^{15}	3-05-00-01676	bz 4
423	Anthracene, 1-methyl- 1-Methylanthracene	$C_{15}H_{12}$ 192.26	610-48-0 85.5	199.5	4-05-00-02310 1.0471^{99}	H_2O 1; EtOH 3; eth 3; bz 3 1.6802^{99}
424	Anthracene, 2-methyl- 2-Methylanthracene	$C_{15}H_{12}$ 192.26	613-12-7 209	sub	4-05-00-02311 1.80^{0}	H_2O 1; EtOH 2; eth 2; ace 1
425	Anthracene, 9-methyl- 9-Methylanthracene	$C_{15}H_{12}$ 192.26	779-02-2 81.5	$196\text{-}7^{12}$	4-05-00-02312 1.065^{99}	EtOH 3; eth 3; ace 3; bz 3 1.6959^{99}
426	Anthracene, 9-nitro- 9-Nitroanthracene	$C_{14}H_9NO_2$ 223.23	602-60-8 146	275^{17}	4-05-00-02296	H_2O 1; EtOH 2; ace 4; chl 2
427	Anthracene, 1,2,3,4,5,6,7,8-octahydro-	$C_{14}H_{18}$ 186.30	1079-71-6 78	294	4-05-00-01584 0.9703^{80}	H_2O 1; EtOH 3; bz 4; ctc 2 1.5372^{80}
428	Anthracene, 9-phenyl- 9-Phenylanthracene	$C_{20}H_{14}$ 254.33	602-55-1 156	417	4-05-00-02628	H_2O 1; EtOH 3; eth 3; bz 3
429	1,2,10-Anthracenetriol Anthrarobin	$C_{14}H_{10}O_3$ 226.23	577-33-3 208	718	4-06-00-07601	H_2O 2; EtOH 4; eth 4; ace 4
430	1,8,9-Anthracenetriol Anthralin	$C_{14}H_{10}O_3$ 226.23	1143-38-0 179	714	4-06-00-07602	H_2O 1; EtOH 3; eth 2; ace 3
431	1-Anthracenol	$C_{14}H_{10}O$ 194.23	610-50-4 158	234^{13}	4-06-00-04928	H_2O 1; EtOH 4; eth 4; NaOH 3
432	9-Anthracenol Anthranol	$C_{14}H_{10}O$ 194.23	529-86-2 152	716	4-06-00-04930	
433	9(10H)-Anthracenone Anthrone	$C_{14}H_{10}O$ 194.23	90-44-8 155	721	4-06-00-04930	ace 3; bz 3; con sulf 3; dil alk 3
434	5,9,14,18-Anthrazinetetrone, 6,15-dihydro- Indanthrene	$C_{28}H_{14}N_2O_4$ 442.43	81-77-6 470-500 dec	4846	5-24-09-00504	H_2O 1; EtOH 1; eth 1; ace 1
435	8'-Apo-β,ψ-carotenal, 3-hydroxy-, (3*R*)- β-Citraurin	$C_{30}H_{40}O_2$ 432.65	650-69-1 147	2326	3-08-00-01599	H_2O 1; EtOH 4; eth 4; ace 4
436	β-*D*-Arabinopyranose Arabinose (*D*)(β)	$C_5H_{10}O_5$ 150.13	6748-95-4 155.5		4-01-00-04215 1.625^{25}	H_2O 4
437	*DL*-Arginine	$C_6H_{14}N_4O_2$ 174.20	7200-25-1 217 dec	805	3-04-00-01359	H_2O 1; EtOH 1; eth 1; bz 1
438	*L*-Arginine	$C_6H_{14}N_4O_2$ 174.20	74-79-3 244 dec	805	4-04-00-02648	H_2O 3; EtOH 2; eth 1

No.	Name Synonym	Mol. Form. Mol. Wt.	CAS RN mp/°C	Merck No. bp/°C	Beil. Ref. den/g cm^{-3}	Solubility n_D
439	Arsenic acid, triethyl ester Ethyl arsenate	$C_6H_{15}AsO_4$ 226.10	15606-95-8	 236.5	4-01-00-01359 1.3021^{20}	 1.4343^{20}
440	Arsenous acid, triethyl ester Ethyl arsenite	$C_6H_{15}AsO_3$ 210.10	3141-12-6	 165.5	4-01-00-01358 1.2239^{20}	 1.4369^{13}
441	Arsine, diethyl- Diethylarsine	$C_4H_{11}As$ 134.05	692-42-2	 105	3-04-00-01797 1.1338^{24}	ace 4; bz 4; eth 4; EtOH 4 1.4709
442	Arsine, dimethyl- Dimethylarsine	C_2H_7As 106.00	593-57-7 -136.1	 36	4-04-00-03665 1.208^{29}	ace 4; bz 4; eth 4; EtOH 4
443	Arsine, diphenyl- Diphenylarsine	$C_{12}H_{11}As$ 230.14	829-83-4	 174^{25}	4-16-00-01138 1.30^{25}	eth 4; EtOH 4
444	Arsine, ethyl- Ethylarsine	C_2H_7As 106.00	593-59-9	 36	3-04-00-01797 1.214^{22}	eth 4; EtOH 4
445	Arsine, ethyldifluoro-	$C_2H_5AsF_2$ 141.98	430-40-0 -38.7	 94.3	3-04-00-01799 1.708^{17}	
446	Arsine, methyl- Methylarsine	CH_5As 91.97	593-52-2 -143	 2	3-04-00-01795	ace 4; eth 4; EtOH 4
447	Arsine, oxophenyl- Oxophenylarsine	C_6H_5AsO 168.03	637-03-6 145	6903	4-16-00-01160	H_2O 1; EtOH 2; eth 1; bz 4
448	Arsine, triethyl- Triethylarsine	$C_6H_{15}As$ 162.11	617-75-4	 138.5	4-04-00-03666 1.150^{20}	ace 4; eth 4; EtOH 4 1.467^{20}
449	Arsine, trimethyl- Trimethylarsine	C_3H_9As 120.03	593-88-4 -87.3	 52	4-04-00-03665 1.144^{15}	bz 4; eth 4; EtOH 4
450	Arsine, triphenyl- Triphenylarsine	$C_{18}H_{15}As$ 306.24	603-32-7 61	 360	4-16-00-01139 1.2634^{18}	H_2O 1; EtOH 2; eth 4; bz 4 1.6888^{21}
451	Arsinic acid, dimethyl- Cacodylic acid	$C_2H_7AsO_2$ 138.00	75-60-5 195	1603	4-04-00-03681	H_2O 4; EtOH 3; eth 1
452	Arsinous acid, dimethyl-, anhydride Cacodyl oxide	$C_4H_{12}As_2O$ 225.98	503-80-0 -25	 150	3-04-00-01814 1.4816^{15}	eth 4; EtOH 4 1.5225^{9}
453	Arsinous chloride, dimethyl- Dimethylchloroarsine	C_2H_6AsCl 140.44	557-89-1 <-45	 109; 77^{37}	4-04-00-03670 1.505^{12}	EtOH 4 1.5203^{12}
454	Arsinous chloride, diphenyl- Chlorodiphenylarsine	$C_{12}H_{10}AsCl$ 264.59	712-48-1 44	 337	3-16-00-00944 1.4820^{16}	ace 4; bz 4; eth 4; EtOH 4 1.6332^{56}
455	Arsonic acid, [4-[(aminocarbonyl)amino]phenyl]- Carbarsone	$C_7H_9AsN_2O_4$ 260.08	121-59-5 174	1788	3-16-00-01112	H_2O 2; EtOH 2; eth 1; chl 1
456	Arsonic acid, (4-aminophenyl)- Arsanilic acid	$C_6H_8AsNO_3$ 217.06	98-50-0 232	818	4-16-00-01190 1.9571^{10}	H_2O 3; EtOH 2; eth 3; ace 1
457	Arsonic acid, ethyl- Ethanearsonic acid	$C_2H_7AsO_3$ 154.00	507-32-4 99.5	3677 $209\text{-}11^{12}$	4-04-00-03682	H_2O 4; EtOH 4
458	Arsonic acid, methyl- Methanearsonic acid	CH_5AsO_3 139.97	124-58-3 160.5	5864	4-04-00-03682	H_2O 3; EtOH 3
459	Arsonic acid, (4-nitrophenyl)- Nitarsone	$C_6H_6AsNO_5$ 247.04	98-72-6 >310 dec	6483	4-16-00-01185	H_2O 2; EtOH 2; DMSO-d_6 2
460	Arsonic acid, phenyl- Benzenearsonic acid	$C_6H_7AsO_3$ 202.04	98-05-5 158- dec	1075	4-16-00-01183	H_2O 4; EtOH 4
461	Arsonic acid, propyl- 1-Propanearsonic acid	$C_3H_9AsO_3$ 168.02	107-34-6 134.5	7810	4-04-00-03682	H_2O 4; EtOH 4; eth 1
462	Arsonous dichloride, phenyl- Dichlorophenylarsine	$C_6H_5AsCl_2$ 222.93	696-28-6 -19	 255	4-16-00-01161 1.6516^{20}	bz 4; eth 4; EtOH 4 1.6386^{15}
463	Arsonous difluoride, methyl- Difluoromethylarsine	CH_3AsF_2 127.95	420-24-6 -29.7	 76.5	3-04-00-01796 1.924^{18}	
464	Arsonous diiodide, phenyl-	$C_6H_5AsI_2$ 405.84	6380-34-3 15	 205^{14} 185^{10}	3-16-00-00958 1.6264^{15}	
465	*DL*-Ascorbic acid	$C_6H_8O_6$ 176.13	62624-30-0 168.5		5-18-05-00026	H_2O 4; EtOH 3; eth 1; bz 1
466	*L*-Ascorbic acid Vitamin C	$C_6H_8O_6$ 176.13	50-81-7 190-2 dec	855	5-18-05-00026 1.65^{25}	H_2O 4; EtOH 3; eth 1; bz 1
467	*L*-Ascorbic acid, 6-deoxy- 6-Desoxy-*L*-ascorbic acid	$C_6H_8O_5$ 160.13	528-81-4 168	2880 sub 160	4-18-00-02301	H_2O 4; ace 4; EtOH 4
468	*L*-Asparagine α-Aminosuccinamic acid	$C_4H_8N_2O_3$ 132.12	70-47-3 235	859	4-04-00-03004 1.543^{15}	H_2O 3; EtOH 1; eth 1; MeOH 1
469	*DL*-Asparagine, monohydrate	$C_4H_{10}N_2O_4$ 150.13	69833-18-7 102.5	859 214 dec	2-04-00-00900 1.4540^{15}	H_2O 2; EtOH 1; eth 1; bz 1
470	*DL*-Aspartic acid	$C_4H_7NO_4$ 133.10	617-45-8 275		2-04-00-00900 1.6622^{13}	H_2O 2; EtOH 1; eth 1; bz 1
471	*L*-Aspartic acid *L*-Aminosuccinic acid	$C_4H_7NO_4$ 133.10	56-84-8 270	862	4-04-00-02998 1.6603^{13}	H_2O 2; EtOH 1; eth 1; bz 1
472	*L*-Aspartic acid, *N*-[2-[(2-amino-2-oxoethyl)amino]-2-carboxyethyl]- Lycomarasmine	$C_9H_{15}N_3O_7$ 277.23	7611-43-0 227-9 dec	5491	3-04-00-01521	
473	Aspidospermidine, 1-acetyl-17-methoxy- Aspidospermine	$C_{22}H_{30}N_2O_2$ 354.49	466-49-9 208	872 220^{2}	5-23-12-00210	H_2O 2; EtOH 3; eth 2; bz 3
474	Asulam Methyl [(4-aminophenyl)sulfonyl]carbamate	$C_8H_{10}N_2O_4S$ 230.24	3337-71-1 144			
475	Atisine Anthorine	$C_{22}H_{33}NO_2$ 343.51	466-43-3 58.5	881		eth 4; EtOH 4; chl 4
476	Atrazine 6-Chloro-*N*-ethyl-*N*′-(1-methylethyl)-1,3,5-triazine-2,4-diamine	$C_8H_{14}ClN_5$ 215.69	1912-24-9 173	886		

No.	Name Synonym	Mol. Form. Mol. Wt.	CAS RN mp/°C	Merck No. bp/°C	Beil. Ref. den/g cm^{-3}	Solubility n_D
477	Avermectin B1a Abamectin	$C_{48}H_{72}O_{14}$ 873.09	71751-41-2 152	1		
478	7-Azabicyclo[4.1.0]heptane 2,3-Tetramethyleneaziridine	$C_6H_{11}N$ 97.16	286-18-0 21	 150; 50^{22}	5-20-04-00304 0.9480^{27}	bz 4; eth 4; EtOH 4
479	9-Azabicyclo[3.3.1]nonan-3-one, 9-methyl- Pseudopelletierine	$C_9H_{15}NO$ 153.22	552-70-5 54	7936 246	5-21-07-00059 1.001^{100}	H_2O 4; eth 4; EtOH 4 1.4760^{100}
480	1-Azabicyclo[2.2.2]octane Quinuclidine	$C_7H_{13}N$ 111.19	100-76-5 158	8109	5-20-04-00335	H_2O 4; ace 4; eth 4; EtOH 4
481	8-Azabicyclo[3.2.1]octane-2-carboxylic acid, 3-(benzoyloxy)-8-methyl-, [1*R*-(*exo,exo*)]- Benzoylecgonine	$C_{16}H_{19}NO_4$ 289.33	519-09-5 195	1125	5-22-05-00053	bz 4; EtOH 4
482	8-Azabicyclo[3.2.1]octane-2-carboxylic acid, 3-(benzoyloxy)-8-methyl-, methyl ester, [1*R*-(*exo,exo*)]- Cocaine	$C_{17}H_{21}NO_4$ 303.36	50-36-2 98	2450 $187^{0.1}$	5-22-05-00054	H_2O 2; EtOH 4; eth 4; ace 3 1.5022^{98}
483	8-Azabicyclo[3.2.1]octane-2-carboxylic acid, 3-hydroxy-8-methyl-, [1*R*-(*exo,exo*)]- Ecgonine	$C_9H_{15}NO_3$ 185.22	481-37-8 205	3467	5-22-05-00053	H_2O 4; EtOH 4
484	8-Azabicyclo[3.2.1]octane, 8-methyl- Tropane	$C_8H_{15}N$ 125.21	529-17-9	9689 166	5-20-04-00331 0.9251^{15}	
485	1-Azabicyclo[2.2.2]octan-3-ol 3-Quinuclidinol	$C_7H_{13}NO$ 127.19	1619-34-7 221	8110 sub 120	5-21-01-00271	ace 3
486	8-Azabicyclo[3.2.1]octan-3-ol, 8-methyl-, *exo*- Pseudotropine	$C_8H_{15}NO$ 141.21	135-97-7 109	7937 241	5-21-01-00219	H_2O 4; EtOH 4; eth 2; bz 3
487	8-Azabicyclo[3.2.1]octan-3-ol, 8-methyl-, benzoate (ester), *endo*-	$C_{15}H_{19}NO_2$ 245.32	19145-60-9 41.5	 177.5	5-21-01-00226	H_2O 4; eth 4; EtOH 4
488	8-Azabicyclo[3.2.1]octan-3-ol, 8-methyl-, benzoate (ester), *exo*- Tropacocaine	$C_{15}H_{19}NO_2$ 245.32	537-26-8 49	9686 dec	5-21-01-00226 1.0426^{100}	bz 4; eth 4; EtOH 4; peth 4 1.5080^{100}
489	8-Azabicyclo[3.2.1]octan-3-ol, 8-methyl-, *endo*- Tropine	$C_8H_{15}NO$ 141.21	120-29-6 64	9693 233	5-21-01-00219 1.016^{100}	H_2O 4; eth 4; EtOH 4 1.4811^{100}
490	8-Azabicyclo[3.2.1]octan-3-one, 8-methyl-	$C_8H_{13}NO$ 139.20	532-24-1 43	 227; 113^{25}	5-21-07-00032 1.9872^{100}	EtOH 3; eth 3; ace 3; bz 3 1.4598^{100}
491	8-Azabicyclo[3.2.1]oct-2-ene-2-carboxylic acid, 8-methyl-, (1*R*)- Ecgonidine	$C_9H_{13}NO_2$ 167.21	484-93-5 226-30 dec	3466	5-22-01-00446	H_2O 4
492	1H-Azepine, hexahydro- Hexamethylenimine	$C_6H_{13}N$ 99.18	111-49-9	 138	5-20-04-00003 0.8643^{22}	H_2O 3; EtOH 4; eth 4 1.4631^{20}
493	2H-Azepin-2-one, hexahydro- Caprolactam	$C_6H_{11}NO$ 113.16	105-60-2 69.3	1762 270	5-21-06-00444	H_2O 4; bz 4; EtOH 4; chl 4
494	Azetidine	C_3H_7N 57.10	503-29-7	 63	5-20-01-00136 0.8436^{20}	ace 4; bz 4; eth 4; EtOH 4 1.4287^{25}
495	2-Azetidinecarboxylic acid	$C_4H_7NO_2$ 101.11	2517-04-6 217 dec	923	5-22-01-00015	
496	2-Azetidinone	C_3H_5NO 71.08	930-21-2 73.5	 106^{15}	5-21-06-00312	eth 4; EtOH 4; chl 4
497	Azinphos-methyl *O,O*-Dimethyl *S*-[(4-oxo-1,2,3-benzotriazin-3(4H)-yl)methyl]phosphorodithioate	$C_{10}H_{12}N_3O_3PS_2$ 317.33	86-50-0 73	926	 1.44^{20}	
498	Aziridine Ethyleneimine	C_2H_5N 43.07	151-56-4 -77.9	3760 56	5-20-01-00003 0.832^{25}	H_2O 5; EtOH 3; eth 4; chl 2
499	1-Aziridineethanol	C_4H_9NO 87.12	1072-52-2	 168	5-20-01-00014 1.088^{25}	 1.4560^{20}
500	Aziridine, 1-methyl-	C_3H_7N 57.10	1072-44-2	 27.5	5-20-01-00007 0.7572^{19}	H_2O 5 1.3885^{19}
501	Aziridine, 2-methyl- 1,2-Propylenimine	C_3H_7N 57.10	75-55-8	 67	5-20-01-00150 0.812^{16}	
502	Azocine, octahydro-	$C_7H_{15}N$ 113.20	1121-92-2 29	 $51\text{-}3^{15}$	5-20-04-00156 0.896^{25}	 1.4720^{20}
503	1H-Azonine, octahydro-	$C_8H_{17}N$ 127.23	5661-71-2 -19	 91^{24}	5-20-04-00203 0.83^{25}	 1.4760^{20}
504	Azulene Bicyclo[5.3.0]decapentaene	$C_{10}H_8$ 128.17	275-51-4 99	939 270 dec; 125^{10}	4-05-00-01636	H_2O 1; EtOH 3; eth 3; ace 3
505	Azulene, 1,4-dimethyl-7-(1-methylethyl)- Guaiazulene	$C_{15}H_{18}$ 198.31	489-84-9 31.5	4464 $167\text{-}8^{12}$	4-05-00-01751 0.973^{20}	EtOH 3; eth 3; AcOEt 3
506	Azulene, 4,8-dimethyl-2-(1-methylethyl)-	$C_{15}H_{18}$ 198.31	529-08-8 32.3	 $122^{1.0}$	4-05-00-01754 0.9735^{19}	eth 4; EtOH 4
507	5-Azulenemethanol, 1,2,3,4,5,6,7,8-octahydro-α,α,3,8-tetramethyl-, Guaiol	$C_{15}H_{26}O$ 222.37	489-86-1 91	4466 288 dec; 165^{17}	3-06-00-00412 0.9074^{100}	H_2O 1; EtOH 3; eth 3 1.4716^{100}
508	Bacitracin	$C_{66}H_{103}N_{17}O_{16}S$ 1422.71	1405-87-4 223	948		H_2O 4; EtOH 4
509	Balan Benzenamine, *N*-butyl-*N*-ethyl-2,6-dinitro-4-(trifluoromethyl)-	$C_{13}H_{16}F_3N_3O_4$ 335.28	1861-40-1 66	1048 $121^{0.5}$; 148^7		

No.	Name Synonym	Mol. Form. Mol. Wt.	CAS RN mp/°C	Merck No. bp/°C	Beil. Ref. den/g cm^{-3}	Solubility n_D
510	Barban Carbamic acid, (3-chlorophenyl)-, 4-chloro-2-butynyl ester	$C_{11}H_9Cl_2NO_2$ 258.10	101-27-9 75	969		
511	Basic copper carbonate Copper, [μ-[carbonato(2-)-O:O']]dihydroxydi-	$CH_2Cu_2O_5$ 221.12	12069-69-1 200 dec	2634	3.8^{20}	
512	Bayleton 2-Butanone, 1-(4-chlorophenoxy)-3,3-dimethyl-1-(1H-1,2,4-triazol-1-yl)-	$C_{14}H_{16}ClN_3O_2$ 293.75	43121-43-3 82	9507	1.22^{20}	
513	Benalaxyl Methyl *N*-(2,6-dimethylphenyl)-*N*-(phenylacetyl)-*DL*-alaninate	$C_{20}H_{23}NO_3$ 325.41	71626-11-4 79		1.27^{25}	
514	Bendiocarb 1,3-Benzodioxol-4-ol, 2,2-dimethyl-, methylcarbamate	$C_{11}H_{13}NO_4$ 223.23	22781-23-3 130	1044	1.25^{20}	
515	Benomyl Carbamic acid, [1-[(butylamino)carbonyl]-1H-benzimidazol-2-yl]-, methyl ester	$C_{14}H_{18}N_4O_3$ 290.32	17804-35-2 dec	1053		
516	Bensulfuron-methyl 2-[[[[[(4,6-Dimethoxypyrimidin-2-yl)amino]carbonyl]amino]sulfonyl]methyl]benzoic acid, methyl ester	$C_{16}H_{18}N_4O_7S$ 410.41	83055-99-6 187			
517	Bensulide Phosphorodithioic acid, *O,O*-bis(1-methylethyl) *S*-[2-[(phenylsulfonyl)amino]ethyl] ester	$C_{14}H_{244}NO_4PS_3$ 397.52	741-58-2 34.4		1.224^{20}	
518	Bentazon 1H-2,1,3-Benzothiadiazin-4(3H)-one, 3-(1-methylethyl)-, 2,2-dioxide	$C_{10}H_{12}N_2O_3S$ 240.28	25057-89-0 138	1060		
519	Benz[a]anthracene 1,2-Benzanthracene	$C_{18}H_{12}$ 228.29	56-55-3 84	1069	4-05-00-02549	EtOH 4
520	Benz[a]anthracene, 7,12-dimethyl- 9,10-Dimethyl-1,2-benzanthracene	$C_{20}H_{16}$ 256.35	57-97-6 122.5	3224	4-05-00-02587	ace 4; bz 4
521	Benz[a]anthracene, 11-methyl-	$C_{19}H_{14}$ 242.32	6111-78-0 118	360 dec	4-05-00-02568	H_2O 1; EtOH 3; HOAc 3
522	Benz[a]anthracene, 3-methyl- 3-Methylbenz[a]anthracene	$C_{19}H_{14}$ 242.32	2498-75-1 156.5	$160^{0.1}$	4-05-00-02562	H_2O 1; EtOH 3; ace 3; bz 3
523	Benz[a]anthracene, 8-methyl-	$C_{19}H_{14}$ 242.32	2381-31-9 158.5	272^{3}; $160^{0.1}$	4-05-00-02566 1.2310^{0}	H_2O 1; EtOH 3; eth 3; bz 3
524	Benzadox Acetic acid, [(benzoylamino)oxy]-	$C_9H_9NO_4$ 195.17	5251-93-4 140			
525	Benzaldehyde Benzenecarboxaldehyde	C_7H_6O 106.12	100-52-7 -26	1065 179.0	4-07-00-00505 1.0415^{10}	H_2O 2; EtOH 5; eth 5; ace 4 1.5463^{20}
526	Benzaldehyde, 2-bromo-	C_7H_5BrO 185.02	6630-33-7 21.5	230	4-07-00-00578	H_2O 1; EtOH 4; bz 4; ctc 2 1.5925^{20}
527	Benzaldehyde, 3-bromo-	C_7H_5BrO 185.02	3132-99-8	234.5	4-07-00-00579	H_2O 1; EtOH 4; eth 4; ctc 2 1.5935^{20}
528	Benzaldehyde, 4-bromo-	C_7H_5BrO 185.02	1122-91-4 67	$66-8^{2}$	4-07-00-00579	H_2O 1; EtOH 4; bz 4; chl 2
529	Benzaldehyde, 2-chloro-	C_7H_5ClO 140.57	89-98-5 12.4	211.9	4-07-00-00561 1.2483^{20}	H_2O 2; EtOH 3; eth 3; ace 3 1.5662^{20}
530	Benzaldehyde, 3-chloro-	C_7H_5ClO 140.57	587-04-2 17.5	213.5	4-07-00-00566 1.2410^{20}	H_2O 2; EtOH 3; eth 3; ace 3 1.5650^{20}
531	Benzaldehyde, 4-chloro-	C_7H_5ClO 140.57	104-88-1 47.5	213.5	4-07-00-00568 1.196^{61}	H_2O 3; EtOH 4; eth 4; ace 3 1.555^{61}
532	Benzaldehyde, 3-chloro-4-hydroxy-	$C_7H_5ClO_2$ 156.57	2420-16-8 139	150^{14}	4-08-00-00296	eth 4; EtOH 4
533	Benzaldehyde, 5-chloro-2-hydroxy-	$C_7H_5ClO_2$ 156.57	635-93-8 100.3	105^{12}	4-08-00-00224	H_2O 1; EtOH 4; eth 3; alk 3
534	Benzaldehyde, 3,5-dibromo-2-hydroxy- 3,5-Dibromosalicylaldehyde	$C_7H_4Br_2O_2$ 279.92	90-59-5 86	3011 sub	3-08-00-00188	bz 4; eth 4; chl 4
535	Benzaldehyde, 2,5-dichloro- 2,5-Dichlorobenzaldehyde	$C_7H_4Cl_2O$ 175.01	6361-23-5 58	232	4-07-00-00576	bz 4; eth 4; EtOH 4; chl 4
536	Benzaldehyde, 3,4-dichloro- 3,4-Dichlorobenzaldehyde	$C_7H_4Cl_2O$ 175.01	6287-38-3 44	247.5	4-07-00-00576	H_2O 1; EtOH 3; eth 3; ctc 2
537	Benzaldehyde, 3,5-dichloro- 3,5-Dichlorobenzaldehyde	$C_7H_4Cl_2O$ 175.01	10203-08-4 65	240	4-07-00-00577	ace 4; bz 4; eth 4; EtOH 4
538	Benzaldehyde, 3,4-diethoxy- 3,4-Diethoxybenzaldehyde	$C_{11}H_{14}O_3$ 194.23	2029-94-9 22	279; 200^{50}	4-08-00-01766 1.0100^{22}	EtOH 4
539	Benzaldehyde, 4-(diethylamino)-	$C_{11}H_{15}NO$ 177.25	120-21-8 41	172^{10}	4-14-00-00066	H_2O 4; EtOH 3; eth 3; bz 3
540	Benzaldehyde, 2,3-dihydroxy-	$C_7H_6O_3$ 138.12	24677-78-9 108	235; 120^{16}	3-08-00-01979	ace 4; EtOH 4; HOAc 4
541	Benzaldehyde, 2,4-dihydroxy- β-Resorcylaldehyde	$C_7H_6O_3$ 138.12	95-01-2 135	8159 226^{22}	4-08-00-01753	H_2O 3; EtOH 4; eth 4; bz 2
542	Benzaldehyde, 3,4-dihydroxy- Protocatechualdehyde	$C_7H_6O_3$ 138.12	139-85-5 153 dec	7908	4-08-00-01762	H_2O 3; EtOH 4; eth 4
543	Benzaldehyde, 2,4-dimethoxy- 2,4-Dimethoxybenzaldehyde	$C_9H_{10}O_3$ 166.18	613-45-6 72	290; 165^{10}	4-08-00-01754	H_2O 1; EtOH 3; eth 3; bz 3
544	Benzaldehyde, 2,5-dimethoxy- 2,5-Dimethoxybenzaldehyde	$C_9H_{10}O_3$ 166.18	93-02-7 52	270; 146^{10}	4-08-00-01759	H_2O 2; EtOH 3; eth 3

No.	Name Synonym	Mol. Form. Mol. Wt.	CAS RN mp/°C	Merck No. bp/°C	Beil. Ref. den/g cm^{-3}	Solubility n_D
545	Benzaldehyde, 3,4-dimethoxy- Veratraldehyde	$C_9H_{10}O_3$ 166.18	120-14-9 43	9852 281; 155^{10}	4-08-00-01765	H_2O 2; EtOH 4; eth 4; chl 2
546	Benzaldehyde, 3,5-dimethoxy- 3,5-Dimethoxybenzaldehyde	$C_9H_{10}O_3$ 166.18	7311-34-4 46.3	 151^{16}	4-08-00-01786	H_2O 2; EtOH 3; bz 3; peth 2
547	Benzaldehyde, 2,4-dimethyl- 2,4-Dimethylbenzaldehyde	$C_9H_{10}O$ 134.18	15764-16-6 -9	 218	4-07-00-00706	EtOH 3; eth 3; ace 3; bz 3
548	Benzaldehyde, 2,5-dimethyl- Isoxylaldehyde	$C_9H_{10}O$ 134.18	5779-94-2	 220	4-07-00-00707 0.9500^{20}	EtOH 4; eth 3; ace 3; bz 3
549	Benzaldehyde, 3,5-dimethyl-	$C_9H_{10}O$ 134.18	5779-95-3 9	 221	4-07-00-00707	ace 4; bz 4; eth 4; EtOH 4
550	Benzaldehyde, 4-(dimethylamino)- Ehrlich's reagent	$C_9H_{11}NO$ 149.19	100-10-7 74.5	3219 $176\text{-}7^{17}$	4-14-00-00051 1.0254^{100}	H_2O 2; EtOH 3; eth 3; ace 3
551	Benzaldehyde, 2,4-dinitro- 2,4-Dinitrobenzaldehyde	$C_7H_4N_2O_5$ 196.12	528-75-6 72	3265 200^{15}	3-07-00-00923	H_2O 2; EtOH 3; eth 3; bz 3
552	Benzaldehyde, 2-ethoxy-	$C_9H_{10}O_2$ 150.18	613-69-4 21	 248	4-08-00-00180	EtOH 5; eth 5; chl 2
553	Benzaldehyde, 3-ethoxy-	$C_9H_{10}O_2$ 150.18	22924-15-8	 245.5	0-08-00-00060 1.0768^{20}	EtOH 3; eth 3; bz 3 1.5408^{20}
554	Benzaldehyde, 4-ethoxy-	$C_9H_{10}O_2$ 150.18	10031-82-0 13.5	 249	4-08-00-00254 1.08^{21}	EtOH 4; eth 4; bz 4
555	Benzaldehyde, 3-ethoxy-4-hydroxy- Ethylvanillin	$C_9H_{10}O_3$ 166.18	121-32-4 77.5	3815	4-08-00-01765	H_2O 2; EtOH 3; eth 3; bz 3
556	Benzaldehyde, 5-ethoxy-2-hydroxy- 5-Ethoxy-2-hydroxybenzaldehyde	$C_9H_{10}O_3$ 166.18	80832-54-8 51.5	 230	3-08-00-02004	eth 4; EtOH 4; chl 4
557	Benzaldehyde, 4-ethoxy-3-methoxy-	$C_{10}H_{12}O_3$ 180.20	120-25-2 64.5	 168^{13}	4-08-00-01766	H_2O 2; EtOH 3; eth 3; bz 3
558	Benzaldehyde, 2-ethyl-	$C_9H_{10}O$ 134.18	22927-13-5	 209	4-07-00-00696 1.0216^{17}	 1.5425^{17}
559	Benzaldehyde, 2-fluoro-	C_7H_5FO 124.11	446-52-6 -44.5	 175	4-07-00-00559 1.178^{25}	 1.5234^{20}
560	Benzaldehyde, 3-fluoro-	C_7H_5FO 124.11	456-48-4	 173	4-07-00-00559 1.17^{25}	 1.5206^{20}
561	Benzaldehyde, 4-fluoro-	C_7H_5FO 124.11	459-57-4 -10	 181.5	4-07-00-00559 1.1810^{19}	
562	Benzaldehyde, 2-(β-*D*-glucopyranosyloxy)- Helicin	$C_{13}H_{16}O_7$ 284.27	618-65-5 175	4544	5-17-07-00150	H_2O 4; EtOH 4
563	Benzaldehyde, hydrazone Benzylidene hydrazine	$C_7H_8N_2$ 120.15	5281-18-5 16	 140^{14}	4-07-00-00530	EtOH 3
564	Benzaldehyde, 2-hydroxy- Salicylaldehyde	$C_7H_6O_2$ 122.12	90-02-8 -7	8295 197	4-08-00-00176 1.1674^{20}	H_2O 2; EtOH 5; eth 5; ace 4 1.5740^{20}
565	Benzaldehyde, 3-hydroxy-	$C_7H_6O_2$ 122.12	100-83-4 108	 240	4-08-00-00240 1.1179^{130}	H_2O 2; EtOH 3; eth 3; ace 3
566	Benzaldehyde, 4-hydroxy- *p*-Hydroxybenzaldehyde	$C_7H_6O_2$ 122.12	123-08-0 117	4741	4-08-00-00251 1.129^{130}	H_2O 2; EtOH 4; eth 4; ace 2 1.5705^{130}
567	Benzaldehyde, 2-hydroxy-4,6-dimethoxy-	$C_9H_{10}O_4$ 182.18	708-76-9 70	 193^{25}; 165^{10}	4-08-00-02717	H_2O 1; EtOH 4; eth 4; bz 4
568	Benzaldehyde, 3-hydroxy-4,5-dimethoxy-	$C_9H_{10}O_4$ 182.18	29865-90-5 62.5	 179^{12}	4-08-00-02718	bz 4; EtOH 4; lig 4
569	Benzaldehyde, 4-hydroxy-3,5-dimethoxy- Syringaldehyde	$C_9H_{10}O_4$ 182.18	134-96-3 113	8996 $192\text{-}3^{14}$	4-08-00-02718	H_2O 2; EtOH 4; eth 4; bz 4
570	Benzaldehyde, 2-hydroxy-3-methoxy-	$C_8H_8O_3$ 152.15	148-53-8 44.5	 265.5	4-08-00-01747	H_2O 2; EtOH 4; eth 4; ctc 4
571	Benzaldehyde, 2-hydroxy-5-methoxy-	$C_8H_8O_3$ 152.15	672-13-9 4	 247.5	4-08-00-01759	eth 4; EtOH 4
572	Benzaldehyde, 3-hydroxy-4-methoxy-	$C_8H_8O_3$ 152.15	621-59-0 114	 179^{15}	4-08-00-01764 1.196^{25}	H_2O 2; EtOH 3; eth 3; bz 3
573	Benzaldehyde, 3-hydroxy-5-methoxy- 3-Hydroxy-5-methoxybenzaldehyde	$C_8H_8O_3$ 152.15	57179-35-8 130.5	 sub 110	3-08-00-02073	bz 4; EtOH 4; HOAc 4
574	Benzaldehyde, 4-hydroxy-3-methoxy- Vanillin	$C_8H_8O_3$ 152.15	121-33-5 81.5	9839 285	4-08-00-01763 1.056^{25}	H_2O 2; EtOH 4; eth 4; ace 4
575	Benzaldehyde, 2-hydroxy-3-methyl- *m*-Tolualdehyde, 2-hydroxy	$C_8H_8O_2$ 136.15	824-42-0 17	 208.5	4-08-00-00430	eth 4; EtOH 4; chl 4
576	Benzaldehyde, 2-hydroxy-4-methyl-	$C_8H_8O_2$ 136.15	698-27-1 60.5	 223	4-08-00-00434	eth 4; EtOH 4; chl 4
577	Benzaldehyde, 2-hydroxy-5-methyl-	$C_8H_8O_2$ 136.15	613-84-3 56	 217.5	4-08-00-00432 1.0913^{59}	eth 4; EtOH 4; chl 4 1.547^{59}
578	Benzaldehyde, 2-hydroxy-6-methyl-	$C_8H_8O_2$ 136.15	18362-36-2 32.3	 230	4-08-00-00430	bz 4
579	Benzaldehyde, 2-hydroxy-, oxime Salicylaldoxime	$C_7H_7NO_2$ 137.14	94-67-7 57	8296	1-08-00-00520	H_2O 2; EtOH 4; eth 4; bz 4
580	Benzaldehyde, 2-iodo-	C_7H_5IO 232.02	26260-02-6 37	 129^{14}	4-07-00-00582	H_2O 2; ace 3
581	Benzaldehyde, 3-iodo-	C_7H_5IO 232.02	696-41-3 57	 125^{13}	4-07-00-00583	EtOH 3; ace 3; bz 3
582	Benzaldehyde, 4-iodo-	C_7H_5IO 232.02	15164-44-0 77.5	 265	4-07-00-00584	H_2O 2; EtOH 3; bz 3
583	Benzaldehyde, 2-methoxy-	$C_8H_8O_2$ 136.15	135-02-4 37.5	 243.5	4-08-00-00180 1.1326^{20}	H_2O 1; EtOH 3; eth 4; ace 4 1.5600^{20}
584	Benzaldehyde, 3-methoxy-	$C_8H_8O_2$ 136.15	591-31-1	 231	4-08-00-00241 1.1187^{20}	H_2O 1; EtOH 3; eth 4; ace 4 1.5530^{20}

No.	Name Synonym	Mol. Form. Mol. Wt.	CAS RN mp/°C	Merck No. bp/°C	Beil. Ref. den/g cm^{-3}	Solubility n_D
585	Benzaldehyde, 4-methoxy- *p*-Anisaldehyde	$C_8H_8O_2$ 136.15	123-11-5 0	693 248; 134^{12}	4-08-00-00252 1.119^{15}	H_2O 1; EtOH 5; eth 5; ace 4 1.5730^{20}
586	Benzaldehyde, 4-methoxy-3-methyl-	$C_9H_{10}O_2$ 150.18	32723-67-4	 80-5^{1}	4-08-00-00432 1.025^{25}	 1.5670^{20}
587	Benzaldehyde, 2-methyl- *o*-Tolualdehyde	C_8H_8O 120.15	529-20-4	9453 200; 94^{10}	4-07-00-00667 1.0328^{20}	H_2O 2; EtOH 3; eth 3; ace 4 1.5462^{20}
588	Benzaldehyde, 3-methyl-	C_8H_8O 120.15	620-23-5	 199	4-07-00-00669 1.0189^{21}	H_2O 2; EtOH 5; eth 5; ace 4 1.5413^{21}
589	Benzaldehyde, 4-methyl- *p*-Tolualdehyde	C_8H_8O 120.15	104-87-0	 204-5.106^{10}	4-07-00-00672 1.0194^{17}	H_2O 2; EtOH 5; eth 5; ace 5 1.5454^{20}
590	Benzaldehyde, 4-(1-methylethyl)- Cuminaldehyde	$C_{10}H_{12}O$ 148.20	122-03-2	2623 235.5	4-07-00-00723 0.9755^{20}	H_2O I; EtOH 3; eth 3; ctc 2 1.5301^{20}
591	Benzaldehyde, 2-nitro-	$C_7H_5NO_3$ 151.12	552-89-6 43.5	 153^{23}	4-07-00-00584 1.2844^{20}	H_2O 2; EtOH 4; eth 4; ace 4
592	Benzaldehyde, 3-nitro-	$C_7H_5NO_3$ 151.12	99-61-6 58.5	 164^{23}	4-07-00-00591 1.2792^{20}	H_2O 2; EtOH 3; eth 3; ace 4
593	Benzaldehyde, 4-nitro-	$C_7H_5NO_3$ 151.12	555-16-8 107	 sub	4-07-00-00598 1.496^{25}	H_2O 2; EtOH 4; eth 2; bz 3
594	Benzaldehyde, oxime, (*E*)-	C_7H_7NO 121.14	622-31-1 35	 119^{10}	4-07-00-00527 1.145^{20}	H_2O 3; EtOH 4; eth 4
595	Benzaldehyde, oxime, (*Z*)-	C_7H_7NO 121.14	622-32-2 36.5	 200	4-07-00-00527 1.1111^{20}	bz 4; eth 4; EtOH 4 1.5908^{20}
596	Benzaldehyde, pentafluoro-	C_7HF_5O 196.08	653-37-2 20	 167	4-07-00-00561	 1.4506^{20}
597	Benzaldehyde, 3-phenoxy-	$C_{13}H_{10}O_2$ 198.22	39515-51-0	 140$^{0.1}$	4-08-00-00242 1.147^{25}	 1.5954^{20}
598	Benzaldehyde, 2,3,5,6-tetramethyl- Durene aldehyde	$C_{11}H_{14}O$ 162.23	17432-37-0 20	 135^{11}	3-07-00-01148	 1.5560^{30}
599	Benzaldehyde, 2,4,6-trimethyl-	$C_{10}H_{12}O$ 148.20	487-68-3 14	 238.5	4-07-00-00730 1.0154^{25}	H_2O 1; EtOH 3; eth 3; ace 3
600	Benzamide Benzoic acid amide	C_7H_7NO 121.14	55-21-0 129.1	1067 290	4-09-00-00725 1.0792^{130}	H_2O 2; EtOH 4; eth 2; bz 2
601	Benzamide, 2-amino-	$C_7H_8N_2O$ 136.15	88-68-6 110.5 dec	 300	4-14-00-01010	H_2O 3; EtOH 3; eth 2; bz 2
602	Benzamide, 4-amino-*N*-[2-(diethylamino)ethyl]-, monohydrochloride Procainamide hydrochloride	$C_{13}H_{22}ClN_3O$ 271.79	614-39-1 167	7762	4-14-00-01154	H_2O 4; EtOH 3; eth 1; bz 1
603	Benzamide, *N*-[(4-aminophenyl)sulfonyl]- Sulfabenzamide	$C_{13}H_{12}N_2O_3S$ 276.32	127-71-9 181.5	8868	4-14-00-02664	
604	Benzamide, 2-bromo-	C_7H_6BrNO 200.03	4001-73-4 160.5	 sub	4-09-00-01013	H_2O 2; EtOH 3; eth 2; tfa 2
605	Benzamide, 3-bromo-	C_7H_6BrNO 200.03	22726-00-7 155.3	 sub	4-09-00-01015	H_2O 2; EtOH 3
606	Benzamide, 5-bromo-*N*,2-dihydroxy- 5-Bromosalicylhydroxamic acid	$C_7H_6BrNO_3$ 232.03	5798-94-7 232 dec	1424	4-10-00-00221	
607	Benzamide, *N*-butyl-4-chloro-2-hydroxy- Buclosamide	$C_{11}H_{14}ClNO_2$ 227.69	575-74-6 91.5	1450		
608	Benzamide, 5-chloro-*N*-(3,4-dichlorophenyl)-2-hydroxy- 3',4',5-Trichlorosalicylanilide	$C_{13}H_8Cl_3NO_2$ 316.57	642-84-2 247	9558	4-12-00-01273	
609	Benzamide, 3,5-dibromo-*N*-(4-bromophenyl)-2-hydroxy- Tribromsalan	$C_{13}H_8Br_3NO_2$ 449.92	87-10-5 227	9529		
610	Benzamide, 3,5-dichloro-*N*-(3,4-dichlorophenyl)-2-hydroxy- 3,3',4',5-Tetrachlorosalicylanilide	$C_{13}H_7Cl_4NO_2$ 351.01	1154-59-2 161	9127		
611	Benzamide, *N*,*N*-diethyl-4-hydroxy-3-methoxy- Ethamivan	$C_{12}H_{17}NO_3$ 223.27	304-84-7 95	3673	4-10-00-01481	chl 3
612	Benzamide, *N*,*N*-diethyl-3-methyl- DEET	$C_{12}H_{17}NO$ 191.27	134-62-3	2848 160^{19}; 111^{1}	4-09-00-01716 0.996^{20}	H_2O 4; bz 4; eth 4; EtOH 4 1.5212^{20}
613	Benzamide, *N*,2-dihydroxy- Salicylhydroxamic acid	$C_7H_7NO_3$ 153.14	89-73-6 168	8300 sub	4-10-00-00192	H_2O 2; EtOH 4; eth 4; HOAc 3
614	Benzamide, 3,5-dinitro- Nitromide	$C_7H_5N_3O_5$ 211.13	121-81-3 184	6533	4-09-00-01351	H_2O 4
615	Benzamide, 2-ethoxy- Ethenzamide	$C_9H_{11}NO_2$ 165.19	938-73-8 133	3685	4-10-00-00175	H_2O 2; EtOH 4; eth 4; chl 2
616	Benzamide, 2-hydroxy- Salicylamide	$C_7H_7NO_2$ 137.14	65-45-2 142	8297 181.5^{14}	4-10-00-00169 1.175^{140}	H_2O 2; EtOH 3; eth 2
617	Benzamide, 2-hydroxy-*N*-phenyl- Salicylanilide	$C_{13}H_{11}NO_2$ 213.24	87-17-2 136.5	8299	4-12-00-00906	H_2O 3; EtOH 2; eth 2; bz 2
618	Benzamide, 4-methoxy-	$C_8H_9NO_2$ 151.17	3424-93-9 166.5	 295	4-10-00-00433	H_2O 4; EtOH 4
619	Benzamide, 2-methyl- *o*-Toluamide	C_8H_9NO 135.17	527-85-5 147	9454	4-09-00-01701	H_2O 2; EtOH 4; eth 2; bz 2
620	Benzamide, *N*-methyl-	C_8H_9NO 135.17	613-93-4 82	 291; 167^{12}	4-09-00-00727	EtOH 3; ace 3
621	Benzamide, *N*-(2-methylphenyl)-	$C_{14}H_{13}NO$ 211.26	584-70-3 145.5		4-12-00-01757 1.205^{15}	H_2O 2; EtOH 3; eth 3

PHYSICAL CONSTANTS OF ORGANIC COMPOUNDS (continued)

No.	Name Synonym	Mol. Form. Mol. Wt.	CAS RN mp/°C	Merck No. bp/°C	Beil. Ref. den/g cm^{-3}	Solubility n_D
622	Benzamide, *N*-(3-methylphenyl)- *m*-Benzotoluidide	$C_{14}H_{13}NO$ 211.26	582-77-4 125		4-12-00-01825 1.170^{15}	EtOH 4
623	Benzamide, *N*-(4-methylphenyl)-	$C_{14}H_{13}NO$ 211.26	582-78-5 158		4-12-00-01910 1.202^{15}	eth 4; EtOH 4
624	Benzamide, 2-nitro-	$C_7H_6N_2O_3$ 166.14	610-15-1 176.6	 317	4-09-00-01049	H_2O 3; EtOH 3; eth 3
625	Benzamide, 3-nitro-	$C_7H_6N_2O_3$ 166.14	645-09-0 142.7	 312.5	4-09-00-01061	H_2O 3; EtOH 3; eth 3
626	Benzamide, *N*-phenyl- Benzanilide	$C_{13}H_{11}NO$ 197.24	93-98-1 163	1068 sub 117	4-12-00-00417 1.315^{25}	H_2O 1; EtOH 2; eth 2; HOAc 2
627	Benzamide, *N,N*,4-trimethyl-	$C_{10}H_{13}NO$ 163.22	14062-78-3 41	 156^{10}	0-09-00-00486	H_2O 2; EtOH 2
628	Benz[b]indeno[1,2-d]pyran-9(6H)-one, 6a,7-dihydro-3,4,6a,10-tetrahydroxy- Hematein	$C_{16}H_{12}O_6$ 300.27	475-25-2 250 dec	4553	4-18-00-03343	H_2O 1; EtOH 2; eth 1; bz 1
629	Benz[b]indeno[1,2-d]pyran-3,4,6a,9,10(6H)-pentol, 7,11b-dihydro-, *cis*-(+)- Hematoxylin	$C_{16}H_{14}O_6$ 302.28	517-28-2 140	4556	2-17-00-00273	H_2O 2; EtOH 3; eth 2; alk 3
630	Benz[c]acridine 12-Azabenz[a]anthracene	$C_{17}H_{11}N$ 229.28	225-51-4 132		5-20-08-00519	bz 4; eth 4; EtOH 4
631	Benz[e]acephenanthrylene Benzo[b]fluoranthene	$C_{20}H_{12}$ 252.32	205-99-2 168		4-05-00-02686	H_2O 1; bz 5
632	1H-Benz[e]indene, 2,3-dihydro-	$C_{13}H_{12}$ 168.24	4944-94-9	 294.5	4-05-00-01865 1.066^{20}	H_2O 1; HOAc 3 1.6290^{20}
633	Benzenamine Aniline	C_6H_7N 93.13	62-53-3 -6.0	687 184.1	4-12-00-00223 1.0217^{20}	H_2O 3; EtOH 5; eth 5; ace 5 1.5863^{20}
634	Benzenamine, 4-[(4-aminophenyl)(4-imino-2,5-cyclohexadien-1-ylidene)methyl]-2-methyl-, monohydrochloride Magenta I	$C_{20}H_{20}ClN_3$ 337.85	632-99-5 200 dec	5528		H_2O 2; EtOH 2; eth 1
635	Benzenamine, 4,4'-azobis- *p*-Diaminoazobenzene	$C_{12}H_{12}N_4$ 212.25	538-41-0 250.5	2954	4-16-00-00515	H_2O 2; EtOH 3; bz 4; chl 4
636	Benzenamine, 2,6-bis(1-methylethyl)-	$C_{12}H_{19}N$ 177.29	24544-04-5 -45	 257	3-12-00-02764 0.94^{25}	 1.5332^{20}
637	Benzenamine, 3,5-bis(trifluoromethyl)-	$C_8H_5F_6N$ 229.12	328-74-5	 85^{15}	3-12-00-02498 1.487^{25}	 1.4335^{20}
638	Benzenamine, 2-bromo- *o*-Bromoaniline	C_6H_6BrN 172.02	615-36-1 32	 229	4-12-00-01487 1.578^{20}	H_2O 1; EtOH 3; eth 3 1.6113^{20}
639	Benzenamine, 3-bromo- *m*-Bromoaniline	C_6H_6BrN 172.02	591-19-5 18.5	 251	4-12-00-01491 1.5793^{20}	H_2O 2; EtOH 3; eth 3 1.6260^{20}
640	Benzenamine, 4-bromo- *p*-Bromoaniline	C_6H_6BrN 172.02	106-40-1 66.4	1392 dec	4-12-00-01497 1.4970^{100}	H_2O 1; EtOH 3; eth 3; chl 2
641	Benzenamine, 2-bromo-4,6-dichloro-	$C_6H_4BrCl_2N$ 240.91	697-86-9 83.5	 273	2-12-00-00355	bz 4; EtOH 4; chl 4
642	Benzenamine, 4-bromo-*N,N*-diethyl-	$C_{10}H_{14}BrN$ 228.13	2052-06-4 38	 270	4-12-00-01500	H_2O 1; EtOH 4; eth 4
643	Benzenamine, 2-bromo-*N,N*-dimethyl-	$C_8H_{10}BrN$ 200.08	698-00-0	 $107\text{-}8^{14}$	4-12-00-01487 1.3839^{25}	EtOH 4 1.5748^{25}
644	Benzenamine, 3-bromo-*N,N*-dimethyl-	$C_8H_{10}BrN$ 200.08	16518-62-0 11	 259	4-12-00-01492 1.3651^{20}	EtOH 4; HOAc 4; con acid 3
645	Benzenamine, 4-bromo-*N,N*-dimethyl-	$C_8H_{10}BrN$ 200.08	586-77-6 55	 264	4-12-00-01499 1.3220^{100}	H_2O 1; EtOH 3; eth 4; os 3
646	Benzenamine, 2-bromo-4,6-dinitro-	$C_6H_4BrN_3O_4$ 262.02	1817-73-8 153.5	 sub	4-12-00-01734	EtOH 4; ace 4; HOAc 3
647	Benzenamine, 2-bromo-4-methyl-	C_7H_8BrN 186.05	583-68-6 26	 240	4-12-00-01992 1.510^{20}	H_2O 1; EtOH 3; eth 3 1.5999^{20}
648	Benzenamine, 2-bromo-5-methyl- 2-Bromo-5-methylaniline	C_7H_8BrN 186.05	53078-85-6 46	 129^{15}	3-12-00-01998 1.470^{25}	eth 4; EtOH 4
649	Benzenamine, 3-bromo-4-methyl-	C_7H_8BrN 186.05	7745-91-7 26	 255.5	4-12-00-01991	eth 4
650	Benzenamine, 4-bromo-2-methyl-	C_7H_8BrN 186.05	583-75-5 59.5	 240	4-12-00-01804	H_2O 2; EtOH 3; eth 4; chl 2
651	Benzenamine, 4-bromo-3-methyl-	C_7H_8BrN 186.05	6933-10-4 81	 240	4-12-00-01856	H_2O 1; EtOH 4; chl 2
652	Benzenamine, 5-bromo-2-methyl-	C_7H_8BrN 186.05	39478-78-9 33	 255 dec; 165^{25}	3-12-00-01923	eth 4; EtOH 4
653	Benzenamine, 2-bromo-6-nitro- Aniline, 2-bromo-6-nitro-	$C_6H_5BrN_2O_2$ 217.02	59255-95-7 74.5		2-12-00-00402 1.988^{25}	EtOH 4
654	Benzenamine, 4-bromo-2-nitro-	$C_6H_5BrN_2O_2$ 217.02	875-51-4 111.5	 sub	3-12-00-01670	EtOH 4
655	Benzenamine, 4-butyl-	$C_{10}H_{15}N$ 149.24	104-13-2	 261	4-12-00-02807 0.945^{20}	ctc 2
656	Benzenamine, *N*-butyl- *N*-Butylaniline	$C_{10}H_{15}N$ 149.24	1126-78-9 -14.4	 243.5	4-12-00-00256 0.9323^{20}	eth 4; EtOH 4 1.5341^{20}
657	Benzenamine, 4,4'-carbonimidoylbis[*N,N*-dimethyl- 4,4'-Dimethylaminobenzophenonimide	$C_{17}H_{21}N_3$ 267.37	492-80-8 136		4-14-00-00256	H_2O 1; EtOH 3; eth 2

No.	Name Synonym	Mol. Form. Mol. Wt.	CAS RN mp/°C	Merck No. bp/°C	Beil. Ref. den/g cm^{-3}	Solubility n_D
658	Benzenamine, 2-chloro- *o*-Chloroaniline	C_6H_6ClN 127.57	95-51-2 -14	2118 208.8	4-12-00-01115	H_2O 1; EtOH 5; eth 3; ace 3 1.5895^{20}
659	Benzenamine, 3-chloro- *m*-Chloroaniline	C_6H_6ClN 127.57	108-42-9 -10.4	2118 230.5	4-12-00-01137 1.2161^{20}	H_2O 1; EtOH 5; eth 5; ace 5 1.5941^{20}
660	Benzenamine, 4-chloro- *p*-Chloroaniline	C_6H_6ClN 127.57	106-47-8 72.5	2118 232	4-12-00-01166 1.429^{19}	H_2O 3; EtOH 3; eth 3; chl 3 1.5546^{87}
661	Benzenamine, 4-chloro-*N*,*N*-diethyl-	$C_{10}H_{14}ClN$ 183.68	2873-89-4 45.5	 252; 113^{6}	4-12-00-01170	EtOH 3
662	Benzenamine, 2-chloro-*N*,*N*-dimethyl-	$C_8H_{10}ClN$ 155.63	698-01-1	 205	4-12-00-01116 1.1067^{20}	bz 4; EtOH 4 1.5578^{20}
663	Benzenamine, 4-chloro-*N*,*N*-dimethyl-	$C_8H_{10}ClN$ 155.63	698-69-1 35.5	 231	4-12-00-01169 1.0480^{100}	EtOH 3
664	Benzenamine, 2-chloro-6-fluoro- Aniline, 2-chloro-6-fluoro	C_6H_5ClFN 145.56	363-51-9	 91^{30}	4-12-00-01238 1.316^{23}	 1.5511^{23}
665	Benzenamine, 2-chloro-, hydrochloride Aniline, 2-chloro, hydrochloride	$C_6H_7Cl_2N$ 164.03	137-04-2 235		3-12-00-01281 1.505^{18}	H_2O 4
666	Benzenamine, 2-chloro-5-methoxy-	C_7H_8ClNO 157.60	2401-24-3 27	 142^{18}	4-13-00-00999	 1.5848^{25}
667	Benzenamine, 4-chloro-2-methoxy- 4-Chloro-*o*-anisidine	C_7H_8ClNO 157.60	93-50-5 52	 260	4-13-00-00885	EtOH 3; eth 3; bz 3; chl 3
668	Benzenamine, 2-chloro-4-methyl- 2-Chloro-*p*-toluidine	C_7H_8ClN 141.60	615-65-6 7	 220	3-12-00-02152 1.151^{20}	EtOH 2; bz 2 1.5748^{22}
669	Benzenamine, 2-chloro-5-methyl- 6-Chloro-*m*-toluidine	C_7H_8ClN 141.60	95-81-8 29.5	 229	4-12-00-01849	EtOH 4
670	Benzenamine, 2-chloro-*N*-methyl-	C_7H_8ClN 141.60	932-32-1	 218	4-12-00-01116 1.1735^{11}	EtOH 3; ace 3; bz 3 1.5780^{25}
671	Benzenamine, 3-chloro-2-methyl- 3-Chloro-*o*-toluidine	C_7H_8ClN 141.60	87-60-5 1	 245	4-12-00-01800	H_2O 3; EtOH 3; eth 1; bz 1 1.5880^{20}
672	Benzenamine, 3-chloro-4-methyl- 3-Chloro-*p*-toluidine	C_7H_8ClN 141.60	95-74-9 26	 243	4-12-00-01985	EtOH 3; ctc 2
673	Benzenamine, 4-chloro-2-methyl-	C_7H_8ClN 141.60	95-69-2 30.3	 244	4-12-00-01796	EtOH 3; ctc 2
674	Benzenamine, 4-chloro-3-methyl- 4-Chloro-*m*-toluidine	C_7H_8ClN 141.60	7149-75-9 83.5	 241	4-12-00-01850	ace 4; bz 4; EtOH 4
675	Benzenamine, 4-chloro-*N*-methyl-	C_7H_8ClN 141.60	932-96-7	 240	4-12-00-01168 1.169^{11}	EtOH 3; ace 3; bz 3 1.5835^{20}
676	Benzenamine, 5-chloro-2-methyl- 5-Chloro-*o*-toluidine	C_7H_8ClN 141.60	95-79-4 26	 239; 140^{38}	4-12-00-01794	EtOH 4
677	Benzenamine, 5-chloro-2-nitro-	$C_6H_5ClN_2O_2$ 172.57	1635-61-6 127.8	 sub	4-12-00-01673	eth 4; EtOH 4
678	Benzenamine, 3-chloro-*N*-phenyl-	$C_{12}H_{10}ClN$ 203.67	101-17-7	 338	4-12-00-01141 1.2^{25}	H_2O 1; EtOH 3; eth 3; bz 3 1.6513^{20}
679	Benzenamine, 4-chloro-*N*-phenyl-	$C_{12}H_{10}ClN$ 203.67	1205-71-6 74	 335	4-12-00-01170	bz 4; eth 4; EtOH 4; lig 4
680	Benzenamine, *N*-cyclohexyl-	$C_{12}H_{17}N$ 175.27	1821-36-9 16	 279; 192^{73}	4-12-00-00268 1.0155^{20}	H_2O 1; EtOH 3; eth 3; bz 3 1.5610^{20}
681	Benzenamine, 2,4-dibromo-	$C_6H_5Br_2N$ 250.92	615-57-6 79.5	 156^{74}	4-12-00-01532 2.260^{20}	EtOH 3; eth 3; chl 3; HOAc 3
682	Benzenamine, 2,6-dibromo-	$C_6H_5Br_2N$ 250.92	608-30-0 87.5	 263	4-12-00-01535	EtOH 4; eth 4; bz 4; chl 4
683	Benzenamine, 3,4-dibromo- 3,4-Dibromoaniline	$C_6H_5Br_2N$ 250.92	615-55-4 80.5	 100	4-12-00-01535	EtOH 4
684	Benzenamine, *N*,*N*-dibutyl-	$C_{14}H_{23}N$ 205.34	613-29-6 -32.2	 274.8	4-12-00-00256 0.9037^{20}	H_2O 1; EtOH 5; eth 5; ace 4 1.5186^{20}
685	Benzenamine, 2,3-dichloro- 2,3-Dichloroaniline	$C_6H_5Cl_2N$ 162.02	608-27-5 24	 252	0-12-00-00621	EtOH 3; eth 4; ace 3; bz 2
686	Benzenamine, 2,4-dichloro- 2,4-Dichloroaniline	$C_6H_5Cl_2N$ 162.02	554-00-7 63.5	 245	4-12-00-01241 1.567^{20}	H_2O 2; EtOH 3; eth 3; chl 2
687	Benzenamine, 2,5-dichloro- 2,5-Dichloroaniline	$C_6H_5Cl_2N$ 162.02	95-82-9 50	 251	4-12-00-01250	H_2O 2; EtOH 3; eth 3; bz 3
688	Benzenamine, 2,6-dichloro- 2,6-Dichloroaniline	$C_6H_5Cl_2N$ 162.02	608-31-1 39		2-12-00-00337	H_2O 2; EtOH 3; eth 3
689	Benzenamine, 3,4-dichloro- 3,4-Dichloroaniline	$C_6H_5Cl_2N$ 162.02	95-76-1 72	3041 272	4-12-00-01257	EtOH 3; eth 3; bz 2; chl 2
690	Benzenamine, 3,5-dichloro- 3,5-Dichloroaniline	$C_6H_5Cl_2N$ 162.02	626-43-7 52	 261	4-12-00-01274	H_2O 1; EtOH 3; eth 3; ctc 3
691	Benzenamine, 2,6-dichloro-4-ethoxy- Aniline, 2,6-dichloro-4-ethoxy	$C_8H_9Cl_2NO$ 206.07	51225-20-8 46	 275	2-13-00-00276	bz 4; eth 4; EtOH 4
692	Benzenamine, 2,6-dichloro-4-nitro- 2,6-Dichloro-4-nitroaniline	$C_6H_4Cl_2N_2O_2$ 207.02	99-30-9 191		4-12-00-01681	EtOH 3; DMSO-d_6 2; acid 3
693	Benzenamine, 3,4-diethoxy- 3,4-Diethoxyaniline	$C_{10}H_{15}NO_2$ 181.23	39052-12-5 48	 $132\text{-}5^{0.45}$	4-13-00-02507	bz 4; eth 4; EtOH 4
694	Benzenamine, 2,6-diethyl-	$C_{10}H_{15}N$ 149.24	579-66-8 3	 243	4-12-00-02841 0.906^{25}	 1.5452^{20}
695	Benzenamine, *N*,*N*-diethyl- *N*,*N*-Diethylaniline	$C_{10}H_{15}N$ 149.24	91-66-7 -38.8	3102 216.3	4-12-00-00252 0.9307^{20}	H_2O 2; EtOH 3; eth 4; ace 3 1.5409^{20}
696	Benzenamine, *N*,*N*-diethyl-2-methyl-	$C_{11}H_{17}N$ 163.26	606-46-2 -60	 209	4-12-00-01748 0.9286^{20}	H_2O 2; EtOH 5; eth 5; ctc 3 1.5153^{20}
697	Benzenamine, *N*,*N*-diethyl-4-methyl-	$C_{11}H_{17}N$ 163.26	613-48-9	 229	4-12-00-01875 0.9242^{18}	H_2O 2; EtOH 5; eth 5

No.	Name Synonym	Mol. Form. Mol. Wt.	CAS RN mp/°C	Merck No. bp/°C	Beil. Ref. den/g cm^{-3}	Solubility n_D
698	Benzenamine, *N,N*-diethyl-4-nitro-	$C_{10}H_{14}N_2O_2$ 194.23	2216-15-1 77.5		4-12-00-01618 1.225^{25}	EtOH 3; lig 2
699	Benzenamine, *N,N*-diethyl-4-nitroso- Aniline, *N,N*-diethyl-4-nitroso	$C_{10}H_{14}N_2O$ 178.23	120-22-9 87.5		4-12-00-01559 1.24^{15}	H_2O 2; EtOH 3; eth 3; ace 3
700	Benzenamine, 2,4-difluoro- 2,4-Difluoroaniline	$C_6H_5F_2N$ 129.11	367-25-9 -7.5	3131 170	4-12-00-01112 1.268^{25}	 1.5063^{20}
701	Benzenamine, 2,4-diiodo- 2,4-Diiodoaniline	$C_6H_5I_2N$ 344.92	533-70-0 95.5	3173	3-12-00-01508 2.748^{25}	ace 4; bz 4; eth 4; EtOH 4
702	Benzenamine, 2,5-dimethoxy-	$C_8H_{11}NO_2$ 153.18	102-56-7 82.5	 270	4-13-00-02548	H_2O 3; EtOH 3; chl 3; lig 3
703	Benzenamine, 2,6-dimethoxy- 2,6-Dimethoxyaniline	$C_8H_{11}NO_2$ 153.18	2734-70-5 76.5	 146^{23}	4-13-00-02526	bz 4; eth 4; EtOH 4; lig 4
704	Benzenamine, 2,3-dimethyl- 2,3-Xylidine	$C_8H_{11}N$ 121.18	87-59-2 <-15	 221.5	4-12-00-02497 0.9931^{20}	H_2O 2; EtOH 4; eth 4; ctc 3 1.5684^{20}
705	Benzenamine, 2,4-dimethyl- 2,4-Xylidine	$C_8H_{11}N$ 121.18	95-68-1 -14.3	 214	4-12-00-02545 0.9723^{20}	H_2O 2; EtOH 3; eth 3; bz 3 1.5569^{20}
706	Benzenamine, 2,5-dimethyl- 2,5-Xylidine	$C_8H_{11}N$ 121.18	95-78-3 15.5	 214	4-12-00-02567 0.9790^{21}	H_2O 2; eth 3; ctc 3 1.5591^{21}
707	Benzenamine, 2,6-dimethyl- 2,6-Xylidine	$C_8H_{11}N$ 121.18	87-62-7 11.2	 215	4-12-00-02521 0.9842^{20}	eth 4; EtOH 4 1.5610^{20}
708	Benzenamine, 3,4-dimethyl- 3,4-Xylidine	$C_8H_{11}N$ 121.18	95-64-7 51	 228	4-12-00-02502 1.076^{18}	H_2O 2; eth 3; chl 2; lig 4
709	Benzenamine, 3,5-dimethyl- 3,5-Xylidine	$C_8H_{11}N$ 121.18	108-69-0 9.8	 220.5	4-12-00-02561 0.9706^{20}	H_2O 2; eth 3; ctc 3 1.5581^{20}
710	Benzenamine, *N*,2-dimethyl-	$C_8H_{11}N$ 121.18	611-21-2	 207.5	4-12-00-01746 0.9709^{20}	H_2O 1; EtOH 5; eth 5; ace 3 1.5649^{20}
711	Benzenamine, *N*,3-dimethyl-	$C_8H_{11}N$ 121.18	696-44-6	 206.5	4-12-00-01815	H_2O 1; EtOH 5; eth 5; ace 3 1.5557^{25}
712	Benzenamine, *N*,4-dimethyl-	$C_8H_{11}N$ 121.18	623-08-5	 210	4-12-00-01874 0.9348^{55}	H_2O 1; EtOH 5; eth 5; ace 3 1.5568^{20}
713	Benzenamine, *N,N*-dimethyl- *N,N*-Dimethylaniline	$C_8H_{11}N$ 121.18	121-69-7 2.4	3223 194.1	4-12-00-00243 0.9557^{20}	H_2O 2; EtOH 3; eth 3; ace 3 1.5582^{20}
714	Benzenamine, 2-(1,1-dimethylethyl)-	$C_{10}H_{15}N$ 149.24	6310-21-0	 234	4-12-00-02823 0.977^{15}	EtOH 4; eth 4; bz 4; acid 3 1.5453^{20}
715	Benzenamine, 4-(1,1-dimethylethyl)-	$C_{10}H_{15}N$ 149.24	769-92-6 17	 241	4-12-00-02828 0.9525^{15}	H_2O 2; EtOH 5; eth 5; bz 4 1.5380^{20}
716	Benzenamine, *N*-(1,1-dimethylethyl)- *N-tert*-Butylaniline	$C_{10}H_{15}N$ 149.24	937-33-7	 215; 95^{19}	4-12-00-00257	EtOH 3; ace 4; bz 4; chl 4 1.5270^{20}
717	Benzenamine, *N,N*-dimethyl-, hydrochloride	$C_8H_{12}ClN$ 157.64	5882-44-0 90		4-12-00-00243 1.1156^{19}	H_2O 4; EtOH 4; chl 4
718	Benzenamine, *N,N*-dimethyl-2-nitro-	$C_8H_{10}N_2O_2$ 166.18	610-17-3 -20	 146^{20}	4-12-00-01564 1.1794^{20}	H_2O 3; EtOH 4; eth 3; chl 4 1.6102^{20}
719	Benzenamine, *N,N*-dimethyl-3-nitro-	$C_8H_{10}N_2O_2$ 166.18	619-31-8 60.5	 282.5	4-12-00-01591 1.313^{17}	H_2O 1; EtOH 3; eth 3
720	Benzenamine, *N,N*-dimethyl-4-nitroso- *p*-Nitroso-*N,N*-dimethylaniline	$C_8H_{10}N_2O$ 150.18	138-89-6 92.5	6559	4-12-00-01558 1.145^{20}	H_2O 2; EtOH 3; eth 3; chl 3
721	Benzenamine, *N,N*-dimethyl-4-(phenylazo)- 4-(Dimethylamino)azobenzene	$C_{14}H_{15}N_3$ 225.29	60-11-7 117	3218 dec	3-16-00-00340	H_2O 1; EtOH 4; eth 3; chl 2
722	Benzenamine, *N,N*-dimethyl-4-(2-phenylethenyl)-, (*E*)-	$C_{16}H_{17}N$ 223.32	838-95-9 150	 $145^{0.1}$	4-12-00-03400	H_2O 1; EtOH 3; chl 3
723	Benzenamine, *N,N*-dimethyl-4-[[3-(trifluoromethyl)phenyl]azo]- Azobenzene, 4-dimethylamino-3'-trifluoromethyl	$C_{15}H_{14}F_3N_3$ 293.29	328-96-1 80	 76^{8}	4-16-00-00487	chl 3
724	Benzenamine, 2,3-dinitro- 2,3-Dinitroaniline	$C_6H_5N_3O_4$ 183.12	602-03-9 128		4-12-00-01689 1.646^{50}	H_2O 1; EtOH 3; eth 2
725	Benzenamine, 2,4-dinitro- 2,4-Dinitroaniline	$C_6H_5N_3O_4$ 183.12	97-02-9 180	3263	4-12-00-01689 1.615^{14}	H_2O 1; EtOH 2; ace 2; HCl 2
726	Benzenamine, 2,6-dinitro- 2,6-Dinitroaniline	$C_6H_5N_3O_4$ 183.12	606-22-4 141.5	3264	4-12-00-01729	H_2O 1; EtOH 2; eth 3; bz 3
727	Benzenamine, 3,5-dinitro- 3,5-Dinitroaniline	$C_6H_5N_3O_4$ 183.12	618-87-1 163		4-12-00-01729 1.601^{50}	H_2O 1; EtOH 3; eth 3; ace 2
728	Benzenamine, *N,N*-diphenyl- Triphenylamine	$C_{18}H_{15}N$ 245.32	603-34-9 127	 365	4-12-00-00276 0.774^{0}	H_2O 1; EtOH 2; eth 3; bz 3 1.353^{16}
729	Benzenamine, 4-(diphenylmethyl)-	$C_{19}H_{17}N$ 259.35	603-38-3 84.5	 248^{12}	4-12-00-03470	bz 4; eth 4; lig 4
730	Benzenamine, *N,N*-dipropyl-	$C_{12}H_{19}N$ 177.29	2217-07-4	 242	4-12-00-00255 0.9104^{20}	H_2O 1; EtOH 3; eth 3; ace 3 1.5271^{20}
731	Benzenamine, 4,4'-(1,2-ethanediyl)bis-	$C_{14}H_{16}N_2$ 212.29	621-95-4 137	 sub	4-13-00-00401	H_2O 1; EtOH 4
732	Benzenamine, 4,4'-(1,2-ethenediyl)bis-, (*E*)-	$C_{14}H_{14}N_2$ 210.28	7314-06-9 231	 sub	4-13-00-00453	H_2O 2; bz 2; MeOH 3; CS_2 2
733	Benzenamine, 2-ethenyl-	C_8H_9N 119.17	3867-18-3	 112^{20}; 97^{8}	4-12-00-02914 1.0181^{20}	ace 3; bz 3 1.6124^{20}
734	Benzenamine, 3-ethenyl-	C_8H_9N 119.17	15411-43-5	 $112\text{-}5^{12}$	4-12-00-02915 1.0198^{20}	ace 3; bz 3 1.6069^{26}
735	Benzenamine, 4-ethenyl-	C_8H_9N 119.17	1520-21-4 23.5	 116^{9}	4-12-00-02915 1.010^{20}	ace 3; bz 3 1.6250^{22}

No.	Name Synonym	Mol. Form. Mol. Wt.	CAS RN mp/°C	Merck No. bp/°C	Beil. Ref. den/g cm^{-3}	Solubility n_D
736	Benzenamine, 2-ethoxy- *o*-Phenetidine	$C_8H_{11}NO$ 137.18	94-70-2 <-21	7187 232.5	4-13-00-00807	H_2O 2; EtOH 3; eth 3; ctc 2 1.5560^{20}
737	Benzenamine, 4-ethoxy- *p*-Phenetidine	$C_8H_{11}NO$ 137.18	156-43-4 2.4	7188 254	4-13-00-01017 1.0652^{16}	H_2O 2; EtOH 3; eth 3; chl 3 1.5528^{20}
738	Benzenamine, 3-ethoxy-*N*,*N*-diethyl-	$C_{12}H_{19}NO$ 193.29	1864-92-2	 286; $97^{0.6}$	4-13-00-00970	EtOH 3; bz 3; HOAc 3 1.5325^{25}
739	Benzenamine, 2-ethoxy-5-nitro- *o*-Phenetidine, 5-nitro	$C_8H_{10}N_2O_3$ 182.18	136-79-8 96.5	6538 $205\text{-}6^{14}$	4-13-00-00896	eth 4; EtOH 4
740	Benzenamine, 2-ethyl- 2-Ethylaniline	$C_8H_{11}N$ 121.18	578-54-1 -43	 209.5	4-12-00-02411 0.983^{22}	H_2O 2; EtOH 4; eth 4; chl 2 1.5584^{22}
741	Benzenamine, 3-ethyl-	$C_8H_{11}N$ 121.18	587-02-0 -64	 214; 94^{6}	4-12-00-02417 0.9896^{25}	eth 4; EtOH 4
742	Benzenamine, 4-ethyl- *p*-Ethylaniline	$C_8H_{11}N$ 121.18	589-16-2 -4.8	 217.5	4-12-00-02419 0.9679^{20}	H_2O 2; EtOH 4; eth 4; ctc 2 1.5554^{20}
743	Benzenamine, *N*-ethyl- *N*-Ethylaniline	$C_8H_{11}N$ 121.18	103-69-5 -63.5	3722 203.0	4-12-00-00250 0.9625^{20}	H_2O 1; EtOH 5; eth 5; ace 4 1.5559^{20}
744	Benzenamine, 2-ethyl-*N*-(2-ethylphenyl)-	$C_{16}H_{19}N$ 225.33	64653-59-4 29	 336^{603}; 173^{10}	4-12-00-02413	H_2O 1; EtOH 4; eth 4; chl 2 1.5550^{25}
745	Benzenamine, 2-ethyl-6-methyl-	$C_9H_{13}N$ 135.21	24549-06-2 -33	 231	4-12-00-02638 0.968^{25}	 1.5525^{20}
746	Benzenamine, *N*-ethyl-2-methyl-	$C_9H_{13}N$ 135.21	94-68-8 <-15	 216	4-12-00-01747 0.948^{25}	EtOH 3; eth 3 1.5456^{20}
747	Benzenamine, *N*-ethyl-3-methyl-	$C_9H_{13}N$ 135.21	102-27-2	 221	4-12-00-01816	EtOH 3; eth 3 1.5451^{20}
748	Benzenamine, *N*-ethyl-4-methyl- *N*-Ethyl-4-toluidine	$C_9H_{13}N$ 135.21	622-57-1	 217	4-12-00-01875 0.9391^{16}	EtOH 3; eth 3
749	Benzenamine, *N*-ethyl-*N*-methyl-	$C_9H_{13}N$ 135.21	613-97-8	 204	4-12-00-00252 0.92^{55}	H_2O 1; EtOH 5; eth 5; ctc 3
750	Benzenamine, *N*-ethyl-*N*-nitroso-	$C_8H_{10}N_2O$ 150.18	612-64-6	 $119\text{-}20^{15}$	4-16-00-00865 1.0874^{20}	H_2O 1; HOAc 3
751	Benzenamine, *N*-ethyl-*N*-phenyl-	$C_{14}H_{15}N$ 197.28	606-99-5	 295.5; 147^{10}	4-12-00-00276 1.0377^{20}	eth 4; EtOH 4 1.6095^{20}
752	Benzenamine, 2-fluoro- *o*-Fluoroaniline	C_6H_6FN 111.12	348-54-9 -34.6	 175; 55^{12}	4-12-00-01099 1.1513^{21}	H_2O 1; EtOH 3; eth 3; ctc 2 1.5421^{20}
753	Benzenamine, 3-fluoro- *m*-Fluoroaniline	C_6H_6FN 111.12	372-19-0	 188	4-12-00-01101 1.1561^{19}	H_2O 2; EtOH 3; eth 3; chl 2 1.5436^{20}
754	Benzenamine, 4-fluoro- *p*-Fluoroaniline	C_6H_6FN 111.12	371-40-4 -0.8	4098 182; 85^{19}	4-12-00-01104 1.1725^{20}	H_2O 2; EtOH 3; eth 3; ctc 2 1.5195^{20}
755	Benzenamine, 4-fluoro-2-methyl-	C_7H_8FN 125.15	452-71-1 14.2	 94^{16}	4-12-00-01786 1.1263^{18}	eth 3; ace 3; bz 3; ctc 3 1.5363^{18}
756	Benzenamine, hydrochloride Aniline, hydrochloride	C_6H_8ClN 129.59	142-04-1 198	687	4-12-00-00223 1.2215^{4}	H_2O 4; EtOH 4; eth 1; chl 1
757	Benzenamine, *N*-hydroxy- Phenylhydroxylamine	C_6H_7NO 109.13	100-65-2 83.5	7266	4-15-00-00004	bz 4; eth 4; EtOH 4; chl 4
758	Benzenamine, *N*-hydroxy-4-methyl-	C_7H_9NO 123.15	623-10-9 96	 117 dec	4-15-00-00019	eth 4; EtOH 4; chl 4
759	Benzenamine, *N*-hydroxy-*N*-nitroso-, ammonium salt Cupferron	$C_6H_9N_3O_2$ 155.16	135-20-6 163.5	2625	4-16-00-00889	DMSO 2
760	Benzenamine, 3-iodo-	C_6H_6IN 219.02	626-01-7 33	 145^{15}	4-12-00-01543	H_2O 1; EtOH 3; chl 3 1.6811^{20}
761	Benzenamine, 4-iodo- *p*-Iodoaniline	C_6H_6IN 219.02	540-37-4 67.5	4920	4-12-00-01544	H_2O 2; EtOH 3; eth 3; os 3
762	Benzenamine, 2-iodo-4-methyl- 2-Iodo-4-methylaniline	C_7H_8IN 233.05	29289-13-2 40	 dec	4-12-00-01996	ace 4; bz 4; eth 4; EtOH 4
763	Benzenamine, 5-iodo-2-methyl- *o*-Toluidine, 5-iodo	C_7H_8IN 233.05	83863-33-6 48.5	 273 dec	1-12-00-00391	EtOH 4
764	Benzenamine, *N*,*N*′-methanetetraylbis-	$C_{13}H_{10}N_2$ 194.24	622-16-2 169	 331; 175^{20}	4-12-00-00866	H_2O 2; EtOH 2; eth 2; bz 3
765	Benzenamine, *N*,*N*′-methanetetraylbis[4-methyl-	$C_{15}H_{14}N_2$ 222.29	726-42-1 58.5	 221^{20}	4-12-00-01938 1.1500^{20}	
766	Benzenamine, 2-methoxy- *o*-Anisidine	C_7H_9NO 123.15	90-04-0 6.2	 224	4-10-00-00000 1.0923^{20}	H_2O 2; EtOH 3; eth 3; ace 3 1.5715^{10}
767	Benzenamine, 3-methoxy- *m*-Anisidine	C_7H_9NO 123.15	536-90-3 -1	 251	4-13-00-00953 1.096^{20}	H_2O 2; EtOH 3; eth 3; ace 3 1.5794^{20}
768	Benzenamine, 4-methoxy- *p*-Anisidine	C_7H_9NO 123.15	104-94-9 57.2	 243	4-13-00-01015 1.071^{57}	H_2O 3; EtOH 4; eth 4; ace 3 1.5559^{60}
769	Benzenamine, 2-methoxy-*N*,*N*-dimethyl-	$C_9H_{13}NO$ 151.21	700-75-4	 210	4-13-00-00816 1.0160^{23}	chl 4
770	Benzenamine, 2-methoxy-5-methyl- 5-Methyl-*o*-anisidine	$C_8H_{11}NO$ 137.18	120-71-8 53	 235	3-13-00-01577	H_2O 2; EtOH 3; eth 3; bz 3
771	Benzenamine, 3-methoxy-4-methyl-	$C_8H_{11}NO$ 137.18	16452-01-0 56.5	 251	4-13-00-01666	bz 4; eth 4; EtOH 4; lig 4
772	Benzenamine, 4-methoxy-2-methyl-	$C_8H_{11}NO$ 137.18	102-50-1 29.5	 248.5	4-13-00-01698 1.065^{25}	EtOH 4 1.5647^{20}
773	Benzenamine, 5-methoxy-2-methyl-	$C_8H_{11}NO$ 137.18	50868-72-9 47	 253; 140^{20}	4-13-00-01711	eth 4

PHYSICAL CONSTANTS OF ORGANIC COMPOUNDS (continued)

No.	Name Synonym	Mol. Form. Mol. Wt.	CAS RN mp/°C	Merck No. bp/°C	Beil. Ref. den/g cm^{-3}	Solubility n_D
774	Benzenamine, 2-methoxy-6-methyl-	$C_8H_{11}NO$ 137.18	50868-73-0 31	 119^{16}	3-13-00-01549	EtOH 4
775	Benzenamine, 2-methoxy-5-nitro- 5-Nitro-*o*-anisidine	$C_7H_8N_2O_3$ 168.15	99-59-2 118		4-13-00-00896 1.2068^{15}	H_2O 3; EtOH 4; eth 3; ace 4
776	Benzenamine, 3-methoxy-4-nitro- *m*-Anisidine, 4-nitro-	$C_7H_8N_2O_3$ 168.15	16292-88-9 169	 sub	4-13-00-01005	ace 4; EtOH 4
777	Benzenamine, 5-methoxy-2-nitro-	$C_7H_8N_2O_3$ 168.15	16133-49-6 131	 sub	4-13-00-01003	EtOH 4
778	Benzenamine, 2-methyl- *o*-Toluidine	C_7H_9N 107.16	95-53-4 -16.3	9462 200.3	4-12-00-01744 0.9984^{20}	H_2O 2; EtOH 5; eth 5; ctc 5 1.5725^{20}
779	Benzenamine, 3-methyl- *m*-Toluidine	C_7H_9N 107.16	108-44-1 -31.2	9462 203.3	4-12-00-01813 0.9889^{20}	ace 4; bz 4; eth 4; EtOH 4 1.5681^{20}
780	Benzenamine, 4-methyl- *p*-Toluidine	C_7H_9N 107.16	106-49-0 43.7	9462 200.4	4-12-00-01866 0.9619^{20}	H_2O 2; EtOH 4; eth 3; ace 3 1.5534^{45}
781	Benzenamine, *N*-methyl- *N*-Methylaniline	C_7H_9N 107.16	100-61-8 -57	5941 196.2	4-12-00-00241 0.9891^{20}	H_2O 1; EtOH 3; eth 3; ctc 3 1.5684^{20}
782	Benzenamine, 4-(6-methyl-2-benzothiazolyl)-	$C_{14}H_{12}N_2S$ 240.33	92-36-4 194.8	 434	4-27-00-05052	EtOH 2; eth 2; bz 2; HOAc 2
783	Benzenamine, *N*-(3-methylbutyl)-	$C_{11}H_{17}N$ 163.26	2051-84-5	 254.5	4-12-00-00259 0.8912^{55}	eth 4; EtOH 4 1.5305^{20}
784	Benzenamine, 4,4'-methylenebis- *p,p'*-Diaminodiphenylmethane	$C_{13}H_{14}N_2$ 198.27	101-77-9 92.5	2958 398; 257^{18}	4-13-00-00390	H_2O 2; EtOH 4; eth 4; bz 4
785	Benzenamine, 4,4'-methylenebis[2-chloro- 4,4'-Methylenebis[2-chloroaniline]	$C_{13}H_{12}Cl_2N_2$ 267.16	101-14-4	5978	1-13-00-00074	ctc 3
786	Benzenamine, 4,4'-methylenebis[*N,N*-dimethyl- Michler's Base	$C_{17}H_{22}N_2$ 254.38	101-61-1 91.5	6099 390 dec; 183^{3}	4-13-00-00390	H_2O 1; EtOH 2; eth 4; bz 4
787	Benzenamine, 2-(1-methylethenyl)-	$C_9H_{11}N$ 133.19	52562-19-3	 95^{13}	4-12-00-02925 0.976^{25}	 1.5722^{20}
788	Benzenamine, 2-(1-methylethyl)- 2-Isopropylaniline	$C_9H_{13}N$ 135.21	643-28-7	 221; 95^{13}	4-12-00-02630 0.9760^{12}	H_2O 1; eth 3; bz 3; ctc 3
789	Benzenamine, 4-(1-methylethyl)- Cumidine	$C_9H_{13}N$ 135.21	99-88-7	2622 225	4-12-00-02632 0.953^{20}	
790	Benzenamine, *N*-(1-methylethyl)- *N*-Isopropylaniline	$C_9H_{13}N$ 135.21	768-52-5	 203	4-12-00-00255 0.9526^{25}	EtOH 3; eth 3; ace 3; bz 3 1.5380^{20}
791	Benzenamine, 2-methyl-, hydrochloride *o*-Toluidine hydrochloride	$C_7H_{10}ClN$ 143.62	636-21-5 215		4-12-00-01744	H_2O 4; EtOH 4
792	Benzenamine, *N*-methyl-, hydrochloride	$C_7H_{10}ClN$ 143.62	2739-12-0 122.7		4-12-00-00241 1.0660^{131}	H_2O 4; EtOH 3; eth 1; bz 1
793	Benzenamine, 2-methyl-5-(1-methylethyl)-	$C_{10}H_{15}N$ 149.24	2051-53-8 -16	 241	3-12-00-02733 0.9942^{20}	ctc 3; CS_2 3 1.5387^{20}
794	Benzenamine, 4-methyl-*N*-(1-methylethyl)-	$C_{10}H_{15}N$ 149.24	10436-75-6	 220	4-12-00-01876 0.9226^{20}	 1.5332
795	Benzenamine, 2-methyl-*N*-(2-methylphenyl)-	$C_{14}H_{15}N$ 197.28	617-00-5 52.5	 316; 192^{23}	4-12-00-01750	H_2O 2
796	Benzenamine, 3-methyl-*N*-(3-methylphenyl)-	$C_{14}H_{15}N$ 197.28	626-13-1 -12-	 319.5	2-12-00-00467	H_2O 2; EtOH 3; eth 3; peth 3
797	Benzenamine, 4-methyl-*N*-(4-methylphenyl)-	$C_{14}H_{15}N$ 197.28	620-93-9 79.8	 330.5	4-12-00-01879	eth 4; peth 4
798	Benzenamine, 2-methyl-4-[(2-methylphenyl)azo]- 2',3-Dimethyl-4-aminoazobenzene	$C_{14}H_{15}N_3$ 225.29	97-56-3 102		4-16-00-00528	eth 4; EtOH 4
799	Benzenamine, 2-methyl-3-nitro-	$C_7H_8N_2O_2$ 152.15	603-83-8 92	 305	4-12-00-01811 1.3780^{15}	H_2O 2; EtOH 3; eth 3; bz 3
800	Benzenamine, 2-methyl-4-nitro-	$C_7H_8N_2O_2$ 152.15	99-52-5 133.5		4-12-00-01809 1.1586^{140}	H_2O 2; EtOH 3; bz 3; HOAc 3
801	Benzenamine, 2-methyl-5-nitro- 5-Nitro-*o*-toluidine	$C_7H_8N_2O_2$ 152.15	99-55-8 105.5		4-12-00-01807	H_2O 2; EtOH 3; eth 3; ace 3
802	Benzenamine, 2-methyl-6-nitro-	$C_7H_8N_2O_2$ 152.15	570-24-1 96		4-12-00-01807 1.1900^{100}	H_2O 2; EtOH 3; eth 3; bz 3
803	Benzenamine, 4-methyl-2-nitro-	$C_7H_8N_2O_2$ 152.15	89-62-3 116.3		4-12-00-02000 1.16^{121}	H_2O 2; EtOH 3; chl 3
804	Benzenamine, *N*-methyl-2-nitro-	$C_7H_8N_2O_2$ 152.15	612-28-2 38	 158^{18}	4-12-00-01564	H_2O 2; EtOH 3; eth 3; ace 3
805	Benzenamine, *N*-methyl-4-nitro-	$C_7H_8N_2O_2$ 152.15	100-15-2 152	 dec	4-12-00-01616 1.201^{155}	H_2O 1; EtOH 3; eth 2; bz 3
806	Benzenamine, *N*-methyl-*N*-nitroso-	$C_7H_8N_2O$ 136.15	614-00-6 14.7	 225 dec; 121^{13}	4-16-00-00864 1.1240^{20}	H_2O 1; EtOH 3; eth 3 1.5769^{20}
807	Benzenamine, *N*-methyl-*N*-phenyl- Methyldiphenylamine	$C_{13}H_{13}N$ 183.25	552-82-9 -7.5	5973 293.5	4-12-00-00275 1.0476^{20}	H_2O 1; EtOH 2; ctc 3; MeOH 2 1.6193^{20}
808	Benzenamine, 3-methyl-*N*-phenyl-	$C_{13}H_{13}N$ 183.25	1205-64-7 30	 316; 183^{17}	4-12-00-01818	bz 4; eth 4; EtOH 4 1.6350^{20}
809	Benzenamine, 2-methyl-*N*-(phenylmethylene)-	$C_{14}H_{13}N$ 195.26	5877-55-4	 314	4-12-00-01751 1.041^{20}	H_2O 1; ace 3 1.6310^{25}
810	Benzenamine, 3-methyl-*N*-(phenylmethylene)-	$C_{14}H_{13}N$ 195.26	5877-58-7 31	 315	3-12-00-01958 1.0390^{25}	ace 3; os 3 1.6353^{25}

PHYSICAL CONSTANTS OF ORGANIC COMPOUNDS (continued)

No.	Name Synonym	Mol. Form. Mol. Wt.	CAS RN mp/°C	Merck No. bp/°C	Beil. Ref. den/g cm^{-3}	Solubility n_D
811	Benzenamine, 4-methyl-*N*-(phenylmethylene)-	$C_{14}H_{13}N$ 195.26	2272-45-9 35	 318; 178^{11}	2-12-00-00496	ace 4
812	Benzenamine, 2-(1-methylpropyl)-	$C_{10}H_{15}N$ 149.24	55751-54-7	 $120\text{-}2^{16}$	4-12-00-02814 0.9574^{20}	EtOH 3; ace 3; bz 3; ctc 2
813	Benzenamine, 4-(1-methylpropyl)-	$C_{10}H_{15}N$ 149.24	30273-11-1	 238; 118^{15}	4-12-00-02814 0.949^{15}	bz 4; eth 4 1.5360^{29}
814	Benzenamine, 4-(2-methylpropyl)-	$C_{10}H_{15}N$ 149.24	30090-17-6	 238; 112^{11}	4-12-00-02818 0.949^{15}	ace 4; bz 4; EtOH 4
815	Benzenamine, 4-methyl-*N*-propyl- *N*-Propyl-*p*-toluidine	$C_{10}H_{15}N$ 149.24	54837-90-0	 235	4-12-00-01876 0.9243^{20}	 1.5367
816	Benzenamine, *N*-(1-methylpropyl)- *N*-*sec*-Butylaniline	$C_{10}H_{15}N$ 149.24	6068-69-5	 225	4-12-00-00257	chl 2; os 3 1.5333^{20}
817	Benzenamine, *N*-(2-methylpropyl)-	$C_{10}H_{15}N$ 149.24	588-47-6	 232; 113^{11}	4-12-00-00257 0.940^{18}	H_2O 1; eth 4; bz 4 1.5328^{20}
818	Benzenamine, *N*-methyl-*N*,2,4,6-tetranitro- Nitramine	$C_7H_5N_5O_8$ 287.15	479-45-8 131.5	6488 exp 180	4-16-00-00895 1.57^{10}	H_2O 1; EtOH 2; eth 2; ace 3
819	Benzenamine, 2-(methylthio)- Aniline, 2-methylthio	C_7H_9NS 139.22	2987-53-3	 234	4-13-00-00910 1.111^{25}	 1.6239^{20}
820	Benzenamine, 4-(methylthio)-	C_7H_9NS 139.22	104-96-1	 272.5	4-13-00-01289 1.1379^{20}	EtOH 3; eth 3; ace 3; bz 3 1.6395^{20}
821	Benzenamine, nitrate Aniline nitrate	$C_6H_8N_2O_3$ 156.14	542-15-4 190 dec	687	4-12-00-00223 1.356^{4}	H_2O 4; eth 4; EtOH 4
822	Benzenamine, 2-nitro- *o*-Nitroaniline	$C_6H_6N_2O_2$ 138.13	88-74-4 71.2	6504 284	4-12-00-01563 1.442^{15}	H_2O 2; EtOH 3; eth 4; ace 4
823	Benzenamine, 3-nitro- *m*-Nitroaniline	$C_6H_6N_2O_2$ 138.13	99-09-2 114	6503 306 dec	4-12-00-01589 0.991^{25}	H_2O 2; EtOH 3; eth 3; ace 3
824	Benzenamine, 4-nitro- *p*-Nitroaniline	$C_6H_6N_2O_2$ 138.13	100-01-6 147	6505 332	4-12-00-01613 1.424^{20}	H_2O 1; EtOH 3; eth 3; ace 3
825	Benzenamine, *N*-nitro-	$C_6H_6N_2O_2$ 138.13	645-55-6 47.3	 exp	4-16-00-00889	H_2O 4; bz 4; eth 4; EtOH 4
826	Benzenamine, 4-nitro-*N*-phenyl-	$C_{12}H_{10}N_2O_2$ 214.22	836-30-6 135.3	 211^{30}	4-12-00-01619	H_2O 1; EtOH 4; ace 2; HOAc 4
827	Benzenamine, 5-nitro-2-propoxy- Aniline, 5-nitro-2-propoxy	$C_9H_{12}N_2O_3$ 196.21	553-79-7 49	6551	4-13-00-00897	EtOH 4
828	Benzenamine, 4-nitroso-	$C_6H_6N_2O$ 122.13	659-49-4 173.5	 dec	4-12-00-01558	H_2O 3; EtOH 3; bz 2
829	Benzenamine, 4-nitroso-*N*-phenyl- *p*-Nitrosodiphenylamine	$C_{12}H_{10}N_2O$ 198.22	156-10-5 143	6560	4-12-00-01560	H_2O 2; EtOH 4; eth 4; bz 4
830	Benzenamine, *N*-nitroso-*N*-phenyl- *N*-Nitroso-*N*-diphenylamine	$C_{12}H_{10}N_2O$ 198.22	86-30-6 66.5		4-16-00-00865	EtOH 2; bz 3; chl 2
831	Benzenamine, 4-octyl-	$C_{14}H_{23}N$ 205.34	16245-79-7 20	 310; 138^{5}	4-12-00-02887 0.9128^{20}	eth 4
832	Benzenamine, 4,4'-oxybis- Bis(*p*-aminophenyl) ether	$C_{12}H_{12}N_2O$ 200.24	101-80-4 189 dec		4-13-00-01038	
833	Benzenamine, 2,3,4,5,6-pentamethyl-	$C_{11}H_{17}N$ 163.26	2243-30-3 152.5	 277.5	3-12-00-02758	eth 4; EtOH 4
834	Benzenamine, 2-phenoxy-	$C_{12}H_{11}NO$ 185.23	2688-84-8 45.8	 308; 172^{14}	4-13-00-00809	EtOH 3; eth 3; ace 3; bz 3
835	Benzenamine, 3-phenoxy-	$C_{12}H_{11}NO$ 185.23	3586-12-7 37	 315; 180^{10}	4-13-00-00955 1.1583^{25}	EtOH 3; eth 3; ace 3; bz 3
836	Benzenamine, *N*-phenyl- Diphenylamine	$C_{12}H_{11}N$ 169.23	122-39-4 52.9	3317 302	4-12-00-00271 1.158^{22}	H_2O 1; EtOH 4; eth 3; ace 4
837	Benzenamine, 4-(phenylazo)- *p*-Aminoazobenzene	$C_{12}H_{11}N_3$ 197.24	60-09-3 127	430 >360	2-16-00-00149	H_2O 2; EtOH 3; eth 3; bz 3
838	Benzenamine, 2-(2-phenylethyl)-	$C_{14}H_{15}N$ 197.28	5697-85-8 33	 193^{19}	4-12-00-03311 1.0430^{20}	
839	Benzenamine, 4-(phenylmethyl)-	$C_{13}H_{13}N$ 183.25	1135-12-2 34.5	 300	4-12-00-03281 1.038^{25}	eth 4; EtOH 4; lig 4
840	Benzenamine, *N*-(phenylmethylene)- Benzylideneaniline	$C_{13}H_{11}N$ 181.24	538-51-2 54	1152 310	2-12-00-00113 1.038^{55}	H_2O 1; EtOH 3; eth 3; chl 2 1.600^{100}
841	Benzenamine, *N*-2-propenyl-	$C_9H_{11}N$ 133.19	589-09-3	 219; 106^{12}	4-12-00-00265 0.982^{25}	H_2O 2; EtOH 3; eth 5; ace 3 1.563^{20}
842	Benzenamine, 2-propyl-	$C_9H_{13}N$ 135.21	1821-39-2	 226; 116^{15}	4-12-00-02590 0.9602^{20}	H_2O 1; EtOH 5; eth 5 1.5427^{20}
843	Benzenamine, *N*-propyl-	$C_9H_{13}N$ 135.21	622-80-0	 222; 98^{11}	4-12-00-00254 0.9443^{20}	eth 4; EtOH 4 1.5428^{20}
844	Benzenamine, *N*-sulfinyl-	C_6H_5NOS 139.18	1122-83-4	 200	4-12-00-01075 1.236^{25}	 1.6270^{20}
845	Benzenamine, 4,4'-sulfinylbis- 4,4'-Sulfinyldianiline	$C_{12}H_{12}N_2OS$ 232.31	119-59-5 175 dec	8927	4-13-00-01306	H_2O 3; EtOH 3
846	Benzenamine, 4,4'-sulfonylbis- Dapsone	$C_{12}H_{12}N_2O_2S$ 248.31	80-08-0 175.5	2820	4-13-00-01306	EtOH 3; DMSO-d_6 2
847	Benzenamine, 2,3,4,5-tetramethyl- Aniline, 2,3,4,5-tetramethyl	$C_{10}H_{15}N$ 149.24	2217-45-0 69	 259.5	3-12-00-02743	eth 4; EtOH 4; lig 4
848	Benzenamine, 2,3,4,6-tetramethyl-	$C_{10}H_{15}N$ 149.24	488-71-1 23.5	 255	4-12-00-02849 0.978^{24}	EtOH 4
849	Benzenamine, *N*,*N*,2,6-tetramethyl-	$C_{10}H_{15}N$ 149.24	769-06-2 -36	 196; 88^{20}	4-12-00-02522 0.9147^{20}	

No.	Name Synonym	Mol. Form. Mol. Wt.	CAS RN mp/°C	Merck No. bp/°C	Beil. Ref. den/g cm^{-3}	Solubility n_D
850	Benzenamine, 4,4'-thiobis- Bis(4-aminophenyl) sulfide	$C_{12}H_{12}N_2S$ 216.31	139-65-1 108.5		4-13-00-01306	H_2O 2; EtOH 4; eth 4; bz 4
851	Benzenamine, 2,4,6-tribromo- 2,4,6-Tribromoaniline	$C_6H_4Br_3N$ 329.82	147-82-0 122	9522 300	4-12-00-01538 2.35^{20}	H_2O 1; EtOH 2; eth 3; chl 3
852	Benzenamine, 2,3,4-trichloro- 2,3,4-Trichloroaniline	$C_6H_4Cl_3N$ 196.46	634-67-3 73	 292	0-12-00-00626	EtOH 4
853	Benzenamine, 2,4,5-trichloro-	$C_6H_4Cl_3N$ 196.46	636-30-6 96.5	 270	4-12-00-01277	EtOH 3; eth 3; CS_2 4; lig 2
854	Benzenamine, 2,4,6-trichloro-	$C_6H_4Cl_3N$ 196.46	634-93-5 78.5	 262	4-12-00-01281	H_2O 1; EtOH 3; eth 3; chl 3
855	Benzenamine, 2-(trifluoromethyl)- *o*-(Trifluoromethyl)aniline	$C_7H_6F_3N$ 161.13	88-17-5 35.5	 68^{15}	4-12-00-01786 1.282^{25}	 1.4810^{20}
856	Benzenamine, 3-(trifluoromethyl)- *m*-(Trifluoromethyl)aniline	$C_7H_6F_3N$ 161.13	98-16-8 5.5	 187; $74\text{-}5^{10}$	4-12-00-01843 1.3047^{12}	H_2O 2; EtOH 3; eth 3 1.4787^{20}
857	Benzenamine, 4-(trifluoromethyl)-	$C_7H_6F_3N$ 161.13	455-14-1 38	 117.5^{60}	4-12-00-01982 1.283^{27}	 1.4815^{25}
858	Benzenamine, 2,4,5-trimethyl-	$C_9H_{13}N$ 135.21	137-17-7 68	 234.5	4-12-00-02645 0.957^{25}	EtOH 4
859	Benzenamine, 2,4,6-trimethyl- Mesitylamine	$C_9H_{13}N$ 135.21	88-05-1 -5	 232.5	4-12-00-02648 0.9633^{25}	ctc 2 1.5495^{20}
860	Benzenamine, *N,N*,2-trimethyl- *N,N*-Dimethyl-*o*-toluidine	$C_9H_{13}N$ 135.21	609-72-3 -60	 194.1	4-12-00-01747 0.9286^{20}	eth 4; EtOH 4 1.5152^{20}
861	Benzenamine, *N,N*,3-trimethyl-	$C_9H_{13}N$ 135.21	121-72-2	 212	4-12-00-01815 0.9410^{20}	EtOH 5; eth 5 1.5492^{20}
862	Benzenamine, *N,N*,4-trimethyl-	$C_9H_{13}N$ 135.21	99-97-8	 211	4-12-00-01874 0.9366^{20}	H_2O 1; EtOH 5; eth 5; ctc 3 1.5366^{20}
863	Benzenamine, 2,4,6-trinitro-	$C_6H_4N_4O_6$ 228.12	489-98-5 193.5	 exp	4-12-00-01735 1.762^{10}	H_2O 1; EtOH 2; eth 2; ace 3
864	Benzenamine, 2,4,6-trinitro-*N*-(2,4,6-trinitrophenyl)- Dipicrylamine	$C_{12}H_5N_7O_{12}$ 439.21	131-73-7 244 dec	3340	4-12-00-01737	H_2O 1; EtOH 1; eth 2; ace 2
865	Benzenaminium, *N*-ethyl-3-hydroxy-*N,N*-dimethyl-, chloride Ammonium, dimethyl ethyl 3-hydroxyphenyl,chloride	$C_{10}H_{16}ClNO$ 201.70	116-38-1	3487	4-13-00-00969	
866	Benzenaminium, *N,N,N*-trimethyl-, iodide Phenyl trimethyl ammonium iodide	$C_9H_{14}IN$ 263.12	98-04-4 224	7289	4-12-00-00249	H_2O 4; EtOH 3; ace 2; chl 1
867	Benzene [6]Annulene	C_6H_6 78.11	71-43-2 5.5	1074 80.0	4-05-00-00583 0.8765^{20}	H_2O 2; EtOH 5; eth 5; ace 5 1.5011^{20}
868	Benzeneacetaldehyde Phenylacetaldehyde	C_8H_8O 120.15	122-78-1 33.5	7236 195	4-07-00-00664 1.0272^{20}	H_2O 2; EtOH 5; eth 5; ace 3 1.5255^{20}
869	Benzeneacetaldehyde, 3,4-dimethoxy-	$C_{10}H_{12}O_3$ 180.20	5703-21-9	 $125^{1.5}$	4-08-00-01828 1.148^{20}	 1.5426^{20}
870	Benzeneacetaldehyde, 2-methoxy-	$C_9H_{10}O_2$ 150.18	33567-59-8	 245; 117^{17}	1-08-00-00543 1.0897^{20}	ace 4; EtOH 4 1.5393^{20}
871	Benzeneacetaldehyde, 4-methoxy-	$C_9H_{10}O_2$ 150.18	5703-26-4	 255.5	4-08-00-00426 1.096^{20}	 1.5359^{20}
872	Benzeneacetaldehyde, α-methyl-	$C_9H_{10}O$ 134.18	93-53-8	 203.5	 1.0089^{20}	EtOH 4 1.5176^{20}
873	Benzeneacetaldehyde, 2-methyl-	$C_9H_{10}O$ 134.18	10166-08-2	 221; 92^{10}	1-07-00-00163 1.0241^{10}	eth 4; EtOH 4; chl 4
874	Benzeneacetaldehyde, 4-methyl-	$C_9H_{10}O$ 134.18	104-09-6 40	 221.5	4-07-00-00705 1.0052^{20}	eth 4; EtOH 4; chl 4 1.5255^{20}
875	Benzeneacetaldehyde, 2-nitro-	$C_8H_7NO_3$ 165.15	1969-73-9 22.5	 133^{5}	2-07-00-00229	bz 4; eth 4; EtOH 4
876	Benzeneacetaldehyde, α-oxo-, aldoxime Isonitrosoacetophenone	$C_8H_7NO_2$ 149.15	532-54-7 129	5077	4-07-00-02132	H_2O 2; chl 3
877	Benzeneacetaldehyde, α-phenyl-	$C_{14}H_{12}O$ 196.25	947-91-1	 315 dec; 157^{7}	4-07-00-01400 1.1061^{21}	H_2O 1; EtOH 4; eth 4; bz 4 1.5920^{21}
878	Benzeneacetamide α-Phenylacetamide	C_8H_9NO 135.17	103-81-1 157	7237	4-09-00-01632	H_2O 2; EtOH 3; eth 2; bz 2
879	Benzeneacetamide, *N*-(aminocarbonyl)- Phenacemide	$C_9H_{10}N_2O_2$ 178.19	63-98-9 215	7154	4-09-00-01636	bz 4; eth 4; EtOH 4
880	Benzeneacetamide, 4-butoxy-*N*-hydroxy- Bufexamac	$C_{12}H_{17}NO_3$ 223.27	2438-72-4 154	1462		
881	Benzeneacetamide, α-ethyl- α-Phenylbutyramide	$C_{10}H_{13}NO$ 163.22	90-26-6 86	7249 185^{16}	4-09-00-01818	H_2O 3; ace 2; ctc 3
882	Benzeneacetamide, α-(1-methylethyl)-	$C_{11}H_{15}NO$ 177.25	5470-47-3 111.5	 180^{14}	3-09-00-02518	eth 4; EtOH 4
883	Benzeneacetic acid Phenylacetic acid	$C_8H_8O_2$ 136.15	103-82-2 76.7	7239 265.5	4-09-00-01614 1.228^{6}	H_2O 2; EtOH 4; eth 4; ace 3
884	Benzeneacetic acid, α-acetyl-, ethyl ester	$C_{12}H_{14}O_3$ 206.24	5413-05-8	 156^{22}	4-10-00-02776 1.0855^{20}	eth 4; EtOH 4 1.5176^{20}
885	Benzeneacetic acid, α-amino- α-Phenylglycine	$C_8H_9NO_2$ 151.17	69-91-0 292 dec	7263 sub 255	4-14-00-01317	alk 3; os 2
886	Benzeneacetic acid, 4-amino- *p*-Aminophenylacetic acid	$C_8H_9NO_2$ 151.17	1197-55-3 200 dec	475	3-14-00-01182	H_2O 1; EtOH 2; DMSO-d_6 2
887	Benzeneacetic acid, anhydride	$C_{16}H_{14}O_3$ 254.29	1555-80-2 73.3	 195^{12}	4-09-00-01629	eth 4; chl 4

PHYSICAL CONSTANTS OF ORGANIC COMPOUNDS (continued)

No.	Name Synonym	Mol. Form. Mol. Wt.	CAS RN mp/°C	Merck No. bp/°C	Beil. Ref. den/g cm^{-3}	Solubility n_D
888	Benzeneacetic acid, 4-bromo-	$C_8H_7BrO_2$ 215.05	1878-68-8 116	 sub	3-09-00-02275	H_2O 2; EtOH 4; eth 4; CS_2 4
889	Benzeneacetic acid, 4-bromo-α-hydroxy-, (±)- Acetic acid, (4-bromophenyl)-hydroxy(*DL*)	$C_8H_7BrO_3$ 231.05	7021-04-7 119	1412	3-10-00-00480	H_2O 4; EtOH 4; eth 4; bz 4
890	Benzeneacetic acid, α-butyl-	$C_{12}H_{16}O_2$ 192.26	24716-09-4	 182^{20}; $118^{0.1}$	4-09-00-01897 1.0225^{19}	 1.5071^{19}
891	Benzeneacetic acid, α-chloro-, ethyl ester, (*S*)- Acetic acid, chlorophenyl-, ethyl ester, *L*-(+)-	$C_{10}H_{11}ClO_2$ 198.65	10606-73-2	 162^{45}; 133^{12}	3-09-00-02266 1.1594^{20}	 1.5152^{20}
892	Benzeneacetic acid, α-chloro-α-phenyl-, ethyl ester	$C_{16}H_{15}ClO_2$ 274.75	52460-86-3 43.5	 185^{14}	4-09-00-02508	eth 4; EtOH 4
893	Benzeneacetic acid, α-cyano-, ethyl ester	$C_{11}H_{11}NO_2$ 189.21	4553-07-5	 275 dec; 165^{20}	3-09-00-04262 1.091^{20}	ace 4; bz 4; eth 4; EtOH 4 1.5012^{25}
894	Benzeneacetic acid, 4-[2-(diethylamino)-2-oxoethoxy]-3-methoxy-, propyl ester Acetic acid, amide,(4(carboxy(propyl ester)methyl)-2-methoxyphenoxy)-*N,N*-diethyl	$C_{18}H_{27}NO_5$ 337.42	1421-14-3	7813 $210-2^{0.7}$		H_2O 1; EtOH 3; chl 3
895	Benzeneacetic acid, 2,5-dihydroxy- Homogentisic acid	$C_8H_8O_4$ 168.15	451-13-8 153	4658	4-10-00-01506	H_2O 4; EtOH 4; eth 4; bz 1
896	Benzeneacetic acid, α,4-dihydroxy-3-methoxy- Vanilmandelic acid	$C_9H_{10}O_5$ 198.18	55-10-7 131-3 dec	9840	3-10-00-02100	H_2O 4; ace 4; eth 4
897	Benzeneacetic acid, α-ethyl-	$C_{10}H_{12}O_2$ 164.20	90-27-7 47.5	 271	2-09-00-00356	eth 3; bz 3; ctc 3
898	Benzeneacetic acid, α-ethyl-, 2-(diethylamino)ethyl ester Butethamate	$C_{16}H_{25}NO_2$ 263.38	14007-64-8	1516 $167-9^{11}$	3-09-00-02463	 1.4909^{20}
899	Benzeneacetic acid, ethyl ester Ethyl phenylacetate	$C_{10}H_{12}O_2$ 164.20	101-97-3 -29.4	3794 227	4-09-00-01618 1.0333^{20}	eth 4; EtOH 4 1.4980^{20}
900	Benzeneacetic acid, α-ethyl-, methyl ester	$C_{11}H_{14}O_2$ 178.23	2294-71-5 77.5	 228	3-09-00-02462	eth 4; EtOH 4
901	Benzeneacetic acid, 4-fluoro- (*p*-Fluorophenyl)acetic acid	$C_8H_7FO_2$ 154.14	405-50-5 86	4105 164^2	4-09-00-01672	
902	Benzeneacetic acid, α-(hydroxymethyl)- 8-methyl-8-azabicyclo[3.2.1]oct-3-yl ester *endo*-(±)- Atropine	$C_{17}H_{23}NO_3$ 289.37	51-55-8 118.5	891 sub 95	5-21-01-00235	H_2O 4; EtOH 4; eth 1; chl 2
903	Benzeneacetic acid, α-hydroxy-, (*S*)-	$C_8H_8O_3$ 152.15	17199-29-0 134		4-10-00-00564 1.341^{25}	H_2O 4; eth 4; EtOH 4; chl 4
904	Benzeneacetic acid, 2-hydroxy-	$C_8H_8O_3$ 152.15	614-75-5 148	 240	4-10-00-00536	H_2O 2; eth 3; chl 2
905	Benzeneacetic acid, 3-hydroxy-	$C_8H_8O_3$ 152.15	621-37-4 132	 190^{11}	4-10-00-00541	H_2O 4; EtOH 4; eth 4; bz 3
906	Benzeneacetic acid, 4-hydroxy-	$C_8H_8O_3$ 152.15	156-38-7 152	 sub	4-10-00-00543	H_2O 2; EtOH 4; eth 4
907	Benzeneacetic acid, α-hydroxy-, ethyl ester, (±)-	$C_{10}H_{12}O_3$ 180.20	4358-88-7 35.5	 254	4-10-00-00569 1.1258^{20}	EtOH 3; eth 3; ctc 2; CS_2 3
908	Benzeneacetic acid, α-(hydroxymethyl)-, 9-methyl-3-oxa-9-azatricyclo[3.3.1.02,4]non-7-yl ester, [(7S)-(1α,2β,4β,5α,7β)]- Scopolamine	$C_{17}H_{21}NO_4$ 303.36	51-34-3	8361	2-27-00-00064	
909	Benzeneacetic acid, α-hydroxy-, ethyl ester	$C_{10}H_{12}O_3$ 180.20	774-40-3 35	 150^{21}	2-10-00-00121 1.1270^{20}	eth 4; EtOH 4
910	Benzeneacetic acid, α-hydroxy-, methyl ester, (±)-	$C_9H_{10}O_3$ 166.18	4358-87-6 58	 250 dec; 144^{20}	2-10-00-00120 1.1756^{20}	EtOH 4; chl 4
911	Benzeneacetic acid, α-hydroxy-, 8-methyl-8-azabicyclo[3.2.1]oct-3-yl ester, *endo*-(±)- Homatropine	$C_{16}H_{21}NO_3$ 275.35	87-00-3 99.5	4649	5-21-01-00234	H_2O 2; EtOH 3; eth 3; ace 3
912	Benzeneacetic acid, α-hydroxy-, 3-methylbutyl ester Mandelic acid isoamyl ester	$C_{13}H_{18}O_3$ 222.28	5421-04-5	5600 172^{11}	4-10-00-00571	
913	Benzeneacetic acid, 4-hydroxy-3-methoxy- Homovanillic acid	$C_9H_{10}O_4$ 182.18	306-08-1 143.5	4662	4-10-00-01509	
914	Benzeneacetic acid, α-hydroxy-α-methyl- Atrolactic acid	$C_9H_{10}O_3$ 166.18	515-30-0 116.5	889	2-10-00-00157	H_2O 4; ace 4; bz 4; EtOH 4
915	Benzeneacetic acid, α-(hydroxymethyl)- Tropic acid	$C_9H_{10}O_3$ 166.18	529-64-6 118	9691	3-10-00-00564	eth 4; EtOH 4
916	Benzeneacetic acid, α-(hydroxymethyl)-, 8-azabicyclo[3.2.1]oct-3-yl ester, [3(*S*)-*endo*]- Norhyoscyamine	$C_{16}H_{21}NO_3$ 275.35	537-29-1 140.5	6623	5-21-01-00219	EtOH 4; chl 4
917	Benzeneacetic acid, α-(hydroxymethyl)-, 8-methyl-8-azabicyclo[3.2.1]oct-3-yl ester, [3(*S*)-*endo*]- Hyoscyamine	$C_{17}H_{23}NO_3$ 289.37	101-31-5 108.5	4795	5-21-01-00235	H_2O 2; EtOH 4; eth 2; bz 2

No.	Name Synonym	Mol. Form. Mol. Wt.	CAS RN mp/°C	Merck No. bp/°C	Beil. Ref. den/g cm^{-3}	Solubility n_D
918	Benzeneacetic acid, α-hydroxy-α-phenyl- Benzilic acid	$C_{14}H_{12}O_3$ 228.25	76-93-7 151	1089 180 dec	4-10-00-01256	H_2O 2; EtOH 4; eth 4; ace 2
919	Benzeneacetic acid, α-hydroxy-α-phenyl-, ethyl ester	$C_{16}H_{16}O_3$ 256.30	52182-15-7 34	 201^{21}	4-10-00-01258 1.5620^{20}	eth 4; EtOH 4
920	Benzeneacetic acid, α-hydroxy-α-phenyl-, methyl ester	$C_{15}H_{14}O_3$ 242.27	76-89-1 75.8	 187^{13}	4-10-00-01258	eth 4; EtOH 4
921	Benzeneacetic acid, 2-methoxy-	$C_9H_{10}O_3$ 166.18	93-25-4 124	 $100\text{-}1^{2}$	4-10-00-00536	H_2O 3; EtOH 4; eth 4; ace 4
922	Benzeneacetic acid, 4-methoxy-	$C_9H_{10}O_3$ 166.18	104-01-8 87	 138^{2}	4-10-00-00544	H_2O 1; EtOH 4; eth 3; bz 3
923	Benzeneacetic acid, 4-methoxy-, ethyl ester	$C_{11}H_{14}O_3$ 194.23	14062-18-1	 $138\text{-}40^{70}$	4-10-00-00549 1.097^{25}	 1.5075^{20}
924	Benzeneacetic acid, α-methyl-, (±)-	$C_9H_{10}O_2$ 150.18	2328-24-7 <-20	 263	4-09-00-01779 1.1^{0}	 1.5237^{20}
925	Benzeneacetic acid, 3-methyl-	$C_9H_{10}O_2$ 150.18	621-36-3 62	 $120\text{-}3^{26}$	4-09-00-01792	H_2O 3; chl 3
926	Benzeneacetic acid, 4-methyl-	$C_9H_{10}O_2$ 150.18	622-47-9 93	 265	4-09-00-01795	bz 4; eth 4; EtOH 4
927	Benzeneacetic acid, α-methylene- Atropic acid	$C_9H_8O_2$ 148.16	492-38-6 106.5	890 267 dec	4-09-00-02050	H_2O 2; EtOH 3; eth 3; bz 3
928	Benzeneacetic acid, α-methylene-, 8-methyl-8-azabicyclo[3.2.1]oct-3-yl ester, *endo*- Apoatropine	$C_{17}H_{21}NO_2$ 271.36	500-55-0 62	770	5-21-01-00228	H_2O 2; EtOH 4; eth 4; ace 4
929	Benzeneacetic acid, methyl ester	$C_9H_{10}O_2$ 150.18	101-41-7	 216.5	4-09-00-01617 1.0622^{16}	H_2O 1; EtOH 5; eth 5; ace 3 1.5075^{20}
930	Benzeneacetic acid, α-(1-methylethyl)-	$C_{11}H_{14}O_2$ 178.23	3508-94-9 63	 159^{14}	3-09-00-02518	eth 4; EtOH 4
931	Benzeneacetic acid, α-methyl-α-phenyl-	$C_{15}H_{14}O_2$ 226.27	5558-66-7 176	 sub	4-09-00-02535	H_2O 2; EtOH 3; eth 4; bz 3
932	Benzeneacetic acid, 2-methylpropyl ester	$C_{12}H_{16}O_2$ 192.26	102-13-6	 247	4-09-00-01619 0.999^{18}	H_2O 1; EtOH 3; eth 3
933	Benzeneacetic acid, 4-nitro- *p*-Nitrophenylacetic acid	$C_8H_7NO_4$ 181.15	104-03-0 154	6543	4-09-00-01687	H_2O 2; EtOH 3; eth 3; bz 3
934	Benzeneacetic acid, α-oxo-	$C_8H_6O_3$ 150.13	611-73-4 66	 163^{15}	4-10-00-02737	H_2O 4; EtOH 3; eth 3; ctc 2
935	Benzeneacetic acid, α-oxo-, ethyl ester	$C_{10}H_{10}O_3$ 178.19	1603-79-8	 256.5	4-10-00-02738 1.1222^{25}	 1.5190^{25}
936	Benzeneacetic acid, α-phenyl- Diphenylacetic acid	$C_{14}H_{12}O_2$ 212.25	117-34-0 147	3316 194^{25}	4-09-00-02492 1.257^{15}	H_2O 2; EtOH 4; eth 3; chl 3
937	Benzeneacetic acid, α-phenyl-, anhydride	$C_{28}H_{22}O_3$ 406.48	1760-46-9 98	 222^{15}	4-09-00-02497	bz 4; chl 4
938	Benzeneacetic acid, α-phenyl-, 2-(diethylamino)ethyl ester, hydrochloride Adiphenine hydrochloride	$C_{20}H_{26}ClNO_2$ 347.88	50-42-0 113.5	151	3-09-00-03297	H_2O 4; EtOH 2; eth 2
939	Benzeneacetic acid, α-phenyl-, ethyl ester	$C_{16}H_{16}O_2$ 240.30	3468-99-3 60	 195^{25}	4-09-00-02493 1.1860^{20}	H_2O 1; EtOH 4; eth 4; CS_2 4
940	Benzeneacetic acid, 2-phenylethyl ester	$C_{16}H_{16}O_2$ 240.30	102-20-5 26.5	 $177\text{-}8^{4.5}$	3-09-00-02183 1.077^{25}	EtOH 4
941	Benzeneacetic acid, α-(phenylmethylene)- α-Phenylcinnamic acid	$C_{15}H_{12}O_2$ 224.26	3368-16-9 172.5	7253 sub	3-09-00-03438	eth 4; EtOH 4
942	Benzeneacetic acid, α-(phenylmethylene)-, ethyl ester	$C_{17}H_{16}O_2$ 252.31	24446-63-7 33.5	 $214\text{-}5^{28}$	3-09-00-03416 1.0971^{18}	 1.5972^{18}
943	Benzeneacetic acid, α-propyl-, (±)- Valeric acid, 2-phenyl-, (±)-	$C_{11}H_{14}O_2$ 178.23	7782-21-0 58	 280	4-09-00-01868	eth 4; EtOH 4
944	Benzeneacetic acid, 2,3,6-trichloro- Chlorfenac	$C_8H_5Cl_3O_2$ 239.48	85-34-7 161	2085	4-09-00-01681	
945	Benzeneacetonitrile Benzyl cyanide	C_8H_7N 117.15	140-29-4 -23.8	1145 233.5	4-09-00-01663 1.0205^{15}	 1.5211^{25}
946	Benzeneacetonitrile, 4-amino-	$C_8H_8N_2$ 132.17	3544-25-0 46	 312	4-14-00-01312	H_2O 2; EtOH 3; CS_2 2; os 3
947	Benzeneacetonitrile, α-bromo- α-Bromobenzyl cyanide	C_8H_6BrN 196.05	5798-79-8 29	1400 242 dec; 133^{12}	3-09-00-02278 1.539^{29}	H_2O 1; EtOH 4; eth 4; ace 4
948	Benzeneacetonitrile, 2-bromo-	C_8H_6BrN 196.05	19472-74-3 1	 140^{13}	4-09-00-01682	H_2O 1; EtOH 4
949	Benzeneacetonitrile, 2-chloro-	C_8H_6ClN 151.60	2856-63-5 24	 251	4-09-00-01674 1.1737^{18}	
950	Benzeneacetonitrile, 3-chloro-	C_8H_6ClN 151.60	1529-41-5	 $134\text{-}6^{10}$	4-09-00-01675 1.283^{25}	 1.5437^{20}
951	Benzeneacetonitrile, α-ethyl-	$C_{10}H_{11}N$ 145.20	769-68-6	 241	4-09-00-01822 0.977^{14}	H_2O 1; EtOH 3; eth 3; bz 3
952	Benzeneacetonitrile, 2-fluoro-	C_8H_6FN 135.14	326-62-5	 232; 102^{10}	4-09-00-01671 1.059^{25}	 1.5009^{20}
953	Benzeneacetonitrile, 4-fluoro-	C_8H_6FN 135.14	459-22-3	 $119\text{-}20^{18}$	4-09-00-01673 1.126^{25}	 1.5002^{20}

No.	Name Synonym	Mol. Form. Mol. Wt.	CAS RN mp/°C	Merck No. bp/°C	Beil. Ref. den/g cm^{-3}	Solubility n_D
954	Benzeneacetonitrile, α-[(6-*O*-β-*D*-glucopyranosyl-β-*D*-glucopyranosyl)oxy]-, (*R*)- Amygdalin	$C_{20}H_{27}NO_{11}$ 457.43	29883-15-6 224.5	629	5-17-08-00118	H_2O 4; EtOH 2; eth 1; chl 1
955	Benzeneacetonitrile, α-(β-*D*-glucopyranosyloxy)- Mandelonitrile glucoside	$C_{14}H_{17}NO_6$ 295.29	138-53-4 122	5602		H_2O 4; EtOH 4
956	Benzeneacetonitrile, α-hydroxy-, (±)-	C_8H_7NO 133.15	613-88-7 22	170	3-10-00-00473 1.1165^{20}	eth 4; EtOH 4; chl 4
957	Benzeneacetonitrile, 2-methoxy-	C_9H_9NO 147.18	7035-03-2 69.8	141^{15}	4-10-00-00538	bz 4
958	Benzeneacetonitrile, 4-methoxy-	C_9H_9NO 147.18	104-47-2	286.5	4-10-00-00555 1.0845^{20}	EtOH 3; eth 3; chl 3 1.5309^{20}
959	Benzeneacetonitrile, α-methyl-	C_9H_9N 131.18	1823-91-2	231	4-09-00-01783 0.9854^{20}	eth 4; EtOH 4 1.5095^{25}
960	Benzeneacetonitrile, 2-methyl-	C_9H_9N 131.18	22364-68-7	244	4-09-00-01791 1.0156^{22}	bz 4; eth 4; EtOH 4 1.5252^{20}
961	Benzeneacetonitrile, 3-methyl-	C_9H_9N 131.18	2947-60-6	248; 133^{15}	4-09-00-01792 1.0022^{22}	H_2O 1; EtOH 3; eth 3; bz 3 1.5233^{20}
962	Benzeneacetonitrile, 4-methyl-	C_9H_9N 131.18	2947-61-7 18	242.5	4-09-00-01796 0.992^{25}	H_2O 1; EtOH 3; eth 3; bz 3 1.5190^{20}
963	Benzeneacetonitrile, α-(1-methylethyl)-	$C_{11}H_{13}N$ 159.23	5558-29-2	247	4-09-00-01879 0.967^{15}	bz 4; EtOH 4 1.5038^{25}
964	Benzeneacetonitrile, α-(2-methylpropyl)- Valeronitrile, 4-methyl-2-phenyl-	$C_{12}H_{15}N$ 173.26	5558-31-6	264; 137^{15}	4-09-00-01903 0.942^{16}	bz 4; eth 4; EtOH 4
965	Benzeneacetonitrile, 2-nitro-	$C_8H_6N_2O_2$ 162.15	610-66-2 84	178^{12}; 138^{1}	3-09-00-02283	ace 4; bz 4; eth 4; EtOH 4
966	Benzeneacetonitrile, 3-nitro-	$C_8H_6N_2O_2$ 162.15	621-50-1 63	180^{15}	4-09-00-01687	bz 4; eth 4; EtOH 4; chl 4
967	Benzeneacetonitrile, 4-nitro- *p*-Nitrobenzyl cyanide	$C_8H_6N_2O_2$ 162.15	555-21-5 117	6512 $195\text{-}7^{12}$	4-09-00-01689	H_2O 2; EtOH 3; eth 3; bz 3
968	Benzeneacetonitrile, α-oxo-	C_8H_5NO 131.13	613-90-1 32.5	206	4-10-00-02743	H_2O 1; EtOH 4; eth 4; chl 2
969	Benzeneacetonitrile, α-phenyl-	$C_{14}H_{11}N$ 193.25	86-29-3 74.3	184^{16}	4-09-00-02505	EtOH 3; eth 4; chl 3; lig 2
970	Benzeneacetonitrile, α-propyl-	$C_{11}H_{13}N$ 159.23	5558-78-1	255; 126^{13}	4-09-00-01869 0.9425^{20}	bz 4; EtOH 4 1.5000^{20}
971	Benzeneacetyl chloride Phenylacetyl chloride	C_8H_7ClO 154.60	103-80-0	170^{250}; 105^{24}	4-09-00-01631 1.1682^{20}	eth 4 1.5325^{20}
972	Benzeneacetyl chloride, α-(acetyloxy)-	$C_{10}H_9ClO_3$ 212.63	1638-63-7 36	152^{33}; 125^{10}	1-10-00-00089	eth 4; bz 4; chl 4; lig 4
973	Benzeneacetyl chloride, α-chloro-	$C_8H_6Cl_2O$ 189.04	2912-62-1	120^{23}	2-09-00-00308 1.196^{25}	1.5440^{20}
974	Benzeneacetyl chloride, α-ethyl-	$C_{10}H_{11}ClO$ 182.65	36854-57-6	$110\text{-}2^{10}$	4-09-00-01817	ctc 3 1.5169^{20}
975	Benzeneacetyl chloride, α-phenyl-	$C_{14}H_{11}ClO$ 230.69	1871-76-7 56.5	170^{16}	3-09-00-03300	lig 3
976	Benzene, azido-	$C_6H_5N_3$ 119.13	622-37-7 -27.5	70^{11}	4-05-00-00759 1.0860^{20}	H_2O 1; EtOH 2; eth 2 1.5589^{25}
977	Benzene, 1-azido-4-bromo-	$C_6H_4BrN_3$ 198.02	2101-88-4 20	105^{10}; 69^{2}	4-05-00-00761 1.5827^{25}	H_2O 1; EtOH 2; eth 4; bz 4
978	Benzene, 1-azido-4-chloro-	$C_6H_4ClN_3$ 153.57	3296-05-7 20	96^{20}	4-05-00-00760 1.2634^{25}	H_2O 1; eth 3
979	Benzene, 1-azido-2-iodo-	$C_6H_4IN_3$ 245.02	54467-95-7	$90\text{-}100^{0.9}$	0-05-00-00278 1.8893^{25}	ace 4; EtOH 4 1.6631^{25}
980	Benzene, (azidomethyl)-	$C_7H_7N_3$ 133.15	622-79-7	108^{23}; 74^{11}	4-05-00-00877 1.0655^{25}	H_2O 1; EtOH 5; eth 5 1.5341^{25}
981	Benzene, 1-azido-2-methyl-	$C_7H_7N_3$ 133.15	31656-92-5 <-10	90.5^{30}; 76^{12}	4-05-00-00876 1.0648^{25}	eth 4
982	Benzene, 1-azido-4-methyl-	$C_7H_7N_3$ 133.15	2101-86-2	180 dec; 80^{10}	4-05-00-00876 1.0527^{23}	eth 4; EtOH 4
983	Benzene, 1,2-bis(bromomethyl)-	$C_8H_8Br_2$ 263.96	91-13-4 95	$128\text{-}30^{4.5}$	4-05-00-00939 1.988^{25}	H_2O 1; EtOH 3; eth 3; ctc 3
984	Benzene, 1,3-bis(bromomethyl)-	$C_8H_8Br_2$ 263.96	626-15-3 77	$135\text{-}40^{20}$	4-05-00-00946 1.959^{25}	H_2O 1; EtOH 3; eth 3; chl 3
985	Benzene, 1,4-bis(bromomethyl)-	$C_8H_8Br_2$ 263.96	623-24-5 144.5	245	4-05-00-00970 2.012^{25}	H_2O 1; EtOH 4; eth 2; bz 3
986	Benzene, 1,2-bis(chloromethyl)-	$C_8H_8Cl_2$ 175.06	612-12-4 55	239.5	4-05-00-00927 1.393^{25}	H_2O 1; EtOH 4; eth 4; ctc 3
987	Benzene, 1,3-bis(chloromethyl)-	$C_8H_8Cl_2$ 175.06	626-16-4 34.2	251.5	4-05-00-00944 1.302^{20}	H_2O 1; EtOH 4; eth 4; chl 2
988	Benzene, 1,4-bis(chloromethyl)-	$C_8H_8Cl_2$ 175.06	623-25-6 100	245 dec; 135^{16}	4-05-00-00967 1.417^{25}	H_2O 1; EtOH 4; eth 4; ace 4
989	Benzene, 1,4-bis(dichloromethyl)-	$C_8H_6Cl_4$ 243.95	7398-82-5 9.5		3-05-00-00857 1.606^{25}	H_2O 1; EtOH 3; eth 3; bz 3
990	Benzene, 1,4-bis(1,1-dimethylethyl)- Benzene, 1,4-di-*tert*-butyl	$C_{14}H_{22}$ 190.33	1012-72-2 79.5	238; 109^{15}	4-05-00-01163 0.9850^{20}	H_2O 1; EtOH 3; eth 3

No.	Name Synonym	Mol. Form. Mol. Wt.	CAS RN mp/°C	Merck No. bp/°C	Beil. Ref. den/g cm^{-3}	Solubility n_D
991	Benzene, 1,4-bis(2-hydroxybenzylideneamino)-2-nitro Ethionine	$C_{20}H_{15}N_3O_4$ 361.36	13073-35-3 195	3693		
992	Benzene, 1,2-bis(1-methylethyl)-	$C_{12}H_{18}$ 162.27	577-55-9 -57	 204	4-05-00-01125 0.8701^{20}	H_2O 1; EtOH 5; eth 5; ace 5 1.4960^{20}
993	Benzene, 1,3-bis(1-methylethyl)-	$C_{12}H_{18}$ 162.27	99-62-7 -63.1	 203.2	4-05-00-01125 0.8559^{20}	H_2O 1; EtOH 5; eth 5; ace 5 1.4883^{20}
994	Benzene, 1,4-bis(1-methylethyl)-	$C_{12}H_{18}$ 162.27	100-18-5 -17	 210.3	4-05-00-01126 0.8568^{20}	H_2O 1; EtOH 5; eth 5; ace 5 1.4898^{20}
995	Benzene, 1,4-bis(phenylmethyl)-	$C_{20}H_{18}$ 258.36	793-23-7 87.5	 225^{18}	4-05-00-02512	EtOH 3; eth 2; bz 3; chl 3
996	Benzene, 1,4-bis(trichloromethyl)-	$C_8H_4Cl_6$ 312.84	68-36-0 109	1313	4-05-00-00968	chl 3
997	Benzene, 1,3-bis(trifluoromethyl)-	$C_8H_4F_6$ 214.11	402-31-3	 116	4-05-00-00943 1.3790^{25}	 1.3916^{25}
998	Benzene, bromo- Bromobenzene	C_6H_5Br 157.01	108-86-1 -30.6	1394 156.0	4-05-00-00670 1.4950^{20}	H_2O 1; EtOH 4; eth 4; bz 4 1.5597^{20}
999	Benzene, 1-bromo-2-(bromomethyl)-	$C_7H_6Br_2$ 249.93	3433-80-5 31	 129^{19}	4-05-00-00836	eth 4; EtOH 4; HOAc 4
1000	Benzene, 1-bromo-4-(bromomethyl)- *p*-Bromobenzyl bromide	$C_7H_6Br_2$ 249.93	589-15-1 63	1397	4-05-00-00836	H_2O 2; EtOH 3; eth 4; bz 3
1001	Benzene, (3-bromo-3-butenyl)- 1-Butene, 2-bromo-4-phenyl	$C_{10}H_{11}Br$ 211.10	62692-40-4	 118^{21}; 90^{5}	3-05-00-01209 1.2907^{20}	ace 4 1.5450^{20}
1002	Benzene, (4-bromo-2-butenyl)- 2-Butene, 1-bromo-4-phenyl-	$C_{10}H_{11}Br$ 211.10	40734-75-6	 $126\text{-}30^{10.5}$	4-05-00-01377 1.2660^{20}	eth 4 1.5678^{20}
1003	Benzene, (4-bromobutoxy)-	$C_{10}H_{13}BrO$ 229.12	1200-03-9 41	 154^{18}	4-06-00-00558	EtOH 2; ctc 2
1004	Benzene, 1-bromo-2-chloro-	C_6H_4BrCl 191.45	694-80-4 -12.3	 204	4-05-00-00680 1.6387^{25}	H_2O 1; bz 4; ctc 2 1.5809^{20}
1005	Benzene, 1-bromo-3-chloro-	C_6H_4BrCl 191.45	108-37-2 -21.5	 196	4-05-00-00680 1.6302^{20}	H_2O 1; EtOH 4; eth 4 1.5771^{20}
1006	Benzene, 1-bromo-4-chloro-	C_6H_4BrCl 191.45	106-39-8 68	 196	4-05-00-00681 1.576^{71}	H_2O 1; EtOH 2; eth 3; bz 3 1.5531^{70}
1007	Benzene, 1-bromo-3-(chloromethyl)-	C_7H_6BrCl 205.48	932-77-4 32.5	 119^{18}	4-05-00-00832	EtOH 4
1008	Benzene, 1-bromo-4-(chloromethyl)- *p*-Bromobenzyl chloride	C_7H_6BrCl 205.48	589-17-3 50	1398 236	4-05-00-00832	H_2O 1; EtOH 4; eth 4; peth 3
1009	Benzene, 1-bromo-4-cyclohexyl-	$C_{12}H_{15}Br$ 239.16	25109-28-8	 160^{25}	3-05-00-01258 1.283^{25}	bz 4; chl 4 1.5584^{20}
1010	Benzene, 1-bromo-2,3-dichloro-	$C_6H_3BrCl_2$ 225.90	56961-77-4 60	 245	0-05-00-00209	bz 4; eth 4; chl 4
1011	Benzene, 1-bromo-2,4-dichloro-	$C_6H_3BrCl_2$ 225.90	1193-72-2 26	 235	3-05-00-00564	bz 4; eth 4; chl 4
1012	Benzene, 1-bromo-3,5-dichloro-	$C_6H_3BrCl_2$ 225.90	19752-55-7 83	 232	2-05-00-00162	H_2O 1; EtOH 3; eth 3; bz 4
1013	Benzene, 2-bromo-1,3-dichloro-	$C_6H_3BrCl_2$ 225.90	19393-92-1 65	 242	0-05-00-00210	bz 4; eth 4; chl 4
1014	Benzene, 2-bromo-1,4-dichloro-	$C_6H_3BrCl_2$ 225.90	1435-50-3 35	 226; 119^{20}	3-05-00-00564	bz 4; eth 4; EtOH 4; lig 4
1015	Benzene, 4-bromo-1,2-dichloro-	$C_6H_3BrCl_2$ 225.90	18282-59-2 25	 237	3-05-00-00564	H_2O 1; EtOH 2; eth 4; bz 4
1016	Benzene, 4-bromo-1,2-dimethoxy-	$C_8H_9BrO_2$ 217.06	2859-78-1	 254.5	4-06-00-05621 1.702^{25}	 1.5743^{20}
1017	Benzene, 1-bromo-2,3-dimethyl-	C_8H_9Br 185.06	576-23-8	 214	4-05-00-00928 1.365^{25}	ace 4; bz 4; eth 4 1.5587^{20}
1018	Benzene, 1-bromo-2,4-dimethyl-	C_8H_9Br 185.06	583-70-0 -17	 205	4-05-00-00945 1.3419^{20}	H_2O 1; EtOH 4; eth 4; ace 4 1.5501^{20}
1019	Benzene, 1-bromo-3,5-dimethyl-	C_8H_9Br 185.06	556-96-7	 204	4-05-00-00945 1.362^{20}	eth 4; ace 3; bz 3 1.5462^{22}
1020	Benzene, 2-bromo-1,3-dimethyl-	C_8H_9Br 185.06	576-22-7	 203.5; 100^{20}	4-05-00-00945	eth 4; ace 3; bz 3 1.5552^{20}
1021	Benzene, 2-bromo-1,4-dimethyl-	C_8H_9Br 185.06	553-94-6 9	 199; $88\text{-}9^{13}$	4-05-00-00969 1.3582^{18}	H_2O 1; EtOH 4; bz 3 1.5514^{18}
1022	Benzene, 4-bromo-1,2-dimethyl-	C_8H_9Br 185.06	583-71-1 -0.2	 214.5	4-05-00-00928 1.3708^{20}	H_2O 1; EtOH 4; eth 4 1.5530^{20}
1023	Benzene, 1-bromo-4-(1,1-dimethylethyl)-	$C_{10}H_{13}Br$ 213.12	3972-65-4 19	 231.5	4-05-00-01050 1.2286^{20}	H_2O 1; eth 3; bz 3; chl 3 1.5436^{20}
1024	Benzene, 1-bromo-2,3-dinitro- 1-Bromo-2,3-dinitrobenzene	$C_6H_3BrN_2O_4$ 247.00	19613-76-4 101.5	 320	4-05-00-00749	EtOH 4
1025	Benzene, (1-bromoethenyl)-	C_8H_7Br 183.05	98-81-7 -44	 86^{14}; 71^{3}	4-05-00-01349 1.4025^{23}	 1.5881^{20}
1026	Benzene, 1-bromo-2-ethenyl-	C_8H_7Br 183.05	2039-88-5 -52.8	 209.2; 98^{20}	4-05-00-01349 1.4160^{20}	 1.5927^{20}
1027	Benzene, 1-bromo-3-ethenyl-	C_8H_7Br 183.05	2039-86-3	 $90\text{-}4^{20}$	3-05-00-01176 1.4059^{20}	 1.5933^{20}
1028	Benzene, 1-bromo-4-ethenyl-	C_8H_7Br 183.05	2039-82-9 7.7	 212; 103^{20}	4-05-00-01349 1.3984^{20}	H_2O 1; chl 4; HOAc 3 1.5947^{20}
1029	Benzene, (2-bromoethenyl)-, (*E*)-	C_8H_7Br 183.05	588-72-7 7	 219 dec; 108^{20}	4-05-00-01349 1.4269^{16}	H_2O 1; EtOH 5; eth 5; chl 3 1.6093^{20}

PHYSICAL CONSTANTS OF ORGANIC COMPOUNDS (continued)

No.	Name Synonym	Mol. Form. Mol. Wt.	CAS RN mp/°C	Merck No. bp/°C	Beil. Ref. den/g cm^{-3}	Solubility n_D
1030	Benzene, 1,1'-(bromoethenylidene)bis-	$C_{14}H_{11}Br$ 259.15	13249-58-6 41.5	 $175^{11.5}$	4-05-00-02177	EtOH 2; eth 3; ace 3; CS_2 3
1031	Benzene, (2-bromoethenyl)-, (*Z*)-	C_8H_7Br 183.05	588-73-8 -7	 $55\text{-}6^{2}$	4-05-00-01349 1.4322^{10}	 1.5990^{22}
1032	Benzene, 1-bromo-4-ethoxy-	C_8H_9BrO 201.06	588-96-5 4	 231	4-06-00-01045 1.4071^{25}	H_2O 1; EtOH 4; eth 4; chl 3 1.5517^{20}
1033	Benzene, (2-bromoethoxy)-	C_8H_9BrO 201.06	589-10-6 39	 240 dec; 128^{20}	4-06-00-00556 1.3555^{20}	H_2O 1; EtOH 4; eth 4
1034	Benzene, (1-bromoethyl)-, (±)-	C_8H_9Br 185.06	38661-81-3	 $202\text{-}3.\ 85^{13}$	4-05-00-00906 1.3605^{20}	EtOH 3; eth 3; bz 3; ctc 2 1.5612^{20}
1035	Benzene, 1-bromo-2-ethyl-	C_8H_9Br 185.06	1973-22-4 -67.9	 199.3	4-05-00-00905 1.3548^{20}	ace 4; bz 4; eth 4; EtOH 4 1.5472^{20}
1036	Benzene, 1-bromo-3-ethyl-	C_8H_9Br 185.06	2725-82-8	 202	4-05-00-00906 1.3493^{20}	 1.5465^{20}
1037	Benzene, 1-bromo-4-ethyl-	C_8H_9Br 185.06	1585-07-5 -43.5	 204	4-05-00-00906 1.3423^{20}	ace 4; bz 4; eth 4; EtOH 4 1.5445^{20}
1038	Benzene, (2-bromoethyl)-	C_8H_9Br 185.06	103-63-9	 219	4-05-00-00907 1.3587^{20}	H_2O 1; eth 3; bz 3; ctc 2 1.5372^{20}
1039	Benzene, (1-bromoethyl)-, (*R*)-	C_8H_9Br 185.06	1459-14-9	 203	4-05-00-00906 1.3108^{23}	H_2O 1; EtOH 3; eth 3 1.5612^{20}
1040	Benzene, 1-bromo-2-ethynyl- *o*-Bromophenylacetylene	C_8H_5Br 181.03	766-46-1	 $92\text{-}3^{20}$	3-05-00-01351 1.4434^{25}	 1.5962^{25}
1041	Benzene, 1-bromo-3-ethynyl- *m*-Bromophenylacetylene	C_8H_5Br 181.03	766-81-4	 $85\text{-}6^{16}$	3-05-00-01351 1.4466^{25}	 1.5896^{25}
1042	Benzene, 1-bromo-4-ethynyl-	C_8H_5Br 181.03	766-96-1 64.5	 89^{16}	4-05-00-01529	chl 3
1043	Benzene, 1-bromo-3-fluoro-	C_6H_4BrF 175.00	1073-06-9	 150	4-05-00-00678 1.7081^{20}	ctc 3 1.5257^{20}
1044	Benzene, 1-bromo-4-fluoro-	C_6H_4BrF 175.00	460-00-4 -17.4	 151.5	4-05-00-00678 1.593^{15}	H_2O 1; EtOH 3; eth 3; chl 3 1.5310^{15}
1045	Benzene, 1-bromo-4-(hexyloxy)-	$C_{12}H_{17}BrO$ 257.17	30752-19-3	 $156\text{-}7^{13}$	4-06-00-01047 1.2306^{20}	 1.5262^{20}
1046	Benzene, 1-bromo-2-iodo-	C_6H_4BrI 282.91	583-55-1 9.5	 257; 120^{15}	4-05-00-00698 2.2570^{25}	H_2O 1; EtOH 2; ace 3; HOAc 2 1.6618^{25}
1047	Benzene, 1-bromo-3-iodo- *m*-Bromoiodobenzene	C_6H_4BrI 282.91	591-18-4 -9.3	 252; 120^{18}	4-05-00-00698	H_2O 1; EtOH 2; HOAc 2
1048	Benzene, 1-bromo-4-iodo-	C_6H_4BrI 282.91	589-87-7 92	 252	4-05-00-00698	H_2O 1; EtOH 2; eth 3; chl 2
1049	Benzene, 1-bromo-4-isocyanato- *p*-Bromophenyl isocyanate	C_7H_4BrNO 198.02	2493-02-9	1418 226	1-12-00-00321	eth 4
1050	Benzene, 1-bromo-2-methoxy-	C_7H_7BrO 187.04	578-57-4 2.5	 216	4-06-00-01037 1.5018^{20}	H_2O 1; EtOH 4; eth 4 1.5727^{20}
1051	Benzene, 1-bromo-3-methoxy-	C_7H_7BrO 187.04	2398-37-0	 211; 105^{16}	4-06-00-01043	H_2O 1; EtOH 3; eth 3; bz 3 1.5635^{20}
1052	Benzene, 1-bromo-4-methoxy-	C_7H_7BrO 187.04	104-92-7 13.5	 215	4-06-00-01044 1.4564^{20}	H_2O 2; EtOH 4; eth 4; ctc 3 1.5642^{20}
1053	Benzene, (bromomethyl)- Benzyl bromide	C_7H_7Br 171.04	100-39-0 -3	1142 201	4-05-00-00829 1.4380^{25}	H_2O 1; EtOH 5; eth 5; ctc 3 1.5752^{20}
1054	Benzene, 1-bromo-2-methyl- *o*-Bromotoluene	C_7H_7Br 171.04	95-46-5 -27.8	 181.7	4-05-00-00825 1.4232^{20}	H_2O 1; EtOH 4; eth 4; bz 4 1.5565^{20}
1055	Benzene, 1-bromo-3-methyl- *m*-Bromotoluene	C_7H_7Br 171.04	591-17-3 -39.8	 183.7	4-05-00-00827 1.4099^{20}	H_2O 1; EtOH 3; eth 5; ace 3 1.5510^{20}
1056	Benzene, 1-bromo-4-methyl- *p*-Bromotoluene	C_7H_7Br 171.04	106-38-7 28.5	1429 184.3	4-05-00-00827 1.3959^{35}	H_2O 1; EtOH 3; eth 3; ace 3 1.5477^{20}
1057	Benzene, 1-(bromomethyl)-3,5-dimethyl- Mesitylene, α-bromo-	$C_9H_{11}Br$ 199.09	27129-86-8 40	 230; 118^{22}	4-05-00-01027	bz 4; eth 4; EtOH 4; chl 4
1058	Benzene, 1,1'-(bromomethylene)bis- Bromodiphenylmethane	$C_{13}H_{11}Br$ 247.13	776-74-9 45	 184^{20}; 152^{2}	4-05-00-01850	EtOH 3; bz 4; chl 3
1059	Benzene, 1-bromo-4-(1-methylethyl)-	$C_9H_{11}Br$ 199.09	586-61-8 -22.5	 218.7	4-05-00-00994 1.3145^{20}	H_2O 1; eth 3; bz 3; ctc 2 1.5569^{20}
1060	Benzene, (2-bromo-1-methylethyl)-	$C_9H_{11}Br$ 199.09	1459-00-3 -59.3	 188.5	4-05-00-00995 1.3020^{20}	
1061	Benzene, 1,1',1''-(bromomethylidyne)tris-	$C_{19}H_{15}Br$ 323.23	590-43-0 153	 230^{15}	4-05-00-02500 1.5500^{20}	
1062	Benzene, 1-(bromomethyl)-2-methyl-	C_8H_9Br 185.06	89-92-9 21	 217; 108^{16}	4-05-00-00928 1.3811^{23}	H_2O 1; EtOH 3; eth 3; ace 3 1.5730^{20}
1063	Benzene, 1-(bromomethyl)-3-methyl-	C_8H_9Br 185.06	620-13-3	 212.5	4-05-00-00945 1.3711^{23}	H_2O 1; EtOH 4; eth 4; os 3 1.5660^{20}
1064	Benzene, 1-(bromomethyl)-4-methyl-	C_8H_9Br 185.06	104-81-4 35	 220	4-05-00-00969 1.324^{25}	H_2O 1; EtOH 3; eth 4; chl 4
1065	Benzene, 2-bromo-1-methyl-4-(1-methylethyl)-	$C_{10}H_{13}Br$ 213.12	2437-76-5	 234.3	4-05-00-01063 1.2689^{18}	eth 4; EtOH 4; chl 4 1.5360^{20}
1066	Benzene, 1-bromo-2-methyl-4-nitro-	$C_7H_6BrNO_2$ 216.03	7149-70-4 78	 141^{17}	4-05-00-00860	eth 4
1067	Benzene, 1-(bromomethyl)-3-nitro-	$C_7H_6BrNO_2$ 216.03	3958-57-4 59.3	 162^{13}	4-05-00-00860	H_2O 1; EtOH 3

No.	Name Synonym	Mol. Form. Mol. Wt.	CAS RN mp/°C	Merck No. bp/°C	Beil. Ref. den/g cm^{-3}	Solubility n_D
1068	Benzene, 1-bromo-3-methyl-2-nitro- 3-Bromo-2-nitrotoluene	$C_7H_6BrNO_2$ 216.03	52414-97-8 28.5	 153^{30}; 129^{10}	4-05-00-00860	eth 4; EtOH 4
1069	Benzene, 1-bromo-3-methyl-5-nitro- Toluene, 3-bromo-5-nitro-	$C_7H_6BrNO_2$ 216.03	52488-28-5 83	 269.5	3-05-00-00752	eth 4; EtOH 4
1070	Benzene, 1-bromo-2-methyl-3-nitro- Toluene, 2-bromo-6-nitro	$C_7H_6BrNO_2$ 216.03	55289-35-5 42	 143^{22}	3-05-00-00751	EtOH 4
1071	Benzene, 2-bromo-1-methyl-3-nitro-	$C_7H_6BrNO_2$ 216.03	41085-43-2 41.5	 161^{30}	4-05-00-00860	EtOH 4
1072	Benzene, 2-bromo-1-methyl-4-nitro-	$C_7H_6BrNO_2$ 216.03	7745-93-9 78	 175^{38}; 150^{20}	4-05-00-00861	eth 4
1073	Benzene, 2-bromo-4-methyl-1-nitro- 2-Bromo-4-methyl-1-nitrobenzene	$C_7H_6BrNO_2$ 216.03	40385-54-4 37.8	 269; 154^{20}	4-05-00-00861	eth 4; EtOH 4
1074	Benzene, 4-bromo-1-methyl-2-nitro- 4-Bromo-2-nitrotoluene	$C_7H_6BrNO_2$ 216.03	60956-26-5 47	 256.5	4-05-00-00860	eth 4; EtOH 4
1075	Benzene, 1-bromo-2-nitro-	$C_6H_4BrNO_2$ 202.01	577-19-5 43	 258	4-05-00-00728 1.6245^{80}	H_2O 1; EtOH 4; eth 3; ace 3
1076	Benzene, 1-bromo-3-nitro-	$C_6H_4BrNO_2$ 202.01	585-79-5 56	 265	4-05-00-00729 1.7036^{20}	H_2O 2; EtOH 3; eth 3; bz 3 1.5979^{20}
1077	Benzene, 1-bromo-4-nitro- *p*-Nitrobromobenzene	$C_6H_4BrNO_2$ 202.01	586-78-7 127	 256	4-05-00-00729 1.948^{25}	H_2O 1; EtOH 3; eth 3; bz 3
1078	Benzene, bromopentafluoro- Bromopentafluorobenzene	C_6BrF_5 246.96	344-04-7 -31	 137	4-05-00-00680 1.981^{25}	 1.4490^{20}
1079	Benzene, 1-bromo-4-phenoxy- 4-Bromophenyl phenyl ether	$C_{12}H_9BrO$ 249.11	101-55-3 18.72	 $124\text{-}8^{3\text{-}4}$	4-06-00-01047 1.6088^{20}	H_2O 1; eth 3; ctc 3 1.6084^{20}
1080	Benzene, 1-bromo-2-(2-phenylethenyl)-, (*E*)- *trans*-2-Bromostilbene	$C_{14}H_{11}Br$ 259.15	54737-45-0 34	 $143^{0.2}$	4-05-00-02164	H_2O I; EtOH 3; peth 3 1.6822^{25}
1081	Benzene, 1-bromo-2-(2-phenylethenyl)-, (*Z*)- Stilbene, 2-bromo (*cis*)	$C_{14}H_{11}Br$ 259.15	4877-77-4	 $121^{0.5}$	4-05-00-02164	H_2O 1; EtOH 3; peth 3 1.6404^{25}
1082	Benzene, 1-bromo-4-(1-propenyl)-	C_9H_9Br 197.07	4489-23-0 35	 238.5	2-05-00-00372 1.332^{36}	 1.590^{36}
1083	Benzene, 1-bromo-4-(2-propenyl)-	C_9H_9Br 197.07	2294-43-1	 224; 96^{12}	4-05-00-01363 1.324^{15}	 1.559^{15}
1084	Benzene, (3-bromo-1-propenyl)-	C_9H_9Br 197.07	4392-24-9 34	 130^{10}	3-05-00-01188 1.3428^{30}	EtOH 4 1.613^{20}
1085	Benzene, (3-bromopropoxy)-	$C_9H_{11}BrO$ 215.09	588-63-6 10.7	 127^{18}	4-06-00-00557 1.364^{16}	eth 4
1086	Benzene, (1-bromopropyl)-, (±)- Benzene, (1-bromopropyl) (*DL*)	$C_9H_{11}Br$ 199.09	63790-14-7	 105^{17}	2-05-00-00305 1.3098^{19}	 1.5517^{19}
1087	Benzene, (2-bromopropyl)-	$C_9H_{11}Br$ 199.09	2114-39-8	 $107\text{-}9^{16}$	4-05-00-00981 1.2908^{16}	chl 4 1.5450^{20}
1088	Benzene, (3-bromopropyl)-	$C_9H_{11}Br$ 199.09	637-59-2	 219.5; 117^{25}	4-05-00-00982 1.3106^{25}	H_2O 1; eth 4 1.5440^{25}
1089	Benzene, 1-bromo-2-(trifluoromethyl)-	$C_7H_4BrF_3$ 225.01	392-83-6	 167.5	4-05-00-00831 1.652^{25}	 1.4817^{20}
1090	Benzene, 1-bromo-3-(trifluoromethyl)-	$C_7H_4BrF_3$ 225.01	401-78-5 1	 151.5	4-05-00-00831 1.613^{25}	 1.4716^{20}
1091	Benzene, 1-bromo-4-(trifluoromethyl)-	$C_7H_4BrF_3$ 225.01	402-43-7	 160	4-05-00-00831 1.607^{25}	 1.4705^{25}
1092	Benzene, 1-bromo-2,3,5-trimethyl- 1-Bromo-2,3,5-trimethylbenzene	$C_9H_{11}Br$ 199.09	31053-99-3 <-15	 238	4-05-00-01014	bz 4 1.5516^{20}
1093	Benzene, 1-bromo-2,4,5-trimethyl-	$C_9H_{11}Br$ 199.09	5469-19-2 73	 234	4-05-00-01014	H_2O 1; EtOH 4
1094	Benzene, 2-bromo-1,3,5-trimethyl-	$C_9H_{11}Br$ 199.09	576-83-0 -1	 225	4-05-00-01027 1.3191^{10}	H_2O 1; eth 4; bz 3; ctc 2 1.5510^{20}
1095	Benzene, 1,1'-(1,3-butadiene-1,4-diyl)bis-, (*E*,*E*)-	$C_{16}H_{14}$ 206.29	538-81-8 154.3	 352	4-05-00-02319	bz 4; eth 4; EtOH 4; peth 4
1096	Benzene, 1,1'-(1,3-butadiene-1,4-diyl)bis-, (*Z*,*Z*)-	$C_{16}H_{14}$ 206.29	5807-76-1 70.5		4-05-00-02318 0.9697^{100}	H_2O 1; EtOH 2; eth 4; bz 4 1.6183^{100}
1097	Benzene, 1,3-butadienyl-, (*E*)-	$C_{10}H_{10}$ 130.19	16939-57-4 4.5	 76^{11}	4-05-00-01536 0.9286^{20}	H_2O I; EtOH 3; eth 3; ace 3 1.6089^{25}
1098	Benzene, 1,3-butadienyl-, (*Z*)-	$C_{10}H_{10}$ 130.19	31915-94-3 -57	 86^{11}	 0.9334^{20}	 1.6095^{20}
1099	Benzene, 1,1'-(1,4-butanediyl)bis- 1,4-Diphenylbutane	$C_{16}H_{18}$ 210.32	1083-56-3 52.5	 317	4-05-00-01937 0.9880^{20}	H_2O 1; EtOH 3; eth 3; chl 3
1100	Benzenebutanoic acid	$C_{10}H_{12}O_2$ 164.20	1821-12-1 52	 290	4-09-00-01811	H_2O 3; EtOH 3; eth 3
1101	Benzenebutanoic acid, 4-[bis(2-chloroethyl)amino]- Chlorambucil	$C_{14}H_{19}Cl_2NO_2$ 304.22	305-03-3 65	2064	4-14-00-01715	
1102	Benzenebutanoic acid, α,2-diamino-γ-oxo- Kynurenine	$C_{10}H_{12}N_2O_3$ 208.22	343-65-7 191 dec	5207	4-14-00-02562	
1103	Benzenebutanoic acid, 4-hydroxy-γ-(4-hydroxyphenyl)-γ-methyl- Diphenolic acid	$C_{17}H_{18}O_4$ 286.33	126-00-1 171.5	3312	4-10-00-01890	H_2O 4; ace 4; EtOH 4

No.	Name Synonym	Mol. Form. Mol. Wt.	CAS RN mp/°C	Merck No. bp/°C	Beil. Ref. den/g cm^{-3}	Solubility n_D
1104	Benzenebutanoic acid, γ-methyl-	$C_{11}H_{14}O_2$ 178.23	16433-43-5 13	 210^{85}; 165^{12}	4-09-00-01866 1.0554^{15}	eth 4; EtOH 4 1.5167^{20}
1105	Benzenebutanol, 4-methoxy-	$C_{11}H_{16}O_2$ 180.25	52244-70-9 3.5	 160^{8}; $125^{1.5}$	4-06-00-06005	 1.5267^{20}
1106	Benzenebutanol, α-methyl-	$C_{11}H_{16}O$ 164.25	2344-71-0	 $134\text{-}5^{16}$	3-06-00-01954 0.9643^{25}	 1.518^{19}
1107	Benzenebutanol, β-methyl-, (±)- 1-Butanol, 2-methyl-4-phenyl (*DL*)	$C_{11}H_{16}O$ 164.25	116783-11-0	 135^{11}	4-06-00-03380 0.9719^{20}	ace 4; bz 4; eth 4; EtOH 4 1.5173^{16}
1108	Benzenebutanol, β-phenyl-, (*R*)- 1-Butanol, 2,4-diphenyl-, (-)-	$C_{16}H_{18}O$ 226.32	17297-04-0 51.5	 178^{1}; $145^{0.1}$		chl 3 1.5686^{25}
1109	Benzene, 1-butenyl-, (*E*)-	$C_{10}H_{12}$ 132.21	1005-64-7 -43.1	 198.7	4-05-00-01374 0.9019^{20}	H_2O 1; EtOH 3; eth 3; bz 3 1.5420^{20}
1110	Benzene, 1-butenyl-, (*Z*)-	$C_{10}H_{12}$ 132.21	1560-09-4	 189	3-05-00-01205 0.9065^{25}	H_2O 1; EtOH 3; eth 3; bz 3 1.5390^{20}
1111	Benzene, 2-butenyl-	$C_{10}H_{12}$ 132.21	1560-06-1	 176	3-05-00-01207 0.8831^{20}	 1.5101^{20}
1112	Benzene, 3-butenyl-	$C_{10}H_{12}$ 132.21	768-56-9 -70	 177	4-05-00-01378 0.8831^{20}	H_2O 1; eth 3; bz 3 1.5059^{20}
1113	Benzene, 1,1'-(1-butenylidene)bis-	$C_{16}H_{16}$ 208.30	1726-14-3 -1.2	 296	4-05-00-02209 0.9934^{20}	 1.5604^{20}
1114	Benzene, butoxy-	$C_{10}H_{14}O$ 150.22	1126-79-0 -19.4	 210	4-06-00-00558 0.9351^{20}	eth 3; ace 3 1.4969^{20}
1115	Benzene, (butoxymethyl)-	$C_{11}H_{16}O$ 164.25	588-67-0	 223	4-06-00-02231 0.9227^{20}	ace 4; eth 4; EtOH 4 1.4833^{20}
1116	Benzene, 1-butoxy-3-methyl-	$C_{11}H_{16}O$ 164.25	23079-65-4	 229.2	4-06-00-02040 0.9406^{70}	eth 4 1.4970^{20}
1117	Benzene, 1-butoxy-4-methyl-	$C_{11}H_{16}O$ 164.25	10519-06-9	 229.5	3-06-00-01354 0.9205^{25}	eth 3 1.4970^{20}
1118	Benzene, butyl- Butylbenzene	$C_{10}H_{14}$ 134.22	104-51-8 -87.9	1549 183.3	4-05-00-01033 0.8601^{20}	H_2O 1; EtOH 5; eth 5; ace 5 1.4898^{20}
1119	Benzene, (1-butylhexadecyl)- Eicosane, 5-phenyl-	$C_{26}H_{46}$ 358.65	2400-04-6 30.2	 197^{1}	4-05-00-01223 0.8549^{20}	 1.4796^{20}
1120	Benzene, 1,1'-butylidenebis- 1,1-Diphenylbutane	$C_{16}H_{18}$ 210.32	719-79-9 27	 287	4-05-00-01944 0.9928^{20}	H_2O 1; EtOH 3; eth 3; bz 3 1.5664^{20}
1121	Benzene, 1-butyl-2-methyl-	$C_{11}H_{16}$ 148.25	1595-11-5	 208	3-05-00-00999 0.8710^{20}	ace 4; bz 4; eth 4 1.4960^{20}
1122	Benzene, 1-butyl-3-methyl-	$C_{11}H_{16}$ 148.25	1595-04-6	 205	3-05-00-00999 0.8590^{20}	ace 4; bz 4; eth 4 1.4910^{20}
1123	Benzene, 1-butyl-4-methyl-	$C_{11}H_{16}$ 148.25	1595-05-7 -85	 207	4-05-00-01094 0.8586^{20}	ace 4; bz 4; eth 4 1.4916^{20}
1124	Benzene, 3-butynyl-	$C_{10}H_{10}$ 130.19	16520-62-0	 190	4-05-00-01535 0.9258^{20}	 1.5208^{20}
1125	Benzenecarbodithioic acid, 2-hydroxy- Dithiosalicylic acid	$C_7H_6OS_2$ 170.26	527-89-9 49	3381	2-10-00-00078	bz 4; eth 4; EtOH 4
1126	Benzenecarboperoxoic acid Perbenzoic acid	$C_7H_6O_3$ 138.12	93-59-4 42	7109 100^{14}	4-09-00-00715	ace 4; bz 4; eth 4; EtOH 4
1127	Benzenecarboperoxoic acid, 3-chloro- 3-Chloroperoxybenzoic acid	$C_7H_5ClO_3$ 172.57	937-14-4 92-4 dec		4-09-00-00972	
1128	Benzenecarboperoxoic acid, 1,1-dimethylethyl ester Benzoyl *tert*-butyl peroxide	$C_{11}H_{14}O_3$ 194.23	614-45-9	 $75\text{-}6^{0.2}$	4-09-00-00715 1.021^{25}	 1.4990^{20}
1129	Benzenecarbothioamide, *N*-phenyl-	$C_{13}H_{11}NS$ 213.30	636-04-4 102	 dec	4-12-00-00423	H_2O 1; EtOH 4; eth 3; bz 3
1130	Benzenecarbothioic acid	C_7H_6OS 138.19	98-91-9 24	 $85\text{-}7^{10}$	4-09-00-01364 1.28^{20}	ace 4; bz 4; eth 4; EtOH 4 1.6040^{20}
1131	Benzene, chloro- Chlorobenzene	C_6H_5Cl 112.56	108-90-7 -45.2	2121 131.7	4-05-00-00640 1.1058^{20}	H_2O 1; EtOH 5; eth 5; bz 4 1.5241^{20}
1132	Benzene, 1-chloro-2-(chloromethyl)-	$C_7H_6Cl_2$ 161.03	611-19-8 -17	 217	4-05-00-00816 1.2699^{0}	H_2O 1; EtOH 2; eth 4; bz 4 1.5530^{20}
1133	Benzene, 1-chloro-3-(chloromethyl)-	$C_7H_6Cl_2$ 161.03	620-20-2	 216; 110^{25}	4-05-00-00816 1.2695^{15}	EtOH 4 1.5554^{20}
1134	Benzene, 1-chloro-4-(chloromethyl)- 4-Chlorobenzyl chloride	$C_7H_6Cl_2$ 161.03	104-83-6 31	 223	4-05-00-00816	ctc 2
1135	Benzene, 1-chloro-4-(chloromethyl)-2-nitro-	$C_7H_5Cl_2NO_2$ 206.03	57403-35-7	 122^{10}	4-05-00-00858 1.32^{25}	 1.5855^{20}
1136	Benzene, 1-chloro-4-[(chloromethyl)thio]-	$C_7H_6Cl_2S$ 193.10	7205-90-5	 $128\text{-}9^{12}$	4-06-00-01593 1.346^{25}	 1.6055^{20}
1137	Benzene, 1-chloro-2,3-dimethyl-	C_8H_9Cl 140.61	608-23-1	 188	4-05-00-00926 1.053^{20}	bz 4; chl 4 1.526^{20}
1138	Benzene, 1-chloro-2,4-dimethyl-	C_8H_9Cl 140.61	95-66-9 -32.5	 184	4-05-00-00943 1.0579^{20}	H_2O 1; ace 3; bz 4 1.5230^{25}
1139	Benzene, 2-chloro-1,3-dimethyl-	C_8H_9Cl 140.61	6781-98-2 -35	 186	4-05-00-00943 1.053^{20}	H_2O 1; ace 3; bz 4; chl 3 1.526^{20}
1140	Benzene, 2-chloro-1,4-dimethyl-	C_8H_9Cl 140.61	95-72-7 1.6	 187	4-05-00-00965 1.0589^{15}	H_2O 1; ace 3; bz 4; ctc 3
1141	Benzene, 4-chloro-1,2-dimethyl-	C_8H_9Cl 140.61	615-60-1 -6	 194	3-05-00-00816 1.0682^{15}	H_2O 1; ace 3; bz 4; ctc 3
1142	Benzene, 1-chloro-4-(1,1-dimethylethyl)-	$C_{10}H_{13}Cl$ 168.67	3972-56-3	 213	4-05-00-01048 1.0075^{18}	 1.5123^{20}

No.	Name	Synonym	Mol. Form.	Mol. Wt.	CAS RN	mp/°C	Merck No.	bp/°C	Beil. Ref.	den/g cm^{-3}	Solubility	n_D
1143	Benzene, (2-chloro-1,1-dimethylethyl)-	Neophyl chloride	$C_{10}H_{13}Cl$	168.67	515-40-2		6375	223; 105^{18}	4-05-00-01048	1.047^{20}	ace 4; bz 4; eth 4; EtOH 4	1.5247^{20}
1144	Benzene, 1-chloro-2,4-dinitro-	1-Chloro-2,4-dinitrobenzene	$C_6H_3ClN_2O_4$	202.55	97-00-7	53	2135	315	4-05-00-00744	1.4982^{75}	H_2O 1; EtOH 2; eth 3; bz 3	1.5857^{60}
1145	Benzene, 2-chloro-1,3-dinitro-	2-Chloro-1,3-dinitrobenzene	$C_6H_3ClN_2O_4$	202.55	606-21-3	88	2136	315	4-05-00-00744	1.6867^{16}	H_2O 1; EtOH 3; eth 3; chl 2	
1146	Benzene, 1,1'-(1-chloro-1,2-ethenediyl)bis-, (*E*)-		$C_{14}H_{11}Cl$	214.69	948-98-1	53.5		322; 128^{2}	4-05-00-02161		H_2O 1; EtOH 3; eth 3; bz 3	
1147	Benzene, 1,1'-(1-chloro-1,2-ethenediyl)bis-, (*Z*)-		$C_{14}H_{11}Cl$	214.69	948-99-2			$160\text{-}2^{12}$	4-05-00-02162		H_2O 1; EtOH 3; eth 3; bz 3	1.6281^{19}
1148	Benzene, (1-chloroethenyl)-		C_8H_7Cl	138.60	618-34-8	-23		199	4-05-00-01345	1.1016^{18}	eth 4; EtOH 4	1.5612^{20}
1149	Benzene, 1-chloro-2-ethenyl-	*o*-Chlorostyrene	C_8H_7Cl	138.60	2039-87-4	-63.1		188.7	4-05-00-01345	1.1000^{20}	EtOH 3; eth 3; ace 3; ctc 3	1.5649^{20}
1150	Benzene, 1-chloro-3-ethenyl-		C_8H_7Cl	138.60	2039-85-2			$62\text{-}3^{6}$	4-05-00-01345	1.1168^{20}	H_2O 1; EtOH 3; eth 3	1.5625^{20}
1151	Benzene, 1-chloro-4-ethenyl-		C_8H_7Cl	138.60	1073-67-2	15.9		192	4-05-00-01345	1.0868^{20}	H_2O 1; EtOH 3; eth 3; ace 5	1.5660^{20}
1152	Benzene, 1-(1-chloroethenyl)-2,4-dimethyl-	Styrene, α-chloro-2,4-dimethyl	$C_{10}H_{11}Cl$	166.65	74346-30-8			$104\text{-}5^{19}$	2-05-00-00381	1.044^{13}		1.5446^{13}
1153	Benzene, (2-chloroethenyl)-, (*E*)-		C_8H_7Cl	138.60	4110-77-4			199	4-05-00-01346	1.1095^{18}	H_2O 1; EtOH 3; eth 3; ace 3	1.5648^{20}
1154	Benzene, 1,1',1''-(1-chloro-1-ethenyl-2-ylidene)tris[4-methoxy-	Chlorotrianisene	$C_{23}H_{21}ClO_3$	380.87	569-57-3	115	2173		4-06-00-07650			
1155	Benzene, (2-chloroethenyl)-, (*Z*)-		C_8H_7Cl	138.60	4604-28-8			$59^{3.5}$	4-05-00-01346	1.1046^{25}		1.5762^{25}
1156	Benzene, 1-chloro-2-ethoxy-		C_8H_9ClO	156.61	614-72-2			210	4-06-00-00785	1.1288^{15}	bz 4; eth 4; EtOH 4	1.5284^{25}
1157	Benzene, 1-chloro-3-ethoxy-	*m*-Chlorophenyl ethyl ether	C_8H_9ClO	156.61	2655-83-6			207	4-06-00-00811	1.1712^{20}	H_2O 1; EtOH 4; eth 4; bz 4	
1158	Benzene, 1-chloro-4-ethoxy-		C_8H_9ClO	156.61	622-61-7	21		213	4-06-00-00823	1.1254^{20}	EtOH 3; eth 3; bz 4; ctc 2	1.5252^{20}
1159	Benzene, (2-chloroethoxy)-		C_8H_9ClO	156.61	622-86-6	28		218.5	4-06-00-00556		H_2O 1; EtOH 4; eth 4; ace 4	
1160	Benzene, (1-chloroethyl)-, (±)-		C_8H_9Cl	140.61	38661-82-4			81^{17}	4-05-00-00898	1.0620^{20}	EtOH 3; eth 4; bz 3	1.5276^{20}
1161	Benzene, 1-chloro-2-ethyl-		C_8H_9Cl	140.61	89-96-3	-82.7		178.4	4-05-00-00897	1.0569^{20}	H_2O 1; ace 3; bz 3; ctc 3	1.5218^{20}
1162	Benzene, 1-chloro-3-ethyl-		C_8H_9Cl	140.61	620-16-6	-55		183.8	4-05-00-00897	1.0529^{20}	ace 4; bz 4; eth 4; EtOH 4	1.5195^{20}
1163	Benzene, 1-chloro-4-ethyl-		C_8H_9Cl	140.61	622-98-0	-62.6		184.4	4-05-00-00897	1.0455^{20}	H_2O 1; EtOH 5; eth 5; ace 5	1.5175^{20}
1164	Benzene, (2-chloroethyl)-		C_8H_9Cl	140.61	622-24-2			197.5; 92^{20}	4-05-00-00899	1.069^{25}	H_2O 1; EtOH 3; eth 3; ace 3	1.5276^{20}
1165	Benzene, [(2-chloroethyl)thio]-		C_8H_9ClS	172.68	5535-49-9			$117\text{-}8^{11}$	4-06-00-01469	1.1769^{25}	eth 4; chl 4	1.5828^{20}
1166	Benzene, 1-chloro-2-fluoro-		C_6H_4ClF	130.55	348-51-6	-43		137.6	4-05-00-00652	1.2233^{30}	H_2O 1; ace 3; bz 3	1.4918^{30}
1167	Benzene, 1-chloro-3-fluoro-		C_6H_4ClF	130.55	625-98-9			127.6	4-05-00-00652	1.221^{25}		1.4911
1168	Benzene, 1-chloro-4-fluoro-		C_6H_4ClF	130.55	352-33-0	-26.8		130	4-05-00-00652	1.4990^{15}	H_2O 1; EtOH 3; eth 3; bz 3	1.4990^{15}
1169	Benzene, 1-chloro-3-fluoro-2-methyl-		C_7H_6ClF	144.58	443-83-4			154	3-05-00-00691	1.191^{25}		1.5026^{20}
1170	Benzene, 2-chloro-4-fluoro-1-methyl-		C_7H_6ClF	144.58	452-73-3			152.5	3-05-00-00691	1.1972^{20}		1.4985^{25}
1171	Benzene, 1-chloro-2-iodo-		C_6H_4ClI	238.45	615-41-8	0.7		234.5	4-05-00-00695	1.9515^{25}	H_2O 1; ace 3; ctc 2	1.6331^{25}
1172	Benzene, 1-chloro-4-iodo-		C_6H_4ClI	238.45	637-87-6	57		227	4-05-00-00696	1.886^{27}	H_2O 1; EtOH 3; chl 2	
1173	Benzene, 1-chloro-2-iodo-4-methyl-	Toluene, 4-chloro-3-iodo-	C_7H_6ClI	252.48	2401-22-1	-10		249	3-05-00-00726	1.7940^{15}		1.614^{23}
1174	Benzene, 4-chloro-1-iodo-2-methyl-	Toluene, 5-chloro-2-iodo	C_7H_6ClI	252.48	23399-70-4	-21		240	3-05-00-00726	1.702^{17}		1.616^{23}
1175	Benzene, 4-chloro-2-iodo-1-methyl-		C_7H_6ClI	252.48	33184-48-4	-25		242.5	3-05-00-00726	1.8358^{15}		1.620^{23}
1176	Benzene, 1-chloro-2-isocyanato-		C_7H_4ClNO	153.57	3320-83-0	30.5		200; 115^{43}	4-12-00-01129		ctc 2	
1177	Benzene, 1-chloro-4-isothiocyanato-		C_7H_4ClNS	169.63	2131-55-7	46		249.5	4-12-00-01214		H_2O 1; EtOH 3	
1178	Benzene, 1-chloro-2-methoxy-		C_7H_7ClO	142.58	766-51-8	-26.8		198.5	4-06-00-00785	1.1911^{20}	H_2O 1; EtOH 3; eth 3; chl 2	1.5480^{20}
1179	Benzene, 1-chloro-3-methoxy-		C_7H_7ClO	142.58	2845-89-8			193.5	4-06-00-00811	1.1759^{12}	H_2O 1; EtOH 3; eth 3	1.5365^{20}
1180	Benzene, 1-chloro-4-methoxy-		C_7H_7ClO	142.58	623-12-1	<-18		197.5	4-06-00-00822	1.201^{20}	H_2O 1; EtOH 4; eth 4; ctc 3	1.5390^{20}
1181	Benzene, [(chloromethoxy)methyl]-		C_8H_9ClO	156.61	3587-60-8			103^{13}	4-06-00-02252	1.1350^{20}		1.5192^{20}

No.	Name	Synonym	Mol. Form.	Mol. Wt.	CAS RN	mp/°C	Merck No.	bp/°C	Beil. Ref.	den/g cm^{-3}	Solubility	n_D
1182	Benzene, (chloromethyl)-	Benzyl chloride	C_7H_7Cl	126.59	100-44-7	-45	1143	179	4-05-00-00809	1.1004^{20}	H_2O 1; EtOH 5; eth 5; ctc 2	1.5391^{20}
1183	Benzene, 1-chloro-2-methyl-	*o*-Chlorotoluene	C_7H_7Cl	126.59	95-49-8	-35.6	2172	159.0	4-05-00-00805	1.0825^{20}	H_2O 1; EtOH 3; eth 5; ace 5	1.5268^{20}
1184	Benzene, 1-chloro-3-methyl-	*m*-Chlorotoluene	C_7H_7Cl	126.59	108-41-8	-47.8	2172	161.8	4-05-00-00806	1.075^{20}	H_2O 1; EtOH 3; eth 5; bz 3	1.5214^{19}
1185	Benzene, 1-chloro-4-methyl-	*p*-Chlorotoluene	C_7H_7Cl	126.59	106-43-4	7.5	2172	162.4	4-05-00-00806	1.0697^{20}	H_2O 1; EtOH 3; eth 5; ctc 3	1.5150^{20}
1186	Benzene, 1-(chloromethyl)-2,4-dimethyl-		$C_9H_{11}Cl$	154.64	824-55-5			215.5; 110^{20}	4-05-00-01012	1.0580^{19}	bz 4; eth 4; EtOH 4	
1187	Benzene, 1,1'-(chloromethylene)bis-		$C_{13}H_{11}Cl$	202.68	90-99-3	16		140^{3}	4-05-00-01847	1.140^{25}	chl 3	1.5951^{20}
1188	Benzene, 1-(2-chloro-1-methylethenyl)-4-methyl-	Styrene, β-chloro, α-4-dimethyl	$C_{10}H_{11}Cl$	166.65	30926-60-4			$107\text{-}8^{10}$	2-05-00-00381	1.058^{20}		1.5549^{23}
1189	Benzene, (1-chloro-1-methylethyl)-		$C_9H_{11}Cl$	154.64	934-53-2			98^{1}	4-05-00-00992	1.192^{25}		1.5290^{25}
1190	Benzene, 1-chloro-2-(1-methylethyl)-		$C_9H_{11}Cl$	154.64	2077-13-6	-74.4		191.1	4-05-00-00992	1.0341^{20}	ace 4; bz 4; eth 4; EtOH 4	1.5168^{20}
1191	Benzene, 1-(chloromethyl)-4-ethyl-		$C_9H_{11}Cl$	154.64	1467-05-6			$95\text{-}6^{15}$	4-05-00-01004		bz 4; EtOH 4; chl 4	1.5290^{25}
1192	Benzene, 1-chloro-4-(1-methylethyl)-		$C_9H_{11}Cl$	154.64	2621-46-7	-12.3		198.3	4-05-00-00992	1.0208^{20}	H_2O 1; EtOH 5; eth 5; ace 5	1.5117^{20}
1193	Benzene, (2-chloro-1-methylethyl)-		$C_9H_{11}Cl$	154.64	824-47-5			85^{13}	4-05-00-00992	1.047^{19}		1.5245^{19}
1194	Benzene, 1-(chloromethyl)-2-fluoro-		C_7H_6ClF	144.58	345-35-7			86^{40}	4-05-00-00813	1.216^{25}		1.5150^{20}
1195	Benzene, 1-(chloromethyl)-4-fluoro-		C_7H_6ClF	144.58	352-11-4			82^{26}	4-05-00-00813	1.207^{25}		1.5130
1196	Benzene, 1,1',1''-(chloromethylidyne)tris-		$C_{19}H_{15}Cl$	278.78	76-83-5	113.5		310	4-05-00-02497		H_2O 1; EtOH 2; eth 4; ace 3	
1197	Benzene, 1-(chloromethyl)-4-methoxy-		C_8H_9ClO	156.61	824-94-2	24.5		262.5	4-06-00-02137	1.261^{20}	ace 4; bz 4; eth 4	1.580^{20}
1198	Benzene, 1-(chloromethyl)-2-methyl-		C_8H_9Cl	140.61	552-45-4			198; 90^{20}	4-05-00-00926	1.063^{25}	eth 4; EtOH 4	1.5410^{25}
1199	Benzene, 1-(chloromethyl)-3-methyl-		C_8H_9Cl	140.61	620-19-9			195.5	4-05-00-00943	1.064^{20}	H_2O 1; EtOH 3; eth 3	1.5345^{20}
1200	Benzene, 1-(chloromethyl)-4-methyl-		C_8H_9Cl	140.61	104-82-5			201; 90^{20}	4-05-00-00966	1.0512^{20}	H_2O 1; EtOH 3; eth 5	1.5380
1201	Benzene, 2-chloro-1-methyl-4-(1-methylethyl)-		$C_{10}H_{13}Cl$	168.67	4395-79-3			216.5	4-05-00-01062	1.0104^{25}	H_2O 1; ace 3; bz 3	1.5078^{20}
1202	Benzene, 2-chloro-4-methyl-1-(1-methylethyl)-		$C_{10}H_{13}Cl$	168.67	4395-80-6			215; 113^{35}	3-05-00-00956	1.018^{15}		1.5180^{20}
1203	Benzene, 1-(chloromethyl)-2-nitro-		$C_7H_6ClNO_2$	171.58	612-23-7	51			4-05-00-00854		H_2O 1; EtOH 3; eth 3; ace 4	1.5557^{62}
1204	Benzene, 1-chloro-2-methyl-3-nitro-		$C_7H_6ClNO_2$	171.58	83-42-1	37.8		238	4-05-00-00854		H_2O 1; EtOH 3	1.5377^{69}
1205	Benzene, 1-chloro-2-methyl-4-nitro-		$C_7H_6ClNO_2$	171.58	13290-74-9	42.5		249	4-05-00-00855		eth 4	
1206	Benzene, 1-(chloromethyl)-3-nitro-		$C_7H_6ClNO_2$	171.58	619-23-8	46		173^{34}	4-05-00-00855		ace 4; bz 4; eth 4; EtOH 4	1.5577^{62}
1207	Benzene, 1-chloro-3-methyl-5-nitro-	3-Chloro-5-nitrotoluene	$C_7H_6ClNO_2$	171.58	16582-38-0	61			4-05-00-00855		EtOH 4	1.5404^{69}
1208	Benzene, 1-(chloromethyl)-4-nitro-	4-Nitrobenzyl chloride	$C_7H_6ClNO_2$	171.58	100-14-1	71			4-05-00-00856		H_2O 1; EtOH 3; eth 3; ace 4	1.5647^{62}
1209	Benzene, 1-chloro-4-methyl-2-nitro-		$C_7H_6ClNO_2$	171.58	89-60-1	7		261; 118^{11}	4-05-00-00855		H_2O 1; ctc 3	1.5572^{20}
1210	Benzene, 2-chloro-1-methyl-4-nitro-		$C_7H_6ClNO_2$	171.58	121-86-8	66.5		260	4-05-00-00855		H_2O 2; EtOH 3; eth 3; chl 2	1.5470^{69}
1211	Benzene, 2-chloro-4-methyl-1-nitro-		$C_7H_6ClNO_2$	171.58	38939-88-7	24		146^{15}	4-05-00-00856		H_2O 1; ctc 3	
1212	Benzene, 4-chloro-1-methyl-2-nitro-		$C_7H_6ClNO_2$	171.58	89-59-8	38		242; 115.5^{11}	4-05-00-00853	1.2559^{80}	H_2O 1; EtOH 3; eth 3; chl 2	
1213	Benzene, 4-chloro-2-methyl-1-nitro-		$C_7H_6ClNO_2$	171.58	5367-28-2	25.5			4-05-00-00854		H_2O 1; ctc 3	1.5100^{65}
1214	Benzene, 1-chloro-2-nitro-	*o*-Chloronitrobenzene	$C_6H_4ClNO_2$	157.56	88-73-3	32.5	2151	245.5	4-05-00-00721	1.368^{242}	H_2O 1; EtOH 3; eth 3; ace 4	
1215	Benzene, 1-chloro-3-nitro-	*m*-Chloronitrobenzene	$C_6H_4ClNO_2$	157.56	121-73-3	44.4	2151	235.5	4-05-00-00722	1.343^{50}	H_2O 1; EtOH 3; eth 3; bz 3	1.5374^{80}
1216	Benzene, 1-chloro-4-nitro-	*p*-Chloronitrobenzene	$C_6H_4ClNO_2$	157.56	100-00-5	83.5	2151	242	4-05-00-00723	1.2979^{90}	H_2O I; EtOH 2; eth 3; chl 3	1.5376^{100}
1217	Benzene, 1-chloro-2-nitro-4-(trifluoromethyl)-	2-Nitro-4-(trifluoromethyl)-1-chlorobenzene	$C_7H_3ClF_3NO_2$	225.55	121-17-5	-2.5		222; 95^{10}	4-05-00-00857	1.511^{25}		1.4893^{20}
1218	Benzene, 1-chloro-4-nitro-2-(trifluoromethyl)-	4-Chloro-3-(trifluoromethyl)nitrobenzene	$C_7H_3ClF_3NO_2$	225.55	777-37-7	22		232	4-05-00-00857	1.527^{25}		1.5083^{28}
1219	Benzene, 4-chloro-1-nitro-2-(trifluoromethyl)-	4-Chloro-2-(trifluoromethyl)nitrobenzene	$C_7H_3ClF_3NO_2$	225.55	118-83-2				3-05-00-00748	1.526^{25}		1.4980^{20}
1220	Benzene, chloropentafluoro-	Chloropentafluorobenzene	C_6ClF_5	202.51	344-07-0			117.9	4-05-00-00654	1.568^{25}		1.4256^{20}

No.	Name	Synonym	Mol. Form.	Mol. Wt.	CAS RN	mp/°C	Merck No.	bp/°C	Beil. Ref.	den/g cm^{-3}	Solubility	n_D
1221	Benzene, 1-chloro-4-phenoxy-	4-Chlorophenyl phenyl ether	$C_{12}H_9ClO$	204.66	7005-72-3			284.5	4-06-00-00826	1.2026^{15}		1.599
1222	Benzene, 1-chloro-2-(2-phenylethenyl)-, (*E*)-		$C_{14}H_{11}Cl$	214.69	1657-52-9	39.5		209^{30}; 139^{2}	4-05-00-02161		H_2O 1; EtOH 3; chl 3	
1223	Benzene, 1-chloro-3-(2-phenylethenyl)-, (*E*)-		$C_{14}H_{11}Cl$	214.69	14064-43-8	73.5		$177^{0.2}$	4-05-00-02161		H_2O 1; EtOH 3; chl 3	
1224	Benzene, 1-chloro-4-(phenylmethyl)-		$C_{13}H_{11}Cl$	202.68	831-81-2	7.5		299; $147\text{-}8^{8}$	4-05-00-01847	1.1247^{20}	ace 4	
1225	Benzene, 1-chloro-4-(phenylsulfonyl)-	Sulphenone	$C_{12}H_9ClO_2S$	252.72	80-00-2	94	8969		4-06-00-01587		H_2O 1; EtOH 2; eth 3; ace 4	
1226	Benzene, (1-chloro-1-propenyl)-		C_9H_9Cl	152.62	35673-03-1			$90.5^{0.9}$	2-05-00-00372	1.085^{20}	ace 4; bz 4	1.5635^{15}
1227	Benzene, (2-chloro-1-propenyl)-		C_9H_9Cl	152.62	13099-50-8			120^{20}; 61^{1}	3-05-00-01186	1.0738^{19}	ace 4; bz 4; chl 4	1.5565^{19}
1228	Benzene, (3-chloro-2-propenyl)-		C_9H_9Cl	152.62	6268-37-7			211	4-05-00-01363	1.073^{13}	H_2O 1; eth 3; ace 3; bz 3	1.545^{14}
1229	Benzene, 1-chloro-3-(1-propenyl)-, (*E*)-		C_9H_9Cl	152.62	23204-80-0	8.5		106.7^{13}		1.0926^{20}	H_2O 1; EtOH 5; eth 5; ace 4	1.5851^{20}
1230	Benzene, (3-chloro-1-propenyl)-, (*E*)-		C_9H_9Cl	152.62	21087-29-6	8.5		$106\text{-}7^{13}$	4-05-00-01360	1.0926^{20}	ace 4; bz 4; eth 4; EtOH 4	1.5851^{20}
1231	Benzene, 1-(3-chloro-2-propenyl)-4-methoxy-	Benzene, 1(3-chloroallyl)-4-methoxy	$C_{10}H_{11}ClO$	182.65	54644-23-4			120^{11}	3-06-00-02416	1.155^{8}		1.553^{8}
1232	Benzene, 1-chloro-4-(2-propenyloxy)-		C_9H_9ClO	168.62	13997-70-1			$106\text{-}7^{12}$	4-06-00-00825	1.131^{15}	bz 4; eth 4; EtOH 4	1.5348^{25}
1233	Benzene, (3-chloropropoxy)-		$C_9H_{11}ClO$	170.64	3384-04-1	12		250	4-06-00-00557	1.1167^{20}	ace 4; eth 4	1.5235^{25}
1234	Benzene, (2-chloropropyl)-, (*R*)-	*R*-(-)-2-Chloro-1-phenylpropane	$C_9H_{11}Cl$	154.64	55449-46-2			94^{17}	4-05-00-00980	1.038^{19}	H_2O 1; ace 3; bz 3; chl 3	1.5198^{22}
1235	Benzene, (3-chloropropyl)-		$C_9H_{11}Cl$	154.64	104-52-9			219.5	4-05-00-00980	1.056^{21}	ctc 2	1.5160^{25}
1236	Benzene, [(3-chloropropyl)thio]-		$C_9H_{11}ClS$	186.71	4911-65-3			$116\text{-}7^{4}$	3-06-00-00981	1.1536^{20}	ace 4; py 4	1.5752^{20}
1237	Benzene, 1-chloro-2-(trichloromethyl)-	1-Chloro-2-(trichloromethyl)benzene	$C_7H_4Cl_4$	229.92	2136-89-2	29.4		264.3	4-05-00-00823	1.5187^{20}	H_2O 1; eth 3; ace 3; ctc 2	1.5836^{20}
1238	Benzene, 1-chloro-3-(trichloromethyl)-	1-Chloro-3-(trichloromethyl)benzene	$C_7H_4Cl_4$	229.92	2136-81-4	0.6		255	4-05-00-00823	1.495^{14}	H_2O 1; eth 3; ace 3; ctc 2	1.4461^{20}
1239	Benzene, 1-chloro-4-(trichloromethyl)-	1-Chloro-4-(trichloromethyl)benzene	$C_7H_4Cl_4$	229.92	5216-25-1			245	4-05-00-00823	1.4463^{20}	ace 4; eth 4	
1240	Benzene, 1-chloro-2-(trifluoromethyl)-		$C_7H_4ClF_3$	180.56	88-16-4	-6		152.2	4-05-00-00814	1.2540^{30}	chl 3	1.4513^{25}
1241	Benzene, 1-chloro-3-(trifluoromethyl)-	Benzotrifluoride, 3-chloro	$C_7H_4ClF_3$	180.56	98-15-7	-56		137.5	4-05-00-00814	1.3311^{25}		1.4438^{25}
1242	Benzene, 1-chloro-4-(trifluoromethyl)-		$C_7H_4ClF_3$	180.56	98-56-6	-33		138.5	4-05-00-00815	1.3340^{25}		1.4431^{30}
1243	Benzene, 2-chloro-1,3,5-trimethyl-		$C_9H_{11}Cl$	154.64	1667-04-5	<-20		205	4-05-00-01026	1.0337^{30}	eth 4; EtOH 4	1.5212^{30}
1244	Benzene, 2-chloro-1,3,5-trinitro-	Picryl chloride	$C_6H_2ClN_3O_6$	247.55	88-88-0	83	7390		4-05-00-00757	1.797^{20}	H_2O 1; EtOH 3; eth 2; ace 4	
1245	Benzene, cyclobutyl-		$C_{10}H_{12}$	132.21	4392-30-7			190^{155}; 101^{41}	4-05-00-01387	0.9378^{20}		1.5277^{20}
1246	Benzene, 1-cyclohexen-1-yl-		$C_{12}H_{14}$	158.24	771-98-2	-11		252	4-05-00-01557	0.9939^{20}	MeOH 4	1.5718^{20}
1247	Benzene, 2-cyclohexen-1-yl-		$C_{12}H_{14}$	158.24	15232-96-9			$115\text{-}7^{16}$	4-05-00-01558	0.9800^{20}		1.5530^{20}
1248	Benzene, cyclohexyl-	Cyclohexylbenzene	$C_{12}H_{16}$	160.26	827-52-1	7.3		240.1	4-05-00-01424	0.9427^{20}	H_2O 1; EtOH 4; eth 3; ctc 2	1.5329^{20}
1249	Benzene, (1-cyclohexylethyl)-		$C_{14}H_{20}$	188.31	4413-16-5			265	4-05-00-01454	0.9773^{17}		1.549^{17}
1250	Benzene, 1-cyclohexyl-2-methoxy-		$C_{13}H_{18}O$	190.29	2206-48-6			268	4-06-00-03899	1.007^{18}		1.5365^{18}
1251	Benzene, 1-cyclohexyl-2-nitro-	Cyclohexane, 2-nitrophenyl	$C_{12}H_{15}NO_2$	205.26	7137-56-6	45		174^{16}; $113^{0.5}$	3-05-00-01259	1.111^{23}		1.5472^{25}
1252	Benzene, 1-cyclohexyl-4-nitro-		$C_{12}H_{15}NO_2$	205.26	5458-48-0	58.5		210^{25}	4-05-00-01426		bz 4; eth 4	
1253	Benzene, (cyclohexyloxy)-		$C_{12}H_{16}O$	176.26	2206-38-4			261	4-06-00-00565	1.0077^{20}	H_2O 1; ace 3; bz 3	1.520^{22}
1254	Benzene, cyclopentyl-		$C_{11}H_{14}$	146.23	700-88-9			219	4-05-00-01409	0.9462^{20}	eth 4	1.5280^{20}
1255	Benzene, (3-cyclopentylpropyl)-		$C_{14}H_{20}$	188.31	2883-12-7			272	4-05-00-01455	0.9233^{19}		1.5130^{19}
1256	Benzene, cyclopropyl-	Cyclopropylbenzene	C_9H_{10}	118.18	873-49-4	-31		173.6; 80^{37}	4-05-00-01370	0.9317^{20}	H_2O 1; eth 3; ace 3; chl 3	1.5285^{20}
1257	Benzenedecanoic acid, ι-octyl-		$C_{24}H_{40}O_2$	360.58	1938-17-6	40.5		$199\text{-}205^{0.09}$	2-09-00-00376	0.9310^{25}		1.4894^{20}
1258	Benzene, decyl-	Decylbenzene	$C_{16}H_{26}$	218.38	104-72-3	-14.4		298	4-05-00-01193	0.8555^{20}	ace 4; bz 4; eth 4; EtOH 4	1.4832^{20}
1259	1,4-Benzenediacetic acid, dimethyl ester		$C_{12}H_{14}O_4$	222.24	36076-25-2	51.5		189^{15}	2-09-00-00624		EtOH 3; eth 3	

No.	Name Synonym	Mol. Form. Mol. Wt.	CAS RN mp/°C	Merck No. bp/°C	Beil. Ref. den/g cm^{-3}	Solubility n_D
1260	1,2-Benzenediamine *o*-Phenylenediamine	$C_6H_8N_2$ 108.14	95-54-5 102.5	7255 257	4-13-00-00038	H_2O 3; EtOH 4; eth 3; bz 3
1261	1,3-Benzenediamine *m*-Phenylenediamine	$C_6H_8N_2$ 108.14	108-45-2 63.5	7254 285	4-13-00-00079 1.0096^{58}	H_2O 4; EtOH 3; eth 3; bz 3 1.6339^{58}
1262	1,4-Benzenediamine *p*-Phenylenediamine	$C_6H_8N_2$ 108.14	106-50-3 146	7256 267	4-13-00-00104	H_2O 2; EtOH 3; eth 3; bz 3
1263	1,4-Benzenediamine, *N*-(4-aminophenyl)- 4,4'-Diaminodiphenylamine	$C_{12}H_{13}N_3$ 199.26	537-65-5 158	2957 dec	4-13-00-00203	eth 4; EtOH 4
1264	1,2-Benzenediamine, 4-chloro-	$C_6H_7ClN_2$ 142.59	95-83-0 76		4-13-00-00068	H_2O 2; EtOH 4; eth 4; bz 3
1265	1,3-Benzenediamine, *N*,*N*-diethyl-	$C_{10}H_{16}N_2$ 164.25	26513-20-2 152	 277	2-13-00-00026	EtOH 4
1266	1,4-Benzenediamine, *N*,*N*-diethyl- *N*,*N*-Diethyl-*p*-phenylenediamine	$C_{10}H_{16}N_2$ 164.25	93-05-0	 261	4-13-00-00109	bz 4
1267	1,2-Benzenediamine, *N*,*N*-dimethyl-	$C_8H_{12}N_2$ 136.20	2836-03-5	 218; 90^{22}	4-13-00-00042 0.995^{22}	H_2O 2; EtOH 4; eth 4; ace 4
1268	1,3-Benzenediamine, *N*,*N*-dimethyl-	$C_8H_{12}N_2$ 136.20	2836-04-6 <-20	 270; 138^{10}	4-13-00-00080 0.995^{25}	H_2O 2; EtOH 4; eth 4
1269	1,2-Benzenediamine, *N*,*N*'-dimethyl- *o*-Phenylenediamine, *N*,*N*'-dimethyl	$C_8H_{12}N_2$ 136.20	3213-79-4 34.5	 247	4-13-00-00042	 1.5914^{25}
1270	1,4-Benzenediamine, *N*,*N*-dimethyl- Dimethyl-*p*-phenylenediamine	$C_8H_{12}N_2$ 136.20	99-98-9 53	3242 263	4-13-00-00106 1.036^{20}	H_2O 3; EtOH 4; eth 4; bz 4
1271	1,4-Benzenediamine, *N*,*N*'-diphenyl- *N*,*N*'-Diphenyl-*p*-phenylenediamine	$C_{18}H_{16}N_2$ 260.34	74-31-7 144	3331	4-13-00-00116	EtOH 2; eth 2; bz 2; chl 2
1272	1,2-Benzenediamine, 4-ethoxy-	$C_8H_{12}N_2O$ 152.20	1197-37-1 71.5	 295	3-13-00-01362	H_2O 4; EtOH 3; eth 3; chl 3
1273	1,2-Benzenediamine, 4-methoxy-	$C_7H_{10}N_2O$ 138.17	102-51-2 51	 200^{21}; 168^{11}	4-13-00-01448	eth 4
1274	1,3-Benzenediamine, 4-methoxy- 4-Methoxy-*m*-phenylenediamine	$C_7H_{10}N_2O$ 138.17	615-05-4 67.5	5920	4-13-00-01425	EtOH 3; eth 3; DMSO-d_6 2
1275	1,2-Benzenediamine, 3-methyl-	$C_7H_{10}N_2$ 122.17	2687-25-4 63.5	 255	3-13-00-00277	ace 4; bz 4; EtOH 4
1276	1,2-Benzenediamine, 4-methyl- Toluene-3,4-diamine	$C_7H_{10}N_2$ 122.17	496-72-0 89.5	 265	4-13-00-00260	H_2O 4; lig 3
1277	1,3-Benzenediamine, 2-methyl- 2,6-Diaminotoluene	$C_7H_{10}N_2$ 122.17	823-40-5 106		3-13-00-00291	H_2O 3; EtOH 3; bz 3
1278	1,3-Benzenediamine, 4-methyl- Toluene-2,4-diamine	$C_7H_{10}N_2$ 122.17	95-80-7 99	 292	4-13-00-00235	H_2O 4; EtOH 4; eth 4; bz 4
1279	1,4-Benzenediamine, 2-methyl-	$C_7H_{10}N_2$ 122.17	95-70-5 64	 273.5	4-13-00-00246	H_2O 3; EtOH 3; eth 3; bz 2
1280	1,4-Benzenediamine, *N*-methyl-	$C_7H_{10}N_2$ 122.17	623-09-6 36	 258	4-13-00-00106	H_2O 4; EtOH 4; eth 4; ace 4
1281	1,2-Benzenediamine, 4-nitro- 4-Nitro-*o*-phenylenediamine	$C_6H_7N_3O_2$ 153.14	99-56-9 199.5	6544	4-13-00-00075	acid 3
1282	1,2-Benzenediamine, *N*-phenyl-	$C_{12}H_{12}N_2$ 184.24	534-85-0 79.5	 313	4-13-00-00043	H_2O 2; ace 3; bz 3; chl 3
1283	1,4-Benzenediamine, *N*-phenyl- *p*-Aminodiphenylamine	$C_{12}H_{12}N_2$ 184.24	101-54-2 66	 354	4-13-00-00113	H_2O 2; EtOH 4; eth 3; chl 2
1284	1,3-Benzenediamine, 4-(phenylazo)-, monohydrochloride Chrysoidine hydrochloride	$C_{12}H_{13}ClN_4$ 248.71	532-82-1 118.5	2262	2-16-00-00203	ace 4
1285	1,3-Benzenediamine, *N*,*N*,*N*',*N*'-tetraethyl- *N*,*N*,*N*',*N*'-Tetraethyl-*m*-phenylenediamine	$C_{14}H_{24}N_2$ 220.36	64287-26-9	 148^9	2-13-00-00026 0.9522^{12}	eth 4; EtOH 4 1.5537^{12}
1286	1,2-Benzenediamine, *N*,*N*,*N*',*N*'-tetramethyl-	$C_{10}H_{16}N_2$ 164.25	704-01-8 8.9	 215.5	4-13-00-00042 0.9560^{20}	
1287	1,4-Benzenediamine, *N*,*N*,*N*',*N*'-tetramethyl- Tetramethyl-*p*-phenylenediamine	$C_{10}H_{16}N_2$ 164.25	100-22-1 51	9159 260	4-13-00-00107	H_2O 2; EtOH 4; eth 4; bz 4
1288	Benzene, 1,2-dibromo- *o*-Dibromobenzene	$C_6H_4Br_2$ 235.91	583-53-9 7.1	 225	4-05-00-00682 1.9843^{20}	H_2O 1; EtOH 3; eth 5; ace 5 1.6155^{20}
1289	Benzene, 1,3-dibromo- *m*-Dibromobenzene	$C_6H_4Br_2$ 235.91	108-36-1 -7	 218	4-05-00-00682 1.9523^{20}	H_2O 1; EtOH 3; eth 5 1.6083^{17}
1290	Benzene, 1,4-dibromo- *p*-Dibromobenzene	$C_6H_4Br_2$ 235.91	106-37-6 87.3	3001 218.5	4-05-00-00683 2.261^{17}	H_2O 1; EtOH 3; eth 4; ace 4 1.5742
1291	Benzene, 1,3-dibromo-5-(bromomethyl)- 3,5-Dibromobenzyl bromide	$C_7H_5Br_3$ 328.83	56908-88-4 96	 173^{19}	3-05-00-00719	EtOH 4
1292	Benzene, 1,2-dibromo-3-chloro- 1,2-Dibromo-3-chlorobenzene	$C_6H_3Br_2Cl$ 270.35	104514-49-0 73.5	 264	1-05-00-00117	bz 4; eth 4; chl 4
1293	Benzene, 1,2-dibromo-4-chloro- 1,2-Dibromo-4-chlorobenzene	$C_6H_3Br_2Cl$ 270.35	60956-24-3 35.5	 256	0-05-00-00212	bz 4; eth 4; chl 4
1294	Benzene, 1,3-dibromo-2-chloro- 1,3-Dibromo-2-chlorobenzene	$C_6H_3Br_2Cl$ 270.35	19230-27-4 72	 265	3-05-00-00569	bz 4; eth 4; chl 4
1295	Benzene, 1,4-dibromo-2-chloro- 1,4-Dibromo-2-chlorobenzene	$C_6H_3Br_2Cl$ 270.35	3460-24-0 40.5	 259	0-05-00-00212	bz 4; eth 4; chl 4
1296	Benzene, 2,4-dibromo-1-chloro-	$C_6H_3Br_2Cl$ 270.35	29604-75-9 27	 258	0-05-00-00212	bz 4; eth 4; chl 4
1297	Benzene, 1,2-dibromo-3,4-dimethyl- 3,4-Dibromo-*o*-xylene	$C_8H_8Br_2$ 263.96	24932-49-8 7	 277	2-05-00-00285 1.782^{15}	EtOH 4
1298	Benzene, 1,5-dibromo-2,4-dimethyl-	$C_8H_8Br_2$ 263.96	615-87-2 70	 255.5	3-05-00-00839	EtOH 4

PHYSICAL CONSTANTS OF ORGANIC COMPOUNDS (continued)

No.	Name	Synonym	Mol. Form.	Mol. Wt.	CAS RN	mp/°C	Merck No.	bp/°C	Beil. Ref.	den/g cm^{-3}	Solubility	n_D
1299	Benzene, (1,2-dibromoethyl)-		$C_8H_8Br_2$	263.96	93-52-7	75		133^{19}	4-05-00-00909		EtOH 3; eth 3; bz 3; chl 3	
1300	Benzene, 2,4-dibromo-1-fluoro-		$C_6H_3Br_2F$	253.90	1435-53-6	69		$103\text{-}5^{32}$	3-05-00-00569	2.047^{20}		1.5840^{20}
1301	Benzene, (dibromomethyl)-		$C_7H_6Br_2$	249.93	618-31-5			156^{23}	4-05-00-00836	1.8365^{28}	H_2O 1; EtOH 5; eth 5	1.6147^{20}
1302	Benzene, 1,2-dibromo-3-methyl-	2,3-Dibromotoluene	$C_7H_6Br_2$	249.93	61563-25-5	30.5		$126\text{-}8^{19}$	4-05-00-00835	1.8234^{15}		1.5984^{25}
1303	Benzene, 1,2-dibromo-4-methyl-	3,4-Dibromotoluene	$C_7H_6Br_2$	249.93	60956-23-2	-10		240; 123^{19}	4-05-00-00836	1.8197^{25}		1.5979^{25}
1304	Benzene, 1,3-dibromo-2-methyl-	2,6-Dibromotoluene	$C_7H_6Br_2$	249.93	69321-60-4	5.5		246	4-05-00-00836	1.812^{22}		
1305	Benzene, 1,4-dibromo-2-methyl-	2,5-Dibromotoluene	$C_7H_6Br_2$	249.93	615-59-8	5.6		236	4-05-00-00835	1.8127^{17}	H_2O 1	1.5982^{18}
1306	Benzene, 2,4-dibromo-1-methyl-		$C_7H_6Br_2$	249.93	31543-75-6	-9.7		$103\text{-}4^{11}$	4-05-00-00835	1.8176^{25}		1.5964^{25}
1307	Benzene, 1,2-dibromo-4-nitro-	3,4-Dibromonitrobenzene	$C_6H_3Br_2NO_2$	280.90	5411-50-7	58.5		296	4-05-00-00732	2.354^{8}	bz 4; EtOH 4; HOAc 4	1.9835^{111}
1308	Benzene, 1,3-dibromo-2-nitro-	2,6-Dibromonitrobenzene	$C_6H_3Br_2NO_2$	280.90	13402-32-9	84		sub	3-05-00-00621	2.211^{8}	ace 4; bz 4; EtOH 4	
1309	Benzene, 1,3-dibromo-5-nitro-		$C_6H_3Br_2NO_2$	280.90	6311-60-0	106			3-05-00-00621	2.36^{8}	bz 4; eth 4; EtOH 4	
1310	Benzene, 1,4-dibromo-2-nitro-		$C_6H_3Br_2NO_2$	280.90	3460-18-2	85.5			4-05-00-00732	2.368^{8}	H_2O 1; ace 3; bz 3; chl 2	
1311	Benzene, 2,4-dibromo-1-nitro-		$C_6H_3Br_2NO_2$	280.90	51686-78-3	62			3-05-00-00620	2.356^{8}	ace 4; bz 4	
1312	Benzene, 1,2-dibromo-3,4,5,6-tetrafluoro-		$C_6Br_2F_4$	307.87	827-08-7	12		198		2.260^{20}		1.5151^{25}
1313	Benzene, 2,4-dibromo-1,3,5-trimethyl-	Mesitylene, 2,4-dibromo-	$C_9H_{10}Br_2$	277.99	6942-99-0	65.5		285	3-05-00-00921		ctc 2	
1314	Benzene, 1,4-dibutoxy-		$C_{14}H_{22}O_2$	222.33	104-36-9	45.5		158^{15}	4-06-00-05721		ctc 3	
1315	1,3-Benzenedicarbonitrile	1,3-Dicyanobenzene	$C_8H_4N_2$	128.13	626-17-5	162		sub	4-09-00-03297	0.992^{40}	H_2O 2; EtOH 4; eth 3; bz 3	
1316	1,4-Benzenedicarbonitrile		$C_8H_4N_2$	128.13	623-26-7	224		sub	4-09-00-03328		H_2O 1; EtOH 2; eth 2; bz 3	
1317	1,3-Benzenedicarbonitrile, 2,4,5,6-tetrachloro-	Chlorothalonil	$C_8Cl_4N_2$	265.91	1897-45-6	250	2167	350		1.7^{25}	H_2O 1; ace 2; cyhex 2	
1318	1,2-Benzenedicarbonyl dichloride	Phthaloyl chloride	$C_8H_4Cl_2O_2$	203.02	88-95-9	15.5	7349	281.1	4-09-00-03261	1.4089^{20}		1.5684^{20}
1319	1,3-Benzenedicarbonyl dichloride		$C_8H_4Cl_2O_2$	203.02	99-63-8	43.5		276	4-09-00-03295	1.3880^{17}	H_2O 2; EtOH 2; eth 3	1.570^{47}
1320	1,4-Benzenedicarbonyl dichloride		$C_8H_4Cl_2O_2$	203.02	100-20-9	83.5		258; 125^{9}	4-09-00-03318		eth 3	
1321	1,2-Benzenedicarbonyl difluoride		$C_8H_4F_2O_2$	170.12	445-69-2	42.5		227	4-09-00-03260	1.3066^{50}	H_2O 1; peth 3	
1322	1,3-Benzenedicarboxaldehyde		$C_8H_6O_2$	134.13	626-19-7	89.5		246; 136^{13}	4-07-00-02139		H_2O 2; EtOH 4; eth 2; ace 3	
1323	1,4-Benzenedicarboxaldehyde		$C_8H_6O_2$	134.13	623-27-8	117		246	4-07-00-02140		H_2O 2; EtOH 4; eth 3; chl 3	
1324	1,2-Benzenedicarboxamide	Phthalamide	$C_8H_8N_2O_2$	164.16	88-96-0	222	7343	dec	4-09-00-03265		H_2O 2; EtOH 2; eth 1	
1325	1,2-Benzenedicarboxamide, *N,N,N',N'*-tetraethyl-	*N,N,N',N'*-Tetraethylphthalamide	$C_{16}H_{24}N_2O_2$	276.38	83-81-8	36	9137	204^{16}	4-09-00-03266			
1326	1,3-Benzenedicarboxamide, *N,N,N',N'*-tetraethyl-	Isophthalamide, *N,N,N',N'*-tetraethyl	$C_{16}H_{24}N_2O_2$	276.38	13698-87-8	86		242^{12}	4-09-00-03296		ace 4; bz 4; eth 4; EtOH 4	
1327	1,2-Benzenedicarboxylic acid	Phthalic acid	$C_8H_6O_4$	166.13	88-99-3	230 dec	7345	dec	4-09-00-03167	2.18^{191}	H_2O 2; EtOH 3; eth 2; chl 1	
1328	1,3-Benzenedicarboxylic acid	Isophthalic acid	$C_8H_6O_4$	166.13	121-91-5	347	5083	sub	4-09-00-03292		H_2O 2; EtOH 3; eth 1; bz 1	
1329	1,4-Benzenedicarboxylic acid	Terephthalic acid	$C_8H_6O_4$	166.13	100-21-0		9093	sub 300	4-09-00-03301		H_2O 1; EtOH 1; eth 1; ctc 2	
1330	1,3-Benzenedicarboxylic acid, 2-amino-		$C_8H_7NO_4$	181.15	39622-79-2	340		sub 267	4-14-00-01899		eth 4; EtOH 4	
1331	1,3-Benzenedicarboxylic acid, 5-amino-		$C_8H_7NO_4$	181.15	99-31-0	360		sub	1-14-00-00636		H_2O 1; EtOH 2	
1332	1,2-Benzenedicarboxylic acid, 3-benzoyl-		$C_{15}H_{10}O_5$	270.24	602-82-4	140.5		dec	3-10-00-04007		bz 4; EtOH 4	
1333	1,2-Benzenedicarboxylic acid, bis(2-ethoxyethyl) ester		$C_{16}H_{22}O_6$	310.35	605-54-9	34		345	2-09-00-00597	1.1229^{21}		
1334	1,2-Benzenedicarboxylic acid, bis(2-ethylhexyl) ester	Bis(2-ethylhexyl) phthalate	$C_{24}H_{38}O_4$	390.56	117-81-7	-55	1262	384	4-09-00-03181	0.981^{25}	ctc 2	1.4853^{20}
1335	1,2-Benzenedicarboxylic acid, bis(3-methylbutyl) ester	Diisoamyl phthalate	$C_{18}H_{26}O_4$	306.40	605-50-5		5006	334 dec	4-09-00-03179	1.0209^{16}	EtOH 4	1.4871^{20}

No.	Name Synonym	Mol. Form. Mol. Wt.	CAS RN mp/°C	Merck No. bp/°C	Beil. Ref. den/g cm^{-3}	Solubility n_D
1336	1,2-Benzenedicarboxylic acid, bis(2-methylpropyl) ester	$C_{16}H_{22}O_4$ 278.35	84-69-5	 296.5; 159^{4}	4-09-00-03177 1.0490^{15}	ctc 3
1337	1,2-Benzenedicarboxylic acid, bis(phenylmethyl) ester	$C_{22}H_{18}O_4$ 346.38	523-31-9 44.0	 277^{15}	4-09-00-03219	H_2O 1; EtOH 4; eth 4; lig 1
1338	1,4-Benzenedicarboxylic acid, bis(2-methylpropyl) ester	$C_{16}H_{22}O_4$ 278.35	18699-48-4 55	 180^{6}	4-09-00-03305	eth 3
1339	1,2-Benzenedicarboxylic acid, dibutyl ester Dibutyl phthalate	$C_{16}H_{22}O_4$ 278.35	84-74-2 -35	1586 340	4-09-00-03175 1.0465^{20}	H_2O 1; EtOH 5; eth 5; bz 5 1.4911^{20}
1340	1,2-Benzenedicarboxylic acid, 3,5-dichloro- Phthalic acid, 3,5-dichloro-	$C_8H_4Cl_2O_4$ 235.02	25641-98-9 164 dec	 sub	4-09-00-03270	ace 4; eth 4; EtOH 4
1341	1,2-Benzenedicarboxylic acid, dicyclohexyl ester	$C_{20}H_{26}O_4$ 330.42	84-61-7 66		4-09-00-03189 1.383^{20}	H_2O 1; EtOH 3; eth 3; chl 2 1.431^{20}
1342	1,2-Benzenedicarboxylic acid, diethyl ester Diethyl phthalate	$C_{12}H_{14}O_4$ 222.24	84-66-2 -40.5	7345 295	4-09-00-03172 1.232^{14}	H_2O I; EtOH 5; eth 5; ace 3 1.5000^{21}
1343	1,2-Benzenedicarboxylic acid, dimethyl ester Dimethyl phthalate	$C_{10}H_{10}O_4$ 194.19	131-11-3 5.5	3243 283.7	4-09-00-03170 1.1905^{20}	H_2O 1; EtOH 5; eth 5; bz 3 1.5138^{20}
1344	1,2-Benzenedicarboxylic acid, dioctyl ester Dioctyl phthalate	$C_{24}H_{38}O_4$ 390.56	117-84-0 25		4-09-00-03180	
1345	1,2-Benzenedicarboxylic acid, diphenyl ester Phenyl phthalate	$C_{20}H_{14}O_4$ 318.33	84-62-8 73	7278 253^{14}	4-09-00-03217	H_2O 1; EtOH 2; eth 2; ctc 2
1346	1,3-Benzenedicarboxylic acid, diethyl ester Diethyl isophthalate	$C_{12}H_{14}O_4$ 222.24	636-53-3 11.5	 302	4-09-00-03294 1.1239^{17}	H_2O 1 1.508^{18}
1347	1,3-Benzenedicarboxylic acid, dimethyl ester Dimethyl isophthalate	$C_{10}H_{10}O_4$ 194.19	1459-93-4 67.5	 282	4-09-00-03293 1.194^{20}	H_2O 2 1.5168^{20}
1348	1,3-Benzenedicarboxylic acid, 4,6-dimethyl-	$C_{10}H_{10}O_4$ 194.19	2790-09-2 266	 sub	3-09-00-04298	EtOH 4
1349	1,4-Benzenedicarboxylic acid, 2,5-dichloro- 2,5-Dichloroterephthalic acid	$C_8H_4Cl_2O_4$ 235.02	13799-90-1 306	 sub	4-09-00-03335	H_2O 2; EtOH 3; eth 4
1350	1,4-Benzenedicarboxylic acid, diethyl ester Diethyl terephthalate	$C_{12}H_{14}O_4$ 222.24	636-09-9 44	 302	4-09-00-03304 1.0989^{45}	H_2O 1; EtOH 4; eth 3
1351	1,4-Benzenedicarboxylic acid, dimethyl ester Dimethyl terephthalate	$C_{10}H_{10}O_4$ 194.19	120-61-6 141	 288	4-09-00-03303 1.075^{141}	H_2O 2; EtOH 2; eth 3; chl 3
1352	1,3-Benzenedicarboxylic acid, 4-hydroxy- 4-Hydroxyisophthalic acid	$C_8H_6O_5$ 182.13	636-46-4 310	4758	4-10-00-02091	H_2O 1; EtOH 4; eth 4; chl 1
1353	1,2-Benzenedicarboxylic acid, 3,4,5,6-tetrachloro-, monoethyl ester Phthalic acid, tetrachloro-, monoethyl ester	$C_{10}H_6Cl_4O_4$ 331.97	602-21-1 94.5	 250 dec	3-09-00-04206	eth 4; EtOH 4
1354	Benzene, 1,2-dichloro- *o*-Dichlorobenzene	$C_6H_4Cl_2$ 147.00	95-50-1 -16.7	3044 180	4-05-00-00654 1.3059^{20}	H_2O 1; EtOH 3; eth 3; ace 5 1.5515^{20}
1355	Benzene, 1,3-dichloro- *m*-Dichlorobenzene	$C_6H_4Cl_2$ 147.00	541-73-1 -24.8	3043 173	4-05-00-00657 1.2884^{20}	H_2O 1; EtOH 3; eth 3; ace 5 1.5459^{20}
1356	Benzene, 1,4-dichloro- *p*-Dichlorobenzene	$C_6H_4Cl_2$ 147.00	106-46-7 52.7	3045 174	4-05-00-00658 1.2475^{55}	H_2O 1; EtOH 5; eth 3; ace 5 1.5285^{20}
1357	Benzene, 1,2-dichloro-4-(chloromethyl)-	$C_7H_5Cl_3$ 195.47	102-47-6 37.5	 241	4-05-00-00820	H_2O 1; EtOH 3; ctc 3; os 3
1358	Benzene, 1,3-dichloro-5-(chloromethyl)-	$C_7H_5Cl_3$ 195.47	3290-06-0 36	 $60^{0.35}$	3-05-00-00699	EtOH 4
1359	Benzene, 1,2-dichloro-4-(dichloromethyl)- Benzylidene chloride, 3,4-dichloro	$C_7H_4Cl_4$ 229.92	56961-84-3	 257	3-05-00-00702 1.515^{22}	bz 4; eth 4; EtOH 4
1360	Benzene, 1,2-dichloro-3,4-dimethyl- *o*-Xylene, 3,4-dichloro	$C_8H_8Cl_2$ 175.06	68266-67-1 9	 231	4-05-00-00926 1.2240^{20}	
1361	Benzene, 1,3-dichloro-2,5-dimethyl-	$C_8H_8Cl_2$ 175.06	38204-89-6 15	 222	3-05-00-00855 1.2156^{20}	
1362	Benzene, 1,4-dichloro-2,5-dimethyl-	$C_8H_8Cl_2$ 175.06	1124-05-6 71	 222	3-05-00-00855	chl 3
1363	Benzene, 1,5-dichloro-2,4-dimethyl-	$C_8H_8Cl_2$ 175.06	2084-45-9 68.5	 223	3-05-00-00835	bz 4; eth 4; chl 4
1364	Benzene, 2,3-dichloro-1,4-dimethyl-	$C_8H_8Cl_2$ 175.06	34840-79-4 -2	 230	3-05-00-00855 1.2310^{20}	
1365	Benzene, 1,1'-(1,2-dichloro-1,2-ethenediyl)bis-, (*E*)-	$C_{14}H_{10}Cl_2$ 249.14	951-86-0 143.5	 180^{18}	1-05-00-00304	H_2O 1; EtOH 4; eth 4
1366	Benzene, 1,1'-(1,2-dichloro-1,2-ethenediyl)bis-, (*Z*)-	$C_{14}H_{10}Cl_2$ 249.14	5216-32-0 67.5	 179^{18}	1-05-00-00304	eth 4; EtOH 4
1367	Benzene, 1,2-dichloro-3-ethenyl- 2,3-Dichlorostyrene	$C_8H_6Cl_2$ 173.04	2123-28-6	 $92\text{-}4^{4.5}$	3-05-00-01174 1.2826^{20}	 1.5848^{20}
1368	Benzene, 1,2-dichloro-4-ethenyl-	$C_8H_6Cl_2$ 173.04	2039-83-0	 95^{5}	4-05-00-01347 1.256^{20}	 1.5857^{20}
1369	Benzene, 1,3-dichloro-2-ethenyl- 2,6-Dichlorostyrene	$C_8H_6Cl_2$ 173.04	28469-92-3 89	 $64\text{-}5^{3}$	3-05-00-01174 1.2631^{20}	 1.5752^{20}
1370	Benzene, 1,3-dichloro-5-ethenyl- 3,5-Dichlorostyrene	$C_8H_6Cl_2$ 173.04	2155-42-2	 59^{1}	3-05-00-01175 1.225^{25}	 1.5745^{25}
1371	Benzene, 1,4-dichloro-2-ethenyl- 2,5-Dichlorostyrene	$C_8H_6Cl_2$ 173.04	1123-84-8	 $72\text{-}3^{2}$	4-05-00-01347 1.4045^{20}	 1.5798^{20}
1372	Benzene, (2,2-dichloroethenyl)-	$C_8H_6Cl_2$ 173.04	698-88-4	 225	4-05-00-01348 1.2531^{20}	ace 3; chl 3 1.5852^{20}

No.	Name Synonym	Mol. Form. Mol. Wt.	CAS RN mp/°C	Merck No. bp/°C	Beil. Ref. den/g cm^{-3}	Solubility n_D
1373	Benzene, 2,4-dichloro-1-ethenyl- 2,4-Dichlorostyrene	$C_8H_6Cl_2$ 173.04	2123-27-5	81^{6}	4-05-00-01347 1.243^{25}	1.5828^{20}
1374	Benzene, 1,1'-(dichloroethenylidene)bis[4-chloro- 2,2-Dichloro-1,1-bis(4-chlorophenyl)ethylene	$C_{14}H_8Cl_4$ 318.03	72-55-9 89		4-05-00-02177	
1375	Benzene, 2,4-dichloro-1-ethoxy-	$C_8H_8Cl_2O$ 191.06	5392-86-9 32	237	4-06-00-00885	H_2O 1; EtOH 3; eth 3; bz 3
1376	Benzene, (1,2-dichloroethyl)-	$C_8H_8Cl_2$ 175.06	1074-11-9	233.5	4-05-00-00901 1.240^{15}	H_2O 1; ace 3; bz 3 1.5544^{15}
1377	Benzene, 1,4-dichloro-2-ethyl- 2,5-Dichloroethylbenzene	$C_8H_8Cl_2$ 175.06	54484-63-8	213.5	4-05-00-00901 1.239^{25}	eth 2; bz 3; chl 2; peth 2
1378	Benzene, 1,1'-(2,2-dichloroethylidene)bis[4-chloro- 1,1-Dichloro-2,2-bis(*p*-chlorophenyl)ethane	$C_{14}H_{10}Cl_4$ 320.04	72-54-8 109.5	3049	4-05-00-01884	chl 2
1379	Benzene, (dichlorofluoromethyl)-	$C_7H_5Cl_2F$ 179.02	498-67-9 -26.8	179	4-05-00-00818 1.3138^{11}	EtOH 4 1.5180^{11}
1380	Benzene, 1,4-dichloro-2-iodo-	$C_6H_3Cl_2I$ 272.90	29682-41-5 22	253	3-05-00-00580	bz 4; eth 4; EtOH 4; chl 4
1381	Benzene, 2,4-dichloro-1-methoxy-	$C_7H_6Cl_2O$ 177.03	553-82-2 28.5	232; 125^{10}	4-06-00-00885	chl 2
1382	Benzene, (dichloromethyl)- Benzal chloride	$C_7H_6Cl_2$ 161.03	98-87-3 -17	1064 205	4-05-00-00817 1.26^{25}	eth 4; EtOH 4 1.5502^{20}
1383	Benzene, 1,2-dichloro-3-methyl- 2,3-Dichlorotoluene	$C_7H_6Cl_2$ 161.03	32768-54-0 6	207.5	4-05-00-00815 1.2458^{20}	bz 4 1.5511^{20}
1384	Benzene, 1,2-dichloro-4-methyl- 3,4-Dichlorotoluene	$C_7H_6Cl_2$ 161.03	95-75-0 -15.2	208.9	4-05-00-00815 1.2564^{20}	H_2O 1; EtOH 5; eth 5; ace 5 1.5471^{20}
1385	Benzene, 1,3-dichloro-2-methyl- 2,6-Dichlorotoluene	$C_7H_6Cl_2$ 161.03	118-69-4 25.8	198	4-05-00-00815 1.2686^{20}	H_2O 1; chl 3 1.5507^{20}
1386	Benzene, 1,3-dichloro-5-methyl- 3,5-Dichlorotoluene	$C_7H_6Cl_2$ 161.03	25186-47-4 26	201.5	4-05-00-00816	H_2O 1
1387	Benzene, 1,4-dichloro-2-methyl- 2,5-Dichlorotoluene	$C_7H_6Cl_2$ 161.03	19398-61-9 5	200	3-05-00-00694 1.2535^{20}	H_2O 1; bz 3 1.5449^{20}
1388	Benzene, 2,4-dichloro-1-methyl- 2,4-Dichlorotoluene	$C_7H_6Cl_2$ 161.03	95-73-8 -13.5	201	4-05-00-00815 1.2476^{20}	H_2O 1; ctc 3 1.5511^{20}
1389	Benzene, 1,1'-(dichloromethylene)bis-	$C_{13}H_{10}Cl_2$ 237.13	2051-90-3	305 dec; 190^{21}	4-05-00-01848 1.235^{18}	eth 3; bz 3; ctc 3
1390	Benzene, 1-(dichloromethyl)-4-methyl-	$C_8H_8Cl_2$ 175.06	23063-36-7 48.5	105^{18}	3-05-00-00855	EtOH 2; bz 2; chl 2
1391	Benzene, 1,2-dichloro-3-nitro-	$C_6H_3Cl_2NO_2$ 192.00	3209-22-1 61.5	257.5	4-05-00-00725 1.721^{14}	H_2O 1; EtOH 3; eth 3; ace 3
1392	Benzene, 1,2-dichloro-4-nitro-	$C_6H_3Cl_2NO_2$ 192.00	99-54-7 43	255.5	4-05-00-00726 1.4558^{75}	H_2O 1; EtOH 3; eth 3; ctc 2
1393	Benzene, 1,3-dichloro-2-nitro- 2,6-Dichloronitrobenzene	$C_6H_3Cl_2NO_2$ 192.00	601-88-7 72.5	130^{8}	4-05-00-00726 1.603^{17}	eth 4; EtOH 4
1394	Benzene, 1,3-dichloro-5-nitro-	$C_6H_3Cl_2NO_2$ 192.00	618-62-2 65.4		4-05-00-00727 1.4000^{100}	H_2O 1; EtOH 3; eth 3 1.4000^{100}
1395	Benzene, 1,4-dichloro-2-nitro-	$C_6H_3Cl_2NO_2$ 192.00	89-61-2 56	267	4-05-00-00726 1.439^{75}	H_2O 1; EtOH 3; eth 3; bz 3 1.4390^{75}
1396	Benzene, 2,4-dichloro-1-nitro-	$C_6H_3Cl_2NO_2$ 192.00	611-06-3 34	258.5	4-05-00-00726 1.4790^{80}	H_2O 1; EtOH 3; eth 3; chl 2 1.5512^{70}
1397	Benzene, (3,3-dichloro-2-propenyl)-	$C_9H_8Cl_2$ 187.07	38862-78-1 59	242; 142^{30}	4-05-00-01363 1.2028^{20}	eth 3; bz 3; chl 3; peth 3
1398	Benzene, 1,2-dichloro-3,4,5,6-tetrafluoro 1,2-Dichlorotetrafluorobenzene	$C_6Cl_2F_4$ 218.97	1198-59-0			
1399	Benzene, 1,3-dichloro-2,4,5,6-tetrafluoro- 1,3-Dichlorotetrafluorobenzene	$C_6Cl_2F_4$ 218.97	1198-61-4			
1400	Benzene, 1,4-dichloro-2,3,5,6-tetrafluoro- 1,4-Dichlorotetrafluorobenzene	$C_6Cl_2F_4$ 218.97	1198-62-5			
1401	Benzene, 1,2-dichloro-4-(trichloromethyl)-	$C_7H_3Cl_5$ 264.36	13014-24-9 25.8	283.1	4-05-00-00824 1.5913^{20}	1.5886^{20}
1402	Benzene, 2,4-dichloro-1,3,5-trimethyl- Mesitylene, 2,4-dichloro-	$C_9H_{10}Cl_2$ 189.08	57386-83-1 59	243.5	3-05-00-00919	EtOH 3; eth 2; bz 2
1403	Benzene, 1,2-diethenyl- *o*-Divinylbenzene	$C_{10}H_{10}$ 130.19	91-14-5	82^{14}	4-05-00-01540 0.9325^{22}	ace 3; bz 3 1.5767^{20}
1404	Benzene, 1,3-diethenyl- *m*-Divinylbenzene	$C_{10}H_{10}$ 130.19	108-57-6 -52.3	121^{76}; 52^{3}	3-05-00-01367 0.9294^{20}	ace 3; bz 3 1.5760^{20}
1405	Benzene, 1,4-diethenyl- *p*-Divinylbenzene	$C_{10}H_{10}$ 130.19	105-06-6 31	95^{18}; $34^{0.2}$	4-05-00-01541 0.913^{40}	ace 3; bz 3 1.5835^{25}
1406	Benzene, 1,2-diethoxy-	$C_{10}H_{14}O_2$ 166.22	2050-46-6 44	219	4-06-00-05565 1.0075^{20}	EtOH 3; eth 4; ctc 3 1.5083^{25}
1407	Benzene, 1,3-diethoxy-	$C_{10}H_{14}O_2$ 166.22	2049-73-2 12.4	235	4-06-00-05664 1.0170^{15}	H_2O 1; EtOH 3; eth 3
1408	Benzene, 1,4-diethoxy-	$C_{10}H_{14}O_2$ 166.22	122-95-2 72	246	4-06-00-05720	EtOH 4; eth 3; bz 3; ctc 3
1409	Benzene, (diethoxymethoxy)-	$C_{11}H_{16}O_3$ 196.25	14444-77-0	$103\text{-}4^{10}$	4-06-00-00610 1.014^{25}	1.4822^{20}
1410	Benzene, 1,4-diethoxy-2-methyl- 2,5-Diethoxytoluene	$C_{11}H_{16}O_2$ 180.25	41901-72-8 24.5	248	3-06-00-04499 1.0134^{15}	bz 4; eth 4; EtOH 4

No.	Name Synonym	Mol. Form. Mol. Wt.	CAS RN mp/°C	Merck No. bp/°C	Beil. Ref. den/g cm^{-3}	Solubility n_D
1411	Benzene, 1-(diethoxymethyl)-4-(1-methylethyl)-	$C_{14}H_{22}O_2$ 222.33	35364-90-0	 265; $142^{18.5}$	2-07-00-00247 0.944^{19}	 1.484^{19}
1412	Benzene, (3,3-diethoxy-1-propynyl)-	$C_{13}H_{16}O_2$ 204.27	6142-95-6	 $99\text{-}100^{2}$	4-07-00-01173 0.991^{25}	 1.5170^{20}
1413	Benzene, 1,2-diethyl- *o*-Diethylbenzene	$C_{10}H_{14}$ 134.22	135-01-3 -31.2	 184	4-05-00-01065 0.8800^{20}	H_2O 1; EtOH 5; eth 5; ace 5 1.5035^{20}
1414	Benzene, 1,3-diethyl- *m*-Diethylbenzene	$C_{10}H_{14}$ 134.22	141-93-5 -83.9	 181.1	4-05-00-01066 0.8602^{20}	H_2O 1; EtOH 5; eth 5; ace 5 1.4955^{20}
1415	Benzene, 1,4-diethyl- *p*-Diethylbenzene	$C_{10}H_{14}$ 134.22	105-05-5 -42.83	 183.7	4-05-00-01067 0.8620^{20}	H_2O 1; EtOH 5; eth 5; ace 5 1.4967^{20}
1416	Benzene, 1,2-diethyl-3-methyl-	$C_{11}H_{16}$ 148.25	13632-93-4	 206.6	4-05-00-01105 0.8910^{20}	ace 4; bz 4; eth 4; EtOH 4 1.5105^{20}
1417	Benzene, 1,2-diethyl-4-methyl- Toluene, 3,4-diethyl-	$C_{11}H_{16}$ 148.25	13732-80-4	 203.6	4-05-00-01106 0.8762^{20}	ace 4; bz 4; eth 4; EtOH 4 1.5039^{20}
1418	Benzene, 1,3-diethyl-2-methyl- Toluene, 2,6-diethyl-	$C_{11}H_{16}$ 148.25	13632-95-6	 208.8	4-05-00-01105 0.8907^{20}	ace 4; bz 4; eth 4; EtOH 4 1.5106^{20}
1419	Benzene, 1,3-diethyl-5-methyl-	$C_{11}H_{16}$ 148.25	2050-24-0 -74.1	 205	4-05-00-01106 0.8748^{20}	H_2O 1; EtOH 5; eth 5; ace 5 1.5027^{20}
1420	Benzene, 1,4-diethyl-2-methyl-	$C_{11}H_{16}$ 148.25	13632-94-5	 207.1	4-05-00-01106 0.8758^{20}	ace 4; bz 4; eth 4; EtOH 4 1.5034^{20}
1421	Benzene, 2,4-diethyl-1-methyl-	$C_{11}H_{16}$ 148.25	1758-85-6	 205	4-05-00-01105 0.8748^{20}	ace 4; bz 4; eth 4; EtOH 4 1.5027^{20}
1422	Benzene, 1,2-difluoro- *o*-Difluorobenzene	$C_6H_4F_2$ 114.09	367-11-3 -34	 94	4-05-00-00637 1.1599^{18}	H_2O 1; ace 3; bz 3; chl 3 1.4451^{18}
1423	Benzene, 1,3-difluoro- *m*-Difluorobenzene	$C_6H_4F_2$ 114.09	372-18-9 -69.0	 82.6	4-05-00-00637 1.1572^{20}	H_2O 1; ace 3; bz 3 1.4374^{20}
1424	Benzene, 1,4-difluoro- *p*-Difluorobenzene	$C_6H_4F_2$ 114.09	540-36-3 -13	3132 89	4-05-00-00637 1.1701^{20}	H_2O 1; ace 3; bz 3; ctc 2 1.4422^{20}
1425	Benzene, (difluoromethyl)-	$C_7H_6F_2$ 128.12	455-31-2	 140	4-05-00-00801 1.1370^{19}	H_2O 1; EtOH 3 1.4577^{20}
1426	Benzene, 2,4-difluoro-1-nitro-	$C_6H_3F_2NO_2$ 159.09	446-35-5 9.8	 207	4-05-00-00720 1.4571^{14}	chl 2 1.5149^{14}
1427	Benzene, 1,2-diiodo-	$C_6H_4I_2$ 329.91	615-42-9 27	 287; 100^{3}	4-05-00-00700 2.54^{20}	H_2O 2; EtOH 2 1.7179^{20}
1428	Benzene, 1,3-diiodo-	$C_6H_4I_2$ 329.91	626-00-6 40.4	 285	4-05-00-00700 2.47^{25}	eth 4; EtOH 4; chl 4
1429	Benzene, 1,4-diiodo-	$C_6H_4I_2$ 329.91	624-38-4 131.5	 285	4-05-00-00700	H_2O 1; EtOH 3; eth 4; chl 2
1430	Benzene, 1,3-diisocyanato-2-methyl- Toluene-2,6-diisocyanate	$C_9H_6N_2O_2$ 174.16	91-08-7 18.3			H_2O 6; ace 3; bz 3
1431	Benzene, 2,4-diisocyanato-1-methyl- Toluene 2,4-diisocyanate	$C_9H_6N_2O_2$ 174.16	584-84-9 20.5	9456 251	4-13-00-00243 1.2244^{20}	ace 4; bz 4; eth 4
1432	Benzene, 1,4-diisothiocyanato- Bitoscanate	$C_8H_4N_2S_2$ 192.27	4044-65-9 132	1318	4-13-00-00174	
1433	1,3-Benzenedimethanamine *m*-Xylenediamine	$C_8H_{12}N_2$ 136.20	1477-55-0	 247	4-13-00-00293 1.052^{20}	H_2O 4; eth 4; EtOH 4
1434	1,3-Benzenedimethanol	$C_8H_{10}O_2$ 138.17	626-18-6 57	 156^{13}	4-06-00-05966 1.1610^{18}	H_2O 4; eth 4; EtOH 4
1435	1,4-Benzenedimethanol	$C_8H_{10}O_2$ 138.17	589-29-7 117.5	 $138\text{-}43^{1}$	4-06-00-05971	H_2O 4; ace 4; eth 4; EtOH 4
1436	Benzene, 1,2-dimethoxy- Veratrole	$C_8H_{10}O_2$ 138.17	91-16-7 22.5	9857 206	4-06-00-05564 1.0810^{25}	H_2O 2; EtOH 3; eth 3; ctc 3 1.5827^{21}
1437	Benzene, 1,3-dimethoxy-	$C_8H_{10}O_2$ 138.17	151-10-0 -52	 217.5	4-06-00-05663 1.0521^{25}	H_2O 2; EtOH 3; eth 3; bz 3 1.5231^{20}
1438	Benzene, 1,4-dimethoxy-	$C_8H_{10}O_2$ 138.17	150-78-7 59	 212.6	4-06-00-05718 1.0375^{55}	H_2O 2; EtOH 3; eth 4; bz 4
1439	Benzene, 1,2-dimethoxy-3-methyl-	$C_9H_{12}O_2$ 152.19	4463-33-6	 204	4-06-00-05860 1.0335^{20}	eth 4; EtOH 4 1.5121^{25}
1440	Benzene, 1,2-dimethoxy-4-methyl-	$C_9H_{12}O_2$ 152.19	494-99-5 24	 220	4-06-00-05879 1.0509^{25}	H_2O 1; ctc 2; os 4 1.5257^{25}
1441	Benzene, 1,3-dimethoxy-5-methyl-	$C_9H_{12}O_2$ 152.19	4179-19-5	 244	3-06-00-04533 1.0478^{15}	bz 4; eth 4; EtOH 4 1.5234^{20}
1442	Benzene, 1,4-dimethoxy-2-(1-methylethyl)-	$C_{11}H_{16}O_2$ 180.25	4132-71-2	 $114\text{-}6^{15}$	4-06-00-05990 1.0129^{17}	bz 4; eth 4 1.5100^{17}
1443	Benzene, 1-(dimethoxymethyl)-3-nitro-	$C_9H_{11}NO_4$ 197.19	3395-79-7	 $162\text{-}4^{19}$	4-07-00-00592 1.209^{15}	bz 4
1444	Benzene, 1-(dimethoxymethyl)-4-nitro-	$C_9H_{11}NO_4$ 197.19	881-67-4 24	 294	4-07-00-00600	bz 4
1445	Benzene, 1,2-dimethoxy-4-nitro-	$C_8H_9NO_4$ 183.16	709-09-1 98	 230^{15}	4-06-00-05627 1.1888^{133}	H_2O 1; EtOH 4; eth 4; chl 3
1446	Benzene, 1,3-dimethoxy-5-nitro-	$C_8H_9NO_4$ 183.16	16147-07-2 89		3-06-00-04347 1.1693^{133}	bz 4; EtOH 4
1447	Benzene, 1,4-dimethoxy-2-nitro-	$C_8H_9NO_4$ 183.16	89-39-4 72.5		4-06-00-05786 1.1666^{132}	H_2O 1; EtOH 3; bz 3; chl 3
1448	Benzene, 2,4-dimethoxy-1-nitro- 1,3-Dimethoxy-4-nitrobenzene	$C_8H_9NO_4$ 183.16	4920-84-7 76.5		4-06-00-05691 1.1876^{132}	H_2O 1; EtOH 3; os 3; lig 1
1449	Benzene, 1,2-dimethoxy-4-(1-propenyl)-	$C_{11}H_{14}O_2$ 178.23	93-16-3 18	 270.5	2-06-00-00918 1.0521^{20}	 1.5616^{20}

No.	Name Synonym	Mol. Form. Mol. Wt.	CAS RN mp/°C	Merck No. bp/°C	Beil. Ref. den/g cm^{-3}	Solubility n_D
1450	Benzene, 1,2-dimethoxy-4-(2-propenyl)-	$C_{11}H_{14}O_2$ 178.23	93-15-2 -4	 254.7	4-06-00-06337 1.0396^{20}	H_2O 1; EtOH 3; eth 3 1.5340^{20}
1451	Benzene, 1,2-dimethoxy-4-(1-propenyl)-, (*Z*)-	$C_{11}H_{14}O_2$ 178.23	6380-24-1 70	 270.5	4-06-00-06325 1.0521^{20}	H_2O 1; ace 3; bz 3 1.5616^{20}
1452	Benzene, 1,2-dimethyl- *o*-Xylene	C_8H_{10} 106.17	95-47-6 -25.2	9988 144.5	4-05-00-00917 0.8802^{10}	H_2O 1; EtOH 5; eth 5; ace 5 1.5055^{20}
1453	Benzene, 1,3-dimethyl- *m*-Xylene	C_8H_{10} 106.17	108-38-3 -47.8	9988 139.1	4-05-00-00932 0.8642^{20}	H_2O 1; EtOH 5; eth 5; ace 5 1.4972^{10}
1454	Benzene, 1,4-dimethyl- *p*-Xylene	C_8H_{10} 106.17	106-42-3 13.2	9988 138.3	4-05-00-00951 0.8611^{20}	H_2O 1; EtOH 5; eth 5; ace 5 1.4958^{20}
1455	Benzene, (1,1-dimethylbutyl)- Pentane, 2-methyl-2-phenyl	$C_{12}H_{18}$ 162.27	1985-57-5	 205.5	4-05-00-01118 0.8796^{10}	 1.4955^{16}
1456	Benzene, 1,1'-(1,2-dimethyl-1,2-ethenediyl)bis-, (*E*)-	$C_{16}H_{16}$ 208.30	782-06-9 107	 156^{16}	4-05-00-02207 0.987^{20}	H_2O 1; EtOH 2; eth 3; MeOH 2
1457	Benzene, 1,1'-(1,2-dimethyl-1,2-ethenediyl)bis-, (*Z*)-	$C_{16}H_{16}$ 208.30	782-05-8 67.5		4-05-00-02207 1.004^{20}	H_2O 1; EtOH 3; HOAc 3; MeOH 3 1.5612^{78}
1458	Benzene, (1,1-dimethylethoxy)-	$C_{10}H_{14}O$ 150.22	6669-13-2 -24	 185.5	4-06-00-00559 0.9214^{20}	
1459	Benzene, (1,1-dimethylethyl)- *tert*-Butylbenzene	$C_{10}H_{14}$ 134.22	98-06-6 -57.8	1551 169.1	4-05-00-01045 0.8665^{20}	H_2O 1; EtOH 4; eth 4; ace 5 1.4927^{20}
1460	Benzene, 1-(1,1-dimethylethyl)-3,5-dimethyl-	$C_{12}H_{18}$ 162.27	98-19-1 -18	 207	4-05-00-01130 0.8668^{20}	ctc 3
1461	Benzene, 1-(1,1-dimethylethyl)-3,5-dimethyl-2,4,6-trinitro- 2,4,6-Trinitro-3,5-dimethyl-*tert*-butylbenzene	$C_{12}H_{15}N_3O_6$ 297.27	81-15-2 110		4-05-00-01132	H_2O 1; EtOH 2; eth 3; chl 3
1462	Benzene, 1-(1,1-dimethylethyl)-4-methoxy-	$C_{11}H_{16}O$ 164.25	5396-38-3	 238	4-06-00-03297 0.9436^{20}	 1.5039^{20}
1463	Benzene, 1-(1,1-dimethylethyl)-2-methoxy-4-methyl-3,5-dinitro-	$C_{12}H_{16}N_2O_5$ 268.27	83-66-9 85	 185^{16}	4-06-00-03402	H_2O 1; EtOH 2; eth 3; chl 3
1464	Benzene, 1-(1,1-dimethylethyl)-2-methyl-	$C_{11}H_{16}$ 148.25	1074-92-6 -50.3	 200.4	4-05-00-01096 0.8897^{20}	ace 4; bz 4; eth 4; EtOH 4 1.5076^{20}
1465	Benzene, 1-(1,1-dimethylethyl)-3-methyl-	$C_{11}H_{16}$ 148.25	1075-38-3 -41.4	 189.3	4-05-00-01096 0.8657^{20}	ace 4; bz 4; eth 4; EtOH 4 1.4944^{20}
1466	Benzene, 1-(1,1-dimethylethyl)-4-methyl- *p*-*tert*-Butyltoluene	$C_{11}H_{16}$ 148.25	98-51-1 -52	 190	4-05-00-01097 0.8612^{20}	H_2O 1; EtOH 2; eth 4; ace 3 1.4918^{20}
1467	Benzene, 1-(1,1-dimethylethyl)-3-nitro-	$C_{10}H_{13}NO_2$ 179.22	23132-52-7 2	 254; 136^{16}	4-05-00-01052 1.0643^{25}	
1468	Benzene, 1-(1,5-dimethyl-4-hexenyl)-4-methyl- ar-Curcumene	$C_{15}H_{22}$ 202.34	644-30-4	 140^{19}	4-05-00-01465 0.8805^{20}	H_2O 1; bz 3 1.4989^{20}
1469	Benzene, 1,2-dimethyl-3-(1-methylethyl)-	$C_{11}H_{16}$ 148.25	22539-65-7	 202.6	4-05-00-01103 0.888^{20}	ace 4; bz 4; eth 4; EtOH 4 1.508^{20}
1470	Benzene, 1,2-dimethyl-4-(1-methylethyl)-	$C_{11}H_{16}$ 148.25	4132-77-8	 200	4-05-00-01103 0.8710^{20}	 1.4951^{20}
1471	Benzene, 1,3-dimethyl-2-(1-methylethyl)-	$C_{11}H_{16}$ 148.25	14411-75-7	 199	4-05-00-01103 0.890^{20}	ace 4; bz 4; eth 4; EtOH 4 1.509^{20}
1472	Benzene, 2,4-dimethyl-1-(1-methylethyl)-	$C_{11}H_{16}$ 148.25	4706-89-2 -82	 197	4-05-00-01104 0.8734^{20}	 1.4998^{25}
1473	Benzene, 2,4-dimethyl-1-(1-methylpropyl)-	$C_{12}H_{18}$ 162.27	1483-60-9	 84^{8}	4-05-00-01129 0.8664^{20}	 1.4939^{25}
1474	Benzene, 1,2-dimethyl-3-nitro-	$C_8H_9NO_2$ 151.17	83-41-0 15	 240	4-05-00-00930 1.1402^{20}	H_2O 1; EtOH 3; ctc 3 1.5441^{20}
1475	Benzene, 1,2-dimethyl-4-nitro- 4-Nitro-*o*-xylene	$C_8H_9NO_2$ 151.17	99-51-4 30.5	 251; 143^{21}	4-05-00-00930 1.112^{15}	H_2O 1; EtOH 5 1.5202^{20}
1476	Benzene, 1,3-dimethyl-2-nitro-	$C_8H_9NO_2$ 151.17	81-20-9 15	 226	4-05-00-00948 1.112^{15}	H_2O 1; EtOH 4; ctc 3 1.5202^{20}
1477	Benzene, 1,3-dimethyl-5-nitro-	$C_8H_9NO_2$ 151.17	99-12-7 75	 274	4-05-00-00948	H_2O 1; EtOH 4; eth 4
1478	Benzene, 1,4-dimethyl-2-nitro-	$C_8H_9NO_2$ 151.17	89-58-7 -25	 240.5	4-05-00-00971 1.132^{15}	H_2O 1; EtOH 3 1.5413^{20}
1479	Benzene, 2,4-dimethyl-1-nitro-	$C_8H_9NO_2$ 151.17	89-87-2 9	 247; 122^{18}	4-05-00-00948 1.135^{15}	H_2O 1; eth 3; ace 3; bz 3 1.5473^{25}
1480	Benzene, (1,1-dimethylpropyl)-	$C_{11}H_{16}$ 148.25	2049-95-8	 192.4	4-05-00-01090 0.8748^{20}	 1.4958^{20}
1481	Benzene, (1,2-dimethylpropyl)-	$C_{11}H_{16}$ 148.25	4481-30-5	 187	4-05-00-01092 0.8672^{16}	 1.4972^{16}
1482	Benzene, 1,3-dimethyl-2-propyl-	$C_{11}H_{16}$ 148.25	17059-45-9	 207.6	4-05-00-01102 0.8856^{20}	ace 4; bz 4; eth 4; EtOH 4 1.5063^{20}
1483	Benzene, 1,3-dimethyl-5-propyl-	$C_{11}H_{16}$ 148.25	3982-64-7 -59.1	 202.2	4-05-00-01102 0.8607^{20}	ace 4; bz 4; eth 4; EtOH 4 1.4952^{20}
1484	Benzene, (2,2-dimethylpropyl)-	$C_{11}H_{16}$ 148.25	1007-26-7	 186	4-05-00-01092 0.8581^{18}	 1.4884^{18}
1485	Benzene, 1,2-dimethyl-3-propyl- *o*-Xylene, 3-propyl	$C_{11}H_{16}$ 148.25	17059-44-8	 210.7	4-05-00-01101 0.8864^{20}	ace 4; bz 4; eth 4; EtOH 4 1.5075^{20}
1486	Benzene, 1,4-dimethyl-2-propyl-	$C_{11}H_{16}$ 148.25	3042-50-0	 204.3	4-05-00-01102 0.8717^{20}	ace 4; bz 4; eth 4; EtOH 4 1.4999^{20}
1487	Benzene, 2,4-dimethyl-1-propyl- *m*-Xylene, 4-propyl-	$C_{11}H_{16}$ 148.25	61827-85-8	 206.6	4-05-00-01102 0.8723^{20}	ace 4; bz 4; eth 4; EtOH 4 1.4998^{20}

PHYSICAL CONSTANTS OF ORGANIC COMPOUNDS (continued)

No.	Name Synonym	Mol. Form. Mol. Wt.	CAS RN mp/°C	Merck No. bp/°C	Beil. Ref. den/g cm^{-3}	Solubility n_D
1488	Benzene, 1-(1,1-dimethylpropyl)-4-methoxy-	$C_{12}H_{18}O$ 178.27	2050-03-5	 238	4-06-00-03383 0.9436^{20}	 1.5039^{20}
1489	Benzene, 1,4-dimethyl-2,3,5-trinitro- *p*-Xylene, 2,3,5-trinitro-	$C_8H_7N_3O_6$ 241.16	602-27-7 139.5	 exp 410	3-05-00-00864 1.59^{19}	EtOH 4
1490	Benzene, 2,4-dimethyl-1,3,5-trinitro-	$C_8H_7N_3O_6$ 241.16	632-92-8 184		4-05-00-00950 1.604^{19}	H_2O 1; EtOH 2; eth 2; ace 3
1491	Benzene, 1,2-dinitro- *o*-Dinitrobenzene	$C_6H_4N_2O_4$ 168.11	528-29-0 118.5	 318; 194^{30}	4-05-00-00738 1.3119^{120}	H_2O 1; EtOH 3; bz 3; chl 3 1.565^{17}
1492	Benzene, 1,3-dinitro- *m*-Dinitrobenzene	$C_6H_4N_2O_4$ 168.11	99-65-0 90	 291; 167^{14}	4-05-00-00739 1.5751^{18}	H_2O 2; EtOH 4; eth 3; ace 4
1493	Benzene, 1,4-dinitro- *p*-Dinitrobenzene	$C_6H_4N_2O_4$ 168.11	100-25-4 174	 297; 183^{34}	4-05-00-00741 1.625^{18}	H_2O 1; EtOH 2; ace 3; bz 3
1494	Benzene, 1,1'-(1,2-dinitro-1,2-ethenediyl)bis-, (*Z*)-	$C_{14}H_{10}N_2O_4$ 270.24	1796-05-0 108.5	 dec	4-05-00-02171	H_2O 1; EtOH 3; eth 3; ace 4
1495	Benzene, 2,4-dinitro-1-phenoxy-	$C_{12}H_8N_2O_5$ 260.21	2486-07-9 71	 230^{27}	4-06-00-01375	eth 4; EtOH 4
1496	Benzene, 2,4-dinitro-1-(2-phenylethenyl)-, (*E*)- Stilbene, 2,4-dinitro (*trans*)	$C_{14}H_{10}N_2O_4$ 270.24	56456-42-9 144	 exp 412	4-05-00-02170	EtOH 1; bz 2; chl 3; CS_2 3
1497	1,2-Benzenediol Pyrocatechol	$C_6H_6O_2$ 110.11	120-80-9 105	8009 245	4-06-00-05557 1.1493^{22}	H_2O 4; bz 4; eth 4; EtOH 4 1.604
1498	1,3-Benzenediol Resorcinol	$C_6H_6O_2$ 110.11	108-46-3 111	8158 178^{16}	4-06-00-05658 1.2717^{25}	H_2O 3; EtOH 3; eth 3; bz 2
1499	1,4-Benzenediol *p*-Hydroquinone	$C_6H_6O_2$ 110.11	123-31-9 172.3	4738 287	4-06-00-05712 1.328^{15}	H_2O 3; EtOH 4; eth 3; ace 4
1500	1,2-Benzenediol, 4-(2-amino-1-hydroxyethyl)-, (*R*)- Norepinephrine	$C_8H_{11}NO_3$ 169.18	51-41-2 217 dec	6612		H_2O 2; EtOH 2; eth 2; alk 4
1501	1,4-Benzenediol, 2,5-bis(1,1-dimethylpropyl)- 2,5-Di-*tert*-pentylhydroquinone	$C_{16}H_{26}O_2$ 250.38	79-74-3 180	3300	3-06-00-04748	
1502	1,3-Benzenediol, 4-bromo-	$C_6H_5BrO_2$ 189.01	6626-15-9 103	 140^{12}	4-06-00-05687	H_2O 4; EtOH 2; eth 4; bz 2
1503	1,4-Benzenediol, 2-bromo-	$C_6H_5BrO_2$ 189.01	583-69-7 111.5	 sub	4-06-00-05780	H_2O 4; EtOH 4; eth 4; bz 4
1504	1,2-Benzenediol, 3-chloro-	$C_6H_5ClO_2$ 144.56	4018-65-9 48.5	 110^{11}	4-06-00-05613	lig 4
1505	1,2-Benzenediol, 4-chloro-	$C_6H_5ClO_2$ 144.56	2138-22-9 90.5	 $139^{10.5}$	4-06-00-05614	H_2O 4; ace 4; eth 4; EtOH 4
1506	1,3-Benzenediol, 5-chloro-	$C_6H_5ClO_2$ 144.56	52780-23-1 118	 sub	4-06-00-05684	H_2O 3; eth 3; ace 3; os 3
1507	1,4-Benzenediol, 2-chloro-	$C_6H_5ClO_2$ 144.56	615-67-8 108	 263	4-06-00-05767	H_2O 4; EtOH 3; eth 3; bz 4
1508	1,2-Benzenediol, diacetate	$C_{10}H_{10}O_4$ 194.19	635-67-6 64.5	 $142\text{-}3^{9}$	4-06-00-05582	H_2O 1; EtOH 4; eth 4; chl 4
1509	1,4-Benzenediol, diacetate	$C_{10}H_{10}O_4$ 194.19	1205-91-0 123.5		4-06-00-05741 0.8731^{25}	H_2O 3; EtOH 4; eth 4; chl 4
1510	1,3-Benzenediol, 4,6-dichloro-	$C_6H_4Cl_2O_2$ 179.00	137-19-9 113	 254	4-06-00-05685	H_2O 4; EtOH 4; eth 4; ace 4
1511	1,2-Benzenediol, 4,5-dimethyl-	$C_8H_{10}O_2$ 138.17	2785-74-2 87.5	 160^{28}	4-06-00-05950	H_2O 4; EtOH 4; eth 4; bz 2
1512	1,3-Benzenediol, 2,5-dimethyl-	$C_8H_{10}O_2$ 138.17	488-87-9 163	 278.5	4-06-00-05970	H_2O 3; EtOH 3; eth 3
1513	1,3-Benzenediol, 4,5-dimethyl-	$C_8H_{10}O_2$ 138.17	527-55-9 136	 284	4-06-00-05948 1.1094^{20}	H_2O 3; EtOH 4; eth 3; bz 2
1514	1,3-Benzenediol, 4,6-dimethyl-	$C_8H_{10}O_2$ 138.17	615-89-4 126.5	 276	4-06-00-05963	H_2O 3; EtOH 3; eth 3
1515	1,2-Benzenediol, 4,4'-(2,3-dimethyl-1,4-butanediyl)bis- Nordihydroguaiaretic acid	$C_{18}H_{22}O_4$ 302.37	500-38-9 185.5	6610	4-06-00-07771	H_2O 2; EtOH 3; eth 3; ace 3
1516	1,2-Benzenediol, 4-(1,1-dimethylethyl)-	$C_{10}H_{14}O_2$ 166.22	98-29-3 54.3	 285; 160^{22}	4-06-00-06014	tfa 3
1517	1,3-Benzenediol, 2,4-dinitro- 2,4-Dinitroresorcinol	$C_6H_4N_2O_6$ 200.11	519-44-8 147.5	3278	4-06-00-05696	H_2O 2; EtOH 2
1518	1,3-Benzenediol, 4-ethyl-	$C_8H_{10}O_2$ 138.17	2896-60-8 98.5	 160^{24}; 131^{15}	4-06-00-05924	H_2O 2; EtOH 2; eth 2
1519	1,3-Benzenediol, 4-hexyl- 4-Hexylresorcinol	$C_{12}H_{18}O_2$ 194.27	136-77-6 68	4633 334	4-06-00-06048	ace 4; eth 4; EtOH 4; chl 4
1520	1,4-Benzenediol, 2-(hydroxymethyl)- Gentisyl alcohol	$C_7H_8O_3$ 140.14	495-08-9 100	4292 sub 75	4-06-00-07380	H_2O 4; EtOH 4; chl 4
1521	1,2-Benzenediol, 4-[1-hydroxy-2-(methylamino)ethyl]-, (*R*)- Epinephrine	$C_9H_{13}NO_3$ 183.21	51-43-4 211.5	3569	4-13-00-02927	H_2O 2; EtOH 1; HOAc 3; min acid 3
1522	1,2-Benzenediol, 4-[1-hydroxy-2-[(1-methylethyl)amino]ethyl]- Ethanol, 1(3,4-dihydroxyphenyl)-2(isopropylamino)	$C_{11}H_{17}NO_3$ 211.26	7683-59-2 170.5	5105	3-13-00-02387	
1523	1,2-Benzenediol, 4-iodo- 4-Iodocatechol	$C_6H_5IO_2$ 236.01	76149-14-9 92	 sub	4-06-00-05625	ace 4; bz 4; eth 4; EtOH 4
1524	1,3-Benzenediol, 5-iodo- Resorcinol, 5-iodo	$C_6H_5IO_2$ 236.01	64339-43-1 92.5	 sub	2-06-00-00821	EtOH 4

PHYSICAL CONSTANTS OF ORGANIC COMPOUNDS (continued)

No.	Name	Synonym	Mol. Form.	Mol. Wt.	CAS RN	mp/°C	Merck No.	bp/°C	Beil. Ref.	den/g cm^{-3}	Solubility	n_D
1525	1,2-Benzenediol, 3-methoxy-		$C_7H_8O_3$	140.14	934-00-9	42.8		163^{48}; 129^{10}	4-06-00-07329		chl 3	
1526	1,3-Benzenediol, 5-methoxy-		$C_7H_8O_3$	140.14	2174-64-3	80.3		213^{16}	4-06-00-07362		H_2O 2; EtOH 4; eth 4; bz 2	
1527	1,2-Benzenediol, 3-methyl-		$C_7H_8O_2$	124.14	488-17-5	68		241	4-06-00-05860		H_2O 3; EtOH 3; bz 3; chl 3	
1528	1,2-Benzenediol, 4-methyl-		$C_7H_8O_2$	124.14	452-86-8	65		251	4-06-00-05878	1.1287^{74}	H_2O 3; EtOH 3; eth 3; ace 3	1.5425^{74}
1529	1,3-Benzenediol, 2-methyl-		$C_7H_8O_2$	124.14	608-25-3	120		264	4-06-00-05877		H_2O 4; bz 4; eth 4; EtOH 4	
1530	1,3-Benzenediol, 4-methyl-		$C_7H_8O_2$	124.14	496-73-1	106		268.5	4-06-00-05864		H_2O 3; EtOH 3; eth 3; bz 2	
1531	1,3-Benzenediol, 5-methyl-	Orcinol	$C_7H_8O_2$	124.14	504-15-4	107.5	6819	289.5	4-06-00-05892	1.290^{4}	H_2O 3; EtOH 3; eth 3; bz 3	
1532	1,4-Benzenediol, 2-methyl-		$C_7H_8O_2$	124.14	95-71-6	128		163^{11}	4-06-00-05866		H_2O 4; EtOH 4; eth 4; ace 3	
1533	1,2-Benzenediol, 4-[2-(methylamino)ethyl]-	Deoxyepinephrine	$C_9H_{13}NO_2$	167.21	501-15-5	188.5	2885		4-13-00-02606			
1534	1,3-Benzenediol, 4-(3-methylbutyl)-		$C_{11}H_{16}O_2$	180.25	15116-17-3	67		177^{7}	3-06-00-04700		bz 4; eth 4; EtOH 4	
1535	1,2-Benzenediol, 4-(1-methylethyl)-		$C_9H_{12}O_2$	152.19	2138-43-4	78		271	3-06-00-04632		peth 3	
1536	1,2-Benzenediol, 3-methyl-6-(1-methylethyl)-		$C_{10}H_{14}O_2$	166.22	490-06-2	48		267	4-06-00-06019		ace 4; eth 4	
1537	1,3-Benzenediol, 2-[3-methyl-6-(1-methylethenyl)-2-cyclohexen-1-yl]-5-pentyl-, (1*R*-*trans*)-	Cannabidiol	$C_{21}H_{30}O_2$	314.47	13956-29-1	67	1750	$187\text{-}90^{2}$	3-06-00-05362	1.040^{40}	H_2O 1; EtOH 3; eth 3; bz 3	1.5404^{20}
1538	1,4-Benzenediol, 2-methyl-5-(1-methylethyl)-		$C_{10}H_{14}O_2$	166.22	2217-60-9	148		290	4-06-00-06019		eth 4; EtOH 4	
1539	1,3-Benzenediol, 4-(2-methylpropyl)-	Resorcinol, 4-isobutyl-	$C_{10}H_{14}O_2$	166.22	18979-62-9	62.5		166^{6}	3-06-00-04667		eth 4; EtOH 4	
1540	1,2-Benzenediol, monoacetate		$C_8H_8O_3$	152.15	2848-25-1	57.5		189^{102}; 148^{25}	4-06-00-05581		H_2O 4; ace 4; EtOH 4; chl 4	
1541	1,3-Benzenediol, 4-nitro-		$C_6H_5NO_4$	155.11	3163-07-3	122		178^{11}	4-06-00-05691		EtOH 3; eth 3; bz 3; ctc 2	
1542	1,3-Benzenediol, 4-[(4-nitrophenyl)azo]-	Magneson	$C_{12}H_9N_3O_4$	259.22	74-39-5	200	5578		4-16-00-00266		H_2O 1; EtOH 2; bz 2; HOAc 2	
1543	1,2-Benzenediol, 3-pentadecyl-	3-Pentadecylcatechol	$C_{21}H_{36}O_2$	320.52	492-89-7	59.5	7061		4-06-00-06109		bz 4; eth 4; EtOH 4	
1544	1,3-Benzenediol, 4-pentyl-		$C_{11}H_{16}O_2$	180.25	533-24-4	72.5		168^{6}	3-06-00-04693		bz 4; eth 4; EtOH 4	
1545	1,3-Benzenediol, 5-(2-phenylethenyl)-, (*E*)-	Pinosylvin	$C_{14}H_{12}O_2$	212.25	22139-77-1	156	7419		3-06-00-05577		ace 4; bz 4; chl 4; HOAc 4	
1546	1,4-Benzenediol, 2-(phenylmethyl)-		$C_{13}H_{12}O_2$	200.24	1706-73-6	105.8		230^{13}	3-06-00-05404		eth 4; EtOH 4	
1547	1,2-Benzenediol, 3-(2-propenyl)-		$C_9H_{10}O_2$	150.18	1125-74-2			$141\text{-}5^{16}$	4-06-00-06334	1.1241^{20}		1.5656^{20}
1548	1,2-Benzenediol, 4-propyl-		$C_9H_{12}O_2$	152.19	2525-02-2	60		177^{30}; $111^{0.2}$	3-06-00-04613	1.100^{18}		1.4440^{18}
1549	1,3-Benzenediol, 4-propyl-		$C_9H_{12}O_2$	152.19	18979-60-7	82.5		172^{14}	4-06-00-05975		H_2O 3; EtOH 2; eth 3	
1550	1,3-Benzenediol, 5-propyl-		$C_9H_{12}O_2$	152.19	500-49-2	82.8		169^{8}; 148^{3}	3-06-00-04622		ace 4; bz 4; eth 4; EtOH 4	
1551	1,4-Benzenediol, 2,3,5,6-tetrabromo-		$C_6H_2Br_4O_2$	425.70	2641-89-6	244			3-06-00-04440	3.023^{21}	eth 4; EtOH 4	
1552	1,4-Benzenediol, 2,3,5,6-tetramethyl-	Durohydroquinone	$C_{10}H_{14}O_2$	166.22	527-18-4	233	3451		4-06-00-06028		EtOH 3; eth 2	
1553	1,3-Benzenediol, 2,4,6-trimethyl-		$C_9H_{12}O_2$	152.19	608-98-0	150.8		274.5	4-06-00-06000		bz 4; eth 4; EtOH 4	
1554	1,3-Benzenediol, 2,4,6-trinitro-	Styphnic acid	$C_6H_3N_3O_8$	245.11	82-71-3	175.5	8828	sub	4-06-00-05699		eth 4; EtOH 4	
1555	Benzene, 1,2-dipropoxy-		$C_{12}H_{18}O_2$	194.27	6280-98-4			235.5	4-06-00-05566	0.9554^{33}	chl 2	1.4950^{27}
1556	Benzene, 1,3-dipropoxy-	Resorcinol dipropyl ether	$C_{12}H_{18}O_2$	194.27	56106-37-7			253	3-06-00-04308	1.033^{20}		1.5138^{83}
1557	Benzene, 1,4-dipropyl-		$C_{12}H_{18}$	162.27	4815-57-0			109^{23}	4-05-00-01124	0.8563^{19}		1.4917^{19}
1558	1,3-Benzenedisulfonamide, 4-amino-6-chloro-	Chloraminophenamide	$C_6H_8ClN_3O_4S_2$	285.73	121-30-2	254.5	2067		4-14-00-02810			
1559	1,3-Benzenedisulfonamide, 4,5-dichloro-	1,3-Benzenedisulfonic acid, diamide,4,5-dichloro	$C_6H_6ClN_2O_4$	205.58	120-97-8	228.7	3067		4-11-00-00555			
1560	1,3-Benzenedisulfonic acid, 4-hydroxy-	Phenoldisulfonic acid	$C_6H_6O_7S_2$	254.24	96-77-5	>100 dec	7207		2-11-00-00139		H_2O 4; EtOH 4	
1561	1,2-Benzenedithiol		$C_6H_6S_2$	142.25	17534-15-5	28.5		238.5	4-06-00-05651		EtOH 4; eth 4; bz 4; AcOEt 3	

No.	Name Synonym	Mol. Form. Mol. Wt.	CAS RN mp/°C	Merck No. bp/°C	Beil. Ref. den/g cm^{-3}	Solubility n_D
1562	1,3-Benzenedithiol	$C_6H_6S_2$ 142.25	626-04-0 27	 245	4-06-00-05705	bz 4; eth 4; EtOH 4
1563	1,3-Benzenedithiol, 4-chloro-	$C_6H_5ClS_2$ 176.69	58593-78-5	 145-6^{13}	4-06-00-05708 1.393^{25}	 1.6704^{20}
1564	1,2-Benzenedithiol, 4-methyl- Toluene-3,4-dithiol	$C_7H_8S_2$ 156.27	496-74-2 29	9457	4-06-00-05890	chl 3
1565	Benzene, docosyl- Docosane, 1-phenyl-	$C_{28}H_{50}$ 386.71	5634-22-0 51	 448	3-05-00-01142 0.8544^{20}	
1566	Benzene, dodecyl- 1-Phenyldodecane	$C_{18}H_{30}$ 246.44	123-01-3 3	 328	4-05-00-01200 0.8551^{20}	H_2O 1 1.4824^{20}
1567	Benzene, 1,1'-dodecylidenebis[4-methyl- Dodecane, 1,1-di-(4-tolyl)	$C_{26}H_{38}$ 350.59	55268-62-7	 197$^{1.0}$	3-05-00-01925 0.9117^{20}	 1.5001^{20}
1568	Benzene, eicosyl-	$C_{26}H_{46}$ 358.65	2398-68-7 44	 429	4-05-00-01222 0.8545^{20}	 1.4805^{20}
1569	Benzeneethanamine Phenylethylamine	$C_8H_{11}N$ 121.18	64-04-0	7186 197.5	4-12-00-02453 0.9580^{24}	H_2O 3; eth 4; eth 4; ctc 3 1.5290^{25}
1570	Benzeneethanamine, 4-bromo- Phenethylamine, 4-bromo	$C_8H_{10}BrN$ 200.08	73918-56-6	 63-72$^{0.2}$	4-12-00-02492 1.29^{25}	 1.5750^{20}
1571	Benzeneethanamine, 3,4-dimethoxy-	$C_{10}H_{15}NO_2$ 181.23	120-20-7	 163-5^{14}	4-13-00-02604	ctc 3 1.5464^{20}
1572	Benzeneethanamine, α,α-dimethyl- Phentermine	$C_{10}H_{15}N$ 149.24	122-09-8	7232 205; 100^{21}	4-12-00-02820	
1573	Benzeneethanamine, *N*,α-dimethyl-, (*S*)- Methamphetamine	$C_{10}H_{15}N$ 149.24	537-46-2	5859	4-12-00-02589	
1574	Benzeneethanamine, *N*,β-dimethyl- Phenylpropylmethylamine	$C_{10}H_{15}N$ 149.24	93-88-9	7280 207.5	3-12-00-02689 0.915^{25}	bz 4; eth 4; EtOH 4
1575	Benzeneethanamine, *N*,α-dimethyl-*N*-(phenylmethyl)-, (+)- Benzphetamine	$C_{17}H_{21}N$ 239.36	156-08-1	1130 127$^{0.02}$	4-12-00-02594	eth 4; EtOH 4; MeOH 4; chl 4 1.5515^{19}
1576	Benzeneethanamine, *N*-ethyl-α-methyl- *N*-Ethylamphetamine	$C_{11}H_{17}N$ 163.26	457-87-4	3720 105-6^{14}	3-12-00-02668	 1.4986^{25}
1577	Benzeneethanamine, 2-methoxy- Phenethylamine, 2-methoxy	$C_9H_{13}NO$ 151.21	2045-79-6	 236.5	3-13-00-01625 1.0400^{25}	 1.5422^{20}
1578	Benzeneethanamine, α-methyl-, (±)- Amphetamine	$C_9H_{13}N$ 135.21	300-62-9	616 203	4-12-00-02587 0.9306^{25}	H_2O 2; EtOH 3; eth 2; chl 3 1.518^{26}
1579	Benzeneethanamine, β-methyl-	$C_9H_{13}N$ 135.21	582-22-9	 210	4-12-00-02634 0.9433^{4}	bz 4; eth 4; EtOH 4 1.5255^{20}
1580	Benzeneethanamine, 4-methyl-	$C_9H_{13}N$ 135.21	3261-62-9	 214	4-12-00-02641 0.93^{25}	 1.5257^{20}
1581	Benzeneethanamine, *N*-methyl-	$C_9H_{13}N$ 135.21	589-08-2	 206	4-12-00-02453 0.93^{25}	 1.5162^{20}
1582	Benzeneethanamine, α-methyl-, (*S*)-, sulfate (2:1) Dextroamphetamine sulfate	$C_{18}H_{28}N_2O_4S$ 368.50	51-63-8 >300	2932	3-12-00-02665 1.15^{25}	H_2O 4
1583	Benzeneethanamine, α-methyl-*N*-(2,2,2-trichloroethylidene)- Amphecloral	$C_{11}H_{12}Cl_3N$ 264.58	5581-35-1	613 96$^{0.5}$		 1.530
1584	Benzeneethanamine, α-phenyl-	$C_{14}H_{15}N$ 197.28	25611-78-3	 311; 175^{15}	4-12-00-03313 1.031^{15}	eth 4; EtOH 4
1585	Benzeneethanamine, *N*-(2-phenylethyl)-	$C_{16}H_{19}N$ 225.33	6308-98-1 29	 360; 190^{15}	4-12-00-02459	eth 4; EtOH 4 1.5550^{25}
1586	Benzeneethanamine, 3,4,5-trimethoxy- Mescaline	$C_{11}H_{17}NO_3$ 211.26	54-04-6 35.5	5808 180^{12}	4-13-00-02919	H_2O 3; EtOH 3; eth 1; bz 3
1587	Benzene, 1,1'-(1,2-ethanediyl)bis- Dibenzyl	$C_{14}H_{14}$ 182.27	103-29-7 52.5	1219 284	4-05-00-01868 0.9780^{25}	H_2O 1; EtOH 3; eth 3; CS_2 3 1.5476^{60}
1588	Benzene, 1,1'-[1,2-ethanediylbis(oxy)]bis-	$C_{14}H_{14}O_2$ 214.26	104-66-5 98	 182^{12}	4-06-00-00573	H_2O 1; EtOH 2; eth 3; chl 3
1589	Benzene, 1,1'-(1,2-ethanediyl)bis[3-methyl-	$C_{16}H_{18}$ 210.32	4662-96-8	 298	4-05-00-01943 0.9703^{22}	 1.5566^{22}
1590	Benzene, 1,1'-(1,2-ethanediyl)bis[4-methyl- 1,2-Di-*p*-tolylethane	$C_{16}H_{18}$ 210.32	538-39-6 85	3386 178^{18}	4-05-00-01943	H_2O 1; EtOH 2; bz 3; peth 3
1591	Benzene, 1,1',1'',1'''-(1,2-ethanediylidene)tetrakis-	$C_{26}H_{22}$ 334.46	632-50-8 214.5	 360	4-05-00-02746	EtOH 2; bz 3; HOAc 3
1592	Benzeneethanol Phenethyl alcohol	$C_8H_{10}O$ 122.17	60-12-8 -27	7183 218.2	4-06-00-03067 1.0202^{20}	H_2O 2; EtOH 5; eth 5 1.5325^{20}
1593	Benzeneethanol, β-amino-	$C_8H_{11}NO$ 137.18	7568-92-5 58.3	 261; 160^{17}	4-13-00-01835	H_2O 4; EtOH 4
1594	Benzeneethanol, 2-amino-	$C_8H_{11}NO$ 137.18	5339-85-5	 147-8$^{3.5}$	3-13-00-01679 1.045^{25}	H_2O 4 1.5849^{20}
1595	Benzeneethanol, 4-bromo-	C_8H_9BrO 201.06	4654-39-1	 138^{9}	4-06-00-03082 1.436^{25}	 1.5735^{20}
1596	Benzeneethanol, 3-chloro-	C_8H_9ClO 156.61	5182-44-5	 133-7^{13}	4-06-00-03079 1.181^{25}	 1.5491^{20}
1597	Benzeneethanol, 4-chloro-	C_8H_9ClO 156.61	1875-88-3	 259	4-06-00-03079 1.1804^{20}	 1.5487^{20}
1598	Benzeneethanol, 4-chloro-α,α-dimethyl-	$C_{10}H_{13}ClO$ 184.67	5468-97-3 34	 126^{12}	4-06-00-03291	 1.5310^{20}
1599	Benzeneethanol, 2,5-dimethoxy-	$C_{10}H_{14}O_3$ 182.22	7417-19-8	 176^{18}	4-06-00-07393 1.055^{25}	 1.5395^{20}

No.	Name Synonym	Mol. Form. Mol. Wt.	CAS RN mp/°C	Merck No. bp/°C	Beil. Ref. den/g cm^{-3}	Solubility n_D
1600	Benzeneethanol, α,α-dimethyl-	$C_{10}H_{14}O$ 150.22	100-86-7 24	215	4-06-00-03290 0.9787^{16}	1.5173^{16}
1601	Benzeneethanol, α,α-diphenyl-	$C_{20}H_{18}O$ 274.36	4428-13-1 89.5	222^{11}	4-06-00-05057	H_2O 1; EtOH 4; eth 2; chl 2
1602	Benzeneethanol, β-ethyl-	$C_{10}H_{14}O$ 150.22	2035-94-1	125^{25}; 81^{2}	2-06-00-00488 0.989^{16}	1.43^{16}
1603	Benzeneethanol, 2-ethyl- Phenethyl alcohol, 2-ethyl	$C_{10}H_{14}O$ 150.22	22545-12-6	249.7; 127^{20}	4-06-00-03351 0.9972^{20}	1.5304^{20}
1604	Benzeneethanol, 4-ethyl- Phenethyl alcohol, 4-ethyl	$C_{10}H_{14}O$ 150.22	22545-13-7 7.9	250	4-06-00-03355 0.9907^{20}	1.5229^{20}
1605	Benzeneethanol, α-ethyl-α-methyl-, (±)- 2-Butanol, 3-chloro (*erythro*, *DL*)	C_4H_9ClO 108.57	116783-12-1	235; 56.1^{30}	3-06-00-01958 1.0610^{25}	eth 4; EtOH 4; chl 4 1.4397^{25}
1606	Benzeneethanol, 3-hydroxy-	$C_8H_{10}O_2$ 138.17	13398-94-2	$168\text{-}73^{4}$	4-06-00-05936 1.082^{25}	1.5643^{20}
1607	Benzeneethanol, α-methyl-	$C_9H_{12}O$ 136.19	698-87-3	125^{25}	4-06-00-03192 0.991^{20}	1.5190^{20}
1608	Benzeneethanol, β-methyl-	$C_9H_{12}O$ 136.19	1123-85-9	$105\text{-}6^{11}$	2-06-00-00477 0.975^{25}	H_2O 1; EtOH 3 1.5582^{2}
1609	Benzeneethanol, 2-methyl-	$C_9H_{12}O$ 136.19	19819-98-8 2	243.5	4-06-00-03234 1.016^{25}	1.5355^{20}
1610	Benzeneethanol, 4-methyl-	$C_9H_{12}O$ 136.19	699-02-5	244.5; 94^{6}	4-06-00-03244 1.0028^{20}	1.5267^{20}
1611	Benzeneethanol, β-(1-methylethyl)- 1-Butanol, 3-methyl-2-phenyl	$C_{11}H_{16}O$ 164.25	90499-41-5	130^{15}	4-06-00-03390 0.9694^{25}	ace 4; bz 4; eth 4; EtOH 4 1.5137^{20}
1612	Benzeneethanol, 2-nitro-	$C_8H_9NO_3$ 167.16	15121-84-3 2	267	3-06-00-01714 1.19^{25}	1.5637^{20}
1613	Benzeneethanol, α-phenyl-	$C_{14}H_{14}O$ 198.26	614-29-9 67	177^{15}	2-06-00-00637 1.0360^{70}	
1614	Benzeneethanol, α-phenyl-, (*S*)-	$C_{14}H_{14}O$ 198.26	5773-56-8 67.5	168^{10}	4-06-00-04701 1.0358^{70}	H_2O 1; EtOH 4; eth 4
1615	Benzeneethanol, β-phenyl- 2,2-Diphenylethanol	$C_{14}H_{14}O$ 198.26	1883-32-5 64.5	193^{20}; 144^{1}	4-06-00-04724	EtOH 2; eth 3; ace 3; chl 3
1616	Benzeneethanol, α,α,β-trimethyl- 2-Butanol, 2-methyl-3-phenyl	$C_{11}H_{16}O$ 164.25	3280-08-8 46	197; 106^{12}	4-06-00-03389 0.9794^{20}	ace 4; bz 4; eth 4; EtOH 4 1.5193^{20}
1617	Benzeneethanol, α,β,β-trimethyl-	$C_{11}H_{16}O$ 164.25	2977-31-3	197	4-06-00-03387	ace 4; bz 4; eth 4; EtOH 4 1.5161^{13}
1618	Benzene, 1,1',1'',1'''-(1-ethanyl-2-ylidyne)tetrakis- 1,1,1,2-Tetraphenylethane	$C_{26}H_{22}$ 334.46	2294-94-2 144.8	278^{21}	4-05-00-02747	bz 4
1619	Benzene, 1,1'-(1,2-ethenediyl)bis-, (*E*)- *trans*-Stilbene	$C_{14}H_{12}$ 180.25	103-30-0 123	8774 307; 166^{12}	4-05-00-02156 0.9707^{20}	H_2O 1; EtOH 2; eth 4; bz 4 1.6264^{17}
1620	Benzene, 1,1'-(1,2-ethenediyl)bis-, (*Z*)- *cis*-Stilbene	$C_{14}H_{12}$ 180.25	645-49-8 -5	8774 141^{12}	4-05-00-02155 1.0143^{20}	H_2O 1; EtOH 3; eth 3; ace 3 1.6130^{20}
1621	Benzene, 1,1'-(1,2-ethenediyl)bis[4-methoxy-	$C_{16}H_{16}O_2$ 240.30	4705-34-4 214.5	sub	4-06-00-06823	H_2O 1; ace 3; bz 3; HOAc 3
1622	Benzene, 1,1'-(1,2-ethenediyl)bis[2-methyl- 2,2'-Dimethylstilbene	$C_{16}H_{16}$ 208.30	10311-74-7 83	$176\text{-}80^{10}$	4-05-00-02208	eth 4; EtOH 4; MeOH 4
1623	Benzene, 1,1',1'',1'''-(1,2-ethenediylidene)tetrakis-	$C_{26}H_{20}$ 332.44	632-51-9 225	420	4-05-00-02780 1.155^{0}	H_2O 1; EtOH 2; eth 2; bz 4
1624	Benzene, ethenyl- Styrene	C_8H_8 104.15	100-42-5 -31	8830 145	4-05-00-01334 0.9060^{20}	H_2O 1; EtOH 3; eth 3; ace 3 1.5468^{20}
1625	Benzene, (1-ethenyl-3-butenyl)- 1,5-Hexadiene, 3-phenyl	$C_{12}H_{14}$ 158.24	1076-66-0	78^{8}	4-05-00-01555 0.8911^{25}	1.5141^{25}
1626	Benzene, 1-ethenyl-2,4-dimethyl-	$C_{10}H_{12}$ 132.21	2234-20-0	$79\text{-}80^{12}$	4-05-00-01386 0.905^{20}	1.539^{20}
1627	Benzene, 1-ethenyl-3,5-dimethyl-	$C_{10}H_{12}$ 132.21	5379-20-4	$57\text{-}8^{4}$	4-05-00-01386 0.894^{25}	1.5382^{20}
1628	Benzene, 2-ethenyl-1,4-dimethyl-	$C_{10}H_{12}$ 132.21	2039-89-6	69^{10}	4-05-00-01385 0.9072^{17}	1.5236^{17}
1629	Benzene, 4-ethenyl-1,2-dimethyl-	$C_{10}H_{12}$ 132.21	27831-13-6	$94\text{-}6^{20}$	4-05-00-01385 0.906^{25}	1.5463^{20}
1630	Benzene, 1-ethenyl-2-ethyl-	$C_{10}H_{12}$ 132.21	7564-63-8 -75.5	187.3; 68^{12}	3-05-00-01216 0.9017^{25}	1.5380^{20}
1631	Benzene, 1-ethenyl-3-ethyl-	$C_{10}H_{12}$ 132.21	7525-62-4 -101	190.0	3-05-00-01217 0.8945^{20}	1.5351^{20}
1632	Benzene, 1-ethenyl-4-ethyl-	$C_{10}H_{12}$ 132.21	3454-07-7 -49.7	192.3; 86^{20}	4-05-00-01384 0.8884^{25}	1.5376^{20}
1633	Benzene, 1-ethenyl-2-fluoro-	C_8H_7F 122.14	394-46-7	46^{32}	4-05-00-01342 1.0282^{20}	bz 4; eth 4; EtOH 4 1.5200^{20}
1634	Benzene, 1-ethenyl-3-fluoro-	C_8H_7F 122.14	350-51-6	$30\text{-}1^{4}$	4-05-00-01343 1.0177^{20}	H_2O 1; EtOH 3; eth 3; bz 3 1.5170^{20}
1635	Benzene, 1-ethenyl-4-fluoro-	C_8H_7F 122.14	405-99-2 -34.5	67.4^{50}; 30^{4}	4-05-00-01343 1.0220^{20}	H_2O 1; EtOH 3; eth 3; bz 3 1.5150^{20}
1636	Benzene, 1,1'-ethenylidenebis- 1,1-Diphenylethene	$C_{14}H_{12}$ 180.25	530-48-3 8.2	3323 277	4-05-00-02173 1.0232^{20}	H_2O 1; eth 3; chl 3 1.6085^{20}
1637	Benzene, 1-ethenyl-2-methoxy-	$C_9H_{10}O$ 134.18	612-15-7 29	197; 83^{12}	4-06-00-03771 1.0049^{17}	ace 4; bz 4; eth 4; EtOH 4 1.5388^{20}
1638	Benzene, 1-ethenyl-3-methoxy-	$C_9H_{10}O$ 134.18	626-20-0	$90\text{-}3^{15}$	4-06-00-03773 0.999^{16}	H_2O 1; EtOH 3; eth 3; bz 3 1.5586^{23}

No.	Name Synonym	Mol. Form. Mol. Wt.	CAS RN mp/°C	Merck No. bp/°C	Beil. Ref. den/g cm^{-3}	Solubility n_D
1639	Benzene, 1-ethenyl-4-methoxy-	$C_9H_{10}O$ 134.18	637-69-4	 205; 91^{13}	4-06-00-03775 1.0001^{13}	H_2O 1; EtOH 3; eth 3; bz 3 1.5642^{13}
1640	Benzene, 1-ethenyl-2-methyl- *o*-Methylstyrene	C_9H_{10} 118.18	611-15-4 -68.5	 169.8	4-05-00-01367 0.9077^{25}	H_2O 1; bz 3; chl 3 1.5437^{20}
1641	Benzene, 1-ethenyl-3-methyl- *m*-Methylstyrene	C_9H_{10} 118.18	100-80-1 -86.3	 164	4-05-00-01367 0.9076^{25}	H_2O 1; EtOH 3; eth 3; bz 3 1.5411^{20}
1642	Benzene, 1-ethenyl-4-methyl- *p*-Methylstyrene	C_9H_{10} 118.18	622-97-9 -34.1	 172.8	4-05-00-01369 0.9173^{25}	H_2O 1; bz 3 1.5420
1643	Benzene, 1-ethenyl-4-(1-methylethyl)-	$C_{11}H_{14}$ 146.23	2055-40-5 -44.7	 204.1	4-05-00-01408 0.8850^{20}	ace 4; bz 4; eth 4; EtOH 4 1.5289^{20}
1644	Benzene, 1-ethenyl-3-nitro-	$C_8H_7NO_2$ 149.15	586-39-0 -10	 $120\text{-}1^{11}$	4-05-00-01351 1.1552^{32}	H_2O 1; EtOH 3; eth 3; bz 3 1.5836^{20}
1645	Benzene, 1-ethenyl-4-nitro-	$C_8H_7NO_2$ 149.15	100-13-0 29	 dec	4-05-00-01351	EtOH 4; eth 4; chl 3; HOAc 3
1646	Benzene, (ethenyloxy)-	C_8H_8O 120.15	766-94-9	 155.5	4-06-00-00561 0.9770^{20}	H_2O 1; eth 4 1.5224^{20}
1647	Benzene, 2-ethenyl-1,3,5-trimethyl-	$C_{11}H_{14}$ 146.23	769-25-5 -37	 209	4-05-00-01408 0.9057^{20}	ctc 3 1.5296^{20}
1648	Benzene, 1,1',1''-(1-ethenyl-2-ylidene)tris-	$C_{20}H_{16}$ 256.35	58-72-0 72.5	 $220\text{-}1^{14}$	4-05-00-02575 1.0373^{78}	H_2O 1; EtOH 3; eth 4; chl 3 1.6292^{78}
1649	Benzene, ethoxy- Phenetole	$C_8H_{10}O$ 122.17	103-73-1 -29.5	7189 169.8	4-06-00-00554 0.9651^{20}	H_2O 1; EtOH 3; eth 3; ctc 3 1.5076^{20}
1650	Benzene, 2-ethoxy-1,3-dinitro- Phenetole, 2,6-dinitro-	$C_8H_8N_2O_5$ 212.16	13027-43-5 60.5	 137^{3}	3-06-00-00868	eth 4
1651	Benzene, (2-ethoxyethenyl)-	$C_{10}H_{12}O$ 148.20	17655-74-2 0	 224.5	 0.979^{20}	 1.5496^{20}
1652	Benzene, (2-ethoxyethoxy)- Ethane, 1-ethoxy-2-phenoxy-	$C_{10}H_{14}O_2$ 166.22	19594-02-6	 230	4-06-00-00572 1.018^{11}	H_2O 1
1653	Benzene, 1-ethoxy-4-ethyl-	$C_{10}H_{14}O$ 150.22	1585-06-4	 210	3-06-00-01665 0.9385^{17}	ace 4; bz 4; EtOH 4
1654	Benzene, 1-ethoxy-2-fluoro-	C_8H_9FO 140.16	451-80-9 -16.7	 171.4	4-06-00-00771 1.0874^{17}	H_2O 1; ace 3; bz 3; ctc 2 1.4932^{17}
1655	Benzene, 1-ethoxy-3-fluoro-	C_8H_9FO 140.16	458-03-7 -27.5	 172; 65.2^{15}	3-06-00-00669 1.0716^{16}	H_2O 1; bz 3 1.4847^{17}
1656	Benzene, 1-ethoxy-4-fluoro-	C_8H_9FO 140.16	459-26-7 -8.5	 173; 54^{7}	4-06-00-00774 1.0715^{18}	H_2O 1; bz 3; ctc 2; chl 3 1.4826^{18}
1657	Benzene, 1-ethoxy-4-iodo-	C_8H_9IO 248.06	699-08-1 29	 250	4-06-00-01077	bz 4; eth 4; EtOH 4; chl 4
1658	Benzene, (ethoxymethyl)- Benzyl ethyl ether	$C_9H_{12}O$ 136.19	539-30-0	1147 186	4-06-00-02229 0.9478^{20}	H_2O 1; EtOH 5; eth 5 1.4955^{20}
1659	Benzene, 1-ethoxy-2-methyl-	$C_9H_{12}O$ 136.19	614-71-1	 184	4-06-00-01944 0.9592^{13}	eth 4; EtOH 4 1.508^{13}
1660	Benzene, 1-ethoxy-3-methyl-	$C_9H_{12}O$ 136.19	621-32-9	 192	4-06-00-02039 0.949^{20}	H_2O 1; EtOH 3; eth 3 1.513^{20}
1661	Benzene, 1-ethoxy-4-methyl-	$C_9H_{12}O$ 136.19	622-60-6	 188.5	4-06-00-02099 0.9509^{18}	H_2O 1; EtOH 3; eth 3; ctc 2 1.5058^{18}
1662	Benzene, 1-ethoxy-2-nitro-	$C_8H_9NO_3$ 167.16	610-67-3 2.1	 267	4-06-00-01250 1.1903^{15}	eth 4; EtOH 4 1.5425^{20}
1663	Benzene, 1-ethoxy-3-nitro-	$C_8H_9NO_3$ 167.16	621-52-3 36	 264; 169^{70}	4-06-00-01271	H_2O 1; EtOH 3; eth 3
1664	Benzene, 1-ethoxy-4-nitro-	$C_8H_9NO_3$ 167.16	100-29-8 60	 283	4-06-00-01283 1.1176^{100}	H_2O 2; EtOH 2; eth 4; ace 5
1665	Benzene, 1-ethoxy-2-(1-propenyl)- Phenetole, 2-propenyl	$C_{11}H_{14}O$ 162.23	67191-37-1	 231; 120^{17}	0-06-00-00565 0.9731^{24}	 1.544^{24}
1666	Benzene, (3-ethoxy-1-propenyl)-	$C_{11}H_{14}O$ 162.23	1476-07-9	 $127\text{-}8^{22}$	3-06-00-02404 0.970^{15}	 1.547^{15}
1667	Benzene, 1-ethoxy-2-propyl- Phenetole, 2-propyl	$C_{11}H_{16}O$ 164.25	101144-90-5	 213; 100^{16}	0-06-00-00499 0.9240^{20}	 1.494^{20}
1668	Benzene, ethyl- Ethylbenzene	C_8H_{10} 106.17	100-41-4 -94.9	3723 136.1	4-05-00-00885 0.8670^{20}	H_2O 1; EtOH 5; eth 5; chl 2 1.4959^{20}
1669	Benzene, (1-ethylbutyl)-	$C_{12}H_{18}$ 162.27	4468-42-2 -55.4	 209	4-05-00-01117 0.8239^{25}	 1.4859^{20}
1670	Benzene, 1-ethyl-2,3-dimethyl- 3-Ethyl-*o*-xylene	$C_{10}H_{14}$ 134.22	933-98-2 -49.5	 194	4-05-00-01069 0.8881^{25}	 1.5117^{20}
1671	Benzene, 1-ethyl-2,4-dimethyl-	$C_{10}H_{14}$ 104.22	874-41-9 -62.9	 188.4	4-05-00-01070 0.8763^{20}	ace 4; bz 4; eth 4; EtOH 4 1.5038^{20}
1672	Benzene, 1-ethyl-3,5-dimethyl-	$C_{10}H_{14}$ 134.22	934-74-7 -84.3	 183.6	4-05-00-01071 0.8608^{25}	H_2O 1; EtOH 5; eth 5; ace 5 1.4981^{20}
1673	Benzene, 2-ethyl-1,3-dimethyl-	$C_{10}H_{14}$ 134.22	2870-04-4 -16.2	 190	4-05-00-01069 0.8864^{25}	 1.5107^{20}
1674	Benzene, 2-ethyl-1,4-dimethyl-	$C_{10}H_{14}$ 134.22	1758-88-9 -53.7	 186.9	4-05-00-01070 0.8732^{25}	H_2O 1; EtOH 5; eth 5; ace 5 1.5043^{20}
1675	Benzene, 4-ethyl-1,2-dimethyl-	$C_{10}H_{14}$ 134.22	934-80-5 -66.9	 189.5	4-05-00-01069 0.8706^{25}	H_2O 1; EtOH 5; eth 5; ace 5 1.5031^{20}
1676	Benzene, 1,1'-(1-ethyl-1,2-ethanediyl)bis-	$C_{16}H_{18}$ 210.32	5223-59-6	 290; 152^{11}	4-05-00-01939 0.9777^{20}	H_2O 1; EtOH 3; eth 3; bz 3 1.5554^{20}
1677	Benzene, 1,1'-ethylidenebis- 1,1-Diphenylethane	$C_{14}H_{14}$ 182.27	612-00-0 -17.9	 272.6	4-05-00-01880 0.9997^{20}	H_2O 1; EtOH 5; eth 5; bz 3 1.5756^{20}
1678	Benzene, 1,1'-ethylidenebis[4-chloro- 1,1-Bis(4-chlorophenyl)ethane	$C_{14}H_{12}Cl_2$ 251.15	3547-04-4 56	 320	4-05-00-01881	

No.	Name Synonym	Mol. Form. Mol. Wt.	CAS RN mp/°C	Merck No. bp/°C	Beil. Ref. den/g cm^{-3}	Solubility n_D
1679	Benzene, 1,1',1''-ethylidynetris-	$C_{20}H_{18}$ 258.36	5271-39-6 95	 207^{18}	4-05-00-02510	H_2O 1; EtOH 2; eth 3; bz 3
1680	Benzene, 1-ethyl-2-iodo- *o*-Ethyliodobenzene	C_8H_9I 232.06	18282-40-1	 226	4-05-00-00910 1.6189^{10}	ace 4; bz 4 1.5941^{22}
1681	Benzene, 1-ethyl-4-iodo-	C_8H_9I 232.06	25309-64-2 -17	 209	3-05-00-00801 1.6095^{10}	ace 4; bz 4 1.5909^{22}
1682	Benzene, 1-ethyl-2-methoxy-	$C_9H_{12}O$ 136.19	14804-32-1	 187; 80^{14}	3-06-00-01656 0.9636^{19}	bz 4; eth 4 1.5142^{20}
1683	Benzene, 1-ethyl-3-methoxy-	$C_9H_{12}O$ 136.19	10568-38-4	 197; 74^{10}	3-06-00-01662 0.9575^{18}	H_2O 1; eth 4; bz 3; ctc 3 1.5102
1684	Benzene, 1-ethyl-4-methoxy-	$C_9H_{12}O$ 136.19	1515-95-3	 198	4-06-00-03021 0.9624^{15}	bz 4; eth 4 1.5120^{20}
1685	Benzene, 1-ethyl-2-methyl- *o*-Ethyltoluene	C_9H_{12} 120.19	611-14-3 -80.8	 165.2	4-05-00-00999 0.8807^{20}	H_2O 1; EtOH 5; eth 5; ace 5 1.5046^{20}
1686	Benzene, 1-ethyl-3-methyl- *m*-Ethyltoluene	C_9H_{12} 120.19	620-14-4 -95.5	 161.3	4-05-00-01001 0.8645^{20}	H_2O 1; EtOH 4; eth 4; ace 5 1.4966^{20}
1687	Benzene, 1-ethyl-4-methyl- *p*-Ethyltoluene	C_9H_{12} 120.19	622-96-8 -62.3	 162	4-05-00-01003 0.8614^{20}	H_2O 1; EtOH 4; eth 4; ace 5 1.4959^{20}
1688	Benzene, 1-ethyl-2-(1-methylethyl)- Benzene, 1-ethyl-2-isopropyl	$C_{11}H_{16}$ 148.25	18970-44-0	 193	4-05-00-01100 0.888^{20}	ace 4; bz 4; eth 4; EtOH 4 1.508^{20}
1689	Benzene, 1-ethyl-3-(1-methylethyl)- Benzene, 1-ethyl-3-isopropyl	$C_{11}H_{16}$ 148.25	4920-99-4 <-20	 192	3-05-00-01005 0.859^{20}	H_2O 1; EtOH 3; eth 3; bz 4 1.4921^{20}
1690	Benzene, 1-ethyl-4-(1-methylethyl)- Benzene, 1-ethyl-4-isopropyl	$C_{11}H_{16}$ 148.25	4218-48-8 <-20	 196.6	4-05-00-01100 0.8585^{20}	H_2O 1; eth 3; bz 4 1.4923^{20}
1691	Benzene, (1-ethyl-1-methylpropyl)- Pentane, 3-methyl-3-phenyl	$C_{12}H_{18}$ 162.27	1985-97-3	 205	4-05-00-01120 0.8773^{15}	 1.4972^{16}
1692	Benzene, 1-ethyl-2-nitro-	$C_8H_9NO_2$ 151.17	612-22-6 -12.3	 232.5	4-05-00-00911 1.1207^{20}	H_2O 1; EtOH 4; eth 4; ace 3 1.5356^{20}
1693	Benzene, 1-ethyl-3-nitro-	$C_8H_9NO_2$ 151.17	7369-50-8 -37.9	 242.5	4-05-00-00911 1.1345^{25}	ace 4; eth 4; EtOH 4
1694	Benzene, 1-ethyl-4-nitro-	$C_8H_9NO_2$ 151.17	100-12-9 -12.3	 245.5	4-05-00-00912 1.1192^{20}	H_2O 1; EtOH 4; eth 4; ace 3 1.5455^{20}
1695	Benzene, (1-ethyloctadecyl)- 3-Phenyleicosane	$C_{26}H_{46}$ 358.65	2400-02-4 29.3	 202^{1}	4-05-00-01222 0.8546^{20}	 1.4796^{20}
1696	Benzene, 1-ethyl-4-(phenylmethyl)-	$C_{15}H_{16}$ 196.29	620-85-9 -24	 297	4-05-00-01926 0.9777^{20}	EtOH 3; eth 3; chl 3 1.5616^{20}
1697	Benzene, 1,1'-(1-ethyl-1,3-propanediyl)bis- Pentane, 1,3-diphenyl-	$C_{17}H_{20}$ 224.35	838-45-9	 304; $113^{0.8}$	4-05-00-01963 0.9734^{21}	 1.553^{21}
1698	Benzene, 1,1'-(2-ethyl-1,3-propanediyl)bis- Propane, 1,1-dibenzyl	$C_{17}H_{20}$ 224.35	1520-45-2	 304.5	4-05-00-01964 0.9734^{21}	 1.553^{21}
1699	Benzene, (1-ethyl-1-propenyl)- 3-Phenyl-2-pentene	$C_{11}H_{14}$ 146.23	4701-36-4	 198; 92^{18}	4-05-00-01404 0.9173^{14}	 1.5266^{15}
1700	Benzene, (1-ethyl-2-propenyl)-	$C_{11}H_{14}$ 146.23	19947-22-9	 191.5; 71^{12}	3-05-00-01240 0.8458^{23}	 1.5030^{21}
1701	Benzene, (1-ethylpropyl)-	$C_{11}H_{16}$ 148.25	1196-58-3	 187.5	4-05-00-01090 0.8649^{20}	 1.4880^{20}
1702	Benzene, 1-ethyl-2-propyl-	$C_{11}H_{16}$ 148.25	16021-20-8	 203	4-05-00-01099 0.8744^{20}	 1.4992^{20}
1703	Benzene, 1-ethyl-4-propyl- 1-Ethyl-4-propylbenzene	$C_{11}H_{16}$ 148.25	20024-90-2	 205	4-05-00-01100 0.8594^{20}	ace 4; bz 4; eth 4; EtOH 4 1.4921^{21}
1704	Benzene, (ethylsulfonyl)-	$C_8H_{10}O_2S$ 170.23	599-70-2 42	 160^{12}	4-06-00-01469 1.1410^{20}	bz 4; eth 4; EtOH 4; chl 4
1705	Benzene, (ethylthio)-	$C_8H_{10}S$ 138.23	622-38-8	 205	4-06-00-01468 1.0211^{20}	EtOH 3 1.5670^{20}
1706	Benzene, 1-(ethylthio)-3-methyl- Ethyl *m*-tolyl sulfide	$C_9H_{12}S$ 152.26	34786-24-8	 219	3-06-00-01332 0.9987^{20}	 1.5590^{20}
1707	Benzene, 1-(ethylthio)-4-methyl-	$C_9H_{12}S$ 152.26	622-63-9	 220	4-06-00-02156 0.9996^{20}	 1.555^{20}
1708	Benzene, 1-ethyl-2,3,4-trimethyl- Benzene, 4-ethyl-1,2,3-trimethyl	$C_{11}H_{16}$ 148.25	61827-86-9	 220.4	4-05-00-01107 0.9019^{20}	ace 4; bz 4; eth 4; EtOH 4 1.5180^{20}
1709	Benzene, 1-ethyl-2,3,5-trimethyl- 1-Ethyl-2,3,5-trimethylbenzene	$C_{11}H_{16}$ 148.25	18262-85-6	 213	4-05-00-01107 0.8897^{20}	ace 4; bz 4; eth 4; EtOH 4 1.5118^{20}
1710	Benzene, 1-ethyl-2,4,5-trimethyl- 1-Ethyl-2,4,5-trimethylbenzene	$C_{11}H_{16}$ 148.25	17851-27-3 -13.5	 213	4-05-00-01108 0.883^{20}	ace 4; bz 4; eth 4; EtOH 4 1.5075^{20}
1711	Benzene, 2-ethyl-1,3,4-trimethyl- Benzene, 3-ethyl-1,2,4-trimethyl	$C_{11}H_{16}$ 148.25	61827-87-0	 216.6	4-05-00-01107 0.895^{20}	ace 4; bz 4; eth 4; EtOH 4 1.5133^{20}
1712	Benzene, 2-ethyl-1,3,5-trimethyl-	$C_{11}H_{16}$ 148.25	3982-67-0 -15.5	 212.4	4-05-00-01107 0.883^{20}	ace 4; bz 4; eth 4; EtOH 4 1.5074^{20}
1713	Benzene, 5-ethyl-1,2,3-trimethyl- 5-Ethyl-1,2,3-trimethylbenzene	$C_{11}H_{16}$ 148.25	31366-00-4	 215.8	4-05-00-01107 0.8863^{20}	ace 4; eth 4; EtOH 4 1.5101^{20}
1714	Benzene, 1,1'-(1,2-ethynediyl)bis- Diphenylacetylene	$C_{14}H_{10}$ 178.23	501-65-5 62.5	9428 300	4-05-00-02276 0.9657^{100}	H_2O 1; EtOH 2; eth 4; ctc 2
1715	Benzene, ethynyl- Ethynylbenzene	C_8H_6 102.14	536-74-3 -44.8	3817 143	4-05-00-01525 0.9300^{20}	H_2O 1; EtOH 5; eth 5; ace 3 1.5470^{20}
1716	Benzene, (ethynyloxy)-	C_8H_6O 118.14	4279-76-9 -36	 $61\text{-}2^{25}$	4-06-00-00565 1.0614^{20}	eth 4; EtOH 4 1.5125^{20}
1717	Benzene, fluoro- Fluorobenzene	C_6H_5F 96.10	462-06-6 -42.2	4099 84.7	4-05-00-00632 1.0225^{20}	bz 4; eth 4; EtOH 4; lig 4 1.4684^{30}
1718	Benzene, 1-fluoro-2,4-dinitro- 2,4-Dinitrophenyl fluoride	$C_6H_3FN_2O_4$ 186.10	70-34-8 25.8	4101 296	4-05-00-00742 1.4718^{54}	EtOH 3; chl 2 1.5690^{20}

PHYSICAL CONSTANTS OF ORGANIC COMPOUNDS (continued)

No.	Name / Synonym	Mol. Form. / Mol. Wt.	CAS RN / mp/°C	Merck No. / bp/°C	Beil. Ref. / den/g cm^{-3}	Solubility / n_D
1719	Benzene, 1-fluoro-2-iodo-	C_6H_4FI 222.00	348-52-7 -41.5	188.6	4-05-00-00693	ace 3; bz 3; chl 3 1.5910^{20}
1720	Benzene, 1-fluoro-4-iodo-	C_6H_4FI 222.00	352-34-1 -27	183	4-05-00-00694 1.9523^{15}	H_2O 1; EtOH 3; eth 3; ace 3 1.5270^{22}
1721	Benzene, 1-fluoro-3-isothiocyanato-	C_7H_4FNS 153.18	404-72-8	227	3-12-00-01276 1.27^{25}	1.6186^{20}
1722	Benzene, 1-fluoro-2-methoxy-	C_7H_7FO 126.13	321-28-8 -39	154.5	4-06-00-00771 1.5489^{17}	H_2O 1; eth 3; ctc 3 1.4969^{17}
1723	Benzene, 1-fluoro-3-methoxy-	C_7H_7FO 126.13	456-49-5 -35	159; 51^{14}	4-06-00-00772 1.104^{25}	1.4876^{20}
1724	Benzene, 1-fluoro-4-methoxy-	C_7H_7FO 126.13	459-60-9 -45	157	4-06-00-00773 1.1781^{18}	eth 3 1.4886^{18}
1725	Benzene, (fluoromethyl)-	C_7H_7F 110.13	350-50-5 -35	140; 40^{14}	4-05-00-00800 1.0228^{25}	ctc 3 1.4892^{25}
1726	Benzene, 1-fluoro-2-methyl- *o*-Fluorotoluene	C_7H_7F 110.13	95-52-3 -62	4108 115	4-05-00-00799 1.0041^{13}	H_2O 1; EtOH 4; eth 4 1.4704^{20}
1727	Benzene, 1-fluoro-3-methyl- *m*-Fluorotoluene	C_7H_7F 110.13	352-70-5 -87	4108 115	4-05-00-00799 0.9974^{20}	H_2O 1; EtOH 4; eth 4 1.4691^{20}
1728	Benzene, 1-fluoro-4-methyl- *p*-Fluorotoluene	C_7H_7F 110.13	352-32-9 -56	4108 116.6	4-05-00-00800 0.9975^{20}	H_2O 1; EtOH 4; eth 4 1.4699^{20}
1729	Benzene, 1-fluoro-4-methyl-2-nitro- Toluene, 4-fluoro-3-nitro	$C_7H_6FNO_2$ 155.13	446-11-7 26.5	241	3-05-00-00743 1.2619^{28}	1.5237^{28}
1730	Benzene, 2-fluoro-4-methyl-1-nitro- 3-Fluoro-4-nitrotoluene	$C_7H_6FNO_2$ 155.13	446-34-4 53.2	97^{3}	4-05-00-00852 1.4380^{25}	
1731	Benzene, 4-fluoro-1-methyl-2-nitro-	$C_7H_6FNO_2$ 155.13	446-10-6 27	213; 138^{83}	4-05-00-00851 1.2686^{20}	1.5218^{20}
1732	Benzene, 4-fluoro-2-methyl-1-nitro-	$C_7H_6FNO_2$ 155.13	446-33-3 27.5	217; $97\text{-}8^{10}$	4-05-00-00851 1.272^{28}	1.5271^{20}
1733	Benzene, 1-fluoro-2-nitro- *o*-Fluoronitrobenzene	$C_6H_4FNO_2$ 141.10	1493-27-2 -6	215 dec	4-05-00-00718 1.3285^{18}	eth 4; EtOH 4 1.5489^{17}
1734	Benzene, 1-fluoro-3-nitro- *m*-Fluoronitrobenzene	$C_6H_4FNO_2$ 141.10	402-67-5 41	199; 86^{19}	4-05-00-00719 1.3254^{19}	H_2O 1; EtOH 3; eth 3; bz 2 1.5262^{15}
1735	Benzene, 1-fluoro-4-nitro- *p*-Fluoronitrobenzene	$C_6H_4FNO_2$ 141.10	350-46-9 21	205	4-05-00-00719 1.3300^{20}	H_2O 1; EtOH 3; eth 3; ctc 2 1.5316^{20}
1736	Benzene, 1-fluoro-2-(trichloromethyl)-	$C_7H_4Cl_3F$ 213.47	488-98-2	75^{5}	3-05-00-00701 1.453^{25}	1.5432^{20}
1737	Benzene, 1-fluoro-2-(trifluoromethyl)-	$C_7H_4F_4$ 164.10	392-85-8	114.5	4-05-00-00804 1.293^{25}	1.4040^{25}
1738	Benzene, 1-fluoro-3-(trifluoromethyl)-	$C_7H_4F_4$ 164.10	401-80-9 -81.5	101.5	4-05-00-00804 1.3021^{17}	
1739	Benzene, 1-fluoro-4-(trifluoromethyl)-	$C_7H_4F_4$ 164.10	402-44-8 -41.7	103.5	4-05-00-00804 1.293^{25}	1.4025^{20}
1740	Benzene, 2-fluoro-1,3,5-trimethyl-	$C_9H_{11}F$ 138.18	392-69-8 -36.7	171.5	4-05-00-01025 0.9745^{25}	1.4809^{25}
1741	Benzene, heneicosyl- Heneicosylbenzene	$C_{27}H_{48}$ 372.68	40775-09-5 48	439	4-05-00-01226 0.8545^{20}	1.4804^{20}
1742	Benzene, heptacosyl- Heptacosylbenzene	$C_{33}H_{60}$ 456.84	61828-25-9 64	397^{100}; 262^{1}	4-05-00-01243 0.8543^{20}	1.4797^{20}
1743	Benzene, heptadecyl- Heptadecane, 1-phenyl-	$C_{23}H_{40}$ 316.57	14752-75-1 32	397	3-05-00-01123 0.8546^{20}	1.4810^{20}
1744	Benzene, 1,1'-(1-heptenylidene)bis- 1,1-Diphenyl-1-heptene	$C_{19}H_{22}$ 250.38	1530-20-7		4-05-00-02252 0.9673^{18}	EtOH 2 1.5648^{18}
1745	Benzene, heptyl-	$C_{13}H_{20}$ 176.30	1078-71-3 -48	240; 109^{10}	4-05-00-01143 0.8567^{20}	H_2O 1; bz 3; chl 3 1.4865^{20}
1746	Benzene, 1,1'-heptylidenebis- 1,1-Diphenylheptane	$C_{19}H_{24}$ 252.40	1530-05-8 14	333.5; 137^{1}	4-05-00-01992 0.9497^{20}	
1747	Benzene, (heptyloxy)-	$C_{13}H_{20}O$ 192.30	32395-96-3 -33.5	267	4-06-00-00560 0.9170^{15}	ace 4; eth 4; EtOH 4 1.4912^{20}
1748	Benzenehexacarboxylic acid Mellitic acid	$C_{12}H_6O_{12}$ 342.17	517-60-2 286-8 dec	5706	4-09-00-03825	H_2O 4; EtOH 3; sulf 3
1749	Benzene, hexachloro- Hexachlorobenzene	C_6Cl_6 284.78	118-74-1 231.8	4600 325	4-05-00-00670 2.044^{23}	H_2O 1; EtOH 2; eth 3; bz 4 1.5691^{23}
1750	Benzene, hexacosyl- Hexacosane, 1-phenyl-	$C_{32}H_{58}$ 442.81	13024-80-1 62	388^{100}; 250^{1}	4-05-00-01242 0.8543^{20}	1.4798^{20}
1751	Benzene, hexadecyl-	$C_{22}H_{38}$ 302.54	1459-09-2 27	385	4-05-00-01216 0.8547^{20}	H_2O 1; EtOH 2; eth 4; bz 4 1.4813^{20}
1752	Benzene, (hexadecyloxy)-	$C_{22}H_{38}O$ 318.54	35021-70-6 41.8	200^{1}	3-06-00-00555 0.8434^{82}	1.4556^{82}
1753	Benzene, 1,3-hexadienyl- 1-Phenyl-1,3-hexadiene	$C_{12}H_{14}$ 158.24	41635-77-2	128^{16}	3-05-00-01378 0.9253^{12}	1.6025^{12}
1754	Benzene, 1,5-hexadienyl- 1,5-Hexadiene, 1-phenyl-	$C_{12}H_{14}$ 158.24	1009-81-0	$102\text{-}3^{8}$	4-05-00-01554 0.9005^{25}	1.5421^{25}
1755	Benzene, 3,5-hexadienyl- 6-Phenyl-1,3-hexadiene	$C_{12}H_{14}$ 158.24	39669-95-9	100.5^{11}	3-05-00-01378 0.9304^{13}	1.5446^{13}
1756	Benzene, hexaethyl- Hexaethylbenzene	$C_{18}H_{30}$ 246.44	604-88-6 129	298	4-05-00-01208 0.8305^{130}	H_2O 1; EtOH 3; eth 4; bz 4 1.4736^{130}
1757	Benzene, hexafluoro- Perfluorobenzene	C_6F_6 186.06	392-56-3 5.3	80.2	4-05-00-00640 1.6184^{20}	1.3777^{20}

No.	Name / Synonym	Mol. Form. / Mol. Wt.	CAS RN / mp/°C	Merck No. / bp/°C	Beil. Ref. / den/g cm^{-3}	Solubility / n_D
1758	Benzene, hexamethyl- Mellitene	$C_{12}H_{18}$ 162.27	87-85-4 166.5	263.4	4-05-00-01137 1.0630^{25}	H_2O 1; EtOH 3; eth 3; ace 3
1759	Benzenehexanoic acid	$C_{12}H_{16}O_2$ 192.26	5581-75-9 23	206^{30}	4-09-00-01895	1.5164^{21}
1760	Benzene, 2-hexenyl- 2-Hexene, 1-phenyl	$C_{12}H_{16}$ 160.26	67590-77-6	108^{16}	0-05-00-00501 0.8898^{16}	1.5058^{16}
1761	Benzene, 3-hexenyl-	$C_{12}H_{16}$ 160.26	35008-86-7	222	4-05-00-01418 0.8790^{20}	1.5039^{20}
1762	Benzene, 5-hexenyl-	$C_{12}H_{16}$ 160.26	1588-44-9	94-5^{10}	3-05-00-01251 0.8839^{20}	1.5033^{20}
1763	Benzene, hexyl-	$C_{12}H_{18}$ 162.27	1077-16-3 -61	226.1	4-05-00-01115 0.8575^{20}	H_2O 1; eth 5; bz 3; peth 3 1.4864^{20}
1764	Benzene, (1-hexylheptyl)- 7-Phenyltridecane	$C_{19}H_{32}$ 260.46	2400-01-3	183-4^{20}	4-05-00-01210 0.8723^{20}	1.4931^{18}
1765	Benzene, (hexyloxy)-	$C_{12}H_{18}O$ 178.27	1132-66-7 -19	240	4-06-00-00560 0.9174^{20}	H_2O 1; eth 3 1.4921^{20}
1766	Benzene, iodo- Iodobenzene	C_6H_5I 204.01	591-50-4 -31.3	4922 188.4	4-05-00-00688 1.8308^{20}	H_2O 1; EtOH 3; eth 5; ace 5 1.6200^{20}
1767	Benzene, 1-iodo-2,3-dimethoxy- 3-Iodoveratrole	$C_8H_9IO_2$ 264.06	25245-33-4 45.5	144-5^{14}	3-06-00-04262 1.7799^{20}	1.6127^{20}
1768	Benzene, 4-iodo-1,2-dimethoxy-	$C_8H_9IO_2$ 264.06	5460-32-2 35	174^{30}	4-06-00-05625	bz 4; EtOH 4
1769	Benzene, 1-iodo-2,3-dimethyl-	C_8H_9I 232.06	31599-60-7	229	3-05-00-00820 1.6395^{20}	ace 4 1.6074^{20}
1770	Benzene, 1-iodo-2,4-dimethyl-	C_8H_9I 232.06	4214-28-2	231 dec; 111^{14}	4-05-00-00947 1.6282^{16}	H_2O 1; ace 3; bz 3 1.6008^{16}
1771	Benzene, 1-iodo-3,5-dimethyl- *m*-Xylene, 5-iodo-	C_8H_9I 232.06	22445-41-6	230.5	3-05-00-00840 1.6085^{18}	ace 4 1.5967^{18}
1772	Benzene, 2-iodo-1,3-dimethyl-	C_8H_9I 232.06	608-28-6 11.2	229.5	4-05-00-00947 1.6158^{20}	H_2O 1; ace 3; bz 3 1.6035^{20}
1773	Benzene, 2-iodo-1,4-dimethyl-	C_8H_9I 232.06	1122-42-5	227 dec	4-05-00-00970 1.6168^{17}	H_2O 1; ace 3; bz 3 1.5992^{17}
1774	Benzene, 4-iodo-1,2-dimethyl- 4-Iodo-*o*-xylene	C_8H_9I 232.06	31599-61-8	231.5	3-05-00-00821 1.6334^{18}	ace 4 1.6049^{18}
1775	Benzene, 1-iodo-2-methoxy- *o*-Iodoanisole	C_7H_7IO 234.04	529-28-2	4921 241; 91^{2}	4-06-00-01070 1.8^{20}	EtOH 4; eth 4; ace 4; bz 4
1776	Benzene, 1-iodo-4-methoxy-	C_7H_7IO 234.04	696-62-8 53	238; 138^{25}	4-06-00-01075	EtOH 3; eth 3; chl 3
1777	Benzene, (iodomethyl)-	C_7H_7I 218.04	620-05-3 24.5	93^{10}	4-05-00-00842 1.7335^{25}	bz 4; eth 4; EtOH 4 1.6334^{25}
1778	Benzene, 1-iodo-2-methyl-	C_7H_7I 218.04	615-37-2	211.5	4-05-00-00838 1.713^{20}	H_2O 1; EtOH 5; eth 5 1.6079^{20}
1779	Benzene, 1-iodo-3-methyl-	C_7H_7I 218.04	625-95-6 -27.2	213	4-05-00-00839 1.705^{20}	H_2O 1; EtOH 5; eth 5 1.6053^{20}
1780	Benzene, 1-iodo-4-methyl-	C_7H_7I 218.04	624-31-7 36.5	211	4-05-00-00840 1.678^{20}	H_2O 1; EtOH 3; eth 3; chl 2
1781	Benzene, 2-iodo-1-methyl-4-(1-methylethyl)- 2-Iodo-*p*-cymene	$C_{10}H_{13}I$ 260.12	56739-95-8	139^{23}	3-05-00-00960 1.4205^{17}	1.5800^{17}
1782	Benzene, 2-iodo-4-methyl-1-(1-methylethyl)- *p*-Cymene, 3-iodo-	$C_{10}H_{13}I$ 260.12	4395-81-7	123^{13}; 80^{5}	3-05-00-00960 1.4113^{17}	1.5690^{17}
1783	Benzene, 1-iodo-2-nitro-	$C_6H_4INO_2$ 249.01	609-73-4 54	290; 162^{18}	4-05-00-00732 1.9186^{75}	H_2O 1; EtOH 3; eth 3
1784	Benzene, 1-iodo-3-nitro-	$C_6H_4INO_2$ 249.01	645-00-1 38.5	280	4-05-00-00733 1.9477^{50}	H_2O 1; EtOH 3; eth 3
1785	Benzene, 1-iodo-4-nitro-	$C_6H_4INO_2$ 249.01	636-98-6 174.7	288	4-05-00-00734 1.8090^{155}	H_2O 1; EtOH 3; HOAc 3
1786	Benzene, isocyanato- Phenyl isocyanate	C_7H_5NO 119.12	103-71-9	7267 163; 55^{13}	4-12-00-00864 1.0956^{20}	eth 4; chl 2 1.5368^{20}
1787	Benzene, 1-isocyanato-2-methyl-	C_8H_7NO 133.15	614-68-6	185	4-12-00-01768	H_2O 1; eth 3 1.5282^{20}
1788	Benzene, 1-isocyanato-2-nitro-	$C_7H_4N_2O_3$ 164.12	3320-86-3 41	137^{18}	3-12-00-01535	bz 4; eth 4; chl 4
1789	Benzene, 1-isocyanato-3-nitro-	$C_7H_4N_2O_3$ 164.12	3320-87-4 51	130^{11}	4-12-00-01604	bz 4; eth 4; chl 4
1790	Benzene, 1-isocyanato-4-nitro-	$C_7H_4N_2O_3$ 164.12	100-28-7 57	162^{20}; 137^{11}	4-12-00-01653	bz 4; eth 4; chl 4
1791	Benzene, (isocyanomethyl)-	C_8H_7N 117.15	10340-91-7	199 dec; 93^{55}	4-12-00-02219 0.972^{15}	1.5193^{20}
1792	Benzene, isothiocyanato- Phenyl isothiocyanate	C_7H_5NS 135.19	103-72-0 -21	7268 221	4-12-00-00867 1.1303^{20}	H_2O 1; EtOH 3; eth 3; ctc 3 1.6492^{23}
1793	Benzene, 1-isothiocyanato-2-methoxy-	C_8H_7NOS 165.22	3288-04-8	131-2^{11}	4-13-00-00855 1.1878^{20}	1.6458^{20}
1794	Benzene, (isothiocyanatomethyl)-	C_8H_7NS 149.22	622-78-6	243	4-12-00-02276 1.1246^{16}	H_2O 1; EtOH 5; eth 3 1.6049^{15}
1795	Benzenemethanamine Benzylamine	C_7H_9N 107.16	100-46-9	1139 185; 90^{12}	4-12-00-02155 0.9813^{20}	H_2O 5; EtOH 5; eth 5; ace 4 1.5401^{20}
1796	Benzenemethanamine, 2-amino-	$C_7H_{10}N_2$ 122.17	4403-69-4 61	269	4-13-00-00265	EtOH 4

No.	Name Synonym	Mol. Form. Mol. Wt.	CAS RN mp/°C	Merck No. bp/°C	Beil. Ref. den/g cm^{-3}	Solubility n_D
1797	Benzenemethanamine, *N*-[bis(4-methoxyphenyl)methylene]- Benzylimidobis(*p*-methoxyphenyl)methane	$C_{22}H_{21}NO_2$ 331.41	524-96-9 90	1153	3-12-00-02254	eth 4; chl 4
1798	Benzenemethanamine, *N*,*N*-bis(phenylmethyl)-	$C_{21}H_{21}N$ 287.40	620-40-6 91.5	 385	4-12-00-02183 0.9912^{95}	H_2O 2; EtOH 2; eth 3; ctc 3
1799	Benzenemethanamine, 4-bromo- Benzyl amine, 4-bromo	C_7H_8BrN 186.05	3959-07-7 20	 250	3-12-00-02350	eth 4
1800	Benzenemethanamine, *N*-3-butynyl-*N*-methyl-	$C_{12}H_{15}N$ 173.26	15240-91-2	 127^{16}	4-12-00-02171 0.9372^{20}	 1.5202^{20}
1801	Benzenemethanamine, *N*-(2-chloroethyl)-*N*-(phenylmethyl)-, hydrochloride *N*-(2-Chloroethyl)dibenzylamine hydrochloride	$C_{16}H_{19}Cl_2N$ 296.24	55-43-6	2138	4-12-00-02180	
1802	Benzenemethanamine, 3-chloro-*N*-methyl-	$C_8H_{10}ClN$ 155.63	39191-07-6	 88^4	3-12-00-02342	chl 3 1.5350^{25}
1803	Benzenemethanamine, 2,4-dichloro-	$C_7H_7Cl_2N$ 176.04	95-00-1	 $124\text{-}6^{13}$	4-12-00-02376	chl 3 1.5762^{25}
1804	Benzenemethanamine, *N*,*N*-diethyl-α-phenyl- *N*,*N*-Diethylbenzhydrylamine	$C_{17}H_{21}N$ 239.36	519-72-2 58.5	3103 170^{17}	4-12-00-03283	
1805	Benzenemethanamine, 3,4-dimethoxy-	$C_9H_{13}NO_2$ 167.21	5763-61-1	 156^{12}; 120^3	4-13-00-02582 1.143^{25}	chl 3
1806	Benzenemethanamine, α,α-dimethyl-	$C_9H_{13}N$ 135.21	585-32-0	 196.5	4-12-00-02633 0.9423^{20}	 1.5181^{25}
1807	Benzenemethanamine, *N*,*N*-dimethyl- Dimethylbenzylamine	$C_9H_{13}N$ 135.21	103-83-3	 181	4-12-00-02161 0.915^0	H_2O 2; EtOH 5; eth 5 1.5011^{20}
1808	Benzenemethanamine, 2-ethoxy- 2-Ethoxybenzylamine	$C_9H_{13}NO$ 151.21	37806-29-4	 $69\text{-}74^{0.2}$	4-13-00-01676 1.015^{25}	 1.5326^{20}
1809	Benzenemethanamine, α-ethyl-	$C_9H_{13}N$ 135.21	2941-20-0	 206; 100^{16}	4-12-00-02582 0.9346^{25}	 1.5173^{23}
1810	Benzenemethanamine, *N*-ethyl-	$C_9H_{13}N$ 135.21	14321-27-8	 194	4-12-00-02163 0.9342^{17}	H_2O 2; EtOH 3; eth 3; bz 3 1.5117^{20}
1811	Benzenemethanamine, *N*-ethyl-*N*-phenyl- Ethylbenzylaniline	$C_{15}H_{17}N$ 211.31	92-59-1 35	3728 288; 185^{22}	4-12-00-02176 1.001^{55}	H_2O 1; EtOH 3; eth 3; chl 3 1.5943^{23}
1812	Benzenemethanamine, 2-methoxy-	$C_8H_{11}NO$ 137.18	6850-57-3	 229	4-13-00-01675 1.051^{25}	 1.5475^{20}
1813	Benzenemethanamine, 4-methoxy-	$C_8H_{11}NO$ 137.18	2393-23-9	 236.5	4-13-00-01719 1.050^{15}	H_2O 2; EtOH 2; eth 2 1.5462^{20}
1814	Benzenemethanamine, α-methyl-, (±)-	$C_8H_{11}N$ 121.18	618-36-0 32	 187	4-12-00-02424 0.9395^{15}	H_2O 3; EtOH 5; eth 5; chl 3 1.5238^{25}
1815	Benzenemethanamine, 2-methyl-	$C_8H_{11}N$ 121.18	89-93-0 -30	 206; 81^{15}	4-12-00-02518 0.9766^{19}	 1.5436^{19}
1816	Benzenemethanamine, 3-methyl-	$C_8H_{11}N$ 121.18	100-81-2	 203.5	4-12-00-02565 0.966^{25}	 1.5360^{20}
1817	Benzenemethanamine, 4-methyl-	$C_8H_{11}N$ 121.18	104-84-7 12.5	 195	4-12-00-02574 0.952^{20}	 1.5340^{20}
1818	Benzenemethanamine, *N*-methyl-	$C_8H_{11}N$ 121.18	103-67-3	 180.5	4-12-00-02161 0.9442^{18}	H_2O 4
1819	Benzenemethanamine, 4-(1-methylethyl)-	$C_{10}H_{15}N$ 149.24	4395-73-7	 227	4-12-00-02836	eth 4; EtOH 4 1.5182^{17}
1820	Benzenemethanamine, *N*-(1-methylethyl)-	$C_{10}H_{15}N$ 149.24	102-97-6	 200; 93^{10}	4-12-00-02165 0.892^{25}	 1.5025^{20}
1821	Benzenemethanamine, *N*-(2-methylphenyl)-	$C_{14}H_{15}N$ 197.28	5405-13-0 60	 302.5	4-12-00-02177 1.0142^{65}	ace 4; EtOH 4; chl 4 1.5861^{65}
1822	Benzenemethanamine, α-methyl-*N*-(1-phenylethyl)-	$C_{16}H_{19}N$ 225.33	10024-74-5	 296.5	4-12-00-02428 1.018^{15}	 1.573
1823	Benzenemethanamine, *N*-methyl-*N*-2-propynyl- Pargyline	$C_{11}H_{13}N$ 159.23	555-57-7	6988 $96\text{-}7^{11}$	2-12-00-00548 0.944^{25}	 1.5213^{20}
1824	Benzenemethanamine, *N*-(3-methylphenyl)- *m*-Toluidine, *N*-benzyl	$C_{14}H_{15}N$ 197.28	5405-17-4	 312	4-12-00-02178 1.0083^{65}	ace 4; bz 4; eth 4 1.5845^{65}
1825	Benzenemethanamine, *N*-(4-methylphenyl)- *p*-Toluidine, *N*-benzyl	$C_{14}H_{15}N$ 197.28	5405-15-2 19.5	 319; 181^{10}	4-12-00-02178 1.0064^{65}	bz 4; eth 4; EtOH 4; chl 4 1.5832^{65}
1826	Benzenemethanamine, α-phenyl- Benzhydrylamine	$C_{13}H_{13}N$ 183.25	91-00-9 34	1085 304; 176^{20}	4-12-00-03282 1.0633^{20}	H_2O 2; ace 3 1.5963
1827	Benzenemethanamine, *N*-phenyl- Benzylaniline	$C_{13}H_{13}N$ 183.25	103-32-2 37.5	1140 306.5	4-12-00-02172 1.0298^{65}	eth 4; EtOH 4 1.6118^{25}
1828	Benzenemethanamine, *N*-(phenylmethyl)- Dibenzylamine	$C_{14}H_{15}N$ 197.28	103-49-1 -26	2993 300 dec; 270^{250}	4-12-00-02179 1.0256^{22}	H_2O 1; EtOH 4; eth 4; ctc 3 1.5781^{20}
1829	Benzenemethanamine, *N*-phenyl-*N*-(phenylmethyl)-	$C_{20}H_{19}N$ 273.38	91-73-6 69	 226^{10}	3-12-00-02225 1.0444^{80}	H_2O 1; EtOH 2; eth 3; bz 3 1.6065^{80}
1830	Benzenemethanamine, α-propyl-	$C_{10}H_{15}N$ 149.24	2941-19-7	 221; 108^{16}	3-12-00-02716 0.9366^{20}	EtOH 4
1831	Benzene, 1,1',1'',1'''-methanetetrayltetrakis- Tetraphenylmethane	$C_{25}H_{20}$ 320.43	630-76-2 282	 431	4-05-00-02741	H_2O 1; EtOH 1; eth 1; bz 3
1832	Benzenemethanethiol Thiobenzyl alcohol	C_7H_8S 124.21	100-53-8 -30	9249 194.5	4-06-00-02632 1.058^{20}	H_2O 1; EtOH 4; eth 4; ctc 2 1.5151^{20}

PHYSICAL CONSTANTS OF ORGANIC COMPOUNDS (continued)

No.	Name Synonym	Mol. Form. Mol. Wt.	CAS RN mp/°C	Merck No. bp/°C	Beil. Ref. den/g cm^{-3}	Solubility n_D
1833	Benzenemethanethiol, 4-chloro-	C_7H_7ClS 158.65	6258-66-8 19.5	 113^{17}	4-06-00-02770 1.202^{25}	 1.5893^{20}
1834	Benzenemethanethiol, α-methyl-, (*S*)- Ethanethiol, 1-phenyl (1)	$C_8H_{10}S$ 138.23	33877-11-1	 199; 83^{10}	4-06-00-03062 1.022^{20}	bz 4; eth 4; EtOH 4 1.5593^{20}
1835	Benzenemethanimine, α-phenyl-	$C_{13}H_{11}N$ 181.24	1013-88-3	 282	4-07-00-01365 1.0847^{19}	eth 4 1.6191^{19}
1836	Benzenemethanol Benzyl alcohol	C_7H_8O 108.14	100-51-6 -15.2	1138 205.3	4-06-00-02222 1.0419^{24}	H_2O 3; EtOH 3; eth 3; ace 3 1.5396^{20}
1837	Benzenemethanol, 2-amino-	C_7H_9NO 123.15	5344-90-1 83.5	 273	3-13-00-01615	H_2O 3; EtOH 3; eth 3; bz 4
1838	Benzenemethanol, α-(1-aminoethyl)-, hydrochloride, (*R**,*S**)-(±)- Phenylpropanolamine hydrochloride	$C_9H_{14}ClNO$ 187.67	154-41-6 194	7279	2-13-00-00371	H_2O 4; EtOH 3; eth 1; bz 1
1839	Benzenemethanol, α-(aminomethyl)- Phenylethanolamine	$C_8H_{11}NO$ 137.18	7568-93-6 56.5	7258 160^{17}	4-13-00-01801	H_2O 4; EtOH 3
1840	Benzenemethanol, α-(1-aminopropyl)- α-(α-Aminopropyl)benzyl alcohol	$C_{10}H_{15}NO$ 165.24	5897-76-7 79.5	483	3-13-00-01791	
1841	Benzenemethanol, 2-[bis(4-hydroxyphenyl)methyl]- Phenolphthalol	$C_{20}H_{18}O_3$ 306.36	81-92-5 201.5	7211	4-06-00-07623	
1842	Benzenemethanol, α,α-bis(1-methylethyl)-	$C_{13}H_{20}O$ 192.30	4397-05-1	 229; 158^{60}	4-06-00-03462 0.9755^{20}	eth 4 1.5239^{20}
1843	Benzenemethanol, 5-bromo-2-hydroxy- Bromosaligenin	$C_7H_7BrO_2$ 203.04	2316-64-5 113	1426	3-06-00-04541	bz 4; eth 4; EtOH 4; chl 4
1844	Benzenemethanol, α-(bromomethyl)-	C_8H_9BrO 201.06	2425-28-7	 $109\text{-}10^{2}$	4-06-00-03053 1.4994^{20}	H_2O 1 1.5800^{17}
1845	Benzenemethanol, 2-bromo-α-methyl-	C_8H_9BrO 201.06	5411-56-3	 $109\text{-}10^{2}$	4-06-00-03053 1.4994^{20}	 1.5800^{17}
1846	Benzenemethanol, α-butyl- 1-Phenyl-1-pentanol	$C_{11}H_{16}O$ 164.25	583-03-9	3921 141^{25}; 102^{3}	2-06-00-00503 0.9655^{20}	ace 4; eth 4; EtOH 4 1.4086^{25}
1847	Benzenemethanol, α-butyl-α-methyl- 2-Hexanol, 2-phenyl-	$C_{12}H_{18}O$ 178.27	4396-98-9	 $129\text{-}30^{4}$	3-06-00-01995 0.9522^{20}	 1.5091^{20}
1848	Benzenemethanol, 2-chloro-	C_7H_7ClO 142.58	17849-38-6 73	 230	4-06-00-02589	H_2O 2; EtOH 4; eth 4; lig 4
1849	Benzenemethanol, 4-chloro-	C_7H_7ClO 142.58	873-76-7 75	 235	4-06-00-02593	bz 4; eth 4; EtOH 4
1850	Benzenemethanol, 4-chloro-α-(4-chlorophenyl)-α-methyl- 1,1-Bis(4-chlorophenyl)ethanol	$C_{14}H_{12}Cl_2O$ 267.15	80-06-8 70	2086	4-06-00-04718	H_2O 1; EtOH 1; eth 3; bz 3
1851	Benzenemethanol, 4-chloro-α-(4-chlorophenyl)-α-(trichloromethyl)- 1,1-Bis(*p*-chlorophenyl)-2,2,2-trichloroethanol	$C_{14}H_9Cl_5O$ 370.49	115-32-2 77.5	3075	4-06-00-04722	H_2O 1; os 1
1852	Benzenemethanol, α-(chloromethyl)-	C_8H_9ClO 156.61	1674-30-2	 128^{17}	4-06-00-03045 1.1926^{20}	EtOH 3; eth 4 1.5523^{20}
1853	Benzenemethanol, 2,4-dichloro- 2,4-Dichlorobenzyl alcohol	$C_7H_6Cl_2O$ 177.03	1777-82-8 59.5	3048 150^{25}	4-06-00-02597	chl 3
1854	Benzenemethanol, 2,4-dichloro-α-methyl-	$C_8H_8Cl_2O$ 191.06	1475-13-4	 $125\text{-}6^{7}$	4-06-00-03048 1.293^{25}	 1.5605^{20}
1855	Benzenemethanol, α,α-diethyl-	$C_{11}H_{16}O$ 164.25	1565-71-5 <-17	 223.5	4-06-00-03381 0.9831^{20}	EtOH 4 1.5165^{20}
1856	Benzenemethanol, 3,4-dimethoxy-	$C_9H_{12}O_3$ 168.19	93-03-8	 298; 172^{12}	4-06-00-07381 1.178^{17}	H_2O 3; EtOH 3 1.555^{17}
1857	Benzenemethanol, α,α-dimethyl- α-Cumyl alcohol	$C_9H_{12}O$ 136.19	617-94-7 36	 202	4-06-00-03219 0.9735^{20}	H_2O 1; EtOH 3; eth 3; bz 3 1.5325^{20}
1858	Benzenemethanol, α,3-dimethyl- 1-(3-Methylphenyl)ethyl alcohol	$C_9H_{12}O$ 136.19	7287-81-2	 112^{12}	3-06-00-01823 0.9974^{15}	eth 4; EtOH 4 1.5240^{20}
1859	Benzenemethanol, α,4-dimethyl- 1-(4-Methylphenyl)ethanol	$C_9H_{12}O$ 136.19	536-50-5	3226 219	4-06-00-03242 0.9668^{25}	H_2O 1; EtOH 4; eth 4 1.5246^{20}
1860	Benzenemethanol, 2,4-dimethyl-	$C_9H_{12}O$ 136.19	16308-92-2 22	 232; 150^{44}	4-06-00-03250 1.0310^{20}	 1.5339^{20}
1861	Benzenemethanol, 3,5-dimethyl-	$C_9H_{12}O$ 136.19	27129-87-9	 219.5	0-06-00-00521 0.927^{25}	 1.5312^{20}
1862	Benzenemethanol, 4-(dimethylamino)-	$C_9H_{13}NO$ 151.21	1703-46-4 69	 123^{1}	4-13-00-01771 1.059^{14}	 1.5727^{14}
1863	Benzenemethanol, α-[1-(dimethylamino)ethyl]-, [*R*-(*R**,*S**)]- (1*R*,2*S*)-*N*-Methylephedrine	$C_{11}H_{17}NO$ 179.26	552-79-4 87.5	5987		H_2O 1; EtOH 3; eth 3; MeOH 3
1864	Benzenemethanol, α-(1,1-dimethylethyl)-	$C_{11}H_{16}O$ 164.25	3835-64-1 45	 111^{15}	1-06-00-00270	eth 4; EtOH 4
1865	Benzenemethanol, 4-(1,1-dimethylethyl)-	$C_{11}H_{16}O$ 164.25	877-65-6	 140^{20}	4-06-00-03402 0.928^{25}	 1.5179^{20}
1866	Benzenemethanol, α,α-diphenyl- Triphenylmethanol	$C_{19}H_{16}O$ 260.34	76-84-6 164.2	9653 380	4-06-00-05014 1.199^{0}	H_2O 1; EtOH 4; eth 4; ace 3
1867	Benzenemethanol, α,α-dipropyl- 4-Heptanol, 4-phenyl-	$C_{13}H_{20}O$ 192.30	4436-96-8	 134^{20}	2-06-00-00513 0.9470^{15}	 1.516^{10}
1868	Benzenemethanol, α-ethenyl- 1-Phenylallyl alcohol	$C_9H_{10}O$ 134.18	4393-06-0		2-06-00-00529 1.0249^{21}	H_2O 2; EtOH 3; eth 3; bz 3 1.5406^{20}
1869	Benzenemethanol, α-ethenyl-α-methyl- 3-Buten-2-ol, 2-phenyl-	$C_{10}H_{12}O$ 148.20	6051-52-1	 $92\text{-}5^{7}$	4-06-00-03839 0.9965^{21}	 1.5277^{21}

No.	Name	Synonym	Mol. Form.	Mol. Wt.	CAS RN	mp/°C	Merck No.	bp/°C	Beil. Ref.	den/g cm^{-3}	Solubility	n_D
1870	Benzenemethanol, α-ethyl-	α-Ethylbenzyl alcohol	$C_9H_{12}O$	136.19	93-54-9		3727	219	4-06-00-03183	0.9915^{25}	bz 4; eth 4; EtOH 4; MeOH 4	1.5169^{23}
1871	Benzenemethanol, α-[1-(ethylamino)ethyl]-, (*R**,*S**)-	*N*-Ethylnorephedrine	$C_{11}H_{17}NO$	179.26	37025-57-3	51.5		143^{18}	4-13-00-01887		bz 4	
1872	Benzenemethanol, α-ethyl-3-hydroxy-	1-(*m*-Hydroxyphenyl)-1-propanol	$C_9H_{12}O_2$	152.19	55789-02-1	107		177^{13}	3-06-00-04623		eth 4; EtOH 4	
1873	Benzenemethanol, α-ethyl-4-methoxy-		$C_{10}H_{14}O_2$	166.22	5349-60-0			143^{20}	3-06-00-04624		ctc 3	1.5277^{20}
1874	Benzenemethanol, α-ethyl-α-methyl-, (*R*)-		$C_{10}H_{14}O$	150.22	1006-06-0	-13		211; 113^{23}		0.984^{25}	H_2O 1; EtOH 3; eth 3	1.5185^{20}
1875	Benzenemethanol, α-ethyl-α-propyl-	3-Hexanol, 3-phenyl	$C_{12}H_{18}O$	178.27	20731-93-5			134^{27}	3-06-00-02000	0.964^{20}		1.5100^{20}
1876	Benzenemethanol, α-ethynyl-	1-Phenylpropargyl alcohol	C_9H_8O	132.16	4187-87-5	22		114^{12}	4-06-00-04066	1.0655^{20}		1.5508^{20}
1877	Benzenemethanol, α-ethynyl-, carbamate	Carfimate	$C_{10}H_9NO_2$	175.19	3567-38-2	86.5	1845		4-06-00-04067			
1878	Benzenemethanol, α-ethynyl-α-methyl-		$C_{10}H_{10}O$	146.19	127-66-2	52.3		217.5; 102^{12}	4-06-00-04073	1.0314^{20}		
1879	Benzenemethanol, 3-fluoro-		C_7H_7FO	126.13	456-47-3			201; 104^{22}	4-06-00-02589	1.164^{25}		1.5095^{20}
1880	Benzenemethanol, 4-fluoro-		C_7H_7FO	126.13	459-56-3	23		210	4-06-00-02589			1.5080^{20}
1881	Benzenemethanol, α-hexyl-	1-Phenyl-1-heptanol	$C_{13}H_{20}O$	192.30	614-54-0			275	4-06-00-03454	0.946^{25}		1.5024^{20}
1882	Benzenemethanol, 2-hydroxy-	Salicyl alcohol	$C_7H_8O_2$	124.14	90-01-7	87	8294	sub	4-06-00-05896	1.1613^{25}	H_2O 3; EtOH 3; eth 3; bz 3	
1883	Benzenemethanol, 3-hydroxy-		$C_7H_8O_2$	124.14	620-24-6	73		300 dec	4-06-00-05907	1.161^{25}	H_2O 4; EtOH 4; eth 4; chl 2	
1884	Benzenemethanol, 4-hydroxy-	*p*-Hydroxybenzyl alcohol	$C_7H_8O_2$	124.14	623-05-2	124.5		252	4-06-00-05909		H_2O 4; EtOH 4; eth 3; bz 4	
1885	Benzenemethanol, 2-hydroxy-3-methoxy-		$C_8H_{10}O_3$	154.17	4383-05-5	61.5		162^{12}	4-06-00-07378		chl 2	
1886	Benzenemethanol, 4-hydroxy-3-methoxy-		$C_8H_{10}O_3$	154.17	498-00-0	115		dec	4-06-00-07381		H_2O 3; EtOH 3; eth 3; bz 3	
1887	Benzenemethanol, 4-hydroxy-α-[(methylamino)methyl]-	Synephrine	$C_9H_{13}NO_2$	167.21	94-07-5	184.5	8994		2-13-00-00491			
1888	Benzenemethanol, 2-iodo-		C_7H_7IO	234.04	5159-41-1	92		$147\text{-}9^{32}$	4-06-00-02605			1.6349^{20}
1889	Benzenemethanol, 2-methoxy-		$C_8H_{10}O_2$	138.17	612-16-8			249	4-06-00-05896	1.0386^{25}	H_2O 1; EtOH 3; eth 5	1.5455^{20}
1890	Benzenemethanol, 3-methoxy-		$C_8H_{10}O_2$	138.17	6971-51-3			252	4-06-00-05907	1.112^{25}		1.5440^{20}
1891	Benzenemethanol, 4-methoxy-	Anise alcohol	$C_8H_{10}O_2$	138.17	105-13-5	25	695	259.1	4-06-00-05909	1.109^{26}	H_2O 3; EtOH 4; eth 4; ctc 3	1.5420^{25}
1892	Benzenemethanol, 4-methoxy-, acetate		$C_{10}H_{12}O_3$	180.20	104-21-2	84		270; 150^{23}	4-06-00-05914	1.105^{25}	ctc 3	
1893	Benzenemethanol, 2-methoxy-α-methyl-		$C_9H_{12}O_2$	152.19	13513-82-1	-85.1		125	3-06-00-04563	0.9647^{20}	ace 4; bz 4; eth 4; EtOH 4	1.4024^{20}
1894	Benzenemethanol, 3-methoxy-α-methyl-		$C_9H_{12}O_2$	152.19	23308-82-9			133^{15}	4-06-00-05929	1.0781^{19}	eth 4; EtOH 4	1.5325^{20}
1895	Benzenemethanol, 4-methoxy-α-methyl-		$C_9H_{12}O_2$	152.19	3319-15-1			310 dec; 140^{17}	4-06-00-05930	1.0794^{20}	ctc 3	1.5310^{25}
1896	Benzenemethanol, α-methyl-	α-Methylbenzyl alcohol	$C_8H_{10}O$	122.17	98-85-1	20		205	2-06-00-00444	1.013^{25}	H_2O 1; EtOH 4; eth 4	1.5265^{20}
1897	Benzenemethanol, 2-methyl-	*o*-Methylbenzyl alcohol	$C_8H_{10}O$	122.17	89-95-2	38		224; 118^{20}	4-06-00-03109	1.023^{40}	eth 4; EtOH 4; chl 4	
1898	Benzenemethanol, 3-methyl-	*m*-Methylbenzyl alcohol	$C_8H_{10}O$	122.17	587-03-1	<-20		215.5	4-06-00-03162	0.9157^{17}	H_2O 2; EtOH 4; eth 4; chl 3	
1899	Benzenemethanol, 4-methyl-	*p*-Methylbenzyl alcohol	$C_8H_{10}O$	122.17	589-18-4	61.5		217	4-06-00-03171	0.978^{22}	eth 4; EtOH 4	
1900	Benzenemethanol, α-[1-(methylamino)ethyl]-, (*R**,*S**)-(±)-	(±)-Ephedrine	$C_{10}H_{15}NO$	165.24	90-81-3	76.5	3561	135^{12}	4-13-00-01881	1.1220^{20}	H_2O 3; EtOH 3; eth 3; bz 0	
1901	Benzenemethanol, α-(1-methylethyl)-	1-Phenyl-2-methylpropyl alcohol	$C_{10}H_{14}O$	150.22	611-69-8			223	4-06-00-03289	0.9869^{14}	H_2O 1; EtOH 3; ace 3	1.5193^{14}
1902	Benzenemethanol, 4-(1-methylethyl)-	Cumic alcohol	$C_{10}H_{14}O$	150.22	536-60-7	28	2621	249	4-06-00-03348	0.9818^{20}	H_2O 1; EtOH 5; eth 5; bz 4	1.5210^{20}
1903	Benzenemethanol, α-[1-[methyl(3-phenyl-2-propenyl)amino]ethyl]-	Cinnamedrine	$C_{19}H_{23}NO$	281.40	90-86-8	75	2299		3-13-00-01734			
1904	Benzenemethanol, α-methyl-α-propyl-		$C_{11}H_{16}O$	164.25	4383-18-0			216	2-06-00-00505	0.9723^{22}	EtOH 4	
1905	Benzenemethanol, α-(2-methylpropyl)-		$C_{11}H_{16}O$	164.25	1565-86-2			236; 112^{9}	4-06-00-03380	0.9537^{19}	ace 4; bz 4; eth 4; EtOH 4	1.5080^{18}
1906	Benzenemethanol, 2-nitro-		$C_7H_7NO_3$	153.14	612-25-9	74		270; 168^{20}	4-06-00-02608		H_2O 2; EtOH 3; eth 3	

No.	Name Synonym	Mol. Form. Mol. Wt.	CAS RN mp/°C	Merck No. bp/°C	Beil. Ref. den/g cm^{-3}	Solubility n_D
1907	Benzenemethanol, 3-nitro-	$C_7H_7NO_3$ 153.14	619-25-0 30.5	 $175\text{-}80^{3}$	4-06-00-02609 1.296^{19}	H_2O 3; EtOH 3; eth 3; chl 2
1908	Benzenemethanol, 4-nitro- *p*-Nitrobenzyl alcohol	$C_7H_7NO_3$ 153.14	619-73-8 96.5	 255 dec; 185^{12}	4-06-00-02611	H_2O 2; EtOH 3; eth 3; ace 2
1909	Benzenemethanol, α-pentyl-	$C_{12}H_{18}O$ 178.27	4471-05-0	 170^{50}; 128^{5}	4-06-00-03417 0.9497^{25}	EtOH 3; eth 3 1.5105^{20}
1910	Benzenemethanol, α-phenyl- Benzohydrol	$C_{13}H_{12}O$ 184.24	91-01-0 69	1100 298; 180^{20}	4-06-00-04648	H_2O 2; EtOH 4; eth 4; ctc 4
1911	Benzenemethanol, α-2-propenyl-	$C_{10}H_{12}O$ 148.20	936-58-3	 228.5	4-06-00-03838 1.004^{18}	 1.5289^{21}
1912	Benzenemethanol, α-propyl-, (*R*)-	$C_{10}H_{14}O$ 150.22	22144-60-1 16	 232	3-06-00-01845 0.9740^{20}	eth 4; EtOH 4 1.5139^{20}
1913	Benzenemethanol, α-[(2-pyridinylamino)methyl]- Phenyramidol	$C_{13}H_{14}N_2O$ 214.27	553-69-5 83.5	7292	5-22-08-00301	
1914	Benzenemethanol, 3,4,5-trimethoxy- 3,4,5-Trimethoxybenzyl alcohol	$C_{10}H_{14}O_4$ 198.22	3840-31-1	 228^{25}	4-06-00-07695 1.232^{25}	 1.5439^{20}
1915	Benzene, methoxy- Anisole	C_7H_8O 108.14	100-66-3 -37.5	699 153.7	4-06-00-00548 0.9940^{20}	H_2O 1; EtOH 3; eth 3; ace 4 1.5174^{20}
1916	Benzene, 1-methoxy-2,3-dimethyl-	$C_9H_{12}O$ 136.19	2944-49-2 29	 199	4-06-00-03097 0.9596^{40}	H_2O 1; EtOH 4; eth 4; ace 3 1.5120^{40}
1917	Benzene, 1-methoxy-2,4-dimethyl-	$C_9H_{12}O$ 136.19	6738-23-4	 192	4-06-00-03127 0.9740^{16}	H_2O 1; EtOH 3; eth 3; bz 3 1.5190^{16}
1918	Benzene, 1-methoxy-3,5-dimethyl-	$C_9H_{12}O$ 136.19	874-63-5	 194; 89^{15}	4-06-00-03142 0.9627^{15}	H_2O 1; EtOH 3; eth 3; bz 3 1.5110^{20}
1919	Benzene, 2-methoxy-1,3-dimethyl-	$C_9H_{12}O$ 136.19	1004-66-6	 182.5	4-06-00-03114 0.9619^{14}	H_2O 1; EtOH 3; eth 3; bz 3 1.5053^{14}
1920	Benzene, 2-methoxy-1,4-dimethyl-	$C_9H_{12}O$ 136.19	1706-11-2	 194	4-06-00-03165 0.9693^{13}	H_2O 1; EtOH 3; eth 3; bz 3 1.5182^{15}
1921	Benzene, 4-methoxy-1,2-dimethyl-	$C_9H_{12}O$ 136.19	4685-47-6	 202	4-06-00-03100 0.9744^{14}	H_2O 1; EtOH 3; eth 3; bz 3 1.5198^{14}
1922	Benzene, 1-methoxy-2,3-dinitro- Anisole, 2,3-dinitro	$C_7H_6N_2O_5$ 198.14	16315-07-4 119		4-06-00-01369 1.23^{137}	EtOH 4; lig 4
1923	Benzene, 1-methoxy-2,4-dinitro-	$C_7H_6N_2O_5$ 198.14	119-27-7 94.5	 206^{12}	4-06-00-01372 1.3364^{131}	H_2O 2; EtOH 3; eth 3; ace 3 1.546^{15}
1924	Benzene, 1-methoxy-3,5-dinitro- 3,5-Dinitroanisole	$C_7H_6N_2O_5$ 198.14	5327-44-6 105.3		4-06-00-01385 1.558^{12}	ace 4; bz 4; MeOH 4
1925	Benzene, 2-methoxy-1,3-dinitro-	$C_7H_6N_2O_5$ 198.14	3535-67-9 118		3-06-00-00868 1.3000^{128}	EtOH 4
1926	Benzene, 2-methoxy-1,4-dinitro- Anisole, 2,5-dinitro-	$C_7H_6N_2O_5$ 198.14	3962-77-4 97	 $136\text{-}8^{2}$	4-06-00-01383 1.476^{18}	H_2O 2; EtOH 3; ace 3; bz 3
1927	Benzene, 4-methoxy-1,2-dinitro-	$C_7H_6N_2O_5$ 198.14	4280-28-8 71	 168^{3}	4-06-00-01384 1.3332^{110}	EtOH 4
1928	Benzene, (2-methoxyethenyl)-	$C_9H_{10}O$ 134.18	4747-15-3	 211.5	2-06-00-00521 0.9894^{23}	 1.5620^{24}
1929	Benzene, 1-methoxy-2-(3-methoxyphenoxy)-	$C_{14}H_{14}O_3$ 230.26	1655-71-6 54.8	 327.5	4-06-00-05670	bz 4; eth 4; EtOH 4
1930	Benzene, (methoxymethyl)- Benzyl methyl ether	$C_8H_{10}O$ 122.17	538-86-3 -52.6	1154 170	4-06-00-02229 0.9634^{20}	H_2O 1; EtOH 4; eth 4; bz 3 1.5008^{20}
1931	Benzene, 1-methoxy-2-methyl-	$C_8H_{10}O$ 122.17	578-58-5 -34.1	 171	4-06-00-01943 0.985^{25}	H_2O 1; EtOH 3; eth 3; ace 3 1.5161^{20}
1932	Benzene, 1-methoxy-3-methyl-	$C_8H_{10}O$ 122.17	100-84-5 -47	 175.5	4-06-00-02039 0.969^{25}	H_2O 1; EtOH 3; eth 3; ace 3 1.5130^{20}
1933	Benzene, 1-methoxy-4-methyl-	$C_8H_{10}O$ 122.17	104-93-8 -32	 175.5	4-06-00-02098 0.969^{25}	H_2O 1; EtOH 3; eth 3; chl 3 1.5112^{20}
1934	Benzene, 4-methoxy-1-methyl-2-nitro-	$C_8H_9NO_3$ 167.16	17484-36-5 17	 266.5	3-06-00-01384 1.207^{25}	 1.5525^{20}
1935	Benzene, 1-methoxy-2-nitro- 2-Nitroanisole	$C_7H_7NO_3$ 153.14	91-23-6 10.5	 277; 144^{4}	4-06-00-01249 1.2540^{20}	H_2O 1; EtOH 5; eth 5; ctc 3 1.5161^{20}
1936	Benzene, 1-methoxy-3-nitro-	$C_7H_7NO_3$ 153.14	555-03-3 38.5	 258	4-06-00-01270 1.373^{18}	H_2O 1; EtOH 3; eth 4
1937	Benzene, 1-methoxy-4-nitro- Anisole, 4-nitro	$C_7H_7NO_3$ 153.14	100-17-4 54	 274	4-06-00-01282 1.2192^{60}	H_2O 1; EtOH 4; eth 4; ctc 3 1.5070^{60}
1938	Benzene, 1-methoxy-2-phenoxy-	$C_{13}H_{12}O_2$ 200.24	1695-04-1 79	 288; 91^{7}	4-06-00-05571	bz 4; eth 4; EtOH 4
1939	Benzene, 1-methoxy-4-(2-phenylethenyl)-, (*E*)-	$C_{15}H_{14}O$ 210.28	1694-19-5 136.5	 142.5^{15}	4-06-00-04855	H_2O 1; EtOH 4; eth 4; ace 4
1940	Benzene, 1-methoxy-4-(1-phenylethyl)-	$C_{15}H_{16}O$ 212.29	2605-18-7	 $180\text{-}2^{19}$	4-06-00-04711 1.0473^{20}	bz 4; eth 4; HOAc 4 1.5725^{20}
1941	Benzene, 1-methoxy-2-(1-propenyl)-	$C_{10}H_{12}O$ 148.20	10577-44-3	 224; 104^{13}	4-06-00-03793 0.9962^{15}	 1.560^{15}
1942	Benzene, 1-methoxy-4-(1-propenyl)- Anethole	$C_{10}H_{12}O$ 148.20	104-46-1 21.3	675 234; 115^{12}	2-06-00-00523 0.9882^{20}	ace 4; bz 4; eth 4; EtOH 4 1.5615^{20}
1943	Benzene, 1-methoxy-4-(2-propenyl)- Estragole	$C_{10}H_{12}O$ 148.20	140-67-0	3657 215.5	4-06-00-03817 0.965^{25}	EtOH 4; chl 4 1.5195^{20}
1944	Benzene, 1-methoxy-4-(1-propenyl)-, (*E*)-	$C_{10}H_{12}O$ 148.20	4180-23-8 21.35	 234; 115^{12}	4-06-00-03796 0.9882^{20}	H_2O 2; EtOH 5; eth 5; ace 3 1.5615^{20}
1945	Benzene, 1-methoxy-4-propyl-	$C_{10}H_{14}O$ 150.22	104-45-0	 211.5	4-06-00-03181 0.9472^{20}	H_2O 2; EtOH 3; eth 4; ace 3 1.5045^{20}

No.	Name Synonym	Mol. Form. Mol. Wt.	CAS RN mp/°C	Merck No. bp/°C	Beil. Ref. den/g cm^{-3}	Solubility n_D
1946	Benzene, 2-methoxy-1,3,5-trinitro- Methyl picrate	$C_7H_5N_3O_7$ 243.13	606-35-9 69		4-06-00-01456 1.4947^{80}	H_2O 1; EtOH 4; eth 3; bz 4
1947	Benzene, methyl- Toluene	C_7H_8 92.14	108-88-3 -94.9	9455 110.6	4-05-00-00766 0.8669^{20}	H_2O 1; EtOH 5; eth 5; ace 3 1.4961^{20}
1948	Benzene, 1-methyl-2,4-bis(1-methylethyl)- Toluene, 2,4-diisopropyl-	$C_{13}H_{20}$ 176.30	1460-98-6	 82^{7}	4-05-00-01152 0.8636^{20}	ctc 3; CS_2 3 1.4912^{20}
1949	Benzene, 1-methyl-3,5-bis(1-methylethyl)-	$C_{13}H_{20}$ 176.30	3055-14-9 -60.8	 216.5	4-05-00-01152 0.8668^{20}	 1.4950^{20}
1950	Benzene, (1-methyl-1-butenyl)- 2-Phenyl-2-pentene	$C_{11}H_{14}$ 146.23	53172-84-2	 202	4-05-00-01402 0.8950^{26}	 1.5196^{26}
1951	Benzene, [(3-methylbutoxy)methyl]-	$C_{12}H_{18}O$ 178.27	122-73-6	 237; 118^{19}	3-06-00-01457 0.9098^{20}	eth 4; EtOH 4 1.4792^{20}
1952	Benzene, (1-methylbutyl)-	$C_{11}H_{16}$ 148.25	2719-52-0	 199	4-05-00-01088 0.8594^{21}	H_2O 1; EtOH 3; eth 3 1.4875^{25}
1953	Benzene, (2-methylbutyl)-	$C_{11}H_{16}$ 148.25	3968-85-2	 195.4	2-05-00-00332 0.8584^{20}	 1.4873^{20}
1954	Benzene, (3-methylbutyl)-	$C_{11}H_{16}$ 148.25	2049-94-7	 195	4-05-00-01089 0.856^{20}	H_2O 1; EtOH 3; eth 3; bz 4 1.4867^{10}
1955	Benzene, (1-methylcyclopropyl)-	$C_{10}H_{12}$ 132.21	2214-14-4	 91^{50}	4-05-00-01388	ace 3; bz 3; chl 3 1.5160^{20}
1956	Benzene, 1-methyl-2,4-dinitro- 2,4-Dinitrotoluene	$C_7H_6N_2O_4$ 182.14	121-14-2 71	 300 dec	4-05-00-00865 1.3208^{71}	H_2O 1; EtOH 3; eth 3; ace 4 1.442
1957	Benzene, 1-methyl-3,5-dinitro-	$C_7H_6N_2O_4$ 182.14	618-85-9 93	 sub	4-05-00-00867 1.2772^{111}	H_2O 2; EtOH 3; eth 3; bz 3
1958	Benzene, 2-methyl-1,3-dinitro- 2,6-Dinitrotoluene	$C_7H_6N_2O_4$ 182.14	606-20-2 66		4-05-00-00866 1.2833^{111}	EtOH 3; chl 3 1.479
1959	Benzene, 2-methyl-1,4-dinitro-	$C_7H_6N_2O_4$ 182.14	619-15-8 52.5		4-05-00-00866 1.282^{111}	EtOH 3; bz 3; CS_2 4
1960	Benzene, 4-methyl-1,2-dinitro- 3,4-Dinitrotoluene	$C_7H_6N_2O_4$ 182.14	610-39-9 58.3		4-05-00-00866 1.2594^{111}	H_2O 1; EtOH 3; chl 2; CS_2 3
1961	Benzene, 1,1'-methylenebis- Diphenylmethane	$C_{13}H_{12}$ 168.24	101-81-5 25.2	3329 265.0	4-05-00-01841 1.001^{26}	H_2O 1; EtOH 3; eth 3; chl 3 1.5753^{20}
1962	Benzene, 1,1'-methylenebis[2-bromo- Diphenylmethane, 2,2'-dibromo	$C_{13}H_{10}Br_2$ 326.03	61592-89-0	 $234\text{-}5^{40}$	3-05-00-01795 1.6197^{20}	 1.6300^{20}
1963	Benzene, 1,1'-methylenebis[4-chloro-	$C_{13}H_{10}Cl_2$ 237.13	101-76-8 55.5	 $186\text{-}90^{18}$	4-05-00-01848 1.365^{17}	EtOH 3
1964	Benzene, 1,1'-methylenebis[4-methyl-	$C_{15}H_{16}$ 196.29	4957-14-6 28	 286	4-05-00-01929 0.9800^{20}	eth 4; EtOH 4
1965	Benzene, 1,1'-[methylenebis(oxy)]bis[4-chloro- Bis(*p*-chlorophenoxy)methane	$C_{13}H_{10}Cl_2O_2$ 269.13	555-89-5 70.5	1258 $189\text{-}94^{6}$	4-06-00-00833	ace 4; bz 4
1966	Benzene, 1,1'-(1-methylene-1,2-ethanediyl)bis-	$C_{15}H_{14}$ 194.28	948-97-0 48	 289	4-05-00-02191 1.1014^{20}	 1.5903^{20}
1967	Benzene, (1-methylene-2-propenyl)-	$C_{10}H_{10}$ 130.19	2288-18-8	 $60\text{-}1^{17}$	4-05-00-01539 0.9266^{20}	H_2O 1; eth 3; bz 3; chl 3 1.5489^{20}
1968	Benzene, (1-methylenepropyl)-	$C_{10}H_{12}$ 132.21	2039-93-2	 182	4-05-00-01380 0.887^{25}	 1.5288^{20}
1969	Benzene, 1,1'-(1-methyl-1,2-ethanediyl)bis-	$C_{15}H_{16}$ 196.29	5814-85-7 52	 280.5	4-05-00-01915 0.9807^{20}	 1.5700^{20}
1970	Benzene, 1,1'-[(1-methyl-1,2-ethanediyl)bis(oxy)]bis- Propane, 1,2-diphenoxy	$C_{15}H_{16}O_2$ 228.29	69813-63-4 32	 $175\text{-}8^{12}$	2-06-00-00151 1.0748^{33}	ace 4; bz 4; eth 4; EtOH 4 1.5542^{33}
1971	Benzene, 1,1'-(1-methyl-1,2-ethenediyl)bis-	$C_{15}H_{14}$ 194.28	779-51-1 82	 285.5	4-05-00-02189 0.9857^{17}	EtOH 3; eth 3; bz 3; lig 2 1.5635^{17}
1972	Benzene, (1-methylethenyl)- α-Methylstyrene	C_9H_{10} 118.18	98-83-9 -23.2	 165.4	4-05-00-01364 0.9106^{20}	H_2O 1; EtOH 3; eth 3; ace 5 1.5386^{20}
1973	Benzene, 1-(1-methylethenyl)-4-(1-methylethyl)-	$C_{12}H_{16}$ 160.26	2388-14-9 -30.6	 220.8	4-05-00-01422 0.8936^{20}	ace 4; bz 4; eth 4; EtOH 4 1.5238^{20}
1974	Benzene, (1-methylethoxy)-	$C_9H_{12}O$ 136.19	2741-16-4 -33	 176.8	4-06-00-00557 0.9408^{25}	H_2O 3; EtOH 3; ace 3; bz 3 1.4975^{20}
1975	Benzene, (1-methylethyl)- Cumene	C_9H_{12} 120.19	98-82-8 -96.0	2619 152.4	4-05-00-00985 0.8618^{20}	H_2O 1; EtOH 5; eth 5; ace 5 1.4915^{20}
1976	Benzene, 1,1'-(1-methylethylidene)bis-	$C_{15}H_{16}$ 196.29	778-22-3 29	 202.5	4-05-00-01925 0.9980^{20}	
1977	Benzene, 1-(1-methylethyl)-2-nitro-	$C_9H_{11}NO_2$ 165.19	6526-72-3	 103^{9}	4-05-00-00997 1.101^{12}	H_2O 1; ace 3; bz 3; HOAc 3 1.5259^{20}
1978	Benzene, 1-(1-methylethyl)-4-nitro-	$C_9H_{11}NO_2$ 165.19	1817-47-6	 122^{9}	4-05-00-00997 1.0830^{20}	H_2O 1; ace 3; bz 3; lig 3 1.5367^{20}
1979	Benzene, 1-(1-methylethyl)-4-(phenylmethyl)-	$C_{16}H_{18}$ 210.32	886-58-8 13.4	 310	4-05-00-01951 1.007^{18}	
1980	Benzene, [(1-methylethyl)thio]-	$C_9H_{12}S$ 152.26	3019-20-3	 208	4-06-00-01471 0.9852^{20}	 1.5464^{20}
1981	Benzene, (1-methylheptyl)- 2-Phenyloctane	$C_{14}H_{22}$ 190.33	777-22-0	 $123\text{-}5^{20}$	4-05-00-01157 0.8611^{20}	 1.4837^{20}
1982	Benzene, 1,1',1''-methylidynetris- Triphenylmethane	$C_{19}H_{16}$ 244.34	519-73-3 94	9655 359; 200^{10}	4-05-00-02495 1.014^{99}	H_2O 1; EtOH 2; eth 4; bz 3 1.5839^{99}
1983	Benzene, 1,1',1''-[methylidynetris(oxy)]tris-	$C_{19}H_{16}O_3$ 292.33	16737-44-3 76.5	 269^{50} dec	4-06-00-00611	eth 4; EtOH 4
1984	Benzene, 1-methyl-2-(1-methylethenyl)-	$C_{10}H_{12}$ 132.21	7399-49-7	 175	3-05-00-01214 0.9180^{15}	 1.5112^{30}

No.	Name Synonym	Mol. Form. Mol. Wt.	CAS RN mp/°C	Merck No. bp/°C	Beil. Ref. den/g cm^{-3}	Solubility n_D
1985	Benzene, 1-methyl-3-(1-methylethenyl)-	$C_{10}H_{12}$ 132.21	1124-20-5	185	4-05-00-01383 0.9010^{25}	1.5335^{20}
1986	Benzene, 1-methyl-4-(1-methylethenyl)-	$C_{10}H_{12}$ 132.21	1195-32-0 -20	185.3	4-05-00-01383 0.8936^{23}	1.5283^{23}
1987	Benzene, 1-methyl-3-(1-methylethoxy)-	$C_{10}H_{14}O$ 150.22	19177-04-9	195	4-06-00-02040 0.931^{20}	1.4959^{20}
1988	Benzene, 1-methyl-2-(1-methylethyl)- *o*-Cymene	$C_{10}H_{14}$ 134.22	527-84-4 -71.5	2770 178.1	4-05-00-01057 0.8766^{20}	H_2O 1; EtOH 5; eth 5; ace 5 1.5006^{20}
1989	Benzene, 1-methyl-3-(1-methylethyl)- *m*-Cymene	$C_{10}H_{14}$ 134.22	535-77-3 -63.7	2770 175.1	4-05-00-01058 0.8610^{20}	H_2O 1; EtOH 5; eth 5; ace 5 1.4930^{20}
1990	Benzene, 1-methyl-4-(1-methylethyl)- *p*-Cymene	$C_{10}H_{14}$ 134.22	99-87-6 -68.9	2770 177.1	4-05-00-01060 0.8573^{20}	H_2O 1; EtOH 5; eth 5; ace 5 1.4909^{20}
1991	Benzene, 1-methyl-4-(1-methylethyl)-2-nitro-	$C_{10}H_{13}NO_2$ 179.22	943-15-7	126^{10}	4-05-00-01064 1.0744^{20}	eth 4; EtOH 4 1.5301^{20}
1992	Benzene, [2-methyl-1-(1-methylethyl)propyl]-	$C_{13}H_{20}$ 176.30	21777-84-4 -24.5	220.7	4-05-00-01146 0.8821^{10}	1.512
1993	Benzene, 1-methyl-2-[(4-methylphenyl)thio]- 2,4'-Ditolyl sulfide	$C_{14}H_{14}S$ 214.33	4279-70-3	173^{11}	4-06-00-02172 1.0774^{15}	eth 4; EtOH 4
1994	Benzene, 1-methyl-2-(1-methylpropyl)-	$C_{11}H_{16}$ 148.25	1595-06-8	196	4-05-00-01094 0.873^{20}	ace 4; bz 4; eth 4; EtOH 4 1.497^{20}
1995	Benzene, 1-methyl-2-(2-methylpropyl)-	$C_{11}H_{16}$ 148.25	36301-29-8 -73.3	196	4-05-00-01095 0.8649^{20}	ace 4; bz 4; eth 4; EtOH 4 1.4935^{20}
1996	Benzene, 1-methyl-3-(1-methylpropyl)-	$C_{11}H_{16}$ 148.25	1772-10-7	194	4-05-00-01095 0.858^{20}	ace 4; bz 4; eth 4; EtOH 4 1.490^{20}
1997	Benzene, 1-methyl-3-(2-methylpropyl)-	$C_{11}H_{16}$ 148.25	5160-99-6	194	3-05-00-01001 0.8536^{20}	ace 4; bz 4; eth 4; EtOH 4 1.4888^{20}
1998	Benzene, 1-methyl-4-(1-methylpropyl)-	$C_{11}H_{16}$ 148.25	1595-16-0	197	4-05-00-01095 0.8640^{19}	bz 4; eth 4; chl 4 1.493^{20}
1999	Benzene, 1-methyl-4-(2-methylpropyl)-	$C_{11}H_{16}$ 148.25	5161-04-6	196	4-05-00-01096 0.8517^{20}	ace 4; bz 4; eth 4; EtOH 4 1.4874^{20}
2000	Benzene, 1-methyl-2-nitro- *o*-Nitrotoluene	$C_7H_7NO_2$ 137.14	88-72-2 -10	6572 222	4-05-00-00845 1.1611^{19}	H_2O 1; EtOH 5; eth 5; ctc 3 1.5450^{20}
2001	Benzene, 1-methyl-3-nitro- *m*-Nitrotoluene	$C_7H_7NO_2$ 137.14	99-08-1 15.5	6572 232	4-05-00-00847 1.1581^{20}	H_2O 1; EtOH 3; eth 5; bz 3 1.5466^{20}
2002	Benzene, 1-methyl-4-nitro- *p*-Nitrotoluene	$C_7H_7NO_2$ 137.14	99-99-0 51.6	6572 238.3	4-05-00-00848 1.1038^{75}	H_2O 1; EtOH 3; eth 4; ace 4
2003	Benzene, (1-methyl-1-nitroethyl)-	$C_9H_{11}NO_2$ 165.19	3457-58-7	224 dec; 126^{15}	4-05-00-00998 1.1025^{20}	ace 4; bz 4 1.5209^{20}
2004	Benzene, 1-methyl-4-(nitromethyl)-	$C_8H_9NO_2$ 151.17	29559-27-1 11.5	$150\text{-}1^{35}$	4-05-00-00972 1.1234^{20}	H_2O 1; bz 3 1.5278^{20}
2005	Benzene, 1-methyl-2-(2-nitrophenoxy)- 2-Nitrophenyl 2-tolyl ether	$C_{13}H_{11}NO_3$ 229.24	54106-40-0 39.5	$194\text{-}6^{14}$	4-06-00-01946 1.195^{20}	
2006	Benzene, 1-methyl-4-(4-nitrophenoxy)-	$C_{13}H_{11}NO_3$ 229.24	3402-74-2 69	224^{25}	4-06-00-02102	bz 4; HOAc 4
2007	Benzene, (1-methylnonadecyl)- Eicosane, 2-phenyl-	$C_{26}H_{46}$ 358.65	2398-66-5 29.0	204.5^{1}	4-05-00-01222 0.8547^{20}	1.4795^{20}
2008	Benzene, (1-methyl-1-pentenyl)- 2-Hexene, 2-phenyl	$C_{12}H_{16}$ 160.26	20247-89-6	$66\text{-}8^{1}$	4-05-00-01419 0.910^{20}	1.5200^{20}
2009	Benzene, (1-methylpentyl)-	$C_{12}H_{18}$ 162.27	6031-02-3	208	4-05-00-01116 0.869^{15}	1.492^{15}
2010	Benzene, (3-methylpentyl)- (3-Methylpentyl)benzene	$C_{12}H_{18}$ 162.27	54410-69-4	220	0-05-00-00444 0.8644^{14}	1.4896
2011	Benzene, (4-methylpentyl)-	$C_{12}H_{18}$ 162.27	4215-86-5	215	4-05-00-01116 0.8568^{16}	bz 4; eth 4; EtOH 4
2012	Benzene, 1-methyl-2-phenoxy-	$C_{13}H_{12}O$ 184.24	3991-61-5 21.7	263.5	4-06-00-01946 1.0480^{25}	
2013	Benzene, 1-methyl-3-phenoxy-	$C_{13}H_{12}O$ 184.24	3586-14-9	272	4-06-00-02041 1.051^{25}	1.5727^{20}
2014	Benzene, 1-methyl-2-(phenylmethyl)-	$C_{14}H_{14}$ 182.27	713-36-0 6.6	280.5	4-05-00-01892 1.0020^{20}	1.5763^{20}
2015	Benzene, 1-methyl-3-(phenylmethyl)-	$C_{14}H_{14}$ 182.27	620-47-3 -28	279	4-05-00-01893 0.9913^{20}	H_2O 1; EtOH 3; eth 3; bz 3 1.5712^{20}
2016	Benzene, 1-methyl-4-(phenylmethyl)-	$C_{14}H_{14}$ 182.27	620-83-7 -30	286	4-05-00-01894 0.9976^{20}	eth 4; bz 4; EtOH 4; chl 4 1.5712^{20}
2017	Benzene, 1-methyl-2-(phenylthio)-	$C_{13}H_{12}S$ 200.30	13963-35-4	308; 164^{12}	4-06-00-02018 1.0893^{20}	H_2O 1; ace 3; bz 3
2018	Benzene, 1-methyl-3-(phenylthio)-	$C_{13}H_{12}S$ 200.30	13865-48-0 -6.5	309.5	4-06-00-02080 1.0937^{15}	H_2O 1; ace 3; bz 3
2019	Benzene, 1-methyl-4-(phenylthio)-	$C_{13}H_{12}S$ 200.30	3699-01-2 15.7	317	4-06-00-02169 1.0986^{25}	H_2O 1; ace 3; bz 3 1.6225^{25}
2020	Benzene, 1,1'-(1-methyl-1,3-propanediyl)bis-, (±)- Butane, 1,3-diphenyl (*DL*)	$C_{16}H_{18}$ 210.32	116783-21-2	293	4-05-00-01938 0.9722^{20}	1.5525^{20}
2021	Benzene, 1,1'-(2-methyl-1,3-propanediyl)bis-	$C_{16}H_{18}$ 210.32	1520-46-3 -33.7	303	4-05-00-01939 0.9669^{20}	
2022	Benzene, 1,1'-(3-methyl-1-propene-1,3-diyl)bis- 1,3-Diphenyl-1-butene	$C_{16}H_{16}$ 208.30	7614-93-9 47.5	311	4-05-00-02205 0.9996^{20}	1.590^{15}
2023	Benzene, 1-methyl-2-(2-propenyl)-	$C_{10}H_{12}$ 132.21	1587-04-8	182.5	3-05-00-01212 0.9005^{20}	1.5187^{20}

No.	Name	Synonym	Mol. Form.	Mol. Wt.	CAS RN	mp/°C	Merck No.	bp/°C	Beil. Ref.	den/g cm^{-3}	Solubility	n_D
2024	Benzene, (2-methyl-1-propenyl)-		$C_{10}H_{12}$	132.21	768-49-0			$99^{43.5}$	4-05-00-01380	0.9029^{20}		1.5388^{20}
2025	Benzene, (1-methyl-1-propenyl)-, (*E*)-		$C_{10}H_{12}$	132.21	768-00-3	-23.5		194.7	4-05-00-01378	0.9138^{25}	H_2O 1; bz 3; chl 3	1.5425^{20}
2026	Benzene, 1-methyl-2-(2-propenyloxy)-		$C_{10}H_{12}O$	148.20	936-72-1			206.5	4-06-00-01946	0.9698^{15}	H_2O 1; ctc 2	1.5188^{15}
2027	Benzene, 1-methyl-3-(2-propenyloxy)-		$C_{10}H_{12}O$	148.20	1758-10-7			212.5	4-06-00-02041	0.9564^{20}	bz 4	1.5179^{20}
2028	Benzene, 1-methyl-4-(2-propenyloxy)-		$C_{10}H_{12}O$	148.20	23431-48-3			214.5	4-06-00-02101	0.9719^{15}	H_2O 1; ace 3; ctc 2	1.5157^{24}
2029	Benzene, (1-methyl-1-propenyl)-, (*Z*)-		$C_{10}H_{12}$	132.21	767-99-7			194.7; 177^{500}	3-05-00-01209	0.9191^{25}	H_2O 1; bz 3; chl 3	1.5402^{25}
2030	Benzene, (1-methylpropoxy)-		$C_{10}H_{14}O$	150.22	10574-17-1			194.5	4-06-00-00559	0.9415^{20}	H_2O 1; eth 3; ctc 2	1.4926^{25}
2031	Benzene, (2-methylpropoxy)-		$C_{10}H_{14}O$	150.22	1126-75-6			197.5	4-06-00-00559	0.9232^{24}		1.4932^{14}
2032	Benzene, 1-methyl-4-propoxy-	Toluene, 4-propoxy	$C_{10}H_{14}O$	150.22	5349-18-8	0		210.4	3-06-00-01354	0.9496^{0}		
2033	Benzene, [(2-methylpropoxy)methyl]-	Benzyl isobutyl ether	$C_{11}H_{16}O$	164.25	940-49-8			212	3-06-00-01457	0.9233^{20}	H_2O 1; eth 4; chl 4	1.4826^{20}
2034	Benzene, 1-(2-methylpropoxy)-2-nitro-	1-Isobutoxy-2-nitrobenzene	$C_{10}H_{13}NO_3$	195.22	56245-02-4			277.5	0-06-00-00218	1.1361^{20}	eth 4	
2035	Benzene, 1-methylpropyl	*sec*-Butylbenzene	$C_{10}H_{14}$	134.22	135-98-8	-82.7	1550	173.5		0.8580^{25}	H_2O 1; EtOH 5; eth 5; ace 5	1.4895^{20}
2036	Benzene, 1-methyl-2-propyl-		$C_{10}H_{14}$	134.22	1074-17-5	-60.3		185	4-05-00-01056	0.8697^{25}		1.4996^{20}
2037	Benzene, 1-methyl-3-propyl-		$C_{10}H_{14}$	134.22	1074-43-7	-82.5		182	4-05-00-01056	0.8569^{25}		1.4935^{20}
2038	Benzene, 1-methyl-4-propyl-		$C_{10}H_{14}$	134.22	1074-55-1	-63.6		183.4	4-05-00-01057	0.8544^{25}	H_2O 1; EtOH 3; eth 3	1.4922^{20}
2039	Benzene, (2-methylpropyl)-	Isobutylbenzene	$C_{10}H_{14}$	134.22	538-93-2	-51.4	5018	172.7	4-05-00-01042	0.8532^{20}	H_2O 1; EtOH 5; eth 5; ace 5	1.4866^{20}
2040	Benzene, (methylsulfinyl)-		C_7H_8OS	140.21	1193-82-4	32.0		263.5; 140^{13}	4-06-00-01467			1.5885^{20}
2041	Benzene, (methylthio)-		C_7H_8S	124.21	100-68-5			193	4-06-00-01466	1.0579^{20}	H_2O 1; EtOH 3; ace 4; os 3	1.5868^{20}
2042	Benzene, [(methylthio)methyl]-		$C_8H_{10}S$	138.23	766-92-7	-30		210; 120^{48}	4-06-00-02633	1.0274^{20}		1.5620^{20}
2043	Benzene, 1-(methylthio)-2-nitro-		$C_7H_7NO_2S$	169.20	3058-47-7	64.5			4-06-00-01661	1.2628^{78}	H_2O 3; EtOH 3; bz 3; chl 3	1.6246^{78}
2044	Benzene, 1-(methylthio)-4-nitro-		$C_7H_7NO_2S$	169.20	701-57-5	72		$135\text{-}40^{2}$	4-06-00-01687	1.2391^{80}	H_2O 1; ace 3; bz 3	1.6401^{20}
2045	Benzene, 1-methyl-4-(trifluoromethyl)-		$C_8H_7F_3$	160.14	6140-17-6			129		1.4320^{26}	eth 3; ace 3; bz 3; chl 3	1.4320^{26}
2046	Benzene, 1-methyl-2,3,4-trinitro-	2,3,4-Trinitrotoluene	$C_7H_5N_3O_6$	227.13	602-29-9	112		301	4-05-00-00872	1.62^{25}	H_2O 1; EtOH 2; eth 3; ace 3	
2047	Benzene, 1-methyl-2,4,5-trinitro-	2,4,5-Trinitrotoluene	$C_7H_5N_3O_6$	227.13	610-25-3	104		exp 300	4-05-00-00872		H_2O 1; EtOH 2; eth 3; ace 3	
2048	Benzene, 2-methyl-1,3,5-trinitro-	2,4,6-Trinitrotoluene	$C_7H_5N_3O_6$	227.13	118-96-7	80.1	9643	exp 240	4-05-00-00873	1.654^{25}	H_2O 1; EtOH 2; eth 3; ace 4	
2049	Benzene, nitro-	Nitrobenzene	$C_6H_5NO_2$	123.11	98-95-3	5.7	6509	210.8	4-05-00-00708	1.2037^{20}	H_2O 2; EtOH 4; eth 4; ace 4	1.5562^{20}
2050	Benzene, (2-nitroethenyl)-, (*E*)-		$C_8H_7NO_2$	149.15	5153-67-3	60		255	4-05-00-01352		H_2O 1; EtOH 3; eth 4; ace 3	
2051	Benzene, (2-nitroethyl)-		$C_8H_9NO_2$	151.17	6125-24-2	-23		250; 137^{16}	4-05-00-00912	1.126^{24}		1.5407^{19}
2052	Benzene, (nitromethyl)-		$C_7H_7NO_2$	137.14	622-42-4			226; 135^{25}	4-05-00-00850	1.1596^{20}	ace 4; eth 4	1.5323^{20}
2053	Benzene, 1-nitro-2-phenoxy-		$C_{12}H_9NO_3$	215.21	2216-12-8	<-20		235^{60}; 184^{8}	4-06-00-01252	1.2539^{22}	bz 4; eth 4; EtOH 4; chl 4	1.575^{20}
2054	Benzene, 1-nitro-4-phenoxy-		$C_{12}H_9NO_3$	215.21	620-88-2	61		320; 225^{30}	4-06-00-01287		H_2O 1; EtOH 2; eth 3; bz 3	
2055	Benzene, 1-nitro-2-(2-phenylethenyl)-, (*E*)-		$C_{14}H_{11}NO_2$	225.25	4264-29-3	73		209^{11}	4-05-00-02166		H_2O 1; EtOH 3; eth 4	
2056	Benzene, 1-nitro-2-(2-phenylethenyl)-, (*Z*)-		$C_{14}H_{11}NO_2$	225.25	52208-62-5	64.8		187^{11}	4-05-00-02166		EtOH 3; eth 4; AcOEt 4; HOAc 4	
2057	Benzene, 1-nitro-2-(phenylthio)-		$C_{12}H_9NO_2S$	231.28	4171-83-9	82		210^{15}	4-06-00-01663		eth 4; EtOH 4	
2058	Benzene, 1-nitro-4-(phenylthio)-		$C_{12}H_9NO_2S$	231.28	952-97-6	56		288^{100}; 240^{25}	4-06-00-01694		eth 4; EtOH 4	
2059	Benzene, nitroso-	Nitrosobenzene	C_6H_5NO	107.11	586-96-9	68.5		58^{18}	4-05-00-00702		H_2O 1; EtOH 3; eth 3; bz 3	
2060	Benzene, 1-nitro-2-(trifluoromethyl)-		$C_7H_4F_3NO_2$	191.11	384-22-5	32.5		217; 105^{20}	4-05-00-00852		H_2O 1; EtOH 4; bz 4; ctc 2	
2061	Benzene, 1-nitro-3-(trifluoromethyl)-	3-(Trifluoromethyl)-1-nitrobenzene	$C_7H_4F_3NO_2$	191.11	98-46-4	-2.4		202.8; 103^{40}	4-05-00-00853	1.4357^{15}	H_2O 1; EtOH 3; eth 3; ctc 2	1.4719^{20}
2062	Benzene, nonadecyl-		$C_{25}H_{44}$	344.62	29136-19-4	40		419	3-05-00-01126	0.8545^{20}		1.4807^{20}

No.	Name	Synonym	Mol. Form.	Mol. Wt.	CAS RN	mp/°C	Merck No.	bp/°C	Beil. Ref.	den/g cm^{-3}	Solubility	n_D
2063	Benzenenonanoic acid, θ-nonyl-		$C_{24}H_{40}O_2$	360.58	1938-22-3	37		200-4$^{0.09}$	2-09-00-00376	0.9312^{25}		1.4891^{20}
2064	Benzene, nonyl-		$C_{15}H_{24}$	204.36	1081-77-2	-24		280.5	3-05-00-01075	0.8584^{20}		1.4816^{20}
2065	Benzene, octacosyl-	Octacosylbenzene	$C_{34}H_{62}$	470.87	61828-26-0	66		405^{100}; 268^{1}	4-05-00-01243	0.8543^{20}		1.4796^{20}
2066	Benzene, octadecyl-	Octadecylbenzene	$C_{24}H_{42}$	330.60	4445-07-2	29		408	4-05-00-01219	0.8546^{20}		1.4809^{20}
2067	Benzene, octyl-		$C_{14}H_{22}$	190.33	2189-60-8	-36		264.5	4-05-00-01157	0.8562^{20}	H_2O 1; eth 5; bz 5	1.4845^{20}
2068	Benzene, (1-octyldodecyl)-	Eicosane, 9-phenyl-	$C_{26}H_{46}$	358.65	2398-65-4	17.9		196$^{1.0}$	4-05-00-01223	0.8534^{20}		1.4790^{20}
2069	Benzene, (octyloxy)-		$C_{14}H_{22}O$	206.33	1818-07-1	8		285	4-06-00-00560	0.9131^{15}	H_2O 1; EtOH 3; eth 3	1.4875^{20}
2070	Benzene, (3-octylundecyl)-	9-(2-Phenylethyl)heptadecane	$C_{25}H_{44}$	344.62	5637-96-7	-26.7		189$^{1.0}$	4-05-00-01221	0.8560^{20}		1.4806^{20}
2071	Benzene, 1,1'-oxybis-	Diphenyl ether	$C_{12}H_{10}O$	170.21	101-84-8	26.8	7259	258.0	4-06-00-00568	1.0661^{30}	H_2O 1; EtOH 3; eth 3; bz 3	1.5787^{25}
2072	Benzene, 1,1'-oxybis[4-bromo-		$C_{12}H_8Br_2O$	328.00	2050-47-7	60.5		339	4-06-00-01048	1.8^{25}	H_2O 1; EtOH 3; eth 4; bz 3	
2073	Benzene, 1,1'-oxybis[4-chloro-		$C_{12}H_8Cl_2O$	239.10	2444-89-5	30		313	4-06-00-00826	1.1231^{20}	H_2O 1	1.611^{20}
2074	Benzene, 1,1'-oxybis[2-ethyl-		$C_{16}H_{18}O$	226.32	56911-77-4			318.5		1.0141^{18}	H_2O 1; eth 3; chl 3	1.5488^{18}
2075	Benzene, 1,1'-oxybis[2-methoxy-		$C_{14}H_{14}O_3$	230.26	1655-70-5	79.5		330.5	3-06-00-04223		H_2O 1; EtOH 4; eth 4; lig 1	
2076	Benzene, 1,1'-oxybis[2-methyl-		$C_{14}H_{14}O$	198.26	4731-34-4			271	4-06-00-01947	1.047^{24}	bz 4; eth 4; EtOH 4	
2077	Benzene, 1,1'-oxybis[3-methyl-		$C_{14}H_{14}O$	198.26	19814-71-2			284; 136^{14}	4-06-00-02043	1.0323^{21}	H_2O 1; EtOH 3; eth 3; bz 3	
2078	Benzene, 1,1'-oxybis[4-methyl-		$C_{14}H_{14}O$	198.26	1579-40-4	51		285	4-06-00-02103		bz 4; eth 4; EtOH 4	
2079	Benzene, 1,1'-[oxybis(methylene)]bis-	Benzyl ether	$C_{14}H_{14}O$	198.26	103-50-4	3.6	1146	298	4-06-00-02240	1.0428^{20}	H_2O 1; EtOH 5; eth 5; ctc 3	1.5168^{20}
2080	Benzene, 1,1'-(oxydi-2,1-ethanediyl)bis-		$C_{16}H_{18}O$	226.32	2396-53-4			318.5; 86$^{0.5}$	4-06-00-03071	1.0141^{18}	eth 4; chl 4	1.5488^{18}
2081	Benzene, 1,1'-(oxydiethylidene)bis-, (*R**,*R**)-(±)-	Bis-(1-phenylethyl)ether (*DL*)	$C_{16}H_{18}O$	226.32	53776-69-5			280.2	4-06-00-03034	1.0058^{15}	H_2O 1; eth 3; chl 3	1.5454^{21}
2082	Benzene, pentabromo-	Pentabromobenzene	C_6HBr_5	472.59	608-90-2	160.5		sub	4-05-00-00687		bz 4; chl 4	
2083	Benzene, pentabromomethyl-		$C_7H_3Br_5$	486.62	87-83-2	288			3-05-00-00720	2.97^{17}	H_2O 1; EtOH 2; bz 3; HOAc 2	
2084	Benzene, pentachloro-	Pentachlorobenzene	C_6HCl_5	250.34	608-93-5	86		277	4-05-00-00669	1.8342^{16}	H_2O 1; EtOH 1; eth 2; bz 2	
2085	Benzene, pentachloro(dichloromethyl)-		C_7HCl_7	333.25	2136-95-0	119.5		334	4-05-00-00824		EtOH 2	
2086	Benzene, pentachloromethyl-		$C_7H_3Cl_5$	264.36	877-11-2	224.8		301	4-05-00-00823		EtOH 2; eth 2; bz 3; tol 3	
2087	Benzene, pentachloronitro-	Quintozene	$C_6Cl_5NO_2$	295.34	82-68-8	144	8108	328 dec	4-05-00-00728	1.718^{25}	H_2O 1; EtOH 2; bz 3; chl 3	
2088	Benzene, pentacosyl-	Pentacosylbenzene	$C_{31}H_{56}$	428.79	61828-06-6	59		380^{100}; 244^{1}	4-05-00-01241	0.8544^{20}		1.4799^{20}
2089	Benzene, pentadecyl-		$C_{21}H_{36}$	288.52	2131-18-2	22		373	3-05-00-01119	0.8548^{20}		1.4815^{20}
2090	Benzene, pentaethyl-	Pentaethylbenzene	$C_{16}H_{26}$	218.38	605-01-6	<-20		277	4-05-00-01197	0.8971^{19}		1.5127^{20}
2091	Benzene, pentafluoro-	Pentafluorobenzene	C_6HF_5	168.07	363-72-4	-47.3		85.7	4-05-00-00639	1.514^{25}		1.3905^{20}
2092	Benzene, pentafluoroiodo-		C_6F_5I	293.96	827-15-6	-29		166	4-05-00-00695	2.212^{20}		1.4950^{25}
2093	Benzene, pentafluoromethoxy-		$C_7H_3F_5O$	198.09	389-40-2	-37		138.5	4-06-00-00782	1.493^{20}		1.4087^{20}
2094	Benzene, pentafluoromethyl-	2,3,4,5,6-Pentafluorotoluene	$C_7H_3F_5$	182.09	771-56-2	-29.8		117.5		1.440^{20}		1.4016^{25}
2095	Benzene, pentafluoro(trifluoromethyl)-	Perfluorotoluene	C_7F_8	236.06	434-64-0	-65.6		104.5	4-05-00-00804			1.3670^{20}
2096	Benzene, pentaiodo-	Pentaiodobenzene	C_6HI_5	707.60	608-96-8	172		sub	4-05-00-00702		bz 4; chl 4	
2097	Benzene, pentamethyl-	Pentamethylbenzene	$C_{11}H_{16}$	148.25	700-12-9	54.5		232	4-05-00-01109	0.917^{20}	H_2O 1; EtOH 4; bz 4; chl 3	1.527^{20}
2098	Benzene, 1,1'-(1,5-pentanediyl)bis-	1,5-Diphenylpentane	$C_{17}H_{20}$	224.35	1718-50-9	-8		327	4-05-00-01963	0.9812^{19}	H_2O 1	1.559^{19}
2099	Benzenepentanoic acid	5-Phenylvaleric acid	$C_{11}H_{14}O_2$	178.23	2270-20-4	57.5		190^{30}	4-09-00-01864		H_2O 2; EtOH 4; os 3	
2100	Benzenepentanoic acid, δ-oxo-, methyl ester		$C_{12}H_{14}O_3$	206.24	1501-04-8	-2		186^{19}; 147^{8}	4-10-00-02783		H_2O 4	
2101	Benzenepentanol		$C_{11}H_{16}O$	164.25	10521-91-2			155^{20}	4-06-00-03373	0.9725^{20}	eth 4; EtOH 4	1.5156^{20}

No.	Name Synonym	Mol. Form. Mol. Wt.	CAS RN mp/°C	Merck No. bp/°C	Beil. Ref. den/g cm^{-3}	Solubility n_D
2102	Benzenepentanol, α-ethyl- 7-Phenylheptan-3-ol	$C_{13}H_{20}O$ 192.30	60012-63-7	277.5; 150^{20}	3-06-00-02027 0.968^{20}	1.5115^{20}
2103	Benzenepentanol, α-methyl- 6-Phenyl-2-hexanol	$C_{12}H_{18}O$ 178.27	38487-94-4	148^{18}	3-06-00-01995 0.9567^{20}	1.5079^{20}
2104	Benzene, 1-pentenyl-	$C_{11}H_{14}$ 146.23	826-18-6	217	4-05-00-01401 0.8782^{22}	1.5158^{20}
2105	Benzene, 3-pentenyl-	$C_{11}H_{14}$ 146.23	1745-16-0	201	4-05-00-01401 0.8884^{16}	1.5089^{16}
2106	Benzene, 4-pentenyl- 1-Pentene, 5-phenyl	$C_{11}H_{14}$ 146.23	1075-74-7	201; 77^{10}	3-05-00-01238 0.8889^{20}	1.5065^{20}
2107	Benzene, pentyl- Amylbenzene	$C_{11}H_{16}$ 148.25	538-68-1 -75	635 205.4	4-05-00-01085 0.8585^{20}	H_2O 1; EtOH 5; eth 5; ace 5 1.4878^{20}
2108	Benzene, 1,1'-pentylidenebis- Pentane, 1,1-diphenyl-	$C_{17}H_{20}$ 224.35	1726-12-1 -12	308	4-05-00-01966 0.9659^{20}	H_2O 1 1.5511^{20}
2109	Benzene, (pentyloxy)- Pentyl phenyl ether	$C_{11}H_{16}O$ 164.25	2050-04-6	200; 111^{17}	3-06-00-00552 0.9270^{20}	ace 4; EtOH 4 1.4947^{20}
2110	Benzene, 1,1'-(phenylmethylene)bis[4-methyl-	$C_{21}H_{20}$ 272.39	603-39-4 56	218^{12}	4-05-00-02520	H_2O 1; EtOH 3; eth 4; bz 4
2111	Benzene, 1-(phenylmethyl)-4-propyl-	$C_{16}H_{18}$ 210.32	62155-41-3	$152\text{-}5^{10}$	4-05-00-01950 0.9739^{18}	1.3552^{20}
2112	Benzene, [(phenylmethyl)sulfonyl]-	$C_{13}H_{12}O_2S$ 232.30	3112-88-7 146		4-06-00-02647 1.1261^{153}	H_2O 1; EtOH 2; eth 2; bz 2
2113	Benzene, [(phenylmethyl)thio]-	$C_{13}H_{12}S$ 200.30	831-91-4 43.5	197^{27}	4-06-00-02644	H_2O 1; EtOH 3; eth 3; con sulf 3
2114	Benzenepropanal Hydrocinnamic aldehyde	$C_9H_{10}O$ 134.18	104-53-0 47	224; 117^{28}	4-07-00-00692 1.0190^{20}	H_2O 1; EtOH 4; eth 5
2115	Benzenepropanal, 2-methyl- Hydrocinnamaldehyde, *o*-methyl-	$C_{10}H_{12}O$ 148.20	19564-40-0	120^{13}	2-07-00-00246 0.998^{19}	1.522^{19}
2116	Benzenepropanal, 4-methyl-	$C_{10}H_{12}O$ 148.20	5406-12-2	223	2-07-00-00247 0.999^{14}	1.525^{14}
2117	Benzenepropanal, α-methyl-4-(1-methylethyl)- 3-*p*-Cumenyl-2-methylpropionaldehyde	$C_{13}H_{18}O$ 190.29	103-95-7	135^{99}; 115^{9}	4-07-00-00788 0.951^{15}	bz 4; eth 4; EtOH 4 1.5068^{20}
2118	Benzenepropanal, α-methyl-β-oxo-	$C_{10}H_{10}O_2$ 162.19	16837-43-7 119.3	155^{125}	3-07-00-03497	bz 4; EtOH 4; MeOH 4
2119	Benzenepropanamine, α-methyl- 1-Methyl-3-phenylpropylamine	$C_{10}H_{15}N$ 149.24	22374-89-6 143	223; 101^{14}	4-12-00-02811 0.9289^{15}	EtOH 4 1.5152^{20}
2120	Benzenepropanaminium, γ-(aminocarbonyl)-*N*-methyl-*N*,*N*-bis(1-methylethyl)-γ-phenyl-, iod Ammonium, (3-carbamoyl-3,3-diphenylpropyl) diisopropyl methyl,iodide	$C_{23}H_{33}IN_2O$ 480.43	71-81-8	5090	4-14-00-01853	
2121	Benzene, 1,1'-(1,3-propanediyl)bis- 1,3-Diphenylpropane	$C_{15}H_{16}$ 196.29	1081-75-0 6	300.3	4-05-00-01913 1.007^{20}	1.5760^{20}
2122	Benzene, 1,1'-[1,3-propanediylbis(oxy)]bis-	$C_{15}H_{16}O_2$ 228.29	726-44-3 62	339	4-06-00-00585	H_2O 1; EtOH 3; eth 3
2123	Benzenepropanenitrile Hydrocinnamonitrile	C_9H_9N 131.18	645-59-0 -1	261; 141^{25}	4-09-00-01764 1.0016^{20}	EtOH 3; eth 3; chl 2 1.5266^{28}
2124	Benzenepropanenitrile, β-oxo- Benzoylacetonitrile	C_9H_7NO 145.16	614-16-4 80.5	160^{10}	4-10-00-02757	H_2O 2; EtOH 3; eth 3; bz 3
2125	Benzenepropanethiol	$C_9H_{12}S$ 152.26	24734-68-7	109^{10}	4-06-00-03206 1.01^{25}	1.5494^{20}
2126	Benzenepropanoic acid Hydrocinnamic acid	$C_9H_{10}O_2$ 150.18	501-52-0 107	4707 279.8	4-09-00-01752 1.0712^{49}	H_2O 3; EtOH 3; eth 3; bz 4
2127	Benzenepropanoic acid, α-acetyl-α-(phenylmethyl)-, ethyl ester	$C_{20}H_{22}O_3$ 310.39	42597-26-2 57	233	3-10-00-03350	EtOH 2; eth 2
2128	Benzenepropanoic acid, β-(aminomethyl)- 4-Amino-3-phenylbutyric acid	$C_{10}H_{13}NO_2$ 179.22	1078-21-3 250-5 dec	476	4-14-00-01723	
2129	Benzenepropanoic acid, α,α-dimethyl-, methyl ester Methyl α,α-dimethylhydrocinnamate	$C_{12}H_{16}O_2$ 192.26	14248-22-7	$121\text{-}2^{23}$	4-09-00-01879 1.0043^{25}	1.4945^{25}
2130	Benzenepropanoic acid, β,β-dimethyl-	$C_{11}H_{14}O_2$ 178.23	1010-48-6 60	167^{10}	4-09-00-01874	H_2O 3; ace 3; peth 3 1.5182^{20}
2131	Benzenepropanoic acid, ethyl ester	$C_{11}H_{14}O_2$ 178.23	2021-28-5	247.2	4-09-00-01753 1.0147^{20}	eth 4; EtOH 4 1.4954^{20}
2132	Benzenepropanoic acid, α-ethyl-β-oxo-, ethyl ester	$C_{13}H_{16}O_3$ 220.27	24346-56-3	152^{7}	3-10-00-03065 1.0706^{15}	eth 4 1.509^{15}
2133	Benzenepropanoic acid, α-hydroxy-, (±)- (±)-3-Phenyllactic acid	$C_9H_{10}O_3$ 166.18	828-01-3 98	149^{15}	2-10-00-00154	H_2O 4; ace 4; eth 4; EtOH 4
2134	Benzenepropanoic acid, 4-hydroxy- *p*-Hydroxyhydrocinnamic acid	$C_9H_{10}O_3$ 166.18	501-97-3 130.8	$208\text{-}10^{14}$	4-10-00-00631	H_2O 3; EtOH 3; eth 3; bz 3
2135	Benzenepropanoic acid, 4-hydroxy-3,5-diiodo-α-phenyl- Iodoalphionic acid	$C_{15}H_{12}I_2O_3$ 494.07	577-91-3 164	4919	4-10-00-01291	H_2O 1; EtOH 3; eth 3; bz 2
2136	Benzenepropanoic acid, 4-(4-hydroxy-3-iodophenoxy)-3,5-diiodo- 3,3',5-Triiodothyropropionic acid	$C_{15}H_{11}I_3O_4$ 635.96	51-26-3	9346	4-10-00-00641	
2137	Benzenepropanoic acid, α-methyl- α-Methyldihydrocinnamic acid	$C_{10}H_{12}O_2$ 164.20	1009-67-2 36.5	272; 152^{12}	2-09-00-00357	eth 4; EtOH 4

No.	Name Synonym	Mol. Form. Mol. Wt.	CAS RN mp/°C	Merck No. bp/°C	Beil. Ref. den/g cm^{-3}	Solubility n_D
2138	Benzenepropanoic acid, β-methyl-, (±)-	$C_{10}H_{12}O_2$ 164.20	772-17-8 46.5	 140-5^{3}	4-09-00-01814 1.0701^{20}	H_2O 2; peth 3 1.5155^{20}
2139	Benzenepropanoic acid, methyl ester Methyl dihydrocinnamate	$C_{10}H_{12}O_2$ 164.20	103-25-3	 238.5; 91^{4}	4-09-00-01753 1.0455^{25}	H_2O 1; EtOH 3; eth 3; bz 3
2140	Benzenepropanoic acid, 1-methylethyl ester	$C_{12}H_{16}O_2$ 192.26	22767-95-9	 126^{11}	2-09-00-00339 0.9860^{25}	eth 4; EtOH 4
2141	Benzenepropanoic acid, β-oxo-, ethyl ester Ethyl benzoylacetate	$C_{11}H_{12}O_3$ 192.21	94-02-0 <0	3726 267 dec; 167^{20}	4-10-00-02756 1.1202^{15}	H_2O 2; EtOH 3; eth 3 1.5317^{15}
2142	Benzenepropanoic acid, β-oxo-, methyl ester	$C_{10}H_{10}O_3$ 178.19	614-27-7	 265 dec; 151^{12}	4-10-00-02755 1.158^{29}	ace 4; eth 4; EtOH 4 1.537^{20}
2143	Benzenepropanoic acid, α-phenyl-, (±)- Propionic acid, 2,3-diphenyl (*DL*)	$C_{15}H_{14}O_2$ 226.27	94942-89-9 88(form a); 95(form b)	 337.5	4-09-00-02528 1.1495^{96}	bz 4; eth 4; EtOH 4
2144	Benzenepropanoic acid, α-(phenylmethyl)-	$C_{16}H_{16}O_2$ 240.30	618-68-8 90	 235^{18}	4-09-00-02546	bz 4; eth 4; EtOH 4
2145	Benzenepropanoic acid, phenylmethyl ester	$C_{16}H_{16}O_2$ 240.30	22767-96-0	 325; 198^{20}	4-09-00-01755 1.090^{15}	eth 4
2146	Benzenepropanoic acid, β-propyl-	$C_{12}H_{16}O_2$ 192.26	5703-52-6	 152^{4}	4-09-00-01900 1.025^{25}	 1.5078
2147	Benzenepropanoic-β-14C acid, α-bromo- Hydrocinnamic acid, α-bromo-	$C_9H_9BrO_2$ 229.07	16503-53-0 52	 150^{13}	4-09-00-01771 1.4790^{25}	ace 4; bz 4; EtOH 4
2148	Benzenepropanol Hydrocinnamyl alcohol	$C_9H_{12}O$ 136.19	122-97-4 <-18	 235	4-06-00-03198 0.995^{25}	H_2O 3; EtOH 5; eth 5; ctc 3 1.5357^{25}
2149	Benzenepropanol, carbamate Phenprobamate	$C_{10}H_{13}NO_2$ 179.22	673-31-4 102	7229		H_2O 1; EtOH 3; chl 3
2150	Benzenepropanol, α,α-dimethyl- Benzyl-*tert*-butanol	$C_{11}H_{16}O$ 164.25	103-05-9 24.5	 121^{13}	4-06-00-03380 0.9626^{21}	H_2O 1; EtOH 4; eth 4; ace 4 1.5077^{21}
2151	Benzenepropanol, β,β-dimethyl-	$C_{11}H_{16}O$ 164.25	13351-61-6 34.5	 125$^{14.5}$	2-06-00-00507	eth 4; EtOH 4
2152	Benzenepropanol, α-ethyl-, (*S*)- 3-Pentanol, 1-phenyl (*d*)	$C_{11}H_{16}O$ 164.25	71747-37-0 38	 143^{19}	3-06-00-01953 0.9687^{20}	EtOH 4
2153	Benzenepropanol, 4-hydroxy-3-methoxy-	$C_{10}H_{14}O_3$ 182.22	2305-13-7 65	 197^{15}	4-06-00-07405	eth 4; EtOH 4 1.5545^{25}
2154	Benzenepropanol, 4-methoxy-	$C_{10}H_{14}O_2$ 166.22	5406-18-8 26	 166^{18}	4-06-00-05982	 1.5315^{20}
2155	Benzenepropanol, α-methyl-	$C_{10}H_{14}O$ 150.22	2344-70-9	 239; 123^{15}	4-06-00-03275 0.9899^{16}	 1.517^{16}
2156	Benzenepropanol, γ-methyl-	$C_{10}H_{14}O$ 150.22	2722-36-3	 139^{33}; 110^{6}	4-06-00-03286 0.9832^{20}	 1.4965^{20}
2157	Benzenepropanol, γ-methylene- 3-Buten-1-ol, 3-phenyl-	$C_{10}H_{12}O$ 148.20	3174-83-2	 123^{10}	4-06-00-03840 1.0272^{20}	 1.5577^{20}
2158	Benzenepropanol, β-phenyl-	$C_{15}H_{16}O$ 212.29	3536-29-6 51	 310; 190^{15}	4-06-00-04754 1.0585^{24}	 1.5742^{19}
2159	Benzenepropanol, γ-propyl- 1-Hexanol, 3-phenyl	$C_{12}H_{18}O$ 178.27	67700-22-5	 127^{5}	3-06-00-02001 0.955^{25}	 1.5101^{25}
2160	Benzenepropanoyl chloride	C_9H_9ClO 168.62	645-45-4	 225 dec; 105^{10}	4-09-00-01762 1.135^{21}	eth 3; CS_2 3
2161	Benzene, 1,1'-(1-propene-1,3-diyl)bis-	$C_{15}H_{14}$ 194.28	5209-18-7 16	 302; 178^{15}	4-05-00-02187 1.0120^{20}	
2162	Benzene, 1-propenyl-	C_9H_{10} 118.18	637-50-3 -27.1	 175.5	3-05-00-01185 0.9019^{25}	ace 4; bz 4; eth 4; EtOH 4 1.5508^{20}
2163	Benzene, 1-propenyl-, (*E*)-	C_9H_{10} 118.18	873-66-5 -29.3	 178.3	4-05-00-01359 0.9023^{25}	H_2O 1; EtOH 5; eth 5; ace 5 1.5506^{20}
2164	Benzene, 1-propenyl-, (*Z*)-	C_9H_{10} 118.18	766-90-5 -61.6	 167.5	4-05-00-01359 0.9088^{20}	H_2O 1; EtOH 5; eth 5; ace 5 1.5420^{20}
2165	Benzene, 2-propenyl-	C_9H_{10} 118.18	300-57-2 -40	 156	4-05-00-01362 0.8920^{20}	H_2O 1; EtOH 3; eth 3; bz 3 1.5131^{20}
2166	Benzene, 1,1'-(1-propenylidene)bis-	$C_{15}H_{14}$ 194.28	778-66-5 52	 280-1 149^{11}	4-05-00-02192 1.0250^{20}	H_2O 1; EtOH 3; bz 3 1.5880^{20}
2167	Benzene, (2-propenyloxy)-	$C_9H_{10}O$ 134.18	1746-13-0	 191.7	4-06-00-00562 0.9811^{20}	H_2O 1; EtOH 3; eth 5; ctc 2 1.5223^{20}
2168	Benzene, propoxy-	$C_9H_{12}O$ 136.19	622-85-5 -27	 189.9	4-06-00-00556 0.9474^{20}	EtOH 3; eth 3 1.5014^{20}
2169	Benzene, propyl- Propylbenzene	C_9H_{12} 120.19	103-65-1 -99.5	7856 159.2	4-05-00-00977 0.8620^{20}	H_2O 1; EtOH 5; eth 5; ace 5 1.4920^{20}
2170	Benzene, (1-propylheptadecyl)- Eicosane, 4-phenyl-	$C_{26}H_{46}$ 358.65	2400-03-5 31.4	 199^{1}	4-05-00-01223 0.8546^{20}	 1.4794^{20}
2171	Benzene, 1,1'-propylidenebis- 1,1-Diphenylpropane	$C_{15}H_{16}$ 196.29	1530-03-6 13.3	 281.6	4-05-00-01920 0.9951^{14}	 1.5681^{14}
2172	Benzene, (propylthio)-	$C_9H_{12}S$ 152.26	874-79-3 -45	 220	4-06-00-01470 0.9995^{20}	 1.5571^{20}
2173	Benzene, 1-propynyl-	C_9H_8 116.16	673-32-5	 183	4-05-00-01530 0.942^{15}	 1.563^{15}
2174	Benzeneseleninic acid Phenylseleninic acid	$C_6H_6O_2Se$ 189.07	6996-92-5 124.5		4-11-00-00708 1.93^{20}	H_2O 2; bz 1; alk 4
2175	Benzene, 1,1'-selenobis-	$C_{12}H_{10}Se$ 233.17	1132-39-4 2.5	 301.5	4-06-00-01779 1.351^{20}	H_2O 1; EtOH 5; eth 5; bz 3 1.5500^{20}

PHYSICAL CONSTANTS OF ORGANIC COMPOUNDS (continued)

No.	Name Synonym	Mol. Form. Mol. Wt.	CAS RN mp/°C	Merck No. bp/°C	Beil. Ref. den/g cm^{-3}	Solubility n_D
2176	Benzene, 1,1'-selenobis[4-methyl-	$C_{14}H_{14}Se$ 261.23	22077-55-0 69.5	 196^{16}	4-06-00-02218	H_2O 1; EtOH 3; eth 2; chl 2
2177	Benzeneselenol	C_6H_6Se 157.07	645-96-5	 183.6; 84^{25}	4-06-00-01777 1.4865^{15}	H_2O 1; EtOH 3; eth 4; ctc 4
2178	Benzenesulfenyl chloride, 2,4-dinitro- 2,4-Dinitrobenzenesulfenyl chloride	$C_6H_3ClN_2O_4S$ 234.62	528-76-7 99	3267	4-06-00-01772	bz 4; chl 4; HOAc 4; peth 2
2179	Benzenesulfenyl chloride, 4-nitro-	$C_6H_4ClNO_2S$ 189.62	937-32-6 52	 $125^{0.1}$	4-06-00-01717	bz 4
2180	Benzenesulfinic acid	$C_6H_6O_2S$ 142.18	618-41-7 84	 dec	4-11-00-00003	H_2O 2; EtOH 3; eth 3; bz 3
2181	Benzenesulfinic acid, 2-methyl- *o*-Toluenesulfinic acid	$C_7H_8O_2S$ 156.21	13165-77-0 80	 dec	4-11-00-00008	H_2O 4; ace 4; eth 4; EtOH 4
2182	Benzenesulfinic acid, 4-methyl- *p*-Toluenesulfinic acid	$C_7H_8O_2S$ 156.21	536-57-2 86.5	9458	4-11-00-00009	H_2O 3; EtOH 4; eth 4; bz 2
2183	Benzene, 1,1'-sulfinylbis-	$C_{12}H_{10}OS$ 202.28	945-51-7 71.2	 340^{16}	4-06-00-01489	EtOH 4; eth 4; bz 4; chl 2
2184	Benzene, 1,1'-[sulfinylbis(methylene)]bis-	$C_{14}H_{14}OS$ 230.33	621-08-9 134	 210 dec	4-06-00-02651	H_2O 1; EtOH 4; eth 3
2185	Benzenesulfinyl chloride	C_6H_5ClOS 160.62	4972-29-6 38	 71-$2^{1.5}$	4-11-00-00005 1.3469^{25}	eth 3; chl 3 1.3470^{25}
2186	Benzenesulfinyl chloride, 4-methyl-	C_7H_7ClOS 174.65	10439-23-3 57	 $113^{3.5}$	4-11-00-00010	chl 4
2187	Benzenesulfonamide, 4-amino- Sulfanilamide	$C_6H_8N_2O_2S$ 172.21	63-74-1 165.5	8898	4-14-00-02658 1.08^{25}	H_2O 3; EtOH 3; eth 3; ace 3
2188	Benzenesulfonamide, 4-amino-*N*-[4-(aminosulfonyl)phenyl]- *N*4-Sulfanilylsulfanilamide	$C_{12}H_{13}N_3O_4S_2$ 327.39	547-52-4 137	8905	4-14-00-02760	H_2O 2; EtOH 3; eth 3; ace 3
2189	Benzenesulfonamide, 4-amino-*N*-(aminothioxomethyl)- Sulfathiourea	$C_7H_9N_3O_2S_2$ 231.30	515-49-1 182	8921	4-14-00-02671	H_2O 1; EtOH 2
2190	Benzenesulfonamide, 4-amino-*N*-(aminocarbonyl)- Sulfanilylurea	$C_7H_9N_3O_3S$ 215.23	547-44-4 147 dec	8906	4-14-00-02667	
2191	Benzenesulfonamide, 4-amino-*N*-(aminoiminomethyl)- Sulfaguanidine	$C_7H_{10}N_4O_2S$ 214.25	57-67-0 191.5	8879	4-14-00-02668	
2192	Benzenesulfonamide, 4-amino-*N*-[(butylamino)carbonyl]- Carbutamide	$C_{11}H_{17}N_3O_3S$ 271.34	339-43-5 144.5	1839	4-14-00-02667	
2193	Benzenesulfonamide, 4-amino-*N*-(6-chloro-3-pyridazinyl)- Sulfachlorpyridazine	$C_{10}H_9ClN_4O_2S$ 284.73	80-32-0 187	8871	4-25-00-02069	
2194	Benzenesulfonamide, 4-amino-*N*-(2,6-dimethoxy-4-pyrimidinyl)- Sulfadimethoxine	$C_{12}H_{14}N_4O_4S$ 310.33	122-11-2 203.5	8876	4-25-00-03483	
2195	Benzenesulfonamide, 4-amino-*N*-(3,4-dimethyl-5-isoxazolyl)- Sulfisoxazole	$C_{11}H_{13}N_3O_3S$ 267.31	127-69-5 191	8930	4-27-00-04747	
2196	Benzenesulfonamide, 4-amino-*N*-(4,6-dimethyl-2-pyrimidinyl)- Sulfamethazine	$C_{12}H_{14}N_4O_2S$ 278.33	57-68-1 198.5	8886	4-25-00-02215	H_2O 3; acid 3; alk 3
2197	Benzenesulfonamide, 4-amino-*N*-(6-methoxy-3-pyridazinyl)- Sulfamethoxypyridazine	$C_{11}H_{12}N_4O_3S$ 280.31	80-35-3 182.5	8890	4-25-00-03332	
2198	Benzenesulfonamide, 4-amino-*N*-(5-methyl-3-isoxazolyl)- Sulfamethoxazole	$C_{10}H_{11}N_3O_3S$ 253.28	723-46-6	8889	4-27-00-04685	
2199	Benzenesulfonamide, 4-amino-*N*-(4-methyl-2-pyrimidinyl)- Sulfamerazine	$C_{11}H_{12}N_4O_2S$ 264.31	127-79-7 236	8884	4-25-00-02159	H_2O 2; EtOH 2; eth 1; ace 2
2200	Benzenesulfonamide, 4-amino-*N*-(4-methyl-2-thiazolyl)- Sulfamethylthiazole	$C_{10}H_{11}N_3O_2S_2$ 269.35	515-59-3 237	8891	4-27-00-04715	EtOH 4
2201	Benzenesulfonamide, 4-amino-*N*-(1-phenyl-1H-pyrazol-5-yl)- Sulfaphenazole	$C_{15}H_{14}N_4O_2S$ 314.37	526-08-9	8910	4-25-00-02029	
2202	Benzenesulfonamide, 4-amino-*N*-pyrazinyl- Sulfapyrazine	$C_{10}H_{10}N_4O_2S$ 250.28	116-44-9 251	8912	4-25-00-02143	H_2O 1; EtOH 1; eth 1; ace 2
2203	Benzenesulfonamide, 4-amino-*N*-2-pyridinyl- Sulfapyridine	$C_{11}H_{11}N_3O_2S$ 249.29	144-83-2 192	8913	5-22-08-00424	H_2O 1; EtOH 3; bz 1; ctc 1
2204	Benzenesulfonamide, 4-amino-*N*-2-pyrimidinyl- Sulfadiazine	$C_{10}H_{10}N_4O_2S$ 250.28	68-35-9 255-6 dec	8874	4-25-00-02097	H_2O 2; EtOH 2; ace 2
2205	Benzenesulfonamide, 4-amino-*N*-2-quinoxalinyl- Sulfaquinoxaline	$C_{14}H_{12}N_4O_2S$ 300.34	59-40-5 247.5	8914	4-25-00-02606	H_2O 2; EtOH 2; ace 2; aq alk 3
2206	Benzenesulfonamide, 4-amino-*N*-2-thiazolyl- Sulfathiazole	$C_9H_9N_3O_2S_2$ 255.32	72-14-0 175(form a); 202(form b)	8920	4-27-00-04623	H_2O 2; EtOH 2; DMSO-d_6 2

PHYSICAL CONSTANTS OF ORGANIC COMPOUNDS (continued)

No.	Name Synonym	Mol. Form. Mol. Wt.	CAS RN mp/°C	Merck No. bp/°C	Beil. Ref. den/g cm^{-3}	Solubility n_D
2207	Benzenesulfonamide, *N*-[(butylamino)carbonyl]-4-methyl- Tolbutamide	$C_{12}H_{18}N_2O_3S$ 270.35	64-77-7 128.5	9432	4-11-00-00396 1.245^{25}	H_2O 2; EtOH 3; eth 3; chl 3
2208	Benzenesulfonamide, 4-chloro-*N*-[(propylamino)carbonyl]- Chloropropamide	$C_{10}H_{13}ClN_2O_3S$ 276.74	94-20-2	2187	4-11-00-00119	
2209	Benzenesulfonamide, 4-[(2,4-diaminophenyl)azo]- Prontosil	$C_{12}H_{14}ClN_5O_2S$ 327.79	103-12-8 249.5	8895	4-16-00-00563	H_2O 2; EtOH 3; ace 3; oils 3
2210	Benzenesulfonamide, *N*,*N*-dichloro-4-methyl- Dichloramine-T	$C_7H_7Cl_2NO_2S$ 240.11	473-34-7 83	3034	3-11-00-00301	H_2O 1; EtOH 3; eth 3; bz 3
2211	Benzenesulfonamide, *N*,4-dimethyl-	$C_8H_{11}NO_2S$ 185.25	640-61-9 78.5		4-11-00-00377 1.340^{25}	eth 4; EtOH 4
2212	Benzenesulfonamide, *N*,4-dimethyl-*N*-nitroso- *p*-Tolylsulfonylmethylnitrosamide	$C_8H_{10}N_2O_3S$ 214.25	80-11-5 60	9469	4-11-00-00478	H_2O 1; EtOH 4; eth 4
2213	Benzenesulfonamide, 4-[(phenylmethyl)amino]- *N*4-Benzylsulfanilamide	$C_{13}H_{14}N_2O_2S$ 262.33	104-22-3 171	1161	3-14-00-02026	
2214	Benzenesulfonic acid Besylic acid	$C_6H_6O_3S$ 158.18	98-11-3 65	1078	4-11-00-00027	H_2O 4; EtOH 4; eth 1; bz 2
2215	Benzenesulfonic acid, 2-amino- Orthanilic acid	$C_6H_7NO_3S$ 173.19	88-21-1 >320 dec	6835	4-14-00-02638	H_2O 2; EtOH 1; eth 1
2216	Benzenesulfonic acid, 3-amino- Metanilic acid	$C_6H_7NO_3S$ 173.19	121-47-1 dec	5832	4-14-00-02647	H_2O 2; EtOH 2; eth 1
2217	Benzenesulfonic acid, 4-amino- Sulfanilic acid	$C_6H_7NO_3S$ 173.19	121-57-3 288	8901	4-14-00-02655 1.485^{25}	H_2O 2; EtOH 1; eth 1
2218	Benzenesulfonic acid, 2-amino-5-[(4-amino-3-sulfophenyl)(4-imino-3-sulfo-2,5-cyclohexadien-1-ylidene)methyl]-3 Fuchsin, acid	$C_{20}H_{17}N_3Na_2O_9S_3$ 585.55	3244-88-0	103		H_2O 2; EtOH 2
2219	Benzenesulfonic acid, *sec*-butyl ester Benzenesulfonic acid, chloride	$C_{10}H_{14}O_3S$ 214.29	103275-78-1 14.5	251 dec; 120^{10}	4-11-00-00032 1.3830^{15}	H_2O 1; EtOH 4; eth 3; ctc 3
2220	Benzenesulfonic acid, 4-chloro- *p*-Chlorobenzenesulfonic acid	$C_6H_5ClO_3S$ 192.62	98-66-8 67	2122 $147\text{-}8^{25}$	4-11-00-00107	H_2O 3; EtOH 3; eth 1; bz 1
2221	Benzenesulfonic acid, 4-chloro-, 4-chlorophenyl ester Ovex	$C_{12}H_8Cl_2O_3S$ 303.17	80-33-1 86.5	6856	4-11-00-00109	H_2O 1; EtOH 2; ace 3
2222	Benzenesulfonic acid, 4-[[4-(dimethylamino)phenyl]azo]-, sodium salt Methyl orange	$C_{14}H_{14}N_3NaO_3S$ 327.34	547-58-0 dec	6019		H_2O 2; EtOH 2; eth 1; py 2
2223	Benzenesulfonic acid, ethyl ester Ethyl benzenesulfonate	$C_8H_{10}O_3S$ 186.23	515-46-8	1078 156^{15}	4-11-00-00030 1.2167^{20}	H_2O 2; EtOH 3; eth 4; chl 4 1.5081^{20}
2224	Benzenesulfonic acid, 4-hydrazino- *p*-Sulfophenylhydrazine	$C_6H_8N_2O_3S$ 188.21	98-71-5 286	4695	3-15-00-00865	H_2O 2; EtOH 2
2225	Benzenesulfonic acid, 2-hydroxy- *o*-Phenolsulfonic acid	$C_6H_6O_4S$ 174.18	609-46-1 145 dec		4-11-00-00574	H_2O 4; EtOH 4
2226	Benzenesulfonic acid, 4-hydroxy- *p*-Phenolsulfonic acid	$C_6H_6O_4S$ 174.18	98-67-9	7212	4-11-00-00582	H_2O 4; EtOH 4
2227	Benzenesulfonic acid, methyl ester Methyl benzenesulfonate	$C_7H_8O_3S$ 172.20	80-18-2	154^{20}	4-11-00-00029 1.2730^{17}	H_2O 2; EtOH 4; eth 4; chl 4 1.5151^{20}
2228	Benzenesulfonic acid, 2-methyl-	$C_7H_8O_3S$ 172.20	88-20-0 67.5	128.8^{25}	4-11-00-00228	H_2O 4; EtOH 3; eth 1
2229	Benzenesulfonic acid, 4-methyl- *p*-Toluenesulfonic acid	$C_7H_8O_3S$ 172.20	104-15-4 104.5	9459 140^{20}	4-11-00-00241	H_2O 4; EtOH 3; eth 3
2230	Benzenesulfonic acid, 4-methyl-, butyl ester	$C_{11}H_{16}O_3S$ 228.31	778-28-9	$164\text{-}6^{6}$	4-11-00-00250 1.1319^{20}	H_2O 1; eth 3; ctc 2 1.5050^{20}
2231	Benzenesulfonic acid, 4-methyl-, ethyl ester Ethyl *p*-toluenesulfonate	$C_9H_{12}O_3S$ 200.26	80-40-0 34.5	3814 173^{15}	4-11-00-00248 1.166^{48}	H_2O 1; EtOH 3; eth 3; ctc 2
2232	Benzenesulfonic acid, 4-methyl-, methyl ester	$C_8H_{10}O_3S$ 186.23	80-48-8 28.5	292; 186^{22}	4-11-00-00247 1.2087^{40}	H_2O 1; EtOH 4; eth 3; bz 4
2233	Benzenesulfonic acid, 4-methyl-, propyl ester	$C_{10}H_{14}O_3S$ 214.29	599-91-7 <-20	189^{9}	4-11-00-00249 1.144^{20}	1.4998^{20}
2234	Benzenesulfonic acid, 4-(phenylamino)- *N*-Phenylsulfanilic acid	$C_{12}H_{11}NO_3S$ 249.29	101-57-5 206	7284	3-14-00-02025	H_2O 4; EtOH 4
2235	Benzenesulfonic acid, propyl ester Propyl benzenesulfonate	$C_9H_{12}O_3S$ 200.26	80-42-2	$162\text{-}3^{15}$	4-11-00-00031 1.1804^{17}	H_2O 2; EtOH 3; eth 4; chl 4 1.5035^{25}
2236	Benzenesulfonothioic acid, 4-chloro-, *S*-phenyl ester Benzenesulfonic acid, *p*-chlorothio-, *S*-phenyl ester	$C_{12}H_9ClO_2S_2$ 284.79	1142-97-8 -86.8	118.5	4-11-00-00223 0.8341^{20}	1.4087^{14}
2237	Benzene, 1,1'-sulfonylbis- Diphenyl sulfone	$C_{12}H_{10}O_2S$ 218.28	127-63-9 128.5	3336 379	4-06-00-01490 1.252^{20}	H_2O 1; EtOH 3; eth 3; bz 3
2238	Benzene, 1,1'-sulfonylbis[4-bromo-	$C_{12}H_8Br_2O_2S$ 376.07	2050-48-8 172		4-06-00-01652 1.88^{25}	H_2O 1; EtOH 2; HOAc 3
2239	Benzene, 1,1'-sulfonylbis[4-chloro-	$C_{12}H_8Cl_2O_2S$ 287.17	80-07-9 147.9	250^{10}	4-06-00-01587	H_2O 2; EtOH 3; chl 3
2240	Benzene, 1,1'-sulfonylbis[4-methyl-	$C_{14}H_{14}O_2S$ 246.33	599-66-6 159	406	4-06-00-02174	H_2O 2; EtOH 3; eth 2; bz 3
2241	Benzene, 1,1'-[sulfonylbis(methylene)]bis-	$C_{14}H_{14}O_2S$ 246.33	620-32-6 152	290 dec	4-06-00-02651	H_2O 1; EtOH 2; ace 4; bz 3

PHYSICAL CONSTANTS OF ORGANIC COMPOUNDS (continued)

No.	Name Synonym	Mol. Form. Mol. Wt.	CAS RN mp/°C	Merck No. bp/°C	Beil. Ref. den/g cm^{-3}	Solubility n_D
2242	Benzenesulfonyl chloride Phenylsulfonyl chloride	$C_6H_5ClO_2S$ 176.62	98-09-9 14.5	1080 251 dec	4-11-00-00049 1.3470^{15}	H_2O 1; EtOH 4; eth 3; ctc 3
2243	Benzenesulfonyl chloride, 4-(acetylamino)- Acetylsulfanilyl chloride	$C_8H_8ClNO_3S$ 233.68	121-60-8 149	99	4-14-00-02703	EtOH 4; eth 4; bz 3; chl 3
2244	Benzenesulfonyl chloride, 4-bromo- *p*-Brosyl chloride	$C_6H_4BrClO_2S$ 255.52	98-58-8 76	1395 153^{15}	4-11-00-00162	H_2O 1; eth 4; chl 3
2245	Benzenesulfonyl chloride, 4-chloro-	$C_6H_4Cl_2O_2S$ 211.07	98-60-2 51	141^{15}	4-11-00-00114	eth 4; bz 4
2246	Benzenesulfonyl chloride, 4-fluoro-	$C_6H_4ClFO_2S$ 194.61	349-88-2 30	106^{9}	4-11-00-00102	bz 4; eth 4; chl 4
2247	Benzenesulfonyl chloride, 4-iodo- Pipsyl chloride	$C_6H_4ClIO_2S$ 302.52	98-61-3 85	7458	4-11-00-00169	
2248	Benzenesulfonyl chloride, 2-methoxy-	$C_7H_7ClO_3S$ 206.65	10130-87-7 56	$126^{0.3}$	4-11-00-00575	eth 3; peth 3
2249	Benzenesulfonyl chloride, 2-methyl- *o*-Toluenesulfonyl chloride	$C_7H_7ClO_2S$ 190.65	133-59-5 10.2	154^{36}	4-11-00-00229 1.3383^{20}	H_2O 1; EtOH 3; eth 3; bz 3 1.5565^{20}
2250	Benzenesulfonyl chloride, 4-methyl- *p*-Toluenesulfonyl chloride	$C_7H_7ClO_2S$ 190.65	98-59-9 71	9460 $145\text{-}6^{15}$	4-11-00-00375	H_2O 1; EtOH 3; eth 3; bz 4
2251	Benzenesulfonyl fluoride Phenylsulfonyl fluoride	$C_6H_5FO_2S$ 160.17	368-43-4	203.5	4-11-00-00049 1.3286^{20}	EtOH 3; eth 3 1.4932^{18}
2252	Benzenesulfonyl fluoride, 4-amino- *p*-Sulfanilyl fluoride	$C_6H_6FNO_2S$ 175.18	98-62-4 68.5	8904	4-14-00-02658	
2253	Benzene, 1,2,3,5-tetrabromo- 1,2,3,5-Tetrabromobenzene	$C_6H_2Br_4$ 393.70	634-89-9 99.5	329	4-05-00-00686	bz 4; eth 4; EtOH 4
2254	Benzene, 1,2,4,5-tetrabromo- 2,3,5,6-Tetrabromobenzene	$C_6H_2Br_4$ 393.70	636-28-2 182	3.1^{20}	4-05-00-00687 3.072^{20}	eth 4
2255	Benzene, 1,2,3,4-tetrabromo-5,6-dimethyl-	$C_8H_6Br_4$ 421.75	2810-69-7 262	374.5	3-05-00-00820	bz 4
2256	Benzene, 1,2,3,5-tetrabromo-4-methoxy- 2,3,4,6-Tetrabromoanisole	$C_7H_4Br_4O$ 423.72	95970-07-3 113.5	340	3-06-00-00766	EtOH 4
2257	1,2,4,5-Benzenetetracarboxylic acid Pyromellitic acid	$C_{10}H_6O_8$ 254.15	89-05-4 276	8015	4-09-00-03804	H_2O 2; EtOH 3
2258	1,2,4,5-Benzenetetracarboxylic acid, tetramethyl ester	$C_{14}H_{14}O_8$ 310.26	635-10-9 144	sub	4-09-00-03804	EtOH 4
2259	Benzene, 1,2,3,4-tetrachloro- 1,2,3,4-Tetrachlorobenzene	$C_6H_2Cl_4$ 215.89	634-66-2 47.5	254	4-05-00-00667	H_2O 1; EtOH 2; eth 4; CS_2 4
2260	Benzene, 1,2,4,5-tetrachloro- s-Tetrachlorobenzene	$C_6H_2Cl_4$ 215.89	95-94-3 139.5	244.5	4-05-00-00668 1.858^{22}	H_2O 1; EtOH 2; eth 3; bz 3
2261	Benzene, 1,2,3,5-tetrachloro-4,6-dimethyl-	$C_8H_6Cl_4$ 243.95	877-09-8 223		4-05-00-00944 1.703^{25}	H_2O 1; EtOH 1; eth 3; bz 3
2262	Benzene, 1,2,3,5-tetrachloro-4-methyl- 2,3,4,6-Tetrachlorotoluene	$C_7H_4Cl_4$ 229.92	875-40-1 94	276.5	3-05-00-00702	bz 4; EtOH 4
2263	Benzene, 1,2,4,5-tetrachloro-3-methyl- 2,3,5,6-Tetrachlorotoluene	$C_7H_4Cl_4$ 229.92	1006-31-1 93.5	sub	3-05-00-00702	eth 4; EtOH 4
2264	Benzene, 1,2,4,5-tetrachloro-3-nitro-	$C_6HCl_4NO_2$ 260.89	117-18-0 99.5	304	4-05-00-00728 1.744^{25}	H_2O 1; EtOH 3; bz 3; chl 3
2265	Benzene, tetracosyl- Tetracosylbenzene	$C_{30}H_{54}$ 414.76	61828-05-5 57	373^{100}; 238^{1}	4-05-00-01239 0.8544^{20}	1.4800^{20}
2266	Benzene, tetradecyl-	$C_{20}H_{34}$ 274.49	1459-10-5 16	359	4-05-00-01212 0.8549^{20}	1.4818^{20}
2267	Benzene, 1,1'-tetradecylidenebis-	$C_{26}H_{38}$ 350.59	55268-63-8 17.9	207^{1}	4-05-00-02017 0.9187^{20}	1.5202^{20}
2268	Benzene, 1,2,3,4-tetraethyl- 1,2,3,4-Tetraethylbenzene	$C_{14}H_{22}$ 190.33	642-32-0 11.8	252; 121.7^{14}	4-05-00-01168 0.8875^{20}	eth 4; EtOH 4 1.5125^{20}
2269	Benzene, 1,2,4,5-tetraethyl- 1,2,4,5-Tetraethylbenzene	$C_{14}H_{22}$ 190.33	635-81-4 10	250	4-05-00-01168 0.8788^{20}	eth 4; EtOH 4 1.5054^{20}
2270	Benzene, 1,2,3,4-tetrafluoro- 1,2,3,4-Tetrafluorobenzene	$C_6H_2F_4$ 150.08	551-62-2	94.3	4-05-00-00639	1.4054^{20}
2271	Benzene, 1,2,3,5-tetrafluoro- 1,2,3,5-Tetrafluorobenzene	$C_6H_2F_4$ 150.08	2367-82-0 -48	84.4	4-05-00-00639 1.319^{25}	1.4035^{20}
2272	Benzene, 1,2,4,5-tetrafluoro- 1,2,4,5-Tetrafluorobenzene	$C_6H_2F_4$ 150.08	327-54-8 4.5	90.2	4-05-00-00639 1.4255^{20}	1.4075^{20}
2273	Benzene, 1,2,3,4-tetraiodo- 1,2,3,4-Tetraiodobenzene	$C_6H_2I_4$ 581.70	034-00-4 136	sub	0-05-00-00229	eth 4; EtOH 4
2274	Benzene, 1,2,3,5-tetraiodo- 1,2,3,5-Tetraiodobenzene	$C_6H_2I_4$ 581.70	634-92-4 148	sub	0-05-00-00229	HOAc 4
2275	Benzene, 1,2,4,5-tetraiodo- 1,2,4,5-Tetraiodobenzene	$C_6H_2I_4$ 581.70	636-31-7 254	sub	0-05-00-00229	HOAc 4
2276	Benzene, 1,2,4,5-tetrakis(1-methylethyl)- 1,2,4,5-Tetraisopropylbenzene	$C_{18}H_{30}$ 246.44	635-11-0 118.4	259; 133^{17}	4-05-00-01207 0.758^{150}	
2277	Benzene, 1,2,3,4-tetramethoxy-5-(2-propenyl)-	$C_{13}H_{18}O_4$ 238.28	15361-99-6 25	145^{12}	3-06-00-06691 1.087^{25}	1.5146^{25}
2278	Benzene, 1,2,3,4-tetramethyl-	$C_{10}H_{14}$ 134.22	488-23-3 -6.2	205	4-05-00-01072 0.9052^{20}	H_2O 1; EtOH 5; eth 5; ace 5 1.5203^{20}
2279	Benzene, 1,2,3,5-tetramethyl- Isodurene	$C_{10}H_{14}$ 134.22	527-53-7 -23.7	5050 198	4-05-00-01073 0.8903^{20}	H_2O 1; EtOH 5; eth 5; ace 5 1.5130^{20}
2280	Benzene, 1,2,4,5-tetramethyl- Durene	$C_{10}H_{14}$ 134.22	95-93-2 79.3	3450 196.8	4-05-00-01076 0.8380^{81}	H_2O 1; EtOH 5; eth 5; ace 5 1.4790^{81}

No.	Name Synonym	Mol. Form. Mol. Wt.	CAS RN mp/°C	Merck No. bp/°C	Beil. Ref. den/g cm^{-3}	Solubility n_D
2281	Benzene, 1,1'-(1,1,4,4-tetramethyl-1,4-butanediyl)bis- Bineophyl	$C_{20}H_{26}$ 266.43	17648-05-4 61.8	 $146\text{-}50^{0.5}$	4-05-00-01997	chl 3
2282	Benzene, 1,2,4,5-tetramethyl-3,6-dinitro-	$C_{10}H_{12}N_2O_4$ 224.22	5465-13-4 211.5	 sub	4-05-00-01080	H_2O 1; EtOH 2; eth 4; bz 3
2283	Benzene, 1,1'-thiobis- Phenyl sulfide	$C_{12}H_{10}S$ 186.28	139-66-2 -25.9	7285 296	4-06-00-01488 1.1136^{20}	H_2O 1; EtOH 3; eth 5; bz 5 1.6334^{20}
2284	Benzene, 1,1'-thiobis[4-bromo-	$C_{12}H_8Br_2S$ 344.07	3393-78-0 115	 268.5^{40}	4-06-00-01651 1.84^{25}	H_2O 1; EtOH 2; eth 3; ctc 3
2285	Benzene, 1,1'-thiobis[2-methyl-	$C_{14}H_{14}S$ 214.33	4537-05-7 64	 285	4-06-00-02019	eth 4; EtOH 4; chl 4
2286	Benzene, 1,1'-thiobis[4-methyl-	$C_{14}H_{14}S$ 214.33	620-94-0 57.3	 >300; 175^{16}	4-06-00-02173	H_2O 1; EtOH 3; eth 4; ace 3
2287	Benzene, 1,1'-[thiobis(methylene)]bis- Benzyl sulfide	$C_{14}H_{14}S$ 214.33	538-74-9 49.5	1162 dec	4-06-00-02649 1.0583^{50}	H_2O 1; EtOH 3; eth 3; CS_2 3
2288	Benzene, 1,1'-thiobis[2,4,6-trinitro- Picryl sulfide	$C_{12}H_4N_6O_{12}S$ 456.26	2217-06-3		4-06-00-01775	
2289	Benzenethiol Phenyl mercaptan	C_6H_6S 110.18	108-98-5 -14.9	9285 169.1	4-06-00-01463 1.0775^{20}	H_2O 1; EtOH 3; eth 3; bz 3 1.5893^{20}
2290	Benzenethiol, 2-amino-	C_6H_7NS 125.19	137-07-5 26	 234	4-13-00-00910	EtOH 3; eth 3 1.4606^{20}
2291	Benzenethiol, 4-amino-	C_6H_7NS 125.19	1193-02-8 46	 143^{17}	4-13-00-01289	H_2O 3; EtOH 3
2292	Benzenethiol, 4-bromo-	C_6H_5BrS 189.08	106-53-6 73	 230.5	4-06-00-01650 1.5260^{83}	H_2O 2; EtOH 2; eth 4; ctc 4
2293	Benzenethiol, 2-chloro-	C_6H_5ClS 144.62	6320-03-2	 205.5	4-06-00-01570 1.2752^{10}	H_2O 2; EtOH 2
2294	Benzenethiol, 3-chloro-	C_6H_5ClS 144.62	2037-31-2	 206	4-06-00-01576 1.2637^{13}	H_2O 1; EtOH 3; eth 3; chl 3
2295	Benzenethiol, 4-chloro-	C_6H_5ClS 144.62	106-54-7 61	 206	4-06-00-01581 1.1911^{20}	H_2O 1; EtOH 4; eth 4; bz 4 1.5480^{20}
2296	Benzenethiol, 4-ethoxy-	$C_8H_{10}OS$ 154.23	699-09-2 1.6	 238	4-06-00-05792 1.1070^{17}	ace 4; bz 4; eth 4; EtOH 4
2297	Benzenethiol, 3-ethyl- 3-Ethylthiophenol	$C_8H_{10}S$ 138.23	62154-77-2	 211	3-06-00-01663 1.038^{20}	 1.572^{30}
2298	Benzenethiol, 4-ethyl-	$C_8H_{10}S$ 138.23	4946-13-8	 211	4-06-00-03027 1.038^{20}	 1.572^{20}
2299	Benzenethiol, 3-methoxy-	C_7H_8OS 140.21	15570-12-4	 224.5; 114^{20}	4-06-00-05701	chl 3 1.5874^{20}
2300	Benzenethiol, 4-methoxy-	C_7H_8OS 140.21	696-63-9	 228	4-06-00-05790 1.1313^{25}	EtOH 3; eth 3; bz 3; chl 2 1.5801^{25}
2301	Benzenethiol, 2-methyl-	C_7H_8S 124.21	137-06-4 15	 195	4-06-00-02014 1.041^{20}	H_2O 1; EtOH 3; eth 4 1.570^{20}
2302	Benzenethiol, 3-methyl-	C_7H_8S 124.21	108-40-7 -20	 195	4-06-00-02079 1.044^{20}	H_2O 1; EtOH 3; eth 5 1.572^{20}
2303	Benzenethiol, 4-methyl-	C_7H_8S 124.21	106-45-6 43	 195	4-06-00-02153 1.0220^{51}	H_2O 1; EtOH 3; eth 4; chl 3
2304	Benzenethiol, pentafluoro-	C_6HF_5S 200.13	771-62-0 -24	 143	 1.501^{25}	 1.4645^{20}
2305	1,2,3-Benzenetriamine 1,2,3-Triaminobenzene	$C_6H_9N_3$ 123.16	608-32-2 103	 336	3-13-00-00551	H_2O 4; eth 4; EtOH 4
2306	1,2,4-Benzenetriamine	$C_6H_9N_3$ 123.16	615-71-4 97.3	 340	4-13-00-00506	H_2O 4; EtOH 4; chl 4
2307	Benzene, 1,2,3-tribromo- 1,2,3-Tribromobenzene	$C_6H_3Br_3$ 314.80	608-21-9 87.5		3-05-00-00569 2.658^{25}	eth 4
2308	Benzene, 1,2,4-tribromo- 1,2,4-Tribromobenzene	$C_6H_3Br_3$ 314.80	615-54-3 44.5	 275	4-05-00-00685	H_2O 1; EtOH 3; eth 4; ace 4
2309	Benzene, 1,3,5-tribromo- 1,3,5-Tribromobenzene	$C_6H_3Br_3$ 314.80	626-39-1 122.8	 271	4-05-00-00685	H_2O 1; EtOH 2; eth 3; bz 3
2310	Benzene, 1,2,3-tribromo-5-methoxy- Anisole, 3,4,5-tribromo	$C_7H_5Br_3O$ 344.83	73557-60-5 93.3	 303	2-06-00-00195	ace 4; bz 4; EtOH 4
2311	Benzene, 1,2,4-tribromo-5-methoxy- 2,4,5-Tribromoanisole	$C_7H_5Br_3O$ 344.83	95970-10-8 105	 307	4-06-00-01067	ace 4; bz 4; EtOH 4
2312	Benzene, 1,2,5-tribromo-3-methoxy- Anisole, 2,3,5-tribromo	$C_7H_5Br_3O$ 344.83	73931-44-9 82	 308.5	2-06-00-00192	ace 4; bz 4; EtOH 4
2313	Benzene, 1,3,5-tribromo-2-methoxy-	$C_7H_5Br_3O$ 344.83	607-99-8 88	 298	4-06-00-01067 2.491^{25}	H_2O 2; EtOH 2; ace 4; bz 4
2314	Benzene, 1,3,5-tribromo-2-methyl- 2,4,6-Tribromotoluene	$C_7H_5Br_3$ 328.83	6320-40-7 70	 290	3-05-00-00719 2.479^{17}	
2315	Benzene, 1,2,3-tribromo-5-nitro- 3,4,5-Tribromonitrobenzene	$C_6H_2Br_3NO_2$ 359.80	3460-20-6 112	 sub	4-05-00-00732 2.645^{25}	
2316	Benzene, 1,3,5-tribromo-2-nitro-	$C_6H_2Br_3NO_2$ 359.80	3463-40-9 125	 177^{11}	4-05-00-00732	eth 4; chl 4
2317	1,3,5-Benzenetricarbonyl trichloride	$C_9H_3Cl_3O_3$ 265.48	4422-95-1 36.3	 180^{16}	4-09-00-03748	chl 3
2318	1,2,3-Benzenetricarboxylic acid Hemimellitic acid	$C_9H_6O_6$ 210.14	569-51-7 200		4-09-00-03745 1.546^{20}	eth 4; EtOH 4
2319	1,2,4-Benzenetricarboxylic acid Trimellitic acid	$C_9H_6O_6$ 210.14	528-44-9 219	9616	4-09-00-03746	H_2O 4; eth 4; EtOH 4
2320	1,3,5-Benzenetricarboxylic acid, 2-chloro- 2-Chloro-1,3,5-benzenetricarboxylic acid	$C_9H_5ClO_6$ 244.59	56961-24-1 285	 sub	4-09-00-03749	H_2O 4; eth 4; EtOH 4

No.	Name Synonym	Mol. Form. Mol. Wt.	CAS RN mp/°C	Merck No. bp/°C	Beil. Ref. den/g cm^{-3}	Solubility n_D
2321	Benzene, 1,2,3-trichloro- 1,2,3-Trichlorobenzene	$C_6H_3Cl_3$ 181.45	87-61-6 53.5	9542 218.5	4-05-00-00664 1.4533^{25}	H_2O 1; EtOH 2; eth 4; bz 4
2322	Benzene, 1,2,4-trichloro- 1,2,4-Trichlorobenzene	$C_6H_3Cl_3$ 181.45	120-82-1 17	9543 213.5	4-05-00-00664 1.459^{25}	H_2O 1; EtOH 2; eth 4; ctc 2 1.5717^{20}
2323	Benzene, 1,3,5-trichloro- 1,3,5-Trichlorobenzene	$C_6H_3Cl_3$ 181.45	108-70-3 63.5	9544 208	4-05-00-00666	H_2O 1; EtOH 2; eth 4; bz 4
2324	Benzene, 1,2,4-trichloro-5-(chloromethyl)-	$C_7H_4Cl_4$ 229.92	3955-26-8	273	4-05-00-00822 1.547^{20}	ace 4; eth 4; EtOH 4
2325	Benzene, 1,2,4-trichloro-5-[(4-chlorophenyl)sulfonyl]- Tetradifon	$C_{12}H_6Cl_4O_2S$ 356.06	116-29-0 146	9132	4-06-00-01636 1.151^{20}	
2326	Benzene, 1,2,3-trichloro-4-(dichloromethyl)- Benzylidene chloride, 2,3,4-trichloro	$C_7H_3Cl_5$ 264.36	56961-82-1 84	278	1-05-00-00153	bz 4
2327	Benzene, 1,2,4-trichloro-5-(dichloromethyl)- Benzylidene chloride, 2,4,5-trichloro	$C_7H_3Cl_5$ 264.36	33429-70-8	280.5	4-05-00-00824 1.5956^{20}	bz 4 1.5992^{20}
2328	Benzene, 1,1'-(2,2,2-trichloroethylidene)bis[4-chloro- DDT	$C_{14}H_9Cl_5$ 354.49	50-29-3 108.5	2832 260; $186^{0.05}$	4-05-00-01885	H_2O 1; EtOH 2; eth 4; ace 4
2329	Benzene, 1,1'-(2,2,2-trichloroethylidene)bis[4-methoxy- Methoxychlor	$C_{16}H_{15}Cl_3O_2$ 345.65	72-43-5 87	5913	4-06-00-06691 1.41^{25}	H_2O 1; EtOH 3; eth 4; bz 4
2330	Benzene, 1,2,4-trichloro-3-methoxy-	$C_7H_5Cl_3O$ 211.47	50375-10-5 45	227	3-06-00-00717	ace 4; EtOH 4
2331	Benzene, 1,2,4-trichloro-5-methoxy-	$C_7H_5Cl_3O$ 211.47	6130-75-2 77.5	254	4-06-00-00963	EtOH 4; ace 4; os 3
2332	Benzene, 1,3,5-trichloro-2-methoxy- 2,4,6-Trichloroanisole	$C_7H_5Cl_3O$ 211.47	87-40-1 61.5	9541 241	4-06-00-01005 1.640^{25}	EtOH 3; ace 4; bz 3; chl 3
2333	Benzene, (trichloromethyl)- (Trichloromethyl)benzene	$C_7H_5Cl_3$ 195.47	98-07-7 -5	1120 221	4-05-00-00820 1.3723^{20}	H_2O 1; EtOH 3; eth 3; bz 3 1.5580^{20}
2334	Benzene, 1,2,3-trichloro-4-methyl- 2,3,4-Trichlorotoluene	$C_7H_5Cl_3$ 195.47	7359-72-0 43.5	244	4-05-00-00819	H_2O 1; EtOH 3; eth 3; ace 3
2335	Benzene, 1,2,3-trichloro-5-methyl- Toluene, 3,4,5-trichloro-	$C_7H_5Cl_3$ 195.47	21472-86-6 45.5	246	0-05-00-00299	H_2O 1; EtOH 3; ace 3; os 3
2336	Benzene, 1,2,4-trichloro-5-methyl- 2,4,5-Trichlorotoluene	$C_7H_5Cl_3$ 195.47	6639-30-1 82.4	231	4-05-00-00819	H_2O 1; EtOH 3; ace 3; os 3
2337	Benzene, 1,2,5-trichloro-3-methyl- 2,3,5-Trichlorotoluene	$C_7H_5Cl_3$ 195.47	56961-86-5 45.5	230	0-05-00-00299	H_2O 1; EtOH 3; ace 3; os 3
2338	Benzene, 1,2,3-trichloro-5-nitro-	$C_6H_2Cl_3NO_2$ 226.45	20098-48-0 72.5		2-05-00-00187 1.807^{25}	EtOH 4
2339	Benzene, 1,2,4-trichloro-5-nitro-	$C_6H_2Cl_3NO_2$ 226.45	89-69-0 57.5	288	4-05-00-00728 1.790^{23}	H_2O 1; EtOH 2; eth 3; bz 3
2340	Benzene, 1,3,5-trichloro-2,4,6-trifluoro- 1,3,5-Trichloro-2,4,6-trifluorobenzene	$C_6Cl_3F_3$ 235.42	319-88-0	198.4		
2341	Benzene, tricosyl- Tricosylbenzene	$C_{29}H_{52}$ 400.73	61828-04-4 54	365^{100}; 231^{1}	4-05-00-01237 0.8544^{20}	1.4801^{20}
2342	Benzene, tridecyl- 1-Phenyltridecane	$C_{19}H_{32}$ 260.46	123-02-4 10	9575 346	3-05-00-01108 0.8550^{20}	1.4821^{20}
2343	Benzene, 1,3,5-triethoxy- 1,3,5-Triethoxybenzene	$C_{12}H_{18}O_3$ 210.27	2437-88-9 43.5	170^{24}	4-06-00-07362	eth 4; EtOH 4
2344	Benzene, 1,2,4-triethyl- 1,2,4-Triethylbenzene	$C_{12}H_{18}$ 162.27	877-44-1	218; 99^{15}	4-05-00-01133 0.8738^{20}	H_2O 1; EtOH 3; eth 3 1.5024^{20}
2345	Benzene, 1,3,5-triethyl- 1,3,5-Triethylbenzene	$C_{12}H_{18}$ 162.27	102-25-0 -66.5	215.9	4-05-00-01133 0.8631^{20}	H_2O 1; EtOH 4; eth 4 1.4969^{20}
2346	Benzene, 1,2,4-trifluoro-	$C_6H_3F_3$ 132.09	367-23-7	90	4-05-00-00638 1.264^{25}	1.4171^{20}
2347	Benzene, 1,3,5-trifluoro- 1,3,5-Trifluorobenzene	$C_6H_3F_3$ 132.09	372-38-3 -5.5	75.5	4-05-00-00638 1.277^{25}	1.4140^{20}
2348	Benzene, (trifluoromethyl)- (Trifluoromethyl)benzene	$C_7H_5F_3$ 146.11	98-08-8 -29.1	1121 102.1	4-05-00-00802 1.1884^{20}	EtOH 5; eth 5; ace 5; bz 5 1.4146^{20}
2349	Benzene, 1,2,3-triiodo- 1,2,3-Triiodobenzene	$C_6H_3I_3$ 455.80	608-29-7 116	sub	1-05-00-00122	eth 4; EtOH 4; chl 4
2350	Benzene, 1,2,4-triiodo- 1,2,4-Triiodobenzene	$C_6H_3I_3$ 455.80	615-68-9 91.5	sub	3-05-00-00585	EtOH 4; chl 4
2351	Benzene, 1,3,5-triiodo- 1,3,5-Triiodobenzene	$C_6H_3I_3$ 455.80	626-44-8 184.2	sub	4-05-00-00702	H_2O 1; EtOH 2; eth 2; bz 2
2352	Benzene, 1,3,5-triiodo-2-methyl- 2,4,6-Triiodotoluene	$C_7H_5I_3$ 469.83	36994-79-3 118.5	300 dec	1-05-00-00158	EtOH 4
2353	Benzene, 1,2,3-triiodo-5-nitro-	$C_6H_2I_3NO_2$ 500.80	53663-23-3 167		3-05-00-00626 3.256^{25}	bz 4; eth 4; EtOH 4; chl 4
2354	Benzene, 1,2,3-trimethoxy-	$C_9H_{12}O_3$ 168.19	634-36-6 48.5	235	4-06-00-07329 1.1009^{45}	H_2O 1; EtOH 3; eth 3; bz 3
2355	Benzene, 1,3,5-trimethoxy-	$C_9H_{12}O_3$ 168.19	621-23-8 54.5	255.5	4-06-00-07362	H_2O 1; EtOH 3; eth 3; bz 3
2356	Benzene, 1,2,4-trimethoxy-5-(1-propenyl)-	$C_{12}H_{16}O_3$ 208.26	494-40-6 67	296	4-06-00-07476 1.165^{20}	eth 4; EtOH 4; peth 4 1.5683^{20}
2357	Benzene, 1,2,3-trimethyl- 1,2,3-Trimethylbenzene	C_9H_{12} 120.19	526-73-8 -25.4	176.1	4-05-00-01007 0.8944^{20}	H_2O 1; EtOH 5; eth 5; ace 5 1.5139^{20}
2358	Benzene, 1,2,4-trimethyl- Pseudocumene	C_9H_{12} 120.19	95-63-6 -43.8	7929 169.3	4-05-00-01010 0.8758^{20}	H_2O 1; EtOH 5; eth 5; ace 5 1.5048^{20}

No.	Name Synonym	Mol. Form. Mol. Wt.	CAS RN mp/°C	Merck No. bp/°C	Beil. Ref. den/g cm^{-3}	Solubility n_D
2359	Benzene, 1,3,5-trimethyl- Mesitylene	C_9H_{12} 120.19	108-67-8 -44.7	5810 164.7	4-05-00-01016 0.8652^{20}	H_2O 1; EtOH 5; eth 5; ace 5 1.4994^{20}
2360	Benzene, 1,3,5-trimethyl-2,4-dinitro-	$C_9H_{10}N_2O_4$ 210.19	608-50-4 85.3	 exp 418	4-05-00-01029	H_2O 1; EtOH 3
2361	Benzene, 1,2,4-trimethyl-5-(1-methylethyl)-	$C_{12}H_{18}$ 162.27	10222-95-4 21	 221	3-05-00-01038 0.8795^{21}	 1.5065^{21}
2362	Benzene, 1,2,4-trimethyl-5-nitro-	$C_9H_{11}NO_2$ 165.19	610-91-3 71	 265	4-05-00-01015	EtOH 3; chl 3; peth 3
2363	Benzene, 1,3,5-trimethyl-2-nitro-	$C_9H_{11}NO_2$ 165.19	603-71-4 44	 255	4-05-00-01028 1.51^{25}	EtOH 4
2364	Benzene, 1,3,5-trimethyl-2,4,6-trinitro-	$C_9H_9N_3O_6$ 255.19	602-96-0 238.2	 exp 415	4-05-00-01029	H_2O 1; EtOH 2; eth 2; ace 3
2365	Benzene, 1,3,5-trinitro- *sym*-Trinitrobenzene	$C_6H_3N_3O_6$ 213.11	99-35-4 121.5	9639 315	4-05-00-00755 1.4775^{152}	H_2O 2; EtOH 2; eth 2; ace 4
2366	1,2,3-Benzenetriol Pyrogallol	$C_6H_6O_3$ 126.11	87-66-1 133	8010 309	4-06-00-07327 1.453^{4}	H_2O 4; EtOH 4; eth 4; ace 3 1.561^{134}
2367	1,2,4-Benzenetriol Hydroxyhydroquinone	$C_6H_6O_3$ 126.11	533-73-3 140.5	1081	4-06-00-07338	H_2O 4; EtOH 4; eth 4; bz 1
2368	1,3,5-Benzenetriol Phloroglucinol	$C_6H_6O_3$ 126.11	108-73-6 218.5	7301 sub	4-06-00-07361 1.46^{25}	H_2O 2; EtOH 4; eth 4; ace 3
2369	1,2,3-Benzenetriol, 5-methyl-	$C_7H_8O_3$ 140.14	609-25-6 129	 sub	4-06-00-07376	bz 4
2370	1,2,4-Benzenetriol, triacetate	$C_{12}H_{12}O_6$ 252.22	613-03-6 99	 300	4-06-00-07342	EtOH 3; chl 3; MeOH 3
2371	1,3,5-Benzenetriol, 2,4,6-trichloro- Trichloro-1,3,5-trihydroxybenzene	$C_6H_3Cl_3O_3$ 229.45	56961-23-0 136	 sub	0-06-00-01104	EtOH 4
2372	Benzene, 1,2,4-tris(1-methylethyl)- 1,2,4-Triisopropylbenzene	$C_{15}H_{24}$ 204.36	948-32-3	 244	4-05-00-01178 0.8574^{25}	 1.4896^{25}
2373	Benzene, 1,3,5-tris(1-methylethyl)-	$C_{15}H_{24}$ 204.36	717-74-8 -7.4	 238	4-05-00-01178 0.8545^{20}	ace 3; bz 3; chl 3 1.4882^{20}
2374	Benzene, undecyl-	$C_{17}H_{28}$ 232.41	6742-54-7 -5	 316	3-05-00-01100 0.8553^{20}	 1.4828^{20}
2375	9,10[1',2']-Benzenoanthracene, 9,10-dihydro- Triptycene	$C_{20}H_{14}$ 254.33	477-75-8	9663	4-05-00-02639	
2376	1H-Benzimidazole *N,N'*-Methenyl-o-phenylenediamine	$C_7H_6N_2$ 118.14	51-17-2 170.5	1091 >360	5-23-06-00196	H_2O 2; EtOH 4; eth 2; bz 1
2377	1H-Benzimidazole, 4,5-dichloro-2-(trifluoromethyl)- Chloroflurazole	$C_8H_3Cl_2F_3N_2$ 255.03	3615-21-2 213.5			
2378	1H-Benzimidazole, 1,5-dimethyl- 1,5-Dimethylbenzimidazole	$C_9H_{10}N_2$ 146.19	10394-35-1 95	 300	5-23-06-00380	bz 4; eth 4; EtOH 4
2379	1H-Benzimidazole, 2,5-dimethyl-	$C_9H_{10}N_2$ 146.19	1792-41-2 203	 350	5-23-06-00448	eth 4; EtOH 4
2380	1H-Benzimidazole, 5,6-dimethyl- Dimedazole	$C_9H_{10}N_2$ 146.19	582-60-5 205.5	3225 sub	5-23-06-00454	H_2O 3; EtOH 3; eth 3; chl 3
2381	1H-Benzimidazole, 1-ethyl-2-methyl- 1-Ethyl-2-methylbenzimidazole	$C_{10}H_{12}N_2$ 160.22	5805-76-5 51	 296	5-23-06-00326 1.073^{25}	
2382	1H-Benzimidazole, 1-methyl-	$C_8H_8N_2$ 132.17	1632-83-3 66	 286	5-23-06-00215 1.1254^{20}	peth 3 1.6013^{7}
2383	1H-Benzimidazole, 1-phenyl-	$C_{13}H_{10}N_2$ 194.24	2622-60-8 97	 221^{25}	5-23-06-00222	H_2O 2; EtOH 3; MeOH 3
2384	1H-Benzimidazole, 2-phenyl- Phenzidole	$C_{13}H_{10}N_2$ 194.24	716-79-0 293	7245	5-23-08-00437	H_2O 2; EtOH 3; bz 2; chl 2
2385	1H-Benzimidazole, 2-(phenylmethyl)- Bendazol	$C_{14}H_{12}N_2$ 208.26	621-72-7 187	1043	5-23-08-00498	bz 4; EtOH 4; gl HOAc 4
2386	2H-Benzimidazole-2-thione, 1,3-dihydro- 2-Benzimidazolethiol	$C_7H_6N_2S$ 150.20	583-39-1 298	1092	5-24-02-00413	EtOH 4
2387	1,2-Benzisothiazole 1-Thia-2-azaindene	C_7H_5NS 135.19	272-16-2 38	 220	2-27-00-00016	ctc 3
2388	1,2-Benzisothiazol-3(2H)-one, 1,1-dioxide Saccharin	$C_7H_5NO_3S$ 183.19	81-07-2 228 dec	8282 sub	4-27-00-02649 0.828^{25}	H_2O 2; EtOH 3; eth 2; ace 3
2389	1,2-Benzisoxazole Indoxazene	C_7H_5NO 119.12	271-95-4	 100^{26}	4-27-00-01067 1.1727^{21}	eth 4 1.5570^{20}
2390	2,1-Benzisoxazole Anthranil	C_7H_5NO 119.12	271-58-9 <-18	 215	4-27-00-01068 1.1827^{20}	ace 4; EtOH 4 1.5845^{20}
2391	Benz[j]aceanthrylene, 1,2-dihydro- Cholanthrene	$C_{20}H_{14}$ 254.33	479-23-2 174	2200	4-05-00-02638	H_2O 1; EtOH 3; bz 3; HOAc 3
2392	Benz[j]aceanthrylene, 1,2-dihydro-3-methyl- 3-Methylcholanthrene	$C_{21}H_{16}$ 268.36	56-49-5 180	5967 280^{80}	4-05-00-02648 1.28^{20}	H_2O 1
2393	11H-Benzo[a]fluorene	$C_{17}H_{12}$ 216.28	238-84-6 189.5	 405	4-05-00-02473	EtOH 2; eth 3; bz 3; chl 3
2394	Benzo[a]phenazine 7,12-Diazabenz[a]anthracene	$C_{16}H_{10}N_2$ 230.27	225-61-6 142.5	 >360	5-23-09-00494	H_2O 1; EtOH 3; eth 3; ace 3
2395	Benzo[a]pyrene 2,3-Benzopyrene	$C_{20}H_{12}$ 252.32	50-32-8 176.5	1113	4-05-00-02687	chl 4
2396	2H-Benzo[a]quinolizin-2-one, 1,3,4,6,7,11b-hexahydro-9,10-dimethoxy-3-(2-methylpropyl)- 1,2-Benzoquinolizine-2-one, 9,10-dimethoxy-1,2,3,4,6,7,11b-heptahydro-3-isobutyl	$C_{19}H_{27}NO_3$ 317.43	58-46-8 128	9118	5-21-13-00178	chl 3

No.	Name Synonym	Mol. Form. Mol. Wt.	CAS RN mp/°C	Merck No. bp/°C	Beil. Ref. den/g cm^{-3}	Solubility n_D
2397	2H,8H-Benzo[1,2-b:5,4-b']dipyran-2-one, 8,8-dimethyl- Xanthyletin	$C_{14}H_{12}O_3$ 228.25	553-19-5 131.5	9978 $140-5^{0.1}$	5-19-04-00470	EtOH 3; peth 3
2398	2H,8H-Benzo[1,2-b:5,4-b']dipyran-2-one, 5-methoxy-8,8-dimethyl- Xanthoxyletin	$C_{15}H_{14}O_4$ 258.27	84-99-1 133	9975	5-19-06-00039	H_2O 1; EtOH 3; eth 2; ace 3
2399	Benzo[b]thiophene Thianaphthene	C_8H_6S 134.20	95-15-8 32	9232 221	5-17-02-00006 1.1484^{32}	H_2O 1; EtOH 4; eth 3; ace 3 1.6374^{37}
2400	Benzo[b]thiophene-2-carboxylic acid Thionaphthene-2-carboxylic acid	$C_9H_6O_2S$ 178.21	6314-28-9 240.5	9274	5-18-06-00427	eth 4
2401	Benzo[b]thiophene-2,3-dione	$C_8H_4O_2S$ 164.18	493-57-2 121	 247	5-17-11-00251	H_2O 1; EtOH 3; bz 3; HOAc 3
2402	Benzo[b]thiophene, 5-methyl-	C_9H_8S 148.23	14315-14-1 20.5	 $111-5^{12}$	5-17-02-00031 1.111^{22}	 1.615^{22}
2403	Benzo[b]thiophene-4-ol	C_8H_6OS 150.20	3610-02-4 79.8	 sub	5-17-04-00199	EtOH 4
2404	Benzo[b]thiophen-3(2H)-one, 2-(3-oxobenzo[b]thien-2(3H)-ylidene)-	$C_{16}H_8O_2S_2$ 296.37	522-75-8 359	 sub	5-19-05-00281	H_2O 1; EtOH 1; bz 3; chl 2
2405	7H-Benzo[c]carbazole 7-Aza-7H-benzo[c]fluorene	$C_{16}H_{11}N$ 217.27	205-25-4 134	 448	5-20-08-00455	bz 4; EtOH 4; HOAc 4
2406	Benzo[c]phenanthrene Tetrahelicene	$C_{18}H_{12}$ 228.29	195-19-7 68	1129	4-05-00-02552	H_2O 1; EtOH 2; lig 2
2407	5H-Benzocyclohepten-5-one, 6,7,8,9-tetrahydro-	$C_{11}H_{12}O$ 160.22	826-73-3	 124^7; 108^1	4-07-00-01029 1.0780^{20}	EtOH 3 1.5698^{20}
2408	5H-Benzocyclohepten-5-one, 2,3,4,6-tetrahydroxy- Purpurogallin	$C_{11}H_8O_5$ 220.18	569-77-7 274-5 dec	7963	4-08-00-03456	
2409	3H-1,4-Benzodiazepin-2-amine, 7-chloro-*N*-methyl-5-phenyl-, 4-oxide Chlorodiazepoxide	$C_{16}H_{14}ClN_3O$ 299.76	58-25-3 236.2	2082		
2410	2H-1,4-Benzodiazepin-2-one, 7-chloro-1,3-dihydro-1-methyl-5-phenyl- Diazepam	$C_{16}H_{13}ClN_2O$ 284.74	439-14-5 132	2977	5-24-04-00300	
2411	2H-1,4-Benzodiazepin-2-one, 7-chloro-1,3-dihydro-5-phenyl- Nordazepam	$C_{15}H_{11}ClN_2O$ 270.72	1088-11-5 216.5	6608	5-24-04-00291	
2412	1,3,2-Benzodioxaborole	$C_6H_5BO_2$ 119.92	274-07-7 12	 88^{156}; 50^{50}	 1.2700^{20}	 1.5070^{20}
2413	1,3,2-Benzodioxaphosphole, 2-chloro-	$C_6H_4ClO_2P$ 174.52	1641-40-3 30	 80^{20}	4-06-00-05598 1.4650^{20}	 1.5712^{20}
2414	1,4-Benzodioxin, 2,3-dihydro-	$C_8H_8O_2$ 136.15	493-09-4	 103^6	5-19-01-00439 1.142^{25}	 1.5485^{20}
2415	1,3-Benzodioxole	$C_7H_6O_2$ 122.12	274-09-9	 172.5; 77^{27}	5-19-01-00434 1.064^{25}	 1.5398^{20}
2416	1,3-Benzodioxole-5-acetic acid, methyl ester Phenylacetic acid, 3,4-methylenedioxy, methyl ester	$C_{10}H_{10}O_4$ 194.19	326-59-0	 $153-5^{10}$	2-19-00-00295 1.246^{20}	 1.534
2417	1,3-Benzodioxole-5-carboxaldehyde Piperonal	$C_8H_6O_3$ 150.13	120-57-0 37	7445 263	5-19-04-00225	H_2O 2; EtOH 4; eth 5; ace 3
2418	1,3-Benzodioxole-5-carboxylic acid Piperonylic acid	$C_8H_6O_4$ 166.13	94-53-1 229	7447	5-19-07-00300	
2419	1,3-Benzodioxole-5-carboxylic acid, ethyl ester	$C_{10}H_{10}O_4$ 194.19	6951-08-2 18.5	 285.5; 135^6	5-19-07-00300	eth 4; EtOH 4; peth 4
2420	1,3-Benzodioxole-5-carboxylic acid, methyl ester	$C_9H_8O_4$ 180.16	326-56-7 53	 273 dec	5-19-07-00300	eth 4; EtOH 4
2421	1,3-Benzodioxole, 5-(chloromethyl)-	$C_8H_7ClO_2$ 170.60	20850-43-5 20.5	 $134-5^{14}$	5-19-01-00443 1.312^{25}	 1.5660^{20}
2422	1,3-Benzodioxole, 4,7-dimethoxy-5-(1-propenyl)-, (*E*)-	$C_{12}H_{14}O_4$ 222.24	17672-88-7 56	 303.5	4-19-00-01033	ace 4; bz 4; eth 4; EtOH 4
2423	1,3-Benzodioxole, 4,7-dimethoxy-5-(2-propenyl)- Apiole	$C_{12}H_{14}O_4$ 222.24	523-80-8 29.5	767 294. 179^{35}	5-19-03-00307 1.015^{20}	ace 4; bz 4; EtOH 4; lig 4 1.5360^{20}
2424	1,3-Benzodioxole-5-ethanamine	$C_9H_{11}NO_2$ 165.19	1484-85-1	 166^{20}	5-19-08-00407 1.225^{20}	 1.5620^{20}
2425	1,3-Benzodioxole, 5-ethenyl-	$C_9H_8O_2$ 148.16	7315-32-4	 224	5-19-01-00544 1.1488^{18}	eth 4; EtOH 4; chl 4 1.5802
2426	1,3-Benzodioxole-5-methanamine	$C_8H_9NO_2$ 151.17	2620-50-0	 $138-9^{13}$	5-19-08-00395 1.214^{25}	 1.5635^{20}
2427	1,3-Benzodioxole-5-methanol	$C_8H_8O_3$ 152.15	495-76-1 58	 157^{16}	5-19-02-00543	H_2O 2; EtOH 3; eth 3; bz 3
2428	1,3-Benzodioxole, 4-methoxy-6-(2-propenyl)- Myristicin	$C_{11}H_{12}O_3$ 192.21	607-91-0 <-20	6247 276.5	5-19-02-00631 1.1416^{20}	H_2O 1; EtOH 2; eth 3; bz 3 1.5403^{20}
2429	1,3-Benzodioxole, 5-(2-propenyl)- Safrole	$C_{10}H_{10}O_2$ 162.19	94-59-7 11.2	8287 234.5	5-19-01-00553 1.1000^{20}	H_2O 1; EtOH 4; eth 5; chl 5 1.5381^{20}
2430	1,3-Benzodioxole, 5-(1-propenyl)-, (*E*)-	$C_{10}H_{10}O_2$ 162.19	4043-71-4 6.8	 253	5-19-01-00552 1.1224^{20}	H_2O 1; EtOH 5; eth 5; ace 4 1.5782^{20}

No.	Name Synonym	Mol. Form. Mol. Wt.	CAS RN mp/°C	Merck No. bp/°C	Beil. Ref. den/g cm^{-3}	Solubility n_D
2431	1,3-Benzodioxole, 5-propyl- Dihydrosafrole	$C_{10}H_{12}O_2$ 164.20	94-58-6		5-19-01-00473	ctc 3
2432	[1,3]Benzodioxolo[5,6-c]-1,3-dioxolo[4,5-i]phenanthridinium, 13-methyl- Sanguinarine	$C_{20}H_{15}NO_5$ 349.34	2447-54-3 266	8320		ace 4; bz 4; eth 4; EtOH 4
2433	[1,3]Benzodioxolo[5,6-c]phenanthridinium, 1,2-dimethoxy-12-methyl- Chelerythrine	$C_{21}H_{19}NO_5$ 365.39	34316-15-9	2041 207		chl 4
2434	1,4-Benzodithiin	$C_8H_6S_2$ 166.27	255-50-5	 $67\text{-}70^{0.1}$	5-19-01-00536 1.2799^{20}	 1.6754^{25}
2435	Benzo[e]-1,3-dioxolo[4,5-l][2]benzazecin-12(5H)-one, 4,6,7,13-tetrahydro-9,10-dimethoxy-5-methyl- Cryptopine	$C_{21}H_{23}NO_5$ 369.42	482-74-6 223	2611	4-27-00-06652 1.315^{20}	H_2O 1; EtOH 2; eth 2; bz 2
2436	Benzo[e]pyrene 1,2-Benzpyrene	$C_{20}H_{12}$ 252.32	192-97-2 177.5	1114 311	4-05-00-02689	
2437	Benzo[f]quinoline β-Naphthoquinoline	$C_{13}H_9N$ 179.22	85-02-9 94	1115 352; $202\text{-}5^{8}$	5-20-08-00220	H_2O 2; EtOH 4; eth 4; ace 3
2438	Benzofuran Coumarone	C_8H_6O 118.14	271-89-6 <-18	1098 174	5-17-02-00003 1.0913^{25}	H_2O 1; EtOH 3; eth 3 1.5615^{17}
2439	2-Benzofurancarboxylic acid Coumarilic acid	$C_9H_6O_3$ 162.14	496-41-3 192.5	2562 312.5	5-18-06-00419	EtOH 4
2440	2-Benzofurancarboxylic acid, ethyl ester	$C_{11}H_{10}O_3$ 190.20	3199-61-9 30.5	 276; 161^{15}	5-18-06-00420 1.1656^{28}	 1.564^{27}
2441	Benzofuran, 2-(chloromethyl)-2,3-dihydro- 2-(Chloromethyl)-2,3-dihydrobenzofuran	C_9H_9ClO 168.62	53491-32-0 41.5	 $118\text{-}9^{11}$	4-17-00-00421 1.2183^{7}	ace 4; eth 4 1.5620^{7}
2442	Benzofuran, 2,3-dihydro- Coumaran	C_8H_8O 120.15	496-16-2 -21.5	2560 188.5	5-17-01-00581 1.058^{25}	eth 4; EtOH 4; chl 4 1.5497^{20}
2443	Benzofuran, 2,3-dihydro-2-methyl-	$C_9H_{10}O$ 134.18	1746-11-8	 197.5	5-17-01-00597 1.061^{25}	 1.5308
2444	Benzofuran, 2,3-dihydro-5-methyl- 2,3-Dihydro-5-methylbenzofuran	$C_9H_{10}O$ 134.18	76429-68-0	 210.5	4-17-00-00423 1.0463^{19}	 1.54^{20}
2445	Benzofuran, 2,5-dimethyl- 2,5-Dimethylbenzofuran	$C_{10}H_{10}O$ 146.19	29040-46-8	 220	5-17-02-00045 1.031^{20}	eth 4 1.5534^{20}
2446	Benzofuran, 2,6-dimethyl- 2,6-Dimethylbenzofuran	$C_{10}H_{10}O$ 146.19	24410-51-3	 217.5	5-17-02-00046 1.051^{12}	eth 4 1.554^{15}
2447	Benzofuran, 2,7-dimethyl- 2,7-Dimethylfuran	$C_{10}H_{10}O$ 146.19	59020-74-5	 216	4-17-00-00512 1.036^{20}	eth 4 1.5546^{20}
2448	Benzofuran, 3,5-dimethyl- 3,5-Dimethylbenzofuran	$C_{10}H_{10}O$ 146.19	10410-35-2	 220.5	5-17-02-00047 1.036^{20}	eth 4 1.550^{20}
2449	Benzofuran, 3,6-dimethyl- 3,6-DimethylfuranBenzofuran	$C_{10}H_{10}O$ 146.19	24410-50-2	 222	4-17-00-00512 1.0456^{20}	 1.5505^{20}
2450	Benzofuran, 4,6-dimethyl-	$C_{10}H_{10}O$ 146.19	116668-34-9	 219	0-17-00-00063 1.037^{20}	eth 4 1.5485^{21}
2451	Benzofuran, 4,7-dimethyl- 4,7-Dimethylbenzofuran	$C_{10}H_{10}O$ 146.19	28715-26-6	 216	0-17-00-00063 1.041^{17}	eth 4 1.549^{17}
2452	Benzofuran, 5,6-dimethyl- 5,6-Dimethylbenzofuran	$C_{10}H_{10}O$ 146.19	24410-52-4	 221	5-17-02-00048 1.060^{15}	eth 4 1.5516^{15}
2453	Benzofuran, 5,7-dimethyl- 5,7-Dimethylbenzofuran	$C_{10}H_{10}O$ 146.19	64965-91-9	 222	0-17-00-00063 1.0262^{18}	eth 4 1.5358^{20}
2454	Benzofuran, 6,7-dimethyl- 6,7-Dimethylfuran	$C_{10}H_{10}O$ 146.19	35355-36-3	 218	0-17-00-00063 1.038^{20}	eth 4 1.5478^{20}
2455	2-Benzofuranmethanol 2-Hydroxymethylbenzofuran	$C_9H_8O_2$ 148.16	55038-01-2	 $149\text{-}50^{14}$	5-17-04-00224 1.177^{13}	 1.5550^{11}
2456	Benzofuran, 2-methyl-	C_9H_8O 132.16	4265-25-2	 197.5	5-17-02-00022 1.0540^{20}	eth 4; EtOH 4 1.5495^{22}
2457	Benzofuran, 3-methyl- 3-Methylbenzo[b]furan	C_9H_8O 132.16	21535-97-7	 197; 86^{20}	5-17-02-00025 1.0540^{25}	H_2O 1; EtOH 3; eth 3 1.5536^{16}
2458	Benzofuran, 5-methyl- 5-Methylbenzo[b]furan	C_9H_8O 132.16	18441-43-5	 198	5-17-02-00030 1.0603^{19}	H_2O 1; EtOH 3; eth 3; sulf 3 1.5570^{19}
2459	Benzofuran, 7-methyl- 7-Methylbenzo[b]furan	C_9H_8O 132.16	17059-52-8 109	 190.5	5-17-02-00032 1.0490^{19}	H_2O 1; EtOH 3; eth 3; sulf 3 1.5525^{19}
2460	Benzofuran, 4-methyl-7-(1-methylethyl)- Benzofuran, 7-isopropyl-4-methyl	$C_{12}H_{14}O$ 174.24	95835-77-1	 241.5	0-17-00-00065 1.0145^{18}	 1.5363^{16}
2461	2(3H)-Benzofuranone	$C_8H_6O_2$ 134.13	553-86-6 50	 249	5-17-10-00006 1.2236^{14}	
2462	3(2H)-Benzofuranone	$C_8H_6O_2$ 134.13	7169-34-8 102.5	 152^{15}	5-17-04-00185	bz 4
2463	2(3H)-Benzofuranone, 3-[2-(diethylamino)ethyl]-3-phenyl- Amolanone	$C_{20}H_{23}NO_2$ 309.41	76-65-3 43.4	603 $192\text{-}4^{2.0}$	4-18-00-07995	 1.5614^{25}
2464	5-Benzofuranpropanol, 2-(1,3-benzodioxol-5-yl)-7-methoxy- Egonol	$C_{19}H_{18}O_5$ 326.35	530-22-3 118	 $228\text{-}30^{15}$	5-19-09-00583	chl 4
2465	Benzofuran, 4,5,6,7-tetrahydro-3,6-dimethyl-	$C_{10}H_{14}O$ 150.22	494-90-6 86	 80^{18}	5-17-01-00518 0.972^{15}	
2466	6H-Benzofuro[3,2-c][1]benzopyran-6-one, 3,9-dihydroxy- Coumestrol	$C_{15}H_8O_5$ 268.23	479-13-0 385 dec	2565	5-19-06-00405	H_2O 1; EtOH 2; eth 1; ace 2

No.	Name Synonym	Mol. Form. Mol. Wt.	CAS RN mp/°C	Merck No. bp/°C	Beil. Ref. den/g cm^{-3}	Solubility n_D
2467	Benzo[g]-1,3-benzodioxolo[5,6-a]quinolizinium, 5,6-dihydro-9,10-dimethoxy- Berberine	$C_{20}H_{19}NO_5$ 353.37	2086-83-1 145	1169		eth 4; EtOH 4
2468	5H-Benzo[g]-1,3-benzodioxolo[6,5,4-de]quinoline, 6,7,7a,8-tetrahydro-10,11-dimethoxy-7-methyl-, (*S*)- Dicentrine	$C_{20}H_{21}NO_4$ 339.39	517-66-8	3028	4-27-00-06486	chl 3
2469	5H-Benzo[g]-1,3-benzodioxolo[6,5,4-de]quinoline, 6,7,7a,8-tetrahydro-11-methoxy-7-methyl-, (*R*)- Laureline	$C_{19}H_{19}NO_3$ 309.36	81-38-9 114	5251	1-27-00-00461	H_2O 1; EtOH 3; eth 3; dil acid 3
2470	5H-Benzo[g]-1,3-benzodioxolo[6,5,4-de]quinolin-12-ol, 6,7,7a,8-tetrahydro-11-methoxy-7-methyl-, (*S*)- Bulbocapnine	$C_{19}H_{19}NO_4$ 325.36	298-45-3 199.5	1471	2-27-00-00554	H_2O 1; EtOH 3; chl 4; os 3
2471	Benzo[ghi]perylene 1,12-Benzperylene	$C_{22}H_{12}$ 276.34	191-24-2		4-05-00-02766	
2472	Benzo[g]pteridine-2,4(1H,3H)-dione, 7,8-dimethyl- Lumichrome	$C_{12}H_{10}N_4O_2$ 242.24	1086-80-2	5468	4-26-00-02538	
2473	Benzo[g]quinoline 6,7-Benzoquinoline	$C_{13}H_9N$ 179.22	260-36-6 114	 202^{14}	5-20-08-00195	eth 4; EtOH 4
2474	Benzo[h]quinoline	$C_{13}H_9N$ 179.22	230-27-3 52	 339; 233^{47}	5-20-08-00215 1.2340^{20}	H_2O 2; EtOH 3; eth 3; ace 3
2475	Benzoic acid Benzenecarboxylic acid	$C_7H_6O_2$ 122.12	65-85-0 122.4	1101 249.2	4-09-00-00273 1.2659^{15}	H_2O 2; EtOH 4; eth 4; ace 3 1.504^{132}
2476	Benzoic acid, 2-acetyl-	$C_9H_8O_3$ 164.16	577-56-0 114.5	 $110\text{-}2^{2}$	4-10-00-02766	H_2O 4; eth 4; EtOH 4
2477	Benzoic acid, 3-acetyl-	$C_9H_8O_3$ 164.16	586-42-5 172	 $110\text{-}2^{2}$	4-10-00-02769	H_2O 3; EtOH 5
2478	Benzoic acid, 4-acetyl-	$C_9H_8O_3$ 164.16	586-89-0 208	 sub	4-10-00-02769	H_2O 4
2479	Benzoic acid, 4-acetyl-, methyl ester	$C_{10}H_{10}O_3$ 178.19	3609-53-8 95	 142^{4}	3-10-00-03030	H_2O 4
2480	Benzoic acid, 2-(acetyloxy)- Aspirin	$C_9H_8O_4$ 180.16	50-78-2 135	873	4-10-00-00138	H_2O 3; EtOH 4; eth 3; bz 2
2481	Benzoic acid, 2-(acetyloxy)-5-bromo- 5-Bromoacetylsalicylic acid	$C_9H_7BrO_4$ 259.06	1503-53-3 60	1425	2-10-00-00064	H_2O 1; EtOH 4; eth 3
2482	Benzoic acid, 2-(acetyloxy)-, methyl ester Methyl acetylsalicylate	$C_{10}H_{10}O_4$ 194.19	580-02-9 51.5	5934 $134\text{-}6^{9}$	4-10-00-00147	eth 4; EtOH 4; chl 4
2483	Benzoic acid, 2-(acetyloxy)-, phenyl ester Phenyl acetylsalicylate	$C_{15}H_{12}O_4$ 256.26	134-55-4 96	7241	4-10-00-00156	
2484	Benzoic acid, 2-amino- *o*-Anthranilic acid	$C_7H_7NO_2$ 137.14	118-92-3 146.5	433 sub	4-14-00-01004 1.412^{20}	H_2O 3; EtOH 3; eth 3; bz 2
2485	Benzoic acid, 3-amino- Aniline-3-carboxylic acid	$C_7H_7NO_2$ 137.14	99-05-8 173	432	4-14-00-01092 1.51^{25}	H_2O 2; EtOH 2; eth 3; ace 4
2486	Benzoic acid, 4-amino- Aniline-4-carboxylic acid	$C_7H_7NO_2$ 137.14	150-13-0 188.5	434	4-14-00-01126 1.374^{20}	H_2O 3; EtOH 3; eth 3; ace 2
2487	Benzoic acid, 2-amino-5-bromo- 5-Bromoanthranilic acid	$C_7H_6BrNO_2$ 216.03	5794-88-7 219.5	1393	4-14-00-01081	DMSO 3
2488	Benzoic acid, 4-amino-, butyl ester Butamben	$C_{11}H_{15}NO_2$ 193.25	94-25-7 58	1504 $173\text{-}4^{8}$	4-14-00-01130	H_2O 1; EtOH 3; eth 3; bz 3
2489	Benzoic acid, 4-(aminocarbonyl)-	$C_8H_7NO_3$ 165.15	6051-43-0 300	 sub	4-09-00-03318	H_2O 1; EtOH 1; eth 1; bz 1
2490	Benzoic acid, 5-amino-2-chloro- 5-Amino-2-chlorobenzoic acid	$C_7H_6ClNO_2$ 171.58	89-54-3 188		4-14-00-01116 1.519^{15}	EtOH 4
2491	Benzoic acid, 2-amino-3,6-dichloro-	$C_7H_5Cl_2NO_2$ 206.03	3032-32-4 155	 sub	4-14-00-01078	ace 4; bz 4; eth 4; EtOH 4
2492	Benzoic acid, 3-amino-2,5-dichloro- Chloramben	$C_7H_5Cl_2NO_2$ 206.03	133-90-4 200	2063		DMSO 2
2493	Benzoic acid, 4-amino-, 2-(diethylamino)ethyl ester Procaine	$C_{13}H_{20}N_2O_2$ 236.31	59-46-1 61	7763	4-14-00-01138	H_2O 2; EtOH 3; eth 3; bz 3
2494	Benzoic acid, 4-[[(2-amino-1,4-dihydro-4-oxo-6-pteridinyl)methyl]formylamino]- Rhizopterin	$C_{15}H_{12}N_6O_4$ 340.30	119-20-0 >300	8181	4-26-00-03949	H_2O 1; EtOH 1; eth 1; aq alk 3
2495	Benzoic acid, 2-amino-, ethyl ester	$C_9H_{11}NO_2$ 165.19	87-25-2 13	 268	4-14-00-01008 1.1174^{20}	eth 4; EtOH 4 1.5646^{20}
2496	Benzoic acid, 3-amino-, ethyl ester	$C_9H_{11}NO_2$ 165.19	582-33-2	 294; 160^{5}	4-14-00-01093 1.1248^{22}	H_2O 2; EtOH 4; eth 4; ctc 3 1.5600^{22}
2497	Benzoic acid, 4-amino-, ethyl ester Ethyl aminobenzoate	$C_9H_{11}NO_2$ 165.19	94-09-7 92	3719 310	4-14-00-01129	H_2O 1; EtOH 4; eth 4; chl 3
2498	Benzoic acid, 4-amino-2-hydroxy- *p*-Aminosalicylic acid	$C_7H_7NO_3$ 153.14	65-49-6 150-1 dec	491	4-14-00-01967	H_2O 3; EtOH 3; eth 3; ace 3
2499	Benzoic acid, 5-amino-2-hydroxy- Mesalamine	$C_7H_7NO_3$ 153.14	89-57-6 283	5807	4-14-00-02058	H_2O 2; EtOH 1
2500	Benzoic acid, 4-amino-2-hydroxy-, hydrazide *p*-Aminosalicylic acid hydrazide	$C_7H_9N_3O_2$ 167.17	6946-29-8 195	492	4-14-00-01995	EtOH 4
2501	Benzoic acid, 3-amino-4-hydroxy-, methyl ester Orthocaine	$C_8H_9NO_3$ 167.16	536-25-4 143	6836	3-14-00-01477	H_2O 1; EtOH 4; eth 3; bz 2

No.	Name	Synonym	Mol. Form.	Mol. Wt.	CAS RN	mp/°C	Merck No.	bp/°C	Beil. Ref.	den/g cm^{-3}	Solubility	n_D
2502	Benzoic acid, 4-amino-2-hydroxy-, phenyl ester	Phenyl *p*-aminosalicylate	$C_{13}H_{11}NO_3$	229.24	133-11-9	153	7243		4-14-00-01979			
2503	Benzoic acid, 2-amino-, methyl ester	Methyl anthranilate	$C_8H_9NO_2$	151.17	134-20-3	24.5	5942	256	4-14-00-01008	1.1682^{10}	H_2O 2; EtOH 4; eth 4	1.5810
2504	Benzoic acid, 3-amino-, methyl ester		$C_8H_9NO_2$	151.17	4518-10-9	39		$152\text{-}3^{11}$	4-14-00-01093	1.232^{20}	EtOH 4; eth 4; bz 4; chl 4	
2505	Benzoic acid, 4-amino-, 2-methylpropyl ester	Isobutyl *p*-aminobenzoate	$C_{11}H_{15}NO_2$	193.25	94-14-4	64.5	5017		4-14-00-01130			
2506	Benzoic acid, 2-amino-3-nitro-	3-Nitroanthranilic acid	$C_7H_6N_2O_4$	182.14	606-18-8	208.5			4-14-00-01087	1.558^{15}	eth 4; EtOH 4	
2507	Benzoic acid, 4-amino-, propyl ester	Risocaine	$C_{10}H_{13}NO_2$	179.22	94-12-2	75	8226		4-14-00-01130		bz 4; eth 4; EtOH 4; chl 4	
2508	Benzoic acid, 4-(aminosulfonyl)-	Carzenide	$C_7H_7NO_4S$	201.20	138-41-0	290-2 dec	1885		4-11-00-00690		H_2O 1; EtOH 4; eth 2; bz 1	
2509	Benzoic acid, ammonium salt		$C_7H_9NO_2$	139.15	1863-63-4	194.5	515		4-09-00-00273		H_2O 3	
2510	Benzoic acid, anhydride	Benzoic anhydride	$C_{14}H_{10}O_3$	226.23	93-97-0	42.5	1102	360	4-09-00-00550	1.989^{15}	H_2O 1; EtOH 3; eth 3; chl 2	1.5767^{15}
2511	Benzoic acid, 3-benzoyl-		$C_{14}H_{10}O_3$	226.23	579-18-0	163.3		sub	4-10-00-02982		eth 4; EtOH 4	
2512	Benzoic acid, 4-benzoyl-		$C_{14}H_{10}O_3$	226.23	611-95-0	199		sub	3-10-00-03305		H_2O 2; EtOH 3; eth 3; bz 2	
2513	Benzoic acid, 3-(benzoylamino)-		$C_{14}H_{11}NO_3$	241.25	587-54-2	252.5			4-14-00-01105	1.510^4	EtOH 4	
2514	Benzoic acid, 4-(benzoylamino)-2-hydroxy-	Benzoyl-PAS	$C_{14}H_{11}NO_4$	257.25	13898-58-3	260.5	1127		4-14-00-02029			
2515	Benzoic acid, 2-benzoyl-, ethyl ester		$C_{16}H_{14}O_3$	254.29	604-61-5	59.5			3-10-00-03291	1.221^{64}	H_2O 1; EtOH 4; eth 4; chl 2	1.560^{64}
2516	Benzoic acid, 2-benzoyl-, methyl ester		$C_{15}H_{12}O_3$	240.26	606-28-0	52		351	4-10-00-02977	1.1903^{19}	H_2O 1; EtOH 4; eth 4; sulf 3	1.591^{20}
2517	Benzoic acid, 2-[bis(4-hydroxyphenyl)methyl]-	Phenolphthalin	$C_{20}H_{16}O_4$	320.34	81-90-3	230.5	7210		4-10-00-01932		eth 4; EtOH 4	
2518	Benzoic acid, 2-bromo-		$C_7H_5BrO_2$	201.02	88-65-3	150		sub	4-09-00-01011	1.929^{25}	H_2O 2; EtOH 3; eth 3; ace 3	
2519	Benzoic acid, 3-bromo-		$C_7H_5BrO_2$	201.02	585-76-2	155		>280	4-09-00-01013	1.845^{20}	H_2O 1; EtOH 3; eth 3	
2520	Benzoic acid, 4-bromo-	*p*-Bromobenzoic acid	$C_7H_5BrO_2$	201.02	586-76-5	254.5	1396		4-09-00-01017	1.894^{20}	H_2O 2; EtOH 3; eth 3	
2521	Benzoic acid, 2-bromo-, ethyl ester		$C_9H_9BrO_2$	229.07	6091-64-1			254.5	4-09-00-01012	1.4438^{15}	ace 4; bz 4; eth 4; EtOH 4	1.5455^{15}
2522	Benzoic acid, 3-bromo-, ethyl ester		$C_9H_9BrO_2$	229.07	24398-88-7			261	4-09-00-01014	1.4308^{19}	ace 4; bz 4; eth 4; EtOH 4	1.5430^{19}
2523	Benzoic acid, 4-bromo-, ethyl ester		$C_9H_9BrO_2$	229.07	5798-75-4			263; 125^{15}	3-09-00-01405	1.4332^{17}	H_2O 2; EtOH 3; eth 3; ace 3	1.5438^{17}
2524	Benzoic acid, 5-bromo-2-hydroxy-		$C_7H_5BrO_3$	217.02	89-55-4	169.8		sub 100	4-10-00-00217		H_2O 2; EtOH 4; eth 4; ace 2	
2525	Benzoic acid, 3-bromo-, methyl ester		$C_8H_7BrO_2$	215.05	618-89-3	32		125^{15}	4-09-00-01014		H_2O 2; EtOH 3; eth 3	
2526	Benzoic acid, 4-bromo-, methyl ester		$C_8H_7BrO_2$	215.05	619-42-1	81			4-09-00-01017	1.689^{25}	EtOH 3; eth 3; ace 3; bz 4	
2527	Benzoic acid, 2-bromo-4-nitro-	2-Bromo-4-nitrobenzoic acid	$C_7H_4BrNO_4$	246.02	16426-64-5	166.5		sub 150	3-09-00-01771		eth 4; EtOH 4	
2528	Benzoic acid, 2-bromo-5-nitro-	2-Bromo-5-nitrobenzoic acid	$C_7H_4BrNO_4$	246.02	943-14-6	180.5		sub	3-09-00-01771		eth 4; EtOH 4	
2529	Benzoic acid, 4-bromo-3-nitro-	4-Bromo-3-nitrobenzoic acid	$C_7H_4BrNO_4$	246.02	6319-40-0	203.5		sub	4-09-00-01236		EtOH 4	
2530	Benzoic acid, 5-bromo-2-nitro-	2-Nitro-5-bromobenzoic acid	$C_7H_4BrNO_4$	246.02	6950-43-2	140			1-09-00-00165	1.920^{18}	bz 4; EtOH 4	
2531	Benzoic acid, 4-(butylamino)-, 2-(dimethylamino)ethyl ester, monohydrochloride	Tetracaine hydrochloride	$C_{15}H_{25}ClN_2O_2$	300.83	136-47-0	147	9123		4-14-00-01172			
2532	Benzoic acid, butyl ester	Butyl benzoate	$C_{11}H_{14}O_2$	178.23	136-60-7	-22.4	1552	250.3	4-09-00-00290	1.000^{20}	H_2O 1; EtOH 5; eth 5; ace 3	1.4940^{25}
2533	Benzoic acid, 4,4'-carbonylbis-		$C_{15}H_{10}O_5$	270.24	964-68-1	365		sub	4-10-00-03497		HOAc 4	
2534	Benzoic acid, 2-chloro-	*o*-Chlorobenzoic acid	$C_7H_5ClO_2$	156.57	118-91-2	140.2	2125	sub	4-09-00-00956	1.544^{20}	H_2O 3; EtOH 4; eth 4; ace 4	
2535	Benzoic acid, 3-chloro-	*m*-Chlorobenzoic acid	$C_7H_5ClO_2$	156.57	535-80-8	158	2124	sub	4-09-00-00969	1.496^{25}	H_2O 4; EtOH 3; eth 4; bz 2	
2536	Benzoic acid, 4-chloro-	*p*-Chlorobenzoic acid	$C_7H_5ClO_2$	156.57	74-11-3	243	2126		4-09-00-00973		H_2O 1; EtOH 4; eth 2; ace 2	
2537	Benzoic acid, 2-chloro-, ethyl ester		$C_9H_9ClO_2$	184.62	7335-25-3			243	4-09-00-00957	1.1942^{15}	eth 4; EtOH 4	1.5242^{15}
2538	Benzoic acid, 3-chloro-, ethyl ester		$C_9H_9ClO_2$	184.62	1128-76-3			243	4-09-00-00970	1.1859^{15}	H_2O 1; EtOH 3; eth 3	1.5223^{20}
2539	Benzoic acid, 4-chloro-, ethyl ester		$C_9H_9ClO_2$	184.62	7335-27-5			237.5	4-09-00-00975	1.1873^{14}	EtOH 4	
2540	Benzoic acid, 3-chloro-4-hydroxy-		$C_7H_5ClO_3$	172.57	3964-58-7	171		sub	4-10-00-00463		H_2O 2; EtOH 4; eth 4; ace 4	

No.	Name Synonym	Mol. Form. Mol. Wt.	CAS RN mp/°C	Merck No. bp/°C	Beil. Ref. den/g cm^{-3}	Solubility n_D
2541	Benzoic acid, 3-chloro-2-hydroxy-, ethyl ester Salicylic acid, 3-chloro, ethyl ester	$C_9H_9ClO_3$ 200.62	56961-32-1 21	 269.5	0-10-00-00101	EtOH 4
2542	Benzoic acid, 5-chloro-2-hydroxy-, methyl ester	$C_8H_7ClO_3$ 186.59	4068-78-4 50	 249 dec; 120^{12}	4-10-00-00209	EtOH 4
2543	Benzoic acid, 4-chloro-, methyl ester	$C_8H_7ClO_2$ 170.60	1126-46-1 43.5		4-09-00-00974 1.382^{20}	EtOH 4
2544	Benzoic acid, 2-chloro-3-nitro- 2-Chloro-3-nitrobenzoic acid	$C_7H_4ClNO_4$ 201.57	3970-35-2 183.5		4-09-00-01225 1.662^{18}	H_2O 3; EtOH 3
2545	Benzoic acid, 2-chloro-5-nitro-	$C_7H_4ClNO_4$ 201.57	2516-96-3 166.5		4-09-00-01228 1.608^{18}	H_2O 2; EtOH 3; eth 3; ace 2
2546	Benzoic acid, 3-chloro-2-nitro- 2-Nitro-3-chlorobenzoic acid	$C_7H_4ClNO_4$ 201.57	4771-47-5 238		4-09-00-01225 1.566^{18}	eth 4; EtOH 4
2547	Benzoic acid, 4-chloro-3-nitro-	$C_7H_4ClNO_4$ 201.57	96-99-1 182.8		4-09-00-01226 1.645^{18}	H_2O 1; EtOH 2; ace 2
2548	Benzoic acid, 4-chloro-3-nitro-, methyl ester Methyl 4-chloro-3-nitrobenzoate	$C_8H_6ClNO_4$ 215.59	14719-83-6 83		4-09-00-01226 1.522^{18}	EtOH 4
2549	Benzoic acid, 5-chloro-2-nitro-, methyl ester Methyl 5-chloro-2-nitrobenzoate	$C_8H_6ClNO_4$ 215.59	51282-49-6 48.5		0-09-00-00401 1.453^{18}	MeOH 4
2550	Benzoic acid, 3-cyano-	$C_8H_5NO_2$ 147.13	1877-72-1 219	 sub	4-09-00-03296	H_2O 2; EtOH 3; eth 3
2551	Benzoic acid, cyclohexyl ester	$C_{13}H_{16}O_2$ 204.27	2412-73-9 <-10	 285	4-09-00-00296 1.0429^{20}	H_2O 1; EtOH 3; eth 3 1.5200^{20}
2552	Benzoic acid, 2,3-diamino- 2,3-Diaminobenzoic acid	$C_7H_8N_2O_2$ 152.15	603-81-6 201	 dec	3-14-00-01172	EtOH 4; HOAc 4
2553	Benzoic acid, 2,4-diamino- 4-Aminoanthranilic acid	$C_7H_8N_2O_2$ 152.15	611-03-0 140	 200 dec	4-14-00-01299	EtOH 4; HOAc 4
2554	Benzoic acid, 3,5-diamino- 3,5-Diaminobenzoic acid	$C_7H_8N_2O_2$ 152.15	535-87-5 228	2955	4-14-00-01304	H_2O 2; EtOH 3; eth 4; tfa 2
2555	Benzoic acid, 2,4-dibromo- 2,4-Dibromobenzoic acid	$C_7H_4Br_2O_2$ 279.92	611-00-7 174	 sub	4-09-00-01027	eth 4; EtOH 4
2556	Benzoic acid, 2,5-dibromo- 2,5-Dibromobenzoic acid	$C_7H_4Br_2O_2$ 279.92	610-71-9 157	 sub	4-09-00-01027	eth 4; EtOH 4; chl 4
2557	Benzoic acid, 2,6-dibromo- 2,6-Dibromobenzoic acid	$C_7H_4Br_2O_2$ 279.92	601-84-3 150.5	 209^{16}	4-09-00-01028	H_2O 3; EtOH 4; eth 4; ace 3
2558	Benzoic acid, 3,5-dibromo-2-hydroxy- 3,5-Dibromosalicylic acid	$C_7H_4Br_2O_3$ 295.91	3147-55-5 228	3012	4-10-00-00222	ace 3
2559	Benzoic acid, 3,5-dibromo-2-hydroxy-, methyl ester Methyl 3,5-dibromosalicylate	$C_8H_6Br_2O_3$ 309.94	21702-79-4 149	 181^{12}	4-10-00-00223	eth 4
2560	Benzoic acid, 2,6-dibromo-3,4,5-trihydroxy- Dibromogallic acid	$C_7H_4Br_2O_5$ 327.91	602-92-6 150	3006	2-10-00-00347	H_2O 4; eth 4; EtOH 4
2561	Benzoic acid, 2,4-dichloro- 2,4-Dichlorobenzoic acid	$C_7H_4Cl_2O_2$ 191.01	50-84-0 164.2	 sub	4-09-00-00998	H_2O 3; EtOH 3; eth 3; ace 2
2562	Benzoic acid, 2,5-dichloro- 2,5-Dichlorobenzoic acid	$C_7H_4Cl_2O_2$ 191.01	50-79-3 154.4	 301	4-09-00-01005	H_2O 2; EtOH 3; eth 3
2563	Benzoic acid, 2,6-dichloro- 2,6-Dichlorobenzoic acid	$C_7H_4Cl_2O_2$ 191.01	50-30-6 144	 sub	4-09-00-01005	H_2O 3; EtOH 3; eth 3; bz 3
2564	Benzoic acid, 3,5-dichloro- 3,5-Dichlorobenzoic acid	$C_7H_4Cl_2O_2$ 191.01	51-36-5 188	 sub	4-09-00-01008	H_2O 2; EtOH 3; eth 3; lig2
2565	Benzoic acid, 4-[(dichloroamino)sulfonyl]- Halazone	$C_7H_5Cl_2NO_4S$ 270.09	80-13-7 195 dec	4503	2-11-00-00220	H_2O 2; chl 2; HOAc 4; peth 1
2566	Benzoic acid, 3,5-dichloro-2-hydroxy-	$C_7H_4Cl_2O_3$ 207.01	320-72-9 220.5	 sub	4-10-00-00213	H_2O 2; EtOH 4; eth 3
2567	Benzoic acid, 3,6-dichloro-2-methoxy- Dicamba	$C_8H_6Cl_2O_3$ 221.04	1918-00-9 115	3026	 1.57^{25}	
2568	Benzoic acid, 2,3-dihydroxy-	$C_7H_6O_4$ 154.12	303-38-8 204		4-10-00-01414 1.542^{20}	H_2O 3; EtOH 3; eth 3; ace 2
2569	Benzoic acid, 2,4-dihydroxy- β-Resorcylic acid	$C_7H_6O_4$ 154.12	89-86-1 225-7 dec	8160	4-10-00-01420	H_2O 3; EtOH 3; eth 3; bz 3
2570	Benzoic acid, 2,5-dihydroxy- Gentisic acid	$C_7H_6O_4$ 154.12	490-79-9 199.5	4290	4-10-00-01441	H_2O 4; EtOH 4; eth 4; ace 3
2571	Benzoic acid, 3,4-dihydroxy- Protocatechuic acid	$C_7H_6O_4$ 154.12	99-50-3 200-2 dec	7909	4-10-00-01459 1.524^{4}	H_2O 2; EtOH 4; eth 3; bz 1
2572	Benzoic acid, 2-(3,6-dihydroxy-9H-xanthen-9-yl)- Fluorescin	$C_{20}H_{14}O_5$ 334.33	518-44-6 126	4007	2-18-00-00307	H_2O 1; EtOH 3; eth 3; ace 3
2573	Benzoic acid, 2,4-dihydroxy-6-methyl- *o*-Orsellinic acid	$C_8H_8O_4$ 168.15	480-64-8 176 dec	6834	4-10-00-01526	EtOH 3; eth 3
2574	Benzoic acid, 2,4-dihydroxy-6-methyl-, ethyl ester	$C_{10}H_{12}O_4$ 196.20	2524-37-0 132	 sub	4-10-00-01526	eth 4; EtOH 4
2575	Benzoic acid, 3,4-dihydroxy-5-[(3,4,5-trihydroxybenzoyl)oxy]- Digallic acid	$C_{14}H_{10}O_9$ 322.23	536-08-3 268-70 dec	3135	3-10-00-02086	ace 4; EtOH 4
2576	Benzoic acid, 3,4-dimethoxy- Veratric acid	$C_9H_{10}O_4$ 182.18	93-07-2 181	9854 sub	4-10-00-01460	H_2O 1; EtOH 4; eth 4; chl 2
2577	Benzoic acid, 3,5-dimethoxy- 3,5-Dimethoxybenzoic acid	$C_9H_{10}O_4$ 182.18	1132-21-4 185.5	 sub	4-10-00-01501	eth 4; EtOH 4

No.	Name	Synonym	Mol. Form.	Mol. Wt.	CAS RN	mp/°C	Merck No.	bp/°C	Beil. Ref.	den/g cm^{-3}	Solubility	n_D
2578	Benzoic acid, 3,4-dimethoxy-, ethyl ester		$C_{11}H_{14}O_4$	210.23	3943-77-9	43.5		295.5	4-10-00-01469		eth 4; EtOH 4	
2579	Benzoic acid, 3,4-dimethoxy-, methyl ester		$C_{10}H_{12}O_4$	196.20	2150-38-1	60.8		283	4-10-00-01467		bz 4; eth 4; EtOH 4	
2580	Benzoic acid, 2,4-dimethyl-	4-Carboxy-1,3-dimethylbenzene	$C_9H_{10}O_2$	150.18	611-01-8	90		268	4-09-00-01801		H_2O 2; EtOH 3; ace 3; bz 3	
2581	Benzoic acid, 2,5-dimethyl-		$C_9H_{10}O_2$	150.18	610-72-0	132		sub	4-09-00-01802	1.069^{21}	H_2O 1; EtOH 3; eth 3; ace 3	
2582	Benzoic acid, 2,6-dimethyl-	2,6-Dimethylbenzoic acid	$C_9H_{10}O_2$	150.18	632-46-2	116		274.5; 155^{17}	4-09-00-01798		H_2O 2; EtOH 3; eth 3; lig 2	
2583	Benzoic acid, 3,5-dimethyl-	Mesitylenic acid	$C_9H_{10}O_2$	150.18	499-06-9	171.1		sub	4-09-00-01806		H_2O 2; EtOH 4; eth 4	
2584	Benzoic acid, 2-(dimethylamino)-		$C_9H_{11}NO_2$	165.19	610-16-2	72		sub	4-14-00-01017		H_2O 4; eth 4; EtOH 4	
2585	Benzoic acid, 4-(dimethylamino)-	*p*-(Dimethylamino)benzoic acid	$C_9H_{11}NO_2$	165.19	619-84-1	242.5	3221		4-14-00-01164		EtOH 3; eth 2	
2586	Benzoic acid, 6-[[6-[2-(dimethylamino)ethyl]-4-methoxy-1,3-benzodioxol-5-yl]acetyl]-2,3-dimethoxy-	Narceine	$C_{23}H_{27}NO_8$	445.47	131-28-2	138	6341		5-19-09-00051			
2587	Benzoic acid, 2-[[4-(dimethylamino)phenyl]azo]-	Methyl red	$C_{15}H_{15}N_3O_2$	269.30	493-52-7	183	6037		4-16-00-00504		H_2O 2; EtOH 3; ace 4; bz 4	
2588	Benzoic acid, 4-(1,1-dimethylethyl)-	*p-tert*-Butylbenzoic acid	$C_{11}H_{14}O_2$	178.23	98-73-7	164.5			4-09-00-01884		H_2O 1; EtOH 4; bz 4; chl 3	
2589	Benzoic acid, 2,4-dimethyl-, methyl ester	Methyl 2,4-dimethylbenzoate	$C_{10}H_{12}O_2$	164.20	23617-71-2	-2		232.5	2-09-00-00350			1.5052^{20}
2590	Benzoic acid, 3,4-dinitro-	3,4-Dinitrobenzoic acid	$C_7H_4N_2O_6$	212.12	528-45-0	166	3268		4-09-00-01242		H_2O 2; EtOH 4; eth 4	
2591	Benzoic acid, 3,5-dinitro-	3,5-Dinitrobenzoic acid	$C_7H_4N_2O_6$	212.12	99-34-3	205	3269		4-09-00-01242		EtOH 4; HOAc 4	
2592	Benzoic acid, 3,5-dinitro-, butyl ester	Butyl 3,5-dinitrobenzoate	$C_{11}H_{12}N_2O_6$	268.23	10478-02-1	62.5			4-09-00-01246		EtOH 3; DMSO-d_6 2	1.488
2593	Benzoic acid, 3,5-dinitro-, ethyl ester	Ethyl 3,5-dinitrobenzoate	$C_9H_8N_2O_6$	240.17	618-71-3	94			4-09-00-01243	1.295^{111}	EtOH 3	1.560
2594	Benzoic acid, 3,3'-[(1,6-dioxo-1,6-hexanediyl)diimino]bis[2,4,6-triiodo-	Iodipamide	$C_{20}H_{14}I_6N_2O_6$	1139.77	606-17-7	307 dec	4916		4-14-00-01122		H_2O 1; EtOH 2; eth 2; ace 2	
2595	Benzoic acid, 4-[(dipropylamino)sulfonyl]-	Probenecid	$C_{13}H_{19}NO_4S$	285.36	57-66-9	195	7760		4-11-00-00691			
2596	Benzoic acid, 2-ethoxy-		$C_9H_{10}O_3$	166.18	134-11-2	20.7		$211\text{-}2^{39}$	3-10-00-00098		H_2O 2; EtOH 2; ctc 2	
2597	Benzoic acid, 3-ethoxy-		$C_9H_{10}O_3$	166.18	621-51-2	137		sub	3-10-00-00245		H_2O 3; EtOH 3; eth 3; bz 3	
2598	Benzoic acid, 2-ethoxy-, ethyl ester	Ethyl 2-ethoxybenzoate	$C_{11}H_{14}O_3$	194.23	6290-24-0			251	3-10-00-00117	1.1005^{20}	eth 4; EtOH 4	
2599	Benzoic acid, 4-ethoxy-, ethyl ester	Ethyl 4-ethoxybenzoate	$C_{11}H_{14}O_3$	194.23	23676-09-7			275	4-10-00-00368	1.076^{12}	H_2O 1; EtOH 4; eth 4; ctc 2	
2600	Benzoic acid, 2-ethyl-		$C_9H_{10}O_2$	150.18	612-19-1	68		259	4-09-00-01789	1.0431^{100}	eth 4; EtOH 4	1.5099^{100}
2601	Benzoic acid, 3-ethyl-		$C_9H_{10}O_2$	150.18	619-20-5	47			4-09-00-01791	1.042^{100}	eth 4; EtOH 4	1.5345^{100}
2602	Benzoic acid, ethyl ester	Ethyl benzoate	$C_9H_{10}O_2$	150.18	93-89-0	-34	3725	212	4-09-00-00285	1.0511^{15}	H_2O 1; EtOH 3; eth 5; ace 3	1.5007^{20}
2603	Benzoic acid, 2-fluoro-		$C_7H_5FO_2$	140.11	445-29-4	126.5			4-09-00-00950	1.460^{25}	H_2O 2; EtOH 4; eth 4; bz 1	
2604	Benzoic acid, 3-fluoro-		$C_7H_5FO_2$	140.11	455-38-9	124			4-09-00-00952	1.474^{25}	H_2O 2; eth 3	
2605	Benzoic acid, 4-fluoro-	*p*-Fluorobenzoic acid	$C_7H_5FO_2$	140.11	456-22-4	185	4100		4-09-00-00953	1.479^{25}	H_2O 2; EtOH 3; eth 3; ace 2	
2606	Benzoic acid, 4-fluoro-, ethyl ester		$C_9H_9FO_2$	168.17	451-46-7	26		210	4-09-00-00953	1.146^{25}	eth 4; EtOH 4	1.4864^{20}
2607	Benzoic acid, 2-formyl-		$C_8H_6O_3$	150.13	119-67-5	98			4-10-00-02748	1.404^{25}	H_2O 3; EtOH 4; eth 4	
2608	Benzoic acid, 6-formyl-2,3-dimethoxy-	Opianic acid	$C_{10}H_{10}O_5$	210.19	519-05-1	150	6806		4-10-00-03863		EtOH 3; eth 3	
2609	Benzoic acid, 3-formyl-4-hydroxy-		$C_8H_6O_4$	166.13	584-87-2	244		sub	4-10-00-03622		eth 4; EtOH 4	
2610	Benzoic acid, hexyl ester		$C_{13}H_{18}O_2$	206.28	6789-88-4			272; 139^{8}	4-09-00-00293	0.9793^{20}	H_2O 1; EtOH 3; ace 3	
2611	Benzoic acid, hydrazide		$C_7H_8N_2O$	136.15	613-94-5	115		267 dec	4-09-00-00922		H_2O 3; EtOH 3; eth 2; ace 2	
2612	Benzoic acid, 2-hydroxy-	Salicylic acid	$C_7H_6O_3$	138.12	69-72-7	158	8301	211^{20}	4-10-00-00125	1.443^{20}	H_2O 2; EtOH 4; eth 4; ace 4	1.565
2613	Benzoic acid, 4-hydroxy-	*p*-Salicylic acid	$C_7H_6O_3$	138.12	99-96-7	214.5	4742		4-10-00-00345	1.46^{25}	H_2O 2; EtOH 4; eth 3; ace 3	
2614	Benzoic acid, 2-hydroxy-, butyl ester		$C_{11}H_{14}O_3$	194.23	2052-14-4	-5.9		271	4-10-00-00153	1.0728^{20}	ctc 2	1.5115^{20}
2615	Benzoic acid, 4-hydroxy-, butyl ester	Butylparaben	$C_{11}H_{14}O_3$	194.23	94-26-8	68.5	1583		4-10-00-00375		H_2O 2; EtOH 3; ctc 2	

No.	Name	Synonym	Mol. Form.	Mol. Wt.	CAS RN	mp/°C	Merck No.	bp/°C	Beil. Ref.	den/g cm^{-3}	Solubility	n_D
2616	Benzoic acid, 2-hydroxy-, 2-carboxyphenyl ester	Salsalate	$C_{14}H_{10}O_5$	258.23	552-94-3	147	8307		4-10-00-00165		ace 2	
2617	Benzoic acid, 2-hydroxy-3,5-diiodo-	3,5-Diiodosalicylic acid	$C_7H_4I_2O_3$	389.92	133-91-5	235.5	3175		4-10-00-00226		H_2O 2; EtOH 4; eth 4; bz 1	
2618	Benzoic acid, 4-hydroxy-3,5-diiodo-		$C_7H_4I_2O_3$	389.92	618-76-8	237		260 dec	4-10-00-00479		H_2O 1; EtOH 4; eth 4; bz 2	
2619	Benzoic acid, 2-hydroxy-, ethyl ester	Ethyl salicylate	$C_9H_{10}O_3$	166.18	118-61-6	45	3804	$150\text{-}1^{10}$	4-10-00-00149	1.1326^{20}	H_2O 1; EtOH 5; eth 4; ctc 3	1.5296^{20}
2620	Benzoic acid, 3-hydroxy-, ethyl ester		$C_9H_{10}O_3$	166.18	7781-98-8	74			4-10-00-00321	1.0680^{131}	H_2O 2; EtOH 3; eth 3; chl 2	
2621	Benzoic acid, 4-hydroxy-, ethyl ester	Ethylparaben	$C_9H_{10}O_3$	166.18	120-47-8	117	3792	297.5	4-10-00-00367		H_2O 2; EtOH 4; eth 4; chl 2	
2622	Benzoic acid, 2-hydroxy-, 2-hydroxyethyl ester	Glycol salicylate	$C_9H_{10}O_4$	182.18	87-28-5	37	4395	173^{15}	3-10-00-00138	1.2526^{15}	H_2O 2; EtOH 4; eth 4; bz 4	
2623	Benzoic acid, 4-hydroxy-3-iodo-	4-Hydroxybenzoic acid, 3-iodo	$C_7H_5IO_3$	264.02	37470-46-5	174.3		sub	4-10-00-00478		eth 4; EtOH 4	
2624	Benzoic acid, 5-hydroxy-2-iodo-	3-Hydroxybenzoic acid, 6-iodo	$C_7H_5IO_3$	264.02	57772-57-3	198		sub 160	3-10-00-00261		eth 4; EtOH 4	
2625	Benzoic acid, 4-hydroxy-3-methoxy-	Vanillic acid	$C_8H_8O_4$	168.15	121-34-6	211.5	9838	sub	4-10-00-01459		H_2O 2; EtOH 4; eth 3; DMSO 3	
2626	Benzoic acid, 4-hydroxy-3-methoxy-, ethyl ester		$C_{10}H_{12}O_4$	196.20	617-05-0	44		292	3-10-00-01413		H_2O 1; EtOH 4; eth 4; chl 3	
2627	Benzoic acid, 4-hydroxy-3-methoxy-, methyl ester		$C_9H_{10}O_4$	182.18	3943-74-6	64		286	3-10-00-01410		EtOH 3; chl 2; peth 3	
2628	Benzoic acid, 4-hydroxy-3-methoxy-, 2-methylpropyl ester		$C_{12}H_{16}O_4$	224.26	7152-88-7	56.5		125^{2}	3-10-00-01415		chl 2	
2629	Benzoic acid, 2-hydroxy-3-methyl-	*o*-Cresotic acid	$C_8H_8O_3$	152.15	83-40-9	165.5	2585		4-10-00-00601		H_2O 2; EtOH 3; eth 3; bz 3	
2630	Benzoic acid, 2-hydroxy-4-methyl-	*m*-Cresotic acid	$C_8H_8O_3$	152.15	50-85-1	177	2584		4-10-00-00617		H_2O 2; EtOH 3; eth 4; bz 3	
2631	Benzoic acid, 2-hydroxy-5-methyl-	*p*-Cresotic acid	$C_8H_8O_3$	152.15	89-56-5	151	2586		4-10-00-00610		H_2O 2; EtOH 3; eth 3; bz 3	
2632	Benzoic acid, 3-hydroxy-5-methyl-		$C_8H_8O_3$	152.15	585-81-9	210		sub	4-10-00-00609		H_2O 3; EtOH 4; eth 4; chl 2	
2633	Benzoic acid, 2-hydroxy-, 3-methylbutyl ester	Isopentyl salicylate	$C_{12}H_{16}O_3$	208.26	87-20-7		5007	278; 151^{15}	4-10-00-00153	1.0535^{20}	H_2O 1; EtOH 4; eth 3; ctc 2	1.5080^{20}
2634	Benzoic acid, 2-hydroxy-, methyl ester	Methyl salicylate	$C_8H_8O_3$	152.15	119-36-8	-8	6038	222.9	4-10-00-00143	1.181^{25}	eth 4; EtOH 4; chl 4	1.535^{20}
2635	Benzoic acid, 2-hydroxy-, 1-methylethyl ester	Isopropyl salicylate	$C_{10}H_{12}O_3$	180.20	607-85-2			238	4-10-00-00152	1.0729^{20}	H_2O 1; EtOH 5; eth 5	1.5065^{20}
2636	Benzoic acid, 3-hydroxy-, methyl ester		$C_8H_8O_3$	152.15	19438-10-9	73		281; 178^{17}	4-10-00-00319	1.1528^{100}	EtOH 3; bz 3; chl 2; peth 3	
2637	Benzoic acid, 4-hydroxy-, methyl ester	Methylparaben	$C_8H_8O_3$	152.15	99-76-3	131	6021	275 dec	4-10-00-00360		H_2O 2; EtOH 4; eth 4; ace 4	
2638	Benzoic acid, 2-hydroxy-3-methyl-, methyl ester		$C_9H_{10}O_3$	166.18	23287-26-5	29		235	4-10-00-00602	1.1683^{25}		1.5354^{16}
2639	Benzoic acid, 2-hydroxy-4-methyl-, methyl ester		$C_9H_{10}O_3$	166.18	4670-56-8	27.5		243	2-10-00-00139	1.1483^{15}		1.5378^{15}
2640	Benzoic acid, 2-hydroxy-5-methyl-, methyl ester		$C_9H_{10}O_3$	166.18	22717-57-3	-1		244.5	4-10-00-00611	1.1673^{25}		1.5351^{15}
2641	Benzoic acid, 2-hydroxy-3-methyl-6-(1-methylethyl)-		$C_{11}H_{14}O_3$	194.23	4389-53-1	136		sub	4-10-00-00738		eth 4; EtOH 4	
2642	Benzoic acid, 2-hydroxy-6-methyl-3-(1-methylethyl)-	*o*-Thymotic acid	$C_{11}H_{14}O_3$	194.23	548-51-6	127	9342	sub	3-10-00-00629		bz 4; eth 4; EtOH 4	
2643	Benzoic acid, 2-hydroxy-, 2-methylpropyl ester	Isobutyl salicylate	$C_{11}H_{14}O_3$	194.23	87-19-4	5.9		261	3-10-00-00121	1.0630^{20}	H_2O 1; EtOH 3; eth 3; ctc 3	1.5087^{20}
2644	Benzoic acid, 2-hydroxy-, 1-naphthalenyl ester	1-Naphthyl salicylate	$C_{17}H_{12}O_3$	264.28	550-97-0	83	6334		2-10-00-00052		eth 4	
2645	Benzoic acid, 2-hydroxy-, 2-naphthalenyl ester	2-Naphthyl salicylate	$C_{17}H_{12}O_3$	264.28	613-78-5	95.5	6335		4-10-00-00158	1.11^{116}	H_2O 1; EtOH 2; eth 3; bz 3	
2646	Benzoic acid, 2-hydroxy-3-nitro-	3-Nitrosalicylic acid	$C_7H_5NO_5$	183.12	85-38-1	148	6553		4-10-00-00228		H_2O 2; EtOH 4; eth 4; ace 3	
2647	Benzoic acid, 2-hydroxy-5-nitro-	5-Nitrosalicylic acid	$C_7H_5NO_5$	183.12	96-97-9	229.5	6554		4-10-00-00255	1.650^{20}	H_2O 2; EtOH 4; eth 4; ace 4	
2648	Benzoic acid, 2-hydroxy-5-[(4-nitrophenyl)azo]-	Alizarine Yellow R	$C_{13}H_9N_3O_5$	287.23	2243-76-7	253-4 dec	241		4-16-00-00372		H_2O 4; EtOH 4	
2649	Benzoic acid, 2-hydroxy-, pentyl ester		$C_{12}H_{16}O_3$	208.26	2050-08-0			270	4-10-00-00153	1.064^{15}	H_2O 2; EtOH 5; eth 5	1.506^{20}
2650	Benzoic acid, 2-hydroxy-, phenyl ester	Phenyl salicylate	$C_{13}H_{10}O_3$	214.22	118-55-8	130.5	7282	173^{12}	4-10-00-00154	1.2614^{30}	H_2O 1; EtOH 4; eth 3; ace 4	

No.	Name Synonym	Mol. Form. Mol. Wt.	CAS RN mp/°C	Merck No. bp/°C	Beil. Ref. den/g cm^{-3}	Solubility n_D
2651	Benzoic acid, 2-hydroxy-, phenylmethyl ester Benzyl salicylate	$C_{14}H_{12}O_3$ 228.25	118-58-1	1160 320	4-10-00-00157 1.1799^{20}	H_2O 2; EtOH 3; eth 3; ctc 3 1.5805^{20}
2652	Benzoic acid, 2-hydroxy-, propyl ester	$C_{10}H_{12}O_3$ 180.20	607-90-9 97	 239	4-10-00-00152 1.0979^{20}	ctc 3; CS_2 3 1.5161^{20}
2653	Benzoic acid, 4-hydroxy-, propyl ester Propylparaben	$C_{10}H_{12}O_3$ 180.20	94-13-3 97	7879	4-10-00-00374 1.0630^{102}	H_2O 1; EtOH 3; eth 3; chl 2 1.5050^{102}
2654	Benzoic acid, 2-hydroxy-5-[[4-[(2-pyridinylamino)sulfonyl]phenyl]azo]- Sulfasalazine	$C_{18}H_{14}N_4O_5S$ 398.40	599-79-1 220 dec	8917	5-22-08-00433	
2655	Benzoic acid, 2-hydroxy-5-sulfo- Sulfosalicylic acid	$C_7H_6O_6S$ 218.19	97-05-2 120	8944	4-11-00-00702	H_2O 5; EtOH 5; eth 4
2656	Benzoic acid, 2,2'-iminobis- Diphenylamine-2,2'-dicarboxylic acid	$C_{14}H_{11}NO_4$ 257.25	579-92-0 296-7 dec	3318	4-14-00-01058	
2657	Benzoic acid, 2-iodo- *o*-Iodobenzoic acid	$C_7H_5IO_2$ 248.02	88-67-5 163	4923 exp 233	4-09-00-01030 2.25^{25}	H_2O 2; EtOH 4; eth 4; ace 2
2658	Benzoic acid, 3-iodo-	$C_7H_5IO_2$ 248.02	618-51-9 188.3	 sub	4-09-00-01033	H_2O 2; EtOH 4; eth 2
2659	Benzoic acid, 4-iodo-	$C_7H_5IO_2$ 248.02	619-58-9 270	 sub	4-09-00-01035 2.184^{20}	H_2O 2; EtOH 2; eth 3
2660	Benzoic acid, 2-iodo-, methyl ester	$C_8H_7IO_2$ 262.05	610-97-9	 280; 146^{18}	4-09-00-01031	EtOH 3 1.6052^{20}
2661	Benzoic acid, 3-iodo-, methyl ester	$C_8H_7IO_2$ 262.05	618-91-7 54.5	 277; 150^{18}	4-09-00-01034	H_2O 1; EtOH 3; eth 4; ace 4
2662	Benzoic acid, 4-iodo-, methyl ester	$C_8H_7IO_2$ 262.05	619-44-3 114.8	 sub	4-09-00-01036 2.0200^{10}	EtOH 3; eth 3
2663	Benzoic acid, 2-mercapto- *o*-Thiosalicylic acid	$C_7H_6O_2S$ 154.19	147-93-3 168.5	9291 sub	4-10-00-00272	H_2O 3; EtOH 3; eth 3; lig 2
2664	Benzoic acid, 2-methoxy-	$C_8H_8O_3$ 152.15	579-75-9 101	 200	4-10-00-00130	H_2O 2; EtOH 4; eth 4; bz 3
2665	Benzoic acid, 3-methoxy-	$C_8H_8O_3$ 152.15	586-38-9 107	 170^{10}	4-10-00-00316	H_2O 2; EtOH 3; eth 3; bz 3
2666	Benzoic acid, 4-methoxy- *p*-Anisic acid	$C_8H_8O_3$ 152.15	100-09-4 185	696 276.5	4-10-00-00346	H_2O 1; EtOH 4; eth 4; chl 3
2667	Benzoic acid, 2-(4-methoxybenzoyl)- *o*-(*p*-Anisoyl)benzoic acid	$C_{15}H_{12}O_4$ 256.26	1151-15-1 146	702	3-10-00-04429	eth 4; EtOH 4; tol 4
2668	Benzoic acid, 4-methoxy-, butyl ester	$C_{12}H_{16}O_3$ 208.26	6946-35-6	 183^{40}	3-10-00-00307 1.054^{16}	 1.5141^{16}
2669	Benzoic acid, 2-methoxy-, ethyl ester	$C_{10}H_{12}O_3$ 180.20	7335-26-4	 261	4-10-00-00150 1.1124^{20}	eth 4; EtOH 4 1.5224^{20}
2670	Benzoic acid, 3-methoxy-, ethyl ester	$C_{10}H_{12}O_3$ 180.20	10259-22-0	 260.5	4-10-00-00321 1.0993^{20}	eth 4; EtOH 4 1.5161^{20}
2671	Benzoic acid, 4-methoxy-, ethyl ester	$C_{10}H_{12}O_3$ 180.20	94-30-4 7.5	 269.5	4-10-00-00368 1.1038^{20}	H_2O 1; EtOH 3; eth 3 1.5254^{20}
2672	Benzoic acid, 3-methoxy-2-(methylamino)-, methyl ester Damascenine	$C_{10}H_{13}NO_3$ 195.22	483-64-7 28	2809 271; 147^{10}	4-14-00-02073	bz 4; eth 4; EtOH 4; lig 4
2673	Benzoic acid, 2-methoxy-, methyl ester	$C_9H_{10}O_3$ 166.18	606-45-1	 246.5	4-10-00-00144 1.1571^{19}	H_2O 1; EtOH 3 1.534^{19}
2674	Benzoic acid, 3-methoxy-, methyl ester	$C_9H_{10}O_3$ 166.18	5368-81-0	 248	4-10-00-00319 1.1310^{20}	H_2O 1; EtOH 3 1.5224^{20}
2675	Benzoic acid, 4-methoxy-, methyl ester	$C_9H_{10}O_3$ 166.18	121-98-2 49	 244	4-10-00-00360	H_2O 1; EtOH 3; eth 3; chl 3
2676	Benzoic acid, 2-methyl- *o*-Toluic acid	$C_8H_8O_2$ 136.15	118-90-1 103.7	9461 259	4-09-00-01697 1.062^{115}	H_2O 1; EtOH 4; eth 4; chl 3 1.512^{115}
2677	Benzoic acid, 3-methyl- *m*-Toluic acid	$C_8H_8O_2$ 136.15	99-04-7 108.7	9461	4-09-00-01712 1.054^{112}	H_2O 2; EtOH 4; eth 4; chl 2 1.509
2678	Benzoic acid, 4-methyl- *p*-Toluic acid	$C_8H_8O_2$ 136.15	99-94-5 179.6	9461	4-09-00-01724	H_2O 1; EtOH 4; eth 4; tfa 2
2679	Benzoic acid, 2-(methylamino)-	$C_8H_9NO_2$ 151.17	119-68-6 180.5	 $80^{0.01}$	4-14-00-01015	H_2O 2; EtOH 4; eth 4; bz 4
2680	Benzoic acid, 2-(methylamino)-, ethyl ester Anthranilic acid, *N*-methyl-, ethyl ester	$C_{10}H_{13}NO_2$ 179.22	35472-56-1 39	 266; 172^{45}	2-14-00-00213	eth 4
2681	Benzoic acid, 2-(methylamino)-, methyl ester	$C_9H_{11}NO_2$ 165.19	85-91-6 19	 255	4-14-00-01016 1.120^{15}	H_2O 1; EtOH 3; eth 3 1.5839^{15}
2682	Benzoic acid, 2-methyl-, anhydride	$C_{16}H_{14}O_3$ 254.29	607-86-3 38.5	 >325; 220^{11}	4-09-00-01700	eth 4
2683	Benzoic acid, 3-methyl-, anhydride	$C_{16}H_{14}O_3$ 254.29	21436-44-2 71	 230^{17}	4-09-00-01715	ace 4; bz 4; eth 4; EtOH 4
2684	Benzoic acid, 2-(4-methylbenzoyl)- 2-(*p*-Toluyl)benzoic acid	$C_{15}H_{12}O_3$ 240.26	85-55-2 146	9465	4-10-00-02991	H_2O 2; EtOH 4; eth 4; ace 4
2685	Benzoic acid, 3-methyl-, butyl ester Butyl *m*-toluate	$C_{12}H_{16}O_2$ 192.26	6640-77-3	 262.3	3-09-00-00397 1.0040^{20}	H_2O 1; EtOH 3; eth 5 1.4950^{20}
2686	Benzoic acid, 3,3'-methylenebis[6-hydroxy- 5,5'-Methylenedisalicylic acid	$C_{15}H_{12}O_6$ 288.26	122-25-8 243.5	5984	4-10-00-02337	ace 4; eth 4; EtOH 4
2687	Benzoic acid, methyl ester Methyl benzoate	$C_8H_8O_2$ 136.15	93-58-3 -15	5947 199	4-09-00-00283 1.0933^{15}	H_2O 1; EtOH 3; eth 5; ctc 3 1.5164^{20}
2688	Benzoic acid, 2-(1-methylethyl)-	$C_{10}H_{12}O_2$ 164.20	2438-04-2 64	 160-1^{25}	4-09-00-01839	H_2O 2; EtOH 4; eth 4; bz 3
2689	Benzoic acid, 4-(1-methylethyl)- Cumic acid	$C_{10}H_{12}O_2$ 164.20	536-66-3 117.5	2620 sub	4-09-00-01843 1.162^4	H_2O 2; EtOH 4; eth 4; peth 3

No.	Name Synonym	Mol. Form. Mol. Wt.	CAS RN mp/°C	Merck No. bp/°C	Beil. Ref. den/g cm^{-3}	Solubility n_D
2690	Benzoic acid, 2-methyl-, ethyl ester	$C_{10}H_{12}O_2$ 164.20	87-24-1 <-10	 227; 113^{18}	4-09-00-01699 1.0325^{21}	H_2O 1; EtOH 5; eth 5 1.507^{22}
2691	Benzoic acid, 3-methyl-, ethyl ester	$C_{10}H_{12}O_2$ 164.20	120-33-2	 234	4-09-00-01713 1.0265^{21}	eth 4; EtOH 4 1.5052^{22}
2692	Benzoic acid, 4-methyl-, ethyl ester	$C_{10}H_{12}O_2$ 164.20	94-08-6	 232	4-09-00-01726 1.0269^{18}	H_2O 1; EtOH 5; eth 5 1.5089^{18}
2693	Benzoic acid, 1-methylethyl ester	$C_{10}H_{12}O_2$ 164.20	939-48-0	 216	4-09-00-00289 1.0163^{15}	H_2O 1; EtOH 3; eth 3; ace 3 1.4890^{20}
2694	Benzoic acid, 4-(1-methylethyl)-, methyl ester	$C_{11}H_{14}O_2$ 178.23	20185-55-1	 126^{14}	4-09-00-01843 1.018^{19}	 1.515^{19}
2695	Benzoic acid, 4-(1-methylethyl)-, 2-methylpropyl ester Benzoic acid, 4-isopropyl, isobutyl ester	$C_{14}H_{20}O_2$ 220.31	6315-03-3	 155^{14}	2-09-00-00359 0.966^{19}	 1.497^{19}
2696	Benzoic acid, 2-methyl-, methyl ester	$C_9H_{10}O_2$ 150.18	89-71-4 <-50	 215	4-09-00-01699 1.068^{20}	H_2O 1; EtOH 5; eth 5
2697	Benzoic acid, 3-methyl-, methyl ester	$C_9H_{10}O_2$ 150.18	99-36-5	 221	4-09-00-01713 1.061^{20}	H_2O 1; EtOH 3; ctc 2
2698	Benzoic acid, 4-methyl-, methyl ester	$C_9H_{10}O_2$ 150.18	99-75-2 33.2	 220	4-09-00-01726	H_2O 1; EtOH 4; eth 4
2699	Benzoic acid, 2-methylphenyl ester	$C_{14}H_{12}O_2$ 212.25	617-02-7	 309; $155^{8.5}$	4-09-00-00305 1.114^{19}	eth 4; EtOH 4
2700	Benzoic acid, 4-methylphenyl ester	$C_{14}H_{12}O_2$ 212.25	614-34-6 71.5	 316	4-09-00-00306	eth 4; EtOH 4
2701	Benzoic acid, 2-methylpropyl ester	$C_{11}H_{14}O_2$ 178.23	120-50-3	 242	4-09-00-00291 0.9990^{20}	H_2O 1; EtOH 5; eth 5; ace 3
2702	Benzoic acid, 2-[(1-naphthalenylamino)carbonyl]- Naptalam	$C_{18}H_{13}NO_3$ 291.31	132-66-1 185	6338	3-12-00-02876 1.4^{20}	H_2O 1; EtOH 2; ace 2; bz 2
2703	Benzoic acid, 2-nitro- *o*-Nitrobenzoic acid	$C_7H_5NO_4$ 167.12	552-16-9 147.5		4-09-00-01046 1.575^{20}	H_2O 3; EtOH 4; eth 3; ace 4
2704	Benzoic acid, 3-nitro- *m*-Nitrobenzoic acid	$C_7H_5NO_4$ 167.12	121-92-6 141		4-09-00-01055 1.494^{20}	H_2O 2; EtOH 4; eth 4; ace 4
2705	Benzoic acid, 4-nitro- *p*-Nitrobenzoic acid	$C_7H_5NO_4$ 167.12	62-23-7 242	 sub	4-09-00-01072 1.610^{20}	ace 4; eth 4; EtOH 4; chl 4
2706	Benzoic acid, 4-nitro-, butyl ester	$C_{11}H_{13}NO_4$ 223.23	120-48-9 35.3	 160^8	3-09-00-01544	bz 4; eth 4
2707	Benzoic acid, 2-nitro-, ethyl ester	$C_9H_9NO_4$ 195.17	610-34-4 30	 275; 178^{23}	4-09-00-01047	eth 4; EtOH 4
2708	Benzoic acid, 3-nitro-, ethyl ester	$C_9H_9NO_4$ 195.17	618-98-4 47	 297	4-09-00-01056	H_2O 1; EtOH 4; eth 4
2709	Benzoic acid, 4-nitro-, ethyl ester	$C_9H_9NO_4$ 195.17	99-77-4 57	 186.3; 153^8	4-09-00-01074	H_2O 1; EtOH 3; eth 3
2710	Benzoic acid, 2-nitro-, methyl ester	$C_8H_7NO_4$ 181.15	606-27-9 -13	 275	4-09-00-01047 1.2855^{20}	H_2O 1; EtOH 3; eth 3; bz 3
2711	Benzoic acid, 3-nitro-, methyl ester	$C_8H_7NO_4$ 181.15	618-95-1 78	 279^{60}	4-09-00-01056	H_2O 1; EtOH 2; eth 2; MeOH 2
2712	Benzoic acid, pentachloro- Perchlorobenzoic acid	$C_7HCl_5O_2$ 294.35	1012-84-6 209.3	 sub	4-09-00-01011	EtOH 4; tol 4; HOAc 4
2713	Benzoic acid, pentamethyl- Permethylbenzoic acid	$C_{12}H_{16}O_2$ 192.26	2243-32-5 210.5	 sub	4-09-00-01920	H_2O 1; EtOH 4
2714	Benzoic acid, 2-phenoxy-	$C_{13}H_{10}O_3$ 214.22	2243-42-7 113	 355	4-10-00-00132 1.1553^{50}	H_2O 1; EtOH 4; eth 4; chl 3
2715	Benzoic acid, 2-(phenylamino)- *N*-Phenylanthranilic acid	$C_{13}H_{11}NO_2$ 213.24	91-40-7 183.5	7244	4-14-00-01019	H_2O 1; EtOH 4; eth 2; bz 2
2716	Benzoic acid, phenyl ester Phenyl benzoate	$C_{13}H_{10}O_2$ 198.22	93-99-2 71	7246 314	4-09-00-00303 1.235^{20}	H_2O 1; EtOH 3; eth 3; chl 3
2717	Benzoic acid, 2-phenylhydrazide	$C_{13}H_{12}N_2O$ 212.25	532-96-7 168	 314	4-15-00-00169	H_2O 2; EtOH 3; eth 2; bz 3
2718	Benzoic acid, 2-(phenylmethyl)-	$C_{14}H_{12}O_2$ 212.25	612-35-1 118	 sub	4-09-00-02516	H_2O 2; EtOH 3; eth 3; bz 3
2719	Benzoic acid, 4-(phenylmethyl)-	$C_{14}H_{12}O_2$ 212.25	620-86-0 158.8	 sub	4-09-00-02518	bz 4; eth 4; EtOH 4; chl 4
2720	Benzoic acid, phenylmethyl ester Benzyl benzoate	$C_{14}H_{12}O_2$ 212.25	120-51-4 21	1141 323.5	4-09-00-00307 1.1121^{25}	H_2O 1; EtOH 3; eth 3; ace 3 1.5680^{20}
2721	Benzoic acid, 4-[[(phenylmethyl)sulfonyl]amino]- Carinamide	$C_{14}H_{13}NO_4S$ 291.33	536-95-8 229.5	1163	4-14-00-01264	EtOH 4
2722	Benzoic acid, 2-propenyl ester	$C_{10}H_{10}O_2$ 162.19	583-04-0		4-09-00-00296 1.0569^{15}	H_2O 1; EtOH 3; eth 3; ace 3 1.5178^{20}
2723	Benzoic acid, *o*-propyl- Benzoic acid, 2-propyl	$C_{10}H_{12}O_2$ 164.20	2438-03-1 58	 273; 164^{20}	4-09-00-01836 1.0020^{15}	eth 4; EtOH 4
2724	Benzoic acid, propyl ester	$C_{10}H_{12}O_2$ 164.20	2315-68-6 -51.6	 211	4-09-00-00288 1.0230^{20}	H_2O 1; EtOH 5; eth 5 1.5000^{20}
2725	Benzoic acid, sodium salt	$C_7H_5NaO_2$ 144.11	532-32-1 >300	8527	4-09-00-00273	H_2O 3
2726	Benzoic acid, 2-[[[4-[(2-thiazolylamino)sulfonyl]phenyl]amino]carbonyl]- Phthalylsulphathiazole	$C_{17}H_{13}N_3O_5S_2$ 403.44	85-73-4 273	7351	4-27-00-04639	H_2O 1; EtOH 2; eth 1; chl 1

No.	Name	Synonym	Mol. Form.	Mol. Wt.	CAS RN	mp/°C	Merck No.	bp/°C	Beil. Ref.	den/g cm^{-3}	Solubility	n_D
2727	Benzoic acid, 2,3,6-trichloro-	2,3,6-Trichlorobenzoic acid	$C_7H_3Cl_3O_2$	225.46	50-31-7	124.5			4-09-00-01009		H_2O 2; eth 3	
2728	Benzoic acid, 2,4,5-trichloro-	2,4,5-Trichlorobenzoic acid	$C_7H_3Cl_3O_2$	225.46	50-82-8	166.5		sub	4-09-00-01010		H_2O 1; EtOH 3; eth 3	
2729	Benzoic acid, 2-[[3-(trifluoromethyl)phenyl]amino]-	Flufenamic acid	$C_{14}H_{10}F_3NO_2$	281.23	530-78-9	133.5	4060		3-14-00-00905		DMSO 3	
2730	Benzoic acid, 2,3,4-trihydroxy-		$C_7H_6O_5$	170.12	610-02-6	221		sub	4-10-00-01971		H_2O 2; EtOH 3; eth 3; ace 3	
2731	Benzoic acid, 3,4,5-trihydroxy-	Gallic acid	$C_7H_6O_5$	170.12	149-91-7	253 dec	4251		3-10-00-02070	1.694^{6}	H_2O 2; EtOH 4; eth 2; ace 3	
2732	Benzoic acid, 3,4,5-trihydroxy-, propyl ester	Propyl gallate	$C_{10}H_{12}O_5$	212.20	121-79-9	130	7872		4-10-00-02003		H_2O 2	
2733	Benzoic acid, 2,4,5-trimethoxy-		$C_{10}H_{12}O_5$	212.20	490-64-2	145		300	4-10-00-01980		H_2O 4; bz 4; EtOH 4; peth 4	
2734	Benzoic acid, 3,4,5-trimethoxy-		$C_{10}H_{12}O_5$	212.20	118-41-2	172.3		$225-7^{10}$	4-10-00-01996		H_2O 2; EtOH 4; eth 4; chl 4	
2735	Benzoic acid, 2,4,6-trinitro-	2,4,6-Trinitrobenzoic acid	$C_7H_3N_3O_8$	257.12	129-66-8	228 dec	9640		4-09-00-01362		H_2O 2; EtOH 4; eth 3; ace 3	
2736	1H,5H-Benzo[ij]quinolizine-1,6(7H)-dione, 2,3-dihydro-	Julolidine, 1,6-dioxo	$C_{12}H_{11}NO_2$	201.22	39052-57-8	145.5		$190-210^{0.3}$	4-21-00-05525		EtOH 4	
2737	1H,5H-Benzo[ij]quinolizine, 2,3,6,7-tetrahydro-	Julolidine	$C_{12}H_{15}N$	173.26	479-59-4	40		280 dec; 155^{17}	5-20-07-00189	1.003^{20}		1.568^{25}
2738	Benzo[j]fluoranthene	Dibenzo[a,jk]fluorene	$C_{20}H_{12}$	252.32	205-82-3	166			4-05-00-02687		H_2O 1; EtOH 2; HOAc 2	
2739	Benzo[k]fluoranthene	2,3,1',8'-Binaphthylene	$C_{20}H_{12}$	252.32	207-08-9	217		480	4-05-00-02686		H_2O 1; EtOH 3; bz 3; HOAc 3	
2740	Benzonitrile	Phenyl cyanide	C_7H_5N	103.12	100-47-0	-12.7	1107	191.1	4-09-00-00892	1.0093^{15}	H_2O 2; EtOH 5; eth 5; ace 4	1.5289^{20}
2741	Benzonitrile, 2-amino-		$C_7H_6N_2$	118.14	1885-29-6	51		263	4-14-00-01013		H_2O 2; EtOH 4; eth 4; ace 4	
2742	Benzonitrile, 3-amino-		$C_7H_6N_2$	118.14	2237-30-1	54.3		289	4-14-00-01095		H_2O 2; EtOH 4; eth 4; ace 4	
2743	Benzonitrile, 5-amino-2-methyl-		$C_8H_8N_2$	132.17	50670-64-9	88		$100-10^{22}$	3-14-00-01201		EtOH 4	
2744	Benzonitrile, 2-bromo-		C_7H_4BrN	182.02	2042-37-7	55.5		252	4-09-00-01013		H_2O 3; EtOH 4; chl 2	
2745	Benzonitrile, 3-bromo-		C_7H_4BrN	182.02	6952-59-6	39.5		225	4-09-00-01016		EtOH 4; eth 4; chl 2	
2746	Benzonitrile, 4-bromo-		C_7H_4BrN	182.02	623-00-7	114		236	4-09-00-01025		H_2O 3; EtOH 3; eth 3; chl 3	
2747	Benzonitrile, 2-chloro-		C_7H_4ClN	137.57	873-32-5	46.3		232	4-09-00-00965		H_2O 2; EtOH 3; eth 3; chl 3	
2748	Benzonitrile, 3-chloro-		C_7H_4ClN	137.57	766-84-7	41		100^{15}	4-09-00-00973		H_2O 1; EtOH 3; eth 3	
2749	Benzonitrile, 4-chloro-		C_7H_4ClN	137.57	623-03-0	95		223; 95^{5}	4-09-00-00992	1.1133^{17}	H_2O 2; EtOH 3; eth 3; bz 3	
2750	Benzonitrile, 2-(chloromethyl)-		C_8H_6ClN	151.60	612-13-5	61		252	4-09-00-01707		EtOH 4	
2751	Benzonitrile, 3,5-dibromo-4-hydroxy-	Bromoxynil	$C_7H_3Br_2NO$	276.91	1689-84-5	190	1431		4-10-00-00475			
2752	Benzonitrile, 2,6-dichloro-	Dichlobenil	$C_7H_3Cl_2N$	172.01	1194-65-6	144.5	3029	270	4-09-00-01006			
2753	Benzonitrile, 2,6-dimethoxy-	2,6-Dimethoxybenzonitrile	$C_9H_9NO_2$	163.18	16932-49-3	119		310	2-10-00-00260		H_2O 2; EtOH 4; eth 2; ace 3	
2754	Benzonitrile, 2-ethoxy-	*o*-Ethoxybenzonitrile	C_9H_9NO	147.18	6609-57-0	5		260.5	0-10-00-00097	1.0650^{15}	eth 4; EtOH 4; lig 4	
2755	Benzonitrile, 4-ethoxy-		C_9H_9NO	147.18	25117-74-2	61.5		258	3-10-00-00344		eth 4; EtOH 4; lig 4	
2756	Benzonitrile, 4-fluoro-		C_7H_4FN	121.11	1194-02-1	34.8		188.8	4-09-00-00954	1.1070^{55}	chl 2; peth 3	1.4925^{55}
2757	Benzonitrile, 3-formyl-		C_8H_5NO	131.13	24964-64-5	76.5		210	4-10-00-02752		H_2O 4; EtOH 4; eth 4; chl 4	
2758	Benzonitrile, 4-formyl-		C_8H_5NO	131.13	105-07-7	100.5		133^{12}	4-10-00-02754		H_2O 3; EtOH 4; eth 4; chl 4	
2759	Benzonitrile, 2-hydroxy-		C_7H_5NO	119.12	611-20-1	98		149^{14}	4-10-00-00191	1.1052^{100}	H_2O 2; EtOH 4; eth 4; bz 4	1.5372^{100}
2760	Benzonitrile, 2-methoxy-		C_8H_7NO	133.15	6609-56-9	24.5		255.5	4-10-00-00191	1.1063^{20}	EtOH 3; eth 4	
2761	Benzonitrile, 3-methoxy-		C_8H_7NO	133.15	1527-89-5			$111-2^{13}$	4-10-00-00329	1.089^{25}		1.5402^{20}
2762	Benzonitrile, 4-methoxy-		C_8H_7NO	133.15	874-90-8	61.5		256.5	4-10-00-00441		H_2O 1; EtOH 4; eth 4; bz 3	
2763	Benzonitrile, 2-methyl-	*o*-Tolunitrile	C_8H_7N	117.15	529-19-1	-13.5	9463	205	4-09-00-01703	0.9955^{20}	H_2O 1; EtOH 5; eth 5; ctc 2	1.5279^{20}
2764	Benzonitrile, 3-methyl-	*m*-Tolunitrile	C_8H_7N	117.15	620-22-4	-23		213	4-09-00-01717	1.0316^{20}	H_2O 1; EtOH 5; eth 5; ctc 2	1.5252^{20}
2765	Benzonitrile, 4-methyl-	*p*-Tolunitrile	C_8H_7N	117.15	104-85-8	29.5	9464	217.0	4-09-00-01738	0.9762^{30}	H_2O 1; EtOH 4; eth 4; ctc 2	

No.	Name Synonym	Mol. Form. Mol. Wt.	CAS RN mp/°C	Merck No. bp/°C	Beil. Ref. den/g cm^{-3}	Solubility n_D
2766	Benzonitrile, 2-methyl-5-nitro-	$C_8H_6N_2O_2$ 162.15	939-83-3 106	174^{18}	3-09-00-02314	ace 4; bz 4; eth 4; EtOH 4
2767	Benzonitrile, pentafluoro-	C_7F_5N 193.08	773-82-0 2.4	162	4-09-00-00956 1.563^{20}	1.4402^{25}
2768	Benzonitrile, 4-(2-phenylethenyl)-, (*Z*)- 4-Stilbenecarbonitrile (*cis*)	$C_{15}H_{11}N$ 205.26	14064-68-7 44.5	$130^{0.1}$	4-09-00-02615	H_2O 2; EtOH 3; chl 3
2769	Benzonitrile, 2-(trifluoromethyl)-	$C_8H_4F_3N$ 171.12	447-60-9 18	205	4-09-00-01705	ctc 3
2770	Benzonitrile, 3-(trifluoromethyl)-	$C_8H_4F_3N$ 171.12	368-77-4 14.5	189	3-09-00-02327 1.2813^{20}	1.4508^{20}
2771	Benzophenone, 3-(dimethylamino)- 3-(Dimethylamino)benzophenone	$C_{15}H_{15}NO$ 225.29	31766-07-1 47	216^{15}	4-14-00-00247	EtOH 3
2772	2H-1-Benzopyran 1,2-Chromene	C_9H_8O 132.16	254-04-6	1112 132^{102}; 91^{13}	5-17-02-00019 1.0993^{16}	H_2O 1 1.5869^{24}
2773	1H-2-Benzopyran-3-carboxylic acid, 6-[2-(4,9-dihydro-8-hydroxy-5,7-dimethoxy-4,9-dioxonaphtho[2,3-b]furan-2-yl Collinomycin	$C_{27}H_{20}O_{12}$ 536.45	27267-69-2 281	2478	5-19-08-00331	ace 4; diox 4; chl 4
2774	3H-2-Benzopyran-7-carboxylic acid, 4,6-dihydro-8-hydroxy-3,4,5-trimethyl-6-oxo-, (3*R*-*trans*)- Citrinin	$C_{13}H_{14}O_5$ 250.25	518-75-2 178-9 dec	2329	5-18-09-00061	H_2O 1; EtOH 2; eth 2; ace 3
2775	2H-1-Benzopyran-3-carboxylic acid, 2-oxo- Coumarin-3-carboxylic acid	$C_9H_6O_3$ 162.14	531-81-7 190 dec	2564	5-18-08-00323	EtOH 4
2776	1H-2-Benzopyran, 3,4-dihydro- Isochroman	$C_9H_{10}O$ 134.18	493-05-0	90^{12}	5-17-01-00595 1.067^{25}	1.5444^{20}
2777	2H-1-Benzopyran, 3,4-dihydro-	$C_9H_{10}O$ 134.18	493-08-3	215; 98^{18}	5-17-01-00593 1.0610^{20}	H_2O 3; os 5 1.5444^{20}
2778	2H-1-Benzopyran, 3,4-dihydro-2-methyl- 2-Methylchroman	$C_{10}H_{12}O$ 148.20	13030-26-7	224.5	5-17-01-00609 1.0340^{20}	bz 4; eth 4; EtOH 4 1.532^{13}
2779	2H-1-Benzopyran, 3,4-dihydro-3-methyl- Chroman, 3-methyl	$C_{10}H_{12}O$ 148.20	70401-56-8	$102\text{-}4^{15}$	4-17-00-00430 1.002^{22}	1.5335^{20}
2780	2H-1-Benzopyran, 3,4-dihydro-6-methyl- Chroman, 6-methyl-	$C_{10}H_{12}O$ 148.20	3722-74-5	234	4-17-00-00431 1.0374^{14}	1.5392^{25}
2781	2H-1-Benzopyran, 3,4-dihydro-7-methyl- Chroman, 7-methyl-	$C_{10}H_{12}O$ 148.20	3722-73-4	$141\text{-}3^{60}$	4-17-00-00431 1.028^{26}	1.5380^{20}
2782	2H-1-Benzopyran, 3,4-dihydro-2-phenyl- Flavan	$C_{15}H_{14}O$ 210.28	494-12-2 46.3	332; 190^{16}	5-17-02-00283	bz 4; eth 4; EtOH 4
2783	[1]Benzopyrano[3,4-b]furo[2,3-h][1]benzopyran-6(6aH)-one, 1,2,12,12a-tetrahydro-8,9-dimethoxy-2-(1-methylethen Rotenone	$C_{23}H_{22}O_6$ 394.42	83-79-4 176	8248 $210\text{-}20^{0.5}$	5-19-10-00581	H_2O 1; EtOH 3; eth 2; ace 3
2784	[2]Benzopyrano[6,5,4-def][2]benzopyran-1,3,6,8-tetrone	$C_{14}H_4O_6$ 268.18	81-30-1 450	sub 320	5-19-05-00423	H_2O 1; $Na2CO_3$ 3; HOAc 3
2785	2H-1-Benzopyran-6-ol, 3,4-dihydro-2,8-dimethyl-2-(4,8,12-trimethyltridecyl)-, [2*R*-[2*R**(4*R**,8*R**)]]- δ-Tocopherol	$C_{27}H_{46}O_2$ 402.66	119-13-1	9419 $150^{0.001}$		H_2O 1; EtOH 4; eth 4; ace 4
2786	2H-1-Benzopyran-7-ol, 3,4-dihydro-3-(4-hydroxyphenyl)-, (*S*)- Equol	$C_{14}H_{14}O_3$ 230.26	531-95-3 189.5	3583	5-17-05-00413	
2787	2H-1-Benzopyran-6-ol, 3,4-dihydro-2,5,7,8-tetramethyl-2-(4,8,12-trimethyltridecyl)-, [2*R*-[2*R**(4*R**,8*R**)]]- Vitamin E	$C_{29}H_{50}O_2$ 430.71	59-02-9 3.0	9931 $210^{0.1}$	5-17-04-00168 0.950^{25}	H_2O 1; EtOH 3; eth 3; ace 3 1.5045^{25}
2788	2H-1-Benzopyran-6-ol, 3,4-dihydro-2,5,7,8-tetramethyl-2-(4,8,12-trimethyltridecyl)-, acetate, [2*R*-[2*R**(4*R**,8*R** Vitamin E acetate	$C_{31}H_{52}O_3$ 472.75	58-95-7 -27.5	9932 $184^{0.01}$	5-17-04-00168 0.9533^{21}	H_2O 1; EtOH 2; eth 3; ace 3 1.497^{20}
2789	2H-1-Benzopyran-6-ol, 3,4-dihydro-2,5,8-trimethyl-2-(4,8,12-trimethyltridecyl)- β-Tocopherol	$C_{28}H_{48}O_2$ 416.69	148-03-8	9417 $200\text{-}10^{0.1}$		ace 4; eth 4; EtOH 4; chl 4
2790	2H-1-Benzopyran-6-ol, 3,4-dihydro-2,7,8-trimethyl-2-(4,8,12-trimethyltridecyl)- γ-Tocopherol	$C_{28}H_{48}O_2$ 416.69	7616-22-0 3	9418 $200\text{-}10^{0.1}$		H_2O 1; EtOH 5; eth 5; ace 5
2791	1H-2-Benzopyran-1-one Isocoumarin	$C_9H_6O_2$ 146.15	491-31-6 47	286	5-17-10-00149	H_2O 1; EtOH 4; eth 4; bz 4
2792	2H-1-Benzopyran-2-one Coumarin	$C_9H_6O_2$ 146.15	91-64-5 71	2563 301.7	5-17-10-00143 0.935^{20}	H_2O 3; EtOH 3; eth 4; chl 4
2793	4H-1-Benzopyran-4-one	$C_9H_6O_2$ 146.15	491-38-3 59	sub	5-17-10-00139 1.2900^{20}	H_2O 2; EtOH 3; eth 3; bz 3
2794	4H-1-Benzopyran-4-one, 7-[[6-*O*-(6-deoxy-α-*L*-mannopyranosyl)-β-*D*-glucopyranosyl]oxy]-2,3-dih Hesperidin	$C_{28}H_{34}O_{15}$ 610.57	520-26-3 262	4591	5-18-05-00218	py 4; EtOH 4; HOAc 4
2795	4H-1-Benzopyran-4-one, 7-[[2-*O*-(6-deoxy-α-*L*-mannopyranosyl)-β-*D*-glucopyranosyl]oxy]-2,3-dihydro-5-hy Naringin	$C_{27}H_{32}O_{14}$ 580.54	10236-47-2	6345	5-18-04-00528	H_2O 2; EtOH 2; eth 1; bz 1

No.	Name Synonym	Mol. Form. Mol. Wt.	CAS RN mp/°C	Merck No. bp/°C	Beil. Ref. den/g cm^{-3}	Solubility n_D
2796	4H-1-Benzopyran-4-one, 3-[(6-deoxy-α-*L*-mannopyranosyl)oxy]-2-(3,4-dihydroxyphenyl)-5,7-dihydroxy- Quercitrin	$C_{21}H_{20}O_{11}$ 448.38	522-12-3 170	8047	5-18-05-00514	H_2O 1; EtOH 3; eth 1; HOAc 3
2797	1H-2-Benzopyran-1-one, 3,4-dihydro-	$C_9H_8O_2$ 148.16	4702-34-5	176^{20}, $112^{0.4}$	5-17-10-00018 1.197^{25}	1.5629^{25}
2798	2H-1-Benzopyran-2-one, 3,4-dihydro-	$C_9H_8O_2$ 148.16	119-84-6 25	272	5-17-10-00013 1.169^{18}	H_2O 1; EtOH 2; eth 2; ctc 2 1.5563^{20}
2799	4H-1-Benzopyran-4-one, 2,3-dihydro- 4-Chromanone	$C_9H_8O_2$ 148.16	491-37-2 36.5	160^{50}, 127^{13}	5-17-10-00014 1.1291^{100}	EtOH 3; eth 4; ace 4; bz 4 1.5750
2800	4H-1-Benzopyran-4-one, 2,3-dihydro-5,7-dihydroxy-2-(3-hydroxy-4-methoxyphenyl)-, (*S*)- Hesperetin	$C_{16}H_{14}O_6$ 302.28	520-33-2 227.5	4590 sub 205	4-18-00-03215	eth 4; EtOH 4
2801	4H-1-Benzopyran-4-one, 2,3-dihydro-5,7-dihydroxy-3-(3-hydroxy-4-methoxyphenyl)- Flavanone, 4'-methoxy-3',5,7-trihydroxy	$C_{16}H_{14}O_6$ 302.28	99365-26-1 227.5	sub 205	2-18-00-00204	H_2O 1; EtOH 4; eth 3; bz 2
2802	4H-1-Benzopyran-4-one, 2,3-dihydro-5,7-dihydroxy-2-(4-hydroxyphenyl)-, (*S*)- Naringenin	$C_{15}H_{12}O_5$ 272.26	480-41-1 251	6344	5-18-04-00524	bz 4; eth 4; EtOH 4
2803	4H-1-Benzopyran-4-one, 2,3-dihydro-6-methyl-	$C_{10}H_{10}O_2$ 162.19	39513-75-2 35	$141\text{-}3^{13.5}$	5-17-10-00036 1.1245^{57}	1.555^{57}
2804	2H-1-Benzopyran-2-one, 6,7-dihydroxy- Esculetin	$C_9H_6O_4$ 178.14	305-01-1 276	3645 sub	5-18-03-00202	H_2O 2; EtOH 3; eth 2; ace 3
2805	2H-1-Benzopyran-2-one, 7,8-dihydroxy- Daphnetin	$C_9H_6O_4$ 178.14	486-35-1 262	2816 sub	5-18-03-00211	H_2O 3; EtOH 3; eth 2; bz 2
2806	4H-1-Benzopyran-4-one, 5,7-dihydroxy-2-(4-hydroxyphenyl)- Apigenin	$C_{15}H_{10}O_5$ 270.24	520-36-5 347.5	763	5-18-04-00574	H_2O 1; EtOH 3; py 3; dil alk 4
2807	4H-1-Benzopyran-4-one, 5,7-dihydroxy-3-(4-hydroxyphenyl)- Genistein	$C_{15}H_{10}O_5$ 270.24	446-72-0 301-2 dec	4281	5-18-04-00594	
2808	2H-1-Benzopyran-2-one, 7,8-dihydroxy-6-methoxy- Fraxetin	$C_{10}H_8O_5$ 208.17	574-84-5 231	4178	5-18-04-00332	EtOH 4
2809	4H-1-Benzopyran-4-one, 5,7-dihydroxy-2-(4-methoxyphenyl)- Acacetin	$C_{16}H_{12}O_5$ 284.27	480-44-4 263	9	5-18-04-00575	EtOH 4
2810	4H-1-Benzopyran-4-one, 5,7-dihydroxy-6-methoxy-2-phenyl- Oroxylin A	$C_{16}H_{12}O_5$ 284.27	480-11-5 231.5	6830	5-18-04-00569	ace 4; eth 4; EtOH 4
2811	4H-1-Benzopyran-4-one, 5,7-dihydroxy-2-phenyl- Chrysin	$C_{15}H_{10}O_4$ 254.24	480-40-0 285.5	2261	5-18-04-00076	H_2O 1; EtOH 3; eth 2; ace 3
2812	4H-1-Benzopyran-4-one, 2-(3,4-dihydroxyphenyl)-2,3-dihydro-5,7-dihydroxy-, (*S*)- Eriodictyol	$C_{15}H_{12}O_6$ 288.26	552-58-9 267 dec	3616	5-18-05-00214	EtOH 4; HOAc 4
2813	4H-1-Benzopyran-4-one, 2-(3,4-dihydroxyphenyl)-3,7-dihydroxy- Fisetin	$C_{15}H_{10}O_6$ 286.24	528-48-3 330	4026	5-18-05-00291	H_2O 1; EtOH 3; eth 2; ace 3
2814	4H-1-Benzopyran-4-one, 2-(3,4-dihydroxyphenyl)-5,7-dihydroxy- Luteolin	$C_{15}H_{10}O_6$ 286.24	491-70-3 329-30 dec	5483	5-18-05-00296	H_2O 2; EtOH 3; eth 3; alk 3
2815	4H-1-Benzopyran-4-one, 2-(3,4-dihydroxyphenyl)-3,5-dihydroxy-7-methoxy- Rhamnetin	$C_{16}H_{12}O_7$ 316.27	90-19-7 295	8170	5-18-05-00495	H_2O 2; EtOH 3; ace 3; dil alk 4
2816	4H-1-Benzopyran-4-one, 2-(2,4-dihydroxyphenyl)-3,5,7-trihydroxy- Morin	$C_{15}H_{10}O_7$ 302.24	480-16-0 303.5	6179	5-18-05-00492	H_2O 2; EtOH 4; eth 2; bz 3
2817	4H-1-Benzopyran-4-one, 2-(3,4-dihydroxyphenyl)-3,5,7-trihydroxy- Quercetin	$C_{15}H_{10}O_7$ 302.24	117-39-5 316.5	8044 sub	5-18-05-00494	H_2O 2; EtOH 3; eth 2; ace 3
2818	2H-1-Benzopyran-2-one, 5,7-dimethoxy- Limettin	$C_{11}H_{10}O_4$ 206.20	487-06-9 149	5370 200 dec	5-18-03-00199	H_2O 2; EtOH 4; eth 1; ace 4
2819	4H-1-Benzopyran-4-one, 8-[(dimethylamino)methyl]-7-methoxy-3-methyl-2-phenyl- Dimefline	$C_{20}H_{21}NO_3$ 323.39	1165-48-6 109.5	3190	5-18-12-00083	chl 3
2820	2H-1-Benzopyran-2-one, 3-ethyl- Coumarin, 3-ethyl-	$C_{11}H_{10}O_2$ 174.20	66898-39-3 72.3	299	5-17-10-00185	eth 4; EtOH 4
2821	2H-1-Benzopyran-2-one, 7-(β-*D*-glucopyranosyloxy)- Skimmin	$C_{15}H_{16}O_8$ 324.29	93-39-0 220	8506	5-18-01-00409	H_2O 3; EtOH 3; eth 1; chl 1
2822	2H-1-Benzopyran-2-one, 6-(β-*D*-glucopyranosyloxy)-7-hydroxy- Esculin	$C_{15}H_{16}O_9$ 340.29	531-75-9 205 dec	3646 230 dec	5-18-03-00208	H_2O 2; EtOH 2; eth 2; chl 3

No.	Name Synonym	Mol. Form. Mol. Wt.	CAS RN mp/°C	Merck No. bp/°C	Beil. Ref. den/g cm^{-3}	Solubility n_D
2823	2H-1-Benzopyran-2-one, 8-(β-*D*-glucopyranosyloxy)-7-hydroxy-6-methoxy- Fraxin	$C_{16}H_{18}O_{10}$ 370.31	524-30-1 205	4179	5-18-04-00334	
2824	4H-1-Benzopyran-4-one, 3-[4-(β-*D*-glucopyranosyloxy)phenyl]-5,7-dihydroxy- Sophoricoside	$C_{21}H_{20}O_{10}$ 432.38	152-95-4 274	8675	5-18-04-00597	
2825	2H-1-Benzopyran-2-one, 6-hydroxy-	$C_9H_6O_3$ 162.14	6093-68-1 250		5-18-01-00385 1.25^{25}	EtOH 4
2826	2H-1-Benzopyran-2-one, 7-hydroxy- Umbelliferone	$C_9H_6O_3$ 162.14	93-35-6 230.5	9758 sub	5-18-01-00386	EtOH 4; HOAc 4; chl 4
2827	4H-1-Benzopyran-4-one, 7-hydroxy-3-(4-hydroxyphenyl)- Daidzein	$C_{15}H_{10}O_4$ 254.24	486-66-8 323 dec	2805 sub	5-18-04-00089	EtOH 3; eth 3
2828	4H-1-Benzopyran-4-one, 5-hydroxy-3-(4-hydroxyphenyl)-7-methoxy- Prunetin	$C_{16}H_{12}O_5$ 284.27	552-59-0 239.5	7923	5-18-04-00595	
2829	2H-1-Benzopyran-2-one, 7-hydroxy-6-methoxy- Scopoletin	$C_{10}H_8O_4$ 192.17	92-61-5 204	8363	5-18-03-00203	H_2O 2; EtOH 2; eth 2; bz 1
2830	4H-1-Benzopyran-4-one, 7-hydroxy-3-(4-methoxyphenyl)- Formononetin	$C_{16}H_{12}O_4$ 268.27	485-72-3 256.5	4157	5-18-04-00089	
2831	2H-1-Benzopyran-2-one, 7-hydroxy-4-methyl- Hymecromone	$C_{10}H_8O_3$ 176.17	90-33-5 194.5	4792	5-18-01-00439	H_2O 2; EtOH 3; eth 2; chl 2
2832	2H-1-Benzopyran-2-one, 4-hydroxy-3-(3-oxo-1-phenylbutyl)- Warfarin	$C_{19}H_{16}O_4$ 308.33	81-81-2 161	9950	5-18-04-00162	H_2O 1; EtOH 3; ace 3; diox 3
2833	2H-1-Benzopyran-2-one, 3-methyl- 3-Methylcoumarin	$C_{10}H_8O_2$ 160.17	2445-82-1 91	 292.5	5-17-10-00165	H_2O 2; EtOH 3; chl 3; KOH 1
2834	2H-1-Benzopyran-2-one, 5-methyl- 5-Methylcoumarin	$C_{10}H_8O_2$ 160.17	42286-84-0 65.8	 173^{12}	5-17-10-00167	H_2O 2; EtOH 4; eth 4; bz 4
2835	2H-1-Benzopyran-2-one, 6-methyl-	$C_{10}H_8O_2$ 160.17	92-48-8 76.5	 304; 174^{14}	5-17-10-00168	EtOH 4; eth 4; bz 4; chl 2
2836	2H-1-Benzopyran-2-one, 7-methyl- 7-Methylcoumarin	$C_{10}H_8O_2$ 160.17	2445-83-2 128	 171.5^{11}	5-17-10-00169	H_2O 2; EtOH 4; eth 3; HOAc 4
2837	2H-1-Benzopyran-2-one, 8-methyl- 8-Methylcoumarin	$C_{10}H_8O_2$ 160.17	1807-36-9 110.5	 178^{20}	5-17-10-00170	EtOH 4; eth 4; bz 4; ctc 4
2838	4H-1-Benzopyran-4-one, 3-methyl- Tricromyl	$C_{10}H_8O_2$ 160.17	85-90-5	9574	5-17-10-00164	chl 3
2839	2H-1-Benzopyran-2-one, 3,3'-methylenebis[4-hydroxy- Dicumarol	$C_{19}H_{12}O_6$ 336.30	66-76-2 290	3080	5-19-06-00682	
2840	4H-1-Benzopyran-4-one, 2-phenyl- Flavone	$C_{15}H_{10}O_2$ 222.24	525-82-6 100	4030	5-17-10-00552	H_2O 1; EtOH 3; eth 3; ace 3
2841	4H-1-Benzopyran-4-one, 3-phenyl- Isoflavone	$C_{15}H_{10}O_2$ 222.24	574-12-9 148	5057	5-17-10-00562	
2842	4H-1-Benzopyran-4-one, 3,5,7-trihydroxy-2-(2-hydroxyphenyl)- Datiscetin	$C_{15}H_{10}O_6$ 286.24	480-15-9 277.5	2822	4-18-00-03281	ace 4; eth 4; EtOH 4
2843	4H-1-Benzopyran-4-one, 3,5,7-trihydroxy-2-(4-hydroxyphenyl)- Kaempferol	$C_{15}H_{10}O_6$ 286.24	520-18-3 277	5156	5-18-05-00251	H_2O 2; EtOH 4; eth 4; ace 4
2844	4H-1-Benzopyran-4-one, 5,6,7-trihydroxy-2-phenyl- Baicalein	$C_{15}H_{10}O_5$ 270.24	491-67-8 264-5 dec	954	5-18-04-00569	H_2O 2; EtOH 3; eth 3; ace 3
2845	2H-1-Benzopyran-2-one, 6,7,8-trimethoxy-	$C_{12}H_{12}O_5$ 236.22	6035-49-0 104.2	 $90\text{-}100^{0.2}$	5-18-04-00333	EtOH 3; eth 3
2846	2H-1-Benzopyran-3,5,7-triol, 2-(3,4-dihydroxyphenyl)-3,4-dihydro-, (2*S*-*cis*)-	$C_{15}H_{14}O_6$ 290.27	35323-91-2 242		5-17-08-00447 1.344^4	ace 4; EtOH 4
2847	1-Benzopyrylium, 3,5,7-trihydroxy-2-(4-hydroxy-3,5-dimethoxyphenyl)-, chloride Malvidin chloride	$C_{17}H_{15}ClO_7$ 366.75	643-84-5 >300	5596	5-17-08-00514	H_2O 2; EtOH 3; MeOH 3
2848	1-Benzopyrylium, 3,5,7-trihydroxy-2-(4-hydroxyphenyl)-, chloride Pelargonidin chloride	$C_{15}H_{11}ClO_5$ 306.70	134-04-3 >350	7014	1-18-00-00410	H_2O 3; EtOH 4; chl 2; con sulf 3
2849	1-Benzopyrylium, 3,5,7-trihydroxy-2-(3,4,5-trihydroxyphenyl)-, chloride Delphinidin	$C_{15}H_{11}ClO_7$ 338.70	528-53-0 >350	2866	5-17-08-00514	H_2O 4; EtOH 4; MeOH 4; AcOEt 3
2850	Benzo[rst]pentaphene 1,2:7,8-Dibenzopyrene	$C_{24}H_{14}$ 302.38	189-55-9 281.5		4-05-00-02803	
2851	2H-1,2,4-Benzothiadiazine-7-sulfonamide, 3-bicyclo[2.2.1]hept-5-en-2-yl-6-chloro-3,4-dihydro-, 1,1-dioxide Cyclothiazide	$C_{14}H_{16}ClN_3O_4S_2$ 389.88	2259-96-3 234	2760		
2852	2H-1,2,4-Benzothiadiazine-7-sulfonamide, 6-chloro-, 1,1-dioxide Chlorothiazide	$C_7H_6ClN_3O_4S_2$ 295.73	58-94-6 350 dec	2169	4-27-00-08036	

No.	Name Synonym	Mol. Form. Mol. Wt.	CAS RN mp/°C	Merck No. bp/°C	Beil. Ref. den/g cm^{-3}	Solubility n_D
2853	2H-1,2,4-Benzothiadiazine-7-sulfonamide, 6-chloro-3-(cyclopentylmethyl)-3,4-dihydro-, 1,1-dioxide Cyclopenthiazide	$C_{13}H_{18}ClN_3O_4S_2$ 379.89	742-20-1 238	2750		
2854	2H-1,2,4-Benzothiadiazine-7-sulfonamide, 6-chloro-3-(dichloromethyl)-3,4-dihydro-, 1,1-dioxide Trichloromethiazide	$C_8H_8Cl_3N_3O_4S_2$ 380.66	133-67-5 270 dec	9537		H_2O 2; EtOH 3
2855	2H-1,2,4-Benzothiadiazine-7-sulfonamide, 6-chloro-3,4-dihydro-, 1,1-dioxide Hydrochlorothiazide	$C_7H_8ClN_3O_4S_2$ 297.74	58-93-5 274	4704		
2856	2H-1,2,4-Benzothiadiazine-7-sulfonamide, 6-chloro-3,4-dihydro-3-(2-methylpropyl)-, 1,1-dioxide Buthiazide	$C_{11}H_{16}ClN_3O_4S_2$ 353.85	2043-38-1 221.5	1519		
2857	2H-1,2,4-Benzothiadiazine-7-sulfonamide, 6-chloro-3,4-dihydro-2-methyl-3-[[(2,2,2-trifluoroethyl)thio]methyl]- Polythiazide	$C_{11}H_{13}ClF_3N_3O_4S_3$ 439.89	346-18-9 214	7561		
2858	2H-1,2,4-Benzothiadiazine-7-sulfonamide, 3,4-dihydro-6-(trifluoromethyl)-, 1,1-dioxide Hydroflumethiazide	$C_8H_8F_3N_3O_4S_2$ 331.30	135-09-1 270.5	4716	4-27-00-08035	
2859	2-Benzothiazolamine 2-Aminobenzothiazole	$C_7H_6N_2S$ 150.20	136-95-8 132	435	4-27-00-04824	H_2O 2; EtOH 3; eth 3; chl 3
2860	6-Benzothiazolamine 6-Aminobenzothiazole	$C_7H_6N_2S$ 150.20	533-30-2 87	436	4-27-00-04884	H_2O 1; EtOH 3; eth 1
2861	Benzothiazole Benzosulfonazole	C_7H_5NS 135.19	95-16-9 2	1118 231	4-27-00-01069 1.2460^{20}	H_2O 2; EtOH 4; eth 4; ace 3 1.6379^{20}
2862	Benzothiazole, 2-chloro- 2-Chlorobenzothiazole	C_7H_4ClNS 169.63	615-20-3 24	 248	4-27-00-01072 1.3715^{10}	ace 4; eth 4; EtOH 4 1.6338^{10}
2863	Benzothiazole, 2-methyl- 2-Methylbenzothiazole	C_8H_7NS 149.22	120-75-2 14	 238	4-27-00-01080 1.1763^{19}	H_2O 1; EtOH 3; chl 3 1.6092^{19}
2864	Benzothiazole, 2-phenyl- 2-Phenylbenzothiazole	$C_{13}H_9NS$ 211.29	883-93-2 115	 371	4-27-00-01385	H_2O 1; EtOH 3; eth 3; CS_2 3
2865	2-Benzothiazolesulfonamide, 6-ethoxy- Ethoxzolamide	$C_9H_{10}N_2O_3S_2$ 258.32	452-35-7 189	3711	4-27-00-04404	
2866	2(3H)-Benzothiazolethione 2-Mercaptobenzothiazole	$C_7H_5NS_2$ 167.26	149-30-4 181	5759	4-27-00-02709 1.42^{20}	H_2O 1; EtOH 3; eth 2; bz 2
2867	2(3H)-Benzothiazolethione, 3-methyl-	$C_8H_7NS_2$ 181.28	2254-94-6 90	 335	4-27-00-02714	H_2O 1; EtOH 2; eth 2; bz 4
2868	2(3H)-Benzothiazolone	C_7H_5NOS 151.19	934-34-9 139	 360	4-27-00-02693	H_2O 1; EtOH 4; eth 4
2869	4H-1-Benzothiopyran-4-one, 2,3-dihydro-	C_9H_8OS 164.23	3528-17-4 29	 154^{12}	5-17-10-00016 1.2487^{14}	 1.6395^{20}
2870	1,2,4-Benzotriazine 1,2,4-Triazanaphthalene	$C_7H_5N_3$ 131.14	254-87-5 75.8	 237.5	4-26-00-00166	H_2O 4; eth 4; EtOH 4
2871	1H-Benzotriazole 1,2,3-Triaza-1H-indene	$C_6H_5N_3$ 119.13	95-14-7 100	1119 204^{15}	4-26-00-00093	H_2O 2; EtOH 3; bz 3; chl 3
2872	1H-Benzotriazole, 1-methyl-	$C_7H_7N_3$ 133.15	13351-73-0 64.5	 270.5	4-26-00-00095	bz 4; EtOH 4; HOAc 4
2873	1H-Benzotriazole, 5-nitro- 5-Nitrobenzotriazole	$C_6H_4N_4O_2$ 164.12	2338-12-7 217		4-26-00-00129	
2874	1,2,3-Benzoxadiazole, 5,7-dinitro- 2,3-Benzoxadiazole, 5,7-dinitro	$C_6H_2N_4O_5$ 210.11	87-31-0 158			EtOH 4
2875	3H-2,1-Benzoxathiol-3-one, 1,1-dioxide	$C_7H_4O_4S$ 184.17	81-08-3 129.5	 184^{18}	5-19-04-00215	bz 4; chl 4
2876	2-Benzoxazolamine, 5-chloro- Zoxazolamine	$C_7H_5ClN_2O$ 168.58	61-80-3 184.5	10098	4-27-00-04819	EtOH 4
2877	Benzoxazole 1-Oxa-3-azaindene	C_7H_5NO 119.12	273-53-0 31	 182.5	4-27-00-01069 1.1754^{20}	H_2O 1; EtOH 3; sulf 3 1.5594^{20}
2878	Benzoxazole, 2-chloro- 2-Chlorobenzoxazole	C_7H_4ClNO 153.57	615-18-9 7	 201.5	2-27-00-00017 1.3453^{18}	 1.5678^{20}
2879	Benzoxazole, 2,5-dimethyl- 2,5-Dimethylbenzoxazole	C_9H_9NO 147.18	5676-58-4	 218.5	4-27-00-01101 1.0880^{18}	ctc 3 1.5412^{20}
2880	Benzoxazole, 2-methyl- 2-Methylbenzoxazole	C_8H_7NO 133.15	95-21-6 9.5	 200.5	4-27-00-01078 1.1211^{20}	H_2O 1; EtOH 4; eth 5 1.5497^{20}
2881	2-Benzoxazolol 2-Hydroxybenzoxazole	$C_7H_5NO_2$ 135.12	69564-68-7 141	 230^{30}	2-27-00-00233	eth 4; EtOH 4
2882	2(3H)-Benzoxazolone	$C_7H_5NO_2$ 135.12	59-49-4 138	 335; 230^{30}	4-27-00-02677	H_2O 2; EtOH 3; eth 3; tfa 3
2883	2(3H)-Benzoxazolone, 5-chloro- Chlorzoxazone	$C_7H_4ClNO_2$ 169.57	95-25-0 191.5	2197	4-27-00-02683	EtOH 4; MeOH 4
2884	Benzoyl azide Benzazide	$C_7H_5N_3O$ 147.14	582-61-6 32	 exp	4-09-00-00945 1.1680^{35}	eth 4; EtOH 4
2885	Benzoyl bromide Benzoic acid, bromide	C_7H_5BrO 185.02	618-32-6 -24	 218.5	4-09-00-00724 1.570^{15}	eth 5 1.5868^{25}
2886	Benzoyl chloride Benzenecarbonyl chloride	C_7H_5ClO 140.57	98-88-4 -1.0	1124 197.2; 71^{9}	4-09-00-00721 1.2120^{20}	eth 5; bz 3; ctc 3; CS_2 3 1.5537^{20}
2887	Benzoyl chloride, 2-bromo-	C_7H_4BrClO 219.46	7154-66-7 11	 243	4-09-00-01013	ctc 2 1.5963^{20}
2888	Benzoyl chloride, 4-bromo-	C_7H_4BrClO 219.46	586-75-4 42	 246; 181^{125}	4-09-00-01023	EtOH 4; eth 4; bz 4; lig 4

PHYSICAL CONSTANTS OF ORGANIC COMPOUNDS (continued)

No.	Name Synonym	Mol. Form. Mol. Wt.	CAS RN mp/°C	Merck No. bp/°C	Beil. Ref. den/g cm^{-3}	Solubility n_D
2889	Benzoyl chloride, 4-butyl-	$C_{11}H_{13}ClO$ 196.68	28788-62-7	 $155\text{-}6^{22}$	3-09-00-02520 1.051^{25}	 1.5351^{20}
2890	Benzoyl chloride, 2-chloro-	$C_7H_4Cl_2O$ 175.01	609-65-4 -4	 238	4-09-00-00962	ctc 3 1.5726^{16}
2891	Benzoyl chloride, 3-chloro- *m*-Chlorobenzoyl chloride	$C_7H_4Cl_2O$ 175.01	618-46-2	 225	4-09-00-00972	 1.5677^{20}
2892	Benzoyl chloride, 4-chloro-	$C_7H_4Cl_2O$ 175.01	122-01-0 16	 222	4-09-00-00984 1.3770^{20}	chl 2 1.5758^{20}
2893	Benzoyl chloride, 2,4-dichloro- 2,4-Dichlorobenzoyl chloride	$C_7H_3Cl_3O$ 209.46	89-75-8 16.5	 150^{34}; $111^{7.5}$	4-09-00-01002	ctc 3 1.5895^{20}
2894	Benzoyl chloride, 3,4-dichloro- 3,4-Dichlorobenzoyl chloride	$C_7H_3Cl_3O$ 209.46	3024-72-4 25	 242	3-09-00-01379	ctc 2
2895	Benzoyl chloride, 4-(1,1-dimethylethyl)-	$C_{11}H_{13}ClO$ 196.68	1710-98-1	 135^{20}	4-09-00-01887 1.007^{25}	 1.5364^{20}
2896	Benzoyl chloride, 3,5-dinitro- 3,5-Dinitrobenzoyl chloride	$C_7H_3ClN_2O_5$ 230.56	99-33-2 74	3270 196^{12}	4-09-00-01350	eth 3; chl 3
2897	Benzoyl chloride, 2-fluoro-	C_7H_4ClFO 158.56	393-52-2 4	 $90\text{-}2^{15}$	4-09-00-00951 1.328^{25}	 1.5365^{20}
2898	Benzoyl chloride, 3-fluoro-	C_7H_4ClFO 158.56	1711-07-5 -30	 189	4-09-00-00953 1.304^{25}	 1.5285^{20}
2899	Benzoyl chloride, 4-fluoro-	C_7H_4ClFO 158.56	403-43-0 9	 82^{20}	4-09-00-00954 1.342^{25}	 1.5296^{20}
2900	Benzoyl chloride, 2-hydroxy-	$C_7H_5ClO_2$ 156.57	1441-87-8 19	 92^{15}	3-10-00-00150 1.3112^{20}	eth 4 1.5812^{20}
2901	Benzoyl chloride, 4-methoxy- *p*-Anisoyl chloride	$C_8H_7ClO_2$ 170.60	100-07-2 24.5	703 262.5	4-10-00-00430 1.261^{20}	eth 3; ace 3; bz 4; ctc 2 1.580^{20}
2902	Benzoyl chloride, 2-methyl-	C_8H_7ClO 154.60	933-88-0	 213.5	3-09-00-02304	eth 4; EtOH 4 1.5549^{20}
2903	Benzoyl chloride, 3-methyl-	C_8H_7ClO 154.60	1711-06-4 -23	 219.5	4-09-00-01716 1.0265^{21}	eth 4; EtOH 4 1.505^{22}
2904	Benzoyl chloride, 4-methyl-	C_8H_7ClO 154.60	874-60-2 -3	 226	4-09-00-01733 1.1686^{20}	ctc 3 1.5547^{20}
2905	Benzoyl chloride, 3-nitro-	$C_7H_4ClNO_3$ 185.57	121-90-4 36	 276.5	4-09-00-01061	eth 4
2906	Benzoyl chloride, 4-nitro- *p*-Nitrobenzoyl chloride	$C_7H_4ClNO_3$ 185.57	122-04-3 75	6511 203^{105}; 151^{15}	4-09-00-01191	eth 3
2907	Benzoyl chloride, 4-pentyl-	$C_{12}H_{15}ClO$ 210.70	49763-65-7	 121^{8}	4-09-00-01910 1.036^{25}	 1.5300^{20}
2908	Benzoyl chloride, 4-(pentyloxy)-	$C_{12}H_{15}ClO_2$ 226.70	36823-84-4	 $196\text{-}200^{30}$	4-10-00-00431 1.087^{25}	 1.5434^{20}
2909	Benzoyl fluoride Benzoic acid, fluoride	C_7H_5FO 124.11	455-32-3 -28	 154.5	4-09-00-00721 1.1400^{20}	EtOH 4; eth 4; ctc 3
2910	Benzoyl iodide	C_7H_5IO 232.02	618-38-2 3	 128^{20}	4-09-00-00725 1.746^{18}	eth 4; EtOH 4
2911	Berbaman-12-ol, 6,6',7-trimethoxy-2,2'-dimethyl- Berbamine	$C_{37}H_{40}N_2O_6$ 608.73	478-61-5 198.5	1168	2-27-00-00891	H_2O 2; EtOH 3; eth 3; chl 3
2912	Beryllium, bis(2,4-pentanedionato-O,O')-, (T-4)- Beryllium acetylacetonate	$C_{10}H_{14}BeO_4$ 207.23	10210-64-7 108	1180 270		
2913	Bicyclo[2.2.1]hepta-2,5-diene	C_7H_8 92.14	121-46-0 -19.1	 89.5	4-05-00-00879 0.9064^{20}	H_2O 1; EtOH 3; eth 3; ace 3 1.4702^{20}
2914	Bicyclo[2.2.1]heptane	C_7H_{12} 96.17	279-23-2 87.5	 105.3	4-05-00-00258	ace 4; bz 4; eth 4; EtOH 4
2915	Bicyclo[4.1.0]heptane Norcarane	C_7H_{12} 96.17	286-08-8	6605 116.5	4-05-00-00257 0.853^{25}	 1.4564^{20}
2916	Bicyclo[2.2.1]heptane-2-carboxaldehyde 2-Norbornanecarboxaldehyde	$C_8H_{10}O$ 122.17	19396-83-9	 $70\text{-}2^{22}$	4-07-00-00157 1.0227^{19}	eth 4 1.4760^{25}
2917	Bicyclo[2.2.1]heptane-2-carboxylic acid, 4,7,7-trimethyl-3-oxo- *d*-Camphocarboxylic acid	$C_{11}H_{16}O_3$ 196.25	18530-30-8 127.5	1737	3-10-00-02925	bz 4; eth 4; EtOH 4
2918	Bicyclo[2.2.1]heptane, 2-chloro-	$C_7H_{11}Cl$ 130.62	29342-53-8 -5	 162; 52^{11}	4-05-00-00259 1.0603^{20}	chl 4 1.4849^{20}
2919	Bicyclo[2.2.1]heptane, 2-chloro-1,7,7-trimethyl- Bornane, 2-chloro-	$C_{10}H_{17}Cl$ 172.70	6120-13-4 132.8	 207	4-05-00-00319	eth 4; EtOH 4
2920	Bicyclo[2.2.1]heptane, 2-chloro-1,7,7-trimethyl-, *endo*- Bornyl chloride	$C_{10}H_{17}Cl$ 172.70	464-41-5 132	1341 207.5	4-05-00-00319	bz 4; eth 4; EtOH 4; peth 4
2921	Bicyclo[2.2.1]heptane, 2,2-dimethyl-3-methylene-, (±)-	$C_{10}H_{16}$ 136.24	565-00-4 51.5	 158.5	4-05-00-00462 0.879^{20}	H_2O 1; EtOH 4; eth 4 1.4551^{54}
2922	Bicyclo[2.2.1]heptane, 2,2-dimethyl-5-methylene-	$C_{10}H_{16}$ 136.24	497-32-5	 152	4-05-00-00465 0.8591^{20}	 1.4645^{25}
2923	Bicyclo[2.2.1]heptane, 7,7-dimethyl-2-methylene-, (±)- α-Fenchene (*DL*)	$C_{10}H_{16}$ 136.24	2623-54-3	 155	3-05-00-00389 0.8660^{20}	ace 4; eth 4; EtOH 4 1.4705^{20}

No.	Name Synonym	Mol. Form. Mol. Wt.	CAS RN mp/°C	Merck No. bp/°C	Beil. Ref. den/g cm^{-3}	Solubility n_D
2924	Bicyclo[3.1.1]heptane, 6,6-dimethyl-2-methylene-, (1*R*)-	$C_{10}H_{16}$ 136.24	19902-08-0 -50	7415 164.6	4-05-00-00456 0.8654^{20}	bz 4; eth 4; EtOH 4; chl 4 1.4789^{20}
2925	Bicyclo[3.1.1]heptane, 6,6-dimethyl-2-methylene- β-Pinene	$C_{10}H_{16}$ 136.24	127-91-3 -61.5	7415 166	 0.860^{25}	H_2O 1; bz 3; EtOH 3; eth 3 1.4768^{25}
2926	Bicyclo[2.2.1]heptane-2,3-dione, 1,7,7-trimethyl-, (1*R*)-	$C_{10}H_{14}O_2$ 166.22	10334-26-6 199	 sub	4-07-00-02039	H_2O 3; EtOH 4; eth 4; bz 3
2927	Bicyclo[2.2.1]heptane-1-methanamine, 7,7-dimethyl- α-Camphylamine	$C_{10}H_{19}N$ 153.27	54131-44-1	 195; 95^{12}	1-12-00-00129 0.8688^{20}	 1.4728^{18}
2928	Bicyclo[3.1.1]heptane-2-methanamine, 6,6-dimethyl-, [1*S*-(1α,2β,5α)]-	$C_{10}H_{19}N$ 153.27	73522-42-6	 $94\text{-}9^{27}$	4-06-00-00384 0.915^{25}	 1.4877^{20}
2929	Bicyclo[2.2.1]heptane-1-methanesulfonic acid, 7,7-dimethyl-2-oxo-, (1*S*)- *D*-Camphorsulfonic acid	$C_{10}H_{16}O_4S$ 232.30	3144-16-9 195 dec	1740	4-11-00-00642	H_2O 4; eth 1; HOAc 2
2930	Bicyclo[3.1.1]heptane-2-methanol, 6,6-dimethyl-, [1*S*-(1α,2β,5α)]-	$C_{10}H_{18}O$ 154.25	51152-12-6	 $70\text{-}2^{1}$	4-06-00-00278 0.986^{25}	 1.4890^{20}
2931	Bicyclo[2.2.1]heptane, 1,3,3-trimethyl-, (+)- Fenchane(*d*)	$C_{10}H_{18}$ 138.25	116435-30-4 <-15	 151	3-05-00-00256 0.8345^{20}	H_2O 1; EtOH 3; eth 3; HOAc 2 1.44714^{20}
2932	Bicyclo[2.2.1]heptane, 1,7,7-trimethyl-	$C_{10}H_{18}$ 138.25	464-15-3	 161	4-05-00-00319	H_2O 1; EtOH 3; eth 3; AcOEt 3
2933	Bicyclo[2.2.1]heptane, 2,2,3-trimethyl-	$C_{10}H_{18}$ 138.25	473-19-8 62.5	 166	3-05-00-00263 0.8276^{67}	ace 4; EtOH 4
2934	Bicyclo[3.1.1]heptane, 2,6,6-trimethyl-, (1α,2α,5α)-(±)- Pinane (*DL*)	$C_{10}H_{18}$ 138.25	112456-46-9	 164.5	1-05-00-00048 0.8551^{20}	bz 4; eth 4 1.4609^{20}
2935	Bicyclo[3.1.1]heptane, 2,6,6-trimethyl-, [1*R*-(1α,2β,5α)]-	$C_{10}H_{18}$ 138.25	4795-86-2 -53	 169	4-05-00-00318 0.8560^{20}	bz 4; eth 4 1.4629^{20}
2936	Bicyclo[4.1.0]heptane, 3,7,7-trimethyl-, [1*S*-(1α,3β,6α)]-	$C_{10}H_{18}$ 138.25	2778-68-9	 169	4-05-00-00316 0.8410^{20}	 1.4569
2937	Bicyclo[4.1.0]heptane, 3,7,7-trimethyl-, [1*S*-(1α,3α,6α)]-	$C_{10}H_{18}$ 138.25	6069-97-2	 169	4-05-00-00316 0.841^{20}	 1.456
2938	Bicyclo[2.2.1]heptan-2-ol, 1,7-dimethyl-, (*exo*,*syn*)- β-Santenol	$C_9H_{16}O$ 140.23	509-12-6 101.5	 192; 60^{15}	4-06-00-00242	eth 4
2939	Bicyclo[2.2.1]heptan-2-ol, 1-phenyl-, acetate	$C_{15}H_{18}O_2$ 230.31	71173-15-4 44	 159^{13}; $128^{1.0}$	3-06-00-02755	ctc 2
2940	Bicyclo[2.2.1]heptan-2-ol, 1,3,3-trimethyl-, *endo*-(±)-	$C_{10}H_{18}O$ 154.25	36386-49-9 39	 199.5	3-06-00-00288 0.9420^{40}	eth 4; EtOH 4
2941	Bicyclo[2.2.1]heptan-2-ol, 1,3,3-trimethyl-, (1*S*-*endo*)- *l*-α-Fenchyl alcohol	$C_{10}H_{18}O$ 154.25	512-13-0 48	 94^{20}	4-06-00-00278 0.9034^{84}	
2942	Bicyclo[2.2.1]heptan-2-ol, 1,7,7-trimethyl-, *endo*-(±)-	$C_{10}H_{18}O$ 154.25	6627-72-1 210.5	 sub	4-06-00-00281 1.011^{20}	H_2O 1; EtOH 4; eth 4; bz 4
2943	Bicyclo[2.2.1]heptan-2-ol, 1,7,7-trimethyl-, *exo*-(±)-	$C_{10}H_{18}O$ 154.25	24393-70-2 212	 sub	4-06-00-00282 1.10^{20}	H_2O 1; EtOH 4; eth 4; bz 2
2944	Bicyclo[3.1.1]heptan-2-ol, 4,6,6-trimethyl-, [1*R*-(1α,2α,4α,5α)]-	$C_{10}H_{18}O$ 154.25	515-88-8 58	 218	3-06-00-00284 0.940^{20}	 1.4702^{20}
2945	Bicyclo[2.2.1]heptan-2-ol, 1,7,7-trimethyl-, acetate, *endo*-(±)-	$C_{12}H_{20}O_2$ 196.29	36386-52-4 <-17	 223.5	4-06-00-00283 1.4630^{20}	ctc 3 1.4630^{20}
2946	Bicyclo[2.2.1]heptan-2-ol, 1,7,7-trimethyl-, acetate, *exo*-(±)-	$C_{12}H_{20}O_2$ 196.29	17283-45-3	 116^{21}	4-06-00-00283 0.9841^{20}	ace 4; EtOH 4 1.4640^{20}
2947	Bicyclo[2.2.1]heptan-2-ol, 4,7,7-trimethyl-, acetate, (1*R*-*endo*)- Epiborneol, acetate	$C_{12}H_{20}O_2$ 196.29	22621-74-5 <-15	 101^{11}	2-06-00-00092 0.9872^{14}	 1.4651^{14}
2948	Bicyclo[2.2.1]heptan-2-ol, 1,7,7-trimethyl-, acetate, *endo*- Borneol, acetate	$C_{12}H_{22}O_2$ 198.31	76-49-3 29	1339 221	3-06-00-00303	
2949	Bicyclo[2.2.1]heptan-2-ol, 1,7,7-trimethyl-, formate, (1*R*-*endo*)- Bornyl formate (*d*)	$C_{11}H_{18}O_2$ 182.26	74219-20-8	 90^{10}	3-06-00-00301 1.009^{22}	 1.4708^{15}
2950	Bicyclo[2.2.1]heptan-2-one, 3-(acetyl-17*O*)-1,7,7-trimethyl- Camphor, 3-acetyl	$C_{12}H_{18}O_2$ 194.27	15068-90-3	 $127\text{-}8^{15}$	2-07-00-00563 1.0314^{19}	 1.4949^{17}

No.	Name Synonym	Mol. Form. Mol. Wt.	CAS RN mp/°C	Merck No. bp/°C	Beil. Ref. den/g cm^{-3}	Solubility n_D
2951	Bicyclo[2.2.1]heptan-2-one, 1-(bromomethyl)-7,7-dimethyl-, (1*S*)- *d*-Camphor, 10-bromo-	$C_{10}H_{15}BrO$ 231.13	64161-50-8 78	 265 dec	4-07-00-00218	ace 4; bz 4; eth 4; EtOH 4
2952	Bicyclo[2.2.1]heptan-2-one, 3-bromo-1,7,7-trimethyl-, (1*R*-*endo*)-	$C_{10}H_{15}BrO$ 231.13	10293-06-8 76	 274 dec	 1.449^{25}	bz 4; eth 4; EtOH 4
2953	Bicyclo[2.2.1]heptan-2-one, 3,3-dibromo-1,7,7-trimethyl-, (1*R*)- α,α'-Dibromo-*d*-camphor	$C_{10}H_{14}Br_2O$ 310.03	514-12-5 61	3002	4-07-00-00220 1.854^{21}	H_2O 1; EtOH 4; eth 4; bz 4
2954	Bicyclo[2.2.1]heptan-2-one, 3,3-dimethyl-, (1*S*)-	$C_9H_{14}O$ 138.21	6069-71-2 41.9	 193; 76^{12}	4-07-00-00177 0.807^{20}	eth 4
2955	Bicyclo[3.1.1]heptan-2-one, 6,6-dimethyl-, (1*R*)-	$C_9H_{14}O$ 138.21	38651-65-9 -1	 209	4-07-00-00176 0.9807^{20}	eth 4; EtOH 4 1.4787^{20}
2956	Bicyclo[3.1.1]heptan-3-one, 6,6-dimethyl-2-methylene-, (1*S*)-	$C_{10}H_{14}O$ 150.22	19890-00-7 -1.8	 $67\text{-}9^{4}$	 0.9875^{15}	 1.4950^{20}
2957	Bicyclo[2.2.1]heptan-2-one, 3-hydroxy-1,7,7-trimethyl- 3-Hydroxycamphor	$C_{10}H_{16}O_2$ 168.24	10373-81-6 205.5	4747	4-08-00-00081	eth 4; EtOH 4; chl 4
2958	Bicyclo[2.2.1]heptan-2-one, 3-methylene- 3-Methylene-2-norbornanone	$C_8H_{10}O$ 122.17	5597-27-3	 $69\text{-}71^{11}$	4-07-00-00301 0.997^{25}	 1.4891^{20}
2959	Bicyclo[2.2.1]heptan-2-one, 1,4,7,7-tetramethyl-	$C_{11}H_{18}O$ 166.26	10309-50-9 168	 213	2-07-00-00113	eth 4; EtOH 4
2960	Bicyclo[2.2.1]heptan-2-one, 1,3,3-trimethyl-, (±)-	$C_{10}H_{16}O$ 152.24	18492-37-0 -18	3911 193.5	3-07-00-00393 0.9492^{15}	H_2O 1; EtOH 4; eth 3; ace 3 1.4702^{20}
2961	Bicyclo[2.2.1]heptan-2-one, 1,7,7-trimethyl-, (1*R*)- (+)-Camphor	$C_{10}H_{16}O$ 152.24	464-49-3 178.8	1738 207.4	4-07-00-00213 0.990^{25}	H_2O 1; EtOH 4; eth 4; ace 3 1.5462
2962	Bicyclo[2.2.1]heptan-2-one, 1,7,7-trimethyl-, oxime, (1*S*)- (-)-Camphor oxime	$C_{10}H_{17}NO$ 167.25	36065-15-3 118	 252 dec	4-07-00-00215 1.01^{116}	eth 4; EtOH 4
2963	Bicyclo[2.2.1]heptan-2-one, 4,7,7-trimethyl-, (1*S*)-	$C_{10}H_{16}O$ 152.24	10292-98-5 186.5	 213	3-07-00-00421	H_2O 2; EtOH 4; eth 4; peth 4
2964	Bicyclo[3.1.1]heptan-2-one, 4,6,6-trimethyl-, [1*R*-(1α,4α,5α)]-	$C_{10}H_{16}O$ 152.24	515-90-2	 222	4-07-00-00210 0.961^{20}	 1.4752^{20}
2965	Bicyclo[3.1.1]heptan-3-one, 2,6,6-trimethyl-, [1*S*-(1α,2β,5α)]-	$C_{10}H_{16}O$ 152.24	14575-93-0	 212	4-07-00-00209 0.964^{25}	 1.4735^{25}
2966	Bicyclo[2.2.1]hept-5-ene-2-carbonitrile	C_8H_9N 119.17	95-11-4 13	 $82\text{-}6^{10}$	 0.999^{25}	 1.4885^{20}
2967	Bicyclo[2.2.1]hept-5-ene-2-carboxaldehyde	$C_8H_{10}O$ 122.17	5453-80-5	 $67\text{-}70^{12}$	2-07-00-00125 1.018^{25}	 1.4893^{20}
2968	Bicyclo[2.2.1]hept-2-ene, 2,3-dimethyl- 2-Norbornene, 2,3-dimethyl-	C_9H_{14} 122.21	529-16-8	 140.5	4-05-00-00427 0.8698^{17}	eth 3; ace 3; bz 3 1.4688^{17}
2969	Bicyclo[3.1.1]hept-2-ene-2-ethanol, 6,6-dimethyl-	$C_{11}H_{18}O$ 166.26	128-50-7	 235; 110^{10}	4-06-00-00398 0.973^{25}	chl 3 1.4930^{20}
2970	Bicyclo[2.2.1]hept-2-ene, 1,7,7-trimethyl-	$C_{10}H_{16}$ 136.24	464-17-5 113	 146	2-05-00-00105	bz 4; eth 4; EtOH 4
2971	Bicyclo[3.1.1]hept-2-ene, 2,6,6-trimethyl-, (1*S*)- (-)-α-Pinene	$C_{10}H_{16}$ 136.24	7785-26-4 -64	7414 155	4-05-00-00452 0.859^{20}	 1.4660^{20}
2972	Bicyclo[3.1.1]hept-2-ene, 2,6,6-trimethyl-, (±)- (±)-α-Pinene	$C_{10}H_{16}$ 136.24	2437-95-8 -55	7414 156.2	2-05-00-00097 0.8582^{20}	H_2O 1; EtOH 5; eth 5; chl 5 1.4658^{20}
2973	Bicyclo[3.1.1]hept-2-ene, 2,6,6-trimethyl 2-Pinene	$C_{10}H_{16}$ 136.24	80-56-8 -64	 155.9	 0.8539^{25}	 1.4632^{25}
2974	Bicyclo[4.1.0]hept-3-ene, 3,7,7-trimethyl-, (1*S*)-	$C_{10}H_{16}$ 136.24	498-15-7	 171; 123^{200}	4-05-00-00450 0.8549^{30}	ace 4; bz 4; eth 4 1.469^{3}
2975	Bicyclo[3.1.1]hept-3-en-2-ol, 4,6,6-trimethyl-, (1α,2β,5α)-	$C_{10}H_{16}O$ 152.24	1845-30-3 15.5	 90^{10}	 0.9684^{25}	 1.4912^{25}
2976	Bicyclo[3.1.1]hept-3-en-2-ol, 4,6,6-trimethyl-, (1α,2α,5α)-	$C_{10}H_{16}O$ 152.24	1820-09-3 24	 92^{10}	4-06-00-00383 0.9657^{25}	 1.4908^{25}
2977	Bicyclo[3.1.1]hept-2-en-6-one, 2,7,7-trimethyl- Chrysanthenone	$C_{10}H_{14}O$ 150.22	473-06-3	2256 $88\text{-}9^{12}$	4-07-00-00327	EtOH 4 1.4720^{22}
2978	Bicyclo[3.1.1]hept-3-en-2-one, 4,6,6-trimethyl-, (1*R*)- *d*-Verbenone	$C_{10}H_{14}O$ 150.22	18309-32-5 9.8	9862 227.5	4-07-00-00327 0.9978^{20}	H_2O 3; EtOH 3; ace 3; bz 3 1.4993^{18}
2979	[Bi-1,4-cyclohexadien-1-yl]-3,3',6,6'-tetrone, 2,2'-dihydroxy-4,4'-dimethyl- Phenicin	$C_{14}H_{10}O_6$ 274.23	128-68-7 230.5	7194	4-08-00-03660	H_2O 2; EtOH 4; chl 4; HOAc 4
2980	Bicyclo[3,1,0]hexane, 1-methyl- 1-Methylbicyclo[3,1,0]hexane	C_7H_{12} 96.17	4625-24-5	 93.1		

No.	Name Synonym	Mol. Form. Mol. Wt.	CAS RN mp/°C	Merck No. bp/°C	Beil. Ref. den/g cm^{-3}	Solubility n_D
2981	Bicyclo[3.1.0]hexane, 4-methylene-1-(1-methylethyl)-, (±)- 4(10)-Thujene, (±)-	$C_{10}H_{16}$ 136.24	15826-80-9	 164; 69^{30}	3-05-00-00365 0.8437^{20}	H_2O 1; EtOH 4; eth 3; ace 3 1.4676^{20}
2982	Bicyclo[3.1.0]hexane, 4-methyl-1-(1-methylethyl)-	$C_{10}H_{18}$ 138.25	471-12-5	 157	4-05-00-00317 0.8139^{20}	 1.4376^{20}
2983	Bicyclo[3.1.0]hexan-3-ol, 4-methylene-1-(1-methylethyl)- 4(10)-Thujene-3-ol	$C_{10}H_{16}O$ 152.24	59905-55-4	 208	4-06-00-00382 0.9461^{25}	 1.4871^{25}
2984	Bicyclo[3.1.0]hexan-3-ol, 4-methylene-1-(1-methylethyl)-, [1*S*-(1α,3β,5α)]-	$C_{10}H_{16}O$ 152.24	471-16-9	 208	4-06-00-00382 0.9488^{19}	eth 3 1.4871^{25}
2985	Bicyclo[3.1.0]hexan-3-ol, 4-methyl-1-(1-methylethyl)-	$C_{10}H_{18}O$ 154.25	513-23-5	 209	3-06-00-00278 0.9210^{20}	 1.4621^{25}
2986	Bicyclo[3.1.0]hex-2-ene, 2-methyl-5-(1-methylethyl)-	$C_{10}H_{16}$ 136.24	2867-05-2	 151	4-05-00-00451 0.8301^{20}	 1.4515^{20}
2987	Bicyclo[3.1.0]hex-3-en-2-one, 4-methyl-1-(1-methylethyl)-	$C_{10}H_{14}O$ 150.22	24545-81-1	 220	0-07-00-00159 0.9572^{15}	 1.48325
2988	1,1'-Bicyclohexyl Cyclohexylcyclohexane	$C_{12}H_{22}$ 166.31	92-51-3 4	 238		H_2O 2; EtOH 3; eth 3
2989	[1,1'-Bicyclohexyl]-2-one	$C_{12}H_{20}O$ 180.29	90-42-6 -32	 264	4-07-00-00248 0.9698^{25}	 1.4877^{25}
2990	Bicyclo[3.3.1]nonane	C_9H_{16} 124.23	280-65-9 145.5	 169.5	4-05-00-00293	EtOH 4
2991	Bicyclo[4.2.0]octane	C_8H_{14} 110.20	278-30-8	 136	4-05-00-00276 0.8573^{20}	 1.4613^{20}
2992	Bicyclo[2.2.2]octane, 2-methyl- 2-Methylbicyclo[2.2.2]octane	C_9H_{16} 124.23	766-53-0 33.5	 159	4-05-00-00295 0.8664^{40}	 1.4608^{40}
2993	Bicyclo[4.2.0]oct-2-ene	C_8H_{12} 108.18	13367-29-8	 138	4-05-00-00413 0.8948^{20}	 1.4810^{20}
2994	[1,1'-Bicyclopentyl]-2-ol Bicyclopentyl, 2-hydroxy	$C_{10}H_{18}O$ 154.25	4884-25-7 20	 235.5	3-06-00-00261 0.9785^{15}	 1.4884^{17}
2995	[1,1'-Bicyclopentyl]-2-one	$C_{10}H_{16}O$ 152.24	4884-24-6 -13	 232.5	4-07-00-00198 0.9745^{21}	 1.4763
2996	Bicyclo[7.2.0]undec-4-ene, 4,11,11-trimethyl-8-methylene-, [1*R*-(1*R**,4E,9*S**)]- Caryophyllene	$C_{15}H_{24}$ 204.36	87-44-5	1884 $122^{13.5}$	4-05-00-01182 0.9075^{20}	bz 4 1.4986^{20}
2997	Bifenox Benzoic acid, 5-(2,4-dichlorophenoxy)-2-nitro-, methyl ester	$C_{14}H_9Cl_2NO_5$ 342.13	42576-02-3 85	1228		
2998	Bifenthrin 2-Methylbiphenyl-3-ylmethyl (*Z*)-(1RS,3RS)-3-(2-chloro-3,3,3-trifluoroprop-1-enyl)- 2,2-dimethylcyclopropanecarboxylate	$C_{23}H_{22}ClF_3O_2$ 422.87	82657-04-3 69	1229	 1.2^{125}	
2999	[2,2'-Bi-1H-indene]-1,1',3,3'(2H,2'H)-tetrone, 2,2'-dihydroxy- Hydrindantin	$C_{18}H_{10}O_6$ 322.27	5103-42-4 250 dec	4698	4-08-00-03701	
3000	21H-Biline-8,12-dipropanoic acid, 2,17-diethenyl-1,10,19,22,23,24-hexahydro-3,7,13,18-tetramethyl-1,19-dioxo- Bilirubin	$C_{33}H_{36}N_4O_6$ 584.67	635-65-4	1235	4-26-00-03268	H_2O 1; EtOH 2; eth 2; bz 3
3001	21H-Biline-8,12-dipropanoic acid, 3,18-diethenyl-1,19,22,24-tetrahydro-2,7,13,17-tetramethyl-1,19-dioxo- Biliverdine	$C_{33}H_{34}N_4O_6$ 582.66	114-25-0 >300	1236	4-26-00-03272	H_2O 1; EtOH 3; eth 2; bz 3
3002	[2,2'-Bimorphinan]-3,3',6,6'-tetrol, 7,7',8,8'-tetradehydro-4,5:4',5'-diepoxy-17,17'-dimethyl- Pseudomorphine	$C_{34}H_{36}N_2O_6$ 568.67	125-24-6 282.5	7934	4-27-00-08724	H_2O 1; EtOH 1; eth 1; chl 1
3003	Binapacryl 2-Butenoic acid, 3-methyl-, 2-(1-methylpropyl)-4,6-dinitrophenyl ester	$C_{15}H_{18}N_2O_6$ 322.32	485-31-4 70	1237	 1.27^{20}	
3004	1,1'-Binaphthalene 1,1'-Binaphthyl	$C_{20}H_{14}$ 254.33	604-53-5 160	 >360; 240^{12}	4-05-00-02634 1.3000^{20}	H_2O 1; EtOH 2; eth 3; ace 3
3005	2,2'-Binaphthalene	$C_{20}H_{14}$ 254.33	612-78-2 188.3	 452	4-05-00-02636	H_2O 1; EtOH 2; eth 3; bz 3
3006	[1,1'-Binaphthalene]-4,4'-diol	$C_{20}H_{14}O_2$ 286.33	1446-34-0 300	 sub	4-06-00-07021	eth 4; EtOH 4
3007	2,2'-Bioxirane, (*R**,*S**)- Erythritol anhydride	$C_4H_6O_2$ 86.09	564-00-1 -19	3621 138	5-19-01-00184 1.113^{18}	H_2O 5; EtOH 4 1.4330^{20}
3008	1,1'-Biphenyl Diphenyl	$C_{12}H_{10}$ 154.21	92-52-4 69	3314 256.1	4-05-00-01807 1.04^{20}	H_2O 1; EtOH 3; eth 3; bz 4 1.588^{77}
3009	[1,1'-Biphenyl]-4-acetic acid Felbinac	$C_{14}H_{12}O_2$ 212.25	5728-52-9 160.5	3893	4-09-00-02520	
3010	[1,1'-Biphenyl]-2-amine	$C_{12}H_{11}N$ 169.23	90-41-5 51	 299	4-12-00-03223	H_2O 1; EtOH 3; eth 3; bz 3
3011	[1,1'-Biphenyl]-4-amine *p*-Biphenylamine	$C_{12}H_{11}N$ 169.23	92-67-1 53.5	1248 302	4-12-00-03241	H_2O 2; EtOH 3; eth 3; ace 3
3012	[1,1'-Biphenyl]-4-amine, 4'-chloro- 4-Amino-4'-chlorodiphenyl	$C_{12}H_{10}ClN$ 203.67	135-68-2 134	445	4-12-00-03269	ace 4; bz 4; eth 4

No.	Name Synonym	Mol. Form. Mol. Wt.	CAS RN mp/°C	Merck No. bp/°C	Beil. Ref. den/g cm^{-3}	Solubility n_D
3013	[1,1'-Biphenyl]-2-amine, 6-methyl- Biphenyl, 2-amino-6-methyl	$C_{13}H_{13}N$ 183.25	76472-83-8 43.5	 355; 144^{2}	3-12-00-03234 1.9060^{0}	bz 4; eth 4; EtOH 4
3014	[1,1'-Biphenyl]-4-amine, 3-methyl-	$C_{13}H_{13}N$ 183.25	63019-98-7 43	 190^{15}	3-12-00-03238	bz 4; eth 4; EtOH 4
3015	[1,1'-Biphenyl]-4-amine, 4'-methyl- Biphenyl, 4-amino-4'-methyl	$C_{13}H_{13}N$ 183.25	1204-78-0 99	 190^{18}	2-12-00-00771	ace 4; bz 4; eth 4; EtOH 4
3016	1,1'-Biphenyl, 2-bromo-	$C_{12}H_9Br$ 233.11	2052-07-5 1.5	 297	4-05-00-01818 1.2175^{26}	eth 4; EtOH 4 1.6248^{25}
3017	1,1'-Biphenyl, 3-bromo-	$C_{12}H_9Br$ 233.11	2113-57-7	 300; 171^{17}	4-05-00-01818	H_2O 1 1.6411^{20}
3018	1,1'-Biphenyl, 4-bromo-	$C_{12}H_9Br$ 233.11	92-66-0 91.5	 310	4-05-00-01819 0.9327^{25}	H_2O 1; EtOH 3; eth 3; bz 3
3019	[1,1'-Biphenyl]-4-butanoic acid	$C_{16}H_{16}O_2$ 240.30	6057-60-9 119.5	 $190^{0.3}$	4-09-00-02558	eth 4; EtOH 4
3020	[1,1'-Biphenyl]-2-carbonitrile	$C_{13}H_9N$ 179.22	24973-49-7 41	 175^{13}	3-09-00-03269	eth 4; EtOH 4
3021	[1,1'-Biphenyl]-2-carboxylic acid	$C_{13}H_{10}O_2$ 198.22	947-84-2 114.3	 343.5	4-09-00-02472	H_2O 1; EtOH 4; bz 4; HOAc 4
3022	[1,1'-Biphenyl]-4-carboxylic acid	$C_{13}H_{10}O_2$ 198.22	92-92-2 228	 sub	4-09-00-02479	H_2O 1; EtOH 3; eth 3; bz 3
3023	1,1'-Biphenyl, 2-chloro-	$C_{12}H_9Cl$ 188.66	2051-60-7 34	 274	4-05-00-01816 1.1499^{32}	eth 4; EtOH 4; lig 4
3024	1,1'-Biphenyl, 3-chloro-	$C_{12}H_9Cl$ 188.66	2051-61-8 16	 284.5	4-05-00-01816 1.1579^{25}	ace 4; eth 4; EtOH 4 1.6181^{25}
3025	1,1'-Biphenyl, 4-chloro-	$C_{12}H_9Cl$ 188.66	2051-62-9 78.8	 292.9; 146^{10}	4-05-00-01816	H_2O 1; EtOH 3; eth 3; lig 3
3026	Biphenyl, 2-chloro-2'-methyl- 2-Chloro-2'-methylbiphenyl	$C_{13}H_{11}Cl$ 202.68	19493-31-3 17	 276; 174^{50}	3-05-00-01799	 1.588^{25}
3027	1,1'-Biphenyl, 2,2',3,3',4,4',5,5',6,6'-decafluoro-	$C_{12}F_{10}$ 334.12	434-90-2 67.5	 206	4-05-00-01816 1.785^{20}	
3028	[1,1'-Biphenyl]-2,2'-diamine	$C_{12}H_{12}N_2$ 184.24	1454-80-4 81	 162^{4}	4-13-00-00356 1.3090^{20}	H_2O 3; ace 3; bz 3
3029	[1,1'-Biphenyl]-2,4'-diamine 2,4'-Biphenyldiamine	$C_{12}H_{12}N_2$ 184.24	492-17-1 54.5	1249 363	4-13-00-00360	H_2O 1; EtOH 3; eth 3
3030	[1,1'-Biphenyl]-3,3'-diamine	$C_{12}H_{12}N_2$ 184.24	2050-89-7 93.5	 $205^{0.001}$	4-13-00-00363	H_2O 2; eth 4; bz 3
3031	[1,1'-Biphenyl]-4,4'-diamine Benzidine	$C_{12}H_{12}N_2$ 184.24	92-87-5 120	1086 401	4-13-00-00364	H_2O 2; EtOH 3; eth 2; DMSO 2
3032	[1,1'-Biphenyl]-4,4'-diamine, 2,2'-dichloro- *o*-Dichlorobenzidine	$C_{12}H_{10}Cl_2N_2$ 253.13	84-68-4 165	3046	4-13-00-00384	eth 4; EtOH 4
3033	[1,1'-Biphenyl]-4,4'-diamine, 3,3'-dimethoxy- *o*-Dianisidine	$C_{14}H_{16}N_2O_2$ 244.29	119-90-4 137	2970	4-13-00-02834	H_2O 1; EtOH 3; eth 3; ace 3
3034	[1,1'-Biphenyl]-4,4'-diamine, 3,3'-dimethyl- *o*-Tolidine	$C_{14}H_{16}N_2$ 212.29	119-93-7 131.5	9437	4-13-00-00419	H_2O 2; EtOH 4; eth 4; chl 2
3035	[1,1'-Biphenyl]-4,4'-diamine, *N,N'*-diphenyl- *N,N'*-Diphenylbenzidine	$C_{24}H_{20}N_2$ 336.44	531-91-9 247	3319	4-13-00-00368	H_2O 1; EtOH 2; eth 2; bz 2
3036	[1,1'-Biphenyl]-4,4'-diamine, tetramethyl-	$C_{16}H_{20}N_2$ 240.35	34314-06-2 198	 >360	3-13-00-00429	bz 4; chl 4
3037	1,1'-Biphenyl, 4,4'-dibromo-	$C_{12}H_8Br_2$ 312.00	92-86-4 164	 357.5	4-05-00-01820	H_2O 1; EtOH 2; bz 3
3038	[1,1'-Biphenyl]-2,2'-dicarboxylic acid *o,o'*-Diphenic acid	$C_{14}H_{10}O_4$ 242.23	482-05-3 233.5	3309 sub	4-09-00-03552	H_2O 1; EtOH 3; eth 3; os 3
3039	[1,1'-Biphenyl]-2,2'-dicarboxylic acid, dimethyl ester	$C_{16}H_{14}O_4$ 270.28	5807-64-7 74	 $204\text{-}6^{14}$	4-09-00-03553	H_2O 1; EtOH 4; eth 3; bz 3
3040	1,1'-Biphenyl, 3,3'-dichloro-	$C_{12}H_8Cl_2$ 223.10	2050-67-1 29	 320	3-05-00-01739	bz 4; eth 4; EtOH 4
3041	1,1'-Biphenyl, 4,4'-dichloro-	$C_{12}H_8Cl_2$ 223.10	2050-68-2 149.3	 317	4-05-00-01817 1.4420^{0}	H_2O 1; EtOH 2; bz 3; chl 2
3042	1,1'-Biphenyl, 2,2'-difluoro-	$C_{12}H_8F_2$ 190.19	388-82-9 118.5		4-05-00-01816 1.393^{20}	H_2O 1; EtOH 3; xyl 3
3043	1,1'-Biphenyl, 3,3'-difluoro-	$C_{12}H_8F_2$ 190.19	396-64-5 8	 130^{14}	4-05-00-01815 1.192^{25}	ctc 2 1.5678^{20}
3044	1,1'-Biphenyl, 4,4'-difluoro- 4,4'-Difluorodiphenyl	$C_{12}H_8F_2$ 190.19	398-23-2 94.5	3133 254.5	4-05-00-01816	H_2O 1; EtOH 4; eth 3; ace 3
3045	1,1'-Biphenyl, 2,2'-dimethoxy- 2,2'-Dimethoxybiphenyl	$C_{14}H_{14}O_2$ 214.26	4877-93-4 155	 307.5	4-06-00-06645 1.200^{25}	H_2O 1; EtOH 4; eth 2; bz 4
3046	1,1'-Biphenyl, 3,3'-dimethoxy- 3,3'-Dimethoxybiphenyl	$C_{14}H_{14}O_2$ 214.26	6161-50-8 36	 328	4-06-00-06650	H_2O 1; EtOH 4; eth 4; ace 3
3047	1,1'-Biphenyl, 4,4'-dimethoxy-	$C_{14}H_{14}O_2$ 214.26	2132-80-1 175	 sub	4-06-00-06651	H_2O 1; EtOH 4; eth 2; bz 4
3048	1,1'-Biphenyl, 2,2'-dimethoxy-5,5'-dimethyl- Biphenyl, 2,2'-dimethoxy-5,5'-dimethyl	$C_{16}H_{18}O_2$ 242.32	7168-55-0 71	 188^{12}	3-06-00-05447	ace 4; bz 4; eth 4; EtOH 4
3049	1,1'-Biphenyl, 2,5'-dimethoxy-2',5-dimethyl- Biphenyl, 2,5'-dimethoxy-2',5-dimethyl	$C_{16}H_{18}O_2$ 242.32	72935-12-7 66	 168^{4}	4-06-00-06704	bz 4; peth 4
3050	1,1'-Biphenyl, 2,2'-dimethyl-	$C_{14}H_{14}$ 182.27	605-39-0 19.5	 256	4-05-00-01897 0.9906^{20}	H_2O 1; EtOH 4; eth 4; ace 3 1.5752^{20}
3051	1,1'-Biphenyl, 2,3'-dimethyl-	$C_{14}H_{14}$ 182.27	611-43-8	 270	4-05-00-01902 0.9924^{20}	H_2O 1; EtOH 4; eth 4; ace 3 1.5810^{20}

No.	Name Synonym	Mol. Form. Mol. Wt.	CAS RN mp/°C	Merck No. bp/°C	Beil. Ref. den/g cm^{-3}	Solubility n_D
3052	1,1'-Biphenyl, 2,3-dimethyl- 2,3-Dimethylbiphenyl	$C_{14}H_{14}$ 182.27	3864-18-4 42	 141^{14}; 98^{2}	4-05-00-01897	H_2O 1; eth 3; peth 3 1.5845^{23}
3053	1,1'-Biphenyl, 2,4'-dimethyl-	$C_{14}H_{14}$ 182.27	611-61-0	 277.2	4-05-00-01903 0.9924^{20}	H_2O 1; EtOH 4; eth 4; ace 3 1.5826^{20}
3054	1,1'-Biphenyl, 2,4-dimethyl- 2,4-Dimethylbiphenyl	$C_{14}H_{14}$ 182.27	4433-10-7	 273; 144^{20}	4-05-00-01897 0.9947^{20}	 1.5844^{20}
3055	1,1'-Biphenyl, 2,5-dimethyl-	$C_{14}H_{14}$ 182.27	7372-85-2	 $140^{14.5}$	4-05-00-01897 0.9931^{20}	 1.5819^{20}
3056	1,1'-Biphenyl, 2,6-dimethyl-	$C_{14}H_{14}$ 182.27	3976-34-9 -5	 262.5	4-05-00-01897 0.9907^{20}	 1.5745^{20}
3057	1,1'-Biphenyl, 3,3'-dimethyl-	$C_{14}H_{14}$ 182.27	612-75-9 9	 280	4-05-00-01903 0.9995^{20}	H_2O 1; EtOH 4; eth 4; ace 3 1.5946^{20}
3058	1,1'-Biphenyl, 3,4'-dimethyl-	$C_{14}H_{14}$ 182.27	7383-90-6 14.5	 289; 153^{15}	4-05-00-01905 0.9978^{20}	H_2O 1; bz 3 1.5968^{20}
3059	1,1'-Biphenyl, 3,4-dimethyl- 3,4-Dimethylbiphenyl	$C_{14}H_{14}$ 182.27	4433-11-8 29.5	 282	4-05-00-01903 1.0087^{20}	H_2O 1; bz 3; chl 3 1.6036^{20}
3060	1,1'-Biphenyl, 3,5-dimethyl- 3,5-Dimethylbiphenyl	$C_{14}H_{14}$ 182.27	17057-88-4 22.5	 274.5	4-05-00-01903 0.9990^{20}	 1.5952^{20}
3061	1,1'-Biphenyl, 4,4'-dimethyl-	$C_{14}H_{14}$ 182.27	613-33-2 125	 295	4-05-00-01906 0.917^{121}	H_2O 1; EtOH 2; eth 3; ace 3
3062	1,1'-Biphenyl, 2,2'-dinitro-	$C_{12}H_8N_2O_4$ 244.21	2436-96-6 126	 305	4-05-00-01826 1.45^{25}	H_2O 1; EtOH 4; eth 3; ace 2
3063	1,1'-Biphenyl, 2,4'-dinitro-	$C_{12}H_8N_2O_4$ 244.21	606-81-5 93.5		4-05-00-01827 1.474^{25}	H_2O 1; EtOH 4; eth 3; bz 3
3064	[1,1'-Biphenyl]-2,2'-diol	$C_{12}H_{10}O_2$ 186.21	1806-29-7 109	 320	4-06-00-06645 1.3420^{20}	H_2O 3; EtOH 3; eth 3; ace 3
3065	[1,1'-Biphenyl]-2,4'-diol	$C_{12}H_{10}O_2$ 186.21	611-62-1 162.5	 342	3-06-00-05387	eth 4
3066	[1,1'-Biphenyl]-3,3'-diol	$C_{12}H_{10}O_2$ 186.21	612-76-0 124.8	 247^{18}	3-06-00-05388	H_2O 2; EtOH 3; eth 3; bz 3
3067	[1,1'-Biphenyl]-2,2'-diol, 3,3'-dimethyl-	$C_{14}H_{14}O_2$ 214.26	32750-14-4 113	 sub	3-06-00-05445	bz 4; eth 4; EtOH 4
3068	[1,1'-Biphenyl]-2,2'-diol, 5,5'-dimethyl-	$C_{14}H_{14}O_2$ 214.26	15519-73-0 153.5	 sub	4-06-00-06705	H_2O 2; EtOH 3; eth 4; ace 3
3069	[1,1'-Biphenyl]-4,4'-diol, 3,3',5,5'-tetramethyl-	$C_{16}H_{18}O_2$ 242.32	2417-04-1 221.8	 sub	4-06-00-06744	EtOH 2; bz 2; gl HOAc 2; tol 2
3070	[1,1'-Biphenyl]-4,4'-disulfonic acid 4,4'-Biphenyldisulfonic acid	$C_{12}H_{10}O_6S_2$ 314.34	5314-37-4 72.5	 >200	3-11-00-00472	H_2O 4
3071	1,1'-Biphenyl, 3-ethoxy- 3-Ethoxybiphenyl	$C_{14}H_{14}O$ 198.26	54852-73-2 35	 305	3-06-00-03313	EtOH 3; eth 3; ace 3; bz 3
3072	1,1'-Biphenyl, 2-ethyl-	$C_{14}H_{14}$ 182.27	1812-51-7 -6	 266; 128^{11}	4-05-00-01896 0.9966^{20}	 1.5758^{25}
3073	1,1'-Biphenyl, 3-ethyl-	$C_{14}H_{14}$ 182.27	5668-93-9 -27.5	 283.5	4-05-00-01896 1.048^{25}	 1.5859^{25}
3074	1,1'-Biphenyl, 2-fluoro-	$C_{12}H_9F$ 172.20	321-60-8 73.5	 248	4-05-00-01815 1.2452^{25}	EtOH 3; eth 3; chl 3; lig 2
3075	1,1'-Biphenyl, 4-fluoro-	$C_{12}H_9F$ 172.20	324-74-3 74.2	 253	4-05-00-01815 1.247^{25}	EtOH 2; eth 3; gl HOAc 3
3076	1,1'-Biphenyl, 2,2',4,4',6,6'-hexamethyl-	$C_{18}H_{22}$ 238.37	4482-03-5 103.5	 297	4-05-00-01989 1.023^{50}	H_2O 1; EtOH 2; eth 3; bz 3
3077	1,1'-Biphenyl, 2-iodo-	$C_{12}H_9I$ 280.11	2113-51-1	 190^{36}; 158^{6}	4-05-00-01820 1.5991^{25}	H_2O 1; EtOH 3; eth 3; bz 3 1.6620^{20}
3078	1,1'-Biphenyl, 3-iodo-	$C_{12}H_9I$ 280.11	20442-79-9 26.5	 188^{16}	4-05-00-01821 1.5967^{25}	
3079	1,1'-Biphenyl, 4-iodo-	$C_{12}H_9I$ 280.11	1591-31-7 113.5	 320; 183^{11}	4-05-00-01821	H_2O 1; EtOH 3; eth 3; bz 3
3080	1,1'-Biphenyl, 2-methoxy-	$C_{13}H_{12}O$ 184.24	86-26-0 29	 274	4-06-00-04580 1.0233^{99}	H_2O 1; EtOH 3; ctc 2; peth 3 1.5641^{99}
3081	1,1'-Biphenyl, 4-methoxy-	$C_{13}H_{12}O$ 184.24	613-37-6 90	 157^{10}	4-06-00-04600 1.0278^{100}	H_2O 1; EtOH 3; eth 3 1.5744^{100}
3082	1,1'-Biphenyl, 2-methyl-	$C_{13}H_{12}$ 168.24	643-58-3 -0.2	 255.5	4-05-00-01855 1.010^{22}	H_2O 1; EtOH 3; eth 3 1.5914^{20}
3083	1,1'-Biphenyl, 3-methyl-	$C_{13}H_{12}$ 168.24	643-93-6 4.5	 272.7	4-05-00-01858 1.0182^{17}	H_2O 1; EtOH 3; eth 3; ctc 3 1.5972^{20}
3084	1,1'-Biphenyl, 4-methyl-	$C_{13}H_{12}$ 168.24	644-08-6 49.5	 267.5	4-05-00-01860 1.015^{27}	H_2O 1; EtOH 3; eth 3; ctc 2
3085	1,1'-Biphenyl, 2-nitro- *o*-Nitrobiphenyl	$C_{12}H_9NO_2$ 199.21	86-00-0 37.2	6513 320	4-05-00-01823 1.44^{25}	H_2O 1; EtOH 3; eth 3; chl 3
3086	1,1'-Biphenyl, 3-nitro-	$C_{12}H_9NO_2$ 199.21	2113-58-8 62	 227^{35}; 143^{9}	4-05-00-01823	H_2O 1; EtOH 3; eth 3; HOAc 3
3087	1,1'-Biphenyl, 4-nitro- *p*-Nitrobiphenyl	$C_{12}H_9NO_2$ 199.21	92-93-3 114	6514 340	4-05-00-01823	H_2O 1; EtOH 2; eth 3; bz 3
3088	[1,1'-Biphenyl]-2-ol *o*-Phenylphenol	$C_{12}H_{10}O$ 170.21	90-43-7 59	7276 286	4-06-00-04579 1.213^{25}	H_2O 1; EtOH 3; eth 4; ace 3
3089	[1,1'-Biphenyl]-3-ol *m*-Phenylphenol	$C_{12}H_{10}O$ 170.21	580-51-8 78	 >300	4-06-00-04597	H_2O 2; EtOH 4; eth 4; bz 4
3090	[1,1'-Biphenyl]-4-ol *p*-Phenylphenol	$C_{12}H_{10}O$ 170.21	92-69-3 166	7277 305	4-06-00-04600	H_2O 2; EtOH 4; eth 4; chl 4

No.	Name Synonym	Mol. Form. Mol. Wt.	CAS RN mp/°C	Merck No. bp/°C	Beil. Ref. den/g cm^{-3}	Solubility n_D
3091	[1,1'-Biphenyl]-4-ol, acetate	$C_{14}H_{12}O_2$ 212.25	148-86-7 89	 196^{13}	3-06-00-03326	H_2O 2; EtOH 2; eth 3; bz 3
3092	[1,1'-Biphenyl]-2-ol, 3-chloro- 2-Phenyl-6-chlorophenol	$C_{12}H_9ClO$ 204.66	85-97-2 6	7251 317 dec	4-06-00-04586 1.24^{25}	H_2O 1; EtOH 3; eth 3; ace 3 1.6237^{30}
3093	1,1'-Biphenyl, 2-(phenylmethyl)-	$C_{19}H_{16}$ 244.34	606-97-3 55	 283^{110}	3-05-00-02323	bz 4; eth 4; EtOH 4
3094	1,1'-Biphenyl, 4-(phenylmethyl)-	$C_{19}H_{16}$ 244.34	613-42-3 85	 $285\text{-}6^{110}$	4-05-00-02502 1.171^{0}	H_2O 1; EtOH 3; eth 4; bz 4
3095	1,1'-Biphenyl, 2,2',4,4'-tetramethyl-	$C_{16}H_{18}$ 210.32	3976-36-1 41	 287	3-05-00-01891	EtOH 3; eth 2; bz 2; chl 3
3096	1,1'-Biphenyl, 2,2',5,5'-tetramethyl-	$C_{16}H_{18}$ 210.32	3075-84-1 51.5	 285	3-05-00-01892	eth 3; bz 3
3097	[1,1'-Biphenyl]-3,3',5,5'-tetrol Diresorcinol	$C_{12}H_{10}O_4$ 218.21	531-02-2 310	3361	2-06-00-01129	H_2O 4; eth 4; EtOH 4
3098	2,3'-Bipiperidine 2,3'-Bipiperidyl	$C_{10}H_{20}N_2$ 168.28	2467-09-6 68.5	 269.5	4-23-00-00512	H_2O 4; EtOH 4
3099	2,2'-Bipyridine α,α'-Dipyridyl	$C_{10}H_8N_2$ 156.19	366-18-7 72	3355 273.5	5-23-08-00016	H_2O 2; EtOH 4; eth 4; bz 4
3100	2,3'-Bipyridine 2,3'-Bipyridyl	$C_{10}H_8N_2$ 156.19	581-50-0	 295.5	5-23-08-00026 1.140^{20}	H_2O 1; EtOH 4; eth 4; bz 4 1.6223^{20}
3101	2,4'-Bipyridine	$C_{10}H_8N_2$ 156.19	581-47-5 61.5	 281	5-23-08-00027	H_2O 2; EtOH 4; eth 4; chl 4
3102	3,3'-Bipyridine 3,3'-Dipyridyl	$C_{10}H_8N_2$ 156.19	581-46-4 68	 291.5	5-23-08-00027 1.1614^{20}	H_2O 4; EtOH 4; eth 2
3103	3,4'-Bipyridine 3,4'-Bipyridyl	$C_{10}H_8N_2$ 156.19	4394-11-0 62	 297; 144^{15}	5-23-08-00028	EtOH 4; peth 4
3104	4,4'-Bipyridine γ,γ'-Dipyridyl	$C_{10}H_8N_2$ 156.19	553-26-4 111	3356 305	5-23-08-00028	H_2O 2; EtOH 4; eth 3; bz 4
3105	2,3'-Bipyridine, 1,2,3,6-tetrahydro-, (*S*)- Anatabine	$C_{10}H_{12}N_2$ 160.22	581-49-7	661 $145\text{-}6^{10}$	5-23-06-00507 1.091^{19}	H_2O 5; EtOH 3; eth 3; bz 3 1.5676^{20}
3106	[5,5'-Bipyrimidine]-2,2',4,4',6,6'(1H,1'H,3H,3'H,5H,5'H)-hexone-5,5'-dihydroxy Alloxantin	$C_8H_6N_4O_8$ 286.16	76-24-4 253-5 dec	282	4-26-00-02782	H_2O 2; EtOH 2; eth 2
3107	2,3'-Biquinoline 2,3'-Biquinolyl	$C_{18}H_{12}N_2$ 256.31	612-81-7 175.8	 >400	5-23-10-00009	bz 4; eth 4; EtOH 4; chl 4
3108	Bis[1,3]benzodioxolo[4,5-c:5',6'-g]azecin-13(5H)-one, 4,6,7,14-tetrahydro-5,14-dimethyl-, (*R*)- Corycavamine	$C_{21}H_{21}NO_5$ 367.40	521-85-7 149	2540	2-27-00-00621	EtOH 4; chl 4
3109	Bis[1,3]benzodioxolo[4,5-c:5',6'-g]azecin-13(5H)-one, 4,6,7,14-tetrahydro-5-methyl- Protopine	$C_{20}H_{19}NO_5$ 353.37	130-86-9 208	7912	4-27-00-06881	H_2O 1; EtOH 2; eth 2; bz 2
3110	3H-Bis[1]benzopyrano[3,4-b:6',5'-e]pyran-7(7aH)-one, 13,13a-dihydro-7a-hydroxy-9,10-dimethoxy-3,3-dimethyl Tephrosin	$C_{23}H_{22}O_7$ 410.42	76-80-2 198	9082	4-19-00-05271	ace 4; eth 4; chl 4
3111	Bismuthine, triphenyl- Triphenylbismuth	$C_{18}H_{15}Bi$ 440.30	603-33-8 77.6	 242^{14}	4-16-00-01226 1.715^{75}	EtOH 2; eth 3; ace 3; bz 3 1.7040^{75}
3112	2,2'-Bithiophene	$C_8H_6S_2$ 166.27	492-97-7 33	 260	5-19-01-00537	H_2O 1; EtOH 4; eth 3; ctc 3
3113	Boranamine, *N,N*,1,1-tetramethyl-	$C_4H_{12}BN$ 84.96	1113-30-0 -92	 65	4-04-00-04367	eth 4; ace 4; os 5
3114	Boranediamine, 1-fluoro-*N,N,N',N'*-tetramethyl-	$C_4H_{12}BFN_2$ 117.96	383-90-4 -44.3	 106	4-04-00-00303	ace 4; eth 4
3115	Borane, difluoro(4-methylphenyl)-	$C_7H_7BF_2$ 139.94	768-39-8	 128	4-16-00-01674 1.055^{25}	bz 4; eth 4 1.4535^{25}
3116	Borane, difluorophenyl-	$C_6H_5BF_2$ 125.91	368-98-9 -36.2	 98	4-16-00-01661 1.087^{25}	bz 4; eth 4 1.4441^{25}
3117	Borane, dimethoxy- Dimethoxyborane	$C_2H_7BO_2$ 73.89	4542-61-4 -130.6	 25.9	4-01-00-01269	H_2O 6
3118	Borane, triethyl Triethylborane	$C_6H_{15}B$ 98.00	97-94-9 -93	 95	 0.70^{23}	EtOH 3; eth 3 1.3971
3119	Borane, trimethyl- Trimethylborane	C_3H_9B 55.92	593-90-8 -161.5	 -20.2		
3120	Borane, trimethyl-, monoammoniate Aminetrimethylboron	$C_3H_{12}BN$ 72.95	1830-95-1 73.5			
3121	Borane, tripropyl-	$C_9H_{21}B$ 140.08	1116-61-6 -56	 159	4-04-00-04360 0.7204^{25}	 1.4135^{22}
3122	Borane, tris(3-methylbutyl)- Borane, triisopentyl-	$C_{15}H_{33}B$ 224.24	3062-81-5	 119^{14}	2-04-00-01023 0.7600^{25}	ace 4; eth 4; EtOH 4 1.4321
3123	Borane, tris(2-methylpropyl)-	$C_{12}H_{27}B$ 182.16	1116-39-8	 188; 86^{20}	4-04-00-04361 0.7380^{25}	bz 4; eth 4; EtOH 4 1.4188^{23}
3124	Boric acid, tributyl ester	$C_{12}H_{27}BO_3$ 230.16	688-74-4	 234	4-01-00-01544 0.8567^{20}	EtOH 3; eth 4; bz 3; MeOH 4 1.4106^{18}
3125	Boric acid, triethyl ester	$C_6H_{15}BO_3$ 145.99	150-46-9 -84.8	 120	4-01-00-01365 0.8546^{20}	EtOH 5; eth 5 1.3749^{20}
3126	Boric acid, trimethyl ester Trimethyl borate	$C_3H_9BO_3$ 103.91	121-43-7 -29.3	9626 67.5	4-01-00-01269 0.915^{25}	eth 4; MeOH 4 1.3568^{20}

No.	Name Synonym	Mol. Form. Mol. Wt.	CAS RN mp/°C	Merck No. bp/°C	Beil. Ref. den/g cm^{-3}	Solubility n_D
3127	Boric acid, tripropyl ester	$C_9H_{21}BO_3$ 188.07	688-71-1	 179.5	4-01-00-01436 0.8576^{20}	EtOH 4; eth 5; PrOH 3 1.3948^{20}
3128	Boric acid, tris(1-methylethyl) ester Isopropyl borate	$C_9H_{21}BO_3$ 188.07	5419-55-6	 140; 75^{76}	4-01-00-01488 0.8251^{20}	EtOH 4; eth 4; bz 4; PrOH 4 1.3777^{20}
3129	Borinic acid, dimethylthio-, methyl ester	C_3H_9BS 87.98	19163-05-4 -84	 71	4-04-00-04368	ace 4; eth 4
3130	Borinic acid, diphenyl-	$C_{12}H_{11}BO$ 182.03	2622-89-1 57.5	 $150-5^{20}$	4-16-00-01639 1.0740^{20}	EtOH 3; bz 3; peth 3
3131	3-Bornanecarboxylic acid Camphane-3-carboxylic acid	$C_{11}H_{18}O_2$ 182.26	91965-23-0 90.5	 153^{13}	3-09-00-00244	EtOH 4; HOAc 4
3132	Boron, (*N,N*-dimethylmethanamine)trihydro-, (T-4)- Trimethylamineborane	$C_3H_{12}BN$ 72.95	75-22-9 94	 172	4-04-00-00134 0.792^{25}	eth 4; EtOH 4
3133	Boronic acid, ethyl-	$C_2H_7BO_2$ 73.89	4433-63-0 40	 dec	4-04-00-04379	H_2O 4; eth 4; EtOH 4
3134	Boronic acid, (2-methylphenyl)-	$C_7H_9BO_2$ 135.96	16419-60-6 166.5	 dec	3-16-00-01277	bz 4; eth 4; EtOH 4
3135	Boronic acid, phenyl- Benzeneboronic acid	$C_6H_7BO_2$ 121.93	98-80-6 219	1076	4-16-00-01654	H_2O 2; EtOH 3; eth 3; bz 3
3136	Boronic acid, propyl-	$C_3H_9BO_2$ 87.91	17745-45-8 107	 dec	3-04-00-01964	H_2O 4; eth 4; EtOH 4
3137	Boron, trifluoro[1,1'-oxybis[ethane]]-, (T-4)- Boron trifluoride etherate	$C_4H_{10}BF_3O$ 141.93	109-63-7 -60.4	1352 125.5	4-01-00-01314 1.3572^{20}	eth 4; EtOH 4 1.4447^{20}
3138	Boron, trifluoro[oxybis[methane]]-, (T-4)- Boron trifluoride compd. with methyl ether	$C_2H_6BF_3O$ 113.88	353-42-4 -14	 127 dec	4-01-00-01245 1.2410^{20}	 1.302^{20}
3139	Bromacil 2,4(1H,3H)-Pyrimidinedione, 5-bromo-6-methyl-3-(1-methylpropyl)-	$C_9H_{13}BrN_2O_2$ 261.12	314-40-9 158	1370	 1.55^{25}	
3140	Bromopropylate Benzeneacetic acid, 4-bromo-α-(4-bromophenyl)-α-hydroxy-, 1-methylethyl ester	$C_{17}H_{16}Br_2O_3$ 428.12	18181-80-1 77	1422	 1.59^{20}	
3141	Bufa-20,22-dienolide, 16-(acetyloxy)-3,14-dihydroxy-, (3β,5β,16β)- Bufotalin	$C_{26}H_{36}O_6$ 444.57	471-95-4 223 dec	1466	5-18-04-00443	H_2O 1; EtOH 3; chl 3
3142	Bufa-4,20,22-trienolide, 6-(acetyloxy)-3-(β-*D*-glucopyranosyloxy)-8,14-dihydroxy-, (3β,6β)- Scilliroside	$C_{32}H_{44}O_{12}$ 620.69	507-60-8 169	8356 dec	4-18-00-03178	H_2O 2; EtOH 4; eth 1; ace 2
3143	*sec*-Bumeton 1,3,5-Triazine-2,4-diamine, *N*-ethyl-6-methoxy-*N*'-(1-methylpropyl)-	$C_{10}H_{19}N_5O$ 225.29	26259-45-0 87			
3144	Butachlor Acetamide, *N*-(butoxymethyl)-2-chloro-*N*-(2,6-diethylphenyl)-	$C_{17}H_{26}ClNO_2$ 311.85	23184-66-9 <-5	1498 $156^{.5}$	 1.070^{25}	
3145	1,2-Butadiene Methylallene	C_4H_6 54.09	590-19-2 -136.2	 10.9	4-01-00-00975 0.676^{0}	H_2O 1; EtOH 5; eth 5; bz 4 1.4205^{1}
3146	1,3-Butadiene Divinyl	C_4H_6 54.09	106-99-0 -108.9	1500 -4.4	4-01-00-00976 $*0.6149^{25}$	H_2O 1; EtOH 3; eth 3; ace 4 1.4292^{-25}
3147	1,2-Butadiene, 4-bromo- 4-Bromo-1,2-butadiene	C_4H_5Br 132.99	18668-68-3	 110	3-01-00-00929 1.4255^{20}	ace 4 1.5248^{20}
3148	1,3-Butadiene, 1-bromo- 1-Bromo-1,3-butadiene	C_4H_5Br 132.99	21890-35-7	 93	4-01-00-00989 1.416^{0}	eth 4; EtOH 4; peth 4
3149	1,3-Butadiene, 2-bromo-	C_4H_5Br 132.99	1822-86-2	 $42-3^{165}$	4-01-00-00989 1.397^{20}	eth 4; EtOH 4 1.4988^{20}
3150	1,2-Butadiene, 4-chloro-	C_4H_5Cl 88.54	25790-55-0	 88	4-01-00-00975 0.9891^{20}	ace 4; bz 4; eth 4 1.4775^{20}
3151	1,3-Butadiene, 1-chloro-	C_4H_5Cl 88.54	627-22-5	 68	3-01-00-00949 0.9606^{20}	eth 4; EtOH 4; chl 4 1.4712^{20}
3152	1,3-Butadiene, 2-chloro- 2-Chloro-1,3-butadiene	C_4H_5Cl 88.54	126-99-8 -130	 59.4	4-01-00-00984 0.956^{20}	H_2O 2; eth 5; ace 5; bz 5 1.4583^{20}
3153	1,3-Butadiene, 1-chloro-2-methyl- Chloromethylbutadiene	C_5H_7Cl 102.56	35383-51-8	 107	3-01-00-00974 0.9710^{20}	ace 4; EtOH 4 1.4702^{20}
3154	1,3-Butadiene, 1-chloro-3-methyl- 4-Chloroisoprene	C_5H_7Cl 102.56	51034-46-9	 100.7	4-01-00-01005 0.9543^{20}	ace 4; eth 4; EtOH 4; chl 4 1.4719^{20}
3155	1,3-Butadiene, 2-chloro-3-methyl-	C_5H_7Cl 102.56	1809-02-5	 93	4-01-00-01004 0.9593^{20}	H_2O 1; EtOH 3; eth 4; ace 3 1.4686^{20}
3156	1,3-Butadiene, 1,1-dichloro- 1,1-Dichloro-1,3-butadiene	$C_4H_4Cl_2$ 122.98	6061-06-9	 $42-3^{90}$	4-01-00-00985 1.1831^{20}	H_2O 1; eth 4; bz 4; chl 4 1.5022^{20}
3157	1,3-Butadiene, 1,2-dichloro- 1,2-Dichloro-1,3-butadiene	$C_4H_4Cl_2$ 122.98	3574-40-1	 62^{105}; 35^{40}	4-01-00-00985 1.1991^{20}	ctc 4 1.4960^{20}
3158	1,3-Butadiene, 2,3-dichloro-	$C_4H_4Cl_2$ 122.98	1653-19-6	 98	4-01-00-00986 1.1829^{20}	chl 4 1.4890^{20}
3159	1,3-Butadiene, 2,3-dimethyl- Diisopropenyl	C_6H_{10} 82.15	513-81-5 -76	3228 68.8	4-01-00-01023 0.7222^{25}	ctc 3 1.4394^{20}
3160	1,3-Butadiene, 1-ethoxy-	$C_6H_{10}O$ 98.14	5614-32-4	 110.5	4-01-00-02222 0.8154^{20}	ace 4; bz 4; eth 4; EtOH 4 1.4529^{20}
3161	1,3-Butadiene, 2-ethoxy-	$C_6H_{10}O$ 98.14	4747-05-1	 94.5	4-01-00-02223 0.8177^{20}	ace 4; bz 4; eth 4; EtOH 4 1.4400^{20}
3162	1,3-Butadiene, 2-fluoro- Fluoroprene	C_4H_5F 72.08	381-61-3	 12	4-01-00-00962 0.843^{4}	 1.400^{4}
3163	1,3-Butadiene, 1,1,2,3,4,4-hexachloro- Hexachloro-1,3-butadiene	C_4Cl_6 260.76	87-68-3 -21	 215	4-01-00-00988 1.556^{25}	H_2O 1; EtOH 3; eth 3 1.5542^{20}

PHYSICAL CONSTANTS OF ORGANIC COMPOUNDS (continued)

No.	Name	Synonym	Mol. Form.	Mol. Wt.	CAS RN	mp/°C	Merck No.	bp/°C	Beil. Ref.	den/g cm^{-3}	Solubility	n_D
3164	1,3-Butadiene, 1,1,2,3,4,4-hexafluoro-	1,3-Butadiene, hexafluoro	C_4F_6	162.03	685-63-2	-132		6	4-01-00-00983	1.553^{-20}		1.378^{-20}
3165	1,3-Butadiene, 2-iodo-		C_4H_5I	179.99	19221-28-4			112	3-01-00-00956	1.7278^{20}		1.5616
3166	1,2-Butadiene, 4-iodo-	4-Iodo-1,2-butadiene	C_4H_5I	179.99	67885-08-9			130	3-01-00-00929	1.7129^{20}		1.5709^{20}
3167	1,2-Butadiene, 4-methoxy-		C_5H_8O	84.12	36678-06-5			88	4-01-00-02221	0.8286^{20}	EtOH 4	1.435^{20}
3168	1,3-Butadiene, 1-methoxy-		C_5H_8O	84.12	3036-66-6			91.5	3-01-00-01975	0.8296^{20}	H_2O 3; EtOH 3	1.4594^{20}
3169	1,3-Butadiene, 2-methoxy-		C_5H_8O	84.12	3588-30-5			75	4-01-00-02223	0.8272^{20}	ace 4; bz 4; eth 4; EtOH 4	1.4442^{20}
3170	1,2-Butadiene, 3-methyl-		C_5H_8	68.12	598-25-4	-113.6		40.83	4-01-00-01006	0.6806^{25}	ace 4; bz 4; eth 4; EtOH 4	1.4203^{20}
3171	1,3-Butadiene, 2-methyl-	Isoprene	C_5H_8	68.12	78-79-5	-145.9	5087	34.0	4-01-00-01001	0.679^{20}	H_2O 1; EtOH 5; eth 5; ace 5	1.4219^{20}
3172	1,3-Butadiene, 1,1,2,4,4-pentafluoro-3-(trifluoromethyl)-	1,3-Butadiene, pentafluoro-2-(trifluoromethyl)-	C_5F_8	212.04	384-04-3			39		1.527^{0}	ace 4; bz 4; eth 4	1.3000^{0}
3173	1,3-Butadiene, 1,2,3,4-tetrachloro-		$C_4H_2Cl_4$	191.87	1637-31-6	-4		188	4-01-00-00987	1.515^{15}	H_2O 1; EtOH 3; eth 3; ace 3	1.5455^{20}
3174	1,3-Butadiene, 1,2,3-trichloro-	1,2,3-Trichloro-1,3-butadiene	$C_4H_3Cl_3$	157.43	1573-58-6			$33\text{-}4^{7}$	3-01-00-00954	1.4060^{20}	eth 4; chl 4	1.5262^{20}
3175	2,3-Butadien-1-ol		C_4H_6O	70.09	18913-31-0			128.5	4-01-00-02221	0.9164^{20}	ace 4; bz 4; eth 4; EtOH 4	1.4759^{20}
3176	1,3-Butadien-1-ol, acetate		$C_6H_8O_2$	112.13	1515-76-0			$60\text{-}1^{40}$	3-02-00-00295	0.945^{25}		1.4690^{20}
3177	1,3-Butadiyne		C_4H_2	50.06	460-12-8	-36.4		10.3	4-01-00-01116	0.7364^{0}	H_2O 4; EtOH 3; eth 4; ace 4	1.4189^{5}
3178	Butanal	Butyraldehyde	C_4H_8O	72.11	123-72-8	-99	1591	74.8	4-01-00-03229	0.8016^{20}	H_2O 3; EtOH 5; eth 5; ace 4	1.3843^{20}
3179	Butanal, 2-bromo-		C_4H_7BrO	151.00	24764-97-4			33^{17}	4-01-00-03241	1.469^{20}	eth 3; ace 3; bz 3; chl 3	1.4683^{20}
3180	Butanal, 2-chloro-		C_4H_7ClO	106.55	28832-55-5			107	4-01-00-03238	1.1072^{17}		1.4262^{17}
3181	Butanal, 4-chloro-		C_4H_7ClO	106.55	6139-84-0			$50\text{-}1^{13}$	4-01-00-03240	1.106^{8}	ace 4; eth 4; EtOH 4	1.4466^{8}
3182	Butanal, 2,3-dichloro-		$C_4H_6Cl_2O$	141.00	55775-41-2			$58\text{-}60^{20}$	4-01-00-03240	1.2666^{21}	ace 4; eth 4; EtOH 4; chl 4	1.4618^{21}
3183	Butanal, 2-ethyl-	Diethylacetaldehyde	$C_6H_{12}O$	100.16	97-96-1			$117\text{-}9^{160}$	4-01-00-03310	0.8110^{20}	H_2O 2; EtOH 5; eth 5; ctc 2	1.4025^{20}
3184	Butanal, 3-hydroxy-	Aldol	$C_4H_8O_2$	88.11	107-89-1		217	83^{20}	4-01-00-03984	1.103^{20}	H_2O 5; EtOH 5; eth 3; ace 4	1.4238^{20}
3185	Butanal, 2-methyl-, (±)-		$C_5H_{10}O$	86.13	57456-98-1			92.5	4-01-00-03285	0.8029^{20}	ace 4; eth 4; EtOH 4	1.3869^{20}
3186	Butanal, 3-methyl-	Isovaleraldehyde	$C_5H_{10}O$	86.13	590-86-3	-51	5118	92.5	4-01-00-03291	0.7977^{20}	H_2O 2; EtOH 3; eth 3	1.3902^{20}
3187	Butanal, 3-methyl-, oxime	Butyraldehyde, 3-methyl, oxime	$C_5H_{11}NO$	101.15	626-90-4	48.5		161.3	4-01-00-03293	0.8934^{20}	ace 4; eth 4; EtOH 4	1.4367^{20}
3188	Butanal, oxime	Butyraldehyde, oxime	C_4H_9NO	87.12	110-69-0	-29.5		154	4-01-00-03234	0.923^{20}	H_2O 4; EtOH 5; eth 5; ace 4	
3189	Butanal, phenylhydrazone	Butyraldehyde, phenylhydrazone	$C_{10}H_{14}N_2$	162.23	940-54-5	94		192^{80}, 152^{14}	3-15-00-00080		ctc 2	
3190	Butanal, 2-(phenylmethylene)-		$C_{11}H_{12}O$	160.22	28467-92-7			$157\text{-}8^{5}$	4-07-00-01024	1.0201^{22}		1.578^{20}
3191	Butanal, 2,2,3-trichloro-	2,2,3-Trichlorobutyraldehyde	$C_4H_5Cl_3O$	175.44	76-36-8		9545	164	4-01-00-03241	1.3956^{20}	H_2O 4; eth 4; EtOH 4	1.4755^{20}
3192	Butanal, 2,3,4-trihydroxy-, [*R*-(*R**,*R**)]-	*D*-Erythrose	$C_4H_8O_4$	120.11	583-50-6		3637		4-01-00-04172		H_2O 3; EtOH 4	
3193	Butanal, 2,3,4-trihydroxy-, [*S*-(*R**,*S**)]-	Threose (*D*-)	$C_4H_8O_4$	120.11	95-43-2	129	9317		4-01-00-04173			
3194	Butanamide	Butyramide	C_4H_9NO	87.12	541-35-5	114.8	1592	216	4-02-00-00804	0.8850^{120}	H_2O 2; EtOH 3; eth 2; bz 1	1.4087^{130}
3195	Butanamide, *N*-(aminocarbonyl)-2-bromo-2-ethyl-	Carbromal	$C_7H_{13}BrN_2O_2$	237.10	77-65-6	118	1837		4-03-00-00117	1.544^{25}	H_2O 2; ace 3; bz 3; chl 2	
3196	Butanamide, *N*-(aminocarbonyl)-2-bromo-3-methyl-	Bromisovalum	$C_6H_{11}BrN_2O_2$	223.07	496-67-3	154	1385	sub	3-03-00-00123	1.56^{15}	ace 4; bz 4; eth 4; EtOH 4	
3197	Butanamide, *N*-[5-(aminosulfonyl)-1,3,4-thiadiazol-2-yl]-	Butazolamide	$C_6H_{10}N_4O_3S_2$	250.30	16790-49-1		1511		4-27-00-08223			
3198	Butanamide, 2-bromo-2-ethyl-	Diethylbromoacetamide	$C_6H_{12}BrNO$	194.07	511-70-6	67	3105		3-02-00-00755		H_2O 2; EtOH 4; eth 4; bz 4	
3199	Butanamide, 2-bromo-2-ethyl-3-methyl-	Ibrotamide	$C_7H_{14}BrNO$	208.10	466-14-8	50.5	4809	sub	4-02-00-00979		ace 4; bz 4; eth 4; EtOH 4	
3200	Butanamide, *N*,*N*-diethyl-		$C_8H_{17}NO$	143.23	1114-76-7			206	4-04-00-00354		H_2O 4; EtOH 4	1.4403^{25}
3201	Butanamide, *N*,*N*-diethyl-3-methyl-	Isovaleryl diethylamide	$C_9H_{19}NO$	157.26	533-32-4		5122	211	4-04-00-00356	0.8764^{20}	eth 4; EtOH 4	1.4422^{20}

PHYSICAL CONSTANTS OF ORGANIC COMPOUNDS (continued)

No.	Name Synonym	Mol. Form. Mol. Wt.	CAS RN mp/°C	Merck No. bp/°C	Beil. Ref. den/g cm^{-3}	Solubility n_D
3202	Butanamide, 2,4-dihydroxy-*N*-(3-hydroxypropyl)-3,3-dimethyl-, (*R*)- Dexpanthenol	$C_9H_{19}NO_4$ 205.25	81-13-0	2924 dec	4-04-00-01652 1.20^{20}	H_2O 4; EtOH 4; eth 2; MeOH 4 1.497^{20}
3203	Butanamide, *N*,*N*-dimethyl-	$C_6H_{13}NO$ 115.18	760-79-2 -40	 186; 125^{100}	4-04-00-00185 0.9064^{25}	ace 4; bz 4; eth 4; EtOH 4 1.4391^{25}
3204	Butanamide, *N*,*N*'-(3,3'-dimethyl[1,1'-biphenyl]-4,4'-diyl)bis[3-oxo- *N*,*N*'-Bis(acetoacetyl)-3,3'-dimethylbenzidine	$C_{22}H_{24}N_2O_4$ 380.44	91-96-3 212		3-13-00-00490	DMSO 2
3205	Butanamide, *N*-(4-hydroxyphenyl)- 4'-Hydroxybutyranilide	$C_{10}H_{13}NO_2$ 179.22	101-91-7 139.5	4745	4-13-00-01109	H_2O 4; EtOH 4
3206	Butanamide, 3-methyl- Isovaleramide	$C_5H_{11}NO$ 101.15	541-46-8 137	5119 226	4-02-00-00902	H_2O 3; EtOH 3; eth 3; peth 4
3207	Butanamide, 3-oxo-*N*-phenyl- Acetoacetanilide	$C_{10}H_{11}NO_2$ 177.20	102-01-2 86	50	4-12-00-00955	H_2O 2; EtOH 3; eth 3; bz 3
3208	Butanamide, *N*-phenyl-	$C_{10}H_{13}NO$ 163.22	1129-50-6 97	 189^{15}	4-12-00-00387 1.134^{25}	H_2O 1; EtOH 4; eth 4; chl 2
3209	1-Butanamine Butylamine	$C_4H_{11}N$ 73.14	109-73-9 -49.1	1543 77.0	4-04-00-00540 0.7414^{20}	H_2O 5; EtOH 3; eth 3 1.4031^{20}
3210	2-Butanamine, (±)- *sec*-Butylamine, (±)-	$C_4H_{11}N$ 73.14	33966-50-6 <-72	1544 63.5	4-04-00-00618 0.7246^{20}	H_2O 3; EtOH 5; eth 5; ace 4 1.3932^{20}
3211	1-Butanamine, *N*-butyl- Dibutylamine	$C_8H_{19}N$ 129.25	111-92-2 -62	3019 159.6	4-04-00-00550 0.7670^{20}	H_2O 3; EtOH 4; eth 4; ace 3 1.4177^{20}
3212	1-Butanamine, *N*-butyl-*N*-nitroso- Dibutylnitrosamine	$C_8H_{18}N_2O$ 158.24	924-16-3	 105^8	4-04-00-03389	
3213	1-Butanamine, *N*,*N*-dibutyl- Tributylamine	$C_{12}H_{27}N$ 185.35	102-82-9 -70	9530 216.5	4-04-00-00554 0.7770^{20}	H_2O 2; EtOH 4; eth 4; ace 3 1.4299^{20}
3214	1-Butanamine, 4,4-diethoxy-	$C_8H_{19}NO_2$ 161.24	6346-09-4	 196	4-04-00-01928 0.933^{25}	 1.4275^{20}
3215	1-Butanamine, *N*,*N*-dimethyl-	$C_6H_{15}N$ 101.19	927-62-8	 95	4-04-00-00546 0.7206^{20}	H_2O 5; EtOH 5; eth 5; ace 5 1.3970^{20}
3216	2-Butanamine, 2,3-dimethyl-	$C_6H_{15}N$ 101.19	4358-75-2 -70.5	 104.5	4-04-00-00733 0.7683^0	 1.4096^{17}
3217	2-Butanamine, 3,3-dimethyl-	$C_6H_{15}N$ 101.19	3850-30-4 -20	 102	4-04-00-00730 0.7668^{20}	H_2O 4 1.4105^{25}
3218	1-Butanamine, *N*-ethyl- Butylethylamine	$C_6H_{15}N$ 101.19	13360-63-9	 107.5	4-04-00-00547 0.7398^{20}	EtOH 5; eth 5; ace 5; bz 5 1.4040^{20}
3219	2-Butanamine, *N*-ethyl-, (±)- *sec*-Butyl ethyl amine (*DL*)	$C_6H_{15}N$ 101.19	116724-10-8 -104.3	 98	2-04-00-00636 0.7358^{20}	ace 4; bz 4; eth 4; EtOH 4
3220	1-Butanamine, 3-methyl- Isoamylamine	$C_5H_{13}N$ 87.16	107-85-7	4994 96	4-04-00-00696 0.7505^{20}	H_2O 5; EtOH 5; eth 5; ace 3 1.4083^{20}
3221	1-Butanamine, *N*-methyl- Butylmethylamine	$C_5H_{13}N$ 87.16	110-68-9	 91	4-04-00-00546 0.7637^{15}	
3222	2-Butanamine, 2-methyl-	$C_5H_{13}N$ 87.16	594-39-8 -105	 77	4-04-00-00694 0.731^{25}	H_2O 4; ace 4; eth 4; EtOH 4 1.3954^{25}
3223	2-Butanamine, 3-methyl-	$C_5H_{13}N$ 87.16	598-74-3 -50	 85.5	2-04-00-00644 0.7574^{19}	H_2O 4; EtOH 3 1.4096^{18}
3224	1-Butanamine, 2-methyl-*N*,*N*-bis(2-methylbutyl)-	$C_{15}H_{33}N$ 227.43	620-43-9	 232	0-04-00-00179 0.9^{13}	ace 4; bz 4; eth 4; EtOH 4 1.4330^{20}
3225	1-Butanamine, 3-methyl-*N*,*N*-bis(3-methylbutyl)-	$C_{15}H_{33}N$ 227.43	645-41-0	 235	4-04-00-00700 0.7848^{20}	H_2O 1; EtOH 4; eth 5; bz 5 1.4331^{20}
3226	1-Butanamine, 3-methyl-*N*-(3-methylbutyl)- Diisoamylamine	$C_{10}H_{23}N$ 157.30	544-00-3 -44	3178 188	4-04-00-00699 0.7672^{21}	H_2O 1; EtOH 3; eth 5 1.4235^{20}
3227	2-Butanamine, *N*-(1-methylpropyl)-	$C_8H_{19}N$ 129.25	626-23-3	 135	4-04-00-00620 0.7534^{20}	H_2O 4; EtOH 3; os 3 1.4162^{20}
3228	1-Butanamine, 1,1,2,2,3,3,4,4,4-nonafluoro-*N*,*N*-bis(nonafluorobutyl)-	$C_{12}F_{27}N$ 671.10	311-89-7	 178	4-02-00-00819 1.884^{25}	ace 3 1.291^{25}
3229	Butane	C_4H_{10} 58.12	106-97-8 -138.2	1507 -0.5	4-01-00-00236 *0.573^{25}	H_2O 3; EtOH 4; eth 4; chl 4 1.3326^{20}
3230	Butane, 2,2-bis(ethylsulfonyl)- Sulfonethylmethane	$C_8H_{18}O_4S_2$ 242.36	76-20-0 76	8937 dec	3-01-00-02790 1.199^{85}	chl 3
3231	Butane, 1-bromo- Butyl bromide	C_4H_9Br 137.02	109-65-9 -112.4	1553 101.6	4-01-00-00258 1.2758^{20}	H_2O 1; EtOH 5; eth 5; ace 5 1.4401^{20}
3232	Butane, 2-bromo-, (±)- (±)-*sec*-Butyl bromide	C_4H_9Br 137.02	5787-31-5 -111.9	1554 91.2	4-01-00-00261 1.2585^{20}	ace 4; eth 4; chl 4 1.4366^{20}
3233	Butane, 1-bromo-4-chloro-	C_4H_8BrCl 171.46	6940-78-9	 175; 63^{10}	4-01-00-00264 1.488^{20}	H_2O 1; EtOH 3; eth 3; ctc 2 1.4885^{20}
3234	Butane, 2-bromo-1-chloro- 2-Bromo-1-chlorobutane	C_4H_8BrCl 171.46	79504-01-1	 146.5	4-01-00-00264 1.468^{20}	bz 4; eth 4; EtOH 4; chl 4 1.4880^{20}
3235	Butane, 1-bromo-3,3-dimethyl-	$C_6H_{13}Br$ 165.07	1647-23-0	 138	3-01-00-00409 1.1556^{20}	eth 4; EtOH 4; chl 4 1.4440^{20}
3236	Butane, 2-bromo-2,3-dimethyl- 2-Bromo-2,3-dimethylbutane	$C_6H_{13}Br$ 165.07	594-52-5 24.5	 133; 87^{180}	4-01-00-00374 1.1772^{10}	eth 4; chl 4 1.4517
3237	Butane, 1-bromo-4-fluoro- 1-Bromo-4-fluorobutane	C_4H_8BrF 155.01	462-72-6	 135	4-01-00-00263	eth 4; EtOH 4 1.4370^{25}
3238	Butane, 1-bromo-2-methyl-, (*S*)- *d*-Amyl bromide	$C_5H_{11}Br$ 151.05	534-00-9	636 121.6	4-01-00-00327 1.2234^{20}	H_2O 1; EtOH 3; eth 3; chl 4 1.4451^{20}
3239	Butane, 1-bromo-3-methyl- Isoamyl bromide	$C_5H_{11}Br$ 151.05	107-82-4 -112	4996 120.4	4-01-00-00328 1.2071^{20}	H_2O 1; EtOH 3; eth 3; ctc 2 1.4420^{20}

No.	Name Synonym	Mol. Form. Mol. Wt.	CAS RN mp/°C	Merck No. bp/°C	Beil. Ref. den/g cm^{-3}	Solubility n_D
3240	Butane, 2-bromo-2-methyl- *tert*-Amyl bromide	$C_5H_{11}Br$ 151.05	507-36-8	638 108	4-01-00-00327 1.197^{18}	 -1.4421
3241	Butane, 1-chloro- Butyl chloride	C_4H_9Cl 92.57	109-69-3 -123.1	1560 78.6	4-01-00-00246 0.8862^{20}	H_2O 1; EtOH 5; eth 5; ctc 2 1.4021^{20}
3242	Butane, 2-chloro-, (±)- *sec*-Butyl chloride (*DL*)	C_4H_9Cl 92.57	53178-20-4 -131.3	1561 68.2	4-01-00-00248 0.8732^{20}	bz 4; eth 4; EtOH 4; chl 4 1.3971^{20}
3243	Butane, 3-chloro-1,1-diethoxy- Butyraldehyde, 3-chloro-, diethyl acetal	$C_8H_{17}ClO_2$ 180.67	51786-70-0	 $70\text{-}1^{12}$	4-01-00-03239 0.9709^{20}	bz 4; eth 4; EtOH 4 1.4195^{25}
3244	Butane, 3-chloro-1,1-dimethoxy- Butyraldehyde, 3-chloro-, dimethyl acetal	$C_6H_{13}ClO_2$ 152.62	50710-40-2	 $40\text{-}4^{7}$	4-01-00-03239 1.0024^{20}	 1.4160^{25}
3245	Butane, 1-chloro-2,3-dimethyl- 1-Chloro-2,3-dimethylbutane	$C_6H_{13}Cl$ 120.62	600-06-6	 124	4-01-00-00373 0.887^{22}	eth 4; EtOH 4; chl 4
3246	Butane, 1-chloro-3,3-dimethyl-	$C_6H_{13}Cl$ 120.62	2855-08-5	 115	4-01-00-00369 0.8670^{20}	eth 4; EtOH 4; chl 4 1.4161^{20}
3247	Butane, 2-chloro-2,3-dimethyl-	$C_6H_{13}Cl$ 120.62	594-57-0 -10.4	 112	4-01-00-00373 0.8780^{20}	ace 4; EtOH 4 1.4191^{20}
3248	Butane, 3-chloro-2,2-dimethyl-	$C_6H_{13}Cl$ 120.62	5750-00-5 0.9	 111	4-01-00-00369 0.8767^{20}	eth 4 1.4182^{20}
3249	Butane, 1-(2-chloroethoxy)-	$C_6H_{13}ClO$ 136.62	10503-96-5	154 dec; 50^{11}	4-01-00-01519 0.9335^{20}	eth 4 1.4155^{20}
3250	Butane, 1-chloro-4-fluoro- 1-Chloro-4-fluorobutane	C_4H_8ClF 110.56	462-73-7	 114.7	4-01-00-00249 1.0627^{25}	eth 4; EtOH 4 1.4020^{25}
3251	Butane, 1-chloro-2-methyl-, (±)- Butane, 1-chloro-2-methyl (*DL*)	$C_5H_{11}Cl$ 106.60	114180-21-1	 99.9	4-01-00-00324 0.8810^{15}	EtOH 3; eth 3 1.4102^{25}
3252	Butane, 1-chloro-3-methyl- Isoamyl chloride	$C_5H_{11}Cl$ 106.60	107-84-6 -104.4	4998 98.9	4-01-00-00325 0.8750^{20}	H_2O 2; EtOH 5; eth 5; chl 4 1.4084^{20}
3253	Butane, 2-chloro-2-methyl-	$C_5H_{11}Cl$ 106.60	594-36-5 -73.5	 85.6	4-01-00-00324 0.8653^{20}	H_2O 2; EtOH 3; eth 3; ctc 3 1.4055^{20}
3254	Butane, 2-chloro-2,3,3-trimethyl-	$C_7H_{15}Cl$ 134.65	918-07-0 136	 sub	4-01-00-00411	eth 4
3255	Butane, decafluoro- Perfluorobutane	C_4F_{10} 238.03	355-25-9 -128.2	 -1.9	4-01-00-00245 1.6484^{25}	bz 3; chl 3
3256	Butanedial	$C_4H_6O_2$ 86.09	638-37-9	170 dec; 58^{9}	4-01-00-03642 1.064^{20}	H_2O 4; ace 4; eth 4; EtOH 4 1.4262^{18}
3257	Butanediamide Succinamide	$C_4H_8N_2O_2$ 116.12	110-14-5 268 dec	8837 sub 125	4-02-00-01922	
3258	1,4-Butanediamine Putrescine	$C_4H_{12}N_2$ 88.15	110-60-1 27.5	7964 158.5	4-04-00-01283 0.877^{25}	H_2O 3 1.4969^{20}
3259	1,4-Butanediamine, *N,N'*-bis(3-aminopropyl)- Spermine	$C_{10}H_{26}N_4$ 202.34	71-44-3 29	8699 150^{5}	4-04-00-01301	
3260	1,4-Butanediamine, dihydrochloride Butane, 1,4-diamino, dihydrochloride	$C_4H_{14}Cl_2N_2$ 161.07	333-93-7 280 dec	 sub	4-04-00-01283	H_2O 4; EtOH 4; eth 1; bz 1
3261	2,3-Butanediamine, (*R**,*R**)-(±)-	$C_4H_{12}N_2$ 88.15	20699-48-3 -22	 $57\text{-}8^{60}$	4-04-00-01302 0.8499^{25}	 1.4408^{25}
3262	1,4-Butanediamine, *N,N,N',N'*-tetraethyl- *N,N,N',N'*-Tetraethyl-1,4-butanediamine	$C_{12}H_{28}N_2$ 200.37	69704-44-5	 $85\text{-}6^{5}$	4-04-00-01286	H_2O 4; eth 4; EtOH 4 1.4383^{25}
3263	1,4-Butanediamine, *N,N,N',N'*-tetramethyl- *N,N,N',N'*-Tetramethyl-1,4-diaminobutane	$C_8H_{20}N_2$ 144.26	111-51-3	9157 168	4-04-00-01284 0.7942^{15}	H_2O 5; EtOH 3; eth 3 1.4621^{25}
3264	Butane, 1,1-dibromo- 1,1-Dibromobutane	$C_4H_8Br_2$ 215.92	62168-25-6	 91^{101}; 72^{43}	4-01-00-00266 1.80^{25}	 1.4980^{25}
3265	Butane, 1,2-dibromo- α-Butylene dibromide	$C_4H_8Br_2$ 215.92	533-98-2 -65.4	1565 166.3	4-01-00-00266 1.7915^{20}	H_2O 1; eth 3; chl 3 1.4025^{20}
3266	Butane, 1,3-dibromo- 1,3-Dibromobutane	$C_4H_8Br_2$ 215.92	107-80-2	 174	4-01-00-00266 1.800^{20}	H_2O 1; eth 3; ctc 2; chl 3 1.507^{20}
3267	Butane, 1,4-dibromo- 1,4-Dibromobutane	$C_4H_8Br_2$ 215.92	110-52-1 -16.5	 197	4-01-00-00267 1.8270^{20}	H_2O 1; ctc 2; chl 3 1.5190^{20}
3268	Butane, 2,3-dibromo- 2,3-Dibromobutane	$C_4H_8Br_2$ 215.92	5408-86-6 -24	 161	3-01-00-00296 1.7893^{22}	H_2O 1; eth 3 1.5133^{22}
3269	Butane, 1,2-dibromo-2,3-dimethyl- 1,2-Dibromo-2,3-dimethylbutane	$C_6H_{12}Br_2$ 243.97	29916-45-8	 80^{17}	3-01-00-00414 1.6033^{20}	 1.5105^{20}
3270	Butane, 1,4-dibromo-2,3-dimethyl- 1,4-Dibromo-2,3-dimethylbutane	$C_6H_{12}Br_2$ 243.97	54462-70-3	 $106\text{-}8^{22}$	4-01-00-00375 1.620^{20}	 1.5128^{20}
3271	Butane, 2,3-dibromo-2,3-dimethyl- 1,2-Dibromo-1,1,2,2-tetramethylethane	$C_6H_{12}Br_2$ 243.97	594-81-0 171 dec	 78^{10}	4-01-00-00375 1.8110^{21}	
3272	Butane, 1,2-dibromo-2-methyl- 1,2-Dibromo-2-methylbutane	$C_5H_{10}Br_2$ 229.94	10428-64-5	 $47\text{-}8^{9}$	4-01-00-00330 1.6711^{20}	 1.5088^{20}
3273	Butane, 1,4-dibromo-2-methyl-, (*R*)- Butane, 1,4-dibromo-2-methyl (*d*)	$C_5H_{10}Br_2$ 229.94	60109-20-0	 $78\text{-}9^{12}$	4-01-00-00330 1.695^{17}	 1.5128^{20}
3274	Butane, 2,3-dibromo-2-methyl- 1,2-Dibromo-1,1,2-trimethylethane	$C_5H_{10}Br_2$ 229.94	594-51-4 7	 $61\text{-}4^{17}$	4-01-00-00330 1.6717^{20}	 1.5729^{25}
3275	Butane, 1,4-dibromo-1,1,2,2,3,3,4,4-octafluoro- 1,4-Dibromooctafluorobutane	$C_4Br_2F_8$ 359.84	335-48-8	 97		
3276	Butane, 1,1-dichloro- Butylidene chloride	$C_4H_8Cl_2$ 127.01	541-33-3	1571 113.8	4-01-00-00250 1.0863^{20}	H_2O 1; chl 3 1.4355^{20}
3277	Butane, 1,2-dichloro- 1,2-Dichlorobutane	$C_4H_8Cl_2$ 127.01	616-21-7	 124.1	4-01-00-00250 1.1116^{25}	H_2O 1; eth 3; ctc 2; chl 3 1.4450^{20}
3278	Butane, 1,3-dichloro- 1,3-Dichlorobutane	$C_4H_8Cl_2$ 127.01	1190-22-3	 134	4-01-00-00250 1.1158^{20}	H_2O 1; eth 3; ctc 2; chl 3 1.4445^{20}

PHYSICAL CONSTANTS OF ORGANIC COMPOUNDS (continued)

No.	Name	Synonym	Mol. Form.	Mol. Wt.	CAS RN	mp/°C	Merck No.	bp/°C	Beil. Ref.	den/g cm^{-3}	Solubility	n_D
3279	Butane, 1,4-dichloro-	1,4-Dichlorobutane	$C_4H_8Cl_2$	127.01	110-56-5	-37.3		161	4-01-00-00250	1.1408^{20}	chl 4	1.4542^{20}
3280	Butane, 2,2-dichloro-	2,2-Dichlorobutane	$C_4H_8Cl_2$	127.01	4279-22-5	-74		104	4-01-00-00251	1.1048^{25}	H_2O 1; chl 3	1.4295
3281	Butane, 2,3-dichloro-, (*R**,*R**)-(±)-		$C_4H_8Cl_2$	127.01	2211-67-8	-80		118; 53.2^{80}	4-01-00-00251	1.063^{25}		1.4409^{25}
3282	Butane, 1,1-dichloro-3-methyl-	1,1-Dichloro-3-methylbutane	$C_5H_{10}Cl_2$	141.04	625-66-1			130	4-01-00-00326	1.0473^{20}	eth 4; EtOH 4	1.4344^{20}
3283	Butane, 1,2-dichloro-2-methyl-	1,2-Dichloro-2-methylbutane	$C_5H_{10}Cl_2$	141.04	23010-04-0			134	3-01-00-00360	1.0785^{20}	eth 4; EtOH 4; chl 4	1.4432^{21}
3284	Butane, 1,3-dichloro-3-methyl-		$C_5H_{10}Cl_2$	141.04	624-96-4			146.7	4-01-00-00326	1.0654^{20}	eth 4; chl 4	1.4455^{20}
3285	Butane, 1,4-dichloro-2-methyl-	1,4-Dichloro-2-methylbutane	$C_5H_{10}Cl_2$	141.04	623-34-7			168.5	3-01-00-00360	1.1003^{25}	chl 4	1.4562^{21}
3286	Butane, 2,3-dichloro-2-methyl-	Amylene dichloride	$C_5H_{10}Cl_2$	141.04	507-45-9		645	129	4-01-00-00325	1.0696^{15}	eth 4; EtOH 4	1.4450^{18}
3287	Butane, 2,3-dichloro-1,1,1,2,3,4,4,4-octafluoro-		$C_4Cl_2F_8$	270.94	355-20-4	-68		63	4-01-00-00252	1.6801^{20}		1.3100^{20}
3288	Butane, 1,4-diiodo-		$C_4H_8I_2$	309.92	628-21-7	5.8		125^{15} dec	4-01-00-00276	2.3659^{15}	H_2O 1; ctc 2; os 3	1.6239^{15}
3289	Butane, 2,2-dimethyl-	Neohexane	C_6H_{14}	86.18	75-83-2	-99		49.7	4-01-00-00367	0.6444^{25}	H_2O 1; EtOH 3; eth 3; ace 4	1.3688^{20}
3290	Butane, 2,3-dimethyl-	2,3-Dimethylbutane	C_6H_{14}	86.18	79-29-8	-128.8		57.9	4-01-00-00371	0.6616^{20}	H_2O 1; EtOH 3; eth 3; ace 4	1.3750^{20}
3291	Butanedinitrile	Succinonitrile	$C_4H_4N_2$	80.09	110-61-2	54.5	8843	266	4-02-00-01923	0.9867^{60}	H_2O 4; EtOH 3; eth 2; ace 3	1.4173^{60}
3292	Butanedinitrile, tetramethyl-	Tetramethylsuccinonitrile	$C_8H_{12}N_2$	136.20	3333-52-6	170.5			4-02-00-02054	1.070^{25}	EtOH 3	
3293	Butane, 1,4-dinitro-	1,4-Dinitrobutane	$C_4H_8N_2O_4$	148.12	4286-49-1	33.5		176^{13}	4-01-00-00280		H_2O 1; EtOH 2; eth 3; bz 3	
3294	Butanedioic acid	Succinic acid	$C_4H_6O_4$	118.09	110-15-6	188	8840	235 dec	4-02-00-01908	1.572^{25}	H_2O 2; EtOH 3; eth 3; ace 3	1.450
3295	Butanedioic acid, acetyl-, diethyl ester		$C_{10}H_{16}O_5$	216.23	1115-30-6			255; 133^{17}	4-03-00-01826	1.081^{20}	H_2O 1; EtOH 3; eth 3; bz 3	1.4346^{20}
3296	Butanedioic acid, bis(1,1-dimethylethyl) ester	Di-*tert*-butyl succinate	$C_{12}H_{22}O_4$	230.30	926-26-1	36.5	3024	$109\text{-}10^{9}$	4-02-00-01916			
3297	Butanedioic acid, bis(1-methylpropyl) ester	Di-*sec*-butyl succinate	$C_{12}H_{22}O_4$	230.30	626-31-3			256	3-02-00-01665	0.9735^{20}	bz 4; eth 4; EtOH 4	1.4238^{25}
3298	Butanedioic acid, bis(phenylmethyl) ester		$C_{18}H_{18}O_4$	298.34	103-43-5	49.5		245^{15}	4-06-00-02271	1.256^{25}	H_2O 1; EtOH 3; eth 3; bz 3	1.596
3299	Butanedioic acid, bromo-, (±)-		$C_4H_5BrO_4$	196.99	584-98-5	161			4-02-00-01929	2.073^{25}	H_2O 3; EtOH 3; HOAc 2	
3300	Butanedioic acid, bromo-	Bromosuccinic acid	$C_4H_5BrO_4$	196.99	923-06-8	161	1427		1-02-00-00268	2.073^{25}	EtOH 4	
3301	Butanedioic acid, dibutyl ester	Dibutyl succinate	$C_{12}H_{22}O_4$	230.30	141-03-7	-29.2	3023	274.5	4-02-00-01916	0.9752^{20}	H_2O 1; EtOH 3; eth 3; bz 3	1.4299^{20}
3302	Butanedioic acid, 2,3-dichloro-, (±)-	Succinic acid, 2,3-dichloro-, (±)-	$C_4H_4Cl_2O_4$	186.98	1114-09-6	175 dec			4-02-00-01928	1.844^{15}	ace 4; eth 4; chl 4	
3303	Butanedioic acid, diethyl ester	Diethyl succinate	$C_8H_{14}O_4$	174.20	123-25-1	-21	8840	217.7	4-02-00-01914	1.0402^{20}	H_2O 1; EtOH 5; eth 5; ace 3	1.4201^{20}
3304	Butanedioic acid, 2,3-dihydroxy-, (*R**,*S**)-	*meso*-Tartaric acid	$C_4H_6O_6$	150.09	147-73-9	147	9040		4-03-00-01218	1.666^{20}	H_2O 4; EtOH 4	
3305	Butanedioic acid, 2,3-dihydroxy-, bis(1-methylethyl) ester, (*R**,*R**)-(±)-	Tartaric acid, diisopropyl ester (*DL*)	$C_{10}H_{18}O_6$	234.25	58167-01-4	34		275; 154^{12}	2-03-00-00337	1.1166^{20}	ace 4; eth 4; EtOH 4	
3306	Butanedioic acid, 2,3-dihydroxy- [*R*-(*R**,*R**)]-, bis(phenylmethyl) ester		$C_{18}H_{18}O_6$	330.34	622-00-4	50		$250\text{-}70^{4}$	3-06-00-01537	1.2036^{72}	py 4; EtOH 4	
3307	Butanedioic acid, 2,3-dihydroxy-, diethyl ester, (*R**,*R**)-(±)-	Diethyl *DL*-tartrate	$C_8H_{14}O_6$	206.20	57968-71-5	18.7		281; 158^{14}	4-03-00-01232	1.2046^{20}	H_2O 2; EtOH 5; eth 5; ace 3	1.4438^{20}
3308	Butanedioic acid, 2,3-dihydroxy-, dimethyl ester, (*R**,*R**)-(±)-		$C_6H_{10}O_6$	178.14	608-69-5	90		280	4-03-00-01232	1.2604^{90}	H_2O 4; ace 4; eth 4; EtOH 4	
3309	Butanedioic acid, 2,3-dihydroxy- [*R*-(*R**,*R**)]-	*L*-Tartaric acid	$C_4H_6O_6$	150.09	87-69-4	169	9039		4-03-00-01219			
3310	Butanedioic acid, 2,3-dihydroxy- [*R*-(*R**,*R**)]-, dipropyl ester	Tartaric acid, dipropyl ester, (+)-	$C_{10}H_{18}O_6$	234.25	2217-14-3			303	2-03-00-00337	1.1390^{20}	H_2O 4; ace 4; eth 4; EtOH 4	
3311	Butanedioic acid, 2,3-dihydroxy-, (*R**,*R**)-(±)-	*DL*-Tartaric acid	$C_4H_6O_6$	150.09	133-37-9	206	9038		4-03-00-01229	1.788^{25}	H_2O 3; EtOH 3; eth 2; bz 1	
3312	Butanedioic acid, 2,3-dihydroxy-, [*S*-(*R**,*R**)]-	*D*-Tartaric acid	$C_4H_6O_6$	150.09	147-71-7	172.5	9037		4-03-00-01229	1.7598^{20}	DMSO 2	1.4955^{20}
3313	Butanedioic acid, 2,3-dihydroxy- [*R*-(*R**,*R**)]-, dibutyl ester		$C_{12}H_{22}O_6$	262.30	87-92-3	22		320	4-03-00-01232	1.0909^{20}	H_2O 4; ace 4; EtOH 4	1.4451^{20}
3314	Butanedioic acid, 2,3-dihydroxy- [*R*-(*R**,*R**)]-, disodium salt	Tartaric acid, disodium salt	$C_4H_4Na_2O_6$	194.05	868-18-8		8640				H_2O 3	

No.	Name Synonym	Mol. Form. Mol. Wt.	CAS RN mp/°C	Merck No. bp/°C	Beil. Ref. den/g cm^{-3}	Solubility n_D
3315	Butanedioic acid, 2,3-dihydroxy- [*R*-(*R**,*R**)]-, monoethyl ester Tartaric acid, ethyl ester	$C_6H_{10}O_6$ 178.14	608-89-9 90	3812	3-03-00-01028	H_2O 4; EtOH 4
3316	Butanedioic acid, [(dimethoxyphosphinothioyl)thio]-, diethyl ester Malathion	$C_{10}H_{19}O_6PS_2$ 330.36	121-75-5 2.8	5582 $156^{0.7}$ dec	4-03-00-01136 1.2076^{20}	H_2O 2; EtOH 3; eth 3; bz 3 1.4960^{20}
3317	Butanedioic acid, 2,3-dimethyl-	$C_6H_{10}O_4$ 146.14	13545-04-5 132 dec	 $87\text{-}90^{2}$	4-02-00-01998	ace 4; eth 4
3318	Butanedioic acid, dimethyl ester	$C_6H_{10}O_4$ 146.14	106-65-0 19	 196.4	4-02-00-01913 1.1198^{20}	H_2O 2; EtOH 3; eth 4; ace 3 1.4197^{20}
3319	Butanedioic acid, dipentyl ester	$C_{14}H_{26}O_4$ 258.36	645-69-2 -10.8	 171.5^{16}	3-02-00-01665 0.9616^{20}	
3320	Butanedioic acid, diphenyl ester Diphenyl succinate	$C_{16}H_{14}O_4$ 270.28	621-14-7 121	 330; 222.5^{15}	3-06-00-00605	H_2O 1; EtOH 3; eth 3; ace 3
3321	Butanedioic acid, dipropyl ester	$C_{10}H_{18}O_4$ 202.25	925-15-5 -5.9	 250.8	4-02-00-01916 1.0020^{20}	ace 4; bz 4; eth 4 1.4250^{20}
3322	Butanedioic acid, ethyl-, (*R*)-	$C_6H_{10}O_4$ 146.14	4074-24-2 96	 181.5	4-02-00-01995 1.0017^{20}	H_2O 4; eth 4; EtOH 4
3323	Butanedioic acid, ethyl methyl ester	$C_7H_{12}O_4$ 160.17	627-73-6 <-20	 208.2	0-02-00-00609 1.076^{20}	eth 4; EtOH 4
3324	Butanedioic acid, formyl-, diethyl ester	$C_9H_{14}O_5$ 202.21	5472-38-8	 $130\text{-}4^{15}$	4-03-00-01819	eth 4; EtOH 4 1.4486^{25}
3325	Butanedioic acid, methyl-, (±)-	$C_5H_8O_4$ 132.12	636-60-2 115	 dec	4-02-00-01948 1.4200^{0}	H_2O 4; EtOH 4; eth 3; chl 2 1.4303
3326	Butanedioic acid, methyl-, diethyl ester	$C_9H_{16}O_4$ 188.22	4676-51-1	 217.5	3-02-00-01696 1.012^{25}	 1.4199^{20}
3327	Butanedioic acid, methyl-, dimethyl ester	$C_7H_{12}O_4$ 160.17	1604-11-1	 196	3-02-00-01696 1.076^{25}	 1.4200^{20}
3328	Butanedioic acid, methylene- Itaconic acid	$C_5H_6O_4$ 130.10	97-65-4 175	5130 dec	4-02-00-02228 1.632^{25}	H_2O 3; EtOH 3; eth 2; ace 3
3329	Butanedioic acid, methylene-, diethyl ester	$C_9H_{14}O_4$ 186.21	2409-52-1 58.5	 228	4-02-00-02230 1.0467^{20}	EtOH 5; eth 3; ace 4; bz 3 1.4377^{20}
3330	Butanedioic acid, methylene-, dimethyl ester	$C_7H_{10}O_4$ 158.15	617-52-7 38	 208	4-02-00-02229 1.1241^{18}	EtOH 3; eth 3; ace 4; MeOH 3 1.4457^{20}
3331	Butanedioic acid, (1-methylethylidene)-	$C_7H_{10}O_4$ 158.15	584-27-0 160.5	 174 dec	4-02-00-02251	eth 4; EtOH 4
3332	Butanedioic acid, (1-methylpropyl)-, diethyl ester Succinic acid, *sec*-butyl-, diethyl ester	$C_{12}H_{22}O_4$ 230.30	69248-35-7	 $75\text{-}6^{0.2}$	3-02-00-01779 0.9745^{25}	 1.4293^{25}
3333	Butanedioic acid, monobutyl ester	$C_8H_{14}O_4$ 174.20	5150-93-6 8.6	 136.5^{3}	3-02-00-01665 1.0732^{20}	 1.4360^{20}
3334	Butanedioic acid, mono(2,2-dimethylhydrazide) Daminozide	$C_6H_{12}N_2O_3$ 160.17	1596-84-5 154.5	2810		
3335	Butanedioic acid, monoethyl ester Succinic acid, ethyl ester	$C_6H_{10}O_4$ 146.14	1070-34-4 8	 172^{42}; 119^{3}	4-02-00-01914 1.1466^{20}	H_2O 4; eth 4; EtOH 4 1.4327^{20}
3336	Butanedioic acid, monomethyl ester	$C_5H_8O_4$ 132.12	3878-55-5 58	 151^{20}; 122^{4}	4-02-00-01913	H_2O 3
3337	Butanedioic acid, monopentyl ester Succinic acid, pentyl ester	$C_9H_{16}O_4$ 188.22	97479-78-2 17.2	 147^{3}	2-02-00-00551 1.0460^{20}	 1.4378^{20}
3338	Butanedioic acid, monopropyl ester Succinic acid, monopropyl ester	$C_7H_{12}O_4$ 160.17	6946-88-9 15	 126^{3}	3-02-00-01664 1.1071^{20}	 1.4343^{20}
3339	Butanedioic acid, oxo- Oxalacetic acid	$C_4H_4O_5$ 132.07	328-42-7 161 dec	6863	4-03-00-01808	
3340	Butanedioic acid, oxo-, diethyl ester Diethyl oxalacetate	$C_8H_{12}O_5$ 188.18	108-56-5	3791 $131\text{-}2^{24}$	4-03-00-01809 1.131^{20}	H_2O 1; EtOH 5; eth 5; ace 4 1.4561^{17}
3341	Butanedioic acid, phenyl-, (±)-	$C_{10}H_{10}O_4$ 194.19	10424-29-0 168	 dec	4-09-00-03352	H_2O 2; EtOH 4; eth 4; ace 4
3342	Butanedioic acid, tetrahydroxy- Dihydroxytartaric acid	$C_4H_6O_8$ 182.09	76-30-2 114.5	3172	3-03-00-01419	
3343	Butanedioic acid, tetramethyl-	$C_8H_{14}O_4$ 174.20	630-51-3 200	 sub	4-02-00-02054 1.30^{25}	bz 4; EtOH 4; chl 4
3344	1,2-Butanediol, (±)-	$C_4H_{10}O_2$ 90.12	26171-83-5	 190.5	4-01-00-02507 1.0024^{20}	H_2O 3; EtOH 3; ace 3 1.4378^{20}
3345	1,3-Butanediol 1,3-Butylene glycol	$C_4H_{10}O_2$ 90.12	107-88-0 <-50	1566 207.5	0-01-00-00477 1.0053^{20}	 1.4401^{20}
3346	1,4-Butanediol Tetramethylene glycol	$C_4H_{10}O_2$ 90.12	110-63-4 20.1	 235	4-01-00-02515 1.0171^{20}	H_2O 5; EtOH 3; eth 2; DMSO 2 1.4460^{20}
3347	2,3-Butanediol, (*R**,*R**)-(±)-	$C_4H_{10}O_2$ 90.12	6982-25-8 7.6	 182.5	4-01-00-02525 1.0033^{20}	H_2O 5; EtOH 5; eth 3; ace 3 1.4310^{25}
3348	1,4-Butanediol, diacetate	$C_8H_{14}O_4$ 174.20	628-67-1 12	 229	4-02-00-00224 1.0479^{15}	 1.4251^{15}
3349	2,3-Butanediol, 1,4-dichloro- 1,4-Dichloro-2,3-butanediol	$C_4H_8Cl_2O_2$ 159.01	2419-73-0 126.5	 150^{30}	4-01-00-02531	EtOH 4
3350	2,3-Butanediol, 2,3-dimethyl- Pinacol	$C_6H_{14}O_2$ 118.18	76-09-5 43	7408 174.4	4-01-00-02575	H_2O 2; EtOH 4; eth 4; CS_2 2
3351	2,3-Butanediol, 2,3-dimethyl-, hexahydrate	$C_6H_{26}O_8$ 226.27	6091-58-3 45.4	 172	2-01-00-00553 0.967^{15}	eth 4; EtOH 4

No.	Name	Synonym	Mol. Form.	Mol. Wt.	CAS RN	mp/°C	Merck No.	bp/°C	Beil. Ref.	den/g cm^{-3}	Solubility	n_D
3352	1,2-Butanediol, 3-methyl-	3-Methyl-1,2-butanediol	$C_5H_{12}O_2$	104.15	50468-22-9			206	4-01-00-02549	0.9987^{0}	eth 4; EtOH 4	
3353	1,3-Butanediol, 2-methyl-	2-Methyl-1,3-butanediol	$C_5H_{12}O_2$	104.15	684-84-4			200; 115^{20}	4-01-00-02546	0.9912^{20}	H_2O 3; EtOH 3; eth 3	
3354	1,3-Butanediol, 3-methyl-		$C_5H_{12}O_2$	104.15	2568-33-4			202.5	4-01-00-02549	0.9448^{20}	H_2O 3; EtOH 3	1.4452^{20}
3355	2,3-Butanediol, 2-methyl-	2,3-Dihydroxy-2-methylbutane	$C_5H_{12}O_2$	104.15	5396-58-7			175	4-01-00-02547	0.9920^{25}	H_2O 4; eth 4; EtOH 4	1.4375^{20}
3356	1,2-Butanediol, 2-phenyl-	2-Phenyl-1,2-butanediol	$C_{10}H_{14}O_2$	166.22	90925-48-7	56		165^{23}	3-06-00-04667		H_2O 4; eth 4; EtOH 4	
3357	1,3-Butanediol, 1-phenyl-	1-Phenylbutane-1,3-diol	$C_{10}H_{14}O_2$	166.22	65469-88-7	73.5		175^{21}; 130^{2}	4-06-00-06006	1.0720^{25}	bz 4; EtOH 4; lig 4	
3358	1,3-Butanediol, 3-phenyl-		$C_{10}H_{14}O_2$	166.22	7133-68-8			150^{2}; $109^{0.5}$	4-06-00-06009	1.0865^{20}		1.5341^{20}
3359	2,3-Butanedione	Diacetyl	$C_4H_6O_2$	86.09	431-03-8	-2.4	2946	88	4-01-00-03644	0.9808^{18}	H_2O 4; EtOH 5; eth 5; ace 4	1.3951^{20}
3360	2,3-Butanedione, dioxime	Dimethylglyoxime	$C_4H_8N_2O_2$	116.12	95-45-4	245.5	3235	sub 234	3-01-00-03105		H_2O 1; EtOH 4; eth 4; bz 2	
3361	2,3-Butanedione, monooxime		$C_4H_7NO_2$	101.11	57-71-6	76.8		185.5	4-01-00-03646		H_2O 2; EtOH 4; eth 4; chl 4	
3362	1,3-Butanedione, 1-phenyl-		$C_{10}H_{10}O_2$	162.19	93-91-4	56		261.5	4-07-00-02151	1.0599^{74}	H_2O 1; eth 3; chl 2	1.5678^{78}
3363	1,3-Butanedione, 1-(2-pyridinyl)-		$C_9H_9NO_2$	163.18	40614-52-6	50		140^{15}	4-21-00-04750		bz 4; eth 4; EtOH 4; chl 4	
3364	1,3-Butanedione, 1-(3-pyridinyl)-		$C_9H_9NO_2$	163.18	3594-37-4	85		171^{15}	5-21-10-00180		H_2O 2; EtOH 3; eth 3; ace 3	
3365	1,3-Butanedione, 1-(4-pyridinyl)-		$C_9H_9NO_2$	163.18	75055-73-1	62		145^{18}	5-21-10-00180		bz 4; eth 4; EtOH 4	
3366	Butanedioyl dichloride	Succinyl chloride	$C_4H_4Cl_2O_2$	154.98	543-20-4	20	8844	193.3	4-02-00-01921	1.3748^{20}	eth 3; ace 3; bz 3	1.4683^{20}
3367	Butanedioyl dichloride, methylene-		$C_5H_4Cl_2O_2$	166.99	1931-60-8			89^{17}	3-02-00-01934		ace 3	1.4919^{20}
3368	1,4-Butanedithiol	Tetramethylenedithiol	$C_4H_{10}S_2$	122.26	1191-08-8	-53.9		195.5	4-01-00-02523	1.0021^{0}	H_2O 1; EtOH 4; ctc 2	1.5290^{20}
3369	Butane, 1-(ethenyloxy)-	Butyl vinyl ether	$C_6H_{12}O$	100.16	111-34-2	-92		94	4-01-00-02052	0.7888^{20}	H_2O 1; EtOH 4; eth 5; ace 4	1.4026^{20}
3370	Butane, 2-(ethenyloxy)-		$C_6H_{12}O$	100.16	1888-85-3			81	4-01-00-02054	0.7715^{20}		1.3970^{20}
3371	Butane, 1-(ethenyloxy)-3-methyl-	Isopentyl vinyl ether	$C_7H_{14}O$	114.19	39782-38-2			112.5	4-01-00-02055	0.7826^{20}	eth 4; EtOH 4	1.4072^{20}
3372	Butane, 1-ethoxy-	Butyl ethyl ether	$C_6H_{14}O$	102.18	628-81-9	-124		92.3	4-01-00-01518	0.7495^{20}	H_2O 1; EtOH 5; eth 5; ace 4	1.3818^{20}
3373	Butane, 2-ethoxy-	*sec*-Butyl ethyl ether	$C_6H_{14}O$	102.18	2679-87-0			81	4-01-00-01572	0.7503^{20}	H_2O 1; EtOH 4; eth 4	1.3802^{20}
3374	Butane, 1-ethoxy-3-methyl-	Ethyl isopentyl ether	$C_7H_{16}O$	116.20	628-04-6			112.5	4-01-00-01681	0.7688^{21}	eth 4; EtOH 4	
3375	Butane, 2-ethoxy-2-methyl-	Ethyl *tert*-amyl ether	$C_7H_{16}O$	116.20	919-94-8			102	3-01-00-01626	0.751^{18}	eth 4; EtOH 4	
3376	Butane, 1-(ethylthio)-	Butyl ethyl sulfide	$C_6H_{14}S$	118.24	638-46-0	-95.1		144.3	4-01-00-01558	0.8376^{20}	EtOH 4; chl 3	1.4492^{10}
3377	Butane, 1-(ethynyloxy)-		$C_6H_{10}O$	98.14	3329-56-4			104	4-01-00-02213	0.8200^{20}	eth 4; EtOH 4	1.4067
3378	Butane, 1-fluoro-	Butyl fluoride	C_4H_9F	76.11	2366-52-1	-134		32.5	4-01-00-00244	0.7789^{20}	EtOH 4	1.3396^{20}
3379	Butane, 2-fluoro-	*sec*-Butyl fluoride	C_4H_9F	76.11	359-01-3	-121.4		25.1	4-01-00-00244	0.7559^{25}		
3380	Butane, 1,1,2,2,3,4,4-heptachloro-		$C_4H_3Cl_7$	299.24	34973-41-6			$137.5^{13.5}$	4-01-00-00257	1.739^{20}	ace 4; ctc 4	1.5407^{20}
3381	Butane, 1-iodo-	Butyl iodide	C_4H_9I	184.02	542-69-8	-103	1572	130.6	4-01-00-00271	1.6154^{20}	H_2O 1; EtOH 5; eth 5; chl 4	1.5001^{20}
3382	Butane, 2-iodo-, (±)-	Butane, 2-iodo (*DL*)	C_4H_9I	184.02	52152-71-3	-104.2	1573	120	4-01-00-00272	1.5920^{20}	H_2O 1; EtOH 5; eth 5; chl 4	1.4991^{20}
3383	Butane, 1-iodo-2-methyl-, (*S*)-	Butane, 1-iodo-2-methyl (*d*)	$C_5H_{11}I$	198.05	29394-58-9			148; 132.6^{500}	3-01-00-00366	1.5253^{20}	eth 4; EtOH 4	1.4977^{20}
3384	Butane, 1-iodo-3-methyl-	Isoamyl iodide	$C_5H_{11}I$	198.05	541-28-6		5002	147	4-01-00-00331	1.5118^{20}	H_2O 2; EtOH 5; eth 5; ctc 2	1.4939^{20}
3385	Butane, 2-iodo-2-methyl-		$C_5H_{11}I$	198.05	594-38-7			124.5	4-01-00-00331	1.4937^{20}	H_2O 1; EtOH 5; eth 5	1.4981^{20}
3386	Butane, 1-isocyanato-3-methyl-		$C_6H_{11}NO$	113.16	1611-65-0			137	0-04-00-00186	0.806^{20}	H_2O 1; EtOH 5; eth 5	1.406^{20}
3387	Butane, 1-isocyano-	Butyl isocyanide	C_5H_9N	83.13	2769-64-4			120	4-04-00-00562	0.78^{20}	eth 4; EtOH 4	
3388	Butane, 1-isothiocyanato-		C_5H_9NS	115.20	592-82-5			168	4-04-00-00596	0.9546^{20}	eth 4; EtOH 4	1.501^{20}
3389	Butane, 2-isothiocyanato-, (±)-	*sec*-Butyl isothiocyanate (*DL*)	C_5H_9NS	115.20	116724-11-9			159.5	4-04-00-00624	0.944^{12}	eth 4; EtOH 4	
3390	Butane, 1-isothiocyanato-3-methyl-		$C_6H_{11}NS$	129.23	628-03-5			183	4-04-00-00707	0.9419^{17}	eth 4; EtOH 4	
3391	Butane, 1-methoxy-	Butyl methyl ether	$C_5H_{12}O$	88.15	628-28-4	-115.5		70.1	4-01-00-01518	0.7443^{20}	H_2O 1; EtOH 5; eth 5; ace 3	1.3736^{20}

No.	Name Synonym	Mol. Form. Mol. Wt.	CAS RN mp/°C	Merck No. bp/°C	Beil. Ref. den/g cm^{-3}	Solubility n_D
3392	Butane, 2-methoxy-, (*S*)- *sec*-Butyl methyl ether (*d*)	$C_5H_{12}O$ 88.15	66610-39-7	 60.7	4-01-00-01571 0.7400^{25}	
3393	Butane, 2-methoxy-, (±)- *sec*-Butyl methyl ether (*DL*)	$C_5H_{12}O$ 88.15	116783-23-4	 60	4-01-00-01571 0.7415^{20}	ace 4; eth 4; EtOH 4 1.3680^{25}
3394	Butane, 1-methoxy-2-methyl- Methyl (2-methylbutyl) ether	$C_6H_{14}O$ 102.18	62016-48-2	 90	4-01-00-01667 0.754^{18}	 1.3849^{20}
3395	Butane, 1-methoxy-3-methyl-	$C_6H_{14}O$ 102.18	626-91-5	 91	4-01-00-01681 0.7517^{20}	eth 4; EtOH 4 1.3830^{20}
3396	Butane, 2-methoxy-2-methyl-	$C_6H_{14}O$ 102.18	994-05-8	 86.3	4-01-00-01671 0.7703^{20}	eth 4; EtOH 4 1.3855^{20}
3397	Butane, 2-methyl- Isopentane	C_5H_{12} 72.15	78-78-4 -159.9	 27.8	4-01-00-00320 0.6201^{20}	H_2O 1; EtOH 5; eth 5 1.3537^{20}
3398	Butane, 1,1'-[methylenebis(oxy)]bis-	$C_9H_{20}O_2$ 160.26	2568-90-3 -58.1	 179.2	4-01-00-03029 0.8339^{20}	 1.4072^{17}
3399	Butane, 1-(1-methylethoxy)- Butyl isopropyl ether	$C_7H_{16}O$ 116.20	1860-27-1	 109	4-01-00-01519 0.7594^{15}	H_2O 1; EtOH 3; eth 3; ace 3 1.3870^{15}
3400	Butane, 3-methyl-1-nitro- 3-Methyl-1-nitrobutane	$C_5H_{11}NO_2$ 117.15	627-67-8 97	 163.5; 72^{27}	4-01-00-00332 0.9458^{25}	
3401	Butane, 1-(2-methylpropoxy)- Butyl isobutyl ether	$C_8H_{18}O$ 130.23	17071-47-5	 151	4-01-00-01594 0.7980^{22}	ace 4; eth 4; EtOH 4 1.4077^{21}
3402	Butane, 1-(methylthio)- Butyl methyl sulfide	$C_5H_{12}S$ 104.22	628-29-5 -97.8	 123.5	4-01-00-01557 0.8426^{20}	EtOH 4; MeOH 4 1.4477^{20}
3403	Butanenitrile Propyl cyanide	C_4H_7N 69.11	109-74-0 -111.9	1597 117.6	4-02-00-00806 0.7936^{20}	H_2O 2; EtOH 5; eth 5; bz 3 1.3842^{20}
3404	Butanenitrile, 4-bromo-	C_4H_6BrN 148.00	5332-06-9	 206	4-02-00-00836 1.4967^{20}	EtOH 3; eth 3; chl 3 1.4818^{20}
3405	Butanenitrile, 4-chloro-	C_4H_6ClN 103.55	628-20-6	 192	4-02-00-00827 1.0934^{15}	H_2O 1; EtOH 3; eth 3; ctc 2 1.4413^{20}
3406	Butanenitrile, 4,4-diethoxy-	$C_8H_{15}NO_2$ 157.21	18381-45-8	 $104\text{-}6^{10}$	3-03-00-01210 0.937^{25}	H_2O 2; EtOH 5; eth 5 1.4186^{20}
3407	Butanenitrile, 2-ethyl-	$C_6H_{11}N$ 97.16	617-80-1	 145.5	4-02-00-00953	eth 4; EtOH 4 1.3891^{24}
3408	Butanenitrile, 2-hydroxy-3-methyl-2-(1-methylethyl)- Diisopropyl ketone, cyanohydrin	$C_8H_{15}NO$ 141.21	4390-75-4 59	 111^{18}	4-03-00-00884	ace 4; bz 4; eth 4; EtOH 4
3409	Butanenitrile, 2-methyl- 2-Methylbutanenitrile	C_5H_9N 83.13	18936-17-9	 125	4-02-00-00892 0.7913^{15}	eth 4; EtOH 4 1.3933^{20}
3410	Butanenitrile, 3-methyl- Isobutyl cyanide	C_5H_9N 83.13	625-28-5 -101	 127.5	4-02-00-00902 0.7914^{20}	H_2O 2; EtOH 5; eth 5; ace 4 1.3927^{20}
3411	Butanenitrile, 4-phenoxy-	$C_{10}H_{11}NO$ 161.20	2243-43-8 45.5	 288	4-06-00-00646	eth 3
3412	Butane, 1-nitro- 1-Nitrobutane	$C_4H_9NO_2$ 103.12	627-05-4	 153	4-01-00-00277 0.970^{25}	H_2O 2; EtOH 5; eth 5; alk 3 1.4303^{20}
3413	Butane, 2-nitro-, (±)-	$C_4H_9NO_2$ 103.12	116781-85-2 -132	 140	4-01-00-00278 0.9854^{17}	chl 3 1.4044^{20}
3414	Butane, 1,1,1,2,3,4,4,4-octafluoro-2,3-bis(trifluoromethyl)- Perfluoro-2,3-dimethylbutane	C_6F_{14} 338.04	354-96-1 -15	 59.8		
3415	Butane, 1,1'-oxybis- Dibutyl ether	$C_8H_{18}O$ 130.23	142-96-1 -95.2	1568 140.2	4-01-00-01520 0.7684^{20}	H_2O 1; EtOH 5; eth 5; ace 4 1.3992^{20}
3416	Butane, 2,2'-oxybis-, (±)- *sec*-Butyl ether, (±)-	$C_8H_{18}O$ 130.23	17226-28-7	 120.5	3-01-00-01533 0.756^{25}	ace 4; eth 4; EtOH 4 1.393^{25}
3417	Butane, 1,1'-[oxybis(2,1-ethanediyloxy)]bis-	$C_{12}H_{26}O_3$ 218.34	112-73-2 -60	 256	4-01-00-02395 0.885^{25}	 1.4235^{20}
3418	Butane, 1,1'-oxybis[3-methyl- Diisoamyl ether	$C_{10}H_{22}O$ 158.28	544-01-4	5000 172.5	4-01-00-01682 0.7777^{20}	ace 4; EtOH 4; chl 4 1.4085^{20}
3419	Butane, 1,1,2,3,4-pentachloro- Butane, 1,1,2,3,4-pentachloro (liquid)	$C_4H_5Cl_5$ 230.35	77753-24-3	 95.5^{11}	4-01-00-00256 1.561^{18}	EtOH 4; ctc 4 1.6141^{18}
3420	Butane, 1,2,2,3,4-pentachloro- 1,2,2,3,4-Pentachlorobutane	$C_4H_5Cl_5$ 230.35	2431-52-9	 85^{10}	4-01-00-00256 1.5543^{20}	ace 4; chl 4 1.5157^{20}
3421	Butane, 1-propoxy- Butyl propyl ether	$C_7H_{16}O$ 116.20	3073-92-5	 118.1	4-01-00-01519 0.7772^{0}	H_2O 1; EtOH 4; eth 4
3422	Butane, 1,1'-sulfinylbis-	$C_8H_{18}OS$ 162.30	2168-93-6 32.6	 dec	4-01-00-01561 0.8317^{23}	H_2O 1; EtOH 3; eth 3 1.4669^{20}
3423	Butane, 1,1'-sulfonylbis-	$C_8H_{18}O_2S$ 178.30	598-04-9 45	 291	4-01-00-01561 0.9885^{47}	H_2O 1; EtOH 3; eth 3
3424	Butane, 1,1,4,4-tetrabromo-	$C_4H_6Br_4$ 373.71	116779-77-2	 $138\text{-}45^{10}$	3-01-00-00298 2.529^{20}	bz 4; eth 4; chl 4 1.6077^{20}
3425	Butane, 1,2,2,3-tetrabromo-	$C_4H_6Br_4$ 373.71	116779-78-3 -2	 97.5^{7}	3-01-00-00298 2.510^{20}	
3426	Butane, 1,2,3,4-tetrabromo-, *DL*- Butane, 1,2,3,4-tetrabromo (*DL*)	$C_4H_6Br_4$ 373.71	2657-65-0 118	 260; 180^{60}	4-01-00-00271	ace 4; eth 4; EtOH 4; lig 4
3427	Butane, 1,1,1,2-tetrachloro- 1,1,1,2-Tetrachlorobutane	$C_4H_6Cl_4$ 195.90	39966-95-5	 135	4-01-00-00253 1.3907^{20}	 1.4920^{25}
3428	Butane, 1,2,2,3-tetrachloro- 1,2,2,3-Tetrachlorobutane	$C_4H_6Cl_4$ 195.90	79630-70-9 -48	 182	3-01-00-00286 1.4276^{18}	ace 4; chl 4 1.491^{20}
3429	Butane, 1,2,3,3-tetrachloro- 1,2,3,3-Tetrachlorobutane	$C_4H_6Cl_4$ 195.90	13138-51-7	 90^{32}; 56^{10}	4-01-00-00254 1.4204^{20}	ace 4; eth 4; EtOH 4 1.4958^{20}
3430	Butane, 2,2,3,3-tetramethyl- 2,2,3,3-Tetramethylbutane	C_8H_{18} 114.23	594-82-1 100.7	 106.4	4-01-00-00447 0.8242^{20}	H_2O 1; eth 3; chl 3 1.4695^{20}

No.	Name Synonym	Mol. Form. Mol. Wt.	CAS RN mp/°C	Merck No. bp/°C	Beil. Ref. den/g cm^{-3}	Solubility n_D
3431	1,2,3,4-Butanetetrol, (*R**,*S**)- Erythritol	$C_4H_{10}O_4$ 122.12	149-32-6 121.5	3620 330.5	4-01-00-02807 1.451^{20}	H_2O 3; eth 1; bz 1
3432	1,2,3,4-Butanetetrol, tetranitrate, (*R**,*S**)- Erythrityl tetranitrate	$C_4H_6N_4O_{12}$ 302.11	7297-25-8 61	3622	4-01-00-02809	EtOH 4
3433	Butane, 1,1'-thiobis- Dibutyl sulfide	$C_8H_{18}S$ 146.30	544-40-1 -79.7	1590 185	4-01-00-01559 0.8386^{20}	eth 4; EtOH 4; chl 4 1.4530^{20}
3434	Butane, 2,2'-thiobis-	$C_8H_{18}S$ 146.30	626-26-6	 165	4-01-00-01586 0.8348^{20}	H_2O 1; EtOH 4; eth 4 1.4506^{20}
3435	Butane, 1,1'-thiobis[2-methyl- Bis(2-methylbutyl) sulfide	$C_{10}H_{22}S$ 174.35	96034-00-3	 165; 95^{13}	0-01-00-00387 0.8348^{20}	eth 4; EtOH 4 1.4506^{20}
3436	Butane, 1,1'-thiobis[3-methyl- Isoamyl sulfide	$C_{10}H_{22}S$ 174.35	544-02-5 -74.6	 211	4-01-00-01689 0.8323^{20}	H_2O 1; EtOH 5; eth 4 1.4520^{20}
3437	1-Butanethiol Butyl mercaptan	$C_4H_{10}S$ 90.19	109-79-5 -115.7	1575 98.5	4-01-00-01555 0.8416^{20}	H_2O 2; EtOH 4; eth 4; chl 2 1.4440^{20}
3438	2-Butanethiol, (±)- 2-Butanethiol (*DL*)	$C_4H_{10}S$ 90.19	91840-99-2 -165	1576 85	4-01-00-01584 0.8295^{20}	EtOH 3; eth 3; bz 3; ctc 2 1.4366^{20}
3439	1-Butanethiol, 2-methyl-, (*S*)- (+)-2-Methyl-1-butanethiol	$C_5H_{12}S$ 104.22	20089-07-0	 118.2	4-01-00-01668 0.8420^{20}	 1.4440^{20}
3440	1-Butanethiol, 3-methyl- Isopentyl mercaptan	$C_5H_{12}S$ 104.22	541-31-1	 120	4-01-00-01688 0.8350^{20}	H_2O 1; EtOH 5; eth 5; ctc 3 1.4418^{20}
3441	2-Butanethiol, 2-methyl- 2-Methyl-2-butanethiol	$C_5H_{12}S$ 104.22	1679-09-0	 99.1	4-01-00-01674 0.8120^{20}	 1.4385^{20}
3442	Butane, 1,1,2-tribromo- 1,1,2-Tribromocyclobutane	$C_4H_7Br_3$ 294.81	3675-68-1	 216.2	2-01-00-00084 2.1835^{20}	eth 4; EtOH 4; chl 4 1.5626^{17}
3443	Butane, 1,2,2-tribromo- 1,2,2-Tribromobutane	$C_4H_7Br_3$ 294.81	3675-69-2	 213.8	2-01-00-00085 2.1691^{20}	eth 4; EtOH 4; chl 4 1.5624^{10}
3444	Butane, 1,2,3-tribromo- 1,2,3-Tribromobutane	$C_4H_7Br_3$ 294.81	632-05-3 -19	 220	4-01-00-00271 2.1907^{20}	eth 4; EtOH 4; chl 4 1.5680^{20}
3445	Butane, 1,2,4-tribromo- 1,2,4-Tribromobutane	$C_4H_7Br_3$ 294.81	38300-67-3 -18	 215	4-01-00-00271 2.170^{20}	eth 4; EtOH 4; chl 4 1.5608^{20}
3446	Butane, 1,3,3-tribromo- 1,3,3-Tribromobutane	$C_4H_7Br_3$ 294.81	62127-46-2	 202.5	3-01-00-00298 2.1445^{20}	eth 4; EtOH 4; chl 4 1.5564^{20}
3447	Butane, 2,2,3-tribromo- 2,2,3-Tribromobutane	$C_4H_7Br_3$ 294.81	62127-47-3 1.8	 206	3-01-00-00298 2.1723^{20}	H_2O 1; EtOH 3; eth 3; ctc 2 1.5602^{20}
3448	Butane, 1,1,3-trichloro- 1,1,3-Trichlorobutane	$C_4H_7Cl_3$ 161.46	13279-87-3	 152	4-01-00-00252 1.317^{15}	H_2O 1; EtOH 3; eth 3; chl 4 1.4600^{15}
3449	Butane, 1,2,3-trichloro- 1,2,3-Trichlorobutane	$C_4H_7Cl_3$ 161.46	18338-40-4	 168; 63^{20}	3-01-00-00285 1.3164^{20}	H_2O 1; EtOH 3; eth 3; chl 4 1.4790^{20}
3450	Butane, 1,2,4-trichloro- 1,2,4-Trichlorobutane	$C_4H_7Cl_3$ 161.46	1790-22-3	 61-3^{10}	4-01-00-00253 1.3175^{20}	H_2O 1; bz 3; ctc 3; chl 4 1.4820^{20}
3451	Butane, 2,2,3-trichloro- 2,2,3-Trichlorobutane	$C_4H_7Cl_3$ 161.46	10403-60-8	 144	3-01-00-00285 1.2699^{20}	H_2O 1; chl 4 1.4645^{20}
3452	Butane, 2,2,3-trichloro-1,1,1,3,4,4,4-heptafluoro-	$C_4Cl_3F_7$ 287.39	335-44-4 4	 98	4-01-00-00253 1.7484^{20}	 1.3530^{20}
3453	Butane, 1,2,3-trichloro-2-methyl- Butane, 2-methyl-1,2,3-trichloro	$C_5H_9Cl_3$ 175.48	62521-69-1	 184	3-01-00-00361 1.2527^{20}	chl 4
3454	Butane, 2,2,3-trichloro-3-methyl- Butane, 2-methyl-2,3,3-trichloro	$C_5H_9Cl_3$ 175.48	98070-91-8	 182.5	3-01-00-00361 1.215^{15}	chl 4; HOAc 4 1.473^{21}
3455	Butane, 1,1,3-trimethoxy-	$C_7H_{16}O_3$ 148.20	10138-89-3	 157	4-01-00-03985 0.935^{25}	 1.4032^{20}
3456	Butane, 1,3,3-trimethoxy-	$C_7H_{16}O_3$ 148.20	6607-66-5	 61-3^{20}	4-01-00-03995 0.94^{25}	 1.4096^{20}
3457	Butane, 2,2,3-trimethyl- 2,2,3-Trimethylbutane	C_7H_{16} 100.20	464-06-2 -25	 80.8	4-01-00-00410 0.6901^{20}	H_2O 1; EtOH 3; eth 3; ace 4 1.3864^{20}
3458	1,2,3-Butanetriol 1-Methylglycerol	$C_4H_{10}O_3$ 106.12	4435-50-1	 170^{20}	4-01-00-02774	H_2O 4; EtOH 4 1.4462^{20}
3459	1,2,4-Butanetriol	$C_4H_{10}O_3$ 106.12	3068-00-6	 172-4^{12}	2-01-00-00596 1.018^{20}	H_2O 4; EtOH 4 1.4688^{20}
3460	Butanoic acid Butyric acid	$C_4H_8O_2$ 88.11	107-92-6 -5.7	1593 163.7	4-02-00-00779 0.9577^{20}	H_2O 5; EtOH 5; eth 5; ctc 2 1.3980^{20}
3461	Butanoic acid, 2-acetyl-3-methyl-, ethyl ester	$C_9H_{16}O_3$ 172.22	1522-46-9	 201; 97-8^{20}	4-03-00-01608 0.9648^{18}	H_2O 1; EtOH 5; eth 5 1.4256^{18}
3462	Butanoic acid, 2-acetyl-3-oxo-, ethyl ester	$C_8H_{12}O_4$ 172.18	603-69-0	 210	4-03-00-01781 1.1045^{20}	bz 4; eth 4; EtOH 4 1.4690^{20}
3463	Butanoic acid, 2-amino-, (±)-	$C_4H_9NO_2$ 103.12	2835-81-6 305 dec	 sub	3-04-00-01296 1.2300^{20}	H_2O 4; EtOH 2; eth 1; bz 1
3464	Butanoic acid, 4-amino- γ-Aminobutyric acid	$C_4H_9NO_2$ 103.12	56-12-2 203 dec	441	4-04-00-02600	H_2O 4; EtOH 2; eth 1; ace 2
3465	Butanoic acid, 3-amino-4-hydroxy- γ-Hydroxy-β-aminobutyric acid	$C_4H_9NO_3$ 119.12	589-44-6 216	455	4-04-00-03216	
3466	Butanoic acid, 4-amino-2-hydroxy-4-oxo-	$C_4H_7NO_4$ 133.10	66398-52-5 147.5		3-03-00-00918 1.577^{18}	H_2O 3; EtOH 2; eth 1; MeOH 2
3467	Butanoic acid, anhydride Butyric anhydride	$C_8H_{14}O_3$ 158.20	106-31-0 -75	1594 200	4-02-00-00802 0.9668^{20}	eth 3; ctc 2 1.4070^{20}
3468	Butanoic acid, 2-bromo-, (±)- *DL*-α-Bromobutyric acid	$C_4H_7BrO_2$ 167.00	2385-70-8 -4	1401 217 dec; 181^{250}	4-02-00-00833 1.5641^{20}	H_2O 3; EtOH 3; eth 3
3469	Butanoic acid, 2-bromo-, ethyl ester	$C_6H_{11}BrO_2$ 195.06	533-68-6	 177; 43^{5}	4-02-00-00834 1.3273^{20}	H_2O 1; EtOH 5; eth 5; chl 3 1.4475^{20}

No.	Name Synonym	Mol. Form. Mol. Wt.	CAS RN mp/°C	Merck No. bp/°C	Beil. Ref. den/g cm^{-3}	Solubility n_D
3470	Butanoic acid, 4-bromo-, ethyl ester	$C_6H_{11}BrO_2$ 195.06	2969-81-5	82^{10}	4-02-00-00836 1.363^{25}	1.4559^{20}
3471	Butanoic acid, 2-bromo-3-methyl-, (±)-	$C_5H_9BrO_2$ 181.03	10323-40-7 44	230 dec; 138^{25}	3-02-00-00706 1.459^{20}	H_2O 2; EtOH 4; eth 3; ace 3
3472	Butanoic acid, 3-bromo-3-methyl- β-Bromoisovaleric acid	$C_5H_9BrO_2$ 181.03	5798-88-9 74	1410	4-02-00-00905	bz 4; eth 4; EtOH 4
3473	Butanoic acid, 2-bromo-, methyl ester	$C_5H_9BrO_2$ 181.03	3196-15-4	168	4-02-00-00834 1.4528^{20}	EtOH 4 1.4029^{25}
3474	Butanoic acid, 4-bromo-, methyl ester	$C_5H_9BrO_2$ 181.03	4897-84-1	186.5	4-02-00-00835 1.4^{25}	EtOH 4 1.4567^{25}
3475	Butanoic acid, 2-bromo-3-methyl-, ethyl ester	$C_7H_{13}BrO_2$ 209.08	609-12-1	186	4-02-00-00906 1.2760^{20}	eth 4; EtOH 4 1.4496^{20}
3476	Butanoic acid, 2-bromo-3-methyl-, methyl ester	$C_6H_{11}BrO_2$ 195.06	26330-51-8	177	3-02-00-00706 1.352^{13}	eth 4; EtOH 4 1.4530^{20}
3477	Butanoic acid, 2-bromo-3-oxo-, ethyl ester	$C_6H_9BrO_3$ 209.04	609-13-2	210 dec; 107^{15}	4-03-00-01551 1.4294^{16}	eth 4; EtOH 4 1.463^{14}
3478	Butanoic acid, 4-bromo-3-oxo-, ethyl ester	$C_6H_9BrO_3$ 209.04	13176-46-0	115^{14}	4-03-00-01551 1.4840^{20}	eth 4; EtOH 4 1.5281^{20}
3479	Butanoic acid, butyl ester Butyl butanoate	$C_8H_{16}O_2$ 144.21	109-21-7 -91.5	1556 166	4-02-00-00789 0.8700^{20}	H_2O 1; EtOH 5; eth 5; ctc 3 1.4075^{20}
3480	Butanoic acid, 2-chloro-	$C_4H_7ClO_2$ 122.55	4170-24-5	189^{627}; 101^{15}	4-02-00-00821 1.1796^{20}	H_2O 2; EtOH 4; eth 4 1.441^{20}
3481	Butanoic acid, 3-chloro-, (±)-	$C_4H_7ClO_2$ 122.55	625-68-3 16	116^{22}	4-02-00-00823 1.1898^{20}	EtOH 3; eth 4; ctc 2 1.4221^{20}
3482	Butanoic acid, 4-chloro-	$C_4H_7ClO_2$ 122.55	627-00-9 16	196^{22}; $68^{0.2}$	4-02-00-00825 1.2236^{20}	EtOH 4 1.4642^{20}
3483	Butanoic acid, 2-chloro-, ethyl ester	$C_6H_{11}ClO_2$ 150.60	7425-45-8	163.5	4-02-00-00822 1.056^{18}	eth 4; EtOH 4 1.4180^{25}
3484	Butanoic acid, 3-chloro-, ethyl ester	$C_6H_{11}ClO_2$ 150.60	7425-48-1		4-02-00-00824 1.0517^{20}	EtOH 4 1.4246^{20}
3485	Butanoic acid, 4-chloro-, ethyl ester	$C_6H_{11}ClO_2$ 150.60	3153-36-4	184	4-02-00-00825 1.0756^{20}	ace 4; eth 4; EtOH 4 1.4311^{20}
3486	Butanoic acid, 2-chloro-2-methyl- Butyric acid, 2-chloro-2-methyl-	$C_5H_9ClO_2$ 136.58	73758-54-0	200 dec	3-02-00-00688 1.1204^{20}	eth 4; EtOH 4 1.4445^{20}
3487	Butanoic acid, 2-chloro-3-methyl-	$C_5H_9ClO_2$ 136.58	921-08-4 21	210; 126^{32}	4-02-00-00904 1.135^{13}	H_2O 1; EtOH 3; eth 3 1.4450^{11}
3488	Butanoic acid, 2-chloro-, methyl ester	$C_5H_9ClO_2$ 136.58	26464-32-4	146	3-02-00-00622 1.0979^{14}	eth 4; EtOH 4 1.4247^{20}
3489	Butanoic acid, 4-chloro-, methyl ester	$C_5H_9ClO_2$ 136.58	3153-37-5	175; 55^{4}	4-02-00-00825 1.1201^{20}	H_2O 1; EtOH 4; eth 4; ace 3 1.4321^{20}
3490	Butanoic acid, 2-chloro-3-methyl-, ethyl ester Ethyl 2-chloro-3-methylbutanoate	$C_7H_{13}ClO_2$ 164.63	91913-99-4	179	0-02-00-00316 1.021^{13}	eth 4; EtOH 4
3491	Butanoic acid, 2-chloro-2-methyl-, ethyl ester Butyric acid, 2-chloro-2-methyl, ethyl ester	$C_7H_{13}ClO_2$ 164.63	58190-94-6	176	3-02-00-00688 1.069^{14}	eth 4; EtOH 4 1.4388^{11}
3492	Butanoic acid, 2-chloro-2-methyl-3-oxo-, ethyl ester Butyric acid, 2-chloro-2-methyl-3-oxo, ethyl ester	$C_7H_{11}ClO_3$ 178.62	37935-39-0	$116\text{-}7^{75}$	2-03-00-00433 1.157^{18}	EtOH 4 1.4382^{18}
3493	Butanoic acid, 4-(4-chloro-2-methylphenoxy)- 4-(4-Chloro-2-methylphenoxy)butyric acid	$C_{11}H_{13}ClO_3$ 228.68	94-81-5 100		4-06-00-01996	
3494	Butanoic acid, 4-chloro-3-oxo-, ethyl ester	$C_6H_9ClO_3$ 164.59	638-07-3 -8	220 dec; 115^{14}	4-03-00-01550 1.218^{25}	1.4520^{20}
3495	Butanoic acid, 2-cyano-2-ethyl-, ethyl ester Cyanoacetic acid, diethyl, ethyl ester	$C_9H_{15}NO_2$ 169.22	1619-56-3	214.5	3-02-00-01761	eth 4; EtOH 4 1.4200^{27}
3496	Butanoic acid, cyclohexyl ester	$C_{10}H_{18}O_2$ 170.25	1551-44-6	213	4-06-00-00037 0.9572^{0}	H_2O 1; EtOH 3; ctc 2; os 3
3497	Butanoic acid, 2,3-dibromo-	$C_4H_6Br_2O_2$ 245.90	600-30-6 87	$100\text{-}10^{20}$	4-02-00-00837	H_2O 2; EtOH 4; eth 4; bz 4
3498	Butanoic acid, 2,3-dibromo-, ethyl ester	$C_6H_{10}Br_2O_2$ 273.95	609-11-0 58.5	113^{30}	3-02-00-00634 1.6800^{20}	H_2O 2; EtOH 3; eth 3; ctc 2
3499	Butanoic acid, 2,4-dibromo-, ethyl ester	$C_6H_{10}Br_2O_2$ 273.95	36847-51-5	$149\text{-}50^{52}$	4-02-00-00838 1.6987^{20}	H_2O 1; EtOH 3; eth 3 1.4960^{20}
3500	Butanoic acid, 2,2-dichloro- Butyric acid, 2,2-dichloro-	$C_4H_6Cl_2O_2$ 157.00	13023-00-2	$107\text{-}10^{14}$	2-02-00-00254 1.389^{20}	H_2O 2; EtOH 4; eth 4
3501	Butanoic acid, 3,4-dichloro-, butyl ester Butyric acid, 3,4-dichloro, butyl ester	$C_8H_{14}Cl_2O_2$ 213.10	116723-94-5	$127\text{-}8^{15}$	4-02-00-00829 1.143^{25}	1.454^{24}
3502	Butanoic acid, 3,4-dichloro-, methyl ester	$C_5H_8Cl_2O_2$ 171.02	819-93-2	90^{15}	4-02-00-00829 1.278^{23}	1.459^{23}
3503	Butanoic acid, 2,2-dichloro-3-oxo-, ethyl ester	$C_6H_8Cl_2O_3$ 199.03	6134-66-3	206; 91^{12}	2-03-00-00427 1.291^{16}	EtOH 4 1.4492^{17}
3504	Butanoic acid, 2,2-diethyl-3-oxo-, ethyl ester	$C_{10}H_{18}O_3$ 186.25	1619-57-4	216; 64^{3}	4-03-00-01625 0.9717^{18}	H_2O 1; EtOH 5; eth 5; ctc 3 1.4326^{17}
3505	Butanoic acid, 2,3-dihydroxypropyl ester	$C_7H_{14}O_4$ 162.19	557-25-5	280; 117^{10}	4-02-00-00798 1.129^{18}	H_2O 4; EtOH 4 1.4531^{20}
3506	Butanoic acid, 2,2-dimethyl-	$C_6H_{12}O_2$ 116.16	595-37-9 -14	186	4-02-00-00954 0.9276^{20}	H_2O 2; EtOH 3; eth 3 1.4145^{20}

No.	Name Synonym	Mol. Form. Mol. Wt.	CAS RN mp/°C	Merck No. bp/°C	Beil. Ref. den/g cm^{-3}	Solubility n_D
3507	Butanoic acid, 2,3-dimethyl- Butyric acid, 2,3-dimethyl	$C_6H_{12}O_2$ 116.16	14287-61-7 -1.5	 191.7	4-02-00-00958 0.9275^{20}	eth 4; EtOH 4 1.4146^{20}
3508	Butanoic acid, 3,3-dimethyl- *tert*-Butylacetic acid	$C_6H_{12}O_2$ 116.16	1070-83-3 6.5	1538 190	4-02-00-00955 0.9124^{20}	EtOH 3; eth 3 1.4096^{20}
3509	Butanoic acid, 1,1-dimethylethyl ester	$C_8H_{16}O_2$ 144.21	2308-38-5	 146	4-02-00-00790	ace 4; eth 4; EtOH 4 1.4007^{17}
3510	Butanoic acid, 3,3-dimethyl-2-oxo-	$C_6H_{10}O_3$ 130.14	815-17-8 90.5	 189; 80^{15}	4-03-00-01595	H_2O 2; eth 3; bz 3; chl 3
3511	Butanoic acid, 2,2-dimethyl-3-oxo-, ethyl ester	$C_8H_{14}O_3$ 158.20	597-04-6	 182	4-03-00-01594 0.9759^{20}	eth 4; EtOH 4 1.4180^{20}
3512	Butanoic acid, 1,2-ethanediyl ester	$C_{10}H_{18}O_4$ 202.25	105-72-6 -80	 240	4-02-00-00796 1.0005^{20}	eth 4; EtOH 4 1.4262^{20}
3513	Butanoic acid, ethenyl ester Vinyl butyrate	$C_6H_{10}O_2$ 114.14	123-20-6	 116.7; 64^{130}	4-02-00-00792 0.9006^{20}	
3514	Butanoic acid, 2-ethyl- Diethylacetic acid	$C_6H_{12}O_2$ 116.16	88-09-5 -31.8	3099 194	4-02-00-00950 0.9239^{20}	H_2O 2; EtOH 5; eth 5; ctc 2 1.4132^{20}
3515	Butanoic acid, ethyl ester Ethyl butanoate	$C_6H_{12}O_2$ 116.16	105-54-4 -98	3733 121.5	4-02-00-00787 0.8844^{15}	H_2O 2; EtOH 3; eth 3; ctc 2 1.4000^{20}
3516	Butanoic acid, 2-ethyl-2-methyl-	$C_7H_{14}O_2$ 130.19	19889-37-3 <-20	 207	4-02-00-00981	EtOH 4 1.4250^{20}
3517	Butanoic acid, 2-ethyl-, methyl ester	$C_7H_{14}O_2$ 130.19	816-11-5	 137	4-02-00-00950 0.8886^{12}	 1.4067^{12}
3518	Butanoic acid, 2-ethyl-3-oxo-, ethyl ester	$C_8H_{14}O_3$ 158.20	607-97-6	 198.0; 80^{10}	4-03-00-01592 0.9847^{16}	EtOH 5; eth 5 1.4214^{25}
3519	Butanoic acid, 2-ethyl-3-oxo-, methyl ester	$C_7H_{12}O_3$ 144.17	51756-08-2	 182	4-03-00-01592 0.995^{14}	ace 4; eth 4; EtOH 4
3520	Butanoic acid, 2-furanylmethyl ester	$C_9H_{12}O_3$ 168.19	623-21-2	 212.5	4-17-00-01247 1.0530^{20}	H_2O 2; EtOH 3; eth 5
3521	Butanoic acid, heptafluoro-	$C_4HF_7O_2$ 214.04	375-22-4 -17.5	 121	4-02-00-00810 1.651^{20}	H_2O 3; eth 3; tol 3; peth 1 1.295^{25}
3522	Butanoic acid, heptafluoro-, anhydride	$C_8F_{14}O_3$ 410.06	336-59-4 -43	 106.5	4-02-00-00817 1.665^{20}	 1.285^{20}
3523	Butanoic acid, heptafluoro-, ethenyl ester	$C_6H_3F_7O_2$ 240.08	356-28-5	 79	4-02-00-00815 1.418^{20}	 1.309
3524	Butanoic acid, heptafluoro-, ethyl ester	$C_6H_5F_7O_2$ 242.09	356-27-4	 95	4-02-00-00813 1.394^{20}	H_2O 2; eth 3; ace 3 1.3011^{20}
3525	Butanoic acid, heptafluoro-, methyl ester	$C_5H_3F_7O_2$ 228.07	356-24-1 -86	 80	4-02-00-00812 1.483^{20}	H_2O 2; eth 3; ace 3 1.295^{20}
3526	Butanoic acid, heptafluoro-, 1-methylethyl ester Isopropyl heptafluorobutyrate	$C_7H_7F_7O_2$ 256.12	425-23-0	 106	4-02-00-00813 1.324^{20}	 1.310^{20}
3527	Butanoic acid, heptyl ester	$C_{11}H_{22}O_2$ 186.29	5870-93-9 -57.5	 225.8	4-02-00-00791 0.8637^{20}	EtOH 4 1.4231^{20}
3528	Butanoic acid, hexyl ester	$C_{10}H_{20}O_2$ 172.27	2639-63-6 -78	 208	4-02-00-00791 0.8652^{20}	H_2O 1; EtOH 3; chl 2; os 3 1.4160^{15}
3529	Butanoic acid, hydrazide Butyrhydrazide	$C_4H_{10}N_2O$ 102.14	3538-65-6 45.5	 138^{20}; 120^{10}	4-02-00-00807	H_2O 4; eth 4; EtOH 4
3530	Butanoic acid, 2-hydroxy-, (±)-	$C_4H_8O_3$ 104.11	600-15-7 44.2	 260 dec; 140^{14}	3-03-00-00561 1.125^{20}	H_2O 3; EtOH 3; eth 3
3531	Butanoic acid, 3-hydroxy-, (±)-	$C_4H_8O_3$ 104.11	625-71-8 49	 130^{12}; $94^{0.1}$	4-03-00-00761	H_2O 4; EtOH 4; eth 4; bz 1 1.4424^{20}
3532	Butanoic acid, 2-hydroxy-, ethyl ester, (±)- Butyric acid, 2-hydroxy, ethyl ester (*DL*)	$C_6H_{12}O_3$ 132.16	68057-83-0	 167	4-03-00-00756 1.0069^{20}	EtOH 4 1.4179^{20}
3533	Butanoic acid, 3-hydroxy-, ethyl ester, (±)-	$C_6H_{12}O_3$ 132.16	35608-64-1	 185; 76^{15}	4-03-00-00762 1.017^{20}	H_2O 3; EtOH 3; ctc 2 1.4182^{20}
3534	Butanoic acid, 3-hydroxy-, ethyl ester	$C_6H_{12}O_3$ 132.16	5405-41-4	 185; 76^{15}	4-03-00-00762 1.017^{20}	H_2O 4; EtOH 4 1.4182^{20}
3535	Butanoic acid, 3-hydroxy-3-methyl-	$C_5H_{10}O_3$ 118.13	625-08-1 <-32	 162^{12}	4-03-00-00827 0.9384^{20}	H_2O 4; eth 4; EtOH 4 1.5081^{20}
3536	Butanoic acid, 2-hydroxy-, 2-methylpropyl ester Butyric acid, 2-hydroxy, isobutyl ester	$C_8H_{16}O_3$ 160.21	116723-95-6	 196	0-03-00-00302 0.944^{15}	 1.4182
3537	Butanoic acid, 3-mercapto-	$C_4H_8O_2S$ 120.17	26473-49-4	 110^{10}; $88^{2.5}$	4-03-00-00769 1.1371^{20}	 1.4782^{20}
3538	Butanoic acid, 2-methyl-, (±)- (±)-2-Methylbutyric acid	$C_5H_{10}O_2$ 102.13	600-07-7 <-80	 177	4-02-00-00889 0.934^{20}	H_2O 2; EtOH 5; eth 5; chl 3 1.4051^{20}
3539	Butanoic acid, 3-methyl- Isovaleric acid	$C_5H_{10}O_2$ 102.13	503-74-2 -29.3	5120 176.5	4-02-00-00895 0.931^{20}	H_2O 3; EtOH 5; eth 5; chl 5 1.4033^{20}
3540	Butanoic acid, 2-methyl-, 10-(acetyloxy)-9,10-dihydro-8,8-dimethyl-2-oxo-2H,8H-benzo[1,2-b:3,4-b']dipyran-9-yl Visnadine	$C_{21}H_{24}O_7$ 388.42	477-32-7 85.5	9915	5-19-06-00272	H_2O 1; EtOH 3; eth 3
3541	Butanoic acid, 3-(methylamino)-, ethyl ester Butyric acid, 3-methylamino-, ethyl ester	$C_7H_{15}NO_2$ 145.20	68384-70-3	 $72^{12.5}$	3-04-00-01313 0.9282^{20}	 1.4250^{20}
3542	Butanoic acid, 3-methyl-, anhydride	$C_{10}H_{18}O_3$ 186.25	1468-39-9	 215	4-02-00-00901 0.9327^{20}	eth 4 1.4043^{20}
3543	Butanoic acid, 3-methyl-, butyl ester Butyl isovalerate	$C_9H_{18}O_2$ 158.24	109-19-3		3-02-00-00698	 1.4058^{25}

No.	Name	Synonym	Mol. Form.	Mol. Wt.	CAS RN	mp/°C	Merck No.	bp/°C	Beil. Ref.	den/g cm^{-3}	Solubility	n_D
3544	Butanoic acid, 3-methylbutyl ester	Isopentyl butanoate	$C_9H_{18}O_2$	158.24	106-27-4		4997	179	4-02-00-00791	0.865^{19}	H_2O 1; EtOH 4; eth 4	1.4110^{20}
3545	Butanoic acid, 2-methyl-, butyl ester	Butyric acid, (2-methylbutyl) ester	$C_9H_{18}O_2$	158.24	15706-73-7			179	0-02-00-00304	0.8620^{20}		1.4135^{20}
3546	Butanoic acid, 3-methyl-, 1,1-dimethylpropyl ester	*tert*-Amyl isovalerate	$C_{10}H_{20}O_2$	172.27	542-37-0		647	173.5	0-02-00-00312	0.8729^{0}	EtOH 4	
3547	Butanoic acid, methyl ester	Methyl butanoate	$C_5H_{10}O_2$	102.13	623-42-7	-85.8	5957	102.8	4-02-00-00786	0.8984^{20}	H_2O 2; EtOH 5; eth 5; ctc 2	1.3878^{20}
3548	Butanoic acid, 3-methyl-, ethyl ester	Ethyl isovalerate	$C_7H_{14}O_2$	130.19	108-64-5	-99.3	3772	135.0	4-02-00-00898	0.8656^{20}	EtOH 4; eth 4	1.3962^{20}
3549	Butanoic acid, 1-methylethyl ester	Isopropyl butyrate	$C_7H_{14}O_2$	130.19	638-11-9			130.5	4-02-00-00789	0.8588^{20}	H_2O 1; EtOH 3	1.3936^{20}
3550	Butanoic acid, 2-methyl-, ethyl ester, (*S*)-		$C_7H_{14}O_2$	130.19	10307-61-6			133; 35^{16}	4-02-00-00890	0.8689^{25}	H_2O 1; EtOH 3; bz 3	1.3964^{20}
3551	Butanoic acid, 3-methyl-, 3-methylbutyl ester	Isoamyl isovalerate	$C_{10}H_{20}O_2$	172.27	659-70-1		5003	190.4	4-02-00-00899	0.8583^{19}		1.4130^{19}
3552	Butanoic acid, 3-methyl-, methyl ester	Methyl isovalerate	$C_6H_{12}O_2$	116.16	556-24-1		6006	116.5	4-02-00-00897	0.8808^{20}	H_2O 1; EtOH 4; eth 4; ace 4	1.3927^{20}
3553	Butanoic acid, 3-methyl-, 1-methylethyl ester	Butyric acid, 3-methyl, isopropyl ester	$C_8H_{16}O_2$	144.21	32665-23-9			142; 70^{55}	3-02-00-00698	0.8538^{17}	ace 4; eth 4; EtOH 4	1.3960^{20}
3554	Butanoic acid, 3-methyl-, 5-methyl-2-(1-methylethyl)cyclohexyl ester		$C_{15}H_{28}O_2$	240.39	16409-46-4			129^{9}	2-06-00-00043	0.9089^{15}	H_2O 1; EtOH 3; ace 3	1.4486^{20}
3555	Butanoic acid, 3-methyl-, 2-methylpropyl ester	Isobutyl isovalerate	$C_9H_{18}O_2$	158.24	589-59-3		5029	168.5	4-02-00-00898	0.853^{20}	H_2O 1; EtOH 5; eth 5; ace 4	1.4057^{20}
3556	Butanoic acid, 3-methyl-2-oxo-		$C_5H_8O_3$	116.12	759-05-7	31.5		170.5	4-03-00-01577	0.9968^{20}	H_2O 3; EtOH 3; eth 3	1.3850^{16}
3557	Butanoic acid, 2-methyl-3-oxo-, ethyl ester		$C_7H_{12}O_3$	144.17	609-14-3			187	4-03-00-01573	0.9941^{20}	H_2O 2; EtOH 3; eth 3; ace 4	1.4185^{20}
3558	Butanoic acid, 2-methyl-3-oxo-, methyl ester		$C_6H_{10}O_3$	130.14	17094-21-2			177.4	3-03-00-01225	1.0217^{25}	eth 4; EtOH 4	1.416^{24}
3559	Butanoic acid, 1-methylpentyl ester, (*S*)-	Butyric acid (1-methylpentyl) ester (*d*)	$C_{10}H_{20}O_2$	172.27	116723-96-7			85^{20}	1-02-00-00120	0.8744^{21}	eth 4; EtOH 4	
3560	Butanoic acid, 3-methyl-, phenylmethyl ester		$C_{12}H_{16}O_2$	192.26	103-38-8			245; 136^{25}	4-06-00-02266	0.9983^{15}		1.4884^{20}
3561	Butanoic acid, 3-methyl-, 1,2,3-propanetriyl ester		$C_{18}H_{32}O_6$	344.45	620-63-3			332.5	4-02-00-00900	0.9984^{20}	eth 4; EtOH 4	1.4354^{20}
3562	Butanoic acid, 3-methyl-, propyl ester	Propyl isovalerate	$C_8H_{16}O_2$	144.21	557-00-6			155.9	4-02-00-00898	0.8617^{20}	eth 4; EtOH 4	1.4031^{20}
3563	Butanoic acid, 2-methylpropyl ester	Isobutyl butanoate	$C_8H_{16}O_2$	144.21	539-90-2		5020	156.9	4-02-00-00790	0.8364^{18}	H_2O 2; EtOH 5; eth 5	1.4032^{20}
3564	Butanoic acid, 1-methylpropyl ester, (*S*)-	Butyric acid, *sec*-butyl ester (*d*)	$C_8H_{16}O_2$	144.21	116836-55-6	-91.5		152; 54^{18}	3-02-00-00600	0.8737^{13}	EtOH 3; bz 3; CS_2 3; py 3	1.4011^{20}
3565	Butanoic acid, 3-methyl-, 1,7,7-trimethylbicyclo[2.2.1]hept-2-yl ester, (1*R*-*endo*)-	*d*-Bornyl isovalerate	$C_{15}H_{26}O_2$	238.37	53022-14-3		1342	257.5	2-06-00-08488	0.955^{25}	eth 4; EtOH 4	
3566	Butanoic acid, octyl ester		$C_{12}H_{24}O_2$	200.32	110-39-4	-55.6		244.1	4-02-00-00791	0.8629^{20}	EtOH 4	1.4267^{15}
3567	Butanoic acid, 2-oxo-		$C_4H_6O_3$	102.09	600-18-0	33		$80\text{-}2^{16}$	4-03-00-01524	1.200^{17}	H_2O 4; EtOH 4; eth 2	1.3972^{20}
3568	Butanoic acid, 3-oxo-	Acetoacetic acid	$C_4H_6O_3$	102.09	541-50-4	36.5	51	100 dec	4-03-00-01527		H_2O 4; eth 4; EtOH 4	
3569	Butanoic acid, 3-oxo-, butyl ester		$C_8H_{14}O_3$	158.20	591-60-6	-35.6		127^{50}; 85^{8}	4-03-00-01536	0.9671^{25}	H_2O 2; EtOH 5; bz 5; lig 5	1.4137^{20}
3570	Butanoic acid, 3-oxo-, 2-chloroethyl ester		$C_6H_9ClO_3$	164.59	54527-68-3			198; 107^{14}	4-03-00-01534	1.190^{20}	bz 4; eth 4; EtOH 4	1.4430^{20}
3571	Butanoic acid, 3-oxo-, ethyl ester	Ethyl acetoacetate	$C_6H_{10}O_3$	130.14	141-97-9	-45	3714	180.8	4-03-00-01528	1.0368^{10}	H_2O 4; EtOH 5; eth 5; bz 3	1.4171^{20}
3572	Butanoic acid, 4-oxo-, ethyl ester		$C_6H_{10}O_3$	130.14	10138-10-0			162; 85^{12}	4-03-00-01555	1.0490^{22}	chl 2	1.4287^{20}
3573	Butanoic acid, 3-oxo-, methyl ester	Methyl acetoacetate	$C_5H_8O_3$	116.12	105-45-3	27.5	5933	171.7	4-03-00-01527	1.0762^{20}	H_2O 4; EtOH 5; eth 5; ctc 3	1.4184^{20}
3574	Butanoic acid, 3-oxo-, 1-methylethyl ester	Isopropyl acetoacetate	$C_7H_{12}O_3$	144.17	542-08-5	-27.3	5094	186	4-03-00-01535	0.9835^{20}	eth 4; EtOH 4; lig 4	1.4173^{20}
3575	Butanoic acid, 4-oxo-4-(phenylamino)-	Succinanilic acid	$C_{10}H_{11}NO_3$	193.20	102-14-7	148.5	8839		4-12-00-00468		H_2O 2; EtOH 3; eth 4	
3576	Butanoic acid, 3-oxo-2-(phenylmethylene)-, ethyl ester		$C_{13}H_{14}O_3$	218.25	620-80-4	60.5		296; 180^{17}	3-10-00-03158		H_2O 1; EtOH 2; eth 2; bz 2	
3577	Butanoic acid, 3-oxo-, 2-propenyl ester		$C_7H_{10}O_3$	142.15	1118-84-9	-85		196; 66.5^{14}	4-03-00-01538	1.0366^{20}	H_2O 3; EtOH 5; bz 5; lig 3	1.4398^{20}
3578	Butanoic acid, 4-oxo-4-[[4-[(2-thiazolylamino)sulfonyl]phenyl]amino]-	Succinylsulphathiazole	$C_{13}H_{13}N_3O_5S_2$	355.40	116-43-8	193.5	8850		4-27-00-04637		H_2O 1; EtOH 2; eth 1; ace 2	
3579	Butanoic acid, pentyl ester	Amyl butyrate	$C_9H_{18}O_2$	158.24	540-18-1	-73.2	639	186.4	4-02-00-00790	0.8713^{15}	H_2O 1; EtOH 4; eth 4	1.4123^{20}
3580	Butanoic acid, 2-phenoxy-		$C_{10}H_{12}O_3$	180.20	13794-14-4	98		258	3-06-00-00616		H_2O 2; EtOH 3; eth 3; ace 3	

No.	Name Synonym	Mol. Form. Mol. Wt.	CAS RN mp/°C	Merck No. bp/°C	Beil. Ref. den/g cm^{-3}	Solubility n_D
3581	Butanoic acid, 4-phenoxy-, ethyl ester Ethyl 4-phenoxybutyrate	$C_{12}H_{16}O_3$ 208.26	2364-59-2	$170\text{-}3^{25}$	2-06-00-00159 1.042^{35}	eth 4; EtOH 4 1.491^{35}
3582	Butanoic acid, 2-phenoxyethyl ester	$C_{12}H_{16}O_3$ 208.26	23511-70-8	$251; 88^{2}$	3-06-00-00571 1.0388^{21}	ace 4; eth 4; EtOH 4
3583	Butanoic acid, 2-phenoxy-, phenyl ester 2-Phenoxybutyric acid, phenyl ester	$C_{16}H_{16}O_3$ 256.30	116836-19-2 48.5	$202\text{-}3^{25}$	4-06-00-00645 1.135^{15}	
3584	Butanoic acid, phenyl ester Phenyl butyrate	$C_{10}H_{12}O_2$ 164.20	4346-18-3	225	4-06-00-00615 1.0382^{15}	H_2O 1; EtOH 3; eth 3
3585	Butanoic acid, phenylmethyl ester	$C_{11}H_{14}O_2$ 178.23	103-37-7	239	4-06-00-02266 1.0111^{20}	H_2O 1; EtOH 4; eth 4; ctc 3 1.4920^{20}
3586	Butanoic acid, 1,2,3-propanetriyl ester Tributyrin	$C_{15}H_{26}O_6$ 302.37	60-01-5 -75	9532 307.5	4-02-00-00799 1.0350^{20}	H_2O 1; EtOH 3; eth 4; ace 3 1.4359^{20}
3587	Butanoic acid, 2-propenyl ester	$C_7H_{12}O_2$ 128.17	2051-78-7	$142; 44.5^{15}$	4-02-00-00793 0.9017^{20}	H_2O 1; EtOH 5; eth 5; ctc 2 1.4158^{20}
3588	Butanoic acid, propyl ester Propyl butanoate	$C_7H_{14}O_2$ 130.19	105-66-8 -95.2	7858 143.0	4-02-00-00788 0.8730^{20}	H_2O 2; EtOH 5; eth 5 1.4001^{20}
3589	Butanoic acid, 3-thioxo-, ethyl ester Butyric acid, 3-thioxo, ethyl ester	$C_6H_{10}O_2S$ 146.21	7740-33-2	75^{15}	4-03-00-01553 1.0554^{31}	eth 4; EtOH 4 1.4712^{26}
3590	Butanoic acid, 2,2,3-trichloro- 2,2,3-Trichlorobutyric acid	$C_4H_5Cl_3O_2$ 191.44	5344-55-8 60	237	3-02-00-00629	eth 4
3591	Butanoic acid, 2,2,3-trichloro-, ethyl ester Butyric acid, 2,2,3-trichloro, ethyl ester	$C_6H_9Cl_3O_2$ 219.49	116723-97-8	212	0-02-00-00281 1.3114^{20}	eth 4; EtOH 4
3592	Butanoic acid, 4,4,4-trifluoro-3-oxo-, ethyl ester	$C_6H_7F_3O_3$ 184.11	372-31-6	133	4-03-00-01548 1.2586^{15}	EtOH 3; eth 3 1.3783^{15}
3593	1-Butanol Butyl alcohol	$C_4H_{10}O$ 74.12	71-36-3 -89.8	1540 117.7	4-01-00-01506 0.8098^{20}	H_2O 3; EtOH 5; eth 5; ace 4 1.3993^{20}
3594	2-Butanol *sec*-Butyl alcohol	$C_4H_{10}O$ 74.12	78-92-2 -114.7	1541 99.5	2-01-00-00400 0.8063^{20}	H_2O 4; EtOH 5; eth 5; ace 4 1.3978^{20}
3595	1-Butanol, 2-amino-, (±)-	$C_4H_{11}NO$ 89.14	13054-87-0 -2	178	4-04-00-01705 0.9162^{20}	H_2O 5; EtOH 5; eth 5; chl 2 1.4489^{25}
3596	1-Butanol, 3-amino-	$C_4H_{11}NO$ 89.14	2867-59-6	$82\text{-}5^{19}$	4-04-00-01710	H_2O 4; EtOH 4 1.4534^{25}
3597	1-Butanol, 4-amino-	$C_4H_{11}NO$ 89.14	13325-10-5	$205; 125^{34}$	4-04-00-01711 0.967^{12}	H_2O 3; EtOH 3; eth 1 1.4625^{20}
3598	2-Butanol, 3-amino-	$C_4H_{11}NO$ 89.14	42551-55-3 19	$160; 79^{20}$	3-04-00-00779 0.9399^{20}	H_2O 4; eth 4; EtOH 4 1.4445^{25}
3599	1-Butanol, 2-amino-3-methyl-, (±)-	$C_5H_{13}NO$ 103.16	16369-05-4	$75\text{-}7^{8}$	4-04-00-01777 0.93^{25}	1.4543^{20}
3600	1-Butanol, 3-bromo-	C_4H_9BrO 153.02	6089-12-9	$69\text{-}71^{20}$	4-01-00-01553 1.7107^{18}	1.5010^{21}
3601	2-Butanol, 3-bromo-, (*R**,*S**)-(±)- 2-Butanol, 3-bromo-, *erythro*-(±)-	C_4H_9BrO 153.02	19246-39-0	$154; 53^{13}$	4-01-00-01580 1.4550^{20}	eth 4; EtOH 4 1.4780^{20}
3602	1-Butanol, 2-chloro-	C_4H_9ClO 108.57	26106-95-6	$74\text{-}6^{25}$	4-01-00-01549 1.062^{25}	eth 4; EtOH 4 1.4438^{20}
3603	1-Butanol, 3-chloro-	C_5H_9ClO 120.58	2203-35-2	$175; 66^{15}$	3-01-00-01517 1.0883^{20}	EtOH 3; eth 3; ace 3; ctc 2 1.4518^{20}
3604	1-Butanol, 4-chloro-	C_4H_9ClO 108.57	928-51-8	$84\text{-}5^{16}$	4-01-00-01550 1.0883^{20}	eth 4; EtOH 4 1.4518^{20}
3605	2-Butanol, 1-chloro- α-Butylene chlorohydrin	C_4H_9ClO 108.57	1873-25-2	141	4-01-00-01578 1.068^{25}	EtOH 3; eth 3 1.4400^{20}
3606	2-Butanol, 3-chloro- β-Butylene chlorohydrin	C_4H_9ClO 108.57	563-84-8	139	3-01-00-01538 1.0669^{20}	EtOH 4; eth 4 1.4432^{20}
3607	2-Butanol, 4-chloro-	C_4H_9ClO 108.57	2203-34-1	67^{20}	4-01-00-01578	eth 4; EtOH 4 1.4408^{20}
3608	2-Butanol, 1-chloro-2,3-dimethyl- 1-Chloro-2,3-dimethyl-2-butanol	$C_6H_{13}ClO$ 136.62	66235-62-9	163	4-01-00-01730 1.049^{20}	1.4459^{20}
3609	2-Butanol, 1-chloro-3,3-dimethyl- 1-Chloro-3,3-dimethyl-2-butanol	$C_6H_{13}ClO$ 136.62	36402-31-0 13	157.5	4-01-00-01728 1.0063^{25}	1.4432^{25}
3610	2-Butanol, 1-chloro-2-methyl-	$C_5H_{11}ClO$ 122.59	74283-48-0	151	4-01-00-01673 1.0161^{20}	EtOH 4 1.4469^{20}
3611	2-Butanol, 3-chloro-2-methyl-	$C_5H_{11}ClO$ 122.59	21326-62-5	141.5	3-01-00-01627 1.0295^{20}	eth 4; EtOH 4 1.4436^{20}
3612	1-Butanol, 2,3-dibromo- 2,3-Dibromo-1-butanol	$C_4H_8Br_2O$ 231.91	4021-75-4 32	99.5^{10}	4-01-00-01554 1.9475^{20}	1.5442^{20}
3613	1-Butanol, 3,4-dibromo- 3,4-Dibromobutanol	$C_4H_8Br_2O$ 231.91	87018-30-2	114^{11}	3-01-00-01518 1.98^{15}	1.548^{15}
3614	2-Butanol, 1,4-dibromo- 2-Hydroxy-1,4-dibromobutane	$C_4H_8Br_2O$ 231.91	19398-47-1	$114\text{-}5^{15}$	4-01-00-01581 1.953^{20}	1.544^{20}
3615	1-Butanol, 2,2-dimethyl- 2,2-Dimethyl-1-butanol	$C_6H_{14}O$ 102.18	1185-33-7 <-15	136.5	4-01-00-01726 0.8283^{20}	H_2O 2; EtOH 3; eth 3 1.4208^{20}
3616	1-Butanol, 2,3-dimethyl-, (±)-	$C_6H_{14}O$ 102.18	20281-85-0	144.5	4-01-00-01729 0.8297^{20}	EtOH 3; eth 3; ace 3 1.4195^{20}
3617	1-Butanol, 3,3-dimethyl-	$C_6H_{14}O$ 102.18	624-95-3 -60	143	4-01-00-01729 0.844^{15}	H_2O 2; EtOH 3; eth 3; ace 3 1.4323^{15}
3618	2-Butanol, 2,3-dimethyl-	$C_6H_{14}O$ 102.18	594-60-5 -14	118.4	4-01-00-01729 0.8236^{20}	H_2O 3; EtOH 5; eth 5 1.4176^{20}
3619	2-Butanol, 3,3-dimethyl-, (±)- 3,3-Dimethylbutan-2-ol (*DL*-)	$C_6H_{14}O$ 102.18	20281-91-8 5.6	120.4	4-01-00-01727 0.8122^{25}	H_2O 2; EtOH 3; eth 5 1.4148^{20}
3620	1-Butanol, 2-ethyl- 2-Ethyl-1-butanol	$C_6H_{14}O$ 102.18	97-95-0 <-15	147	4-01-00-01725 0.8326^{20}	H_2O 2; EtOH 3; eth 3; chl 3 1.4220^{20}

PHYSICAL CONSTANTS OF ORGANIC COMPOUNDS (continued)

No.	Name Synonym	Mol. Form. Mol. Wt.	CAS RN mp/°C	Merck No. bp/°C	Beil. Ref. den/g cm^{-3}	Solubility n_D
3621	1-Butanol, 4-fluoro- 4-Fluoro-1-butanol	C_4H_9FO 92.11	372-93-0	 58^{15}	4-01-00-01546	ace 4; eth 4; EtOH 4 1.3942^{15}
3622	1-Butanol, 2,2,3,3,4,4,4-heptafluoro-	$C_4H_3F_7O$ 200.06	375-01-9	 95	4-01-00-01547 1.600^{20}	EtOH 3; ace 3 1.294^{20}
3623	2-Butanol, 1-[(2-hydroxyethyl)amino]- 2-Hydroxyethyl 2-hydroxybutyl amine	$C_6H_{15}NO_2$ 133.19	6967-43-7	 137^{9}	4-04-00-01725 1.0310^{20}	H_2O 4; ace 4; EtOH 4 1.4690^{30}
3624	2-Butanol, 3,3'-iminobis- Bis-(3-hydroxy-2-butyl)amine	$C_8H_{19}NO_2$ 161.24	6959-06-4	 $112\text{-}5^{3}$	3-04-00-00781 0.9775^{20}	H_2O 4; ace 4; EtOH 4 1.4162^{20}
3625	1-Butanol, 3-methoxy- 3-Methoxy-1-butanol	$C_5H_{12}O_2$ 104.15	2517-43-3	 157	4-01-00-02509 0.923^{23}	EtOH 4; eth 3; ace 4; chl 2 1.4148^{25}
3626	1-Butanol, 2-methyl- 2-Methyl-1-butanol	$C_5H_{12}O$ 88.15	137-32-6	5952 128	4-01-00-01666 0.8150^{25}	H_2O 2; EtOH 5; eth 5; ace 4 1.4092^{20}
3627	1-Butanol, 3-methyl- Isopentyl alcohol	$C_5H_{12}O$ 88.15	123-51-3 -117.2	5081 131.1	4-01-00-01677 0.8104^{20}	ace 4; eth 4; EtOH 4 1.4053^{20}
3628	2-Butanol, 2-methyl- *tert*-Amyl alcohol	$C_5H_{12}O$ 88.15	75-85-4 -8.8	7096 102.4	4-01-00-01668 0.8096^{20}	H_2O 3; EtOH 5; eth 5; ace 4 1.4052^{20}
3629	2-Butanol, 3-methyl-, (±)- (±)-3-Methyl-2-butanol	$C_5H_{12}O$ 88.15	70116-68-6	5953 112.9	4-01-00-01675 0.8180^{20}	H_2O 2; EtOH 5; eth 5; ace 4 1.4089^{20}
3630	1-Butanol, 2-methyl-, acetate	$C_7H_{14}O_2$ 130.19	624-41-9	 140	4-02-00-00156 0.8740^{20}	ace 4; eth 4; EtOH 4 1.4040^{20}
3631	1-Butanol, 3-methyl-, acetate Isopentyl acetate	$C_7H_{14}O_2$ 130.19	123-92-2 -78.5	4993 142.5	4-02-00-00157 0.876^{15}	H_2O 2; EtOH 5; eth 5; ace 3 1.4000^{20}
3632	1-Butanol, 3-methyl-, benzoate Isoamyl benzoate	$C_{12}H_{16}O_2$ 192.26	94-46-2	4995 261	4-09-00-00292 0.993^{15}	EtOH 4
3633	1-Butanol, 3-methyl-, carbamate	$C_6H_{13}NO_2$ 131.17	543-86-2 59	 220	4-03-00-00058 0.9438^{71}	eth 4; EtOH 4 1.4175^{20}
3634	2-Butanol, 2-methyl-, carbamate *tert*-Amyl carbamate	$C_6H_{13}NO_2$ 131.17	590-60-3 86	641	1-03-00-00014	ace 4; bz 4
3635	1-Butanol, 3-methyl-, carbonate (2:1)	$C_{11}H_{22}O_3$ 202.29	2050-95-5	 234; 122^{16}	3-03-00-00011 0.9067^{20}	H_2O 1 1.4174^{20}
3636	1-Butanol, 3-methyl-, formate Isopentyl formate	$C_6H_{12}O_2$ 116.16	110-45-2 -93.5	5001 123.5	4-02-00-00030 0.877^{20}	H_2O 2; EtOH 3; eth 5; ctc 2 1.3967^{20}
3637	1-Butanol, 3-methyl-, nitrate Isoamyl nitrate	$C_5H_{11}NO_3$ 133.15	543-87-3	5004 148	4-01-00-01683 0.996^{22}	 1.4122^{21}
3638	1-Butanol, 3-methyl-2-nitro-	$C_5H_{11}NO_3$ 133.15	77392-54-2	 111^{10}	3-01-00-01645 1.0889^{25}	 1.4430^{20}
3639	1-Butanol, 3-methyl-, propanoate Isopentyl propanoate	$C_8H_{16}O_2$ 144.21	105-68-0	 160.2	4-02-00-00709 0.8697^{20}	eth 4; EtOH 4 1.4069^{20}
3640	1-Butanol, 2-nitro- 2-Nitro-1-butanol	$C_4H_9NO_3$ 119.12	609-31-4 -47	 105^{10}	4-01-00-01555 1.1332^{25}	H_2O 3; EtOH 5; eth 5; ace 3 1.4390^{20}
3641	2-Butanol, 1-nitro- 2-Hydroxy-1-nitrobutane	$C_4H_9NO_3$ 119.12	3156-74-9	 204; 75^{2}	4-01-00-01582 1.1353^{20}	H_2O 3; EtOH 4; eth 4; ace 3 1.4435^{20}
3642	2-Butanol, 3-nitro- 3-Nitro-2-butanol	$C_4H_9NO_3$ 119.12	6270-16-2	 112^{38}; $55^{0.5}$	4-01-00-01583 1.1166^{7}	 1.4414^{20}
3643	2-Butanol, 1,1,1-trichloro- Ethyl(trichloromethyl)carbinol	$C_4H_7Cl_3O$ 177.46	6111-61-1	 171; 83^{22}	4-01-00-01579 1.3630^{25}	ace 4; bz 4; eth 4; EtOH 4 1.4800^{20}
3644	2-Butanol, 2,3,3-trimethyl-	$C_7H_{16}O$ 116.20	594-83-2 17	 131	4-01-00-01755 0.8380^{25}	ace 4; eth 4; EtOH 4 1.4233^{22}
3645	2-Butanone Methyl ethyl ketone	C_4H_8O 72.11	78-93-3 -86.65	5991 79.59	4-01-00-03243 0.7999^{25}	H_2O 4; EtOH 5; eth 5; ace 5 1.3788^{20}
3646	2-Butanone, 1-chloro- Chloromethyl ethyl ketone	C_4H_7ClO 106.55	616-27-3	 137.5	4-01-00-03255 1.0850^{20}	MeOH 4 1.4372^{20}
3647	2-Butanone, 3-chloro-	C_4H_7ClO 106.55	4091-39-8	 115	4-01-00-03256 1.0554^{25}	 1.4219^{20}
3648	2-Butanone, 4-chloro- 1-Chloro-3-butanone	C_4H_7ClO 106.55	6322-49-2	 120 dec	4-01-00-03256 1.0680^{23}	eth 4; EtOH 4 1.4284^{23}
3649	2-Butanone, 3-chloro-3-methyl- 3-Chloro-3-methyl-2-butanone	C_5H_9ClO 120.58	5950-19-6	 117.2	3-01-00-02818 1.0083^{20}	eth 4; EtOH 4 1.4204^{20}
3650	1-Butanone, 4-chloro-1-phenyl-	$C_{10}H_{11}ClO$ 182.65	939-52-6 19.5	 $130\text{-}3^{4}$	4-07-00-00711 1.137^{25}	 1.5459^{20}
3651	1-Butanone, 4-[4-(4-chlorophenyl)-4-hydroxy- 1-piperidinyl]-1-(4-fluorophenyl)- Haloperidol	$C_{21}H_{23}ClFNO_2$ 375.87	52-86-8 151.5	4511	5-21-02-00377	
3652	1-Butanone, 1-cyclohexyl-	$C_{10}H_{18}O$ 154.25	1462-27-7	 94^{13}	4-07-00-00080 0.903^{20}	 1.4537^{20}
3653	2-Butanone, 3,4-dibromo- 1,2-Dibromoethyl methyl ketone	$C_4H_6Br_2O$ 229.90	25109-57-3	 $80\text{-}1^{10}$	3-01-00-02788 1.9662^{15}	 1.5314^{15}
3654	2-Butanone, 1,3-dichloro- 1,3-Dichloro-2-butanone	$C_4H_6Cl_2O$ 141.00	16714-77-5	 100.5	3-01-00-02786 1.3116^{20}	ace 4; bz 4; eth 4; EtOH 4 1.4686^{20}
3655	2-Butanone, 3,4-dichloro-	$C_4H_6Cl_2O$ 141.00	58625-77-7	 $65\text{-}70^{16}$	3-01-00-02787 1.293^{18}	 1.4628^{18}
3656	2-Butanone, 4-(diethylamino)-	$C_8H_{17}NO$ 143.23	3299-38-5	 84^{30}	4-04-00-01930 0.8630^{20}	ace 4; bz 4; eth 4; EtOH 4 1.4333^{24}
3657	2-Butanone, 3,3-dimethyl- Pinacolone	$C_6H_{12}O$ 100.16	75-97-8 -52.5	7409 106.1	4-01-00-03310 0.7229^{25}	H_2O 2; EtOH 3; eth 3; ace 3 1.3952^{20}
3658	2-Butanone, 3,3-diphenyl- 2,2-Diphenyl-3-butanone	$C_{16}H_{16}O$ 224.30	2575-20-4 41	 310.5	4-07-00-01451 1.069^{20}	H_2O 1; EtOH 3; eth 4; chl 4 1.5748^{20}
3659	2-Butanone, 4-(2-furanyl)-	$C_8H_{10}O_2$ 138.17	699-17-2	 203	4-17-00-04559 1.0361^{19}	 1.4696^{17}
3660	2-Butanone, 1-hydroxy-	$C_4H_8O_2$ 88.11	5077-67-8	 160; 78^{60}	4-01-00-03989 1.0272^{20}	H_2O 4; EtOH 4; eth 4 1.4189^{20}

No.	Name Synonym	Mol. Form. Mol. Wt.	CAS RN mp/°C	Merck No. bp/°C	Beil. Ref. den/g cm^{-3}	Solubility n_D
3661	2-Butanone, 3-hydroxy- Acetoin	$C_4H_8O_2$ 88.11	513-86-0 15	55 143	2-01-00-00870 1.0044^{20}	H_2O 4; ace 4; EtOH 4 1.4171^{20}
3662	2-Butanone, 4-hydroxy-	$C_4H_8O_2$ 88.11	590-90-9	 109^{30}; 90^{11}	4-01-00-03994 1.1089^{24}	H_2O 5; EtOH 5; eth 5; ace 4 1.4585^{14}
3663	2-Butanone, 4-(4-hydroxy-3-methoxyphenyl)- Zingerone	$C_{11}H_{14}O_3$ 194.23	122-48-5 40.5	10072 $187\text{-}8^{14}$	4-08-00-01866	eth 4
3664	2-Butanone, 4-hydroxy-3-methyl-	$C_5H_{10}O_2$ 102.13	3393-64-4	 $90\text{-}5^{15}$	4-01-00-04009 0.993^{25}	 1.4340^{20}
3665	1-Butanone, 1-(2-hydroxyphenyl)-	$C_{10}H_{12}O_2$ 164.20	2887-61-8 10	 $124\text{-}6^{14}$	4-08-00-00497 1.0683^{24}	 1.5379^{25}
3666	1-Butanone, 2-hydroxy-1-phenyl- Butyrophenone, 2-hydroxy-	$C_{10}H_{12}O_2$ 164.20	16183-46-3	 $132\text{-}3^{12}$	4-08-00-00502 1.077^{19}	 1.529^{19}
3667	1-Butanone, 4-hydroxy-1-phenyl-	$C_{10}H_{12}O_2$ 164.20	39755-03-8 30	 $187\text{-}8^{9}$	3-08-00-00449	EtOH 4
3668	2-Butanone, 3-methyl- Methyl isopropyl ketone	$C_5H_{10}O$ 86.13	563-80-4 -92	 94.3	4-01-00-03287 0.8051^{20}	H_2O 2; EtOH 5; eth 5; ace 4 1.3880^{20}
3669	1-Butanone, 3-methyl-1-(2-methylphenyl)- Isobutyl 2-tolyl ketone	$C_{12}H_{16}O$ 176.26	58138-81-1	 248; $85\text{-}6^{3}$	4-07-00-00765 0.9578^{20}	 1.5104^{20}
3670	1-Butanone, 3-methyl-1-(4-methylphenyl)- Isobutyl 4-tolyl ketone	$C_{12}H_{16}O$ 176.26	61971-91-3	 259; 137^{18}	4-07-00-00765 0.9557^{20}	 1.5081^{20}
3671	1-Butanone, 3-methyl-1-phenyl-	$C_{11}H_{14}O$ 162.23	582-62-7	 236.5	4-07-00-00739 0.9701^{16}	H_2O 1; EtOH 5; eth 5; ace 4 1.5139^{15}
3672	1-Butanone, 1-(4-methylphenyl)- Butyrophenone, 4-methyl	$C_{11}H_{14}O$ 162.23	4160-52-5 12	 251.5	4-07-00-00743 0.9745^{20}	eth 4; EtOH 4 1.5232^{20}
3673	2-Butanone, (1-methylpropylidene)hydrazone	$C_8H_{16}N_2$ 140.23	5921-54-0	 171.5	4-01-00-03252 0.8404^{20}	 1.4511^{20}
3674	1-Butanone, 1-(1-naphthalenyl)- 1-Butyronaphthone	$C_{14}H_{14}O$ 198.26	2876-62-2	 317	4-07-00-01305 1.0861^{25}	bz 4; eth 4; EtOH 4 1.596^{27}
3675	2-Butanone, oxime	C_4H_9NO 87.12	96-29-7 -29.5	 152.5	4-01-00-03250 0.9232^{20}	H_2O 3; EtOH 5; eth 5; chl 3 1.4410^{20}
3676	1-Butanone, 1-phenyl-	$C_{10}H_{12}O$ 148.20	495-40-9 12	 228.5	4-07-00-00709 0.988^{20}	H_2O 1; EtOH 5; eth 5; ace 4 1.5203^{20}
3677	2-Butanone, 1-phenyl-	$C_{10}H_{12}O$ 148.20	1007-32-5	 230; 111^{16}	4-07-00-00712 1.002^{20}	H_2O 1; EtOH 3; eth 5; ace 4
3678	2-Butanone, 4-phenyl-	$C_{10}H_{12}O$ 148.20	2550-26-7 -13	 233.5	4-07-00-00713 0.9849^{22}	H_2O 1; EtOH 3; eth 3; ace 4 1.511^{22}
3679	2-Butanone, 1,3,4-trihydroxy-, (*S*)- *L*-Erythrulose	$C_4H_8O_4$ 120.11	533-50-6 dec	3641 dec	4-01-00-04176	H_2O 4; EtOH 4
3680	Butanoyl bromide Butyryl bromide	C_4H_7BrO 151.00	5856-82-6	 129	4-02-00-00804 1.4162^{17}	 1.596^{17}
3681	Butanoyl chloride Butyryl chloride	C_4H_7ClO 106.55	141-75-3 -89	1598 102	4-02-00-00803 1.0277^{20}	eth 5 1.4121^{20}
3682	Butanoyl chloride, 2-bromo-	C_4H_6BrClO 185.45	22118-12-3	 151	4-02-00-00834 1.5320^{20}	eth 4 1.5320^{20}
3683	Butanoyl chloride, 2-chloro-	$C_4H_6Cl_2O$ 141.00	7623-11-2	 130.5	3-02-00-00623 1.2360^{17}	eth 4 1.4475^{20}
3684	Butanoyl chloride, 3-chloro- 3-Chlorobutyryl chloride	$C_4H_6Cl_2O$ 141.00	1951-11-7	 $40\text{-}1^{12}$	4-02-00-00824 1.2163^{20}	 1.4509^{20}
3685	Butanoyl chloride, 4-chloro-	$C_4H_6Cl_2O$ 141.00	4635-59-0	 173.5	4-02-00-00826 1.2581^{20}	eth 3 1.4616^{20}
3686	Butanoyl chloride, 2-chloro-3-methyl- Butyryl chloride, 2-chloro-3-methyl	$C_5H_8Cl_2O$ 155.02	35383-59-8	 148.5; 56^{20}	4-02-00-00904 1.135^{13}	eth 4
3687	Butanoyl chloride, 4-chloro-3-oxo- Butyryl chloride, 4-chloro-3-oxo	$C_4H_4Cl_2O_2$ 154.98	41295-64-1	 $117\text{-}9^{17}$	3-03-00-01207 1.4397^{20}	bz 4 1.4860^{20}
3688	Butanoyl chloride, 2,2-dimethyl-	$C_6H_{11}ClO$ 134.61	5856-77-9	 132	4-02-00-00955 0.9810^{20}	eth 4 1.4245^{20}
3689	Butanoyl chloride, 2,3-dimethyl- Butyryl chloride, 2,3-dimethyl	$C_6H_{11}ClO$ 134.61	51760-90-8	 136; 38^{18}	3-02-00-00761 0.9795^{20}	eth 4
3690	Butanoyl chloride, 3,3-dimethyl-	$C_6H_{11}ClO$ 134.61	7065-46-5	 130; 68^{100}	4-02-00-00956 0.9696^{20}	eth 4 1.4210^{20}
3691	Butanoyl chloride, 2-ethyl-	$C_6H_{11}ClO$ 134.61	2736-40-5	 140	4-02-00-00952 0.9825^{20}	eth 4 1.4234^{20}
3692	Butanoyl chloride, heptafluoro-	C_4ClF_7O 232.48	375-16-6	 38.5	4-02-00-00818 1.55^{20}	 1.288^{20}
3693	Butanoyl chloride, 3-methyl- Isovaleryl chloride	C_5H_9ClO 120.58	108-12-3	5121 114	4-02-00-00901 0.9844^{20}	eth 3 1.4149^{20}
3694	Butanoyl chloride, 2-methyl-, (±)- Butyryl chloride, 2-methyl (*DL*)	C_5H_9ClO 120.58	57526-28-0	 116	4-02-00-00891 0.9917^{20}	 1.4170^{20}
3695	2-Butenal, (*E*)- *trans*-Crotonaldehyde	C_4H_6O 70.09	123-73-9 -76	2599 102.2	4-01-00-03447 0.8516^{20}	H_2O 3; EtOH 4; eth 4; ace 4 1.4366^{20}
3696	2-Butenal, 2-bromo-	C_4H_5BrO 148.99	24247-53-8	 63^{15}	3-01-00-02981 1.5797^{13}	 1.5184^{20}
3697	2-Butenal, 2-chloro- α-Chlorocrotonaldehyde	C_4H_5ClO 104.54	53175-28-3	 147.5	3-01-00-02981 1.1404^{23}	H_2O 2; EtOH 3; eth 3; ctc 3 1.4780^{25}
3698	2-Butenal, 2-methyl-	C_5H_8O 84.12	1115-11-3	 117; 64^{110}	3-01-00-02990 0.8710^{20}	H_2O 3; EtOH 5; eth 5 1.4475^{20}
3699	2-Butenal, 2-methyl-, (*E*)-	C_5H_8O 84.12	497-03-0	 117; 64^{119}	4-01-00-03464 0.8710^{20}	H_2O 4; EtOH 4 1.4475^{20}
3700	2-Butenal, 3-methyl- Senecialdehyde	C_5H_8O 84.12	107-86-8	8399 134	4-01-00-03464 0.8722^{20}	H_2O 3; EtOH 3; eth 3 1.4528^{20}

No.	Name Synonym	Mol. Form. Mol. Wt.	CAS RN mp/°C	Merck No. bp/°C	Beil. Ref. den/g cm^{-3}	Solubility n_D
3701	2-Butenal, 3-phenyl-	$C_{10}H_{10}O$ 146.19	1196-67-4 95	124^{12}; 109^{4}	4-07-00-01011	1.5876^{23}
3702	2-Butenamide, (*E*)-	C_4H_7NO 85.11	625-37-6 161.5	sub 140	2-02-00-00392 0.9461^{120}	H_2O 2; EtOH 3; eth 2; bz 3 1.4420^{165}
3703	3-Butenamide, 3-methyl-4-phenyl- β-Benzalbutyramide	$C_{11}H_{13}NO$ 175.23	7236-47-7 133	1063		
3704	1-Butene 1-Butylene	C_4H_8 56.11	106-98-9 -185.3	1513 -6.2	4-01-00-00765 $*0.588^{25}$	H_2O 1; EtOH 4; eth 4; bz 3 1.3962^{20}
3705	2-Butene, (*E*)- *trans*-2-Butene	C_4H_8 56.11	624-64-6 -105.5	1514 0.8	4-01-00-00781 $*0.599^{25}$	bz 3 1.3848^{-25}
3706	2-Butene, (*Z*)- *cis*-2-Butene	C_4H_8 56.11	590-18-1 -138.9	1514 3.7	4-01-00-00778 $*0.616^{25}$	H_2O 1; EtOH 4; eth 4; bz 3 1.3931^{-25}
3707	1-Butene, 1-bromo-, (*E*)- *trans*-1-Butenyl bromide	C_4H_7Br 135.00	32620-08-9 -100.3	94.7	1.3209^{15}	H_2O 1; eth 3; ace 3; bz 3 1.4527^{20}
3708	1-Butene, 1-bromo-, (*Z*)-	C_4H_7Br 135.00	31849-78-2	86.1	1.3265^{15}	H_2O 1; eth 3; ace 3; bz 3 1.4536^{20}
3709	1-Butene, 2-bromo- 2-Bromo-1-butene	C_4H_7Br 135.00	23074-36-4 -133.4	88	4-01-00-00775 1.3209^{15}	H_2O 1; eth 3; ace 3; bz 3 1.4527^{20}
3710	1-Butene, 3-bromo- α-Methallyl bromide	C_4H_7Br 135.00	22037-73-6	7^{10}	4-01-00-00775 1.2998^{25}	1.4618^{20}
3711	1-Butene, 4-bromo-	C_4H_7Br 135.00	5162-44-7	98.5	4-01-00-00775 1.3230^{20}	bz 4; eth 4; EtOH 4 1.4622^{20}
3712	2-Butene, 1-bromo-	C_4H_7Br 135.00	4784-77-4	104.5	4-01-00-00789 1.3371^{25}	H_2O 1; EtOH 3; eth 3; bz 4 1.4822^{20}
3713	2-Butene, 2-bromo-, (*E*)-	C_4H_7Br 135.00	3017-71-8 -114.6	85.6	4-01-00-00790 1.3323^{15}	H_2O 1; EtOH 3; eth 3; bz 4 1.4602^{16}
3714	2-Butene, 2-bromo-, (*Z*)-	C_4H_7Br 135.00	3017-68-3 -111.5	93.9	4-01-00-00790 1.3416^{15}	H_2O 1; EtOH 3; eth 3; bz 4 1.4631^{19}
3715	1-Butene, 4-bromo-3-chloro-3,4,4-trifluoro- 4-Bromo-3-chloro-3,4,4-trifluoro-1-butene	$C_4H_3BrClF_3$ 223.42	374-25-4	99.5	4-01-00-00776 1.678^{25}	1.4092^{25}
3716	1-Butene, 2-bromo-3-methyl- 2-Bromo-3-methyl-1-butene	C_5H_9Br 149.03	31844-96-9	105	3-01-00-00800 1.2328^{20}	bz 4; eth 4; chl 4 1.4504^{20}
3717	2-Butene, 1-bromo-3-methyl-	C_5H_9Br 149.03	870-63-3	131 dec; 50^{40}	4-01-00-00824 1.2817^{20}	ace 4; bz 4; eth 4; EtOH 4 1.4930^{15}
3718	2-Butene, 2-bromo-3-methyl-	C_5H_9Br 149.03	3017-70-7	119.5	4-01-00-00824 1.2773^{20}	ace 4; eth 4; chl 4 1.4738^{20}
3719	2-Butene, 2-bromo-3-phenyl- 2-Bromo-3-phenyl-2-butene	$C_{10}H_{11}Br$ 211.10	90841-14-8	$120\text{-}30^{11}$	4-05-00-01379 1.3348^{20}	bz 4; eth 4; chl 4 1.5811^{20}
3720	1-Butene, 1-chloro-, (*E*)- *trans*-1-Chloro-1-butene	C_4H_7Cl 90.55	7611-87-2	68	3-01-00-00723 0.9205^{15}	H_2O 1; EtOH 5; eth 4; ace 3 1.4223^{15}
3721	1-Butene, 1-chloro-, (*Z*)- *cis*-1-Chloro-1-butene	C_4H_7Cl 90.55	7611-86-1	63.5	3-01-00-00723 0.9153^{15}	ace 4; eth 4; EtOH 4; chl 4 1.4194^{15}
3722	1-Butene, 2-chloro- 2-Chloro-1-butene	C_4H_7Cl 90.55	2211-70-3	58.5	4-01-00-00769 0.9107^{15}	ace 4; bz 4; eth 4; EtOH 4 1.4165^{21}
3723	1-Butene, 3-chloro- 3-Chloro-1-butene	C_4H_7Cl 90.55	563-52-0	2131 64.5	4-01-00-00769 0.8978^{20}	eth 4; ace 4; chl 3 1.4149^{20}
3724	1-Butene, 4-chloro-	C_4H_7Cl 90.55	927-73-1	75	4-01-00-00771 0.9211^{20}	ace 4; eth 4; chl 4 1.4233^{20}
3725	2-Butene, 1-chloro-, (*E*)-	C_4H_7Cl 90.55	4894-61-5	85	4-01-00-00784 0.9295^{20}	H_2O 1; ace 3; chl 3 1.4350^{20}
3726	2-Butene, 1-chloro-, (*Z*)-	C_4H_7Cl 90.55	4628-21-1	84.1	4-01-00-00783 0.9426^{20}	H_2O 1; EtOH 3; ace 3; chl 3 1.4390^{20}
3727	2-Butene, 2-chloro-, (*E*)-	C_4H_7Cl 90.55	2211-68-9 -105.8	62.8	4-01-00-00785 0.9138^{20}	H_2O 1; EtOH 5; ace 3; chl 3 1.4190^{20}
3728	2-Butene, 2-chloro-, (*Z*)-	C_4H_7Cl 90.55	2211-69-0 -117.3	70.6	4-01-00-00785 0.9230^{20}	H_2O 1; EtOH 5; ace 3; chl 3 1.4240^{20}
3729	1-Butene, 3-chloro-2-(chloromethyl)- 3-Chloro-2-(chloromethyl)-1-butene	$C_5H_8Cl_2$ 139.02	69295-21-2	155	3-01-00-00787 1.1233^{20}	ace 4; chl 4 1.4724^{20}
3730	2-Butene, 2-chloro-1,1-dimethoxy- Crotonaldehyde-2-chloro, dimethylacetal	$C_6H_{11}ClO_2$ 150.60	108365-83-9	58^{13}	3-01-00-02981 1.074^{18}	1.4466^{18}
3731	2-Butene, 1-chloro-2,3-dimethyl- 1-Chloro-2,3-dimethyl-2-butene	$C_6H_{11}Cl$ 118.61	37866-06-1	112	4-01-00-00856 0.9355^{20}	ace 4; eth 4; chl 4 1.4605^{20}
3732	1-Butene, 1-chloro-2-methyl- 1-Chloro-2-methyl-1-butene	C_5H_9Cl 104.58	23378-11-2	96.5	3-01-00-00787 0.9170^{20}	ace 4; eth 4 1.4141^{20}
3733	1-Butene, 1-chloro-3-methyl- 1-Chloro-3-methyl-1-butene	C_5H_9Cl 104.58	23010-00-6	87	3-01-00-00799	ace 4; eth 4; chl 4 1.4229^{20}
3734	1-Butene, 3-chloro-2-methyl- 3-Chloro-2-methyl-1-butene	C_5H_9Cl 104.58	5166-35-8	94	4-01-00-00810 0.9088^{20}	ace 4; eth 4; chl 4 1.4304^{20}
3735	2-Butene, 1-chloro-2-methyl- 1-Chloro-2-methyl-2-butene	C_5H_9Cl 104.58	13417-43-1	110	4-01-00-00822 0.9327^{20}	ace 4; eth 4; EtOH 4 1.4481^{20}
3736	2-Butene, 1-chloro-3-methyl-	C_5H_9Cl 104.58	503-60-6	109	4-01-00-00823 0.9273^{20}	ace 4; eth 4; EtOH 4; chl 4 1.4485^{20}
3737	2-Butene, 2-chloro-3-methyl- 2-Chloro-3-methyl-2-butene	C_5H_9Cl 104.58	17773-65-8	96	3-01-00-00794 0.9324^{20}	ace 4; eth 4; EtOH 4; chl 4 1.4320^{20}
3738	1-Butene, 1,1-dibromo- 1,1-Dibromo-1-butene	$C_4H_6Br_2$ 213.90	73383-24-1	$53\text{-}5^{22}$	3-01-00-00727 1.8315^{20}	1.5168^{20}
3739	1-Butene, 1,2-dibromo- 1,2-dibromo-1-butene	$C_4H_6Br_2$ 213.90	55030-56-3 -49.5	150	0-01-00-00204 1.887^{25}	eth 4; EtOH 4
3740	1-Butene, 3,4-dibromo- 1,2-Dibromo-3-butene	$C_4H_6Br_2$ 213.90	10463-48-6	52^{10}	4-01-00-00776 1.865^{24}	1.541^{21}

No.	Name	Synonym	Mol. Form.	Mol. Wt.	CAS RN	mp/°C	Merck No.	bp/°C	Beil. Ref.	den/g cm^{-3}	Solubility	n_D
3741	2-Butene, 1,3-dibromo-	2,4-Dibromo-2-butene	$C_4H_6Br_2$	213.90	64930-16-1			168 dec	3-01-00-00746	1.8768^{20}		1.5485^{20}
3742	2-Butene, 1,4-dibromo-, (*E*)-		$C_4H_6Br_2$	213.90	821-06-7	53.4		203; 74^{14}	4-01-00-00791		H_2O 2; EtOH 4; ace 3; chl 2	
3743	1-Butene, 1,3-dichloro-	1,3-Dichloro-1-butene	$C_4H_6Cl_2$	125.00	52497-07-1			125	4-01-00-00772	1.1341^{24}	ace 4; eth 4; EtOH 4; chl 4	1.4647^{20}
3744	1-Butene, 2,3-dichloro-	2,3-Dichloro-1-butene	$C_4H_6Cl_2$	125.00	7013-11-8			112	4-01-00-00772	1.1340^{20}	ace 4; eth 4; chl 4	1.4580^{20}
3745	1-Butene, 3,4-dichloro-	1,2-Dichloro-3-butene	$C_4H_6Cl_2$	125.00	760-23-6	-61		116	4-01-00-00772	1.1170^{20}	H_2O 1; EtOH 3; eth 3; bz 4	1.4641^{20}
3746	2-Butene, 1,1-dichloro-	1,1-Dichloro-2-butene	$C_4H_6Cl_2$	125.00	56800-09-0			126	3-01-00-00741	1.1310^{20}		1.466^{18}
3747	2-Butene, 1,2-dichloro-	1,2-Dichloro-2-butene	$C_4H_6Cl_2$	125.00	13602-13-6			130.5	3-01-00-00741	1.1601^{20}	ace 4; eth 4; chl 4	1.4734^{20}
3748	2-Butene, 1,3-dichloro-, (*E*)-		$C_4H_6Cl_2$	125.00	7415-31-8			131; 53^{50}	4-01-00-00786	1.1585^{20}	ace 4; bz 4; eth 4; EtOH 4	1.4719^{20}
3749	2-Butene, 1,3-dichloro-, (*Z*)-		$C_4H_6Cl_2$	125.00	10075-38-4			131; 34^{20}	4-01-00-00786	1.1605^{20}	ace 4; bz 4; eth 4; EtOH 4	1.4735^{20}
3750	2-Butene, 1,4-dichloro-, (*E*)-	1,4-Dichloro-*trans*-2-butene	$C_4H_6Cl_2$	125.00	110-57-6	2.0		155.4	4-01-00-00787	1.183^{25}	ace 4; bz 4; eth 4; EtOH 4	1.4871^{25}
3751	2-Butene, 1,4-dichloro-, (*Z*)-		$C_4H_6Cl_2$	125.00	1476-11-5	-48		152.5	4-01-00-00787	1.188^{25}	ace 4; bz 4; eth 4; EtOH 4	1.4887^{25}
3752	2-Butene, 2,3-dichloro-, (*E*)-		$C_4H_6Cl_2$	125.00	1587-29-7			102	4-01-00-00787	1.1416^{20}	H_2O 1; EtOH 3; eth 3; ace 3	1.4582^{20}
3753	2-Butene, 2,3-dichloro-, (*Z*)-		$C_4H_6Cl_2$	125.00	1587-26-4			125.5	4-01-00-00787	1.1618^{20}	H_2O 1; EtOH 3; eth 3; ace 3	1.4590^{20}
3754	2-Butene, 2,3-dichloro-1,1,1,4,4,4-hexafluoro-	2,3-Dichlorohexafluoro-2-butene	$C_4Cl_2F_6$	232.94	303-04-8	-67.3		68.5	4-01-00-00787	1.6233^{20}		1.3459^{20}
3755	1-Butene, 3,3-dichloro-2-methyl-	3,3-Dichloro-2-methyl-1-butene	$C_5H_8Cl_2$	139.02	42101-38-2				3-01-00-00787	1.1276^{18}	ace 4; eth 4; chl 4	1.4737^{18}
3756	2-Butene, 1,3-dichloro-2-methyl-	1,3-Dichloro-2-methyl-2-butene	$C_5H_8Cl_2$	139.02	25148-87-2			150	3-01-00-00795	1.1293^{20}	ace 4; chl 4	
3757	2-Butene, 1,4-dichloro-2-methyl-	1,4-Dichloro-2-methyl-2-butene	$C_5H_8Cl_2$	139.02	29843-58-1			93^{50}; 56^{10}	3-01-00-00795	1.1526^{20}	ace 4; chl 4	1.4932^{20}
3758	2-Butene, 1,1-diethoxy-	Crotonaldehyde, diethyl acetal	$C_8H_{16}O_2$	144.21	10602-34-3			147.5	3-01-00-02980	0.8473^{18}	H_2O 1; EtOH 5; eth 5; ace 5	1.4097^{20}
3759	1-Butene, 2,3-dimethyl-	2,3-Dimethyl-1-butene	C_6H_{12}	84.16	563-78-0	-157.3		55.6	4-01-00-00852	0.6803^{20}	H_2O 1; EtOH 3; eth 3; ace 3	1.3995^{20}
3760	1-Butene, 3,3-dimethyl-		C_6H_{12}	84.16	558-37-2	-115.2		41.2	4-01-00-00850	0.6529^{20}	H_2O 1; EtOH 3; eth 3; ctc 3	1.3763^{20}
3761	2-Butene, 2,3-dimethyl-	2,3-Dimethyl-2-butene	C_6H_{12}	84.16	563-79-1	-74.6		73.3	4-01-00-00853	0.7080^{20}	H_2O 1; EtOH 3; eth 3; ace 3	1.4122^{20}
3762	2-Butenedinitrile, (*E*)-		$C_4H_2N_2$	78.07	764-42-1	96.8		186	4-02-00-02219	0.9416^{111}	H_2O 3; EtOH 3; eth 3; ace 3	1.4349^{111}
3763	2-Butenedinitrile, 2-chloro-, (*Z*)-	Fumaronitrile, chloro-	C_4HClN_2	112.52	71200-79-8	60.5		172; 68^{25}	3-02-00-01909	1.2499^{20}		1.4957^{20}
3764	2-Butenedioic acid (*E*)-	Fumaric acid	$C_4H_4O_4$	116.07	110-17-8	287 dec	4200	sub 165	3-02-00-01890	1.635^{20}	H_2O 2; EtOH 3; eth 2; ace 2	
3765	2-Butenedioic acid (*Z*)-	Maleic acid	$C_4H_4O_4$	116.07	110-16-7	130.5	5585		4-02-00-02199	1.590^{20}	H_2O 4; EtOH 4; eth 3; ace 4	
3766	2-Butenedioic acid (*E*)-, bis(1-methylethyl) ester	Diisopropyl fumarate	$C_{10}H_{16}O_4$	200.23	7283-70-7	2.1		280	4-02-00-02209		ace 4; eth 4; EtOH 4	
3767	2-Butenedioic acid (*E*)-, bis(2-methylpropyl) ester	Diisobutyl fumarate	$C_{12}H_{20}O_4$	228.29	7283-69-4			170^{160}; 122^5	4-02-00-02211	0.9760^{20}	H_2O 1; EtOH 3; eth 3; ace 3	1.4432^{20}
3768	2-Butenedioic acid (*E*)-, bis(phenylmethyl) ester	Benzyl fumarate	$C_{18}H_{16}O_4$	296.32	538-64-7	59	1149	210-11^5	4-06-00-02275		eth 4; EtOH 4; chl 4	
3769	2-Butenedioic acid, 2-bromo-, (*E*)-		$C_4H_3BrO_4$	194.97	584-99-6	141		dec	4-02-00-02224		H_2O 4; EtOH 4; eth 4	
3770	2-Butenedioic acid, 2-bromo-, (*Z*)-	2-Bromofumaric acid	$C_4H_3BrO_4$	194.97	644-80-4	184.3		200 dec	4-02-00-02224		H_2O 4; EtOH 4	
3771	2-Butenedioic acid, 2-chloro-, diethyl ester, (*E*)-	Diethyl chloromaleate	$C_8H_{11}ClO_4$	206.63	626-10-8			235 dec; 125^{19}	3-02-00-01928	1.1741^{20}	ace 4; eth 4; EtOH 4	
3772	2-Butenedioic acid, 2-chloro-, diethyl ester, (*Z*)-		$C_8H_{11}ClO_4$	206.63	10302-94-0			250 dec; 127^{10}	3-02-00-01909	1.1880^{20}	eth 4; EtOH 4	1.4571^{20}
3773	2-Butenedioic acid, 2-chloro-, dimethyl ester, (*E*)-	Dimethyl chloromaleate	$C_6H_7ClO_4$	178.57	19393-45-4			224; 100^{17}	3-02-00-01928	1.2899^{25}	eth 4; EtOH 4	1.4720^{18}
3774	2-Butenedioic acid (*E*)-, dibutyl ester		$C_{12}H_{20}O_4$	228.29	105-75-9			150^4	4-02-00-02210	0.9869^{20}	H_2O 1; ace 3; chl 3	1.4469^{20}
3775	2-Butenedioic acid (*E*)-, diethyl ester		$C_8H_{12}O_4$	172.18	623-91-6	1.5		214	4-02-00-02207	1.0452^{20}	H_2O 1; ace 3; chl 3	1.4412^{20}
3776	2-Butenedioic acid (*Z*)-, diethyl ester	Diethyl maleate	$C_8H_{12}O_4$	172.18	141-05-9	-8.8	3113	223	4-02-00-02207	1.0662^{20}	H_2O 1; EtOH 3; eth 3; chl 2	1.4416^{20}

PHYSICAL CONSTANTS OF ORGANIC COMPOUNDS (continued)

No.	Name Synonym	Mol. Form. Mol. Wt.	CAS RN mp/°C	Merck No. bp/°C	Beil. Ref. den/g cm^{-3}	Solubility n_D
3777	2-Butenedioic acid, 2,3-dihydroxy-, (*Z*)- Dihydroxymaleic acid	$C_4H_4O_6$ 148.07	526-84-1 155 dec	3170	4-03-00-01975	H_2O 2; EtOH 3; eth 2; MeOH 2
3778	2-Butenedioic acid (*E*)-, dimethyl ester	$C_6H_8O_4$ 144.13	624-49-7 103.5	 193	4-02-00-02205 1.37^{20}	H_2O 1; ace 3; chl 3 1.4062^{111}
3779	2-Butenedioic acid (*Z*)-, dimethyl ester	$C_6H_8O_4$ 144.13	624-48-6 -19	 202	4-02-00-02204 1.1606^{20}	H_2O 1; eth 3; ctc 3; lig 1 1.4416^{20}
3780	2-Butenedioic acid (*Z*)-, diphenyl ester	$C_{16}H_{12}O_4$ 268.27	7242-17-3 73	 226^{15}	4-06-00-00628	ace 4; bz 4; eth 4; EtOH 4
3781	2-Butenedioic acid (*E*)-, di-2-propenyl ester Diallyl fumarate	$C_{10}H_{12}O_4$ 196.20	2807-54-7	 140^{3}	3-02-00-01906 1.060^{25}	ace 4; bz 4; eth 4; EtOH 4 1.4670^{25}
3782	2-Butenedioic acid (*Z*)-, di-2-propenyl ester Diallyl maleate	$C_{10}H_{12}O_4$ 196.20	999-21-3	 $108\text{-}10^{3}$	4-02-00-02214 1.0773^{20}	chl 3 1.4699^{20}
3783	2-Butenedioic acid (*E*)-, dipropyl ester Dipropyl fumarate	$C_{10}H_{16}O_4$ 200.23	14595-35-8	 110^{5}	4-02-00-02209 1.0129^{20}	EtOH 3; eth 3 1.4435^{20}
3784	2-Butenedioic acid (*Z*)-, dipropyl ester	$C_{10}H_{16}O_4$ 200.23	2432-63-5	 126^{12}	4-02-00-02209 1.0245^{20}	H_2O 1; EtOH 3; eth 3; ace 3 1.4434^{20}
3785	2-Butenedioic acid, 2-methyl-, (*E*)- Mesaconic acid	$C_5H_6O_4$ 130.10	498-24-8 204.5	5806 sub	4-02-00-02231 1.466^{20}	H_2O 2; EtOH 4; eth 3; bz 2
3786	2-Butenedioic acid, 2-methyl-, (*Z*)- Citraconic acid	$C_5H_6O_4$ 130.10	498-23-7 93.5	2323	4-02-00-02230 1.617^{25}	H_2O 4; eth 2; bz 1; chl 2
3787	2-Butenedioic acid, 2-methyl-, diethyl ester, (*E*)- Diethyl mesaconate	$C_9H_{14}O_4$ 186.21	2418-31-7	 229	4-02-00-02232 1.0434^{20}	ace 4; bz 4; eth 4; EtOH 4 1.4488^{20}
3788	2-Butenedioic acid, 2-methyl-, diethyl ester, (*Z*)-	$C_9H_{14}O_4$ 186.21	691-83-8	 228; 120^{20}	4-02-00-02232 1.0491^{20}	eth 4; EtOH 4; HOAc 4 1.4467^{20}
3789	2-Butenedioic acid, 2-methyl-, dimethyl ester, (*E*)-	$C_7H_{10}O_4$ 158.15	617-53-8	 203.5	4-02-00-02232 1.0914^{20}	ace 4; eth 4; EtOH 4 1.4512^{20}
3790	2-Butenedioic acid, 2-methyl-, dimethyl ester, (*Z*)-	$C_7H_{10}O_4$ 158.15	617-54-9	 210.5	4-02-00-02232 1.1153^{20}	ace 4; eth 4; EtOH 4 1.4473^{20}
3791	2-Butenedioic acid (*E*)-, monoethyl ester	$C_6H_8O_4$ 144.13	2459-05-4 70	 147^{16}	4-02-00-02206 1.1109^{87}	EtOH 3; ace 3; chl 2
3792	2-Butene-1,4-diol, (*E*)- *trans*-2-Butene-1,4-diol	$C_4H_8O_2$ 88.11	821-11-4 25	 131^{13}	4-01-00-02660 1.0700^{20}	H_2O 4; EtOH 4 1.4755^{20}
3793	2-Butene-1,4-diol, (*Z*)- *cis*-2-Butene-1,4-diol	$C_4H_8O_2$ 88.11	6117-80-2 4	 235	4-01-00-02660 1.0698^{20}	H_2O 3; EtOH 4 1.4782^{20}
3794	3-Butene-1,2-diol	$C_4H_8O_2$ 88.11	497-06-3	 196.5	4-01-00-02658 1.0470^{20}	H_2O 3; EtOH 3 1.4628^{21}
3795	2-Butenedioyl dichloride, (*E*)- Fumaric acid dichloride	$C_4H_2Cl_2O_2$ 152.96	627-63-4	 159	4-02-00-02217 1.408^{20}	 1.5004^{18}
3796	2-Butenedioyl dichloride, 2-chloro-, (*Z*)-	$C_4HCl_3O_2$ 187.41	17096-37-6	 185 dec	3-02-00-01909 1.564^{20}	eth 4; HOAc 4 1.5206^{20}
3797	1-Butene, 2-ethyl-3-methyl- 2-Ethyl-3-methylbut-1-ene	C_7H_{14} 98.19	7357-93-9	 89	4-01-00-00871 0.7150^{20}	H_2O 1; eth 3; ace 3; bz 3 1.410^{20}
3798	2-Butene, 1,1,2,3,4,4-hexachloro- 2-Butene, 1,1,2,3,4,4-hexachloro (liquid)	$C_4H_2Cl_6$ 262.78	13045-99-3 -19	 $97\text{-}8^{10}$	4-01-00-00789 1.650^{15}	H_2O 1; chl 3 1.5331
3799	1-Butene, 2-methyl- 2-Methyl-1-butene	C_5H_{10} 70.13	563-46-2 -137.5	 31.2	4-01-00-00818 0.6504^{20}	H_2O 1; EtOH 3; eth 3; bz 3 1.3778^{20}
3800	1-Butene, 3-methyl- 3-Methyl-1-butene	C_5H_{10} 70.13	563-45-1 -168.5	 20.1	4-01-00-00825 0.6213^{25}	H_2O 1; EtOH 5; eth 5; bz 3 1.3643^{20}
3801	2-Butene, 2-methyl- 2-Methyl-2-butene	C_5H_{10} 70.13	513-35-9 -133.7	644 38.5	4-01-00-00820 0.6623^{20}	H_2O 1; EtOH 3; eth 3; bz 3 1.3874^{20}
3802	2-Butenenitrile, (*E*)-	C_4H_5N 67.09	627-26-9 -51.5	 120	4-02-00-01507 0.8239^{20}	eth 3; ace 3 1.4225^{20}
3803	3-Butenenitrile Allyl cyanide	C_4H_5N 67.09	109-75-1 -87	288 119	4-02-00-01491 0.8341^{20}	H_2O 2; EtOH 5; eth 5 1.4060^{20}
3804	2-Butene, 1,1,1,2,3,4,4,4-octafluoro- Perfluoro-2-butene	C_4F_8 200.03	360-89-4 -129	 1.5	4-01-00-00783 1.5297^{25}	
3805	2-Butene, 1,1,1,4,4-pentachloro- 1,1,1,4,4-Pentachloro-2-butene	$C_4H_3Cl_5$ 228.33	77753-21-0	 $78\text{-}80^{11}$	3-01-00-00744 1.609^{21}	EtOH 4; chl 4 1.5538^{21}
3806	1-Butene, 1,3,4,4-tetrachloro- 1,3,4,4-Tetrachloro-1-butene	$C_4H_4Cl_4$ 193.89	2984-40-9	 88^{20}	4-01-00-00774 1.476^{20}	chl 4 1.4773^{20}
3807	1-Butene, 2,3,3,4-tetrachloro- 2,3,3,4-Tetrachloro-1-butene	$C_4H_4Cl_4$ 193.89	64346-45-8	 $41\text{-}2^{7}$	3-01-00-00726 1.4602^{20}	ace 4; chl 4 1.5135^{20}
3808	1-Butene, 1,3,4,4-tetrachloro-1,2,3,4-tetrafluoro- 1-Butene, 1,3,4,4-tetrachloro-tetrafluoro	$C_4Cl_4F_4$ 265.85	357-20-0	 140.8	4-01-00-00774 1.6902^{25}	 1.429^{20}
3809	1-Butene, 2,3,4-trichloro- 2,3,4-Trichloro-1-butene	$C_4H_5Cl_3$ 159.44	2431-50-7	 60^{20}	3-01-00-00725 1.3430^{20}	ace 4; chl 4 1.4944^{20}
3810	2-Butene, 1,2,4-trichloro- 1,2,4-Trichloro-2-butene	$C_4H_5Cl_3$ 159.44	2431-54-1	 $67\text{-}9^{10}$	3-01-00-00744 1.3843^{20}	bz 4; eth 4; EtOH 4; chl 4 1.5175^{20}
3811	2-Butene, 1,1,3-triethoxy- 2-Butenal, 3-ethoxy, diethyl acetal	$C_{10}H_{20}O_3$ 188.27	69190-65-4	 192.5	4-01-00-04082 0.908^{21}	EtOH 4 1.430^{21}
3812	1-Butene, 2,3,3-trimethyl-	C_7H_{14} 98.19	594-56-9 -109.9	 77.9	4-01-00-00873 0.7050^{20}	H_2O 1; eth 3; bz 3; chl 3 1.4025^{20}
3813	3-Butenoic acid	$C_4H_6O_2$ 86.09	625-38-7 -35	 169	4-02-00-01491 1.0091^{20}	H_2O 3; EtOH 5; eth 5 1.4239^{20}

No.	Name Synonym	Mol. Form. Mol. Wt.	CAS RN mp/°C	Merck No. bp/°C	Beil. Ref. den/g cm^{-3}	Solubility n_D
3814	2-Butenoic acid, (*E*)- *trans*-Crotonic acid	$C_4H_6O_2$ 86.09	107-93-7 72	2600 184.7	4-02-00-01498 0.9604^{77}	H_2O 4; EtOH 4; eth 3; ace 3 1.4249^{77}
3815	2-Butenoic acid, (*Z*)- *cis*-Crotonic acid	$C_4H_6O_2$ 86.09	503-64-0 15	5048 169	4-02-00-01497 1.0267^{20}	H_2O 4; EtOH 3 1.4483^{14}
3816	2-Butenoic acid, 3-amino-, ethyl ester, (*E*)-	$C_6H_{11}NO_2$ 129.16	41867-20-3 34	 212 dec	3-03-00-01199 1.0219^{19}	H_2O 1; EtOH 3; eth 3; bz 3 1.4988^{22}
3817	2-Butenoic acid, 4-amino-4-oxo-, (*Z*)- Maleamic acid	$C_4H_5NO_3$ 115.09	557-24-4 172.5	5583	4-02-00-02218	H_2O 4; EtOH 4
3818	2-Butenoic acid, anhydride Crotonic acid anhydride	$C_8H_{10}O_3$ 154.17	623-68-7	 247; 129^{19}	4-02-00-01505 1.0397^{20}	eth 4 1.4745^{20}
3819	2-Butenoic acid, 4-bromo-, ethyl ester, (*E*)-	$C_6H_9BrO_2$ 193.04	37746-78-4	 $97\text{-}8^{15}$	4-02-00-01517 1.402^{16}	EtOH 4 1.4925^{20}
3820	2-Butenoic acid, 3-chloro-, (*E*)-	$C_4H_5ClO_2$ 120.54	6214-28-4 61	 195	4-02-00-01510 1.1969^{69}	H_2O 2; EtOH 4; chl 3; CS_2 4
3821	2-Butenoic acid, 3-chloro-, (*Z*)-	$C_4H_5ClO_2$ 120.54	6213-90-7 61		4-02-00-01510 1.1905^{66}	H_2O 2; EtOH 4; peth 4 1.4704^{66}
3822	2-Butenoic acid, 4-chloro- γ-Chlorocrotonic acid	$C_4H_5ClO_2$ 120.54	16197-90-3 83	 117^{13}	4-02-00-01511	eth 4; HOAc 4
3823	2-Butenoic acid, 2-chloro-, ethyl ester Ethyl β-chlorocrotonate	$C_6H_9ClO_2$ 148.59	56216-12-7	 176	4-02-00-01510 1.1133^{20}	 1.453^{20}
3824	2-Butenoic acid, 2-chloro-, ethyl ester, (*E*)- 2-Butenoic acid, 2-chloro, ethyl ester (*trans*)	$C_6H_9ClO_2$ 148.59	77825-53-7	 177; 75^{30}	4-02-00-01510 1.1135^{20}	eth 4; EtOH 4 1.4538^{20}
3825	2-Butenoic acid, 2-chloro-, ethyl ester, (*Z*)- 2-Butenoic acid, 2-chloro, ethyl ester (*cis*)	$C_6H_9ClO_2$ 148.59	77825-54-8	 75^{30}	4-02-00-01510 1.1021^{18}	eth 4; EtOH 4
3826	2-Butenoic acid, 3-chloro-, ethyl ester 3-Chlorocrotonic acid ethyl ester	$C_6H_9ClO_2$ 148.59	38624-62-3	 180; 75^{14}	4-02-00-01511 1.1062^{20}	 1.459^{20}
3827	2-Butenoic acid, 3-chloro-, ethyl ester, (*E*)-	$C_6H_9ClO_2$ 148.59	6127-92-0	 184	4-02-00-01511 1.1062^{20}	eth 4; EtOH 4 1.4592^{20}
3828	2-Butenoic acid, 3-chloro-, ethyl ester, (*Z*)-	$C_6H_9ClO_2$ 148.59	6127-93-1	 161.4	4-02-00-01511 1.0860^{20}	eth 4; EtOH 4 1.4542^{19}
3829	2-Butenoic acid, 4-chloro-, ethyl ester Crotonic acid, 4-chloro-, ethyl ester	$C_6H_9ClO_2$ 148.59	15333-22-9	 93^{10}	4-02-00-01511 1.090^{18}	 1.462^{18}
3830	2-Butenoic acid, 2-chloro-, methyl ester, (*E*)-	$C_5H_7ClO_2$ 134.56	22038-57-9	 161.5; 63^{17}	4-02-00-01510 1.160^{20}	eth 4 1.4569^{23}
3831	2-Butenoic acid, 3-chloro-, methyl ester, (*E*)-	$C_5H_7ClO_2$ 134.56	6372-01-6	 $64\text{-}7^{14}$	4-02-00-01510 1.157^{20}	H_2O 2; EtOH 4; eth 4 1.4630^{20}
3832	2-Butenoic acid, 3-chloro-, methyl ester, (*Z*)-	$C_5H_7ClO_2$ 134.56	6214-25-1	 142.4; 70^{24}	4-02-00-01500 1.138^{20}	H_2O 2; eth 4; MeOH 4 1.4573^{19}
3833	2-Butenoic acid, 4-chloro-, methyl ester Methyl γ-chlorocrotonate	$C_5H_7ClO_2$ 134.56	15320-72-6	 68^{11}	4-02-00-01511 1.169^{18}	 1.467^{18}
3834	2-Butenoic acid, 3-cyclohexyl- Cicrotoic acid	$C_{10}H_{16}O_2$ 168.24	25229-42-9 85.5	2272		
3835	3-Butenoic acid, 2,2-dimethyl- 2,2-Dimethyl-3-butenoic acid	$C_6H_{10}O_2$ 114.14	10276-09-2 -6	 185	4-02-00-01574 0.9567^{15}	 1.4305^{15}
3836	2-Butenoic acid, 2,3-dimethyl-, ethyl ester Ethyl 2,3-dimethyl-2-butenoate	$C_8H_{14}O_2$ 142.20	13979-28-7	 155	4-02-00-01576 0.9072^{19}	 1.430^{19}
3837	2-Butenoic acid, 2-ethyl-, (*E*)- 2-Ethylcrotonic acid	$C_6H_{10}O_2$ 114.14	1187-13-9 45.5	 209	3-02-00-01329 0.9578^{50}	eth 4; EtOH 4 1.4475^{50}
3838	2-Butenoic acid, ethyl ester	$C_6H_{10}O_2$ 114.14	10544-63-5	 136.5	2-02-00-00394 0.9175^{20}	eth 4; EtOH 4 1.4243^{20}
3839	2-Butenoic acid, ethyl ester, (*E*)- Ethyl *trans*-crotonate	$C_6H_{10}O_2$ 114.14	623-70-1	2600 138	4-02-00-01500 0.9175^{20}	H_2O 1; EtOH 3; eth 3 1.4243^{20}
3840	2-Butenoic acid, ethyl ester, (*Z*)- *cis*-Ethyl crotonate	$C_6H_{10}O_2$ 114.14	6776-19-8	2600 136	4-02-00-01497 0.9182^{20}	ace 4; eth 4; EtOH 4 1.4242^{20}
3841	3-Butenoic acid, ethyl ester Ethyl 3-butenoate	$C_6H_{10}O_2$ 114.14	1617-18-1	 119	4-02-00-01491 0.9122^{20}	EtOH 3 1.4105^{20}
3842	3-Butenoic acid, 2-hydroxy- Vinylglycolic acid	$C_4H_6O_3$ 102.09	600-17-9 43	 124^{9}	3-03-00-00685 1.2020^{21}	H_2O 4; eth 4; EtOH 4; chl 4
3843	3-Butenoic acid, 2-hydroxy-, ethyl ester Ethyl 2-hydroxy-3-butenoate	$C_6H_{10}O_3$ 130.14	91890-87-8	 173 dec; 68^{15}	3-03-00-00685 1.0470^{15}	H_2O 4; eth 4; EtOH 4 1.436^{13}
3844	2-Butenoic acid, 2-methyl-, (*E*)- Tiglic acid	$C_5H_8O_2$ 100.12	80-59-1 64.5	9365 198.5	4-02-00-01552 0.9641^{76}	eth 4; EtOH 4 1.4330^{76}
3845	2-Butenoic acid, 2-methyl-, (*Z*)- Angelic acid	$C_5H_8O_2$ 100.12	565-63-9 45.5	678 185	4-02-00-01551 0.9834^{49}	H_2O 2; EtOH 3; eth 4 1.4434^{47}
3846	2-Butenoic acid, 3-methyl-	$C_5H_8O_2$ 100.12	541-47-9 69.5	 197	4-02-00-01555 1.0062^{24}	
3847	2-Butenoic acid, methyl ester, (*E*)- Methyl *trans*-crotonate	$C_5H_8O_2$ 100.12	623-43-8 -42	2600 121	4-02-00-01500 0.9444^{20}	H_2O 1; EtOH 4; eth 4 1.4242^{20}
3848	2-Butenoic acid, methyl ester, (*Z*)- Methyl *cis*-crotonate	$C_5H_8O_2$ 100.12	4358-59-2	5048 118	4-02-00-01500	 1.4175^{20}
3849	2-Butenoic acid, 2-methyl-, ethyl ester, (*E*)-	$C_7H_{12}O_2$ 128.17	5837-78-5	 156	4-02-00-01553 0.9200^{20}	 1.4340^{20}
3850	2-Butenoic acid, 3-methyl-, ethyl ester	$C_7H_{12}O_2$ 128.17	638-10-8	 153.5	4-02-00-01556 0.9199^{21}	 1.4345^{20}
3851	2-Butenoic acid, 2-methyl-, methyl ester, (*E*)-	$C_6H_{10}O_2$ 114.14	6622-76-0	 139	4-02-00-01553 0.9498^{20}	 1.4370^{20}
3852	2-Butenoic acid, 3-methyl-, methyl ester	$C_6H_{10}O_2$ 114.14	924-50-5 114	 136.5	3-02-00-01313 0.9337^{20}	 1.432^{20}

PHYSICAL CONSTANTS OF ORGANIC COMPOUNDS (continued)

No.	Name Synonym	Mol. Form. Mol. Wt.	CAS RN mp/°C	Merck No. bp/°C	Beil. Ref. den/g cm^{-3}	Solubility n_D
3853	2-Butenoic acid, 4-oxo-4-(phenylamino)-, (*Z*)- Maleanilic acid	$C_{10}H_9NO_3$ 191.19	555-59-9 192 dec	5584	3-12-00-00571 1.418^{30}	
3854	2-Butenoic acid, 3-phenyl-, (*Z*)-	$C_{10}H_{10}O_2$ 162.19	704-79-0 131.5	170^{14}	4-09-00-02057	bz 3; CS_2 3; peth 2
3855	3-Butenoic acid, 4-phenyl- Styrylacetic acid	$C_{10}H_{10}O_2$ 162.19	2243-53-0 87	302	4-09-00-02054	H_2O 1; EtOH 4; eth 4; CS_2 4
3856	2-Butenoic acid, 2-propenyl ester Allyl crotonate	$C_7H_{10}O_2$ 126.16	20474-93-5	$88\text{-}9^{70}$	4-02-00-01503 0.9440^{20}	1.4465^{20}
3857	2-Butenoic acid, 2-propenyl ester, (*E*)-	$C_7H_{10}O_2$ 126.16	5453-44-1	$88\text{-}9^{70}$	4-02-00-01503 0.9440^{20}	1.4465^{20}
3858	3-Buten-2-ol, (±)-	C_4H_8O 72.11	6118-14-5 <-100	97.3	4-01-00-02102 0.8318^{20}	1.4137^{20}
3859	2-Buten-1-ol, (*E*)- *trans*-Crotyl alcohol	C_4H_8O 72.11	504-61-0 <-30	121.2	4-01-00-02107 0.8521^{20}	H_2O 4; EtOH 5; eth 5; chl 3 1.4288^{20}
3860	2-Buten-1-ol, (*Z*)- *cis*-Crotyl alcohol	C_4H_8O 72.11	4088-60-2	123	4-01-00-02107 0.8662^{20}	1.4342^{25}
3861	3-Buten-1-ol	C_4H_8O 72.11	627-27-0	113.5	4-01-00-02105 0.8424^{20}	H_2O 3; EtOH 5; eth 5; ace 3 1.4224^{20}
3862	2-Buten-1-ol, acetate	$C_6H_{10}O_2$ 114.14	628-08-0	132	3-02-00-00283 0.9192^{20}	bz 4; eth 4; EtOH 4 1.4181^{20}
3863	2-Buten-1-ol, 2-chloro-	C_4H_7ClO 106.55	116723-93-4	159	3-01-00-01900 1.1180^{20}	H_2O 4; EtOH 4 1.4682^{20}
3864	2-Buten-1-ol, 3-chloro- γ-Chlorocrotyl alcohol	C_4H_7ClO 106.55	40605-42-3 -42.9	161.5	4-01-00-02108 1.095^{23}	1.4652^{20}
3865	2-Buten-1-ol, 4-chloro- 4-Chloro-2-buten-1-ol	C_4H_7ClO 106.55	7523-44-6	$64\text{-}5^{2}$	4-01-00-02110	eth 4; EtOH 4 1.4845^{20}
3866	3-Buten-2-ol, 1-chloro- 1-Chloro-3-buten-2-ol	C_4H_7ClO 106.55	671-56-7	145.5	3-01-00-01893 1.111^{20}	chl 4 1.4643^{20}
3867	3-Buten-1-ol, 2-chloro-	C_4H_7ClO 106.55	75455-41-3	$66\text{-}7^{30}$	4-01-00-02106 1.1044^{20}	eth 4; EtOH 4 1.4665^{20}
3868	3-Buten-2-ol, 3-chloro- 3-Chloro-3-buten-2-ol	C_4H_7ClO 106.55	6498-47-1	$53\text{-}7^{19}$	3-01-00-01893 1.1138^{23}	eth 4; EtOH 4
3869	2-Buten-1-ol, 3-methyl-	$C_5H_{10}O$ 86.13	556-82-1	140	4-01-00-02129 0.848^{25}	1.4412^{20}
3870	3-Buten-2-ol, 2-methyl-	$C_5H_{10}O$ 86.13	115-18-4 -28	97	4-01-00-02132 0.82^{20}	
3871	3-Buten-2-ol, 3-methyl- Methyl isopropenyl carbinol	$C_5H_{10}O$ 86.13	10473-14-0	114	4-01-00-02127 0.8531^{17}	1.4288^{17}
3872	2-Buten-1-ol, 3-phenyl- 2-Butene-1-ol, 3-phenyl	$C_{10}H_{12}O$ 148.20	1504-54-7	145^{15}	4-06-00-03839 1.035^{20}	ctc 3 1.5678^{20}
3873	3-Buten-2-ol, 4-(2,6,6-trimethyl-1-cyclohexen-1-yl)- β-Ionol	$C_{13}H_{22}O$ 194.32	22029-76-1	131^{15}; $80^{0.7}$	3-06-00-00401 0.9243^{20}	EtOH 3; eth 3; ace 3 1.4969^{20}
3874	3-Buten-2-ol, 4-(2,6,6-trimethyl-2-cyclohexen-1-yl)- α-Ionol	$C_{13}H_{22}O$ 194.32	25312-34-9	127^{15}	3-06-00-00402 0.9474^{20}	1.4735^{20}
3875	3-Buten-2-one Methyl vinyl ketone	C_4H_6O 70.09	78-94-4	6052 81.4	4-01-00-03444 0.864^{20}	H_2O 3; EtOH 3; eth 4; ace 4 1.4081^{20}
3876	2-Buten-1-one, 1,3-diphenyl- Dypnone	$C_{16}H_{14}O$ 222.29	495-45-4	3460 342.5	4-07-00-01682 1.1080^{15}	eth 4; EtOH 4 1.6343^{20}
3877	1-Buten-1-one, 2-ethyl- Diethylketene	$C_6H_{10}O$ 98.14	24264-08-2	88.5	4-01-00-03477 0.831^{20}	1.4135^{11}
3878	3-Buten-2-one, 4-(2-furanyl)-	$C_8H_8O_2$ 136.15	623-15-4 39.5	229 dec; 113^{10}	5-17-09-00578 1.0496^{57}	H_2O 1; EtOH 4; eth 4; chl 4 1.5788^{45}
3879	3-Buten-2-one, 3-methyl- Isopropenyl methyl ketone	C_5H_8O 84.12	814-78-8 -54	98	4-01-00-03462 0.8527^{20}	EtOH 4 1.4220^{20}
3880	2-Buten-1-one, 1-phenyl-	$C_{10}H_{10}O$ 146.19	495-41-0 20.5	$111\text{-}2^{9}$	3-07-00-01410 1.025^{15}	1.5626^{18}
3881	3-Buten-2-one, 4-phenyl- Benzylideneacetone	$C_{10}H_{10}O$ 146.19	122-57-6 41.5	1151 261	2-07-00-00287 1.0368^{15}	bz 4; eth 4; EtOH 4; chl 4
3882	3-Buten-2-one, 4-phenyl-, (*E*)-	$C_{10}H_{10}O$ 146.19	1896-62-4 42	262	4-07-00-01003 1.0097^{45}	H_2O 1; EtOH 4; eth 3; ace 3 1.5836^{45}
3883	3-Buten-2-one, 4-(2,5,6,6-tetramethyl-1-cyclohexen-1-yl)- β-Irone	$C_{14}H_{22}O$ 206.33	79-70-9	4978 $85\text{-}90^{0.1}$	4-07-00-00377 0.9434^{21}	H_2O 2; EtOH 4; eth 4; bz 4 1.5017^{20}
3884	3-Buten-2-one, 4-(2,5,6,6-tetramethyl-2-cyclohexen-1-yl)- α-Irone	$C_{14}H_{22}O$ 206.33	79-69-6	4977 $07\text{-}92^{0.4}$	0.9362^{20}	1.5002^{20}
3885	3-Buten-2-one, 4-(2,6,6-trimethyl-1-cyclohexen-1-yl)-, (*E*)-	$C_{13}H_{20}O$ 192.30	79-77-6	140^{18}; $73^{0.1}$	4-07-00-00361 0.9462^{20}	H_2O 2; EtOH 5; eth 5; chl 3 1.5198^{20}
3886	3-Buten-2-one, 4-(2,6,6-trimethyl-2-cyclohexen-1-yl)-, (*E*)-(±)-	$C_{13}H_{20}O$ 192.30	30685-95-1	146^{28}; $75^{0.15}$	4-07-00-00363 0.9298^{21}	ace 4; eth 4; EtOH 4 1.5041^{20}
3887	2-Butenoyl chloride	C_4H_5ClO 104.54	10487-71-5	124.5	2-02-00-00394 1.0905^{20}	ace 4 1.460^{18}
3888	2-Butenoyl chloride, 3-methyl-	C_5H_7ClO 118.56	3350-78-5	146	4-02-00-01558 1.065^{25}	1.4770^{20}

No.	Name Synonym	Mol. Form. Mol. Wt.	CAS RN mp/°C	Merck No. bp/°C	Beil. Ref. den/g cm^{-3}	Solubility n_D
3889	1-Buten-3-yne Vinylacetylene	C_4H_4 52.08	689-97-4	 5.1	4-01-00-01083 0.7094^{0}	H_2O 1; bz 3 1.4161^{1}
3890	1-Buten-3-yne, 4-chloro-	C_4H_3Cl 86.52	40589-38-6	 56	4-01-00-01085 1.0022^{20}	chl 4 1.4656^{20}
3891	1-Buten-3-yne, 1-methoxy-	C_5H_6O 82.10	2798-73-4	 123 dec	4-01-00-02301 0.906^{20}	H_2O 1; chl 3; os 4 1.4818^{20}
3892	1-Buten-3-yne, 2-methyl- Isopropenylacetylene	C_5H_6 66.10	78-80-8 -113	 32	4-01-00-01089 0.6801^{11}	chl 3 1.4140^{20}
3893	Butralin Benzenamine, 4-(1,1-dimethylethyl)-*N*-(1-methylpropyl)-2,6-dinitro-	$C_{14}H_{21}N_3O_4$ 295.34	33629-47-9 60	1532 135$^{0.5}$		
3894	*N-sec*-Butyl-2,6-dinitrobenzene Benzeneamine, *N*-(1-methylpropyl)-2,6-dinitro	$C_{10}H_{13}N_3O_4$ 239.23	55702-49-3			
3895	2-Butynal Tetrolaldehyde	C_4H_4O 68.08	1119-19-3 -26	 106.5	4-01-00-03540 0.9264^{17}	eth 3; ace 3; chl 3 1.446^{19}
3896	3-Butyn-2-amine, 2-methyl-	C_5H_9N 83.13	2978-58-7 18	 79.5	4-04-00-01145 0.79^{25}	 1.4235^{20}
3897	1-Butyne Ethylacetylene	C_4H_6 54.09	107-00-6 -125.7	 8.0	4-01-00-00969 0.6783^{0}	H_2O 1; EtOH 3; eth 3 1.3962^{20}
3898	2-Butyne Dimethylacetylene	C_4H_6 54.09	503-17-3 -32.3	 26.9	4-01-00-00971 0.6910^{20}	H_2O 1; EtOH 3; eth 3; ctc 3 1.3921^{20}
3899	1-Butyne, 3-chloro-	C_4H_5Cl 88.54	21020-24-6	 68.5	4-01-00-00970 1.4218^{25}	 1.4218^{25}
3900	2-Butyne, 1-chloro-	C_4H_5Cl 88.54	3355-17-7	 103	4-01-00-00973 1.0152^{20}	ace 4; eth 4; EtOH 4 1.4581^{20}
3901	1-Butyne, 3-chloro-3-methyl-	C_5H_7Cl 102.56	1111-97-3 -61	 76	4-01-00-01000 0.9061^{20}	
3902	2-Butynediamide Cellocidin	$C_4H_4N_2O_2$ 112.09	543-21-5 217 dec	1958	4-02-00-02295	H_2O 2; EtOH 2; eth 2; chl 2
3903	2-Butyne-1,4-diamine	$C_4H_8N_2$ 84.12	53878-96-9 44	 96^{16}	4-04-00-01392	H_2O 4; EtOH 4; MeOH 4
3904	2-Butyne, 1,4-dibromo- 1,4-Dibromo-2-butyne	$C_4H_4Br_2$ 211.88	2219-66-1	 92^{15}	4-01-00-00974 2.014^{18}	eth 3; ace 3; chl 4 1.588^{18}
3905	2-Butyne, 1,4-dichloro-	$C_4H_4Cl_2$ 122.98	821-10-3	 165.5	4-01-00-00973 1.258^{20}	eth 3; ace 3; ctc 2; chl 4 1.5058^{20}
3906	2-Butyne, 1,4-diiodo-	$C_4H_4I_2$ 305.88	116529-73-8 54.5	 70$^{0.1}$	4-01-00-00975	H_2O 2; EtOH 3; eth 3; ace 3
3907	1-Butyne, 3,3-dimethyl- *tert*-Butylacetylene	C_6H_{10} 82.15	917-92-0 -78.2	 37.7	4-01-00-01022 0.6623^{25}	 1.3736^{20}
3908	2-Butynedinitrile	C_4N_2 76.06	1071-98-3 20.5	 76.5	4-02-00-02295 0.9708^{25}	 1.4647^{25}
3909	2-Butynedioic acid, diethyl ester	$C_8H_{10}O_4$ 170.17	762-21-0 1.5	 184^{200}	4-02-00-02294 1.0075^{20}	EtOH 3; eth 3; ctc 3 1.4425^{20}
3910	2-Butynedioic acid, dimethyl ester	$C_6H_6O_4$ 142.11	762-42-5	 197 dec; 98^{20}	4-02-00-02291 1.1564^{20}	EtOH 3; eth 3; ctc 3 1.4434^{20}
3911	2-Butyne-1,4-diol Bis(hydroxymethyl)acetylene	$C_4H_6O_2$ 86.09	110-65-6 50	 238	4-01-00-02687	H_2O 4; EtOH 4; eth 2; ace 4 1.4804^{20}
3912	3-Butyne-1,2-diol 1-Butyne-3,4-diol	$C_4H_6O_2$ 86.09	616-28-4 40	 64$^{0.2}$	4-01-00-02689	H_2O 4; EtOH 4
3913	2-Butyne-1,4-diol, diacetate 1,4-Diacetoxy-2-butyne	$C_8H_{10}O_4$ 170.17	1573-17-7	 122-3^{10}	4-02-00-00244	ctc 3 1.4611^{20}
3914	2-Butyne, 1,1,1,4,4,4-hexafluoro-	C_4F_6 162.03	692-50-2 -117.4	 -24.6	4-01-00-00972	EtOH 3; eth 3; ace 3; ctc 3
3915	2-Butyne, 1-methoxy-	C_5H_8O 84.12	2768-41-4	 99.5	3-01-00-01973 0.8496^{20}	bz 4; eth 4; EtOH 4 1.4262^{20}
3916	1-Butyne, 3-methyl-	C_5H_8 68.12	598-23-2 -89.7	 26.3	4-01-00-00999 0.6660^{20}	H_2O 1; EtOH 5; eth 5 1.3723^{20}
3917	2-Butynoic acid	$C_4H_4O_2$ 84.07	590-93-2 78	 203	4-02-00-01690 0.9641^{20}	H_2O 4; eth 4; EtOH 4; chl 4
3918	2-Butynoic acid, ethyl ester	$C_6H_8O_2$ 112.13	4341-76-8	 163	4-02-00-01691 0.9641^{20}	 1.4372^{20}
3919	2-Butyn-1-ol	C_4H_6O 70.09	764-01-2 -2.2	 148	4-01-00-02220 0.9370^{20}	eth 4; EtOH 4 1.4530^{20}
3920	3-Butyn-1-ol	C_4H_6O 70.09	927-74-2 -63.6	 129	4-01-00-02219 0.9257^{20}	H_2O 4; EtOH 4 1.4409^{20}
3921	3-Butyn-2-ol	C_4H_6O 70.09	2028-63-9 -3	 106.5	4-01-00-02218 0.8618^{20}	H_2O 4; eth 4; EtOH 4 1.4207^{20}
3922	3-Butyn-2-ol, 2-methyl- 1,1-Dimethylpropargyl alcohol	C_5H_8O 84.12	115-19-5 3	5956 104	4-01-00-02229 0.8618^{20}	 1.4207^{20}
3923	3-Butyn-2-one Ethynyl methyl ketone	C_4H_4O 68.08	1423-60-5	 84	4-01-00-03539 0.8860^{15}	 1.4070^{20}
3924	3-Butyn-2-one, 4-phenyl-	$C_{10}H_8O$ 144.17	1817-57-8	 75-6$^{0.8}$	4-07-00-01175 0.99^{25}	 1.5762^{20}
3925	Butyrac 118 Butanoic acid, 4-(2,4-dichlorophenoxy)-	$C_{10}H_{10}Cl_2O_3$ 249.09	94-82-6 118	2828		
3926	Butyric acid, 3,4-dichloro- 3,4-Dichlorobutyric acid	$C_4H_6Cl_2O_2$ 157.00	29653-38-1 49	 123^{8}	4-02-00-00828	eth 4; chl 4; HOAc 4
3927	Butyric acid, 4-fluoro- 4-Fluorobutyric acid	$C_4H_7FO_2$ 106.10	462-23-7	 76-8^{5}	4-02-00-00809	eth 4; EtOH 4 1.3993^{25}

PHYSICAL CONSTANTS OF ORGANIC COMPOUNDS (continued)

No.	Name Synonym	Mol. Form. Mol. Wt.	CAS RN mp/°C	Merck No. bp/°C	Beil. Ref. den/g cm^{-3}	Solubility n_D
3928	Butyric acid, 2-iodo-, ethyl ester	$C_6H_{11}IO_2$ 242.06	7425-47-0	 100^{21}; 46^{1}	4-02-00-00839 1.570^{17}	 1.4923^{20}
3929	Cadmium, dimethyl- Dimethyl cadmium	C_2H_6Cd 142.48	506-82-1 -4.5	3229 105.5; exp 150	 1.9846^{18}	peth 3 1.5488
3930	Calcium cyanide Calcyan	C_2CaN_2 92.11	592-01-8	1664		
3931	Camphoric acid, *trans*-(±)- Isocamphoric acid, (*DL*)	$C_{10}H_{16}O_4$ 200.23	560-08-7 195		3-09-00-03879 1.249^{25}	eth 4
3932	Carbamic acid, (aminocarbonyl)-, ethyl ester	$C_4H_8N_2O_3$ 132.12	626-36-8 196.5	 dec	4-03-00-00127	H_2O 1; EtOH 2; eth 1; bz 2
3933	Carbamic acid, butyl-, 2-[[(aminocarbonyl)oxy]methyl]-2-methylpentyl ester 2-Methyl-2-propyltrimethylene butylcarbamate	$C_{13}H_{26}N_2O_4$ 274.36	4268-36-4	9737		
3934	Carbamic acid, butyl-, butyl ester	$C_9H_{19}NO_2$ 173.26	13105-52-7	 88^{3}	4-04-00-00577 0.9221^{20}	eth 4; EtOH 4 1.4359^{20}
3935	Carbamic acid, butyl ester Butyl carbamate	$C_5H_{11}NO_2$ 117.15	592-35-8 53	 204 dec; 108^{14}	4-03-00-00054	EtOH 4; chl 2
3936	Carbamic acid, butyl-, ethyl ester	$C_7H_{15}NO_2$ 145.20	591-62-8 -22	 202; 100^{15}	4-04-00-00577 0.9434^{26}	 1.4278^{26}
3937	Carbamic acid, butyl-, methyl ester	$C_6H_{13}NO_2$ 131.17	2594-21-0 -18	 92^{15}	0-04-00-00158 0.9689^{23}	 1.4289^{23}
3938	Carbamic acid, *N*-butyl-*N*-nitro,butyl ester	$C_9H_{18}N_2O_4$ 218.25	110795-27-2	 98^{3}	 1.046^{20}	EtOH 5; eth 5 1.448^{20}
3939	Carbamic acid, (3-chlorophenyl)-, 1-methylethyl ester Chloropropham	$C_{10}H_{12}ClNO_2$ 213.66	101-21-3 41	2188 149^{2}	4-12-00-01149 1.18^{30}	 1.5388^{20}
3940	Carbamic acid, dichloro-, ethyl ester	$C_3H_5Cl_2NO_2$ 157.98	13698-16-3	 $55\text{-}6^{15}$	3-03-00-00050 1.349^{25}	 1.4595^{20}
3941	Carbamic acid, diethyl- Carbamic acid, *N,N*-diethyl	$C_5H_{11}NO_2$ 117.15	24579-70-2 -15	 171 dec	3-04-00-00222 0.9276^{20}	H_2O 4; eth 4; EtOH 4 1.4206^{20}
3942	Carbamic acid, (1,1-dimethylethyl)-, ethyl ester	$C_7H_{15}NO_2$ 145.20	1611-50-3 21.5	 72^{16}	4-04-00-00665 0.943^{15}	
3943	Carbamic acid, (1,1-dimethylethyl)-, methyl ester	$C_6H_{13}NO_2$ 131.17	27701-01-5 27	 63.3^{17}	3-04-00-00325 0.966^{15}	
3944	Carbamic acid, (1,1-dimethylethyl)nitro-, ethyl ester	$C_7H_{14}N_2O_4$ 190.20	55696-02-1	 56^{2} dec	4-04-00-03409 1.049^{20}	eth 4; EtOH 4 1.4331^{20}
3945	Carbamic acid, dimethyl-, 3-methyl-1-phenyl-1H-pyrazol-5-yl ester Pyrolan	$C_{13}H_{15}N_3O_2$ 245.28	87-47-8 50	8013 $160\text{-}2^{0.2}$	5-23-10-00530	ctc 3; CS_2 3
3946	Carbamic acid, diphenyl-, ethyl ester	$C_{15}H_{15}NO_2$ 241.29	603-52-1 72	 360	4-12-00-00850	H_2O 4; eth 4; bz 4
3947	Carbamic acid, ethyl-, butyl ester Carbamic acid, *N*-ethyl, butyl ester	$C_7H_{15}NO_2$ 145.20	16246-07-4	 66^{3}	 0.9396^{20}	eth 4; EtOH 4 1.4301^{20}
3948	Carbamic acid, ethyl ester Urethan	$C_3H_7NO_2$ 89.09	51-79-6 49	9789 185	4-03-00-00040 0.9862^{21}	H_2O 4; EtOH 4; eth 4; bz 4 1.4144^{51}
3949	Carbamic acid, ethyl-, ethyl ester	$C_5H_{11}NO_2$ 117.15	623-78-9	 176	4-04-00-00365 0.9813^{20}	H_2O 4; eth 4; EtOH 4 1.4215^{20}
3950	Carbamic acid, ethylnitro-, ethyl ester Carbamic acid, *N*-ethyl-*N*-nitro, ethyl ester	$C_5H_{10}N_2O_4$ 162.15	6274-16-4	 107^{31}	4-04-00-03404 1.161^{20}	eth 4; EtOH 4 1.4432^{20}
3951	Carbamic acid, ethylnitro-, methyl ester Carbamic acid, *N*-ethyl-*N*-nitro, butyl ester	$C_7H_{14}N_2O_4$ 190.20	6162-79-4	 79^{3}	4-04-00-03403 1.089^{20}	H_2O 4; eth 4; EtOH 4 1.4455^{20}
3952	Carbamic acid, (3-methylbutyl)-, ethyl ester Urethane, isopentyl	$C_8H_{17}NO_2$ 159.23	1611-52-5	 218	2-04-00-00648 0.9322^{20}	 1.4326^{20}
3953	Carbamic acid, methyl ester Methyl carbamate	$C_2H_5NO_2$ 75.07	598-55-0 54	5958 177	4-03-00-00037 1.1361^{56}	H_2O 4; EtOH 4; eth 3 1.4125^{56}
3954	Carbamic acid, methyl-, ethyl ester	$C_4H_9NO_2$ 103.12	105-40-8	 170	4-04-00-00200 1.0115^{20}	H_2O 4; EtOH 4 1.4183^{20}
3955	Carbamic acid, 1-methylethyl ester Isopropyl carbamate	$C_4H_9NO_2$ 103.12	1746-77-6 93	 183	4-03-00-00053 0.9951^{66}	
3956	Carbamic acid, (1-methylethyl)-, ethyl ester	$C_6H_{13}NO_2$ 131.17	2594-20-9	 79^{15}	4-04-00-00520 0.9531^{20}	eth 4; EtOH 4 1.4229^{20}
3957	Carbamic acid, (1-methylethyl)nitro-, ethyl ester Carbamic acid, *N*-isopropyl-*N*-nitro, ethyl ester	$C_6H_{12}N_2O_4$ 176.17	62261-05-6	 72^{7}	4-04-00-03407 1.110^{20}	eth 4; EtOH 4 1.4381^{20}
3958	Carbamic acid, methylphenyl-, methyl ester	$C_9H_{11}NO_2$ 165.19	28685-60-1 44	 235	2-12-00-00235 1.2960^{19}	eth 4; lig 4
3959	Carbamic acid, 2-methylpropyl ester Isobutyl carbamate	$C_5H_{11}NO_2$ 117.15	543-28-2 67	5021 207	4-03-00-00056	eth 4; EtOH 4 1.4098^{76}
3960	Carbamic acid, (1-methylpropyl)-, ethyl ester Urethane, *sec*-butyl	$C_7H_{15}NO_2$ 145.20	10212-74-5 -14	 194	4-04-00-00622 0.9404^{26}	 1.4267^{26}
3961	Carbamic acid, (2-methylpropyl)-, ethyl ester Isobutyl urethane	$C_7H_{15}NO_2$ 145.20	539-89-9 <-65	5037 110^{30}	4-04-00-00647 0.9432^{20}	eth 4; EtOH 4 1.4288^{20}

No.	Name Synonym	Mol. Form. Mol. Wt.	CAS RN mp/°C	Merck No. bp/°C	Beil. Ref. den/g cm^{-3}	Solubility n_D
3962	Carbamic acid, (1-methylpropyl)-, methyl ester Carbamic acid, *N*-*sec*-butyl, methyl ester	$C_6H_{13}NO_2$ 131.17	39076-02-3	 83^{16}	0-04-00-00162 0.9651^{25}	 1.4263^{25}
3963	Carbamic acid, nitro-, ethyl ester	$C_3H_6N_2O_4$ 134.09	626-37-9 64	 140 dec	4-03-00-00247 1.0074^{20}	H_2O 4; EtOH 4; eth 4; lig 2
3964	Carbamic acid, nitropropyl-, ethyl ester Carbamic acid, *N*-propyl-*N*-nitro, ethyl ester	$C_6H_{12}N_2O_4$ 176.17	13855-77-1	 66^{3}	4-04-00-03406 1.121^{20}	eth 4; EtOH 4 1.4431^{20}
3965	Carbamic acid, phenyl-, ethyl ester Phenylurethane	$C_9H_{11}NO_2$ 165.19	101-99-5 53	7291 237 dec	4-12-00-00619 1.1064^{30}	H_2O 1; EtOH 4; eth 4; bz 3 1.5376^{30}
3966	Carbamic acid, phenyl-, 1-methylethyl ester Propham	$C_{10}H_{13}NO_2$ 179.22	122-42-9 90	7828	4-12-00-00620 1.09^{20}	bz 4; EtOH 4 1.4989^{91}
3967	Carbamic acid, (phenylmethyl)-, ethyl ester	$C_{10}H_{13}NO_2$ 179.22	2621-78-5 49	 230	4-12-00-02250	bz 4; eth 4; EtOH 4; chl 4
3968	Carbamic acid, phenyl-, 2-methylpropyl ester	$C_{11}H_{15}NO_2$ 193.25	2291-80-7 86	 216	3-12-00-00614	bz 4; eth 4; EtOH 4
3969	Carbamic acid, propyl ester	$C_4H_9NO_2$ 103.12	627-12-3 60	 196	4-03-00-00052	ace 4; eth 4; EtOH 4
3970	Carbamic acid, propyl-, ethyl ester	$C_6H_{13}NO_2$ 131.17	623-85-8	 192	4-04-00-00480 0.9921^{15}	EtOH 4
3971	Carbamic chloride Carbamyl chloride	CH_2ClNO 79.49	463-72-9	1785 62 dec	3-03-00-00065	
3972	Carbamic chloride, diethyl- Carbamoyl chloride, *N,N*-diethyl	$C_5H_{10}ClNO$ 135.59	88-10-8	 186	4-04-00-00379	
3973	Carbamic chloride, dimethyl- Dimethylcarbamoyl chloride	C_3H_6ClNO 107.54	79-44-7 -33	 167	4-04-00-00224 1.168^{25}	 1.4540^{20}
3974	Carbamic chloride, methylphenyl-	C_8H_8ClNO 169.61	4285-42-1 88.5	 280	4-12-00-00839	eth 4; EtOH 4
3975	Carbamodithioic acid, dimethyl-, 2,4-dinitrophenyl ester	$C_9H_9N_3O_4S_2$ 287.32	89-37-2 152.5		4-06-00-01760 1.54^{20}	H_2O 1; EtOH 3; ace 3; bz 3
3976	Carbamothioic acid, butylethyl-, *S*-propyl ester Pebulate	$C_{10}H_{21}NOS$ 203.35	1114-71-2	7007 142^{20}	 0.9458^{20}	ace 4; bz 4; MeOH 4 1.4752^{20}
3977	Carbamothioic acid, diphenyl-, *S*-[2-(diethylamino)ethyl] ester Phencarbamide	$C_{19}H_{24}N_2OS$ 328.48	3735-90-8 48.5	7178 $120\text{-}6^{0.01}$		eth 4; chl 4; MeOH 4; peth 4
3978	Carbamothioic acid, *O*-ethyl ester	C_3H_7NOS 105.16	625-57-0 41	 dec	4-03-00-00294 1.069^{20}	eth 4; EtOH 4; chl 4 1.520^{20}
3979	Carbamothioic acid, *S*-ethyl ester	C_3H_7NOS 105.16	637-98-9 109	 dec	4-03-00-00294	eth 4; EtOH 4
3980	Carbamothioic chloride, dimethyl-	C_3H_6ClNS 123.61	16420-13-6 42.5	 98^{10}	3-04-00-00147	eth 4; chl 3; peth 3
3981	9H-Carbazole Dibenzo[b,d]pyrrole	$C_{12}H_9N$ 167.21	86-74-8 246.2	1792 354.69	5-20-08-00009	H_2O 1; EtOH 2; eth 2; ace 3
3982	9H-Carbazole-9-acetic acid (*N*-Carbazolyl)acetic acid	$C_{14}H_{11}NO_2$ 225.25	524-80-1 215	1793	5-20-08-00033	eth 4; EtOH 4; chl 4; HOAc 4
3983	9H-Carbazole, 9-acetyl-	$C_{14}H_{11}NO$ 209.25	574-39-0 69	 190^{6}	5-20-08-00030 1.158^{100}	 1.640^{100}
3984	9H-Carbazole, 9-butyl-	$C_{16}H_{17}N$ 223.32	1484-08-8 58	 $218\text{-}9^{19}$	5-20-08-00018	eth 4
3985	9H-Carbazole, 9-ethyl-	$C_{14}H_{13}N$ 195.26	86-28-2 68	 190^{10}	5-20-08-00016 1.059^{80}	H_2O 1; EtOH 4; eth 4 1.6394^{80}
3986	9H-Carbazole, 3-methyl-	$C_{13}H_{11}N$ 181.24	4630-20-0 208.5	 365	5-20-08-00076	bz 4; eth 4
3987	9H-Carbazole, 9-methyl-	$C_{13}H_{11}N$ 181.24	1484-12-4 89.34	343.64; 195^{12}	5-20-08-00015	eth 4
3988	9H-Carbazole, 9-(phenylmethyl)-	$C_{19}H_{15}N$ 257.33	19402-87-0 119	 267^{24}	5-20-08-00023	bz 4
3989	1H-Carbazole, 2,3,4,9-tetrahydro-	$C_{12}H_{13}N$ 171.24	942-01-8 120	 327.5	5-20-07-00468	H_2O 1; EtOH 3; eth 4; bz 4
3990	Carbendazim Carbamic acid, 1H-benzimidazol-2-yl-, methyl ester	$C_9H_9N_3O_2$ 191.19	10605-21-7 300 dec	1794	 1.45	
3991	Carbofuran 7-Benzofuranol, 2,3-dihydro-2,2-dimethyl-, methylcarbamate	$C_{12}H_{15}NO_3$ 221.26	1563-66-2 151	1810	 1.18	
3992	Carbon dioxide Carbonic anhydride	CO_2 44.01	124-38-9 -56.57 (5.11 atm)	1816 sub -78.5 (1 atm)	0-03-00-00004 *0.720^{25}	
3993	Carbon disulfide Carbon bisulfide	CS_2 76.14	75-15-0 -111.5	1818 46	4-03-00-00395 1.2632^{20}	H_2O 3; EtOH 5; eth 5; chl 3 1.6319^{20}
3994	Carbonic acid, bis(1,1-dimethylethyl) ester	$C_9H_{18}O_3$ 174.24	34619-03-9 40	 158	4-03-00-00009	EtOH 4
3995	Carbonic acid, bis(1-methylethyl) ester	$C_7H_{14}O_3$ 146.19	6482-34-4	 147	3-03-00-00009 0.9162^{20}	EtOH 4 1.3932^{20}
3996	Carbonic acid, bis(2-methylphenyl) ester	$C_{15}H_{14}O_3$ 242.27	617-09-4 60	 $144^{0.5}$	3-06-00-01256	HOAc 4
3997	Carbonic acid, bis(2-methylpropyl) ester	$C_9H_{18}O_3$ 174.24	539-92-4	 190	4-03-00-00009 0.9138^{20}	H_2O 1; EtOH 5; eth 5 1.4072^{20}
3998	Carbonic acid, dibutyl ester Dibutyl carbonate	$C_9H_{18}O_3$ 174.24	542-52-9	1558 207	4-03-00-00008 0.9251^{20}	H_2O 1; EtOH 3; eth 3 1.4117^{20}

No.	Name Synonym	Mol. Form. Mol. Wt.	CAS RN mp/°C	Merck No. bp/°C	Beil. Ref. den/g cm^{-3}	Solubility n_D
3999	Carbonic acid, diethyl ester Diethyl carbonate	$C_5H_{10}O_3$ 118.13	105-58-8 -43	3738 126	4-03-00-00005 0.9692^{25}	H_2O 1; EtOH 3; eth 3; chl 3 1.3845^{20}
4000	Carbonic acid, dimethyl ester Methyl carbonate	$C_3H_6O_3$ 90.08	616-38-6 0.5	5960 90.5	4-03-00-00003 1.0636^{25}	H_2O 1; EtOH 3; eth 3; ctc 2 1.3687^{20}
4001	Carbonic acid, diphenyl ester Phenyl carbonate	$C_{13}H_{10}O_3$ 214.22	102-09-0 83	7250 306	4-06-00-00629 1.1215^{87}	H_2O 1; EtOH 3; eth 3; ctc 3
4002	Carbonic acid, dipropyl ester Dipropyl carbonate	$C_7H_{14}O_3$ 146.19	623-96-1	 168	4-03-00-00006 0.9435^{20}	H_2O 2; EtOH 5; eth 5 1.4008^{20}
4003	Carbonic acid, ethyl methyl ester Ethyl methyl carbonate	$C_4H_8O_3$ 104.11	623-53-0 -14	 107.5	4-03-00-00004 1.012^{20}	eth 4; EtOH 4 1.3778^{20}
4004	Carbonic acid monosodium salt	$CHNaO_3$ 84.01	144-55-8	8528		H_2O 3; EtOH 1
4005	Carbonic chloride fluoride Carbonyl chloride fluoride	CClFO 82.46	353-49-1		4-03-00-00031	
4006	Carbonic dibromide Carbonyl bromide	CBr_2O 187.82	593-95-3	 64.5	 2.52^{15}	
4007	Carbonic dichloride Phosgene	CCl_2O 98.92	75-44-5 -127.9	7310 8	4-03-00-00031 $*1.3719^{25}$	bz 3; ctc 3; chl 3; tol 3
4008	Carbonic difluoride Carbonyl fluoride	CF_2O 66.01	353-50-4 -111.2	1826 -84.5	4-03-00-00021 1.139^{25}	
4009	Carbonic dihydrazide Carbohydrazide	CH_6N_4O 90.09	497-18-7 154	1811	4-03-00-00240 1.616^{20}	H_2O 4; EtOH 4
4010	Carbonic dihydrazide, 2,2'-diphenyl- *sym*-Diphenylcarbazide	$C_{13}H_{14}N_4O$ 242.28	140-22-7 170	3321 dec	4-15-00-00182	H_2O 2; EtOH 3; eth 2; ace 3
4011	Carbon monoxide Carbon oxide	CO 28.01	630-08-0 -205	1820 -191.5	0-01-00-00720 0.7909^{-19}	H_2O 2; bz 3; HOAc 3
4012	Carbonochloridic acid, butyl ester Butyl chloroformate	$C_5H_9ClO_2$ 136.58	592-34-7	 142	4-03-00-00025 1.074^{25}	eth 5; ace 3; ctc 2 1.4114^{20}
4013	Carbonochloridic acid, 2-chloroethyl ester 2-Chloroethyl chloroformate	$C_3H_4Cl_2O_2$ 142.97	627-11-2	 155	4-03-00-00024 1.3847^{20}	H_2O 1; EtOH 3; eth 3; ace 3 1.4483^{20}
4014	Carbonochloridic acid, chloromethyl ester Chloromethyl chloroformate	$C_2H_2Cl_2O_2$ 128.94	22128-62-7	 107	4-03-00-00030 1.465^{15}	ace 4; eth 4 1.4286^{22}
4015	Carbonochloridic acid, 3-chloropropyl ester	$C_4H_6Cl_2O_2$ 157.00	628-11-5	 177	3-03-00-00025 1.2926^{25}	H_2O 1 1.4456^{20}
4016	Carbonochloridic acid, dichloromethyl ester	$C_2HCl_3O_2$ 163.39	22128-63-8	 110.5	4-03-00-00031 1.5600^{15}	ace 4
4017	Carbonochloridic acid, 2-ethoxyethyl ester	$C_5H_9ClO_3$ 152.58	628-64-8	 67.2^{14}	3-03-00-00029 1.1341^{25}	EtOH 4 1.4169^{25}
4018	Carbonochloridic acid, ethyl ester Ethyl chloroformate	$C_3H_5ClO_2$ 108.52	541-41-3 -80.6	3742 95	4-03-00-00023 1.1352^{20}	bz 4; eth 4; chl 4 1.3974^{20}
4019	Carbonochloridic acid, 2-methoxyethyl ester	$C_4H_7ClO_3$ 138.55	628-12-6	 58.7^{13}	3-03-00-00029 1.1905^{25}	eth 4 1.4163^{20}
4020	Carbonochloridic acid, methyl ester Methyl chlorocarbonate	$C_2H_3ClO_2$ 94.50	79-22-1	5966 70.5	4-03-00-00023 1.2231^{20}	EtOH 5; eth 5; bz 3; ctc 3 1.3868^{20}
4021	Carbonochloridic acid, 3-methylbutyl ester	$C_6H_{11}ClO_2$ 150.60	628-50-2	 154.3	4-03-00-00026 1.0288^{17}	eth 4; EtOH 4 1.4176^{20}
4022	Carbonochloridic acid, 1-methylethenyl ester Isopropenyl chloroformate	$C_4H_5ClO_2$ 120.54	57933-83-2	 102	3-03-00-00028 1.101^{20}	eth 4
4023	Carbonochloridic acid, 1-methylethyl ester Isopropyl chloroformate	$C_4H_7ClO_2$ 122.55	108-23-6	 105	4-03-00-00024	eth 4 1.4013^{20}
4024	Carbonochloridic acid, 2-methylpropyl ester Isobutyl chlorocarbonate	$C_5H_9ClO_2$ 136.58	543-27-1	5023 128.8	4-03-00-00026 1.0426^{18}	EtOH 3; eth 5; bz 3; chl 3 1.4071^{18}
4025	Carbonochloridic acid, pentyl ester	$C_6H_{11}ClO_2$ 150.60	638-41-5	 $60\text{-}2^{15}$	4-03-00-00026	eth 3 1.4181^{18}
4026	Carbonochloridic acid, phenyl ester Phenyl chloroformate	$C_7H_5ClO_2$ 156.57	1885-14-9	 71^{9}	4-06-00-00629	
4027	Carbonochloridic acid, phenylmethyl ester Carbobenzoxy chloride	$C_8H_7ClO_2$ 170.60	501-53-1	1807 103^{20}	4-06-00-02278 1.195^{25}	eth 3; ace 3; bz 3 1.5190^{20}
4028	Carbonochloridic acid, propyl ester Propyl chlorocarbonate	$C_4H_7ClO_2$ 122.55	109-61-5	7860 115.2	4-03-00-00024 1.0901^{20}	EtOH 5; eth 5 1.4035^{20}
4029	Carbonochloridic acid, trichloromethyl ester Diphosgene	$C_2Cl_4O_2$ 197.83	503-38-8 -57	3339 128	4-03-00-00033 1.6525^{14}	eth 4; EtOH 4 1.4566^{22}
4030	Carbonocyanidic acid, ethyl ester	$C_4H_5NO_2$ 99.09	623-49-4	 115.5	4-02-00-01862 1.003^{25}	H_2O 1; EtOH 3; eth 3; ctc 3 1.3820^{20}
4031	Carbonodithioic acid, *S,S*-diethyl ester	$C_5H_{10}OS_2$ 150.27	623-80-3	 197	3-03-00-00339 1.085^{20}	eth 4; EtOH 4 1.5237^{18}
4032	Carbonothioic dichloride Thiophosgene	CCl_2S 114.98	463-71-8	 73	4-03-00-00281 1.508^{15}	H_2O 1; EtOH 3; eth 3 1.5442^{20}
4033	Carbonothioic dihydrazide 1,3-Diamino-2-thiourea	CH_6N_4S 106.15	2231-57-4 170 dec		4-03-00-00388	H_2O 4
4034	Carbonotrithioic acid Trithiocarbonic acid	CH_2S_3 110.22	594-08-1 -30	9671 57 dec	4-03-00-00428 1.47^{17}	EtOH 4; tol 4; chl 4
4035	Carbon oxide sulfide Carbonyl sulfide	COS 60.08	463-58-1 -138.8	 -50	4-03-00-00271 1.028^{17}	H_2O 2; EtOH 3; KOH 4 1.24^{-87}
4036	Carbon selenide Carbon diselenide	CSe_2 169.93	506-80-9 -45.5	1817 125.5	4-03-00-00436 2.6823^{20}	 1.8454^{20}
4037	Carbophenothion Phosphorodithioic acid, *S*-[[(4-chlorophenyl)thio]methyl] *O,O*-diethyl ester	$C_{11}H_{16}ClO_2PS_3$ 342.87	786-19-6	1827 $82^{0.01}$	 1.271^{20}	

No.	Name Synonym	Mol. Form. Mol. Wt.	CAS RN mp/°C	Merck No. bp/°C	Beil. Ref. den/g cm^{-3}	Solubility n_D
4038	Carbosulfan 2,3-Dihydro-2,2-dimethyl-7-benzofuranyl (dibutylamino)thio methylcarbamate	$C_{20}H_{32}N_2O_3S$ 380.55	55285-14-8	 126	 1.056^{20}	
4039	Carboxin 1,4-Oxathiil*N*-3-carboxamide, 5,6-dihydro-2-methyl-*N*-phenyl-	$C_{12}H_{13}NO_2S$ 235.31	5234-68-4 94	1832		
4040	Card-20(22)-enolide, 3-[(6-deoxy-α-*L*-mannopyranosyl)oxy]-1,5,11,14,19-pentahydroxy-, (1β,3β Ouabain	$C_{29}H_{44}O_{12}$ 584.66	630-60-4 200	6854	5-18-05-00625	H_2O 2; EtOH 4
4041	Card-20(22)-enolide, 3-[(*O*-2,6-dideoxy-β-*D*-ribo-hexopyranosyl-(1-4)-*O*-2,6-dideoxy-β-*D*-ribo-hexopyranosyl-(1-4)-2,6-dideoxy-β-*D*, 12,14-dihydroxy Digoxin	$C_{41}H_{64}O_{14}$ 780.95	20830-75-5 248-50 dec	3150	5-18-04-00381	EtOH 4
4042	Card-20(22)-enolide, 3-[(*O*-2,6-dideoxy-β-*D*-ribo-hexopyranosyl-(1-4)-*O*-2,6-dideoxy-β-*D*-ribo-hexopyranosyl-(1-4)-2,6-dideoxy-β-*D*, 14,16-dihydroxy Gitoxin	$C_{41}H_{64}O_{14}$ 780.95	4562-36-1 285 dec	4328	5-18-04-00388	
4043	Card-20(22)-enolide, 3-[(*O*-2,6-dideoxy-β-*D*-ribo-hexopyranosyl-(1-4)-*O*-2,6-dideoxy-β-*D*-ribo-hexopyranosyl-(1-4)-2,6-dideoxy-β-*D*, 14-dihydroxy Digitoxin	$C_{41}H_{64}O_{13}$ 764.95	71-63-6 255.5	3146	5-18-03-00354	H_2O 2; EtOH 4; eth 3; chl 3
4044	Card-20(22)-enolide, 3,14-dihydroxy-, (3β,5β)- Digitoxigenin	$C_{23}H_{34}O_4$ 374.52	143-62-4 253	3145	5-18-03-00348	EtOH 3; MeOH 4
4045	Card-20(22)-enolide, 3-[(6-*O*-β-*D*-glucopyranosyl-β-D-glucopyranosyl)oxy]-14-hydroxy-, Uzarin	$C_{35}H_{54}O_{14}$ 698.81	20231-81-6 269	9809	4-18-00-01497	
4046	Card-20(22)-enolide, 3,11,14-trihydroxy-, (3β,5β,11α)- Sarmentogenin	$C_{23}H_{34}O_5$ 390.52	76-28-8 280	8334	5-18-04-00378	H_2O 1; EtOH 3; eth 1; ace 2
4047	Card-20(22)-enolide, 3,12,14-trihydroxy-, (3β,5β,12β)- Digoxigenin	$C_{23}H_{34}O_5$ 390.52	1672-46-4 222	3149	5-18-04-00380	EtOH 4; chl 2; MeOH 4
4048	Card-20(22)-enolide, 3,14,16-trihydroxy-, (3β,5β,16β)- Gitoxigenin	$C_{23}H_{34}O_5$ 390.52	545-26-6 234	4327	5-18-04-00385	H_2O 1; eth 2; chl 3
4049	Card-20(22)-enolide, 3,5,14-trihydroxy-19-oxo-, (3β,5β)- Strophanthidin	$C_{23}H_{32}O_6$ 404.50	66-28-4 173 dec	8818	5-18-05-00126	H_2O 1; EtOH 3; eth 1; ace 3
4050	4-Carene, (1*S*,3*R*,6*R*)-(-)- Δ4-Carene (*l*)	$C_{10}H_{16}$ 136.24	5208-49-1	 169; 64^{20}	4-05-00-00451 0.8441^{30}	ace 4; bz 4; eth 4 1.4740^{30}
4051	β,β-Carotene β-Carotene, all-*trans*-	$C_{40}H_{56}$ 536.88	7235-40-7 183	1860	4-05-00-02617 1.00^{20}	H_2O 1; EtOH 2; eth 3; ace 3
4052	β,ψ-Carotene γ-Carotene	$C_{40}H_{56}$ 536.88	472-93-5 178	1861	4-05-00-02616	H_2O 1; EtOH 1; eth 2; bz 3
4053	ψ,ψ-Carotene *trans*-Lycopene	$C_{40}H_{56}$ 536.88	502-65-8 175	5492	4-01-00-01165	EtOH 2; eth 3; bz 4; chl 4
4054	β,ε-Carotene, (6'*R*)- α-Carotene	$C_{40}H_{56}$ 536.88	7488-99-5 187.5	1859	4-05-00-02620 1.00^{20}	bz 4; eth 4; chl 4
4055	β,β-Carotene, 3'-(acetyloxy)-6',7'-didehydro-5,6-epoxy-5,5',6,6',7,8-hexahydro-3,5'-dihydroxy-8-oxo-, Fucoxanthin	$C_{40}H_{60}O_6$ 636.91	3351-86-8 168	4195	5-18-04-00673	eth 4; EtOH 4
4056	β,β-Carotene-3,3'-diol, 5,6:5',6'-diepoxy-5,5',6,6'-tetrahydro- Violaxanthin	$C_{40}H_{56}O_4$ 600.88	126-29-4 208	9902	5-19-03-00464	EtOH 3; eth 3; CS_2 3; peth 1
4057	β,β-Carotene-3,3'-diol, (3*R*,3'*R*)- Zeaxanthin	$C_{40}H_{56}O_2$ 568.88	144-68-3 215.5	10019 $226\text{-}9^{0.06}$	4-06-00-07017	H_2O 1; EtOH 2; eth 3; ace 3
4058	β,ε-Carotene-3,3'-diol, (3*R*,3'*R*,6'*R*)- Xanthophyll	$C_{40}H_{56}O_2$ 568.88	127-40-2 196	9972	3-06-00-05871	bz 4; eth 4; EtOH 4; peth 4
4059	β,ψ-Caroten-16'-oic acid, 3',4'-didehydro- Torularhodin	$C_{40}H_{52}O_2$ 564.85	514-92-1 211	9476	4-09-00-02767	py 4; chl 4; CS_2 4
4060	ψ,ψ-Caroten-16-ol Lycoxanthin	$C_{40}H_{56}O$ 552.88	19891-74-8 168	5499	4-01-00-02368	H_2O 1; EtOH 2; bz 3; CS_2 3
4061	β,β-Caroten-3-ol, (3*R*)- Cryptoxanthin	$C_{40}H_{56}O$ 552.88	472-70-8 160	2612	4-06-00-05111	bz 4; chl 4
4062	β,ψ-Caroten-3-ol, (3*R*)- Rubixanthin	$C_{40}H_{56}O$ 552.88	3763-55-1 160	8265	4-06-00-05110	EtOH 2; bz 3; chl 3; peth 2
4063	β,κ-Caroten-6'-one, 3,3'-dihydroxy-, (3*R*,3'*S*,5'*R*)- Capsanthin	$C_{40}H_{56}O_3$ 584.88	465-42-9 176	1768	4-08-00-02657	
4064	Carpaine 2,15-Dioxa-12,25-diazatricyclo[22.2.2.211,14]triacontane, carpaine deriv.	$C_{28}H_{50}N_2O_4$ 478.72	3463-92-1 121	1866	2-27-00-00209	ace 4; bz 4; eth 4; EtOH 4

No.	Name Synonym	Mol. Form. Mol. Wt.	CAS RN mp/°C	Merck No. bp/°C	Beil. Ref. den/g cm^{-3}	Solubility n_D
4065	CDEC Carbamodithioic acid, diethyl-, 2-chloro-2-propenyl ester	$C_8H_{14}ClNS_2$ 223.79	95-06-7	8882 129^{1}	1.088	
4066	Cedrene 1H-3a,7-Methanoazulene, cedrene deriv.	$C_{15}H_{24}$ 204.36	11028-42-5	262.5	3-05-00-01095 0.9342^{20}	H_2O 1; bz 3; lig 3 1.5034^{20}
4067	Cevane-3,4,12,14,16,17,20-heptol, 3-acetate, (3β,4α,16β)- Sabadine	$C_{29}H_{47}NO_8$ 537.69	124-80-1 258	8281	4-21-00-02883	ace 4; EtOH 4
4068	Cevane-3,4,7,14,15,16,20-heptol, 4,9-epoxy-, (3β,4α,7α,15α,16β)- Germine	$C_{27}H_{43}NO_8$ 509.64	508-65-6 220	4307	5-21-13-00706	bz 3; MeOH 3; alk 3; acid 3
4069	Cevane-3,4,12,14,16,17,20-heptol, 4,9-epoxy-, 3-(2-methyl-2-butenoate), [3β(*Z*),4α,16β]- Cevadine	$C_{32}H_{49}NO_9$ 591.74	62-59-9 213-4 dec	2025	4-21-00-06820	
4070	Cevane-3,4,6,7,14,15,16,20-octol, 4,9-epoxy-, (3β,4α,6α,7α,15α,16β Protoverine	$C_{27}H_{43}NO_9$ 525.64	76-45-9 221	7916	4-21-00-06841	H_2O 1; EtOH 3; bz 3; aq acid 3
4071	Chelidonine Stylophorine	$C_{20}H_{19}NO_5$ 353.37	476-32-4 135.5	2043 $220^{0.002}$	2-27-00-00615	H_2O 1; EtOH 3; eth 3; chl 3
4072	Chinomethionat Carbonic acid, dithio-, cyclic *S,S*-(6-methyl-2,3-quinoxalinediyl) ester	$C_{10}H_6N_2OS_2$ 234.30	2439-01-2 170	6933		
4073	*DL*-chiro-Inositol	$C_6H_{12}O_6$ 180.16	18685-70-6 253		4-06-00-07920 1.752^{15}	H_2O 4; EtOH 2; eth 1; bz 1
4074	*D*-chiro-Inositol, 2-deoxy- *D*-Quercitol	$C_6H_{12}O_5$ 164.16	488-73-3 236	8046	4-06-00-07883 1.5845^{13}	H_2O 4
4075	Chlorbenside Benzene, 1-chloro-4-[[(4-chlorophenyl)methyl]thio]-	$C_{13}H_{10}Cl_2S$ 269.19	103-17-3 75	2074	1.4210^{20}	
4076	Chlorbromuron Urea, *N'*-(4-bromo-3-chlorophenyl)-*N*-methoxy-*N*-methyl-	$C_9H_{10}BrClN_2O_2$ 293.55	13360-45-7 96		1.69^{20}	
4077	Chlordane 4,7-Methano-1H-indene, 1,2,4,5,6,7,8,8-octachloro-2,3,3a,4,7,7a-hexahydro-	$C_{10}H_6Cl_8$ 409.78	57-74-9 106	2079 175^{1}	1.60^{25}	
4078	Chlorfenvinphos Phosphoric acid, 2-chloro-1-(2,4-dichlorophenyl)ethenyl diethyl ester	$C_{12}H_{14}Cl_3O_4P$ 359.57	470-90-6	2087 $170^{0.05}$		
4079	Chlorflurecol 9H-Fluorene-9-carboxylic acid, 2-chloro-9-hydroxy-	$C_{14}H_9ClO_3$ 260.68	2464-37-1		1.496^{20}	
4080	Chloridazon 3(2H)-Pyridazinone, 5-amino-4-chloro-2-phenyl-	$C_{10}H_8ClN_3O$ 221.65	1698-60-8 205			
4081	Chlorimuron 2-(4-Chloro-6-methoxypyrimidin-2-ylcarbamoylsulfamoyl) benzoic acid, ethyl ester	$C_{15}H_{15}ClN_4O_6S$ 414.83	90982-32-4 186	2092		
4082	Chlorobenzilate Benzeneacetic acid, 4-chloro-α-(4-chlorophenyl)-α-hydroxy-, ethyl ester	$C_{16}H_{14}Cl_2O_3$ 325.19	510-15-6 37	$157^{0.07}$	1.2816^{20}	
4083	5-Chloro-3-methyl-4-nitro-1H-pyrazole 1H-Pyrazole, 3-chloro-5-methyl-4-nitro-	$C_4H_4ClN_3O_2$ 161.55	6814-58-0 111			
4084	Chloroneb Benzene, 1,4-dichloro-2,5-dimethoxy-	$C_8H_8Cl_2O_2$ 207.06	2675-77-6 134	268		
4085	Chlorosulfuric acid, ethyl ester	$C_2H_5ClO_3S$ 144.58	625-01-4	152.5; 93^{100}	4-01-00-01326 1.3502^{25}	eth 4; chl 4; lig 4 1.416^{20}
4086	Chlorosulfuric acid, methyl ester	CH_3ClO_3S 130.55	812-01-1 -70	134	4-01-00-01252 1.4805^{25}	ace 4; bz 4; eth 4 1.4138^{18}
4087	Chlorosulfurous acid, ethyl ester	$C_2H_5ClO_2S$ 128.58	6378-11-6	52.5^{44}; 32^{16}	3-01-00-01316 1.2766^{25}	eth 4 1.4550^{25}
4088	Chloroxuron Urea, *N'*-[4-(4-chlorophenoxy)phenyl]-*N,N*-dimethyl-	$C_{15}H_{15}ClN_2O_2$ 290.75	1982-47-4 151			
4089	Chlorpyrifos Phosphorothioic acid, *O,O*-diethyl *O*-(3,5,6-trichloro-2-pyridinyl) ester	$C_9H_{11}Cl_3NO_3PS$ 350.59	2921-88-2 42	2190		
4090	Chlorpyrifos-methyl Phosphorothioic acid, *O,O*-dimethyl *O*-(3,5,6-trichloro-2-pyridinyl) ester	$C_7H_7Cl_3NO_3PS$ 322.54	5598-13-0 43			
4091	Chlorsulfuron 2-Chloro-*N*-[[(4-methoxy-6-methyl-1,3,5-triazi*N*-2-yl)amino]carbonyl]benzenesulfonamide	$C_{12}H_{12}ClN_5O_4S$ 357.78	64902-72-3 176	2192		
4092	Chlorthiophos Phosphorothioic acid, *O*-[2,5-dichloro-4-(methylthio)phenyl] *O,O*-diethyl ester	$C_{11}H_{15}Cl_2O_3PS_2$ 361.25	21923-23-9	$150^{0.001}$		

No.	Name Synonym	Mol. Form. Mol. Wt.	CAS RN mp/°C	Merck No. bp/°C	Beil. Ref. den/g cm^{-3}	Solubility n_D
4093	Cholane 1H-Cyclopenta[a]phenanthrene, hexadecahydro-10,13-dimethyl-17-(1-methylbutyl)-, [8*R*-[8α,9β,10α,1	$C_{24}H_{42}$ 330.60	548-98-1 90	2198 $190^{0.001}$	4-05-00-01220	
4094	Cholan-24-oic acid Cholanic acid	$C_{24}H_{40}O_2$ 360.58	25312-65-6 163.5	2199	4-09-00-01992	EtOH 3; chl 3; HOAc 3
4095	Cholan-24-oic acid, 3,12-dihydroxy-, (3α,5β,12α)- Deoxycholic acid	$C_{24}H_{40}O_4$ 392.58	83-44-3 177	2881	4-10-00-01608	
4096	Cholan-24-oic acid, 3,6-dihydroxy-, (3α,5β,6α)- Hyodeoxycholic acid	$C_{24}H_{40}O_4$ 392.58	83-49-8 198.5	4794	4-10-00-01600	H_2O 2; EtOH 3; eth 2; ace 2
4097	Cholan-24-oic acid, 3,7-dihydroxy-, (3α,5β,7α)- Chenodiol	$C_{24}H_{40}O_4$ 392.58	474-25-9 119	2044	4-10-00-01604	H_2O 1; EtOH 4; eth 3; ace 4
4098	Cholan-24-oic acid, 3,7-dihydroxy-, (3α,5β,7β)- Ursodiol	$C_{24}H_{40}O_4$ 392.58	128-13-2 203	9801	4-10-00-01604	EtOH 4; eth 2
4099	Cholan-24-oic acid, 3-hydroxy-, (3α,5β)- Lithocholic acid	$C_{24}H_{40}O_3$ 376.58	434-13-9 186	5423	4-10-00-00785	H_2O 1; EtOH 3; eth 2; chl 3
4100	Cholan-24-oic acid, 3,7,12-trihydroxy-, (3α,5β,7α,12α)- Cholic acid	$C_{24}H_{40}O_5$ 408.58	81-25-4 198	2206	4-10-00-02072	H_2O 2; EtOH 3; eth 4; ace 3
4101	Cholan-24-oic acid, 3,7,12-trioxo-, (5β)- Dehydrocholic acid	$C_{24}H_{34}O_5$ 402.53	81-23-2 237	2858	4-10-00-03478	H_2O 1; EtOH 2; eth 1; ace 3
4102	Cholesta-3,5-diene	$C_{27}H_{44}$ 368.65	747-90-0 80	260^{13}	4-05-00-01620 0.925^{100}	H_2O 1; EtOH 3; eth 5; bz 5
4103	Cholesta-5,7-dien-3-ol, (3β)- 7-Dehydrocholesterol	$C_{27}H_{44}O$ 384.65	434-16-2 150.5	2857	4-06-00-04153	H_2O 1; EtOH 2; eth 3; ace 3
4104	Cholesta-8,24-dien-3-ol, (3β,5α)-	$C_{27}H_{44}O$ 384.65	128-33-6 110	$160^{0.001}$	3-06-00-02828	ace 3; chl 3; MeOH 3
4105	Cholestane, (5α)- 28,29,30-Trinorlanostane	$C_{27}H_{48}$ 372.68	481-21-0 80	2202 250^{1}	4-05-00-01227 0.9090^{88}	H_2O 1; EtOH 2; eth 4; bz 4 1.4887^{88}
4106	Cholestane, (5β)- Coprostane	$C_{27}H_{48}$ 372.68	481-20-9 72	2518	4-05-00-01226 0.9119^{87}	eth 4; chl 4 1.4884^{88}
4107	Cholestanol Dihydrocholesterol	$C_{27}H_{48}O$ 388.68	80-97-7 141.5	2203	4-06-00-03577	eth 4; chl 4
4108	Cholestan-3-ol, (3α,5α)- Epicholestanol	$C_{27}H_{48}O$ 388.68	516-95-0 185.5	3564	4-06-00-03577	chl 3
4109	Cholestan-3-ol, (3β,5β)- Coprosterol	$C_{27}H_{48}O$ 388.68	360-68-9 102	2519	4-06-00-03576	H_2O 1; EtOH 4; eth 4; bz 4
4110	Cholest-5-en-3-ol (3β)- Cholesterol	$C_{27}H_{46}O$ 386.66	57-88-5 148.5	2204 360 dec; $233^{0.5}$	4-06-00-04000 1.067^{20}	H_2O 1; EtOH 2; eth 4; ace 2
4111	Cholest-4-en-3-ol, (3β)- Allocholesterol	$C_{27}H_{46}O$ 386.66	517-10-2 132	253	3-06-00-02604	H_2O 1; EtOH 3; eth 4; ace 4
4112	Cholest-5-en-3-ol, (3α)- Epicholesterol	$C_{27}H_{46}O$ 386.66	474-77-1 141.5	3565	4-06-00-04001	EtOH 2
4113	6-Chrysenamine 6-Aminochrysene	$C_{18}H_{13}N$ 243.31	2642-98-0 210.5	2258	4-12-00-03492	
4114	Chrysene 1,2-Benzophenanthrene	$C_{18}H_{12}$ 228.29	218-01-9 258.2	2259 448	4-05-00-02554 1.274^{20}	H_2O 1; EtOH 2; eth 2; ace 2
4115	Chrysene, 5,6-dimethyl- 5,6-Dimethylchrysene	$C_{20}H_{16}$ 256.35	3697-27-6 129.3	$200^{0.5}$	4-05-00-02590	H_2O 1; EtOH 3; CS_2 3; HOAc 3
4116	5,6-Chrysenedione	$C_{18}H_{10}O_2$ 258.28	2051-10-7 239.5	sub	4-07-00-02646	H_2O 1; EtOH 3; eth 2; bz 3
4117	Chrysene, 1-methyl- 1-Methylchrysene	$C_{19}H_{14}$ 242.32	3351-28-8 256.5	sub 140	4-05-00-02572	H_2O 1; EtOH 3
4118	Chrysene, octadecahydro-	$C_{18}H_{30}$ 246.44	2090-14-4 115	353	3-05-00-01107	EtOH 4
4119	Cinchonan-6',9-diol, (8α,9*R*)- Cupreine	$C_{19}H_{22}N_2O_2$ 310.40	524-63-0 202	2626	4-23-00-03260	EtOH 4
4120	Cinchonan-9-ol, (9*S*)- Cinchonine	$C_{19}H_{22}N_2O$ 294.40	118-10-5 265	2289	5-23-12-00406	
4121	Cinchonan-9-ol, 10,11-dihydro-, (8α,9*R*)- Hydrocinchonidine	$C_{19}H_{24}N_2O$ 296.41	485-64-3 229	4705		EtOH 4
4122	Cinchonan-9-ol, 10,11-dihydro-, (9*S*)- Hydrocinchonine	$C_{19}H_{24}N_2O$ 296.41	485-65-4 268.5	4706		H_2O 3; EtOH 2; eth 1
4123	Cinchonan-9-ol, 10,11-dihydro-6'-methoxy-, (8α,9*R*)- Hydroquinine	$C_{20}H_{26}N_2O_2$ 326.44	522-66-7 172.5	4737	5-23-13-00340	ace 4; eth 4; EtOH 4; chl 4
4124	Cinchonan-9-ol, 10,11-dihydro-6'-methoxy-, (9*S*)- Hydroquinidine	$C_{20}H_{26}N_2O_2$ 326.44	1435-55-8 168.5	4736	5-23-13-00340	EtOH 3; eth 3; ace 3; chl 3
4125	Cinchonan-9-ol, 6'-methoxy-, (9*R*)- Epiquinidine	$C_{20}H_{24}N_2O_2$ 324.42	572-59-8 113	3570	0-23-00-00505	EtOH 4; eth 3
4126	Cinchonan-9-ol, 6'-methoxy-, (9*S*)- Quinidine	$C_{20}H_{24}N_2O_2$ 324.42	56-54-2 174	8072	5-23-13-00395	H_2O 2; EtOH 3; eth 2; bz 3

No.	Name Synonym	Mol. Form. Mol. Wt.	CAS RN mp/°C	Merck No. bp/°C	Beil. Ref. den/g cm^{-3}	Solubility n_D
4127	Cinchonan-9-ol, 6'-methoxy-, (8α,9*R*)- Quinine	$C_{20}H_{24}N_2O_2$ 324.42	130-95-0 57	8075	5-23-13-00395	H_2O 2; EtOH 4; eth 3; ace 2 1.625^{15}
4128	Cinchonan-9-ol, 6'-methoxy-, carbonate (2:1) (ester*R*), (8α,9*R*)- Quinine carbonate	$C_{41}H_{46}N_4O_5$ 674.84	146-06-5 189	8077	4-23-00-03282	
4129	Cinchonan-9-ol, 6'-methoxy-, (8α,9*R*)-, monoformate (salt) Quinine, formate	$C_{21}H_{26}N_2O_4$ 370.45	130-90-5 149.5	8081	4-23-00-03274	H_2O 4; EtOH 4; chl 4
4130	Cinchonan-9-ol, 6'-methoxy-, monohydrochloride, (8α,9*R*)- Quinine hydrochloride	$C_{20}H_{25}ClN_2O_2$ 360.88	130-89-2 159	8085		H_2O 4; EtOH 4; chl 4
4131	Cinchonan-9-ol, 6'-methoxy-, (8α,9*R*)-, mono(2-hydroxybenzoate) (salt) Quinine salicylate	$C_{27}H_{32}N_2O_6$ 480.56	750-90-3 195	8088	4-23-00-03277	EtOH 4; chl 4
4132	Cinchonan-9-ol, 6'-methoxy-, (8α,9*R*)-, sulfate (2:1) (salt) Quinine sulfate	$C_{40}H_{50}N_4O_8S$ 746.93	804-63-7 235.2	8089		EtOH 4
4133	Cinchonan-9-ol, (8α,9*R*)- Cinchonidine	$C_{19}H_{22}N_2O$ 294.40	485-71-2 210.5	2288 sub	5-23-12-00406	H_2O 1; EtOH 3; eth 2; bz 1
4134	Cinchonan-9-one, 6'-methoxy-, (8α)- Quininone	$C_{20}H_{22}N_2O_2$ 322.41	84-31-1 108	8093	4-25-00-00204	bz 4; eth 4; EtOH 4
4135	Cinnamic acid, β-propyl- 2-Hexenoic acid, 3-phenyl-	$C_{12}H_{14}O_2$ 190.24	4362-01-0 94	$183\text{-}4^{14}$	3-09-00-02811	bz 4
4136	Cinnoline 1,2-Benzodiazine	$C_8H_6N_2$ 130.15	253-66-7 38	2309 $114^{0.3}$	5-23-07-00124	eth 4; EtOH 4
4137	Ciodrin 2-Butenoic acid, 3-[(dimethoxyphosphinyl)oxy]-, 1-phenylethyl ester, (*E*)-	$C_{14}H_{19}O_6P$ 314.28	7700-17-6	2603 $135^{0.03}$	1.19^{25}	
4138	Clethodim	$C_{17}H_{26}ClNO_3S$ 359.92	99129-21-2		1.14^{20}	
4139	Clofentezine 3,6-Bis(2-chlorophenyl)-1,2,4,5-tetrazine	$C_{14}H_8Cl_2N_4$ 303.15	74115-24-5 182	2373		
4140	Clomazone 2-(2-Chlorobenzyl)-4,4-dimethyl-1,2-oxazolidin-3-one	$C_{12}H_{14}ClNO_2$ 239.70	81777-89-1	3203	1.192^{20}	
4141	Cloprop Propanoic acid, 2-(3-chlorophenoxy)-	$C_9H_9ClO_3$ 200.62	101-10-0			
4142	Clopyralid 3,6-Dichloro-2-pyridinecarboxylic acid	$C_6H_3Cl_2NO_2$ 192.00	1702-17-6 151	2398		
4143	Con-5-enin-3-amine, *N,N*-dimethyl-, (3β)- Conessine	$C_{24}H_{40}N_2$ 356.60	546-06-5 125.5	2492 $165\text{-}7^{0.1}$		H_2O 2; chl 3; HOAc 3
4144	Conhydrine, *N*-methyl- *N*-Methylconhydrine	$C_9H_{19}NO$ 157.26	94826-08-1	$94\text{-}5^{11}$	2-21-00-00011 0.9400^{20}	1.4708^{19}
4145	Coronene	$C_{24}H_{12}$ 300.36	191-07-1 437.3	525	4-05-00-02830 1.371^{25}	H_2O 1; bz 2; con sulf 1
4146	Corynan-16-carboxylic acid, 16,17,18,19-tetradehydro-17-methoxy-, methyl ester, (16E)- Corynantheine	$C_{22}H_{26}N_2O_3$ 366.46	18904-54-6 165.5	2546	4-25-00-01255	EtOH 4
4147	Coumaphos Coumarin, 3-chloro-7-hydroxy-4-methyl, *O*-ester with *O,O*-diethylphosphorothioate	$C_{14}H_{16}ClO_5PS$ 362.77	56-72-4 93	2559	1.474	
4148	*o*-Cresol, 4-(3-methylpentyl)- Phenol, 2-methyl-4-(3-methyl pentyl)	$C_{13}H_{20}O$ 192.30	882-25-7 35	$145\text{-}6^{11}$		ace 4 1.5200^{25}
4149	Crotonaldehyde, 2-bromo-, diethyl acetal 2-Butenal, 2-bromo, diethyl acetal	$C_8H_{15}BrO_2$ 223.11	98551-14-5	86^{15}	1-01-00-00380 1.2255^{21}	1.4565^{21}
4150	Crotonic acid, 2-chloro- 2-Chlorocrotonic acid	$C_4H_5ClO_2$ 120.54	600-13-5 99.8	212; 111^{14}	4-02-00-01509	eth 4; EtOH 4
4151	Crotonic acid, 4-chloro-, (*E*)- 2-Butenoic acid, 4-chloro (*trans*)	$C_4H_5ClO_2$ 120.54	26340-58-9 83	$117\text{-}8^{13}$	4-02-00-01511	eth 4
4152	Crotonic acid, 4,4-dichloro- 4,4-Dichlorocrotonic acid	$C_4H_4Cl_2O_2$ 154.98	99980-00-4 100.5	130^{18}	4-02-00-01512 1.3331^{99}	eth 4; EtOH 4; chl 4 1.4597^{99}
4153	Crufomate Phosphoramidic acid, methyl-, 2-chloro-4-(1,1-dimethylethyl)phenyl methyl ester	$C_{12}H_{19}ClNO_3P$ 291.71	299-86-5 60	2607 $118^{0.01}$		
4154	Curan-17-ol, (16α)- Geissoschizoline	$C_{19}H_{26}N_2O$ 298.43	18397-07-4 135 dec	4274	5-23-12-00211	H_2O 1; EtOH 4; eth 4; chl 4
4155	Cyanamide Cyanogenamide	CH_2N_2 42.04	420-04-2 44	2691 140^{19}	4-03-00-00145 1.282^{20}	H_2O 4; EtOH 4; eth 3; ace 3 1.4418^{48}
4156	Cyanamide, diethyl-	$C_5H_{10}N_2$ 98.15	617-83-4 -80.6	188	4-04-00-00381 0.8542^{20}	H_2O 1; EtOH 3; eth 3 1.4126^{25}
4157	Cyanamide, dimethyl-	$C_3H_6N_2$ 70.09	1467-79-4	163.5	4-04-00-00226	ace 4; eth 4; EtOH 4 1.4089^{19}
4158	Cyanamide, diphenyl- Diphenylcyanamide	$C_{13}H_{10}N_2$ 194.24	27779-01-7 73.5	237^{60}	4-12-00-00857	EtOH 4; lig 4
4159	Cyanamide, di-2-propenyl- Diallylcyanamide	$C_7H_{10}N_2$ 122.17	538-08-9	2952 142^{90}; 95^{9}	4-04-00-01078	EtOH 3; eth 2; ctc 2
4160	Cyanamide, methylphenyl-	$C_8H_8N_2$ 132.17	18773-77-8 32	137^{12}	4-12-00-00840 1.0400^{55}	

No.	Name Synonym	Mol. Form. Mol. Wt.	CAS RN mp/°C	Merck No. bp/°C	Beil. Ref. den/g cm^{-3}	Solubility n_D
4161	Cyanazine Propanenitrile, 2-[[4-chloro-6-(ethylamino)-1,3,5-triazin-2-yl]amino]-2-methyl-	$C_9H_{13}ClN_6$ 240.70	21725-46-2 168	2692		
4162	Cyanic acid Hydrogen cyanate (HOCN)	CHNO 43.03	420-05-3 -86	2693 23	4-03-00-00080 1.140^{20}	H_2O 4; bz 4; eth 4; chl 4
4163	Cyanic acid, ethyl ester Ethyl cyanate	C_3H_5NO 71.08	627-48-5	162 dec; 30^{12}	0.89^{20}	EtOH 4 1.3788^{25}
4164	Cyanic acid, sodium salt Sodium cyanate	CNNaO 65.01	917-61-3 550	8552	1.893^{20}	H_2O 3; EtOH 2; eth 1
4165	Cyanogen bromide Bromine cyanide	CBrN 105.92	506-68-3 52	2700 61.5	4-03-00-00092 2.015^{20}	H_2O 3; EtOH 3; eth 3
4166	Cyanogen chloride Chlorine cyanide	CClN 61.47	506-77-4 -6.5	2701 13	4-03-00-00090 1.186^{20}	H_2O 3; EtOH 3; eth 4
4167	Cyanogen fluoride Fluorine cyanide	CFN 45.02	1495-50-7 -82	-46		
4168	Cycloate Carbamothioic acid, cyclohexylethyl-, *S*-ethyl ester	$C_{11}H_{21}NOS$ 215.36	1134-23-2 11.5	145^{10}	1.0156^{30}	
4169	Cyclobutanamine Aminocyclobutane	C_4H_9N 71.12	2516-34-9	82	4-12-00-00003 0.8328^{20}	1.4363^{19}
4170	Cyclobutane Tetramethylene	C_4H_8 56.11	287-23-0 -90.6	2720 12.6	4-05-00-00006 0.6890^{25}	H_2O 1; EtOH 4; eth 5; ace 4 1.365^{290}
4171	Cyclobutaneacetic acid, 3-(ethoxycarbonyl)-2,2-dimethyl-, ethyl ester Pinic acid diethyl ester	$C_{13}H_{22}O_4$ 242.32	28664-03-1	$142\text{-}6^{10}$	2-09-00-00529 1.0104^{20}	1.4496^{70}
4172	Cyclobutane, 1,2-bis(methylene)-	C_6H_8 80.13	14296-80-1	64	4-05-00-00389 0.7698^{20}	1.4232^{20}
4173	Cyclobutanecarbonitrile Cyanocyclobutane	C_5H_7N 81.12	4426-11-3	149.6		
4174	Cyclobutanecarboxylic acid Cyclobutylcarboxylic acid	$C_5H_8O_2$ 100.12	3721-95-7 -2	190; $74\text{-}5^{2}$	4-09-00-00006 1.0599^{20}	H_2O 2; EtOH 5; eth 5 1.4400^{20}
4175	Cyclobutanecarboxylic acid, ethyl ester Ethyl cyclobutanecarboxylate	$C_7H_{12}O_2$ 128.17	14924-53-9	159	3-09-00-00007 0.928^{25}	ctc 2 1.4261^{20}
4176	Cyclobutanecarboxylic acid, methyl ester Methyl cyclobutanecarboxylate	$C_6H_{10}O_2$ 114.14	765-85-5	135.5		
4177	Cyclobutane, 1-chloro-1,2,2,3,3,4,4-heptafluoro- Chloroheptafluorocyclobutane	C_4ClF_7 216.49	377-41-3			
4178	Cyclobutane, 1,1-dibromo- 1,1-Dibromocyclobutane	$C_4H_6Br_2$ 213.90	33742-81-3	160	0-05-00-00017 1.930^{20}	bz 4; EtOH 4; lig 4 1.5362^{20}
4179	Cyclobutane, 1,2-dibromo- 1,2-Dibromocyclobutane	$C_4H_6Br_2$ 213.90	89033-70-5 2.5	172.5	0-05-00-00017 1.972^{0}	
4180	1,1-Cyclobutanedicarboxylic acid, diethyl ester Diethyl 1,1-cyclobutanedicarboxylate	$C_{10}H_{16}O_4$ 200.23	3779-29-1	224	4-09-00-02788 1.0456^{20}	EtOH 4; ctc 2 1.4330^{26}
4181	1,2-Cyclobutanedicarboxylic acid, dimethyl ester, *trans*- Dimethyl *trans*-1,2-cyclobutanedicarboxylate	$C_8H_{12}O_4$ 172.18	7371-67-7	225; 118^{24}	3-09-00-03799 1.1276^{20}	1.4450^{20}
4182	1,3-Cyclobutanedicarboxylic acid, *cis*-	$C_6H_8O_4$ 144.13	2398-16-5 143.5	252	3-09-00-03801	H_2O 4; EtOH 4
4183	1,3-Cyclobutanedicarboxylic acid, 2,2-dimethyl-, *cis*- *cis*-Norpinic acid	$C_8H_{12}O_4$ 172.18	3211-48-1 175	sub 100	4-09-00-02813	eth 4; chl 4
4184	1,3-Cyclobutanedicarboxylic acid, 2,2-dimethyl-, dimethyl ester Norpinic acid, dimethyl ester	$C_{10}H_{16}O_4$ 200.23	91057-79-3	229; 113^{11}	3-09-00-03829 1.0700^{14}	1.4459^{17}
4185	1,3-Cyclobutanedicarboxylic acid, *trans*-	$C_6H_8O_4$ 144.13	7439-33-0 171	sub	3-09-00-03802	H_2O 4; EtOH 4
4186	Cyclobutane, 1,2-diethenyl-, *cis*-	C_8H_{12} 108.18	16177-46-1	38^{38}	4-05-00-00412 0.8010^{20}	1.4563^{20}
4187	Cyclobutane, 1,2-diethenyl-, *trans*- *trans*-1,2-Divinylcyclobutane	C_8H_{12} 108.18	6553-48-6	112.5	4-05-00-00412 0.7817^{20}	1.4451^{20}
4188	Cyclobutane, 1,2-dimethyl-, *trans*- *trans*-1,2-Dimethylcyclobutane	C_6H_{12} 84.16	15679-02-4 -122.5	58	4-05-00-00088 0.703^{20}	1.395^{20}
4189	Cyclobutane, 1,3-dimethyl-, *cis*- *cis*-1,3-Dimethylcyclobutane	C_6H_{12} 84.16	2398-09-6	60.5	4-05-00-00089 0.7060^{25}	1.3933^{20}
4190	Cyclobutane, 1,3-dimethyl-, *trans*- *trans*-1,3-Dimethylcyclobutane	C_6H_{12} 84.16	2398-10-9	57.5	4-05-00-00089 0.6970^{25}	1.3896^{20}
4191	Cyclobutane, ethyl- Ethylcyclobutane	C_6H_{12} 84.16	4806-61-5 -142.9	70.8	4-05-00-00087 0.7284^{20}	H_2O 1; EtOH 5; eth 5; ace 3 1.4020^{20}
4192	Cyclobutane, methyl- Methylcyclobutane	C_5H_{10} 70.13	598-61-8 -161.5	36.3	4-05-00-00021 0.6884^{20}	H_2O 1; EtOH 5; eth 5; ace 3 1.3866^{20}
4193	Cyclobutane, methylene- *exo*-Methylenecyclobutane	C_5H_8 68.12	1120-56-5 -134.7	42.2	4-05-00-00216 0.7401^{20}	1.4210^{20}
4194	Cyclobutane, octafluoro- Perfluorocyclobutane	C_4F_8 200.03	115-25-3 -40.1	6666 -5.9	4-05-00-00008 $*1.500^{25}$	eth 3
4195	Cyclobutane, 1,2,3,4-tetrabromo- 1,2,3,4-Tetrabromocyclobutane	$C_4H_4Br_4$ 371.69	101257-79-8	$110^{1.5}$	4-05-00-00012 2.5672^{20}	1.6303^{20}
4196	Cyclobutanol Hydroxycyclobutane	C_4H_8O 72.11	2919-23-5	124	4-06-00-00003 0.9218^{15}	1.4371^{20}

No.	Name Synonym	Mol. Form. Mol. Wt.	CAS RN mp/°C	Merck No. bp/°C	Beil. Ref. den/g cm^{-3}	Solubility n_D
4197	Cyclobutanone	C_4H_6O 70.09	1191-95-3 -50.9	 99	4-07-00-00003 0.9547^{0}	H_2O 3; EtOH 4; eth 3; bz 3 1.4215^{20}
4198	Cyclobutene	C_4H_6 54.09	822-35-5	 2	4-05-00-00207 0.733^{0}	ace 4; bz 3; peth 3
4199	Cyclobutene, 1,2-dichloro-3,3,4,4-tetrafluoro-	$C_4Cl_2F_4$ 194.94	377-93-5 -43.4	 67	4-05-00-00209 1.534^{25}	 1.3699^{25}
4200	Cyclobutene, hexafluoro-	C_4F_6 162.03	697-11-0 -60	 5.5	4-05-00-00208 1.602^{-20}	 1.298^{-20}
4201	Cyclobutene, 1-methyl- 1-Methylcyclobutene	C_5H_8 68.12	1489-60-7	 37	4-05-00-00216 0.7244^{20}	 1.4088^{18}
4202	Cyclodecane	$C_{10}H_{20}$ 140.27	293-96-9 10	 202	4-05-00-00144 0.8538^{25}	 1.4716^{20}
4203	1,2-Cyclodecanedione Sebacil	$C_{10}H_{16}O_2$ 168.24	96-01-5 40.5	8370 $104\text{-}5^{10}$	4-07-00-02009	
4204	Cyclodecanol	$C_{10}H_{20}O$ 156.27	1502-05-2 40.5	 125^{12}	4-06-00-00138 0.9606^{20}	EtOH 3 1.4926^{20}
4205	Cyclodecanone	$C_{10}H_{18}O$ 154.25	1502-06-3 28	 $106\text{-}7^{13}$	4-07-00-00076 0.9654^{20}	bz 4; eth 4; chl 4 1.4806^{20}
4206	Cyclodecanone, 2-hydroxy- Sebacoin	$C_{10}H_{18}O_2$ 170.25	96-00-4 38.5	8371 $134\text{-}8^{14}$	4-08-00-00037	
4207	Cyclodecene, (*E*)-	$C_{10}H_{18}$ 138.25	2198-20-1	 $68\text{-}70^{10}$	4-05-00-00296 0.8672^{20}	H_2O 1; bz 3; chl 3 1.4822^{20}
4208	Cyclodecene, (*Z*)-	$C_{10}H_{18}$ 138.25	935-31-9	 195	4-05-00-00295 0.8770^{20}	 1.4854^{20}
4209	Cyclodecyne	$C_{10}H_{16}$ 136.24	3022-41-1	 204	4-05-00-00430 0.8975^{20}	 1.4903^{20}
4210	Cyclododecane	$C_{12}H_{24}$ 168.32	294-62-2 60.4	 247	4-05-00-00169 0.82^{80}	
4211	Cyclododecanone Cyclodoecanone	$C_{12}H_{22}O$ 182.31	830-13-7 59	 $126\text{-}8^{12}$	4-07-00-00100 0.9059^{66}	 1.4571^{60}
4212	1,5,9-Cyclododecatriene CDT	$C_{12}H_{18}$ 162.27	4904-61-4 -17	 240	4-05-00-01114 0.84^{100}	
4213	1,5,9-Cyclododecatriene, (*Z,E,E*)- 1,5,9-Cyclododecatriene, (*cis, trans, trans*)	$C_{12}H_{18}$ 162.27	53859-78-2 -18	 $100\text{-}1^{11}$	4-05-00-01115 0.8910^{20}	bz 4; chl 4 1.5058^{20}
4214	1,5,9-Cyclododecatriene, (*Z,Z,Z*)-	$C_{12}H_{18}$ 162.27	4736-48-5 -1.5	 108^{17}		bz 4; chl 4 1.5100^{25}
4215	Cyclododecene, (*E*)-	$C_{12}H_{22}$ 166.31	1486-75-5	 $112\text{-}5^{17}$	4-05-00-00332	bz 4; chl 4 1.4850^{20}
4216	Cyclododecene, (*Z*)-	$C_{12}H_{22}$ 166.31	1129-89-1	 133^{35}; 71^{2}	4-05-00-00332	bz 4; chl 4 1.4840^{20}
4217	9-Cycloheptadecen-1-one 9-Cycloheptadecene-1-one	$C_{17}H_{30}O$ 250.42	74244-64-7 32.5	 343; 159^{2}	3-07-00-00524 0.9170^{33}	H_2O 2; EtOH 3; bz 3 1.4830^{33}
4218	9-Cycloheptadecen-1-one, (*Z*)- Civetone	$C_{17}H_{30}O$ 250.42	542-46-1 32.5	2337 343; 159^{2}	3-07-00-00524	
4219	1,3-Cycloheptadiene	C_7H_{10} 94.16	4054-38-0 -110.4	 120.5	4-05-00-00390 0.868^{25}	 1.4978^{20}
4220	2,4-Cycloheptadien-1-one, 2,6,6-trimethyl- Eucarvone	$C_{10}H_{14}O$ 150.22	503-93-5	 105^{22}; 88^{16}	4-07-00-00311 0.9490^{20}	eth 3; ace 3 1.5087^{20}
4221	Cycloheptane	C_7H_{14} 98.19	291-64-5 -8.0	 118.4	4-05-00-00092 0.8098^{20}	H_2O 1; EtOH 4; eth 4; bz 3 1.4436^{20}
4222	Cycloheptane, bromo- Cycloheptyl bromide	$C_7H_{13}Br$ 177.08	2404-35-5	 101.5^{40}; 75^{12}	4-05-00-00093 1.2887^{22}	H_2O 1; eth 4; chl 4 1.4996^{20}
4223	Cycloheptanecarboxylic acid Cycloheptylcarboxylic acid	$C_8H_{14}O_2$ 142.20	1460-16-8	 259; 130^{8}	4-09-00-00038 1.0423^{20}	ace 3 1.4753^{20}
4224	Cycloheptanecarboxylic acid, ethyl ester Ethyl cycloheptanecarboxylate	$C_{10}H_{18}O_2$ 170.25	32777-26-7	 74^{3}	4-09-00-00038 0.9515^{20}	 1.4482^{20}
4225	1,2-Cycloheptanedione	$C_7H_{10}O_2$ 126.16	3008-39-7 -40	 $107\text{-}9^{17}$	4-07-00-01989 1.0583^{22}	EtOH 3 1.4689^{22}
4226	Cycloheptane, methyl- Methylcycloheptane	C_8H_{16} 112.22	4126-78-7	 134	4-05-00-00114 0.800^{20}	bz 4; eth 4; EtOH 4; peth 4 1.439^{18}
4227	Cycloheptane, 1,1,2-trimethyl- 1,1,2-Trimethylcycloheptane	$C_{10}H_{20}$ 140.27	35099-89-9	 $104\text{-}5^{100}$	3-05-00-00127 0.8243^{20}	 1.4527^{20}
4228	Cycloheptane, 1,1,4-trimethyl- 1,1,4-Trimethylcycloheptane	$C_{10}H_{20}$ 140.27	2158-55-6	 164	4-05-00-00146 0.8011^{20}	 1.4420^{20}
4229	Cycloheptanol	$C_7H_{14}O$ 114.19	502-41-0 2	 185	4-06-00-00094 0.9554^{20}	H_2O 2; EtOH 4; eth 4 1.40705^{20}
4230	Cycloheptanol, 1-methyl- 1-Methyl-1-cycloheptanol	$C_8H_{16}O$ 128.21	3761-94-2	 100.6	4-06-00-00114 0.9382^{21}	 1.4677^{22}
4231	Cycloheptanol, 2-methyl- 2-Methylcycloheptanol	$C_8H_{16}O$ 128.21	59777-92-3	 191	3-06-00-00083 0.9492^{15}	 1.4762^{15}
4232	Cycloheptanol, 4-methyl-	$C_8H_{16}O$ 128.21	90200-61-6	 $105\text{-}6^{39}$	3-06-00-00084 0.9134^{31}	 1.4574
4233	Cycloheptanol, 2,6,6-trimethyl- 3,3,7-Trimethylcycloheptanol	$C_{10}H_{20}O$ 156.27	33515-82-1	 216	1-06-00-00018 0.9096^{24}	 1.4369
4234	Cycloheptanone Suberone	$C_7H_{12}O$ 112.17	502-42-1	2728 178.5	4-07-00-00039 0.9508^{20}	H_2O 1; EtOH 4; eth 3 1.4608^{20}
4235	Cycloheptanone, 2-methyl- 2-Methylcycloheptanone	$C_8H_{14}O$ 126.20	932-56-9	 185	4-07-00-00051 0.9395^{18}	 1.461^{18}
4236	Cycloheptanone, 2-phenyl- 2-Phenylcycloheptanone	$C_{13}H_{16}O$ 188.27	14996-78-2 22.2	 141^{5}; 125^{2}	4-07-00-01061	 1.5395^{20}

No.	Name	Synonym	Mol. Form.	Mol. Wt.	CAS RN	mp/°C	Merck No.	bp/°C	Beil. Ref.	den/g cm^{-3}	Solubility	n_D
4237	Cycloheptanone, 2,2,6-trimethyl-		$C_{10}H_{18}O$	154.25	1686-41-5			191	3-07-00-00138	0.878^{14}		1.4421^{16}
4238	Cycloheptanone, 2,6,6-trimethyl-	Tetrahydroeucarvone	$C_{10}H_{18}O$	154.25	4436-59-3			208.5	4-07-00-00079	0.9095^{18}		1.4568^{18}
4239	Cycloheptasiloxane, tetradecamethyl-	Tetradecamethylcycloheptasiloxane	$C_{14}H_{42}O_7Si_7$	519.08	107-50-6	-26		154^{20}	4-04-00-04130	0.9703^{20}		1.4040^{20}
4240	1,3,5-Cycloheptatriene	Tropilidene	C_7H_8	92.14	544-25-2	-79.5		117; 60.5^{122}	4-05-00-00765	0.8875^{19}	H_2O 1; EtOH 3; eth 3; bz 4	1.5343^{20}
4241	1,3,6-Cycloheptatriene-1-carboxylic acid, 5,5-dimethyl-	Thujic acid	$C_{10}H_{12}O_2$	164.20	499-89-8		9325		4-09-00-01810			
4242	2,4,6-Cycloheptatrien-1-one		C_7H_6O	106.12	539-80-0	-7		113^{15}; 84^{6}	4-07-00-00501	1.095^{22}	bz 4; chl 4	1.6172^{22}
4243	2,4,6-Cycloheptatrien-1-one, 2-hydroxy-		$C_7H_6O_2$	122.12	533-75-5	50.8		sub 40	4-08-00-00159		H_2O 3; eth 3; ace 3; os 3	
4244	Cycloheptene		C_7H_{12}	96.17	628-92-2	-56		115	4-05-00-00244	0.8228^{20}	H_2O 1; EtOH 3; eth 3; bz 3	1.4552^{20}
4245	Cycloheptene, 1-methyl-	1-Methylcycloheptene	C_8H_{14}	110.20	1453-25-4			136	4-05-00-00265	0.8243^{22}		1.4581^{20}
4246	Cycloheptene, 5-methyl-	5-Methylcycloheptene	C_8H_{14}	110.20	2505-06-8			$69\text{-}70^{38}$	3-05-00-00211	0.7606^{31}		1.4202^{31}
4247	Cycloheptene, 1,4,4-trimethyl-	1,4,4-Trimethyl-1-cycloheptene	$C_{10}H_{18}$	138.25	4755-36-6			165	3-05-00-00232	0.8185^{20}		1.4561^{20}
4248	1,3-Cyclohexadiene		C_6H_8	80.13	592-57-4	-89		80.5	4-05-00-00382	0.8405^{20}	H_2O 1; EtOH 3; eth 4; bz 3	1.4755^{20}
4249	1,4-Cyclohexadiene	1,4-Dihydrobenzene	C_6H_8	80.13	628-41-1	-49.2		85.5	4-05-00-00385	0.8471^{20}	H_2O 1; EtOH 5; eth 5; bz 3	1.4725^{20}
4250	1,3-Cyclohexadiene-1-carboxaldehyde, 2,6,6-trimethyl-	Safranal	$C_{10}H_{14}O$	150.22	116-26-7		8286	70^{1}	4-07-00-00318	0.9734^{19}	EtOH 4; peth 4	1.5281^{19}
4251	1,4-Cyclohexadiene-1,2-dicarbonitrile, 4,5-dichloro-3,6-dioxo-	2,3-Dichloro-5,6-dicyanobenzoquinone	$C_8Cl_2N_2O_2$	227.01	84-58-2	214.5	3052		0-10-00-00902		bz 4; HOAc 4; diox 4	
4252	1,3-Cyclohexadiene, 1,3-dimethyl-	1,3-Dimethyl-1,3-cyclohexadiene	C_8H_{12}	108.18	4573-05-1			135.5	3-05-00-00325	0.8373^{20}		1.4776^{20}
4253	1,3-Cyclohexadiene, 1,4-dimethyl-	1,4-Dimethyl-1,3-cyclohexadiene	C_8H_{12}	108.18	26120-52-5			136.5	4-05-00-00409	0.830^{20}		1.4792^{19}
4254	1,3-Cyclohexadiene, 1,5-dimethyl-	1,5-Dimethyl-1,3-cyclohexadiene	C_8H_{12}	108.18	1453-17-4			130	1-05-00-00063	0.821^{20}		1.471^{20}
4255	1,3-Cyclohexadiene, 2,5-dimethyl-	2,5-Dimethyl-1,3-cyclohexadiene	C_8H_{12}	108.18	2050-33-1			127	3-05-00-00325	0.8245^{18}		1.4631^{18}
4256	1,3-Cyclohexadiene, 2,6-dimethyl-	1,3-Cyclohexadiene, 3,5-dimethyl	C_8H_{12}	108.18	2050-32-0			131	0-05-00-00119	0.8225^{20}		1.4675^{20}
4257	2,5-Cyclohexadiene-1,4-dione	*p*-Benzoquinone	$C_6H_4O_2$	108.10	106-51-4	115.7	8103	sub	4-07-00-02065	1.318^{20}	H_2O 2; EtOH 3; eth 3; chl 3	
4258	2,5-Cyclohexadiene-1,4-dione, 2,5-dichloro-3,6-dihydroxy-	Chloranilic acid	$C_6H_2Cl_2O_4$	208.98	87-88-7	283.5	2072		4-08-00-02707		H_2O 3	
4259	2,5-Cyclohexadiene-1,4-dione, 2,5-dihydroxy-3,6-dinitro-	Nitranilic acid	$C_6H_2N_2O_8$	230.09	479-22-1	100 dec	6489	exp 170	3-08-00-03351		H_2O 4; EtOH 4; eth 1	
4260	2,5-Cyclohexadiene-1,4-dione, 2,5-dihydroxy-3-methoxy-6-methyl-	Spinulosin	$C_8H_8O_5$	184.15	85-23-4	202.5	8706	sub 120	3-08-00-03983		H_2O 2; alk 3	
4261	2,5-Cyclohexadiene-1,4-dione, 2,5-dihydroxy-3-undecyl-	Embelin	$C_{17}H_{26}O_4$	294.39	550-24-3	142.5	3513		4-08-00-02769		bz 4; eth 4; EtOH 4	
4262	2,5-Cyclohexadiene-1,4-dione, 2,6-dimethoxy-	2,6-Dimethoxy-*p*-quinone	$C_8H_8O_4$	168.15	530-55-2	256	3214	sub	4-08-00-02710		H_2O 2; EtOH 2; eth 2; tfa 3	
4263	2,5-Cyclohexadiene-1,4-dione, 2,3-dimethyl-		$C_8H_8O_2$	136.15	526-86-3	55		sub	4-07-00-02093		H_2O 2; EtOH 3; eth 3; chl 3	
4264	2,5-Cyclohexadiene-1,4-dione, 2,6-dimethyl-		$C_8H_8O_2$	136.15	527-61-7	72.5		sub	4-07-00-02095	1.0479^{28}	chl 3	
4265	2,5-Cyclohexadiene-1,4-dione, 3-hydroxy-2-methoxy-5-methyl-	Fumigatin	$C_8H_8O_4$	168.15	484-89-9	116	4201		3-08-00-03374		ace 4; bz 4; eth 4; EtOH 4	
4266	2,5-Cyclohexadiene-1,4-dione, 2-methyl-		$C_7H_6O_2$	122.12	553-97-9	69		sub	4-07-00-02088	1.08^{75}	H_2O 2; EtOH 3; eth 3	
4267	2,5-Cyclohexadiene-1,4-dione, 2-methyl-5-(1-methylethyl)-		$C_{10}H_{12}O_2$	164.20	490-91-5	45.5		232	4-07-00-02100		chl 3	
4268	2,5-Cyclohexadiene-1,4-dione, 2,3,5,6-tetrachloro-	Chloranil	$C_6Cl_4O_2$	245.88	118-75-2	290	2071	sub	4-07-00-02083		H_2O 1; EtOH 2; eth 3; chl 2	
4269	2,5-Cyclohexadiene-1,4-dione, 2,3,5,6-tetrahydroxy-	Tetroquinone	$C_6H_4O_6$	172.09	319-89-1		9177		4-08-00-03604		H_2O 2; EtOH 4; eth 2; ctc 2	
4270	2,5-Cyclohexadiene-1,4-dione, 2,3,5,6-tetramethyl-	Duroquinone	$C_{10}H_{12}O_2$	164.20	527-17-3	111.5	3452		4-07-00-02102		H_2O 1; EtOH 3; eth 3; ace 3	
4271	1,3-Cyclohexadiene, 2-methyl-	4,5-Dihydrotoluene	C_7H_{10}	94.16	1489-57-2			107.5	3-06-00-00211	0.8260^{18}		1.4662^{18}

No.	Name Synonym	Mol. Form. Mol. Wt.	CAS RN mp/°C	Merck No. bp/°C	Beil. Ref. den/g cm^{-3}	Solubility n_D
4272	1,3-Cyclohexadiene, 5-methyl-, (±)- 1,3-Cyclohexadiene, 5-methyl (*DL*)	C_7H_{10} 94.16	116781-86-3	 101.5	3-05-00-00318 0.8354^{20}	bz 4; eth 4; EtOH 4; lig 4 1.4763^{20}
4273	1,3-Cyclohexadiene, 1-methyl-4-(1-methylethyl)-	$C_{10}H_{16}$ 136.24	99-86-5	 175	4-05-00-00435 0.8375^{19}	H_2O 1; EtOH 5; eth 5 1.477^{19}
4274	1,3-Cyclohexadiene, 2-methyl-5-(1-methylethyl)-, (±)-	$C_{10}H_{16}$ 136.24	13811-01-3	 175.5	3-05-00-00342 0.8410^{22}	H_2O 1; EtOH 1; eth 3 1.4772^{19}
4275	1,4-Cyclohexadiene, 1-methyl-4-(1-methylethyl)-	$C_{10}H_{16}$ 136.24	99-85-4	 183	4-05-00-00436 0.849^{20}	 1.4765^{14}
4276	1,3-Cyclohexadiene, 1,2,3,4,5,5,6,6-octafluoro-	C_6F_8 224.05	377-70-8	 63.5	4-05-00-00384 1.601^{20}	 1.3149^{20}
4277	1,4-Cyclohexadiene, 1,2,3,3,4,5,6,6-octafluoro-	C_6F_8 224.05	775-51-9	 57.5	4-05-00-00386 1.6010^{25}	 1.318^{18}
4278	2,5-Cyclohexadien-1-one, 4-[bis(4-hydroxyphenyl)methylene]- Aurin	$C_{19}H_{14}O_3$ 290.32	603-45-2 308-10 dec	899	4-08-00-02646	H_2O 1; EtOH 3; eth 2; bz 1
4279	2,5-Cyclohexadien-1-one, 2,6-dibromo-4-(chloroimino)- 2,6-Dibromoquinone-4-chlorimide	$C_6H_2Br_2ClNO$ 299.35	537-45-1 83	3010	4-07-00-02086	EtOH 4
4280	2,5-Cyclohexadien-1-one, 2,6-dichloro-4-(chloroimino)- Gibbs' reagent	$C_6H_2Cl_3NO$ 210.45	101-38-2 66	4315	3-07-00-03377	
4281	2,4-Cyclohexadien-1-one, 3,5-dihydroxy-2,6,6-tris(3-methyl-2-butenyl)-4-(3-methyl-1-oxobutyl)- Lupulon	$C_{26}H_{38}O_4$ 414.59	468-28-0 93	5482	3-07-00-04753	H_2O 1; EtOH 3; peth 3; hx 3
4282	2,4-Cyclohexadien-1-one, 3,5,6-trihydroxy-4,6-bis(3-methyl-2-butenyl)-2-(3-methyl-1-oxobutyl)-, (*R*)- Humulon	$C_{21}H_{30}O_5$ 362.47	26472-41-3 66.5	4673	4-08-00-03410	H_2O 2; EtOH 3; eth 3; ace 3
4283	Cyclohexanamine Cyclohexylamine	$C_6H_{13}N$ 99.18	108-91-8 -17.7	2735 134	4-12-00-00008 0.8191^{20}	H_2O 3; EtOH 4; eth 5; ace 5 1.4372^{20}
4284	Cyclohexanamine, *N*-cyclohexyl- Dicyclohexylamine	$C_{12}H_{23}N$ 181.32	101-83-7 -0.1	3085 256 dec; 114^{9}	4-12-00-00022 0.9123^{20}	H_2O 2; EtOH 3; eth 3; bz 3 1.4842^{20}
4285	Cyclohexanamine, *N,N*-diethyl-	$C_{10}H_{21}N$ 155.28	91-65-6	 193; 85^{20}	4-12-00-00019 0.872^{0}	EtOH 3; ctc 2
4286	Cyclohexanamine, *N,N*-dimethyl- Cyclohexyldimethylamine	$C_8H_{17}N$ 127.23	98-94-2	 162	4-12-00-00018	
4287	Cyclohexanamine, 2-ethyl- 2-Ethylcyclohexylamine	$C_8H_{17}N$ 127.23	6850-36-8	 170.5	3-12-00-00093 0.8744^{20}	 1.4682^{20}
4288	Cyclohexanamine, *N*-ethyl-	$C_8H_{17}N$ 127.23	5459-93-8	 164	4-12-00-00018 0.868^{0}	H_2O 2; EtOH 5; eth 5; ctc 2
4289	Cyclohexanamine, *N*-(2-ethylhexyl)-	$C_{14}H_{29}N$ 211.39	5432-61-1	 270; 140^{14}	3-12-00-00018 0.8473^{20}	ctc 2
4290	Cyclohexanamine, 1-ethynyl-	$C_8H_{13}N$ 123.20	30389-18-5	 $65\text{-}6^{20}$	4-12-00-00214 0.913^{25}	 1.4817^{20}
4291	Cyclohexanamine, *N,N'*-methanetetraylbis- Dicyclohexylcarbodiimide	$C_{13}H_{22}N_2$ 206.33	538-75-0 34.5	3086 123^{6}; $99^{0.5}$	4-12-00-00072	
4292	Cyclohexanamine, 2-methyl-, *cis*-	$C_7H_{15}N$ 113.20	2164-19-4	 153.5	4-12-00-00118 0.8778^{20}	 1.4688^{20}
4293	Cyclohexanamine, 2-methyl-, *trans*- *trans*-2-Methylcyclohexylamine	$C_7H_{15}N$ 113.20	931-10-2	 150	4-12-00-00118 0.8685^{20}	 1.4650^{20}
4294	Cyclohexanamine, 3-methyl-, *cis*-	$C_7H_{15}N$ 113.20	1193-16-4	 153	4-12-00-00121 0.8552^{20}	 1.4538^{20}
4295	Cyclohexanamine, 3-methyl-, *trans*-	$C_7H_{15}N$ 113.20	1193-17-5	 151.5	4-12-00-00121 0.8572^{20}	 1.4547^{20}
4296	Cyclohexanamine, 4-methyl-, *cis*-	$C_7H_{15}N$ 113.20	2523-56-0	 153.5	4-12-00-00122 0.8567^{20}	 1.4559^{20}
4297	Cyclohexanamine, 4-methyl-, *trans*-	$C_7H_{15}N$ 113.20	2523-55-9	 152	4-12-00-00122 0.8543^{20}	 1.4550^{20}
4298	Cyclohexanamine, *N*-methyl-	$C_7H_{15}N$ 113.20	100-60-7	 147	4-12-00-00017 0.8660^{23}	H_2O 2; EtOH 4; eth 5; chl 3 1.4560^{20}
4299	Cyclohexanamine, 4,4'-methylenebis-	$C_{13}H_{26}N_2$ 210.36	1761-71-3 15	 320	 0.92^{75}	
4300	Cyclohexanamine, 4,4' methylenebis-, [cis(*cis*)]-	$C_{13}H_{26}N_2$ 210.36	6693-31-8 61	 141^{2}	4-13-00-00031	H_2O 1 1.5014^{27}
4301	Cyclohexanamine, 4,4'-methylenebis-, [*trans*(*cis*)]-	$C_{13}H_{26}N_2$ 210.36	6693-30-7 36.5	 $127\text{-}8^{1.2}$	4-13-00-00031 0.9608^{25}	H_2O 1 1.5046^{25}
4302	Cyclohexanamine, 4,4'-methylenebis-, [trans(*trans*)]-	$C_{13}H_{26}N_2$ 210.36	6693-29-4 64.7	 $131^{0.8}$	4-13-00-00031	H_2O 1 1.5032^{25}
4303	Cyclohexanamine, *N*-(1-methylethyl)-	$C_9H_{19}N$ 141.26	1195-42-2	 $60\text{-}5^{12}$	4-12-00-00019 0.859^{25}	 1.4480^{20}

No.	Name Synonym	Mol. Form. Mol. Wt.	CAS RN mp/°C	Merck No. bp/°C	Beil. Ref. den/g cm^{-3}	Solubility n_D
4304	Cyclohexanamine, *N*-2-propenyl-	$C_9H_{17}N$ 139.24	6628-00-8	 65-6^{12}	4-12-00-00021 0.962^{25}	 1.4664^{20}
4305	Cyclohexane Hexahydrobenzene	C_6H_{12} 84.16	110-82-7 6.59	2729 80.73	4-05-00-00027 0.7739^{25}	H_2O 1; EtOH 5; eth 5; ace 5 1.4266^{20}
4306	Cyclohexaneacetic acid Cyclohexylacetic acid	$C_8H_{14}O_2$ 142.20	5292-21-7 33	 245	4-09-00-00038 1.0423^{18}	H_2O 2; eth 3; ace 3; os 3 1.4775^{20}
4307	Δ1,α-Cyclohexaneacetic acid, 3,3-dimethyl- Allocyclogeranic acid	$C_{10}H_{16}O_2$ 168.24	6790-47-2	 145^{14}	4-09-00-00161 0.9921^{17}	 1.4765^{17}
4308	Cyclohexaneacetic acid, ethyl ester	$C_{10}H_{18}O_2$ 170.25	5452-75-5	 211	4-09-00-00038 0.9537^{14}	 1.451^{14}
4309	Cyclohexaneacetic acid, α-ethyl-1-hydroxy- Cyclobutyrol	$C_{10}H_{18}O_3$ 186.25	512-16-3 81.5	2721 164^{24}	4-10-00-00049 1.0010^{18}	ace 4; eth 4; EtOH 4; chl 4 1.4680^{18}
4310	Cyclohexaneacetic acid, methyl ester Methyl cyclohexaneacetate	$C_9H_{16}O_2$ 156.22	14352-61-5	 201	4-09-00-00038 0.9896^{24}	 1.459^{14}
4311	Cyclohexaneacetic acid, 2-methyl- Cyclohexanacetic acid, 2-methyl	$C_9H_{16}O_2$ 156.22	6617-04-5	 145-7^{13}	2-09-00-00013 1.012^{25}	 1.4656^{25}
4312	Cyclohexaneacetic acid, 3-methyl- Cyclohexanacetic acid, 3-methyl	$C_9H_{16}O_2$ 156.22	67451-76-7	 148^{18}	2-09-00-00013 0.9847^{20}	 1.495^{20}
4313	Cyclohexaneacetonitrile	$C_8H_{13}N$ 123.20	4435-14-7	 105^{22}	4-09-00-00040 0.9473^{21}	eth 4; EtOH 4 1.4787^{21}
4314	Cyclohexane, 1,2-bis(methylene)-	C_8H_{12} 108.18	2819-48-9	 125; 60-1^{90}	4-05-00-00409 0.8229^{25}	H_2O 1; EtOH 3; eth 3; ace 4 1.4718^{25}
4315	Cyclohexane, bromo- Cyclohexyl bromide	$C_6H_{11}Br$ 163.06	108-85-0 -56.5	2736 166.2	4-05-00-00067 1.3359^{20}	H_2O 1; EtOH 5; eth 5; ace 5 1.4957^{20}
4316	Cyclohexane, (2-bromoethyl)-	$C_8H_{15}Br$ 191.11	1647-26-3 -57	 212	4-05-00-00116 1.2357^{20}	 1.4899^{20}
4317	Cyclohexane, 1-bromo-2-methoxy-	$C_7H_{13}BrO$ 193.08	24618-31-3	 72-4^{8}	3-06-00-00043 1.3257^{20}	 1.4871^{20}
4318	Cyclohexane, (bromomethyl)-	$C_7H_{13}Br$ 177.08	2550-36-9	 76-7^{26}	4-05-00-00100 1.2763^{25}	bz 4; eth 4; chl 4 1.4907^{30}
4319	Cyclohexane, 1-bromo-1-methyl-	$C_7H_{13}Br$ 177.08	931-77-1	 158	4-05-00-00100 1.2510^{20}	H_2O 1; EtOH 3; chl 3 1.4866^{20}
4320	Cyclohexane, 1-bromo-3-methyl- 3-Methylcyclohexyl bromide	$C_7H_{13}Br$ 177.08	13905-48-1	 181; 60^{11}	3-05-00-00076 1.275^{25}	H_2O 1; eth 4; bz 3 1.4979^{20}
4321	Cyclohexane, 1-bromo-4-methyl-1-(1-methylethyl)- *p*-Menthane, 4-bromo	$C_{10}H_{19}Br$ 219.16	116836-10-3	 98-9^{11}	0-05-00-00051 1.165^{20}	 1.4872^{20}
4322	Cyclohexane, (2-bromo-2-propenyl)- Propylene, 2-bromo-3-cyclohexyl	$C_9H_{15}Br$ 203.12	53608-85-8	 90^{15}	3-05-00-00224 1.215^{17}	eth 4 1.495^{17}
4323	Cyclohexane, 1,1'-(1,4-butanediyl)bis- 1,4-Dicyclohexylbutane	$C_{16}H_{30}$ 222.41	6165-44-2 12	 305	4-05-00-00357 0.8771^{21}	EtOH 2; eth 2; bz 2; chl 2 1.475^{21}
4324	Cyclohexanebutanol	$C_{10}H_{20}O$ 156.27	4441-57-0	 103-4^{4}	4-06-00-00139 0.902^{25}	 1.4660^{20}
4325	Cyclohexane, butyl- Butylcyclohexane	$C_{10}H_{20}$ 140.27	1678-93-9 -74.7	 180.9	4-05-00-00146 0.7902^{20}	H_2O 1 1.4408^{20}
4326	Cyclohexane, 1,1'-butylidenebis- Butane, 1,1-dicyclohexyl-	$C_{16}H_{30}$ 222.41	54890-00-5 -10.5	 293	4-05-00-00359 0.8841^{16}	 1.485^{16}
4327	Cyclohexane, 2-butyl-1,1,3-trimethyl- 2-Butyl-1,1,3-trimethylcyclohexane	$C_{13}H_{26}$ 182.35	54676-39-0	 94-5^{10}	4-05-00-00176 0.8292^{19}	 1.4563^{19}
4328	Cyclohexanecarbonitrile	$C_7H_{11}N$ 109.17	766-05-2			
4329	Cyclohexanecarbonitrile, 1-hydroxy-	$C_7H_{11}NO$ 125.17	931-97-5 35	 132^{20}	4-10-00-00021 1.0172^{20}	H_2O 4; eth 4 1.4693^{20}
4330	Cyclohexanecarbonyl chloride	$C_7H_{11}ClO$ 146.62	2719-27-9	 180	4-09-00-00026 1.0962^{15}	 1.4711^{29}
4331	Cyclohexanecarboxaldehyde	$C_7H_{12}O$ 112.17	2043-61-0	 159.3	4-07-00-00045 0.9035^{20}	H_2O 3; eth 3 1.4496^{20}
4332	Cyclohexanecarboxylic acid Hexahydrobenzoic acid	$C_7H_{12}O_2$ 128.17	98-89-5 31.5	2730 232.5	4-09-00-00016 1.0334^{22}	H_2O 2; EtOH 4; bz 4; ctc 2 1.4530^{20}
4333	Cyclohexanecarboxylic acid, 4-(aminomethyl)-, *trans*- Tranexamic acid	$C_8H_{15}NO_2$ 157.21	1197-18-8 >300	9487	3-14-00-00868	H_2O 4
4334	Cyclohexanecarboxylic acid, ethyl ester	$C_9H_{16}O_2$ 156.22	3289-28-9	 196	4-09-00-00018 0.9362^{20}	ace 4; eth 4; EtOH 4; chl 4 1.4501^{15}
4335	Cyclohexanecarboxylic acid, methyl ester	$C_8H_{14}O_2$ 142.20	4630-82-4	 183	4-09-00-00017 0.9954^{15}	H_2O 1; EtOH 3; eth 3; ace 3 1.4433^{20}
4336	Cyclohexanecarboxylic acid, 3-methyl-4-oxo- Cyclohexanone-4-carboxylic acid, 2-methyl	$C_8H_{12}O_3$ 156.18	101567-36-6 93.5	 195^{20}	4-10-00-02617	bz 4; EtOH 4; chl 4
4337	Cyclohexanecarboxylic acid, propyl ester Propyl cyclohexanecarboxylate	$C_{10}H_{18}O_2$ 170.25	6739-34-0	 215.5	4-09-00-00018 0.9530^{15}	ace 4; eth 4; EtOH 4; chl 4 1.4486^{15}
4338	Cyclohexanecarboxylic acid, 1,3,4,5-tetrahydroxy-, [1*R*-(1α,3α,4α,5β)]- Quinic acid	$C_7H_{12}O_6$ 192.17	77-95-2 162.5	8071	4-10-00-02257 1.64^{25}	H_2O 4; EtOH 4; HOAc 4
4339	Cyclohexane, chloro- Cyclohexyl chloride	$C_6H_{11}Cl$ 118.61	542-18-7 -44	2738 142	4-05-00-00048 1.000^{20}	H_2O 1; EtOH 5; eth 5; ace 5 1.4626^{20}
4340	Cyclohexane, cyclopentyl-	$C_{11}H_{20}$ 152.28	1606-08-2	 215.1	4-05-00-00328 0.8758^{20}	 1.4725^{20}
4341	Cyclohexane, (3-cyclopentylpropyl)- Propane, 1-cyclohexyl-3-cyclopentyl-	$C_{14}H_{26}$ 194.36	2883-07-0	 270	4-05-00-00347 0.8751^{20}	 1.4765^{20}

No.	Name Synonym	Mol. Form. Mol. Wt.	CAS RN mp/°C	Merck No. bp/°C	Beil. Ref. den/g cm^{-3}	Solubility n_D
4342	Cyclohexane, 1,1,2,2,3,3,4,4,5,6-decafluoro-5,6-bis(trifluoromethyl)- Cyclohexane, 1,2-dimethyl, perfluoro	C_8F_{16} 400.06	306-98-9 -56	 101.5	3-05-00-00098 1.829^{25}	 1.283^{25}
4343	Cyclohexane, 1,1,2,2,3,4,4,5,5,6-decafluoro-3,6-bis(trifluoromethyl)-	C_8F_{16} 400.06	374-77-6	 100.5	4-05-00-00124 1.8503^{20}	 1.2897^{20}
4344	Cyclohexane, decyl-	$C_{16}H_{32}$ 224.43	1795-16-0 -1.7	 299	4-05-00-00187 0.8186^{20}	 1.4534^{20}
4345	Cyclohexane, 1,1-dibromo- 1,1-Dibromocyclohexane	$C_6H_{10}Br_2$ 241.95	10489-97-1	 87^{14}	4-05-00-00070 1.7349^{20}	 1.5392^{25}
4346	Cyclohexane, 1,2-dibromo-, *cis*- *cis*-1,2-Dibromocyclohexane	$C_6H_{10}Br_2$ 241.95	19246-38-9 9.7	 115^{14}	4-05-00-00070 1.798^{25}	ace 4; bz 4; eth 4; lig 4 1.5514^{25}
4347	Cyclohexane, 1,2-dibromo-, *trans*-(±)-	$C_6H_{10}Br_2$ 241.95	5183-77-7 -4	 145^{100}; 105^{20}	4-05-00-00071 1.7759^{20}	ace 4; bz 4; eth 4; EtOH 4 1.5445^{19}
4348	Cyclohexane, 1,3-dibromo-, *trans*- *trans*-1,3-Dibromocyclohexane	$C_6H_{10}Br_2$ 241.95	29624-17-7 1	 116^{16}	4-05-00-00072	bz 4; EtOH 4 1.5480^{20}
4349	Cyclohexane, 1,4-dibromo-, *cis*- *cis*-1,4-Dibromocyclohexane	$C_6H_{10}Br_2$ 241.95	16661-99-7	 $137\text{-}8^{25}$	4-05-00-00072 1.7834^{20}	eth 4 1.5531^{20}
4350	1,2-Cyclohexanedicarboxylic acid, diethyl ester, *trans*-(±)- Cyclohexane-1,2-dicarboxylic acid, diethyl ester (*trans, DL*)	$C_{12}H_{20}O_4$ 228.29	96836-97-4	 133^{11}	4-09-00-02803 1.040^{20}	eth 4 1.4522^{13}
4351	1,3-Cyclohexanedicarboxylic acid, diethyl ester, *cis*- Cyclohexane-1,3-dicarboxylic acid, diethyl ester (*cis*)	$C_{12}H_{20}O_4$ 228.29	62059-56-7	 288; 151^{15}	3-09-00-03817 1.045^{20}	 1.452^{20}
4352	1,3-Cyclohexanedicarboxylic acid, dimethyl ester, *cis*- Dimethyl *cis*-1,3-cyclohexanedicarboxylate	$C_{10}H_{16}O_4$ 200.23	6998-82-9	 263	4-09-00-02806 1.0997^{20}	 1.4568^{20}
4353	1,3-Cyclohexanedicarboxylic acid, dimethyl ester, *trans*- Cyclohexane-1,3-dicarboxylic acid, dimethyl ester (*trans*)	$C_{10}H_{16}O_4$ 200.23	10021-92-8	 139^{20}	4-09-00-02806 1.1095^{20}	 1.4577^{25}
4354	1,4-Cyclohexanedicarboxylic acid, *trans*-	$C_8H_{12}O_4$ 172.18	619-82-9 312.5	 sub 300	4-09-00-02807	H_2O 2; EtOH 4; eth 2; ace 3
4355	1,4-Cyclohexanedicarboxylic acid, diethyl ester, *cis*- Cyclohexane-1,4-dicarboxylic acid, diethyl ester (*cis*)	$C_{12}H_{20}O_4$ 228.29	116724-15-3	 151^{13}	2-09-00-00524 1.015^{20}	eth 4
4356	1,4-Cyclohexanedicarboxylic acid, diethyl ester, *trans*- Cyclohexane-1,4-dicarboxylic acid, diethyl ester (*trans*)	$C_{12}H_{20}O_4$ 228.29	19145-96-1 43.5		2-09-00-00524 1.0110^{20}	eth 4 1.4337^{64}
4357	1,4-Cyclohexanedicarboxylic acid, dimethyl ester, *cis*- *cis*-Dimethyl 1,4-cyclohexanedicarboxylate	$C_{10}H_{16}O_4$ 200.23	3399-21-1 14	 132.5^{10}	4-09-00-02808 1.1112^{20}	
4358	Cyclohexane, 1,1-dichloro- 1,1-Dichlorocyclohexane	$C_6H_{10}Cl_2$ 153.05	2108-92-1 -47	 171	4-05-00-00050 1.1559^{20}	 1.4803^{20}
4359	Cyclohexane, 1,2-dichloro-, *cis*- *cis*-1,2-Dichlorocyclohexane	$C_6H_{10}Cl_2$ 153.05	10498-35-8 -1.5	 206.9	4-05-00-00050 1.2021^{20}	bz 4 1.4967^{20}
4360	Cyclohexane, 1,2-dichloro-, *trans*-(±)-	$C_6H_{10}Cl_2$ 153.05	5183-79-9 -6.3	 189	4-05-00-00051 1.1839^{20}	 1.4902^{20}
4361	Cyclohexane, 1,4-dichloro-, *cis*- *cis*-1,4-Dichlorocyclohexane	$C_6H_{10}Cl_2$ 153.05	16749-11-4 18	 80.3^{25}	4-05-00-00051 1.1900^{20}	 1.4942^{20}
4362	1,3-Cyclohexanedimethanamine Cyclohexane, 1,3-bis(aminomethyl)	$C_8H_{18}N_2$ 142.24	2579-20-6 <-70	 220	 0.945^{20}	H_2O 4; eth 4; EtOH 4
4363	1,3-Cyclohexanedimethanol Cyclohexane, 1,3-bis-(hydroxymethyl)	$C_8H_{16}O_2$ 144.21	3971-28-6	 105.5	4-06-00-05236 1.036^{25}	 1.4912^{20}
4364	Cyclohexane, 1,1-dimethyl- 1,1-Dimethylcyclohexane	C_8H_{16} 112.22	590-66-9 -33.3	 119.6	4-05-00-00117 0.7809^{20}	H_2O 1; EtOH 3; eth 3; ace 3 1.4290^{20}
4365	Cyclohexane, 1,2-dimethyl-, *cis*- *cis*-1,2-Dimethylcyclohexane	C_8H_{16} 112.22	2207-01-4 -49.9	 129.8	4-05-00-00118 0.7963^{20}	H_2O 1; EtOH 3; eth 5; ace 5 1.4360^{20}
4366	Cyclohexane, 1,2-dimethyl-, *trans*- *trans*-1,2-Dimethylcyclohexane	C_8H_{16} 112.22	6876-23-9 -90	 123.5	4-05-00-00118 0.7760^{20}	H_2O 1; EtOH 3; eth 3; ace 5 1.4270^{20}
4367	Cyclohexane, 1,3-dimethyl-, *cis*- *cis*-1,3-Dimethylcyclohexane	C_8H_{16} 112.22	638-04-0 -75.6	 120.1	4-05-00-00121 0.7660^{20}	H_2O 1; EtOH 5; eth 5; ace 5 1.4229^{20}
4368	Cyclohexane, 1,4-dimethyl-, *cis*- *cis*-1,4-Dimethylcyclohexane	C_8H_{16} 112.22	624-29-3 -87.4	 124.4	4-05-00-00122 0.7820^{20}	H_2O 1; EtOH 5; eth 5; ace 5 1.4230^{20}
4369	Cyclohexane, (1,1-dimethylethyl)-	$C_{10}H_{20}$ 140.27	3178-22-1 -41.2	 171.5	4-05-00-00147 0.8127^{20}	H_2O 1 1.4469^{20}
4370	Cyclohexane, 1,3-dimethyl-, *trans*- *trans*-1,3-Dimethylcyclohexane	C_8H_{16} 112.22	2207-03-6 -90.1	 124.5	 0.79^{15}	 1.4284^{25}
4371	Cyclohexane, 1,4-dimethyl-, *trans*- *trans*-1,4-Dimethylcyclohexane	C_8H_{16} 112.22	2207-04-7 -36.9	 119.4	 0.77^{15}	 1.4185^{25}
4372	1,2-Cyclohexanediol, *cis*-	$C_6H_{12}O_2$ 116.16	1792-81-0 100	 120^{15}	4-06-00-05193 1.0297^{101}	EtOH 3; ace 3; bz 3; chl 2
4373	1,2-Cyclohexanediol, *trans*-(±)-	$C_6H_{12}O_2$ 116.16	54383-22-1 105	 117^{13}	4-06-00-05194 1.147^{24}	H_2O 4; bz 4; eth 4; EtOH 4
4374	1,4-Cyclohexanediol, *trans*- *trans*-1,4-Dihydroxycyclohexane	$C_6H_{12}O_2$ 116.16	6995-79-5 143		4-06-00-05209 1.18^{20}	H_2O 3; EtOH 3; eth 1; ace 2

No.	Name Synonym	Mol. Form. Mol. Wt.	CAS RN mp/°C	Merck No. bp/°C	Beil. Ref. den/g cm^{-3}	Solubility n_D
4375	1,2-Cyclohexanedione 1,2-Dioxocyclohexane	$C_6H_8O_2$ 112.13	765-87-7 40	194	4-07-00-01982 1.1187^{21}	H_2O 3; EtOH 3; eth 3; bz 3 1.4995^{20}
4376	1,3-Cyclohexanedione	$C_6H_8O_2$ 112.13	504-02-9 105.5	3161	4-07-00-01985 1.0861^{91}	H_2O 3; EtOH 3; eth 2; ace 3 1.4576^{102}
4377	1,4-Cyclohexanedione Tetrahydroquinone	$C_6H_8O_2$ 112.13	637-88-7 78	132^{20}	4-07-00-01986 1.0861^{91}	H_2O 3; EtOH 3; eth 3; ace 3
4378	1,3-Cyclohexanedione, 5,5-dimethyl- 5,5-Dimethyldihydroresorcinol	$C_8H_{12}O_2$ 140.18	126-81-8 149-51 dec	3231	4-07-00-01999	H_2O 2; eth 2; ace 3; ctc 3
4379	1,2-Cyclohexanedione, dioxime Nioxime	$C_6H_{10}N_2O_2$ 142.16	492-99-9 192	6477	4-07-00-01982	H_2O 3; ace 3; chl 3; tfa 2
4380	Cyclohexane, docosyl- Docosane, 1-cyclohexyl-	$C_{28}H_{56}$ 392.75	61828-07-7 54.4	449	3-05-00-00167 0.8334^{20}	1.4632^{20}
4381	Cyclohexane, dodecafluoro- Perfluorocyclohexane	C_6F_{12} 300.05	355-68-0 62.5 (trip. pt.)	52.8 (subl. pt.)	4-05-00-00048	
4382	Cyclohexane, dodecyl-	$C_{18}H_{36}$ 252.48	1795-17-1 12.5	331	4-05-00-00191 0.8223^{20}	1.4559^{20}
4383	Cyclohexane, dotriacontyl- Dotriacontylcyclohexane	$C_{38}H_{76}$ 533.02	61828-17-9 74.8	527	4-05-00-00204 0.8388^{20}	1.4668^{20}
4384	Cyclohexane, eicosyl- 1-Cyclohexyleicosane	$C_{26}H_{52}$ 364.70	4443-55-4 48.5	430	4-05-00-00198 0.8318^{20}	1.4622^{20}
4385	Cyclohexaneethanamine, *N*,α-dimethyl- Propylhexedrine	$C_{10}H_{21}N$ 155.28	101-40-6	7873 205; 82^{10}	3-12-00-00108 0.8501^{20}	EtOH 4 1.4600^{20}
4386	Cyclohexane, 1,1'-(1,2-ethanediyl)bis- 1,2-Dicyclohexylethane	$C_{14}H_{26}$ 194.36	3321-50-4 11.5	272.5	4-05-00-00344 0.8728^{24}	1.4745^{24}
4387	Cyclohexaneethanol	$C_8H_{16}O$ 128.21	4442-79-9	208	4-06-00-00119 0.9229^{20}	EtOH 3; eth 3; bz 3 1.4641^{20}
4388	Cyclohexane, ethenyl-	C_8H_{14} 110.20	695-12-5	128	4-05-00-00268 0.8166^{19}	1.455^{19}
4389	Cyclohexane, 1-ethenyl-1-methyl-2,4-bis(1-methylethenyl)-, (1α,2β,4β)- β-Elemene	$C_{15}H_{24}$ 204.36	33880-83-0	$117-24^{15.5}$	4-05-00-01180 0.8749^{20}	1.4935^{20}
4390	Cyclohexane, ethyl- Ethylcyclohexane	C_8H_{16} 112.22	1678-91-7 -111.3	131.9	4-05-00-00115 0.7880^{20}	H_2O 1; EtOH 3; eth 3; ace 3 1.4330^{20}
4391	Cyclohexane, ethylidene- Ethylidenecyclohexane	C_8H_{14} 110.20	1003-64-1	136	4-05-00-00267 0.822^{25}	1.4618^{20}
4392	Cyclohexane, 1,1'-ethylidenebis- 1,1-Dicyclohexylethane	$C_{14}H_{26}$ 194.36	2319-61-1 -20.9	256; 112^7	4-05-00-00345 0.9070^{25}	1.4887^{25}
4393	Cyclohexane, 1-ethylidene-2-methyl- Cyclohexane, 2-methyl-1-ethylidene	C_9H_{16} 124.23	40514-70-3	158	0-05-00-00078 0.823^{25}	1.47
4394	Cyclohexane, 1-ethyl-1-methyl-	C_9H_{18} 126.24	4926-90-3	152.2	4-05-00-00135 0.8025^{25}	1.4419^{20}
4395	Cyclohexane, 1-ethyl-2-methyl-, *cis*-	C_9H_{18} 126.24	4923-77-7	156	4-05-00-00136 0.8059^{25}	1.4456^{20}
4396	Cyclohexane, 1-ethyl-2-methyl-, *trans*- Cyclohexane, 1-methyl-2-ethyl (*trans*)	C_9H_{18} 126.24	4923-78-8	151.7	4-05-00-00136 0.790^{25}	1.4381^{20}
4397	Cyclohexane, 1-ethyl-3-methyl-, *cis*- *cis*-1-Ethyl-3-methylcyclohexane	C_9H_{18} 126.24	19489-10-2	156	4-05-00-00136 0.8094^{20}	1.4432^{20}
4398	Cyclohexane, 1-ethyl-3-methyl-, *trans*-	C_9H_{18} 126.24	4926-76-5	150	4-05-00-00136 0.7972^{20}	1.4382^{20}
4399	Cyclohexane, 1-ethyl-4-methyl-, *cis*- *cis*-1-Ethyl-4-methylcyclohexane	C_9H_{18} 126.24	4926-78-7	152.6	4-05-00-00136 0.7969^{20}	1.4374^{20}
4400	Cyclohexane, 1-ethyl-4-methyl-, *trans*- *trans*-1-Methyl-4-ethylcyclohexane	C_9H_{18} 126.24	6236-88-0 -80.8	149	4-05-00-00136 0.7798^{20}	1.4304^{20}
4401	Cyclohexane, 1-ethyl-1-methyl-2,4-bis(1-methylethyl)-, [1*R*-(1α,2β,4β)]-	$C_{15}H_{30}$ 210.40	515-12-8	$115-9^{10}$	4-05-00-00185 0.8509^{20}	bz 4; eth 4; peth 4 1.4640^{20}
4402	Cyclohexane, 1,1'-(2-ethyl-1,3-propanediyl)bis- Propane, 1,3-dicyclohexyl-2-ethyl	$C_{17}H_{32}$ 236.44	54833-34-0	306	4-05-00-00363 0.8845^{20}	1.483^{21}
4403	Cyclohexane, fluoro- Cyclohexyl fluoride	$C_6H_{11}F$ 102.15	372-46-3 13	101	4-05-00-00044 0.9279^{20}	H_2O 1; py 3 1.4146^{20}
4404	Cyclohexane, heptadecyl- Heptadecane, 1-cyclohexyl-	$C_{23}H_{46}$ 322.62	19781-73-8 38	397	3-05-00-00161 0.8290^{20}	1.4603^{20}
4405	Cyclohexane, heptyl-	$C_{13}H_{26}$ 182.35	5617-41-4 -30	244	4-05-00-00174 0.8109^{20}	1.4484^{20}
4406	Cyclohexane, 1,2,3,4,5,6-hexachloro-, (1α,2β,3α,4β,5α,6β)- β-Lindane	$C_6H_6Cl_6$ 290.83	319-85-7	$60^{0.50}$	4-05-00-00061 1.89^{19}	H_2O 1; EtOH 2; bz 2; chl 2
4407	Cyclohexane, 1,2,3,4,5,6-hexachloro-, (1α,2α,3α,4β,5α,6β)- δ-Lindane	$C_6H_6Cl_6$ 290.83	319-86-8 141.5	$60^{0.36}$	4-05-00-00056	
4408	Cyclohexane, 1,2,3,4,5,6-hexachloro-, (1α,2α,3β,4α,5β,6β)-(±)- Cyclohexane, 1,2,3,4,5,6-hexachloro, (α, *DL*)	$C_6H_6Cl_6$ 290.83	60291-32-9 159.5	288	4-05-00-00060 1.87^{20}	H_2O 1; EtOH 4; bz 3; chl 3
4409	Cyclohexane, 1,2,3,4,5,6-hexachloro-, (1α,2α,3β,4α,5α,6β)- Lindane	$C_6H_6Cl_6$ 290.83	58-89-9 112.5	5379 323.4	4-05-00-00058	ace 4; bz 4
4410	Cyclohexane, hexacosyl- Hexacosylcyclohexane	$C_{32}H_{64}$ 448.86	61828-11-3 64	484	4-05-00-00203 0.8359^{20}	1.4649^{20}

No.	Name Synonym	Mol. Form. Mol. Wt.	CAS RN mp/°C	Merck No. bp/°C	Beil. Ref. den/g cm^{-3}	Solubility n_D
4411	Cyclohexane, hexadecyl- Hexadecane, 1-cyclohexyl-	$C_{22}H_{44}$ 308.59	6812-38-0 33.6	 385	4-05-00-00196 0.8279^{20}	 1.4596^{20}
4412	Cyclohexanehexanoic acid Caproic acid, 6-cyclohexyl	$C_{12}H_{22}O_2$ 198.31	4354-56-7 34	 180^{11}	4-09-00-00095 0.9626^{20}	eth 3; chl 2 1.4750^{20}
4413	Cyclohexane, hexatriacontyl- Hexatriacontylcyclohexane	$C_{42}H_{84}$ 589.13	61828-21-5 80	 551	4-05-00-00205 0.8402^{20}	 1.4678^{20}
4414	Cyclohexane, hexyl-	$C_{12}H_{24}$ 168.32	4292-75-5 -43	 224	4-05-00-00170 0.8076^{20}	 1.4462^{20}
4415	Cyclohexane, iodo- Cyclohexyl iodide	$C_6H_{11}I$ 210.06	626-62-0	 180 dec; 81^{20}	4-05-00-00078 1.6244^{20}	H_2O 1; EtOH 3; eth 3; ace 3 1.5477^{20}
4416	Cyclohexane, isocyanato- Cyclohexyl isocyanate	$C_7H_{11}NO$ 125.17	3173-53-3	 172	4-12-00-00071 0.98^{25}	 1.4551^{20}
4417	Cyclohexane, isothiocyanato-	$C_7H_{11}NS$ 141.24	1122-82-3	 220	4-12-00-00073 0.996^{25}	H_2O 1; EtOH 3; eth 3; ctc 2 1.5375^{20}
4418	Cyclohexanemethanamine	$C_7H_{15}N$ 113.20	3218-02-8	 160	4-12-00-00123 0.87^{25}	 1.4630^{20}
4419	Cyclohexanemethanol Cyclohexylcarbinol	$C_7H_{14}O$ 114.19	100-49-2 -43	2737 183	4-06-00-00106 0.9297^{20}	eth 4; EtOH 4 1.4644^{20}
4420	Cyclohexanemethanol, α-methyl-	$C_8H_{16}O$ 128.21	1193-81-3	 189	1-06-00-00012 0.928^{25}	EtOH 4; eth 4; ctc 2 1.4656^{20}
4421	Cyclohexanemethanol, α-methyl-, (S)-	$C_8H_{16}O$ 128.21	3113-98-2	 105^{35}; 82^{12}	4-06-00-00117 0.9254^{20}	 1.4635^{25}
4422	Cyclohexanemethanol, 2-methyl-, *cis*- Cyclohexane, 1-hydroxymethyl-2-methyl (*cis*)	$C_8H_{16}O$ 128.21	3937-45-9	 188.5	4-06-00-00122 0.9342^{20}	 1.4689^{20}
4423	Cyclohexanemethanol, 2-methyl-, *trans*- Cyclohexane, 1-hydroxymethyl-2-methyl (*trans*)	$C_8H_{16}O$ 128.21	3937-46-0	 192	4-06-00-00122 0.9224^{20}	 1.4665^{20}
4424	Cyclohexanemethanol, 4-methyl-	$C_8H_{16}O$ 128.21	34885-03-5	 75$^{2.5}$	2-06-00-00030 0.9074^{20}	 1.4617^{20}
4425	Cyclohexane, methoxy-	$C_7H_{14}O$ 114.19	931-56-6 -74.4	 133	4-06-00-00026 0.8756^{20}	eth 4; EtOH 4 1.4355^{20}
4426	Cyclohexane, methyl- Methylcyclohexane	C_7H_{14} 98.19	108-87-2 -126.6	 100.9	4-05-00-00094 0.7694^{20}	H_2O 1; EtOH 3; eth 3; ace 5 1.4231^{20}
4427	Cyclohexane, (3-methylbutyl)-	$C_{11}H_{22}$ 154.30	54105-76-9	 196.5	3-05-00-00143 0.8023^{20}	bz 4; lig 4 1.4420^{20}
4428	Cyclohexane, methylene- Methylenecyclohexane	C_7H_{12} 96.17	1192-37-6 -106.7	 102.5	4-05-00-00250 0.8074^{20}	H_2O 1; eth 3; bz 3; chl 3 1.4523^{20}
4429	Cyclohexane, 1,1'-methylenebis- Dicyclohexylmethane	$C_{13}H_{24}$ 180.33	3178-23-2	 110-5^{18}	4-05-00-00340 0.8750^{20}	 1.4752^{20}
4430	Cyclohexane, 1,1'-(1-methyl-1,2-ethanediyl)bis- 1,2-Dicyclohexylpropane	$C_{15}H_{28}$ 208.39	41851-34-7 -22	 284.5	4-05-00-00350 0.8724^{21}	 1.479^{21}
4431	Cyclohexane, (1-methylethyl)- Isopropylcyclohexane	C_9H_{18} 126.24	696-29-7 -89.4	 154.8	4-05-00-00134 0.8023^{20}	H_2O 1; EtOH 4; eth 4; ace 5 1.4410^{20}
4432	Cyclohexane, 1,1'-(1-methylethylidene)bis- Propane, 2,2-dicyclohexyl	$C_{15}H_{28}$ 208.39	54934-90-6 15.6	 286	4-05-00-00350 0.9001^{23}	 1.490^{23}
4433	Cyclohexane, 1,1',1''-methylidynetris- Methane, tricyclohexyl-	$C_{19}H_{34}$ 262.48	1610-24-8 48	 325.5	4-05-00-00504 0.9273^{50}	H_2O 1; EtOH 2; eth 3; bz 3 1.4986^{40}
4434	Cyclohexane, 1-methyl-4-methylene-	C_8H_{14} 110.20	2808-80-2	 122	4-05-00-00271 0.7910^{19}	 1.4465^{18}
4435	Cyclohexane, 1-methyl-3-(1-methylethenyl)- *m*-Menth-8-ene	$C_{10}H_{18}$ 138.25	16605-36-0	 170	3-05-00-00235 0.8178^{20}	 1.4546
4436	Cyclohexane, 1-methyl-2-(1-methylethyl)-	$C_{10}H_{20}$ 140.27	16580-23-7	 171	4-05-00-00150 0.8134^{22}	 1.447^{21}
4437	Cyclohexane, 1-methyl-3-(1-methylethyl)-	$C_{10}H_{20}$ 140.27	16580-24-8	 166.5	3-05-00-00132 0.7963^{20}	 1.44^{24}
4438	Cyclohexane, 1-methyl-4-(1-methylethyl)-	$C_{10}H_{20}$ 140.27	99-82-1 -87.6	 170.7	4-05-00-00151 0.797^{25}	 1.4373^{20}
4439	Cyclohexane, 1-methyl-4-(1-methylethyl)-, *cis*-	$C_{10}H_{20}$ 140.27	6069-98-3 -89.9	 172	4-05-00-00151 0.8039^{20}	H_2O 1; EtOH 4; eth 4; bz 3 1.4431^{20}
4440	Cyclohexane, 1-methyl-4-(1-methylethyl)-, *trans*-	$C_{10}H_{20}$ 140.27	1678-82-6 -86.3	 170.6	4-05-00-00151 0.7928^{20}	bz 4; eth 4; EtOH 4; lig 4 1.4366^{20}
4441	Cyclohexane, 1-methyl-3-(1-methylethylidene)-	$C_{10}H_{18}$ 138.25	13828-34-7	 174	1-05-00-00043 0.8214^{20}	 1.4670^{20}
4442	Cyclohexane, 1-methyl-4-(1-methylethylidene)-	$C_{10}H_{18}$ 100.20	1124-27-2	 173	4-05-00-00302 0.819^{21}	 1.4568^{21}
4443	Cyclohexane, 1-methyl-1-nitro- 1-Methyl-1-nitrocyclohexane	$C_7H_{13}NO_2$ 143.19	59368-15-9 -71	 118-20^{50}	4-05-00-00103 1.0384^{20}	 1.4598^{20}
4444	Cyclohexane, 1-methyl-2-nitro- 2-Methylnitrocyclohexane	$C_7H_{13}NO_2$ 143.19	74221-86-6	 100^{20}	4-05-00-00103 1.046^{25}	 1.4608^{25}
4445	Cyclohexane, 1-methyl-4-nitro- 1-Methyl-4-nitrocyclohexane	$C_7H_{13}NO_2$ 143.19	89895-45-4	 102^{20}	4-05-00-00103 1.037^{25}	 1.4567^{23}
4446	Cyclohexane, 1-methyl-2-pentyl- 1-Methyl-2-pentylcyclohexane	$C_{12}H_{24}$ 168.32	54411-01-7	 217.5	4-05-00-00171 0.815^{20}	 1.4487^{20}
4447	Cyclohexane, 1,1'-(2-methyl-1,3-propanediyl)bis- 1,3-Dicyclohexyl-2-methylpropane	$C_{16}H_{30}$ 222.41	2883-08-1 0.6	 295.2	4-05-00-00358 0.8715^{20}	 1.4756^{20}

No.	Name Synonym	Mol. Form. Mol. Wt.	CAS RN mp/°C	Merck No. bp/°C	Beil. Ref. den/g cm^{-3}	Solubility n_D
4448	Cyclohexane, (1-methylpropyl)- *sec*-Butylcyclohexane	$C_{10}H_{20}$ 140.27	7058-01-7	 179.3	4-05-00-00146 0.8131^{20}	H_2O 1; ace 3 1.4467^{20}
4449	Cyclohexane, 1-methyl-1-propyl- 1-Methyl-1-propylcyclohexane	$C_{10}H_{20}$ 140.27	4258-93-9	 172	4-05-00-00148 0.8101^{20}	 1.4440^{20}
4450	Cyclohexane, 1-methyl-2-propyl- 1-Methyl-2-propylcyclohexane	$C_{10}H_{20}$ 140.27	4291-79-6	 176; 56^{13}	4-05-00-00148 0.8130^{19}	 1.4468^{19}
4451	Cyclohexane, 1-methyl-3-propyl- 1-Methyl-3-propylcyclohexane	$C_{10}H_{20}$ 140.27	4291-80-9	 164.5	4-05-00-00149 0.7895^{21}	 1.4377^{20}
4452	Cyclohexane, 1-methyl-4-propyl- 1-Methyl-4-propylcyclohexane	$C_{10}H_{20}$ 140.27	4291-81-0	 175.5	4-05-00-00149 0.797^{20}	 1.4393^{20}
4453	Cyclohexane, (2-methylpropyl)-	$C_{10}H_{20}$ 140.27	1678-98-4 -95	 171.3	4-05-00-00147 0.7952^{20}	H_2O 1; EtOH 3; eth 4; ace 3 1.4386^{20}
4454	Cyclohexane, nitro- Nitrocyclohexane	$C_6H_{11}NO_2$ 129.16	1122-60-7 -34	 205; 95^{22}	4-05-00-00081 1.0610^{20}	H_2O 1; EtOH 3; lig 3 1.4612^{19}
4455	Cyclohexane, (nitromethyl)- (Nitromethyl)cyclohexane	$C_7H_{13}NO_2$ 143.19	2625-30-1	 123^{40}; 98^{10}	4-05-00-00103 1.0482^{20}	 1.4705^{20}
4456	Cyclohexane, nonadecyl-	$C_{25}H_{50}$ 350.67	22349-03-7 45	 420	4-05-00-00197 0.8310^{20}	 1.4616^{20}
4457	Cyclohexane, nonyl-	$C_{15}H_{30}$ 210.40	2883-02-5 -10	 282	4-05-00-00183 0.8163^{20}	 1.4519^{20}
4458	Cyclohexane, octacosyl- Octacosylcyclohexane	$C_{34}H_{68}$ 476.91	61828-13-5 68	 499	4-05-00-00203 0.8370^{20}	 1.4656^{20}
4459	Cyclohexane, octadecyl- Octadecane, 1-cyclohexyl-	$C_{24}H_{48}$ 336.65	4445-06-1 41.6	 409; 175^1	4-05-00-00197 0.8300^{20}	 1.4610^{20}
4460	Cyclohexane, octyl-	$C_{14}H_{28}$ 196.38	1795-15-9 -20	 264	4-05-00-00178 0.8138^{20}	 1.4503^{20}
4461	Cyclohexane, 1,1'-oxybis-	$C_{12}H_{22}O$ 182.31	4645-15-2 -36	 242.5	3-06-00-00019 0.9227^{20}	 1.4741^{20}
4462	Cyclohexane, pentacosyl- Pentacosylcyclohexane	$C_{31}H_{62}$ 434.83	61828-10-2 62	 476	4-05-00-00202 0.8353^{20}	 1.4645^{20}
4463	Cyclohexane, pentadecyl-	$C_{21}H_{42}$ 294.56	6006-95-7 29	 373	4-05-00-00195 0.8267^{20}	 1.4588^{20}
4464	Cyclohexane, 1,1'-(1,5-pentanediyl)bis- Pentane, 1,5-dicyclohexyl-	$C_{17}H_{32}$ 236.44	54833-31-7 -13.6	 325	4-05-00-00362 0.8718^{21}	 1.479^{21}
4465	Cyclohexane, pentyl- Pentylcyclohexane	$C_{11}H_{22}$ 154.30	4292-92-6 -57.5	 203.7	4-05-00-00164 0.8037^{20}	ace 4; bz 4; eth 4; EtOH 4 1.4437^{20}
4466	Cyclohexane, 1,1'-(1,3-propanediyl)bis- 1,3-Dicyclohexylpropane	$C_{15}H_{28}$ 208.39	3178-24-3 -17	 291.5	4-05-00-00349 0.8728^{24}	 1.4736^{24}
4467	Cyclohexanepropanenitrile, 2-oxo-	$C_9H_{13}NO$ 151.21	4594-78-9	 $138\text{-}42^{10}$	4-10-00-02623 1.0181^{20}	ctc 2 1.4755^{20}
4468	Cyclohexanepropanoic acid	$C_9H_{16}O_2$ 156.22	701-97-3 16	 276.5	4-09-00-00056 0.912^{25}	H_2O 3; eth 3; ctc 2 1.4638^{20}
4469	Cyclohexanepropanol, α-methyl- 3-Cyclohexyl-1-methylpropanol	$C_{10}H_{20}O$ 156.27	10528-67-3	 112^{16}	3-06-00-00123 0.903^{21}	 1.464^{21}
4470	Cyclohexane, 2-propenyl-	C_9H_{16} 124.23	2114-42-3	 132	4-05-00-00283 0.8135^{20}	EtOH 4; eth 3; ace 3; bz 3 1.4500^{20}
4471	Cyclohexane, propyl- Propylcyclohexane	C_9H_{18} 126.24	1678-92-8 -94.9	 156.7	4-05-00-00134 0.7936^{20}	H_2O 1; EtOH 5; eth 3; ace 5 1.4370^{20}
4472	Cyclohexane, 1,1'-propylidenebis- Propane, 1,1-dicyclohexyl-	$C_{15}H_{28}$ 208.39	54934-91-7 -23.5	 283.6	4-05-00-00350 0.8886^{23}	 1.485^{23}
4473	Cyclohexane, 2-propynyl-	C_9H_{14} 122.21	17715-00-3	 157.5	4-05-00-00419 0.8449^{20}	eth 3 1.4605^{20}
4474	Cyclohexane, tetracosyl- Tetracosylcyclohexane	$C_{30}H_{60}$ 420.81	61828-09-9 59.5	 467	4-05-00-00202 0.8347^{20}	 1.4641^{20}
4475	Cyclohexane, tetradecyl-	$C_{20}H_{40}$ 280.54	1795-18-2 24	 360	4-05-00-00194 0.8254^{20}	 1.4579^{20}
4476	Cyclohexane, 1,1'-tetradecylidenebis- 1,1-Dicyclohexyltetradecane	$C_{26}H_{50}$ 362.68	55334-08-2 37.6	 406	4-05-00-00374 0.8735^{20}	 1.4799^{20}
4477	Cyclohexane, 1,1,3,4-tetramethyl- 1,1,3,4-Tetramethylcyclohexane	$C_{10}H_{20}$ 140.27	24612-75-7	 160	4-05-00-00159 0.7976^{20}	 1.4380^{20}
4478	Cyclohexane, 1,1,3,5-tetramethyl-, *cis*- *cis*-1,1,3,5-Tetramethylcyclohexane	$C_{10}H_{20}$ 140.27	50876-32-9	 152.5	4-05-00-00159 0.7813^{20}	 1.4319^{20}
4479	Cyclohexane, tetratriacontyl- Tetratriacontylcyclohexane	$C_{40}H_{80}$ 561.08	61828-19-1 78	 540	4-05-00-00205 0.8395^{20}	 1.4673^{20}
4480	Cyclohexanethiol Cyclohexyl mercaptan	$C_6H_{12}S$ 116.23	1569-69-3	 158.9	4-06-00-00072 0.9782^{20}	ace 4; bz 4; eth 4; EtOH 4 1.4921^{20}
4481	Cyclohexanethione Thiocyclohexanone	$C_6H_{10}S$ 114.21	2720-41-4	 82^{17}; 74^{11}	3-07-00-00039	EtOH 3; eth 3; ace 3 1.5375^{20}
4482	Cyclohexane, triacontyl- Triacontylcyclohexane	$C_{36}H_{72}$ 504.97	61828-15-7 71.6	 514	4-05-00-00204 0.8379^{20}	 1.4662^{20}
4483	Cyclohexane, tricosyl- Tricosylcyclohexane	$C_{29}H_{58}$ 406.78	61828-08-8 57	 459	4-05-00-00202 0.8341^{20}	 1.4637^{20}
4484	Cyclohexane, tridecyl- Tridecane, 1-cyclohexyl-	$C_{19}H_{38}$ 266.51	6006-33-3 18.5	 346	4-05-00-00192 0.8239^{20}	 1.4570^{20}
4485	Cyclohexane, 1,1,2-trimethyl- 1,1,2-Trimethylcyclohexane	C_9H_{18} 126.24	7094-26-0 -29	 145.2	4-05-00-00137 0.7963^{25}	 1.4382^{20}
4486	Cyclohexane, 1,1,3-trimethyl- Cyclogeraniolane	C_9H_{18} 126.24	3073-66-3 -65.7	 136.6	4-05-00-00137 0.7749^{25}	 1.4295^{20}
4487	Cyclohexane, 1,1,4-trimethyl- 1,1,4-Trimethylcyclohexane	C_9H_{18} 126.24	7094-27-1	 135	4-05-00-00137 0.7685^{25}	 1.4251^{20}

No.	Name Synonym	Mol. Form. Mol. Wt.	CAS RN mp/°C	Merck No. bp/°C	Beil. Ref. den/g cm^{-3}	Solubility n_D
4488	Cyclohexane, 1,2,3-trimethyl- 1,2,3-Trimethylcyclohexane	C_9H_{18} 126.24	1678-97-3	142	4-05-00-00137 0.7898^{20}	1.4346^{20}
4489	Cyclohexane, 1,2,3-trimethyl-, (1α,2β,3α)-	C_9H_{18} 126.24	1678-81-5 -66.9	144	4-05-00-00137 0.777^{25}	1.430^{20}
4490	Cyclohexane, 1,2,3-trimethyl-, (1α,2α,3α)-	C_9H_{18} 126.24	1839-88-9 -85	151.7	4-05-00-00137 0.7989^{25}	1.4405^{20}
4491	Cyclohexane, 1,2,3-trimethyl-, (1α,2α,3β)-	C_9H_{18} 126.24	7667-55-2 -85.7	151.2	4-05-00-00137 0.7995^{25}	1.4399^{20}
4492	Cyclohexane, 1,2,4-trimethyl-, (1α,2α,4α)-	C_9H_{18} 126.24	1678-80-4 -77.5	146.6	4-05-00-00138 0.783^{25}	1.4340^{20}
4493	Cyclohexane, 1,2,4-trimethyl-, (1α,2β,4α)-	C_9H_{18} 126.24	7667-59-6 -86	142.9	4-05-00-00138 0.7683^{25}	1.4266^{20}
4494	Cyclohexane, 1,2,4-trimethyl-, (1α,2α,4β)-	C_9H_{18} 126.24	7667-58-5 -91.8	146.7	4-05-00-00138 0.7870^{25}	1.4345^{20}
4495	Cyclohexane, 1,2,4-trimethyl-, (1α,2β,4β)-	C_9H_{18} 126.24	7667-60-9 -83.5	142.9	4-05-00-00138 0.7870^{25}	1.4341^{20}
4496	Cyclohexane, 1,3,5-trimethyl-, (1α,3α,5α)-	C_9H_{18} 126.24	1795-27-3 -49.7	138.5	4-05-00-00138 0.7708^{20}	H_2O 1; eth 3; bz 3; ctc 3 1.4269^{20}
4497	Cyclohexane, 1,3,5-trimethyl-, (1α,3α,5β)- *trans*-1,3,5-Trimethylcyclohexane	C_9H_{18} 126.24	1795-26-2 -107.4	140.5	4-05-00-00138 0.7794^{20}	bz 4; eth 4; lig 4 1.4307^{20}
4498	Cyclohexane, undecafluoro- Undecafluorocyclohexane	C_6HF_{11} 282.06	308-24-7	62.0		
4499	Cyclohexane, undecafluoro(trifluoromethyl)- Perfluoro(methylcyclohexane)	C_7F_{14} 350.05	355-02-2 -44.7	76.3	4-05-00-00097 1.7878^{25}	ace 3; bz 3; ctc 3; tol 3 1.285^{17}
4500	Cyclohexanol Cyclohexyl alcohol	$C_6H_{12}O$ 100.16	108-93-0 25.4	2731 160.8	4-06-00-00020 0.9624^{20}	H_2O 3; EtOH 3; eth 3; ace 3 1.4641^{20}
4501	Cyclohexanol, 2-amino-, *trans*-(±)- Cyclohexanol, 2-amino (*trans, DL*)	$C_6H_{13}NO$ 115.18	33092-82-9 68	219; 105^{10}	4-13-00-00540	bz 4; chl 4; HOAc 4
4502	Cyclohexanol, 3-(aminomethyl)-3,5,5-trimethyl- 1-Hydroxy-3-aminomethyl-3,5,5-trimethylcyclohexane	$C_{10}H_{21}NO$ 171.28	15647-11-7 45.5	265	0.969^{25}	1.4904^{20}
4503	Cyclohexanol, 2-bromo- 2-Bromo-1-cyclohexanol	$C_6H_{11}BrO$ 179.06	24796-87-0	85^{10}; 60^{1}	4-06-00-00063 1.4604^{20}	1.5169^{20}
4504	Cyclohexanol, 2-butyl-, *trans*- *trans*-2-Butylcyclohexanol	$C_{10}H_{20}O$ 156.27	35242-05-8	$111\text{-}2^{16}$	3-06-00-00122 0.9020^{20}	H_2O 1; eth 3; ace 3; bz 3 1.4641^{20}
4505	Cyclohexanol, 2-chloro-, *trans*- *trans*-2-Chlorocyclohexanol	$C_6H_{11}ClO$ 134.61	6628-80-4 29	93^{20}	4-06-00-00064 1.146^{16}	bz 4; eth 4; EtOH 4; chl 4 1.4899^{20}
4506	Cyclohexanol, 2-chloro-, *cis*-(±)-	$C_6H_{11}ClO$ 134.61	116783-28-9 36.5	$93\text{-}4^{26}$	4-06-00-00064 1.1261^{25}	bz 4; EtOH 4; chl 4 1.4894^{25}
4507	Cyclohexanol, 4-chloro-, *trans*- *trans*-4-Chlorocyclohexanol	$C_6H_{11}ClO$ 134.61	29538-77-0 82.5	106^{14}	4-06-00-00068 1.1435^{13}	EtOH 3; eth 3; bz 3; chl 3 1.4930^{17}
4508	Cyclohexanol, 1,2-dimethyl-, *cis*- Cyclohexanol, 1,2-dimethyl-, stereoisomer	$C_8H_{16}O$ 128.21	19879-11-9 23.2	82.8^{25}	4-06-00-00121 0.9250^{20}	1.4625^{20}
4509	Cyclohexanol, 1,2-dimethyl-, *trans*- Cyclohexanol, 1,2-dimethyl-, stereoisomer	$C_8H_{16}O$ 128.21	19879-12-0 13.2	74^{25}	4-06-00-00121 0.9187^{20}	1.4590^{20}
4510	Cyclohexanol, 1,3-dimethyl-, *cis*- *cis*-1,3-Dimethylcyclohexanol	$C_8H_{16}O$ 128.21	15466-94-1 27.5	84^{25}	3-06-00-00092 0.9022^{20}	1.4575^{20}
4511	Cyclohexanol, 1,3-dimethyl-, *trans*- *trans*-1,3-Dimethylcyclohexanol	$C_8H_{16}O$ 128.21	15466-93-0 14.5	77.6^{25}	3-06-00-00093 0.8894^{30}	1.4507^{20}
4512	Cyclohexanol, 1,4-dimethyl-, *cis*-	$C_8H_{16}O$ 128.21	16980-61-3 24	83.5^{25}	3-06-00-00098 0.9011^{30}	1.4564^{20}
4513	Cyclohexanol, 2,2-dimethyl- 2,2-Dimethylcyclohexanol	$C_8H_{16}O$ 128.21	1193-46-0 8	177	0.9225^{20}	1.4648^{20}
4514	Cyclohexanol, 2,4-dimethyl- 2,4-Dimethylcyclohexanol	$C_8H_{16}O$ 128.21	69542-91-2	177.5	4-06-00-00124 0.900^{20}	1.4500^{20}
4515	Cyclohexanol, 2,6-dimethyl- 2,6-Dimethyl-1-cyclohexanol	$C_8H_{16}O$ 128.21	5337-72-4 32.5	176	2-06-00-00028 0.944^{25}	1.4600^{20}
4516	Cyclohexanol, 3,3-dimethyl- 3,3-Dimethylcyclohexanol	$C_8H_{16}O$ 128.21	767-12-4 11.5	185; 99.5^{35}	3-06-00-00089 0.9128^{14}	1.4606^{15}
4517	Cyclohexanol, 3,4-dimethyl- 3,4-Dimethylcyclohexyl alcohol	$C_8H_{16}O$ 128.21	5715-23-1	189	3-06-00-00091 0.9073^{16}	1.458^{16}
4518	Cyclohexanol, 3,5-dimethyl- 3,5-Dimethylcyclohexanol	$C_8H_{16}O$ 128.21	5441-52-1 11.6	187	2-06-00-00029 0.898^{20}	1.4550^{20}
4519	Cyclohexanol, 4,4-dimethyl- 4,4-Dimethylcyclohexanol	$C_8H_{16}O$ 128.21	932-01-4 16	186	4-06-00-00120 0.9250^{20}	1.4613^{20}
4520	Cyclohexanol, 3-(dimethylamino)-	$C_8H_{17}NO$ 143.23	6890-03-5 73	231	3-13-00-00719 0.9737^{25}	EtOH 4 1.4852^{20}
4521	Cyclohexanol, 2-(1,1-dimethylethyl)	$C_{10}H_{20}O$ 156.27	13491-79-7 45	139^{95}	3-06-00-00125 0.902^{25}	
4522	Cyclohexanol, 1-ethyl- 1-Ethylcyclohexanol	$C_8H_{16}O$ 128.21	1940-18-7 34.5	166	4-06-00-00115 0.9227^{25}	H_2O 2; bz 3; ctc 2; peth 3 1.4633^{20}
4523	Cyclohexanol, 2-ethyl-, *trans*-(±)- Cyclohexanol, 2-ethyl (*trans, DL*)	$C_8H_{16}O$ 128.21	89886-25-9	79^{12}	4-06-00-00117 0.9193^{21}	ace 4; bz 4; eth 4; peth 4 1.4640^{21}
4524	Cyclohexanol, 2-ethyl-, *cis*-(±)-	$C_8H_{16}O$ 128.21	116697-35-9	181	3-06-00-00085 0.9274^{21}	ace 4; bz 4; eth 4; peth 4 1.4655^{21}
4525	Cyclohexanol, 4-ethyl- 4-Ethylcyclohexanol	$C_8H_{16}O$ 128.21	4534-74-1	84^{10}	2-06-00-00026 0.889^{25}	1.4625^{20}
4526	Cyclohexanol, 1-ethyl-2-methyl- 1-Ethyl-2-methylcyclohexanol	$C_9H_{18}O$ 142.24	102370-18-3	182; 82^{73}	4-06-00-00133 0.9235^{20}	1.458^{20}

No.	Name Synonym	Mol. Form. Mol. Wt.	CAS RN mp/°C	Merck No. bp/°C	Beil. Ref. den/g cm^{-3}	Solubility n_D
4527	Cyclohexanol, 1-ethyl-3-methyl- 1-Ethyl-3-methylcyclohexanol	$C_9H_{18}O$ 142.24	62067-44-1	88^{20}	2-06-00-00035 0.9013^{20}	1.459
4528	Cyclohexanol, 1,1'-(1,2-ethynediyl)bis-	$C_{14}H_{22}O_2$ 222.33	78-54-6 112.5	182^{13}	4-06-00-06076	H_2O 1; EtOH 4; eth 4; ace 4
4529	Cyclohexanol, 1-ethynyl-	$C_8H_{12}O$ 124.18	78-27-3 31.5	174	4-06-00-00348 0.9873^{20}	H_2O 1; EtOH 3; bz 3; chl 2 1.4822^{20}
4530	Cyclohexanol, 1-methyl- 1-Methylcyclohexanol	$C_7H_{14}O$ 114.19	590-67-0 25	155; 70^{25}	4-06-00-00095 0.9194^{20}	H_2O 1; EtOH 3; bz 3; chl 3 1.4595^{20}
4531	Cyclohexanol, 2-methyl-, *cis*-(±)- (±)-*cis*-2-Methylcyclohexanol	$C_7H_{14}O$ 114.19	615-38-3 7	165	4-06-00-00100 0.9360^{20}	EtOH 4 1.4640^{20}
4532	Cyclohexanol, 2-methyl-, *trans*-(±)- (±)-*trans*-2-Methylcyclohexanol	$C_7H_{14}O$ 114.19	615-39-4 -4	167.5	4-06-00-00100 0.9247^{20}	eth 4; EtOH 4 1.4616^{20}
4533	Cyclohexanol, 3-methyl-, *cis*-(±)- *cis*-3-Methylcyclohexanol	$C_7H_{14}O$ 114.19	5454-79-5 -5.5	168; 94^{12}	0.9155^{20}	eth 4; EtOH 4 1.4752^{20}
4534	Cyclohexanol, 3-methyl-, *trans*-(±)- *trans*-3-Methylcyclohexanol	$C_7H_{14}O$ 114.19	7443-55-2 -0.5	167; 84^{13}	0.9214^{30}	eth 4; EtOH 4 1.4580^{20}
4535	Cyclohexanol, 4-methyl-, *cis*- *cis*-4-Methylcyclohexanol	$C_7H_{14}O$ 114.19	7731-28-4 -9.2	173	4-06-00-00105 0.9170^{20}	eth 4; EtOH 4 1.4614^{20}
4536	Cyclohexanol, 4-methyl-, *trans*- *trans*-4-Methylcyclohexanol	$C_7H_{14}O$ 114.19	7731-29-5	174	4-06-00-00105 0.9118^{21}	H_2O 2; EtOH 5; eth 3 1.4561^{20}
4537	Cyclohexanol, 2-methylene-	$C_7H_{12}O$ 112.17	4065-80-9	$82\text{-}4^{13}$	4-06-00-00208 0.955^{20}	1.4843^{20}
4538	Cyclohexanol, 1-(1-methylethyl)-	$C_9H_{18}O$ 142.24	3552-01-0	$76\text{-}9^{10}$	4-06-00-00131 0.928^{19}	1.4683^{19}
4539	Cyclohexanol, 2-(1-methylethyl)-, *cis*- *cis*-2-Isopropylcyclohexanol	$C_9H_{18}O$ 142.24	10488-25-2 52.5	77^{13}	4-06-00-00131 0.9223^{25}	1.4665^{25}
4540	Cyclohexanol, 4-(1-methylethyl)-, *cis*- *cis*-4-Isopropylcyclohexanol	$C_9H_{18}O$ 142.24	22900-08-9	$72^{1.9}$	3-06-00-00111 0.9195^{20}	1.4671^{20}
4541	Cyclohexanol, 4-(1-methylethyl)-, *trans*- *trans*-4-Isopropylcyclohexanol	$C_9H_{18}O$ 142.24	15890-36-5	$79.5^{1.8}$	3-06-00-00112 0.9140^{20}	1.4658^{20}
4542	Cyclohexanol, 1-methyl-4-(1-methylethenyl)- β-Terpineol	$C_{10}H_{18}O$ 154.25	138-87-4 32.5	210; 90^{10}	4-06-00-00254 0.917^{20}	1.4747^{20}
4543	Cyclohexanol, 2-methyl-5-(1-methylethenyl)-, [1*S*-(1α,2β,5α)]-	$C_{10}H_{18}O$ 154.25	22567-21-1	225; 107^{14}	3-06-00-00256 0.9274^{20}	1.4780^{20}
4544	Cyclohexanol, 5-methyl-2-(1-methylethenyl)-, [1*R*-(1α,2β,5α)]-	$C_{10}H_{18}O$ 154.25	89-79-2 78	93^{14}	3-06-00-00257 0.911^{20}	H_2O 2; EtOH 3; eth 3 1.4723^{20}
4545	Cyclohexanol, 5-methyl-2-(1-methylethenyl)-, acetate, [1*R*-(1α,2β,5α)]-	$C_{12}H_{20}O_2$ 196.29	57576-09-7 85	$112\text{-}4^{8}$	4-06-00-00255 0.925^{25}	1.4566^{20}
4546	Cyclohexanol, 1-methyl-4-(1-methylethyl)-	$C_{10}H_{20}O$ 156.27	21129-27-1	208.5	4-06-00-00147 0.90^{20}	1.4619^{20}
4547	Cyclohexanol, 2-methyl-5-(1-methylethyl)-, [1*R*-(1α,2α,5β)]-	$C_{10}H_{20}O$ 156.27	5563-78-0	217.5; 102^{18}	4-06-00-00148 0.9012^{20}	eth 4; EtOH 4 1.4632^{20}
4548	Cyclohexanol, 5-methyl-2-(1-methylethyl)-, (1α,2β,5α)-(±)- (±)-Menthol	$C_{10}H_{20}O$ 156.27	15356-70-4 38	5723 216	4-06-00-00151 0.903^{15}	H_2O 1; EtOH 4; eth 4; ace 4 1.4615^{20}
4549	Cyclohexanol, 5-methyl-2-(1-methylethyl)-, acetate, [1*R*-(1α,2β,5α)]-	$C_{12}H_{22}O_2$ 198.31	2623-23-6	109^{10}	4-06-00-00153 0.9185^{20}	1.4469^{20}
4550	Cyclohexanol, 2-methyl-1-propyl- 2-Methyl-1-propylcyclohexanol	$C_{10}H_{20}O$ 156.27	24580-48-1	$97\text{-}8^{34}$	3-06-00-00127 0.919^{20}	1.48^{20}
4551	Cyclohexanol, 3-methyl-1-propyl- 3-Methyl-1-propylcyclohexanol	$C_{10}H_{20}O$ 156.27	24580-52-7	199; $96\text{-}8^{20}$	1-06-00-00018 0.8903^{21}	1.4566^{24}
4552	Cyclohexanol, 1-phenyl- 1-Phenylcyclohexan-1-ol	$C_{12}H_{16}O$ 176.26	1589-60-2 63.5	157^{28}; 112^{5}	4-06-00-03908 1.035^{16}	bz 4; eth 4; EtOH 4; lig 4 1.5415^{16}
4553	Cyclohexanol, 2-phenyl-, *cis*-(±)- Cyclohexanol, 2-phenyl (*cis*, *DL*)	$C_{12}H_{16}O$ 176.26	40960-73-4 41.5	$140\text{-}1^{16}$	4-06-00-03909 1.035^{16}	1.5415^{16}
4554	Cyclohexanol, 2-phenyl-, *trans*-(±)- Cyclohexanol, 2-phenyl (*trans*, *DL*)	$C_{12}H_{16}O$ 176.26	40960-69-8 56.5	153^{16}	4-06-00-03909	EtOH 4; chl 4
4555	Cyclohexanol, 1-(2-propenyl)-	$C_9H_{16}O$ 140.23	1123-34-8	190	4-06-00-00235 0.9341^{22}	1.4756^{22}
4556	Cyclohexanol, 2-(2-propenyl)-, *trans*- Cyclohexanol, 2-allyl-, *trans*-	$C_9H_{16}O$ 140.23	24844-28-8	$94\text{-}6^{15}$	4-06-00-00235 0.947^{20}	HOAc 4 1.4778^{20}
4557	Cyclohexanol, 1-propyl- 1-Hydroxy-1-propylcyclohexane	$C_9H_{18}O$ 142.24	5445-24-9	180 dec; 85^{20}	4-06-00-00129 0.934^{12}	1.468^{12}
4558	Cyclohexanol, 2-propyl-, *cis*- *cis*-2-Propylcyclohexanol	$C_9H_{18}O$ 142.24	5857-86-3	84^{10}	4-06-00-00130 0.9247^{11}	1.4688^{11}
4559	Cyclohexanol, 2-propyl-, *trans*- *trans*-2-Propylcyclohexanol	$C_9H_{18}O$ 142.24	5846-43-5	90^{14}	4-06-00-00130 0.9162^{11}	1.4668^{11}
4560	Cyclohexanol, 4-propyl-, *trans*- *trans*-4-Propylcyclohexanol	$C_9H_{18}O$ 142.24	77866-58-1	208; 103^{14}	3-06-00-00106 0.9072^{17}	1.4613^{17}
4561	Cyclohexanol, 4-propyl-, *cis*-	$C_9H_{18}O$ 142.24	98790-22-8	99^{17}	3-06-00-00105 0.9102^{16}	1.4652^{16}
4562	Cyclohexanol, 1,2,2-trimethyl- 1,2,2-Trimethylcyclohexanol	$C_9H_{18}O$ 142.24	31720-69-1	81^{20}; 76^{16}	3-06-00-00114 0.923^{20}	bz 4; eth 4; EtOH 4 1.4680^{20}

No.	Name Synonym	Mol. Form. Mol. Wt.	CAS RN mp/°C	Merck No. bp/°C	Beil. Ref. den/g cm^{-3}	Solubility n_D
4563	Cyclohexanol, 1,2,6-trimethyl- 1,2,6-Trimethylcyclohexanol	$C_9H_{18}O$ 142.24	96244-13-2	 78^{22}	1-06-00-00017 0.9126^{15}	ace 4; bz 4; eth 4; EtOH 4 1.4598^{15}
4564	Cyclohexanol, 1,3,5-trimethyl- 1,3,5-Trimethylcyclohexanol	$C_9H_{18}O$ 142.24	90760-75-1	 181	3-06-00-00117 0.8876^{17}	eth 4; EtOH 4; chl 4 1.454^{16}
4565	Cyclohexanol, 2,2,5-trimethyl- 2,2,5-Trimethylcyclohexanol	$C_9H_{18}O$ 142.24	73210-25-0	 188	0-06-00-00022 0.8955^{22}	EtOH 4 1.4569^{20}
4566	Cyclohexanol, 2,2,6-trimethyl- 1,1,3-Trimethyl-2-cyclohexanol	$C_9H_{18}O$ 142.24	10130-91-3 51	 187; 87^{28}	4-06-00-00135 0.9128^{20}	eth 4; EtOH 4; chl 4 1.4600^{20}
4567	Cyclohexanol, 2,3,3-trimethyl-	$C_9H_{18}O$ 142.24	116724-17-5 28	 197	1-06-00-00016 0.9002^{25}	ace 4; EtOH 4 1.4572^{25}
4568	Cyclohexanol, 2,3,6-trimethyl- 2,3,6-Trimethylcyclohexanol	$C_9H_{18}O$ 142.24	58210-03-0	 195	0-06-00-00022 0.9117^{17}	EtOH 4; chl 4
4569	Cyclohexanol, 3,3,5-trimethyl-, *cis-* *cis*-3,3,5-Trimethylcyclohexanol	$C_9H_{18}O$ 142.24	933-48-2 37.3	 202; 92^{12}	4-06-00-00135 0.9006^{16}	H_2O 1; EtOH 3; eth 3; chl 3 1.4550^{16}
4570	Cyclohexanol, 3,3,5-trimethyl-, *trans-* *trans*-3,3,5-Trimethylcyclohexanol	$C_9H_{18}O$ 142.24	767-54-4 55.8	 189.2	4-06-00-00135 0.8631^{60}	H_2O 1; EtOH 3; eth 3; chl 3
4571	Cyclohexanone Pimelic ketone	$C_6H_{10}O$ 98.14	108-94-1 -31	2732 155.4	4-07-00-00015 0.9478^{20}	H_2O 3; EtOH 3; eth 3; ace 3 1.4507^{20}
4572	Cyclohexanone, 2-acetyl-	$C_8H_{12}O_2$ 140.18	874-23-7	 $111\text{-}2^{18}$	4-07-00-01997 1.0782^{25}	ctc 3 1.5138^{20}
4573	Cyclohexanone, 2,6-bis(phenylmethylene)-	$C_{20}H_{18}O$ 274.36	897-78-9 117.5	 $185\text{-}95^{20}$	3-07-00-02661	EtOH 2; bz 3; HOAc 3
4574	Cyclohexanone, 2-bromo-	C_6H_9BrO 177.04	822-85-5	 $89\text{-}90^{14}$	4-07-00-00034 1.340^{25}	 1.5085^{25}
4575	Cyclohexanone, 2-butyl- 2-Butylcyclohexanone	$C_{10}H_{18}O$ 154.25	1126-18-7	 70^{2}	4-07-00-00079 0.905^{20}	H_2O 1; ctc 2; os 4 1.4545^{20}
4576	Cyclohexanone, 2-butylidene- 2-Butylidenecyclohexanone	$C_{10}H_{16}O$ 152.24	7153-14-2	 $98\text{-}100^{10}$	4-07-00-00181 0.935^{20}	ace 4; bz 4; eth 4; EtOH 4 1.4800^{20}
4577	Cyclohexanone, 2-chloro- 2-Chlorocyclohexanone	C_6H_9ClO 132.59	822-87-7 23	 82^{15}	4-07-00-00032 1.160^{20}	eth 3; bz 3; ctc 2; diox 3 1.4825^{20}
4578	Cyclohexanone, 4-chloro- 4-Chlorocyclohexanone	C_6H_9ClO 132.59	21299-26-3	 95^{17}	4-07-00-00032	eth 4 1.4867^{20}
4579	Cyclohexanone, 2-(2-chlorophenyl)-2-(methylamino)-, (±)- Ketamine	$C_{13}H_{16}ClNO$ 237.73	6740-88-1 92.5	5174		
4580	Cyclohexanone, 2,2-dimethyl- 2,2-Dimethyl-1-cyclohexanone	$C_8H_{14}O$ 126.20	1193-47-1 -20.5	 172	4-07-00-00055 0.9145^{20}	 1.4486^{20}
4581	Cyclohexanone, 2,3-dimethyl-, *cis-* *cis*-2,3-Dimethylcyclohexanone	$C_8H_{14}O$ 126.20	1551-88-8	 178.5; 99^{57}	3-07-00-00088 0.9159^{20}	 1.4505^{20}
4582	Cyclohexanone, 2,3-dimethyl-, *trans-* *trans*-2,3-Dimethylcyclohexanone	$C_8H_{14}O$ 126.20	1551-89-9	 96^{57}	4-07-00-00057 0.908^{22}	 1.4511^{25}
4583	Cyclohexanone, 2,4-dimethyl-, (2*S-trans*)- Cyclohexanone, 2,4-dimethyl (*trans*, d)	$C_8H_{14}O$ 126.20	93921-42-7	 179; 69^{17}	3-07-00-00093 0.9004^{16}	ace 4; bz 4; eth 4 1.4488^{22}
4584	Cyclohexanone, 2,4-dimethyl-, *cis-*	$C_8H_{14}O$ 126.20	116783-29-0	 177	3-07-00-00093 0.910^{15}	 1.4493^{15}
4585	Cyclohexanone, 2,5-dimethyl-, *trans*-(±)- *trans*-(±)-2,5-Dimethylcyclohexanone	$C_8H_{14}O$ 126.20	66395-23-1	 172	4-07-00-00059 0.9025^{20}	eth 4; EtOH 4 1.4446^{20}
4586	Cyclohexanone, 2,6-dimethyl- 2,6-Dimethylcyclohexanone	$C_8H_{14}O$ 126.20	2816-57-1	 175	2-07-00-00025 0.925^{25}	 1.4460^{20}
4587	Cyclohexanone, 3,3-dimethyl- 3,3-Dimethylcyclohexanone	$C_8H_{14}O$ 126.20	2979-19-3	 180; $72\text{-}3^{25}$	4-07-00-00056 0.909^{15}	 1.44821^{7}
4588	Cyclohexanone, 3,4-dimethyl- 3,4-Dimethylcyclohexanone	$C_8H_{14}O$ 126.20	5465-09-8	 187	3-07-00-00089 0.906^{20}	 1.4507^{20}
4589	Cyclohexanone, 3,5-dimethyl- 3,5-Dimethylcyclohexanone	$C_8H_{14}O$ 126.20	2320-30-1	 182.5	4-07-00-00058 0.896^{20}	 1.4407^{20}
4590	Cyclohexanone, 4,4-dimethyl- 4,4-Dimethylcyclohexanone	$C_8H_{14}O$ 126.20	4255-62-3 39	 73^{14}	4-07-00-00056 0.932^{20}	 1.4537^{24}
4591	Cyclohexanone, 2-[(dimethylamino)methyl]- 2-[(Dimethylamino)methyl]cyclohexanone	$C_9H_{17}NO$ 155.24	15409-60-6	 $92^{10.5}$	4-14-00-00008 0.9504^{20}	H_2O 2; EtOH 3; eth 3 1.4672^{20}
4592	Cyclohexanone, 4-(1,1-dimethylpropyl)-	$C_{11}H_{20}O$ 168.28	16587-71-6	 $124\text{-}5^{16}$	4-07-00-00096 0.920^{25}	 1.4677^{20}
4593	Cyclohexanone, 4-ethyl- 4-Ethylcyclohexanone	$C_8H_{14}O$ 126.20	5441-51-0	 193	3-07-00-00083 0.895^{25}	 1.4515^{20}
4594	Cyclohexanone, 2-ethylidene- 2-Ethylidenecyclohexanone	$C_8H_{12}O$ 124.18	1122-25-4	 92^{20}	4-07-00-00146 0.962^{20}	H_2O 1; os 4 1.4882^{20}
4595	Cyclohexanone, 2-ethyl-2-methyl- 2-Ethyl-2-methylcyclohexanone	$C_9H_{16}O$ 140.23	17206-52-9	 195	2-07-00-00031 0.9037^{17}	 1.4515^{17}
4596	Cyclohexanone, 2-methyl-, (±)- (±)-2-Methylcyclohexanone	$C_7H_{12}O$ 112.17	24965-84-2 -13.9	 165	4-07-00-00041 0.9250^{20}	H_2O 1; EtOH 3; eth 3 1.4483^{25}
4597	Cyclohexanone, 3-methyl-, (±)-	$C_7H_{12}O$ 112.17	626 00-7 -73.5	 169; 65^{15}	4-07-00-00043 0.9136^{20}	H_2O 1; EtOH 3; eth 3 1.4456^{20}
4598	Cyclohexanone, 4-methyl- 4-Methylcyclohexanone	$C_7H_{12}O$ 112.17	589-92-4 -40.6	 170	4-07-00-00044 0.9138^{20}	H_2O 1; EtOH 3; eth 3; ctc 2 1.4451^{20}
4599	Cyclohexanone, 2-(1-methylethyl)- 2-Isopropylcyclohexanone	$C_9H_{16}O$ 140.23	1004-77-9 72.5	 83^{17}	4-07-00-00066 0.922^{16}	H_2O 1; EtOH 3; eth 3; ace 3 1.4564^{15}
4600	Cyclohexanone, 4-(1-methylethyl)- 4-Isopropylcyclohexanone	$C_9H_{16}O$ 140.23	5432-85-9	 $139\text{-}40^{100}$	4-07-00-00066 0.9158^{25}	 1.4552^{25}
4601	Cyclohexanone, 2-methyl-5-(1-methylethenyl)-, (2*S-trans*)-	$C_{10}H_{16}O$ 152.24	619-02-3 -11	 221.5	 0.9253^{20}	 1.4717^{20}

No.	Name Synonym	Mol. Form. Mol. Wt.	CAS RN mp/°C	Merck No. bp/°C	Beil. Ref. den/g cm^{-3}	Solubility n_D
4602	Cyclohexanone, 5-methyl-2-(1-methylethenyl)-	$C_{10}H_{16}O$ 152.24	529-00-0	 100^{18}	4-07-00-00190 0.9198^{20}	 1.4675^{20}
4603	Cyclohexanone, 5-methyl-2-(1-methylethenyl)-, *trans*-	$C_{10}H_{16}O$ 152.24	29606-79-9	 100^{18}	4-07-00-00190 0.9198^{20}	 1.4675^{20}
4604	Cyclohexanone, 2-methyl-5-(1-methylethyl)-, (2*S*-*trans*)-	$C_{10}H_{18}O$ 154.25	13163-73-0	 220; 96^{15}	4-07-00-00085 0.904^{20}	ace 4; EtOH 4; chl 4 1.4553^{20}
4605	Cyclohexanone, 5-methyl-2-(1-methylethyl)-, (2*R*-*cis*)- Menthone(*d*)	$C_{10}H_{18}O$ 154.25	1196-31-2	 204; 85^{14}	4-07-00-00087 0.8947^{20}	H_2O 2; EtOH 3; eth 3; ace 3 1.4503^{20}
4606	Cyclohexanone, 5-methyl-2-(1-methylethyl)-, (2*S*-*trans*)- (-)-Menthone	$C_{10}H_{18}O$ 154.25	14073-97-3 -6	5724 207	 0.8954^{20}	H_2O 2; EtOH 5; eth 5; ace 3 1.4505^{20}
4607	Cyclohexanone, 5-methyl-2-(1-methylethylidene)-	$C_{10}H_{16}O$ 152.24	15932-80-6	 95-6^{11}	4-07-00-00188 0.9367^{20}	 1.4869^{20}
4608	Cyclohexanone, 5-methyl-2-(1-methylethylidene)-, (*R*)- Pulegone	$C_{10}H_{16}O$ 152.24	89-82-7	7955 224	4-07-00-00188 0.9346^{45}	H_2O 1; EtOH 5; eth 5; ctc 3 1.4894^{20}
4609	Cyclohexanone, 4-(2-methylpropyl)-	$C_{10}H_{18}O$ 154.25	42061-72-3	 218; 98^{12}	4-07-00-00082 0.8914^{24}	 1.4502^{24}
4610	Cyclohexanone, oxime (Hydroxyimino)cyclohexane	$C_6H_{11}NO$ 113.16	100-64-1 90	 206	4-07-00-00021	H_2O 3; EtOH 3; eth 3; chl 2
4611	Cyclohexanone, 2-(1-piperidinylmethyl)- Pimeclone	$C_{12}H_{21}NO$ 195.30	534-84-9	7399 118-20^{14}	5-20-02-00330	
4612	Cyclohexanone, 2-propyl- 2-Propylcyclohexanone	$C_9H_{16}O$ 140.23	94-65-5	 197	4-07-00-00064 0.927^{20}	H_2O 1; EtOH 3; eth 4; ace 3 1.4538^{20}
4613	Cyclohexanone, 4-propyl-	$C_9H_{16}O$ 140.23	40649-36-3	 213; 97.5^{15}	3-07-00-00116 0.9072^{19}	 1.4530^{19}
4614	Cyclohexanone, 2,2,3-trimethyl- 2,2,3-Trimethylcyclohexanone	$C_9H_{16}O$ 140.23	39257-08-4	 66-7^{10}	4-07-00-00069 0.9234^{15}	 1.4569^{15}
4615	Cyclohexanone, 2,2,4-trimethyl- 2,2,4-Trimethylcyclohexanone	$C_9H_{16}O$ 140.23	35413-38-8	 86-7^{12}	3-07-00-00125 0.8914^{20}	 1.4520^{20}
4616	Cyclohexanone, 2,2,5-trimethyl- 2,2,5-Trimethylcyclohexanone	$C_9H_{16}O$ 140.23	933-36-8	 183	3-07-00-00126 0.8871^{24}	 1.4432^{24}
4617	Cyclohexanone, 2,2,6-trimethyl-	$C_9H_{16}O$ 140.23	2408-37-9 -31.8	 178.5	4-07-00-00069 0.9043^{18}	 1.4470^{20}
4618	Cyclohexanone, 2,3,6-trimethyl- 2,3,6-Trimethylcyclohexanone	$C_9H_{16}O$ 140.23	42185-47-7	 190	3-07-00-00126 0.9129^{18}	 1.4464^{21}
4619	Cyclohexanone, 2,4,4-trimethyl- 2,4,4-Trimethylcyclohexanone	$C_9H_{16}O$ 140.23	2230-70-8	 191	4-07-00-00070 0.902^{20}	 1.4493^{20}
4620	Cyclohexanone, 2,4,5-trimethyl- 2,4,5-Trimethylcyclohexanone	$C_9H_{16}O$ 140.23	51299-55-9	 193	4-07-00-00072 0.897^{20}	 1.4479^{20}
4621	Cyclohexanone, 2,4,6-trimethyl- 2,4,6-Trimethylcyclohexanone	$C_9H_{16}O$ 140.23	90645-54-8	 185	3-07-00-00127 0.8992^{20}	 1.4458^{20}
4622	Cyclohexanone, 3,3,5-trimethyl- Dihydroisophorone	$C_9H_{16}O$ 140.23	873-94-9	 189	4-07-00-00070 0.8919^{19}	 1.4454^{15}
4623	Cyclohexanone, 3,4,4-trimethyl- 3,4,4-Trimethylcyclohexanone	$C_9H_{16}O$ 140.23	40441-35-8	 80-1^{13}	3-07-00-00122 0.911^{25}	 1.4552^{20}
4624	Cyclohexasiloxane, dodecamethyl- Dodecamethylcyclohexasiloxane	$C_{12}H_{36}O_6Si_6$ 444.93	540-97-6 -3	3400 245	4-04-00-04129 0.9672^{25}	H_2O 1 1.4015^{20}
4625	Cyclohexasiloxane, 2,4,6,8,10,12-hexamethyl- Hexamethylcyclohexasiloxane	$C_6H_{24}O_6Si_6$ 360.77	6166-87-6 -79		4-04-00-04100 1.006^{20}	H_2O 1 1.3944^{20}
4626	Cyclohexene Tetrahydrobenzene	C_6H_{10} 82.15	110-83-8 -103.5	2733 82.9	4-05-00-00218 0.8110^{20}	H_2O 1; EtOH 5; eth 5; ace 5 1.4465^{20}
4627	Cyclohexene, 1-bromo-	C_6H_9Br 161.04	2044-08-8	 165	4-05-00-00236 1.3901^{20}	eth 3; ace 4; bz 3; chl 4 1.5134^{20}
4628	Cyclohexene, 3-bromo-	C_6H_9Br 161.04	1521-51-3	 80-2^{40}	4-05-00-00237 1.3890^{20}	H_2O 1; eth 3; bz 3; chl 3 1.5320^{20}
4629	3-Cyclohexene-1-carbonitrile	C_7H_9N 107.16	100-45-8	 $39^{1.5}$	4-09-00-00116	ctc 3 1.4716^{20}
4630	1-Cyclohexenecarbonitrile 1-Cyanocyclohexene	C_7H_9N 107.16	1855-63-6	 81^{12}		
4631	1-Cyclohexene-1-carboxaldehyde	$C_7H_{10}O$ 110.16	1192-88-7	 72^{15}	4-07-00-00132 0.9694^{20}	EtOH 3; eth 3 1.5005^{20}
4632	3-Cyclohexene-1-carboxaldehyde	$C_7H_{10}O$ 110.16	100-50-5 2	 105	4-07-00-00134 0.9692^{20}	ace 3; ctc 2; MeOH 3
4633	1-Cyclohexene-1-carboxaldehyde, 4-(1-methylethenyl)- Perillaldehyde	$C_{10}H_{14}O$ 150.22	2111-75-3	7119	2-07-00-00130	
4634	1-Cyclohexene-1-carboxaldehyde, 4-(1-methylethenyl)-, (*R*)-	$C_{10}H_{14}O$ 150.22	5503-12-8	 238; 99^{7}	4-07-00-00316 0.953^{20}	 1.5058^{20}
4635	1-Cyclohexene-1-carboxaldehyde, 4-(1-methylethyl)- Phellandral	$C_{10}H_{16}O$ 152.24	21391-98-0	 223	4-07-00-00187 0.93^{20}	 1.4911^{20}
4636	1-Cyclohexene-1-carboxaldehyde, 2,6,6-trimethyl- β-Cyclocitral	$C_{10}H_{16}O$ 152.24	432-25-7	 95-100^{15}	4-07-00-00192 0.959^{15}	 1.4971^{15}

No.	Name Synonym	Mol. Form. Mol. Wt.	CAS RN mp/°C	Merck No. bp/°C	Beil. Ref. den/g cm^{-3}	Solubility n_D
4637	1-Cyclohexene-1-carboxylic acid Cyclohexene-1-carboxylic acid	$C_7H_{10}O_2$ 126.16	636-82-8 38	 241	4-09-00-00111 1.109^{20}	H_2O 2; EtOH 3; ace 3 1.4902^{20}
4638	3-Cyclohexene-1-carboxylic acid	$C_7H_{10}O_2$ 126.16	4771-80-6 17	 234.5	4-09-00-00114 1.0820^{20}	H_2O 4; EtOH 3; ace 3 1.4814^{20}
4639	3-Cyclohexene-1-carboxylic acid, ethyl ester	$C_9H_{14}O_2$ 154.21	15111-56-5	 194.5	4-09-00-00114 0.9688^{20}	 1.4578^{20}
4640	3-Cyclohexene-1-carboxylic acid, methyl ester	$C_8H_{12}O_2$ 140.18	6493-77-2	 80^{20}	4-09-00-00114 0.995^{25}	 1.4610^{20}
4641	2-Cyclohexene-1-carboxylic acid, 2,6-dimethyl-4-oxo-, ethyl ester	$C_{11}H_{16}O_3$ 196.25	6102-15-4	 157-8^{18}	4-10-00-02674 1.0493^{20}	ace 3 1.4773^{20}
4642	Cyclohexene, 1-chloro- 1-Chlorocyclohexene	C_6H_9Cl 116.59	930-66-5	 142.5	4-05-00-00230 1.0361^{19}	eth 3; ace 3; ctc 3; chl 3 1.4797^{20}
4643	Cyclohexene, decafluoro- Perfluorocyclohexene	C_6F_{10} 262.05	355-75-9	 52.0	4-05-00-00229 1.6650^{25}	 1.293^{20}
4644	1-Cyclohexene-1,2-dicarboxylic acid, diethyl ester Cyclohexene-1,2-dicarboxylic acid, diethyl ester	$C_{12}H_{18}O_4$ 226.27	92687-41-7	 160^{14}	3-09-00-03940 1.0803^{19}	 1.4747^{19}
4645	1-Cyclohexene-1,3-dicarboxylic acid, diethyl ester Diethyl 1-cyclohexene-1,3-dicarboxylate	$C_{12}H_{18}O_4$ 226.27	38511-07-8	 150^{12}	3-09-00-03950 1.0772^{20}	 1.4722^{20}
4646	1-Cyclohexene-1,4-dicarboxylic acid, dimethyl ester Cyclohexene-1,4-dicarboxylic acid, dimethyl ester	$C_{10}H_{14}O_4$ 198.22	22646-79-3 39	 153^{20}	4-09-00-02897	ace 4; eth 4; EtOH 4
4647	Cyclohexene, 1,2-dichloro-3,3,4,4,5,5,6,6-octafluoro-	$C_6Cl_2F_8$ 294.96	336-19-6 -70	 113	4-05-00-00231 1.719^{20}	 1.375^{20}
4648	Cyclohexene, 1,2-dimethyl-	C_8H_{14} 110.20	1674-10-8 -84.1	 138	4-05-00-00268 0.8220^{25}	 1.4620^{20}
4649	Cyclohexene, 1,3-dimethyl- 1,3-Dimethyl-1-cyclohexene	C_8H_{14} 110.20	2808-76-6	 127	4-05-00-00270 0.799^{25}	 1.449^{20}
4650	Cyclohexene, 1,4-dimethyl- 1,4-Dimethyl-1-cyclohexene	C_8H_{14} 110.20	2808-79-9 -59.4	 128.5	4-05-00-00270 0.8005^{20}	 1.4457^{20}
4651	Cyclohexene, 1,5-dimethyl- Cyclohexene, 2,4-dimethyl-	C_8H_{14} 110.20	2808-77-7	 126	3-05-00-00216 0.8009^{25}	 1.450^{20}
4652	Cyclohexene, 1,6-dimethyl-	C_8H_{14} 110.20	1759-64-4	 131	4-05-00-00269 0.815^{20}	 1.454^{20}
4653	Cyclohexene, 3,3-dimethyl- 3,3-Dimethylcyclohexene	C_8H_{14} 110.20	695-28-3	 119	4-05-00-00268 0.804^{20}	 1.445^{20}
4654	Cyclohexene, 3,5-dimethyl- 3,5-Dimethylcyclohexene	C_8H_{14} 110.20	823-17-6	 124.5	3-05-00-00216 0.8005^{18}	 1.443^{18}
4655	Cyclohexene, 3,6-dimethyl- 3,6-Dimethylcyclohexene	C_8H_{14} 110.20	19550-40-4	 125	4-05-00-00271 0.802^{25}	 1.4433^{25}
4656	Cyclohexene, 4,4-dimethyl-	C_8H_{14} 110.20	14072-86-7 -74.4	 117	4-05-00-00268 0.7996^{20}	 1.4420^{20}
4657	Cyclohexene, 1-ethenyl-	C_8H_{12} 108.18	2622-21-1	 145	4-05-00-00405 0.8623^{15}	H_2O 1; eth 3; bz 3; MeOH 4 1.4915^{20}
4658	Cyclohexene, 4-ethenyl-	C_8H_{12} 108.18	100-40-3 -108.9	 128	4-05-00-00406 0.8299^{20}	H_2O 1; eth 3; bz 3; peth 3 1.4639^{20}
4659	Cyclohexene, 6-ethenyl-6-methyl-1-(1-methylethyl)-3-(1-methylethylidene)-, (*S*)-	$C_{15}H_{24}$ 204.36	5951-67-7	 120-30^{8}	3-05-00-01083 0.8782^{20}	ace 4; bz 4 1.5130^{26}
4660	Cyclohexene, 1-ethyl-	C_8H_{14} 110.20	1453-24-3 -109.9	 137	4-05-00-00266 0.8176^{25}	 1.4567^{20}
4661	Cyclohexene, 3-ethyl- 3-Ethyl-1-cyclohexene	C_8H_{14} 110.20	2808-71-1	 134	4-05-00-00266 0.814^{20}	 1.451^{20}
4662	Cyclohexene, 4-ethyl- 4-Ethyl-1-cyclohexene	C_8H_{14} 110.20	3742-42-5	 133	4-05-00-00267 0.810^{20}	 1.449^{20}
4663	Cyclohexene, 1-ethyl-4-methyl- 1-Ethyl-4-methylcyclohexene	C_9H_{16} 124.23	62088-36-2	 149	4-05-00-00286 0.8169^{16}	 1.453^{16}
4664	1-Cyclohexene-1-methanol, 4-(1-methylethenyl)-	$C_{10}H_{16}O$ 152.24	536-59-4	 244; 12.5^{12}	2-06-00-00102 0.9690^{20}	 1.5005^{20}
4665	1-Cyclohexene-1-methanol, 2,4,4-trimethyl- β-Cyclolavandulol	$C_{10}H_{18}O$ 154.25	103985-40-6	 105-6^{15}	4-06-00-00259 0.9153^{20}	 1.4814^{20}
4666	1-Cyclohexene-1-methanol, 2,6,6-trimethyl- Cyclogeranyl alcohol	$C_{10}H_{18}O$ 154.25	472-20-8 43.5		3-06-00-00259 0.945^{19}	 1.487^{19}
4667	2-Cyclohexene-1-methanol, 2,6,6-trimethyl- α-Cyclogeraniol	$C_{10}H_{18}O$ 154.25	6627-74-3	 215	4-06-00-00257 0.9382^{20}	 1.4843^{20}
4668	3-Cyclohexene-1-methanol, α,α,4-trimethyl-, (±)-	$C_{10}H_{18}O$ 154.25	2438-12-2 40.5	 220	4-06-00-00252 0.9337^{20}	ace 4; bz 4; eth 4; EtOH 4 1.4831^{20}
4669	3-Cyclohexene-1-methanol, α,α,4-trimethyl-, acetate	$C_{12}H_{20}O_2$ 196.29	80-26-2	 140^{40}; 105^{11}	4-06-00-00253 0.9659^{21}	H_2O 1; EtOH 3; eth 3; bz 3 1.4689^{21}
4670	Cyclohexene, 1-methyl-	C_7H_{12} 96.17	591-49-1 -120.4	 110.3	4-05-00-00245 0.8102^{20}	H_2O 1; eth 3; bz 3; ctc 3 1.4503^{20}

No.	Name Synonym	Mol. Form. Mol. Wt.	CAS RN mp/°C	Merck No. bp/°C	Beil. Ref. den/g cm^{-3}	Solubility n_D
4671	Cyclohexene, 3-methyl-, (±)- (±)-3-Methylcyclohexene	C_7H_{12} 96.17	56688-75-6 -115.5	 104	4-05-00-00247 0.7990^{20}	bz 4; eth 4; chl 4; peth 4 1.4414^{20}
4672	Cyclohexene, 4-methyl-	C_7H_{12} 96.17	591-47-9 -115.5	 102.7	4-05-00-00248 0.7991^{20}	H_2O 1; EtOH 3; eth 3 1.4414^{20}
4673	Cyclohexene, 3-methylene-6-(1-methylethyl)- β-Phellandrene	$C_{10}H_{16}$ 136.24	555-10-2	7152 171.5	2-05-00-00087 0.8520^{20}	H_2O 1; EtOH 1; eth 3 1.4788^{20}
4674	Cyclohexene, 3-methylene-6-(1-methylethyl)-, (+)-	$C_{10}H_{16}$ 136.24	6153-16-8	 171.5	4-05-00-00436 0.8520^{20}	eth 4 1.4788^{20}
4675	Cyclohexene, 4-methylene-1-(1-methylethyl)-	$C_{10}H_{16}$ 136.24	99-84-3	 173.5	4-05-00-00437 0.838^{22}	 1.4754^{22}
4676	Cyclohexene, 1-methyl-3-(1-methylethenyl)-	$C_{10}H_{16}$ 136.24	38738-60-2	 175	4-05-00-00434 0.8479^{18}	 1.4760^{18}
4677	Cyclohexene, 1-methyl-4-(1-methylethenyl)- Limonene	$C_{10}H_{16}$ 136.24	138-86-3 -95	5371 178	4-05-00-00440 0.8402^{21}	H_2O 1; EtOH 5; eth 5; ctc 3 1.4727^{19}
4678	Cyclohexene, 1-methyl-5-(1-methylethenyl)-	$C_{10}H_{16}$ 136.24	13898-73-2	 176; 95^{20}	4-05-00-00435 0.8481^{20}	eth 4; EtOH 4 1.4804^{20}
4679	Cyclohexene, 5-methyl-1-(1-methylethenyl)-, (±)- Cyclohexene, 1-isopropenyl-5-methyl (*DL*)	$C_{10}H_{16}$ 136.24	105065-24-5	 184	3-05-00-00336 0.8594^{20}	 1.4975
4680	Cyclohexene, 1-methyl-4-(1-methylethyl)-	$C_{10}H_{18}$ 138.25	5502-88-5	 174.5	4-05-00-00300 0.8457^{15}	 1.4735^{20}
4681	Cyclohexene, 1-methyl-4-(1-methylethyl)-, (*R*)-	$C_{10}H_{18}$ 138.25	1195-31-9	 173	4-05-00-00299 0.8246^{18}	bz 4; EtOH 4; peth 4 1.4563^{18}
4682	Cyclohexene, 3-methyl-6-(1-methylethyl)-, (3*R-trans*)-	$C_{10}H_{18}$ 138.25	5113-93-9	 167; 55-6^{12}	4-05-00-00300 0.824^{20}	 1.461^{20}
4683	Cyclohexene, 4-methyl-1-(1-methylethyl)-, (*R*)-	$C_{10}H_{18}$ 138.25	619-52-3	 167.5	4-05-00-00301 0.8078^{20}	H_2O 1; EtOH 3; eth 3; bz 3 1.4503^{15}
4684	Cyclohexene, 1-methyl-4-(1-methylethylidene)- Terpinolene	$C_{10}H_{16}$ 136.24	586-62-9	 186	4-05-00-00437 0.8632^{15}	H_2O 1; EtOH 5; eth 5; bz 3 1.4883^{20}
4685	Cyclohexene, 1-methyl-4-(5-methyl-1-methylene-4-hexenyl)-, (*S*)-	$C_{15}H_{24}$ 204.36	495-61-4	 129-$30^{10.5}$	4-05-00-01174 0.8673^{20}	 1.4880^{20}
4686	5-Cyclohexene-1,2,3,4-tetrone, 5,6-dihydroxy- Rhodizonic acid	$C_6H_2O_6$ 170.08	118-76-3 248 dec	 sub	4-08-00-03609	EtOH 3
4687	Cyclohexene, 1,2,3-trimethyl- 1,2,3-Trimethylcyclohexene	C_9H_{16} 124.23	72312-48-2	 150	1-05-00-00040 0.8347^{11}	 1.463^{11}
4688	Cyclohexene, 1,3,5-trimethyl- 1,3,5-Trimethylcyclohexene	C_9H_{16} 124.23	3643-64-9	 141	3-05-00-00226 0.8025^{11}	 1.449^{13}
4689	Cyclohexene, 1,4,4-trimethyl- 1,4,4-Trimethylcyclohexene	C_9H_{16} 124.23	3419-71-4	 140	1-05-00-00039 0.8032^{19}	 1.444^{23}
4690	Cyclohexene, 1,4,5-trimethyl- 1,4,5-Trimethylcyclohexene	C_9H_{16} 124.23	20030-32-4	 145	2-05-00-00050 0.805^{20}	 1.4482^{20}
4691	Cyclohexene, 1,5,5-trimethyl- α-Cyclogeraniolene	C_9H_{16} 124.23	503-46-8	 140	4-05-00-00286 0.7981^{23}	 1.4461^{21}
4692	Cyclohexene, 1,6,6-trimethyl- 1,6,6-Trimethylcyclohexene	C_9H_{16} 124.23	69745-49-9	 147	3-05-00-00225 0.8217^{20}	 1.456^{20}
4693	Cyclohexene, 1,5,6-trimethyl-	C_9H_{16} 124.23	116724-18-6	 140	3-05-00-00225 0.829^{25}	 1.4572^{25}
4694	2-Cyclohexen-1-ol	$C_6H_{10}O$ 98.14	822-67-3	 164	4-06-00-00196 0.9923^{15}	EtOH 3; ace 3 1.4790^{25}
4695	3-Cyclohexen-1-ol	$C_6H_{10}O$ 98.14	822-66-2	 164; 68^{16}	4-06-00-00200 0.9845^{23}	H_2O 2; eth 3; ace 3 1.4851^{22}
4696	2-Cyclohexen-1-ol, 2-methyl-5-(1-methylethenyl)-	$C_{10}H_{16}O$ 152.24	99-48-9	 228	2-06-00-00102 0.9484^{25}	 1.4942^{25}
4697	2-Cyclohexen-1-ol, 3-methyl-6-(1-methylethyl)-	$C_{10}H_{18}O$ 154.25	491-04-3	 97-$8^{15.5}$	2-06-00-00065 0.9119^{25}	 1.4729^{25}
4698	2-Cyclohexen-1-ol, 6-methyl-3-(1-methylethyl)- Carvenol	$C_{10}H_{18}O$ 154.25	586-27-6	 220	4-06-00-00253 0.925^{22}	 1.479
4699	3-Cyclohexen-1-ol, 4-methyl-1-(1-methylethyl)-	$C_{10}H_{18}O$ 154.25	562-74-3	 209	4-06-00-00250 0.926^{20}	 1.4785^{19}
4700	2-Cyclohexen-1-ol, 1,3,5-trimethyl- Cyclohexen-3-ol, 1,3,5-trimethyl	$C_9H_{16}O$ 140.23	33843-55-9 46	 87-90^{17}	1-06-00-00036 0.9132^{20}	 1.4735^{19}
4701	2-Cyclohexen-1-ol, 2,4,4-trimethyl- 2,4,4-Trimethyl-2-cyclohexen-1-ol	$C_9H_{16}O$ 140.23	73741-61-4	 193	1-06-00-00036 0.9310^{25}	bz 4; EtOH 4
4702	2-Cyclohexen-1-ol, 3,5,5-trimethyl- Isophorol	$C_9H_{16}O$ 140.23	470-99-5	 69^{5}	3-06-00-00230 0.9128^{20}	 1.4717^{20}
4703	2-Cyclohexen-1-one	C_6H_8O 96.13	930-68-7 -53	 170	4-07-00-00124 0.9620^{25}	EtOH 4; ace 3 1.4883^{20}
4704	3-Cyclohexen-1-one	C_6H_8O 96.13	4096-34-8	 170	4-07-00-00126 0.9620^{25}	ace 4; EtOH 4 1.4842^{18}
4705	2-Cyclohexen-1-one, 2,3-dimethyl-	$C_8H_{12}O$ 124.18	1122-20-9	 93-6^{20}	4-07-00-00148 0.9695^{20}	EtOH 3; eth 3 1.4995^{20}
4706	2-Cyclohexen-1-one, 2,5-dimethyl- 2,5-Dimethyl-2-cyclohexenone	$C_8H_{12}O$ 124.18	14845-35-3	 189.5	1-07-00-00051 0.938^{22}	eth 4; EtOH 4 1.4753^{22}
4707	2-Cyclohexen-1-one, 3,5-dimethyl-	$C_8H_{12}O$ 124.18	1123-09-7	 208.5	4-07-00-00149 0.9400^{20}	EtOH 3; eth 3 1.4812^{20}

PHYSICAL CONSTANTS OF ORGANIC COMPOUNDS (continued)

No.	Name Synonym	Mol. Form. Mol. Wt.	CAS RN mp/°C	Merck No. bp/°C	Beil. Ref. den/g cm^{-3}	Solubility n_D
4708	2-Cyclohexen-1-one, 3,6-dimethyl-	$C_8H_{12}O$ 124.18	15329-10-9	 75^{19}	4-07-00-00150 1.007^{18}	H_2O 1; EtOH 3; eth 3 1.4805^{18}
4709	2-Cyclohexen-1-one, 2-hydroxy-3-methyl-6-(1-methylethyl)- Diosphenol	$C_{10}H_{16}O_2$ 168.24	490-03-9 83	3292 109^{10}		
4710	2-Cyclohexen-1-one, 2-methyl-	$C_7H_{10}O$ 110.16	1121-18-2	 178.5	4-07-00-00131 0.966^{20}	bz 3 1.4833^{20}
4711	2-Cyclohexen-1-one, 3-methyl-	$C_7H_{10}O$ 110.16	1193-18-6 -21	 201	4-07-00-00130 0.9693^{20}	H_2O 5; bz 3 1.49475^{20}
4712	2-Cyclohexen-1-one, 5-methyl- 5-Methyl-2-cyclohexenone	$C_7H_{10}O$ 110.16	7214-50-8	 185	4-07-00-00133 0.947^{25}	ace 4; bz 4; EtOH 4 1.4739^{25}
4713	3-Cyclohexen-1-one, 3-methyl-	$C_7H_{10}O$ 110.16	31883-98-4 -21	 201	4-07-00-00131 0.9693^{20}	bz 4 1.4947^{20}
4714	3-Cyclohexen-1-one, 4-methyl-	$C_7H_{10}O$ 110.16	5259-65-4	 171; 74^{17}	4-07-00-00131 0.9551^{20}	EtOH 3; ace 3; bz 3 1.4652^{20}
4715	2-Cyclohexen-1-one, 2-methyl-5-(1-methylethenyl)-, (*S*)- (+)-Carvone	$C_{10}H_{14}O$ 150.22	2244-16-8 <15	1883 231	4-07-00-00315 0.965^{20}	H_2O 2; EtOH 4; eth 3; chl 3 1.4989^{20}
4716	2-Cyclohexen-1-one, 3-methyl-6-(1-methylethyl)-, (±)- (±)-Piperitone	$C_{10}H_{16}O$ 152.24	6091-52-7 -19	7443 232.5	4-07-00-00186 0.9331^{20}	ace 4; EtOH 4 1.4845^{20}
4717	2-Cyclohexen-1-one, 6-methyl-3-(1-methylethyl)-, (±)- Carvenone (*DL*)	$C_{10}H_{16}O$ 152.24	23733-68-8	 235.5	4-07-00-00187 0.9263^{20}	ace 4 1.4826^{20}
4718	2-Cyclohexen-1-one, 3-methyl-6-(1-methylethylidene)- Piperitenone	$C_{10}H_{14}O$ 150.22	491-09-8	 $120\text{-}2^{14}$	4-07-00-00314 0.9774^{20}	EtOH 4; eth 4 1.5294^{20}
4719	2-Cyclohexen-1-one, 3,4,4-trimethyl- Cyclohexen-3-one, 1,6,6-trimethyl	$C_9H_{14}O$ 138.21	17299-41-1	 217; 98^{13}	4-07-00-00164 0.944^{25}	 1.4889^{25}
4720	2-Cyclohexen-1-one, 3,4,6-trimethyl- Cyclohexen-3-one, 1,4,6-trimethyl	$C_9H_{14}O$ 138.21	7474-10-4	 213	4-07-00-00170 0.943^{17}	 1.484^{17}
4721	2-Cyclohexen-1-one, 3,5,5-trimethyl- Isophorone	$C_9H_{14}O$ 138.21	78-59-1 -8.1	 215.2	4-07-00-00165 0.9255^{20}	 1.4766^{18}
4722	2-Cyclohexen-1-one, 3,6,6-trimethyl- 3,6,6-Trimethyl-2-cyclohexenone	$C_9H_{14}O$ 138.21	23438-77-9	 208; $86\text{-}8^{15}$	3-07-00-00286 0.927^{24}	 1.4798^{20}
4723	Cyclohexylamine, *N*,4-dimethyl- Cyclohexane, 1-methylamino-4-methyl	$C_8H_{17}N$ 127.23	90226-23-6 -15	 196	 0.8188^{22}	 1.4175^{22}
4724	Cyclononane	C_9H_{18} 126.24	293-55-0 11	 178.4	4-05-00-00133 0.8463^{25}	 1.4666^{20}
4725	Cyclononanone	$C_9H_{16}O$ 140.23	3350-30-9 34	 148^{24}; 94^{12}	4-07-00-00062 0.9560^{20}	EtOH 3 1.4729^{20}
4726	Cyclononasiloxane, octadecamethyl- Octadecamethylcyclononasiloxane	$C_{18}H_{54}O_9Si_9$ 667.39	556-71-8	 188^{20}	3-04-00-01886	bz 4; lig 4 1.4070^{20}
4727	Cyclononene, (*E*)-	C_9H_{16} 124.23	3958-38-1	 $94\text{-}6^{30}$	4-05-00-00280 0.8615^{20}	 1.4799^{20}
4728	Cyclononene, (*Z*)-	C_9H_{16} 124.23	933-21-1	 168	4-05-00-00280 0.8671^{20}	bz 4 1.4805^{20}
4729	Cyclononyne	C_9H_{14} 122.21	6573-52-0	 178	4-05-00-00419 0.8972^{20}	 1.4890^{20}
4730	2,4(1H)-Cyclo-3,4-secoakuammilanium, 3,17-dihydroxy-16-(methoxycarbonyl)-4-methyl-, (3β,16*R*)- Echitamine	$C_{22}H_{30}N_2O_5$ 402.49	6871-44-9 206	3474		H_2O 3; EtOH 3; eth 3; chl 3
4731	1,3-Cyclooctadecadiene	$C_{18}H_{32}$ 248.45	6568-58-7	 115^{3}	4-05-00-00503 0.8814^{25}	 1.4899^{20}
4732	1,4-Cyclooctadiene	C_8H_{12} 108.18	1073-07-0 -53	 145	 0.8754^{20}	
4733	1,5-Cyclooctadiene	C_8H_{12} 108.18	111-78-4 -56.4	 150.8	3-05-00-00321 0.8818^{25}	bz 4 1.4905^{25}
4734	1,5-Cyclooctadiene, (*Z,Z*)-	C_8H_{12} 108.18	1552-12-1 -70	 151; $51\text{-}2^{25}$	4-05-00-00403 0.8818^{25}	bz 4 1.4905^{25}
4735	Cyclooctanamine Aminocyclooctane	$C_8H_{17}N$ 127.23	5452-37-9 -48	 190	4-12-00-00128 0.928^{25}	 1.4804^{20}
4736	Cyclooctane	C_8H_{16} 112.22	292-64-8 14.8	 149	4-05-00-00111 0.8349^{20}	H_2O 1; bz 3; lig 3 1.4586^{20}
4737	Cyclooctane, butyl- Butylcyclooctane	$C_{12}H_{24}$ 168.32	16538-93-5	 $90\text{-}1^{10}$	4-05-00-00170 0.8260^{25}	 1.4585^{25}
4738	Cyclooctanecarboxaldehyde Formylcyclooctane	$C_9H_{16}O$ 140.23	6688-11-5	 96^{15}	3-07-00-00112 0.94^{25}	 1.4748^{20}
4739	Cyclooctanol	$C_8H_{16}O$ 128.21	696-71-9 25.1	 99^{16}	4-06-00-00113 0.9740^{20}	EtOH 3 1.4871^{20}
4740	Cyclooctanone	$C_8H_{14}O$ 126.20	502-49-8 29	 196	4-07-00-00049 0.9581^{20}	H_2O 1; EtOH 3; ace 3; bz 3 1.4694^{20}
4741	Cyclooctasiloxane, hexadecamethyl- Hexadecamethylcyclooctasiloxane	$C_{16}H_{48}O_8Si_8$ 593.24	556-68-3 31.5	 290	3-04-00-01886 1.177^{25}	bz 4; lig 4 1.4060^{20}
4742	1,3,5,7-Cyclooctatetraene [8]Annulene	C_8H_8 104.15	629-20-9 -4.7	 140.5	4-05-00-01331 0.9206^{20}	EtOH 3; eth 3; ace 3; bz 3 1.5381^{20}
4743	1,3,5,7-Cyclooctatetraene, 1-bromo-	C_8H_7Br 183.05	7567-22-8	 $52\text{-}3^{18}$	4-05-00-01334 1.4206^{25}	ace 3 1.5870^{25}
4744	1,3,5,7-Cyclooctatetraene, 1-butyl-	$C_{12}H_{16}$ 160.26	13402-37-4	 98^{20}	4-05-00-01417 0.8876^{25}	 1.5083^{25}

No.	Name Synonym	Mol. Form. Mol. Wt.	CAS RN mp/°C	Merck No. bp/°C	Beil. Ref. den/g cm^{-3}	Solubility n_D
4745	1,3,5,7-Cyclooctatetraene, 1-chloro- Chlorocyclooctatetraene	C_8H_7Cl 138.60	29554-49-2	 $50-1^{5.5}$	4-05-00-01334 1.1199^{25}	H_2O 1; ace 3; bz 3; chl 3 1.5542^{25}
4746	1,3,5,7-Cyclooctatetraene, 1-ethyl-	$C_{10}H_{12}$ 132.21	13402-35-2	 81^{37}	4-05-00-01373 0.8996^{25}	 1.5187^{25}
4747	1,3,5,7-Cyclooctatetraene, 1-methyl-	C_9H_{10} 118.18	2570-12-9	 84.5^{67}	4-05-00-01358 0.8978^{25}	H_2O 1; EtOH 3; eth 3; bz 3 1.5249^{25}
4748	1,3,5,7-Cyclooctatetraene, propyl- Cyclooctatetraene, propyl	$C_{11}H_{14}$ 146.23	13402-36-3	 73^{9}	4-05-00-01400 0.8870^{25}	 1.5131^{25}
4749	1,3,5-Cyclooctatriene	C_8H_{10} 106.17	1871-52-9 -83	 145.5	4-05-00-00882 0.8971^{25}	 1.5035^{25}
4750	1,3,6-Cyclooctatriene	C_8H_{10} 106.17	3725-30-2 -62	 68^{60}	4-05-00-00883 0.8940^{25}	
4751	Cyclooctene, (*E*)-	C_8H_{14} 110.20	931-89-5 -59	 143	4-05-00-00263 0.8483^{20}	EtOH 3; ctc 2; chl 3 1.4741^{25}
4752	Cyclooctene, (*Z*)-	C_8H_{14} 110.20	931-87-3 -12	 138	3-05-00-00209 0.8472^{20}	EtOH 3; eth 3; ctc 3 1.4698^{20}
4753	2-Cycloocten-1-ol	$C_8H_{14}O$ 126.20	3212-75-7	 74^{2}	4-06-00-00217 0.9756^{25}	 1.4959^{25}
4754	3-Cycloocten-1-one	$C_8H_{12}O$ 124.18	4734-90-1	 200.5	4-07-00-00142 0.990^{20}	ace 4; eth 4; chl 4 1.4953^{25}
4755	Cyclooctyne	C_8H_{12} 108.18	1781-78-8	 158	4-05-00-00401 0.868^{20}	 1.4850^{20}
4756	15H-Cyclopenta[a]phenanthrene, 16,17-dihydro- 1,2-Cyclopentenophenanthrene	$C_{17}H_{14}$ 218.30	482-66-6 135.5	2749	4-05-00-02423	H_2O 1; EtOH 3; peth 3
4757	Cyclopenta[c]furo[3',2':4,5]furo[2,3-h][1]benzopyran-1,11-dione, 2,3,6a,8,9,9a-hexahydro-4-methoxy-, (6a*R*- Aflatoxin B2	$C_{17}H_{14}O_6$ 314.29	7220-81-7 287.5	168	5-19-10-00552	
4758	Cyclopenta[c]pyran-1(4aH)-one, 5,6,7,7a-tetrahydro-4,7-dimethyl- Nepetalactone	$C_{10}H_{14}O_2$ 166.22	490-10-8	6383 $71-2^{0.05}$	4-17-00-04622 1.0663^{25}	 1.4859^{25}
4759	Cyclopentadecane	$C_{15}H_{30}$ 210.40	295-48-7 61.3		4-05-00-00181 0.8364^{61}	 1.4592^{61}
4760	Cyclopentadecanol Exaltol	$C_{15}H_{30}O$ 226.40	4727-17-7 80.5	 177^{11}; $145^{0.3}$	4-06-00-00185 0.930^{20}	 1.4555^{98}
4761	Cyclopentadecanone	$C_{15}H_{28}O$ 224.39	502-72-7 63	 $120^{0.3}$	4-07-00-00113 0.8895^{25}	H_2O 2; EtOH 3; ace 3 1.4637^{60}
4762	Cyclopentadecanone, 2-methyl- 2-Methylcyclopentadecanone	$C_{16}H_{30}O$ 238.41	52914-66-6	 $171-3^{12}$	3-07-00-00208 0.9213^{16}	 1.4812^{16}
4763	Cyclopentadecanone, 3-methyl- Muscone	$C_{16}H_{30}O$ 238.41	541-91-3	6222 329; $130^{0.5}$	4-07-00-00118 0.9221^{17}	ace 4; eth 4; EtOH 4 1.4802^{17}
4764	1,3-Cyclopentadiene Pyropentylene	C_5H_6 66.10	542-92-7 -85	2744 41	4-05-00-00377 0.8021^{20}	H_2O 1; EtOH 5; eth 5; ace 3 1.4440^{20}
4765	1,3-Cyclopentadiene, 5-ethylidene-	C_7H_8 92.14	3839-50-7	 32^{17}	4-05-00-00878 0.865^{18}	 1.526^{20}
4766	1,3-Cyclopentadiene, 1,2,3,4,5,5-hexachloro- Perchlorocyclopentadiene	C_5Cl_6 272.77	77-47-4 -9	 239; $48^{0.3}$	4-05-00-00381 1.7019^{25}	 1.5658^{20}
4767	1,3-Cyclopentadiene, 5-methylene- Fulvene	C_6H_6 78.11	497-20-1	 $7-8^{56}$	4-05-00-00764 0.8241^{20}	H_2O 1; bz 3; chl 3 1.4920^{20}
4768	1,3-Cyclopentadiene, 5-(1-methylethylidene)-	C_8H_{10} 106.17	2175-91-9	 $49-50^{11}$	4-05-00-00974 0.881^{20}	 1.5474^{20}
4769	1,3-Cyclopentadiene, 5-(2-propenylidene)-, (*E*)- Fulvene, 6-vinyl(*trans*)	C_8H_8 104.15	116862-65-8 -35	 45^{12} $20^{0.15}$	 0.898^{20}	H_2O 1; bz 3; chl 3
4770	1,3-Cyclopentadiene, 1,2,3,4-tetrachloro-5,5-dimethoxy-	$C_7H_6Cl_4O_2$ 263.93	2207-27-4	 109	4-07-00-00286 1.501^{25}	 1.5282^{20}
4771	Cyclopentanamine Cyclopentylamine	$C_5H_{11}N$ 85.15	1003-03-8 -82.7	 108	4-12-00-00004 0.8689^{20}	ace 3; bz 3; chl 3 1.4728^{25}
4772	Cyclopentanamine, 1-methyl-	$C_6H_{13}N$ 99.18	40571-45-7	 121	4-12-00-00113 0.8025^{25}	H_2O 4 1.4408^{25}
4773	Cyclopentanamine, 2-methyl-	$C_6H_{13}N$ 99.18	41223-14-7	 123	0-12-00-00007 0.801^{20}	H_2O 4
4774	Cyclopentane Pentamethylene	C_5H_{10} 70.13	287-92-3 -93.8	2746 49.3	4-05-00-00014 0.7457^{20}	H_2O l; EtOH 5; eth 5; ace 5 1.4065^{20}
4775	Cyclopentaneacetic acid 2-Cyclopentylacetic acid	$C_7H_{12}O_2$ 128.17	1123-00-8 13.5	 228	4-09-00-00033 1.0216^{18}	 1.4523^{18}
4776	Cyclopentaneacetic acid, 3-methyl- 3-Methylcyclopentaneacetic acid	$C_8H_{14}O_2$ 142.20	63370-69-4	 237	3-09-00-00060 0.9818^{20}	 1.4504^{20}
4777	Cyclopentane, bromo- Cyclopentyl bromide	C_5H_9Br 149.03	137-43-9	 137.5	4-05-00-00019 1.3873^{20}	ctc 2 1.4886^{20}
4778	Cyclopentane, butyl- Butylcyclopentane	C_9H_{18} 126.24	2040-95-1 -108	 156.6	4-05-00-00139 0.7846^{20}	ace 4; bz 4; eth 4; EtOH 4 1.4316^{20}
4779	Cyclopentanecarbonitrile Cyanocyclopentane	C_6H_9N 95.14	4254-02-8			
4780	Cyclopentanecarboxaldehyde	$C_6H_{10}O$ 98.14	872-53-7	 133.5	4-07-00-00037 0.9371^{20}	H_2O 4; eth 4; EtOH 4 1.4432^{20}
4781	Cyclopentanecarboxylic acid Cyclopentanoic acid	$C_6H_{10}O_2$ 114.14	3400-45-1 -7	 212; 104^{11}	4-09-00-00011 1.0527^{20}	H_2O 2; ctc 2; MeOH 3 1.4532^{20}

No.	Name Synonym	Mol. Form. Mol. Wt.	CAS RN mp/°C	Merck No. bp/°C	Beil. Ref. den/g cm^{-3}	Solubility n_D
4782	Cyclopentanecarboxylic acid, 1-amino- Cycloleucine	$C_6H_{11}NO_2$ 129.16	52-52-8 330 dec	2740	4-14-00-00974	
4783	Cyclopentanecarboxylic acid, 1-methyl- 1-Methylcyclopentanecarboxylic acid	$C_7H_{12}O_2$ 128.17	5217-05-0	 219	4-09-00-00035 1.0218^{20}	eth 4; EtOH 4 1.4529^{20}
4784	Cyclopentanecarboxylic acid, 2-methyl-5-(1-methylethyl)- Puleganic acid	$C_{10}H_{18}O_2$ 170.25	528-23-4 -18	 152^{25}; 139^{12}	4-09-00-00080 0.9642^{20}	 1.4524^{24}
4785	Cyclopentanecarboxylic acid, 2-oxo-, ethyl ester	$C_8H_{12}O_3$ 156.18	611-10-9	 221; 110^{16}	4-10-00-02602 1.0781^{21}	eth 3; bz 3 1.4519^{20}
4786	Cyclopentanecarboxylic acid, 1,2,2,3-tetramethyl- Campholic acid	$C_{10}H_{18}O_2$ 170.25	464-88-0 106	 254; 146^{12}	2-09-00-00019	H_2O 2; EtOH 3; eth 3
4787	Cyclopentanecarboxylic acid, 3-isopropyl-1-methyl-	$C_{12}H_{22}O_2$ 198.31	104068-42-0 18.5	 257	4-09-00-00079 0.9698^{19}	 1.4563^{20}
4788	Cyclopentanecarboxylic acid, 2-methyl-5-(1-methylethyl)-, ethyl ester Cyclopentanecarboxylic acid, 2-isopropyl-5-methyl, ethyl ester	$C_{12}H_{22}O_2$ 198.31	116530-99-5	 145^{4}	1-09-00-00017 0.9178^{12}	 1.4405^{12}
4789	Cyclopentane, chloro- Cyclopentyl chloride	C_5H_9Cl 104.58	930-28-9	 114	4-05-00-00018 1.0051^{20}	H_2O 1; eth 3; ace 3; bz 3 1.4510^{20}
4790	Cyclopentane, decyl- Decane, 1-cyclopentyl-	$C_{15}H_{30}$ 210.40	1795-21-7 -22	 279	4-05-00-00186 0.8110^{20}	ace 4; bz 4; eth 4; EtOH 4 1.4486^{20}
4791	1,1-Cyclopentanedicarboxylic acid, diethyl ester	$C_{11}H_{18}O_4$ 214.26	4167-77-5	 210	4-09-00-02793 1.0186^{20}	 1.4370^{25}
4792	1,2-Cyclopentanedicarboxylic acid, dimethyl ester, *trans*-(±)- Cyclopentane-1,2-dicarboxylic acid, dimethyl ester (*DL*, *trans*)	$C_9H_{14}O_4$ 186.21	80656-12-8	 $118\text{-}9^{17}$	4-09-00-02794 1.1130^{19}	 1.4498^{17}
4793	1,3-Cyclopentanedicarboxylic acid, *cis*-	$C_7H_{10}O_4$ 158.15	876-05-1 121	 300 dec	4-09-00-02795	H_2O 3; EtOH 3; eth 3; ace 3
4794	1,3-Cyclopentanedicarboxylic acid, 1,2,2-trimethyl-, (1*R*-*cis*)-	$C_{10}H_{16}O_4$ 200.23	124-83-4 189.1		4-09-00-02851 1.186^{20}	H_2O 2; EtOH 4; eth 4; ace 3
4795	1,3-Cyclopentanedicarboxylic acid, 1,2,2-trimethyl-, dimethyl ester, (1*R*-*cis*)- Camphoric acid, dimethyl ester, (+)-	$C_{12}H_{20}O_4$ 228.29	15797-21-4 <-16	 264	4-09-00-02851 1.015^{20}	eth 4; EtOH 4 1.4627^{19}
4796	Cyclopentane, 1,1-diethyl- 1,1-Diethylcyclopentane	C_9H_{18} 126.24	2721-38-2 -95.6	 150.5	4-05-00-00141 0.7988^{25}	 1.4388^{20}
4797	Cyclopentane, 1,2-diethyl-, *cis*- *cis*-1,2-Diethylcyclopentane	C_9H_{18} 126.24	932-39-8 -118	 153.6	4-05-00-00141 0.7920^{25}	 1.4355^{20}
4798	Cyclopentane, 1,2-diethyl-, *trans*- *trans*-1,2-Diethylcyclopentane	C_9H_{18} 126.24	932-40-1 -95	 147.5	4-05-00-00141 0.7792^{25}	bz 4; eth 4; peth 4 1.4295^{20}
4799	Cyclopentane, 1,3-diethyl- 1,3-Diethylcyclopentane	C_9H_{18} 126.24	19398-75-5	 150	0-05-00-00045 0.783^{25}	 1.430^{20}
4800	Cyclopentane, 1,1-dimethyl- 1,1-Dimethylcyclopentane	C_7H_{14} 98.19	1638-26-2 -69.8	 87.5	4-05-00-00105 0.7499^{25}	 1.4136^{20}
4801	Cyclopentane, 1,2-dimethyl-, *cis*- *cis*-1,2-Dimethylcyclopentane	C_7H_{14} 98.19	1192-18-3 -54	 99.5	4-05-00-00106 0.7680^{25}	 1.4222^{20}
4802	Cyclopentane, 1,2-dimethyl-, *trans*- *trans*-1,2-Dimethylcyclopentane	C_7H_{14} 98.19	822-50-4 -117.6	 91.9	4-05-00-00107 0.7468^{25}	 1.4120^{20}
4803	Cyclopentane, 1,3-dimethyl-, *cis*- *cis*-1,3-Dimethylcyclopentane	C_7H_{14} 98.19	2532-58-3 -133.7	 90.8	4-05-00-00107 0.7402^{25}	 1.4089^{20}
4804	Cyclopentane, 1,3-dimethyl-, *trans*- *trans*-1,3-Dimethylcyclopentane	C_7H_{14} 98.19	1759-58-6 -134	 91.7	4-05-00-00108 0.7443^{25}	 1.4107^{20}
4805	Cyclopentane, (1,1-dimethylethyl)-	C_9H_{18} 126.24	3875-52-3 -95.8	 144.9	3-05-00-00124 0.7870^{25}	 1.4338^{20}
4806	1,3-Cyclopentanediol, *cis*-	$C_5H_{10}O_2$ 102.13	16326-97-9 31	 105^{5}; 90^{1}	3-06-00-04057 1.099^{15}	 1.4792^{20}
4807	1,3-Cyclopentanediol, *trans*- Cyclopentane-1,3-diol (*trans*)	$C_5H_{10}O_2$ 102.13	16326-98-0 40	 $80\text{-}5^{0.1}$	3-06-00-04057 1.094^{25}	 1.4830^{20}
4808	Cyclopentane, docosyl- Docosylcyclopentane	$C_{27}H_{54}$ 378.73	62016-55-1 45	 433	4-05-00-00201 0.8295^{20}	bz 4; eth 4 1.4608^{20}
4809	Cyclopentane, dodecyl-	$C_{17}H_{34}$ 238.46	5634-30-0 -5	 312	3-05-00-00156 0.8158^{20}	ace 4; bz 4; eth 4; EtOH 4 1.4518^{20}
4810	Cyclopentaneethanamine, *N*,α-dimethyl- Cyclopentamine	$C_9H_{19}N$ 141.26	102-45-4	2745 171	4-12-00-00137	 1.4500^{20}
4811	Cyclopentaneethanol	$C_7H_{14}O$ 114.19	766-00-7	 183; 96^{24}	4-06-00-00110 0.9180^{20}	eth 4 1.4577^{20}
4812	Cyclopentane, ethyl- Ethylcyclopentane	C_7H_{14} 98.19	1640-89-7 -138.4	 103.5	4-05-00-00104 0.7665^{20}	H_2O 1; EtOH 5; eth 5; ace 5 1.4198^{20}
4813	Cyclopentane, ethylidene- Ethylidenecyclopentane	C_7H_{12} 96.17	2146-37-4 -129.5	 115	4-05-00-00253 0.8020^{20}	 1.4481^{20}
4814	Cyclopentane, 1-ethyl-1-methyl- 1-Ethyl-1-methylcyclopentane	C_8H_{16} 112.22	16747-50-5 -143.8	 121.6	4-05-00-00126 0.7767^{25}	ace 4; bz 4; eth 4; EtOH 4 1.4272^{20}
4815	Cyclopentane, 1-ethyl-2-methyl-, *cis*- *cis*-1-Methyl-2-ethylcyclopentane	C_8H_{16} 112.22	930-89-2 -106	 128	4-05-00-00126 0.7852^{20}	 1.4293^{20}

No.	Name Synonym	Mol. Form. Mol. Wt.	CAS RN mp/°C	Merck No. bp/°C	Beil. Ref. den/g cm^{-3}	Solubility n_D
4816	Cyclopentane, 1-ethyl-2-methyl-, *trans-*	C_8H_{16} 112.22	930-90-5 -105.9	 121.2	4-05-00-00126 0.7649^{25}	 1.4219^{20}
4817	Cyclopentane, 1-ethyl-3-methyl-, *cis-* 1-Methyl-*cis*-3-ethylcyclopentane	C_8H_{16} 112.22	2613-66-3	 121	4-05-00-00127 0.7724^{20}	ace 4; bz 4; eth 4; EtOH 4 1.4203^{20}
4818	Cyclopentane, 1-ethyl-3-methyl-, *trans-* 1-Methyl-*trans*-3-ethylcyclopentane	C_8H_{16} 112.22	2613-65-2 -108	 121	4-05-00-00127 0.7619^{20}	 1.4186^{20}
4819	Cyclopentane, heneicosyl-	$C_{26}H_{52}$ 364.70	6703-82-8 42	 420	4-05-00-00200 0.8286^{20}	 1.4602^{20}
4820	Cyclopentane, heptyl- Heptane, 1-cyclopentyl-	$C_{12}H_{24}$ 168.32	5617-42-5 -53	 224	4-05-00-00173 0.8010^{20}	ace 4; bz 4; eth 4; EtOH 4 1.4421^{20}
4821	Cyclopentane, hexacosyl- Hexacosylcyclopentane	$C_{31}H_{62}$ 434.83	62016-58-4 56	 468	4-05-00-00203 0.8326^{20}	 1.4628^{20}
4822	Cyclopentane, hexadecyl- Hexadecane, 1-cyclopentyl-	$C_{21}H_{42}$ 294.56	6812-39-1 21	 368	4-05-00-00195 0.8228^{20}	ace 4; bz 4; eth 4; EtOH 4 1.4564^{20}
4823	Cyclopentane, hexyl-	$C_{11}H_{22}$ 154.30	4457-00-5 -73	 203	4-05-00-00168 0.7965^{20}	ace 4; bz 4; eth 4; EtOH 4 1.4392^{20}
4824	Cyclopentane, iodo- Cyclopentyl iodide	C_5H_9I 196.03	1556-18-9	 166.5	4-05-00-00020 1.7096^{20}	H_2O 1; eth 3; bz 3; ctc 2 1.5447^{20}
4825	Cyclopentanemethanol	$C_6H_{12}O$ 100.16	3637-61-4	 163	4-06-00-00092 0.9332^{20}	 1.4579^{20}
4826	Cyclopentane, methyl- Methylcyclopentane	C_6H_{12} 84.16	96-37-7 -142.5	 71.8	4-05-00-00084 0.7486^{20}	H_2O 1; EtOH 5; eth 5; ace 5 1.4097^{20}
4827	Cyclopentane, methylene- Methylenecyclopentane	C_6H_{10} 82.15	1528-30-9	 75.5	4-05-00-00241 0.7787^{20}	bz 3; chl 3 1.4355^{20}
4828	Cyclopentane, (1-methylethyl)- Isopropylcyclopentane	C_8H_{16} 112.22	3875-51-2 -111.4	 126.5	4-05-00-00125 0.7765^{20}	H_2O 1; EtOH 5; eth 3; ace 5 1.4258^{20}
4829	Cyclopentane, 1-methyl-2-(1-methylethyl)- Cyclopentane, 1-isopropyl-2-methyl-	C_9H_{18} 126.24	89223-57-4	 143	4-05-00-00141 0.7832^{15}	 1.4279^{20}
4830	Cyclopentane, 1-methyl-1-nitro- 1-Methyl-1-nitrocyclopentane	$C_6H_{11}NO_2$ 129.16	30168-50-4	 181; 92^{40}	4-05-00-00087 1.0395^{20}	 1.4504^{20}
4831	Cyclopentane, 1-methyl-2-nitro- 1-Methyl-2-nitrocyclopentane	$C_6H_{11}NO_2$ 129.16	102153-88-8	 185 dec; 98^{40}	1-05-00-00011 1.0381^{22}	 1.4488^{22}
4832	Cyclopentane, (1-methylpropyl)-	C_9H_{18} 126.24	4850-32-2	 154.4	3-05-00-00124 0.7905^{25}	 1.4357^{20}
4833	Cyclopentane, 1-methyl-1-propyl- 1-Methyl-1-propylcyclopentane	C_9H_{18} 126.24	16631-63-3	 146	4-05-00-00140 0.795^{25}	 1.437^{20}
4834	Cyclopentane, 1-methyl-2-propyl-, *cis-* *cis*-1-Methyl-2-propylcyclopentane	C_9H_{18} 126.24	932-43-4 -104	 152.6	4-05-00-00140 0.7881^{25}	 1.4343^{20}
4835	Cyclopentane, 1-methyl-2-propyl-, *trans-* *trans*-1-Methyl-2-propylcyclopentane	C_9H_{18} 126.24	932-44-5 -123	 146.4	4-05-00-00140 0.7735^{25}	 1.4274^{20}
4836	Cyclopentane, 1-methyl-3-propyl-, *cis-* *cis*-1-Methyl-3-propylcyclopentane	C_9H_{18} 126.24	2443-04-1	 148	3-05-00-00125 0.776^{25}	 1.426^{20}
4837	Cyclopentane, 1-methyl-3-propyl-, *trans-* *trans*-1-Methyl-3-propylcyclopentane	C_9H_{18} 126.24	2443-03-0	 148	3-05-00-00125 0.776^{25}	 1.426^{20}
4838	Cyclopentane, (2-methylpropyl)- Isobutylcyclopentane	C_9H_{18} 126.24	3788-32-7 -115.2	 148	4-05-00-00140 0.7769^{25}	 1.4298^{20}
4839	Cyclopentane, nitro- Nitrocyclopentane	$C_5H_9NO_2$ 115.13	2562-38-1	 $90\text{-}1^{40}$	4-05-00-00020 1.0776^{23}	H_2O l; bz 3; chl 3 1.4538^{20}
4840	Cyclopentane, (nitromethyl)- (Nitromethyl)cyclopentane	$C_6H_{11}NO_2$ 129.16	2625-31-2	 110^{35}	1-05-00-00011 1.0713^{20}	 1.4587^{20}
4841	Cyclopentane, nonyl-	$C_{14}H_{28}$ 196.38	2882-98-6 -29	 262	4-05-00-00181 0.8081^{20}	ace 4; bz 4; eth 4; EtOH 4 1.4467^{20}
4842	Cyclopentane, octadecyl- Octadecane, 1-cyclopentyl-	$C_{23}H_{46}$ 322.62	62016-53-9 30	 391	4-05-00-00196 0.8254^{20}	ace 4; bz 4; eth 4; EtOH 4 1.4581^{20}
4843	Cyclopentane, octyl- Octane, 1-cyclopentyl-	$C_{13}H_{26}$ 182.35	1795-20-6 -44	 243	4-05-00-00176 0.8048^{20}	 1.4446^{20}
4844	Cyclopentane, pentadecyl- Pentadecane, 1-cyclopentyl-	$C_{20}H_{40}$ 280.54	4669-01-6 17	 355	4-05-00-00195 0.8213^{20}	ace 4; bz 4; eth 4; EtOH 4 1.4554^{20}
4845	Cyclopentane, pentatriacontyl- Pentatriacontylcyclopentane	$C_{40}H_{80}$ 561.08	61827-98-3 74	 531	4-05-00-00205 0.8374^{20}	 1.4660^{20}
4846	Cyclopentane, pentyl-	$C_{10}H_{20}$ 140.27	3741-00-2 -83	 180	4-05-00-00160 0.7912^{20}	ace 4; bz 4; eth 4; EtOH 4 1.4356^{20}
4847	Cyclopentanepropanoic acid	$C_8H_{14}O_2$ 142.20	140-77-2	 $130\text{-}2^{12}$	4-09-00-00046 0.996^{25}	 1.4570^{20}
4848	Cyclopentane, 2-propenyl-	C_8H_{14} 110.20	3524-75-2 -110.7	 125	4-05-00-00272 0.793^{25}	chl 3 1.4412^{20}
4849	Cyclopentane, propyl- Propylcyclopentane	C_8H_{16} 112.22	2040-96-2 -117.3	 131	4-05-00-00125 0.7763^{20}	H_2O 1; EtOH 5; eth 5; ace 5 1.4266^{20}
4850	Cyclopentane, tetradecyl- Tetradecane, 1-cyclopentyl-	$C_{19}H_{38}$ 266.51	1795-22-8 9	 341	3-05-00-00158 0.8196^{20}	 1.4543^{20}
4851	Cyclopentane, 1,1,3,3-tetramethyl- 1,1,3,3-Tetramethylcyclopentane	C_9H_{18} 126.24	50876-33-0 -88.4	 118	4-05-00-00143 0.7469^{25}	 1.4125^{20}
4852	Cyclopentanethiol Cyclopentyl mercaptan	$C_5H_{10}S$ 102.20	1679-07-8	 132.1	4-06-00-00015 0.9550^{20}	
4853	Cyclopentane, triacontyl- Triacontylcyclopentane	$C_{35}H_{70}$ 490.94	61827-93-8 65	 498	4-05-00-00204 0.8350^{20}	 1.4644^{20}
4854	Cyclopentane, tridecyl- Tridecane, 1-cyclopentyl-	$C_{18}H_{36}$ 252.48	6006-34-4 5	 327	4-05-00-00192 0.8178^{20}	ace 4; bz 4; eth 4; EtOH 4 1.4531^{20}
4855	Cyclopentane, 1,1,2-trimethyl- Camphocean	C_8H_{16} 112.22	4259-00-1 -21.6	 114; 53^{100}	4-05-00-00127 0.7660^{20}	 1.4199^{20}

No.	Name Synonym	Mol. Form. Mol. Wt.	CAS RN mp/°C	Merck No. bp/°C	Beil. Ref. den/g cm^{-3}	Solubility n_D
4856	Cyclopentane, 1,1,3-trimethyl- 1,1,3-Trimethylcyclopentane	C_8H_{16} 112.22	4516-69-2 -142.4	 104.9	3-05-00-00111 0.7439^{25}	 1.4112^{20}
4857	Cyclopentane, 1,2,3-trimethyl-, (1α,2β,3α)- Cyclopentane, 1,2,3-trimethyl-, *trans*-1,2,*cis*-1,3-	C_8H_{16} 112.22	19374-46-0 -112.7	 110.4	4-05-00-00128 0.7492^{25}	 1.4138^{20}
4858	Cyclopentane, 1,2,3-trimethyl-, (1α,2α,3α)- Cyclopentane, 1,2,3-trimethyl-, *cis*-1,2,*cis*-1,3-	C_8H_{16} 112.22	2613-69-6 -116.4	 123	4-05-00-00128 0.7751^{25}	 1.4262^{20}
4859	Cyclopentane, 1,2,3-trimethyl-, (1α,2α,3β)-	C_8H_{16} 112.22	15890-40-1 -112	 117.5	4-05-00-00128 0.7661^{25}	 1.4218^{20}
4860	Cyclopentane, 1,2,4-trimethyl-, (1α,2β,4α)-	C_8H_{16} 112.22	16883-48-0 -130.8	 109.3	4-05-00-00129 0.7430^{25}	 1.4106^{20}
4861	Cyclopentane, 1,2,4-trimethyl-, (1α,2α,4α)- Cyclopentane, 1,2,4-trimethyl-, *cis*-1,2,*cis*-1,4-	C_8H_{16} 112.22	2613-72-1 -132.3	 116.8	4-05-00-00129 0.758^{25}	 1.4186^{20}
4862	Cyclopentane, 1,2,4-trimethyl-, (1α,2α,4β)-	C_8H_{16} 112.22	4850-28-6 -132.6	 116.7	4-05-00-00129 0.7592^{25}	 1.4186^{20}
4863	Cyclopentane, undecyl-	$C_{16}H_{32}$ 224.43	6785-23-5 -10	 296	4-05-00-00189 0.8135^{20}	ace 4; bz 4; eth 4; EtOH 4 1.4503^{20}
4864	Cyclopentanol Cyclopentyl alcohol	$C_5H_{10}O$ 86.13	96-41-3 -17.5	2747 140.4	4-06-00-00005 0.9488^{20}	H_2O 2; EtOH 3; eth 3; ace 3 1.4530^{20}
4865	Cyclopentanol, acetate Cyclopentyl acetate	$C_7H_{12}O_2$ 128.17	933-05-1	 $51\text{-}2^{12}$	4-06-00-00007 0.9522^{16}	
4866	Cyclopentanol, 1-ethynyl- 1-Ethynyl-1-cyclopentanol	$C_7H_{10}O$ 110.16	17356-19-3 27	 157.5	4-06-00-00339 0.962^{25}	 1.4751^{20}
4867	Cyclopentanol, 1-methyl-	$C_6H_{12}O$ 100.16	1462-03-9 36	 136; 53^{30}	4-06-00-00086 0.9044^{23}	 1.4429^{23}
4868	Cyclopentanol, 2-methyl-, *cis*- *cis*-2-Methylcyclopentanol	$C_6H_{12}O$ 100.16	25144-05-2	 148.5	3-06-00-00055 0.9379^{16}	 1.4504^{16}
4869	Cyclopentanol, 2-methyl-, *trans*- *trans*-2-Methylcyclopentanol	$C_6H_{12}O$ 100.16	25144-04-1	 150.5	4-06-00-00091 0.9248^{16}	 1.4499^{16}
4870	Cyclopentanol, 3-methyl- 3-Methylcyclopentanol	$C_6H_{12}O$ 100.16	18729-48-1	 148.5	2-06-00-00015 0.9158^{16}	eth 4; EtOH 4 1.4487^{16}
4871	Cyclopentanol, 1-phenyl-	$C_{11}H_{14}O$ 162.23	10487-96-4	 $132\text{-}3^{18}$	4-06-00-03884 1.0530^{20}	 1.5479^{20}
4872	Cyclopentanol, 1-propyl-	$C_8H_{16}O$ 128.21	1604-02-0 -37.5	 173.5	3-06-00-00101 0.9040^{25}	 1.4502^{25}
4873	Cyclopentanol, 2-propyl-, *cis*- *cis*-2-Propylcyclopentanol	$C_8H_{16}O$ 128.21	25172-42-3	 $79\text{-}80^{12}$	3-06-00-00101 0.9163^{9}	 1.4600^{9}
4874	Cyclopentanol, 2-propyl-, *trans*- *trans*-2-Propylcyclopentanol	$C_8H_{16}O$ 128.21	25172-43-4	 $78\text{-}9^{10}$	3-06-00-00101 0.9016^{9}	 1.4565^{9}
4875	Cyclopentanol, 1,2,2-trimethyl- 1,2,2-Trimethylcyclopentanol	$C_8H_{16}O$ 128.21	1121-95-5	 156; 60^{15}	1-06-00-00014 0.9101^{20}	 1.4513^{20}
4876	Cyclopentanol, 1,2,4-trimethyl- 1,2,4-Trimethylcyclopentanol	$C_8H_{16}O$ 128.21	34103-97-4	 158	1-06-00-00014 0.8850^{21}	 1.4424^{21}
4877	Cyclopentanol, 1,2,5-trimethyl- 1,2,5-Trimethylcyclopentanol	$C_8H_{16}O$ 128.21	33840-38-9	 $56\text{-}60^{8}$	1-06-00-00014 0.9121^{15}	 1.4554^{16}
4878	Cyclopentanone Adipic ketone	C_5H_8O 84.12	120-92-3 -51.3	2748 130.5	4-07-00-00005 0.9487^{20}	H_2O 1; EtOH 3; eth 5; ace 3 1.4366^{20}
4879	Cyclopentanone, 2-acetyl- 2-Acetylcyclopentanone	$C_7H_{10}O_2$ 126.16	1670-46-8	 $72\text{-}5^{20}$	4-07-00-01993 1.0431^{25}	 1.4906^{20}
4880	Cyclopentanone, 2-chloro-	C_5H_7ClO 118.56	694-28-0	 $72\text{-}4^{12}$	4-07-00-00012 1.185^{25}	 1.4750^{20}
4881	Cyclopentanone, 2-cyclopentylidene-	$C_{10}H_{14}O$ 150.22	825-25-2	 135^{25}	4-07-00-00320 1.0179^{18}	 1.5215^{18}
4882	Cyclopentanone, 2,5-dimethyl- 2,5-Dimethylcyclopentanone	$C_7H_{12}O$ 112.17	4041-09-2	 146.5	4-07-00-00048 0.882^{25}	 1.4310^{20}
4883	Cyclopentanone, 2-ethyl-5-methyl- 2-Ethyl-5-methylcyclopentanone	$C_8H_{14}O$ 126.20	51686-60-3	 164.5	3-06-00-02552 0.900^{12}	 1.4360^{22}
4884	Cyclopentanone, 4-isopropyl-2-methyl- Cyclopentanone, 3-isopropyl-5-methyl	$C_9H_{16}O$ 140.23	90645-61-7	 191.5	2-07-00-00034 0.8862^{20}	 1.4413^{19}
4885	Cyclopentanone, 2-methyl-	$C_6H_{10}O$ 98.14	1120-72-5 -75	 139.5	4-07-00-00036 0.9139^{20}	H_2O 3; EtOH 4; eth 4; ace 4 1.4364^{20}
4886	Cyclopentanone, 3-methyl-, (±)- (±)-3-Methylcyclopentanone	$C_6H_{10}O$ 98.14	6195-92-2 -58.4	 144	4-07-00-00037 0.913^{22}	H_2O 3; EtOH 4; eth 4; ace 4 1.4329^{20}
4887	Cyclopentanone, 2-(1-methylethyl)- 2-Isopropylcyclopentanone	$C_8H_{14}O$ 126.20	14845-55-7	 175	4-07-00-00060 0.9105^{20}	 1.4425^{15}
4888	Cyclopentanone, 3-methyl-2-(3-methylbutyl)- Cyclopentanone, 2-isopentyl-3-methyl	$C_{11}H_{20}O$ 168.28	52033-97-3	 $98\text{-}9^{8}$	3-07-00-00181 0.8938^{20}	 1.4537^{20}
4889	Cyclopentanone, 2-methyl-2-(1-methylethyl)- Cyclopentanone, 2-isopropyl-2-methyl-	$C_9H_{16}O$ 140.23	32116-65-7	 97.5^{45}	3-07-00-00130 0.9067^{16}	 1.4495^{16}
4890	Cyclopentanone, 2-methyl-5-(1-methylethyl)-	$C_9H_{16}O$ 140.23	6784-18-5	 182	3-07-00-00130 0.889^{20}	 1.4402^{20}
4891	Cyclopentanone, 4-methyl-2-(1-methylethyl)- Cyclopentanone, 2-isopropyl-4-methyl	$C_9H_{16}O$ 140.23	69770-98-5	 183	1-07-00-00026 0.8850^{20}	 1.4392^{20}
4892	Cyclopentanone, oxime	C_5H_9NO 99.13	1192-28-5 57.8	 196	4-07-00-00008	H_2O 4; bz 4
4893	Cyclopentanone, 2-(phenylmethyl)- 2-Benzylcyclopentanone	$C_{12}H_{14}O$ 174.24	2867-63-2	 152^{18}; $84^{0.3}$	4-07-00-01043 1.038^{20}	 1.534^{13}
4894	Cyclopentanone, 2-propyl- 2-Propylcyclopentanone	$C_8H_{14}O$ 126.20	1193-70-0 -68.3	 183	4-07-00-00060 0.9017^{20}	 1.4429^{20}

No.	Name	Synonym	Mol. Form.	Mol. Wt.	CAS RN	mp/°C	Merck No.	bp/°C	Beil. Ref.	den/g cm^{-3}	Solubility	n_D
4895	Cyclopentanone, 3-propyl-	3-Propylcyclopentanone	$C_8H_{14}O$	126.20	82322-93-8			190.5	3-07-00-00099	0.9041^{20}		1.4456^{12}
4896	Cyclopentanone, 2,2,4-trimethyl-	2,2,4-Trimethylcyclopentanone	$C_8H_{14}O$	126.20	28056-54-4	-40.6		158	4-07-00-00061	0.877^{25}		1.4300^{20}
4897	Cyclopentanone, 2,2,5-trimethyl-	2,2,5-Trimethylcyclopentanone	$C_8H_{14}O$	126.20	4573-09-5			152	3-07-00-00107	0.8781^{20}		1.4306^{20}
4898	Cyclopentanone, 2,3,5-trimethyl-	2,3,5-Trimethylcyclopentanol	$C_8H_{14}O$	126.20	90112-26-8			158	1-07-00-00021	0.8778^{19}		1.4316^{19}
4899	Cyclopentanone, 2,4,4-trimethyl-	2,4,4-Trimethylcyclopentanone	$C_8H_{14}O$	126.20	4694-12-6	-25.6		160.5	4-07-00-00061	0.8785^{18}		1.433^{18}
4900	1H-Cyclopentapyrimidine-2,4(3H,5H)-dione, 3-cyclohexyl-6,7-dihydro-	Lenacil	$C_{13}H_{18}N_2O_2$	234.30	2164-08-1	290	5318		5-24-07-00375	1.32^{25}	py 4	
4901	Cyclopentasiloxane, decamethyl-	Decamethylcyclopentasiloxane	$C_{10}H_{30}O_5Si_5$	370.77	541-02-6	-38	2841	210	4-04-00-04128	0.9593^{20}	H_2O 1	1.3982^{20}
4902	Cyclopentasiloxane, 2,4,6,8,10-pentamethyl-		$C_5H_{20}O_5Si_5$	300.64	6166-86-5	-108		169	4-04-00-04099	0.9985^{20}		1.3912^{20}
4903	Cyclopentene		C_5H_8	68.12	142-29-0	-135.1		44.2	4-05-00-00209	0.7720^{20}	H_2O 1; EtOH 3; eth 3; bz 3	1.4225^{20}
4904	3-Cyclopentene-1-acetic acid		$C_7H_{10}O_2$	126.16	767-03-3	19		$93-4^{2.5}$		1.047^{25}		1.4676^{20}
4905	1-Cyclopentene-1-acetonitrile		C_7H_9N	107.16	22734-04-9			124^{100}	3-09-00-00151	0.951^{25}		1.4670^{20}
4906	Cyclopentene, 1-butyl-	1-Butyl-1-cyclopentene	C_9H_{16}	124.23	2423-01-0	-88.4		157	4-05-00-00289	0.8035^{25}		1.4486^{20}
4907	1-Cyclopentenecarbonitrile	1-Cyanocyclopentene	C_6H_7N	93.13	3047-38-9							
4908	1-Cyclopentene-1-carboxaldehyde		C_6H_8O	96.13	6140-65-4	-32		146	4-07-00-00127	0.970^{21}		1.4872^{17}
4909	1-Cyclopentene-1-carboxylic acid		$C_6H_8O_2$	112.13	1560-11-8	123.5		210	4-09-00-00109		H_2O 3; EtOH 3; eth 3; peth 3	
4910	1-Cyclopentene-1-carboxylic acid, 2,3,3-trimethyl-		$C_9H_{14}O_2$	154.21	5587-63-3	135		255.5	4-09-00-00150		eth 4; EtOH 4	
4911	1-Cyclopentene-1-carboxylic acid, 2,3,3-trimethyl-, methyl ester		$C_{10}H_{16}O_2$	168.24	1460-88-4			203.5	0-09-00-00058	0.9656^{20}		1.4571^{20}
4912	2-Cyclopentene-1-carboxylic acid, 1,2,3-trimethyl-	Laurolenic acid	$C_9H_{14}O_2$	154.21	6894-69-5	6.5		192^{100}	2-09-00-00040	1.0318^{25}		1.4766
4913	3-Cyclopentene-1-carboxylic acid, 2,2,3-trimethyl-	α-Campholytic acid	$C_9H_{14}O_2$	154.21	6709-22-4	40.5		241.5; 162^{45}	4-09-00-00150	1.0145^{18}		1.4712^{17}
4914	Cyclopentene, 3-chloro-		C_5H_7Cl	102.56	96-40-2			$25-31^{30}$	4-05-00-00212	1.0577^{15}	eth 4; EtOH 4; chl 4	1.4708^{26}
4915	1-Cyclopentene-1,2-dicarboxylic acid, diethyl ester	Diethyl 1-cyclopentene-1,2-dicarboxylate	$C_{11}H_{16}O_4$	212.25	70202-92-5			140^{16}	3-09-00-03938	1.0805^{20}		1.4652^{20}
4916	1-Cyclopentene-1,3-dicarboxylic acid, diethyl ester	Cyclopentene-1,3-dicarboxylic acid, diethyl ester	$C_{11}H_{16}O_4$	212.25	30689-41-9			168^{21}	3-09-00-03939	1.1121^{24}		1.4564
4917	Cyclopentene, 1,2-dichloro-3,3,4,4,5,5-hexafluoro-		$C_5Cl_2F_6$	244.95	706-79-6	-105.8		90.7	4-05-00-00213	1.6546^{20}		1.3676^{20}
4918	Cyclopentene, 1,2-dimethyl-		C_7H_{12}	96.17	765-47-9	-90.4		105.8	4-05-00-00254	0.7928^{25}		1.4448^{20}
4919	Cyclopentene, 1,3-dimethyl-	1,3-Dimethylcyclopentene	C_7H_{12}	96.17	62184-82-1			92	4-05-00-00255	0.766^{20}		1.428^{20}
4920	Cyclopentene, 1,4-dimethyl-	1,4-Dimethylcyclopentene	C_7H_{12}	96.17	19550-48-2			93.2	4-05-00-00255	0.779^{20}		1.4283^{20}
4921	Cyclopentene, 1,5-dimethyl-		C_7H_{12}	96.17	16491-15-9	-118		99	4-05-00-00254	0.780^{20}		1.4331^{20}
4922	Cyclopentene, 4,4-dimethyl-		C_7H_{12}	96.17	19037-72-0			88	4-05-00-00254	0.766^{25}		1.423^{20}
4923	Cyclopentene, 1,5-dimethyl-4-(1-methylethyl)-	Cyclopentene, 2,3-dimethyl-4-isopropyl	$C_{10}H_{18}$	138.25	6912-05-6			163	1-05-00-00046	0.8085^{22}		1.4503^{22}
4924	Cyclopentene, 1-ethyl-		C_7H_{12}	96.17	2146-38-5	-118.5		106.3	4-05-00-00252	0.7936^{25}		1.4412^{20}
4925	Cyclopentene, 3-ethyl-		C_7H_{12}	96.17	694-35-9			97.8	4-05-00-00252	0.7784^{25}		1.4315^{20}
4926	Cyclopentene, 4-ethyl-	4-Ethylcyclopentene	C_7H_{12}	96.17	3742-38-9			98.3	4-05-00-00253	0.778^{25}		1.431^{20}
4927	Cyclopentene, 4-isopropyl-1-methyl-	4-Isopropyl-1-methylcyclopentene	C_9H_{16}	124.23	90769-70-3			142.5	1-05-00-00041	0.7945^{21}		1.4403^{21}
4928	Cyclopentene, 1-methyl-		C_6H_{10}	82.15	693-89-0	-126.5		75.5	4-05-00-00239	0.7748^{25}		1.4322^{20}
4929	Cyclopentene, 3-methyl-	3-Methylcyclopentene	C_6H_{10}	82.15	1120-62-3			64.9	4-05-00-00240	0.7572^{25}		1.4216^{20}
4930	Cyclopentene, 4-methyl-		C_6H_{10}	82.15	1759-81-5	-160.8		65.7	4-05-00-00240	0.7634^{25}		1.4209^{20}

No.	Name Synonym	Mol. Form. Mol. Wt.	CAS RN mp/°C	Merck No. bp/°C	Beil. Ref. den/g cm^{-3}	Solubility n_D
4931	Cyclopentene, 3-methyl-1-(1-methylethyl)- Pulegene	C_9H_{16} 124.23	51115-02-7	 138.5	1-05-00-00040 0.791^{22}	 1.4380^{23}
4932	Cyclopentene, octachloro- Perchlorocyclopentene	C_5Cl_8 343.68	706-78-5 40	 283	4-05-00-00214 1.8200^{50}	H_2O 1; EtOH 4 1.5660^{50}
4933	1-Cyclopentene-1-pentanoic acid, δ-hydroxy-3,5-bis(1-methylpropyl)-β-oxo-	$C_{18}H_{30}O_4$ 310.43	53109-18-5 183		3-10-00-04194 1.269^{20}	eth 4; EtOH 4
4934	Cyclopentene, 1-pentyl-	$C_{10}H_{18}$ 138.25	4291-98-9 -83	 179	4-05-00-00306 0.8085^{25}	 1.4516^{20}
4935	Cyclopentene, 1-propyl- 1-Propyl-1-cyclopentene	C_8H_{14} 110.20	3074-61-1 -100.3	 131.2	4-05-00-00271 0.7978^{25}	 1.4452^{20}
4936	2-Cyclopentene-1-tridecanoic acid, (*S*)- Chaulmoogric acid	$C_{18}H_{32}O_2$ 280.45	29106-32-9 68.5	2037 $247\text{-}8^{20}$	4-09-00-00216	eth 4; chl 4
4937	Cyclopentene, 1,2,3-trimethyl-	C_8H_{14} 110.20	473-91-6	 121.6	3-05-00-00219 0.8039^{15}	 1.4464^{16}
4938	Cyclopentene, 1,5,5-trimethyl- 1,5,5-Trimethylcyclopentene	C_8H_{14} 110.20	62184-83-2	 109	3-05-00-00219 0.7824^{20}	 1.4324^{20}
4939	4-Cyclopentene-1,2,3-trione, 4,5-dihydroxy- Croconic acid	$C_5H_2O_5$ 142.07	488-86-8 150	 dec	4-08-00-03342	H_2O 4; EtOH 3
4940	2-Cyclopentene-1-undecanoic acid, (*R*)- Hydnocarpic acid	$C_{16}H_{28}O_2$ 252.40	459-67-6 60.5	4679	4-09-00-00215	EtOH 4; chl 4; peth 4
4941	2-Cyclopenten-1-one Cyclopenten-3-one	C_5H_6O 82.10	930-30-3	 136; 40^{12}	4-07-00-00122 0.989^{15}	eth 4; EtOH 4 1.4629^{15}
4942	2-Cyclopenten-1-one, 2,3-dihydroxy- Reductic acid	$C_5H_6O_3$ 114.10	80-72-8 212	8134	4-08-00-01714	H_2O 3; EtOH 3; eth 2; ace 2
4943	2-Cyclopenten-1-one, 2,4-dimethyl-	$C_7H_{10}O$ 110.16	23048-13-7	 165	4-07-00-00138 0.9423^{20}	 1.4655^{20}
4944	2-Cyclopenten-1-one, 2-methyl-	C_6H_8O 96.13	1120-73-6	 157	4-07-00-00127 0.9808^{16}	 1.4762^{15}
4945	2-Cyclopenten-1-one, 3-methyl- 3-Methyl-2-cyclopentenone	C_6H_8O 96.13	2758-18-1	 157.5	4-07-00-00126 0.9712^{20}	 1.4714^{20}
4946	2-Cyclopenten-1-one, 5-methyl- Cyclopenten-3-one, 4-methyl	C_6H_8O 96.13	14963-40-7	 140	3-07-00-00229 0.942^{18}	 1.4460^{18}
4947	2-Cyclopenten-1-one, 3-(1-methylethyl)-	$C_8H_{12}O$ 124.18	1619-28-9	 214	2-07-00-00061 0.9378^{20}	 1.4788^{20}
4948	2-Cyclopenten-1-one, 5-methyl-2-(1-methylethyl)-	$C_9H_{14}O$ 138.21	5587-79-1	 189	3-07-00-00291 0.9143^{20}	 1.4660^{20}
4949	2-Cyclopenten-1-one, 3-methyl-2-(2-pentenyl)-, (*Z*)- Jasmone	$C_{11}H_{16}O$ 164.25	488-10-8	5141 258; 134^{12}	4-07-00-00337 0.9437^{22}	H_2O 2; EtOH 3; eth 3; ctc 3 1.4979^{22}
4950	2-Cyclopenten-1-one, 3-methyl-2-pentyl-	$C_{11}H_{18}O$ 166.26	1128-08-1	 143^{22}, 116^{12}	4-07-00-00230 0.9165^{18}	 1.4767^{20}
4951	2-Cyclopenten-1-one, 3-phenyl-	$C_{11}H_{10}O$ 158.20	3810-26-2 -23	 234.2	4-07-00-01180 0.9711^{20}	EtOH 3; eth 2; ace 3; chl 3 1.5440^{20}
4952	Cycloprate Cyclopropanecarboxylic acid, hexadecyl ester	$C_{20}H_{38}O_2$ 310.52	54460-46-7	 $154^{0.05}$		
4953	Cyclopropanamine Cyclopropylamine	C_3H_7N 57.10	765-30-0 -35.4	 50.5	4-12-00-00003 0.8240^{20}	H_2O 5; EtOH 3; eth 3; chl 3 1.4210^{20}
4954	Cyclopropa[d]naphthalene, 1,1a,4,4a,5,6,7,8-octahydro-2,4a,8,8-tetramethyl-, [1aS-(1aα,4aβ,8a*R**)]- Thujopsene	$C_{15}H_{24}$ 204.36	470-40-6	9327	4-05-00-01189	
4955	Cyclopropane Trimethylene	C_3H_6 42.08	75-19-4 -127.4	2755 -32.8	4-05-00-00003 $*0.617^{25}$	H_2O 3; EtOH 4; eth 4; bz 3 1.3799^{-42}
4956	Cyclopropanecarbonitrile Cyclopropyl cyanide	C_4H_5N 67.09	5500-21-0	 135.1	4-09-00-00005 0.8946^{20}	eth 3; ctc 2; hx 3 1.4229^{20}
4957	Cyclopropanecarbonyl chloride, 2-phenyl-, *trans*- *trans*-2-Phenylcyclopropanecarbonyl chloride	$C_{10}H_9ClO$ 180.63	939-87-7	 $108\text{-}10^{2}$	3-09-00-02774 1.163^{25}	 1.5560^{20}
4958	Cyclopropanecarboxylic acid	$C_4H_6O_2$ 86.09	1759-53-1 18.5	 183	4-09-00-00003 1.0885^{20}	H_2O 3; EtOH 3; eth 3; ctc 2 1.4390^{20}
4959	Cyclopropanecarboxylic acid, 2,2-dimethyl-3-(2-methyl-1-propenyl)-, (1*R-cis*)-	$C_{10}H_{16}O_2$ 168.24	26771-11-9 41	 135^{12}; $95^{0.1}$	4-09-00-00168	eth 4; EtOH 4; chl 4
4960	Cyclopropanecarboxylic acid, 2,2-dimethyl-3-(2-methyl-1-propenyl)-, 2-methyl-4-oxo-3-(2,4-pentadienyl)-2-cyclopenten-1-yl ester, [1R-[1α[S*(Z)],3β]]- Pyrethrin 1	$C_{21}H_{28}O_3$ 328.45	121-21-1	7978 $170^{0.1}$ dec	3-09-00-00215 1.5192^{18}	H_2O 1; EtOH 3; eth 3; ctc 3 1.5192^{18}
4961	Cyclopropanecarboxylic acid, ethyl ester Ethyl cyclopropylcarboxylate	$C_6H_{10}O_2$ 114.14	4606-07-9	 134	4-09-00-00003 0.9608^{15}	 1.4190^{20}
4962	Cyclopropanecarboxylic acid, 3-(3-methoxy-2-methyl-3-oxo-1-propenyl)-2,2-dimethyl-, 2-methyl-4-oxo-3-(2,4-pentadienyl)-2-cyclopenten-1-yl ester [1R-[1α[S*(Z)],3β(E)]]- Pyrethrin 2	$C_{22}H_{28}O_5$ 372.46	121-29-9	7978 $200^{0.1}$ dec	3-09-00-03988	H_2O 1; EtOH 3; eth 3; ctc 3 1.5258^{20}
4963	Cyclopropanecarboxylic acid, methyl ester Methyl cyclopropanecarboxylate	$C_5H_8O_2$ 100.12	2868-37-3	 114.9	4-09-00-00003 0.9848^{20}	ace 3; chl 3 1.4144^{19}

No.	Name Synonym	Mol. Form. Mol. Wt.	CAS RN mp/°C	Merck No. bp/°C	Beil. Ref. den/g cm^{-3}	Solubility n_D
4964	Cyclopropanecarboxylic acid, 2-methyl- 2-Methylcyclopropanecarboxylic acid	$C_5H_8O_2$ 100.12	29555-02-0	191; 96.5^{14}	2-09-00-00006 1.0267^{20}	1.4441^{20}
4965	Cyclopropanecarboxylic acid, 2,2-dimethyl-3-(2-methyl-1-propenyl)-, methyl ester, *trans*-(±)- Chrysanthemum monocarboxylic acid, methyl ester (*DL, trans*)	$C_{11}H_{18}O_2$ 182.26	15543-70-1	108^{20}; 95^{13}	4-09-00-00169 0.9274^{25}	1.4614^{25}
4966	Cyclopropane, (chloromethyl)-	C_4H_7Cl 90.55	5911-08-0 -90.9	88	4-05-00-00013 0.98^{25}	1.4350^{20}
4967	1,1-Cyclopropanedicarboxylic acid, diethyl ester	$C_9H_{14}O_4$ 186.21	1559-02-0	216; 100^{12}	4-09-00-02786 1.0535^{25}	EtOH 4; eth 4 1.4345^{18}
4968	1,2-Cyclopropanedicarboxylic acid, diethyl ester, *cis*- Cyclopropane-1,2-dicarboxylic acid, diethyl ester (*cis*)	$C_9H_{14}O_4$ 186.21	710-43-0		2-09-00-00513 1.062^{12}	eth 4; EtOH 4 1.4450^{20}
4969	1,2-Cyclopropanedicarboxylic acid, dimethyl ester Cyclopropane-1,2-dicarboxylic acid, dimethyl ester	$C_7H_{10}O_4$ 158.15	702-28-3	219; 110^{3}	1.1584^{16}	eth 4; EtOH 4 1.4472^{14}
4970	1,2-Cyclopropanedicarboxylic acid, *trans*-(±)- Cyclopropane-1,2-dicarboxylic acid (*trans, DL*)	$C_5H_6O_4$ 130.10	58616-95-8 175	210^{30}	4-09-00-02786 1.4600^{20}	H_2O 4; eth 4; EtOH 4
4971	Cyclopropane, 1,1-dimethyl- gem-Dimethylcyclopropane	C_5H_{10} 70.13	1630-94-0 -109	20.6	4-05-00-00024 0.6604^{20}	H_2O 1; EtOH 3; eth 4; sulf 4 1.3668^{20}
4972	Cyclopropane, 1,2-dimethyl-, *cis*- *cis*-1,2-Dimethylcyclopropane	C_5H_{10} 70.13	930-18-7 -140.9	37.0	4-05-00-00025 0.6889^{25}	H_2O 1; EtOH 3; eth 4; ctc 2 1.3829^{20}
4973	Cyclopropane, 1,2-dimethyl-, *trans*- *trans*-1,2-Dimethylcyclopropane	C_5H_{10} 70.13	2402-06-4 -149.6	28.2	4-05-00-00026 0.6648^{25}	eth 4; EtOH 4 1.3713^{20}
4974	Cyclopropane, ethyl- Ethylcyclopropane	C_5H_{10} 70.13	1191-96-4 -149.2	35.9	4-05-00-00023 0.6790^{25}	1.3786^{20}
4975	Cyclopropane, 1-ethyl-1-methyl- Cyclopropane, 1-methyl-1-ethyl	C_6H_{12} 84.16	53778-43-1 -130.2	56.8	4-05-00-00091 0.6968^{25}	1.3887^{20}
4976	Cyclopropane, 1-ethyl-2-methyl-, *cis*- *cis*-1-Methyl-2-ethylcyclopropane	C_6H_{12} 84.16	19781-68-1	58.7	3-05-00-00063 0.6935^{20}	1.3846^{20}
4977	Cyclopropane, 1-ethyl-2-methyl-, *trans*- *trans*-1-Methyl-2-ethylcyclopropane	C_6H_{12} 84.16	19781-69-2	59	3-05-00-00063 0.7146^{20}	ctc 3 1.3953^{20}
4978	Cyclopropanemethanol, 2,2-dimethyl-3-(2-methyl-1-propenyl)-	$C_{10}H_{18}O$ 154.25	5617-92-5	66-9$^{0.07}$	4-06-00-00264 0.888^{25}	1.4757^{20}
4979	Cyclopropanemethanol, α-methyl-	$C_5H_{10}O$ 86.13	765-42-4 -32.1	123.5	4-06-00-00020 0.8805^{20}	1.4316^{20}
4980	Cyclopropane, methoxy- Cyclopropyl methyl ether	C_4H_8O 72.11	540-47-6 -119	2756 44.7	3-06-00-00003 0.8100^{20}	H_2O 4; bz 4; eth 4; EtOH 4 1.3802^{20}
4981	Cyclopropane, methyl- Methylcyclopropane	C_4H_8 56.11	594-11-6 -177.3	0.7	4-05-00-00013 0.6912^{-20}	eth 4; EtOH 4
4982	Cyclopropane, (1-methylethenyl)-	C_6H_{10} 82.15	4663-22-3	70	4-05-00-00243 0.7500^{20}	1.4252^{20}
4983	Cyclopropane, (1-methylethyl)-	C_6H_{12} 84.16	3638-35-5 -112.9	58.3	4-05-00-00090 0.6936^{25}	1.3865^{20}
4984	Cyclopropane, propyl- Propane, 1-cyclopropyl-	C_6H_{12} 84.16	2415-72-7	69.1	4-05-00-00090 0.7062^{25}	1.3930^{20}
4985	Cyclopropane, 1,1,2-trimethyl- 1,1,2-Trimethylcyclopropane	C_6H_{12} 84.16	4127-45-1 -138.2	54	4-05-00-00091 0.6897^{25}	1.3864^{20}
4986	Cyclopropane, 1,2,3-trimethyl-, (1α,2α,3α)-	C_6H_{12} 84.16	4806-58-0	66	3-05-00-00063 0.7130^{25}	1.3970^{20}
4987	Cyclopropane, 1,2,3-trimethyl-, (1α,2α,3β)-	C_6H_{12} 84.16	4806-59-1	59.7	3-05-00-00063 0.6929^{25}	1.3873^{20}
4988	Cyclopropanone	C_3H_4O 56.06	5009-27-8 stable only at low temp.			
4989	1H-Cycloprop[e]azulen-4-ol, decahydro-1,1,4,7-tetramethyl-, (1a*R*,4*R*,4a*S*,7*R*,7a*S*,7b*S*)- Ledol	$C_{15}H_{26}O$ 222.37	577-27-5 105	5313 292	0.9078^{100}	ace 4; eth 4; EtOH 4 1.4667^{110}
4990	Cyclopropene	C_3H_4 40.06	2781-85-3	-36 dec		
4991	Cyclotetrasiloxane, ethenylheptamethyl-	$C_9H_{24}O_4Si_4$ 308.63	3763-39-1	84^{20}	0.9505^{20}	ctc 3; CS_2 3 1.4034^{20}
4992	Cyclotetrasiloxane, heptamethyl- Heptamethylcyclotetrasiloxane	$C_7H_{22}O_4Si_4$ 282.59	15721-05-8 -27	165	4-04-00-04125 0.9583^{20}	H_2O I 1.3965^{20}
4993	Cyclotetrasiloxane, octamethyl- Octamethylcyclotetrasiloxane	$C_8H_{24}O_4Si_4$ 296.62	556-67-2 17.5	6668 175.8	4-04-00-04125 0.9561^{20}	H_2O I; ctc 3 1.3968^{20}
4994	Cyclotetrasiloxane, octaphenyl-	$C_{48}H_{40}O_4Si_4$ 793.19	546-56-5 200.5	330^{1}	4-16-00-01530	H_2O I; EtOH 2; bz 3; chl 3
4995	Cyclotetrasiloxane, 2,4,6,8-tetrabutyl-2,4,6,8-tetramethyl- Siloxane, butylmethyl(cyclic tetramer)	$C_{20}H_{48}O_4Si_4$ 464.94	14685-29-1	291	4-04-00-04173 0.9230^{20}	1.4300^{20}
4996	Cyclotetrasiloxane, 2,4,6,8-tetraethenyl-2,4,6,8-tetramethyl-	$C_{12}H_{24}O_4Si_4$ 344.66	2554-06-5 -43.5	224; 111^{12}	4-04-00-04184 0.9875^{20}	ctc 3; CS_2 3

No.	Name Synonym	Mol. Form. Mol. Wt.	CAS RN mp/°C	Merck No. bp/°C	Beil. Ref. den/g cm^{-3}	Solubility n_D
4997	Cyclotetrasiloxane, 2,4,6,8-tetraethyl-2,4,6,8-tetramethyl- Cyclotetrasiloxane, 2,4,6,8-tetramethyl-2,4,6,8-tetraethyl	$C_{12}H_{32}O_4Si_4$ 352.73	7623-01-0 -43.5	 245; 111^{12}	4-04-00-04154 0.9600^{20}	ctc 3; CS_2 3
4998	Cyclotetrasiloxane, 2,4,6,8-tetraethyl-2,4,6,8-tetraphenyl- Cyclotetrasiloxane, 1,3,5,7-tetraethyl-1,3,5,7-tetraphenyl	$C_{32}H_{40}O_4Si_4$ 601.01	18758-34-4 106	 $212^{0.1}$	3-16-00-01211 1.1000^{20}	 1.5430^{25}
4999	Cyclotetrasiloxane, 2,4,6,8-tetramethyl-	$C_4H_{16}O_4Si_4$ 240.51	2370-88-9 -65	 134.5	4-04-00-04099 0.9912^{20}	H_2O I 1.3870^{20}
5000	Cyclotetrasiloxane, 2,4,6,8-tetramethyl-2,4,6,8-tetraphenyl-	$C_{28}H_{32}O_4Si_4$ 544.90	77-63-4 99	 237^{115}	3-16-00-01211 1.1183^{20}	H_2O 1; ace 5; hp 5 1.5461^{20}
5001	Cyclotrisilazane, 2,2,4,4,6,6-hexamethyl- 1,1,3,3,5,5-Hexamethylcyclotrisilazane	$C_6H_{21}N_3Si_3$ 219.51	1009-93-4 -10	 188	4-04-00-04114 0.9196^{20}	 1.448^{20}
5002	Cyclotrisiloxane, hexamethyl- Dimethylsiloxane cyclic trimer	$C_6H_{18}O_3Si_3$ 222.46	541-05-9 64.5	 134	4-04-00-04123 1.1200^{20}	H_2O 1
5003	Cyclotrisiloxane, 2,4,6-triethyl-2,4,6-trimethyl- 2,4,6-Triethyl-2,4,6-trimethylcyclotrisiloxane	$C_9H_{24}O_3Si_3$ 264.54	15901-49-2 -3	 199	4-04-00-04154	ctc 3; CS_2 3
5004	Cyclotrisiloxane, 2,4,6-triethyl-2,4,6-triphenyl- 2,4,6-Triethyl-2,4,6-triphenylcyclotrisiloxane	$C_{24}H_{30}O_3Si_3$ 450.76	546-33-8 177.5	 $166^{0.025}$	3-16-00-01211 1.0952^{25}	H_2O 1 1.5402^{25}
5005	Cyclotrisiloxane, 2,4,6-trimethyl-2,4,6-triphenyl-	$C_{21}H_{24}O_3Si_3$ 408.68	546-45-2 100	 $190^{1.5}$	3-16-00-01211 1.1062^{20}	 1.5397^{20}
5006	1,4,8-Cycloundecatriene, 2,6,6,9-tetramethyl-, (*E,E,E*)- Humulene	$C_{15}H_{24}$ 204.36	6753-98-6	4672 123^{10}	4-05-00-01171 0.8905^{20}	 1.5038^{20}
5007	7,20-Cycloveatchane-1,12,15-triol, 21-ethyl-4-methyl-16-methylene-, (1α,12α,15β)- Luciculine	$C_{22}H_{35}NO_3$ 361.52	5008-52-6 149	6286 $165^{0.02}$	4-21-00-02584	EtOH 4
5008	Cyfluthrin Cyano(4-fluoro-3-phenoxyphenyl)methyl 3-(2,2-dichloroethenyl)-2,2-dimethylcyclopropanecarboxylate	$C_{22}H_{18}Cl_2FNO_3$ 434.29	68359-37-5 60	2764		
5009	Cygon Phosphorodithioic acid, *O,O*-dimethyl *S*-[2-(methylamino)-2-oxoethyl] ester	$C_5H_{12}NO_3PS_2$ 229.26	60-51-5 52	3209 $117^{0.1}$	 1.277^{65}	
5010	Cyhalothrin	$C_{23}H_{19}ClF_3NO_3$ 449.86	91465-08-6 49.2			
5011	Cyhexatin Stannane, tricyclohexylhydroxy-	$C_{18}H_{34}OSn$ 385.18	13121-70-5 196	2767		
5012	Cypermethrin Cyclopropanecarboxylic acid, 3-(2,2-dichloroethenyl)-2,2-dimethyl-, cyano(3-phenoxy-phenyl)methyl ester	$C_{22}H_{19}Cl_2NO_3$ 416.30	52315-07-8 70	2775	 1.25^{20}	
5013	Cyprazine 1,3,5-Triazine-2,4-diamine, 6-chloro-*N*-cyclopropyl-*N'*-(1-methylethyl)-	$C_9H_{14}ClN_5$ 227.70	22936-86-3 167			
5014	*L*-Cysteine Propanoic acid, 2-amino-3-mercapto-, (*R*)-	$C_3H_7NO_2S$ 121.16	52-90-4 240 dec	2787	4-04-00-03144	H_2O 4; ace 4; EtOH 4
5015	*L*-Cysteine, *N*-acetyl- Acetylcysteine	$C_5H_9NO_3S$ 163.20	616-91-1 109.5	82	4-04-00-03160	
5016	*L*-Cysteine, *S*-(2-amino-2-carboxyethyl)-, (*R*)- *L*-Lanthionine	$C_6H_{12}N_2O_4S$ 208.24	922-55-4 293-5 dec	5234	3-04-00-01593	
5017	*L*-Cysteine, *S*-(carboxymethyl)- Carbocysteine	$C_5H_9NO_4S$ 179.20	638-23-3 206	1809	4-04-00-03150	
5018	*L*-Cysteine, *S,S'*-methylenebis- Djenkolic acid	$C_7H_{14}N_2O_4S_2$ 254.33	498-59-9 300-50 dec	3393	3-04-00-01591	
5019	*L*-Cysteine, methyl ester, hydrochloride Methylcysteine hydrochloride	$C_4H_{10}ClNO_2S$ 171.65	18598-63-5 140.5	6668	3-04-00-01601	
5020	*L*-Cystine 3,3'-Dithiobis(2-aminopropanoic acid)	$C_6H_{12}N_2O_4S_2$ 240.30	56-89-3 260-1 dec	2788	4-04-00-03155 1.677^{25}	H_2O 2; EtOH 1; eth 1; bz 1
5021	Cytidine 4-Amino-1-β-*D*-ribofuranosyl-2(1H)-pyrimidinone	$C_9H_{13}N_3O_5$ 243.22	65-46-3 230-1 dec	2792	4-25-00-03667	H_2O 4; EtOH 2
5022	3'-Cytidylic acid Cytidine 3'-monophosphate	$C_9H_{14}N_3O_8P$ 323.20	84-52-6 233 dec	2794	4-25-00-03669	H_2O 3; EtOH 3
5023	1,3-Decadiene 1-Hexyl-1,3-butadiene	$C_{10}H_{18}$ 138.25	2051-25-4	 169	4-01-00-01056 0.752^{30}	bz 4
5024	1,9-Decadiene α,ω-Decadiene	$C_{10}H_{18}$ 138.25	1647 16 1	 167	4-01-00-01056 0.75^{25}	 1.4325^{20}
5025	2,4-Decadienoic acid, (*E,E*)- 2,4-Decadienoic acid (*trans*-2, *trans*-4)	$C_{10}H_{16}O_2$ 168.24	30361-33-2 49.5	 $121^{0.3}$	4-02-00-01730	 1.5058^{31}
5026	2,4-Decadienoic acid, methyl ester, (*E,E*)-	$C_{11}H_{18}O_2$ 182.26	7328-33-8	 $70^{0.2}$	4-02-00-01731 0.982^{22}	 1.4918^{22}
5027	2,4-Decadienoic acid, methyl ester, (*E,Z*)-	$C_{11}H_{18}O_2$ 182.26	4493-42-9	 $71^{0.15}$	4-02-00-01731 0.9128^{22}	 1.4874^{22}

No.	Name Synonym	Mol. Form. Mol. Wt.	CAS RN mp/°C	Merck No. bp/°C	Beil. Ref. den/g cm^{-3}	Solubility n_D
5028	2,4-Decadienoic acid, methyl ester, (*Z,E*)- 2,4-Decadienoic acid, methyl ester (*cis*-2, *trans*-4)	$C_{11}H_{18}O_2$ 182.26	108965-84-0	76-80$^{0.5}$	4-02-00-01731 0.9131^{23}	1.4876^{23}
5029	2,4-Decadienoic acid, methyl ester, (*Z,Z*)- 2,4-Decadienoic acid, methyl ester (*cis*-2, *cis*-4)	$C_{11}H_{18}O_2$ 182.26	108965-86-2	78$^{0.6}$	4-02-00-01731 0.9095^{23}	1.4830^{23}
5030	Decanal Capraldehyde	$C_{10}H_{20}O$ 156.27	112-31-2 -5	208.5	4-01-00-03366 0.830^{15}	H_2O 1; EtOH 3; eth 3; ace 3 1.4287^{20}
5031	Decanamide Capramide	$C_{10}H_{21}NO$ 171.28	2319-29-1 108		4-02-00-01050 0.822^{130}	ace 4; eth 4; EtOH 4 1.4621^{110}
5032	1-Decanamine	$C_{10}H_{23}N$ 157.30	2016-57-1 17	220.5	4-04-00-00783 0.7936^{20}	H_2O 2; EtOH 5; eth 5; ace 5 1.4369^{20}
5033	Decane *n*-Decane	$C_{10}H_{22}$ 142.28	124-18-5 -29.7	174.1	4-01-00-00464 0.7300^{20}	H_2O 1; EtOH 5; eth 3; ctc 2 1.4102^{20}
5034	Decane, 1-bromo-	$C_{10}H_{21}Br$ 221.18	112-29-8 -29.2	240.6	4-01-00-00470 1.0702^{20}	H_2O 1; eth 4; ctc 3; chl 4 1.4557^{20}
5035	Decane, 1-bromo-10-fluoro- 1-Bromo-10-fluorodecane	$C_{10}H_{20}BrF$ 239.17	334-61-2	131-2^{11}	4-01-00-00471 1.152^{20}	eth 4; EtOH 4 1.4512^{25}
5036	Decane, 1-chloro-	$C_{10}H_{21}Cl$ 176.73	1002-69-3 -31.3	223.4	4-01-00-00469 0.8705^{20}	H_2O 1; eth 4; ctc 3; chl 4 1.4379^{20}
5037	Decane, 1-chloro-10-fluoro- 1-Chloro-10-fluorodecane	$C_{10}H_{20}ClF$ 194.72	334-62-3	115^{9}	4-01-00-00469 0.957^{20}	eth 4; EtOH 4 1.4333^{25}
5038	Decane, 1,10-dibromo- Decamethylene dibromide	$C_{10}H_{20}Br_2$ 300.08	4101-68-2 28	161^{9}; 128^{4}	4-01-00-00471 1.335^{30}	H_2O 1; EtOH 2; eth 3 1.4927^{25}
5039	Decane, 5,6-dibromo- 5,6-Dibromodecane	$C_{10}H_{20}Br_2$ 300.08	77928-86-0	119^{9}	4-01-00-00472 1.3484^{20}	1.4912^{20}
5040	Decane, 1,10-dichloro- Decamethylene dichloride	$C_{10}H_{20}Cl_2$ 211.17	2162-98-3 15.6	167-8^{28}	4-01-00-00469 0.9945^{25}	1.4586^{25}
5041	Decanedinitrile	$C_{10}H_{16}N_2$ 164.25	1871-96-1	204^{16}	4-02-00-02089 0.9313^{20}	H_2O 1; chl 3 1.4474^{20}
5042	Decanedioic acid Sebacic acid	$C_{10}H_{18}O_4$ 202.25	111-20-6 130.8	8369 295^{100}; 232^{10}	4-02-00-02078 1.2705^{20}	H_2O 2; EtOH 3; eth 3; bz 1 1.422^{133}
5043	Decanedioic acid, bis(2-ethylhexyl) ester Bis(2-ethylhexyl) sebacate	$C_{26}H_{50}O_4$ 426.68	122-62-3 -48	1263 256^{5}	4-02-00-02083 0.912^{25}	ace 4; bz 4; EtOH 4 1.451^{25}
5044	Decanedioic acid, dibutyl ester Dibutyl sebacate	$C_{18}H_{34}O_4$ 314.47	109-43-3 -10	344.5	4-02-00-02081 0.9405^{15}	H_2O 1; eth 3; ctc 3 1.4433^{15}
5045	Decanedioic acid, diethyl ester	$C_{14}H_{26}O_4$ 258.36	110-40-7 5	305; 188^{19}	4-02-00-02080 0.9646^{20}	H_2O 2; EtOH 3; ace 3; bz 1 1.4306^{20}
5046	Decanedioic acid, dimethyl ester	$C_{12}H_{22}O_4$ 230.30	106-79-6 38	175^{20}; 144^{5}	4-02-00-02080 0.9882^{28}	H_2O 1; EtOH 3; eth 3; ace 3 1.4355^{28}
5047	1,10-Decanediol Decamethylene glycol	$C_{10}H_{22}O_2$ 174.28	112-47-0 74	2842 192^{20}	4-01-00-02613	H_2O 2; EtOH 4; eth 2; DMSO 3
5048	2,9-Decanediol	$C_{10}H_{22}O_2$ 174.28	14021-92-2 33	114^{5}	3-01-00-02232	1.4505^{21}
5049	Decanedioyl dichloride	$C_{10}H_{16}Cl_2O_2$ 239.14	111-19-3 -2.5	220^{75}; 165^{11}	4-02-00-02088 1.1212^{20}	1.4684^{18}
5050	Decane, docosafluoro- Perfluorodecane	$C_{10}F_{22}$ 538.07	307-45-9	144.2		
5051	Decane, 1-(ethenyloxy)- Decyl vinyl ether	$C_{12}H_{24}O$ 184.32	765-05-9 -41	101^{10}	4-01-00-02057 0.812^{20}	1.4346^{20}
5052	Decane, 1-fluoro- Decyl fluoride	$C_{10}H_{21}F$ 160.28	334-56-5 -35	186.2	4-01-00-00468 0.8194^{20}	eth 4 1.4085
5053	Decane, 1-iodo-	$C_{10}H_{21}I$ 268.18	2050-77-3 -16.3	132^{15}	4-01-00-00472 1.2546^{20}	H_2O 1; EtOH 3; eth 3; ctc 3 1.4858^{20}
5054	Decane, 2-methyl- 2-Methyldecane	$C_{11}H_{24}$ 156.31	6975-98-0 -48.9	189.3	4-01-00-00491 0.7368^{20}	1.4154^{20}
5055	Decane, 3-methyl- 3-Methyldecane	$C_{11}H_{24}$ 156.31	13151-34-3 -92.9	188.1	4-01-00-00491 0.7422^{20}	1.4177^{20}
5056	Decane, 4-methyl- 4-Methyldecane	$C_{11}H_{24}$ 156.31	2847-72-5 -77.5	187		1.4352^{20}
5057	Decanenitrile Caprinitrile	$C_{10}H_{19}N$ 153.27	1975-78-6 -17.9	243. 106^{10}	4-02-00-01051 0.8199^{20}	ace 4; eth 4; EtOH 4; chl 4 1.4296^{20}
5058	1-Decanethiol Decyl mercaptan	$C_{10}H_{22}S$ 174.35	143-10-2 -26	240.6	4-01-00-01821 0.8443^{20}	H_2O 1; EtOH 3; eth 3 1.4509^{20}
5059	Decanoic acid Capric acid	$C_{10}H_{20}O_2$ 172.27	334-48-5 31.9	1759 268.7	4-02-00-01041 0.8858^{40}	ace 4; bz 4; eth 4; EtOH 4 1.4288^{40}
5060	Decanoic acid, anhydride	$C_{20}H_{38}O_3$ 326.52	2082-76-0 24.7		4-02-00-01049 0.8865^{25}	eth 4; EtOH 4 1.400^{25}
5061	Decanoic acid, 2-bromo-	$C_{10}H_{19}BrO_2$ 251.16	2623-95-2 4	140-1^{2}	4-02-00-01054 1.1912^{24}	eth 4 1.4595^{24}
5062	Decanoic acid, decyl ester	$C_{20}H_{40}O_2$ 312.54	1654-86-0 9.7	219^{15}	4-02-00-01045 0.8586^{20}	eth 4 1.4423^{20}
5063	Decanoic acid, ethyl ester Ethyl caprate	$C_{12}H_{24}O_2$ 200.32	110-38-3 -20	3734 241.5	4-02-00-01044 0.8650^{20}	eth 4; EtOH 4; chl 4 1.4256^{20}
5064	Decanoic acid, 10-fluoro- Capric acid, 10-fluoro	$C_{10}H_{19}FO_2$ 190.26	334-59-8 49	135-8^{10}	4-02-00-01051	eth 4; EtOH 4; lig 4
5065	Decanoic acid, 2-hexyl- 2-Hexyldecanoic acid	$C_{16}H_{32}O_2$ 256.43	25354-97-6	4630 140-50$^{0.02}$	4-02-00-01189	1.4432^{24}

No.	Name Synonym	Mol. Form. Mol. Wt.	CAS RN mp/°C	Merck No. bp/°C	Beil. Ref. den/g cm^{-3}	Solubility n_D
5066	Decanoic acid, methyl ester	$C_{11}H_{22}O_2$ 186.29	110-42-9 -18	224	4-02-00-01044 0.8730^{20}	H_2O 1; EtOH 4; eth 4; ctc 2 1.4259^{20}
5067	Decanoic acid, 1-methylethyl ester	$C_{13}H_{26}O_2$ 214.35	2311-59-3	121^{10}	4-02-00-01045 0.8543^{20}	1.4221^{25}
5068	Decanoic acid, 2-octyl-	$C_{18}H_{36}O_2$ 284.48	619-39-6 38.5	215^{13}	4-02-00-01254 0.8447^{70}	eth 4; EtOH 4
5069	Decanoic acid, propyl ester	$C_{13}H_{26}O_2$ 214.35	30673-60-0	128.5^{10}	4-02-00-01045 0.8623^{20}	1.4280^{20}
5070	1-Decanol Capric alcohol	$C_{10}H_{22}O$ 158.28	112-30-1 6.9	2847 231.1	4-01-00-01815 0.8297^{20}	H_2O 1; EtOH 5; eth 5; ace 5 1.4372^{20}
5071	2-Decanol, (±)- 2-Decanol (*DL*)	$C_{10}H_{22}O$ 158.28	74742-10-2 -2.4	211	4-01-00-01823 0.8250^{20}	EtOH 3; eth 5; ace 5; bz 3 1.4326^{25}
5072	4-Decanol 1-Propylheptyl alcohol	$C_{10}H_{22}O$ 158.28	2051-31-2 -11	210.5	4-01-00-01824 0.8261^{20}	H_2O 1; EtOH 3; ctc 3 1.4320^{20}
5073	1-Decanol, 10-chloro-	$C_{10}H_{21}ClO$ 192.73	51309-10-5 12.5	$185\text{-}9^{15}$	4-01-00-01821 0.9630^{25}	eth 4; EtOH 4 1.4578^{20}
5074	1-Decanol, 10-fluoro- 10-Fluoro-1-decanol	$C_{10}H_{21}FO$ 176.27	334-64-5 22	$136\text{-}7^{15}$	4-01-00-01821 0.919^{20}	eth 4; EtOH 4 1.4322^{25}
5075	2-Decanone Methyl octyl ketone	$C_{10}H_{20}O$ 156.27	693-54-9 14	210; 96^{12}	4-01-00-03367 0.8248^{20}	H_2O 1; EtOH 3; eth 3; ctc 2 1.4255^{20}
5076	3-Decanone Ethyl heptyl ketone	$C_{10}H_{20}O$ 156.27	928-80-3 2.5	203	4-01-00-03368 0.8251^{20}	EtOH 3; eth 3; ctc 3 1.4252^{20}
5077	4-Decanone Hexyl propyl ketone	$C_{10}H_{20}O$ 156.27	624-16-8 -9	206.5	4-01-00-03368 0.824^{20}	H_2O 1; EtOH 5; eth 5 1.4240^{21}
5078	Decanoyl chloride Caprinoyl chloride	$C_{10}H_{19}ClO$ 190.71	112-13-0 -34.5	95	4-02-00-01050 0.919^{25}	eth 3; ctc 3 1.4410^{20}
5079	Decasiloxane, docosamethyl- Decasiloxane, dicosamethyl	$C_{22}H_{66}O_9Si_{10}$ 755.62	556-70-7	$183.^4$	3-04-00-01881 0.925^{20}	bz 4; lig 4 1.3988^{20}
5080	2-Decenal	$C_{10}H_{18}O$ 154.25	3913-71-1	230	4-01-00-03511 0.845^{17}	1.4533^{17}
5081	3-Decenal	$C_{10}H_{18}O$ 154.25	58474-80-9	$93\text{-}4^{14}$	4-01-00-03512 0.850^{15}	1.4462^{15}
5082	1-Decene	$C_{10}H_{20}$ 140.27	872-05-9 -66.3	170.5	3-01-00-00858 0.7408^{20}	H_2O 1; EtOH 5; eth 5 1.4215^{20}
5083	4-Decene	$C_{10}H_{20}$ 140.27	19689-18-0	170.6	4-01-00-00902 0.7404^{20}	1.4243^{20}
5084	5-Decene, (*E*)-	$C_{10}H_{20}$ 140.27	7433-56-9 -73	171	4-01-00-00902 0.7401^{20}	H_2O 1; EtOH 5; eth 5; ctc 2 1.4243^{20}
5085	5-Decene, (*Z*)-	$C_{10}H_{20}$ 140.27	7433-78-5 -112	171; 73^{20}	3-01-00-00859 0.7445^{20}	H_2O 1; EtOH 5; eth 5; ctc 2 1.4258^{20}
5086	1-Decene, 2-bromo- 2-Bromo-1-decene	$C_{10}H_{19}Br$ 219.16	3017-67-2	$115\text{-}6^{22}$	3-01-00-00859 1.0844^{20}	1.4629^{20}
5087	2-Decene, 1-bromo- 1-Bromo-2-decene	$C_{10}H_{19}Br$ 219.16	14304-30-4	121^{17}	4-01-00-00902 1.074^{18}	lig 4 1.4716^{18}
5088	2-Decenoic acid Δ 2-Decenoic acid	$C_{10}H_{18}O_2$ 170.25	3913-85-7 12	165^{15}	4-02-00-01606 0.9280^{18}	1.4616^{20}
5089	3-Decenoic acid	$C_{10}H_{18}O_2$ 170.25	15469-77-9 18	$154\text{-}63^{11}$	4-02-00-01606 0.914^{15}	1.4510^{18}
5090	4-Decenoic acid Deconic acid telomer	$C_{10}H_{18}O_2$ 170.25	26303-90-2	149^{13}	4-02-00-01607 0.9197^{20}	bz 4; eth 4 1.4497^{20}
5091	9-Decenoic acid Caproleic acid	$C_{10}H_{18}O_2$ 170.25	14436-32-9	158^{21}; 142^4	4-02-00-01605 0.9238^{15}	eth 4; EtOH 4 1.4507^{15}
5092	9-Decen-1-ol Decylenic alcohol	$C_{10}H_{20}O$ 156.27	13019-22-2	236	4-01-00-02184 0.876^{25}	1.4480^{20}
5093	3-Decen-2-one Heptylidene acetone	$C_{10}H_{18}O$ 154.25	10519-33-2	$102\text{-}3^{15.3}$	4-01-00-03512 0.8473^{20}	1.4480^{20}
5094	1-Decen-3-yne	$C_{10}H_{16}$ 136.24	33622-26-3	76^{20}	4-01-00-01105 0.7873^{20}	1.4620^{20}
5095	1-Decen-4-yne	$C_{10}H_{16}$ 136.24	24948-66-1	$73\text{-}4^{22}$	3-01-00-01049 0.7880^{20}	1.445^{20}
5096	2-Decen-4-yne	$C_{10}H_{16}$ 136.24	116668-40-7	55^5	3-01-00-01049 0.7850^{25}	1.4609^{25}
5097	1-Decyne Octylacetylene	$C_{10}H_{18}$ 138.25	764-93-2 -44	174	4-01-00-01054 0.7655^{20}	H_2O 1; EtOH 3; eth 3; os 3 1.4265^{20}
5098	3-Decyne	$C_{10}H_{18}$ 138.25	2384-85-2	177	4-01-00-01055 0.7619^{25}	1.4315^{20}
5099	4-Decyne	$C_{10}H_{18}$ 138.25	2384-86-3	74.5^{19}	3-01-00-01017 0.772^{17}	1.436^{17}
5100	5-Decyne Dibutylacetylene	$C_{10}H_{18}$ 138.25	1942-46-7 73	177; 78.6^{26}	4-01-00-01055 0.7690^{20}	H_2O 1; EtOH 3; eth 3 1.4331^{20}
5101	4-Decyne, 3,3-dimethyl- 3,3-Dimethyl-4-decyne	$C_{12}H_{22}$ 166.31	70732-45-5	86^{20}	3-01-00-01026 0.7731^{20}	1.4399^{20}
5102	Deltamethrin Cyano(3-phenoxyphenyl)methyl-3-(2,2-dibromoethenyl)-2,2-dimethylcyclopropanecarboxylate	$C_{22}H_{19}Br_2NO_3$ 505.21	52918-63-5 99	2869		
5103	Demeton S methyl Phosphorothioic acid, *S*-[2-(ethylthio)ethyl] *O*,*O*-dimethyl ester	$C_6H_{15}O_3PS_2$ 230.29	919-86-8	$89^{0.15}$; 118^1	1.207^{20}	

No.	Name Synonym	Mol. Form. Mol. Wt.	CAS RN mp/°C	Merck No. bp/°C	Beil. Ref. den/g cm^{-3}	Solubility n_D
5104	Desmedipham Carbamic acid, [3-[[(phenylamino)carbonyl]oxy]phenyl]-, ethyl ester	$C_{16}H_{16}N_2O_4$ 300.31	13684-56-5 120			
5105	Dialifor Phosphorodithioic acid, *S*-[2-chloro-1-(1,3-dihydro-1,3-dioxo-2H-isoindol-2-yl)ethyl]	$C_{14}H_{17}ClNO_4PS_2$ 393.85	10311-84-9 68	2949		
5106	Diallate Carbamothioic acid, bis(1-methylethyl)-, *S*-(2,3-dichloro-2-propenyl) ester	$C_{10}H_{17}Cl_2NOS$ 270.22	2303-16-4	2950 150^{9}		
5107	8,8'-Diapo-ψ,ψ-carotenedioic acid Crocetin (*trans*)	$C_{20}H_{24}O_4$ 328.41	27876-94-4 286	2592	3-02-00-02018	H_2O 2; EtOH 2; eth 1; bz 1
5108	6,6'-Diapo-ψ,ψ-carotenedioic acid, monomethyl ester, 9-*cis*- Bixin	$C_{25}H_{30}O_4$ 394.51	6983-79-5 198	1320	3-02-00-02021	H_2O 1; EtOH 3; eth 2; ace 3
5109	Diarsine, tetraethyl- Tetraethyldiarsenic	$C_8H_{20}As_2$ 266.09	612-08-8	 186	3-04-00-01832 1.1388^{24}	eth 4; EtOH 4 1.4709
5110	Diarsine, tetramethyl- Cacodyl	$C_4H_{12}As_2$ 209.98	471-35-2 -6	1602 165	4-04-00-03683 1.447^{15}	eth 4; EtOH 4
5111	1,4-Diazabicyclo[2.2.2]octane Triethylenediamine	$C_6H_{12}N_2$ 112.17	280-57-9 159	9584	5-23-03-00487	chl 3
5112	Diazene, bis(2-ethoxyphenyl)- Azobenzene, 2,2'-diethoxy-	$C_{16}H_{18}N_2O_2$ 270.33	613-43-4 131	 240 dec	4-16-00-00138	H_2O 1; EtOH 3; HCl 3
5113	Diazene, bis(4-ethoxyphenyl)-	$C_{16}H_{18}N_2O_2$ 270.33	588-52-3 162	 dec	4-16-00-00174	H_2O 1; EtOH 2; eth 3; bz 3
5114	Diazene, bis(2-methylphenyl)-	$C_{14}H_{14}N_2$ 210.28	584-90-7 55.5		2-16-00-00019 1.0215^{65}	H_2O 1; EtOH 3; eth 4; bz 4 1.6180^{65}
5115	Diazene, bis(3-methylphenyl)-, (*E*)-	$C_{14}H_{14}N_2$ 210.28	51437-67-3 54		4-16-00-00070 1.0123^{66}	EtOH 4; eth 4; bz 4 1.6152^{66}
5116	Diazene, bis(2-methylphenyl)-, 1-oxide, (*Z*)- Azoxybenzene, 2,2'-dimethyl (*trans*)	$C_{14}H_{14}N_2O$ 226.28	51284-68-5 60		4-16-00-00064 1.0215^{65}	bz 4; eth 4; EtOH 4 1.6180^{65}
5117	Diazene, bis(2-methylphenyl)-, 1-oxide, (*E*)- Azoxybenzene, 2,2'-dimethyl(*trans*)	$C_{14}H_{14}N_2O$ 226.28	116723-89-8		2-16-00-00318 1.0215^{65}	H_2O 1; EtOH 3; eth 3; bz 3 1.6180^{65}
5118	Diazene, bis(3-methylphenyl)-, 1-oxide, (*Z*)- Azoxybenzene, 3,3'-dimethyl (*trans*)	$C_{14}H_{14}N_2O$ 226.28	71297-97-7 39		4-16-00-00071 1.0123^{66}	bz 4; eth 4; EtOH 4; lig 4 1.6152^{66}
5119	Diazene, bis(3-methylphenyl)-, 1-oxide, (*E*)- Azoxybenzene, 3,3'-dimethyl(*trans*)	$C_{14}H_{14}N_2O$ 226.28	116723-90-1 39		2-16-00-00319 1.0123^{66}	H_2O 1; EtOH 3; eth 3; bz 3 1.6152^{66}
5120	Diazene, (4-bromophenyl)phenyl-, 1-oxide Azoxybenzene, 4-bromo-	$C_{12}H_9BrN_2O$ 277.12	16109-68-5 93.5		3-16-00-00584 1.4138^{100}	H_2O 1; EtOH 3; bz 3; MeOH 3
5121	Diazenecarbothioic acid, phenyl-, 2-phenylhydrazide Dithizone	$C_{13}H_{12}N_4S$ 256.33	60-10-6 165-9 dec	3383	4-16-00-00018	H_2O 1; EtOH 2; eth 2; chl 3
5122	Diazenecarboxylic acid, phenyl-, 2-phenylhydrazide Diphenylcarbazone	$C_{13}H_{12}N_4O$ 240.26	538-62-5 157 dec	3322	4-16-00-00017	H_2O 1; EtOH 4; bz 4; chl 4
5123	Diazene, dibutyl Azobutane	$C_8H_{18}N_2$ 142.24	2159-75-3	 60^{18}		
5124	Diazenedicarboxamide Azodicarbonamide	$C_2H_4N_4O_2$ 116.08	123-77-3 212 dec	932	3-03-00-00234	
5125	Diazene, dimethyl- Azomethane	$C_2H_6N_2$ 58.08	503-28-6 -78	 1.5	3-04-00-01747 0.743^{0}	ace 4; eth 4; EtOH 4 1.4199^{19}
5126	Diazene, dimethyl-, (*E*)-	$C_2H_6N_2$ 58.08	4143-41-3 -78	 1.5	4-04-00-03366 0.743^{0}	ctc 3; hp 3
5127	Diazene, di-1-naphthalenyl- 1,1'-Azobisnaphthalene	$C_{20}H_{14}N_2$ 282.34	487-10-5 190	 sub	2-16-00-00026	H_2O 1; EtOH 2; ace 3; bz 4
5128	Diazene, di-2-naphthalenyl- 2,2'-Azonaphthalene	$C_{20}H_{14}N_2$ 282.34	582-08-1 208	 sub	2-16-00-00026	H_2O 1; EtOH 2; eth 2; bz 3
5129	Diazene, diphenyl-, (*E*)- *trans*-Azobenzene	$C_{12}H_{10}N_2$ 182.22	17082-12-1 68.5	930 293	4-16-00-00007 1.203^{20}	H_2O 2; EtOH 3; eth 3; bz 3 1.6266^{78}
5130	Diazene, diphenyl-, (*Z*)- *cis*-Azobenzene	$C_{12}H_{10}N_2$ 182.22	1080-16-6 71	930	4-16-00-00007	H_2O 2; EtOH 3; eth 3; bz 3
5131	Diazene, diphenyl-, 1-oxide, (*Z*)- *cis*-Azoxybenzene	$C_{12}H_{10}N_2O$ 198.22	20972-43-4 87	937	4-16-00-00010 1.166^{20}	 1.633^{20}
5132	Diazene, dipropyl Azopropane	$C_6H_{14}N_2$ 114.19	821-67-0	 114		
5133	Diazene, (4-ethoxyphenyl)phenyl-	$C_{14}H_{14}N_2O$ 226.28	7466-38-8 85	 339.5	2-16-00-00040 1.0400^{100}	EtOH 4; eth 4; ace 3; bz 4 1.6419^{100}
5134	Diazene, ethylphenyl-	$C_8H_{10}N_2$ 134.18	935-08-0	 177	4-16-00-00007 0.9628^{22}	bz 4; eth 4; EtOH 4 1.5579
5135	Diazene, (2-methoxyphenyl)phenyl-	$C_{13}H_{12}N_2O$ 212.25	6319-21-7 41	 $195\text{-}7^{14}$	3-16-00-00081 1.0728^{100}	H_2O 1; EtOH 3; ace 3; os 4
5136	Diazene, (3-methoxyphenyl)phenyl-	$C_{13}H_{12}N_2O$ 212.25	34238-81-8 33.3	 193^{15}	4-16-00-00155 1.1023^{53}	H_2O 1; EtOH 3; ace 3; os 4
5137	Diazene, (4-methoxyphenyl)phenyl-	$C_{13}H_{12}N_2O$ 212.25	2396-60-3 56	 340	2-16-00-00040 1.12^{75}	H_2O 1; EtOH 3; eth 3; ace 3
5138	Diazene, (3-methylphenyl)phenyl- Azobenzene, 3-methyl	$C_{13}H_{12}N_2$ 196.25	17478-66-9 18.5	 175^{19}	3-16-00-00047 1.065^{20}	
5139	Diazene, (4-methylphenyl)phenyl- *p*-Methylazobenzene	$C_{13}H_{12}N_2$ 196.25	949-87-1 71.5	 312	3-16-00-00049	bz 4; eth 4; chl 4

No.	Name Synonym	Mol. Form. Mol. Wt.	CAS RN mp/°C	Merck No. bp/°C	Beil. Ref. den/g cm^{-3}	Solubility n_D
5140	1H-1,4-Diazepine, 1-[(4-chlorophenyl)phenylmethyl]hexahydro-4-methyl- Homochlorocyclizine	$C_{19}H_{23}ClN_2$ 314.86	848-53-3	4653	5-23-03-00248	
5141	1H-1,4-Diazepine, hexahydro-1-methyl-	$C_6H_{14}N_2$ 114.19	4318-37-0	 154	5-23-03-00241 0.9111^{20}	 1.4769^{20}
5142	Dibenz[a,h]anthracene 1,2:5,6-Dibenzanthracene	$C_{22}H_{14}$ 278.35	53-70-3 269.5	2989	4-05-00-02722	H_2O 1; EtOH 2; ace 3; bz 3
5143	Dibenz[a,j]acridine 7-Azadibenz[a,j]anthracene	$C_{21}H_{13}N$ 279.34	224-42-0 216		5-20-08-00656	
5144	5H-Dibenz[b,f]azepine-5-carboxamide Carbamazepine	$C_{15}H_{12}N_2O$ 236.27	298-46-4 190.2	1783	5-20-08-00247	
5145	Dibenz[c,e]oxepin-5,7-dione	$C_{14}H_8O_3$ 224.22	6050-13-1 217	 sub	5-17-11-00525	H_2O 1; eth 2
5146	10H-Dibenzo[a,d]cyclohepten-10-one, 5,11-dihydro-	$C_{15}H_{12}O$ 208.26	6374-70-5 30	 $203\text{-}4^7$	4-07-00-01675 1.1635^{20}	 1.6324^{20}
5147	5H-Dibenzo[a,d]cyclohepten-5-one, 10,11-dihydro-	$C_{15}H_{12}O$ 208.26	1210-35-1 30	 $203\text{-}4^7$	4-07-00-01674 1.1635^{20}	 1.6324^{20}
5148	13H-Dibenzo[a,g]fluorene	$C_{21}H_{14}$ 266.34	207-83-0 174.5	 $195\text{-}200^{0.5}$	4-05-00-02693	H_2O 1; EtOH 2; eth 2; bz 3
5149	6H-Dibenzo[a,g]quinolizine, 5,8,13,13a-tetrahydro-2,3,9,10-tetramethoxy-13-methyl-, (13*S*-*trans*)- Corydaline (*d*)	$C_{22}H_{27}NO_4$ 369.46	518-69-4 136	2543	5-21-06-00173	bz 4; eth 4; EtOH 4; chl 4
5150	6H-Dibenzo[a,g]quinolizin-2-ol, 5,8,13,13a-tetrahydro-3,9,10-trimethoxy-13-methyl-, (13S-*trans*)- Isocorybulbine	$C_{21}H_{25}NO_4$ 355.43	22672-74-8 187.5	5045	5-21-06-00173 1.045^{20}	H_2O 1; EtOH 3; chl 3; acid 3
5151	6H-Dibenzo[a,g]quinolizin-3-ol, 5,8,13,13a-tetrahydro-2,9,10-trimethoxy-13-methyl-, (13S-*trans*)- Corybulbine	$C_{21}H_{25}NO_4$ 355.43	518-77-4 237.5	2539	5-21-06-00173	H_2O 1; EtOH 2; eth 2; ace 3
5152	13H-Dibenzo[a,h]fluoren-13-one 1,2:6,7-Dibenzo-9-fluorenone	$C_{21}H_{12}O$ 280.33	4599-94-4 214	 sub 200	3-07-00-02897	
5153	11H-Dibenzo[b,e][1,4]dioxepin-11-one, 2,4,7,9-tetrachloro-3-hydroxy-8-methoxy-1,6-dimethyl- Diploicin	$C_{16}H_{10}Cl_4O_5$ 424.06	527-93-5 232	3345	5-19-06-00364	
5154	6H-Dibenzo[b,d]pyran-1-ol, 6,6,9-trimethyl-3-pentyl- Cannabinol	$C_{21}H_{26}O_2$ 310.44	521-35-7 77	1751 $185^{0.05}$	5-17-04-00567	H_2O 1; EtOH 3; eth 3; ace 3
5155	4H-Dibenzo[de,g]quinoline-10,11-diol, 5,6,6a,7-tetrahydro-6-methyl-, (*R*)- Apomorphine	$C_{17}H_{17}NO_2$ 267.33	58-00-4 195 dec	776	5-21-05-00333	H_2O 2; EtOH 3; eth 3; ace 3
5156	4H-Dibenzo[de,g]quinoline-2,9-diol, 5,6,6a,7-tetrahydro-1,10-dimethoxy-6-methyl-, (*S*)- Boldine	$C_{19}H_{21}NO_4$ 327.38	476-70-0 163	1328	5-21-06-00118	EtOH 4; chl 4
5157	4H-Dibenzo[de,g]quinoline, 5,6,6a,7-tetrahydro-1,2,9,10-tetramethoxy-6-methyl-, (*S*)- *d*-Glaucine	$C_{21}H_{25}NO_4$ 355.43	475-81-0 120	4332	5-21-06-00122	ace 4; EtOH 4; chl 4
5158	4H-Dibenzo[de,g]quinolin-1-ol, 5,6,6a,7-tetrahydro-2,11-dimethoxy-6-methyl-, (*S*)- Isothebaine	$C_{19}H_{21}NO_3$ 311.38	568-21-8 203.5	5116	5-21-05-00658	H_2O 1; EtOH 5; eth 2; chl 5
5159	4H-Dibenzo[de,g]quinolin-10-ol, 5,6,6a,7-tetrahydro-1,2-dimethoxy-, (*R*)- Tsuduranine	$C_{18}H_{19}NO_3$ 297.35	517-97-5 204	9710	5-21-05-00655	ace 4; eth 4; EtOH 4
5160	4H-Dibenzo[de,g]quinolin-11-ol, 5,6,6a,7-tetrahydro-10-methoxy-6-methyl-, (*R*)- Apocodeine	$C_{18}H_{19}NO_2$ 281.35	641-36-1 123.5	771	5-21-05-00334	EtOH 2; eth 3; ace 3; bz 3
5161	4H-Dibenzo[de,g]quinolin-11-ol, 5,6,6a,7-tetrahydro-1,2,10-trimethoxy-6-methyl-, (*S*)- Isocorydine	$C_{20}H_{23}NO_4$ 341.41	475-67-2 185	5046	5-21-06-00132	chl 4
5162	4H-Dibenzo[de,g]quinolin-1-ol, 5,6,6a,7-tetrahydro-2,10,11-trimethoxy-6-methyl-, (*S*)- Corydine	$C_{20}H_{23}NO_4$ 341.41	476-69-7 149	2545	5-21-06-00134	eth 4; EtOH 4; chl 4
5163	Dibenzofuran 2,2'-Biphenylene oxide	$C_{12}H_8O$ 168.19	132-64-9 86.5	 287	5-17-02-00234 1.0886^{99}	H_2O 2; EtOH 3; eth 4; ace 3 1.6079^{99}
5164	Dibenzofuran, 2-bromo- 2-Bromodibenzofuran	$C_{12}H_7BrO$ 247.09	86-76-0 110	 220^{40}	5-17-02-00238	EtOH 3; HOAc 2
5165	Dibenzofuran, 3-bromo- 3-Bromodibenzofuran	$C_{12}H_7BrO$ 247.09	26608-06-0 121	 220^{40}	4-17-00-00588	H_2O 1; EtOH 3; eth 3
5166	Dibenzofuran, 3-nitro- 3-Nitrodibenzofuran	$C_{12}H_7NO_3$ 213.19	5410-97-9 181.5	 $180\text{-}5^3$	5-17-02-00239	H_2O 1; EtOH 2; eth 2; HOAc 3
5167	Dibenzothiophene	$C_{12}H_8S$ 184.26	132-65-0 99.5	 332.5	5-17-02-00239	H_2O 3; EtOH 4; bz 4; chl 3
5168	Diborane(6), mu-(methylthio)- Diborane, methylthio	CH_8B_2S 73.76	91572-15-5 -101.5	 53	4-01-00-01287	ace 4; bz 4

No.	Name Synonym	Mol. Form. Mol. Wt.	CAS RN mp/°C	Merck No. bp/°C	Beil. Ref. den/g cm^{-3}	Solubility n_D
5169	Dicarbonic acid, diethyl ester Pyrocarbonic acid diethyl ester	$C_6H_{10}O_5$ 162.14	1609-47-8	8008 $93\text{-}4^{18}$	4-03-00-00018 1.120^{20}	ace 4; EtOH 4; lig 4 1.3960^{20}
5170	Dichlorvos Phosphoric acid, 2,2-dichloroethenyl dimethyl ester	$C_4H_7Cl_2O_4P$ 220.98	62-73-7	3069 140^{20}; 84^{1}	1.415^{25}	
5171	Diclofop-methyl Propanoic acid, 2-[4-(2,4-dichlorophenoxy)phenoxy]-, methyl ester	$C_{16}H_{14}Cl_2O_4$ 341.19	51338-27-3 40	3072 $176^{0.1}$		
5172	Dicrotophos Phosphoric acid, 3-(dimethylamino)-1-methyl-3-oxo-1-propenyl dimethyl ester, (*E*)-	$C_8H_{16}NO_5P$ 237.19	141-66-2	3077 400; $130^{0.1}$	1.216^{15}	
5173	Diethatyl-ethyl Glycine, *N*-(chloroacetyl)-*N*-(2,6-diethylphenyl)-, ethyl ester	$C_{16}H_{22}ClNO_3$ 311.81	38727-55-8			
5174	Difenoconazole *cis,trans*-3-Chloro-4-[4-methyl-2-(1H-1,2,4-triazol-1-ylmethyl)-1,3-dioxolan-2-yl]phenyl 4-chlorophenyl ether	$C_{19}H_{17}Cl_2N_3O_3$ 406.27	119446-68-3 76	$220^{0.03}$		
5175	Difenzoquat methyl sulfate 1H-Pyrazolium, 1,2-dimethyl-3,5-diphenyl-, methyl sulfate	$C_{17}H_{17}N_2 \cdot CH_3O_4S$ 360.43	43222-48-6 157			
5176	Diflubenzuron Benzamide, *N*-[[(4-chlorophenyl)amino]carbonyl]-2,6-difluoro-	$C_{14}H_9ClF_2N_2O_2$ 310.69	35367-38-5 239	3128		
5177	1,4:5,8-Dimethanonaphthalene, 1,2,3,4,10,10-hexachloro-1,4,4a,5,8,8a-hexahydro-, Aldrin	$C_{12}H_8Cl_6$ 364.91	309-00-2 104	219	3-05-00-01385	H_2O 1; EtOH 3; eth 3; ace 3
5178	2,7:3,6-Dimethanonaphth[2,3-b]oxirene, 3,4,5,6,9,9-hexachloro-1a,2,2a,3,6,6a,7,7a-octahydro-,(1aα,2β,2aα,3β,6β,6aα,7β,7aα)- Dieldrin	$C_{12}H_8Cl_6O$ 380.91	60-57-1 175.5	3093	5-17-02-00087 1.75^{25}	H_2O 1; EtOH 2; ace 3; bz 3
5179	Dimethipin 2,3-Dihydro-5,6-dimethyl-1,4-dithiin, 1,1,4,4-tetraoxide	$C_6H_{10}O_4S_2$ 210.28	55290-64-7 165			
5180	Dimethyl tetrachloroterephthalate 1,4-Benzenedicarboxylic acid, 2,3,5,6-tetrachloro-, dimethyl ester	$C_{10}H_6Cl_4O_4$ 331.97	1861-32-1 155	2830		
5181	Dinitramine 1,3-Benzenediamine, *N3,N3*-diethyl-2,4-dinitro-6-(trifluoromethyl)-	$C_{11}H_{13}F_3N_4O_4$ 322.24	29091-05-2 98			
5182	Dinocap 2-Butenoic acid, 2-(1-methylheptyl)-4,6-dinitrophenyl ester	$C_{18}H_{24}N_2O_6$ 364.40	6119-92-2	$136^{0.01}$		
5183	Dinoseb Phenol, 2-(1-methylpropyl)-4,6-dinitro-	$C_{10}H_{12}N_2O_5$ 240.22	88-85-7 40	3282	1.265^{45}	
5184	2,3-Dioxabicyclo[2.2.2]oct-5-ene, 1-methyl-4-(1-methylethyl)- Ascaridole	$C_{10}H_{16}O_2$ 168.24	512-85-6 3.3	852 115^{15}; $39^{0.2}$	5-19-01-00319 1.0103^{20}	H_2O 1; EtOH 3; ace 3; bz 3 1.4769^{20}
5185	1,7-Dioxadispiro[4.0.4.2]dodeca-3,9-diene-2,8-dione, *trans*- Anemonin	$C_{10}H_8O_4$ 192.17	508-44-1 158	674	5-19-05-00101	chl 4
5186	1,3-Dioxane 1,3-Dioxacyclohexane	$C_4H_8O_2$ 88.11	505-22-6 -45	106.1	5-19-01-00011 1.0342^{20}	H_2O 5; EtOH 5; eth 5; ace 5 1.4165^{20}
5187	1,4-Dioxane 1,4-Dioxacyclohexane	$C_4H_8O_2$ 88.11	123-91-1 11.8	3294 101.5	5-19-01-00016 1.0337^{20}	H_2O 5; EtOH 5; eth 5; ace 5 1.4224^{20}
5188	1,4-Dioxane, 2,3-dichloro-	$C_4H_6Cl_2O_2$ 157.00	95-59-0 30	$80\text{-}2^{10}$	4-19-00-00030 1.468^{20}	H_2O 1; eth 4; ace 4; bz 4 1.4928^{20}
5189	1,3-Dioxane, 2,4-dimethyl-	$C_6H_{12}O_2$ 116.16	766-20-1	116.5	5-19-01-00072 0.9392^{20}	H_2O 2; acid 4 1.4136^{20}
5190	1,3-Dioxane-4,6-dione, 2,2-dimethyl- Meldrum's acid	$C_6H_8O_4$ 144.13	2033-24-1 94	5696	5-19-05-00008	
5191	1,3-Dioxane-2-ethanol, 4-hydroxy-α,6-dimethyl- Paraldol	$C_8H_{11}O_4$ 171.17	19404-07-0 90	90^{15}	4-01-00-03987 1.116^{20}	H_2O 4; eth 4; EtOH 4 1.4610^{20}
5192	1,4-Dioxane, 2,2,3,3,5,5,6-heptachloro- 1,4-Dioxane, heptachloro	$C_4HCl_7O_2$ 329.22	6629-96-5 55	$123\text{-}8^{8}$	4-19-00-00032	H_2O 1; eth 4; ace 4; bz 4
5193	1,3-Dioxane, 4-methyl-	$C_5H_{10}O_2$ 102.13	1120-97-4 -44.5	114	5-19-01-00054 0.9758^{20}	H_2O 2; os 4 1.4159^{20}
5194	1,3-Dioxane, 4-methyl-4-phenyl- *m*-Dioxane, 4-methyl-4-phenyl-	$C_{11}H_{14}O_2$ 178.23	1200-73-3 37.5	256	5-19-01-00481 1.0864^{20}	H_2O 1; chl 2; os 3 1.5240^{20}
5195	1,3-Dioxane, 2-phenyl-	$C_{10}H_{12}O_2$ 164.20	772-01-0 41	253	5-19-01-00459 1.6053^{60}	EtOH 4; eth 4
5196	1,3-Dioxane, 4-phenyl-	$C_{10}H_{12}O_2$ 164.20	772-00-9	247	5-19-01-00464 1.1038^{20}	H_2O 1; os 3 1.5306^{18}
5197	1,3-Dioxan-5-ol, 2-methyl-	$C_5H_{10}O_3$ 118.13	3774-03-6	176; 79^{18}	4-19-00-00624 1.0705^{17}	H_2O 4 1.4375^{17}
5198	1,3,2-Dioxathiane, 4-methyl-, 2-oxide 1,3-Butanediol, sulfite	$C_4H_8O_3S$ 136.17	4426-51-1 5	185	4-01-00-02510 1.2352^{20}	EtOH 4; eth 4; ace 4; bz 4 1.4661^{20}

No.	Name Synonym	Mol. Form. Mol. Wt.	CAS RN mp/°C	Merck No. bp/°C	Beil. Ref. den/g cm^{-3}	Solubility n_D
5199	1,3,2-Dioxathiolane, 2,2-dioxide	$C_2H_4O_4S$ 124.12	1072-53-3 99	 sub	3-01-00-02110	ace 4; bz 4; eth 4; EtOH 4
5200	1,3,2-Dioxathiolane, 2-oxide	$C_2H_4O_3S$ 108.12	3741-38-6 -11	 173	4-01-00-02409 1.4402^{20}	H_2O 4; EtOH 4; eth 4; ace 4 1.4463^{20}
5201	Dioxathion Phosphorodithioic acid, *S,S'*-1,4-dioxane-2,3-diyl *O,O,O',O'*-tetraethyl ester	$C_{12}H_{26}O_6P_2S_4$ 456.55	78-34-2 -20	3296	 1.257^{26}	
5202	1,4-Dioxin	$C_4H_4O_2$ 84.07	290-67-5	 76	5-19-01-00275 1.115^{20}	eth 3; ace 3; bz 3 1.4350^{20}
5203	1,4-Dioxin, 2,3-dihydro-	$C_4H_6O_2$ 86.09	543-75-9	 94.1	5-19-01-00181 1.0836^{20}	ctc 3 1.4372^{20}
5204	1,3-Dioxolane 1,3-Dioxacyclopentane	$C_3H_6O_2$ 74.08	646-06-0 -95	 78	5-19-01-00006 1.060^{20}	H_2O 5; EtOH 3; eth 3; ace 3 1.3974^{20}
5205	1,3-Dioxolane, 2-(3-bromophenyl)-	$C_9H_9BrO_2$ 229.07	17789-14-9	 132.5	5-19-01-00448 1.514^{25}	 1.5627^{20}
5206	1,3-Dioxolane-4-carboxaldehyde, 2,2-dimethyl-	$C_6H_{10}O_3$ 130.14	5736-03-8	 74^{50}	5-19-04-00041	H_2O 4 1.4189^{25}
5207	1,3-Dioxolane-4-methanol, 2,2-dimethyl- Isopropylidene glycerol	$C_6H_{12}O_3$ 132.16	100-79-8	5101 82^{10}	5-19-02-00362 1.064^{20}	 1.4383^{20}
5208	1,3-Dioxolane-4-methanol, 2-methyl-	$C_5H_{10}O_3$ 118.13	3773-93-1	 187	5-19-02-00350 1.1243^{17}	EtOH 4 1.4413^{17}
5209	1,3-Dioxolane-4-methanol, 2-phenyl-	$C_{10}H_{12}O_3$ 180.20	1708-39-0	 $143\text{-}4^{2}$	5-19-02-00562 1.1916^{17}	 1.5389^{17}
5210	1,3-Dioxolane, 2-methyl-	$C_4H_8O_2$ 88.11	497-26-7	 81.5	5-19-01-00042 0.9811^{20}	H_2O 4; EtOH 5; eth 5 1.4035^{17}
5211	1,3-Dioxolan-2-one Ethylene glycol carbonate	$C_3H_4O_3$ 88.06	96-49-1 36.4	 248	5-19-04-00006 1.3214^{39}	H_2O 5; EtOH 5; eth 5; bz 5 1.4148^{50}
5212	1,3-Dioxolan-4-one, 2,5-bis(trichloromethyl)- Chloralide	$C_5H_2Cl_6O_3$ 322.79	554-21-2 116	 272.5	4-19-00-01571	eth 4; HOAc 4
5213	1,3-Dioxolan-2-one, 4-methyl-	$C_4H_6O_3$ 102.09	108-32-7 -48.8	 242	5-19-04-00021 1.2047^{20}	H_2O 4; EtOH 4; eth 4; ace 4 1.4189^{20}
5214	1,3-Dioxolo[4,5-g]isoquinoline, 5,6,7,8-tetrahydro-4-methoxy-6-methyl- Hydrocotarnine	$C_{12}H_{15}NO_3$ 221.26	550-10-7 56	4715	4-27-00-06395	H_2O 1; EtOH 3; eth 3; ace 3
5215	1,3-Dioxolo[4,5-g]isoquinoline, 5,6,7,8-tetrahydro-6-methyl- Hydrohydrastinine	$C_{11}H_{13}NO_2$ 191.23	494-55-3 66	4732 303	4-27-00-06306	ace 4; bz 4; eth 4; EtOH 4
5216	1,3-Dioxolo[4,5-g]isoquinolin-5-ol, 5,6,7,8-tetrahydro-4-methoxy-6-methyl- Cotarnine	$C_{12}H_{15}NO_4$ 237.26	82-54-2 132-3 dec	2551	4-27-00-06448	H_2O 2; EtOH 3; eth 3; bz 3
5217	1,3-Dioxolo[4,5-g]isoquinolin-5-ol, 5,6,7,8-tetrahydro-6-methyl- Hydrastinine	$C_{11}H_{13}NO_3$ 207.23	6592-85-4 116.5	4689	4-27-00-06396	H_2O 3; EtOH 4; eth 4; chl 4
5218	1,3-Dioxolo[4,5-h]quinolin-8(9H)-one, 6-methoxy-9-methyl- Casimiroin	$C_{12}H_{11}NO_4$ 233.22	477-89-4	1894	4-27-00-06618	chl 2
5219	Diphenamid Benzeneacetamide, *N,N*-dimethyl-α-phenyl-	$C_{16}H_{17}NO$ 239.32	957-51-7 135	3305	 $1.17^{23.3}$	
5220	Diphosphoramide, octamethyl- Schradan	$C_8H_{24}N_4O_3P_2$ 286.25	152-16-9 17	8351 $188\text{-}22^{0.3}$	4-04-00-00288 1.1343^{25}	H_2O 4; EtOH 4; chl 4 1.462^{25}
5221	Diphosphoric acid, tetraethyl ester Tetraethyl pyrophosphate	$C_8H_{20}O_7P_2$ 290.19	107-49-3 170 dec	9138 155^{3}	4-01-00-01342 1.1847^{20}	H_2O 5; EtOH 5; eth 5; ace 5 1.4180^{20}
5222	Dipropetryn 1,3,5-Triazine-2,4-diamine, 6-(ethylthio)-*N,N'*-bis(1-methylethyl)-	$C_{11}H_{21}N_5S$ 255.39	4147-51-7 105	3349		
5223	5H,10H-Dipyrrolo[1,2-a:1',2'-d]pyrazine-5,10-dione Pyrocoll	$C_{10}H_6N_2O_2$ 186.17	484-73-1 268	 sub	5-24-08-00263	eth 4; EtOH 4; chl 4
5224	Diquat dibromide Dipyrido[1,2-a:2',1'-c]pyrazinediium, 6,7-dihydro-, dibromide	$C_{12}H_{12}N_2Br$ 344.05	85-00-7 337	3359	 1.24^{20}	
5225	Diselenide, diphenyl Diphenyl diselenide	$C_{12}H_{10}Se_2$ 312.13	1666-13-3 63.5	 202^{11}	4-06-00-01781 1.557^{80}	EtOH 3; eth 3; xyl 3; MeOH 3 1.743^{20}
5226	Disilane, 1,2-dichloro-1,1,2,2-tetramethyl-	$C_4H_{12}Cl_2Si_2$ 187.22	4342-61-4	 $49\text{-}50^{18}$	4-04-00-04280 1.010^{20}	 1.4548^{20}
5227	Disilane, 1,2-difluoro-1,1,2,2-tetramethyl- 1,2-Difluorotetramethyldisilane	$C_4H_{12}F_2Si_2$ 154.31	661-68-7	 94	4-04-00-04279 0.9120^{20}	 1.3837^{20}
5228	Disilane, hexamethyl- Permethyldisilane	$C_6H_{18}Si_2$ 146.38	1450-14-2 13.5	 113.5	4-04-00-04277 0.7247^{22}	H_2O 1; eth 3; ace 3; bz 3 1.4229^{20}
5229	Disiloxane, 1,3-bis(bromomethyl)-1,1,3,3-tetramethyl- 1,3-Bis(bromomethyl)tetramethyldisiloxane	$C_6H_{16}Br_2OSi_2$ 320.17	2351-13-5	 $102\text{-}4^{15}$	4-04-00-04027 1.3918^{25}	 1.4719^{25}
5230	Disiloxane, 1,3-bis(chloromethyl)-1,1,3,3-tetramethyl-	$C_6H_{16}Cl_2OSi_2$ 231.27	2362-10-9 -90	 $204; 92^{21}$	4-04-00-04026 1.0381^{20}	 1.4398^{20}
5231	Disiloxane, 1,3-bis(dichloromethyl)-1,1,3,3-tetramethyl- 1,3-Bis-(dichloromethyl)tetramethyl disiloxane	$C_6H_{14}Cl_4OSi_2$ 300.16	2943-70-6	 $149\text{-}50^{50}$	4-04-00-04031 1.2213^{20}	 1.4660^{20}
5232	Disiloxane, 1,3-dichloro-1,1,3,3-tetramethyl-	$C_4H_{12}Cl_2OSi_2$ 203.22	2401-73-2 -37.5	 138	4-04-00-04119 1.038^{20}	

No.	Name Synonym	Mol. Form. Mol. Wt.	CAS RN mp/°C	Merck No. bp/°C	Beil. Ref. den/g cm^{-3}	Solubility n_D
5233	Disiloxane, 1,3-diethenyl-1,1,3,3-tetramethyl-	$C_8H_{18}OSi_2$ 186.40	2627-95-4 -99.7	 39	4-04-00-04080 0.811^{20}	 1.4123^{20}
5234	Disiloxane, 1,3-dimethoxy-1,1,3,3-tetramethyl- Bis(methoxydimethylsilyl) oxide	$C_6H_{18}O_3Si_2$ 194.38	18187-24-1	 139	4-04-00-04117 0.9048^{20}	 1.3835^{20}
5235	Disiloxane, hexaethyl- 1,1,1,3,3,3-Hexaethyldisiloxane	$C_{12}H_{30}OSi_2$ 246.54	994-49-0	 233; 129^{30}	4-04-00-04055 0.8589^{0}	 1.4340^{20}
5236	Disiloxane, hexamethyl- Hexamethyldisiloxane	$C_6H_{18}OSi_2$ 162.38	107-46-0 -66	 99	4-04-00-04018 0.7638^{20}	H_2O 1 1.3774^{20}
5237	Disiloxane, 1,1,3,3-tetramethyl-	$C_4H_{14}OSi_2$ 134.33	3277-26-7	 72	4-04-00-03991 0.7572^{20}	 1.3700^{20}
5238	Disiloxane, 1,1,3,3-tetramethyl-1,3-diphenyl-	$C_{16}H_{22}OSi_2$ 286.52	56-33-7	 155-8^{13}	4-16-00-01475 0.9763^{20}	ctc 3 1.5176^{20}
5239	Disiloxane, 1,1,1-trimethyl-3,3,3-triphenyl- 1,1,1-Trimethyl-3,3,3-triphenyldisiloxane	$C_{21}H_{24}OSi_2$ 348.59	799-53-1 51	 349	4-16-00-01486 1.0320^{25}	ctc 3; CS_2 3
5240	Dispiro[5.1.5.1]tetradecane Dispiro[5,1,5,1]tetradecane	$C_{14}H_{24}$ 192.34	184-97-4 10	 131^{10}	4-05-00-00491 0.872^{26}	 1.4735^{27}
5241	Distannoxane, hexabutyl- Bis(tributyltin) oxide	$C_{24}H_{54}OSn_2$ 596.11	56-35-9 45	 180^{2}		
5242	Disulfide, bis(1,1-dimethylethyl)	$C_8H_{18}S_2$ 178.36	110-06-5 -5	 88^{21}	4-01-00-01638 0.9226^{20}	 1.4899^{20}
5243	Disulfide, bis(3-methylbutyl)	$C_{10}H_{22}S_2$ 206.42	2051-04-9	 247	4-01-00-01689 0.9192^{20}	 1.4864^{20}
5244	Disulfide, bis(1-methylethyl)	$C_6H_{14}S_2$ 150.31	4253-89-8 -69	 177	4-01-00-01503 0.9435^{20}	 1.4916^{20}
5245	Disulfide, bis(3-methylphenyl)	$C_{14}H_{14}S_2$ 246.40	20333-41-9 -21	 177^{3}	4-06-00-02085	EtOH 3; eth 3; ace 3
5246	Disulfide, bis(4-methylphenyl)	$C_{14}H_{14}S_2$ 246.40	103-19-5 47.5	 210-15^{20}	4-06-00-02206 1.114^{51}	H_2O 1; EtOH 3; eth 4; ace 3
5247	Disulfide, bis(3-nitrophenyl) Nitrophenide	$C_{12}H_8N_2O_4S_2$ 308.34	537-91-7 84	6539	4-06-00-01686	EtOH 2; eth 3; chl 2
5248	Disulfide, bis(phenylmethyl) Dibenzyl disulfide	$C_{14}H_{14}S_2$ 246.40	150-60-7 71.5	2995	4-06-00-02760	H_2O 2; EtOH 3; eth 3; bz 3
5249	Disulfide, bis(trifluoromethyl)	$C_2F_6S_2$ 202.14	372-64-5	 34.6	4-03-00-00278	EtOH 4; peth 4
5250	Disulfide, diacetyl Acetyl disulfide	$C_4H_6O_2S_2$ 150.22	592-22-3 20	 105^{18}	4-02-00-00564	eth 4; EtOH 4
5251	Disulfide, dibenzoyl	$C_{14}H_{10}O_2S_2$ 274.36	644-32-6 134.5	 dec	3-09-00-01977	H_2O 1; EtOH 2; eth 2; CS_2 3
5252	Disulfide, dibutyl	$C_8H_{18}S_2$ 178.36	629-45-8	 117^{20}; 85^{3}	4-01-00-01562 0.9371^{20}	H_2O 1; EtOH 5; eth 5 1.4923^{20}
5253	Disulfide, diethyl Diethyl disulfide	$C_4H_{10}S_2$ 122.26	110-81-6 -101.5	 154.1	4-01-00-01397 0.9931^{20}	H_2O 2; EtOH 5; eth 5 1.5073^{20}
5254	Disulfide, dimethyl Dimethyl disulfide	$C_2H_6S_2$ 94.20	624-92-0 -85	 109.8	4-01-00-01281 1.0625^{20}	H_2O 1; EtOH 5; eth 5 1.5289^{20}
5255	Disulfide, di-1-naphthalenyl	$C_{20}H_{14}S_2$ 318.46	39178-11-5 91		4-06-00-04245 1.144^{20}	H_2O 1; EtOH 2; eth 4; chl 2
5256	Disulfide, di-2-naphthalenyl	$C_{20}H_{14}S_2$ 318.46	5586-15-2 139.5		4-06-00-04320 1.144^{145}	H_2O 1; EtOH 4; eth 4; lig 1 1.4555^{20}
5257	Disulfide, dipentyl	$C_{10}H_{22}S_2$ 206.42	112-51-6	 119^{7}	4-01-00-01654 0.9221^{20}	H_2O 1; eth 3 1.4889^{20}
5258	Disulfide, diphenyl	$C_{12}H_{10}S_2$ 218.34	882-33-7 62	 310	4-06-00-01560 1.353^{20}	H_2O 1; EtOH 3; eth 3; bz 3
5259	Disulfide, dipropyl	$C_6H_{14}S_2$ 150.31	629-19-6 -85.6	 193.5	4-01-00-01454 0.9599^{20}	 1.4981^{20}
5260	Disulfoton Phosphorodithioic acid, *O,O*-diethyl *S*-[2-(ethylthio)ethyl] ester	$C_8H_{19}O_2PS_3$ 274.41	298-04-4 -25	3371 $108^{0.01}$; 128^{1}	 1.144^{20}	
5261	1,2-Dithiane	$C_4H_8S_2$ 120.24	505-20-4 32.5	 80^{14}; 60^{5}	5-19-01-00010	eth 3; bz 3; chl 3 1.5981^{25}
5262	1,3-Dithiane	$C_4H_8S_2$ 120.24	505-23-7 54	 89^{14}	5-19-01-00013	bz 4; eth 4; chl 4 1.5981^{25}
5263	1,4-Dithiane	$C_4H_8S_2$ 120.24	505-29-3 112.3	 199.5	5-19-01-00036	H_2O 2; EtOH 3; eth 3; ctc 3
5264	4H-1,3,5-Dithiazine, dihydro-5-methyl- 1,3,5-Dithiazine, 4,5-dihydro-5-methyl	$C_4H_9NS_2$ 135.25	6302-94-9 65	 185	4-27-00-06285	eth 4; EtOH 4; HOAc 4
5265	4H-1,3,5-Dithiazine, dihydro-2,4,6-trimethyl-, (2α,4α,6α)-	$C_6H_{13}NS_2$ 163.31	638-17-5 46	 dec	2-27-00-00525 1.0613^{50}	eth 4; EtOH 4
5266	1,3-Dithiolane 1,3-Dithiacyclopentane	$C_3H_6S_2$ 106.21	4829-04-3 -50	 175	5-19-01-00008 1.259^{17}	EtOH 3; eth 3; xyl 3 1.5975^{15}
5267	Dithiopyr *S,S'*-Dimethyl-2-difluoromethyl-4-isobutyl-6-trifluoromethylpyridine-3,5-dicarbothioate	$C_{15}H_{16}F_5NO_2S_2$ 401.42	97886-45-8 65			
5268	Docosane *n*-Docosane	$C_{22}H_{46}$ 310.61	629-97-0 44.4	 368.6	4-01-00-00572 0.7944^{20}	H_2O 1; EtOH 3; eth 4; chl 3 1.4455^{20}
5269	Docosane, 5-butyl- 5-*n*-Butyldocosane	$C_{26}H_{54}$ 366.71	55282-16-1 208	 244^{10}	4-01-00-00584 0.8058^{20}	 1.4503^{20}
5270	Docosane, 7-butyl- 7-Butyldocosane	$C_{26}H_{54}$ 366.71	55282-15-0 3.2	 241^{10}	4-01-00-00584 0.8046^{20}	 1.4499^{20}

No.	Name	Synonym	Mol. Form.	Mol. Wt.	CAS RN	mp/°C	Merck No.	bp/°C	Beil. Ref.	den/g cm^{-3}	Solubility	n_D
5271	Docosane, 9-butyl-	9-Butyldocosane	$C_{26}H_{54}$	366.71	55282-14-9	1.3		242.5^{10}	4-01-00-00584	0.8044^{20}		1.4498^{20}
5272	Docosane, 11-butyl-	11-Butyldocosane	$C_{26}H_{54}$	366.71	13475-76-8			242.5^{10}	4-01-00-00584	0.8041^{20}		1.4499^{20}
5273	Docosane, decyl-	Docosane, 11-decyl	$C_{32}H_{66}$	450.88	28261-97-4	1.0		286^{10}	4-01-00-00597	0.8131^{20}		1.4546^{20}
5274	Docosane, 7-hexyl-	7-Hexyldocosane	$C_{28}H_{58}$	394.77	55373-86-9	19.3		258^{10}	4-01-00-00590	0.8080^{20}		1.4517^{20}
5275	Docosane, 9-octyl-	9-Octyldocosane	$C_{30}H_{62}$	422.82	55319-83-0	8.6		272.5^{10}	4-01-00-00594	0.8114^{20}		1.4537^{20}
5276	Docosanoic acid	Behenic acid	$C_{22}H_{44}O_2$	340.59	112-85-6	81	1033	306^{60}	4-02-00-01290	0.8223^{90}	H_2O 2; EtOH 2; eth 2	1.4270^{100}
5277	Docosanoic acid, ethyl ester		$C_{24}H_{48}O_2$	368.64	5908-87-2	50		240^{10}	4-02-00-01292		eth 4; EtOH 4	
5278	Docosanoic acid, methyl ester		$C_{23}H_{46}O_2$	354.62	929-77-1	54			4-02-00-01291		eth 4; EtOH 4	1.4339^{60}
5279	1-Docosanol		$C_{22}H_{46}O$	326.61	661-19-8	72.5		$180^{0.22}$	4-01-00-01906		H_2O 2; EtOH 4; eth 2; chl 3	
5280	1-Docosene		$C_{22}H_{44}$	308.59	1599-67-3	38		367	3-01-00-00883	0.794^{25}		
5281	13-Docosenoic acid		$C_{22}H_{42}O_2$	338.57	1072-39-5	34.7		242^{5}	4-02-00-01676	0.8532^{20}		1.4444^{70}
5282	13-Docosenoic acid, (*E*)-	Brassidic acid	$C_{22}H_{42}O_2$	338.57	506-33-2	61.9	1359	282^{30}; 256^{10}	4-02-00-01677	0.8585^{57}		1.4347^{100}
5283	13-Docosenoic acid, (*Z*)-	Erucic acid	$C_{22}H_{42}O_2$	338.57	112-86-7	33.5	3619	265^{15}	4-02-00-01676	0.860^{55}	H_2O 1; EtOH 3; eth 4; ctc 3	1.4758^{20}
5284	13-Docosen-1-ol, (*Z*)-	Erucyl alcohol	$C_{22}H_{44}O$	324.59	629-98-1	34.7		240^{10}	4-01-00-02209	0.8416^{33}	bz 4; EtOH 4; peth 4	
5285	1,11-Dodecadiene		$C_{12}H_{22}$	166.31	5876-87-9			208.5	4-01-00-01067	0.7702^{20}		1.4400^{20}
5286	3,9-Dodecadiyne		$C_{12}H_{18}$	162.27	61827-89-2							
5287	5,7-Dodecadiyne	Dibutylbutadiyne	$C_{12}H_{18}$	162.27	1120-29-2			103^{8}	4-01-00-01130			
5288	Dodecanal	Lauraldehyde	$C_{12}H_{24}O$	184.32	112-54-9	44.5		185^{100}; $100^{2.5}$	4-01-00-03380	0.8352^{15}	H_2O 1; EtOH 2; eth 3	1.435^{22}
5289	Dodecanamide		$C_{12}H_{25}NO$	199.34	1120-16-7	110		199^{12}	4-02-00-01103		H_2O 1; EtOH 3; eth 2; ace 3	1.4287^{110}
5290	Dodecanamide, *N,N*-diethyl-		$C_{16}H_{33}NO$	255.44	3352-87-2			$166\text{-}7^{2}$	4-04-00-00357	0.847^{25}	chl 3	1.4545^{20}
5291	1-Dodecanamine		$C_{12}H_{27}N$	185.35	124-22-1	28.3		259	4-04-00-00794	0.8015^{20}	H_2O 2; EtOH 5; eth 5; bz 5	1.4421^{20}
5292	Dodecane		$C_{12}H_{26}$	170.34	112-40-3	-9.6		216.3	4-01-00-00498	0.7487^{20}	H_2O 1; EtOH 4; eth 4; ace 4	1.4216^{20}
5293	Dodecane, 1-bromo-	Lauryl bromide	$C_{12}H_{25}Br$	249.23	143-15-7	-9.5	5260	276	4-01-00-00502	1.0399^{20}	H_2O 1; EtOH 3; eth 3; ace 5	1.4583^{20}
5294	Dodecane, 1-bromo-12-fluoro-	1-Bromo-12-fluorododecane	$C_{12}H_{24}BrF$	267.22	353-29-7			$85.8^{0.15}$	4-01-00-00502		ace 4; eth EtOH 4	1.4524^{25}
5295	Dodecane, 1-chloro-		$C_{12}H_{25}Cl$	204.78	112-52-7	-9.3		260	4-01-00-00501	0.8687^{20}	H_2O 1; EtOH 4; ace 5; bz 3	1.4433^{20}
5296	Dodecane, 1,12-dibromo-		$C_{12}H_{24}Br_2$	328.13	3344-70-5	41		215^{15}	4-01-00-00503		H_2O 1; EtOH 4; eth 3; chl 4	
5297	Dodecane, 1,1-dimethoxy-	Lauraldehyde, dimethyl acetal	$C_{14}H_{30}O_2$	230.39	14620-52-1			$132\text{-}4^{5}$	4-01-00-03381		eth 4; EtOH 4	1.4310^{25}
5298	Dodecane, 1,12-dimethoxy-	1,12-Dodecanediol, dimethyl ether	$C_{14}H_{30}O_2$	230.39	73120-52-2	11.5		$155\text{-}6^{15}$	4-01-00-02627	0.8563^{22}		1.436^{22}
5299	Dodecanedioic acid		$C_{12}H_{22}O_4$	230.30	693-23-2	128		222^{25}	4-02-00-02126	1.15^{25}	tfa 3	
5300	Dodecanedioic acid, dimethyl ester		$C_{14}H_{26}O_4$	258.36	1731-79-9	31.9		170^{10}	4-02-00-02126		H_2O 1; AcOEt 3	
5301	1,12-Dodecanediol	1,12-Dihydroxydodecane	$C_{12}H_{26}O_2$	202.34	5675-51-4	81.3		189^{12}	4-01-00-02627		tfa 3	
5302	Dodecane, 1-iodo-		$C_{12}H_{25}I$	296.23	4292-19-7	0.3		298.2	4-01-00-00503	1.1999^{20}	H_2O 1; EtOH 3; eth 5; ace 5	1.4840^{20}
5303	Dodecanenitrile	Lauronitrile	$C_{12}H_{23}N$	181.32	2437-25-4	4		277; 198^{100}	4-02-00-01104	0.8240^{20}	H_2O 1; EtOH 5; eth 5; ace 5	1.4361^{20}
5304	1-Dodecanethiol		$C_{12}H_{26}S$	202.40	112-55-0			$142\text{-}5^{15}$	4-01-00-01851	0.8435^{20}	H_2O 1; EtOH 3; eth 3; chl 3	1.4589^{20}
5305	Dodecanoic acid	Lauric acid	$C_{12}H_{24}O_2$	200.32	143-07-7	43.2	5254	91.4	4-02-00-01082	0.8679^{50}	H_2O 1; EtOH 4; eth 4; ace 3	1.4183^{82}
5306	Dodecanoic acid, anhydride		$C_{24}H_{46}O_3$	382.63	645-66-9	41.8			4-02-00-01101	0.8533^{70}	EtOH 4	1.4292^{70}
5307	Dodecanoic acid, 2-bromo-		$C_{12}H_{23}BrO_2$	279.22	111-56-8	32		$157\text{-}9^{2}$	4-02-00-01106	1.1474^{74}	bz 4; eth 4; EtOH 4; lig 4	1.4585^{24}
5308	Dodecanoic acid, 2-bromoethyl ester		$C_{14}H_{27}BrO_2$	307.27	6309-50-8			176^{12}; 145^{3}	4-02-00-01092	1.088^{25}		1.4547^{25}
5309	Dodecanoic acid, 2-bromo-, methyl ester		$C_{13}H_{25}BrO_2$	293.24	617-60-7			$169\text{-}71^{12}$	4-02-00-01107	1.113^{15}		1.4572^{25}

No.	Name Synonym	Mol. Form. Mol. Wt.	CAS RN mp/°C	Merck No. bp/°C	Beil. Ref. den/g cm^{-3}	Solubility n_D
5310	Dodecanoic acid, 2,3-dihydroxypropyl ester, (±)- Glycerol, 1-laurate (*DL*)	$C_{15}H_{30}O_4$ 274.40	40738-26-9 63	 186^{2}	4-02-00-01096 0.9248^{97}	EtOH 2; eth 4; ace 4; bz 3 1.4350^{86}
5311	Dodecanoic acid, 1,2-ethanediyl ester Ethylene glycol dilaurate	$C_{26}H_{50}O_4$ 426.68	624-04-4 56.6	4393 188^{20}	4-02-00-01094	eth 4; EtOH 4
5312	Dodecanoic acid, ethyl ester Ethyl laurate	$C_{14}H_{28}O_2$ 228.38	106-33-2 -10	3774 271; 154^{15}	4-02-00-01092 0.8618^{20}	H_2O 1; EtOH 4; eth 5; ctc 2 1.4311^{20}
5313	Dodecanoic acid, 2-(2-hydroxyethoxy)ethyl ester Diethylene glycol monolaurate	$C_{16}H_{32}O_4$ 288.43	141-20-8 17.5	3110 >270	 0.96^{25}	EtOH 5; eth 5; ace 5; bz 3
5314	Dodecanoic acid, 2-methyl-	$C_{13}H_{26}O_2$ 214.35	2874-74-0 22	 153^{1}	4-02-00-01121 0.890^{18}	
5315	Dodecanoic acid, methyl ester Methyl laurate	$C_{13}H_{26}O_2$ 214.35	111-82-0 5.2	 267	4-02-00-01090 0.8702^{20}	H_2O 1; EtOH 5; eth 5; ace 5 1.4319^{20}
5316	Dodecanoic acid, 1-methylethyl ester	$C_{15}H_{30}O_2$ 242.40	10233-13-3	 196^{60}; 117^{2}	4-02-00-01092 0.8536^{20}	eth 4; EtOH 4 1.4280^{25}
5317	Dodecanoic acid, phenyl ester Phenyl laurate	$C_{18}H_{28}O_2$ 276.42	4228-00-6 24.5	 210^{15}	4-06-00-00618 0.9354^{30}	ace 4; eth 4; EtOH 4
5318	Dodecanoic acid, phenylmethyl ester	$C_{19}H_{30}O_2$ 290.45	140-25-0 8.5	 $209\text{-}11^{12}$	4-06-00-02267 0.9429^{25}	bz 4; eth 4; EtOH 4; peth 4 1.4812^{24}
5319	Dodecanoic acid, 1,2,3-propanetriyl ester	$C_{39}H_{74}O_6$ 639.01	538-24-9		4-02-00-01098 0.8986^{55}	H_2O 1; EtOH 3; eth 3; ace 4 1.4404^{60}
5320	Dodecanoic acid, propyl ester Propyl laurate	$C_{15}H_{30}O_2$ 242.40	3681-78-5	 205^{60}; 124^{2}	4-02-00-01092 0.8600^{20}	 1.4335^{20}
5321	1-Dodecanol Lauryl alcohol	$C_{12}H_{26}O$ 186.34	112-53-8 24	3402 259	4-01-00-01844 0.8309^{24}	H_2O 1; EtOH 3; eth 3; bz 2
5322	2-Dodecanol	$C_{12}H_{26}O$ 186.34	10203-28-8 19	 252	3-01-00-01793 0.8286^{20}	 1.4400^{20}
5323	3-Dodecanol	$C_{12}H_{26}O$ 186.34	10203-30-2 25	 130^{15}	4-01-00-01854 0.8223^{32}	
5324	6-Dodecanol	$C_{12}H_{26}O$ 186.34	6836-38-0 30	 225; 119^{9}	3-01-00-01794 0.8201^{40}	eth 4; EtOH 4
5325	2-Dodecanone Decyl methyl ketone	$C_{12}H_{24}O$ 184.32	6175-49-1 21	 246.5	4-01-00-03382 0.8198^{20}	H_2O 1; EtOH 3; eth 3; ace 3 1.4330^{20}
5326	6-Dodecanone Amyl hexyl ketone	$C_{12}H_{24}O$ 184.32	6064-27-3 10	 112^{9}	4-01-00-03383	 1.4302^{20}
5327	1-Dodecanone, 1-phenyl-	$C_{18}H_{28}O$ 260.42	1674-38-0 47	 201^{9}; 181^{5}	4-07-00-00847 0.8794^{18}	H_2O 1; ace 3; ctc 2 1.4700^{18}
5328	Dodecanoyl chloride	$C_{12}H_{23}ClO$ 218.77	112-16-3 -17	 145^{18}	4-02-00-01103 0.9169^{25}	eth 4 1.4458^{20}
5329	1,3,6,10-Dodecatetraene, 3,7,11-trimethyl-, (*E,E*)- α-Farnesene	$C_{15}H_{24}$ 204.36	502-61-4	3883 $129\text{-}32^{12}$	3-01-00-01067 0.8410^{20}	H_2O 1; eth 3; ace 3; peth 5 1.4836^{20}
5330	2,6,10-Dodecatrienal, 3,7,11-trimethyl-	$C_{15}H_{24}O$ 220.35	19317-11-4	 $172\text{-}4^{14}$	4-01-00-03603 0.893^{18}	 1.4995
5331	1,6,10-Dodecatriene, 7,11-dimethyl-3-methylene-, (*E*)- β-Farnesene	$C_{15}H_{24}$ 204.36	18794-84-8	3884 $121\text{-}2^{9}$	4-01-00-01133 0.8363^{20}	ace 4; eth 4; chl 4 1.4899^{20}
5332	1,6,10-Dodecatrien-3-ol, 3,7,11-trimethyl-, [*S*-(*Z*)]-	$C_{15}H_{26}O$ 222.37	142-50-7	6388 276; $70^{0.1}$	4-01-00-02336 0.8778^{20}	EtOH 4; eth 3; ace 3; os 3 1.4898^{20}
5333	2,6,10-Dodecatrien-1-ol, 3,7,11-trimethyl-, (*E,E*)-	$C_{15}H_{26}O$ 222.37	106-28-5	 160^{10}	4-01-00-02335 0.8846^{20}	H_2O 1; EtOH 4; eth 3; ace 3 1.4877^{20}
5334	2,6,10-Dodecatrien-1-ol, 3,7,11-trimethyl-, (*Z,E*)-	$C_{15}H_{26}O$ 222.37	3790-71-4	 156^{12}; $120^{0.3}$	4-01-00-02335 0.8908^{20}	ace 4; eth 4; EtOH 4 1.4877^{20}
5335	1-Dodecene	$C_{12}H_{24}$ 168.32	112-41-4 -35.2	 213.8	4-01-00-00914 0.7584^{20}	H_2O 1; EtOH 3; eth 3; ace 3 1.4300^{20}
5336	2-Dodecenedioic acid, (*E*)- Traumatic acid	$C_{12}H_{20}O_4$ 228.29	6402-36-4 165.5	9493	4-02-00-02279	eth 4; EtOH 4; chl 4
5337	2-Dodecenoic acid	$C_{12}H_{22}O_2$ 198.31	4412-16-2 17.1	 155^{3}; $127^{0.15}$	4-02-00-01619 0.9265^{20}	 1.4629^{25}
5338	4-Dodecenoic acid Linderic acid	$C_{12}H_{22}O_2$ 198.31	505-92-0 1.3	 171^{13}	4-02-00-01619 0.9081^{15}	bz 4; eth 4; chl 4 1.4529^{20}
5339	5-Dodecenoic acid	$C_{12}H_{22}O_2$ 198.31	2761-84-4 1.3	 $170\text{-}2^{13}$	4-02-00-01619 0.9081^{20}	
5340	11-Dodecenoic acid	$C_{12}H_{22}O_2$ 198.31	65423-25-8 20	 171^{13}; 144^{3}	4-02-00-01618 0.9014^{20}	 1.4510^{20}
5341	11-Dodecenoic acid, methyl ester Methyl 11-dodecenoate	$C_{13}H_{24}O_2$ 212.33	29972-79-0	 $138\text{-}9^{13}$	4-02-00-01618 0.8789^{22}	 1.4414^{20}
5342	1-Dodecen-3-yne	$C_{12}H_{20}$ 164.29	74744-36-8	 78^{4}	4-01-00-01112 0.7858^{25}	 1.4510^{25}
5343	1-Dodecyne Decylacetylene	$C_{12}H_{22}$ 166.31	765-03-7 -19	 215	4-01-00-01066 0.7788^{20}	 1.4340^{20}
5344	2-Dodecyne	$C_{12}H_{22}$ 166.31	629-49-2 -9	 105^{15}	0-01-00-00261 0.7917^{15}	 1.4828^{20}
5345	3-Dodecyne	$C_{12}H_{22}$ 166.31	6790-27-8	 95^{12}	3-01-00-01025 0.7871^{20}	ace 4; eth 4 1.4442^{20}

No.	Name	Synonym	Mol. Form.	Mol. Wt.	CAS RN	mp/°C	Merck No.	bp/°C	Beil. Ref.	den/g cm^{-3}	Solubility	n_D
5346	6-Dodecyne		$C_{12}H_{22}$	166.31	6975-99-1			$210; 100^{14}$	4-01-00-01067	0.7871^{20}	ace 4; eth 4; EtOH 4	1.4442^{20}
5347	Dodine	Guanidine, dodecyl-, monoacetate	$C_{13}H_{29}N_3 \cdot C_2H_4O_2$	287.45	2439-10-3	136	3406					
5348	Dotriacontane	Bicetyl	$C_{32}H_{66}$	450.88	544-85-4	69.7		467	4-01-00-00595	0.8124^{20}	H_2O 1; EtOH 2; eth 3; bz 4	1.4550^{20}
5349	Eburnamenine-14-carboxylic acid, 14,15-dihydro-14-hydroxy-, methyl ester, (3α,14β,16α)-	Vincamine	$C_{21}H_{26}N_2O_3$	354.45	1617-90-9	231.5	9888		4-25-00-01252			
5350	Eicosane		$C_{20}H_{42}$	282.55	112-95-8	36.8		343	4-01-00-00563	0.7886^{20}	H_2O 1; eth 3; ace 4; bz 3	1.4425^{20}
5351	Eicosane, 1-cyclopentyl-	Cyclopentane, eicosyl	$C_{25}H_{50}$	350.67	22331-38-0	38		413	4-05-00-00198	0.8276^{20}	ace 4; bz 4; eth 4; EtOH 4	1.4595^{20}
5352	Eicosanedioic acid	1,18-Octadecanedicarboxylic acid	$C_{20}H_{38}O_4$	342.52	2424-92-2	125.5		$233\text{-}4^{2}$	4-02-00-02185		eth 3	
5353	Eicosanedioic acid, diethyl ester		$C_{24}H_{46}O_4$	398.63	42235-39-2	54.5		$240^{12}; 230^{2}$	3-02-00-01881		eth 4; EtOH 4	
5354	Eicosane, 9-octyl-	Eicosane, 9-ocytl	$C_{28}H_{58}$	394.77	13475-77-9	0.5		$257^{10}; 199^{0.5}$	4-01-00-00590	0.8075^{20}		1.4515^{20}
5355	Eicosanoic acid	Arachidic acid	$C_{20}H_{40}O_2$	312.54	506-30-9	76.5	791	328 dec; 204^{1}	4-02-00-01275	0.8240^{100}	H_2O 1; EtOH 2; eth 4; bz 3	1.425^{100}
5356	Eicosanoic acid, ethyl ester		$C_{22}H_{44}O_2$	340.59	18281-05-5	50		$295^{100}; 186^{2}$	3-02-00-01067		H_2O 1; EtOH 3; eth 3; bz 3	
5357	Eicosanoic acid, methyl ester		$C_{21}H_{41}O_2$	325.56	1120-28-1	54.5		215^{10}	4-02-00-01276		bz 4; eth 4; EtOH 4; chl 4	1.4317^{60}
5358	1-Eicosanol	Arachic alcohol	$C_{20}H_{42}O$	298.55	629-96-9	66.1		$309; 222^{3}$	4-01-00-01900	0.8405^{20}	H_2O 1; EtOH 2; ace 4; bz 3	1.4350^{20}
5359	2-Eicosanol	1-Methyl-1-nonadecanol	$C_{20}H_{42}O$	298.55	4340-76-5	63.5		357	4-01-00-01902	0.8378^{20}	ace 4; bz 4	1.4912^{80}
5360	5,8,11,14-Eicosatetraenoic acid	5,8,11,14-Eicosotetraenoic acid	$C_{20}H_{32}O_2$	304.47	7771-44-0	-49.5		$169^{0.15}$	4-02-00-01802	0.9219^{20}	H_2O 1; EtOH 3; eth 3; ace 3	1.4824^{20}
5361	5,8,11,14-Eicosatetraenoic acid, (all-Z)-	Arachidonic acid	$C_{20}H_{32}O_2$	304.47	506-32-1	-49.5	792	163^{1}	4-02-00-01802	0.9082^{20}	ace 4; eth 4; EtOH 4; peth 4	1.4824^{20}
5362	1-Eicosene		$C_{20}H_{40}$	280.54	3452-07-1	28.5		$341; 151^{2}$	4-01-00-00934	0.7882^{30}	H_2O 1; bz 3; peth 3	1.4440^{30}
5363	11-Eicosenoic acid	Δ 11-Eicosenoic acid	$C_{20}H_{38}O_2$	310.52	2462-94-4	24		267^{15}	4-02-00-01673	0.8826^{25}	EtOH 4; MeOH 4	
5364	9-Eicosenoic acid		$C_{20}H_{38}O_2$	310.52	506-31-0	23		$220^{6}; 170^{0.1}$	4-02-00-01672	0.8882^{25}		1.4597^{25}
5365	9-Eicosenoic acid, (*Z*)-		$C_{20}H_{38}O_2$	310.52	29204-02-2	24.5		220^{6}	4-02-00-01672	0.8882^{25}		
5366	1-Eicosyne		$C_{20}H_{38}$	278.52	765-27-5	36		340	4-01-00-01077	0.8073^{20}	bz 4; peth 4	1.4501^{20}
5367	Emetan-6'-ol, 7',10,11-trimethoxy-	Cephaeline	$C_{28}H_{38}N_2O_4$	466.62	483-17-0	115.5	1970		5-23-13-00611		ace 4; EtOH 4; MeOH 4; chl 4	
5368	Emetan, 6',7',10,11-tetramethoxy-	Emetine	$C_{29}H_{40}N_2O_4$	480.65	483-18-1	74	3517		4-23-00-03422		H_2O 1; EtOH 3; eth 3; ace 3	
5369	Endosulfan	5,9-Methano-2,4,3-benzodioxathiepin, 6,7,8,9,10,10-hexachloro-1,5,5a,6,9,9a-hexahydro-, 3-oxide	$C_9H_6Cl_6O_3S$	406.93	115-29-7	106	3529	$106^{0.7}$		1.745^{20}		
5370	Endothall, disodium salt	7-Oxabicyclo[2.2.1]heptane-2,3-dicarboxylic acid, disodium salt	$C_8H_{10}Na_2O_5$	232.14	145-73-3	144	3530			1.431^{20}		
5371	4,7-Epoxyisobenzofuran-1,3-dione, hexahydro-3a,7a-dimethyl-, (3aα,4β,7β,7aα)-	Cantharidin	$C_{10}H_{12}O_4$	196.20	56-25-7	218	1755	sub 84	5-19-05-00051		H_2O 1; EtOH 2; eth 2; ace 2	
5372	EPTC	Carbamothioic acid, dipropyl-, *S*-ethyl ester	$C_9H_{19}NOS$	189.32	759-94-4		3580	127^{20}		0.9546^{30}		
5373	Ergoline-8-carboxamide, 9,10-didehydro-*N*-(2-hydroxy-1-methylethyl)-6-methyl-, [8α(*S*)]-	Ergometrinine	$C_{19}H_{23}N_3O_2$	325.41	479-00-5	195-7 dec	3599		4-25-00-00950		chl 4	
5374	Ergoline-8-carboxamide, 9,10-didehydro-6-methyl-, (8β)	Lysergamide	$C_{16}H_{17}N_3O$	267.33	478-94-4	137.5	5505		4-26-00-00007		EtOH 2; ace 2; os 2	
5375	Ergoline-8-carboxylic acid, 9,10-didehydro-6-methyl-, (8α)-	Isolysergic acid	$C_{16}H_{16}N_2O_2$	268.32	478-95-5	218 dec	5065		4-25-00-00935		H_2O 2; EtOH 2; py 3	
5376	Ergoline-8-carboxylic acid, 9,10-didehydro-6-methyl-, (8β)-	Lysergic acid	$C_{16}H_{16}N_2O_2$	268.32	82-58-6	240 dec	5506		4-25-00-00934		H_2O 2; EtOH 3; eth 2; bz 2	
5377	Ergostane, (5α)-		$C_{28}H_{50}$	386.71	511-20-6	85	3602		4-05-00-01234		ace 4; eth 4; chl 4	
5378	Ergostane, (5β)-	Coproergostane	$C_{28}H_{50}$	386.71	511-21-7	64	2516		3-05-00-01143		eth 4; chl 4	

PHYSICAL CONSTANTS OF ORGANIC COMPOUNDS (continued)

No.	Name Synonym	Mol. Form. Mol. Wt.	CAS RN mp/°C	Merck No. bp/°C	Beil. Ref. den/g cm^{-3}	Solubility n_D
5379	Ergostan-3-ol, (3β,5α)- Ergostanol	$C_{28}H_{50}O$ 402.70	6538-02-9 144.5	3603	4-06-00-03602	H_2O 1; eth 3; chl 3
5380	Ergosta-5,7,9(11),22-tetraen-3-ol, (3β,22E)- Dehydroergosterol	$C_{28}H_{42}O$ 394.64	516-85-8 146	2861 $230^{0.5}$	4-06-00-04838	ace 4; bz 4; eth 4; EtOH 4
5381	Ergosta-5,7,22-trien-3-ol, (3β,10α,22E)- Pyrocalciferol	$C_{28}H_{44}O$ 396.66	128-27-8 94	8007	3-06-00-03098	H_2O 1; EtOH 3; chl 3; MeOH 3
5382	Ergosta-5,7,22-trien-3-ol, (3β,9β,10α,22E)- Lumisterol	$C_{28}H_{44}O$ 396.66	474-69-1 118	5471	4-06-00-04408	H_2O 1; EtOH 3; eth 4; ace 4
5383	Ergosta-5,7,22-trien-3-ol, (3β,22E)- Ergosterol	$C_{28}H_{44}O$ 396.66	57-87-4 170	3607 $250^{0.01}$	4-06-00-04407	H_2O 1; EtOH 2; eth 2; bz 3
5384	Ergost-7-en-3-ol, (3β,5α)- γ-Ergostenol	$C_{28}H_{48}O$ 400.69	516-78-9 146	3606	4-06-00-04029	eth 3
5385	Ergost-8(14)-en-3-ol, (3β,5α)- α-Ergostenol	$C_{28}H_{48}O$ 400.69	632-32-6 131	3604	4-06-00-04031	EtOH 2; eth 3; bz 3; chl 3
5386	Ergost-5-en-3-ol, (3β,24*R*)- Campesterol	$C_{28}H_{48}O$ 400.69	474-62-4 157.5	1735	3-06-00-02680	
5387	Ergotaman-3',6',18-trione, 12'-hydroxy-2',5'-bis(1-methylethyl)-, Ergocornine	$C_{31}H_{39}N_5O_5$ 561.68	564-36-3 182-4 dec	3591	4-25-00-00963	H_2O 1; EtOH 3; ace 3; bz 3
5388	Ergotaman-3',6',18-trione, 12'-hydroxy-2',5'-bis(1-methylethyl)-, Ergocorninine	$C_{31}H_{39}N_5O_5$ 561.68	564-37-4 228 dec	3592	4-25-00-00963	ace 4; bz 4; EtOH 4; chl 4
5389	Ergotaman-3',6',18-trione, 12'-hydroxy-2'-(1-methylethyl)-5'-(2-methylpropyl)- Ergocryptine	$C_{32}H_{41}N_5O_5$ 575.71	511-09-1 212-4 dec	3595	4-25-00-00964	H_2O 1; EtOH 3; chl 3
5390	Ergotaman-3',6',18-trione, 12'-hydroxy-2'-(1-methylethyl)-5'-(2-methylpropyl)-, (5'α,8α)- Ergocryptinine	$C_{33}H_{41}N_5O_5$ 587.72	511-10-4 245 dec	3596	4-25-00-00964	ace 4; chl 4
5391	Ergotaman-3',6',18-trione, 12'-hydroxy-2'-(1-methylethyl)-5'-(phenylmethyl)- Ergocristine	$C_{35}H_{39}N_5O_5$ 609.73	511-08-0 175 dec	3593	4-25-00-00966	H_2O 1; EtOH 3; ace 3; chl 3
5392	Ergotaman-3',6',18-trione, 12'-hydroxy-2'-(1-methylethyl)-5'-(phenylmethyl)- Ergocristinine	$C_{35}H_{39}N_5O_5$ 609.73	511-07-9 237-8 dec	3594	4-25-00-00967	H_2O 1; EtOH 2; ace 2; chl 2
5393	Ergotaman-3',6',18-trione, 12'-hydroxy-2'-methyl-5'-(2-methylpropyl)-, Ergosine	$C_{30}H_{37}N_5O_5$ 547.65	561-94-4 228 dec	3601	4-25-00-00961	ace 3; chl 3; MeOH 2
5394	Ergotaman-3',6',18-trione, 12'-hydroxy-2'-methyl-5'-(phenylmethyl)-, (5'α)- Ergotamine	$C_{33}H_{55}N_5O_5$ 601.83	113-15-5 213-4 dec	3609	4-25-00-00964	bz 4; eth 4; chl 4
5395	Ergotaman-3',6',18-trione, 12'-hydroxy-2'-methyl-5'-(phenylmethyl)-, Ergotaminine	$C_{33}H_{35}N_5O_5$ 581.67	639-81-6 252 dec	3610	4-25-00-00966	H_2O 1; EtOH 2; ace 2; bz 2
5396	Erythromycin Propiocine	$C_{37}H_{67}NO_{13}$ 733.94	114-07-8 191	3626	5-18-10-00398	ace 4; eth 4; EtOH 4; chl 4
5397	Estra-1,3,5,7,9-pentaen-17-one, 3-hydroxy- Equilenin	$C_{18}H_{18}O_2$ 266.34	517-09-9 258.5	3581 sub 170	4-08-00-01420	EtOH 2; ace 2; chl 2
5398	Estra-1,3,5(10),7-tetraen-17-one, 3-hydroxy- Equilin	$C_{18}H_{20}O_2$ 268.36	474-86-2 239	3582 sub 170	4-08-00-01366	H_2O 2; EtOH 3; ace 3; diox 3
5399	Estra-1,3,5(10)-triene-3,17-diol (17β)- Estradiol	$C_{18}H_{24}O_2$ 272.39	50-28-2 178.5	3653	4-06-00-06611	ace 4; EtOH 4; Diox 4
5400	Estra-1,3,5(10)-triene-3,17-diol, (17α)- α-Estradiol	$C_{18}H_{24}O_2$ 272.39	57-91-0 221.5	3654	4-06-00-06611	H_2O 1; EtOH 3; eth 2; ace 3
5401	Estra-1,3,5(10)-triene-3,17-diol, (8α,17β)- Isoestradiol	$C_{18}H_{24}O_2$ 272.39	517-04-4 181	5051	4-06-00-06611	EtOH 3; diox 3
5402	Estra-5,7,9-triene-3,17-diol, (3β,17β)- Hexahydroequilenin	$C_{18}H_{24}O_2$ 272.39	517-07-7 168.5	4607 sub 150	4-06-00-06618	
5403	Estra-1,3,5(10)-triene-3,17-diol (17β)-, 3-benzoate Estradiol benzoate	$C_{25}H_{28}O_3$ 376.50	50-50-0 196	3655	4-09-00-00406	
5404	Estra-1,3,5(10)-triene-3,16,17-triol, (16α,17β)- Estriol	$C_{18}H_{24}O_3$ 288.39	50-27-1 288 dec	3659	4-06-00-07550 1.27^{25}	EtOH 3; eth 2; bz 2; tfa 2
5405	Estra-1,3,5(10)-triene-3,16,17-triol, (16β,17β)- 16-Epiestriol	$C_{18}H_{24}O_3$ 288.39	547-81-9 290	3567	4-06-00-07549	
5406	Estra-1,3,5(10)-trien-17-one, 3-hydroxy- Estrone	$C_{18}H_{22}O_2$ 270.37	53-16-7 260.2	3660	3-08-00-01171 1.236^{25}	H_2O 1; EtOH 2; eth 2; ace 3
5407	Estra-1,3,5(10)-trien-17-one, 3-hydroxy-, (8α)- 8-Isoestrone	$C_{18}H_{22}O_2$ 270.37	517-06-6 247	5052	3-08-00-01170	eth 4; Diox 4
5408	Estr-4-en-3-one, 17-(1-oxo-3-phenylpropoxy)-, (17β)- Nandrolone phenpropionate	$C_{27}H_{34}O_3$ 406.57	62-90-8	6283	4-09-00-01758	
5409	Ethalfluralin Benzenamine, *N*-ethyl-*N*-(2-methyl-2-propenyl)-2,6-dinitro-4-(trifluoromethyl)-	$C_{13}H_{14}F_3N_3O_4$ 333.27	55283-68-6 57	3671 256 dec		
5410	Ethanamine Ethylamine	C_2H_7N 45.08	75-04-7 -80.5	3718 16.5	4-04-00-00307 $*0.677^{25}$	H_2O 5; EtOH 5; eth 5 1.3663^{20}
5411	Ethanamine, 2-[(3-butyl-1-isoquinolinyl)oxy]-*N*,*N*-dimethyl- Dimethisoquin	$C_{17}H_{24}N_2O$ 272.39	86-80-6 146	3207 $155\text{-}7^{3}$	5-21-03-00414	H_2O 3; EtOH 3 1.5486^{20}

No.	Name	Synonym	Mol. Form.	Mol. Wt.	CAS RN	mp/°C	Merck No.	bp/°C	Beil. Ref.	den/g cm^{-3}	Solubility	n_D
5412	Ethanamine, 2-chloro-*N*,*N*-bis(2-chloroethyl)-	2,2',2"-Trichlorotriethylamine	$C_6H_{12}Cl_3N$	204.53	555-77-1	-4	9560	143-4^{15}	4-04-00-00447		bz 4; eth 4; EtOH 4	
5413	Ethanamine, 2-chloro-*N*-(2-chloroethyl)-*N*-methyl-	Mechlorethamine	$C_5H_{11}Cl_2N$	156.05	51-75-2	-60	5655	87^{18}; 64^{5}			H_2O 2; ctc 5; DMF 5	
5414	Ethanamine, 2,2-diethoxy-		$C_6H_{15}NO_2$	133.19	645-36-3	-78		163	4-04-00-01918	0.9159^{25}	H_2O 4; eth 4; EtOH 4; chl 4	1.4123^{25}
5415	Ethanamine, 2,2-diethoxy-*N*,*N*-diethyl-		$C_{10}H_{23}NO_2$	189.30	3616-57-7			194.5	4-04-00-01919	0.863^{25}	H_2O 3; EtOH 3; eth 3; ctc 2	1.4189^{20}
5416	Ethanamine, 2,2-diethoxy-*N*,*N*-dimethyl-		$C_8H_{19}NO_2$	161.24	3616-56-6			170.5	3-04-00-00870	0.885^{7}	H_2O 4; EtOH 3; eth 3; ace 3	1.4129^{20}
5417	Ethanamine, *N*,*N*-diethyl-	Triethylamine	$C_6H_{15}N$	101.19	121-44-8	-114.7	9582	89	4-04-00-00322	0.7275^{20}	H_2O 3; EtOH 3; eth 3; ace 4	1.4010^{20}
5418	Ethanamine, *N*,*N*-diethyl-, hydrochloride		$C_6H_{16}ClN$	137.65	554-68-7	260 dec		sub 245	4-04-00-00322	1.0689^{21}	H_2O 4; EtOH 4; chl 4	
5419	Ethanamine, 2,2-dimethoxy-		$C_4H_{11}NO_2$	105.14	22483-09-6			135-9^{95}	4-04-00-01918	0.965^{25}		1.4170^{20}
5420	Ethanamine, 2,2-dimethoxy-*N*-methyl-		$C_5H_{13}NO_2$	119.16	122-07-6			140	2-04-00-00759	0.928^{25}		1.4115^{20}
5421	Ethanamine, *N*,*N*-dimethyl-		$C_4H_{11}N$	73.14	598-56-1	-140		36.5	4-04-00-00312	0.675^{20}		1.3705^{25}
5422	Ethanamine, 2-(diphenylmethoxy)-*N*,*N*-dimethyl-	Diphenhydramine	$C_{17}H_{21}NO$	255.36	58-73-1		3308		4-06-00-04658			
5423	Ethanamine, 2-ethoxy-		$C_4H_{11}NO$	89.14	110-76-9			108	4-04-00-01411	0.8512^{20}	H_2O 5; EtOH 5; eth 5; ace 3	1.4101^{20}
5424	Ethanamine, 2-ethoxy-*N*,*N*-dimethyl-		$C_6H_{15}NO$	117.19	26311-17-1			121	4-04-00-01425	0.806^{20}		1.406^{20}
5425	Ethanamine, 2-ethoxy-*N*-(2-ethoxyethyl)-	Bis-(2-ethoxyethyl) amine	$C_8H_{19}NO_2$	161.24	124-21-0	-50		194	4-04-00-01516	0.883^{25}		1.4213^{20}
5426	Ethanamine, 2-ethoxy-*N*-methyl-	Methyl(2-ethoxyethyl)amine	$C_5H_{13}NO$	103.16	38256-94-9			115	3-04-00-00647	0.8363^{20}	ace 4; bz 4; eth 4; EtOH 4	1.4147^{20}
5427	Ethanamine, *N*-ethyl-	Diethylamine	$C_4H_{11}N$	73.14	109-89-7	-49.8	3100	55.5	4-04-00-00313	0.7056^{20}	H_2O 4; EtOH 5; eth 3; ctc 3	1.3864^{20}
5428	Ethanamine, *N*-ethyl-, hydrochloride		$C_4H_{12}ClN$	109.60	660-68-4	228.5			4-04-00-00313	1.0477^{22}	H_2O 4; EtOH 4	
5429	Ethanamine, *N*-ethyl-*N*-hydroxy-		$C_4H_{11}NO$	89.14	3710-84-7	10		133	4-04-00-03304	0.8669^{20}		1.4195^{20}
5430	Ethanamine, *N*-ethyl-*N*-methyl-		$C_5H_{13}N$	87.16	616-39-7	-196		66	4-04-00-00321	0.703^{25}	H_2O 4; EtOH 4; eth 4	1.3879^{25}
5431	Ethanamine, *N*-ethyl-*N*-nitro-		$C_4H_{10}N_2O_2$	118.14	7119-92-8			206.5	4-04-00-03403	1.057^{15}	eth 4; EtOH 4	
5432	Ethanamine, *N*-ethyl-*N*-nitroso-	*N*-Nitrosodiethylamine	$C_4H_{10}N_2O$	102.14	55-18-5		6557	176.9	4-04-00-03386	0.9422^{20}	H_2O 3; EtOH 3; eth 3; chl 2	1.4386^{20}
5433	Ethanamine, hydrochloride		C_2H_8ClN	81.54	557-66-4	109.5	3718		4-04-00-00307	1.2160^{20}	H_2O 4; EtOH 4	
5434	Ethanamine, *N*-hydroxy-	*N*-Ethylhydroxylamine	C_2H_7NO	61.08	624-81-7	59-60 dec			4-04-00-03304	0.9079^{20}	H_2O 4; EtOH 4	1.4152^{66}
5435	Ethanamine, *N*-methyl-, hydrochloride	Ethyl methyl amine, hydrochloride	$C_3H_{10}ClN$	95.57	624-60-2	128			2-04-00-00589	1.0874^{20}	H_2O 4; EtOH 4; eth 1; chl 3	
5436	Ethanamine, 1,1,2,2,2-pentafluoro-*N*,*N*-bis(pentafluoroethyl)-		$C_6F_{15}N$	371.05	359-70-6			70.3	4-02-00-00471	1.736^{20}		1.262^{25}
5437	Ethanamine, *N*-(phenylmethylene)-		$C_9H_{11}N$	133.19	6852-54-6			196	4-07-00-00518	0.937^{20}	H_2O 1; EtOH 3; eth 3	1.5378^{15}
5438	Ethanaminium, 2-(acetyloxy)-*N*,*N*,*N*-trimethyl-, bromide	Acetyl choline bromide	$C_7H_{16}BrNO_2$	226.11	66-23-9	146	80		4-04-00-01446		H_2O 4	
5439	Ethanaminium, 2-(acetyloxy)-*N*,*N*,*N*-trimethyl-, chloride	Acetyl choline chloride	$C_7H_{16}ClNO_2$	181.66	60-31-1	150	81		4-04-00-01446		H_2O 3; EtOH 3; eth 1	
5440	Ethanaminium, 2-[(aminocarbonyl)oxy]-*N*,*N*,*N*-trimethyl-, chloride	Carbamoylcholine chloride	$C_6H_{15}ClN_2O_2$	182.65	51-83-2	210 dec	1780		4-04-00-01455		H_2O 4; EtOH 2; eth 1; chl 1	
5441	Ethanaminium, 2,2',2"-[1,2,3-benzenetriyltris(oxy)]tris[*N*,*N*,*N*-triethyl-, tr	Gallamine triethiodide	$C_{30}H_{60}I_3N_3O_3$	891.54	65-29-2	147.5	4249		4-06-00-07333		H_2O 4; EtOH 4; eth 2; ace 2	
5442	Ethanaminium, 2-chloro-*N*,*N*,*N*-trimethyl-, chloride		$C_6H_{10}Cl_2N$	158.07	999-81-5	239 dec	2100		4-04-00-00445			
5443	Ethanaminium, 2-[(cyclohexylhydroxyphenylacetyl)oxy]-*N*,*N*-diethyl-*N*-methyl-, bromide	Oxyphenonium bromide	$C_{21}H_{34}BrNO_3$	428.41	50-10-2	191.5	6928		3-10-00-00911		H_2O 4; EtOH 2	
5444	Ethanaminium, *N*-[4-[[4-(diethylamino)phenyl]phenylmethylene]-2,5-cyclohexadien-1-ylidene]-*N*-ethyl-, sulfate (1	Brilliant green	$C_{27}H_{34}N_2O_4S$	482.64	633-03-4		1367		4-13-00-02281		H_2O 4; EtOH 4	
5445	Ethanaminium, *N*,*N*-diethyl-*N*-methyl-2-[(9H-xanthen-9-ylcarbonyl)oxy]-, bromide	Methantheline bromide	$C_{21}H_{26}BrNO_3$	420.35	53-46-3	174.5	5869		5-18-06-00590		H_2O 3; EtOH 3; eth 1; chl 3	

No.	Name Synonym	Mol. Form. Mol. Wt.	CAS RN mp/°C	Merck No. bp/°C	Beil. Ref. den/g cm^{-3}	Solubility n_D
5446	Ethanaminium, 2-hydroxy-*N,N,N*-trimethyl-, chloride Choline chloride	$C_5H_{14}ClNO$ 139.63	67-48-1 305-5 dec	2208	4-04-00-01443	H_2O 4; EtOH 4
5447	Ethanaminium, *N,N,N*-triethyl-, bromide Tetraethylammonium bromide	$C_8H_{20}BrN$ 210.16	71-91-0 285-7 dec	9133	4-04-00-00331 1.3970^{20}	H_2O 4; EtOH 4; chl 4; MeOH 4
5448	Ethane	C_2H_6 30.07	74-84-0 -182.8	3676 -88.6	4-01-00-00108 0.5446^{-89}	bz 4
5449	Ethane, 1,2-bis(2-chloroethoxy)-	$C_6H_{12}Cl_2O_2$ 187.07	112-26-5	 232	4-01-00-02379 1.195^{20}	ctc 3 1.4592^{25}
5450	Ethane, 1,1-bis(ethylthio)-	$C_6H_{14}S_2$ 150.31	14252-42-7	 184	4-01-00-03162 0.9706^{20}	eth 3; ace 3; chl 3 1.5025^{20}
5451	Ethane, 1,2-bis(ethylthio)-	$C_6H_{14}S_2$ 150.31	5395-75-5 180	 217; 108^{27}	4-01-00-02452 0.9815^{20}	EtOH 3; eth 3 1.5118^{20}
5452	Ethane, 1,2-bis(methylthio)-	$C_4H_{10}S_2$ 122.26	6628-18-8	 183; 79^{11}	4-01-00-02451 1.0371^{20}	H_2O 4; ace 4; eth 4; EtOH 4 1.5292^{20}
5453	Ethanebis(thioic) acid, *S,S*-diethyl ester	$C_6H_{10}O_2S_2$ 178.28	615-85-0 27	 235	1-02-00-00244 1.0565^{21}	eth 4
5454	Ethane, bromo- Ethyl bromide	C_2H_5Br 108.97	74-96-4 -118.6	3730 38.5	4-01-00-00150 1.4604^{20}	H_2O 2; EtOH 5; eth 5; chl 5 1.4239^{20}
5455	Ethane, 1-bromo-1-chloro- 1-Bromo-1-chloroethane	C_2H_4BrCl 143.41	593-96-4	 83	4-01-00-00155 1.667^{10}	 1.4660^{2-}
5456	Ethane, 1-bromo-2-chloro-	C_2H_4BrCl 143.41	107-04-0 -16.7	 107	4-01-00-00155 1.7392^{20}	H_2O 2; EtOH 3; eth 3; chl 3 1.4908^{20}
5457	Ethane, 1-bromo-2-chloro-1,1,2-trifluoro- 1-Bromo-2-chloro-1,1,2-trifluoroethane	$C_2HBrClF_3$ 197.38	354-06-3	 52.5	4-01-00-00156 1.8636^{25}	 1.3738^{20}
5458	Ethane, 2-bromo-2-chloro-1,1,1-trifluoro- Halothane	$C_2HBrClF_3$ 197.38	151-67-7	4517 50.2; 20^{243}	 1.871^{20}	H_2O 2; peth 3 1.3697^{0}
5459	Ethane, 2-bromo-1,1-diethoxy-	$C_6H_{13}BrO_2$ 197.07	2032-35-1	 $66\text{-}7^{18}$	4-01-00-03151 1.310^{25}	EtOH 3; eth 3 1.4387^{20}
5460	Ethane, 2-bromo-1,1-dimethoxy-	$C_4H_9BrO_2$ 169.02	7252-83-7	 149	3-01-00-02672 1.430^{20}	eth 3; ace 3; chl 3 1.4450^{20}
5461	Ethane, 1-bromo-2-ethoxy- 2-Bromoethyl ethyl ether	C_4H_9BrO 153.02	592-55-2	 127.5	4-01-00-01386 1.3852^{0}	H_2O 2; EtOH 5; eth 5 1.4447^{20}
5462	Ethane, 1-bromo-2-fluoro-	C_2H_4BrF 126.96	762-49-2	 71.5	4-01-00-00154 1.7044^{25}	eth 4; EtOH 4 1.4236^{20}
5463	Ethane, 1-bromo-2-iodo- 1-Bromo-2-iodoethane	C_2H_4BrI 234.86	590-16-9 28	 163	4-01-00-00168 2.516^{29}	
5464	Ethane, (bromomethoxy)-	C_3H_7BrO 138.99	53588-92-4	 110	3-01-00-02594 1.4402^{20}	eth 4 1.4515^{20}
5465	Ethane, 1-bromo-2-methoxy-	C_3H_7BrO 138.99	6482-24-2	 110 dec	4-01-00-01386 1.4623^{20}	 1.44753^{20}
5466	Ethane, 2-bromo-1,1,1-trifluoro-	$C_2H_2BrF_3$ 162.94	421-06-7 -93.9	 26	4-01-00-00154 1.7881^{20}	 1.3331^{20}
5467	Ethane, chloro- Ethyl chloride	C_2H_5Cl 64.51	75-00-3 -138.7	3740 12.3	4-01-00-00124 *0.8902^{25}	H_2O 2; EtOH 4; eth 5; chl 2 1.3676^{20}
5468	Ethane, 1-chloro-1-(2-chloroethoxy)- 1-Chloroethyl 2-chloroethyl ether	$C_4H_8Cl_2O$ 143.01	1462-34-6	 56^{17}	3-01-00-02655 1.1867^{20}	eth 4; EtOH 4; chl 4 1.4473^{20}
5469	Ethane, 2-chloro-1,1-diethoxy-	$C_6H_{13}ClO_2$ 152.62	621-62-5	 157.4	4-01-00-03134 1.0180^{20}	H_2O 2; EtOH 5; eth 5; ctc 2 1.4170^{20}
5470	Ethane, 1-chloro-1,1-difluoro 1-Chloro-1,1-difluoroethane	$C_2H_3ClF_2$ 100.50	75-68-3 -130.8	 -9.7	 1.107^{25}	H_2O 1; bz 3
5471	Ethane, 2-chloro-1,1-difluoro 1-Chloro-2,2-difluoroethane	$C_2H_3ClF_2$ 100.50	338-65-8	 35.1		
5472	Ethane, 2-chloro-1,1-dimethoxy-	$C_4H_9ClO_2$ 124.57	97-97-2	 127.5	4-01-00-03134 1.068^{20}	EtOH 2; eth 2; bz 2; ctc 2 1.4150^{20}
5473	Ethane, 1-chloro-1-ethoxy-	C_4H_9ClO 108.57	7081-78-9	 93.5	4-01-00-03119 0.9655^{20}	 1.4053^{20}
5474	Ethane, 1-chloro-2-ethoxy-	C_4H_9ClO 108.57	628-34-2	 107.5	4-01-00-01375 0.9895^{20}	H_2O 2; eth 5; chl 3 1.4113^{20}
5475	Ethane, 1-chloro-1-fluoro- 1-Chloro-1-fluoroethane	C_2H_4ClF 82.50	1615-75-4	 16.2	4-01-00-00127	
5476	Ethane, 1-chloro-2-fluoro-	C_2H_4ClF 82.50	762-50-5	 59	4-01-00-00127 1.1747^{20}	eth 4; EtOH 4 1.3775^{20}
5477	Ethane, 1-chloro-2-iodo-	C_2H_4ClI 190.41	624-70-4 -15.6	 140	4-01-00-00167 2.1644^{25}	
5478	Ethane, (chloromethoxy)- Chloromethyl ethyl ether	C_3H_7ClO 94.54	3188-13-4	 83	4-01-00-03047 1.0188^{15}	 1.4040^{20}
5479	Ethane, 1-chloro-1-methoxy-	C_3H_7ClO 94.54	1538-87-0	 73	4-01-00-03119 0.9902^{20}	eth 4 1.4004^{20}
5480	Ethane, 1-chloro-2-methoxy-	C_3H_7ClO 94.54	627-42-9	 92.5	4-01-00-01375 1.0345^{20}	H_2O 4; eth 4 1.4111^{20}
5481	Ethane, 1-chloro-2-(methylthio)-	C_3H_7ClS 110.61	542-81-4	 60^{30}; 44^{20}	4-01-00-01406 1.1225^{20}	EtOH 3; eth 3; ace 3 1.4902^{20}
5482	Ethane, 1-chloro-1-nitro- 1-Chloro-1-nitroethane	$C_2H_4ClNO_2$ 109.51	598-92-5	 124.5	4-01-00-00172 1.2837^{20}	H_2O 1; EtOH 3; ctc 3; alk 3 1.4224^{20}
5483	Ethane, 1-chloro-2-nitro- 2-Chloro-1-nitroethane	$C_2H_4ClNO_2$ 109.51	625-47-8	 173.5	4-01-00-00173 1.3550^{20}	 1.407^{7}
5484	Ethane, chloropentafluoro- Chloropentafluoroethane	C_2ClF_5 154.47	76-15-3 -99.4	 -37.9	4-01-00-00129 1.5678^{-42}	H_2O 1; EtOH 3; eth 3 1.2678^{-42}

No.	Name Synonym	Mol. Form. Mol. Wt.	CAS RN mp/°C	Merck No. bp/°C	Beil. Ref. den/g cm^{-3}	Solubility n_D
5485	Ethane, 1-chloro-1,1,2,2-tetrafluoro- 1-Chloro-1,1,2,2-tetrafluoroethane	C_2HClF_4 136.48	354-25-6 -117	-10	4-01-00-00129	
5486	Ethane, 1-chloro-1,2,2,2-tetrafluoro- 1-Chloro-1,2,2,2-tetrafluoroethane	C_2HClF_4 136.48	2837-89-0	-12		
5487	Ethane, 1-chloro-1,1,2-trifluoro 1-Chloro-1,1,2-trifluoroethane	$C_2H_2ClF_3$ 118.49	421-04-5	12		
5488	Ethane, 1-chloro-1,2,2-trifluoro 1-Chloro-1,2,2-trifluoroethane	$C_2H_2ClF_3$ 118.49	431-07-2	17		
5489	Ethane, 2-chloro-1,1,1-trifluoro- 2-Chloro-1,1,1-trifluoroethane	$C_2H_2ClF_3$ 118.49	75-88-7 -105.5	6.1	4-01-00-00128 1.389^{0}	1.3090^{0}
5490	Ethane, 1-chloro-1,1,2-trifluoro-2-iodo- Ethane, 1-chloro-2-iodo-1,1,2-trifluoro	C_2HClF_3I 244.38	354-26-7	77^{630}	4-01-00-00167 2.181^{20}	1.4320^{20}
5491	Ethanedial Glyoxal	$C_2H_2O_2$ 58.04	107-22-2 15	4405 50.4	4-01-00-03625 1.14^{20}	H_2O 4; EtOH 3; eth 3 1.3826^{20}
5492	Ethanedial, dioxime	$C_2H_4N_2O_2$ 88.07	557-30-2 178 dec	sub	3-01-00-03084	H_2O 4; EtOH 4; eth 4
5493	Ethanediamide Oxamide	$C_2H_4N_2O_2$ 88.07	471-46-5 350 dec	6871	4-02-00-01860 1.667^{20}	H_2O 2; EtOH 2; eth 1
5494	Ethanediamide, *N,N'*-diethyl-	$C_6H_{12}N_2O_2$ 144.17	615-84-9 175		4-04-00-00359 1.169^{4}	EtOH 4
5495	Ethanediamide, *N,N'*-diphenyl-	$C_{14}H_{12}N_2O_2$ 240.26	620-81-5 254	>360	4-12-00-00463	bz 4
5496	1,2-Ethanediamine Ethylenediamine	$C_2H_8N_2$ 60.10	107-15-3 11.1	3752 117	4-04-00-01166 0.8979^{20}	H_2O 4; EtOH 5; eth 1; bz 1 1.4565^{20}
5497	1,2-Ethanediamine, *N*-(2-aminoethyl)- Diethylenetriamine	$C_4H_{13}N_3$ 103.17	111-40-0 -39	207	4-04-00-01238 0.9569^{20}	H_2O 5; EtOH 5; eth 1; lig 3 1.4810^{25}
5498	1,2-Ethanediamine, *N*-(2-aminoethyl)-*N'*-[2-[(2-aminoethyl)amino]ethyl]- Tetraethylenepentamine	$C_8H_{23}N_5$ 189.30	112-57-2	341.5	4-04-00-01244	H_2O 3 1.5042^{20}
5499	1,2-Ethanediamine, *N,N'*-bis(2-aminoethyl)- Trientine	$C_6H_{18}N_4$ 146.24	112-24-3 12	9579 266.5	4-04-00-01242	H_2O 3; EtOH 3; acid 3 1.4971^{20}
5500	1,2-Ethanediamine, *N,N'*-bis(phenylmethyl)- Benzathine	$C_{16}H_{20}N_2$ 240.35	140-28-3 26	1072 195^{4}	4-12-00-02321 1.024^{20}	bz 4; eth 4; EtOH 4 1.5635^{20}
5501	1,2-Ethanediamine, *N,N'*-diethyl-	$C_6H_{16}N_2$ 116.21	111-74-0	146	4-04-00-01174 0.8280^{20}	H_2O 4; eth 4; EtOH 4; tol 4 1.4340^{20}
5502	1,2-Ethanediamine, *N,N*-diethyl- *N,N*-Diethylethylenediamine	$C_6H_{16}N_2$ 116.21	100-36-7	144	4-04-00-01175 0.8280^{20}	H_2O 5; EtOH 3; eth 3; ctc 3 1.4340^{20}
5503	1,2-Ethanediamine, *N,N'*-dimethyl-	$C_4H_{12}N_2$ 88.15	110-70-3	120	4-04-00-01171 0.828^{15}	EtOH 3; eth 3; dil HCl 3
5504	1,2-Ethanediamine, *N,N*-dimethyl-*N'*-(phenylmethyl)-	$C_{11}H_{18}N_2$ 178.28	103-55-9	$122\text{-}4^{11}$	4-12-00-02320 0.922^{25}	1.5089^{20}
5505	1,2-Ethanediamine, *N,N*-dimethyl-*N'*-(phenylmethyl)-*N'*-2-pyridinyl- Tripelennamine	$C_{16}H_{21}N_3$ 255.36	91-81-6	9651 $167\text{-}72^{0.1}$	5-22-08-00378	
5506	1,2-Ethanediamine, *N,N*-dimethyl-*N'*-phenyl-*N'*-(phenylmethyl)- Phenbenzamine	$C_{17}H_{22}N_2$ 254.38	961-71-7	7176 $179\text{-}80^{7}$	4-12-00-02324 1.016^{25}	1.5794^{25}
5507	1,2-Ethanediamine, *N,N*-dimethyl-*N'*-2-pyridinyl-*N'*-(2-thienylmethyl)- Methapyrilene	$C_{14}H_{19}N_3S$ 261.39	91-80-5	5871 $173\text{-}5^{3}$	5-22-08-00401	1.5915^{20}
5508	1,2-Ethanediamine, *N,N'*-diphenyl- 1,2-Dianilinoethane	$C_{14}H_{16}N_2$ 212.29	150-61-8 74	2969 229^{12}; 178^{2}	4-12-00-00986	H_2O 1; EtOH 3; eth 3; tfa 2
5509	1,2-Ethanediamine, *N*-ethyl-	$C_4H_{12}N_2$ 88.15	110-72-5	129	4-04-00-01174 0.837^{25}	1.4385^{20}
5510	1,2-Ethanediamine, *N'*-ethyl-*N,N*-dimethyl- Ethylenediamine, *N,N*-dimethyl-*N'*-ethyl	$C_6H_{15}N_2$ 115.20	123-83-1	134.5	4-04-00-01174 0.738^{25}	1.4222^{20}
5511	1,2-Ethanediamine, *N*-[(4-methoxyphenyl)methyl]-*N',N'*-dimethyl-*N*-2-pyridinyl- Pyrilamine	$C_{17}H_{23}N_3O$ 285.39	91-84-9	7996 201^{5}	5-22-08-00381	
5512	1,2-Ethanediamine, *N*-methyl-	$C_3H_{10}N_2$ 74.13	109-81-9	115	4-04-00-01170 0.841^{25}	1.4395^{20}
5513	1,2-Ethanediamine, monohydrate	$C_2H_{10}N_2O$ 78.11	6780-13-8 10	118	4-04-00-01166 0.964^{20}	H_2O 4 1.4500^{20}
5514	1,2-Ethanediamine, *N*-1-naphthalenyl- *N*-(1-Naphthyl)ethylenediamine	$C_{12}H_{14}N_2$ 186.26	551-09-7	6329 204^{9}	4-12-00-03101 1.114^{25}	1.6648^{25}
5515	1,2-Ethanediamine, *N,N,N',N'*-tetraethyl-	$C_{10}H_{24}N_2$ 172.31	150-77-6	192	4-04-00-01177 0.808^{25}	1.4343^{20}
5516	1,2-Ethanediamine, *N,N,N',N'*-tetramethyl- *N,N,N',N'*-Tetramethyldiaminoethane	$C_6H_{16}N_2$ 116.21	110-18-9 -55	121	4-04-00-01172 0.77^{25}	1.4179^{20}
5517	Ethane, 1,1-dibromo- Ethylidene dibromide	$C_2H_4Br_2$ 187.86	557-91-5 -63	109	4-01-00-00157 2.0554^{20}	H_2O 1; EtOH 3; eth 4; ace 3 1.5128^{20}
5518	Ethane, 1,2-dibromo- Ethylene dibromide	$C_2H_4Br_2$ 187.86	106-93-4 9.9	3753 131.6	4-01-00-00158 2.1791^{20}	ace 4; bz 4; eth 4; EtOH 4 1.5387^{20}
5519	Ethane, 1,2-dibromo-1-chloro-1,2,2-trifluoro- 1,2-Dibromo-1-chloro-1,2,2-trifluoroethane	$C_2Br_2ClF_3$ 276.28	354-51-8 50	93		
5520	Ethane, 1,2-dibromo-1,1-dichloro-	$C_2H_2Br_2Cl_2$ 256.75	75-81-0 -26	195	4-01-00-00161 2.135^{20}	ace 4; bz 4; eth 4; EtOH 4 1.5662^{20}
5521	Ethane, 1,2-dibromo-1,2-dichloro-	$C_2H_2Br_2Cl_2$ 256.75	683-68-1 -26	195	4-01-00-00161 2.135^{20}	H_2O 1; EtOH 3; eth 3; ace 3 1.5662^{20}

No.	Name	Synonym	Mol. Form.	Mol. Wt.	CAS RN	mp/°C	Merck No.	bp/°C	Beil. Ref.	den/g cm^{-3}	Solubility	n_D
5522	Ethane, 1,2-dibromo-1,1-difluoro-	Genetron 132b-B2	$C_2H_2Br_2F_2$	223.84	75-82-1	-61.3		92.5	4-01-00-00160	2.2238^{20}		1.4456^{20}
5523	Ethane, 1,2-dibromo-1-ethoxy-		$C_4H_8Br_2O$	231.91	2983-26-8			80^{20}	4-01-00-03152	1.7320^{20}	EtOH 4; chl 4	1.5044^{20}
5524	Ethane, 1,2-dibromo-1,1,2,2-tetrachloro-		$C_2Br_2Cl_4$	325.64	630-25-1	202.5			3-01-00-00190	2.713^{25}		
5525	Ethane, 1,2-dibromo-1,1,2,2-tetrafluoro-	1,2-Dibromotetrafluoroethane	$C_2Br_2F_4$	259.82	124-73-2	-110.4		47.3	4-01-00-00160	2.149^{25}		1.361^{25}
5526	Ethane, 1,2-dibromo-1,1,2-trifluoro-	Halon 2302	$C_2HBr_2F_3$	241.83	354-04-1			76	4-01-00-00160	2.274^{27}		1.4191^{24}
5527	Ethane, 1,2-dibutoxy	Ethylene glycol dibutyl ether	$C_{10}H_{22}O_2$	174.28	112-48-1	-69.1		203.3		0.8319^{25}		
5528	Ethane, 1,1-dichloro-	Ethylidene dichloride	$C_2H_4Cl_2$	98.96	75-34-3	-96.9	3766	57.4	4-01-00-00130	1.1757^{20}	H_2O 2; EtOH 4; eth 4; ace 3	1.4164^{20}
5529	Ethane, 1,2-dichloro-	Ethylene dichloride	$C_2H_4Cl_2$	98.96	107-06-2	-35.5	3754	83.5	4-01-00-00131	1.2454^{25}	H_2O 2; EtOH 4; eth 5; ace 3	1.4448^{20}
5530	Ethane, 1,1-dichloro-2,2-diethoxy-		$C_6H_{12}Cl_2O_2$	187.07	619-33-0			183.5	4-01-00-03140	1.1383^{14}	EtOH 3; eth 3; ctc 2	
5531	Ethane, 1,2-dichloro-1,1-difluoro-		$C_2H_2Cl_2F_2$	134.94	1649-08-7	-101.2		46.8	4-01-00-00135	1.4163^{20}		1.36193^{20}
5532	Ethane, 1,2-dichloro-1,2-difluoro-		$C_2H_2Cl_2F_2$	134.94	431-06-1	-101.2		46.8	4-01-00-00135	1.4163^{20}		1.3619^{20}
5533	Ethane, 1,2-dichloro-1-ethoxy-		$C_4H_8Cl_2O$	143.01	623-46-1			145	4-01-00-03136	1.1370^{20}	chl 2	1.4435^{20}
5534	Ethane, 1,1-dichloro-1-fluoro-	1,1-Dichloro-1-fluoroethane	$C_2H_3Cl_2F$	116.95	1717-00-6	-103.5		32	4-01-00-00134	1.250^{10}		1.3600^{10}
5535	Ethane, 1,2-dichloro-1-fluoro-		$C_2H_3Cl_2F$	116.95	430-57-9	-60		73.7	4-01-00-00134	1.3814^{20}		1.4132^{20}
5536	Ethane, 1,1-dichloro-1-nitro-	Ethide	$C_2H_3Cl_2NO_2$	143.96	594-72-9			123.5	4-01-00-00173		ctc 3	
5537	Ethane, 1,1-dichloro-1,2,2,2-tetrafluoro-	1,1-Dichlorotetrafluoroethane	$C_2Cl_2F_4$	170.92	374-07-2	-56.6		4	4-01-00-00136	*1.455^{25}	bz 4; eth 4; EtOH 4	1.3092^{0}
5538	Ethane, 1,2-dichloro-1,1,2,2-tetrafluoro-	Refrigerant 114	$C_2Cl_2F_4$	170.92	76-14-2	-94	2608	3.8	4-01-00-00137	*1.455^{25}	eth 4; EtOH 4	1.3092^{0}
5539	Ethane, 1,2-dichloro-1,1,2-trifluoro-		$C_2HCl_2F_3$	152.93	354-23-4	-78		28	4-01-00-00136	1.50^{25}		
5540	Ethane, 2,2-dichloro-1,1,1-trifluoro-	2,2-Dichloro-1,1,1-trifluoroethane	$C_2HCl_2F_3$	152.93	306-83-2	-107		27.82	4-01-00-00135	1.4638^{25}		
5541	Ethane, 1,1-diethoxy-	Acetal	$C_6H_{14}O_2$	118.18	105-57-7	-100	31	102.2	4-01-00-03103	0.8254^{20}	H_2O 3; EtOH 5; eth 5; ace 4	1.3834^{20}
5542	Ethane, 1,2-diethoxy-	Ethylene glycol diethyl ether	$C_6H_{14}O_2$	118.18	629-14-1			119.4	4-01-00-02379	0.8484^{20}	ace 4; bz 4; eth 4; EtOH 4	1.3860^{20}
5543	Ethane, 1,1-diethoxy-2-iodo-		$C_6H_{13}IO_2$	244.07	51806-20-3			132^{90}; 100^{10}	4-01-00-03156	1.489^{22}		1.4734^{22}
5544	Ethane, 1,1-difluoro-		$C_2H_4F_2$	66.05	75-37-6	-117		-24.9	4-01-00-00120	*0.896^{20}		1.3011^{-72}
5545	Ethane, 1,2-difluoro-	Freon 152	$C_2H_4F_2$	66.05	624-72-6			30.7	4-01-00-00121		bz 4; eth 4; chl 4	
5546	Ethane, 1,1-diiodo-	1,1-Diiodoethane	$C_2H_4I_2$	281.86	594-02-5			179.5	4-01-00-00169	2.84^{25}	ace 4; eth 4; EtOH 4; chl 4	1.673^{20}
5547	Ethane, 1,2-diiodo-		$C_2H_4I_2$	281.86	624-73-7	83		200	4-01-00-00169	3.325^{20}	H_2O 2; EtOH 3; eth 3; ace 3	1.871^{20}
5548	Ethane, 1,1-dimethoxy-	Dimethyl acetal	$C_4H_{10}O_2$	90.12	534-15-6	-113.2	3215	64.5	4-01-00-03103	0.8501^{20}	H_2O 3; EtOH 3; eth 3; ace 4	1.3668^{20}
5549	Ethane, 1,2-dimethoxy-	Ethylene glycol dimethyl ether	$C_4H_{10}O_2$	90.12	110-71-4	-69.20	3213	85	4-01-00-02376	0.8691^{20}	H_2O 3; EtOH 3; eth 3; ace 3	1.3796^{20}
5550	Ethanedinitrile	Cyanogen	C_2N_2	52.04	460-19-5	-27.9	2698	-21.1	4-02-00-01863	0.9537^{-21}	H_2O 3; EtOH 3; eth 3	
5551	Ethane, 1,1-dinitro-	1,1-Dinitroethane	$C_2H_4N_2O_4$	120.06	600-40-8			185.5	4-01-00-00174	1.349^{24}	H_2O 2; EtOH 3; eth 3	
5552	Ethane, 1,2-dinitro-	1,2-Dinitroethane	$C_2H_4N_2O_4$	120.06	7570-26-5	39.5		$94\text{-}6^{5}$	4-01-00-00175	1.4597^{20}	eth 4; EtOH 4	1.4468^{20}
5553	Ethanedioic acid	Oxalic acid	$C_2H_2O_4$	90.04	144-62-7	189.5 dec	6865	sub 157	4-02-00-01819	1.900^{17}	H_2O 3; EtOH 4; eth 2; bz 1	
5554	Ethanedioic acid, bis(2-chloroethyl) ester		$C_6H_8Cl_2O_4$	215.03	7208-92-6	45		150^{4}	4-02-00-01849		bz 4; EtOH 4	
5555	Ethanedioic acid, bis(3-methylbutyl) ester		$C_{12}H_{22}O_4$	230.30	2051-00-5	-9		267.5	4-02-00-01851	0.968^{11}	eth 4; EtOH 4	
5556	Ethanedioic acid, bis(1-methylethyl) ester		$C_8H_{14}O_4$	174.20	615-81-6			191	4-02-00-01850	1.001^{20}	eth 4; EtOH 4	1.4100^{20}
5557	Ethanedioic acid, bis(2-methylpropyl) ester	Diisobutyl oxalate	$C_{10}H_{18}O_4$	202.25	2050-61-5			229	4-02-00-01850	0.9737^{20}	ace 4; eth 4; EtOH 4	1.4180^{20}
5558	Ethanedioic acid, dibutyl ester		$C_{10}H_{18}O_4$	202.25	2050-60-4	-30.5		241; 96^{2}	4-02-00-01850	0.9873^{20}	H_2O 1; EtOH 3; eth 3	1.4234^{20}
5559	Ethanedioic acid, dicyclohexyl ester	Dicyclohexyl oxalate	$C_{14}H_{22}O_4$	254.33	620-82-6	43.5		190^{73}	3-06-00-00026		H_2O 1; EtOH 4; eth 4; MeOH 3	
5560	Ethanedioic acid, diethyl ester	Diethyl oxalate	$C_6H_{10}O_4$	146.14	95-92-1	-40.6	3115	185.7	4-02-00-01848	1.0785^{20}	H_2O 2; EtOH 5; eth 5; ace 5	1.4101^{20}

No.	Name Synonym	Mol. Form. Mol. Wt.	CAS RN mp/°C	Merck No. bp/°C	Beil. Ref. den/g cm^{-3}	Solubility n_D
5561	Ethanedioic acid, dihydrate Oxalic acid dihydrate	$C_2H_6O_6$ 126.07	6153-56-6 101.5		4-02-00-01819 1.653^{18}	H_2O 3; EtOH 3; eth 2
5562	Ethanedioic acid, dihydrazide	$C_2H_6N_4O_2$ 118.10	996-98-5 243 dec		4-02-00-01868 1.458^{22}	H_2O 3; EtOH 2; eth 2; bz 2
5563	Ethanedioic acid, dimethyl ester Dimethyl oxalate	$C_4H_6O_4$ 118.09	553-90-2 54.3	6020 163.5	4-02-00-01847 1.1716^{60}	H_2O 2; EtOH 3; eth 3; ace 3 1.379^{82}
5564	Ethanedioic acid, di-2-propenyl ester Diallyl oxalate	$C_8H_{10}O_4$ 170.17	615-99-6	 217	4-02-00-01851 1.1582^{20}	H_2O 1; EtOH 3; ace 3; bz 3 1.4481^{20}
5565	Ethanedioic acid, dipropyl ester	$C_8H_{14}O_4$ 174.20	615-98-5 -44.3	 211	4-02-00-01849 1.0188^{20}	H_2O 2; EtOH 5; eth 3 1.4158^{20}
5566	Ethanedioic acid, disodium salt Oxalic acid, disodium salt	$C_2Na_2O_4$ 134.00	62-76-0 260 dec	8601	4-02-00-01819 2.34^{25}	H_2O 2; EtOH 1; eth 1
5567	Ethanedioic acid, ethyl methyl ester	$C_5H_8O_4$ 132.12	615-52-1	 173.7	0-02-00-00535 1.1555^{0}	eth 4; EtOH 4
5568	Ethanedioic acid, lead (II) salt Lead oxalate	C_2O_4Pb 295.22	814-93-7 300 dec	5294	4-02-00-01819 5.28	H_2O i; EtOH 1; HNO_3 3
5569	1,2-Ethanediol Ethylene glycol	$C_2H_6O_2$ 62.07	107-21-1 -13	3755 197.3	4-01-00-02369 1.1135^{20}	H_2O 5; EtOH 5; eth 3; ace 5 1.4318^{20}
5570	1,1-Ethanediol, diacetate Ethylidene diacetate	$C_6H_{10}O_4$ 146.14	542-10-9 18.9	3767 169	4-02-00-00282 1.070^{25}	eth 4; EtOH 4 1.3985^{25}
5571	1,2-Ethanediol, diacetate Ethylene glycol diacetate	$C_6H_{10}O_4$ 146.14	111-55-7 -31	3756 190	4-02-00-00217 1.1043^{20}	H_2O 4; EtOH 5; eth 5; ace 5 1.4159^{20}
5572	1,2-Ethanediol, dibenzoate	$C_{16}H_{14}O_4$ 270.28	94-49-5 73.5	 360 dec	4-09-00-00356	H_2O 1; eth 3; chl 3
5573	1,2-Ethanediol, diformate	$C_4H_6O_4$ 118.09	629-15-2	 174	4-02-00-00037 1.193^{0}	H_2O 2; EtOH 3; eth 3 1.3580
5574	1,2-Ethanediol, dinitrate Ethylene glycol dinitrate	$C_2H_4N_2O_6$ 152.06	628-96-6 -22.3	 198.5	4-01-00-02413 1.4918^{20}	eth 4; EtOH 4
5575	1,2-Ethanediol, 1,2-diphenyl-, (*R**,*R**)-(±)-	$C_{14}H_{14}O_2$ 214.26	655-48-1 122.5	 >300	4-06-00-06682	H_2O 1; EtOH 4; eth 4; ace 3
5576	1,2-Ethanediol, dipropanoate	$C_8H_{14}O_4$ 174.20	123-80-8	 211	4-02-00-00715 1.020^{15}	H_2O 2; EtOH 5; eth 5; ace 4
5577	1,2-Ethanediol, monoacetate Ethylene glycol, monoacetate	$C_4H_8O_3$ 104.11	542-59-6	3757 188	4-02-00-00214 1.108^{15}	H_2O 5; EtOH 5; eth 5
5578	1,2-Ethanediol, monobenzoate	$C_9H_{10}O_3$ 166.18	94-33-7 45	 150^{10}	4-09-00-00354 1.1101^{30}	EtOH 4
5579	1,2-Ethanediol, 1-phenyl- Styrene glycol	$C_8H_{10}O_2$ 138.17	93-56-1 67.5	8831 273	4-06-00-05939	H_2O 4; eth 4; bz 4; EtOH 4
5580	1,2-Ethanediol, 1,1,2,2-tetraphenyl- Benzopinacol	$C_{26}H_{22}O_2$ 366.46	464-72-2 182	1110	4-06-00-07053	H_2O 1; EtOH 2; eth 3; ace 3
5581	1,1-Ethanediol, 2,2,2-tribromo- Acetaldehyde, tribromo, hydrate	$C_2H_3Br_3O_2$ 298.76	507-42-6 53.5	1373 dec	4-01-00-03155 2.5661^{40}	eth 4; EtOH 4
5582	1,1-Ethanediol, 2,2,2-trichloro- Chloral hydrate	$C_2H_3Cl_3O_2$ 165.40	302-17-0 57	2061 96 dec	4-01-00-03143 1.9081^{20}	H_2O 4; bz 4; eth 4; EtOH 4
5583	Ethanedione, di-2-furanyl-, dioxime α-Furildioxime	$C_{10}H_8N_2O_4$ 220.18	522-27-0 167	4218	4-19-00-02008	EtOH 2; eth 2; bz 2; lig 2
5584	Ethanedione, diphenyl- Benzil	$C_{14}H_{10}O_2$ 210.23	134-81-6 94.8	1087 347	4-07-00-02502 1.084^{102}	H_2O 1; EtOH 4; eth 4; ace 3
5585	Ethanedioyl dichloride Oxalyl chloride	$C_2Cl_2O_2$ 126.93	79-37-8 -16	6867 63.5	4-02-00-01853 1.4785^{20}	eth 3 1.4316^{20}
5586	Ethane, 1,1-diproxy- 1,2-Dipropoxyethane	$C_8H_{18}O_2$ 146.23	18854-56-3			
5587	1,2-Ethanedisulfonic acid Ethylenedisulfonic acid	$C_2H_6O_6S_2$ 190.20	110-04-3 173	3678	4-04-00-00078	Diox 4
5588	Ethanedithioamide Rubeanic acid	$C_2H_4N_2S_2$ 120.20	79-40-3 170 dec	8256	4-02-00-01871	H_2O 2; EtOH 2; con sulf 3
5589	Ethane(dithioic) acid	$C_2H_4S_2$ 92.19	594-03-6	 66^{85}; 37^{15}	4-02-00-00572 1.24^{20}	H_2O 3; EtOH 4; eth 4; ace 4
5590	1,2-Ethanedithiol Ethylenedimercaptan	$C_2H_6S_2$ 94.20	540-63-6 -41.2	3679 146	4-01-00-02450 1.234^{20}	H_2O 1; EtOH 3; eth 3; ace 3 1.5590^{20}
5591	Ethane, 1-ethoxy-2-methoxy- 1-Ethoxy-2-methoxyethane	$C_5H_{12}O_2$ 104.15	5137-45-1	 102.1	3-01-00-02078 0.8529^{20}	 1.3868^{20}
5592	Ethane, 1,1'-[ethylidenebis(oxy)]bis[2-chloro-	$C_6H_{12}Cl_2O_2$ 187.07	14689-97-5	 195	4-01-00-03104 1.1737^{20}	eth 4; EtOH 4 1.4527^{20}
5593	Ethane, fluoro- Ethyl fluoride	C_2H_5F 48.06	353-36-6 -143.2	 -37.6	4-01-00-00120 0.7182^{20}	H_2O 3; EtOH 4; eth 4 1.2656^{20}
5594	Ethane, hexabromo- Hexabromoethane	C_2Br_6 503.45	594-73-0	 200 dec	3-01-00-00193 3.823^{20}	EtOH 2; eth 2; CS_2 2 1.863
5595	Ethane, hexachloro- Hexachloroethane	C_2Cl_6 236.74	07-72-1 187 (triple point)	4601	4-01-00-00148 2.091^{20}	H_2O 1; EtOH 4; eth 4; bz 3
5596	Ethane, hexafluoro- Perfluoroethane	C_2F_6 138.01	76-16-4 -100.7	 -78.1	4-01-00-00123 1.590^{-78}	H_2O 1; EtOH 2; eth 2
5597	Ethane, iodo- Ethyl iodide	C_2H_5I 155.97	75-03-6 -111.1	3769 72.5	4-01-00-00163 1.9358^{20}	H_2O 2; EtOH 5; eth 3; chl 3 1.5133^{20}
5598	Ethane, isocyanato- Ethyl isocyanate	C_3H_5NO 71.08	109-90-0	 60	4-04-00-00402 0.9031^{20}	H_2O 1; EtOH 5; eth 5 1.3808^{20}
5599	Ethane, isocyano-	C_3H_5N 55.08	624-79-3 <-66	 79	4-04-00-00342 0.7402^{20}	H_2O 4; EtOH 5; eth 5; ace 3 1.3622^{20}
5600	Ethane, isothiocyanato- Ethyl isothiocyanate	C_3H_5NS 87.15	542-85-8 -5.9	3771 131.5	4-04-00-00403 0.9990^{20}	H_2O 1; EtOH 5; eth 5 1.5130^{20}

No.	Name Synonym	Mol. Form. Mol. Wt.	CAS RN mp/°C	Merck No. bp/°C	Beil. Ref. den/g cm^{-3}	Solubility n_D
5601	Ethane, 1,1',1'',1'''-[methanetetrayltetrakis(oxy)]tetrakis-	$C_9H_{20}O_4$ 192.26	78-09-1	 159.5	4-03-00-00006 0.9186^{20}	EtOH 5; eth 5; ctc 3 1.3905^{25}
5602	Ethane, methoxy- Ethyl methyl ether	C_3H_8O 60.10	540-67-0 -113	3783 7.4	4-01-00-01314 0.7251^{0}	H_2O 3; EtOH 5; eth 5; ace 3 1.3420^{4}
5603	Ethane, 1,1'-[methylenebis(oxy)]bis- Diethoxymethane	$C_5H_{12}O_2$ 104.15	462-95-3 -66.5	 88	4-01-00-03027 0.8319^{20}	H_2O 3; EtOH 5; eth 5; ace 4 1.3748^{18}
5604	Ethane, 1,1',1''-[methylidynetris(oxy)]tris- Ethyl orthoformate	$C_7H_{16}O_3$ 148.20	122-51-0	 143; 60^{20}	4-02-00-00025 0.8909^{20}	EtOH 3; eth 3 1.3922^{20}
5605	Ethane, 1,1',1''-[methylidynetris(thio)]tris-	$C_7H_{16}S_3$ 196.40	6267-24-9	 235 dec; 127^{12}	4-02-00-00093 1.053^{20}	eth 4; EtOH 4 1.5410^{15}
5606	Ethane, (methylthio)- Ethyl methyl sulfide	C_3H_8S 76.16	624-89-5 -105.9	 66.7	4-01-00-01392 0.8422^{20}	H_2O 1; EtOH 5; eth 3; chl 3 1.4404^{20}
5607	Ethane, nitro- Nitroethane	$C_2H_5NO_2$ 75.07	79-24-3 -89.5	6518 114.0	4-01-00-00170 1.0448^{25}	H_2O 2; EtOH 5; eth 5; ace 3 1.3917^{20}
5608	Ethane, 1,1'-oxybis- Diethyl ether	$C_4H_{10}O$ 74.12	60-29-7 -116.3	3762 34.5	4-01-00-01314 0.7138^{20}	H_2O 2; EtOH 5; eth 5; ace 4 1.3526^{20}
5609	Ethane, 1,1'-oxybis[2-bromo-	$C_4H_8Br_2O$ 231.91	5414-19-7	 115^{32}	4-01-00-01386 1.8222^{27}	 1.5131^{27}
5610	Ethane, 1,1'-oxybis[1-chloro-	$C_4H_8Cl_2O$ 143.01	6986-48-7	 116.5	4-01-00-03120 1.1060^{25}	eth 4; EtOH 4; chl 4 1.4186^{25}
5611	Ethane, 1,1'-oxybis[2-chloro- Bis(2-chloroethyl) ether	$C_4H_8Cl_2O$ 143.01	111-44-4 -51.9	3055 178.5	4-01-00-01375 1.22^{20}	H_2O 1; EtOH 3; eth 3; ace 3 1.451^{20}
5612	Ethane, 1,1'-oxybis[2-ethoxy- Diethylene glycol diethyl ether	$C_8H_{18}O_3$ 162.23	112-36-7 -45	3108 188	4-01-00-02394 0.9063^{20}	H_2O 4; EtOH 4; eth 3; os 4 1.4115^{20}
5613	Ethane, 1,1'-oxybis[2-methoxy- Diethylene glycol dimethyl ether	$C_6H_{14}O_3$ 134.18	111-96-6 -68	3148 162	4-01-00-02393 0.9434^{20}	H_2O 5; EtOH 5; eth 5 1.4097^{20}
5614	Ethane, 1,1'-[oxybis(methyleneoxy)]bis- Bis(ethoxymethyl)ether	$C_6H_{14}O_3$ 134.18	5648-29-3	 140.6		
5615	Ethane, 1,1'-oxybis[1,2,2,2-tetrachloro- Bis(1,2,2,2-tetrachloroethyl) ether	$C_4H_2Cl_8O$ 349.68	41284-12-2 40.2	 188; 130^{11}	2-01-00-00681	bz 4; peth 4
5616	Ethane, pentabromo- 1,1,1,2,2-Pentabromoethane	C_2HBr_5 424.55	75-95-6 56.5	 210^{300}	3-01-00-00193 3.312^{20}	H_2O 1; EtOH 3; eth 4; chl 3
5617	Ethane, pentachloro- Pentachloroethane	C_2HCl_5 202.29	76-01-7 -29	7058 159.8	4-01-00-00147 1.6796^{20}	H_2O 1; EtOH 5; eth 5 1.5025^{20}
5618	Ethane, pentachlorofluoro-	C_2Cl_5F 220.28	354-56-3 101.3	 135	4-01-00-00148 1.74^{25}	H_2O 1; EtOH 3; eth 3
5619	Ethaneperoxoic acid Peroxyacetic acid	$C_2H_4O_3$ 76.05	79-21-0 -0.2	7107 110	4-02-00-00390 1.226^{15}	H_2O 4; EtOH 3; eth 4; sulf 4 1.3974^{20}
5620	Ethane, 1,1'-selenobis-	$C_4H_{10}Se$ 137.08	627-53-2 55	 108	4-01-00-01411 1.2300^{20}	 1.4768^{20}
5621	Ethane, 1,1'-sulfinylbis-	$C_4H_{10}OS$ 106.19	70-29-1 14	 104^{25}; 90^{15}	4-01-00-01395 1.0092^{22}	H_2O 4; eth 4; EtOH 4
5622	Ethanesulfonic acid Ethylsulfonic acid	$C_2H_6O_3S$ 110.13	594-45-6 -17	 123^{1}	4-04-00-00033 1.3341^{25}	H_2O 4; EtOH 4 1.4335^{20}
5623	Ethanesulfonic acid, 2-amino- Taurine	$C_2H_7NO_3S$ 125.15	107-35-7 328	9043	4-04-00-03289	H_2O 4
5624	Ethanesulfonic acid, 2-(methylamino)- *N*-Methyltaurine	$C_3H_9NO_3S$ 139.18	107-68-6 241.5	6043	3-04-00-01699	H_2O 4; EtOH 1; eth 1
5625	Ethanesulfonic acid, 2-[[(3α,5β,7α,12α)-3,7,12-trihydroxy-24-oxocholan-24-yl]amino]- Taurocholic acid	$C_{26}H_{45}NO_7S$ 515.71	81-24-3 125 dec	9044	4-10-00-02078	H_2O 4; EtOH 4; eth 2; AcOEt 2
5626	Ethane, 1,1'-sulfonylbis-	$C_4H_{10}O_2S$ 122.19	597-35-3 73.5	 248	4-01-00-01396 1.357^{20}	H_2O 3; eth 3; bz 4; peth 1
5627	Ethanesulfonyl chloride	$C_2H_5ClO_2S$ 128.58	594-44-5	 174	4-04-00-00034 1.357^{22}	eth 4; CS_2 3 1.4531^{20}
5628	Ethanesulfonyl chloride, 2-bromo- 2-Bromoethanesulfonyl chloride	$C_2H_4BrClO_2S$ 207.48	54429-56-0	 102^{13}	4-04-00-00037 1.921^{20}	 1.5242^{20}
5629	Ethanesulfonyl chloride, 2-chloro-	$C_2H_4Cl_2O_2S$ 163.02	1622-32-8	 201.5	4-04-00-00036 1.555^{20}	 1.4920^{20}
5630	Ethane, 1,1'-tellurobis-	$C_4H_{10}Te$ 185.72	627-54-3	 137.5	4-01-00-01412 1.599^{15}	EtOH 4 1.5182^{15}
5631	Ethane, 1,1,1,2-tetrabromo- 1,2,2,2-Tetrabromoethane	$C_2H_2Br_4$ 345.65	630-16-0 0	 112^{18} dec	4-01-00-00162 2.8747^{20}	ace 4; bz 4; eth 4; EtOH 4 1.6277^{20}
5632	Ethane, 1,1,2,2-tetrabromo- Acetylene tetrabromide	$C_2H_2Br_4$ 345.65	79-27-6 0	9121 243.5; 151^{54}	4-01-00-00162 2.9655^{20}	H_2O 1; EtOH 5; eth 5; ace 3 1.6353^{20}
5633	1,1,2,2-Ethanetetracarboxylic acid, tetraethyl ester	$C_{14}H_{22}O_8$ 318.32	632-56-4 77	 305 dec	4-02-00-02415 1.064^{80}	EtOH 3 1.4105^{80}
5634	Ethane, 1,1,1,2-tetrachloro- 1,1,1,2-Tetrachloroethane	$C_2H_2Cl_4$ 167.85	630-20-6 -70.2	 130.5	4-01-00-00143 1.5406^{20}	H_2O 2; EtOH 5; eth 5; ace 3 1.4821^{20}
5635	Ethane, 1,1,2,2-tetrachloro- Acetylene tetrachloride	$C_2H_2Cl_4$ 167.85	79-34-5 -43.8	9125 146.5	4-01-00-00144 1.5953^{20}	H_2O 2; EtOH 5; eth 5; ace 3 1.4940^{20}
5636	Ethane, 1,1,1,2-tetrachloro-2,2-difluoro- 1,1-Difluoroperchloroethane	$C_2Cl_4F_2$ 203.83	76-11-9 40.6	 91.5	4-01-00-00146 1.649^{25}	H_2O 1; EtOH 3; eth 3; chl 3
5637	Ethane, 1,1,2,2-tetrachloro-1,2-difluoro- Tetrachloro-1,2-difluoroethane	$C_2Cl_4F_2$ 203.83	76-12-0 26	 93	4-01-00-00146 1.6447^{25}	H_2O 1; EtOH 3; eth 3; chl 3 1.4130^{25}
5638	Ethane, 1,1,2,2-tetrachloro-1-fluoro-	C_2HCl_4F 185.84	354-14-3 -82.6	 116	4-01-00-00145 1.5497^{17}	 1.4390^{20}

No.	Name Synonym	Mol. Form. Mol. Wt.	CAS RN mp/°C	Merck No. bp/°C	Beil. Ref. den/g cm^{-3}	Solubility n_D
5639	Ethane, 1,1,1,2-tetrachloro-2-fluoro- 1,1,1,2-Tetrachloro-2-fluoroethane	C_2HCl_4F 185.84	354-11-0 -95.3	 116.5		
5640	Ethane, 1,1,1,2-tetrafluoro- 1,1,1,2-Tetrafluoroethane	$C_2H_2F_4$ 102.03	811-97-2 -103.3	 -26.08	4-01-00-00123 1.2072^{25}	H_2O 1; eth 3
5641	Ethane, 1,1,2,2-tetrafluoro- 1,1,2,2-Tetrafluoroethane	$C_2H_2F_4$ 102.03	359-35-3 -89	 -19.9		
5642	Ethane, 1,1,2,2-tetrafluoro-1,2-dinitro- Ethane, 1,2-dinitro-1,1,2,2-tetrafluoro	$C_2F_4N_2O_4$ 192.03	356-16-1 -41.5	 58.5	3-01-00-00203 1.6024^{25}	H_2O 1; ace 3 1.3265^{25}
5643	Ethanethioamide Thioacetamide	C_2H_5NS 75.13	62-55-5 115.5	9246	4-02-00-00565	H_2O 4; EtOH 4; eth 2; bz 2
5644	Ethane, 1,1'-thiobis- Diethyl sulfide	$C_4H_{10}S$ 90.19	352-93-2 -103.9	3809 92.1	4-01-00-01394 0.8362^{20}	H_2O 2; EtOH 3; eth 3; ctc 2 1.4430^{20}
5645	Ethane, 1,1'-thiobis[2-chloro- Mustard gas	$C_4H_8Cl_2S$ 159.08	505-60-2 13.5	6225 216	4-01-00-01407 1.2741^{20}	 1.5313^{20}
5646	Ethanethioic acid Thioacetic acid	C_2H_4OS 76.12	507-09-5 <-17	9247 93; 26^{35}	4-02-00-00542 1.064^{20}	H_2O 3; EtOH 4; eth 5; ace 4 1.4648^{20}
5647	Ethanethioic acid, *S*-ethyl ester	C_4H_8OS 104.17	625-60-5	 116.4	4-02-00-00543 0.9792^{20}	H_2O 1; EtOH 4; eth 4 1.4583^{21}
5648	Ethanethiol Ethyl mercaptan	C_2H_6S 62.14	75-08-1 -147.8	3680 35.1	4-01-00-01390 0.8315^{25}	H_2O 2; EtOH 3; eth 3; ace 3 1.4310^{20}
5649	Ethanethiol, 2-amino- Cysteamine	C_2H_7NS 77.15	60-23-1 99.5	2785 dec	4-04-00-01570	H_2O 4; EtOH 4
5650	Ethanethiol, 2-(butylamino)- 2-(Butylamino)ethanethiol	$C_6H_{15}NS$ 133.26	5842-00-2	 112-5^{10}	4-04-00-01602 0.901^{25}	 1.4711^{20}
5651	Ethanethiol, 2-chloro- 2-Chloroethanethiol	C_2H_5ClS 96.58	4325-97-7	 113	4-01-00-01406 1.1826^{20}	eth 4; EtOH 4; Diox 4 1.4929^{20}
5652	Ethanethiol, 2,2'-thiobis-	$C_4H_{10}S_3$ 154.32	3570-55-6	 135-6^{10}	4-01-00-02455 1.183^{25}	 1.5982^{20}
5653	Ethane, 1,1,2-tribromo- 1,1,2-Tribromoethane	$C_2H_3Br_3$ 266.76	78-74-0 -29.3	 188.93	4-01-00-00161 2.6210^{20}	H_2O 1; EtOH 3; eth 3; bz 3 1.5933^{20}
5654	1,1,2-Ethanetricarboxylic acid, triethyl ester	$C_{11}H_{18}O_6$ 246.26	7459-46-3	 99^{05}	4-02-00-02364 1.074^{25}	 1.4290^{20}
5655	Ethane, 1,1,1-trichloro- Methylchloroform	$C_2H_3Cl_3$ 133.40	71-55-6 -30.4	9549 74.0	4-01-00-00138 1.3390^{20}	H_2O 2; EtOH 3; eth 5; chl 3 1.4379^{20}
5656	Ethane, 1,1,2-trichloro- 1,1,2-Trichloroethane	$C_2H_3Cl_3$ 133.40	79-00-5 -36.6	9550 113.8	4-01-00-00139 1.4397^{20}	H_2O 1; EtOH 3; eth 3; chl 3 1.4714^{20}
5657	Ethane, 1,1,1-trichloro-2,2-diethoxy- Chloral, diethyl acetal	$C_6H_{11}Cl_3O_2$ 221.51	599-97-3	 197	4-01-00-03144 1.266^{20}	H_2O 2; EtOH 5; eth 5; glycerol 5 1.4586^{25}
5658	Ethane, 1,1,1-trichloro-2,2-difluoro 1,1,1-Trichloro-2,2-difluoroethane	$C_2HCl_3F_2$ 169.38	354-12-1	 73		
5659	Ethane, 1,2,2-trichloro-1,1-difluoro-	$C_2HCl_3F_2$ 169.38	354-21-2	 71.9	4-01-00-00141 1.5447^{20}	 1.3889^{20}
5660	Ethane, 1,2,2-trichloro-1,2-difluoro- 1,2,2-Trichloro-1,2-difluoroethane	$C_2HCl_3F_2$ 169.38	354-15-4	 72.5		
5661	Ethane, 1,1,2-trichloro-2-fluoro-	$C_2H_2Cl_3F$ 151.39	359-28-4	 102	4-01-00-00141 1.5393^{20}	H_2O 1 1.4390^{20}
5662	Ethane, 1,1,1-trichloro-2,2,2-trifluoro- 1,1,1-Trichlorotrifluoroethane	$C_2Cl_3F_3$ 187.38	354-58-5 14.2	 46.1	4-01-00-00142 1.5790^{20}	H_2O 1; EtOH 3; eth 3; chl 3 1.3610^{35}
5663	Ethane, 1,1,2-trichloro-1,2,2-trifluoro- 1,1,2-Trichlorotrifluoroethane	$C_2Cl_3F_3$ 187.38	76-13-1 -35	 47.7	4-01-00-00142 1.5635^{25}	H_2O 1; EtOH 3; eth 5; bz 5 1.3557^{25}
5664	Ethane, 1,1,1-triethoxy-	$C_8H_{18}O_3$ 162.23	78-39-7	 145	4-02-00-00137 0.8847^{25}	H_2O 1; EtOH 5; eth 5; ctc 5 1.3980^{20}
5665	Ethane, 1,1,1-trifluoro- Methylfluoroform	$C_2H_3F_3$ 84.04	420-46-2 -111.3	 -47.5	4-01-00-00122	eth 3; chl 3
5666	Ethane, 1,1,2-trifluoro- 1,1,2-Trifluoroethane	$C_2H_3F_3$ 84.04	430-66-0 -84	 5	4-01-00-00123	
5667	Ethane, 1,1,1-trifluoro-2-iodo-	$C_2H_2F_3I$ 209.94	353-83-3	 54.5	4-01-00-00166 2.13^{25}	 1.4009^{20}
5668	Ethane, 1,1,1-trifluoro-2-nitro-	$C_2H_2F_3NO_2$ 129.04	819-07-8	 96	4-01-00-00172 1.3914^{20}	eth 4 1.3394^{20}
5669	Ethane, 1,1,1-trimethoxy-	$C_5H_{12}O_3$ 120.15	1445-45-0	 108	4-02-00-00127 0.9438^{25}	eth 4; EtOH 4 1.3859^{25}
5670	Ethanimidamide, *N,N'*-bis(4-ethoxyphenyl)-, monohydrochloride Holocaine, hydrochloride	$C_{18}H_{23}ClN_2O_2$ 334.85	620-99-5 191	7153	3-13-00-01069	H_2O 4; EtOH 4; chl 4
5671	Ethanimidamide, monohydrochloride Acetamidine, hydrochloride	$C_2H_7ClN_2$ 94.54	124-42-5 177.5	37	4-02-00-00428	H_2O 4; EtOH 4
5672	Ethanol Ethyl alcohol	C_2H_6O 46.07	64-17-5 -114.1	3716 78.2	4-01-00-01289 0.7893^{20}	H_2O 5; EtOH 5; eth 5; ace 5 1.3611^{20}
5673	Ethanol, 2-[2-(acetyloxy)ethoxy]-	$C_6H_{12}O_4$ 148.16	2093-20-1	 114-5^{7}	2-02-00-00155 1.1208^{20}	 1.4320^{20}
5674	Ethanol, 1-amino- Acetaldehyde ammonia	C_2H_7NO 61.08	75-39-8 97	33 110 dec		H_2O 3; eth 2
5675	Ethanol, 2-amino- Ethanolamine	C_2H_7NO 61.08	141-43-5 10.5	3681 171	4-04-00-01406 1.0180^{20}	H_2O 5; EtOH 5; eth 2; bz 2 1.4541^{20}
5676	Ethanol, 2-[(2-aminoethyl)amino]- *N*-(Aminoethyl)ethanolamine	$C_4H_{12}N_2O$ 104.15	111-41-1	 239; 105^{10}	4-04-00-01558 1.0286^{20}	H_2O 5; EtOH 5; ace 3; bz 2 1.4863^{20}
5677	Ethanol, 2-[(2-amino-2-methylpropyl)amino]-	$C_6H_{16}N_2O$ 132.21	68750-16-3 5	 236		chl 3 1.4698^{25}

No.	Name Synonym	Mol. Form. Mol. Wt.	CAS RN mp/°C	Merck No. bp/°C	Beil. Ref. den/g cm^{-3}	Solubility n_D
5678	Ethanol, 2-[2-[2-(3-aminopropoxy)ethoxy]ethoxy]-	$C_9H_{21}NO_4$ 207.27	49542-66-7 -50	 184^{10}	 1.0663^{20}	H_2O 4; EtOH 4 1.4668^{20}
5679	Ethanol, 2,2'-[(3-aminopropyl)imino]bis- *N*-(3-Aminopropyl)diethanolamine	$C_7H_{18}N_2O_2$ 162.23	4985-85-7	 160^{1}	3-04-00-00720	
5680	Ethanol, 1-amino-2,2,2-trichloro- Chloral ammonia	$C_2H_4Cl_3NO$ 164.42	507-47-1 73	2057 100 dec	4-01-00-03147	bz 4; eth 4; EtOH 4
5681	Ethanol, 2-azido-	$C_2H_5N_3O$ 87.08	1517-05-1	 75^{40}	4-01-00-01389 1.146^{24}	H_2O 4
5682	Ethanol, 2-[bis(3-methylbutyl)amino]- Diisopentyl 2-hydroxyethylamine	$C_{12}H_{27}NO$ 201.35	3574-43-4	 248	3-04-00-00685 0.8492^{20}	eth 4; EtOH 4; chl 4 1.4435^{20}
5683	Ethanol, 2-[bis(1-methylethyl)amino]- *N,N*-Diisopropyl-2-aminoethanol	$C_8H_{19}NO$ 145.25	96-80-0	 190	4-04-00-01504 0.826^{25}	 1.4417^{20}
5684	Ethanol, 2-[bis(2-methylpropyl)amino]- Ethanolamine, *N,N*-diisobutyl	$C_{10}H_{23}NO$ 173.30	4535-66-4	 214	4-04-00-01509 0.8407^{20}	eth 4; EtOH 4; lig 4; chl 4 1.4355^{20}
5685	Ethanol, 2-bromo- Ethylene bromohydrin	C_2H_5BrO 124.97	540-51-2	3749 150; 51^{4}	4-01-00-01385 1.7629^{20}	H_2O 5; EtOH 5; eth 5; os 4 1.4915^{20}
5686	Ethanol, 2-bromo-, acetate	$C_4H_7BrO_2$ 167.00	927-68-4 -13.8	 162.5	4-02-00-00136 1.514^{20}	H_2O 4; EtOH 5; eth 5; chl 4 1.457^{23}
5687	Ethanol, 2-butoxy- Ethylene glycol monobutyl ether	$C_6H_{14}O_2$ 118.18	111-76-2 -74.8	1559 168.4	4-01-00-02380 0.9015^{20}	H_2O 5; EtOH 5; eth 5; ctc 2 1.4198^{20}
5688	Ethanol, 2-(2-butoxyethoxy)- Diethylene glycol monobutyl ether	$C_8H_{18}O_3$ 162.23	112-34-5 -68	1557 231	4-01-00-02394 0.9553^{20}	H_2O 5; EtOH 4; eth 4; ace 4 1.4306^{20}
5689	Ethanol, 2-(2-butoxyethoxy)-, acetate	$C_{10}H_{20}O_4$ 204.27	124-17-4 -32	 245	3-02-00-00308 0.985^{20}	ace 4; eth 4; EtOH 4 1.4262^{20}
5690	Ethanol, 2-[2-(2-butoxyethoxy)ethoxy]-	$C_{10}H_{22}O_4$ 206.28	143-22-6	 278	4-01-00-02402 0.9890^{20}	EtOH 4; MeOH 4 1.4389^{20}
5691	Ethanol, 2-(butylamino)-	$C_6H_{15}NO$ 117.19	111-75-1	 199; $91\text{-}2^{11}$	3-04-00-00682 0.8907^{20}	H_2O 4; EtOH 4; eth 4 1.4437^{20}
5692	Ethanol, 2,2'-(butylimino)bis-	$C_8H_{19}NO_2$ 161.24	102-79-4	 275; 80^{35}	4-04-00-01520 0.9592^{20}	chl 3 1.4625^{20}
5693	Ethanol, 2-chloro- Ethylene chlorohydrin	C_2H_5ClO 80.51	107-07-3 -67.5	3750 128.6	4-01-00-01372 1.2019^{20}	H_2O 5; EtOH 5; eth 2; chl 3 1.4419^{20}
5694	Ethanol, 1-chloro-, acetate	$C_4H_7ClO_2$ 122.55	5912-58-3	 123	4-02-00-00282 1.110^{20}	eth 4 1.409^{20}
5695	Ethanol, 2-chloro-, acetate β-Chloroethyl acetate	$C_4H_7ClO_2$ 122.55	542-58-5	2137 145	4-02-00-00136 1.178^{20}	H_2O 1; EtOH 5; eth 5; ctc 3 1.4234^{20}
5696	Ethanol, 2-chloro-, benzoate	$C_9H_9ClO_2$ 184.62	939-55-9 118	 255	4-09-00-00288 1.1789^{20}	eth 4; EtOH 4
5697	Ethanol, 2-chloro-, carbonate (2:1) Bis(2-chloroethyl) carbonate	$C_5H_8Cl_2O_3$ 187.02	623-97-2 8	 241	4-03-00-00006 1.3506^{20}	H_2O 1 1.461^{20}
5698	Ethanol, 2-(2-chloroethoxy)-	$C_4H_9ClO_2$ 124.57	628-89-7	 $79\text{-}81^{5}$	3-01-00-02078 1.18^{25}	H_2O 4; EtOH 5; eth 5 1.4529^{20}
5699	Ethanol, 2-chloro-, methanesulfonate	$C_3H_7ClO_3S$ 158.61	3570-58-9	 $125\text{-}6^{9}$	4-04-00-00012 1.39^{25}	chl 3 1.4562^{20}
5700	Ethanol, 2-[2-[4-[(4-chlorophenyl)phenylmethyl]-1-piperazinyl]ethoxy]- Hydroxyzine	$C_{21}H_{27}ClN_2O_2$ 374.91	68-88-2	4786	5-23-01-00462	
5701	Ethanol, 2-chloro-, phosphate (3:1)	$C_6H_{12}Cl_3O_4P$ 285.49	115-96-8	 330; 194^{10}	4-01-00-01379 1.39^{25}	ctc 3 1.4721^{20}
5702	Ethanol, 2-chloro-, phosphite (3:1)	$C_6H_{12}Cl_3O_3P$ 269.49	140-08-9	 $112\text{-}5^{2}$	4-01-00-01378 1.328^{25}	 1.4868^{20}
5703	Ethanol, 2-(dibutylamino)-	$C_{10}H_{23}NO$ 173.30	102-81-8	 114^{18}	4-04-00-01506	
5704	Ethanol, 2,2-dichloro- 2,2-Dichloroethanol	$C_2H_4Cl_2O$ 114.96	598-38-9	 146	4-01-00-01383 1.4040^{25}	H_2O 2; EtOH 3; eth 3; ctc 2 1.4626^{25}
5705	Ethanol, 2,2-diethoxy-	$C_6H_{14}O_3$ 134.18	621-63-6	 167	4-01-00-03958 0.9^{25}	eth 4; EtOH 4 1.4073^{20}
5706	Ethanol, 2-(diethylamino)- 2-(Diethylamino)ethanol	$C_6H_{15}NO$ 117.19	100-37-8	3101 163	4-04-00-01471 0.8921^{20}	H_2O 5; EtOH 3; eth 3; ace 3 1.4412^{20}
5707	Ethanol, 2-[2-(diethylamino)ethoxy]-	$C_8H_{19}NO_2$ 161.24	140-82-9	 221.5; 92^{7}	4-04-00-01474 0.9421^{25}	 1.4480^{20}
5708	Ethanol, 2,2-difluoro-	$C_2H_4F_2O$ 82.05	359-13-7 -28.2	 95.5	4-01-00-01369 1.3084^{17}	ace 4; eth 4; EtOH 4 1.3345^{11}
5709	Ethanol, 2-(dimethylamino)- Deanol	$C_4H_{11}NO$ 89.14	108-01-0 -59	2834 134	4-04-00-01424 0.8866^{20}	H_2O 5; EtOH 5; eth 5; chl 3 1.4300^{20}
5710	Ethanol, 2-[(1,1-dimethylethyl)amino]-	$C_6H_{15}NO$ 117.19	4620-70-6 44	 176.5; 72^{14}	4-04-00-01509 0.8818^{20}	
5711	Ethanol, 2,2',2'',2'''-[(4,8-di-1-piperidinylpyrimido[5,4-d]pyrimidine- Dipyridamole	$C_{24}H_{40}N_8O_4$ 504.63	58-32-2 163	3354	4-26-00-03840	
5712	Ethanol, 2-(dipropylamino)-	$C_8H_{19}NO$ 145.25	3238-75-3	 196	4-04-00-01502 0.8576^{20}	eth 4; EtOH 4 1.4402^{20}
5713	Ethanol, 2,2'-[1,2-ethanediylbis(oxy)]bis- Triethylene glycol	$C_6H_{14}O_4$ 150.17	112-27-6 -7	9585 285	4-01-00-02400 1.1274^{15}	H_2O 5; EtOH 5; eth 2; bz 5 1.4531^{20}
5714	Ethanol, 2,2'-[1,2-ethanediylbis(oxy)]bis-, diacetate Triethylene glycol, diacetate	$C_{10}H_{18}O_6$ 234.25	111-21-7 -50	 286	4-02-00-00215 1.1153^{20}	H_2O 4; eth 4; EtOH 4
5715	Ethanol, 2,2'-(1,2-ethanediyldiimino)bis-	$C_6H_{16}N_2O_2$ 148.21	4439-20-7 97.5	 $135\text{-}7^{1}$	4-04-00-01559	H_2O 3

PHYSICAL CONSTANTS OF ORGANIC COMPOUNDS (continued)

No.	Name Synonym	Mol. Form. Mol. Wt.	CAS RN mp/°C	Merck No. bp/°C	Beil. Ref. den/g cm^{-3}	Solubility n_D
5716	Ethanol, 2-(ethenyloxy)-	$C_4H_8O_2$ 88.11	764-48-7	 141.6	4-01-00-02387 0.9821^{20}	H_2O 3; EtOH 3; eth 3; bz 3 1.4564^{17}
5717	Ethanol, 2-ethoxy- Ethylene glycol monoethyl ether	$C_4H_{10}O_2$ 90.12	110-80-5 -70	3707 135	4-01-00-02377 0.9297^{20}	H_2O 4; ace 4; eth 4; EtOH 4 1.4080^{20}
5718	Ethanol, 2-ethoxy-, acetate Ethylene glycol monoethyl ether acetate	$C_6H_{12}O_3$ 132.16	111-15-9 -61.7	3708 156.4	4-02-00-00214 0.9740^{20}	H_2O 4; ace 4; eth 4; EtOH 4 1.4054^{20}
5719	Ethanol, 2-ethoxy-, carbonate (2:1)	$C_9H_{18}O_5$ 206.24	2049-74-3	 245.5	3-03-00-00017 1.0439^{20}	ace 4; eth 4; EtOH 4 1.4227^{20}
5720	Ethanol, 2-(2-ethoxyethoxy)- Diethylene glycol monoethyl ether	$C_6H_{14}O_3$ 134.18	111-90-0	1806 196	4-01-00-02393 0.9885^{20}	H_2O 5; EtOH 5; eth 4; ace 5 1.4300^{20}
5721	Ethanol, 2-(2-ethoxyethoxy)-, acetate Diethylene glycol monoethyl ether acetate	$C_8H_{16}O_4$ 176.21	112-15-2 -25	 218.5	3-02-00-00308 1.0096^{20}	H_2O 4; ace 4; eth 4; EtOH 4 1.4213^{20}
5722	Ethanol, 2-(ethylamino)-	$C_4H_{11}NO$ 89.14	110-73-6	 169.5	4-04-00-01465 0.914^{20}	H_2O 4; EtOH 4; eth 4; chl 3 1.444^{20}
5723	Ethanol, 2,2'-(ethylimino)bis-	$C_6H_{15}NO_2$ 133.19	139-87-7 -50	 247	3-04-00-00693 1.0135^{20}	H_2O 4; EtOH 4; eth 2 1.4663^{20}
5724	Ethanol, 2-[ethyl(3-methylphenyl)amino]-	$C_{11}H_{17}NO$ 179.26	91-88-3	 $118^{1.5}$	4-12-00-01819	ctc 3 1.5540^{20}
5725	Ethanol, 2-(ethylsulfonyl)- Ethylsulfonylethyl alcohol	$C_4H_{10}O_3S$ 138.19	513-12-2	3810	4-01-00-02430	chl 2
5726	Ethanol, 2-(ethylthio)- 2-(Ethylthio)ethanol	$C_4H_{10}OS$ 106.19	110-77-0 -100	3813 184	4-01-00-02430 1.0166^{20}	H_2O 2; EtOH 3; ace 4 1.4867^{20}
5727	Ethanol, 2-fluoro- Ethylene fluorohydrin	C_2H_5FO 64.06	371-62-0 -26.4	 103.5	4-01-00-01366 1.1040^{20}	H_2O 5; EtOH 5; eth 5; ace 4 1.3647^{18}
5728	Ethanol, 2-(hexyloxy)-	$C_8H_{18}O_2$ 146.23	112-25-4 -45.1	 208	4-01-00-02383 0.8878^{20}	H_2O 2; EtOH 4; eth 4 1.4291^{20}
5729	Ethanol, 2-hydrazino- 2-Hydrazinoethanol	$C_2H_8N_2O$ 76.10	109-84-2 -70	4696 219; $120^{17.5}$	4-04-00-03348 1.119^{25}	H_2O 4; EtOH 4; MeOH 4
5730	Ethanol, 2,2'-iminobis- Diethanolamine	$C_4H_{11}NO_2$ 105.14	111-42-2 28	3097 268.8	4-04-00-01514 1.0966^{20}	H_2O 4; EtOH 4; eth 2; bz 2 1.4776^{20}
5731	Ethanol, 2-iodo-	C_2H_5IO 171.97	624-76-0	 176 dec	4-01-00-01387 2.1967^{20}	H_2O 4; eth 4; EtOH 4 1.5713^{20}
5732	Ethanol, 2-mercapto- 2-Mercaptoethanol	C_2H_6OS 78.14	60-24-2	5760 158; 55^{13}	4-01-00-02428 1.1143^{20}	H_2O 3; EtOH 3; eth 3; bz 3 1.4996^{20}
5733	Ethanol, 2-methoxy- Ethylene glycol monomethyl ether	$C_3H_8O_2$ 76.10	109-86-4 -85.1	5961 124.1	4-01-00-02375 0.9647^{20}	H_2O 5; EtOH 4; eth 5; ace 3 1.4024^{20}
5734	Ethanol, methoxy-, acetate Acetic acid, 2-methoxyethyl ester	$C_5H_{10}O_3$ 118.13	32718-56-2	 145; 40^{12}	2-02-00-00154 1.0074^{19}	H_2O 3; EtOH 3; eth 3; ctc 2 1.4002^{20}
5735	Ethanol, 2-methoxy-, acetate 2-Methoxyethyl acetate	$C_5H_{10}O_3$ 118.13	110-49-6 -70	5962 143	4-02-00-00214 1.0074^{19}	H_2O 4; EtOH 4 1.4002^{20}
5736	Ethanol, 2-methoxy-, carbonate (2:1) Bis(2-methoxyethyl) carbonate	$C_7H_{14}O_5$ 178.19	626-84-6	 231	3-03-00-00017 1.0988^{20}	EtOH 3; eth 3; ace 3 1.4204^{20}
5737	Ethanol, 2-(2-methoxyethoxy)- 2-(2-Methoxyethoxy)ethanol	$C_5H_{12}O_3$ 120.15	111-77-3	5959 193	4-01-00-02392 1.035^{20}	H_2O 5; EtOH 4; eth 4; ace 5 1.4264^{20}
5738	Ethanol, 2-(methylamino)- *N*-Methyl-2-ethanolamine	C_3H_9NO 75.11	109-83-1	5939 158	4-04-00-01422 0.937^{20}	H_2O 5; EtOH 5; eth 5 1.4385^{20}
5739	Ethanol, 2-[(3-methylbutyl)amino]- *N*-Isopentylethanolamine	$C_7H_{17}NO$ 131.22	34240-76-1	 210; $82^{5.5}$	2-04-00-00728 0.8822^{20}	H_2O 4; eth 4; EtOH 4 1.4447^{20}
5740	Ethanol, 2-(1-methylethoxy)- 2-Isopropoxyethanol	$C_5H_{12}O_2$ 104.15	109-59-1	 145	4-01-00-02380 0.9030^{20}	H_2O 5; EtOH 5; eth 5; ace 3 1.4095^{20}
5741	Ethanol, 2-[(1-methylethyl)amino]-	$C_5H_{13}NO$ 103.16	109-56-8 128.5	 173	3-04-00-00681 0.8970^{20}	H_2O 5; EtOH 5; eth 5 1.4395^{20}
5742	Ethanol, 2,2'-(methylimino)bis-	$C_5H_{13}NO_2$ 119.16	105-59-9 -21	 247	4-04-00-01517 1.038^{25}	H_2O 4 1.4685^{20}
5743	Ethanol, 2-(methylphenylamino)-	$C_9H_{13}NO$ 151.21	93-90-3	 150^{14}	4-12-00-00280 0.9993^{15}	H_2O 3; EtOH 4; eth 4; ace 4
5744	Ethanol, 2-[(2-methylphenyl)amino]-	$C_9H_{13}NO$ 151.21	136-80-1	 285.5	4-12-00-01750 1.0794^{20}	eth 4; EtOH 4 1.5675^{20}
5745	Ethanol, 2-[(4-methylphenyl)amino]-	$C_9H_{13}NO$ 151.21	2933-74-6 42.5	 287	4-12-00-01880	bz 4; eth 4; EtOH 4; chl 4
5746	Ethanol, 2-(2-methylpropoxy)-	$C_6H_{14}O_2$ 118.18	4439-24-1	 160	4-01-00-02382 0.8900^{20}	 1.4143^{20}
5747	Ethanol, 2-[(2-methylpropyl)amino]- Ethanolamine, *N*-isobutyl	$C_6H_{15}NO$ 117.19	17091-40-6	 199; 90^{16}	3-04-00-00683 0.8818^{20}	H_2O 4; EtOH 4; eth 4 1.4402^{20}
5748	Ethanol, 2-[(2-methylpropyl)amino]-, 4-aminobenzoate (ester) Butethamine	$C_{13}H_{20}N_2O_2$ 236.31	2090-89-3	1517		
5749	Ethanol, 2-(methylthio)-	C_3H_8OS 92.16	5271-38-5	 $68\text{-}70^{20}$	4-01-00-02429 1.063^{20}	H_2O 4; eth 4; EtOH 4 1.4861^{30}
5750	Ethanol, 2-(2-naphthalenylamino)- Ethanolamine, *N*-(β-naphthyl)	$C_{12}H_{13}NO$ 187.24	36190-77-9 52	 $197\text{-}8^{3}$	4-12-00-03131	eth 4
5751	Ethanol, 2,2',2''-nitrilotris- Triethanolamine	$C_6H_{15}NO_3$ 149.19	102-71-6 20.5	9581 335.4	4-04-00-01524 1.1242^{20}	H_2O 5; EtOH 5; eth 2; bz 2 1.4852^{20}
5752	Ethanol, 2-nitro- 2-Nitroethanol	$C_2H_5NO_3$ 91.07	625-48-9 -80	 194; 102^{10}	4-01-00-01388 1.270^{15}	H_2O 5; EtOH 5; eth 5; bz 1 1.4438^{19}
5753	Ethanol, 2,2'-oxybis- Diethylene glycol	$C_4H_{10}O_3$ 106.12	111-46-6 -10.4	3109 245.8	4-01-00-02390 1.1197^{15}	H_2O 3; EtOH 3; eth 3; chl 3 1.4472^{20}
5754	Ethanol, 2,2'-oxybis-, diacetate Diethylene glycol, diacetate	$C_8H_{14}O_5$ 190.20	628-68-2 18	 200	4-02-00-00216 1.1068^{15}	EtOH 4 1.4348^{20}
5755	Ethanol, 2,2'-oxybis-, dibenzoate	$C_{18}H_{18}O_5$ 314.34	120-55-8 33.5	 280^{24}; 250^{1}	4-09-00-00356 1.1690^{15}	H_2O 4; EtOH 4

No.	Name Synonym	Mol. Form. Mol. Wt.	CAS RN mp/°C	Merck No. bp/°C	Beil. Ref. den/g cm^{-3}	Solubility n_D
5756	Ethanol, 2,2'-[oxybis(2,1-ethanediyloxy)]bis- Tetraethylene glycol	$C_8H_{18}O_5$ 194.23	112-60-7 -6.2	 328	4-01-00-02403 1.1285^{15}	H_2O 4; EtOH 3; eth 3; ctc 3 1.4577^{20}
5757	Ethanol, 2-(pentyloxy)-	$C_7H_{16}O_2$ 132.20	6196-58-3	 182	4-01-00-02383 0.8918^{15}	 1.4213^{25}
5758	Ethanol, 2-phenoxy- 2-Phenoxyethanol	$C_8H_{10}O_2$ 138.17	122-99-6 14	7226 245	4-06-00-00571 1.102^{22}	H_2O 1; EtOH 3; eth 3; chl 3 1.534^{20}
5759	Ethanol, 2-(phenylamino)-	$C_8H_{11}NO$ 137.18	122-98-5	 279.5; 150^{10}	4-12-00-00277 1.0945^{20}	H_2O 2; EtOH 4; eth 4; chl 4 1.5760^{20}
5760	Ethanol, 2,2'-(phenylimino)bis-	$C_{10}H_{15}NO_2$ 181.23	120-07-0 57	 200^{10}	4-12-00-00285 1.201^{60}	ace 4; bz 4; eth 4; EtOH 4
5761	Ethanol, 2-(phenylmethoxy)-	$C_9H_{12}O_2$ 152.19	622-08-2 <-75	 256	4-06-00-02241 1.0640^{20}	H_2O 4; eth 4; EtOH 4 1.5233^{20}
5762	Ethanol, 2-[(phenylmethyl)amino]-	$C_9H_{13}NO$ 151.21	104-63-2	 $153\text{-}6^{12}$	4-12-00-02184 1.065^{25}	 1.5430^{20}
5763	Ethanol, 2-(2-propenyloxy)-	$C_5H_{10}O_2$ 102.13	111-45-5	 158.5	4-01-00-02388 0.9580^{20}	H_2O 5; EtOH 4; bz 3; ctc 3 1.4358^{20}
5764	Ethanol, 2-propoxy- 2-Propoxyethanol	$C_5H_{12}O_2$ 104.15	2807-30-9	 149.8	4-01-00-02379 0.9112^{20}	H_2O 3; EtOH 4; eth 4 1.4133^{20}
5765	Ethanol, 2-(propylamino)-	$C_5H_{13}NO$ 103.16	16369-21-4	 183	3-04-00-00681 0.9005^{20}	 1.4428^{20}
5766	Ethanol, 2,2'-thiobis- 2,2'-Thiodiethanol	$C_4H_{10}O_2S$ 122.19	111-48-8 -10.2	9259 282	4-01-00-02437 1.1793^{25}	H_2O 5; EtOH 5; eth 3; bz 2 1.5211^{20}
5767	Ethanol, 2,2,2-tribromo- Tribromoethanol	$C_2H_3Br_3O$ 282.76	75-80-9 81	9525 $92\text{-}3^{10}$	3-01-00-01362	bz 4; eth 4; EtOH 4
5768	Ethanol, 2,2,2-trichloro- 2,2,2-Trichloroethanol	$C_2H_3Cl_3O$ 149.40	115-20-8 19	9551 152; 52^{11}	4-01-00-01383	H_2O 2; EtOH 5; eth 5; ctc 2 1.4861^{20}
5769	Ethanol, 2,2,2-trichloro-1-ethoxy- Chloral alcoholate	$C_4H_7Cl_3O_2$ 193.46	515-83-3 56.5	2056 115.5	4-01-00-03144 1.143^{40}	H_2O 3; EtOH 3; eth 3
5770	Ethanol, 2,2,2-trifluoro-	$C_2H_3F_3O$ 100.04	75-89-8 -43.5	 74	4-01-00-01370 1.3842^{20}	EtOH 4; eth 3; ace 3; bz 3 1.2907^{22}
5771	Ethanone, 2-(acetyloxy)-1-phenyl-	$C_{10}H_{10}O_3$ 178.19	2243-35-8 49	 270	4-08-00-00368 1.1169^{65}	H_2O 1; EtOH 4; eth 4; bz 2 1.5036^{65}
5772	Ethanone, 1-(3-aminophenyl)-	C_8H_9NO 135.17	99-03-6 98.5	 289.5	4-14-00-00096	H_2O 2; EtOH 3
5773	Ethanone, 1-(4-aminophenyl)-	C_8H_9NO 135.17	99-92-3 106	 294; 195^{15}	4-14-00-00100	eth 4; EtOH 4
5774	Ethanone, 2-amino-1-phenyl- Phenacylamine	C_8H_9NO 135.17	613-89-8 20	7159 251	4-14-00-00114	H_2O 1; eth 3; ctc 2 1.6160^{20}
5775	Ethanone, 1-(2-benzofuranyl)-	$C_{10}H_8O_2$ 160.17	1646-26-0 76	 126^{11}	5-17-10-00172	H_2O 3
5776	Ethanone, 1-[1,1'-biphenyl]-4-yl-	$C_{14}H_{12}O$ 196.25	92-91-1 121	 326	4-07-00-01407 1.2510^{0}	H_2O 1; EtOH 4; ace 4; chl 2
5777	Ethanone, 2-bromo-1-(4-bromophenyl)- *p*-Bromophenacyl bromide	$C_8H_6Br_2O$ 277.94	99-73-0 111	1415	4-07-00-00652	H_2O 1; EtOH 3; eth 3; chl 3
5778	Ethanone, 2-bromo-1-(4-chlorophenyl)- *p*-Chlorophenacyl bromide	C_8H_6BrClO 233.49	536-38-9 96.5	2153	4-07-00-00651	
5779	Ethanone, 2-bromo-1-(4-methylphenyl)-	C_9H_9BrO 213.07	619-41-0 51	 157^{14}	4-07-00-00705	eth 4; EtOH 4
5780	Ethanone, 1-(3-bromophenyl)-	C_8H_7BrO 199.05	2142-63-4 7.5	 133^{19}	4-07-00-00646	H_2O 1; ace 3; bz 3 1.5755^{20}
5781	Ethanone, 1-(4-bromophenyl)- *p*-Bromoacetophenone	C_8H_7BrO 199.05	99-90-1 50.5	1390 257; 130^{11}	4-07-00-00647 1.647^{25}	H_2O 1; EtOH 3; eth 3; bz 3 1.647
5782	Ethanone, 2-bromo-1-phenyl- ω-Bromacetophenone	C_8H_7BrO 199.05	70-11-1 50.5	1391 135^{18}	4-07-00-00649 1.647^{20}	H_2O 1; EtOH 3; eth 4; bz 4
5783	Ethanone, 1-(5-bromo-2-thienyl)-	C_6H_5BrOS 205.08	5370-25-2 94.5	 103^{4}	5-17-09-00392	EtOH 2; ctc 3
5784	Ethanone, 2-chloro-1-(4-chlorophenyl)-	$C_8H_6Cl_2O$ 189.04	937-20-2 101.5	 270	4-07-00-00644	EtOH 3; bz 3; MeOH 3
5785	Ethanone, 2-chloro-1,2-diphenyl-	$C_{14}H_{11}ClO$ 230.69	447-31-4 68.5	 dec	3-07-00-02106	EtOH 3; chl 2; alk 1
5786	Ethanone, 2-chloro-1-(4-hydroxy-3-methoxyphenyl)-	$C_9H_9ClO_3$ 200.62	6344-28-1 102	 208^{22}	3-08-00-02114	H_2O 2; EtOH 3
5787	Ethanone, 2-chloro-1-(4-methylphenyl)-	C_9H_9ClO 168.62	4209-24-9 57.5	 261.5	4-07-00-00704	eth 4; EtOH 4
5788	Ethanone, 1-(3-chlorophenyl)-	C_8H_7ClO 154.60	99-02-5	 244; 129^{30}	4-07-00-00638 1.2130^{40}	EtOH 3; eth 3; ace 3 1.5494^{20}
5789	Ethanone, 1-(4-chlorophenyl)- *p*-Chloroacetophenone	C_8H_7ClO 154.60	99-91-2 20	2114 232	4-07-00-00639 1.1922^{20}	H_2O 1; EtOH 5; eth 5; chl 3 1.5550^{20}
5790	Ethanone, 2-chloro-1-phenyl- ω-Chloroacetophenone	C_8H_7ClO 154.60	532-27-4 56.5	2115 247	4-07-00-00641 1.324^{15}	H_2O 1; EtOH 4; eth 4; ace 3
5791	Ethanone, 1-(5-chloro-2-thienyl)-	C_6H_5ClOS 160.62	6310-09-4 52	 117^{17}; $88^{4.5}$	5-17-09-00391	H_2O 4; eth 4; EtOH 4
5792	Ethanone, 1-cyclobutyl-	$C_6H_{10}O$ 98.14	3019-25-8	 138	4-07-00-00037 0.9020^{20}	H_2O 2 1.4322^{19}
5793	Ethanone, 1-(1-cyclohexen-1-yl)-	$C_8H_{12}O$ 124.18	932-66-1 73	 201.5	4-07-00-00143 0.9655^{20}	EtOH 3; eth 3 1.4881^{20}
5794	Ethanone, 1-cyclohexyl-	$C_8H_{14}O$ 126.20	823-76-7	 180.5	4-07-00-00053 0.9176^{20}	H_2O 1; eth 3 1.4565^{16}
5795	Ethanone, 1-cyclopentyl-	$C_7H_{12}O$ 112.17	6004-60-0	 158.5	4-07-00-00047 0.916^{20}	eth 3 1.4409^{20}

No.	Name Synonym	Mol. Form. Mol. Wt.	CAS RN mp/°C	Merck No. bp/°C	Beil. Ref. den/g cm^{-3}	Solubility n_D
5796	Ethanone, 1-cyclopropyl- Cyclopropyl methyl ketone	C_5H_8O 84.12	765-43-5 -68.3	 111.3	4-07-00-00014 0.8984^{20}	H_2O 4; eth 4; EtOH 4 1.4251^{20}
5797	Ethanone, 1-(3,5-dibromophenyl)-	$C_8H_6Br_2O$ 277.94	14401-73-1 68	 198^{15}	3-07-00-00984	bz 4; eth 4; EtOH 4
5798	Ethanone, 2,2-dibromo-1-phenyl-	$C_8H_6Br_2O$ 277.94	13665-04-8 36.5	 159^{13}	4-07-00-00653	EtOH 3; eth 3; chl 3
5799	Ethanone, 1-(2,4-dichlorophenyl)-	$C_8H_6Cl_2O$ 189.04	2234-16-4 33.5		4-07-00-00643	H_2O 1 1.5640^{20}
5800	Ethanone, 1-(2,5-dichlorophenyl)-	$C_8H_6Cl_2O$ 189.04	2476-37-1 12	 118^{12}	4-07-00-00643 1.321^{30}	 1.5595^{30}
5801	Ethanone, 1-(3,4-dichlorophenyl)-	$C_8H_6Cl_2O$ 189.04	2642-63-9 76	 135^{12}	4-07-00-00643	H_2O 1; ctc 3; lig 3
5802	Ethanone, 2,2-dichloro-1-phenyl-	$C_8H_6Cl_2O$ 189.04	2648-61-5 20.5	 249	4-07-00-00644 1.340^{16}	EtOH 3; bz 3; ctc 3 1.5686^{20}
5803	Ethanone, 1-(2,4-dihydroxyphenyl)- Resacetophenone	$C_8H_8O_3$ 152.15	89-84-9 146	8143	4-08-00-01792 1.18^{141}	H_2O 1; EtOH 3; eth 2; bz 2
5804	Ethanone, 1-(3,4-dihydroxyphenyl)-2-(methylamino)- Adrenalone	$C_9H_{11}NO_3$ 181.19	99-45-6 235-6 dec	161	4-14-00-00832	H_2O 2; EtOH 2; eth 2
5805	Ethanone, 1-(3,4-dimethoxyphenyl)-	$C_{10}H_{12}O_3$ 180.20	1131-62-0 51	 287	4-08-00-01815	H_2O 4; bz 4; EtOH 4; chl 4
5806	Ethanone, 1-[3-(dimethylamino)phenyl]-	$C_{10}H_{13}NO$ 163.22	18992-80-8 43	 148^{13}	4-14-00-00096	ace 4; eth 4
5807	Ethanone, 1-[10-[3-(dimethylamino)propyl]-10H-phenothiazin-2-yl]- Acepromazine	$C_{19}H_{22}N_2OS$ 326.46	61-00-7	27 $230^{0.5}$	4-27-00-02911	
5808	Ethanone, 1-[4-(1,1-dimethylethyl)phenyl]-	$C_{12}H_{16}O$ 176.26	943-27-1	 $136\text{-}8^{20}$	4-07-00-00769 0.9705^{25}	 1.518^{15}
5809	Ethanone, 1-(2,4-dimethylphenyl)-	$C_{10}H_{12}O$ 148.20	89-74-7	 228	4-07-00-00729 1.0121^{15}	eth 4; EtOH 4 1.5340^{20}
5810	Ethanone, 1-(2,5-dimethylphenyl)-	$C_{10}H_{12}O$ 148.20	2142-73-6 -18.1	 232.5	4-07-00-00728 0.9963^{19}	H_2O 1; EtOH 4; eth 4; bz 4 1.5291^{20}
5811	Ethanone, 1-(3,4-dimethylphenyl)-	$C_{10}H_{12}O$ 148.20	3637-01-2 -3	 246.5	4-07-00-00728 1.0090^{14}	H_2O 1; EtOH 4; eth 4; bz 4 1.5413^{15}
5812	Ethanone, 1,2-diphenyl-	$C_{14}H_{12}O$ 196.25	451-40-1 60	 320	4-07-00-01393 1.201^{0}	H_2O 2; EtOH 3; eth 3; ctc 3
5813	Ethanone, 2-ethoxy-1,2-diphenyl-	$C_{16}H_{16}O_2$ 240.30	574-09-4 62	 $194\text{-}5^{20}$	3-08-00-01279 1.1016^{17}	bz 4; eth 4; EtOH 4; lig 4 1.5727^{17}
5814	Ethanone, 2-ethoxy-1-phenyl- 2-Ethoxyacetophenone	$C_{10}H_{12}O_2$ 164.20	14869-39-7 43	 243.5	4-08-00-00365 1.0036^{78}	eth 4; EtOH 4; lig 4
5815	Ethanone, 1-(4-fluorophenyl)-	C_8H_7FO 138.14	403-42-9 -45	 196	4-07-00-00635 1.1382^{25}	H_2O 1; bz 3; chl 3 1.5081^{25}
5816	Ethanone, 2-fluoro-1-phenyl-	C_8H_7FO 138.14	450-95-3 29	 $90\text{-}1^{12}$	4-07-00-00636 1.152^{20}	 1.5200^{20}
5817	Ethanone, 1-(2-furanyl)-	$C_6H_6O_2$ 110.11	1192-62-7 33	 175	5-17-09-00381 1.098^{20}	H_2O 1; EtOH 3; eth 3 1.5017^{20}
5818	Ethanone, 1-[4-(β-*D*-glucopyranosyloxy)phenyl]- Picein	$C_{14}H_{18}O_7$ 298.29	530-14-3 195.5	7367	5-17-07-00151	H_2O 2; EtOH 3; eth 3; chl 1
5819	Ethanone, 1-(1H-indol-3-yl)-	$C_{10}H_9NO$ 159.19	703-80-0 192.3	 144^{10}	5-21-08-00297	EtOH 4
5820	Ethanone, 1-(1H-pyrrol-2-yl)-	C_6H_7NO 109.13	1072-83-9 90	 220	5-21-07-00204	H_2O 3; EtOH 3; eth 3
5821	Ethanone, 1-(1-hydroxycyclohexyl)-	$C_8H_{14}O_2$ 142.20	1123-27-9	 125.5; 91^{11}	4-08-00-00023 1.0248^{25}	eth 4; EtOH 4 1.4670^{25}
5822	Ethanone, 2-hydroxy-1,2-diphenyl-, (±)- (±)-Benzoin	$C_{14}H_{12}O_2$ 212.25	579-44-2 137	1103 344; 194^{12}	4-08-00-01279 1.310^{20}	EtOH 4; chl 4
5823	Ethanone, 1-(2-hydroxy-4-methoxyphenyl)-	$C_9H_{10}O_3$ 166.18	552-41-0 52.5	 158^{20}	4-08-00-01793 1.3102^{81}	bz 4; eth 4; EtOH 4; chl 4 1.5452^{81}
5824	Ethanone, 1-(4-hydroxy-3-methoxyphenyl)- Apocynin	$C_9H_{10}O_3$ 166.18	498-02-2 115	772 297; 234^{15}	4-08-00-01814	H_2O 2; EtOH 3; eth 4; ace 3
5825	Ethanone, 1-[6-hydroxy-2-(1-methylethenyl)-5-benzofuranyl]- Euparin	$C_{13}H_{12}O_3$ 216.24	532-48-9 121.5	3857	5-18-02-00029	eth 3; bz 3; chl 3; NaOH 2
5826	Ethanone, 1-(2-hydroxy-4-methylphenyl)-	$C_9H_{10}O_2$ 150.18	6921-64-8 21	 245	4-08-00-00480 1.1012^{10}	 1.5527^{13}
5827	Ethanone, 1-(2-hydroxy-5-methylphenyl)-	$C_9H_{10}O_2$ 150.18	1450-72-2 50	 210; 120^{20}	4-08-00-00472 1.0797^{53}	bz 4; eth 4; EtOH 4; chl 4
5828	Ethanone, 1-(4-hydroxy-2-methylphenyl)-	$C_9H_{10}O_2$ 150.18	875-59-2 128	 313	4-08-00-00400 1.0592^{135}	eth 4; EtOH 4 1.5369^{135}
5829	Ethanone, 1-(1-hydroxy-2-naphthalenyl)-	$C_{12}H_{10}O_2$ 186.21	711-79-5 101	 325 dec	4-08-00-01175	bz 4; HOAc 4
5830	Ethanone, 1-(2-hydroxyphenyl)-	$C_8H_8O_2$ 136.15	118-93-4 5	 218	4-08-00-00320 1.1307^{20}	eth 4; EtOH 4; HOAc 4 1.5584^{20}
5831	Ethanone, 1-(3-hydroxyphenyl)-	$C_8H_8O_2$ 136.15	121-71-1 96	 296; 153^{5}	4-08-00-00334 1.0992^{109}	H_2O 2; EtOH 4; eth 4; bz 4 1.5348^{109}
5832	Ethanone, 1-(4-hydroxyphenyl)-	$C_8H_8O_2$ 136.15	99-93-4 109.5	 $147\text{-}8^{3}$	4-08-00-00339 1.1090^{109}	H_2O 2; EtOH 4; eth 4 1.5577^{109}
5833	Ethanone, 2-hydroxy-1-phenyl-	$C_8H_8O_2$ 136.15	582-24-1 90	 125^{12}; 56^{1}	4-08-00-00365 1.0963^{99}	H_2O 3; EtOH 3; eth 3; chl 3

No.	Name	Synonym	Mol. Form.	Mol. Wt.	CAS RN	mp/°C	Merck No.	bp/°C	Beil. Ref.	den/g cm^{-3}	Solubility	n_D
5834	Ethanone, 2-[6-(2-hydroxy-2-phenylethyl)-1-methyl-2-piperidinyl]-1-phenyl-, [2*R*-[2α,6α(*S**)]]-	Lobeline	$C_{22}H_{27}NO_2$	337.46	90-69-7	130.5	5432		5-21-12-00627		H_2O 2; EtOH 3; eth 3; ace 4	
5835	Ethanone, 1-(3-hydroxy-2-thienyl)-		$C_6H_6O_2S$	142.18	5556-07-0	51.5		$47^{0.2}$	5-18-01-00121	1.5000^{20}	EtOH 3	1.5795^{20}
5836	Ethanone, 1-(3-iodophenyl)-		C_8H_7IO	246.05	14452-30-3			129^{8}; 117^{4}	4-07-00-00654		bz 3	1.622^{20}
5837	Ethanone, 1-(4-iodophenyl)-		C_8H_7IO	246.05	13329-40-3	86		153^{18}	4-07-00-00654		EtOH 3; eth 2; bz 3; HOAc 3	
5838	Ethanone, 2-iodo-1-phenyl-		C_8H_7IO	246.05	4636-16-2	34.4		154^{15}	4-07-00-00654		ace 3	
5839	Ethanone, 2-methoxy-1,2-diphenyl-		$C_{15}H_{14}O_2$	226.27	3524-62-7	49.5		$188\text{-}9^{15}$	4-08-00-01280	1.1278^{14}	bz 4; eth 4; EtOH 4	
5840	Ethanone, 1-(3-methoxyphenyl)-		$C_9H_{10}O_2$	150.18	586-37-8	95.5		240	4-08-00-00334	1.0343^{19}	H_2O 3; EtOH 3; ace 3; ctc 3	1.5410^{20}
5841	Ethanone, 1-(4-methoxyphenyl)-		$C_9H_{10}O_2$	150.18	100-06-1	38.5		258	4-08-00-00340	1.0818^{41}	H_2O 2; EtOH 3; eth 3; ace 3	1.547^{41}
5842	Ethanone, 2-methoxy-1-phenyl-		$C_9H_{10}O_2$	150.18	4079-52-1	8		245; 125^{19}	4-08-00-00365	1.0897^{20}	H_2O 2; EtOH 3; ace 3	1.5393^{20}
5843	Ethanone, 1-(1-methylcyclohexyl)-		$C_9H_{16}O$	140.23	2890-62-2			186.2	4-07-00-00067	0.9504^{20}		1.4543^{20}
5844	Ethanone, 1-(3-methylcyclohexyl)-		$C_9H_{16}O$	140.23	7193-78-4			197	3-07-00-00120	0.912^{19}		1.4517^{19}
5845	Ethanone, 1-(4-methylcyclohexyl)-		$C_9H_{16}O$	140.23	1879-06-7			196	3-07-00-00121	0.9055^{18}		1.4509^{18}
5846	Ethanone, 1-(1-methylcyclopentyl)-		$C_8H_{14}O$	126.20	13388-93-7			48.4^{10}	3-07-00-00102	0.9104^{20}		1.4430^{20}
5847	Ethanone, 1-(2-methylcyclopentyl)-		$C_8H_{14}O$	126.20	1601-00-9			170.5; 59.6^{15}	3-07-00-00103	0.9222^{4}		1.4434^{16}
5848	Ethanone, 1-[4-(1-methylethyl)phenyl]-		$C_{11}H_{14}O$	162.23	645-13-6			254	4-07-00-00747	0.9753^{15}		1.5235^{20}
5849	Ethanone, 1-(1-methyl-1H-pyrrol-2-yl)-		C_7H_9NO	123.15	932-16-1			201^{252}; 93^{22}	5-21-07-00205	1.0445^{15}	EtOH 3; bz 3; chl 3	1.5403^{15}
5850	Ethanone, 1-[2-methyl-5-(1-methylethyl)phenyl]-		$C_{12}H_{16}O$	176.26	1202-08-0	<-20		249.5	4-07-00-00772	0.956^{20}		1.5181^{20}
5851	Ethanone, 1-(2-methylphenyl)-		$C_9H_{10}O$	134.18	577-16-2			214	4-07-00-00697	1.026^{20}		1.5276^{20}
5852	Ethanone, 1-(3-methylphenyl)-		$C_9H_{10}O$	134.18	585-74-0			220	4-07-00-00699	1.0165^{0}	EtOH 3; eth 3; ace 3; ctc 2	1.533^{15}
5853	Ethanone, 1-(4-methylphenyl)-		$C_9H_{10}O$	134.18	122-00-9	28		226; 93.5^{7}	4-07-00-00701	1.0051^{20}	bz 4; eth 4; EtOH 4; chl 4	1.5335^{20}
5854	Ethanone, 2,2'-(1-methyl-2,6-piperidinediyl)bis[1-phenyl-, *cis*-	Lobelanine	$C_{22}H_{25}NO_2$	335.45	579-21-5	99	5430		5-21-11-00432		ace 4; bz 4; EtOH 4; chl 4	
5855	Ethanone, 1-(6-methyl-3-pyridinyl)-		C_8H_9NO	135.17	36357-38-7	17.6		144^{50}	5-21-07-00425	1.0168^{25}	H_2O 4	1.5302^{25}
5856	Ethanone, 1-(5-methyl-2-thienyl)-		C_7H_8OS	140.21	13679-74-8	27.5		232.5	5-17-09-00425	1.1185^{25}	eth 3; ace 3; bz 3; chl 2	1.5604
5857	Ethanone, 1-(1-naphthalenyl)-		$C_{12}H_{10}O$	170.21	941-98-0	34		297	4-07-00-01292	1.1171^{21}	H_2O 1; EtOH 3; eth 3; ace 3	1.6280^{22}
5858	Ethanone, 1-(2-naphthalenyl)-		$C_{12}H_{10}O$	170.21	93-08-3	56		302	4-07-00-01294		EtOH 2; ctc 2	
5859	Ethanone, 1-(1-naphthalenyl)-2-phenyl-	Benzyl 1-naphthyl ketone	$C_{18}H_{14}O$	246.31	605-85-6	66.5		$194\text{-}6^{0.05}$	4-07-00-01790		eth 4	
5860	Ethanone, 1-(3-nitrophenyl)-		$C_8H_7NO_3$	165.15	121-89-1	81		202; 167^{18}	4-07-00-00656		H_2O 4; EtOH 2; eth 4; chl 2	
5861	Ethanone, 1-(4-nitrophenyl)-		$C_8H_7NO_3$	165.15	100-19-6	81.8		165^{5}	4-07-00-00657		eth 4; EtOH 4	
5862	Ethanone, 2-nitro-1-phenyl-		$C_8H_7NO_3$	165.15	614-21-1	106		158^{16}; 142^{10}	4-07-00-00659		eth 4; EtOH 4	1.5468^{30}
5863	Ethanone, 1-(2-nitrophenyl)-	Acetophene,2-nitro	$C_8H_7N_3O$	161.16	577-59-3	28.5		178^{32}; 158^{16}	4-07-00-00655	1.2370^{25}	H_2O 1; EtOH 4; eth 4; chl 4	1.5468^{20}
5864	Ethanone, 1-phenyl-	Acetophenone	C_8H_8O	120.15	98-86-2	20	65	202	4-07-00-00619	1.0281^{20}	H_2O 1; EtOH 3; eth 3; ace 3	1.5372^{20}
5865	Ethanone, 1,1'-(1,3-phenylene)bis-		$C_{10}H_{10}O_2$	162.19	6781-42-6	32		152^{15}	4-07-00-02156		H_2O 2; EtOH 3; bz 3; chl 3	
5866	Ethanone, 1-phenyl-, oxime		C_8H_9NO	135.17	613-91-2	60		245	3-07-00-00954	1.0515^{78}	H_2O 2; EtOH 4; eth 4; ace 4	
5867	Ethanone, 1-(2-pyridinyl)-		C_7H_7NO	121.14	1122-62-9			192	5-21-07-00385	1.077^{25}	EtOH 3; eth 3; ctc 2; HOAc 3	1.5203^{20}
5868	Ethanone, 1-(3-pyridinyl)-	Methyl 3-pyridyl ketone	C_7H_7NO	121.14	350-03-8	13.5	6034	220	5-21-07-00394		H_2O 3; EtOH 3; eth 3; acid 3	1.5341^{20}
5869	Ethanone, 1-(4-pyridinyl)-		C_7H_7NO	121.14	1122-54-9	16		212	5-21-07-00400	1.097^{25}	EtOH 2; eth 2; acid 2	1.5282^{25}

PHYSICAL CONSTANTS OF ORGANIC COMPOUNDS (continued)

No.	Name Synonym	Mol. Form. Mol. Wt.	CAS RN mp/°C	Merck No. bp/°C	Beil. Ref. den/g cm^{-3}	Solubility n_D
5870	Ethanone, 1-(2-thienyl)-	C_6H_6OS 126.18	88-15-3 10.5	 213.5	5-17-09-00387 1.1679^{20}	H_2O 2; EtOH 5; eth 5; ctc 3 1.5667^{20}
5871	Ethanone, 2,2,2-trichloro-1-phenyl-	$C_8H_5Cl_3O$ 223.49	2902-69-4	 256.5	4-07-00-00645 1.425^{16}	eth 4; EtOH 4
5872	Ethanone, 2,2,2-trifluoro-1-phenyl-	$C_8H_5F_3O$ 174.12	434-45-7 -40	 153	4-07-00-00637 1.279^{20}	 1.4583^{20}
5873	Ethanone, 1-(2,3,4-trihydroxyphenyl)- Gallacetophenone	$C_8H_8O_4$ 168.15	528-21-2 173	4248	4-08-00-02721	H_2O 3; EtOH 4; eth 3; ace 4
5874	Ethanone, 1-(2,3,4-trimethoxyphenyl)-	$C_{11}H_{14}O_4$ 210.23	13909-73-4 15.8	 296	4-08-00-02721	 1.5384^{20}
5875	Ethanone, 1-(2,4,5-trimethylphenyl)-	$C_{11}H_{14}O$ 162.23	2040-07-5 10.5	 246.5	4-07-00-00751 1.0039^{15}	H_2O 1; EtOH 4; eth 4; bz 4 1.541^{15}
5876	Ethanone, 1-(2,4,6-trimethylphenyl)-	$C_{11}H_{14}O$ 162.23	1667-01-2	 241; 120^{12}	4-07-00-00749 0.9754^{20}	H_2O 1; EtOH 3; eth 3; ace 3 1.5175^{20}
5877	Ethanone, 1-(3,4,5-trimethylphenyl)-	$C_{11}H_{14}O$ 162.23	2047-21-4 4.7	 101.5^3	4-07-00-00749 1.0037^{25}	 1.5420^{25}
5878	Ethenaminium, *N,N,N*-trimethyl-, hydroxide Neurine	$C_5H_{13}NO$ 103.16	463-88-7	6393	3-04-00-00442	H_2O 4; eth 4; EtOH 4
5879	Ethene Ethylene	C_2H_4 28.05	74-85-1 -169	3748 -103.7	4-01-00-00677 0.5678^{-104}	H_2O 1; EtOH 2; eth 3; ace 2 1.363^{100}
5880	Ethene, bromo- Vinyl bromide	C_2H_3Br 106.95	593-60-2 -137.8	 15.8	4-01-00-00718 1.4933^{20}	H_2O 1; EtOH 3; eth 3; ace 3 1.4380^{20}
5881	Ethene, 1-bromo-2-chloro- Ethylene, 1-bromo-2-chloro-	C_2H_2BrCl 141.39	3018-09-5 -86.7	 84.6	3-01-00-00671 1.7972^{15}	 1.4982
5882	Ethene, 2-bromo-1,1-dichloro-	C_2HBrCl_2 175.84	5870-61-1 -88.5	 107.5	4-01-00-00720 1.9053^{15}	
5883	Ethene, 1-bromo-1,2-difluoro- 1-Bromo-1,2-difluoroethylene	C_2HBrF_2 142.93	358-99-6	 19	0-01-00-00189 1.8434^{25}	 1.3846^0
5884	Ethene, chloro- Vinyl chloride	C_2H_3Cl 62.50	75-01-4 -153.7	9898 -13.3	4-01-00-00700 0.9106^{20}	H_2O 2; EtOH 3; eth 4 1.3700^{20}
5885	Ethene, 2-chloro-1,1-difluoro- 1,1-Difluoro-2-chloroethylene	C_2HClF_2 98.48	359-10-4 -138.5	 -18.5	4-01-00-00703	
5886	Ethene, 1-chloro-2-ethoxy- 2-Chlorovinyl ethyl ether	C_4H_7ClO 106.55	928-56-3	 120	4-01-00-02061 1.0386^{20}	 1.4385^{20}
5887	Ethene, (2-chloroethoxy)- 2-Chloroethyl vinyl ether	C_4H_7ClO 106.55	110-75-8 -70	2139 108	4-01-00-02051 1.0495^{20}	EtOH 4; eth 4; chl 2 1.4378^{20}
5888	Ethene, 1-chloro-2-iodo-	C_2H_2ClI 188.40	20244-71-7 -38.2	 119	3-01-00-00674 2.2298^{25}	
5889	Ethene, chlorotrifluoro- Chlorotrifluoroethylene	C_2ClF_3 116.47	79-38-9 -158	 -27.8	4-01-00-00704 1.54^{-60}	bz 3; chl 3 1.38^0
5890	Ethene, 1,1-dibromo-	$C_2H_2Br_2$ 185.85	593-92-0	 92	4-01-00-00720 2.1779^{21}	EtOH 3; eth 3; ace 3; bz 3
5891	Ethene, 1,2-dibromo-, (*E*)- *trans*-1,2-Dibromoethylene	$C_2H_2Br_2$ 185.85	590-12-5 -6.5	 108	4-01-00-00721 2.2308^{20}	H_2O 1; EtOH 4; eth 4; ace 3 1.5505^{18}
5892	Ethene, 1,2-dibromo-, (*Z*)- *cis*-1,2-Dibromoethylene	$C_2H_2Br_2$ 185.85	590-11-4 -53	 112.5	4-01-00-00720 2.2464^{20}	H_2O 1; EtOH 4; eth 4; ace 3 1.5428^{20}
5893	Ethene, 1,1-dibromo-2-ethoxy- Ethylene, 1,1-dibromo-2-ethoxy	$C_4H_6Br_2O$ 229.90	77295-79-5	 172; 74^{15}	2-01-00-00473 1.7697^{18}	eth 4
5894	Ethene, 1,1-dichloro- Vinylidene chloride	$C_2H_2Cl_2$ 96.94	75-35-4 -122.5	9900 31.6	4-01-00-00706 1.213^{20}	H_2O 1; EtOH 3; eth 4; ace 3 1.4249^{20}
5895	Ethene, 1,2-dichloro-, (*E*)- *trans*-1,2-Dichloroethylene	$C_2H_2Cl_2$ 96.94	156-60-5 -49.8	86 48.7	4-01-00-00709 1.2565^{20}	H_2O 2; EtOH 5; eth 5; ace 5 1.4454^{20}
5896	Ethene, 1,2-dichloro-, (*Z*)- *cis*-1,2-Dichloroethylene	$C_2H_2Cl_2$ 96.94	156-59-2 -80	86 60.1	4-01-00-00707 1.2837^{20}	H_2O 2; EtOH 5; eth 5; ace 5 1.4490^{20}
5897	Ethene, 1,1-dichloro-2,2-difluoro-	$C_2Cl_2F_2$ 132.92	79-35-6 -116	 19	4-01-00-00711 1.555^{-20}	 1.383^{-20}
5898	Ethene, 1,2-dichloro-1,2-difluoro-	$C_2Cl_2F_2$ 132.92	598-88-9 -130.5	 21.1	 1.4950^0	 1.3777^0
5899	Ethene, 1,2-dichloro-1-ethoxy-	$C_4H_6Cl_2O$ 141.00	42345-82-4	 128.2	3-01-00-02950 1.1972^{25}	 1.4558^{17}
5900	Ethene, 1,1-dichloro-2-fluoro- 1,1-Dichloro-2-fluoroethylene	C_2HCl_2F 114.93	359-02-4 -108.8	 37.5	4-01-00-00711 1.3732^{16}	 1.4031^{16}
5901	Ethene, 1,1-diethoxy-	$C_6H_{12}O_2$ 116.16	2678-54-8	 68^{100}	4-01-00-03420 0.7932^{20}	 1.3643^{21}
5902	Ethene, 1,1-difluoro- Vinylidene fluoride	$C_2H_2F_2$ 64.03	75-38-7 -144	 -85.7	4-01-00-00696	eth 4; EtOH 4
5903	Ethene, 1,2-difluoro-, (*Z*)- *cis*-1,2-Difluoroethylene	$C_2H_2F_2$ 64.03	1630-77-9			
5904	Ethene, 1,2-diiodo-, (*Z*)-	$C_2H_2I_2$ 279.85	600-20-1 -14	 72.5^{16}	4-01-00-00724 3.0625^{20}	eth 3; chl 3
5905	Ethene, ethoxy- Ethyl vinyl ether	C_4H_8O 72.11	109-92-2 -115.8	 35.5	4-01-00-02049 0.7589^{20}	H_2O 2; EtOH 3; eth 5; ctc 2 1.3767^{20}
5906	Ethene, fluoro- Vinyl fluoride	C_2H_3F 46.04	75-02-5 -160.5	 -72	4-01-00-00694	H_2O 1; EtOH 3; ace 3
5907	Ethene, iodo- Iodoethylene	C_2H_3I 153.95	593-66-8	 56	4-01-00-00722 2.037^{20}	eth 4; EtOH 4 1.5385^{20}
5908	Ethene, methoxy- Methyl vinyl ether	C_3H_6O 58.08	107-25-5 -122	 5.5	4-01-00-02049 0.7725^0	H_2O 2; EtOH 4; eth 4; ace 4 1.3730^0
5909	Ethene, (methylsulfonyl)-	$C_3H_6O_2S$ 106.15	3680-02-2	 $122\text{-}4^{24}$	4-01-00-02065 1.2117^{20}	eth 3; ace 3 1.4636^{20}

No.	Name	Synonym	Mol. Form.	Mol. Wt.	CAS RN	mp/°C	Merck No.	bp/°C	Beil. Ref.	den/g cm^{-3}	Solubility	n_D
5910	Ethene, (methylthio)-		C_3H_6S	74.15	1822-74-8			69.5	4-01-00-02065	0.9026^{20}	eth 3; ace 3; chl 3	1.4837^{20}
5911	Ethene, nitro-		$C_2H_3NO_2$	73.05	3638-64-0	-55.5		98.5	4-01-00-00725	1.2212^{14}	EtOH 4; eth 4; ace 4; bz 4	1.4282^{20}
5912	Ethene, 1,1'-oxybis-	Divinyl ether	C_4H_6O	70.09	109-93-3	-101	9899	28.3	4-01-00-02058	0.773^{20}	H_2O 1; EtOH 5; eth 5; ace 5	1.3989^{20}
5913	Ethene, 1,1'-sulfonylbis-		$C_4H_6O_2S$	118.16	77-77-0	-26		234.5	4-01-00-02068	1.177^{25}		1.4765^{20}
5914	Ethene, tetrabromo-	Tetrabromoethylene	C_2Br_4	343.64	79-28-7	56.5		226	4-01-00-00722		H_2O 1; EtOH 3; eth 3; ace 3	
5915	Ethenetetracarbonitrile	Tetracyanoethylene	C_6N_4	128.09	670-54-2	199	9129	223	4-02-00-02450	1.348^{25}	eth 2; ace 3; bz 2; ctc 2	1.560^{25}
5916	Ethenetetracarboxylic acid, tetraethyl ester		$C_{14}H_{20}O_8$	316.31	6174-95-4	58		328 dec; 203^{13}	4-02-00-02450		H_2O 1; EtOH 4; eth 4	
5917	Ethene, tetrachloro-	Tetrachloroethylene	C_2Cl_4	165.83	127-18-4	-22.3	9126	121.3	4-01-00-00715	1.6227^{20}	H_2O 1; EtOH 5; eth 5; bz 5	1.5053^{20}
5918	Ethene, tetrafluoro-	Tetrafluoroethylene	C_2F_4	100.02	116-14-3	-142.5		-75.9	4-01-00-00698	1.519^{-76}	H_2O 1	
5919	Ethene, tetraiodo-	Tetraiodoethylene	C_2I_4	531.64	513-92-8	187	9151	sub	4-01-00-00724	2.983^{20}	bz 4; chl 4	
5920	Ethene, 1,1'-thiobis-		C_4H_6S	86.16	627-51-0	20		84	4-01-00-02068	0.9174^{15}	H_2O 2; EtOH 5; eth 5; ace 3	
5921	Ethene, tribromo-		C_2HBr_3	264.74	598-16-3			164	4-01-00-00722	2.708^{20}	H_2O 2; EtOH 4; eth 3; ace 3	1.6045^{16}
5922	Ethene, trichloro-	Trichloroethylene	C_2HCl_3	131.39	79-01-6	-84.7	9552	87.2	4-01-00-00712	1.4642^{20}	H_2O 2; EtOH 5; eth 5; ace 3	1.4773^{20}
5923	Ethene, trichlorofluoro-		C_2Cl_3F	149.38	359-29-5	-108.9		71	4-01-00-00715	1.5460^{20}	chl 3	1.4379^{20}
5924	Ethene, trifluoro-	Trifluoroethylene	C_2HF_3	82.03	359-11-5			-51	4-01-00-00697	1.26^{-70}	H_2O 1; EtOH 2; eth 3	
5925	Ethene, trifluoroiodo-		C_2F_3I	207.92	359-37-5			30	4-01-00-00723	2.284^{25}		1.4143^{0}
5926	1,4-Etheno-3H,7H-benzo[1,2-c:3,4-c']dipyran-3,7-dione, 9-(3-furanyl)-1,4,4a,5,6,6a,9,10,10a,10b-decahydro-4-hydroxy-4a,10a-dimethyl-, [1R-(1α,4β,4aα,6aβ,9β,10aα,10b)]	Columbin	$C_{20}H_{22}O_6$	358.39	546-97-4		2489		5-19-10-00546			
5927	Ethenone	Ketene	C_2H_2O	42.04	463-51-4	-151	5177	-49.8	4-01-00-03418		eth 2; ace 2	
5928	Ethenone, diphenyl-	Diphenylketene	$C_{14}H_{10}O$	194.23	525-06-4		3327	267.5	4-07-00-01653	1.1107^{13}		1.615^{14}
5929	Ethephon	Phosphonic acid, (2-chloroethyl)-	$C_2H_6ClO_3P$	144.49	16672-87-0	74	3686			1.2		
5930	Ethiolate	Carbamothioic acid, diethyl-, *S*-ethyl ester	$C_7H_{15}NOS$	161.27	2941-55-1			$52^{0.5}$; 142^{87}		0.9791^{30}		
5931	Ethion	Phosphorodithioic acid, *S,S'*-methylene *O,O,O',O'*-tetraethyl ester	$C_9H_{22}O_4P_2S_4$	384.48	563-12-2	-13	3691	$165^{0.3}$		1.22^{20}		
5932	Ethirimol	4(1H)-Pyrimidinone, 5-butyl-2-(ethylamino)-6-methyl-	$C_{11}H_{19}N_3O$	209.29	23947-60-6	160	3695			1.21^{25}		
5933	Ethofumesate	2-Ethoxy-2,3-dihydro-3,3-dimethyl-5-benzofuranyl methanesulfonate	$C_{13}H_{18}O_5S$	286.35	26225-79-6	71	3697			1.14		
5934	Ethoprop	Phosphorodithioic acid, *O*-ethyl *S,S*-dipropyl ester	$C_8H_{19}O_2PS_2$	242.34	13194-48-4		3702	$88^{0.2}$		1.094^{20}		
5935	Ethylene, 1-bromo-1,2-dichloro-, *trans*-	Ethylene, 1-bromo-1,2-dichloro (dis)	C_2HBrCl_2	175.84	6795-75-1	-83.5		113.8	3-01-00-00671	1.9133^{15}		1.5218^{15}
5936	Ethyne	Acetylene	C_2H_2	26.04	74-86-2	-80.7 (trip. pt.)	84	-84.7 (subl. point)	4-01-00-00939	*0.377^{25}	H_2O 2; EtOH 2; ace 3; bz 3	
5937	Ethyne, bromo-	Bromoacetylene	C_2HBr	104.93	593-61-3			4.7	3-01-00-00919		eth 4	
5938	Ethyne, chloro-	Chloroacetylene	C_2HCl	60.48	593-63-5	-126		-30	4-01-00-00957		EtOH 2	
5939	Ethyne, dibromo-		C_2Br_2	183.83	624-61-3	-24		exp 76	3-01-00-00919		ace 4; bz 4; eth 4; EtOH 4	
5940	Ethyne, dichloro-	Dichloroacetylene	C_2Cl_2	94.93	7572-29-4	-66		33	4-01-00-00957	1.261^{20}	EtOH 3; eth 3; ace 3	1.42790^{20}
5941	Ethyne, diiodo-	Diiodoacetylene	C_2I_2	277.83	624-74-8	81.5		exp	4-01-00-00958		ace 4; bz 4; eth 4; EtOH 4	
5942	Ethyne, ethoxy-		C_4H_6O	70.09	927-80-0			50	4-01-00-02211	0.8000^{20}		1.3796^{20}
5943	Ethyne, fluoro-	Fluoroacetylene	C_2HF	44.03	2713-09-9				4-01-00-00957			
5944	Ethyne, methoxy-		C_3H_4O	56.06	6443-91-0				4-01-00-02211	0.8001^{20}	eth 4; EtOH 4	1.3812^{20}

No.	Name Synonym	Mol. Form. Mol. Wt.	CAS RN mp/°C	Merck No. bp/°C	Beil. Ref. den/g cm^{-3}	Solubility n_D
5945	Etrimfos Phosphorothioic acid, *O*-(6-ethoxy-2-ethyl-4-pyrimidinyl) *O,O*-dimethyl ester	$C_{10}H_{17}N_2O_4PS$ 292.30	38260-54-7 -3.35	3847	 1.195^{20}	
5946	Famphur Phosphorothioic acid, *O*-[4-[(dimethylamino)sulfonyl]phenyl] *O,O*-dimethyl ester	$C_{10}H_{16}NO_5PS_2$ 325.35	52-85-7 53	3882		
5947	Fenamiphos Phosphoramidic acid, (1-methylethyl)-, ethyl 3-methyl-4-(methylthio)phenyl ester	$C_{13}H_{22}NO_3PS$ 303.36	22224-92-6 49	3901	 1.15^{20}	
5948	Fenarimol 5-Pyrimidinemethanol, α-(2-chlorophenyl)-α-(4-chlorophenyl)-	$C_{17}H_{12}Cl_2N_2O$ 331.20	60168-88-9 118	3903		
5949	Fenbuconazole 4-(4-Chlorophenyl)-2-phenyl-2-(1H-1,2,4-triazol-1-ylmethyl)butyronitrile	$C_{20}H_{19}ClN_4$ 350.85	114369-43-6 125			
5950	Fenbutatin oxide Distannoxane, hexakis(2-methyl-2-phenylpropyl)-	$C_{60}H_{78}OSn_2$ 1052.70	13356-08-6 138	3907		
5951	Fenitrothion Phosphorothioic acid, *O,O*-dimethyl *O*-(3-methyl-4-nitrophenyl) ester	$C_9H_{12}NO_5PS$ 277.24	122-14-5	3922 $118^{0.05}$, 164^{1}	 1.3227^{25}	
5952	Fenoxycarb Ethyl 2-(4-phenoxyphenoxy)ethylcarbamate	$C_{17}H_{19}NO_4$ 301.34	79127-80-3 53			
5953	Fenpropathrin Cyano(3-phenoxyphenyl)methyl 2,2,3,3-tetramethylcyclopropanecarboxylate	$C_{22}H_{23}NO_3$ 349.43	64257-84-7 47		 1.15^{25}	
5954	Fensulfothion Phosphorothioic acid, *O,O*-diethyl *O*-[4-(methylsulfinyl)phenyl] ester	$C_{11}H_{17}O_4PS_2$ 308.36	115-90-2	3943 $140^{0.01}$	 1.202^{20}	
5955	Fenthion Phosphorothioic acid, *O,O*-dimethyl *O*-[3-methyl-4-(methylthio)phenyl] ester	$C_{10}H_{15}O_3PS_2$ 278.33	55-38-9 7.5	3945 $87^{0.01}$	 1.246^{20}	
5956	Fenvalerate Benzeneacetic acid, 4-chloro-α-(1-methylethyl)-, cyano(3-phenoxyphenyl)methyl ester	$C_{25}H_{22}ClNO_3$ 419.91	51630-58-1	3952 dec	 1.15^{25}	
5957	Ferbam Iron, tris(dimethylcarbamodithioato-*S,S'*)-, (OC-6-11)-	$C_9H_{18}FeN_3S_6$ 416.51	14484-64-1 180 dec	3954		
5958	Ferrate(2-), chloro[7,12-diethenyl-3,8,13,17-tetramethyl-21H,23H-porphine-2,18-dipropanoato(4-)-*N*21,*N*22,*N*23, Hemin	$C_{34}H_{32}ClFeN_4O_4$ 651.95	16009-13-5 >300	4563		
5959	Ferrate(2-), [7,12-diethenyl-3,8,13,17-tetramethyl-21H,23H-porphine-2,18-dipropanoato(4-)-*N*21,*N*22,*N*23,*N*24 Hematin	$C_{34}H_{33}FeN_4O_5$ 633.51	15489-90-4 >200	4554	4-26-00-03047	H_2O 1; EtOH 3; eth 1; alk 3
5960	Ferrocene Bis(cyclopentadienyl)iron	$C_{10}H_{10}Fe$ 186.04	102-54-5 172.5	3985 249		H_2O 1
5961	Ferrocene, 1,1'-bis(1,1,3,3-tetramethyl-3-phenyldisiloxanyl)- Disiloxane, 1-(cyclopentadienyl)-1,1,3,3-tetramethyl-3-phenyl-, Fe deriv.	$C_{30}H_{42}FeO_2Si_4$ 602.85	12321-04-9 18.7	 $200\text{-}5^{0.03}$	 1.1063^{25}	 1.5473^{25}
5962	Fluazipop-butyl (±)-2-[4-[[5-(Trifluoromethyl)-2-pyridinyl]oxy]phenoxy]propanoic acid, butyl ester	$C_{19}H_{20}F_3NO_4$ 383.37	69806-50-4 13	4049 $170^{0.5}$	 1.21^{20}	
5963	Fluazipop-P-butyl (*R*)-2-[4-(5-Trifluoromethyl-2-pyridyloxy)phenoxy]propionic acid, butyl ester	$C_{19}H_{20}F_3NO_4$ 383.37	79241-46-6 5			
5964	Fluchloralin Benzenamine, *N*-(2-chloroethyl)-2,6-dinitro-*N*-propyl-4-(trifluoromethyl)-	$C_{12}H_{13}ClF_3N_3O_4$ 355.70	33245-39-5 42	4052		
5965	Flucythrinate Cythrin	$C_{26}H_{23}F_2NO_4$ 451.47	70124-77-5	4055 $108^{0.35}$	 1.189^{22}	
5966	Fluoranthene 1,2-(1,8-Naphthylene)benzene	$C_{16}H_{10}$ 202.26	206-44-0 107.8	 384	4-05-00-02463 1.252^{0}	H_2O 1; EtOH 3; eth 3; bz 3
5967	9H-Fluorene 2,2'-Methylenebiphenyl	$C_{13}H_{10}$ 166.22	86-73-7 114.8	4081 295	4-05-00-02142 1.203^{0}	H_2O 1; EtOH 2; eth 3; ace 3
5968	9H-Fluorene, 2-bromo- 2-Bromofluorene	$C_{13}H_9Br$ 245.12	1133-80-8 113.5	 185^{135}	4-05-00-02147	H_2O 1; EtOH 3; chl 4; HOAc 3
5969	9H-Fluorene-2-carboxylic acid, 9-oxo-	$C_{14}H_8O_3$ 224.22	784-50-9 341	 sub 340	4-10-00-03040	H_2O 1; EtOH 3; HOAc 3
5970	9H-Fluorene, 9-[(2-chlorophenyl)methylene]- Fluorene, 9-(2-chlorobenzylidene)	$C_{20}H_{13}Cl$ 288.78	1643-49-8 69.5	 $180^{0.7}$	4-05-00-02631	EtOH 4; HOAc 4
5971	9H-Fluorene-2,7-diamine 2,7-Diaminofluorene	$C_{13}H_{12}N_2$ 196.25	525-64-4 166	4082	4-13-00-00449	H_2O 1; EtOH 3; chl 3
5972	9H-Fluorene, 2,7-dichloro- 2,7-Dichlorofluorene	$C_{13}H_8Cl_2$ 235.11	7012-16-0 128	 sub	4-05-00-02147	bz 4; chl 4

No.	Name	Synonym	Mol. Form.	Mol. Wt.	CAS RN	mp/°C	Merck No.	bp/°C	Beil. Ref.	den/g cm^{-3}	Solubility	n_D
5973	1H-Fluorene, dodecahydro-		$C_{13}H_{22}$	178.32	5744-03-6			260	4-05-00-00489	0.92^{25}		1.5012^{20}
5974	9H-Fluorene, 9-methyl-	9-Methylfluorene	$C_{14}H_{12}$	180.25	2523-37-7	46.5		$154\text{-}6^{15}$	4-05-00-02185	1.0263^{66}	H_2O 1; EtOH 3; eth 3; ace 3	1.610^{66}
5975	9H-Fluoren-9-one		$C_{13}H_8O$	180.21	486-25-9	84		341.5	4-07-00-01629	1.1300^{99}	H_2O 1; EtOH 3; eth 4; ace 3	1.6309^{99}
5976	9H-Fluoren-9-one, 2-chloro-		$C_{13}H_7ClO$	214.65	3096-47-7	125.5		sub	4-07-00-01634		EtOH 4	
5977	9H-Fluoren-9-one, 2-nitro-		$C_{13}H_7NO_3$	225.20	3096-52-4	224.3		sub	4-07-00-01636		EtOH 2; ace 3; sulf 3; HOAc 3	
5978	9H-Fluoren-9-one, 1,2,3,4-tetrahydro-	Phentydrone	$C_{13}H_{12}O$	184.24	634-19-5	81.5	7235	$139\text{-}40^{0.05}$	4-07-00-01302			
5979	9H-Fluoren-9-one, 2,4,7-trinitro-	2,4,7-Trinitrofluorenone	$C_{13}H_5N_3O_7$	315.20	129-79-3	176	9641		4-07-00-01638		H_2O 2; ace 4; bz 4; chl 4	
5980	Fluorodifen	Benzene, 2-nitro-1-(4-nitrophenoxy)-4-(trifluoromethyl)-	$C_{13}H_7F_3N_2O_5$	328.20	15457-05-3	94						
5981	Fluridone	4(1H)-Pyridinone, 1-methyl-3-phenyl-5-[3-(trifluoromethyl)phenyl]-	$C_{19}H_{14}F_3NO$	329.32	59756-60-4	155						
5982	Fluroxypyr	[(4-Amino-3,5-dichloro-6-fluoro-2-pyridyl)oxy]acetic acid	$C_7H_5Cl_2FN_2O_3$	255.03	69377-81-7	232	4128					
5983	Fluvalinate	2-[[2-Chloro-4-(trifluoromethyl)phenyl]amino]-3-methylbutanoic acid, (±)-cyano(3-phenoxy phenyl)methyl ester	$C_{26}H_{22}ClF_3N_2O_3$	502.92	102851-06-9			>450		1.29^{25}		
5984	Folpet	1H-Isoindole-1,3(2H)-dione, 2-[(trichloromethyl)thio]-	$C_9H_4Cl_3NO_2S$	296.56	133-07-3	177	4142					
5985	Fomesafen	5-[2-Chloro-4-(trifluoromethyl)phenoxy]-*N*-(methylsulfonyl)-2-nitrobenzamide	$C_{15}H_{10}ClF_3N_2O_6S$	438.77	72178-02-0	220				1.28^{20}		
5986	Fonofos	Phosphonodithioic acid, ethyl-, *O*-ethyl *S*-phenyl ester	$C_{10}H_{15}OPS_2$	246.33	944-22-9		4147	$130^{0.1}$		1.16^{25}		
5987	Formaldehyde	Methanal	CH_2O	30.03	50-00-0	-92	4148	-19.1	4-01-00-03017	0.815^{-20}	H_2O 3; EtOH 3; eth 5; ace 5	
5988	Formaldehyde, oxime		CH_3NO	45.04	75-17-2	2.5		109^{15}	4-01-00-03055	1.133^{25}	H_2O 3; EtOH 4; eth 4	
5989	Formamide	Methanamide	CH_3NO	45.04	75-12-7	2.55	4151	220	4-02-00-00045	1.1334^{20}	H_2O 5; EtOH 5; eth 2; ace 3	1.4472^{20}
5990	Formamide, *N*-cyclohexyl-		$C_7H_{13}NO$	127.19	766-93-8	39.5		260; $113^{0.7}$	4-12-00-00038	1.0123^{17}		
5991	Formamide, *N,N*-diethyl-		$C_5H_{11}NO$	101.15	617-84-5			177.5	4-04-00-00346	0.9080^{19}	H_2O 5; EtOH 4; eth 4; ace 5	1.4321^{25}
5992	Formamide, *N,N*-dimethyl-	*N,N*-Dimethylformamide	C_3H_7NO	73.09	68-12-2	-60.4	3232	153	4-04-00-00171	0.944^{25}	H_2O 5; EtOH 5; eth 5; ace 5	1.4305^{20}
5993	Formamide, *N,N*-diphenyl-		$C_{13}H_{11}NO$	197.24	607-00-1	73.5		337.5; 189^{13}	4-12-00-00370		H_2O 1; EtOH 3; eth 3; bz 3	
5994	Formamide, *N*-ethyl-	*N*-Ethylformamide	C_3H_7NO	73.09	627-45-2			198	4-04-00-00346	0.9552^{20}	H_2O 5; EtOH 5; eth 5	1.4320^{20}
5995	Formamide, *N*-methyl-	*N*-Methylformamide	C_2H_5NO	59.07	123-39-7	-3.8		199.5	4-04-00-00170	1.011^{19}	H_2O 4; ace 4; EtOH 4	1.4319^{20}
5996	Formamide, *N*-(2-methylphenyl)-		C_8H_9NO	135.17	94-69-9	62		288	3-12-00-01852	1.086^{55}	H_2O 3; EtOH 4	
5997	Formamide, *N*-(3-methylphenyl)-		C_8H_9NO	135.17	3085-53-8	<-18		290 dec; 176^{17}	3-12-00-01962		H_2O 3; NaOH 3	
5998	Formamide, *N*-methyl-*N*-phenyl-		C_8H_9NO	135.17	93-61-8	14.5		243	4-12-00-00370	1.0948^{20}	H_2O 2; EtOH 3; ace 3; ctc 2	1.5589^{20}
5999	Formamide, *N*-phenyl-	Formanilide	C_7H_7NO	121.14	103-70-8	46	4152	271	4-12-00-00368	1.1186^{50}	H_2O 3; EtOH 4; eth 3; bz 3	
6000	Formamide, *N*-(2,2,2-trichloro-1-hydroxyethyl)-	Chloral formamide	$C_3H_4Cl_3NO_2$	192.43	515-82-2	120	2060		3-02-00-00059		ace 4; eth 4; EtOH 4	
6001	Formetanate hydrochloride	Methanimidamide, *N,N*-dimethyl-*N*′-[3-[[(methylamino)carbonyl]oxy]phenyl]-, monohydrochloride	$C_{11}H_{16}ClN_3O_2$	257.72	23422-53-9							
6002	Formic acid	Methanoic acid	CH_2O_2	46.03	64-18-6	8.3	4153	101	4-02-00-00003	1.220^{20}	H_2O 5; EtOH 5; eth 5; ace 4	1.3714^{20}
6003	Formic acid, ammonium salt	Ammonium formate	CH_5NO_2	63.06	540-69-2		548		4-02-00-00003			
6004	Formic acid, butyl ester	Butyl formate	$C_5H_{10}O_2$	102.13	592-84-7	-91.5		106.1	4-02-00-00028	0.8885^{20}	H_2O 2; EtOH 5; eth 5; ace 3	1.3912^{20}
6005	Formic acid, cyclohexyl ester		$C_7H_{12}O_2$	128.17	4351-54-6			163	4-06-00-00035	1.0057^{0}	H_2O 1; EtOH 3; eth 4; HCOOH 3	1.4430^{20}

No.	Name Synonym	Mol. Form. Mol. Wt.	CAS RN mp/°C	Merck No. bp/°C	Beil. Ref. den/g cm^{-3}	Solubility n_D
6006	Formic acid, ethenyl ester Vinyl formate	$C_3H_4O_2$ 72.06	692-45-5		4-02-00-00033	
6007	Formic acid, ethyl ester Ethyl formate	$C_3H_6O_2$ 74.08	109-94-4 -79.6	3763 54.4	4-02-00-00023 0.9168^{20}	H_2O 3; EtOH 5; eth 5; ace 4 1.3598^{10}
6008	Formic acid, heptyl ester Heptyl formate	$C_8H_{16}O_2$ 144.21	112-23-2	 178.1	4-02-00-00031 0.8784^{20}	H_2O 1; EtOH 5; eth 5 1.4140^{20}
6009	Formic acid, hexyl ester Hexyl formate	$C_7H_{14}O_2$ 130.19	629-33-4 -62.6	 155.5	4-02-00-00031 0.8813^{20}	H_2O 1; EtOH 5; eth 5 1.4071^{20}
6010	Formic acid, methyl ester Methyl formate	$C_2H_4O_2$ 60.05	107-31-3 -99	5994 31.7	4-02-00-00020 0.9742^{20}	H_2O 4; EtOH 5; eth 3; chl 3 1.3433^{20}
6011	Formic acid, 1-methylethyl ester Isopropyl formate	$C_4H_8O_2$ 88.11	625-55-8	 68.2	4-02-00-00027 0.8728^{20}	H_2O 2; EtOH 5; eth 5; ace 4 1.3678^{20}
6012	Formic acid, 1-methylpropyl ester	$C_5H_{10}O_2$ 102.13	589-40-2	 97	4-02-00-00029 0.8846^{20}	H_2O 2; EtOH 5; eth 5; ace 3 1.3865^{20}
6013	Formic acid, 2-methylpropyl ester Isobutyl formate	$C_5H_{10}O_2$ 102.13	542-55-2 -95.8	5026 98.2	4-02-00-00029 0.8776^{20}	H_2O 2; EtOH 5; eth 5; ace 4 1.3857^{20}
6014	Formic acid, octyl ester	$C_9H_{18}O_2$ 158.24	112-32-3 -39.1	 198.8	4-02-00-00031 0.8744^{20}	H_2O 1; EtOH 3; eth 5; ctc 2 1.4208^{15}
6015	Formic acid, pentyl ester Pentyl formate	$C_6H_{12}O_2$ 116.16	638-49-3 -73.5	 130.4	4-02-00-00030 0.8853^{20}	H_2O 2; EtOH 5; eth 5 1.3992^{20}
6016	Formic acid, phenylmethyl ester Benzyl formate	$C_8H_8O_2$ 136.15	104-57-4	1148 203; 84^{10}	4-06-00-02262 1.081^{20}	H_2O 1; EtOH 3; eth 5; ace 3 1.5154^{20}
6017	Formic acid, 2-propenyl ester Allyl formate	$C_4H_6O_2$ 86.09	1838-59-1	 83.6	3-02-00-00046 0.9460^{20}	H_2O 2; EtOH 3; eth 5
6018	Formic acid, propyl ester Propyl formate	$C_4H_8O_2$ 88.11	110-74-7 -92.9	7871 80.9	4-02-00-00026 0.9058^{20}	H_2O 2; EtOH 5; eth 5; ctc 2 1.3779^{20}
6019	Formyl fluoride Fluoroformaldehyde	CHFO 48.02	1493-02-3 -142.2	4162 -26.5	4-02-00-00042 1.1950^{-30}	
6020	Fosetyl-Al Ethyl hydrogen phosphonate, aluminum salt	$C_6H_{18}AlO_9P_3$ 354.11	39148-24-8 >300	4167		
6021	Fosthiazate (*RS*)-*S*-*sec*-Butyl *O*-ethyl 2-oxo-1,3-thiazolidin-3-ylphosphonothioate	$C_9H_{18}NO_3PS_2$ 283.35	98886-44-3			
6022	*D*-Fructopyranose, 3-*O*-α-*D*-glucopyranosyl- *D*-Turanose	$C_{12}H_{22}O_{11}$ 342.30	547-25-1 168	9728	5-17-07-00213	H_2O 4; EtOH 3; MeOH 3
6023	β-*D*-Fructose β-Levulose	$C_6H_{12}O_6$ 180.16	53188-23-1 103 dec	4185	4-01-00-04401 1.60^{20}	H_2O 4; EtOH 3; ace 4; MeOH 3
6024	Fulminic acid Carbyloxime	CHNO 43.03	506-85-4 unstable in pure form			eth 3
6025	Furan Oxacyclopentadiene	C_4H_4O 68.08	110-00-9 -85.6	4206 31.5	5-17-01-00291 0.9514^{20}	H_2O 1; EtOH 4; eth 4; ace 3 1.4214^{20}
6026	2-Furanacetic acid	$C_6H_6O_3$ 126.11	2745-26-8 68.5	 $102^{0.4}$	5-18-06-00206	H_2O 3; bz 3; MeOH 3; peth 3
6027	2-Furanacetic acid, butyl ester 2-Furylacetic acid, butyl ester	$C_{10}H_{14}O_3$ 182.22	4915-23-5	 110-11^{13}	4-18-00-04062 1.0232^{25}	 1.4558^{25}
6028	2-Furanacetic acid, ethyl ester	$C_8H_{10}O_3$ 154.17	4915-21-3	 88^{15}	5-18-06-00206 1.0763^{25}	 1.4571^{25}
6029	2-Furanacetic acid, methyl ester	$C_7H_8O_3$ 140.14	4915-22-4	 87-82^{21}	5-18-06-00206 1.1250^{20}	 1.4638^{25}
6030	3-Furanacetic acid, tetrahydro-2,2-dimethyl-5-oxo- Terpenylic acid	$C_8H_{12}O_4$ 172.18	26754-48-3 90	9100	5-18-08-00032	H_2O 4
6031	2-Furanacetonitrile 2-Furylacetonitrile	C_6H_5NO 107.11	2745-25-7	 75-80^{20}	5-18-06-00207 1.0854^{25}	eth 4; EtOH 4 1.4693^{20}
6032	Furan, 2,5-bis(1,1-dimethylethyl)-	$C_{12}H_{20}O$ 180.29	4789-40-6	 210	5-17-01-00439 0.837^{20}	H_2O 1; EtOH 3; eth 3 1.4369^{20}
6033	Furan, 2-bromo-	C_4H_3BrO 146.97	584-12-3	 103	5-17-01-00295 1.6500^{20}	H_2O 2; EtOH 3; eth 3; ace 3 1.4980^{20}
6034	Furan, 3-bromo- 3-Bromofuran	C_4H_3BrO 146.97	22037-28-1	 103	5-17-01-00295 1.6606^{20}	ace 4; bz 4; eth 4; EtOH 4 1.4958^{20}
6035	Furan, 2-(bromomethyl)-	C_5H_5BrO 161.00	4437-18-7	 32-4^{2}	4-17-00-00268 1.557^{20}	 1.5380^{20}
6036	Furan, 2-(bromomethyl)tetrahydro-	C_5H_9BrO 165.03	1192-30-9	 170; 70^{22}	5-17-01-00080 1.3653^{20}	EtOH 3; eth 3 1.4850^{20}
6037	Furan, 2-(butoxymethyl)-	$C_9H_{14}O_2$ 154.21	56920-82-2	 190	2-17-00-00114 0.9516^{20}	eth 4; EtOH 4 1.4522^{20}
6038	Furan, 2,2'-*sec*-butylidenedi- Butane, 2,2-di(2-furyl)	$C_{12}H_{14}O_2$ 190.24	100121-73-1	 64-6^{1}	4-19-00-00207 1.0330^{20}	eth 4; EtOH 4 1.4970^{20}
6039	2-Furancarbonitrile	C_5H_3NO 93.08	617-90-3	 147	5-18-06-00122 1.0822^{20}	EtOH 3; eth 3 1.4798^{20}
6040	2-Furancarbonyl chloride	$C_5H_3ClO_2$ 130.53	527-69-5 -2	 173	5-18-06-00110 1.324^{25}	H_2O 1; eth 3; ctc 2; chl 3 1.5310^{20}
6041	2-Furancarboxaldehyde Furfural	$C_5H_4O_2$ 96.09	98-01-1 -36.5	4214 161.7	5-17-09-00292 1.1594^{20}	H_2O 3; EtOH 4; eth 5; ace 4 1.5261^{20}
6042	3-Furancarboxaldehyde	$C_5H_4O_2$ 96.09	498-60-2	 145; 70-2^{43}	5-17-09-00372 1.109^{20}	 1.4945^{20}
6043	2-Furancarboxaldehyde, 5-bromo-	$C_5H_3BrO_2$ 174.98	1899-24-7 83.5	 201; 112^{18}	5-17-09-00320	eth 4; EtOH 4

No.	Name Synonym	Mol. Form. Mol. Wt.	CAS RN mp/°C	Merck No. bp/°C	Beil. Ref. den/g cm^{-3}	Solubility n_D
6044	2-Furancarboxaldehyde, 5-(hydroxymethyl)- 5-(Hydroxymethyl)-2-furaldehyde	$C_6H_6O_3$ 126.11	67-47-0 31.5	4764 $114-6^{1}$	5-18-01-00130 1.2062^{25}	H_2O 3; EtOH 3; eth 2; bz 3 1.5627^{18}
6045	2-Furancarboxaldehyde, 5-methyl-	$C_6H_6O_2$ 110.11	620-02-0	 187; 89^{26}	5-17-09-00404 1.1072^{18}	H_2O 3; EtOH 4; eth 5; ctc 2 1.5264^{20}
6046	2-Furancarboxaldehyde, 5-nitro-	$C_5H_3NO_4$ 141.08	698-63-5 35.5	 $128-32^{10}$	5-17-09-00323	H_2O 2; peth 3
6047	2-Furancarboxaldehyde, oxime, (*E*)-	$C_5H_5NO_2$ 111.10	620-03-1	 205 dec; 98^{9}	5-17-09-00306 1.1550^{80}	H_2O 3; EtOH 4; eth 4; bz 4
6048	2-Furancarboxaldehyde, oxime, (*Z*)-	$C_5H_5NO_2$ 111.10	1450-58-4 91	 100^{12}	5-17-09-00306 1.3800^{20}	H_2O 2; EtOH 4; eth 4; bz 4
6049	2-Furancarboxaldehyde, tetrahydro-	$C_5H_8O_2$ 100.12	7681-84-7	 142; 46^{29}	5-17-09-00027 1.0727^{20}	H_2O 4; eth 4 1.4366^{20}
6050	2-Furancarboxylic acid 2-Furoic acid	$C_5H_4O_3$ 112.08	88-14-2 133.5	4219 231	5-18-06-00102	H_2O 3; EtOH 3; eth 4; ace 2
6051	3-Furancarboxylic acid	$C_5H_4O_3$ 112.08	488-93-7 122.5	 sub 105	5-18-06-00196	H_2O 2; EtOH 3; eth 4; AcOEt 3
6052	2-Furancarboxylic acid, butyl ester	$C_9H_{12}O_3$ 168.19	583-33-5	 233	4-18-00-03918 1.0555^{20}	H_2O 1; EtOH 3; eth 3; bz 3 1.4740
6053	2-Furancarboxylic acid, 5-chloro-, ethyl ester Ethyl 5-chloro-2-furancarboxylate	$C_7H_7ClO_3$ 174.58	4301-39-7 2	 217; 108^{23}	5-18-06-00131 1.2418^{25}	
6054	3-Furancarboxylic acid, 4,5-dihydro-5-oxo-	$C_5H_4O_4$ 128.08	585-68-2 164	 dec	4-18-00-05333	H_2O 4; EtOH 3; MeOH 3
6055	3-Furancarboxylic acid, 2,5-dimethyl-, ethyl ester	$C_9H_{12}O_3$ 168.19	29113-63-1	 208.5; 96^{19}	5-18-06-00233 1.0478^{23}	 1.4686^{20}
6056	2-Furancarboxylic acid, ethyl ester	$C_7H_8O_3$ 140.14	614-99-3 34.5	 196.8	5-18-06-00104 1.1174^{21}	H_2O 1; EtOH 5; eth 5; ace 5 1.4797^{21}
6057	2-Furancarboxylic acid, 2-furanylmethyl ester	$C_{10}H_8O_4$ 192.17	615-11-2 27.5	 122^{2}	4-18-00-03936 1.2347^{25}	ace 4; bz 4; eth 4; EtOH 4 1.5280^{20}
6058	2-Furancarboxylic acid, hexyl ester α-Furoic acid, hexyl ester	$C_{11}H_{18}O_3$ 196.25	39251-86-0	 $105-7^{1}$	2-18-00-00267 1.0170^{20}	H_2O 1; EtOH 3; eth 3; bz 5
6059	2-Furancarboxylic acid, methyl ester	$C_6H_6O_3$ 126.11	611-13-2	 181.3	5-18-06-00103 1.1786^{21}	H_2O 1; EtOH 3; eth 3; bz 3 1.4860^{20}
6060	2-Furancarboxylic acid, 5-methyl-	$C_6H_6O_3$ 126.11	1917-15-3 109.5	 105^{1}	5-18-06-00217	H_2O 2; EtOH 3; eth 3; bz 2
6061	2-Furancarboxylic acid, 1-methylethyl ester	$C_8H_{10}O_3$ 154.17	6270-34-4	 198.5	0-18-00-00275 1.0655^{24}	H_2O 1; EtOH 3; eth 3; ace 3 1.4682^{24}
6062	2-Furancarboxylic acid, 2-methylpropyl ester	$C_9H_{12}O_3$ 168.19	20279-53-2	 222	2-18-00-00266 1.0388^{20}	ace 4; bz 4; eth 4; EtOH 4 1.4676^{20}
6063	3-Furancarboxylic acid, methyl ester	$C_6H_6O_3$ 126.11	13129-23-2	 160	5-18-06-00196 1.1733^{15}	H_2O 1; EtOH 3; ace 3 1.4676^{20}
6064	3-Furancarboxylic acid, 2-methyl-, ethyl ester	$C_8H_{10}O_3$ 154.17	28921-35-9	 $85-7^{20}$	5-18-06-00214 1.0102^{25}	eth 4 1.4620^{25}
6065	3-Furancarboxylic acid, 5-methyl- 3-Furoic acid, 5-methyl-	$C_6H_6O_3$ 126.11	21984-93-0 119	 sub	5-18-06-00216	H_2O 4; eth 4; EtOH 4
6066	2-Furancarboxylic acid, 5-nitro-	$C_5H_3NO_5$ 157.08	645-12-5 186	 sub	5-18-06-00139	H_2O 3; EtOH 3; eth 3; ace 2
6067	2-Furancarboxylic acid, octyl ester α-Furoic acid, octyl ester	$C_{13}H_{20}O_3$ 224.30	39251-88-2	 $126-7^{1}$	2-18-00-00267 0.9885^{20}	H_2O 1; EtOH 3
6068	2-Furancarboxylic acid, pentyl ester 2-Furoic acid, pentyl ester	$C_{10}H_{14}O_3$ 182.22	4996-48-9	 $95-7^{1}$	2-18-00-00266 1.0335^{20}	H_2O 1; EtOH 5
6069	2-Furancarboxylic acid, phenylmethyl ester	$C_{12}H_{10}O_3$ 202.21	5380-40-5	 $179-81^{18}$	4-18-00-03921 1.1623^{22}	ace 4; eth 4 1.5550^{20}
6070	2-Furancarboxylic acid, 2-propenyl ester	$C_8H_8O_3$ 152.15	4208-49-5	 207.5	5-18-06-00105 1.115^{25}	eth 3; ace 3; ctc 2 1.4945^{20}
6071	2-Furancarboxylic acid, propyl ester	$C_8H_{10}O_3$ 154.17	615-10-1	 210.9; 112^{20}	5-18-06-00104 1.0745^{20}	H_2O 1; EtOH 3; eth 3; ace 3 1.4787^{20}
6072	2-Furancarboxylic acid, tetrahydro-	$C_5H_8O_3$ 116.12	16874-33-2 21	 145^{25}	5-18-06-00010 1.1912^{20}	H_2O 4 1.4612^{20}
6073	Furan, 2-chloro- 2-Chlorofuran	C_4H_3ClO 102.52	3187-94-8	 78	5-17-01-00294 1.1923^{20}	ace 4; eth 4; EtOH 4 1.4569^{20}
6074	Furan, 3-chloro- 3-Chlorofuran	C_4H_3ClO 102.52	50689-17-3	 80	5-17-01-00295 1.2094^{20}	ace 4; eth 4 1.4601^{20}
6075	Furan, 2-(chloromethyl)-	C_5H_5ClO 116.55	617-88-9	 49^{26}	5-17-01-00323 1.1783^{20}	bz 4; eth 4; EtOH 4 1.4941^{20}
6076	Furan, 2-(cyclohexyloxy)- Cyclohexyl 2-furyl ether	$C_{10}H_{14}O_2$ 166.22	99172-57-3	 $118-9^{28}$	4-17-00-01219 1.0200^{28}	ace 4; eth 4; EtOH 4 1.4861^{28}
6077	Furan, 2,3-dibromo- 2,3-Dibromofuran	$C_4H_2Br_2O$ 225.87	30544-34-4	 167	5-17-01-00295 1.9938^{25}	 1.5458^{25}
6078	Furan, 2,5-dibromo- 2,5-Dibromofuran	$C_4H_2Br_2O$ 225.87	32460-00-7 9.5	 164.5	5-17-01-00295 2.27^{20}	 1.5455^{20}
6079	2,3-Furandicarboxylic acid Isocitrate lactone	$C_6H_4O_5$ 156.09	4282-24-0 226	 sub	5-18-07-00043	H_2O 4; EtOH 4; eth 2; AcOEt 4
6080	2,4-Furandicarboxylic acid	$C_6H_4O_5$ 156.09	4282-28-4 266	 sub	5-18-07-00045	H_2O 2; EtOH 4; ace 4; chl 2
6081	2,5-Furandicarboxylic acid Dehydromucic acid	$C_6H_4O_5$ 156.09	3238-40-2 342	 sub	5-18-07-00047 1.7400^{20}	H_2O 2; EtOH 2

PHYSICAL CONSTANTS OF ORGANIC COMPOUNDS (continued)

No.	Name Synonym	Mol. Form. Mol. Wt.	CAS RN mp/°C	Merck No. bp/°C	Beil. Ref. den/g cm^{-3}	Solubility n_D
6082	3,4-Furandicarboxylic acid, diethyl ester 2,3-Bis(ethoxycarbonyl)furan	$C_{10}H_{12}O_5$ 212.20	30614-77-8	 155^{13}	5-18-07-00053 1.14^{25}	 1.4717^{20}
6083	2,5-Furandicarboxylic acid, dimethyl ester Dimethyl 2,5-furandicarboxylate	$C_8H_8O_5$ 184.15	4282-32-0 112	 154^{15}	5-18-07-00048	eth 4; EtOH 4; chl 4
6084	Furan, 2-(diethoxymethyl)-	$C_9H_{14}O_3$ 170.21	13529-27-6	 191.5	5-17-09-00299 0.9976^{20}	EtOH 4 1.4451^{20}
6085	Furan, 2,2-diethyltetrahydro- 2,2-Diethyltetrahydrofuran	$C_8H_{16}O$ 128.21	1193-35-7	 146	4-17-00-00107 0.8703^{20}	ace 4; bz 4; eth 4; EtOH 4 1.4317^{20}
6086	Furan, 2,3-dihydro- 2,3-Dihydrofuran	C_4H_6O 70.09	1191-99-7	 54.5	5-17-01-00172 0.927^{25}	 1.4239^{20}
6087	Furan, 2,5-dihydro-2,5-dimethoxy- 2,5-Dimethoxy-2,5-dihydrofuran	$C_6H_{10}O_3$ 130.14	332-77-4	 161	5-17-05-00073 1.073^{25}	 1.4339^{20}
6088	Furan, 2,4-dimethyl- 3,5-Dimethylfuran	C_6H_8O 96.13	3710-43-8	 94	5-17-01-00343 0.8993^{20}	 1.4371^{20}
6089	Furan, 2,5-dimethyl- 2,5-Dimethylfuran	C_6H_8O 96.13	625-86-5 -62.8	 93.5	5-17-01-00344 0.8883^{20}	H_2O 1; EtOH 3; eth 3; ace 3 1.4363^{20}
6090	Furan, 2-(1,1-dimethylethyl)- 2-*tert*-Butylfuran	$C_8H_{12}O$ 124.18	7040-43-9	 119.5	5-17-01-00371 0.869^{20}	ace 4; eth 4; EtOH 4 1.4373^{20}
6091	2,5-Furandione Maleic anhydride	$C_4H_2O_3$ 98.06	108-31-6 52.8	5586 202	5-17-11-00055 1.314^{60}	H_2O 3; eth 3; ace 3; chl 3
6092	2,5-Furandione, 3-chloro-	C_4HClO_3 132.50	96-02-6 33	 196	5-17-11-00063 1.5375^{25}	 1.4980^{20}
6093	2,5-Furandione, dihydro- Succinic anhydride	$C_4H_4O_3$ 100.07	108-30-5 119	8841 261	5-17-11-00006 1.2^{20}	H_2O 1; EtOH 3; eth 2; chl 3
6094	2,5-Furandione, dihydro-3-methyl-	$C_5H_6O_3$ 114.10	4100-80-5 34	 239	5-17-11-00011 1.22^{25}	
6095	2,5-Furandione, dihydro-3-methylene-	$C_5H_4O_3$ 112.08	2170-03-8 69	 139^{30}, 114^{18}	5-17-11-00066	eth 2; chl 4
6096	2,5-Furandione, dihydro-3-phenyl-, (±)- Succinic anhydride, phenyl (*DL*)	$C_{10}H_8O_3$ 176.17	112489-85-7 54	 204^{22}	5-17-11-00275	ace 4; bz 4; eth 4; EtOH 4
6097	2,5-Furandione, 3,4-dimethyl-	$C_6H_6O_3$ 126.11	766-39-2 96	 223	5-17-11-00069 1.107^{100}	H_2O 2; EtOH 4; eth 4; bz 4
6098	2,5-Furandione, 3-methyl-	$C_5H_4O_3$ 112.08	616-02-4 7.5	 213.5	5-17-11-00065 1.2469^{16}	ace 4; eth 4; EtOH 4 1.4710^{21}
6099	2,5-Furandione, 3,3,4,4-tetrafluorodihydro-	$C_4F_4O_3$ 172.04	699-30-9	 54.5	5-17-11-00007 1.6209^{20}	 1.3240^{20}
6100	Furan, 2,5-diphenyl-	$C_{16}H_{12}O$ 220.27	955-83-9 91	 344	5-17-02-00398	H_2O 1; EtOH 4; eth 4; ace 3
6101	Furan, 2,2'-[dithiobis(methylene)]bis-	$C_{10}H_{10}O_2S_2$ 226.32	4437-20-1 10	 167^{13}, $112^{0.5}$	4-17-00-01258	EtOH 4
6102	Furan, 2-ethenyl-	C_6H_6O 94.11	1487-18-9 -94	 99.5	5-17-01-00477 0.9445^{19}	 1.4992^{19}
6103	Furan, 2-ethoxy- Ethyl 2-furyl ether	$C_6H_8O_2$ 112.13	5809-07-4	 125.5	4-17-00-01219 0.9849^{23}	 1.4500^{23}
6104	Furan, 2-(ethoxymethyl)-	$C_7H_{10}O_2$ 126.16	6270-56-0	 149	5-17-03-00340 0.9844^{20}	H_2O 1; EtOH 3; eth 3 1.4523^{20}
6105	Furan, 2-ethyl- 2-Ethylfuran	C_6H_8O 96.13	3208-16-0	 92	5-17-01-00338 0.911^{15}	EtOH 3; eth 3; ace 3; bz 3 1.4466^{23}
6106	Furan, 2-ethyl-5-methyl- 2-Methyl-5-ethylfuran	$C_7H_{10}O$ 110.16	1703-52-2	 118.5	5-17-01-00357 0.8883^{20}	bz 4; eth 4; EtOH 4 1.4473^{20}
6107	Furan, 2-ethyltetrahydro- 2-Ethyltetrahydrofuran	$C_6H_{12}O$ 100.16	1003-30-1	 109	5-17-01-00095 0.8570^{19}	ace 4; bz 4; eth 4; EtOH 4 1.4147^{19}
6108	Furan, 2-ethynyl-	C_6H_4O 92.10	18649-64-4	 105.5	4-17-00-00396 0.9919^{20}	 1.5055^{20}
6109	Furan, 2,2,3,3,4,4,5-heptafluorotetrahydro-5-(nonafluorobutyl)- Perfluoro-2-butyltetrahydrofuran	$C_8F_{16}O$ 416.06	335-36-4	 102.6		
6110	Furan, 2-iodo- 2-Furyl iodide	C_4H_3IO 193.97	54829-48-0	 $43\text{-}5^{15}$	5-17-01-00295 2.024^{20}	eth 4 1.5661^{20}
6111	Furan, 3-iodo- 3-Iodofuran	C_4H_3IO 193.97	29172-20-1	 133; $37\text{-}8^{22}$	5-17-01-00296 2.045^{20}	H_2O 1; eth 3 1.5610^{20}
6112	Furan, 2-isopropoxy- 2-Furyl isopropyl ether	$C_7H_{10}O_2$ 126.16	98272-34-5	 135.5	4-17-00-01219 0.9689^{20}	 1.4419^{20}
6113	2-Furanmethanamine Furfurylamine	C_5H_7NO 97.12	617-89-0	 145.5	5-18-09-00541 1.0995^{20}	H_2O 5; EtOH 5; eth 3; chl 3 1.4908^{20}
6114	2-Furanmethanamine, *N*-(2-furanylmethyl)- Difurfurylamine	$C_{10}H_{11}NO_2$ 177.20	18240-50-1	 $105\text{-}42^{15}$	4-18-00-07092 1.1045^{20}	eth 4 1.5168^{20}
6115	2-Furanmethanamine, *N*-methyl-	C_6H_9NO 111.14	4753-75-7	 149	5-18-09-00541 0.989^{25}	 1.4729^{20}
6116	2-Furanmethanamine, tetrahydro- Tetrahydrofurfurylamine	$C_5H_{11}NO$ 101.15	4795-29-3	 153	5-18-09-00488 0.9752^{20}	H_2O 4; eth 4; EtOH 4 1.4551^{20}
6117	2-Furanmethanethiol	C_5H_6OS 114.17	98-02-2	 157	5-17-03-00351 1.1319^{20}	H_2O 1; chl 2 1.5329^{20}
6118	2-Furanmethanol Furfuryl alcohol	$C_5H_6O_2$ 98.10	98-00-0 -31	4215 171	5-17-03-00338 1.1296^{20}	H_2O 5; EtOH 4; eth 4; chl 3 1.4869^{20}
6119	2-Furanmethanol, acetate	$C_7H_8O_3$ 140.14	623-17-6	 179	5-17-03-00344 1.1175^{20}	H_2O 1; EtOH 3; eth 3 1.4327^{20}
6120	2-Furanmethanol, 5-(aminomethyl)tetrahydro-	$C_6H_{13}NO_2$ 131.17	589-14-0 <-70	 $103^{0.3}$	 1.1021^{25}	 1.4870^{25}

No.	Name	Synonym	Mol. Form.	Mol. Wt.	CAS RN	mp/°C	Merck No.	bp/°C	Beil. Ref.	den/g cm^{-3}	Solubility	n_D
6121	2-Furanmethanol, α-methyl-		$C_6H_8O_2$	112.13	4208-64-4			162.5	5-17-03-00374	1.0739^{25}		1.4827^{15}
6122	2-Furanmethanol, 5-methyl-		$C_6H_8O_2$	112.13	3857-25-8			195 dec; 81^{23}	5-17-03-00387	1.0769^{20}	eth 4; EtOH 4	1.4853^{20}
6123	2-Furanmethanol, propanoate		$C_8H_{10}O_3$	154.17	623-19-8			195	2-17-00-00115	1.1085^{20}	H_2O 2; EtOH 3; eth 5; ace 3	
6124	2-Furanmethanol, tetrahydro-	Tetrahydrofurfuryl alcohol	$C_5H_{10}O_2$	102.13	97-99-4	<-80	9146	178	5-17-03-00115	1.0524^{20}	ace 4; eth 4	1.4520^{20}
6125	2-Furanmethanol, tetrahydro-, acetate		$C_7H_{12}O_3$	144.17	637-64-9			193; 89^{18}	5-17-03-00119	1.0624^{20}	H_2O 4; eth 4; EtOH 4; chl 4	1.4350^{25}
6126	2-Furanmethanol, tetrahydro-, propanoate		$C_8H_{14}O_3$	158.20	637-65-0			205.5	5-17-03-00119	1.044^{20}	eth 4; EtOH 4; chl 4	
6127	Furan, 2-methoxy-	2-Methoxyfuran	$C_5H_6O_2$	98.10	25414-22-6			110.5	5-17-03-00283	1.0646^{25}		1.4468^{25}
6128	Furan, 2-(methoxymethyl)-		$C_6H_8O_2$	112.13	13679-46-4			132	5-17-03-00340	1.0163^{20}	H_2O 1; EtOH 3; eth 4	1.4570^{20}
6129	Furan, 2-(methoxymethyl)-5-nitro-	2-(Methoxymethyl)-5-nitrofuran	$C_6H_7NO_4$	157.13	586-84-5		5917	$104\text{-}5^{3}$	4-17-00-01254	1.281^{20}	EtOH 4	1.5325^{20}
6130	Furan, 2-methyl-	2-Methylfuran	C_5H_6O	82.10	534-22-5	-87.5		65	5-17-01-00322	0.9132^{20}	H_2O 2; EtOH 3; eth 3; ctc 2	1.4342^{20}
6131	Furan, 3-methyl-	3-Methylfuran	C_5H_6O	82.10	930-27-8			66	5-17-01-00330	0.923^{18}	H_2O 1; EtOH 3; eth 3	1.4330^{19}
6132	Furan, 2,2'-methylenebis-		$C_9H_8O_2$	148.16	1197-40-6			94^{22}	5-19-01-00545	1.102^{20}	H_2O 1; EtOH 4; eth 4; ace 4	1.5049^{20}
6133	Furan, 3-(4-methyl-3-pentenyl)-		$C_{10}H_{14}O$	150.22	539-52-6			185.5	5-17-01-00515	0.9017^{20}		1.4705^{21}
6134	Furan, 2-nitro-		$C_4H_3NO_3$	113.07	609-39-2	30		134^{123}; 84^{13}	5-17-01-00296		H_2O 3; EtOH 3; eth 3	
6135	Furan, 2-(octyloxy)-	2-Furyl octyl ether	$C_{12}H_{20}O_2$	196.29	100314-90-7			$129\text{-}30^{18}$	4-17-00-01219	0.9214^{28}	eth 4	1.4520^{20}
6136	2-Furanol, 3,4-bis(1,3-benzodioxol-5-ylmethyl)tetrahydro-	Cubebin	$C_{20}H_{20}O_6$	356.38	18423-69-3	131.5	2616		5-19-12-00048		eth 4; EtOH 4; chl 4	
6137	3-Furanol, tetrahydro-		$C_4H_8O_2$	88.11	453-20-3			181	5-17-03-00064	1.09^{25}		1.4500^{20}
6138	2-Furanol, tetrahydro-, benzoate		$C_{11}H_{12}O_3$	192.21	3333-44-6			302; 139^{2}	5-17-03-00060	1.137^{20}	eth 4; EtOH 4	
6139	2(3H)-Furanone, 3-acetyldihydro-	α-Acetylbutyrolactone	$C_6H_8O_3$	128.13	517-23-7		77	$107\text{-}8^{5}$	5-17-11-00016	1.1846^{20}	H_2O 4	1.4585^{20}
6140	2(3H)-Furanone, 5-butyldihydro-		$C_8H_{14}O_2$	142.20	104-50-7			$132\text{-}3^{20}$	5-17-09-00073	0.9796^{19}	EtOH 3; ctc 2	1.4451^{19}
6141	2(3H)-Furanone, 5-(chloromethyl)dihydro-	Valeric acid, 5-chloro-4-hydroxy-, γ-lactone	$C_5H_7ClO_2$	134.56	39928-72-8			$132\text{-}5^{12}$	5-17-09-00025	1.625^{18}	EtOH 2; eth 2; bz 2	
6142	2(3H)-Furanone, dihydro-	γ-Butyrolactone	$C_4H_6O_2$	86.09	96-48-0	-43.3	1596	204	5-17-09-00007	1.1284^{16}	ace 4; bz 4; eth 4; EtOH 4	1.4341^{20}
6143	2(3H)-Furanone, dihydro-3-hydroxy-4,4-dimethyl-, (*R*)-	Pantolactone	$C_6H_{10}O_3$	130.14	599-04-2	92	6963		5-18-01-00022			
6144	2(3H)-Furanone, dihydro-3-(hydroxyphenylmethyl)-4-[(1-methyl-1H-imidazol-5-yl)methyl]-,	Isopilosine	$C_{16}H_{18}N_2O_3$	286.33	491-88-3	187	5085		1-27-00-00612		EtOH 4	
6145	2(3H)-Furanone, dihydro-3-methyl-		$C_5H_8O_2$	100.12	1679-47-6			$78\text{-}81^{10}$	5-17-09-00028	1.063^{25}		1.4325^{20}
6146	2(3H)-Furanone, dihydro-5-methyl-, (±)-		$C_5H_8O_2$	100.12	57129-69-8	-31		206	5-17-09-00024	1.0465^{25}	H_2O 5; EtOH 3; ace 3; ctc 2	1.4328^{20}
6147	2(3H)-Furanone, dihydro-3-methylene-	α-Methylene butyrolactone	$C_5H_6O_2$	98.10	547-65-9		5981	$85\text{-}6^{10}$	5-17-09-00127	1.1206^{20}	H_2O 3; EtOH 4; eth 3; ace 3	1.4650^{20}
6148	2(3H)-Furanone, dihydro-5-methylene-		$C_5H_6O_2$	98.10	10008-73-8			$85\text{-}6^{10}$	5-17-09-00124	1.1206^{20}	ace 4; bz 4; eth 4; EtOH 4	1.4650^{20}
6149	2(3H)-Furanone, dihydro-5-pentyl-	4-Hydroxynonanoic acid lactone	$C_9H_{16}O_2$	156.22	104-61-0				5-17-09-00084			
6150	2(3H)-Furanone, 5-ethyldihydro-		$C_6H_{10}O_2$	114.14	695-06-7	-18		215.5	5-17-09-00040	1.0261^{20}	H_2O 4; EtOH 4	1.4495^{20}
6151	2(3H)-Furanone, 3-ethyldihydro-4-[(1-methyl-1H-imidazol-5-yl)methyl]-, (3*S*-*cis*)-	Pilocarpine	$C_{11}H_{16}N_2O_2$	208.26	92-13-7	34	7395	260^{5}	2-27-00-00694		H_2O 3; EtOH 3; eth 2; bz 2	
6152	2(3H)-Furanone, 5-heptyldihydro-	4-Hydroxyundecanoic acid lactone	$C_{11}H_{20}O_2$	184.28	104-67-6			286	5-17-09-00098	0.9494^{20}	EtOH 4	1.4512^{20}
6153	2(3H)-Furanone, 5-methyl-		$C_5H_6O_2$	98.10	591-12-8	18		56^{12}	5-17-09-00120	1.084^{20}	H_2O 3; EtOH 3; eth 3; ctc 2	1.4476^{20}
6154	2(5H)-Furanone, 5-methyl-		$C_5H_6O_2$	98.10	591-11-7	<-17		209; 98^{15}	5-17-09-00121	1.0810^{20}	H_2O 5; EtOH 3; eth 3	1.4454^{20}
6155	2(5H)-Furanone, 5-methylene-	Protoanemonin	$C_5H_4O_2$	96.09	108-28-1		7907	73^{11}	5-17-09-00371		H_2O 2; chl 3	
6156	Furan, 2,2'-[oxybis(methylene)]bis-		$C_{10}H_{10}O_3$	178.19	4437-22-3			101^{2}	5-17-03-00347	1.1405^{20}	H_2O 1	1.5088^{20}
6157	Furan, 2-phenoxy-	2-Furyl phenyl ether	$C_{10}H_8O_2$	160.17	60698-31-9			$105\text{-}6^{18}$	4-17-00-01219	1.1010^{23}	eth 3	1.5418^{23}

No.	Name	Synonym	Mol. Form.	Mol. Wt.	CAS RN	mp/°C	Merck No.	bp/°C	Beil. Ref.	den/g cm^{-3}	Solubility	n_D
6158	Furan, 2-(phenoxymethyl)-	Furfuryl phenyl ether	$C_{11}H_{10}O_2$	174.20	4437-23-4			$139-40^{20}$	5-17-03-00340	1.123^{17}		1.5535^{17}
6159	Furan, 2-phenyl-	2-Phenylfuran	$C_{10}H_8O$	144.17	17113-33-6			$107-8^{18}$	5-17-02-00139	1.083^{20}	ace 4; bz 4	1.5920^{20}
6160	Furan, 2-(phenylthio)-	2-Furyl phenyl sulfide	$C_{10}H_8OS$	176.24	16003-14-8			$119-20^{8}$	5-17-03-00285	1.1341^{26}	eth 4; EtOH 4	1.5976^{20}
6161	2-Furanpropanoic acid		$C_7H_8O_3$	140.14	935-13-7	58		229	5-18-06-00224		H_2O 4; eth 4; chl 4	
6162	2-Furanpropanoic acid, methyl ester		$C_8H_{10}O_3$	154.17	37493-31-5			89^{15}	5-18-06-00224	1.088^{20}		1.4662^{20}
6163	2-Furanpropanoic acid, β-oxo-, ethyl ester		$C_9H_{10}O_4$	182.18	615-09-8			$142-3^{10}$	5-18-08-00143	1.165^{17}	H_2O 1; EtOH 3; eth 3	
6164	2-Furanpropanoic acid, tetrahydro-	Propionic acid, 3-(2-tetrahydrofuryl)	$C_7H_{12}O_3$	144.17	935-12-6			263	5-18-06-00033	1.1135^{20}	ace 4; EtOH 4	1.4578^{25}
6165	2-Furanpropanoic acid, tetrahydro-, ethyl ester		$C_9H_{16}O_3$	172.22	4525-36-4			$222; 73^{2}$	5-18-06-00034	1.023^{7}	ace 4; EtOH 4	1.440^{20}
6166	2-Furanpropanol, tetrahydro-α-[2-(tetrahydro-2-furanyl)ethyl]-	1,5-Bis(2-tetrahydrofuryl)-3-pentanol	$C_{13}H_{24}O_3$	228.33	6265-26-5			$116^{0.05}$	4-19-00-00704		chl 2	1.4541^{25}
6167	Furan, 2-propyl-	2-Propylfuran	$C_7H_{10}O$	110.16	4229-91-8			115	5-17-01-00354	0.8876^{20}	EtOH 3; eth 3; ace 3	1.4549^{20}
6168	Furan, tetrahydro-	Tetrahydrofuran	C_4H_8O	72.11	109-99-9	-108.3	9144	65	5-17-01-00027	0.8892^{20}	H_2O 3; EtOH 4; eth 4; ace 4	1.4050^{20}
6169	Furan, tetrahydro-2,5-dimethoxy-		$C_6H_{12}O_3$	132.16	696-59-3			145.7	5-17-05-00004	1.02^{25}		1.4180^{20}
6170	Furan, tetrahydro-2-(methoxymethyl)-		$C_6H_{12}O_2$	116.16	19354-27-9			140	5-17-03-00116	0.952^{25}		1.4270^{20}
6171	Furan, tetrahydro-2-methyl-	2-Methyltetrahydrofuran	$C_5H_{10}O$	86.13	96-47-9			78	5-17-01-00078	0.8552^{20}	H_2O 3; EtOH 4; eth 4; ace 4	1.4059^{21}
6172	Furan, tetrahydro-3-methyl-	3-Methyltetrahydrofuran	$C_5H_{10}O$	86.13	13423-15-9			86.5	5-17-01-00081	0.8642^{20}	ace 4; bz 4; eth 4; EtOH 4	1.4122^{20}
6173	Furan, tetrahydro-2-propyl-		$C_7H_{14}O$	114.19	3208-22-8			136	5-17-01-00116	0.8547^{20}		1.4242^{20}
6174	Furan, tetrahydro-2,2,4,4-tetramethyl-	2,2,4,4-Tetramethyltetrahydrofuran	$C_8H_{16}O$	128.21	3358-28-9			121	4-17-00-00109		H_2O 1; EtOH 5; eth 5	1.4074^{20}
6175	Furan, tetraphenyl-	2,3,4,5-Tetraphenylfuran	$C_{28}H_{20}O$	372.47	1056-77-5	175		220	4-17-00-00810		H_2O 1; EtOH 3; eth 3; ace 3	
6176	Furazan, dimethyl-		$C_4H_6N_2O$	98.10	4975-21-7	-7		156	4-27-00-07096	1.0528^{14}	eth 4; EtOH 4	1.4327^{20}
6177	3aH-Furo[2,3-b]indol-3a-ol, 8a-(5-ethenyl-1-azabicyclo[2.2.2]oct-2-yl)-2,3,8,8a-tetrahydro-	Conquinamine	$C_{19}H_{24}N_2O_2$	312.41	464-86-8	123	2505		2-27-00-00667		H_2O 2; EtOH 3; eth 3; chl 3	
6178	3aH-Furo[2,3-b]indol-3a-ol, 8a-(5-ethenyl-1-azabicyclo[2.2.2]oct-2-yl)-2,3,8,8a-tetrahydro-	Quinamine	$C_{19}H_{24}N_2O_2$	312.41	464-85-7	185.5	8060		2-27-00-00667		H_2O 1; EtOH 4; eth 3; ace 3	
6179	Furo[2,3-b]quinoline, 4,8-dimethoxy-	Fagarine	$C_{13}H_{11}NO_3$	229.24	524-15-2	142	3879		4-27-00-02211		H_2O 2; EtOH 3; eth 3; bz 3	
6180	Furo[2,3-b]quinoline, 4-methoxy-	Dictamnine	$C_{12}H_9NO_2$	199.21	484-29-7	133.5	3079		4-27-00-02030		H_2O 2; EtOH 4; eth 3; chl 3	
6181	Furo[2,3-b]quinoline, 4,7,8-trimethoxy-	Skimmianine	$C_{14}H_{13}NO_4$	259.26	83-95-4	177	8505		4-27-00-02296		H_2O 1; EtOH 3; eth 2; chl 3	
6182	Furo[2,3-b]quinolin-4(2H)-one, 3,9-dihydro-8-methoxy-9-methyl-2-(1-methylethyl)-, (*R*)-	Lunacrine	$C_{16}H_{19}NO_3$	273.33	82-40-6		5473		4-27-00-03566		chl 3	
6183	4H-Furo[3,2-c]pyran-2(6H)-one, 4-hydroxy-	Patulin	$C_7H_6O_4$	154.12	149-29-1	111	7002		5-18-03-00005		H_2O 3; EtOH 3; eth 3; ace 3	
6184	1H,12H-Furo[3',2':4,5]furo[2,3-h]pyrano[3,4-c][1]benzopyran-1,12-dione, 3,4,7a,10a-tetrahydro-5-meth	Aflatoxin G1	$C_{17}H_{12}O_7$	328.28	1165-39-5	245			5-19-12-00081			
6185	5H-Furo[3,2-g][1]benzopyran-5-one, 4,9-dimethoxy-7-methyl-	Khellin	$C_{14}H_{12}O_5$	260.25	82-02-0	154-5 dec	5189	$180-200^{0.05}$	5-19-06-00320		H_2O 1; EtOH 3; eth 2; ace 3	
6186	5H-Furo[3,2-g][1]benzopyran-5-one, 4-methoxy-7-methyl-	Visnagin	$C_{13}H_{10}O_4$	230.22	82-57-5	144.5	9916		5-19-06-00030		H_2O 2; EtOH 2; chl 4	
6187	7H-Furo[3,2-g][1]benzopyran-7-one, 9-methoxy-	Methoxsalen	$C_{12}H_8O_4$	216.19	298-81-7	148	5911		5-19-06-00015		H_2O 2; EtOH 4; eth 2; ace 2	
6188	7H-Furo[3,2-g][1]benzopyran-7-one, 9-methoxy-2-(1-methylethyl)-	Peucedanin	$C_{15}H_{14}O_4$	258.27	133-26-6	85	7144	$276-81^{17}$	5-19-06-00042		H_2O 2; EtOH 3; eth 3; bz 2	
6189	7H-Furo[3,2-g][1]benzopyran-7-one, 9-[(3-methyl-2-butenyl)oxy]-	Imperatorin	$C_{16}H_{14}O_4$	270.28	482-44-0	102	4839		5-19-06-00016		H_2O 2; EtOH 3; eth 3; bz 3	
6190	3-Furoic acid, 5-bromo-, ethyl ester	β-Furoic acid, 5-bromo, ethyl ester	$C_7H_7BrO_3$	219.03	32460-20-1	17		$235; 135^{34}$	5-18-06-00198	1.528^{20}	eth 4; EtOH 4	
6191	Furo[3',4':6,7]naphtho[2,3-d]-1,3-dioxol-6(5aH)-one, 5,8,8a,9-tetrahydro-9-hydroxy-5-(3,4,5-trimethoxyphenyl)-	Picropodophyllin	$C_{22}H_{22}O_8$	414.41	477-47-4	228	7385		5-19-10-00665		ace 4; bz 4; eth 4; EtOH 4	

No.	Name Synonym	Mol. Form. Mol. Wt.	CAS RN mp/°C	Merck No. bp/°C	Beil. Ref. den/g cm^{-3}	Solubility n_D
6192	Furo[3',4':6,7]naphtho[2,3-d]-1,3-dioxol-6(5aH)-one, 5,8,8a,9-tetrahydro-9-hydroxy-5-(3,4,5-trimetho Podophyllotoxin	$C_{22}H_{22}O_8$ 414.41	518-28-5 183	7520	5-19-10-00666	H_2O 2; EtOH 4; eth 1; ace 3
6193	Galactaric acid Mucic acid	$C_6H_{10}O_8$ 210.14	526-99-8 255 dec	4237	4-03-00-01292	
6194	Galactitol Dulcose	$C_6H_{14}O_6$ 182.17	608-66-2 189.5	4238 $275\text{-}80^{1}$	4-01-00-02844 1.47^{20}	H_2O 3; EtOH 2; eth 1; bz 1
6195	β-*D*-Galactopyranoside, (2α,3β,5α,15β,25*R*)-2,15-dihydroxyspirostan-3-yl Digitonin	$C_{56}H_{92}O_{29}$ 1229.33	11024-24-1 237.5	3144	5-19-03-00603	
6196	β-*D*-Galactopyranoside, (3β,5α,22β,25*S*)-spirosolan-3-yl *O*-β- Tomatine	$C_{50}H_{83}NO_{21}$ 1034.20	17406-45-0 270	9471	4-27-00-01954	EtOH 4; diox 4
6197	*D*-Galactose	$C_6H_{12}O_6$ 180.16	59-23-4 170	4241	4-01-00-04336	H_2O 4; EtOH 2; eth 1; bz 1
6198	Galactose, 6-deoxy-3-*O*-methyl- Digitalose	$C_7H_{14}O_5$ 178.19	4481-08-7 119	3142	4-01-00-04270	H_2O 4
6199	*D*-Galacturonic acid	$C_6H_{10}O_7$ 194.14	685-73-4 166	4242	1-03-00-00306	H_2O 3; EtOH 3; eth 1
6200	Galanthamine Lycoremine	$C_{17}H_{21}NO_3$ 287.36	357-70-0 126.5	4245	4-27-00-02184	ace 4; EtOH 4; chl 4
6201	Galanthan-1,2-diol, 3,12-didehydro-9,10-[methylenebis(oxy)]-, (1α,2β)- Lycorine	$C_{16}H_{17}NO_4$ 287.32	476-28-8 280	5498 sub	2-27-00-00547	H_2O 1; EtOH 2; eth 2; chl 2
6202	Gardona Phosphoric acid, 2-chloro-1-(2,4,5-trichlorophenyl)ethenyl dimethyl ester	$C_{10}H_9Cl_4O_4P$ 365.96	961-11-5			
6203	Gelsemine Spiro[3,5,8-ethanylylidene-1H-pyrano[3,4-c]pyridine-10,3'-[3H]indol]-2'(1'H)-one, 5-ethenyl	$C_{20}H_{22}N_2O_2$ 322.41	509-15-9 178	4277	4-27-00-07526	ace 4; bz 4; eth 4; EtOH 4
6204	Gibb-3-ene-1,10-dicarboxylic acid, 2,4a,7-trihydroxy-1-methyl-8-methylene-, 1,4a-lactone, Gibberellic acid	$C_{19}H_{22}O_6$ 346.38	77-06-5 234	4313	5-18-09-00269	ace 4; EtOH 4; MeOH 4
6205	*D*-Glucaric acid *D*-Tetrahydroxyadipic acid	$C_6H_{10}O_8$ 210.14	87-73-0 125.5	4344	4-03-00-01291	H_2O 4; EtOH 4; eth 2; chl 2
6206	*D*-Glucitol Sorbitol	$C_6H_{14}O_6$ 182.17	50-70-4 111	8680 $295^{3.5}$	4-01-00-02839 1.489^{20}	H_2O 4; ace 4 1.3330^{20}
6207	*D*-Glucitol, 1-amino-1-deoxy- Glucamine	$C_6H_{15}NO_5$ 181.19	488-43-7 127	4343	4-04-00-01913	H_2O 4; EtOH 4
6208	*D*-Glucitol, 1-deoxy-1-(methylamino)- *N*-Methylglucamine	$C_7H_{17}NO_5$ 195.22	6284-40-8 128.5	5995	4-04-00-01914	H_2O 3
6209	*D*-Gluconic acid Gluconic acid	$C_6H_{12}O_7$ 196.16	526-95-4 131	4350	4-03-00-01255	H_2O 3; EtOH 2; eth 1; bz 1
6210	*D*-Gluconic acid, 4-*O*-β-*D*-galactopyranosyl- Lactobionic acid	$C_{12}H_{22}O_{12}$ 358.30	96-82-2	5219	5-17-07-00436	H_2O 4; EtOH 2; eth 1; HOAc 2
6211	Gluconic acid, β-lactone Oxetane, gluconic acid deriv.	$C_6H_{10}O_6$ 178.14	3087-62-5 153		2-18-00-00190 1.610^{-5}	H_2O 2; EtOH 2; eth 1; ace 1
6212	*D*-Gluconic acid, δ-lactone δ-*D*-Gluconolactone	$C_6H_{10}O_6$ 178.14	90-80-2	4351	5-18-05-00011	
6213	*D*-Gluconic acid, monosodium salt Gluconic acid, sodium salt	$C_6H_{11}NaO_7$ 218.14	527-07-1	8569	4-03-00-01255	H_2O 3
6214	β-*D*-Glucopyranose, 4-*O*-β-*D*-galactopyranosyl-	$C_{12}H_{22}O_{11}$ 342.30	5965-66-2 254		5-17-07-00196 1.59^{20}	H_2O 4; EtOH 2; eth 1; chl 1
6215	α-*L*-Glucopyranose, 2-(methylamino)-2-deoxy- *N*-Methyl-α-*L*-glucosamine	$C_7H_{15}NO_5$ 193.20	42852-95-9	5996		
6216	α-*D*-Glucopyranose, pentaacetate	$C_{16}H_{22}O_{11}$ 390.34	604-68-2 113.3	sub	5-17-07-00318	H_2O 2; EtOH 2; eth 3; chl 3
6217	β-*D*-Glucopyranose, pentaacetate	$C_{16}H_{22}O_{11}$ 390.34	604-69-3 134	sub	5-17-07-00319 1.2740^{20}	H_2O 1; EtOH 2; eth 2; bz 3
6218	α-*D*-Glucopyranoside, β-*D*-fructofuranosyl Sucrose	$C_{12}H_{22}O_{11}$ 342.30	57-50-1 185.5	8855	5-17-08-00399 1.5805^{17}	H_2O 3; EtOH 2; eth 1; py 3 1.5376
6219	α-*D*-Glucopyranoside, β-*D*-fructofuranosyl *O*-α-*D*-galactopyranosyl-(1 Raffinose	$C_{18}H_{32}O_{16}$ 504.44	512-69-6 80	8120	4-17-00-03801 1.465^{25}	H_2O 4; EtOH 2; eth 1; MeOH 4
6220	α-*D*-Glucopyranoside, α-*D*-glucopyranosyl Trehalose	$C_{12}H_{22}O_{11}$ 342.30	99-20-7 203	9496	1.58^{24}	H_2O 4; EtOH 3; eth 1; bz 1
6221	α-*D*-Glucopyranoside, *O*-α-*D*-glucopyranosyl-(13)-β-*D*-fructofuranosyl Melezitose	$C_{18}H_{32}O_{16}$ 504.44	597-12-6 153	5698	5-17-08-00414 1.5565^{25}	H_2O 4
6222	β-*D*-Glucopyranoside, 2-(hydroxymethyl)phenyl Salicin	$C_{13}H_{18}O_7$ 286.28	138-52-3 207	8293 240 dec	5-17-07-00113 1.434^{20}	H_2O 4; EtOH 4; HOAc 4
6223	β-*D*-Glucopyranoside, 2-(hydroxymethyl)phenyl, 6-benzoate Populin	$C_{20}H_{22}O_8$ 390.39	99-17-2 180	7571	5-17-07-00343	

No.	Name Synonym	Mol. Form. Mol. Wt.	CAS RN mp/°C	Merck No. bp/°C	Beil. Ref. den/g cm^{-3}	Solubility n_D
6224	β-*D*-Glucopyranoside, 4-hydroxyphenyl Arbutin	$C_{12}H_{16}O_7$ 272.25	497-76-7 199.5	799	5-17-07-00110	H_2O 4; EtOH 3; eth 2; bz 1
6225	β-*D*-Glucopyranoside, 4-(3-hydroxy-1-propenyl)-2,6-dimethoxyphenyl Syringin	$C_{17}H_{24}O_9$ 372.37	118-34-3 192	8997	4-17-00-03002	EtOH 4
6226	β-*D*-Glucopyranoside, 4-(3-hydroxy-1-propenyl)-2-methoxyphenyl Coniferin	$C_{16}H_{22}O_8$ 342.35	531-29-3 186	2498	4-17-00-02999	H_2O 3; EtOH 2; eth 1; py 3
6227	α-*D*-Glucopyranoside, methyl α-Methylglucoside	$C_7H_{14}O_6$ 194.18	97-30-3 168	5997 $200^{0.2}$	5-17-07-00010 1.46^{30}	H_2O 4
6228	α-*D*-Glucopyranoside, methyl 2,3,4,6-tetra-*O*-methyl- α-*D*-Glucose, pentamethyl ether	$C_{11}H_{22}O_6$ 250.29	605-81-2	 $180^{0.4}$	5-17-07-00029 1.0944^{20}	H_2O 4; ace 4; eth 4; EtOH 4 1.4466^{20}
6229	α-*D*-Glucopyranoside, 1,3,4,6-tetra-*O*-acetyl-β-*D*-fructofuranosyl, tetraacetate Octaacetyl sucrose	$C_{28}H_{38}O_{19}$ 678.60	126-14-7 86.5	8856 250^{1}	5-17-08-00410 1.27^{16}	H_2O 2; EtOH 3; eth 3; ace 3 1.4660
6230	α-*D*-Glucopyranosiduronic acid, (3β,20β)-20-carboxy-11-oxo-30-norolean-12-en-3-yl 2-*O*- Glycyrrhizic acid	$C_{42}H_{62}O_{16}$ 822.94	1405-86-3 220 dec	4401	4-18-00-05156	H_2O 4
6231	β-*D*-Glucopyranosiduronic acid, 2,2,2-trichloroethyl Urochloralic acid	$C_8H_{11}Cl_3O_7$ 325.53	97-25-6 142	9798	4-18-00-05113	H_2O 4; EtOH 4
6232	β-*D*-Glucose	$C_6H_{12}O_6$ 180.16	28905-12-6 150		4-01-00-04306 1.5620^{18}	H_2O 4; EtOH 2; eth 1; py 3
6233	*D*-Glucose, 2-amino-2-deoxy- *D*-Glucosamine	$C_6H_{13}NO_5$ 179.17	3416-24-8	4352	4-04-00-02017	H_2O 4
6234	*D*-Glucose, 6-*O*-α-*L*-arabinopyranosyl- Vicianose	$C_{11}H_{20}O_{10}$ 312.27	14116-69-9 210 dec	9879	5-17-06-00112	H_2O 4
6235	*D*-Glucose, 6-deoxy- Quinovose	$C_6H_{12}O_5$ 164.16	7658-08-4 139.5	8106	4-01-00-04260	H_2O 4; EtOH 4
6236	*D*-Glucose, 6-*O*-(6-deoxy-α-*L*-mannopyranosyl)- Rutinose	$C_{12}H_{22}O_{10}$ 326.30	90-74-4 189-92 dec	8277	5-17-06-00269	H_2O 4; EtOH 4
6237	*D*-Glucose, 6-*O*-β-*D*-xylopyranosyl- Primeverose	$C_{11}H_{20}O_{10}$ 312.27	26531-85-1 210	7752	4-17-00-02447	H_2O 4; MeOH 4
6238	*D*-Glucuronic acid	$C_6H_{10}O_7$ 194.14	6556-12-3 165	4360	4-03-00-01996	H_2O 4; EtOH 4
6239	*D*-Glucuronic acid, γ-lactone *D*-Glucuronolactone	$C_6H_8O_6$ 176.13	32449-92-6 177.5	4362	5-18-05-00033 1.76^{20}	H_2O 3; EtOH 2; bz 1; MeOH 2
6240	*DL*-Glutamic acid	$C_5H_9NO_4$ 147.13	617-65-2 199 dec		4-04-00-03028 1.4601^{20}	H_2O 2; EtOH 1; eth 2; CS_2 1
6241	*L*-Glutamic acid (*S*)-2-Aminopentanedioic acid	$C_5H_9NO_4$ 147.13	56-86-0 160 dec	4363 sub 175	4-04-00-03028 1.538^{20}	
6242	*L*-Glutamic acid, *N*-(4-aminobenzoyl)- *N*-(*p*-Aminobenzoyl)-L(+)-glutamic acid	$C_{12}H_{14}N_2O_5$ 266.25	4271-30-1 173	437	4-14-00-01153	
6243	*L*-Glutamic acid, *N*-[4-[[(2-amino-1,4-dihydro-4-oxo-6-pteridinyl)methyl]amino]benzoyl]- Folic acid	$C_{19}H_{19}N_7O_6$ 441.40	59-30-3 250 dec	4140	4-26-00-03944	py 4; EtOH 4; HOAc 4
6244	*L*-Glutamic acid, *N*-[4-[[(2-amino-5-formyl-1,4,5,6,7,8-hexahydro-4-oxo-6-pteridinyl)methyl]amino]benzoyl]- Folinic acid	$C_{20}H_{23}N_7O_7$ 473.45	58-05-9 245 dec	4141	4-26-00-03881	H_2O 2
6245	*L*-Glutamic acid, 5-ethyl ester γ-Ethyl *L*-glutamate	$C_7H_{13}NO_4$ 175.18	1119-33-1 191	4364	4-04-00-03033	H_2O 2
6246	*DL*-Glutamic acid, 3-hydroxy-	$C_5H_9NO_5$ 163.13	5985-23-9 209	 dec	4-04-00-03253	H_2O 4; EtOH 1; eth 1; bz 1
6247	*L*-Glutamic acid, 5-[2-[4-(hydroxymethyl)phenyl]hydrazide] Agaritine	$C_{12}H_{17}N_3O_4$ 267.28	2757-90-6 205-9 dec	175		H_2O 4
6248	*L*-Glutamic acid, monosodium salt Glutamic acid, monosodium salt	$C_5H_8NNaO_4$ 169.11	142-47-2	6165		H_2O 3
6249	*L*-Glutamine 2-Aminoglutaramic acid	$C_5H_{10}N_2O_3$ 146.15	56-85-9 185-6 dec	4365	4-04-00-03038	H_2O 3; EtOH 1; eth 1; bz 1
6250	Glycine Aminoacetic acid	$C_2H_5NO_2$ 75.07	56-40-6 290 dec	4386	4-04-00-02349 1.161^{20}	H_2O 4; EtOH 1; eth 1; ace 2
6251	Glycine, *N*-acetyl- Aceturic acid	$C_4H_7NO_3$ 117.10	543-24-8 206	74	4-04-00-02399	H_2O 4; ace 4; EtOH 4
6252	Glycine, *N*-(4-aminobenzoyl)- *p*-Aminohippuric acid	$C_9H_{10}N_2O_3$ 194.19	61-78-9 198.5	454	4-14-00-01152	ace 4; bz 4; EtOH 4
6253	Glycine, *N*-(aminoiminomethyl)- Glycocyamine	$C_3H_7N_3O_2$ 117.11	352-97-6 282	4391	4-04-00-02414	H_2O 2; EtOH 2; eth 2
6254	Glycine, *N*-(aminoiminomethyl)-*N*-methyl- Creatine	$C_4H_9N_3O_2$ 131.13	57-00-1 303 dec	2570	4-04-00-02426 1.33^{25}	H_2O 3; EtOH 2; eth 1
6255	Glycine, *N*-benzoyl- Hippuric acid	$C_9H_9NO_3$ 179.18	495-69-2 191.5	4636	4-09-00-00778 1.371^{20}	H_2O 3; EtOH 3; eth 2; bz 2
6256	Glycine, *N*,*N*-bis(carboxymethyl)- Nitrilotriacetic acid	$C_6H_9NO_6$ 191.14	139-13-9 242 dec	6499	4-04-00-02441	H_2O 2; EtOH 3; DMSO-d_6 2

No.	Name Synonym	Mol. Form. Mol. Wt.	CAS RN mp/°C	Merck No. bp/°C	Beil. Ref. den/g cm^{-3}	Solubility n_D
6257	Glycine, *N,N*-bis[2-[bis(carboxymethyl)amino]ethyl]- Pentetic acid	$C_{14}H_{23}N_3O_{10}$ 393.35	67-43-6	7083	4-04-00-02454	
6258	Glycine, *N,N*-bis(ethoxycarbonyl)-, ethyl ester Aminoacetic acid, *N,N*-(dicarboethoxy), ethyl ester	$C_{10}H_{17}NO_6$ 247.25	127665-32-1 36.5	 152^{10}	0-04-00-00365	bz 4; eth 4; EtOH 4
6259	Glycine, *N,N*-bis(2-hydroxyethyl)- Bicine	$C_6H_{13}NO_4$ 163.17	150-25-4 193-5 dec	1221	4-04-00-02390	H_2O 4; EtOH 1
6260	Glycine, *N*-(carboxymethyl)- Iminodiacetic acid	$C_4H_7NO_4$ 133.10	142-73-4 247.5	4832	4-04-00-02428	H_2O 2; EtOH 1; eth 1
6261	Glycine, *N,N*-dimethyl- *N,N*-Dimethylglycine	$C_4H_9NO_2$ 103.12	1118-68-9 185.5	3233	4-04-00-02365	H_2O 4; EtOH 3; eth 3; ace 3
6262	Glycine, *N,N'*-1,2-ethanediylbis[*N*-(carboxymethyl)- Edetic acid	$C_{10}H_{16}N_2O_8$ 292.25	60-00-4 245 dec	3484	4-04-00-02449	
6263	Glycine, ethyl ester	$C_4H_9NO_2$ 103.12	459-73-4	 149; 58^{18}	4-04-00-02355 1.0275^{10}	H_2O 5; EtOH 5; eth 5; ace 5 1.4242^{10}
6264	Glycine, *N*-(*N*-*L*-γ-glutamyl-*L*-cysteinyl)- Glutatione	$C_{10}H_{17}N_3O_6S$ 307.33	70-18-8 195	4369	4-04-00-03165	H_2O 4; EtOH 1; eth 1; DMF 3
6265	Glycine, *N*-glycyl- 2-(Aminoacetamido)acetic acid	$C_4H_8N_2O_3$ 132.12	556-50-3 215 dec	4399	4-04-00-02459	H_2O 3
6266	Glycine, hydrazide Aminoacetic acid hydrazide	$C_2H_7N_3O$ 89.10	14379-80-7 80.5	 150 dec	3-04-00-01121	EtOH 2; eth 1; chl 3
6267	Glycine, *N*-(4-hydroxyphenyl)- *N*-(4-Hydroxyphenyl)glycine	$C_8H_9NO_3$ 167.16	122-87-2 245-7 dec	4771	4-13-00-01210	H_2O 2; EtOH 2; eth 1; ace 2
6268	Glycine, *N*-methyl- Sarcosine	$C_3H_7NO_2$ 89.09	107-97-1 212-3 dec	8331	4-04-00-02363	
6269	Glycine, *N*-methyl-*N*-(1-oxododecyl)-, sodium salt Gardol	$C_{15}H_{28}NNaO_3$ 293.38	137-16-6	4267		H_2O 2
6270	Glycine, *N*-phenyl- Phenylaminoacetic acid	$C_8H_9NO_2$ 151.17	103-01-5 127.5	7262	4-12-00-00872	H_2O 4; EtOH 4
6271	Glycine, *N*-phenyl-, ethyl ester	$C_{10}H_{13}NO_2$ 179.22	2216-92-4 58	 273.5	4-12-00-00873	eth 4; EtOH 4
6272	Glycine, *N*-(phenylmethyl)-, ethyl ester	$C_{11}H_{15}NO_2$ 193.25	6436-90-4	 $175\text{-}9^{50}$	4-12-00-02277	EtOH 4; eth 4; bz 4 1.5041^{20}
6273	Glycine, *N*-[(3α,5β,7α,12α)-3,7,12-trihydroxy-24-oxocholan-24-yl]- Glycocholic acid	$C_{26}H_{43}NO_6$ 465.63	475-31-0 166.5	4390	4-10-00-02077	H_2O 2; EtOH 4; eth 2
6274	Glycogen	$(C_6H_{10}O_5)_x$ 162.14	9005-79-2	4392		H_2O 4; EtOH 1; eth 1
6275	Glyodin 1H-Imidazole, 2-heptadecyl-4,5-dihydro-, monoacetate	$C_{20}H_{40}N_2{\cdot}C_2H_4O_2$ 368.60	556-22-9	4404	 1.035^{20}	
6276	Glyphosate Glycine, *N*-(phosphonomethyl)-	$C_3H_8NO_5P$ 169.07	1071-83-6 230 dec	4408		
6277	Glyphosine Glycine, *N,N*-bis(phosphonomethyl)-	$C_4H_{11}NO_8P_2$ 263.08	2439-99-8	4409		
6278	Guanidine Aminomethanamidine	CH_5N_3 59.07	113-00-8 50	4475	4-03-00-00148	H_2O 4; EtOH 4
6279	Guanidine, *N,N'*-bis(2-methylphenyl)-	$C_{15}H_{17}N_3$ 239.32	97-39-2 179		4-12-00-01764 1.10^{20}	H_2O 2; EtOH 2; eth 4; chl 3
6280	Guanidine, cyano- Dicyanodiamide	$C_2H_4N_4$ 84.08	461-58-5 211	3082	4-03-00-00160 1.404^{14}	H_2O 3; EtOH 3; eth 1; ace 3
6281	Guanidine, *N,N'*-dimethyl-*N''*-(phenylmethyl)- Bethanidine	$C_{10}H_{15}N_3$ 177.25	55-73-2 196	1208		
6282	Guanidine, *N,N'*-diphenyl- 1,3-Diphenylguanidine	$C_{13}H_{13}N_3$ 211.27	102-06-7 150	3325 170 dec	4-12-00-00769 1.13^{20}	H_2O 2; EtOH 3; eth 4; ctc 3
6283	Guanidine, (3-methyl-2-butenyl)- Galegine	$C_6H_{13}N_3$ 127.19	543-83-9 62.5	4246 dec	3-04-00-00465	H_2O 4; EtOH 4
6284	Guanidine, *N*-methyl-*N'*-nitro-*N*-nitroso- *N*-Methyl-*N'*-nitro-*N*-nitrosoguanidine	$C_2H_5N_5O_3$ 147.09	70-25-7	6017	4-04-00-03386	DMSO 3
6285	Guanidine, monohydrochloride	CH_6ClN_3 95.53	50-01-1 182.3		4-03-00-00148 1.354^{20}	H_2O 4; EtOH 4
6286	Guanidine, mononitrate	$CH_6N_4O_3$ 122.08	506-93-4 217	 dec	4-03-00-00148	H_2O 4; EtOH 4
6287	Guanidine, nitro- Nitroguanidine	$CH_4N_4O_2$ 104.07	556-88-7 239 dec	6529	4-03-00-00249	H_2O 2; EtOH 2; eth 1; alk 4
6288	Guanidine, *N,N',N''*-triphenyl-	$C_{19}H_{17}N_3$ 287.36	101-01-9 146.5	 dec	4-12-00-00866 1.163^{20}	H_2O 2; EtOH 3
6289	Guanosine 2-Amino-1,9-dihydro-9-β-*D*-ribofuranosyl-6H-purin-6-one	$C_{10}H_{13}N_5O_5$ 283.24	118-00-3 239 dec	4480	4-26-00-03901	H_2O 2; EtOH 1; eth 1; HOAc 4
6290	*D*-Gulose Gulose, *D*-	$C_6H_{12}O_6$ 180.16	4205-23-6	4490 dec	4-01-00-04333	H_2O 4
6291	Haloxyfop-(2-ethoxyethyl) (*RS*)-2-[4-(3-Chloro-5-trifluoromethyl-2-pyridyloxy)phenoxy]propionic acid, ethyl ester	$C_{17}H_{15}ClF_3NO_4$ 389.76	87237-48-7 60		 1.34	

No.	Name Synonym	Mol. Form. Mol. Wt.	CAS RN mp/°C	Merck No. bp/°C	Beil. Ref. den/g cm^{-3}	Solubility n_D
6292	7H-Benz[de]anthracen-7-one Benzanthrone	$C_{17}H_{10}O$ 230.27	82-05-3 170	1070		bz 2
6293	Helenine	$C_{15}H_{20}O_2$ 232.32	1407-14-3 78	275; 192^{10}	4-17-00-05030	bz 4; eth 4; EtOH 4
6294	Heneicosane	$C_{21}H_{44}$ 296.58	629-94-7 40.5	356.5	4-01-00-00569 0.7919^{20}	H_2O 1; EtOH 2; peth 3 1.4441^{20}
6295	Heneicosane, 11-decyl-	$C_{31}H_{64}$ 436.85	55320-06-4 10.0	282.0^{10}	4-01-00-00595 0.8116^{20}	1.4540^{20}
6296	Heneicosane, 11-(2,2-dimethylpropyl)-	$C_{26}H_{54}$ 366.71	55282-10-5 -21	238.0^{10}	4-01-00-00585 0.8031^{20}	1.4491^{20}
6297	Heneicosane, 11-phenyl- Benzene, (1-decylundecyl)-	$C_{27}H_{48}$ 372.68	6703-80-6 20.8	205$^{1.0}$	4-05-00-01226 0.8531^{20}	1.4788^{20}
6298	10-Heneicosene, 11-phenyl-	$C_{27}H_{46}$ 370.66	6703-78-2 48.2	203$^{1.0}$ 190$^{0.5}$	4-05-00-01506 0.8638^{20}	1.4922^{20}
6299	Hentriacontane Untriacontane	$C_{31}H_{64}$ 436.85	630-04-6 67.9	458	4-01-00-00594 0.781^{68}	EtOH 2; eth 2; bz 2; chl 2 1.4278^{90}
6300	16-Hentriacontanone	$C_{31}H_{62}O$ 450.83	502-73-8 83.8	250$^{0.1}$	4-01-00-03413 0.7947^{91}	H_2O 1; EtOH 2; eth 3; ace 2 1.4297^{94}
6301	Heptachlor epoxide 2,5-Methano-2H-indeno[1,2-b]oxirene, 2,3,4,5,6,7,7-heptachloro-1a,1b,5,5a,6,6a-hexahydro-, (1aα,1bβ,2α,5α,5aβ,6β,6aα)-	$C_{10}H_5Cl_7O$ 389.32	1024-57-3 160			
6302	Heptacosane	$C_{27}H_{56}$ 380.74	593-49-7 59.5	442	4-01-00-00586 0.7796^{60}	H_2O 1; EtOH 1; eth 2 1.4345^{65}
6303	14-Heptacosanone	$C_{27}H_{54}O$ 394.73	542-50-7 77.5	238$^{0.1}$	4-01-00-03411 0.7986^{81}	
6304	Heptadecanal Margaric aldehyde	$C_{17}H_{34}O$ 254.46	629-90-3 36	204^{26}	4-01-00-03395	bz 4; eth 4
6305	1-Heptadecanamine	$C_{17}H_{37}N$ 255.49	4200-95-7 49	336	4-04-00-00824 0.8510^{20}	H_2O 1; EtOH 3; eth 3 1.4510^{20}
6306	Heptadecane	$C_{17}H_{36}$ 240.47	629-78-7 22	302.0	4-01-00-00548 0.7780^{20}	H_2O 1; EtOH 2; eth 3; ctc 2 1.4369^{20}
6307	Heptadecane, 1-bromo- 1-Bromoheptadecane	$C_{17}H_{35}Br$ 319.37	3508-00-7 32	349	4-01-00-00549 0.9916^{20}	H_2O 1; chl 4 1.4625^{20}
6308	Heptadecane, 1,17-dibromo-	$C_{17}H_{34}Br_2$ 398.26	81726-82-1 38	208^{3}	3-01-00-00564	chl 4
6309	Heptadecane, 9-hexyl- 9-Hexylheptadecane	$C_{23}H_{48}$ 324.63	55124-79-3 -19.4	213^{10}, 157$^{0.5}$	4-01-00-00578 0.7976^{20}	1.4465^{20}
6310	Heptadecane, 2-methyl- 2-Methylheptadecane	$C_{18}H_{38}$ 254.50	1560-89-0 5.7	311	4-01-00-00556 0.7838^{15}	1.4394^{15}
6311	Heptadecane, 9-methyl- 9-Methylheptadecane	$C_{18}H_{38}$ 254.50	18869-72-2	172-4^{10}	3-01-00-00568 0.7810^{20}	1.4388^{20}
6312	Heptadecanenitrile	$C_{17}H_{33}N$ 251.46	5399-02-0 34	349	4-02-00-01195 0.8315^{20}	H_2O 1; EtOH 2; eth 4; chl 2 1.4467^{20}
6313	Heptadecane, 9-octyl-	$C_{25}H_{52}$ 352.69	7225-64-1 -13.8	231.5^{10}	4-01-00-00583 0.8020^{20}	1.4487^{20}
6314	Heptadecanoic acid Margaric acid	$C_{17}H_{34}O_2$ 270.46	506-12-7 61.3	5634 227^{100}	4-02-00-01193 0.8532^{60}	EtOH 2; eth 3; ace 3; bz 3 1.4342^{60}
6315	Heptadecanoic acid, ethyl ester	$C_{19}H_{38}O_2$ 298.51	14010-23-2 28	185^{5}	4-02-00-01194 0.8517^{30}	ace 4; bz 4; eth 4; EtOH 4
6316	Heptadecanoic acid, methyl ester	$C_{18}H_{36}O_2$ 284.48	1731-92-6 30	185^{9}, 152$^{0.05}$	4-02-00-01194	H_2O 1; EtOH 3; eth 4; ace 3
6317	1-Heptadecanol Heptadecyl alcohol	$C_{17}H_{36}O$ 256.47	1454-85-9 53.8	333	4-01-00-01884 0.8475^{20}	H_2O 1; EtOH 3; eth 3
6318	2-Heptadecanol	$C_{17}H_{36}O$ 256.47	16813-18-6 54	155^{1}	4-01-00-01885	eth 4; EtOH 4 1.4407^{37}
6319	9-Heptadecanol	$C_{17}H_{36}O$ 256.47	624-08-8 61	174^{9}	4-01-00-01886	H_2O 1; EtOH 3; eth 3; ace 3 1.4262^{80}
6320	2-Heptadecanone Pentadecyl methyl ketone	$C_{17}H_{34}O$ 254.46	2922-51-2 48	320	4-01-00-03395 0.8049^{48}	H_2O 1; EtOH 2; eth 4; ace 3
6321	9-Heptadecanone	$C_{17}H_{34}O$ 254.46	540-08-9 53	251.5; 142$^{1.5}$	4-01-00-03396 0.8140^{48}	EtOH 2; MeOH 3
6322	1-Heptadecene Hexahydroaplotaxene	$C_{17}H_{34}$ 238.46	6765-39-5 11.5	300	4-01-00-00929 0.7852^{20}	H_2O 1; eth 4; bz 3; lig 5 1.4432^{20}
6323	8-Heptadecene, 9-octyl-	$C_{25}H_{50}$ 350.67	24306-18-1	227^{10}	4-01-00-00936 0.8086^{20}	1.4554^{20}
6324	9-Heptadecenoic acid	$C_{17}H_{32}O_2$ 268.44	10136-52-4 12.5	175$^{0.5}$	4-02-00-01632 0.8942^{20}	1.4598^{20}
6325	2,4-Heptadienal, (*E*,*E*)-	$C_7H_{10}O$ 110.16	4313-03-5	84.5	4-01-00-03549 0.881^{25}	1.5315^{20}
6326	1,2-Heptadiene Butylallene	C_7H_{12} 96.17	2384-90-9	105.5	4-01-00-01027 0.7306^{18}	1.432^{18}
6327	1,4-Heptadiene	C_7H_{12} 96.17	5675-22-9	93	3-01-00-00999 0.7270^{20}	bz 4; eth 4; peth 4 1.4370^{20}
6328	1,5-Heptadiene 1-Methyldiallyl	C_7H_{12} 96.17	1541-23-7	94	3-01-00-00999 0.7186^{20}	ace 4; bz 4; eth 4; EtOH 4 1.4200^{20}

No.	Name Synonym	Mol. Form. Mol. Wt.	CAS RN mp/°C	Merck No. bp/°C	Beil. Ref. den/g cm^{-3}	Solubility n_D
6329	2,4-Heptadiene	C_7H_{12} 96.17	628-72-8	 108	4-01-00-01029 0.7384^{20}	H_2O 1; EtOH 3; eth 3; ace 3 1.4578^{20}
6330	1,3-Heptadiene, 2,6-dimethyl-	C_9H_{16} 124.23	2436-84-2	 141	4-01-00-01051 0.7648^{10}	 1.4606^{22}
6331	1,5-Heptadiene, 2,6-dimethyl-	C_9H_{16} 124.23	6709-39-3 -70	 143	4-01-00-01051 0.7648^{25}	
6332	2,4-Heptadiene, 2,4-dimethyl-	C_9H_{16} 124.23	74421-05-9	 138	4-01-00-01051 0.7750^{4}	 1.4587^{4}
6333	2,4-Heptadiene, 3,5-dimethyl-	C_9H_{16} 124.23	101935-28-8	 142	4-01-00-01052 0.7728^{20}	 1.4487^{20}
6334	1,6-Heptadiene-3,5-dione, 1,7-bis(4-hydroxy-3-methoxyphenyl)-, (*E,E*)- Curcumin	$C_{21}H_{20}O_6$ 368.39	458-37-7 183	2681	4-08-00-03697	EtOH 4; HOAc 4
6335	1,5-Heptadiene, 3-methyl- 3-Methyl-1,5-heptadiene	C_8H_{14} 110.20	4894-62-6 -57	 111.	4-01-00-01040 0.7250^{25}	
6336	1,5-Heptadien-4-ol	$C_7H_{12}O$ 112.17	5638-26-6	 156; 68^{24}	4-01-00-02248 0.8598^{20}	ace 4 1.4510^{25}
6337	1,6-Heptadien-4-ol 4-Hydroxy-1,6-heptadiene	$C_7H_{12}O$ 112.17	2883-45-6	 151	4-01-00-02249 0.864^{25}	 1.4505^{20}
6338	1,6-Heptadien-4-ol, 4-methyl- 1,6-Heptadiene-4-ol, 4-methyl	$C_8H_{14}O$ 126.20	25201-40-5	 158.4	4-01-00-02260 0.8611^{20}	ace 4; EtOH 4 1.4500^{23}
6339	3,5-Heptadien-2-one	$C_7H_{10}O$ 110.16	3916-64-1	 88^{28}	4-01-00-03549 0.8946^{19}	eth 4 1.5177^{19}
6340	2,5-Heptadien-4-one, 2,6-dimethyl- Phorone	$C_9H_{14}O$ 138.21	504-20-1 28	7307 197.5	4-01-00-03564 0.8850^{20}	H_2O 2; EtOH 3; eth 3; ace 3 1.4998^{20}
6341	1,6-Heptadien-3-yne 1,6-Heptadiene-3-yne	C_7H_8 92.14	5150-80-1	 110	3-01-00-01061 0.787^{25}	bz 3; peth 3 1.4694^{25}
6342	1,5-Heptadiyne	C_7H_8 92.14	764-56-7	 26^{30}	2-01-00-00247 0.8100^{21}	bz 4; peth 4 1.4521^{21}
6343	1,6-Heptadiyne	C_7H_8 92.14	2396-63-6 -85	 112	4-01-00-01121 0.8164^{17}	H_2O 1; bz 3; HOAc 3 1.451^{17}
6344	Heptanal Heptaldehyde	$C_7H_{14}O$ 114.19	111-71-7 -43.3	4578 152.8	4-01-00-03314 0.8132^{25}	H_2O 2; EtOH 5; eth 5; ctc 2 1.4113^{20}
6345	Heptanal, 5-methyl-	$C_8H_{16}O$ 128.21	75579-88-3	 72^{25}	4-01-00-03345 0.8164^{25}	 1.4144^{25}
6346	Heptanal, oxime Enanthaldoxime	$C_7H_{15}NO$ 129.20	629-31-2 57.5	 195	4-01-00-03316 0.8583^{55}	H_2O 2; EtOH 3; eth 3 1.4210^{20}
6347	Heptanal, 2-(phenylmethylene)-	$C_{14}H_{18}O$ 202.30	122-40-7 80	 174-5^{20}	2-07-00-00310 0.9711^{20}	H_2O 1; ace 3; ctc 3 1.5381^{20}
6348	Heptanamide	$C_7H_{15}NO$ 129.20	628-62-6 96	 254	4-02-00-00963 0.8521^{110}	H_2O 3; EtOH 3; eth 3; chl 2 1.4217^{110}
6349	1-Heptanamine Heptylamine	$C_7H_{17}N$ 115.22	111-68-2 -18	 156	4-04-00-00734 0.7754^{20}	H_2O 2; EtOH 5; eth 5; chl 2 1.4251^{20}
6350	2-Heptanamine Tuaminoheptane	$C_7H_{17}N$ 115.22	123-82-0	9712 142	3-04-00-00374 0.7665^{19}	H_2O 2; EtOH 3; eth 3; chl 2 1.4199^{19}
6351	4-Heptanamine	$C_7H_{17}N$ 115.22	16751-59-0	 139.5	4-04-00-00745 0.767^{20}	 1.4172^{20}
6352	1-Heptanamine, *N*-heptyl- Diheptylamine	$C_{14}H_{31}N$ 213.41	2470-68-0 31.5	 271; 135^{9}	4-04-00-00736 0.7956^{21}	H_2O 2; EtOH 3; eth 4
6353	2-Heptanamine, 6-methyl-, (±)-	$C_8H_{19}N$ 129.25	5984-58-7	6678 155	4-04-00-00764 0.767^{25}	 1.4209^{20}
6354	2-Heptanamine, 6-methyl-*N*-(3-methylbutyl)- Octamylamine	$C_{13}H_{29}N$ 199.38	502-59-0	6671 100-1^{7}	3-04-00-00387	
6355	Heptane	C_7H_{16} 100.20	142-82-5 -90.6	4580 98.5	4-01-00-00376 0.6837^{20}	H_2O 1; EtOH 4; eth 5; ace 5 1.3878^{20}
6356	Heptane, 1-bromo- Heptyl bromide	$C_7H_{15}Br$ 179.10	629-04-9 -58	 179	4-01-00-00391 1.1400^{20}	H_2O 1; EtOH 4; eth 4; ctc 2 1.4502^{20}
6357	Heptane, 2-bromo- 2-Heptyl bromide	$C_7H_{15}Br$ 179.10	1974-04-5 47	 166	3-01-00-00431 1.1277^{20}	H_2O 1; bz 4; ctc 3; chl 3 1.4503^{20}
6358	Heptane, 4-bromo- 4-Heptyl bromide	$C_7H_{15}Br$ 179.10	998-93-6	 84.6^{72}; 60^{18}	4-01-00-00392 1.1351^{20}	H_2O 1; bz 3; ctc 3; chl 3 1.4495^{20}
6359	Heptane, 1-bromo-7-fluoro-	$C_7H_{14}BrF$ 197.09	334-42-9	 85^{11}	4-01-00-00392	eth 4; EtOH 4 1.4463^{20}
6360	Heptane, 1-chloro- Heptyl chloride	$C_7H_{15}Cl$ 134.65	629-06-1 -69.5	 159	4-01-00-00389 0.8758^{20}	H_2O 1; EtOH 5; eth 5; ctc 2 1.4256^{20}
6361	Heptane, 2-chloro- 2-Chloroheptane	$C_7H_{15}Cl$ 134.65	1001-89-4	 46^{19}	4-01-00-00390 0.8672^{20}	H_2O 1; eth 4; bz 3; chl 3 1.4221^{20}
6362	Heptane, 3-chloro- 3-Chloroheptane	$C_7H_{15}Cl$ 134.65	999-52-0	 144; 48.3^{20}	4-01-00-00390 0.8960^{20}	bz 4; eth 4 1.4228^{20}
6363	Heptane, 4-chloro- 4-Chloroheptane	$C_7H_{15}Cl$ 134.65	998-95-8	 144	4-01-00-00390 0.8710^{20}	bz 4; eth 4 1.4237^{20}
6364	Heptane, 3-chloro-3-ethyl-	$C_9H_{19}Cl$ 162.70	28320-89-0	 46^{3}	3-01-00-00510 0.8856^{20}	EtOH 4; chl 4 1.4400^{20}
6365	Heptane, 1-chloro-7-fluoro-	$C_7H_{14}ClF$ 152.64	334-43-0	 70^{10}	4-01-00-00390 0.993^{20}	bz 4; eth 4; EtOH 4; chl 4 1.4222^{25}
6366	Heptane, 2-chloro-2-methyl- 2-Chloro-2-methylheptane	$C_8H_{17}Cl$ 148.68	4325-49-9	 50^{15}	4-01-00-00428 0.8568^{25}	bz 4; eth 4; EtOH 4; chl 4 1.4240^{25}
6367	Heptane, 2-chloro-6-methyl- 2-Chloro-6-methylheptane	$C_8H_{17}Cl$ 148.68	2350-19-8	 74^{35}	3-01-00-00472	bz 4; eth 4; EtOH 4; chl 4 1.4260^{15}
6368	Heptane, 3-(chloromethyl)-	$C_8H_{17}Cl$ 148.68	123-04-6	 172	4-01-00-00430 0.8769^{20}	H_2O 1; EtOH 3; eth 3; ace 3 1.4319^{20}

No.	Name	Synonym	Mol. Form.	Mol. Wt.	CAS RN	mp/°C	Merck No.	bp/°C	Beil. Ref.	den/g cm^{-3}	Solubility	n_D
6369	Heptane, 3-chloro-3-methyl-	3-Methyl-3-chloroheptane	$C_8H_{17}Cl$	148.68	5272-02-6			64^{27}	4-01-00-00430	0.8764^{20}	bz 4; eth 4; EtOH 4; chl 4	1.4317^{20}
6370	Heptane, 4-chloro-4-methyl-	2-Chloro-2-propylpentane	$C_8H_{17}Cl$	148.68	61764-94-1			50^{12}	3-01-00-00477	0.8690^{20}	bz 4; eth 4; EtOH 4; chl 4	1.4310^{15}
6371	1,7-Heptanediamine		$C_7H_{18}N_2$	130.23	646-19-5	28.9		224	4-04-00-01354		EtOH 3; eth 3; ace 3	
6372	Heptane, 1,2-dibromo-	1,2-Dibromoheptane	$C_7H_{14}Br_2$	258.00	42474-21-5			228	3-01-00-00432	1.5086^{20}		1.4986^{20}
6373	Heptane, 1,5-dibromo-	1,5-Dibromoheptane	$C_7H_{14}Br_2$	258.00	1622-10-2			$113\text{-}5^{11}$	4-01-00-00393	1.536^{15}		1.5041^{15}
6374	Heptane, 1,7-dibromo-	Heptamethylene dibromide	$C_7H_{14}Br_2$	258.00	4549-31-9	41.7		263	4-01-00-00393	1.5306^{20}	H_2O 1; eth 3; ace 3; bz 3	1.5034^{20}
6375	Heptane, 2,3-dibromo-	2,3-Dibromoheptane	$C_7H_{14}Br_2$	258.00	21266-88-6			101^{17}	3-01-00-00432	1.5139^{20}		1.4992^{20}
6376	Heptane, 3,4-dibromo-	3,4-Dibromoheptane	$C_7H_{14}Br_2$	258.00	21266-90-0			107^{24}	3-01-00-00433	1.5182^{20}		1.5010^{20}
6377	Heptane, 1,1-dichloro-	1,1-Dichloroheptane	$C_7H_{14}Cl_2$	169.09	821-25-0			191	4-01-00-00390	1.011^{20}	H_2O 1; eth 3; bz 3; ctc 3	1.4440^{20}
6378	Heptane, 1,2-dichloro-	1,2-Dichloroheptane	$C_7H_{14}Cl_2$	169.09	10575-87-8			$68\text{-}72^{7}$	3-01-00-00430	1.064^{20}	bz 4; eth 4; chl 4	1.4490^{20}
6379	Heptane, 1,7-dichloro-	1,7-Dichloroheptane	$C_7H_{14}Cl_2$	169.09	821-76-1			$124\text{-}5^{35}$	4-01-00-00390	1.0408^{25}		1.4565^{25}
6380	Heptane, 2,2-dichloro-	2,2-Dichloroheptane	$C_7H_{14}Cl_2$	169.09	65786-09-6			77^{25}	3-01-00-00430	1.012^{20}	bz 4; eth 4; chl 4	1.4440^{20}
6381	Heptane, 4,4-dichloro-		$C_7H_{14}Cl_2$	169.09	89796-76-9			86^{27}	2-01-00-00117	1.008^{17}	eth 4; chl 4	1.448^{17}
6382	Heptane, 2,6-dichloro-2,6-dimethyl-		$C_9H_{18}Cl_2$	197.15	35951-36-1	43		93^{16}	3-01-00-00513		bz 4; chl 4	
6383	Heptane, 2,2-dimethyl-	2,2-Dimethylheptane	C_9H_{20}	128.26	1071-26-7	-113		132.7	4-01-00-00457	0.7105^{20}	H_2O 1; eth 3; ace 4; bz 5	1.4016^{20}
6384	Heptane, 2,3-dimethyl-	2,3-Dimethylheptane	C_9H_{20}	128.26	3074-71-3	-116		140.5	4-01-00-00457	0.7260^{20}	H_2O 1; EtOH 5; eth 5; ace 5	1.4088^{20}
6385	Heptane, 2,4-dimethyl-	2,4-Dimethylheptane	C_9H_{20}	128.26	2213-23-2			132.9	4-01-00-00457	0.7115^{25}	H_2O 1; EtOH 5; eth 5; ace 5	1.4034^{20}
6386	Heptane, 2,5-dimethyl-	3,6-Dimethylheptane	C_9H_{20}	128.26	2216-30-0			136	4-01-00-00457	0.7198^{20}	ace 4; bz 4; eth 4; EtOH 4	1.4033^{20}
6387	Heptane, 2,6-dimethyl-	2,6-Dimethylheptane	C_9H_{20}	128.26	1072-05-5	-102.9		135.2	4-01-00-00458	0.7089^{20}	chl 2	1.4011^{20}
6388	Heptane, 3,3-dimethyl-	3,3-Dimethylheptane	C_9H_{20}	128.26	4032-86-4			137.3	4-01-00-00458	0.7254^{20}	H_2O 1; EtOH 5; eth 3; ace 4	1.4087^{20}
6389	Heptane, 3,4-dimethyl-	3,4-Dimethylheptane	C_9H_{20}	128.26	922-28-1			140.6	3-01-00-00514	0.7314^{20}	H_2O 1; eth 3; ace 4; bz 5	1.4108^{20}
6390	Heptane, 3,5-dimethyl-	3,5-Dimethylheptane	C_9H_{20}	128.26	926-82-9			136	4-01-00-00458	0.7225^{20}	H_2O 1; eth 3; ace 4; bz 5	1.4083^{20}
6391	Heptane, 4,4-dimethyl-	4,4-Dimethylheptane	C_9H_{20}	128.26	1068-19-5			135.2	4-01-00-00458	0.7221^{20}	H_2O 1; eth 3; ace 4; bz 5	1.4076^{20}
6392	Heptanedinitrile		$C_7H_{10}N_2$	122.17	646-20-8	-31.4		155^{14}	4-02-00-02006	0.949^{18}	H_2O 1; EtOH 5; eth 5; chl 5	1.4472^{20}
6393	Heptanedioic acid	Pimelic acid	$C_7H_{12}O_4$	160.17	111-16-0	106	7401	272^{100}; 212^{10}	4-02-00-02003	1.329^{15}	H_2O 3; EtOH 3; eth 3; bz 1	
6394	Heptanedioic acid, diethyl ester	Diethyl pimelate	$C_{11}H_{20}O_4$	216.28	2050-20-6	-24		254; 140^{15}	4-02-00-02004	0.9945^{20}	eth 4; EtOH 4	1.4305^{20}
6395	Heptanedioic acid, dimethyl ester		$C_9H_{16}O_4$	188.22	1732-08-7	-21		120^{10}; 80^{1}	4-02-00-02004	1.0625^{20}	H_2O 2; EtOH 3; eth 3; bz 3	1.4309^{20}
6396	Heptanedioic acid, monoethyl ester		$C_9H_{16}O_4$	188.22	33018-91-6	10		182^{18}; 160^{4}	3-02-00-01742			1.4415^{20}
6397	1,4-Heptanediol		$C_7H_{16}O_2$	132.20	40646-07-9			243; 127^{4}	4-01-00-02579	0.9542^{20}		1.4510^{25}
6398	1,5-Heptanediol		$C_7H_{16}O_2$	132.20	60096-09-5			$135\text{-}6^{11}$	3-01-00-02213	0.9705^{20}	eth 4; EtOH 4; chl 4	1.4571^{22}
6399	1,7-Heptanediol		$C_7H_{16}O_2$	132.20	629-30-1	22.5		262	4-01-00-02580	0.9569^{25}	eth 4; EtOH 4	1.4520^{25}
6400	2,4-Heptanediol		$C_7H_{16}O_2$	132.20	20748-86-1			$107\text{-}8^{8}$	4-01-00-02580	0.926^{20}		1.4386^{25}
6401	2,4-Heptanediol, 3-methyl-	3-Methylheptan-2,4-diol	$C_8H_{18}O_2$	146.23	6964-04-1			115^{3}	0-01-00-00491	0.928^{20}	H_2O 1; EtOH 3; ctc 2	1.4459^{20}
6402	2,3-Heptanedione	Acetyl valeryl	$C_7H_{12}O_2$	128.17	96-04-8			46^{13}	4-01-00-03697	0.919^{18}		1.4160^{18}
6403	2,6-Heptanedione		$C_7H_{12}O_2$	128.17	13505-34-5	33.5		222; 97^{11}	4-01-00-03699	0.9399^{37}	bz 4; eth 4; EtOH 4	1.4277^{37}
6404	3,5-Heptanedione, 2,2,6,6-tetramethyl-	Dipivaloylmethane	$C_{11}H_{20}O_2$	184.28	1118-71-4			$72\text{-}3^{6}$	4-01-00-03729	0.883^{25}	ctc 2	1.4589^{20}
6405	Heptane, 3-[(ethenyloxy)methyl]-		$C_{10}H_{20}O$	156.27	103-44-6	-100		76^{32}	4-01-00-02056	0.8108^{20}		1.4247^{25}
6406	Heptane, 1-ethoxy-		$C_9H_{20}O$	144.26	1969-43-3	-68.3		166	4-01-00-01733	0.790^{16}		
6407	Heptane, 3-ethyl-	3-Ethylheptane	C_9H_{20}	128.26	15869-80-4	-114.9		143.0	4-01-00-00456	0.7225^{25}		1.4093^{20}
6408	Heptane, 4-ethyl-	4-Ethylheptane	C_9H_{20}	128.26	2216-32-2			141.2	4-01-00-00457	0.7241^{25}	H_2O 1; EtOH 5; eth 3; ace 5	1.4096^{20}

No.	Name Synonym	Mol. Form. Mol. Wt.	CAS RN mp/°C	Merck No. bp/°C	Beil. Ref. den/g cm^{-3}	Solubility n_D
6409	Heptane, 3-ethyl-2-methyl- Heptane, 2-methyl-3-ethyl	$C_{10}H_{22}$ 142.28	14676-29-0	163	4-01-00-00480 0.7398^{25}	1.4174^{20}
6410	Heptane, 3-ethyl-3-methyl-	$C_{10}H_{22}$ 142.28	17302-01-1	163.8	4-01-00-00481 0.7463^{25}	1.4208^{20}
6411	Heptane, 3-ethyl-4-methyl- 3-Ethyl-4-methylheptane	$C_{10}H_{22}$ 142.28	52896-91-0	165	4-01-00-00481 0.7468^{25}	1.4207^{20}
6412	Heptane, 3-ethyl-5-methyl- 3-Ethyl-5-methylheptane	$C_{10}H_{22}$ 142.28	52896-90-9	158.2	4-01-00-00481 0.7368^{25}	1.4164^{20}
6413	Heptane, 4-ethyl-2-methyl- 4-Ethyl-2-methylheptane	$C_{10}H_{22}$ 142.28	52896-88-5	158	4-01-00-00481 0.7322^{25}	1.4137^{20}
6414	Heptane, 4-ethyl-3-methyl- 4-Ethyl-3-methylheptane	$C_{10}H_{22}$ 142.28	52896-89-6	165	4-01-00-00481 0.7466^{25}	1.4206^{20}
6415	Heptane, 4-ethyl-4-methyl- 4-Ethyl-4-methylheptane	$C_{10}H_{22}$ 142.28	17302-04-4	160.8	4-01-00-00481 0.7472^{25}	1.4210^{25}
6416	Heptane, 5-ethyl-2-methyl-	$C_{10}H_{22}$ 142.28	13475-78-0	159.7	4-01-00-00481 0.7318^{25}	1.4134^{20}
6417	Heptane, 1-fluoro-	$C_7H_{15}F$ 118.19	661-11-0 -73	117.9	4-01-00-00387 0.8062^{20}	H_2O 1; eth 3; ace 3; bz 3 1.3854^{20}
6418	Heptane, hexadecafluoro- Perfluoroheptane	C_7F_{16} 388.05	335-57-9 -78	82.5	4-01-00-00388 1.7333^{20}	ace 4; eth 4; EtOH 4; chl 4 1.2618^{20}
6419	Heptane, 1-iodo-	$C_7H_{15}I$ 226.10	4282-40-0 -48.2	204	4-01-00-00393 1.3791^{20}	H_2O 1; EtOH 3; eth 3; ace 3 1.4904^{20}
6420	Heptane, 2-iodo- 2-Iodoheptane	$C_7H_{15}I$ 226.10	18589-29-2	98^{50}	4-01-00-00393 1.304^{20}	H_2O 1; ace 3; bz 3; ctc 3 1.4826
6421	Heptane, 1-methoxy-	$C_8H_{18}O$ 130.23	629-32-3	151	3-01-00-01682 0.7862^{15}	ace 4; eth 4; EtOH 4 1.4073^{20}
6422	Heptane, 2-methyl- 2-Methylheptane	C_8H_{18} 114.23	592-27-8 -108.9	117.6	4-01-00-00428 0.6980^{20}	H_2O 1; EtOH 5; eth 3; ace 5 1.3949^{20}
6423	Heptane, 3-methyl-, (*S*)-	C_8H_{18} 114.23	6131-25-5 -120	118.9	4-01-00-00429 0.7075^{16}	H_2O 1; EtOH 3; eth 3; ace 5 1.4002^{18}
6424	Heptane, 4-methyl- 4-Methylheptane	C_8H_{18} 114.23	589-53-7 -121	117.7	4-01-00-00431 0.7046^{20}	H_2O 1; EtOH 5; eth 3; ace 5 1.3979^{20}
6425	Heptane, 3-methylene-	C_8H_{16} 112.22	1632-16-2	120	4-01-00-00884 0.7270^{20}	H_2O 1; eth 4; bz 4; peth 4 1.4157^{20}
6426	Heptane, 4-methylene- 2-Propyl-1-pentene	C_8H_{16} 112.22	15918-08-8	117.7	4-01-00-00885 0.7198^{25}	1.4136^{20}
6427	Heptane, 4-(1-methylethyl)- 4-Isopropylheptane	$C_{10}H_{22}$ 142.28	52896-87-4	158.9	4-01-00-00481 0.7354^{25}	1.4153^{20}
6428	Heptanenitrile	$C_7H_{13}N$ 111.19	629-08-3	183; $70\text{-}2^{10}$	4-02-00-00963 0.8106^{20}	H_2O 1; eth 3; ace 3; bz 3 1.4104^{30}
6429	Heptane, 1-nitro- 1-Nitroheptane	$C_7H_{15}NO_2$ 145.20	693-39-0	194	4-01-00-00395 0.9476^{17}	eth 4; EtOH 4
6430	Heptane, 1,1'-oxybis-	$C_{14}H_{30}O$ 214.39	629-64-1	258.5	4-01-00-01733 0.8008^{20}	eth 4; EtOH 4 1.4275^{20}
6431	Heptane, 3,3'-[oxybis(methylene)]bis- Ether, dihexyl,2,2'-diethyl	$C_{16}H_{34}O$ 242.45	10143-60-9	269.4	4-01-00-01785	ctc 2 1.4325^{20}
6432	Heptane, 1,1,1,2,2,3,3,4,4,5,5,6,6,7,7-pentadecafluoro- 1H-Pentadecafluoroheptane	C_7HF_{15} 370.06	375-83-7	96.0	4-01-00-00388 1.725^{25}	1.2690^{25}
6433	Heptane, 2,2,4,6,6-pentamethyl- 2,2,4,6,6-Pentamethylheptane	$C_{12}H_{26}$ 170.34	13475-82-6 -67	177.8	4-01-00-00510 0.7463^{20}	1.4440^{20}
6434	Heptane, 4-propyl- 4-Propylheptane	$C_{10}H_{22}$ 142.28	3178-29-8	157.5	4-01-00-00480 0.7321^{25}	1.4135^{20}
6435	Heptane, 1,1'-thiobis-	$C_{14}H_{30}S$ 230.46	629-65-2 70	298	4-01-00-01739 0.8416^{20}	H_2O 1; eth 3 1.4606^{20}
6436	1-Heptanethiol Heptyl mercaptan	$C_7H_{16}S$ 132.27	1639-09-4 -43	177	4-01-00-01738 0.8427^{20}	H_2O 1; EtOH 5; eth 5; chl 3 1.4521^{20}
6437	Heptane, 2,2,3-trimethyl- 2,2,3-Trimethylheptane	$C_{10}H_{22}$ 142.28	52896-92-1	157.8	4-01-00-00481 0.7385^{25}	1.4168^{20}
6438	Heptane, 2,2,4-trimethyl- 2,2,4-Trimethylheptane	$C_{10}H_{22}$ 142.28	14720-74-2	148.3	4-01-00-00481 0.7237^{25}	H_2O 1; bz 3; ctc 3; chl 3 1.4092^{20}
6439	Heptane, 2,2,5-trimethyl- 2,2,5-Trimethylheptane	$C_{10}H_{22}$ 142.28	20291-95-6	150.8	4-01-00-00482 0.7243^{25}	1.4101^{20}
6440	Heptane, 2,2,6-trimethyl- 2,2,6-Trimethylheptane	$C_{10}H_{22}$ 142.28	1190-83-6 -105	148.9	4-01-00-00482 0.7200^{25}	1.4078^{20}
6441	Heptane, 2,3,3-trimethyl- 2,3,3-Trimethylheptane	$C_{10}H_{22}$ 142.28	52896-93-2	160.2	4-01-00-00482 0.7450^{25}	1.4202^{20}
6442	Heptane, 2,3,4-trimethyl- 2,3,4-Trimethylheptane	$C_{10}H_{22}$ 142.28	52896-95-4	161	4-01-00-00482 0.7447^{25}	1.4195^{20}
6443	Heptane, 2,3,5-trimethyl- 2,3,5-Trimethylheptane	$C_{10}H_{22}$ 142.28	20278-85-7	160.7	4-01-00-00482 0.7413^{25}	1.4169^{20}
6444	Heptane, 2,3,6-trimethyl- 2,3,6-Trimethylheptane	$C_{10}H_{22}$ 142.28	4032-93-3	156.0	4-01-00-00482 0.7305^{25}	1.4131^{20}
6445	Heptane, 2,4,4-trimethyl- 2,4,4-Trimethylheptane	$C_{10}H_{22}$ 142.28	4032-92-2	151.0	4-01-00-00482 0.7308^{25}	1.4142^{20}
6446	Heptane, 2,4,5-trimethyl- 2,4,5-Trimethylheptane	$C_{10}H_{22}$ 142.28	20278-84-6	156.5	4-01-00-00483 0.7373^{25}	1.4160^{20}
6447	Heptane, 2,4,6-trimethyl- 2,4,6-Trimethylheptane	$C_{10}H_{22}$ 142.28	2613-61-8	147.6	4-01-00-00483 0.7190^{25}	1.4071^{20}
6448	Heptane, 2,5,5-trimethyl- 2,5,5-Trimethylheptane	$C_{10}H_{22}$ 142.28	1189-99-7	152.8	4-01-00-00483 0.7362^{25}	1.4149^{20}

No.	Name Synonym	Mol. Form. Mol. Wt.	CAS RN mp/°C	Merck No. bp/°C	Beil. Ref. den/g cm^{-3}	Solubility n_D
6449	Heptane, 3,3,4-trimethyl- 3,3,4-Trimethylheptane	$C_{10}H_{22}$ 142.28	20278-87-9	 161.9	4-01-00-00483 0.7527^{25}	 1.4236^{20}
6450	Heptane, 3,3,5-trimethyl- 3,3,5-Trimethylheptane	$C_{10}H_{22}$ 142.28	7154-80-5	 155.7	4-01-00-00483 0.7248^{20}	H_2O 1; bz 3; ctc 3; chl 3 1.4170^{20}
6451	Heptane, 3,4,4-trimethyl- 3,4,4-Trimethylheptane	$C_{10}H_{22}$ 142.28	20278-88-0	 161.1	4-01-00-00484 0.7535^{25}	 1.4235^{20}
6452	Heptane, 3,4,5-trimethyl- 3,4,5-Trimethylheptane	$C_{10}H_{22}$ 142.28	20278-89-1	 162.5	4-01-00-00484 0.7519^{25}	 1.4229^{20}
6453	1,4,7-Heptanetriol	$C_7H_{16}O_3$ 148.20	3920-53-4 -35	 231^{25}; 146^{1}	4-01-00-02787 1.075^{18}	H_2O 3; EtOH 3; ace 3 1.4725^{20}
6454	2,4,6-Heptanetrione	$C_7H_{10}O_3$ 142.15	626-53-9 49	 121^{10}	4-01-00-03783 1.0599^{40}	H_2O 4; eth 4; EtOH 4 1.4930^{20}
6455	Heptanoic acid Enanthylic acid	$C_7H_{14}O_2$ 130.19	111-14-8 -7.5	4581 222.2	4-02-00-00958 0.9181^{20}	H_2O 2; EtOH 3; eth 3; ace 3 1.4170^{20}
6456	Heptanoic acid, anhydride	$C_{14}H_{26}O_3$ 242.36	626-27-7 -12.4	 269.5	4-02-00-00962 0.9321^{20}	H_2O 1; EtOH 3; eth 3 1.4335^{15}
6457	Heptanoic acid, 2-bromo- 2-Bromoheptanoic acid	$C_7H_{13}BrO_2$ 209.08	2624-01-3	 250 dec; 147^{12}	4-02-00-00967 1.319^{15}	ace 4; eth 4 1.471^{18}
6458	Heptanoic acid, 7-bromo-	$C_7H_{13}BrO_2$ 209.08	30515-28-7 31	 280; 165^{12}	4-02-00-00968	ace 4; bz 4; eth 4; EtOH 4
6459	Heptanoic acid, butyl ester Butyl enanthate	$C_{11}H_{22}O_2$ 186.29	5454-28-4 -67.5	 226.2	3-02-00-00768 0.8638^{20}	ace 4; bz 4; eth 4; EtOH 4 1.4204^{20}
6460	Heptanoic acid, 2,2-dimethyl-6-oxo- Geronic acid	$C_9H_{16}O_3$ 172.22	461-11-0	 278; 169^{12}	3-03-00-01264 1.0211^{20}	eth 4; EtOH 4 1.4488^{20}
6461	Heptanoic acid, 4,4-dimethyl-6-oxo- Isogeronic acid	$C_9H_{16}O_3$ 172.22	471-04-5 50	 178^{20}; 162^{10}	3-03-00-01266	H_2O 4; bz 4; eth 4; EtOH 4
6462	Heptanoic acid, ethyl ester Ethyl oenanthate	$C_9H_{18}O_2$ 158.24	106-30-9 -66.1	3790 187	4-02-00-00960 0.8817^{20}	H_2O 2; EtOH 3; eth 3; ctc 2 1.4100^{20}
6463	Heptanoic acid, 7-fluoro-	$C_7H_{13}FO_2$ 148.18	334-28-1	 133^{10}	4-02-00-00964 1.039^{20}	 1.4207^{25}
6464	Heptanoic acid, heptyl ester Heptyl heptanoate	$C_{14}H_{28}O_2$ 228.38	624-09-9 -33	 277	4-02-00-00961 0.8649^{20}	H_2O 1; EtOH 3; eth 3 1.4320^{20}
6465	Heptanoic acid, hexyl ester Hexyl heptanoate	$C_{13}H_{26}O_2$ 214.35	1119-06-8 -48	 261	3-02-00-00768 0.8611^{20}	ace 4; bz 4; eth 4; EtOH 4 1.429^{15}
6466	Heptanoic acid, methyl ester Methyl heptanoate	$C_8H_{16}O_2$ 144.21	106-73-0 -56	4581 174	4-02-00-00960 0.8815^{20}	H_2O 2; EtOH 3; eth 3; ace 2 1.4152^{20}
6467	Heptanoic acid, 2-methylpropyl ester	$C_{11}H_{22}O_2$ 186.29	7779-80-8	 208	1-02-00-00145 0.8593^{20}	ace 4; bz 4; eth 4; EtOH 4
6468	Heptanoic acid, octyl ester Octyl enanthate	$C_{15}H_{30}O_2$ 242.40	5132-75-2 -22.5	 290	3-02-00-00768 0.8596^{20}	ace 4; bz 4; eth 4; EtOH 4 1.4349^{15}
6469	Heptanoic acid, 6-oxo-	$C_7H_{12}O_3$ 144.17	3128-07-2 40.2	 251^{280}; 135^{1}	4-03-00-01598	H_2O 4; ace 4; eth 4; EtOH 4 1.4306^{25}
6470	Heptanoic acid, pentyl ester Amyl heptanoate	$C_{12}H_{24}O_2$ 200.32	7493-82-5 -50	 245.4	4-02-00-00960 0.8623^{20}	ace 4; bz 4; eth 4; EtOH 4 1.4263^{15}
6471	Heptanoic acid, propyl ester Propyl heptanoate	$C_{10}H_{20}O_2$ 172.27	7778-87-2 -63.5	 208	3-02-00-00767 0.8641^{15}	ace 4; bz 4; eth 4; EtOH 4 1.4183^{15}
6472	1-Heptanol Heptyl alcohol	$C_7H_{16}O$ 116.20	111-70-6 -34	4582 176.4	4-01-00-01731 0.8219^{20}	H_2O 2; EtOH 5; eth 5; ctc 2 1.4249^{20}
6473	2-Heptanol, (±)-	$C_7H_{16}O$ 116.20	52390-72-4	4583 159	4-01-00-01740 0.8167^{20}	H_2O 2; EtOH 3; eth 3; ctc 2 1.4210^{20}
6474	3-Heptanol, (*S*)-	$C_7H_{16}O$ 116.20	26549-25-7 -70	 157; 66^{18}	2-01-00-00444 0.8227^{20}	H_2O 2; EtOH 3; eth 3; ctc 2 1.4201^{20}
6475	4-Heptanol Dipropylcarbinol	$C_7H_{16}O$ 116.20	589-55-9 -41.2	 156	4-01-00-01743 0.8183^{20}	H_2O 1; EtOH 3; eth 3 1.4205^{20}
6476	4-Heptanol, acetate	$C_9H_{18}O_2$ 158.24	5921-84-6	 171	4-02-00-00163 0.8741^{0}	eth 4 1.4105^{19}
6477	1-Heptanol, 7-chloro- Heptamethylene chlorohydrin	$C_7H_{15}ClO$ 150.65	55944-70-2 11	 150^{20}	4-01-00-01738 0.9998^{15}	EtOH 4; peth 4 1.4537^{25}
6478	2-Heptanol, 1-chloro- 1-Chloro-2-heptanol	$C_7H_{15}ClO$ 150.65	53660-21-2	 93^{13}	4-01-00-01741 0.9885^{20}	EtOH 3; eth 3; ace 3 1.4499^{20}
6479	2-Heptanol, 2,6-dimethyl- 2,6-Dimethyl-2-heptanol	$C_9H_{20}O$ 144.26	13254-34-7	 173	2-01-00-00457 0.8186^{20}	 1.4242^{20}
6480	3-Heptanol, 2,3-dimethyl- Butylisopropylmethylcarbinol	$C_9H_{20}O$ 144.26	19549-71-4	 173; 75^{16}	3-01-00-01752 0.8395^{20}	 1.4355^{20}
6481	3-Heptanol, 2,6-dimethyl- 2,6-Dimethyl-3-heptanol	$C_9H_{20}O$ 144.26	19549-73-6	 175	3-01-00-01753 0.8212^{20}	 1.4246^{20}
6482	3-Heptanol, 3,5-dimethyl-	$C_9H_{20}O$ 144.26	19549-74-7	 $95-7^{50}$	4-01-00-01811 0.8177^{25}	 1.4251^{20}
6483	3-Heptanol, 3,6-dimethyl- 3,6-Dimethylheptan-3-ol	$C_9H_{20}O$ 144.26	1573-28-0	 173	4-01-00-01810 0.8285^{16}	eth 4; EtOH 4; lig 4 1.4326^{16}
6484	4-Heptanol, 2,4-dimethyl- 2,4-Dimethyl-4-heptanol	$C_9H_{20}O$ 144.26	19549-77-0	 171; 70^{26}	4-01-00-01810 0.8215^{20}	 1.4292^{20}
6485	4-Heptanol, 2,6-dimethyl- Diisobutylcarbinol	$C_9H_{20}O$ 144.26	108-82-7	 174.5	4-01-00-01810 0.8114^{20}	H_2O 1; EtOH 3; eth 3; ctc 2 1.4242^{20}
6486	4-Heptanol, 3,5-dimethyl- 3,5-Dimethyl-4-heptanol	$C_9H_{20}O$ 144.26	19549-79-2	 186	4-01-00-01811 0.836^{18}	 1.4283^{20}
6487	4-Heptanol, 4-ethyl-	$C_9H_{20}O$ 144.26	597-90-0	 182	4-01-00-01809 0.8350^{20}	eth 4; EtOH 4 1.4332^{20}

No.	Name Synonym	Mol. Form. Mol. Wt.	CAS RN mp/°C	Merck No. bp/°C	Beil. Ref. den/g cm^{-3}	Solubility n_D
6488	3-Heptanol, 3-ethyl-2-methyl-	$C_{10}H_{22}O$ 158.28	66719-37-7	 193	1-01-00-00214 0.8455^{20}	 1.4378
6489	3-Heptanol, 3-ethyl-6-methyl-	$C_{10}H_{22}O$ 158.28	66719-40-2	 $83-6^{15}$	0-01-00-00427 0.852^{25}	 1.4409
6490	3-Heptanol, 5-ethyl-4-methyl-	$C_{10}H_{22}O$ 158.28	66731-94-0	 $86-8^{14}$	3-01-00-01770 0.865^{17}	 1.4565^{17}
6491	1-Heptanol, 7-fluoro-	$C_7H_{15}FO$ 134.19	408-16-2	 $98-9^{12}$	4-01-00-01736 0.956^{20}	eth 4; EtOH 4 1.4197^{25}
6492	1-Heptanol, 3-methyl- 3-Methyl-1-heptanol	$C_8H_{18}O$ 130.23	1070-32-2	 101^{20}	4-01-00-01782 0.824^{24}	 1.4295^{25}
6493	1-Heptanol, 4-methyl- 4-Methyl-1-heptanol	$C_8H_{18}O$ 130.23	817-91-4	 188	4-01-00-01789 0.8065^{25}	EtOH 4 1.4253^{25}
6494	1-Heptanol, 6-methyl- 6-Methyl-1-heptanol	$C_8H_{18}O$ 130.23	1653-40-3 -106	 188; 95.8^{20}	4-01-00-01782 0.8176^{25}	H_2O 1; EtOH 3; eth 3 1.4251^{25}
6495	2-Heptanol, 2-methyl- 2-Methyl-2-heptanol	$C_8H_{18}O$ 130.23	625-25-2 -50.4	 156	4-01-00-01780 0.8142^{20}	H_2O 1; EtOH 3; eth 3 1.4250^{20}
6496	2-Heptanol, 3-methyl- 3-Methyl-2-heptanol	$C_8H_{18}O$ 130.23	31367-46-1	 166.1	 0.8177^{25}	H_2O 1; EtOH 3; eth 3; ctc 3 1.4199^{25}
6497	2-Heptanol, 5-methyl- 5-Methyl-2-heptanol	$C_8H_{18}O$ 130.23	54630-50-1 -61	 170	4-01-00-01783 0.8174^{21}	
6498	2-Heptanol, 6-methyl-	$C_8H_{18}O$ 130.23	4730-22-7 -105	 174	3-01-00-01728 0.8218^{20}	 1.4238^{10}
6499	3-Heptanol, 2-methyl-, (±)-	$C_8H_{18}O$ 130.23	100296-26-2 -85	 167.5	4-01-00-01781 0.8235^{20}	H_2O 2; EtOH 3; eth 3; ctc 3 1.4265^{20}
6500	3-Heptanol, 3-methyl- 2-Ethyl-2-hexanol	$C_8H_{18}O$ 130.23	5582-82-1 -83	 163	4-01-00-01783 0.8282^{20}	H_2O 1; EtOH 3; eth 3; ctc 3 1.4279^{20}
6501	3-Heptanol, 4-methyl- 4-Methyl-3-heptanol	$C_8H_{18}O$ 130.23	14979-39-6 -123	 170	4-01-00-01789 0.827^{25}	 1.4300^{20}
6502	3-Heptanol, 5-methyl-	$C_8H_{18}O$ 130.23	18720-65-5 -91.2	 172	4-01-00-01783 0.8425^{25}	 1.433^{24}
6503	4-Heptanol, 2-methyl- 6-Methyl-4-heptanol	$C_8H_{18}O$ 130.23	21570-35-4 -81	 164	4-01-00-01781 0.8207^{20}	eth 4; EtOH 4 1.4203
6504	4-Heptanol, 3-methyl-, (*R**,*S**)-(±)- 4-Heptanol, 3-methyl(*DL*)	$C_8H_{18}O$ 130.23	92737-91-2	 162	3-01-00-01731 0.8335^{25}	H_2O 2; EtOH 3; eth 3; ctc 3 1.4211^{25}
6505	4-Heptanol, 4-methyl- 4-Hydroxy-4-methylheptane	$C_8H_{18}O$ 130.23	598-01-6 -82	 161	4-01-00-01789 0.8248^{20}	H_2O 1; EtOH 3; eth 3; ctc 3 1.4258^{20}
6506	2-Heptanol, 6-methyl-, acetate Acetic acid, (6-methyl-2-heptyl) ester	$C_{10}H_{20}O_2$ 172.27	67952-57-2	 187	0-02-00-00135 0.8474^{20}	EtOH 4 1.413^{20}
6507	3-Heptanol, 6-methyl-, acetate Acetic acid, (6-methyl-3-heptyl) ester	$C_{10}H_{20}O_2$ 172.27	32764-34-4	 184.5	0-02-00-00135 0.8554^{20}	EtOH 4 1.4160^{20}
6508	4-Heptanol, 4-propyl-	$C_{10}H_{22}O$ 158.28	2198-72-3	 191	4-01-00-01831 0.8337^{21}	bz 4; eth 4; EtOH 4 1.4355^{21}
6509	4-Heptanol, 2,4,6-trimethyl-	$C_{10}H_{22}O$ 158.28	60836-07-9	 181; 79^{12}	4-01-00-01833 0.823^{21}	 1.4334^{18}
6510	2-Heptanone Methyl pentyl ketone	$C_7H_{14}O$ 114.19	110-43-0 -35	4584 151.0	4-01-00-03318 0.8111^{20}	H_2O 4; EtOH 3; eth 3 1.4088^{20}
6511	3-Heptanone Ethyl butyl ketone	$C_7H_{14}O$ 114.19	106-35-4 -39	 147	4-01-00-03321 0.8183^{20}	H_2O 1; EtOH 5; eth 5; ctc 2 1.4057^{20}
6512	4-Heptanone Dipropyl ketone	$C_7H_{14}O$ 114.19	123-19-3 -33	3351 144	4-01-00-03323 0.8174^{20}	H_2O 1; EtOH 5; eth 5; ctc 3 1.4069^{20}
6513	2-Heptanone, 1-chloro- 1-Chloro-2-heptanone	$C_7H_{13}ClO$ 148.63	41055-92-9	 83^{16}	4-01-00-03320 0.802^{20}	eth 4; EtOH 4 1.4371^{20}
6514	4-Heptanone, 2,6-dimethyl- Diisobutyl ketone	$C_9H_{18}O$ 142.24	108-83-8 -41.5	 169.4	4-01-00-03360 0.8062^{20}	H_2O 1; EtOH 5; eth 5; ctc 3 1.412^{21}
6515	4-Heptanone, 3,5-dimethyl- Di-*sec*-butyl ketone	$C_9H_{18}O$ 142.24	19549-84-9	 162	3-01-00-02893 0.826^{14}	 1.4193^{14}
6516	3-Heptanone, 6-(dimethylamino)-4,4-diphenyl-, hydrochloride Methadone hydrochloride	$C_{21}H_{28}ClNO$ 345.91	1095-90-5 235	5852	3-14-00-00279	H_2O 4; EtOH 4
6517	4-Heptanone, 2-hydroxy-	$C_7H_{14}O_2$ 130.19	54862-92-9	 95^{12}	4-01-00-04034 0.9296^{19}	 1.4357^{19}
6518	2-Heptanone, 4-hydroxy-3-methyl- 4-Heptanol-2-one, 3-methyl	$C_8H_{16}O_2$ 144.21	56072-27-6	 110^{16}	3-01-00-03249 0.9238^{20}	 1.442^{20}
6519	3-Heptanone, 5-hydroxy-5-methyl- 3-Heptanol-5-one, 3-methyl	$C_8H_{16}O_2$ 144.21	39121-37-4	 85^{15}	4-01-00-04045 0.9315^{15}	 1.4367^{15}
6520	2-Heptanone, 3-methyl- 3-Methyl-2-heptanone	$C_8H_{16}O$ 128.21	2371-19-9	 164	3-01-00-02878 0.8218^{20}	H_2O 2; EtOH 3; eth 3; ace 3 1.4172^{20}
6521	2-Heptanone, 6-methyl-	$C_8H_{16}O$ 128.21	928-68-7	 167	4-01-00-03344 0.8151^{20}	H_2O 2; EtOH 4; eth 4; ace 5 1.4162^{20}
6522	3-Heptanone, 2-methyl- Butyl isopropyl ketone	$C_8H_{16}O$ 128.21	13019-20-0	 158	4-01-00-03343 0.8163^{20}	H_2O 1; EtOH 3; eth 3; ace 4 1.4115^{20}
6523	3-Heptanone, 6-methyl-	$C_8H_{16}O$ 128.21	624-42-0	 164	4-01-00-03344 0.8304^{20}	H_2O 1; EtOH 3; eth 3; bz 3 1.4209^{20}
6524	4-Heptanone, 2-methyl- Isobutyl propyl ketone	$C_8H_{16}O$ 128.21	626-33-5	 155	4-01-00-03343 0.813^{22}	H_2O 1; EtOH 3; eth 3
6525	4-Heptanone, 3-methyl- 3-Methyl-4-heptanone	$C_8H_{16}O$ 128.21	15726-15-5	 154	4-01-00-03344 0.817^{20}	 1.4103^{25}
6526	1-Heptanone, 1-phenyl-	$C_{13}H_{18}O$ 190.29	1671-75-6 16.4	 283.3	4-07-00-00780 0.9516^{20}	ace 4; eth 4; EtOH 4 1.5060^{20}
6527	Heptanoyl chloride	$C_7H_{13}ClO$ 148.63	2528-61-2 -83.8	 125.2	4-02-00-00963 0.9590^{20}	eth 3; ctc 2; lig 4 1.4345^{18}

No.	Name / Synonym	Mol. Form. / Mol. Wt.	CAS RN / mp/°C	Merck No. / bp/°C	Beil. Ref. / den/g cm^{-3}	Solubility / n_D
6528	Heptasiloxane, hexadecamethyl- / Hexadecamethylheptasiloxane	$C_{16}H_{48}O_6Si_7$ / 533.15	541-01-5 / -78	270	4-04-00-04129 / 0.9012^{20}	bz 4; lig 4 / 1.3965^{20}
6529	1,3,5-Heptatriene	C_7H_{10} / 94.16	2196-23-8 / 23	113.5; 31^{25}	4-01-00-01097 / 0.764^{20}	1.5079
6530	2-Heptenal / Butylacrolein	$C_7H_{12}O$ / 112.17	2463-63-0	166	4-01-00-03478 / 0.864^{17}	1.4468^{17}
6531	3-Heptenal	$C_7H_{12}O$ / 112.17	89896-73-1	151	4-01-00-03480 / 0.851^{15}	eth 3; ace 3; bz 3 / 1.4348^{15}
6532	5-Hepten-2-amine, *N*,6-dimethyl- / Isometheptene	$C_9H_{19}N$ / 141.26	503-01-5	5070 / 177	3-04-00-00467	eth 4; EtOH 4
6533	1-Heptene	C_7H_{14} / 98.19	592-76-7 / -119.7	93.6	4-01-00-00857 / 0.6970^{20}	H_2O 1; EtOH 3; eth 3; ctc 2 / 1.3998^{20}
6534	2-Heptene, (*E*)-	C_7H_{14} / 98.19	14686-13-6 / -109.5	98	4-01-00-00860 / 0.7012^{20}	H_2O 1; EtOH 3; eth 3; ace 3 / 1.4045^{20}
6535	2-Heptene, (*Z*)- / *cis*-2-Heptene	C_7H_{14} / 98.19	6443-92-1	98.4	3-01-00-00824 / 0.708^{20}	H_2O 1; EtOH 3; eth 3; ace 3 / 1.406^{20}
6536	3-Heptene, (*E*)- / *trans*-3-Heptene	C_7H_{14} / 98.19	14686-14-7 / -136.6	95.7	4-01-00-00861 / 0.6981^{20}	H_2O 1; EtOH 3; eth 3; ace 3 / 1.4043^{20}
6537	3-Heptene, (*Z*)- / *cis*-3-Heptene	C_7H_{14} / 98.19	7642-10-6 / -136.6	95.8	4-01-00-00861 / 0.7030^{20}	H_2O 1; EtOH 3; eth 3; ace 3 / 1.4059^{20}
6538	1-Heptene, 1-chloro- / 1-Chloro-1-heptene	$C_7H_{13}Cl$ / 132.63	2384-75-0	155	3-01-00-00823 / 0.8948^{20}	ace 4; bz 4; eth 4; chl 4 / 1.4380^{20}
6539	1-Heptene, 2-chloro-	$C_7H_{13}Cl$ / 132.63	65786-11-0	139	3-01-00-00823 / 0.8895^{20}	eth 4; EtOH 4; chl 4 / 1.4349^{20}
6540	3-Heptene, 4-chloro-	$C_7H_{13}Cl$ / 132.63	2431-24-5	139	3-01-00-00827 / 0.883^{14}	eth 4; EtOH 4; chl 4 / 1.437^{14}
6541	2-Heptene, 6-chloro-2-methyl-	$C_8H_{15}Cl$ / 146.66	80325-37-7	$60\text{-}1^{15}$	2-01-00-00200 / 0.8931^{18}	ace 4; bz 4; eth 4; EtOH 4 / 1.4458^{18}
6542	2-Heptene, 2,3-dimethyl- / 2,3-Dimethyl-2-heptene	C_9H_{18} / 126.24	3074-64-4 / -108.5	145.2	4-01-00-00897 / 0.731^{25}	1.4519^{20}
6543	1-Heptene, 2-methyl- / 2-Methyl-1-heptene	C_8H_{16} / 112.22	15870-10-7 / -90	119.3	4-01-00-00881 / 0.7104^{25}	1.4123^{20}
6544	1-Heptene, 3-methyl- / 3-Methyl-1-heptene	C_8H_{16} / 112.22	4810-09-7	111	4-01-00-00883 / 0.707^{25}	1.406^{20}
6545	1-Heptene, 4-methyl- / 4-Methyl-1-heptene	C_8H_{16} / 112.22	13151-05-8	112.8	4-01-00-00884 / 0.713^{25}	1.410^{20}
6546	1-Heptene, 5-methyl- / 5-Methyl-1-heptene	C_8H_{16} / 112.22	13151-04-7	113.3	4-01-00-00884 / 0.7125^{25}	1.4094^{20}
6547	1-Heptene, 6-methyl- / 6-Methyl-1-heptene	C_8H_{16} / 112.22	5026-76-6	113.2	4-01-00-00883 / 0.7079^{25}	1.4070^{20}
6548	2-Heptene, 2-methyl- / 2-Methyl-2-heptene	C_8H_{16} / 112.22	627-97-4	122.6	4-01-00-00882 / 0.7200^{25}	H_2O 1; eth 3; bz 3; ctc 3 / 1.4170^{20}
6549	2-Heptene, 3-methyl-, (*E*)- / 3-Methyl-*trans*-2-heptene	C_8H_{16} / 112.22	22768-20-3	122	4-01-00-00883 / 1.419^{20}	
6550	2-Heptene, 3-methyl-, (*Z*)- / 2-Heptene, 3-methyl (*cis*)	C_8H_{16} / 112.22	22768-19-0	122	3-01-00-00843 / 0.725^{25}	1.419^{20}
6551	2-Heptene, 4-methyl-, (*E*)- / 4-Methyl-*trans*-2-heptene	C_8H_{16} / 112.22	66225-17-0	114	4-01-00-00885 / 0.712^{25}	1.410^{20}
6552	2-Heptene, 4-methyl-, (*Z*)- / 2-Heptene, 4-methyl (*cis*)	C_8H_{16} / 112.22	66225-16-9	114	4-01-00-00885 / 0.712^{25}	1.408^{20}
6553	2-Heptene, 6-methyl- / 2-Methyl-5-heptene	C_8H_{16} / 112.22	73548-72-8	117	3-01-00-00843 / 0.714^{25}	1.410^{20}
6554	3-Heptene, 2-methyl- / 2-Methyl-3-heptene	C_8H_{16} / 112.22	17618-76-7 / -107.5	112	4-01-00-00882 / 0.702^{25}	1.407^{20}
6555	3-Heptene, 3-methyl- / 2-Ethyl-2-hexene	C_8H_{16} / 112.22	7300-03-0	121	3-01-00-00843 / 0.724^{25}	1.418^{20}
6556	3-Heptene, 4-methyl- / 4-Methyl-3-heptene	C_8H_{16} / 112.22	4485-16-9	121	1-01-00-00094 / 0.721^{25}	1.417^{20}
6557	3-Heptene, 5-methyl-	C_8H_{16} / 112.22	13172-91-3	112	4-01-00-00884 / 0.709^{25}	1.410^{20}
6558	3-Heptene, 6-methyl- / 2-Methyl-4-heptene	C_8H_{16} / 112.22	3404-57-7	115	4-01-00-00882 / 0.709^{25}	1.410^{20}
6559	3-Heptene, 4-propyl- / 4-Propyl-3-heptene	$C_{10}H_{20}$ / 140.27	4485-13-6	160.5	4-01-00-00906 / 0.7518^{17}	1.4302^{18}
6560	1-Heptene, 1,1,2,3,3,4,4,5,5,6,6,7,7,7-tetradecafluoro- / Perfluoro-1-heptene	C_7F_{14} / 350.05	355-63-5	81.0	4-01-00-00859	
6561	2-Heptenoic acid	$C_7H_{12}O_2$ / 128.17	18999-28-5 / -19	226.5	4-02-00-01577 / 0.9575^{20}	1.4488^{20}
6562	4-Heptenoic acid	$C_7H_{12}O_2$ / 128.17	35194-37-7	$98\text{-}100^{5}$	4-02-00-01578 / 0.949^{15}	1.4418^{15}
6563	5-Heptenoic acid	$C_7H_{12}O_2$ / 128.17	3593-00-8	223	4-02-00-01578 / 0.9496^{20}	1.4444^{20}
6564	6-Heptenoic acid	$C_7H_{12}O_2$ / 128.17	1119-60-4 / -6.5	226	4-02-00-01576 / 0.9515^{14}	1.4404^{14}
6565	2-Heptenoic acid, 6-methyl-	$C_8H_{14}O_2$ / 142.20	90112-75-7 / 3	239.5	1-02-00-00194 / 0.938^{20}	1.4511^{10}
6566	1-Hepten-4-ol	$C_7H_{14}O$ / 114.19	3521-91-3	152.1	3-01-00-01936 / 0.8384^{22}	1.4347^{20}
6567	2-Hepten-1-ol, (*E*)-	$C_7H_{14}O$ / 114.19	33467-76-4	178; 75^{10}	3-01-00-01936 / 0.8516^{20}	EtOH 3; ace 3 / 1.4460^{20}

No.	Name Synonym	Mol. Form. Mol. Wt.	CAS RN mp/°C	Merck No. bp/°C	Beil. Ref. den/g cm^{-3}	Solubility n_D
6568	2-Hepten-1-ol, (*Z*)- *cis*-2-Hepten-1-ol	$C_7H_{14}O$ 114.19	55454-22-3	178; 82^{20}	4-01-00-02155 0.8421^{20}	1.4410^{20}
6569	2-Hepten-4-ol, (±)-	$C_7H_{14}O$ 114.19	115113-98-9	153	4-01-00-02155 0.8445^{20}	EtOH 3; eth 3 1.4373^{20}
6570	3-Hepten-2-ol	$C_7H_{14}O$ 114.19	98991-54-9	$66\text{-}7^{16}$	4-01-00-02156 0.834^{17}	1.4391^{18}
6571	6-Hepten-2-ol	$C_7H_{14}O$ 114.19	24395-10-6	$64\text{-}5^{13}$	2-01-00-00488 0.8484^{18}	1.4387^{20}
6572	6-Hepten-3-ol 1-Heptene-4-ol	$C_7H_{14}O$ 114.19	19781-77-2	$60\text{-}2^{11}$	4-01-00-02154 0.8447^{18}	1.4369^{18}
6573	1-Hepten-4-ol, 4-methyl- 1-Heptene-4-ol, 4-methyl	$C_8H_{16}O$ 128.21	1186-31-8	159.5	3-01-00-01945 0.8345^{20}	eth 4; EtOH 4 1.4479^{18}
6574	2-Hepten-4-ol, 5-methyl-	$C_8H_{16}O$ 128.21	4048-31-1	$69\text{-}70^{18}$	2-01-00-00490 0.8473^{20}	1.4411^{20}
6575	2-Hepten-4-ol, 6-methyl- 6-Methyl-2-hepten-4-ol	$C_8H_{16}O$ 128.21	4798-62-3	75^{21}; 67^{11}	2-01-00-00490 0.8354^{15}	1.4378^{20}
6576	3-Hepten-2-ol, 2-methyl-	$C_8H_{16}O$ 128.21	116668-45-2	$62\text{-}3^{14}$	2-01-00-00490 0.8398^{10}	1.4416^{10}
6577	4-Hepten-2-ol, 2-methyl-	$C_8H_{16}O$ 128.21	116668-44-1 -28	$63\text{-}4^{20}$	3-01-00-01944 0.8424^{14}	1.4407^{14}
6578	4-Hepten-3-ol, 3-methyl-	$C_8H_{16}O$ 128.21	4048-30-0	103^{75}; 62^{15}	2-01-00-00490 0.8477^{17}	1.4465^{20}
6579	4-Hepten-3-ol, 4-methyl-	$C_8H_{16}O$ 128.21	81280-12-8	66^{17}	4-01-00-02170 0.8525^{18}	1.4479^{18}
6580	5-Hepten-2-ol, 6-methyl-	$C_8H_{16}O$ 128.21	1569-60-4	175	1-01-00-00230 0.8545^{20}	1.4505^{20}
6581	6-Hepten-1-ol, 3-methyl-	$C_8H_{16}O$ 128.21	4048-32-2	$97\text{-}9^{22}$	2-01-00-00491 0.8562^{19}	1.4470^{14}
6582	6-Hepten-2-ol, 2-methyl- 1,1-Dimethyl-5-hexen-1-ol	$C_8H_{16}O$ 128.21	77437-98-0	$65\text{-}6^{16}$	3-01-00-01945 0.8393^{16}	1.4393^{16}
6583	1-Hepten-3-one Butyl vinyl ketone	$C_7H_{12}O$ 112.17	2918-13-0	$42\text{-}3^{15}$	4-01-00-03477 0.8434^{20}	1.4305^{19}
6584	4-Hepten-2-one	$C_7H_{12}O$ 112.17	24332-22-7	$61\text{-}2^{20}$	3-01-00-03004 0.8618^{21}	1.4290^{21}
6585	6-Hepten-2-one Methyl 4-pentenyl ketone	$C_7H_{12}O$ 112.17	21889-88-3	$41\text{-}3^{10}$	4-01-00-03478 0.8673^{18}	1.4350^{18}
6586	6-Hepten-3-one 1-Hepten-5-one	$C_7H_{12}O$ 112.17	2565-39-1	142	4-01-00-03478 0.8487^{18}	1.4254^{18}
6587	3-Hepten-2-one, (*E*)-	$C_7H_{12}O$ 112.17	5609-09-6	62^{15}	4-01-00-03481 0.8496^{20}	H_2O 1; EtOH 4; eth 4 1.4436^{20}
6588	3-Hepten-2-one, (*Z*)- 3-Hepten-2-one (*cis*)	$C_7H_{12}O$ 112.17	69668-88-8	$34\text{-}9^{11}$	4-01-00-03481 0.8440^{20}	1.4325^{20}
6589	5-Hepten-2-one, (*E*)-	$C_7H_{12}O$ 112.17	1071-94-9	153	4-01-00-03480 0.8445^{20}	1.4309^{20}
6590	3-Hepten-2-one, 3-ethyl-4-methyl-	$C_{10}H_{18}O$ 154.25	54244-90-5	190; 83^{14}	3-01-00-03028 0.8559^{20}	1.4535^{20}
6591	4-Hepten-2-one, 3-ethyl-4-methyl-	$C_{10}H_{18}O$ 154.25	53252-08-7	69^{11}	3-01-00-03028 0.8450^{20}	1.4405^{20}
6592	4-Hepten-3-one, 5-ethyl-4-methyl-	$C_{10}H_{18}O$ 154.25	22319-28-4	80^{10}	4-01-00-03517 0.8622^{10}	1.4545^{19}
6593	5-Hepten-2-one, 4-ethyl-6-methyl- 2-Hepten-6-one, 4-ethyl-2-methyl	$C_{10}H_{18}O$ 154.25	104035-66-7	$74\text{-}5^{13}$	4-01-00-03517 0.8430^{20}	1.4546^{20}
6594	5-Hepten-3-one, 5-ethyl-4-methyl-	$C_{10}H_{18}O$ 154.25	74764-56-0 74.5	$135\text{-}9^{3}$	4-01-00-03517 0.8564^{21}	1.4452^{21}
6595	1-Hepten-3-one, 5-methyl-	$C_8H_{14}O$ 126.20	44829-76-7	$40\text{-}1^{7}$	3-01-00-03014 0.8479^{19}	1.4360^{19}
6596	2-Hepten-4-one, 5-methyl- 5-Methyl-2-hepten-4-one	$C_8H_{14}O$ 126.20	81925-81-7	171	4-01-00-03496 0.848^{25}	1.4400^{25}
6597	2-Hepten-4-one, 6-methyl-	$C_8H_{14}O$ 126.20	49852-35-9	170	4-01-00-03494 0.841^{25}	1.4388^{25}
6598	3-Hepten-2-one, 6-methyl-	$C_8H_{14}O$ 126.20	2009-74-7	179	3-01-00-03011 0.8443^{17}	1.4427^{10}
6599	4-Hepten-2-one, 6-methyl- 2-Methyl-3-hepten-6-one	$C_8H_{14}O$ 126.20	39273-81-9	163	1-01-00-00385 0.8345^{20}	1.4315
6600	4-Hepten-3-one, 5-methyl- Homomesityl oxide	$C_8H_{14}O$ 126.20	1447-26-3	166.5	3-01-00-03012 0.8591^{20}	1.4476^{20}
6601	5-Hepten-2-one, 3-methyl- 3-Methyl-5-hepten-2-one	$C_8H_{14}O$ 126.20	38552-72-6	$62\text{-}4^{20}$	4-01-00-03496 0.8463^{18}	1.4345^{18}
6602	5-Hepten-2-one, 5-methyl-	$C_8H_{14}O$ 126.20	10339-67-0	84^{45}	4-01-00-03494 0.8647^{16}	1.4434^{16}
6603	5-Hepten-2-one, 6-methyl-	$C_8H_{14}O$ 126.20	110-93-0 173.5		4-01-00-03493 0.8546^{16}	eth 4; EtOH 4 1.4445^{20}
6604	5-Hepten-3-one, 2-methyl-	$C_8H_{14}O$ 126.20	77958-21-5	161.5	1-01-00-00385 0.842^{20}	1.4310^{20}
6605	5-Hepten-3-one, 5-methyl- 5-Methylhept-5-en-3-one	$C_8H_{14}O$ 126.20	1190-34-7	63^{19}	3-01-00-03011 0.8524^{21}	1.4367^{21}
6606	1-Hepten-3-yne	C_7H_{10} 94.16	2384-73-8	110	4-01-00-01097 0.7603^{20}	ace 4; bz 4; EtOH 4; peth 4 1.4520^{25}
6607	1-Hepten-4-yne, 6,6-dimethyl-	C_9H_{14} 122.21	31508-08-4	125	0.758^{20}	ace 4; bz 4; EtOH 4; peth 4 1.4312^{20}

No.	Name Synonym	Mol. Form. Mol. Wt.	CAS RN mp/°C	Merck No. bp/°C	Beil. Ref. den/g cm^{-3}	Solubility n_D
6608	6-Hepten-4-yn-3-ol, 3-ethyl- Diethylvinylethynylcarbinol	$C_9H_{14}O$ 138.21	3142-84-5	 62^{4}	3-01-00-02034 0.8875^{20}	eth 4; EtOH 4 1.4800^{20}
6609	1-Heptyne	C_7H_{12} 96.17	628-71-7 -81	 99.7	4-01-00-01025 0.7328^{20}	H_2O 2; EtOH 5; eth 5; bz 3 1.4087^{20}
6610	2-Heptyne 1-Methyl-2-butylacetylene	C_7H_{12} 96.17	1119-65-9	 112	4-01-00-01026 0.744^{25}	H_2O 1; EtOH 5; eth 5; bz 3 1.4230^{20}
6611	3-Heptyne 1-Ethyl-2-propylacetylene	C_7H_{12} 96.17	2586-89-2 -130.5	 107.2	4-01-00-01027 0.7336^{25}	H_2O 1; EtOH 5; eth 5; bz 3 1.4189^{20}
6612	1-Heptyne, 1-bromo- 1-Bromo-1-heptyne	$C_7H_{11}Br$ 175.07	19821-84-2	 164; 69^{25}	3-01-00-00998 1.2120^{22}	ace 4; eth 4; EtOH 4; chl 4 1.4678^{22}
6613	2-Heptyne, 1-bromo- 2-Heptynyl bromide	$C_7H_{11}Br$ 175.07	18495-26-6	 104^{55}; 84^{20}	4-01-00-01026	ace 4; eth 4; EtOH 4 1.4878^{25}
6614	1-Heptyne, 1-chloro- 1-Chloro-1-heptyne	$C_7H_{11}Cl$ 130.62	51556-10-6	 141	3-01-00-00997 0.9250^{24}	eth 4; EtOH 4 1.4411^{24}
6615	2-Heptyne, 7-chloro- 1-Chloro-5-heptyne	$C_7H_{11}Cl$ 130.62	70396-13-3	 166	3-01-00-00998	eth 4; EtOH 4 1.4507^{25}
6616	3-Heptyne, 7-chloro- 4-Heptynyl chloride	$C_7H_{11}Cl$ 130.62	51575-85-0	 164	3-01-00-00999	eth 4; EtOH 4 1.4517^{20}
6617	3-Heptyne, 2,6-dimethyl-	C_9H_{16} 124.23	19549-97-4	 130.6	3-01-00-01015 0.785^{20}	ace 4; eth 4
6618	3-Heptyne, 5,5-dimethyl- 5,5-Dimethyl-3-heptyne	C_9H_{16} 124.23	23097-98-5	 69^{100}	3-01-00-01015 0.7610^{20}	ace 4; eth 4 1.4360^{20}
6619	3-Heptyne, 5-ethyl-5-methyl- 5-Ethyl-5-methyl-3-heptyne	$C_{10}H_{18}$ 138.25	61228-10-2	 88^{100}	3-01-00-01021 0.7714^{20}	ace 4; eth 4 1.4386^{20}
6620	1-Heptyne, 1-iodo- 1-Iodo-1-heptyne	$C_7H_{11}I$ 222.07	54573-13-6	 $90\text{-}2^{17}$	3-01-00-00998 1.4701^{19}	 1.5123^{19}
6621	Hexacene	$C_{26}H_{16}$ 328.41	258-31-1 380	 sub	4-05-00-02834	H_2O 1; EtOH 1
6622	Hexacosane	$C_{26}H_{54}$ 366.71	630-01-3 56.4	 412.2	4-01-00-00583 0.7783^{60}	bz 4; lig 4; chl 4 1.4357^{60}
6623	Hexacosane, 13-dodecyl-	$C_{38}H_{78}$ 535.04	55517-73-2 13.7	 272^{1}	4-01-00-00600 0.8188^{20}	 1.4577^{20}
6624	Hexacosanoic acid Cerotic acid	$C_{26}H_{52}O_2$ 396.70	506-46-7 88.5		4-02-00-01310 0.8198^{100}	H_2O 1; EtOH 4; eth 4 1.4301^{100}
6625	Hexacosanoic acid, methyl ester Methyl hexacosanoate	$C_{27}H_{54}O_2$ 410.72	5802-82-4 63.8	 286^{15}; 237^{2}	4-02-00-01311	chl 3 1.4301^{80}
6626	1-Hexacosanol	$C_{26}H_{54}O$ 382.71	506-52-5 80	 305^{20} dec	4-01-00-01912	H_2O 1; EtOH 3; eth 3
6627	6,10-Hexadecadiyne	$C_{16}H_{26}$ 218.38	10160-99-3	 157^{10}	3-01-00-01067 0.7907^{20}	 1.4523^{20}
6628	Hexadecanal	$C_{16}H_{32}O$ 240.43	629-80-1 35	 200^{29}	4-01-00-03393	H_2O 1; EtOH 3; eth 3; ace 3
6629	Hexadecanamide	$C_{16}H_{33}NO$ 255.44	629-54-9 107	 236^{12}	4-02-00-01182 1.0000^{20}	H_2O 1; EtOH 2; eth 2; ace 2
6630	Hexadecanamide, *N*-phenyl-	$C_{22}H_{37}NO$ 331.54	6832-98-0 91.8	 283^{17}; 132^{10}	4-12-00-00400	ace 4; bz 4; EtOH 4; chl 4
6631	1-Hexadecanamine	$C_{16}H_{35}N$ 241.46	143-27-1 46.8	 322.5	4-04-00-00818 0.8129^{20}	H_2O 1; EtOH 4; eth 4; ace 3 1.4496^{20}
6632	Hexadecane Cetane	$C_{16}H_{34}$ 226.45	544-76-3 18.1	 286.8	4-01-00-00537 0.7733^{20}	H_2O 1; EtOH 2; eth 5; ctc 3 1.4345
6633	Hexadecane, 1-bromo-	$C_{16}H_{33}Br$ 305.34	112-82-3 18	 336	4-01-00-00542 0.9991^{20}	H_2O 1; eth 3 1.4618^{25}
6634	Hexadecane, 1-chloro-	$C_{16}H_{33}Cl$ 260.89	4860-03-1 17.9	 322	4-01-00-00542 0.8652^{20}	H_2O 1 1.4505^{20}
6635	Hexadecane, 1,16-dibromo- Hexadecamethylene dibromide	$C_{16}H_{32}Br_2$ 384.24	45223-18-5 56	 204^{4}	2-01-00-00138	chl 4
6636	Hexadecane, 1,1-dimethoxy- Palmitaldehyde, dimethyl acetal	$C_{18}H_{38}O_2$ 286.50	2791-29-9 10	 144^{2}	4-01-00-03393 0.8542^{20}	ace 4; eth 4; EtOH 4 1.4382^{25}
6637	Hexadecane, 6,11-dipentyl-	$C_{26}H_{54}$ 366.71	15874-03-0 -16.2	 231^{10}	4-01-00-00586 0.8072^{20}	 1.4502^{20}
6638	Hexadecane, 1-(ethenyloxy)-	$C_{18}H_{36}O$ 268.48	822-28-6 16	 160^{2}	4-01-00-02057 0.821^{27}	 1.4444^{25}
6639	Hexadecane, 1-fluoro- 1-Fluorohexadecane	$C_{16}H_{33}F$ 244.44	408-38-8 18	 289	4-01-00-00542 0.8321^{20}	eth 4; lig 4 1.4317^{20}
6640	Hexadecane, 1-iodo-	$C_{16}H_{33}I$ 352.34	544-77-4 22	 212^{15}	4-01-00-00543 1.123^{25}	H_2O 1; EtOH 2; eth 3; ace 3 1.4806^{20}
6641	Hexadecanenitrile	$C_{16}H_{31}N$ 237.43	629-79-8 31	 333	4-02-00-01183 0.8303^{20}	H_2O 1; EtOH 4; eth 4; ace 4 1.4450^{20}
6642	Hexadecane, 1,1'-oxybis-	$C_{32}H_{66}O$ 466.88	4113-12-6 55	 270 dec	4-01-00-01878 0.978^{19}	H_2O 1; EtOH 3; eth 3; chl 3
6643	1-Hexadecanethiol Cetyl mercaptan	$C_{16}H_{34}S$ 258.51	2917-26-2 19	 $123\text{-}8^{0.5}$	4-01-00-01881	H_2O 1; EtOH 2; eth 3; ctc 2
6644	Hexadecanoic acid Palmitic acid	$C_{16}H_{32}O_2$ 256.43	57-10-3 63.1	6947 351.5	4-02-00-01157 0.8527^{62}	H_2O 1; EtOH 3; eth 5; ace 3 1.43345^{60}
6645	Hexadecanoic acid, aluminum salt	$C_{48}H_{93}AlO_6$ 793.24	555-35-1	361	4-02-00-01157	H_2O 1; EtOH 3; peth 3; bz 3
6646	Hexadecanoic acid, anhydride	$C_{32}H_{62}O_3$ 494.84	623-65-4 64		4-02-00-01181 0.8388^{83}	eth 4 1.4364^{68}
6647	Hexadecanoic acid, butyl ester	$C_{20}H_{40}O_2$ 312.54	111-06-8 16.9		4-02-00-01167	H_2O 1; EtOH 3; eth 3 1.4312^{50}

No.	Name Synonym	Mol. Form. Mol. Wt.	CAS RN mp/°C	Merck No. bp/°C	Beil. Ref. den/g cm^{-3}	Solubility n_D
6648	Hexadecanoic acid, 2-[(dichloroacetyl)amino]-3-hydroxy-3-(4-nitrophenyl)propyl ester, [*R*-(*R**,*R**)]- Chloramphenicol palmitate	$C_{27}H_{42}Cl_2N_2O_6$ 561.55	530-43-8 90	2069	4-13-00-02753	bz 4; eth 4; EtOH 4
6649	Hexadecanoic acid, 1,2-ethanediyl ester	$C_{34}H_{66}O_4$ 538.90	624-03-3 72		4-02-00-01169 0.8594^{78}	H_2O 1; EtOH 1; eth 3; ace 4
6650	Hexadecanoic acid, ethyl ester	$C_{18}H_{36}O_2$ 284.48	628-97-7 24	191^{10}	4-02-00-01165 0.8577^{25}	H_2O 1; EtOH 3; eth 3; ace 3 1.4347^{34}
6651	Hexadecanoic acid, hexadecyl ester Cetyl palmitate	$C_{32}H_{64}O_2$ 480.86	540-10-3 54	2023	4-02-00-01168 0.989^{20}	eth 4; EtOH 4 1.4398^{70}
6652	Hexadecanoic acid, 2-hydroxyethyl ester	$C_{18}H_{36}O_3$ 300.48	4219-49-2 51	$173\text{-}4^{3}$	4-02-00-01168 0.8768^{60}	EtOH 4
6653	Hexadecanoic acid, methyl ester	$C_{17}H_{34}O_2$ 270.46	112-39-0 30	417; 148^{2}	4-02-00-01165 0.8247^{75}	H_2O 1; EtOH 4; eth 3; ace 4
6654	Hexadecanoic acid, 1-methylethyl ester	$C_{19}H_{38}O_2$ 298.51	142-91-6 13.5	160^{2}	4-02-00-01167 0.8404^{38}	ace 4; bz 4; eth 4; EtOH 4 1.4364^{25}
6655	Hexadecanoic acid, phenylmethyl ester Benzyl palmitate	$C_{23}H_{38}O_2$ 346.55	41755-60-6 36		3-06-00-01481 0.9109^{35}	bz 4; eth 4; EtOH 4; chl 4 1.4689^{50}
6656	Hexadecanoic acid, 1,2,3-propanetriyl ester Tripalmitin	$C_{51}H_{98}O_6$ 807.34	555-44-2 66.5	9648 315	4-02-00-01176 0.8752^{70}	H_2O 1; EtOH 2; eth 4; bz 3 1.4381^{80}
6657	Hexadecanoic acid, propyl ester	$C_{19}H_{38}O_2$ 298.51	2239-78-3 20.4	190^{12}	4-02-00-01167 0.8455^{33}	1.4392^{25}
6658	Hexadecanoic acid, 3,7,11,15-tetramethyl- Phytanic acid	$C_{20}H_{40}O_2$ 312.54	14721-66-5 -65			
6659	1-Hexadecanol Cetyl alcohol	$C_{16}H_{34}O$ 242.45	36653-82-4 49.3	2020 334	4-01-00-01876 0.8187^{50}	H_2O 1; EtOH 2; eth 4; ace 3 1.4283^{79}
6660	2-Hexadecanol	$C_{16}H_{34}O$ 242.45	14852-31-4 44	314	4-01-00-01882 0.8338^{20}	1.4479^{20}
6661	3-Hexadecanol	$C_{16}H_{34}O$ 242.45	593-03-3 50	152^{4}	4-01-00-01882 0.8000^{67}	
6662	1-Hexadecanol, acetate	$C_{18}H_{36}O_2$ 284.48	629-70-9 -18.5	$220\text{-}5^{205}$	4-02-00-00171 0.8574^{25}	H_2O 1; EtOH 2; ctc 3 1.4438^{20}
6663	3-Hexadecanone	$C_{16}H_{32}O$ 240.43	18787-64-9 43	184^{17}; 140^{2}	4-01-00-03394	chl 3
6664	1-Hexadecanone, 1-phenyl-	$C_{22}H_{36}O$ 316.53	6697-12-7 59	251^{15}	4-07-00-00860 0.8692^{76}	1.4675^{76}
6665	Hexadecanoyl chloride	$C_{16}H_{31}ClO$ 274.87	112-67-4 12	199^{20}	4-02-00-01182 0.9016^{25}	eth 4 1.4514^{20}
6666	6,10,14-Hexadecatrienoic acid Hiragonic acid	$C_{16}H_{26}O_2$ 250.38	4444-12-6	$180\text{-}90^{15}$	3-02-00-01505 0.9296^{20}	eth 4; EtOH 4 1.4850^{50}
6667	1-Hexadecene 1-Cetene	$C_{16}H_{32}$ 224.43	629-73-2 4.1	284.9	4-01-00-00927 0.7811^{20}	H_2O 1; EtOH 3; eth 3; ctc 3 1.4412^{20}
6668	7-Hexadecenoic acid Dinoroleic acid	$C_{16}H_{30}O_2$ 254.41	1191-75-9 33	230^{10}	0-02-00-00460	EtOH 4; eth 4
6669	9-Hexadecenoic acid	$C_{16}H_{30}O_2$ 254.41	2091-29-4 0.5	181^{1}	4-02-00-01629	bz 4; eth 4; peth 4
6670	9-Hexadecenoic acid, (*Z*)- Palmitoleic acid	$C_{16}H_{30}O_2$ 254.41	373-49-9 -0.1	$131^{0.06}$	4-02-00-01629	
6671	1-Hexadecen-3-ol, 3,7,11,15-tetramethyl- Isophytol	$C_{20}H_{40}O$ 296.54	505-32-8	5084 $107\text{-}10^{0.01}$	2-01-00-00504 0.8519^{20}	bz 4; eth 4; EtOH 4 1.4571^{20}
6672	2-Hexadecen-1-ol, 3,7,11,15-tetramethyl-, [*R*-[*R**,*R**-(*E*)]]- Phytol	$C_{20}H_{40}O$ 296.54	150-86-7	7362 203^{10}	4-01-00-02208 0.8497^{25}	1.4595^{25}
6673	1-Hexadecyne	$C_{16}H_{30}$ 222.41	629-74-3 15	284	4-01-00-01073 0.7965^{20}	bz 4 1.4440^{20}
6674	2-Hexadecyne	$C_{16}H_{30}$ 222.41	629-75-4 20	160^{15}	3-01-00-01028 0.8039^{20}	
6675	2,4-Hexadienal, (*E*,*E*)-	C_6H_8O 96.13	142-83-6	174; 76^{30}	4-01-00-03545 0.898^{20}	1.5384^{20}
6676	1,2-Hexadiene Propylallene	C_6H_{10} 82.15	592-44-9	76	4-01-00-01011 0.7149^{20}	eth 4; chl 4 1.4282^{20}
6677	1,4-Hexadiene	C_6H_{10} 82.15	592-45-0	65	4-01-00-01013 0.7000^{20}	H_2O 1; eth 4 1.4150^{20}
6678	1,5-Hexadiene Biallyl	C_6H_{10} 82.15	592-42-7 -140.7	59.4	4-01-00-01013 0.6878^{25}	H_2O 1; EtOH 3; eth 3; bz 3 1.4042^{20}
6679	2,3-Hexadiene Methylethylallene	C_6H_{10} 82.15	592-49-4	78	4-01-00-01016 0.684^{15}	1.395^{20}
6680	2,4-Hexadiene	C_6H_{10} 82.15	592-46-1 -79	80	3-01-00-00985 0.7196^{20}	H_2O 1; EtOH 3; eth 3; chl 3 1.4500^{20}
6681	1,3-Hexadiene, (*E*)-	C_6H_{10} 82.15	20237-34-7 -102.4	73.2	4-01-00-01011 0.6995^{25}	1.4406^{20}
6682	1,3-Hexadiene, (*Z*)-	C_6H_{10} 82.15	14596-92-0	73.1	4-01-00-01011 0.7033^{25}	1.4379^{20}
6683	1,4-Hexadiene, (*E*)-	C_6H_{10} 82.15	7319-00-8 -138.7	65.0	0.695^{25}	1.4104^{20}
6684	1,4-Hexadiene, (*Z*)-	C_6H_{10} 82.15	7318-67-4	66.3	0.695^{25}	eth 4 1.4049^{20}
6685	2,4-Hexadiene, (*E*,*E*)-	C_6H_{10} 82.15	5194-51-4 -44.9	82.2	4-01-00-01016 0.7101^{25}	1.4510^{20}
6686	2,4-Hexadiene, (*E*,*Z*)-	C_6H_{10} 82.15	5194-50-3 -96.1	83.5	4-01-00-01016 0.7185^{25}	1.4560^{20}

No.	Name	Synonym	Mol. Form.	Mol. Wt.	CAS RN	mp/°C	Merck No.	bp/°C	Beil. Ref.	den/g cm^{-3}	Solubility	n_D
6687	2,4-Hexadiene, (*Z,Z*)-		C_6H_{10}	82.15	6108-61-8			85		0.7298^{25}		1.4606^{20}
6688	1,3-Hexadiene, 3-chloro-		C_6H_9Cl	116.59	101870-06-8			68^{117}	4-01-00-01012	0.9390^{20}	eth 4; chl 4	1.4770^{20}
6689	1,5-Hexadiene, 1,1,2,3,3,4,4,5,6,6-decachloro-		C_6Cl_{10}	426.59	29030-84-0	49		$121^{0.03}$	4-01-00-01016	1.905^{52}	ace 4; eth 4; chl 4	1.6012^{51}
6690	2,4-Hexadiene, 1,3-dichloro-		$C_6H_8Cl_2$	151.03	73454-83-8			$80\text{-}2^{17}$	4-01-00-01017	1.1456^{20}	bz 4; chl 4	1.5271^{20}
6691	1,5-Hexadiene, 2,5-dimethyl-		C_8H_{14}	110.20	627-58-7	-75.6		114.3	4-01-00-01042	0.743^{20}	H_2O 1; ace 3; chl 3	1.43995^{21}
6692	2,4-Hexadiene, 2,5-dimethyl-		C_8H_{14}	110.20	764-13-6	14		134.5	4-01-00-01043	0.7577^{25}	H_2O 1; EtOH 3; eth 3; bz 3	1.4785^{20}
6693	1,5-Hexadiene-3,4-diol, (*R*,R**)-(±)-		$C_6H_{10}O_2$	114.14	19700-97-1	21.7		$90\text{-}1^{8}$		1.017^{19}	H_2O 3; EtOH 3; eth 3; chl 3	1.4790^{19}
6694	1,5-Hexadiene, 3-ethyl-2-methyl-	1,5-Hexadiene, 2-methyl-3-ethyl	C_9H_{16}	124.23	73398-15-9	-70		145	0-01-00-00260	0.7594^{25}		
6695	1,5-Hexadiene, 2-methyl-		C_7H_{12}	96.17	4049-81-4	-128.8		88.1	4-01-00-01030	0.7153^{25}		1.4183^{20}
6696	2,4-Hexadiene, 2-methyl-		C_7H_{12}	96.17	28823-41-8	-74.2		111.5	4-01-00-01030	0.7411^{25}		
6697	2,4-Hexadiene, 1,3,4,6-tetrachloro-		$C_6H_6Cl_4$	219.92	100367-45-1			$84\text{-}9^{2}$	3-01-00-00987	1.4013^{20}	MeOH 4; chl 4	1.5465^{20}
6698	2,4-Hexadienoic acid	Sorbic acid	$C_6H_8O_2$	112.13	110-44-1	134.5	8677	228 dec; 153^{50}	2-02-00-00452	1.204^{19}	H_2O 3; EtOH 3; eth 4; chl 3	
6699	2,4-Hexadienoic acid, ethyl ester		$C_8H_{12}O_2$	140.18	110318-09-7			195.5 85^{20}	2-02-00-00452	0.9506^{20}	H_2O 1; EtOH 3; eth 3; chl 3	1.4951^{20}
6700	2,4-Hexadienoic acid, ethyl ester, (*E,E*)-		$C_8H_{12}O_2$	140.18	2396-84-1			195.5	4-02-00-01703	0.9506^{20}	eth 4; EtOH 4; chl 4	1.4951^{20}
6701	2,5-Hexadienoic acid, 3-methoxy-5-methyl-4-oxo-	Penicillic acid	$C_8H_{10}O_4$	170.17	90-65-3	83	7033		3-03-00-01467		H_2O 3; EtOH 4; eth 4; ace 3	
6702	2,4-Hexadienoic acid, methyl ester	Methyl 2,4-hexadienoate	$C_7H_{10}O_2$	126.16	1515-80-6	15		180; 70^{20}	2-02-00-00452	0.9777^{20}	H_2O 1; EtOH 3; eth 3	1.5025^{22}
6703	2,4-Hexadienoic acid, methyl ester, (*E,E*)-		$C_7H_{10}O_2$	126.16	689-89-4	15		180	4-02-00-01703	0.9777^{20}	eth 4; EtOH 4	1.5025^{22}
6704	2,4-Hexadienoic acid, potassium salt, (*E,E*)-		$C_6H_7KO_2$	150.22	24634-61-5	>270 dec	7661			1.361^{25}	H_2O 4; EtOH 3	
6705	1,4-Hexadien-3-ol		$C_6H_{10}O$	98.14	1070-14-0			$50\text{-}2^{15}$	4-01-00-02238	0.859^{23}		1.4502^{23}
6706	1,5-Hexadien-3-ol	3-Hydroxy-1,5-hexadiene	$C_6H_{10}O$	98.14	924-41-4			133.5	4-01-00-02239	0.8596^{25}		1.4464^{25}
6707	2,4-Hexadien-1-ol	1-Hydroxy-2,4-hexadiene	$C_6H_{10}O$	98.14	111-28-4	30.5	8678	76^{12}	4-01-00-02239	0.8967^{23}	H_2O 1; EtOH 3; eth 3	1.4981^{20}
6708	3,5-Hexadien-2-ol	1,3-Hexadien-5-ol	$C_6H_{10}O$	98.14	3280-51-1			$77\text{-}8^{26}$	4-01-00-02237	0.8678^{20}	EtOH 4	1.4896^{20}
6709	2,4-Hexadien-1-ol, (*E,E*)-		$C_6H_{10}O$	98.14	17102-64-6	30.5		76^{12}	4-01-00-02239	0.8967^{20}	eth 4; EtOH 4	1.4981^{20}
6710	3,5-Hexadien-2-one, 6-phenyl-		$C_{12}H_{12}O$	172.23	4173-44-8	68		171^{15}; 155^{12}	4-07-00-01183		bz 4; eth 4; EtOH 4; chl 4	
6711	2,4-Hexadienoyl chloride, (*E,E*)-		C_6H_7ClO	130.57	2614-88-2			78^{15}	3-02-00-01460	1.0666^{19}	ace 4	1.5545^{20}
6712	1,3-Hexadien-5-yne		C_6H_6	78.11	10420-90-3	-81		83.5	4-01-00-01120	0.7806^{20}	bz 4	1.5095^{20}
6713	1,5-Hexadien-3-yne	Divinylacetylene	C_6H_6	78.11	821-08-9	-88		85	4-01-00-01120	0.7851^{20}	H_2O 1; bz 3	1.5035^{20}
6714	1,5-Hexadien-3-yne, 2,5-dimethyl-	Diisopropenylacetylene	C_8H_{10}	106.17	3725-05-1			123	4-01-00-01124	0.7863^{25}	bz 4; chl 4	1.4845^{20}
6715	1,4-Hexadiyne		C_6H_6	78.11	10420-91-4	<-80		81.7	3-01-00-01057	0.825^{0}	bz 4; chl 4	
6716	1,5-Hexadiyne	Bipropargyl	C_6H_6	78.11	628-16-0	-6		86	4-01-00-01118	0.8049^{20}	H_2O 1; EtOH 3; eth 3; ace 3	1.4380^{23}
6717	2,4-Hexadiyne	Dimethyldiacetylene	C_6H_6	78.11	2809-69-0	67.8		129.5	4-01-00-01119		EtOH 4; eth 4	
6718	Hexanal	Caproaldehyde	$C_6H_{12}O$	100.16	66-25-1	-56	1761	131	4-01-00-03296	0.8335^{20}	H_2O 2; EtOH 4; eth 4; ace 3	1.4039^{20}
6719	Hexanal, 2-ethyl-	2-Ethylhexanal	$C_8H_{16}O$	128.21	123-05-7	<-100		163	4-01-00 03345	0.8540^{20}	H_2O 1; EtOH 3; eth 3; ctc 2	1.4142^{20}
6720	Hexanal, 3-methyl-	Caproaldehyde, 3-methyl	$C_7H_{14}O$	114.19	19269-28-4			143	2-01-00-00756	0.8203^{20}	H_2O 1; EtOH 3; eth 3	1.4122^{20}
6721	Hexanamide		$C_6H_{13}NO$	115.18	628-02-4	101		255	4-02-00-00929	0.999^{20}	bz 4; eth 4; EtOH 4; chl 4	1.4200^{110}
6722	Hexanamide, *N*-phenyl-		$C_{12}H_{17}NO$	191.27	621-15-8	95			4-12-00-00392	1.112^{25}	eth 4; EtOH 4	
6723	1-Hexanamine	Hexylamine	$C_6H_{15}N$	101.19	111-26-2	-22.9		132.8	4-04-00-00709	0.7660^{20}	H_2O 2; EtOH 5; eth 5; chl 3	1.4180^{20}
6724	2-Hexanamine, (±)-		$C_6H_{15}N$	101.19	68107-05-1	-19		117.5	4-04-00-00721	0.7533^{20}	EtOH 3; eth 3	1.4080^{25}

PHYSICAL CONSTANTS OF ORGANIC COMPOUNDS (continued)

No.	Name Synonym	Mol. Form. Mol. Wt.	CAS RN mp/°C	Merck No. bp/°C	Beil. Ref. den/g cm^{-3}	Solubility n_D
6725	1-Hexanamine, *N,N*-dihexyl-	$C_{18}H_{39}N$ 269.51	102-86-3	261.7	4-04-00-00711 0.7976^{21}	H_2O 1; EtOH 4; eth 4; acid 3
6726	1-Hexanamine, 2-ethyl- 2-Ethyl-1-hexylamine	$C_8H_{19}N$ 129.25	104-75-6		4-04-00-00766	
6727	1-Hexanamine, *N*-hexyl-	$C_{12}H_{27}N$ 185.35	143-16-8 -13.1	236; 75^{1}	4-04-00-00711 0.7889^{20}	EtOH 3; eth 3 1.4339^{20}
6728	1-Hexanamine, 3-methyl- Hexylamine, 3-methyl-	$C_7H_{17}N$ 115.22	65530-93-0	149; 67^{45}	2-04-00-00653 0.772^{26}	1.4249^{25}
6729	1-Hexanamine, 4-methyl- Hexylamine, 4-methyl-	$C_7H_{17}N$ 115.22	34263-68-8	153	2-04-00-00654 0.7802^{20}	1.4238^{20}
6730	2-Hexanamine, 4-methyl- Methylhexaneamine	$C_7H_{17}N$ 115.22	105-41-9	6000 132.5	4-04-00-00747 0.7655^{20}	H_2O 2; EtOH 4; eth 4; chl 4 1.4150^{25}
6731	Hexane	C_6H_{14} 86.18	110-54-3 -95.3	4613 68.7	4-01-00-00338 0.6548^{25}	H_2O 1; EtOH 4; eth 3; chl 3 1.3749^{20}
6732	Hexane, 1-bromo- Hexyl bromide	$C_6H_{13}Br$ 165.07	111-25-1 -84.7	155.3	4-01-00-00352 1.1744^{20}	H_2O 1; EtOH 5; eth 5; ace 3 1.4478^{20}
6733	Hexane, 2-bromo- 2-Bromohexane	$C_6H_{13}Br$ 165.07	3377-86-4	144; 78^{90}	4-01-00-00353 1.1658^{20}	H_2O 1; EtOH 4; eth 3; ace 3 1.4832^{25}
6734	Hexane, 3-bromo- 3-Bromohexane	$C_6H_{13}Br$ 165.07	3377-87-5	142	4-01-00-00353 1.1799^{20}	ace 4; eth 4; EtOH 4; chl 4 1.4472^{20}
6735	Hexane, 1-bromo-6-fluoro- 1-Bromo-6-fluorohexane	$C_6H_{12}BrF$ 183.06	373-28-4	$67\text{-}8^{11}$	4-01-00-00353 1.293^{20}	ace 4; eth 4; EtOH 4; chl 4 1.4436^{25}
6736	Hexane, 1-bromo-6-methoxy-	$C_7H_{15}BrO$ 195.10	50592-87-5	112^{35}; 98^{19}	4-01-00-01704 1.1887^{25}	1.4469^{25}
6737	Hexane, 1-chloro- Hexyl chloride	$C_6H_{13}Cl$ 120.62	544-10-5 -94	2144 135	4-01-00-00349 0.8785^{20}	H_2O 1; EtOH 3; eth 3; ace 3 1.4199^{20}
6738	Hexane, 2-chloro- 2-Hexyl chloride	$C_6H_{13}Cl$ 120.62	638-28-8	122.5	4-01-00-00349 0.8694^{21}	ace 4; bz 4; eth 4; EtOH 4 1.4142^{22}
6739	Hexane, 3-chloro- 3-Chlorohexane	$C_6H_{13}Cl$ 120.62	2346-81-8	123	4-01-00-00349 0.8684^{20}	ace 4; bz 4; eth 4; EtOH 4 1.4163^{20}
6740	Hexane, 2-chloro-2,5-dimethyl- 2-Chloro-2,5-dimethylhexane	$C_8H_{17}Cl$ 148.68	29342-44-7	86^{100}	4-01-00-00434 0.8476^{18}	ace 4; bz 4; eth 4; EtOH 4 1.4232^{20}
6741	Hexane, 3-chloro-2,3-dimethyl-	$C_8H_{17}Cl$ 148.68	101654-30-2	$41\text{-}3^{12}$	3-01-00-00481 0.8869^{20}	EtOH 4 1.4333^{25}
6742	Hexane, 1-chloro-6-fluoro- 1-Chloro-6-fluorohexane	$C_6H_{12}ClF$ 138.61	1550-09-0	168; 62^{15}	4-01-00-00350 1.015^{20}	eth 4; EtOH 4; chl 4 1.4168^{25}
6743	Hexane, 1-chloro-3-methyl-	$C_7H_{15}Cl$ 134.65	101257-63-0	151	2-01-00-00119 0.8766^{20}	eth 4; EtOH 4; chl 4 1.4274^{20}
6744	Hexane, 2-chloro-2-methyl-	$C_7H_{15}Cl$ 134.65	4398-65-6	135 dec; 60^{52}	4-01-00-00398 0.8635^{20}	eth 4; EtOH 4; chl 4 1.4200^{20}
6745	Hexane, 2-chloro-5-methyl- 2-Chloro-5-methylhexane	$C_7H_{15}Cl$ 134.65	58766-17-9	138 dec	3-01-00-00436 0.863^{20}	eth 4; EtOH 4; chl 4
6746	Hexane, 3-chloro-3-methyl-	$C_7H_{15}Cl$ 134.65	43197-78-0	135	3-01-00-00438 0.8787^{20}	eth 4; EtOH 4; chl 4 1.4250^{20}
6747	Hexane, 3-chloro-2,2,3-trimethyl-	$C_9H_{19}Cl$ 162.70	102449-95-6	$64\text{-}5^{13}$	3-01-00-00515 0.9010^{20}	eth 4; EtOH 4; chl 4 1.4465^{20}
6748	Hexanedial	$C_6H_{10}O_2$ 114.14	1072-21-5 -8	$92\text{-}4^{9}$	4-01-00-03686 1.003^{19}	bz 4; eth 4; EtOH 4 1.4350^{20}
6749	1,6-Hexanediamine Hexamethylenediamine	$C_6H_{16}N_2$ 116.21	124-09-4 41.5	4614 205	4-04-00-01320	H_2O 4; EtOH 3; bz 3
6750	1,6-Hexanediamine, *N,N'*-dibutyl- *N,N'*-Dibutylhexamethylenediamine	$C_{14}H_{32}N_2$ 228.42	4835-11-4	$137\text{-}40^{3\text{-}4}$	4-04-00-01327	1.4470^{25}
6751	2,5-Hexanediamine, 2,5-dimethyl-	$C_8H_{20}N_2$ 144.26	23578-35-0	$63\text{-}4^{8}$	4-04-00-01363 0.832^{25}	1.4459^{20}
6752	1,6-Hexanediamine, *N,N,N',N'*-tetramethyl-	$C_{10}H_{24}N_2$ 172.31	111-18-2	209.5	4-04-00-01322 0.806^{25}	1.4359^{20}
6753	1,6-Hexanediaminium, *N,N'*-di-9H-fluoren-9-yl-*N,N,N',N'*-tetramethyl-dibromide Hexafluorenium bromide	$C_{36}H_{42}Br_2N_2$ 662.55	317-52-2 153.5	4606	4-12-00-03397	
6754	Hexane, 1,2-dibromo- 1,2-Dibromohexane	$C_6H_{12}Br_2$ 243.97	624-20-4	104^{36}; 82^{12}	3-01-00-00392 1.5774^{20}	bz 4; eth 4; chl 4 1.5024^{20}
6755	Hexane, 1,4-dibromo- 1,4-Dibromohexane	$C_6H_{12}Br_2$ 243.97	25118-28-9	$106\text{-}8^{15}$	3-01-00-00392 1.602^{15}	1.5084^{15}
6756	Hexane, 1,5-dibromo- 1,5-Dibromohexane	$C_6H_{12}Br_2$ 243.97	627-96-3	153^{100}; 103^{9}	4-01-00-00354 1.5989^{20}	1.5072^{15}
6757	Hexane, 1,6-dibromo-	$C_6H_{12}Br_2$ 243.97	629-03-8 -2.3	245.5	4-01-00-00354 1.5948^{15}	H_2O 1; eth 3; ace 3; ctc 2 1.5037^{20}
6758	Hexane, 2,3-dibromo- 2,3-Dibromohexane	$C_6H_{12}Br_2$ 243.97	6423-02-5	90^{16}	3-01-00-00393 1.5812^{20}	1.5025^{20}
6759	Hexane, 2,5-dibromo-, (*R**,*R**)-(±)- Hexane, 2,5-dibromo(*DL*)	$C_6H_{12}Br_2$ 243.97	54462-68-9 -44.64	108^{30} 94^{13}	3-01-00-00393 1.5788^{20}	H_2O 1; eth 3; ace 3; ctc 3 1.5007^{20}
6760	Hexane, 3,4-dibromo-	$C_6H_{12}Br_2$ 243.97	89583-12-0	$80\text{-}1^{13}$	3-01-00-00394 1.6027^{20}	1.5043^{20}
6761	Hexane, 1,2-dichloro- 1,2-Dichlorohexane	$C_6H_{12}Cl_2$ 155.07	2162-92-7	173	4-01-00-00350 1.085^{15}	eth 4; chl 4
6762	Hexane, 1,6-dichloro- 1,6-Dichlorohexane	$C_6H_{12}Cl_2$ 155.07	2163-00-0	204	4-01-00-00350 1.0677^{20}	H_2O 1; eth 3; ctc 3; chl 3 1.4572^{20}
6763	Hexane, 2,2-dichloro- 2,2-Dichlorohexane	$C_6H_{12}Cl_2$ 155.07	42131-89-5	68^{49}	3-01-00-00390 1.0150^{25}	eth 4; chl 4 1.4353^{25}

No.	Name	Synonym	Mol. Form.	Mol. Wt.	CAS RN	mp/°C	Merck No.	bp/°C	Beil. Ref.	den/g cm^{-3}	Solubility	n_D
6764	Hexane, 2,3-dichloro-	2,3-Dichlorohexane	$C_6H_{12}Cl_2$	155.07	54305-87-2			163.5	4-01-00-00350	1.0527^{11}	eth 4; chl 4	
6765	Hexane, 2,5-dichloro-2,5-dimethyl-	2,5-Dimethyl-2,5-dichlorohexane	$C_8H_{16}Cl_2$	183.12	6223-78-5	67.5			4-01-00-00435	0.9543^{20}	bz 4; eth 4; EtOH 4; chl 4	
6766	Hexane, 2,5-dichloro-, (*R**,*R**)-(±)-	(±)-2,5-Dichlorohexane	$C_6H_{12}Cl_2$	155.07	41761-12-0	-38		177; 106^{91}	4-01-00-00350	1.0474^{20}	chl 4	1.4491^{20}
6767	Hexane, 3,3-diethyl-		$C_{10}H_{22}$	142.28	17302-02-2			166.3	4-01-00-00484	0.7575^{25}		1.4258^{20}
6768	Hexane, 3,4-diethyl-	3,4-Diethylhexane	$C_{10}H_{22}$	142.28	19398-77-7			163.9	4-01-00-00484	0.7472^{25}		1.4190^{20}
6769	Hexane, 1,6-diiodo-	Hexamethylene diiodide	$C_6H_{12}I_2$	337.97	629-09-4	9.5		163^{17}; 141^{10}	4-01-00-00356	2.0416^{15}	H_2O 1; EtOH 4; eth 4	1.5899^{15}
6770	Hexane, 1,6-diisocyanato-	Hexamethylene diisocyanate	$C_8H_{12}N_2O_2$	168.20	822-06-0			122^{10}; 94^{1}	4-04-00-01349	1.0528^{20}		1.4585^{20}
6771	Hexane, 2,2-dimethyl-	2,2-Dimethylhexane	C_8H_{18}	114.23	590-73-8	-121.1		106.8	4-01-00-00432	0.6953^{20}	ace 4; bz 4; eth 4; EtOH 4	1.3935^{20}
6772	Hexane, 2,3-dimethyl-	2,3-Dimethylhexane	C_8H_{18}	114.23	584-94-1			115.6	4-01-00-00432	0.6912^{25}	ace 4; bz 4; EtOH 4; lig 4	1.4011^{20}
6773	Hexane, 2,4-dimethyl-	2,4-Dimethylhexane	C_8H_{18}	114.23	589-43-5			109.5		0.6962^{25}		1.3929^{25}
6774	Hexane, 2,5-dimethyl-	2,5-Dimethylhexane	C_8H_{18}	114.23	592-13-2	-91		109.1	4-01-00-00434	0.6901^{25}	H_2O 1; EtOH 5; eth 3; ace 5	1.3925^{20}
6775	Hexane, 3,3-dimethyl-	3,3-Dimethylhexane	C_8H_{18}	114.23	563-16-6	-126.1		111.9	4-01-00-00435	0.7100^{20}	H_2O 1; EtOH 5; eth 4; ace 4	1.4001^{20}
6776	Hexane, 3,4-dimethyl-	3,4-Dimethylhexane	C_8H_{18}	114.23	583-48-2			117.7	4-01-00-00436	0.7151^{25}	H_2O 1; EtOH 5; eth 3; ace 5	1.4041^{20}
6777	Hexanedinitrile	Adiponitrile	$C_6H_8N_2$	108.14	111-69-3	1		295	4-02-00-01975	0.9676^{20}	H_2O 2; EtOH 3; eth 2; chl 3	1.4380^{20}
6778	Hexane, 1,6-dinitro-	1,6-Dinitrohexane	$C_6H_{12}N_2O_4$	176.17	4286-47-9	37.5		$100\text{-}3^{0.3}$	4-01-00-00357		H_2O 1; HOAc 3	
6779	Hexanedioic acid	Adipic acid	$C_6H_{10}O_4$	146.14	124-04-9	153.2	152	337.5	4-02-00-01956	1.360^{25}	H_2O 2; EtOH 4; eth 3; HOAc 1	
6780	Hexanedioic acid, bis(2-ethylbutyl) ester	Adipic acid, di-2-ethylbutyl ester	$C_{18}H_{34}O_4$	314.47	10022-60-3	-15		200^{19}	4-02-00-01964	0.934^{25}	ace 4; EtOH 4	1.4434^{20}
6781	Hexanedioic acid, bis(2-ethylhexyl) ester	Bis(2-ethylhexyl) adipate	$C_{22}H_{42}O_4$	370.57	103-23-1	-67.8		214^{5}	4-02-00-01964	0.922^{25}	ace 4; eth 4; EtOH 4	1.4474^{20}
6782	Hexanedioic acid, bis(1-methylethyl) ester		$C_{12}H_{22}O_4$	230.30	6938-94-9	-1.1		$120^{6.5}$	4-02-00-01961	0.9569^{20}	ace 4; eth 4; EtOH 4	1.4247^{20}
6783	Hexanedioic acid, bis(2-methylpropyl) ester		$C_{14}H_{26}O_4$	258.36	141-04-8			$186\text{-}8^{15}$	4-02-00-01962	0.9543^{19}		1.4301^{20}
6784	Hexanedioic acid, dibutyl ester		$C_{14}H_{26}O_4$	258.36	105-99-7	-32.4		165^{10}	4-02-00-01961	0.9613^{20}	H_2O 1; EtOH 5; eth 5	1.4369^{20}
6785	Hexanedioic acid, diethyl ester		$C_{10}H_{18}O_4$	202.25	141-28-6	-19.8		245	4-02-00-01960	1.0076^{20}	H_2O 1; EtOH 3; eth 3	1.4272^{20}
6786	Hexanedioic acid, dimethyl ester		$C_8H_{14}O_4$	174.20	627-93-0	10.3		115^{13}	4-02-00-01959	1.0600^{20}	H_2O 1; EtOH 3; eth 3; ctc 3	1.4283^{20}
6787	Hexanedioic acid, dipropyl ester		$C_{12}H_{22}O_4$	230.30	106-19-4	-15.7		151^{11}	4-02-00-01961	0.9790^{20}	eth 4; EtOH 4; chl 4	1.4314^{20}
6788	Hexanedioic acid, 2-methyl-		$C_7H_{12}O_4$	160.17	626-70-0	64		209^{13}	4-02-00-02010		H_2O 4; eth 4; EtOH 4; chl 4	
6789	Hexanedioic acid, 3-methyl-, (±)-	Adipic acid, 3-methyl (*DL*)	$C_7H_{12}O_4$	160.17	81177-02-8	97		230^{30}	4-02-00-02014		H_2O 4; EtOH 4; eth 4; ace 4	
6790	Hexanedioic acid, 3-methyl-, diethyl ester	Diethyl 3-methyladipate	$C_{11}H_{20}O_4$	216.28	55877-01-5			258; 126.5^{10}	4-02-00-02015	0.9948^{20}		1.4335
6791	Hexanedioic acid, monoethyl ester		$C_8H_{14}O_4$	174.20	626-86-8	29		285	4-02-00-01960	0.9796^{20}	EtOH 3; eth 3; peth 3	1.4311^{20}
6792	Hexanedioic acid, monomethyl ester		$C_7H_{12}O_4$	160.17	627-91-8	9		158^{10}	3-02-00-01711	1.0623^{20}	EtOH 3	1.4283^{20}
6793	1,3-Hexanediol		$C_6H_{14}O_2$	118.18	21531-91-9			123^{13}	4-01-00-02555	0.958^{22}		1.4461^{22}
6794	1,4-Hexanediol		$C_6H_{14}O_2$	118.18	16432-53-4			$134\text{-}5^{18.5}$	4-01-00-02556	0.982^{16}	eth 4; EtOH 4	1.4530^{16}
6795	1,5-Hexanediol		$C_6H_{14}O_2$	118.18	928-40-5			237; $90^{0.5}$	4-01-00-02556	0.981^{25}		1.4511^{20}
6796	1,6-Hexanediol	Hexamethylene glycol	$C_6H_{14}O_2$	118.18	629-11-8	42.8	4610	208	4-01-00-02556		H_2O 3; EtOH 3; eth 2; ace 3	1.4579^{25}
6797	2,3-Hexanediol		$C_6H_{14}O_2$	118.18	617-30-1	60		205	4-01-00-02561	0.9900^{15}	H_2O 4; eth 4; EtOH 4	1.4510^{15}
6798	2,4-Hexanediol		$C_6H_{14}O_2$	118.18	19780-90-6			211; 105^{9}	4-01-00-02562	0.9516^{21}		1.4418^{21}
6799	2,5-Hexanediol	Diisopropanol	$C_6H_{14}O_2$	118.18	2935-44-6	43		218; 88^{1}	4-01-00-02562	0.9610^{20}	H_2O 3; EtOH 3; eth 3; ctc 2	1.4475^{20}
6800	3,4-Hexanediol, 3,4-diethyl-	3,4-Diethyl-3,4-hexanediol	$C_{10}H_{22}O_2$	174.28	6931-71-1	28		230	4-01-00-02621	0.9602^{13}	eth 4; EtOH 4	1.467^{13}
6801	2,5-Hexanediol, 2,5-dimethyl-	1,1,4,4-Tetramethyl-1,4-butanediol	$C_8H_{18}O_2$	146.23	110-03-2	92		214	4-01-00-02600	0.898^{20}	H_2O 3; EtOH 4; bz 4; chl 4	
6802	1,3-Hexanediol, 2-ethyl-	Ethohexadiol	$C_8H_{18}O_2$	146.23	94-96-2	-40	3699	244	4-01-00-02597	0.9325^{22}	H_2O 2; EtOH 3; eth 3	1.4497^{20}

No.	Name Synonym	Mol. Form. Mol. Wt.	CAS RN mp/°C	Merck No. bp/°C	Beil. Ref. den/g cm^{-3}	Solubility n_D
6803	2,4-Hexanediol, 2-methyl-	$C_7H_{16}O_2$ 132.20	66225-35-2	121^{30}	2-01-00-00554 0.9321^{18}	1.4407^{18}
6804	2,4-Hexanedione Propionylacetone	$C_6H_{10}O_2$ 114.14	3002-24-2	160	4-01-00-03687 0.959^{20}	1.4516^{20}
6805	2,5-Hexanedione Acetonylacetone	$C_6H_{10}O_2$ 114.14	110-13-4 -5.5	63 194	4-01-00-03688 0.7370^{20}	H_2O 4; bz 4; eth 4; EtOH 4 1.4232^{20}
6806	3,4-Hexanedione Bipropionyl	$C_6H_{10}O_2$ 114.14	4437-51-8 -10	130	4-01-00-03690 0.941^{21}	1.4130^{21}
6807	3,4-Hexanedione, 2,5-dimethyl- Biisobutyryl	$C_8H_{14}O_2$ 142.20	4388-87-8	144.5	4-01-00-03713 0.9232^{20}	1.4206^{20}
6808	2,5-Hexanedione, 3-hydroxy- 3-Hydroxy-2,5-hexanedione	$C_6H_{10}O_3$ 130.14	61892-85-1	$62\text{-}7^{0.5}$	3-01-00-03317	bz 4; eth 4 1.4497^{25}
6809	2,3-Hexanedione, 5-methyl- 2-Methylhexa-4,5-dione	$C_7H_{12}O_2$ 128.17	13706-86-0	138	4-01-00-03701 0.908^{22}	1.4119^{20}
6810	2,5-Hexanedione, 3-methyl- 3-Methyl-2,5-hexanedione	$C_7H_{12}O_2$ 128.17	4437-50-7	196; 71^{10}	3-01-00-03137 0.9527^{20}	eth 4; EtOH 4 1.4260^{20}
6811	3,4-Hexanedione, 2,2,5,5-tetramethyl- Pivalil	$C_{10}H_{18}O_2$ 170.25	4388-88-9 -2	169	4-01-00-03726 0.8776^{20}	H_2O 1; eth 5 1.4157^{20}
6812	1,6-Hexanedithiol	$C_6H_{14}S_2$ 150.31	1191-43-1	$118\text{-}9^{15}$	4-01-00-02559 0.983^{25}	1.5110^{20}
6813	Hexane, 1-(ethenyloxy)- Hexyl vinyl ether	$C_8H_{16}O$ 128.21	5363-64-4	143	4-01-00-02055 0.7966^{20}	1.4171^{20}
6814	Hexane, 1-ethoxy-	$C_8H_{18}O$ 130.23	5756-43-4	143	4-01-00-01697 0.7722^{20}	eth 4; EtOH 4 1.4008^{20}
6815	Hexane, 3-ethyl- 3-Ethylhexane	C_8H_{18} 114.23	619-99-8	118.6	4-01-00-00431 0.7136^{20}	H_2O 1; EtOH 5; eth 5; ace 5 1.4018^{20}
6816	Hexane, 3-ethyl-2,2-dimethyl- 3-Ethyl-2,2-dimethylhexane	$C_{10}H_{22}$ 142.28	20291-91-2	156.1	4-01-00-00484 0.7447^{25}	1.4197^{20}
6817	Hexane, 3-ethyl-2,3-dimethyl- 3-Ethyl-2,3-dimethylhexane	$C_{10}H_{22}$ 142.28	52897-00-4	166	4-01-00-00484 0.7599^{25}	1.4270^{207}
6818	Hexane, 3-ethyl-2,4-dimethyl- 3-Ethyl-2,4-dimethylhexane	$C_{10}H_{22}$ 142.28	7220-26-0	162	4-01-00-00484 0.7514^{25}	1.4225^{20}
6819	Hexane, 3-ethyl-2,5-dimethyl- 3-Ethyl-2,5-dimethylhexane	$C_{10}H_{22}$ 142.28	52897-04-8	154.1	4-01-00-00484 0.7368^{25}	1.4157^{20}
6820	Hexane, 3-ethyl-3,4-dimethyl- 3-Ethyl-3,4-dimethylhexane	$C_{10}H_{22}$ 142.28	52897-06-0	162.1	4-01-00-00485 0.7596^{25}	1.4267^{207}
6821	Hexane, 4-ethyl-2,2-dimethyl- Hexane, 2,2-dimethyl-4-ethyl	$C_{10}H_{22}$ 142.28	52896-99-8	147	4-01-00-00485 0.7330^{20}	1.4131^{20}
6822	Hexane, 4-ethyl-2,3-dimethyl- 4-Ethyl-2,3-dimethylhexane	$C_{10}H_{22}$ 142.28	52897-01-5	162	4-01-00-00485 0.7516^{25}	1.4226^{20}
6823	Hexane, 4-ethyl-2,4-dimethyl- 4-Ethyl-2,4-dimethylhexane	$C_{10}H_{22}$ 142.28	52897-03-7	161.1	4-01-00-00485 0.7525^{25}	1.4235^{20}
6824	Hexane, 4-ethyl-3,3-dimethyl- 4-Ethyl-3,3-dimethylhexane	$C_{10}H_{22}$ 142.28	52897-05-9	162.9	4-01-00-00485 0.7598^{25}	1.4269^{20}
6825	Hexane, 3-ethyl-2-methyl- 3-Ethyl-2-methylhexane	C_9H_{20} 128.26	16789-46-1	138	4-01-00-00459 0.7310^{20}	1.4106^{20}
6826	Hexane, 3-ethyl-3-methyl- 3-Ethyl-3-methylhexane	C_9H_{20} 128.26	3074-76-8	140.6	4-01-00-00459 0.7371^{25}	1.4140^{20}
6827	Hexane, 3-ethyl-4-methyl- 4-Ethyl-3-methylhexane	C_9H_{20} 128.26	3074-77-9	140	3-01-00-00514 0.7420^{20}	1.4134^{20}
6828	Hexane, 4-ethyl-2-methyl- 4-Ethyl-2-methylhexane	C_9H_{20} 128.26	3074-75-7	133.8	4-01-00-00459 0.7195^{25}	1.4063^{20}
6829	Hexane, 1-fluoro- Hexyl fluoride	$C_6H_{13}F$ 104.17	373-14-8 -103	91.5	4-01-00-00348 0.7995^{20}	eth 3; bz 3 1.3738^{20}
6830	Hexane, 2-fluoro- 2-Fluorohexane	$C_6H_{13}F$ 104.17	372-54-3	86	4-01-00-00348 0.7916^{20}	1.3693^{20}
6831	Hexane, 1-iodo- Hexyl iodide	$C_6H_{13}I$ 212.07	638-45-9 -74.2	181	4-01-00-00355 1.4397^{20}	H_2O 1 1.4929^{20}
6832	Hexane, 2-iodo-, (*R*)- Hexane, 2-iodo (*l*)	$C_6H_{13}I$ 212.07	59654-13-6	90^{70}; 45^{9}	4-01-00-00355 1.4354^{17}	H_2O 1; ace 3; chl 3 1.4878^{25}
6833	Hexane, 1-methoxy- Hexyl methyl ether	$C_7H_{16}O$ 116.20	4747-07-3	126.1		
6834	Hexane, 2-methyl- 2-Methylhexane	C_7H_{16} 100.20	591-76-4 -118.2	90.0	4-01-00-00397 0.6787^{20}	H_2O 1; EtOH 3; eth 5; ace 5 1.3848^{20}
6835	Hexane, 3-methyl-, (*S*)- (+)-3-Methylhexane	C_7H_{16} 100.20	6131-24-4 -119	92	4-01-00-00399 0.6860^{20}	H_2O 1; EtOH 3; eth 5; ace 5 1.3887^{20}
6836	Hexane, 2-methyl-3-methylene- 2-Isopropyl-1-pentene	C_8H_{16} 112.22	16746-02-4	113	4-01-00-00887 0.721^{25}	1.414^{20}
6837	Hexane, 2-methyl-4-methylene- 1-Pentene, 2-ethyl-4-methyl-	C_8H_{16} 112.22	3404-80-6	110.3	4-01-00-00888 0.7152^{25}	1.4105^{20}
6838	Hexane, 3-methyl-4-methylene-	C_8H_{16} 112.22	3404-67-9	112.5	4-01-00-00890 0.725^{25}	1.4142^{20}
6839	Hexane, 2-methyl-3-(1-methylethyl)- Hexane, 3-isopropyl-2-methyl-	$C_{10}H_{22}$ 142.28	62016-13-1	165	4-01-00-00484 0.7436^{25}	1.4195^{20}
6840	Hexanenitrile Capronitrile	$C_6H_{11}N$ 97.16	628-73-9 -80.3	163.6	4-02-00-00930 0.8051^{20}	H_2O 1; EtOH 3; eth 3; chl 2 1.4068^{20}
6841	Hexane, 1-nitro- 1-Nitrohexane	$C_6H_{13}NO_2$ 131.17	646-14-0	193; 84^{21}	4-01-00-00356 0.9396^{20}	H_2O 1; EtOH 3; eth 3; ace 3 1.4270^{20}
6842	Hexane, 1,1'-oxybis-	$C_{12}H_{26}O$ 186.34	112-58-3	223	4-01-00-01698 0.7936^{20}	H_2O 1; eth 3; ctc 2 1.4204^{20}

No.	Name Synonym	Mol. Form. Mol. Wt.	CAS RN mp/°C	Merck No. bp/°C	Beil. Ref. den/g cm^{-3}	Solubility n_D
6843	Hexane, tetradecafluoro- Perfluorohexane	C_6F_{14} 338.04	355-42-0 -87.1	 56.6	4-01-00-00348 1.6995^{20}	H_2O 1; eth 3; bz 3; chl 3 1.2515^{20}
6844	Hexane, 2,2,3,3-tetramethyl- 2,2,3,3-Tetramethylhexane	$C_{10}H_{22}$ 142.28	13475-81-5 -54	 160.3	4-01-00-00485 0.7609^{25}	 1.4282^{20}
6845	Hexane, 2,2,3,4-tetramethyl- 2,2,3,4-Tetramethylhexane	$C_{10}H_{22}$ 142.28	52897-08-2	 157	4-01-00-00485 0.7513^{25}	 1.4216^{20}
6846	Hexane, 2,2,3,5-tetramethyl- 2,2,3,5-Tetramethylhexane	$C_{10}H_{22}$ 142.28	52897-09-3	 148.4	4-01-00-00485 0.7336^{25}	 1.4142^{20}
6847	Hexane, 2,2,4,4-tetramethyl- 2,2,4,4-Tetramethylhexane	$C_{10}H_{22}$ 142.28	51750-65-3	 153.8	4-01-00-00486 0.7424^{25}	 1.4208^{20}
6848	Hexane, 2,2,4,5-tetramethyl- 2,2,4,5-Tetramethylhexane	$C_{10}H_{22}$ 142.28	16747-42-5	 147.9	4-01-00-00486 0.7316^{25}	 1.4132^{20}
6849	Hexane, 2,2,5,5-tetramethyl- 2,2,5,5-Tetramethylhexane	$C_{10}H_{22}$ 142.28	1071-81-4 -12.6	 137.4	4-01-00-00486 0.7148^{25}	 1.4055^{20}
6850	Hexane, 2,3,3,4-tetramethyl- 2,3,3,4-Tetramethylhexane	$C_{10}H_{22}$ 142.28	52897-10-6	 164.6	4-01-00-00486 0.7656^{25}	 1.4298^{20}
6851	Hexane, 2,3,3,5-tetramethyl- 2,3,3,5-Tetramethylhexane	$C_{10}H_{22}$ 142.28	52897-11-7	 153.1	4-01-00-00486 0.7449^{25}	 1.4196^{20}
6852	Hexane, 2,3,4,4-tetramethyl- 2,3,4,4-Tetramethylhexane	$C_{10}H_{22}$ 142.28	52897-12-8	 161.6	4-01-00-00487 0.7586^{25}	 1.4267^{20}
6853	Hexane, 2,3,4,5-tetramethyl- 2,3,4,5-Tetramethylhexane	$C_{10}H_{22}$ 142.28	52897-15-1	 156.2	4-01-00-00487 0.7456^{25}	 1.4204^{20}
6854	Hexane, 3,3,4,4-tetramethyl- 3,3,4,4-Tetramethylhexane	$C_{10}H_{22}$ 142.28	5171-84-6	 170.0	3-01-00-00534 0.7789^{25}	 1.4368^{20}
6855	Hexane, 1,1'-thiobis-	$C_{12}H_{26}S$ 202.40	6294-31-1	 230; 136^{20}	4-01-00-01706 0.8411^{20}	 1.4586^{20}
6856	1-Hexanethiol Hexyl mercaptan	$C_6H_{14}S$ 118.24	111-31-9 -81	 151	4-01-00-01705 0.8424^{20}	H_2O 1; EtOH 4; eth 4 1.4496^{20}
6857	2-Hexanethiol	$C_6H_{14}S$ 118.24	1679-06-7 -147	 142	4-01-00-01711 0.8345^{20}	H_2O 1; EtOH 3; eth 3; bz 3 1.4451^{20}
6858	3-Hexanethiol 3-Mercaptohexane	$C_6H_{14}S$ 118.24	1633-90-5	 57^{25}	4-01-00-01713 0.9206^{20}	EtOH 3 1.4496^{20}
6859	Hexane, 2,2,3-trimethyl- 2,2,3-Trimethylhexane	C_9H_{20} 128.26	16747-25-4	 133.6	4-01-00-00459 0.7257^{25}	 1.4106^{20}
6860	Hexane, 2,2,4-trimethyl- 2,2,4-Trimethylhexane	C_9H_{20} 128.26	16747-26-5 -120	 126.5	4-01-00-00459 0.711^{20}	 1.4033^{20}
6861	Hexane, 2,2,5-trimethyl- 2,2,5-Trimethylhexane	C_9H_{20} 128.26	3522-94-9 -105.7	 124.0	4-01-00-00460 0.7072^{20}	H_2O 1; EtOH 4; eth 4; ace 4 1.3997^{20}
6862	Hexane, 2,3,3-trimethyl- 2,3,3-Trimethylhexane	C_9H_{20} 128.26	16747-28-7 -116.8	 137.7	4-01-00-00461 0.7345^{25}	 1.4141^{20}
6863	Hexane, 2,3,4-trimethyl- 2,3,4-Trimethylhexane	C_9H_{20} 128.26	921-47-1	 139.1	4-01-00-00461 0.7354^{25}	 1.4144^{20}
6864	Hexane, 2,3,5-trimethyl- 2,3,5-Trimethylhexane	C_9H_{20} 128.26	1069-53-0 -127.9	 131.4	4-01-00-00461 0.7218^{20}	 1.4051^{20}
6865	Hexane, 2,4,4-trimethyl- 2,4,4-Trimethylhexane	C_9H_{20} 128.26	16747-30-1 -113.4	 130.7	4-01-00-00461 0.7201^{25}	 1.4074^{20}
6866	Hexane, 3,3,4-trimethyl- 3,3,4-Trimethylhexane	C_9H_{20} 128.26	16747-31-2 -101.2	 140.5	4-01-00-00462 0.7414^{25}	 1.4178^{20}
6867	1,2,5-Hexanetriol	$C_6H_{14}O_3$ 134.18	10299-30-6	 181^{10}; 149^{3}	4-01-00-02784 1.1012^{20}	H_2O 4; EtOH 4
6868	1,2,6-Hexanetriol 1,2,6-Trihydroxyhexane	$C_6H_{14}O_3$ 134.18	106-69-4	 170^{3}	4-01-00-02784 1.03^{20}	 1.58^{20}
6869	2,3,4-Hexanetriol	$C_6H_{14}O_3$ 134.18	93972-93-1 47	 256.5	3-01-00-02349	H_2O 4; EtOH 4
6870	Hexanoic acid Caproic acid	$C_6H_{12}O_2$ 116.16	142-62-1 -3	1760 205.2	4-02-00-00917 0.9274^{20}	H_2O 1; EtOH 3; eth 3; chl 3 1.4163^{20}
6871	Hexanoic acid, 6-(acetylamino)- ε-Acetamidocaproic acid	$C_8H_{15}NO_3$ 173.21	57-08-9 104.5	38	4-04-00-02700	
6872	Hexanoic acid, 2-acetyl-, ethyl ester	$C_{10}H_{18}O_3$ 186.25	1540-29-0	 221.5	4-03-00-01616 0.9523^{20}	ace 4; eth 4 1.4301^{20}
6873	Hexanoic acid, 6-amino- ε-Aminocaproic acid	$C_6H_{13}NO_2$ 131.17	60-32-2 205	442	4-04-00-02695	H_2O 4; EtOH 1; MeOH 2
6874	Hexanoic acid, anhydride	$C_{12}H_{22}O_3$ 214.30	2051-49-2 -41	 255 dec	4-02-00-00928 0.9240^{15}	eth 4; EtOH 4 1.4297^{20}
6875	Hexanoic acid, 2-bromo-, (±)-	$C_6H_{11}BrO_2$ 195.06	2681-83-6 4	 240; 140^{23}	3-02-00-00736 1.2810^{33}	EtOH 3; eth 3
6876	Hexanoic acid, 6-bromo-	$C_6H_{11}BrO_2$ 195.06	4224-70-8 35	 $165\text{-}70^{20}$	4-02-00-00940	peth 4
6877	Hexanoic acid, 2-bromo-, ethyl ester, (±)-	$C_8H_{15}BrO_2$ 223.11	63927-44-6	 208.7	4-02-00-00938 1.2210^{25}	EtOH 3
6878	Hexanoic acid, 5-bromo-, ethyl ester Caproic acid, 5-bromo,ethyl ester	$C_8H_{15}BrO_2$ 223.11	90202-08-7	 121^{21}; 70^{3}	3-02-00-00737 1.1908^{25}	 1.4525^{25}
6879	Hexanoic acid, 6-bromo-, ethyl ester	$C_8H_{15}BrO_2$ 223.11	25542-62-5 33	 $126\text{-}7^{21}$	4-02-00-00940 1.238^{23}	 1.4566^{21}
6880	Hexanoic acid, butyl ester	$C_{10}H_{20}O_2$ 172.27	626-82-4 -64.3	 208	4-02-00-00922 0.8653^{20}	H_2O 1; EtOH 3; eth 5 1.4152^{20}
6881	Hexanoic acid, 2-cyano-, ethyl ester 2-Cyanocaproic acid, ethyl ester	$C_9H_{15}NO_2$ 169.22	7391-39-1	 105^{9}	3-02-00-01746 0.988^{15}	 1.4248^{20}
6882	Hexanoic acid, 2-ethyl- 2-Ethylhexanoic acid	$C_8H_{16}O_2$ 144.21	149-57-5	 228; 120^{13}	3-02-00-00804 0.9031^{25}	H_2O 3; EtOH 2; eth 3; ctc 3 1.4241^{20}

No.	Name	Synonym	Mol. Form.	Mol. Wt.	CAS RN	mp/°C	Merck No.	bp/°C	Beil. Ref.	den/g cm^{-3}	Solubility	n_D
6883	Hexanoic acid, 3-ethyl-	3-Ethylhexanoic acid	$C_8H_{16}O_2$	144.21	41065-91-2			106^{5}	4-02-00-01008	0.911^{30}		1.4287^{25}
6884	Hexanoic acid, ethyl ester	Ethyl hexanoate	$C_8H_{16}O_2$	144.21	123-66-0	-67	3735	167	4-02-00-00921	0.873^{20}	eth 4; EtOH 4	1.4073^{20}
6885	Hexanoic acid, 2-ethyl-, ethyl ester		$C_{10}H_{20}O_2$	172.27	2983-37-1			90^{28}	4-02-00-01004	0.8586^{25}		1.4123^{25}
6886	Hexanoic acid, 2-ethyl-3-methyl-	2-Ethyl-3-methylhexanoic acid	$C_9H_{18}O_2$	158.24	74581-94-5			231	3-02-00-00832	0.9060^{25}		1.4302^{25}
6887	Hexanoic acid, 3-ethyl-3-methyl-	3-Ethyl-3-methylhexanoic acid	$C_9H_{18}O_2$	158.24	50902-82-4			$139\text{-}40^{22}$	4-02-00-01036	0.9205^{25}		1.4375^{25}
6888	Hexanoic acid, 6-fluoro-	Caproic acid, 6-fluoro	$C_6H_{11}FO_2$	134.15	373-05-7			138^{28}; $67^{0.6}$	4-02-00-00932		eth 4; EtOH 4	1.4166^{25}
6889	Hexanoic acid, heptyl ester		$C_{13}H_{26}O_2$	214.35	6976-72-3	-34.4		261	4-02-00-00923	0.8611^{20}	ace 4; bz 4; eth 4; EtOH 4	1.4293^{15}
6890	Hexanoic acid, hexyl ester		$C_{12}H_{24}O_2$	200.32	6378-65-0	-55		246	4-02-00-00922	0.865^{18}	ace 4; bz 4; eth 4; EtOH 4	1.4264^{15}
6891	Hexanoic acid, 2-hydroxy-, (±)-		$C_6H_{12}O_3$	132.16	636-36-2	61.3		270	4-03-00-00838		H_2O 4; EtOH 4; eth 3; chl 4	
6892	Hexanoic acid, 3-hydroxy-2,2-dimethyl-, β-lactone	Caproic acid, 2,2-dimethyl-3-hydroxy, β-lactone	$C_8H_{14}O_2$	142.20	90112-82-6	143.5		$93\text{-}100^{14}$			DMSO 2	1.4260^{20}
6893	Hexanoic acid, 2-methyl-, (±)-	Caproic acid, 2-methyl (*DL*)	$C_7H_{14}O_2$	130.19	22160-12-9			215.5	4-02-00-00969	0.9612^{20}	H_2O 5; EtOH 5; eth 5; ace 5	1.4195^{20}
6894	Hexanoic acid, 3-methyl-		$C_7H_{14}O_2$	130.19	3780-58-3			213; 116^{15}	2-02-00-00298	0.9187^{20}		1.4222^{20}
6895	Hexanoic acid, 4-methyl-, (±)-		$C_7H_{14}O_2$	130.19	22160-41-4	-80		217.5	4-02-00-00974	0.9215^{20}	H_2O 2; EtOH 3; eth 3; ace 3	1.4211^{20}
6896	Hexanoic acid, 5-methyl-		$C_7H_{14}O_2$	130.19	628-46-6	<-25		216	4-02-00-00970	0.9138^{21}	ace 4; bz 4; eth 4; EtOH 4	1.4220^{20}
6897	Hexanoic acid, 3-methylbutyl ester		$C_{11}H_{22}O_2$	186.29	2198-61-0			225.5	3-02-00-00727	0.861^{20}	H_2O 1; EtOH 3; eth 3	
6898	Hexanoic acid, methyl ester	Methyl caproate	$C_7H_{14}O_2$	130.19	106-70-7	-71		149.5	4-02-00-00921	0.8846^{20}	H_2O 1; EtOH 4; eth 4; ace 3	1.4049^{20}
6899	Hexanoic acid, 3-methyl-, ethyl ester	Ethyl 3-methylhexanoate	$C_9H_{18}O_2$	158.24	41692-47-1			177	2-02-00-00298	0.8679^{20}		1.4119^{20}
6900	Hexanoic acid, 4-methyl-, ethyl ester	Ethyl 4-methylhexanoate	$C_9H_{18}O_2$	158.24	1561-10-0			180	4-02-00-00974	0.8708^{20}		1.4051^{20}
6901	Hexanoic acid, 4-methyl-5-oxo-		$C_7H_{12}O_3$	144.17	6818-07-1			160^{15}; 140^{5}	4-03-00-01601	0.9803^{20}		1.4320^{20}
6902	Hexanoic acid, 4-methyl-5-oxo-, ethyl ester	Valeric acid, 4-acetyl, ethyl ester	$C_9H_{16}O_3$	172.22	53068-88-5			114^{15}	4-03-00-01601	0.9803^{20}		1.4282^{25}
6903	Hexanoic acid, 4-methyl-5-oxo-, methyl ester	Valeric acid, 4-acetyl, methyl ester	$C_8H_{14}O_3$	158.20	36045-56-4			$113\text{-}7^{16}$	4-03-00-01601	1.0988^{20}		1.4288^{20}
6904	Hexanoic acid, octyl ester		$C_{14}H_{28}O_2$	228.38	4887-30-3	-28		275	3-02-00-00727	0.8603^{20}	ace 4; bz 4; eth 4; EtOH 4	1.4326^{15}
6905	Hexanoic acid, 5-oxo-		$C_6H_{10}O_3$	130.14	3128-06-1	13.5		274.5	4-03-00-01583	1.09^{25}	H_2O 3; EtOH 3; eth 3; ctc 2	1.4451^{20}
6906	Hexanoic acid, 5-oxo-, ethyl ester	Butyric acid, 4-acetyl, ethyl ester	$C_8H_{14}O_3$	158.20	13984-57-1			221.5	4-03-00-01583	0.989^{25}		1.4277^{20}
6907	Hexanoic acid, pentyl ester	Amyl caproate	$C_{11}H_{22}O_2$	186.29	540-07-8	-47	640	226	4-02-00-00922	0.8612^{25}	EtOH 3; eth 3; ace 3; ctc 2	1.4202^{25}
6908	Hexanoic acid, 1,2,3-propanetriyl ester		$C_{21}H_{38}O_6$	386.53	621-70-5	-60		>200	4-02-00-00926	0.9867^{20}	H_2O 1; EtOH 5; eth 5; ace 4	1.4427^{20}
6909	Hexanoic acid, propyl ester		$C_9H_{18}O_2$	158.24	626-77-7	-68.7		187	4-02-00-00922	0.8672^{20}	eth 4; EtOH 4	1.4170^{20}
6910	Hexanoic acid, 3,5,6-trichloro-2,2,3,4,4,5,6,6-octafluoro-	Hexanoic acid, 2,2,3,4,4,5,6,6-octafluoro-3,5,6-trichloro	$C_6HCl_3F_8O_2$	363.42	2106-54-9			$105\text{-}6^{5}$	4-02-00-00937	1.860^{20}		1.3903^{20}
6911	1-Hexanol	Caproyl alcohol	$C_6H_{14}O$	102.18	111-27-3	-44.6	4615	157.6	4-01-00-01694	0.8136^{20}	H_2O 2; EtOH 3; eth 5; ace 3	1.4178^{20}
6912	2-Hexanol, (±)-		$C_6H_{14}O$	102.18	20281-86-1			140	4-01-00-01708	0.8159^{20}	H_2O 2; EtOH 3; eth 3; ctc 2	1.4144^{20}
6913	2-Hexanol, (*R*)-		$C_6H_{14}O$	102.18	26549-24-6			137	3-01-00-01663	0.8178^{18}	H_2O 2; EtOH 3; eth 3	
6914	3-Hexanol, (±)-		$C_6H_{14}O$	102.18	17015-11-1			135	4-01-00-01711	0.8182^{20}	H_2O 2; EtOH 3; eth 5; ace 3	1.4167^{20}
6915	3-Hexanol, (*R*)-	3-Hexanol (*l*)	$C_6H_{14}O$	102.18	13471-42-6			135		0.8213^{20}	H_2O 2; EtOH 3; eth 5; ace 3	1.4140^{20}
6916	1-Hexanol, 6-chloro-		$C_6H_{13}ClO$	136.62	2009-83-8			107^{12}	4-01-00-01704	1.0241^{20}	H_2O 2; EtOH 4; eth 4	1.4550^{20}
6917	2-Hexanol, 1-chloro-	1-Chloro-2-hexanol	$C_6H_{13}ClO$	136.62	52802-07-0			$73\text{-}5^{12}$	4-01-00-01710	1.0139^{20}	ace 4; eth 4; EtOH 4	1.4478^{20}
6918	3-Hexanol, 1-chloro-	1-Chloro-3-hexanol	$C_6H_{13}ClO$	136.62	52418-81-2			120^{35}; 78^{6}	4-01-00-01712	1.003^{25}	ace 4; bz 4; eth 4; EtOH 4	1.446^{25}
6919	3-Hexanol, 5-chloro-		$C_6H_{13}ClO$	136.62	58588-28-6			$78\text{-}9^{13}$	4-01-00-01712	1.0012^{15}	ace 4; bz 4; eth 4; EtOH 4	1.4433^{19}
6920	1-Hexanol, 3,5-dimethoxy-		$C_8H_{18}O_3$	162.23	90952-10-6			114^{13}	4-01-00-02785	0.9631^{25}		1.4329^{25}

No.	Name Synonym	Mol. Form. Mol. Wt.	CAS RN mp/°C	Merck No. bp/°C	Beil. Ref. den/g cm^{-3}	Solubility n_D
6921	1-Hexanol, 2,5-dimethyl-	$C_8H_{18}O$ 130.23	6886-16-4	 179.5	0-01-00-00422 0.8280^{20}	 1.5095^{17}
6922	2-Hexanol, 2,3-dimethyl- 2,3-Dimethyl-2-hexanol	$C_8H_{18}O$ 130.23	19550-03-9	 160; 62^{10}	3-01-00-01737 0.8365^{20}	 1.4335^{20}
6923	2-Hexanol, 2,5-dimethyl- 2,5-Dimethyl-2-hexanol	$C_8H_{18}O$ 130.23	3730-60-7	 154.5	4-01-00-01792 0.8227^{20}	 1.4208^{20}
6924	2-Hexanol, 3,4-dimethyl-	$C_8H_{18}O$ 130.23	19550-05-1	 165.5	3-01-00-01740 0.840^{15}	 1.4325^{15}
6925	3-Hexanol, 2,2-dimethyl- Propyl-*tert*-butylcarbinol	$C_8H_{18}O$ 130.23	4209-90-9	 156	3-01-00-01737 0.8342^{20}	 1.4275^{20}
6926	3-Hexanol, 2,3-dimethyl-	$C_8H_{18}O$ 130.23	4166-46-5	 61-2^{18}	4-01-00-01791 0.8371^{20}	 1.4309^{20}
6927	3-Hexanol, 2,5-dimethyl- 2,5-Dimethyl-3-hexanol	$C_8H_{18}O$ 130.23	19550-07-3	 157.5	4-01-00-01793 0.8212^{20}	 1.4246^{20}
6928	3-Hexanol, 3,5-dimethyl-, (±)- 3-Hexanol, 3,5-dimethyl (*DL*)	$C_8H_{18}O$ 130.23	19113-78-1	 152	4-01-00-01792 0.8373^{20}	bz 4; eth 4; EtOH 4 1.4300^{20}
6929	3-Hexanol, 4,4-dimethyl-	$C_8H_{18}O$ 130.23	19550-09-5	 160.4	4-01-00-01793 0.8341^{20}	 1.4345^{20}
6930	1-Hexanol, 2-ethyl- 2-Ethyl-1-hexanol	$C_8H_{18}O$ 130.23	104-76-7 -70	3764 184.6	4-01-00-01783 0.8319^{25}	H_2O 1; EtOH 3; eth 3; ace 3 1.4300^{20}
6931	3-Hexanol, 3-ethyl-	$C_8H_{18}O$ 130.23	597-76-2	 160	4-01-00-01790 0.8373^{20}	H_2O 1; EtOH 3; eth 3; ace 3 1.4300^{20}
6932	1-Hexanol, 2-ethyl-3-methyl-	$C_9H_{20}O$ 144.26	66794-04-5	 193; 83^{10}	4-02-00-02107 0.8486^{20}	 1.4455^{20}
6933	2-Hexanol, 3-ethyl-2-methyl-	$C_9H_{20}O$ 144.26	66794-02-3	 71.5^{14}	4-01-00-01812 0.8334^{25}	 1.4331^{25}
6934	3-Hexanol, 3-ethyl-2-methyl-	$C_9H_{20}O$ 144.26	66794-03-4	 177.5	4-01-00-01811 0.8445^{25}	 1.4369^{25}
6935	3-Hexanol, 3-ethyl-5-methyl-	$C_9H_{20}O$ 144.26	597-77-3	 172; 68^{18}	3-01-00-01755 0.8396^{22}	eth 4; EtOH 4 1.4346^{13}
6936	1-Hexanol, 6-fluoro-	$C_6H_{13}FO$ 120.17	373-32-0	 85-6^{14}	4-01-00-01703 0.975^{20}	eth 4; EtOH 4 1.4141^{25}
6937	1-Hexanol, 2-methyl-, (±)- 1-Hexanol, 2-methyl (*DL*)	$C_7H_{16}O$ 116.20	111768-04-8	 164; 71^{15}	4-01-00-01745 0.8270^{20}	eth 4; EtOH 4 1.4226^{20}
6938	1-Hexanol, 2-methyl-, (*R*)- (*R*)-2-Methyl-1-hexanol	$C_7H_{16}O$ 116.20	66050-98-4	 164; 71^{15}	 0.8313^{13}	eth 4; EtOH 4 1.4245^{17}
6939	1-Hexanol, 3-methyl-, (±)-	$C_7H_{16}O$ 116.20	111768-08-2	 169; 91^{13}	4-01-00-01748 0.8258^{20}	H_2O 1; EtOH 4; eth 4; ace 3 1.4245^{20}
6940	1-Hexanol, 4-methyl-, (±)- 1-Hexanol, 4-methyl (*DL*)	$C_7H_{16}O$ 116.20	111768-05-9	 173	4-01-00-01749 0.8239^{20}	ace 4; bz 4; eth 4; EtOH 4 1.4219^{20}
6941	1-Hexanol, 5-methyl- 5-Methyl-1-hexanol	$C_7H_{16}O$ 116.20	627-98-5	 170; 53-5^{15}	4-01-00-01748 0.8119^{20}	eth 4; EtOH 4 1.4175^{20}
6942	2-Hexanol, 2-methyl-	$C_7H_{16}O$ 116.20	625-23-0	 143	4-01-00-01745 0.8119^{20}	H_2O 2; EtOH 5; eth 5 1.4175^{20}
6943	2-Hexanol, 3-methyl- 3-Methyl-2-hexanol	$C_7H_{16}O$ 116.20	2313-65-7	 79-81^{52}	 0.8820^{25}	H_2O 1; EtOH 4; eth 4; ace 3 1.4198^{18}
6944	2-Hexanol, 5-methyl- 5-Methyl-2-hexanol	$C_7H_{16}O$ 116.20	627-59-8	 151; 78^{28}	4-01-00-01747 0.814^{20}	H_2O 1; EtOH 3; eth 3 1.4180^{20}
6945	3-Hexanol, 2-methyl-, (±)- (±)-2-Methyl-3-hexanol	$C_7H_{16}O$ 116.20	100295-82-7	 143	4-01-00-01746 0.8407^{20}	H_2O 1; EtOH 3; eth 3; ctc 3 1.4149^{20}
6946	3-Hexanol, 3-methyl- 2-Ethyl-2-pentanol	$C_7H_{16}O$ 116.20	597-96-6	 143	4-01-00-01749 0.8233^{20}	H_2O 2; EtOH 3; eth 3; ctc 3 1.4231^{20}
6947	3-Hexanol, 5-methyl-, (±)- (±)-5-Methyl-3-hexanol	$C_7H_{16}O$ 116.20	100295-83-8	 148	4-01-00-01747 0.827^{25}	H_2O 1; EtOH 3; eth 3; ctc 2 1.4128^{20}
6948	1-Hexanol, 5-methyl-2-(1-methylethyl)- 2-Isopropyl-5-methyl-1-hexanol	$C_{10}H_{22}O$ 158.28	2051-33-4	 211	4-01-00-01833 0.8345^{20}	H_2O 1; EtOH 3; eth 4 1.4369^{20}
6949	3-Hexanol, 1-phenyl-	$C_{12}H_{18}O$ 178.27	2180-43-0 34	 170^{50}; 146^{16}	4-06-00-03418 0.9525^{20}	
6950	3-Hexanol, 2,2,5,5-tetramethyl-	$C_{10}H_{22}O$ 158.28	55073-86-4 52.5	 170	3-01-00-01772	ace 4; eth 4; EtOH 4; peth 4
6951	1-Hexanol, 3,5,5-trimethyl-	$C_9H_{20}O$ 144.26	3452-97-9	 194	4-01-00-01813 0.8236^{25}	 1.4300^{25}
6952	2-Hexanol, 2,3,4-trimethyl-	$C_9H_{20}O$ 144.26	21102-13-6	 57^{5}	3-01-00-01756 0.853^{15}	ace 4; eth 4; EtOH 4; peth 4 1.4415^{15}
6953	3-Hexanol, 2,2,3-trimethyl-	$C_9H_{20}O$ 144.26	5340-41-0	 170	3-01-00-01755 0.8474^{20}	EtOH 3; eth 3; ace 3 1.4402^{20}
6954	3-Hexanol, 2,3,5-trimethyl-	$C_9H_{20}O$ 144.26	65927-60-8	 72^{21}	4-01-00-01813 0.8256^{20}	ace 4; eth 4; EtOH 4 1.4321^{20}
6955	3-Hexanol, 2,4,4-trimethyl-	$C_9H_{20}O$ 144.26	66793-92-8	 170	3-01-00-01756 0.8489^{20}	EtOH 3; eth 3; ace 3 1.4395^{20}
6956	3-Hexanol, 2,5,5-trimethyl-	$C_9H_{20}O$ 144.26	66793-72-4	 77^{32}	3-01-00-01756 0.8250^{20}	ace 4; eth 4; EtOH 4 1.4286^{20}
6957	3-Hexanol, 3,4,4-trimethyl-	$C_9H_{20}O$ 144.26	66793-74-6	 165.5	0-01-00-00425 0.8322^{21}	EtOH 3; eth 3; ace 3 1.4341^{21}
6958	3-Hexanol, 3,5,5-trimethyl- 3,5,5-Trimethylhexan-3-ol	$C_9H_{20}O$ 144.26	66810-87-5	 62^{14}	3-01-00-01755 0.8350^{20}	ace 4; eth 4; EtOH 4 1.4352^{20}
6959	2-Hexanone Butyl methyl ketone	$C_6H_{12}O$ 100.16	591-78-6 -55.5	5955 127.6	4-01-00-03298 0.8113^{20}	H_2O 2; EtOH 5; eth 5; ace 3 1.4007^{20}
6960	3-Hexanone Ethyl propyl ketone	$C_6H_{12}O$ 100.16	589-38-8 -55.5	 123.5	4-01-00-03301 0.8118^{20}	H_2O 2; EtOH 5; eth 5; ace 3 1.4004^{20}

No.	Name Synonym	Mol. Form. Mol. Wt.	CAS RN mp/°C	Merck No. bp/°C	Beil. Ref. den/g cm^{-3}	Solubility n_D
6961	2-Hexanone, 6-bromo-	$C_6H_{11}BrO$ 179.06	10226-29-6	216; 136^{90}	3-01-00-02829 1.3494^{0}	eth 4; EtOH 4
6962	1-Hexanone, 1-(2,4-dihydroxyphenyl)-	$C_{12}H_{16}O_3$ 208.26	3144-54-5 57	343	4-08-00-01888	H_2O 1; EtOH 3; eth 3; ace 3
6963	2-Hexanone, 3,3-dimethyl- 3,3-Dimethyl-2-hexanone	$C_8H_{16}O$ 128.21	26118-38-7	149	4-01-00-03350 0.838^{0}	ace 4; eth 4; EtOH 4 1.4098^{20}
6964	2-Hexanone, 3,4-dimethyl- 3,4-Dimethyl-2-hexanone	$C_8H_{16}O$ 128.21	19550-10-8	158	3-01-00-02882 0.8295^{22}	EtOH 3; eth 3; ace 3 1.4193^{20}
6965	3-Hexanone, 2,2-dimethyl- *tert*-Butyl propyl ketone	$C_8H_{16}O$ 128.21	5405-79-8	147	4-01-00-03347 0.8105^{25}	EtOH 3; eth 3; ace 3 1.4119^{20}
6966	3-Hexanone, 2,5-dimethyl-	$C_8H_{16}O$ 128.21	1888-57-9	147.5	4-01-00-03349 0.8269^{0}	ace 4; eth 4; EtOH 4 1.4049^{20}
6967	3-Hexanone, 4,4-dimethyl-	$C_8H_{16}O$ 128.21	19550-14-2	148	4-01-00-03350 0.8285^{20}	bz 4; EtOH 4; chl 4 1.4208^{25}
6968	3-Hexanone, 6-(dimethylamino)-4,4-diphenyl- Normethadone	$C_{20}H_{25}NO$ 295.42	467-85-6	6628 $164\text{-}7^{3}$	4-14-00-00297	
6969	2-Hexanone, 5-hydroxy- 5-Hydroxy-2-hexanone	$C_6H_{12}O_2$ 116.16	56745-61-0	203^{270}; 80^{10}	4-01-00-04019 0.9626^{25}	H_2O 4; eth 4; EtOH 4 1.4312^{25}
6970	2-Hexanone, 6-hydroxy-	$C_6H_{12}O_2$ 116.16	21856-89-3	227	4-01-00-04020 0.9886^{15}	eth 4; EtOH 4 1.4494
6971	3-Hexanone, 4-hydroxy-	$C_6H_{12}O_2$ 116.16	4984-85-4	133^{227}; 73^{20}	4-01-00-04021 0.956^{21}	ace 4; EtOH 4 1.4340^{21}
6972	3-Hexanone, 5-hydroxy- 5-Hydroxy-3-hexanone	$C_6H_{12}O_2$ 116.16	33683-44-2	$75\text{-}8^{12}$	4-01-00-04021 0.950^{20}	1.4280^{25}
6973	3-Hexanone, 2-hydroxy-2-methyl- 2-Hexanol-3-one, 2-methyl	$C_7H_{14}O_2$ 130.19	18905-91-4	$60\text{-}2^{15}$	4-01-00-04035 0.899^{18}	1.419^{18}
6974	3-Hexanone, 5-hydroxy-2-methyl- 2-Hexanol-4-one, 5-methyl	$C_7H_{14}O_2$ 130.19	59357-07-2	$72\text{-}3^{9}$	4-01-00-04035 0.929^{20}	1.4278^{20}
6975	3-Hexanone, 4-hydroxy-2,2,5,5-tetramethyl- Pivaloin	$C_{10}H_{20}O_2$ 172.27	815-66-7 81	80^{10}	4-01-00-04060	eth 4
6976	2-Hexanone, 3-methyl- 3-Methyl-2-hexanone	$C_7H_{14}O$ 114.19	2550-21-2	143.5	3-01-00-02863 0.828^{25}	ace 4; bz 4; eth 4; EtOH 4 1.4035^{20}
6977	2-Hexanone, 4-methyl- Methyl 2-methylbutyl ketone	$C_7H_{14}O$ 114.19	105-42-0	144.5	4-01-00-03329 0.8130^{20}	H_2O 2; EtOH 4; eth 4; ace 4 1.4081^{24}
6978	2-Hexanone, 5-methyl- Methyl isopentyl ketone	$C_7H_{14}O$ 114.19	110-12-3	144	4-01-00-03329 0.888^{20}	H_2O 2; EtOH 5; eth 5; ace 4 1.4062^{20}
6979	3-Hexanone, 2-methyl-	$C_7H_{14}O$ 114.19	7379-12-6	135	4-01-00-03328 0.8091^{20}	EtOH 3; eth 3; ace 4; chl 3 1.4042^{20}
6980	3-Hexanone, 4-methyl-	$C_7H_{14}O$ 114.19	17042-16-9	134.5	4-01-00-03329 0.8162^{20}	ace 4; bz 4; eth 4; EtOH 4 1.4069^{20}
6981	3-Hexanone, 5-methyl-	$C_7H_{14}O$ 114.19	623-56-3	135	4-01-00-03328 0.8090^{20}	H_2O 1; EtOH 5; eth 5; ctc 3 1.4047^{20}
6982	2-Hexanone, 5-methyl-, oxime	$C_7H_{15}NO$ 129.20	624-44-2	195.5	0-01-00-00701 0.8881^{20}	chl 2 1.4448^{20}
6983	1-Hexanone, 1-phenyl-	$C_{12}H_{16}O$ 176.26	942-92-7 27	265	4-07-00-00754 0.9576^{20}	H_2O 2; EtOH 3; eth 3; ace 3 1.5027^{25}
6984	2-Hexanone, 3-phenyl-	$C_{12}H_{16}O$ 176.26	6306-30-5	235.5; 114^{13}	3-07-00-01160 0.970^{0}	1.5020^{20}
6985	3-Hexanone, 2-phenyl- 1-Phenylethyl propyl ketone	$C_{12}H_{16}O$ 176.26	65248-43-3	241	2-07-00-00258 0.941^{25}	1.4988^{25}
6986	Hexanoyl chloride Caproyl chloride	$C_6H_{11}ClO$ 134.61	142-61-0 -87	1763 153	4-02-00-00928 0.9784^{20}	eth 3; ace 3 1.4264^{20}
6987	Hexanoyl chloride, 2-ethyl-	$C_8H_{15}ClO$ 162.66	760-67-8	$67\text{-}8^{11}$	3-02-00-00805 0.939^{25}	1.4335^{20}
6988	Hexanoyl chloride, 3-methyl- Caproyl chloride, 3-methyl	$C_7H_{13}ClO$ 148.63	57323-93-0	163; 82^{50}	3-02-00-00777 0.967^{20}	bz 4 1.4293^{25}
6989	Hexanoyl chloride, 4-methyl- Caproyl chloride, 4-methyl	$C_7H_{13}ClO$ 148.63	50599-73-0	167	2-02-00-00299 0.9677^{20}	eth 4
6990	Hexasiloxane, tetradecamethyl- Tetradecamethylhexasiloxane	$C_{14}H_{42}O_5Si_6$ 459.00	107-52-8 -59	9131 245.5	4-04-00-04127 0.8910^{20}	bz 4 1.3948^{20}
6991	Hexatriacontane	$C_{36}H_{74}$ 506.98	630-06-8 76.5	298.4^{3}	4-01-00-00599 0.7803^{80}	1.4397^{80}
6992	1,3,5-Hexatriene, (*E*)-	C_6H_8 80.13	821-07-8 -12	78.5	4-01-00-01093 0.7369^{15}	H_2O 1; EtOH 3; ace 3; chl 3 1.5135^{20}
6993	1,3,5-Hexatriene, (*Z*)-	C_6H_8 80.13	2612-46-6 -12	78	0.7175^{20}	H_2O 1; EtOH 3; ace 3; chl 3 1.4577^{20}
6994	1,3,5-Hexatriene, 2,5-dimethyl-	C_8H_{12} 108.18	4916-63-6 -9	146	4-01-00-01102 0.7822^{20}	H_2O 1; ace 3; MeOH 3; lig 3 1.5122^{20}
6995	Hexazinone 1,3,5-Triazine-2,4(1H,3H)-dione, 3-cyclohexyl-6-(dimethylamino)-1-methyl-	$C_{12}H_{20}N_4O_2$ 252.32	51235-04-2 99	4617 dec	1.25	
6996	2-Hexenal, (*E*)-	$C_6H_{10}O$ 98.14	6728-26-3	146.5; 50^{20}	4-01-00-03468 0.8491^{20}	1.4480^{20}
6997	3-Hexenal, (*E*)-	$C_6H_{10}O$ 98.14	69112-21-6	$42\text{-}3^{28}$	4-01-00-03469 0.8455^{22}	H_2O 1; eth 3; ace 3 1.4275^{21}
6998	3-Hexenal, (*Z*)-	$C_6H_{10}O$ 98.14	6789-80-6	121	4-01-00-03469 0.8533^{22}	1.4300^{21}
6999	1-Hexene	C_6H_{12} 84.16	592-41-6 -139.7	63.4	4-01-00-00828 0.6731^{20}	bz 4; eth 4; EtOH 4; peth 4 1.3837^{20}

No.	Name	Synonym	Mol. Form.	Mol. Wt.	CAS RN	mp/°C	Merck No.	bp/°C	Beil. Ref.	den/g cm^{-3}	Solubility	n_D
7000	2-Hexene, (*E*)-	*trans*-2-Hexene	C_6H_{12}	84.16	4050-45-7	-133		67.9	4-01-00-00834	0.6732^{25}	H_2O 1; EtOH 3; eth 3; bz 3	1.3936^{20}
7001	2-Hexene, (*Z*)-	*cis*-2-Hexene	C_6H_{12}	84.16	7688-21-3	-141.1		68.8	4-01-00-00833	0.6869^{20}	H_2O 1; EtOH 3; eth 3; bz 3	1.3979^{20}
7002	3-Hexene, (*E*)-		C_6H_{12}	84.16	13269-52-8	-115.4		67.1	4-01-00-00837	0.6772^{20}	H_2O 1; EtOH 3; eth 3; bz 3	1.3943^{20}
7003	3-Hexene, (*Z*)-		C_6H_{12}	84.16	7642-09-3	-137.8		66.4	4-01-00-00837	0.6796^{20}	H_2O 1; EtOH 3; eth 3; bz 3	1.3947^{20}
7004	1-Hexene, 1-chloro-	1-Chloro-1-hexene	$C_6H_{11}Cl$	118.61	22922-67-4			121	4-01-00-00831	0.8872^{22}	ace 4; bz 4; eth 4; chl 4	1.4300^{22}
7005	1-Hexene, 2-chloro-	2-Chloro-1-hexene	$C_6H_{11}Cl$	118.61	10124-73-9			113	3-01-00-00803	0.8886^{25}	ace 4; bz 4; chl 4	1.4278^{25}
7006	1-Hexene, 5-chloro-	5-Chloro-1-hexene	$C_6H_{11}Cl$	118.61	927-54-8			120.7	4-01-00-00832	0.8891^{25}	ace 4; bz 4; eth 4; chl 4	1.4305^{20}
7007	2-Hexene, 4-chloro-	4-Chloro-2-hexene	$C_6H_{11}Cl$	118.61	6734-98-1			123	4-01-00-00835	0.8934^{20}	ace 4; bz 4; eth 4; chl 4	1.4400^{20}
7008	3-Hexene, 1-chloro-	3-Hexen-1-yl chloride	$C_6H_{11}Cl$	118.61	62706-16-5			61^{60}	4-01-00-00838	0.900^{24}	ace 4; bz 4; eth 4; chl 4	1.435^{24}
7009	3-Hexene, 3-chloro-, (*Z*)-	3-Hexene, 3-chloro (*cis*)	$C_6H_{11}Cl$	118.61	17226-34-5			117	4-01-00-00838	0.9009^{20}	ace 4; bz 4; eth 4; chl 4	1.4360^{20}
7010	3-Hexene, 1-chloro-4-ethyl-		$C_8H_{15}Cl$	146.66	82507-04-8			173	2-01-00-00201	0.9102^{20}	bz 4; chl 4	1.4524^{20}
7011	1-Hexene, 1,2-dichloro-, (*E*)-		$C_6H_{10}Cl_2$	153.05	59697-51-7			$63\text{-}5^{22}$	3-01-00-00803	1.1167^{25}	bz 4; chl 4	1.4576^{25}
7012	1-Hexene, 1,2-dichloro-, (*Z*)-		$C_6H_{10}Cl_2$	153.05	59697-55-1			88^{30}	3-01-00-00803	1.0812^{25}	bz 4; chl 4	1.4631^{25}
7013	1-Hexene, 2,3-dimethyl-	2,3-Dimethyl-1-hexene	C_8H_{16}	112.22	16746-86-4			110.5	3-01-00-00845	0.7172^{25}		1.4113^{20}
7014	1-Hexene, 2,4-dimethyl-	2,4-Dimethyl-1-hexene	C_8H_{16}	112.22	16746-87-5			111.2	4-01-00-00887	0.716^{25}		1.409^{20}
7015	1-Hexene, 2,5-dimethyl-	2,5-Dimethyl-1-hexene	C_8H_{16}	112.22	6975-92-4			111.6	3-01-00-00846	0.7129^{25}		1.4105^{20}
7016	1-Hexene, 3,3-dimethyl-	3,3-Dimethyl-1-hexene	C_8H_{16}	112.22	3404-77-1			104	4-01-00-00889	0.7099^{25}		1.4070^{20}
7017	1-Hexene, 3,4-dimethyl-	3,4-Dimethyl-1-hexene	C_8H_{16}	112.22	16745-94-1			112	4-01-00-00889	0.720^{25}		1.413^{20}
7018	1-Hexene, 3,5-dimethyl-		C_8H_{16}	112.22	7423-69-0			104	3-01-00-00846	0.712^{25}		1.404^{20}
7019	1-Hexene, 4,4-dimethyl-	4,4-Dimethyl-1-hexene	C_8H_{16}	112.22	1647-08-1			107.2	3-01-00-00847	0.7157^{25}		1.4102^{20}
7020	1-Hexene, 4,5-dimethyl-	4,5-Dimethyl-1-hexene	C_8H_{16}	112.22	16106-59-5			109	4-01-00-00887	0.724^{25}		1.414^{20}
7021	1-Hexene, 5,5-dimethyl-	5,5-Dimethyl-1-hexene	C_8H_{16}	112.22	7116-86-1			104	4-01-00-00886	0.705^{25}		1.4049^{20}
7022	2-Hexene, 2,3-dimethyl-	2,3-Dimethyl-2-hexene	C_8H_{16}	112.22	7145-20-2	-115.1		121.8	4-01-00-00887	0.7366^{25}		1.4268^{20}
7023	2-Hexene, 2,4-dimethyl-	2,4-Dimethyl-2-hexene	C_8H_{16}	112.22	14255-23-3			110.6	4-01-00-00887	0.7171^{25}		1.4118^{20}
7024	2-Hexene, 2,5-dimethyl-	2,5-Dimethyl-2-hexene	C_8H_{16}	112.22	3404-78-2			112.2	4-01-00-00888	0.7182^{20}		1.4140^{20}
7025	2-Hexene, 4,5-dimethyl-	2,3-Dimethyl-4-hexene	C_8H_{16}	112.22	73548-71-7			110	3-01-00-00846	0.721^{25}		1.413^{20}
7026	2-Hexene, 5,5-dimethyl-, (*E*)-		C_8H_{16}	112.22	39782-43-9			104.1	4-01-00-00886	0.7023^{25}		1.4055^{20}
7027	2-Hexene, 5,5-dimethyl-, (*Z*)-	*cis*-5,5-Dimethyl-2-hexene	C_8H_{16}	112.22	39761-61-0			106.9	4-01-00-00886	0.7125^{25}		1.4113^{20}
7028	3-Hexene, 2,2-dimethyl-, (*E*)-		C_8H_{16}	112.22	690-93-7			100.8	4-01-00-00885	0.6995^{25}		1.4063^{20}
7029	3-Hexene, 2,2-dimethyl-, (*Z*)-		C_8H_{16}	112.22	690-92-6	-137.4		105.5	4-01-00-00885	0.7086^{25}		1.4099^{20}
7030	3-Hexene, 2,4-dimethyl-, (*E*)-	2,4-Dimethyl-*trans*-3-hexene	C_8H_{16}	112.22	61847-78-7			107.6	4-01-00-00887	0.7101^{25}		1.4126^{20}
7031	3-Hexene, 2,4-dimethyl-, (*Z*)-	3-Hexene, 2,4-dimethyl (*cis*)	C_8H_{16}	112.22	37549-89-6			109	4-01-00-00887	0.7135^{25}		1.4140^{20}
7032	1-Hexene, 1,1,2,3,3,4,4,5,5,6,6,6-dodecafluoro-	Perfluoro-1-hexene	C_6F_{12}	300.05	755-25-9			57.0	4-01-00-00831		chl 4	
7033	1-Hexene, 3-ethyl-	3-Propyl-1-pentene	C_8H_{16}	112.22	3404-58-8			110.3	4-01-00-00885	0.711^{25}		1.407^{20}
7034	3-Hexene, 3-ethyl-		C_8H_{16}	112.22	16789-51-8			116	3-01-00-00845	0.725^{25}		1.418^{20}
7035	3-Hexene, 1,2,3,4,5,6-hexachloro-	1,2,3,4,5,6-Hexachloro-3-hexene	$C_6H_6Cl_6$	290.83	1725-74-2	58.5		112^{2}	3-01-00-00808		MeOH 4; chl 4	
7036	1-Hexene, 2-methyl-	2-Methyl-1-hexene	C_7H_{14}	98.19	6094-02-6	-102.8		92	4-01-00-00863	0.7000^{20}		1.4035^{20}
7037	1-Hexene, 3-methyl-	3-Methyl-1-hexene	C_7H_{14}	98.19	3404-61-3			83.9	4-01-00-00865	0.6871^{25}		1.3965^{20}
7038	1-Hexene, 4-methyl-	4-Methyl-1-hexene	C_7H_{14}	98.19	3769-23-1	-141.5		86.7	3-01-00-00830	0.6942^{25}		1.4000^{20}
7039	1-Hexene, 5-methyl-	5-Methyl-1-hexene	C_7H_{14}	98.19	3524-73-0			85.3	4-01-00-00865	0.6877^{25}		1.3967^{20}

No.	Name Synonym	Mol. Form.	Mol. Wt.	CAS RN	mp/°C	Merck No.	bp/°C	Beil. Ref.	den/g cm^{-3}	Solubility	n_D
7040	2-Hexene, 2-methyl- 2-Methyl-2-hexene	C_7H_{14}	98.19	2738-19-4	-130.4		95.4	4-01-00-00863	0.7038^{25}		1.4106^{20}
7041	2-Hexene, 3-methyl-, (*Z*)-	C_7H_{14}	98.19	10574-36-4	-118.5		95.6	4-01-00-00866	0.712^{20}		1.4126^{20}
7042	2-Hexene, 4-methyl-, (*E*)-	C_7H_{14}	98.19	3683-22-5	-125.7		87.6	4-01-00-00867	0.6925^{25}		1.4025^{20}
7043	2-Hexene, 4-methyl-, (*Z*)-	C_7H_{14}	98.19	3683-19-0			86.3	4-01-00-00867	0.6952^{25}		1.4026^{20}
7044	2-Hexene, 5-methyl-, (*E*)-	C_7H_{14}	98.19	7385-82-2	-124.3		88.1	4-01-00-00865	0.6883^{25}		1.4006^{20}
7045	2-Hexene, 5-methyl-, (*Z*)-	C_7H_{14}	98.19	13151-17-2			89.5	4-01-00-00865	0.697^{25}		1.404^{20}
7046	3-Hexene, 2-methyl-, (*E*)-	C_7H_{14}	98.19	692-24-0	-141.6		85.9	4-01-00-00864	0.6853^{25}		1.4001^{20}
7047	3-Hexene, 2-methyl-, (*Z*)-	C_7H_{14}	98.19	15840-60-5			86	4-01-00-00864	0.690^{25}		1.401^{20}
7048	1-Hexene, 1,1,2-trichloro-	$C_6H_9Cl_3$	187.50	53977-99-4			$90-3^{10}$	3-01-00-00803	1.125^{25}	eth 4	1.4760^{25}
7049	2-Hexenoic acid	$C_6H_{10}O_2$	114.14	1191-04-4	36.5		216.5	4-02-00-01563	0.965^{20}	eth 4	1.4460^{40}
7050	3-Hexenoic acid Hydrosorbic acid	$C_6H_{10}O_2$	114.14	4219-24-3	12		208	4-02-00-01566	0.9640^{23}		1.4935^{20}
7051	5-Hexenoic acid 5-Hexanoic acid	$C_6H_{10}O_2$	114.14	1577-22-6	-37		203	4-02-00-01562	0.9610^{20}	eth 4; EtOH 4	1.4343^{20}
7052	4-Hexenoic acid, 6-(1,3-dihydro-4-hydroxy-6-methoxy-7-methyl-3-oxo-5-isobenzofuranyl)-4-methyl-, (*E*)- Mycophenolic acid	$C_{17}H_{20}O_6$	320.34	24280-93-1	141	6238		4-18-00-06513		H_2O 1; EtOH 4; eth 4; bz 2	
7053	2-Hexenoic acid, ethyl ester Ethyl 2-hexenoate	$C_8H_{14}O_2$	142.20	1552-67-6			$174.5; 80^{22}$	3-02-00-01318	0.8986^{20}		1.4348^{20}
7054	3-Hexenoic acid, ethyl ester Hydrosorbic acid, ethyl ester	$C_8H_{14}O_2$	142.20	2396-83-0			166.5	4-02-00-01567	0.8957^{20}		1.4255^{20}
7055	2-Hexenoic acid, 2-methyl-, (*E*)-	$C_7H_{12}O_2$	128.17	97961-66-5			$205; 118^{11}$	4-02-00-01581	0.9627^{20}	eth 3; ace 3	1.4601^{20}
7056	2-Hexenoic acid, 4-methyl- 4-Methyl-2-hexenoic acid	$C_7H_{12}O_2$	128.17	37549-83-0			125^{13}	3-02-00-01336	0.9441^{20}		1.4526^{20}
7057	2-Hexenoic acid, 5-methyl- 5-Methyl-2-hexenoic acid	$C_7H_{12}O_2$	128.17	41653-96-7	16.5		226.5	2-02-00-00411	0.942^{20}		1.4425^{17}
7058	3-Hexenoic acid, 2-methyl- Hydrosorbic acid, 2-methyl-	$C_7H_{12}O_2$	128.17	73513-50-5			122^{24}	4-02-00-01582	0.9353^{20}		1.4379^{20}
7059	3-Hexenoic acid, 3-methyl-	$C_7H_{12}O_2$	128.17	35205-71-1			113^{10}	3-02-00-01335	0.9549^{20}		1.4469^{20}
7060	3-Hexenoic acid, 4-methyl- 4-Ethyl-3-pentenoic acid	$C_7H_{12}O_2$	128.17	55665-79-7			118^{12}	3-02-00-01336	0.9644^{16}		1.4512^{16}
7061	4-Hexenoic acid, 5-methyl-	$C_7H_{12}O_2$	128.17	5636-65-7	-28		217	4-02-00-01581	0.9862^{0}		1.4504
7062	3-Hexenoic acid, methyl ester Methyl 3-hexenoate	$C_7H_{12}O_2$	128.17	2396-78-3			$67-8^{34}$	4-02-00-01566	0.9132^{25}		1.4240^{23}
7063	2-Hexenoic acid, 2-methyl-, ethyl ester	$C_9H_{16}O_2$	156.22	26311-33-1			72^{10}	3-02-00-01333	0.9031^{20}		1.4407^{20}
7064	3-Hexenoic acid, 2-methyl-, ethyl ester	$C_9H_{16}O_2$	156.22	21994-78-5			78^{25}	3-02-00-01334	0.8778^{20}		1.4237^{20}
7065	3-Hexenoic acid, 3-methyl-, ethyl ester	$C_9H_{16}O_2$	156.22	21994-77-4			$84-5^{26}$	3-02-00-01335	0.8961^{20}		1.4309^{20}
7066	1-Hexen-3-ol	$C_6H_{12}O$	100.16	4798-44-1			134	4-01-00-02136	0.834^{22}	ace 4; eth 4; EtOH 4	1.4297^{18}
7067	4-Hexen-2-ol	$C_6H_{12}O$	100.16	52387-50-5			137.5	4-01-00-02139	0.8405^{18}		1.4392^{20}
7068	5-Hexen-2-ol	$C_6H_{12}O$	100.16	626-94-8			139	3-01-00-01926	0.842^{16}	H_2O 2	
7069	2-Hexen-1-ol, (*E*)-	$C_6H_{12}O$	100.16	928-95-0			157	4-01-00-02138	0.8490^{16}		1.4340^{20}
7070	2-Hexen-1-ol, (*Z*)-	$C_6H_{12}O$	100.16	928-94-9			157	4-01-00-02138	0.8472^{20}	H_2O 3; EtOH 4; eth 3; ace 3	1.4397^{20}
7071	3-Hexen-1-ol, (*Z*)-	$C_6H_{12}O$	100.16	928-96-1			156.5	4-01-00-02141	0.8478^{22}	H_2O 3; EtOH 4; eth 3	1.4380^{20}
7072	4-Hexen-1-ol, (*E*)-	$C_6H_{12}O$	100.16	928-92-7			159	4-01-00-02140	0.8513^{20}		1.4402^{20}
7073	1-Hexen-3-ol, 3,5-dimethyl-	$C_8H_{16}O$	128.21	3329-48-4			146.5	4-01-00-02173	0.8382^{20}		1.4342^{20}
7074	4-Hexen-3-ol, 2,2-dimethyl- 2-Hexen-4-ol, 5,5-dimethyl	$C_8H_{16}O$	128.21	37674-67-2			$75-6^{50}$	4-01-00-02171	0.8308^{25}		1.4369^{25}
7075	4-Hexen-3-ol, 2,5-dimethyl- 2,5-Dimethyl-4-hexen-3-ol	$C_8H_{16}O$	128.21	60703-31-3			162	2-01-00-00491	0.8444^{20}		1.4449^{20}
7076	4-Hexen-3-ol, 3,5-dimethyl- 2-Hexen-4-ol, 2,4-dimethyl	$C_8H_{16}O$	128.21	1569-43-3			$48-50^{3-6}$	4-01-00-02172	0.8600^{17}		1.4460^{17}
7077	1-Hexen-3-ol, 5-methyl-	$C_7H_{14}O$	114.19	4798-46-3			$125; 58^{15}$	3-01-00-01939	0.8306^{15}		1.4263^{23}
7078	2-Hexen-1-ol, 5-methyl- 5-Methyl-2-hexenol	$C_7H_{14}O$	114.19	77053-92-0			169	4-01-00-02158	0.8355^{20}		1.4390^{20}

PHYSICAL CONSTANTS OF ORGANIC COMPOUNDS (continued)

No.	Name Synonym	Mol. Form. Mol. Wt.	CAS RN mp/°C	Merck No. bp/°C	Beil. Ref. den/g cm^{-3}	Solubility n_D
7079	3-Hexen-2-ol, 2-methyl-	$C_7H_{14}O$ 114.19	18812-62-9	49^{11}	4-01-00-02157 0.8536^{18}	1.443^{18}
7080	3-Hexen-2-ol, 3-methyl-	$C_7H_{14}O$ 114.19	110383-31-8	89^{55}	0-01-00-00447 0.8678^{9}	1.4487^{9}
7081	4-Hexen-3-ol, 2-methyl-	$C_7H_{14}O$ 114.19	4798-60-1	139.5	3-01-00-01939 0.8426^{20}	1.438^{20}
7082	4-Hexen-3-ol, 3-methyl-	$C_7H_{14}O$ 114.19	60111-14-0	$72\text{-}3^{60}$	4-01-00-02160 0.8471^{25}	1.4268^{16}
7083	5-Hexen-2-ol, 2-methyl- Dimethyl(vinylethyl)carbinol	$C_7H_{14}O$ 114.19	16744-89-1 -22	142.5	4-01-00-02158 0.8397^{15}	1.4349^{17}
7084	5-Hexen-3-ol, 3-methyl- Methylethylallylcarbinol	$C_7H_{14}O$ 114.19	1569-44-4	140; 61^{35}	4-01-00-02160 0.8432^{20}	1.4370^{20}
7085	5-Hexen-3-ol, 4-methyl- 4-Methyl-5-hexen-3-ol	$C_7H_{14}O$ 114.19	1838-77-3	140.5	3-01-00-01940 0.8452^{22}	1.4365^{22}
7086	3-Hexen-2-one 1-Butenyl methyl ketone	$C_6H_{10}O$ 98.14	763-93-9	140	4-01-00-03468 0.8655^{20}	ace 4; eth 4; EtOH 4 1.4418^{20}
7087	4-Hexen-2-one 2-Hexen-5-one	$C_6H_{10}O$ 98.14	25659-22-7	127	3-01-00-02994 0.8520^{16}	1.4300^{16}
7088	4-Hexen-3-one	$C_6H_{10}O$ 98.14	2497-21-4	138.5	4-01-00-03468 0.8559^{20}	EtOH 3; eth 3; ace 4 1.4388^{20}
7089	5-Hexen-2-one	$C_6H_{10}O$ 98.14	109-49-9	129.5	4-01-00-03467 0.833^{27}	1.4178^{27}
7090	3-Hexen-2-one, (*E*)-	$C_6H_{10}O$ 98.14	4376-23-2	140	4-01-00-03469 0.8665^{20}	EtOH 3; eth 3; ace 4 1.4418^{20}
7091	3-Hexen-2-one, 3,4-dimethyl-	$C_8H_{14}O$ 126.20	1635-02-5	158; 65^{20}	4-01-00-03500 0.8585^{20}	1.4418^{20}
7092	4-Hexen-2-one, 3,4-dimethyl- 2-Hexen-5-one, 3,4-dimethyl	$C_8H_{14}O$ 126.20	53252-21-4	154	4-01-00-03500 0.8538^{19}	1.4377^{19}
7093	1-Hexen-3-one, 5-methyl- Isobutyl vinyl ketone	$C_7H_{12}O$ 112.17	2177-32-4	41^{22}; 32^{10}	4-01-00-03483 0.8400^{15}	1.4293^{15}
7094	3-Hexen-2-one, 5-methyl- 2-Oxo-5-methylhex-3-ene	$C_7H_{12}O$ 112.17	5166-53-0	77.5^{50}	3-01-00-03005 0.8407^{22}	1.4395^{22}
7095	4-Hexen-2-one, 5-methyl-	$C_7H_{12}O$ 112.17	28332-44-7	76^{50}	4-01-00-03482 0.9012^{21}	1.4385^{20}
7096	4-Hexen-3-one, 2-methyl-	$C_7H_{12}O$ 112.17	53252-19-0	148	4-01-00-03483 0.843^{20}	EtOH 3; eth 3; ace 4 1.4345^{20}
7097	5-Hexen-2-one, 3-methyl- 1-Methyl-1-allyl acetone	$C_7H_{12}O$ 112.17	2550-22-3	137	4-01-00-03484 0.845^{15}	1.4215^{25}
7098	5-Hexen-2-one, 4-methyl- 4-Methyl-5-hexen-2-one	$C_7H_{12}O$ 112.17	61675-14-7	137.5	3-01-00-03006 0.8273^{25}	1.4193^{22}
7099	5-Hexen-2-one, 5-methyl- 5-Hexene-2-one,5-methyl	$C_7H_{12}O$ 112.17	3240-09-3	150	4-01-00-03482 0.8460^{20}	ace 4; eth 4; EtOH 4 1.4348^{20}
7100	1-Hexen-3-one, 5-methyl-1-phenyl-	$C_{13}H_{16}O$ 188.27	2892-18-4 43	154^{25}	4-07-00-01058 0.9509^{46}	H_2O 2; EtOH 3; bz 3; chl 3 1.5523^{25}
7101	5-Hexen-2-one, 4-phenyl-	$C_{12}H_{14}O$ 174.24	50552-30-2	$85\text{-}6^{1}$	3-07-00-01460 0.9848^{25}	1.5193^{25}
7102	5-Hexen-2-one, 6-phenyl- 6-Phenyl-5-hexen-2-one	$C_{12}H_{14}O$ 174.24	69371-59-1	154^{10}, $98^{0.3}$	3-07-00-01455 0.9967^{25}	1.5458^{25}
7103	1-Hexen-3-yne Ethylvinylacetylene	C_6H_8 80.13	13721-54-5	85	4-01-00-01091 0.7492^{20}	bz 4; eth 4; peth 4; chl 4 1.4522^{20}
7104	1-Hexen-4-yne	C_6H_8 80.13	5009-11-0	87	4-01-00-01092 0.767^{14}	1.446^{14}
7105	1-Hexen-5-yne Hex-5-en-1-yne	C_6H_8 80.13	14548-31-3	70	4-01-00-01092 0.7650^{20}	bz 4; eth 4; peth 4; chl 4 1.4318^{20}
7106	2-Hexen-4-yne 2-Hexene-4-yne	C_6H_8 80.13	14092-20-7	88.5	4-01-00-01092 0.7710^{20}	1.4918^{20}
7107	1-Hexen-3-yne, 5-chloro-5-methyl- Dimethylvinylethynylchloromethane	C_7H_9Cl 128.60	819-44-3	48^{28}	4-01-00-01098 0.9375^{15}	ace 4; bz 4; eth 4; EtOH 4 1.4778^{20}
7108	3-Hexen-1-yne, 3-propyl-	C_9H_{14} 122.21	688-52-8	136	3-01-00-01049 0.7799^{25}	bz 4; peth 4; chl 4 1.4432^{25}
7109	Hexitol, 2,5-anhydro-3,4-dideoxy- Tetrahydro-2,5-furandimethanol	$C_6H_{12}O_3$ 132.16	104-80-3 <-50	9145 265	4-17-00-02052 1.154^{20}	H_2O 4; ace 4; bz 4; EtOH 4
7110	*D-arabino*-Hexose, 2-deoxy- 2-Deoxy-*D*-glucose	$C_6H_{12}O_5$ 164.16	154-17-6 146.5	2886	4-01-00-04282	
7111	*D-ribo*-Hexose, 2,6-dideoxy- Digitoxose	$C_6H_{12}O_4$ 148.16	527-52-6 112	3147	4-01-00-04191	H_2O 4; ace 4; py 3; AcOEt 3
7112	*ribo*-Hexose, 2,6-dideoxy-3-*O*-methyl- Cymarose	$C_7H_{14}O_4$ 162.19	579-04-4 101	2769	4-01-00-04193	H_2O 4; ace 4; EtOH 4
7113	3-Hexynal	C_6H_8O 96.13	89533-67-5	$42\text{-}3^{17}$	4-01-00-03544 0.9036^{23}	1.4498^{22}
7114	1-Hexyne Butylacetylene	C_6H_{10} 82.15	693-02-7 -131.9	71.3	4-01-00-01006 0.7155^{20}	H_2O 1; EtOH 3; eth 3; bz 3 1.3989^{20}
7115	2-Hexyne 1-Methyl-2-propylacetylene	C_6H_{10} 82.15	764-35-2 -89.6	84.5	4-01-00-01009 0.7315^{20}	H_2O 1; EtOH 5; eth 5; bz 3 1.4138^{20}
7116	3-Hexyne Diethylacetylene	C_6H_{10} 82.15	928-49-4 -103	81	4-01-00-01009 0.7231^{20}	H_2O 1; EtOH 3; eth 3; bz 3 1.4115^{20}
7117	3-Hexyne, 2,5-dichloro-2,5-dimethyl- 3-Hexyne, 2,5-dimethyl-2,5-dichloro	$C_8H_{12}Cl_2$ 179.09	2431-30-3 29	178; 62^{14}	4-01-00-01042 1.0118^{20}	chl 4
7118	3-Hexyne-2,5-diol	$C_6H_{10}O_2$ 114.14	3031-66-1	120^{11}	3-01-00-02271 1.0180^{20}	1.4691^{20}

PHYSICAL CONSTANTS OF ORGANIC COMPOUNDS (continued)

No.	Name Synonym	Mol. Form. Mol. Wt.	CAS RN mp/°C	Merck No. bp/°C	Beil. Ref. den/g cm^{-3}	Solubility n_D
7119	3-Hexyne-2,5-diol, 2,5-dimethyl-	$C_8H_{14}O_2$ 142.20	142-30-3 95	205	4-01-00-02699 0.947^{20}	H_2O 3; EtOH 4; eth 4; ace 4
7120	1-Hexyne, 5-methyl- 5-Methyl-1-hexyne	C_7H_{12} 96.17	2203-80-7 -125	92	3-01-00-01000 0.7274^{20}	H_2O 1; EtOH 3; eth 3; bz 3 1.4059^{-20}
7121	2-Hexyne, 5-methyl-	C_7H_{12} 96.17	53566-37-3 -92.9	102.5	4-01-00-01030 0.7378^{20}	H_2O 1; eth 3; ace 3; bz 3 1.4176^{20}
7122	3-Hexyne, 2-methyl- 1-Ethyl-2-isopropylacetylene	C_7H_{12} 96.17	36566-80-0 -116.7	95.2	4-01-00-01029 0.7263^{20}	bz 4; eth 4; chl 4; peth 4 1.4120^{20}
7123	2-Hexynoic acid Propiolic acid, propyl-	$C_6H_8O_2$ 112.13	764-33-0 27	126^{24}	4-02-00-01699 0.9820^{20}	eth 4; EtOH 4; lig 4
7124	1-Hexyn-3-ol	$C_6H_{10}O$ 98.14	105-31-7 -80	142	4-01-00-02234 0.8704^{20}	ctc 3 1.4340^{25}
7125	2-Hexyn-1-ol	$C_6H_{10}O$ 98.14	764-60-3	100^{53}; 80^{21}	4-01-00-02235 0.8472^{20}	1.4341^{20}
7126	3-Hexyn-1-ol 3-Hexynol	$C_6H_{10}O$ 98.14	1002-28-4	163; 65^{12}	4-01-00-02236 0.8982^{20}	1.4530^{20}
7127	5-Hexyn-2-ol	$C_6H_{10}O$ 98.14	23470-12-4	152	4-01-00-02235 0.899^{20}	1.4481^{20}
7128	1-Hexyn-3-ol, 3-methyl- Ethynylmethylpropylcarbinol	$C_7H_{12}O$ 112.17	4339-05-3	137	4-01-00-02252 0.8620^{20}	H_2O 4; eth 4; EtOH 4 1.4338^{20}
7129	3-Hexyn-2-ol, 2-methyl- 2-Methyl-3-hexyn-2-ol	$C_7H_{12}O$ 112.17	5075-33-2 -6	146	4-01-00-02250 0.962^{25}	eth 4; EtOH 4 1.4392^{25}
7130	3-Hexyn-2-one 3-Hexyne-2-one	C_6H_8O 96.13	1679-36-3	$77\text{-}8^{75}$	4-01-00-03544 0.8804^{20}	1.4400^{20}
7131	5-Hexyn-2-one	C_6H_8O 96.13	2550-28-9	149	4-01-00-03543 0.9065^{20}	1.4366^{20}
7132	*L*-Histidine Glyoxaline-5-alanine	$C_6H_9N_3O_2$ 155.16	71-00-1 287 dec	4642	4-25-00-04343	H_2O 3; EtOH 2; eth 1; ace 1
7133	*L*-Histidine, *N*-β-alanyl- Carnosine	$C_9H_{14}N_4O_3$ 226.24	305-84-0 260	1857	4-25-00-04381	H_2O 4
7134	Homoserine (*S*)-2-Amino-4-hydroxybutanoic acid	$C_4H_9NO_3$ 119.12	672-15-1 203 dec	4661	4-04-00-03187	H_2O 4; EtOH 2; eth 1; bz 1
7135	*L*-Homoserine, *O*-[(aminoiminomethyl)amino]- Canavanine	$C_5H_{12}N_4O_3$ 176.18	543-38-4	1745		H_2O 4
7136	Hydramethylnon Tetrahydro-5,5-dimethyl-2-(1H)-pyrimidinone[3[4-(trifluoromethyl-styryl)-cinnamylidene- hydrazone	$C_{25}H_{24}F_6N_4$ 494.48	67485-29-4 190	4684		
7137	Hydrazine, 1,2-bis(1-methylethyl)-	$C_6H_{16}N_2$ 116.21	3711-34-0	125	4-04-00-03339 0.7894^{20}	ace 4; bz 4; eth 4; EtOH 4 1.4173^{20}
7138	Hydrazine, 1,2-bis(3-methylphenyl)-	$C_{14}H_{16}N_2$ 212.29	621-26-1 38	224	4-15-00-01220	H_2O 1; EtOH 3; eth 3; bz 3
7139	Hydrazine, 1,2-bis(4-methylphenyl)-	$C_{14}H_{16}N_2$ 212.29	637-47-8 135		4-15-00-01230 0.957^{20}	bz 4; eth 4; EtOH 4
7140	Hydrazine, 1,2-bis(2-methylpropyl)-	$C_8H_{20}N_2$ 144.26	3711-37-3	171; 63^{18}	2-04-00-00962 0.8002^{20}	ace 4; bz 4; eth 4; EtOH 4 1.4276
7141	Hydrazine, (4-bromophenyl)- (*p*-Bromophenyl)hydrazine	$C_6H_7BrN_2$ 187.04	589-21-9 108	1417	4-15-00-00282	eth 4; EtOH 4; lig 4
7142	Hydrazinecarbothioamide Thiosemicarbazide	CH_5N_3S 91.14	79-19-6 183	9292	4-03-00-00374	H_2O 4; EtOH 4
7143	Hydrazinecarboxamide, monohydrochloride Semicarbazide hydrochloride	CH_6ClN_3O 111.53	563-41-7 175-7 dec	8396	4-03-00-00177	H_2O 4
7144	Hydrazinecarboxamide, 2-[(5-nitro-2-furanyl)methylene]- Nitrofurazone	$C_6H_6N_4O_4$ 198.14	59-87-0 236-40 dec	6521	5-17-09-00335	H_2O 1; EtOH 2; eth 1; DMSO 2
7145	Hydrazinecarboxamide, 2-phenyl- Phenicarbazide	$C_7H_9N_3O$ 151.17	103-03-7 172	7193	4-15-00-00180	H_2O 2; EtOH 3; eth 2; ace 3
7146	Hydrazinecarboxamide, *N*-phenyl- 4-Phenylsemicarbazide	$C_7H_9N_3O$ 151.17	537-47-3 128	7283	3-12-00-00822	H_2O 2; EtOH 4; eth 1; chl 4
7147	Hydrazinecarboximidamide Aminoguanidine	CH_6N_4 74.09	79-17-4 dec	453	4-03-00-00236	H_2O 4; EtOH 4
7148	Hydrazinecarboxylic acid, ethyl ester	$C_3H_8N_2O_2$ 104.11	4114-31-2 46	198 dec; 93^9	4-03-00-00174	EtOH 3; eth 3; chl 2
7149	Hydrazinecarboxylic acid, methyl ester	$C_2H_6N_2O_2$ 90.08	6294-89-9 73	108^{12}	2-03-00-00078	H_2O 3; EtOH 3; bz 2; peth 1
7150	1,1-Hydrazinedicarboxylic acid, diethyl ester	$C_6H_{12}N_2O_4$ 176.17	5311-96-6 30.5	138^{12}	2-03-00-00079	EtOH 4
7151	1,2-Hydrazinedicarboxylic acid, diethyl ester	$C_6H_{12}N_2O_4$ 176.17	4114-28-7 135	250 dec	4-03-00-00175 1.324^8	eth 4; EtOH 4
7152	Hydrazine, 1,1-diethyl- 1,1-Diethylhydrazine	$C_4H_{12}N_2$ 88.15	616-40-0	99	4-04-00-03335 0.8804^{20}	bz 4; eth 4; EtOH 4; chl 4 1.4214^{20}
7153	Hydrazine, 1,2-diethyl- 1,2-Diethylhydrazine	$C_4H_{12}N_2$ 88.15	1615-80-1	85.5	4-04-00-03336 0.797^{26}	bz 4; eth 4; EtOH 4 1.4204^{20}
7154	Hydrazine, 1,1-dimethyl- 1,1-Dimethylhydrazine	$C_2H_8N_2$ 60.10	57-14-7 -58	3236 63.9	4-04-00-03322 0.791^{22}	H_2O 4; EtOH 4; eth 4; MeOH 4 1.4075^{22}
7155	Hydrazine, 1,2-dimethyl- 1,2-Dimethylhydrazine	$C_2H_8N_2$ 60.10	540-73-8	3237 81	4-04-00-03323 0.8274^{20}	H_2O 5; EtOH 5; eth 5 1.4209^{20}
7156	Hydrazine, (2,4-dinitrophenyl)- (2,4-Dinitrophenyl)hydrazine	$C_6H_6N_4O_4$ 198.14	119-26-6 194	3277	4-15-00-00380	H_2O 1; EtOH 3; eth 2; bz 2

No.	Name Synonym	Mol. Form. Mol. Wt.	CAS RN mp/°C	Merck No. bp/°C	Beil. Ref. den/g cm^{-3}	Solubility n_D
7157	Hydrazine, 1,1-diphenyl- 1,1-Diphenylhydrazine	$C_{12}H_{12}N_2$ 184.24	530-50-7 50.5	3326 220^{40}	4-15-00-00055 1.190^{16}	bz 4; eth 4; EtOH 4; chl 4
7158	Hydrazine, 1,2-diphenyl- *N,N'*-Diphenylhydrazine	$C_{12}H_{12}N_2$ 184.24	122-66-7 131		4-15-00-00056 1.158^{16}	EtOH 4; bz 2; DMSO-d_6 2; HOAc 1
7159	Hydrazine, 1-ethyl-1-phenyl- 1-Phenyl-1-ethylhydrazine	$C_8H_{12}N_2$ 136.20	644-21-3	 237	3-15-00-00074 1.0181^{21}	ace 4; bz 4; eth 4; EtOH 4 1.5711^{21}
7160	Hydrazine, 1-ethyl-2-phenyl- 1-Ethyl-2-phenylhydrazine	$C_8H_{12}N_2$ 136.20	622-82-2	 241; 110^{14}	4-15-00-00054 1.0150^{20}	bz 4; eth 4; EtOH 4; chl 4 1.5676^{20}
7161	Hydrazine, methyl- Methylhydrazine	CH_6N_2 46.07	60-34-4 -52.4	6001 87.5	4-04-00-03322	H_2O 3; EtOH 5; eth 3; ctc 3 1.4325^{20}
7162	Hydrazine, 1-(3-methylbutyl)-1-phenyl- Hydrazine, 1-isopentyl-1-phenyl	$C_{11}H_{18}N_2$ 178.28	636-10-2 236	 260	0-15-00-00121 0.9588^{15}	
7163	Hydrazine, 1-methyl-2-(3-methylphenyl)- Hydrazine, 1-methyl-2-(3-tolyl)	$C_8H_{12}N_2$ 136.20	116836-06-7 60		2-15-00-00229 1.0265^{100}	bz 4; EtOH 4
7164	Hydrazine, (3-methylphenyl)-	$C_7H_{10}N_2$ 122.17	536-89-0	 244 dec	4-15-00-01220 1.057^{20}	H_2O 2; EtOH 4; eth 4; bz 4
7165	Hydrazine, 1-methyl-1-phenyl-	$C_7H_{10}N_2$ 122.17	618-40-6	 228; 131^{35}	4-15-00-00053 1.0404^{20}	H_2O 2; EtOH 5; eth 5; bz 5 1.5691^{20}
7166	Hydrazine, 1-methyl-2-phenyl- 1-Methyl-2-phenylhydrazine	$C_7H_{10}N_2$ 122.17	622-36-6	 231; 112^{14}	4-15-00-00054 1.0320^{20}	bz 4; eth 4; EtOH 4; chl 4 1.5733^{20}
7167	Hydrazine, 1-naphthalenyl-	$C_{10}H_{10}N_2$ 158.20	2243-55-2 117	 203^{20}	3-15-00-00728	bz 4; eth 4; EtOH 4; chl 4
7168	Hydrazine, (4-nitrophenyl)- (4-Nitrophenyl)hydrazine	$C_6H_7N_3O_2$ 153.14	100-16-3 158 dec	6545	4-15-00-00317	H_2O 2; EtOH 3; eth 3; bz 3
7169	Hydrazine, phenyl- Phenylhydrazine	$C_6H_8N_2$ 108.14	100-63-0 19.6	7264 243.5	4-15-00-00050 1.0986^{20}	H_2O 3; EtOH 5; eth 5; ace 4 1.6084^{10}
7170	Hydrazine, (1-phenylethyl)-	$C_8H_{12}N_2$ 136.20	65-64-5	 $69^{0.65}$	4-15-00-01267 0.9672^{25}	 1.5436^{25}
7171	Hydrazine, phenyl-, hemihydrate	$C_{12}H_{18}N_4O$ 234.30	6152-31-4 24	 120^{12}	3-15-00-00067 1.0938^{25}	 1.6081^{20}
7172	Hydrazine, phenyl-, monohydrochloride	$C_6H_9ClN_2$ 144.60	59-88-1 243-6 dec	7265 sub		H_2O 4; EtOH 4
7173	Hydrazine, triphenyl- Triphenylhydrazine	$C_{18}H_{16}N_2$ 260.34	606-88-2 142 dec		4-15-00-00058 0.869^{70}	EtOH 2; bz 2
7174	Hydrazobenzene, 3-methyl- Hydrazine, 1-phenyl-2-(3-tolyl)	$C_{13}H_{14}N_2$ 198.27	621-25-0 61		2-15-00-00229 1.0265^{100}	bz 4; EtOH 4; lig 4
7175	Hydrocinnamic acid, β-ethyl-β-methyl- Valeric acid, 3-methyl-3-phenyl	$C_{12}H_{16}O_2$ 192.26	105401-59-0 284	 174^{14}	4-09-00-01905 1.050^{25}	H_2O 3 1.5197^{25}
7176	Hydrocyanic acid Hydrogen cyanide	CHN 27.03	74-90-8 -13.4	4722 26	4-02-00-00050 0.6876^{20}	H_2O 5; EtOH 5; eth 5 1.2614^{20}
7177	Hydroperoxide, cyclohexyl Cyclohexyl hydroperoxide	$C_6H_{12}O_2$ 116.16	766-07-4 -20	 $42^{0.1}$	4-06-00-00053 1.019^{20}	eth 4; EtOH 4; HOAc 4 1.4645^{25}
7178	Hydroperoxide, 1,1-dimethylethyl *tert*-Butyl hydroperoxide	$C_4H_{10}O_2$ 90.12	75-91-2 6	1569 89 dec; 36^{17}	4-01-00-01616 0.8960^{20}	H_2O 3; EtOH 3; eth 3; ctc 3 1.4015^{20}
7179	Hydroperoxide, ethyl Ethyl hydrogen peroxide	$C_2H_6O_2$ 62.07	3031-74-1 -100	 95	4-01-00-01323 0.9332^{20}	H_2O 4; bz 4; eth 4; EtOH 4 1.3800^{20}
7180	Hydroperoxide, methyl	CH_4O_2 48.04	3031-73-0	 $38\text{-}40^{65}$	4-01-00-01249 1.9967^{15}	H_2O 4; bz 4; eth 4; EtOH 4 1.3641^{15}
7181	Hydroperoxide, 1-methyl-1-phenylethyl Cumene hydroperoxide	$C_9H_{12}O_2$ 152.19	80-15-9		4-06-00-03221	
7182	Hydroprene 2,4-Dodecadienoic acid, 3,7,11-trimethyl-, ethyl ester, *(E,E)*-	$C_{17}H_{30}O_2$ 266.42	41096-46-2	 174^{19}	 0.8955^{20}	
7183	Hypochlorous acid, 1,1-dimethylethyl ester *tert*-Butyl hypochlorite	C_4H_9ClO 108.57	507-40-4	1570 77.5	4-01-00-01621 0.9583^{18}	H_2O 1; eth 4; ace 3; bz 4 1.403^{20}
7184	Hypochlorous acid, ethyl ester Ethyl hypochlorite	C_2H_5ClO 80.51	624-85-1	 37	4-01-00-01324 1.013^{-6}	H_2O 1; EtOH 3; eth 5; bz 5
7185	Hypophosphoric acid, tetraethyl ester	$C_8H_{20}O_6P_2$ 274.19	679-37-8	 $116\text{-}7^{2}$	4-01-00-01358 1.1283^{18}	bz 3; ctc 3 1.4284^{20}
7186	Ibogamine, 12-methoxy- Ibogaine	$C_{20}H_{26}N_2O$ 310.44	83-74-9 148	4806	5-23-12-00284	chl 3
7187	Imazalil 1H-Imidazole, 1-[2-(2,4-dichlorophenyl)-2-(2-propenyloxy)ethyl]-	$C_{14}H_{14}Cl_2N_2O$ 297.18	35554-44-0 50	3537 dec	 1.243^{23}	
7188	Imazapyr 2-(4-Isopropyl-4-methyl-5-oxo-2-imidazolin-2-yl)nicotinic acid	$C_{13}H_{15}N_3O_3$ 261.28	81334-34-1 171			
7189	Imazaquin *(RS)*-2-(4-Isopropyl-4-methyl-5-oxo-2-imidazolin-2-yl)quinoline-3-carboxylic acid	$C_{17}H_{17}N_3O_3$ 311.34	81335-37-7 221	4826		
7190	Imazethapyr *(RS)*-5-Ethyl-2-(4-isopropyl-4-methyl-5-oxo-2-imidazolin-2-yl)nicotinic acid	$C_{15}H_{19}N_3O_3$ 289.33	81335-77-5 173			
7191	Imidazo[4,5-d]imidazole-2,5(1H,3H)-dione, tetrahydro- Acetyleneurea	$C_4H_6N_4O_2$ 142.12	496-46-8 300 dec	87	2-26-00-00260	H_2O 2; EtOH 1; eth 3; HCl 3
7192	1H-Imidazole 1,3-Diazole	$C_3H_4N_2$ 68.08	288-32-4 90.5	4828 257	5-23-04-00191 1.0303^{101}	H_2O 4; EtOH 4; eth 3; ace 3 1.4801^{101}
7193	1H-Imidazole, 1,1'-carbonylbis- *N,N'*-Carbonyldiimidazole	$C_7H_6N_4O$ 162.15	530-62-1 119	1825	5-23-04-00245	

PHYSICAL CONSTANTS OF ORGANIC COMPOUNDS (continued)

No.	Name Synonym	Mol. Form. Mol. Wt.	CAS RN mp/°C	Merck No. bp/°C	Beil. Ref. den/g cm^{-3}	Solubility n_D
7194	1H-Imidazole-1-carboxylic acid, 2,3-dihydro-3-methyl-2-thioxo-, ethyl ester Carbimazole	$C_7H_{10}N_2O_2S$ 186.23	22232-54-8 123.5	1803	4-24-00-00064	ace 4; chl 4
7195	1H-Imidazole-4,5-dicarboxylic acid	$C_5H_4N_2O_4$ 156.10	570-22-9 290 dec		4-25-00-01051 1.749^{25}	H_2O 2; EtOH 1; eth 1; bz 1
7196	1H-Imidazole, 1,2-diethyl- 1,2-Diethylimidazole	$C_7H_{12}N_2$ 124.19	51807-53-5	219	5-23-05-00165 0.9813^{25}	H_2O 4; eth 4; EtOH 4
7197	1H-Imidazole, 4,5-dihydro-2-methyl- Lysidine	$C_4H_8N_2$ 84.12	534-26-9 107	5508 196.5	5-23-03-00385	H_2O 4; EtOH 4; eth 1; chl 3
7198	1H-Imidazole, 4,5-dihydro-2-(phenylmethyl)- 2-Imidazolidine, 2-benzyl	$C_{10}H_{12}N_2$ 160.22	59-98-3	9430	5-23-06-00488	
7199	1H-Imidazole, 4,5-dihydro-2,4,5-triphenyl-, *cis*-	$C_{21}H_{18}N_2$ 298.39	573-33-1 134.5	198 dec	5-23-10-00053	bz 4; eth 4; EtOH 4
7200	1H-Imidazole, 1,2-dimethyl-	$C_5H_8N_2$ 96.13	1739-84-0	206	5-23-05-00039 1.0051^{11}	H_2O 4; eth 4; EtOH 4
7201	1H-Imidazole, 1,4-dimethyl- 1,4-Dimethylimidazole	$C_5H_8N_2$ 96.13	6338-45-0 25.8	199	5-23-05-00094 0.9960^{16}	
7202	1H-Imidazole, 1,2-dimethyl-5-nitro- Dimetridazole	$C_5H_7N_3O_2$ 141.13	551-92-8 138.5	3255	5-23-05-00058	eth 4; EtOH 4
7203	1H-Imidazole-4-ethanamine Histamine	$C_5H_9N_3$ 111.15	51-45-6 83	4640 209^{18}	4-25-00-02049	H_2O 3; EtOH 3; eth 2; chl 3
7204	1H-Imidazole-4-ethanaminium, α-carboxy-2,3-dihydro-*N,N,N*-trimethyl-2-thioxo-, hydroxide, in Ergothioneine	$C_9H_{15}N_3O_2S$ 229.30	497-30-3 290 dec	3611	4-25-00-04487	H_2O 4; EtOH 2; eth 1; ace 2
7205	1H-Imidazole-1-ethanol	$C_5H_8N_2O$ 112.13	1615-14-1 39	203^{20}	5-23-04-00309	H_2O 4; EtOH 4
7206	1H-Imidazole-1-ethanol, 2-methyl-5-nitro- Metronidazole	$C_6H_9N_3O_3$ 171.16	443-48-1 160.5	6079	5-23-05-00063	
7207	1H-Imidazole, 1-ethyl-	$C_5H_8N_2$ 96.13	7098-07-9	208	5-23-04-00261 0.999^{25}	H_2O 5
7208	1H-Imidazole, 1-methyl-	$C_4H_6N_2$ 82.11	616-47-7 -6	195.5	5-23-04-00256 1.0325^{20}	H_2O 4; ace 4; eth 4; EtOH 4 1.4970^{20}
7209	1H-Imidazole, 2-methyl-	$C_4H_6N_2$ 82.11	693-98-1 144	267	5-23-05-00035	H_2O 4; EtOH 4
7210	1H-Imidazole, 4-methyl-	$C_4H_6N_2$ 82.11	822-36-6 56	263	5-23-05-00089 1.0416^{14}	H_2O 4; EtOH 4 1.5037^{14}
7211	1H-Imidazole, 2-nitro- Azomycin	$C_3H_3N_3O_2$ 113.08	527-73-1 287 dec	934	5-23-04-00471	
7212	1H-Imidazole, 1-phenyl-	$C_9H_8N_2$ 144.18	7164-98-9 13	276	5-23-04-00267 1.1397^{15}	H_2O 1; eth 4; ace 4; chl 4 1.6025^{25}
7213	1H-Imidazole, 2-phenyl-	$C_9H_8N_2$ 144.18	670-96-2 149.3	340	5-23-07-00194	EtOH 4
7214	2H-Imidazole-2-thione, 1,3-dihydro-1-methyl- Methimazole	$C_4H_6N_2S$ 114.17	60-56-0 146	5892 280 dec	5-24-01-00305	H_2O 4; EtOH 3; eth 2; bz 2
7215	1H-Imidazole, 2,4,5-triphenyl-	$C_{21}H_{16}N_2$ 296.37	484-47-9 275	sub	5-23-10-00107	H_2O 1; EtOH 3; eth 3
7216	2,4-Imidazolidinedione Hydantoin	$C_3H_4N_2O_2$ 100.08	461-72-3 220	4678	5-24-05-00188	H_2O 4; EtOH 3; eth 2; alk 3
7217	2,4-Imidazolidinedione, 1,3-dibromo-5,5-dimethyl- Dibromantine	$C_5H_6Br_2N_2O_2$ 285.92	77-48-5 197-9 dec		5-24-05-00374	
7218	2,4-Imidazolidinedione, 5,5-dimethyl-	$C_5H_8N_2O_2$ 128.13	77-71-4 178	sub	5-24-05-00348	H_2O 4; EtOH 4; eth 4; ace 4
7219	2,4-Imidazolidinedione, 5,5-diphenyl- Phenytoin	$C_{15}H_{12}N_2O_2$ 252.27	57-41-0 286	7293	5-24-08-00376	H_2O 1; EtOH 3; eth 2; ace 3
7220	2,4-Imidazolidinedione, 5-(1-ethylpentyl)-3-[(trichloromethyl)thio]- Chlordantoin	$C_{11}H_{17}Cl_3N_2O_2S$ 347.69	5588-20-5	2080		CS_2 3
7221	2,4-Imidazolidinedione, 3-methyl-5-phenyl- 3-Methyl-5-phenylhydantoin	$C_{10}H_{10}N_2O_2$ 190.20	6846-11-3 164.5	6026	5-24-08-00005	chl 3
7222	2,4-Imidazolidinedione, 1-[[(5-nitro-2-furanyl)methylene]amino]- Nitrofurantoin	$C_8H_6N_4O_5$ 238.16	67-20-9 263	6520	5-24-05-00224	
7223	4-Imidazolidinehexanoic acid, 5-methyl-2-oxo-, (4*R*-*cis*)- Desthiobiotin	$C_{10}H_{18}N_2O_3$ 214.26	533-48-2 157	2914	4-25-00-01543	H_2O 3
7224	4-Imidazolidinehexanoic acid, 5-methyl-2-oxo-, methyl ester, (4*R*-*cis*)- Desthiobiotin, methyl ester	$C_{11}H_{20}N_2O_3$ 228.29	6020-51-5 69.5	$194\text{-}7^{0.03}$	4-25-00-01544	EtOH 4
7225	2-Imidazolidinethione Ethylenethiourea	$C_3H_6N_2S$ 102.16	96-45-7 203	3759	5-24-01-00165	H_2O 4; EtOH 3; eth 1; bz 1
7226	Imidazolidinetrione Parabanic acid	$C_3H_2N_2O_3$ 114.06	120-89-8 244 dec	6970 sub 100	5-24-09-00009	H_2O 3; EtOH 4
7227	2-Imidazolidinone Ethyleneurea	$C_3H_6N_2O$ 86.09	120-93-4 131	4830	5-24-01-00022	H_2O 4; EtOH 4; eth 2; chl 2
7228	4-Imidazolidinone, 5,5-diphenyl- Doxenitoin	$C_{15}H_{14}N_2O$ 238.29	3254-93-1 183	3424	5-24-04-00145	
7229	4-Imidazolidinone, 5-(2-methylpropyl)-3-(2-propenyl)-2-thioxo- Albutoin	$C_{10}H_{16}N_2OS$ 212.32	830-89-7 210.5	210	4-24-00-01139	

No.	Name Synonym	Mol. Form. Mol. Wt.	CAS RN mp/°C	Merck No. bp/°C	Beil. Ref. den/g cm^{-3}	Solubility n_D
7230	4H-Imidazol-4-one, 2-amino-1,5-dihydro-1-methyl- Creatinine	$C_4H_7N_3O$ 113.12	60-27-5 300 dec	2571	4-25-00-03543	H_2O 3; EtOH 2; eth 1; ace 1
7231	Imidodicarbonic acid, diethyl ester	$C_6H_{11}NO_4$ 161.16	19617-44-8 50	 226	3-03-00-00050	H_2O 4
7232	Imidodicarbonic diamide Biuret	$C_2H_5N_3O_2$ 103.08	108-19-0 190 dec	1319	4-03-00-00141	H_2O 2; EtOH 4; eth 1
7233	Imidodicarbonimidic diamide Biguanide	$C_2H_7N_5$ 101.11	56-03-1 136	1233 142 dec	4-03-00-00162	H_2O 4; EtOH 3; bz 1; chl 1
7234	Imidodicarbonimidic diamide, *N*-phenyl- Phenyl biguanide	$C_8H_{11}N_5$ 177.21	102-02-3 143	7247	4-12-00-00771	
7235	1H-Indazole 1H-Benzopyrazole	$C_7H_6N_2$ 118.14	271-44-3 148	4848 269	5-23-06-00156	H_2O 3; EtOH 3; eth 3
7236	1H-Indazole, 3-chloro- 3-Chloroindazole	$C_7H_5ClN_2$ 152.58	29110-74-5 149	 sub	5-23-06-00173	H_2O 3; EtOH 4; eth 4; bz 4
7237	1H-Indazole, 1,5-dimethyl- 1,5-Dimethylindazole	$C_9H_{10}N_2$ 146.19	70127-93-4 62.5	 260	2-23-00-00157	ace 4; eth 4; EtOH 4; MeOH 4
7238	2H-Indazole, 2,3-dimethyl- 2,3-Dimethylindazole	$C_9H_{10}N_2$ 146.19	50407-18-6 79.5	 286	5-23-06-00317	eth 4; EtOH 4
7239	1H-Indazole, 1-methyl- 1-Methylindazole	$C_8H_8N_2$ 132.17	13436-48-1 60.5	 231	5-23-06-00159 1.0315^{99}	H_2O 1; EtOH 3; ace 3; chl 3
7240	1H-Indazole, 3-methyl- 3-Methylindazole	$C_8H_8N_2$ 132.17	3176-62-3 113	 280.5	5-23-06-00312	H_2O 3; EtOH 3; eth 3; chl 3
7241	1H-Indazole, 5-methyl- 5-Methylindazole	$C_8H_8N_2$ 132.17	1776-37-0 117	 294	5-23-06-00319	H_2O 3; EtOH 3; eth 3; chl 3
7242	2H-Indazole, 2-methyl- 2-Methyl-2H-indazole	$C_8H_8N_2$ 132.17	4838-00-0 56	 261	5-23-06-00165 1.0450^{99}	H_2O 1; EtOH 3; eth 3; ace 3
7243	1H-Inden-1-amine, 2,3-dihydro-, (±)- (*RS*)-1-Aminoindan	$C_9H_{11}N$ 133.19	61949-83-5	 221; 96^{8}	3-12-00-02798 1.038^{15}	H_2O 2; eth 3; ace 3; bz 3 1.5613^{20}
7244	1H-Inden-5-amine, 2,3-dihydro-	$C_9H_{11}N$ 133.19	24425-40-9 37.5	 248; 131^{15}	4-12-00-02929	H_2O 2; eth 3; ace 3; bz 3
7245	1H-Indene Indonaphthene	C_9H_8 116.16	95-13-6 -1.8	4851 182	4-05-00-01532 0.9960^{25}	H_2O 1; EtOH 5; eth 5; ace 3 1.5768^{20}
7246	1H-Indene-1-carboxylic acid 1-Indenecarboxylic acid	$C_{10}H_8O_2$ 160.17	5020-21-3 74	 $193\text{-}5^{12}$	3-09-00-03068	H_2O 2; EtOH 4; eth 2; bz 2
7247	1H-Indene, 1,2-dibromo-2,3-dihydro- 1,2-Dibromoindan	$C_9H_8Br_2$ 275.97	20357-79-3 32	 140^{11}	2-05-00-00377 1.7470^{25}	
7248	2H-Indene-2,2-dicarboxylic acid, 1,3-dihydro-1,3-dioxo-, diethyl ester Phthaloylmalonic acid, diethyl ester	$C_{15}H_{14}O_6$ 290.27	116836-20-5		1-10-00-00441 1.1896^{83}	eth 4; EtOH 4 1.541^{84}
7249	1H-Indene, 1,2-dichloro-2,3-dihydro- 1,2-Dichloroindan	$C_9H_8Cl_2$ 187.07	74925-48-7	 $87\text{-}90^{2}$	3-05-00-01202 1.254^{25}	 1.5715^{23}
7250	1H-Indene, 2,3-dihydro- Indan	C_9H_{10} 118.18	496-11-7 -51.4	4844 177.9	4-05-00-01371 0.9639^{20}	H_2O 1; EtOH 5; eth 5; chl 2 1.5378^{20}
7251	1H-Indene, 2,3-dihydro-1,1-dimethyl- 1,1-Dimethylindan	$C_{11}H_{14}$ 146.23	4912-92-9	 191	4-05-00-01415 0.919^{20}	 1.5135^{25}
7252	1H-Indene, 2,3-dihydro-1,2-dimethyl- 1,2-Dimethylindan	$C_{11}H_{14}$ 146.23	17057-82-8	 $79\text{-}80^{10}$	4-05-00-01415 0.927^{20}	 1.5186^{20}
7253	1H-Indene, 2,3-dihydro-4,7-dimethyl- 4,7-Dimethylindan	$C_{11}H_{14}$ 146.23	6682-71-9	 $94\text{-}7^{10}$	4-05-00-01416 0.949^{20}	 1.5342^{20}
7254	1H-Indene, 2,3-dihydro-5,6-dimethyl- 5,6-Dimethylindan	$C_{11}H_{14}$ 146.23	1075-22-5	 94^{10}	4-05-00-01416 0.9449^{20}	 1.5360^{20}
7255	1H-Indene, 2,3-dihydro-1-methyl- 1-Methylindan	$C_{10}H_{12}$ 132.21	767-58-8	 190.6	4-05-00-01397 0.938^{25}	H_2O 1 1.5266^{20}
7256	1H-Indene, 2,3-dihydro-2-methyl- 2-Methylindan	$C_{10}H_{12}$ 132.21	824-63-5	 195	4-05-00-01397 0.940^{25}	H_2O 1 1.5220^{20}
7257	1H-Indene, 2,3-dihydro-4-methyl- 4-Methylindan	$C_{10}H_{12}$ 132.21	824-22-6	 205.3	4-05-00-01397 0.9577^{20}	 1.5356^{20}
7258	1H-Indene, 2,3-dihydro-5-methyl- 5-Methylindan	$C_{10}H_{12}$ 132.21	874-35-1	 202	4-05-00-01398 0.944^{25}	 1.5336^{20}
7259	1H-Indene, 2,3-dihydro-2-phenyl- 2-Phenylindan	$C_{15}H_{14}$ 194.28	22253-11-8	 $162\text{-}3^{10}$	4-05-00-02198 1.0821^{16}	 1.5955^{15}
7260	1H-Indene, 2,3-dihydro-1,1,4,7-tetramethyl- 1,1,4,7-Tetramethylindan	$C_{13}H_{18}$ 174.29	1078-04-2	 114.2^{15}	3-05-00-01282 0.934^{25}	 1.5216^{25}
7261	1H-Indene, 2,3-dihydro-1,1,5-trimethyl- 1,1,5-Trimethylindan	$C_{12}H_{16}$ 160.26	40650-41-7	 86^{10}	4-05-00-01433 0.9119^{20}	 1.5126^{20}
7262	1H-Indene, 2,3-dihydro-1,4,7-trimethyl- 1,4,7-Trimethylindan	$C_{12}H_{16}$ 160.26	54340-87-3	 95^{10}	4-05-00-01434 0.938^{20}	 1.5252^{20}
7263	1H-Indene, 2,3-dihydro-1,1,3-trimethyl-3-phenyl- Indan, 1-phenyl-1,3,3-trimethyl	$C_{18}H_{20}$ 236.36	3910-35-8 52.5	 308.5	4-05-00-02246 1.0009^{20}	H_2O 1; EtOH 3; bz 3; MeOH 3 1.5681^{20}
7264	1H-Indene-1,3(2H)-dione	$C_9H_6O_2$ 146.15	606-23-5 131-2 dec		4-07-00-02344 1.37^{21}	H_2O 2; EtOH 4; eth 3; bz 3
7265	1H-Indene-1,3(2H)-dione, 2-(4-bromophenyl)- Bromindione	$C_{15}H_9BrO_2$ 301.14	1146-98-1 138	1381	4-07-00-02572	
7266	1H-Indene-1,3(2H)-dione, 2-(4-chlorophenyl)- Clorindione	$C_{15}H_9ClO_2$ 256.69	1146-99-2 145.5	2402	4-07-00-02571	bz 4; eth 4; EtOH 4
7267	1H-Indene-1,3(2H)-dione, 2,2-dihydroxy- Ninhydrin	$C_9H_6O_4$ 178.14	485-47-2 241-3 dec	6470	4-07-00-02786	H_2O 4; EtOH 3; eth 2; alk 3

PHYSICAL CONSTANTS OF ORGANIC COMPOUNDS (continued)

No.	Name Synonym	Mol. Form. Mol. Wt.	CAS RN mp/°C	Merck No. bp/°C	Beil. Ref. den/g cm^{-3}	Solubility n_D
7268	1H-Indene-1,3(2H)-dione, 2-(diphenylacetyl)- Diphenadione	$C_{23}H_{16}O_3$ 340.38	82-66-6 146.5	3304	4-07-00-02838	ace 4; HOAc 4 1.670
7269	1H-Indene-1,3(2H)-dione, 2-(4-methoxyphenyl)- Anisindione	$C_{16}H_{12}O_3$ 252.27	117-37-3 156.5	698	3-08-00-02931	
7270	1H-Indene-1,3(2H)-dione, 2-phenyl- Phenindione	$C_{15}H_{10}O_2$ 222.24	83-12-5 150	7196	4-07-00-02570	H_2O 1; EtOH 3; eth 3; ace 3
7271	1H-Indene, 1,3-diphenyl- 1,3-Diphenylindene	$C_{21}H_{16}$ 268.36	4467-88-3 85	230^{15}	3-05-00-02474	ace 4; eth 4
7272	1H-Indene, 2,3-diphenyl- 2,3-Diphenylindene	$C_{21}H_{16}$ 268.36	5324-00-5 109.3	237^{12}	4-05-00-02641	H_2O 1; EtOH 3; eth 3; ace 3
7273	1H-Indene, 1-ethyl-2,3-dihydro- 1-Ethylindan	$C_{11}H_{14}$ 146.23	4830-99-3	212	4-05-00-01414 0.9348^{25}	1.5121^{25}
7274	1H-Indene, 5-hexyl-2,3-dihydro- 5-Hexylindan	$C_{15}H_{22}$ 202.34	54889-55-3 -34	292.1	4-05-00-01474 0.9114^{20}	1.5122^{20}
7275	1H-Indene, 1-methyl- 1-Methylindene	$C_{10}H_{10}$ 130.19	767-59-9	199; 82^{15}	3-05-00-01371 0.970^{25}	1.5616^{20}
7276	1H-Indene, 2-methyl- 2-Methylindene	$C_{10}H_{10}$ 130.19	2177-47-1 80	208	4-05-00-01545 0.974^{25}	H_2O 1; eth 3; ace 3; bz 3 1.5652^{20}
7277	1H-Indene, 3-methyl- 3-Methylindene	$C_{10}H_{10}$ 130.19	767-60-2	198	4-05-00-01545 0.972^{25}	H_2O 1; eth 3; ace 3; bz 3 1.5621^{20}
7278	1H-Indene, 4-methyl- 4-Methylindene	$C_{10}H_{10}$ 130.19	7344-34-5	209; 88^{13}	2-05-00-00418 0.989^{25}	1.568^{20}
7279	1H-Indene, 6-methyl- 6-Methylindene	$C_{10}H_{10}$ 130.19	20232-11-5	207; 90^{17}	4-05-00-01545 0.977^{25}	1.566^{20}
7280	1H-Indene, 7-methyl- 7-Methylindene	$C_{10}H_{10}$ 130.19	7372-92-1	209; 90^{15}	2-05-00-00418 0.989^{25}	1.568^{20}
7281	1H-Indene, octahydro- Hexahydroindan	C_9H_{16} 124.23	496-10-6 -53	167	4-05-00-00292 0.876^{25}	1.4702^{20}
7282	6H-Indeno[1,2-b]oxirene, 1a,6a-dihydro-	C_9H_8O 132.16	768-22-9 24.5	113^{20}; 98^6	5-17-02-00033 1.1255^{24}	chl 3
7283	1H-Inden-4-ol, 7-chloro-2,3-dihydro- Chlorindanol	C_9H_9ClO 168.62	145-94-8 92	2094	4-06-00-03828	
7284	1H-Inden-1-ol, 2,3-dihydro-	$C_9H_{10}O$ 134.18	6351-10-6 54.8	220; 128^{12}	4-06-00-03824	bz 4; EtOH 4; chl 4
7285	1H-Inden-5-ol, 2,3-dihydro-	$C_9H_{10}O$ 134.18	1470-94-6 58	253	4-06-00-03829	H_2O 2; EtOH 4; eth 4; sulf 3
7286	1H-Inden-1-one, 2,3-dihydro-	C_9H_8O 132.16	83-33-0 42	243; 129^{12}	4-07-00-00999 1.0943^{40}	H_2O 2; EtOH 4; eth 4; ace 4 1.561^{25}
7287	2H-Inden-2-one, 1,3-dihydro- 2-Indanone	C_9H_8O 132.16	615-13-4 59	218 dec	4-07-00-01002 1.0712^{69}	H_2O 1; EtOH 4; eth 4; ace 4 1.538^{67}
7288	1H-Inden-1-one, 2,3-dihydro-3,3-dimethyl-	$C_{11}H_{12}O$ 160.22	26465-81-6	$130\text{-}1^{18}$	4-07-00-01035 1.0320^{14}	1.5453^{14}
7289	1H-Inden-1-one, 2,3-dihydro-2-methyl-	$C_{10}H_{10}O$ 146.19	17496-14-9	250; 125^{18}	4-07-00-01019 1.0651^{21}	1.553^{22}
7290	1H-Inden-1-one, 2,3-dihydro-4-methyl- 4-Methyl-1-indanone	$C_{10}H_{10}O$ 146.19	24644-78-8 95	144^{13}; 120^5	4-07-00-01020	bz 4; eth 4; EtOH 4
7291	1H-Inden-1-one, 2,3-dihydro-2-phenyl- 2-Phenyl-1-indanone	$C_{15}H_{12}O$ 208.26	16619-12-8 77.5	344	4-07-00-01673	ace 4; bz 4; EtOH 4
7292	2H-Inden-2-one, octahydro-, *cis*-	$C_9H_{14}O$ 138.21	5689-04-3 10	225; 113^{25}	4-07-00-00174 0.9970^{20}	EtOH 3; bz 3; lig 3 1.4830^{20}
7293	2H-Inden-2-one, octahydro-, *trans*-	$C_9H_{14}O$ 138.21	16484-17-6 -12	218	4-07-00-00174 0.9807^{17}	EtOH 3; bz 3; lig 3 1.4769^{17}
7294	1H-Indole 2,3-Benzopyrrole	C_8H_7N 117.15	120-72-9 52.5	4869 253.6	5-20-07-00005 1.22^{25}	H_2O 3; EtOH 4; eth 4; bz 3
7295	1H-Indole-3-acetic acid Indoleacetic acid	$C_{10}H_9NO_2$ 175.19	87-51-4 168.5	4870	5-22-03-00065	H_2O 1; EtOH 4; eth 3; ace 3
7296	1H-Indole-3-butanoic acid Indolebutyric acid	$C_{12}H_{13}NO_2$ 203.24	133-32-4 124.5	4871	5-22-03-00140	bz 4; DMSO 3; peth 1
7297	1H-Indole, 2,3-dihydro-	C_8H_9N 119.17	496-15-1	229	5-20-06-00238 1.069^{20}	H_2O 2; eth 3; ace 3; bz 3 1.5923^{20}
7298	1H-Indole, 1,3-dimethyl-	$C_{10}H_{11}N$ 145.20	875-30-9 142	258.5	5-20-07-00071	eth 3
7299	1H-Indole, 2,5-dimethyl- 2,5-Dimethylindole	$C_{10}H_{11}N$ 145.20	1196-79-8 115.3	188^{40}	5-20-07-00111	H_2O 1; EtOH 2; eth 3; bz 3
7300	1H-Indole-2,3-dione Isatin	$C_8H_5NO_2$ 147.13	91-56-5 203 dec	4985	5-21-10-00221	H_2O 3; EtOH 4; eth 2; ace 3
7301	1H-Indole-5,6-dione, 2,3-dihydro-3-hydroxy-1-methyl- Adrenochrome	$C_9H_9NO_3$ 179.18	54-06-8 125 dec	162	5-21-13-00079	H_2O 4; EtOH 4; eth 1; bz 1
7302	1H-Indole-3-ethanamine Tryptamine	$C_{10}H_{12}N_2$ 160.22	61-54-1 118	9705 $137^{0.15}$	5-22-10-00045	H_2O 1; EtOH 3; eth 1; ace 3
7303	1H-Indole-3-ethanamine, *N,N*-dimethyl- *N,N*-Dimethyltryptamine	$C_{12}H_{16}N_2$ 188.27	61-50-7 46	3251	5-22-10-00048	
7304	1H-Indole-3-ethanamine, 5-methoxy- 5-Methoxytryptamine	$C_{11}H_{14}N_2O$ 190.25	608-07-1 121.5	5926	5-22-12-00018	
7305	1H-Indole-3-ethanol Tryptophol	$C_{10}H_{11}NO$ 161.20	526-55-6 59	9708 174^2	5-21-03-00061	ace 4; eth 4; EtOH 4; chl 4
7306	1H-Indole-3-ethanol, 2-(5-ethenyl-1-azabicyclo[2.2.2]oct-2-yl)- Cinchonamine	$C_{19}H_{24}N_2O$ 296.41	482-28-0 186	2287	4-23-00-02735	H_2O 1; EtOH 4; eth 4; bz 3

No.	Name Synonym	Mol. Form. Mol. Wt.	CAS RN mp/°C	Merck No. bp/°C	Beil. Ref. den/g cm^{-3}	Solubility n_D
7307	1H-Indole, 1-ethyl-	$C_{10}H_{11}N$ 145.20	10604-59-8 105	 252.5; 122^{14}	5-20-07-00014 1.2563^{15}	
7308	1H-Indole, 3-ethyl-	$C_{10}H_{11}N$ 145.20	1484-19-1 37	 279; 138^{6}	5-20-07-00099	bz 4; eth 4; EtOH 4
7309	1H-Indole, 3-ethyl-2-methyl-	$C_{11}H_{13}N$ 159.23	35246-18-5 44.5	 292; 156^{12}	5-20-07-00136 1.0414^{0}	eth 4; EtOH 4
7310	1H-Indole-3-methanamine, *N*,*N*-dimethyl- Gramine	$C_{11}H_{14}N_2$ 174.25	87-52-5 138.5	4440	5-22-10-00025	H_2O 1; EtOH 3; eth 3; chl 3
7311	1H-Indole, 7-methoxy-	C_9H_9NO 147.18	3189-22-8	 $108\text{-}10^{0.3}$	5-21-03-00024 1.126^{25}	 1.6120^{20}
7312	1H-Indole, 1-methyl-	C_9H_9N 131.18	603-76-9	 237	5-20-07-00012 1.0707^{25}	H_2O 1; EtOH 3; eth 3; bz 3
7313	1H-Indole, 2-methyl-	C_9H_9N 131.18	95-20-5 61	 272	5-20-07-00059 1.07^{20}	H_2O 2; EtOH 4; eth 4; ace 3
7314	1H-Indole, 3-methyl- Skatole	C_9H_9N 131.18	83-34-1 97.5	8503 266	5-20-07-00069	H_2O 3; EtOH 3; eth 3; ace 3
7315	1H-Indole, 4-methyl-	C_9H_9N 131.18	16096-32-5 5	 267	5-20-07-00079 1.062^{20}	
7316	1H-Indole, 5-methyl-	C_9H_9N 131.18	614-96-0 60	 267	5-20-07-00080 1.0202^{78}	H_2O 3; EtOH 3; eth 3; bz 3
7317	1H-Indole, 7-methyl- 7-Methylindole	C_9H_9N 131.18	933-67-5 85	 266	5-20-07-00081 1.0202^{100}	
7318	1H-Indole, 3-methyl-2-phenyl-	$C_{15}H_{13}N$ 207.27	10257-92-8 91.5	 $280\text{-}90^{120}$	5-20-08-00274	H_2O 1; EtOH 3; bz 3
7319	1H-Indole, octahydro-	$C_8H_{15}N$ 125.21	4375-14-8 143	 185.5	5-20-04-00358 0.9472^{20}	bz 4; eth 4; EtOH 4 1.4892^{20}
7320	1H-Indole, 2-phenyl-	$C_{14}H_{11}N$ 193.25	948-65-2 190.5	 250^{10}	5-20-08-00232	H_2O 2; eth 3; bz 3; chl 3
7321	1H-Indole, 1-(phenylmethyl)-	$C_{15}H_{13}N$ 207.27	3377-71-7 45.8	 172^{2}	5-20-07-00016	bz 4; eth 4; EtOH 4; lig 4
7322	1H-Indole, 1,2,3-trimethyl- 1,2,3-Trimethylindole	$C_{11}H_{13}N$ 159.23	1971-46-6 18	 283	5-20-07-00102	H_2O 2; EtOH 2; eth 2; bz 2
7323	1H-Indole, 2,3,5-trimethyl- 2,3,5-Trimethylindole	$C_{11}H_{13}N$ 159.23	21296-92-4 121.5	 297	5-20-07-00143	EtOH 4; chl 4; lig 4
7324	Indolizine	C_8H_7N 117.15	274-40-8 75	 205	5-20-07-00047	H_2O 1; EtOH 3
7325	Indolizine, octahydro-	$C_8H_{15}N$ 125.21	13618-93-4	 75^{43}	5-20-04-00360 0.9074^{10}	eth 4; EtOH 4 1.4748
7326	1H-Indol-5-ol, 3-[2-(dimethylamino)ethyl]- Bufotenine	$C_{12}H_{16}N_2O$ 204.27	487-93-4 146.5	1467 $320^{0.1}$	5-22-12-00026	eth 4; EtOH 4
7327	2H-Indol-2-one, 1,3-dihydro-	C_8H_7NO 133.15	59-48-3 128	 227^{23}; 195^{17}	5-21-08-00007	H_2O 3; EtOH 3; eth 3
7328	2H-Indol-2-one, 1,3-dihydro-1,3-dimethyl-	$C_{10}H_{11}NO$ 161.20	24438-17-3 55	 273; 137^{11}	5-21-08-00077	eth 4; EtOH 4
7329	2H-Indol-2-one, 1,3-dihydro-3,3-dimethyl-	$C_{10}H_{11}NO$ 161.20	19155-24-9 154.3	 300	5-21-08-00113	bz 4; eth 4; EtOH 4
7330	2H-Indol-2-one, 1,3-dihydro-4,7-dimethyl- 4,7-Dimethyl-2-indolinone	$C_{10}H_{11}NO$ 161.20	59022-71-8 159	 $205\text{-}10^{15}$	4-21-00-03676	ace 4; EtOH 4; chl 4
7331	3H-Indol-3-one, 2-(1,3-dihydro-3-oxo-2H-indol-2-ylidene)-1,2-dihydro- Indigo	$C_{16}H_{10}N_2O_2$ 262.27	482-89-3 390 dec	4855 sub 300	5-24-08-00503	
7332	2H-Indol-2-one, 1-ethyl-1,3-dihydro-3-methyl- 2-Indolinone, 1-ethyl-3-methyl-	$C_{11}H_{13}NO$ 175.23	84258-49-1	 283; $105^{0.5}$	2-21-00-00258	eth 4; EtOH 4 1.557^{25}
7333	Indolo[2',3':3,4]pyrido[2,1-b]quinazolin-5(7H)-one, 8,13,13b,14-tetrahydro-14-methyl-, (*S*)- Evodiamine	$C_{19}H_{17}N_3O$ 303.36	518-17-2 28	3865	4-26-00-00525	
7334	Indolo[2',3':3,4]pyrido[2,1-b]quinazolin-5(7H)-one, 8,13-dihydro- Rutecarpine	$C_{18}H_{13}N_3O$ 287.32	84-26-4 259.5	8271	4-26-00-00528	EtOH 2; ace 2; bz 2
7335	Inosine Hypoxanthine riboside	$C_{10}H_{12}N_4O_5$ 268.23	58-63-9 218 dec	4880	4-26-00-02086	EtOH 4
7336	*scyllo*-Inositol	$C_6H_{12}O_6$ 180.16	488-59-5 353 dec		4-06-00-07920 1.659^{19}	H_2O 2; EtOH 1; eth 1; bz 1
7337	Iodine cyanide Cyanogen iodide	CIN 152.92	506-78-5 146.7	2702 sub 45	4-03-00-00093 2.84^{18}	eth 4; EtOH 4
7338	Iprodione 1-Imidazolidinecarboxamide, 3-(3,5-dichlorophenyl)-*N*-(1-methylethyl)-2,4-dioxo-	$C_{13}H_{13}Cl_2N_3O_3$ 330.17	36734-19-7 136	4964		
7339	Iron carbonyl Iron pentacarbonyl	C_5FeO_5 195.90	13463-40-6 -20	4980 103	 1.5^{20}	H_2O 1; EtOH 2; bz 3; ace 3 1.453^{22}
7340	Isazophos Phosphorothioic acid, *O*-[2-chloro-l-(l-methylethyl)-lH-imidazol-4-yl] *O*,*O*-diethyl ester	$C_9H_{17}ClN_3O_3PS$ 313.74	67329-04-8	4988 170; $100^{0.001}$	 1.22^{20}	
7341	5-Isobenzofurancarboxylic acid, 1,3-dihydro-1,3-dioxo- Trimellitic anhydride	$C_9H_4O_5$ 192.13	552-30-7 162	9617 $240\text{-}3^{14}$	5-18-08-00562	
7342	1,3-Isobenzofurandione Phthalic anhydride	$C_8H_4O_3$ 148.12	85-44-9 130.8	7346 295	5-17-11-00253 1.527^{4}	H_2O 2; EtOH 3; eth 2; ace 3

No.	Name Synonym	Mol. Form. Mol. Wt.	CAS RN mp/°C	Merck No. bp/°C	Beil. Ref. den/g cm^{-3}	Solubility n_D
7343	1,3-Isobenzofurandione, 4,5-dichloro- Phthalic anhydride, 3,4-dichloro	$C_8H_2Cl_2O_3$ 217.01	56962-07-3 121	 329	4-17-00-06142	EtOH 4; chl 4
7344	1,3-Isobenzofurandione, 5,6-dichloro- 4,5-Dichlorophthalic anhydride	$C_8H_2Cl_2O_3$ 217.01	942-06-3 188	 313	5-17-11-00260	eth 4; EtOH 4; tol 4
7345	1,3-Isobenzofurandione, hexahydro- Hexahydrophthalic anhydride	$C_8H_{10}O_3$ 154.17	85-42-7 32	 145^{18}	5-17-11-00076	
7346	1,3-Isobenzofurandione, 4,5,6,7-tetrachloro- Tetrachlorophthalic anhydride	$C_8Cl_4O_3$ 285.90	117-08-8 254.5	 sub	5-17-11-00260 1.49^{275}	eth 2
7347	1,3-Isobenzofurandione, 4,5,6,7-tetrahydro- 3,4,5,6-Tetrahydrophthalic anhydride	$C_8H_8O_3$ 152.15	2426-02-0 74		5-17-11-00133 1.2^{105}	EtOH 3; eth 4; ace 3; chl 3
7348	1,3-Isobenzofurandione, 4,5,6,7-tetraiodo-	$C_8I_4O_3$ 651.70	632-80-4 327.5	 sub	5-17-11-00266	H_2O 1; EtOH 1; bz 1; HOAc 2
7349	1(3H)-Isobenzofuranone	$C_8H_6O_2$ 134.13	87-41-2 75	 290	5-17-10-00007 1.1636^{99}	H_2O 3; EtOH 4; eth 4; chl 2 1.536^{99}
7350	1(3H)-Isobenzofuranone, 3,3-bis(3,5-dibromo-4-hydroxyphenyl)- 3',3'',5',5''-Tetrabromophenolphthalein	$C_{20}H_{10}Br_4O_4$ 633.91	76-62-0 296	9122	5-18-04-00190	H_2O 1; EtOH 2; eth 4; alk 3
7351	1(3H)-Isobenzofuranone, 3,3-bis(4-hydroxy-3,5-diiodophenyl)-	$C_{20}H_{10}I_4O_4$ 821.91	386-17-4 308		4-18-00-01949 2.0201^{22}	H_2O 1; EtOH 1; eth 1; chl 2
7352	1(3H)-Isobenzofuranone, 3,3-bis[4-hydroxy-2-methyl-5-(1-methylethyl)phenyl]- Thymolphthalein	$C_{28}H_{30}O_4$ 430.54	125-20-2 253	9336	5-18-04-00194	H_2O 1; EtOH 3; eth 3; ace 3
7353	1(3H)-Isobenzofuranone, 3,3-bis(4-hydroxy-3-methylphenyl)- *o*-Cresolphthalein	$C_{22}H_{18}O_4$ 346.38	596-27-0 223	2582	5-18-04-00193	EtOH 4
7354	1(3H)-Isobenzofuranone, 3,3-bis(4-hydroxyphenyl)- Phenolphthalein	$C_{20}H_{14}O_4$ 318.33	77-09-8 262.5	7208	5-18-04-00188 1.277^{32}	H_2O 1; EtOH 4; eth 3; ace 4
7355	1(3H)-Isobenzofuranone, 6,7-dimethoxy- Meconin	$C_{10}H_{10}O_4$ 194.19	569-31-3 102.5	5665	5-18-03-00068	H_2O 2; EtOH 3; eth 3; ace 3
7356	1(3H)-Isobenzofuranone, 6,7-dimethoxy-3-(5,6,7,8-tetrahydro-4-methoxy-6-methyl-1,3-dioxolo[4,5-g]isoquinolin-5-yl)- Noscapine	$C_{22}H_{23}NO_7$ 413.43	128-62-1 176	6638	4-27-00-06838	H_2O 1; EtOH 3; eth 2; ace 4
7357	1(3H)-Isobenzofuranone, 6,7-dimethoxy-3-(5,6,7,8-tetrahydro-6-methyl-1,3-dioxolo[4,5-g]isoquinolin-5-yl)-, [*S*-(R*, S*)]- Hydrastine	$C_{21}H_{21}NO_6$ 383.40	118-08-1 132	4688	4-27-00-06833	H_2O 1; ace 3; bz 3
7358	1(3H)-Isobenzofuranone, 3,3-diphenyl-	$C_{20}H_{14}O_2$ 286.33	596-29-2 120	 235^{15}	5-17-10-00735	chl 2
7359	Isofenphos Benzoic acid, 2-[[ethoxy[(1-methylethyl)amino]phosphinothioyl]oxy]-, 1-methylethyl ester	$C_{15}H_{24}NO_4PS$ 345.40	25311-71-1 <-12	5055 $120^{0.01}$	 1.134^{20}	
7360	Isofulminic acid	CHNO 43.03	51060-05-0 stable only at low temp.			
7361	2H-Isoindole-2-acetic acid, 1,3-dihydro-1,3-dioxo-, ethyl ester	$C_{12}H_{11}NO_4$ 233.22	6974-10-3 112.5	 300	5-21-10-00431	bz 4; eth 4; EtOH 4; chl 4
7362	1H-Isoindole-1,3(2H)-dione Phthalimide	$C_8H_5NO_2$ 147.13	85-41-6 238	7347	5-21-10-00270	bz 4
7363	1H-Isoindole-1,3(2H)-dione, 2-(2,6-dioxo-3-piperidinyl)- Thalidomide	$C_{13}H_{10}N_2O_4$ 258.23	50-35-1 270	9182	5-22-13-00224	py 4; diox 4
7364	1H-Isoindole-1,3(2H)-dione, *N*-ethyl-	$C_{10}H_9NO_2$ 175.19	5022-29-7 79	 285.5	5-21-10-00274	EtOH 3; eth 3
7365	1H-Isoindole-1,3(2H)-dione, 2-methyl-	$C_9H_7NO_2$ 161.16	550-44-7 134	 286	5-21-10-00273	H_2O 1; EtOH 2
7366	1H-Isoindole-1,3(2H)-dione, 2-phenyl-	$C_{14}H_9NO_2$ 223.23	520-03-6 210	 sub	5-21-10-00292	H_2O 1; EtOH 2; chl 5
7367	1H-Isoindole-1,3(2H)-dione, 2-(phenylmethyl)-	$C_{15}H_{11}NO_2$ 237.26	2142-01-0 116		5-21-10-00299 1.343^{18}	EtOH 3; DMSO-d_6 2; HOAc 3
7368	1H-Isoindole-1,3(2H)-dione, 3a,4,7,7a-tetrahydro-2-[(trichloromethyl)thio]- Captan	$C_9H_8Cl_3NO_2S$ 300.59	133-06-2 172.5	1771	5-21-10-00136 1.74^{25}	chl 4
7369	1H-Isoindol-1-one, 2,3-dihydro-	C_8H_7NO 133.15	480-91-1 151	 338; 103^{18}	5-21-08-00018	eth 4; EtOH 4; chl 4
7370	*L*-Isoleucine 2-Amino-3-methylpentanoic acid	$C_6H_{13}NO_2$ 131.17	73-32-5 284 dec	5064	4-04-00-02775	H_2O 3; EtOH 1
7371	Isopropalin Benzenamine, 4-(1-methylethyl)-2,6-dinitro-*N,N*-dipropyl-	$C_{15}H_{23}N_3O_4$ 309.37	33820-53-0	5089		
7372	5-Isoquinolinamine 5-Aminoisoquinoline	$C_9H_8N_2$ 144.18	1125-60-6 128	 196^{18}	5-22-10-00342	chl 2; lig 3
7373	Isoquinoline Benzopyridine	C_9H_7N 129.16	119-65-3 26.47	5110 243.2	5-20-07-00333 1.0910^{30}	H_2O 1; EtOH 4; eth 5; ace 4 1.6148^{20}
7374	Isoquinoline, 4-bromo- 4-Bromoisoquinoline	C_9H_6BrN 208.06	1532-97-4 41.5	 282.5	5-20-07-00363	eth 4
7375	5-Isoquinolinecarbonitrile	$C_{10}H_6N_2$ 154.17	27655-41-0 137	 sub 100	4-22-00-01208	eth 4; EtOH 4

No.	Name Synonym	Mol. Form. Mol. Wt.	CAS RN mp/°C	Merck No. bp/°C	Beil. Ref. den/g cm^{-3}	Solubility n_D
7376	Isoquinoline, 1-chloro- 1-Chloroisoquinoline	C_9H_6ClN 163.61	19493-44-8 37.5	 274.5	5-20-07-00361	H_2O 1; bz 3
7377	Isoquinoline, 1-[(3,4-diethoxyphenyl)methyl]-6,7-diethoxy- Isoquinoline, 6,7-diethoxy-1(3,4-diethoxybenzyl)	$C_{24}H_{29}NO_4$ 395.50	486-47-5 100	3682	5-21-06-00183	H_2O 1; EtOH 3; eth 2; chl 2
7378	Isoquinoline, 1-[(3,4-dimethoxyphenyl)methyl]-6,7-dimethoxy- Papaverine	$C_{20}H_{21}NO_4$ 339.39	58-74-2 147.5	6968 sub 135	5-21-06-00182 1.337^{20}	H_2O 2; EtOH 4; eth 2; ace 3 1.625
7379	Isoquinoline, 1-[(3,4-dimethoxyphenyl)methyl]-1,2,3,4-tetrahydro-6,7-dimethoxy-2-methyl-, (*S*)- Laudanosine	$C_{21}H_{27}NO_4$ 357.45	2688-77-9 89	5249	5-21-06-00049	ace 4; eth 4; EtOH 4; chl 4
7380	Isoquinoline, 7-methoxy- 7-Methoxyisoquinoline	$C_{10}H_9NO$ 159.19	39989-39-4 49	 194^{50}, 184^{34}	5-21-03-00326	H_2O 1; EtOH 3; lig 3
7381	Isoquinoline, 1-methyl- Isoquinaldine	$C_{10}H_9N$ 143.19	1721-93-3 10	 248	5-20-07-00407 1.0777^{20}	H_2O 2; eth 3; ace 3; bz 3 1.6095^{20}
7382	Isoquinoline, 3-methyl- 3-Methylisoquinoline	$C_{10}H_9N$ 143.19	1125-80-0 68	 249	5-20-07-00410	H_2O 2; eth 3; ace 3; chl 2
7383	Isoquinoline, 4-methyl- 4-Methylisoquinoline	$C_{10}H_9N$ 143.19	1196-39-0 <-75	 256	5-20-07-00413	H_2O 2; eth 3; ace 3; bz 3
7384	Isoquinoline, 6-methyl- 6-Methylisoquinoline	$C_{10}H_9N$ 143.19	42398-73-2 85.5	 265.5	5-20-07-00414	ace 4; bz 4; eth 4; EtOH 4
7385	Isoquinoline, 8-methyl- 8-Methylisoquinoline	$C_{10}H_9N$ 143.19	62882-00-2 205	 258	5-20-07-00414	ace 4; bz 4; eth 4
7386	Isoquinoline, 5-nitro- 5-Nitroisoquinoline	$C_9H_6N_2O_2$ 174.16	607-32-9 110	 sub	5-20-07-00365	H_2O 2; EtOH 4; eth 4; chl 4
7387	Isoquinoline, 3-(phenylmethyl)- Isoquinoline, 3-benzyl	$C_{16}H_{13}N$ 219.29	90210-56-3 104	 311^{23}	5-20-08-00383	bz 4; chl 4
7388	Isoquinoline, 1,2,3,4-tetrahydro-	$C_9H_{11}N$ 133.19	91-21-4 <-15	 232.5	5-20-06-00320 1.0642^{24}	H_2O 1; EtOH 3; bz 3; acid 3 1.5668^{20}
7389	Isoquinoline, 5,6,7,8-tetrahydro- 5,6,7,8-Tetrahydroisoquinoline	$C_9H_{11}N$ 133.19	36556-06-6 146.5	 218; 106^{13}	5-20-06-00319 1.0504^{10}	H_2O 1; ace 3; bz 3; chl 3 1.5276^{10}
7390	Isoquinoline, 1,2,3,4-tetrahydro-6,7-dimethoxy-1,2-dimethyl-, (±)- Carnegine	$C_{13}H_{19}NO_2$ 221.30	490-53-9	1854 170^{1}	5-21-04-00501	H_2O 4; eth 4; EtOH 4
7391	Isoquinoline, 1,2,3,4-tetrahydro-6,7-dimethoxy-1,2-dimethyl- Isoquinoline, 6,7-dimethoxy-1,2-dimethyl-1,2,3,4-tetrahydro	$C_{13}H_{19}NO_2$ 221.30	71783-56-7 262.5	 170^{1}; $100^{0.01}$	5-21-04-00501	H_2O 3; EtOH 3; eth 3; chl 3
7392	7-Isoquinolinol, 1-[(3,4-dimethoxyphenyl)methyl]-1,2,3,4-tetrahydro-6-methoxy-2-methyl-, (*S*)- Codamine	$C_{20}H_{25}NO_4$ 343.42	21040-59-5 127	2458	4-21-00-02704	eth 4; EtOH 4; chl 4
7393	8-Isoquinolinol, 1,2,3,4-tetrahydro-6,7-dimethoxy- Anhalamine	$C_{11}H_{15}NO_3$ 209.25	643-60-7 187.5	682	5-21-05-00513	eth 4; EtOH 4
7394	8-Isoquinolinol, 1,2,3,4-tetrahydro-6,7-dimethoxy-1,2-dimethyl-, (*S*)- Pellotine	$C_{13}H_{19}NO_3$ 237.30	83-14-7 111.5	7018	4-21-00-02525	ace 4; eth 4; EtOH 4; peth 4
7395	8-Isoquinolinol, 1,2,3,4-tetrahydro-6,7-dimethoxy-1-methyl-, (*S*)- Anhalonidine	$C_{12}H_{17}NO_3$ 223.27	17627-77-9 160.5	683	4-21-00-02524	H_2O 4; EtOH 4
7396	7-Isoquinolinol, 1,2,3,4-tetrahydro-1-[(4-hydroxyphenyl)methyl]-6-methoxy-, (*S*)- Coclaurine	$C_{17}H_{19}NO_3$ 285.34	486-39-5 220.5	2455	4-21-00-02605	
7397	6-Isoquinolinol, 1,2,3,4-tetrahydro-7-methoxy-1-methyl-, (*S*)- Salsoline	$C_{11}H_{15}NO_2$ 193.25	89-31-6 221.5	8308	5-21-04-00499	H_2O 2; EtOH 2; eth 1; bz 2
7398	1(2H)-Isoquinolinone, 3,4-dihydro-6,7-dimethoxy- Corydaldine	$C_{11}H_{13}NO_3$ 207.23	493-49-2 175	2542	5-21-13-00086	H_2O 4; bz 4; eth 4; EtOH 4
7399	Isothiocyanic acid	CHNS 59.09	3129-90-6			
7400	*DL*-Isovaline	$C_5H_{11}NO_2$ 117.15	595-39-1 316.5	 sub 300	4-04-00-02655	H_2O 3; EtOH 3; eth 2; bz 1
7401	*L*-Isovaline 2-Amino-2-methylbutyric acid	$C_5H_{11}NO_2$ 117.15	595-40-4	5124	4-04-00-02655	
7402	Isoxazole 1-Oxa-2-azacyclopentadiene	C_3H_3NO 69.06	288-14-2	 95	4-27-00-00959 1.078^{20}	H_2O 3 1.4298^{17}
7403	Isoxazole, 3,5-dimethyl- 3,5-Dimethylisoxazole	C_5H_7NO 97.12	300-87-8	 143	4-27-00-00978 0.99^{25}	 1.4421^{20}
7404	5-Isoxazolemethanol, 3-methyl-	$C_5H_7NO_2$ 113.12	14716-89-3	 130^{20}	4-27-00-01745 1.132^{25}	 1.4790^{20}
7405	Isoxazole, 5-methyl- 5-Methylisoxazole	C_4H_5NO 83.09	5765-44-6	 122	4-27-00-00966 1.023^{20}	DMSO 3 1.4386^{20}
7406	3-Isoxazolidinone, 4-amino-, (*R*)- Cycloserine	$C_3H_6N_2O_2$ 102.09	68-41-7 155-6 dec	2758	4-27-00-05549	H_2O 3; MeOH 2
7407	Isoxazolium, 2-ethyl-5-(3-sulfophenyl)-, hydroxide, inner salt Woodward's Reagent K	$C_{11}H_{11}NO_4S$ 253.28	4156-16-5 dec 207	9962		

No.	Name Synonym	Mol. Form. Mol. Wt.	CAS RN mp/°C	Merck No. bp/°C	Beil. Ref. den/g cm^{-3}	Solubility n_D
7408	Kepone 1,3,4-Metheno-2H-cyclobuta[cd]pentalen-2-one, 1,1a,3,3a,4,5,5,5a,5b,6-decachloro-octahydro-	$C_{10}Cl_{10}O$ 490.64	143-50-0 350 dec	2081	1.61^{25}	
7409	Ketone, 1-methyl-4-piperidyl phenyl Piperidine, 4-benzoyl-*N*-methyl	$C_{13}H_{17}NO$ 203.28	92040-00-1 36	190^{21}; 133^{2}	4-21-00-03689	H_2O 3; EtOH 3; eth 4; ace 3 1.5430^{23}
7410	Lactofen (±)-2-Ethoxy-1-ethyl-2-oxoethyl 5-[2-chloro-4-(trifluoromethyl)phenoxy]-2-nitrobenzoate	$C_{19}H_{15}ClF_3NO_7$ 461.78	77501-63-4			
7411	Lanosta-8,24-dien-3-ol, (3β)- Lanosterol	$C_{30}H_{50}O$ 426.73	79-63-0 140.5	5232	4-06-00-04188	eth 4; EtOH 4; chl 4
7412	*DL*-Leucine	$C_6H_{13}NO_2$ 131.17	328-39-2 293	sub	2-04-00-00870 1.293^{18}	H_2O 3; EtOH 2; eth 1
7413	*L*-Leucine 2-Amino-4-methylpentanoic acid	$C_6H_{13}NO_2$ 131.17	61-90-5 293	5331 sub	4-04-00-02738 1.293^{18}	H_2O 2; EtOH 1; eth 1
7414	*DL*-Leucine, *N*-glycyl-	$C_8H_{16}N_2O_3$ 188.23	688-14-2 242 dec		4-04-00-02766 1.181^{25}	H_2O 4; EtOH 1
7415	Linuron Urea, *N*'-(3,4-dichlorophenyl)-*N*-methoxy-*N*-methyl-	$C_9H_{10}Cl_2N_2O_2$ 249.10	330-55-2 93	5387		
7416	Lup-20(29)-ene-3,28-diol, (3β)- Betulin	$C_{30}H_{50}O_2$ 442.73	473-98-3 250	1212 sub 240	4-06-00-06534	H_2O 1; EtOH 2; eth 3; bz 2
7417	Lup-20(29)-en-3-ol, (3β)- Lupeol	$C_{30}H_{50}O$ 426.73	545-47-1 216	5478	4-06-00-04200 0.9457^{218}	H_2O 1; EtOH 4; eth 4; ace 4 1.4910^{218}
7418	*DL*-Lysine 2,6-Diaminohexanoic acid	$C_6H_{14}N_2O_2$ 146.19	70-54-2 224		2-04-00-00858	H_2O 2
7419	*L*-Lysine 2,6-Diaminohexanoic acid	$C_6H_{14}N_2O_2$ 146.19	56-87-1 224-5 dec	5509	4-04-00-02717	H_2O 4; EtOH 1; eth 1; ace 1
7420	Maneb Manganese, [[1,2-ethanediylbis[carbamodithioato]](2-)]-	$C_4H_6MnN_2S_4$ 265.31	12427-38-2 dec 200	5603		
7421	Manganous dimethyldithiocarbamate Carbamic acid, dimethyldithio-, manganese(2+) salt	$C_6H_{12}MnN_2S_4$ 295.38	15339-36-3			
7422	*D*-Mannitol Cordycepic acid	$C_6H_{14}O_6$ 182.17	69-65-8 168	5629 $295^{3.5}$	4-01-00-02841 1.489^{20}	H_2O 4; EtOH 2; eth 1; py 2 1.3330
7423	*D*-Mannitol, 1,4:3,6-dianhydro-	$C_6H_{10}O_4$ 146.14	641-74-7 88	274 dec	5-19-03-00201	H_2O 4
7424	*D*-Mannitol, hexanitrate Mannitol, hexanitrate	$C_6H_8N_6O_{18}$ 452.16	15825-70-4 107	5630 exp	4-01-00-02849 1.8^{20}	bz 4; eth 4; EtOH 4
7425	*D*-Mannose Seminose	$C_6H_{12}O_6$ 180.16	3458-28-4 132 dec	5632	4-01-00-04328 1.539^{20}	H_2O 4; EtOH 2; eth 1; bz 1
7426	Matridin-15-one Matrine	$C_{15}H_{24}N_2O$ 248.37	519-02-8	5641 223^{6}	5-24-02-00301	H_2O 3; EtOH 4; eth 3; ace 3 1.5286^{25}
7427	Mefluidide Acetamide, *N*-[2,4-dimethyl-5-[[(trifluoromethyl)sulfonyl]amino]phenyl]-	$C_{11}H_{13}F_3N_2O_3S$ 310.30	53780-34-0 184	5684		
7428	*m*-Menthan-6-ol Cyclohexanol-4-isopropyl-2-methyl	$C_{10}H_{20}O$ 156.27	1490-05-7	119.21^{28}	2-06-00-00038 0.9140^{20}	eth 4; EtOH 4 1.4666^{20}
7429	*p*-Menth-3-ene, (*S*)-(-)- Cyclohexene, 1-isopropyl-4-methyl (*l*)	$C_{10}H_{18}$ 138.25	22564-83-6	167	4-05-00-00301 0.8099^{19}	H_2O 1; EtOH 3; eth 3; bz 3 1.4511^{20}
7430	Mepiquat chloride Piperidinium, 1,1-dimethyl-, chloride	$C_7H_{16}NCl$ 149.66	24307-26-4 223	5746		
7431	Mercury, (acetato-O)phenyl- Phenylmercuric acetate	$C_8H_8HgO_2$ 336.74	62-38-4 153	7271	4-16-00-01720	chl 3
7432	Mercury, bis(4-methylphenyl)- Bis(*p*-tolyl)mercury	$C_{14}H_{14}Hg$ 382.86	537-64-4 245.7	3387	4-16-00-01705	
7433	Mercury, dibutyl- Dibutylmercury	$C_8H_{18}Hg$ 314.82	629-35-6	105^{10}	4-04-00-04429 1.7779^{20}	1.5057^{20}
7434	Mercury, diethyl- Diethyl mercury	$C_4H_{10}Hg$ 258.71	627-44-1	159; 57^{16}	2.43^{20}	eth 3; EtOH 2
7435	Mercury, dimethyl- Dimethyl mercury	C_2H_6Hg 230.66	593-74-8	3238 93	3.17^{25}	H_2O 1; EtOH 4; eth 4 1.5452^{20}
7436	Mercury, diphenyl- Benzene, mercuriodi-	$C_{12}H_{10}Hg$ 354.80	587-85-9	204^{10}	4-16-00-01702 2.318^{25}	H_2O 1; EtOH 2; eth 2; bz 3
7437	Mercury, [2-methyl-5-nitrophenolato(2-)-C6,O1]- Nitromersol	$C_7H_5HgNO_3$ 351.71	133-58-4	6531		H_2O 1; ace 2; EtOH 2; alk 3
7438	Merphos Phosphorotrithious acid, *S,S,S*-tributyl ester	$C_{12}H_{27}PS_3$ 298.52	150-50-5 100	$137^{0.7}$; 176^{15}	1.02^{20}	
7439	Metalaxyl *N*-(2,6-Dimethylphenyl)-*N*-(methoxyacetyl)alanine, methyl ester	$C_{15}H_{21}NO_4$ 279.34	57837-19-1 71	5826		
7440	Metaldehyde Acetaldehyde polymer	$(C_2H_4O)_x$	37273-91-9 246	5827 sub 115	4-19-00-05643	H_2O 1; EtOH 2; eth 2; ace 1
7441	Methamidophos Phosphoramidothioic acid, *O,S*-dimethyl ester	$C_2H_8NO_2PS$ 141.13	10265-92-6 46	5858	1.31^{20}	

No.	Name Synonym	Mol. Form. Mol. Wt.	CAS RN mp/°C	Merck No. bp/°C	Beil. Ref. den/g cm^{-3}	Solubility n_D
7442	Methanamine Methylamine	CH_5N 31.06	74-89-5 -93.4	5938 -6.3	4-04-00-00118 *0.656^{25}	H_2O 4; EtOH 3; eth 5; ace 3
7443	Methanamine, 1,1-diethoxy-*N*,*N*-dimethyl-	$C_7H_{17}NO_2$ 147.22	1188-33-6	 129	4-04-00-00174 0.859^{25}	 1.4007^{20}
7444	Methanamine, *N*,*N*-dimethyl- Trimethylamine	C_3H_9N 59.11	75-50-3 -117.1	9625 2.8	4-04-00-00134 *0.627^{25}	H_2O 4; EtOH 3; eth 3; bz 3 1.3631^{0}
7445	Methanamine, *N*,*N*-dimethyl-, hydrochloride	$C_3H_{10}ClN$ 95.57	593-81-7 277.5	 sub 200	4-04-00-00134	H_2O 4; EtOH 4; chl 4
7446	Methanamine, hydrochloride	CH_6ClN 67.52	593-51-1 227.5	5938 227^{15}	4-04-00-00118	H_2O 4; EtOH 4
7447	Methanamine, *N*-hydroxy- Hydroxylamine, *N*-methyl	CH_5NO 47.06	593-77-1 87.5	 62.5^{15}	4-04-00-03299 1.0003^{20}	H_2O 4; EtOH 4 1.4164^{20}
7448	Methanamine, *N*-methyl- Dimethylamine	C_2H_7N 45.08	124-40-3 -92.2	3217 6.8	4-04-00-00128 0.6804^{0}	H_2O 4; EtOH 3; eth 3 1.350^{17}
7449	Methanamine, *N*-methyl-*N*-nitro-	$C_2H_6N_2O_2$ 90.08	4164-28-7 58	 187	4-04-00-03401 1.1090^{72}	H_2O 4; ace 4; eth 4; EtOH 4 1.4462^{72}
7450	Methanamine, *N*-methyl-*N*-nitroso- *N*-Nitrosodimethylamine	$C_2H_6N_2O$ 74.08	62-75-9	6558 154	4-04-00-03384 1.0059^{20}	H_2O 3; EtOH 3; eth 3; chl 3 1.4358^{20}
7451	Methanamine, *N*-nitro-	$CH_4N_2O_2$ 76.06	598-57-2 38	 $80\text{-}5^{10}$	4-04-00-03400 1.2433^{49}	H_2O 4; EtOH 4; eth 3; bz 4 1.4616^{49}
7452	Methanamine, *N*-(phenylmethylene)-	C_8H_9N 119.17	622-29-7	 185; 92^{34}	4-07-00-00518 0.9671^{14}	EtOH 3; eth 3; ace 3; chl 3 1.5526^{20}
7453	Methanaminium, *N*-[4-[bis[4-(dimethylamino)phenyl]methylene]-2,5-cyclohexadien-1-ylidene]-*N*-methyl-, chloride Crystal violet	$C_{25}H_{30}ClN_3$ 407.99	548-62-9 215 dec	4287		H_2O 4; chl 4
7454	Methanaminium, 1-carboxy-*N*,*N*,*N*-trimethyl-, hydroxide, inner salt Betaine	$C_5H_{11}NO_2$ 117.15	107-43-7 293 dec	1201	4-04-00-02369	H_2O 4; EtOH 3; eth 2; chl 2
7455	Methanaminium, *N*-[4-[[4-(dimethylamino)phenyl]phenylmethylene]-2,5-cyclohexadien-1-ylidene]-*N*-methyl-, chlorid Malachite green	$C_{23}H_{25}ClN_2$ 364.92	569-64-2	5581		H_2O 4; EtOH 4; MeOH 4
7456	Methanaminium, *N*,*N*,*N*-trimethyl-, bromide Tetramethylammonium bromide	$C_4H_{12}BrN$ 154.05	64-20-0 230 dec		4-04-00-00145 1.56^{25}	H_2O 4; EtOH 2; eth 1; bz 1
7457	Methanaminium, *N*,*N*,*N*-trimethyl-, chloride Tetramethylammonium chloride	$C_4H_{12}ClN$ 109.60	75-57-0 420 dec		4-04-00-00145 1.169^{20}	H_2O 3; EtOH 2; eth 1; bz 1
7458	Methanaminium, *N*,*N*,*N*-trimethyl-, iodide Tetramethylammonium iodide	$C_4H_{12}IN$ 201.05	75-58-1 >230 dec	9156	4-04-00-00145 1.829^{25}	H_2O 2; EtOH 2; eth 1; ace 2
7459	Methane	CH_4 16.04	74-82-8 -182.4	5863 -161.5	4-01-00-00003 0.4228^{-162}	H_2O 3; EtOH 3; eth 3; ace 2
7460	Methane, azido Methyl azide	CH_3N_3 57.06	624-90-8	 20.5; exp 500	 0.869^{15}	
7461	Methane, bis(methylthio)- Bis[methylmercapto]methane	$C_3H_8S_2$ 108.23	1618-26-4	1267 148	4-01-00-03088	
7462	Methane, bromo- Methyl bromide	CH_3Br 94.94	74-83-9 -93.7	5951 3.5	4-01-00-00068 1.6755^{20}	H_2O 2; EtOH 5; eth 5; chl 5 1.4218^{20}
7463	Methane, bromochloro- Bromochloromethane	CH_2BrCl 129.38	74-97-5 -87.9	 68.0	4-01-00-00074 1.9344^{20}	H_2O 1; EtOH 3; eth 3; ace 3 1.4838^{20}
7464	Methane, bromochlorodifluoro- Bromochlorodifluoromethane	$CBrClF_2$ 165.36	353-59-3 -159.5	 -3.7	4-01-00-00075	
7465	Methane, bromochlorodinitro- Bromochlorodinitromethane	$CBrClN_2O_4$ 219.38	33829-48-0 9.3	 $75\text{-}6^{15}$	2-01-00-00044 2.0393^{20}	EtOH 4 1.4793
7466	Methane, bromochlorofluoro- Bromochlorofluoromethane	$CHBrClF$ 147.37	593-98-6 -115	 36	4-01-00-00075 1.9771^{0}	H_2O 1; eth 3; ace 3; chl 3 1.4144^{25}
7467	Methane, bromodichloro- Bromodichloromethane	$CHBrCl_2$ 163.83	75-27-4 -57	 90	4-01-00-00076 1.980^{20}	H_2O 1; EtOH 4; eth 4; ace 4 1.4964^{20}
7468	Methane, bromodichlorofluoro- Bromodichlorofluoromethane	$CBrCl_2F$ 181.82	353-58-2		4-01-00-00076	
7469	Methane, bromodifluoro- Bromodifluoromethane	$CHBrF_2$ 130.92	1511-62-2 -14.5		4-01-00-00072 1.55^{16}	H_2O 3; EtOH 4
7470	Methane, bromofluoro- Bromofluoromethane	CH_2BrF 112.93	373-52-4	 19	4-01-00-00072	EtOH 3; chl 4
7471	Methane, bromoiodo- Bromoiodomethane	CH_2BrI 220.84	557-68-6	 139.5	4-01-00-00095 2.926^{17}	chl 4 1.6410^{20}
7472	Methane, bromomethoxy-	C_2H_5BrO 124.97	13057-17-5	 87	4-01-00-03051 1.5976^{20}	 1.4562^{20}
7473	Methane, bromonitro- Bromonitromethane	CH_2BrNO_2 139.94	563-70-2	 149; 71^{40}	4-01-00-00106	EtOH 4 1.4880^{20}
7474	Methane, bromotrichloro- Bromotrichloromethane	$CBrCl_3$ 198.27	75-62-7 -5.7	 105	4-01-00-00077 2.012^{25}	eth 4; EtOH 4 1.5065^{20}
7475	Methane, bromotrifluoro- Bromotrifluoromethane	$CBrF_3$ 148.91	75-63-8 -172	 -57.8	4-01-00-00073 1.5800^{20}	chl 4
7476	Methane, bromotrinitro- Bromotrinitromethane	$CBrN_3O_6$ 229.93	560-95-2 17.5	 56^{10}	3-01-00-00116 2.0312^{20}	EtOH 4; chl 4 1.4808^{20}
7477	Methane, chloro- Methyl chloride	CH_3Cl 50.49	74-87-3 -97.7	5964 -24.0	4-01-00-00028 *0.911^{25}	H_2O 3; EtOH 3; eth 5; ace 5 1.3389^{20}
7478	Methane, chlorodifluoro- Chlorodifluoromethane	$CHClF_2$ 86.47	75-45-6 -157.4	 -40.7	4-01-00-00032 1.4909^{-69}	H_2O 4; eth 3; ace 3; chl 3

No.	Name	Synonym	Mol. Form.	Mol. Wt.	CAS RN	mp/°C	Merck No.	bp/°C	Beil. Ref.	den/g cm^{-3}	Solubility	n_D
7479	Methane, chlorodiiodo-		$CHClI_2$	302.28	638-73-3	-4		200 dec; 101^{33}	4-01-00-00097	3.1700^{0}	ace 4; eth 4; chl 4	
7480	Methane, chlorodinitro-	Chlorodinitromethane	$CHClN_2O_4$	140.48	921-13-1			$34\text{-}6^{13}$	3-01-00-00115	1.6125^{20}		1.4575^{20}
7481	Methane, chlorofluoro-	Chlorofluoromethane	CH_2ClF	68.48	593-70-4	-133		-9.1	4-01-00-00032		chl 4	
7482	Methane, chloroiodo-	Chloroiodomethane	CH_2ClI	176.38	593-71-5			109	4-01-00-00094	2.422^{20}	ace 4; bz 4; eth 4; EtOH 4	1.5822^{20}
7483	Methane, chloromethoxy-	Chloromethyl methyl ether	C_2H_5ClO	80.51	107-30-2	-103.5	2146	59.5	4-01-00-03046	1.063^{10}	EtOH 3; eth 3; ace 3; chl 3	1.397^{20}
7484	Methane, chloro(methylthio)-		C_2H_5ClS	96.58	2373-51-5			105	4-01-00-03079	1.153^{25}		1.4963^{20}
7485	Methane, chloronitro-	Chloronitromethane	CH_2ClNO_2	95.49	1794-84-9			122.5	4-01-00-00106	1.466^{15}	H_2O 4	
7486	Methane, chlorotrifluoro-	Chlorotrifluoromethane	$CClF_3$	104.46	75-72-9	-181		-81.4	4-01-00-00034			
7487	Methane, chlorotrinitro-		$CClN_3O_6$	185.48	1943-16-4	4.5		134 dec; 56^{40}	3-01-00-00116	1.6769^{20}	eth 4; EtOH 4; chl 4	1.4500^{20}
7488	Methanediamine, 1-(2-furanyl)-*N*,*N*'-bis(2-furanylmethylene)-	Hydrofuramide	$C_{15}H_{12}N_2O_3$	268.27	494-47-3	117	4718		5-17-09-00306		H_2O 1; EtOH 4; eth 4	
7489	Methanediamine, *N*,*N*,*N*',*N*'-tetraethyl-		$C_9H_{22}N_2$	158.29	102-53-4			165.8	4-04-00-00338	0.8000^{20}		1.4420^{25}
7490	Methanediamine, *N*,*N*,*N*',*N*'-tetramethyl-	Tetramethylmethanediamine	$C_5H_{14}N_2$	102.18	51-80-9			83	4-04-00-00153	0.7491^{18}	H_2O 3	
7491	Methane, diazo-	Diazomethane	CH_2N_2	42.04	334-88-3	-145	2983	-23	4-01-00-03056		eth 4; diox 4	
7492	Methane, dibromo-	Methylene bromide	CH_2Br_2	173.83	74-95-3	-52.5	5980	97	4-01-00-00078	2.4969^{20}	H_2O 2; EtOH 5; eth 5; ace 5	1.5420^{20}
7493	Methane, dibromochloro-	Chlorodibromomethane	$CHBr_2Cl$	208.28	124-48-1	-20		120	4-01-00-00081	2.451^{20}	H_2O 1; EtOH 3; eth 3; ace 3	1.5482^{20}
7494	Methane, dibromochlorofluoro-	Dibromochlorofluoromethane	CBr_2ClF	226.27	353-55-9			80.3	4-01-00-00082	2.3173^{22}		1.4570^{20}
7495	Methane, dibromodichloro-	Dibromodichloromethane	CBr_2Cl_2	242.72	594-18-3	38		150.2	4-01-00-00082	2.42^{25}	H_2O 1; EtOH 3; eth 3; ace 3	
7496	Methane, dibromodifluoro-	Dibromodifluoromethane	CBr_2F_2	209.82	75-61-6	-110.1		25	4-01-00-00080		H_2O 3; eth 3; ace 3; bz 3	
7497	Methane, dibromodinitro-	Dibromodinitromethane	$CBr_2N_2O_4$	263.83	2973-00-4	5.5		158 dec; 77^{21}	3-01-00-00115	2.4439^{20}	EtOH 4	1.5280^{25}
7498	Methane, dibromofluoro-	Dibromofluoromethane	$CHBr_2F$	191.83	1868-53-7	-78		64.9	4-01-00-00080	2.421^{20}	H_2O 1; EtOH 3; eth 3; ace 3	1.4685^{20}
7499	Methane, dichloro-	Methylene chloride	CH_2Cl_2	84.93	75-09-2	-95.1	5982	40	4-01-00-00035	1.3266^{20}	H_2O 2; EtOH 5; eth 5; ctc 3	1.4242^{20}
7500	Methane, dichlorodifluoro-	Dichlorodifluoromethane	CCl_2F_2	120.91	75-71-8	-158	3053	-29.8	4-01-00-00040		H_2O 3; EtOH 3; eth 3; HOAc 3	
7501	Methane, dichlorodinitro-	Dichlorodinitromethane	$CCl_2N_2O_4$	174.93	1587-41-3			121.5	3-01-00-00115	1.6124^{20}	bz 4; eth 4; EtOH 4; chl 4	1.4575^{20}
7502	Methane, dichlorofluoro-	Dichlorofluoromethane	$CHCl_2F$	102.92	75-43-4	-135		8.9	4-01-00-00039	1.405^{9}	H_2O 1; EtOH 3; eth 3; ctc 3	1.3724^{9}
7503	Methane, dichloroiodo-	Dichloroiodomethane	$CHCl_2I$	210.83	594-04-7			132	4-01-00-00095	2.392^{20}	ace 4; bz 4; eth 4; EtOH 4	1.5840^{20}
7504	Methane, difluoro-	Methylene difluoride	CH_2F_2	52.02	75-10-5	-136		-51.6	4-01-00-00024	$*0.960^{25}$	H_2O 1; EtOH 3	
7505	Methane, difluoroiodo-	Difluoroiodomethane	CHF_2I	177.92	1493-03-4	-122		21.6	4-01-00-00092	3.238^{-19}		
7506	Methane, diiodo-	Methylene iodide	CH_2I_2	267.84	75-11-6	6.1	5985	182	4-01-00-00096	3.3212^{20}	H_2O 2; EtOH 3; eth 3; bz 3	1.7425^{20}
7507	Methane, dimethoxy-	Methylal	$C_3H_8O_2$	76.10	109-87-5	-104.8	5936	42	4-01-00-03026	0.8593^{20}	ace 4; bz 4; eth 4; EtOH 4	1.3513^{20}
7508	Methane, dinitro-	Dinitromethane	$CH_2N_2O_4$	106.04	625-76-3	<-15		exp 100	4-01-00-00107		H_2O 1; EtOH 3; eth 3	
7509	Methanediol, dibenzoate		$C_{15}H_{12}O_4$	256.26	5342-31-4	99		225 dec	4-09-00-00448	1.275^{22}	ace 4; bz 4; eth 4	
7510	Methanediol, 2-furanyl-, diacetate		$C_9H_{10}O_5$	198.18	613-75-2	53.3		220	5-17-09-00300		bz 4; eth 4; EtOH 4	
7511	Methanediol, (3-nitrophenyl)-, diacetate (ester)	Benzaldehyde, 3-nitro, diacetate	$C_{11}H_{11}NO_6$	253.21	29949-19-7	72			3-07-00-00900	1.393^{25}	ace 4; bz 4; eth 4; EtOH 4	
7512	Methanediol, phenyl-, diacetate		$C_{11}H_{12}O_4$	208.21	581-55-5	46		220	4-07-00-00516	1.11^{20}	bz 4; eth 4; EtOH 4	
7513	Methane, diphenyl-*m*-tolyl-	Triphenylmethane, 3-methyl	$C_{20}H_{18}$	258.36	603-26-9	62		357	3-05-00-02332	1.07^{16}	bz 4; eth 4; chl 4; lig 4	
7514	Methanedisulfonic acid	Methionic acid	$CH_4O_6S_2$	176.17	503-40-2	98	5895		4-01-00-03054		H_2O 1; HNO_3 3	
7515	Methane, fluoro-	Methyl fluoride	CH_3F	34.03	593-53-3	-141.8	4103	-78.4	4-01-00-00022	$*0.566^{25}$	H_2O 4; EtOH 4; eth 4; bz 3	
7516	Methane, fluorodiiodo-		$CHFI_2$	285.83	1493-01-2	-34.5		100	4-01-00-00097	3.1969^{22}	eth 4; EtOH 4	

No.	Name Synonym	Mol. Form. Mol. Wt.	CAS RN mp/°C	Merck No. bp/°C	Beil. Ref. den/g cm^{-3}	Solubility n_D
7517	Methane, fluoroiodo-	CH_2FI	373-53-5		4-01-00-00092	ace 4; bz 4; eth 4; chl 4
	Fluoroiodomethane	159.93		53.4	2.366^{20}	1.5256^{20}
7518	Methane, iodo-	CH_3I	74-88-4	6002	4-01-00-00087	H_2O 2; EtOH 5; eth 5; ace 3
	Methyl iodide	141.94	-66.4	42.5	2.279^{20}	1.5380^{20}
7519	Methane, iodomethoxy-	C_2H_5IO	13057-19-7		4-01-00-03052	
	(Iodomethyl) methyl ether	171.97		122	2.030^{20}	1.5472^{20}
7520	Methane, isocyanato-	C_2H_3NO	624-83-9	6004	4-04-00-00247	H_2O 4
	Methyl isocyanate	57.05	-45	39.5	0.9230^{27}	1.3419^{18}
7521	Methane, isothiocyanato-	C_2H_3NS	556-61-6	6005	4-04-00-00248	H_2O 2; EtOH 5; eth 4
	Methyl isothiocyanate	73.12	36	119	1.0691^{37}	1.5258
7522	Methane, nitro-	CH_3NO_2	75-52-5	6532	4-01-00-00100	H_2O 3; EtOH 3; eth 3; ace 3
	Nitromethane	61.04	-28.5	101.1	1.1371^{20}	1.3817^{20}
7523	Methane, oxybis-	C_2H_6O	115-10-6	5990	4-01-00-01245	H_2O 3; EtOH 3; eth 3; ace 3
	Dimethyl ether	46.07	-141.5	-24.8		
7524	Methane, oxybis[bromo-	$C_2H_4Br_2O$	4497-29-4		4-01-00-03052	
		203.86	-34	154.5	2.2013^{20}	
7525	Methane, oxybis[chloro-	$C_2H_4Cl_2O$	542-88-1	3058	4-01-00-03051	EtOH 5; eth 5
	Bis(chloromethyl) ether	114.96	-41.5	106	1.323^{15}	1.435^{21}
7526	Methane, selenobis-	C_2H_6Se	593-79-3		4-01-00-01288	eth 4; EtOH 4; chl 4
		109.03		55	1.4077^{15}	
7527	Methanesulfenyl chloride, trichloro-	CCl_4S	594-42-3		4-03-00-00290	eth 3
	Perchloromethyl mercaptan	185.89		147.5	1.6947^{20}	1.5484^{20}
7528	Methanesulfenyl fluoride, trifluoro-	CF_4S	17742-04-0			eth 4
	Methyl mercaptan, perfluoro	120.07		147.5	1.6947^{20}	1.5484^{20}
7529	Methane, sulfinylbis-	C_2H_6OS	67-68-5	3247	4-01-00-01277	H_2O 3; EtOH 3; eth 3; ace 3
	Dimethyl sulfoxide	78.14	18.5	189	1.1014^{20}	1.4170^{20}
7530	Methanesulfonic acid	CH_4O_3S	75-75-2	5865	4-04-00-00010	H_2O 3
	Methylsulfonic acid	96.11	20	167^{10}	1.4812^{18}	1.4317^{18}
7531	Methanesulfonic acid, ethyl ester	$C_3H_8O_3S$	62-50-0	3782	4-04-00-00012	
	Ethyl methanesulfonate	124.16		86^{10}		
7532	Methanesulfonic acid, methyl ester	$C_2H_6O_3S$	66-27-3	6010	4-04-00-00011	
	Methyl methanesulfonate	110.13	20	202.5	1.2943^{20}	1.4138^{20}
7533	Methanesulfonic acid, trifluoro-, ethyl ester	$C_3H_5F_3O_3S$	425-75-2		4-03-00-00034	eth 3
		178.13		115	1.3740^{0}	
7534	Methane, sulfonylbis-	$C_2H_6O_2S$	67-71-0	3246	4-01-00-01279	H_2O 3; EtOH 3; bz 3
	Dimethyl sulfone	94.13	109	238	1.1700^{110}	1.4226
7535	Methanesulfonyl chloride	CH_3ClO_2S	124-63-0	5866	4-04-00-00027	H_2O 1; EtOH 3; eth 3
	Methanesulfonic acid chloride	114.55		162; 55^{11}	1.4805^{18}	1.4573^{20}
7536	Methanesulfonyl chloride, trichloro-	CCl_4O_2S	2547-61-7		4-03-00-00036	H_2O 1; EtOH 3; eth 3; CS_2 3
		217.89	140.5	170		
7537	Methanesulfonyl chloride, trifluoro-	$CClF_3O_2S$	421-83-0		4-03-00-00035	H_2O 1
		168.52		162; 62^{18}		1.3344^{20}
7538	Methanesulfonyl fluoride	CH_3FO_2S	558-25-8		4-04-00-00027	
	Methanesulfonic acid fluoride	98.10		123.5		
7539	Methane, tetrabromo-	CBr_4	558-13-4		4-01-00-00085	H_2O 1; EtOH 3; eth 3; chl 3
	Carbon tetrabromide	331.63	90.1	189.5	2.9608^{100}	1.5942^{100}
7540	Methane, tetrachloro-	CCl_4	56-23-5	1822	4-01-00-00056	H_2O 1; EtOH 3; eth 5; ace 3
	Carbon tetrachloride	153.82	-23	76.8	1.5940^{20}	1.4601^{20}
7541	Methane, tetrafluoro-	CF_4	75-73-0	1823	4-01-00-00026	H_2O 2; bz 3; chl 3
	Carbon tetrafluoride	88.00	-183.5	-128.0	3.034^{25}	
7542	Methane, tetraiodo-	CI_4	507-25-5	1824	4-01-00-00098	py 4; chl 4
	Carbon tetraiodide	519.63	171	130-40^{1-2}	4.23^{20}	
7543	Methane, tetramethoxy-	$C_5H_{12}O_4$	1850-14-2		4-03-00-00004	
		136.15	-5	114	1.023^{25}	1.3845^{20}
7544	Methane, tetranitro-	CN_4O_8	509-14-8	9164	4-01-00-00107	H_2O 1; EtOH 3; eth 3
	Tetranitromethane	196.03	13.8	126.1	1.6380^{20}	1.4384^{20}
7545	Methane, thiobis-	C_2H_6S	75-18-3	6042	4-01-00-01275	H_2O 1; EtOH 3; eth 3
	Dimethyl sulfide	62.14	-98.3	37.3	0.8483^{20}	1.4438^{20}
7546	Methanethiol	CH_4S	74-93-1	5867	4-01-00-01273	H_2O 2; EtOH 4; eth 4; chl 2
	Methyl mercaptan	48.11	-123	5.9	0.8665^{20}	
7547	Methanethione, diphenyl-	$C_{13}H_{10}S$	1450-31-3		4-07-00-01385	EtOH 2; eth 2; bz 4; chl 4
		198.29	53.5	174^{14}		
7548	Methane, tribromo-	$CHBr_3$	75-25-2	1407	4-01-00-00082	H_2O 2; EtOH 5; eth 5; bz 3
	Bromoform	252.73	8.0	149.1	2.899^{15}	1.6005^{15}
7549	Methane, tribromochloro-	CBr_3Cl	594-15-0		4-01-00-00085	eth 4
	Tribromochloromethane	287.18	55	158.5	2.71^{15}	
7550	Methane, tribromofluoro-	CBr_3F	353-54-8			EtOH 3
	Tribromofluoromethane	270.72	-73.6	108		
7551	Methane, tribromonitro-	CBr_3NO_2	464-10-8		4-01-00-00106	H_2O 1; EtOH 3; eth 3; ace 4
		297.73	10	127^{18}	2.811^{12}	1.5790^{20}
7552	Methanetricarboxylic acid, triethyl ester	$C_{10}H_{16}O_6$	6279-86-3		4-02-00-02360	
	Triethyl methanetricarboxylate	232.23	29	253	1.1084^{16}	1.4243^{20}
7553	Methanetricarboxylic acid, trimethyl ester	$C_7H_{10}O_6$	1186-73-8		3-02-00-02023	bz 4; eth 4; EtOH 4; chl 4
	Trimethyl methanetricarboxylate	190.15	46.5	242.7; 128^{15}		
7554	Methane, trichloro-	$CHCl_3$	67-66-3	2141	4-01-00-00042	H_2O 2; EtOH 5; eth 5; ace 3
	Chloroform	119.38	-63.6	61.1	1.4832^{20}	1.4459^{20}
7555	Methane, trichlorofluoro-	CCl_3F	75-69-4	9553	4-01-00-00054	
	Trichlorofluoromethane	137.37	-111.1	23.7		
7556	Methane, trichloroiodo-	CCl_3I	594-22-9		4-01-00-00095	ace 4; bz 4; eth 4; chl 4
	Trichloroiodomethane	245.27	-7.8	142	2.355^{20}	1.5854^{20}

No.	Name Synonym	Mol. Form. Mol. Wt.	CAS RN mp/°C	Merck No. bp/°C	Beil. Ref. den/g cm^{-3}	Solubility n_D
7557	Methane, trichloronitro- Chloropicrin	CCl_3NO_2 164.37	76-06-2 -64	2156 112	4-01-00-00106 1.6558^{20}	H_2O 3; EtOH 5; ace 5; bz 5 1.4611^{20}
7558	Methane, trifluoro- Fluoroform	CHF_3 70.01	75-46-7 -155.1	4102 -82.1	4-01-00-00024 *0.673^{25}	H_2O 3; EtOH 4; ace 3; bz 3
7559	Methane, trifluoroiodo- Trifluoroiodomethane	CF_3I 195.91	2314-97-8	 -22.5	4-01-00-00092 2.3607^{-32}	 1.3790^{-32}
7560	Methane, triiodo- Iodoform	CHI_3 393.73	75-47-8 119	4926 218	4-01-00-00097 4.008^{25}	H_2O 1; EtOH 3; eth 3; ace 3
7561	Methane, trimethoxy-	$C_4H_{10}O_3$ 106.12	149-73-5 15	 104	4-02-00-00022 0.9676^{20}	EtOH 3; eth 3 1.3793^{20}
7562	Methane, trinitro- Trinitromethane	CHN_3O_6 151.04	517-25-9 15	9642 exp	4-01-00-00107 1.479^{20}	ace 4; EtOH 4 1.4451^{24}
7563	Methanimidamide Formamidine	CH_4N_2 44.06	463-52-5 81	 dec	4-02-00-00082	H_2O 4; EtOH 4
7564	Methanimidamide, *N*'-(4-chloro-2-methylphenyl)-*N*,*N*-dimethyl- Chlordimeform	$C_{10}H_{13}ClN_2$ 196.68	6164-98-3 35	2083 $156\text{-}7^{0.4}$	 1.105^{25}	bz 4; eth 4; EtOH 4 1.5885^{25}
7565	Methanimidamide, *N*,*N*'-diphenyl-	$C_{13}H_{12}N_2$ 196.25	622-15-1 142	 >250	4-12-00-00372	H_2O 2; EtOH 3; eth 4; ace 3
7566	Methanimidamide, *N*-hydroxy-	CH_4N_2O 60.06	624-82-8 114.5	 dec	4-02-00-00084	EtOH 4
7567	Methanimidic acid, *N*-phenyl-, ethyl ester	$C_9H_{11}NO$ 149.19	6780-49-0	 214; 87^{10}	4-12-00-00371 1.0051^{20}	eth 3; bz 3 1.5279^{20}
7568	1,4-Methanoazulene, decahydro-4,8,8-trimethyl-9-methylene-, [1*S*-(1α,3aβ,4α,8aβ)]- Longifolene	$C_{15}H_{24}$ 204.36	475-20-7	5446 258; 126^{15}	4-05-00-01192 0.9319^{18}	H_2O 1; bz 3 1.5040^{20}
7569	1,4-Methanoazulen-9-ol, decahydro-1,5,5,8a-tetramethyl-, [1*R*-(1α,3aβ,4α,8aβ,9*S**)]	$C_{15}H_{26}O$ 222.37	465-24-7 112	 287 dec	3-06-00-00426 1.0441^{20}	 1.519
7570	1H-3a,7-Methanoazulen-6-ol, octahydro-3,6,8,8-tetramethyl-, Cedrol	$C_{15}H_{26}O$ 222.37	77-53-2 86	1919	3-06-00-00424 0.9479^{90}	 1.4824^{90}
7571	7,14-Methano-2H,6H-dipyrido[1,2-a:1',2'-e][1,5]diazocine, dodecahydro-, Sparteine	$C_{15}H_{26}N_2$ 234.38	90-39-1 30.5	8692 325; 173^{8}	5-23-05-00497 1.0196^{20}	eth 4; EtOH 4; chl 4 1.5312^{20}
7572	7,14-Methano-2H,6H-dipyrido[1,2-a:1',2'-e][1,5]diazocin-6-one, dodecahydro-, [7*R*-(7α,7aβ,14α,14aalp Aphylline	$C_{15}H_{24}N_2O$ 248.37	577-37-7 54.5	761 200 dec	5-24-02-00298	ace 4; bz 4; eth 4; EtOH 4
7573	7,14-Methano-4H,6H-dipyrido[1,2-a:1',2'-e][1,5]diazocin-4-one, 7,7a,8,9,10,11,13,14-octahydro-, Anagyrine	$C_{15}H_{20}N_2O$ 244.34	486-89-5	660 265^{12}; 212^{4}	5-24-03-00410	H_2O 3; EtOH 4; eth 3; bz 3
7574	7,14-Methano-4H,6H-dipyrido[1,2-a:1',2'-e][1,5]diazocin-4-one, dodecahydro-9-hydroxy-, Hydroxylupanine	$C_{15}H_{24}N_2O_2$ 264.37	15358-48-2 169.5	4760	4-25-00-00082	H_2O 4; EtOH 4; chl 4
7575	2,5-Methano-2H-furo[3,2-b]pyrrol-6-ol, hexahydro-4-methyl-, (2α,3aβ,5α,6β,6aβ Scopoline	$C_8H_{13}NO_2$ 155.20	487-27-4 108.5	8365 248	2-27-00-00061 1.0891^{134}	H_2O 3
7576	4,7-Methano-1H-indene, 1,4,5,6,7,8,8-heptachloro-3a,4,7,7a-tetrahydro- Heptachlor	$C_{10}H_5Cl_7$ 373.32	76-44-8 95.5	4576	 1.57^{9}	bz 4; eth 4; EtOH 4; lig 4
7577	4,7-Methano-1H-indene, 3a,4,7,7a-tetrahydro-, (3aα,4α,7α,7aα)-	$C_{10}H_{12}$ 132.21	1755-01-7 32	 170 dec; 65^{14}	4-05-00-01399 0.9302^{35}	eth 4; EtOH 4 1.5050^{35}
7578	1,5-Methano-8H-pyrido[1,2-a][1,5]diazocin-8-one, 1,2,3,4,5,6-hexahydro-3-methyl-, (1*R*)- Caulophylline	$C_{12}H_{16}N_2O$ 204.27	486-86-2 137	1914	5-24-02-00538	H_2O 4; ace 4; bz 4; EtOH 4
7579	3,6-Methano-8H-1,5,7-trioxacyclopenta[ij]cycloprop[a]azulene-4,8(3H)-dione, hexahydro-2a-hydroxy Picrotoxin	$C_{30}H_{34}O_{13}$ 602.59	124-87-8 203.5	7388	4-19-00-05245	py 4; EtOH 4
7580	4,7-Methanoisobenzofuran-1,3-dione, 3a,4,7,7a-tetrahydro-, (3aα,4α,7α,7aα)- Carbic anhydride	$C_9H_8O_3$ 164.16	129-64-6 164.5	1801	5-17-11-00192 1.417^{25}	ace 4; bz 4; EtOH 4; chl 4
7581	Methanol Methyl alcohol	CH_4O 32.04	67-56-1 -97.6	5868 64.6	4-01-00-01227 0.7914^{20}	H_2O 5; EtOH 5; eth 5; ace 5 1.3288^{20}
7582	Methanol, chloro-, acetate	$C_3H_5ClO_2$ 108.52	625-56-9	 116	4-02-00-00280 1.194^{20}	eth 4; EtOH 4 1.409^{20}
7583	Methanol, (1,3,5-triazine-2,4,6-triyltriimino)tris- Trimethylolmelamine	$C_6H_{12}N_6O_3$ 216.20	1017-56-7	9631	4-26-00-01273	
7584	Methanol, (1,3,5-triazine-2,4,6-triyltrinitrilo)hexakis- Hexamethylolmelamine	$C_9H_{18}N_6O_6$ 306.28	531-18-0 137	4611	4-26-00-01274	H_2O 4
7585	Methanol, trichloro-, carbonate (2:1)	$C_3Cl_6O_3$ 296.75	32315-10-9 79	 203	4-03-00-00033 1.6290^{80}	
7586	1,6-Methanonaphthalen-1(2H)-ol, octahydro-4,8a,9,9-tetramethyl-, Patchouli alcohol	$C_{15}H_{26}O$ 222.37	5986-55-0 56	7001	3-06-00-00426 0.9906^{65}	H_2O 1; EtOH 3; eth 3 1.5029^{65}

No.	Name Synonym	Mol. Form. Mol. Wt.	CAS RN mp/°C	Merck No. bp/°C	Beil. Ref. den/g cm^{-3}	Solubility n_D
7587	Methanone, 1,3,5-benzenetriyltris[phenyl- 1,3,5-Tribenzoylbenzene	$C_{27}H_{18}O_3$ 390.44	25871-69-6 120	 260^{1}	4-07-00-02847	chl 3
7588	Methanone, 2-benzofuranylphenyl-	$C_{15}H_{10}O_2$ 222.24	6272-40-8 91	 360	5-17-10-00568	H_2O 1
7589	Methanone, bis(3-aminophenyl)-	$C_{13}H_{12}N_2O$ 212.25	611-79-0 173.5	 285^{11}	4-14-00-00255	eth 4; EtOH 4
7590	Methanone, bis(4-bromophenyl)-	$C_{13}H_8Br_2O$ 340.01	3988-03-2 177	 395	4-07-00-01380	bz 4; HOAc 4; chl 4
7591	Methanone, bis(3-chlorophenyl)-	$C_{13}H_8Cl_2O$ 251.11	7094-34-0 124	 163^{2}	4-07-00-01376	H_2O 1; EtOH 3; eth 3
7592	Methanone, bis(4-chlorophenyl)-	$C_{13}H_8Cl_2O$ 251.11	90-98-2 147.5	 353	4-07-00-01376 1.4500^{20}	H_2O 1; EtOH 3; eth 4; ace 2
7593	Methanone, bis[4-(dimethylamino)phenyl]- Michler's ketone	$C_{17}H_{20}N_2O$ 268.36	90-94-8 179	6100 360 dec	4-14-00-00255	H_2O 1; EtOH 2; eth 1; bz 4
7594	Methanone, bis(2,4-dimethylphenyl)- 2,2',4,4'-Tetramethylbenzophenone	$C_{17}H_{18}O$ 238.33	3478-88-4	 190^{10}	4-07-00-01470 1.043^{15}	 1.5790^{25}
7595	Methanone, bis(2-hydroxy-4-methoxyphenyl)- Benzophenone 6	$C_{15}H_{14}O_5$ 274.27	131-54-4 139.5	1109	4-08-00-03505	
7596	Methanone, bis(2-hydroxyphenyl)-	$C_{13}H_{10}O_3$ 214.22	835-11-0 59.5	 333	4-08-00-02446	H_2O 1; EtOH 3; eth 3; chl 3
7597	Methanone, bis(4-hydroxyphenyl)-	$C_{13}H_{10}O_3$ 214.22	611-99-4 210		4-08-00-02452 1.133^{131}	H_2O 2; EtOH 3; eth 3; ace 3
7598	Methanone, bis(4-iodophenyl)- 4,4'-Diiodobenzophenone	$C_{13}H_8I_2O$ 434.01	5630-56-8 238.5	 281^{12}	4-07-00-01381 2.2600^{20}	
7599	Methanone, bis(4-methylphenyl)-	$C_{15}H_{14}O$ 210.28	611-97-2 96.5	 334	4-07-00-01434	ace 4; bz 4; eth 4; EtOH 4
7600	Methanone, (2-bromophenyl)phenyl- 2-Bromobenzophenone	$C_{13}H_9BrO$ 261.12	13047-06-8 42	 345	3-07-00-02079	ace 4; EtOH 4; lig 4
7601	Methanone, (3-bromophenyl)phenyl-	$C_{13}H_9BrO$ 261.12	1016-77-9 81	 186^{6}	4-07-00-01378	EtOH 4
7602	Methanone, (4-bromophenyl)phenyl-	$C_{13}H_9BrO$ 261.12	90-90-4 82.5	 350	4-07-00-01378	H_2O 1; EtOH 2; eth 2; bz 2
7603	Methanone, (2-chlorophenyl)(4-chlorophenyl)-	$C_{13}H_8Cl_2O$ 251.11	85-29-0 67	 $214\text{-}5^{22}$	4-07-00-01376 1.303^{14}	EtOH 3; chl 2
7604	Methanone, (4-chlorophenyl)phenyl-	$C_{13}H_9ClO$ 216.67	134-85-0 77.5	 332	4-07-00-01375	EtOH 3; eth 3; ace 3; ctc 2
7605	Methanone, cyclobutylphenyl-	$C_{11}H_{12}O$ 160.22	5407-98-7	 260	4-07-00-01027 1.0426^{25}	 1.5472^{20}
7606	Methanone, dicyclohexyl-	$C_{13}H_{22}O$ 194.32	119-60-8	 159^{20}	4-07-00-00262 0.986^{0}	eth 3; ace 3; ctc 3 1.4860^{20}
7607	Methanone, dicyclopropyl- Dicyclopropyl ketone	$C_7H_{10}O$ 110.16	1121-37-5	 161	4-07-00-00139 0.977^{25}	 1.4670^{20}
7608	Methanone, (2,6-dihydroxy-4-methoxyphenyl)phenyl- Cotoin	$C_{14}H_{12}O_4$ 244.25	479-21-0 130.5	2553	4-08-00-03163	ace 4; bz 4; eth 4; EtOH 4
7609	Methanone, (2,4-dihydroxyphenyl)phenyl- Benzoresorcinol	$C_{13}H_{10}O_3$ 214.22	131-56-6 144	1117	4-08-00-02442	H_2O 1; EtOH 3; eth 4; bz 2
7610	Methanone, (3,4-dihydroxyphenyl)(2,4,6-trihydroxyphenyl)- Maclurin	$C_{13}H_{10}O_6$ 262.22	519-34-6 222.5	5519	3-08-00-04226	eth 4; EtOH 4
7611	Methanone, (6,7-dimethoxy-1-isoquinolinyl)(3,4-dimethoxyphenyl)- Papaveraldine	$C_{20}H_{19}NO_5$ 353.37	522-57-6 210.5	6966	5-21-13-00658	H_2O 1; EtOH 2; eth 2; bz 3
7612	Methanone, (2,4-dimethoxyphenyl)phenyl- 2,4-Dimethoxybenzophenone	$C_{15}H_{14}O_3$ 242.27	3555-84-8 87.5	 218^{10}	4-08-00-02442	H_2O 1; EtOH 4; eth 2; chl 4
7613	Methanone, [4-(dimethylamino)phenyl]phenyl- 4-(Dimethylamino)benzophenone	$C_{15}H_{15}NO$ 225.29	530-44-9 92.5	3222	4-14-00-00248	H_2O 1; EtOH 2; eth 4; chl 3
7614	Methanone, (2,4-dimethylphenyl)phenyl-	$C_{15}H_{14}O$ 210.28	1140-14-3	 321.5	4-07-00-01429 1.071^{20}	bz 4; eth 4; EtOH 4; chl 4
7615	Methanone, (3,4-dimethylphenyl)phenyl- 3,4-Dimethylbenzophenone	$C_{15}H_{14}O$ 210.28	2571-39-3 47.5	 341	4-07-00-01433	bz 4; eth 4; EtOH 4; chl 4
7616	Methanone, diphenyl- Benzophenone	$C_{13}H_{10}O$ 182.22	119-61-9 47.8	1108 305.4	4-07-00-01357	H_2O 1; EtOH 4; eth 4; ace 4 1.6077^{19}
7617	Methanone, di-2-thienyl- Di-(2-thienyl) ketone	$C_9H_6OS_2$ 194.28	704-38-1 90	 326	5-19-04-00318	ace 4; eth 4
7618	Methanone, 2-furanylphenyl-	$C_{11}H_8O_2$ 172.18	2689-59-0 <-15	 285	5-17-10-00289 1.1732^{20}	eth 4; EtOH 4 1.6055^{20}
7619	Methanone, (2-hydroxy-4-methoxyphenyl)(2-hydroxyphenyl)- Dioxybenzone	$C_{14}H_{12}O_4$ 244.25	131-53-3	3298 $170\text{-}5^{1}$	4-08-00-03163	
7620	Methanone, (2-hydroxy-4-methoxyphenyl)phenyl- Oxybenzone	$C_{14}H_{12}O_3$ 228.25	131-57-7 65.5	6907	4-08-00-02442	ctc 3
7621	Methanone, [2-hydroxy-4-(octyloxy)phenyl]phenyl- Octabenzone	$C_{21}H_{26}O_3$ 326.44	1843-05-6 48.5	6662		
7622	Methanone, (2-hydroxyphenyl)phenyl-	$C_{13}H_{10}O_2$ 198.22	117-99-7 40	 250^{560}	4-08-00-01246	H_2O 1; EtOH 4; eth 4; bz 4
7623	Methanone, (2-methoxyphenyl)phenyl-	$C_{14}H_{12}O_2$ 212.25	2553-04-0 41	 210^{27}; 195^{18}	4-08-00-01246	H_2O 1; EtOH 3; bz 3; HOAc 3

No.	Name Synonym	Mol. Form. Mol. Wt.	CAS RN mp/°C	Merck No. bp/°C	Beil. Ref. den/g cm^{-3}	Solubility n_D
7624	Methanone, (3-methoxyphenyl)phenyl-	$C_{14}H_{12}O_2$ 212.25	6136-67-0 44	 343; 201^{17}	4-08-00-01262	H_2O 1; EtOH 4; bz 4; HOAc 4
7625	Methanone, (4-methoxyphenyl)phenyl-	$C_{14}H_{12}O_2$ 212.25	611-94-9 61.5	 355; 168^{12}	4-08-00-01263	H_2O 1; EtOH 4; eth 4; ace 3
7626	Methanone, [2-(methylamino)phenyl]phenyl-	$C_{14}H_{13}NO$ 211.26	1859-76-3 69	 282	4-14-00-00244	bz 4; EtOH 4; HOAc 4
7627	Methanone, (4-methyl-2-nitrophenyl)phenyl- 4-Methyl-2-nitrobenzophenone	$C_{14}H_{11}NO_3$ 241.25	100224-75-7 126.5	 sub	0-07-00-00442	bz 4; EtOH 4; chl 4
7628	Methanone, (2-methylphenyl)(4-methylphenyl)- 2,4'-Dimethylbenzophenone	$C_{15}H_{14}O$ 210.28	1140-16-5	 317	4-07-00-01432 1.074^{19}	eth 4; EtOH 4; HOAc 4
7629	Methanone, (2-methylphenyl)-1-naphthalenyl-	$C_{18}H_{14}O$ 246.31	68723-25-1 64	 365	3-07-00-02639	eth 4; EtOH 4
7630	Methanone, (4-methylphenyl)(4-nitrophenyl)-	$C_{14}H_{11}NO_3$ 241.25	5350-47-0 124	 sub	4-07-00-01406	bz 4; eth 4; EtOH 4; chl 4
7631	Methanone, (2-methylphenyl)phenyl-	$C_{14}H_{12}O$ 196.25	131-58-8 <-18	 308; 128^{12}	4-07-00-01401 1.1098^{20}	H_2O 1; EtOH 4; os 3
7632	Methanone, (3-methylphenyl)phenyl-	$C_{14}H_{12}O$ 196.25	643-65-2	 317; 170^{9}	4-07-00-01403 1.088^{17}	H_2O 1; EtOH 3; eth 3; bz 3
7633	Methanone, (4-methylphenyl)phenyl-	$C_{14}H_{12}O$ 196.25	134-84-9 59.5	 228^{70}	4-07-00-01403 0.9926^{0}	H_2O 1; EtOH 2; eth 3; bz 3
7634	Methanone, 1-naphthalenyl-2-naphthalenyl- 1,2'-Dinaphthyl ketone	$C_{21}H_{14}O$ 282.34	605-79-8 136.5	 235$^{0.6}$	4-07-00-01886	chl 4
7635	Methanone, (3-nitrophenyl)phenyl-	$C_{13}H_9NO_3$ 227.22	2243-80-3 95	 234^{18}	4-07-00-01382	H_2O 1; EtOH 3
7636	Methanone, (4-nitrophenyl)phenyl-	$C_{13}H_9NO_3$ 227.22	1144-74-7 138		4-07-00-01382 1.406^{9}	bz 4
7637	Methanone, phenyl-1H-pyrrol-2-yl- Pyrrole, 2-benzoyl	$C_{11}H_9NO$ 171.20	7697-46-3 79	 305	5-21-08-00461	bz 4; EtOH 4; peth 4; HOAc 4
7638	Methanone, phenyl-2-pyridinyl-	$C_{12}H_9NO$ 183.21	91-02-1 42	 317	5-21-08-00566 1.1556^{20}	chl 3
7639	Methanone, phenyl-4-pyridinyl-	$C_{12}H_9NO$ 183.21	14548-46-0 72	 315; 170^{10}	5-21-08-00576	H_2O 2; EtOH 3; eth 3; bz 3
7640	Methanone, phenyl-4-quinolinyl-	$C_{16}H_{11}NO$ 233.27	54885-00-6 60	 222^{15}	1-21-00-00320	bz 4; eth 4; EtOH 4
7641	Methanone, phenyl-2-thienyl-	$C_{11}H_8OS$ 188.25	135-00-2 56.5	 300	5-17-10-00291 1.1890^{54}	H_2O 1; EtOH 3; eth 3 1.6181^{54}
7642	Methanone, phenyl-3-thienyl-	$C_{11}H_8OS$ 188.25	6453-99-2 63.5	 129^{3}	5-17-10-00295	H_2O 1; EtOH 3; eth 3
7643	Methazole 1,2,4-Oxadiazolidine-3,5-dione, 2-(3,4-dichlorophenyl)-4-methyl-	$C_9H_6Cl_2N_2O_3$ 261.06	20354-26-1 123	5876	 1.24^{25}	
7644	1,3,4-Metheno-1H-cyclobuta[cd]pentalene, 1,1a,2,2,3,3a,4,5,5,5a,5b,6-dodecachlorooctahydro- Mirex	$C_{10}Cl_{12}$ 545.54	2385-85-5 485 dec	6126	3-05-00-01237	bz 4; diox 4
7645	Methidathion Phosphorodithioic acid, S-[(5-methoxy-2-oxo-1,3,4-thiadiazol-3(2H)-yl)methyl], O,O-dimethyl ester	$C_6H_{11}N_2O_4PS_3$ 302.34	950-37-8 39	5891		
7646	Methiocarb Phenol, 3,5-dimethyl-4-(methylthio)-, methylcarbamate	$C_{11}H_{15}NO_2S$ 225.31	2032-65-7 120	5893		
7647	*DL*-Methionine	$C_5H_{11}NO_2S$ 149.21	59-51-8 281 dec		4-04-00-03190 1.340^{25}	H_2O 4; EtOH 2; eth 1; bz 1
7648	*L*-Methionine 2-Amino-4-methylthiobutanoic acid	$C_5H_{11}NO_2S$ 149.21	63-68-3 281 dec	5896	4-04-00-03189	H_2O 3; EtOH 1; eth 1; ace 1
7649	*L*-Methionine, *N*-acetyl- Methionamine	$C_7H_{13}NO_3S$ 191.25	65-82-7 105.5	90	4-04-00-03205	
7650	Methomyl Acetimidic acid, *N*-[(methylcarbamoyl)oxy]thio-, methyl ester	$C_5H_{10}N_2O_2S$ 162.21	16752-77-5 78	5905	 1.2946^{24}	
7651	Methoprene 2,4-Dodecadienoic acid, 11-methoxy-3,7,11-trimethyl-, 1-methylethyl ester, (*E,E*)-	$C_{19}H_{34}O_3$ 310.48	40596-69-8	5906 100$^{0.05}$	 0.926^{20}	
7652	Methylamine, *N*-(*o*-chlorobenzylidene)- Toluene, 2-chloro-α-methylimino	C_8H_8ClN 153.61	17972-08-6 7	 78-80^{2}	3-07-00-00866	 1.5660^{20}
7653	Methyl 3,6-dichloro-*o*-anisate Benzoic acid, 3,6-dichloro-2-methoxy-, methyl ester	$C_9H_8Cl_2O_3$ 235.07	6597-78-0	 115$^{2.4}$		
7654	Methylene bispropionate Propanoic acid, methylene ester	$C_7H_{12}O_4$ 160.17	7044-96-4	 98^{1}; 191^{745}	 1.053^{20}	
7655	Methylisonitrile Methyl isocyanate	C_2H_3N 41.05	593-75-9 -45	 59.6 exp	 0.756^{4}	
7656	Metobromuron Urea, 3-(*p*-bromophenyl)-1-methoxy-1-methyl-	$C_9H_{11}BrN_2O_2$ 259.10	3060-89-7 95	6061	 1.60^{20}	
7657	Metolachlor Acetamide, 2-chloro-*N*-(2-ethyl-6-methylphenyl)-*N*-(2-methoxy-1-methylethyl)-	$C_{15}H_{22}ClNO_2$ 283.80	51218-45-2	6067 100$^{0.001}$	 1.12^{20}	

No.	Name Synonym	Mol. Form. Mol. Wt.	CAS RN mp/°C	Merck No. bp/°C	Beil. Ref. den/g cm^{-3}	Solubility n_D
7658	Metribuzin 1,2,4-Triazin-5(4H)-one, 4-amino-6-(1,1-dimethylethyl)-3-(methylthio)-	$C_8H_{14}N_4OS$ 214.29	21087-64-9 126	6076	 1.31^{20}	
7659	Mevinphos 2-Butenoic acid, 3-[(dimethoxyphosphinyl)oxy]-, methyl ester,(*E*)-	$C_7H_{13}O_6P$ 224.15	7786-34-7 21 (E), 6.9 (Z)	6089 $101^{0.3}$		
7660	MGK 264 4,7-Methano-1H-isoindole-1,3(2H)-dione, 2-(2-ethylhexyl)-3a,4,7,7a-tetrahydro-	$C_{17}H_{25}NO_2$ 275.39	113-48-4 <-20	 157	 1.04	
7661	Molinate 1H-Azepine-1-carbothioic acid, hexahydro-, *S*-ethyl ester	$C_9H_{17}NOS$ 187.31	2212-67-1	 202^{10}	 1.063^{20}	
7662	Molybdenum carbonyl Molybdenum hexacarbonyl	C_6MoO_6 264.00	13939-06-5 dec 150	 701		os 3
7663	Monocrotophos (*E*)-Dimethyl 1-methyl-3-(methylamino)-3-oxo-1-propenyl phosphate	$C_7H_{14}NO_5P$ 223.17	6923-22-4 55	6161 $125^{0.0005}$	 1.33^{20}	
7664	Morphinan, 6,7-didehydro-4,5-epoxy-3,6-dimethoxy-17-methyl-, (5α)- Dihydrothebaine	$C_{19}H_{23}NO_3$ 313.40	561-25-1 162.5	3164	2-27-00-00110	H_2O 1; EtOH 3; bz 3; AcOEt 3
7665	Morphinan-3,6-diol, 7,8-didehydro-4,5-epoxy-, (5α,6α)- Normorphine	$C_{16}H_{17}NO_3$ 271.32	466-97-7 273	6630		
7666	Morphinan-3,6-diol, 7,8-didehydro-4,5-epoxy-17-methyl- (5α,6α)- Morphine	$C_{17}H_{19}NO_3$ 285.34	57-27-2 255	6186 sub 190	4-27-00-02223	H_2O 1; EtOH 2; eth 1; ace 1
7667	Morphinan-3,6-diol, 7,8-didehydro-4,5-epoxy-17-methyl- (5α,6α)-, diacetate (ester) Diacetylmorphine	$C_{21}H_{23}NO_5$ 369.42	561-27-3 173	2948 $272\text{-}4^{12}$	4-27-00-02236 1.56^{25}	bz 4; chl 4
7668	Morphinan-3,6-diol, 7,8-didehydro-4,5-epoxy-17-methyl- (5α,6α)-, 17-oxide Morphine, *N*-oxide	$C_{17}H_{19}NO_4$ 301.34	639-46-3 274.5	6192	4-27-00-02249	H_2O 2; EtOH 2; ace 1; bz 1
7669	Morphinan-6-ol, 7,8-didehydro-4,5-epoxy-3-methoxy-17-methyl-, (5α,6α)- Codeine	$C_{18}H_{21}NO_3$ 299.37	76-57-3 157.5	2459 250^{22}; $140^{1.5}$	2-27-00-00136 1.32^{25}	H_2O 3; EtOH 4; eth 3; bz 3
7670	Morphinan-6-ol, 7,8-didehydro-4,5-epoxy-3-methoxy-17-methyl-, (5α,6β)-	$C_{18}H_{21}NO_3$ 299.37	509-64-8 171.5	 dec	2-27-00-00175 1.87^4	 1.675
7671	Morphinan-6-ol, 8,14-didehydro-4,5-epoxy-3-methoxy-17-methyl-, (5α,6α)- Neopine	$C_{18}H_{21}NO_3$ 299.37	467-14-1 127.5	6376		H_2O 3; EtOH 3; eth 3; bz 3
7672	Morphinan-8-ol, 6,7-didehydro-4,5-epoxy-3-methoxy-17-methyl-, (5α,8β)- Pseudocodeine	$C_{18}H_{21}NO_3$ 299.37	466-96-6 181.5	7927	2-27-00-00112 1.290^{80}	 1.574
7673	Morphinan-6-ol, 7,8-didehydro-4,5-epoxy-3-methoxy-17-methyl-, (5α,6α)-, phosphate (1:1) (salt) Codeine, phosphate	$C_{18}H_{24}NO_7P$ 397.36	52-28-8 220-35 dec	2462	4-27-00-02228	EtOH 4; chl 4
7674	Morphinan-6-ol, 4,5-epoxy-3-methoxy-17-methyl-, (5α,6α)- Dihydrocodeine	$C_{18}H_{23}NO_3$ 301.39	125-28-0 112.5	3154 248^{15}	2-27-00-00103	
7675	Morphinan-6-one, 7,8-didehydro-4,5-epoxy-14-hydroxy-3-methoxy-17-methyl-, (5α)- Hydroxycodeinone	$C_{18}H_{19}NO_4$ 313.35	508-54-3 275 dec	4752		
7676	Morphinan-6-one, 7,8-didehydro-4-hydroxy-3,7-dimethoxy-17-methyl-, (9α,13α,14α)- Sinomenine	$C_{19}H_{23}NO_4$ 329.40	115-53-7 162	8496	5-21-13-00516	H_2O 2; EtOH 3; eth 2; ace 3
7677	Morphinan-6-one, 7,8-didehydro-4-hydroxy-3-methoxy-17-methyl- Thebainone	$C_{18}H_{21}NO_3$ 299.37	467-98-1 151.5	9203	5-21-13-00268	H_2O 2; EtOH 2; eth 2; ace 3
7678	Morphinan-6-one, 4,5-epoxy-3-methoxy-17-methyl-, (5α)- Hydrocodone	$C_{18}H_{21}NO_3$ 299.37	125-29-1 198	4708	4-27-00-03580	H_2O 1; EtOH 3
7679	Morphinan, 6,7,8,14-tetradehydro-4,5-epoxy-3,6-dimethoxy-17-methyl-, (5α)- Thebaine	$C_{19}H_{21}NO_3$ 311.38	115-37-7 193	9202 sub 91	4-27-00-02271 1.305^{20}	H_2O 1; EtOH 4; eth 2; bz 3
7680	4-Morpholinamine	$C_4H_{10}N_2O$ 102.14	4319-49-7	 166	4-27-00-00623 1.059^{25}	 1.4772^{20}
7681	Morpholine Tetrahydro-1,4-oxazine	C_4H_9NO 87.12	110-91-8 -4.9	6194 128	4-27-00-00015 1.0005^{20}	H_2O 5; EtOH 3; eth 3; ace 3 1.4548^{20}
7682	Morpholine, 4-acetyl-	$C_6H_{11}NO_2$ 129.16	1696-20-4 14.5	152^{50}; 118^{12}	4-27-00-00274 1.1145^{20}	H_2O 5; EtOH 3; ace 3; ctc 3 1.4827^{20}
7683	Morpholine, 4-[3-(4-butoxyphenoxy)propyl]- Pramoxine	$C_{17}H_{27}NO_3$ 293.41	140-65-8	7707 196^6	4-27-00-00127	
7684	Morpholine, 4-butyl-	$C_8H_{17}NO$ 143.23	1005-67-0 -57.1	 213.5	4-27-00-00026 0.9068^{20}	H_2O 4; ace 4; bz 4; EtOH 4 1.4451^{20}
7685	4-Morpholinecarboxaldehyde	$C_5H_9NO_2$ 115.13	4394-85-8 21	 239	4-27-00-00274 1.1520^{20}	 1.4845^{20}

No.	Name Synonym	Mol. Form. Mol. Wt.	CAS RN mp/°C	Merck No. bp/°C	Beil. Ref. den/g cm^{-3}	Solubility n_D
7686	4-Morpholinecarboxamide, *N,N'*-1,2-ethanediylbis[*N*-butyl- Dimorpholamine	$C_{20}H_{38}N_4O_4$ 398.55	119-48-2 41.5	3257 $229^{0.4}$	4-27-00-00298	H_2O 4
7687	Morpholine, 2,6-dimethyl- 2,6-Dimethylmorpholine	$C_6H_{13}NO$ 115.18	141-91-3 -88	 146.6	4-27-00-00674 0.9329^{20}	H_2O 5; EtOH 5; ace 3; bz 5 1.4460^{20}
7688	Morpholine, 4,4'-dithiobis- 4,4'-Dithiodimorpholine	$C_8H_{16}N_2O_2S_2$ 236.36	103-34-4 124.5	3379	4-27-00-00613	chl 3
7689	4-Morpholineethanamine	$C_6H_{14}N_2O$ 130.19	2038-03-1 25.6	 205	4-27-00-00370 0.9897^{20}	H_2O 5; EtOH 5; ace 3; bz 5 1.4715^{20}
7690	Morpholine, 4,4'-(1,2-ethanediyl)bis-	$C_{10}H_{20}N_2O_2$ 200.28	1723-94-0 75	 285; 160^{25}	4-27-00-00375	H_2O 4; ace 4; bz 4; EtOH 4
7691	4-Morpholineethanol	$C_6H_{13}NO_2$ 131.17	622-40-2 -1.6	 227	4-27-00-00056 1.0710^{20}	H_2O 3; EtOH 3; ctc 2 1.4763^{20}
7692	4-Morpholineethanol, α-methyl-	$C_7H_{15}NO_2$ 145.20	2109-66-2	 $92\text{-}4^{13}$	4-27-00-00119 1.0174^{20}	H_2O 4; ace 4; bz 4; EtOH 4 1.4638^{20}
7693	Morpholine, 4-(2-ethoxyethyl)- Morpholine, *N*-(2-ethoxyethyl)	$C_8H_{17}NO_2$ 159.23	622-09-3 -100	 206	4-27-00-00057 0.963^{20}	H_2O 4; ace 4; bz 4; eth 4
7694	Morpholine, 4-ethyl- *N*-Ethylmorpholine	$C_6H_{13}NO$ 115.18	100-74-3	 138.5	4-27-00-00023 0.8996^{20}	H_2O 5; EtOH 5; eth 5; ace 3 1.4400^{20}
7695	Morpholine, 4-(2-hydroxybenzoyl)-	$C_{11}H_{10}NO_3$ 204.21	3202-84-4	8302	4-27-00-00346	DMSO 3
7696	Morpholine, 4-methyl- *N*-Methylmorpholine	$C_5H_{11}NO$ 101.15	109-02-4	 116	4-27-00-00022 0.9051^{20}	H_2O 3; EtOH 3; eth 3 1.4332^{20}
7697	Morpholine, 3-methyl-2-phenyl- Phenmetrazine	$C_{11}H_{15}NO$ 177.25	134-49-6	7200 139^{12}; 104^{1}	4-27-00-01050	
7698	Morpholine, 4-(4-methylphenyl)-	$C_{11}H_{15}NO$ 177.25	3077-16-5 51	 167^{30}	2-27-00-00004	eth 4; EtOH 4
7699	Morpholine, 4-nitroso- *N*-Nitrosomorpholine	$C_4H_8N_2O_2$ 116.12	59-89-2 29	6561 225; 140^{25}		H_2O 3
7700	Morpholine, 4-(phenylmethyl)-	$C_{11}H_{15}NO$ 177.25	10316-00-4 194	 260.5	4-27-00-00039 1.0387^{20}	ace 4; bz 4 1.5302^{20}
7701	4-Morpholinepropanamine 4-(3-Aminopropyl)morpholine	$C_7H_{16}N_2O$ 144.22	123-00-2 -15	 220; 134^{50}	4-27-00-00411 0.9854^{20}	H_2O 5; EtOH 5; ace 3; bz 5 1.4762^{20}
7702	Naled Phosphoric acid, 1,2-dibromo-2,2-dichloroethyl dimethyl ester	$C_4H_7Br_2Cl_2O_4P$ 380.78	300-76-5 27	6272 $110^{0.5}$	 1.96^{20}	
7703	Naphthacene 2,3-Benzanthracene	$C_{18}H_{12}$ 228.29	92-24-0 357	6288 sub	4-05-00-02545	H_2O 2; bz 2; con sulf 3; os 2
7704	2-Naphthacenecarboxamide, 7-chloro-4-(dimethylamino)-1,4,4a,5,5a,6,11,12a-octahydro-3,6,10,12,12a-pentahydroxy-6-methyl-1,11-dioxo, [4S-(4α,4aα,5aα,6β,12aα)]- Chlortetracycline	$C_{22}H_{23}ClN_2O_8$ 478.89	57-62-5 168.5	2193	4-14-00-02629	H_2O 1; EtOH 2; eth 1; ace 2
7705	2-Naphthacenecarboxamide, 4-(dimethylamino)-1,4,4a,5,5a,6,11,12a-octahydro-3,5,6,10,12,12a-hexahydroxy-6-methyl-1,11-dioxo-,[4S-(4α,4aα,5aα,6β,12aα)]- Oxytetracycline	$C_{22}H_{24}N_2O_9$ 460.44	79-57-2 184.5	6931	4-14-00-02633 1.634^{20}	
7706	2-Naphthacenecarboxamide, 4-(dimethylamino)-1,4,4a,5,5a,6,11,12a-octahydro-3,6,10,12,12a-pentahydroxy-6-methyl-1,11-dioxo-,[4S-(4α,4aα,5aα,6β,12aα)]- Tetracycline	$C_{22}H_{24}N_2O_8$ 444.44	60-54-8 170-5 dec	9130	4-14-00-02625	
7707	1,2-Naphthacenedione 9,10-Naphthacenequinone	$C_{18}H_{10}O_2$ 258.28	29276-40-2 294	 sub	3-07-00-04273	EtOH 4
7708	1-Naphthalenamine 1-Naphthylamine	$C_{10}H_9N$ 143.19	134-32-7 49.2	6318 300.7	4-12-00-03009 1.0228^{20}	chl 3 1.6140^{20}
7709	2-Naphthalenamine 2-Naphthylamine	$C_{10}H_9N$ 143.19	91-59-8 113	6319 306.2	4-12-00-03122 1.6414^{98}	H_2O 3; EtOH 3; eth 3 1.6493^{98}
7710	2-Naphthalenamine, *N,N*-bis(2-chloroethyl)- Chlornaphazine	$C_{14}H_{15}Cl_2N$ 268.19	494-03-1 55	2107 210^{5}	4-12-00-03126	ace 4; bz 4; eth 4; EtOH 4
7711	1-Naphthalenamine, 5-bromo- 1-Amino-5-bromonaphthalene	$C_{10}H_8BrN$ 222.08	4766-33-0 70	 sub	4-12-00-03111	bz 4; eth 4; EtOH 4; lig 4
7712	1-Naphthalenamine, decahydro-	$C_{10}H_{19}N$ 153.27	7250-95-5 13.5	 227.6	3-12-00-00183 0.9508^{21}	
7713	1-Naphthalenamine, *N,N*-diethyl-	$C_{14}H_{17}N$ 199.30	84-95-7	 285	3-12-00-02855 1.013^{20}	EtOH 3; eth 3; bz 3; ctc 2 1.5961^{20}
7714	1-Naphthalenamine, 5,8-dihydro- 1-Naphthylamine, 5,8-dihydro-	$C_{10}H_{11}N$ 145.20	32666-56-1 37.5	 247^{408}	4-12-00-02995	EtOH 4; chl 4
7715	1-Naphthalenamine, *N,N*-dimethyl- *N,N*-Dimethyl-1-naphthylamine	$C_{12}H_{13}N$ 171.24	86-56-6	3240 250; 140^{13}	4-12-00-03013 1.0423^{20}	H_2O 1; EtOH 3; eth 3; ctc 3 1.624^{15}
7716	2-Naphthalenamine, *N,N*-dimethyl-	$C_{12}H_{13}N$ 171.24	2436-85-3 52.5	 305	4-12-00-03125 1.0279^{60}	H_2O 1; EtOH 3; eth 3 1.6443^{53}
7717	1-Naphthalenamine, *N,N*-diphenyl- 1-Naphthylamine, *N,N*-diphenyl-	$C_{22}H_{17}N$ 295.38	61231-45-6 142	 $335\text{-}40^{80}$	0-12-00-01225	ace 4; bz 4; eth 4; EtOH 4
7718	1-Naphthalenamine, *N*-ethyl-	$C_{12}H_{13}N$ 171.24	118-44-5	 305; 191^{16}	4-12-00-03013 1.060^{20}	eth 4; EtOH 4 1.6477^{15}

No.	Name Synonym	Mol. Form. Mol. Wt.	CAS RN mp/°C	Merck No. bp/°C	Beil. Ref. den/g cm^{-3}	Solubility n_D
7719	2-Naphthalenamine, *N*-ethyl-	$C_{12}H_{13}N$ 171.24	2437-03-8 236	 316.5	4-12-00-03126 1.0545^{21}	 1.6544^{21}
7720	1-Naphthalenamine, 4-methyl-	$C_{11}H_{11}N$ 157.22	4523-45-9 51.5	 176^{12}	4-12-00-03190	eth 4
7721	1-Naphthalenamine, *N*-methyl-	$C_{11}H_{11}N$ 157.22	2216-68-4 174	 294.5	4-12-00-03013	eth 4; EtOH 4 1.6722^{20}
7722	2-Naphthalenamine, *N*-methyl-	$C_{11}H_{11}N$ 157.22	2216-67-3	 317	4-12-00-03125	eth 4; EtOH 4 1.6722^{20}
7723	1-Naphthalenamine, *N*-(2-methylphenyl)- 1-Naphthyl(2-tolyl)amine	$C_{17}H_{15}N$ 233.31	634-41-3 94.5	 200^{9}	3-12-00-02857	bz 4; eth 4; EtOH 4
7724	1-Naphthalenamine, *N*-(4-methylphenyl)-	$C_{17}H_{15}N$ 233.31	634-43-5 79	 360^{528}; 230^{10}	4-12-00-03016	bz 4; eth 4
7725	2-Naphthalenamine, *N*-(2-methylphenyl)- 2-Naphthyl(2-tolyl)amine	$C_{17}H_{15}N$ 233.31	644-15-5 95.5	 402.5	1-12-00-00535	ace 4; bz 4; eth 4; EtOH 4
7726	1-Naphthalenamine, *N*-1-naphthalenyl-	$C_{20}H_{15}N$ 269.35	737-89-3 115	 311^{15}	4-12-00-03018	H_2O 1; EtOH 3; eth 3; ace 3
7727	2-Naphthalenamine, *N*-2-naphthalenyl- β,β'-Dinaphthylamine	$C_{20}H_{15}N$ 269.35	532-18-3 172.2	3260 471	4-12-00-03131	H_2O 1; EtOH 2; eth 3; bz 2
7728	2-Naphthalenamine, *N*-phenyl- *N*-Phenyl-β-naphthylamine	$C_{16}H_{13}N$ 219.29	135-88-6 108	 395.5	4-12-00-03128	H_2O 1; EtOH 3; eth 3; bz 3
7729	2-Naphthalenamine, 1-(phenylazo)- Yellow AB	$C_{16}H_{13}N_3$ 247.30	85-84-7 103	10003		EtOH 4; HOAc 4
7730	2-Naphthalenamine, *N*-(phenylmethyl)-	$C_{17}H_{15}N$ 233.31	13672-18-9 68	 405	4-12-00-03130	bz 4; lig 4
7731	1-Naphthalenamine, 5,6,7,8-tetrahydro-	$C_{10}H_{13}N$ 147.22	2217-41-6 38	 279	4-12-00-02934 1.0625^{16}	H_2O 2; EtOH 3; eth 3; acid 3 1.5900^{20}
7732	Naphthalene	$C_{10}H_8$ 128.17	91-20-3 80.2	6289 217.9	4-05-00-01640 1.0253^{20}	H_2O 1; EtOH 3; eth 4; ace 4 1.5898^{25}
7733	1-Naphthaleneacetamide Acetic acid, amide,1-naphthyl	$C_{12}H_{11}NO$ 185.23	86-86-2	 sub 180	4-09-00-02426	H_2O 1; eth 3; bz 3; CS_2 3
7734	1-Naphthaleneacetic acid 1-Naphthylacetic acid	$C_{12}H_{10}O_2$ 186.21	86-87-3 135	6290 dec	4-09-00-02424	H_2O 2; EtOH 2; eth 4; ace 4
7735	1-Naphthaleneacetic acid, ethyl ester Ethyl 1-naphthylacetate	$C_{14}H_{14}O_2$ 214.26	2122-70-5 88.5	 222^{20}; 118^{13}	4-09-00-02425	EtOH 3; eth 3
7736	1-Naphthaleneacetonitrile	$C_{12}H_9N$ 167.21	132-75-2 32.5	 192^{18}; 163^{12}	4-09-00-02428	EtOH 3 1.6192^{20}
7737	2-Naphthaleneacetonitrile	$C_{12}H_9N$ 167.21	7498-57-9 83	 303	3-09-00-03212 1.092^{25}	
7738	Naphthalene, 1-azido-	$C_{10}H_7N_3$ 169.19	6921-40-0 12	 $75^{0.001}$	 1.1713^{25}	EtOH 4; eth 5; ace 5 1.6550^{25}
7739	Naphthalene, 1-bromo- 1-Naphthyl bromide	$C_{10}H_7Br$ 207.07	90-11-9 -1.8	1413 281	4-05-00-01665 1.4785^{20}	H_2O 3; EtOH 5; eth 5; ace 3 1.658^{20}
7740	Naphthalene, 2-bromo- 2-Bromonaphthalene	$C_{10}H_7Br$ 207.07	580-13-2 59	1414 281.5	4-05-00-01667 1.605^{25}	H_2O 1; EtOH 3; eth 3; bz 3 1.6382^{60}
7741	Naphthalene, 1-bromo-4-chloro- 1-Bromo-4-chloronaphthalene	$C_{10}H_6BrCl$ 241.51	53220-82-9 66.5	 303	3-05-00-01585	eth 4; HOAc 4
7742	Naphthalene, 1-(bromomethyl)-	$C_{11}H_9Br$ 221.10	3163-27-7 56	 183^{18}; 167^{10}	4-05-00-01693	ace 4; bz 4; eth 4; EtOH 4
7743	Naphthalene, 2-(bromomethyl)-	$C_{11}H_9Br$ 221.10	939-26-4 56	 213^{100}; 167^{14}	4-05-00-01698	EtOH 3; eth 3; chl 3; HOAc 3
7744	Naphthalene, 1-bromo-3-nitro-	$C_{10}H_6BrNO_2$ 252.07	7499-65-2 132.3	 sub	4-05-00-01678	eth 4; EtOH 4
7745	Naphthalene, 2-bromo-1-nitro- 2-Bromo-1-nitronaphthalene	$C_{10}H_6BrNO_2$ 252.07	4185-62-0 104.3	 >360	4-05-00-01677	bz 4; eth 4; EtOH 4
7746	Naphthalene, 1-butyl-	$C_{14}H_{16}$ 184.28	1634-09-9 -19.8	 289.3	4-05-00-01737 0.9738^{20}	H_2O 1; EtOH 3; eth 3; ace 3 1.5819^{20}
7747	Naphthalene, 2-butyl- 2-Butylnaphthalene	$C_{14}H_{16}$ 186.30	1134-62-9 -5	 292	4-05-00-01738 0.9673^{20}	ace 4; bz 4; EtOH 4 1.5777^{20}
7748	1-Naphthalenecarbonitrile	$C_{11}H_7N$ 153.18	86-53-3 37.5	 299	4-09-00-02407 1.1080^{25}	H_2O 1; EtOH 4; eth 4; lig 3 1.6298^{18}
7749	2-Naphthalenecarbonitrile	$C_{11}H_7N$ 153.18	613-46-7 66	 306.5	4-09-00-02419 1.0755^{60}	H_2O 2; EtOH 3; eth 3; chl 2
7750	2-Naphthalenecarbonyl chloride	$C_{11}H_7ClO$ 190.63	2243-83-6 51	 305	4-09-00-02417	bz 4; eth 4; chl 4
7751	1-Naphthalenecarboxaldehyde	$C_{11}H_8O$ 156.18	66-77-3 33.5	 292	4-07-00-01286 1.1503^{20}	H_2O 1; EtOH 3; eth 3; ace 3 1.6507^{20}
7752	2-Naphthalenecarboxaldehyde	$C_{11}H_8O$ 156.18	66-99-9 62	 160^{19}	4-07-00-01288 1.0775^{99}	H_2O 2; EtOH 4; eth 4; ace 3 1.6211^{99}
7753	1-Naphthalenecarboxaldehyde, 2-ethoxy-	$C_{13}H_{12}O_2$ 200.24	19523-57-0 115	 $185\text{-}7^{25}$	3-08-00-01110	EtOH 3; HOAc 3
7754	1-Naphthalenecarboxaldehyde, 2-hydroxy-	$C_{11}H_8O_2$ 172.18	708-06-5 83	 192^{27}	4-08-00-01160	H_2O 1; EtOH 3; eth 3; aq alk 3
7755	1-Naphthalenecarboxaldehyde, 4-methoxy-	$C_{12}H_{10}O_2$ 186.21	15971-29-6 34.8	 212^{40}	4-08-00-01170 1.1879^{39}	

No.	Name	Synonym	Mol. Form.	Mol. Wt.	CAS RN	mp/°C	Merck No.	bp/°C	Beil. Ref.	den/g cm^{-3}	Solubility	n_D
7756	1-Naphthalenecarboxaldehyde, 4-methyl-	1-Naphthaldehyde, 4-methyl-	$C_{12}H_{10}O$	170.21	33738-48-6	33.5		175^{13}	3-07-00-01972	1.1252^{38}		
7757	1-Naphthalenecarboxamide		$C_{11}H_9NO$	171.20	2243-81-4	205.8		sub	4-09-00-02405		H_2O 2; EtOH 2; con HCl 3; HOAc 3	
7758	1-Naphthalenecarboxylic acid	1-Naphthoic acid	$C_{11}H_8O_2$	172.18	86-55-5	161	6301	>300	4-09-00-02402	1.398^{25}	eth 4; EtOH 4; chl 4	1.46
7759	2-Naphthalenecarboxylic acid	2-Naphthoic acid	$C_{11}H_8O_2$	172.18	93-09-4	185.5	6302	>300	4-09-00-02414	1.077^{100}	H_2O 2; EtOH 3; eth 3; chl 3	
7760	1-Naphthalenecarboxylic acid, 5-amino-		$C_{11}H_9NO_2$	187.20	32018-88-5	212.3		sub	4-14-00-01812		EtOH 4; HOAc 4	
7761	1-Naphthalenecarboxylic acid, 5-bromo-		$C_{11}H_7BrO_2$	251.08	16726-67-3	261		sub	3-09-00-03153		H_2O 2; EtOH 2; eth 2; bz 4	
7762	1-Naphthalenecarboxylic acid, 5-chloro-	1-Naphthoic acid, 5-chloro-	$C_{11}H_7ClO_2$	206.63	16650-52-5	245		sub	3-09-00-03149		H_2O 2; EtOH 4; bz 2; HOAc 2	
7763	1-Naphthalenecarboxylic acid, 8-chloro-		$C_{11}H_7ClO_2$	206.63	4537-00-2	171.5		sub	3-09-00-03151		H_2O 3; EtOH 3; ace 4; bz 3	
7764	1-Naphthalenecarboxylic acid, 3,4-dihydro-	3,4-Dihydro-1-naphthoic acid	$C_{11}H_{10}O_2$	174.20	3333-23-1	125		306	4-09-00-02338		H_2O 2; EtOH 3; chl 3; MeOH 3	
7765	1-Naphthalenecarboxylic acid, ethyl ester		$C_{13}H_{12}O_2$	200.24	3007-97-4			310	4-09-00-02403	1.1264^{15}	H_2O 1; EtOH 3	1.5966^{15}
7766	1-Naphthalenecarboxylic acid, methyl ester		$C_{12}H_{10}O_2$	186.21	2459-24-7	59.5		168^{20}; $101^{0.04}$	4-09-00-02402	1.1290^{20}	bz 4; EtOH 4	1.6086^{20}
7767	1-Naphthalenecarboxylic acid, 5-nitro-		$C_{11}H_7NO_4$	217.18	1975-44-6	241.5		sub	4-09-00-02411		H_2O 1; EtOH 3; eth 2; bz 2	
7768	2-Naphthalenecarboxylic acid, 3-amino-	3-Amino-2-naphthoic acid	$C_{11}H_9NO_2$	187.20	5959-52-4	216.5	464		3-14-00-01341		EtOH 3; eth 3	
7769	2-Naphthalenecarboxylic acid, 5-bromo-	5-Bromo-2-naphthoic acid	$C_{11}H_7BrO_2$	251.08	1013-83-8	270		sub	3-09-00-03196		bz 4; eth 4; EtOH 4; HOAc 4	
7770	2-Naphthalenecarboxylic acid, ethyl ester		$C_{13}H_{12}O_2$	200.24	3007-91-8	32		308.5	4-09-00-02415	1.1143^{23}	eth 4; EtOH 4; chl 4; HOAc 4	1.5951^{23}
7771	2-Naphthalenecarboxylic acid, 1-hydroxy-	1-Hydroxy-2-naphthoic acid	$C_{11}H_8O_3$	188.18	86-48-6	195	4766		4-10-00-01181		H_2O 2; EtOH 4; eth 4; bz 3	
7772	2-Naphthalenecarboxylic acid, 3-hydroxy-	3-Hydroxy-2-naphthoic acid	$C_{11}H_8O_3$	188.18	92-70-6	222.5	4767		4-10-00-01184		H_2O 2; EtOH 4; eth 4; bz 3	
7773	2-Naphthalenecarboxylic acid, 3-hydroxy-, ethyl ester		$C_{13}H_{12}O_3$	216.24	7163-25-9	85		291	4-10-00-01186		ace 4; chl 4	
7774	2-Naphthalenecarboxylic acid, 3-hydroxy-, hexadecyl ester	Hexadecyl 3-hydroxy-2-naphthoate	$C_{27}H_{40}O_3$	412.61	531-84-0	72.5	4604		3-10-00-01087		bz 4; HOAc 4	
7775	2-Naphthalenecarboxylic acid, 3-hydroxy-, methyl ester		$C_{12}H_{10}O_3$	202.21	883-99-8	75.5		206	4-10-00-01186		H_2O 1; EtOH 3; os 3	
7776	2-Naphthalenecarboxylic acid, methyl ester		$C_{12}H_{10}O_2$	186.21	2459-25-8	77		290	4-09-00-02414		bz 4; eth 4; EtOH 4; chl 4	
7777	2-Naphthalenecarboxylic acid, 4,4'-methylenebis[3-hydroxy-	Pamoic acid	$C_{23}H_{16}O_6$	388.38	130-85-8	315	6955		4-10-00-02362			
7778	2-Naphthalenecarboxylic acid, 1,2,3,4-tetrahydro-	1,2,3,4-Tetrahydro-2-naphthoic acid	$C_{11}H_{12}O_2$	176.22	53440-12-3	97.7		168^{15}	3-09-00-02805		bz 4; eth 4; EtOH 4; chl 4	
7779	2-Naphthalenecarboxylic acid, 5,6,7,8-tetrahydro-		$C_{11}H_{12}O_2$	176.22	1131-63-1	155.5		216^{14}	4-09-00-02096		EtOH 4; bz 3; HOAc 3; peth 1	
7780	Naphthalene, 1-chloro-	1-Naphthyl chloride	$C_{10}H_7Cl$	162.62	90-13-1	-2.5	2149	259; 106.5^{5}	4-05-00-01658	1.1938^{20}	H_2O 1; EtOH 3; eth 3; bz 3	1.6326^{20}
7781	Naphthalene, 2-chloro-	2-Chloronaphthalene	$C_{10}H_7Cl$	162.62	91-58-7	61	2150	256	4-05-00-01660	1.1377^{71}	H_2O 1; EtOH 3; eth 3; bz 3	1.6079^{13}
7782	Naphthalene, 2-chlorodecahydro-	Decalin, 2-chloro	$C_{10}H_{17}Cl$	172.70	5597-81-9			112^{15}	4-05-00-00313	1.0421^{20}		1.5020^{20}
7783	Naphthalene, 1-(chloromethyl)-		$C_{11}H_9Cl$	176.65	86-52-2	32		291.5	4-05-00-01692	1.1813^{20}	H_2O 1; EtOH 3; ctc 3; peth 3	1.6380^{20}
7784	Naphthalene, 1-chloro-2-methyl-	1-Chloro-2-methylnaphthalene	$C_{11}H_9Cl$	176.65	5859-45-0			$150\text{-}2^{10}$	3-05-00-01632		ctc 3	1.6265^{20}
7785	Naphthalene, 2-(chloromethyl)-		$C_{11}H_9Cl$	176.65	2506-41-4	48.5		169^{20}	4-05-00-01697		H_2O 1; EtOH 3; peth 3	
7786	Naphthalene, 1-chloro-5-nitro-	1-Chloro-5-nitronaphthalene	$C_{10}H_6ClNO_2$	207.62	605-63-0	111		>360	3-05-00-01597		EtOH 4	
7787	Naphthalene, 1-chloro-8-nitro-	1-Chloro-8-nitronaphthalene	$C_{10}H_6ClNO_2$	207.62	602-37-9	95.8		175^{2}	4-05-00-01676		EtOH 3; bz 3; chl 2; HOAc 3	
7788	Naphthalene, 2-chloro-1-nitro-	2-Chloro-1-nitronaphthalene	$C_{10}H_6ClNO_2$	207.62	4185-63-1	99.5		>360	4-05-00-01676	1.4854^{25}	EtOH 4; eth 4; ace 4; bz 4	
7789	Naphthalene, decahydro-	Decalin	$C_{10}H_{18}$	138.25	91-17-8	-43	2839	155.5	3-05-00-00245	0.8965^{22}	ace 4; bz 4; eth 4; EtOH 4	1.4810^{20}
7790	Naphthalene, decahydro-, *cis*-	*cis*-Decalin	$C_{10}H_{18}$	138.25	493-01-6	-42.9	2839	195.8	4-05-00-00310	0.8965^{20}	H_2O 1; EtOH 5; eth 4; ace 4	1.4810^{20}
7791	Naphthalene, decahydro-, *trans*-	*trans*-Decalin	$C_{10}H_{18}$	138.25	493-02-7	-30.3	2839	187.3	4-05-00-00311	0.8699^{20}	H_2O 1; EtOH 4; eth 4; ace 4	1.4695^{20}

No.	Name Synonym	Mol. Form. Mol. Wt.	CAS RN mp/°C	Merck No. bp/°C	Beil. Ref. den/g cm^{-3}	Solubility n_D
7792	Naphthalene, decahydro-2-methylene-, *trans*- Decalin, 2-methylene (*trans*)	$C_{11}H_{18}$ 150.26	7787-72-6	 201; 82^{10}	3-05-00-00397 0.8897^{20}	 1.4841^{22}
7793	Naphthalene, 1-decyl- Decane, 1-(1-naphthyl)-	$C_{20}H_{28}$ 268.44	26438-27-7 15	 379	4-05-00-01786 0.9322^{20}	 1.5435^{20}
7794	Naphthalene, 2-decyl-	$C_{20}H_{28}$ 268.44	14188-79-5 20	 387	4-05-00-01786 0.9253^{20}	 1.5413^{20}
7795	1,2-Naphthalenediamine 1,2-Diaminonaphthalene	$C_{10}H_{10}N_2$ 158.20	938-25-0 98.5	 214^{13}; $150^{0.5}$	4-13-00-00337	H_2O 2; EtOH 4; eth 4; chl 4
7796	1,4-Naphthalenediamine 1,4-Diaminonaphthalene	$C_{10}H_{10}N_2$ 158.20	2243-61-0 120		4-13-00-00338	ace 4; bz 4; EtOH 4; chl 4 1.6441^{18}
7797	1,5-Naphthalenediamine	$C_{10}H_{10}N_2$ 158.20	2243-62-1 190	 sub	4-13-00-00340 1.4^{25}	H_2O 3; EtOH 3; eth 3; chl 4
7798	1,8-Naphthalenediamine Naphthalene, 1,8-diamino	$C_{10}H_{10}N_2$ 158.20	479-27-6 66.5	6291 205^{12}	4-13-00-00344 1.1265^{90}	eth 4; EtOH 4 1.6828^{99}
7799	2,3-Naphthalenediamine 2,3-Diaminonaphthalene	$C_{10}H_{10}N_2$ 158.20	771-97-1 199		4-13-00-00346 1.0968^{26}	H_2O 2; EtOH 4; eth 3 1.6392^{26}
7800	1,4-Naphthalenediamine, 2-methyl- Vitamin K6	$C_{11}H_{12}N_2$ 172.23	83-68-1 113.5	9937	3-13-00-00406	
7801	Naphthalene, 1,3-dibromo- 1,3-Dibromonaphthalene	$C_{10}H_6Br_2$ 285.97	52358-73-3 62.8	 290	3-05-00-01586	EtOH 4
7802	Naphthalene, 1,4-dibromo- 1,4-Dibromonaphthalene	$C_{10}H_6Br_2$ 285.97	83-53-4 83	 310	4-05-00-01669	H_2O 1; EtOH 3; eth 3; HOAc 2
7803	Naphthalene, 1,5-dibromo- 1,5-Dibromonaphthalene	$C_{10}H_6Br_2$ 285.97	7351-74-8 131.3	 326	4-05-00-01669	eth 4; EtOH 4
7804	1,8-Naphthalenedicarbonyl dichloride Naphthaloyl chloride	$C_{12}H_6Cl_2O_2$ 253.08	6423-29-6 85	 $197^{0.2}$	3-09-00-04468	bz 4; chl 4
7805	1,8-Naphthalenedicarboxylic acid Naphthalic acid	$C_{12}H_8O_4$ 216.19	518-05-8 260	6299	4-09-00-03534	H_2O 1; EtOH 2; eth 2
7806	1,8-Naphthalenedicarboxylic acid, diethyl ester Naphthalic acid, diethyl ester	$C_{16}H_{16}O_4$ 272.30	58618-39-6 59.5	 238^{19}; 202^{11}	4-09-00-03534 1.1399^{70}	
7807	Naphthalene, 1,2-dichloro- 1,2-Dichloronaphthalene	$C_{10}H_6Cl_2$ 197.06	2050-69-3 36	 296.5	4-05-00-01661 1.3147^{49}	EtOH 3; eth 3 1.5338^{49}
7808	Naphthalene, 1,3-dichloro- 1,3-Dichloronaphthalene	$C_{10}H_6Cl_2$ 197.06	2198-75-6 62.3	 291	4-05-00-01661	EtOH 3
7809	Naphthalene, 1,4-dichloro- 1,4-Dichloronaphthalene	$C_{10}H_6Cl_2$ 197.06	1825-31-6 67.5	 288; 147^{12}	4-05-00-01661 1.2997^{76}	H_2O 1; EtOH 2; eth 3; ace 4 1.6228^{76}
7810	Naphthalene, 1,5-dichloro- 1,5-Dichloronaphthalene	$C_{10}H_6Cl_2$ 197.06	1825-30-5 107	 sub	4-05-00-01662 1.4900^{20}	H_2O 1; EtOH 2; eth 3
7811	Naphthalene, 1,7-dichloro- 1,7-Dichloronaphthalene	$C_{10}H_6Cl_2$ 197.06	2050-73-9 63.5	 285.5	4-05-00-01662 1.2611^{100}	EtOH 3; eth 3; bz 3; HOAc 3 1.6092^{100}
7812	Naphthalene, 1,8-dichloro- 1,8-Dichloronaphthalene	$C_{10}H_6Cl_2$ 197.06	2050-74-0 89	 sub	4-05-00-01662 1.2924^{100}	EtOH 3; peth 3 1.6236^{100}
7813	Naphthalene, 2,6-dichloro- 2,6-Dichloronaphthalene	$C_{10}H_6Cl_2$ 197.06	2065-70-5 140.5	 285	4-05-00-01662	EtOH 2; eth 3; bz 3; chl 3
7814	Naphthalene, 1,2-dihydro-	$C_{10}H_{10}$ 130.19	447-53-0 -8	 206.5	4-05-00-01543 0.9974^{20}	 1.5814^{20}
7815	Naphthalene, 1,4-dihydro- Δ 2-Dialin	$C_{10}H_{10}$ 130.19	612-17-9 25	 211.5	4-05-00-01544 0.9928^{33}	 1.5577^{20}
7816	Naphthalene, 1,2-dihydro-3-methyl-	$C_{11}H_{12}$ 144.22	2717-44-4	 105^{13}; $64^{0.6}$	4-05-00-01552 0.9819^{20}	H_2O 1; eth 3; bz 3 1.5751^{20}
7817	Naphthalene, 1,2-dihydro-4-methyl-	$C_{11}H_{12}$ 144.22	4373-13-1	 112^{18}; 84^{5}	4-05-00-01551 0.9895^{20}	H_2O 1; eth 3; bz 3 1.5758^{20}
7818	Naphthalene, 1,2-dimethyl- 1,2-Dimethylnaphthalene	$C_{12}H_{12}$ 156.23	573-98-8 1.6	 266.5	4-05-00-01708 1.0179^{20}	H_2O 1; eth 3; bz 3 1.6166^{20}
7819	Naphthalene, 1,3-dimethyl- 1,3-Dimethylnaphthalene	$C_{12}H_{12}$ 156.23	575-41-7 -6	 263	4-05-00-01708 1.0144^{20}	H_2O 1; eth 3; bz 3 1.6140^{20}
7820	Naphthalene, 1,4-dimethyl- 1,4-Dimethylnaphthalene	$C_{12}H_{12}$ 156.23	571-58-4 7.6	 268	4-05-00-01709 1.0166^{20}	H_2O 1; EtOH 4; eth 5; ace 5 1.6127^{20}
7821	Naphthalene, 1,5-dimethyl- 1,5-Dimethylnaphthalene	$C_{12}H_{12}$ 156.23	571-61-9 82	 265	4-05-00-01710	bz 4; eth 4
7822	Naphthalene, 1,6-dimethyl- 1,6-Dimethylnaphthalene	$C_{12}H_{12}$ 156.23	575-43-9 -16.9	 264	4-05-00-01711 1.0021^{20}	H_2O 1; eth 3; bz 3 1.6166^{20}
7823	Naphthalene, 1,7-dimethyl- 1,7-Dimethylnaphthalene	$C_{12}H_{12}$ 156.23	575-37-1 -13.9	 263	4-05-00-01711 1.0115^{20}	H_2O 1; eth 3; bz 3 1.6083^{20}
7824	Naphthalene, 1,8-dimethyl- 1,8-Dimethylnaphthalene	$C_{12}H_{12}$ 156.23	569-41-5 65	 270	4-05-00-01712 1.003^{20}	H_2O 1; eth 3; bz 3
7825	Naphthalene, 2,3-dimethyl- Guajen	$C_{12}H_{12}$ 156.23	581-40-8 105	 268	4-05-00-01713 1.003^{20}	bz 4; eth 4 1.5060^{20}
7826	Naphthalene, 2,6-dimethyl- 2,6-Dimethylnaphthalene	$C_{12}H_{12}$ 156.23	581-42-0 112	 262	4-05-00-01714 1.003^{20}	
7827	Naphthalene, 2,7-dimethyl- 2,7-Dimethylnaphthalene	$C_{12}H_{12}$ 156.23	582-16-1 97	 265	4-05-00-01715 1.003^{20}	
7828	Naphthalene, 1-(1,1-dimethylethyl)-	$C_{14}H_{16}$ 184.28	17085-91-5 -11.3	 279	4-05-00-01739 0.9629^{20}	 1.5726^{20}
7829	Naphthalene, 2-(1,1-dimethylethyl)-	$C_{14}H_{16}$ 184.28	2876-35-9 -4	 276; 125^{4}	4-05-00-01740 0.9674^{20}	ace 4; bz 4; EtOH 4 1.5685^{20}

No.	Name Synonym	Mol. Form. Mol. Wt.	CAS RN mp/°C	Merck No. bp/°C	Beil. Ref. den/g cm^{-3}	Solubility n_D
7830	Naphthalene, 1,6-dimethyl-4-(1-methylethyl)- Cadalene	$C_{15}H_{18}$ 198.31	483-78-3	1607 294; 149^{10}	4-05-00-01758 0.9667^{25}	oils 4 1.5785^{25}
7831	Naphthalene, 1,3-dinitro- 1,3-Dinitronaphthalene	$C_{10}H_6N_2O_4$ 218.17	606-37-1 148	sub	4-05-00-01680	H_2O 1; EtOH 3; ace 3
7832	Naphthalene, 1,5-dinitro- 1,5-Dinitronaphthalene	$C_{10}H_6N_2O_4$ 218.17	605-71-0 219	sub	4-05-00-01680 1.5860^{20}	H_2O 1; EtOH 2; eth 4; ace 2
7833	Naphthalene, 1,8-dinitro- 1,8-Dinitronaphthalene	$C_{10}H_6N_2O_4$ 218.17	602-38-0 173	445 dec	4-05-00-01681	H_2O 1; EtOH 2; ace 3; bz 2
7834	1,3-Naphthalenediol Naphthoresorcinol	$C_{10}H_8O_2$ 160.17	132-86-5 123.5	6316	4-06-00-06543	H_2O 3; EtOH 3; eth 3; ace 2
7835	1,5-Naphthalenediol	$C_{10}H_8O_2$ 160.17	83-56-7 262 dec	sub	4-06-00-06554	H_2O 2; EtOH 2; eth 4; ace 4
7836	1,6-Naphthalenediol	$C_{10}H_8O_2$ 160.17	575-44-0 138	sub	4-06-00-06557	H_2O 2; EtOH 2; eth 3; ace 3
7837	1,7-Naphthalenediol	$C_{10}H_8O_2$ 160.17	575-38-2 180.5	sub	4-06-00-06559	H_2O 2; EtOH 4; eth 4; bz 3
7838	2,6-Naphthalenediol	$C_{10}H_8O_2$ 160.17	581-43-1 220	sub	4-06-00-06566	H_2O 2; EtOH 3; eth 3; ace 3
7839	2,7-Naphthalenediol	$C_{10}H_8O_2$ 160.17	582-17-2 193	sub	4-06-00-06570	H_2O 3; EtOH 3; eth 3; ace 2
7840	1,8-Naphthalenediol, cyclic sulfite 1-Naphthalenesulfonic acid, 8-hydroxy, sulfone	$C_{10}H_6O_3S$ 206.22	21849-97-8 97	>360	2-06-00-00954	bz 4; chl 4
7841	1,4-Naphthalenediol, 2-methyl-, diacetate Menadiol diacetate	$C_{15}H_{14}O_4$ 258.27	573-20-6 113	5710	4-06-00-06581	EtOH 4
7842	1,2-Naphthalenedione 1,2-Naphthoquinone	$C_{10}H_6O_2$ 158.16	524-42-5 146	6314	4-07-00-02417 1.450^{25}	H_2O 3; EtOH 3; eth 3; sulf 3
7843	1,4-Naphthalenedione 1,4-Naphthoquinone	$C_{10}H_6O_2$ 158.16	130-15-4 128.5	6315 sub	4-07-00-02422	H_2O 2; EtOH 4; eth 3; bz 3
7844	1,2-Naphthalenedione, 3-bromo- 3-Bromo-1,2-naphthoquinone	$C_{10}H_5BrO_2$ 237.05	7474-83-1 178	sub	4-07-00-02422	bz 4; EtOH 4
7845	1,4-Naphthalenedione, 2-bromo-3-hydroxy-	$C_{10}H_5BrO_3$ 253.05	1203-39-0 202	sub	4-08-00-02364	ace 4
7846	1,4-Naphthalenedione, 2-bromo-3-methyl-	$C_{11}H_7BrO_2$ 251.08	3129-39-3 152.8	sub 100	4-07-00-02431	ace 4; bz 4; eth 4; EtOH 4
7847	1,4-Naphthalenedione, 5-chloro- 5-Chloro-1,4-naphthoquinone	$C_{10}H_5ClO_2$ 192.60	40242-15-7 163	sub	4-07-00-02425	EtOH 4; lig 4
7848	1,4-Naphthalenedione, 2-chloro-3-hydroxy-	$C_{10}H_5ClO_3$ 208.60	1526-73-4 216.5	sub	4-08-00-02363	bz 4; eth 4; EtOH 4
7849	1,2-Naphthalenedione, 3,4-dichloro-	$C_{10}H_4Cl_2O_2$ 227.05	18398-36-2 184	sub	4-07-00-02422	bz 4; chl 4
7850	1,4-Naphthalenedione, 2,3-dichloro- Dichlone	$C_{10}H_4Cl_2O_2$ 227.05	117-80-6 195	3032	4-07-00-02426	H_2O 1; EtOH 2; eth 2; bz 2
7851	1,4-Naphthalenedione, 5,6-dichloro- 1,4-Naphthoquinone, 5,6-dichloro	$C_{10}H_4Cl_2O_2$ 227.05	56961-93-4 181	sub	0-07-00-00730	eth 4
7852	1,4-Naphthalenedione, 2,3-dihydroxy-	$C_{10}H_6O_4$ 190.16	605-37-8 282	sub	4-08-00-02953	H_2O 2; EtOH 2; eth 2; ace 3
7853	1,4-Naphthalenedione, 5,8-dihydroxy-	$C_{10}H_6O_4$ 190.16	475-38-7 232	sub	4-08-00-02946	H_2O 2; EtOH 2; eth 2; HOAc 3
7854	1,4-Naphthalenedione, 5,8-dihydroxy-6-(1-hydroxy-4-methyl-3-pentenyl)-, (*S*)- Alkannin	$C_{16}H_{16}O_5$ 288.30	23444-65-7 149	243 sub 140	3-08-00-04089	EtOH 4
7855	1,4-Naphthalenedione, 5,8-dihydroxy-6-methoxy-2-methyl-3-(2-oxopropyl)- Javanicin	$C_{15}H_{14}O_6$ 290.27	476-45-9 208 dec	5143	4-08-00-03646	alk 3
7856	1,4-Naphthalenedione, 3,5-dihydroxy-2-methyl-	$C_{11}H_8O_4$ 204.18	478-40-0 181	sub 100	4-08-00-02958	H_2O 3; EtOH 3; eth 3; peth 3
7857	1,4-Naphthalenedione, 2-ethyl-3,5,6,7,8-pentahydroxy- Echinochrome A	$C_{12}H_{10}O_7$ 266.21	517-82-8 220 dec	3470 sub 120	4-08-00-03714	H_2O 2; EtOH 3; eth 4; ace 3
7858	1,4-Naphthalenedione, 6-ethyl-2,3,5,7,8-pentahydroxy-	$C_{12}H_{10}O_7$ 266.21	1471-96-1 220 dec	120$^{0.0004}$	4-08-00-03714	ace 4; eth 4; EtOH 4
7859	1,4-Naphthalenedione, 2-hydroxy- Lawsone	$C_{10}H_6O_3$ 174.16	83-72-7 195-6 dec	5263	4-08-00-02360	EtOH 4; eth 1; bz 1; chl 1
7860	1,4-Naphthalenedione, 5-hydroxy- Juglone	$C_{10}H_6O_3$ 174.16	481-39-0 155	5150 sub	4-08-00-02368	H_2O 1; EtOH 3; eth 3; bz 3
7861	1,4-Naphthalenedione, 2-hydroxy-3-methyl- Phthiocol	$C_{11}H_8O_3$ 188.18	483-55-6 173.5	7352 sub	4-08-00-02375	ace 4; eth 4
7862	1,4-Naphthalenedione, 2-hydroxy-3-(3-methyl-2-butenyl)- Lapachol	$C_{15}H_{14}O_3$ 242.27	84-79-7 139.5	5235	4-08-00-02487	H_2O 1; EtOH 3; eth 3; bz 3
7863	1,4-Naphthalenedione, 5-hydroxy-2-methyl- 1,4-Naphthoquinone, 5-hydroxy-2-methyl	$C_{11}H_8O_3$ 188.18	481-42-5 78.5	7511 sub	4-08-00-02376	ace 4; bz 4; eth 4; EtOH 4
7864	1,4-Naphthalenedione, 2-methyl- Menadione	$C_{11}H_8O_2$ 172.18	58-27-5 107	5714	4-07-00-02430	H_2O 1; EtOH 2; eth 3; bz 3
7865	1,4-Naphthalenedione, 2-methyl-3-(3,7,11,15-tetramethyl-2-hexadecenyl)-, [*R*-[*R**,*R**-(*E*)]]- Vitamin K1	$C_{31}H_{46}O_2$ 450.71	84-80-0 -20	9933 140-5$^{0.001}$	4-07-00-02496 0.964^{25}	ace 4; bz 4; eth 4; EtOH 4 1.5250^{25}

No.	Name Synonym	Mol. Form. Mol. Wt.	CAS RN mp/°C	Merck No. bp/°C	Beil. Ref. den/g cm^{-3}	Solubility n_D
7000	1,4-Naphthalenedione, 2-(phenylamino)-	$C_{16}H_{11}NO_2$ 249.27	6628-97-3 193	 sub	4-14-00-00399	bz 4; eth 4
7867	1,5-Naphthalenedisulfonic acid Armstrong's acid	$C_{10}H_8O_6S_2$ 288.30	81-04-9 240-5 dec	812	4-11-00-00561 1.493^{25}	H_2O 4; EtOH 3; eth 1
7868	1,6-Naphthalenedisulfonic acid Eiver-pick acid	$C_{10}H_8O_6S_2$ 288.30	525-37-1 125 dec	6292	4-11-00-00562	H_2O 4; EtOH 3; eth 1
7869	2,7-Naphthalenedisulfonic acid Naphthalene-2,7-disulfonic acid	$C_{10}H_8O_6S_2$ 288.30	92-41-1 199	6294	4-11-00-00563	H_2O 3; con HCl 2
7870	1,3-Naphthalenedisulfonic acid, 7-amino- Amido-G-Acid	$C_{10}H_9NO_6S_2$ 303.32	86-65-7 274	412	4-14-00-02811	H_2O 4; EtOH 4
7871	1,3-Naphthalenedisulfonic acid, 7-hydroxy- 2-Naphthol-6,8-disulfonic acid	$C_{10}H_8O_7S_2$ 304.30	118-32-1	6308	4-11-00-00625	H_2O 3
7872	1,6-Naphthalenedisulfonic acid, 4-amino- 1-Naphthylamine-4,7-disulfonic acid	$C_{10}H_9NO_6S_2$ 303.32	85-75-6	6322	2-14-00-00476	H_2O 4
7873	1,7-Naphthalenedisulfonic acid, 4-amino- 1-Naphthylamine-4,6-disulfonic acid	$C_{10}H_9NO_6S_2$ 303.32	85-74-5	6321	2-14-00-00477	H_2O 4; EtOH 4
7874	2,7-Naphthalenedisulfonic acid, 4-amino-5-hydroxy- 1-Amino-8-naphthol-3,6-disulfonic acid	$C_{10}H_9NO_7S_2$ 319.32	90-20-0	6305	4-14-00-02824	H_2O 2; EtOH 2; eth 2
7875	2,7-Naphthalenedisulfonic acid, 4,5-dihydroxy- Chromotropic acid	$C_{10}H_8O_8S_2$ 320.30	148-25-4	2243	4-11-00-00636	H_2O 3; EtOH 1; eth 1; alk 3
7876	2,7-Naphthalenedisulfonic acid, 3-hydroxy- 2-Naphthol-3,6-disulfonic acid	$C_{10}H_8O_7S_2$ 304.30	148-75-4 dec	6307	4-11-00-00624	H_2O 4; EtOH 4
7877	1,3-Naphthalenedisulfonic acid, 6,6'-[(3,3'-dimethyl[1,1'-biphenyl]-4,4'-diyl)bis(azo)]bis[4-amino-5-hydroxy-, C.I. Direct blue 53	$C_{34}H_{24}N_6Na_4O_{14}S_4$ 960.82	314-13-6	3863		H_2O 3; EtOH 3; acid 3
7878	2,7-Naphthalenedisulfonic acid, 4,5-dihydroxy-3-[(4-nitrophenyl)azo]-, disodium salt	$C_{16}H_9N_3Na_2O_{10}S_2$ 513.37	548-80-1	2242		
7879	Naphthalene, 1-dodecyl- Dodecane, 1-(1-naphthyl)-	$C_{22}H_{32}$ 296.50	26438-28-8 27	 415	4-05-00-01790 0.9240^{20}	 1.5364^{20}
7880	Naphthalene, 2-dodecyl- Dodecane, 1-(2-naphthyl)-	$C_{22}H_{32}$ 296.50	60899-39-0 26	 414	4-05-00-01791 0.9177^{20}	 1.5343^{20}
7881	Naphthalene, 1-ethenyl-	$C_{12}H_{10}$ 154.21	826-74-4	 $124\text{-}5^{15}$	4-05-00-01833 1.036^{18}	 1.644^{20}
7882	Naphthalene, 2-ethenyl- 2-Vinylnaphthalene	$C_{12}H_{10}$ 154.21	827-54-3 66	 135^{18}; 95^{2}	4-05-00-01833	H_2O 1; EtOH 3; ace 3; bz 3
7883	Naphthalene, 1-ethoxy-	$C_{12}H_{12}O$ 172.23	5328-01-8 5.5	 280.5	4-06-00-04212 1.060^{20}	H_2O 1; EtOH 4; eth 4 1.5953^{25}
7884	Naphthalene, 2-ethoxy- 2-Ethoxynaphthalene	$C_{12}H_{12}O$ 172.23	93-18-5 37.5	3709 282	4-06-00-04257 1.0640^{20}	H_2O 1; EtOH 3; eth 3; tol 3 1.5975^{36}
7885	Naphthalene, 1-ethyl- 1-Ethylnaphthalene	$C_{12}H_{12}$ 156.23	1127-76-0 -13.9	 258.6	4-05-00-01705 1.0082^{20}	H_2O 1; EtOH 5; eth 5 1.6062^{20}
7886	Naphthalene, 2-ethyl- 2-Ethylnaphthalene	$C_{12}H_{12}$ 156.23	939-27-5 -7.4	 258	4-05-00-01707 0.9922^{20}	H_2O 1; EtOH 5; eth 5; chl 2 1.5999^{20}
7887	Naphthalene, 1-ethyl-5-methyl- 1-Methyl-5-ethylnaphthalene	$C_{13}H_{14}$ 170.25	17057-92-0 40	 133^{10}	3-05-00-01659	 1.600^{30}
7888	Naphthalene, 1-ethyl-1,2,3,4-tetrahydro- 1-Ethyltetralin	$C_{12}H_{16}$ 160.26	13556-58-6	 241	4-05-00-01429 0.9528^{20}	 1.5318^{20}
7889	Naphthalene, 2-ethyl-1,2,3,4-tetrahydro- 2-Ethyltetralin	$C_{12}H_{16}$ 160.26	32367-54-7 -42.8	 235	4-05-00-01429 0.9401^{15}	 1.5250^{15}
7890	Naphthalene, 5-ethyl-1,2,3,4-tetrahydro-	$C_{12}H_{16}$ 160.26	42775-75-7 -44.5	 245	4-05-00-01428 0.973^{20}	 1.540^{20}
7891	Naphthalene, 6-ethyl-1,2,3,4-tetrahydro- 6-Ethyltetralin	$C_{12}H_{16}$ 160.26	22531-20-0 -70	 244	4-05-00-01429 0.9632^{17}	 1.5414^{16}
7892	Naphthalene, 1-ethynyl-	$C_{12}H_8$ 152.20	15727-65-8 2	 $143\text{-}4^{25}$	3-05-00-01935 1.0513^{20}	 1.6360^{20}
7893	Naphthalene, 1-fluoro- 1-Fluoronaphthalene	$C_{10}H_7F$ 146.16	321-38-0 -9	 215; 80^{11}	4-05-00-01657 1.1322^{20}	H_2O 1; EtOH 3; eth 3; bz 3 1.5939^{20}
7894	Naphthalene, 2-fluoro- 2-Fluoronaphthalene	$C_{10}H_7F$ 146.16	323-09-1 61	 212; 90^{16}	4-05-00-01658	H_2O 1; EtOH 3; eth 3; bz 3
7895	Naphthalene, 1-heptyl- 1-Heptylnaphthalene	$C_{17}H_{22}$ 226.36	2876-52-0 -8	 340	4-05-00-01775 0.9491^{20}	 1.5582^{20}
7896	Naphthalene, 2-heptyl- 2-Heptylnaphthalene	$C_{17}H_{22}$ 226.36	2876-45-1 1	 341	4-05-00-01775 0.9410^{20}	 1.5556^{20}
7897	Naphthalene, 1,2,3,4,4a,8a-hexahydro- Naphthalene, 1,2,3,4,9,10-hexahydro	$C_{10}H_{14}$ 134.22	62690-62-4	 197	4-05-00-01082 0.934^{23}	bz 4; eth 4 1.5260^{16}
7898	Naphthalene, 1,2,4a,5,8,8a-hexahydro-4,7-dimethyl-1-(1-methylethyl)-, [1*S*-(1α,4aβ,8aα)]-	$C_{15}H_{24}$ 204.36	523-47-7	 274; 136^{11}	3-05-00-01086 0.9230^{20}	eth 4; lig 4 1.5059^{20}
7899	Naphthalene, 1,2,3,5,6,8a-hexahydro-4,7-dimethyl-1-(1-methylethyl)-, (1*S*-*cis*)-	$C_{15}H_{24}$ 204.36	483-76-1	 $125\text{-}6^{12}$	2-05-00-00348 0.9160^{15}	 1.5089^{15}
7900	Naphthalene, 1-hexyl- 1-Hexylnaphthalene	$C_{16}H_{20}$ 212.33	2876-53-1 -18	 322	4-05-00-01767 0.9566^{20}	 1.5647^{20}
7901	Naphthalene, 2-hexyl-	$C_{16}H_{20}$ 212.33	2876-46-2 -3	 324	4-05-00-01767 0.9479^{20}	 1.5620^{20}
7902	Naphthalene, 1-iodo-	$C_{10}H_7I$ 254.07	90-14-2 4.2	 302	4-05-00-01670 1.7399^{20}	H_2O 1; EtOH 5; eth 5; bz 5 1.7026^{20}

No.	Name Synonym	Mol. Form. Mol. Wt.	CAS RN mp/°C	Merck No. bp/°C	Beil. Ref. den/g cm^{-3}	Solubility n_D
7903	Naphthalene, 2-iodo-	$C_{10}H_7I$ 254.07	612-55-5 54.5	 308	4-05-00-01671 1.6319^{99}	H_2O 1; EtOH 4; eth 4; HOAc 4 1.6662^{99}
7904	Naphthalene, 1-isocyanato- 1-Naphthyl isocyanate	$C_{11}H_7NO$ 169.18	86-84-0	6330 269	4-12-00-03094 1.1774^{20}	eth 3; bz 3
7905	Naphthalene, 1-isopropyl-2,4,7-trimethyl- Cadalene, 3-methyl	$C_{18}H_{20}$ 212.33	6995-30-8	 $129\text{-}31^{1.5}$	3-05-00-01689 0.9721^{30}	 1.5840^{30}
7906	Naphthalene, 1-isothiocyanato- 1-Naphthyl isothiocyanate	$C_{11}H_7NS$ 185.25	551-06-4 58	6331	4-12-00-03094	bz 4; eth 4; EtOH 4; chl 4
7907	2-Naphthalenemethanamine	$C_{11}H_{11}N$ 157.22	2018-90-8 59.5	 180^{24}, 148^{12}	4-12-00-03207	eth 4; EtOH 4
7908	1-Naphthalenemethanol	$C_{11}H_{10}O$ 158.20	4780-79-4 64	 304; 163^{12}	4-06-00-04332 1.1039^{80}	H_2O 2; EtOH 4; eth 4
7909	2-Naphthalenemethanol	$C_{11}H_{10}O$ 158.20	1592-38-7 81.3	 178^{12}	4-06-00-04340	H_2O 2; EtOH 3; eth 3
7910	1-Naphthalenemethanol, α,α-diphenyl- Benzhydrol, α-(1-naphthyl)	$C_{23}H_{18}O$ 310.40	630-95-5 136.5	 dec	4-06-00-05134	bz 4; eth 4; EtOH 4
7911	1-Naphthalenemethanol, α-methyl-, (±)- (±)-1-(α-Naphthyl)ethanol	$C_{12}H_{12}O$ 172.23	57605-95-5 66	 178^{15}	4-06-00-04346 1.1190^{14}	ace 4; bz 4; EtOH 4; chl 4 1.6188^{25}
7912	1-Naphthalenemethanol, α-phenyl-	$C_{17}H_{14}O$ 234.30	642-28-4 87.8	 360	4-06-00-04989	H_2O 1; EtOH 3; eth 3; bz 3
7913	Naphthalene, 1-methoxy-	$C_{11}H_{10}O$ 158.20	2216-69-5 <-10	 269	4-06-00-04211 1.0963^{14}	H_2O 1; EtOH 3; eth 3; bz 3 1.6940^{25}
7914	Naphthalene, 2-methoxy- 2-Methoxynaphthalene	$C_{11}H_{10}O$ 158.20	93-04-9 73.5	5918 274	4-06-00-04257	bz 4; eth 4; chl 4
7915	Naphthalene, 1-methyl- 1-Methylnaphthalene	$C_{11}H_{10}$ 142.20	90-12-0 -30.4	 244.7	4-05-00-01687 1.0202^{20}	H_2O 1; EtOH 4; eth 4; bz 3 1.6170^{20}
7916	Naphthalene, 2-methyl- 2-Methylnaphthalene	$C_{11}H_{10}$ 142.20	91-57-6 34.4	 241.1	4-05-00-01693 1.0058^{20}	H_2O 1; EtOH 4; eth 4; bz 3 1.6015^{40}
7917	Naphthalene, 1-(3-methylbutoxy)-	$C_{15}H_{18}O$ 214.31	20213-30-3	 319; 150^{3}	3-06-00-02925 1.0069^{14}	 1.5705^{16}
7918	Naphthalene, 2-(3-methylbutoxy)- Isopentyl β-Naphthyl ether	$C_{15}H_{18}O$ 214.31	635-88-1 26.5	 325 dec	4-06-00-04259 1.0155^{12}	eth 4; EtOH 4 1.5768^{12}
7919	Naphthalene, 1,1'-methylenebis-	$C_{21}H_{16}$ 268.36	607-50-1 109	 >360; 270^{14}	4-05-00-02646	H_2O 2; EtOH 2; eth 3; bz 3
7920	Naphthalene, 1-(1-methylethyl)-	$C_{13}H_{14}$ 170.25	6158-45-8 -16	 268	4-05-00-01722 0.9956^{20}	 1.5952^{20}
7921	Naphthalene, 2-(1-methylethyl)-	$C_{13}H_{14}$ 170.25	2027-17-0 14.5	 268.2	4-05-00-01723 0.9753^{20}	H_2O 1; EtOH 4; eth 4; bz 3 1.5848^{20}
7922	Naphthalene, 1-methyl-4-(1-methylethyl)-	$C_{14}H_{16}$ 184.28	1680-53-1	 $145\text{-}8^{12}$	3-05-00-01671 0.9934^{14}	 1.5907^{14}
7923	Naphthalene, 1-methyl-7-(1-methylethyl)-	$C_{14}H_{16}$ 184.28	490-65-3	 140^{11}	4-05-00-01741 0.9740^{20}	H_2O 1; eth 3; ace 3; bz 3 1.5833^{20}
7924	Naphthalene, 7-methyl-1-(1-methylethyl)- Naphthalene, 1-isopropyl-7-methyl	$C_{14}H_{16}$ 184.28	66577-17-1	 282	3-05-00-01672 0.9833^{20}	 1.5884^{20}
7925	Naphthalene, 1-methyl-4-nitro- 4-Methyl-1-nitronaphthalene	$C_{11}H_9NO_2$ 187.20	880-93-3 71.5	 182^{18}	3-05-00-01624	ace 4; eth 4; EtOH 4
7926	Naphthalene, 2-methyl-1-nitro- 2-Methyl-1-nitronaphthalene	$C_{11}H_9NO_2$ 187.20	881-03-8 81.5	 188^{20}	4-05-00-01698	H_2O 1; EtOH 3; ace 4; os 4
7927	Naphthalene, 1-(1-methylpropyl)-	$C_{14}H_{16}$ 184.28	1680-58-6	 273	4-05-00-01738 0.9742^{20}	 1.5701^{20}
7928	Naphthalene, 1-(2-naphthalenyloxy)- 1,2'-Bis-naphthyl ether	$C_{20}H_{14}O$ 270.33	611-49-4 81	 264^{15}	2-06-00-00600	bz 4; eth 4
7929	Naphthalene, 1-nitro- 1-Nitronaphthalene	$C_{10}H_7NO_2$ 173.17	86-57-7 61	6535 180^{14}	4-05-00-01673 1.332^{20}	H_2O 1; EtOH 4; eth 4; bz 4
7930	Naphthalene, 2-nitro- 2-Nitronaphthalene	$C_{10}H_7NO_2$ 173.17	581-89-5 79	 314; 165^{15}	4-05-00-01675	H_2O 1; EtOH 4; eth 4
7931	Naphthalene, 1-nonyl-	$C_{19}H_{26}$ 254.42	26438-26-6 8	 366	4-05-00-01784 0.9371^{20}	 1.5477^{20}
7932	Naphthalene, 2-nonyl-	$C_{19}H_{26}$ 254.42	61886-67-7 12	 372	4-05-00-01784 0.9298^{20}	 1.5454^{20}
7933	Naphthalene, octachloro- Perchloronaphthalene	$C_{10}Cl_8$ 403.73	2234-13-1 197.5	 441^{7}; $248^{0.5}$	4-05-00-01665	EtOH 2; bz 4; chl 4; lig 4
7934	Naphthalene, octadecafluorodecahydro- Perfluorodecalin	$C_{10}F_{18}$ 462.08	306-94-5 -10	 142	3-05-00-00249	
7935	Naphthalene, octafluoro- Perfluoronaphthalene	$C_{10}F_8$ 272.10	313-72-4 87.5	 209	4-05-00-01658	
7936	Naphthalene, 1,2,3,4,4a,5,6,8a-octahydro-7-methyl-4-methylene-1-(1-methylethyl)-, (1α,4aβ,8aα)- γ-Cadinene	$C_{15}H_{24}$ 204.36	39029-41-9	 $124\text{-}8^{12}$	3-05-00-01088 0.9182^{15}	 1.3166^{20}
7937	Naphthalene, 1-octyl- 1-Octylnaphthalene	$C_{18}H_{24}$ 240.39	2876-51-9 -2	 356	4-05-00-01779 0.9427^{20}	 1.5526^{20}
7938	Naphthalene, 2-octyl-	$C_{18}H_{24}$ 240.39	2876-44-0 12	 357	4-05-00-01779 0.9350^{20}	 1.5501^{20}
7939	Naphthalene, 1,1'-oxybis- 1-Naphthyl ether	$C_{20}H_{14}O$ 270.33	607-52-3 110	 $280\text{-}5^{22}$	4-06-00-04215	bz 4; eth 4
7940	Naphthalene, 2,2'-oxybis-	$C_{20}H_{14}O$ 270.33	613-80-9 105	 250^{19} dec	4-06-00-04260	H_2O 1; EtOH 2; eth 4; bz 4

No.	Name Synonym	Mol. Form. Mol. Wt.	CAS RN mp/°C	Merck No. bp/°C	Beil. Ref. den/g cm^{-3}	Solubility n_D
7941	Naphthalene, 1-pentyl-	$C_{15}H_{18}$ 198.31	86-89-5 -22	 307	4-05-00-01755 0.9656^{20}	 1.5725^{20}
7942	Naphthalene, 2-pentyl- 2-Pentylnaphthalene	$C_{15}H_{18}$ 198.31	93-22-1 -4	 310	4-05-00-01755 0.9561^{20}	 1.5694^{20}
7943	Naphthalene, 1-(pentyloxy)- 1-Naphthyl pentyl ether	$C_{15}H_{18}O$ 214.31	108438-40-0 30	 322	4-06-00-04213	bz 4; eth 4; EtOH 4
7944	Naphthalene, 2-(pentyloxy)- 2-Naphthyl pentyl ether	$C_{15}H_{18}O$ 214.31	31059-19-5 24.5	 335	3-06-00-02973	bz 4; eth 4; EtOH 4; chl 4 1.5587^{30}
7945	Naphthalene, 1-phenoxy-	$C_{16}H_{12}O$ 220.27	3402-76-4 55.5	 349.5	4-06-00-04214	H_2O 1; EtOH 2; eth 3; ace 3
7946	Naphthalene, 2-phenoxy-	$C_{16}H_{12}O$ 220.27	19420-29-2 46	 335	3-06-00-02975	H_2O 1; eth 3; chl 3; HOAc 3
7947	Naphthalene, 1-phenyl- 1-Phenylnaphthalene	$C_{16}H_{12}$ 204.27	605-02-7 45	 334	4-05-00-02411 1.096^{20}	H_2O 1; EtOH 4; eth 4; bz 4 1.6664^{20}
7948	Naphthalene, 2-phenyl-	$C_{16}H_{12}$ 204.27	612-94-2 103.5	 345.5	4-05-00-02412 1.2180^{20}	EtOH 3; eth 4; bz 3; chl 3
7949	Naphthalene, 1-(phenylmethyl)-	$C_{17}H_{14}$ 218.30	611-45-0 59.5	 350	4-05-00-02420 1.166^{17}	H_2O 1; EtOH 2; eth 4; bz 3
7950	Naphthalene, 2-(phenylmethyl)- 2-Benzylnaphthalene	$C_{17}H_{14}$ 218.30	613-59-2 58	 350	4-05-00-02420 1.176^{12}	bz 4; eth 4; chl 4
7951	Naphthalene, 1-(phenylthio)-	$C_{16}H_{12}S$ 236.34	7570-98-1 41.8	 255^{43}; 222^{11}	4-06-00-04242 1.167^{15}	eth 4; EtOH 4
7952	1-Naphthalenepropanoic acid	$C_{13}H_{12}O_2$ 200.24	3243-42-3 156.5	 179^{11}	4-09-00-02435	H_2O 3; EtOH 3; bz 3
7953	2-Naphthalenepropanoic acid, 6-hydroxy- Allenolic acid	$C_{13}H_{12}O_3$ 216.24	553-39-9 180.5	247	3-10-00-01113	py 4; EtOH 4; MeOH 4
7954	Naphthalene, 1-(2-propenyl)-	$C_{13}H_{12}$ 168.24	2489-86-3	 266	3-05-00-01805 1.0228^{20}	EtOH 3; bz 4; chl 3 1.6140^{20}
7955	Naphthalene, 1-propoxy-	$C_{13}H_{14}O$ 186.25	20009-26-1	 293.5	4-06-00-04213 1.0447^{18}	 1.5928^{18}
7956	Naphthalene, 2-propoxy-	$C_{13}H_{14}O$ 186.25	19718-45-7 41	 305	3-06-00-02973	EtOH 4
7957	Naphthalene, 1-propyl-	$C_{13}H_{14}$ 170.25	2765-18-6 -8.6	 274.5	4-05-00-01721 0.9897^{20}	 1.5923^{20}
7958	Naphthalene, 2-propyl-	$C_{13}H_{14}$ 170.25	2027-19-2 -3	 273.5	4-05-00-01722 0.9770^{20}	 1.5872^{20}
7959	1-Naphthalenesulfonic acid	$C_{10}H_8O_3S$ 208.24	85-47-2 140	6295	4-11-00-00521	H_2O 3; EtOH 3; eth 2
7960	2-Naphthalenesulfonic acid β-Naphthylsulfonic acid	$C_{10}H_8O_3S$ 208.24	120-18-3 91	6296 dec	4-11-00-00527 1.441^{25}	H_2O 4; EtOH 4; eth 3; bz 2
7961	1-Naphthalenesulfonic acid, 2-amino- 2-Naphthylamine-1-sulfonic acid	$C_{10}H_9NO_3S$ 223.25	81-16-3	6326	4-14-00-02792	DMSO 3
7962	1-Naphthalenesulfonic acid, 4-amino- 1-Naphthylamine-4-sulfonic acid	$C_{10}H_9NO_3S$ 223.25	84-86-6 dec	6323	4-14-00-02793 1.6703^{25}	H_2O 1; EtOH 2; MeOH 3; py 3
7963	1-Naphthalenesulfonic acid, 5-amino- 1-Naphthylamine-5-sulfonic acid	$C_{10}H_9NO_3S$ 223.25	84-89-9	6324	4-14-00-02800	H_2O 3; eth 1
7964	1-Naphthalenesulfonic acid, 7-amino- Badische acid	$C_{10}H_9NO_3S$ 223.25	86-60-2	952	4-14-00-02802	HOAc 4
7965	1-Naphthalenesulfonic acid, 8-amino- 1-Naphthylamine-8-sulfonic acid	$C_{10}H_9NO_3S$ 223.25	82-75-7	6325	4-14-00-02802	gl HOAc 4
7966	1-Naphthalenesulfonic acid, 4-amino-3-hydroxy- 1-Amino-2-naphthol-4-sulfonic acid	$C_{10}H_9NO_4S$ 239.25	116-63-2	466	4-14-00-02825	
7967	2-Naphthalenesulfonic acid, 8-amino- 1,7-Cleve's acid	$C_{10}H_9NO_3S$ 223.25	119-28-8	2349	4-14-00-02805	
7968	1-Naphthalenesulfonic acid, 3,6-bis(1,1-dimethylethyl)-, ethyl ester Ethyl dibunate	$C_{20}H_{28}O_3S$ 348.51	5560-69-0	3745		chl 3
7969	1-Naphthalenesulfonic acid, 3,3'-[[1,1'-biphenyl]-4,4'-diylbis(azo)]bis[4-amino-, disodium salt Congo red	$C_{32}H_{22}N_6Na_2O_6S_2$ 696.68	573-58-0 >360	2493	1-16-00-00342	H_2O 3; EtOH 2; eth 1
7970	1-Naphthalenesulfonic acid, 3,4-dihydro-3,4-dioxo-, sodium salt 1,2-Naphthoquinone-4-sulfonic acid,sodium salt	$C_{10}H_5NaO_5S$ 260.20	521-24-4 287 dec	8597		
7971	1-Naphthalenesulfonic acid, 4-hydroxy- 1-Naphthol-4-sulfonic acid	$C_{10}H_8O_4S$ 224.24	84-87-7 170 dec	6311	4-11-00-00614	H_2O 4; eth 1
7972	1-Naphthalenesulfonic acid, 7-hydroxy- Croceic acid	$C_{10}H_8O_4S$ 224.24	132-57-0	2591	4-11-00-00623	H_2O 3
7973	2-Naphthalenesulfonic acid, 1-hydroxy- 1-Naphthol-2-sulfonic acid	$C_{10}H_8O_4S$ 224.24	567-18-0 >250	6310	4-11-00-00613	H_2O 2; EtOH 3; eth 1; dil HCl 2
7974	2-Naphthalenesulfonic acid, 6-hydroxy- 2-Naphthol-6-sulfonic acid	$C_{10}H_8O_4S$ 224.24	93-01-6 125	6312	4-11-00-00622	H_2O 4; EtOH 4; eth 1; HOAc 3
7975	1-Naphthalenesulfonyl chloride	$C_{10}H_7ClO_2S$ 226.68	85-46-1 68	 209^{20}; $147^{0.9}$	3-11-00-00383	bz 4; eth 4; EtOH 4

No.	Name Synonym	Mol. Form. Mol. Wt.	CAS RN mp/°C	Merck No. bp/°C	Beil. Ref. den/g cm^{-3}	Solubility n_D
7976	2-Naphthalenesulfonyl chloride	$C_{10}H_7ClO_2S$ 226.68	93-11-8 81	201^{13}; $148^{0.5}$	4-11-00-00529	H_2O 1; EtOH 3; eth 4; bz 3
7977	1-Naphthalenesulfonyl chloride, 5-(dimethylamino)- Dansyl chloride	$C_{12}H_{12}ClNO_2S$ 269.75	605-65-2 70	2812	4-14-00-02801	
7978	2-Naphthalenesulfonyl chloride, 5,6,7,8-tetrahydro- 6-Tetralinsulfonyl chloride	$C_{10}H_{11}ClO_2S$ 230.71	61551-49-3 58	197^{18}	3-11-00-00375	eth 4
7979	Naphthalene, 1,2,3,4-tetrahydro- Tetralin	$C_{10}H_{12}$ 132.21	119-64-2 -35.7	9152 207.6	4-05-00-01388 0.9660^{25}	H_2O 1; EtOH 4; eth 4; chl 3 1.5413^{20}
7980	Naphthalene, 1,2,3,4-tetrahydro-1,1-dimethyl- 1,1-Dimethyltetralin	$C_{12}H_{16}$ 160.26	1985-59-7	221	4-05-00-01430 0.950^{20}	1.5292^{20}
7981	Naphthalene, 1,2,3,4-tetrahydro-1,2-dimethyl- 1,2,3,4-Tetrahydro-1,2-dimethylnaphthalene	$C_{12}H_{16}$ 160.26	5195-40-4	235	3-05-00-01262 0.9470^{20}	1.5286^{20}
7982	Naphthalene, 1,2,3,4-tetrahydro-1,3-dimethyl- Tetralin, 1,3-dimethyl	$C_{12}H_{16}$ 160.26	5195-37-9	234; 78^{1}	3-05-00-01263 0.940^{20}	1.525^{20}
7983	Naphthalene, 1,2,3,4-tetrahydro-1,4-dimethyl- Tetralin, 1,4-dimethyl	$C_{12}H_{16}$ 160.26	4175-54-6	226	4-05-00-01430 0.940^{20}	1.525^{20}
7984	Naphthalene, 1,2,3,4-tetrahydro-1,5-dimethyl-	$C_{12}H_{16}$ 160.26	21564-91-0	239	3-05-00-01263 0.941^{20}	1.526^{20}
7985	Naphthalene, 1,2,3,4-tetrahydro-2,2-dimethyl- Tetralin, 2,2-dimethyl	$C_{12}H_{16}$ 160.26	13556-55-3	230	4-05-00-01431 0.935^{20}	1.5200^{20}
7986	Naphthalene, 1,2,3,4-tetrahydro-2,3-dimethyl- Tetralin, 2,3-dimethyl	$C_{12}H_{16}$ 160.26	21564-92-1 -5		3-05-00-01265 0.940^{20}	1.523^{20}
7987	Naphthalene, 1,2,3,4-tetrahydro-2,5-dimethyl- Tetralin, 2,5-dimethyl	$C_{12}H_{16}$ 160.26	25419-37-8	236	3-05-00-01263 0.946^{20}	1.526^{20}
7988	Naphthalene, 1,2,3,4-tetrahydro-2,6-dimethyl-	$C_{12}H_{16}$ 160.26	7524-63-2 20	240; 117^{15}	4-05-00-01432 0.941^{20}	1.526^{20}
7989	Naphthalene, 1,2,3,4-tetrahydro-2,7-dimethyl- Tetralin, 2,7-dimethyl	$C_{12}H_{16}$ 160.26	13065-07-1	237	3-05-00-01265 0.941^{20}	1.526^{20}
7990	Naphthalene, 1,2,3,4-tetrahydro-2,8-dimethyl- Tetralin, 2,8-dimethyl	$C_{12}H_{16}$ 160.26	25419-36-7	236	3-05-00-01264 0.941^{20}	1.526^{20}
7991	Naphthalene, 1,2,3,4-tetrahydro-5,6-dimethyl- Tetralin, 5,6-dimethyl	$C_{12}H_{16}$ 160.26	20027-77-4	252	4-05-00-01430 0.975^{20}	1.552^{20}
7992	Naphthalene, 1,2,3,4-tetrahydro-5,7-dimethyl- Tetralin, 5,7-dimethyl	$C_{12}H_{16}$ 160.26	21693-54-9 -6	253	4-05-00-01430 0.9583^{20}	1.5405^{20}
7993	Naphthalene, 1,2,3,4-tetrahydro-5,8-dimethyl-	$C_{12}H_{16}$ 160.26	14108-88-4 -4.1	254	4-05-00-01430 0.967^{20}	1.547^{20}
7994	Naphthalene, 1,2,3,4-tetrahydro-6,7-dimethyl-	$C_{12}H_{16}$ 160.26	1076-61-5 10.	252	4-05-00-01431 0.954^{20}	1.538^{20}
7995	Naphthalene, 1,2,3,4-tetrahydro-1-methyl-	$C_{11}H_{14}$ 146.23	1559-81-5	220.6	4-05-00-01412 0.9583^{20}	1.5353^{20}
7996	Naphthalene, 1,2,3,4-tetrahydro-2-methyl-	$C_{11}H_{14}$ 146.23	3877-19-8 -43.1	221	4-05-00-01414 0.952^{20}	1.531^{20}
7997	Naphthalene, 1,2,3,4-tetrahydro-5-methyl-	$C_{11}H_{14}$ 146.23	2809-64-5 -23	234	4-05-00-01412 0.9720^{20}	1.5439^{20}
7998	Naphthalene, 1,2,3,4-tetrahydro-6-methyl-	$C_{11}H_{14}$ 146.23	1680-51-9 -40	229	4-05-00-01413 0.9537^{20}	1.5357^{20}
7999	Naphthalene, 1,2,3,4-tetrahydro-1,1,2,6-tetramethyl- Irene	$C_{14}H_{20}$ 188.31	1681-22-7	$120\text{-}5^{10}$	4-05-00-01460 0.9332^{20}	H_2O 1; ace 3; bz 3 1.5217^{20}
8000	Naphthalene, 1,2,3,4-tetrahydro-1,1,6-trimethyl-	$C_{13}H_{18}$ 174.29	475-03-6	240; 90^{4}	4-05-00-01445 0.9356^{20}	EtOH 3; eth 3; bz 3; chl 3 1.5257^{20}
8001	Naphthalene, 1,2,4,7-tetramethyl-	$C_{14}H_{16}$ 184.28	16020-17-0 -3	$174\text{-}5^{12}$	4-05-00-01746 1.011^{20}	1.6032^{20}
8002	Naphthalene, 1,2,5,6-tetramethyl-	$C_{14}H_{16}$ 184.28	2131-43-3 118	152^{12}	4-05-00-01746	H_2O 1; EtOH 3; bz 3; peth 3
8003	Naphthalene, 1,3,6,8-tetramethyl- 1,3,6,8-Tetramethylnaphthalene	$C_{14}H_{16}$ 184.28	14558-14-6 84.5	$115\text{-}6^{2}$	4-05-00-01746	H_2O 1; EtOH 3
8004	Naphthalene, 1,3,6,8-tetranitro-	$C_{10}H_4N_4O_8$ 308.16	28995-89-3 207	exp	4-05-00-01683 1.6400^{0}	bz 4; HOAc 4
8005	Naphthalene, 1,1'-thiobis-	$C_{20}H_{14}S$ 286.40	607-53-4 110	289^{15}	4-06-00-04243	bz 4; HOAc 4
8006	Naphthalene, 2,2'-thiobis-	$C_{20}H_{14}S$ 286.40	613-81-0 151	296^{15}	4-06-00-04314	EtOH 1; bz 4; CS_2 4; HOAc 4
8007	1-Naphthalenethiol 1-Naphthyl mercaptan	$C_{10}H_8S$ 160.24	529-36-2	6297 285 dec; 161^{20}	4-06-00-04241 1.1607^{20}	H_2O 2; EtOH 4; eth 4; dil alk 2 1.6802^{20}
8008	2-Naphthalenethiol 2-Naphthyl mercaptan	$C_{10}H_8S$ 160.24	91-60-1 81	6298 288	4-06-00-04312 1.550^{25}	H_2O 2; EtOH 4; eth 4; lig 4
8009	Naphthalene, 1,2,4-trimethyl- 1,2,4-Trimethylnaphthalene	$C_{13}H_{14}$ 170.25	2717-42-2 55.5	146^{12}	4-05-00-01725	EtOH 3; eth 3; bz 3; chl 3
8010	Naphthalene, 1,2,5-trimethyl- 1,2,5-Trimethylnaphthalene	$C_{13}H_{14}$ 170.25	641-91-8 33.5	140^{12}	4-05-00-01726 1.0103^{22}	H_2O 1; EtOH 2; eth 4; bz 4 1.6093^{22}
8011	Naphthalene, 1,2,6-trimethyl- 1,2,6-Trimethylnaphthalene	$C_{13}H_{14}$ 170.25	3031-05-8 14	146^{10}	4-05-00-01726	H_2O 1; eth 3; bz 3 1.6010^{20}
8012	Naphthalene, 1,2,7-trimethyl-	$C_{13}H_{14}$ 170.25	486-34-0	$147\text{-}8^{18}$	4-05-00-01727 1.0087^{20}	H_2O 1; eth 3; bz 3 1.6097^{20}

No.	Name Synonym	Mol. Form. Mol. Wt.	CAS RN mp/°C	Merck No. bp/°C	Beil. Ref. den/g cm^{-3}	Solubility n_D
8013	Naphthalene, 1,3,7-trimethyl- 2,4,6-Trimethylnaphthalene	$C_{13}H_{14}$ 170.25	2131-38-6 12	 131-3^{9}	4-05-00-01728 0.9801^{21}	 1.5972^{15}
8014	Naphthalene, 1,2,5-trimethyl-8-(1-methylethyl)- Cadalene, 5-methyl	$C_{16}H_{20}$ 212.33	6897-76-3	 144-5$^{1.5}$	3-05-00-01689 0.9847^{30}	 1.5926^{30}
8015	Naphthalene, 1,3,8-trimethyl-5-(1-methylethyl)- Cadalene, 8-methyl	$C_{16}H_{20}$ 212.33	6897-88-7	 135-6$^{1.5}$	4-05-00-01770 0.9800^{30}	 1.5876^{30}
8016	Naphthalene, 1,3,5-trinitro-	$C_{10}H_5N_3O_6$ 263.17	2243-94-9 124	 exp 364	4-05-00-01682	ace 4; EtOH 4; chl 4
8017	1-Naphthalenol 1-Naphthol	$C_{10}H_8O$ 144.17	90-15-3 95	6303 288; 184^{40}	4-06-00-04208 1.0989^{99}	H_2O 1; EtOH 4; eth 4; ace 3 1.6224^{99}
8018	2-Naphthalenol 2-Naphthol	$C_{10}H_8O$ 144.17	135-19-3 123	6304 285	4-06-00-04253 1.28^{20}	H_2O 1; EtOH 4; eth 4; bz 3
8019	2-Naphthalenol, 6-amino-	$C_{10}H_9NO$ 159.19	4363-04-6 193	 212 dec	3-13-00-01902	H_2O 4; EtOH 4
8020	2-Naphthalenol, 8-amino-	$C_{10}H_9NO$ 159.19	118-46-7 206	 sub	3-13-00-01907	H_2O 3; EtOH 4; eth 3; bz 2
8021	2-Naphthalenol, benzoate 2-Naphthyl benzoate	$C_{17}H_{12}O_2$ 248.28	93-44-7 107	6328	4-09-00-00339	H_2O 1; EtOH 3; eth 2; HOAc 2
8022	2-Naphthalenol, 1-bromo-	$C_{10}H_7BrO$ 223.07	573-97-7 84	 130	4-06-00-04301	H_2O 1; EtOH 3; eth 3; bz 3
8023	2-Naphthalenol, 6-chloro-	$C_{10}H_7ClO$ 178.62	40604-49-7 116.5	 sub	4-06-00-04294	bz 4; eth 4; EtOH 4; chl 4
8024	2-Naphthalenol, decahydro-	$C_{10}H_{18}O$ 154.25	825-51-4	 109^{14}	1-06-00-00044 0.996^{25}	 1.4992^{20}
8025	1-Naphthalenol, 2,3-dichloro- 2,3-Dichloro-1-naphthol	$C_{10}H_6Cl_2O$ 213.06	71284-96-3 101	 sub	0-06-00-00612	bz 4; eth 4; EtOH 4; chl 4
8026	1-Naphthalenol, 2,4-dichloro-	$C_{10}H_6Cl_2O$ 213.06	2050-76-2 107.5	 180	4-06-00-04233	bz 4; eth 4; EtOH 4
8027	2-Naphthalenol, 1,4-dimethyl- 2-Naphthol, 1,4-dimethyl-	$C_{12}H_{12}O$ 172.23	4705-94-6 135.5	 315.5	4-06-00-04351	eth 4; EtOH 4
8028	2-Naphthalenol, 1-[(2,4-dimethylphenyl)azo]- C.I. Solvent Orange 7	$C_{18}H_{16}N_2O$ 276.34	3118-97-6 166		4-16-00-00234	eth 4; EtOH 4
8029	2-Naphthalenol, 1-[(2,5-dimethylphenyl)azo]- 1-(2,5-Xylylazo)-2-naphthol	$C_{18}H_{16}N_2O$ 276.34	85-82-5 153	9997	4-16-00-00234	
8030	1-Naphthalenol, 4-methyl-	$C_{11}H_{10}O$ 158.20	10240-08-1 85.8	 166^{13}	4-06-00-04330	H_2O 3; EtOH 3; eth 3; ace 3
8031	2-Naphthalenol, 1-methyl-	$C_{11}H_{10}O$ 158.20	1076-26-2 112	 160^{10}	4-06-00-04329	H_2O 2; EtOH 4; eth 4; ace 4
8032	1-Naphthalenol, methylcarbamate Carbaryl	$C_{12}H_{11}NO_2$ 201.22	63-25-2 145	1789	4-06-00-04219 1.228^{25}	ace 4; DMF 4
8033	2-Naphthalenol, 1-[[2-methyl-4-[(2-methylphenyl)azo]phenyl]azo]- Scarlet red	$C_{24}H_{20}N_4O$ 380.45	85-83-6 186 dec	8349 260 dec	4-16-00-00249	chl 4; peth 4
8034	2-Naphthalenol, 1-nitro- 1-Nitro-2-naphthol	$C_{10}H_7NO_3$ 189.17	550-60-7 104	6536 115$^{0.05}$	4-06-00-04307	H_2O 3; EtOH 3; eth 4; chl 2
8035	2-Naphthalenol, 1-nitroso- 1-Nitroso-2-naphthol	$C_{10}H_7NO_2$ 173.17	131-91-9 109.5	6562	4-07-00-02419	bz 4; eth 4
8036	1-Naphthalenol, 2-(phenylazo)-	$C_{16}H_{12}N_2O$ 248.28	3375-23-3 138	 sub	4-16-00-00219	EtOH 4; HOAc 4
8037	2-Naphthalenol, 1-(phenylazo)- Azo, benzene 1'-naphthalene,2'-hydroxy	$C_{16}H_{12}N_2O$ 248.28	842-07-9		4-16-00-00228	
8038	2-Naphthalenol, 1-[[4-(phenylazo)phenyl]azo]- Sudan III	$C_{22}H_{16}N_4O$ 352.40	85-86-9 195	8858	4-16-00-00248	H_2O 1; EtOH 3; eth 3; ace 3
8039	1-Naphthalenol, 2-(phenylmethyl)-	$C_{17}H_{14}O$ 234.30	36441-32-4 73.5	 237^{12}	4-06-00-04990	H_2O 1; bz 3; lig 2
8040	2-Naphthalenol, 1-(phenylmethyl)- 2-Naphthol, 1-benzyl	$C_{17}H_{14}O$ 234.30	36441-31-3 114	 247^{14}	4-06-00-04988	ace 4; bz 4; eth 4; EtOH 4
8041	1-Naphthalenol, 1,2,3,4-tetrahydro-	$C_{10}H_{12}O$ 148.20	529-33-9	 102-4^{2}	3-06-00-02457 1.09^{25}	 1.5638^{20}
8042	1-Naphthalenol, 5,6,7,8-tetrahydro-	$C_{10}H_{12}O$ 148.20	529-35-1 70	 266; 143^{11}	4-06-00-03851 1.0556^{75}	
8043	2-Naphthalenol, 1,2,3,4-tetrahydro- 2-Tetralol	$C_{10}H_{12}O$ 148.20	530-91-6 15.5	9153 140^{12}	0-06-00-00580	
8044	2-Naphthalenol, 5,6,7,8-tetrahydro-	$C_{10}H_{12}O$ 148.20	1125-78-6 57	 275.5	4-06-00-03854 1.0552^{65}	
8045	1(2H)-Naphthalenone, 2-acetyl-3,4-dihydro-	$C_{12}H_{12}O_2$ 188.23	17216-08-9 56	 135-42^{1}	4-07-00-02358	chl 3
8046	1(2H)-Naphthalenone, 3,4-dihydro-	$C_{10}H_{10}O$ 146.19	529-34-0 8	 115^{6}	4-07-00-01015 1.0988^{16}	 1.5672^{20}
8047	2(1H)-Naphthalenone, 3,4-dihydro-	$C_{10}H_{10}O$ 146.19	530-93-8 18	 237	4-07-00-01018 1.1055^{27}	H_2O 1; eth 3; bz 3 1.5598^{20}
8048	1(2H)-Naphthalenone, 3,4-dihydro-6-methoxy- 6-Methoxy-α-tetralone	$C_{11}H_{12}O_2$ 176.22	1078-19-9 78	5924 171^{11}	4-08-00-00904	
8049	1(2H)-Naphthalenone, 3,4-dihydro-2-methyl-	$C_{11}H_{12}O$ 160.22	1590-08-5	 127-31^{12}	4-07-00-01033 1.057^{25}	 1.5535^{20}
8050	1(2H)-Naphthalenone, 3,4-dihydro-4-methyl-	$C_{11}H_{12}O$ 160.22	19832-98-5	 133-4^{12}	4-07-00-01031 1.0779^{19}	 1.5620^{19}

No.	Name Synonym	Mol. Form. Mol. Wt.	CAS RN mp/°C	Merck No. bp/°C	Beil. Ref. den/g cm^{-3}	Solubility n_D
8051	1(2H)-Naphthalenone, 7-ethyl-3,4-dihydro- 1-Tetralone, 7-ethyl	$C_{12}H_{14}O$ 174.24	22531-06-2	 152-3^{12}	4-07-00-01049 1.0556^{17}	bz 4; eth 4; EtOH 4 1.5599^{17}
8052	2(3H)-Naphthalenone, 4,4a,5,6,7,8-hexahydro-4,4a-dimethyl-6-(1-methylethylidene)-, (*4R-cis*)-	$C_{15}H_{22}O$ 218.34	15764-04-2 51.5	 144^{2}	 1.0035^{20}	ace 4 1.5370^{20}
8053	1(2H)-Naphthalenone, octahydro-, *cis*- 1-Decalone (*cis*)	$C_{10}H_{16}O$ 152.24	32166-40-8 2	 126^{20}	4-07-00-00201 1.008^{20}	 1.4936^{20}
8054	1(2H)-Naphthalenone, octahydro-, *trans*-	$C_{10}H_{16}O$ 152.24	21370-71-8 33	 122^{20}	4-07-00-00201 0.986^{20}	 1.4849^{21}
8055	2(1H)-Naphthalenone, octahydro-	$C_{10}H_{16}O$ 152.24	4832-17-1	 96$^{2.5}$	4-07-00-00203 0.979^{25}	 1.4900^{20}
8056	Naphtho[2,3-b]-1,4-dithiin-2,3-dicarbonitrile, 5,10-dihydro-5,10-dioxo- Dithianone	$C_{14}H_4N_2O_2S_2$ 296.33	3347-22-6 220	3375	5-19-08-00189	
8057	Naphtho[1,2-b]furan-2,8(3H,4H)-dione, 3a,5,5a,9b-tetrahydro-3,5a,9-trimethyl-, α-Santonin	$C_{15}H_{18}O_3$ 246.31	481-06-1 175	8325	5-17-11-00314 1.590^{25}	H_2O 2; EtOH 2; eth 2; bz 3
8058	Naphtho[1,2-b]furan-2,8(3H,4H)-dione, 3a,5,5a,9b-tetrahydro-4,5-dihydroxy-9-methoxy-3,5a-dimethyl- Cedrin	$C_{15}H_{18}O_6$ 294.30	6040-62-6 266	1918		EtOH 4; MeOH 4; chl 4
8059	Naphtho[2,3-b]furan-2(3H)-one, 3a,5,6,7,8,8a,9,9a-octahydro-5,8a-dimethyl-3-methylene-, Alantolactone	$C_{15}H_{20}O_2$ 232.32	546-43-0 76	198 275	5-17-10-00097	bz 4; eth 4; EtOH 4; chl 4
8060	Naphtho[1,2-c]furan-1,3-dione, 4,5-dihydro-	$C_{12}H_8O_3$ 200.19	37845-14-0 126.5	 227-30^{23}	5-17-11-00444	bz 4; MeOH 4
8061	Naphtho[1,2,3,4-def]chrysene Dibenzo[a,e]pyrene	$C_{24}H_{14}$ 302.38	192-65-4 233.5		4-05-00-02803	EtOH 2; ace 2; bz 2; tol 3
8062	Naphtho[2,3-f]quinoline-7,12-dione, 5,6-dihydroxy- Alizarine Blue	$C_{17}H_9NO_4$ 291.26	568-02-5 269	239	5-21-13-00590	bz 4; gl HOAc 4
8063	2-Naphthol, 8-chloro-	$C_{10}H_7ClO$ 178.62	29921-50-4 101	 307.5	3-06-00-02992	bz 4; eth 4; EtOH 4; chl 4
8064	2-Naphthylamine, 1,4-dimethyl- Naphthalene, 2-amino-1,4-dimethyl	$C_{12}H_{13}N$ 171.24	878-93-3 75	 333	0-12-00-01317	eth 4; EtOH 4
8065	1,5-Naphthyridine 1,5-Diazanaphthalene	$C_8H_6N_2$ 130.15	254-79-5 75	 112^{12}	5-23-07-00156 1.2100^{20}	
8066	1,8-Naphthyridine-3-carboxylic acid, 1-ethyl-1,4-dihydro-7-methyl-4-oxo-	$C_{12}H_{12}N_2O_3$ 232.24	389-08-2 229.5	6273		EtOH 2; eth 2; chl 3
8067	Napropamide Propanamide, *N,N*-diethyl-2-(1-naphthalenyloxy)-	$C_{17}H_{21}NO_2$ 271.36	15299-99-7 75	6336		
8068	Nickel carbonyl Nickel tetracarbonyl	C_4NiO_4 170.74	13463-39-3 -19.3	6411 43	 1.31^{25}	
8069	Nicosulfuron 2-(4,6-Dimethoxypyrimidi*N*-2-ylcarbamoylsulfamoyl)-*N,N*-dimethylnicotinamide	$C_{15}H_{18}N_6O_6S$ 410.41	111991-09-4 172			
8070	Nitralin Benzenamine, 4-(methylsulfonyl)-2,6-dinitro-*N,N*-dipropyl-	$C_{13}H_{19}N_3O_6S$ 345.38	4726-14-1 150	6486		
8071	Nitrapyrin Pyridine, 2-chloro-6-(trichloromethyl)-	$C_6H_3Cl_4N$ 230.91	1929-82-4 63	6490 136^{11}		
8072	Nitric acid, butyl ester Butyl nitrate	$C_4H_9NO_3$ 119.12	928-45-0	 133	4-01-00-01524 1.0228^{30}	H_2O 1; EtOH 3; eth 3; ctc 2 1.4013^{23}
8073	Nitric acid, decyl ester Decyl nitrate	$C_{10}H_{21}NO_3$ 203.28	2050-78-4	 128^{11}; 88^{1}	4-01-00-01819 0.951^{0}	EtOH 3; eth 3
8074	Nitric acid, ethyl ester Ethyl nitrate	$C_2H_5NO_3$ 91.07	625-58-1 -94.6	 87.2	4-01-00-01327 1.1084^{20}	H_2O 3; EtOH 5; eth 5 1.3852^{20}
8075	Nitric acid, methyl ester Methyl nitrate	CH_3NO_3 77.04	598-58-3 -82.5	6015 64.6 exp	4-01-00-01254 1.2075^{20}	H_2O 2; EtOH 3; eth 3 1.3748^{20}
8076	Nitric acid, 1-methylethyl ester Isopropyl nitrate	$C_3H_7NO_3$ 105.09	1712-64-7	 100	4-01-00-01475 1.034^{19}	EtOH 3; eth 3 1.3912^{16}
8077	Nitric acid, 1-methylpropyl ester	$C_4H_9NO_3$ 119.12	924-52-7	 124	4-01-00-01573 1.0264^{20}	eth 4; EtOH 4 1.4015^{20}
8078	Nitric acid, 2-methylpropyl ester Isobutyl nitrate	$C_4H_9NO_3$ 119.12	543-29-3	5031 123.4	4-01-00-01595 1.0152^{20}	 1.4028^{20}
8079	Nitric acid, octyl ester	$C_8H_{17}NO_3$ 175.23	629-39-0	 110-2^{20}	4-01-00-01762 0.8409^{17}	H_2O 2; EtOH 3; eth 3
8080	Nitric acid, propyl ester Propyl nitrate	$C_3H_7NO_3$ 105.09	627-13-4	7877 110	4-01-00-01424 1.0538^{20}	H_2O 2; EtOH 3; eth 3; ctc 3 1.3973^{20}
8081	Nitrofen Benzene, 2,4-dichloro-1-(4-nitrophenoxy)-	$C_{12}H_7Cl_2NO_3$ 284.10	1836-75-5 70	6519		
8082	Nitrofluorfen Benzene, 2-chloro-1-(4-nitrophenoxy)-4-(trifluoromethyl)-	$C_{13}H_7ClF_3NO_3$ 317.65	42874-01-1 68			
8083	Nitrous acid, butyl ester Butyl nitrite	$C_4H_9NO_2$ 103.12	544-16-1	1581 78	4-01-00-01523 0.9114^{25}	EtOH 5; eth 5 1.3762^{20}

No.	Name Synonym	Mol. Form. Mol. Wt.	CAS RN mp/°C	Merck No. bp/°C	Beil. Ref. den/g cm^{-3}	Solubility n_D
8084	Nitrous acid, decyl ester Decyl nitrite	$C_{10}H_{21}NO_2$ 187.28	1653-57-2	105-8^{12}	4-01-00-01819	EtOH 3; eth 3 1.4247^{20}
8085	Nitrous acid, 1,1-dimethylethyl ester *tert*-Butyl nitrite	$C_4H_9NO_2$ 103.12	540-80-7	1582 63	4-01-00-01622 0.8670^{20}	H_2O 2; EtOH 3; eth 3; chl 3 1.368^{20}
8086	Nitrous acid, 1,2-ethanediyl ester Ethylene glycol, dinitrite	$C_2H_4N_2O_4$ 120.06	629-16-3 <-15	98	4-01-00-02411 1.2156^{0}	eth 4; EtOH 4
8087	Nitrous acid, ethyl ester Ethyl nitrite	$C_2H_5NO_2$ 75.07	109-95-5	3786 18	4-01-00-01327 0.899^{15}	EtOH 5; eth 5 1.3418^{10}
8088	Nitrous acid, heptyl ester Heptyl nitrite	$C_7H_{15}NO_2$ 145.20	629-43-6	155.8	4-01-00-01735 0.8939^{0}	eth 4 1.4032^{20}
8089	Nitrous acid, hexyl ester Hexyl nitrite	$C_6H_{13}NO_2$ 131.17	638-51-7	129; 52^{44}	4-01-00-01699 0.8778^{20}	H_2O 1; EtOH 3; eth 3 1.3987^{20}
8090	Nitrous acid, 3-methylbutyl ester Isoamyl nitrite	$C_5H_{11}NO_2$ 117.15	110-46-3	5005 99.2	4-01-00-01683 0.8828^{20}	H_2O 2; EtOH 5; eth 5 1.3918^{20}
8091	Nitrous acid, methyl ester Methyl nitrite	CH_3NO_2 61.04	624-91-9 -16	-12	4-01-00-01253 0.991^{15}	EtOH 3; eth 3
8092	Nitrous acid, 1-methylethyl ester Isopropyl nitrite	$C_3H_7NO_2$ 89.09	541-42-4	5104 40	4-01-00-01474 0.8684^{15}	H_2O 1; EtOH 3; eth 3
8093	Nitrous acid, 1-methylpropyl ester	$C_4H_9NO_2$ 103.12	924-43-6	68.5	4-01-00-01573 0.8726^{20}	eth 4; EtOH 4; chl 4 1.3710^{20}
8094	Nitrous acid, 2-methylpropyl ester Isobutyl nitrite	$C_4H_9NO_2$ 103.12	542-56-3	5032 67	4-01-00-01595 0.8699^{22}	H_2O 2; EtOH 3; eth 3 1.3715^{22}
8095	Nitrous acid, octyl ester	$C_8H_{17}NO_2$ 159.23	629-46-9	174.5	4-01-00-01762 0.862^{17}	H_2O 2; EtOH 4; eth 4 1.4127^{20}
8096	Nitrous acid, pentyl ester	$C_5H_{11}NO_2$ 117.15	463-04-7	104.5	4-01-00-01644 0.8817^{20}	H_2O 2; EtOH 5; eth 5 1.3851^{20}
8097	Nitrous acid, phenylmethyl ester Benzyl nitrite	$C_7H_7NO_2$ 137.14	935-05-7	80-3^{35}	4-06-00-02568 1.075^{25}	1.4989^{25}
8098	Nitrous acid, propyl ester Propyl nitrite	$C_3H_7NO_2$ 89.09	543-67-9	7878 48	4-01-00-01424 0.886^{20}	H_2O 2; EtOH 3; eth 3 1.3604^{20}
8099	Nonacosane Celidoniol, deoxy-	$C_{29}H_{60}$ 408.80	630-03-5 63.7	440.8	4-01-00-00591 0.8083^{20}	H_2O 1; EtOH 4; eth 4; ace 4 1.4529^{20}
8100	1-Nonacosanol Nonacosanyl alcohol	$C_{29}H_{60}O$ 424.79	6624-76-6 84.5	200$^{1.0}$	4-01-00-01916	EtOH 3
8101	Nonadecane	$C_{19}H_{40}$ 268.53	629-92-5 32.1	329.9	4-01-00-00560 0.7855^{20}	H_2O 1; EtOH 2; eth 3; ace 3 1.4409^{20}
8102	1,2,3-Nonadecanetricarboxylic acid, 2-hydroxy- Agaricic acid	$C_{22}H_{40}O_7$ 416.56	666-99-9 142 dec	174	4-03-00-01284	H_2O 3; EtOH 2; eth 2; bz 1
8103	Nonadecanoic acid	$C_{19}H_{38}O_2$ 298.51	646-30-0 69.4	297^{100}; 228^{10}	4-02-00-01256 0.8468^{70}	H_2O 1; EtOH 4; eth 4; bz 4
8104	1-Nonadecanol	$C_{19}H_{40}O$ 284.53	1454-84-8 63.3	166$^{0.3}$	4-01-00-01898	eth 3; ace 3 1.4328^{75}
8105	10-Nonadecanone	$C_{19}H_{38}O$ 282.51	504-57-4 65.5	>350; 156$^{1.1}$	4-01-00-03402	H_2O 1; EtOH 2; eth 3; ace 3
8106	2-Nonadecanone	$C_{19}H_{38}O$ 282.51	629-66-3 57	266^{110}; 165^{2}	4-01-00-03400 0.8108^{56}	H_2O 1; EtOH 2; eth 4; ace 3
8107	1-Nonadecyne	$C_{19}H_{36}$ 264.49	26186-01-6 37.5	327	3-01-00-01030 0.8054^{20}	ace 4; bz 4; eth 4 1.4488^{20}
8108	2,6-Nonadienal Leaf aldehyde violet	$C_9H_{14}O$ 138.21	26370-28-5	85-7^{11}	4-01-00-03560 0.8678^{20}	1.4460^{20}
8109	2,4-Nonadienal, (*E*,*E*)-	$C_9H_{14}O$ 138.21	5910-87-2	97-8^{10}	4-01-00-03560 0.862^{25}	1.5207^{20}
8110	1,8-Nonadiene	C_9H_{16} 124.23	4900-30-5	142.5	4-01-00-01048 0.7511^{20}	1.4302^{20}
8111	2,7-Nonadiene	C_9H_{16} 124.23	51333-70-1 -72.5	151.5	3-01-00-01013 0.7499^{20}	1.4358^{20}
8112	2,6-Nonadien-1-ol	$C_9H_{16}O$ 140.23	7786-44-9	96-100^{11}	4-01-00-02266 0.8604^{25}	1.4598^{25}
8113	6,8-Nonadien-2-one, 8-methyl-5-(1-methylethyl)-, [*S*-(*E*)]- Solanone	$C_{13}H_{22}O$ 194.32	1937-54-8	8663 60^{1}	0.870^{20}	1.4755^{20}
8114	1,4-Nonadiyne	C_9H_{12} 120.19	6088-94-4	83-4^{41}	4-01-00-01125 0.8112^{25}	1.4518^{25}
8115	1,8-Nonadiyne	C_9H_{12} 120.19	2396-65-8 -27.3	162	4-01-00-01125 0.8158^{20}	eth 3; ace 3 1.4490^{20}
8116	2,7-Nonadiyne	C_9H_{12} 120.19	31699-35-1 4.3	180	3-01-00-01063 0.8332^{20}	1.4674^{20}
8117	Nonanal Nonaldehyde	$C_9H_{18}O$ 142.24	124-19-6	191	4-01-00-03352 0.8264^{22}	eth 3; chl 3 1.4273^{20}
8118	Nonanamide	$C_9H_{19}NO$ 157.26	1120-07-6 99.5	sub	4-02-00-01023 0.8394^{110}	H_2O 1; EtOH 2; eth 2 1.4248^{110}
8119	1-Nonanamine	$C_9H_{21}N$ 143.27	112-20-9 -1	202.2	4-04-00-00777 0.7886^{20}	H_2O 2; EtOH 3; eth 3; chl 2 1.4336^{20}
8120	Nonane	C_9H_{20} 128.26	111-84-2 -53.5	150.8	4-01-00-00447 0.7176^{20}	H_2O 1; EtOH 4; eth 4; ace 5 1.4054^{20}
8121	Nonane, 1-bromo-	$C_9H_{19}Br$ 207.15	693-58-3	88^{4}	4-01-00-00451 1.0165^{20}	1.4533^{20}

No.	Name Synonym	Mol. Form. Mol. Wt.	CAS RN mp/°C	Merck No. bp/°C	Beil. Ref. den/g cm^{-3}	Solubility n_D
8122	Nonane, 5-butyl-	$C_{13}H_{28}$ 184.37	17312-63-9	217.5	4-01-00-00517 0.7635^{18}	1.4273^{18}
8123	Nonane, 1-chloro-	$C_9H_{19}Cl$ 162.70	2473-01-0 -39.4	203.5	4-01-00-00450 0.8720^{20}	H_2O 1; eth 3; chl 3 1.4345^{20}
8124	Nonane, 2-chloro- 2-Chlorononane	$C_9H_{19}Cl$ 162.70	2216-36-8	190	0-01-00-00166 0.8790^{20}	chl 4 1.4420^{20}
8125	Nonane, 5-chloro- 5-Chlorononane	$C_9H_{19}Cl$ 162.70	28123-70-8	$85\text{-}7^{14}$	2-01-00-00128 0.8639^{15}	eth 4 1.4314^{15}
8126	Nonane, 1-chloro-9-fluoro-	$C_9H_{18}ClF$ 180.69	463-23-0	102^{11}	4-01-00-00450 0.966^{20}	eth 4; EtOH 4 1.4301^{25}
8127	Nonane, 1,2-dibromo- 1,2-Dibromononane	$C_9H_{18}Br_2$ 286.05	73642-91-8	141.5^{20}	3-01-00-00506 1.3980^{20}	1.4942^{20}
8128	Nonane, 1,9-dichloro- 1,9-Dichlorononane	$C_9H_{18}Cl_2$ 197.15	821-99-8	260; 138^{17}	4-01-00-00451 1.0173^{25}	1.4586^{25}
8129	Nonanedinitrile	$C_9H_{14}N_2$ 150.22	1675-69-0	$198\text{-}9^{25}$	4-02-00-02059 0.9200^{19}	eth 4; EtOH 4; bz 4 1.4518^{19}
8130	Nonanedioic acid Azelaic acid	$C_9H_{16}O_4$ 188.22	123-99-9 106.5	921 287^{100}; 225^{10}	4-02-00-02055 1.225^{25}	H_2O 2; EtOH 3; eth 2; bz 2 1.4303^{111}
8131	Nonanedioic acid, bis(2-ethylbutyl) ester	$C_{21}H_{40}O_4$ 356.55	105-03-3 -45	230^5	4-02-00-02057 0.928^{25}	ace 4; bz 4; EtOH 4 1.443^{25}
8132	Nonanedioic acid, bis(2-ethylhexyl) ester	$C_{25}H_{48}O_4$ 412.65	103-24-2 -78	237^5	3-02-00-01787 0.915^{25}	H_2O 1; EtOH 3; ace 3; bz 3 1.446^{25}
8133	Nonanedioic acid, diethyl ester	$C_{13}H_{24}O_4$ 244.33	624-17-9 -18.5	291.5	4-02-00-02056 0.9729^{20}	H_2O 1; EtOH 3; eth 3 1.4351^{20}
8134	Nonanedioic acid, dimethyl ester	$C_{11}H_{20}O_4$ 216.28	1732-10-1	156^{20}	4-02-00-02056 1.0082^{20}	H_2O 1; EtOH 3; ace 3; bz 3 1.4367^{20}
8135	Nonanedioic acid, 3-methyl- Azelaic acid, 3-methyl	$C_{10}H_{18}O_4$ 202.25	76078-85-8 44	$168^{0.2}$	3-02-00-01819	1.4670^{19}
8136	1,9-Nonanediol 1,9-Dihydroxynonane	$C_9H_{20}O_2$ 160.26	3937-56-2 45.8	173^{20}; 150^3	4-01-00-02607	H_2O 2; EtOH 4; eth 4; bz 3
8137	2,4-Nonanedione Methyl β-oxoheptyl ketone	$C_9H_{16}O_2$ 156.22	6175-23-1 -18	$105\text{-}7^{22}$	4-01-00-03715 0.938^{25}	
8138	Nonanedioyl dichloride	$C_9H_{14}Cl_2O_2$ 225.11	123-98-8	166^{18}	3-02-00-01789 1.4680^{20}	eth 3; bz 4 1.4680^{20}
8139	Nonane, icosofluoro Perfluorononane	C_9F_{20} 488.07	375-96-2	125.3	1.800^{25}	
8140	Nonane, 2-methyl- 2-Methylnonane	$C_{10}H_{22}$ 142.28	871-83-0 -74.6	167.1	4-01-00-00473 0.7281^{20}	H_2O 1; eth 3; bz 3; chl 3 1.4099^{20}
8141	Nonane, 3-methyl- 3-Methylnonane	$C_{10}H_{22}$ 142.28	5911-04-6 -84.8	167.9	2-01-00-00130 0.7354^{20}	bz 4; eth 4; chl 4 1.4125^{20}
8142	Nonane, 4-methyl- 4-Methylnonane	$C_{10}H_{22}$ 142.28	17301-94-9 -99	165.7	4-01-00-00475 0.7323^{20}	bz 4; eth 4; chl 4 1.4123^{20}
8143	Nonane, 5-methyl- 5-Methylnonane	$C_{10}H_{22}$ 142.28	15869-85-9 -87.7	165.1	4-01-00-00475 0.7326^{20}	H_2O 1; eth 3; bz 3; chl 3 1.4116^{20}
8144	Nonanenitrile	$C_9H_{17}N$ 139.24	2243-27-8 -34.2	224.4	4-02-00-01024 0.8178^{20}	H_2O 1; EtOH 3; eth 3; ctc 2 1.4255^{20}
8145	1-Nonanethiol Nonyl mercaptan	$C_9H_{20}S$ 160.32	1455-21-6 -20.1	220	4-01-00-01802 0.842^{25}	1.4548^{20}
8146	Nonanoic acid Pelargonic acid	$C_9H_{18}O_2$ 158.24	112-05-0 12.3	7013 254.5	4-02-00-01018 0.9052^{20}	H_2O 1; EtOH 3; eth 3; chl 3 1.4343^{19}
8147	Nonanoic acid, butyl ester Pelargonic acid, butyl ester	$C_{13}H_{26}O_2$ 214.35	50623-57-9 -38	$122\text{-}4^{20}$	3-02-00-00820 0.8520^{25}	1.4262^{25}
8148	Nonanoic acid, 9-cyano-, methyl ester Sebacic acid, monomitrile, methyl ester	$C_{11}H_{19}NO_2$ 197.28	53663-26-6 3.5	178^{16}	4-02-00-02089 0.934^{20}	eth 4 1.4398^{25}
8149	Nonanoic acid, ethenyl ester	$C_{11}H_{20}O_2$ 184.28	6280-03-1	133.5^{50}	4-02-00-01020 0.8689^{30}	1.4447^{30}
8150	Nonanoic acid, ethyl ester Ethyl nonanoate	$C_{11}H_{22}O_2$ 186.29	123-29-5 -36.7	3793 227.0	4-02-00-01019 0.8657^{20}	H_2O 1; EtOH 3; eth 3; ace 3 1.4220^{20}
8151	Nonanoic acid, 9-fluoro- Pelargonic acid, 9-fluoro	$C_9H_{17}FO_2$ 176.23	463-16-1 18	$89^{0.2}$	4-02-00-01024	eth 4; EtOH 4 1.4289^{25}
8152	Nonanoic acid, heptyl ester Pelargonic acid, heptyl ester	$C_{16}H_{32}O_2$ 256.43	71605-85-1 -15.5	210^{75}	3-02-00-00821 0.8553^{25}	1.4350^{20}
8153	Nonanoic acid, 3-methyl- Pelargonic acid, 3-methyl	$C_{10}H_{20}O_2$ 172.27	35205-79-9	$147\text{-}8^{12}$	4-02-00-01057 0.9012^{20}	1.4342^{20}
8154	Nonanoic acid, methyl ester	$C_{10}H_{20}O_2$ 172.27	1731-84-6	213.5	4-02-00-01019 0.8799^{15}	H_2O 1; EtOH 3; eth 3; ctc 2 1.4214^{20}
8155	Nonanoic acid, 3-methyl-, ethyl ester Pelargonic acid 3-methyl, ethyl ester	$C_{12}H_{24}O_2$ 200.32	86051-37-8	115^{13}	3-02-00-00849 0.8653^{20}	1.4240^{20}
8156	Nonanoic acid, pentyl ester Pelargonic acid, pentyl ester	$C_{14}H_{28}O_2$ 228.38	61531-45-1 -27	$130\text{-}2^{20}$	3-02-00-00820 0.8506^{25}	1.4318^{20}
8157	1-Nonanol Nonyl alcohol	$C_9H_{20}O$ 144.26	143-08-8 -5	6598 213.3	4-01-00-01798 0.8273^{20}	H_2O 1; EtOH 3; eth 3; ctc 2 1.4333^{20}
8158	2-Nonanol, (±)-	$C_9H_{20}O$ 144.26	74683-66-2 -35	193.5	4-01-00-01803 0.8471^{20}	eth 4; EtOH 4 1.4353^{20}
8159	3-Nonanol, (±)-	$C_9H_{20}O$ 144.26	74742-08-8 22	195; 93^{18}	4-01-00-01803 0.8250^{20}	H_2O 1; EtOH 3; eth 3 1.4289^{20}
8160	4-Nonanol, (S)-	$C_9H_{20}O$ 144.26	52708-03-9	192.5; 94^{18}	3-01-00-01747 0.8282^{20}	H_2O 1; EtOH 3; eth 3 1.4197^{20}
8161	5-Nonanol Dibutylcarbinol	$C_9H_{20}O$ 144.26	623-93-8	193; 97^{20}	4-01-00-01804 0.8356^{20}	H_2O 1; EtOH 3 1.4289^{20}

No.	Name Synonym	Mol. Form. Mol. Wt.	CAS RN mp/°C	Merck No. bp/°C	Beil. Ref. den/g cm^{-3}	Solubility n_D
8162	5-Nonanol, 5-butyl-	$C_{13}H_{28}O$ 200.36	597-93-3 20	 232 dec	4-01-00-01863 0.8408^{20}	EtOH 4 1.4445^{20}
8163	1-Nonanol, 9-chloro- 9-Chlorononan-1-ol	$C_9H_{19}ClO$ 178.70	51308-99-7 28	 147^{14}	4-01-00-01802	eth 4; EtOH 4 1.4575^{20}
8164	5-Nonanol, 2,8-dimethyl- 2,8-Dimethylnonan-5-ol	$C_{11}H_{24}O$ 172.31	19780-96-2	 105^{9}	0-01-00-00428 0.8305^{12}	 1.4380^{12}
8165	1-Nonanol, 9-fluoro-	$C_9H_{19}FO$ 162.25	463-24-1	 $125\text{-}6^{15}$	4-01-00-01801 0.928^{20}	eth 4; EtOH 4 1.4279^{25}
8166	1-Nonanol, 3-methyl- 3-Hexylbutanol-1	$C_{10}H_{22}O$ 158.28	22663-64-5	 114^{14}; 103^{9}	4-01-00-01825 0.8342^{20}	 1.4361^{20}
8167	1-Nonanol, 4-methyl-	$C_{10}H_{22}O$ 158.28	1489-47-0	 120^{17}	4-01-00-01826 0.826^{27}	 1.4364^{20}
8168	1-Nonanol, 8-methyl 8-Methyl-1-nonanol	$C_{10}H_{22}O$ 158.28	55505-26-5	 108^{10}		
8169	3-Nonanol, 2-methyl-	$C_{10}H_{22}O$ 158.28	26533-33-5	 115^{28}	4-01-00-01825 0.8290^{20}	 1.4346^{20}
8170	4-Nonanol, 4-methyl- 1-Methyl-1-propylhexanol	$C_{10}H_{22}O$ 158.28	23418-38-4	 $92\text{-}3^{15}$	4-01-00-01826 0.8245^{25}	 1.4338^{20}
8171	5-Nonanol, 5-methyl- 2-Butyl-2-hexanol	$C_{10}H_{22}O$ 158.28	33933-78-7	 $84\text{-}5^{10}$	4-01-00-01827 0.8290^{20}	 1.4341^{20}
8172	2-Nonanone Heptyl methyl ketone	$C_9H_{18}O$ 142.24	821-55-6 -7.5	 195.3	4-01-00-03353 0.8208^{20}	H_2O 1; EtOH 3; eth 3; ace 4 1.4210^{20}
8173	3-Nonanone Ethyl hexyl ketone	$C_9H_{18}O$ 142.24	925-78-0 -8	 190; 86^{20}	4-01-00-03354 0.8241^{20}	H_2O 1; EtOH 3; eth 3; ace 4 1.4208^{20}
8174	4-Nonanone Propyl amyl ketone	$C_9H_{18}O$ 142.24	4485-09-0	 187.5	4-01-00-03354 0.8190^{25}	H_2O 1; EtOH 3; eth 3; ace 4 1.4189^{20}
8175	5-Nonanone Dibutyl ketone	$C_9H_{18}O$ 142.24	502-56-7 -5.9	 188.4	4-01-00-03355 0.8217^{20}	H_2O 1; EtOH 3; eth 4; chl 4 1.4195^{20}
8176	5-Nonanone, 2-methyl- 2-Methyl-5-nonanone	$C_{10}H_{20}O$ 156.27	22287-02-1	 203.5	3-01-00-02900 0.8213^{20}	 1.4239^{20}
8177	Nonanoyl chloride	$C_9H_{17}ClO$ 176.69	764-85-2 -60.5	 215.3	4-02-00-01023 0.9463^{15}	eth 3; ace 3
8178	Nonasiloxane, eicosamethyl- Eicosamethylnonasiloxane	$C_{20}H_{60}O_8Si_9$ 681.46	2652-13-3	3494 307.5; 198^{16}	4-04-00-04130 0.9173^{20}	bz 4 1.3980^{20}
8179	2-Nonenal 2-Nonen-1-al	$C_9H_{16}O$ 140.23	2463-53-8	 $56\text{-}8^{0.1}$	4-01-00-03501 0.8418^{25}	 1.4502^{25}
8180	2-Nonenal, (*E*)-	$C_9H_{16}O$ 140.23	18829-56-6	6595	4-01-00-03502	
8181	6-Nonenamide, *N*-[(4-hydroxy-3-methoxyphenyl)methyl]-8-methyl-, (*E*)- Capsaicin	$C_{18}H_{27}NO_3$ 305.42	404-86-4 65	1767 $210\text{-}20^{0.01}$	4-13-00-02588	H_2O 1; EtOH 4; eth 3; bz 3
8182	1-Nonene	C_9H_{18} 126.24	124-11-8 -81.3	 146.9	4-01-00-00894 0.7253^{25}	 1.4257^{20}
8183	3-Nonene, (*E*)-	C_9H_{18} 126.24	20063-92-7	 147.5	3-01-00-00853 0.732^{21}	H_2O 1; eth 3; bz 3; ctc 3 1.4181^{21}
8184	4-Nonene, 5-butyl- 5-Butyl-4-nonene	$C_{13}H_{26}$ 182.35	7367-38-6	 215.5	4-01-00-00922 0.7745^{20}	 1.4375^{20}
8185	1-Nonene, 2-methyl- 2-Methylnon-1-ene	$C_{10}H_{20}$ 140.27	2980-71-4 -64.2	 168.4	4-01-00-00903 0.7412^{25}	 1.4241^{20}
8186	3-Nonene, 2-methyl-	$C_{10}H_{20}$ 140.27	53966-53-3	 161.0	4-01-00-00903 0.7340^{20}	 1.4202^{20}
8187	3-Nonenoic acid	$C_9H_{16}O_2$ 156.22	4124-88-3	 106^{1}	4-02-00-01598 0.9254^{20}	 1.4454^{25}
8188	8-Nonenoic acid	$C_9H_{16}O_2$ 156.22	31642-67-8 5	 $116\text{-}8^{1}$	4-02-00-01597 0.9146^{16}	 1.4492^{15}
8189	1-Nonen-3-ol 1-Vinylheptanol	$C_9H_{18}O$ 142.24	21964-44-3	 193.5	3-01-00-01947 0.824^{21}	 1.4382^{15}
8190	8-Nonen-1-ol	$C_9H_{18}O$ 142.24	13038-21-6	 135^{20}	3-01-00-01947 0.8394^{25}	 1.4450^{23}
8191	5-Nonen-4-one, 6-methyl-	$C_{10}H_{18}O$ 154.25	7036-98-8	 $90\text{-}2^{16}$	3-01-00-03022 0.8608^{20}	 1.4518^{20}
8192	6-Nonen-4-one, 6-methyl-	$C_{10}H_{18}O$ 154.25	53252-20-3	 94^{18}	3-01-00-03022 0.8413^{21}	 1.4429^{21}
8193	1-Nonen-3-yne	C_9H_{14} 122.21	57223-18-4	 $27\text{-}8^{4}$	4-01-00-01103 0.7602^{25}	ace 4; eth 4 1.4487^{25}
8194	1-Nonen-4-yne	C_9H_{14} 122.21	31508-12-0	 58^{22}	4-01-00-01103 0.777^{25}	ace 4; eth 4 1.4418^{25}
8195	2-Nonen-4-yne	C_9H_{14} 122.21	90644-60-3	 70^{20}	3-01-00-01048 0.7832^{25}	ace 4; eth 4 1.4590^{25}
8196	1-Nonyne Heptylacetylene	C_9H_{16} 124.23	3452-09-3 -50	 150.8	4-01-00-01047 0.7658^{20}	H_2O 1; eth 3; bz 3; ctc 3 1.4217^{20}
8197	2-Nonyne	C_9H_{16} 124.23	19447-29-1	 161.9	3-01-00-01013 0.7690^{20}	H_2O 1; eth 3; lig 3 1.4337^{20}
8198	3-Nonyne	C_9H_{16} 124.23	20184-89-8	 157.1	3-01-00-01013 0.7616^{20}	H_2O 1; eth 3; lig 3 1.4299^{20}
8199	4-Nonyne	C_9H_{16} 124.23	20184-91-2	 152	4-01-00-01047 0.757^{25}	H_2O 1; eth 3; ace 3 1.4296^{25}
8200	1-Nonyne, 1-chloro-	$C_9H_{15}Cl$ 158.67	90722-14-8	 $75\text{-}7^{15}$	3-01-00-01012 0.906^{20}	eth 4 1.450^{20}
8201	4-Nonyne, 3,3-dimethyl- 3,3-Dimethyl-4-nonyne	$C_{11}H_{20}$ 152.28	29022-31-9	 82^{40}	3-01-00-01024 0.7667^{20}	bz 4; eth 4 1.4317^{20}

No.	Name Synonym	Mol. Form. Mol. Wt.	CAS RN mp/°C	Merck No. bp/°C	Beil. Ref. den/g cm^{-3}	Solubility n_D
8202	4-Nonyne, 8-methyl- 8-Methyl-4-nonyne	$C_{10}H_{18}$ 138.25	70732-43-3	104.5^{97}	3-01-00-01018 0.7681^{20}	bz 4; eth 4 1.4311^{20}
8203	2-Nonynoic acid	$C_9H_{14}O_2$ 154.21	1846-70-4 -8	$154\text{-}6^{18}$	4-02-00-01721 0.9525^{12}	1.4605^{25}
8204	3-Nonynoic acid	$C_9H_{14}O_2$ 154.21	56630-33-2 17	118^{2}	4-02-00-01721	1.4603^{25}
8205	5-Nonynoic acid	$C_9H_{14}O_2$ 154.21	56630-34-3 -3	115^{1}	4-02-00-01723	1.4558^{25}
8206	6-Nonynoic acid	$C_9H_{14}O_2$ 154.21	56630-31-0 14	130^{3}	4-02-00-01723	1.4578^{25}
8207	8-Nonynoic acid	$C_9H_{14}O_2$ 154.21	30964-01-3 19	100^{6}	4-02-00-01721	1.4524^{25}
8208	1-Nonyn-3-ol	$C_9H_{16}O$ 140.23	7383-20-2	85^{10}; 63^{2}	4-01-00-02265 0.8627^{20}	1.4440^{20}
8209	2-Nonyn-1-ol	$C_9H_{16}O$ 140.23	5921-73-3	109^{11}	4-01-00-02265 0.878^{16}	1.4576^{16}
8210	24-Norcholan-23-oic acid, (5β)- Norcholanic acid	$C_{23}H_{38}O_2$ 346.55	511-18-2 177	6606	3-09-00-02652	
8211	29-Nordammara-1,17(20),24-trien-21-oic acid, 6,16-bis(acetyloxy)-3,7-dioxo-, Helvolic acid	$C_{33}H_{44}O_8$ 568.71	29400-42-8 212 dec	4552	4-10-00-04083	H_2O 2; EtOH 2; eth 3; ace 3
8212	Norea Urea, *N*,*N*-dimethyl-*N*'-(octahydro-4,7-methano-1H-inde*N*-5-yl)-, (3aα,4α, 5α,7α,7aα)-	$C_{13}H_{22}N_2O$ 222.33	18530-56-8 177	6611		
8213	Norflurazon 3(2H)-Pyridazinone, 4-chloro-5-(methylamino)-2-[3-(trifluoromethyl)phenyl]-	$C_{12}H_9ClF_3N_3O$ 303.67	27314-13-2 184	6618		
8214	Norleucine 2-Aminohexanoic acid	$C_6H_{13}NO_2$ 131.17	327-57-1 301 dec	6624	4-04-00-02686	H_2O 4
8215	*DL*-Norleucine Hexanoic acid, 2-amino(*DL*)	$C_6H_{13}NO_2$ 131.17	616-06-8 327 dec	6624	3-04-00-01386 1.172^{25}	H_2O 3; EtOH 2; eth 1
8216	19-Norpregna-1,3,5(10)-trien-20-yne-3,17-diol, (17α)- 17α-Ethinyl-17β-estradiol	$C_{20}H_{24}O_2$ 296.41	57-63-6	3689	4-06-00-06877	chl 2
8217	*DL*-Norvaline	$C_5H_{11}NO_2$ 117.15	760-78-1 303	6636 sub	4-04-00-02629	H_2O 3; EtOH 1; eth 1; chl 1
8218	*L*-Norvaline (*S*)-2-Aminopentanoic acid	$C_5H_{11}NO_2$ 117.15	6600-40-4 307	6636	4-04-00-02629	H_2O 4
8219	Nuarimol 5-Pyrimidinemethanol, α-(2-chlorophenyl)-α-(4-fluorophenyl)-	$C_{17}H_{12}ClFN_2O$ 314.75	63284-71-9 126			
8220	Octacosane	$C_{28}H_{58}$ 394.77	630-02-4 64.5	431.6	4-01-00-00588 0.8067^{20}	H_2O 1; ace 5; bz 3; chl 3 1.4330^{70}
8221	Octacosanoic acid Montanic acid	$C_{28}H_{56}O_2$ 424.75	506-48-9 90.9		4-02-00-01318 0.8191^{100}	bz 4; chl 4 1.4313^{100}
8222	1-Octacosanol Montanyl alcohol	$C_{28}H_{58}O$ 410.77	557-61-9 83.3	6664 200^{1}	4-01-00-01915	
8223	9,12-Octadecadienoic acid (*Z*,*Z*)- Linoleic acid	$C_{18}H_{32}O_2$ 280.45	60-33-3 -12	5382 $229\text{-}30^{16}$	4-02-00-01754 0.9022^{20}	ace 4; bz 4; eth 4; EtOH 4 1.4699^{20}
8224	10,12-Octadecadienoic acid, (*E*,*E*)-	$C_{18}H_{32}O_2$ 280.45	1072-36-2 56.5		4-02-00-01752 0.8686^{70}	1.4689^{40}
8225	9,12-Octadecadienoic acid	$C_{18}H_{32}I_2$ 502.26	2197-37-7 -5	$229\text{-}30^{16}$	4-02-00-01754 0.9022^{20}	ace 4; bz 4; eth 4; EtOH 4 1.4699^{20}
8226	9,12-Octadecadienoic acid (*Z*,*Z*)-, ethyl ester Ethyl linoleate	$C_{20}H_{36}O_2$ 308.50	544-35-4	3776 272^{180}, 212^{12}	4-02-00-01757 0.8865^{20}	eth 4; EtOH 4
8227	9,12-Octadecadienoic acid, ethyl ester	$C_{20}H_{36}O_2$ 308.50	7619-08-1	272^{180} 212^{12}	2-02-00-00461 0.8865^{20}	H_2O 1; EtOH 3; eth 3
8228	9,12-Octadecadienoic acid (*Z*,*Z*)-, methyl ester Methyl linoleate	$C_{19}H_{34}O_2$ 294.48	112-63-0 -35	6008 215^{20}	4-02-00-01756 0.8886^{10}	eth 4; EtOH 4 1.4638^{20}
8229	9,12-Octadecadienoic acid, methyl ester Methyl 9,12-octadecadienoate	$C_{19}H_{34}O_2$ 294.48	2462-85-3 -35	215^{20} 169^{1}	2-02-00-00461 0.8886^{18}	H_2O 1; EtOH 3; eth 3; ctc 2 1.4638^{20}
8230	7,11-Octadecadiyne	$C_{18}H_{30}$ 246.44	103697-14-9	$167\text{-}8^{7}$	3-01-00-01068 0.841^{19}	1.4698^{19}
8231	Octadecanamide	$C_{18}H_{37}NO$ 283.50	124-26-5 109	250^{12}	4-02-00-01240	eth 4; chl 4
8232	Octadecanamide, *N*-(4-hydroxyphenyl)-	$C_{24}H_{41}NO_2$ 375.60	103-99-1 133.8	239.5^{10}	4-13-00-01113	H_2O 1; eth 2; ace 3; bz 2
8233	1-Octadecanamine	$C_{18}H_{39}N$ 269.51	124-30-1 52.9	346.8	4-04-00-00825 0.8618^{20}	H_2O 1; EtOH 3; eth 3; ace 2 1.4522^{20}
8234	Octadecane	$C_{18}H_{38}$ 254.50	593-45-3 28.2	316.3	4-01-00-00553 0.7768^{28}	H_2O 1; EtOH 2; eth 3; ace 3 1.4390^{20}
8235	Octadecane, 1-bromo-	$C_{18}H_{37}Br$ 333.40	112-89-0 28.2	210^{10}	4-01-00-00555 0.9848^{20}	H_2O 1; EtOH 3; eth 3; ctc 2 1.4631^{20}
8236	Octadecane, 1-chloro-	$C_{18}H_{37}Cl$ 288.94	3386-33-2 28.6	348	4-01-00-00554 0.8641^{20}	H_2O 1; ctc 2 1.4531^{20}
8237	Octadecane, 1,1-dimethoxy- Octadecanal, dimethyl acetal	$C_{20}H_{42}O_2$ 314.55	14620-55-4	$168\text{-}70^{3}$	4-01-00-03397	eth 4; EtOH 4 1.4410^{25}

No.	Name	Synonym	Mol. Form.	Mol. Wt.	CAS RN	mp/°C	Merck No.	bp/°C	Beil. Ref.	den/g cm^{-3}	Solubility	n_D
8238	Octadecanedioic acid, diethyl ester		$C_{22}H_{42}O_4$	370.57	1472-90-8	54.5		240^{12}	4-02-00-02176		eth 4; EtOH 4	
8239	Octadecanedioic acid, 9,10-dihydroxy-, (*R**,*R**)-(±)-	Phloionic acid	$C_{18}H_{34}O_6$	346.46	23843-52-9	126	7297		4-03-00-01250			
8240	Octadecane, 1-(ethenyloxy)-		$C_{20}H_{40}O$	296.54	930-02-9	30		182^{3}	4-01-00-02057	0.8138^{40}	chl 2	
8241	Octadecane, 3-ethyl-5-(2-ethylbutyl)-		$C_{26}H_{54}$	366.71	55282-12-7			229.5^{10}	4-01-00-00586	0.8115^{20}		1.4523^{20}
8242	Octadecane, 1-iodo-		$C_{18}H_{37}I$	380.40	629-93-6	34		383	4-01-00-00556	1.0994^{20}	H_2O 1; EtOH 2; eth 2	1.4810^{20}
8243	Octadecanenitrile		$C_{18}H_{35}N$	265.48	638-65-3	41		362	4-02-00-01242	0.8325^{20}	H_2O 1; EtOH 3; eth 4; ace 4	1.4389^{45}
8244	1-Octadecanethiol	Stearyl mercaptan	$C_{18}H_{38}S$	286.57	2885-00-9	30		$204\text{-}10^{11}$	4-01-00-01894	0.8475^{20}	eth 4	1.4645^{20}
8245	Octadecane, 9-*p*-tolyl-	Toluene, *p*-(1-octyldecyl)-	$C_{25}H_{44}$	344.62	4445-08-3			185^{10}	4-05-00-01221	0.8549^{20}		1.4811^{20}
8246	9,11,13-Octadecanetrienoic acid (*Z,Z,Z*)	Eleostearic acid	$C_{18}H_{30}O_2$	278.44	3884-88-6	48.5						
8247	Octadecanoic acid	Stearic acid	$C_{18}H_{36}O_2$	284.48	57-11-4	68.8	8761	350 dec; 232^{15}	4-02-00-01206	0.9408^{20}	H_2O 1; EtOH 2; eth 4; ace 3	1.4299^{80}
8248	Octadecanoic acid, aluminum salt		$C_{54}H_{105}AlO_6$	877.41	637-12-7	118	370		4-02-00-01206		H_2O 1; EtOH 3; peth 3	
8249	Octadecanoic acid, anhydride		$C_{36}H_{70}O_3$	550.95	638-08-4	72			4-02-00-01239	0.8365^{82}	H_2O 1; EtOH 1; eth 2; bz 2	1.4362^{80}
8250	Octadecanoic acid, 18-bromo-	Stearic acid, 18-bromo	$C_{18}H_{35}BrO_2$	363.38	2536-38-1	75.5		240^{4}	2-02-00-00361		bz 4; eth 4; EtOH 4	
8251	Octadecanoic acid, butyl ester	Butyl stearate	$C_{22}H_{44}O_2$	340.59	123-95-5	27	1589	343	4-02-00-01219	0.854^{25}	H_2O 1; EtOH 3; ace 4	1.4328^{50}
8252	Octadecanoic acid, calcium salt		$C_{36}H_{70}CaO_4$	607.03	1592-23-0	179.5	1710		4-02-00-01206		H_2O 1; EtOH 1; eth 1	
8253	Octadecanoic acid, cyclohexyl ester	Stearic acid, cyclohexyl ester	$C_{24}H_{46}O_2$	366.63	104-07-4	44			4-06-00-00038	0.880^{15}	eth 4	
8254	Octadecanoic acid, 9,10-dihydroxy-	9,10-Dihydroxystearic acid	$C_{18}H_{36}O_4$	316.48	120-87-6	90	3171		4-03-00-01092		H_2O 1; EtOH 2; eth 2	
8255	Octadecanoic acid, 2,3-dihydroxypropyl ester, (±)-		$C_{21}H_{42}O_4$	358.56	22610-63-5	74			4-02-00-01225	0.9841^{20}	H_2O 1; EtOH 2; eth 2; lig 3	1.4400^{86}
8256	Octadecanoic acid, 1,2-ethanediyl ester		$C_{38}H_{74}O_4$	595.00	627-83-8	79		241^{20}	4-02-00-01223	0.8581^{78}	H_2O 1; EtOH 1; eth 4; ace 4	
8257	Octadecanoic acid, ethyl ester		$C_{20}H_{40}O_2$	312.54	111-61-5	33		199^{10}	4-02-00-01218	1.057^{20}	H_2O 1; EtOH 3; eth 3; ace 4	1.4349^{40}
8258	Octadecanoic acid, hexadecyl ester		$C_{34}H_{68}O_2$	508.91	1190-63-2	57			4-02-00-01220		ace 4; eth 4; chl 4	1.4410^{70}
8259	Octadecanoic acid, 2-[2-[2-(2-hydroxyethoxy)ethoxy]ethoxy]ethyl ester		$C_{26}H_{52}O_6$	460.70	106-07-0	40		328		1.1285^{15}		1.4593^{20}
8260	Octadecanoic acid, 2-hydroxyethyl ester		$C_{20}H_{40}O_3$	328.54	111-60-4	60.5		$189\text{-}91^{3}$	4-02-00-01222	0.8780^{60}	EtOH 2; eth 3	1.4310^{60}
8261	Octadecanoic acid, lead (II) salt	Lead stearate	$C_{36}H_{70}O_4Pb$	774.15	7428-48-0	125			4-02-00-01206	1.4	H_2O 1; Hot EtOH 3; eth 1	
8262	Octadecanoic acid, 14-methyl-	Stearic acid, 14-methyl	$C_{19}H_{38}O_2$	298.51	94434-64-7	37.5		$182^{0.4}$	4-02-00-01265	0.9400^{20}		
8263	Octadecanoic acid, 17-methyl-		$C_{19}H_{38}O_2$	298.51	2724-59-6	67.5		$180^{0.3}$	4-02-00-01260	0.8420^{70}		1.4336^{70}
8264	Octadecanoic acid, 9-methyl-	Stearic acid, 9-methyl	$C_{19}H_{38}O_2$	298.51	86073-38-3	40		$171^{0.1}$	4-02-00-01271	0.9980^{20}		
8265	Octadecanoic acid, 3-methylbutyl ester	Stearic acid, isopentyl ester	$C_{23}H_{46}O_2$	354.62	627-88-3	25.5		192^{2}	2-02-00-00353	0.855^{20}	H_2O 1; EtOH 2; eth 3; ace 3	1.433^{50}
8266	Octadecanoic acid, methyl ester		$C_{19}H_{38}O_2$	298.51	112-61-8	39.1		443; 215^{15}	4-02-00-01216	0.8498^{40}	eth 4; chl 4	1.4367^{40}
8267	Octadecanoic acid, 1-methylethyl ester		$C_{21}H_{42}O_2$	326.56	112-10-7	28		207^{6}	4-02-00-01219	0.8403^{38}	ace 4; eth 4; EtOH 4; chl 4	
8268	Octadecanoic acid, 2-methylpropyl ester	Isobutyl stearate	$C_{22}H_{44}O_2$	340.59	646-13-9	28.9	5034	223^{15}	3-02-00-01017	0.8498^{20}	eth 4	
8269	Octadecanoic acid, 12-oxo-, ethyl ester	Stearic acid, 12-oxo, ethyl ester	$C_{20}H_{38}O_3$	326.52	88472-61-1	38		199^{3}	3-03-00-01294		EtOH 4	
8270	Octadecanoic acid, pentyl ester	Stearic acid, pentyl ester	$C_{23}H_{46}O_2$	354.62	6382-13-4	30			4-02-00-01220		eth 4; EtOH 4	1.4342^{50}
8271	Octadecanoic acid, phenyl ester	Stearic acid, phenyl ester	$C_{24}H_{40}O_2$	360.58	637-55-8	52		267^{15}	4-06-00-00618		H_2O 1; EtOH 3; eth 3	
8272	Octadecanoic acid, 1,2,3-propanetriyl ester	Tristearin	$C_{57}H_{110}O_6$	891.50	555-43-1		9669		4-02-00-01233	0.8559^{90}	H_2O 1; EtOH 1; ace 3; bz 2	1.4395^{80}
8273	Octadecanoic acid, propyl ester		$C_{21}H_{42}O_2$	326.56	3634-92-2	28.9		186.8^{2}	4-02-00-01219	0.8452^{38}	ace 4; eth 4; EtOH 4	1.4400^{30}
8274	Octadecanoic acid, 9,10,12,13-tetrabromo-, methyl ester	Stearic acid, 9,10,12,13-tetrabromo, methyl ester	$C_{19}H_{34}Br_4O_2$	614.09	62080-86-8	63		215^{15}	3-02-00-01049		eth 4; EtOH 4; chl 4	1.4346^{45}

No.	Name Synonym	Mol. Form. Mol. Wt.	CAS RN mp/°C	Merck No. bp/°C	Beil. Ref. den/g cm^{-3}	Solubility n_D
8275	Octadecanoic acid, (tetrahydro-2-furanyl)methyl ester Stearic acid, tetrahydrofurfuryl ester	$C_{23}H_{44}O_3$ 368.60	6940-09-6 22		4-02-00-01248 0.914^{25}	eth 4; EtOH 4
8276	Octadecanoic acid, 9,10,18-trihydroxy-, (*R**,*R**)- Phloionolic acid	$C_{18}H_{36}O_5$ 332.48	583-86-8 101.5	7298	4-03-00-01119	
8277	Octadecanoic acid, zinc salt	$C_{36}H_{70}O_4Zn$ 632.34	557-05-1 130	10063	4-02-00-01206	H_2O 1; EtOH 1; eth 1
8278	1-Octadecanol Stearyl alcohol	$C_{18}H_{38}O$ 270.50	112-92-5 59.5	210.5^{15}	4-01-00-01888 0.8124^{59}	H_2O 1; EtOH 3; eth 3; ace 2
8279	3-Octadecanol, (±)- 3-Octadecanol (*DL*)	$C_{18}H_{38}O$ 270.50	111897-18-8 51	202^{13}	4-01-00-01896 0.7858^{80}	1.4290^{80}
8280	3-Octadecanone	$C_{18}H_{36}O$ 268.48	18261-92-2 51	197^{11}	4-01-00-03398	ctc 3
8281	Octadecanoyl chloride	$C_{18}H_{35}ClO$ 302.93	112-76-5 23	215^{15}	4-02-00-01240 0.8969^{0}	EtOH 2 1.4523^{24}
8282	9,11,13-Octadecatrienoic acid, (*E*,*E*,*E*)-	$C_{18}H_{30}O_2$ 278.44	544-73-0 71.5	188^{1}	4-02-00-01787 0.8839^{80}	EtOH 4 1.5000^{80}
8283	9,12,15-Octadecatrienoic acid, ethyl ester, (*Z*,*Z*,*Z*)-	$C_{20}H_{34}O_2$ 306.49	1191-41-9	218^{15}	4-02-00-01782 0.8919^{20}	eth 4; EtOH 4 1.4694^{20}
8284	9,11,13-Octadecatrienoic acid, (*E*,*Z*,*E*)-	$C_{18}H_{30}O_2$ 278.44	506-23-0 49	235^{12} dec; 170^{1}	4-02-00-01787 0.9028^{50}	eth 4; EtOH 4 1.5112^{50}
8285	9,12,15-Octadecatrienoic acid, methyl ester, (*Z*,*Z*,*Z*)-	$C_{19}H_{32}O_2$ 292.46	301-00-8	182^{3}	4-02-00-01782 0.896^{25}	1.4709^{20}
8286	9,11,13-Octadecatrienoic acid, (*Z*,*Z*,*E*)-	$C_{18}H_{30}O_2$ 278.44	544-72-9 62		4-02-00-01787 0.9027^{20}	EtOH 4; peth 4 1.5114^{50}
8287	9,12,15-Octadecatrienoic acid, (*Z*,*Z*,*Z*)- Linolenic acid	$C_{18}H_{30}O_2$ 278.44	463-40-1 -11	5383 231^{17}; $129^{0.05}$	4-02-00-01781 0.9164^{20}	H_2O 1; EtOH 3; eth 3; bz 2 1.4800^{20}
8288	9-Octadecenal Octadecenyl aldehyde	$C_{18}H_{34}O$ 266.47	5090-41-5	$168\text{-}9^{3}$	4-01-00-03533 0.8509^{20}	1.4558^{20}
8289	9-Octadecenal, (*Z*)-	$C_{18}H_{34}O$ 266.47	2423-10-1	$168\text{-}9^{3}$	4-01-00-03533 0.8509^{20}	1.4558^{20}
8290	9-Octadecenamide, *N*-phenyl-, (*Z*)-	$C_{24}H_{39}NO$ 357.58	5429-85-6 41	143.5^{10}	4-12-00-00411	bz 4; eth 4; HOAc 4
8291	1-Octadecene	$C_{18}H_{36}$ 252.48	112-88-9 17.5	179^{15}; 145^{8}	4-01-00-00930 0.7891^{20}	H_2O 1; ace 3; ctc 3 1.4448^{20}
8292	9-Octadecene	$C_{18}H_{36}$ 252.48	5557-31-3 -30.5	162^{9}	4-01-00-00932 0.7916^{20}	1.4470^{20}
8293	9-Octadecenenitrile, (*Z*)-	$C_{18}H_{33}N$ 263.47	112-91-4 -1	332 dec	4-02-00-01668 0.847^{17}	EtOH 4 1.4566^{20}
8294	9-Octadecenoic acid (*Z*)- Oleic acid	$C_{18}H_{34}O_2$ 282.47	112-80-1 13.4	6788 286^{100}	4-02-00-01641 0.8935^{20}	H_2O 1; EtOH 5; eth 5; ace 5 1.4582^{20}
8295	11-Octadecenoic acid, (*E*)- Vaccenic acid	$C_{18}H_{34}O_2$ 282.47	693-72-1 44	9810	4-02-00-01640	ace 3 1.4499^{60}
8296	2-Octadecenoic acid, (*E*)-	$C_{18}H_{34}O_2$ 282.47	2825-79-8 58.5		4-02-00-01632 0.8484^{90}	H_2O 1; EtOH 2; eth 3; bz 3
8297	9-Octadecenoic acid, (*E*)- Elaidic acid	$C_{18}H_{34}O_2$ 282.47	112-79-8 45	3497 288^{100}; 234^{15}	4-02-00-01647 0.8734^{45}	H_2O 1; EtOH 3; eth 3; bz 3 1.4499^{45}
8298	9-Octadecenoic acid (*Z*)-, butyl ester	$C_{22}H_{42}O_2$ 338.57	142-77-8 -26.4	$227\text{-}8^{15}$	4-02-00-01653 0.8704^{15}	EtOH 4 1.4480^{25}
8299	9-Octadecenoic acid (*Z*)-, 2,3-dihydroxypropyl ester	$C_{21}H_{40}O_4$ 356.55	111-03-5 35	$238\text{-}40^{3}$	2-02-00-00439 0.9420^{20}	H_2O 1; EtOH 3; eth 3; chl 3 1.4626^{20}
8300	6-Octadecenoic acid, 6,7-diiodo- Diiodotariric acid	$C_{18}H_{32}I_2O_2$ 534.26	533-86-8 50.3	dec	4-02-00-01638	bz 4; eth 4; chl 4
8301	9-Octadecenoic acid (*Z*)-, ethyl ester	$C_{20}H_{38}O_2$ 310.52	111-62-6	207^{13}	4-02-00-01651 0.8720^{20}	eth 4; EtOH 4 1.4515^{20}
8302	9-Octadecenoic acid, ethyl ester, (*E*)-	$C_{20}H_{38}O_2$ 310.52	6114-18-7 5.8	$217\text{-}9^{15}$	4-02-00-01652 0.8664^{25}	eth 4; EtOH 4 1.4480^{25}
8303	9-Octadecenoic acid, 12-hydroxy-, [*R*-(*Z*)]- Ricinoleic acid	$C_{18}H_{34}O_3$ 298.47	141-22-0	8213 227^{10}	4-03-00-01026 0.9450^{21}	eth 4; EtOH 4 1.4716^{21}
8304	9-Octadecenoic acid, 12-hydroxy-, butyl ester, [*R*-(*Z*)]-	$C_{22}H_{42}O_3$ 354.57	151-13-3	278^{12}	0-03-00-00388 0.9058^{22}	eth 4 1.4566^{22}
8305	9-Octadecenoic acid, 12-hydroxy-, ethyl ester, [*R*-(*Z*)]-	$C_{20}H_{38}O_3$ 326.52	55066-53-0	258^{13}	4-03-00-01029 0.9045^{22}	1.4618^{22}
8306	9-Octadecenoic acid (*Z*)-, 3-methylbutyl ester Oleic acid, isopentyl ester	$C_{23}H_{44}O_2$ 352.60	627-89-4	$223\text{-}4^{10}$	3-02-00-01414 0.897^{15}	eth 4; EtOH 4
8307	9-Octadecenoic acid (*Z*)-, methyl ester Methyl oleate	$C_{19}H_{36}O_2$ 296.49	112-62-9 -19.9	218.5^{20}	4-02-00-01649 0.8739^{20}	H_2O 1; EtOH 5; eth 5; chl 3 1.4522^{20}
8308	9-Octadecenoic acid, methyl ester, (*E*)-	$C_{19}H_{36}O_2$ 296.49	1937-62-8	$213\text{-}5^{15}$	4-02-00-01651 0.8730^{20}	eth 4; EtOH 4 1.4513^{20}
8309	9-Octadecenoic acid (*Z*)-, phenylmethyl ester Oleic acid, benzyl ester	$C_{25}H_{40}O_2$ 372.59	55130-16-0	237^{1}	4-06-00-02269 0.9302^{25}	eth 4; EtOH 4 1.4875^{25}

PHYSICAL CONSTANTS OF ORGANIC COMPOUNDS (continued)

No.	Name Synonym	Mol. Form. Mol. Wt.	CAS RN mp/°C	Merck No. bp/°C	Beil. Ref. den/g cm^{-3}	Solubility n_D
8310	9-Octadecenoic acid (*Z*)-, 1,2,3-propanetriyl ester Triolein	$C_{57}H_{104}O_6$ 885.45	122-32-7 -32	9644 $235-40^{18}$	4-02-00-01664 0.8988^{40}	H_2O 1; EtOH 2; eth 4; chl 3 1.4621^{40}
8311	9-Octadecen-1-ol, (*E*)-	$C_{18}H_{36}O$ 268.48	506-42-3 36.5	 333	4-01-00-02204 0.8338^{40}	ace 4; eth 4; EtOH 4 1.4552^{40}
8312	9-Octadecen-1-ol, (*Z*)- Oleyl alcohol	$C_{18}H_{36}O$ 268.48	143-28-2 6.5	6791 $205-10^{15}$	4-01-00-02204 0.8489^{20}	H_2O 1; EtOH 3; eth 3; ctc 2 1.4606^{20}
8313	1-Octadecyne	$C_{18}H_{34}$ 250.47	629-89-0 22.5	 313	4-01-00-01075 0.8025^{20}	 1.4774^{20}
8314	2-Octadecyne	$C_{18}H_{34}$ 250.47	61847-97-0 30	 184^{15}	4-01-00-01075 0.8016^{30}	
8315	9-Octadecyne	$C_{18}H_{34}$ 250.47	35365-59-4 3	 $163-4^{7}$	4-01-00-01076 0.8012^{20}	 1.4488^{25}
8316	9-Octadecynoic acid 9-Stearolic acid	$C_{18}H_{32}O_2$ 280.45	506-24-1 48	 $189^{1.8}$	4-02-00-01751	H_2O 1; EtOH 3; eth 4 1.4510^{54}
8317	2,6-Octadienal, 3,7-dimethyl-, (*E*)-	$C_{10}H_{16}O$ 152.24	141-27-5	 229	4-01-00-03569 0.8888^{20}	H_2O 1; EtOH 5; eth 5 1.4898^{20}
8318	2,6-Octadienal, 3,7-dimethyl-, (*Z*)-	$C_{10}H_{16}O$ 152.24	106-26-3	 120^{20}	4-01-00-03569 0.8869^{20}	H_2O 1; EtOH 5; eth 5 1.4869^{20}
8319	1,7-Octadiene α,ω-Octadiene	C_8H_{14} 110.20	3710-30-3	 115.5	4-01-00-01038 0.734^{20}	 1.4245^{20}
8320	2,6-Octadiene 1,6-Dimethyldiallyl	C_8H_{14} 110.20	4974-27-0 -76	 124.5	4-01-00-01039 0.7406^{25}	 1.4292^{17}
8321	2,6-Octadiene, 1-bromo-3,7-dimethyl-, (*E*)- *trans*-Geranyl bromide	$C_{10}H_{17}Br$ 217.15	6138-90-5	 101^{12}, $47^{0.005}$	4-01-00-01059 1.0940^{22}	 1.5027^{20}
8322	2,4-Octadiene, 3,7-dimethyl- 3,7-Dimethylocta-2,4-diene	$C_{10}H_{18}$ 138.25	56523-26-3	 165.5	3-01-00-01020 0.7933^{20}	 1.456^{20}
8323	2,5-Octadiene, 2,6-dimethyl-	$C_{10}H_{18}$ 138.25	116668-48-5	 59.5^{12}	3-01-00-01018 0.773^{20}	 1.4500^{20}
8324	2,6-Octadiene, 2,6-dimethyl-, (*Z*)- *cis*-2,6-Dimethyl-2,6-octadiene	$C_{10}H_{18}$ 138.25	2492-22-0	 168	4-01-00-01058 0.775^{21}	ace 4; eth 4; EtOH 4 1.4498^{20}
8325	2,6-Octadiene, 2,7-dimethyl- 2,7-Dimethyl-2,6-octadiene	$C_{10}H_{18}$ 138.25	16736-42-8 -74.4	 166	4-01-00-01061 0.7849^{20}	 1.4481^{20}
8326	2,6-Octadiene, 3,6-dimethyl-	$C_{10}H_{18}$ 138.25	116668-49-6	 154	4-01-00-01061 0.7767^{20}	 1.4445^{20}
8327	2,6-Octadiene, 4,5-dimethyl- 4,5-Dimethyl-2,6-octadiene	$C_{10}H_{18}$ 138.25	18476-57-8 -64.8	 153.5	3-01-00-01021 0.7611^{25}	 1.4375^{25}
8328	2,4-Octadiene, 7-methyl-	C_9H_{16} 124.23	2216-70-8	 149	2-01-00-00239 0.7521^{18}	 1.4543^{18}
8329	3,5-Octadiene, 4-methyl- 4-Methyl-3,5-octadiene	C_9H_{16} 124.23	36903-95-4	 149.5	2-01-00-00239 0.7640^{25}	 1.4628^{25}
8330	1,6-Octadiene, 7-methyl-3-methylene- β-Myrcene	$C_{10}H_{16}$ 136.24	123-35-3	6243 167	4-01-00-01108 0.8013^{15}	H_2O 1; EtOH 3; eth 3; bz 3 1.4722^{20}
8331	2,4-Octadienoic acid, 3,7-dimethyl-	$C_{10}H_{16}O_2$ 168.24	101715-60-0	 $151-4^{14}$	4-02-00-01736 0.959^{17}	 1.4919
8332	1,6-Octadien-3-ol, 3,7-dimethyl-, (±)- Linalol	$C_{10}H_{18}O$ 154.25	22564-99-4	 198; 86^{13}	4-01-00-02278 0.870^{15}	 1.4627
8333	2,6-Octadien-1-ol, 3,7-dimethyl-, (*E*)- *trans*-Geraniol	$C_{10}H_{18}O$ 154.25	106-24-1 <-15	4298 230	4-01-00-02277 0.8894^{20}	H_2O 1; EtOH 3; eth 3; ace 3 1.4766^{20}
8334	2,6-Octadien-1-ol, 3,7-dimethyl-, (*Z*)- Nerol	$C_{10}H_{18}O$ 154.25	106-25-2 <-15	6387 225; 125^{25}	4-01-00-02276 0.8756^{20}	EtOH 4 1.4746^{20}
8335	1,6-Octadien-3-ol, 3,7-dimethyl-, acetate, (*R*)-	$C_{12}H_{20}O_2$ 196.29	16509-46-9	 220; 115^{25}	4-02-00-00204 0.8951^{20}	eth 4; EtOH 4 1.4544^{21}
8336	2,6-Octadien-1-ol, 3,7-dimethyl-, acetate, (*Z*)-	$C_{12}H_{20}O_2$ 196.29	141-12-8	 134^{25}; 93^{3}	2-02-00-00153 0.905^{15}	 1.452^{20}
8337	1,6-Octadien-3-ol, 3,7-dimethyl-, formate Linalyl formate	$C_{11}H_{18}O_2$ 182.26	115-99-1	 $100-3^{10}$	4-02-00-00035 0.915^{25}	EtOH 4 1.456^{20}
8338	2,6-Octadien-1-ol, 3,7-dimethyl-, formate, (*E*)-	$C_{11}H_{18}O_2$ 182.26	105-86-2	 229 dec; 113^{25}	4-02-00-00035 0.9086^{25}	H_2O 1; EtOH 4; eth 3; ace 3 1.4659^{20}
8339	2,6-Octadien-4-yne, 3,6-diethyl-	$C_{12}H_{18}$ 162.27	100319-48-0	 170	3-01-00-01066 0.8196^{20}	ace 4; eth 4 1.4965^{20}
8340	2,6-Octadien-4-yne, 3,6-dimethyl- 3,6-Dimethyl-2,6-octadiene-4-yne	$C_{10}H_{14}$ 134.22	3725-07-3 -45	 170	4-01-00-01129 0.8071^{22}	ace 4; eth 4 1.4998^{20}
8341	1,5-Octadien-3-yne, 5-propyl- 5-Propyl-1,5-octadien-3-yne	$C_{11}H_{16}$ 148.25	70732-44-4	 $57-8^{6}$	4-01-00-01129 0.8047^{20}	ace 4; eth 4 1.4949^{20}
8342	1,7-Octadiyne	C_8H_{10} 106.17	871-84-1	 135.5; 59^{35}	4-01-00-01122 0.8169^{21}	eth 3 1.4521^{18}
8343	2,6-Octadiyne	C_8H_{10} 106.17	764-73-8 27	 62^{19}	4-01-00-01123 0.828^{80}	eth 4 1.4658^{30}
8344	3,5-Octadiyne	C_8H_{10} 106.17	16387-70-5	 163.5	4-01-00-01123 0.826^{0}	eth 3 1.4968^{0}
8345	3,5-Octadiyne, 2,7-dimethyl- Diisopropylbutadiyne	$C_{10}H_{14}$ 134.22	14813-68-4	 74^{12}	1-01-00-00128 0.8090^{20}	eth 4
8346	Octanal Caprylic aldehyde	$C_8H_{16}O$ 128.21	124-13-0	1766 171	4-01-00-03337 0.8211^{20}	ace 4; bz 4; eth 4; EtOH 4 1.4217^{20}
8347	Octanal, 7-hydroxy-3,7-dimethyl-	$C_{10}H_{20}O_2$ 172.27	107-75-5	 103^{3}	4-01-00-04058 0.9220^{20}	H_2O 2; EtOH 3; ace 3 1.4494^{20}
8348	Octanal, oxime	$C_8H_{17}NO$ 143.23	929-55-5 60	 112^{9}	2-01-00-00758	ace 4; EtOH 4

No.	Name Synonym	Mol. Form. Mol. Wt.	CAS RN mp/°C	Merck No. bp/°C	Beil. Ref. den/g cm^{-3}	Solubility n_D
8349	Octanamide	$C_8H_{17}NO$ 143.23	629-01-6 108	239	4-02-00-00992 0.8450^{110}	H_2O 2; EtOH 4; eth 3; ace 3
8350	1-Octanamine	$C_8H_{19}N$ 129.25	111-86-4 0	179.6	4-04-00-00751 0.7826^{20}	H_2O 2; EtOH 4; eth 4; ctc 3 1.4292^{20}
8351	2-Octanamine, (±)-	$C_8H_{19}N$ 129.25	44855-57-4 97	164	4-04-00-00762 0.7744^{20}	eth 4; EtOH 4 1.4232^{25}
8352	2-Octanamine, *N*-(1-methylheptyl)-	$C_{16}H_{35}N$ 241.46	5412-92-0	283	4-04-00-00764 0.7948^{20}	H_2O 1; EtOH 4; eth 4
8353	1-Octanamine, *N*-methyl-*N*-octyl-	$C_{17}H_{37}N$ 255.49	4455-26-9 -30.1	158^{10}	4-04-00-00753	1.4424^{20}
8354	1-Octanamine, *N*-octyl-	$C_{16}H_{35}N$ 241.46	1120-48-5 35.5	297.5	4-04-00-00753 0.7963^{26}	eth 4; EtOH 4 1.4415^{26}
8355	Octane	C_8H_{18} 114.23	111-65-9 -56.8	6672 125.6	4-01-00-00412 0.6986^{25}	H_2O 1; EtOH 5; eth 3; ace 5 1.3974^{20}
8356	Octane, 1-bromo- Octyl bromide	$C_8H_{17}Br$ 193.13	111-83-1 -55	6684 200	4-01-00-00422 1.108^{25}	H_2O 1; EtOH 5; eth 5; ctc 2 1.4524^{20}
8357	Octane, 2-bromo-, (±)- Octane, 2-bromo (*DL*)	$C_8H_{17}Br$ 193.13	60251-57-2	188.5	4-01-00-00423 1.0878^{25}	H_2O 1; EtOH 5; eth 5 1.4442^{25}
8358	Octane, 1-bromo-8-fluoro-	$C_8H_{16}BrF$ 211.12	593-12-4	$118\text{-}20^{22.5}$	4-01-00-00424	eth 4; EtOH 4 1.4500^{20}
8359	Octane, 1-chloro- Octyl chloride	$C_8H_{17}Cl$ 148.68	111-85-3 -57.8	181.5	4-01-00-00419 0.8738^{20}	H_2O 1; EtOH 4; eth 4; ctc 2 1.4305^{20}
8360	Octane, 2-chloro-	$C_8H_{17}Cl$ 148.68	628-61-5	172; 75^{28}	1-01-00-00060 0.8658^{17}	H_2O 1; EtOH 4; eth 3 1.4273^{21}
8361	Octane, 1-chloro-8-fluoro-	$C_8H_{16}ClF$ 166.67	593-14-6	87^{10}	4-01-00-00420 0.978^{20}	ace 4; eth 4; EtOH 4 1.4266^{25}
8362	Octane, 3-chloro-3-methyl-	$C_9H_{19}Cl$ 162.70	28320-88-9	$73\text{-}4^{15}$	3-01-00-00508 0.8680^{25}	ace 4; bz 4; eth 4; chl 4 1.4351^{20}
8363	Octane, 4-chloro-4-methyl- 4-Chloro-4-methyloctane	$C_9H_{19}Cl$ 162.70	36903-89-6	$71^{14.5}$	3-01-00-00509 0.8723^{20}	eth 4; chl 4 1.4360^{20}
8364	Octanedial	$C_8H_{14}O_2$ 142.20	638-54-0	235 dec; 97^{3}	4-01-00-03706	H_2O 4; EtOH 4 1.4439^{20}
8365	1,8-Octanediamine	$C_8H_{20}N_2$ 144.26	373-44-4 52	225	3-04-00-00612	H_2O 4; eth 4; EtOH 4
8366	Octane, 1,2-dibromo- 1,2-Dibromooctane	$C_8H_{16}Br_2$ 272.02	6269-92-7	241	3-01-00-00468 1.4580^{20}	1.4970^{20}
8367	Octane, 1,4-dibromo-	$C_8H_{16}Br_2$ 272.02	70690-24-3	$125\text{-}6^{11}$	4-01-00-00424 1.466^{11}	1.5003^{11}
8368	Octane, 1,5-dibromo- 1,5-Dibromooctane	$C_8H_{16}Br_2$ 272.02	17912-17-3	$127\text{-}8^{11}$	4-01-00-00424 1.447^{16}	1.4968^{16}
8369	Octane, 1,8-dibromo- Octamethylene dibromide	$C_8H_{16}Br_2$ 272.02	4549-32-0 15.5	271	4-01-00-00424 1.4594^{25}	H_2O 1; eth 3; ctc 3; chl 3 1.4971^{25}
8370	Octane, 4,5-dibromo-	$C_8H_{16}Br_2$ 272.02	61539-75-1	$84^{4.3}$	3-01-00-00469 1.4569^{20}	1.4981^{20}
8371	Octane, 1,8-dichloro- 1,8-Dichlorooctane	$C_8H_{16}Cl_2$ 183.12	2162-99-4	241	4-01-00-00420 1.0248^{25}	1.4572^{25}
8372	Octane, 2,3-dichloro-	$C_8H_{16}Cl_2$ 183.12	21948-47-0 -55.1	206	4-01-00-00425 0.9712^{20}	1.4523^{20}
8373	Octane, 2,2-dimethyl- 2,2-Dimethyloctane	$C_{10}H_{22}$ 142.28	15869-87-1	155	4-01-00-00476 0.7208^{25}	1.4082^{20}
8374	Octane, 2,3-dimethyl- 2,3-Dimethyloctane	$C_{10}H_{22}$ 142.28	7146-60-3	164.3	4-01-00-00476 0.7377^{20}	1.4146^{20}
8375	Octane, 2,4-dimethyl- 2,4-Dimethyloctane	$C_{10}H_{22}$ 142.28	4032-94-4	156	4-01-00-00477 0.7226^{25}	1.4091^{20}
8376	Octane, 2,5-dimethyl- 2,5-Dimethyloctane	$C_{10}H_{22}$ 142.28	15869-89-3	158.5	4-01-00-00477 0.7264^{25}	1.4112^{20}
8377	Octane, 2,6-dimethyl- 2,6-Dimethyloctane	$C_{10}H_{22}$ 142.28	2051-30-1	160.4	2-01-00-00131 0.7313^{20}	1.4097^{20}
8378	Octane, 2,7-dimethyl-	$C_{10}H_{22}$ 142.28	1072-16-8 -54.9	159.9	4-01-00-00478 0.7202^{25}	eth 3; HOAc 3 1.4086^{20}
8379	Octane, 3,3-dimethyl- 3,3-Dimethyloctane	$C_{10}H_{22}$ 142.28	4110-44-5	161.2	4-01-00-00479 0.7351^{25}	1.4165^{20}
8380	Octane, 3,4-dimethyl- 3,4-Dimethyloctane	$C_{10}H_{22}$ 142.28	15869-92-8	163.4	4-01-00-00479 0.7410^{25}	1.4182^{20}
8381	Octane, 3,5-dimethyl- 3,5-Dimethyloctane	$C_{10}H_{22}$ 142.28	15869-93-9	159.4	4-01-00-00479 0.7329^{25}	1.4139^{20}
8382	Octane, 3,6-dimethyl- 3,6-Dimethyloctane	$C_{10}H_{22}$ 142.28	15869-94-0	160.8	4-01-00-00479 0.7324^{25}	1.4139^{20}
8383	Octane, 4,4-dimethyl- 4,4-Dimethyloctane	$C_{10}H_{22}$ 142.28	15869-95-1	157.5	4-01-00-00480 0.7312^{25}	1.4144^{20}
8384	Octane, 4,5-dimethyl- 4,5-Dimethyloctane	$C_{10}H_{22}$ 142.28	15869-96-2	162.15	4-01-00-00480 0.7432^{25}	1.4190^{20}
8385	Octanedinitrile Suberonitrile	$C_8H_{12}N_2$ 136.20	629-40-3 -3.5	185^{15}	3-02-00-01768 0.954^{25}	1.4436^{20}
8386	Octanedioic acid Suberic acid	$C_8H_{14}O_4$ 174.20	505-48-6 144	8833 219^{20}	4-02-00-02028	H_2O 1; eth 5; bz 5
8387	Octanedioic acid, diethyl ester	$C_{12}H_{22}O_4$ 230.30	2050-23-9 5.9	282.6	4-02-00-02029 0.9811^{20}	H_2O 1; EtOH 3; eth 3; ctc 2 1.4328^{20}
8388	Octanedioic acid, dimethyl ester	$C_{10}H_{18}O_4$ 202.25	1732-09-8 -3.1	268	4-02-00-02029 1.0217^{20}	H_2O 1; EtOH 3; eth 3; ace 3 1.4341^{20}

No.	Name	Synonym	Mol. Form.	Mol. Wt.	CAS RN	mp/°C	Merck No.	bp/°C	Beil. Ref.	den/g cm^{-3}	Solubility	n_D
8389	1,8-Octanediol		$C_8H_{18}O_2$	146.23	629-41-4	63		172^{20}	4-01-00-02592		H_2O 2; EtOH 4; eth 2; bz 3	
8390	2,4-Octanediol		$C_8H_{18}O_2$	146.23	90162-24-6			$117\text{-}8^{8}$	3-01-00-02219	0.918^{25}		1.4422^{25}
8391	4,5-Octanediol, (±)-		$C_8H_{18}O_2$	146.23	22520-40-7	30		110^{8}	4-01-00-02593		ctc 2	1.4419^{25}
8392	1,7-Octanediol, 3,7-dimethyl-		$C_{10}H_{22}O_2$	174.28	107-74-4			265	3-01-00-02233	0.937^{20}	bz 2; tol 2	1.4599^{20}
8393	2,4-Octanedione	Valerylacetone	$C_8H_{14}O_2$	142.20	14090-87-0			$79\text{-}83^{20}$	4-01-00-03706	0.9233^{20}		1.4559^{20}
8394	2,7-Octanedione		$C_8H_{14}O_2$	142.20	1626-09-1	44		110^{13}	4-01-00-03707		EtOH 4	
8395	3,6-Octanedione		$C_8H_{14}O_2$	142.20	2955-65-9	35.5		98^{14}	4-01-00-03708		EtOH 4	
8396	4,5-Octanedione		$C_8H_{14}O_2$	142.20	5455-24-3			168	4-01-00-03708	0.934^{0}	EtOH 4; eth 4; ace 4	
8397	4,5-Octanedione, dioxime	Dipropylglyoxime	$C_8H_{16}N_2O_2$	172.23	61050-68-8	186.5		sub	2-01-00-00846		eth 4; EtOH 4	
8398	3,5-Octanedione, 6,6,7,7,8,8,8-heptafluoro-2,2-dimethyl-		$C_{10}H_{11}F_7O_2$	296.18	17587-22-3	38		$46\text{-}7^{5}$		1.273^{25}		1.3766^{20}
8399	2,3-Octanedione, 3-oxime	3-Oximino-2-octanone	$C_8H_{15}NO_2$	157.21	584-92-9	59		133^{13}	0-01-00-00795		eth 4	
8400	3,6-Octanedione, 2,2,7,7-tetramethyl-	2,2,7,7-Tetramethyl-3,6-octanedione	$C_{12}H_{22}O_2$	198.31	27610-88-4	2.5		116^{17}; $57^{0.5}$	4-01-00-03734	0.900^{27}	EtOH 4	1.4400^{20}
8401	Octane, 1-(ethenyloxy)-		$C_{10}H_{20}O$	156.27	929-62-4			58^{4}	4-01-00-02056	0.8020^{20}		1.4268^{20}
8402	Octane, 1-ethoxy-		$C_{10}H_{22}O$	158.28	929-61-3	12.5		186.3	4-01-00-01759	0.7847^{20}	EtOH 4	1.4127^{20}
8403	Octane, 3-ethyl-	3-Ethyloctane	$C_{10}H_{22}$	142.28	5881-17-4			166.5	4-01-00-00476	0.7359^{25}		1.4156^{20}
8404	Octane, 4-ethyl-	4-Ethyloctane	$C_{10}H_{22}$	142.28	15869-86-0			163.7	4-01-00-00476	0.7343^{25}		1.4151^{20}
8405	Octane, 1-fluoro-	Octyl fluoride	$C_8H_{17}F$	132.22	463-11-6			142.4	4-01-00-00418	0.8108^{20}		1.3935^{20}
8406	Octane, 1-iodo-		$C_8H_{17}I$	240.13	629-27-6	-45.7		225.5	4-01-00-00425	1.3297^{20}	EtOH 3; eth 3	1.4889^{20}
8407	Octane, 2-iodo-, (±)-	(±)-2-Octyl iodide	$C_8H_{17}I$	240.13	36049-78-2			210; 95^{16}	4-01-00-00425	1.3251^{20}	H_2O 1; EtOH 3; eth 3; lig 3	1.4896^{20}
8408	Octane, 2-methyl-	2-Methyloctane	C_9H_{20}	128.26	3221-61-2	-80.3		143.2	4-01-00-00454	0.7095^{25}	H_2O 1; EtOH 3; eth 3; ctc 2	1.4031^{20}
8409	Octane, 3-methyl-	3-Methyloctane	C_9H_{20}	128.26	2216-33-3	-107.6		144.2		0.717^{25}		1.4040^{25}
8410	Octane, 4-methyl-	4-Methyloctane	C_9H_{20}	128.26	2216-34-4	-113.3		142.4		0.716^{25}		1.4039^{25}
8411	Octanenitrile	Caprylnitrile	$C_8H_{15}N$	125.21	124-12-9	-45.6		205.2	4-02-00-00993	0.8136^{20}	eth 4	1.4203^{20}
8412	Octane, 1-nitro-	1-Nitrooctane	$C_8H_{17}NO_2$	159.23	629-37-8	15		208.5	4-01-00-00426	0.9346^{20}		1.4322^{20}
8413	Octane, 2-nitro-, (+)-		$C_8H_{17}NO_2$	159.23	116836-12-5			$102\text{-}5^{23}$	4-01-00-00426	0.9166^{20}		1.4280^{20}
8414	Octane, 3-nitro-		$C_8H_7NO_2$	149.15	4609-92-1			91^{25}	4-01-00-00427	0.917^{25}		1.4210^{25}
8415	Octane, octadecafluoro-	Perfluorooctane	C_8F_{18}	438.06	307-34-6			105.9	4-01-00-00418	1.73^{20}		1.282^{20}
8416	Octane, 1,1'-oxybis-		$C_{16}H_{34}O$	242.45	629-82-3	-7.6		283	4-01-00-01760	0.8063^{20}	H_2O 2; EtOH 3; eth 3; ctc 3	1.4327^{20}
8417	Octane, 1,1'-thiobis-		$C_{16}H_{34}S$	258.51	2690-08-6			180^{10}	4-01-00-01768	0.842^{25}		1.4610^{20}
8418	1-Octanethiol	Octyl mercaptan	$C_8H_{18}S$	146.30	111-88-6	-49.2		199.1	4-01-00-01767	0.8433^{20}	EtOH 3; ctc 2	1.4540^{20}
8419	2-Octanethiol, (±)-	2-Octanethiol (*DL*)	$C_8H_{18}S$	146.30	10435-81-1	-79		186.4	4-01-00-01777	0.8366^{20}	bz 4; eth 4; EtOH 4	1.4504^{20}
8420	Octane, 2,2,4-trimethyl-	Octane, 2,2,4-trimethyl	$C_{11}H_{24}$	156.31	18932-14-4			168.5	4-01-00-00494	0.7381^{20}		1.4155^{20}
8421	Octane, 2,4,7-trimethyl	2,4,7-Trimethyloctane	$C_{11}H_{24}$	156.31	62016-38-0			168.1				
8422	Octanoic acid	Caprylic acid	$C_8H_{16}O_2$	144.21	124-07-2	16.3	1765	239	4-02-00-00982	0.9106^{20}	H_2O 2; EtOH 5; chl 5; CH_3CN 5	1.4285^{20}
8423	Octanoic acid, 2-amino-, (±)-		$C_8H_{17}NO_2$	159.23	644-90-6	270		sub	4-04-00-02799		H_2O 2; EtOH 2; eth 2; bz 2	
8424	Octanoic acid, anhydride		$C_{16}H_{30}O_3$	270.41	623-66-5	-1		282.5	3-02-00-00796	0.9065^{18}	ace 4; eth 4; EtOH 4	1.4358^{18}
8425	Octanoic acid, 2-bromo-		$C_8H_{15}BrO_2$	223.11	2623-82-7			140^{5}; $118^{0.5}$	4-02-00-01000	1.2785^{24}		1.4613^{24}
8426	Octanoic acid, 8-bromo-		$C_8H_{15}BrO_2$	223.11	17696-11-6	38.5		147^{2}	4-02-00-01000		bz 4; eth 4; EtOH 4	
8427	Octanoic acid, butyl ester		$C_{12}H_{24}O_2$	200.32	589-75-3	-42.9		240.5	4-02-00-00987	0.8628^{20}	ace 4; eth 4; EtOH 4	1.4232^{25}

No.	Name Synonym	Mol. Form. Mol. Wt.	CAS RN mp/°C	Merck No. bp/°C	Beil. Ref. den/g cm^{-3}	Solubility n_D
8428	Octanoic acid, ethyl ester Ethyl octanoate	$C_{10}H_{20}O_2$ 172.27	106-32-1 -43.1	3736 208.5	4-02-00-00987 0.866^{18}	H_2O 1; EtOH 4; eth 4; ctc 2 1.4178^{20}
8429	Octanoic acid, 8-fluoro- Caprylic acid, 8-fluoro	$C_8H_{15}FO_2$ 162.20	353-25-3 35	 132^{4}	4-02-00-00994	eth 4; EtOH 4
8430	Octanoic acid, heptyl ester	$C_{15}H_{30}O_2$ 242.40	4265-97-8 -10.6	 290.5	3-02-00-00794 0.8596^{20}	ace 4; eth 4; EtOH 4 1.4340^{20}
8431	Octanoic acid, hexyl ester	$C_{14}H_{28}O_2$ 228.38	1117-55-1 -30.6	 277.4	3-02-00-00794 0.8603^{20}	H_2O 1; EtOH 3; eth 3; ace 3 1.4323^{25}
8432	Octanoic acid, 2-hydroxy-	$C_8H_{16}O_3$ 160.21	617-73-2 70	 162^{10}	3-03-00-00642	H_2O 2; EtOH 4; eth 4; chl 2
8433	Octanoic acid, 3-methyl- Caprylic acid, 3-methyl	$C_9H_{18}O_2$ 158.24	6061-10-5	 135^{16}	3-02-00-00826 0.899^{23}	 1.4298^{25}
8434	Octanoic acid, methyl ester Methyl caprylate	$C_9H_{18}O_2$ 158.24	111-11-5 -40	 192.9	4-02-00-00986 0.8775^{20}	H_2O 1; EtOH 4; eth 4; ctc 2 1.4170^{20}
8435	Octanoic acid, 1-methylethyl ester	$C_{11}H_{22}O_2$ 186.29	5458-59-3	 93.8^{10}	4-02-00-00987 0.8555^{20}	 1.4147^{25}
8436	Octanoic acid, 2-methyl-3-oxo-, ethyl ester Caprylic acid, 2-methyl-3-oxo, ethyl ester	$C_{11}H_{20}O_3$ 200.28	10488-94-5	 $128\text{-}9^{12}$	4-03-00-01629 0.963^{0}	EtOH 4
8437	Octanoic acid, octyl ester	$C_{16}H_{32}O_2$ 256.43	2306-88-9 -18.1	 306.8	4-02-00-00988 0.8554^{20}	ace 4; eth 4; EtOH 4 1.4352^{20}
8438	Octanoic acid, pentadecafluoro-, methyl ester	$C_9H_3F_{15}O_2$ 428.10	376-27-2	 158	4-02-00-00995 1.684^{20}	 1.304^{27}
8439	Octanoic acid, pentyl ester	$C_{13}H_{26}O_2$ 214.35	638-25-5 -34.8	 260.2	3-02-00-00794 0.8613^{20}	H_2O 1; EtOH 3; eth 3; ace 3 1.4262^{25}
8440	Octanoic acid, 1,2,3-propanetriyl ester	$C_{27}H_{50}O_6$ 470.69	538-23-8 10	 233	4-02-00-00991 0.9540^{20}	H_2O 1; EtOH 5; eth 4; bz 4 1.4482^{20}
8441	Octanoic acid, propyl ester	$C_{11}H_{22}O_2$ 186.29	624-13-5 -46.2	 226.4	4-02-00-00987 0.8659^{20}	ace 4; eth 4; EtOH 4 1.4191^{25}
8442	Octanoic acid, 3,5,7,8-tetrachloro-2,2,3,4,4,5,6,6,7,8,8-undecafluoro- Octanoic acid, 3,5,7,8-tetrachloro-2,2,3,4,4,5,6,6.7,8,8-undecafluoro	$C_8HCl_4F_{11}O_2$ 479.89	2923-68-4	 $154.0\text{-}6.5^{10}$	4-02-00-00998 1.899^{20}	 1.3980^{20}
8443	1-Octanol Caprylic alcohol	$C_8H_{18}O$ 130.23	111-87-5 -15.5	6674 195.1	4-01-00-01756 0.8262^{25}	H_2O 1; EtOH 5; eth 5; ctc 3 1.4295^{20}
8444	2-Octanol, (±)-	$C_8H_{18}O$ 130.23	4128-31-8 -31.6	6675 180	4-01-00-01770 0.8193^{20}	H_2O 2; EtOH 3; eth 3; ace 3 1.4203^{20}
8445	3-Octanol Ethylamylcarbinol	$C_8H_{18}O$ 130.23	589-98-0 -45	 171	3-01-00-01723 0.8258^{20}	
8446	4-Octanol, (±)-	$C_8H_{18}O$ 130.23	74778-22-6 -40.7	 176.3	4-01-00-01779 0.8186^{20}	H_2O 2; EtOH 3; ctc 2 1.4248^{20}
8447	2-Octanol, acetate, (±)- Acetic acid, 2-octyl ester(*DL*)	$C_{10}H_{20}O_2$ 172.27	74112-36-0	 195	4-02-00-00165 0.8626^{14}	eth 4; EtOH 4 1.4146^{20}
8448	1-Octanol, 8-chloro-	$C_8H_{17}ClO$ 164.68	23144-52-7	 139^{19}	4-01-00-01766	eth 4; EtOH 4 1.4563^{25}
8449	1-Octanol, 3,7-dimethyl-, (±)- Geraniol, tetrahydro (*DL*)	$C_{10}H_{22}O$ 158.28	59204-02-3	 212.5; 106^{13}	4-01-00-01830 0.8308^{10}	H_2O 1; EtOH 3; eth 3; ace 3 1.4367^{20}
8450	2-Octanol, 2,6-dimethyl- Tetrahydromyrcenol	$C_{10}H_{22}O$ 158.28	18479-57-7	 80.5^{10}	4-01-00-01829 0.8023^{25}	 1.4220^{25}
8451	3-Octanol, 2,3-dimethyl- Amylisopropylmethylcarbinol	$C_{10}H_{22}O$ 158.28	19781-10-3	 $69\text{-}70^{5}$	4-01-00-01828 0.8401^{20}	 1.4380^{20}
8452	3-Octanol, 2,7-dimethyl-	$C_{10}H_{22}O$ 158.28	66719-55-9	 193.5	1-01-00-00214 0.8152^{20}	 1.4302^{20}
8453	3-Octanol, 3,6-dimethyl-	$C_{10}H_{22}O$ 158.28	151-19-9 -67.5	 202.2	1-01-00-00214 0.8347^{22}	 1.4370^{20}
8454	3-Octanol, 3,6-dimethyl-(+-)- Linalool, tetrahydro (*DL*)	$C_{10}H_{22}O$ 158.28	57706-88-4 31.5	 196.5	3-01-00-01767 0.8280^{20}	EtOH 4 1.4335^{20}
8455	4-Octanol, 2,4-dimethyl-	$C_{10}H_{22}O$ 158.28	33933-79-8	 97^{27}	4-01-00-01828 0.8238^{20}	 1.4338^{20}
8456	4-Octanol, 2,5-dimethyl-	$C_{10}H_{22}O$ 158.28	66719-53-7	 $102\text{-}4^{34}$	1-01-00-00213 0.8125^{25}	 1.4260^{25}
8457	4-Octanol, 2,6-dimethyl-	$C_{10}H_{22}O$ 158.28	66719-54-8	 195	2-01-00-00460 0.8114^{20}	 1.4270^{20}
8458	4-Octanol, 2,7-dimethyl- 2,7-Dimethyloctan-4-ol	$C_{10}H_{22}O$ 158.28	19781-11-4	 96^{18}	3-01-00-01769 0.8139^{20}	 1.4280^{20}
8459	3-Octanol, 3-ethyl-	$C_{10}H_{22}O$ 158.28	2051-32-3	 199	3-01-00-01766 0.8361^{25}	EtOH 4 1.4390^{20}
8460	1-Octanol, 8-fluoro-	$C_8H_{17}FO$ 148.22	408-27-5	 $106\text{-}7^{10}$	4-01-00-01766 0.945^{20}	eth 4; EtOH 4 1.4248^{25}
8461	1-Octanol, 3-methyl- 3-Methyl-1-octanol	$C_9H_{20}O$ 144.26	38514-02-2	 110^{25}	4-01-00-01806 0.827^{4}	 1.4328^{25}
8462	1-Octanol, 4-methyl- 4-Methyl-1-octanol	$C_9H_{20}O$ 144.26	38514-03-3	 106^{17}	4-01-00-01807 0.820^{27}	 1.4335^{25}
8463	1-Octanol, 7-methyl-	$C_9H_{20}O$ 144.26	2430-22-0 64.5	 206	4-01-00-01806 0.8260^{25}	 1.4316^{25}
8464	2-Octanol, 2-methyl- 2-Methyl-2-octanol	$C_9H_{20}O$ 144.26	628-44-4	 178	4-01-00-01805 0.8210^{20}	H_2O 1; EtOH 3; eth 3 1.4280^{20}
8465	2-Octanol, 3-methyl-	$C_9H_{20}O$ 144.26	27644-49-1	 75^{15}	3-01-00-01748 0.831^{27}	 1.437^{27}
8466	3-Octanol, 3-methyl- Amylethylmethylcarbinol	$C_9H_{20}O$ 144.26	5340-36-3	 97.5^{50}; 36^{3}	4-01-00-01807 0.8258^{25}	 1.4257^{25}

No.	Name Synonym	Mol. Form. Mol. Wt.	CAS RN mp/°C	Merck No. bp/°C	Beil. Ref. den/g cm^{-3}	Solubility n_D
8467	3-Octanol, 4-methyl-	$C_9H_{20}O$ 144.26	66793-80-4	132-3^{20}	3-01-00-01750 0.8437^{25}	1.4372^{20}
8468	3-Octanol, 6-methyl- 6-Methyl-3-octanol	$C_9H_{20}O$ 144.26	40225-75-0	81-3^{15}	3-01-00-01749 0.8320^{28}	1.4372^{20}
8469	4-Octanol, 4-methyl- 4-Methyl-4-octanol	$C_9H_{20}O$ 144.26	23418-37-3	180; 78^{15}	4-01-00-01807 0.8267^{20}	1.4327^{20}
8470	4-Octanol, 5-methyl-	$C_9H_{20}O$ 144.26	59734-23-5	74-6^{9}	1-01-00-00211 0.8156^{25}	1.4262^{25}
8471	2-Octanone Hexyl methyl ketone	$C_8H_{16}O$ 128.21	111-13-7 -16	4632 172.5	4-01-00-03339 0.820^{20}	H_2O 2; EtOH 5; eth 5 1.4151^{20}
8472	3-Octanone Ethyl amyl ketone	$C_8H_{16}O$ 128.21	106-68-3	3721 167.5	4-01-00-03341 0.822^{25}	H_2O 1; EtOH 5; eth 5 1.4150^{20}
8473	4-Octanone Butyl propyl ketone	$C_8H_{16}O$ 128.21	589-63-9	163	4-01-00-03342 0.8146^{25}	H_2O 1; EtOH 5; eth 5; ctc 3 1.4173^{14}
8474	4-Octanone, 5-hydroxy- Butyroin	$C_8H_{16}O_2$ 144.21	496-77-5 -10	1595 185	4-01-00-04042 0.9107^{16}	1.4345^{16}
8475	2-Octanone, 3-methyl- 3-Methyl-2-octanone	$C_9H_{18}O$ 142.24	6137-08-2	64-5^{18}	4-01-00-03356 0.832^{27}	1.424^{27}
8476	3-Octanone, 4-methyl- 4-Methyl-3-octanone	$C_9H_{18}O$ 142.24	6137-15-1	62-4^{10}	4-01-00-03357 0.820^{25}	1.4186^{25}
8477	3-Octanone, 7-methyl- 7-Methyl-3-octanone	$C_9H_{18}O$ 142.24	5408-57-1	182.5	3-01-00-02889 0.8353^{25}	1.4748
8478	4-Octanone, 3-methyl-	$C_9H_{18}O$ 142.24	20754-04-5	174	4-01-00-03357 0.829^{14}	1.4200^{14}
8479	4-Octanone, 7-methyl- 7-Methyl-4-octanone	$C_9H_{18}O$ 142.24	20809-46-5	178	4-01-00-03356 0.8239^{20}	H_2O 1; EtOH 3; eth 3 1.4210^{20}
8480	1-Octanone, 1-phenyl-	$C_{14}H_{20}O$ 204.31	1674-37-9 22.8	285; 164^{15}	4-07-00-00798 0.9360^{30}	EtOH 3; eth 3
8481	Octanoyl chloride	$C_8H_{15}ClO$ 162.66	111-64-8 -63	195.6	4-02-00-00992 0.9535^{15}	eth 3 1.4335^{20}
8482	Octasiloxane, octadecamethyl- Octadecamethyloctasiloxane	$C_{18}H_{54}O_7Si_8$ 607.31	556-69-4	153$^{5.1}$	4-04-00-04130 0.913^{25}	bz 4; peth 4; lig 4 1.3970^{20}
8483	1,3,5,7-Octatetraene Octatetraene	C_8H_{10} 106.17	1482-91-3 50	sub	4-01-00-01124	peth 3; HOAc 3
8484	2,4,6-Octatrienal Octatrienal	$C_8H_{10}O$ 122.17	17609-31-3 56.8	75^{3}	4-01-00-03591 0.8891^{60}	eth 4; EtOH 4
8485	1,3,6-Octatriene, 3,7-dimethyl- Ocimene	$C_{10}H_{16}$ 136.24	13877-91-3	6661 177 dec; 73^{21}	3-01-00-01052 0.800^{20}	bz 4; eth 4; EtOH 4; chl 4 1.4862^{20}
8486	1,3,7-Octatriene, 3,7-dimethyl-	$C_{10}H_{16}$ 136.24	502-99-8	177 dec	3-01-00-01050 0.8000^{20}	H_2O 1; EtOH 3; eth 3; chl 3 1.4862^{20}
8487	2,4,6-Octatriene, 2,6-dimethyl-, (*E,E*)-	$C_{10}H_{16}$ 136.24	3016-19-1 -35.4	188; 91^{20}	4-01-00-01106 0.8118^{20}	1.5446^{20}
8488	2,4,6-Octatriene, 2,6-dimethyl-, (*E,Z*)-	$C_{10}H_{16}$ 136.24	7216-56-0 -20.6	89^{20}	4-01-00-01106 0.8060^{20}	1.5446^{20}
8489	2,4,6-Octatriene, (*E,E,E*)-	C_8H_{12} 108.18	15192-80-0 52	147.5	4-01-00-01101 0.7961^{23}	EtOH 3; chl 3; lig 3 1.5131^{27}
8490	6-Octenal, 3,7-dimethyl- Resorcinol, 4-acetyl	$C_{10}H_{18}O$ 154.25	106-23-0	2331 47^{1}	1-01-00-00387 0.85^{20}	H_2O 2; EtOH 3
8491	1-Octene Caprylene	C_8H_{16} 112.22	111-66-0 -101.7	1764 121.2	4-01-00-00874 0.7149^{20}	H_2O 1; EtOH 5; eth 3; ace 3 1.4087^{20}
8492	2-Octene, (*E*)-	C_8H_{16} 112.22	13389-42-9 -87.7	125	4-01-00-00879 0.7199^{20}	H_2O 1; EtOH 3; eth 3; ace 3 1.4132^{20}
8493	2-Octene, (*Z*)-	C_8H_{16} 112.22	7642-04-8 -100.2	125.6	4-01-00-00878 0.7243^{20}	H_2O 1; EtOH 3; eth 3; ace 3 1.4150^{20}
8494	3-Octene, (*E*)-	C_8H_{16} 112.22	14919-01-8 -110	123.3	4-01-00-00880 0.7152^{20}	H_2O 1; EtOH 3; eth 3; ace 3 1.4126^{20}
8495	3-Octene, (*Z*)-	C_8H_{16} 112.22	14850-22-7 -126	122.9	4-01-00-00880 0.7159^{20}	ace 4; bz 4; eth 4; EtOH 4 1.4135^{20}
8496	4-Octene, (*E*)-	C_8H_{16} 112.22	14850-23-8 -93.8	122.3	4-01-00-00880 0.7141^{20}	H_2O 1; EtOH 3; eth 3; ace 3 1.4114^{20}
8497	4-Octene, (*Z*)-	C_8H_{16} 112.22	7642-15-1 -118.7	122.5	4-01-00-00880 0.7212^{20}	ace 4; bz 4; eth 4; EtOH 4 1.4148^{20}
8498	1-Octene, 2-chloro-	$C_8H_{15}Cl$ 146.66	31283-43-9	169	3-01-00-00840 0.9273^{0}	ace 4; bz 4; eth 4
8499	2-Octene, 2-chloro-	$C_8H_{15}Cl$ 146.66	90202-21-4	167.5; 82^{50}	4-01-00-00879 0.8914^{16}	ace 4; bz 4; eth 4; EtOH 4 1.4424^{16}
8500	4-Octene, 4-chloro-, (*Z*)- 4-Octene, 4-chloro (*cis*)	$C_8H_{15}Cl$ 146.66	7321-48-4	165.3; 75^{50}	4-01-00-00881 0.8912^{20}	eth 4 1.4447^{20}
8501	2-Octene, 4-chloro-	$C_8H_{15}Cl$ 146.66	116668-50-9	153	3-01-00-00840 0.8924^{20}	ace 4; bz 4; eth 4; chl 4 1.4952^{20}
8502	1-Octene, 3,7-dimethyl- 2,6-Dimethyl-7-octene	$C_{10}H_{20}$ 140.27	4984-01-4	154	3-01-00-00862 0.7396^{20}	1.4212^{20}
8503	2-Octene, 2,6-dimethyl- 2,6-Dimethyl-2-octene	$C_{10}H_{20}$ 140.27	4057-42-5	163	3-01-00-00861 0.746^{22}	1.425^{22}
8504	4-Octene-2,7-dione	$C_8H_{12}O_2$ 140.18	2130-23-6 30.5	72$^{0.03}$	4-01-00-03757	eth 4; EtOH 4
8505	1-Octene, 2-methyl- 2-Methyl-1-octene	C_9H_{18} 126.24	4588-18-5 -77.8	144.8	4-01-00-00896 0.7343^{20}	1.4184^{20}
8506	2-Octene, 3-methyl-	C_9H_{18} 126.24	113426-22-5	145	3-01-00-00854 0.7409^{25}	1.4247^{20}

PHYSICAL CONSTANTS OF ORGANIC COMPOUNDS (continued)

No.	Name Synonym	Mol. Form. Mol. Wt.	CAS RN mp/°C	Merck No. bp/°C	Beil. Ref. den/g cm^{-3}	Solubility n_D
8507	3-Octene, 7-methyl-	C_9H_{18} 126.24	86668-33-9	 142	3-01-00-00854 0.7278^{20}	 1.4168^{20}
8508	4-Octene, 2-methyl-	C_9H_{18} 126.24	64501-77-5	 139	3-01-00-00854 0.7379^{20}	 1.4181^{20}
8509	1-Octen-4-ol	$C_8H_{12}O$ 124.18	40575-42-6	 172; 68^{10}	3-01-00-01942 0.8373^{20}	 1.4383^{20}
8510	2-Octen-1-ol	$C_8H_{16}O$ 128.21	22104-78-5	 $87\text{-}9^{11}$	4-01-00-02164 0.850^{20}	 1.4470^{20}
8511	2-Octen-4-ol	$C_8H_{16}O$ 128.21	4798-61-2	 175	3-01-00-01943 0.8393^{20}	 1.4395
8512	4-Octen-1-ol	$C_8H_{16}O$ 128.21	67700-26-9	 95^{17}	4-01-00-02166 0.842^{20}	 1.4462^{20}
8513	5-Octen-1-ol	$C_8H_{16}O$ 128.21	90200-83-2	 91.5^{14}	4-01-00-02166 0.8492^{21}	 1.4478^{21}
8514	6-Octen-1-ol, 3,7-dimethyl-, (±)- (±)-Citronellol	$C_{10}H_{20}O$ 156.27	26489-01-0	2332 224; 99^{10}	4-01-00-02188 0.8560^{20}	H_2O 2; EtOH 5; eth 5 1.4543^{20}
8515	6-Octen-3-ol, 3,7-dimethyl-	$C_{10}H_{20}O$ 156.27	18479-51-1	 94^{14}	2-01-00-00494 0.8695^{15}	 1.4569^{15}
8516	7-Octen-1-ol, 3,7-dimethyl-, (*S*)- (-)-α-Citronellol	$C_{10}H_{20}O$ 156.27	6812-78-8	8186 $114\text{-}5^{12}$	2-01-00-00497 0.8549^{20}	eth 4; EtOH 4 1.4556^{20}
8517	1-Octen-4-ol, 4-methyl-	$C_9H_{18}O$ 142.24	62108-06-9	 178	2-01-00-00492 0.8440^{20}	 1.4399
8518	2-Octen-4-ol, 7-methyl-	$C_9H_{18}O$ 142.24	4798-64-5	 85^{12}	2-01-00-00492 0.8402^{14}	 1.4449^{14}
8519	3-Octen-2-ol, 2-methyl-, (*Z*)- 3-Octen-2-ol, 2-methyl (*cis*)	$C_9H_{18}O$ 142.24	18521-07-8	 177	4-01-00-02177 0.8378^{20}	 1.4426^{20}
8520	4-Octen-3-ol, 3-methyl-	$C_9H_{18}O$ 142.24	90676-55-4	 $69\text{-}70^{10}$	4-01-00-02178 0.812^{14}	 1.4414^{14}
8521	5-Octen-4-ol, 5-methyl- 3-Octen-5-ol, 4-methyl	$C_9H_{18}O$ 142.24	36903-92-1	 89^{16}	2-01-00-00492 0.8468^{25}	 1.4446^{25}
8522	1-Octen-3-one, 5-methyl-	$C_9H_{16}O$ 140.23	116836-13-6	 $72\text{-}3^{16}$	3-01-00-03018 0.8450^{19}	 1.4380^{19}
8523	5-Octen-4-one, 2-methyl- 2-Methyl-5-octen-4-one	$C_9H_{16}O$ 140.23	17577-93-4	 183	4-01-00-03504 0.839^{25}	 1.4387^{25}
8524	5-Octen-4-one, 7-methyl-	$C_9H_{16}O$ 140.23	32064-78-1	 $68\text{-}78^{24}$	3-01-00-03018 0.9011^{12}	 1.4748
8525	7-Octen-4-one, 2-methyl- 1-Octen-5-one, 7-methyl	$C_9H_{16}O$ 140.23	54298-97-4	 $62\text{-}3^{14}$	2-01-00-00801 0.8361^{12}	 1.4288^{12}
8526	1-Octen-3-yne	C_8H_{12} 108.18	17679-92-4	 62^{80}	4-01-00-01100 0.7830^{20}	eth 4 1.4592^{20}
8527	7-Octen-5-yn-4-ol, 4-methyl- 1-Octen-3-yne-5-ol, 5-methyl	$C_9H_{14}O$ 138.21	39118-35-9	 80^{13}	3-01-00-02034 0.8851^{15}	EtOH 4 1.4785^{20}
8528	Octhilinone 3(2H)-Isothiazolone, 2-octyl-	$C_{11}H_{19}NOS$ 213.34	26530-20-1	6677 $120^{0.01}$		
8529	1-Octyne Hexylacetylene	C_8H_{14} 110.20	629-05-0 -79.3	 126.3	4-01-00-01034 0.7461^{20}	H_2O 1; EtOH 3; eth 3 1.4159^{20}
8530	2-Octyne Methylpentylacetylene	C_8H_{14} 110.20	2809-67-8 -61.6	 137.6	4-01-00-01035 0.7596^{20}	H_2O 1; EtOH 3; eth 3 1.4278^{20}
8531	3-Octyne	C_8H_{14} 110.20	15232-76-5 -103.9	 133.1	4-01-00-01036 0.7529^{20}	H_2O 1; EtOH 3; eth 3 1.4250^{20}
8532	4-Octyne Dipropylacetylene	C_8H_{14} 110.20	1942-45-6 -101	 131.6	4-01-00-01037 0.7509^{20}	H_2O 1; EtOH 3; eth 3 1.4248^{20}
8533	1-Octyne, 1-chloro- 1-Chloro-1-octyne	$C_8H_{13}Cl$ 144.64	64531-26-6	 $61\text{-}2^{17}$	3-01-00-01005 0.912^{20}	eth 4; EtOH 4 1.445^{20}
8534	3-Octyne, 2-chloro-2-methyl- 2-Chloro-2-methyloct-3-yne	$C_9H_{15}Cl$ 158.67	20599-21-7	 68^{15}	4-01-00-01048 0.8929^{20}	H_2O 1; eth 4 1.4480^{20}
8535	3-Octyne, 2,2-dimethyl- 2,2-Dimethyl-3-octyne	$C_{10}H_{18}$ 138.25	19482-57-6	 79^{60}	3-01-00-01018 0.7491^{20}	eth 4 1.4270^{20}
8536	2-Octyne, 1-methoxy- Methyl (2-octyn-1-yl) ether	$C_9H_{16}O$ 140.23	18495-23-3	 77^{19}	3-01-00-01996 0.8370^{25}	eth 4 1.4380^{20}
8537	3-Octyne, 7-methyl- 7-Methyl-3-octyne	C_9H_{16} 124.23	37050-06-9	 87^{99}	3-01-00-01014 0.7599^{20}	eth 4 1.4280^{20}
8538	2-Octynoic acid Pentylpropiolic acid	$C_8H_{12}O_2$ 140.18	5663-96-7 4	 148^{19}; 114^{2}	4-02-00-01714 0.9623^{13}	 1.4595^{20}
8539	3-Octynoic acid	$C_8H_{12}O_2$ 140.18	57074-96-1 18	 110^{2}	4-02-00-01715	 1.4577^{25}
8540	5-Octynoic acid	$C_8H_{12}O_2$ 140.18	76469-08-4 8	 111^{2}	4-02-00-01715 0.9787^{27}	 1.4540^{25}
8541	7-Octynoic acid	$C_8H_{12}O_2$ 140.18	10297-09-3 19	 123^{2}	4-02-00-01714	 1.4502^{25}
8542	2-Octynoic acid, methyl ester	$C_9H_{14}O_2$ 154.21	111-12-6	 107^{20}; 94^{10}	4-02-00-01715 0.926^{20}	 1.4464^{20}
8543	2-Octyn-1-ol 2-Octynol	$C_8H_{14}O$ 126.20	20739-58-6 -18	 $98\text{-}9^{15}$	4-01-00-02256 0.8805^{20}	eth 4 1.4556^{20}
8544	1-Octyn-3-ol, 3-methyl- 1-Octyne-3-ol, 3-methyl	$C_9H_{16}O$ 140.23	23580-51-0	 75^{10}	4-01-00-02268 0.863^{10}	 1.443^{10}
8545	Olean-12-en-28-oic acid, 3,23-dihydroxy-, (3β,4α)- Hederagenin	$C_{30}H_{48}O_4$ 472.71	465-99-6 333	4540		

No.	Name Synonym	Mol. Form. Mol. Wt.	CAS RN mp/°C	Merck No. bp/°C	Beil. Ref. den/g cm^{-3}	Solubility n_D
8546	Olean-12-en-28-oic acid, 3,16-dihydroxy-23-oxo-, (3β,4α,16α)- Quillaic acid	$C_{30}H_{46}O_5$ 486.69	631-01-6 294	8049	3-10-00-04652	ace 4; eth 4; py 4; EtOH 4
8547	Olean-12-en-28-oic acid, 3-hydroxy-, (3β)- Oleanolic acid	$C_{30}H_{48}O_3$ 456.71	508-02-1 310 dec	6787 sub 280	4-10-00-01164	H_2O 1; EtOH 2; eth 2; ace 2
8548	Olean-12-en-3-ol, (3β)- β-Amyrin	$C_{30}H_{50}O$ 426.73	559-70-6 197	654 260^{05}	4-06-00-04195	H_2O 1; EtOH 2; eth 3; bz 3
8549	Orcein	$C_{28}H_{24}N_2O_7$ 500.51	1400-62-0	6818	2-06-00-00876	
8550	*L*-Ornithine (*S*)-2,5-Diaminopentanoic acid	$C_5H_{12}N_2O_2$ 132.16	70-26-8 140	6826	4-04-00-02644	H_2O 4; EtOH 4
8551	*L*-Ornithine, *N*5-(aminocarbonyl)- Citrulline	$C_6H_{13}N_3O_3$ 175.19	372-75-8 235.5	2333	4-04-00-02647	H_2O 3; EtOH 1; MeOH 1
8552	Orthoformic acid, triisopentyl ester Isopentyl alcohol, orthoformate (3:1)	$C_{16}H_{34}O_3$ 274.44	5337-70-2	 268 dec	4-02-00-00030 0.8628^{20}	eth 4; EtOH 4 1.4238^{20}
8553	Oryzalin Benzenesulfonamide, 4-(dipropylamino)-3,5-dinitro-	$C_{12}H_{18}N_4O_6S$ 346.36	19044-88-3 141	6840		
8554	7-Oxabicyclo[4.1.0]heptane	$C_6H_{10}O$ 98.14	286-20-4 <-10	 131.5	5-17-01-00203 0.9663^{20}	H_2O 1; EtOH 4; eth 4; ace 4 1.4519^{20}
8555	7-Oxabicyclo[4.1.0]heptane, 3-ethenyl- Cyclohexane, 1,2-epoxy-4-vinyl	$C_8H_{12}O$ 124.18	106-86-5 <-100	 169; 70^{20}	4-17-00-00314 0.9581^{20}	 1.4700^{20}
8556	7-Oxabicyclo[2.2.1]heptane, 1-methyl-4-(1-methylethyl)-	$C_{10}H_{18}O$ 154.25	470-67-7 1	 173.5	5-17-01-00273 0.8997^{20}	H_2O 2; EtOH 5; eth 5; bz 3 1.4562^{20}
8557	7-Oxabicyclo[4.1.0]heptane, 3-oxiranyl- 4-Vinyl-1-cyclohexene dioxide	$C_8H_{12}O_2$ 140.18	106-87-6 <-55	 227	5-19-01-00295 1.0966^{20}	H_2O 4 1.4787^{20}
8558	6-Oxabicyclo[3.1.0]hexane	C_5H_8O 84.12	285-67-6	 102	5-17-01-00190 0.964^{25}	 1.4336^{20}
8559	3-Oxabicyclo[3.2.1]octane-2,4-dione, 1,8,8-trimethyl-, (±)-	$C_{10}H_{14}O_3$ 182.22	595-30-2 221	 270	5-17-11-00089 1.194^{20}	bz 4
8560	2-Oxabicyclo[2.2.2]octane, 1,3,3-trimethyl- Eucalyptol	$C_{10}H_{18}O$ 154.25	470-82-6 1.5	3851 176.4	5-17-01-00273 0.9267^{20}	H_2O 1; EtOH 3; eth 3; ctc 2 1.4586^{20}
8561	6-Oxabicyclo[3.2.1]oct-3-ene, 4,7,7-trimethyl-, (±)- (±)-Pinol	$C_{10}H_{16}O$ 152.24	60761-04-8	 183.5	4-17-00-00327 0.9515^{20}	eth 4; EtOH 4 1.4695^{20}
8562	Oxacyclohexadecan-2-one Exaltolide	$C_{15}H_{28}O_2$ 240.39	106-02-5	3867	5-17-09-00106	
8563	1,3,4-Oxadiazole 1-Oxa-3,4-diazacyclopentadiene	$C_2H_2N_2O$ 70.05	288-99-3	6859 150		 1.4300^{25}
8564	1,2,4-Oxadiazole, 3,5-diphenyl- 3,5-Diphenyl-1,2,4-oxadiazole	$C_{14}H_{10}N_2O$ 222.25	888-71-1 109	 290; 210^{17}	4-27-00-07210	bz 4; eth 4; EtOH 4
8565	Oxadiazon 1,2,4-Oxadiazolidin-3,5-dione, 2-*tert*-butyl-4-(2,4-dichloro-5-isopropoxyphenyl)-	$C_{15}H_{18}Cl_2N_2O_3$ 345.23	19666-30-9 90	6860		
8566	Oxadixyl *N*-(2,6-Dimethylphenyl)-2-methoxy-*N*-(2-oxo-3-oxazolidinyl)acetamide	$C_{14}H_{18}N_2O_4$ 278.31	77732-09-3 104			
8567	16(15H)-Oxaerythrinan-15-one, 1,2,6,7-tetradehydro-14,17-dihydro-3-methoxy-, (3β)- β-Erythroidine	$C_{16}H_{19}NO_3$ 273.33	466-81-9 99.5	3625	4-27-00-03568	H_2O 3; EtOH 4; eth 3; bz 4
8568	Oxamyl Ethanimidothioic acid, 2-(dimethylamino)-*N*-[[(methylamino)carbonyl]oxy]-2-oxo-, methyl ester	$C_7H_{13}N_3O_3S$ 219.26	23135-22-0 109	6873 dec	 0.97^{25}	
8569	1,4-Oxathiane	C_4H_8OS 104.17	15980-15-1	 147	5-19-01-00033 1.1174^{20}	H_2O 2
8570	1,2-Oxathiolane, 2,2-dioxide Propane sultone	$C_3H_6O_3S$ 122.14	1120-71-4		5-19-01-00005	chl 3
8571	Oxayohimban-16-carboxylic acid, 16,17-didehydro-19-methyl-, methyl ester, (19α)- Raubasine	$C_{21}H_{24}N_2O_3$ 352.43	483-04-5 258 dec	8129	4-27-00-07927	H_2O 1; MeOH 3
8572	Oxayohimbanium, 3,4,5,6,16,17-hexadehydro-16-(methoxycarbonyl)-19-methyl-, (19α)- Serpentine	$C_{21}H_{20}N_2O_3$ 348.40	18786-24-8 175	8415		H_2O 1; EtOH 3; eth 3; ace 3
8573	2H-1,3-Oxazine-2,4(3H)-dione, 5,5-diethyldihydro- Diethadione	$C_8H_{13}NO_3$ 171.20	702-54-5 97.5	3096	4-27-00-03263	
8574	2-Oxazolamine, 4,5-dihydro-5-phenyl- Aminorex	$C_9H_{10}N_2O$ 162.19	2207-50-3 137	490		
8575	Oxazole 1,3-Oxazole	C_3H_3NO 69.06	288-42-6	 69.5	4-27-00-00960	 1.4285^{17}
8576	Oxazole, 4,5-dihydro-2-methyl-	C_4H_7NO 85.11	1120-64-5	 111	4-27-00-00921 1.005^{25}	 1.4340^{20}
8577	Oxazole, 2,4-dimethyl- 2,4-Dimethyloxazole	C_5H_7NO 97.12	7208-05-1	 108	4-27-00-00980 0.9352^{15}	H_2O 4; eth 4; EtOH 4 1.4166^{15}
8578	Oxazole, 2,5-dimethyl- 2,5-Dimethyloxazole	C_5H_7NO 97.12	23012-11-5	 117.5	4-27-00-00985 0.9958^{21}	H_2O 4 1.4385^{21}

No.	Name	Synonym	Mol. Form.	Mol. Wt.	CAS RN	mp/°C	Merck No.	bp/°C	Beil. Ref.	den/g cm^{-3}	Solubility	n_D
8579	Oxazole, 2,4-diphenyl-	2,4-Diphenyloxazole	$C_{15}H_{11}NO$	221.26	838-41-5	103		339	4-27-00-01435		EtOH 4; eth 4; bz 4	
8580	Oxazole, 2,5-diphenyl-		$C_{15}H_{11}NO$	221.26	92-71-7	74		360	4-27-00-01435	1.0940^{100}	H_2O 1; EtOH 4; eth 4; chl 2	1.6231^{100}
8581	Oxazole, 4,5-diphenyl-	4,5-Diphenyloxazole	$C_{15}H_{11}NO$	221.26	4675-18-7	44		193^{15}; $132^{0.05}$	4-27-00-01438		H_2O 2; EtOH 4; eth 4; con acid 3	1.6283^{100}
8582	Oxazole, 5-methoxy-2-phenyl-	5-Methoxy-2-phenyloxazole	$C_{10}H_9NO_2$	175.19	40527-16-0			140^{10}	4-27-00-01958		ctc 3	1.584^{23}
8583	Oxazole, 4-methyl-	4-Methyloxazole	C_4H_5NO	83.09	693-93-6			88	4-27-00-00968	1.015^{25}		1.4317^{20}
8584	2,4-Oxazolidinedione, 5,5-dimethyl-	Dimethadione	$C_5H_7NO_3$	129.12	695-53-4	76.5	3201		4-27-00-03237			
8585	2,4-Oxazolidinedione, 5,5-dipropyl-	5,5-Dipropyl-2,4-oxazolidinedione	$C_9H_{15}NO_3$	185.22	512-12-9	42.5	3352	$148\text{-}50^{3}$	4-27-00-03269			
8586	2,4-Oxazolidinedione, 5-methyl-3-(2-propenyl)-	Aloxidone	$C_7H_9NO_3$	155.15	526-35-2		305	138^{35}; $86^{0.5}$	4-27-00-03214			1.4688^{25}
8587	2,4-Oxazolidinedione, 3,5,5-trimethyl-	Trimethadione	$C_6H_9NO_3$	143.14	127-48-0	46	9620	$78\text{-}80^{5}$	4-27-00-03237		H_2O 3; EtOH 4; eth 4; ace 4	
8588	2-Oxazolidinone, 5-[(2-methoxyphenoxy)methyl]-	Mephenoxalone	$C_{11}H_{13}NO_4$	223.23	70-07-5	144	5739					
8589	2-Oxazolidinone, 3-[[(5-nitro-2-furanyl)methylene]amino]-	Furazolidone	$C_8H_7N_3O_5$	225.16	67-45-8	255	4210		4-27-00-02530			
8590	Oxepane		$C_6H_{12}O$	100.16	592-90-5			119	5-17-01-00090	0.89^{25}		1.4400^{20}
8591	2-Oxepanone	Caprolactone	$C_6H_{10}O_2$	114.14	502-44-3	-18		215	5-17-09-00034	1.0693^{20}	EtOH 3; eth 3; ace 3	1.4611^{20}
8592	Oxetane	Trimethylene oxide	C_3H_6O	58.08	503-30-0	-97	9630	47.6	5-17-01-00011	0.8930^{25}	H_2O 5; EtOH 5; eth 3; ace 4	1.3961^{20}
8593	Oxetane, 3,3-dimethyl-	3,3-Dimethyloxetane	$C_5H_{10}O$	86.13	6921-35-3			80.6	5-17-01-00085	0.834^{25}		1.3965^{20}
8594	Oxetane, 2-methyl-		C_4H_8O	72.11	2167-39-7			59	5-17-01-00054	0.841^{25}		1.3885^{20}
8595	2-Oxetanone	β-Propiolactone	$C_3H_4O_2$	72.06	57-57-8	-33.4	7832	162	5-17-09-00003	1.1460^{20}	eth 5; chl 3	1.4105^{20}
8596	3-Oxetanone		$C_3H_4O_2$	72.06	6704-31-0							
8597	2-Oxetanone, 3,3-dimethyl-4-(1-methylethylidene)-		$C_8H_{12}O_2$	140.18	3173-79-3	-18		170	5-17-09-00181	0.947^{25}		1.4380^{20}
8598	2-Oxetanone, 4-methylene-	Diketene	$C_4H_4O_2$	84.07	674-82-8	-6.5		126.1	5-17-09-00115	1.0877^{20}		1.4379^{20}
8599	Oxirane	Ethylene oxide	C_2H_4O	44.05	75-21-8	-111.7	3758	10.6	5-17-01-00003	0.8821^{10}	H_2O 3; EtOH 3; eth 3; ace 3	1.3597^{7}
8600	Oxirane, (bromomethyl)-, (±)-	Propane, 3-bromo-1,2-epoxy (*DL*)	C_3H_5BrO	136.98	82584-73-4	-40		137	5-17-01-00023	1.615^{14}	H_2O 1; EtOH 3; eth 3; bz 3	1.4841^{20}
8601	Oxirane, 2,2'-[1,4-butanediylbis(oxymethylene)]bis-		$C_{10}H_{18}O_4$	202.25	2425-79-8			266; 155^{11}	5-17-03-00023	1.1^{25}		1.4611^{20}
8602	Oxirane, (butoxymethyl)-	Butyl glycidyl ether	$C_7H_{14}O_2$	130.19	2426-08-6			169; 75^{26}	5-17-03-00011	0.918^{20}		
8603	Oxiranecarboxaldehyde	Glycidaldehyde	$C_3H_4O_2$	72.06	765-34-4	-62		112.5	5-17-09-00006			1.4265^{20}
8604	Oxiranecarboxylic acid, 3-methyl-3-phenyl-, ethyl ester		$C_{12}H_{14}O_3$	206.24	77-83-8			273.5	5-18-06-00385	1.044^{20}		1.5182^{20}
8605	Oxirane, (chloromethyl)-, (±)-	*DL*-α-Epichlorohydrin	C_3H_5ClO	92.52	13403-37-7	-26	3563	117; 62^{100}	5-17-01-00020	1.1812^{20}	H_2O 1; EtOH 5; eth 5; bz 3	1.4361^{20}
8606	Oxirane, 2-(chloromethyl)-2-methyl-		C_4H_7ClO	106.55	598-09-4			122	5-17-01-00059	1.1011^{20}	H_2O 4; eth 4	1.4310^{20}
8607	Oxirane, 2,2-dimethyl-		C_4H_8O	72.11	558-30-5			52	5-17-01-00058	0.8112^{20}	EtOH 3; eth 3	1.3712^{22}
8608	Oxirane, 2,3-dimethyl-, *cis*-		C_4H_8O	72.11	1758-33-4	-80		60	5-17-01-00061	0.8226^{25}	eth 4; ace 4; bz 4; os 4	1.3802^{20}
8609	Oxirane, 2,3-dimethyl-, *trans*-		C_4H_8O	72.11	21490-63-1	-85		56.5	5-17-01-00061	0.8010^{25}	eth 4; ace 4; bz 4; os 4	1.3736^{20}
8610	Oxirane, [(1,1-dimethylethoxy)methyl]-		$C_7H_{14}O_2$	130.19	7665-72-7	-70		152	5-17-03-00011	0.898^{20}		
8611	Oxirane, [[4-(1,1-dimethylethyl)phenoxy]methyl]-		$C_{13}H_{18}O_2$	206.28	3101-60-8			$165\text{-}70^{14}$	5-17-03-00017	1.038^{25}		1.5145^{20}
8612	Oxirane, 2-(2,2-dimethylpropyl)-2-methyl-		$C_8H_{16}O$	128.21	107-48-2	-64		140.9	4-17-00-00112	0.8272^{20}	bz 4; eth 4	1.4097^{20}
8613	Oxirane, ethenyl-		C_4H_6O	70.09	930-22-3			68	5-17-01-00180	0.9006^{25}	EtOH 3; eth 3; bz 3; os 3	1.4168^{20}
8614	Oxirane, (ethoxymethyl)-	2,3-Epoxypropyl ethyl ether	$C_5H_{10}O_2$	102.13	4016-11-9			128	5-17-03-00010	0.9700^{20}	H_2O 3; EtOH 3; eth 3; ctc 2	1.4320^{20}

No.	Name Synonym	Mol. Form. Mol. Wt.	CAS RN mp/°C	Merck No. bp/°C	Beil. Ref. den/g cm^{-3}	Solubility n_D
8615	Oxirane, ethyl- 1,2-Epoxybutane	$C_4H_8O_2$ 88.11	106-88-7 -150	 63.3	5-17-01-00056 0.8297^{20}	EtOH 4; eth 5; ace 4; os 4 1.3851^{20}
8616	Oxirane, (fluoromethyl)-	C_3H_5FO 76.07	503-09-3	 85.5	5-17-01-00019 1.067^{25}	 1.3715^{20}
8617	Oxirane, (iodomethyl)-	C_3H_5IO 183.98	624-57-7	 161; 62^{24}	5-17-01-00024 1.982^{24}	eth 4; EtOH 4
8618	Oxiranemethanol Glycidol	$C_3H_6O_2$ 74.08	556-52-5	4385 167 dec; $66^{2.5}$	5-17-03-00009 1.1143^{25}	H_2O 5; EtOH 5; eth 3; ace 3 1.4287^{20}
8619	Oxiranemethanol, 3-phenyl-	$C_9H_{10}O_2$ 150.18	21915-53-7 26.5	 134^{4}	4-17-00-01349 1.1512^{27}	 1.5427^{27}
8620	Oxirane, (methoxymethyl)-	$C_4H_8O_2$ 88.11	930-37-0	 113	5-17-03-00010 0.9890^{20}	H_2O 4; ace 4; eth 4; EtOH 4 1.4320^{20}
8621	Oxirane, methyl- Propylene oxide	C_3H_6O 58.08	75-56-9 -111.9	7869 35	5-17-01-00017 0.859^{0}	H_2O 4; eth 4; EtOH 4 1.3670^{20}
8622	Oxirane, 2-methyl-2-phenyl-	$C_9H_{10}O$ 134.18	2085-88-3	 83-5^{15}	5-17-01-00591 1.0280^{20}	 1.5232^{20}
8623	Oxirane, (phenoxymethyl)- Phenyl glycidyl ether	$C_9H_{10}O_2$ 150.18	122-60-1	 243	5-17-03-00013 1.1109^{21}	 1.5307^{21}
8624	Oxirane, phenyl- Phenylethylene oxide	C_8H_8O 120.15	96-09-3 -35.6	 194.1	5-17-01-00577 1.0523^{16}	H_2O 1; EtOH 3; eth 3; chl 3 1.5342^{20}
8625	Oxirane, 2,2'-[1,3-phenylenebis(oxymethylene)]bis- Diglycidyl resorcinol ether	$C_{12}H_{14}O_4$ 222.24	101-90-6 42.5	 143-$51^{0.4}$	5-17-03-00024 1.2183^{30}	 1.5408^{20}
8626	Oxirane, [(2-propenyloxy)methyl]- Allyl glycidyl ether	$C_6H_{10}O_2$ 114.14	106-92-3	 154	5-17-03-00012 0.9698^{20}	 1.4332^{20}
8627	Oxirane, tetramethyl-	$C_6H_{12}O$ 100.16	5076-20-0	 90-$4^{74.5}$	5-17-01-00108 0.8156^{16}	H_2O 3 1.3984^{16}
8628	Oxirane, (trichloromethyl)-	$C_3H_3Cl_3O$ 161.41	3083-23-6	 149; 44^{13}	5-17-01-00023 1.495^{20}	eth 4; chl 3 1.4737^{25}
8629	Oxirane, trifluoro(trifluoromethyl)- Perfluoropropylene oxide	C_3F_6O 166.02	428-59-1		5-17-01-00020	
8630	Oxyacanthan, 6',12'-dimethoxy-2,2'-dimethyl-6,7-[methylenebis(oxy)]- Cepharanthine	$C_{37}H_{38}N_2O_6$ 606.72	481-49-2 150	1981	4-27-00-09061	
8631	Oxyacanthan-12'-ol, 6,6',7-trimethoxy-2,2'-dimethyl- Oxyacanthine	$C_{37}H_{40}N_2O_6$ 608.73	548-40-3 216.5	6906	2-27-00-00894	H_2O 1; EtOH 3; eth 3; bz 3
8632	Oxydemeton-methyl Phosphorothioic acid, *S*-[2-(ethylsulfinyl)ethyl] *O,O*-dimethyl ester	$C_6H_{15}O_4PS_2$ 246.29	301-12-2 <-20	$106^{0.01}$	1.289^{20}	
8633	Oxyfluorfen Benzene, 2-chloro-1-(3-ethoxy-4-nitrophenoxy)-4-(trifluoromethyl)-	$C_{15}H_{11}ClF_3NO_4$ 361.70	42874-03-3 84	6916 358.2 dec	1.35^{73}	
8634	Paraformaldehyde Formaldehyde polymer	$(CH_2O)_x$	30525-89-4 164 dec	6974	5-19-09-00103	H_2O 2; EtOH 1; eth 1; alk 3
8635	Pendimethalin Benzenamine, *N*-(1-ethylpropyl)-3,4-dimethyl-2,6-dinitro-	$C_{13}H_{19}N_3O_4$ 281.31	40487-42-1 56	7026 dec	1.19^{25}	
8636	Pentacene Benzo[b]naphthacene	$C_{22}H_{14}$ 278.35	135-48-8 257	7057	4-05-00-02721	H_2O 1; bz 2; $PhNO_2$ 3; os 2
8637	Pentacosane	$C_{25}H_{52}$ 352.69	629-99-2 54	 401.9; 282^{40}	4-01-00-00582 0.8012^{20}	bz 3; chl 3 1.4491^{20}
8638	Pentacosane, 13-phenyl- Benzene, (1-dodecyltridecyl)-	$C_{31}H_{56}$ 428.79	6006-90-2 31.7	 235^{10}	4-05-00-01241 0.8537^{20}	 1.4787^{20}
8639	Pentacosane, 13-undecyl-	$C_{36}H_{74}$ 506.98	55517-89-0 9.7	 307^{10}	4-01-00-00600 0.8168^{20}	 1.4567^{20}
8640	Pentadecanal Pentadecaldehyde	$C_{15}H_{30}O$ 226.40	2765-11-9 24.5	 185^{25}	4-01-00-03391	ace 4; eth 4; EtOH 4
8641	1-Pentadecanamine	$C_{15}H_{33}N$ 227.43	2570-26-5 37.3	 307.6	4-04-00-00817 0.8104^{20}	eth 4; EtOH 4 1.4480^{20}
8642	Pentadecane	$C_{15}H_{32}$ 212.42	629-62-9 9.9	 270.6	4-01-00-00529 0.7685^{20}	H_2O 1; EtOH 4; eth 4 1.4315^{20}
8643	Pentadecane, 1-bromo-	$C_{15}H_{31}Br$ 291.32	629-72-1 19	 322	4-01-00-00531 1.0675^{20}	H_2O 1; ace 3; chl 4 1.4611^{20}
8644	Pentadecane, 2,6,10,14-tetramethyl- Pristane	$C_{19}H_{40}$ 268.53	1921-70-6	7757 296	3-01-00-00570 0.783^{20}	bz 4; eth 4; chl 4; peth 4 1.4379^{20}
8645	Pentadecanoic acid Pentadecylic acid	$C_{15}H_{30}O_2$ 242.40	1002-84-2 52.3	257^{100}; 158^{1}	4-02-00-01147 0.8423^{80}	H_2O 1; EtOH 4; eth 3; ace 4 1.4254^{80}
8646	Pentadecanoic acid, methyl ester Methyl pentadecanoate	$C_{16}H_{32}O_2$ 256.43	7132-64-1 18.5	 153.5	3-02-00-00936 0.8618^{25}	EtOH 3; eth 3; os 3 1.4390^{25}
8647	2-Pentadecanol *sec*-Pentadecyl alcohol	$C_{15}H_{32}O$ 228.42	1653-34-5 35	 299	4-01-00-01871 0.8328^{20}	 1.4463^{20}
8648	3-Pentadecanol	$C_{15}H_{32}O$ 228.42	53346-71-7 38.8	 388; 163^{12}	4-01-00-01871	 1.4227^{80}
8649	1-Pentadecanol, 2-methyl- 2-Methyl-1-pentadecanol	$C_{16}H_{34}O$ 242.45	25354-98-7 12.5	 185-7^{15}	3-01-00-01827 0.8320^{25}	 1.4453^{25}
8650	6-Pentadecanol, 6-methyl-	$C_{16}H_{34}O$ 242.45	108836-86-8	 199-200^{50}	3-01-00-01828 0.8316^{25}	 1.4446^{25}

No.	Name	Synonym	Mol. Form.	Mol. Wt.	CAS RN	mp/°C	Merck No.	bp/°C	Beil. Ref.	den/g cm^{-3}	Solubility	n_D
8651	2-Pentadecanone		$C_{15}H_{30}O$	226.40	2345-28-0	39.5		294	4-01-00-03391	0.8182^{39}		
8652	8-Pentadecanone		$C_{15}H_{30}O$	226.40	818-23-5	43		291	4-01-00-03392	0.8180^{39}	EtOH 3; eth 3; bz 3; ctc 3	
8653	1-Pentadecene		$C_{15}H_{30}$	210.40	13360-61-7	-2.8		268.2	4-01-00-00926	0.7764^{20}	H_2O 1; ace 3	1.4389^{20}
8654	7-Pentadecene		$C_{15}H_{30}$	210.40	15430-98-5			$114^{3.2}$	3-01-00-00875	0.7765^{20}		1.4420^{20}
8655	1-Pentadecen-3-ol		$C_{15}H_{30}O$	226.40	99814-65-0	27.5		$150\text{-}2^{6}$	3-01-00-01961	0.834^{32}		1.4481^{32}
8656	1-Pentadecyne		$C_{15}H_{28}$	208.39	765-13-9	10		268	4-01-00-01072	0.7928^{20}	ace 4	1.4419^{20}
8657	2,4-Pentadienal, 5-phenyl-	5-Phenyl-2,4-pentadienal	$C_{11}H_{10}O$	158.20	13466-40-5	42.5		160^{3}; $133^{1.0}$	4-07-00-01179		H_2O 1; EtOH 5; eth 4; bz 5	
8658	1,2-Pentadiene	Ethylallene	C_5H_8	68.12	591-95-7	-137.3		44.9	4-01-00-00993	0.6926^{20}	EtOH 5; eth 5; ace 5; bz 5	1.4209^{20}
8659	1,4-Pentadiene		C_5H_8	68.12	591-93-5	-148.8		26	4-01-00-00998	0.6608^{20}	H_2O 1; EtOH 4; eth 4; ace 4	1.3888^{20}
8660	2,3-Pentadiene	1,3-Dimethylallene	C_5H_8	68.12	591-96-8	-125.6		48.2	4-01-00-00999	0.6950^{20}	H_2O 1; EtOH 5; eth 5; ace 5	1.4284^{20}
8661	1,3-Pentadiene, (*E*)-	*trans*-Piperylene	C_5H_8	68.12	2004-70-8	-87.4		42	4-01-00-00994	0.6710^{25}		1.4301^{20}
8662	1,3-Pentadiene, (*Z*)-	*cis*-Piperylene	C_5H_8	68.12	1574-41-0	-140.8		44.1	4-01-00-00994	0.6910^{20}	EtOH 5; eth 5; ace 5; bz 5	1.4363^{20}
8663	1,3-Pentadiene, 3-chloro-	3-Chloro-1,3-pentadiene	C_5H_7Cl	102.56	37710-49-9			100	3-01-00-00962	0.9576^{20}	ace 4; bz 4; eth 4	1.4785^{20}
8664	1,2-Pentadiene, 1-chloro-3-methyl-		C_6H_9Cl	116.59	32337-74-9			$68\text{-}70^{100}$	3-01-00-00990	0.9562^{20}	eth 4	
8665	1,3-Pentadiene, 2,4-dimethyl-		C_7H_{12}	96.17	1000-86-8	-114		93.2	4-01-00-01034	0.7343^{23}		1.4390^{23}
8666	1,2-Pentadiene, 3-methyl-	1-Ethyl-1-methylallene	C_6H_{10}	82.15	7417-48-3			70	4-01-00-01021	0.710^{25}		1.425^{20}
8667	1,2-Pentadiene, 4-methyl-		C_6H_{10}	82.15	13643-05-5			70	2-01-00-00232	0.703^{25}		1.424^{20}
8668	1,3-Pentadiene, 2-methyl-, (*Z*)-	*cis*-2-Methyl-1,3-pentadiene	C_6H_{10}	82.15	1501-60-6	-117.6		76	4-01-00-01019	0.714^{25}		1.446^{20}
8669	1,3-Pentadiene, 3-methyl-		C_6H_{10}	82.15	4549-74-0			77	4-01-00-01021	0.730^{25}		1.452^{20}
8670	1,3-Pentadiene, 4-methyl-	1,1-Dimethylbutadiene	C_6H_{10}	82.15	926-56-7			76.5	4-01-00-01020	0.7181^{20}		1.4532^{20}
8671	1,4-Pentadiene, 2-methyl-	2-Methyl-1,4-pentadiene	C_6H_{10}	82.15	763-30-4			56	4-01-00-01020	0.689^{25}		1.405^{20}
8672	2,3-Pentadiene, 2-methyl-		C_6H_{10}	82.15	3043-33-2			72	4-01-00-01020	0.706^{25}		1.425^{20}
8673	2,4-Pentadienenitrile		C_5H_5N	79.10	1615-70-9	-60		136.5	4-02-00-01692	0.8444^{20}		1.4880
8674	2,4-Pentadienenitrile, (*E*)-		C_5H_5N	79.10	2180-68-9	-43		41^{13}	4-02-00-01695	0.8576^{20}	eth 3; ace 3	1.4986^{20}
8675	2,4-Pentadienenitrile, (*Z*)-		C_5H_5N	79.10	2180-69-0	-64		49^{32}; 32^{13}	4-02-00-01695	0.8541^{26}	eth 3; ace 3	1.4855^{20}
8676	2,4-Pentadienoic acid		$C_5H_6O_2$	98.10	626-99-3	79		112 dec	3-02-00-01451		H_2O 4; EtOH 4; eth 4; bz 4	
8677	2,4-Pentadienoic acid, 5-(1,3-benzodioxol-5-yl)-	2,4-Pentadienoic acid, 5(3,4-metylenedioxyphenyl)	$C_{12}H_{10}O_4$	218.21	5285-18-7	215	7437		2-19-00-00300		H_2O 1; EtOH 3; eth 2; bz 2	
8678	2,4-Pentadienoic acid, 5-(1,3-benzodioxol-5-yl)-, (*E*,*E*)-		$C_{12}H_{10}O_4$	218.21	136-72-1	215.8		sub	5-19-07-00382		EtOH 4	
8679	2,4-Pentadienoic acid, 5-(1-hydroxy-2,6,6-trimethyl-4-oxo-2-cyclohexen-1-yl)-3-methyl-, [*S*-(*Z*,*E*)]-	Abscisic acid	$C_{15}H_{20}O_4$	264.32	21293-29-8	160	6	sub 120			ace 4; eth 4; chl 4	
8680	2,4-Pentadienoic acid, 5-phenyl-, ethyl ester	Ethyl cinnamylideneacetate	$C_{13}H_{14}O_2$	202.25	1552-95-0	26		$149\text{-}50^{4}$	2-09-00-00441	1.0467^{20}	eth 4; EtOH 4	1.5768^{80}
8681	1,4-Pentadien-3-ol		C_5H_8O	84.12	922-65-6			115.5	4-01-00-02229	0.860^{23}		1.4400^{17}
8682	1,4-Pentadien-3-one, 1,5-di-2-furanyl-		$C_{13}H_{10}O_3$	214.22	886-77-1	60.5		181^{4}	5-19-04-00459		eth 4; EtOH 4; chl 4	
8683	1,4-Pentadien-3-one, 1,5-diphenyl-	Dibenzalacetone	$C_{17}H_{14}O$	234.30	538-58-9	113 dec	2988	dec	4-07-00-01747		H_2O 1; EtOH 2; eth 2; ace 3	
8684	1,3-Pentadiyne	Methyldiacetylene	C_5H_4	64.09	4911-55-1	-38.5		55	4-01-00-01117	0.7909^{20}	H_2O 1; eth 3; bz 3; chl 3	1.4431^{21}
8685	1,5-Pentalenedione, hexahydro-	Bicyclo[3.3.0]octane-2,6-dione	$C_8H_{10}O_2$	138.17	62353-69-9	45		$86\text{-}8^{0.2}$	3-07-00-03280	1.1290^{60}		1.4877^{54}
8686	Pentalene, octahydro-, *cis*-		C_8H_{14}	110.20	1755-05-1	<-80		139	4-05-00-00277	0.8638^{25}	H_2O 1; EtOH 3	1.4595^{25}
8687	Pentalene, octahydro-, *trans*-		C_8H_{14}	110.20	5597-89-7	-30		132	4-05-00-00277	0.8624^{18}	H_2O 1; EtOH 3	1.4625^{18}
8688	2(1H)-Pentalenone, hexahydro-	Bicyclo[3.3.0]octane-2-one	$C_8H_{12}O$	124.18	56180-61-1			72^{13}	4-07-00-00156	1.0097^{20}	ace 4; EtOH 4	1.4790^{20}

No.	Name Synonym	Mol. Form. Mol. Wt.	CAS RN mp/°C	Merck No. bp/°C	Beil. Ref. den/g cm^{-3}	Solubility n_D
8689	2(1H)-Pentalenone, hexahydro-, *cis*- *cis*-Bicyclo[3.3.0]octan-3-one	$C_8H_{12}O$ 124.18	19915-11-8	$50^{2.3}$	4-07-00-00156 1.0660^{25}	1.4766^{25}
8690	Pentanal Valeraldehyde	$C_5H_{10}O$ 86.13	110-62-3 -91.5	9813 103	4-01-00-03268 0.8095^{20}	H_2O 2; EtOH 3; eth 3 1.3944^{20}
8691	Pentanal, 3-hydroxy-2-methyl-	$C_6H_{12}O_2$ 116.16	615-30-5	$94\text{-}6^{23}$	4-01-00-04022 0.986^{25}	H_2O 4; ace 4; eth 4; EtOH 4 1.4502^{20}
8692	Pentanal, 3-hydroxy-2,2,4-trimethyl-	$C_8H_{16}O_2$ 144.21	918-79-6	$118\text{-}20^{14}$	4-01-00-04049 0.9482^{20}	1.4501^{20}
8693	Pentanal, 2-methyl- 2-Methylvaleraldehyde	$C_6H_{12}O$ 100.16	123-15-9	117	4-01-00-03304	H_2O 3; eth 3; ace 3; ctc 2
8694	Pentanal, 3-methyl-	$C_6H_{12}O$ 100.16	15877-57-3	121	3-01-00-02836 0.811^{25}	1.4026^{20}
8695	Pentanal, 2-oxo-	$C_5H_8O_2$ 100.12	7332-93-6	112	4-01-00-03658	H_2O 4; eth 4; EtOH 4 1.4043^{25}
8696	Pentanal, 4-oxo-	$C_5H_8O_2$ 100.12	626-96-0 $<$-21	187 dec	4-01-00-03659 1.0134^{21}	H_2O 4; ace 4; eth 4; EtOH 4 1.4257^{22}
8697	Pentanal, 2-propyl- Valeraldehyde, 2-propyl-	$C_8H_{16}O$ 128.21	18295-59-5	160	3-01-00-02880 0.8347^{15}	1.4142^{15}
8698	Pentanamide	$C_5H_{11}NO$ 101.15	626-97-1 106	225	4-02-00-00874 0.8735^{110}	H_2O 4; EtOH 4; eth 4; chl 2 1.4183^{110}
8699	Pentanamide, *N*,*N*-dimethyl-	$C_7H_{15}NO$ 129.20	6225-06-5 -51	141^{100}	3-04-00-00127 0.8962^{25}	H_2O 4; eth 4; EtOH 4 1.4419^{25}
8700	1-Pentanamine Amylamine	$C_5H_{13}N$ 87.16	110-58-7 -55	630 104.3	4-04-00-00674 0.7544^{20}	H_2O 5; EtOH 5; eth 5; ace 4 1.448^{20}
8701	2-Pentanamine	$C_5H_{13}N$ 87.16	625-30-9	92	4-04-00-00689 0.7384^{20}	ace 4; bz 4; eth 4; EtOH 4 1.4027^{20}
8702	3-Pentanamine	$C_5H_{13}N$ 87.16	616-24-0	89	4-04-00-00692 0.7487^{20}	EtOH 3; chl 2 1.4063^{20}
8703	1-Pentanamine, *N*,*N*-dimethyl-	$C_7H_{17}N$ 115.22	26153-88-8	123	4-04-00-00675 0.743^{20}	eth 4 1.4083^{20}
8704	2-Pentanamine, 2,4-dimethyl-	$C_7H_{17}N$ 115.22	64379-30-2	122	4-04-00-00749 0.7719^{20}	1.4009^{20}
8705	1-Pentanamine, *N*,*N*-dipentyl-	$C_{15}H_{33}N$ 227.43	621-77-2	242.5	4-04-00-00676 0.7907^{20}	H_2O 1; EtOH 3; eth 3; acid 3 1.4366^{20}
8706	1-Pentanamine, 3-methyl- 3-Methylpentylamine	$C_6H_{15}N$ 101.19	42245-37-4	122	3-04-00-00366 0.767^{20}	1.4196^{25}
8707	1-Pentanamine, 4-methyl-	$C_6H_{15}N$ 101.19	5344-20-7 -94.4	125	4-04-00-00727 0.758^{52}	
8708	1-Pentanamine, *N*-pentyl- Diamylamine	$C_{10}H_{23}N$ 157.30	2050-92-2	202.5	4-04-00-00676 0.7771^{20}	H_2O 2; EtOH 4; eth 5; ace 3 1.4272^{20}
8709	Pentane	C_5H_{12} 72.15	109-66-0 -129.7	7072 36.0	4-01-00-00303 0.6262^{20}	H_2O 2; EtOH 5; eth 5; ace 5 1.3575^{20}
8710	Pentane, 1-bromo- Amyl bromide	$C_5H_{11}Br$ 151.05	110-53-2 -95	637 129.8	4-01-00-00312 1.2182^{20}	H_2O 1; EtOH 3; eth 5; bz 3 1.4447^{20}
8711	Pentane, 2-bromo- 2-Pentyl bromide	$C_5H_{11}Br$ 151.05	107-81-3 -95.5	117.4	4-01-00-00312 1.2075^{20}	bz 4; eth 4; EtOH 4; chl 4 1.4413^{20}
8712	Pentane, 3-bromo- 3-Pentyl bromide	$C_5H_{11}Br$ 151.05	1809-10-5 -126.2	118.6	4-01-00-00313 1.214^{20}	H_2O 1; EtOH 3; eth 3; bz 3 1.4441^{20}
8713	Pentane, 1-bromo-5-fluoro- 1-Bromo-5-fluoropentane	$C_5H_{10}BrF$ 169.04	407-97-6	162	4-01-00-00313 1.3604^{25}	eth 4; EtOH 4 1.4406^{25}
8714	Pentane, 1-bromo-2-methyl- 2-Methylpentyl bromide	$C_6H_{13}Br$ 165.07	25346-33-2	141	4-01-00-00361 1.1624^{20}	eth 4; chl 4 1.4495^{20}
8715	Pentane, 1-bromo-3-methyl- 3-Methylpentyl bromide	$C_6H_{13}Br$ 165.07	51116-73-5	148	4-01-00-00365 1.1829^{20}	H_2O 1; eth 3; chl 4 1.4496^{20}
8716	Pentane, 1-bromo-4-methyl-	$C_6H_{13}Br$ 165.07	626-88-0	145	4-01-00-00361 1.1683^{20}	eth 4; chl 4 1.4490
8717	Pentane, 2-bromo-2-methyl- 1-Bromo-1,1-dimethylbutane	$C_6H_{13}Br$ 165.07	4283-80-1	142.5	4-01-00-00361	eth 4; chl 4 1.442^{23}
8718	Pentane, 3-bromo-3-methyl- 3-Bromo-3-methylpentane	$C_6H_{13}Br$ 165.07	25346-31-0	130; 76^{100}	4-01-00-00365 1.1835^{20}	eth 4; chl 4 1.4525^{20}
8719	Pentane, 1-chloro- Amyl chloride	$C_5H_{11}Cl$ 106.60	543-59-9 -99	642 107.8	4-01-00-00309 0.8820^{20}	H_2O 1; EtOH 5; eth 5; bz 3 1.4127^{20}
8720	Pentane, 2-chloro-, (+)- Pentane, 2-chloro (*d*)	$C_5H_{11}Cl$ 106.60	29882-57-3 -137	96.9	4-01-00-00309 0.8698^{20}	H_2O 1; EtOH 3; eth 3; bz 3 1.4069^{20}
8721	Pentane, 3-chloro- 1-Ethylpropyl chloride	$C_5H_{11}Cl$ 106.60	616-20-6 -105	97.5	4-01-00-00309 0.8731^{20}	H_2O 1; EtOH 3; eth 3; ace 2 1.4082^{20}
8722	Pentane, 2-chloro-2,3-dimethyl-	$C_7H_{15}Cl$ 134.65	59889-45-1	$38\text{-}9^{20}$	4-01-00-00406	bz 4; eth 4 1.4264^{20}
8723	Pentane, 2-chloro-2,4-dimethyl-	$C_7H_{15}Cl$ 134.65	35951-33-8	129; 33^{20}	4-01-00-00408 0.861^{20}	eth 4 1.4180^{20}
8724	Pentane, 3-chloro-2,3-dimethyl-	$C_7H_{15}Cl$ 134.65	595-38-0	136 dec; 41^{20}	4-01-00-00406 0.882^{22}	eth 4; chl 4 1.4318^{20}
8725	Pentane, 4-chloro-2,2-dimethyl-	$C_7H_{15}Cl$ 134.65	33429-72-0	93^{250}	4-01-00-00404 0.855^{20}	eth 4; chl 4 1.4180^{20}
8726	Pentane, 3-chloro-3-ethyl-	$C_7H_{15}Cl$ 134.65	994-25-2	143.5	4-01-00-00403 0.8856^{20}	eth 4 1.4400^{20}
8727	Pentane, 3-chloro-3-ethyl-2,2-dimethyl- Pentane 3-chloro-2,2-dimethyl-3-ethyl	$C_9H_{19}Cl$ 162.70	86661-53-2	53^{6} dec	4-01-00-00462	eth 4; chl 4 1.4528^{25}
8728	Pentane, 1-chloro-5-fluoro-	$C_5H_{10}ClF$ 124.59	407-98-7	143.2	4-01-00-00309 1.0325^{25}	eth 4; EtOH 4 1.4120^{23}

No.	Name Synonym	Mol. Form. Mol. Wt.	CAS RN mp/°C	Merck No. bp/°C	Beil. Ref. den/g cm^{-3}	Solubility n_D
8729	Pentane, 2-chloro-2-methyl-	$C_6H_{13}Cl$ 120.62	4325-48-8	110 dec; 36^{15}	4-01-00-00360 0.863^{20}	eth 4 1.4126^{20}
8730	Pentane, 2-chloro-4-methyl- 2-Chloro-4-methylpentane	$C_6H_{13}Cl$ 120.62	25346-32-1	113	3-01-00-00399 0.8610^{20}	eth 4 1.4113^{20}
8731	Pentane, 3-(chloromethyl)-	$C_6H_{13}Cl$ 120.62	4737-41-1	126; 83^{202}	3-01-00-00403 0.8914^{20}	bz 4; eth 4; chl 4 1.4222^{20}
8732	Pentane, 3-chloro-3-methyl-	$C_6H_{13}Cl$ 120.62	918-84-3	116	4-01-00-00365 0.8900^{20}	bz 4; eth 4; chl 4 1.4210^{20}
8733	Pentane, 2-chloro-2,4,4-trimethyl-	$C_8H_{17}Cl$ 148.68	6111-88-2 -26	147 dec; 44^{16}	4-01-00-00444 0.8746^{20}	EtOH 4 1.4308^{20}
8734	Pentanedial Glutaraldehyde	$C_5H_8O_2$ 100.12	111-30-8	4366 188 dec	4-01-00-03659	H_2O 5; EtOH 5; bz 3
8735	1,5-Pentanediamine Cadaverine	$C_5H_{14}N_2$ 102.18	462-94-2 9	1608 179	4-04-00-01310 0.873^{25}	H_2O 3; EtOH 3; eth 2 1.463^{20}
8736	1,4-Pentanediamine, *N*1,*N*1-diethyl- Novoldiamine	$C_9H_{22}N_2$ 158.29	140-80-7	6642 201	4-04-00-01308 0.814^{20}	1.4429^{20}
8737	1,5-Pentanediamine, *N*-(6-methoxy-8-quinolinyl)-*N*'-(1-methylethyl)-	$C_{18}H_{27}N_3O$ 301.43	86-78-2	$165\text{-}70^{0.02}$	5-22-12-00126	dil HCl 4 1.5785^{25}
8738	1,5-Pentanediaminium, *N,N,N,N',N',N'*-hexamethyl-, dibromide Ammonium, pentamethylenebis trimethyl, bromide	$C_{11}H_{28}Br_2N_2$ 348.16	541-20-8 301	7070	4-04-00-01311	H_2O 2
8739	Pentane, 1,3-dibromo- 1,3-Dibromopentane	$C_5H_{10}Br_2$ 229.94	42474-20-4	193; 64^{10}	4-01-00-00314 1.6721^{20}	1.5048^{207}
8740	Pentane, 1,4-dibromo- 1,4-Dibromopentane	$C_5H_{10}Br_2$ 229.94	626-87-9 -34.4	146^{150}; 99^{14}	4-01-00-00314 1.6222^{20}	1.5086^{20}
8741	Pentane, 1,5-dibromo-	$C_5H_{10}Br_2$ 229.94	111-24-0 -39.5	222.3	4-01-00-00314 1.7018^{20}	H_2O 1; bz 3; ctc 2; chl 3 1.5126^{20}
8742	Pentane, 2,3-dibromo- 2,3-Dibromopentane	$C_5H_{10}Br_2$ 229.94	5398-25-4 -56	178; 91^{50}	3-01-00-00347 1.6865^{14}	1.5087^{20}
8743	Pentane, 2,4-dibromo- 2,4-Dibromopentane	$C_5H_{10}Br_2$ 229.94	19398-53-9	60^{12}	3-01-00-00348 1.6659^{20}	1.4987^{20}
8744	Pentane, 1,2-dichloro- 1,2-Dichloropentane	$C_5H_{10}Cl_2$ 141.04	1674-33-5	148.3	3-01-00-00341 1.0872^{20}	H_2O 1; EtOH 3; chl 4 1.4485^{20}
8745	Pentane, 1,3-dichloro- 1,3-Dichloropentane	$C_5H_{10}Cl_2$ 141.04	30122-12-4	80.4^{60}	3-01-00-00342 1.0834^{20}	ace 4; chl 4 1.4485^{20}
8746	Pentane, 1,4-dichloro- 1,4-Dichloropentane	$C_5H_{10}Cl_2$ 141.04	626-92-6	162; 88^{60}	4-01-00-00309 1.0840^{20}	ace 3; chl 3 1.4503^{20}
8747	Pentane, 1,5-dichloro- 1,5-Dichloropentane	$C_5H_{10}Cl_2$ 141.04	628-76-2 -72.8	179	4-01-00-00310 1.1006^{22}	H_2O 1; EtOH 3; eth 3; bz 3 1.4564^{20}
8748	Pentane, 2,2-dichloro- 2,2-Dichloropentane	$C_5H_{10}Cl_2$ 141.04	34887-14-4	128.5	2-01-00-00095 1.040^{20}	H_2O 1; eth 3; bz 3; chl 3 1.434^{20}
8749	Pentane, 2,3-dichloro- 2,3-Dichloropentane	$C_5H_{10}Cl_2$ 141.04	600-11-3 -77.3	139	4-01-00-00310 1.0789^{20}	1.4464^{20}
8750	Pentane, 2,4-dichloro- 2,4-Dichloropentane	$C_5H_{10}Cl_2$ 141.04	625-67-2	145	3-01-00-00343 1.0634^{15}	H_2O 1; eth 4; bz 3; chl 4 1.447^{18}
8751	Pentane, 3,3-dichloro- 3,3-Dichloropentane	$C_5H_{10}Cl_2$ 141.04	21571-91-5	132; 32^{14}	3-01-00-00343 1.053^{20}	bz 4; chl 4 1.442^{20}
8752	Pentane, 1,2-dichloro-4,4-dimethyl- 1,2-Dichloro-4,4-dimethylpentane	$C_7H_{14}Cl_2$ 169.09	6065-90-3	175; 58^{12}	3-01-00-00445 1.0259^{20}	bz 4; chl 4 1.4489^{20}
8753	Pentane, 1,5-dichloro-3,3-dimethyl- 1,5-Dichloro-3,3-dimethylpentane	$C_7H_{14}Cl_2$ 169.09	62496-53-1	135^{80}; 58^{8}	4-01-00-00409 1.0563^{20}	chl 4 1.4652^{20}
8754	Pentane, 2,4-dichloro-2,4-dimethyl- 2,4-Dichloro-2,4-dimethylpentane	$C_7H_{14}Cl_2$ 169.09	33553-93-4 23.5	$51\text{-}7^{8}$	4-01-00-00408 1.0292^{20}	1.4537^{20}
8755	Pentane, 1,1-diethoxy-	$C_9H_{20}O_2$ 160.26	3658-79-5	59^{12}	4-01-00-03269 0.829^{22}	1.4029^{22}
8756	Pentane, 3,3-diethyl- 3,3-Diethylpentane	C_9H_{20} 128.26	1067-20-5 -33.1	146.3	4-01-00-00462 0.7536^{20}	H_2O 1; eth 3; bz 3; os 3 1.4206^{20}
8757	Pentane, 3,3-diethyl-2-methyl- 3,3-Diethyl-2-methylpentane	$C_{10}H_{22}$ 142.28	52897-16-2	172	4-01-00-00487 0.7755^{25}	1.4343^{20}
8758	Pentane, 1,5-diiodo- Pentamethylene diiodide	$C_5H_{10}I_2$ 323.94	628-77-3 9	149^{20}; 101^{3}	4-01-00-00317 2.1903^{15}	H_2O 1; eth 3; chl 3 1.6046^{15}
8759	Pentane, 2,4-diiodo-	$C_5H_{10}I_2$ 323.94	66719-29-7	147^{34}	0-01-00-00133 2.195^{20}	1.600^{20}
8760	Pentane, 2,2-dimethyl- 2,2-Dimethylpentane	C_7H_{16} 100.20	590-35-2 -123.8	79.2	4-01-00-00403 0.6739^{20}	H_2O 1; EtOH 3; eth 3; ace 5 1.3822^{20}
8761	Pentane, 2,3-dimethyl- 2,3-Dimethylpentane	C_7H_{16} 100.20	565-59-3	89.7	3-01-00-00445 0.6951^{20}	H_2O 1; EtOH 3; eth 3; ace 5 1.3919^{20}
8762	Pentane, 2,4-dimethyl- 2,4-Dimethylpentane	C_7H_{16} 100.20	108-08-7 -119.9	80.4	4-01-00-00406 0.6727^{20}	H_2O 1; EtOH 3; eth 3; ace 5 1.3815^{20}
8763	Pentane, 3,3-dimethyl- 3,3-Dimethylpentane	C_7H_{16} 100.20	562-49-2 -134.9	86.0	4-01-00-00409 0.6936^{20}	H_2O 1; EtOH 3; eth 3; ace 5 1.3909^{20}
8764	Pentane, 2,2-dimethyl-3-methylene-	C_8H_{16} 112.22	18231-53-3	110	3-01-00-00848 0.724^{25}	1.4159^{20}
8765	Pentane, 2,4-dimethyl-3-methylene- 1-Butene, 2-isopropyl-3-methyl-	C_8H_{16} 112.22	111823-35-9	104	3-01-00-00851 0.718^{25}	1.4086^{20}

No.	Name Synonym	Mol. Form. Mol. Wt.	CAS RN mp/°C	Merck No. bp/°C	Beil. Ref. den/g cm^{-3}	Solubility n_D
8766	Pentane, 2,4-dimethyl-3-(1-methylethyl)-	$C_{10}H_{22}$ 142.28	13475-79-1 -81.7	 157.1	3-01-00-00534 0.7545^{25}	 1.4246^{20}
8767	Pentanedinitrile Glutaronitrile	$C_5H_6N_2$ 94.12	544-13-8 -29	4368 286	4-02-00-01941 0.9911^{15}	EtOH 4; chl 4 1.4295^{20}
8768	Pentanedinitrile, 2-bromo-2-(bromomethyl)- 1,2-Dibromo-2,4-dicyanobutane	$C_6H_6Br_2N_2$ 265.94	35691-65-7 52	3004		H_2O 1; ace 4; bz 4; DMF 4
8769	Pentane, 1,5-dinitro- 1,5-Dinitropentane	$C_5H_{10}N_2O_4$ 162.15	6848-84-6	 $134^{1.2}$	4-01-00-00319	H_2O 1; bz 3 1.461^{20}
8770	Pentanedioic acid Glutaric acid	$C_5H_8O_4$ 132.12	110-94-1 97.8	4367 303 dec	4-02-00-01934 1.429^{15}	H_2O 4; EtOH 4; eth 4; bz 1 1.4188^{106}
8771	Pentanedioic acid, 2-acetyl-, diethyl ester	$C_{11}H_{18}O_5$ 230.26	1501-06-0	 271 dec	4-03-00-01835 1.0712^{20}	H_2O 2; EtOH 3; eth 3; ctc 2 1.4420^{15}
8772	Pentanedioic acid, diethyl ester	$C_9H_{16}O_4$ 188.22	818-38-2 -24.1	 236.5	4-02-00-01937 1.0220^{20}	eth 4 1.4241^{20}
8773	Pentanedioic acid, 3,3-dimethyl-	$C_7H_{12}O_4$ 160.17	4839-46-7 103.5	 126^{415}; 89^2	4-02-00-02023 1.4278^{20}	H_2O 4; EtOH 4; eth 4; bz 2
8774	Pentanedioic acid, 3,3-dimethyl-, dimethyl ester	$C_9H_{16}O_4$ 188.22	19184-67-9	 $103\text{-}4^{15}$	3-02-00-01758 1.0366^{20}	ctc 3
8775	Pentanedioic acid, dimethyl ester	$C_7H_{12}O_4$ 160.17	1119-40-0 -42.5	 214; 109^{21}	4-02-00-01937 1.0876^{20}	EtOH 4; eth 4; chl 3 1.4242^{20}
8776	Pentanedioic acid, diphenyl ester Diphenyl glutarate	$C_{17}H_{16}O_4$ 284.31	47172-89-4 54	 236.5^{15}	4-06-00-00626	ace 4; EtOH 4; lig 4
8777	Pentanedioic acid, 3-ethyl-3-methyl- 3-Ethyl-3-methylglutaric acid	$C_8H_{14}O_4$ 174.20	5345-01-7 87	 261	3-02-00-01781	H_2O 3; EtOH 3; eth 3; bz 3
8778	Pentanedioic acid, 3-hydroxy-, diethyl ester Diethyl 3-hydroxyglutarate	$C_9H_{16}O_5$ 204.22	32328-03-3	 $156\text{-}7^{23}$	4-03-00-01149 1.103^{25}	 1.4368^{20}
8779	Pentanedioic acid, 3-methyl- Glutaric acid, 3-methyl	$C_6H_{10}O_4$ 146.14	626-51-7 87	 $165\text{-}7^{0.5}$	4-02-00-01992	H_2O 3; EtOH 3; eth 3; bz 2
8780	Pentanedioic acid, monomethyl ester	$C_6H_{10}O_4$ 146.14	1501-27-5	 $150\text{-}1^{10}$	4-02-00-01937 1.169^{25}	 1.4381^{20}
8781	Pentanedioic acid, 2-oxo- α-Ketoglutaric acid	$C_5H_6O_5$ 146.10	328-50-7 115.5	5182	4-03-00-01813	H_2O 4; EtOH 4; eth 4; ace 3
8782	Pentanedioic acid, 3-oxo- Acetonedicarboxylic acid	$C_5H_6O_5$ 146.10	542-05-2 138 dec	60	4-03-00-01816	H_2O 3; EtOH 3; eth 2; bz 1
8783	Pentanedioic acid, 3-oxo-, diethyl ester	$C_9H_{14}O_5$ 202.21	105-50-0	 250	4-03-00-01817 1.113^{20}	H_2O 2; EtOH 5
8784	Pentanedioic acid, 3-oxo-, dimethyl ester	$C_7H_{10}O_5$ 174.15	1830-54-2	 150^{25}	4-03-00-01817 1.185^{25}	 1.4434^{20}
8785	1,3-Pentanediol	$C_5H_{12}O_2$ 104.15	3174-67-2	 $120\text{-}2^{16}$	4-01-00-02538 0.9863^{20}	 1.4659^{20}
8786	1,4-Pentanediol 4-Hydroxy-1-pentanol	$C_5H_{12}O_2$ 104.15	626-95-9	 202; 125^{10}	3-01-00-02191 0.9883^{20}	H_2O 4; EtOH 4; chl 4 1.4452^{23}
8787	1,5-Pentanediol Pentamethylene glycol	$C_5H_{12}O_2$ 104.15	111-29-5 -18	7073 239	4-01-00-02540 0.9914^{20}	H_2O 3; EtOH 3; eth 2; bz 2 1.4494^{20}
8788	2,3-Pentanediol	$C_5H_{12}O_2$ 104.15	42027-23-6	 187.5; 100^{17}	4-01-00-02543 0.9798^{19}	H_2O 3; EtOH 3; eth 2 1.4412^{25}
8789	2,4-Pentanediol 2,4-Amylene glycol	$C_5H_{12}O_2$ 104.15	625-69-4	 199; $97\text{-}8^{13}$	3-01-00-02195 0.9635^{20}	H_2O 4; EtOH 4 1.4349^{20}
8790	1,2-Pentanediol, (±)- 1,2-Pentanediol (*DL*)	$C_5H_{12}O_2$ 104.15	91049-43-3	 211	4-01-00-02538 0.9784^{20}	 1.4412^{19}
8791	1,5-Pentanediol, diacetate Pentamethylene acetate	$C_9H_{16}O_4$ 188.22	6963-44-6	 $122\text{-}3^3$	4-02-00-00226 1.0296^{20}	 1.4261^{19}
8792	1,3-Pentanediol, 2,2-dimethyl- 2,2-Dimethyl-1,3-dihydroxypentane	$C_7H_{16}O_2$ 132.20	2157-31-5 62.3	 213	0-01-00-00490	EtOH 4
8793	2,4-Pentanediol, 2-methyl- Hexylene glycol	$C_6H_{14}O_2$ 118.18	107-41-5 -50	4631 197.1	4-01-00-02565 0.923^{15}	H_2O 3; EtOH 3; eth 3; ctc 2 1.4276^{20}
8794	2,4-Pentanediol, 3-methyl- 3-Methyl-2,4-pentanediol	$C_6H_{14}O_2$ 118.18	5683-44-3	 211.5	4-01-00-02572 0.9640^{20}	H_2O 3; EtOH 3 1.4433^{20}
8795	1,3-Pentanediol, 2,2,4-trimethyl- 2,2,4-Trimethyl-1,3-pentanediol	$C_8H_{18}O_2$ 146.23	144-19-4 51.5	 235; $81\text{-}2^1$	4-01-00-02604 0.936^{15}	H_2O 2; EtOH 4; eth 4; bz 3 1.4513^{15}
8796	1,4-Pentanediol, 2,2,4-trimethyl-	$C_8H_{18}O_2$ 146.23	80864-10-4 86	 210	0-01-00-00493	eth 4; EtOH 4
8797	2,3-Pentanedione Acetylpropionyl	$C_5H_8O_2$ 100.12	600-14-6	 108	4-01-00-03660 0.9565^{19}	H_2O 3; EtOH 5; eth 5; ace 5 1.4014^{19}
8798	2,4-Pentanedione Acetylacetone	$C_5H_8O_2$ 100.12	123-54-6 -23	75 138	4-01-00-03662 0.9721^{25}	H_2O 4; EtOH 5; eth 5; ace 5 1.4494^{20}
8799	2,4-Pentanedione, 3,3-dimethyl- 3,3-Dimethylacetylacetone	$C_7H_{12}O_2$ 128.17	3142-58-3 19	 173	4-01-00-03705 0.9575^{20}	eth 4 1.4306^{20}
8800	2,3-Pentanedione, dioxime	$C_5H_{10}N_2O_2$ 130.15	4775-86-4 172.5	 sub	4-01-00-03661	EtOH 4
8801	2,4-Pentanedione, 3-ethyl-	$C_7H_{12}O_2$ 128.17	1540-34-7	 178.5	4-01-00-03703 0.9531^{19}	eth 4; EtOH 4; chl 4 1.4408^{19}
8802	2,4-Pentanedione, 1,1,1,5,5,5-hexafluoro- Acetylacetone, hexafluoro	$C_5H_2F_6O_2$ 208.06	1522-22-1	 54.1	4-01-00-03681 1.485^{20}	 1.3333^{20}
8803	2,3-Pentanedione, 3-oxime 3-Oximino-2-pentanone	$C_5H_9NO_2$ 115.13	609-29-0 58.5	 183	4-01-00-03661	eth 4; EtOH 4; chl 4
8804	1,4-Pentanedione, 1-phenyl-	$C_{11}H_{12}O_2$ 176.22	583-05-1	 162^{12}	4-07-00-02160	ace 4 1.5250^{30}
8805	2,4-Pentanedione, 3-(2-propenyl)-	$C_8H_{12}O_2$ 140.18	3508-78-9	 199; 92^{16}	4-01-00-03759 0.974^{15}	 1.4698^{14}

PHYSICAL CONSTANTS OF ORGANIC COMPOUNDS (continued)

No.	Name Synonym	Mol. Form. Mol. Wt.	CAS RN mp/°C	Merck No. bp/°C	Beil. Ref. den/g cm^{-3}	Solubility n_D
8806	Pentanedioyl dichloride	$C_5H_6Cl_2O_2$ 169.01	2873-74-7	 217	4-02-00-01939 1.324^{20}	eth 3; chl 2 1.4728^{20}
8807	Pentane, dodecafluoro- Perfluoropentane	C_5F_{12} 288.04	678-26-2 -10	 29.2	4-01-00-00308	
8808	Pentane, 1-ethoxy- Ethyl pentyl ether	$C_7H_{16}O$ 116.20	17952-11-3	 117.6	3-01-00-01602 0.7622^{20}	eth 4; EtOH 4 1.3927^{20}
8809	Pentane, 3-ethyl- 3-Ethylpentane	C_7H_{16} 100.20	617-78-7 -118.6	 93.5	4-01-00-00402 0.6982^{20}	H_2O 1; EtOH 3; eth 3; ace 5 1.3934^{20}
8810	Pentane, 3-ethyl-2,2-dimethyl-	C_9H_{20} 128.26	16747-32-3 -99.3	 133.8	4-01-00-00462 0.7438^{20}	 1.4123^{20}
8811	Pentane, 3-ethyl-2,3-dimethyl-	C_9H_{20} 128.26	16747-33-4	 144.7	4-01-00-00462 0.7508^{25}	 1.4221^{20}
8812	Pentane, 3-ethyl-2,4-dimethyl-	C_9H_{20} 128.26	1068-87-7 -122.4	 136.7	4-01-00-00462 0.7365^{20}	 1.4131^{20}
8813	Pentane, 3-ethyl-2-methyl- 3-Ethyl-2-methylpentane	C_8H_{18} 114.23	609-26-7 -114.9	 115.6	4-01-00-00437 0.7193^{20}	H_2O 1; EtOH 5; eth 3; ace 5 1.4040^{20}
8814	Pentane, 3-ethyl-3-methyl- 3-Ethyl-3-methylpentane	C_8H_{18} 114.23	1067-08-9 -90.9	 118.2	4-01-00-00438 0.7274^{20}	H_2O 1; EtOH 5; eth 3; ace 5 1.4078^{20}
8815	Pentane, 3-ethyl-2,2,3-trimethyl- 3-Ethyl-2,2,3-trimethylpentane	$C_{10}H_{22}$ 142.28	52897-17-3 -42.5	 169.5	4-01-00-00487 0.7780^{25}	 1.4420^{20}
8816	Pentane, 3-ethyl-2,3,4-trimethyl- 3-Ethyl-2,3,4-trimethylpentane	$C_{10}H_{22}$ 142.28	52897-19-5	 169.47	4-01-00-00487 0.7735^{25}	 1.4333^{20}
8817	Pentane, 1-fluoro- Pentyl fluoride	$C_5H_{11}F$ 90.14	592-50-7 -120	 62.8	4-01-00-00308 0.7907^{20}	eth 4; EtOH 4 1.3591^{2-}
8818	Pentane, 1-iodo- Amyl iodide	$C_5H_{11}I$ 198.05	628-17-1 -85.6	 155	4-01-00-00315 1.5161^{20}	chl 3 1.4959^{20}
8819	Pentane, 2-iodo-, (±)- Pentane, 2-iodo (*DL*)	$C_5H_{11}I$ 198.05	52152-72-4	 141	4-01-00-00316 1.5096^{20}	H_2O 1; eth 3; ace 3; bz 3 1.4961^{20}
8820	Pentane, 3-iodo- 3-Iodopentane	$C_5H_{11}I$ 198.05	1809-05-8	 145.5	3-01-00-00349 1.5176^{20}	ace 4; bz 4; eth 4 1.4974^{20}
8821	Pentane, 1-isocyano- Pentyl isocyanide	$C_6H_{11}N$ 97.16	18971-59-0 -51.1	 155.5	3-04-00-00331 0.806^{20}	H_2O 1; EtOH 3
8822	Pentane, 1-methoxy- Methyl pentyl ether	$C_6H_{14}O$ 102.18	628-80-8	 99	4-01-00-01643 0.759^{22}	ace 4; eth 4; EtOH 4 1.3862^{22}
8823	Pentane, 2-methyl- Isohexane	C_6H_{14} 86.18	107-83-5 -153.7	 60.2	4-01-00-00358 0.650^{25}	H_2O 1; EtOH 3; eth 3; ace 5 1.3715^{20}
8824	Pentane, 3-methyl- 3-Methylpentane	C_6H_{14} 86.18	96-14-0 -162.9	 63.2	4-01-00-00363 0.6598^{25}	H_2O 1; EtOH 3; eth 5; ace 5 1.3765^{20}
8825	Pentane, 3-methylene- 2-Ethyl-1-butene	C_6H_{12} 84.16	760-21-4 -131.5	 64.7	4-01-00-00850 0.6894^{20}	H_2O 1; eth 3; ace 3; bz 3 1.3969^{20}
8826	Pentane, 1-(methylthio)- Methyl pentyl sulfide	$C_6H_{14}S$ 118.24	1741-83-9 -94	 145.1	3-01-00-01608 0.8431^{20}	EtOH 3; eth 3; ace 3; bz 3 1.4506^{20}
8827	Pentanenitrile Valeronitrile	C_5H_9N 83.13	110-59-8 -96.2	 141.3	4-02-00-00875 0.8008^{20}	eth 3; ace 3; bz 3; ctc 2 1.3971^{20}
8828	Pentanenitrile, 5-bromo-	C_5H_8BrN 162.03	5414-21-1	 111^{12}	4-02-00-00884 1.386^{25}	 1.4780^{20}
8829	Pentanenitrile, 2-hydroxy-2-propyl-	$C_8H_{15}NO$ 141.21	5699-74-1	 $119\text{-}20^{21}$	4-03-00-00880 0.9077^{18}	EtOH 4 1.4337^{18}
8830	Pentanenitrile, 3-imino-2-methyl- Valeronitrile, 3-imino-2-methyl	$C_6H_{10}N_2$ 110.16	95642-49-2 46.3	 257.5	 0.9525^{19}	eth 4; EtOH 4
8831	Pentanenitrile, 4-methyl- Isoamyl cyanide	$C_6H_{11}N$ 97.16	542-54-1 -51	4999 156.5	4-02-00-00946 0.8030^{20}	H_2O 1; EtOH 3; eth 5; ctc 2 1.4059^{20}
8832	Pentane, 1-nitro- 1-Nitropentane	$C_5H_{11}NO_2$ 117.15	628-05-7	 172.5	4-01-00-00317 0.9525^{20}	EtOH 3; eth 3; bz 3 1.4175^{20}
8833	Pentane, 2-nitro- 2-Nitropentane	$C_5H_{11}NO_2$ 117.15	4609-89-6	 149; 88^{100}	4-01-00-00318 0.938^{25}	 1.4092^{25}
8834	Pentane, 3-nitro- 3-Nitropentane	$C_5H_{11}NO_2$ 117.15	551-88-2	6537 154	4-01-00-00318 0.957^{0}	ace 4; eth 4; EtOH 4
8835	Pentane, 1,1'-oxybis- Amyl ether	$C_{10}H_{22}O$ 158.28	693-65-2 -69	646 190	4-01-00-01643 0.7833^{20}	H_2O 1; EtOH 5; eth 5; chl 3 1.4119^{20}
8836	Pentane, 2,2,3,3,4-pentamethyl- 2,2,3,3,4-Pentamethylpentane	$C_{10}H_{22}$ 142.28	16747-44-7 -36.4	 166.1	4-01-00-00487 0.7767^{25}	 1.4361^{20}
8837	Pentane, 2,2,3,4,4-pentamethyl- 2,2,3,4,4-Pentamethylpentane	$C_{10}H_{22}$ 142.28	16747-45-8 -38.7	 159.3	4-01-00-00487 0.7636^{25}	 1.4307^{20}
8838	Pentane, 1,1,1,5-tetrachloro- 1,1,1,5-Tetrachloropentane	$C_5H_8Cl_4$ 209.93	2467-10-9	 112^{24}	4-01-00-00311 1.3416^{25}	 1.4859^{25}
8839	Pentane, 2,2,3,3-tetramethyl- 2,2,3,3-Tetramethylpentane	C_9H_{20} 128.26	7154-79-2 -9.8	 140.2	4-01-00-00463 0.7530^{25}	 1.4236^{20}
8840	Pentane, 2,2,3,4-tetramethyl- 2,2,3,4-Tetramethylpentane	C_9H_{20} 128.26	1186-53-4 -121.0	 133.0	4-01-00-00463 0.7389^{20}	 1.4147^{20}
8841	Pentane, 2,2,4,4-tetramethyl- Di-*tert*-butylmethane	C_9H_{20} 128.26	1070-87-7 -66.5	 122.2	4-01-00-00464 0.7195^{20}	H_2O 1; EtOH 4; bz 4; os 4 1.4069^{20}
8842	Pentane, 2,3,3,4-tetramethyl- 2,3,3,4-Tetramethylpentane	C_9H_{20} 128.26	16747-38-9 -102.1	 141.5	4-01-00-00464 0.7547^{20}	 1.4222^{20}
8843	Pentane, 1,1'-thiobis-	$C_{10}H_{22}S$ 174.35	872-10-6 -51.3	 $86^{3.7}$	4-01-00-01654 0.8407^{20}	H_2O 1; eth 3 1.4561^{20}
8844	1-Pentanethiol Pentyl mercaptan	$C_5H_{12}S$ 104.22	110-66-7 -75.7	648 126.6	4-01-00-01653 0.850^{20}	H_2O 1; EtOH 5; eth 5 1.4469^{20}
8845	2-Pentanethiol *sec*-Amylmercaptan	$C_5H_{12}S$ 104.22	2084-19-7 -169	 112.9	4-01-00-01662 0.8327^{20}	EtOH 3; lig 3 1.4412^{20}

No.	Name Synonym	Mol. Form. Mol. Wt.	CAS RN mp/°C	Merck No. bp/°C	Beil. Ref. den/g cm^{-3}	Solubility n_D
8846	3-Pentanethiol 3-Pentyl mercaptan	$C_5H_{12}S$ 104.22	616-31-9 -110.8	 105	4-01-00-01665 0.8410^{20}	EtOH 3; DMSO-d_6 2 1.4447^{20}
8847	Pentane, 2,2,3-trimethyl- 2-*tert*-Butylbutane	C_8H_{18} 114.23	564-02-3 -112.2	 110	4-01-00-00438 0.7161^{20}	H_2O 1; EtOH 5; eth 5; ace 5 1.4030^{20}
8848	Pentane, 2,2,4-trimethyl- Isooctane	C_8H_{18} 114.23	540-84-1 -107.3	5079 99.2	4-01-00-00439 0.6877^{25}	H_2O 1; EtOH 5; eth 3; ace 5 1.3915^{20}
8849	Pentane, 2,3,3-trimethyl- 2,3,3-Trimethylpentane	C_8H_{18} 114.23	560-21-4 -100.9	 114.8	4-01-00-00445 0.7262^{20}	H_2O 1; EtOH 4; eth 5; ace 5 1.4075^{20}
8850	Pentane, 2,3,4-trimethyl- 2,3,4-Trimethylpentane	C_8H_{18} 114.23	565-75-3 -109.2	 113.5	4-01-00-00446 0.7191^{20}	H_2O 1; EtOH 4; eth 5; ace 5 1.4042^{20}
8851	1,2,5-Pentanetriol 1,4,5-Pentanetriol	$C_5H_{12}O_3$ 120.15	14697-46-2	 $190\text{-}1^{13}$	4-01-00-02779 1.136^{20}	H_2O 4; EtOH 4 1.4730^{20}
8852	1,3,5-Pentanetriol	$C_5H_{12}O_3$ 120.15	4328-94-3	188^{11}; $139^{0.1}$	4-01-00-02779 1.1291^{20}	H_2O 4; EtOH 4; eth 3; ace 3 1.4785^{20}
8853	Pentane, 1,1,1,2,2,3,4,4,5,5,5-undecafluoro-3-(trifluoromethyl) Perfluoro-3-methylpentane	C_6F_{14} 338.04	865-71-4 -115	 58.4		bz 3
8854	Pentane, 1,1,1,2,2,3,3,4,5,5,5-undecafluoro-4-(trifluoromethyl)- Perfluoro(2-methylpentane)	C_6F_{14} 338.04	355-04-4	 57.6	4-01-00-00360 1.7326^{20}	H_2O 1; bz 3 1.2564^{22}
8855	Pentanoic acid Valeric acid	$C_5H_{10}O_2$ 102.13	109-52-4 -34	9815 186.1	4-02-00-00868 0.9391^{20}	H_2O 3; EtOH 3; eth 3; ctc 2 1.4085^{20}
8856	Pentanoic acid, 2-acetyl-, ethyl ester	$C_9H_{16}O_3$ 172.22	1540-28-9	 224; 90^{15}	4-03-00-01602 0.9661^{20}	eth 4; EtOH 4 1.4255^{20}
8857	Pentanoic acid, 5-amino-	$C_5H_{11}NO_2$ 117.15	660-88-8 157-8 dec	 dec	4-04-00-02636	H_2O 3; EtOH 2; eth 1; bz 1
8858	Pentanoic acid, anhydride	$C_{10}H_{18}O_3$ 186.25	2082-59-9 -56.1	 227	4-02-00-00874 0.924^{20}	eth 4; EtOH 4 1.4171^{26}
8859	Pentanoic acid, 2-bromo-, (±)-	$C_5H_9BrO_2$ 181.03	2681-92-7	 118^{12}	4-02-00-00883 1.381^{20}	eth 4; EtOH 4
8860	Pentanoic acid, 2-bromo-, ethyl ester	$C_7H_{13}BrO_2$ 209.08	615-83-8	 191	3-02-00-00681 1.226^{18}	H_2O 1; EtOH 3; eth 3 1.4496^{20}
8861	Pentanoic acid, 5-bromo-, ethyl ester	$C_7H_{13}BrO_2$ 209.08	14660-52-7	 $107\text{-}8^{20}$	4-02-00-00884 1.2804^{20}	ctc 2 1.4543^{20}
8862	Pentanoic acid, butyl ester	$C_9H_{18}O_2$ 158.24	591-68-4 -92.8	 185.8	3-02-00-00671 0.8710^{15}	H_2O 2; EtOH 3; eth 3 1.4128^{20}
8863	Pentanoic acid, 2-chloro-, (±)-	$C_5H_9ClO_2$ 136.58	94347-45-2 -15	 222	4-02-00-00878 1.141^{13}	eth 4; EtOH 4 1.4481^{11}
8864	Pentanoic acid, 3-chloro- Valeric acid, 3-chloro	$C_5H_9ClO_2$ 136.58	51637-47-9 33	 112^{10}	4-02-00-00879 1.1484^{20}	eth 4; EtOH 4 1.4462^{20}
8865	Pentanoic acid, 4-chloro- 4-Chlorovaleric acid	$C_5H_9ClO_2$ 136.58	32607-54-8	 116^{10}	4-02-00-00879 1.1514^{20}	eth 4; EtOH 4 1.4458^{20}
8866	Pentanoic acid, 5-chloro-	$C_5H_9ClO_2$ 136.58	1119-46-6 18	 230	4-02-00-00880 1.3416^{25}	eth 4; EtOH 4 1.4555^{20}
8867	Pentanoic acid, 4-chloro-, ethyl ester Ethyl 4-chlorovalerate	$C_7H_{13}ClO_2$ 164.63	70786-82-2	 196	4-02-00-00879 1.0393^{20}	eth 4; EtOH 4 1.4310^{20}
8868	Pentanoic acid, 5-chloro-, ethyl ester	$C_7H_{13}ClO_2$ 164.63	2323-81-1	 205.5	4-02-00-00880 1.0561^{20}	H_2O 1; EtOH 3; eth 3; chl 2 1.4355^{20}
8869	Pentanoic acid, 2-chloroethyl ester	$C_7H_{13}ClO_2$ 164.63	7735-33-3	 185	3-02-00-00670 1.040^{12}	eth 4 1.4307^{11}
8870	Pentanoic acid, 2,3-dibromo- Valeric acid, 2,3-dibromo-	$C_5H_8Br_2O_2$ 259.93	79912-57-5 57	 $104\text{-}6^{13}$	3-02-00-00683	bz 4; CS_2 4
8871	Pentanoic acid, 2,5-dibromo- Valeric acid, 2,5-dibromo-	$C_5H_8Br_2O_2$ 259.93	1450-81-3	 173^{15}; 151^{5}	3-02-00-00683 1.8629^{25}	 1.5347^{25}
8872	Pentanoic acid, 2,3-dibromo-, ethyl ester Valeric acid, 2,3-dibromo, ethyl ester	$C_7H_{12}Br_2O_2$ 287.98	79912-55-3	 117^{14}	2-02-00-00269 1.6199^{15}	 1.4953^{15}
8873	Pentanoic acid, 2,5-dibromo-, ethyl ester Valeric acid, 2,5-dibromo-, ethyl ester	$C_7H_{12}Br_2O_2$ 287.98	29823-16-3	 $133\text{-}5^{14}$	3-02-00-00683 1.6289^{25}	 1.4947^{25}
8874	Pentanoic acid, 2,2-dimethyl-4-oxo-	$C_7H_{12}O_3$ 144.17	470-49-5 76.8	 138^{15}	3-03-00-01247	H_2O 4; bz 4; eth 4; EtOH 4
8875	Pentanoic acid, 2,4-dioxo-	$C_5H_6O_4$ 130.10	5699-58-1 99.5	 130^{37}	3-03-00-01331	H_2O 3; EtOH 3; eth 3; ace 3
8876	Pentanoic acid, 2,4-dioxo-, ethyl ester	$C_7H_{10}O_4$ 158.15	615-79-2 18	 214	4-03-00-01777 1.1251^{20}	eth 4; EtOH 4 1.4757^{17}
8877	Pentanoic acid, 2-ethyl-, (±)- Valeric acid, 2-ethyl (*DL*)	$C_7H_{14}O_2$ 130.19	116908-83-9	 209.2	4-02-00-00975 0.9311^{33}	eth 4; EtOH 4
8878	Pentanoic acid, ethyl ester Ethyl valerate	$C_7H_{14}O_2$ 130.19	539-82-2 -91.2	 146.1	4-02-00-00872 0.8770^{20}	H_2O 1; EtOH 5; eth 5; ctc 2 1.4120^{20}
8879	Pentanoic acid, 5-fluoro- Valeric acid, 5-fluoro-	$C_5H_9FO_2$ 120.12	407-75-0	 83^{2}	4-02-00-00876	eth 4; EtOH 4 1.4080^{25}
8880	Pentanoic acid, 2-furanylmethyl ester Furfuryl valerate	$C_{10}H_{14}O_3$ 182.22	36701-01-6	 228; 82^{1}	2-17-00-00115 1.0284^{20}	eth 4; EtOH 4
8881	Pentanoic acid, heptyl ester	$C_{12}H_{24}O_2$ 200.32	5451-80-9 -46.4	 245.2	4-02-00-00872 0.8623^{20}	ace 4; eth 4; EtOH 4 1.4254^{15}
8882	Pentanoic acid, hexyl ester	$C_{11}H_{22}O_2$ 186.29	1117-59-5 -63.1	 226.3	3-02-00-00671 0.8635^{20}	ace 4; eth 4; EtOH 4 1.4228^{15}
8883	Pentanoic acid, 2-hydroxy-	$C_5H_{10}O_3$ 118.13	617-31-2 34	 sub	4-03-00-00807	H_2O 3; EtOH 3; eth 3
8884	Pentanoic acid, 4-hydroxy-, ethyl ester Valeric acid, 4-hydroxy-, ethyl ester	$C_7H_{14}O_3$ 146.19	6149-46-8	 $85\text{-}6^{2}$	3-03-00-00612 0.9504^{25}	eth 4; EtOH 4 1.4265^{25}

No.	Name Synonym	Mol. Form. Mol. Wt.	CAS RN mp/°C	Merck No. bp/°C	Beil. Ref. den/g cm^{-3}	Solubility n_D
8885	Pentanoic acid, 2-hydroxyethyl ester Valeric acid, 2-hydroxyethyl ester	$C_7H_{14}O_3$ 146.19	16179-36-5	 190	3-02-00-00672 0.9839^{20}	 1.4220^{20}
8886	Pentanoic acid, 2-methoxyphenyl ester Guaiacol valerate	$C_{12}H_{16}O_3$ 208.26	531-39-5	4461 265	 1.05^{25}	bz 4; eth 4; EtOH 4
8887	Pentanoic acid, 2-methyl-, (±)-	$C_6H_{12}O_2$ 116.16	22160-39-0	 195.6	4-02-00-00942 0.9230^{20}	H_2O 3; EtOH 3; eth 3; ctc 2 1.413^{20}
8888	Pentanoic acid, 3-methyl-, (±)- Valeric acid, 3-methyl (*DL*)	$C_6H_{12}O_2$ 116.16	22160-40-3 -41.6	 197.5	4-02-00-00948 0.9262^{20}	eth 4; EtOH 4 1.4159^{20}
8889	Pentanoic acid, 4-methyl-	$C_6H_{12}O_2$ 116.16	646-07-1 -33	 200.5	4-02-00-00944 0.9225^{20}	H_2O 2; EtOH 3; eth 3; chl 3 1.4144^{20}
8890	Pentanoic acid, 2-methylene- Valeric acid, 2-methylene	$C_6H_{10}O_2$ 114.14	5650-75-9 24.4	 214	0-02-00-00437 0.9783^{25}	bz 4; eth 4; chl 4
8891	Pentanoic acid, methyl ester Methyl valerate	$C_6H_{12}O_2$ 116.16	624-24-8	 127.4	4-02-00-00871 0.8947^{20}	H_2O 2; EtOH 5; eth 5; ace 3 1.4003^{20}
8892	Pentanoic acid, 4-methyl-, ethyl ester	$C_8H_{16}O_2$ 144.21	25415-67-2	 163; 52^{10}	4-02-00-00945 0.8705^{20}	 1.4050^{20}
8893	Pentanoic acid, 1-methylethyl ester	$C_8H_{16}O_2$ 144.21	18362-97-5		3-02-00-00671 0.8579^{20}	H_2O 1; EtOH 3; eth 3; ace 3 1.4061^{20}
8894	Pentanoic acid, 4-methyl-, 2-methylpropyl ester	$C_{10}H_{20}O_2$ 172.27	25415-70-7	 172; 76^{10}	4-02-00-00945 0.8536^{20}	 1.4135^{20}
8895	Pentanoic acid, 2-methyl-4-oxo-	$C_6H_{10}O_3$ 130.14	6641-83-4	 165^{40}	4-03-00-01587 1.11^{25}	 1.4431^{20}
8896	Pentanoic acid, 3-methyl-4-oxo-	$C_6H_{10}O_3$ 130.14	6628-79-1 31.5	 241.5	3-03-00-01238 1.0932^{17}	H_2O 4; EtOH 4 1.4443^{17}
8897	Pentanoic acid, 4-methyl-3-oxo-, ethyl ester	$C_8H_{14}O_3$ 158.20	7152-15-0 -9	 173	4-03-00-01589 0.98^{25}	 1.250^{20}
8898	Pentanoic acid, 3-methyl-4-oxo-, methyl ester Levulinic acid, 3-methyl-, methyl ester	$C_7H_{12}O_3$ 144.17	25234-83-7	 200; 82^{10}	1-03-00-00240 1.022^{20}	 1.4052^{73}
8899	Pentanoic acid, 1-methylpropyl ester, (+)- Valeric acid, *sec*-butyl ester (*d*)	$C_9H_{18}O_2$ 158.24	116836-32-9	 174.5	3-02-00-00671 0.8605^{20}	bz 4; eth 4; py 4; EtOH 4 1.4070^{20}
8900	Pentanoic acid, 2-methylpropyl ester	$C_9H_{18}O_2$ 158.24	10588-10-0	 179	3-02-00-00671 0.8625^{25}	H_2O 1; EtOH 5; eth 3; ace 3 1.4046^{20}
8901	Pentanoic acid, octyl ester Valeric acid, octyl ester	$C_{13}H_{26}O_2$ 214.35	5451-85-4 -42.3	 261.6	3-02-00-00672 0.8615^{20}	ace 4; eth 4; EtOH 4 1.4273^{15}
8902	Pentanoic acid, 2-oxo-	$C_5H_8O_3$ 116.12	1821-02-9 6.5	 179	4-03-00-01558 1.0970^{14}	H_2O 2; eth 3; bz 3; chl 3
8903	Pentanoic acid, 4-oxo- Levulinic acid	$C_5H_8O_3$ 116.12	123-76-2 33	5352 245 dec	4-03-00-01560 1.1335^{20}	H_2O 4; EtOH 4; eth 4; chl 3 1.4396^{20}
8904	Pentanoic acid, 4-oxo-, butyl ester	$C_9H_{16}O_3$ 172.22	2052-15-5	 237.5	4-03-00-01563 0.9735^{20}	chl 2 1.4290^{20}
8905	Pentanoic acid, 2-oxo-, ethyl ester Valeric acid, 2-oxo-, ethyl ester	$C_7H_{12}O_3$ 144.17	50461-74-0	 182.5; 70^{11}	4-03-00-01558 0.9985^{18}	 1.4170^{18}
8906	Pentanoic acid, 3-oxo-, ethyl ester	$C_7H_{12}O_3$ 144.17	4949-44-4	 191	4-03-00-01559 1.0120^{20}	bz 4; eth 4; EtOH 4 1.4230^{20}
8907	Pentanoic acid, 4-oxo-, ethyl ester Ethyl levulinate	$C_7H_{12}O_3$ 144.17	539-88-8	3775 205.8	4-03-00-01562 1.0111^{20}	H_2O 4; EtOH 4 1.4229^{20}
8908	Pentanoic acid, 4-oxo-, methyl ester	$C_6H_{10}O_3$ 130.14	624-45-3	 196	4-03-00-01562 1.0511^{20}	H_2O 2; EtOH 3; eth 5; ace 3 1.4233^{20}
8909	Pentanoic acid, 4-oxo-, 1-methylethyl ester Levulinic acid, isopropyl ester	$C_8H_{14}O_3$ 158.20	21884-26-4	 209.3	4-03-00-01563 0.9842^{20}	ace 4; bz 4; eth 4; EtOH 4 1.4420^{20}
8910	Pentanoic acid, 4-oxo-, 2-methylpropyl ester	$C_9H_{16}O_3$ 172.22	3757-32-2	 228	4-03-00-01563 0.9669^{20}	ace 4; eth 4; EtOH 4 1.4249^{20}
8911	Pentanoic acid, 4-oxo-, phenylmethyl ester Levulinic acid, benzyl ester	$C_{12}H_{14}O_3$ 206.24	6939-75-9	 $132\text{-}4^{2}$	4-06-00-02481 1.0935^{20}	tol 4 1.5090^{20}
8912	Pentanoic acid, 4-oxo-, propyl ester Levulinic acid, propyl ester	$C_8H_{14}O_3$ 158.20	645-67-0	 218	4-03-00-01563 0.9896^{20}	H_2O 2; EtOH 3; eth 3; ace 3 1.4258^{20}
8913	Pentanoic acid, pentyl ester	$C_{10}H_{20}O_2$ 172.27	2173-56-0 -78.8	 203.7	4-02-00-00872 0.8638^{20}	H_2O 2; EtOH 5; eth 5 1.4164^{20}
8914	Pentanoic acid, propyl ester	$C_8H_{16}O_2$ 144.21	141-06-0 -70.7	 167.5	4-02-00-00872 0.8699^{20}	H_2O 1; EtOH 3; eth 3; chl 3 1.4065^{20}
8915	1-Pentanol Amyl alcohol	$C_5H_{12}O$ 88.15	71-41-0 -78.9	7074 137.9	4-01-00-01640 0.8144^{20}	H_2O 1; EtOH 5; eth 5; ace 3 1.4101^{20}
8916	2-Pentanol *sec*-Amyl alcohol	$C_5H_{12}O$ 88.15	6032-29-7 -73	7075 119.3	3-01-00-01609 0.8094^{20}	H_2O 4; EtOH 3; eth 3; ctc 3 1.4053^{20}
8917	3-Pentanol Diethyl carbinol	$C_5H_{12}O$ 88.15	584-02-1 -69	7076 116.2	4-01-00-01662 0.8203^{20}	H_2O 2; EtOH 3; eth 3; ace 3 1.4104^{20}
8918	2-Pentanol, acetate, (*R*)- *sec*-Amyl acetate (*R*)	$C_7H_{14}O_2$ 130.19	54638-10-7	 142	0-02-00-00132 0.8803^{18}	eth 4; EtOH 4 1.4012^{20}
8919	3-Pentanol, acetate	$C_7H_{14}O_2$ 130.19	620-11-1	 133	4-02-00-00155 0.8712^{20}	eth 4; EtOH 4 1.4005^{20}
8920	1-Pentanol, 5-amino-	$C_5H_{13}NO$ 103.16	2508-29-4 38.5	 221.5	4-04-00-01750 0.9488^{17}	H_2O 5; EtOH 5; ace 5 1.4618^{17}
8921	3-Pentanol, 2-amino-	$C_5H_{13}NO$ 103.16	116836-16-9	 171	1-04-00-00442 0.9289^{23}	 1.4458
8922	1-Pentanol, 2-amino-4-methyl-, (±)-	$C_6H_{15}NO$ 117.19	16369-17-8	 199	2-04-00-00748 0.9173^{13}	H_2O 4; EtOH 4
8923	1-Pentanol, 5-chloro-	$C_5H_{11}ClO$ 122.59	5259-98-3	 112^{12}	4-01-00-01650	eth 4; EtOH 4 1.4518^{20}
8924	2-Pentanol, 1-chloro-	$C_5H_{11}ClO$ 122.59	17658-32-1	 160; 60^{14}	3-01-00-01613 1.035^{20}	eth 4; EtOH 4 1.4404^{25}

No.	Name Synonym	Mol. Form. Mol. Wt.	CAS RN mp/°C	Merck No. bp/°C	Beil. Ref. den/g cm^{-3}	Solubility n_D
8925	3-Pentanol, 1-chloro- 1-Chloro-3-hydroxypentane	$C_5H_{11}ClO$ 122.59	32541-33-6	 173	4-01-00-01665 1.0327^{25}	eth 4; EtOH 4 1.448^{25}
8926	1-Pentanol, 2-(diethylamino)-4-methyl-, 4-aminobenzoate (ester), monomethanesulfonate (salt) Panthesin	$C_{18}H_{32}N_2O_5S$ 388.53	135-44-4 158	5332	4-14-00-01143	H_2O 4; EtOH 4
8927	1-Pentanol, 2,4-dimethyl-, (±)- 1-Pentanol, 2,4-dimethyl (*DL*)	$C_7H_{16}O$ 116.20	111768-02-6	 159	4-01-00-01753 0.7932^{20}	eth 4; EtOH 4 1.427^{20}
8928	2-Pentanol, 2,4-dimethyl- Isobutyldimethylcarbinol	$C_7H_{16}O$ 116.20	625-06-9 <-20	 133.1	4-01-00-01753 0.8103^{20}	H_2O 1; EtOH 3; eth 3; ctc 3 1.4172^{20}
8929	3-Pentanol, 2,2-dimethyl- 2,2-Dimethyl-3-pentanol	$C_7H_{16}O$ 116.20	3970-62-5 -5	 135	3-01-00-01697 0.8253^{20}	H_2O 1; EtOH 3; eth 3 1.4223^{20}
8930	3-Pentanol, 2,3-dimethyl- 2,3-Dimethyl-3-pentanol	$C_7H_{16}O$ 116.20	595-41-5 <-30	 139.7	4-01-00-01752 0.833^{20}	H_2O 1; EtOH 3; eth 3; bz 2 1.4287^{20}
8931	3-Pentanol, 2,4-dimethyl-	$C_7H_{16}O$ 116.20	600-36-2 <-70	 138.7	4-01-00-01754 0.8288^{20}	H_2O 2; EtOH 3; eth 3 1.4250^{20}
8932	3-Pentanol, 3-ethyl-	$C_7H_{16}O$ 116.20	597-49-9 -12.5	 142	4-01-00-01750 0.8407^{22}	H_2O 1; EtOH 3; eth 3 1.4294^{20}
8933	2-Pentanol, 3-ethyl-2-methyl-	$C_8H_{18}O$ 130.23	19780-63-3	 156	4-01-00-01794 0.8382^{20}	bz 4; eth 4; EtOH 4; lig 4 1.4325^{20}
8934	3-Pentanol, 3-ethyl-2-methyl- Diethylisopropylcarbinol	$C_8H_{18}O$ 130.23	597-05-7	 160; 56^{48}	4-01-00-01794 0.8280^{20}	H_2O 1; EtOH 3; eth 3 1.4372^{10}
8935	1-Pentanol, 5-fluoro-	$C_5H_{11}FO$ 106.14	592-80-3	 $70\text{-}1^{11}$	4-01-00-01647	eth 4; EtOH 4 1.4057^{25}
8936	2-Pentanol, 1,1'-iminobis[2-methyl- Dipentylamine, 2,2'-dimethyl,2,2'-dihydroxy	$C_{12}H_{27}NO_2$ 217.35	85733-97-7	 $165\text{-}70^{15}$	 0.9264^{20}	ace 4; eth 4; EtOH 4 1.4584^{20}
8937	1-Pentanol, 2-methyl- 2-Methyl-1-pentanol	$C_6H_{14}O$ 102.18	105-30-6	 149	3-01-00-01665 0.8263^{20}	EtOH 3; eth 3; ace 3; ctc 3 1.4182^{20}
8938	1-Pentanol, 3-methyl-, (±)- 1-Pentanol, 3-methyl (*DL*)	$C_6H_{14}O$ 102.18	20281-83-8	 153	4-01-00-01722 0.8242^{20}	H_2O 1; EtOH 3; eth 3 1.4112^{23}
8939	1-Pentanol, 4-methyl- Isohexyl alcohol	$C_6H_{14}O$ 102.18	626-89-1	 151.9	4-01-00-01721 0.8131^{20}	H_2O 1; EtOH 3; eth 3 1.4134^{25}
8940	2-Pentanol, 2-methyl- 2-Methyl-2-pentanol	$C_6H_{14}O$ 102.18	590-36-3 -103	 121.1	4-01-00-01714 0.8350^{16}	H_2O 2; EtOH 3; eth 3 1.4100^{20}
8941	2-Pentanol, 3-methyl- 3-Methyl-2-pentanol	$C_6H_{14}O$ 102.18	565-60-6	 134.3	3-01-00-01672 0.8307^{20}	H_2O 2; EtOH 3; eth 3 1.4182^{20}
8942	2-Pentanol, 4-methyl- 4-Methyl-2-pentanol	$C_6H_{14}O$ 102.18	108-11-2 -90	 131.6	4-01-00-01717 0.8075^{20}	H_2O 2; EtOH 3; eth 3; ctc 2 1.4100^{20}
8943	3-Pentanol, 2-methyl- 2-Methyl-3-pentanol	$C_6H_{14}O$ 102.18	565-67-3	 126.5	4-01-00-01716 0.8243^{20}	H_2O 2; EtOH 5; eth 5 1.4175^{20}
8944	3-Pentanol, 3-methyl-	$C_6H_{14}O$ 102.18	77-74-7 -23.6	 122.4	4-01-00-01723 0.8286^{20}	H_2O 2; EtOH 5; eth 5; ctc 2 1.4186^{20}
8945	1-Pentanol, 2-methyl-, acetate Acetic acid, (2-methylpentyl) ester	$C_8H_{16}O_2$ 144.21	7789-99-3	 163	4-02-00-00160 0.8691^{25}	eth 4; EtOH 4
8946	2-Pentanol, 2-methyl-, acetate 1,1-Dimethylbutyl acetate	$C_8H_{16}O_2$ 144.21	34859-98-8	 153; 34^{10}	3-02-00-00257 0.8798^{18}	eth 4; EtOH 4 1.4068^{10}
8947	2-Pentanol, 4-methyl-, acetate 1,3-Dimethylbutyl acetate	$C_8H_{16}O_2$ 144.21	108-84-9	 147.5	4-02-00-00161 0.8805^{25}	eth 4; EtOH 4 1.3980^{20}
8948	3-Pentanol, 2-methyl-, acetate Acetic acid, (2-methyl-3-pentyl) ester	$C_8H_{16}O_2$ 144.21	35897-16-6	 149	4-02-00-00160 0.8688^{20}	eth 4; EtOH 4
8949	3-Pentanol, 3-methyl-, acetate Acetic acid, (3-methyl-3-pentyl) ester	$C_8H_{16}O_2$ 144.21	10250-47-2	 148	4-02-00-00161 0.8818^{20}	eth 4; EtOH 4 1.4169^{18}
8950	1-Pentanol, 2-nitro- 2-Nitro-1-pentanol	$C_5H_{11}NO_3$ 133.15	2899-90-3	133^{28}; 117^{10}	3-01-00-01606 1.0818^{25}	 1.4405^{25}
8951	2-Pentanol, 1-nitro-	$C_5H_{11}NO_3$ 133.15	2224-37-5	 $87\text{-}8^{1}$	4-01-00-01661 1.0968^{20}	 1.4493^{20}
8952	2-Pentanol, 3-nitro- 3-Nitro-2-pentanol	$C_5H_{11}NO_3$ 133.15	5447-99-4	 $60^{0.5}$	3-01-00-01614 1.075^{25}	 1.4430^{20}
8953	1-Pentanol, 2,2,3,3,4,4,5,5-octafluoro- 2,2,3,3,4,4,5,5-Octafluoro-1-pentanol	$C_5H_4F_8O$ 232.07	355-80-6	6667 140.5	4-01-00-01648 1.6647^{20}	 1.3178^{20}
8954	2-Pentanol, 2,4,4-trimethyl- 2,4,4-Trimethyl-2-pentanol	$C_8H_{18}O$ 130.23	690-37-9 -20	 147.5	4-01-00-01796 0.8225^{20}	H_2O 1; EtOH 2; eth 3 1.4284^{20}
8955	3-Pentanol, 2,2,4-trimethyl- Isopropyl-*tert*-butylcarbinol	$C_8H_{18}O$ 130.23	5162-48-1 -13	 150.5	4-01-00-01795 0.8297^{20}	 1.4288^{20}
8956	3-Pentanol, 2,3,4-trimethyl-	$C_8H_{18}O$ 130.23	3054-92-0	 156.5	4-01-00-01798 0.8492^{20}	 1.4353^{20}
8957	2-Pentanone Methyl propyl ketone	$C_5H_{10}O$ 86.13	107-87-9 -76.9	6032 102.2	4-01-00-03271 0.809^{20}	H_2O 2; EtOH 5; eth 5; ctc 2 1.3895^{20}
8958	3-Pentanone Diethyl ketone	$C_5H_{10}O$ 86.13	96-22-0 -39	3111 101.96	4-01-00-03279 0.8098^{25}	H_2O 4; EtOH 5; eth 5; ctc 3 1.3905^{25}
8959	2-Pentanone, 4-amino-4-methyl- Diacetonamine	$C_6H_{13}NO$ 115.18	625-04-7	2942 $25^{0.14}$	3-04-00-00894	H_2O 3; EtOH 5; eth 5
8960	1-Pentanone, 1-(4-aminophenyl)-	$C_{11}H_{15}NO$ 177.25	38237-74-0 74.5	 $160\text{-}3^{3}$	4-14-00-00178	H_2O 1; EtOH 3; eth 3
8961	2-Pentanone, 5-chloro-	C_5H_9ClO 120.58	5891-21-4	 76^{34}	4-01-00-03277 1.0523^{20}	eth 3; ace 3; ctc 2 1.4375^{20}
8962	3-Pentanone, 1-chloro-	C_5H_9ClO 120.58	32830-97-0	 68^{20}	4-01-00-03284	eth 4; EtOH 4 1.4361^{20}
8963	2-Pentanone, 3,5-dichloro- 3,5-Dichloro-2-pentanone	$C_5H_8Cl_2O$ 155.02	58371-98-5	 99^{15}; 58^{3}	4-01-00-03277 1.239^{15}	 1.4632^{20}

PHYSICAL CONSTANTS OF ORGANIC COMPOUNDS (continued)

No.	Name Synonym	Mol. Form. Mol. Wt.	CAS RN mp/°C	Merck No. bp/°C	Beil. Ref. den/g cm^{-3}	Solubility n_D
8964	2-Pentanone, 5-(diethylamino)-	$C_9H_{19}NO$ 157.26	105-14-6	 $83\text{-}5^{15}$	4-04-00-01935 0.861^{25}	 1.4350^{20}
8965	2-Pentanone, 4,4-dimethyl-	$C_7H_{14}O$ 114.19	590-50-1 -64	 126	4-01-00-03331 0.809^{25}	 1.4036^{20}
8966	3-Pentanone, 2,2-dimethyl- 2,2-Dimethyl-3-pentanone	$C_7H_{14}O$ 114.19	564-04-5 -45	 125.6	4-01-00-03331 0.8125^{20}	H_2O 2; EtOH 3; eth 3; ace 3 1.4065^{20}
8967	3-Pentanone, 2,4-dimethyl- Diisopropyl ketone	$C_7H_{14}O$ 114.19	565-80-0 -69	 125.4	4-01-00-03334 0.8108^{20}	H_2O 1; EtOH 5; eth 5; bz 3 1.3999^{20}
8968	3-Pentanone, 2,4-dimethyl-, oxime Diisopropyl ketone oxime	$C_7H_{15}NO$ 129.20	1113-74-2 34	 181; 91.5^{21}	4-01-00-03335 0.9022^{17}	
8969	3-Pentanone, 1,5-diphenyl- Diphenethyl ketone	$C_{17}H_{18}O$ 238.33	5396-91-8 13.5	 352	4-07-00-01462 1.0356^{0}	
8970	2-Pentanone, 3-ethyl-4-methyl-	$C_8H_{16}O$ 128.21	71172-57-1	 154.5	3-01-00-02883 0.812^{20}	H_2O 3; EtOH 5; eth 5; bz 5 1.4105^{20}
8971	2-Pentanone, 1-hydroxy- 1-Pentanol-2-one	$C_5H_{10}O_2$ 102.13	64502-89-2	 152	4-01-00-04004 0.9860^{20}	H_2O 4; eth 4; EtOH 4 1.4234^{12}
8972	2-Pentanone, 3-hydroxy-	$C_5H_{10}O_2$ 102.13	3142-66-3	 147.5	3-01-00-03220 0.9500^{20}	H_2O 4; ace 4; bz 4; eth 4 1.4350^{10}
8973	2-Pentanone, 4-hydroxy-	$C_5H_{10}O_2$ 102.13	4161-60-8	 177	4-01-00-04005 1.0071^{20}	H_2O 3; EtOH 3; eth 3 1.4415^{18}
8974	2-Pentanone, 5-hydroxy-	$C_5H_{10}O_2$ 102.13	1071-73-4	 209; 117^{33}	4-01-00-04006 1.0071^{20}	H_2O 5; EtOH 3; eth 3 1.4390^{20}
8975	3-Pentanone, 2-hydroxy- 2-Pentanol-3-one	$C_5H_{10}O_2$ 102.13	5704-20-1	 152.5	4-01-00-04008 0.9742^{20}	eth 4; EtOH 4 1.4128^{20}
8976	2-Pentanone, 4-hydroxy-4-methyl- Diacetone alcohol	$C_6H_{12}O_2$ 116.16	123-42-2 -44	2944 167.9	4-01-00-04023 0.9387^{20}	H_2O 5; EtOH 5; eth 5; chl 3 1.4213^{20}
8977	2-Pentanone, 4-imino-	C_5H_9NO 99.13	870-74-6 43	 209	4-01-00-03678 0.9427^{0}	H_2O 4; eth 4
8978	2-Pentanone, 4-methoxy-4-methyl- Pentoxone	$C_7H_{14}O_2$ 130.19	107-70-0	 160	4-01-00-04024 0.8980^{25}	 1.418^{20}
8979	1-Pentanone, 1-(3-methoxyphenyl)-	$C_{12}H_{16}O_2$ 192.26	20359-55-1	 196^{30}	4-08-00-00533 1.012^{20}	 1.5242^{31}
8980	2-Pentanone, 3-methyl-, (±)- 2-Pentanone, 3-methyl (*DL*)	$C_6H_{12}O$ 100.16	55156-16-6	 118	4-01-00-03309 0.8130^{20}	H_2O 2; EtOH 5; eth 5; chl 3 1.4002^{20}
8981	2-Pentanone, 4-methyl- Isobutyl methyl ketone	$C_6H_{12}O$ 100.16	108-10-1 -84	5095 116.5	4-01-00-03305 0.7965^{25}	H_2O 2; EtOH 5; eth 5; ace 5 1.3962^{20}
8982	3-Pentanone, 2-methyl- Ethyl isopropyl ketone	$C_6H_{12}O$ 100.16	565-69-5	 113.5	4-01-00-03304 0.814^{18}	H_2O 2; EtOH 4; eth 5; ace 5 1.3975^{20}
8983	1-Pentanone, 4-methyl-1-phenyl-	$C_{12}H_{16}O$ 176.26	2050-07-9 -2	 255.5	3-07-00-01158 0.9623^{15}	ace 4; bz 4; eth 4; EtOH 4 1.533^{20}
8984	2-Pentanone, 3-methyl-1-phenyl- Benzyl *sec*-butyl ketone	$C_{12}H_{16}O$ 176.26	27993-42-6	 $127\text{-}8^{17}$	4-07-00-00758 0.964^{20}	 1.5019^{20}
8985	2-Pentanone, oxime Methyl propyl ketone oxime	$C_5H_{11}NO$ 101.15	623-40-5	 168	4-01-00-03274 0.9095^{20}	H_2O 4; eth 4; EtOH 4 1.4450^{20}
8986	1-Pentanone, 1-phenyl-	$C_{11}H_{14}O$ 162.23	1009-14-9 -9.4	 245	4-07-00-00733 0.986^{20}	H_2O 1; EtOH 4; eth 4; ctc 2 1.5158^{20}
8987	2-Pentanone, 3-(phenylmethylene)- Methyl α-ethylstyryl ketone	$C_{12}H_{14}O$ 174.24	3437-89-6	 $136\text{-}8^{12}$	3-07-00-01461 1.0005^{22}	 1.5650^{22}
8988	3-Pentanone, 2,2,4,4-tetramethyl-	$C_9H_{18}O$ 142.24	815-24-7 -25.2	 152	4-01-00-03365 0.8240^{18}	H_2O 1; EtOH 3; eth 3; ace 3 1.4194^{20}
8989	3-Pentanone, 2,2,4-trimethyl- *tert*-Butyl isopropyl ketone	$C_8H_{16}O$ 128.21	5857-36-3	 135.1	4-01-00-03351 0.8065^{20}	H_2O 1; eth 3; ace 3 1.4060
8990	Pentanoyl chloride Valeroyl chloride	C_5H_9ClO 120.58	638-29-9 -110	 109	4-02-00-00874 1.0155^{15}	 1.4200^{20}
8991	Pentanoyl chloride, 2-chloro-	$C_5H_8Cl_2O$ 155.02	61589-68-2	 156	4-02-00-00879 1.1765^{20}	eth 4 1.4465^{20}
8992	Pentanoyl chloride, 5-chloro-	$C_5H_8Cl_2O$ 155.02	1575-61-7	 83^{12}	4-02-00-00881 1.210^{18}	eth 4 1.4639^{20}
8993	Pentanoyl chloride, 2-methyl-, (±)- Valeryl chloride, 2-methyl (*DL*)	$C_6H_{11}ClO$ 134.61	116908-84-0	 141	3-02-00-00741 0.9781^{20}	eth 4 1.4330^{27}
8994	Pentanoyl chloride, 3-methyl-, (±)- Valeryl chloride, 3-methyl (*DL*)	$C_6H_{11}ClO$ 134.61	116908-85-1	 141	4-02-00-00949 0.9781^{20}	eth 4
8995	2,5,8,11,14-Pentaoxapentadecane Tetraethylene glycol dimethyl ether	$C_{10}H_{22}O_5$ 222.28	143-24-8	9141 275.3	4-01-00-02404 1.0114^{20}	H_2O 5; EtOH 3; eth 3; ctc 3
8996	Pentaphene 2,3:6,7-Dibenzphenanthrene	$C_{22}H_{14}$ 278.35	222-93-5 257	2992	4-05-00-02721	H_2O 1; EtOH 2; eth 2; bz 3
8997	Pentasiloxane, dodecamethyl- Dodecamethylpentasiloxane	$C_{12}H_{36}O_4Si_5$ 384.84	141-63-9 -80	3401 232; 105^{12}	4-04-00-04124 0.8755^{20}	ctc 3; CS_2 3 1.3925^{20}
8998	1,2,3,5,6-Pentathiepane Lenthionine	$C_2H_4S_5$ 188.38	292-46-6 60.5	5321	5-19-12-00251	
8999	Pentatriacontane	$C_{35}H_{72}$ 492.96	630-07-9 75	 490	4-01-00-00598 0.8157^{20}	H_2O 1; eth 2; ace 3 1.4568^{20}
9000	18-Pentatriacontanone	$C_{35}H_{70}O$ 506.94	504-53-0 89.0	 $270^{0.1}$	4-01-00-03413 0.793^{95}	H_2O 1; EtOH 2; eth 2; ace 2
9001	4-Pentenal Pent-4-en-1-al	C_5H_8O 84.12	2100-17-6	 99	4-01-00-03459 0.852^{20}	H_2O 1; eth 3; ace 3 1.4191^{20}
9002	2-Pentenal, 2-methyl-	$C_6H_{10}O$ 98.14	623-36-9	 136.5	4-01-00-03471 0.8581^{20}	H_2O 1; EtOH 3; eth 3; bz 3 1.4488^{20}
9003	4-Pentenamide, 2,2-diethyl- Novonal	$C_9H_{17}NO$ 155.24	512-48-1 75.5	6643	3-02-00-01351	eth 4; EtOH 4

No.	Name Synonym	Mol. Form. Mol. Wt.	CAS RN mp/°C	Merck No. bp/°C	Beil. Ref. den/g cm^{-3}	Solubility n_D
9004	1-Pentene α-Amylene	C_5H_{10} 70.13	109-67-1 -165.2	7079 29.9	4-01-00-00808 0.6405^{20}	H_2O 1; EtOH 5; eth 5; bz 3 1.3715^{20}
9005	2-Pentene, (*E*)- *trans*-2-Pentene	C_5H_{10} 70.13	646-04-8 -140.2	7080 36.3	4-01-00-00814 0.6431^{25}	H_2O 1; EtOH 5; eth 5; bz 3 1.3793^{20}
9006	2-Pentene, (*Z*)- *cis*-2-Pentene	C_5H_{10} 70.13	627-20-3 -151.4	7080 36.9	4-01-00-00814 0.6556^{20}	H_2O 1; EtOH 5; eth 5; bz 3 1.3830^{20}
9007	1-Pentene, 1-bromo- 1-Bromo-1-pentene	C_5H_9Br 149.03	60468-23-7	 119	3-01-00-00775 1.2606^{20}	bz 4; eth 4; chl 4 1.4572^{20}
9008	1-Pentene, 2-bromo- 2-Bromo-1-pentene	C_5H_9Br 149.03	31844-95-8	 107.5	3-01-00-00775 1.228^{20}	bz 4; eth 4; lig 4 1.4535^{20}
9009	1-Pentene, 3-bromo- 1-Ethylallyl bromide	C_5H_9Br 149.03	53045-71-9	 30.5^{30}	2-01-00-00183 1.2417^{25}	ace 4; bz 4; chl 4 1.4626^{25}
9010	1-Pentene, 5-bromo-	C_5H_9Br 149.03	1119-51-3	 125.5	3-01-00-00775 1.2581^{20}	 1.4640^{20}
9011	2-Pentene, 1-bromo-	C_5H_9Br 149.03	20599-27-3	 123.5	3-01-00-00783 1.2545^{20}	ace 4; bz 4; chl 4 1.4731^{20}
9012	2-Pentene, 2-bromo- 2-Bromo-2-pentene	C_5H_9Br 149.03	80204-19-9	 111	3-01-00-00784 1.275^{20}	ace 4; bz 4; chl 4 1.4580^{20}
9013	2-Pentene, 3-bromo- 3-Bromo-2-pentene	C_5H_9Br 149.03	21964-23-8	 116	3-01-00-00784 1.271^{20}	ace 4; bz 4; chl 4 1.4628^{20}
9014	2-Pentene, 4-bromo-	C_5H_9Br 149.03	1809-26-3	117 dec; 22^{9}	4-01-00-00816 1.2312^{21}	ace 4; bz 4; chl 4 1.4752^{21}
9015	2-Pentene, 5-bromo- 3-Pentenyl bromide	C_5H_9Br 149.03	51952-42-2	 121.7^{621}	4-01-00-00816 1.2715^{20}	ace 4; bz 4; chl 4 1.4695^{20}
9016	1-Pentene, 2-chloro- 2-Chloro-1-pentene	C_5H_9Cl 104.58	42131-85-1	 96	3-01-00-00774 0.872^{5}	eth 4; EtOH 4
9017	1-Pentene, 3-chloro- 1-Ethylallyl chloride	C_5H_9Cl 104.58	24356-00-1	 93	3-01-00-00774 0.8978^{20}	ace 4; eth 4; EtOH 4 1.4254^{20}
9018	1-Pentene, 4-chloro- 4-Chloro-1-pentene	C_5H_9Cl 104.58	10524-08-0	 97	2-01-00-00183 0.934^{15}	H_2O 1; eth 4; chl 4 1.417^{15}
9019	1-Pentene, 5-chloro-	C_5H_9Cl 104.58	928-50-7	 105	4-01-00-00810 0.9125^{20}	ace 4; eth 4 1.4297^{20}
9020	2-Pentene, 1-chloro- γ-Ethylallyl chloride	C_5H_9Cl 104.58	10071-60-0	 108	3-01-00-00782 0.908^{22}	ace 4; eth 4; EtOH 4; chl 4 1.4352^{22}
9021	2-Pentene, 2-chloro- 2-Chloro-2-pentene	C_5H_9Cl 104.58	67747-70-0	 96	3-01-00-00782 0.9067^{20}	ace 4; eth 4 1.4261^{20}
9022	2-Pentene, 3-chloro- 3-Chloro-2-pentene	C_5H_9Cl 104.58	34238-52-3	 91	3-01-00-00782 0.9125^{20}	ace 4; eth 4 1.423^{24}
9023	2-Pentene, 4-chloro-	C_5H_9Cl 104.58	1458-99-7	 103; 47^{25}	4-01-00-00816 0.8988^{20}	ace 4; eth 4; chl 4 1.4322^{20}
9024	2-Pentene, 5-chloro-	C_5H_9Cl 104.58	16435-50-0	 108	4-01-00-00816 0.9043^{20}	ace 4; bz 4; chl 4 1.4310^{20}
9025	1-Pentene, 3-chloro-2-methyl-	$C_6H_{11}Cl$ 118.61	4104-01-2	 122.5	3-01-00-00809	bz 4; eth 4; chl 4 1.4422^{20}
9026	2-Pentene, 5-chloro-2-methyl- 5-Chloro-2-methyl-2-pentene	$C_6H_{11}Cl$ 118.61	7712-60-9	 133	4-01-00-00843 0.9135^{20}	ace 4; chl 4
9027	1-Pentene, 1,1,2,3,3,4,4,5,5,5-decafluoro-	C_5F_{10} 250.04	376-87-4	 30	4-01-00-00810	chl 4 1.2571^{25}
9028	2-Pentene, 2,5-dichloro- 1,4-Dichloro-3-pentene	$C_5H_8Cl_2$ 139.02	20177-02-0	 $40\text{-}1^{8}$	4-01-00-00816 1.1182^{15}	chl 4
9029	1-Pentene, 2,3-dimethyl- 2,3-Dimethyl-1-pentene	C_7H_{14} 98.19	3404-72-6 -134.3	 84.3	3-01-00-00832 0.7051^{20}	H_2O 1; EtOH 5; eth 5; dil sulf 4 1.4033^{20}
9030	1-Pentene, 2,4-dimethyl- 2,4-Dimethyl-1-pentene	C_7H_{14} 98.19	2213-32-3 -124.1	 81.6	4-01-00-00871 0.6943^{20}	H_2O 1; EtOH 5; eth 5; bz 3 1.3986^{20}
9031	1-Pentene, 3,3-dimethyl- 3,3-Dimethyl-1-pentene	C_7H_{14} 98.19	3404-73-7 -134.3	 77.5	4-01-00-00873 0.6974^{20}	H_2O 1; EtOH 5; eth 5; bz 3 1.3984^{20}
9032	1-Pentene, 3,4-dimethyl- 3,4-Dimethyl-1-pentene	C_7H_{14} 98.19	7385-78-6	 80.8	4-01-00-00871 0.6934^{25}	 1.3992^{20}
9033	1-Pentene, 4,4-dimethyl-	C_7H_{14} 98.19	762-62-9 -136.6	 72.5	4-01-00-00869 0.6827^{20}	H_2O 1; EtOH 5; eth 5; bz 3 1.3818^{20}
9034	2-Pentene, 2,3-dimethyl- Ethyltrimethylethylene	C_7H_{14} 98.19	10574-37-5 -118.3	 97.5	4-01-00-00870 0.7277^{20}	H_2O 1; EtOH 3; eth 3; bz 3 1.4208^{20}
9035	2-Pentene, 2,4-dimethyl- 2,4-Dimethyl-2-pentene	C_7H_{14} 98.19	625-65-0 -127.7	 83.4	4-01-00-00872 0.6954^{20}	H_2O 1; EtOH 3; eth 3; bz 3 1.4040^{20}
9036	2-Pentene, 3,4-dimethyl- 3,4-Dimethyl-2-pentene	C_7H_{14} 98.19	24910-63-2 -124.2	 89	3-01-00-00833 0.7126^{20}	H_2O 1; EtOH 3; eth 3; bz 3 1.4070^{20}
9037	2-Pentene, 3,4-dimethyl-, (*E*)-	C_7H_{14} 98.19	4914-92-5 -124.2	 91.5	4-01-00-00871 0.7124^{25}	 1.4128^{20}
9038	2-Pentene, 3,4-dimethyl-, (*Z*)-	C_7H_{14} 98.19	4914-91-4 -113.4	 89.3	4-01-00-00871 0.7092^{25}	 1.4104^{20}
9039	2-Pentene, 4,4-dimethyl-, (*E*)-	C_7H_{14} 98.19	690-08-4 -115.2	 76.7	4-01-00-00868 0.6889^{20}	H_2O 1; EtOH 3; eth 3; bz 3 1.3982^{20}
9040	2-Pentene, 4,4-dimethyl-, (*Z*)-	C_7H_{14} 98.19	762-63-0 -135.4	 80.4	4-01-00-00868 0.6951^{25}	 1.4026^{20}
9041	2-Pentenedioic acid, diethyl ester Diethyl glutaconate	$C_9H_{14}O_4$ 186.21	2049-67-4	 237	 1.0496^{20}	eth 4; EtOH 4 1.4411^{20}
9042	2-Pentenedioic acid, diethyl ester, (*E*)- *trans*-2-Pentenedioic acid diethyl ester	$C_9H_{14}O_4$ 186.21	73178-43-5	 237; 126^{14}	4-02-00-02227 1.0496^{20}	eth 4; EtOH 4 1.4411^{20}

No.	Name	Synonym	Mol. Form.	Mol. Wt.	CAS RN	mp/°C	Merck No.	bp/°C	Beil. Ref.	den/g cm^{-3}	Solubility	n_D
9043	1-Pentene, 2-ethyl-	2-Ethyl-1-pentene	C_7H_{14}	98.19	3404-71-5			94	4-01-00-00867	0.7079^{20}	bz 4; eth 4; EtOH 4	1.405^{20}
9044	1-Pentene, 3-ethyl-	3-Ethyl-1-pentene	C_7H_{14}	98.19	4038-04-4	-127.5		84.1	4-01-00-00867	0.6917^{25}		1.3982^{20}
9045	2-Pentene, 3-ethyl-	3-Ethyl-2-pentene	C_7H_{14}	98.19	816-79-5			96	4-01-00-00868	0.7204^{20}	H_2O 1; EtOH 3; eth 3; bz 3	1.4148^{20}
9046	1-Pentene, 3-ethyl-2-methyl-		C_8H_{16}	112.22	19780-66-6	-112.9		109.5	4-01-00-00891	0.7262^{20}		1.4140^{20}
9047	1-Pentene, 3-ethyl-3-methyl-	3-Ethyl-3-methyl-1-pentene	C_8H_{16}	112.22	6196-60-7			112	4-01-00-00891	0.7264^{20}		1.418^{20}
9048	1-Pentene, 3-ethyl-4-methyl-	1-Pentene, 4-methyl-3-ethyl	C_8H_{16}	112.22	61847-80-1			107.5	4-01-00-00890	0.7158^{25}		1.4097^{20}
9049	2-Pentene, 3-ethyl-2-methyl-	3-Ethyl-2-methyl-2-pentene	C_8H_{16}	112.22	19780-67-7			115	4-01-00-00890	0.735^{25}		1.4247^{20}
9050	2-Pentene, 3-ethyl-4-methyl-, (*E*)-	2-Pentene, 4-methyl-3-ethyl (*trans*)	C_8H_{16}	112.22	42067-49-2			114.3	4-01-00-00890	0.7308^{25}		1.4210^{20}
9051	2-Pentene, 3-ethyl-4-methyl-, (*Z*)-	2-Pentene, 4-methyl-3-ethyl (*cis*)	C_8H_{16}	112.22	42067-48-1			116	4-01-00-00890	0.735^{25}		1.424^{20}
9052	1-Pentene, 2-methyl-	2-Methyl-1-pentene	C_6H_{12}	84.16	763-29-1	-135.7		62.1	4-01-00-00841	0.6799^{20}	H_2O 1; EtOH 3; bz 3; ctc 2	1.3920^{20}
9053	1-Pentene, 3-methyl-	*sec*-Butylethene	C_6H_{12}	84.16	760-20-3	-153		54.2	3-01-00-00812	0.6675^{20}	H_2O 1; EtOH 3; bz 3; chl 3	1.3841^{20}
9054	1-Pentene, 4-methyl-	Isobutylethene	C_6H_{12}	84.16	691-37-2	-153.6		53.9	4-01-00-00846	0.6642^{20}	H_2O 1; EtOH 3; bz 3; chl 3	1.3828^{20}
9055	2-Pentene, 2-methyl-	2-Methyl-2-pentene	C_6H_{12}	84.16	625-27-4	-135		67.3	4-01-00-00842	0.6863^{20}	H_2O 1; EtOH 3; bz 3; ctc 3	1.4004^{20}
9056	2-Pentene, 3-methyl-, (*E*)-		C_6H_{12}	84.16	616-12-6	-138.5		70.4	4-01-00-00848	0.6930^{25}	H_2O 1; EtOH 3; bz 3; ctc 3	1.4045^{20}
9057	2-Pentene, 3-methyl-, (*Z*)-	(*cis*)-3-Methyl-2-pentene	C_6H_{12}	84.16	922-62-3	-134.8		67.7	4-01-00-00848	0.6886^{25}	H_2O 1; EtOH 3; bz 3; chl 3	1.4016^{20}
9058	2-Pentene, 4-methyl-, (*E*)-	4-Methyl-*trans*-2-pentene	C_6H_{12}	84.16	674-76-0	-140.8		58.6	4-01-00-00844	0.6686^{20}	H_2O 1; EtOH 3; bz 3; ctc 2	1.3889^{20}
9059	2-Pentene, 4-methyl-, (*Z*)-	*cis*-4-Methyl-2-pentene	C_6H_{12}	84.16	691-38-3	-134.8		56.3	4-01-00-00844	0.6690^{20}	H_2O 1; EtOH 3; bz 3; chl 3	1.3800^{20}
9060	1-Pentene, 3-methylene-		C_6H_{10}	82.15	3404-63-5			75	4-01-00-01022	0.712^{25}		1.445^{20}
9061	3-Pentenenitrile, (*E*)	*trans*-3-Pentenenitrile	C_5H_7N	81.12	16545-78-1			142.6				
9062	4-Pentenenitrile		C_5H_7N	81.12	592-51-8			140	4-02-00-01543	0.8239^{24}	H_2O 1; EtOH 5; eth 5	1.4213^{14}
9063	3-Pentenenitrile, 2-hydroxy-, (*Z*)-		C_5H_7NO	97.12	116908-78-2			139^{70}	4-03-00-01001	0.9675^{15}	bz 4; eth 4; EtOH 4; chl 4	1.4460^{21}
9064	Pentene, 2,4,4-trimethyl-	Diisobutylene	C_8H_{16}	112.22	25167-70-8			101	4-01-00-00892			
9065	1-Pentene, 2,3,3-trimethyl-	2,3,3-Trimethyl-1-pentene	C_8H_{16}	112.22	560-23-6	-69		108.3	4-01-00-00894	0.7308^{25}		1.4174^{20}
9066	1-Pentene, 2,3,4-trimethyl-	2,3,4-Trimethyl-1-pentene	C_8H_{16}	112.22	565-76-4			108	4-01-00-00894	0.725^{25}		1.415^{20}
9067	1-Pentene, 2,4,4-trimethyl-	2,2,4-Trimethyl-4-pentene	C_8H_{16}	112.22	107-39-1	-93.5		101.4	4-01-00-00892	0.7150^{20}	H_2O 1; eth 3; bz 3; ctc 3	1.4086^{20}
9068	1-Pentene, 3,3,4-trimethyl-		C_8H_{16}	112.22	560-22-5			105	3-01-00-00851	0.725^{25}		1.4144^{20}
9069	1-Pentene, 3,4,4-trimethyl-	3,4,4-Trimethyl-1-pentene	C_8H_{16}	112.22	564-03-4			104	4-01-00-00891	0.715^{25}		1.412^{20}
9070	2-Pentene, 2,3,4-trimethyl-	2,3,4-Trimethyl-2-pentene	C_8H_{16}	112.22	565-77-5	-113.4		116.5	4-01-00-00894	0.7434^{20}		1.4274^{20}
9071	2-Pentene, 2,4,4-trimethyl-	2,4,4-Trimethyl-2-pentene	C_8H_{16}	112.22	107-40-4	-106.3		104.9	4-01-00-00891	0.7218^{20}	H_2O 1; eth 3; bz 3; ctc 3	1.4160^{20}
9072	2-Pentene, 3,4,4-trimethyl-	3,4,4-Trimethyl-2-pentene	C_8H_{16}	112.22	598-96-9			112	4-01-00-00891	0.735^{25}	EtOH 4	1.423^{20}
9073	4-Pentenoic acid	Allylacetic acid	$C_5H_8O_2$	100.12	591-80-0	-22.5		188.5	4-02-00-01542	0.9809^{20}	H_2O 2; EtOH 4; eth 4	1.4281^{20}
9074	4-Pentenoic acid, 2-acetyl-, ethyl ester		$C_9H_{14}O_3$	170.21	610-89-9			208	3-03-00-01314	0.9898^{20}	EtOH 5; eth 5; bz 5	1.4388^{18}
9075	4-Pentenoic acid, 2,2-dimethyl-	2,2-Dimethylpent-4-enoic acid	$C_7H_{12}O_2$	128.17	16386-93-9			116^{30}	4-02-00-01585	0.933^{20}		1.4338^{20}
9076	2-Pentenoic acid, 2-methyl-, (*E*)-	2-Pentenoic acid, 2-methyl (*trans*)	$C_6H_{10}O_2$	114.14	16957-70-3	24.4		214; 112^{12}	4-02-00-01568	0.9751^{20}	H_2O 2; eth 3; chl 3; CS_2 3	1.4513^{20}
9077	2-Pentenoic acid, 3-methyl-, (*Z*)-	2-Pentenoic acid, 3-methyl (*cis*)	$C_6H_{10}O_2$	114.14	19866-51-4	12		96^{5}	4-02-00-01571	0.9830^{20}		1.4650
9078	2-Pentenoic acid, 4-methyl-	4,4-Dimethyl-2-butenoic acid	$C_6H_{10}O_2$	114.14	10321-71-8	35		217	3-02-00-01324	0.9529^{21}	ace 4; eth 4; EtOH 4	1.4489^{21}
9079	3-Pentenoic acid, 2-methyl-		$C_6H_{10}O_2$	114.14	37674-63-8			199	3-02-00-01324	0.966^{15}		1.4402^{25}
9080	2-Pentenoic acid, 4-methyl-, ethyl ester	Ethyl 4,4-dimethylcrotonate	$C_8H_{14}O_2$	142.20	2351-97-5			174; 76-7^{25}	3-02-00-01324	0.9048^{21}		1.4328^{20}
9081	3-Pentenoic acid, 3-methyl-, ethyl ester	Ethyl 3-methyl-3-pentenoate	$C_8H_{14}O_2$	142.20	86623-80-5			61^{13}	3-02-00-01328	0.9183^{20}		1.4298^{20}
9082	3-Pentenoic acid, 4-methyl-, ethyl ester	Ethyl 4-methyl-3-pentenoate	$C_8H_{14}O_2$	142.20	6849-18-9			58^{11}	3-02-00-01324	0.9134^{17}		1.4329^{17}

No.	Name Synonym	Mol. Form. Mol. Wt.	CAS RN mp/°C	Merck No. bp/°C	Beil. Ref. den/g cm^{-3}	Solubility n_D
9083	2-Pentenoic acid, 2-methyl-, methyl ester	$C_7H_{12}O_2$ 128.17	10478-12-3	51^{11}	4-02-00-01568 0.920^{25}	1.4336
9084	2-Pentenoic acid, 3-methyl-, methyl ester, (*E*)-	$C_7H_{12}O_2$ 128.17	17447-01-7	79^{50}	4-02-00-01571 0.9302^{20}	1.4446^{20}
9085	2-Pentenoic acid, 3-methyl-, methyl ester, (*Z*)- *cis*-3-Methyl-2-pentenoic acid methyl ester	$C_7H_{12}O_2$ 128.17	17447-00-6	74^{50}	4-02-00-01571 0.9279^{20}	1.4420^{20}
9086	3-Pentenoic acid, 3-methyl-, methyl ester Methyl 3-methyl-3-pentenoate	$C_7H_{12}O_2$ 128.17	2258-58-4	74^{50}	4-02-00-01573 0.9949^{20}	1.4306^{20}
9087	1-Penten-3-ol	$C_5H_{10}O$ 86.13	616-25-1	115	1-01-00-00227 0.839^{20}	H_2O 2; EtOH 5; eth 5 1.4239^{20}
9088	4-Penten-1-ol	$C_5H_{10}O$ 86.13	821-09-0	141	4-01-00-02119 0.8457^{20}	H_2O 2; eth 3; ctc 2 1.4309^{20}
9089	4-Penten-2-ol	$C_5H_{10}O$ 86.13	625-31-0	116	1-01-00-00227 0.8367^{20}	H_2O 4; EtOH 5; eth 5 1.4225^{20}
9090	2-Penten-1-ol, (*E*)-	$C_5H_{10}O$ 86.13	1576-96-1	138	4-01-00-02121 0.8471^{20}	EtOH 3; eth 3; ace 3 1.4341^{20}
9091	2-Penten-1-ol, (*Z*)-	$C_5H_{10}O$ 86.13	1576-95-0	138	4-01-00-02121 0.8529^{20}	EtOH 3; eth 3; ace 3 1.4354^{20}
9092	3-Penten-2-ol, (±)-	$C_5H_{10}O$ 86.13	42569-16-4	121.6; 65^{70}	4-01-00-02122 0.8328^{25}	ace 4; eth 4; EtOH 4 1.4280^{20}
9093	2-Penten-1-ol, 5-(2,3-dimethyltricyclo[2.2.1.02,6]hept-3-yl)-2-methyl-, stereoisomer α-Santalol	$C_{15}H_{24}O$ 220.35	115-71-9	8321 301.5	4-06-00-03521 0.9679^{20}	H_2O 1; EtOH 3 1.5023^{20}
9094	2-Penten-1-ol, 2-methyl-	$C_6H_{12}O$ 100.16	1610-29-3	167.5	4-01-00-02144 0.8501^{24}	1.444^{24}
9095	2-Penten-1-ol, 4-methyl-	$C_6H_{12}O$ 100.16	5362-55-0	159	4-01-00-02146 0.8489^{16}	1.4403^{16}
9096	3-Penten-1-ol, 4-methyl- 4-Methyl-3-pentenol	$C_6H_{12}O$ 100.16	763-89-3	157	4-01-00-02145 0.8600^{20}	1.4432^{21}
9097	3-Penten-2-ol, 2-methyl- 2-Methyl-3-penten-2-ol	$C_6H_{12}O$ 100.16	63468-05-3	112; 65^{72}	4-01-00-02145 0.8347^{20}	1.4302^{20}
9098	3-Penten-2-ol, 3-methyl- 3-Methyl-3-penten-2-ol	$C_6H_{12}O$ 100.16	2747-53-7	140	4-01-00-02149 0.8792^{0}	1.4428^{17}
9099	3-Penten-2-ol, 4-methyl-	$C_6H_{12}O$ 100.16	4325-82-0	134	4-01-00-02144 0.840^{15}	1.9377^{15}
9100	4-Penten-2-ol, 2-methyl-	$C_6H_{12}O$ 100.16	624-97-5 -73	119.5	4-01-00-02146 0.8300^{20}	H_2O 2; EtOH 3; eth 3; ctc 2 1.4268^{20}
9101	4-Penten-2-ol, 3-methyl-	$C_6H_{12}O$ 100.16	1569-59-1	125.5	3-01-00-01933 0.8429^{22}	1.4326^{22}
9102	4-Penten-2-ol, 4-methyl-	$C_6H_{12}O$ 100.16	2004-67-3	129	4-01-00-02143 0.8436^{20}	1.4297^{17}
9103	2-Penten-1-ol, 2-methyl-5-(2-methyl-3-methylenebicyclo[2.2.1]hept-2-yl)-, [1*S*-[1α,2α(*Z*),4α] β-Santalol	$C_{15}H_{24}O$ 220.35	77-42-9	8322 167-8^{10}	4-06-00-03519 0.9750^{20}	1.5115^{20}
9104	1-Penten-3-one Ethyl vinyl ketone	C_5H_8O 84.12	1629-58-9	103; 44^{90}	4-01-00-03457 0.8468^{20}	H_2O 1; EtOH 3; eth 3; ace 3 1.4195^{20}
9105	3-Penten-2-one, (*E*)-	C_5H_8O 84.12	3102-33-8	122	4-01-00-03460 0.8624^{20}	H_2O 3; eth 3; ace 3; ctc 3 1.4350^{20}
9106	1-Penten-3-one, 4,4-dimethyl-1-phenyl-	$C_{13}H_{16}O$ 188.27	538-44-3 43	154^{25}	3-07-00-01490 0.9508^{46}	1.5523^{25}
9107	1-Penten-3-one, 2-methyl-	$C_6H_{10}O$ 98.14	25044-01-3 -69.5	118.5	4-01-00-03470 1.8530^{20}	ace 4; EtOH 4 1.4289^{20}
9108	3-Penten-2-one, 4-methyl- Mesityl oxide	$C_6H_{10}O$ 98.14	141-79-7 -59	5811 130	4-01-00-03471 0.8653^{20}	H_2O 3; EtOH 5; eth 5; ace 3 1.4440^{20}
9109	4-Penten-2-one, 4-methyl-	$C_6H_{10}O$ 98.14	3744-02-3 -72.6	124.2	4-01-00-03470 0.8411^{20}	
9110	1-Penten-3-one, 1-phenyl-	$C_{11}H_{12}O$ 160.22	3152-68-9 38.5	142^{12}	2-07-00-00298 0.8697^{20}	H_2O 2; EtOH 4; eth 4; bz 4 1.5684^{20}
9111	1-Penten-3-yne Methylvinylacetylene	C_5H_6 66.10	646-05-9	59.5	4-01-00-01087 0.7401^{20}	bz 4; eth 4 1.4496^{20}
9112	1-Penten-4-yne	C_5H_6 66.10	871-28-3	42.5	4-01-00-01088 0.738^{16}	H_2O 1; eth 3; bz 3 1.4125^{16}
9113	3-Penten-1-yne, (*E*)- *trans*-3-Penten-1-yne	C_5H_6 66.10	2004-69-5	52.2	4-01-00-01085	
9114	3-Penten-1-yne, (*Z*)- *cis*-3-Penten-1-yne	C_5H_6 66.10	1574-40-9	44.6		
9115	3-Penten-1-yne, 3-ethyl-	C_7H_{10} 94.16	14272-54-9	96.5	4-01-00-01099 0.7886^{25}	bz 4; eth 4 1.4338^{25}
9116	1-Penten-3-yne, 2-methyl- Methylisopropenylacetylene	C_6H_8 80.13	926-55-6	81-2^{100}	4-01-00-01095	bz 4; eth 4 1.4002^{20}
9117	3-Penten-1-yne, 3-methyl- 3-Methyl-3-penten-1-yne	C_6H_8 80.13	1574-33-0	66.5	4-01-00-01096 0.739^{20}	eth 3; bz 3 1.4332^{20}
9118	4-Penten-2-yn-1-ol 1-Hydroxypent-2-yn-4-ene	C_5H_6O 82.10	2919-05-3	57^{10}; 44$^{2.5}$	4-01-00-02303 0.9581^{20}	1.4986^{20}
9119	Pentitol, 1,2-dideoxy- 1,2,3-Pentanetriol	$C_5H_{12}O_3$ 120.15	5371-48-2	192^{63}	4-01-00-02778 1.0849^{34}	H_2O 4; EtOH 4; eth 4
9120	Pentonic acid, 5-C-[3,5-bis(1-methylpropyl)-1-cyclopenten-1-yl]-4-deoxy-	$C_{18}H_{32}O_5$ 328.45	491-14-5 196		3-10-00-02045 1.292^{14}	EtOH 4

No.	Name	Synonym	Mol. Form.	Mol. Wt.	CAS RN	mp/°C	Merck No.	bp/°C	Beil. Ref.	den/g cm^{-3}	Solubility	n_D
9121	1-Pentyn-3-amine, 3-ethyl-		$C_7H_{13}N$	111.19	3234-64-8			$71-2^{90}$	4-04-00-01154	0.828^{25}		1.4409^{20}
9122	1-Pentyne	Propylacetylene	C_5H_8	68.12	627-19-0	-90		40.1	4-01-00-00990	0.6901^{20}	H_2O 1; EtOH 4; eth 5; bz 3	1.3852^{20}
9123	2-Pentyne		C_5H_8	68.12	627-21-4	-109.3		56.1	4-01-00-00992	0.7058^{25}	H_2O 1; EtOH 4; eth 5; bz 3	1.4039^{20}
9124	1-Pentyne, 1-bromo-	1-Pentynyl bromide	C_5H_7Br	147.01	14752-60-4			$44-6^{52}$	3-01-00-00958	1.281^{13}		1.4579^{13}
9125	1-Pentyne, 3-chloro-3-ethyl-	3-Chloro-3-ethyl-1-pentyne	$C_7H_{11}Cl$	130.62	6080-79-1			$73-6^{100}$	4-01-00-01032	0.9230^{19}	bz 4; eth 4; chl 4	1.4437^{19}
9126	1-Pentyne, 3-chloro-3-methyl-	3-Chloro-3-methyl-1-pentyne	C_6H_9Cl	116.59	14179-94-3			102.3	4-01-00-01021	0.9163^{20}	H_2O 1; eth 3; bz 3; chl 3	1.4330^{20}
9127	2-Pentyne, 4-chloro-4-methyl-		C_6H_9Cl	116.59	999-79-1			55^{70}	4-01-00-01018		ace 4; bz 4; chl 4	1.4143^{20}
9128	1-Pentyne, 3,3-dimethyl-	3,3-Dimethyl-1-pentyne	C_7H_{12}	96.17	918-82-1			70	4-01-00-01034	0.7032^{25}		1.3934^{20}
9129	1-Pentyne, 4,4-dimethyl-	4,4-Dimethyl-1-pentyne	C_7H_{12}	96.17	13361-63-2	-75.7		76.1	4-01-00-01033	0.7142^{20}	bz 4; eth 4; chl 4	1.3983^{20}
9130	2-Pentyne, 4,4-dimethyl-		C_7H_{12}	96.17	999-78-0	-82.4		83	4-01-00-01032	0.7176^{20}	H_2O 1; eth 3; bz 3; ctc 2	1.4071^{20}
9131	1-Pentyne, 3-ethyl-	3-Ethyl-1-pentyne	C_7H_{12}	96.17	21020-26-8			84	3-01-00-01002	0.7246^{25}	bz 4; eth 4; chl 4	1.4043^{25}
9132	1-Pentyne, 3-ethyl-3-methyl-	3-Ethyl-3-methyl-1-pentyne	C_8H_{14}	110.20	919-12-0			101.5	4-01-00-01046	0.7422^{20}	bz 4; eth 4; chl 4	1.4110^{20}
9133	1-Pentyne, 1-iodo-	1-Iodo-1-pentyne	C_5H_7I	194.02	14752-61-5			54^{23}	4-01-00-00992	1.6127^{19}		1.5148^{19}
9134	1-Pentyne, 3-methyl-	3-Methyl-1-pentyne	C_6H_{10}	82.15	922-59-8			57.7	4-01-00-01021	0.6992^{25}		1.3916^{20}
9135	1-Pentyne, 4-methyl-		C_6H_{10}	82.15	7154-75-8	-104.6		61.2	4-01-00-01019	0.7000^{25}	H_2O 1; bz 3; chl 3	1.3936^{20}
9136	2-Pentyne, 4-methyl-		C_6H_{10}	82.15	21020-27-9	-110.3		73.1	4-01-00-01018	0.7112^{25}	bz 4; chl 4	1.4057^{20}
9137	2-Pentynoic acid	Ethylpropiolic acid	$C_5H_6O_2$	98.10	5963-77-9	50		122^{10}; 81^{3}	4-02-00-01693	0.978^{20}	H_2O 4	1.4619^{20}
9138	4-Pentynoic acid	Propargylacetic acid	$C_5H_6O_2$	98.10	6089-09-4	57.7		110^{30}; 102^{17}	4-02-00-01693		eth 4; EtOH 4	
9139	2-Pentynoic acid, methyl ester	Methyl 2-pentynoate	$C_6H_8O_2$	112.13	24342-04-9			72^{10}	4-02-00-01694	0.937^{20}		1.4455^{20}
9140	1-Pentyn-3-ol	Ethylethynylcarbinol	C_5H_8O	84.12	4187-86-4			125	4-01-00-02224	0.8926^{15}		1.4347^{15}
9141	2-Pentyn-1-ol		C_5H_8O	84.12	6261-22-9			$61-2^{15}$	4-01-00-02226	0.913^{17}		1.4518^{17}
9142	3-Pentyn-1-ol		C_5H_8O	84.12	10229-10-4			154	4-01-00-02227	0.9002^{20}		1.4454^{20}
9143	4-Pentyn-1-ol		C_5H_8O	84.12	5390-04-5			154	4-01-00-02225	0.913^{20}		1.4414^{20}
9144	1-Pentyn-3-ol, 3,4-dimethyl-		$C_7H_{12}O$	112.17	1482-15-1			134	4-01-00-02255	0.8691^{20}	H_2O 4; eth 4; EtOH 4	1.4372^{20}
9145	1-Pentyn-3-ol, 3-ethyl-		$C_7H_{12}O$	112.17	6285-06-9			138	4-01-00-02254	0.8691^{20}	H_2O 2	
9146	1-Pentyn-3-ol, 3-methyl-	Meparfynol	$C_6H_{10}O$	98.14	77-75-8	30.5	5731	120.5	2-01-00-00506	0.8688^{20}		1.4310^{20}
9147	2-Pentynoyl chloride, 4,4-dimethyl-	4,4-Dimethyl-2-pentynoyl chloride	C_7H_9ClO	144.60	52324-03-5			52^{20}	2-02-00-00454	0.9743^{20}		1.4443^{20}
9148	Perchloric acid, trichloromethyl ester	Trichloro perchlorate	CCl_4O_4	217.82	67632-66-0	-55		exp 40	3-03-00-00037		eth 4	
9149	Perfluidone	*o*-Toluidine, 4-(phenylsulfonyl)-*N*-[(trifluoromethyl)sulfono]-	$C_{14}H_{12}F_3NO_4S_2$	379.38	37924-13-3	143	7113					
9150	Permethrin	Cyclopropanecarboxylic acid, 3-(2,2-dichloroethenyl)-2,2-dimethyl-, (3-phenoxyphenyl), methyl ester	$C_{21}H_{20}Cl_2O_3$	391.29	52645-53-1	34	7132	$200^{0.01}$		1.23^{20}		
9151	Peroxide, acetyl benzoyl	Acetozone	$C_9H_8O_4$	180.16	644-31-5	37	71	130^{19}	4-09-00-00717		eth 4	
9152	Peroxide, bis(2,4-dichlorobenzoyl)		$C_{14}H_6Cl_4O_4$	380.01	133-14-2	106			4-09-00-01001			
9153	Peroxide, bis(1,1-dimethylethyl)	DTBP	$C_8H_{18}O_2$	146.23	110-05-4	-40	3446	111	4-01-00-01619	0.704^{20}	H_2O 1; ace 5; ctc 3; os 3	1.3890^{20}
9154	Peroxide, bis(1,1-dimethylpropyl)		$C_{10}H_{22}O_2$	174.28	10508-09-5			58.5^{14}; 44^{10}	4-01-00-01671	0.821^{20}		1.4095^{20}
9155	Peroxide, bis(1-oxododecyl)	Lauroyl peroxide	$C_{24}H_{46}O_4$	398.63	105-74-8	49			4-02-00-01102		H_2O 1; chl 3	
9156	Peroxide, diacetyl	Acetyl peroxide	$C_4H_6O_4$	118.09	110-22-5	30		63^{21}	4-02-00-00392		eth 4; EtOH 4	
9157	Peroxide, dibenzoyl	Benzoyl peroxide	$C_{14}H_{10}O_4$	242.23	94-36-0	105	1128	exp	4-09-00-00717		H_2O 2; EtOH 3; eth 3; ace 3	1.543
9158	Peroxide, diethyl		$C_4H_{10}O_2$	90.12	628-37-5	-70		65	4-01-00-01323	0.8240^{19}	H_2O 2; EtOH 5; eth 5	1.3715^{17}

No.	Name Synonym	Mol. Form. Mol. Wt.	CAS RN mp/°C	Merck No. bp/°C	Beil. Ref. den/g cm^{-3}	Solubility n_D
9159	Peroxide, dimethyl	$C_2H_6O_2$ 62.07	690-02-8	 14	4-01-00-01249 0.8677^{0}	EtOH 2; eth 2; tol 3; HOAc 3 1.3503^{0}
9160	Perthane Ethane, 1,1-dichloro-2,2-bis(*p*-ethylphenyl)-	$C_{18}H_{20}Cl_2$ 307.26	72-56-0 56	3050		
9161	Perylene Dibenz[de,kl]anthracene	$C_{20}H_{12}$ 252.32	198-55-0 274	7137	4-05-00-02689 1.35^{25}	H_2O 1; EtOH 2; eth 2; ace 4
9162	9-Phenanthrenamine	$C_{14}H_{11}N$ 193.25	947-73-9 138.3	 sub	4-12-00-03443	eth 2; bz 2; chl 2
9163	Phenanthrene	$C_{14}H_{10}$ 178.23	85-01-8 99.2	7167 340	4-05-00-02297 0.9800^{4}	H_2O 1; EtOH 3; eth 3; ace 3 1.5943
9164	Phenanthrene, 9-bromo- 9-Phenanthryl bromide	$C_{14}H_9Br$ 257.13	573-17-1 64.5	 >360	4-05-00-02303 1.4093^{10}	H_2O 1; EtOH 3; eth 3; chl 2
9165	9-Phenanthrenecarboxylic acid 9-Phenanthroic acid	$C_{15}H_{10}O_2$ 222.24	837-45-6 257.3	 sub	4-09-00-02674	H_2O 1; EtOH 3; eth 3; bz 3
9166	1-Phenanthrenecarboxylic acid, 1,2,3,4,4a,5,6,9,10,10a-decahydro-1,4a-dimethyl-7-(1-methylethyl)-, Palustric acid	$C_{20}H_{30}O_2$ 302.46	1945-53-5 164.5	6950	4-09-00-02174	
9167	1-Phenanthrenecarboxylic acid, 1,2,3,4,4a,4b,5,6,10,10a-decahydro-1,4a-dimethyl-7-(1-methylethyl)-, Abietic acid	$C_{20}H_{30}O_2$ 302.46	514-10-3 173.5	2 250^{9}	4-09-00-02175	ace 4; bz 4; eth 4; EtOH 4
9168	1-Phenanthrenecarboxylic acid, 1,2,3,4,4a,4b,5,6,10,10a-decahydro-1,4a-dimethyl-7-(1-methylethyl)-, methyl ester, [1R-(1α,4aβ,4bα,10aα)]- Methyl abietate	$C_{21}H_{32}O_2$ 316.48	127-25-3	5930 $225-6^{16}$	4-09-00-02176 1.049^{20}	H_2O 1; EtOH 3; HOAc 3 1.5344
9169	1-Phenanthrenecarboxylic acid, 1,2,3,4,4a,4b,5,9,10,10a-decahydro-1,4a-dimethyl-7-(1-methylethyl)-, Levopimaric acid	$C_{20}H_{30}O_2$ 302.46	79-54-9 150	5348	4-09-00-02173	
9170	1-Phenanthrenecarboxylic acid, 7-ethenyl-1,2,3,4,4a,4b,5,6,7,9,10,10a-dodecahydro-1,4a,7-trimethyl-, [1R-(1α,4aβ,4bα,7β,10aα)]- Pimaric acid	$C_{20}H_{30}O_2$ 302.46	127-27-5 218.5	7398 282^{18}	4-09-00-02179	eth 4; py 4; EtOH 4
9171	Phenanthrene, 9,10-dihydro- 9,10-Dihydrophenanthrene	$C_{14}H_{12}$ 180.25	776-35-2 34.5	 $168-9^{15}$	4-05-00-02183 1.0757^{40}	chl 3 1.6415^{20}
9172	Phenanthrene, 9,10-dimethyl- 9,10-Phenanthraquinodimethane	$C_{16}H_{14}$ 206.29	604-83-1 144	 sub	4-05-00-02335	EtOH 2; bz 3; chl 3; HOAc 3
9173	3,4-Phenanthrenediol	$C_{14}H_{10}O_2$ 210.23	478-71-7 143	 sub 130	3-06-00-05689	eth 4; EtOH 4
9174	9,10-Phenanthrenedione Phenanthrenequinone	$C_{14}H_8O_2$ 208.22	84-11-7 209	7168	4-07-00-02565 1.405^{22}	H_2O 1; EtOH 2; eth 3; bz 2
9175	9,10-Phenanthrenedione, 2-hydroxy-	$C_{14}H_8O_3$ 224.22	4088-81-7 284	 sub	3-08-00-02929	H_2O 4
9176	9,10-Phenanthrenedione, 3-hydroxy- 9,10-Phenanthraquinone, 3-hydroxy	$C_{14}H_8O_3$ 224.22	57404-54-3 330 dec	 sub	2-08-00-00398	H_2O 4; EtOH 4
9177	9,10-Phenanthrenedione, 1-methyl-7-(1-methylethyl)-	$C_{18}H_{16}O_2$ 264.32	5398-75-4 197.5	 sub	4-07-00-02595	eth 4; EtOH 4
9178	Phenanthrene, dodecahydro- Dodecahydrophenanthrene	$C_{14}H_{22}$ 190.33	1322-67-4	 $81-2^{1.5}$	3-05-00-01074 0.9674^{20}	ace 4; bz 4 1.5102^{20}
9179	Phenanthrene, 9-ethyl- 9-Ethylphenanthrene	$C_{16}H_{14}$ 206.29	3674-75-7 62.5	 $199-200^{1}$	4-05-00-02332 1.0603^{78}	H_2O 1; EtOH 3; bz 3; peth 2 1.6582^{78}
9180	Phenanthrene, 3-methyl- 3-Methylphenanthrene	$C_{15}H_{12}$ 192.26	832-71-3 65	 350; 145^{6}	4-05-00-02314	H_2O 1; EtOH 3; ace 3; chl 2
9181	Phenanthrene, 4-methyl- 4-Methylphenanthrene	$C_{15}H_{12}$ 192.26	832-64-4 53.5	 $175-80^{10}$	3-05-00-02154	H_2O 1; EtOH 3; ctc 3
9182	Phenanthrene, 9-(1-methylethenyl)- 9-Isopropenylphenanthrene	$C_{17}H_{14}$ 218.30	58873-44-2 38	 163^{20}	3-05-00-02240	 1.6765^{22}
9183	Phenanthrene, 1-methyl-7-(1-methylethyl)- Retene	$C_{18}H_{18}$ 234.34	483-65-8 101	8162 390	4-05-00-02368 1.035^{25}	H_2O 1; EtOH 3; eth 3; bz 3
9184	Phenanthrene, 1,2,3,4,4a,9,10,10a-octahydro- 1,2,3,4,4a,9,10,10-Octahydrophenanthrene	$C_{14}H_{18}$ 186.30	16306-39-1	 159^{15}	4-05-00-01585 0.9973^{32}	 1.5527^{19}
9185	Phenanthrene, 1,2,3,4,5,6,7,8-octahydro- 1,2,3,4,5,6,7,8-Octahydrophenanthrene	$C_{14}H_{18}$ 186.30	5325-97-3 16.7	 295	4-05-00-01585 1.026^{20}	H_2O 1; ace 3; bz 3; CS_2 3 1.5569^{17}
9186	Phenanthrene, tetradecahydro-	$C_{14}H_{24}$ 192.34	5743-97-5	 $86-9^{2}$	4-05-00-00492 0.9447^{20}	H_2O 1; eth 3; ace 3; bz 3 1.5011^{20}
9187	Phenanthrene, 1,2,3,4-tetrahydro- 1,2,3,4-Tetrahydrophenanthrene	$C_{14}H_{14}$ 182.27	1013-08-7 33.5	 173^{11}	4-05-00-01909 1.0601^{40}	H_2O 1; EtOH 3; eth 3; ace 3
9188	Phenanthridine	$C_{13}H_9N$ 179.22	229-87-8 107.4	 349	5-20-08-00223	H_2O 2; EtOH 4; eth 4; ace 3
9189	1,10-Phenanthroline *o*-Phenanthroline	$C_{12}H_8N_2$ 180.21	66-71-7 117	7169 >300	5-23-08-00419	H_2O 4; EtOH 3; ace 3; bz 3
9190	1,7-Phenanthroline	$C_{12}H_8N_2$ 180.21	230-46-6 78	 360	5-23-08-00417	H_2O 3; EtOH 4; eth 1; bz 1
9191	4,7-Phenanthroline	$C_{12}H_8N_2$ 180.21	230-07-9 177	 sub 100	5-23-08-00430	H_2O 3; EtOH 4; eth 2; bz 2
9192	1,10-Phenanthroline, 2,9-dimethyl- Neocuproine	$C_{14}H_{12}N_2$ 208.26	484-11-7 159.5	6364	5-23-08-00527	
9193	Phenazine Dibenzopyrazine	$C_{12}H_8N_2$ 180.21	92-82-0 176.5	7173	5-23-08-00389	H_2O 2; EtOH 3; eth 2; bz 3

No.	Name	Synonym	Mol. Form.	Mol. Wt.	CAS RN	mp/°C	Merck No.	bp/°C	Beil. Ref.	den/g cm^{-3}	Solubility	n_D
9194	2,3-Phenazinediamine	2,3-Diaminophenazine	$C_{12}H_{10}N_4$	210.24	655-86-7	264	2960	sub	4-25-00-03028		bz 4; EtOH 4	
9195	Phenazine, 2-methyl-	Tolazine	$C_{13}H_{10}N_2$	194.24	1016-94-0	118.5		350	5-23-08-00475		eth 4; EtOH 4; chl 4	
9196	Phenazinium, 1-hydroxy-5-methyl-, hydroxide, inner salt	Pyocyanine	$C_{13}H_{10}N_2O$	210.24	85-66-5	133 dec	7965		5-24-04-00100		H_2O 2; EtOH 3; eth 1; ace 3	
9197	1-Phenazinol	Hemipyocyanine	$C_{12}H_8N_2O$	196.21	528-71-2	158	4564	sub	5-23-12-00297		H_2O 2; EtOH 2; bz 3; os 4	
9198	Phenmedipham	Carbamic acid, (3-methylphenyl)-, 3-[(methoxycarbonyl)amino]phenyl ester	$C_{16}H_{16}N_2O_4$	300.31	13684-63-4	143	7199					
9199	Phenol	Hydroxybenzene	C_6H_6O	94.11	108-95-2	40.9	7206	181.8	4-06-00-00531	1.0545^{45}	H_2O 3; EtOH 3; eth 4; ace 5	1.5408^{41}
9200	Phenol, amino-	Hydroxyphenylamine	C_6H_7NO	109.13	27598-85-2	83.5			2-15-00-00004		H_2O 3; EtOH 4; eth 4; bz 4	
9201	Phenol, 2-amino-	*o*-Aminophenol	C_6H_7NO	109.13	95-55-6	174	473	sub 153	4-13-00-00805	1.328^{25}	H_2O 3; EtOH 4; eth 3; bz 2	
9202	Phenol, 3-amino-	*m*-Aminophenol	C_6H_7NO	109.13	591-27-5	123	472	164^{11}	4-13-00-00952		H_2O 3; EtOH 4; eth 4; bz 2	
9203	Phenol, 4-amino-	*p*-Aminophenol	C_6H_7NO	109.13	123-30-8	187.5	474	$110^{0.3}$	4-13-00-01014		H_2O 2; EtOH 4; bz 1; chl 1	
9204	Phenol, 2-amino-4-chloro-	2-Hydroxy-5-chloroaniline	C_6H_6ClNO	143.57	95-85-2	140			4-13-00-00878		DMSO 2	
9205	Phenol, 2-amino-4,6-dichloro-	2-Amino-4,6-dichlorophenol	$C_6H_5Cl_2NO$	178.02	527-62-8	95.5	448	sub 70	4-13-00-00889			
9206	Phenol, 4-amino-2,6-dichloro-		$C_6H_5Cl_2NO$	178.02	5930-28-9	168		sub	2-13-00-00274		H_2O 1; EtOH 4; eth 4; ace 3	
9207	Phenol, 2-amino-4,6-dinitro-	Picramic acid	$C_6H_5N_3O_5$	199.12	96-91-3	169	7379		4-13-00-00909		bz 4; EtOH 4	
9208	Phenol, 4-(2-aminoethyl)-	Tyramine	$C_8H_{11}NO$	137.18	51-67-2	164.5	9743	$205\text{-}7^{25}$	4-13-00-01788		H_2O 2; EtOH 3; bz 2; xzl 3	
9209	Phenol, 2-amino-3-methyl-		C_7H_9NO	123.15	2835-97-4	150		sub	2-13-00-00324		eth 4	
9210	Phenol, 2-amino-4-methyl-		C_7H_9NO	123.15	95-84-1	136		sub	4-13-00-01712		H_2O 2; EtOH 3; eth 3; bz 2	
9211	Phenol, 3-amino-4-methyl-		C_7H_9NO	123.15	2836-00-2	157.8		sub	4-13-00-01710		eth 4	
9212	Phenol, 3-amino-5-methyl-	*m*-Cresol, 5-amino-	C_7H_9NO	123.15	76619-89-1	139		245	4-13-00-01697		EtOH 4	
9213	Phenol, 4-amino-2-methyl-		C_7H_9NO	123.15	2835-96-3	176.5		sub	4-13-00-01672		H_2O 2; EtOH 3; eth 3; bz 2	
9214	Phenol, 2-amino-3-nitro-		$C_6H_6N_2O_3$	154.13	603-85-0	216.5		sub	4-13-00-00895		H_2O 4	
9215	Phenol, 4-(2-aminopropyl)-, (±)-	Hydroxyamphetamine	$C_9H_{13}NO$	151.21	1518-86-1	125.5	4740		4-13-00-01871		H_2O 3; EtOH 3; bz 3; chl 3	
9216	Phenol, 2,2'-azobis-		$C_{12}H_{10}N_2O_2$	214.22	2050-14-8	173		$140^{0.001}$	2-16-00-00033		H_2O 1; EtOH 2; eth 4; bz 2	
9217	Phenol, 2-(2-benzothiazolyl)-		$C_{13}H_9NOS$	227.29	3411-95-8	131		$175\text{-}93^{3}$	4-27-00-02081		EtOH 3	
9218	Phenol, 2-(2-benzoxazolyl)-		$C_{13}H_9NO_2$	211.22	835-64-3	123.5		338	4-27-00-02080		H_2O 2; EtOH 4; eth 3; ace 3	
9219	Phenol, 2,4-bis(1,1-dimethylethyl)-		$C_{14}H_{22}O$	206.33	96-76-4	56.5		263.5	4-06-00-03493		ctc 2; alk 1	1.5080^{20}
9220	Phenol, 2,6-bis(1,1-dimethylethyl)-		$C_{14}H_{22}O$	206.33	128-39-2	39		161^{50}, 133^{20}	4-06-00-03492		EtOH 2; ctc 3; alk 1	1.5001^{20}
9221	Phenol, 2,6-bis(1,1-dimethylethyl)-4-ethyl-		$C_{16}H_{26}O$	234.38	4130-42-1	44		272	4-06-00-03529		alk 1	
9222	Phenol, 2,4-bis(1,1-dimethylethyl)-5-methyl-	DBMC	$C_{15}H_{24}O$	220.35	497-39-2	62.1	2829	282	4-06-00-03510	0.912^{80}	H_2O 1; EtOH 3; eth 3; ace 3	
9223	Phenol, 2,4-bis(1,1-dimethylethyl)-6-methyl-		$C_{15}H_{24}O$	220.35	616-55-7	51		269	4-06-00-03510	0.891^{80}	alk 1	
9224	Phenol, 2,6-bis(1,1-dimethylethyl)-4-methyl-	2,6-Di-*tert*-butyl-4-methylphenol	$C_{15}H_{24}O$	220.35	128-37-0	71	1548	265	4-06-00-03511	0.8937^{75}	H_2O 1; EtOH 3; ace 3; bz 3	1.4859^{75}
9225	Phenol, 2,6-bis(1,1-dimethylpropyl)-4-methyl-	*p*-Cresol, 2,6-di-*tert*-pentyl-	$C_{17}H_{28}O$	248.41	56103-67-4			283	4-06-00-03533	0.931^{25}		1.4950^{20}
9226	Phenol, 2,6-bis(1-methylethyl)-	Propofol	$C_{12}H_{18}O$	178.27	2078-54-8	18	7847	256	4-06-00-03435	1.5140^{20}		1.5140^{20}
9227	Phenol, 2,6-bis(1-methylpropyl)-	2,6-Di-*sec*-butylphenol	$C_{14}H_{22}O$	206.33	5510-99-6	-42		257.5	4-06-00-03490			1.5080^{20}
9228	Phenol, 2-bromo-		C_6H_5BrO	173.01	95-56-7	5.6		194.5	4-06-00-01037	1.4924^{20}	H_2O 2; EtOH 3; eth 3; chl 2	1.589^{20}
9229	Phenol, 3-bromo-		C_6H_5BrO	173.01	591-20-8	33		236.5	4-06-00-01042		H_2O 2; EtOH 4; eth 4; ctc 2	
9230	Phenol, 4-bromo-		C_6H_5BrO	173.01	106-41-2	66.4		238	4-06-00-01043	1.840^{15}	H_2O 3; EtOH 4; eth 4; chl 3	
9231	Phenol, 3-bromo-5-chloro-		C_6H_4BrClO	207.45	56962-04-0	70		256	2-06-00-00187		EtOH 4; HOAc 4	
9232	Phenol, 4-bromo-2-chloro-	4-Bromo-2-chlorophenol	C_6H_4BrClO	207.45	3964-56-5	50.5		233.5	3-06-00-00750	1.6170^{20}	EtOH 4; eth 4; ace 4; bz 4	1.5859^{20}

No.	Name Synonym	Mol. Form. Mol. Wt.	CAS RN mp/°C	Merck No. bp/°C	Beil. Ref. den/g cm^{-3}	Solubility n_D
9233	Phenol, 2-bromo-4,6-dichloro-	$C_6H_3BrCl_2O$ 241.90	4524-77-0 68	 220^{200}	3-06-00-00751	bz 4; eth 4; chl 4
9234	Phenol, 2-bromo-4,6-dinitro-	$C_6H_3BrN_2O_5$ 263.00	2316-50-9 119.3	 sub	3-06-00-00871	H_2O 2; EtOH 3; eth 4; bz 4
9235	Phenol, 4-bromo-2,6-dinitro-	$C_6H_3BrN_2O_5$ 263.00	40466-95-3 78	 sub	3-06-00-00872	bz 4; eth 4; EtOH 4; chl 4
9236	Phenol, 2-bromo-4-methyl-	C_7H_7BrO 187.04	6627-55-0 56.5	 213.5	4-06-00-02143 1.5422^{25}	H_2O 2; EtOH 3; bz 3; chl 3 1.5772^{20}
9237	Phenol, 2-bromo-5-methyl- 6-Bromo-*m*-cresol	C_7H_7BrO 187.04	14847-51-9 38.8	 209	3-06-00-01320	ace 4; eth 4; EtOH 4
9238	Phenol, 3-bromo-2-methyl-	C_7H_7BrO 187.04	7766-23-6 95	 55^{4}	0-06-00-00360	ace 4; bz 4; eth 4; EtOH 4
9239	Phenol, 3-bromo-4-methyl-	C_7H_7BrO 187.04	60710-39-6 56	 246	4-06-00-02143	ace 4; bz 4; eth 4; EtOH 4
9240	Phenol, 3-bromo-5-methyl-	C_7H_7BrO 187.04	74204-00-5 56.5	 161^{28}	2-06-00-00357	eth 4; EtOH 4
9241	Phenol, 4-bromo-2-methyl-	C_7H_7BrO 187.04	2362-12-1 64.7	 235	4-06-00-02006	ace 4; eth 4; EtOH 4
9242	Phenol, 4-bromo-3-methyl-	C_7H_7BrO 187.04	14472-14-1 63.5	 142^{23}	4-06-00-02072	eth 4; py 4
9243	Phenol, 2-bromo-3-nitro-	$C_6H_4BrNO_3$ 218.01	101935-40-4 147.5	 sub	3-06-00-00844	eth 4; EtOH 4
9244	Phenol, 4-bromo-2-nitro- 4-Bromo-2-nitrophenol	$C_6H_4BrNO_3$ 218.01	7693-52-9 92	 sub	4-06-00-01363	bz 4; eth 4; EtOH 4; chl 4
9245	Phenol, 2-butoxy-	$C_{10}H_{14}O_2$ 166.22	39075-90-6	 232.5	4-06-00-05566 1.026^{25}	 1.5113^{25}
9246	Phenol, 2-butyl-	$C_{10}H_{14}O$ 150.22	3180-09-4 -20	 235	4-06-00-03267 0.975^{20}	H_2O 1; EtOH 3; eth 3; alk 3 1.5180^{25}
9247	Phenol, 3-butyl-	$C_{10}H_{14}O$ 150.22	4074-43-5	 248	4-06-00-03269 0.974^{20}	eth 4; EtOH 4
9248	Phenol, 4-butyl-	$C_{10}H_{14}O$ 150.22	1638-22-8 22	 248	4-06-00-03269 0.978^{22}	H_2O 1; EtOH 3; eth 3; ctc 2 1.5165^{25}
9249	Phenol, 4,4'-butylidenebis-	$C_{16}H_{18}O_2$ 242.32	4731-84-4 137	 270^{12}	4-06-00-06736	bz 4; EtOH 4; tol 4
9250	Phenol, 2-butyl-4-methyl-	$C_{11}H_{16}O$ 164.25	6891-45-8 19	 228; 124^{15}	4-06-00-03392 0.9680^{20}	
9251	Phenol, 2-chloro- *o*-Chlorophenol	C_6H_5ClO 128.56	95-57-8 9.8	2154 174.9	4-06-00-00782 1.2634^{20}	H_2O 2; EtOH 3; eth 3; bz 4 1.5524^{20}
9252	Phenol, 3-chloro- *m*-Chlorophenol	C_6H_5ClO 128.56	108-43-0 32.6	2154 214	4-06-00-00810 1.245^{45}	H_2O 2; EtOH 3; eth 3; bz 4 1.5565^{40}
9253	Phenol, 4-chloro- *p*-Chlorophenol	C_6H_5ClO 128.56	106-48-9 42.7	2154 220	4-06-00-00820 1.2651^{40}	H_2O 1; EtOH 4; eth 4; bz 4 1.5579^{40}
9254	Phenol, 2-chloro-4-cyclohexyl- 2-Chloro-4-cyclohexylphenol	$C_{12}H_{15}ClO$ 210.70	3964-61-2 40.5	 160^{12}; 122^{3}	3-06-00-02508 1.1600^{15}	ctc 3; CS_2 3
9255	Phenol, 4-chloro-2-cyclopentyl- Dowicide 9	$C_{11}H_{13}ClO$ 196.68	13347-42-7	3420 $181\text{-}5^{18}$	3-06-00-02478	
9256	Phenol, 2-chloro-3,4-dimethyl-	C_8H_9ClO 156.61	10283-15-5 27	 188	2-06-00-00456 1.5538^{20}	peth 4
9257	Phenol, 4-chloro-3,5-dimethyl- Chloroxylenol	C_8H_9ClO 156.61	88-04-0 115	2176 246	4-06-00-03152	H_2O 2; EtOH 3; eth 3; bz 2
9258	Phenol, 4-chloro-2,6-dinitro- 4-Chloro-2,6-dinitrophenol	$C_6H_3ClN_2O_5$ 218.55	88-87-9 81		4-06-00-01386 1.74^{22}	eth 4; EtOH 4; chl 4
9259	Phenol, 5-chloro-2,4-dinitro- 3-Chloro-4,6-dinitrophenol	$C_6H_3ClN_2O_5$ 218.55	54715-57-0 92		4-06-00-01385 1.74^{22}	eth 4; EtOH 4; chl 4; peth 4
9260	Phenol, 2-chloro-3-methyl-	C_7H_7ClO 142.58	608-26-4 56	 230	4-06-00-02062	H_2O 2; HOAc 4; peth 3
9261	Phenol, 2-chloro-4-methyl-	C_7H_7ClO 142.58	6640-27-3	 195.5	4-06-00-02135 1.1785^{27}	bz 4; eth 4; EtOH 4 1.5200^{27}
9262	Phenol, 2-chloro-5-methyl- 6-Chloro-*m*-cresol	C_7H_7ClO 142.58	615-74-7 45.5	 196	3-06-00-01315 1.215^{15}	H_2O 4; EtOH 4
9263	Phenol, 2-chloro-6-methyl-	C_7H_7ClO 142.58	87-64-9	 189; 80^{20}	4-06-00-01984	eth 3 1.5449^{20}
9264	Phenol, 3-chloro-2-methyl-	C_7H_7ClO 142.58	3260-87-5 86	 225	4-06-00-02000	bz 4; eth 4; EtOH 4
9265	Phenol, 3-chloro-4-methyl-	C_7H_7ClO 142.58	615-62-3 55.5	 228	3-06-00-01374	bz 4; eth 4; EtOH 4
9266	Phenol, 4-chloro-2-methyl- 4-Chloro-*o*-cresol	C_7H_7ClO 142.58	1570-64-5 51	 223	4-06-00-01987	H_2O 2; peth 3
9267	Phenol, 4-chloro-3-methyl- 4-Chloro-*m*-cresol	C_7H_7ClO 142.58	59-50-7 67	2133 235	4-06-00-02064	H_2O 2; EtOH 3; eth 3; chl 2
9268	Phenol, 4-chloro-5-methyl-2-(1-methylethyl)- Chlorothymol	$C_{10}H_{13}ClO$ 184.67	89-68-9 63	2171 258.5	4-06-00-03344	H_2O 4; EtOH 3; eth 3; bz 3
9269	Phenol, 5-chloro-2-nitro- 5-Chloro-2-nitrophenol	$C_6H_4ClNO_3$ 173.56	611-07-4 41	 sub	3-06-00-00836	H_2O 2; EtOH 3; eth 3; HOAc 3
9270	Phenol, 4-chloro-2-(phenylmethyl)- Clorophene	$C_{13}H_{11}ClO$ 218.68	120-32-1 48.5	2403 $160\text{-}2^{3.5}$	4-06-00-04636 1.185^{58}	ctc 3; CS_2 3
9271	Phenol, 4-chloro-2-(2-propenyl)-	C_9H_9ClO 168.62	13997-73-4 48	 $130\text{-}2^{15}$	4-06-00-03813 1.171^{15}	bz 4; EtOH 4
9272	Phenol, 4-cyclohexyl-	$C_{12}H_{16}O$ 176.26	1131-60-8 133	 294; 133^{4}	4-06-00-03904	H_2O 1; EtOH 4; eth 4; bz 3

No.	Name Synonym	Mol. Form. Mol. Wt.	CAS RN mp/°C	Merck No. bp/°C	Beil. Ref. den/g cm^{-3}	Solubility n_D
9273	Phenol, 2-cyclohexyl-4,6-dinitro- 4,6-Dinitro-*o*-cyclohexylphenol	$C_{12}H_{14}N_2O_5$ 266.25	131-89-5	2739	4-06-00-03903	
9274	Phenol, 2,4-diamino- 2,4-Diaminophenol	$C_6H_8N_2O$ 124.14	95-86-3 78-80 dec	2961	4-13-00-01425	H_2O 4; ace 4; EtOH 4
9275	Phenol, 4,4'-(1,2-diarsenediyl)bis[2-amino-, dihydrochloride Salvarsan	$C_{12}H_{20}As_2Cl_2O_5$ 465.04	139-93-5 185-95 dec	839	2-16-00-00560	H_2O 4
9276	Phenol, 2,4-dibromo- 2,4-Dibromophenol	$C_6H_4Br_2O$ 251.91	615-58-7 38	 238.5	4-06-00-01061 2.0700^{20}	H_2O 2; EtOH 4; eth 4; bz 4
9277	Phenol, 2,6-dibromo- 2,6-Dibromophenol	$C_6H_4Br_2O$ 251.91	608-33-3 56.5	 255; 162^{21}	4-06-00-01064	H_2O 3; EtOH 4; eth 4
9278	Phenol, 3,5-dibromo- 3,5-Dibromophenol	$C_6H_4Br_2O$ 251.91	626-41-5 81	 274; 121^{3}	4-06-00-01065	H_2O 2; EtOH 3; eth 3; lig 2
9279	Phenol, 2,4-dibromo-6-methyl-	$C_7H_6Br_2O$ 265.93	609-22-3 58	 265 dec; 105^{4}	3-06-00-01271	chl 3
9280	Phenol, 3,5-dibromo-2-methyl-	$C_7H_6Br_2O$ 265.93	14122-00-0 100.3	 285	2-06-00-00334	peth 4
9281	Phenol, 2,4-dibromo-3-methyl-6-(1-methylethyl)- Thymol, 2,6-dibromo	$C_{10}H_{12}Br_2O$ 308.01	52262-38-1 3.7	 $160\text{-}1^{16}$	2-06-00-00500 1.6568^{17}	
9282	Phenol, 2,3-dichloro- 2,3-Dichlorophenol	$C_6H_4Cl_2O$ 163.00	576-24-9 58		4-06-00-00883	EtOH 3; eth 3; bz 3; lig 3
9283	Phenol, 2,4-dichloro- 2,4-Dichlorophenol	$C_6H_4Cl_2O$ 163.00	120-83-2 45	3061 210	4-06-00-00885	H_2O 2; EtOH 3; eth 3; bz 3
9284	Phenol, 2,5-dichloro- 2,5-Dichlorophenol	$C_6H_4Cl_2O$ 163.00	583-78-8 59	 211	4-06-00-00942	H_2O 2; EtOH 4; eth 4; bz 3
9285	Phenol, 2,6-dichloro- 2,6-Dichlorophenol	$C_6H_4Cl_2O$ 163.00	87-65-0 68.5	3062 220; 82^{4}	4-06-00-00949 1.653^{20}	EtOH 4; eth 4; bz 3; peth 3
9286	Phenol, 3,4-dichloro- 3,4-Dichlorophenol	$C_6H_4Cl_2O$ 163.00	95-77-2 68	 253	4-06-00-00952	H_2O 2; EtOH 4; eth 4; bz 3
9287	Phenol, 3,5-dichloro- 3,5-Dichlorophenol	$C_6H_4Cl_2O$ 163.00	591-35-5 68	 233	4-06-00-00957	H_2O 2; EtOH 4; eth 4; peth 3
9288	Phenol, 2,4-dichloro-, acetate 2,4-Dichlorophenyl acetate	$C_8H_6Cl_2O_2$ 205.04	6341-97-5	 244.5	3-06-00-00704	EtOH 3; eth 3; chl 3
9289	Phenol, 2,4-dichloro-, benzenesulfonate Genite	$C_{12}H_8Cl_2O_3S$ 303.17	97-16-5 45.5	4283	4-11-00-00034	ctc 3; CS_2 3
9290	Phenol, 2,4-dichloro-3,5-dimethyl- Dichloroxylenol	$C_8H_8Cl_2O$ 191.06	133-53-9 83	3066	4-06-00-03156	eth 4
9291	Phenol, 2,6-dichloro-3,5-dimethyl-	$C_8H_8Cl_2O$ 191.06	1943-54-0 87.5	 $105\text{-}10^{1}$	4-06-00-03157	eth 4; chl 4
9292	Phenol, 2,4-dichloro-3-methyl- *m*-Cresol, 2,4-dichloro	$C_7H_6Cl_2O$ 177.03	17788-00-0 58	 236; 77^{4}	3-06-00-01319	eth 4; chl 4
9293	Phenol, 2,4-dichloro-5-methyl-	$C_7H_6Cl_2O$ 177.03	1124-07-8 73	 234	4-06-00-02069	chl 4; peth 3 1.572^{20}
9294	Phenol, 2,6-dichloro-3-methyl- 2,6-Dichloro-*m*-cresol	$C_7H_6Cl_2O$ 177.03	13481-70-4 28	 237; 77^{4}	3-06-00-01319	eth 4; chl 4
9295	Phenol, 2,6-dichloro-4-methyl-	$C_7H_6Cl_2O$ 177.03	2432-12-4 39	 231; 138^{28}	4-06-00-02141	eth 4; EtOH 4; HOAc 4
9296	Phenol, 2,6-dichloro-4-nitro-	$C_6H_3ClNO_3$ 172.55	618-80-4 127 exp		4-06-00-01361 1.822^{25}	eth 4; chl 4
9297	Phenol, 3-(diethylamino)-	$C_{10}H_{15}NO$ 165.24	91-68-9 78	 276; 170^{15}	4-13-00-00969	H_2O 3; EtOH 3; eth 3; CS_2 3
9298	Phenol, 4,4'-(1,2-diethyl-1,2-ethanediyl)bis-, (*R**,*S**)- Hexestrol	$C_{18}H_{22}O_2$ 270.37	84-16-2 186.5	4621	4-06-00-06761	ace 4; eth 4; EtOH 4
9299	Phenol, 4,4'-(1,2-diethyl-1,2-ethanediyl)bis[2-methyl- Methestrol	$C_{20}H_{26}O_2$ 298.43	130-73-4 145	5888	4-06-00-06786	
9300	Phenol, 4,4'-(1,2-diethyl-1,2-ethenediyl)bis-, (*E*)- Diethylstilbestrol	$C_{18}H_{20}O_2$ 268.36	56-53-1 170.5	3118	4-06-00-06856	eth 4; EtOH 4; chl 4
9301	Phenol, 4,4'-(1,2-diethyl-1,2-ethenediyl)bis-, dipropanoate, (*E*)- Diethylstilbestrol dipropionate	$C_{24}H_{28}O_4$ 380.48	130-80-3 104	3119	4-06-00-06857	bz 4; eth 4; EtOH 4
9302	Phenol, 4,4'-(1,2-diethylidene-1,2-ethanediyl)bis- Dienestrol	$C_{18}H_{18}O_2$ 266.34	84-17-3 227.5	3094 sub 130	4-06-00-06916	ace 4; eth 4; EtOH 4
9303	Phenol, 4,4'-(1,2-diethyl-3-methyl-1,3-propanediyl)bis- Benzestrol	$C_{20}H_{26}O_2$ 298.43	85-95-0 164	1082	3-06-00-05538	ace 4; eth 4; EtOH 4; HOAc 4
9304	Phenol, 2,6-diiodo-4-nitro- Disophenol	$C_6H_3I_2NO_3$ 390.90	305-85-1 157	3365	4-06-00-01368	EtOH 4
9305	Phenol, 2,6-dimethoxy-	$C_8H_{10}O_3$ 154.17	91-10-1 56.5	 261	4-06-00-07329	eth 4; EtOH 4
9306	Phenol, 3,5-dimethoxy-	$C_8H_{10}O_3$ 154.17	500-99-2 37	 199^{35}; 170^{10}	4-06-00-07362	eth 3; bz 3; lig 2
9307	Phenol, 2,3-dimethyl- 2,3-Xylenol	$C_8H_{10}O$ 122.17	526-75-0 72.8	9989 216.9	4-06-00-03096	H_2O 2; EtOH 3; eth 3 1.5420^{20}

No.	Name Synonym	Mol. Form. Mol. Wt.	CAS RN mp/°C	Merck No. bp/°C	Beil. Ref. den/g cm^{-3}	Solubility n_D
9308	Phenol, 2,4-dimethyl- 2,4-Xylenol	$C_8H_{10}O$ 122.17	105-67-9 24.5	9989 210.9	4-06-00-03126 0.9650^{20}	H_2O 2; EtOH 5; eth 5; ctc 3 1.5420^{14}
9309	Phenol, 2,5-dimethyl- 2,5-Xylenol	$C_8H_{10}O$ 122.17	95-87-4 74.8	9989 211.1	4-06-00-03164	H_2O 3; EtOH 3; eth 4; chl 2
9310	Phenol, 2,6-dimethyl- 2,6-Xylenol	$C_8H_{10}O$ 122.17	576-26-1 45.7	9989 201.0	4-06-00-03112	H_2O 3; EtOH 3; eth 3; ctc 3
9311	Phenol, 3,4-dimethyl- 3,4-Xylenol	$C_8H_{10}O$ 122.17	95-65-8 60.8	9989 227	4-06-00-03099 0.9830^{20}	H_2O 2; EtOH 3; eth 5; ctc 3
9312	Phenol, 3,5-dimethyl- 3,5-Xylenol	$C_8H_{10}O$ 122.17	108-68-9 63.6	9989 221.7	4-06-00-03141 0.9680^{20}	H_2O 3; EtOH 3; ctc 3
9313	Phenol, 2,4-dimethyl-, acetate	$C_{10}H_{12}O_2$ 164.20	877-53-2	 226	2-06-00-00459 1.0298^{15}	eth 4; EtOH 4 1.4990^{15}
9314	Phenol, 2,5-dimethyl-, acetate	$C_{10}H_{12}O_2$ 164.20	877-48-5 <-20	 237	2-06-00-00467 1.0624^{15}	eth 4; EtOH 4
9315	Phenol, 3,4-dimethyl-, acetate	$C_{10}H_{12}O_2$ 164.20	22618-23-1 22	 235	4-06-00-03102	bz 4; eth 4; EtOH 4
9316	Phenol, 3,5-dimethyl-, acetate	$C_{10}H_{12}O_2$ 164.20	877-82-7 <-20	 237	3-06-00-01757 1.0624^{15}	eth 4; EtOH 4
9317	Phenol, 2-(dimethylamino)-	$C_8H_{11}NO$ 137.18	3743-22-4 45	 199.5	4-13-00-00815	eth 4; EtOH 4; HOAc 4
9318	Phenol, 3-(dimethylamino)-	$C_8H_{11}NO$ 137.18	99-07-0 86	 266.5	4-13-00-00959	H_2O 1; EtOH 3; eth 3; ace 3 1.5895^{26}
9319	Phenol, 4-(dimethylamino)-	$C_8H_{11}NO$ 137.18	619-60-3 77	 165^{30}	4-13-00-01040	H_2O 2; EtOH 3; eth 3
9320	Phenol, 4-[2-(dimethylamino)ethyl]- Hordenine	$C_{10}H_{15}NO$ 165.24	539-15-1 117.5	4666 173^{11}	4-13-00-01790	eth 4; EtOH 4; chl 4
9321	Phenol, 2-(1,1-dimethylethyl)-	$C_{10}H_{14}O$ 150.22	88-18-6 -6.8	 223	4-06-00-03292 0.9783^{20}	EtOH 3; eth 4; ctc 3; alk 3 1.5160^{20}
9322	Phenol, 3-(1,1-dimethylethyl)-	$C_{10}H_{14}O$ 150.22	585-34-2 42.3	 240	4-06-00-03294	EtOH 3; eth 4; alk 3
9323	Phenol, 4-(1,1-dimethylethyl)- *p-tert*-Butylphenol	$C_{10}H_{14}O$ 150.22	98-54-4 98	1584 237	4-06-00-03296 0.908^{80}	H_2O 3; EtOH 3; eth 3; chl 3 1.4787^{114}
9324	Phenol, 2-(1,1-dimethylethyl)-4,5-dimethyl- 3,4-Xylenol, 6-*tert*-butyl-	$C_{12}H_{18}O$ 178.27	1445-23-4 46	 258.5	3-06-00-02019 0.920^{80}	 1.5222^{20}
9325	Phenol, 2-(1,1-dimethylethyl)-4,6-dimethyl-	$C_{12}H_{18}O$ 178.27	1879-09-0 22.3	 249	4-06-00-03442 0.917^{80}	alk 1 1.5183^{20}
9326	Phenol, 4-(1,1-dimethylethyl)-2,5-dimethyl-	$C_{12}H_{18}O$ 178.27	17696-37-6 71.2	 264	3-06-00-02019 0.939^{80}	alk 3 1.5311^{20}
9327	Phenol, 4-(1,1-dimethylethyl)-2,6-dimethyl-	$C_{12}H_{18}O$ 178.27	879-97-0 82.4	 248	3-06-00-02019 0.916^{80}	alk 3
9328	Phenol, 2-(1,1-dimethylethyl)-4,6-dinitro- 2-*tert*-Butyl-4,6-dinitrophenol	$C_{10}H_{12}N_2O_5$ 240.22	1420-07-1		4-06-00-03293	
9329	Phenol, (1,1-dimethylethyl)-4-methoxy- Butylated hydroxyanisole	$C_{11}H_{16}O_2$ 180.25	25013-16-5 51	1547 268		H_2O 1; peth 3; EtOH 3
9330	Phenol, 2-(1,1-dimethylethyl)-4-methyl-	$C_{11}H_{16}O$ 164.25	2409-55-4 51.5	 237	4-06-00-03397 0.9247^{75}	H_2O 2; ace 3; bz 3; chl 3 1.4969^{75}
9331	Phenol, 2-(1,1-dimethylethyl)-5-methyl-	$C_{11}H_{16}O$ 164.25	88-60-8 46.5	 127^{11}	4-06-00-03400 0.922^{80}	H_2O 1; EtOH 3; eth 3; ace 3 1.5250^{20}
9332	Phenol, 2-(1,1-dimethylethyl)-6-methyl-	$C_{11}H_{16}O$ 164.25	2219-82-1 31	 230	4-06-00-03396 0.9240^{80}	 1.5195^{20}
9333	Phenol, 4-(1,1-dimethylethyl)-2-methyl-	$C_{11}H_{16}O$ 164.25	98-27-1 27.5	 237; 132^{20}	4-06-00-03398 0.965^{20}	H_2O 1; eth 3; ace 3; bz 3 1.5230^{20}
9334	Phenol, 3,6-dimethyl-2-nitro-	$C_8H_9NO_3$ 167.16	71608-10-1 34.5	 236 dec; 150^{15}	4-06-00-03169	EtOH 4; eth 3; ace 3; bz 3
9335	Phenol, 2,4-dimethyl-, phosphate (3:1)	$C_{24}H_{27}O_4P$ 410.45	3862-12-2	 233.5	0-06-00-00488 1.142^{38}	H_2O 1; bz 3; chl 3; hx 3 1.5550^{20}
9336	Phenol, 2,5-dimethyl-, phosphate (3:1)	$C_{24}H_{27}O_4P$ 410.45	19074-59-0 79.8	 $260\text{-}5^{8}$	3-06-00-01773 1.197^{25}	H_2O 1; EtOH 2; eth 3; bz 3
9337	Phenol, 2,6-dimethyl-, phosphate (3:1)	$C_{24}H_{27}O_4P$ 410.45	121-06-2 137.8	 $262\text{-}4^{6}$	4-06-00-03121	H_2O 1; EtOH 2; bz 3; hx 2
9338	Phenol, 3,4-dimethyl-, phosphate (3:1)	$C_{24}H_{27}O_4P$ 410.45	3862-11-1 72	 $260\text{-}3^{7}$	0-06-00-00482	H_2O 1; EtOH 2; bz 3; chl 2
9339	Phenol, 4-(1,1-dimethylpropyl)- *p-tert*-Pentylphenol	$C_{11}H_{16}O$ 164.25	80-46-6 95	7098 262.5	4-06-00-03383	
9340	Phenol, 4-(1,1-dimethylpropyl)-2-methyl-	$C_{12}H_{18}O$ 178.27	71745-63-6 15	 260	3-06-00-02010 0.969^{20}	chl 2 1.5240^{20}
9341	Phenol, 2,3-dinitro- 2,3-Dinitrophenol	$C_6H_4N_2O_5$ 184.11	66-56-8 144.5		4-06-00-01369 1.681^{20}	H_2O 2; EtOH 4; eth 4; bz 3
9342	Phenol, 2,4-dinitro- 2,4-Dinitrophenol	$C_6H_4N_2O_5$ 184.11	51-28-5 115.5	3274 sub	4-06-00-01369 1.683^{24}	H_2O 2; EtOH 3; eth 3; ace 3
9343	Phenol, 2,5-dinitro- 2,5-Dinitrophenol	$C_6H_4N_2O_5$ 184.11	329-71-5 108	3275	4-06-00-01383	bz 4; eth 4
9344	Phenol, 2,6-dinitro- 2,6-Dinitrophenol	$C_6H_4N_2O_5$ 184.11	573-56-8 63.5	3276	4-06-00-01383	H_2O 1; EtOH 4; eth 4; ace 4
9345	Phenol, 3,4-dinitro-	$C_6H_4N_2O_5$ 184.11	577-71-9 134		4-06-00-01384 1.672^{25}	bz 4; eth 4; EtOH 4
9346	Phenol, 3,5-dinitro- 3,5-Dinitrophenol	$C_6H_4N_2O_5$ 184.11	586-11-8 125.1		4-06-00-01385 1.702^{25}	bz 4; eth 4; EtOH 4; chl 4
9347	Phenol, 2,6-dipropyl- 2,6-Dipropylphenol	$C_{12}H_{18}O$ 178.27	6626-32-0 27	 256	3-06-00-02013 1.6450^{0}	eth 4

No.	Name Synonym	Mol. Form. Mol. Wt.	CAS RN mp/°C	Merck No. bp/°C	Beil. Ref. den/g cm^{-3}	Solubility n_D
9348	Phenol, 2-ethenyl-	C_8H_8O 120.15	695-84-1 29.5	 101^{14}	4-06-00-03771 1.0609^{18}	eth 4; EtOH 4 1.5851^{20}
9349	Phenol, 3-ethenyl-	C_8H_8O 120.15	620-18-8	 $114\text{-}6^{16}$	4-06-00-03773 1.0353^{21}	 1.5804^{21}
9350	Phenol, 2-ethoxy-	$C_8H_{10}O_2$ 138.17	94-71-3 29	 217	4-06-00-05565 1.0903^{25}	H_2O 2; EtOH 5; eth 5; ctc 2
9351	Phenol, 3-ethoxy-	$C_8H_{10}O_2$ 138.17	621-34-1	 131^{10}	4-06-00-05664 1.0705^{4}	H_2O 1; EtOH 3; eth 3; bz 3
9352	Phenol, 4-ethoxy-	$C_8H_{10}O_2$ 138.17	622-62-8 66.5	 246.5	4-06-00-05719	H_2O 2; EtOH 4; eth 4; chl 3
9353	Phenol, 2-ethyl- *o*-Ethylphenol	$C_8H_{10}O$ 122.17	90-00-6 18	7302 204.5	4-06-00-03011 1.0146^{25}	ace 4; bz 4; eth 4; EtOH 4 1.5367^{20}
9354	Phenol, 3-ethyl- *m*-Ethylphenol	$C_8H_{10}O$ 122.17	620-17-7 -4	 218.4	4-06-00-03016 1.0283^{20}	H_2O 2; EtOH 4; eth 4; chl 2
9355	Phenol, 4-ethyl- *p*-Ethylphenol	$C_8H_{10}O$ 122.17	123-07-9 45.0	 217.9	4-06-00-03020	H_2O 2; EtOH 4; eth 4; ace 3 1.5239^{25}
9356	Phenol, 3-(ethylamino)-	$C_8H_{11}NO$ 137.18	621-31-8 62	 176^{12}	0-13-00-00408	bz 4; eth 4; EtOH 4
9357	Phenol, 4-ethyl-2-methoxy-	$C_9H_{12}O_2$ 152.19	2785-89-9 -7	 236.5	4-06-00-05927 1.0931^{18}	
9358	Phenol, 4-[1-ethyl-2-(4-methoxyphenyl)-1-butenyl]- Mestilbol	$C_{19}H_{22}O_2$ 282.38	7773-60-6 117.5	5818 $185\text{-}95^{0.3}$	4-06-00-06856	ace 4; eth 4; EtOH 4
9359	Phenol, 4-[1-ethyl-2-(4-methoxyphenyl)-1-butenyl]-, (*E*)-	$C_{19}H_{22}O_2$ 282.38	18839-90-2 117	 $185\text{-}95^{0.3}$	4-06-00-06856	EtOH 4
9360	Phenol, 2-fluoro-	C_6H_5FO 112.10	367-12-4 16.1	 151.5	4-06-00-00770 1.120^{25}	H_2O 3 1.5144^{20}
9361	Phenol, 3-fluoro-	C_6H_5FO 112.10	372-20-3 13.7	 178	4-06-00-00772 1.238^{25}	 1.5140^{20}
9362	Phenol, 4-fluoro-	C_6H_5FO 112.10	371-41-5 48	 185.5	4-06-00-00773 1.1889^{56}	H_2O 2; ace 3; peth 3
9363	Phenol, 4,4'-(3H-2,1-benzoxathiol-3-ylidene)bis-, *S,S*-dioxide Phenolsulfonphthalein	$C_{19}H_{14}O_5S$ 354.38	143-74-8 >300	7213	5-19-03-00457	H_2O 2; EtOH 2; eth 1; ace 2
9364	Phenol, 4,4'-(3H-2,1-benzoxathiol-3-ylidene)bis[2-bromo-6-methyl-, *S,S*-dioxide Bromcresol purple	$C_{21}H_{16}Br_2O_5S$ 540.23	115-40-2 241.5	1376	5-19-03-00460	
9365	Phenol, 4,4'-(3H-2,1-benzoxathiol-3-ylidene)bis[2-bromo-3-methyl-6-(1-methylethyl)-, *S,S*-dioxide Bromthymol blue	$C_{27}H_{28}Br_2O_5S$ 624.39	76-59-5 201	1435	5-19-03-00461	eth 4; EtOH 4
9366	Phenol, 4,4'-(3H-2,1-benzoxathiol-3-ylidene)bis[2,6-dibromo-, *S,S*-dioxide Bromphenol blue	$C_{19}H_{10}Br_4O_5S$ 669.97	115-39-9 279 dec	1434	5-19-03-00458	H_2O 2; EtOH 3; bz 3; HOAc 3
9367	Phenol, 4,4'-(3H-2,1-benzoxathiol-3-ylidene)bis[2,6-dibromo-3-methyl-, *S,S*-dioxide Bromcresol green	$C_{21}H_{14}Br_4O_5S$ 698.02	76-60-8 218.5	1375	5-19-03-00460	H_2O 2; EtOH 4; eth 4; bz 3
9368	Phenol, 4,4'-(3H-2,1-benzoxathiol-3-ylidene)bis[2-methyl-, *S,S*-dioxide Cresol red	$C_{21}H_{18}O_5S$ 382.44	1733-12-6 >300	2583	5-19-03-00460	H_2O 4; EtOH 4
9369	Phenol, 4,4'-(3H-2,1-benzoxathiol-3-ylidene)bis[5-methyl-2-(1-methylethyl)-, *S,S*-dioxide Thymol blue	$C_{27}H_{30}O_5S$ 466.60	76-61-9 221-4 dec	9334	5-19-03-00462	H_2O 2; EtOH 3; ace 2; bz 2
9370	Phenol, 4-(3-hydroxy-1-propenyl)-2-methoxy- Coniferyl alcohol	$C_{10}H_{12}O_3$ 180.20	458-35-5 74	2499 $163\text{-}5^{3}$	2-06-00-01093	H_2O 1; EtOH 3; eth 4; alk 3
9371	Phenol, 2-iodo- *o*-Iodophenol	C_6H_5IO 220.01	533-58-4 43	4928 186^{160}; 91^{2}	4-06-00-01070 1.8757^{80}	H_2O 3; EtOH 4; eth 4; CS_2 4
9372	Phenol, 3-iodo-	C_6H_5IO 220.01	626-02-8 118	 186^{100}	4-06-00-01073	H_2O 2; EtOH 3; eth 3
9373	Phenol, 4-iodo- *p*-Iodophenol	C_6H_5IO 220.01	540-38-5 93.5	4929 139^{5} dec	4-06-00-01074 1.8573^{112}	H_2O 2; EtOH 4; eth 4
9374	Phenol, 2-iodo-4-methyl- *p*-Cresol, 2-iodo-	C_7H_7IO 234.04	16188-57-1 35	 117^{12}	4-06-00-02147 1.684^{25}	 1.5331^{25}
9375	Phenol, 2-iodo-6-methyl-	C_7H_7IO 234.04	24885-45-8 15.5	 $70\text{-}5^{2}$	4-06-00-02008 1.661^{25}	 1.610^{25}
9376	Phenol, 2-mercapto-	C_6H_6OS 126.18	1121-24-0 5.5	 217; 89^{8}	4-06-00-05633 1.2371^{0}	bz 4; eth 4; EtOH 4
9377	Phenol, 3-mercapto-	C_6H_6OS 126.18	40248-84-8 16.5	 168^{35}	2-06-00-00827	bz 4; EtOH 4
9378	Phenol, 4-mercapto-	C_6H_6OS 126.18	637-89-8 29.5	 167^{45}; 135^{11}	4-06-00-05790 1.1285^{25}	H_2O 3; EtOH 3; alk 3; con sulf 3 1.5101^{25}
9379	Phenol, 2-methoxy- Guaiacol	$C_7H_8O_2$ 124.14	90-05-1 32	4457 205	4-06-00-05563 1.1287^{21}	H_2O 2; EtOH 3; eth 3; ctc 3 1.5429^{20}
9380	Phenol, 3-methoxy-	$C_7H_8O_2$ 124.14	150-19-6 <-17	 114^{5}	4-06-00-05662 1.131^{25}	H_2O 2; EtOH 5; eth 5; chl 2 1.5510^{20}

No.	Name Synonym	Mol. Form. Mol. Wt.	CAS RN mp/°C	Merck No. bp/°C	Beil. Ref. den/g cm^{-3}	Solubility n_D
9381	Phenol, 4-methoxy- 4-Methoxyphenol	$C_7H_8O_2$ 124.14	150-76-5 57	 243	4-06-00-05717	H_2O 3; EtOH 4; eth 4; bz 3
9382	Phenol, 2-methoxy-, acetate	$C_9H_{10}O_3$ 166.18	613-70-7 31.5	 $123\text{-}4^{13}$	4-06-00-05581 1.1285^{25}	H_2O 1; EtOH 3; eth 3 1.5101^{25}
9383	Phenol, 2-methoxy-, benzoate Guaiacol benzoate	$C_{14}H_{12}O_3$ 228.25	531-37-3 57.5	4458	4-09-00-00371	eth 4; chl 4
9384	Phenol, 2-methoxy-, carbonate (2:1) Guaiacol carbonate	$C_{15}H_{14}O_5$ 274.27	553-17-3 89	4459	3-06-00-04233	H_2O 1; EtOH 2; eth 3; chl 4
9385	Phenol, 2-methoxy-4-methyl- Creosol	$C_8H_{10}O_2$ 138.17	93-51-6 5.5	2573 221	4-06-00-05878 1.098^{20}	eth 4; EtOH 4 1.5353^{25}
9386	Phenol, 2-methoxy-3-nitro-	$C_7H_7NO_4$ 169.14	20734-71-8 70	 sub	3-06-00-04263	H_2O 2; EtOH 4; chl 3; peth 3
9387	Phenol, 2-methoxy-, phosphate (3:1) Guaiacol phosphate	$C_{21}H_{21}O_7P$ 416.37	563-03-1 91	4460 $275\text{-}80^{3}$	3-06-00-04246	ace 4; tol 4; chl 4
9388	Phenol, 2-methoxy-4-(1-propenyl)- Isoeugenol	$C_{10}H_{12}O_2$ 164.20	97-54-1	5054 266	3-06-00-04993 1.080^{25}	eth 4; EtOH 4 1.5739^{19}
9389	Phenol, 2-methoxy-4-(2-propenyl)- Eugenol	$C_{10}H_{12}O_2$ 164.20	97-53-0 -7.5	3855 253.2	4-06-00-06337 1.0652^{20}	H_2O 1; EtOH 5; eth 5; chl 3 1.5405^{20}
9390	Phenol, 2-methoxy-5-(2-propenyl)-	$C_{10}H_{12}O_2$ 164.20	501-19-9 8.5	 253.5	3-06-00-05024 1.0613^{25}	eth 4; EtOH 4 1.5413^{20}
9391	Phenol, 2-methoxy-6-(1-propenyl)-	$C_{10}H_{12}O_2$ 164.20	1076-55-7 81	 267.5	3-06-00-04991	H_2O 2; EtOH 3; eth 3; ace 3
9392	Phenol, 2-methoxy-6-(2-propenyl)-	$C_{10}H_{12}O_2$ 164.20	579-60-2	 250.5	4-06-00-06334 1.062^{20}	ace 3; bz 3 1.5545^{20}
9393	Phenol, 2-methoxy-4-(1-propenyl)-, acetate, (*E*)-	$C_{12}H_{14}O_3$ 206.24	5912-87-8 80.5	 282.5	4-06-00-06330 1.0251^{100}	eth 4 1.5052^{100}
9394	Phenol, 2-methoxy-4-(1-propenyl)-, acetate, (*Z*)- Isoeugenol, acetate (*cis*)	$C_{12}H_{14}O_3$ 206.24	97412-23-2	 $160\text{-}2^{13}$	3-06-00-05007 1.0947^{19}	H_2O 1; eth 3 1.5418^{20}
9395	Phenol, 2-methoxy-4-(2-propenyl)-, acetate	$C_{12}H_{14}O_3$ 206.24	93-28-7 30.5	 281; $127\text{-}8^{6}$	3-06-00-05029 1.0806^{20}	H_2O 1; EtOH 3; ctc 2 1.5205^{20}
9396	Phenol, 2-methoxy-4-(1-propenyl)-, (*E*)-	$C_{10}H_{12}O_2$ 164.20	5932-68-3 33.5	 $141\text{-}2^{13}$	4-06-00-06324 1.0852^{20}	H_2O 2; EtOH 3; eth 3; chl 3 1.5784^{20}
9397	Phenol, 2-methoxy-4-(1-propenyl)-, (*Z*)-	$C_{10}H_{12}O_2$ 164.20	5912-86-7	 133^{13}; $80^{0.5}$	4-06-00-06324 1.0837^{20}	H_2O 2; EtOH 3; eth 3 1.5726^{20}
9398	Phenol, 2-methoxy-5-[(1,2,3,4-tetrahydro-6,7-dimethoxy-2-methyl-1-isoquinolinyl)methyl]-, (±)- Laudanine	$C_{20}H_{25}NO_4$ 343.42	85-64-3 167	5248	5-21-06-00048 1.26^{20}	H_2O 2; EtOH 2; eth 2; bz 3
9399	Phenol, 2-methoxy-5-[(1,2,3,4-tetrahydro-6,7-dimethoxy-2-methyl-1-isoquinolinyl)methyl]-, (*R*)- Laudanidine	$C_{20}H_{25}NO_4$ 343.42	301-21-3 184.5	5247	5-21-06-00048	H_2O 4; bz 4
9400	Phenol, 2-methyl- *o*-Cresol	C_7H_8O 108.14	95-48-7 29.8	2580 191.0	4-06-00-01940 1.135^{25}	H_2O 3; EtOH 4; eth 4; ace 5 1.5361^{20}
9401	Phenol, 3-methyl- *m*-Cresol	C_7H_8O 108.14	108-39-4 11.8	2579 202.2	4-06-00-02035 1.0341^{20}	H_2O 2; EtOH 5; eth 5; ace 5 1.5438^{20}
9402	Phenol, 4-methyl- *p*-Cresol	C_7H_8O 108.14	106-44-5 35.5	2581 201.9	4-06-00-02093 1.0185^{40}	H_2O 2; EtOH 5; eth 5; ace 5 1.5312^{20}
9403	Phenol, 4-[2-(methylamino)propyl]- Pholedrine	$C_{10}H_{15}NO$ 165.24	370-14-9 161	7304	3-13-00-01710	eth 4; EtOH 4
9404	Phenol, 4-(3-methylbutyl)-	$C_{11}H_{16}O$ 164.25	1805-61-4 29	 255; 128^{7}	4-06-00-03378 0.9562^{23}	 1.505^{27}
9405	Phenol, 2-methyl-4,6-dinitro- Dinitrocresol	$C_7H_6N_2O_5$ 198.14	534-52-1 86.5	3272	4-06-00-02014	H_2O 2; EtOH 3; eth 3; ace 3
9406	Phenol, 3,3'-methylenebis- Diphenylmethane, 3,3'-dihydroxy	$C_{13}H_{12}O_2$ 200.24	10193-50-7 102.5	 235^{3}	3-06-00-05411	eth 4; EtOH 4; HOAc 4
9407	Phenol, 4,4'-methylenebis-	$C_{13}H_{12}O_2$ 200.24	620-92-8 162.5	 sub	4-06-00-06664	EtOH 3; eth 3; chl 3; DMSO-d_6 2
9408	Phenol, 2,2'-methylenebis[4-chloro- Dichlorophene	$C_{13}H_{10}Cl_2O_2$ 269.13	97-23-4 177.5	3059	4-06-00-06658	H_2O 1; EtOH 3; ace 3
9409	Phenol, 2,2'-methylenebis[3,4,6-trichloro- Hexachlorophene	$C_{13}H_6Cl_6O_2$ 406.91	70-30-4 166.5	4602	4-06-00-06659	H_2O 1; EtOH 3; eth 3; ace 3
9410	Phenol, 2-(1-methylethyl)-	$C_9H_{12}O$ 136.19	88-69-7 15.5	 213.5	4-06-00-03210 1.012^{20}	H_2O 2; EtOH 3; eth 3; bz 3 1.5315^{20}
9411	Phenol, 3-(1-methylethyl)-	$C_9H_{12}O$ 136.19	618-45-1 26	 228	4-06-00-03214	eth 4 1.5261^{20}
9412	Phenol, 4-(1-methylethyl)-	$C_9H_{12}O$ 136.19	99-89-8 62.3	 230; 110^{10}	4-06-00-03215 0.990^{20}	H_2O 2; EtOH 3; chl 3 1.5228^{20}
9413	Phenol, 4,4'-(1-methylethylidene)bis- Bisphenol A	$C_{15}H_{16}O_2$ 228.29	80-05-7 153	1311 220^{4}; 222^{3}	4-06-00-06717	H_2O 1; EtOH 4; eth 4; bz 4
9414	Phenol, 2-methyl-4-(1-methylethyl)-	$C_{10}H_{14}O$ 150.22	1740-97-2 8.6	 230; 83^{3}	4-06-00-03330 0.9793^{25}	H_2O 3; EtOH 3; bz 3; chl 3 1.5253^{20}
9415	Phenol, 2-methyl-5-(1-methylethyl)- Carvacrol	$C_{10}H_{14}O$ 150.22	499-75-2 1	1881 237.7	4-06-00-03331 0.9772^{20}	H_2O 2; EtOH 3; eth 3; ace 4 1.5230^{20}
9416	Phenol, 2-methyl-6-(1-methylethyl)-	$C_{10}H_{14}O$ 150.22	3228-04-4 -14.5	 225.5	3-06-00-01882 0.9782^{25}	EtOH 3; bz 3; chl 3; os 3 1.5239^{15}
9417	Phenol, 4-methyl-2-(1-methylethyl)-	$C_{10}H_{14}O$ 150.22	4427-56-9 36.5	 228.5	4-06-00-03329 0.9910^{20}	H_2O 2; EtOH 3; bz 3; chl 3 1.5275^{20}

No.	Name Synonym	Mol. Form. Mol. Wt.	CAS RN mp/°C	Merck No. bp/°C	Beil. Ref. den/g cm^{-3}	Solubility n_D
9418	Phenol, 5-methyl-2-(1-methylethyl)- Thymol	$C_{10}H_{14}O$ 150.22	89-83-8 51.5	9333 232.5	4-06-00-03334 0.970^{25}	H_2O 1; EtOH 4; eth 4; chl 4 1.5227^{20}
9419	Phenol, 2-methyl-5-(1-methylethyl)-, acetate Carvacrol, acetate	$C_{12}H_{16}O_2$ 192.26	6380-28-5	 246.5	2-06-00-00494 0.9896^{25}	eth 4; EtOH 4 1.4913^{28}
9420	Phenol, 5-methyl-2-(1-methylethyl)-, acetate Thymol, acetate	$C_{12}H_{16}O_2$ 192.26	528-79-0	 245	4-06-00-03337 1.009^{9}	bz 4; eth 4; EtOH 4; chl 4
9421	Phenol, 5-methyl-2-(1-methylpropyl)- 6-*sec*-Butyl-*m*-cresol	$C_{11}H_{16}O$ 164.25	29472-95-5	 $89\text{-}90^{2}$	4-06-00-03394 0.9612^{25}	 1.5188^{20}
9422	Phenol, 2-methyl-4-nitro-	$C_7H_7NO_3$ 153.14	99-53-6 96	 188^{9}; $132^{0.2}$	4-06-00-02011	H_2O 2; EtOH 4; eth 4; bz 4
9423	Phenol, 2-methyl-6-nitro-	$C_7H_7NO_3$ 153.14	13073-29-5 70	 255 dec; 102^{9}	4-06-00-02009	H_2O 1; EtOH 4; eth 4; peth 3
9424	Phenol, 4-methyl-2-nitro-	$C_7H_7NO_3$ 153.14	119-33-5 36.5	 125^{22}	4-06-00-02149 1.2399^{20}	ace 4; bz 4; eth 4; EtOH 4 1.5744^{40}
9425	Phenol, 5-methyl-2-pentyl- 6-*n*-Amyl-*m*-cresol	$C_{12}H_{18}O$ 178.27	1300-94-3 24	643 $137\text{-}9^{15}$	3-06-00-02005	ace 4; eth 4; EtOH 4
9426	Phenol, 4-methyl-2-(phenylazo)-	$C_{13}H_{12}N_2O$ 212.25	952-47-6 108.5	 sub	4-16-00-00197	EtOH 3; eth 3; bz 3; chl 3
9427	Phenol, 2-methyl-6-(2-propenyl)-	$C_{10}H_{12}O$ 148.20	3354-58-3	 232	4-06-00-03844 0.992^{25}	 1.5381^{20}
9428	Phenol, 4-methyl-2-(1-propenyl)-	$C_{10}H_{12}O$ 148.20	53889-94-4	 $126\text{-}8^{14}$	4-06-00-03844 1.019^{20}	 1.5727^{20}
9429	Phenol, 4-methyl-2-(2-propenyl)-	$C_{10}H_{12}O$ 148.20	6628-06-4	 237	4-06-00-03845 0.96^{25}	 1.5385^{20}
9430	Phenol, 2-(1-methylpropyl)- *o*-*sec*-Butylphenol	$C_{10}H_{14}O$ 150.22	89-72-5 16	 228; 116^{21}	3-06-00-01852 0.9804^{25}	 1.5200^{25}
9431	Phenol, 4-(1-methylpropyl)-	$C_{10}H_{14}O$ 150.22	99-71-8 61.5	 241	4-06-00-03279 0.986^{20}	H_2O 1; EtOH 3; eth 4; alk 3 1.5182^{21}
9432	Phenol, 4-(2-methylpropyl)-	$C_{10}H_{14}O$ 150.22	4167-74-2 51.5	 237	4-06-00-03288 0.9778^{20}	ace 4; eth 4; EtOH 4 1.5319^{25}
9433	Phenol, 4,4'-(1-methylpropylidene)bis- Bisphenol B	$C_{16}H_{18}O_2$ 242.32	77-40-7 120.5	1312	4-06-00-06738	ace 4; MeOH 4
9434	Phenol, 3-methyl-2,4,6-trinitro- 3-Methylpicric acid	$C_7H_5N_3O_7$ 243.13	602-99-3 109.5	 exp 150	4-06-00-02079	H_2O 2; EtOH 3; eth 3; ace 3
9435	Phenol, 2-nitro- *o*-Nitrophenol	$C_6H_5NO_3$ 139.11	88-75-5 44.8	6541 216	4-06-00-01246 1.2942^{40}	H_2O 2; EtOH 4; eth 4; ace 4 1.5723^{50}
9436	Phenol, 3-nitro- *m*-Nitrophenol	$C_6H_5NO_3$ 139.11	554-84-7 96.8	6540 194^{70}	4-06-00-01269 1.2797^{100}	H_2O 2; EtOH 4; eth 4; ace 4
9437	Phenol, 4-nitro- *p*-Nitrophenol	$C_6H_5NO_3$ 139.11	100-02-7 113.8	6542	4-06-00-01279 1.479^{20}	H_2O 2; EtOH 4; eth 4; ace 4
9438	Phenol, 4-nitroso- 4-Nitrosophenol	$C_6H_5NO_2$ 123.11	104-91-6 144 dec	6563	4-07-00-02073	H_2O 2; EtOH 3; eth 3; ace 3
9439	Phenol, pentabromo- Pentabromophenol	C_6HBr_5O 488.59	608-71-9 229.5	 sub	3-06-00-00766	H_2O 1; EtOH 3; eth 2; bz 3
9440	Phenol, pentachloro- Pentachlorophenol	C_6HCl_5O 266.34	87-86-5 174	7059 310 dec	4-06-00-01025 1.978^{22}	H_2O 2; EtOH 4; eth 4; bz 3
9441	Phenol, 3-pentadecyl-	$C_{21}H_{36}O$ 304.52	501-24-6 53.5	 230^{8}; $197^{1.5}$	4-06-00-03556	ace 4; bz 4; EtOH 4
9442	Phenol, pentafluoro- Pentafluorophenol	C_6HF_5O 184.07	771-61-9 32.8	 145.6	4-06-00-00782	 1.4263^{20}
9443	Phenol, pentamethyl- Pentamethylphenol	$C_{11}H_{16}O$ 164.25	2819-86-5 128	 267	4-06-00-03412	H_2O 1; EtOH 3
9444	Phenol, 4-pentyl-	$C_{11}H_{16}O$ 164.25	14938-35-3 23	 250.5	4-06-00-03370 0.960^{20}	eth 4; EtOH 4 1.5272^{25}
9445	Phenol, 2-(phenylamino)-	$C_{12}H_{11}NO$ 185.23	644-71-3 69.5	 185^{20}	4-13-00-00822	H_2O 2; EtOH 3; eth 3; bz 2
9446	Phenol, 3-(phenylamino)-	$C_{12}H_{11}NO$ 185.23	101-18-8 81.5	 340	3-13-00-00944	H_2O 2; EtOH 4; eth 4; ace 4
9447	Phenol, 4-(phenylamino)-	$C_{12}H_{11}NO$ 185.23	122-37-2 73	 330	4-13-00-01052	H_2O 2; EtOH 4; eth 4; bz 4
9448	Phenol, 4-(phenylazo)-	$C_{12}H_{10}N_2O$ 198.22	1689-82-3 155	 225^{20} dec	2-16-00-00038	H_2O 1; EtOH 4; eth 4; bz 3
9449	Phenol, 4-(phenylazo)-, acetate (ester)	$C_{14}H_{12}N_2O_2$ 240.26	13102-31-3 89	 360 dec	1-16-00-00236	EtOH 4; lig 4
9450	Phenol, 2-[(phenylimino)methyl]-	$C_{13}H_{11}NO$ 197.24	779-84-0 49.5		4-12-00-00339 1.087^{25}	H_2O 1; EtOH 3
9451	Phenol, 2-(phenylmethoxy)-	$C_{13}H_{12}O_2$ 200.24	6272-38-4	 $173\text{-}4^{13}$	4-06-00-05572 1.154^{22}	eth 4; EtOH 4 1.5906^{18}
9452	Phenol, 3-(phenylmethoxy)-	$C_{13}H_{12}O_2$ 200.24	3769-41-3 69.2	 200^{5}	4-06-00-05669	H_2O 4
9453	Phenol, 4-(phenylmethoxy)- Monobenzone	$C_{13}H_{12}O_2$ 200.24	103-16-2 122	6159	4-06-00-05728	H_2O 2; EtOH 4; eth 4; ace 3
9454	Phenol, 2-(phenylmethyl)- *o*-Benzylphenol	$C_{13}H_{12}O$ 184.24	28994-41-4 21	1158 312	4-06-00-04628	ace 4; bz 4; EtOH 4 1.5994^{20}
9455	Phenol, 4-(phenylmethyl)- *p*-Benzylphenol	$C_{13}H_{12}O$ 184.24	101-53-1 84	1159 322	4-06-00-04640	H_2O 3; EtOH 3; eth 3; bz 3
9456	Phenol, 2-(1-propenyl)-	$C_9H_{10}O$ 134.18	6380-21-8 -6	 220	3-06-00-02393 1.0246^{15}	eth 4 1.5181^{20}
9457	Phenol, 2-(2-propenyl)-	$C_9H_{10}O$ 134.18	1745-81-9 -6	 220	4-06-00-03807 1.0246^{15}	eth 4 1.5181^{20}

No.	Name Synonym	Mol. Form. Mol. Wt.	CAS RN mp/°C	Merck No. bp/°C	Beil. Ref. den/g cm^{-3}	Solubility n_D
9458	Phenol, 4-(2-propenyl)- Chavicol	$C_9H_{10}O$ 134.18	501-92-8 15.8	2039 238	4-06-00-03817 1.0203^{15}	eth 4; EtOH 4; chl 4 1.5441^{18}
9459	Phenol, 2-(1-propenyl)-, (*E*)-	$C_9H_{10}O$ 134.18	23619-59-2 37.5	 230.5	4-06-00-03793 1.0441^{16}	eth 4; lig 4 1.5823^{21}
9460	Phenol, 2-propoxy-	$C_9H_{12}O_2$ 152.19	6280-96-2	 227	4-06-00-05566 1.0523^{25}	EtOH 4 1.5176^{25}
9461	Phenol, 2-propyl-	$C_9H_{12}O$ 136.19	644-35-9 7	 220	4-06-00-03176 1.015^{20}	eth 4; EtOH 4
9462	Phenol, 3-propyl-	$C_9H_{12}O$ 136.19	621-27-2 26	 228	4-06-00-03180 0.987^{20}	H_2O 2; EtOH 3; eth 3 1.5223^{20}
9463	Phenol, 4-propyl-	$C_9H_{12}O$ 136.19	645-56-7 22	 232.6	4-06-00-03181 1.009^{20}	H_2O 2; EtOH 3; ctc 2 1.5379^{25}
9464	Phenol, 4,4'-propylidenebis-	$C_{15}H_{16}O_2$ 228.29	1576-13-2 132	 275^{20}	3-06-00-05457	eth 4; EtOH 4
9465	Phenol, 4,4'-(2-pyridinylmethylene)bis-, diacetate (ester) Bisacodyl	$C_{22}H_{19}NO_4$ 361.40	603-50-9 133.5	1253	5-21-05-00415	
9466	Phenol, 4,4'-sulfonylbis-	$C_{12}H_{10}O_4S$ 250.28	80-09-1 240.5		4-06-00-05809 1.3663^{15}	H_2O 1; EtOH 3; eth 3; bz 2
9467	Phenol, 2,3,4,6-tetrabromo- 2,3,4,6-Tetrabromophenol	$C_6H_2Br_4O$ 409.70	14400-94-3 113.5	 sub	3-06-00-00766	bz 4; EtOH 4
9468	Phenol, 2,3,4,5-tetrabromo-6-methyl- 3,4,5,6-Tetrabromo-*o*-cresol	$C_7H_4Br_4O$ 423.72	576-55-6 208	9120 dec	3-06-00-01272	H_2O 1; EtOH 3; eth 3; bz 3
9469	Phenol, 2,3,4,5-tetrachloro- 2,3,4,5-Tetrachlorophenol	$C_6H_2Cl_4O$ 231.89	4901-51-3 116.5	 sub	4-06-00-01020	EtOH 4
9470	Phenol, 2,3,4,6-tetrachloro- 2,3,4,6-Tetrachlorophenate	$C_6H_2Cl_4O$ 231.89	58-90-2 70	 150^{15}	4-06-00-01021	H_2O 1; EtOH 3; bz 3; chl 3
9471	Phenol, 2,3,5,6-tetrachloro- 2,3,5,6-Tetrachlorophenol	$C_6H_2Cl_4O$ 231.89	935-95-5 115		4-06-00-01025	H_2O 2; bz 4; lig 3
9472	Phenol, 2,3,5,6-tetrachloro-4-methoxy- Drosophilin A	$C_7H_4Cl_4O_2$ 261.92	484-67-3 116	3441	3-06-00-04436	
9473	Phenol, 2,3,4,5-tetramethyl- Prehnitenol	$C_{10}H_{14}O$ 150.22	488-70-0 85.3	 266	4-06-00-03359	H_2O 2; EtOH 4; eth 4; lig 2
9474	Phenol, 2,3,4,6-tetramethyl-	$C_{10}H_{14}O$ 150.22	3238-38-8 80.5	 240	4-06-00-03360	EtOH 3
9475	Phenol, 2,3,5,6-tetramethyl-	$C_{10}H_{14}O$ 150.22	527-35-5 118.5	 247	4-06-00-03363	chl 3; peth 3; HOAc 3
9476	Phenol, 2,3,4,6-tetranitro- 2,3,4,6-Tetranitrophenol	$C_6H_2N_4O_9$ 274.10	641-16-7 140 dec	 exp	2-06-00-00284	H_2O 4; chl 3
9477	Phenol, 2,2'-thiobis[4-chloro- Fenticlor	$C_{12}H_8Cl_2O_2S$ 287.17	97-24-5 174	3947	4-06-00-05645	H_2O 1; EtOH 3; eth 3; gl HOAc 3
9478	Phenol, 2,2'-thiobis[4,6-dichloro- Bithionol	$C_{12}H_6Cl_4O_2S$ 356.06	97-18-7 188	1316	4-06-00-05646 1.73^{25}	ace 4
9479	Phenol, 2,4,6-tribromo- 2,4,6-Tribromophenol	$C_6H_3Br_3O$ 330.80	118-79-6 95.5	9526 286	4-06-00-01067 2.55^{20}	H_2O 2; EtOH 4; eth 3; bz 3
9480	Phenol, 2,4,6-tribromo-3-methyl- 2,4,6-Tribromo-*m*-cresol	$C_7H_5Br_3O$ 344.83	4619-74-3 84	9524	3-06-00-01324	EtOH 3; eth 3; bz 3; chl 2
9481	Phenol, 2,3,4-trichloro- 2,3,4-Trichlorophenol	$C_6H_3Cl_3O$ 197.45	15950-66-0 83.5	 sub	3-06-00-00716	EtOH 3; eth 3; bz 3; alk 3
9482	Phenol, 2,3,5-trichloro- 2,3,5-Trichlorophenol	$C_6H_3Cl_3O$ 197.45	933-78-8 62	 $248\text{-}9^{250}$	3-06-00-00716	eth 4; EtOH 4
9483	Phenol, 2,3,6-trichloro- 2,3,6-Trichlorophenol	$C_6H_3Cl_3O$ 197.45	933-75-5 58		4-06-00-00962	H_2O 2; EtOH 4; eth 4; bz 4
9484	Phenol, 2,4,5-trichloro- 2,4,5-Trichlorophenol	$C_6H_3Cl_3O$ 197.45	95-95-4 69	9555 247	4-06-00-00962	H_2O 2; EtOH 4; eth 4; bz 4
9485	Phenol, 2,4,6-trichloro- 2,4,6-Trichlorophenol	$C_6H_3Cl_3O$ 197.45	88-06-2 69	9556 246	4-06-00-01005 1.4901^{75}	H_2O 2; EtOH 3; eth 3; HOAc 3
9486	Phenol, 3,4,5-trichloro- 3,4,5-Trichlorophenol	$C_6H_3Cl_3O$ 197.45	609-19-8 101	 275	3-06-00-00729	H_2O 2; eth 3; lig 2
9487	Phenol, 2,3,4-trichloro-6-methyl- 4,5,6-Trichloro-*o*-cresol	$C_7H_5Cl_3O$ 211.47	551-78-0 77	9548	3-06-00-01268	
9488	Phenol, 2,3,6-trichloro-4-methyl- 2,3,6-Trichloro-*p*-cresol	$C_7H_5Cl_3O$ 211.47	551-77-9 66.5	9546	0-06-00-00404	EtOH 4
9489	Phenol, 2,4,6-trichloro-3-methyl- 2,4,6-Trichloro-*m*-cresol	$C_7H_5Cl_3O$ 211.47	551-76-8 46	9547 265	4-06-00-02070	EtOH 4; MeOH 4; chl 4
9490	Phenol, 3,4,6-trichloro-2-nitro- 3,4,6-Trichloro-2-nitrophenol	$C_6H_2Cl_3NO_3$ 242.44	82-62-2 92.5	9554	3-06-00-00842	
9491	Phenol, 3-(trifluoromethyl)-	$C_7H_5F_3O$ 162.11	98-17-9 -1.8	 178	4-06-00-02060 1.3418^{25}	
9492	Phenol, 2,4,6-triiodo- 2,4,6-Triiodophenol	$C_6H_3I_3O$ 471.80	609-23-4 159.8	 sub	4-06-00-01085	H_2O 1; EtOH 2; eth 3; ace 3
9493	Phenol, 2,3,4-trimethyl-	$C_9H_{12}O$ 136.19	526-85-2 81	 236	4-06-00-03245	bz 4; eth 4; EtOH 4
9494	Phenol, 2,4,5-trimethyl-	$C_9H_{12}O$ 136.19	496-78-6 72	 232	4-06-00-03247	H_2O 1; EtOH 4; eth 4
9495	Phenol, 2,4,6-trimethyl-	$C_9H_{12}O$ 136.19	527-60-6 73	 220	4-06-00-03253	eth 4; EtOH 4
9496	Phenol, 2,4,5-trimethyl-, acetate 2,4,5-Trimethylphenyl acetate	$C_{11}H_{14}O_2$ 178.23	69305-42-6 34	 245.5	2-06-00-00482	eth 4; EtOH 4

No.	Name Synonym	Mol. Form. Mol. Wt.	CAS RN mp/°C	Merck No. bp/°C	Beil. Ref. den/g cm^{-3}	Solubility n_D
9497	Phenol, 2,4,6-trinitro- Picric acid	$C_6H_3N_3O_7$ 229.11	88-89-1 122.5	7380 exp 300	4-06-00-01388	H_2O 2; EtOH 3; eth 3; ace 4 1.763
9498	Phenol, 2,4,6-tris(1,1-dimethylethyl)-	$C_{18}H_{30}O$ 262.44	732-26-3 131	 278	4-06-00-03539 0.864^{27}	H_2O 1; EtOH 3; ace 3; ctc 3
9499	10H-Phenothiazine Thiodiphenylamine	$C_{12}H_9NS$ 199.28	92-84-2 187.5	7220 371	4-27-00-01214	ace 4; bz 4; eth 4; EtOH 4
9500	10H-Phenothiazine-2-acetic acid, 10-methyl- Metiazinic acid	$C_{15}H_{13}NO_2S$ 271.34	13993-65-2 144	6057		chl 3
9501	10H-Phenothiazine, 2-chloro-10-[3-(4-methyl-1-piperazinyl)propyl]- Phenothiazine, 2-chloro-10(3(4-methyl-1-piperazinyl)propyl)	$C_{20}H_{24}ClN_3S$ 373.95	58-38-8 228	7768	4-27-00-01307	
9502	10H-Phenothiazine, 10-[(dimethylamino)acetyl]- Ahistan	$C_{16}H_{16}N_2OS$ 284.38	518-61-6 144.5	179	4-27-00-01272	
9503	10H-Phenothiazine-10-ethanamine, *N,N*-diethyl- Diethazine	$C_{18}H_{22}N_2S$ 298.45	60-91-3	3098 $167^{0.5}$	4-27-00-01240	H_2O 1; dil HCl 3
9504	10H-Phenothiazine-10-ethanamine, *N,N*,α-trimethyl- Promethazine	$C_{17}H_{20}N_2S$ 284.43	60-87-7 60	7797 $190\text{-}3^{0.5}$	4-27-00-01253	H_2O 1; dil HCl 4
9505	10H-Phenothiazine-10-propanamine, 2-chloro-*N,N*-dimethyl- Chlorpromazine	$C_{17}H_{19}ClN_2S$ 318.87	50-53-3	2186 $200\text{-}5^{0.8}$	4-27-00-01300	H_2O 1; EtOH 4; eth 4; bz 4
9506	Phenothiazin-5-ium, 3,7-bis(dimethylamino)-, chloride Methylene blue	$C_{16}H_{18}ClN_3S$ 319.86	61-73-4	5979	4-27-00-05152	H_2O 3; EtOH 3; eth 1; chl 3
9507	10H-Phenoxazine Phenoxazine	$C_{12}H_9NO$ 183.21	135-67-1 156	7222 dec	4-27-00-01209	bz 4; eth 4; EtOH 4
9508	3H-Phenoxazin-3-one, 7-hydroxy-, 10-oxide Resazurin	$C_{12}H_7NO_4$ 229.19	550-82-3	8144 sub	4-27-00-03594	H_2O 1; EtOH 2; eth 1; alk 3
9509	*DL*-Phenylalanine	$C_9H_{11}NO_2$ 165.19	150-30-1 284-8 dec	7242 sub	4-14-00-01553	H_2O 4; EtOH 1; eth 1; bz 1
9510	*L*-Phenylalanine (*S*)-α-Aminobenzenepropanoic acid	$C_9H_{11}NO_2$ 165.19	63-91-2 283 dec	7242	4-14-00-01552	H_2O 4; EtOH 1; eth 1; bz 1
9511	*L*-Phenylalanine, *N*-*L*-α-aspartyl-, 1-methyl ester Aspartame	$C_{14}H_{18}N_2O_5$ 294.31	22839-47-0 246.5	861		
9512	*L*-Phenylalanine, ethyl ester	$C_{11}H_{15}NO_2$ 193.25	3081-24-1	 148^{13}	4-14-00-01556 1.065^{15}	H_2O 2
9513	Phorate Phosphorodithioic acid, *O,O*-diethyl *S*-[(ethylthio)methyl] ester	$C_7H_{17}O_2PS_3$ 260.38	298-02-2 <-15	7305 $119^{0.8}$	 1.16^{25}	
9514	Phosalone Phosphorodithioic acid, *S*-[(6-chloro-2-oxo-3(2H)-benzoxazolyl)methyl]*O,O*-diethyl ester	$C_{12}H_{15}ClNO_4PS_2$ 367.81	2310-17-0 46	7308		
9515	Phosmet Phosphorodithioic acid, *S*-[(1,3-dihydro-1,3-dioxo-2H-isoindol-2-yl)methyl] *O,O*-dimethyl ester	$C_{11}H_{12}NO_4PS_2$ 317.33	732-11-6 72	7311 dec		
9516	Phosphine, dicyclohexyl- Dicyclohexylphosphine	$C_{12}H_{23}P$ 198.29	829-84-5	 281; 129^8	4-16-00-00947 0.904^{25}	 1.5163^{20}
9517	Phosphine oxide, triethyl-	$C_6H_{15}OP$ 134.16	597-50-2 48	 243	4-04-00-03463	H_2O 4; eth 4; EtOH 4
9518	Phosphine oxide, triphenyl-	$C_{18}H_{15}OP$ 278.29	791-28-6 156.5	 >360	4-16-00-01011 1.2124^{23}	H_2O 2; EtOH 4; eth 2; bz 4
9519	Phosphine, phenyl- Monophenylphosphine	C_6H_7P 110.10	638-21-1	 160.5	4-16-00-00948 1.001^{15}	 1.5796^{20}
9520	Phosphine, tributyl- Tributylphosphine	$C_{12}H_{27}P$ 202.32	998-40-3	 150^{50}	4-04-00-03436 0.812^{25}	 1.4619^{20}
9521	Phosphine, triethyl- Triethyl phosphine	$C_6H_{15}P$ 118.16	554-70-1 -88	9590 129	4-04-00-03431 0.8006^{19}	H_2O 1; EtOH 5; eth 5 1.458^{15}
9522	Phosphine, trimethyl- Trimethylphosphine	C_3H_9P 76.08	594-09-2 -85	 37.5	4-04-00-03429	H_2O 1; eth 3
9523	Phosphine, triphenyl- Triphenylphosphine	$C_{18}H_{15}P$ 262.29	603-35-0 80	9657 188^1	4-16-00-00951 1.0749^{80}	H_2O 1; EtOH 3; eth 4; bz 3 1.6358^{80}
9524	Phosphine, tris(chloromethyl)- Tris(chloromethyl)phosphine	$C_3H_6Cl_3P$ 179.41	13482-62-7	 100^7	3-01-00-02608 1.414^{20}	ace 4; bz 4; eth 4; EtOH 4
9525	Phosphinic acid, diethyl- Diethylphosphinic acid	$C_4H_{11}O_2P$ 122.10	813-76-3 18.5	 320	4-04-00-03474	H_2O 4; EtOH 4; eth 4
9526	Phosphinic acid, diethyl-, anhydride	$C_8H_{20}O_3P_2$ 226.19	7495-97-8	 188^{14}	4-04-00-03477 1.1053^{20}	 1.4647^{20}
9527	Phosphinic acid, diethyl-, ethyl ester Ethyl diethylphosphinate	$C_6H_{15}O_2P$ 150.16	4775-09-1	 95^{14}	4-04-00-03475 0.9908^{20}	eth 4; EtOH 4 1.4337^{20}
9528	Phosphinic acid, dimethyl- Dimethylphosphinic acid	$C_2H_7O_2P$ 94.05	3283-12-3 92	 377	4-04-00-03470	H_2O 4; EtOH 4; eth 4; bz 3
9529	Phosphinic chloride, diethyl-	$C_4H_{10}ClOP$ 140.55	1112-37-4	 104^{15}	4-04-00-03476 1.1394^{20}	 1.4647^{20}
9530	Phosphonic acid, acetyl-, diethyl ester *O,O*-Diethyl acetylphosphonate	$C_6H_{13}O_4P$ 180.14	919-19-7	 $114\text{-}5^{20}$	4-02-00-00440 1.1005^{20}	 1.4200^{26}

No.	Name Synonym	Mol. Form. Mol. Wt.	CAS RN mp/°C	Merck No. bp/°C	Beil. Ref. den/g cm^{-3}	Solubility n_D
9531	Phosphonic acid, bis(1,1-dimethylethyl) ester	$C_8H_{19}O_3P$ 194.21	13086-84-5	 $70\text{-}2^{10}$	4-01-00-01623 0.975^{25}	 1.4168^{25}
9532	Phosphonic acid, dibutyl ester	$C_8H_{19}O_3P$ 194.21	1809-19-4	 131^{19}; $77^{1.5}$	4-01-00-01525 0.9485^{20}	 1.4220^{20}
9533	Phosphonic acid, (1,1-dimethylethyl)-	$C_4H_{11}O_3P$ 138.10	4923-84-6 192	 dec	4-04-00-03557	H_2O 4; EtOH 4
9534	Phosphonic acid, diphenyl ester	$C_{12}H_{11}O_3P$ 234.19	4712-55-4 12	 $218\text{-}9^{26}$	4-06-00-00703 1.223^{25}	 1.5575^{20}
9535	Phosphonic acid, ethenyl-, diethyl ester	$C_6H_{13}O_3P$ 164.14	682-30-4	 110^{2}	4-04-00-03560 1.068^{25}	 1.4290^{20}
9536	Phosphonic acid, ethyl-, diethyl ester Diethyl ethylphosphonate	$C_6H_{15}O_3P$ 166.16	78-38-6	 198; 90^{16}	4-04-00-03524 1.0259^{20}	H_2O 2; EtOH 3; eth 3 1.4163^{20}
9537	Phosphonic acid, ethyl-, dimethyl ester	$C_4H_{11}O_3P$ 138.10	6163-75-3	 82^{18}	4-04-00-03523 1.1029^{30}	H_2O 4; bz 4; EtOH 4 1.4128^{30}
9538	Phosphonic acid, methyl-	CH_5O_3P 96.02	993-13-5 108.5	 dec	4-04-00-03498	H_2O 4; EtOH 4; eth 4; bz 1
9539	Phosphonic acid, methyl-, diethyl ester	$C_5H_{13}O_3P$ 152.13	683-08-9	 194	4-04-00-03500 1.0406^{30}	H_2O 3; EtOH 3; eth 3; bz 1 1.4101^{30}
9540	Phosphonic acid, methyl-, dimethyl ester	$C_3H_9O_3P$ 124.08	756-79-6	 181; 79.5^{20}	4-04-00-03499 1.1507^{20}	H_2O 3; EtOH 3; eth 3 1.4099^{30}
9541	Phosphonic acid, methyl-, diphenyl ester Diphenyl methylphosphonate	$C_{13}H_{13}O_3P$ 248.22	7526-26-3 35	 205^{13}	4-06-00-00703 1.2051^{20}	H_2O 1
9542	Phosphonic acid, (1-methylethyl)- 2-Propanephosphonic acid	$C_3H_9O_3P$ 124.08	4721-37-3 74.5	 dec	4-04-00-03548	H_2O 4; eth 4; EtOH 4
9543	Phosphonic acid, methyl-, monoethyl ester	$C_3H_9O_3P$ 124.08	1832-53-7	 $106\text{-}7^{0.1}$	4-04-00-03499 1.1800^{20}	 1.4258^{20}
9544	Phosphonic acid, (2-methylpropyl)-	$C_4H_{11}O_3P$ 138.10	4721-34-0 119	 dec	4-04-00-03556	H_2O 4; eth 4; EtOH 4; xyl 4
9545	Phosphonic acid, (phenylmethyl)-, diethyl ester	$C_{11}H_{17}O_3P$ 228.23	1080-32-6	 110^{2}	4-16-00-01093	ctc 3 1.4930^{20}
9546	Phosphonic acid, 1,3-propanediylbis-, tetraethyl ester	$C_{11}H_{26}O_6P_2$ 316.27	22401-25-8	 $185\text{-}6^{3}$	4-04-00-03582 1.1227^{20}	 1.4500^{20}
9547	Phosphonic acid, propyl-	$C_3H_9O_3P$ 124.08	4672-38-2 73.8	 dec	4-04-00-03542	H_2O 4; eth 4; EtOH 4
9548	Phosphonic dichloride, phenyl-	$C_6H_5Cl_2OP$ 194.98	824-72-6 1	 258	4-16-00-01074 1.197^{25}	DMSO-d_6 2 1.5581^{25}
9549	Phosphonothioic acid, methyl-, *O*,*O*-bis(1-methylethyl) ester Phosphonic acid, thiono, methyl, diisopropyl ester	$C_7H_{17}O_2PS$ 196.25	66295-45-2	 42^{1}	4-04-00-03516 0.9885^{25}	 1.4512^{25}
9550	Phosphonothioic acid, methyl-, *O*,*O*-dimethyl ester	$C_3H_9O_2PS$ 140.14	681-06-1	 43^{4}	4-04-00-03514 1.1375^{25}	 1.4738^{25}
9551	Phosphonothioic acid, phenyl-, *O*-ethyl *O*-(4-nitrophenyl) ester *O*-Ethyl *O*-*p*-nitrophenyl benzenethiophosphonate	$C_{14}H_{14}NO_4PS$ 323.31	2104-64-5 36	3576	4-16-00-01089 1.27^{25}	bz 4; eth 4; EtOH 4 1.5978^{30}
9552	Phosphonothioic dichloride, phenyl- Dichlorophenylphosphine sulfide	$C_6H_5Cl_2PS$ 211.05	3497-00-5	 205^{130}	4-16-00-01090 1.376^{13}	
9553	Phosphonous dichloride, (2,5-dimethylphenyl)- Phosphonous dichloride, 2,5-xylyl-	$C_8H_9Cl_2P$ 207.04	57150-66-0 -30	 253.5	3-16-00-00848 1.25^{18}	
9554	Phosphonous dichloride, (4-ethylphenyl)-	$C_8H_9Cl_2P$ 207.04	5274-50-0	 251	4-16-00-00974 1.237^{20}	 1.584^{20}
9555	Phosphonous dichloride, [4-(1-methylethyl)phenyl]- Benzene, 1-chlorophosphino-4-isopropyl	$C_9H_{11}Cl_2P$ 221.07	82906-78-3	 269	3-16-00-00848 1.190^{12}	ace 4 1.5677^{25}
9556	Phosphonous dichloride, phenyl- Dichlorophenylphosphine	$C_6H_5Cl_2P$ 178.98	644-97-3 -51	 225; 142^{57}	4-16-00-00972 1.356^{20}	bz 4 1.6030^{20}
9557	Phosphoramidothioic dichloride, dimethyl- Phosphoric acid, amide,thiono,dichloride,*N*,*N*-dimethyl	$C_2H_6Cl_2NPS$ 178.02	1498-65-3	 77.5^{10}	4-04-00-00293 1.3731^{20}	 1.5390^{20}
9558	Phosphoric acid, diethyl ester	$C_4H_{11}O_4P$ 154.10	598-02-7	 203 dec	4-01-00-01339 1.186^{25}	eth 4 1.4170^{20}
9559	Phosphoric acid, dimethyl ester	$C_2H_7O_4P$ 126.05	813-78-5	 174 dec	4-01-00-01259 1.335^{25}	H_2O 4; ace 4; EtOH 4 1.408^{25}
9560	Phosphoric acid, 2-ethylhexyl diphenyl ester	$C_{20}H_{27}O_4P$ 362.41	1241-94-7	 232^{5}	4-06-00-00718 1.090^{25}	 1.510^{25}
9561	Phosphoric acid, monoethyl ester	$C_2H_7O_4P$ 126.05	1623-14-9	 dec	4-01-00-01338 1.430^{25}	H_2O 4; ace 4; eth 4; EtOH 4 1.427
9562	Phosphoric acid tributyl ester Tributyl phosphate	$C_{12}H_{27}O_4P$ 266.32	126-73-8	9531 289	4-01-00-01531 0.9727^{25}	H_2O 3; EtOH 5; eth 3; bz 3 1.4224^{25}
9563	Phosphoric acid, triethyl ester Ethyl phosphate	$C_6H_{15}O_4P$ 182.16	78-40-0 -56.4	9589 215.5	4-01-00-01339 1.0695^{20}	H_2O 3; EtOH 4; eth 3; bz 3 1.4053^{20}
9564	Phosphoric acid, trimethyl ester	$C_3H_9O_4P$ 140.08	512-56-1 -46	 197.2	4-01-00-01259 1.2144^{20}	H_2O 4; EtOH 2; eth 3 1.3967^{20}
9565	Phosphoric acid, tripentyl ester	$C_{15}H_{33}O_4P$ 308.40	2528-38-3	 225^{50}; 167^{5}	4-01-00-01645 0.9608^{20}	H_2O 1; EtOH 3; eth 3; tol 3 1.4319^{20}
9566	Phosphoric acid, triphenyl ester Triphenyl phosphate	$C_{18}H_{15}O_4P$ 326.29	115-86-6 50.5	9656 245^{11}	4-06-00-00720 1.2055^{50}	H_2O 1; EtOH 3; eth 4; bz 4

No.	Name	Synonym	Mol. Form.	Mol. Wt.	CAS RN	mp/°C	Merck No.	bp/°C	Beil. Ref.	den/g cm^{-3}	Solubility	n_D
9567	Phosphoric acid, tripropyl ester	Tripropyl phosphate	$C_9H_{21}O_4P$	224.24	513-08-6			252	4-01-00-01428	1.0121^{20}	H_2O 2; EtOH 3; eth 3; chl 2	1.4165^{20}
9568	Phosphoric acid, tris(1-methylethyl) ester		$C_9H_{21}O_4P$	224.24	513-02-0			219	4-01-00-01478	0.9867^{20}	EtOH 4	1.4057^{20}
9569	Phosphoric acid, tris(2-methylphenyl) ester	Tri-*o*-tolyl phosphate	$C_{21}H_{21}O_4P$	368.37	78-30-8	11	9676	410	4-06-00-01979	1.1955^{20}	H_2O 1; EtOH 4; eth 4; ctc 4	1.5575^{20}
9570	Phosphoric acid, tris(3-methylphenyl) ester		$C_{21}H_{21}O_4P$	368.37	563-04-2	25.5		260^{15}	4-06-00-02057	1.150^{25}	H_2O 1; EtOH 2; eth 3; ctc 4	1.5575^{20}
9571	Phosphoric acid, tris(4-methylphenyl) ester		$C_{21}H_{21}O_4P$	368.37	78-32-0	77.5		224^{35}	4-06-00-02130	1.247^{25}	EtOH 3; eth 3; bz 3; chl 3	
9572	Phosphoric acid, tris(2-methylpropyl) ester	Triisobutyl phosphate	$C_{12}H_{27}O_4P$	266.32	126-71-6			264	4-01-00-01598	0.9681^{20}	H_2O 4; bz 4; eth 4; EtOH 4	1.4193^{20}
9573	Phosphoric triamide, hexamethyl-	Tris(dimethylamino)phosphine oxide	$C_6H_{18}N_3OP$	179.20	680-31-9		4568	232.5	4-04-00-00284	1.03^{20}	EtOH 3; eth 3	1.4579^{20}
9574	Phosphorochloridic acid, diethyl ester	Diethyl chlorophosphate	$C_4H_{10}ClO_3P$	172.55	814-49-3			93.5	4-01-00-01345	1.205^{19}		1.4170^{20}
9575	Phosphorochloridic acid, diphenyl ester		$C_{12}H_{10}ClO_3P$	268.64	2524-64-3			$314\text{-}6^{272}$	4-06-00-00737	1.296^{25}	tfa 3	1.5500^{20}
9576	Phosphorochloridothioic acid, *O,O*-diethyl ester	Diethyl chlorothiophosphate	$C_4H_{10}ClO_2PS$	188.61	2524-04-1			45^{3}	4-01-00-01352		ctc 3	
9577	Phosphorodichloridic acid, ethyl ester	Ethyl dichlorophosphate	$C_2H_5Cl_2O_2P$	162.94	1498-51-7			$60\text{-}5^{10}$	4-01-00-01345			1.4338^{20}
9578	Phosphorodichloridothioic acid, *O*-(2,4,5-trichlorophenyl) ester	Phenol, 2,4,5-trichloro-, *O*-phosphorodichloridothioate	$C_6H_2Cl_5OPS$	330.38	16805-78-0			110^{1}	4-06-00-00996	1.6653^{20}		1.6084^{20}
9579	Phosphorodithioic acid, *O,O,S*-triethyl ester		$C_6H_{15}O_2PS_2$	214.29	2524-09-6			128^{20}	4-01-00-01402	1.1336^{0}		1.5043
9580	Phosphorofluoridic acid, bis(1-methylethyl) ester	Isoflurophate	$C_6H_{14}FO_3P$	184.15	55-91-4		5060	62^{9}	4-01-00-01480	1.055^{25}	H_2O 2; eth 3; oils 4; lig 2	1.3830^{25}
9581	Phosphorothioic acid, *O*-(2,4-dichlorophenyl) *O,O*-diethyl ester	Dichlofenthion	$C_{10}H_{13}Cl_2O_3PS$	315.16	97-17-6		3030		4-06-00-00939		ctc 3; CS_2 3	
9582	Phosphorothioic acid, *O,O*-diethyl *O*-[6-methyl-2-(1-methylethyl)-4-pyrimidinyl] ester	Diazinon	$C_{12}H_{21}N_2O_3PS$	304.35	333-41-5		2978	$85\text{-}90^{0.05}$	5-23-11-00187	1.1088^{20}		1.4922^{20}
9583	Phosphorothioic acid, *O,O*-diethyl *O*-(4-nitrophenyl) ester	Parathion	$C_{10}H_{14}NO_5PS$	291.26	56-38-2	6.1	6983	375	4-06-00-01337	1.2681^{20}	H_2O 1; EtOH 5; eth 3; ace 3	1.5370^{25}
9584	Phosphorothioic acid, *O,O*-dimethyl *O*-(2,4,5-trichlorophenyl) ester	Ronnel	$C_8H_8Cl_3O_3PS$	321.55	299-84-3	41	8239	$151\text{-}4^{0.4}$	4-06-00-00994	1.44^{32}		1.5335^{35}
9585	Phosphorothioic acid, *O,O,O*-triethyl ester		$C_6H_{15}O_3PS$	198.22	126-68-1			100^{16}	4-01-00-01348	1.082^{25}		1.4480^{20}
9586	Phosphorous acid, bis(phenylmethyl) ester	Dibenzyl phosphite	$C_{14}H_{15}O_3P$	262.25	17176-77-1	-5	2996	$162^{0.1}$	4-06-00-02571			1.5521^{18}
9587	Phosphorous acid, diethyl ester		$C_4H_{11}O_3P$	138.10	123-22-8			87^{20}		1.0720^{20}	EtOH 3; eth 3	1.4101^{20}
9588	Phosphorous acid, dimethyl ester		$C_2H_7O_3P$	110.05	868-85-9			170.5		1.2002^{20}	EtOH 3; ctc 2; py 3	1.4036^{20}
9589	Phosphorous acid, 2-ethylhexyl diphenyl ester		$C_{20}H_{27}O_3P$	346.41	15647-08-2			$148\text{-}56^{0.15}$	4-06-00-00695	1.054^{20}		1.5207^{27}
9590	Phosphorous acid, tributyl ester		$C_{12}H_{27}O_3P$	250.32	102-85-2			122^{12}	4-01-00-01527	0.9259^{20}	EtOH 3; eth 4; ctc 2	1.4321^{19}
9591	Phosphorous acid, triethyl ester	Triethyl phosphite	$C_6H_{15}O_3P$	166.16	122-52-1			157.9	4-01-00-01333	0.9629^{20}	H_2O 1; EtOH 4; eth 4	1.4127^{20}
9592	Phosphorous acid, trimethyl ester	Trimethyl phosphite	$C_3H_9O_3P$	124.08	121-45-9			111.5	4-01-00-01256	1.0518^{20}	EtOH 4; eth 4; ctc 2	1.4095^{20}
9593	Phosphorous acid, triphenyl ester		$C_{18}H_{15}O_3P$	310.29	101-02-0	25		360	4-06-00-00695	1.1842^{20}	H_2O 1; EtOH 4; os 4	1.5900^{20}
9594	Phosphorous acid, tripropyl ester	Propyl phosphite	$C_9H_{21}O_3P$	208.24	923-99-9			206.5	4-01-00-01426	0.9417^{20}	eth 4; EtOH 4	1.4282^{20}
9595	Phosphorous acid, tris(1-methylethyl) ester		$C_9H_{21}O_3P$	208.24	116-17-6			$60\text{-}1^{10}$	4-01-00-01476	0.9685^{18}	EtOH 3; eth 3; chl 3	1.4085^{25}
9596	Phosphorous acid, tris(2-methylphenyl) ester		$C_{21}H_{21}O_3P$	352.37	2622-08-4			238^{11} 198^{2}	4-06-00-01977	1.1423^{20}	eth 3; chl 2	1.5740^{28}
9597	Phosphorous acid, tris(4-methylphenyl) ester		$C_{21}H_{21}O_3P$	352.37	620-42-8			$250\text{-}5^{10}$	4-06-00-02128	1.1280^{25}	eth 4	1.5703^{28}
9598	*p*-Phthalaldehydic acid, 3-hydroxy-	3-Hydroxybenzoic acid, 4-formyl	$C_8H_6O_4$	166.13	619-12-5	234		sub	4-10-00-03623		eth 4; EtOH 4	
9599	Phthalazine	2,3-Benzodiazine	$C_8H_6N_2$	130.15	253-52-1	90.5	7344	316	5-23-07-00149		H_2O 3; EtOH 3; eth 2; bz 3	
9600	1,4-Phthalazinedione, 5-amino-2,3-dihydro-	Luminol	$C_8H_7N_3O_2$	177.16	521-31-3	330.5	5470		4-25-00-04192		H_2O 1; EtOH 2; eth 2; alk 4	
9601	Picene	Benzo[a]chrysene	$C_{22}H_{14}$	278.35	213-46-7	368	7368	519	4-05-00-02724		H_2O 1; EtOH 2; bz 2; chl 2	
9602	Piperazine	Diethylenediamine	$C_4H_{10}N_2$	86.14	110-85-0	106	7431	146	5-23-01-00030		H_2O 4; EtOH 3; eth 1; chl 3	1.446^{113}

No.	Name	Synonym	Mol. Form.	Mol. Wt.	CAS RN	mp/°C	Merck No.	bp/°C	Beil. Ref.	den/g cm^{-3}	Solubility	n_D
9603	1-Piperazinecarboxamide, *N,N*-diethyl-4-methyl-	Diethylcarbamazine	$C_{10}H_{21}N_3O$	199.30	90-89-1	48	3106	109-11^{3}	4-23-00-00225			
9604	1-Piperazinecarboxylic acid, ethyl ester		$C_7H_{14}N_2O_2$	158.20	120-43-4			237	5-23-01-00055		H_2O 4; eth 4; EtOH 4	1.4760^{25}
9605	Piperazine, 1-[(4-chlorophenyl)phenylmethyl]-4-[(3-methylphenyl)methyl]-	Meclizine	$C_{25}H_{27}ClN_2$	390.96	569-65-3		5657	230	5-23-01-00235		CS_2 3	
9606	1,4-Piperazinedicarboxylic acid, diethyl ester	*N,N'*-Piperazinedicarboxylic acid, diethyl ester	$C_{10}H_{18}N_2O_4$	230.26	5470-28-0	47		315	4-23-00-00254		eth 4; EtOH 4	
9607	Piperazine, dihydrochloride	Diethylenediamine dihydrochloride	$C_4H_{12}Cl_2N_2$	159.06	142-64-3				5-23-01-00030		H_2O 2; EtOH 1	
9608	Piperazine, 1,4-dimethyl-		$C_6H_{14}N_2$	114.19	106-58-1			131	5-23-01-00128	0.8600^{20}	H_2O 4; EtOH 4; eth 4	1.4474^{20}
9609	Piperazine, 2,5-dimethyl-, *cis*-	*cis*-2,5-Dimethylpiperazine	$C_6H_{14}N_2$	114.19	6284-84-0	114		162	5-23-03-00304		H_2O 4; EtOH 4; eth 2; bz 2	1.4720^{20}
9610	Piperazine, 2,5-dimethyl-, *trans*-	*trans*-2,5-Dimethylpiperazine	$C_6H_{14}N_2$	114.19	2815-34-1	118.5		162	5-23-03-00304		H_2O 4; EtOH 4; eth 2; bz 2	
9611	Piperazine, 2,6-dimethyl-, *cis*-	*cis*-2,6-Dimethylpiperazine	$C_6H_{14}N_2$	114.19	21655-48-1	114		160	5-23-03-00326		H_2O 3; EtOH 3; eth 1; bz 2	
9612	2,5-Piperazinedione	Glycine anhydride	$C_4H_6N_2O_2$	114.10	106-57-0	312 dec	7434	sub 260	5-24-05-00294		H_2O 2; EtOH 2; HCl 3	
9613	Piperazine, 1,4-diphenyl-		$C_{16}H_{18}N_2$	238.33	613-39-8	165		300 dec; 232^{12}	5-23-01-00173		H_2O 1; EtOH 2	
9614	1,4-Piperazinedipropanamine	1,4-Bis(3-aminopropyl)piperazine	$C_{10}H_{24}N_4$	200.33	7209-38-3	15		150-2^{2}	5-23-01-00356	0.973^{25}		1.5015^{20}
9615	1-Piperazineethanamine	1-(2-Aminoethyl)piperazine	$C_6H_{15}N_3$	129.21	140-31-8			220	5-23-01-00257	0.985^{25}		1.4983^{20}
9616	1-Piperazineethanol		$C_6H_{14}N_2O$	130.19	103-76-4			246	5-23-01-00406	1.061^{25}		1.5065^{20}
9617	1-Piperazineethanol, 4-[3-(2-chloro-10H-phenothiazin-10-yl)propyl]-	Perphenazine	$C_{21}H_{26}ClN_3OS$	403.98	58-39-9	97	7135		4-27-00-01307			
9618	Piperazine, 1-methyl-		$C_5H_{12}N_2$	100.16	109-01-3			138	5-23-01-00082		H_2O 4; eth 4; EtOH 4	1.4378^{20}
9619	Piperazine, 2-methyl-	2-Methylpiperazine	$C_5H_{12}N_2$	100.16	109-07-9	62		153	5-23-03-00267		H_2O 4; EtOH 3; eth 3; bz 3	
9620	Piperazine, 1-phenyl-		$C_{10}H_{14}N_2$	162.23	92-54-6			286.5; 161^{15}	5-23-01-00160	1.0621^{20}	H_2O 1; EtOH 5; eth 5; chl 3	1.5875^{20}
9621	Piperazine, 1-(phenylmethyl)-		$C_{11}H_{16}N_2$	176.26	2759-28-6			145-7^{12}	5-23-01-00193		H_2O 3; EtOH 3; eth 3; chl 2	1.5430^{28}
9622	Piperazine, 1,2,4-trimethyl-	1,2,4-Trimethylpiperazine	$C_7H_{16}N_2$	128.22	120-85-4			149.5	5-23-03-00271		ctc 3	1.4433^{20}
9623	Piperazinium, 4-(2-cyclohexyl-2-hydroxy-2-phenylethyl)-1,1-dimethyl-, methyl sulfate (salt)	Hexocyclium methyl sulfate	$C_{21}H_{36}N_2O_5S$	428.59	115-63-9		4627					
9624	1-Piperidinamine		$C_5H_{12}N_2$	100.16	2213-43-6			147	5-20-03-00516	0.928^{25}		1.4750^{20}
9625	1-Piperidinamine, 2,6-dimethyl-		$C_7H_{16}N_2$	128.22	39135-39-2			65-80^{30}	5-20-04-00187	0.865^{25}		1.4650^{20}
9626	3-Piperidinamine, 1-ethyl-		$C_7H_{16}N_2$	128.22	6789-94-2			155; 82^{40}	5-22-08-00026	0.923^{25}		1.4715^{20}
9627	4-Piperidinamine, 1-methyl-*N*-phenyl-*N*-(phenylmethyl)-	Bamipine	$C_{19}H_{24}N_2$	280.41	4945-47-5	115	966		5-22-08-00036			
9628	4-Piperidinamine, 2,2,6,6-tetramethyl-		$C_9H_{20}N_2$	156.27	36768-62-4	17		188.5	5-22-08-00162	0.912^{25}		1.4706^{20}
9629	Piperidine	Azacyclohexane	$C_5H_{11}N$	85.15	110-89-4	-11.03	7438	106.2	5-20-02-00003	0.8606^{20}	H_2O 5; EtOH 5; eth 3; ace 3	1.4530^{20}
9630	Piperidine, 1-acetyl-		$C_7H_{13}NO$	127.19	618-42-8	-13.4		226.5	5-20-02-00440	1.011^{9}	H_2O 4; EtOH 4	1.4790^{25}
9631	Piperidine, 1-acetyl-3-methyl-		$C_8H_{15}NO$	141.21	4593-16-2	-13.6		239	5-20-04-00107	0.9684^{25}	H_2O 4	1.4731^{25}
9632	Piperidine, 1-[5-(1,3-benzodioxol-5-yl)-1-oxo-2,4-pentadienyl]-, (*E,E*)-	Piperine	$C_{17}H_{19}NO_3$	285.34	94-62-2	131.5	7442		5-20-03-00469		H_2O 1; EtOH 3; eth 2; bz 3	
9633	Piperidine, 1-[5-(1,3-benzodioxol-5-yl)-1-oxo-2,4-pentadienyl]-, (*Z,Z*)-	Chavicine	$C_{17}H_{19}NO_3$	285.34	495-91-0		2038		5-20-03-00469		eth 4; EtOH 4; peth 4	
9634	Piperidine, 1-benzoyl-		$C_{12}H_{15}NO$	189.26	776-75-0	49		320.5	5-20-02-00450		H_2O 1; EtOH 3; eth 3; ctc 2	
9635	Piperidine, 1-butyl-		$C_9H_{19}N$	141.26	4945-48-6			176	5-20-02-00029	0.8245^{20}		1.4467^{20}
9636	Piperidine, 3-butyl-	3-Butylpiperidine	$C_9H_{19}N$	141.26	13603-21-9			196.5	5-20-04-00230	0.8620^{18}	eth 4; EtOH 4; chl 4	
9637	Piperidine, 4-butyl-	4-*n*-Butylpiperidine	$C_9H_{19}N$	141.26	24152-39-4			194	4-20-00-01635	0.879^{23}		1.4472^{23}
9638	1-Piperidinecarboxaldehyde		$C_6H_{11}NO$	113.16	2591-86-8	-30.8		222.5	5-20-02-00432	1.0158^{25}	H_2O 5; EtOH 5; eth 5; bz 5	1.4805^{25}

No.	Name Synonym	Mol. Form. Mol. Wt.	CAS RN mp/°C	Merck No. bp/°C	Beil. Ref. den/g cm^{-3}	Solubility n_D
9639	2-Piperidinecarboxamide, *N*-(2,6-dimethylphenyl)-1-methyl- Mepivacaine	$C_{15}H_{22}N_2O$ 246.35	96-88-8	5748	5-22-01-00223	CS_2 3
9640	2-Piperidinecarboxylic acid Pipecolic acid	$C_6H_{11}NO_2$ 129.16	535-75-1 264	7425	5-22-01-00220	H_2O 3; EtOH 2
9641	3-Piperidinecarboxylic acid Nipecotic acid	$C_6H_{11}NO_2$ 129.16	498-95-3 261 dec	6478	5-22-01-00235	H_2O 4
9642	4-Piperidinecarboxylic acid Isonipecotic acid	$C_6H_{11}NO_2$ 129.16	498-94-2 336	5075	5-22-01-00244	
9643	4-Piperidinecarboxylic acid, ethyl ester	$C_8H_{15}NO_2$ 157.21	1126-09-6	$100\text{-}1^{10}$	4-22-00-00128	H_2O 4; bz 4; eth 4; EtOH 4 1.4591^{20}
9644	4-Piperidinecarboxylic acid, methyl ester Methyl isonipecotate	$C_7H_{13}NO_2$ 143.19	2971-79-1	$107\text{-}10^{22}$	4-22-00-00128	H_2O 4; bz 4; eth 4; EtOH 4 1.4635^{25}
9645	4-Piperidinecarboxylic acid, 1-methyl-, methyl ester Methyl *N*-methylisonipecotinate	$C_8H_{15}NO_2$ 157.21	1690-75-1	98^{20}; 84^{10}	5-22-01-00245	H_2O 3; EtOH 3; eth 3; bz 3 1.4539^{24}
9646	2,6-Piperidinediethanol, 1-methyl-α,α'-diphenyl-, [2α(*R**),6α(*S**)]- Lobelanidine	$C_{22}H_{29}NO_2$ 339.48	552-72-7 150	5429	4-21-00-02386	H_2O 1; EtOH 3; eth 2; ace 4
9647	Piperidine, 2,5-diethyl-	$C_9H_{19}N$ 141.26	116836-21-6	190	0-20-00-00128 0.8722^{25}	eth 4; chl 4
9648	Piperidine, 1,2-dimethyl-, (±)- Piperidine, *N*,2-dimethyl (*DL*)	$C_7H_{15}N$ 113.20	2512-81-4	127.5	5-20-04-00072 0.824^{15}	H_2O 4; eth 4; EtOH 4 1.4395^{20}
9649	Piperidine, 2,5-dimethyl- 2,5-Dimethylpiperidine	$C_7H_{15}N$ 113.20	34893-50-0	137	5-20-04-00179 0.8317^{20}	EtOH 4 1.4452^{25}
9650	Piperidine, 2,6-dimethyl-	$C_7H_{15}N$ 113.20	504-03-0	127	5-20-04-00180 0.8158^{25}	H_2O 5; EtOH 5; eth 5; ctc 2 1.4377^{20}
9651	Piperidine, 3,3-dimethyl- 3,3-Dimethylpiperidine	$C_7H_{15}N$ 113.20	1193-12-0	137	5-20-04-00189	EtOH 4 1.4452^{25}
9652	Piperidine, 3,5-dimethyl- 3,5-Lupetidine	$C_7H_{15}N$ 113.20	35794-11-7	144	5-20-04-00191 0.853^{25}	1.4454^{20}
9653	Piperidine, 1-(1,1-dimethylethyl)- Piperidine, *N*-*tert*-butyl	$C_9H_{19}N$ 141.26	14446-69-6	166	5-20-02-00031 0.8465^{20}	eth 4; EtOH 4 1.4532^{20}
9654	2,4-Piperidinedione, 3,3-diethyl- Piperidione	$C_9H_{15}NO_2$ 169.22	77-03-2 104	7439	5-21-09-00609	H_2O 4; EtOH 4; chl 4; MeOH 4
9655	2,6-Piperidinedione, 4-[2-(3,5-dimethyl-2-oxocyclohexyl)-2-hydroxyethyl]-, Cycloheximide	$C_{15}H_{23}NO_4$ 281.35	66-81-9 119	2734	5-21-13-00434	EtOH 4
9656	2,6-Piperidinedione, 4-ethyl-4-methyl- Bemegride	$C_8H_{13}NO_2$ 155.20	64-65-3 126.5	1036 sub 100	5-21-09-00601	chl 3
9657	Piperidine, 1-[2-[3-(2,2-diphenylethyl)-1,2,4-oxadiazol-5-yl]ethyl]-, monohydrochloride Prenoxdiazine hydrochloride	$C_{23}H_{28}ClN_3O$ 397.95	982-43-4 186.5	7743		
9658	Piperidine, 1-(3,3-diphenylpropyl)- Fenpiprane	$C_{20}H_{25}N$ 279.43	3540-95-2 41.5	3934 $210\text{-}20^{8}$	5-20-02-00091	
9659	Piperidine, 1,1'-dithiobis-	$C_{10}H_{20}N_2S_2$ 232.41	10220-20-9 64.2	160.1^{11}	5-20-03-00487	acid 1
9660	Piperidine, 1-dodecyl-	$C_{17}H_{35}N$ 253.47	5917-47-5	161^{5}; $115^{0.6}$	4-20-00-00321 0.8378^{20}	1.4588^{20}
9661	Piperidine, 1,1'-(1,2-ethanediyl)bis-	$C_{12}H_{24}N_2$ 196.34	1932-04-3 -0.5	265	5-20-03-00148 0.9160^{25}	1.4853^{25}
9662	1-Piperidineethanol	$C_7H_{15}NO$ 129.20	3040-44-6 17.9	202; 90^{12}	5-20-02-00101 0.9703^{25}	H_2O 5; EtOH 4 1.4749^{20}
9663	2-Piperidineethanol 2-(2-Hydroxyethyl)piperidine	$C_7H_{15}NO$ 129.20	1484-84-0 69	202; 145^{36}	5-21-01-00121 1.01^{27}	H_2O 4
9664	4-Piperidineethanol 4-(2-Hydroxyethyl)piperidine	$C_7H_{15}NO$ 129.20	622-26-4 132.5	227.5	4-21-00-00093 1.0059^{15}	H_2O 4; eth 4; EtOH 4 1.4907^{20}
9665	1-Piperidineethanol, 2-(diphenylmethyl)- Diphemethoxidine	$C_{20}H_{25}NO$ 295.42	13862-07-2 106.5	3303 $180\text{-}1^{0.1}$	5-20-08-00173	
9666	Piperidine, 3-ethenyl- Piperidine, 3-vinyl-	$C_7H_{13}N$ 111.19	57502-49-5	153.5	4-20-00-01946 0.9274^{25}	1.4731^{25}
9667	Piperidine, 1-ethyl- *N*-Ethylpiperidine	$C_7H_{15}N$ 113.20	766-09-6	130.8	5-20-02-00025 0.8237^{20}	1.4480^{20}
9668	Piperidine, 2-ethyl-, (±)- Piperidine, 2-ethyl (*DL*)	$C_7H_{15}N$ 113.20	78738-37-1	142.5	5-20-04-00169 0.8649^{0}	1.4494^{21}
9669	Piperidine, 3-ethyl-, (±)- Piperidine, 3-ethyl (*DL*)	$C_7H_{15}N$ 113.20	59433-08-8	152.6	5-20-04-00173 0.8565^{23}	ace 4 1.4531^{20}
9670	Piperidine, 4-ethyl- 4-Ethylpiperidine	$C_7H_{15}N$ 113.20	3230-23-7	154	5-20-04-00174 0.8759^{25}	1.4503^{25}
9671	Piperidine, 1-(1-ethyl-1-methylpropyl)- Piperidine, *N*-(3-methyl-3-pentyl)	$C_{11}H_{23}N$ 169.31	14045-26-2	214	5-20-02-00033 0.8614^{20}	1.4637^{20}
9672	Piperidine, 1-hexyl- 1-Hexylpiperidine	$C_{11}H_{23}N$ 169.31	7335-01-5	219.2	5-20-02-00032 0.8292^{20}	1.4522^{20}
9673	4-Piperidinemethanamine 4-(Aminomethyl)piperidine	$C_6H_{14}N_2$ 114.19	7144-05-0 25	200; 31^{10}	5-22-08-00128	1.4900^{20}
9674	2-Piperidinemethanol	$C_6H_{13}NO$ 115.18	3433-37-2 69	104^{10}; 80^{1}	5-21-01-00098	chl 2
9675	3-Piperidinemethanol	$C_6H_{13}NO$ 115.18	4606-65-9 61	$106\text{-}7^{3.5}$	5-21-01-00107 1.0263^{20}	chl 2 1.4964^{20}

No.	Name Synonym	Mol. Form. Mol. Wt.	CAS RN mp/°C	Merck No. bp/°C	Beil. Ref. den/g cm^{-3}	Solubility n_D
9676	2-Piperidinemethanol, α,α-diphenyl- Pipradrol	$C_{18}H_{21}NO$ 267.37	467-60-7	7455	5-21-04-00070	
9677	2-Piperidinemethanol, α-ethyl-, (*R**,*S**)-(±)- Conhydrine (*DL*) (lower m.p.)	$C_8H_{17}NO$ 143.23	3238-62-8 121	2495 226	5-21-01-00138	H_2O 2; bz 4; eth 4; EtOH 4
9678	3-Piperidinemethanol, 1-methyl-	$C_7H_{15}NO$ 129.20	7583-53-1	 142.5; 114^{20}	5-21-01-00108 1.013^{25}	 1.4772^{20}
9679	Piperidine, 1-methyl- *N*-Methylpiperidine	$C_6H_{13}N$ 99.18	626-67-5	 107	5-20-02-00021 0.8159^{20}	H_2O 4; EtOH 5; eth 5; ctc 3 1.4355^{20}
9680	Piperidine, 2-methyl-, (±)-	$C_6H_{13}N$ 99.18	3000-79-1 -4.9	 118	5-20-04-00071 0.8436^{24}	H_2O 4; EtOH 3; eth 3; chl 2 1.4459^{20}
9681	Piperidine, 3-methyl-, (±)- Piperidine, 3-methyl (*DL*)	$C_6H_{13}N$ 99.18	53152-98-0 -24	 125.5	5-20-04-00099 0.8446^{26}	H_2O 4; chl 2 1.4470^{20}
9682	Piperidine, 4-methyl-	$C_6H_{13}N$ 99.18	626-58-4	 130	5-20-04-00116 0.8674^{25}	H_2O 4; chl 2 1.4458^{20}
9683	Piperidine, 1,1'-methylenebis-	$C_{11}H_{22}N_2$ 182.31	880-09-1	 122-3^{15}	5-20-02-00293 0.915^{25}	 1.4820^{20}
9684	Piperidine, 1-(1-methylethyl)-	$C_8H_{17}N$ 127.23	766-79-0	 149.5	5-20-02-00028 0.8389^{20}	H_2O 4 1.4491^{20}
9685	Piperidine, 1-(2-methylpropyl)-	$C_9H_{19}N$ 141.26	10315-89-6	 160.5	5-20-02-00030 0.8161^{25}	H_2O 4 1.4382^{25}
9686	Piperidine, 1-methyl-2-propyl-, (*S*)- Methylconiine	$C_9H_{19}N$ 141.26	35305-13-6	5968 174	5-20-04-00206 0.8326^{22}	ace 4; EtOH 4 1.4538^{12}
9687	Piperidine, 1-nitro-	$C_5H_{10}N_2O_2$ 130.15	7119-94-0 -5.5	 245; 121^{20}	5-20-03-00531 1.1519^{26}	 1.4954^{26}
9688	Piperidine, 1-nitroso- *N*-Nitrosopiperidine	$C_5H_{10}N_2O$ 114.15	100-75-4	 219; 109^{20}	5-20-03-00530 1.0631^{18}	H_2O 3; HCl 3 1.4933^{18}
9689	Piperidine, 1-nonyl-	$C_{14}H_{29}N$ 211.39	30538-80-8	 135-7^{11}	4-20-00-00319 0.8313^{25}	 1.4538^{25}
9690	Piperidine, 1-octyl-	$C_{13}H_{27}N$ 197.36	7335-02-6	 137^{13}; 89^{1}	5-20-02-00034 0.8324^{20}	 1.4544^{20}
9691	Piperidine, 1,2,2,6,6-pentamethyl- Pempidine	$C_{10}H_{21}N$ 155.28	79-55-0	7022 147	5-20-04-00238 0.8580^{0}	 1.4550^{21}
9692	Piperidine, 1-pentyl-	$C_{10}H_{21}N$ 155.28	10324-58-0	 198.2	5-20-02-00031 0.8282^{20}	 1.4498^{20}
9693	Piperidine, 1-phenyl-	$C_{11}H_{15}N$ 161.25	4096-20-2 4.7	 258	5-20-02-00057 0.9944^{25}	EtOH 4; eth 4; bz 4; chl 4 1.5598^{25}
9694	Piperidine, 3-phenyl-	$C_{11}H_{15}N$ 161.25	3973-62-4 14.5	 257; 140^{19}	5-20-06-00405 1.0010^{25}	 1.5473^{25}
9695	Piperidine, 4-phenyl- 4-Phenylpiperidine	$C_{11}H_{15}N$ 161.25	771-99-3 60.5	 257	5-20-06-00408 0.9998^{16}	chl 3
9696	Piperidine, 1-(1-phenylcyclohexyl)- Phencyclidine	$C_{17}H_{25}N$ 243.39	77-10-1 46.5	7179 135-7$^{1.0}$	5-20-02-00078	
9697	Piperidine, 1-(1-phenylethyl)-	$C_{13}H_{19}N$ 189.30	7529-63-7 -13	 272	5-20-02-00064 0.945^{25}	 1.5218^{25}
9698	Piperidine, 1-(phenylmethyl)-	$C_{12}H_{17}N$ 175.27	2905-56-8	 245	5-20-02-00062 0.9625^{16}	 1.5227^{20}
9699	Piperidine, 2-(phenylmethyl)- 2-Benzylpiperidine	$C_{12}H_{17}N$ 175.27	32838-55-4 32	 267.5	5-20-06-00438 0.9660^{25}	eth 4; EtOH 4
9700	Piperidine, 4-(phenylmethyl)- 4-Benzylpiperidine	$C_{12}H_{17}N$ 175.27	31252-42-3 16.8	 270; 150^{17}	5-20-06-00439 0.9970^{20}	H_2O 1; EtOH 3; eth 3 1.5337^{25}
9701	Piperidine, 4,4'-(1,3-propanediyl)bis-	$C_{13}H_{26}N_2$ 210.36	16898-52-5 67.1	 329	5-23-04-00101	H_2O 4
9702	Piperidine, 4,4'-(1,3-propanediyl)bis[1-methyl- Propane, 1,3-bis(1-methyl-4-piperidyl)	$C_{15}H_{30}N_2$ 238.42	64168-11-2 13.7	 215^{50}	4-23-00-00536 0.8962^{25}	 1.4804^{25}
9703	1-Piperidinepropanenitrile	$C_8H_{14}N_2$ 138.21	3088-41-3 -6.8	 145^{50}	5-20-03-00047 0.9403^{25}	 1.4676^{25}
9704	1-Piperidinepropanoic acid, ethyl ester Piperidine, *N*-carboxyethyl, ethyl ester	$C_{10}H_{19}NO_2$ 185.27	19653-33-9	 139^{50}	5-20-03-00042 0.9627^{25}	H_2O 4 1.4525^{25}
9705	1-Piperidinepropanol, α-cyclohexyl-α-phenyl-, hydrochloride Trihexyphenidyl hydrochloride	$C_{20}H_{32}ClNO$ 337.93	52-49-3 258.5	9607		
9706	Piperidine, 2-(1-propenyl)-, (±)-	$C_8H_{15}N$ 125.21	5913-85-9 6	 168	5-20-04-00346 0.8716^{15}	eth 4; EtOH 4
9707	Piperidine, 1-propyl-	$C_8H_{17}N$ 127.23	5470-02-0	 151.5	5-20-02-00027 0.8231^{20}	H_2O 4; eth 4; EtOH 4 1.4446^{20}
9708	Piperidine, 2-propyl-, (*S*)- Coniine	$C_8H_{17}N$ 127.23	458-88-8 -2	2500 166.5	5-20-04-00205 0.8440^{20}	H_2O 2; EtOH 5; eth 4; bz 3 1.4512^{22}
9709	Piperidine, 4-propyl- 4-Propylpiperidine	$C_8H_{17}N$ 127.23	22398-09-0	 173	5-20-04-00208 0.864^{22}	EtOH 3; chl 2 1.4465^{23}
9710	Piperidine, 2,2,4,6-tetramethyl- 2,2,4,6-Tetramethylpiperidine	$C_9H_{19}N$ 141.26	6292-82-6	 156	4-20-00-01638 0.816^{25}	 1.4374^{25}
9711	Piperidine, 2,2,6,6-tetramethyl- Norpempidine	$C_9H_{19}N$ 141.26	768-66-1 28	 156	5-20-04-00237 0.8367^{16}	eth 4 1.4455^{20}
9712	Piperidine, 2,2,4-trimethyl-	$C_8H_{17}N$ 127.23	101257-71-0	 148	4-20-00-01624 0.832^{15}	eth 4; EtOH 4 1.4458^{20}
9713	Piperidine, 2,3,6-trimethyl- 2,3,6-Trimethylpiperidine	$C_8H_{17}N$ 127.23	50402-72-7	 36^{5}	4-20-00-01625 0.8302^{20}	 1.4434^{20}
9714	Piperidine, 2,4,6-trimethyl- 2,4,6-Trimethylpiperidine	$C_8H_{17}N$ 127.23	21974-48-1 <20	 145	5-20-04-00220 0.8315^{19}	H_2O 2 1.4412^{20}
9715	3-Piperidinol, 1-ethyl- 1-Ethyl-3-piperidinol	$C_7H_{15}NO$ 129.20	13444-24-1	3799 93-5^{15}	5-21-01-00039	 1.4777^{14}

No.	Name Synonym	Mol. Form. Mol. Wt.	CAS RN mp/°C	Merck No. bp/°C	Beil. Ref. den/g cm^{-3}	Solubility n_D
9716	3-Piperidinol, 1-methyl-	$C_6H_{13}NO$ 115.18	3554-74-3	76-8^{11}	5-21-01-00029 0.999^{25}	1.4735^{20}
9717	4-Piperidinol, 1-methyl-	$C_6H_{13}NO$ 115.18	106-52-5 29	200	5-21-01-00052	1.4775^{20}
9718	3-Piperidinol, 6-propyl-, (3*S-trans*)- Pseudoconhydrine	$C_8H_{17}NO$ 143.23	140-55-6 106	7928 236	5-21-01-00138	H_2O 4; eth 4; EtOH 4
9719	2-Piperidinone	C_5H_9NO 99.13	675-20-7 39.5	256	5-21-06-00396	H_2O 4; EtOH 4; eth 4; dil acid 3
9720	4-Piperidinone, 1-acetyl-	$C_7H_{11}NO_2$ 141.17	32161-06-1	218; 124$^{0.2}$	5-21-06-00426 1.146^{25}	1.5026^{20}
9721	2-Piperidinone, 1-methyl-	$C_6H_{11}NO$ 113.16	931-20-4	105^{12}; 94^{9}	5-21-06-00398 1.0263^{25}	1.4820^{20}
9722	2-Piperidinone, 3-methyl-	$C_6H_{11}NO$ 113.16	3768-43-2 55.3	249.5	5-21-06-00479	bz 4; eth 4; EtOH 4
9723	3-Piperidinone, 1-methyl-	$C_6H_{11}NO$ 113.16	5519-50-6	63-4^{13}	5-21-06-00417 0.9655^{25}	1.4559^{25}
9724	4-Piperidinone, 1-methyl-	$C_6H_{11}NO$ 113.16	1445-73-4	56-8^{11}	5-21-06-00419 0.9696^{25}	1.4580^{25}
9725	4-Piperidinone, 1,2,3,6-tetramethyl- 1,2,5,6-Tetramethyl-4-piperidinone	$C_9H_{17}NO$ 155.24	53381-90-1	86-8$^{7.5}$	4-21-00-03272 0.9499^{20}	1.4680^{20}
9726	4-Piperidinone, 2,2,6,6-tetramethyl-	$C_9H_{17}NO$ 155.24	826-36-8 36	205	5-21-06-00538	H_2O 3; EtOH 3; eth 3; chl 2
9727	Piperonyl butoxide 1,3-Benzodioxole, 5-[[2-(2-butoxyethoxy)ethoxy]methyl]-6-propyl-	$C_{19}H_{30}O_5$ 338.44	51-03-6	7446 180^{1}	1.05^{25}	
9728	Pirimicarb Carbamic acid, dimethyl-, 2-(dimethylamino)-5,6-dimethyl-4-pyrimidinyl ester	$C_{11}H_{18}N_4O_2$ 238.29	23103-98-2 90.5	7468		
9729	Pirimiphos-ethyl Phosphorothioic acid, *O*-[2-(diethylamino)-6-methyl-4-pyrimidinyl] *O,O*-diethyl ester	$C_{13}H_{24}N_3O_3PS$ 333.39	23505-41-1	7469 dec >130	1.14^{20}	
9730	Pirimiphos-methyl Phosphorothioic acid, *O*-[2-(diethylamino)-6-methyl-4-pyrimidinyl] *O,O*-dimethyl ester	$C_{11}H_{20}N_3O_3PS$ 305.34	29232-93-7 15	dec	1.17^{20}	
9731	Plumbane, tetraethyl Tetraethyl lead	$C_8H_{20}Pb$ 323.45	78-00-2	9136 200 dec	1.653^{20}	H_2O 1; bz 3; EtOH 2 1.5198^{20}
9732	Plumbane, tetramethyl Tetramethyl lead	$C_4H_{12}Pb$ 267.34	75-74-1 -27.5	110	1.995^{20}	
9733	Plumbane, tetraphenyl-	$C_{24}H_{20}Pb$ 515.62	595-89-1 228.3	126^{13}	4-16-00-01615 1.5298^{20}	chl 3
9734	21H,23H-Porphine	$C_{20}H_{14}N_4$ 310.36	101-60-0 360	7574 sub 300	4-26-00-01900 1.336^{25}	H_2O 1; EtOH 2; eth 1; ace 1
9735	21H,23H-Porphine-2,18-dipropanoic acid, 7,12-bis(1-hydroxyethyl)-3,8,13,17-tetramethyl- Hematoporphyrin	$C_{34}H_{38}N_4O_6$ 598.70	14459-29-1 172.5	4555	4-26-00-03157	H_2O 1; EtOH 3; eth 2; chl 2
9736	Pregna-4,6-diene-21-carboxylic acid, 17-hydroxy-3-oxo-, γ-lactone, (17α)- Canrenone	$C_{22}H_{28}O_3$ 340.46	976-71-6 150	1753	5-17-11-00476	
9737	Pregna-1,4-diene-3,20-dione, 9-fluoro-11,17,21-trihydroxy-16-methyl-, (11β,16α)- Dexamethasone	$C_{22}H_{29}FO_5$ 392.47	50-02-2 262	2922		
9738	Pregna-1,4-diene-3,20-dione, 11,17,21-trihydroxy-, (11β)- Prednisolone	$C_{21}H_{28}O_5$ 360.45	50-24-8 235	7719	4-08-00-03467	
9739	Pregna-1,4-diene-3,11,20-trione, 17,21-dihydroxy- Prednisone	$C_{21}H_{26}O_5$ 358.43	53-03-2 233-5 dec	7727	4-08-00-03531	
9740	Pregnane, (5α)- Allopregnane	$C_{21}H_{36}$ 288.52	641-85-0 84.5	257	4-05-00-01215	
9741	Pregnane, (5β)- 17β-Ethyletiocholane	$C_{21}H_{36}$ 288.52	481-26-5 83.5	7731	4-05-00-01215 1.032^{15}	H_2O 1; chl 3; MeOH 3
9742	Pregnane-3,20-diol, (3α,5β,20*S*)- Pregnanediol	$C_{21}H_{36}O_2$ 320.52	80-92-2 243.5	7732	4-06-00-06111 1.15^{25}	EtOH 2; eth 2; ace 3; os 2
9743	Pregnane-3,20-dione, (5β)- 5β-Pregnane-3,20-dione	$C_{21}H_{32}O_2$ 316.48	128-23-4 123	7733	4-07-00-02198	H_2O 1; EtOH 4; eth 3; ace 3
9744	Pregnane-11,20-dione, 3,17,21-trihydroxy-, (3α,5β)- Tetrahydrocortisone	$C_{21}H_{32}O_5$ 364.48	53-05-4 190	9143	4-08-00-03395	
9745	Pregnan-20-one, 3-hydroxy-, (3α,5β)- Pregnan-3α-ol-20-one	$C_{21}H_{34}O_2$ 318.50	128-20-1 149.5	7734		EtOH 4
9746	Pregn-4-en-18-al, 11,21-dihydroxy-3,20-dioxo-, (11β)- Aldosterone	$C_{21}H_{28}O_5$ 360.45	52-39-1 166.5	218	4-08-00-03491	
9747	Pregn-4-ene-3,20-dione, 21-(acetyloxy)-11,17-dihydroxy-, (11β)- Hydrocortisone acetate	$C_{23}H_{23}O_6$ 395.43	50-03-3 223 dec	4711	4-08-00-03424 1.289^{20}	
9748	Pregn-4-ene-3,20-dione, 11,21-dihydroxy-, (11β)- Corticosterone	$C_{21}H_{30}O_4$ 346.47	50-22-6 181	2532 190$^{0.01}$	4-08-00-02907	H_2O 1; EtOH 3; eth 3; ace 3

No.	Name Synonym	Mol. Form. Mol. Wt.	CAS RN mp/°C	Merck No. bp/°C	Beil. Ref. den/g cm^{-3}	Solubility n_D
9749	Pregn-4-ene-3,20-dione, 17,21-dihydroxy- 11-Deoxy-17-hydroxycorticosterone	$C_{21}H_{30}O_4$ 346.47	152-58-9 215	2912	4-08-00-02913	ace 4; EtOH 4; chl 4
9750	Pregn-4-ene-3,20-dione, 17-hydroxy- 17α-Hydroxyprogesterone	$C_{21}H_{30}O_3$ 330.47	68-96-2	4773	4-08-00-02189	chl 2
9751	Pregn-4-ene-3,20-dione, 21-hydroxy- Deoxycorticosterone	$C_{21}H_{30}O_3$ 330.47	64-85-7 141.5	2882	4-08-00-02195	H_2O 2; EtOH 4; eth 2; ace 4
9752	Pregn-4-ene-3,20-dione, 17-hydroxy-6-methyl-, (6α)- Medroxyprogesterone	$C_{22}H_{32}O_3$ 344.49	520-85-4 214.5	5677	4-08-00-02211	chl 4
9753	Pregn-4-ene-3,20-dione, 11,17,21-trihydroxy-, (11β)- Hydrocortisone	$C_{21}H_{30}O_5$ 362.47	50-23-7 220	4710	4-08-00-03422	H_2O 2; EtOH 3; diox 3; HOAc 3
9754	Pregn-4-ene-3,11,20-trione, 17,21-dihydroxy- Cortisone	$C_{21}H_{28}O_5$ 360.45	53-06-5 222	2533	4-08-00-03480	H_2O 2; EtOH 3; eth 2; ace 3
9755	Pregn-4-ene-3,11,20-trione, 21-hydroxy- 11-Dehydrocorticosterone	$C_{21}H_{28}O_4$ 344.45	72-23-1 183.5	2859	4-08-00-02998	H_2O 1; EtOH 3; ace 3; bz 3
9756	Pregn-5-en-20-one, 3-hydroxy-, (3β)- Pregnenolone	$C_{21}H_{32}O_2$ 316.48	145-13-1 192	7739	4-08-00-01019	
9757	Pregn-4-en-3-one, 11,17,20,21-tetrahydroxy-, (11β,20*R*)- 4-Pregnene-11β,17α,20β,21-tetrol-3-one	$C_{21}H_{32}O_5$ 364.48	116-58-5 125 dec	7736	4-08-00-03384	ace 4; EtOH 4
9758	Pregn-4-en-20-yn-3-one, 17-hydroxy-, (17α)- Ethisterone	$C_{21}H_{28}O_2$ 312.45	434-03-7 272	3696	4-08-00-01225	
9759	Procyazine Propanenitrile, 2-[[4-chloro-6-(cyclopropylamino)-1,3,5-triazin-2-yl]amino]-2-methyl-	$C_{10}H_{13}ClN_6$ 252.71	32889-48-8 170			
9760	Procymidone *N*-(3,5-Dichlorophenyl)-1,2-dimethylcyclopropane-1,2-dicarboximide	$C_{13}H_{11}Cl_2NO_2$ 284.14	32809-16-8 166		 1.452^{25}	
9761	Prodiamine 1,3-Benzenediamine, 2,4-dinitro-*N*3,*N*3-dipropyl-6-(trifluoromethyl)-	$C_{13}H_{17}F_3N_4O_4$ 350.30	29091-21-2 124	7773	 1.47^{25}	
9762	Profenofos Phosphorothioic acid, *O*-(4-bromo-2-chlorophenyl) *O*-ethyl *S*-propyl ester	$C_{11}H_{15}BrClO_3PS$ 373.63	41198-08-7	 $110^{0.001}$	 1.455^{20}	
9763	Profluralin Benzenamine, *N*-(cyclopropylmethyl)-2,6-dinitro-*N*-propyl-4-(trifluoromethyl)-	$C_{14}H_{16}F_3N_3O_4$ 347.29	26399-36-0 34	7781		
9764	*L*-Proline 2-Pyrrolidinecarboxylic acid	$C_5H_9NO_2$ 115.13	147-85-3 221 dec	7790	5-22-01-00031	H_2O 4; EtOH 2; eth 1; ace 2
9765	*DL*-Proline	$C_5H_9NO_2$ 115.13	609-36-9 205 dec		5-22-01-00033	H_2O 4; EtOH 4
9766	*L*-Proline, 1-acetyl-4-hydroxy-, *trans*- Oxaceprol	$C_7H_{11}NO_4$ 173.17	33996-33-7 132	6857	5-22-05-00013	H_2O 4; MeOH 4
9767	*L*-Proline, 4-hydroxy-, *trans*- 4-Hydroxy-*L*-proline	$C_5H_9NO_3$ 131.13	51-35-4 274	4775	5-22-05-00007	H_2O 4; EtOH 2
9768	*L*-Proline, 5-oxo- *L*-Pyroglutamic acid	$C_5H_7NO_3$ 129.12	98-79-3 162	8012	5-22-06-00007	DMSO 3
9769	Promecarb Phenol, 3-methyl-5-(1-methylethyl)-, methylcarbamate	$C_{12}H_{17}NO_2$ 207.27	2631-37-0 87	7794 $117^{0.01}$		
9770	Prometryn 1,3,5-Triazine-2,4-diamine, *N*,*N*′-bis(1-methylethyl)-6-(methylthio)-	$C_{10}H_{19}N_5S$ 241.36	7287-19-6 119	7800	 1.157^{20}	
9771	Propachlor Acetamide, 2-chloro-*N*-(1-methylethyl)-*N*-phenyl-	$C_{11}H_{14}ClNO$ 211.69	1918-16-7 77	7805 $110^{0.03}$	 1.242^{25}	
9772	1,2-Propadiene Allene	C_3H_4 40.06	463-49-0 -136.2	 -34.4	4-01-00-00966 $*0.584^{25}$	bz 4; peth 4 1.4168
9773	1,2-Propadiene-1,3-dione Carbon suboxide	C_3O_2 68.03	504-64-3 -107	1821 6.8	4-01-00-03764 1.114^{0}	eth 3; bz 3; CS_2 3 1.4538^{0}
9774	Propanal Propionaldehyde	C_3H_6O 58.08	123-38-6 -80	7835 48	4-01-00-03165 0.8657^{25}	H_2O 3; EtOH 5; eth 5 1.3636^{20}
9775	Propanal, 2-bromo-	C_3H_5BrO 136.98	19967-57-8	 109.5	4-01-00-03177 1.592^{20}	eth 4 1.4813^{20}
9776	Propanal, 2-bromo-2-methyl-	C_4H_7BrO 151.00	13206-46-7	 113	4-01-00-03267 1.383^{25}	 1.4518^{25}
9777	Propanal, 2-chloro-	C_3H_5ClO 92.52	683-50-1	 86	3-01-00-02691 1.182^{15}	bz 4; eth 4 1.431^{17}
9778	Propanal, 3-chloro-	C_3H_5ClO 92.52	19434-65-2 <20	 130; 45^{10}	4-01-00-03174 1.268^{15}	eth 4; EtOH 4 1.475^{25}
9779	Propanal, 2-chloro-2-methyl-	C_4H_7ClO 106.55	917-93-1	 90	4-01-00-03267 1.053^{15}	eth 4; EtOH 4 1.4160^{16}
9780	Propanal, 2,3-dibromo-	$C_3H_4Br_2O$ 215.87	5221-17-0	 $73\text{-}5^{10}$	4-01-00-03178 2.198^{15}	eth 4 1.5082^{20}
9781	Propanal, 2,3-dichloro-	$C_3H_4Cl_2O$ 126.97	10140-89-3	 73^{50}; 48^{14}	4-01-00-03174 1.400^{20}	 1.4762^{20}
9782	Propanal, 2,3-dihydroxy-, (±)-	$C_3H_6O_3$ 90.08	56-82-6 145	 $140\text{-}50^{0.8}$	3-01-00-03282 1.453^{18}	H_2O 3; EtOH 2; eth 2; bz 1

No.	Name	Synonym	Mol. Form.	Mol. Wt.	CAS RN	mp/°C	Merck No.	bp/°C	Beil. Ref.	den/g cm^{-3}	Solubility	n_D
9783	Propanal, 2,2-dimethyl-	Pivaldehyde	$C_5H_{10}O$	86.13	630-19-3	6		77.5	4-01-00-03295	0.7923^{17}	EtOH 3; eth 3	1.3791^{20}
9784	Propanal, 2-(hydroxyimino)-, oxime		$C_3H_6N_2O_2$	102.09	1804-15-5	157		sub	4-01-00-03633		H_2O 2; EtOH 3; eth 3; bz 3	
9785	Propanal, 2-methyl-	Isobutanal	C_4H_8O	72.11	78-84-2	-65.9	5038	64.5	4-01-00-03262	0.7891^{20}	H_2O 3; eth 3; ace 3; ctc 2	1.3730^{20}
9786	Propanal, 3-(methylthio)-	3-(Methylthio)propionaldehyde	C_4H_8OS	104.17	3268-49-3			62^{11}	4-01-00-03974			
9787	Propanal, oxime		C_3H_7NO	73.09	627-39-4	40		131.5	4-01-00-03170	0.9258^{20}		1.4287^{20}
9788	Propanal, 2-oxo-	Pyruvaldehyde	$C_3H_4O_2$	72.06	78-98-8		8030	72	4-01-00-03631	1.0455^{20}	EtOH 3; eth 3; bz 3	1.4002^{18}
9789	Propanal, 2-oxo-, 1-oxime	Isonitrosoacetone	$C_3H_5NO_2$	87.08	306-44-5	69	5076	sub	3-01-00-03092	1.0744^{67}	H_2O 3; eth 3; bz 2; ctc 2	
9790	Propanal, 2,2,3-trichloro-		$C_3H_3Cl_3O$	161.41	7789-90-4			$63\text{-}5^{45}$	3-01-00-02693	1.470^{25}	eth 4	1.473^{25}
9791	Propanamide	Propionamide	C_3H_7NO	73.09	79-05-0	81.3	7836	213	4-02-00-00725	0.9262^{110}	H_2O 4; EtOH 4; eth 4; chl 4	1.4180^{110}
9792	Propanamide, 2-bromo-2-methyl-	Isobutyramide, 2-bromo	C_4H_8BrNO	166.02	7462-74-0	148		145^{17}	4-02-00-00863		EtOH 4; chl 4	
9793	Propanamide, *N*,*N*-diethyl-		$C_7H_{15}NO$	129.20	1114-51-8			191	4-04-00-00353	0.8972^{20}	EtOH 4	1.4425^{20}
9794	Propanamide, *N*,*N*-diethyl-2,2-dimethyl-	Pivalamide, *N*,*N*-diethyl	$C_9H_{19}NO$	157.26	24331-72-4			206	3-04-00-00211	0.891^{15}	eth 4; EtOH 4	
9795	Propanamide, *N*-(2,3-dihydro-1,5-dimethyl-3-oxo-2-phenyl-1H-pyrazol-4-yl)-2-(dimethylamino)-	Aminopropylon	$C_{16}H_{22}N_4O_2$	302.38	3690-04-8	181	484		4-25-00-03595		H_2O 4	
9796	Propanamide, *N*-(4-ethoxyphenyl)-2-hydroxy-	*p*-Lactophenetide	$C_{11}H_{15}NO_3$	209.25	539-08-2	118	5220		3-13-00-01135		H_2O 3; EtOH 4; eth 2; bz 2	
9797	Propanamide, 2-hydroxy-, (±)-		$C_3H_7NO_2$	89.09	65144-02-7	75.5			4-03-00-00674	1.1381^{80}	H_2O 4; EtOH 4	
9798	Propanamide, 2-methyl-		C_4H_9NO	87.12	563-83-7	128		218	4-02-00-00852		chl 3	
9799	Propanamide, *N*-methyl-		C_4H_9NO	87.12	1187-58-2	-30.9		148	4-04-00-00183	0.9305^{25}		1.4345^{25}
9800	Propanamide, *N*-phenyl-		$C_9H_{11}NO$	149.19	620-71-3	105.5		222.2	4-12-00-00385	1.175^{25}	H_2O 2; EtOH 4; eth 4	
9801	Propanamide, *N*-phenyl-*N*-[1-(2-phenylethyl)-4-piperidinyl]-	Fentanyl	$C_{22}H_{28}N_2O$	336.48	437-38-7	87.5	3944		5-22-08-00049			
9802	Propanamide, *N*-propyl-		$C_6H_{13}NO$	115.18	3217-86-5	154		215; 108^{9}	4-04-00-00477	0.8985^{25}	H_2O 2; eth 2	
9803	1-Propanamine	Propylamine	C_3H_9N	59.11	107-10-8	-83	7855	47.2	4-04-00-00464	0.7173^{20}	H_2O 3; EtOH 4; eth 4; ace 4	1.3870^{20}
9804	2-Propanamine	Isopropylamine	C_3H_9N	59.11	75-31-0	-95.1	5097	31.7	4-04-00-00504	0.6891^{20}	H_2O 5; EtOH 5; eth 5; ace 4	1.3742^{20}
9805	1-Propanamine, *N*,*N*-dichloro-	*N*,*N*-Dichloropropylamine	$C_3H_7Cl_2N$	128.00	69947-01-9			117	0-04-00-00145	1.1454^{23}		1.4525^{23}
9806	1-Propanamine, 2,2-dimethyl-		$C_5H_{13}N$	87.16	5813-64-9			82	4-04-00-00707	0.7455^{20}	eth 4	1.4023^{20}
9807	1-Propanamine, *N*,*N*-dimethyl-	Dimethylpropylamine	$C_5H_{13}N$	87.16	926-63-6			66	4-04-00-00467	0.7152^{20}	bz 4; eth 4; EtOH 4	1.3860^{20}
9808	1-Propanamine, *N*,*N*-dipropyl-	Tripropylamine	$C_9H_{21}N$	143.27	102-69-2	-93.5		156	4-04-00-00470	0.7558^{20}	eth 4; EtOH 4	1.4181^{20}
9809	1-Propanamine, *N*-ethyl-		$C_5H_{13}N$	87.16	20193-20-8			81	4-04-00-00468	0.7204^{17}	H_2O 2; ace 4; EtOH 4	1.3858^{25}
9810	2-Propanamine, *N*-ethyl-	Ethylisopropylamine	$C_5H_{13}N$	87.16	19961-27-4			69.6	4-04-00-00508			1.3872^{25}
9811	2-Propanamine, *N*-ethyl-*N*-(1-methylethyl)-		$C_8H_{19}N$	129.25	7087-68-5			126.5	4-04-00-00511	0.742^{25}	ctc 3	1.4138^{20}
9812	1-Propanamine, *N*-ethyl-*N*-propyl-		$C_8H_{19}N$	129.25	20634-92-8			138	2-04-00-00622	0.807^{24}	ace 4; bz 4; eth 4	
9813	1-Propanamine, 1,1,2,2,3,3,3-heptafluoro-*N*,*N*-bis(heptafluoropropyl)-		$C_9F_{21}N$	521.07	338-83-0			130	4-02-00-00742	1.822^{4}		1.279^{25}
9814	1-Propanamine, *N*-hydroxy-*N*-propyl-	*N*,*N*-Dipropylhydroxylamine	$C_6H_{15}NO$	117.19	7446-43-7	29		154.5	4-04-00-03308		bz 4; eth 4; EtOH 4; chl 4	
9815	2-Propanamine, *N*,*N*'-methanetetraylbis-		$C_7H_{14}N_2$	126.20	693-13-0			147	4-04-00-00531	0.806^{25}		1.4320^{20}
9816	1-Propanamine, 3-methoxy-		$C_4H_{11}NO$	89.14	5332-73-0			117.5	4-04-00-01623	0.8727^{20}	H_2O 3; ace 3; bz 3; ctc 3	1.4391^{20}
9817	1-Propanamine, 2-methyl-	Isobutylamine	$C_4H_{11}N$	73.14	78-81-9	-86.7	5016	67.7	4-04-00-00625	0.724^{25}		1.3988^{19}
9818	2-Propanamine, 2-methyl-	*tert*-Butylamine	$C_4H_{11}N$	73.14	75-64-9	-66.9	1545	44.0	4-04-00-00657	0.6958^{20}	H_2O 5; EtOH 5; eth 5; chl 3	1.3784^{20}
9819	2-Propanamine, *N*-methyl-	Isopropylmethylamine	$C_4H_{11}N$	73.14	4747-21-1			50.4				
9820	1-Propanamine, 2-methyl-*N*,*N*-bis(2-methylpropyl)-	Triisobutylamine	$C_{12}H_{27}N$	185.35	1116-40-1	-21.8		191.5	4-04-00-00631	0.7684^{20}	eth 4; EtOH 4	1.4252^{17}

No.	Name Synonym	Mol. Form. Mol. Wt.	CAS RN mp/°C	Merck No. bp/°C	Beil. Ref. den/g cm^{-3}	Solubility n_D
9821	2-Propanamine, *N*-(1-methylethyl)- Diisopropylamine	$C_6H_{15}N$ 101.19	108-18-9 -61	3181 83.9	4-04-00-00510 0.7153^{20}	ace 4; bz 4; eth 4; EtOH 4 1.3924^{20}
9822	2-Propanamine, *N*-(1-methylethyl)-*N*-nitroso-	$C_6H_{14}N_2O$ 130.19	601-77-4 48	 194.5	4-04-00-03388 0.9422^{20}	H_2O 2; EtOH 3; eth 3; bz 3
9823	1-Propanamine, 2-methyl-*N*-(2-methylpropyl)- Diisobutylamine	$C_8H_{19}N$ 129.25	110-96-3 -73.5	 139.6	4-04-00-00630	H_2O 2; EtOH 3; eth 3; ace 3 1.4090^{20}
9824	1-Propanamine, *N*-nitro-	$C_3H_8N_2O_2$ 104.11	627-07-6 -21	 $128\text{-}9^{40}$	4-04-00-03405 1.1046^{15}	eth 4; EtOH 4 1.4610^{20}
9825	1-Propanamine, *N*-nitroso-*N*-propyl- Dipropylnitrosamine	$C_6H_{14}N_2O$ 130.19	621-64-7	 206; 113^{40}	4-04-00-03388 0.9163^{20}	H_2O 2; EtOH 5; eth 5 1.4437^{20}
9826	1-Propanamine, *N*-propyl- Dipropylamine	$C_6H_{15}N$ 101.19	142-84-7 -63	3350 109.3	4-04-00-00469 0.7400^{20}	H_2O 3; EtOH 3; eth 5; ace 4 1.4050^{20}
9827	2-Propanamine, *N*-propyl Isopropylpropylamine	$C_6H_{15}N$ 101.19	21968-17-2	 96.9		
9828	1-Propanamine, 3-(triethoxysilyl)-	$C_9H_{23}NO_3Si$ 221.37	919-30-2	 $122\text{-}3^{31}$	4-04-00-04273 0.9506^{20}	 1.4225^{20}
9829	1-Propanamine, *N*,*N*,2-trimethyl-	$C_6H_{15}N$ 101.19	7239-24-9	 80.5	4-04-00-00627 0.7097^{20}	H_2O 4 1.3907^{20}
9830	1-Propanaminium, 2-(acetyloxy)-*N*,*N*,*N*-trimethyl-, chloride Methacholine chloride	$C_8H_{18}ClNO_2$ 195.69	62-51-1 172	5847		H_2O 4; EtOH 4; chl 4
9831	1-Propanaminium, 2-hydroxy-*N*,*N*,*N*-trimethyl-, chloride (2-Hydroxypropyl)trimethyl ammonium chloride	$C_6H_{16}ClNO$ 153.65	2382-43-6 165	 dec	3-04-00-00754	H_2O 4; EtOH 4
9832	2-Propanaminium, *N*-methyl-*N*-(1-methylethyl)-*N*-[2-[(9H-xanthen-9-ylcarbonyl)oxy]ethyl]-, bromide Propantheline bromide	$C_{23}H_{30}BrNO_3$ 448.40	50-34-0 159.5	7816	5-18-06-00590	
9833	1-Propanaminium, *N*,*N*,*N*-tripropyl-, iodide Tetrapropylammonium iodide	$C_{12}H_{28}IN$ 313.27	631-40-3 280 dec		4-04-00-00471 1.3138^{25}	H_2O 4; EtOH 3; eth 2; chl 4
9834	Propane	C_3H_8 44.10	74-98-6 -187.6	7809 -42.1	4-01-00-00176 $^*0.493^{25}$	H_2O 3; EtOH 3; eth 4; ace 2
9835	Propane, 2,2-bis(ethylsulfonyl)- Sulfonmethane	$C_7H_{16}O_4S_2$ 228.33	115-24-2 125.8	8939 300 dec	3-01-00-02754	bz 4; EtOH 4; chl 4
9836	Propane, 1-bromo- Propyl bromide	C_3H_7Br 122.99	106-94-5 -110	7857 71.1	4-01-00-00205 1.3537^{20}	H_2O 2; EtOH 3; eth 3; ace 3 1.4343^{20}
9837	Propane, 2-bromo- Isopropyl bromide	C_3H_7Br 122.99	75-26-3 -89	5098 59.5	4-01-00-00208 1.3140^{20}	H_2O 2; EtOH 5; eth 5; ace 3 1.4251^{20}
9838	Propane, 1-bromo-2-chloro- 1-Bromo-2-chloropropane	C_3H_6BrCl 157.44	3017-96-7	 118	4-01-00-00212 1.531^{20}	ace 4; bz 4; eth 4; EtOH 4 1.4745^{20}
9839	Propane, 1-bromo-3-chloro- 1-Bromo-3-chloropropane	C_3H_6BrCl 157.44	109-70-6 -58.9	 143.3	4-01-00-00212 1.5969^{20}	H_2O 1; EtOH 4; eth 4; chl 4 1.4864^{20}
9840	Propane, 2-bromo-1-chloro- 2-Bromo-1-chloropropane	C_3H_6BrCl 157.44	3017-95-6	 118	4-01-00-00212 1.537^{20}	H_2O 1; EtOH 4; eth 4; ace 3 1.4795^{20}
9841	Propane, 2-bromo-2-chloro- 2-Bromo-2-chloropropane	C_3H_6BrCl 157.44	2310-98-7	 95	4-01-00-00213 1.474^{22}	ace 4; bz 4; eth 4; EtOH 4 1.4575^{20}
9842	Propane, 1-bromo-3-chloro-2-methyl-	C_4H_8BrCl 171.46	6974-77-2	 155	4-01-00-00298 1.4839^{20}	 1.4796^{25}
9843	Propane, 1-bromo-2,2-dimethyl-	$C_5H_{11}Br$ 151.05	630-17-1	 106	4-01-00-00337 1.1997^{20}	H_2O 1; EtOH 3; eth 3; ace 3 1.4370^{20}
9844	Propane, 1-bromo-3-fluoro-	C_3H_6BrF 140.98	352-91-0	 101.4	4-01-00-00210 1.542^{25}	bz 4; eth 4; EtOH 4; chl 4 1.4290^{25}
9845	Propane, 1-bromo-2-methyl- Isobutyl bromide	C_4H_9Br 137.02	78-77-3 -119	5019 91.1	4-01-00-00294 1.272^{15}	EtOH 4; eth 4; ace 4; bz 4 1.4348^{20}
9846	Propane, 2-bromo-2-methyl- *tert*-Butyl bromide	C_4H_9Br 137.02	507-19-7 -16.2	1555 73.3	4-01-00-00295 1.4278^{20}	H_2O 1; ctc 2 1.4278^{20}
9847	Propane, 2-bromo-2-nitro-	$C_3H_6BrNO_2$ 167.99	5447-97-2	 153; $73\text{-}5^{50}$	4-01-00-00233 1.6562^{25}	eth 4; EtOH 4
9848	Propane, 1-chloro- Propyl chloride	C_3H_7Cl 78.54	540-54-5 -122.8	7859 46.5	4-01-00-00189 0.8899^{20}	H_2O 2; EtOH 5; eth 5; bz 3 1.3879^{20}
9849	Propane, 2-chloro- Isopropyl chloride	C_3H_7Cl 78.54	75-29-6 -117.2	5099 35.7	4-01-00-00191 0.8617^{20}	H_2O 2; EtOH 5; eth 5; bz 3 1.3777^{20}
9850	Propane, 3-chloro-1,1-diethoxy-	$C_7H_{15}ClO_2$ 166.65	35573-93-4	 84^{25}	3-01-00-02592 0.9951^{19}	ace 4; bz 4 1.4268^{20}
9851	Propane, 1-chloro-2,2-difluoro- 2,2-Difluoropropyl chloride	$C_3H_5ClF_2$ 114.52	420-99-5 -56.2	 55	4-01-00-00193 1.2001^{20}	eth 3; bz 3; chl 3 1.3520^{20}
9852	Propane, 3-chloro-1,1-dimethoxy-	$C_5H_{11}ClO_2$ 138.59	35502-06-8	 86^{100}; 51^{19}	3-01-00-02692 1.064^{15}	 1.4163^{20}
9853	Propane, 1-chloro-2,2-dimethyl-	$C_5H_{11}Cl$ 106.60	753-89-9 -20	 84.3	4-01-00-00336 0.8660^{20}	bz 4; eth 4; EtOH 4; chl 4 1.4044^{20}
9854	Propane, 1-(1-chloroethoxy)- 1-Chloroethyl propyl ether	$C_5H_{11}ClO$ 122.59	692-35-3	 114	3-01-00-02655 0.9322^{20}	eth 4 1.4013^{20}
9855	Propane, 2-[(2-chloroethyl)thio]-2-methyl-	$C_6H_{13}ClS$ 152.69	4303-44-0 -49	 $81\text{-}2^{30}$	3-01-00-01592 1.0001^{25}	
9856	Propane, 1-chloro-3-fluoro- 1-Chloro-3-fluoropropane	C_3H_6ClF 96.53	462-38-4	 80	4-01-00-00193	bz 4; eth 4; EtOH 4; chl 4 1.3871^{25}
9857	Propane, 1-chloro-3-iodo-	C_3H_6ClI 204.44	6940-76-7	 171	4-01-00-00226 1.904^{20}	H_2O 1; eth 3; bz 3; ctc 2 1.5472^{20}
9858	Propane, 1-(chloromethoxy)-	C_4H_9ClO 108.57	3587-57-3	 109	3-01-00-02589 0.9884^{20}	eth 4; EtOH 4 1.4125^{20}

No.	Name Synonym	Mol. Form. Mol. Wt.	CAS RN mp/°C	Merck No. bp/°C	Beil. Ref. den/g cm^{-3}	Solubility n_D
9859	Propane, 1-chloro-2-methoxy-	C_4H_9ClO 108.57	5390-72-7	 103.5	3-01-00-01470 1.009^{20}	 1.4137^{20}
9860	Propane, 1-chloro-3-methoxy-	C_4H_9ClO 108.57	36215-07-3	 111	4-01-00-01441 1.0013^{20}	 1.4131^{20}
9861	Propane, 2-chloro-1-methoxy-	C_4H_9ClO 108.57	5390-71-6	 99	3-01-00-01424 0.9946^{20}	 1.4075^{20}
9862	Propane, 1-chloro-2-methyl- Isobutyl chloride	C_4H_9Cl 92.57	513-36-0 -130.3	5022 68.5	4-01-00-00287 0.8773^{20}	eth 3; ace 3; ctc 2; chl 3 1.3984^{20}
9863	Propane, 2-chloro-2-methyl- *tert*-Butyl chloride	C_4H_9Cl 92.57	507-20-0 -26	1562 50.9	4-01-00-00288 0.8420^{20}	H_2O 2; EtOH 5; eth 5; bz 3 1.3857^{20}
9864	Propane, 1-chloro-3-(methylthio)- 3-Chloropropyl methyl sulfide	C_4H_9ClS 124.63	13012-59-4	 $71\text{-}2^{29}$	4-01-00-01458 1.0844^{20}	 1.4833^{20}
9865	Propane, 2-chloro-1-(methylthio)- 2-Chloropropyl methyl sulfide	C_4H_9ClS 124.63	19987-13-4	 67^{37}	4-01-00-01457 1.076^{20}	 1.4905^{20}
9866	Propane, 1-chloro-1-nitro- 1-Chloro-1-nitropropane	$C_3H_6ClNO_2$ 123.54	600-25-9	 142	4-01-00-00232 1.207^{20}	H_2O 2; EtOH 3; eth 3; chl 2 1.4251^{20}
9867	Propane, 1-chloro-2-nitro-	$C_3H_6ClNO_2$ 123.54	2425-66-3	 172.5	4-01-00-00233 1.245^{22}	eth 4; EtOH 4; chl 4 1.4432^{25}
9868	Propane, 1-chloro-3-nitro- 1-Chloro-3-nitropropane	$C_3H_6ClNO_2$ 123.54	16694-52-3	 197 dec; 115^{40}	3-01-00-00259 1.267^{20}	eth 4; EtOH 4
9869	Propane, 2-chloro-1-nitro-	$C_3H_6ClNO_2$ 123.54	503-76-4	 172	4-01-00-00232 1.2361^{15}	eth 4; EtOH 4 1.4447^{20}
9870	Propane, 2-chloro-2-nitro- 2-Chloro-2-nitropropane	$C_3H_6ClNO_2$ 123.54	594-71-8	 134 dec; 57^{50}	4-01-00-00233 1.230^{19}	H_2O 2; EtOH 3; eth 3; ctc 3 1.4378^{19}
9871	Propane, 3-chloro-1,1,1-trifluoro-	$C_3H_4ClF_3$ 132.51	460-35-5 -106.5	 45.1	4-01-00-00193 1.3253^{20}	 1.3350^{20}
9872	1,2-Propanediamine Propylenediamine	$C_3H_{10}N_2$ 74.13	78-90-0	7865 119.5	2-04-00-00698 0.878^{15}	H_2O 4; eth 1 1.4460^{20}
9873	1,3-Propanediamine 1,3-Diaminopropane	$C_3H_{10}N_2$ 74.13	109-76-2	 139.8	4-04-00-01258 0.884^{25}	H_2O 3; EtOH 5; eth 5 1.4600^{20}
9874	1,3-Propanediamine, *N*-(3-aminopropyl)- Bis(3-aminopropyl)amine	$C_6H_{17}N_3$ 131.22	56-18-8 -14	 151^{50}	4-04-00-01278 0.938^{25}	chl 3 1.4810^{20}
9875	1,3-Propanediamine, *N*-(3-aminopropyl)-*N*-methyl-	$C_7H_{19}N_3$ 145.25	105-83-9	 232.5; 112^{6}	4-04-00-01279 0.9023^{20}	 1.4705^{25}
9876	1,3-Propanediamine, *N,N*-diethyl- 3-(Diethylamino)propylamine	$C_7H_{18}N_2$ 130.23	104-78-9	 168.5	4-04-00-01260 0.822^{20}	 1.443^{20}
9877	1,3-Propanediamine, *N,N*-diethyl-*N'*-(6-methoxy-8-quinolinyl)- Plasmocid	$C_{17}H_{25}N_3O$ 287.41	551-01-9	7495 $182^{1.0}$	5-22-12-00126 1.0569^{24}	 1.5855^{24}
9878	1,2-Propanediamine, 2-methyl-	$C_4H_{12}N_2$ 88.15	811-93-8	 123	4-04-00-01306 0.841^{25}	ctc 3 1.4410^{20}
9879	1,3-Propanediamine, *N,N,N',N'*-tetramethyl-	$C_7H_{18}N_2$ 130.23	110-95-2	 144	4-04-00-01259 0.7837^{18}	H_2O 5; EtOH 5; eth 5
9880	Propane, 1,1-dibromo- 1,1-Dibromopropane	$C_3H_6Br_2$ 201.89	598-17-4	 133.5	4-01-00-00215 1.982^{20}	eth 4; EtOH 4; chl 4 1.5100^{20}
9881	Propane, 1,2-dibromo- Propylene dibromide	$C_3H_6Br_2$ 201.89	78-75-1 -55.2	7866 141.9	4-01-00-00215 1.9324^{20}	EtOH 3; eth 3; ctc 2; chl 3 1.5201^{20}
9882	Propane, 1,3-dibromo- 1,3-Dibromopropane	$C_3H_6Br_2$ 201.89	109-64-8 -34.2	9628 167.3	4-01-00-00216 1.9822^{20}	EtOH 3; eth 3; ctc 2; chl 3 1.5232^{20}
9883	Propane, 2,2-dibromo- 2,2-Dibromopropane	$C_3H_6Br_2$ 201.89	594-16-1	 115	4-01-00-00217 1.7825^{20}	eth 4; EtOH 4; chl 4
9884	Propane, 1,3-dibromo-2,2-bis(bromomethyl)-	$C_5H_8Br_4$ 387.73	3229-00-3 163	 305.5	4-01-00-00337 2.596^{15}	EtOH 3; eth 2; bz 3; chl 2
9885	Propane, 1,2-dibromo-3-chloro- Dibromochloropropane	$C_3H_5Br_2Cl$ 236.33	96-12-8	3003 196	3-01-00-00250 2.093^{14}	 1.553^{14}
9886	Propane, 1,1-dibromo-2,2-dimethyl- Neopentane, 1,1-dibromo	$C_5H_{10}Br_2$ 229.94	2443-91-6	 $64\text{-}5^{43}$	0-01-00-00141 1.7880^{0}	 1.5085
9887	Propane, 1,3-dibromo-2,2-dimethyl-	$C_5H_{10}Br_2$ 229.94	5434-27-5	 187 dec; 80^{26}	4-01-00-00337 1.7049^{0}	 1.5090
9888	Propane, 1,2-dibromo-3-methoxy-	$C_4H_8Br_2O$ 231.91	5836-66-8	 185	4-01-00-01447 1.8320^{12}	eth 4 1.5123^{20}
9889	Propane, 1,2-dibromo-2-methyl-	$C_4H_8Br_2$ 215.92	594-34-3 10.5	 150	4-01-00-00298 1.7827^{20}	EtOH 3; eth 3; chl 3 1.5119^{20}
9890	Propane, 1,3-dibromo-2-methyl-	$C_4H_8Br_2$ 215.92	28148-04-1	 177.5; 62^{13}	4-01-00-00299 1.8204^{20}	 1.5068^{20}
9891	Propane, 1,1-dichloro- Propylidene chloride	$C_3H_6Cl_2$ 112.99	78-99-9	7874 88.1	4-01-00-00195 1.1321^{20}	EtOH 3; eth 3; bz 3; chl 3 1.4289^{20}
9892	Propane, 1,2-dichloro-, (±)- Propane, 1,2-dichloro (*DL*)	$C_3H_6Cl_2$ 112.99	26198-63-0 -100.4	7867 96.4	4-01-00-00195 1.1560^{20}	H_2O 2; EtOH 3; eth 3; bz 3 1.4394^{20}
9893	Propane, 1,3-dichloro- Trimethylene dichloride	$C_3H_6Cl_2$ 112.99	142-28-9 -99.5	 120.9	4-01-00-00196 1.1876^{20}	H_2O 2; EtOH 4; eth 4; bz 3 1.4487^{20}
9894	Propane, 2,2-dichloro-	$C_3H_6Cl_2$ 112.99	594-20-7 -33.8	 69.3	4-01-00-00196 1.1136^{20}	H_2O 1; EtOH 3; eth 5; bz 3 1.4148^{20}
9895	Propane, 1,3-dichloro-2,2-bis(chloromethyl)-	$C_5H_8Cl_4$ 209.93	3228-99-7 97	 110^{12}	4-01-00-00336	H_2O 1; eth 3; ctc 3; chl 3

No.	Name Synonym	Mol. Form. Mol. Wt.	CAS RN mp/°C	Merck No. bp/°C	Beil. Ref. den/g cm^{-3}	Solubility n_D
9896	Propane, 1,3-dichloro-2-(ethoxymethoxy)- Formaldehyde, (2,2′-dichloroisopropyl) ethyl acetal	$C_6H_{12}Cl_2O_2$ 187.07	89583-61-9	 96-8^{16}	3-01-00-02574 1.181^{17}	eth 4 1.4491^{17}
9897	Propane, 1,2-dichloro-2-fluoro-	$C_3H_5Cl_2F$ 130.98	420-97-3 -91.7	 88.6	4-01-00-00197 1.2624^{20}	H_2O 2; ace 3; bz 3 1.4099^{20}
9898	Propane, 1,2-dichloro-1,1,2,3,3,3-hexafluoro- 1,2-Dichlorohexafluoropropane	$C_3Cl_2F_6$ 220.93	661-97-2	 34.1		
9899	Propane, 1,1-dichloro-2-methyl- Isobutane, 1,1-dichloro	$C_4H_8Cl_2$ 127.01	598-76-5	 105.5	4-01-00-00292 1.011^{12}	bz 4; eth 4; EtOH 4; chl 4 1.4330^{25}
9900	Propane, 1,2-dichloro-2-methyl- Isobutane, 1,2-dichloro	$C_4H_8Cl_2$ 127.01	594-37-6	 106.5	4-01-00-00292 1.093^{20}	H_2O 1; EtOH 5; eth 5; ace 5 1.4370^{20}
9901	Propane, 1,3-dichloro-2-methyl-	$C_4H_8Cl_2$ 127.01	616-19-3	 136.4	3-01-00-00319 1.131^{20}	H_2O 1; EtOH 3; eth 4; bz 4 1.4488^{25}
9902	Propane, 1,1-dichloro-1-nitro- 1,1-Dichloro-1-nitropropane	$C_3H_5Cl_2NO_2$ 157.98	595-44-8	 145	3-01-00-00259 1.312^{20}	ctc 3
9903	Propane, 1,1-diethoxy-	$C_7H_{16}O_2$ 132.20	4744-08-5	 123	4-01-00-03168 0.8232^{20}	H_2O 3; EtOH 4; eth 4; ace 3 1.3924^{19}
9904	Propane, 2,2-diethoxy-	$C_7H_{16}O_2$ 132.20	126-84-1	 114	4-01-00-03200 0.8200^{21}	EtOH 3; eth 4; ace 3; bz 3 1.3891^{20}
9905	Propane, 1,3-difluoro- 1,3-Difluoropropane	$C_3H_6F_2$ 80.08	462-39-5	 41.6	3-01-00-00218 1.0057^{25}	bz 4 1.3190^{26}
9906	Propane, 2,2-difluoro- Dimethyldifluoromethane	$C_3H_6F_2$ 80.08	420-45-1 -104.8	 -0.4	4-01-00-00187 0.9205^{20}	
9907	Propane, 1,3-diiodo- Trimethylene diiodide	$C_3H_6I_2$ 295.89	627-31-6 -20	 227 dec; 110^{19}	4-01-00-00228 2.5754^{20}	H_2O 1; eth 3; ctc 3; chl 3 1.6423^{20}
9908	Propane, 2,2-diiodo- 2,2-Diiodopropane	$C_3H_6I_2$ 295.89	630-13-7	 173; 156^{500}	3-01-00-00255 2.5754^{20}	eth 4; chl 4 1.651^{20}
9909	Propane, 2,2-dimethoxy-	$C_5H_{12}O_2$ 104.15	77-76-9 -47	 83	4-01-00-03199 0.847^{25}	 1.3780^{20}
9910	Propane, 2,2-dimethyl- Neopentane	C_5H_{12} 72.15	463-82-1 -16.6	6372 9.4	4-01-00-00333 *0.5852^{25}	H_2O 1; EtOH 3; eth 3; ctc 3 1.3476^{6}
9911	Propanedinitrile Malononitrile	$C_3H_2N_2$ 66.06	109-77-3 32	5592 218.5	4-02-00-01892 1.1910^{20}	H_2O 3; EtOH 4; eth 4; ace 3 1.4146^{34}
9912	Propanedinitrile, dichloro-	$C_3Cl_2N_2$ 134.95	13063-43-9	 97	3-02-00-01638 1.312^{25}	 1.4312^{20}
9913	Propanedinitrile, dimethyl-	$C_5H_6N_2$ 94.12	7321-55-3 31.5	 169.5; 62^{22}	3-02-00-01704	H_2O 3; EtOH 2
9914	Propanedinitrile, (ethoxymethylene)-	$C_6H_6N_2O$ 122.13	123-06-8 66	 160^{12}	4-03-00-01194	EtOH 3; eth 3; chl 2
9915	Propanedinitrile, oxo- Carbonyl cyanide	C_3N_2O 80.05	1115-12-4 -36	 65.5	4-03-00-01806 1.124^{20}	eth 3; ace 3; ctc 3; chl 3 1.3919^{20}
9916	Propanedinitrile, phenyl- Phenylmalononitrile	$C_9H_6N_2$ 142.16	3041-40-5 70.5	 152^{21}	4-09-00-03341	EtOH 4
9917	Propanedinitrile, (phenylmethyl)-	$C_{10}H_8N_2$ 156.19	1867-37-4 91	 174^{23}	4-09-00-03359	bz 4; eth 4; EtOH 4
9918	Propane, 1,1-dinitro- 1,1-Dinitropropane	$C_3H_6N_2O_4$ 134.09	601-76-3 -42	 184	4-01-00-00234 1.2610^{25}	alk 3 1.4339^{20}
9919	Propane, 1,3-dinitro- 1,3-Dinitropropane	$C_3H_6N_2O_4$ 134.09	6125-21-9 -21.4	 103^{1}	4-01-00-00234 1.353^{26}	H_2O 1; eth 3 1.4654^{20}
9920	Propane, 2,2-dinitro- 2,2-Dinitropropane	$C_3H_6N_2O_4$ 134.09	595-49-3 53	 185.5	4-01-00-00234 1.30^{25}	H_2O 2
9921	Propanedioic acid Malonic acid	$C_3H_4O_4$ 104.06	141-82-2 135 dec	5591 sub	4-02-00-01874 1.619^{10}	H_2O 4; EtOH 3; eth 3; bz 1
9922	Propanedioic acid, (acetylamino)-, diethyl ester	$C_9H_{15}NO_5$ 217.22	1068-90-2 96.3	 185^{20}	4-04-00-02993	H_2O 2; EtOH 3; eth 2; tfa 3
9923	Propanedioic acid, acetyl-, diethyl ester	$C_9H_{14}O_5$ 202.21	570-08-1	 232	4-03-00-01819 1.0834^{26}	H_2O 1; ace 3 1.4435^{25}
9924	Propanedioic acid, amino-, diethyl ester Diethyl aminomalonate	$C_7H_{13}NO_4$ 175.18	6829-40-9	 122-3^{16}	4-04-00-02991 1.100^{16}	H_2O 4; EtOH 4; eth 4; ace 3 1.4353^{16}
9925	Propanedioic acid, bis(1,1-dimethylethyl) ester Di-*tert*-butyl malonate	$C_{11}H_{20}O_4$ 216.28	541-16-2 -6	3021 113^{31}; 66^{2}	4-02-00-01884 1.4184^{20}	ace 3; chl 3 1.4184^{29}
9926	Propanedioic acid, bis(phenylmethyl)-, diethyl ester Diethyl dibenzylmalonate	$C_{21}H_{24}O_4$ 340.42	597-55-7 14	 234-5^{23}	3-09-00-04552 1.093^{20}	eth 4; EtOH 4
9927	Propanedioic acid, bis(phenylmethyl) ester	$C_{17}H_{16}O_4$ 284.31	15014-25-2	 188$^{0.2}$	4-06-00-02270 1.137^{25}	 1.5447^{20}
9928	Propanedioic acid, bromo-, diethyl ester	$C_7H_{11}BrO_4$ 239.07	685-87-0 -54	 254 dec	4-02-00-01904 1.4022^{25}	H_2O 1; EtOH 5; eth 5; ace 3 1.4521^{20}
9929	Propanedioic acid, butyl- *n*-Butylmalonic acid	$C_7H_{12}O_4$ 160.17	534-59-8 104.5	1574	4-02-00-02011	H_2O 4; EtOH 3; eth 3
9930	Propanedioic acid, butyl-, diethyl ester	$C_{11}H_{20}O_4$ 216.28	133-08-4	 237.5	4-02-00-02011	EtOH 4; eth 4 1.4250^{20}
9931	Propanedioic acid, chloro-, diethyl ester	$C_7H_{11}ClO_4$ 194.61	14064-10-9	 222	4-02-00-01903 1.2040^{20}	H_2O 1; EtOH 5; eth 5; chl 5 1.4327^{20}
9932	Propanedioic acid, (3-chloropropyl)-, diethyl ester	$C_{10}H_{17}ClO_4$ 236.70	18719-43-2	 147-9^{10}	4-02-00-01992	H_2O 1; EtOH 3; eth 3; ctc 2 1.4429^{20}

No.	Name	Synonym	Mol. Form.	Mol. Wt.	CAS RN	mp/°C	Merck No.	bp/°C	Beil. Ref.	den/g cm^{-3}	Solubility	n_D
9933	Propanedioic acid, (2-cyanoethyl)-, diethyl ester		$C_{10}H_{15}NO_4$	213.23	17216-62-5			$115-6^{2}$	4-02-00-02365	1.078^{25}		1.4368^{20}
9934	Propanedioic acid, cyclohexyl-, diethyl ester		$C_{13}H_{22}O_4$	242.32	2163-44-2			$163-5^{20}$	3-09-00-03834	1.0281^{19}	ace 4; bz 4; eth 4; EtOH 4	1.4478^{25}
9935	Propanedioic acid, 2-cyclopenten-1-yl-, diethyl ester	Malonic acid, 2-cyclopentenyl, diethyl ester	$C_{12}H_{18}O_4$	226.27	53608-93-8			141^{10}	3-09-00-03951	1.0507^{20}	H_2O 1; eth 3; ace 3	1.4536^{20}
9936	Propanedioic acid, cyclopentylidene-, diethyl ester	Malonic acid, cyclopentylidene, diethyl ester	$C_{12}H_{18}O_4$	226.27	41589-42-8			140^{10}	3-09-00-03952	1.0616^{20}	ace 4; eth 4	1.4724^{20}
9937	Propanedioic acid, dibutyl-, diethyl ester	Diethyl dibutylmalonate	$C_{15}H_{28}O_4$	272.38	596-75-8			$153-4^{14}$	4-02-00-02119	0.9457^{20}	H_2O 1; EtOH 3; eth 3; ctc 3	1.4341^{20}
9938	Propanedioic acid, dibutyl ester		$C_{11}H_{20}O_4$	216.28	1190-39-2	-83		251.5	4-02-00-01884	0.9824^{20}	H_2O 1; EtOH 3; eth 3; ace 3	1.4262^{20}
9939	Propanedioic acid, diethyl-	Diethylmalonic acid	$C_7H_{12}O_4$	160.17	510-20-3	127 dec	3114		4-02-00-02026		H_2O 4; EtOH 4; eth 4; bz 2	
9940	Propanedioic acid, diethyl-, diethyl ester	Diethyl 2,2-diethylmalonate	$C_{11}H_{20}O_4$	216.28	77-25-8		3746	230	4-02-00-02026	0.9643^{30}	H_2O 1; EtOH 5; eth 5; ctc 3	1.4240^{20}
9941	Propanedioic acid, diethyl ester	Diethyl malonate	$C_7H_{12}O_4$	160.17	105-53-3	-50	3779	200	4-02-00-01881	1.0551^{20}	H_2O 2; EtOH 5; eth 5; ace 4	1.4139^{20}
9942	Propanedioic acid, dimethyl-, diethyl ester	Diethyl dimethylmalonate	$C_9H_{16}O_4$	188.22	1619-62-1	-30.4		197	4-02-00-01955	0.9964^{20}	H_2O 1; EtOH 5; eth 5; ctc 2	1.4129^{20}
9943	Propanedioic acid, dimethyl ester	Methyl malonate	$C_5H_8O_4$	132.12	108-59-8	-61.9	6009	181.4	4-02-00-01880	1.528^{20}	H_2O 2; EtOH 5; eth 5; ace 4	1.4135^{20}
9944	Propanedioic acid, di-2-propenyl-, diethyl ester		$C_{13}H_{20}O_4$	240.30	3195-24-2			243.5	3-02-00-02003	0.9943^{20}	EtOH 3; eth 3; ace 3; ctc 2	1.4445^{22}
9945	Propanedioic acid, dipropyl ester	Dipropyl malonate	$C_9H_{16}O_4$	188.22	1117-19-7	-77.1		229	3-02-00-01619	1.0097^{20}	ace 4; bz 4; eth 4; EtOH 4	1.4206^{20}
9946	Propanedioic acid, di-2-propynyl-, diethyl ester	Diethyl dipropargylmalonate	$C_{13}H_{16}O_4$	236.27	2689-88-5	45.5		129^{11}	4-02-00-02340		chl 3	
9947	Propanedioic acid, (ethoxymethylene)-, diethyl ester		$C_{10}H_{16}O_5$	216.23	87-13-8			280 dec; 165^{19}	4-03-00-01192		H_2O 1; EtOH 3; eth 3; chl 2	1.4600^{20}
9948	Propanedioic acid, ethyl-		$C_5H_8O_4$	132.12	601-75-2	114		$180^{0.05}$	4-02-00-01952		H_2O 4; EtOH 3; eth 3; ace 1	
9949	Propanedioic acid, ethyl-, diethyl ester		$C_9H_{16}O_4$	188.22	133-13-1			208; 98^{12}	4-02-00-01953	1.0047^{20}	H_2O 2; EtOH 4; eth 4; ace 3	1.4166^{20}
9950	Propanedioic acid, ethylidene-, diethyl ester		$C_9H_{14}O_4$	186.21	1462-12-0			$115-8^{17}$	4-02-00-02234	1.0194^{17}	eth 4; EtOH 4	1.4308^{17}
9951	Propanedioic acid, ethyl(1-methylethyl)-, diethyl ester	Malonic acid, ethylisopropyl-, diethyl ester	$C_{12}H_{22}O_4$	230.30	2049-66-3			234; 109^{11}	4-02-00-02052		ace 4; eth 4; EtOH 4	1.4280^{25}
9952	Propanedioic acid, ethylphenyl-, diethyl ester		$C_{15}H_{20}O_4$	264.32	76-67-5			170^{19}	4-09-00-03385	1.071^{20}	H_2O 1; EtOH 3; eth 3; chl 2	1.4896^{25}
9953	Propanedioic acid, (formylamino)-, diethyl ester		$C_8H_{13}NO_5$	203.19	6326-44-9	48.5		173^{11}	4-04-00-02992		EtOH 4	
9954	Propanedioic acid, hexadecyl-, diethyl ester		$C_{23}H_{44}O_4$	384.60	41433-81-2			238^{14}	4-02-00-02183	0.9118^{20}	EtOH 4	1.4433^{20}
9955	Propanedioic acid, hexyl-, diethyl ester		$C_{13}H_{24}O_4$	244.33	5398-10-7			269; 143^{15}	3-02-00-01791	0.9577^{21}	ace 4; bz 4; eth 4; EtOH 4	1.4278^{21}
9956	Propanedioic acid, hydroxy-	Tartronic acid	$C_3H_4O_5$	120.06	80-69-3	157	9042	sub	4-03-00-01120		H_2O 3; EtOH 3; eth 2	
9957	Propanedioic acid, hydroxy-, diethyl ester		$C_7H_{12}O_5$	176.17	13937-08-1	-2.5		221	3-03-00-00905	1.1520^{15}	ace 4; bz 4; eth 4; EtOH 4	
9958	Propanedioic acid, (hydroxyimino)-, diethyl ester		$C_7H_{11}NO_5$	189.17	6829-41-0			172^{12}	4-03-00-01805	1.1821^{18}	ace 4; bz 4; eth 4; EtOH 4	1.4544^{18}
9959	Propanedioic acid, (hydroxymethyl)-	β-Isomalic acid	$C_4H_6O_5$	134.09	4360-96-7	142 dec		170 dec	0-03-00-00441		H_2O 4; EtOH 4; eth 4	
9960	Propanedioic acid, methyl-		$C_4H_6O_4$	118.09	516-05-2	135 dec			4-02-00-01932	1.455^{20}	H_2O 4; EtOH 4; eth 4; bz 2	
9961	Propanedioic acid, (3-methylbutyl)-, diethyl ester		$C_{12}H_{22}O_4$	230.30	5398-08-3			241	3-02-00-01777	0.9580^{25}	ace 4; eth 4; EtOH 4	1.4255^{25}
9962	Propanedioic acid, methyl-, diethyl ester		$C_8H_{14}O_4$	174.20	609-08-5			201	4-02-00-01932	1.0225^{20}	H_2O 2; EtOH 4; eth 4; ace 3	1.4126^{20}
9963	Propanedioic acid, methyl-, dimethyl ester		$C_6H_{10}O_4$	146.14	609-02-9			174	3-02-00-01681	1.0977^{20}	ace 4; eth 4; EtOH 4; chl 4	1.4128^{20}
9964	Propanedioic acid, (1-methylethyl)-, diethyl ester		$C_{10}H_{18}O_4$	202.25	759-36-4			215	4-02-00-02001	0.9961^{20}	H_2O 2; EtOH 4; eth 4; ctc 2	1.4188^{21}
9965	Propanedioic acid, (1-methylethylidene)-		$C_6H_8O_4$	144.13	4441-90-1	170.5		dec	3-02-00-01948		EtOH 4	
9966	Propanedioic acid, (1-methylethylidene)-, diethyl ester	Diethyl isopropylidenemalonate	$C_{10}H_{16}O_4$	200.23	6802-75-1			176.5; 116^{14}	4-02-00-02244	1.0282^{18}	ace 4; EtOH 4	1.4486^{17}

No.	Name Synonym	Mol. Form. Mol. Wt.	CAS RN mp/°C	Merck No. bp/°C	Beil. Ref. den/g cm^{-3}	Solubility n_D
9967	Propanedioic acid, (1-methylpropyl)-, diethyl ester	$C_{11}H_{20}O_4$ 216.28	83-27-2		4-02-00-02019 0.988^{15}	H_2O 1; EtOH 4; eth 4; ctc 2 1.4248^{20}
9968	Propanedioic acid, (2-methylpropyl)-	$C_7H_{12}O_4$ 160.17	4361-06-2 112 dec	dec	4-02-00-02023	eth 4; EtOH 4
9969	Propanedioic acid, (2-methylpropyl)-, diethyl ester Diethyl isobutylmalonate	$C_{11}H_{20}O_4$ 216.28	10203-58-4		3-02-00-01756 0.9804^{20}	H_2O 1; EtOH 4; eth 4; chl 3 1.4236^{20}
9970	Propanedioic acid, 1-naphthalenyl-, diethyl ester	$C_{17}H_{18}O_4$ 286.33	6341-60-2 61.3	$176^{1.0}$	4-09-00-03536	EtOH 3; bz 3; chl 2
9971	Propanedioic acid, octyl-	$C_{11}H_{20}O_4$ 216.28	760-55-4 116		4-02-00-02114 1.173^{17}	ace 4; EtOH 4
9972	Propanedioic acid, (1-oxobutyl)-, diethyl ester	$C_{11}H_{18}O_5$ 230.26	21633-79-4	249.5	4-03-00-01833 1.056^{20}	1.4451^{20}
9973	Propanedioic acid, oxo-, diethyl ester	$C_7H_{10}O_5$ 174.15	609-09-6 -30	210; 105^{19}	3-03-00-01356 1.1419^{16}	H_2O 4; EtOH 3; eth 3; chl 3 1.4310^{22}
9974	Propanedioic acid, pentyl-, diethyl ester	$C_{12}H_{22}O_4$ 230.30	6065-59-4	$134\text{-}6^{14}$	4-02-00-02034 0.9652^{20}	H_2O 1; EtOH 4; eth 4; ctc 2 1.4253^{20}
9975	Propanedioic acid, phenyl-, diethyl ester	$C_{13}H_{16}O_4$ 236.27	83-13-6 16.5	205 dec; 168^{12}	4-09-00-03339 1.0950^{20}	ace 4; EtOH 4 1.4977^{20}
9976	Propanedioic acid, phenyl-, dimethyl ester	$C_{11}H_{12}O_4$ 208.21	37434-59-6 51.6	147^{13}	4-09-00-03339	eth 4; EtOH 4
9977	Propanedioic acid, (phenylmethyl)-, diethyl ester	$C_{14}H_{18}O_4$ 250.29	607-81-8	300	4-09-00-03357 1.076^{15}	H_2O 1; chl 2 1.4872^{20}
9978	Propanedioic acid, (phenylmethylene)-, diethyl ester	$C_{14}H_{16}O_4$ 248.28	5292-53-5 32	216^{30}, 198^{14}	4-09-00-03461 1.1045^{20}	H_2O 1; EtOH 3; eth 3; ace 3 1.5389^{20}
9979	Propanedioic acid, 2-propenyl-	$C_6H_8O_4$ 144.13	2583-25-7 105	180 dec	3-02-00-01945	H_2O 4; eth 4; EtOH 4
9980	Propanedioic acid, 2-propenyl-, diethyl ester	$C_{10}H_{16}O_4$ 200.23	2049-80-1	222.5; 93^{6}	4-02-00-02240 1.0098^{20}	H_2O 1; EtOH 4; eth 4; ctc 3 1.4305^{20}
9981	Propanedioic acid, propyl-, diethyl ester	$C_{10}H_{18}O_4$ 202.25	2163-48-6	221; 114^{22}	4-02-00-01991 0.9873^{20}	H_2O 2; EtOH 4; eth 4 1.4197^{20}
9982	1,2-Propanediol 1,2-Propylene glycol	$C_3H_8O_2$ 76.10	57-55-6 -60	7868 187.6	3-01-00-02142 1.0361^{20}	H_2O 5; EtOH 5; eth 3; bz 3 1.4324^{20}
9983	1,3-Propanediol Trimethylene glycol	$C_3H_8O_2$ 76.10	504-63-2 -26.7	9629 214.4	4-01-00-02493 1.0538^{20}	H_2O 5; EtOH 5; eth 4; bz 2 1.4398^{20}
9984	1,2-Propanediol, 3-amino-, (±)- 1,2-Propanediol, 3-amino(*DL*)	$C_3H_9NO_2$ 91.11	13552-31-3	265 dec; 145^{9}	3-04-00-00839 1.1752^{20}	H_2O 3; EtOH 3; eth 1; bz 1 1.4910^{25}
9985	1,3-Propanediol, 2-amino-2-ethyl- 2-Amino-2-ethyl-1,3-propanediol	$C_5H_{13}NO_2$ 119.16	115-70-8 37.5	451 $152\text{-}3^{10}$	3-04-00-00850 1.099^{20}	H_2O 5 1.490^{20}
9986	1,3-Propanediol, 2-amino-2-(hydroxymethyl)- Tromethamine	$C_4H_{11}NO_3$ 121.14	77-86-1 171.5	9684 $219\text{-}20^{10}$	4-04-00-01903	H_2O 4; MeOH 3
9987	1,3-Propanediol, 2-amino-2-methyl- 2-Amino-2-methyl-1,3-propanediol	$C_4H_{11}NO_2$ 105.14	115-69-5 110	460 $151\text{-}2^{10}$	4-04-00-01881	H_2O 4; EtOH 3
9988	1,3-Propanediol, 2,2-bis[(acetyloxy)methyl]-, diacetate Pentaerythrityl tetraacetate	$C_{13}H_{20}O_8$ 304.30	597-71-7 83.5	7065	4-02-00-00264 1.273^{18}	H_2O 3; EtOH 4; eth 4
9989	1,3-Propanediol, 2,2-bis(hydroxymethyl)- Pentaerythritol	$C_5H_{12}O_4$ 136.15	115-77-5 260	7062 sub	4-01-00-02812	H_2O 3; eth 1; bz 1 1.548
9990	1,3-Propanediol, 2,2-bis[(nitrooxy)methyl]-, dinitrate (ester) Pentaerythrityl tetranitrate	$C_5H_8N_4O_{12}$ 316.14	78-11-5 140.5	7066	4-01-00-02816 1.773^{20}	H_2O 2; EtOH 2; eth 2; ace 4
9991	1,3-Propanediol, 2-bromo-2-nitro- Bronopol	$C_3H_6BrNO_4$ 199.99	52-51-7 131.5	1437	4-01-00-02501	
9992	1,3-Propanediol, 2-butyl-2-ethyl-	$C_9H_{20}O_2$ 160.26	115-84-4 43.8	262	4-01-00-02611 0.927^{50}	H_2O 2; EtOH 3; ace 2 1.4587^{25}
9993	1,3-Propanediol, 2-butyl-2-methyl-	$C_8H_{18}O_2$ 146.23	3121-83-3 43.8	262	4-01-00-02599 0.927^{50}	EtOH 4 1.4587^{25}
9994	1,2-Propanediol, 3-chloro- α-Chlorohydrin	$C_3H_7ClO_2$ 110.54	96-24-2	2145 213 dec; 116^{11}	4-01-00-02484 1.325^{18}	H_2O 3; EtOH 3; eth 3 1.4809^{20}
9995	1,3-Propanediol, 2-chloro- Glycerol β-chlorohydrin	$C_3H_7ClO_2$ 110.54	497-04-1	146^{18}	4-01-00-02499 1.3219^{20}	H_2O 4; ace 4; EtOH 4 1.4831^{20}
9996	1,2-Propanediol, 3-chloro-, diacetate	$C_7H_{11}ClO_4$ 194.61	869-50-1	245	3-02-00-00313 1.199^{25}	eth 4; EtOH 4 1.4407^{20}
9997	1,2-Propanediol, 3-chloro-, dinitrate Clonitrate	$C_3H_5ClN_2O_6$ 200.54	2612-33-1	2390 192.5	4-01-00-02487 1.5112^{9}	ace 4; EtOH 4; chl 4
9998	1,2-Propanediol, 3-chloro-2-methyl- Isobutylene glycol, 3-chloro	$C_4H_9ClO_2$ 124.57	597-33-1	$114\text{-}7^{20}$	4-01-00-02536 1.2362^{20}	H_2O 4; eth 4; EtOH 4 1.4748^{20}
9999	1,2-Propanediol, 3-(2-chlorophenoxy)-	$C_9H_{11}ClO_3$ 202.64	5112-21-0 71.5	250^{19}	4-06-00-00790	H_2O 3; eth 4; bz 3
10000	1,2-Propanediol, 3-(4-chlorophenoxy)- Chlorphenesin	$C_9H_{11}ClO_3$ 202.64	104-29-0 78	2178 $214\text{-}5^{19}$	4-06-00-00831	H_2O 1; EtOH 4; eth 4; bz 3

No.	Name Synonym	Mol. Form. Mol. Wt.	CAS RN mp/°C	Merck No. bp/°C	Beil. Ref. den/g cm^{-3}	Solubility n_D
10001	1,2-Propanediol, 3-(4-chlorophenoxy)-, 1-carbamate Chlorphenesin carbamate	$C_{10}H_{12}ClNO_4$ 245.66	886-74-8 90	2179		ace 4; EtOH 4; diox 4
10002	1,2-Propanediol, diacetate	$C_7H_{12}O_4$ 160.17	623-84-7	190.5	4-02-00-00220 1.059^{20}	H_2O 4; EtOH 3; eth 3 1.4173^{20}
10003	1,3-Propanediol, diacetate	$C_7H_{12}O_4$ 160.17	628-66-0	209.5	4-02-00-00221 1.070^{14}	H_2O 4; EtOH 3 1.4192
10004	1,3-Propanediol, 2,2-diethyl-	$C_7H_{16}O_2$ 132.20	115-76-4 61.5	240.5	4-01-00-02589 1.050^{20}	H_2O 4; EtOH 4; eth 4; chl 3 1.4574^{25}
10005	1,3-Propanediol, 2,2-dimethyl- Neopentyl glycol	$C_5H_{12}O_2$ 104.15	126-30-7 130	6374 208	4-01-00-02551	H_2O 3; EtOH 4; eth 4; bz 3
10006	1,3-Propanediol, 2-ethyl-2-(hydroxymethyl)- Trimethylolpropane	$C_6H_{14}O_3$ 134.18	77-99-6 58	160^{5}	4-01-00-02786	H_2O 4; EtOH 4
10007	1,3-Propanediol, 2-ethyl-2-methyl-	$C_6H_{14}O_2$ 118.18	77-84-9 40.3	225.8	4-01-00-02574	chl 3
10008	1,3-Propanediol, 2-ethyl-2-nitro- 2-Nitro-2-ethyl-1,3-propanediol	$C_5H_{11}NO_4$ 149.15	597-09-1 57.5	dec	4-01-00-02550	H_2O 4; eth 4; EtOH 4
10009	1,2-Propanediol, 3-(hexadecyloxy)-	$C_{19}H_{40}O_3$ 316.52	53584-29-5 64	$205^{1.0}$	3-01-00-02322	ace 4; chl 4; peth 4
10010	1,2-Propanediol, 3-(hexadecyloxy)-, (S)- Chimyl alcohol	$C_{19}H_{40}O_3$ 316.52	506-03-6 64	2049 $120^{0.005}$	3-01-00-02322	ace 4; peth 4; chl 4
10011	1,3-Propanediol, 2-(hydroxymethyl)-2-methyl-	$C_5H_{12}O_3$ 120.15	77-85-0 204	$135\text{-}7^{15}$	4-01-00-02780	H_2O 5; EtOH 5; eth 1; bz 1
10012	1,3-Propanediol, 2-(hydroxymethyl)-2-nitro- Tris(hydroxymethyl)nitromethane	$C_4H_9NO_5$ 151.12	126-11-4 165	9667 dec	4-01-00-02777	H_2O 4; eth 4; EtOH 4
10013	1,2-Propanediol, 3-mercapto- Thioglycerol	$C_3H_8O_2S$ 108.16	96-27-5	9264 $100\text{-}1^{1}$	3-01-00-02339 1.2455^{20}	H_2O 2; EtOH 5; eth 2; ace 4 1.5268^{20}
10014	1,2-Propanediol, 3-methoxy-	$C_4H_{10}O_3$ 106.12	623-39-2	220	4-01-00-02755 1.114^{20}	H_2O 4; EtOH 4; eth 3; ace 4 1.442^{25}
10015	1,3-Propanediol, 2-methoxy- Glycerol, 2-methyl ether	$C_4H_{10}O_3$ 106.12	761-06-8	232	3-01-00-02317 1.124^{25}	H_2O 4; ace 4; eth 4; EtOH 4 1.4505^{12}
10016	1,2-Propanediol, 3-(2-methoxyphenoxy)- Guaifenesin	$C_{10}H_{14}O_4$ 198.22	93-14-1 78.5	4465 215^{19}; $127^{0.2}$	4-06-00-05576	H_2O 3; EtOH 4; bz 3; chl 3
10017	1,2-Propanediol, 2-methyl-	$C_4H_{10}O_2$ 90.12	558-43-0	176	4-01-00-02533 1.0024^{20}	H_2O 4; eth 4; EtOH 4 1.4350^{20}
10018	1,3-Propanediol, 2-methyl-2-nitro-	$C_4H_9NO_4$ 135.12	77-49-6 150.1	dec	4-01-00-02537	H_2O 4; EtOH 4; DMSO-d_6 2
10019	1,2-Propanediol, 3-(2-methylphenoxy)- Mephenesin	$C_{10}H_{14}O_3$ 182.22	59-47-2 70 dec	5737	4-06-00-01952	H_2O 2; EtOH 3; eth 2
10020	1,3-Propanediol, 2-methyl-2-propyl-	$C_7H_{16}O_2$ 132.20	78-26-2 62.5	234; 121^{10}	4-01-00-02585	H_2O 3; chl 2; os 3; hx 3
10021	1,3-Propanediol, 2-methyl-2-propyl-, dicarbamate Meprobamate	$C_9H_{18}N_2O_4$ 218.25	57-53-4 105	5751	4-03-00-00073	bz 4; eth 4; EtOH 4
10022	1,2-Propanediol, 1-nitrate	$C_3H_7NO_4$ 121.09	20266-65-3	$148\text{-}9^{4}$	4-01-00-02476 1.2417^{20}	1.4368^{20}
10023	1,2-Propanediol, 3-(octadecyloxy)- Batyl alcohol	$C_{21}H_{44}O_3$ 344.58	544-62-7 70.5	1021 217^{2}	4-01-00-02758	eth 4
10024	Propanediol, oxybis-	$C_6H_{14}O_5$ 166.17	59113-36-9	267^{15}; 225^{3}	3-01-00-02327 1.2751^{20}	H_2O 4; EtOH 4 1.4890^{20}
10025	1,2-Propanediol, 3,3'-oxybis-	$C_6H_{14}O_5$ 166.17	627-82-7	267^{15}; 225^{3}	3-01-00-02327 1.2751^{20}	H_2O 4; EtOH 4 1.4890^{20}
10026	1,2-Propanediol, 3-phenoxy- Phenylglyceryl ether	$C_9H_{12}O_3$ 168.19	538-43-2 67.5	7261 200^{22}	4-06-00-00589 1.225^{20}	H_2O 4; bz 4; eth 4; EtOH 4
10027	1,3-Propanedione, 1,3-diphenyl- Dibenzoylmethane	$C_{15}H_{12}O_2$ 224.26	120-46-7 70.5	2991	4-07-00-02512	EtOH 3; eth 3; chl 3; dil NaOH 3
10028	1,2-Propanedione, 1-phenyl-	$C_9H_8O_2$ 148.16	579-07-7 <20	222; 102^{12}	4-07-00-02143 1.1006^{20}	H_2O 3; EtOH 3; eth 3 1.537^{10}
10029	Propanedioyl dichloride	$C_3H_2Cl_2O_2$ 140.95	1663-67-8	58^{26}	4-02-00-01887 1.4509^{20}	eth 3; AcOEt 3 1.4639^{20}
10030	1,2-Propanedithiol	$C_3H_8S_2$ 108.23	814-67-5	152	4-01-00-02492 1.08^{20}	chl 3 1.532^{20}
10031	1,3-Propanedithiol Trimethylene dimercaptan	$C_3H_8S_2$ 108.23	109-80-8 -79	7811 172.9	4-01-00-02503 1.0772^{20}	H_2O 2; EtOH 5; eth 5; bz 5 1.5392^{20}
10032	Propane, 1,1'-[1,2-ethanediylbis(oxy)]bis[2-methyl- Ethylene glycol diisobutyl ether	$C_{10}H_{22}O_2$ 174.28	54922-86-0	171.3	4-01-00-02382 0.8200^{20}	1.4021^{20}
10033	Propane, 1-(ethenyloxy)-	$C_5H_{10}O$ 86.13	764-47-6	65	4-01-00-02052 0.7674^{20}	1.3908^{20}
10034	Propane, 2-(ethenyloxy)-	$C_5H_{10}O$ 86.13	926-65-8 -140	55.5	4-01-00-02052 0.7534^{20}	ace 4; bz 4; eth 4; EtOH 4 1.3840^{20}
10035	Propane, 1-(ethenyloxy)-2-methyl- Isobutyl vinyl ether	$C_6H_{12}O$ 100.16	109-53-5 -112	83	4-01-00-02054 0.7645^{20}	H_2O 2; EtOH 4; eth 5; ace 4 1.3966^{20}
10036	Propane, 2-(ethenyloxy)-2-methyl-	$C_6H_{12}O$ 100.16	926-02-3 -112	75	4-01-00-02055 0.7691^{20}	1.3922^{20}
10037	Propane, 1-ethoxy- Ethyl propyl ether	$C_5H_{12}O$ 88.15	628-32-0 -127.5	63.2	4-01-00-01421 0.7386^{20}	eth 4; EtOH 4; HOAc 4 1.3695^{20}
10038	Propane, 2-ethoxy- Ethyl isopropyl ether	$C_5H_{12}O$ 88.15	625-54-7	54.1	4-01-00-01471 0.720^{25}	H_2O 3; EtOH 5; eth 5; ace 3 1.3698^{25}

No.	Name	Synonym	Mol. Form.	Mol. Wt.	CAS RN	mp/°C	Merck No.	bp/°C	Beil. Ref.	den/g cm^{-3}	Solubility	n_D
10039	Propane, 1-ethoxy-2-methyl-		$C_6H_{14}O$	102.18	627-02-1			81	4-01-00-01593	0.751^{20}	H_2O 1; EtOH 5; eth 5; ace 3	1.3739^{25}
10040	Propane, 2-ethoxy-2-methyl-	Ethyl *tert*-butyl ether	$C_6H_{14}O$	102.18	637-92-3	-94	3732	72.6	4-01-00-01615	0.736^{25}	H_2O 1; EtOH 4; eth 4	1.3756^{20}
10041	Propane, 1,1'-[ethylidenebis(oxy)]bis[2-methyl-		$C_{10}H_{22}O_2$	174.28	5669-09-0			171.3	4-01-00-03107	0.8200^{20}		1.4021^{20}
10042	Propane, 1-(ethylthio)-	Ethyl propyl sulfide	$C_5H_{12}S$	104.22	4110-50-3	-117		118.6	4-01-00-01451	0.8370^{20}	EtOH 3	1.4462^{20}
10043	Propane, 2-(ethylthio)-	Ethyl isopropyl sulfide	$C_5H_{12}S$	104.22	5145-99-3	-122.2		107.5	4-01-00-01501	0.8246^{20}		
10044	Propane, 1-(ethylthio)-2-methyl-	Ethyl isobutyl sulfide	$C_6H_{14}S$	118.24	1613-45-2			134.2	4-01-00-01606	0.8306^{20}	chl 3	1.4450^{20}
10045	Propane, 1-(ethynyloxy)-		C_5H_8O	84.12	42842-08-0			75	4-01-00-02213	0.8080^{20}	eth 4; EtOH 4	1.3935^{20}
10046	Propane, 1-fluoro-	Propyl fluoride	C_3H_7F	62.09	460-13-9	-159		2.5	4-01-00-00187	0.7956^{20}	H_2O 2; EtOH 4; eth 4	1.3115^{20}
10047	Propane, 2-fluoro-	Isopropyl fluoride	C_3H_7F	62.09	420-26-8				4-01-00-00187			
10048	Propane, 1,1,1,2,2,3,3-heptachloro-	1,1,1,2,2,3,3-Heptachloropropane	C_3HCl_7	285.21	594-89-8	29.4		247.5	4-01-00-00205	1.8048^{34}	ctc 2; chl 3	
10049	Propane, 1,1,1,2,3,3,3-heptachloro-	1,1,1,2,3,3,3-Heptachloropropane	C_3HCl_7	285.21	3849-33-0	11		249	4-01-00-00205	1.7921^{34}	chl 4	1.5427^{21}
10050	Propane, 1,1,1,2,3,3,3-heptafluoro-2-(trifluoromethyl)-	Perfluoroisobutane	C_4F_{10}	238.03	354-92-7			0				
10051	Propane, 1,1,1,2,2,3-hexachloro-	1,1,1,2,2,3-Hexachloropropane	$C_3H_2Cl_6$	250.77	24425-97-6	26		216	4-01-00-00204	1.7187^{25}		1.5282^{25}
10052	Propane, 1,1,1,3,3,3-hexachloro-	1,1,1,3,3,3-Hexachloropropane	$C_3H_2Cl_6$	250.77	3607-78-1	-27		206	4-01-00-00205	1.6800^{20}		1.5179^{20}
10053	Propane, 1,1,2,2,3,3-hexachloro-	1,1,2,2,3,3-Hexachloropropane	$C_3H_2Cl_6$	250.77	15600-01-8			218.5	4-01-00-00205	1.7137^{34}		1.5262^{18}
10054	Propane, 1-iodo-	Propyl iodide	C_3H_7I	169.99	107-08-4	-101.3	7875	102.6	4-01-00-00222	1.7489^{20}	H_2O 2; EtOH 5; eth 5; ctc 2	1.5058^{20}
10055	Propane, 2-iodo-	Isopropyl iodide	C_3H_7I	169.99	75-30-9	-90	5102	89.5	4-01-00-00223	1.7042^{20}	H_2O 2; EtOH 5; eth 5; bz 5	1.5028^{20}
10056	Propane, 1-iodo-2,2-dimethyl-		$C_5H_{11}I$	198.05	15501-33-4			128 dec	4-01-00-00338	1.4940^{20}	H_2O 1; EtOH 3; eth 3	1.4890^{20}
10057	Propane, 1-iodo-2-methyl-	Isobutyl iodide	C_4H_9I	184.02	513-38-2		5027	121.1	4-01-00-00299	1.6035^{20}		1.4959^{20}
10058	Propane, 2-iodo-2-methyl-	*tert*-Butyl iodide	C_4H_9I	184.02	558-17-8	-38.2		100.1	4-01-00-00300	1.571^{25}	EtOH 5; eth 5	1.4918^{20}
10059	Propane, 1-isocyanato-	Propyl isocyanate	C_4H_7NO	85.11	110-78-1			83.5	4-04-00-00491	0.908^{25}		1.3970^{20}
10060	Propane, 2-isocyanato-	Isopropyl isocyanate	C_4H_7NO	85.11	1795-48-8			74.5	4-04-00-00530	0.866^{25}		1.3825^{20}
10061	Propane, 1-isocyanato-2-methyl-	Isobutyl isocyanate	C_5H_9NO	99.13	1873-29-6			106	4-04-00-00653			
10062	Propane, 2-isocyanato-2-methyl-	*tert*-Butyl isocyanate	C_5H_9NO	99.13	1609-86-5	85.5			4-04-00-00669	0.8670^{7}		1.4061^{20}
10063	Propane, 2-isocyano-	Isopropyl isocyanide	C_4H_7N	69.11	598-45-8			87	0-04-00-00154	0.7596^{25}	H_2O 1; EtOH 5; eth 5	
10064	Propane, 1-isothiocyanato-		C_4H_7NS	101.17	628-30-8			153	4-04-00-00491	0.9781^{16}	H_2O 2; EtOH 5; eth 5	1.5085^{16}
10065	Propane, 1-isothiocyanato-2-methyl-		C_5H_9NS	115.20	591-82-2			160	4-04-00-00653	0.9631^{14}		1.5005^{14}
10066	Propane, 2-isothiocyanato-2-methyl-	*tert*-Butyl isothiocyanate	C_5H_9NS	115.20	590-42-1	10.5		140	4-04-00-00669	0.9187^{10}		
10067	Propane, 1-isothiocyanato-3-(methylsulfonyl)-	Cheirolin	$C_5H_9NO_2S_2$	179.26	505-34-0	47.5	2040	200^{3}	4-04-00-01664		EtOH 4; chl 4	
10068	Propane, 1,1',1'',1'''-[methanetetrayltetrakis(oxy)]tetrakis-		$C_{13}H_{28}O_4$	248.36	597-72-8			224.2	4-03-00-00007	0.897^{20}	EtOH 4; eth 4	1.4100^{20}
10069	Propane, 1-methoxy-	Methyl propyl ether	$C_4H_{10}O$	74.12	557-17-5		6031	39.1	4-01-00-01421	0.7356^{13}	H_2O 3; EtOH 5; eth 5; ace 3	1.3579^{25}
10070	Propane, 2-methoxy-	Isopropyl methyl ether	$C_4H_{10}O$	74.12	598-53-8			30.7	4-01-00-01471	0.7237^{15}	H_2O 2; EtOH 5; eth 5	1.3576^{20}
10071	Propane, 1-methoxy-2-methyl-	Isobutyl methyl ether	$C_5H_{12}O$	88.15	625-44-5			58.6	4-01-00-01593	0.7311^{20}	eth 4; EtOH 4	
10072	Propane, 2-methoxy-2-methyl-	*tert*-Butyl methyl ether	$C_5H_{12}O$	88.15	1634-04-4	-108.6	5954	55.2	4-01-00-01615	0.7405^{20}	H_2O 3; EtOH 4; eth 4	1.3690^{20}
10073	Propane, 2-methyl-	Isobutane	C_4H_{10}	58.12	75-28-5	-138.3		-11.7	4-01-00-00282	*0.5510^{25}	H_2O 2; EtOH 3; eth 3; chl 3	1.3518^{-25}
10074	Propane, 1,1'-[methylenebis(oxy)]bis-		$C_7H_{16}O_2$	132.20	505-84-0	-97.3		140.5	4-01-00-03029	0.8345^{20}	ace 4; bz 4; eth 4; EtOH 4	1.3939^{19}
10075	Propane, 1,1'-[methylenebis(oxy)]bis[2-methyl-		$C_9H_{20}O_2$	160.26	2568-91-4			165.5	4-01-00-03030	0.8268^{20}		1.4108^{17}
10076	Propane, 1-(1-methylethoxy)-		$C_6H_{14}O$	102.18	627-08-7			83	4-01-00-01471	0.7370^{20}	H_2O 2; EtOH 4; eth 3; ace 3	1.376^{21}
10077	Propane, 1-[(1-methylethyl)thio]-	Isopropyl propyl sulfide	$C_6H_{14}S$	118.24	5008-73-1			132.1	4-01-00-01501	0.8269^{20}		

No.	Name Synonym	Mol. Form. Mol. Wt.	CAS RN mp/°C	Merck No. bp/°C	Beil. Ref. den/g cm^{-3}	Solubility n_D
10078	Propane, 1,1',1"-[methylidynetris(oxy)]tris-	$C_{10}H_{22}O_3$	621-76-1		4-02-00-00027	eth 4; EtOH 4
		190.28		191; 93^{30}	0.8805^{20}	1.4072^{20}
10079	Propane, 2,2',2"-[methylidynetris(oxy)]tris-	$C_{10}H_{22}O_3$	4447-60-3		4-02-00-00028	eth 4; EtOH 4
		190.28		167	0.8621^{20}	1.4000^{20}
10080	Propane, 1,1',1"-[methylidynetris(oxy)]tris[2-methyl-	$C_{13}H_{28}O_3$	16754-49-7		4-02-00-00029	eth 4; EtOH 4
	Orthoformic acid, triisobutyl ester	232.36		223	0.8582^{20}	1.4120^{20}
10081	Propane, 2-methyl-2-(1-methylethoxy)-	$C_7H_{16}O$	17348-59-3		4-01-00-01616	chl 3
		116.20	-88	87.6	0.7365^{25}	
10082	Propane, 2-methyl-1-(methylthio)-	$C_5H_{12}S$	5008-69-5		4-01-00-01600	EtOH 3; eth 3; ace 3
		104.22		112.5	0.8335^{20}	1.4433^{20}
10083	Propane, 2-methyl-2-(methylthio)-	$C_5H_{12}S$	6163-64-0		4-01-00-01636	
	tert-Butyl methyl sulfide	104.22		99		
10084	Propane, 2-methyl-1-nitro-	$C_4H_9NO_2$	625-74-1		4-01-00-00301	H_2O 2; EtOH 5; eth 5
		103.12		140.5	0.9597^{25}	1.4066^{25}
10085	Propane, 2-methyl-2-nitro-	$C_4H_9NO_2$	594-70-7		4-01-00-00301	EtOH 5; eth 5; ace 5; bz 5
		103.12	26.23	127.16	0.9501^{28}	1.4015^{20}
10086	Propane, 2-methyl-1-propoxy-	$C_7H_{16}O$	15268-49-2		4-01-00-01594	eth 4; EtOH 4
		116.20		107	0.7549^{20}	1.3852^{25}
10087	Propane, 2-methyl-2-propoxy-	$C_7H_{16}O$	29072-93-3		3-01-00-01578	
		116.20		100	0.7472^{25}	1.3830^{25}
10088	Propane, 1-(methylthio)-	$C_4H_{10}S$	3877-15-4		4-01-00-01450	H_2O 3; EtOH 3; eth 3; ace 3
	Methyl propyl sulfide	90.19	-113	95.6	0.8424^{20}	1.4442^{20}
10089	Propane, 2-(methylthio)-	$C_4H_{10}S$	1551-21-9		4-01-00-01500	EtOH 3; eth 3; ace 3
	Isopropyl methyl sulfide	90.19	-101.5	84.8	0.8291^{20}	1.4932^{20}
10090	Propanenitrile	C_3H_5N	107-12-0	7839	4-02-00-00728	H_2O 4; EtOH 3; eth 3; ace 3
	Ethyl cyanide	55.08	-92.8	97.1	0.7818^{20}	1.3655^{20}
10091	Propanenitrile, 2-(acetyloxy)-	$C_5H_7NO_2$	15657-96-2		4-03-00-00676	H_2O 4
		113.12	-75	172.5	1.0278^{20}	1.4027^{20}
10092	Propanenitrile, 3-amino-	$C_3H_6N_2$	151-18-8	481	4-04-00-02530	
	3-Aminopropionitrile	70.09		185; 88^{20}	0.9584^{20}	1.4396^{20}
10093	Propanenitrile, 2,2'-azobis[2-methyl-	$C_8H_{12}N_4$	78-67-1	931	4-04-00-03377	H_2O 1; EtOH 2; eth 2
	2,2'-Azobis(isobutyronitrile)	164.21				
10094	Propanenitrile, 2-bromo-	C_3H_4BrN	19481-82-4		4-02-00-00764	ace 4; eth 4
		133.98		59^{24}	1.5505^{20}	1.4585^{20}
10095	Propanenitrile, 3-bromo-	C_3H_4BrN	2417-90-5		4-02-00-00766	EtOH 4; eth 4; ctc 2
		133.98		92^{25}; 69^{7}	1.6152^{20}	1.4800^{20}
10096	Propanenitrile, 2-bromo-2-methyl-	C_4H_6BrN	41658-69-9		4-02-00-00863	ace 4; eth 4; EtOH 4
		148.00		139.5	1.4796^{15}	1.4379^{15}
10097	Propanenitrile, 3-chloro-	C_3H_4ClN	542-76-7	2160	4-02-00-00751	ctc 2
	β-Chloropropionitrile	89.52	-51	175.5	1.1573^{20}	1.4360^{20}
10098	Propanenitrile, 3-chloro-2-hydroxy-2-methyl-	C_4H_6ClNO	33401-05-7		3-03-00-00599	H_2O 4; EtOH 4; ace 4
	Lactonitrile, 3-chloro-2-methyl-	119.55		110^{27}	1.2027^{15}	1.4356^{11}
10099	Propanenitrile, 2,3-dichloro-	$C_3H_3Cl_2N$	2601-89-0		4-02-00-00757	
		123.97		58-9^{8}	1.3500^{20}	1.4640^{20}
10100	Propanenitrile, 2,2-dimethyl-	C_5H_9N	630-18-2		4-02-00-00913	
	tert-Butyl cyanide	83.13	15	106.1	0.7586^{25}	1.3774^{20}
10101	Propanenitrile, 3-(dimethylamino)-	$C_5H_{10}N_2$	1738-25-6		4-04-00-02533	
	3-(Dimethylamino)propionitrile	98.15		173	0.8705^{20}	
10102	Propanenitrile, 2-ethoxy-	C_5H_9NO	14631-45-9		4-03-00-00675	eth 4; EtOH 4
	2-Ethoxypropionitrile	99.13		136	0.8743^{20}	1.3890^{22}
10103	Propanenitrile, 3-ethoxy-	C_5H_9NO	2141-62-0		4-03-00-00708	eth 4; EtOH 4
		99.13		171	0.9285^{15}	1.4068^{20}
10104	Propanenitrile, 2-(β-*D*-glucopyranosyloxy)-2-methyl-	$C_{10}H_{17}NO_6$	554-35-8	5375	5-17-07-00397	ace 4
	Linamarin	247.25	145			
10105	Propanenitrile, 2-hydroxy-	C_3H_5NO	78-97-7		3-03-00-00451	H_2O 5; EtOH 5; eth 3; chl 3
	Acetaldehyde, cyanohydrin	71.08	-40	183	0.9877^{20}	1.4058^{18}
10106	Propanenitrile, 3-hydroxy-	C_3H_5NO	109-78-4	3751	4-03-00-00708	H_2O 5; EtOH 5; eth 2; chl 3
	Hydracrylonitrile	71.08	-46	221	1.0404^{25}	1.4248^{20}
10107	Propanenitrile, 3-[2-(2-hydroxyethoxy)ethoxy]-	$C_7H_{13}NO_3$	10143-54-1		4-03-00-00713	
		159.19		186^{9}	1.089^{32}	1.4452^{22}
10108	Propanenitrile, 2-hydroxy-2-methyl-	C_4H_7NO	75-86-5	59	4-03-00-00785	H_2O 4; EtOH 4; eth 4; ace 3
	Acetone cyanohydrin	85.11	-19	82^{23}	0.932^{19}	1.3992^{20}
10109	Propanenitrile, 2-methoxy-	C_4H_7NO	33695-59-9		4-03-00-00675	EtOH 4
	2-Methoxypropionitrile	85.11		119	0.8928^{20}	1.3818^{20}
10110	Propanenitrile, 3-methoxy-	C_4H_7NO	110-67-8		4-03-00-00708	EtOH 3; eth 3; chl 3
		85.11		163	0.9379^{20}	1.4043^{20}
10111	Propanenitrile, 2-methyl-	C_4H_7N	78-82-0		4-02-00-00853	H_2O 2; EtOH 4; eth 4; ace 3
	Isobutyronitrile	69.11	-71.5	103.9	0.7704^{20}	1.3720^{20}
10112	Propanenitrile, 3-(methylamino)-	$C_4H_8N_2$	693-05-0		3-04-00-01264	H_2O 3; ace 3; bz 3; chl 3
		84.12		101-4^{49}	0.8992^{20}	1.4320^{20}
10113	Propanenitrile, 2-oxo-	C_3H_3NO	631-57-2		4-03-00-01515	eth 3; ace 3; CH_3CN 3
		69.06		92.3	0.9745^{20}	1.3764^{20}
10114	Propanenitrile, 3,3'-oxybis-	$C_6H_8N_2O$	1656-48-0		4-03-00-00720	
		124.14		110-$12^{0.5}$	1.043^{25}	1.4405^{20}
10115	Propanenitrile, 3,3,3-trichloro-2-hydroxy-	$C_3H_2Cl_3NO$	513-96-2	2134	4-03-00-00680	H_2O 4; eth 4; EtOH 4
	Chlorocyanohydrin	174.41	61	217 dec		
10116	Propane, 1-nitro-	$C_3H_7NO_2$	108-03-2	6548	4-01-00-00229	H_2O 2; EtOH 5; eth 5; chl 3
	1-Nitropropane	89.09	-108	131.1	0.9961^{25}	1.4018^{20}

No.	Name Synonym	Mol. Form. Mol. Wt.	CAS RN mp/°C	Merck No. bp/°C	Beil. Ref. den/g cm^{-3}	Solubility n_D
10117	Propane, 2-nitro- 2-Nitropropane	$C_3H_7NO_2$ 89.09	79-46-9 -91.3	6549 120.2	4-01-00-00230 0.9821^{25}	H_2O 2; chl 3 1.3944^{20}
10118	Propane, octafluoro- Perfluoropropane	C_3F_8 188.02	76-19-7 -147.6	 -36.6	4-01-00-00189	
10119	Propane, 1,1'-oxybis- Dipropyl ether	$C_6H_{14}O$ 102.18	111-43-3 -126.1	7870 90.0	4-01-00-01422 0.7466^{20}	eth 4; EtOH 4 1.3809^{20}
10120	Propane, 2,2'-oxybis- Diisopropyl ether	$C_6H_{14}O$ 102.18	108-20-3 -86.8	5100 68.5	4-01-00-01471 0.7241^{20}	H_2O 2; EtOH 5; eth 5; ace 3 1.3679^{20}
10121	Propane, 1,1'-oxybis[2-chloro-	$C_6H_{12}Cl_2O$ 171.07	54460-96-7	 188	2-01-00-00370 1.109^{20}	EtOH 3; eth 3 1.4467^{20}
10122	Propane, 1,1'-oxybis[3-chloro-	$C_6H_{12}Cl_2O$ 171.07	629-36-7	 216; 90.5^{11}	4-01-00-01442 1.138^{20}	EtOH 3; eth 3 1.4158^{20}
10123	Propane, 2,2'-oxybis[1-chloro- 2,2'-Dichlorodiisopropyl ether	$C_6H_{12}Cl_2O$ 171.07	108-60-1	 187	3-01-00-01470 1.103^{20}	H_2O 1; EtOH 5; eth 5; ace 5 1.4505^{20}
10124	Propane, 2,2'-oxybis[2-methyl- Di-*tert*-Butyl ether	$C_8H_{18}O$ 130.23	6163-66-2	3020 107.2	4-01-00-01616 0.7658^{20}	 1.3949^{20}
10125	Propane, 1,1,1,2,3-pentachloro- 1,1,1,2,3-Pentachloropropane	$C_3H_3Cl_5$ 216.32	21700-31-2	 179.5	4-01-00-00203	eth 4; EtOH 4 1.5130^{20}
10126	Propane, 1,1,2,3,3-pentachloro- 1,1,2,3,3-Pentachloropropane	$C_3H_3Cl_5$ 216.32	15104-61-7	 199; 99^{20}	4-01-00-00204 1.6086^{39}	H_2O l; EtOH 4; eth 3; chl 3 1.5131^{17}
10127	Propane, 1,1,1,2,3-pentachloro-2-methyl- Isobutane, 1,1,1,2,3-pentachloro	$C_4H_5Cl_5$ 230.35	4749-31-9 73.5	 215	4-01-00-00294 1.5686^{25}	chl 4 1.5165^{20}
10128	Propane, 1,1,1,2,2-pentafluoro 1,1,1,2,2-Pentafluoropropane	$C_3H_3F_5$ 134.05	1814-88-6	 -17.4		
10129	Propane, 1,1'-sulfinylbis-	$C_6H_{14}OS$ 134.24	4253-91-2 22.5	 80^2	4-01-00-01453 0.9654^{20}	eth 4; EtOH 4 1.4663^{20}
10130	2-Propanesulfonic acid Isopropanesulfonic acid	$C_3H_8O_3S$ 124.16	14159-48-9 -37	 $159^{1.4}$	4-04-00-00042 1.187^{25}	H_2O 4 1.4332^{20}
10131	Propane, 1,1'-sulfonylbis-	$C_6H_{14}O_2S$ 150.24	598-03-8 29.5		4-01-00-01453 1.0278^{50}	H_2O 2; EtOH 3; eth 3 1.4456^{30}
10132	Propane, 1,1'-sulfonylbis[2-methyl-	$C_8H_{18}O_2S$ 178.30	10495-45-1 17	 265	4-01-00-01607 1.0056^{18}	
10133	1-Propanesulfonyl chloride	$C_3H_7ClO_2S$ 142.61	10147-36-1	 180 dec; 77^{12}	4-04-00-00039 1.2826^{15}	 1.452^{20}
10134	1-Propanesulfonyl chloride, 2-methyl-	$C_4H_9ClO_2S$ 156.63	35432-36-1	 190	4-04-00-00049 1.2014^{20}	 1.4520^{25}
10135	Propane, 1,1,1,2-tetrabromo-	$C_3H_4Br_4$ 359.68	62127-49-5	 122^{15}	3-01-00-00251 2.679^{20}	 1.6187^{20}
10136	Propane, 1,1,2,2-tetrabromo- 1,1,2,2-Tetrabromopropane	$C_3H_4Br_4$ 359.68	34570-59-7	 $105\text{-}7^9$	4-01-00-00222 2.687^{20}	 1.6166^{20}
10137	Propane, 1,1,3,3-tetrabromo-	$C_3H_4Br_4$ 359.68	51525-97-4	 $154\text{-}6^{19}$	2-01-00-00077 2.702^{21}	 1.6225^{21}
10138	Propane, 1,2,2,3-tetrabromo- 1,2,2,3-Tetrabromopropane	$C_3H_4Br_4$ 359.68	54268-02-9 10.5	 169^{80}; 115^9	3-01-00-00251 2.703^{20}	 1.6200^{20}
10139	1,1,2,3-Propanetetracarboxylic acid, tetraethyl ester Tetraethyl 1,1,2,3-propanetetracarboxylate	$C_{15}H_{24}O_8$ 332.35	635-03-0	 $203\text{-}4^{18}$	3-02-00-02077 1.1184^{20}	EtOH 4 1.4395^{20}
10140	1,1,3,3-Propanetetracarboxylic acid, tetraethyl ester 1,1,3,3-Tetracarbethoxypropane	$C_{15}H_{24}O_8$ 332.35	2121-66-6 -30	 305 dec; 195^8	4-02-00-02417 1.116^{20}	EtOH 3 1.4398^{20}
10141	Propane, 1,1,1,2-tetrachloro- 1,1,1,2-Tetrachloropropane	$C_3H_4Cl_4$ 181.88	812-03-3 -64	 152.5	4-01-00-00201 1.473^{20}	H_2O 1; EtOH 4; eth 3; chl 3 1.4867^{20}
10142	Propane, 1,1,1,3-tetrachloro- 1,1,1,3-Tetrachloropropane	$C_3H_4Cl_4$ 181.88	1070-78-6	 157	4-01-00-00201 1.4509^{20}	H_2O 1; EtOH 4; eth 4; bz 4 1.4825^{20}
10143	Propane, 1,1,2,2-tetrachloro- 1,1,2,2-Tetrachloropropane	$C_3H_4Cl_4$ 181.88	13116-60-4	 153	4-01-00-00201 1.47^{12}	H_2O 1; EtOH 5; eth 5; chl 3 1.4850^{25}
10144	Propane, 1,1,2,3-tetrachloro- 1,1,2,3-Tetrachloropropane	$C_3H_4Cl_4$ 181.88	18495-30-2	 179.5	4-01-00-00202 1.513^{17}	H_2O 1; EtOH 3; eth 4; chl 3 1.5037^{17}
10145	Propane, 1,2,2,3-tetrachloro- 1,2,2,3-Tetrachloropropane	$C_3H_4Cl_4$ 181.88	13116-53-5	 165	4-01-00-00202 1.500^{18}	H_2O 1; EtOH 4; eth 4; chl 3 1.4940^{18}
10146	Propane, 1,1,2,3-tetrachloro-2-(chloromethyl)- 1,1,2,3-Tetrachloro-2-(chloromethyl)propane	$C_4H_5Cl_5$ 230.35	66997-45-3	 227; 95^9	3-01-00-00321 1.5686^{25}	bz 4; EtOH 4; chl 4 1.5165^{25}
10147	Propane, 1,1,1,2-tetrachloro-2-methyl- Isobutane, 1,1,1,2-tetrachloro	$C_4H_6Cl_4$ 195.90	7086-07-9 178.5	 192^{117}	4-01-00-00293	eth 4; EtOH 4; chl 4
10148	Propane, 1,1,2,3-tetrachloro-2-methyl-	$C_4H_6Cl_4$ 195.90	18963-01-4 -46	 190.5	4-01-00-00294 1.4393^{25}	bz 4; eth 4; chl 4 1.4963^{20}
10149	Propane, 1,2,2,3-tetrachloro-1,1,3,3-tetrafluoro- Propane, 1,2,2,3-tetrachloro-tetrafluoro	$C_3Cl_4F_4$ 253.84	677-68-9 -42.9	 112	4-01-00-00203 1.7199^{20}	 1.3958^{20}
10150	Propane, 1,1,3,3-tetramethoxy-	$C_7H_{16}O_4$ 164.20	102-52-3	 183; 66^{12}	4-01-00-03635 0.997^{25}	 1.4081^{20}
10151	Propane, 1,1'-thiobis- Dipropyl sulfide	$C_6H_{14}S$ 118.24	111-47-7 -102.5	7881 142.9	4-01-00-01452 0.814^{17}	H_2O 1; EtOH 3; eth 3 1.4487^{20}
10152	Propane, 2,2'-thiobis- Diisopropyl sulfide	$C_6H_{14}S$ 118.24	625-80-9 -78.1	 120.1	4-01-00-01502 0.8142^{20}	H_2O 1; EtOH 3; eth 3 1.4438^{20}
10153	Propane, 1,1'-thiobis[3-chloro- Bis(3-chloropropyl) sulfide	$C_6H_{12}Cl_2S$ 187.13	55882-21-8	 162^{43}	3-01-00-01438 1.1774^{25}	eth 4; EtOH 4; tol 4 1.5075^{20}
10154	Propane, 1,1'-thiobis[2-methyl- Diisobutyl sulfide	$C_8H_{18}S$ 146.30	592-65-4 -105.5	5035 171	4-01-00-01607 0.8363^{10}	

No.	Name Synonym	Mol. Form. Mol. Wt.	CAS RN mp/°C	Merck No. bp/°C	Beil. Ref. den/g cm^{-3}	Solubility n_D
10155	Propane, 2,2'-thiobis[2-methyl- Di-*tert*-butyl sulfide	$C_8H_{18}S$ 146.30	107-47-1 -9.0	 149.1	4-01-00-01637 0.815^{25}	 1.4506^{20}
10156	1-Propanethiol Propyl mercaptan	C_3H_8S 76.16	107-03-9 -113.3	 67.8	4-01-00-01449 0.8411^{20}	H_2O 2; EtOH 3; eth 3; ace 3 1.4380^{20}
10157	2-Propanethiol Isopropyl mercaptan	C_3H_8S 76.16	75-33-2 -130.5	 52.6	4-01-00-01498 0.8143^{20}	H_2O 2; EtOH 5; eth 5; ace 4 1.4255^{20}
10158	1-Propanethiol, 2-methyl- Isobutyl mercaptan	$C_4H_{10}S$ 90.19	513-44-0 <-70	5030 88.5	4-01-00-01605 0.8357^{20}	H_2O 2; EtOH 4; eth 4; ace 4 1.4387^{20}
10159	2-Propanethiol, 2-methyl- *tert*-Butyl mercaptan	$C_4H_{10}S$ 90.19	75-66-1 -0.5	1577 64.3	4-01-00-01634 0.7943^{25}	H_2O 1; ctc 3; hp 3 1.4232^{20}
10160	Propane, 1,1,1-tribromo-	$C_3H_5Br_3$ 280.78	62127-61-1	 $77\text{-}83^{25}$	4-01-00-00221 2.35^{25}	 1.5606^{25}
10161	Propane, 1,1,2-tribromo-	$C_3H_5Br_3$ 280.78	14602-62-1	 200.5	3-01-00-00251 2.3547^{20}	H_2O 1; EtOH 3; eth 4; chl 3 1.5790^{20}
10162	Propane, 1,2,2-tribromo-	$C_3H_5Br_3$ 280.78	14476-30-3	 190.5	2-01-00-00077 2.2984^{20}	eth 4; EtOH 4; chl 4 1.5670^{20}
10163	Propane, 1,2,3-tribromo- 1,2,3-Tribromopropane	$C_3H_5Br_3$ 280.78	96-11-7 16.9	9527 222.1	4-01-00-00221 2.4208^{20}	H_2O 1; EtOH 4; eth 4; ctc 2 1.5862^{20}
10164	Propane, 1,2,3-tribromo-2-methyl-	$C_4H_7Br_3$ 294.81	631-28-7	 178^{235}; 108^{18}	4-01-00-00299 2.2105^{14}	H_2O 1; eth 3; chl 3 1.5701^{21}
10165	1,2,3-Propanetricarboxylic acid Tricarballylic acid	$C_6H_8O_6$ 176.13	99-14-9 166	9534	4-02-00-02366	H_2O 4; EtOH 4; eth 2
10166	1,2,3-Propanetricarboxylic acid, 2-(acetyloxy)-, triethyl ester	$C_{14}H_{22}O_8$ 318.32	77-89-4	 214^{40}	3-03-00-01105 1.135^{25}	 1.4380
10167	1,2,3-Propanetricarboxylic acid, 2-hydroxy- Citric acid	$C_6H_8O_7$ 192.13	77-92-9 153	2328 dec	4-03-00-01272 1.665^{20}	H_2O 4; EtOH 4; eth 3; bz 1
10168	1,2,3-Propanetricarboxylic acid, 2-hydroxy-, tributyl ester Butyl citrate	$C_{18}H_{32}O_7$ 360.45	77-94-1 -20	1564 233^{22}	4-03-00-01276 1.043^{20}	 1.4460^{20}
10169	1,2,3-Propanetricarboxylic acid, 2-hydroxy-, triethyl ester	$C_{12}H_{20}O_7$ 276.29	77-93-0	 294	4-03-00-01276 1.1369^{20}	H_2O 1; EtOH 3; eth 3; ctc 2 1.4455^{20}
10170	1,2,3-Propanetricarboxylic acid, 2-hydroxy-, trimethyl ester Trimethyl citrate	$C_9H_{14}O_7$ 234.21	1587-20-8 79.3	 285; 176^{16}	4-03-00-01276	eth 4; EtOH 4
10171	Propane, 1,1,1-trichloro- 1,1,1-Trichloropropane	$C_3H_5Cl_3$ 147.43	7789-89-1	 108	4-01-00-00198 1.287^{23}	H_2O 1; EtOH 3; eth 4; chl 4
10172	Propane, 1,1,2-trichloro- 1,1,2-Trichloropropane	$C_3H_5Cl_3$ 147.43	598-77-6	 132.0; 117^{500}	4-01-00-00199 1.372^{15}	H_2O 1; EtOH 3; eth 4; ctc 2
10173	Propane, 1,1,3-trichloro- 1,1,3-Trichloropropane	$C_3H_5Cl_3$ 147.43	20395-25-9 -59	 145.5	4-01-00-00199 1.3557^{20}	eth 4; EtOH 4; chl 4 1.4718^{20}
10174	Propane, 1,2,2-trichloro- 1,2,2-Trichloropropane	$C_3H_5Cl_3$ 147.43	3175-23-3	 124	4-01-00-00199 1.318^{25}	H_2O 1; EtOH 3; eth 3; chl 4 1.4609^{20}
10175	Propane, 1,2,3-trichloro- 1,2,3-Trichloropropane	$C_3H_5Cl_3$ 147.43	96-18-4 -14.7	 157	4-01-00-00199 1.3889^{20}	H_2O 2; EtOH 3; eth 3; ctc 2 1.4852^{20}
10176	Propane, 1,2,3-trichloro-2-(chloromethyl)-	$C_4H_6Cl_4$ 195.90	18963-00-3	 211; 87^{9}	3-01-00-00321 1.5036^{25}	chl 4 1.508^{20}
10177	Propane, 1,1,2-trichloro-2-methyl- Isobutane, 1,1,2-trichloro	$C_4H_7Cl_3$ 161.46	29559-52-2 -6	 163	4-01-00-00293 1.2588^{20}	eth 4; chl 4 1.4666^{20}
10178	Propane, 1,2,3-trichloro-2-methyl-	$C_4H_7Cl_3$ 161.46	1871-58-5	 162.5	4-01-00-00293 1.3012^{25}	chl 4 1.4765^{20}
10179	Propane, 1,1,1-trichloro-2,2,3,3,3-pentafluoro-	$C_3Cl_3F_5$ 237.38	4259-43-2 -80	 74	4-01-00-00200 1.643^{25}	
10180	Propane, 1,2,2-trichloro-1,1,3,3,3-pentafluoro-	$C_3Cl_3F_5$ 237.38	1599-41-3 -4.3	 72.0	3-01-00-00233 1.6681^{20}	 1.3519^{20}
10181	Propane, 1,2,3-trichloro-1,1,2,3,3-pentafluoro-	$C_3Cl_3F_5$ 237.38	76-17-5 -72	 73.7	4-01-00-00200 1.6631^{20}	 1.3512^{20}
10182	Propane, 1,1,1-triethoxy-	$C_9H_{20}O_3$ 176.26	115-80-0	 171	4-02-00-00707	eth 4; EtOH 4 1.4000^{25}
10183	Propane, 1,1,3-triethoxy-	$C_9H_{20}O_3$ 176.26	7789-92-6	 185 dec; 78^{14}	4-01-00-03971 0.898^{19}	H_2O 2; EtOH 3; ctc 3 1.4067^{20}
10184	Propane, 1,1,1-trifluoro-3-nitro- Propane, 1-nitro-3,3,3-trifluoro	$C_3H_4F_3NO_2$ 143.07	461-35-8	 135	4-01-00-00232 1.4178^{20}	eth 4 1.3549^{20}
10185	Propane, 1,2,3-trimethoxy- Glycerol trimethyl ether	$C_6H_{14}O_3$ 134.18	20637-49-4	 148	4-01-00-02755 0.9460^{15}	H_2O 5; eth 3; ace 4; bz 3 1.4055^{15}
10186	1,2,3-Propanetriol Glycerol	$C_3H_8O_3$ 92.09	56-81-5 18.2	4379 290	4-01-00-02751 1.2613^{20}	H_2O 5; EtOH 5; eth 2; bz 1 1.4746^{20}
10187	1,2,3-Propanetriol, 1-acetate, (±)- Glycerol, 1-monoacetate(*DL*)	$C_5H_{10}O_4$ 134.13	93713-40-7	 158^{165} 130^{3}	4-02-00-00251 1.2060^{20}	H_2O 3; EtOH 3; eth 2; bz 1 1.4157^{20}
10188	1,2,3-Propanetriol, 1-(4-aminobenzoate) Glyceryl *p*-aminobenzoate	$C_{10}H_{13}NO_4$ 211.22	136-44-7	4382		H_2O 1; EtOH 3
10189	1,2,3-Propanetriol, diacetate Diacetin	$C_7H_{12}O_5$ 176.17	25395-31-7	2941 259	4-02-00-00252 1.184^{16}	H_2O 4; eth 4; EtOH 4 1.44^{20}
10190	1,2,3-Propanetriol, 1,3-diacetate 1,3-Diacetin	$C_7H_{12}O_5$ 176.17	105-70-4	2941 173^{40}	4-02-00-00252 1.179^{15}	H_2O 4; EtOH 4; eth 2; CS_2 1 1.4395^{20}

No.	Name Synonym	Mol. Form. Mol. Wt.	CAS RN mp/°C	Merck No. bp/°C	Beil. Ref. den/g cm^{-3}	Solubility n_D
10191	1,2,3-Propanetriol, 1,3-dinitrate	$C_3H_6N_2O_7$ 182.09	623-87-0 26	148^{15}; $116^{0.6}$	2-01-00-00591 1.523^{20}	H_2O 4; eth 4; EtOH 4 1.4715^{20}
10192	1,2,3-Propanetriol, 1-nitrate	$C_3H_7NO_5$ 137.09	624-43-1 61	157.5	4-01-00-02761 1.4164^{20}	H_2O 4; EtOH 4; eth 2 1.4698^{20}
10193	1,2,3-Propanetriol, 2-nitrate	$C_3H_7NO_5$ 137.09	620-12-2 54	157.5	2-01-00-00591 1.40^{22}	H_2O 4; eth 4; EtOH 4
10194	1,2,3-Propanetriol, 1-phenyl- Glycerol, 1-phenyl	$C_9H_{12}O_3$ 168.19	63157-81-3 100.5	185	4-06-00-07407 1.2213^{13}	1.5606^{15}
10195	1,2,3-Propanetriol, triacetate Triacetin	$C_9H_{14}O_6$ 218.21	102-76-1 -78	9504 259	4-02-00-00253 1.1583^{20}	H_2O 2; EtOH 5; eth 5; ace 4 1.4301^{20}
10196	1,2,3-Propanetriol, tribenzoate	$C_{24}H_{20}O_6$ 404.42	614-33-5 76		3-09-00-00666 1.228^{12}	H_2O 1; EtOH 3; eth 4; ace 4
10197	1,2,3-Propanetriol, trinitrate Nitroglycerin	$C_3H_5N_3O_9$ 227.09	55-63-0 13.5	6528 $93^{0.31}$; exp 218	4-01-00-02762 1.5931^{20}	H_2O 2; EtOH 3; eth 5; ace 4 1.4786^{12}
10198	1,2,3-Propanetriol, tripropanoate	$C_{12}H_{20}O_6$ 260.29	139-45-7	$175\text{-}6^{20}$	4-02-00-00717 1.099^{20}	H_2O 1; EtOH 3; eth 4; chl 3 1.4318^{19}
10199	Propanetrione, diphenyl-	$C_{15}H_{10}O_3$ 238.24	643-75-4 70	289^{175}; 179^{4}	4-07-00-02822	eth 4
10200	Propanil Propanamide, *N*-(3,4-dichlorophenyl)-	$C_9H_9Cl_2NO$ 218.08	709-98-8 92	7814	1.25^{25}	
10201	Propanoic acid Propionic acid	$C_3H_6O_2$ 74.08	79-09-4 -20.7	7837 141.1	4-02-00-00695 0.9930^{20}	H_2O 5; EtOH 5; eth 3; chl 2 1.3809^{20}
10202	Propanoic acid, 2-(acetyloxy)-, (±)-	$C_5H_8O_4$ 132.12	3853-80-3	$167\text{-}70^{78}$	4-03-00-00638 1.1758^{20}	bz 4; EtOH 4 1.4240^{20}
10203	Propanoic acid, anhydride Propionic anhydride	$C_6H_{10}O_3$ 130.14	123-62-6 -45	7838 170; 67.5^{18}	4-02-00-00722 1.0110^{20}	eth 5; ctc 2 1.4038^{20}
10204	Propanoic acid, 2-bromo-, (±)-	$C_3H_5BrO_2$ 152.98	10327-08-9 25.7	203.5	1.7000^{20}	H_2O 4; EtOH 4; eth 4; chl 2 1.4753^{20}
10205	Propanoic acid, 3-bromo- β-Bromopropionic acid	$C_3H_5BrO_2$ 152.98	590-92-1 62.5	1421 $140\text{-}2^{45}$	4-02-00-00764 1.48^{25}	H_2O 3; EtOH 3; eth 3; bz 3
10206	Propanoic acid, 3-bromo-, butyl ester	$C_7H_{13}BrO_2$ 209.08	6973-79-1	130^{26}	4-02-00-00765 1.4549^{20}	ace 4; eth 4; EtOH 4 1.3051^{20}
10207	Propanoic acid, 2-bromo-, ethyl ester Ethyl α-bromopropionate	$C_5H_9BrO_2$ 181.03	535-11-5	3731 160 dec; 71^{26}	4-02-00-00762 1.4135^{20}	H_2O 1; EtOH 5; eth 5; chl 3 1.4490^{20}
10208	Propanoic acid, 3-bromo-, ethyl ester	$C_5H_9BrO_2$ 181.03	539-74-2	179; 65^{15}	4-02-00-00765 1.4123^{18}	EtOH 3; eth 3; ace 3; ctc 2 1.4516^{20}
10209	Propanoic acid, 2-bromo-2-methyl- α-Bromoisobutyric acid	$C_4H_7BrO_2$ 167.00	2052-01-9 48.5	1408 199; 115^{24}	4-02-00-00862 1.4969^{60}	
10210	Propanoic acid, 3-bromo-, 3-methylbutyl ester Isopentyl 3-bromopropionate	$C_8H_{15}BrO_2$ 223.11	100983-12-8	$110\text{-}1^{11}$	2-02-00-00231 1.2217^{15}	eth 4; EtOH 4 1.4556^{8}
10211	Propanoic acid, 2-bromo-, methyl ester, (±)-	$C_4H_7BrO_2$ 167.00	57885-43-5	144	4-02-00-00762 1.4966^{25}	eth 4; EtOH 4 1.4451^{22}
10212	Propanoic acid, 3-bromo-, methyl ester	$C_4H_7BrO_2$ 167.00	3395-91-3	105.5^{60}	4-02-00-00765 1.4897^{15}	EtOH 3; eth 3; ace 3; os 4 1.4542^{20}
10213	Propanoic acid, 2-bromo-2-methyl-, ethyl ester	$C_6H_{11}BrO_2$ 195.06	600-00-0	163	4-02-00-00862 1.3263^{20}	H_2O 1; EtOH 3; eth 5 1.4446^{20}
10214	Propanoic acid, butyl ester Butyl propionate	$C_7H_{14}O_2$ 130.19	590-01-2 -89	1587 146.8	4-02-00-00708 0.8754^{20}	H_2O 2; EtOH 5; eth 5; ctc 2 1.4014^{20}
10215	Propanoic acid, 2-chloro-	$C_3H_5ClO_2$ 108.52	598-78-7	185	4-02-00-00745 1.2585^{20}	H_2O 5; EtOH 5; eth 5; ace 3 1.4380^{20}
10216	Propanoic acid, 3-chloro- β-Chloropropionic acid	$C_3H_5ClO_2$ 108.52	107-94-8 41	2159 204 dec	4-02-00-00748	H_2O 3; EtOH 3; eth 5; chl 3
10217	Propanoic acid, 2-chloro-, butyl ester	$C_7H_{13}ClO_2$ 164.63	54819-86-2	184	4-02-00-00746 1.0253^{20}	eth 4 1.4263^{20}
10218	Propanoic acid, 3-chloro-, butyl ester	$C_7H_{13}ClO_2$ 164.63	27387-79-7	104^{22}	4-02-00-00749 1.0370^{20}	H_2O 4; eth 4 1.4321^{20}
10219	Propanoic acid, 3-chloro-2,2-dimethyl-	$C_5H_9ClO_2$ 136.58	13511-38-1 41.5	110^{10}	4-02-00-00914	ctc 4
10220	Propanoic acid, 2-chloro-, ethyl ester Ethyl α-chloropropionate	$C_5H_9ClO_2$ 136.58	535-13-7	3743 147	4-02-00-00746 1.0793^{20}	H_2O 1; EtOH 5; eth 5; ctc 2 1.4178^{20}
10221	Propanoic acid, 3-chloro-, ethyl ester	$C_5H_9ClO_2$ 136.58	623-71-2	162	4-02-00-00749 1.1086^{20}	H_2O 2; EtOH 5; eth 5; os 3 1.4254^{20}
10222	Propanoic acid, 2-chloro-, methyl ester Methyl 2-chloropropionate	$C_4H_7ClO_2$ 122.55	17639-93-9	132.5	4-02-00-00746 1.075^{25}	
10223	Propanoic acid, 2-chloro-, 1-methylethyl ester Isopropyl 2-chloropropionate	$C_6H_{11}ClO_2$ 150.60	40058-87-5	151.5	4-02-00-00746 1.0315^{20}	H_2O 1; EtOH 3; eth 3 1.4149^{20}
10224	Propanoic acid, 2-chloro-, 2-methylpropyl ester, (*R*)- Propionic acid, 2-chloro, isobutyl ester (*d*)	$C_7H_{13}ClO_2$ 164.63	114489-96-2	176	0-02-00-00248 1.0312^{20}	1.4247^{20}
10225	Propanoic acid, 3-chloro-, methyl ester	$C_4H_7ClO_2$ 122.55	6001-87-2	156	4-02-00-00748 1.1861^{15}	EtOH 3; ctc 3 1.4263^{20}
10226	Propanoic acid, 3-chloro-, 3-methylbutyl ester	$C_8H_{15}ClO_2$ 178.66	62108-70-7	208; 87^{12}	4-02-00-00749 1.0171^{20}	eth 4; EtOH 4 1.4343^{20}
10227	Propanoic acid, 3-chloro-, 2-methylpropyl ester	$C_7H_{13}ClO_2$ 164.63	62108-68-3	191.3	4-02-00-00749 1.0323^{20}	eth 4; EtOH 4 1.4295^{20}

No.	Name Synonym	Mol. Form. Mol. Wt.	CAS RN mp/°C	Merck No. bp/°C	Beil. Ref. den/g cm^{-3}	Solubility n_D
10228	Propanoic acid, 2-chloro-2-methyl-	$C_4H_7ClO_2$ 122.55	594-58-1 31	118^{50}; 93^{20}	4-02-00-00858 1.1790^{20}	EtOH 4 1.450^{20}
10229	Propanoic acid, 2-chloro-2-methyl-, methyl ester	$C_5H_9ClO_2$ 136.58	22421-97-2	135	4-02-00-00859 1.0883^{15}	eth 4 1.4122^{21}
10230	Propanoic acid, 3-chloro-2-methyl-	$C_4H_7ClO_2$ 122.55	16674-04-7	$128\text{-}33^{50}$	4-02-00-00860 1.0153^{20}	eth 4; EtOH 4 1.4310^{20}
10231	Propanoic acid, 2-(4-chlorophenoxy)-2-methyl-, ethyl ester Clofibrate	$C_{12}H_{15}ClO_3$ 242.70	637-07-0	2374 $148\text{-}50^{20}$	4-06-00-00851	
10232	Propanoic acid, 3-chloro-, propyl ester	$C_6H_{11}ClO_2$ 150.60	62108-66-1	180	4-02-00-00749 1.0656^{20}	eth 4; EtOH 4 1.4290^{20}
10233	Propanoic acid, 2-cyano-	$C_4H_5NO_2$ 99.09	632-07-5 35	143^{11}	3-02-00-01683 1.1400^{20}	
10234	Propanoic acid, 3-cyano-, methyl ester	$C_5H_7NO_2$ 113.12	4107-62-4	211	3-02-00-01672	chl 3 1.4250^{20}
10235	Propanoic acid, cyclohexyl ester	$C_9H_{16}O_2$ 156.22	6222-35-1	193; 93^{35}	4-06-00-00037 0.9359^{20}	H_2O 1; EtOH 3; eth 3; ace 3 1.4403^{20}
10236	Propanoic acid, 2-diazo-, ethyl ester Ethyl 2-diazopropionate	$C_5H_8N_2O_2$ 128.13	6111-99-5	$65\text{-}8^{41}$	2-03-00-00403 1.086^{12}	1.4472^{18}
10237	Propanoic acid, 2-diazo-, methyl ester Methyl 2-diazopropionate	$C_4H_6N_2O_2$ 114.10	34757-14-7	$43\text{-}5^{11}$	2-03-00-00403 1.1011^{13}	1.4487^{20}
10238	Propanoic acid, 2,3-dibromo-	$C_3H_4Br_2O_2$ 231.87	600-05-5 66.5	160^{20}; 138^{12}	4-02-00-00767	bz 4; eth 4; EtOH 4
10239	Propanoic acid, 2,3-dibromo-, butyl ester Butyl 2,3-dibromopropionate	$C_7H_{12}Br_2O_2$ 287.98	21179-48-6	237.5	4-02-00-00768 1.6107^{20}	1.4890^{20}
10240	Propanoic acid, 2,3-dibromo-, ethyl ester	$C_5H_8Br_2O_2$ 259.93	3674-13-3	214.5	4-02-00-00767 1.7966^{20}	EtOH 3; eth 3 1.5007^{20}
10241	Propanoic acid, 2,3-dibromo-, methyl ester	$C_4H_6Br_2O_2$ 245.90	1729-67-5	206	4-02-00-00767 1.9333^{20}	EtOH 3 1.5127^{20}
10242	Propanoic acid, 2,3-dibromo-2-methyl-, methyl ester	$C_5H_8Br_2O_2$ 259.93	3673-79-8	203	4-02-00-00864 1.805^{20}	1.505^{20}
10243	Propanoic acid, 2,3-dibromo-, propyl ester Propyl 2,3-dibromopropionate	$C_6H_{10}Br_2O_2$ 273.95	79762-76-8	222.5	3-02-00-00572 1.6799^{20}	1.4950^{20}
10244	Propanoic acid, 2,2-dichloro- Dalapon	$C_3H_4Cl_2O_2$ 142.97	75-99-0	2806 187.5; 92^{14}	4-02-00-00753 1.389^{12}	H_2O 4; EtOH 4; eth 3; ctc 3
10245	Propanoic acid, 2,3-dichloro-	$C_3H_4Cl_2O_2$ 142.97	565-64-0 50	210; 130^{28}	4-02-00-00756	H_2O 3; EtOH 3; eth 3; peth 2 1.4650^{20}
10246	Propanoic acid, 2,3-dichloro-, ethyl ester	$C_5H_8Cl_2O_2$ 171.02	6628-21-3	183.5	4-02-00-00756 1.2401^{20}	eth 4; EtOH 4 1.4482^{20}
10247	Propanoic acid, 3,3-dichloro-2-hydroxy-2-methyl-	$C_4H_6Cl_2O_3$ 173.00	31340-44-0 82.5	dec	2-03-00-00224	H_2O 3; EtOH 3; eth 3
10248	Propanoic acid, 2,3-dichloro-, methyl ester	$C_4H_6Cl_2O_2$ 157.00	3674-09-7	92^{50}	4-02-00-00756 1.3282^{20}	ace 4; eth 4; EtOH 4
10249	Propanoic acid, 2,3-dichloro-, 1-methylethyl ester	$C_6H_{10}Cl_2O_2$ 185.05	54774-99-1	$61\text{-}2^{5}$	4-02-00-00756 1.2010^{25}	1.4470
10250	Propanoic acid, 2-(2,4-dichlorophenoxy)- Dichlorprop	$C_9H_8Cl_2O_3$ 235.07	120-36-5 117.5	3068	3-06-00-00707	H_2O 2; EtOH 3; eth 3; lig 2
10251	Propanoic acid, 2,2-dichloro-3,3,3-trifluoro-, methyl ester Propionic acid, 2,2-dichloro-3,3,3-trifluoro-, methyl ester	$C_4H_3Cl_2F_3O_2$ 210.97	378-68-7	$116\text{-}7^{625}$	4-02-00-00758 1.5092^{20}	1.3806^{20}
10252	Propanoic acid, 2,3-dihydroxy- Glyceric acid	$C_3H_6O_4$ 106.08	473-81-4	4378 dec	4-03-00-01050	H_2O 4; ace 4; EtOH 4
10253	Propanoic acid, 2,3-dihydroxy-, ethyl ester, (±)- Glyceric acid, ethyl ester (*DL*)	$C_5H_{10}O_4$ 134.13	89300-69-6	235; 120^{14}	4-03-00-01052 1.1897^{15}	H_2O 4; eth 4; EtOH 4
10254	Propanoic acid, 2,3-dihydroxy-, methyl ester, (±)- Methyl *DL*-glycerate	$C_4H_8O_4$ 120.11	15909-76-9	241.5; 119^{14}	4-03-00-01052 1.2802^{15}	H_2O 4; EtOH 4 1.4502^{20}
10255	Propanoic acid, 2,3-dihydroxy-2-methyl-, ethyl ester Glyceric acid, 2-methyl, ethyl ester	$C_6H_{12}O_4$ 148.16	67535-07-3	95^{10}	3-03-00-00860 1.114^{25}	1.4370^{25}
10256	Propanoic acid, 2,3-dihydroxy-2-methyl-, methyl ester Glyceric acid, 2-methyl-, methyl ester	$C_5H_{10}O_4$ 134.13	19860-56-1	89^{10}	4-03-00-01060 1.185^{25}	1.4438^{25}
10257	Propanoic acid, 2,3-dihydroxy-, propyl ester	$C_6H_{12}O_4$ 148.16	116435-95-1	$132\text{-}4^{3}$	3-03-00-00854 1.1537^{20}	1.4503^{20}
10258	Propanoic acid, 2,2-dimethyl- Pivalic acid	$C_5H_{10}O_2$ 102.13	75-98-9 35	7482 164	4-02-00-00908 0.905^{50}	H_2O 2; EtOH 4; eth 4 1.3931^{30}
10259	Propanoic acid, 2,2-dimethyl-, ethyl ester Ethyl 2,2-dimethylpropionate	$C_7H_{14}O_2$ 130.19	3938-95-2 -89.5	118.4	4-02-00-00910 0.856^{20}	EtOH 3; eth 3 1.3906^{20}
10260	Propanoic acid, 2,2-dimethyl-, methyl ester Methyl 2,2-dimethylpropionate	$C_6H_{12}O_2$ 116.16	598-98-1	101.1	4-02-00-00909 0.891^{0}	eth 4; EtOH 4 1.3905^{20}
10261	Propanoic acid, ethenyl ester Vinyl propionate	$C_5H_8O_2$ 100.12	105-38-4	91.2	4-02-00-00711	

No.	Name	Synonym	Mol. Form.	Mol. Wt.	CAS RN	mp/°C	Merck No.	bp/°C	Beil. Ref.	den/g cm^{-3}	Solubility	n_D
10262	Propanoic acid, 3-ethoxy-, ethyl ester		$C_7H_{14}O_3$	146.19	763-69-9			$48-9^{5}$	4-03-00-00697	0.92^{25}		1.4065^{20}
10263	Propanoic acid, ethyl ester	Ethyl propionate	$C_5H_{10}O_2$	102.13	105-37-3	-73.9	3801	99.1	4-02-00-00705	0.8917^{20}	H_2O 2; EtOH 5; eth 5; ace 3	1.3839^{20}
10264	Propanoic acid, heptyl ester		$C_{10}H_{20}O_2$	172.27	2216-81-1	-50.9		210	4-02-00-00710	0.8679^{20}	H_2O 1; EtOH 3; eth 3; ace 3	1.4201^{15}
10265	Propanoic acid, hexyl ester		$C_9H_{18}O_2$	158.24	2445-76-3	-57.5		190	4-02-00-00709	0.8698^{20}	H_2O 1; EtOH 3; eth 3; ace 3	1.4162^{15}
10266	Propanoic acid, 2-hydroxy-, (±)-	Lactic acid (*DL*)	$C_3H_6O_3$	90.08	598-82-3	18	5215	122^{15}	2-03-00-00192	1.2060^{21}	H_2O 4; eth 2; EtOH 4	1.4392^{20}
10267	Propanoic acid, 3-hydroxy-	Hydracrylic acid	$C_3H_6O_3$	90.08	503-66-2		4681	dec	4-03-00-00689		H_2O 4; EtOH 3; eth 5	1.4489^{20}
10268	Propanoic acid, 2-hydroxy-, butyl ester, (*R*)-	Lactic acid, butyl ester (*d*)	$C_7H_{14}O_3$	146.19	34451-18-8			77^{10}	3-03-00-00450	0.9744^{27}	eth 4; EtOH 4	
10269	Propanoic acid, 2-hydroxy-, ethyl ester, (±)-	Ethyl lactate	$C_5H_{10}O_3$	118.13	2676-33-7	-26	3773	154.5		1.0328^{20}	H_2O 4; eth 4; EtOH 4	1.4124^{20}
10270	Propanoic acid, 3-hydroxy-, ethyl ester		$C_5H_{10}O_3$	118.13	623-72-3			187.5	2-03-00-00213	1.059^{20}		1.4271^{23}
10271	Propanoic acid, 2-hydroxy-, hexadecyl ester	Cetyl lactate	$C_{19}H_{38}O_3$	314.51	35274-05-6	41	2022	219^{10}; 170^{1}	4-03-00-00660			1.4410^{40}
10272	Propanoic acid, 3-hydroxy-2-(hydroxymethyl)-2-methyl-	Dimethylolpropionic acid	$C_5H_{10}O_4$	134.13	4767-03-7	190	3241		4-03-00-01066			
10273	Propanoic acid, 2-hydroxy-, methyl ester, (±)-	Lactic acid, methyl ester (*DL*)	$C_4H_8O_3$	104.11	2155-30-8			144.8	4-03-00-00640	1.0928^{20}	H_2O 4; eth 4; EtOH 4	1.4141^{20}
10274	Propanoic acid, 2-hydroxy-, 3-methylbutyl ester		$C_8H_{16}O_3$	160.21	19329-89-6			202.4	3-03-00-00483	0.9589^{25}	eth 4; EtOH 4	1.4240^{25}
10275	Propanoic acid, 2-hydroxy-, 1-methylethyl ester	Isopropyl lactate	$C_6H_{12}O_3$	132.16	617-51-6			167	3-03-00-00479	0.9980^{20}	H_2O 4; bz 4; eth 4; EtOH 4	1.4082^{25}
10276	Propanoic acid, 3-hydroxy-, methyl ester		$C_4H_8O_3$	104.11	6149-41-3			179	3-03-00-00525	1.105^{16}	H_2O 5; EtOH 5; eth 5	1.43^{23}
10277	Propanoic acid, 3-hydroxy-, 1-methylethyl ester	Hydracrylic acid, isopropyl ester	$C_6H_{12}O_3$	132.16	84098-45-3			128.5^{82}; 95^{12}	1-03-00-00113	1.058^{25}		1.4303^{23}
10278	Propanoic acid, 2-hydroxy-2-methyl-		$C_4H_8O_3$	104.11	594-61-6	82.5		212	4-03-00-00782		H_2O 4; EtOH 4; eth 4; bz 2	
10279	Propanoic acid, 2-hydroxy-2-methyl-, ethyl ester		$C_6H_{12}O_3$	132.16	80-55-7			150	4-03-00-00783	0.987^{20}	H_2O 5; EtOH 5; ctc 3	1.4080^{20}
10280	Propanoic acid, 2-hydroxy-2-methyl-, methyl ester		$C_5H_{10}O_3$	118.13	2110-78-3			137	4-03-00-00783		H_2O 4; EtOH 4	1.4056^{20}
10281	Propanoic acid, 2-hydroxy-, 2-propenyl ester		$C_6H_{10}O_3$	130.14	5349-55-3			$56-60^{8}$	3-03-00-00486	1.0452^{20}	py 3	1.4369^{20}
10282	Propanoic acid, 2-iodo-, (±)-		$C_3H_5IO_2$	199.98	18791-49-6	46		$93-6^{0.2}$	3-02-00-00573	2.073^{18}	H_2O 2; EtOH 4; eth 4	
10283	Propanoic acid, 3-iodo-, ethyl ester		$C_5H_9IO_2$	228.03	6414-69-3			202	4-02-00-00770	1.6226^{30}		1.4961^{25}
10284	Propanoic acid, 3-iodo-, methyl ester		$C_4H_7IO_2$	214.00	5029-66-3			188	3-02-00-00574	1.8408^{7}	EtOH 4	
10285	Propanoic acid, 2-mercapto-	Thiolactic acid	$C_3H_6O_2S$	106.15	79-42-5	12	9269	$106-7^{15}$	4-03-00-00682	1.1938^{20}	H_2O 4; eth 4; EtOH 4	1.4810^{20}
10286	Propanoic acid, 3-mercapto-		$C_3H_6O_2S$	106.15	107-96-0	18		111^{15}; 86^{3}	4-03-00-00726	1.218^{21}	H_2O 3; EtOH 3; eth 3; ctc 3	1.494^{20}
10287	Propanoic acid, 3-mercapto-, methyl ester		$C_4H_8O_2S$	120.17	2935-90-2			$54-5^{14}$	3-03-00-00553	1.085^{25}		1.4640^{20}
10288	Propanoic acid, 2-mercapto-2-methyl-		$C_4H_8O_2S$	120.17	4695-31-2	47		101^{15}	4-03-00-00797		H_2O 4	
10289	Propanoic acid, 3-methoxy-, methyl ester		$C_5H_{10}O_3$	118.13	3852-09-3			142.8	4-03-00-00693	1.0139^{15}		1.4030^{20}
10290	Propanoic acid, 2-methyl-	Isobutyric acid	$C_4H_8O_2$	88.11	79-31-2	-46	5039	154.4	4-02-00-00843	0.9681^{20}	H_2O 4; EtOH 5; eth 5; ctc 2	1.3930^{20}
10291	Propanoic acid, 2-methyl-, anhydride	Isobutyric anhydride	$C_8H_{14}O_3$	158.20	97-72-3	-53.5		183; 89^{32}	4-02-00-00851	0.9535^{20}	eth 5; chl 3	1.4061^{19}
10292	Propanoic acid, 2-methyl-, 2-chloroethyl ester	2-Chloroethyl isobutyrate	$C_6H_{11}ClO_2$	150.60	33662-96-3			170	4-02-00-00847	1.062^{25}	eth 4; EtOH 4	1.4109^{16}
10293	Propanoic acid, 2-methyl-, cyclohexyl ester	Isobutyric acid, cyclohexyl ester	$C_{10}H_{18}O_2$	170.25	1129-47-1			204	3-06-00-00024	0.9489^{0}	eth 4; EtOH 4	
10294	Propanoic acid, 2-methyl-, 1,1-dimethylethyl ester		$C_8H_{16}O_2$	144.21	16889-72-8			126.7	3-02-00-00648		ace 4; eth 4; EtOH 4	1.3921^{20}
10295	Propanoic acid, 2-methyl-, 3,7-dimethyl-2,6-octadienyl ester, (*E*)-		$C_{14}H_{24}O_2$	224.34	2345-26-8			$135-7^{13}$	2-02-00-00261	0.8997^{15}		1.4576^{20}
10296	Propanoic acid, methyl ester	Methyl propionate	$C_4H_8O_2$	88.11	554-12-1	-87.5	6030	79.8	4-02-00-00704	0.9150^{20}	H_2O 2; EtOH 5; eth 5; ace 3	1.3775^{20}
10297	Propanoic acid, 2-methyl-, ethenyl ester		$C_6H_{10}O_2$	114.14	2424-98-8			104.5	4-02-00-00848	0.8932^{20}		1.4061^{20}

PHYSICAL CONSTANTS OF ORGANIC COMPOUNDS (continued)

No.	Name Synonym	Mol. Form. Mol. Wt.	CAS RN mp/°C	Merck No. bp/°C	Beil. Ref. den/g cm^{-3}	Solubility n_D
10298	Propanoic acid, 2-methyl-, ethyl ester Ethyl isobutanoate	$C_6H_{12}O_2$ 116.16	97-62-1 -88.2	3770 110.1	4-02-00-00846 0.868^{20}	H_2O 2; EtOH 5; eth 5; ace 3 1.3869^{18}
10299	Propanoic acid, 1-methylethyl ester Isopropyl propionate	$C_6H_{12}O_2$ 116.16	637-78-5	 109.5	4-02-00-00708 0.8660^{20}	H_2O 2; EtOH 5; eth 5 1.3872^{20}
10300	Propanoic acid, 2-methyl-, 2-furanylmethyl ester	$C_9H_{12}O_3$ 168.19	6270-55-9	 $85\text{-}6^{15}$	2-17-00-00115 1.0313^{20}	eth 4; EtOH 4
10301	Propanoic acid, 2-methyl-, methyl ester Methyl isobutyrate	$C_5H_{10}O_2$ 102.13	547-63-7 -84.7	6003 92.5	4-02-00-00846 0.8906^{20}	H_2O 2; EtOH 5; eth 5; ace 3 1.3840^{20}
10302	Propanoic acid, 2-methyl-, 3-methylbutyl ester	$C_9H_{18}O_2$ 158.24	2050-01-3	 168.5	3-02-00-00649 0.8627^{20}	H_2O 2; EtOH 3; eth 3; ace 3
10303	Propanoic acid, 2-methyl-, 1-methylethyl ester Isopropyl isobutyrate	$C_7H_{14}O_2$ 130.19	617-50-5	 120.7	4-02-00-00847 0.8471^{21}	H_2O 1; EtOH 3; eth 3; ace 3
10304	Propanoic acid, 2-methyl-, 2-methylpropyl ester Isobutyl isobutanoate	$C_8H_{16}O_2$ 144.21	97-85-8 -80.7	5028 148.6	4-02-00-00847 0.8542^{20}	H_2O 2; EtOH 3; eth 5; ace 3 1.3999^{20}
10305	Propanoic acid, 2-methyl-, 2-nitropentyl ester Isobutyric acid, (2-nitropentyl) ester	$C_9H_{17}NO_4$ 203.24	116530-63-3	 249.5; 121^{10}	3-02-00-00648 1.0310^{20}	ace 4; eth 4 1.4315^{20}
10306	Propanoic acid, 2-methyl-, 2-phenylethyl ester	$C_{12}H_{16}O_2$ 192.26	103-48-0	 $122\text{-}4^{15}$	4-06-00-03073 0.9950^{15}	 1.4871^{20}
10307	Propanoic acid, 2-methyl-, phenylmethyl ester	$C_{11}H_{14}O_2$ 178.23	103-28-6	 $114\text{-}5^{20}$	4-06-00-02266 1.0075^{15}	 1.4883^{20}
10308	Propanoic acid, 2-methyl-, propyl ester Propyl isobutyrate	$C_7H_{14}O_2$ 130.19	644-49-5	 135.4	4-02-00-00847 0.8843^{0}	H_2O 2; EtOH 3; eth 4; ace 3 1.3955^{20}
10309	Propanoic acid, 1-methylpropyl ester	$C_7H_{14}O_2$ 130.19	591-34-4	 133	4-02-00-00709 0.8657^{20}	EtOH 3; eth 3 1.3952^{20}
10310	Propanoic acid, 2-methylpropyl ester Isobutyl propionate	$C_7H_{14}O_2$ 130.19	540-42-1 -71.4	5033 137	4-02-00-00709 0.888^{0}	H_2O 2; EtOH 4; eth 4; ace 3 1.3973^{20}
10311	Propanoic acid, 3-(methylthio)-, methyl ester	$C_5H_{10}O_2S$ 134.20	13532-18-8	 $74\text{-}5^{13}$	4-03-00-00737 1.077^{25}	 1.4650^{20}
10312	Propanoic acid, 2-nitro-, ethyl ester	$C_5H_9NO_4$ 147.13	2531-80-8	 190.5	4-02-00-00771	bz 4; eth 4; EtOH 4 1.4210^{20}
10313	Propanoic acid, octyl ester	$C_{11}H_{22}O_2$ 186.29	142-60-9 -42.6	 228	4-02-00-00710 0.8663^{20}	H_2O 1; EtOH 3; eth 3; bz 3 1.4221^{15}
10314	Propanoic acid, 2-oxo- Pyruvic acid	$C_3H_4O_3$ 88.06	127-17-3 13.8	8032 165 dec; 54^{10}	4-03-00-01505 1.2272^{20}	H_2O 5; EtOH 5; eth 5; ace 3 1.4280^{20}
10315	Propanoic acid, 2-oxo-, ethyl ester	$C_5H_8O_3$ 116.12	617-35-6 -50	 155	4-03-00-01513 1.0596^{15}	H_2O 2; EtOH 5; eth 5; ace 3 1.4052^{20}
10316	Propanoic acid, 2-oxo-, methyl ester	$C_4H_6O_3$ 102.09	600-22-6	 135.5	4-01-00-04146 1.154^{0}	H_2O 2; EtOH 5; eth 5; ace 3 1.4046^{25}
10317	Propanoic acid, pentafluoro-, methyl ester Pentafluoropropionic acid, methyl ester	$C_4H_3F_5O_2$ 178.06	378-75-6	 59.5	4-02-00-00740 1.390^{25}	 1.2869^{25}
10318	Propanoic acid, pentyl ester	$C_8H_{16}O_2$ 144.21	624-54-4 -73.1	 168.6	4-02-00-00709 0.8761^{25}	H_2O 1; EtOH 5; eth 5; bz 3 1.4096^{15}
10319	Propanoic acid, 2-phenoxy-, (±)-	$C_9H_{10}O_3$ 166.18	1912-21-6 115.5	 265.5	4-06-00-00642 1.1865^{20}	eth 4; EtOH 4 1.5184^{20}
10320	Propanoic acid, 3-phenoxy-	$C_9H_{10}O_3$ 166.18	7170-38-9 97.5	 240; 188^{26}	4-06-00-00643	H_2O 3; chl 2; lig 3
10321	Propanoic acid, 2-phenoxy-, ethyl ester	$C_{11}H_{14}O_3$ 194.23	42412-84-0	 243.5	4-06-00-00643 1.360^{17}	H_2O 1; EtOH 3; eth 3 1.360^{17}
10322	Propanoic acid, 3-phenoxy-, ethyl ester	$C_{11}H_{14}O_3$ 194.23	22409-91-2 24	 142^{11}	4-06-00-00644 1.0745^{20}	 1.5007^{18}
10323	Propanoic acid, phenyl ester	$C_9H_{10}O_2$ 150.18	637-27-4 20	 211	4-06-00-00615 1.0436^{25}	H_2O 1; EtOH 4; eth 4; bz 3 1.4980^{20}
10324	Propanoic acid, 2-phenylethyl ester 2-Phenethyl propionate	$C_{11}H_{14}O_2$ 178.23	122-70-3		3-06-00-01709	
10325	Propanoic acid, 2-propenyl ester	$C_6H_{10}O_2$ 114.14	2408-20-0	 124	4-02-00-00711 0.9140^{20}	EtOH 3; eth 3; ace 3 1.4105^{20}
10326	Propanoic acid, propyl ester Propyl propionate	$C_6H_{12}O_2$ 116.16	106-36-5 -75.9	7880 122.5	4-02-00-00707 0.8809^{20}	H_2O 2; EtOH 5; eth 5; ace 3 1.3935^{20}
10327	Propanoic acid, sodium salt	$C_3H_5NaO_2$ 96.06	137-40-6	8623	4-02-00-00695	H_2O 2
10328	Propanoic acid, 2,3,3,3-tetrafluoro-, ethyl ester Ethyl 2,3,3,3-tetrafluoropropionate	$C_5H_6F_4O_2$ 174.10	399-92-8	 108.5	4-02-00-00737 1.289^{20}	 1.3260^{20}
10329	Propanoic acid, 3,3'-thiobis- 3,3'-Thiodipropionic acid	$C_6H_{10}O_4S$ 178.21	111-17-1 129	9261	4-03-00-00735	H_2O 4; EtOH 4
10330	Propanoic acid, 3,3'-thiobis-, diethyl ester	$C_{10}H_{18}O_4S$ 234.32	673-79-0	 121^{2}	4-03-00-00741 1.095^{25}	 1.4655^{20}
10331	Propanoic acid, 3,3'-thiobis-, dimethyl ester	$C_8H_{14}O_4S$ 206.26	4131-74-2	 148^{18}	4-03-00-00738 1.198^{25}	 1.4740^{20}
10332	Propanoic acid, 2,2,3-trichloro-	$C_3H_3Cl_3O_2$ 177.41	3278-46-4 65.5	 140^{40}	4-02-00-00759	H_2O 3; EtOH 3; bz 3; CS_2 3
10333	Propanoic acid, 3,3,3-trichloro-2-hydroxy- Lactic acid, 3,3,3-trichloro-	$C_3H_3Cl_3O_3$ 193.41	599-01-9 125	 140^{45}	3-03-00-00501	H_2O 4; eth 4; EtOH 4; chl 4
10334	1-Propanol Propyl alcohol	C_3H_8O 60.10	71-23-8 -126.1	7854 97.2	4-01-00-01413 0.7997^{25}	H_2O 5; EtOH 5; eth 5; ace 3 1.3850^{20}
10335	2-Propanol Isopropyl alcohol	C_3H_8O 60.10	67-63-0 -89.5	5096 82.3	4-01-00-01461 0.7809^{25}	H_2O 5; EtOH 5; eth 5; ace 3 1.3776^{20}

No.	Name Synonym	Mol. Form. Mol. Wt.	CAS RN mp/°C	Merck No. bp/°C	Beil. Ref. den/g cm^{-3}	Solubility n_D
10336	1-Propanol, 2-amino-, (±)-	C_3H_9NO 75.11	6168-72-5	 174.5	3-04-00-00737	H_2O 4; EtOH 4; eth 4; chl 2 1.4502^{20}
10337	1-Propanol, 3-amino- Propanolamine	C_3H_9NO 75.11	156-87-6 11	 187.5	4-04-00-01623 0.9824^{26}	H_2O 3; EtOH 3; eth 3 1.4617^{20}
10338	2-Propanol, 1-amino-, (±)- *DL*-1-Amino-2-propanol	C_3H_9NO 75.11	1674-56-2 1.7	 159.4	4-04-00-01665 0.9611^{20}	H_2O 5; EtOH 5; eth 5; ace 5 1.4479^{20}
10339	2-Propanol, 1-amino-3-(diethylamino)- 3-(Diethylamino)-2-hydroxypropylamine	$C_7H_{18}N_2O$ 146.23	6322-01-6	 223	3-04-00-00767 0.931^{20}	EtOH 3; eth 3; chl 2 1.465^{20}
10340	2-Propanol, 1-[(2-aminoethyl)amino]- *N*-(2-Hydroxypropyl)ethylenediamine	$C_5H_{14}N_2O$ 118.18	123-84-2	450 94^{3}	4-04-00-01684 0.9837^{25}	 1.4738^{20}
10341	1-Propanol, 2-amino-2-methyl- 2-Aminoisobutanol	$C_4H_{11}NO$ 89.14	124-68-5 25.5	461 165.5	4-04-00-01740 0.934^{20}	H_2O 5; ctc 3 1.449^{20}
10342	1-Propanol, 2-[(2-amino-2-methylpropyl)amino]-2-methyl- 1-Propanol, 2-[2-(2-amino-2-methylpropyl)amino]-2-methyl-	$C_8H_{20}N_2O$ 160.26	72622-74-3 46	 231	3-04-00-00788 0.9360^{20}	chl 2 1.4635^{25}
10343	2-Propanol, 1,3-bis(dimethylamino)-	$C_7H_{18}N_2O$ 146.23	5966-51-8	 181.5	4-04-00-01695 0.8788^{20}	H_2O 4 1.4418^{20}
10344	1-Propanol, 3-bromo-	C_3H_7BrO 138.99	627-18-9	 105^{185}; 62^{5}	4-01-00-01446 1.5374^{20}	H_2O 3; EtOH 5; eth 5 1.4834^{25}
10345	2-Propanol, 1-bromo-	C_3H_7BrO 138.99	19686-73-8	 146.5	3-01-00-01474 1.5585^{30}	H_2O 3; EtOH 4; eth 4 1.4801^{20}
10346	2-Propanol, 1-butoxy-	$C_7H_{16}O_2$ 132.20	5131-66-8	 171.5; 71^{20}	4-01-00-02471 0.882^{20}	EtOH 3; eth 3; bz 3; ctc 3 1.4168^{20}
10347	1-Propanol, 2-chloro- Propylene chlorohydrin	C_3H_7ClO 94.54	78-89-7	7863 133.5	4-01-00-01440 1.103^{20}	H_2O 4; eth 4; EtOH 4 1.4390^{20}
10348	1-Propanol, 3-chloro- 3-Chloro-1-propanol	C_3H_7ClO 94.54	627-30-5	 165	4-01-00-01441 1.1309^{20}	H_2O 4; EtOH 3; eth 3; ctc 2 1.4459^{20}
10349	2-Propanol, 1-chloro- *sec*-Propylene chlorohydrin	C_3H_7ClO 94.54	127-00-4	7864 127	4-01-00-01490 1.113^{20}	H_2O 5; EtOH 5; eth 5; ctc 2 1.4392^{20}
10350	1-Propanol, 3-chloro-, acetate	$C_5H_9ClO_2$ 136.58	628-09-1	 164; 62^{10}	4-02-00-00140 1.250^{19}	H_2O 1; EtOH 3; eth 3; ace 3 1.431^{20}
10351	2-Propanol, 1-chloro-, acetate	$C_5H_9ClO_2$ 136.58	623-60-9	 149.5	4-02-00-00142 1.0788^{20}	eth 4; EtOH 4 1.4223^{20}
10352	1-Propanol, 2-chloro-2-methyl- Isobutyl alcohol, 2-chloro	C_4H_9ClO 108.57	558-38-3	 132 dec; 60^{50}	4-01-00-01603 1.0472^{20}	 1.4388^{20}
10353	1-Propanol, 3-chloro-2-methyl- Isobutyl alcohol, 3-chloro	C_4H_9ClO 108.57	10317-10-9	 76-8^{11}	3-01-00-01564 1.083^{25}	eth 4; EtOH 4 1.4460^{25}
10354	2-Propanol, 1-chloro-2-methyl- Isobutylene chlorohydrin	C_4H_9ClO 108.57	558-42-9 -20	 128.5	4-01-00-01628 1.0628^{20}	H_2O 4; EtOH 4 1.4380^{24}
10355	2-Propanol, 1-chloro-3-(1-methylethoxy)- 2-Propanol, 1-chloro-3-isopropoxy-	$C_6H_{13}ClO_2$ 152.62	4288-84-0	 87^{20}	4-01-00-02485 1.0530^{25}	EtOH 3; eth 3 1.4370^{25}
10356	2-Propanol, 1-chloro-3-propoxy-	$C_6H_{13}ClO_2$ 152.62	6943-58-4	 92-5^{15}	3-01-00-02153 1.0526^{25}	ace 4; eth 4 1.4378^{25}
10357	2-Propanol, 1,3-dibromo-	$C_3H_6Br_2O$ 217.89	96-21-9	 219 dec; 105^{16}	4-01-00-01496 2.1201^{25}	ace 4; eth 4; EtOH 4 1.5495^{25}
10358	1-Propanol, 2,3-dibromo-, phosphate (3:1) Tris(2,3-dibromopropyl) phosphate	$C_9H_{15}Br_6O_4P$ 697.61	126-72-7	9665		chl 3
10359	1-Propanol, 2,3-dichloro-	$C_3H_6Cl_2O$ 128.99	616-23-9	 184	4-01-00-01442 1.3607^{20}	H_2O 2; EtOH 5; eth 5; ace 5 1.4819^{20}
10360	1-Propanol, 3,3-dichloro-	$C_3H_6Cl_2O$ 128.99	83682-72-8	 82-3^{20}	3-01-00-01427 1.316^{25}	 1.467^{20}
10361	2-Propanol, 1,1-dichloro-	$C_3H_6Cl_2O$ 128.99	53894-19-2	 147	2-01-00-00383 1.3334^{22}	ace 4; eth 4; EtOH 4
10362	2-Propanol, 1,3-dichloro- 1,3-Dichloro-2-propanol	$C_3H_6Cl_2O$ 128.99	96-23-1	3063 176	4-01-00-01491 1.3506^{17}	H_2O 4; EtOH 4; eth 5; ace 3 1.4837^{20}
10363	2-Propanol, 1,3-dichloro-, acetate	$C_5H_8Cl_2O_2$ 171.02	3674-10-0	 205	3-02-00-00233 1.281^{20}	eth 4; EtOH 4; chl 4 1.4542^{20}
10364	2-Propanol, 1,1-dichloro-2-methyl- 2,2-Dichloro-1,1-dimethylethanol	$C_4H_8Cl_2O$ 143.01	4773-53-9 8	 152.5	4-01-00-01629 1.2363^{19}	bz 4; EtOH 4 1.4598^{19}
10365	2-Propanol, 1,3-dichloro-2-methyl- β-Methylglycerol α,γ-dichlorohydrin	$C_4H_8Cl_2O$ 143.01	597-32-0	 174.5	4-01-00-01629 1.2745^{20}	H_2O 3; EtOH 4; ace 3 1.4744^{21}
10366	1-Propanol, 2-(diethylamino)- 2-Diethylamino-1-propanol	$C_7H_{17}NO$ 131.22	611-12-1	 168; 78^{12}	4-04-00-01618 0.8649^{27}	EtOH 3; eth 3; ace 3; bz 3 1.4332^{20}
10367	1-Propanol, 3-(diethylamino)-	$C_7H_{17}NO$ 131.22	622-93-5	 189.5	4-04-00-01633 0.8600^{20}	EtOH 3; eth 3; ace 3; bz 3 1.4439^{20}
10368	2-Propanol, 1-(diethylamino)-, (±)- 2-Propanol, 1-diethylamino (*DL*)	$C_7H_{17}NO$ 131.22	78738-36-0 13.5	 159; 63^{22}	3-04-00-00759 0.8511^{20}	H_2O 4; ace 4; EtOH 4 1.4255^{20}
10369	2-Propanol, 1,3-difluoro- 1,3-Difluoro-2-propanol	$C_3H_6F_2O$ 96.08	453-13-4	 54-5^{34}	4-01-00-01489 1.24^{25}	 1.3725^{20}
10370	1-Propanol, 2,3-dimercapto- Dimercaprol	$C_3H_8OS_2$ 124.23	59-52-9	3196 82-4$^{0.8}$	4-01-00-02770 1.2463^{20}	EtOH 3; eth 3; chl 2; oils 3 1.5749^{20}
10371	2-Propanol, 1,3-dimercapto-	$C_3H_8OS_2$ 124.23	584-04-3	 94^{12}	4-01-00-02773 1.2386^{20}	 1.5700^{20}
10372	2-Propanol, 1,3-dimethoxy- Glycerol 1,3-dimethyl ether	$C_5H_{12}O_3$ 120.15	623-69-8	 169	3-01-00-02318 1.0085^{20}	H_2O 5; EtOH 4; eth 3 1.4192^{20}
10373	1-Propanol, 2,2-dimethyl- Neopentyl alcohol	$C_5H_{12}O$ 88.15	75-84-3 52.5	6373 113.5	4-01-00-01690 0.812^{20}	H_2O 2; EtOH 4; eth 4; ctc 3

No.	Name Synonym	Mol. Form. Mol. Wt.	CAS RN mp/°C	Merck No. bp/°C	Beil. Ref. den/g cm^{-3}	Solubility n_D
10374	1-Propanol, 2-(dimethylamino)- 2-Dimethylamino-1-propanol	$C_5H_{13}NO$ 103.16	15521-18-3	 150.3	4-04-00-01617 0.8820^{26}	H_2O 3
10375	1-Propanol, 3-(dimethylamino)-	$C_5H_{13}NO$ 103.16	3179-63-3	 163.5	4-04-00-01628 0.872^{25}	ctc 3 1.4360^{20}
10376	2-Propanol, 1-(dimethylamino)-	$C_5H_{13}NO$ 103.16	108-16-7	 124.5	3-04-00-00756 0.837^{25}	ctc 3 1.4193^{20}
10377	2-Propanol, 1,3-diphenoxy-	$C_{15}H_{16}O_3$ 244.29	622-04-8 81.5	 224^{17}; 175^{2}	4-06-00-00590 1.179^{24}	bz 4; eth 4; EtOH 4; chl 4
10378	2-Propanol, 1,3-diphenoxy-, acetate Glycerol, 1,3-diphenyl ether-2-acetate	$C_{17}H_{18}O_4$ 286.33	71159-31-4 33	 190^{160}	3-06-00-00583	bz 4; eth 4; EtOH 4; chl 4
10379	2-Propanol, 1,1',1'',1'''-(1,2-ethanediyldinitrilo)tetrakis- ENTPROL	$C_{14}H_{32}N_2O_4$ 292.42	102-60-3	3551	4-04-00-01685 1.030^{25}	chl 2 1.478^{25}
10380	1-Propanol, 2-ethoxy- Propylene glycol β-monoethyl ether	$C_5H_{12}O_2$ 104.15	19089-47-5	 140.5	3-01-00-02147 0.9044^{20}	H_2O 3; EtOH 3; eth 3 1.4122^{20}
10381	2-Propanol, 1-ethoxy- 1-Ethoxy-2-propanol	$C_5H_{12}O_2$ 104.15	1569-02-4	 131	4-01-00-02471 0.9028^{20}	H_2O 3; EtOH 3; eth 3 1.4075^{20}
10382	1-Propanol, 3-fluoro- 3-Hydroxypropyl fluoride	C_3H_7FO 78.09	462-43-1	 127.8	4-01-00-01437 1.0390^{25}	H_2O 4; eth 4; EtOH 4 1.3771^{25}
10383	2-Propanol, 1-[(2-hydroxyethyl)amino]-	$C_5H_{13}NO_2$ 119.16	6579-55-1	 $120\text{-}3^{1}$	4-04-00-01678 1.0455^{20}	H_2O 3; EtOH 3; ace 3 1.4695^{20}
10384	2-Propanol, 1,1'-iminobis- Diisopropanolamine	$C_6H_{15}NO_2$ 133.19	110-97-4 44.5	 250; 151^{23}	3-04-00-00761 0.989^{20}	H_2O 3; EtOH 3; eth 2
10385	1-Propanol, 3-iodo-	C_3H_7IO 185.99	627-32-7	 115^{38}; 88^{4}	4-01-00-01448 1.9976^{20}	 1.5585^{20}
10386	1-Propanol, 2-methoxy- 2-Methoxy-1-propanol	$C_4H_{10}O_2$ 90.12	1589-47-5	 130	4-01-00-02471 0.938^{20}	 1.4070^{20}
10387	2-Propanol, 1-methoxy- 1-Methoxy-2-propanol	$C_4H_{10}O_2$ 90.12	107-98-2	 119	3-01-00-02146 0.9620^{20}	 1.4034^{20}
10388	1-Propanol, 2-methyl- Isobutyl alcohol	$C_4H_{10}O$ 74.12	78-83-1 -108	5015 107.8	4-01-00-01588 0.8018^{20}	H_2O 2; EtOH 3; eth 3; ace 3 1.3955^{20}
10389	2-Propanol, 2-methyl- *tert*-Butyl alcohol	$C_4H_{10}O$ 74.12	75-65-0 25.4	1542 82.4	4-01-00-01609 0.7887^{20}	H_2O 5; EtOH 5; eth 5; chl 3 1.3878^{20}
10390	2-Propanol, 1-(1-methylethoxy)- 1-Isopropoxy-2-propanol	$C_6H_{14}O_2$ 118.18	3944-36-3	 137.5	4-01-00-02471 0.879^{20}	 1.4070^{20}
10391	1-Propanol, 2-methyl-2-nitro-	$C_4H_9NO_3$ 119.12	76-39-1 89.5	 94^{10}	4-01-00-01604	H_2O 2; EtOH 4; eth 4; chl 3
10392	1-Propanol, 3-[methyl(phenylmethyl)amino]-	$C_{11}H_{17}NO$ 179.26	5814-42-6	 $132\text{-}5^{4}$	4-12-00-02208 1.001^{25}	ctc 2; CS_2 3 1.5230^{20}
10393	2-Propanol, 1,1',1''-nitrilotris-	$C_9H_{21}NO_3$ 191.27	122-20-3 45	 $170\text{-}80^{10}$	4-04-00-01680 1.0^{20}	H_2O 3; EtOH 3; chl 2
10394	1-Propanol, 2-nitro-	$C_3H_7NO_3$ 105.09	2902-96-7	 100^{12}	4-01-00-01448 1.1841^{25}	H_2O 3; EtOH 3; eth 3; chl 2 1.4379^{20}
10395	1-Propanol, 3-nitro-	$C_3H_7NO_3$ 105.09	25182-84-7	 $138\text{-}40^{32}$	3-01-00-01429 1.173^{13}	H_2O 4; eth 4; EtOH 4
10396	2-Propanol, 1-nitro- 1-Nitro-2-propanol	$C_3H_7NO_3$ 105.09	3156-73-8	 112^{13}; 68^{1}	4-01-00-01496 1.1906^{20}	H_2O 3; EtOH 3; eth 3; bz 3 1.4383^{20}
10397	Propanol, oxybis- Dipropylene glycol	$C_6H_{14}O_3$ 134.18	25265-71-8	 230.5	4-01-00-02473 1.0206^{20}	H_2O 5; EtOH 3
10398	1-Propanol, 2-phenoxy-	$C_9H_{12}O_2$ 152.19	4169-04-4	 244	4-06-00-00582 0.9801^{25}	EtOH 3; eth 3 1.4760^{25}
10399	2-Propanol, 1-phenoxy-	$C_9H_{12}O_2$ 152.19	770-35-4	 134.5^{20}	4-06-00-00582 1.0622^{20}	 1.5232^{20}
10400	1-Propanol, 3-(phenylmethoxy)-	$C_{10}H_{14}O_2$ 166.22	4799-68-2	 172^{43}	4-06-00-02243 1.0474^{20}	EtOH 3; eth 3
10401	2-Propanol, 1-propoxy- 1,2-Propylene glycol 1-propyl ether	$C_6H_{14}O_2$ 118.18	1569-01-3	 150	3-01-00-02147 0.8886^{20}	 1.4130^{20}
10402	2-Propanol, 1,1,3,3-tetrachloro- 1,1,3,3-Tetrachloro-2-propanol	$C_3H_4Cl_4O$ 197.87	18992-39-7	 $80\text{-}90^{14}$	3-01-00-01474 1.612^{20}	eth 3 1.5133^{20}
10403	2-Propanol, 1,1,1,3-tetrachloro-	$C_3H_4Cl_4O$ 197.87	88947-16-4	 $95\text{-}6^{17}$	3-01-00-01474 1.610^{20}	eth 4 1.5145^{20}
10404	1-Propanol, 2,2,3,3-tetrafluoro- 2,2,3,3-Tetrafluoro-1-propanol	$C_3H_4F_4O$ 132.06	76-37-9 -15	9139 109.5	4-01-00-01438 1.4853^{20}	EtOH 3; ace 3; chl 3 1.3197^{20}
10405	2-Propanol, 1,1,1-tribromo-2-methyl- Tribromo-*tert*-butyl alcohol	$C_4H_7Br_3O$ 310.81	76-08-4 169	9523 sub	3-01-00-01588	H_2O 2; EtOH 3; eth 3; chl 2
10406	2-Propanol, 1,1,1-trichloro- 1,1,1-Trichloro-2-propanol	$C_3H_5Cl_3O$ 163.43	76-00-6 50.5	9557 163; $53\text{-}5^{12}$	3-01-00-01474	ace 4; bz 4; eth 4; EtOH 4
10407	2-Propanol, 1,1,1-trichloro-2-methyl- Chlorobutanol	$C_4H_7Cl_3O$ 177.46	57-15-8 97	2129 167	4-01-00-01629	H_2O 1; EtOH 3; eth 3; ace 3
10408	2-Propanol, 1,1,1-trichloro-3-nitro- 2-Propanol, 1-nitro-3,3,3-trichloro	$C_3H_4Cl_3NO_3$ 208.43	760-40-7 45.5	 $105^{3.5}$	4-01-00-01497	eth 4; EtOH 4
10409	2-Propanol, 1,1,1-trifluoro-, (±)- 2-Propanol, 1,1,1-trifluoro(*DL*)	$C_3H_5F_3O$ 114.07	17556-48-8 -52	 78	4-01-00-01489 1.2632^{25}	EtOH 4; eth 4; ace 3; bz 3 1.3130^{25}
10410	1-Propanol, 3-(trimethylsilyl)-	$C_6H_{16}OSi$ 132.28	2917-47-7	 141; 82^{24}	4-04-00-03953 0.822^{25}	 1.4298^{20}
10411	2-Propanone Acetone	C_3H_6O 58.08	67-64-1 -94.8	58 56.05	4-01-00-03180 0.7845^{25}	H_2O 5; EtOH 5; eth 5; ace 5 1.3588^{20}
10412	2-Propanone, 1-(acetyloxy)-	$C_5H_8O_3$ 116.12	592-20-1	 171; 63^{11}	4-02-00-00297 1.0757^{20}	H_2O 4; eth 4; EtOH 4 1.4141^{20}
10413	1-Propanone, 1-(4-aminophenyl)- *p*-Aminopropiophenone	$C_9H_{11}NO$ 149.19	70-69-9 140	482	4-14-00-00139	DMSO 3

No.	Name Synonym	Mol. Form. Mol. Wt.	CAS RN mp/°C	Merck No. bp/°C	Beil. Ref. den/g cm^{-3}	Solubility n_D
10414	1-Propanone, 2-amino-1-phenyl-	$C_9H_{11}NO$ 149.19	5265-18-9 46.5	 $93^{0.8}$	3-14-00-00147	ace 4; eth 4; EtOH 4
10415	2-Propanone, 1,3-bis(1-methyl-2-pyrrolidinyl)-, (*R**,*S**)- Cuscohygrine	$C_{13}H_{24}N_2O$ 224.35	454-14-8	2685 169^{23}; 122^2	5-24-01-00458 0.9733^{20}	H_2O 4; bz 4; eth 4; EtOH 4 1.4832^{20}
10416	2-Propanone, 1-bromo- Bromoacetone	C_3H_5BrO 136.98	598-31-2 -36.5	1389 138; 31.5^8	4-01-00-03223 1.634^{23}	H_2O 2; EtOH 3; eth 3; ace 3 1.4697^{15}
10417	2-Propanone, 1-bromo-3-chloro- 1-Bromo-3-chloroacetone	C_3H_4BrClO 171.42	53535-68-5 34.5	 179	4-01-00-03224	eth 4; EtOH 4
10418	1-Propanone, 1-(4-bromophenyl)-	C_9H_9BrO 213.07	10342-83-3 48	 169^{15}; 137^2	4-07-00-00684	H_2O 1; EtOH 3; eth 3; ace 3
10419	1-Propanone, 2-bromo-1-phenyl-	C_9H_9BrO 213.07	2114-00-3	 247.5	4-07-00-00684 1.4298^{20}	H_2O 1; EtOH 3; eth 3; ace 3 1.5720^{20}
10420	2-Propanone, 1-chloro- Chloroacetone	C_3H_5ClO 92.52	78-95-5 -44.5	2113 119	4-01-00-03215 1.15^{20}	H_2O 3; EtOH 3; eth 3; chl 3
10421	2-Propanone, 1-chloro-1,1,3,3,3-pentafluoro- Chloropentafluoroacetone	C_3ClF_5O 182.48	79-53-8 -133	 8		
10422	1-Propanone, 1-(4-chlorophenyl)-	C_9H_9ClO 168.62	6285-05-8 37.3	 135^{31}; 114^2	4-07-00-00683	H_2O 1; EtOH 3; chl 2; CS_2 3
10423	1-Propanone, 1-cyclohexyl- Cyclohexyl ethyl ketone	$C_9H_{16}O$ 140.23	1123-86-0	 196	4-07-00-00064 0.9105^{20}	 1.4530^{20}
10424	2-Propanone, 1,3-dibromo-	$C_3H_4Br_2O$ 215.87	816-39-7 26	 97^{22}	4-01-00-03225 2.1670^{18}	eth 4; CS_2 4
10425	2-Propanone, 1,1-dichloro- 1,1-Dichloroacetone	$C_3H_4Cl_2O$ 126.97	513-88-2	3038 120	4-01-00-03218 1.304^{18}	H_2O 2; EtOH 3; eth 5
10426	2-Propanone, 1,3-dichloro- 1,3-Dichloroacetone	$C_3H_4Cl_2O$ 126.97	534-07-6 45	3039 173.4	4-01-00-03219 1.3826^{46}	H_2O 3; EtOH 3; eth 3 1.4716^{40}
10427	2-Propanone, 1-(diethylamino)-	$C_7H_{15}NO$ 129.20	1620-14-0	 157 dec	4-04-00-01925 0.8620^{20}	H_2O 5; EtOH 5; eth 5; ctc 3 1.4249^{20}
10428	1-Propanone, 2-(diethylamino)-1-phenyl- Diethylpropion	$C_{13}H_{19}NO$ 205.30	90-84-6	3116	4-14-00-00144	
10429	2-Propanone, 1,3-dihydroxy- Dihydroxyacetone	$C_3H_6O_3$ 90.08	96-26-4 90	3166	4-01-00-04119	H_2O 3; EtOH 3; eth 3; ace 3
10430	1-Propanone, 1-(2,4-dihydroxyphenyl)-	$C_9H_{10}O_3$ 166.18	5792-36-9 99	 176^6	4-08-00-01837	bz 4; eth 4; EtOH 4
10431	1-Propanone, 2,2-dimethyl-1-phenyl-	$C_{11}H_{14}O$ 162.23	938-16-9	 220	4-07-00-00740 0.963^{26}	ace 3 1.5086^{19}
10432	2-Propanone, 1,1-diphenyl-	$C_{15}H_{14}O$ 210.28	781-35-1 46	 307; 174^{10}	4-07-00-01426	EtOH 3; eth 3; bz 3; chl 3 1.5361^{16}
10433	2-Propanone, 1,3-diphenyl-	$C_{15}H_{14}O$ 210.28	102-04-5 35	 331	4-07-00-01420 1.195^0	H_2O 1; EtOH 3; eth 3; peth 3
10434	1-Propanone, 3-(3-ethenyl-4-piperidinyl)-1-(6-methoxy-4-quinolinyl)-, (3*R*-*cis*)- Viquidil	$C_{20}H_{24}N_2O_2$ 324.42	84-55-9 60	9908	4-25-00-00181	eth 4; EtOH 4; chl 4
10435	1-Propanone, 3-(3-ethenyl-4-piperidinyl)-1-(4-quinolinyl)-, (3*R*-*cis*)- Cinchotoxine	$C_{19}H_{22}N_2O$ 294.40	69-24-9 59	2291	4-24-00-00651	H_2O 1; EtOH 4; eth 4; ace 4
10436	1-Propanone, 3-ethoxy-1-phenyl-	$C_{11}H_{14}O_2$ 178.23	34008-71-4 11.5	 135^{18}	4-08-00-00453 1.0386^{20}	 1.5190^{20}
10437	2-Propanone, 1-fluoro-	C_3H_5FO 76.07	430-51-3	 77	4-01-00-03213 1.0288^{20}	 1.3700^{20}
10438	1-Propanone, 1-(2-furanyl)-	$C_7H_8O_2$ 124.14	3194-15-8 28	 88^{14}	5-17-09-00415 1.0626^{28}	eth 3; ctc 2 1.4922^{25}
10439	2-Propanone, 1-(2-furanyl)-	$C_7H_8O_2$ 124.14	6975-60-6 29	 179.5	5-17-09-00418 1.104^{20}	 1.5035^{20}
10440	2-Propanone, 1,1,1,3,3,3-hexachloro- Hexachloroacetone	C_3Cl_6O 264.75	116-16-5 -2	 203	4-01-00-03223 1.7434^{12}	H_2O 2; ace 3 1.5112^{20}
10441	2-Propanone, 1,1,1,3,3,3-hexafluoro- Perfluoroacetone	C_3F_6O 166.02	684-16-2 -125	 -27.4	4-01-00-03215	
10442	2-Propanone, 1-(1H-indol-3-yl)- 3-Indolylacetone	$C_{11}H_{11}NO$ 173.21	1201-26-9 116	4873	5-21-08-00327	
10443	2-Propanone, 1-hydroxy-	$C_3H_6O_2$ 74.08	116-09-6 -17	 145.5	4-01-00-03977 1.0805^{20}	H_2O 4; EtOH 4; eth 4 1.4295^{20}
10444	1-Propanone, 1-(2-hydroxy-3-methylphenyl)- 2-Methyl-6-propionylphenol	$C_{10}H_{12}O_2$ 164.20	3338-15-6 22.5	 124^{13}	4-08-00-00517 1.0900^{20}	ctc 3
10445	1-Propanone, 1-(2-hydroxy-4-methylphenyl)- 2'-Hydroxy-4'-methylpropiophenone	$C_{10}H_{12}O_2$ 164.20	2886-52-4 44	 115^{10}	4-08-00-00519	chl 3
10446	1-Propanone, 1-(2-hydroxy-5-methylphenyl)-	$C_{10}H_{12}O_2$ 164.20	938-45-4 2	 $129\text{-}30^{16.5}$	4-08-00-00517 1.0841^{14}	chl 3 1.549^{13}
10447	1-Propanone, 1-(2-hydroxy-5-methylphenyl)-2-methyl-	$C_{11}H_{14}O_2$ 178.23	64207-03-0	 251; 125^{11}	4-08-00-00545 1.0460^{16}	 1.538^{16}
10448	1-Propanone, 1-(4-hydroxy-3-methylphenyl)-2-methyl- *o*-Cresol, 4-isobutyryl	$C_{11}H_{14}O_2$ 178.23	73206-57-2 122	 182^{12}	2-08-00-00130	eth 4; EtOH 4
10449	1-Propanone, 1-(2-hydroxyphenyl)-	$C_9H_{10}O_2$ 150.18	610-99-1	 150^{80}; 115^{15}	4-08-00-00436	H_2O 2; EtOH 3; eth 3; ctc 3 1.5501^{20}
10450	1-Propanone, 1-(4-hydroxyphenyl)- Paroxypropione	$C_9H_{10}O_2$ 150.18	70-70-2 149	6992	4-08-00-00441	H_2O 2; EtOH 3; eth 3; ace 2

PHYSICAL CONSTANTS OF ORGANIC COMPOUNDS (continued)

No.	Name	Synonym	Mol. Form.	Mol. Wt.	CAS RN	mp/°C	Merck No.	bp/°C	Beil. Ref.	den/g cm^{-3}	Solubility	n_D
10451	1-Propanone, 2-hydroxy-1-phenyl-		$C_9H_{10}O_2$	150.18	5650-40-8			251	4-08-00-00450	1.1085^{18}		1.536^{23}
10452	1-Propanone, 3-hydroxy-1-phenyl-		$C_9H_{10}O_2$	150.18	5650-41-9			$98^{0.2}$	4-08-00-00453	1.1170^{20}		1.5450^{20}
10453	1-Propanone, 3-(4-hydroxyphenyl)-1-(2,4,6-trihydroxyphenyl)-	Phloretin	$C_{15}H_{14}O_5$	274.27	60-82-2	263-4 dec	7299		4-08-00-03518		H_2O 2; EtOH 5; eth 1; ace 3	
10454	2-Propanone, 1-iodo-		C_3H_5IO	183.98	3019-04-3			62^{12}	4-01-00-03226	2.17^{15}	EtOH 3	
10455	2-Propanone, 1-methoxy-		$C_4H_8O_2$	88.11	5878-19-3			116	4-01-00-03977	0.957^{25}		1.3970^{20}
10456	1-Propanone, 1-(3-methoxyphenyl)-	3-Methoxypropiophenone	$C_{10}H_{12}O_2$	164.20	37951-49-8			259	4-08-00-00441	1.0812^{25}	eth 4; EtOH 4	1.5230^{25}
10457	1-Propanone, 2-methoxy-1-phenyl-	Propiophenone, 2-methoxy-	$C_{10}H_{12}O_2$	164.20	6493-83-0			$79\text{-}80^{1}$	4-08-00-00451	1.0704^{20}		1.5225^{20}
10458	1-Propanone, 3-methoxy-1-phenyl-		$C_{10}H_{12}O_2$	164.20	55563-72-9			$125\text{-}6^{16}$	4-08-00-00453	1.0602^{20}		1.5250^{20}
10459	2-Propanone, 1-(methoxyphenyl)-	2-Propanone, (methoxyphenyl)-	$C_{10}H_{12}O_2$	164.20	31116-84-4	<-15		267-9 142^{14}	2-08-00-00104	1.0670^{18}	H_2O 2; EtOH 3; eth 3	1.5253^{20}
10460	2-Propanone, 1-(3-methoxyphenyl)-		$C_{10}H_{12}O_2$	164.20	3027-13-2			259	4-08-00-00459	1.0812^{6}	EtOH 3; eth 3	1.5230^{25}
10461	2-Propanone, 1-(4-methoxyphenyl)-		$C_{10}H_{12}O_2$	164.20	122-84-9	<-15		268	4-08-00-00460	1.0694^{17}	eth 4; EtOH 4	1.5253^{20}
10462	1-Propanone, 2-methyl-1,2-di-3-pyridinyl-	Metyrapone	$C_{14}H_{14}N_2O$	226.28	54-36-4	50.5	6083		5-24-04-00048			
10463	2-Propanone, (1-methylethylidene)hydrazone		$C_6H_{12}N_2$	112.17	627-70-3	-12.5		133	4-01-00-03207	0.8390^{20}	H_2O 5; EtOH 5; eth 5; ace 3	1.4535^{20}
10464	1-Propanone, 2-methyl-1-(1-naphthalenyl)-	Isopropyl 1-naphthyl ketone	$C_{14}H_{14}O$	198.26	61838-78-6			307	3-07-00-01985	1.0761^{25}	bz 4; eth 4; EtOH 4	1.5948^{20}
10465	1-Propanone, 2-methyl-1-(2-naphthalenyl)-	2-Isobutyronaphthone	$C_{14}H_{14}O$	198.26	107574-57-2			313	3-07-00-01985	1.0617^{25}	bz 4; eth 4; EtOH 4	
10466	1-Propanone, 1-(2-methylphenyl)-		$C_{10}H_{12}O$	148.20	2040-14-4	-27.6		219.5	4-07-00-00720	1.0119^{0}		
10467	1-Propanone, 1-(3-methylphenyl)-		$C_{10}H_{12}O$	148.20	51772-30-6			235; 132^{31}	4-07-00-00721	1.0059^{0}	H_2O 1; EtOH 3; eth 3; ace 3	
10468	1-Propanone, 1-(4-methylphenyl)-		$C_{10}H_{12}O$	148.20	5337-93-9	7.2		236	4-07-00-00722	0.9926^{20}	H_2O 1; EtOH 3; eth 3; ace 3	1.5278^{20}
10469	1-Propanone, 2-methyl-1-phenyl-		$C_{10}H_{12}O$	148.20	611-70-1	-1.3		220	4-07-00-00716	0.9863^{11}	eth 4; EtOH 4	1.5172^{20}
10470	2-Propanone, 1-(1-methyl-2-piperidinyl)-, (±)-		$C_9H_{17}NO$	155.24	18747-42-7			$96\text{-}8^{13}$	5-21-06-00523	0.9478^{20}	H_2O 4; lig 4	1.4674^{20}
10471	2-Propanone, 1-(1-methyl-2-pyrrolidinyl)-, (*R*)-	Hygrine	$C_8H_{15}NO$	141.21	496-49-1		4788	76.5^{11}	5-21-06-00511		EtOH 4; chl 4	1.4555^{20}
10472	1-Propanone, 2-methyl-1-(2-thienyl)-	Isopropyl 2-thienyl ketone	$C_8H_{10}OS$	154.23	36448-60-9			229	5-17-09-00439	1.0894^{20}		1.5405^{20}
10473	1-Propanone, 1-(1-naphthalenyl)-		$C_{13}H_{12}O$	184.24	2876-63-3			306	4-07-00-01298	1.0971^{20}		1.6108^{20}
10474	1-Propanone, 1-(2-naphthalenyl)-		$C_{13}H_{12}O$	184.24	6315-96-4	60		313	4-07-00-01299		EtOH 4; chl 4	
10475	2-Propanone, 1-nitro-		$C_3H_5NO_3$	103.08	10230-68-9	50.3		103^{24}	4-01-00-03226		bz 4; eth 4; EtOH 4	
10476	2-Propanone, oxime	Acetoxime	C_3H_7NO	73.09	127-06-0	61	68	136; 61^{20}	4-01-00-03202	0.9113^{62}	H_2O 3; EtOH 3; eth 3; chl 3	1.4156^{20}
10477	2-Propanone, 1,1,1,3,3-pentabromo-	Pentabromoacetone	C_3HBr_5O	452.56	79-49-2	79.5	7056	sub	4-01-00-03226		H_2O 1; EtOH 4; eth 4; ace 4	
10478	2-Propanone, 1,1,1,3,3-pentachloro-		C_3HCl_5O	230.30	1768-31-6	2.1		192^{253}; 98^{40}	4-01-00-03222	1.69^{15}		
10479	2-Propanone, 1-phenoxy-		$C_9H_{10}O_2$	150.18	621-87-4			229.5	4-06-00-00604	1.0903^{20}	eth 3; ace 3	1.5228^{20}
10480	1-Propanone, 1-phenyl-	Propiophenone	$C_9H_{10}O$	134.18	93-55-0	18.6	7842	217.5	4-07-00-00680	1.0096^{20}	H_2O 1; EtOH 3; eth 3; chl 3	1.5269^{20}
10481	2-Propanone, 1-phenyl-	Phenylacetone	$C_9H_{10}O$	134.18	103-79-7	-15	7240	216.5	4-07-00-00687	1.0157^{20}	H_2O 1; EtOH 4; eth 4; bz 5	1.5168^{20}
10482	2-Propanone, phenylhydrazone	Acetone, phenylhydrazone	$C_9H_{12}N_2$	148.21	103-02-6	42		163^{50}	4-15-00-00060		EtOH 3; eth 3; dil acid 3	
10483	1-Propanone, 1-phenyl-, oxime		$C_9H_{11}NO$	149.19	2157-50-8	54.8		245; 165^{38}	4-07-00-00682	1.0639^{18}	H_2O 1; EtOH 3; eth 3; chl 2	
10484	1-Propanone, 1-(2-piperidinyl)-	Piperidine, 2(1-propionyl)	$C_8H_{15}NO$	141.21	97073-23-9			94^{18}	5-21-06-00522	0.9380^{20}		1.4621^{20}
10485	2-Propanone, 1-(2-piperidinyl)-, (*R*)-	(-)-Pelletierine	$C_8H_{15}NO$	141.21	2858-66-4		7015	195; 125^{100}	5-21-06-00522	0.988^{20}	eth 4; EtOH 4; chl 4	1.4683^{20}
10486	2-Propanone, 1,1,1,3-tetrachloro-		$C_3H_2Cl_4O$	195.86	16995-35-0	65		183	4-01-00-03222	1.624^{25}	H_2O 2; eth 4; ace 4	1.497^{18}
10487	2-Propanone, 1,1,3,3-tetrachloro-		$C_3H_2Cl_4O$	195.86	632-21-3			183	4-01-00-03222		H_2O 4; EtOH 4; eth 4; ace 4	1.4944^{20}
10488	2-Propanone, 1,1,1-trichloro-		$C_3H_3Cl_3O$	161.41	918-00-3			149; 28^{10}	4-01-00-03221	1.435^{20}	H_2O 1; EtOH 4; eth 4	1.4635^{17}
10489	1-Propanone, 1-(2,4,6-trihydroxyphenyl)-	Flopropione	$C_9H_{10}O_4$	182.18	2295-58-1	175.5	4039		3-08-00-03413		eth 4; EtOH 4	

No.	Name Synonym	Mol. Form. Mol. Wt.	CAS RN mp/°C	Merck No. bp/°C	Beil. Ref. den/g cm^{-3}	Solubility n_D
10490	Propanoyl bromide Propionic acid bromide	C_3H_5BrO 136.98	598-22-1	 102	4-02-00-00724 1.5210^{16}	eth 4 1.4578^{16}
10491	Propanoyl bromide, 2-bromo-2-methyl-	$C_4H_6Br_2O$ 229.90	20769-85-1	 163	3-02-00-00661 1.4067^{14}	ace 4; CS_2 4
10492	Propanoyl bromide, 2-methyl-	C_4H_7BrO 151.00	2736-37-0	 117	4-02-00-00852 1.4067^{15}	 1.4552^{25}
10493	Propanoyl chloride Propionyl chloride	C_3H_5ClO 92.52	79-03-8 -94	7840 80	4-02-00-00724 1.0646^{20}	eth 3 1.4032^{20}
10494	Propanoyl chloride, 2-(acetyloxy)-, (±)- Lactyl chloride, acetate (*DL*)	$C_5H_7ClO_3$ 150.56	55057-45-9	 150 dec; 56^{11}	4-03-00-00674 1.1920^{17}	 1.4241^{17}
10495	Propanoyl chloride, 2-bromo-	C_3H_4BrClO 171.42	7148-74-5	 132	4-02-00-00764 1.697^{11}	eth 3; ctc 2; chl 3 1.4780^{20}
10496	Propanoyl chloride, 2-chloro-, (*S*)- Propanoic acid, chloride, 2-chloro(*d*)	$C_3H_4Cl_2O$ 126.97	70110-24-6	 110	4-02-00-00747 1.2394^{7}	 1.4400^{20}
10497	Propanoyl chloride, 3-chloro-	$C_3H_4Cl_2O$ 126.97	625-36-5	 144	4-02-00-00750 1.3307^{13}	H_2O 2; EtOH 4; eth 4; chl 4 1.4549^{20}
10498	Propanoyl chloride, 2-chloro-2-methyl-	$C_4H_6Cl_2O$ 141.00	13222-26-9	 126.5	4-02-00-00859	eth 4 1.4369^{20}
10499	Propanoyl chloride, 3-chloro-2-methyl- Isobutyryl chloride, 3-chloro	$C_4H_6Cl_2O$ 141.00	7623-10-1 -90	 92	4-02-00-00860 1.0174^{20}	eth 4 1.4079^{20}
10500	Propanoyl chloride, 2,2-dichloro-	$C_3H_3Cl_3O$ 161.41	26073-26-7	 118; 70^{89}	4-02-00-00755 1.4062^{20}	 1.4524^{20}
10501	Propanoyl chloride, 2,3-dichloro-	$C_3H_3Cl_3O$ 161.41	7623-13-4	 $52\text{-}4^{16}$	4-02-00-00757 1.4757^{20}	 1.4764^{20}
10502	Propanoyl chloride, 3,3-dichloro- 3,3-Dichloropropionyl chloride	$C_3H_3Cl_3O$ 161.41	17880-86-3	 $43\text{-}4^{10}$	3-02-00-00562 1.4557^{20}	eth 4; Diox 4 1.4738^{20}
10503	Propanoyl chloride, 2,2-dimethyl- Pivalic acid chloride	C_5H_9ClO 120.58	3282-30-2	 107	4-02-00-00912 1.003^{20}	eth 4 1.4139^{20}
10504	Propanoyl chloride, 2-methyl- Isobutyric acid chloride	C_4H_7ClO 106.55	79-30-1 -90	 92	4-02-00-00852	eth 3 1.4079^{20}
10505	Propanoyl chloride, 2-phenoxy-	$C_9H_9ClO_2$ 184.62	122-35-0	 147; 116^{10}	4-06-00-00643 1.1865^{20}	eth 3 1.5178^{20}
10506	Propanoyl fluoride Propionyl fluoride	C_3H_5FO 76.07	430-71-7	 44	4-02-00-00724 0.972^{15}	 1.329^{13}
10507	Propargite Sulfurous acid, 2-[4-(1,1-dimethylethyl)phenoxy]cyclohexyl 2-propynyl ester	$C_{19}H_{26}O_4S$ 350.48	2312-35-8	7818	 1.10^{25}	
10508	Propazine 1,3,5-Triazine-2,4-diamine, 6-chloro-*N,N'*-bis(1-methylethyl)-	$C_9H_{16}ClN_5$ 229.71	139-40-2 213	7822	 1.162^{20}	
10509	2-Propenal Acrolein	C_3H_4O 56.06	107-02-8 -87.7	122 52.6	4-01-00-03435 0.840^{20}	H_2O 4; EtOH 3; eth 3; ace 3 1.4017^{20}
10510	2-Propenal, 3-bromo-3-phenyl-	C_9H_7BrO 211.06	14804-59-2	 $144\text{-}6^{12}$	3-07-00-01383 1.492^{20}	eth 4
10511	2-Propenal, 2-chloro-	C_3H_3ClO 90.51	683-51-2	 40^{30}	4-01-00-03440 1.199^{20}	eth 4; EtOH 4 1.463^{20}
10512	2-Propenal, 3-(2-furanyl)-	$C_7H_6O_2$ 122.12	623-30-3 54	 135^{14}	5-17-09-00570	H_2O 1; EtOH 5; eth 3; chl 2
10513	2-Propenal, 3-(2-furanyl)-, (*E*)-	$C_7H_6O_2$ 122.12	39511-08-5 50	 143^{37}; 100^{12}	5-17-09-00570	bz 4; eth 4; EtOH 4 1.5286^{20}
10514	2-Propenal, 3-(4-hydroxy-3-methoxyphenyl)-	$C_{10}H_{10}O_3$ 178.19	458-36-6 84		4-08-00-01979 1.1562^{102}	bz 4; eth 4; EtOH 4
10515	2-Propenal, 2-methyl- Methacrolein	C_4H_6O 70.09	78-85-3	 68.4	4-01-00-03455 0.840^{25}	H_2O 5; EtOH 5; eth 5 1.4144^{20}
10516	2-Propenal, 2-methyl-3-phenyl-	$C_{10}H_{10}O$ 146.19	101-39-3	 150^{100}	2-07-00-00291 1.0407^{17}	 1.6057^{17}
10517	2-Propenal, 3-(4-methylphenyl)-	$C_{10}H_{10}O$ 146.19	1504-75-2 41.5	 154^{25}	4-07-00-01013 0.9670^{23}	EtOH 4
10518	2-Propenal, 3-phenyl-, (*E*)- *trans*-Cinnamaldehyde	C_9H_8O 132.16	14371-10-9 -7.5	 253 dec; 127^{16}	4-07-00-00984 1.0497^{20}	H_2O 2; EtOH 3; eth 3; chl 3 1.6195^{20}
10519	2-Propenamide Acrylamide	C_3H_5NO 71.08	79-06-1 84.5	123 192.6	4-02-00-01471	H_2O 4; EtOH 3; eth 3; ace 3
10520	2-Propenamide, *N*-(3,4-dichlorophenyl)-2-methyl- Dicryl	$C_{10}H_9Cl_2NO$ 230.09	2164-09-2 128	3078		ace 4; EtOH 4
10521	2-Propenamide, *N*-(1,1-dimethyl-3-oxobutyl)- Diacetone acrylamide	$C_9H_{15}NO_2$ 169.22	2873-97-4	2943		chl 3
10522	2-Propen-1-amine Allylamine	C_3H_7N 57.10	107-11-9 -88.2	285 53.3	4-04-00-01057 0.758^{20}	H_2O 5; EtOH 5; eth 5; chl 3 1.4205^{20}
10523	2-Propen-1-amine, *N,N*-di-2-propenyl- Triallylamine	$C_9H_{15}N$ 137.22	102-70-5 94	 155.5	4-04-00-01061 0.809^{20}	EtOH 3; eth 3; ace 3; bz 3 1.4502^{20}
10524	2-Propen-1-amine, *N*-methyl-	C_4H_9N 71.12	627-37-2	 65	4-04-00-01058	H_2O 4; ace 4; eth 4; EtOH 4 1.4065^{20}
10525	2-Propen-1-amine, *N*-(1-methylethyl)-	$C_6H_{13}N$ 99.18	35000-22-7	 96.5	4-04-00-01059 0.7400^{30}	 1.4140^{25}
10526	2-Propen-1-amine, *N*-2-propenyl- Diallylamine	$C_6H_{11}N$ 97.16	124-02-7	2951 111	4-04-00-01060	EtOH 3; eth 3 1.4387^{20}

No.	Name	Synonym	Mol. Form.	Mol. Wt.	CAS RN	mp/°C	Merck No.	bp/°C	Beil. Ref.	den/g cm^{-3}	Solubility	n_D
10527	2-Propen-1-amine, *N*-propyl-		$C_6H_{13}N$	99.18	5666-21-7			112	0-04-00-00207	0.7708^{18}	H_2O 4	1.4140^{25}
10528	1-Propene	Propylene	C_3H_6	42.08	115-07-1	-185.2	7862	-47.6	4-01-00-00725	$*0.505^{25}$	H_2O 4; EtOH 4; HOAc 4	1.3567^{-70}
10529	1-Propene, 1-bromo-, (*Z*)-		C_3H_5Br	120.98	590-13-6	-113		57.8	4-01-00-00754	1.4291^{20}	H_2O 1; eth 3; ace 3; chl 3	1.4560^{20}
10530	1-Propene, 2-bromo-		C_3H_5Br	120.98	557-93-7	-126		48.4	4-01-00-00754	1.3965^{16}	H_2O 1; eth 3; ace 3; chl 3	1.4467^{16}
10531	1-Propene, 3-bromo-	Allyl bromide	C_3H_5Br	120.98	106-95-6	-119	286	70.1	4-01-00-00754	1.398^{20}	H_2O 1; EtOH 5; eth 5; ctc 3	1.4697^{20}
10532	1-Propene, 3-bromo-3,3-difluoro-	Propylene, 3-bromo-3,3-difluoro	$C_3H_3BrF_2$	156.96	420-90-6			42	4-01-00-00756	1.543^{25}		1.3773^{25}
10533	1-Propene, 1-chloro-, (*E*)-	*trans*-1-Chloropropene	C_3H_5Cl	76.53	16136-85-9	-99		37.4	4-01-00-00737	0.9349^{20}	H_2O 1; eth 3; ace 3; bz 3	1.4054^{20}
10534	1-Propene, 1-chloro-, (*Z*)-	*cis*-1-Chloropropene	C_3H_5Cl	76.53	16136-84-8	-134.8		32.8	4-01-00-00737	0.9347^{20}	H_2O 1; eth 3; ace 3; bz 3	1.4055^{20}
10535	1-Propene, 2-chloro-	Isopropenyl chloride	C_3H_5Cl	76.53	557-98-2	-137.4		22.6	4-01-00-00737	0.9017^{20}	H_2O 1; eth 3; ace 3; bz 3	1.3973^{20}
10536	1-Propene, 3-chloro-	Allyl chloride	C_3H_5Cl	76.53	107-05-1	-134.5	287	45.1	4-01-00-00738	0.9376^{20}	H_2O 1; EtOH 5; eth 5; ace 5	1.4157^{20}
10537	1-Propene, 3-chloro-2-(chloromethyl)-		$C_4H_6Cl_2$	125.00	1871-57-4	-14		138	4-01-00-00805	1.1782^{20}	EtOH 4; chl 4	1.4753
10538	1-Propene, 2-chloro-3-isothiocyanato-		C_4H_4ClNS	133.60	14214-31-4			182	0-04-00-00219	1.27^{12}	EtOH 2	
10539	1-Propene, 1-chloro-2-methyl-	1-Chloro-2-methylpropene	C_4H_7Cl	90.55	513-37-1		2147	68	4-01-00-00803	0.9186^{20}	chl 3	1.4221^{20}
10540	1-Propene, 3-chloro-2-methyl-	3-Chloro-2-methylpropene	C_4H_7Cl	90.55	563-47-3		2148	71.5	4-01-00-00803	0.9165^{20}	EtOH 5; eth 5; ace 3; chl 4	1.4291^{20}
10541	1-Propene, 2-chloro-1,1,3,3,3-pentafluoro-		C_3ClF_5	166.48	2804-50-4	-130		6.8	4-01-00-00741	1.5920^{20}		
10542	1-Propene, 1,1-dibromo-		$C_3H_4Br_2$	199.87	13195-80-7			125	4-01-00-00759	1.9767^{20}	H_2O 2; bz 3; ctc 3; chl 3	1.5260^{20}
10543	1-Propene, 1,3-dibromo-, (*E*)-	Propylene, 1,3-dibromo (*trans*)	$C_3H_4Br_2$	199.87	32121-07-6			66^{25}	4-01-00-00760	1.9791^{20}		1.5495^{20}
10544	1-Propene, 1,3-dibromo-, (*Z*)-	Propylene, 1,3-dibromo (*cis*)	$C_3H_4Br_2$	199.87	32121-06-5			60^{25}	4-01-00-00760	2.0599^{20}		1.5550^{20}
10545	1-Propene, 2,3-dibromo-	2,3-Dibromopropene	$C_3H_4Br_2$	199.87	513-31-5		3009	141; 37.7^{11}	4-01-00-00760	2.0345^{25}	H_2O 1; eth 3; ace 3; chl 3	1.5416^{25}
10546	1-Propene, 1,3-dibromo-2-methyl-	Isobutylene, 1,3-dibromo	$C_4H_6Br_2$	213.90	35911-17-2			102^{100}	3-01-00-00769	1.8939^{4}		1.5396^{15}
10547	1-Propene, 1,1-dichloro-		$C_3H_4Cl_2$	110.97	563-58-6			76.5	4-01-00-00742	1.1864^{25}	H_2O 1; eth 3; ace 3; chl 3	1.4430^{25}
10548	1-Propene, 1,2-dichloro-, (*E*)-		$C_3H_4Cl_2$	110.97	7069-38-7			77	4-01-00-00742	1.1818^{20}	H_2O 1; EtOH 4; ctc 4; MeOH 4	1.4471^{20}
10549	1-Propene, 1,2-dichloro-, (*Z*)-		$C_3H_4Cl_2$	110.97	6923-20-2			93	4-01-00-00742		H_2O 1; ace 3; bz 3; chl 3	1.4549^{20}
10550	1-Propene, 1,3-dichloro-, (*E*)-	*trans*-1,3-Dichloropropylene	$C_3H_4Cl_2$	110.97	10061-02-6		3064	112	4-01-00-00744	1.217^{20}	H_2O 1; eth 3; bz 3; chl 3	1.4730^{20}
10551	1-Propene, 1,3-dichloro-, (*Z*)-	*cis*-1,3-Dichloropropylene	$C_3H_4Cl_2$	110.97	10061-01-5		3064	104.3	4-01-00-00743	1.224^{20}	H_2O 1; eth 3; bz 3; chl 3	1.4682^{20}
10552	1-Propene, 2,3-dichloro-	2,3-Dichloropropene	$C_3H_4Cl_2$	110.97	78-88-6	10		94	4-01-00-00744	1.211^{20}	H_2O 1; EtOH 5; eth 3; bz 3	1.4603^{20}
10553	1-Propene, 3,3-dichloro-		$C_3H_4Cl_2$	110.97	563-57-5			84.4	4-01-00-00745	1.175^{20}	bz 4; eth 4; EtOH 4	1.4510^{20}
10554	1-Propene, 1,1-dichloro-2-fluoro-	Propylene, 1,1-dichloro-2-fluoro	$C_3H_3Cl_2F$	128.96	430-95-5			78	4-01-00-00745	1.3026^{25}		1.4196^{25}
10555	1-Propene, 1,1-dichloro-2-methyl-		$C_4H_6Cl_2$	125.00	6065-93-6	-43		108.5	4-01-00-00804	1.1447^{20}	bz 4; eth 4; chl 4	1.4580^{20}
10556	1-Propene, 3,3-dichloro-2-methyl-		$C_4H_6Cl_2$	125.00	22227-75-4			108	4-01-00-00805	1.136^{24}	bz 4; eth 4; chl 4	1.4523^{24}
10557	1-Propene, 1,2-dichloro-1,3,3,3-tetrafluoro-		$C_3Cl_2F_4$	182.93	431-53-8	-137		47.3	4-01-00-00747	1.5468^{25}		1.3511^{20}
10558	Propene, 1,2-dichloro-3,3,3-trifluoro-	Propylene, 1,2-dichloro-3,3,3-trifluoro	$C_3HCl_2F_3$	164.94	431-27-6	-109.2		53.7	4-01-00-00746	1.4653^{20}		1.3670^{20}
10559	1-Propene, 1,1-diethoxy-		$C_7H_{14}O_2$	130.19	21504-43-8			133.5	4-01-00-03434	0.8628^{25}		1.4083^{25}
10560	1-Propene, 3,3-diethoxy-	Acrolein, diethyl acetal	$C_7H_{14}O_2$	130.19	3054-95-3			123.5	4-01-00-03437	0.8543^{15}	H_2O 2; EtOH 5; eth 5	1.4000^{20}
10561	1-Propene, 1,1-difluoro	1,1-Difluoro-1-propene	$C_3H_4F_2$	78.06	430-63-7							
10562	1-Propene, 3,3-dimethoxy-		$C_5H_{10}O_2$	102.13	6044-68-4			88	4-01-00-03437	0.862^{25}		1.3954^{20}
10563	2-Propene-1,1-diol, diacetate		$C_7H_{10}O_4$	158.15	869-29-4	-37.6		180	4-02-00-00291	1.0760^{20}	ace 4; bz 4; eth 4; EtOH 4	1.4193^{20}
10564	2-Propene-1,1-diol, 2-methyl-, diacetate	Methacrolein diacetate	$C_8H_{12}O_4$	172.18	10476-95-6			191	4-02-00-00292			1.4241^{20}
10565	1-Propene, 3-(ethenyloxy)-		C_5H_8O	84.12	3917-15-5			66	4-01-00-02085	0.8050^{20}	H_2O 1; eth 3; ace 3; chl 3	1.4062^{20}

No.	Name Synonym	Mol. Form. Mol. Wt.	CAS RN mp/°C	Merck No. bp/°C	Beil. Ref. den/g cm^{-3}	Solubility n_D
10566	1-Propene, 3-ethoxy- Allyl ethyl ether	$C_5H_{10}O$ 86.13	557-31-3	291 67.6	4-01-00-02083 0.7651^{20}	H_2O 1; EtOH 5; eth 5; ace 3 1.3881^{20}
10567	1-Propene, 1-fluoro- (*Z*) *cis*-1-Fluoropropene	C_3H_5F 60.07	19184-10-2			
10568	1-Propene, 1-fluoro- (*E*) *trans*-1-Fluoropropene	C_3H_5F 60.07	20327-65-5			
10569	1-Propene, 2-fluoro- 2-Fluoropropene	C_3H_5F 60.07	1184-60-7	 -24	4-01-00-00733	
10570	1-Propene, 3-fluoro- 3-Fluoropropene	C_3H_5F 60.07	818-92-8	 -3	4-01-00-00733	H_2O 2; EtOH 4; eth 4; chl 3
10571	1-Propene, 1,1,2,3,3,3-hexachloro- Hexachloropropylene	C_3Cl_6 248.75	1888-71-7 -72.9	 209.5	4-01-00-00753 1.7632^{20}	H_2O 1; ctc 3; chl 3 1.5091^{20}
10572	1-Propene, 1,1,2,3,3,3-hexafluoro- Perfluoropropene	C_3F_6 150.02	116-15-4 -156.5	 -29.6	4-01-00-00735	 1.583^{-40}
10573	1-Propene, 3-iodo- Allyl iodide	C_3H_5I 167.98	556-56-9 -99.3	292 103	4-01-00-00761 1.848^{12}	H_2O 1; EtOH 3; eth 3; chl 3 1.5540^{21}
10574	1-Propene, 3-isocyano- Allyl isocyanide	C_4H_5N 67.09	2835-21-4	 98	0-04-00-00208 0.794^{17}	eth 4; EtOH 4
10575	1-Propene, 3-isothiocyanato- Allyl isothiocyanate	C_4H_5NS 99.16	57-06-7 -80	293 152	4-04-00-01081 1.0126^{20}	bz 4; eth 4; EtOH 4 1.5306^{20}
10576	1-Propene, 3-methoxy-	C_4H_8O 72.11	627-40-7	 44	4-01-00-02083 0.77^{11}	H_2O 1; EtOH 5; eth 5; ace 3 1.3778^{20}
10577	1-Propene, 3-methoxy-2-methyl-	$C_5H_{10}O$ 86.13	22418-49-1	 68	3-01-00-01904 0.7698^{20}	 1.3964^{20}
10578	1-Propene, 2-methyl- Isobutene	C_4H_8 56.11	115-11-7 -140.4	5024 -6.9	4-01-00-00796 $*0.589^{25}$	H_2O 1; EtOH 4; eth 4; bz 3 1.3926^{-25}
10579	1-Propene, 3-(1-methylethoxy)-	$C_6H_{12}O$ 100.16	6140-80-3	 83.5	4-01-00-02083 0.7764^{20}	ace 4; eth 4; EtOH 4 1.3946^{20}
10580	1-Propene, 2-methyl-, tetramer Propene, 2-methyl(tetramer)	$C_{16}H_{32}$ 224.43	15220-85-6 -98	 244; 109^{15}	3-01-00-00763 0.7944^{20}	 1.4482^{20}
10581	1-Propene, 3-(methylthio)-	C_4H_8S 88.17	10152-76-8	 92	4-01-00-02095 0.8767^{20}	 1.4714^{20}
10582	1-Propene, 2-methyl-, trimer Isobutylene (trimer)	$C_{12}H_{24}$ 168.32	7756-94-7 -76	 180	3-01-00-00763 0.7590^{20}	 1.4314^{20}
10583	2-Propenenitrile Acrylonitrile	C_3H_3N 53.06	107-13-1 -83.5	125 77.3	4-02-00-01473 0.8060^{20}	ace 4; bz 4; eth 4; EtOH 4 1.3911^{20}
10584	2-Propenenitrile, 2-chloro-	C_3H_2ClN 87.51	920-37-6 -65	 88.5	4-02-00-01484 1.096^{25}	 1.4290^{20}
10585	2-Propenenitrile, 3-(2-furanyl)- 2-Furanacrylonitrile	C_7H_5NO 119.12	7187-01-1 38	4208 96	4-18-00-04152	tol 4 1.5824^{25}
10586	2-Propenenitrile, 2-methyl- Methylacrylonitrile	C_4H_5N 67.09	126-98-7 -35.8	5850 90.3	4-02-00-01539 0.8001^{20}	H_2O 1; EtOH 5; eth 5; ace 5 1.4003^{20}
10587	2-Propenenitrile, 3-phenyl-, (*E*)-	C_9H_7N 129.16	1885-38-7 22	 263.8	4-09-00-02027 1.0304^{20}	H_2O 1; EtOH 3; ace 3; ctc 3 1.6013^{20}
10588	2-Propenenitrile, 3-phenyl-, (*Z*)-	C_9H_7N 129.16	24840-05-9 -4.4	 249; 139^{30}	4-09-00-02027 1.0289^{20}	H_2O 1; EtOH 3; bz 4 1.5843^{20}
10589	1-Propene, 1-nitro-	$C_3H_5NO_2$ 87.08	3156-70-5	 60^{34}; 37^{10}	3-01-00-00715 1.0661^{20}	eth 3; ace 3; chl 3 1.4527^{20}
10590	1-Propene, 2-nitro-	$C_3H_5NO_2$ 87.08	4749-28-4	 57^{100}; 32^{30}	4-01-00-00764 1.0643^{20}	eth 3; ace 3; chl 3 1.4358^{20}
10591	1-Propene, 3-nitro-	$C_3H_5NO_2$ 87.08	625-46-7	 127.5	4-01-00-00764 1.051^{22}	eth 4; EtOH 4
10592	1-Propene, 3,3'-oxybis- Allyl ether	$C_6H_{10}O$ 98.14	557-40-4 -6	290 94	4-01-00-02086 0.8260^{20}	H_2O 1; EtOH 5; eth 5; ace 4 1.4163^{20}
10593	1-Propene, 1,1,2,3,3-pentachloro-	C_3HCl_5 214.30	1600-37-9	 185	4-01-00-00753 1.6317^{34}	eth 4 1.5313^{20}
10594	1-Propene, 3-propoxy-	$C_6H_{12}O$ 100.16	1471-03-0	 91	4-01-00-02083 0.7764^{20}	ace 4; eth 4; EtOH 4 1.3919^{20}
10595	2-Propene-1-sulfinothioic acid, *S*-2-propenyl ester Allicin	$C_6H_{10}OS_2$ 162.28	539-86-6	249 dec	4-04-00-00007 1.112^{20}	H_2O 4 1.561^{20}
10596	1-Propene, 3,3'-sulfinylbis-	$C_6H_{10}OS$ 130.21	14180-63-3 23.5	 114^{12}; $89^{2.6}$	4-01-00-02097 1.0261^{20}	EtOH 3; eth 3; ace 3 1.5115^{20}
10597	1-Propene, 3,3'-sulfonylbis-	$C_6H_{10}O_2S$ 146.21	16841-48-8	 109^{3}	4-01-00-02098 1.1213^{20}	 1.4893^{20}
10598	1-Propene, 1,2,3,3-tetrachloro-	$C_3H_2Cl_4$ 179.86	20589-85-9	 162	3-01-00-00708 1.537^{19}	bz 4; chl 4 1.5121^{19}
10599	1-Propene, 3,3'-thiobis- Diallyl sulfide	$C_6H_{10}S$ 114.21	592-88-1 -85	295 138.6	4-01-00-02097 0.8877^{27}	eth 4; EtOH 4 1.4870^{25}
10600	2-Propene-1-thiol	C_3H_6S 74.15	870-23-5	 65	4-01-00-02095 0.925^{23}	H_2O 1; EtOH 5; eth 5; chl 3 1.4832^{20}
10601	1-Propene-1,2,3-tricarboxylic acid, triethyl ester, (*E*)- Aconitic acid, triethyl ester (*trans*)	$C_{12}H_{18}O_6$ 258.27	68077-28-1	 275 dec; 159^{9}	4-02-00-02405 1.1064^{20}	eth 4; EtOH 4
10602	1-Propene, 1,1,2-trichloro-	$C_3H_3Cl_3$ 145.41	21400-25-9	 118	4-01-00-00747 1.382^{20}	H_2O 1; EtOH 3; eth 3; bz 3 1.4827^{20}
10603	1-Propene, 1,2,3-trichloro-	$C_3H_3Cl_3$ 145.41	96-19-5	 142	4-01-00-00748 1.412^{20}	H_2O 1; EtOH 4; eth 4; bz 3 1.5030^{20}
10604	1-Propene, 3,3,3-trichloro-	$C_3H_3Cl_3$ 145.41	2233-00-3 -30	 114.5	4-01-00-00749 1.367^{20}	H_2O 1; EtOH 3; eth 3; bz 3 1.4827^{20}

PHYSICAL CONSTANTS OF ORGANIC COMPOUNDS (continued)

No.	Name Synonym	Mol. Form. Mol. Wt.	CAS RN mp/°C	Merck No. bp/°C	Beil. Ref. den/g cm^{-3}	Solubility n_D
10605	1-Propene, 1,1,3-trichloro-2-methyl-	$C_4H_5Cl_3$ 159.44	31702-33-7	156	4-01-00-00805 1.346^{20}	ace 4; bz 4; chl 4 1.4990^{20}
10606	1-Propene, 3,3,3-trichloro-2-methyl-	$C_4H_5Cl_3$ 159.44	4749-27-3	133	4-01-00-00805 1.293^{20}	H_2O 1; ace 3; bz 3; chl 4 1.4770^{20}
10607	1-Propene, 1,1,2-trichloro-3,3,3-trifluoro-	$C_3Cl_3F_3$ 199.39	431-52-7 -114.6	88.1	4-01-00-00751 1.617^{20}	
10608	1-Propene, 3,3,3-trifluoro- 3,3,3-Trifluoropropene	$C_3H_3F_3$ 96.05	677-21-4	-17	4-01-00-00734 0	
10609	2-Propenoic acid Acrylic acid	$C_3H_4O_2$ 72.06	79-10-7 12.3	124 141	4-02-00-01455 1.0511^{20}	H_2O 5; EtOH 5; eth 5; ace 3 1.4224^{20}
10610	2-Propenoic acid, 2-bromo-	$C_3H_3BrO_2$ 150.96	10443-65-9 71.5	dec	4-02-00-01486	H_2O 4; EtOH 4
10611	2-Propenoic acid, 2-bromoethyl ester	$C_5H_7BrO_2$ 179.01	4823-47-6	156.5; 53^{5}	4-02-00-01462 1.4581^{25}	1.4660^{25}
10612	2-Propenoic acid, 3-bromo-2-methyl- Methacrylic acid, 3-bromo	$C_4H_5BrO_2$ 164.99	89123-63-7 64	229	0-02-00-00424	H_2O 4
10613	2-Propenoic acid, 2-bromo-3-phenyl-, ethyl ester	$C_{11}H_{11}BrO_2$ 255.11	57296-00-1	295	4-09-00-02042 1.3885^{25}	1.5845^{25}
10614	2-Propenoic acid, 3-bromo-3-phenyl-, ethyl ester, (*Z*)- Cinnamic acid, β-bromo-, ethyl ester	$C_{11}H_{11}BrO_2$ 255.11	1504-70-7	167^{12}	3-09-00-02733 1.4044^{10}	1.5841^{10}
10615	2-Propenoic acid, 2-bromo-3-phenyl-, methyl ester Methyl α-bromocinnamate	$C_{10}H_9BrO_2$ 241.08	24127-62-6 23	$120\text{-}1^{0.6}$	4-09-00-02042 1.4975^{20}	
10616	2-Propenoic acid, butyl ester Butyl acrylate	$C_7H_{12}O_2$ 128.17	141-32-2 -64.6	1539 145	4-02-00-01463 0.8898^{20}	H_2O 1; EtOH 3; eth 3; ace 3 1.4185^{20}
10617	2-Propenoic acid, 3-chloro-, (*E*)-	$C_3H_3ClO_2$ 106.51	2345-61-1 86	94^{18}	4-02-00-01481	EtOH 3; eth 3; chl 2
10618	2-Propenoic acid, 3-chloro-, (*Z*)-	$C_3H_3ClO_2$ 106.51	1609-93-4 63.5	106^{17}	4-02-00-01481	EtOH 3; eth 3
10619	2-Propenoic acid, chloro-, ethyl ester Acrylic acid, chloro-, ethyl ester	$C_5H_7ClO_2$ 134.56	30600-19-2	$51\text{-}3^{18}$	3-02-00-01245 1.1404^{20}	EtOH 4; eth 4 1.4384^{20}
10620	2-Propenoic acid, 2-chloroethyl ester	$C_5H_7ClO_2$ 134.56	2206-89-5	$51\text{-}3^{18}$	3-02-00-01225 1.1604^{20}	eth 4; EtOH 4 1.4384^{20}
10621	2-Propenoic acid, 2-chloro-, methyl ester Methyl 2-chloroacrylate	$C_4H_5ClO_2$ 120.54	80-63-7	52^{51}	4-02-00-01482 1.189^{20}	eth 4 1.4420^{20}
10622	2-Propenoic acid, 2-chloro-3-phenyl-, methyl ester, (*E*)- Cinnamic acid, α-chloro-, methyl ester, (*E*)-	$C_{10}H_9ClO_2$ 196.63	14737-94-1 33	$108\text{-}9^{0.5}$	4-09-00-02035	ace 4
10623	2-Propenoic acid, 3-chloro-3-phenyl-, methyl ester Cinnamic acid, β-chloro-, methyl ester	$C_{10}H_9ClO_2$ 196.63	87541-87-5 29	$113\text{-}4^{0.5}$	1.2248^{21}	ace 4 1.5781^{21}
10624	2-Propenoic acid, 2-cyano-, methyl ester Mecrylate	$C_5H_5NO_2$ 111.10	137-05-3	5667 $47\text{-}8^{2}$	4-02-00-02226 1.1012^{20}	1.4430
10625	2-Propenoic acid, 2-cyano-3-phenyl-, ethyl ester	$C_{12}H_{11}NO_2$ 201.22	2025-40-3 51	188^{15}	2-09-00-00640 1.1076^{25}	ace 4; chl 4 1.5033
10626	2-Propenoic acid, cyclohexyl ester	$C_9H_{14}O_2$ 154.21	3066-71-5	183; 88^{20}	4-06-00-00038 1.0275^{20}	H_2O 1; EtOH 5; eth 5; chl 3 1.4673^{20}
10627	2-Propenoic acid, 2,3-dibromo-, (*Z*)-	$C_3H_2Br_2O_2$ 229.86	24557-10-6 85.5	148^{6}	3-02-00-01249	H_2O 2; EtOH 3; eth 3; bz 2
10628	2-Propenoic acid, 3,3-dibromo- Acrylic acid, 3,3-dibromo-	$C_3H_2Br_2O_2$ 229.86	1578-21-8 86.3	245	4-02-00-01487	eth 4; EtOH 4
10629	2-Propenoic acid, 2,3-dibromo-3-phenyl-, (*Z*)-	$C_9H_6Br_2O_2$ 305.95	708-82-7 100	$124^{0.5}$	4-09-00-02043	H_2O 1; EtOH 3; eth 3; chl 3
10630	2-Propenoic acid, 3,3-dichloro- Acrylic acid, 3,3-dichloro-	$C_3H_2Cl_2O_2$ 140.95	1561-20-2 77.5	sub	4-02-00-01484	eth 4; chl 4
10631	2-Propenoic acid, 2-(diethylamino)ethyl ester	$C_9H_{17}NO_2$ 171.24	2426-54-2 <-60	81^{10}	4-04-00-01477 0.937^{20}	1.4376^{25}
10632	2-Propenoic acid, 3-(3,4-dihydroxyphenyl)- Caffeic acid	$C_9H_8O_4$ 180.16	331-39-5 225 dec	1634	2-10-00-00294	EtOH 4
10633	2-Propenoic acid, 2-(dimethylamino)ethyl ester	$C_7H_{13}NO_2$ 143.19	2439-35-2 <-60	95^{50}	4-04-00-01431 0.938^{20}	
10634	2-Propenoic acid, 2-ethoxyethyl ester	$C_7H_{12}O_3$ 144.17	106-74-1 -47	174	4-02-00-01469 0.983^{20}	1.4274^{20}
10635	2-Propenoic acid, ethyl ester Ethyl acrylate	$C_5H_8O_2$ 100.12	140-88-5 -71.2	3715 99.4	4-02-00-01460 0.9234^{20}	H_2O 2; EtOH 5; eth 5; chl 3 1.4068^{20}
10636	2-Propenoic acid, 2-ethylhexyl ester 2-Ethylhexyl acrylate	$C_{11}H_{20}O_2$ 184.28	103-11-7 -90	$123\text{-}7^{60}$	4-02-00-01467 0.880^{25}	1.4332^{25}
10637	2-Propenoic acid, 2-fluoro-3-phenyl- Cinnamic acid, α-fluoro-	$C_9H_7FO_2$ 166.15	350-90-3 157.6	290	1-09-00-00237	EtOH 3; eth 3
10638	2-Propenoic acid, 3-(2-furanyl)- 2-Furanacrylic acid	$C_7H_6O_3$ 138.12	539-47-9 141	4207 286	5-18-06-00306	eth 4; EtOH 4
10639	2-Propenoic acid, 3-(2-furanyl)-, ethyl ester	$C_9H_{10}O_3$ 166.18	623-20-1 24.5	232	5-18-06-00307 1.0891^{25}	H_2O 1; EtOH 5; eth 5; chl 2 1.5286^{20}
10640	2-Propenoic acid, 3-(2-furanyl)-, methyl ester	$C_8H_8O_3$ 152.15	623-18-7 36	227; 117^{23}	4-18-00-04144	bz 4; eth 4; EtOH 4 1.4447^{20}

No.	Name	Synonym	Mol. Form.	Mol. Wt.	CAS RN	mp/°C	Merck No.	bp/°C	Beil. Ref.	den/g cm^{-3}	Solubility	n_D
10641	2-Propenoic acid, 3-(2-furanyl)-, pentyl ester	Acrylic acid, 3-(2-furyl), pentyl ester	$C_{12}H_{16}O_3$	208.26	2438-19-9			116-8^{2}	4-18-00-04146	1.0322^{20}	eth 4; EtOH 4	1.5289^{14}
10642	2-Propenoic acid, 3-(2-furanyl)-, propyl ester	Acrylic acid, 3-(2-furyl), propyl ester	$C_{10}H_{12}O_3$	180.20	623-22-3			91-4^{3}	4-18-00-04146	1.0744^{20}	bz 4; eth 4; EtOH 4	1.5392^{24}
10643	2-Propenoic acid, hexyl ester		$C_9H_{16}O_2$	156.22	2499-95-8	-45		40^{1}	4-02-00-01466	0.878^{20}		
10644	2-Propenoic acid, 3-(4-hydroxyphenyl)-	*p*-Coumaric acid	$C_9H_8O_3$	164.16	7400-08-0	211.5	2561		4-10-00-01005		eth 4; EtOH 4	
10645	2-Propenoic acid, 2-isocyano-3-phenyl-, ethyl ester	Malonomononitrile, benzylidene,monoethyl ester	$C_{12}H_{11}NO_2$	201.22	52744-85-1	51		188^{15}; 118$^{0.4}$		1.0762^{25}	H_2O 1; EtOH 2; eth 3; bz 4	1.5033
10646	2-Propenoic acid, isodecyl ester		$C_{13}H_{24}O_2$	212.33	1330-61-6	-100		158^{50}		0.885^{20}		1.4416^{20}
10647	2-Propenoic acid, 3-(2-methoxyphenyl)-, methyl ester		$C_{11}H_{12}O_3$	192.21	15854-58-7			304; 161-3^{3}	3-10-00-00835	1.1366^{17}		1.5854^{16}
10648	2-Propenoic acid, 2-methyl-	Methacrylic acid	$C_4H_6O_2$	86.09	79-41-4	16	5849	162.5	4-02-00-01518	1.0153^{20}	H_2O 3; EtOH 5; eth 5; chl 3	1.4314^{20}
10649	2-Propenoic acid, 2-methyl-, anhydride	Methacrylic acid anhydride	$C_8H_{10}O_3$	154.17	760-93-0			89^{5}	4-02-00-01537		EtOH 5; eth 5	1.4540^{20}
10650	2-Propenoic acid, 2-methyl-, butyl ester	Butyl methacrylate	$C_8H_{14}O_2$	142.20	97-88-1			160	4-02-00-01525	0.8936^{20}	eth 4; EtOH 4	1.4240^{20}
10651	2-Propenoic acid, methylbutyl ester	Propenoic acid, 2-methyl,butyl ester	$C_8H_{14}O_2$	142.20	90638-48-5			160; 71^{37}	3-02-00-01286	0.8936^{20}	H_2O 1; EtOH 5; eth 5; ctc 2	1.4240^{20}
10652	2-Propenoic acid, 2-methylbutyl ester	Acrylic acid, (2-methylbutyl) ester	$C_8H_{14}O_2$	142.20	44914-03-6			160; 45^{10}	3-02-00-01228	0.8936^{20}	eth 4; EtOH 4	1.4240^{20}
10653	2-Propenoic acid, 2-methyl-, cyclohexyl ester		$C_{10}H_{16}O_2$	168.24	101-43-9			210	4-06-00-00039	0.9626^{20}		1.4578^{20}
10654	2-Propenoic acid, 2-methyl-, 2-(dimethylamino)ethyl ester	β-Dimethylaminoethyl methacrylate	$C_8H_{15}NO_2$	157.21	2867-47-2			62-5^{6}	4-04-00-01432			
10655	2-Propenoic acid, methyl ester	Methyl acrylate	$C_4H_6O_2$	86.09	96-33-3	<-75	5935	80.7	4-02-00-01457	0.9535^{20}	H_2O 2; EtOH 3; eth 3; ace 3	1.4040^{20}
10656	2-Propenoic acid, 2-methyl-, 1,2-ethanediyl ester		$C_{10}H_{14}O_4$	198.22	97-90-5	-40		260	4-02-00-01532	1.053^{20}	bz 4; EtOH 4; lig 4	1.4532^{25}
10657	2-Propenoic acid, 2-methyl-, 1,2-ethanediylbis(oxy-2,1-ethanediyl) ester		$C_{14}H_{22}O_6$	286.33	109-16-0				4-02-00-01531	1.072^{25}	ace 4; eth 4; EtOH 4; peth 4	1.4595^{25}
10658	2-Propenoic acid, 2-methyl-, ethyl ester	Ethyl methacrylate	$C_6H_{10}O_2$	114.14	97-63-2			117	4-02-00-01523	0.9135^{20}	H_2O 2; EtOH 5; eth 5; chl 2	1.4147^{20}
10659	2-Propenoic acid, 2-methyl-, 2-ethylhexyl ester	2-Ethylhexyl methacrylate	$C_{12}H_{22}O_2$	198.31	688-84-6			110^{14}	4-02-00-01528	0.880^{25}		1.436^{25}
10660	2-Propenoic acid, 2-methyl-, hexyl ester		$C_{10}H_{18}O_2$	170.25	142-09-6			162; 86^{17}	4-02-00-01527	0.880^{25}	ace 4; bz 4; eth 4; EtOH 4	1.429^{25}
10661	2-Propenoic acid, 2-methyl-, 2-hydroxyethyl ester		$C_6H_{10}O_3$	130.14	868-77-9			67$^{3.5}$	4-02-00-01530	1.034^{25}		1.4515^{20}
10662	2-Propenoic acid, 2-methyl-, 2-hydroxypropyl ester		$C_7H_{12}O_3$	144.17	923-26-2			90^{9}	4-02-00-01532	1.066^{25}		1.4458^{20}
10663	2-Propenoic acid, 2-methyl-, methyl ester	Methyl methacrylate	$C_5H_8O_2$	100.12	80-62-6	-48	5849	100.5	4-02-00-01519	0.9440^{20}	H_2O 2; EtOH 5; eth 5; ace 5	1.4142^{20}
10664	2-Propenoic acid, 2-methyl-, 1-methylethyl ester		$C_7H_{12}O_2$	128.17	4655-34-9			125	4-02-00-01525	0.8847^{20}	ace 4; bz 4; eth 4; EtOH 4	1.4122^{20}
10665	2-Propenoic acid, 2-methyl-, 1-methylpropyl ester		$C_8H_{14}O_2$	142.20	2998-18-7			147; 35^{4}	4-02-00-01526	0.886^{20}		1.4179^{20}
10666	2-Propenoic acid, 2-methyl-, 2-methylpropyl ester	Isobutyl methacrylate	$C_8H_{14}O_2$	142.20	97-86-9			155	4-02-00-01526	0.8858^{20}	H_2O 1; EtOH 5; eth 5	1.4199^{20}
10667	2-Propenoic acid, 2-methyl-, octadecyl ester		$C_{22}H_{42}O_2$	338.57	32360-05-7			195^{6}	4-02-00-01529	0.880^{25}		1.429^{25}
10668	2-Propenoic acid, 2-methyl-, oxiranylmethyl ester		$C_7H_{10}O_3$	142.15	106-91-2			75^{10}	5-17-03-00035	1.07^{25}	bz 4; eth 4; EtOH 4	1.448^{25}
10669	2-Propenoic acid, 2-methyl-, oxydi-2,1-ethanediyl ester		$C_{12}H_{18}O_5$	242.27	2358-84-1			>200	4-02-00-01531	1.056^{20}		1.4571^{25}
10670	2-Propenoic acid, 2-methyl-3-phenyl-, (*Z*)-		$C_{10}H_{10}O_2$	162.19	15250-29-0	94		288	4-09-00-02063		H_2O 2; EtOH 4; eth 4; bz 4	
10671	2-Propenoic acid, 2-methyl-, 2-propenyl ester		$C_7H_{10}O_2$	126.16	96-05-9			55^{30}; 60^{43}	4-02-00-01529	0.927^{20}		1.4360^{20}
10672	2-Propenoic acid, 2-methyl-, propyl ester	Propyl methacrylate	$C_7H_{12}O_2$	128.17	2210-28-8			141	4-02-00-01524	0.9022^{20}	H_2O 1; EtOH 5; eth 5	1.4190^{20}
10673	2-Propenoic acid, 2-methylpropyl ester	Isobutyl acrylate	$C_7H_{12}O_2$	128.17	106-63-8	-61		132	4-02-00-01465	0.8896^{20}	H_2O 2; EtOH 3; eth 3; MeOH 3	1.4150^{20}

No.	Name Synonym	Mol. Form. Mol. Wt.	CAS RN mp/°C	Merck No. bp/°C	Beil. Ref. den/g cm^{-3}	Solubility n_D
10674	2-Propenoic acid, 2-methyl-, (tetrahydro-2-furanyl)methyl ester	$C_9H_{14}O_3$ 170.21	2455-24-5	 265; 81^{4}	4-17-00-01105 1.040^{25}	 1.4554^{25}
10675	2-Propenoic acid, 3-(1-naphthalenyl)-, (*E*)- Acrylic acid, 3-(1-naphtyl) (*trans*)	$C_{13}H_{10}O_2$ 198.22	2006-14-6 211.5	 sub	4-09-00-02486	H_2O 2; EtOH 3; eth 3; chl 3
10676	2-Propenoic acid, 3-(2-nitrophenyl)-, (*E*)-	$C_9H_7NO_4$ 193.16	1013-96-3 244.3	 sub	4-09-00-02043	H_2O 1; EtOH 3; bz 1
10677	2-Propenoic acid, 3-(2-nitrophenyl)-, methyl ester, (*E*)- Cinnamic acid, 2-nitro, methyl ester (*trans*)	$C_{10}H_9NO_4$ 207.19	39228-29-0 73	 187^{15}	4-09-00-02044	EtOH 4
10678	2-Propenoic acid, 3-(3-nitrophenyl)-, methyl ester, (*E*)- Cinnamic acid, 3-nitro, methyl ester (*trans*)	$C_{10}H_9NO_4$ 207.19	659-04-1 125.3	 dec	3-09-00-02742	bz 4; eth 4; chl 4
10679	2-Propenoic acid, 3-(4-nitrophenyl)-, methyl ester, (*E*)-	$C_{10}H_9NO_4$ 207.19	637-57-0 162	 283.5	4-09-00-02046	EtOH 4
10680	2-Propenoic acid, oxiranylmethyl ester	$C_6H_8O_3$ 128.13	106-90-1	 53^{10}	4-17-00-01005 1.1^{20}	bz 4 1.4490^{20}
10681	2-Propenoic acid, 2-phenoxyethyl ester	$C_{11}H_{12}O_3$ 192.21	48145-04-6	 110^{2}	3-06-00-00572 1.090^{25}	ace 4; eth 4; chl 4
10682	2-Propenoic acid, 3-phenyl-, (*E*)- *trans*-Cinnamic acid	$C_9H_8O_2$ 148.16	140-10-3 133	2300 300	4-09-00-02002 1.2475^{4}	H_2O 1; EtOH 4; eth 3; ace 3
10683	2-Propenoic acid, 3-phenyl-, (*Z*)-	$C_9H_8O_2$ 148.16	102-94-3 42		4-09-00-02001	EtOH 4; HOAc 4; lig 4
10684	2-Propenoic acid, 3-phenyl-, anhydride Cinnamic anhydride	$C_{18}H_{14}O_3$ 278.31	538-56-7 136	2300		bz 4
10685	2-Propenoic acid, 3-phenyl-, ethyl ester, (*E*)-	$C_{11}H_{12}O_2$ 176.22	4192-77-2 10	2300 271.5	4-09-00-02006 1.0491^{20}	H_2O 1; EtOH 4; eth 4; ace 4 1.5598^{20}
10686	2-Propenoic acid, 3-phenyl-, methyl ester, (*E*)-	$C_{10}H_{10}O_2$ 162.19	1754-62-7 36.5	2300 261.9	4-09-00-02005 1.042^{36}	H_2O 1; EtOH 4; eth 4; ace 3 1.5766^{22}
10687	2-Propenoic acid, 3-phenyl-, 1-methylethyl ester, (*E*)- Cinnamic acid, isopropyl ester (*trans*)	$C_{12}H_{14}O_2$ 190.24	60512-85-8	 269	4-09-00-02008 1.0320^{20}	H_2O 1; EtOH 3; eth 3; ace 3 1.5455^{20}
10688	2-Propenoic acid, phenylmethyl ester	$C_{10}H_{10}O_2$ 162.19	2495-35-4	 228	3-06-00-01481 1.0573^{20}	H_2O 1; EtOH 3; eth 3; ace 3 1.5143^{20}
10689	2-Propenoic acid, 3-phenyl-, phenyl ester, (*E*)- Cinnamic acid, phenyl ester (*trans*)	$C_{15}H_{12}O_2$ 224.26	25695-77-6 74.3	 205^{15}	4-09-00-02011 1.0650^{100}	bz 4; eth 4
10690	2-Propenoic acid, 3-phenyl-, phenylmethyl ester, (*E*)- Cinnamic acid, benzyl ester (*trans*)	$C_{16}H_{14}O_2$ 238.29	78277-23-3 39	 350 dec; 244^{5}	4-09-00-02012 1.109^{15}	H_2O 1; EtOH 3; eth 3; bz 2
10691	2-Propenoic acid, 3-phenyl-, 3-phenyl-2-propenyl ester Cinnamyl cinnamate	$C_{18}H_{16}O_2$ 264.32	122-69-0 44	2307	4-09-00-02013 1.1565^{4}	H_2O 1; EtOH 3; eth 4; chl 3
10692	2-Propenoic acid, 3-phenyl-, 2-propenyl ester Cinnamic acid, allyl ester	$C_{12}H_{12}O_2$ 188.23	1866-31-5	 268 dec; 163^{17}	4-09-00-02010 1.048^{23}	H_2O 1; EtOH 4; eth 5; ctc 2 1.530^{20}
10693	2-Propenoic acid, 3-phenyl-, propyl ester, (*E*)- Cinnamic acid, propyl ester (*trans*)	$C_{12}H_{14}O_2$ 190.24	74513-58-9	 285	4-09-00-02007 1.0433^{0}	H_2O 1
10694	2-Propenoic acid, 2-propenyl ester	$C_6H_8O_2$ 112.13	999-55-3	 121	4-02-00-01468 0.9441^{20}	H_2O 2; EtOH 3; eth 3; acid 3 1.4320^{20}
10695	2-Propenoic acid, (tetrahydro-2-furanyl)methyl ester	$C_8H_{12}O_3$ 156.18	2399-48-6 <-60	 96^{6}	4-17-00-01104 1.061^{20}	
10696	2-Propenoic acid, 2,3,3-trichloro-	$C_3HCl_3O_2$ 175.40	2257-35-4 76	 222	4-02-00-01486	H_2O 3; EtOH 3; eth 3; ctc 3 1.5271^{18}
10697	2-Propen-1-ol Allyl alcohol	C_3H_6O 58.08	107-18-6 -129	284 97.0	4-01-00-02079 0.8540^{20}	H_2O 5; EtOH 5; eth 5; chl 3 1.4135^{20}
10698	1-Propen-2-ol, acetate Isopropenyl acetate	$C_5H_8O_2$ 100.12	108-22-5 -92.9	5091 94	4-02-00-00179 0.9090^{20}	H_2O 2; EtOH 3; eth 4; ace 3 1.4033^{20}
10699	2-Propen-1-ol, 2-bromo-	C_3H_5BrO 136.98	598-19-6	 154; 62^{11}	4-01-00-02094 1.621^{18}	eth 4; chl 4 1.500^{18}
10700	2-Propen-1-ol, 2-chloro-	C_3H_5ClO 92.52	5976-47-6	 135	4-01-00-02091 1.1618^{20}	EtOH 3; ctc 2 1.4588^{20}
10701	2-Propen-1-ol, 3-chloro- Allyl alcohol, 3-chloro	C_3H_5ClO 92.52	29560-84-7	 147	4-01-00-02090 1.1769^{20}	 1.4738^{20}
10702	2-Propen-1-ol, 2-methyl- Methallyl alcohol	C_4H_8O 72.11	513-42-8	 114.5	4-01-00-02114 0.8515^{20}	H_2O 4; EtOH 5; eth 5 1.4255^{20}
10703	2-Propen-1-ol, 2-methyl-, carbonate (2:1) Bis-(2-methallyl)carbonate	$C_9H_{14}O_3$ 170.21	64057-79-0 201.3	 66^{3}	4-03-00-00011 0.943^{25}	 1.4371^{20}
10704	2-Propen-1-ol, 3-phenyl-, (*E*)-	$C_9H_{10}O$ 134.18	4407-36-7 34	 257.5	4-06-00-03799 1.0440^{20}	H_2O 2; EtOH 4; eth 4; chl 2 1.5819^{20}
10705	2-Propen-1-ol, 3-phenyl-, (*Z*)-	$C_9H_{10}O$ 134.18	4510-34-3 34	 257.5	3-06-00-02401 1.0440^{20}	eth 4; EtOH 4 1.5819^{20}
10706	2-Propen-1-ol, 3-phenyl-, acetate, (*E*)-	$C_{11}H_{12}O_2$ 176.22	21040-45-9	 145^{15}; 114^{1}	4-06-00-03801 1.0567^{20}	H_2O 1; EtOH 3; eth 3; ace 3 1.5425^{20}
10707	2-Propen-1-ol, 3-phenyl-, formate	$C_{10}H_{10}O_2$ 162.19	104-65-4 0	 252	0-06-00-00571 1.086^{25}	
10708	2-Propen-1-one, 1,3-diphenyl- Chalcone	$C_{15}H_{12}O$ 208.26	94-41-7 59	2028 346 dec	4-07-00-01658 1.0712^{62}	bz 4; EtOH 4; chl 4

No.	Name	Synonym	Mol. Form.	Mol. Wt.	CAS RN	mp/°C	Merck No.	bp/°C	Beil. Ref.	den/g cm^{-3}	Solubility	n_D
10709	2-Propen-1-one, 1,3-diphenyl-, (*E*)-		$C_{15}H_{12}O$	208.26	614-47-1	59		346 dec	4-07-00-01658	1.0712^{62}	H_2O 1; EtOH 2; eth 3; bz 3	
10710	2-Propen-1-one, 3-(2-furanyl)-1-phenyl-		$C_{13}H_{10}O_2$	198.22	717-21-5	47		317	5-17-10-00371	1.1140^{20}	EtOH 3; eth 3	
10711	2-Propen-1-one, 3-hydroxy-1,3-diphenyl-		$C_{15}H_{12}O_2$	224.26	1704-15-0	79.8		220^{18}, $140^{1.0}$	4-07-00-02512		eth 4; EtOH 4; chl 4	
10712	2-Propen-1-one, 3-(3-methoxyphenyl)-1-phenyl-		$C_{16}H_{14}O_2$	238.29	5470-91-7	65		247^{12}	1-08-00-00579		ace 4; bz 4; eth 4; EtOH 4	
10713	2-Propen-1-one, 3-(4-methoxyphenyl)-1-phenyl-		$C_{16}H_{14}O_2$	238.29	959-33-1	79		187^{19}	2-08-00-00218		H_2O 1; EtOH 4; eth 3; ctc 3	
10714	2-Propenoyl chloride	Acrylic acid chloride	C_3H_3ClO	90.51	814-68-6			75.5	4-02-00-01471	1.1136^{20}	chl 4	1.4343^{20}
10715	2-Propenoyl chloride, 2-methyl-	Methacrylic acid chloride	C_4H_5ClO	104.54	920-46-7	-60		96	4-02-00-01537	1.0871^{20}	eth 3; ace 3; chl 3	1.4435^{20}
10716	2-Propenoyl chloride, 3-phenyl-	Cinnamoyl chloride	C_9H_7ClO	166.61	102-92-1	36	2303	257	2-09-00-00395	1.1617^{45}	lig 4	1.614^{42}
10717	2-Propenoyl chloride, 3-phenyl-, (*E*)-		C_9H_7ClO	166.61	17082-09-6	37.5		257.5	2-09-00-00390	1.1617^{45}	H_2O 1; EtOH 3; ctc 3; lig 3	1.614^{42}
10718	2-Propenoyl chloride, 2,3,3-trichloro-		C_3Cl_4O	193.84	815-58-7			158	4-02-00-01486		bz 4	1.5271^{18}
10719	Propetamphos	2-Butenoic acid, 3-[[(ethylamino)methoxyphosphinothioyl]oxy]-, 1-methylethyl ester, (*E*)-	$C_{10}H_{20}NO_4PS$	281.31	31218-83-4		7827	$88^{0.005}$		1.1294^{20}		
10720	Propiconazole	1-[[2-(2,4-Dichlorophenyl)-4-propyl-1,3-dioxolan-2-yl]methyl]-1H-1,2,4-triazole	$C_{15}H_{17}Cl_2N_3O_2$	342.22	60207-90-1			$180^{0.1}$		1.27^{20}		
10721	Propionic acid, 2,2-dichloro-3,3,3-trifluoro-, propyl ester	Propanoic acid, 2,2-dichloro-3,3,3-trifluoro, propyl ester	$C_6H_7Cl_2F_3O_2$	239.02	357-49-3			144.5^{625}	4-02-00-00758	1.3531^{20}		1.3888^{20}
10722	Propoxur	Phenol, 2-(1-methylethoxy)-, methylcarbamate	$C_{11}H_{15}NO_3$	209.25	114-26-1	87	7849	dec		1.12^{20}		
10723	2-Propynal	Propargyl aldehyde	C_3H_2O	54.05	624-67-9			60	4-01-00-03537	0.9152^{20}	H_2O 5; EtOH 3; eth 3; ace 3	1.4033^{25}
10724	2-Propynal, 3-phenyl-		C_9H_6O	130.15	2579-22-8			127^{28}, 104^{11}	4-07-00-01173	1.0680^{12}		1.6079^{12}
10725	2-Propyn-1-amine		C_3H_5N	55.08	2450-71-7			83	4-04-00-01135	0.803^{25}		1.4480^{20}
10726	2-Propyn-1-amine, *N*-methyl-		C_4H_7N	69.11	35161-71-8			83	0-04-00-00228	0.819^{25}		1.4332^{20}
10727	1-Propyne	Methylacetylene	C_3H_4	40.06	74-99-7	-102.7		-23.2	4-01-00-00958	*0.607^{25}	H_2O 2; EtOH 4; bz 3; chl 3	1.3863^{-40}
10728	1-Propyne, 3-bromo-	Propargyl bromide	C_3H_3Br	118.96	106-96-7			89	4-01-00-00964	1.579^{19}	EtOH 3; eth 3; bz 3; ctc 3	1.4922^{20}
10729	1-Propyne, 3-chloro-	Propargyl chloride	C_3H_3Cl	74.51	624-65-7		7820	58	4-01-00-00963	1.030^{25}	H_2O 1; EtOH 5; eth 5; bz 3	1.4349^{20}
10730	1-Propyne, 1,3-dibromo-	Propyne, 1,3-dibromo-	$C_3H_2Br_2$	197.86	627-16-7			$73\text{-}4^{30}$	4-01-00-00965	2.1894^{20}	eth 4; chl 4	1.5690^{20}
10731	1-Propyne, 3,3-diethoxy-		$C_7H_{12}O_2$	128.17	10160-87-9			139	4-01-00-03538	0.8942^{22}	ace 4; eth 4; EtOH 4; chl 4	1.4140^{20}
10732	1-Propyne, 1-ethoxy-		C_5H_8O	84.12	14273-06-4			84	3-01-00-01969	0.8276^{20}	eth 4; EtOH 4	1.4039^{20}
10733	1-Propyne, 3-ethoxy-		C_5H_8O	84.12	628-33-1			82	4-01-00-02215	0.8326^{20}	eth 4; EtOH 4	1.4039^{20}
10734	1-Propyne, 1-iodo-		C_3H_3I	165.96	624-66-8			110	4-01-00-00965	2.08^{22}	eth 4; EtOH 4	
10735	1-Propyne, 3-iodo-		C_3H_3I	165.96	659-86-9			115	4-01-00-00965	2.0177^{25}	eth 3	
10736	1-Propyne, 3-methoxy-		C_4H_6O	70.09	627-41-8			63	4-01-00-02215	0.83^{12}	eth 4; EtOH 4	1.5035^{20}
10737	2-Propynenitrile	Cyanoacetylene	C_3HN	51.05	1070-71-9	5		42.5	4-02-00-01689	0.8167^{17}	H_2O 2; EtOH 3	1.3868^{25}
10738	2-Propynoic acid	Propiolic acid	$C_3H_2O_2$	70.05	471-25-0	9	7833	144 dec; 72^{50}	4-02-00-01687	1.1380^{20}	H_2O 4; eth 4; EtOH 4; chl 4	1.4306^{20}
10739	2-Propynoic acid, ethyl ester		$C_5H_6O_2$	98.10	623-47-2			120	4-02-00-01688	0.9560^{25}	H_2O 1; EtOH 4; eth 4; chl 4	1.4105^{20}
10740	2-Propynoic acid, 3-(2-nitrophenyl)-	*o*-Nitrophenylpropiolic acid	$C_9H_5NO_4$	191.14	530-85-8	157 dec	6546	exp	4-09-00-02330		eth 4; EtOH 4	
10741	2-Propynoic acid, 3-phenyl-, ethyl ester		$C_{11}H_{10}O_2$	174.20	2216-94-6			265; $128^{1.6}$	4-09-00-02328	1.055^{25}	eth 3	1.5520^{20}
10742	2-Propynoic acid, 3-phenyl-, methyl ester		$C_{10}H_8O_2$	160.17	4891-38-7	26		158^{48}, 132^{16}	4-09-00-02328	1.0830^{25}		1.5618^{25}

No.	Name Synonym	Mol. Form. Mol. Wt.	CAS RN mp/°C	Merck No. bp/°C	Beil. Ref. den/g cm^{-3}	Solubility n_D
10743	2-Propyn-1-ol Propargyl alcohol	C_3H_4O 56.06	107-19-7 -51.8	7819 113.6	4-01-00-02214 0.9478^{20}	H_2O 3; EtOH 5; eth 5; chl 3 1.4322^{20}
10744	2-Propyn-1-ol, acetate Propargyl acetate	$C_5H_6O_2$ 98.10	627-09-8	 121.5	4-02-00-00197 0.9982^{20}	H_2O 2; EtOH 3; eth 3 1.4187^{20}
10745	2-Propyn-1-ol, 3-phenyl-	C_9H_8O 132.16	1504-58-1	 $137\text{-}8^{15}$	4-06-00-04065 1.07^{18}	eth 3; ace 3; bz 3 1.5873^{28}
10746	Propyzamide Benzamide, 3,5-dichloro-*N*-(1,1-dimethyl-2-propynyl)-	$C_{12}H_{11}Cl_2NO$ 256.13	23950-58-5 155	7886		
10747	Pteridine Pyrazino[2,3-d]pyrimidine	$C_6H_4N_4$ 132.12	91-18-9 139.5	7947 sub 125	4-26-00-01770	H_2O 4; EtOH 3; eth 2; bz 2
10748	2,4(1H,3H)-Pteridinedione Lumazine	$C_6H_4N_4O_2$ 164.12	487-21-8 348.5	5467	4-26-00-02489	HOAc 4
10749	4,6-Pteridinedione, 2-amino-1,5-dihydro- Xanthopterin	$C_6H_5N_5O_2$ 179.14	119-44-8 >410 dec	9973 $98\text{-}100^{18}$	4-26-00-04000 1.559^{25}	H_2O 1; EtOH 2; eth 2; acid 4
10750	2,4,7-Pteridinetriamine, 6-phenyl- Triamterene	$C_{12}H_{11}N_7$ 253.27	396-01-0	9515	4-26-00-03837	
10751	1H-Purin-6-amine Adenine	$C_5H_5N_5$ 135.13	73-24-5 360 dec	141 sub 220	4-26-00-03561	H_2O 3; EtOH 2; eth 1; chl 1
10752	1H-Purine 7H-Imidazo[4,5-d]pyrimidine	$C_5H_4N_4$ 120.11	120-73-0 216.5	7959	4-26-00-01736	H_2O 4; EtOH 4; eth 2; ace 3
10753	1H-Purine, 6-chloro- 6-Chloropurine	$C_5H_3ClN_4$ 154.56	87-42-3 175-7 dec	2161	4-26-00-01742	
10754	1H-Purine-2,6-diamine 2,6-Diaminopurine	$C_5H_6N_6$ 150.14	1904-98-9 302	2963	4-26-00-03785	
10755	1H-Purine-2,6-dione, 3,7-dihydro- Xanthine	$C_5H_4N_4O_2$ 152.11	69-89-6 dec	9968 sub	4-26-00-02327	
10756	1H-Purine-2,6-dione, 3,7-dihydro-1,3-dimethyl- Theophylline	$C_7H_8N_4O_2$ 180.17	58-55-9 273	9212	4-26-00-02331	H_2O 4; EtOH 2; eth 2; chl 2
10757	1H-Purine-2,6-dione, 3,7-dihydro-3,7-dimethyl- Theobromine	$C_7H_8N_4O_2$ 180.17	83-67-0 357	9209 sub 290	4-26-00-02336	H_2O 2; EtOH 2; eth 1; bz 1
10758	1H-Purine-2,6-dione, 3,7-dihydro-7-(2-hydroxyethyl)-1,3-dimethyl- Etofylline	$C_9H_{12}N_4O_3$ 224.22	519-37-9 158	3835	4-26-00-02357	H_2O 4; EtOH 3; eth 2; bz 2
10759	1H-Purine-2,6-dione, 3,7-dihydro-8-methoxy-1,3,7-trimethyl-	$C_9H_{12}N_4O_3$ 224.22	569-34-6 176		4-26-00-02727 1.399^{25}	H_2O 3; EtOH 4; bz 3
10760	1H-Purine-2,6-dione, 3,7-dihydro-1,3,7-trimethyl- Caffeine	$C_8H_{10}N_4O_2$ 194.19	58-08-2 238	1635 sub 90	4-26-00-02338 1.23^{19}	H_2O 2; EtOH 2; eth 1; ace 2
10761	1H-Purine-2,6-dione, 7-(2,3-dihydroxypropyl)-3,7-dihydro-1,3-dimethyl- Dyphylline	$C_{10}H_{14}N_4O_4$ 254.25	479-18-5 161.5	3459	4-26-00-02370	
10762	6H-Purine-6-thione, 2-amino-1,7-dihydro- Thioguanine	$C_5H_5N_5S$ 167.19	154-42-7 >360	9266	4-26-00-03926	
10763	6H-Purine-6-thione, 1,7-dihydro- 6-Mercaptopurine	$C_5H_4N_4S$ 152.18	50-44-2 313 dec	5762		H_2O 1; alk 3
10764	1H-Purine-2,6,8(3H)-trione, 7,9-dihydro- Uric acid	$C_5H_4N_4O_3$ 168.11	69-93-2 dec	9790 dec	4-26-00-02619 1.89^{25}	H_2O 1; EtOH 1; eth 1; alk 4
10765	1H-Purine-2,6,8(3H)-trione, 4,9-dihydro-3-methyl- 3-Methyluric acid	$C_6H_6N_4O_3$ 182.14	605-99-2 >350		4-26-00-02621 1.6104^{25}	H_2O 2; EtOH 2; alk 3 1.6334^{25}
10766	1H-Purine-2,6,8(3H)-trione, 7,9-dihydro-7-methyl- 7-Methyluric acid	$C_6H_6N_4O_3$ 182.14	612-37-3 370-80 dec		4-26-00-02622 1.706^{25}	H_2O 2; NaOH 3
10767	2H-Purin-2-one, 6-amino-1,3-dihydro- Isoguanine	$C_5H_5N_5O$ 151.13	3373-53-3 >360			
10768	6H-Purin-6-one, 2-amino-1,7-dihydro- Guanine	$C_5H_5N_5O$ 151.13	73-40-5 360 dec	4477 sub	4-26-00-03890	H_2O 1; EtOH 2; eth 2; alk 3
10769	6H-Purin-6-one, 2-amino-1,7-dihydro-7-methyl 7-Methylguanine	$C_6H_7N_5O$ 165.15	578-76-7 370			
10770	6H-Purin-6-one, 1,7-dihydro- Hypoxanthine	$C_5H_4N_4O$ 136.11	68-94-0 150 dec	4805	4-26-00-02081	H_2O 2; alk 3; dil acid 3
10771	Pyran	C_5H_6O 82.10	33941-07-0	 80	0-17-00-00036	EtOH 3; eth 3; bz 3 1.4559^{20}
10772	2H-Pyran-5-carboxylic acid, 3,4-dihydro-3-(hydroxymethyl)-2-methyl-4-[(9-methyl-9H-pyrido[3,4-b]indol-1-yl)methyl]-, methyl ester, [2S-(2α,3β,4β)]- Alstonidine	$C_{22}H_{24}N_2O_4$ 380.44	25394-75-6 189	313	4-27-00-07962	ace 4; EtOH 4
10773	2H-Pyran-5-carboxylic acid, 2-oxo- Coumalic acid	$C_6H_4O_4$ 140.10	500-05-0 205-10 dec	2558 218^{120}	5-18-08-00120	H_2O 2; EtOH 3; eth 2; ace 2
10774	2H-Pyran-4-carboxylic acid, tetrahydro- Pyran-4-carboxylic acid, tetrahydro-	$C_6H_{10}O_3$ 130.14	5337-03-1 88	 144^{11}	5-18-06-00023	H_2O 3; peth 2
10775	4H-Pyran-2,6-dicarboxylic acid, 3-hydroxy-4-oxo- Meconic acid	$C_7H_4O_7$ 200.10	497-59-6 120 dec	5664 120 dec	5-18-09-00319	H_2O 2; EtOH 3; eth 2; ace 2
10776	4H-Pyran-2,6-dicarboxylic acid, 4-oxo- Chelidonic acid	$C_7H_4O_6$ 184.11	99-32-1 262	2042	5-18-08-00646	H_2O 3; EtOH 2

No.	Name	Synonym	Mol. Form.	Mol. Wt.	CAS RN	mp/°C	Merck No.	bp/°C	Beil. Ref.	den/g cm^{-3}	Solubility	n_D
10777	2H-Pyran, 3,4-dihydro-	3,4-Dihydro-2H-pyran	C_5H_8O	84.12	110-87-2			86	5-17-01-00181	0.921^{19}	H_2O 3; EtOH 3; chl 2	1.4402^{19}
10778	2H-Pyran, 3,6-dihydro-4-methyl-		$C_6H_{10}O$	98.14	16302-35-5			117.5	5-17-01-00194	0.912^{25}		1.4495^{20}
10779	2H-Pyran-2,4(3H)-dione, 3-acetyl-6-methyl-	Dehydroacetic acid	$C_8H_8O_4$	168.15	520-45-6	109	2855	270	4-17-00-06699		H_2O 4; EtOH 2; eth 4; chl 2	
10780	2H-Pyran-2,6(3H)-dione, dihydro-		$C_5H_6O_3$	114.10	108-55-4	56.3		158^{15}	5-17-11-00009	1.4110^{20}		
10781	2H-Pyran-2,6(3H)-dione, dihydro-3-phenyl-		$C_{11}H_{10}O_3$	190.20	2959-96-8	95		224^{13}	5-17-11-00279		EtOH 4	
10782	2H-Pyran-2,6(3H)-dione, dihydro-4-phenyl-		$C_{11}H_{10}O_3$	190.20	4160-80-9	105.6		217^{15}	5-17-11-00279		bz 4; eth 4; chl 4	
10783	2H-Pyran, 2-ethoxy-3,4-dihydro-		$C_7H_{12}O_2$	128.17	103-75-3			42^{16}	5-17-03-00196	0.957^{25}		1.4394^{20}
10784	2H-Pyran-2-methanol, tetrahydro-		$C_6H_{12}O_2$	116.16	100-72-1			185	5-17-03-00136	1.027^{25}		1.458^{20}
10785	4H,5H-Pyrano[4,3-b]pyran-4,5-dione, 2,3-dihydro-3-hydroxy-2-methyl-7-(1-propenyl)-, [2*S*-[2α,3β,7(*E*)]]-	Radicinin	$C_{12}H_{12}O_5$	236.22	10088-95-6	221.5	8117		4-19-00-02746		chl 2	
10786	Pyrano[3,2-c][2]benzopyran-6(2H)-one, 3,4,4a,10b-tetrahydro-3,4,8,10-tetrahydroxy-2-(hydroxymethyl)-9-methoxy-[2R-(2α,3β,4α,4aα,10bβ)]-	Bergenin	$C_{14}H_{16}O_9$	328.28	477-90-7	238	1174		5-19-07-00132		H_2O 4; EtOH 4	
10787	2H-Pyran-4-ol, tetrahydro-		$C_5H_{10}O_2$	102.13	2081-44-9			87^{15}	5-17-03-00111	1.071^{25}		1.4600^{20}
10788	2H-Pyran-2-one		$C_5H_4O_2$	96.09	504-31-4	8.5		207.5	5-17-09-00288	1.200^{20}	H_2O 5; ace 4	1.5270^{25}
10789	4H-Pyran-4-one		$C_5H_4O_2$	96.09	108-97-4	32.5		212.5	5-17-09-00290	1.190^{25}	H_2O 4; EtOH 3; eth 4; bz 3	1.5238
10790	2H-Pyran-2-one, 6-[2-(1,3-benzodioxol-5-yl)ethenyl]-5,6-dihydro-4-methoxy-, [*R*-(*E*)]-	Methysticin	$C_{15}H_{14}O_5$	274.27	495-85-2	137	6056		5-19-10-00456			
10791	2H-Pyran-2-one, 5,6-dihydro-6-methyl-, (*S*)-	Parasorbic acid	$C_6H_8O_2$	112.13	10048-32-5		6981	100^{15}	5-17-09-00130	1.079^{18}	H_2O 4; eth 4; EtOH 4	1.4736^{20}
10792	2H-Pyran-2-one, 4,6-dimethyl-		$C_7H_8O_2$	124.14	675-09-2	51.5		245	5-17-09-00409		H_2O 4; eth 4; EtOH 4	
10793	4H-Pyran-4-one, 2,6-dimethyl-		$C_7H_8O_2$	124.14	1004-36-0	132		251; 140^{25}	5-17-09-00410	0.9953^{137}	H_2O 3; EtOH 3; eth 3; ace 3	
10794	2H-Pyran-2-one, 3-hydroxy-		$C_5H_4O_3$	112.08	496-64-0	93.8		112^{20}; 102^{15}	5-18-01-00095		H_2O 3; EtOH 3; eth 3; bz 3	
10795	4H-Pyran-4-one, 5-hydroxy-2-(hydroxymethyl)-	Kojic acid	$C_6H_6O_4$	142.11	501-30-4	153.5	5197		5-18-02-00516		H_2O 2; EtOH 3; eth 3; ace 3	
10796	4H-Pyran-4-one, 3-hydroxy-2-methyl-	Maltol	$C_6H_6O_3$	126.11	118-71-8	161.5	5594	sub 93	5-18-01-00114		H_2O 2; eth 2; bz 2; chl 4	
10797	4H-Pyran-4-one, 2-methoxy-3,5-dimethyl-6-[tetrahydro-4-[2-methyl-3-(4-nitrophenyl)-2-propenylidene]-2-furanyl]	Aureothin	$C_{22}H_{23}NO_6$	397.43	2825-00-5	158	897		5-19-06-00099		ace 4; EtOH 4; chl 4	
10798	2H-Pyran-2-one, tetrahydro-		$C_5H_8O_2$	100.12	542-28-9	-12.5		219	5-17-09-00017	1.0794^{20}	H_2O 3; EtOH 5; eth 5; ctc 2	1.4503^{20}
10799	4H-Pyran-4-one, tetrahydro-		$C_5H_8O_2$	100.12	29943-42-8			166.5	5-17-09-00021	1.084^{25}		1.4520^{20}
10800	2H-Pyran, tetrahydro-	Oxane	$C_5H_{10}O$	86.13	142-68-7	-45	9149	88	5-17-01-00064	0.8814^{20}	EtOH 3; eth 3; bz 3; ctc 3	1.4200^{20}
10801	Pyrazine	1,4-Diazine	$C_4H_4N_2$	80.09	290-37-9	55	7971	115	5-23-05-00351	1.0311^{61}	H_2O 3; EtOH 3; eth 3; ace 3	1.4953^{61}
10802	Pyrazinecarboxamide	Pyrazinamide	$C_5H_5N_3O$	123.11	98-96-4	192	7970	sub	4-25-00-00772		H_2O 3; EtOH 3	
10803	Pyrazinecarboxylic acid	Pyrazinoic acid	$C_5H_4N_2O_2$	124.10	98-97-5	225 dec	7973	sub	4-25-00-00771			
10804	2,3-Pyrazinedicarboxylic acid	2,3-Dicarboxypyrazine	$C_6H_4N_2O_4$	168.11	89-01-0	193 dec	7972		4-25-00-01064		H_2O 4; EtOH 2; eth 2; ace 3	
10805	Pyrazine, 2,3-dimethyl-	2,3-Dimethylpyrazine	$C_6H_8N_2$	108.14	5910-89-4			156	5-23-05-00402	1.0281^{0}	H_2O 3; EtOH 3; eth 3	
10806	Pyrazine, 2,5-dimethyl-	2,5-Dimethylpyrazine	$C_6H_8N_2$	108.14	123-32-0	15		155	5-23-05-00403	0.9887^{20}	H_2O 5; EtOH 5; eth 5; ace 3	1.4980^{20}
10807	Pyrazine, 2,6-dimethyl-	2,6-Dimethylpyrazine	$C_6H_8N_2$	108.14	108-50-9	47.5		155.6	5-23-05-00406	0.9647^{50}	H_2O 3; EtOH 3; eth 3; ctc 2	
10808	Pyrazineethanol		$C_6H_8N_2O$	124.14	6705-31-3			129^{10}	5-23-11-00173	1.163^{20}		1.5378^{20}
10809	Pyrazine, 3-ethyl-2,5-dimethyl-		$C_8H_{12}N_2$	136.20	13360-65-1			180.5	5-23-05-00440	0.9657^{24}	H_2O 2; EtOH 2; eth 2	1.5014^{24}
10810	Pyrazine, methyl-	2-Methylpyrazine	$C_5H_6N_2$	94.12	109-08-0	-29		137	5-23-05-00386	1.03^{20}	H_2O 5; EtOH 5; eth 5; ace 3	1.5042^{20}
10811	2(1H)-Pyrazinone, 1-hydroxy-6-(1-methylpropyl)-3-(2-methylpropyl)-, (+)-	Aspergillic acid	$C_{12}H_{20}N_2O_2$	224.30	490-02-8	98	863		4-24-00-00235		bz 4; eth 4; EtOH 4	

No.	Name Synonym	Mol. Form. Mol. Wt.	CAS RN mp/°C	Merck No. bp/°C	Beil. Ref. den/g cm^{-3}	Solubility n_D
10812	1H-Pyrazol-4-amine 4-Aminopyrazole	$C_3H_5N_3$ 83.09	28466-26-4 81	 sub	4-25-00-02029	H_2O 4; EtOH 4; chl 4
10813	1H-Pyrazol-4-amine, 3-methyl-1-phenyl- Pyrazole, 4-amino-3-methyl-1-phenyl-	$C_{10}H_{11}N_3$ 173.22	103095-51-8 88.8	 314.5	4-25-00-02033	bz 4; eth 4; EtOH 4; chl 4
10814	1H-Pyrazol-5-amine, 3-methyl-1-phenyl-	$C_{10}H_{11}N_3$ 173.22	1131-18-6 116	 333	4-25-00-02034	H_2O 3; EtOH 3; bz 2; chl 3
10815	1H-Pyrazole 1,2-Diazole	$C_3H_4N_2$ 68.08	288-13-1 68	7974 187	5-23-04-00122	H_2O 3; EtOH 3; eth 3; bz 3 1.4203
10816	1H-Pyrazole, 4-bromo-1,3-dimethyl- 4-Bromo-1,3-dimethylpyrazole	$C_5H_7BrN_2$ 175.03	5775-82-6	 $76-7^{10}$	5-23-05-00022 1.4975^{15}	ace 4; eth 4; EtOH 4 1.5214^{15}
10817	1H-Pyrazole, 4-bromo-3-methyl- 4-Bromo-3-methylpyrazole	$C_4H_5BrN_2$ 161.00	13808-64-5 76.5		5-23-05-00022 1.5638^{100}	ace 4; eth 4; EtOH 4 1.5182^{100}
10818	1H-Pyrazole, 3-chloro-1,5-dimethyl- 3-Chloro-1,5-dimethylpyrazole	$C_5H_7ClN_2$ 130.58	51500-32-4 47.5	 211	2-23-00-00049 1.0823^{100}	ace 4; bz 4; eth 4; EtOH 4 1.4648^{100}
10819	1H-Pyrazole, 4-chloro-3,5-dimethyl-	$C_5H_7ClN_2$ 130.58	15953-73-8 117.5	 221	5-23-05-00151	ace 4; bz 4; eth 4; EtOH 4
10820	1H-Pyrazole, 5-chloro-1,3-dimethyl- 5-Chloro-1,3-dimethylpyrazole	$C_5H_7ClN_2$ 130.58	54454-10-3	 157.5	5-23-05-00019 1.1367^{18}	H_2O 4 1.4877^{18}
10821	1H-Pyrazole, 3-chloro-5-methyl-	$C_4H_5ClN_2$ 116.55	15953-45-4 118.5	 258; 138^{15}	5-23-05-00019	H_2O 4
10822	1H-Pyrazole, 4,5-dihydro- 2-Pyrazoline	$C_3H_6N_2$ 70.09	109-98-8	7975 144	5-23-03-00361 1.0200^{17}	H_2O 4; eth 4; EtOH 4 1.4796^{17}
10823	1H-Pyrazole, 4,5-dihydro-1-phenyl- 1-Phenyl-2-pyrazoline	$C_9H_{10}N_2$ 146.19	936-53-8 52	 273; 151^{17}	5-23-03-00363 1.0689^{58}	H_2O 1; EtOH 3; eth 3; bz 3 1.6015^{58}
10824	1H-Pyrazole, 4,5-dihydro-3-phenyl- 2-Pyrazoline, 3-phenyl	$C_9H_{10}N_2$ 146.19	936-48-1 45.8	 164^{17}	5-23-06-00396 1.0892^{54}	
10825	1H-Pyrazole, 1,3-dimethyl-	$C_5H_8N_2$ 96.13	694-48-4	 137	5-23-05-00012 0.9561^{17}	H_2O 4 1.4734^{15}
10826	1H-Pyrazole, 1,5-dimethyl-	$C_5H_8N_2$ 96.13	694-31-5	 149	5-23-05-00006 0.9813^{17}	H_2O 4; EtOH 4; eth 4 1.4782^{16}
10827	1H-Pyrazole, 3,4-dimethyl-	$C_5H_8N_2$ 96.13	2820-37-3 58	 110^{11}	5-23-05-00108 0.9325^{99}	EtOH 3; eth 3; ace 3; bz 3
10828	1H-Pyrazole, 3,5-dimethyl-	$C_5H_8N_2$ 96.13	67-51-6 107.5	 218	5-23-05-00110 0.8839^{16}	H_2O 3; EtOH 4; eth 4; ace 3
10829	1H-Pyrazole, 3,5-dimethyl-1-phenyl-	$C_{11}H_{12}N_2$ 172.23	1131-16-4	 273; $145^{12.5}$	5-23-05-00127 1.0566^{20}	eth 4; EtOH 4; chl 4 1.5738^{19}
10830	1H-Pyrazole, 1,3-diphenyl- 1,3-Diphenylpyrazole	$C_{15}H_{12}N_2$ 220.27	4492-01-7 85.8	 341^{270}	5-23-07-00178 1.0794^{101}	
10831	1H-Pyrazole, 1,5-diphenyl- 1,5-Diphenylpyrazole	$C_{15}H_{12}N_2$ 220.27	6831-89-6 55.5	 338; 167^{12}	5-23-07-00176 1.0696^{100}	bz 4; eth 4; EtOH 4
10832	1H-Pyrazole, 1-ethyl-	$C_5H_8N_2$ 96.13	2817-71-2	 139	5-23-04-00135 0.9537^{20}	ace 4; bz 4; eth 4; EtOH 4 1.4700^{20}
10833	1H-Pyrazole, 1-ethyl-3-methyl- 1-Ethyl-3-methylpyrazole	$C_6H_{10}N_2$ 110.16	30433-57-9 115	 152	5-23-05-00013 0.936^{20}	
10834	1H-Pyrazole, 1-methyl- 1-Methylpyrazole	$C_4H_6N_2$ 82.11	930-36-9	 127	5-23-04-00134 0.9929^{13}	 1.4787^{13}
10835	1H-Pyrazole, 3-methyl-	$C_4H_6N_2$ 82.11	1453-58-3 36.5	 204; 108^{25}	5-23-05-00004 1.0203^{16}	H_2O 5; EtOH 5; eth 5 1.4915^{20}
10836	1H-Pyrazole, 4-methyl- 4-Methylpyrazole	$C_4H_6N_2$ 82.11	7554-65-6	10146 206; 95^{13}	5-23-05-00031 1.015^{20}	
10837	1H-Pyrazole, 1-methyl-3-phenyl-	$C_{10}H_{10}N_2$ 158.20	3463-26-1 56	 280.5	5-23-07-00178 1.0232^{99}	
10838	1H-Pyrazole, 3-methyl-1-phenyl-	$C_{10}H_{10}N_2$ 158.20	1128-54-7 37	 255	5-23-05-00013 1.076^{20}	
10839	1H-Pyrazole, 1,3,4-trimethyl- 1,3,4-Trimethylpyrazole	$C_6H_{10}N_2$ 110.16	15802-99-0	 160	5-23-05-00109 0.9567^{17}	 1.4866^{17}
10840	1H-Pyrazole, 1,3,5-trimethyl- 1,3,5-Trimethylpyrazole	$C_6H_{10}N_2$ 110.16	1072-91-9	 170	5-23-05-00125 0.9130^{57}	 1.4589^{57}
10841	1H-Pyrazole, 1,4,5-trimethyl- 1,4,5-Trimethylpyrazole	$C_6H_{10}N_2$ 110.16	15802-97-8	 176.5	5-23-05-00108 0.9685^{17}	 1.4849^{17}
10842	1H-Pyrazole, 3,4,5-trimethyl-	$C_6H_{10}N_2$ 110.16	5519-42-6 138.5	 233	5-23-05-00198	bz 4; eth 4; EtOH 4
10843	3,5-Pyrazolidinedione, 1,4-diphenyl- Phenopyrazone	$C_{15}H_{12}N_2O_2$ 252.27	3426-01-5 233.5	7217	5-24-08-00003	
10844	3-Pyrazolidinone, 1-phenyl- 1-Phenyl-3-pyrazolidinone	$C_9H_{10}N_2O$ 162.19	92-43-3 126	7281	5-24-01-00011	eth 1; lig 1
10845	3H-Pyrazol-3-one, 4-amino-1,2-dihydro-1,5-dimethyl-2-phenyl- Ampyrone	$C_{11}H_{13}N_3O$ 203.24	83-07-8 109	625	4-25-00-03554	H_2O 3; EtOH 3; eth 2; bz 3
10846	3H-Pyrazol-3-one, 4-bromo-1,2-dihydro-1,5-dimethyl-2-phenyl- 4-Bromoantipyrine	$C_{11}H_{11}BrN_2O$ 267.13	5426-65-3 122	 300^{9}	5-24-01-00343 1.5900^{20}	H_2O 3; EtOH 4; eth 3; chl 4
10847	3H-Pyrazol-3-one, 1,2-dihydro-1,5-dimethyl-2-phenyl- Antipyrine	$C_{11}H_{12}N_2O$ 188.23	60-80-0 114	748 319	5-24-01-00330	H_2O 4; EtOH 4
10848	3H-Pyrazol-3-one, 2,4-dihydro-5-methyl-4-nitro-2-(4-nitrophenyl)- Picrolonic acid	$C_{10}H_8N_4O_5$ 264.20	550-74-3 116	7383 dec		H_2O 2; EtOH 3; eth 3; MeOH 3

No.	Name	Synonym	Mol. Form.	Mol. Wt.	CAS RN	mp/°C	Merck No.	bp/°C	Beil. Ref.	den/g cm^{-3}	Solubility	n_D
10849	3H-Pyrazol-3-one, 1,2-dihydro-5-methyl-2-phenyl-		$C_{10}H_{10}N_2O$	174.20	19735-89-8	128		287^{105}; 191^{17}	5-24-01-00317	1.2600^{20}	H_2O 3; EtOH 3; bz 2; peth 1	1.637
10850	3H-Pyrazol-3-one, 1,2-dihydro-2-(1-methyl-4-piperidinyl)-5-phenyl-4-(phenylmethyl)-	Benzpiperylon	$C_{22}H_{25}N_3O$	347.46	53-89-4	181-3 dec	1131		5-24-04-00398			
10851	3H-Pyrazol-3-one, 4-(dimethylamino)-1,2-dihydro-1,5-dimethyl-2-phenyl-	Aminopyrine	$C_{13}H_{17}N_3O$	231.30	58-15-1	134.5	488		4-24-00-00096		H_2O 4; bz 4; EtOH 4	
10852	4H-Pyrazolo[3,4-d]pyrimidin-4-one, 1,5-dihydro-	Allopurinol	$C_5H_4N_4O$	136.11	315-30-0	350	278		4-26-00-02076			
10853	Pyrene	Benzo[def]phenanthrene	$C_{16}H_{10}$	202.26	129-00-0	151.2	7977	404	4-05-00-02467	1.271^{23}	H_2O 1; EtOH 3; eth 3; bz 3	
10854	1-Pyrenecarboxylic acid	3-Pyrenecarboxylic acid	$C_{17}H_{10}O_2$	246.27	19694-02-1	274		sub	3-09-00-03575		eth 4; chl 4	
10855	Pyridazine	1,2-Diazabenzene	$C_4H_4N_2$	80.09	289-80-5	-8	7982	208	5-23-05-00321	1.1035^{23}	H_2O 5; EtOH 5; eth 4; ace 4	1.5218^{20}
10856	Pyridazine, 3-methyl-	3-Methyl-1,2-diazine	$C_5H_6N_2$	94.12	1632-76-4	184		214	5-23-05-00371	1.0450^{26}		1.5145^{20}
10857	2-Pyridinamine	2-Aminopyridine	$C_5H_6N_2$	94.12	504-29-0	57.5	486	105^{20}	5-22-08-00280		EtOH 3; eth 3; ace 3; bz 3	
10858	3-Pyridinamine	β-Aminopyridine	$C_5H_6N_2$	94.12	462-08-8	64.5	487	252	5-22-09-00003		H_2O 3; EtOH 3; eth 3; lig 2	
10859	4-Pyridinamine	4-Aminopyridine	$C_5H_6N_2$	94.12	504-24-5	158.5		273	5-22-09-00106		H_2O 3; EtOH 4; eth 3; bz 3	
10860	3-Pyridinamine, 6-butoxy-		$C_9H_{14}N_2O$	166.22	539-23-1			$148\text{-}50^{12}$	4-22-00-05574	1.037^{25}	EtOH 3; eth 3; chl 2; dil acid 3	1.5373^{20}
10861	2-Pyridinamine, 5-chloro-		$C_5H_5ClN_2$	128.56	1072-98-6	137		127^{11}	5-22-08-00451		H_2O 3; EtOH 3; peth 1; lig 1	
10862	3-Pyridinamine, 2,6-dichloro-	Pyridine, 3-amino-2,6-dichloro	$C_5H_4Cl_2N_2$	163.01	62476-56-6	119		$110^{0.3}$	4-22-00-04093		eth 4; EtOH 4	
10863	2-Pyridinamine, *N*,*N*-dimethyl-		$C_7H_{10}N_2$	122.17	5683-33-0	182		196	5-22-08-00287	1.0149^{14}	EtOH 3; eth 3; bz 3	1.5663^{20}
10864	4-Pyridinamine, 2,6-dimethyl-		$C_7H_{10}N_2$	122.17	3512-80-9	192.5		246	5-22-09-00417		H_2O 3; EtOH 3; eth 2; ace 3	
10865	3-Pyridinamine, 6-methoxy-		$C_6H_8N_2O$	124.14	6628-77-9	30		125^{10}; 87^{1}	5-22-11-00408			1.5745^{20}
10866	2-Pyridinamine, 3-methyl-		$C_6H_8N_2$	108.14	1603-40-3	33.5		222; 95^{8}	5-22-09-00272		H_2O 4; EtOH 3; eth 3; ace 3	
10867	2-Pyridinamine, 4-methyl-	2-Amino-4-picoline	$C_6H_8N_2$	108.14	695-34-1	100	478	$115\text{-}7^{11}$	5-22-09-00325		H_2O 4; EtOH 3; eth 3; ace 3	
10868	2-Pyridinamine, 6-methyl-		$C_6H_8N_2$	108.14	1824-81-3	41		208.5	5-22-09-00210		H_2O 4; EtOH 3; eth 3; ace 3	
10869	2-Pyridinamine, *N*-methyl-		$C_6H_8N_2$	108.14	4597-87-9	15		200.5	5-22-08-00286	1.048^{29}	H_2O 3; EtOH 4; eth 4; bz 3	
10870	3-Pyridinamine, 4-methyl-		$C_6H_8N_2$	108.14	3430-27-1	106		255	5-22-09-00344		H_2O 2; EtOH 2; eth 2; ace 2	
10871	2-Pyridinamine, *N*-2-pyridinyl-		$C_{10}H_9N_3$	171.20	1202-34-2	90.5		307.5	5-22-08-00415		H_2O 2; EtOH 4; eth 4; ace 4	
10872	Pyridine	Azine	C_5H_5N	79.10	110-86-1	-41.6	7983	115.2	5-20-05-00160	0.9819^{20}	H_2O 5; EtOH 5; eth 5; ace 5	1.5095^{20}
10873	1(4H)-Pyridineacetic acid, 3,5-diiodo-4-oxo-, propyl ester	Propyliodone	$C_{10}H_{11}I_2NO_3$	447.01	587-61-1	186	7876		5-21-07-00164			
10874	Pyridine, 2,6-bis(1,1-dimethylethyl)-	2,6-Di-*tert*-butylpyridine	$C_{13}H_{21}N$	191.32	585-48-8		3022	$120\text{-}1^{20}$	5-20-06-00181			
10875	Pyridine, 2-bromo-		C_5H_4BrN	158.00	109-04-6			193; 75^{13}	5-20-05-00422	1.657^{15}	H_2O 2; EtOH 3; eth 3; ctc 3	1.5734^{20}
10876	Pyridine, 3-bromo-	3-Bromopyridine	C_5H_4BrN	158.00	626-55-1	-27.3		173; 69^{18}	5-20-05-00426	1.645^{0}	H_2O 3; EtOH 4; eth 4	1.5694^{20}
10877	Pyridine, 4-bromo-	4-Bromopyridine	C_5H_4BrN	158.00	1120-87-2	0.5		$28\text{-}30^{0.4}$	5-20-05-00431	1.6450^{0}	ace 3; bz 3	1.5694^{20}
10878	Pyridine, 2-butyl-	2-Butylpyridine	$C_9H_{13}N$	135.21	5058-19-5	97		192	5-20-06-00118	0.9135^{15}		
10879	Pyridine, 4-butyl-		$C_9H_{13}N$	135.21	5335-75-1			209; 98^{20}	5-20-06-00120	0.9157^{20}		1.4946^{20}
10880	Pyridine, 4-(1-butylpentyl)-		$C_{14}H_{23}N$	205.34	2961-47-9			181^{50}	5-20-06-00189	0.8878^{25}		1.4846^{25}
10881	2-Pyridinecarbonitrile		$C_6H_4N_2$	104.11	100-70-9	29		224.5	5-22-02-00019	1.0810^{25}	H_2O 3; EtOH 4; eth 4; bz 4	1.5242^{25}
10882	3-Pyridinecarbonitrile		$C_6H_4N_2$	104.11	100-54-9	51		206.9; 170^{300}	5-22-02-00115	1.1590^{25}	H_2O 4; EtOH 4; eth 4; bz 4	
10883	3-Pyridinecarbonitrile, 1,2-dihydro-4-methoxy-1-methyl-2-oxo-	Ricinine	$C_8H_8N_2O_2$	164.16	524-40-3	201.5	8212	sub 170	5-22-07-00318		H_2O 3; EtOH 2; bz 2; chl 3	

No.	Name Synonym	Mol. Form. Mol. Wt.	CAS RN mp/°C	Merck No. bp/°C	Beil. Ref. den/g cm^{-3}	Solubility n_D
10884	3-Pyridinecarbonitrile, 1,2-dihydro-1-methyl-2-oxo-	$C_7H_6N_2O$ 134.14	767-88-4 140	 243^{28}	5-22-06-00129	EtOH 4
10885	4-Pyridinecarbothioamide, 2-ethyl- Ethionamide	$C_8H_{10}N_2S$ 166.25	536-33-4 163	3692	5-22-02-00360	
10886	4-Pyridinecarbothioamide, 2-propyl- Protionamide	$C_9H_{12}N_2S$ 180.27	14222-60-7 136.7	7905	5-22-02-00376	
10887	2-Pyridinecarboxaldehyde	C_6H_5NO 107.11	1121-60-4	 180; 62^{13}	5-21-07-00293 1.1255^{18}	H_2O 3; EtOH 3; eth 3; ctc 2 1.5389^{18}
10888	3-Pyridinecarboxaldehyde	C_6H_5NO 107.11	500-22-1	 89.5^{14}	5-21-07-00334 1.1394^{25}	H_2O 3; EtOH 3; eth 2; ace 3
10889	4-Pyridinecarboxaldehyde	C_6H_5NO 107.11	872-85-5	 $77\text{-}8^{12}$	5-21-07-00351	H_2O 3; eth 3; ctc 3 1.5423^{20}
10890	3-Pyridinecarboxamide Niacinamide	$C_6H_6N_2O$ 122.13	98-92-0 130	6398 $157^{0.0005}$	5-22-02-00080 1.400^{25}	H_2O 4; EtOH 4; chl 2; glycerol 4 1.466
10891	3-Pyridinecarboxamide, *N,N*-diethyl- Nikethamide	$C_{10}H_{14}N_2O$ 178.23	59-26-7 25	6459 280 dec; 175^{25}	5-22-02-00084 1.060^{25}	DMSO-d_6 2 1.525^{20}
10892	4-Pyridinecarboxamide, *N,N*-diethyl- Isonicotinic acid diethylamide	$C_{10}H_{14}N_2O$ 178.23	530-40-5	5074 $119\text{-}20^{1}$	5-22-02-00196	H_2O 4; ace 4; eth 4; EtOH 4 1.525^{20}
10893	2-Pyridinecarboxylic acid Picolinic acid	$C_6H_5NO_2$ 123.11	98-98-6 136.5	7375 sub	5-22-02-00003	H_2O 2; EtOH 3; eth 1; bz 2
10894	3-Pyridinecarboxylic acid Nicotinic acid	$C_6H_5NO_2$ 123.11	59-67-6 236.6	6435 sub	5-22-02-00057 1.473^{25}	H_2O 2; EtOH 2; eth 2
10895	4-Pyridinecarboxylic acid Isonicotinic acid	$C_6H_5NO_2$ 123.11	55-22-1 315	5073 sub 260	5-22-02-00188	H_2O 2; EtOH 2; eth 2; bz 2
10896	2-Pyridinecarboxylic acid, 4-amino-3,5,6-trichloro- Picloram	$C_6H_3Cl_3N_2O_2$ 241.46	1918-02-1 218.5	7370	5-22-13-00585	
10897	3-Pyridinecarboxylic acid, 6-amino- 6-Aminonicotinic acid	$C_6H_6N_2O_2$ 138.13	3167-49-5 312	468	4-22-00-06726	
10898	2-Pyridinecarboxylic acid, 5-butyl- Fusaric acid	$C_{10}H_{13}NO_2$ 179.22	536-69-6 97	4227	5-22-02-00384	
10899	3-Pyridinecarboxylic acid, 1,6-dihydro-6-oxo-	$C_6H_5NO_3$ 139.11	5006-66-6 310 dec	 sub	5-22-06-00119	H_2O 2; EtOH 1; eth 1; bz 1
10900	4-Pyridinecarboxylic acid, 1,2-dihydro-6-hydroxy-2-oxo- Citrazinic acid	$C_6H_5NO_4$ 155.11	99-11-6 >300 dec	2327	5-22-07-00024	H_2O 3; alk 3; HCl 2
10901	2-Pyridinecarboxylic acid, ethyl ester	$C_8H_9NO_2$ 151.17	2524-52-9 1	 243	5-22-02-00004 1.1194^{20}	H_2O 4; eth 4; EtOH 4 1.5104^{20}
10902	3-Pyridinecarboxylic acid, ethyl ester	$C_8H_9NO_2$ 151.17	614-18-6 8.5	 224	5-22-02-00060 1.1070^{20}	H_2O 4; EtOH 4; eth 4; bz 4 1.5024^{20}
10903	4-Pyridinecarboxylic acid, ethyl ester	$C_8H_9NO_2$ 151.17	1570-45-2 23	 219.5	5-22-02-00190 1.0091^{15}	H_2O 2; EtOH 3; eth 4; bz 3 1.5017^{20}
10904	4-Pyridinecarboxylic acid, [1-(2-furanyl)ethylidene]hydrazide Furonazide	$C_{12}H_{11}N_3O_2$ 229.24	3460-67-1 202.3	4220	5-22-02-00264	
10905	4-Pyridinecarboxylic acid, hydrazide Isoniazid	$C_6H_7N_3O$ 137.14	54-85-3 171.4	5071	5-22-02-00219	H_2O 4; EtOH 4
10906	3-Pyridinecarboxylic acid, methyl ester Methyl nicotinate	$C_7H_7NO_2$ 137.14	93-60-7 42.5	6014 204	5-22-02-00059	H_2O 3; EtOH 3; bz 3
10907	3-Pyridinecarboxylic acid, 1-methylethyl ester	$C_9H_{11}NO_2$ 165.19	553-60-6	 126^{30}; 92.5^{5}	5-22-02-00060 1.0624^{20}	 1.4926^{20}
10908	4-Pyridinecarboxylic acid, methyl ester	$C_7H_7NO_2$ 137.14	2459-09-8 16.1	 208	5-22-02-00189 1.1599^{20}	H_2O 2; EtOH 3; eth 3; bz 3 1.5135^{20}
10909	3-Pyridinecarboxylic acid, 1-oxide Oxiniacic acid	$C_6H_5NO_3$ 139.11	2398-81-4 254-5 dec	6895	5-22-02-00140	H_2O 4; MeOH 4
10910	4-Pyridinecarboxylic acid, 2-[3-oxo-3-[(phenylmethyl)amino]propyl]hydrazide Nialamide	$C_{16}H_{18}N_4O_2$ 298.34	51-12-7 151.6	6399	5-22-02-00251	
10911	3-Pyridinecarboxylic acid, phenylmethyl ester Nicotinic acid benzyl ester	$C_{13}H_{11}NO_2$ 213.24	94-44-0	6436 170^{3}	5-22-02-00062	
10912	3-Pyridinecarboxylic acid, 1,2,5,6-tetrahydro- Guvacine	$C_6H_9NO_2$ 127.14	498-96-4 295 dec	4495	5-22-01-00322	H_2O 4
10913	3-Pyridinecarboxylic acid, 1,2,5,6-tetrahydro-1-methyl- Arecaidine	$C_7H_{11}NO_2$ 141.17	499-04-7 232 dec	801	5-22-01-00322	H_2O 4; EtOH 1; eth 1; bz 1
10914	3-Pyridinecarboxylic acid, 1,2,5,6-tetrahydro-1-methyl-, methyl ester Arecoline	$C_8H_{13}NO_2$ 155.20	63-75-2	802 209	5-22-01-00322 1.0485^{20}	H_2O 5; EtOH 5; eth 5; chl 3 1.486^{20}
10915	Pyridine, 2-chloro- 2-Chloropyridine	C_5H_4ClN 113.55	109-09-1	 170	5-20-05-00402 1.205^{15}	H_2O 2; EtOH 3; eth 3 1.5320^{20}
10916	Pyridine, 3-chloro- 3-Chloropyridine	C_5H_4ClN 113.55	626-60-8	 148; 86^{100}	5-20-05-00406	H_2O 2 1.5304^{20}
10917	Pyridine, 4-chloro- 4-Chloropyridine	C_5H_4ClN 113.55	626-61-9 -43.5	 147.5	5-20-05-00410 1.2000^{25}	H_2O 3; EtOH 5
10918	Pyridine, 2-chloro-6-methoxy-	C_6H_6ClNO 143.57	17228-64-7	 185.5	5-21-02-00018 1.207^{25}	 1.5263^{20}
10919	Pyridine, 2-chloro-4-methyl-	C_6H_6ClN 127.57	3678-62-4 115	 192	5-20-05-00579 1.1459^{20}	 1.5293^{8}

No.	Name Synonym	Mol. Form. Mol. Wt.	CAS RN mp/°C	Merck No. bp/°C	Beil. Ref. den/g cm^{-3}	Solubility n_D
10920	Pyridine, 2-chloro-6-methyl-	C_6H_6ClN 127.57	18368-63-3	183.5; 64^{10}	5-20-05-00498 1.167^{25}	1.5270^{20}
10921	Pyridine, 2-[(4-chlorophenyl)methyl]-	$C_{12}H_{10}ClN$ 203.67	4350-41-8	$181\text{-}3^{20}$	5-20-07-00558 1.39^{25}	1.5868^{20}
10922	Pyridine, 4-(3-cyclohexen-1-yl)- 4-(Cyclohex-3-enyl)pyridine	$C_{11}H_{13}N$ 159.23	70644-46-1 22.1	226	4-20-00-03239 1.0222^{25}	1.5466^{25}
10923	2,3-Pyridinediamine	$C_5H_7N_3$ 109.13	452-58-4 120.8	$148\text{-}50^{5}$	5-22-11-00241	H_2O 3; EtOH 3; bz 3
10924	2,5-Pyridinediamine 2,5-Diaminopyridine	$C_5H_7N_3$ 109.13	4318-76-7 110.3	182^{12}	5-22-11-00252	H_2O 4; EtOH 4
10925	2,6-Pyridinediamine	$C_5H_7N_3$ 109.13	141-86-6 121.5	285; 148^{5}	5-22-11-00255	H_2O 2; ace 2
10926	Pyridine, 3,5-dibromo- 3,5-Dibromopyridine	$C_5H_3Br_2N$ 236.89	625-92-3 112	222	5-20-05-00436	H_2O 2; EtOH 3; eth 3
10927	2,3-Pyridinedicarboxylic acid Quinolinic acid	$C_7H_5NO_4$ 167.12	89-00-9 228.5	8102	5-22-04-00115	H_2O 2; EtOH 1; eth 1; bz 1
10928	2,4-Pyridinedicarboxylic acid	$C_7H_5NO_4$ 167.12	499-80-9 249		5-22-04-00119 0.942^{25}	H_2O 2; EtOH 3; eth 1; bz 1
10929	2,5-Pyridinedicarboxylic acid Isocinchomeronic acid	$C_7H_5NO_4$ 167.12	100-26-5 254	5043	5-22-04-00124	H_2O 3; EtOH 2; eth 1; bz 1
10930	3,4-Pyridinedicarboxylic acid Cinchomeronic acid	$C_7H_5NO_4$ 167.12	490-11-9 256	2285 sub	5-22-04-00135	H_2O 2; EtOH 2; eth 1; bz 2
10931	3,5-Pyridinedicarboxylic acid	$C_7H_5NO_4$ 167.12	499-81-0 324	sub	5-22-04-00139	H_2O 1; eth 2; HCl 3; HOAc 2
10932	3,5-Pyridinedicarboxylic acid, 2,6-dimethyl-, diethyl ester	$C_{13}H_{17}NO_4$ 251.28	1149-24-2 71	301; 208^{40}	5-22-04-00162	H_2O 1; EtOH 3; eth 3; bz 3
10933	Pyridine, 2,4-diethyl- 2,4-Diethylpyridine	$C_9H_{13}N$ 135.21	626-21-1 112	187.5	4-20-00-02825 0.9338^{25}	
10934	3,4-Pyridinedimethanol, 5-hydroxy-6-methyl-, hydrochloride Pyridoxine hydrochloride	$C_8H_{12}ClNO_3$ 205.64	58-56-0 207	7995 sub	5-21-05-00492	H_2O 4
10935	Pyridine, 2,6-dimethoxy- 2,6-Dimethoxypyridine	$C_7H_9NO_2$ 139.15	6231-18-1	179	5-21-04-00416 1.053^{25}	1.5029^{20}
10936	Pyridine, 2,3-dimethyl- 2,3-Lutidine	C_7H_9N 107.16	583-61-9	161.2	5-20-06-00015 0.9319^{25}	H_2O 3; EtOH 3; eth 3 1.5057^{20}
10937	Pyridine, 2,4-dimethyl- 2,4-Lutidine	C_7H_9N 107.16	108-47-4 -64	158.5	5-20-06-00019 0.9309^{20}	H_2O 4; EtOH 4; eth 4; ace 3 1.5010^{20}
10938	Pyridine, 2,5-dimethyl- 2,5-Lutidine	C_7H_9N 107.16	589-93-5 -16	157.0	5-20-06-00027 0.9297^{20}	H_2O 2; EtOH 4; eth 5; ace 3 1.5006^{20}
10939	Pyridine, 2,6-dimethyl- 2,6-Lutidine	C_7H_9N 107.16	108-48-5 -6.1	5485 144.1	5-20-06-00032 0.9226^{20}	H_2O 5; EtOH 2; eth 3; ace 3 1.4953^{20}
10940	Pyridine, 3,4-dimethyl- 3,4-Lutidine	C_7H_9N 107.16	583-58-4 -11	179.1	5-20-06-00054 0.9281^{20}	H_2O 2; EtOH 3; eth 3; ace 3 1.5096^{20}
10941	Pyridine, 3,5-dimethyl- 3,5-Lutidine	C_7H_9N 107.16	591-22-0 -6.6	172	5-20-06-00060 0.9419^{20}	H_2O 3; EtOH 3; eth 3; ace 3 1.5061^{20}
10942	Pyridine, 2-(1,1-dimethylethyl)- 2-*tert*-Butylpyridine	$C_9H_{13}N$ 135.21	5944-41-2 -33	170	5-20-06-00121	1.4891^{20}
10943	Pyridine, 3-(1,1-dimethylethyl)- 3-*tert*-Butylpyridine	$C_9H_{13}N$ 135.21	38031-78-6 -44	194.8	5-20-06-00122	1.4965^{20}
10944	Pyridine, 4-(1,1-dimethylethyl)-	$C_9H_{13}N$ 135.21	3978-81-2 -41	196.5	5-20-06-00123 0.915^{25}	ctc 3; CS_2 3 1.4958^{20}
10945	Pyridine, 2,6-dimethyl-, 1-oxide	C_7H_9NO 123.15	1073-23-0	$115\text{-}9^{18}$	5-20-06-00040 1.073^{25}	1.5706^{20}
10946	2,4(1H,3H)-Pyridinedione, 3,3-diethyl- Pyrithyldione	$C_9H_{13}NO_2$ 167.21	77-04-3 90.7	8005	5-21-10-00086	
10947	Pyridine, 2,6-diphenyl- 2,6-Diphenylpyridine	$C_{17}H_{13}N$ 231.30	3558-69-8 82	397	5-20-08-00459	eth 4; EtOH 4
10948	2-Pyridineethanamine	$C_7H_{10}N_2$ 122.17	2706-56-1	131.5^{50}	5-22-09-00368 1.0220^{25}	1.5335^{25}
10949	4-Pyridineethanamine	$C_7H_{10}N_2$ 122.17	13258-63-4	121^{10}	4-22-00-04201 1.0302^{25}	H_2O 4 1.5381^{25}
10950	2-Pyridineethanol	C_7H_9NO 123.15	103-74-2 -7.8	190^{200}; 170^{100}	5-21-02-00206 1.091^{25}	H_2O 4; EtOH 4; eth 2; chl 4 1.5366^{20}
10951	2-Pyridineethanol, α-methyl- 2-(2-Hydroxypropyl)pyridine	$C_8H_{11}NO$ 137.18	5307-19-7 33	124^{20}; 110^{10}	5-21-02-00234	H_2O 4; EtOH 4; chl 4
10952	Pyridine, 2-ethenyl-	C_7H_7N 105.14	100-69-6	159.5	5-20-06-00211 0.9983^{20}	H_2O 2; EtOH 4; eth 4; ace 4 1.5495^{20}
10953	Pyridine, 3-ethenyl-	C_7H_7N 105.14	1121-55-7	162	5-20-06-00212	H_2O 2; EtOH 3; eth 3 1.5530^{20}
10954	Pyridine, 4-ethenyl-	C_7H_7N 105.14	100-43-6	121^{150}; 65^{15}	5-20-06-00213 0.9800^{20}	H_2O 3; EtOH 3; eth 2; chl 3 1.5449^{20}
10955	Pyridine, 5-ethenyl-2-methyl- 5-Vinyl-2-picoline	C_8H_9N 119.17	140-76-1		5-20-06-00218	
10956	Pyridine, 2-ethyl- 2-Ethylpyridine	C_7H_9N 107.16	100-71-0 -63.1	148.6	5-20-06-00003 0.9502^{25}	H_2O 3; EtOH 5; eth 4; ace 4 1.4964^{20}
10957	Pyridine, 3-ethyl- 3-Ethylpyridine	C_7H_9N 107.16	536-78-7 -76.9	3802 165	5-20-06-00007 0.9539^{25}	H_2O 3; EtOH 3; eth 3; ace 4 1.5021^{20}

No.	Name	Synonym	Mol. Form.	Mol. Wt.	CAS RN	mp/°C	Merck No.	bp/°C	Beil. Ref.	den/g cm^{-3}	Solubility	n_D
10958	Pyridine, 4-ethyl-	4-Ethylpyridine	C_7H_9N	107.16	536-75-4	-90.5	3803	168.3	5-20-06-00010	0.9417^{20}	H_2O 3; EtOH 3; eth 3; ace 4	1.5009^{20}
10959	Pyridine, 2-ethyl-3,5-dimethyl-		$C_9H_{13}N$	135.21	1123-96-2			198; 86^{15}	5-20-06-00129	0.9338^{25}	H_2O 3; EtOH 4; eth 4; ace 4	
10960	Pyridine, 3-ethyl-2,6-dimethyl-		$C_9H_{13}N$	135.21	23580-52-1			$83\text{-}5^{23}$	5-20-06-00129	0.9120^{18}	H_2O 2	
10961	Pyridine, 4-ethyl-2,6-dimethyl-		$C_9H_{13}N$	135.21	36917-36-9	97		187.5	5-20-06-00130	0.9089^{25}		1.4964^{25}
10962	Pyridine, 2-ethyl-4-methyl-		$C_8H_{11}N$	121.18	2150-18-7			175	5-20-06-00086	0.9238^{20}	ace 4; eth 4; EtOH 4	
10963	Pyridine, 2-ethyl-6-methyl-		$C_8H_{11}N$	121.18	1122-69-6			160.5	5-20-06-00089	0.9207^{25}	H_2O 2; EtOH 3; eth 3; ace 4	1.4920^{25}
10964	Pyridine, 3-ethyl-4-methyl-	3-Ethyl-4-picoline	$C_8H_{11}N$	121.18	529-21-5		3796	198	5-20-06-00090	0.9286^{17}	H_2O 2; EtOH 3; eth 3; ace 4	
10965	Pyridine, 4-ethyl-2-methyl-	4-Ethyl-2-picoline	$C_8H_{11}N$	121.18	536-88-9		3797	179	5-20-06-00085	0.9130^{25}	ace 4; bz 4; eth 4; EtOH 4	
10966	Pyridine, 5-ethyl-2-methyl-	5-Ethyl-2-picoline	$C_8H_{11}N$	121.18	104-90-5		3798	178.3	5-20-06-00086	0.9202^{20}	H_2O 2; EtOH 3; eth 3; ace 4	1.4971^{20}
10967	Pyridine, 2-(1-ethylpropyl)-		$C_{10}H_{15}N$	149.24	7399-50-0			195.4	5-20-06-00145	0.8981^{20}		1.4850^{25}
10968	Pyridine, 4-(1-ethylpropyl)-	4-(3-Pentyl)pyridine	$C_{10}H_{15}N$	149.24	35182-51-5	125.5		217; 80^{12}	5-20-06-00145	0.9085^{25}		1.40905^{25}
10969	Pyridine, hydrochloride	Pyridinium chloride	C_5H_6ClN	115.56	628-13-7	146		222	5-20-05-00160		H_2O 4; EtOH 4; chl 4	
10970	Pyridine, 2-iodo-	2-Iodopyridine	C_5H_4IN	205.00	5029-67-4			93^{13}	5-20-05-00438	1.9735^{20}	EtOH 3; eth 3; ace 3; bz 3	1.6366^{20}
10971	2-Pyridinemethanamine		$C_6H_8N_2$	108.14	3731-51-9			203; 91^{17}	5-22-09-00232	1.0525^{25}	H_2O 4	1.5431^{25}
10972	3-Pyridinemethanamine		$C_6H_8N_2$	108.14	3731-52-0	-21.1		226	5-22-09-00301	1.064^{20}	H_2O 4; eth 4; EtOH 4	1.552^{20}
10973	4-Pyridinemethanamine		$C_6H_8N_2$	108.14	3731-53-1	-7.6		230; 103^{11}	5-22-09-00345	1.072^{20}	H_2O 4	1.5495^{25}
10974	2-Pyridinemethanamine, *N*-(2-pyridinylmethyl)-		$C_{12}H_{13}N_3$	199.26	1539-42-0			200^{10}	5-22-09-00267	1.1074^{25}		1.5757^{25}
10975	2-Pyridinemethanol		C_6H_7NO	109.13	586-98-1			112^{16}; 102.5^{8}	5-21-02-00150	1.1317^{20}	H_2O 5; EtOH 4; eth 4; ace 4	1.5444^{20}
10976	3-Pyridinemethanol	Nicotinyl alcohol	C_6H_7NO	109.13	100-55-0	-6.5	6438	266	5-21-02-00172	1.131^{20}	H_2O 4; eth 4	1.5455^{20}
10977	4-Pyridinemethanol	4-Picolyl alcohol	C_6H_7NO	109.13	586-95-8	53		$140\text{-}2^{12}$	5-21-02-00191		chl 3	
10978	3-Pyridinemethanol, 4-(aminomethyl)-5-hydroxy-6-methyl-, dihydrochloride	Pyridoxamine dihydrochloride	$C_8H_{14}Cl_2N_2O_2$	241.12	524-36-7	226-7 dec	7993		5-22-12-00324		H_2O 4; EtOH 2	
10979	Pyridine, 2-(2-methoxyethyl)-	Metyridine	$C_8H_{11}NO$	137.18	114-91-0		6084	203; 96^{17}	5-21-02-00206	0.988^{20}	H_2O 4; EtOH 4	1.4975^{20}
10980	Pyridine, 2-methyl-	2-Picoline	C_6H_7N	93.13	109-06-8	-66.7	7372	129.3	5-20-05-00464	0.9443^{20}	H_2O 4; EtOH 5; eth 5; ace 4	1.4957^{20}
10981	Pyridine, 3-methyl-	3-Picoline	C_6H_7N	93.13	108-99-6	-18.1	7373	144.1	5-20-05-00506	0.9566^{20}	H_2O 5; EtOH 5; eth 5; ace 4	1.5040^{20}
10982	Pyridine, 4-methyl-	4-Picoline	C_6H_7N	93.13	108-89-4	3.66	7374	145.3	5-20-05-00543	0.9548^{20}	H_2O 5; EtOH 5; eth 5; ace 3	1.5037^{20}
10983	Pyridine, 2-[2-(1-methylethoxy)ethyl]-	Pyridine, 2-(2-isopropoxyethyl)	$C_{10}H_{15}NO$	165.24	70715-19-4			133^{50}		0.9502^{25}	H_2O 4	1.4820^{25}
10984	Pyridine, 2-(1-methylethyl)-		$C_8H_{11}N$	121.18	644-98-4	-141		159.8	5-20-06-00080	0.9342^{25}	H_2O 2; EtOH 5; eth 5; ace 4	1.4915^{20}
10985	Pyridine, 3-(1-methylethyl)-	3-Isopropylpyridine	$C_8H_{11}N$	121.18	6304-18-3	-45.8		177.5	5-20-06-00081	0.9217^{16}		
10986	Pyridine, 4-(1-methylethyl)-		$C_8H_{11}N$	121.18	696-30-0	-54.9		178	5-20-06-00082	0.9382^{25}	H_2O 2; EtOH 5; eth 5; ace 4	1.4962^{20}
10987	Pyridine, 3-(1-methyl-1H-pyrrol-2-yl)-		$C_{10}H_{10}N_2$	158.20	487-19-4			281; 150^{16}	5-23-07-00227	1.241^{20}	EtOH 4	1.6057^{20}
10988	Pyridine, 3-methyl-, 1-oxide		C_6H_7NO	109.13	1003-73-2	39		148^{15}	5-20-05-00517		chl 3	
10989	Pyridine, 2-(4-methylphenyl)-		$C_{12}H_{11}N$	169.23	4467-06-5			$170\text{-}80^{20}$	5-20-07-00569	0.999^{25}		1.6125^{20}
10990	Pyridine, 2-methyl-5-phenyl-		$C_{12}H_{11}N$	169.23	3256-88-0			189^{50}	5-20-07-00573	1.0590^{25}		1.6055^{25}
10991	Pyridine, 3-methyl-2-phenyl-	3-Picoline, 2-phenyl-	$C_{12}H_{11}N$	169.23	10273-90-2			$148\text{-}9^{16}$	5-20-07-00571	1.065^{25}		1.6026^{20}
10992	Pyridine, 3-(1-methyl-2-pyrrolidinyl)-, (*S*)-	Nicotine	$C_{10}H_{14}N_2$	162.23	54-11-5	-79	6434	247; 125^{18}	5-23-06-00064	1.0097^{20}	H_2O 5; EtOH 4; eth 4; chl 4	1.5282^{20}
10993	Pyridine, 4-nitro-, 1-oxide	4-Nitropyridine *N*-oxide	$C_5H_4N_2O_3$	140.10	1124-33-0	160.5			5-20-05-00448			
10994	Pyridine, 1-oxide	Pyridine *N*-oxide	C_5H_5NO	95.10	694-59-7	65.5	7984	$146\text{-}7^{13}$	5-20-05-00217			
10995	Pyridine, pentachloro-		C_5Cl_5N	251.33	2176-62-7	125.5		280	5-20-05-00422		bz 4; EtOH 4; lig 4	
10996	Pyridine, 2-(3-pentenyl)-		$C_{10}H_{13}N$	147.22	2057-43-4			216; 93^{12}	5-20-06-00355	0.9234^{25}		1.5076^{25}

No.	Name Synonym	Mol. Form. Mol. Wt.	CAS RN mp/°C	Merck No. bp/°C	Beil. Ref. den/g cm^{-3}	Solubility n_D
10997	Pyridine, 2-phenyl- 2-Phenylpyridine	$C_{11}H_9N$ 155.20	1008-89-5	 271	5-20-07-00545 1.0833^{25}	H_2O 2; EtOH 5; eth 5 1.6210^{20}
10998	Pyridine, 3-phenyl- 3-Phenylpyridine	$C_{11}H_9N$ 155.20	1008-88-4 164	 272	5-20-07-00548	H_2O 2; EtOH 3; eth 3 1.6123^{25}
10999	Pyridine, 4-phenyl- 4-Phenylpyridine	$C_{11}H_9N$ 155.20	939-23-1 77.5	 281	5-20-07-00549	H_2O 3; EtOH 3; eth 3
11000	Pyridine, 2-(2-phenylethenyl)-, (*E*)-	$C_{13}H_{11}N$ 181.24	538-49-8 91.5	 325; 194^{14}	5-20-08-00057 1.0147^{100}	H_2O 1; EtOH 3; eth 3; bz 3
11001	Pyridine, 4-(2-phenylethenyl)-, (*E*)-	$C_{13}H_{11}N$ 181.24	5097-93-8 131	 208^{33}	5-20-08-00063	H_2O 1; EtOH 3; eth 3; chl 3
11002	Pyridine, 2-(2-phenylethenyl)-, (*Z*)-	$C_{13}H_{11}N$ 181.24	1519-59-1 -50	 141^{10}	4-20-00-03874	EtOH 3; eth 3; bz 3
11003	Pyridine, 2-(2-phenylethyl)-	$C_{13}H_{13}N$ 183.25	2116-62-3 -3	 289	5-20-07-00588 1.0465^{0}	
11004	Pyridine, 2-(phenylmethyl)- 2-Benzylpyridine	$C_{12}H_{11}N$ 169.23	101-82 6 12.5	 277; 149^{16}	5-20-07-00556 1.067^{0}	H_2O 1; EtOH 3; eth 3; chl 3 1.5785^{20}
11005	Pyridine, 3-(phenylmethyl)- 3-Benzylpyridine	$C_{12}H_{11}N$ 169.23	620-95-1 34	 287	5-20-07-00560 1.061^{25}	eth 4; EtOH 4
11006	Pyridine, 4-(phenylmethyl)-	$C_{12}H_{11}N$ 169.23	2116-65-6 12.4	 288; 180^{31}	5-20-07-00561 1.0612^{20}	H_2O 1; EtOH 3; eth 4; ctc 3 1.5818^{20}
11007	Pyridine, 4-(3-phenylpropyl)-	$C_{14}H_{15}N$ 197.28	2057-49-0	 322; 150^{5}	5-20-07-00607 1.024^{25}	bz 4; eth 4; py 4; EtOH 4 1.5616^{25}
11008	Pyridine, 3-(2-piperidinyl)-, (*S*)- Anabasine	$C_{10}H_{14}N_2$ 162.23	494-52-0 9	655 276	5-23-06-00096 1.0436^{20}	H_2O 5; EtOH 3; eth 3; bz 3 1.5430^{20}
11009	Pyridine, 4-(4-piperidinyl)-	$C_{10}H_{14}N_2$ 162.23	581-45-3 79	 292	5-23-06-00115	H_2O 2; EtOH 4; eth 3; bz 3
11010	2-Pyridinepropanamine, *N,N*-dimethyl-γ-phenyl- Pheniramine	$C_{16}H_{20}N_2$ 240.35	86-21-5	7198 181^{13}; $135^{0.5}$	5-22-10-00487 1.0081^{25}	bz 4; eth 4; EtOH 4; chl 4 1.5519^{25}
11011	1(4H)-Pyridinepropanoic acid, α-amino-3-hydroxy-4-oxo-, (*S*)- Mimosine	$C_8H_{10}N_2O_4$ 198.18	500-44-7 228 dec	6118	5-21-12-00107	H_2O 2; EtOH 1; eth 1; ace 1
11012	2-Pyridinepropanol 2-(3-Hydroxypropyl)pyridine	$C_8H_{11}NO$ 137.18	2859-68-9 34	 260.2; 116^{4}	5-21-02-00236 1.060^{25}	H_2O 4 1.5298^{20}
11013	3-Pyridinepropanol	$C_8H_{11}NO$ 137.18	2859-67-8	 284; 130^{3}	5-21-02-00237 1.063^{25}	H_2O 4 1.5313^{20}
11014	Pyridine, 2-(2-propenyl)-	C_8H_9N 119.17	2835-33-8	 190; 63^{15}	5-20-06-00216 0.959^{20}	eth 4; EtOH 4
11015	Pyridine, 2-propyl-	$C_8H_{11}N$ 121.18	622-39-9 2	 167	5-20-06-00076 0.9119^{20}	H_2O 2; EtOH 5; eth 5; ace 4 1.4925^{20}
11016	Pyridine, 4-propyl-	$C_8H_{11}N$ 121.18	1122-81-2	 185	5-20-06-00078 0.9381^{15}	eth 4; EtOH 4 1.4966^{20}
11017	Pyridine, 3-(2-pyrrolidinyl)-, (*S*)- Nornicotine	$C_9H_{12}N_2$ 148.21	494-97-3	6631 270	4-23-00-00998 1.0737^{19}	H_2O 4; ace 4; eth 4; EtOH 4 1.5378^{18}
11018	3-Pyridinesulfonic acid 3-Pyridylsulfonic acid	$C_5H_5NO_3S$ 159.17	636-73-7 357 dec		5-22-07-00552 1.713^{25}	H_2O 4; EtOH 2; eth 1
11019	Pyridine, 2,3,5,6-tetrachloro- 2,3,5,6-Tetrachloropyridine	C_5HCl_4N 216.88	2402-79-1 90.5	 250.5	5-20-05-00421	eth 4; EtOH 4; peth 4
11020	Pyridine, 1,2,3,6-tetrahydro- Δ 3-Piperidine	C_5H_9N 83.13	694-05-3 -48	 108	5-20-04-00282 0.911^{25}	chl 3 1.4800^{20}
11021	Pyridine, 2,3,4,5-tetrahydro-6-propyl- γ-Coniceine	$C_8H_{15}N$ 125.21	1604-01-9	2497 174	5-20-04-00345 0.8753^{15}	 1.4661^{16}
11022	Pyridine, 2,3,4,6-tetramethyl- 2,3,4,6-Tetramethylpyridine	$C_9H_{13}N$ 135.21	20820-82-0	 203	5-20-06-00132 0.9322^{25}	eth 4; EtOH 4 1.5087^{25}
11023	Pyridine, 2-(trichloromethyl)- 2-(Trichloromethyl)pyridine	$C_6H_4Cl_3N$ 196.46	4377-37-1 -10	 $125\text{-}6^{25}$	5-20-05-00499 1.4526^{25}	 1.5596^{25}
11024	Pyridine, 2,3,4-trimethyl- 2,3,4-Collidine	$C_8H_{11}N$ 121.18	2233-29-6	 192.5	5-20-06-00091 0.9127^{15}	H_2O 3; EtOH 3; eth 3 1.5150^{20}
11025	Pyridine, 2,3,5-trimethyl- 2,3,5-Collidine	$C_8H_{11}N$ 121.18	695-98-7	 184	5-20-06-00092 0.9352^{19}	H_2O 2; EtOH 3; eth 3; ace 3 1.5057^{25}
11026	Pyridine, 2,3,6-trimethyl- 2,3,6-Collidine	$C_8H_{11}N$ 121.18	1462-84-6	 171.6	5-20-06-00092 0.9220^{25}	H_2O 3; EtOH 3; eth 3; ace 3 1.5053^{20}
11027	Pyridine, 2,4,5-trimethyl- 2,4,5-Collidine	$C_8H_{11}N$ 121.18	1122-39-0	 188	5-20-06-00093 0.9330^{25}	ace 4; bz 4; eth 4; EtOH 4 1.5054^{25}
11028	Pyridine, 2,4,6-trimethyl- 2,4,6-Collidine	$C_8H_{11}N$ 121.18	108-75-8 -46	9632 170.6	5-20-06-00093 0.9166^{22}	H_2O 3; EtOH 3; eth 3; ace 3 1.4959^{25}
11029	Pyridinium, 3-carboxy-1-methyl-, hydroxide, inner salt Trigonelline	$C_7H_7NO_2$ 137.14	535-83-1	9606	5-22-02-00143	H_2O 4
11030	Pyridinium, 1-hexadecyl-, chloride Hexadecyl pyridinium chloride	$C_{21}H_{38}ClN$ 339.99	123-03-5 80	2024	5-20-05-00233	H_2O 4; chl 4
11031	2-Pyridinol	C_5H_5NO 95.10	72762-00-6 107.8		5-21-07-00106 1.3910^{20}	H_2O 4; bz 4; EtOH 4
11032	4-Pyridinol	C_5H_5NO 95.10	626-64-2 149.8	 >350; 257^{10}	5-21-07-00152	H_2O 3; EtOH 3; eth 1; bz 1
11033	2(1H)-Pyridinone	C_5H_5NO 95.10	142-08-5 107.8	 280	5-21-07-00106 1.3910^{20}	H_2O 3; EtOH 3; eth 2; bz 3
11034	2(1H)-Pyridinone, hydrazone	$C_5H_7N_3$ 109.13	4930-98-7 46.6	 185^{140}; 90^{1}	5-22-14-00486	chl 3

No.	Name Synonym	Mol. Form. Mol. Wt.	CAS RN mp/°C	Merck No. bp/°C	Beil. Ref. den/g cm^{-3}	Solubility n_D
11035	2(1H)-Pyridinone, 1-methyl-	C_6H_7NO 109.13	694-85-9 31	 250	5-21-07-00109 1.1120^{20}	H_2O 5; peth 2; lig 2
11036	2(1H)-Pyridinone, 3-methyl-	C_6H_7NO 109.13	1003-56-1 141.5	 289	5-21-07-00199	H_2O 4; EtOH 4; chl 4
11037	2(1H)-Pyridinone, 4-methyl-	C_6H_7NO 109.13	13466-41-6 130	 308	5-21-07-00201	H_2O 3; EtOH 3; eth 2; bz 2
11038	3H-Pyrido[3,4-b]indole, 4,9-dihydro-7-methoxy-1-methyl- Harmaline	$C_{13}H_{14}N_2O_{13}$ 406.26	304-21-2 250 dec	4528	5-23-12-00148	H_2O 2; EtOH 2; eth 2; chl 3
11039	9H-Pyrido[3,4-b]indole, 7-methoxy-1-methyl- Harmine	$C_{13}H_{12}N_2O$ 212.25	442-51-3 273	4531 sub	5-23-12-00237	H_2O 2; EtOH 2; eth 2; chl 2
11040	9H-Pyrido[3,4-b]indole, 1-methyl- Harman	$C_{12}H_{10}N_2$ 182.22	486-84-0 236.5	4530	5-23-08-00261	
11041	9H-Pyrido[3,4-b]indole, 1-[(2-methylphenyl)methyl]-	$C_{19}H_{16}N_2$ 272.35	525-15-5 217.8	 $150^{0.01}$	5-23-09-00536	EtOH 3; eth 3; bz 3; chl 3
11042	2-Pyrimidinamine	$C_4H_5N_3$ 95.10	109-12-6 127.5	 sub	4-25-00-02071	H_2O 3; chl 2
11043	4-Pyrimidinamine, 2-chloro-*N*,*N*,6-trimethyl- Crimidine	$C_7H_{10}ClN_3$ 171.63	535-89-7 87	2590 $140\text{-}7^{4}$	4-25-00-02184	EtOH 4
11044	2-Pyrimidinamine, 4,5-dimethyl- 2-Amino-4,5-dimethylpyrimidine	$C_6H_9N_3$ 123.16	1193-74-4 215.8	 sub	4-25-00-02203	ace 4; bz 4; EtOH 4; chl 4
11045	4-Pyrimidinamine, 2,6-dimethyl- Kyanmethin	$C_6H_9N_3$ 123.16	461-98-3 183	5205 sub	4-25-00-02192	H_2O 2; EtOH 2; bz 2; chl 2
11046	2-Pyrimidinamine, 4-methyl-	$C_5H_7N_3$ 109.13	108-52-1 160.3	 sub	4-25-00-02152	H_2O 3; EtOH 3; chl 2
11047	2-Pyrimidinamine, 5-methyl-	$C_5H_7N_3$ 109.13	50840-23-8 193.5	 sub	5-25-10-00191	EtOH 4
11048	4-Pyrimidinamine, 6-methyl-	$C_5H_7N_3$ 109.13	3435-28-7 197	 sub	4-25-00-02182	H_2O 3; EtOH 3
11049	Pyrimidine 1,3-Diazine	$C_4H_4N_2$ 80.09	289-95-2 22	7998 123.8	5-23-05-00334	H_2O 5; EtOH 3 1.4998^{20}
11050	4-Pyrimidinecarboxylic acid, 1,2,3,6-tetrahydro-2,6-dioxo- Orotic acid	$C_5H_4N_2O_4$ 156.10	65-86-1 345.5	6828	4-25-00-01759	H_2O 2; os 1
11051	2,4-Pyrimidinediamine, 5-(4-chlorophenyl)-6-ethyl- Pyrimethamine	$C_{12}H_{13}ClN_4$ 248.71	58-14-0 233.5	7997	4-25-00-03014	
11052	Pyrimidine, 2,4-dichloro-5-methyl- 2,4-Dichloro-5-methylpyrimidine	$C_5H_4Cl_2N_2$ 163.01	1780-31-0 26	 235	5-23-05-00385	H_2O 2; EtOH 4; eth 4; bz 4
11053	Pyrimidine, 2,4-dichloro-6-methyl- 2,6-Dichloro-4-methylpyrimidine	$C_5H_4Cl_2N_2$ 163.01	5424-21-5 46.5	 219	5-23-05-00381	bz 4; eth 4; EtOH 4; chl 4
11054	Pyrimidine, 2,4-dichloro-5-nitro- 2,4-Dichloro-5-nitropyrimidine	$C_4HCl_2N_3O_2$ 193.98	49845-33-2 29.3	 154^{58}, 135^{17}	5-23-05-00350	EtOH 2; eth 2
11055	Pyrimidine, 2,4-dimethyl- 2,4-Dimethylpyrimidine	$C_6H_8N_2$ 108.14	14331-54-5 -4	 150.5	5-23-05-00397 1.168^{14}	 1.4880^{25}
11056	Pyrimidine, 2,5-dimethyl- 2,5-Dimethylpyrimidine	$C_6H_8N_2$ 108.14	22868-76-4 19	 59^{13}	5-23-05-00399	H_2O 4
11057	Pyrimidine, 4,5-dimethyl- 4,5-Dimethylpyrimidine	$C_6H_8N_2$ 108.14	694-81-5 3	 177	5-23-05-00399	H_2O 4
11058	Pyrimidine, 4,6-dimethyl- 4,6-Dimethylpyrimidine	$C_6H_8N_2$ 108.14	1558-17-4 25	 159	5-23-05-00400	H_2O 4 1.4880^{20}
11059	2,4(1H,3H)-Pyrimidinedione Uracil	$C_4H_4N_2O_2$ 112.09	66-22-8 338	9761	5-24-06-00043	H_2O 2; EtOH 4; eth 4
11060	2,4(1H,3H)-Pyrimidinedione, 6-amino-3-ethyl-1-(2-propenyl)- Aminometradine	$C_9H_{13}N_3O_2$ 195.22	642-44-4 143	463	4-25-00-04113	
11061	2,4(1H,3H)-Pyrimidinedione, 5-[bis(2-chloroethyl)amino]- Uracil mustard	$C_8H_{11}Cl_2N_3O_2$ 252.10	66-75-1 206 dec	9762	4-25-00-04127	H_2O 2
11062	2,4(1H,3H)-Pyrimidinedione, 5-bromo- 5-Bromouracil	$C_4H_3BrN_2O_2$ 190.98	51-20-7 310	1430	5-24-06-00322	
11063	4,6(1H,5H)-Pyrimidinedione, dihydro-5-(2-methylpropyl)-5-(2-propenyl)-2-thioxo-, monosodium salt Barbituric acid, 5-allyl-5-isobutyl-2-thio, sodium salt	$C_{11}H_{15}N_2NaO_2S$ 262.31	510-90-7	1518		H_2O 4; EtOH 2; eth 1; bz 1
11064	4,6(1H,5H)-Pyrimidinedione, 5-ethyldihydro-5-(2-methyl-2-propenyl)-2-thioxo- Methallatal	$C_{10}H_{14}N_2O_2S$ 226.30	115-56-0 160.5	5855	5-24-09-00224	
11065	4,6(1H,5H)-Pyrimidinedione, 5-ethyldihydro-5-(1-methylpropyl)-2-thioxo- Barbituric acid, 5-*sec*-butyl-5-ethyl-2-thio-	$C_{10}H_{16}N_2O_2S$ 228.32	2095-57-0 169	9250	5-24-09-00164	
11066	4,6(1H,5H)-Pyrimidinedione, 5-ethyldihydro-5-phenyl- Primidone	$C_{12}H_{14}N_2O_2$ 218.26	125-33-7 281.5	7753	5-24-08-00102	
11067	2,4(1H,3H)-Pyrimidinedione, 5-fluoro- 5-Fluorouracil	$C_4H_3FN_2O_2$ 130.08	51-21-8 283	4109 sub 190	5-24-06-00271	
11068	2,4(1H,3H)-Pyrimidinedione, 5-methyl- Thymine	$C_5H_6N_2O_2$ 126.11	65-71-4 316	9332	5-24-07-00055	H_2O 2; EtOH 2; eth 2

No.	Name	Synonym	Mol. Form.	Mol. Wt.	CAS RN	mp/°C	Merck No.	bp/°C	Beil. Ref.	den/g cm^{-3}	Solubility	n_D
11069	2,4(1H,3H)-Pyrimidinedione, 6-methyl-	6-Methyluracil	$C_5H_6N_2O_2$	126.11	626-48-2	270-80 dec	6051		5-24-07-00006		H_2O 3; EtOH 3; eth 2; tfa 2	
11070	Pyrimidine, 2-methyl-	2-Methyl-1,3-diazine	$C_5H_6N_2$	94.12	5053-43-0	-4		138	5-23-05-00376		H_2O 5	
11071	Pyrimidine, 4-methyl-	4-Methyl-1,3-diazine	$C_5H_6N_2$	94.12	3438-46-8	32		142	5-23-05-00379	1.030^{16}	H_2O 5	1.500^{20}
11072	Pyrimidine, 5-methyl-	5-Methyl-1,3-diazine	$C_5H_6N_2$	94.12	2036-41-1	30.5		153	5-23-05-00383		H_2O 4	
11073	2,4,5,6(1H,3H)-Pyrimidinetetrone	Alloxan	$C_4H_2N_2O_4$	142.07	50-71-5	256 dec	281	sub	5-24-09-00382		H_2O 4; EtOH 3; ace 3; bz 3	
11074	2,4,5,6(1H,3H)-Pyrimidinetetrone, 5-oxime	Violuric acid	$C_4H_3N_3O_4$	157.09	87-39-8	203-4 dec	9904		5-24-09-00386		H_2O 2; EtOH 3	
11075	Pyrimidine, 2,4,6-trichloro-	2,4,6-Trichloropyrimidine	$C_4HCl_3N_2$	183.42	3764-01-0	22.5		212.5	5-23-05-00347			1.5700^{20}
11076	2,4,6(1H,3H,5H)-Pyrimidinetrione	Barbituric acid	$C_4H_4N_2O_3$	128.09	67-52-7	248	973	260 dec	5-24-09-00078		H_2O 3; EtOH 2; eth 3	
11077	2,4,6(1H,3H,5H)-Pyrimidinetrione, 5-amino-	Uramil	$C_4H_5N_3O_3$	143.10	118-78-5	>400	9763		4-25-00-04228		H_2O 3; eth 1; bz 1; chl 3	
11078	2,4,6(1H,3H,5H)-Pyrimidinetrione, 5-(2-bromo-2-propenyl)-5-(1-methylethyl)-	Propallylonal	$C_{10}H_{13}BrN_2O_3$	289.13	545-93-7	181	7807		4-24-00-01994		H_2O 2; EtOH 4; eth 2; ace 4	
11079	2,4,6(1H,3H,5H)-Pyrimidinetrione, 5-(2-bromo-2-propenyl)-1-methyl-5-(1-methylethyl)-	Narcobarbital	$C_{11}H_{15}BrN_2O_3$	303.16	125-55-3	115	6342		4-24-00-01995		H_2O 2; EtOH 3; py 3	
11080	2,4,6(1H,3H,5H)-Pyrimidinetrione, 5-(2-bromo-2-propenyl)-5-(1-methylpropyl)-	Butallylonal	$C_{11}H_{15}BrN_2O_3$	303.16	1142-70-7	131.5	1503		5-24-09-00231		eth 4; EtOH 4	
11081	2,4,6(1H,3H,5H)-Pyrimidinetrione, 5-butyl-1-cyclohexyl-	Bucolome	$C_{14}H_{22}N_2O_3$	266.34	841-73-6	84	1452	$185\text{-}7^{0.8}$	5-24-09-00132			
11082	2,4,6(1H,3H,5H)-Pyrimidinetrione, 5-butyl-5-ethyl-	Butethal	$C_{10}H_{16}N_2O_3$	212.25	77-28-1	128.5	1515		5-24-09-00161			
11083	2,4,6(1H,3H,5H)-Pyrimidinetrione, 5-(1-cyclohexen-1-yl)-1,5-dimethyl-	Hexobarbital	$C_{12}H_{16}N_2O_3$	236.27	56-29-1	146.5	4625		5-24-09-00253			
11084	2,4,6(1H,3H,5H)-Pyrimidinetrione, 5-(1-cyclohexen-1-yl)-5-ethyl-	Cyclobarbital	$C_{12}H_{16}N_2O_3$	236.27	52-31-3	173	2717		5-24-09-00258		H_2O 1; EtOH 4; eth 3; dil alk 3	
11085	2,4,6(1H,3H,5H)-Pyrimidinetrione, 5-(2-cyclopenten-1-yl)-5-(2-propenyl)-	Cyclopentobarbital	$C_{12}H_{14}N_2O_3$	234.25	76-68-6	139.5	2751		5-24-09-00269		H_2O 2; EtOH 4	
11086	2,4,6(1H,3H,5H)-Pyrimidinetrione, 5,5-dibromo-		$C_4H_2Br_2N_2O_3$	285.88	511-67-1	235		dec	4-24-00-01881		eth 4; EtOH 4	
11087	2,4,6(1H,3H,5H)-Pyrimidinetrione, 5,5-diethyl-	Barbital	$C_8H_{12}N_2O_3$	184.19	57-44-3	190	972		5-24-09-00137	1.220^{25}	H_2O 2; EtOH 3; eth 3; ace 3	
11088	2,4,6(1H,3H,5H)-Pyrimidinetrione, 5,5-diethyl-1-methyl-	Metharbital	$C_9H_{14}N_2O_3$	198.22	50-11-3	150.5	5873		5-24-09-00144		H_2O 3; chl 2	
11089	2,4,6(1H,3H,5H)-Pyrimidinetrione, 5,5-diethyl-1-phenyl-	Phenetharbital	$C_{14}H_{16}N_2O_3$	260.29	357-67-5	178	7183		4-24-00-01908		EtOH 4	
11090	2,4,6(1H,3H,5H)-Pyrimidinetrione, 5-(2,2-dimethylpropyl)-5-(2-propenyl)-	Nealbarbital	$C_{12}H_{18}N_2O_3$	238.29	561-83-1	156	6350		5-24-09-00238		ace 4; eth 4; EtOH 4	
11091	2,4,6(1H,3H,5H)-Pyrimidinetrione, 5,5-di-2-propenyl-	Allobarbital	$C_{10}H_{12}N_2O_3$	208.22	52-43-7	172	252		5-24-09-00245		H_2O 2; EtOH 3; eth 3; bz 3	
11092	2,4,6(1H,3H,5H)-Pyrimidinetrione, 5-ethyl-5-(3-methylbutyl)-	Amobarbital	$C_{11}H_{18}N_2O_3$	226.28	57-43-2	157	601		5-24-09-00173		bz 4; EtOH 4; chl 4	
11093	2,4,6(1H,3H,5H)-Pyrimidinetrione, 5-ethyl-5-(1-methylethyl)-	Probarbital	$C_9H_{14}N_2O_3$	198.22	76-76-6	203	7759		5-24-09-00159		H_2O 2; EtOH 1; eth 3	
11094	2,4,6(1H,3H,5H)-Pyrimidinetrione, 5-ethyl-1-methyl-5-phenyl-	Mephobarbital	$C_{13}H_{14}N_2O_3$	246.27	115-38-8	176	5742		5-24-09-00294		H_2O 2; EtOH 4; eth 2; chl 3	
11095	2,4,6(1H,3H,5H)-Pyrimidinetrione, 5-ethyl-5-phenyl-	Phenobarbital	$C_{12}H_{12}N_2O_3$	232.24	50-06-6	174	7201		5-24-09-00286		H_2O 1; EtOH 3; eth 3; bz 1	
11096	2,4,6(1H,3H,5H)-Pyrimidinetrione, 5-ethyl-5-(1-piperidinyl)-	5-Ethyl-5-(1-piperidyl)barbituric acid	$C_{11}H_{17}N_3O_3$	239.27	509-87-5	215	3800		4-25-00-04239		ace 4; eth 4; EtOH 4	
11097	2,4,6(1H,3H,5H)-Pyrimidinetrione, 5-[(hexahydro-2,4,6-trioxo-5-pyrimidinyl)imino]-, monoammonium salt	Murexide	$C_8H_{10}N_6O_7$	302.20	3051-09-0		6215		4-25-00-04236		H_2O 2; EtOH 1; eth 1; alk 3	
11098	2,4,6(1H,3H,5H)-Pyrimidinetrione, 5-(1-methylethyl)-5-(2-propenyl)-	Aprobarbital	$C_{10}H_{14}N_2O_3$	210.23	77-02-1	141	782		5-24-09-00224		ace 4; eth 4; EtOH 4; chl 4	

No.	Name Synonym	Mol. Form. Mol. Wt.	CAS RN mp/°C	Merck No. bp/°C	Beil. Ref. den/g cm^{-3}	Solubility n_D
11099	2,4,6(1H,3H,5H)-Pyrimidinetrione, 1-methyl-5-(1-methylethyl)-5-(2-propenyl)- Enallylpropymal	$C_{11}H_{16}N_2O_3$ 224.26	1861-21-8 56.5	3523 $176-8^{12}$	5-24-09-00225	bz 4; eth 4; EtOH 4; chl 4
11100	2,4,6(1H,3H,5H)-Pyrimidinetrione, 5-methyl-5-phenyl- Phenylmethylbarbituric acid	$C_{11}H_{10}N_2O_3$ 218.21	76-94-8 220	7275	5-24-09-00283	H_2O 1; EtOH 3; eth 3; alk 3
11101	2,4,6(1H,3H,5H)-Pyrimidinetrione, 5-(2-methylpropyl)-5-(2-propenyl)- Butalbital	$C_{11}H_{16}N_2O_3$ 224.26	77-26-9 138.5	1502	5-24-09-00231	H_2O 2; EtOH 3; eth 3; ace 3
11102	2,4,6(1H,3H,5H)-Pyrimidinetrione, 5-nitro- 5-Nitrobarbituric acid	$C_4H_3N_3O_5$ 173.09	480-68-2 180.5	6507	5-24-09-00105	H_2O 3; EtOH 3; eth 1
11103	2,4,6(1H,3H,5H)-Pyrimidinetrione, 5-phenyl-5-(2-propenyl)- Phenallymal	$C_{13}H_{12}N_2O_3$ 244.25	115-43-5 156.5	7162	5-24-09-00334	H_2O 2; EtOH 4; eth 4; bz 2
11104	2(1H)-Pyrimidinone, 4-amino- Cytosine	$C_4H_5N_3O$ 111.10	71-30-7 320-5 dec	2801	4-25-00-03654	H_2O 3; EtOH 2; eth 1; chl 2
11105	2(1H)-Pyrimidinone, 4-amino-5-(hydroxymethyl)- 5-Hydroxymethylcytosine	$C_5H_7N_3O_2$ 141.13	1123-95-1 >300 dec			
11106	2(1H)-Pyrimidinone, 4-amino-5-methyl- 5-Methylcytosine	$C_5H_7N_3O$ 125.13	554-01-8 270 dec	5970	4-25-00-03727	H_2O 3; EtOH 2; eth 1; acid 3
11107	4(1H)-Pyrimidinone, 2,6-diamino-5-(β-*D*-glucopyranosyloxy)- Vicine	$C_{10}H_{16}N_4O_7$ 304.26	152-93-2 239-42 dec	9880	4-25-00-04285	H_2O 2; EtOH 2; acid 4; alk 4
11108	4(1H)-Pyrimidinone, 2,3-dihydro-6-methyl-2-thioxo- Methylthiouracil	$C_5H_6N_2OS$ 142.18	56-04-2 330 dec	6048 sub	5-24-07-00049	H_2O 1; EtOH 2; eth 2; bz 2
11109	4(1H)-Pyrimidinone, 2,3-dihydro-6-propyl-2-thioxo- Propylthiouracil	$C_7H_{10}N_2OS$ 170.24	51-52-5 219	7882	5-24-07-00257	H_2O 2; EtOH 2; eth 1; bz 1
11110	4(1H)-Pyrimidinone, 2,3-dihydro-2-thioxo- 2-Thiouracil	$C_4H_4N_2OS$ 128.15	141-90-2 >340 dec	9298	5-24-06-00361	H_2O 2; EtOH 2; anh HF 3
11111	5H-Pyrimido[4,5-d]thiazolo[3,2-a]pyrimidine-8-ethanol, 2,7-dimethyl- Thiochrome	$C_{12}H_{14}N_4OS$ 262.34	92-35-3 228.8	sub	4-27-00-09599	H_2O 3; EtOH 2; eth 2; ace 2
11112	1H-Pyrrole Azole	C_4H_5N 67.09	109-97-7 -23.42	8025 129.79	5-20-05-00003 0.9698^{20}	H_2O 2; EtOH 3; eth 3; ace 3 1.5085^{20}
11113	1H-Pyrrole, 1-butyl-	$C_8H_{13}N$ 123.20	589-33-3	170.5	5-20-05-00010 0.8747^{20}	1.4727^{20}
11114	1H-Pyrrole-2-carboxaldehyde	C_5H_5NO 95.10	1003-29-8 46.5	218	5-21-07-00169	chl 2; os 4; lig 2 1.5939^{16}
11115	1H-Pyrrole-3-carboxylic acid, 2,4-dimethyl-, ethyl ester	$C_9H_{13}NO_2$ 167.21	2199-51-1 78.5	291	5-22-01-00427	eth 4; EtOH 4
11116	1H-Pyrrole-3-carboxylic acid, 2,5-dimethyl-, ethyl ester	$C_9H_{13}NO_2$ 167.21	2199-52-2 117.5	291; 130^{15}	5-22-01-00432	EtOH 4
11117	1H-Pyrrole, 3,4-diethyl-2-methyl- 3,4-Diethyl-2-methylpyrrole	$C_9H_{15}N$ 137.22	34874-30-1	202.5	4-20-00-02171 0.9100^{17}	1.4988^{17}
11118	1H-Pyrrole, 2,5-dihydro- 3-Pyrroline	C_4H_7N 69.11	109-96-6	8028 90.5	5-20-04-00274 0.9097^{20}	H_2O 4; ace 4; eth 4; EtOH 4 1.4664^{20}
11119	1H-Pyrrole, 2,4-dimethyl- 2,4-Dimethylpyrrole	C_6H_9N 95.14	625-82-1	168	5-20-05-00063 0.9236^{20}	H_2O 2; EtOH 4; eth 4; bz 4 1.5048^{20}
11120	1H-Pyrrole, 2,5-dimethyl-	C_6H_9N 95.14	625-84-3	171; 51^{8}	5-20-05-00064 0.9353^{20}	H_2O 1; EtOH 4; eth 4 1.5036^{20}
11121	1H-Pyrrole-2,5-dione	$C_4H_3NO_2$ 97.07	541-59-3 94	sub	5-21-10-00003 1.2493^{106}	H_2O 3; EtOH 3; eth 3
11122	1H-Pyrrole-2,5-dione, 1-ethyl- *N*-Ethylmaleimide	$C_6H_7NO_2$ 125.13	128-53-0 45.5	3778	5-21-10-00006	H_2O 2; EtOH 4; eth 4; chl 3
11123	1H-Pyrrole-2,5-dione, 1-phenyl- *N*-Phenylmaleimide	$C_{10}H_7NO_2$ 173.17	941-69-5 90.5	7270 162^{12}	5-21-10-00009	bz 4; eth 4; EtOH 4
11124	1H-Pyrrole, 1-ethyl-	C_6H_9N 95.14	617-92-5	129.5	5-20-05-00010 0.9009^{20}	EtOH 4 1.4841^{20}
11125	1H-Pyrrole, 2-ethyl-	C_6H_9N 95.14	1551-06-0	164	5-20-05-00061 0.9042^{20}	H_2O 1; EtOH 3; chl 3 1.4942^{20}
11126	1H-Pyrrole, 3-ethyl-2,4-dimethyl-	$C_8H_{13}N$ 123.20	517-22-6 0	199; 96^{16}	5-20-05-00099 0.913^{20}	H_2O 2; EtOH 3; eth 3; bz 3 1.4961^{20}
11127	1H-Pyrrole, 4-ethyl-2,3-dimethyl-	$C_8H_{13}N$ 123.20	491-18-9 16.5	200; 113^{16}	4-20-00-02152 0.915^{20}	H_2O 4
11128	1H-Pyrrole, 3-ethyl-4-methyl-	$C_7H_{11}N$ 109.17	488-92-6 3	70^{11}	5-20-05-00083 0.9059^{20}	eth 4; EtOH 4 1.4913^{20}
11129	Pyrrole, 2-isopropyl-	$C_7H_{11}N$ 109.17	7696-51-7	172	5-20-05-00081 0.908^{25}	eth 4; EtOH 4 1.491^{25}
11130	1H-Pyrrole, 1-methyl-	C_5H_7N 81.12	96-54-8 -56.32	112.81	5-20-05-00008 0.9145^{15}	H_2O 1; EtOH 5; eth 5 1.4875^{20}
11131	1H-Pyrrole, 2-methyl-	C_5H_7N 81.12	636-41-9 -35.6	147.6	5-20-05-00053 0.9446^{15}	H_2O 1; EtOH 5; eth 5 1.5035^{16}
11132	1H-Pyrrole, 3-methyl- 3-Methylpyrrole	C_5H_7N 81.12	616-43-3 -48.4	142.9; 45^{11}	5-20-05-00056	EtOH 5; eth 5 1.4970^{20}
11133	1H-Pyrrole, 2,2'-methylenebis-	$C_9H_{10}N_2$ 146.19	21211-65-4 73	164^{12}	5-23-06-00468	bz 4; eth 4; EtOH 4

No.	Name	Synonym	Mol. Form.	Mol. Wt.	CAS RN	mp/°C	Merck No.	bp/°C	Beil. Ref.	den/g cm^{-3}	Solubility	n_D
11134	1H-Pyrrole, 1-phenyl-		$C_{10}H_9N$	143.19	635-90-5	62		234	5-20-05-00012		H_2O 1; EtOH 3; eth 3; ace 3	
11135	1H-Pyrrole, 2-phenyl-	2-Phenylpyrrole	$C_{10}H_9N$	143.19	3042-22-6	129		272	5-20-07-00369		H_2O 1; EtOH 4; eth 4; bz 4	
11136	1H-Pyrrole, 1-(phenylmethyl)-		$C_{11}H_{11}N$	157.22	2051-97-0	15		247	5-20-05-00013	1.0183^{20}	H_2O 1; EtOH 4; eth 4	1.5655^{24}
11137	1H-Pyrrole-1-propanenitrile		$C_7H_8N_2$	120.15	43036-06-2			$132-3^{10}$	5-20-05-00023	1.048^{25}		1.5103^{20}
11138	1H-Pyrrole, 1-propyl-		$C_7H_{11}N$	109.17	5145-64-2			145.5	5-20-05-00010	0.8833^{20}	eth 4; EtOH 4	
11139	1H-Pyrrole, 3-propyl-	3-Propylpyrrole	$C_7H_{11}N$	109.17	1551-09-3			90^{30}; 49^{2}	5-20-05-00081		eth 3; ace 3; chl 2	1.4900^{25}
11140	1H-Pyrrole, 2,3,4,5-tetraiodo-	Iodopyrrole	C_4HI_4N	570.68	87-58-1	150 dec	4934		5-20-05-00045		ace 4; eth 4; chl 4	
11141	1H-Pyrrole, 2,3,4,5-tetramethyl-		$C_8H_{13}N$	123.20	1003-90-3	111		130^{7}	5-20-05-00099		EtOH 5; eth 5; ace 5; bz 4	
11142	1H-Pyrrole, 1,2,5-trimethyl-	1,2,5-Trimethylpyrrole	$C_7H_{11}N$	109.17	930-87-0			171	5-20-05-00065	0.807^{25}		1.4969^{20}
11143	Pyrrolidine	Azacyclopentane	C_4H_9N	71.12	123-75-1	-57.8	8026	86.5	5-20-01-00162	0.8586^{20}	H_2O 5; EtOH 3; eth 3; bz 2	1.4431^{20}
11144	3-Pyrrolidineacetic acid, 2-carboxy-4-(1-methylethenyl)-, [2*S*-(2α,3β,4β)]-	Kainic acid	$C_{10}H_{15}NO_4$	213.23	487-79-6		5157		4-22-00-01523			
11145	1-Pyrrolidineacetic acid, 2,5-dioxo-, ethyl ester	Aminoacetic acid, *N*-succinyl, ethyl ester	$C_8H_{11}NO_4$	185.18	14181-05-6	67		198^{32}	5-21-09-00474		H_2O 4; EtOH 4	
11146	Pyrrolidine, 1-butyl-		$C_8H_{17}N$	127.23	767-10-2			154.5	5-20-01-00170	0.816^{25}	EtOH 4	1.4373^{25}
11147	Pyrrolidine, 2-butyl-	2-Butylpyrrolidine	$C_8H_{17}N$	127.23	3446-98-8			174; 67^{18}	5-20-04-00222	0.8277^{20}	H_2O 4; EtOH 4	1.4490^{20}
11148	1,2-Pyrrolidinedicarboxylic acid, 1-(phenylmethyl) ester, (*S*)-		$C_{13}H_{15}NO_4$	249.27	1148-11-4	77			5-22-01-00084		chl 2	1.5310^{20}
11149	Pyrrolidine, 2,2-dimethyl-	2,2-Dimethylpyrrolidine	$C_6H_{13}N$	99.18	35018-15-6			106	4-20-00-01529	0.8211^{25}		1.4304^{25}
11150	Pyrrolidine, 2,4-dimethyl-		$C_6H_{13}N$	99.18	13603-04-8			116	4-20-00-01537	0.8297^{20}	H_2O 4; eth 4; EtOH 4	1.4325^{20}
11151	Pyrrolidine, 2,5-dimethyl-, *cis*-	*cis*-2,5-Dimethylpyrrolidine	$C_6H_{13}N$	99.18	39713-71-8			106.5	5-20-04-00146	0.8205^{20}	H_2O 4; eth 4; EtOH 4	1.4299^{20}
11152	2,5-Pyrrolidinedione	Succinimide	$C_4H_5NO_2$	99.09	123-56-8	126.5	8842	287 dec	5-21-09-00438	1.418^{25}	H_2O 3; EtOH 2; eth 2; ace 2	
11153	2,5-Pyrrolidinedione, 1-bromo-	*N*-Bromosuccinimide	$C_4H_4BrNO_2$	177.99	128-08-5	174	1428		5-21-09-00543	2.098^{25}	H_2O 2; eth 2; ace 4	
11154	2,5-Pyrrolidinedione, 1-chloro-	*N*-Chlorosuccinimide	$C_4H_4ClNO_2$	133.53	128-09-6	150	2165		5-21-09-00543	1.65^{25}	H_2O 2; EtOH 2; ace 3; bz 2	
11155	2,5-Pyrrolidinedione, 1-(chloromethyl)-		$C_5H_6ClNO_2$	147.56	54553-14-9	58		158^{12}	5-21-09-00460		ace 4; bz 4	
11156	2,5-Pyrrolidinedione, 1,3-dimethyl-3-phenyl-	Methsuximide	$C_{12}H_{13}NO_2$	203.24	77-41-8	52.5	5928	$121-2^{0.1}$	5-21-11-00209			
11157	2,5-Pyrrolidinedione, 1-ethyl-		$C_6H_9NO_2$	127.14	2314-78-5	26		236	5-21-09-00442		eth 4; EtOH 4	
11158	2,5-Pyrrolidinedione, 3-ethyl-3-methyl-	Ethosuximide	$C_7H_{11}NO_2$	141.17	77-67-8	64.5	3704		5-21-09-00595		H_2O 4	
11159	2,5-Pyrrolidinedione, 1-iodo-	*N*-Iodosuccinimide	$C_4H_4INO_2$	224.99	516-12-1	200.5	4936		5-21-09-00544	2.245^{25}	H_2O 4; EtOH 3; eth 2; ace 3	
11160	2,5-Pyrrolidinedione, 1-methyl-		$C_5H_7NO_2$	113.12	1121-07-9	71		234	5-21-09-00441		H_2O 3; EtOH 3; eth 4	
11161	2,5-Pyrrolidinedione, 1-methyl-3-phenyl-	Phensuximide	$C_{11}H_{11}NO_2$	189.21	86-34-0	72	7231		5-21-11-00188		EtOH 4; MeOH 4	
11162	2,5-Pyrrolidinedione, 1-phenyl-	Succinanil	$C_{10}H_9NO_2$	175.19	83-25-0	156	8838	400	5-21-09-00446	1.356^{25}	H_2O 1; EtOH 3; eth 3	
11163	2,5-Pyrrolidinedione, 1-(phenylmethyl)-	*N*-Benzylsuccinimide	$C_{11}H_{11}NO_2$	189.21	2142-06-5	103		393	5-21-09-00449		bz 4; eth 4; EtOH 4; chl 4	
11164	1-Pyrrolidineethanamine		$C_6H_{14}N_2$	114.19	7154-73-6			$66-70^{23}$	5-20-01-00405	0.901^{25}		1.4687^{20}
11165	1-Pyrrolidineethanol		$C_6H_{13}NO$	115.18	2955-88-6			$79-81^{13}$	4-20-00-00092	0.985^{25}		1.4713^{20}
11166	Pyrrolidine, 1-ethyl-		$C_6H_{13}N$	99.18	7335-06-0			104	5-20-01-00168	0.8156^{20}		1.4113^{15}
11167	2-Pyrrolidinemethanamine, 1-ethyl-		$C_7H_{16}N_2$	128.22	26116-12-1			$58-60^{16}$	5-22-08-00096	0.887^{25}		1.4665^{20}
11168	Pyrrolidine, 1-methyl-	*N*-Methylpyrrolidine	$C_5H_{11}N$	85.15	120-94-5			81	5-20-01-00166	0.8188^{20}	H_2O 4; eth 4	1.4247^{20}
11169	Pyrrolidine, 1-[2-(3-methylbutoxy)-2-phenylethyl]-	Amixetrine	$C_{17}H_{27}NO$	261.41	24622-72-8		507	121^{2}				1.4978^{22}
11170	Pyrrolidine, 1-nitroso-	*N*-Nitrosopyrrolidine	$C_4H_8N_2O$	100.12	930-55-2		6564	214	5-20-01-00521	1.085^{25}		1.4880^{25}
11171	Pyrrolidine, 1-phenyl-		$C_{10}H_{13}N$	147.22	4096-21-3			$119-20^{12}$	5-20-01-00182	1.0260^{25}	eth 3	1.5813^{20}

No.	Name Synonym	Mol. Form. Mol. Wt.	CAS RN mp/°C	Merck No. bp/°C	Beil. Ref. den/g cm^{-3}	Solubility n_D
11172	Pyrrolidine, 1-(phenylmethyl)-	$C_{11}H_{15}N$ 161.25	29897-82-3	$45^{0.7}$	5-20-01-00184 0.965^{25}	1.5270^{20}
11173	Pyrrolidine, 1-propyl-	$C_7H_{15}N$ 113.20	7335-07-1 105	132.5	5-20-01-00169 0.8171^{20}	1.4389^{20}
11174	Pyrrolidine, 3-propyl-	$C_7H_{15}N$ 113.20	116632-47-4	159	4-20-00-01598 0.8450^{20}	1.4469^{20}
11175	Pyrrolidine, 2,2,4-trimethyl- 2,2,4-Trimethylpyrrolidine	$C_7H_{15}N$ 113.20	35018-28-1	119	5-20-04-00198 0.8063^{25}	1.4259^{25}
11176	Pyrrolidine, 2,2,5-trimethyl-	$C_7H_{15}N$ 113.20	6496-48-6	112	4-20-00-01603 0.7980^{25}	1.4223^{25}
11177	Pyrrolidinium, 2-carboxy-1,1-dimethyl-, hydroxide, inner salt, (*S*)- Stachydrine	$C_7H_{13}NO_2$ 143.19	471-87-4 235	8729		H_2O 4; EtOH 4; EtOH 4
11178	Pyrrolidinium, 2-carboxy-4-hydroxy-1,1-dimethyl-, hydroxide, inner salt, (2*S*-*trans*)- Betonicine	$C_7H_{13}NO_3$ 159.19	515-25-3 252 dec	1209		EtOH 4
11179	Pyrrolidinium, 3-[(cyclopentylhydroxyphenylacetyl)oxy]-1,1-dimethyl-, bromide Glycopyrrolate	$C_{19}H_{28}BrNO_3$ 398.34	596-51-0 192.5	4397	5-21-01-00015	
11180	2-Pyrrolidinone 2-Pyrrolidone	C_4H_7NO 85.11	616-45-5 25	8027 251; 133^{12}	5-21-06-00317 1.120^{20}	H_2O 4; EtOH 4; eth 4; bz 4 1.4806^{30}
11181	2-Pyrrolidinone, 1,5-dimethyl-	$C_6H_{11}NO$ 113.16	5075-92-3	217; 87.5^{10}	5-21-06-00431	H_2O 4; eth 4 1.4650^{20}
11182	2-Pyrrolidinone, 3,3-dimethyl- 3,3-Dimethyl-2-pyrrolidone	$C_6H_{11}NO$ 113.16	4831-43-0 66	237	0-21-00-00242	H_2O 4; ace 4; eth 4; EtOH 4
11183	2-Pyrrolidinone, 1-ethenyl-	C_6H_9NO 111.14	88-12-0 13.5	193^{400}, 93^{11}	5-21-06-00330 1.04^{20}	
11184	2-Pyrrolidinone, 1-(2-hydroxyethyl)-	$C_6H_{11}NO_2$ 129.16	3445-11-2 20	295	5-21-06-00341 1.1435^{20}	
11185	2-Pyrrolidinone, 1-methyl- *N*-Methyl-2-pyrrolidone	C_5H_9NO 99.13	872-50-4 -24	202	5-21-06-00321 1.0230^{25}	H_2O 4; eth 3; ace 3; chl 3 1.4684^{20}
11186	2-Pyrrolidinone, 5-methyl-	C_5H_9NO 99.13	108-27-0 43	248	5-21-06-00431 1.0458^{20}	
11187	3-Pyrrolidinone, 1-methyl-	C_5H_9NO 99.13	68165-06-0	46-7^{18}	5-21-06-00389 0.9646^{25}	1.4431^{25}
11188	2-Pyrrolidinone, 1-[4-(1-pyrrolidinyl)-2-butynyl]- Oxotremorine	$C_{12}H_{18}N_2O$ 206.29	70-22-4	6904 $124^{0.1}$	5-21-06-00370 0.991^{25}	1.5160^{20}
11189	2-Pyrroline, 2-methyl-	C_5H_9N 83.13	100791-95-5	96	0-20-00-00135 0.8995^{22}	H_2O 4; eth 4; EtOH 4
11190	Pyrrolo[1,2-a]pyrimidine, 2,3,4,6,7,8-hexahydro-	$C_7H_{12}N_2$ 124.19	3001-72-7	95-8^7	5-23-05-00239 1.005^{25}	1.5190^{20}
11191	Pyrrolo[2,3-b]indol-5-ol, 1,2,3,3a,8,8a-hexahydro-1,3a,8-trimethyl-, methylcarbamate (ester), (3a*S*-*cis*)- Physostigmine	$C_{15}H_{21}N_3O_2$ 275.35	57-47-6 105.5	7357	5-23-11-00401	H_2O 2; EtOH 3; eth 3; bz 3
11192	Pyrrolo[2,1-b]quinazolin-3-ol, 1,2,3,9-tetrahydro-, (*R*)- Vasicine	$C_{11}H_{12}N_2O$ 188.23	6159-55-3 211.5	9842	5-23-12-00040	H_2O 2; EtOH 3; eth 2; ace 3
11193	1,1':2',1'':2'',1'''-Quaterphenyl	$C_{24}H_{18}$ 306.41	641-96-3 118.5	420; $193^{1.0}$	4-05-00-02729	H_2O 1; EtOH 2; eth 3; ace 3
11194	1,1':4',1'':4'',1'''-Quaterphenyl	$C_{24}H_{18}$ 306.41	135-70-6 320	428^{18}	4-05-00-02730	H_2O 1; EtOH 1; eth 1; bz 3
11195	Quinazoline 1,3-Benzodiazine	$C_8H_6N_2$ 130.15	253-82-7 48	8062 241	5-23-07-00129	H_2O 4; EtOH 3; eth 3; ace 3
11196	Quinazoline, 3,4-dihydro-3-phenyl- Orexine	$C_{14}H_{12}N_2$ 208.26	612-97-5 96.5	dec	4-23-00-01108 1.290^4	bz 4; eth 4; EtOH 4; chl 4
11197	6-Quinazolinesulfonamide, 7-chloro-2-ethyl-1,2,3,4-tetrahydro-4-oxo- Quinethazone	$C_{10}H_{12}ClN_3O_3S$ 289.74	73-49-4	8067		tfa 3
11198	4(1H)-Quinazolinone, 2,3-dihydro-1-(4-morpholinylacetyl)-3-phenyl- Moquizone	$C_{20}H_{21}N_3O_3$ 351.41	19395-58-5 136	6174		chl 3
11199	4(3H)-Quinazolinone, 2-methyl-3-(2-methylphenyl)- Methaqualone	$C_{16}H_{14}N_2O$ 250.30	72-44-6 120	5872	5-24-03-00132	eth 4; EtOH 4; chl 4
11200	4(1H)-Quinazolinone, 1-methyl-2-(phenylmethyl)- Glycosine	$C_{16}H_{14}N_2O$ 250.30	6873-15-0 161.5	4398	5-24-04-00363	
11201	Quinclorac 3,7-Dichloroquinoline-8-carboxylic acid	$C_{10}H_5Cl_2NO_2$ 242.06	84087-01-4 274		1.75	
11202	2-Quinolinamine 2-Aminoquinoline	$C_9H_8N_2$ 144.18	580-22-3 131.5	sub	5-22-10-00220	H_2O 4; EtOH 3; eth 3; ace 3
11203	4-Quinolinamine	$C_9H_8N_2$ 144.18	578-68-7 154.8	180^{12}	5-22-10-00241	H_2O 3; EtOH 4; eth 4; bz 3
11204	5-Quinolinamine	$C_9H_8N_2$ 144.18	611-34-7 110	310; 184^{10}	5-22-10-00297	H_2O 2; EtOH 4; eth 4; bz 3
11205	6-Quinolinamine 6-Aminoquinoline	$C_9H_8N_2$ 144.18	580-15-4 114	187^{12}	5-22-10-00303	H_2O 2; EtOH 3; eth 2; NH_3 3

No.	Name	Synonym	Mol. Form.	Mol. Wt.	CAS RN	mp/°C	Merck No.	bp/°C	Beil. Ref.	den/g cm^{-3}	Solubility	n_D
11206	8-Quinolinamine		$C_9H_8N_2$	144.18	578-66-5	70		157^{19}	5-22-10-00316		H_2O 4; EtOH 4	
11207	4-Quinolinamine, 3-bromo-	4-Amino-3-bromoquinoline	$C_9H_7BrN_2$	223.07	36825-36-2	203		sub	4-22-00-04662		eth 4; EtOH 4	
11208	2-Quinolinamine, 4-methyl-	Lepidine, 2-amino-	$C_{10}H_{10}N_2$	158.20	27063-27-0	133		320	5-22-10-00375		eth 4; EtOH 4; chl 4	
11209	3-Quinolinamine, 2-methyl-		$C_{10}H_{10}N_2$	158.20	21352-22-7	160		278; 198^{16}	5-22-10-00347		bz 4; eth 4; EtOH 4	
11210	4-Quinolinamine, 2-methyl-		$C_{10}H_{10}N_2$	158.20	6628-04-2	169		333	5-22-10-00347		H_2O 2; EtOH 4; eth 4; ace 4	
11211	7-Quinolinamine, 2-methyl-	7-Amino-2-methylquinoline	$C_{10}H_{10}N_2$	158.20	64334-96-9	148		304	5-22-10-00364		ace 4; bz 4; eth 4; EtOH 4	
11212	8-Quinolinamine, 6-methyl-	8-Amino-6-methylquinoline	$C_{10}H_{10}N_2$	158.20	68420-93-9	73		sub	4-22-00-04822		ace 4; bz 4; eth 4; EtOH 4	
11213	6-Quinolinamine, *N,N*,2-trimethyl-		$C_{12}H_{14}N_2$	186.26	92-99-9	101		319	2-22-00-00361		ctc 3; CS_2 3	
11214	Quinoline	1-Azanaphthalene	C_9H_7N	129.16	91-22-5	-14.78	8097	237.1	5-20-07-00276	1.0977^{15}	H_2O 1; EtOH 5; eth 5; ace 5	1.6268^{20}
11215	Quinoline, 6-amino-3-bromo-		$C_9H_7BrN_2$	223.07	7101-96-4	106		sub	5-22-10-00312		EtOH 4	
11216	Quinoline, 2-[2-(1,3-benzodioxol-5-yl)ethyl]-4-methoxy-	Cusparine	$C_{19}H_{17}NO_3$	307.35	529-92-0	92(form a); 116(form b)	2687		2-27-00-00545		ace 4; bz 4; eth 4; EtOH 4	
11217	Quinoline, 3-bromo-	3-Bromoquinoline	C_9H_6BrN	208.06	5332-24-1	13.3		275	5-20-07-00318		chl 3; HOAc 4	1.6641^{20}
11218	Quinoline, 5-bromo-	5-Bromoquinoline	C_9H_6BrN	208.06	4964-71-0	52		280; 106^{12}	5-20-07-00319		acid 3	
11219	Quinoline, 6-bromo-	6-Bromoquinoline	C_9H_6BrN	208.06	5332-25-2	24		281	5-20-07-00319		EtOH 3; eth 3; acid 3	
11220	Quinoline, 7-bromo-	7-Bromoquinoline	C_9H_6BrN	208.06	4965-36-0	34		290	4-20-00-03390		EtOH 3; eth 3; acid 4	
11221	Quinoline, 6-bromo-5-nitro-	6-Bromo-5-nitroquinoline	$C_9H_5BrN_2O_2$	253.05	98203-04-4	130		sub	0-20-00-00376		eth 4; EtOH 4	
11222	Quinoline, 2-butyl-	2-Butylquinoline	$C_{13}H_{15}N$	185.27	7661-39-4			153^{14}	5-20-07-00482	1.003^{20}		1.5699^{20}
11223	2-Quinolinecarbonitrile	2-Cyanoquinoline	$C_{10}H_6N_2$	154.17	1436-43-7	95		160^{23}	5-22-03-00190		H_2O 3; EtOH 4; eth 4; bz 4	
11224	4-Quinolinecarbonitrile		$C_{10}H_6N_2$	154.17	2973-27-5	103.5		242.5	5-22-03-00206		ace 4; bz 4; eth 4; EtOH 4	
11225	4-Quinolinecarboxaldehyde	Cinchoninaldehyde	$C_{10}H_7NO$	157.17	4363-93-3	51		122^{4}	5-21-08-00443		eth 4; tol 4	
11226	4-Quinolinecarboxamide, 2-butoxy-*N*-[2-(diethylamino)ethyl]-, monohydrochloride	Dibucaine hydrochloride	$C_{20}H_{30}ClN_3O_2$	379.93	61-12-1	94 dec	3016		5-22-05-00278		chl 3	
11227	2-Quinolinecarboxylic acid	Quinaldic acid	$C_{10}H_7NO_2$	173.17	93-10-7	156	8054		5-22-03-00183		H_2O 3; bz 4	
11228	7-Quinolinecarboxylic acid		$C_{10}H_7NO_2$	173.17	1078-30-4	249.5		sub	5-22-03-00217		EtOH 4	
11229	8-Quinolinecarboxylic acid	8-Carboxyquinoline	$C_{10}H_7NO_2$	173.17	86-59-9	187	8099	sub	5-22-03-00217		EtOH 4	
11230	2-Quinolinecarboxylic acid, 4,8-dihydroxy-	Xanthurenic acid	$C_{10}H_7NO_4$	205.17	59-00-7	289	9977		4-22-00-02513		H_2O 1; EtOH 3; eth 2; bz 2	
11231	1(2H)-Quinolinecarboxylic acid, 2-ethoxy-, ethyl ester	EEDQ	$C_{14}H_{17}NO_3$	247.29	16357-59-8	56.5	3488	$125\text{-}8^{0.1}$	5-21-03-00028		chl 3	
11232	2-Quinolinecarboxylic acid, 4-hydroxy-	Kynurenic acid	$C_{10}H_7NO_3$	189.17	492-27-3	282.5	5206		5-22-06-00280		H_2O 2; EtOH 3; eth 1; os 2	
11233	4-Quinolinecarboxylic acid, 3-hydroxy-2-phenyl-	Oxycinchophen	$C_{16}H_{11}NO_3$	265.27	485-89-2	206-7 dec	6910		4-22-00-02383		bz 4; EtOH 4; HOAc 4	
11234	4-Quinolinecarboxylic acid, 6-methoxy-	Quininic acid	$C_{11}H_9NO_3$	203.20	86-68-0	285 dec	8092	sub	4-22-00-02297		H_2O 2; EtOH 3; eth 2; bz 2	
11235	4-Quinolinecarboxylic acid, 2-phenyl-	Cinchophen	$C_{16}H_{11}NO_2$	249.27	132-60-5	214.5	2290		5-22-03-00484		H_2O 1; EtOH 3; eth 3; ace 2	
11236	4-Quinolinecarboxylic acid, 2-phenyl-, 2-propenyl ester		$C_{19}H_{15}NO_2$	289.33	524-34-5	31.5		262^{15}	4-22-00-01361		ace 4; eth 4; EtOH 4	
11237	Quinoline, 2-chloro-	2-Chloroquinoline	C_9H_6ClN	163.61	612-62-4	38		266; 153^{22}	5-20-07-00312	1.2464^{25}	H_2O 1; EtOH 4; eth 4; bz 3	1.6342^{25}
11238	Quinoline, 4-chloro-	4-Chloroquinoline	C_9H_6ClN	163.61	611-35-8	34.5		262; 130^{15}	5-20-07-00314	1.251^{25}	H_2O 2; EtOH 4; eth 4; dil HCl 3	
11239	Quinoline, 5-chloro-	5-Chloroquinoline	C_9H_6ClN	163.61	635-27-8	45		256	5-20-07-00314		EtOH 3	
11240	Quinoline, 6-chloro-	6-Chloroquinoline	C_9H_6ClN	163.61	612-57-7	43.8		263	5-20-07-00315			1.6110^{56}
11241	Quinoline, 7-chloro-	7-Chloroquinoline	C_9H_6ClN	163.61	612-61-3	31.5		267.5	5-20-07-00315	1.2158^{58}	H_2O 2; EtOH 4; eth 4; ace 4	1.6108^{58}
11242	Quinoline, 8-chloro-	8-Chloroquinoline	C_9H_6ClN	163.61	611-33-6	-20		288.5	5-20-07-00315	1.2834^{14}	H_2O 3; EtOH 4; eth 4; ace 4	1.6408^{14}

No.	Name Synonym	Mol. Form. Mol. Wt.	CAS RN mp/°C	Merck No. bp/°C	Beil. Ref. den/g cm^{-3}	Solubility n_D
11243	Quinoline, 2-chloro-4-methyl-	$C_{10}H_8ClN$ 177.63	634-47-9 59	 296	5-20-07-00396	H_2O 1; EtOH 4; eth 4; chl 4
11244	Quinoline, 2-chloro-8-methyl- 2-Chloro-8-methylquinoline	$C_{10}H_8ClN$ 177.63	4225-85-8 61	 287	5-20-07-00407	bz 4; eth 4; EtOH 4; chl 4
11245	Quinoline, 4-chloro-2-methyl-	$C_{10}H_8ClN$ 177.63	4295-06-1 42.5	 269.5	5-20-07-00383 0.881^{25}	H_2O 2; EtOH 3; eth 3; bz 3 1.6224^{20}
11246	Quinoline, decahydro-, *cis*- *cis*-Decahydroquinoline	$C_9H_{17}N$ 139.24	10343-99-4 -40	 206; 90^{20}	5-20-04-00405 0.9426^{20}	H_2O 2; EtOH 3; eth 3 1.4926^{20}
11247	Quinoline, decahydro-, *trans*-(±)- Quinoline, decahydro (*trans, DL*)	$C_9H_{17}N$ 139.24	105728-23-2 48	 204	5-20-04-00405 0.9610^{22}	ace 4; bz 4; eth 4; EtOH 4 1.4692^{58}
11248	Quinoline, 2,4-dichloro- 2,4-Dichloroquinoline	$C_9H_5Cl_2N$ 198.05	703-61-7 67.5	 281	4-20-00-03382	bz 4; eth 4; EtOH 4; chl 4
11249	Quinoline, 2,7-dichloro- 2,7-Dichloroquinoline	$C_9H_5Cl_2N$ 198.05	613-77-4 120	 100^{2}	4-20-00-03383	eth 4; EtOH 4
11250	Quinoline, 4,5-dichloro-	$C_9H_5Cl_2N$ 198.05	21617-18-5 118	 $134^{5.5}$	4-20-00-03383	eth 4; EtOH 4
11251	Quinoline, 4,7-dichloro- 4,7-Dichloroquinoline	$C_9H_5Cl_2N$ 198.05	86-98-6 93	 148^{10}	5-20-07-00316	chl 2
11252	Quinoline, 5,8-dichloro-	$C_9H_5Cl_2N$ 198.05	703-32-2 97.5	 sub	5-20-07-00317	eth 4; EtOH 4
11253	Quinoline, 2-[2-(3,4-dimethoxyphenyl)ethyl]-4-methoxy- Galipine	$C_{20}H_{21}NO_3$ 323.39	525-68-8 115.5	4247	4-21-00-02656	ace 4; bz 4; eth 4; EtOH 4
11254	Quinoline, 2,3-dimethyl- 2,3-Dimethylquinoline	$C_{11}H_{11}N$ 157.22	1721-89-7 68.5	 263	5-20-07-00431 1.1013^{25}	eth 4; EtOH 4; lig 4
11255	Quinoline, 2,4-dimethyl- 4-Methylquinaldine	$C_{11}H_{11}N$ 157.22	1198-37-4	 265	5-20-07-00431 1.0611^{15}	H_2O 2; EtOH 4; eth 4; chl 2 1.6075^{20}
11256	Quinoline, 2,5-dimethyl-	$C_{11}H_{11}N$ 157.22	26190-82-9 61	 264.5	4-20-00-03524	bz 4; eth 4; EtOH 4
11257	Quinoline, 2,6-dimethyl-	$C_{11}H_{11}N$ 157.22	877-43-0 60	 266.5	5-20-07-00433	H_2O 2; EtOH 2; eth 2; bz 4
11258	Quinoline, 2,7-dimethyl- *m*-Toluquinaldine	$C_{11}H_{11}N$ 157.22	93-37-8 61	 264.5	5-20-07-00435	H_2O 2; EtOH 3; eth 3; chl 3
11259	Quinoline, 2,8-dimethyl-	$C_{11}H_{11}N$ 157.22	1463-17-8 27	 255.3	5-20-07-00436 1.0394^{20}	H_2O 2; EtOH 4; eth 4; ctc 3 1.6022^{20}
11260	Quinoline, 3,4-dimethyl- 3,4-Dimethylquinoline	$C_{11}H_{11}N$ 157.22	2436-92-2 73.5	 291; 114^{4}	5-20-07-00437	eth 4; EtOH 4
11261	Quinoline, 3,8-dimethyl- 3,8-Dimethylquinoline	$C_{11}H_{11}N$ 157.22	20668-29-5	 270	5-20-07-00437 1.049^{20}	 1.6063^{20}
11262	Quinoline, 4,6-dimethyl- 6-Methyllepidine	$C_{11}H_{11}N$ 157.22	826-77-7 22	 273.5; 140^{12}	5-20-07-00437 1.0577^{20}	eth 4; EtOH 4 1.6100^{20}
11263	Quinoline, 5,8-dimethyl- 5,8-Dimethylquinoline	$C_{11}H_{11}N$ 157.22	2623-50-9 4.5	 266	5-20-07-00438 1.070^{21}	eth 4; EtOH 4
11264	Quinoline, 6,8-dimethyl- 6,8-Dimethylquinoline	$C_{11}H_{11}N$ 157.22	2436-93-3	 268.5	5-20-07-00439 1.0665^{4}	eth 4; EtOH 4
11265	Quinoline, 6-ethoxy-1,2-dihydro-2,2,4-trimethyl- Ethoxyquin	$C_{14}H_{19}NO$ 217.31	91-53-2	3710 $123\text{-}5^{2}$	5-21-03-00095 1.026^{25}	 1.569^{25}
11266	Quinoline, 2-ethyl- 2-Ethylquinoline	$C_{11}H_{11}N$ 157.22	1613-34-9	 245.5; 134^{16}	5-20-07-00428 1.050^{17}	eth 4; EtOH 4; chl 4 1.5979^{23}
11267	Quinoline, 3-ethyl- 3-Ethylquinoline	$C_{11}H_{11}N$ 157.22	1873-54-7	 266	5-20-07-00428 1.0508^{20}	 1.6160^{20}
11268	Quinoline, 6-ethyl- 6-Ethylquinoline	$C_{11}H_{11}N$ 157.22	19655-60-8	 $135\text{-}6^{14}$	5-20-07-00429 1.043^{25}	 1.6009^{25}
11269	Quinoline, 2-ethyl-3-methyl- 2-Ethyl-3-methylquinoline	$C_{12}H_{13}N$ 171.24	27356-52-1 57	 269; 140^{12}	5-20-07-00456	bz 4; eth 4; EtOH 4
11270	Quinoline, 2-ethyl-4-methyl- Lepidine, 2-ethyl-	$C_{12}H_{13}N$ 171.24	33357-44-7	 272.5	5-20-07-00456 1.025^{20}	 1.5941^{20}
11271	Quinoline, 3-ethyl-6-methyl-	$C_{12}H_{13}N$ 171.24	105688-74-2	 284.5	4-20-00-03555 1.0280^{20}	 1.5955^{20}
11272	Quinoline, 4-iodo- 4-Iodoquinoline	C_9H_6IN 255.06	16560-43-3 98.5	 sub	5-20-07-00322	eth 4; EtOH 4
11273	Quinoline, 6-iodo-	C_9H_6IN 255.06	13327-31-6 91	 $120^{1.0}$	5-20-07-00322	eth 4; EtOH 4
11274	Quinoline, 4-methoxy- 4-Methoxyquinoline	$C_{10}H_9NO$ 159.19	607-31-8 41	 245	5-21-03-00229	H_2O 1; EtOH 3; eth 3; bz 3
11275	Quinoline, 6-methoxy- 6-Methoxyquinoline	$C_{10}H_9NO$ 159.19	5263-87-6 26.5	 306; 153^{12}	5-21-03-00244 1.152^{20}	EtOH 3; eth 3; chl 3; dil HCl 3
11276	Quinoline, 8-methoxy- 8-Methoxyquinoline	$C_{10}H_9NO$ 159.19	938-33-0 49.5	 283; 164^{14}	5-21-03-00254 1.034^{29}	EtOH 3; eth 3; bz 3; peth 3
11277	Quinoline, 2-methyl- Quinaldine	$C_{10}H_9N$ 143.19	91-63-4 -1.5	8055 246.5	5-20-07-00375 1.06^{25}	H_2O 2; EtOH 3; eth 3; ace 3 1.6116^{20}
11278	Quinoline, 3-methyl- 3-Methylquinoline	$C_{10}H_9N$ 143.19	612-58-8 16.5	 259.8	5-20-07-00388 1.0673^{20}	ace 4; eth 4; EtOH 4 1.6171^{20}
11279	Quinoline, 4-methyl- Lepidine	$C_{10}H_9N$ 143.19	491-35-0 9.5	5324 262	5-20-07-00389 1.083^{20}	H_2O 2; EtOH 3; eth 3; ace 3 1.6200^{20}
11280	Quinoline, 5-methyl- 5-Methylquinoline	$C_{10}H_9N$ 143.19	7661-55-4 19	 262.7	5-20-07-00399 1.0832^{20}	H_2O 2; EtOH 5; eth 5; ace 3 1.6219^{20}
11281	Quinoline, 6-methyl- 6-Methylquinoline	$C_{10}H_9N$ 143.19	91-62-3 -22	 258.6	5-20-07-00400 1.0654^{20}	H_2O 2; EtOH 3; eth 3; ace 3 1.6157^{20}

No.	Name Synonym	Mol. Form. Mol. Wt.	CAS RN mp/°C	Merck No. bp/°C	Beil. Ref. den/g cm^{-3}	Solubility n_D
11282	Quinoline, 7-methyl- *m*-Toluquinoline	$C_{10}H_9N$ 143.19	612-60-2 39	257.6	5-20-07-00402 1.0609^{20}	H_2O 2; EtOH 3; eth 3; ace 3 1.6150^{20}
11283	Quinoline, 8-methyl- 8-Methylquinoline	$C_{10}H_9N$ 143.19	611-32-5 -80	247.5	5-20-07-00405 1.0719^{20}	H_2O 2; EtOH 5; eth 5; ace 3 1.6164^{20}
11284	Quinoline, 5-nitro- 5-Nitroquinoline	$C_9H_6N_2O_2$ 174.16	607-34-1 74	sub	5-20-07-00325	H_2O 2; EtOH 3; bz 3; chl 2
11285	Quinoline, 6-nitro- 6-Nitroquinoline	$C_9H_6N_2O_2$ 174.16	613-50-3 153.5	$170^{0.2}$	5-20-07-00326	H_2O 3; EtOH 3; eth 2; bz 4
11286	Quinoline, 7-nitro- 7-Nitroquinoline	$C_9H_6N_2O_2$ 174.16	613-51-4 132.5	sub	5-20-07-00327	H_2O 3; EtOH 3; eth 3; chl 4
11287	Quinoline, 4-nitro-, 1-oxide 4-Nitroquinoline oxide	$C_9H_6N_2O_3$ 190.16	56-57-5 154		5-20-07-00324	
11288	Quinoline, 2-phenyl- 2-Phenylquinoline	$C_{15}H_{11}N$ 205.26	612-96-4 86	363; 194^{6}	5-20-08-00355	H_2O 2; EtOH 4; eth 4; ace 4
11289	Quinoline, 3-phenyl- 3-Phenylquinoline	$C_{15}H_{11}N$ 205.26	1666-96-2 52	205^{12}	5-20-08-00359	H_2O 2; EtOH 3; eth 3; ace 4
11290	Quinoline, 4-phenyl- 4-Phenylquinoline	$C_{15}H_{11}N$ 205.26	605-03-8 61.5	260^{77}; $156^{0.3}$	5-20-08-00360	H_2O 1; EtOH 4; eth 4; ace 4
11291	Quinoline, 6-phenyl- 6-Phenylquinoline	$C_{15}H_{11}N$ 205.26	612-95-3 110	260^{77}	5-20-08-00361 1.1945^{20}	H_2O 2; EtOH 4; eth 2; ace 4
11292	Quinoline, 2-propyl- 2-Propylquinoline	$C_{12}H_{13}N$ 171.24	1613-32-7	$130\text{-}1^{10}$	5-20-07-00453 1.038^{17}	1.5886^{23}
11293	5-Quinolinesulfonic acid, 8-ethoxy- Actinoquinol	$C_{11}H_{11}NO_4S$ 253.28	15301-40-3 286 dec	134	5-22-07-00581	alk 3
11294	5-Quinolinesulfonic acid, 8-hydroxy- 8-Hydroxy-5-quinolinesulfonic acid	$C_9H_7NO_4S$ 225.22	84-88-8 322.5	4780	5-22-07-00581	H_2O 2
11295	5-Quinolinesulfonic acid, 8-hydroxy-7-iodo- Ferron	$C_9H_6INO_4S$ 351.12	547-91-1 260 dec	4757	5-22-07-00582	H_2O 2; EtOH 2; eth 1; bz 1
11296	Quinoline, 1,2,3,4-tetrahydro-	$C_9H_{11}N$ 133.19	635-46-1 20	251	5-20-06-00293 1.0588^{20}	H_2O 3; EtOH 5; eth 5; chl 3 1.6062^{19}
11297	Quinoline, 5,6,7,8-tetrahydro- 2,3-Cyclohexenopyridine	$C_9H_{11}N$ 133.19	10500-57-9	222	5-20-06-00292 1.0304^{13}	H_2O 2; EtOH 3; eth 3; ace 3 1.5435^{20}
11298	Quinoline, 1,2,3,4-tetrahydro-6-methoxy-	$C_{10}H_{13}NO$ 163.22	120-15-0 42.5	284; 128^{1}	5-21-02-00314	chl 3 1.5718^{20}
11299	Quinoline, 1,2,3,4-tetrahydro-1-methyl-	$C_{10}H_{13}N$ 147.22	491-34-9	246	5-20-06-00294 1.022^{20}	EtOH 4 1.5802^{23}
11300	Quinoline, 2,3,6-trimethyl-	$C_{12}H_{13}N$ 171.24	2437-73-2 86.5	285	4-20-00-03557	eth 4; EtOH 4
11301	Quinoline, 2,3,8-trimethyl-	$C_{12}H_{13}N$ 171.24	4945-28-2	283	4-20-00-03557 1.0069^{20}	H_2O 1; EtOH 3; eth 3; bz 2
11302	Quinoline, 2,4,6-trimethyl- 4,6-Dimethylquinaldine	$C_{12}H_{13}N$ 171.24	2243-89-2 68	280	5-20-07-00458	ace 4; bz 4; eth 4; EtOH 4
11303	Quinoline, 2,4,7-trimethyl- 2,4,7-Trimethylquinoline	$C_{12}H_{13}N$ 171.24	71633-43-7 63.5	280.5	5-20-07-00458 1.0337^{20}	H_2O 2; EtOH 4; eth 4; ace 4 1.5973^{24}
11304	Quinoline, 2,5,7-trimethyl-	$C_{12}H_{13}N$ 171.24	102871-67-0	287; $107^{1.2}$	4-20-00-03559	H_2O 1; EtOH 4; eth 4 1.5980^{20}
11305	5-Quinolinol 5-Hydroxyquinoline	C_9H_7NO 145.16	578-67-6 226 dec	sub	5-21-03-00240	H_2O 3; EtOH 2; eth 2; bz 3
11306	6-Quinolinol	C_9H_7NO 145.16	580-16-5 195	360	5-21-03-00244	H_2O 1; EtOH 2; eth 2; bz 1
11307	7-Quinolinol 7-Hydroxyquinoline	C_9H_7NO 145.16	580-20-1 239	sub	5-21-03-00249	EtOH 4
11308	8-Quinolinol 8-Hydroxyquinoline	C_9H_7NO 145.16	148-24-3 75.5	4778 267	5-21-03-00252 1.034^{20}	H_2O 1; EtOH 4; eth 1; ace 3
11309	8-Quinolinol, benzoate (ester) Benzoxiquine	$C_{16}H_{11}NO_2$ 249.27	86-75-9	1122	5-21-03-00259	chl 2
11310	8-Quinolinol, 5-bromo-	C_9H_6BrNO 224.06	1198-14-7 124	sub	5-21-03-00288	H_2O 4; EtOH 4; bz 4; chl 4
11311	8-Quinolinol, 5-chloro- Cloxyquin	C_9H_6ClNO 179.61	130-16-5 130	2416	5-21-03-00279	
11312	8-Quinolinol, 5-chloro-7-iodo- Iodochlorhydroxyquin	C_9H_5ClINO 305.50	130-26-7 178.5	4924	5-21-03-00294	EtOH 2; HOAc 3
11313	8-Quinolinol, 5,7-dibromo- Broxyquinoline	$C_9H_5Br_2NO$ 302.95	521-74-4 196	1441 sub	5-21-03-00290	H_2O 1; EtOH 3; eth 2; ace 3
11314	8-Quinolinol, 5,7-dichloro- Chloroxine	$C_9H_5Cl_2NO$ 214.05	773-76-2 179.5	2175	5-21-03-00286	EtOH 2; ace 2; bz 3; chl 2
11315	8-Quinolinol, 5,7-diiodo- Iodoquinol	$C_9H_5I_2NO$ 396.95	83-73-8 210	4935	5-21-03-00296	H_2O 2; EtOH 4; eth 2; bz 2
11316	4-Quinolinol, 2,8-dimethyl-	$C_{11}H_{11}NO$ 173.21	15644-80-1 262.3	sub	5-21-08-00321	EtOH 4
11317	6-Quinolinol, 2,4-dimethyl- 2,4-Dimethyl-6-hydroxyquinoline	$C_{11}H_{11}NO$ 173.21	64165-34-0 214	360 dec	2-21-00-00068	ace 4; EtOH 4
11318	8-Quinolinol, 2,4-dimethyl- Quinoline, 2,4-dimethyl-8-hydroxy	$C_{11}H_{11}NO$ 173.21	115310-98-0 65	281	4-21-00-01292	ace 4; bz 4; eth 4; EtOH 4
11319	6-Quinolinol, 2-methyl-	$C_{10}H_9NO$ 159.19	613-21-8 213	304.5	5-21-03-00339 1.1665^{25}	eth 4; EtOH 4
11320	7-Quinolinol, 6-methyl- Quinoline, 7-hydroxy-6-methyl	$C_{10}H_9NO$ 159.19	84583-53-9 244	sub	0-21-00-00111	bz 4; eth 4; EtOH 4
11321	8-Quinolinol, 2-methyl-	$C_{10}H_9NO$ 159.19	826-81-3 73.8	267	5-21-03-00341	H_2O 1; EtOH 3; eth 3; bz 3

No.	Name Synonym	Mol. Form. Mol. Wt.	CAS RN mp/°C	Merck No. bp/°C	Beil. Ref. den/g cm^{-3}	Solubility n_D
11322	8-Quinolinol, 6-methyl-	$C_{10}H_9NO$ 159.19	20984-33-2 95.5	 sub	4-21-00-01275	EtOH 4
11323	8-Quinolinol, 5-nitro- Nitroxoline	$C_9H_6N_2O_3$ 190.16	4008-48-4 180	6577	5-21-03-00297	
11324	8-Quinolinol, sulfate (2:1) (salt) 8-Hydroxyquinoline sulfate	$C_9H_8NO_3S$ 210.23	134-31-6 177.5	4779	5-21-03-00252	H_2O 4; EtOH 3; eth 1
11325	2(1H)-Quinolinone Carbostyril	C_9H_7NO 145.16	59-31-4 199.5	1831 sub	5-21-08-00217	H_2O 2; EtOH 4; eth 4
11326	2(1H)-Quinolinone, 4-bromo-	C_9H_6BrNO 224.06	938-39-6 266.5	 sub	5-21-08-00229	EtOH 4
11327	2(1H)-Quinolinone, 3,4-dihydro- Hydrocarbostyril	C_9H_9NO 147.18	553-03-7 163.5	4702 201^{45}	5-21-08-00057	eth 4; EtOH 4
11328	2(1H)-Quinolinone, 1-methyl-	$C_{10}H_9NO$ 159.19	606-43-9 74	 325	5-21-08-00218	H_2O 2; EtOH 3; eth 3; ace 3
11329	2(1H)-Quinolinone, 6-methyl-	$C_{10}H_9NO$ 159.19	4053-34-3 237	 240^{12}	5-21-08-00289	ace 4; bz 4; eth 4; EtOH 4
11330	4(1H)-Quinolinone, 1-methyl- Echinopsine	$C_{10}H_9NO$ 159.19	83-54-5	3472	5-21-08-00210	H_2O 3; EtOH 4; eth 2; bz 4
11331	2H-Quinolizine-1-methanol, octahydro-, (1*R*-*trans*)- Lupinine	$C_{10}H_{19}NO$ 169.27	486-70-4 70	5479 270	5-21-01-00338	H_2O 3; EtOH 3; eth 3; bz 3
11332	6-Quinoxalinamine 6-Aminoquinoxaline	$C_8H_7N_3$ 145.16	6298-37-9 159	 sub	4-25-00-02611	H_2O 4; EtOH 4; eth 3; bz 2
11333	Quinoxaline 1,4-Benzodiazine	$C_8H_6N_2$ 130.15	91-19-0 28	8107 229.5	5-23-07-00135 1.1334^{48}	H_2O 3; EtOH 5; eth 5; ace 5 1.6231^{48}
11334	Quinoxaline, 6-chloro- 6-Chloroquinoxaline	$C_8H_5ClN_2$ 164.59	5448-43-1 64	 117^{10}	5-23-07-00144	H_2O 2
11335	Quinoxaline, 2,6-dimethyl-	$C_{10}H_{10}N_2$ 158.20	60814-29-1 54	 268	0-23-00-00192	H_2O 4; ace 4; bz 4; EtOH 4
11336	Quinoxaline, 6-methoxy- 6-Methoxyquinoxaline	$C_9H_8N_2O$ 160.18	6639-82-3 58.9	 128^7	5-23-11-00461	ace 4; eth 4
11337	Quinoxaline, 2-methyl- 2-Methylquinoxaline	$C_9H_8N_2$ 144.18	7251-61-8 180.5	 244	5-23-07-00219	H_2O 5; EtOH 4; eth 5; ace 5
11338	Quinoxaline, 6-methyl- 6-Methylquinoxaline	$C_9H_8N_2$ 144.18	6344-72-5 218.5	 249; 141.5^{29}	5-23-07-00223 1.1164^{20}	H_2O 5; EtOH 5; eth 5; ace 5 1.6211^{18}
11339	Quinoxaline, 1,2,3,4-tetrahydro- 1,2,3,4-Tetrahydroquinoxaline	$C_8H_{10}N_2$ 134.18	3476-89-9 99	 289	5-23-06-00024	H_2O 3; EtOH 4; eth 4; bz 4
11340	(-)-*cis*-Resmethrin Cyclopropanecarboxylic acid, 2,2-dimethyl-3-(2-methyl-1-propenyl)-, [5-(phenylmethyl)- 3-furanyl] methyl ester	$C_{22}H_{26}O_3$ 338.45	10453-86-8 75			
11341	Retinol Vitamin A	$C_{20}H_{30}O$ 286.46	68-26-8 63.5	9918 $137\text{-}8^{10\text{-}6}$	4-06-00-04133	H_2O 1; EtOH 3; eth 3; ace 3
11342	Rheadan, 8-methoxy-16-methyl-2,3:10,11-bis[methylenebis(oxy)]-, (8β)- Rheadine	$C_{21}H_{21}NO_6$ 383.40	2718-25-4 257	8174 sub	4-27-00-06897	
11343	Ribitol Adonitol	$C_5H_{12}O_5$ 152.15	488-81-3 104	157	4-01-00-02832	H_2O 3; EtOH 3; eth 1; lig 1
11344	Riboflavin Riboflavine	$C_{17}H_{20}N_4O_6$ 376.37	83-88-5 280 dec	8201	4-26-00-02542	H_2O 1; EtOH 2; eth 1; ace 1
11345	*D*-Ribose	$C_5H_{10}O_5$ 150.13	50-69-1 88	8205	4-01-00-04211	H_2O 3; EtOH 2
11346	Sarpagan-17-al Vellosimine	$C_{19}H_{20}N_2O$ 292.38	6874-98-2 305.5	9846 sub 180		
11347	Sarpagan-10,17-diol Sarpagine	$C_{19}H_{22}N_2O_2$ 310.40	482-68-8 320	8336	5-23-13-00401	H_2O 1; EtOH 3
11348	9,10-Secocholesta-5,7,10(19)-trien-3-ol, (3β,5Z,7E)- Vitamin D3	$C_{27}H_{44}O$ 384.65	67-97-0 84.5	9929	4-06-00-04149	H_2O 1; os 3
11349	9,10-Secoergosta-5(10),6,8,22-tetraen-3-ol, (3β,6E,22E)- Tachysterol	$C_{28}H_{44}O$ 396.66	115-61-7	9002	4-06-00-04404	H_2O 1; EtOH 3; eth 3; ace 3
11350	9,10-Secoergosta-5,7,10(19),22-tetraen-3-ol, (3β,5Z,7E,22E)- Vitamin D2	$C_{28}H_{44}O$ 396.66	50-14-6 116.5	9928 sub	4-06-00-04404	H_2O 1; EtOH 3; eth 3; ace 3
11351	16,19-Secostrychnidine-10,16-dione, 4-hydroxy-19-methyl- Vomicine	$C_{22}H_{24}N_2O_4$ 380.44	125-15-5 282	9946	4-27-00-07873	EtOH 2; eth 2; ace 2; chl 4
11352	9,10-Secostrychnidin-10-oic acid, 2,3-dihydro-4-nitro-2,3-dioxo- Cacotheline	$C_{21}H_{21}N_3O_7$ 427.41	561-20-6 >300	1604	4-27-00-08014	H_2O 2
11353	Selenanthrene Selanthrene	$C_{12}H_8Se_2$ 310.12	262-30-6 181	 223^{11}	5-19-02-00057	EtOH 2; eth 2; CS_2 3
11354	Selenourea Carbamimidoselenoic acid	CH_4N_2Se 123.02	630-10-4	 200 dec	4-03-00-00435	H_2O 4
11355	*D*-Serine	$C_3H_7NO_3$ 105.09	312-84-5 229 dec	 dec	4-04-00-03118	H_2O 4; EtOH 1; eth 1; bz 1
11356	*DL*-Serine	$C_3H_7NO_3$ 105.09	302-84-1 246 dec		4-04-00-03119 1.603^{22}	H_2O 3; EtOH 1; eth 1; bz 1
11357	*L*-Serine (*S*)-2-Amino-3-hydroxypropanoic acid	$C_3H_7NO_3$ 105.09	56-45-1 228 dec	8411 sub 150	4-04-00-03118 1.6^{22}	H_2O 4; EtOH 1; eth 1; bz 1

No.	Name Synonym	Mol. Form. Mol. Wt.	CAS RN mp/°C	Merck No. bp/°C	Beil. Ref. den/g cm^{-3}	Solubility n_D
11358	Sesone Ethanol, 2-(2,4-dichlorophenoxy)-, hydrogen sulfate, sodium salt	$C_8H_8Cl_2NaO_5S$ 310.11	136-78-7 245 dec	3372		
11359	Sethoxydim (±)-2-[1-(Ethoxyimino)butyl]-5-[2-(ethylthio)propyl]-3-hydroxy-2-cyclohexen-1-one	$C_{17}H_{29}NO_3S$ 327.49	74051-80-2	8424 $>90^{0.00003}$	1.043^{25}	
11360	Silacyclohexane, 1,1-diphenyl- 1,1-Diphenylsilacyclohexane	$C_{17}H_{20}Si$ 252.43	18002-79-4	$193\text{-}8^{5}$	4-27-00-09833 1.0319^{25}	1.5779^{20}
11361	Silanamine, pentamethyl-	$C_5H_{15}NSi$ 117.27	2083-91-2	86	4-04-00-04010 0.7400^{20}	1.4379^{24}
11362	Silanamine, 1,1,1-trimethyl-*N*-phenyl-*N*-(trimethylsilyl)-	$C_{12}H_{23}NSi_2$ 237.49	4147-89-1 16.5	$98\text{-}9^{11}$	0.8951^{23}	1.4855^{20}
11363	Silanamine, 1,1,1-trimethyl-*N*-(trimethylsilyl)- Hexamethyldisilazane	$C_6H_{19}NSi_2$ 161.39	999-97-3	10158 125	4-04-00-04014 0.7741^{25}	1.4090^{20}
11364	Silane, [1,1'-biphenyl]-4-yltrichloro-	$C_{12}H_9Cl_3Si$ 287.65	18030-61-0	$209\text{-}11^{30}$	4-16-00-01572 1.3190^{20}	ctc 3; CS_2 3
11365	Silane, (bromomethyl)chlorodimethyl- (Bromomethyl)dimethylchlorosilane	$C_3H_8BrClSi$ 187.54	16532-02-8	131	4-04-00-04024 1.375^{25}	1.4630^{25}
11366	Silane, (bromomethyl)trimethyl- Bromomethyltrimethylsilane	$C_4H_{11}BrSi$ 167.12	18243-41-9	116.5	4-04-00-03878 1.170^{25}	1.4460^{20}
11367	Silane, (4-bromophenoxy)trimethyl-	$C_9H_{13}BrOSi$ 245.19	17878-44-3	126^{25}	4-06-00-01056 1.2619^{20}	1.5145^{20}
11368	Silane, bromotriethyl-	$C_6H_{15}BrSi$ 195.17	1112-48-7	$66\text{-}7^{24}$	4-04-00-04051 1.1403^{20}	1.4561^{20}
11369	Silane, butylchlorodimethyl-	$C_6H_{15}ClSi$ 150.72	1000-50-6	139	4-04-00-04068 0.8751^{20}	1.5145^{20}
11370	Silane, butyltrichloro- Trichlorobutylsilane	$C_4H_9Cl_3Si$ 191.56	7521-80-4	148.5	4-04-00-04246 1.1606^{20}	eth 3; bz 3; tol 3; AcOEt 3 1.4363^{20}
11371	Silane, butyltrimethyl- Trimethylbutylsilane	$C_7H_{18}Si$ 130.31	1000-49-3	115	4-04-00-03907 0.7353^{0}	H_2O 1
11372	Silane, chloro(chloromethyl)dimethyl-	$C_3H_8Cl_2Si$ 143.09	1719-57-9	115.5	4-04-00-04024 1.0865^{20}	1.4360^{20}
11373	Silane, chloro(dichloromethyl)dimethyl- (Dichloromethyl)dimethylchlorosilane	$C_3H_7Cl_3Si$ 177.53	18171-59-0 -48	149	4-04-00-04031 1.2369^{20}	1.461^{20}
11374	Silane, chlorodimethylphenyl-	$C_8H_{11}ClSi$ 170.71	768-33-2	$80\text{-}4^{16}$	4-16-00-01475 1.032^{20}	1.5082^{20}
11375	Silane, chlorodimethyl-2-propenyl-	$C_5H_{11}ClSi$ 134.68	4028-23-3	111	4-04-00-04080 0.8964^{20}	1.4195^{20}
11376	Silane, chloroethenyldimethyl-	C_4H_9ClSi 120.65	1719-58-0	83.5	4-04-00-04080 0.8744^{20}	1.4141^{20}
11377	Silane, chloroethyldimethyl-	$C_4H_{11}ClSi$ 122.67	6917-76-6	89.5	4-04-00-04035 0.8675^{20}	1.4105^{20}
11378	Silane, chloromethyl Chloromethylsilane	CH_5ClSi 80.59	993-00-0 -135	$7; -45^{63}$		
11379	Silane, (chloromethyl)dimethylphenyl-	$C_9H_{13}ClSi$ 184.74	1833-51-8	225	4-16-00-01376 1.0240^{25}	ctc 3; CS_2 3
11380	Silane, chloromethyldiphenyl-	$C_{13}H_{13}ClSi$ 232.78	144-79-6	295	4-16-00-01479 1.1277^{20}	1.5742^{20}
11381	Silane, chloromethylphenyl-	C_7H_9ClSi 156.69	1631-82-9	113^{100}	4-16-00-01473 1.0540^{20}	1.5171^{20}
11382	Silane, (chloromethyl)trimethyl-	$C_4H_{11}ClSi$ 122.67	2344-80-1	98.5	4-04-00-03877 0.879^{25}	1.4175^{20}
11383	Silane, (4-chlorophenoxy)trimethyl-	$C_9H_{13}ClOSi$ 200.74	17005-59-3	$214; 101^{14}$	4-06-00-00878 1.0320^{20}	1.4930^{20}
11384	Silane, chlorophenyl- Phenylchlorosilane	C_6H_7ClSi 142.66	4206-75-1	162.5	4-16-00-01472 1.0683^{20}	1.5340^{20}
11385	Silane, (3-chlorophenyl)trimethyl-	$C_9H_{13}ClSi$ 184.74	4405-42-9	206.5	4-16-00-01386 1.0071^{20}	1.5108^{20}
11386	Silane, (3-chloropropyl)trimethyl-	$C_6H_{15}ClSi$ 150.72	2344-83-4	151	4-04-00-03905 0.8789^{20}	1.4319^{20}
11387	Silane, chlorotriethoxy-	$C_6H_{15}ClO_3Si$ 198.72	4667-99-6 -51	156	4-01-00-01363 1.030^{20}	EtOH 4 1.3999^{20}
11388	Silane, chlorotriethyl-	$C_6H_{15}ClSi$ 150.72	994-30-9	144.5	4-04-00-04051 0.8967^{20}	1.4314^{20}
11389	Silane, chlorotrimethyl- Trimethylchlorosilane	C_3H_9ClSi 108.64	75-77-4 -40	60	4-04-00-04007 0.856^{25}	1.3870^{20}
11390	Silane, dichloro(chloromethyl)methyl- Dichloro(chloromethyl)methylsilane	$C_2H_5Cl_3Si$ 163.51	1558-33-4	121.5	4-04-00-04135 1.2858^{20}	1.4500^{20}
11391	Silane, dichloro(dichloromethyl)methyl- Dichloromethyl methyl dichlorosilane	$C_2H_4Cl_4Si$ 197.95	1558-31-2	149	4-04-00-04144 1.4116^{20}	1.4700^{20}
11392	Silane, dichlorodiethoxy-	$C_4H_{10}Cl_2O_2Si$ 189.11	4667-38-3 -130	135.9	4-01-00-01364 1.1290^{20}	EtOH 4
11393	Silane, dichlorodiethyl- Dichlorodiethylsilane	$C_4H_{10}Cl_2Si$ 157.11	1719-53-5 -96.5	129 dec	4-04-00-04158 1.0504^{20}	1.4309^{20}
11394	Silane, dichlorodimethyl- Dichlorodimethylsilane	$C_2H_6Cl_2Si$ 129.06	75-78-5 -16	70.3	4-04-00-04111 1.064^{25}	H_2O 6; EtOH 6 1.4038^{20}
11395	Silane, dichlorodiphenyl- Dichlorodiphenylsilane	$C_{12}H_{10}Cl_2Si$ 253.20	80-10-4	305	4-16-00-01526 1.204^{25}	EtOH 3; eth 3; ace 3; bz 3 1.5800^{20}

No.	Name Synonym	Mol. Form. Mol. Wt.	CAS RN mp/°C	Merck No. bp/°C	Beil. Ref. den/g cm^{-3}	Solubility n_D
11396	Silane, dichloroethenylmethyl-	$C_3H_6Cl_2Si$ 141.07	124-70-9	 92.5	4-04-00-04184 1.0868^{20}	H_2O 6 1.4270^{20}
11397	Silane, dichloroethylmethyl- Methylethyldichlorosilane	$C_3H_8Cl_2Si$ 143.09	4525-44-4	 101	4-04-00-04152 1.0630^{20}	 1.4197^{20}
11398	Silane, dichloromethyl- Dichloromethylsilane	CH_4Cl_2Si 115.03	75-54-7 -93	 41	4-04-00-04096 1.105^{25}	
11399	Silane, dichloromethyl(1-methylethyl)-	$C_4H_{10}Cl_2Si$ 157.11	18236-89-0	 121.5	4-04-00-04171 1.0385^{20}	 1.4270^{20}
11400	Silane, dichloromethyl(4-methylphenyl)-	$C_8H_{10}Cl_2Si$ 205.16	18236-57-2 57		 1.1619^{20}	 1.5170^{20}
11401	Silane, dichloromethylphenyl- Dichloromethylphenylsilane	$C_7H_8Cl_2Si$ 191.13	149-74-6	 206.5	4-16-00-01517 1.1866^{20}	 1.5180^{20}
11402	Silane, dichloromethyl-2-propenyl-	$C_4H_8Cl_2Si$ 155.10	1873-92-3	 119.5	4-04-00-04188 1.0758^{20}	 1.4419^{20}
11403	Silane, dichlorophenyl- Phenyldichlorosilane	$C_6H_6Cl_2Si$ 177.10	1631-84-1	 181	4-16-00-01514 1.221^{25}	H_2O 6
11404	Silane, diethenyldiphenyl-	$C_{16}H_{16}Si$ 236.39	17937-68-7	 130-1$^{0.05}$	4-16-00-01368 1.0092^{25}	 1.5350^{25}
11405	Silane, diethoxydimethyl- Dimethyldiethoxysilane	$C_6H_{16}O_2Si$ 148.28	78-62-6 -87	 114	4-04-00-04101 0.865^{25}	ctc 3 1.3811^{20}
11406	Silane, diethoxydiphenyl- Diphenyldiethoxysilane	$C_{16}H_{20}O_2Si$ 272.42	2553-19-7	 167^{15}	4-16-00-01524 1.0329^{20}	 1.5269^{20}
11407	Silane, diethoxymethylphenyl-	$C_{11}H_{18}O_2Si$ 210.35	775-56-4	 218	4-16-00-01515 0.9627^{20}	 1.4690^{20}
11408	Silane, diethoxymethyl-2-propenyl- Silane, allyldiethoxymethyl-	$C_8H_{18}O_2Si$ 174.32	18388-45-9	 155	4-04-00-04187 0.8572^{25}	 1.4104^{20}
11409	Silane, diethyl- 3-Silapentane	$C_4H_{12}Si$ 88.22	542-91-6 -134.3	3117 57	4-04-00-03894 0.6843^{20}	H_2O 1 1.3921^{20}
11410	Silane, diethyldifluoro- Diethyldifluorosilane	$C_4H_{10}F_2Si$ 124.21	358-06-5 -78.7	 61	4-04-00-04157 0.9348^{20}	 1.3385^{20}
11411	Silane, difluorodiphenyl-	$C_{12}H_{10}F_2Si$ 220.29	312-40-3	 157^{50}	4-16-00-01525 1.145^{17}	 1.5221^{25}
11412	Silane, dimethoxydimethyl-	$C_4H_{12}O_2Si$ 120.22	1112-39-6	 82	4-04-00-04100 0.8646^{20}	H_2O 6 1.3708^{20}
11413	Silane, dimethoxydiphenyl-	$C_{14}H_{16}O_2Si$ 244.37	6843-66-9	 161^{15}	4-16-00-01524 1.0771^{20}	 1.5447^{20}
11414	Silane, dimethyl- 2-Silapropane	C_2H_8Si 60.17	1111-74-6 -150	 -20	4-04-00-03874 0.68^{-80}	
11415	Silane, dimethyldiphenoxy- Diphenoxydimethylsilane	$C_{14}H_{16}O_2Si$ 244.37	3440-02-6 -23	 130-2^{5}	4-06-00-00764 1.0599^{25}	 1.5330^{20}
11416	Silane, dimethyldiphenyl- Diphenyl(dimethyl)silane	$C_{14}H_{16}Si$ 212.37	778-24-5	 277; 173^{45}	4-16-00-01366 0.9867^{20}	 1.5644^{20}
11417	Silane, dimethyldi-2-propenyl-	$C_8H_{16}Si$ 140.30	1113-12-8	 68^{50}	4-04-00-03929 0.7679^{20}	 1.4420^{20}
11418	Silane, dimethylphenyl-	$C_8H_{12}Si$ 136.27	766-77-8	 156.5	4-16-00-01360 0.8891^{20}	H_2O 1 1.4995^{20}
11419	Silanediol, dimethyl-, diacetate	$C_6H_{12}O_4Si$ 176.24	2182-66-3 -12.5	 165	4-04-00-04108 1.0540^{20}	 1.4030^{20}
11420	Silane, diphenyl- Diphenylsilane	$C_{12}H_{12}Si$ 184.31	775-12-2	 95-7^{13}	4-16-00-01366 0.9969^{20}	ctc 3; CS_2 3 1.5800^{20}
11421	Silane, 1,2-ethenediylbis[trimethyl-, (*E*)-	$C_8H_{20}Si_2$ 172.42	18178-59-1 -18.5	 145.5	4-04-00-03948 0.7589^{20}	H_2O 1 1.4310^{20}
11422	Silane, ethenyldiethoxymethyl-	$C_7H_{16}O_2Si$ 160.29	5507-44-8	 133	4-04-00-04183 0.8620^{20}	 1.4001^{20}
11423	Silane, ethenylethoxydimethyl-	$C_6H_{14}OSi$ 130.26	5356-83-2	 99	4-04-00-04079 0.790^{20}	 1.3983^{20}
11424	Silane, ethenyltriethoxy-	$C_8H_{18}O_3Si$ 190.31	78-08-0	 148^{535}; 62^{20}	4-04-00-04256 0.9027^{20}	chl 3 1.3960^{25}
11425	Silane, ethenyltrimethyl-	$C_5H_{12}Si$ 100.24	754-05-2	 55	4-04-00-03922 0.6910^{20}	H_2O 1 1.3914^{20}
11426	Silane, ethenyltris(1-methylethoxy)-	$C_{11}H_{24}O_3Si$ 232.40	18023-33-1	 179.5; 77^{20}	4-04-00-04257 0.8627^{25}	ctc 3 1.3981^{20}
11427	Silane, ethenyltris(2-propenyloxy)- Tris(allyloxy)vinylsilane	$C_{11}H_{18}O_3Si$ 226.35	17988-31-7	 90.3$^{7.5}$	4-04-00-04257 0.9435^{25}	 1.4380^{20}
11428	Silane, ethoxytriethyl-	$C_8H_{20}OSi$ 160.33	597-67-1	 154.5	4-04-00-04042 0.8160^{20}	eth 4; EtOH 4 1.4140^{20}
11429	Silane, ethoxytrifluoro- Ethoxytrifluorosilicon	$C_2H_5F_3OSi$ 130.14	460-55-9 -122	 -7	3-01-00-01336 1.3200^{-63}	
11430	Silane, ethoxytrimethyl-	$C_5H_{14}OSi$ 118.25	1825-62-3	 76	4-04-00-03994 0.7573^{20}	H_2O 1; EtOH 3; eth 3; ace 3 1.3741^{20}
11431	Silane, ethoxytriphenyl-	$C_{20}H_{20}OSi$ 304.46	1516-80-9 65	 344	4-16-00-01481	chl 3
11432	Silane, ethyltrifluoro- Trifluoroethylsilane	$C_2H_5F_3Si$ 114.14	353-89-9 -105	 -4.4	4-04-00-04226 1.227^{-78}	
11433	Silane, ethyltrimethoxy- Ethyltrimethoxysilane	$C_5H_{14}O_3Si$ 150.25	5314-55-6	 124.3	4-04-00-04223 0.9488^{20}	EtOH 4 1.3838^{20}
11434	Silane, 1,2-ethynediylbis[trimethyl-	$C_8H_{18}Si_2$ 170.40	14630-40-1 26	 134	4-04-00-03950 0.770^{20}	 1.413^{20}
11435	Silane, ethynyl Ethynylsilane	C_2H_4Si 56.14	1066-27-9	 -22.5		

No.	Name Synonym	Mol. Form. Mol. Wt.	CAS RN mp/°C	Merck No. bp/°C	Beil. Ref. den/g cm^{-3}	Solubility n_D
11436	Silane, methoxy Methyl silyl ether	CH_6OSi 62.14	2171-96-2			
11437	Silane, methyl- Methylsilane	CH_6Si 46.14	992-94-9 -156.5	 -57.5	4-04-00-03873	
11438	Silane, methyldiphenyl-	$C_{13}H_{14}Si$ 198.34	776-76-1	 93.5^{1}	4-16-00-01366 0.9973^{20}	ctc 3 1.5694^{20}
11439	Silane, methylenebis-	CH_8Si_2 76.25	1759-88-2	 15	4-01-00-03072 0.6979^{4}	 1.4115^{4}
11440	Silane, methylenebis[trichloro-	$CH_2Cl_6Si_2$ 282.91	4142-85-2	 183; 64^{10}	4-01-00-03077 1.5567^{20}	eth 4; EtOH 4 1.4740^{20}
11441	Silane, (2-methylphenoxy)triphenyl-	$C_{25}H_{22}OSi$ 366.53	18858-65-6	 192	4-16-00-01482 0.9287^{20}	 1.4830^{20}
11442	Silane, methylphenyl-	$C_7H_{10}Si$ 122.24	766-08-5	 140	4-16-00-01360 0.8895^{20}	 1.5058^{20}
11443	Silane, methyltriphenoxy- Methyltriphenoxysilane	$C_{19}H_{18}O_3Si$ 322.44	3439-97-2	 269^{100}; 210^{12}	4-06-00-00765 1.135^{20}	 1.5599^{20}
11444	Silane, methyltriphenyl- Triphenylmethylsilane	$C_{19}H_{18}Si$ 274.44	791-29-7 69.5	 198^{9}	4-16-00-01370 1.0880^{20}	ctc 3; CS_2 3
11445	Silane, methyltri-*p*-tolyl- Silane, methyl tri-(4-tolyl)	$C_{22}H_{24}Si$ 316.52	18752-92-6 92	 198^{4}	4-16-00-01399	chl 2
11446	Silane, phenyl-	C_6H_8Si 108.22	694-53-1	 119	4-16-00-01360 0.8681^{20}	H_2O 1 1.5125^{20}
11447	Silane, [1,3-phenylenebis(oxy)]bis[trimethyl-	$C_{12}H_{22}O_2Si_2$ 254.48	4520-29-0	 240	4-06-00-05683 0.950^{20}	 1.4748^{20}
11448	Silane, phenyltripropyl- Phenyltripropylsilane	$C_{15}H_{26}Si$ 234.46	78938-11-1	 146-7^{18}	4-16-00-01362 0.8799^{20}	 1.4950^{20}
11449	Silane, tetraethenyl- Tetravinylsilane	$C_8H_{12}Si$ 136.27	1112-55-6	 130.2	4-04-00-03925 0.7999^{20}	 1.4625^{20}
11450	Silane, tetraethyl- Tetraethylsilane	$C_8H_{20}Si$ 144.33	631-36-7	 154.7	4-04-00-03895 0.7658^{20}	H_2O 1 1.4268^{20}
11451	Silane, tetramethyl- Tetramethylsilane	$C_4H_{12}Si$ 88.22	75-76-3 -99.0	 26.6	4-04-00-03875 0.648^{19}	H_2O 1; EtOH 4; eth 4; sulf 1 1.3587^{20}
11452	Silane, tetraphenyl- Tetraphenylsilane	$C_{24}H_{20}Si$ 336.51	1048-08-4 236.5	 228^{3}	4-16-00-01372 1.078^{20}	ctc 3; CS_2 3
11453	Silane, tribromomethyl-	CH_3Br_3Si 282.83	4095-09-4 -28.4	 132	4-04-00-04213 2.2130^{25}	 1.5152^{25}
11454	Silane, tributyl- Tributylsilane	$C_{12}H_{28}Si$ 200.44	998-41-4	 221	4-04-00-03908 0.7794^{20}	 1.4380^{20}
11455	Silane, tributylphenyl- Silane, phenyl-tributyl	$C_{18}H_{32}Si$ 276.54	18510-29-7	 140-2^{4}	4-16-00-01362 0.8753^{20}	 1.4915^{20}
11456	Silane, trichloro(chloromethyl)- (Chloromethyl)trichlorosilane	CH_2Cl_4Si 183.92	1558-25-4	 118	4-01-00-03074 1.4650^{20}	 1.4555^{20}
11457	Silane, trichloro(4-chlorophenyl)-	$C_6H_4Cl_4Si$ 245.99	825-94-5	 115-7^{20}	4-16-00-01563 1.4316^{20}	 1.5418^{20}
11458	Silane, trichloro(3-chloropropyl)-	$C_3H_6Cl_4Si$ 211.98	2550-06-3	 181.5	4-04-00-04243 1.3590^{20}	 1.4668^{20}
11459	Silane, trichloro(dichloromethyl)- (Dichloromethyl)trichlorosilane	$CHCl_5Si$ 218.37	1558-24-3	 145	4-02-00-00088 1.5518^{20}	 1.4714^{20}
11460	Silane, trichlorododecyl- Dodecyltrichlorosilane	$C_{12}H_{25}Cl_3Si$ 303.77	4484-72-4	 155^{10}	4-04-00-04255	 1.4581^{20}
11461	Silane, trichloroethenyl- Vinyltrichlorosilane	$C_2H_3Cl_3Si$ 161.49	75-94-5 -95	 91.5	4-04-00-04258 1.2426^{20}	chl 4 1.4295^{20}
11462	Silane, trichloroethoxy-	$C_2H_5Cl_3OSi$ 179.50	1825-82-7 -135	 101.9	4-01-00-01364 1.2274^{20}	EtOH 4 1.4045^{20}
11463	Silane, trichloroethyl- Ethyltrichlorosilane	$C_2H_5Cl_3Si$ 163.51	115-21-9 -105.6	 100.5	4-04-00-04227 1.2373^{20}	ctc 3 1.4256^{20}
11464	Silane, trichlorohexyl- Hexyltrichlorosilane	$C_6H_{13}Cl_3Si$ 219.61	928-65-4	 190	4-04-00-04251 1.1100^{20}	H_2O 6
11465	Silane, trichloromethyl- Methyltrichlorosilane	CH_3Cl_3Si 149.48	75-79-6 -90	 65.6	4-04-00-04212 1.273^{20}	H_2O 6; EtOH 6 1.4106^{20}
11466	Silane, trichloro(1-methylethyl)-	$C_3H_7Cl_3Si$ 177.53	4170-46-1 -87.7	 121	4-04-00-04244 1.1934^{20}	H_2O 6 1.4319^{20}
11467	Silane, trichloro(2-methylphenyl)-	$C_7H_7Cl_3Si$ 225.58	13835-81-9 -25.5	 226.7	4-16-00-01565 1.306^{25}	 1.5336^{25}
11468	Silane, trichloro(3-methylphenyl)-	$C_7H_7Cl_3Si$ 225.58	13688-75-0 -26	 226.5	4-16-00-01565 1.306^{25}	 1.5336^{25}
11469	Silane, trichloro(2-methylpropyl)-	$C_4H_9Cl_3Si$ 191.56	18169-57-8	 143.3	4-04-00-04248 1.154^{20}	H_2O 6
11470	Silane, trichlorooctadecyl- Octadecyltrichlorosilane	$C_{18}H_{37}Cl_3Si$ 387.94	112-04-9	 223^{10}	4-04-00-04256 0.984^{25}	 1.4602^{20}
11471	Silane, trichlorooctyl- Octyltrichlorosilane	$C_8H_{17}Cl_3Si$ 247.67	5283-66-9	 232	4-04-00-04253	H_2O 6; EtOH 6; ctc 3 1.4480^{20}
11472	Silane, trichloropentyl- Amyltrichlorosilane	$C_5H_{11}Cl_3Si$ 205.59	107-72-2	 172; 60.5^{15}	4-04-00-04249 1.1330^{20}	 1.4503^{20}
11473	Silane, trichlorophenyl- Phenyltrichlorosilane	$C_6H_5Cl_3Si$ 211.55	98-13-5	 201	4-16-00-01560 1.321^{20}	ctc 3; chl 3; CS_2 3 1.5230^{20}
11474	Silane, trichloro(2-phenylethyl)-	$C_8H_9Cl_3Si$ 239.60	940-41-0	 242; 98-9^{5}	4-16-00-01568 1.2397^{20}	 1.5185^{20}
11475	Silane, trichloro-2-propenyl- Allyltrichlorosilane	$C_3H_5Cl_3Si$ 175.52	107-37-9 35	 117.5	4-04-00-04261 1.2011^{20}	 1.4460^{20}

No.	Name Synonym	Mol. Form. Mol. Wt.	CAS RN mp/°C	Merck No. bp/°C	Beil. Ref. den/g cm^{-3}	Solubility n_D
11476	Silane, trichloropropyl- Propyltrichlorosilane	$C_3H_7Cl_3Si$ 177.53	141-57-1	 123.5	4-04-00-04239 1.195^{20}	 1.4310^{20}
11477	Silane, triethoxy- Triethoxysilane	$C_6H_{16}O_3Si$ 164.28	998-30-1	 133.5	4-01-00-01359 0.8745^{20}	
11478	Silane, triethoxyethyl-	$C_8H_{20}O_3Si$ 192.33	78-07-9	 158.5	4-04-00-04223 0.8963^{20}	H_2O 1; EtOH 5; eth 5; chl 3 1.3955^{20}
11479	Silane, triethoxymethyl-	$C_7H_{18}O_3Si$ 178.30	2031-67-6	 142	4-04-00-04204 0.8948^{25}	 1.3832^{20}
11480	Silane, triethoxypentyl-	$C_{11}H_{26}O_3Si$ 234.41	2761-24-2	 95^{13}	4-04-00-04248 0.8862^{20}	 1.4059^{20}
11481	Silane, triethoxyphenyl-	$C_{12}H_{20}O_3Si$ 240.37	780-69-8	 $112\text{-}3^{10}$	4-16-00-01556 0.996^{25}	 1.4604^{20}
11482	Silane, triethoxy-2-propenyl-	$C_9H_{20}O_3Si$ 204.34	2550-04-1	 100^{50}	4-04-00-04261 0.9030^{20}	 1.4072^{20}
11483	Silane, triethyl- Triethylsilane	$C_6H_{16}Si$ 116.28	617-86-7	 109	4-04-00-03895 0.7302^{20}	H_2O 1; sulf 1 1.447^{20}
11484	Silane, triethylfluoro-	$C_6H_{15}FSi$ 134.27	358-43-0	 110	4-04-00-04051 0.8354^{25}	peth 4 1.3900^{25}
11485	Silane, triethylphenyl-	$C_{12}H_{20}Si$ 192.38	2987-77-1	 236	4-16-00-01362 0.8915^{20}	 1.4999^{20}
11486	Silane, trifluorophenyl-	$C_6H_5F_3Si$ 162.19	368-47-8 -18	 101.5	4-16-00-01559 1.2169^{20}	bz 4; EtOH 4 1.4110^{20}
11487	Silane, trimethoxymethyl-	$C_4H_{12}O_3Si$ 136.22	1185-55-3	 102.5	4-04-00-04203 0.9548^{20}	chl 3 1.3696^{20}
11488	Silane, trimethoxyphenyl-	$C_9H_{14}O_3Si$ 198.29	2996-92-1	 130^{45}	4-16-00-01556 1.064^{20}	ctc 3; CS_2 3 1.4734^{20}
11489	Silane, trimethyl- Trimethylsilane	$C_3H_{10}Si$ 74.20	993-07-7 -135.9	 6.7		
11490	Silane, trimethyl(4-methylphenyl)-	$C_{10}H_{16}Si$ 164.32	3728-43-6	 192; 73^{10}	4-16-00-01396 0.860^{25}	 1.4900^{20}
11491	Silane, trimethyl(2-methylpropyl)-	$C_7H_{18}Si$ 130.31	1118-09-8	 108.5	1-04-00-00580 0.7330^{0}	
11492	Silane, trimethylphenoxy-	$C_9H_{14}OSi$ 166.30	1529-17-5 -55	 119	4-06-00-00762 0.8681^{20}	 1.5125^{20}
11493	Silane, trimethylphenyl-	$C_9H_{14}Si$ 150.30	768-32-1	 169.5	4-16-00-01361 0.8722^{20}	ctc 3; CS_2 3 1.4907^{20}
11494	Silane, trimethyl(phenylmethyl)-	$C_{10}H_{16}Si$ 164.32	770-09-2	 190.5	4-16-00-01401 0.8933^{20}	 1.4941^{20}
11495	Silane, trimethyl-2-propenyl-	$C_6H_{14}Si$ 114.26	762-72-1	 86	4-04-00-03927 0.7158^{25}	H_2O 1 1.4074^{20}
11496	Silane, trimethylpropyl- Trimethylpropylsilane	$C_6H_{16}Si$ 116.28	3510-70-1	 89	4-04-00-03901 0.7196^{0}	 1.3929^{20}
11497	Silane, trimethyl[4-[(trimethylsilyl)oxy]phenyl]- Silane, (oxy-*p*-phenylene)bis[trimethyl-	$C_{12}H_{22}OSi_2$ 238.48	18036-81-2	 132^{25}	4-16-00-01441 0.900^{25}	 1.4794^{25}
11498	Silanetriol, ethenyl-, triacetate	$C_8H_{12}O_6Si$ 232.27	4130-08-9	 115^{10}	4-04-00-04258 1.167^{20}	 1.4226^{20}
11499	Silanetriol, methyl-, triacetate	$C_7H_{12}O_6Si$ 220.25	4253-34-3 40.5	 $110\text{-}12^{17}$	4-04-00-04208 1.1750^{20}	 1.4083^{20}
11500	Silane, tripropyl- Tripropylsilane	$C_9H_{22}Si$ 158.36	998-29-8	 172	4-04-00-03902 0.7723^{0}	H_2O 1 1.4280^{20}
11501	Silanol, ethyldimethyl- Silanol, dimethyl ethyl	$C_4H_{12}OSi$ 104.22	5906-73-0	 120; 58^{50}	4-04-00-04034 0.8332^{20}	 1.4070^{20}
11502	Silanol, methyldiphenyl-	$C_{13}H_{14}OSi$ 214.34	778-25-6 167	 184^{24}; 148^{3}	4-16-00-01478 1.0840^{25}	ctc 3; CS_2 3
11503	Silanol, triethyl-	$C_6H_{16}OSi$ 132.28	597-52-4	 154	4-04-00-04040 0.8647^{20}	H_2O 1; EtOH 5; eth 5 1.4329^{20}
11504	Silanol, triphenyl-	$C_{18}H_{16}OSi$ 276.41	791-31-1 154.8		4-16-00-01480 1.1777^{20}	ctc 3; CS_2 3
11505	Silicic acid (H4SiO4), tetrabutyl ester	$C_{16}H_{36}O_4Si$ 320.54	4766-57-8	 120^{3}	4-01-00-01541 0.8990^{20}	 1.4128^{20}
11506	Silicic acid (H4SiO4), tetraethyl ester Ethyl silicate	$C_8H_{20}O_4Si$ 208.33	78-10-4 -82.5	3805 168.8	4-01-00-01360 0.9320^{20}	H_2O 6 1.3928^{20}
11507	Silicic acid (H4SiO4), tetrakis(2-ethylbutyl) ester	$C_{24}H_{52}O_4Si$ 432.76	78-13-7		4-01-00-01725 0.8920^{20}	H_2O 1; EtOH 2; eth 3; bz 3 1.4307^{20}
11508	Silicic acid (H4SiO4), tetramethyl ester Methyl silicate	$C_4H_{12}O_4Si$ 152.22	681-84-5 -2	 121	4-01-00-01266 1.0232^{20}	EtOH 4 1.3683^{20}
11509	Silicic acid (H4SiO4), tetraphenyl ester	$C_{24}H_{20}O_4Si$ 400.51	1174-72-7 49	 417; 236^{1}	4-06-00-00766 1.1412^{60}	
11510	Silicic acid (H4SiO4), tetrapropyl ester	$C_{12}H_{28}O_4Si$ 264.44	682-01-9	 226	4-01-00-01435 0.9158^{20}	ctc 3; CS_2 3 1.4012^{20}
11511	Silicic acid (H4SiO4), triethyl phenyl ester	$C_{12}H_{20}O_4Si$ 256.37	18023-36-4	 $120\text{-}2^{10}$	4-06-00-00765 1.0283^{20}	 1.4525^{20}
11512	Silvex Propanoic acid, 2-(2,4,5-trichlorophenoxy)-	$C_9H_7Cl_3O_3$ 269.51	93-72-1 181.6	8483		
11513	Simazine 1,3,5-Triazine-2,4-diamine, 6-chloro-*N,N'*-diethyl-	$C_7H_{12}ClN_5$ 201.66	122-34-9 226	8485	 1.302^{20}	
11514	Sodium dimethyldithiocarbamate Carbamodithioic acid, dimethyl-, sodium salt	$C_3H_7NNaS_2$ 144.22	128-04-1			

No.	Name Synonym	Mol. Form. Mol. Wt.	CAS RN mp/°C	Merck No. bp/°C	Beil. Ref. den/g cm^{-3}	Solubility n_D
11515	Solanid-5-ene-3,12-diol, (3β,12α)- Rubijervine	$C_{27}H_{43}NO_2$ 413.64	79-58-3 242	8264	5-21-05-00208	bz 4; EtOH 4; chl 4
11516	Solanid-5-ene-3,18-diol, (3β)- Isorubijervine	$C_{27}H_{43}NO_2$ 413.64	468-45-1 242.5	5111	5-21-05-00208	bz 4; chl 4
11517	Sonar 2-Naphthalenol, 1,1'-methylenebis-, monosodium salt	$C_{21}H_{16}NaO_2$ 323.35	38827-66-6 132			
11518	*DL*-Sorbose	$C_6H_{12}O_6$ 180.16	65732-90-3 162.5	8681	4-01-00-04410 1.638^{17}	H_2O 4; MeOH 4
11519	*L*-Sorbose *L*-Sorbinose	$C_6H_{12}O_6$ 180.16	87-79-6 165	8681	4-01-00-04412 1.612^{17}	H_2O 3; EtOH 2; eth 2; MeOH 2
11520	Spiro[benzofuran-2(3H),1'-[2]cyclohexene]-3,4'-dione, 7-chloro-2',4,6-trimethoxy- Griseofulvin, (+)-	$C_{17}H_{17}ClO_6$ 352.77	126-07-8 220	4453	5-18-05-00150	H_2O 1; EtOH 2; eth 2; ace 2
11521	Spiro[4.5]dec-6-en-8-one, 6,10-dimethyl-2-(1-methylethylidene)-, (5*R*-*cis*)-	$C_{15}H_{22}O$ 218.34	18444-79-6 44.5	$141-2^{2}$	4-07-00-00822 1.0001^{20}	ace 3 1.5309^{20}
11522	Spiro[isobenzofuran-1(3H),9'-[9H]xanthen]-3-one, 3',6'-dihydroxy- Fluorescein	$C_{20}H_{12}O_5$ 332.31	2321-07-5 315 dec	4085	5-19-06-00456	H_2O 2; EtOH 2; eth 2; ace 4
11523	Spiro[isobenzofuran-1(3H),9'-[9H]xanthen]-3-one, 3',6'-dihydroxy-4',5'-diiodo- Fluorescein, 4',5'-diiodo	$C_{20}H_{10}I_2O_5$ 584.11	38577-97-8	3174	5-19-06-00466	
11524	Spiro[isobenzofuran-1(3H),9'-[9H]xanthen]-3-one, 3',6'-dihydroxy-2',4',5',7'-tetraiodo-, disodium salt Erythrosin	$C_{20}H_8I_4O_5$ 835.90	16423-68-0	3640	4-19-00-02923	eth 4; EtOH 4
11525	Spiro[isobenzofuran-1(3H),9'-[9H]xanthen]-3-one, 2',4',5',7'-tetrabromo-3',6'-dihydroxy-, disodium salt Eosin	$C_{20}H_6Br_4Na_2O_5$ 691.86	17372-87-1 295.5	3555		EtOH 4
11526	Spiro[isobenzofuran-1(3H),9'-[9H]xanthen]-3-one, 3',4',5',6'-tetrahydroxy- Gallein	$C_{20}H_{12}O_5$ 332.31	2103-64-2 >300	4250	5-19-07-00083	ace 4; EtOH 4
11527	Spiropentane Spiro[2.2]pentane	C_5H_8 68.12	157-40-4 -134.6	39	4-05-00-00218 0.7266^{20}	1.4120^{20}
11528	Spirosolan-3-ol, (3β,5α,22β,25*S*)- Tomatidine	$C_{27}H_{45}NO_2$ 415.66	77-59-8 210.5	9470		EtOH 3; eth 3
11529	Spirosol-5-en-3-ol, (3β,22α,25*R*)- Solasodine	$C_{27}H_{43}NO_2$ 413.64	126-17-0 202	8665	4-27-00-02000	EtOH 3; eth 2; ace 3; bz 3
11530	Spirostan-2,3-diol, (2α,3β,5α,25*R*)- Gitogenin	$C_{27}H_{44}O_4$ 432.64	511-96-6 271.5	4325	5-19-03-00338	H_2O 1; EtOH 3; eth 2; chl 3
11531	Spirostan-3-ol, (3β,5β,25*S*)- Sarsasapogenin	$C_{27}H_{44}O_3$ 416.64	126-19-2 200.5	8338	5-19-02-00672	EtOH 3; ace 3; bz 3; chl 3
11532	Spirostan-3-ol, (3β,5α,25*R*)- Tigogenin	$C_{27}H_{44}O_3$ 416.64	77-60-1 205.5	9367		EtOH 3; eth 3; ace 3; ctc 3
11533	Spirostan-3-ol, (3β,5β,25*R*)- Smilagenin	$C_{27}H_{44}O_3$ 416.64	126-18-1 185	8508	5-19-02-00673	ace 4; bz 4; EtOH 4
11534	Spirostan-12-one, 3-hydroxy-, (3β,5α,25*R*)- Hecogenin	$C_{27}H_{42}O_4$ 430.63	467-55-0 266.5	4537	5-19-05-00592	ace 4; eth 4; EtOH 4
11535	Spirostan-2,3,15-triol, (2α,3β,5α,15β,25*R*)- Digitogenin	$C_{27}H_{44}O_5$ 448.64	511-34-2 281.5	3143	5-19-03-00603	chl 4
11536	Spirost-5-en-3-ol, (3β,25*R*)- Diosgenin	$C_{27}H_{42}O_3$ 414.63	512-04-9 205.5	3289	5-19-03-00030	EtOH 4
11537	Spiro[5.5]undecane Spir[5.]undecane	$C_{11}H_{20}$ 152.28	180-43-8	207	4-05-00-00328 0.8783^{20}	1.4731
11538	Stannane, (acetyloxy)tributyl- Acetoxytributylstannane	$C_{14}H_{30}O_2Sn$ 349.10	56-36-0 84.7			
11539	Stannane, (acetyloxy)triphenyl- Triphenyltin acetate	$C_{20}H_{18}O_2Sn$ 409.07	900-95-8 121.5			
11540	Stannane, chlorotrimethyl-	C_3H_9ClSn 199.27	1066-45-1 38.5	148		H_2O 3; chl 3; os 3
11541	Stannane, chlorotriphenyl- Triphenyltin chloride	$C_{18}H_{15}ClSn$ 385.48	639-58-7 103.5			chl 3
11542	Stannane, chlorotripropyl-	$C_9H_{21}ClSn$ 283.43	2279-76-7 -23.5	123^{13}	1.2678^{28}	ctc 3; os 3 1.4910^{28}
11543	Stannane, tetraethenyl-	$C_8H_{12}Sn$ 226.89	1112-56-7	$55-7^{17}$	4-04-00-04316 1.2651^{25}	ctc 3 1.4993^{25}
11544	Stannane, tetramethyl- Tetramethylstannane	$C_4H_{12}Sn$ 178.85	594-27-4 -54.8	78	4-04-00-04307 1.314^{25}	H_2O 1; ctc 3; CS_2 3; os 3 1.4386
11545	Stannane, tetraphenyl-	$C_{24}H_{20}Sn$ 427.13	595-90-4 228	420	4-16-00-01592	chl 2
11546	Stannane, tetrapropyl-	$C_{12}H_{28}Sn$ 291.06	2176-98-9 -109.1	228	4-04-00-04310 1.1065^{20}	1.4745^{20}
11547	Stibine, dichlorophenyl-	$C_6H_5Cl_2Sb$ 269.77	5035-52-9 69.5	$110-5^{10}$	4-16-00-01205	ace 4; EtOH 4; HOAc 4
11548	Stibine oxide, dihydroxyphenyl- Benzenestibonic acid	$C_6H_7O_3Sb$ 248.88	535-46-6 139	1077	3-16-00-01178	

No.	Name Synonym	Mol. Form. Mol. Wt.	CAS RN mp/°C	Merck No. bp/°C	Beil. Ref. den/g cm^{-3}	Solubility n_D
11549	Stibine, triethyl- Triethylstibine	$C_6H_{15}Sb$ 208.94	617-85-6 -98	 161.4	4-04-00-03685 1.3224^{15}	H_2O 1; EtOH 3; eth 3
11550	Stibine, trimethyl- Trimethylstibine	C_3H_9Sb 166.86	594-10-5 -62	 80.6	4-04-00-03684 1.523^{15}	H_2O 1; EtOH 3; eth 3; CS_2 3 1.42^{15}
11551	Stibine, triphenyl-	$C_{18}H_{15}Sb$ 353.07	603-36-1 53.5	 >360	4-16-00-01198 1.4343^{25}	H_2O 1; EtOH 3; eth 4; ace 4 1.6948^{42}
11552	Stibine, tris(3-methylphenyl)-	$C_{21}H_{21}Sb$ 395.15	35569-54-1 72		4-16-00-01199 1.3957^{16}	bz 4; eth 4; EtOH 4; chl 4
11553	Stibine, tris(4-methylphenyl)-	$C_{21}H_{21}Sb$ 395.15	5395-43-7 127.5	 218^{3}	4-16-00-01199 1.3595^{16}	bz 4; eth 4; chl 4
11554	Stigmasta-5,7-dien-3-ol, (3β)- 7-Dehydrositosterol	$C_{29}H_{48}O$ 412.70	521-04-0 144.5	2863	3-06-00-02855	bz 4; eth 4; EtOH 4
11555	Stigmasta-5,22-dien-3-ol, (3β,22E)- Stigmasterol	$C_{29}H_{48}O$ 412.70	83-48-7 170	8771	4-06-00-04170	bz 4; eth 4; EtOH 4
11556	Stigmasta-7,24(28)-dien-3-ol, 4-methyl-, (3β,4α,5α,24Z)- α1-Sitosterol	$C_{30}H_{50}O$ 426.73	474-40-8 166	8499	3-06-00-02876	EtOH 4; chl 4
11557	Stigmastan-3-ol, (3β,5α)- Stigmastanol	$C_{29}H_{52}O$ 416.73	83-45-4 144	8770	3-06-00-02172	
11558	Stigmast-5-en-3-ol, (3β)- β-Sitosterol	$C_{29}H_{50}O$ 414.72	83-46-5 140	8500	4-06-00-04037	EtOH 3; eth 3; HOAc 3
11559	Stigmast-5-en-3-ol, (3β,24S)- γ-Sitosterol	$C_{29}H_{50}O$ 414.72	83-47-6	8501		
11560	Stilbene, α-ethyl- 1-Butene, 1,2-diphenyl	$C_{16}H_{16}$ 208.30	22692-70-2 57	 296.5	4-05-00-02206 1.0124^{18}	 1.593^{18}
11561	*D*-Streptamine, *O*-2-deoxy-2-(methylamino)-α-*L*-glucopyranosyl-(12)-*O*-5-deoxy-3-*C*-formyl-α-L-lyxofuranosyl-(1-4)-*N,N'* 55-bis(aminoiminomethyl)-, sulfate (2:3)(salt) Streptomycin, sulfate	$C_{42}H_{84}N_{14}O_{36}S_3$ 1457.40	3810-74-0		5-18-11-00082	
11562	Streptomycin *N*-Methyl-*L*-glucosamidinostreptosidostreptidine	$C_{21}H_{39}N_7O_{12}$ 581.58	57-92-1	8786		
11563	Strychnidin-10-one Strychnine	$C_{21}H_{22}N_2O_2$ 334.42	57-24-9 287	8822 270^{5}	4-27-00-07537 1.36^{20}	H_2O 2; EtOH 2; eth 1; ace 2
11564	Strychnidin-10-one, 2,3-dimethoxy- Brucine	$C_{23}H_{26}N_2O_4$ 394.47	357-57-3 178	1443	4-27-00-07875	H_2O 2; EtOH 4; eth 2; bz 2
11565	Strychnidin-10-one, mononitrate Strychnine nitrate	$C_{21}H_{23}N_3O_5$ 397.43	66-32-0 295	8823	2-27-00-00723 1.627^{25}	H_2O 4; MeOH 4
11566	Strychnidin-10-one, sulfate (2:1) Strychnine, sulfate	$C_{42}H_{46}N_4O_8S$ 766.92	60-41-3 200 dec	8826		H_2O 3; EtOH 3; eth 1; chl 2
11567	Succinimide, *N*-(ethoxymethyl)-	$C_7H_{11}NO_3$ 157.17	98431-97-1 31.5	 262	2-21-00-00304	H_2O 4; EtOH 4
11568	Sulfamic acid, cyclohexyl- Cyclamic acid	$C_6H_{13}NO_3S$ 179.24	100-88-9 169.5	2707	4-12-00-00102	alk 4
11569	Sulfamide, tetramethyl-	$C_4H_{12}N_2O_2S$ 152.22	3768-63-6 73	 225	4-04-00-00270	EtOH 4
11570	Sulfide, *m*-tolyl *p*-tolyl 3,4'-Ditolyl sulfide	$C_{14}H_{14}S$ 214.33	107770-92-3 28	 179^{11}	4-06-00-02173	eth 4; EtOH 4
11571	Sulfoxylic acid, diethyl ester	$C_4H_{10}O_2S$ 122.19	10297-38-8	 118	3-01-00-01314 0.9940^{20}	H_2O 3; EtOH 4; eth 4; ace 4 1.4234^{20}
11572	Sulfur cyanide	C_2N_2S 84.10	627-52-1 63.5	 sub 30	4-03-00-00339	eth 4; EtOH 4
11573	Sulfuric acid, diethyl ester Diethyl sulfate	$C_4H_{10}O_4S$ 154.19	64-67-5 -24	3120 208	4-01-00-01326 1.172^{25}	H_2O 1; EtOH 5; eth 5 1.3989^{20}
11574	Sulfuric acid, dimethyl ester Dimethyl sulfate	$C_2H_6O_4S$ 126.13	77-78-1 -27	3244 188 dec; 76^{15}	4-01-00-01251 1.3322^{20}	H_2O 3; EtOH 5; eth 3; bz 3 1.3874^{20}
11575	Sulfuric acid, dipropyl ester	$C_6H_{14}O_4S$ 182.24	598-05-0	 121^{30}	4-01-00-01423 1.1064^{20}	peth 4 1.4135^{20}
11576	Sulfuric acid, monoethyl ester Ethyl sulfate	$C_2H_6O_4S$ 126.13	540-82-9	3808 280 dec	4-01-00-01324 1.3657^{20}	H_2O 4 1.4105^{20}
11577	Sulfuric acid, monomethyl ester Methyl sulfate	CH_4O_4S 112.11	75-93-4 <-30	6041 135 dec	4-01-00-01250	H_2O 4; eth 4; EtOH 4
11578	Sulfurous acid, bis(2-methylpropyl) ester	$C_8H_{18}O_3S$ 194.30	18748-27-1	 210; $92\text{-}4^{13}$	4-01-00-01595 0.9862^{20}	 1.4268^{20}
11579	Sulfurous acid, 2-chloroethyl 2-[4-(1,1-dimethylethyl)phenoxy]-1-methylethyl ester Aramite	$C_{15}H_{23}ClO_4S$ 334.86	140-57-8 -37.3	794 195^{2}	4-06-00-03301 1.143^{20}	ace 4; bz 4; eth 4; EtOH 4 1.5100^{20}
11580	Sulfurous acid, dibutyl ester Butyl sulfite	$C_8H_{18}O_3S$ 194.30	626-85-7	 230	4-01-00-01522 0.9957^{20}	EtOH 3; eth 3 1.4310^{20}
11581	Sulfurous acid, diethyl ester Ethyl sulfite	$C_4H_{10}O_3S$ 138.19	623-81-4	 158; 51^{13}	4-01-00-01324 1.1^{20}	EtOH 3; eth 3 1.4310^{20}
11582	Sulfurous acid, dimethyl ester	$C_2H_6O_3S$ 110.13	616-42-2	 126	4-01-00-01250 1.2129^{20}	H_2O 3; EtOH 3; eth 3 1.4083^{20}
11583	Sulfurous diamide, tetramethyl-	$C_4H_{12}N_2OS$ 136.22	3768-60-3 31	 209; 74^{13}	4-04-00-00269	eth 4
11584	Sulfur, pentafluoro(trifluoromethyl)-, (OC-6-21)-	CF_8S 196.06	373-80-8 -87	 -20	4-03-00-00035	H_2O 1; EtOH 1; CS_2 3

No.	Name Synonym	Mol. Form. Mol. Wt.	CAS RN mp/°C	Merck No. bp/°C	Beil. Ref. den/g cm^{-3}	Solubility n_D
11585	Sulfuryl chloride isocyanate	$CClNO_3S$ 141.53	1189-71-5 -44	 107	4-03-00-00086 1.626^{25}	 1.4467^{20}
11586	Sutan Carbamothioic acid, bis(2-methylpropyl)-, *S*-ethyl ester	$C_{11}H_{23}NOS$ 217.38	2008-41-5	1546 138^{21}	 0.9402^{25}	
11587	*D*-Tagatose Tagatose, *D*-	$C_6H_{12}O_6$ 180.16	87-81-0 134.5	9005	4-01-00-04414	H_2O 4
11588	Tartar emetic Antimonate(2-), bis[μ-[2,3-dihydroxybutanedioato(4-)-O1,O2:O3,O4]]di-, dipotassium, trihydrate, stereoisomer	$C_8H_{10}K_2O_{15}Sb_2$ 667.87	28300-74-5	732	 2.6	
11589	Tebuconazole (*RS*)-1-*p*-Chlorophenyl-4,4-dimethyl-3-(1H-1,2,4-triazol-1-ylmethyl)pentan-3-ol	$C_{16}H_{23}ClN_3O$ 308.83	107534-96-3 102.4			
11590	Tebuthiuron Urea, *N*-[5-(1,1-dimethylethyl)-1,3,4-thiadiazol-2-yl]-*N,N'*-dimethyl-	$C_9H_{16}N_4OS$ 228.32	34014-18-1 163 dec	9053		
11591	Terbacil 2,4(1H,3H)-Pyrimidinedione, 5-chloro-3-(1,1-dimethylethyl)-6-methyl-	$C_9H_{13}ClN_2O_2$ 216.67	5902-51-2 176	9085 sub 175	 1.34^{25}	
11592	Terbufos Phosphorodithioic acid, *S*-[[(1,1-dimethylethyl)thio]methyl] *O,O*-diethyl ester	$C_9H_{21}O_2PS_3$ 288.44	13071-79-9 -29.2	9088 $69^{0.01}$	 1.105^{24}	
11593	Terbuthylazine 1,3,5-Triazine-2,4-diamine, 6-chloro-*N*-(1,1-dimethylethyl)-*N'*-ethyl-	$C_9H_{16}ClN_5$ 229.71	5915-41-3 178		 1.188^{20}	
11594	Terbutryn 1,3,5-Triazine-2,4-diamine, *N*-(1,1-dimethylethyl)-*N'*-ethyl-6-(methylthio)-	$C_{10}H_{19}N_5S$ 241.36	886-50-0 104	 $157^{0.06}$	 1.115^{20}	
11595	1,1':2',1''-Terphenyl *o*-Terphenyl	$C_{18}H_{14}$ 230.31	84-15-1 56.2	 332	4-05-00-02478	H_2O 1; ace 3; bz 3; chl 3
11596	1,1':3',1''-Terphenyl *m*-Terphenyl	$C_{18}H_{14}$ 230.31	92-06-8 87	 363	4-05-00-02480	H_2O 1; EtOH 3; eth 3; bz 3
11597	1,1':4',1''-Terphenyl *p*-Terphenyl	$C_{18}H_{14}$ 230.31	92-94-4 210.1	 376	4-05-00-02483	H_2O 1; EtOH 2; eth 3; bz 3
11598	1,1':3',1''-Terphenyl, 5'-phenyl-	$C_{24}H_{18}$ 306.41	612-71-5 176	 462	4-05-00-02732 1.199^{30}	H_2O 1; EtOH 3; eth 3; bz 4
11599	Terrazole 1,2,4-Thiadiazole, 5-ethoxy-3-(trichloromethyl)-	$C_5H_5Cl_3N_2OS$ 247.53	2593-15-9 19.9	 95^{1}	 1.503^{25}	
11600	2,2':5',2''-Terthiophene α-Terthienyl	$C_{12}H_8S_3$ 248.39	1081-34-1	9106	5-19-09-00226	
11601	1,3,5,7-Tetraazatricyclo[3.3.1.13,7]decane Methenamine	$C_6H_{12}N_4$ 140.19	100-97-0 >250	5879 sub	4-26-00-01680 1.331^{-5}	H_2O 4; EtOH 3; eth 2; ace 3
11602	2,6,10,14,18,22-Tetracosahexaene, 2,6,10,15,19,23-hexamethyl-, (all-E)- Squalene	$C_{30}H_{50}$ 410.73	111-02-4 <-20	8727 280^{17}; 241^{4}	3-01-00-01071 0.8584^{20}	H_2O 1; EtOH 2; eth 3; ace 3 1.4990^{20}
11603	Tetracosane	$C_{24}H_{50}$ 338.66	646-31-1 54	 391.3	4-01-00-00578 0.7991^{20}	H_2O 1; EtOH 2; eth 4 1.4283^{70}
11604	Tetracosane, 11-decyl-	$C_{34}H_{70}$ 478.93	55429-84-0 10.8	 301^{10}	4-01-00-00598 0.8161^{20}	 1.4556^{20}
11605	Tetracosane, 3-ethyl-	$C_{26}H_{54}$ 366.71	55282-17-2 30.1	 255.5^{10}	4-01-00-00584 0.7949^{40}	 1.4436^{40}
11606	Tetracosane, 2,6,10,15,19,23-hexamethyl- Squalane	$C_{30}H_{62}$ 422.82	111-01-3 -38	8726 350	4-01-00-00593 0.8115^{15}	H_2O 1; EtOH 2; eth 3; ace 2 1.4530^{15}
11607	Tetracosanoic acid Lignoceric acid	$C_{24}H_{48}O_2$ 368.64	557-59-5 87.5	5364 272^{10}	4-02-00-01301 0.8207^{100}	bz 4; eth 4 1.4287^{100}
11608	15-Tetracosenoic acid	$C_{24}H_{46}O_2$ 366.63	14490-79-0 61		4-02-00-01681	EtOH 3; chl 3; ace 3
11609	5,9-Tetradecadien-7-yne, 6,9-dimethyl-	$C_{16}H_{26}$ 218.38	101427-54-7	 $95\text{-}8^{0.3}$	4-01-00-01133 0.8241^{20}	bz 4; eth 4 1.4866^{20}
11610	6,8-Tetradecadiyne	$C_{14}H_{22}$ 190.33	16387-72-7 2	 $118\text{-}9^{4}$	3-01-00-01067 0.8699^{16}	eth 4
11611	Tetradecanamide	$C_{14}H_{29}NO$ 227.39	638-58-4 104	 217^{12}	4-02-00-01138	EtOH 4
11612	1-Tetradecanamine	$C_{14}H_{31}N$ 213.41	2016-42-4 83.1	 291.2	4-04-00-00812 0.8079^{20}	H_2O 1; EtOH 4; eth 4; ace 3 1.4463^{20}
11613	Tetradecane	$C_{14}H_{30}$ 198.39	629-59-4 5.8	 253.5	4-01-00-00520 0.7628^{20}	H_2O 1; EtOH 4; eth 4; ctc 3 1.4290^{20}
11614	Tetradecane, 1-bromo-	$C_{14}H_{29}Br$ 277.29	112-71-0 5.6	 307	4-01-00-00523 1.0170^{20}	ace 4; bz 4; EtOH 4 1.4603^{20}
11615	Tetradecane, 1-chloro-	$C_{14}H_{29}Cl$ 232.84	2425-54-9 4.9	 292	3-01-00-00550 0.8665^{20}	H_2O 1; EtOH 3; ace 4; bz 4 1.4473^{20}
11616	Tetradecane, 1,14-dibromo- Tetradecamethylene dibromide	$C_{14}H_{28}Br_2$ 356.18	37688-96-3 50.4	 190^{8}	4-01-00-00523	eth 4; EtOH 4; chl 4
11617	Tetradecane, 1,1-dimethoxy- Myristaldehyde, dimethyl acetal	$C_{16}H_{34}O_2$ 258.44	14620-53-2	 $134\text{-}6^{4}$	4-01-00-03389	eth 4; EtOH 4 1.4342^{25}
11618	1,14-Tetradecanediol	$C_{14}H_{30}O_2$ 230.39	19812-64-7 85.8	 200^{9}	4-01-00-02631	eth 4; EtOH 4

No.	Name Synonym	Mol. Form. Mol. Wt.	CAS RN mp/°C	Merck No. bp/°C	Beil. Ref. den/g cm^{-3}	Solubility n_D
11619	Tetradecanenitrile Myristonitrile	$C_{14}H_{27}N$ 209.38	629-63-0 19	 226^{100}; 119^{1}	4-02-00-01139 0.8281^{19}	H_2O 1; EtOH 5; eth 5; ace 5 1.4392^{23}
11620	1-Tetradecanesulfonic acid Tetradecylsulfonic acid	$C_{14}H_{30}O_3S$ 278.46	7314-37-6 65		4-04-00-00066 0.9996^{25}	H_2O 4
11621	1-Tetradecanethiol	$C_{14}H_{30}S$ 230.46	2079-95-0	 $176\text{-}80^{22}$	4-01-00-01867 0.8469^{20}	H_2O 1; EtOH 3; eth 3; ctc 3 1.4597^{20}
11622	Tetradecanoic acid Myristic acid	$C_{14}H_{28}O_2$ 228.38	544-63-8 53.9	6246 250^{100}	4-02-00-01126 0.8622^{54}	H_2O 1; EtOH 3; eth 2; ace 3 1.4723^{70}
11623	Tetradecanoic acid, anhydride	$C_{28}H_{54}O_3$ 438.73	626-29-9 53.4		4-02-00-01138 0.8502^{70}	eth 4; EtOH 4 1.4335^{70}
11624	Tetradecanoic acid, ethyl ester	$C_{16}H_{32}O_2$ 256.43	124-06-1 12.3	 295	4-02-00-01131 0.8573^{25}	H_2O 1; EtOH 3; eth 2; ctc 3 1.4362^{20}
11625	Tetradecanoic acid, methyl ester	$C_{15}H_{30}O_2$ 242.40	124-10-7 19	 295; 155^{7}	4-02-00-01131 0.8671^{20}	H_2O 1; EtOH 5; eth 5; ace 5 1.425^{45}
11626	Tetradecanoic acid, 1-methylethyl ester Isopropyl myristate	$C_{17}H_{34}O_2$ 270.46	110-27-0	5103 193^{20}; 140^{2}	4-02-00-01132 0.8532^{20}	H_2O 1; EtOH 3; eth 3; ace 4 1.4325^{25}
11627	Tetradecanoic acid, phenylmethyl ester	$C_{21}H_{34}O_2$ 318.50	31161-71-4 20.5	 229.3^{11}	2-06-00-00417 0.9293^{25}	bz 4; eth 4; EtOH 4; chl 4
11628	Tetradecanoic acid, 1,2,3-propanetriyl ester Trimyristin	$C_{45}H_{86}O_6$ 723.17	555-45-3 56.5	9638 311	4-02-00-01135 0.8848^{60}	H_2O 1; EtOH 2; eth 3; ace 3 1.4428^{60}
11629	Tetradecanoic acid, propyl ester	$C_{17}H_{34}O_2$ 270.46	14303-70-9	 147^{2}	4-02-00-01132 0.8592^{20}	ace 4; bz 4; eth 4; EtOH 4 1.4356^{25}
11630	1-Tetradecanol Tetradecyl alcohol	$C_{14}H_{30}O$ 214.39	112-72-1 39.5	6248 289	4-01-00-01864 0.8236^{38}	H_2O 1; EtOH 4; eth 4; ace 4
11631	2-Tetradecanol	$C_{14}H_{30}O$ 214.39	4706-81-4 34	 284	4-01-00-01867 0.8315^{20}	 1.4444^{20}
11632	3-Tetradecanol	$C_{14}H_{30}O$ 214.39	1653-32-3 31.5	 173^{25}; 146^{10}	4-01-00-01868 0.8098^{53}	bz 4; eth 4; EtOH 4 1.4340^{45}
11633	2-Tetradecanone Dodecylmethylketone	$C_{14}H_{28}O$ 212.38	2345-27-9 33.5	 205^{100}; 134^{13}	4-01-00-03389	H_2O 1; EtOH 3; ace 3; os 3
11634	3-Tetradecanone	$C_{14}H_{28}O$ 212.38	629-23-2 34	 152^{16}	4-01-00-03389	H_2O 1; EtOH 3; ace 3; os 3
11635	Tetradecanoyl chloride Myristoyl chloride	$C_{14}H_{27}ClO$ 246.82	112-64-1 -1	 171^{16}	4-02-00-01138 0.9078^{25}	eth 3
11636	1-Tetradecene	$C_{14}H_{28}$ 196.38	1120-36-1 -12	 233	4-01-00-00924 0.7745^{25}	H_2O 1; EtOH 4; eth 4; bz 3 1.4351^{20}
11637	4-Tetradecenoic acid Tuduic acid	$C_{14}H_{26}O_2$ 226.36	544-65-0 18.5	 $185\text{-}8^{13}$	4-02-00-01626 0.9024^{20}	bz 4; peth 4 1.4559^{20}
11638	5-Tetradecenoic acid Physoteric acid	$C_{14}H_{26}O_2$ 226.36	544-66-1 20	 $190\text{-}5^{15}$	3-02-00-01373 0.9046^{20}	 1.4552^{20}
11639	9-Tetradecenoic acid	$C_{14}H_{26}O_2$ 226.36	13147-06-3 -4	 $144^{0.6}$	4-02-00-01626 0.9018^{20}	 1.4519^{20}
11640	2-Tetradecyne	$C_{14}H_{26}$ 194.36	638-60-8 6.5	 252.5	0-01-00-00262 0.8000^{20}	eth 4; EtOH 4
11641	7-Tetradecyne	$C_{14}H_{26}$ 194.36	35216-11-6	 144^{30}	3-01-00-01027 0.7991^{20}	eth 4; EtOH 4 1.4330^{25}
11642	2,5,8,11-Tetraoxadodecane Triglyme	$C_8H_{18}O_4$ 178.23	112-49-2 -45	9604 216	4-01-00-02401 0.986^{20}	H_2O 4; bz 4 1.4224^{20}
11643	2,4,8,10-Tetraoxaspiro[5.5]undecane	$C_7H_{12}O_4$ 160.17	126-54-5 48.3	 147^{53}; 68^{1}	5-19-11-00342	H_2O 4; ace 4; eth 4; EtOH 4
11644	Tetraphenylene Tetrabenzocyclooctatetraene	$C_{24}H_{16}$ 304.39	212-74-8 233	 sub 200	4-05-00-02773	EtOH 3; eth 2; AcOEt 3; $PhNO_2$ 3
11645	Tetraphosphoric acid, hexaethyl ester Ethyl tetraphosphate	$C_{12}H_{30}O_{13}P_4$ 506.26	757-58-4 -40	 150 dec	 1.2917^{27}	ace 4; bz 4; EtOH 4 1.4273^{27}
11646	Tetrasiloxane, decamethyl- Decamethyltetrasiloxane	$C_{10}H_{30}O_3Si_4$ 310.69	141-62-8 -76	2843 194	4-04-00-04119 0.8536^{25}	H_2O 1; EtOH 2; bz 3; peth 3 1.3895^{20}
11647	Tetrasiloxane, 1,1,1,3,5,7,7,7-octamethyl- 1,1,1,3,5,7,7,7-Octamethyltetrasiloxane	$C_8H_{26}O_3Si_4$ 282.63	16066-09-4	 170	4-04-00-04098 0.8559^{20}	 1.3854^{20}
11648	Tetrasul *p*-Chlorophenyl 2,4,5-trichlorophenyl sulfide	$C_{12}H_6Cl_4S$ 324.06	2227-13-6			
11649	Tetrasulfide, bis(1,1-dimethylethyl) Di-*tert*-Butyl tetrasulfide	$C_8H_{18}S_4$ 242.49	5943-35-1 2.3	 $70^{0.2}$	4-01-00-01638 1.0690^{20}	 1.5660^{20}
11650	Tetratriacontane	$C_{34}H_{70}$ 478.93	14167-59-0 72.6	 285.4^{3}	4-01-00-00597 0.7728^{90}	 1.4296^{90}
11651	1,2,4,5-Tetrazine *sym*-Tetrazine	$C_2H_2N_4$ 82.06	290-96-0 99	 sub	4-26-00-01710	H_2O 3; EtOH 3; eth 3; sulf 3
11652	1H-Tetrazole	CH_2N_4 70.05	288-94-8 157	 sub	4-26-00-01652 1.4060^{20}	H_2O 2
11653	2H-Tetrazolium, 2,3,5-triphenyl-, chloride Triphenyltetrazolium chloride	$C_{19}H_{15}ClN_4$ 334.81	298-96-4 243 dec	9658	4-26-00-01774	H_2O 3; EtOH 3; eth 1; ace 3
11654	5H-Tetrazolo[1,5-a]azepine, 6,7,8,9-tetrahydro- Pentylenetetrazole	$C_6H_{10}N_4$ 138.17	54-95-5 59.5	7097 194^{12}	4-26-00-01712	H_2O 4; EtOH 4; eth 3; ace 4
11655	4-Thia-1-azabicyclo[3.2.0]heptane-2-carboxylic acid, 3,3-dimethyl-7-oxo-6-[(phenylacetyl)amino]- Penicillin G, sodium salt	$C_{16}H_{17}N_2NaO_4S$ 356.38	69-57-8	1157	4-27-00-05861	

No.	Name Synonym	Mol. Form. Mol. Wt.	CAS RN mp/°C	Merck No. bp/°C	Beil. Ref. den/g cm^{-3}	Solubility n_D
11656	Thiabendazole 1H-Benzimidazole, 2-(4-thiazolyl)-	$C_{10}H_7N_3S$ 201.25	148-79-8	9217 sub 305		
11657	1,2,4-Thiadiazole 1-Thia-2,4-diazacyclopentadiene	$C_2H_2N_2S$ 86.12	288-92-6 -33	 121	4-27-00-07093 1.3298^{20}	
11658	1,2,5-Thiadiazole Piazthiole	$C_2H_2N_2S$ 86.12	288-39-1 -50.1	 94	 1.268^{25}	 1.5150^{25}
11659	1,3,4-Thiadiazole, 2,5-dimethyl- 2,5-Dimethyl-1,3,4-thiadiazole	$C_4H_6N_2S$ 114.17	27464-82-0 65	 202.5	4-27-00-07097	H_2O 2; EtOH 2; eth 2
11660	Thianthrene	$C_{12}H_8S_2$ 216.33	92-85-3 159.3	 365	5-19-02-00049 1.4420^{20}	H_2O 1; EtOH 2; eth 3; bz 3
11661	Thianthrene, 2,7-dimethyl- Mesulphen	$C_{14}H_{12}S_2$ 244.38	135-58-0 123	5821 184^{3}	5-19-02-00072	ace 4; eth 4; peth 4; chl 4
11662	2-Thiazolamine 2-Aminothiazole	$C_3H_4N_2S$ 100.14	96-50-4 93	494 140^{11}	4-27-00-04574	H_2O 2; EtOH 2; eth 2; chl 2
11663	2-Thiazolamine, 5-[(4-aminophenyl)sulfonyl]- Thiazolsulfone	$C_9H_9N_3O_2S_2$ 255.32	473-30-3 219-21 dec	9237	4-27-00-05401	ace 4; eth 4; EtOH 4; diox 4
11664	2-Thiazolamine, 4,5-dihydro-	$C_3H_6N_2S$ 102.16	1779-81-3 85.3	 dec	4-27-00-04555	H_2O 4; EtOH 4; bz 4; chl 4
11665	2-Thiazolamine, 4-methyl- 2-Amino-4-methylthiazole	$C_4H_6N_2S$ 114.17	1603-91-4 45.5	462 125^{20}; $70^{0.4}$	4-27-00-04687	H_2O 4; EtOH 4; eth 4
11666	2-Thiazolamine, 5-nitro- 2-Amino-5-nitrothiazole	$C_3H_3N_3O_2S$ 145.14	121-66-4 202 dec	469	4-27-00-04675	
11667	Thiazole	C_3H_3NS 85.13	288-47-1	9235 118	4-27-00-00960 1.1998^{17}	H_2O 2; EtOH 3; eth 3; ace 3 1.5969^{20}
11668	Thiazole, 2-bromo- 2-Bromo-1,3-thiazole	C_3H_2BrNS 164.03	3034-53-5	 171	4-27-00-00962 1.82^{25}	 1.5927^{20}
11669	5-Thiazolecarboxylic acid, 4-methyl- 4-Methyl-5-thiazolecarboxylic acid	$C_5H_5NO_2S$ 143.17	20485-41-0 280 dec	 sub 250	4-27-00-04008	H_2O 2; EtOH 3; eth 3; bz 2
11670	Thiazole, 5-(2-chloroethyl)-4-methyl- Clomethiazole	C_6H_8ClNS 161.65	533-45-9	2382 92^{7}	4-27-00-00990 1.233^{25}	
11671	2,4-Thiazolediamine, 5-phenyl- Amiphenazole	$C_9H_9N_3S$ 191.26	490-55-1 163-4 dec	498	4-27-00-05139	
11672	Thiazole, 4,5-dihydro-2-methyl-	C_4H_7NS 101.17	2346-00-1 -101	 145	4-27-00-00921 1.067^{25}	 1.5200^{20}
11673	Thiazole, 4,5-dihydro-2,4,4-trimethyl-	$C_6H_{11}NS$ 129.23	4145-94-2	 147	4-27-00-00931 0.969^{25}	 1.4825^{25}
11674	Thiazole, 2,4-dimethyl- 2,4-Dimethylthiazole	C_5H_7NS 113.18	541-58-2	3249 146; 71^{50}	4-27-00-00981 1.0562^{15}	H_2O 2; EtOH 3; eth 3; chl 3 1.5091^{20}
11675	Thiazole, 4,5-dimethyl- 4,5-Dimethylthiazole	C_5H_7NS 113.18	3581-91-7 83.5	 158	4-27-00-00986 1.0699^{20}	eth 4; EtOH 4
11676	Thiazole, 2,4-diphenyl- 2,4-Diphenylthiazole	$C_{15}H_{11}NS$ 237.33	1826-14-8 92.5	 >360	4-27-00-01435 1.1554^{98}	eth 4; EtOH 4
11677	5-Thiazoleethanol, 4-methyl- 4-Methyl-5-thiazoleethanol	C_6H_9NOS 143.21	137-00-8	6046 135^{7}	4-27-00-01754 1.196^{24}	H_2O 4; EtOH 3; eth 3; bz 3
11678	Thiazole, 2-methyl- 2-Methylthiazole	C_4H_5NS 99.16	3581-87-1	 128.5	4-27-00-00967	H_2O 5; EtOH 3; ace 3 1.510
11679	Thiazole, 4-methyl- 4-Methylthiazole	C_4H_5NS 99.16	693-95-8	 133.3	4-27-00-00969 1.112^{25}	H_2O 3; EtOH 3; eth 3
11680	2(3H)-Thiazolethione, 4-methyl-	$C_4H_5NS_2$ 131.22	5685-06-3 89.3	 188^{3}	4-27-00-02584	EtOH 4
11681	Thiazolidine	C_3H_7NS 89.16	504-78-9	 164.5	4-27-00-00009 1.131^{25}	H_2O 5; EtOH 3; eth 4; ace 4 1.551^{20}
11682	4-Thiazolidinecarboxylic acid Timonacic	$C_4H_7NO_2S$ 133.17	444-27-9 196.5	9375		H_2O 4
11683	2,4-Thiazolidinedione	$C_3H_3NO_2S$ 117.13	2295-31-0 128	 179^{19}	4-27-00-03144	eth 4
11684	4-Thiazolidinone, 5-[[4-(dimethylamino)phenyl]methylene]-2-thioxo- *p*-(Dimethylamino)benzal-5-rhodanine	$C_{12}H_{12}N_2OS_2$ 264.37	536-17-4 270 dec	3220	4-27-00-05697	H_2O 1; EtOH 2; eth 4; ace 3
11685	4-Thiazolidinone, 2-thioxo- Rhodanine	$C_3H_3NOS_2$ 133.19	141-84-4 170	8184	4-27-00-03188 0.868^{25}	H_2O 2; EtOH 4; eth 4
11686	Thiazolium, 3-[(4-amino-2-methyl-5-pyrimidinyl)methyl]-5-(2-hydroxyethyl)-4-methyl- chloride, monohydrochlorid Thiamine hydrochloride	$C_{12}H_{18}Cl_2N_4OS$ 337.27	67-03-8 248 dec	9222	4-27-00-01764	H_2O 4; EtOH 2; eth 1; ace 2
11687	Thidiazuron Urea, *N*-phenyl-*N'*-1,2,3-thiadiazol-5-yl-	$C_9H_8N_4OS$ 220.25	51707-55-2 211 dec			
11688	Thieno[2,3-b]thiophene	$C_6H_4S_2$ 140.23	250-84-0 6.5	 225	5 19 01 00431	eth 4
11689	1H-Thieno[3,4-d]imidazole-4-pentanoic acid, hexahydro-2-oxo-, [3a*S*-(3aα,4β,6aα)]- Biotin	$C_{10}H_{16}N_2O_3S$ 244.31	58-85-5 232 dec	1244	4-27-00-07979	H_2O 3; EtOH 3; eth 2; chl 2
11690	Thiepane	$C_6H_{12}S$ 116.23	4753-80-4 0.5	 173.5	5-17-01-00090 0.9883^{20}	H_2O 1; eth 3; ace 3; chl 3 1.5044^{18}
11691	Thietane Trimethylene sulfide	C_3H_6S 74.15	287-27-4 -73.2	 95	5-17-01-00014 1.0200^{20}	H_2O 1; EtOH 4; ace 3; bz 4 1.5102^{20}
11692	Thietane, 1,1-dioxide Trimethylene sulfone	$C_3H_6O_2S$ 106.15	5687-92-3 75.5	 91.2^{14}	5-17-01-00015	H_2O 3; EtOH 3; eth 2; peth 2 1.5156^{20}
11693	Thietane, 2-methyl-	C_4H_8S 88.17	17837-41-1	 107	5-17-01-00055 0.9571^{20}	H_2O 1; EtOH 3; eth 3; ace 3 1.4852^{20}

No.	Name	Synonym	Mol. Form.	Mol. Wt.	CAS RN	mp/°C	Merck No.	bp/°C	Beil. Ref.	den/g cm^{-3}	Solubility	n_D
11694	Thietane, 1-oxide	1,3-Propanesulfone	$C_3H_6O_3S$	122.14	13153-11-2	31.5		180^{30}	5-17-01-00015	1.392^{25}		
11695	Thiirane		C_2H_4S	60.12	420-12-2			57 dec	5-17-01-00009	1.0368^{0}	EtOH 2; eth 2; ace 3; chl 3	1.4935^{20}
11696	Thiirane, methyl-		C_3H_6S	74.15	1072-43-1	-91		72.5	5-17-01-00025	0.941^{20}	chl 3	1.472^{20}
11697	Thiobencarb	Carbamothioic acid, diethyl-, *S*-[(4-chlorophenyl)methyl] ester	$C_{12}H_{16}ClNOS$	257.78	28249-77-6	3.3		$127^{0.008}$		1.16^{20}		
11698	Thioboric acid, trimethyl ester		$C_3H_9BS_3$	152.11	997-49-9	5		218.2	4-01-00-01287	1.126^{20}	ace 4; eth 4	1.5788^{20}
11699	Thiocyanic acid, butyl ester	Butyl thiocyanate	C_5H_9NS	115.20	628-83-1			186	4-03-00-00329	0.9563^{15}	H_2O 1; EtOH 3; eth 3	1.4360^{20}
11700	Thiocyanic acid, chloromethyl ester		C_2H_2ClNS	107.56	3268-79-9			185	2-03-00-00124	1.37^{15}	os 5	
11701	Thiocyanic acid, 1,1-dimethylethyl ester	*tert*-Butyl thiocyanate	C_5H_9NS	115.20	37985-18-5	10.5		140 dec; 40^{10}	4-03-00-00330	0.9187^{10}		
11702	Thiocyanic acid, 1,2-ethanediyl ester		$C_4H_4N_2S_2$	144.22	629-17-4	90		dec	4-03-00-00333	1.4200^{0}	H_2O 2; EtOH 3; eth 3; ace 4	
11703	Thiocyanic acid, ethyl ester	Ethyl thiocyanate	C_3H_5NS	87.15	542-90-5	-85.5		146	4-03-00-00328	1.007^{23}	H_2O 1; EtOH 5; eth 5; chl 3	1.4684^{15}
11704	Thiocyanic acid, heptyl ester	Heptyl thiocyanate	$C_8H_{15}NS$	157.28	5416-94-4			235	4-03-00-00331	0.92^{20}	eth 4; EtOH 4	
11705	Thiocyanic acid, methyl ester	Methyl thiocyanate	C_2H_3NS	73.12	556-64-9	-5	6047	132.9	4-03-00-00327	1.0678^{25}	H_2O 2; EtOH 5; eth 5; ctc 3	1.4669^{25}
11706	Thiocyanic acid, 1-methylethyl ester	Isopropyl thiocyanate	C_4H_7NS	101.17	625-59-2			153	4-03-00-00329	0.9784^{20}	H_2O 1; EtOH 5; eth 5	
11707	Thiocyanic acid, 2-methylpropyl ester	Isobutyl thiocyanate	C_5H_9NS	115.20	591-84-4	-59	5036	175.4	4-03-00-00330		eth 4; EtOH 4	
11708	Thiocyanic acid, octyl ester		$C_9H_{17}NS$	171.31	19942-78-0	105		$141\text{-}2^{19}$	4-03-00-00331	0.9149^{25}	H_2O 1; EtOH 4; eth 4; bz 3	1.4649^{20}
11709	Thiocyanic acid, phenyl ester	Phenyl thiocyanate	C_7H_5NS	135.19	5285-87-0			232.5	4-06-00-01536	1.153^{18}	H_2O 1; EtOH 3; eth 3	
11710	Thiocyanic acid, phenylmethyl ester		C_8H_7NS	149.22	3012-37-1	43		232	4-06-00-02680		H_2O 1; EtOH 3; eth 3; chl 3	
11711	Thiocyanic acid, potassium salt	KSCN	CKNS	97.18	333-20-0		7687		4-03-00-00299			
11712	Thiocyanic acid, 2-propenyl ester		C_4H_5NS	99.16	764-49-8			161	4-03-00-00332	1.056^{15}	eth 4; EtOH 4	
11713	Thiodicarb	Dimethyl *N,N'*-[thiobis[[(methylimino)carbonyl]oxy]]bis [ethanimidothioate]	$C_{10}H_{18}N_4O_4S_3$	354.48	59669-26-0	173	9258			1.4^{20}		
11714	Thiodicarbonic diamide, tetraethyl-	Sulfiram	$C_{10}H_{20}N_2S_3$	264.48	95-05-6		8928	$225\text{-}40^{3}$	4-04-00-00397	1.12^{20}	chl 3	
11715	Thiodicarbonic diamide, tetramethyl-		$C_6H_{12}N_2S_3$	208.37	97-74-5	109.5			4-04-00-00238	1.37^{25}	H_2O 1; EtOH 3; eth 2; ace 3	
11716	Thiodiphosphoric acid, tetraethyl ester	Sulfotep	$C_8H_{20}O_5P_2S_2$	322.32	3689-24-5		8945	$136\text{-}9^{2}$	4-01-00-01351	1.196^{25}	H_2O 1; EtOH 3	1.4753^{25}
11717	Thiofanox	2-Butanone, 3,3-dimethyl-1-(methylthio)-, *O*-[(methylamino)carbonyl]oxime	$C_9H_{18}N_2O_2S$	218.32	39196-18-4	57						
11718	Thioimidodicarbonic diamide	2,4-Dithiobiuret	$C_2H_5N_3S_2$	135.21	541-53-7	181 dec	3378		4-03-00-00356		ace 4	
11719	Thiomorpholine	Thiamorpholine	C_4H_9NS	103.19	123-90-0		9229	175; 110^{100}	4-27-00-00636	1.0882^{20}	H_2O 4; ace 4; eth 4; EtOH 4	1.5386^{20}
11720	Thionazin	Phosphorothioic acid, *O,O*-diethyl *O*-pyrazinyl ester	$C_8H_{13}N_2O_3PS$	248.24	297-97-2	-1.7	9275	80				
11721	Thioperoxydicarbonic diamide ([(H2N)C(*S*)]2S2), tetraethyl-	Disulfiram	$C_{10}H_{20}N_2S_4$	296.55	97-77-8	71.5	3370	117^{17}	4-04-00-00398		H_2O 1; EtOH 3; eth 2; chl 4	
11722	Thioperoxydicarbonic diamide ([(H2N)C(*S*)]2S2), tetramethyl-	Thiram	$C_6H_{12}N_2S_4$	240.44	137-26-8	155.6	9304	129^{20}	4-04-00-00242		chl 4	
11723	Thiophanate-methyl	Carbamic acid, [1,2-phenylenebis(iminocarbonothioyl)]bis-, dimethyl ester	$C_{12}H_{14}N_4O_4S_2$	342.40	23564-05-8	172 dec						
11724	Thiophene	Thiofuran	C_4H_4S	84.14	110-02-1	-39.4	9283	84.0	5-17-01-00297	1.0649^{20}	EtOH 5; eth 5; ace 5; bz 5	1.5289^{20}
11725	2-Thiopheneacetonitrile		C_6H_5NS	123.18	20893-30-5			$115\text{-}20^{22}$	5-18-06-00208	1.157^{25}		1.5425^{20}
11726	Thiophene, 2-bromo-	2-Thienyl bromide	C_4H_3BrS	163.04	1003-09-4			150	5-17-01-00305	1.684^{20}	H_2O 1; eth 4; ace 4; ctc 3	1.5868^{20}
11727	Thiophene, 3-bromo-		C_4H_3BrS	163.04	872-31-1			159.5	5-17-01-00306	1.735^{20}	H_2O 1; ace 3; bz 3; chl 2	1.5919^{20}
11728	Thiophene, 2-bromo-5-chloro-	2-Bromo-5-chlorothiophene	C_4H_2BrClS	197.48	2873-18-9	-20		70^{18}	5-17-01-00307	1.803^{25}	eth 4	1.5925^{25}

No.	Name Synonym	Mol. Form. Mol. Wt.	CAS RN mp/°C	Merck No. bp/°C	Beil. Ref. den/g cm^{-3}	Solubility n_D
11729	Thiophene, 2-bromo-3-methyl- 2-Bromo-3-methylthiophene	C_5H_5BrS 177.06	14282-76-9	 176; $27^{1.8}$	5-17-01-00333 1.5844^{18}	bz 4; eth 4 1.5714^{20}
11730	Thiophene, 2-bromo-5-methyl-	C_5H_5BrS 177.06	765-58-2	 178; $29^{1.8}$	5-17-01-00327 1.5529^{20}	eth 3; bz 3 1.5673^{20}
11731	Thiophene, 2-butyl- 2-Butylthiophene	$C_8H_{12}S$ 140.25	1455-20-5	 181.5	5-17-01-00369 0.9537^{20}	 1.5090^{20}
11732	2-Thiophenecarbonitrile 2-Cyanothiophene	C_5H_3NS 109.15	1003-31-2	 192	5-18-06-00171 1.172^{25}	chl 3 1.5629^{20}
11733	2-Thiophenecarboxaldehyde	C_5H_4OS 112.15	98-03-3	 197; 85^{16}	5-17-09-00349 1.2127^{21}	H_2O 1; EtOH 4; eth 3; chl 2 1.5920^{20}
11734	3-Thiophenecarboxaldehyde 3-Formylthiophene	C_5H_4OS 112.15	498-62-4	 86.7^{20}	5-17-09-00373	H_2O 1; EtOH 4; eth 3 1.5855^{20}
11735	2-Thiophenecarboxaldehyde, 5-chloro-	C_5H_4ClOS 147.60	7283-96-7	 77.5^{5}	5-17-09-00360	chl 2 1.6036^{25}
11736	2-Thiophenecarboxaldehyde, 5-methyl-	C_6H_6OS 126.18	13679-70-4	 114^{25}	5-17-09-00406	chl 3 1.5825^{20}
11737	2-Thiophenecarboxylic acid 2-Carboxythiophene	$C_5H_4O_2S$ 128.15	527-72-0 129.5	9284 260 dec	5-18-06-00158	H_2O 4; EtOH 4; eth 4; chl 3
11738	3-Thiophenecarboxylic acid 3-Thenoic acid	$C_5H_4O_2S$ 128.15	88-13-1 138	9206	5-18-06-00199	H_2O 3
11739	2-Thiophenecarboxylic acid, ethyl ester	$C_7H_8O_2S$ 156.21	2810-04-0	 218	5-18-06-00159 1.1623^{16}	EtOH 3; ace 3; ctc 2; os 3 1.5248^{20}
11740	Thiophene, 2-chloro- 2-Thienyl chloride	C_4H_3ClS 118.59	96-43-5 -71.9	 128.3	5-17-01-00303 1.2863^{20}	H_2O 1; EtOH 5; eth 5; chl 2 1.5487^{20}
11741	Thiophene, 2-chloro-5-methyl- 2-Chloro-5-methylthiophene	C_5H_5ClS 132.61	17249-82-0	 155; 55^{19}	5-17-01-00326 1.2147^{25}	ace 4; bz 4; eth 4; EtOH 4 1.5372^{20}
11742	Thiophene, 2,3-dibromo- 2,3-Dibromothiophene	$C_4H_2Br_2S$ 241.93	3140-93-0 -17.5	 218.5; 89^{13}	5-17-01-00308	 1.6304^{22}
11743	Thiophene, 2,4-dibromo- 2,4-Dibromothiophene	$C_4H_2Br_2S$ 241.93	3140-92-9 -26.3	 210	5-17-01-00308 2.1488^{20}	
11744	Thiophene, 2,5-dibromo- 2,5-Dibromothiophene	$C_4H_2Br_2S$ 241.93	3141-27-3 -6	 210.3	5-17-01-00308 2.142^{23}	H_2O 1; EtOH 4; eth 4; ctc 3 1.6288^{20}
11745	2,5-Thiophenedicarboxylic acid 2,5-Dicarboxythiophene	$C_6H_4O_4S$ 172.16	4282-31-9 359	 sub 150	5-18-07-00049	H_2O 2; EtOH 3; eth 3
11746	Thiophene, 2,3-dichloro- 2,3-Dichlorothiophene	$C_4H_2Cl_2S$ 153.03	17249-79-5 -26.2	 173.5	5-17-01-00304 1.4605^{20}	H_2O 1; EtOH 5; eth 3; ace 3 1.5651^{20}
11747	Thiophene, 2,4-dichloro- 2,4-Dichlorothiophene	$C_4H_2Cl_2S$ 153.03	17249-75-1 -34	 167.6	5-17-01-00304 1.4553^{20}	H_2O 1; EtOH 4; eth 3; ace 3 1.5660^{20}
11748	Thiophene, 2,5-dichloro- 5-Chloro-2-thienyl chloride	$C_4H_2Cl_2S$ 153.03	3172-52-9 -40.5	 162	5-17-01-00304 1.4422^{20}	H_2O 1; EtOH 5; eth 5; ctc 3 1.5626^{20}
11749	Thiophene, 3,4-dichloro- 3,4-Dichlorothiophene	$C_4H_2Cl_2S$ 153.03	17249-76-2 1	 185	5-17-01-00305 1.4867^{20}	H_2O 1; EtOH 4; eth 4; ace 3 1.5821^{19}
11750	Thiophene, 2,3-dihydro- 2,3-Dihydrothiophene	C_4H_6S 86.16	1120-59-8	 112.1		
11751	Thiophene, 2,5-dihydro- 2,5-Dihydrothiophene	C_4H_6S 86.16	1708-32-3	 122.4		
11752	Thiophene, 2,5-dihydro-, 1,1-dioxide 3-Sulfolene	$C_4H_6O_2S$ 118.16	77-79-2 64.5	8935	5-17-01-00177	chl 3
11753	Thiophene, 2,5-diiodo- 2,5-Diiodothiophene	$C_4H_2I_2S$ 335.93	625-88-7 41.5	 139^{15}	5-17-01-00313	H_2O 1; EtOH 4
11754	Thiophene, 2,3-dimethyl- 2,3-Dimethylthiophene	C_6H_8S 112.20	632-16-6 -49	 141.6	5-17-01-00342 1.0021^{20}	H_2O 1; EtOH 4; eth 4; bz 3 1.5192^{20}
11755	Thiophene, 2,4-dimethyl- 2,4-Dimethylthiophene	C_6H_8S 112.20	638-00-6	 140.7	5-17-01-00344 0.9938^{20}	H_2O 1; EtOH 3; eth 3; bz 3 1.5104^{20}
11756	Thiophene, 2,5-dimethyl- 2,5-Dimethylthiophene	C_6H_8S 112.20	638-02-8 -62.6	 136.5	5-17-01-00345 0.9850^{20}	H_2O 1; EtOH 3; eth 3; bz 3 1.5129^{20}
11757	Thiophene, 3,4-dimethyl- 3,4-Dimethylthiophene	C_6H_8S 112.20	632-15-5	 145	5-17-01-00347 0.993^{25}	H_2O 1; EtOH 3; eth 4 1.5206^{20}
11758	Thiophene, 2,5-dinitro- 2,5-Dinitrothiophene	$C_4H_2N_2O_4S$ 174.14	59434-05-8 81	 290	5-17-01-00315	eth 4; EtOH 4
11759	Thiophene, 2-ethyl- 2-Ethylthiophene	C_6H_8S 112.20	872-55-9	 134	5-17-01-00339 0.9930^{20}	H_2O 1; EtOH 4; eth 4 1.5122^{20}
11760	Thiophene, 3-ethyl- 3-Ethylthiophene	C_6H_8S 112.20	1795-01-3 -89.1	 136	5-17-01-00341 0.9980^{20}	H_2O 1; EtOH 3; eth 4 1.5146^{20}
11761	Thiophene, 2-ethyl-3-methyl- 2-Ethyl-3-methylthiophene	$C_7H_{10}S$ 126.22	31805-48-8	 161	5-17-01-00357 0.9815^{20}	 1.5105^{20}
11762	Thiophene, 2-ethyl-5-methyl- 2-Ethyl-5-methylthiophene	$C_7H_{10}S$ 126.22	40323-88-4 -68.4	 160	5-17-01-00357 0.9003^{20}	 1.5073^{20}
11763	Thiophene, 3-ethyl-4-methyl-	$C_7H_{10}S$ 126.22	66577-03-5	 165.5	4-17-00-00301 0.9976^{20}	 1.5186^{20}
11764	Thiophene, 4-ethyl-2-methyl- 5-Methyl-3-ethylthiophene	$C_7H_{10}S$ 126.22	13678-54-1 -60	 163	4-17-00-00299 0.9742^{20}	 1.5098^{20}
11765	Thiophene, 2-ethyltetrahydro-	$C_6H_{12}S$ 116.23	1551-32-2	 157	5-17-01-00096 0.944^{20}	ace 4; bz 4; eth 4; EtOH 4 1.490^{20}
11766	Thiophene, 3-ethyltetrahydro- 3-Ethyltetrahydrothiophene	$C_6H_{12}S$ 116.23	62184-67-2	 165	5-17-01-00096 0.950^{20}	 1.491^{20}
11767	Thiophene, 2-iodo-	C_4H_3IS 210.04	3437-95-4 -40	 181	5-17-01-00309 2.0595^{25}	EtOH 4; eth 4; chl 2 1.6465^{25}
11768	2-Thiophenemethanol	C_5H_6OS 114.17	636-72-6	 207; 86^{10}	5-17-03-00358 1.2053^{16}	EtOH 3; ace 3 1.5280^{20}

No.	Name Synonym	Mol. Form. Mol. Wt.	CAS RN mp/°C	Merck No. bp/°C	Beil. Ref. den/g cm^{-3}	Solubility n_D
11769	Thiophene, 2-methyl- 2-Methylthiophene	C_5H_6S 98.17	554-14-3 -63.4	 112.6	5-17-01-00324 1.0193^{20}	H_2O 1; EtOH 5; eth 5; ace 5 1.5203^{20}
11770	Thiophene, 3-methyl- 3-Thiotolene	C_5H_6S 98.17	616-44-4 -69	 115.5	5-17-01-00331 1.0218^{20}	H_2O 1; EtOH 5; eth 5; ace 5 1.5204^{20}
11771	Thiophene, 2-(3-methylbutyl)- 2-Isopentylthiophene	$C_9H_{14}S$ 154.28	26963-33-7	 $74\text{-}5^{11}$	4-17-00-00316 0.9481^{12}	 1.5014^{12}
11772	Thiophene, 2-(1-methylethyl)- 2-Isopropylthiophene	$C_7H_{10}S$ 126.22	4095-22-1	 153	5-17-01-00356 0.9678^{20}	 1.5038^{20}
11773	Thiophene, 3-(1-methylethyl)- 3-Isopropylthiophene	$C_7H_{10}S$ 126.22	29488-27-5	 157	5-17-01-00356 0.9733^{20}	 1.5052^{20}
11774	Thiophene, 2-methyl-5-phenyl- 2-Methyl-5-phenylthiophene	$C_{11}H_{10}S$ 174.27	5069-26-1 51	 271	5-17-02-00153	EtOH 4; eth 4; lig 4
11775	Thiophene, 2-nitro- 2-Nitrothiophene	$C_4H_3NO_2S$ 129.14	609-40-5 46.5	 224.5	5-17-01-00313 1.3644^{43}	H_2O 1; EtOH 4; alk 3; peth 2
11776	Thiophene, 3-nitro- 3-Nitrothiophene	$C_4H_3NO_2S$ 129.14	822-84-4 78.5	 225	5-17-01-00313	H_2O 1; EtOH 4; ctc 2; peth 3
11777	Thiophene-2-ol	C_4H_4OS 100.14	17236-58-7 9	 218	 1.255^{20}	H_2O 3; ace 3; chl 3 1.5644^{20}
11778	Thiophene, 2-propyl- 2-Propylthiophene	$C_7H_{10}S$ 126.22	1551-27-5	 158	5-17-01-00355 0.9683^{20}	H_2O 1; EtOH 3; eth 3; bz 3 1.5076^{20}
11779	Thiophene, 3-propyl- 3-Propylthiophene	$C_7H_{10}S$ 126.22	1518-75-8	 163.2	5-17-01-00355 0.9738^{20}	H_2O 1; EtOH 3; eth 3; bz 3 1.5057^{20}
11780	2-Thiophenesulfonyl chloride	$C_4H_3ClO_2S_2$ 182.65	16629-19-9 28	 100^{6}	5-18-09-00426	eth 3
11781	Thiophene, tetrabromo- 2,3,4,5-Tetrabromothiophene	C_4Br_4S 399.73	3958-03-0 117.5	 326	5-17-01-00309	H_2O 1; EtOH 3; eth 4
11782	Thiophene, tetrachloro-	C_4Cl_4S 221.92	6012-97-1 30.5	 233.4	5-17-01-00305 1.7036^{30}	H_2O 1; EtOH 4; eth 5 1.5915^{30}
11783	Thiophene, tetrahydro- Thiacyclopentane	C_4H_8S 88.17	110-01-0 -96.1	 121.0	5-17-01-00036 0.9987^{20}	H_2O 1; EtOH 5; eth 5; ace 5 1.4871^{18}
11784	Thiophene, tetrahydro-2,4-dimethyl-, 1,1-dioxide 2,4-Dimethylsulfolane	$C_6H_{12}O_2S$ 148.23	1003-78-7 -3	3245 281	5-17-01-00097 1.1362^{20}	lig 4 1.4732^{20}
11785	Thiophene, tetrahydro-2,5-dimethyl-, *cis*-	$C_6H_{12}S$ 116.23	5161-13-7 -89	 142.3	5-17-01-00098 0.9222^{20}	ace 4; bz 4; eth 4; EtOH 4 1.4799^{20}
11786	Thiophene, tetrahydro-, 1,1-dioxide Sulfolane	$C_4H_8O_2S$ 120.17	126-33-0 27.6	8934 287.3	5-17-01-00039 1.2723^{18}	chl 3 1.4833^{18}
11787	Thiophene, tetrahydro-2-methyl-	$C_5H_{10}S$ 102.20	1795-09-1 -100.7	 134	5-17-01-00080 0.9564^{18}	H_2O 1; EtOH 4; eth 4; ace 4 1.4886^{15}
11788	Thiophene, tetrahydro-3-methyl-	$C_5H_{10}S$ 102.20	4740-00-5	 139	5-17-01-00082 0.9596^{18}	H_2O 1; EtOH 4; eth 4; ace 4 1.4886^{18}
11789	Thiophene, tetrahydro-3-methyl-, 1,1-dioxide 3-Methylsulfolane	$C_5H_{10}O_2S$ 134.20	872-93-5 1	 276	5-17-01-00082 1.188^{25}	 1.4772^{20}
11790	Thiophene, 2,3,5-tribromo- 2,3,5-Tribromothiophene	C_4HBr_3S 320.83	3141-24-0 29	 260	5-17-01-00309	chl 3
11791	Thiophene, 2,3,5-trichloro- 2,3,5-Trichlorothiophene	C_4HCl_3S 187.48	17249-77-3 -16.1	 198.7	5-17-01-00305 1.5856^{20}	H_2O 1; EtOH 5; eth 5 1.5791^{20}
11792	Thiophene, 2,3,4-trimethyl- 2,3,4-Trimethylthiophene	$C_7H_{10}S$ 126.22	1795-04-6 -26.2	 162; 75^{30}	5-17-01-00359 0.995^{20}	 1.5208^{20}
11793	Thiophene, 2,3,5-trimethyl- 2,3,5-Trimethylthiophene	$C_7H_{10}S$ 126.22	1795-05-7	 164.5	5-17-01-00359 0.9753^{20}	H_2O 1; EtOH 4; eth 4; ace 4 1.5112^{20}
11794	2(3H)-Thiophenone, dihydro-	C_4H_6OS 102.16	1003-10-7	 $39\text{-}40^{1}$	5-17-09-00012 1.18^{25}	 1.5230^{20}
11795	2H-Thiopyran, tetrahydro- Thiacyclohexane	$C_5H_{10}S$ 102.20	1613-51-0 19	 141.8	5-17-01-00068 0.9861^{20}	H_2O 1; EtOH 3; eth 3; ace 3 1.5067^{20}
11796	2H-Thiopyran, tetrahydro-2-methyl-	$C_6H_{12}S$ 116.23	5161-16-0 -58	 153	5-17-01-00093 0.9428^{20}	 1.4905^{20}
11797	2H-Thiopyran, tetrahydro-3-methyl-	$C_6H_{12}S$ 116.23	5258-50-4 -60	 158	5-17-01-00094 0.9473^{20}	 1.4922^{20}
11798	2H-Thiopyran, tetrahydro-4-methyl-	$C_6H_{12}S$ 116.23	5161-17-1 -28	 158.6	5-17-01-00094 0.9471^{20}	 1.4923^{20}
11799	Thiourea Thiocarbamide	CH_4N_2S 76.12	62-56-6 182	9299	4-03-00-00342 1.405^{25}	H_2O 3; EtOH 3; eth 1
11800	Thiourea, *N,N'*-bis[4-(3-methylbutoxy) phenyl]- Tiocarlide	$C_{23}H_{32}N_2O_2S$ 400.59	910-86-1 146	9382	4-13-00-01187	
11801	Thiourea, (2-chlorophenyl)- 1-(2-Chlorophenyl)thiourea	$C_7H_7ClN_2S$ 186.66	5344-82-1 146		3-12-00-01296	bz 4; EtOH 4
11802	Thiourea, *N*-(3,5-dichlorophenyl)-*N'*-(4-fluorophenyl)- Loflucarban	$C_{13}H_9Cl_2FN_2S$ 315.20	790-69-2 163.5	5440	4-12-00-01275	
11803	Thiourea, *N,N'*-diethyl-	$C_5H_{12}N_2S$ 132.23	105-55-5 78	 dec	4-04-00-00375	H_2O 3; EtOH 3; eth 4; ctc 2
11804	Thiourea, *N,N'*-dimethyl- *N,N'*-Dimethylthiourea	$C_3H_8N_2S$ 104.18	534-13-4 62	3250	4-04-00-00217	H_2O 4; EtOH 4; eth 2; ace 4
11805	Thiourea, *N,N'*-diphenyl- *sym*-Diphenylthiourea	$C_{13}H_{12}N_2S$ 228.32	102-08-9 154.5	3337	4-12-00-00810 1.32^{25}	H_2O 2; EtOH 4; eth 4; chl 4
11806	Thiourea, (2-methylphenyl)- *o*-Tolylthiourea	$C_8H_{10}N_2S$ 166.25	614-78-8 162		4-12-00-01766	H_2O 4; EtOH 4; eth 2

No.	Name Synonym	Mol. Form. Mol. Wt.	CAS RN mp/°C	Merck No. bp/°C	Beil. Ref. den/g cm^{-3}	Solubility n_D
11807	Thiourea, 1-naphthalenyl- ANTU	$C_{11}H_{10}N_2S$ 202.28	86-88-4 198	755	4-12-00-03086	H_2O 1; EtOH 2; eth 2; ace 2
11808	Thiourea, phenyl- Phenylthiourea	$C_7H_8N_2S$ 152.22	103-85-5 154	7286	4-12-00-00804	H_2O 2; EtOH 3; NaOH 3
11809	Thiourea, 2-propenyl- Thiosinamine	$C_4H_8N_2S$ 116.19	109-57-9 78	9293	4-04-00-01072 1.217^{20}	H_2O 3; EtOH 3; eth 2; bz 1 1.5936^{78}
11810	Thiourea, tetramethyl-	$C_5H_{12}N_2S$ 132.23	2782-91-4 79.3	 245	4-04-00-00232	H_2O 3; EtOH 3; eth 2; chl 3
11811	9H-Thioxanthene Dibenzothiapyran	$C_{13}H_{10}S$ 198.29	261-31-4 128.5	9300 341	5-17-02-00254	chl 3
11812	9H-Thioxanthen-9-one Thioxanthone	$C_{13}H_8OS$ 212.27	492-22-8 209	9301 373	5-17-10-00437	H_2O 1; EtOH 2; bz 3; chl 3
11813	*L*-Threonine [*R*-(*R**,*S**)]-2-Amino-3-hydroxybutanoic acid	$C_4H_9NO_3$ 119.12	72-19-5 256 dec	9316	4-04-00-03171	H_2O 3; EtOH 1; eth 1; chl 1
11814	Thujane, (1*S*,4*R*,5*S*)-(+)- Sabinane (*d*)	$C_{10}H_{18}$ 138.25	7712-66-5	 157	4-05-00-00318 0.8139^{20}	 1.4376^{20}
11815	4(10)-Thujene, (+)- 4(10)-Thujene (*d*)	$C_{10}H_{16}$ 136.24	2514-91-2	 164	4-05-00-00451 0.842^{20}	ace 4; bz 4; eth 4; EtOH 4 1.4678^{20}
11816	Thymidine Thymine 2-desoxyriboside	$C_{10}H_{14}N_2O_5$ 242.23	50-89-5 186.5	9331	5-24-07-00109	H_2O 3; EtOH 3; ace 3; chl 2
11817	Toluene, 3-chloro-2-iodo-	C_7H_6ClI 252.48	5100-98-1 27.5	 123^{14}	3-05-00-00726	 1.608^{23}
11818	*o*-Toluenesulfonic acid, thio-, *S*-phenyl ester Benzenesulfonic acid, thiolo,2-methyl,phenyl ester	$C_{13}H_{12}O_2S_2$ 264.37	96097-69-7 59.5	 241.6	4-11-00-00237	H_2O 2; EtOH 4; eth 3; chl 2 1.53412^{20}
11819	Triacontane	$C_{30}H_{62}$ 422.82	638-68-6 65.8	 449.7	4-01-00-00592 0.8097^{20}	H_2O 1; EtOH 2; eth 3; bz 4 1.4352^{70}
11820	Triacontanoic acid	$C_{30}H_{60}O_2$ 452.81	506-50-3 93.6		4-02-00-01321	bz 4; CS_2 4; chl 4 1.4323^{100}
11821	1-Triacontanol Myricyl alcohol	$C_{30}H_{62}O$ 438.82	593-50-0 88	9506	4-01-00-01918 0.777^{95}	bz 4; eth 4; EtOH 4
11822	Triadimenol Mercury, chloro(2-methoxyethyl)-	C_3H_7ClHgO 295.13	123-88-6			
11823	Triallate Carbamothioic acid, bis(1-methylethyl)-, *S*-(2,3,3-trichloro-2-propenyl) ester	$C_{10}H_{16}Cl_3NOS$ 304.67	2303-17-5 29	9510 $117^{0.0003}$	 1.273^{25}	
11824	Triasulfuron 2-(2-Chloroethoxy)-*N*-[[(4-methoxy-6-methyl-1,3,5-triazi*N*-2-yl)amino]carbonyl]benzene sulfonamide	$C_{14}H_{16}ClN_5O_5S$ 401.83	82097-50-5 186			
11825	3,5,7-Triaza-1-azoniatricyclo[3.3.1.13,7]decane, 1-(2-propenyl)-, iodide Hexamethylenetetramine, allyl iodide	$C_9H_{17}IN_4$ 308.17	36895-62-2	5880		
11826	1-Triazene, 1,3-diphenyl- Diazoaminobenzene	$C_{12}H_{11}N_3$ 197.24	136-35-6 98	2981	4-16-00-00904	H_2O 1; EtOH 4; eth 4; bz 4
11827	1,3,5-Triazine	$C_3H_3N_3$ 81.08	290-87-9 86	 114	4-26-00-00063 1.38^{25}	EtOH 3; eth 3
11828	1,3,5-Triazine-2,4-diamine, 6-chloro-*N*,*N*,*N*',*N*'-tetraethyl-	$C_{11}H_{20}ClN_5$ 257.77	580-48-3 27	 $154\text{-}6^{9}$	4-26-00-01208 1.0956^{20}	bz 4; chl 4; EtOH 4; lig 4 1.5320^{20}
11829	1,3,5-Triazine-2,4-diamine, 6-phenyl- Benzoguanamine	$C_9H_9N_5$ 187.20	91-76-9 226.5	1099	4-26-00-01244	EtOH 3; eth 3; tfa 2
11830	1,3,5-Triazine-2,4-diamine, *N*-phenyl- Amanozine	$C_9H_9N_5$ 187.20	537-17-7 235.5	379	4-26-00-01195	
11831	1,2,4-Triazine-3,5(2H,4H)-dione, 6-methyl- 6-Azathymine	$C_4H_5N_3O_2$ 127.10	932-53-6 211	919	4-26-00-00556	H_2O 3; EtOH 3; ace 3
11832	1,2,4-Triazine-3,5(2H,4H)-dione, 2-β-*D*-ribofuranosyl- 6-Azauridine	$C_8H_{11}N_3O_6$ 245.19	54-25-1 158	920	4-26-00-00554	H_2O 3
11833	1,3,5-Triazine, hexahydro-1,3,5-trinitro- Cyclonite	$C_3H_6N_6O_6$ 222.12	121-82-4 205.5	2742	4-26-00-00022 1.82^{20}	H_2O 1; EtOH 1; eth 2; ace 3
11834	1,3,5-Triazine, hexahydro-1,3,5-triphenyl-	$C_{21}H_{21}N_3$ 315.42	91-78-1 144	 185; 60^{29}	4-26-00-00007	H_2O 1; EtOH 2; eth 3; ace 3
11835	1,3,5-Triazine-2,4,6-triamine Melamine	$C_3H_6N_6$ 126.12	108-78-1 345 dec	5691 sub	4-26-00-01253 1.573^{16}	H_2O 2; EtOH 2; eth 1 1.872^{20}
11836	1,3,5-Triazine-2,4,6-tricarbonitrile	C_6N_6 156.11	7615-57-8 119	 262; $119^{1.0}$	1-26-00-00091	bz 4
11837	1,3,5-Triazine, 2,4,6-trichloro- Cyanuric acid trichloride	$C_3Cl_3N_3$ 184.41	108-77-0 154	 192	4-26-00-00066	EtOH 4
11838	1,3,5-Triazine-2,4,6(1H,3H,5H)-trione, 1,3-dichloro- Dichlorocyanuric acid	$C_3HCl_2N_3O_3$ 197.96	2782-57-2			
11839	1,3,5-Triazine-2,4,6(1H,3H,5H)-trione, dihydrate Cyanuric acid, dihydrate	$C_3H_7N_3O_5$ 165.11	6202-04-6 >360 dec	 dec	4-26-00-00632 2.50^{20}	H_2O 3; EtOH 2; eth 1; ace 1 2.500^{20}
11840	1,3,5-Triazine-2,4,6(1H,3H,5H)-trione, 1,3,5-trichloro- Symclosene	$C_3Cl_3N_3O_3$ 232.41	87-90-1 246.7 dec	8993	4-26-00-00642	

No.	Name Synonym	Mol. Form. Mol. Wt.	CAS RN mp/°C	Merck No. bp/°C	Beil. Ref. den/g cm^{-3}	Solubility n_D
11841	1,3,5-Triazine-2,4,6(1H,3H,5H)-trione, 1,3,5-trimethyl-	$C_6H_9N_3O_3$ 171.16	827-16-7 176.5	 274	4-26-00-00635	H_2O 1; EtOH 3
11842	1,3,5-Triazine-2,4,6(1H,3H,5H)-trione, 1,3,5-tri-2-propenyl-	$C_{12}H_{15}N_3O_3$ 249.27	1025-15-6 20.5	 149^{4}; $105^{0.5}$	4-26-00-00637 1.1590^{20}	
11843	1,3,5-Triazine, 2,4,6-tri-2-pyridinyl- 2,4,6-Tripyridyl-s-triazine	$C_{18}H_{12}N_6$ 312.33	3682-35-7 210	9664	4-26-00-04192	
11844	1,3,5-Triazine, 2,4,6-tris(1-aziridinyl)- Triethylenemelamine	$C_9H_{12}N_6$ 204.23	51-18-3	9586	4-26-00-01268	
11845	1H-1,2,4-Triazol-3-amine Amitrole	$C_2H_4N_4$ 84.08	61-82-5 159	506	4-26-00-01073	H_2O 4; EtOH 4; eth 1; ace 1
11846	1H-1,2,3-Triazole	$C_2H_3N_3$ 69.07	288-36-8 23	 204	4-26-00-00029 1.1861^{25}	H_2O 3; eth 3; ace 3; os 3 1.4854^{25}
11847	1H-1,2,4-Triazole, 1-butyl 1-Butyl-*s*-triazole	$C_6H_{11}N_3$ 125.17	6086-22-2			
11848	4H-1,2,4-Triazole, 4-cyclohexyl-3-ethyl- Hexazole	$C_{10}H_{17}N_3$ 179.27	4671-03-8 89.5	4618 227^{10}	4-26-00-00049	H_2O 4; bz 4; chl 4
11849	1H-1,2,3-Triazole, 1,5-dimethyl- 1,5-Dimethyl-1,2,3-triazole	$C_4H_7N_3$ 97.12	15922-53-9 -4	 255	1-26-00-00005	H_2O 4
11850	1,2,4-Triazolidine-3,5-dione Urazole	$C_2H_3N_3O_2$ 101.06	3232-84-6 249-50 dec	9780	4-26-00-00538	
11851	7H-1,2,3-Triazolo[4,5-d]pyrimidin-7-one, 5-amino-1,4-dihydro- 8-Azaguanine	$C_4H_4N_6O$ 152.12	134-58-7 300	912	4-26-00-04171	
11852	*S,S,S*-Tributyl phosphorotrothioate *S,S,S*-Tributyl trithiophosphate	$C_{12}H_{27}OPS_3$ 314.52	78-48-8 <-25	 $150^{0.3}$	 1.057^{20}	
11853	Trichlorfon Phosphonic acid, (2,2,2-trichloro-1-hydroxyethyl)-, dimethyl ester	$C_4H_8Cl_3O_4P$ 257.44	52-68-6 77	9536 $100^{0.1}$	 1.73^{20}	
11854	Triclopyr Acetic acid, [(3,5,6-trichloro-2-pyridinyl)oxy]-	$C_7H_4Cl_3NO_3$ 256.47	55335-06-3 149	9572 dec 290		
11855	Tricosane	$C_{23}H_{48}$ 324.63	638-67-5 47.6	 380	4-01-00-00576 0.7785^{48}	H_2O 1; EtOH 2; eth 3; ctc 3 1.4468^{20}
11856	Tricosane, 2-methyl- 2-Methyltricosane	$C_{24}H_{50}$ 338.66	1928-30-9 37.6	 205^{3}	3-01-00-00576 0.7539^{90}	 1.4201^{90}
11857	12-Tricosanone Diundecyl ketone	$C_{23}H_{46}O$ 338.62	540-09-0 70.2		4-01-00-03408 0.8086^{69}	bz 4; eth 4; chl 4 1.4283^{80}
11858	Tricyclazole 1,2,4-Triazolo[3,4-b]benzothiazole, 5-methyl-	$C_9H_7N_3S$ 189.24	41814-78-2 187			
11859	Tricyclo[3.3.1.13,7]decan-1-amine Amantadine	$C_{10}H_{17}N$ 151.25	768-94-5 180	380		H_2O 2
11860	Tricyclo[3.3.1.13,7]decane Adamantane	$C_{10}H_{16}$ 136.24	281-23-2 268	140 sub	4-05-00-00469 1.07^{25}	bz 3; ctc 3 1.568
11861	Tricyclo[4.4.0.02,7]dec-3-ene, 1,3-dimethyl-8-(1-methylethyl)-, stereoisomer Copaene	$C_{15}H_{24}$ 204.36	3856-25-5	2510 248.5	4-05-00-01189 0.8996^{20}	H_2O 1; eth 3; ace 3; HOAc 3 1.4894^{20}
11862	Tricyclo[2.2.1.02,6]heptane-1-carboxylic acid, 7,7-dimethyl- Tricyclenic acid	$C_{10}H_{14}O_2$ 166.22	512-60-7 151.3	 263	4-09-00-00246	bz 4; EtOH 4; peth 4
11863	Tricyclo[2.2.1.02,6]heptane-1-carboxylic acid, 7,7-dimethyl-, ethyl ester Tricyclenic acid, ethyl ester	$C_{12}H_{18}O_2$ 194.27	56694-98-5	 $100-1^{10}$	2-09-00-00064 1.0143^{20}	 1.4230^{20}
11864	Tricyclo[2.2.1.02,6]heptane-1-carboxylic acid, 7,7-dimethyl-, methyl ester Tricyclenic acid, methyl ester	$C_{11}H_{16}O_2$ 180.25	56694-97-4 45.5	 99^{14}	2-09-00-00064 1.0255^{42}	 1.4695^{42}
11865	Tricyclo[2.2.1.02,6]heptane, 3,3-dimethyl-	C_9H_{14} 122.21	473-02-9 42.5	 138	4-05-00-00430 0.8710^{40}	bz 4; eth 4; EtOH 4 1.4514^{40}
11866	Tricyclo[2.2.1.02,6]heptane, 1,3,3-trimethyl- Cyclofenchene	$C_{10}H_{16}$ 136.24	488-97-1	 145	3-05-00-00392 0.859^{20}	 1.4503^{22}
11867	Tricyclo[2.2.1.02,6]heptane, 1,7,7-trimethyl-	$C_{10}H_{16}$ 136.24	508-32-7 67.5	 152.5	4-05-00-00468 0.8668^{80}	 1.4296^{80}
11868	1,12-Tridecadiyne	$C_{13}H_{20}$ 176.30	38628-39-6 -3	 115.5^{12}	2-01-00-00248 0.8262^{21}	 1.454^{20}
11869	Tridecanal	$C_{13}H_{26}O$ 198.35	10486-19-8 14	 156^{13}	4-01-00-03386 0.8356^{18}	H_2O 1; EtOH 3; os 4 1.4384^{18}
11870	1-Tridecanamine Tridecane, 1-amino	$C_{13}H_{29}N$ 199.38	2869-34-3 27.4	 275.8	4-04-00-00810 0.8049^{20}	H_2O 2; EtOH 3; eth 3 1.4443^{20}
11871	Tridecane	$C_{13}H_{28}$ 184.37	629-50-5 -5.3	 235.4	4-01-00-00512 0.7564^{20}	H_2O 1; EtOH 4; eth 4; ctc 3 1.4256^{20}
11872	Tridecane, 1-bromo-	$C_{13}H_{27}Br$ 263.26	765-09-3 6.2	 296	4-01-00-00514 1.0177^{20}	H_2O 1; chl 4 1.4593^{20}
11873	Tridecane, 1-bromo-12-methyl- Tridecane, 2-methyl-13-bromo	$C_{14}H_{29}Br$ 277.29	111772-90-8	 $120-2^{3}$	3-01-00-00551 1.0241^{20}	 1.4598^{20}
11874	Tridecane, 1,13-dibromo- Tridecamethylene dibromide	$C_{13}H_{26}Br_2$ 342.16	31772-05-1 9	 $188-92^{12}$	3-01-00-00548 1.276^{15}	eth 4; chl 4 1.4880^{27}
11875	1,12-Tridecanediol	$C_{13}H_{28}O_2$ 216.36	99706-68-0 60.5	 188^{8}	2-01-00-00563	EtOH 4
11876	1,13-Tridecanediol	$C_{13}H_{28}O_2$ 216.36	13362-52-2 76.5	 195^{10}	4-01-00-02630	EtOH 4

No.	Name	Synonym	Mol. Form.	Mol. Wt.	CAS RN	mp/°C	Merck No.	bp/°C	Beil. Ref.	den/g cm^{-3}	Solubility	n_D
11877	Tridecane, 2,2,4,10,12,12-hexamethyl-7-(3,5,5-trimethylhexyl)-		$C_{28}H_{58}$	394.77	3035-75-4			$166^{0.5}$	4-01-00-00590	0.8137^{20}		1.4558^{20}
11878	Tridecane, 7-hexyl-		$C_{19}H_{40}$	268.53	7225-66-3	-28.3		170.5^{10}	4-01-00-00562	0.7877^{20}		1.4409^{20}
11879	Tridecane, 7-methyl-	7-Methyltridecane	$C_{14}H_{30}$	198.39	26730-14-3	-37.2		115.5^{10}	4-01-00-00525	0.7634^{20}		1.4291^{20}
11880	Tridecanenitrile		$C_{13}H_{25}N$	195.35	629-60-7	9.7		293	3-02-00-00906	0.8257^{20}	EtOH 4; eth 4	1.4378^{20}
11881	Tridecanoic acid		$C_{13}H_{26}O_2$	214.35	638-53-9	41.5		236^{100}; 140^{1}	4-02-00-01117	0.8458^{80}	H_2O 1; EtOH 4; eth 4; ace 3	1.4286^{60}
11882	Tridecanoic acid, methyl ester	Methyl tridecanoate	$C_{14}H_{28}O_2$	228.38	1731-88-0	6.5		92^{1}	4-02-00-01118		EtOH 5; ctc 3	1.4405^{20}
11883	1-Tridecanol	Tridecyl alcohol	$C_{13}H_{28}O$	200.36	112-70-9	32.5		152^{14}	4-01-00-01860	0.8223^{31}	H_2O 1; EtOH 3; eth 3	
11884	2-Tridecanol	2-Hydroxytridecane	$C_{13}H_{28}O$	200.36	1653-31-2	23		161^{30}; 145^{12}	4-01-00-01861		bz 4; eth 4; EtOH 4	1.4188^{70}
11885	3-Tridecanol		$C_{13}H_{28}O$	200.36	10289-68-6	32		139^{12}	4-01-00-01861	0.8139^{46}		
11886	1-Tridecanol, 12-methyl-	12-Methyl-1-tridecanol	$C_{14}H_{30}O$	214.39	21987-21-3	10.5		$145\text{-}50^{6}$	4-01-00-01869	0.8414^{20}		1.4437^{20}
11887	3-Tridecanol, 2-methyl-		$C_{14}H_{30}O$	214.39	98930-89-3	20		274	1-01-00-00219	0.8390^{20}		1.4460^{20}
11888	4-Tridecanol, 2-methyl	Acetic acid, trifluoro,(1,2-dimethyl-1-ethylbutyl) ester	$C_{14}H_{30}O$	214.39	36691-77-7			$147\text{-}8^{14}$	4-01-00-01869		chl 2	1.4404^{25}
11889	2-Tridecanone	Methyl undecyl ketone	$C_{13}H_{26}O$	198.35	593-08-8	30.5		263	4-01-00-03386	0.8217^{30}	H_2O 1; EtOH 4; eth 4; ace 4	1.4318^{20}
11890	3-Tridecanone		$C_{13}H_{26}O$	198.35	1534-26-5	31		140^{17}	4-01-00-03387		H_2O 1; ace 3	
11891	7-Tridecanone	Dihexyl ketone	$C_{13}H_{26}O$	198.35	462-18-0	33		261	4-01-00-03315	0.825^{30}	EtOH 3; eth 4; chl 3; lig 3	
11892	3-Tridecanone, 2-methyl-	Decyl isopropyl ketone	$C_{14}H_{28}O$	212.38	40239-35-8	47		267; 142^{17}	4-01-00-03390	0.8314^{20}		
11893	1-Tridecene		$C_{13}H_{26}$	182.35	2437-56-1	-13		232.8	4-01-00-00921	0.7658^{20}	H_2O 1; EtOH 4; eth 4; bz 3	1.4340^{20}
11894	2-Tridecenoic acid		$C_{13}H_{24}O_2$	212.33	6969-16-0			$167\text{-}71^{2}$	4-02-00-01622	0.8995^{30}		1.4612^{20}
11895	12-Tridecenoic acid	(10-Undecenyl)acetic acid	$C_{13}H_{24}O_2$	212.33	6006-06-0	38.5		192^{20}	2-02-00-00422		bz 4; eth 4; EtOH 4; peth 4	
11896	12-Tridecenoic acid, methyl ester		$C_{14}H_{26}O_2$	226.36	29780-00-5			143^{8}	4-02-00-01622	0.8803^{20}		1.4438^{20}
11897	1-Tridecyne		$C_{13}H_{24}$	180.33	26186-02-7			94.5^{25}	4-01-00-01069	0.7729^{20}	bz 4; eth 4	1.4309^{20}
11898	Tridemorph	Morpholine, 2,6-dimethyl-4-tridecyl-	$C_{19}H_{39}NO$	297.52	24602-86-6		9576	$141^{1.3}$		0.86		
11899	Tridiphane	(±)-2-(3,5-Dichlorophenyl)-2-(2,2,2-trichloroethyl)oxirane	$C_{10}H_7Cl_5O$	320.43	58138-08-2	42.8						
11900	Triflumizole	(*E*)-1-[1-[[4-Chloro-2-(trifluoromethyl)phenyl]imino]-2-propoxyethyl]-1H-imidazole	$C_{15}H_{15}ClF_3N_3O$	345.75	68694-11-1	63.5						
11901	Trifluralin	Benzenamine, 2,6-dinitro-*N*,*N*-dipropyl-4-(trifluoromethyl)-	$C_{13}H_{16}F_3N_3O_4$	335.28	1582-09-8	49	9598	$140^{4.2}$				
11902	Triforine	Formamide, *N*,*N*'-[1,4-piperazinediylbis(2,2,2-trichloroethylidene)]bis-	$C_{10}H_{14}Cl_6N_4O_2$	434.96	26644-46-2	155 dec	9602					
11903	2,5,7-Trioxabicyclo[2.2.1]heptane, 1,3,3,4,6,6-hexamethyl-	Bicylo[2.2.1]heptane, 1,3,3,4,6,6-hexamethyl-2,5,7-trioxa	$C_{10}H_{18}O_3$	186.25	7045-89-8			166	5-19-09-00139	0.9702^{17}	H_2O 1	
11904	2H,4aH-1,4,5-Trioxadicyclopent[a,hi]indene-7-carboxylic acid, 3-ethylidene-3,3a,7a,9b-tetrahydro-2-oxo-, methyl ester, [3aS-(3E,3aα,4aβ,7aβ,9aR*,9bβ)]-	Plumericin	$C_{15}H_{14}O_6$	290.27	77-16-7		7512		5-19-11-00097		chl 3	
11905	1,3,5-Trioxane	Formaldehyde, trimer	$C_3H_6O_3$	90.08	110-88-3	60.2	9646	114.5	5-19-09-00103	1.17^{65}	H_2O 4; EtOH 3; eth 3; bz 3	
11906	1,3,5-Trioxane, 2,4,6-triethenyl-	Metacrolein	$C_9H_{12}O_3$	168.19	588-01-2	50		170	5-19-09-00165		eth 4; EtOH 4	
11907	1,3,5-Trioxane-2,4,6-triimine	Cyamelide	$C_3H_3N_3O_3$	129.08	462-02-2	dec	2689	dec	3-03-00-00030	1.127^{15}	eth 4; EtOH 4	
11908	1,3,5-Trioxane, 2,4,6-trimethyl-	Paraldehyde	$C_6H_{12}O_3$	132.16	123-63-7	12.6	6975	124.3	5-19-09-00112	0.9943^{20}	H_2O 2; EtOH 5; eth 5; chl 5	1.4049^{20}

No.	Name Synonym	Mol. Form. Mol. Wt.	CAS RN mp/°C	Merck No. bp/°C	Beil. Ref. den/g cm^{-3}	Solubility n_D
11909	2,6,7-Trioxa-1-phosphabicyclo[2.2.2]octane, 4-ethyl- Trimethylolpropane phosphite	$C_6H_{11}O_3P$ 162.13	824-11-3 53.7			chl 3
11910	Triphenylene 9,10-Benzophenanthrene	$C_{18}H_{12}$ 228.29	217-59-4 199	9654 425	4-05-00-02556	H_2O 1; EtOH 3; bz 4; chl 4
11911	Triphenyltin hydroxide Stannane, hydroxytriphenyl-	$C_{18}H_{16}OSn$ 367.03	76-87-9 119	9659	 1.54^{20}	
11912	Tripropylamine, eicosafluoro-2-(trifluoromethyl)- Dipropylisobutylamine, perfluoro	$C_{10}F_{23}N$ 571.08	307-11-9	 147	4-02-00-00858 1.84^{25}	eth 4; EtOH 4 1.283^{25}
11913	Trisiloxane, 1,5-dichloro-1,1,3,3,5,5-hexamethyl-	$C_6H_{18}Cl_2O_2Si_3$ 277.37	3582-71-6 -53	 184	4-04-00-04123 1.018^{20}	H_2O 6
11914	Trisiloxane, octamethyl- Octamethyltrisiloxane	$C_8H_{24}O_2Si_3$ 236.53	107-51-7 -80	6669 153; $50\text{-}2^{17}$	4-04-00-04115 0.8200^{20}	EtOH 2; bz 3; peth 3 1.3840^{20}
11915	Trisiloxane, 1,1,3,5,5-pentamethyl-1,3,5-triphenyl- 1,1,3,5,5-Pentamethyl-1,3,5-triphenyltrisiloxane	$C_{23}H_{30}O_2Si_3$ 422.75	80-14-8	 $169^{0.7}$	3-16-00-01210 1.0227^{20}	bz 4; lig 4 1.5280^{20}
11916	Trisulfide, diethyl	$C_4H_{10}S_3$ 154.32	3600-24-6	 $96\text{-}7^{26}$	4-01-00-01398 1.114^{20}	 1.5689^{13}
11917	Trisulfide, di-2-propenyl	$C_6H_{10}S_3$ 178.34	2050-87-5	 $112\text{-}22^{16}$	0-01-00-00441 1.0845^{15}	eth 4
11918	Tritetracontane	$C_{43}H_{88}$ 605.17	7098-21-7 85.5	 332^3	3-01-00-00593 0.7812^{90}	 1.4340^{90}
11919	1,3,5-Trithiane	$C_3H_6S_3$ 138.28	291-21-4 220	 sub	5-19-09-00105 1.6374^{24}	H_2O 2; EtOH 2; eth 2; bz 3
11920	1,3,5-Trithiane, 2,4,6-trimethyl- Thioacetaldehyde trimer	$C_6H_{12}S_3$ 180.36	2765-04-0	9245	4-19-00-04718	
11921	1,3,5-Trithiane, 2,4,6-trimethyl-, (2α,4α,6β)-	$C_6H_{12}S_3$ 180.36	23769-39-3 101	 246.5	2-19-00-00399	H_2O 1; EtOH 3; eth 3; ace 3
11922	*L*-Tryptophan *l*-α-Aminoindole-3-propionic acid	$C_{11}H_{12}N_2O_2$ 204.23	73-22-3 289	9707	5-22-14-00014	H_2O 2; EtOH 3; eth 1; chl 1
11923	*L*-Tryptophan, *N*-methyl- *L*-Abrine	$C_{12}H_{14}N_2O_2$ 218.26	526-31-8 295 dec	5	5-22-14-00040	H_2O 2; MeOH 2; eth 1; alk 3
11924	Tubocuraran-7',12'-diol, 6,6'-dimethoxy-2,2'-dimethyl-, (1'α)- Bebeerine	$C_{36}H_{38}N_2O_6$ 594.71	477-60-1 221	1024	2-27-00-00896	EtOH 3; eth 3; ace 4; chl 4
11925	Tubocuraran-7',12'-diol, 6,6'-dimethoxy-2,2'-dimethyl-, (1β)- Curine	$C_{36}H_{38}N_2O_6$ 594.71	436-05-5 221	2682	4-27-00-08725	ace 4; bz 4; py 4
11926	*L*-Tyrosine *L*-Phenylalanine, 4-hydroxy-	$C_9H_{11}NO_3$ 181.19	60-18-4 343 dec	9747 sub	4-14-00-02264	H_2O 2; EtOH 1; eth 1; HOAc 2
11927	*L*-Tyrosine, 3,5-dibromo- Tyrosine, 3,5-dibromo(*L*)	$C_9H_9Br_2NO_3$ 338.98	300-38-9 245	3014	4-14-00-02365	H_2O 2; EtOH 2; eth 1; alk 3
11928	Tyrosine, 3,5-diiodo- 3,5-Diiodotyrosine	$C_9H_9I_2NO_3$ 432.98	66-02-4 213 dec	3177	2-14-00-00384	H_2O 2; EtOH 1; eth 1; bz 1
11929	*L*-Tyrosine, 3-hydroxy- Levodopa	$C_9H_{11}NO_4$ 197.19	59-92-7 277 dec	5344	4-14-00-02492	H_2O 3; EtOH 1; eth 1; ace 1
11930	*L*-Tyrosine, *O*-(4-hydroxy-3,5-diiodophenyl)-3,5-diiodo- Thyroxine	$C_{15}H_{11}I_4NO_4$ 776.87	51-48-9 235.5	9348	4-14-00-02373	H_2O 2; EtOH 1; bz 1
11931	*L*-Tyrosine, 3-hydroxy-α-methyl- Methyldopa	$C_{10}H_{13}NO_4$ 211.22	555-30-6 300 dec	5974		
11932	*L*-Tyrosine, 3-iodo- 3-Iodotyrosine	$C_9H_{10}INO_3$ 307.09	70-78-0 204-6 dec	4937	4-14-00-02369	
11933	*L*-Tyrosine, *N*-methyl- Surinamine	$C_{10}H_{13}NO_3$ 195.22	537-49-5 293	8988	3-14-00-01513	
11934	1,10-Undecadiyne	$C_{11}H_{16}$ 148.25	4117-15-1 -17	 83^{12}	2-01-00-00248 0.8182^{21}	ace 4; bz 4 1.453^{21}
11935	Undecanal	$C_{11}H_{22}O$ 170.30	112-44-7 -4	 117^{18}	4-01-00-03374 0.8251^{23}	H_2O 1; EtOH 3; eth 3 1.4520^{20}
11936	Undecanal, 2-methyl-	$C_{12}H_{24}O$ 184.32	110-41-8	 114^{10}	4-01-00-03383 0.830^{15}	H_2O 2; EtOH 3; eth 3 1.4321^{20}
11937	1-Undecanamine	$C_{11}H_{25}N$ 171.33	7307-55-3 17	 242	4-04-00-00792 0.7979^{20}	H_2O 3; EtOH 3; eth 1; ctc 2 1.4398^{20}
11938	Undecane Hendecane	$C_{11}H_{24}$ 156.31	1120-21-4 -25.6	 195.9	4-01-00-00487 0.7402^{20}	H_2O 1; EtOH 5; eth 5 1.4398^{20}
11939	Undecane, 1-bromo-11-fluoro- 1-Bromo-11-fluoroundecane	$C_{11}H_{22}BrF$ 253.20	463-33-2	 $95^{0.6}$	4-01-00-00490	eth 4; EtOH 4 1.4518^{25}
11940	Undecane, 1,11-dibromo- 1,11-Dibromoundecane	$C_{11}H_{22}Br_2$ 314.10	16696-65-4 166	 $170\text{-}5^{15}$	3-01-00-00536 1.332^{15}	
11941	Undecanenitrile Decyl cyanide	$C_{11}H_{21}N$ 167.29	2244-07-7	 253	3-02-00-00860 0.8254^{30}	H_2O 1; EtOH 3; eth 3; ctc 3 1.4293^{30}
11942	1-Undecanethiol Undecyl mercaptan	$C_{11}H_{24}S$ 188.38	5332-52-5 -3	 257.4	4-01-00-01838 0.8448^{20}	 1.4585^{20}
11943	Undecanoic acid	$C_{11}H_{22}O_2$ 186.29	112-37-8 28.6	 280	4-02-00-01068 0.8907^{20}	H_2O 1; EtOH 4; eth 3; ace 4 1.4294^{45}

No.	Name Synonym	Mol. Form. Mol. Wt.	CAS RN mp/°C	Merck No. bp/°C	Beil. Ref. den/g cm^{-3}	Solubility n_D
11944	Undecanoic acid, 11-bromo- 11-Bromoundecanoic acid	$C_{11}H_{21}BrO_2$ 265.19	2834-05-1 57	 188^{18}	4-02-00-01073	ace 4; bz 4; eth 4; EtOH 4
11945	Undecanoic acid, 2-bromo- 2-Bromoundecanoic acid	$C_{11}H_{21}BrO_2$ 265.19	2623-84-9 10	 180^{14}	4-02-00-01073 1.1586^{24}	
11946	Undecanoic acid, ethyl ester Ethyl undecylate	$C_{13}H_{26}O_2$ 214.35	627-90-7 -15	 131^{14}	3-02-00-00858 0.8633^{20}	H_2O 1; EtOH 3; eth 3; ace 3 1.4285^{20}
11947	Undecanoic acid, 11-fluoro-	$C_{11}H_{21}FO_2$ 204.28	463-17-2 36	 $113^{0.25}$	4-02-00-01071	eth 4; EtOH 4; lig 4
11948	Undecanoic acid, 11-hydroxy- 11-Hydroxyundecanoic acid	$C_{11}H_{22}O_3$ 202.29	3669-80-5 65.5	 $148\text{-}50^{4}$	4-03-00-00903	eth 4; EtOH 4
11949	Undecanoic acid, 11-hydroxy-, methyl ester Methyl 11-hydroxyhendecanoate	$C_{12}H_{24}O_3$ 216.32	24724-07-0 27.5	 168^{8}	2-03-00-00243 0.9542^{20}	bz 4; eth 4; EtOH 4
11950	1-Undecanol	$C_{11}H_{24}O$ 172.31	112-42-5 19	 243	4-01-00-01835 0.8298^{20}	H_2O 1; EtOH 3; eth 4 1.4392^{20}
11951	2-Undecanol, (±)-	$C_{11}H_{24}O$ 172.31	113666-64-1 12	 228	4-01-00-01838 0.8268^{19}	H_2O 1; EtOH 3; eth 3 1.4369^{20}
11952	3-Undecanol, (*R*)-	$C_{11}H_{24}O$ 172.31	107494-37-1 17	 229	2-01-00-00462 0.8295^{20}	bz 4; eth 4; EtOH 4 1.4367^{20}
11953	5-Undecanol 1-Butylheptyl alcohol	$C_{11}H_{24}O$ 172.31	37493-70-2 -3.5	 229	4-01-00-01839 0.8292^{20}	eth 4; EtOH 4 1.4354^{24}
11954	6-Undecanol	$C_{11}H_{24}O$ 172.31	23708-56-7 25	 228	4-01-00-01839 0.8334^{20}	H_2O 1; EtOH 3; ace 3 1.4374^{20}
11955	1-Undecanol, 2-methyl- 2-Methyl-1-undecanol	$C_{12}H_{26}O$ 186.34	10522-26-6	 $129\text{-}31^{12}$	4-01-00-01855 0.8300^{15}	eth 4; EtOH 4 1.4382^{20}
11956	3-Undecanol, 2-methyl- 3-Hydroxy-2-methylundecane	$C_{12}H_{26}O$ 186.34	60671-36-5	 135^{28}	1-01-00-00217 0.8327^{20}	 1.4405^{20}
11957	5-Undecanol, 2-methyl-	$C_{12}H_{26}O$ 186.34	33978-71-1	 $130\text{-}3^{20}$	4-01-00-01855 0.8235^{20}	eth 4; EtOH 4 1.4346^{20}
11958	2-Undecanone Methyl nonyl ketone	$C_{11}H_{22}O$ 170.30	112-12-9 15	 231.5	4-01-00-03374 0.8250^{20}	H_2O 1; EtOH 3; eth 3; ace 3 1.4291^{20}
11959	3-Undecanone Ethyl octyl ketone	$C_{11}H_{22}O$ 170.30	2216-87-7 12	 227	4-01-00-03375 0.8272^{20}	H_2O 1; EtOH 3; eth 3; ctc 2 1.4296^{20}
11960	4-Undecanone Heptyl propyl ketone	$C_{11}H_{22}O$ 170.30	14476-37-0 4.5	 106^{13}	4-01-00-03375 0.8274^{25}	eth 4; EtOH 4 1.4248^{24}
11961	5-Undecanone Butyl hexyl ketone	$C_{11}H_{22}O$ 170.30	33083-83-9 2	 227	4-01-00-03375 0.8278^{19}	H_2O 1; EtOH 3; eth 3; ctc 3 1.4275^{18}
11962	6-Undecanone Diamyl ketone	$C_{11}H_{22}O$ 170.30	927-49-1 14.5	 228	4-01-00-03376 0.8308^{20}	H_2O 1; EtOH 4; eth 4 1.4270^{20}
11963	4-Undecanone, 7-ethyl-2-methyl-	$C_{14}H_{28}O$ 212.38	6976-00-7	 252.5	3-01-00-02920 0.8362^{20}	H_2O 1; EtOH 4; eth 4; ace 4 1.4370^{20}
11964	Undecasiloxane, tetracosamethyl- Tetracosamethylhendecasiloxane	$C_{24}H_{72}O_{10}Si_{11}$ 829.77	107-53-9	9128 322.8; 202^{47}	3-04-00-01881 0.9247^{25}	bz 4 1.3994^{20}
11965	3,5,9-Undecatrien-2-one, 6,10-dimethyl- Pseudoionone	$C_{13}H_{20}O$ 192.30	141-10-6	7931 $143\text{-}5^{12}$	3-01-00-03067 0.8984^{20}	EtOH 3; eth 3; chl 3; MeOH 3 1.5335^{20}
11966	1-Undecene	$C_{11}H_{22}$ 154.30	821-95-4 -49.2	 192.7	4-01-00-00910 0.7503^{20}	H_2O 1; eth 3; chl 3; lig 3 1.4261^{20}
11967	3-Undecene	$C_{11}H_{22}$ 154.30	60669-40-1 -62	 193.5	 0.7516^{20}	 1.4290^{20}
11968	2-Undecene, (*E*)-	$C_{11}H_{22}$ 154.30	693-61-8 -48.3	 192.5	4-01-00-00911 0.7528^{20}	 1.4292^{20}
11969	2-Undecene, (*Z*)-	$C_{11}H_{22}$ 154.30	821-96-5 -66.5	 196.1	 0.7576^{20}	
11970	4-Undecene, (*E*)-	$C_{11}H_{22}$ 154.30	693-62-9 -63.7	 193	 0.7508^{20}	 1.4285^{20}
11971	4-Undecene, (*Z*)-	$C_{11}H_{22}$ 154.30	821-98-7 -97	 192.6	 0.7541^{20}	 1.4302^{20}
11972	5-Undecene, (*E*)-	$C_{11}H_{22}$ 154.30	764-97-6 -61.1	 192	4-01-00-00911 0.7497^{20}	eth 4; chl 4; lig 4 1.4285^{20}
11973	5-Undecene, (*Z*)-	$C_{11}H_{22}$ 154.30	764-96-5 -106.5	 192.3	4-01-00-00911 0.7537^{20}	 1.4302^{20}
11974	10-Undecenoic acid Undecylenic acid	$C_{11}H_{20}O_2$ 184.28	112-38-9 24.5	9760 275	4-02-00-01612 0.9072^{24}	H_2O 1; EtOH 3; eth 3; ctc 2 1.4486^{24}
11975	9-Undecenoic acid, (*E*)- *trans*-9-Undecenoic acid	$C_{11}H_{20}O_2$ 184.28	37973-84-5 19	 273	4-02-00-01616 0.9118^{25}	eth 4 1.4519^{20}
11976	9-Undecenoic acid, (*Z*)-	$C_{11}H_{20}O_2$ 184.28	116836-30-7 2.5	 156^{4}; $119^{0.8}$	4-02-00-01616 0.9150^{20}	 1.4530^{20}
11977	10-Undecenoic acid, ethyl ester	$C_{13}H_{24}O_2$ 212.33	692-86-4 -38	 264.5	4-02-00-01614 0.8827^{15}	H_2O 1; EtOH 3; eth 3; ctc 2 1.4449^{25}
11978	10-Undecenoic acid, methyl ester	$C_{12}H_{22}O_2$ 198.31	111-81-9 -27.5	 248	4-02-00-01613 0.889^{15}	H_2O 1; EtOH 3; eth 3; ctc 2 1.4393^{20}
11979	10-Undecen-1-ol	$C_{11}H_{22}O$ 170.30	112-43-6 -2	 250	4-01-00-02194 0.8495^{15}	H_2O 1; EtOH 3; eth 3; ctc 2 1.4500^{20}
11980	2-Undecen-4-ol	$C_{11}H_{22}O$ 170.30	22381-86-8	 $114\text{-}5^{13}$	4-01-00-02194 0.843^{22}	 1.4485^{22}
11981	9-Undecen-1-ol	$C_{11}H_{22}O$ 170.30	112-46-9 6	 248.5	3-01-00-01957 0.8507^{15}	 1.4535^{19}
11982	10-Undecenoyl chloride	$C_{11}H_{19}ClO$ 202.72	38460-95-6	 $120\text{-}2^{10}$	4-02-00-01615 0.944^{20}	 1.454^{20}
11983	1-Undecen-3-yne	$C_{11}H_{18}$ 150.26	74744-28-8	 74^{9}	3-01-00-01054 0.7962^{20}	 1.4606^{20}

No.	Name Synonym	Mol. Form. Mol. Wt.	CAS RN mp/°C	Merck No. bp/°C	Beil. Ref. den/g cm^{-3}	Solubility n_D
11984	1-Undecyne	$C_{11}H_{20}$ 152.28	2243-98-3 -25	 196	4-01-00-01064 0.7728^{20}	ace 4; bz 4; eth 4; EtOH 4 1.4306^{20}
11985	2-Undecyne	$C_{11}H_{20}$ 152.28	60212-29-5 -30.1	 204.2	0-01-00-00261 0.7827^{20}	 1.4391^{20}
11986	4-Undecyne	$C_{11}H_{20}$ 152.28	60212-31-9 -74.7	 198.5	3-01-00-01022 0.7752^{20}	 1.4369
11987	5-Undecyne	$C_{11}H_{20}$ 152.28	2294-72-6 -74.1	 198	4-01-00-01064 0.7753^{20}	 1.4369^{20}
11988	6-Undecynoic acid	$C_{11}H_{18}O_2$ 182.26	55182-83-7	 $124\text{-}5^{0.17}$	4-02-00-01739 0.9537^{25}	 1.4566^{25}
11989	9-Undecynoic acid	$C_{11}H_{18}O_2$ 182.26	22202-65-9 61	 170^{15}, $132^{0.1}$	3-02-00-01471	H_2O 4; eth 4; EtOH 4
11990	9-Undecynoic acid, methyl ester Methyl 9-undecynoate	$C_{12}H_{20}O_2$ 196.29	18937-76-3	 $93\text{-}4^{1}$	3-02-00-01472 0.9245^{17}	 1.4732^{19}
11991	10-Undecynoic acid, methyl ester Methyl 10-undecynoate	$C_{12}H_{20}O_2$ 196.29	2777-66-4 42	 121^{15}	4-02-00-01738 0.9177^{20}	 1.4421^{21}
11992	Urea Carbamide	CH_4N_2O 60.06	57-13-6 132.7	9781 dec	4-03-00-00094 1.3230^{20}	H_2O 4; EtOH 4; eth 1; bz 1 1.484
11993	Urea, (aminoiminomethyl)-	$C_2H_6N_4O$ 102.10	141-83-3 105	 160 dec	4-03-00-00155	H_2O 3; EtOH 2; eth 1; bz 1
11994	Urea, *N,N*'-bis(2-chloroethyl)-*N*-nitroso- Carmustine	$C_5H_9Cl_2N_3O_2$ 214.05	154-93-8 31	1852		H_2O 4; EtOH 4
11995	Urea, *N,N*'-bis(hydroxymethyl)- Oxymethurea	$C_3H_8N_2O_3$ 120.11	140-95-4 126	6921 149^{25}	4-03-00-00107	H_2O 3; EtOH 3; eth 1; MeOH 3
11996	Urea, *N,N*-bis(2-methylpropyl)- 1,1-Diisobutylurea	$C_9H_{20}N_2O$ 172.27	77464-06-3 73	 180^{25}	0-04-00-00170	eth 4; EtOH 4
11997	Urea, *N,N*'-bis(4-nitrophenyl)- 4,4'-Dinitrocarbanilide	$C_{13}H_{10}N_4O_5$ 302.25	587-90-6 312 dec	3271	4-12-00-01646	
11998	Urea, (4-bromophenyl)-	$C_7H_7BrN_2O$ 215.05	1967-25-5 226	 260	4-12-00-01511	H_2O 2; EtOH 4; eth 4; bz 4
11999	Urea, *N*'-(4-chlorophenyl)-*N,N*-dimethyl- Monuron	$C_9H_{11}ClN_2O$ 198.65	150-68-5 170.5	6169	4-12-00-01191	H_2O 1; EtOH 2; ace 2
12000	Urea, *N*'-(3,4-dichlorophenyl)-*N,N*-dimethyl- Diuron	$C_9H_{10}Cl_2N_2O$ 233.10	330-54-1 158	3388	4-12-00-01263	
12001	Urea, *N,N*'-diethyl-	$C_5H_{12}N_2O$ 116.16	623-76-7 112.5	 263	4-04-00-00370 1.0415^{25}	H_2O 4; EtOH 4; eth 4 1.4616^{40}
12002	Urea, *N,N*-diethyl-	$C_5H_{12}N_2O$ 116.16	634-95-7 75	 $94\text{-}6^{0.02}$	4-04-00-00380	H_2O 4; EtOH 4; eth 3; bz 4
12003	Urea, *N,N*'-diethyl-*N,N*'-diphenyl- *N,N*'-Diethylcarbanilide	$C_{17}H_{20}N_2O$ 268.36	85-98-3 79	3107	4-12-00-00844	H_2O 1; EtOH 4; chl 3
12004	Urea, *N,N*'-dimethyl-	$C_3H_8N_2O$ 88.11	96-31-1 108	 269	4-04-00-00207 1.142^{25}	H_2O 4; EtOH 4; eth 1; chl 2
12005	Urea, *N,N*-dimethyl-	$C_3H_8N_2O$ 88.11	598-94-7 183.5		4-04-00-00224 1.2555^{25}	H_2O 3; EtOH 2; eth 1; tfa 2
12006	Urea, *N,N*'-dimethyl-*N,N*'-diphenyl-	$C_{15}H_{16}N_2O$ 240.30	611-92-7 122	 350	4-12-00-00839	H_2O 4; EtOH 4; eth 2; ace 4
12007	Urea, *N,N*-dimethyl-*N*'-[3-(trifluoromethyl)phenyl]- Fluometuron	$C_{10}H_{11}F_3N_2O$ 232.21	2164-17-2 164	4080		ace 4; EtOH 4
12008	Urea, *N,N*'-di-1-naphthalenyl-	$C_{21}H_{16}N_2O$ 312.37	607-56-7 296	 sub	4-12-00-03078	py 4
12009	Urea, (2,5-dioxo-4-imidazolidinyl)- Allantoin	$C_4H_6N_4O_3$ 158.12	97-59-6 239	246	4-25-00-04071	H_2O 2; EtOH 3; eth 1; MeOH 1
12010	Urea, *N,N*'-diphenyl- Carbanilide	$C_{13}H_{12}N_2O$ 212.25	102-07-8 236	1787 262	4-12-00-00741 1.239^{25}	H_2O 2; EtOH 2; eth 3; ace 2
12011	Urea, *N,N*-diphenyl- 1,1-Diphenylurea	$C_{13}H_{12}N_2O$ 212.25	603-54-3 189	 dec	4-12-00-00852 1.276^{25}	H_2O 2; EtOH 3; eth 3; chl 3
12012	Urea, *N,N*'-dipropyl-	$C_7H_{16}N_2O$ 144.22	623-95-0 105	 255	4-04-00-00482	eth 4; EtOH 4
12013	Urea, (4-ethoxyphenyl)- Dulcin	$C_9H_{12}N_2O_2$ 180.21	150-69-6 173.5	3448 dec	4-13-00-01154	H_2O 2; EtOH 3; AcOEt 4
12014	Urea, ethyl-	$C_3H_8N_2O$ 88.11	625-52-5 92.5	 dec	4-04-00-00369 1.2130^{18}	H_2O 4; EtOH 4; eth 3; bz 4
12015	Urea, *N*-ethyl-*N*-nitroso- *N*-Ethyl-*N*-nitrosourea	$C_3H_7N_3O_2$ 117.11	759-73-9 98-100 dec	3788	3-04-00-00233	chl 3
12016	Urea, hydroxy- Hydroxyurea	$CH_4N_2O_2$ 76.06	127-07-1 141	4785 dec	4-03-00-00170	H_2O 4
12017	Urea, methyl-	$C_2H_6N_2O$ 74.08	598-50-5 103	 dec	4-04-00-00205 1.2040^{0}	H_2O 4; EtOH 4; eth 1; bz 1
12018	Urea, *N*-methyl-*N,N*'-diphenyl-	$C_{14}H_{14}N_2O$ 226.28	612-01-1 106	 203	4-12-00-00839	H_2O 2; EtOH 2; eth 4; bz 3
12019	Urea, *N*-methyl-*N*-nitroso- Methylnitrosourea	$C_2H_5N_3O_2$ 103.08	684-93-5 123-4 dec		4-04-00-03385	H_2O 2; EtOH 2; eth 2
12020	Urea, monohydrochloride Urea hydrochloride	CH_5ClN_2O 96.52	506-89-8 145 dec	9782	4-03-00-00094	H_2O 3
12021	Urea, mononitrate Urea nitrate	$CH_5N_3O_4$ 123.07	124-47-0 152 dec	9784	4-03-00-00094 1.690^{20}	EtOH 4

No.	Name Synonym	Mol. Form. Mol. Wt.	CAS RN mp/°C	Merck No. bp/°C	Beil. Ref. den/g cm^{-3}	Solubility n_D
12022	Urea, 1-naphthalenyl- 1-Naphthylurea	$C_{11}H_{10}N_2O$ 186.21	6950-84-1 219.5		4-12-00-03076	H_2O 1; EtOH 4; eth 3
12023	Urea, 2-naphthalenyl- 2-Naphthylurea	$C_{11}H_{10}N_2O$ 186.21	13114-62-0 219		4-12-00-03149	EtOH 4
12024	Urea, nitro- Nitrourea	$CH_3N_3O_3$ 105.05	556-89-8 158-9 dec	6573	4-03-00-00248	ace 4; EtOH 4
12025	Urea, (4-nitrophenyl)- 4-Nitrophenylurea	$C_7H_7N_3O_3$ 181.15	556-10-5 238	6547	4-12-00-01645	H_2O 4; EtOH 4
12026	Urea, phenyl- Phenylurea	$C_7H_8N_2O$ 136.15	64-10-8 147	7290 238	4-12-00-00734 1.302^{25}	H_2O 2; EtOH 3; eth 2; AcOEt 3
12027	Urea, (phenylmethyl)- Benzylurea	$C_8H_{10}N_2O$ 150.18	538-32-9 148	1165 200 dec	4-12-00-02253	ace 4; EtOH 4
12028	Urea, 2-propenyl- Allylurea	$C_4H_8N_2O$ 100.12	557-11-9 85	296	4-04-00-01070	H_2O 5; EtOH 5; eth 2; chl 2
12029	Urea, tetraethyl- Urea, 1,1,3,3-tetraethyl-	$C_9H_{20}N_2O$ 172.27	1187-03-7	 209	4-04-00-00380 0.919^{20}	H_2O 1; alk 1; acid 1 1.4474^{20}
12030	Urea, tetramethyl- Tetramethylurea	$C_5H_{12}N_2O$ 116.16	632-22-4 -1.2	9160 176.5	4-04-00-00225 0.9687^{20}	EtOH 2; eth 2; ctc 2 1.4496^{23}
12031	Urea, tetraphenyl- Tetraphenylurea	$C_{25}H_{20}N_2O$ 364.45	632-89-3 183		4-12-00-00853 1.222^{25}	H_2O 1; EtOH 2
12032	Urea, trimethyl- Trimethylurea	$C_4H_{10}N_2O$ 102.14	632-14-4 75.5	 232.5	4-04-00-00224 1.1900^{20}	H_2O 3; EtOH 3; eth 2; bz 2
12033	Uridine 1-β-*D*-Ribofuranosyluracil	$C_9H_{12}N_2O_6$ 244.20	58-96-8 165	9792	5-24-06-00132	H_2O 3; EtOH 3; py 3
12034	Uridine, 2'-deoxy-5-fluoro- Floxuridine	$C_9H_{11}FN_2O_5$ 246.20	50-91-9	4045	5-24-06-00291	
12035	5'-Uridylic acid Uridine 5'-phosphoric acid	$C_9H_{13}N_2O_9P$ 324.18	58-97-9 202 dec	9796	5-24-06-00173	H_2O 4; MeOH 3
12036	Urs-12-ene-27,28-dioic acid, 3-hydroxy-, (3β)- Quinovic acid	$C_{30}H_{46}O_5$ 486.69	465-74-7 298 dec	8104	4-10-00-02197	
12037	Urs-12-en-28-oic acid, 3-hydroxy-, (3β)- Ursolic acid	$C_{30}H_{48}O_3$ 456.71	77-52-1 284	9802	4-10-00-01160	ace 4; eth 4; chl 4
12038	Urs-12-en-3-ol, (3β)- α-Amyrin	$C_{30}H_{50}O$ 426.73	638-95-9 186	653 $243^{0.5}$	4-06-00-04191	EtOH 3; eth 3; bz 3; chl 3
12039	Valeric acid, 3-chloro-, ethyl ester	$C_7H_{13}ClO_2$ 164.63	6513-13-9	 189	3-02-00-00678 1.0330^{20}	eth 4; EtOH 4 1.4278^{20}
12040	*DL*-Valine	$C_5H_{11}NO_2$ 117.15	516-06-3 298 dec	9818 sub	3-04-00-01370 1.316^{25}	H_2O 3; EtOH 2; eth 2; bz 2
12041	*L*-Valine 2-Aminoisovaleric acid	$C_5H_{11}NO_2$ 117.15	72-18-4 315	9818 sub	4-04-00-02659 1.23^{25}	H_2O 4
12042	*D*-Valine, 3-[(2-amino-2-carboxyethyl)dithio]-, (*R*)- Penicillamine cysteine disulfide	$C_8H_{16}N_2O_4S_2$ 268.36	18840-45-4 195	7030	4-04-00-03231	
12043	*D*-Valine, 3-mercapto- Penicillamine	$C_5H_{11}NO_2S$ 149.21	52-67-5 198.5	7029	4-04-00-03228	
12044	Veratraman-3,23-diol, 14,15,16,17-tetradehydro-, (3β,23β)- Veratramine	$C_{27}H_{39}NO_2$ 409.61	60-70-8 206	9853	5-21-05-00297	EtOH 3; bz 3; chl 3; dil acid 3
12045	Veratraman-11-one, 17,23-epoxy-3-(β-*D*-glucopyranosyloxy)-, (3β,23β)- Pseudojervine	$C_{33}H_{49}NO_8$ 587.75	36069-05-3 304-5 dec	7932	4-27-00-03592	H_2O 1; EtOH 3; eth 1; bz 1
12046	Veratraman-11-one, 17,23-epoxy-3-hydroxy-, (3β,23β)- Jervine	$C_{27}H_{39}NO_3$ 425.61	469-59-0 243 dec	5145	4-27-00-03590	H_2O 1; EtOH 3; eth 2; ace 3
12047	Vernolate Carbamothioic acid, dipropyl-, *S*-propyl ester	$C_{10}H_{21}NOS$ 203.35	1929-77-7	9866 150^{30}	 0.952^{20}	
12048	Vinclozolin 2,4-Oxazolidinedione, 3-(3,5-dichlorophenyl)-5-ethenyl-5-methyl-	$C_{12}H_9Cl_2NO_3$ 286.11	50471-44-8 108	9890 $131^{0.05}$	 1.51	
12049	9H-Xanthene 10H-9-Oxaanthracene	$C_{13}H_{10}O$ 182.22	92-83-1 100.5	 311	5-17-02-00252	H_2O 1; EtOH 2; eth 3; bz 3
12050	9H-Xanthen-9-one Xanthone	$C_{13}H_8O_2$ 196.21	90-47-1 174	9971 351; 146^3	5-17-10-00430	H_2O 1; EtOH 3; eth 3; bz 3
12051	9H-Xanthen-9-one, 1,3-dihydroxy-	$C_{13}H_8O_4$ 228.20	3875-68-1 259	 sub	5-18-04-00004	EtOH 4
12052	9H-Xanthen-9-one, 2,7-dihydroxy- Xanthone, 2,7-dihydroxy	$C_{13}H_8O_4$ 228.20	64632-72-0 330	 sub	1-18-00-00357	eth 4; EtOH 4
12053	9H-Xanthen-9-one, 3,6-dihydroxy-	$C_{13}H_8O_4$ 228.20	1214-24-0 350 dec	 sub	5-18-04-00010	EtOH 4; HOAc 4
12054	9H-Xanthen-9-one, 1,7-dihydroxy-3-methoxy- Gentisin	$C_{14}H_{10}O_5$ 258.23	437-50-3 266.5	4291	5-18-04-00497	H_2O 1; EtOH 4; ace 1; py 3
12055	*peri*-Xanthenoxanthene Dinaphthalene dioxide	$C_{20}H_{10}O_2$ 282.30	191-28-6 243.5	 400^{20-25}	5-19-02-00242	bz 3; chl 3
12056	Xanthylium, 9-(2-carboxyphenyl)-3,6-bis(diethylamino)-, chloride Rhodamine, B	$C_{28}H_{32}ClN_2O_3$ 480.03	81-88-9 165	8182	5-19-08-00669	H_2O 3; EtOH 3; eth 3; bz 3
12057	*o*-Xylene, 4-propyl-	$C_{11}H_{16}$ 148.25	3982-66-9 -20	 208.9	3-05-00-01005 0.8715^{20}	ace 4; bz 4; eth 4; EtOH 4 1.5000^{20}

No.	Name Synonym	Mol. Form. Mol. Wt.	CAS RN mp/°C	Merck No. bp/°C	Beil. Ref. den/g cm^{-3}	Solubility n_D
12058	Xylitol Xylite	$C_5H_{12}O_5$ 152.15	87-99-0 93.5	9992 216	4-01-00-02832	H_2O 4; py 4; EtOH 4
12059	*D*-Xylose	$C_5H_{10}O_5$ 150.13	58-86-6 90.5	9995	4-01-00-04223 1.525^{20}	H_2O 4; EtOH 3; eth 2
12060	Yohimban-16-carboxylic acid, 11,17-dimethoxy-18-[[1-oxo-3-(3,4,5-trimethoxyphenyl)-2-propenyl]oxy]-, methyl ester, (3β,16β,17α,18β,20α)- Rescinnamine	$C_{35}H_{42}N_2O_9$ 634.73	24815-24-5 238.5	8146	4-25-00-01323	H_2O 1; EtOH 2; ace 3; chl 3
12061	Yohimban-16-carboxylic acid, 11,17-dimethoxy-18-[(3,4,5-trimethoxybenzoyl)oxy]-, methyl ester, Reserpine	$C_{33}H_{40}N_2O_9$ 608.69	50-55-5 264.5	8149	4-25-00-01319	H_2O 2; EtOH 3; eth 2; ace 2
12062	Yohimban-16-carboxylic acid, 17-hydroxy-, methyl ester, (16α,17α)- Yohimbine	$C_{21}H_{26}N_2O_3$ 354.45	146-48-5 241	10011 sub 160	4-25-00-01237	H_2O 2; EtOH 3; eth 3; bz 2
12063	Yohimban-16-carboxylic acid, 18-hydroxy-11,17-dimethoxy-, (3β,16β,17α,18β,20 Reserpic acid	$C_{22}H_{28}N_2O_5$ 400.47	83-60-3 242	8147	4-25-00-01305	
12064	Yohimban-16-carboxylic acid, 17-hydroxy-18-[(3,4,5-trimethoxybenzoyl)oxy]-, methyl ester, (3β,16β Raunescine	$C_{31}H_{36}N_2O_8$ 564.64	117-73-7 165	8130	4-25-00-01281	H_2O 1; EtOH 3; chl 3; HOAc 3
12065	Yohimban-16-carboxylic acid, 17-methoxy-18-[(3,4,5-trimethoxybenzoyl)oxy]-, methyl ester, (3β,16β Deserpidine	$C_{32}H_{38}N_2O_8$ 578.66	131-01-1 230.5	2901	4-25-00-01282	H_2O 1; EtOH 3; chl 3
12066	Zinc, dimethyl Dimethyl zinc	C_2H_6Zn 95.46	544-97-8 -40	3252 46	1.386^{10}	eth 3; peth 5
12067	Zinc, [[1,2-ethanediylbis[carbamodithioato]](2-)]- Zineb	$C_4H_6N_2S_4Zn$ 275.76	12122-67-7 157 dec	10071		
12068	Ziram Zinc, bis(dimethylcarbamodithioato-*S,S'*)-, (T-4)-	$C_6H_{12}N_2S_4Zn$ 305.83	137-30-4			

STRUCTURAL FORMULAS OF ORGANIC COMPOUNDS

In numeric order as they occur in the Table of Physical Constants of Organic Compounds

1

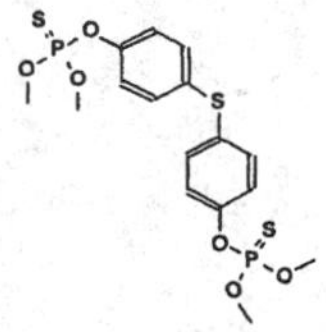

3

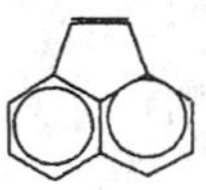

4

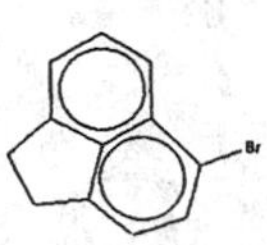

5

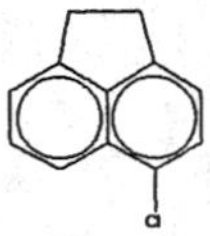

6

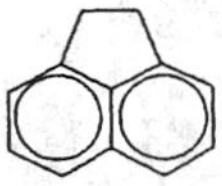

8

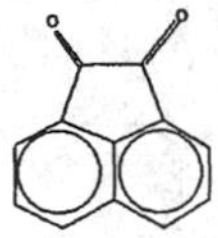

9

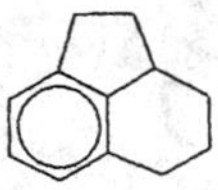

10

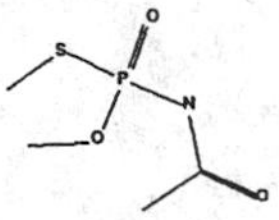

11

12

14

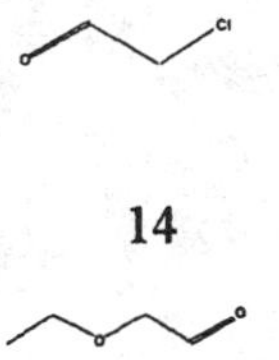

15

16

17

18

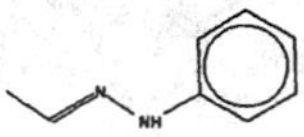

19

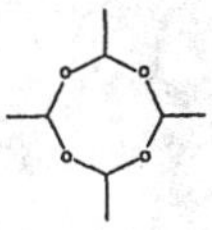

20

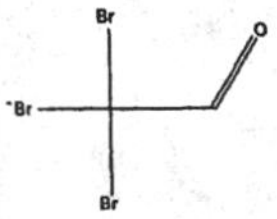

21

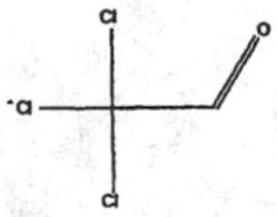

22

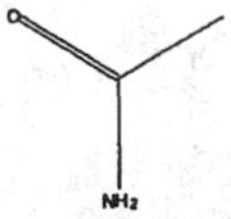

23

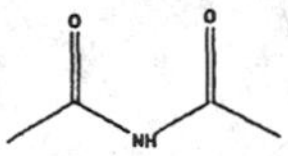

24

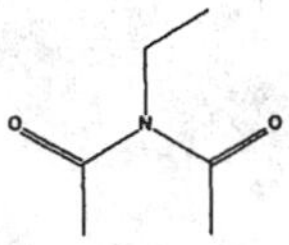

25

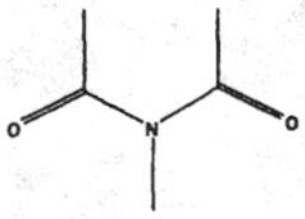

26

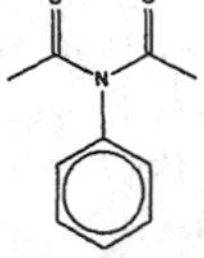

27

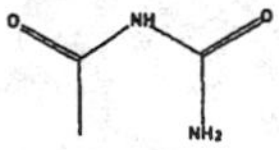

28

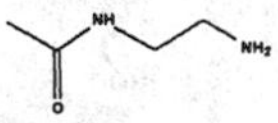

29

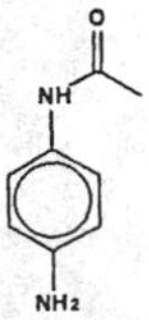

30

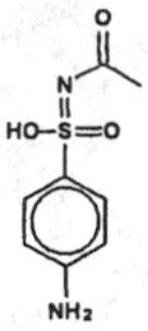

31

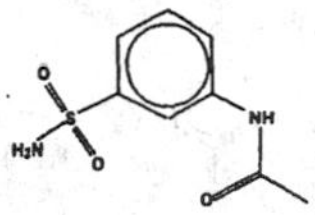

32

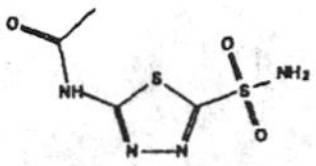

33

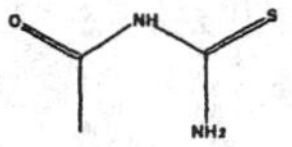

34

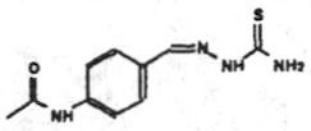

35

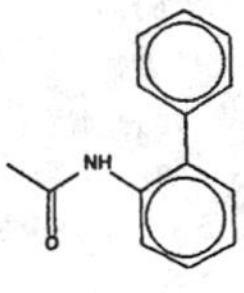

37

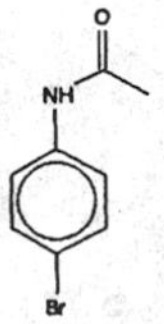

38

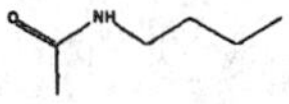

39

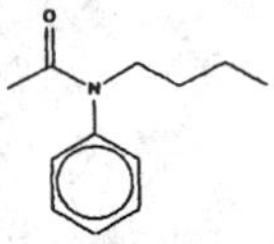

40

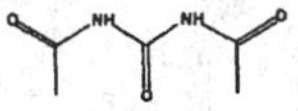

41

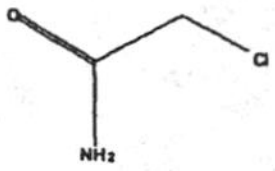

42

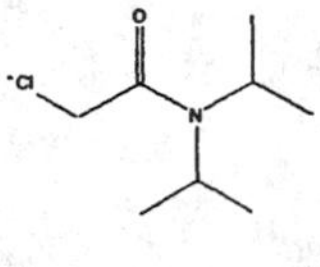

43

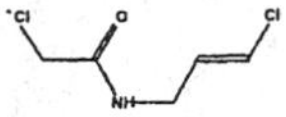

44

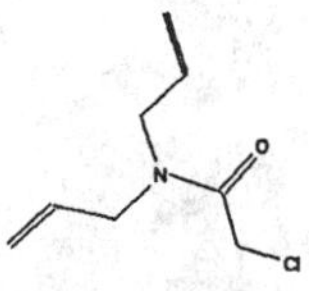

45

46

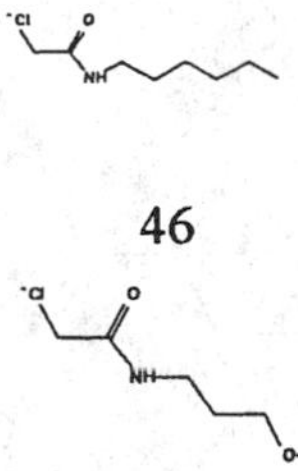

STRUCTURAL FORMULAS OF ORGANIC COMPOUNDS (continued)

In numeric order as they occur in the Table of Physical Constants of Organic Compounds

47

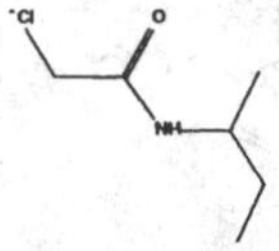

48

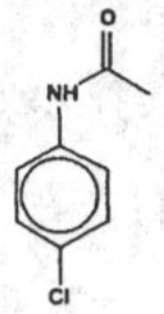

49

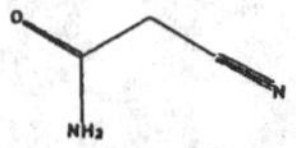

50

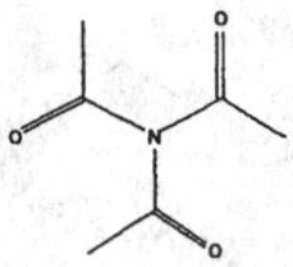

51

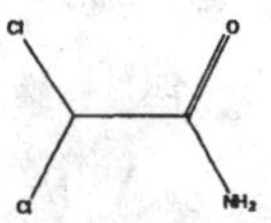

52

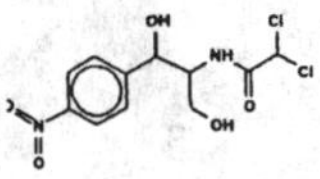

53

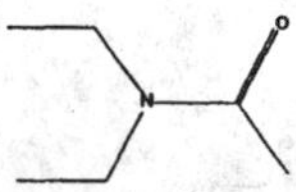

54

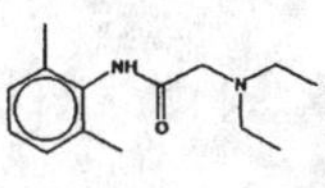

55

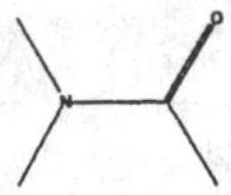

56

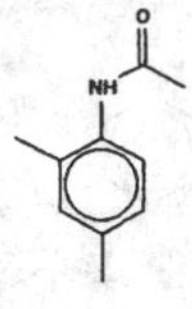

57

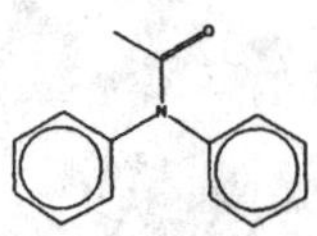

58

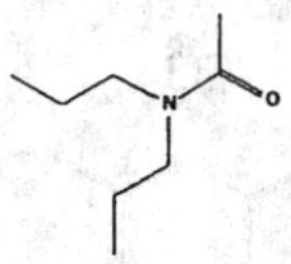

59

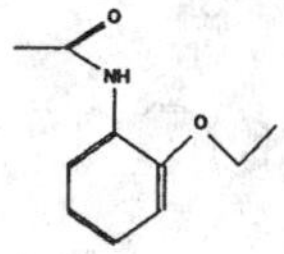

60

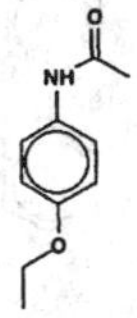

61

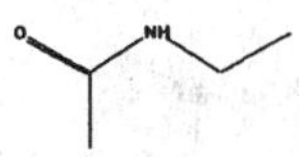

62

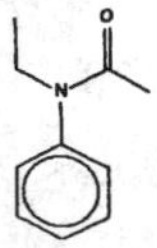

63

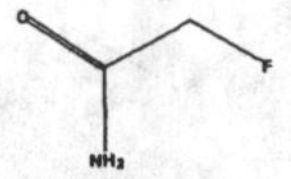

64

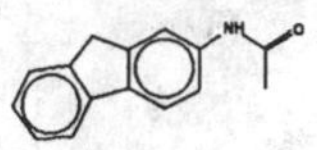

65

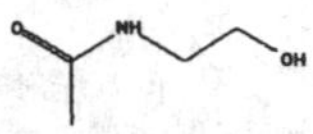

66

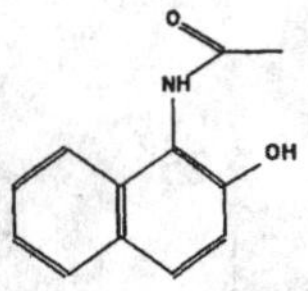

67

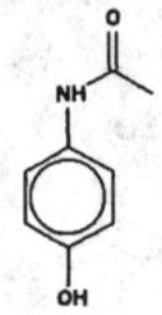

68

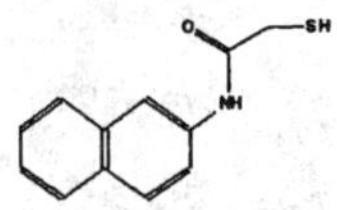

69

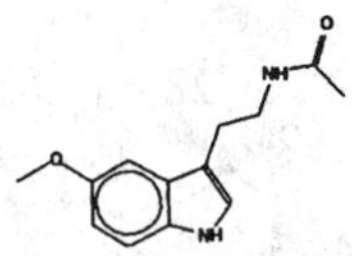

70

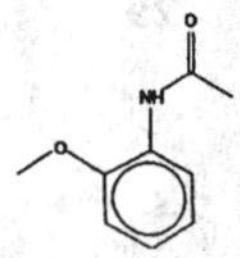

71

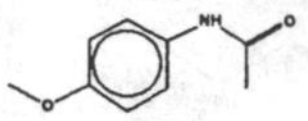

72

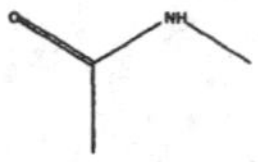

73

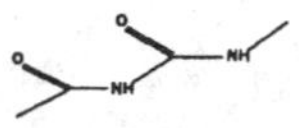

74

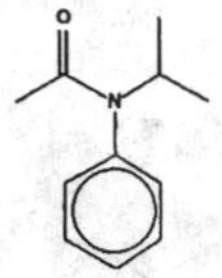

75

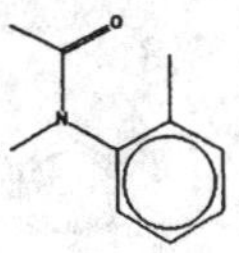

76

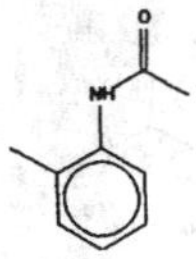

77

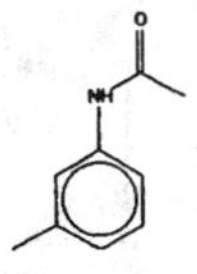

78

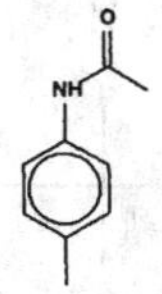

79

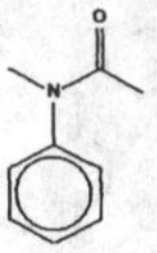

80

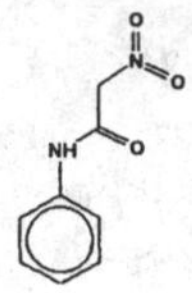

81

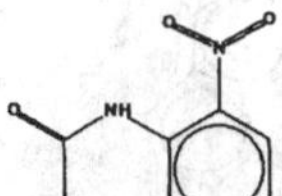

82

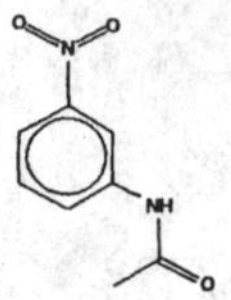

83

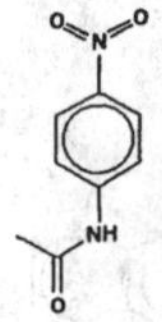

84

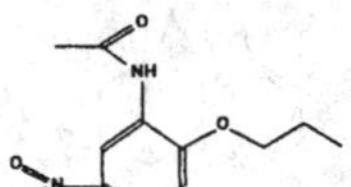

85

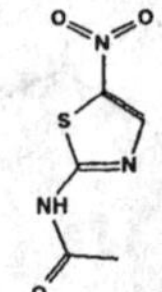

STRUCTURAL FORMULAS OF ORGANIC COMPOUNDS (continued)

In numeric order as they occur in the Table of Physical Constants of Organic Compounds

86

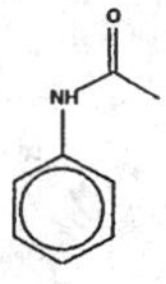

87

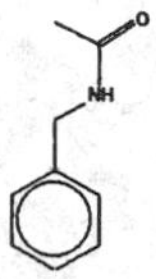

88

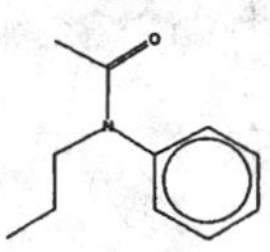

89

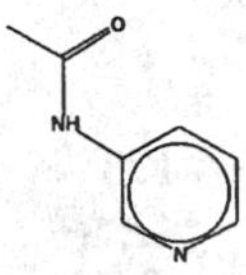

90

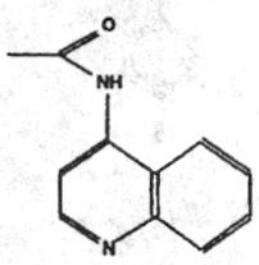

91

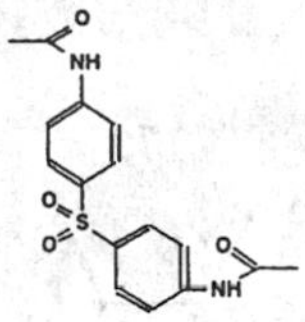

92

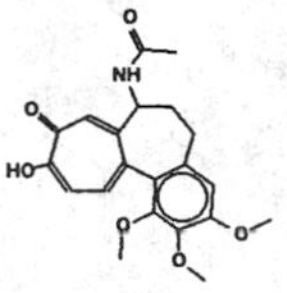

93

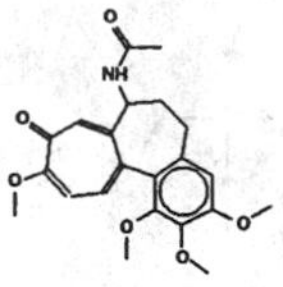

95

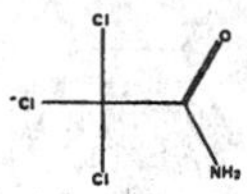

96

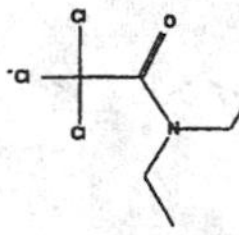

97

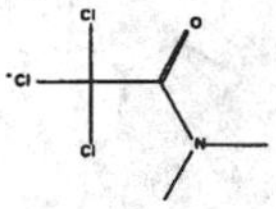

98

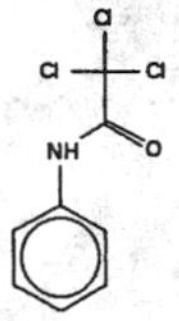

99

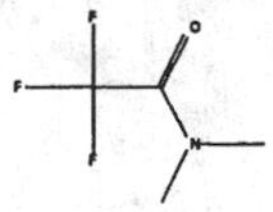

100

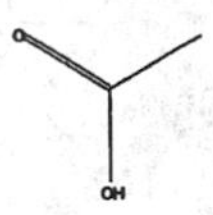

101

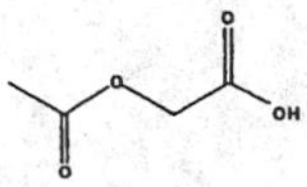

102

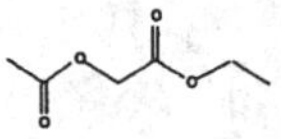

103

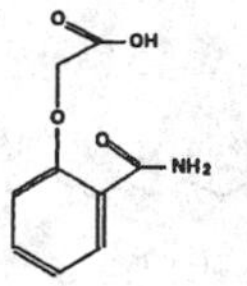

104

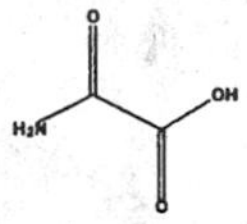

105

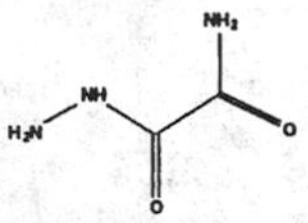

107

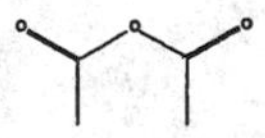

108

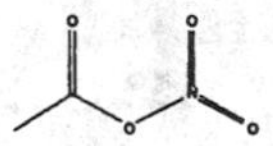

109

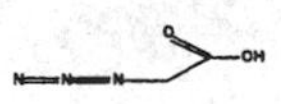

111

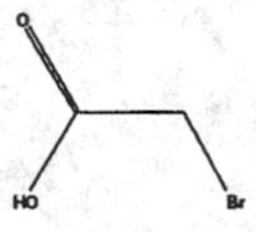

112

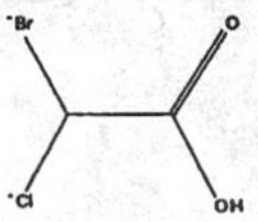

113

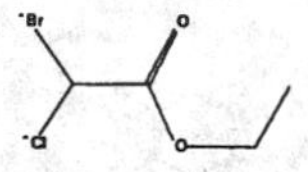

114

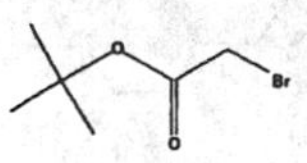

115

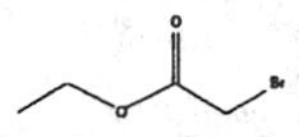

116

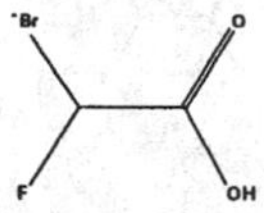

117

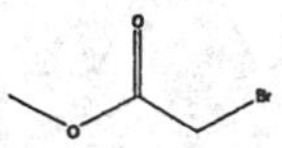

118

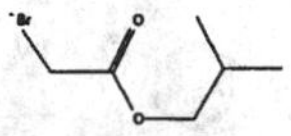

119

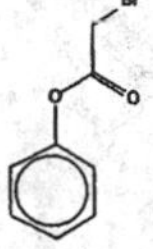

120

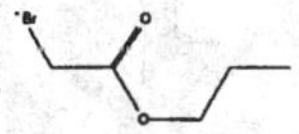

121

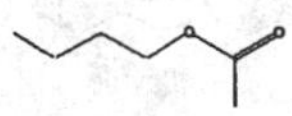

123

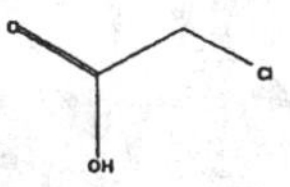

124

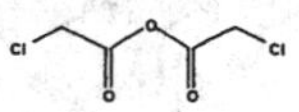

125

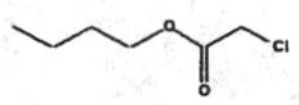

126

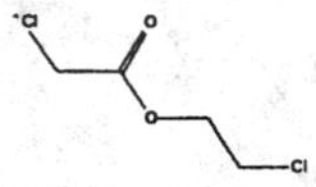

127

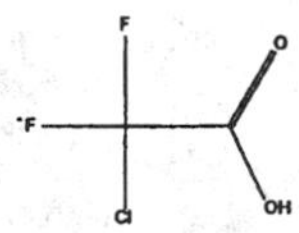

128

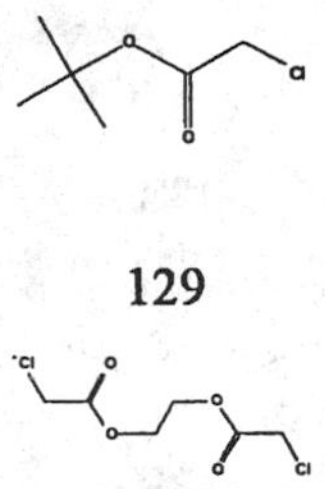

129

130

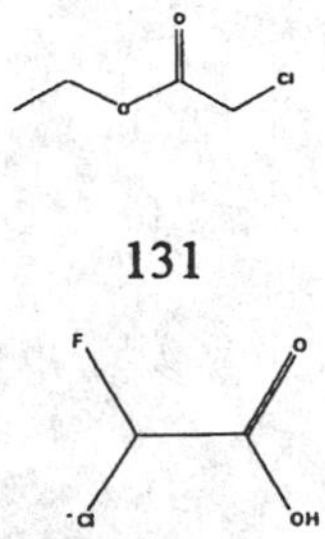

131

STRUCTURAL FORMULAS OF ORGANIC COMPOUNDS (continued)

In numeric order as they occur in the Table of Physical Constants of Organic Compounds

132

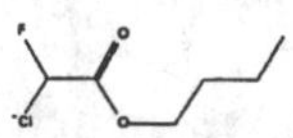

133

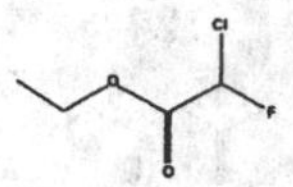

134

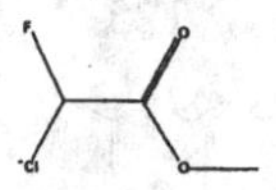

135

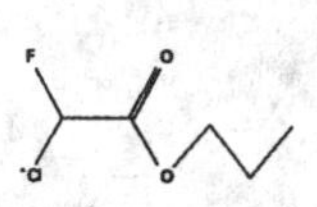

136

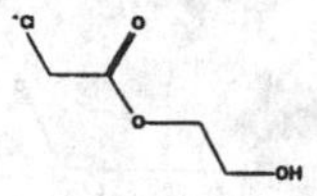

137

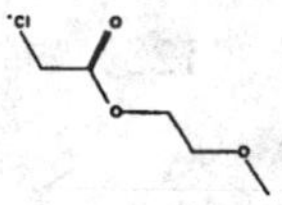

138, 139

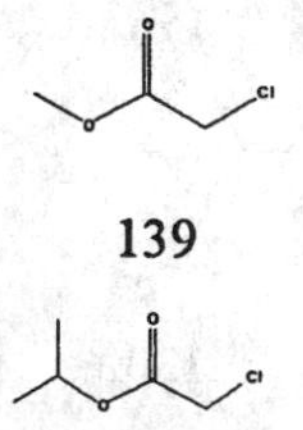

140

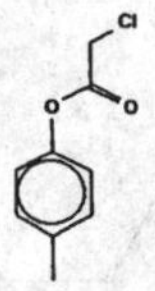

141

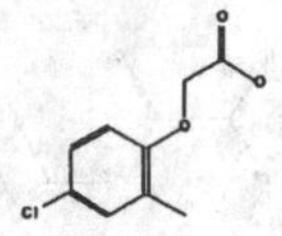

142

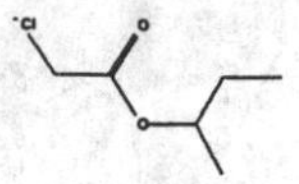

143

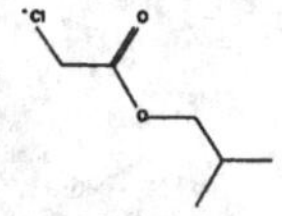

144

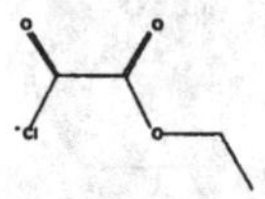

145

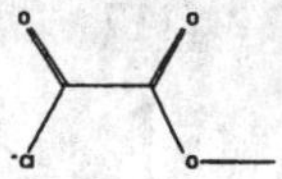

146

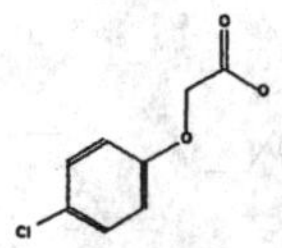

147, 148

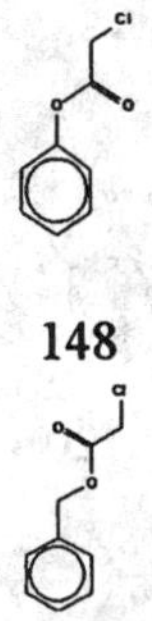

149

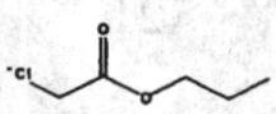

150

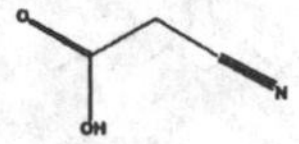

151

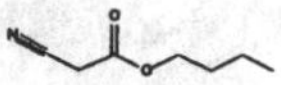

152, 153

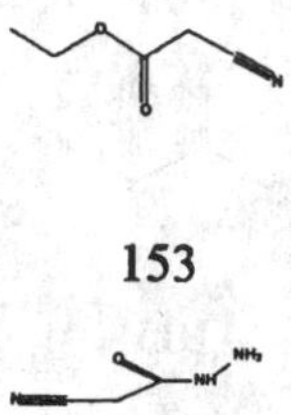

154

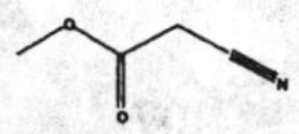

157

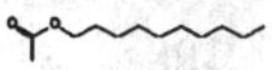

158

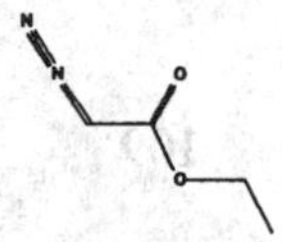

159

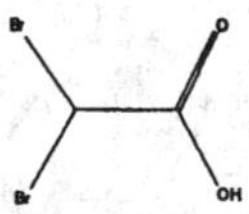

160

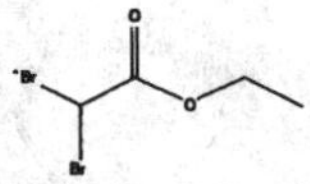

161

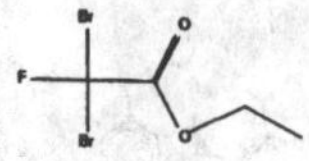

162

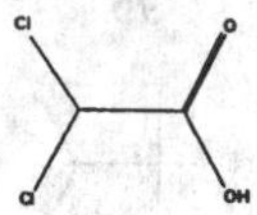

163

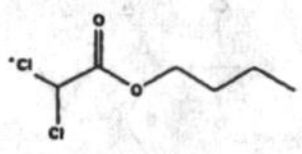

164

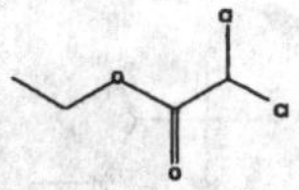

165

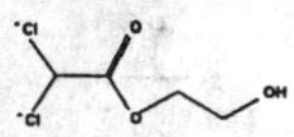

166

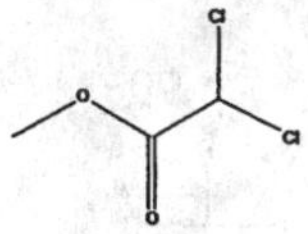

167

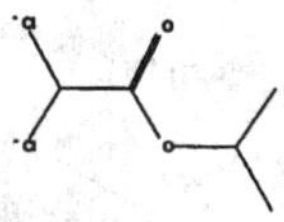

168

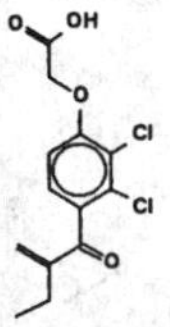

169

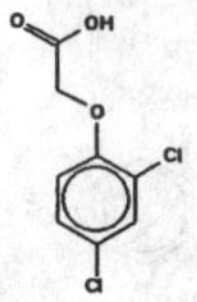

170, 171

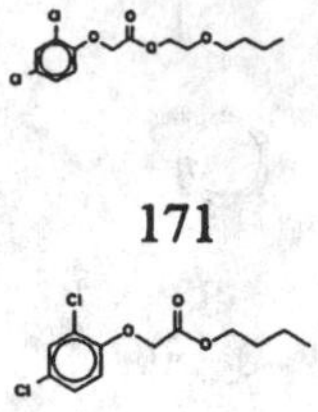

172

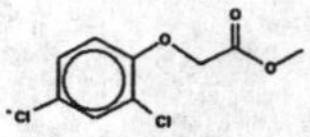

173

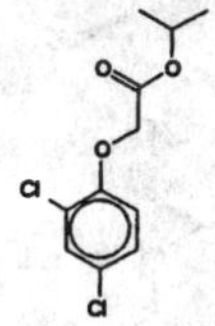

174

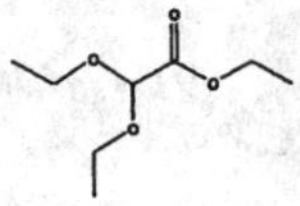

175

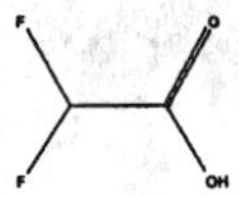

176

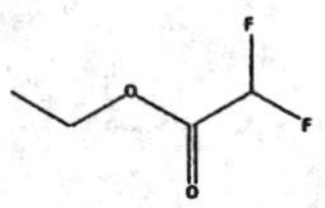

177

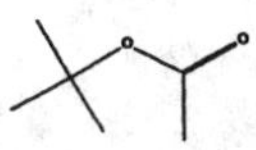

In numeric order as they occur in the Table of Physical Constants of Organic Compounds

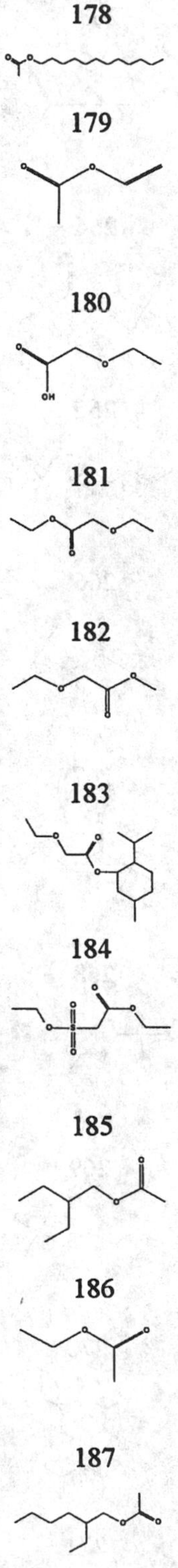

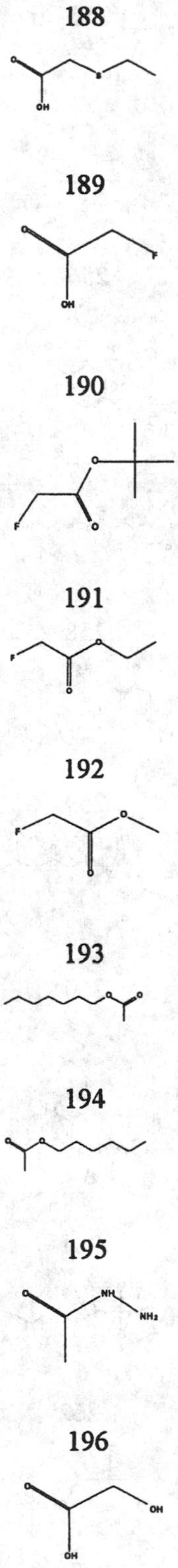

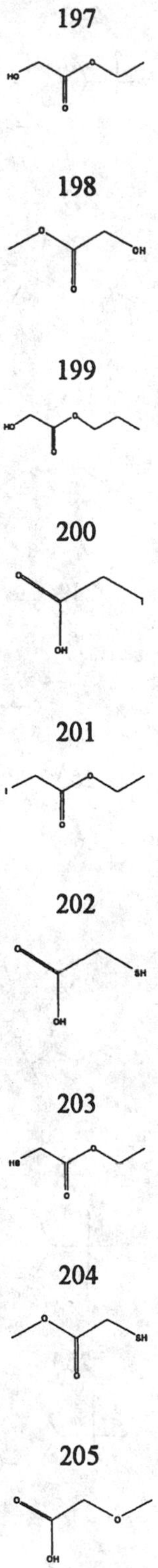

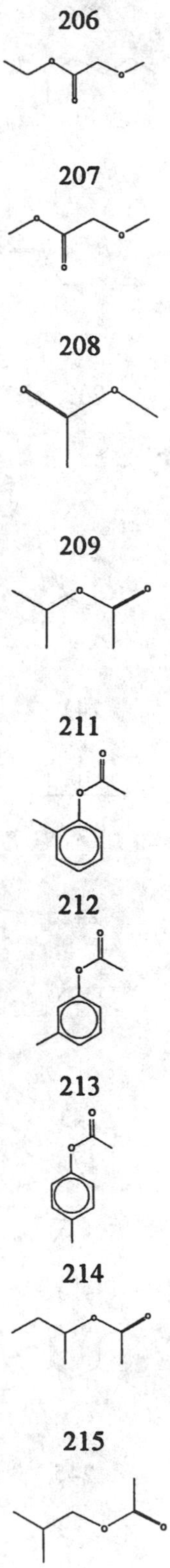

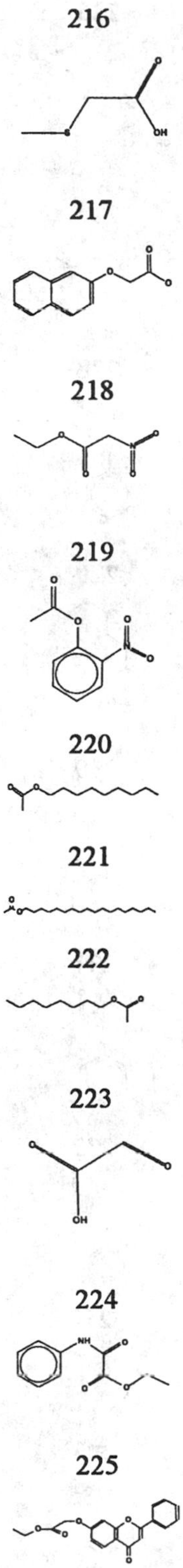

STRUCTURAL FORMULAS OF ORGANIC COMPOUNDS (continued)

In numeric order as they occur in the Table of Physical Constants of Organic Compounds

226

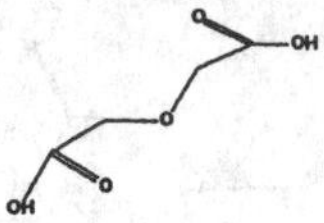

227

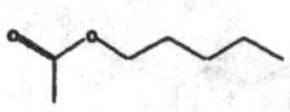

228

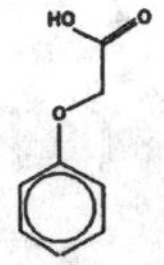

229

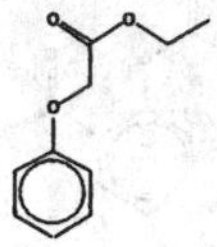

230

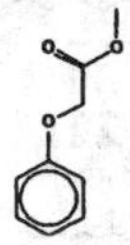

231

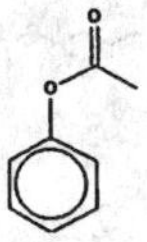

232

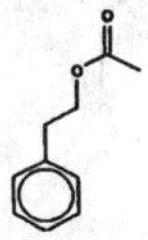

233

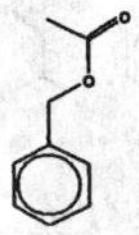

235

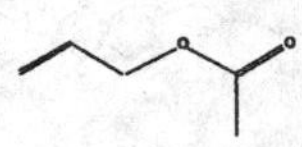

236

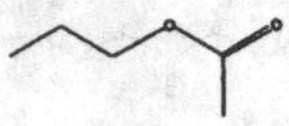

238

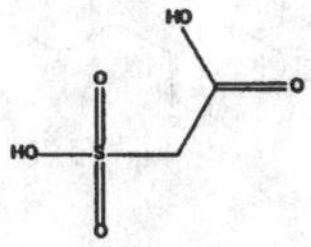

239

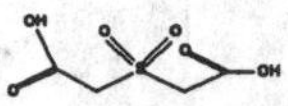

240

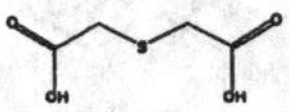

241

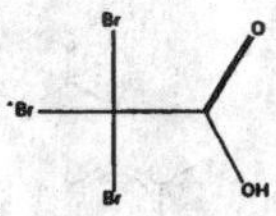

242

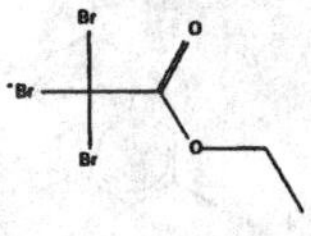

243

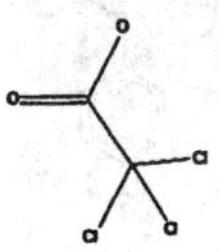

244

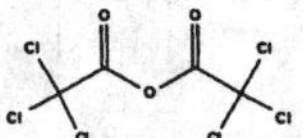

245

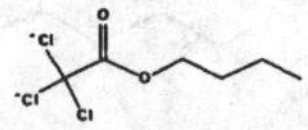

246

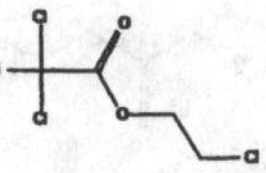

247

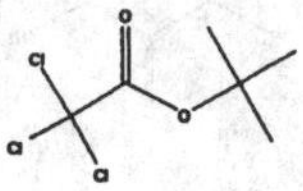

248

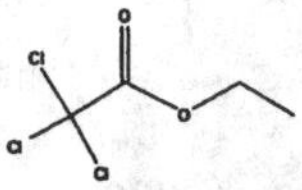

249

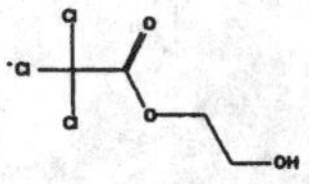

250

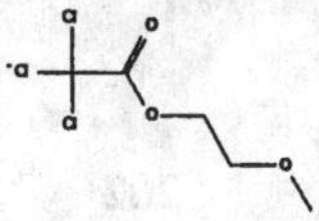

252

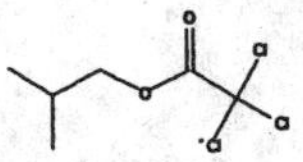

253

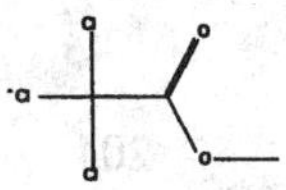

254

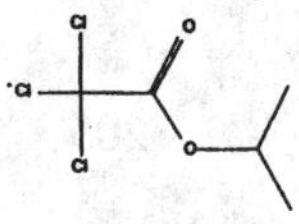

255

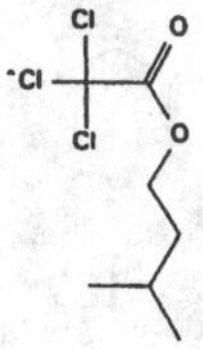

256

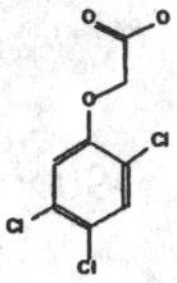

257

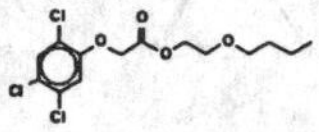

258

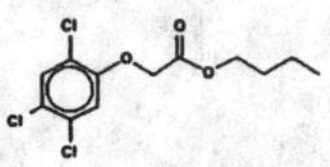

259

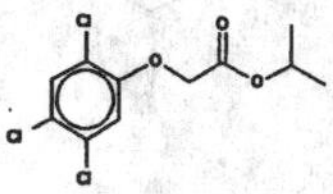

260

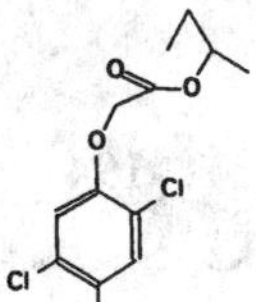

261

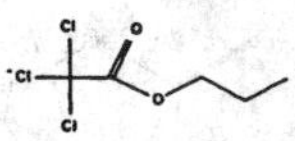

262

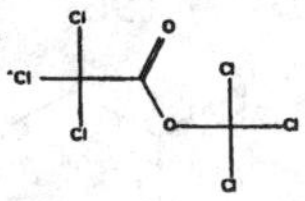

263

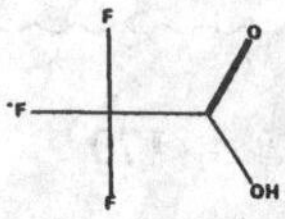

264

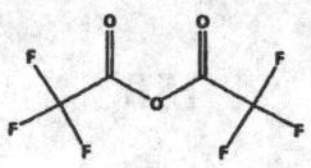

265

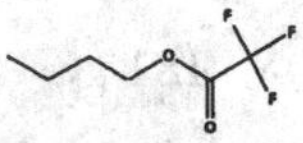

266

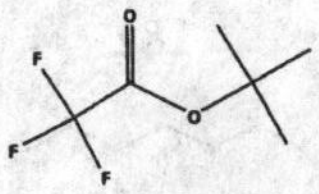

267

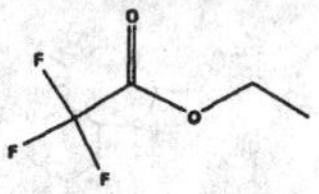

268

269

270

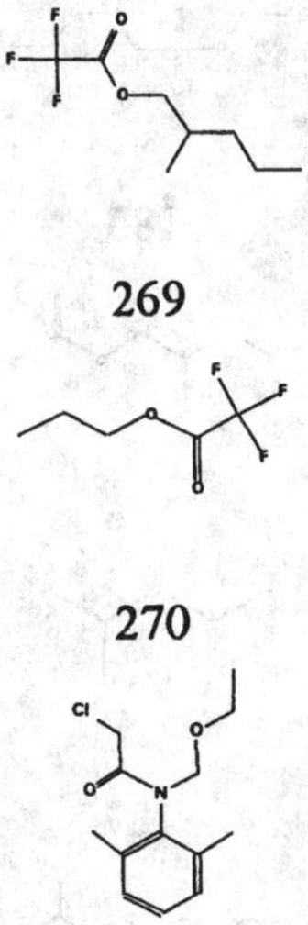

271

272

273

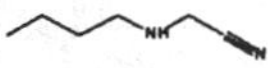

274

277

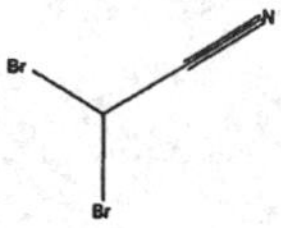

278

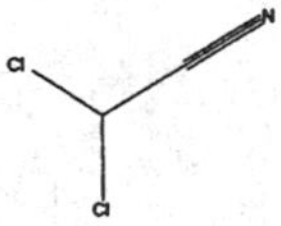

279

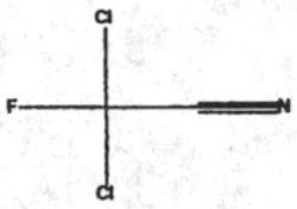

280

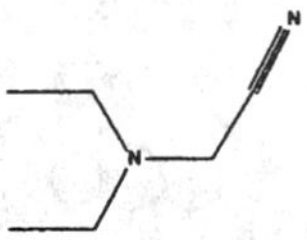

281

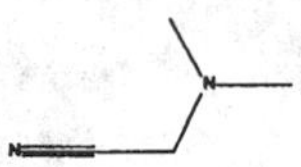

282

283

284

285

286

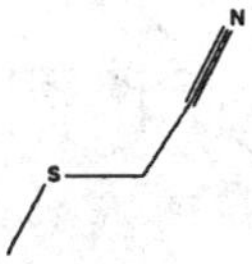

287

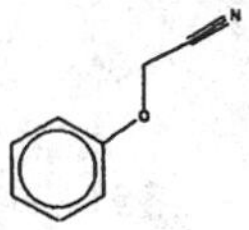

288

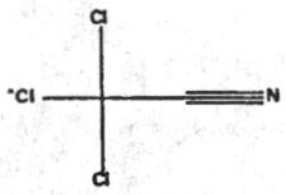

289

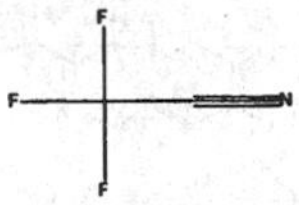

290

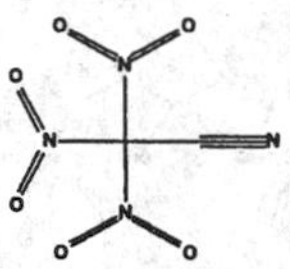

291

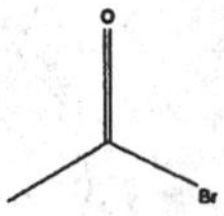

292

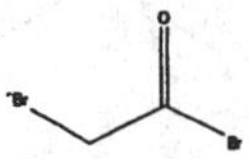

293

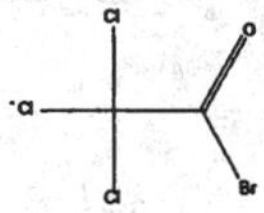

294

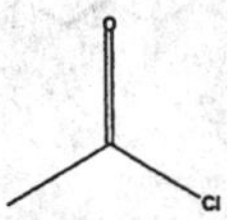

295

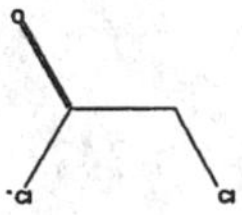

296

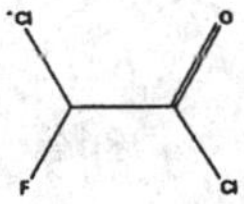

297

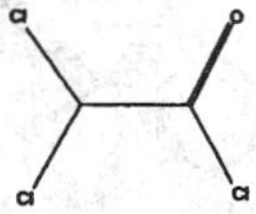

298

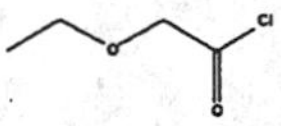

299

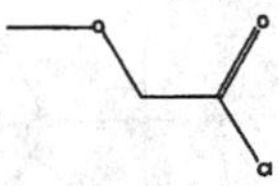

300

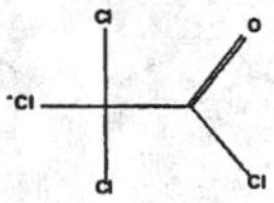

301

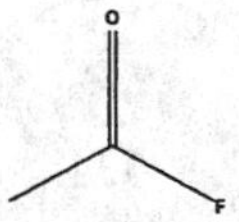

302

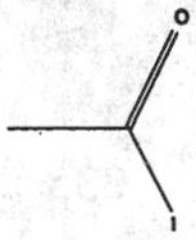

303

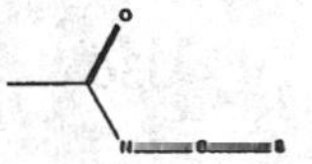

304

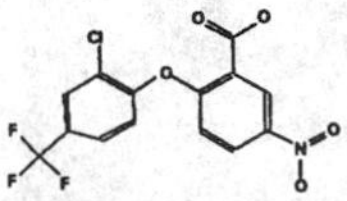

306

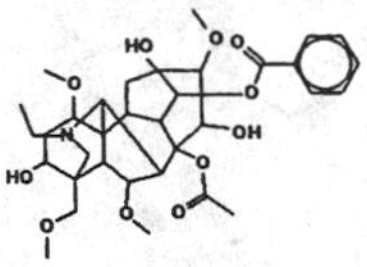

312

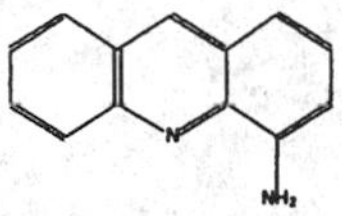

313

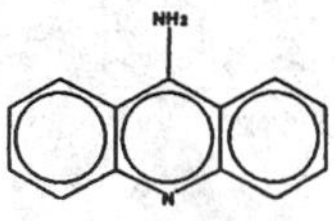

314

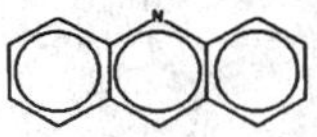

315

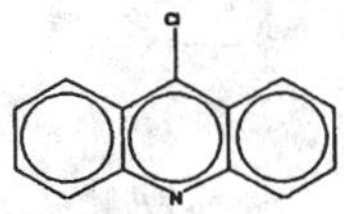

316

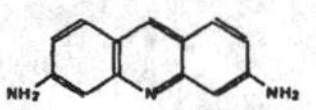

317

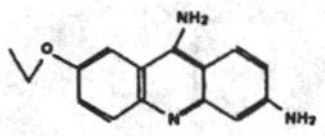

318

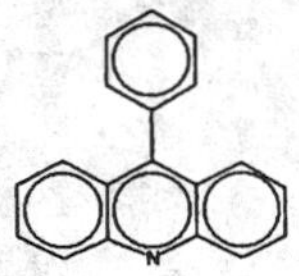

STRUCTURAL FORMULAS OF ORGANIC COMPOUNDS (continued)

In numeric order as they occur in the Table of Physical Constants of Organic Compounds

319

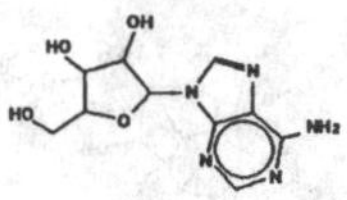

320

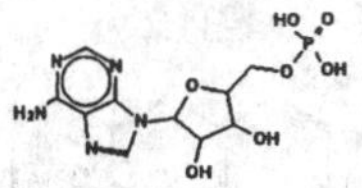

321

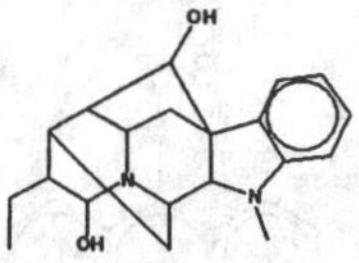

322

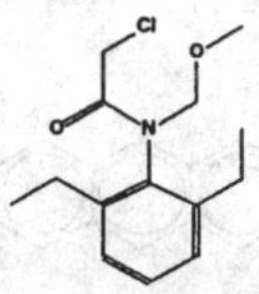

323

324

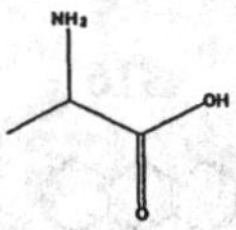

325

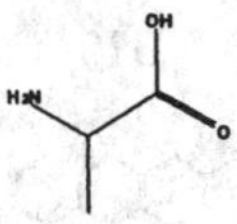

326

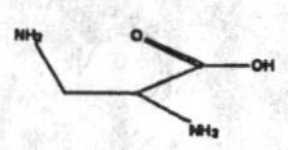

327

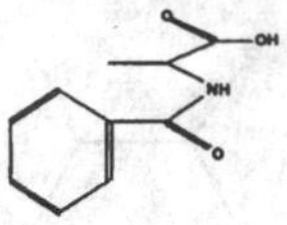

329

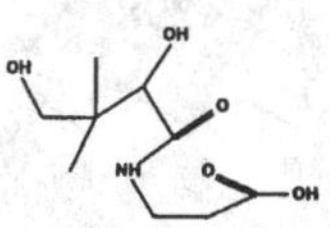

331

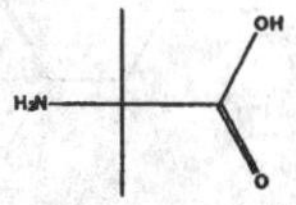

332

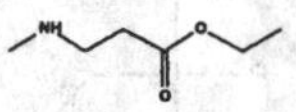

333

334

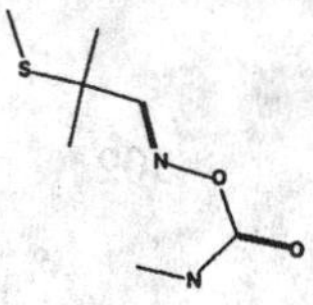

335

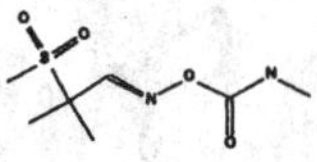

336

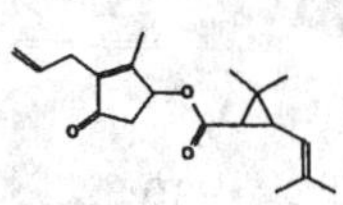

337

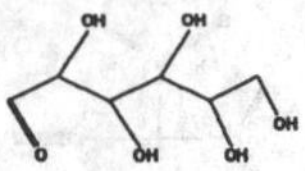

338

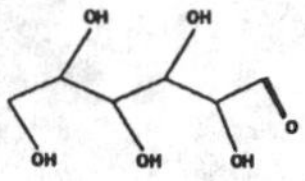

339

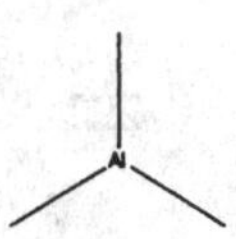

340

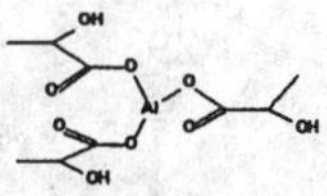

341

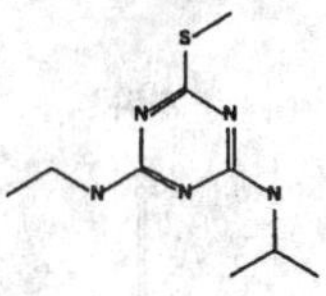

342

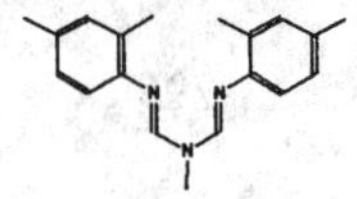

343

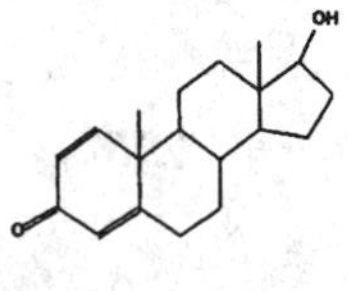

344

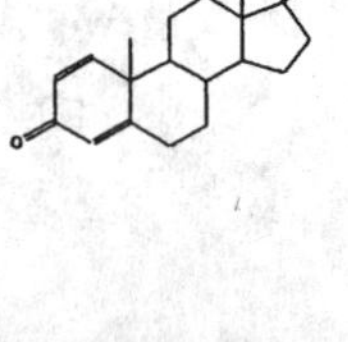

345

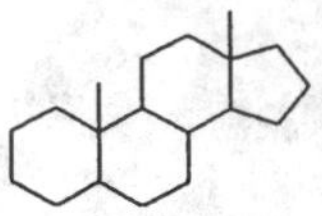

346

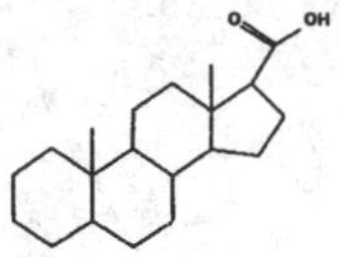

347

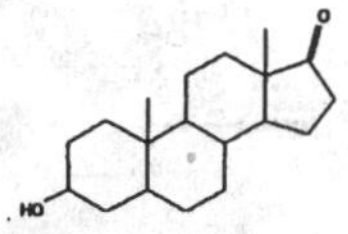

348

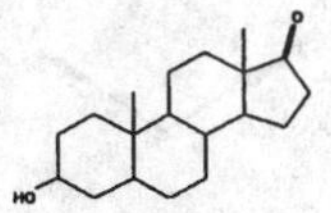

349

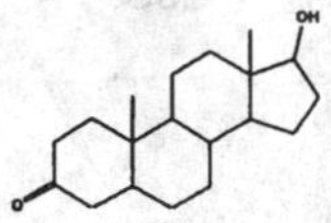

350

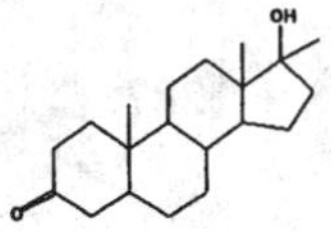

351

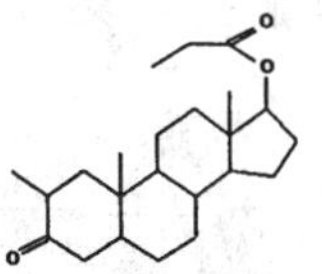

352

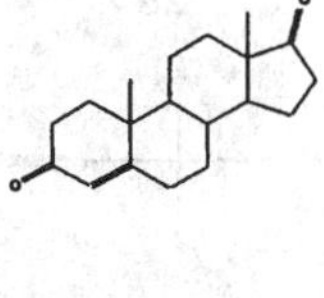

353

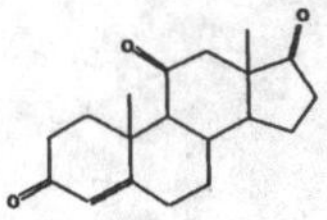

354

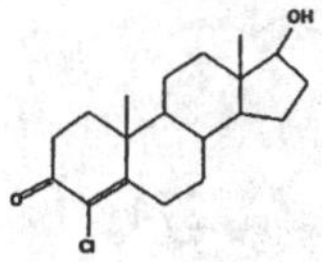

355

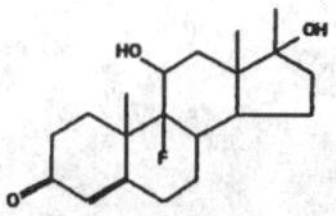

356

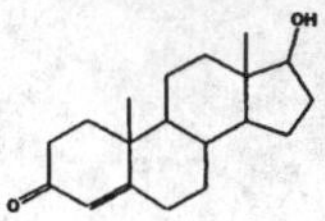

357

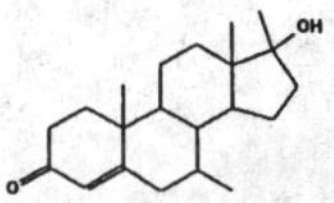

358

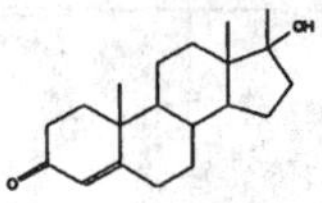

359

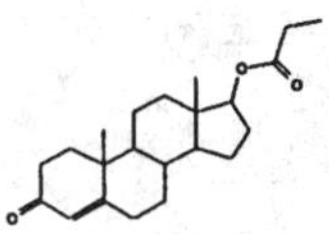

360

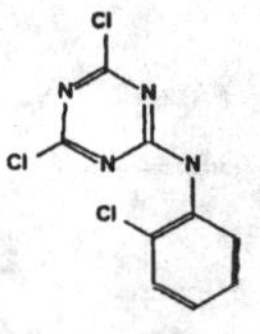

In numeric order as they occur in the Table of Physical Constants of Organic Compounds

362
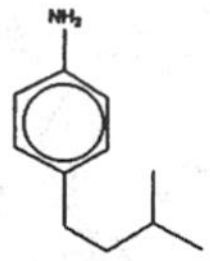

364
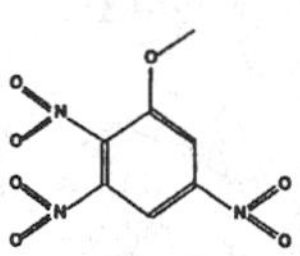

365
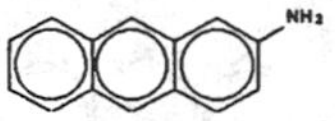

366
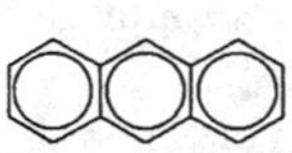

367
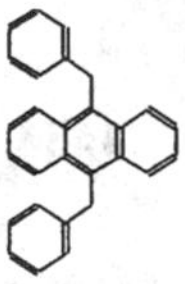

368
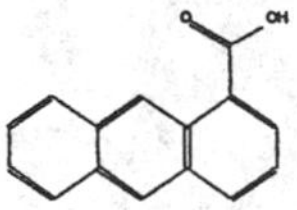

369
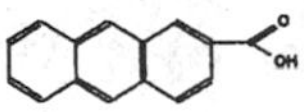

370
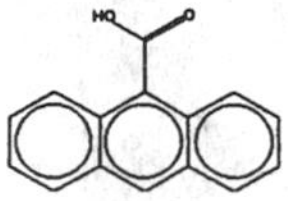

371
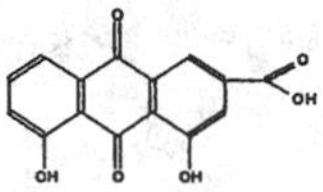

372
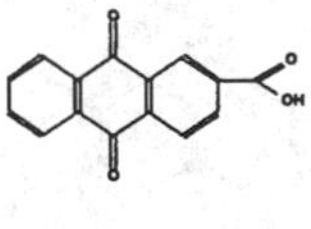

373
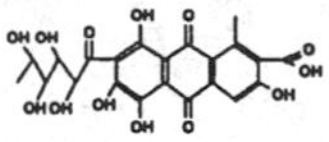

374
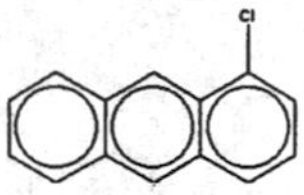

375
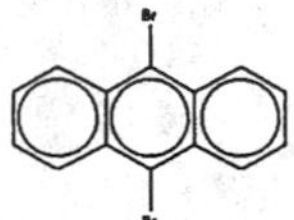

376
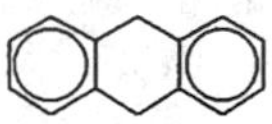

377
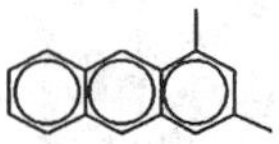

379
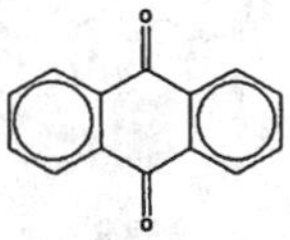

380
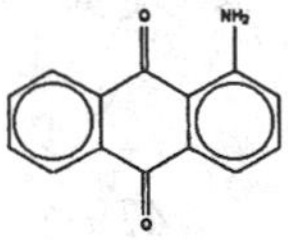

381
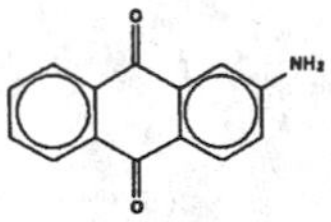

382
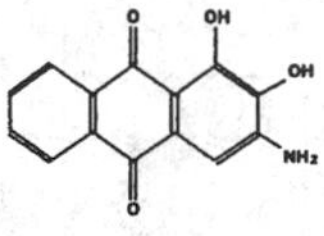

383
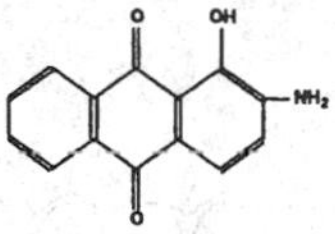

384
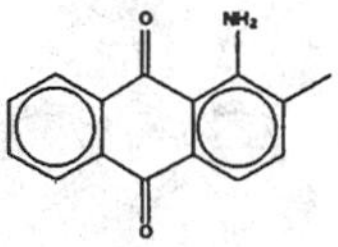

385
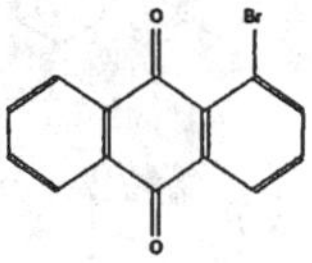

386
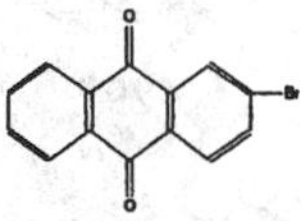

387
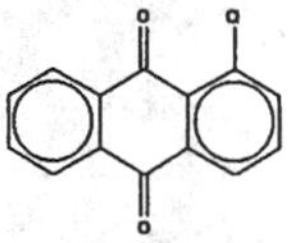

388
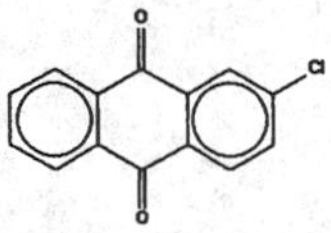

389
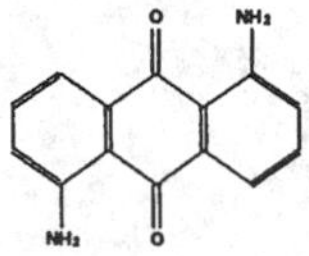

390
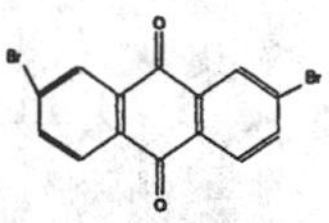

391
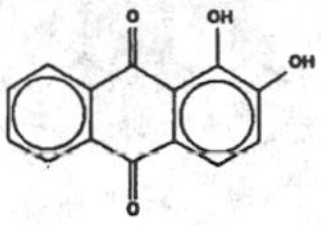

392
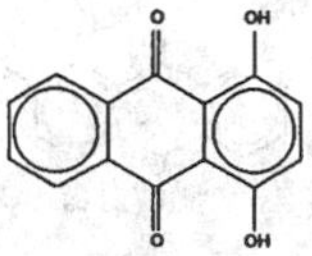

393
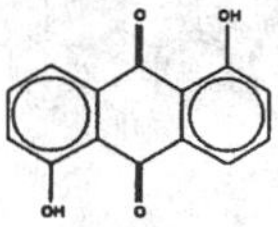

394
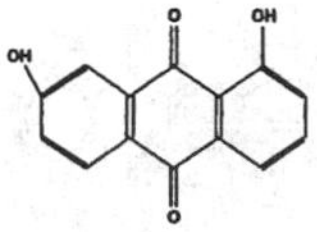

395
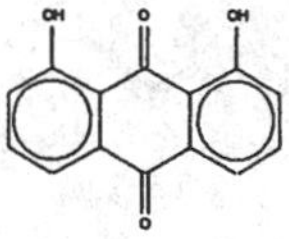

396
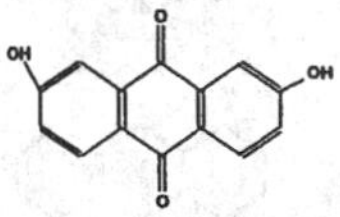

397
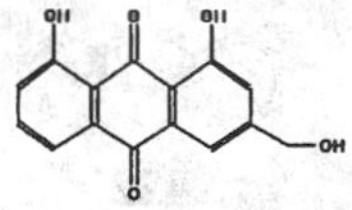

398
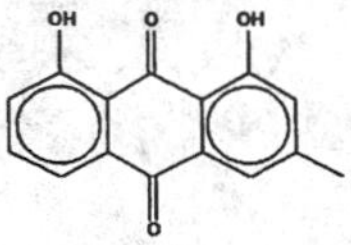

399
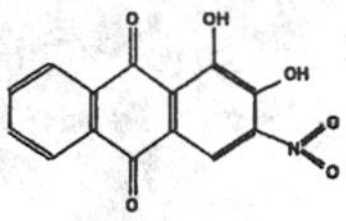

400
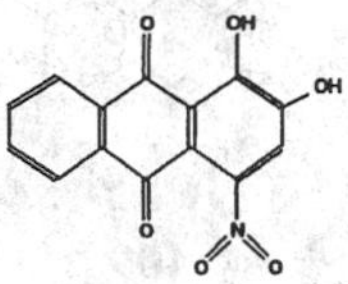

401
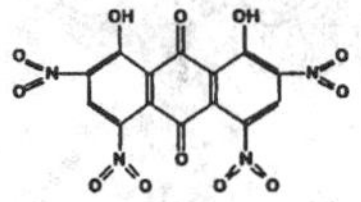

402

403
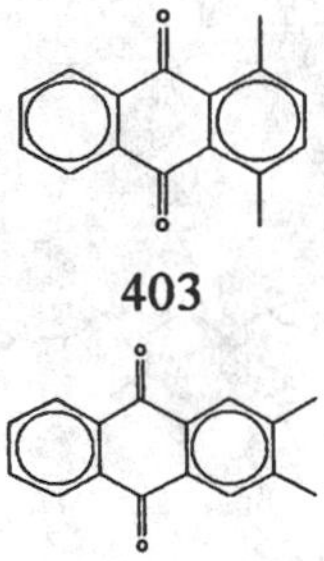

404

405
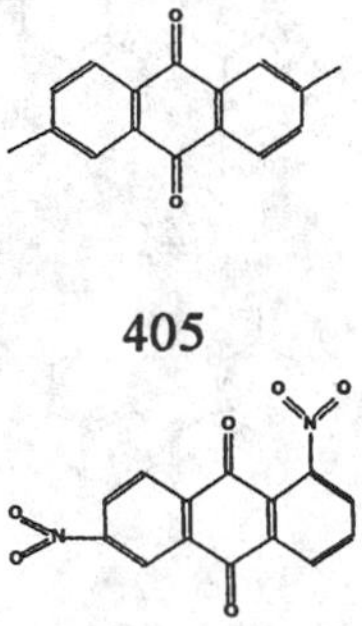

In numeric order as they occur in the Table of Physical Constants of Organic Compounds

406
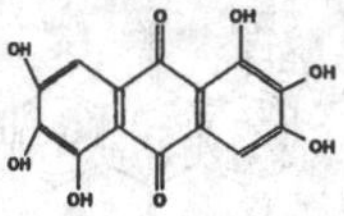

407
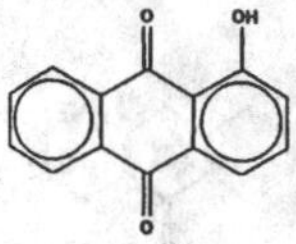

408
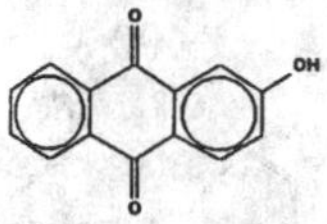

409
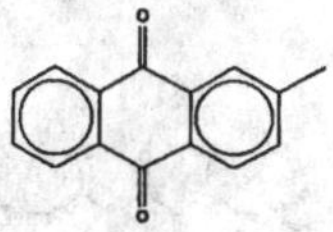

410
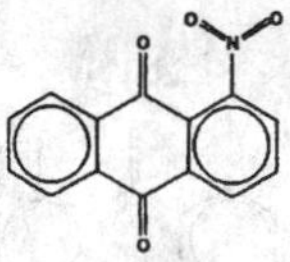

411
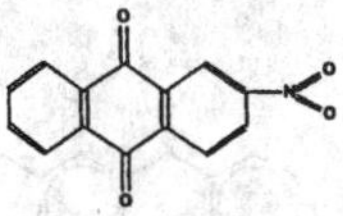

412
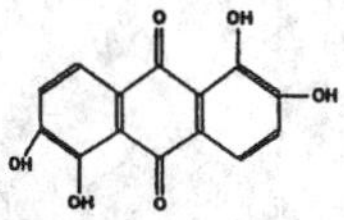

413
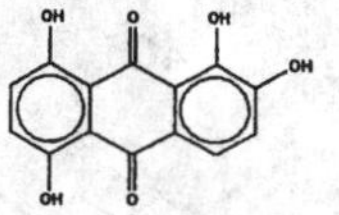

414
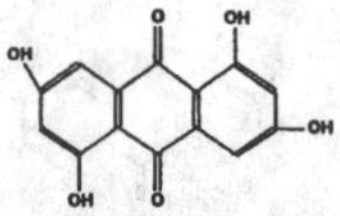

415
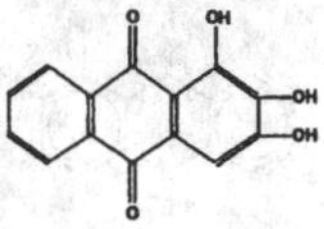

416
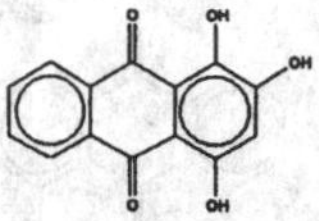

417
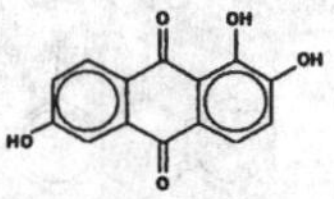

418
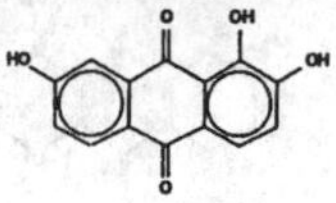

419
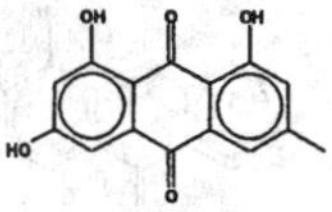

420
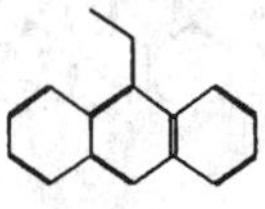

421

422
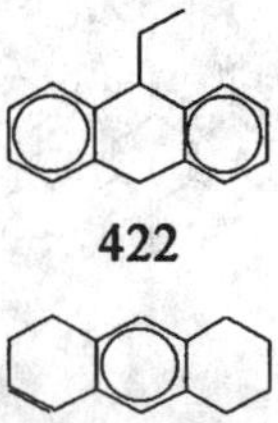

423

424

425
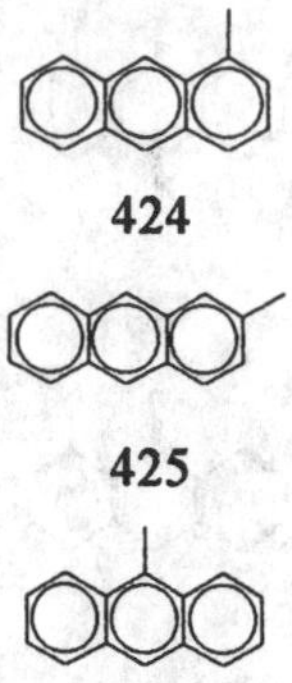

426
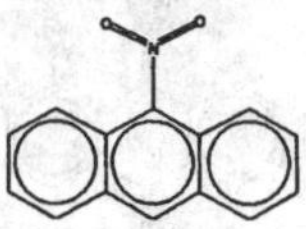

427
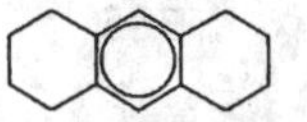

428
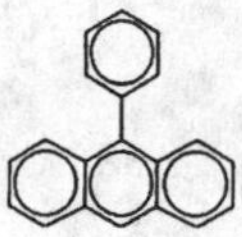

429
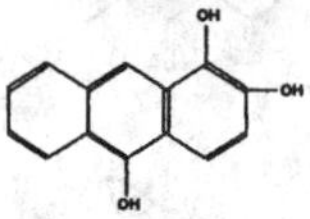

430
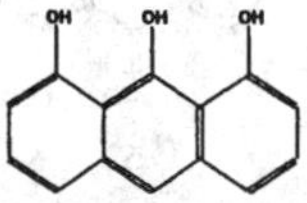

431
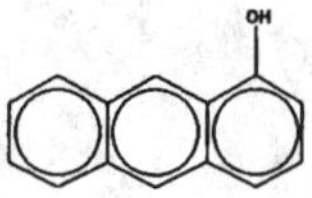

432
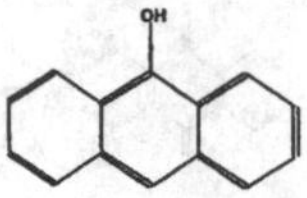

433
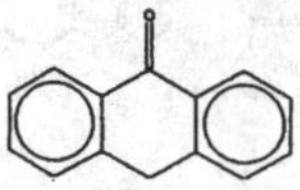

434
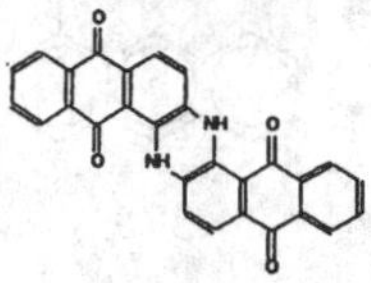

435
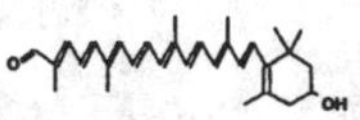

436
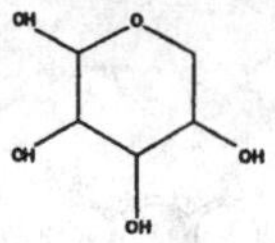

437
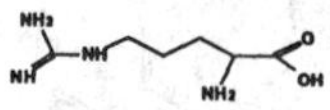

438
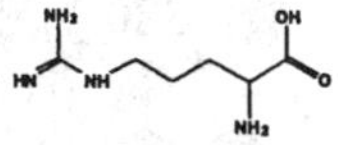

439
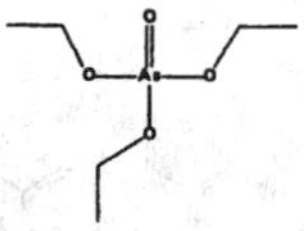

440
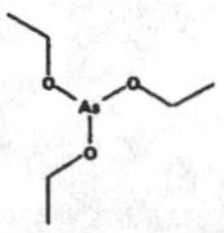

441
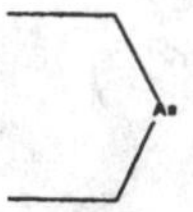

442

443
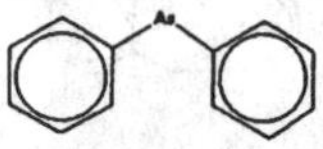

444

445
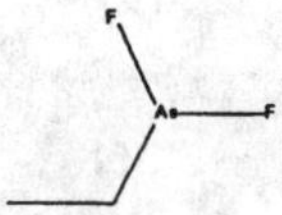

446

447
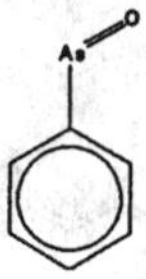

448
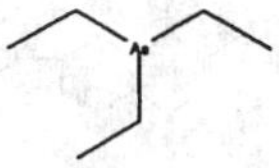

STRUCTURAL FORMULAS OF ORGANIC COMPOUNDS (continued)

In numeric order as they occur in the Table of Physical Constants of Organic Compounds

449

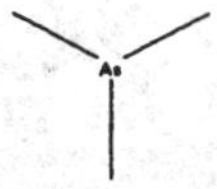

450

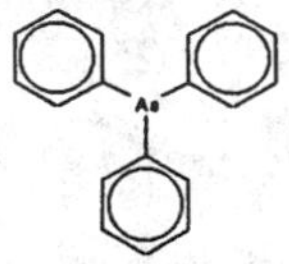

451

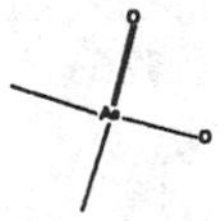

452

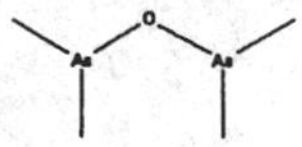

453

454

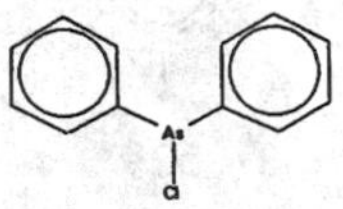

455

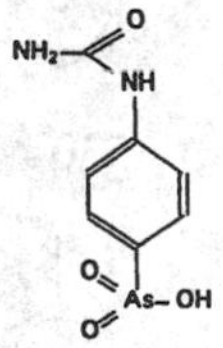

456

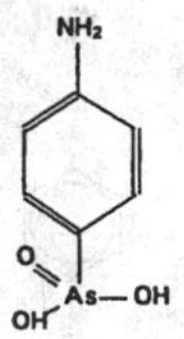

457

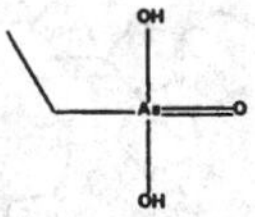

458

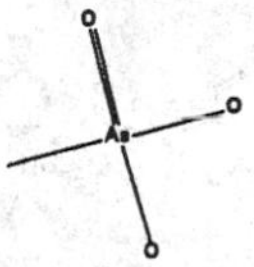

459

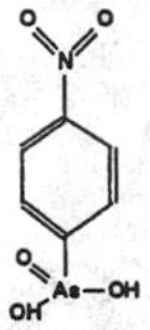

460

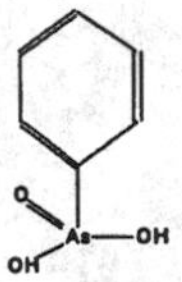

461

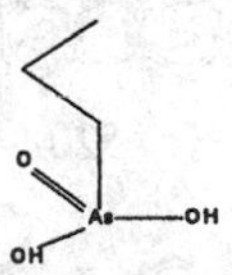

462

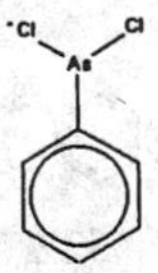

463

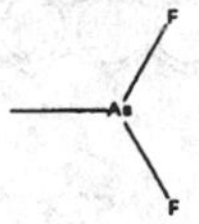

464

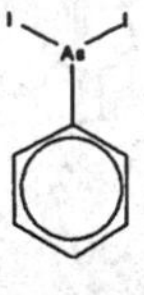

465

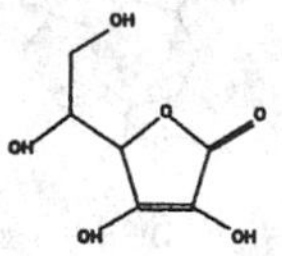

466

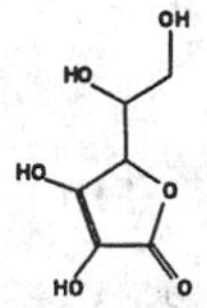

467

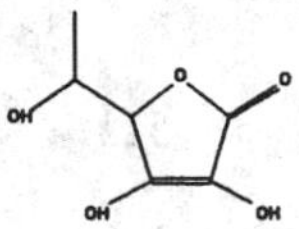

468

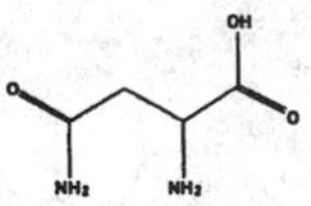

470

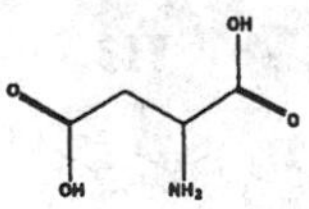

471

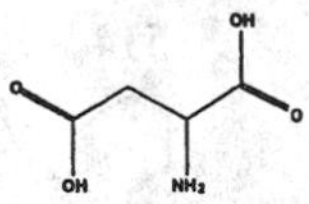

472

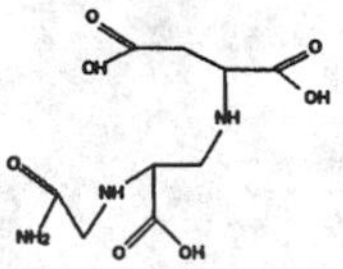

473

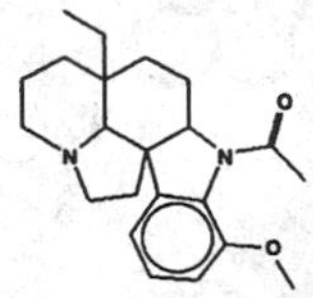

474

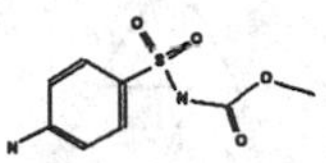

475

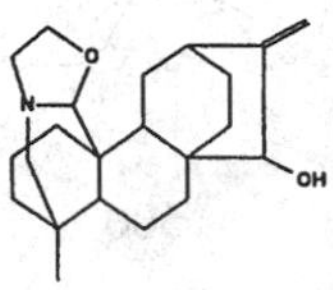

476

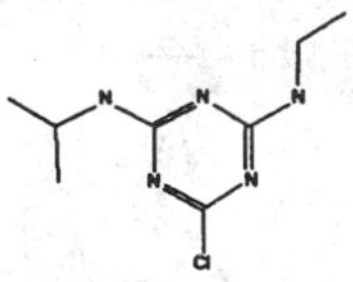

477

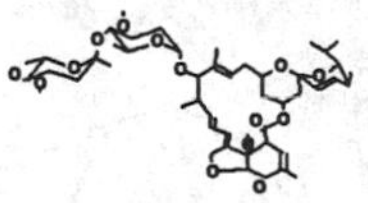

478

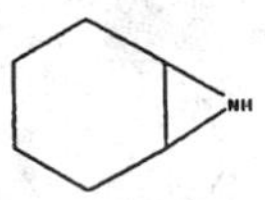

479

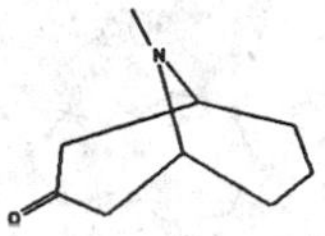

480

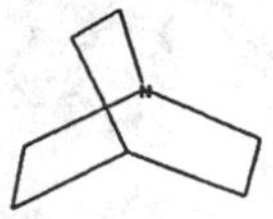

481

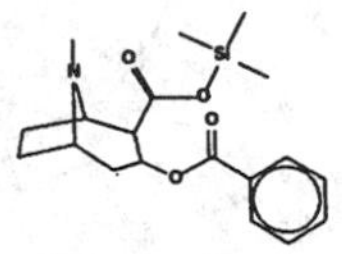

482

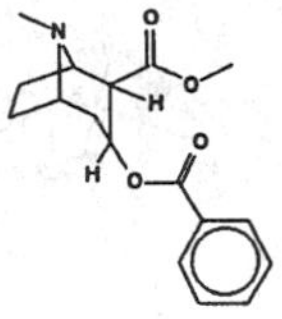

483

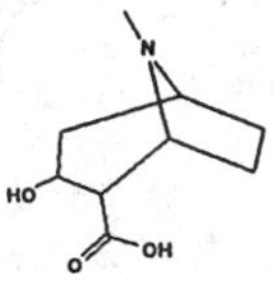

484

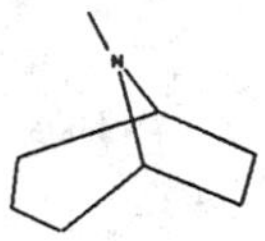

485

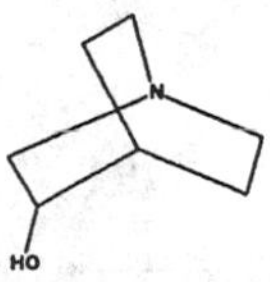

486

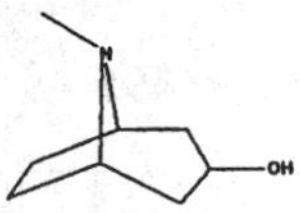

487

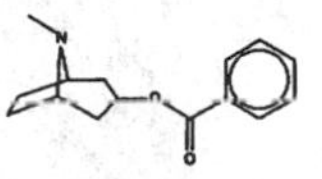

488

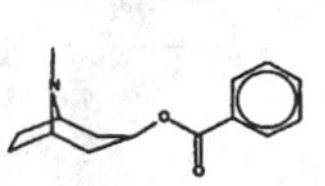

In numeric order as they occur in the Table of Physical Constants of Organic Compounds

489

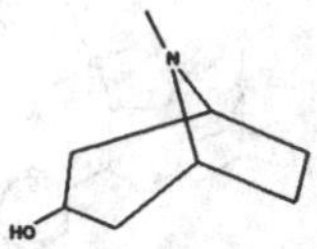

490

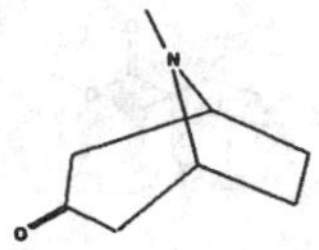

491

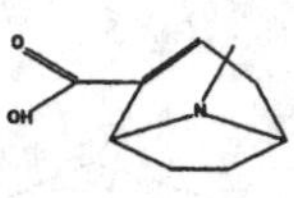

492

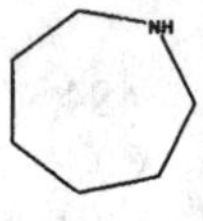

493

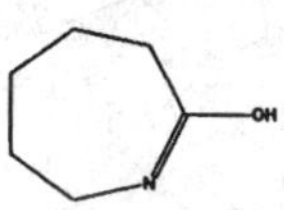

494

495

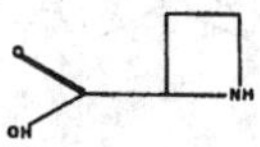

496

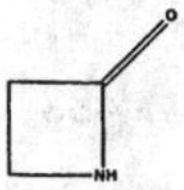

497

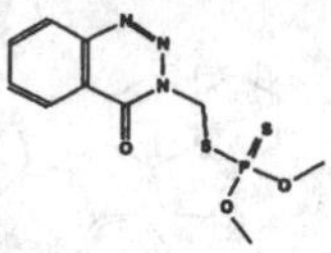

498

499

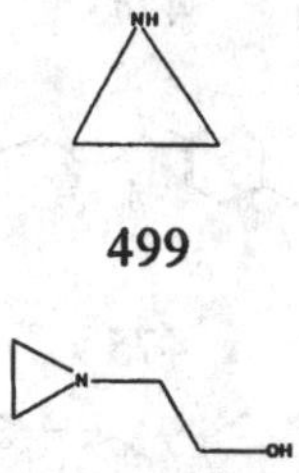

500

501

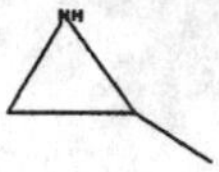

502

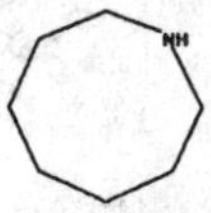

503

504

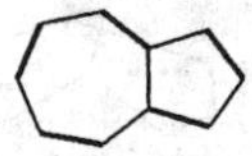

505

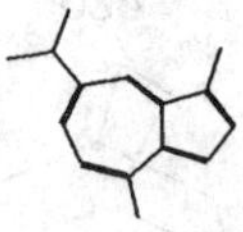

506

507

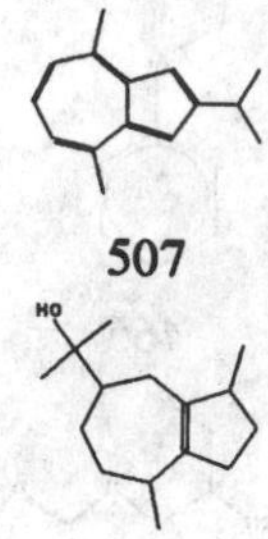

509

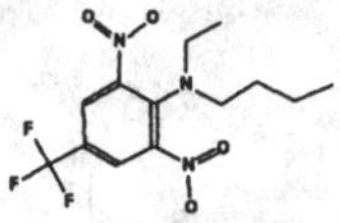

510

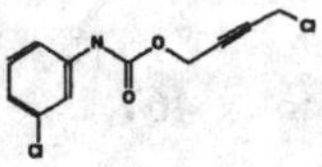

511

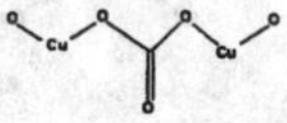

512

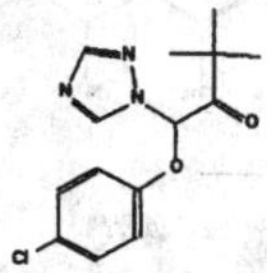

513

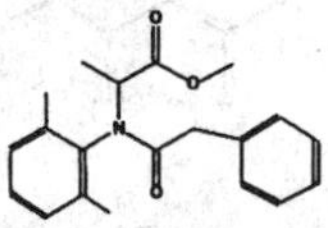

514

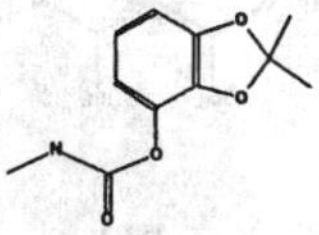

515

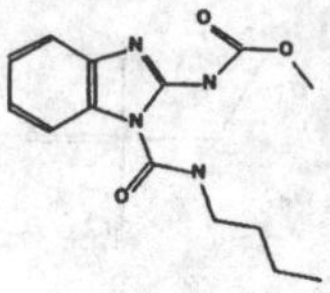

516

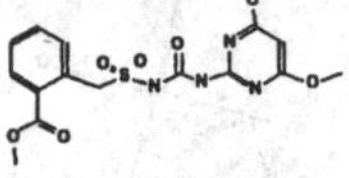

517

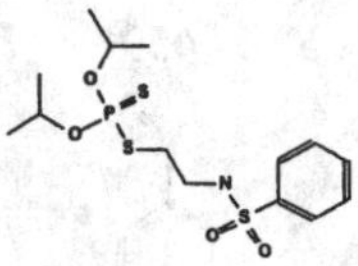

518

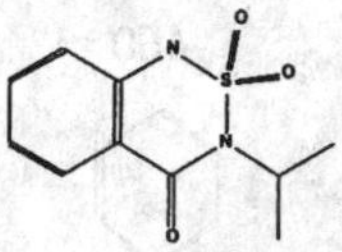

519

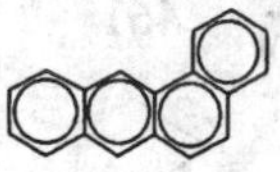

520

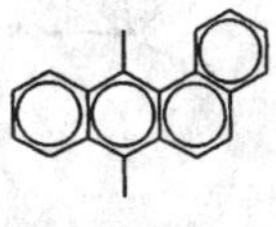

521

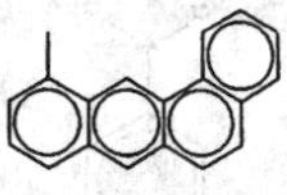

522

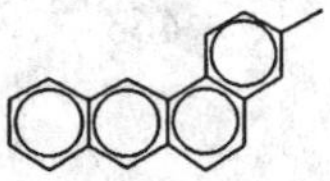

523

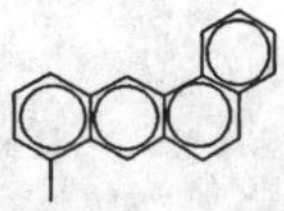

524

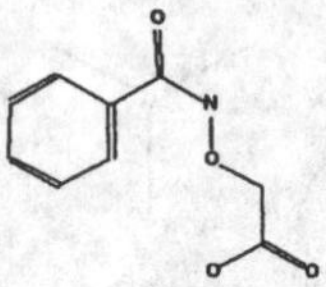

525

526

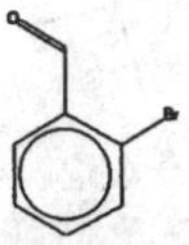

527

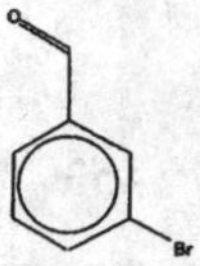

528

529

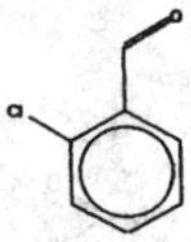

530

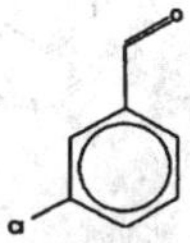

531

In numeric order as they occur in the Table of Physical Constants of Organic Compounds

532

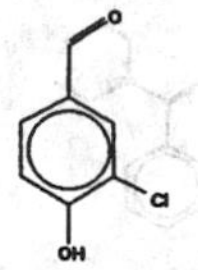

533

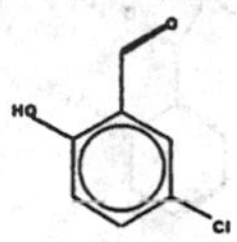

535

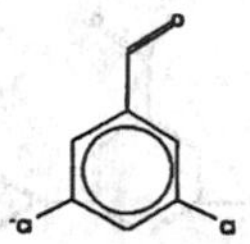

536

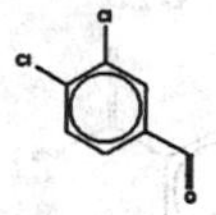

537

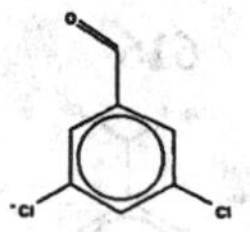

538

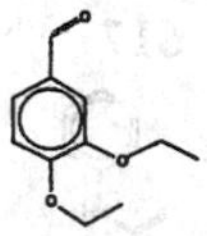

539

540

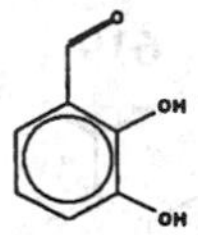

541

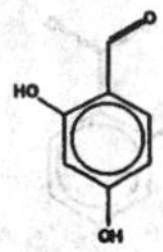

542

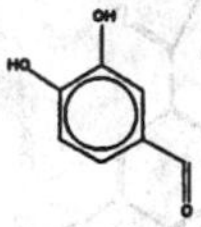

543

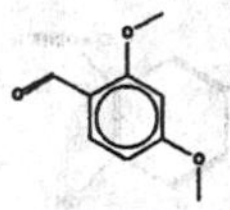

544

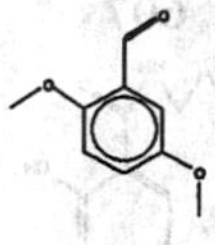

545

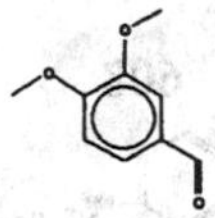

546

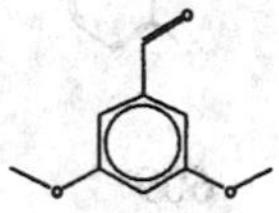

547

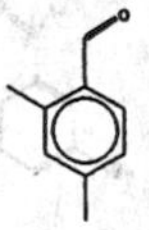

548

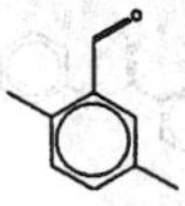

549

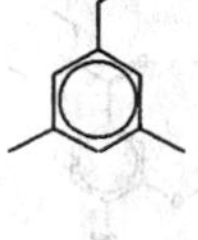

550

551

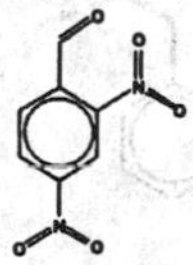

552

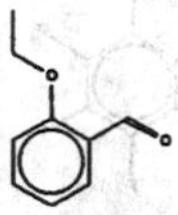

553

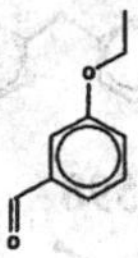

554

555

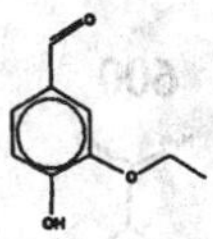

556

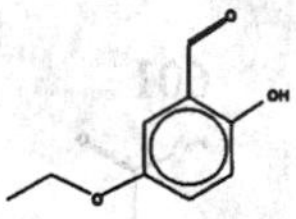

557

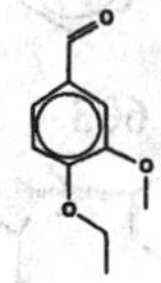

558

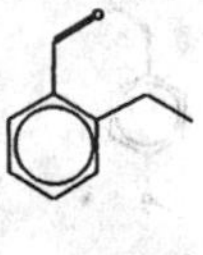

559

560

561

562

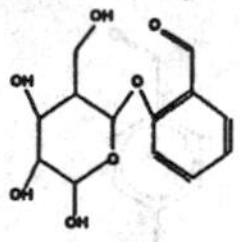

563

564

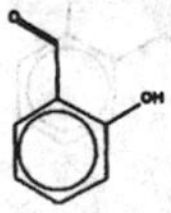

565

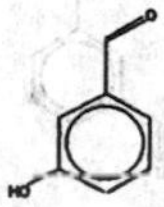

566

567

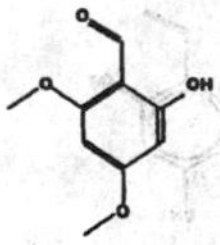

568

569

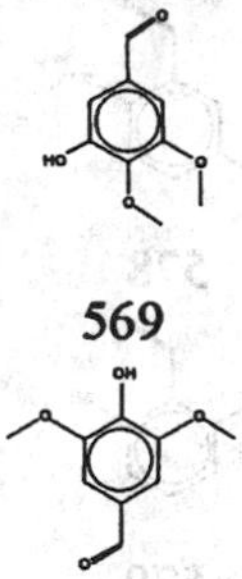

570

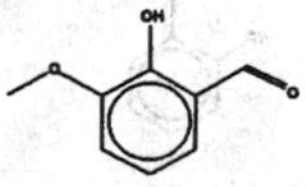

571

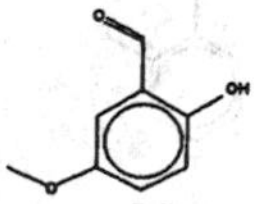

572

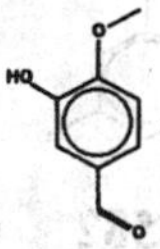

573

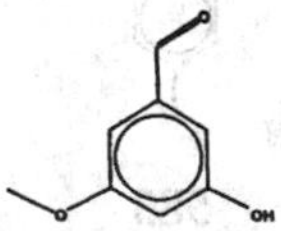

574

575

576

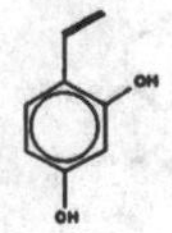

577

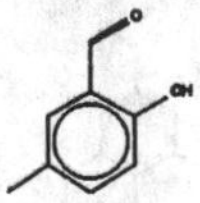

578

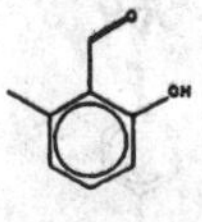

579

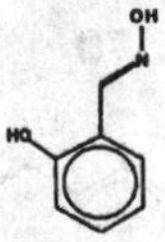

580

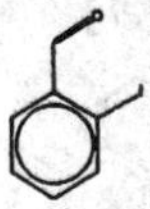

581

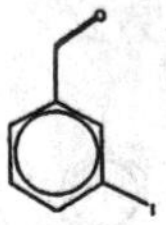

582

583

584

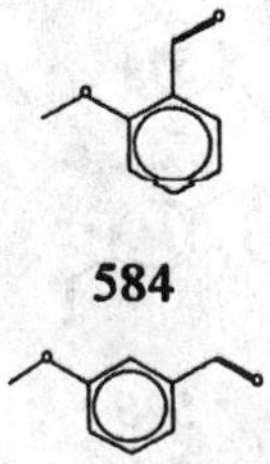

585

586

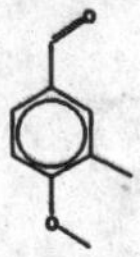

587

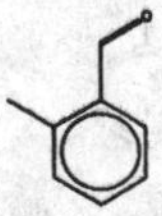

588

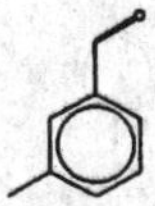

589

590

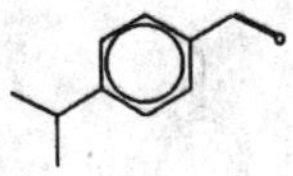

591

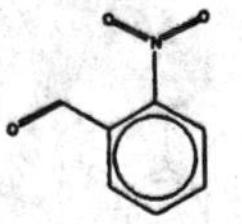

592

593

594

595

596

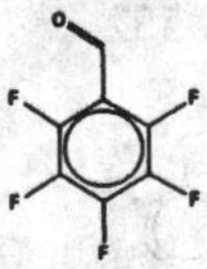

597

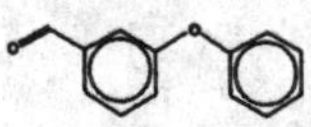

598

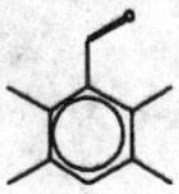

599

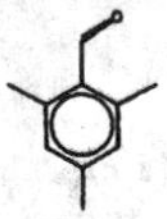

600

601

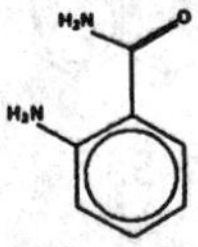

603

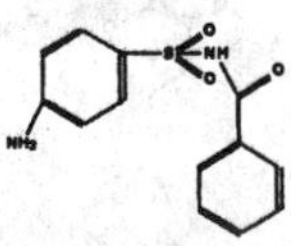

604

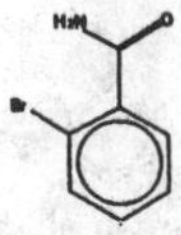

605

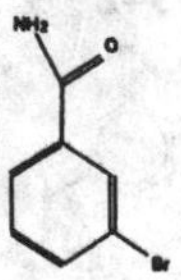

606

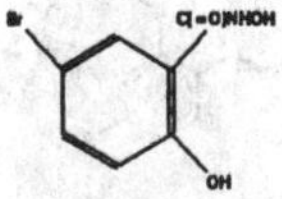

607

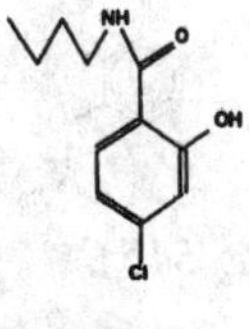

608

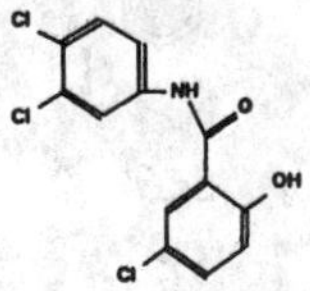

609

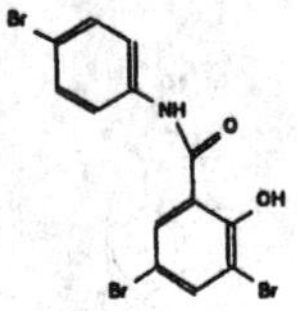

610

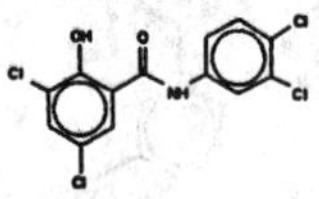

611

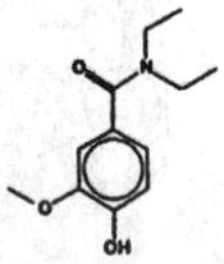

612

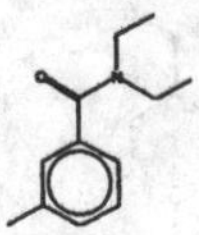

613

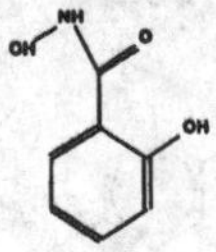

614

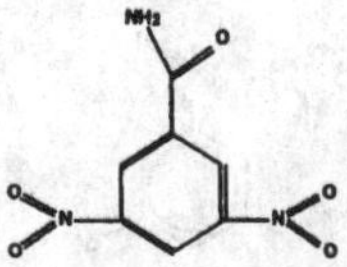

615

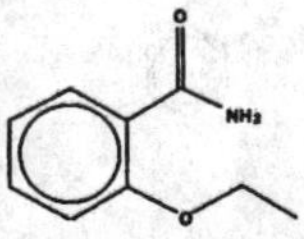

616

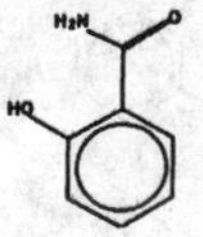

617

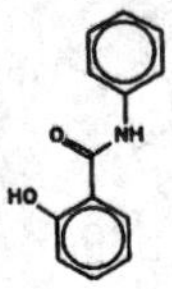

618

619

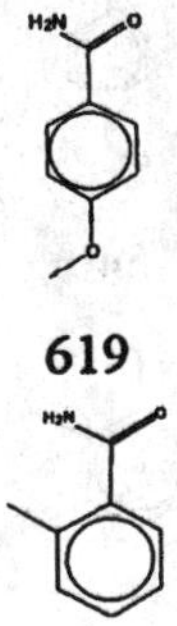

620

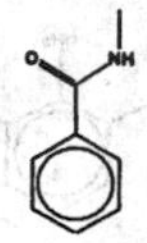

621

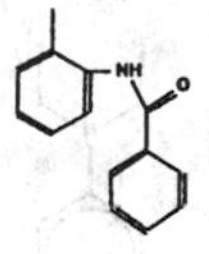

622

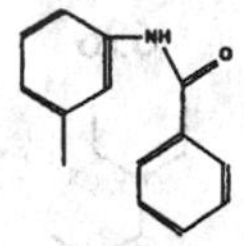

623

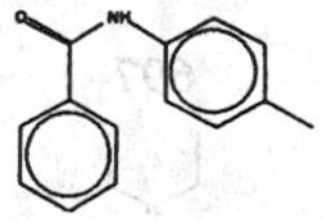

624

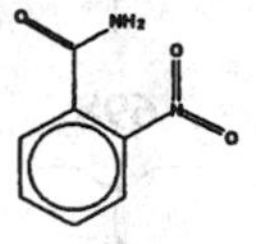

625

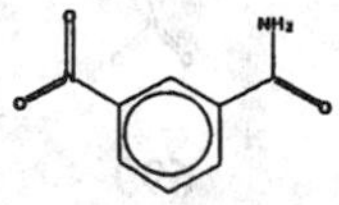

626

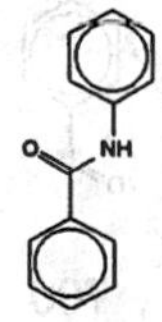

627

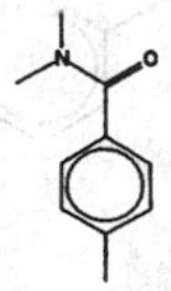

628

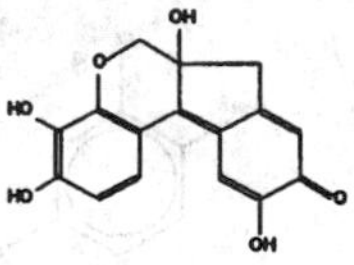

629

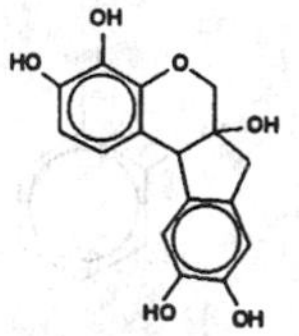

630

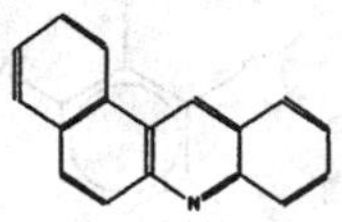

631

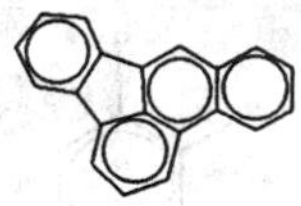

632

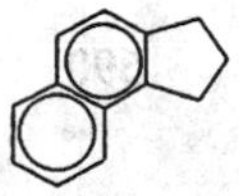

633

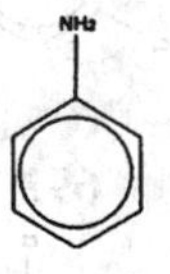

635

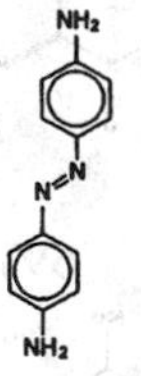

636

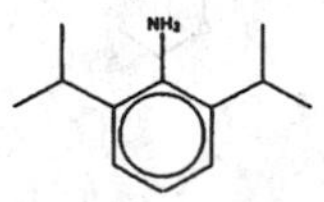

637

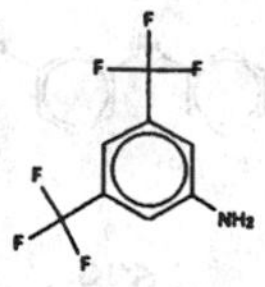

638

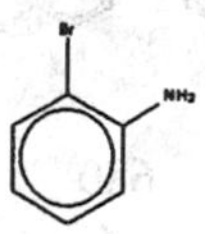

639

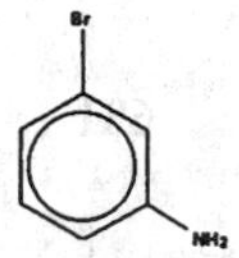

640

641

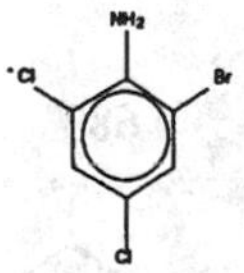

642

643

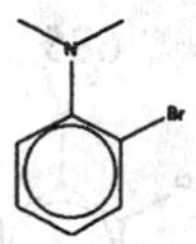

644

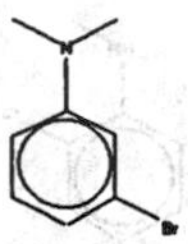

645

646

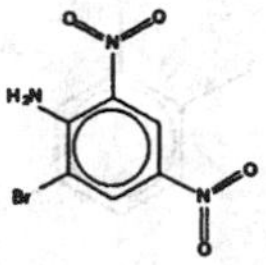

647

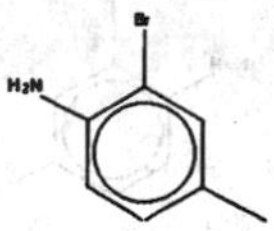

648

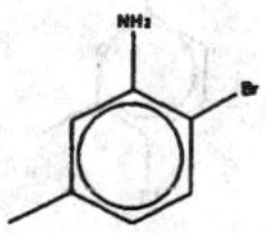

649

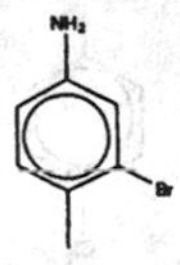

650

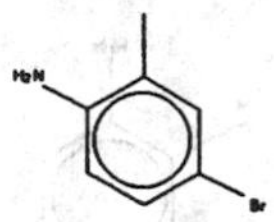

651

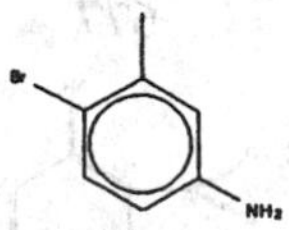

652

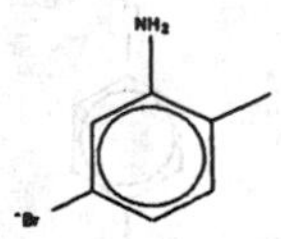

653

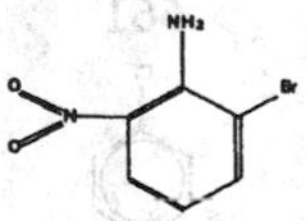

654

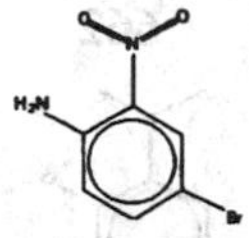

655

656

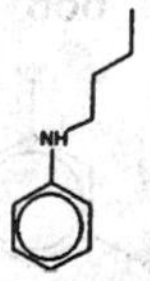

657

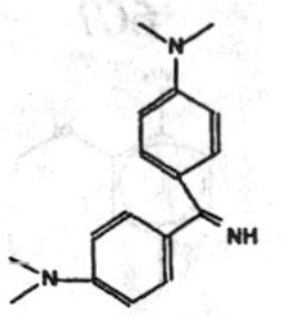

658

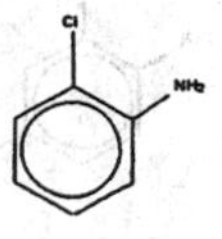

659

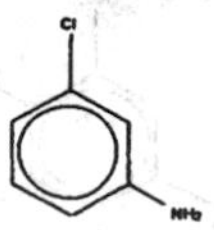

In numeric order as they occur in the Table of Physical Constants of Organic Compounds

660

661

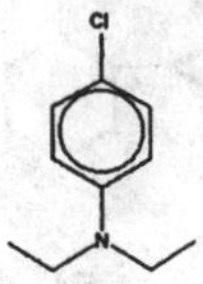

662

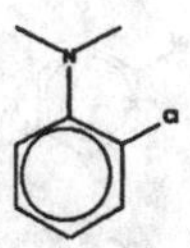

663

666

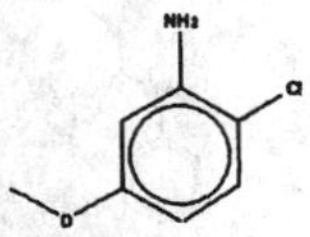

667

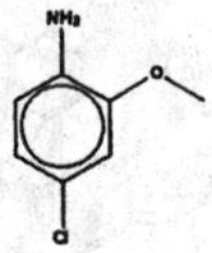

668

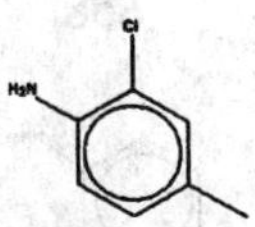

669

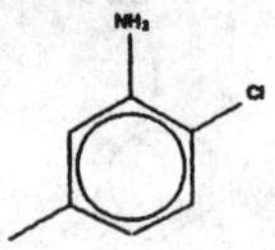

670

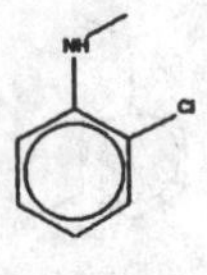

671

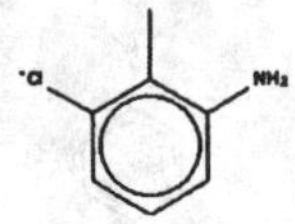

672

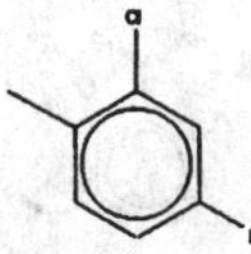

673

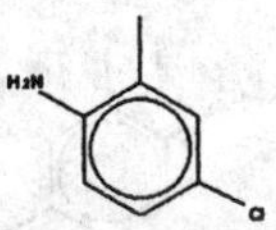

674

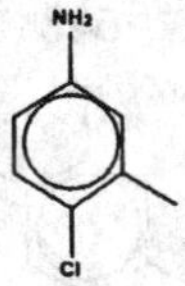

675

676

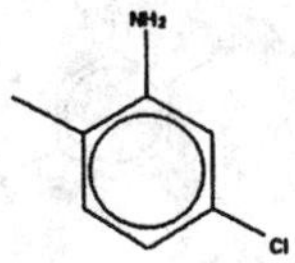

677

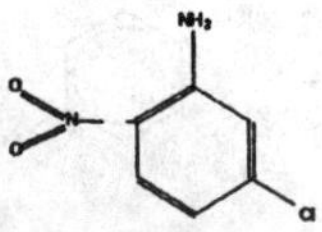

678

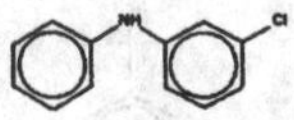

679

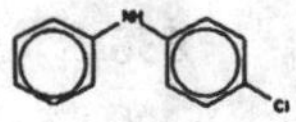

680

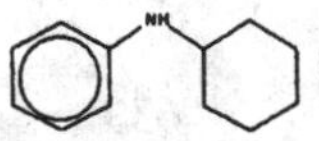

681

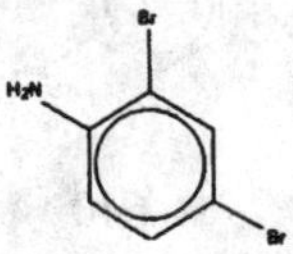

682

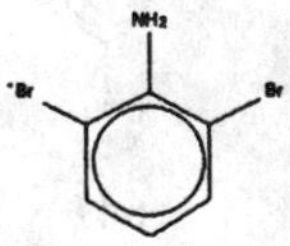

683

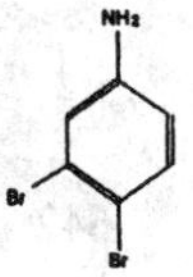

684

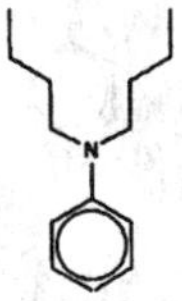

685

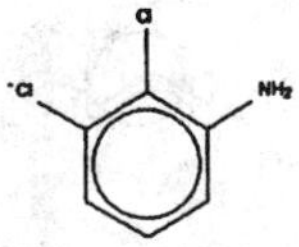

686

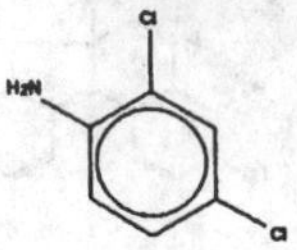

687

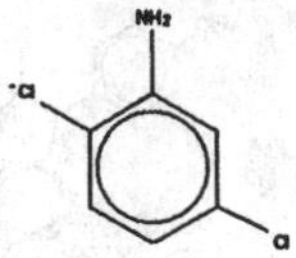

688

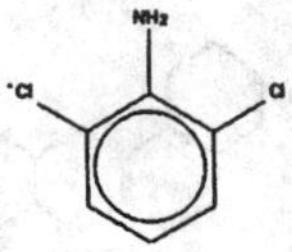

689

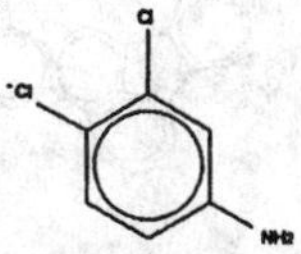

690

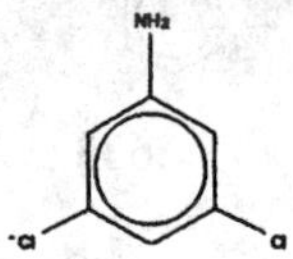

692

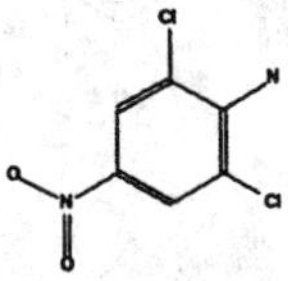

693

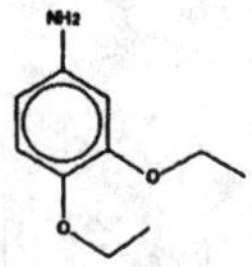

694

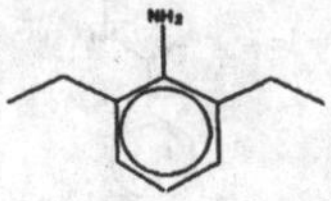

695

696

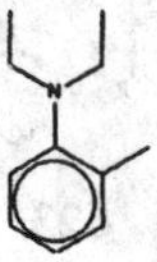

697

698

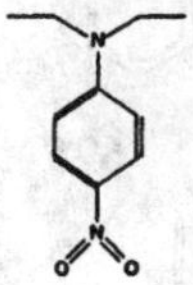

699

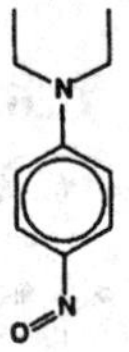

700

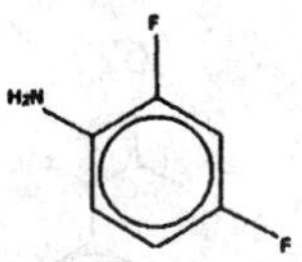

In numeric order as they occur in the Table of Physical Constants of Organic Compounds

701

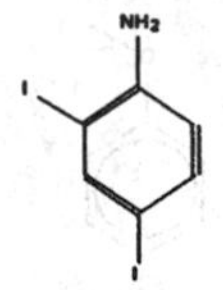

702

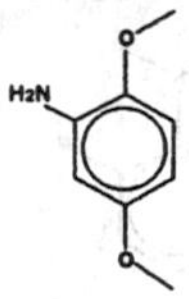

703

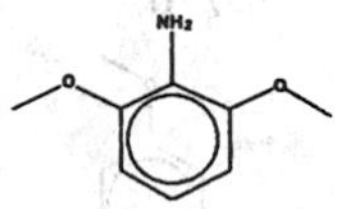

704

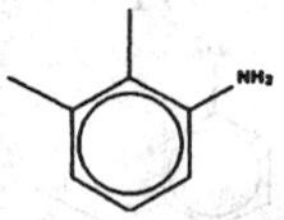

705

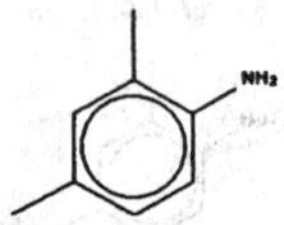

706

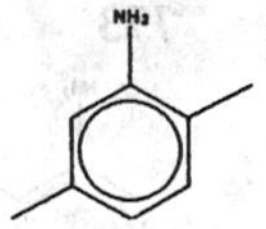

707

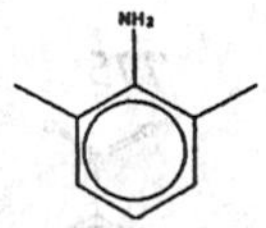

708

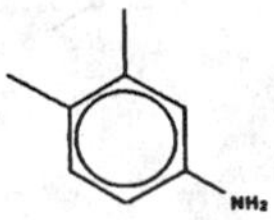

709

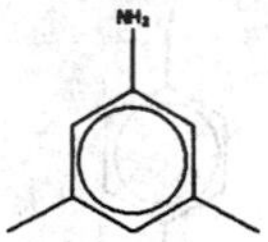

710

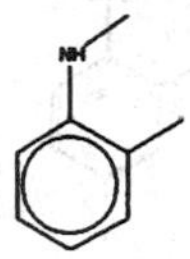

711

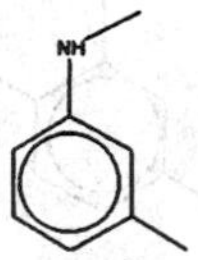

712

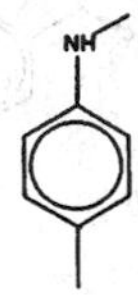

713

714

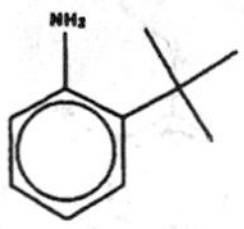

715

716

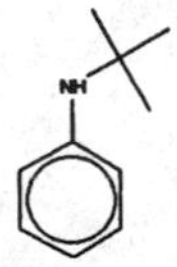

718

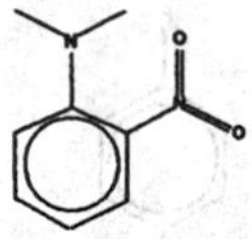

719

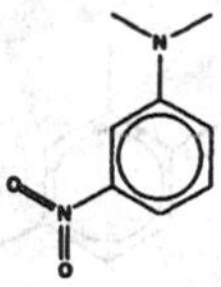

720

721

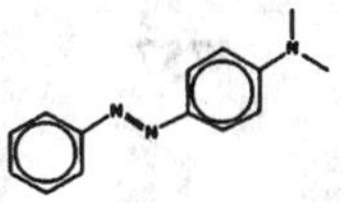

722

723

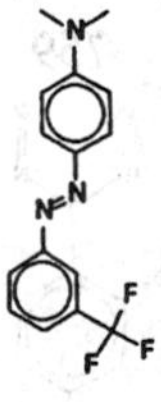

724

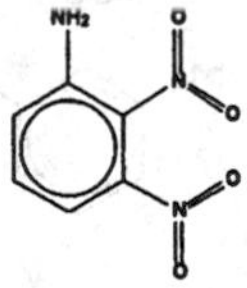

725

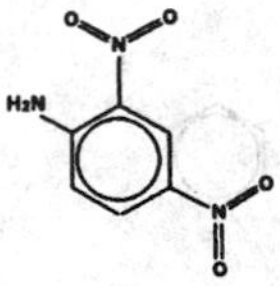

726

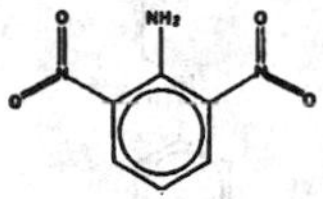

727

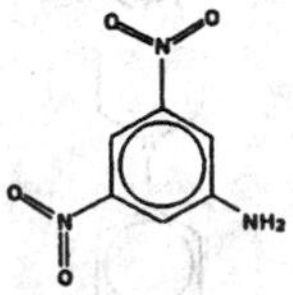

728

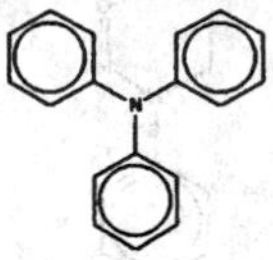

729

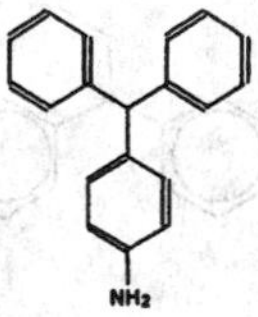

730

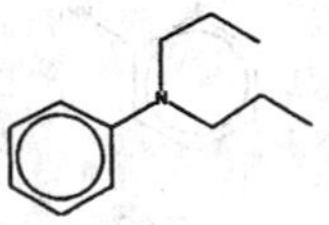

731

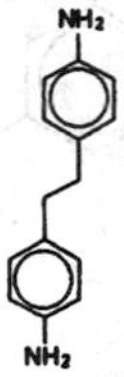

732

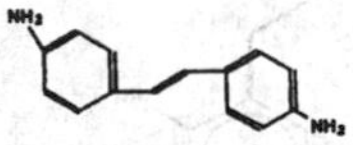

733

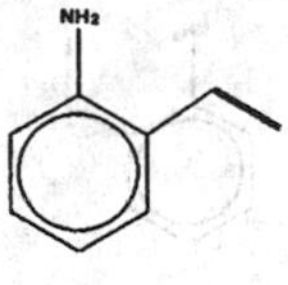

734

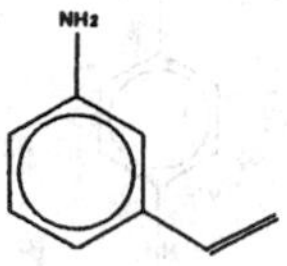

735

736

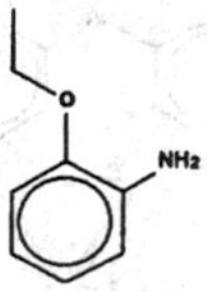

737

738

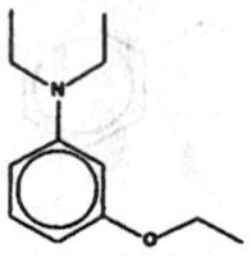

740

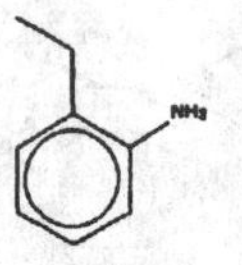

741

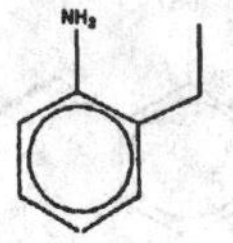

742

743

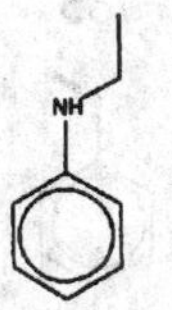

744

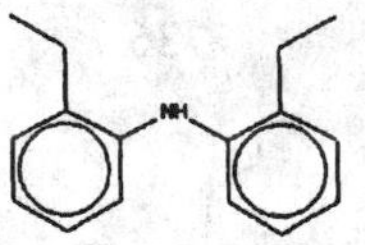

745

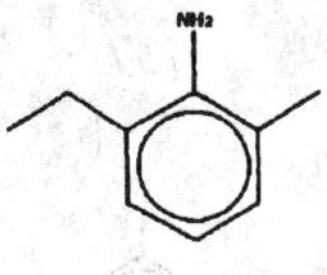

746

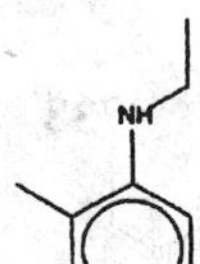

747

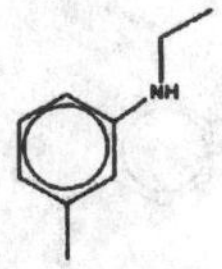

748

749

750

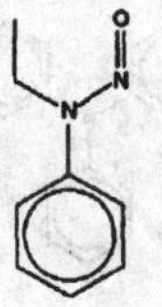

751

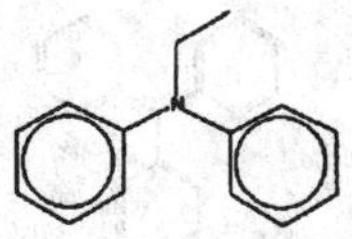

752

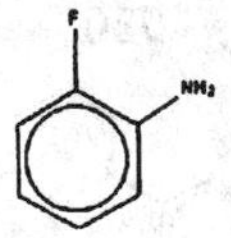

753

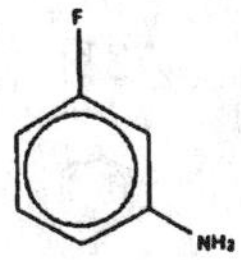

754

755

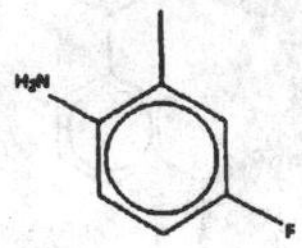

756

757

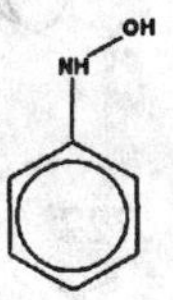

758

759

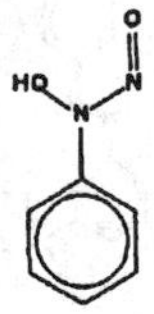

760

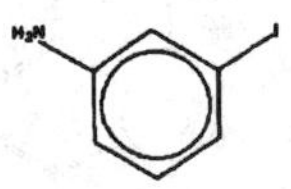

761

762

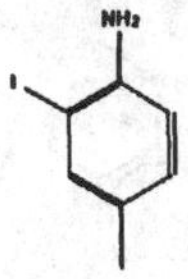

763

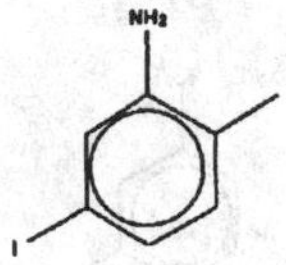

764

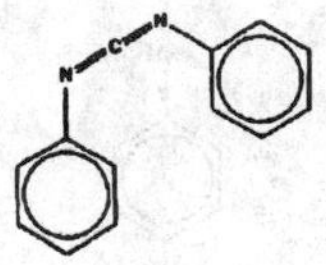

765

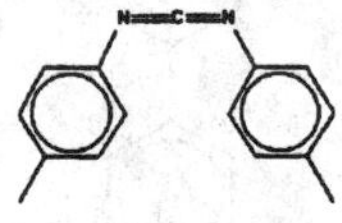

766

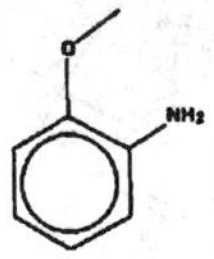

767

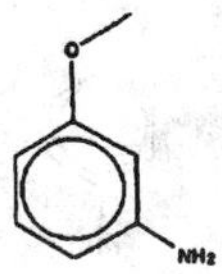

768

769

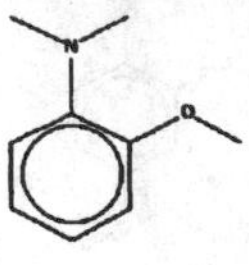

770

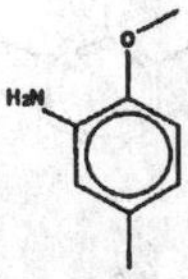

771

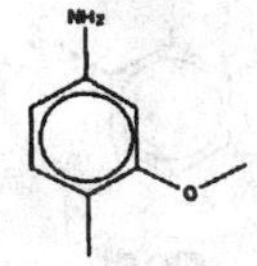

772

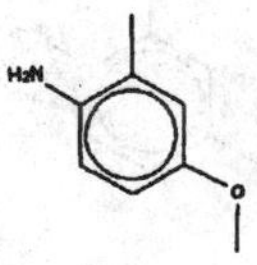

773

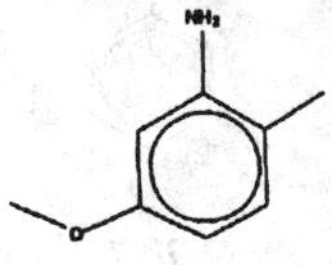

775

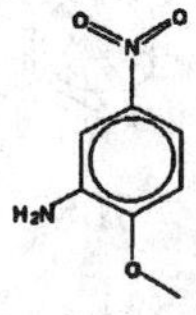

STRUCTURAL FORMULAS OF ORGANIC COMPOUNDS (continued)

In numeric order as they occur in the Table of Physical Constants of Organic Compounds

776

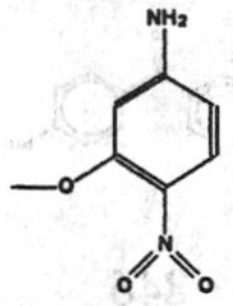

777

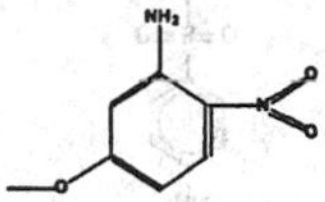

778

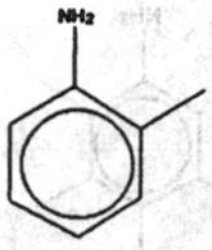

779

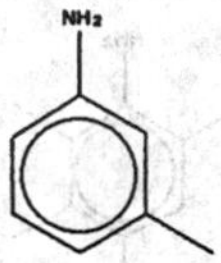

780

781

782

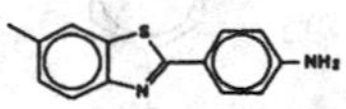

783

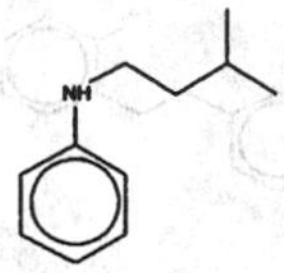

784

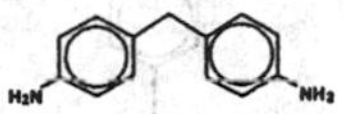

785

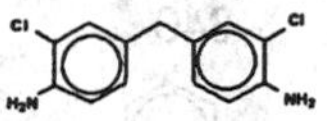

786

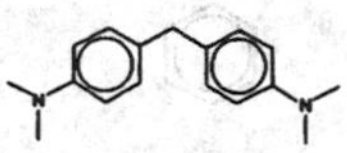

787

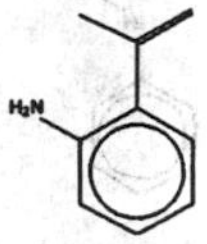

788

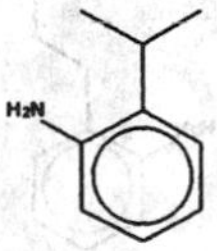

789

790

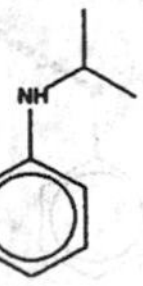

791

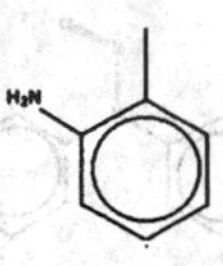

793

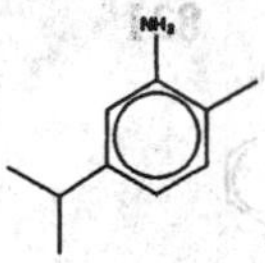

794

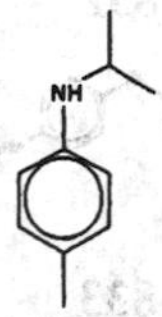

795

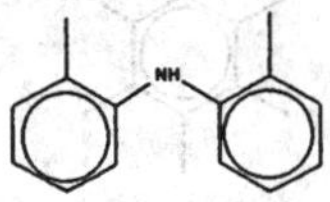

796

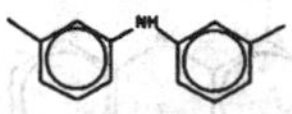

797

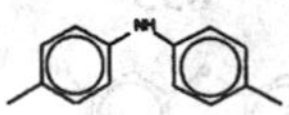

798

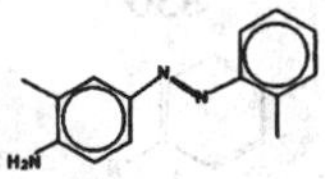

799

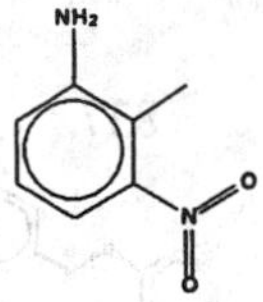

800

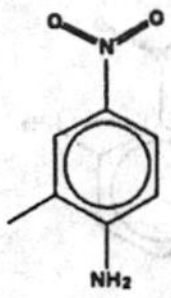

801

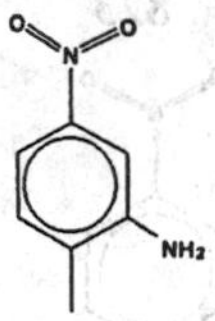

802

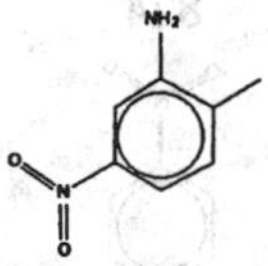

803

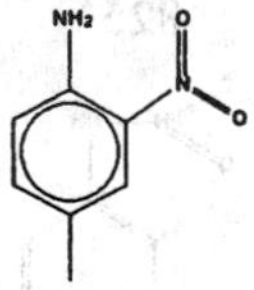

804

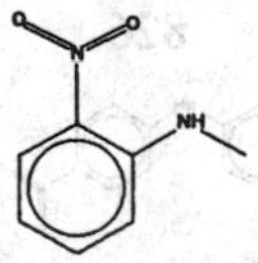

805

806

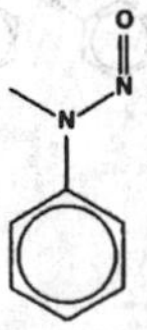

807

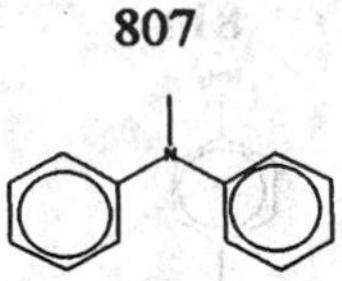

808

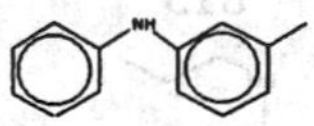

809

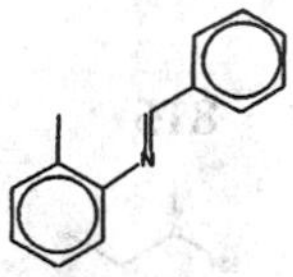

810

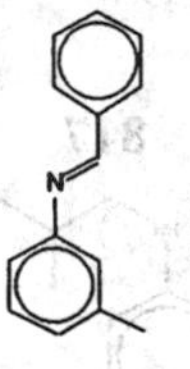

811

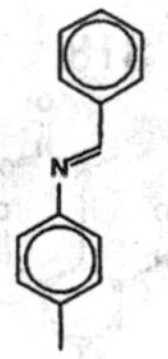

812

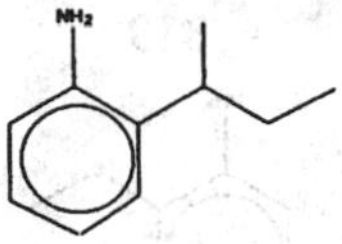

813

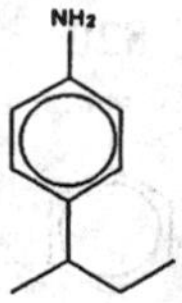

STRUCTURAL FORMULAS OF ORGANIC COMPOUNDS (continued)

In numeric order as they occur in the Table of Physical Constants of Organic Compounds

814

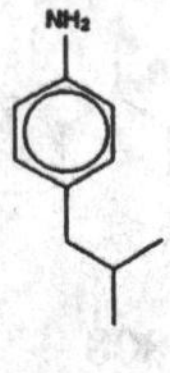

815

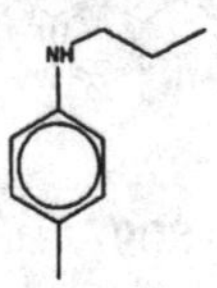

816

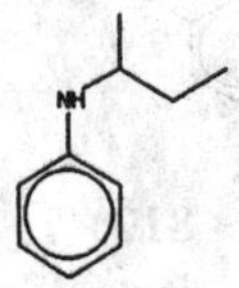

817

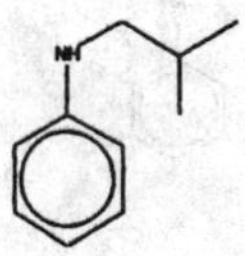

818

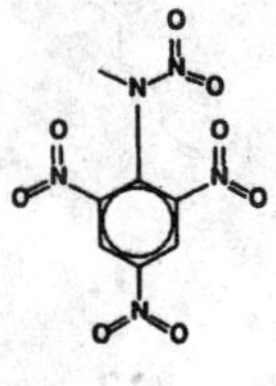

819

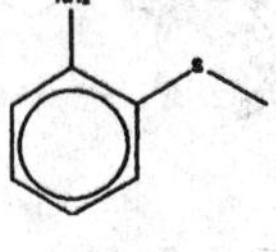

820

822

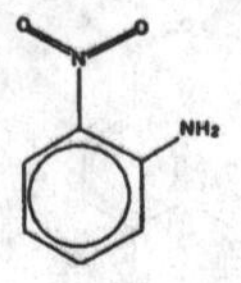

823

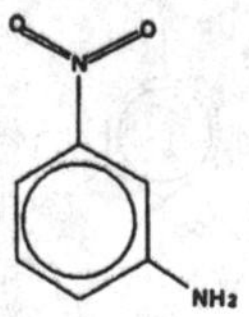

824

825

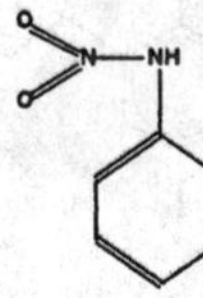

826

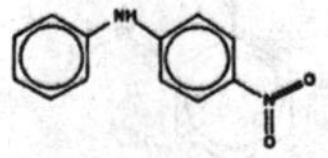

828

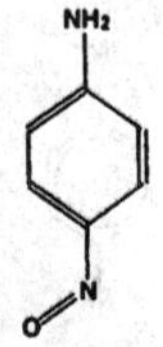

829

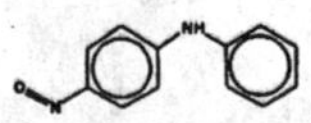

830

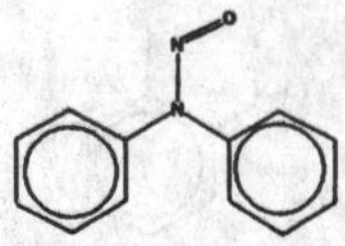

831

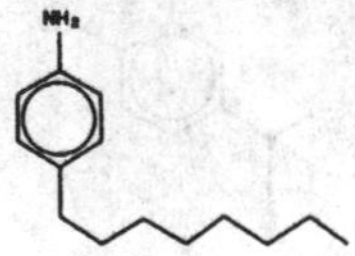

832

833

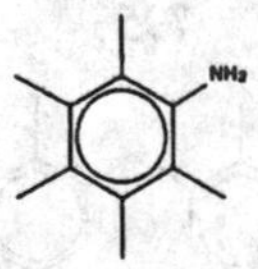

834

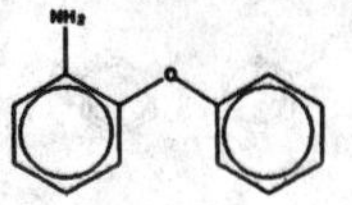

835

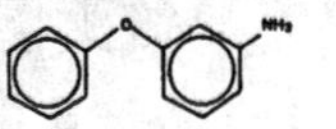

836

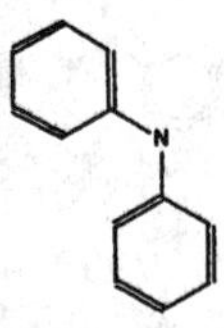

837

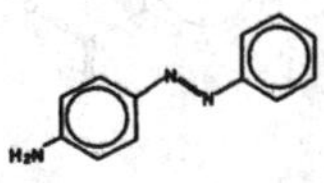

838

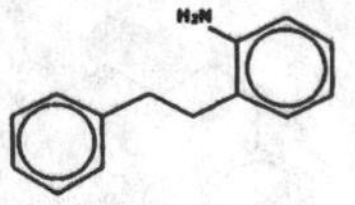

839

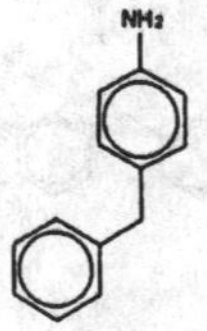

840

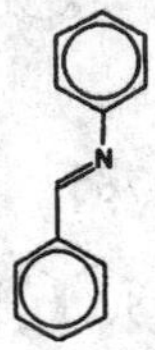

841

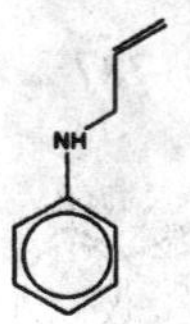

842

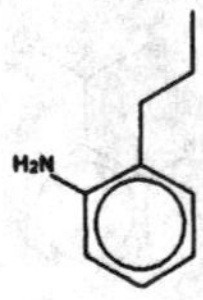

843

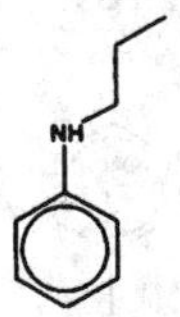

844

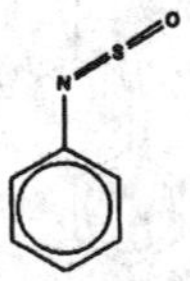

845

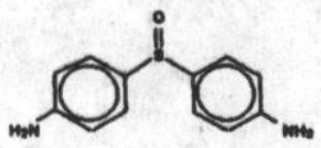

846

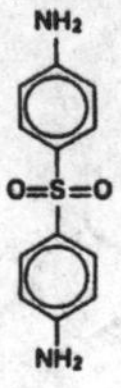

847

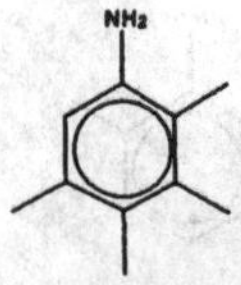

848

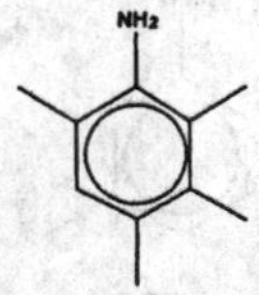

849

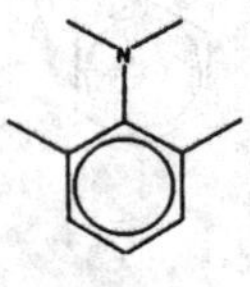

850

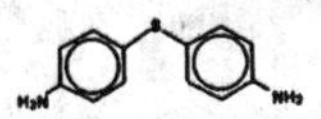

851

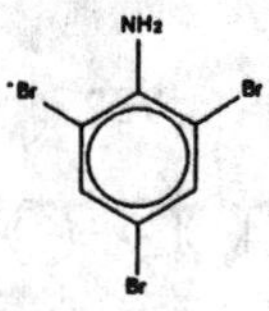

In numeric order as they occur in the Table of Physical Constants of Organic Compounds

852

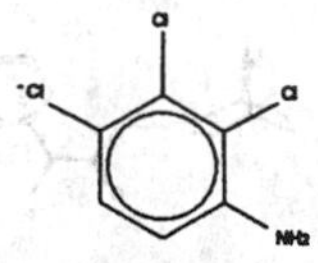

853

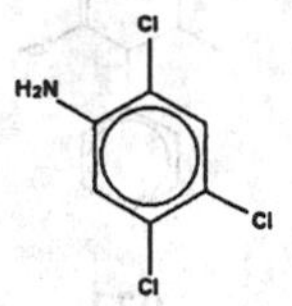

854

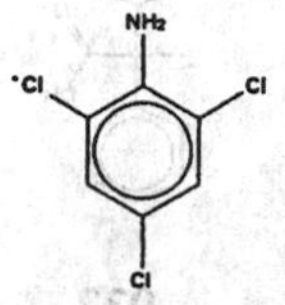

855

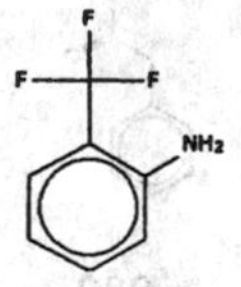

856

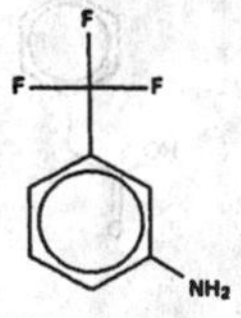

857

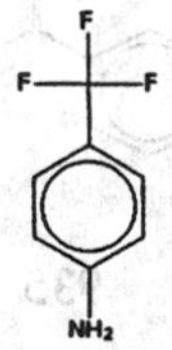

858

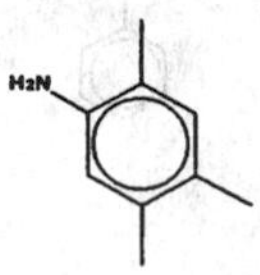

859

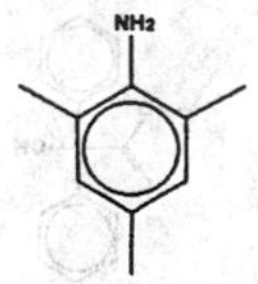

860

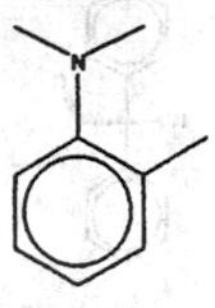

861

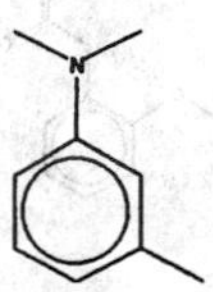

862

863

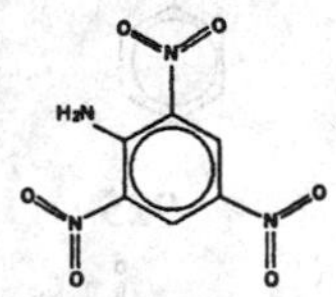

864

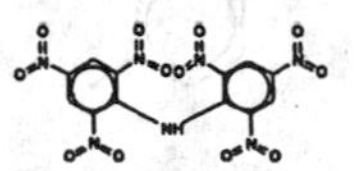

867

868

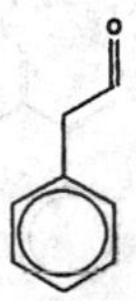

869

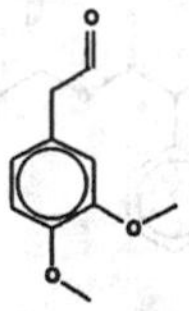

870

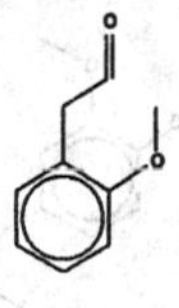

871

872

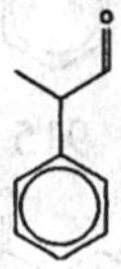

873

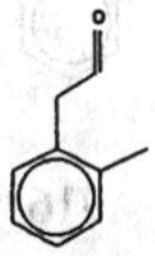

874

875

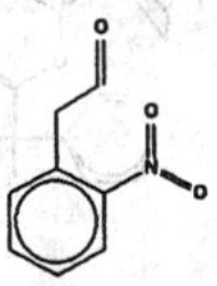

877

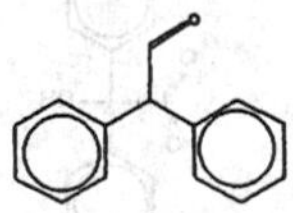

878

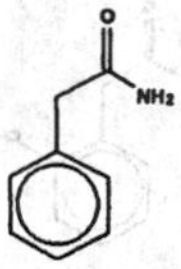

879

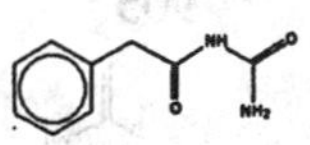

880

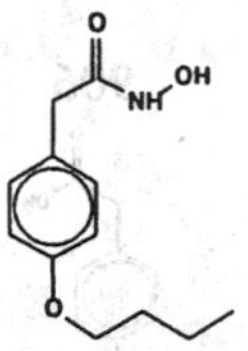

881

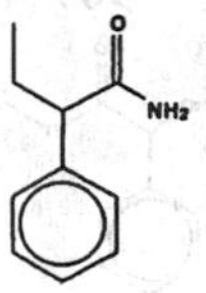

882

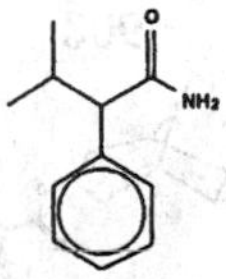

883

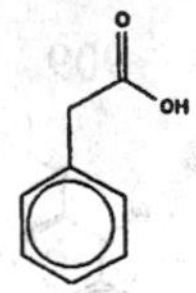

884

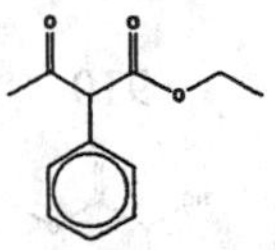

886

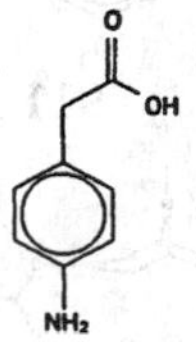

887

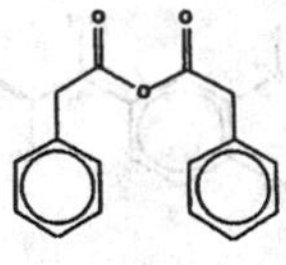

888

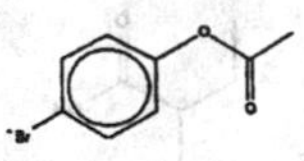

890

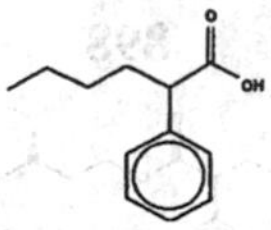

891

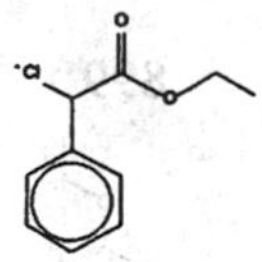

892

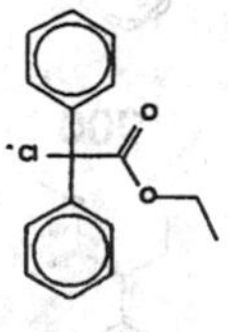

893

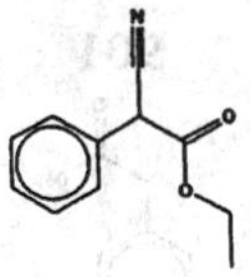

894

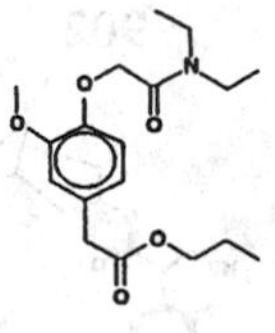

895

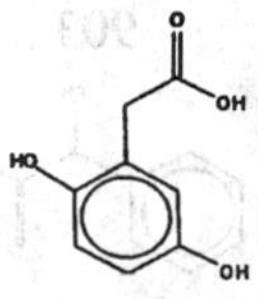

STRUCTURAL FORMULAS OF ORGANIC COMPOUNDS (continued)

In numeric order as they occur in the Table of Physical Constants of Organic Compounds

896

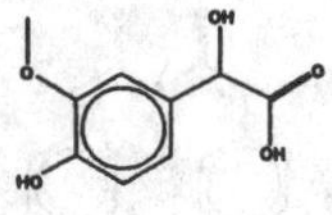

897

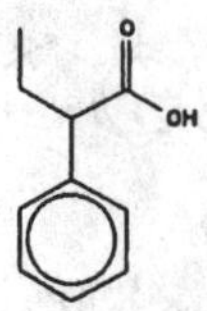

898

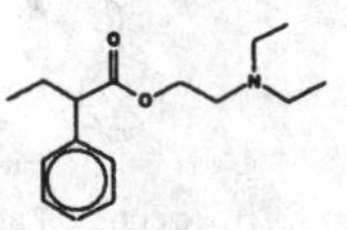

899

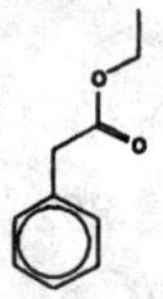

900

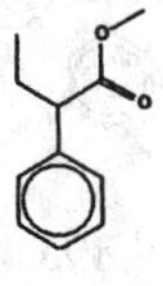

901

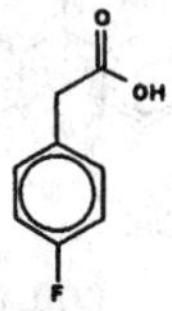

902

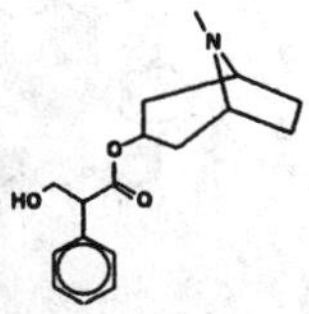

903

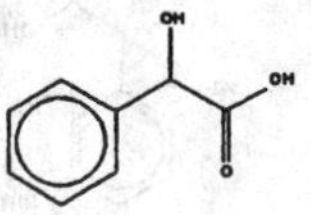

904

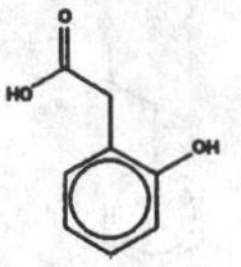

905

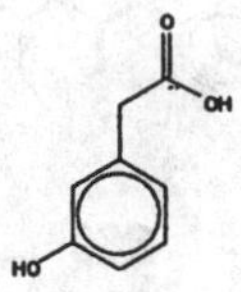

906

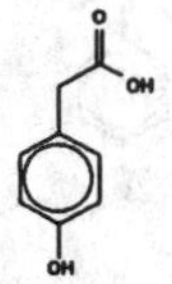

907

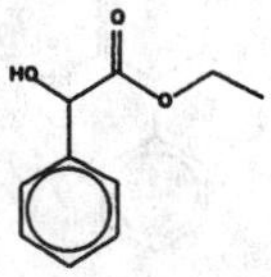

908

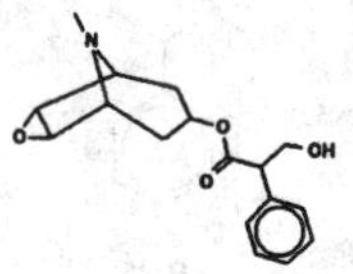

909

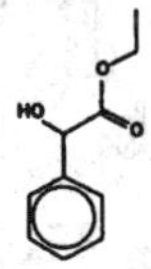

910

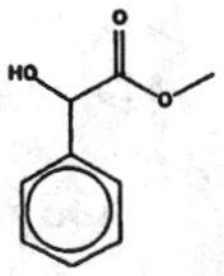

911

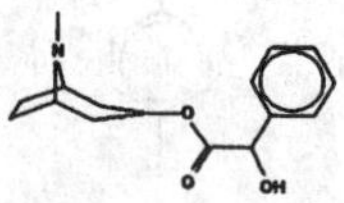

912

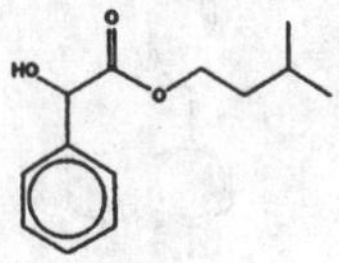

913

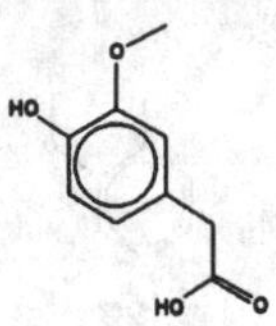

914

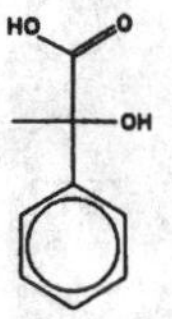

915

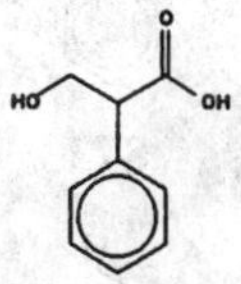

916

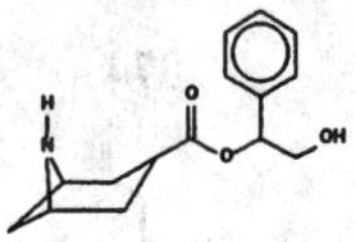

917

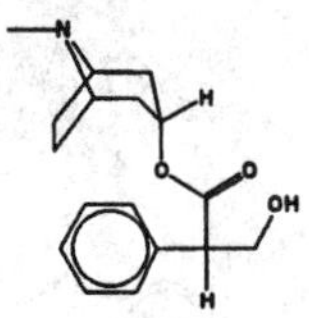

918

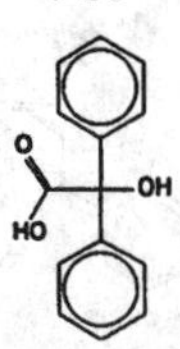

919

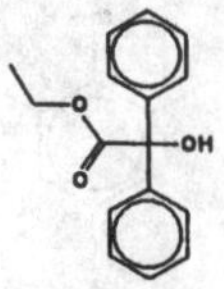

920

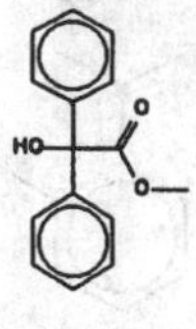

921

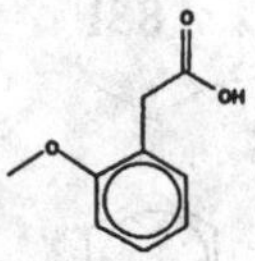

923

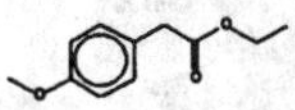

924

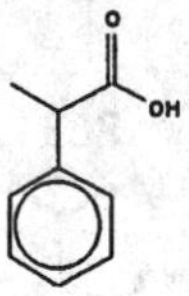

925

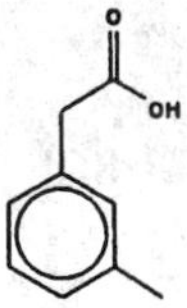

926

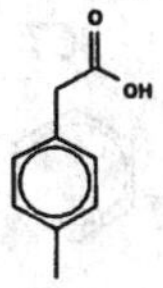

927

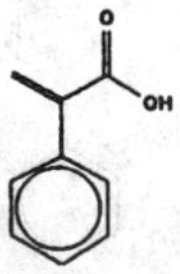

928

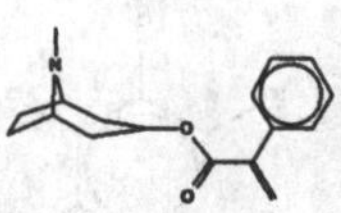

930

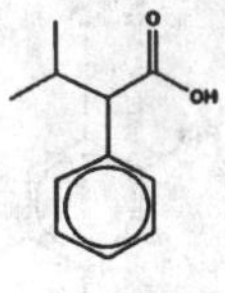

931

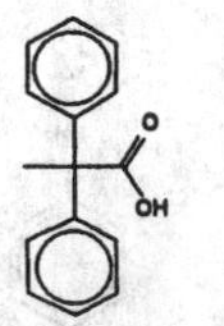

932

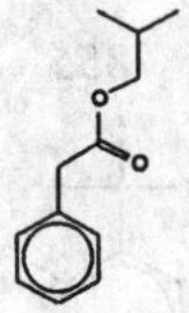

933

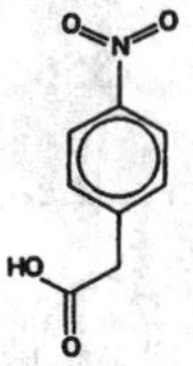

934

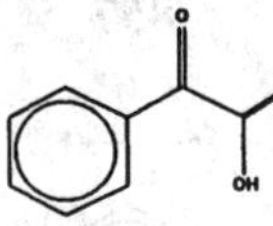

935

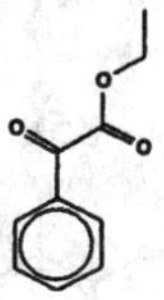

STRUCTURAL FORMULAS OF ORGANIC COMPOUNDS (continued)

In numeric order as they occur in the Table of Physical Constants of Organic Compounds

937

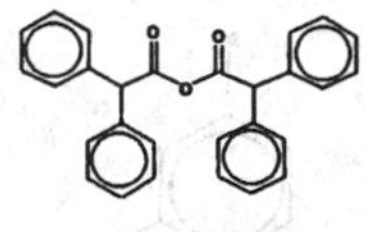

938

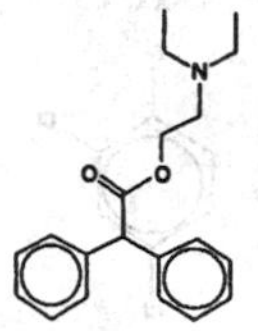

939

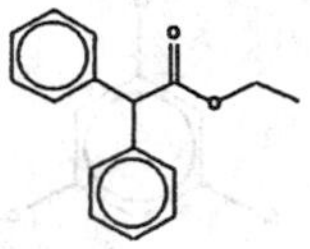

940

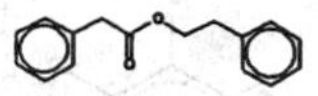

941

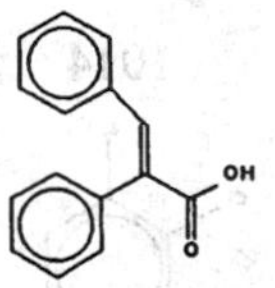

942

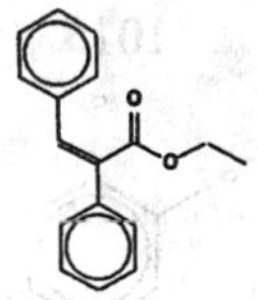

944

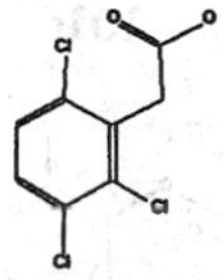

945

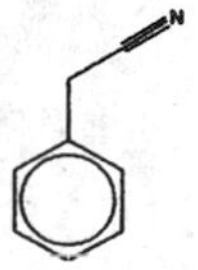

946

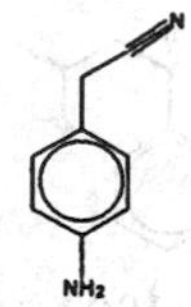

947

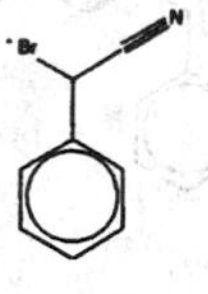

948

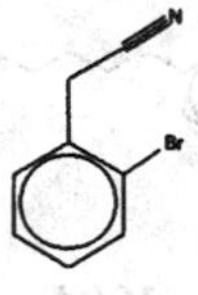

949

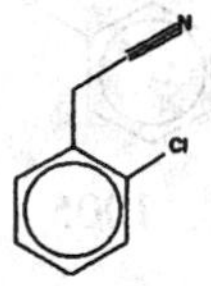

951

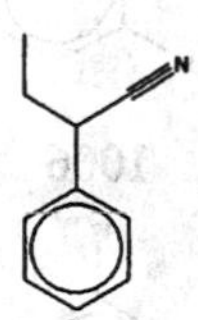

952

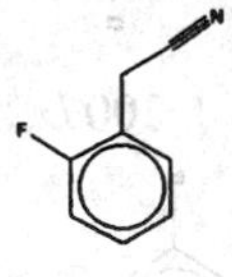

953

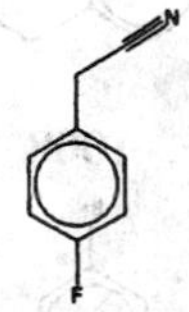

954

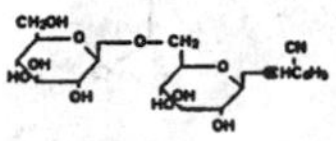

956

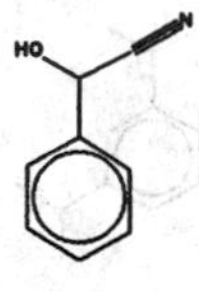

957

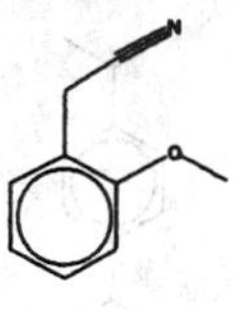

959

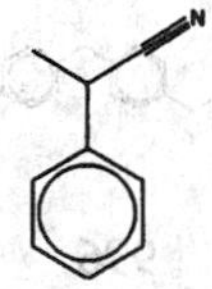

960

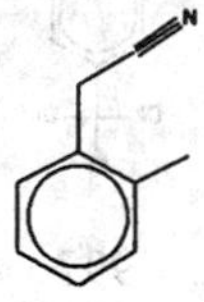

961

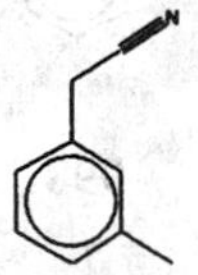

962

963

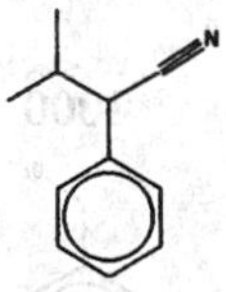

964

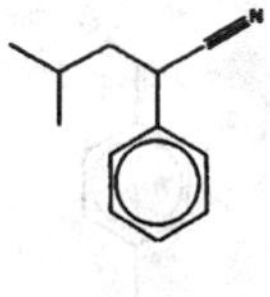

965

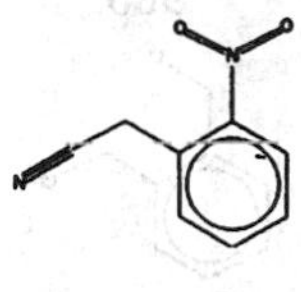

966

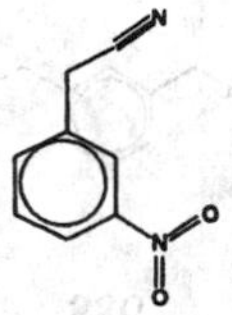

967

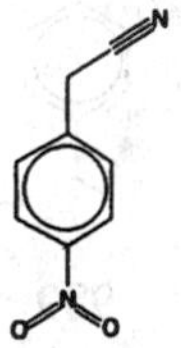

968

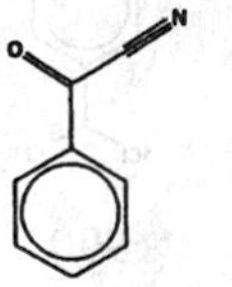

969

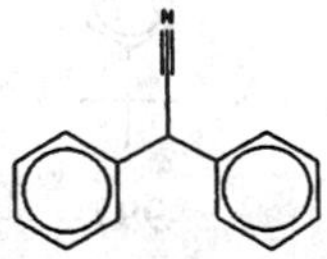

970

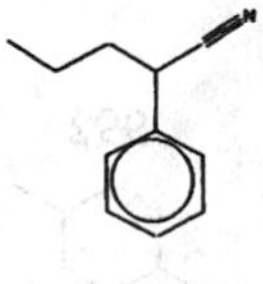

971

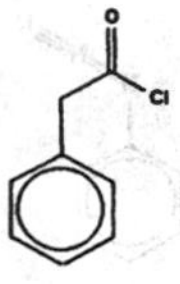

972

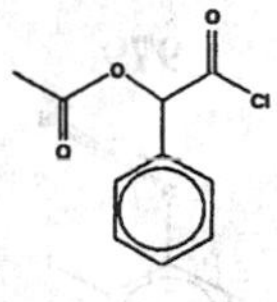

973

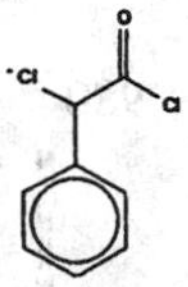

974

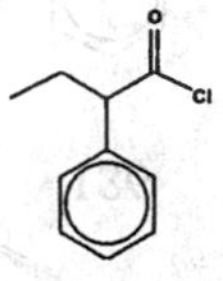

975

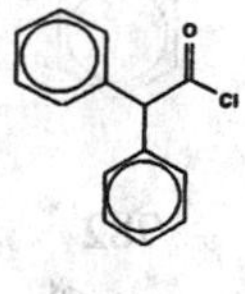

976

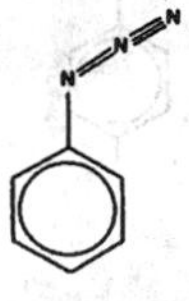

977

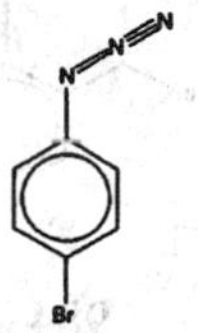

STRUCTURAL FORMULAS OF ORGANIC COMPOUNDS (continued)

In numeric order as they occur in the Table of Physical Constants of Organic Compounds

978

979

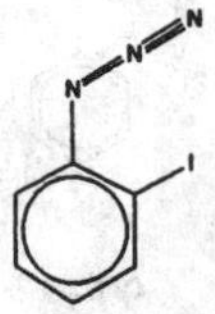

980

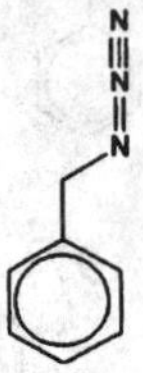

981

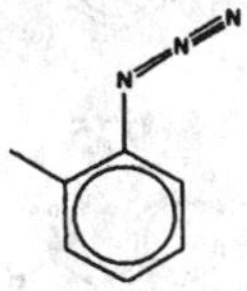

982

983

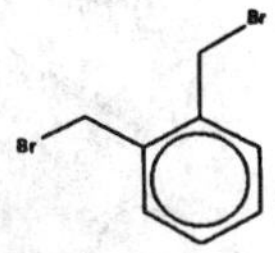

984

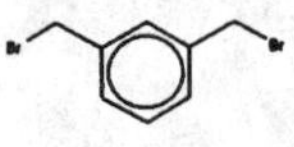

985

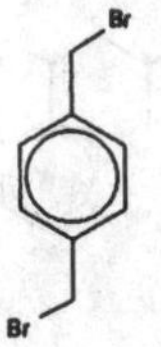

986

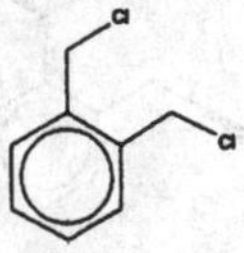

987

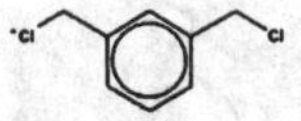

988

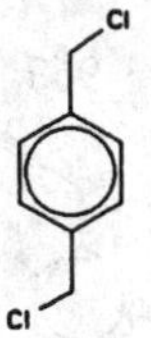

989

990

991

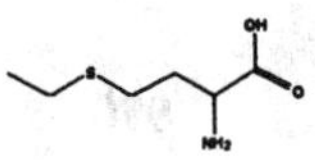

992

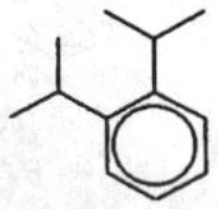

993

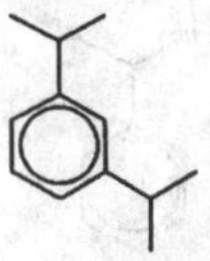

994

995

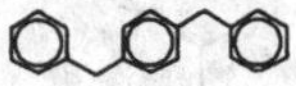

996

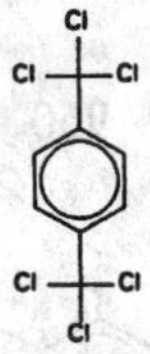

997

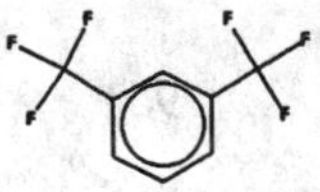

998

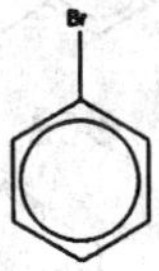

999

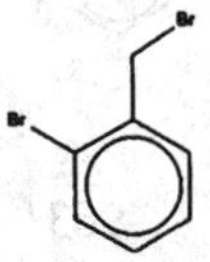

1000

1001

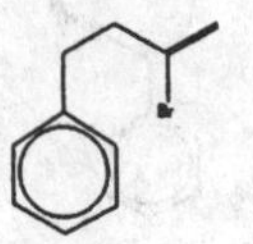

1002

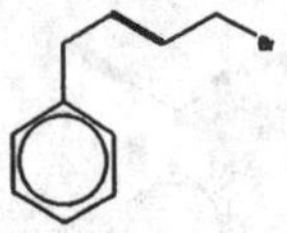

1003

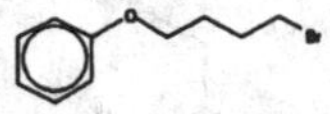

1004

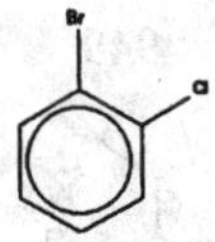

1005

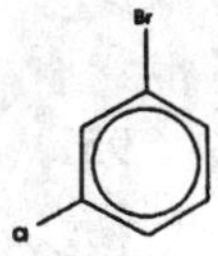

1006

1007

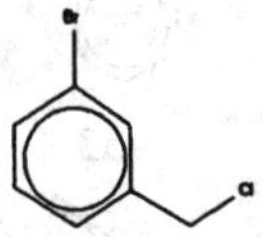

1008

1010

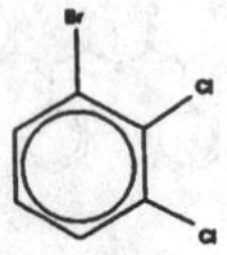

1011

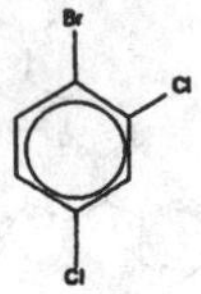

1012

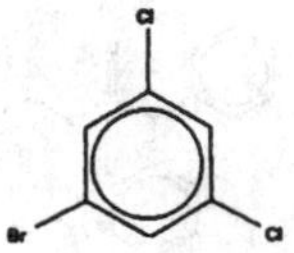

1013

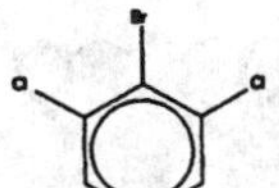

1014

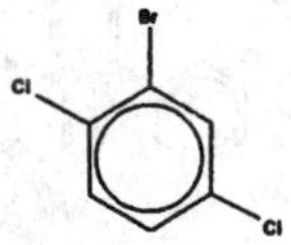

1015

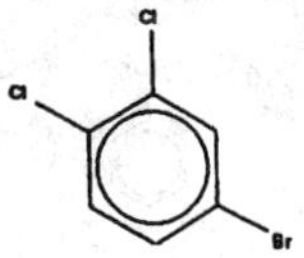

1016

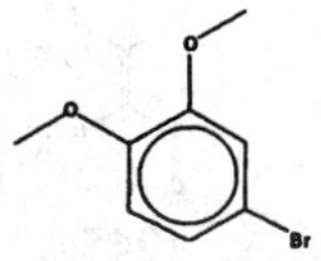

STRUCTURAL FORMULAS OF ORGANIC COMPOUNDS (continued)

In numeric order as they occur in the Table of Physical Constants of Organic Compounds

1017

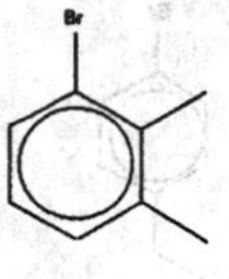

1018

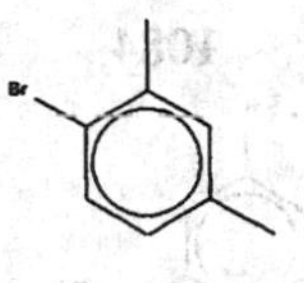

1019

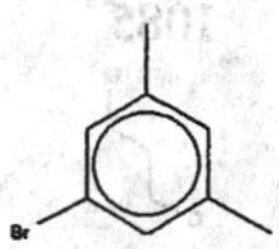

1020

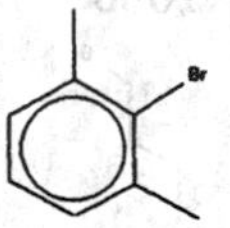

1021

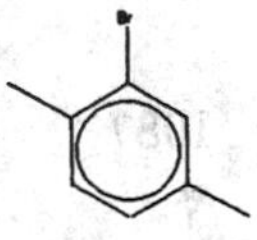

1022

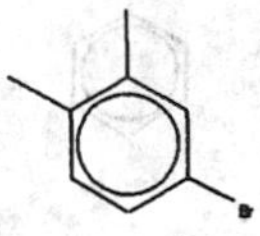

1023

1024

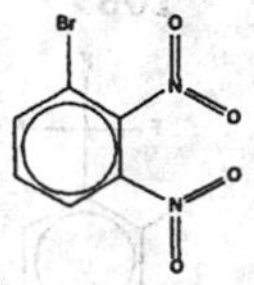

1025

1026

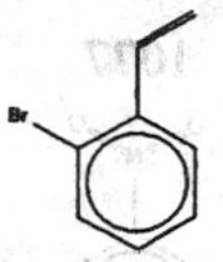

1027

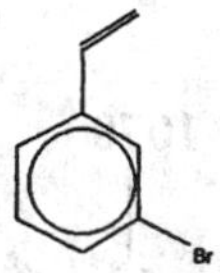

1028

1029

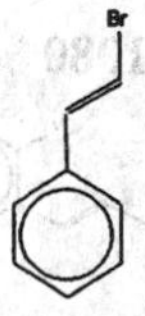

1030

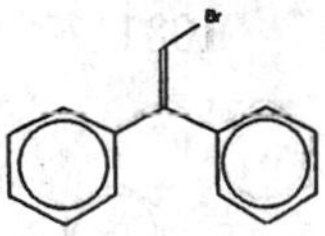

1031

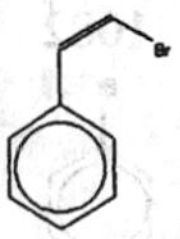

1032

1033

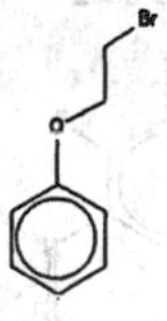

1034

1035

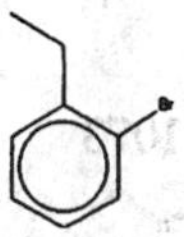

1036

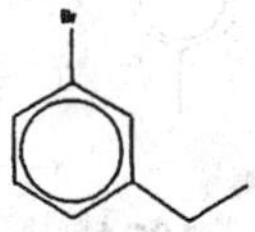

1037

1038

1039

1040

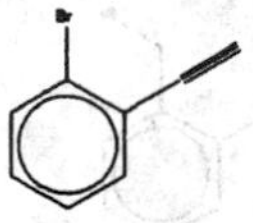

1041

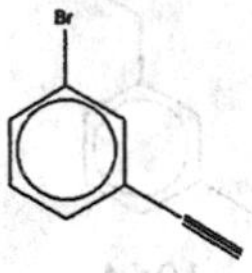

1042

1043

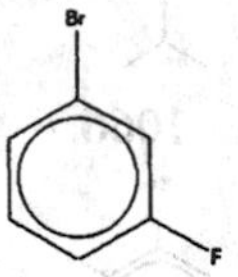

1044

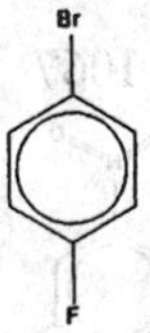

1045

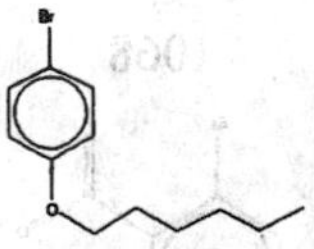

1046

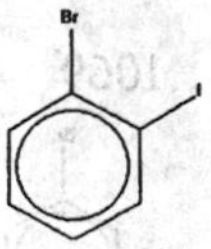

1047

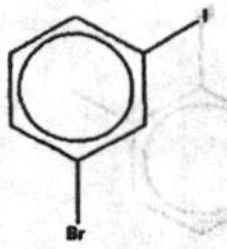

1048

1049

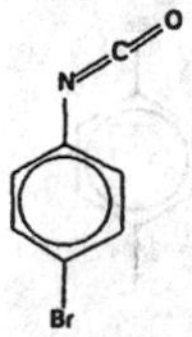

1050

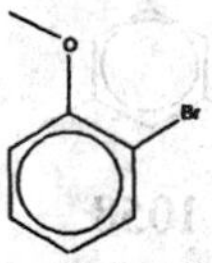

1051

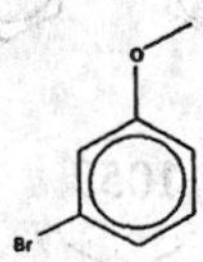

1052

1053

In numeric order as they occur in the Table of Physical Constants of Organic Compounds

1054

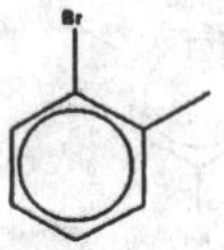

1055

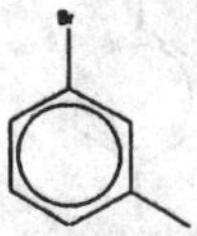

1056

1057

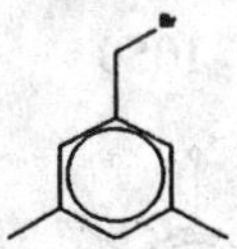

1058

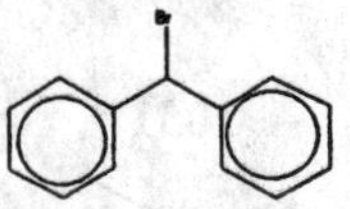

1059

1060

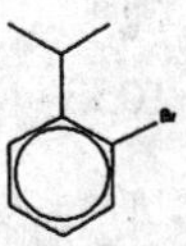

1061

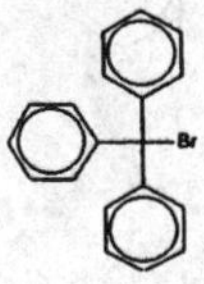

1062

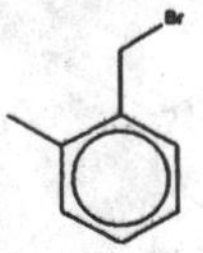

1063

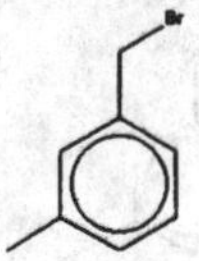

1064

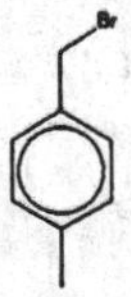

1065

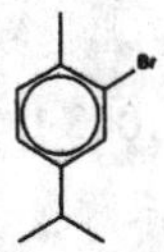

1066

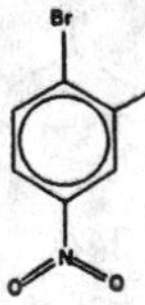

1067

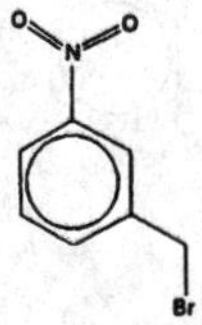

1068

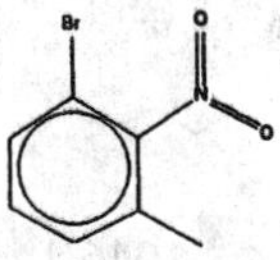

1069

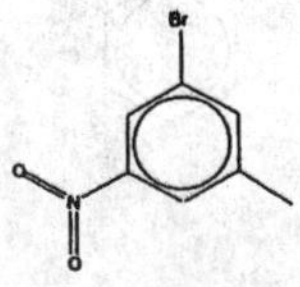

1070

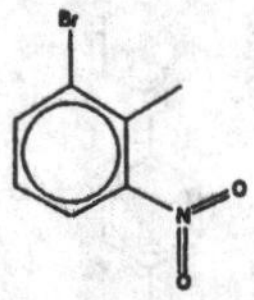

1071

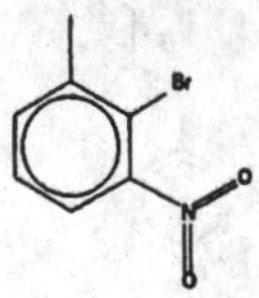

1072

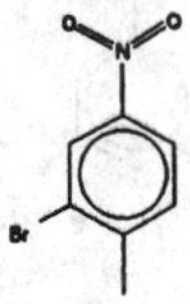

1073

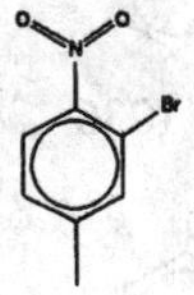

1074

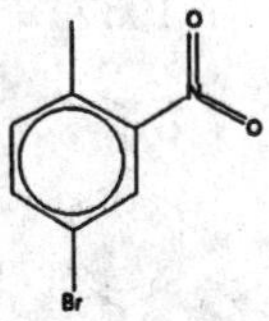

1075

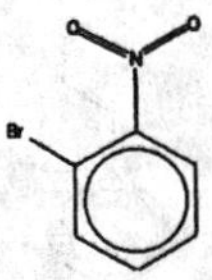

1076

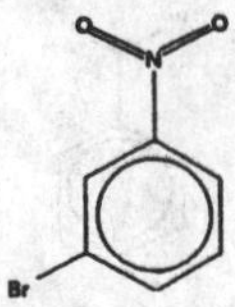

1077

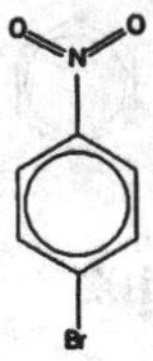

1078

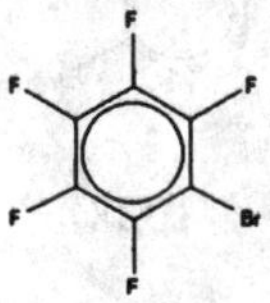

1079

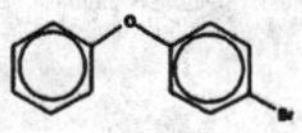

1080

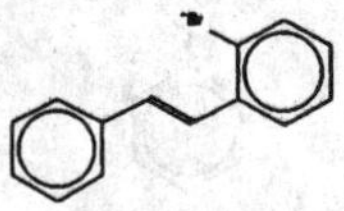

1081

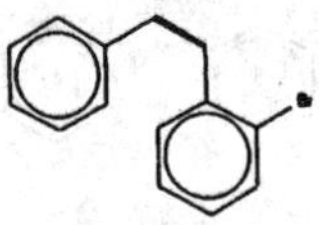

1082

1083

1084

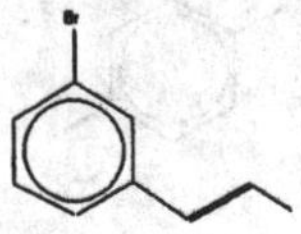

1085

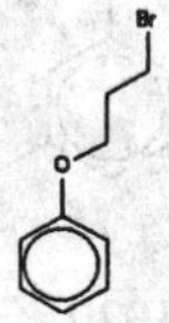

1086

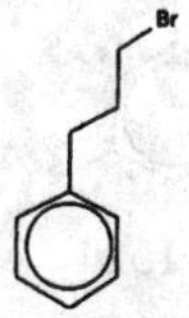

1087

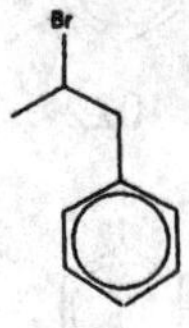

1088

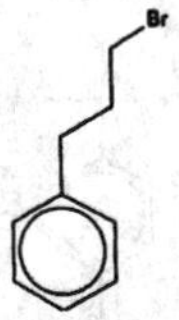

1089

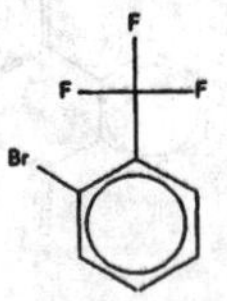

STRUCTURAL FORMULAS OF ORGANIC COMPOUNDS (continued)

In numeric order as they occur in the Table of Physical Constants of Organic Compounds

1090

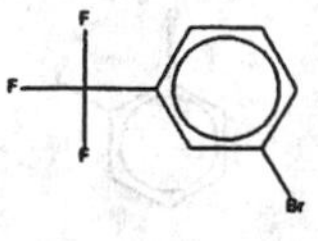

1091

1092

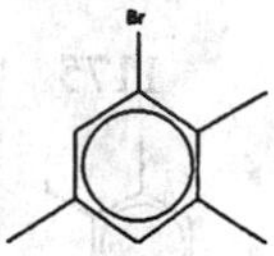

1093

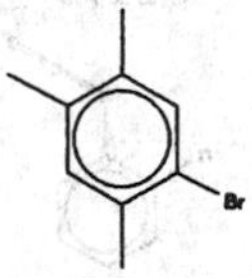

1094

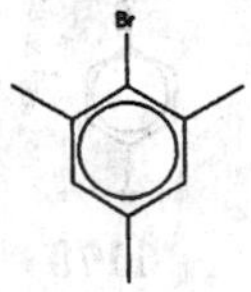

1095

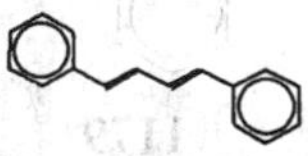

1096

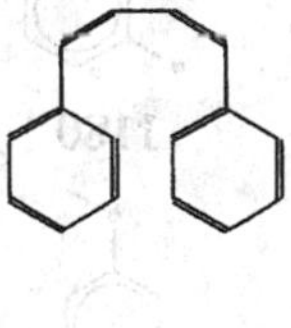

1097

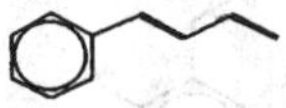

1098

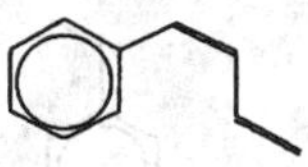

1099

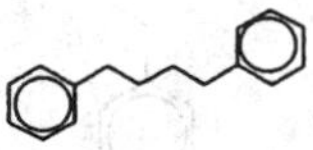

1100

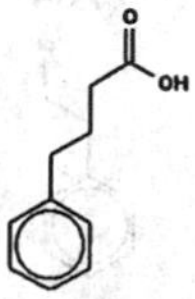

1101

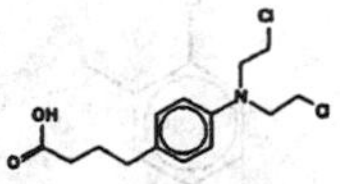

1102

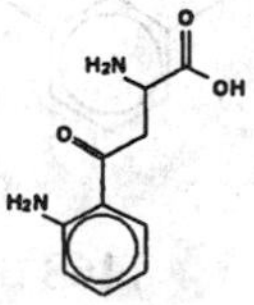

1103

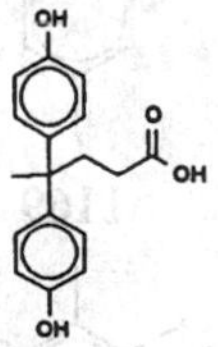

1104

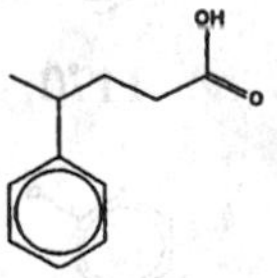

1105

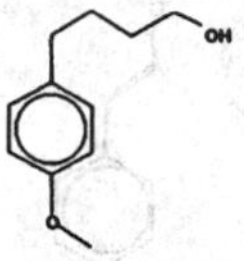

1106

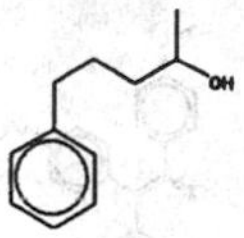

1108

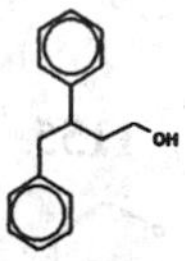

1109

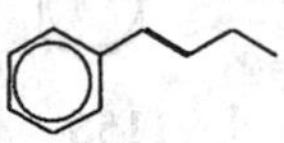

1110

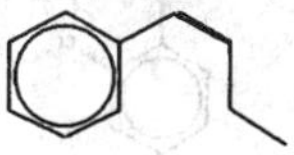

1111

1112

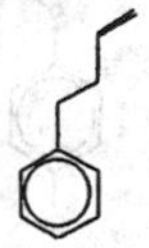

1113

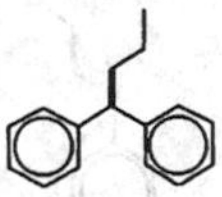

1114

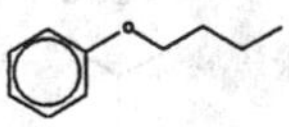

1115

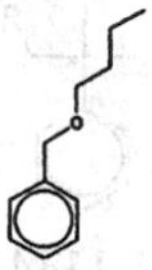

1116

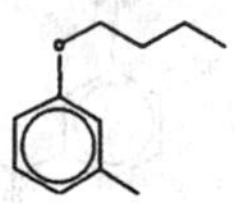

1117

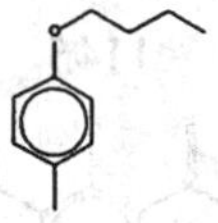

1118

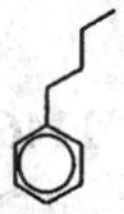

1119

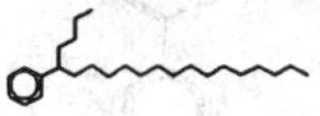

1120

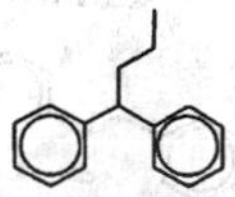

1121

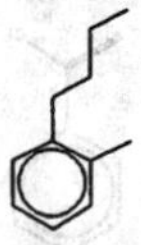

1122

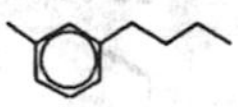

1123

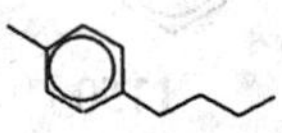

1124

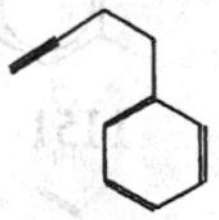

1125

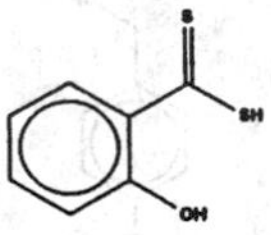

1126

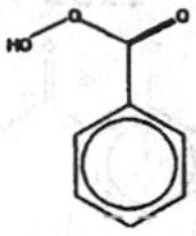

1127

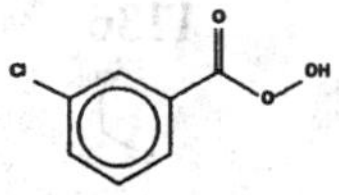

1128

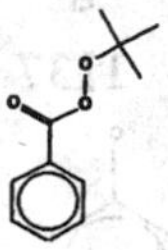

1129

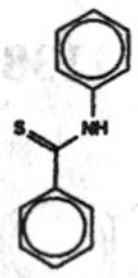

1130

1131

1132

1133

STRUCTURAL FORMULAS OF ORGANIC COMPOUNDS (continued)

In numeric order as they occur in the Table of Physical Constants of Organic Compounds

1134

1135

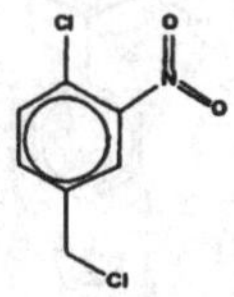

1136

1137

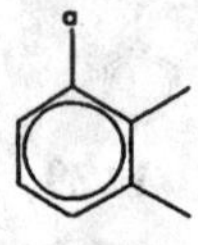

1138

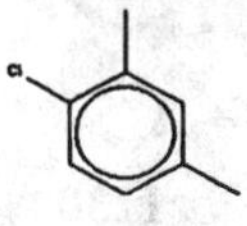

1139

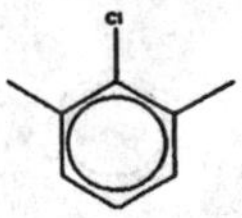

1140

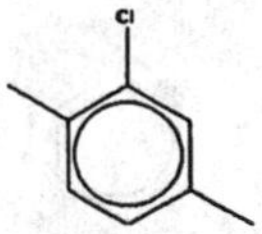

1141

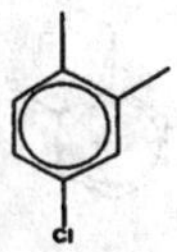

1142

1143

1144

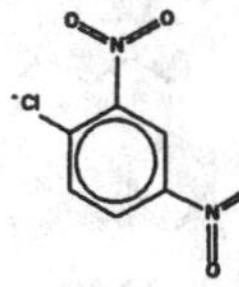

1145

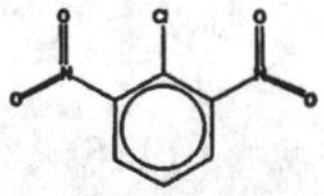

1146

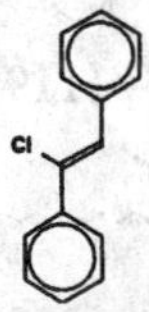

1147

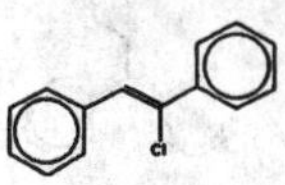

1148

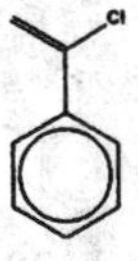

1149

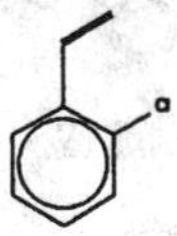

1150

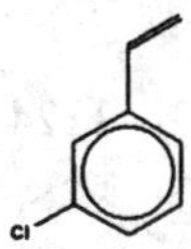

1151

1153

1154

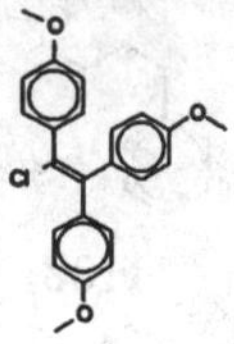

1155

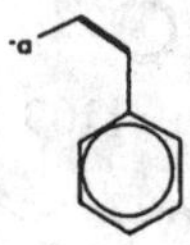

1156

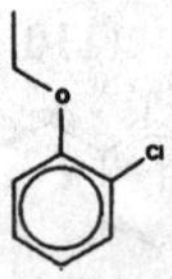

1157

1158

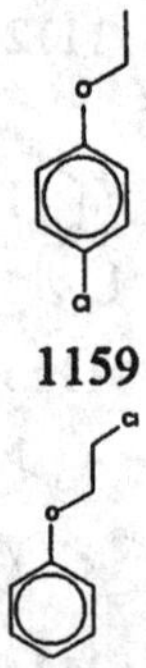

1159

1160

1161

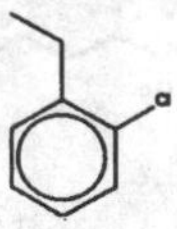

1162

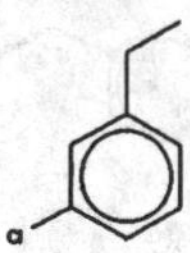

1163

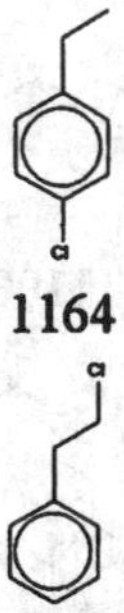

1164

1166

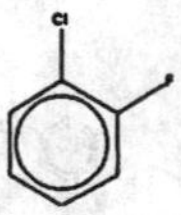

1167

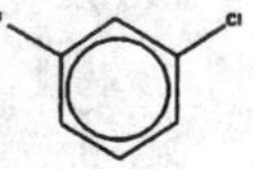

1168

1169

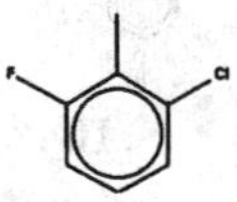

1170

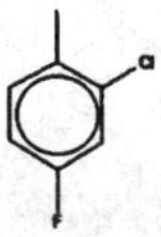

1171

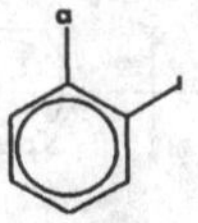

1172

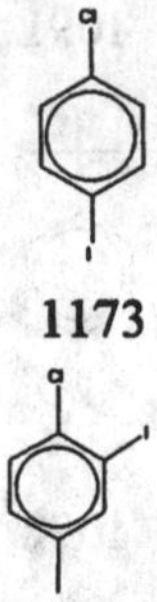

1173

1175

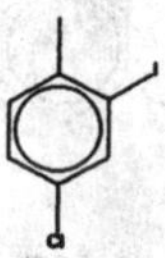

1176

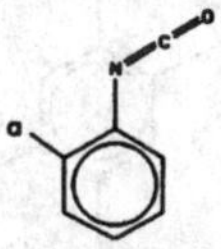

1177

1178

1179

1180

In numeric order as they occur in the Table of Physical Constants of Organic Compounds

1181
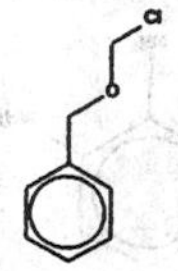
1182

1183
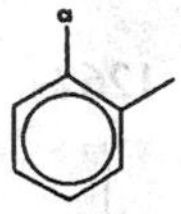
1184
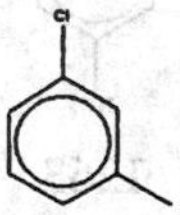
1185

1186
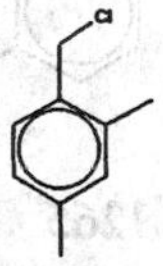
1187
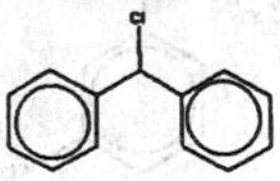
1189
1190

1191
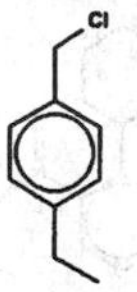
1192

1193
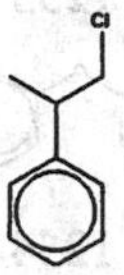
1194
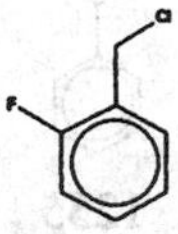
1195

1196
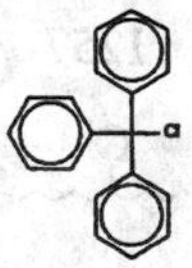
1197

1198
1199

1200

1201
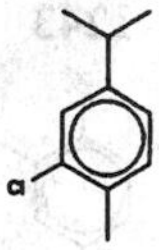
1202
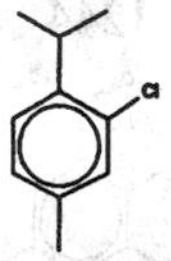
1203
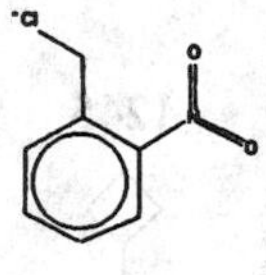
1204
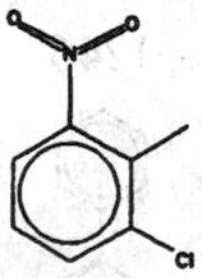
1205
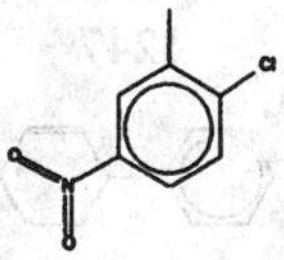
1206
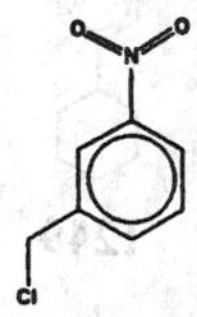
1207
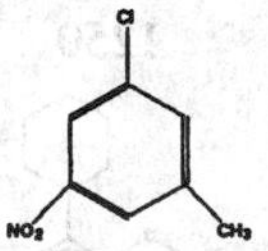

1208

1209
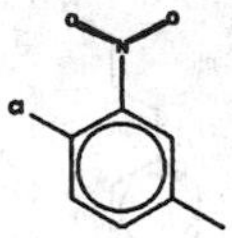
1210
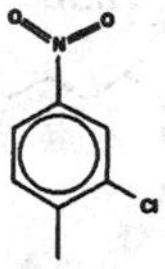
1211
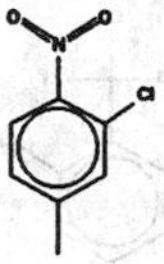
1212
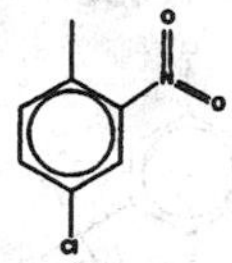
1213
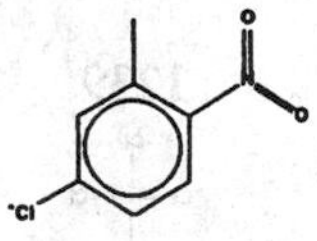
1214
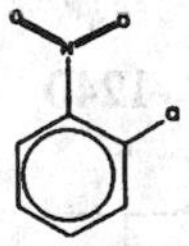
1215
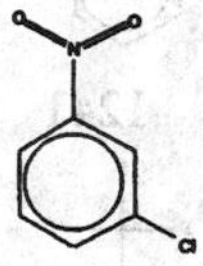

1216

1217
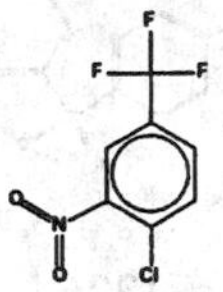
1218
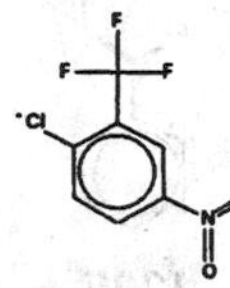
1219
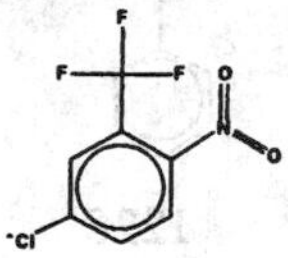
1220
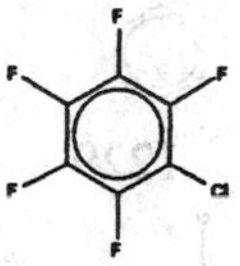
1221
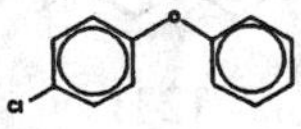
1222
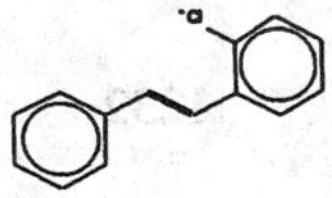
1223
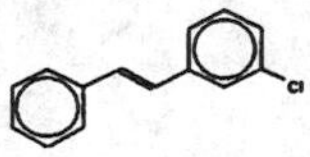

1224

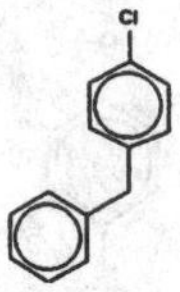

1225

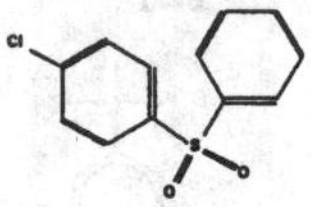

1226

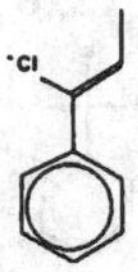

1227

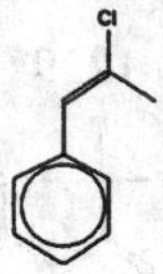

1228

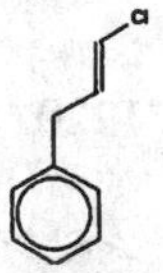

1229

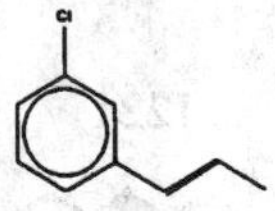

1231

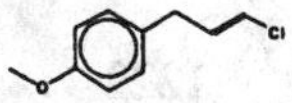

1232

1234

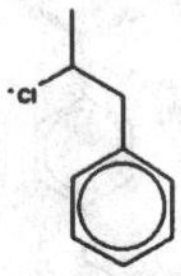

1235

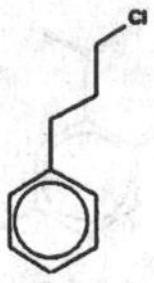

1236

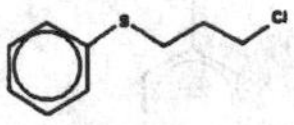

1237

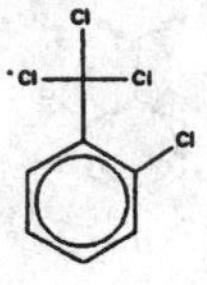

1238

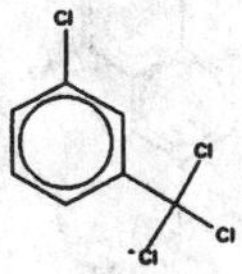

1239

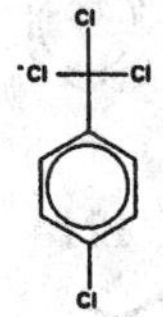

1240

1241

1242

1243

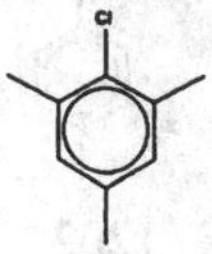

1244

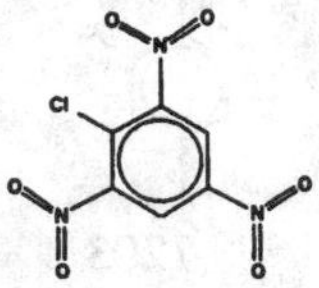

1245

1246

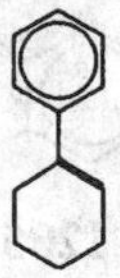

1247

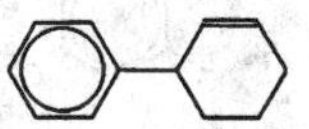

1248

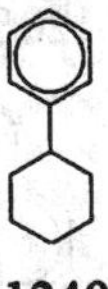

1249

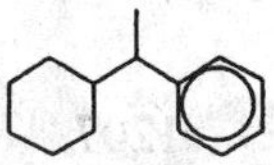

1250

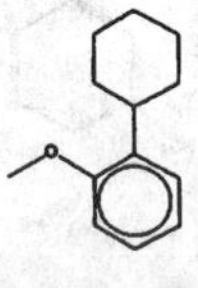

1251

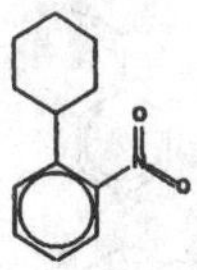

1252

1253

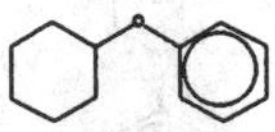

1254

1255

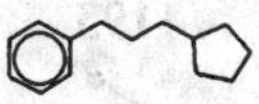

1256

1257

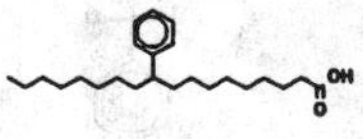

1258

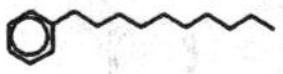

1259

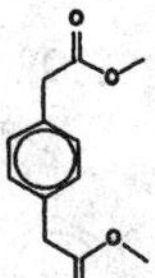

1260

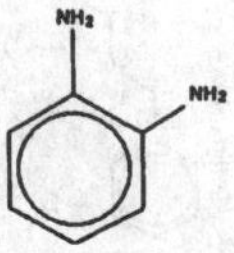

1261

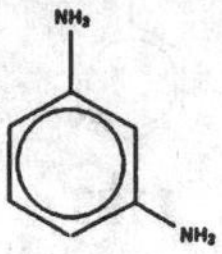

1262

1263

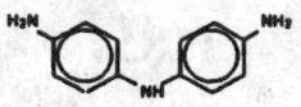

1264

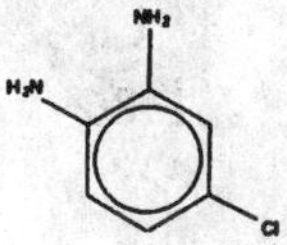

1265

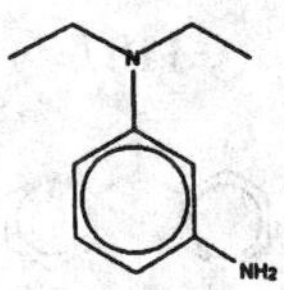

1266

STRUCTURAL FORMULAS OF ORGANIC COMPOUNDS (continued)

In numeric order as they occur in the Table of Physical Constants of Organic Compounds

1267

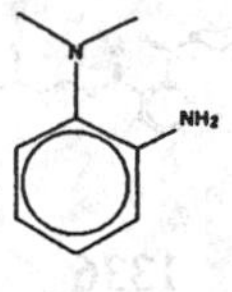

1268

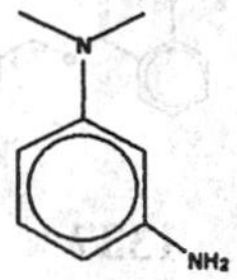

1270

1271

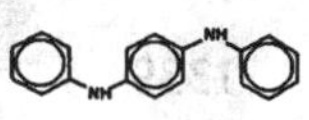

1272

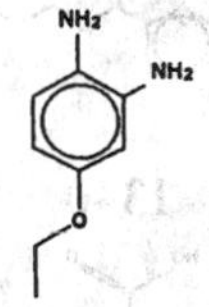

1273

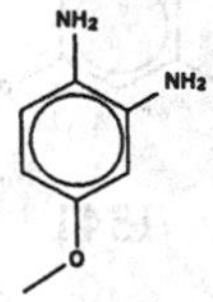

1274

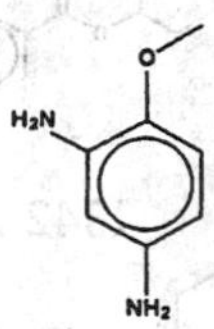

1275

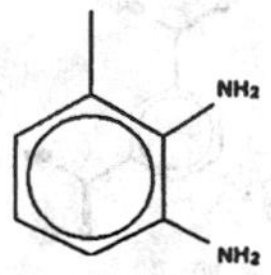

1276

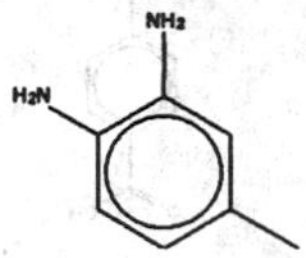

1277

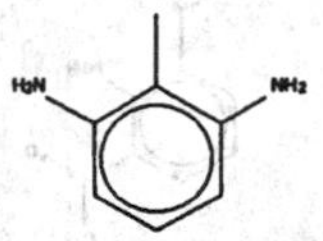

1278

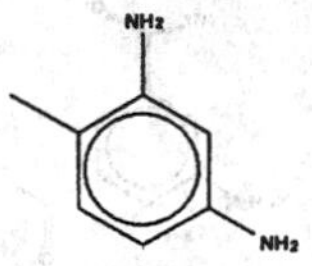

1279

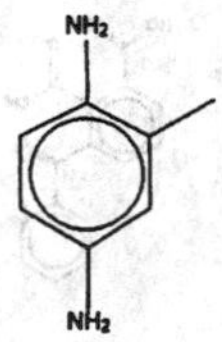

1280

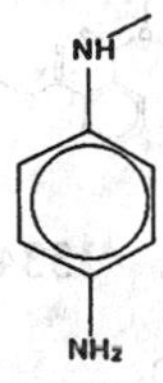

1281

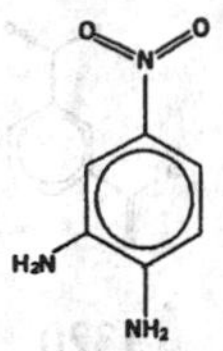

1282

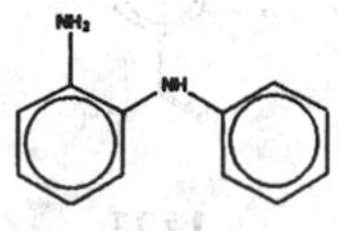

1283

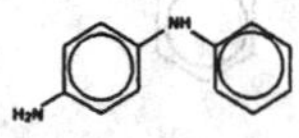

1284

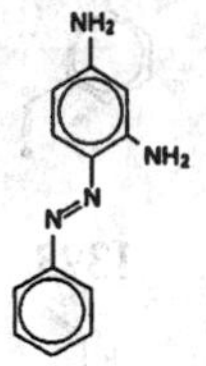

1285

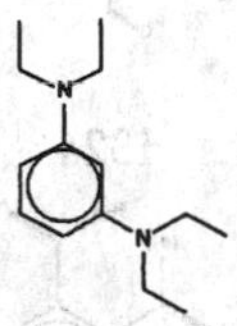

1286

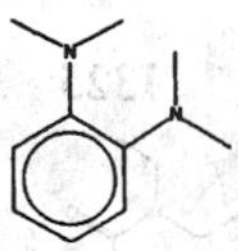

1287

1288

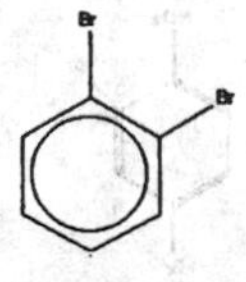

1289

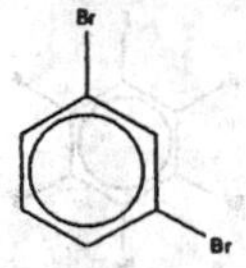

1290

1291

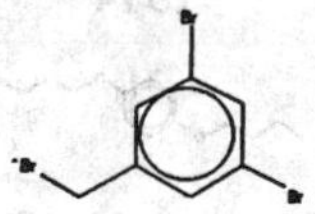

1292

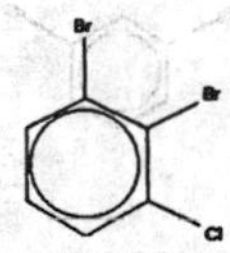

1293

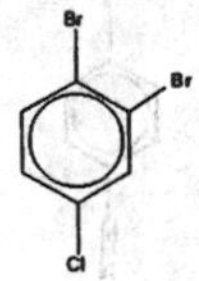

1294

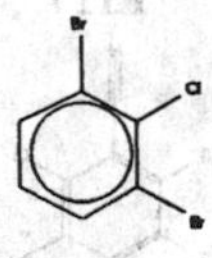

1295

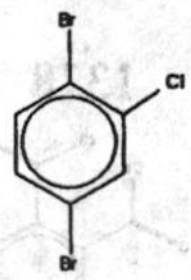

1297

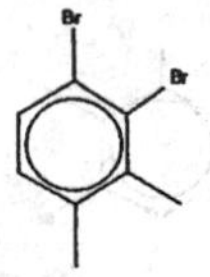

1298

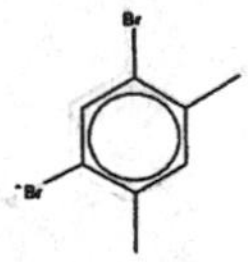

1299

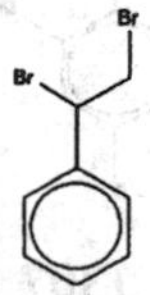

1300

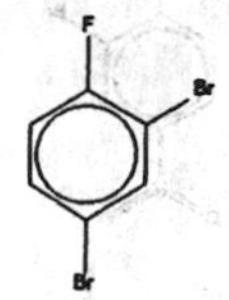

1301

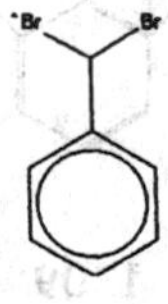

1302

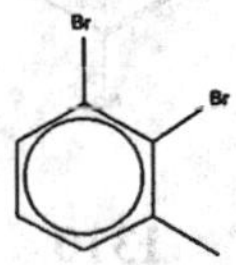

1303

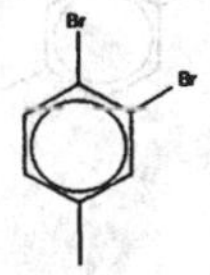

In numeric order as they occur in the Table of Physical Constants of Organic Compounds

1304

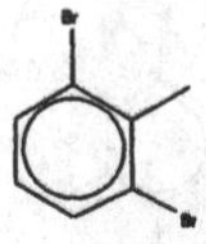

1305

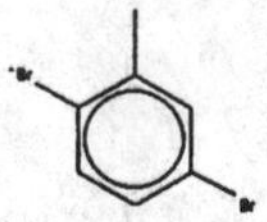

1306

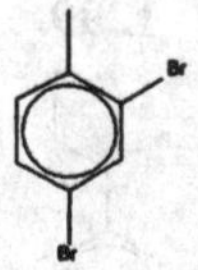

1307

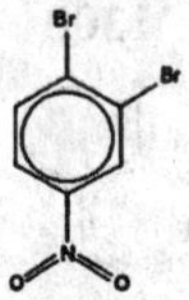

1308

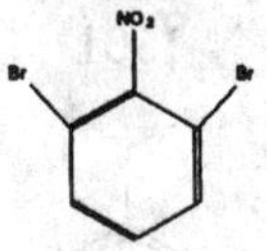

1309

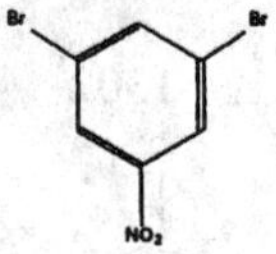

1310

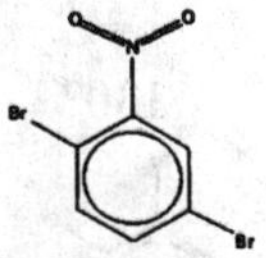

1311

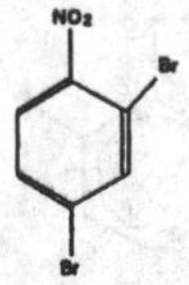

1312

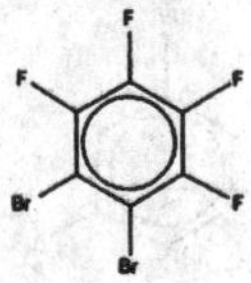

1313

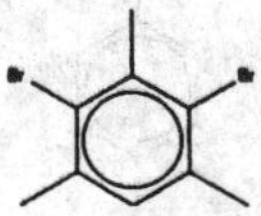

1314

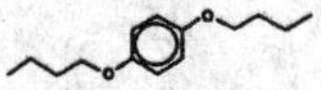

1315

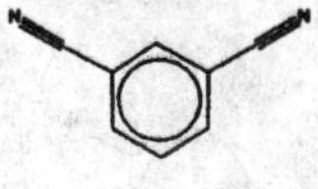

1316

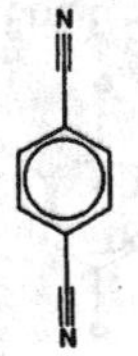

1317

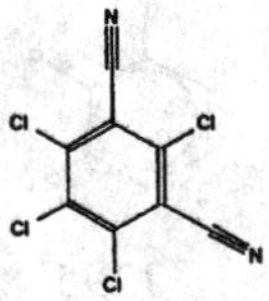

1318

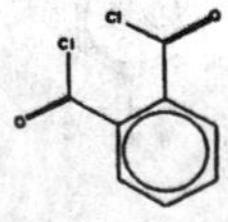

1319

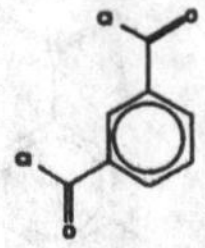

1320

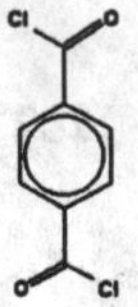

1321

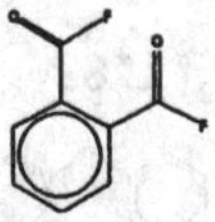

1322

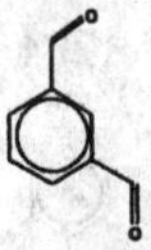

1323

1324

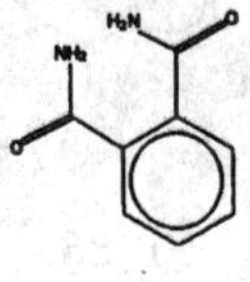

1325

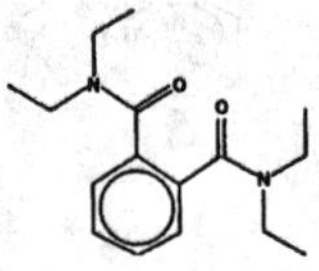

1327

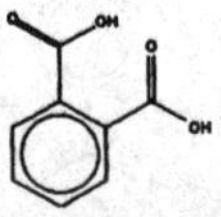

1328

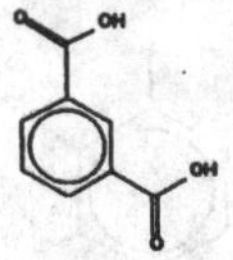

1329

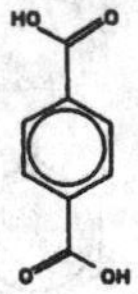

1330

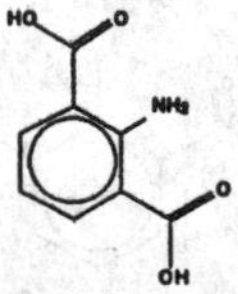

1331

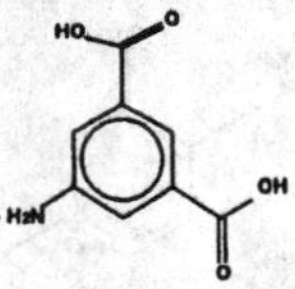

1332

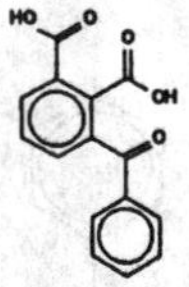

1333

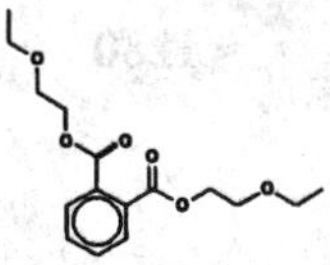

1334

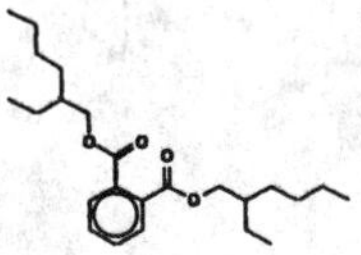

1335

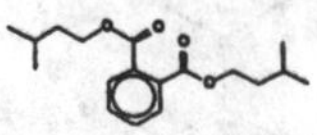

1336

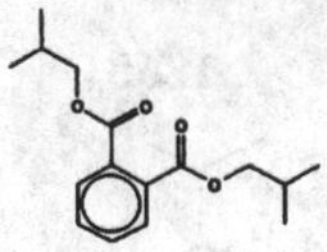

1337

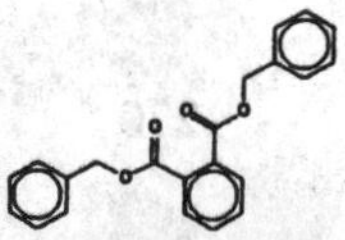

1338

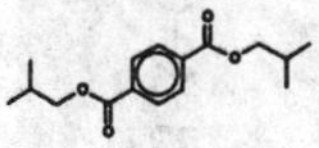

1339

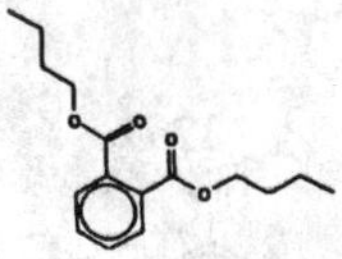

1340

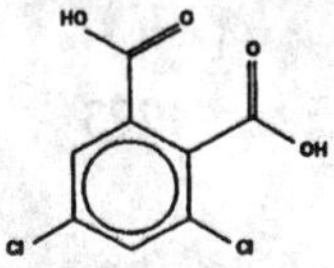

1341

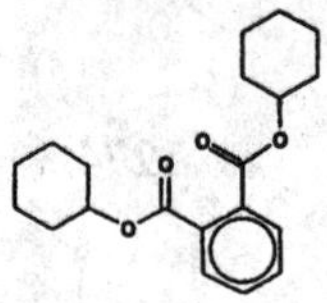

1342

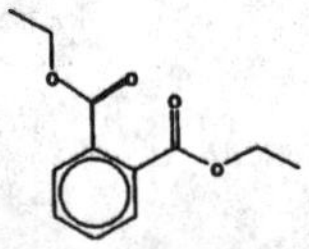

STRUCTURAL FORMULAS OF ORGANIC COMPOUNDS (continued)

In numeric order as they occur in the Table of Physical Constants of Organic Compounds

1343

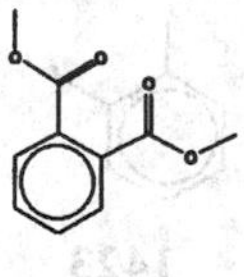

1344

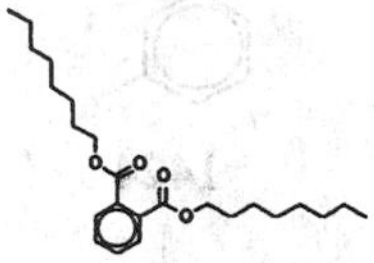

1345

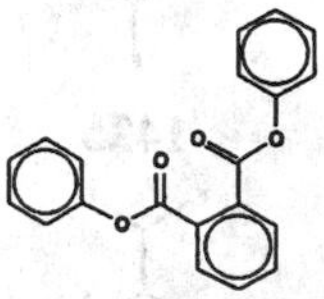

1346

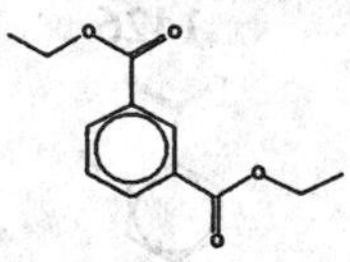

1347

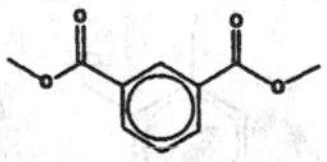

1348

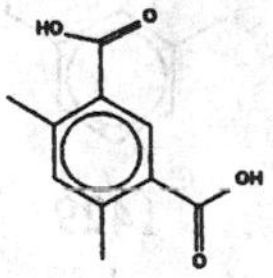

1349

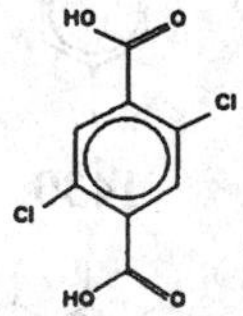

1350

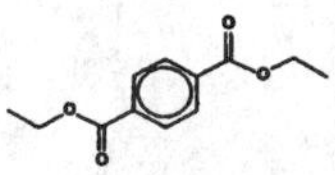

1351

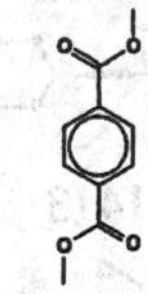

1352

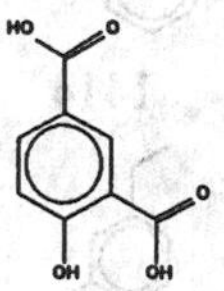

1353

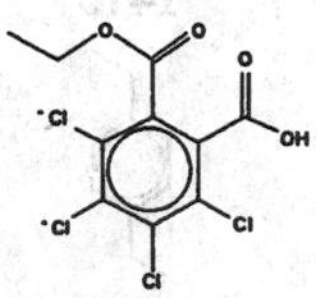

1354

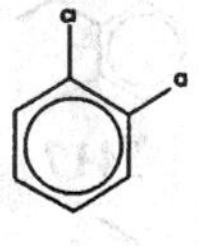

1355

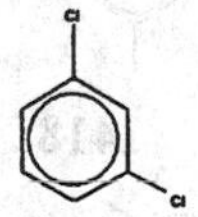

1356

1357

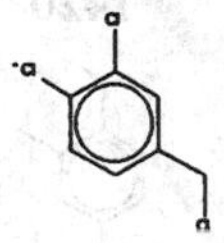

1358

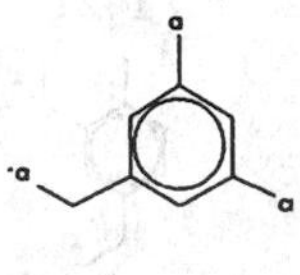

1359

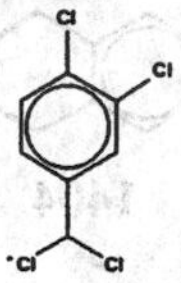

1361

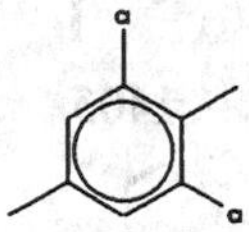

1362

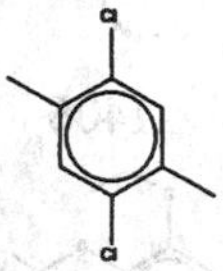

1363

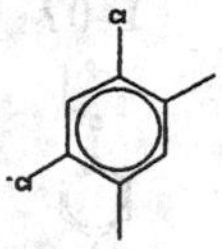

1364

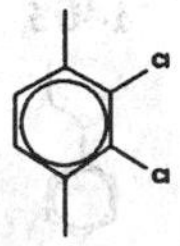

1365

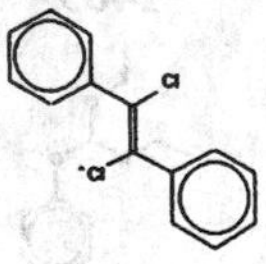

1366

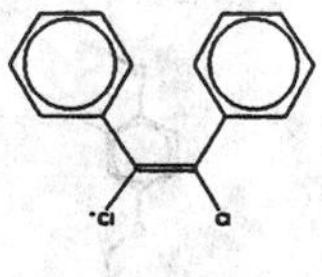

1367

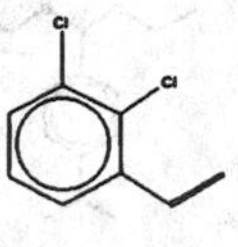

1368

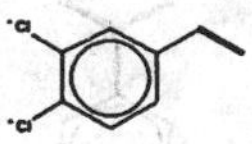

1369

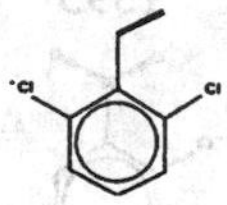

1370

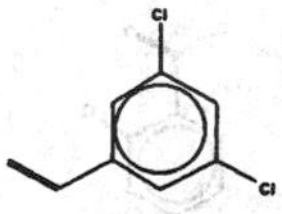

1371

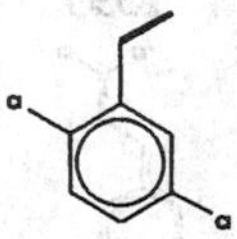

1372

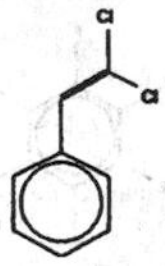

1373

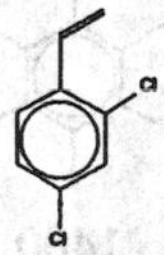

1374

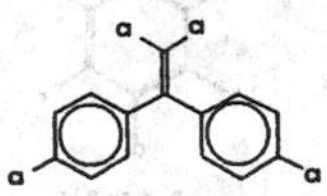

1375

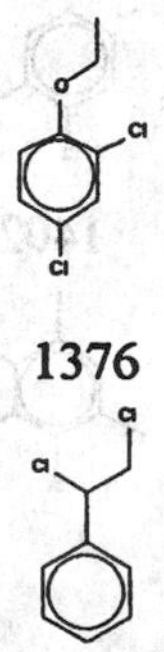

1376

1377

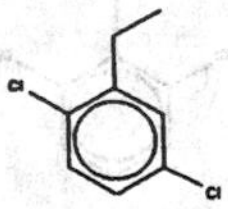

1378

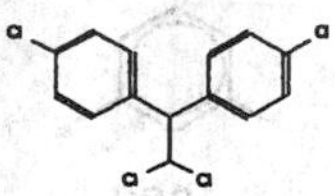

1379

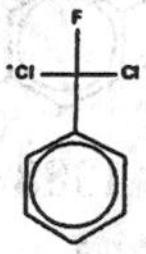

1380

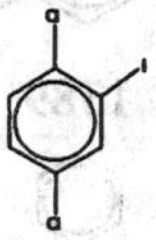

1381

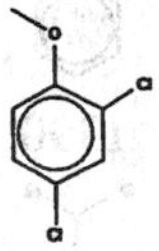

1382

1383

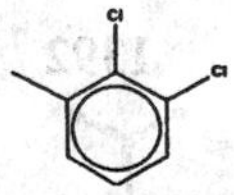

1384

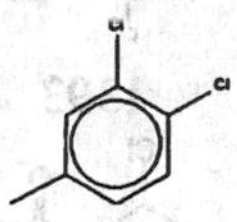

In numeric order as they occur in the Table of Physical Constants of Organic Compounds

1385
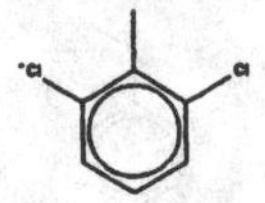
1386
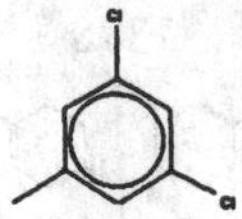
1387
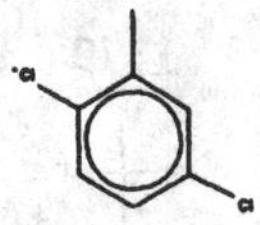
1388
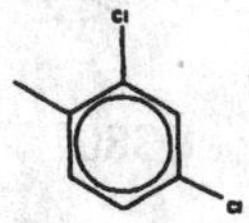
1389
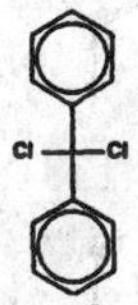
1390

1391
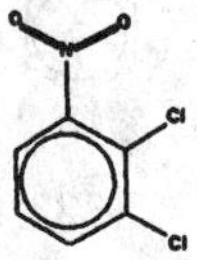
1392
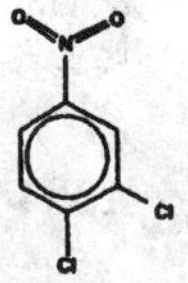
1393
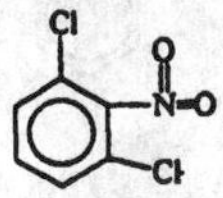
1394
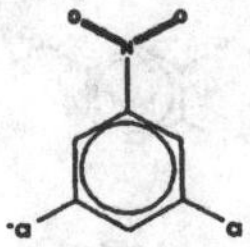
1395
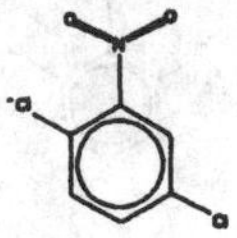
1396
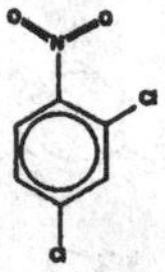
1397

1398
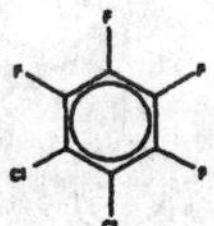
1399
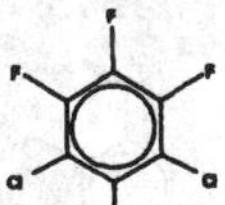
1400
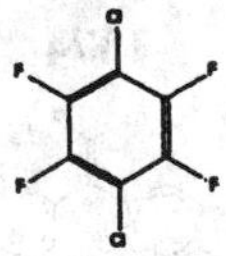
1401
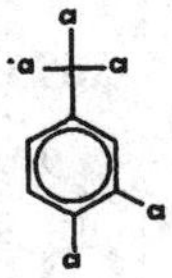
1402
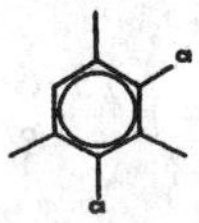
1403
1404
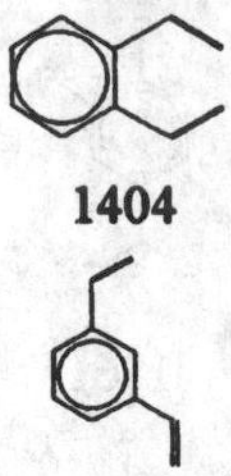
1405
1406
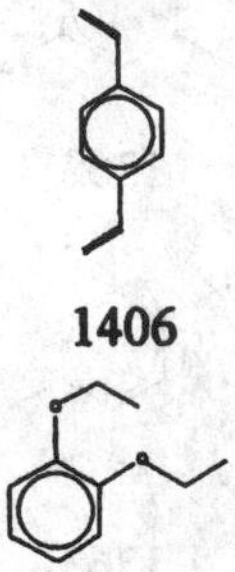
1407
1408

1409
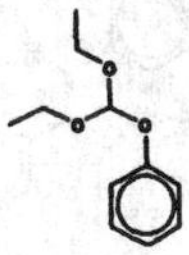
1410
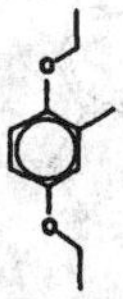
1411
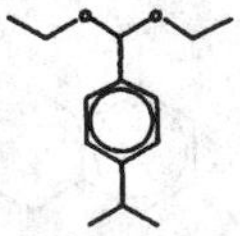
1412
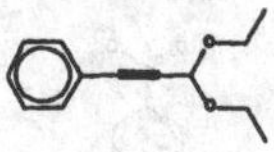
1413
1414

1415
1416
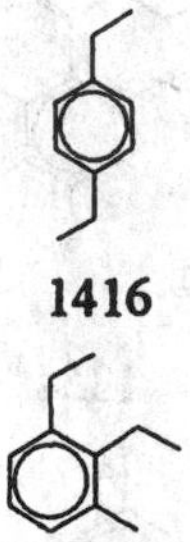
1417
1418

1419
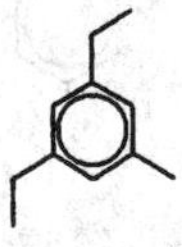
1420
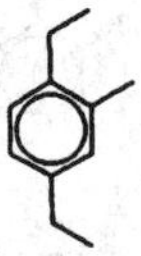
1421
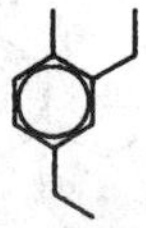
1422
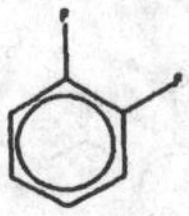
1423
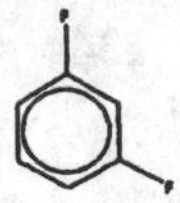
1424

1425

1426
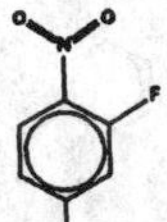
1427
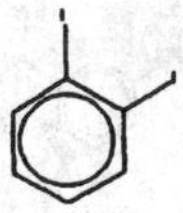
1428
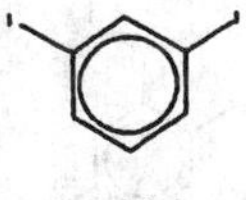
1429

1430
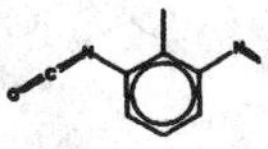

In numeric order as they occur in the Table of Physical Constants of Organic Compounds

1431

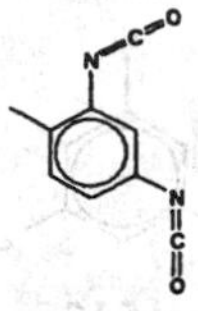

1432

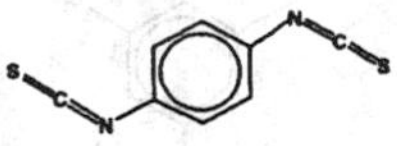

1433

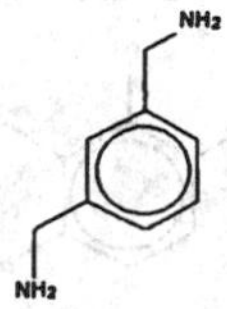

1434

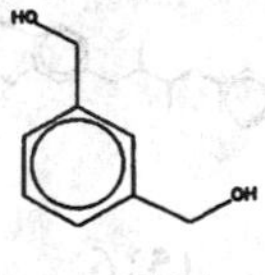

1435

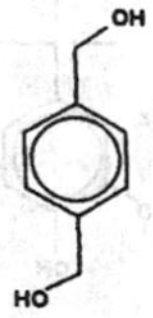

1436

1437

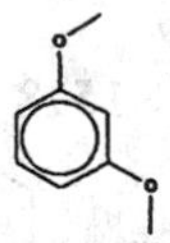

1438

1439

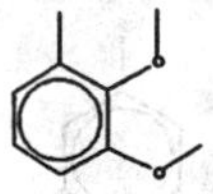

1440

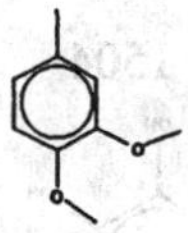

1441

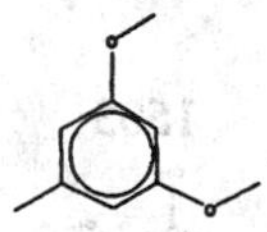

1442

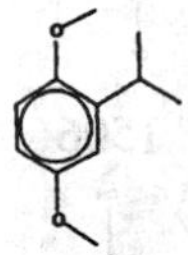

1443

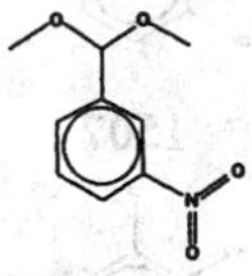

1444

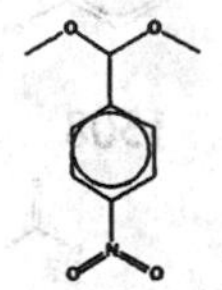

1445

1447

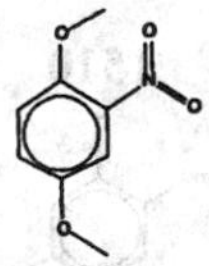

1448

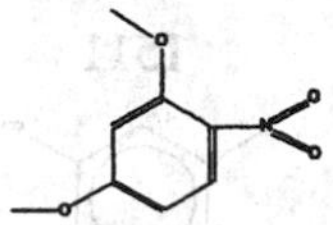

1449

1450

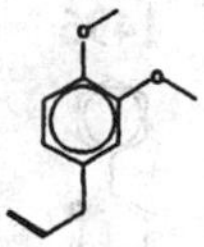

1451

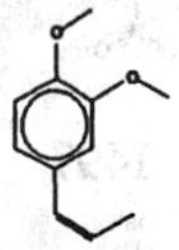

1452

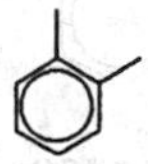

1453

1454

1455

1456

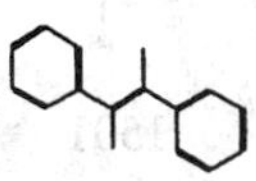

1457

1458

1459

1460

1461

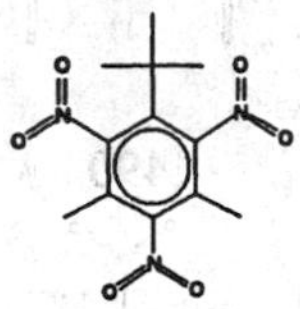

1462

1463

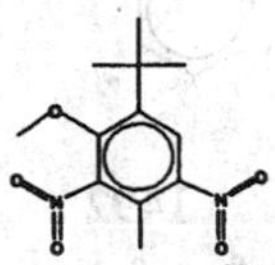

1464

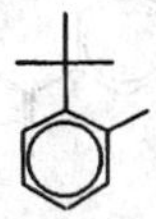

1465

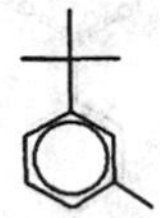

1466

1467

1468

1469

1470

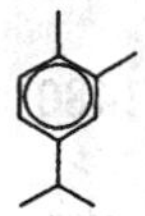

1471

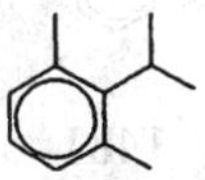

1472

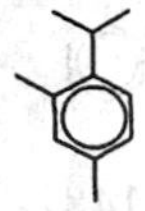

1473

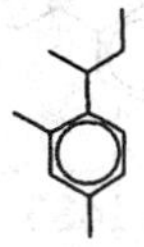

1474

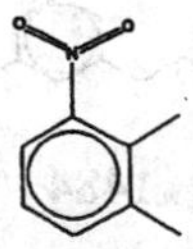

1475

1476

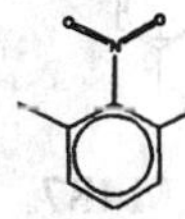

1477

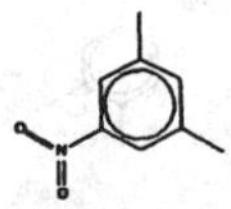

In numeric order as they occur in the Table of Physical Constants of Organic Compounds

1478

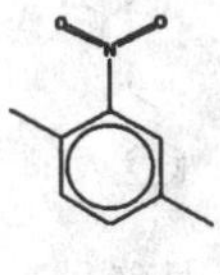

1479

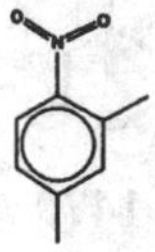

1480

1481

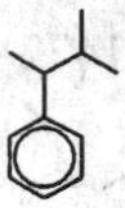

1482

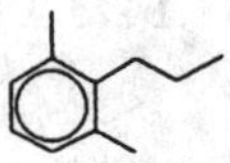

1483

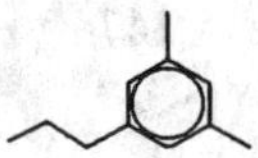

1484

1486

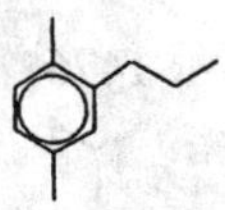

1487

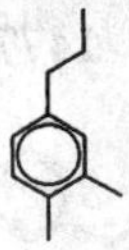

1488

1489

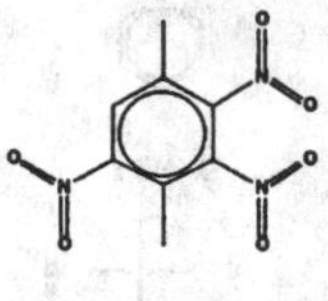

1490

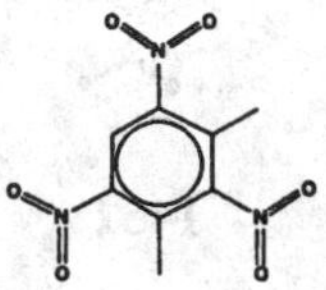

1491

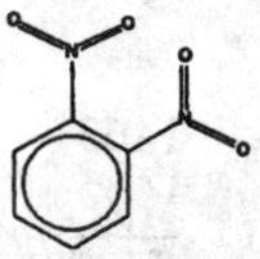

1492

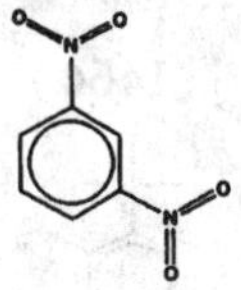

1493

1494

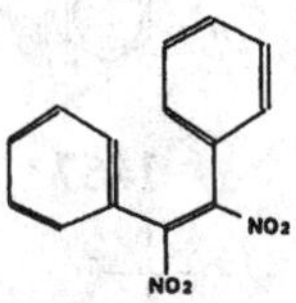

1495

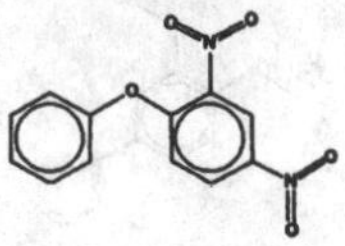

1496

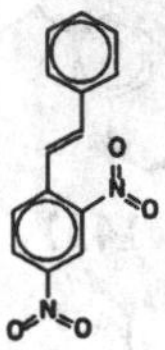

1497

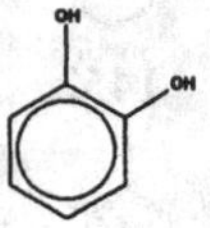

1498

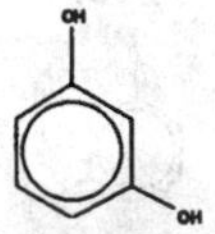

1499

1500

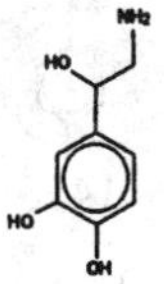

1501

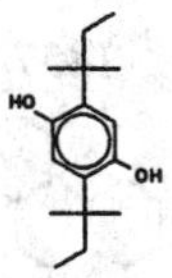

1502

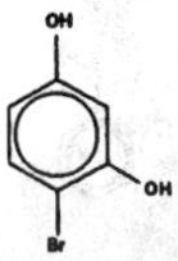

1503

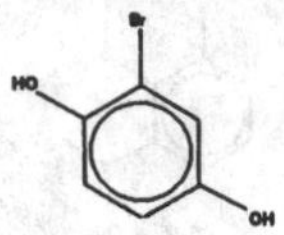

1504

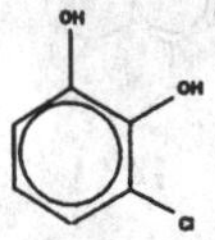

1505

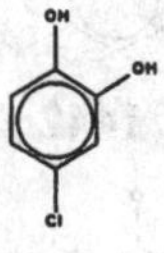

1506

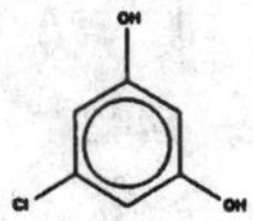

1507

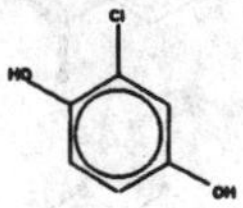

1508

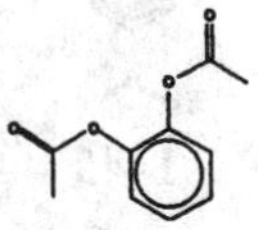

1509

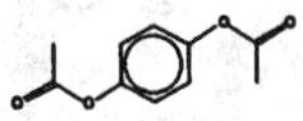

1510

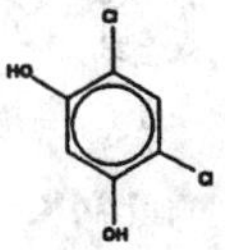

1511

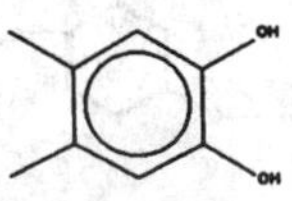

1512

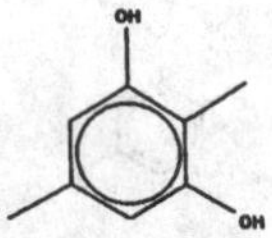

1513

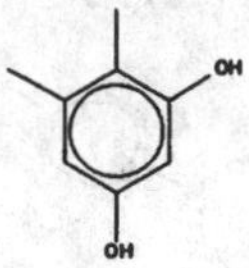

1514

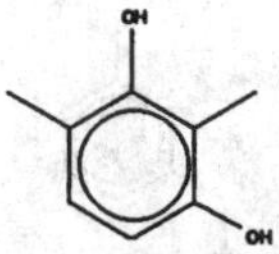

1515

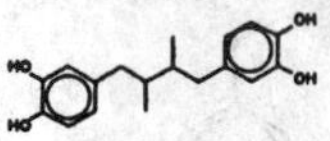

1516

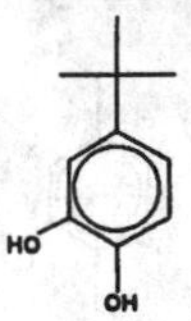

1517

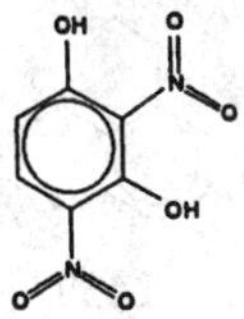

1518

1519

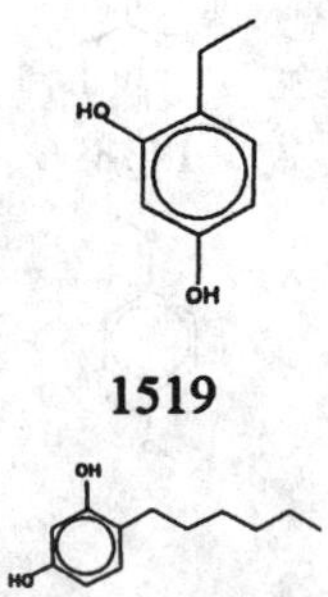

STRUCTURAL FORMULAS OF ORGANIC COMPOUNDS (continued)

In numeric order as they occur in the Table of Physical Constants of Organic Compounds

1520

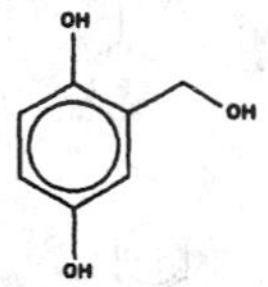

1521

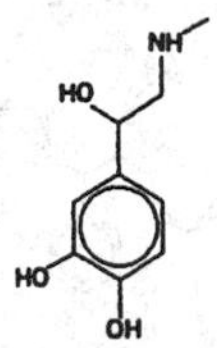

1522

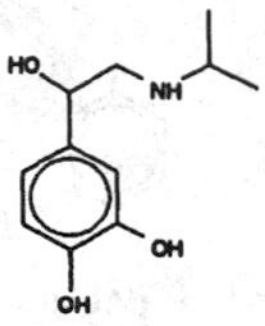

1523

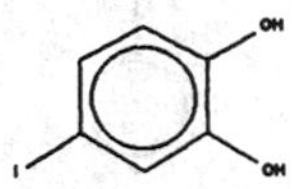

1524

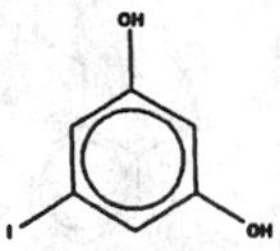

1525

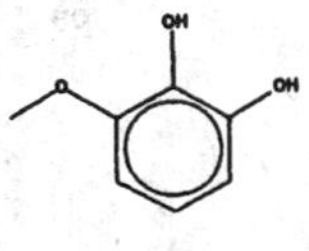

1526

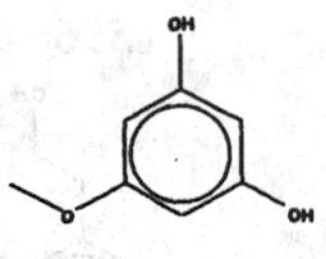

1527

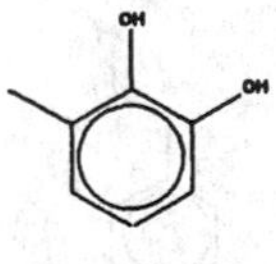

1528

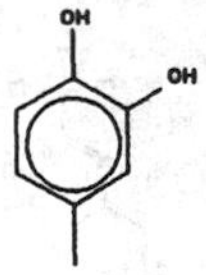

1529

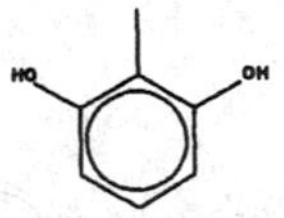

1530

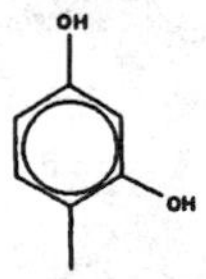

1531

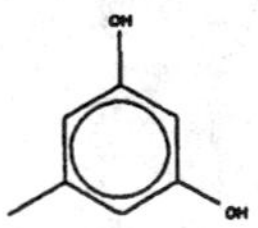

1532

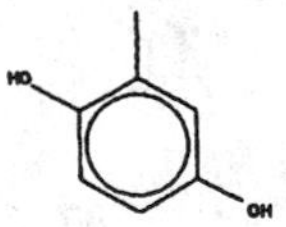

1533

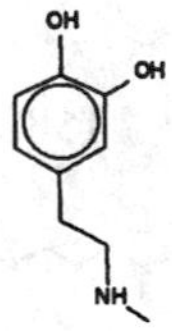

1534

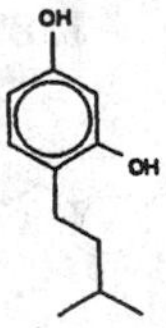

1535

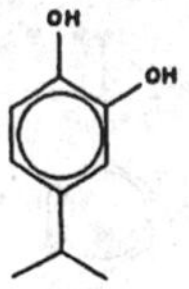

1536

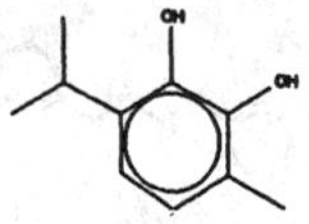

1537

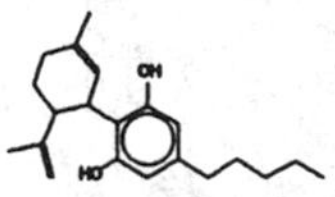

1538

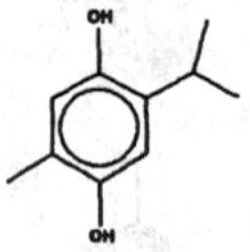

1539

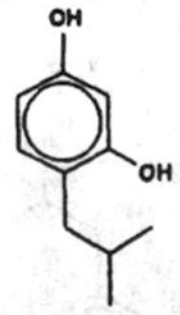

1540

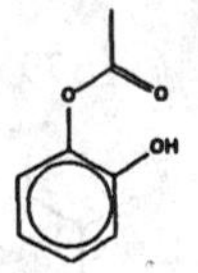

1541

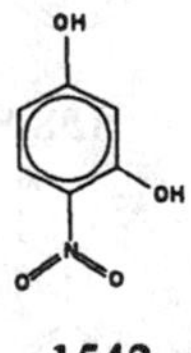

1542

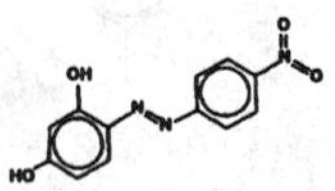

1543

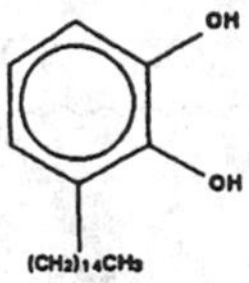

1544

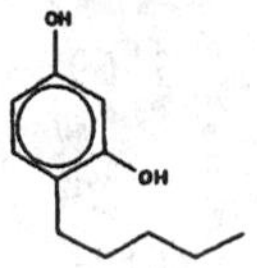

1545

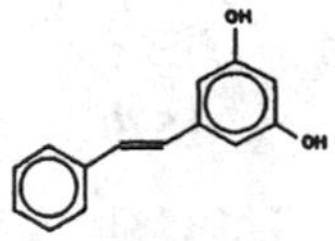

1546

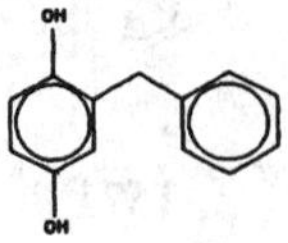

1547

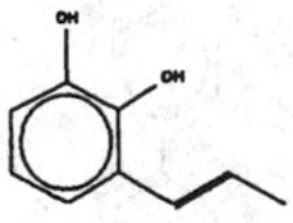

1548

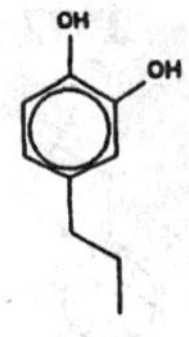

1549

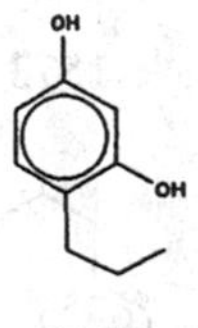

1550

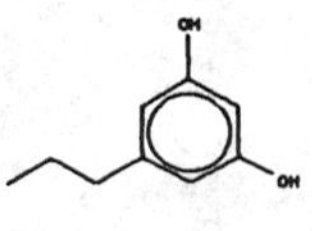

1551

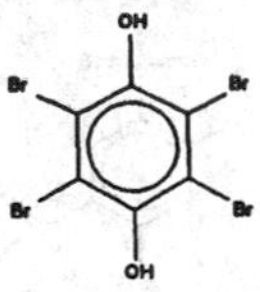

1552

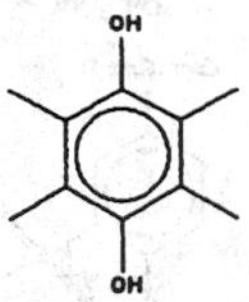

1553

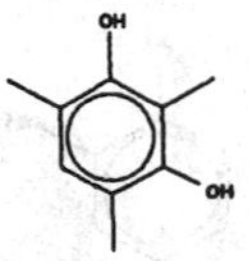

1554

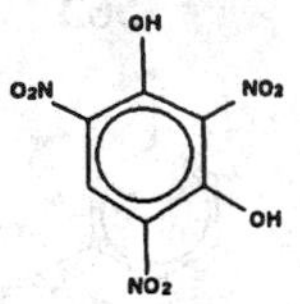

1555

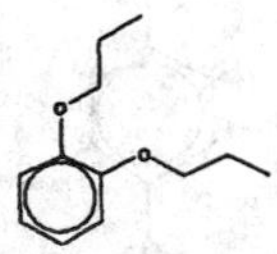

1556

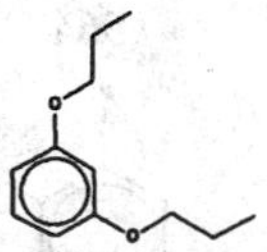

1557

1558

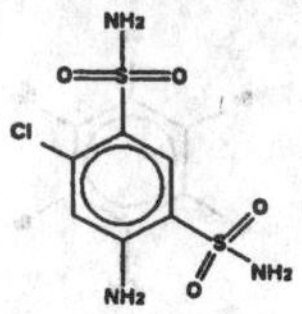

1560

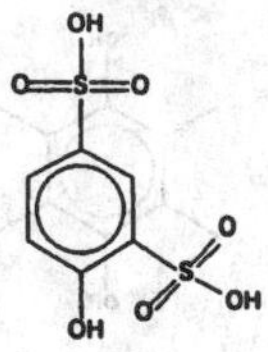

1561

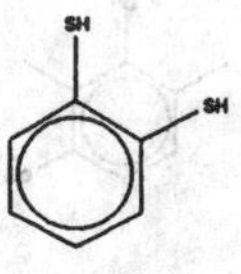

1562

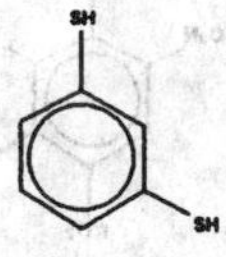

1563

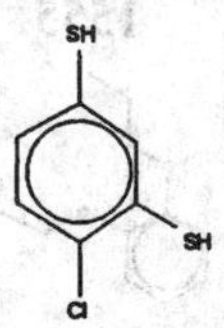

1564

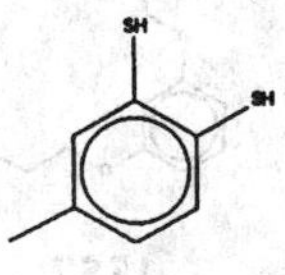

1565

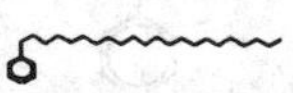

1566

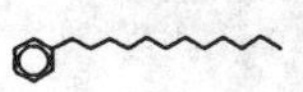

1567

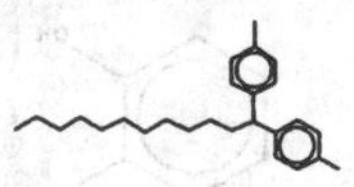

1568

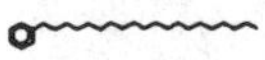

1569

1570

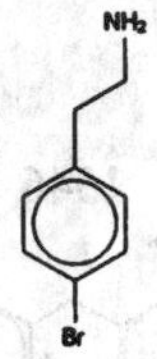

1571

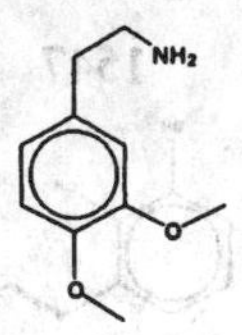

1572

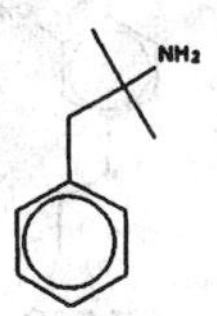

1573

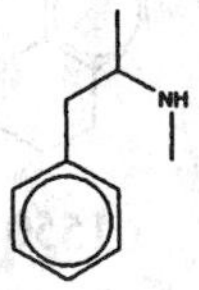

1574

1575

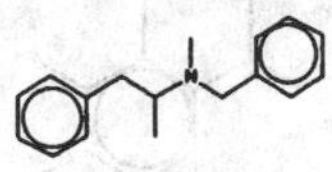

1576

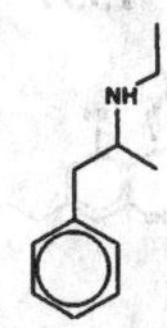

1577

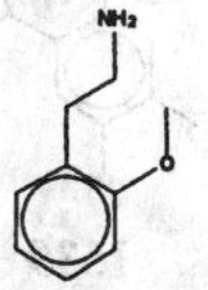

1578

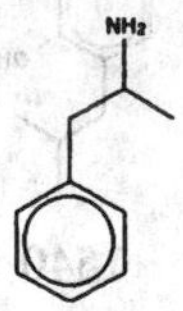

1579

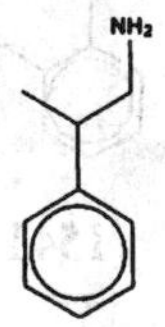

1580

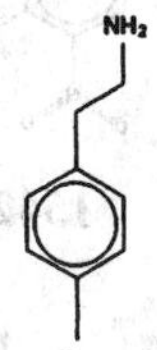

1581

1583

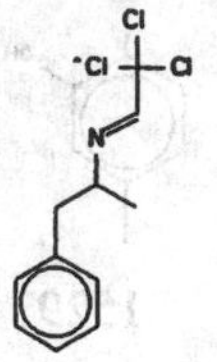

1584

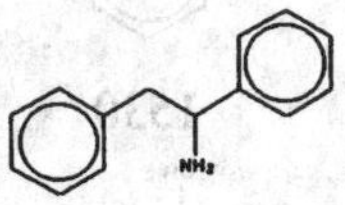

1585

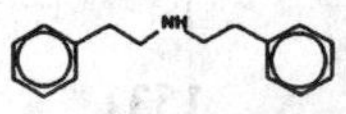

1586

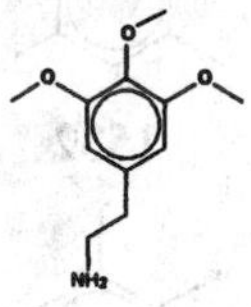

1587

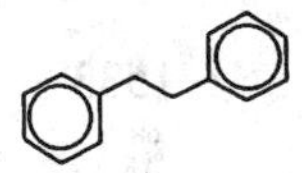

1588

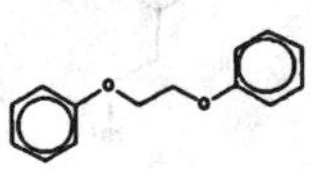

1589

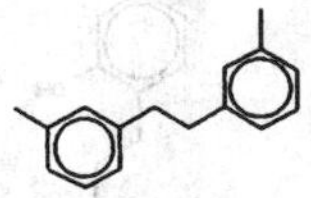

1590

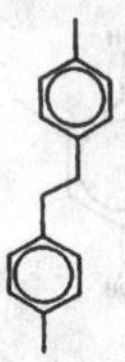

1591

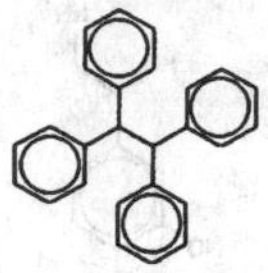

1592

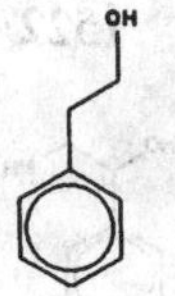

1593

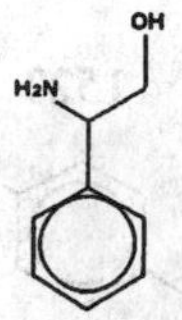

1594

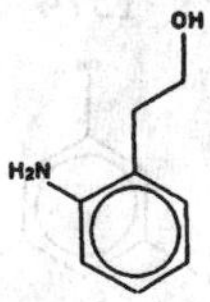

1595

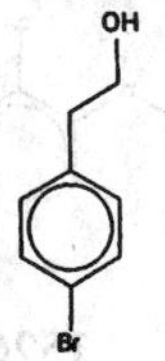

1596

STRUCTURAL FORMULAS OF ORGANIC COMPOUNDS (continued)

In numeric order as they occur in the Table of Physical Constants of Organic Compounds

1597

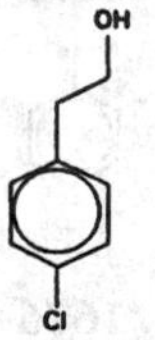

1598

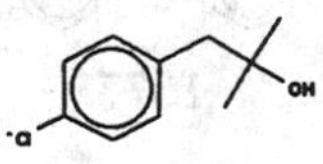

1599

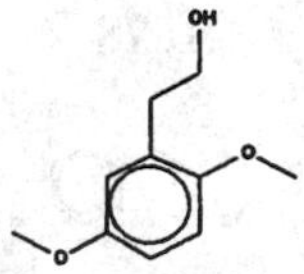

1600

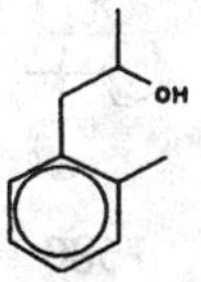

1601

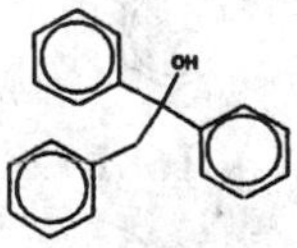

1602

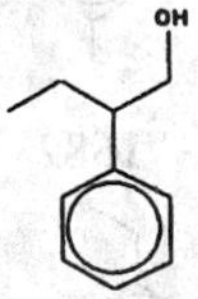

1603

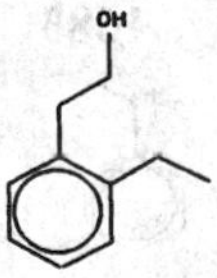

1604

1606

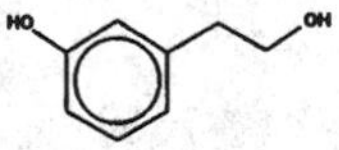

1607

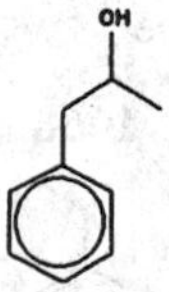

1608

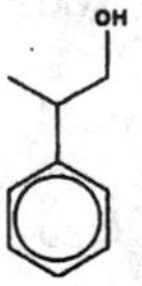

1609

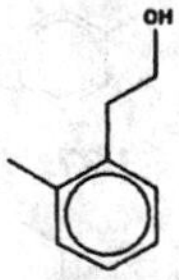

1610

1612

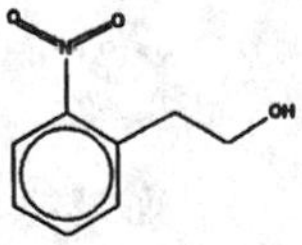

1613

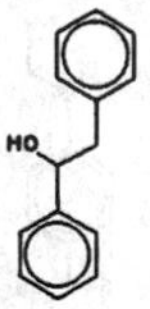

1614

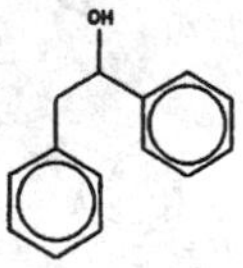

1615

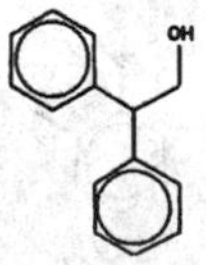

1616

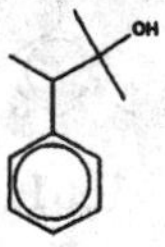

1617

1618

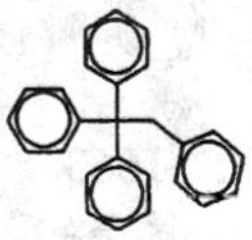

1619

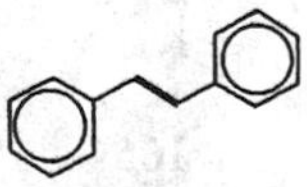

1620

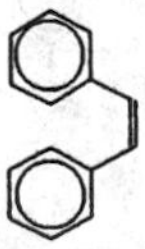

1621

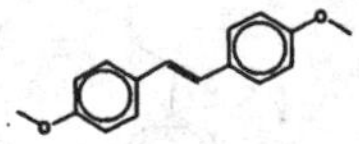

1622

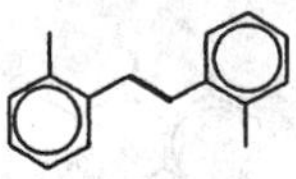

1623

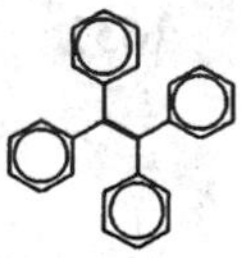

1624

1626

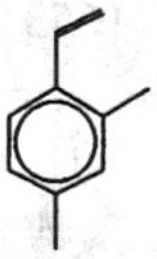

1627

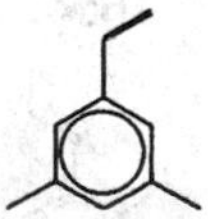

1628

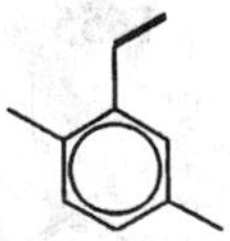

1629

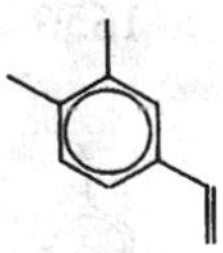

1630

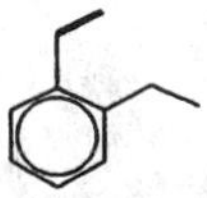

1631

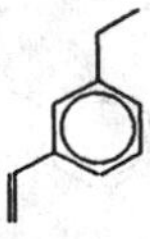

1632

1633

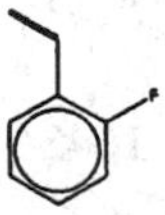

1634

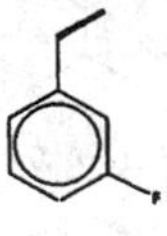

1635

1636

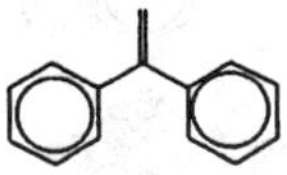

1637

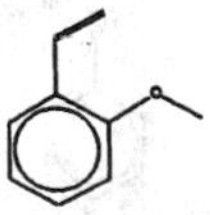

1638

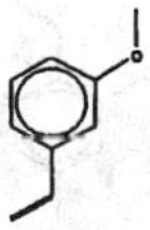

1639

In numeric order as they occur in the Table of Physical Constants of Organic Compounds

1640

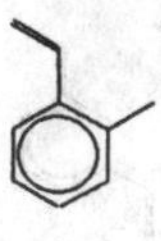

1641

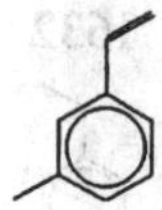

1642

1643

1644

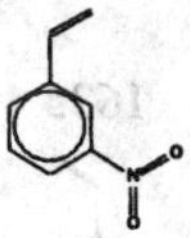

1645

1646

1647

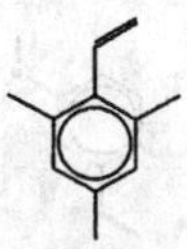

1648

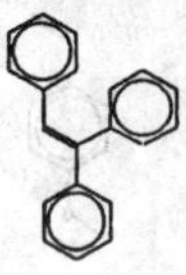

1649

1650

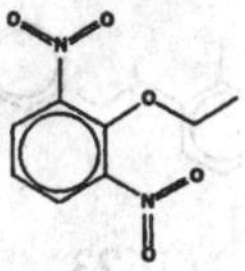

1651

1652

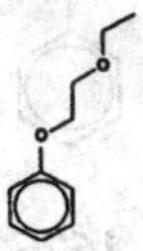

1653

1654

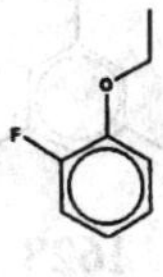

1655

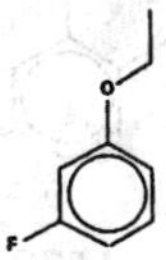

1656

1657

1658

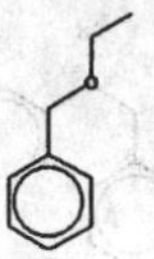

1659

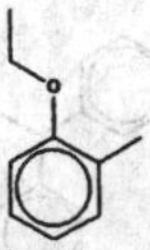

1660

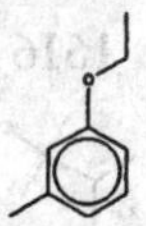

1661

1662

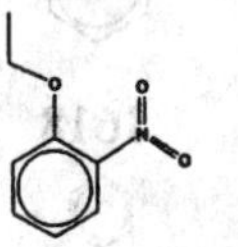

1663

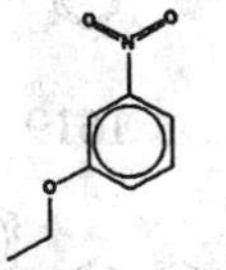

1664

1665

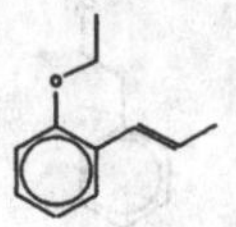

1666

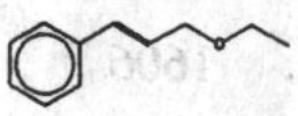

1667

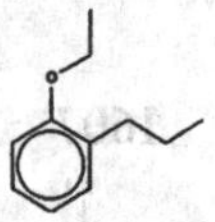

1668

1669

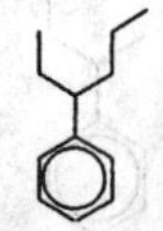

1670

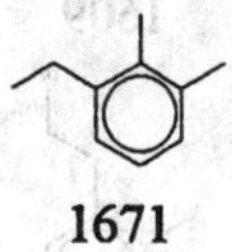

1671

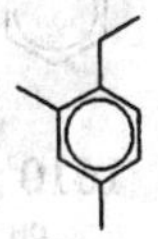

1672

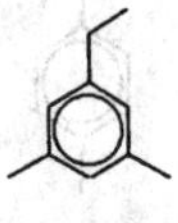

1673

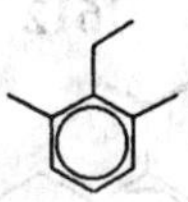

1674

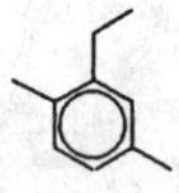

1675

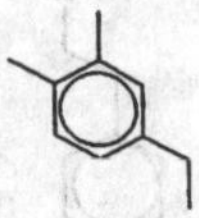

1676

1677

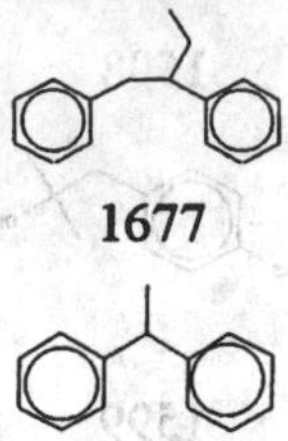

1678

1679

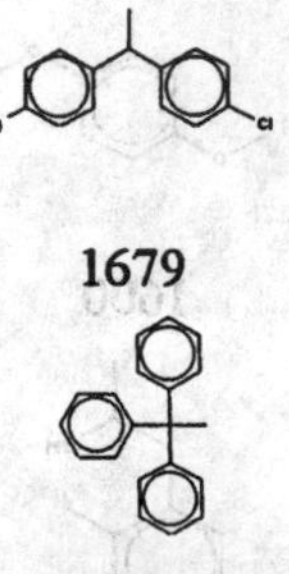

1680

1681

1682

1683

STRUCTURAL FORMULAS OF ORGANIC COMPOUNDS (continued)

In numeric order as they occur in the Table of Physical Constants of Organic Compounds

1684

1685

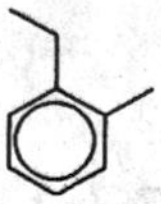

1686

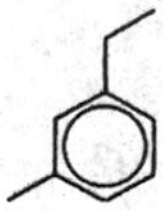

1687

1689

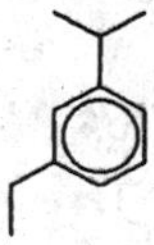

1690

1691

1692

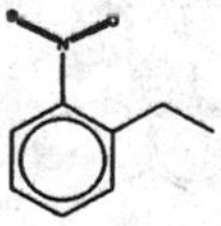

1693

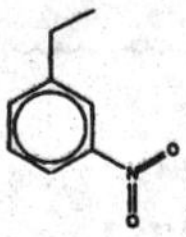

1694

1695

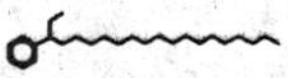

1696

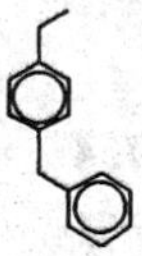

1697

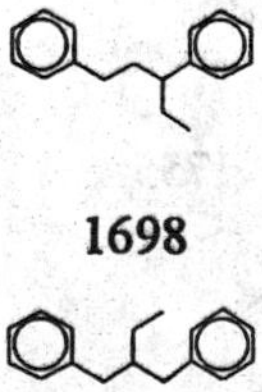

1698

1699

1700

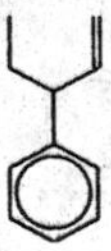

1701

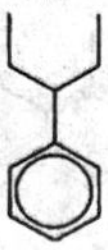

1702

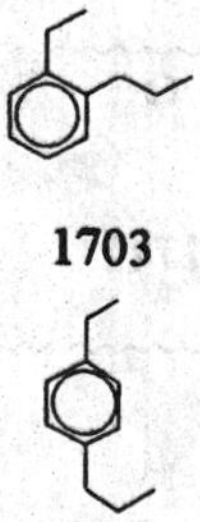

1703

1704

1705

1706

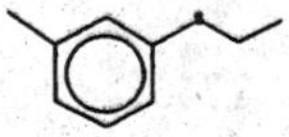

1707

1708

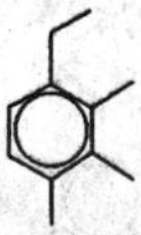

1709

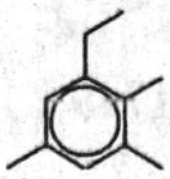

1710

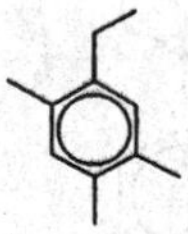

1711

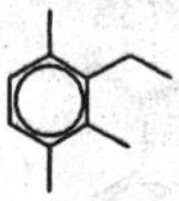

1712

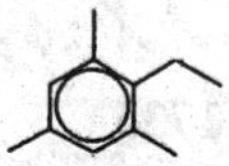

1713

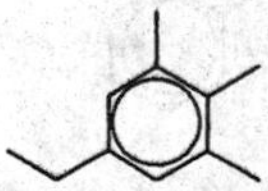

1714

1715

1716

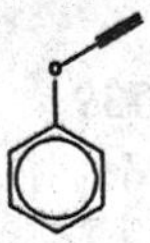

1717

1718

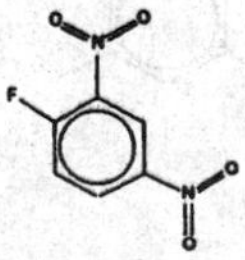

1719

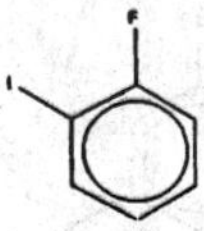

1720

1721

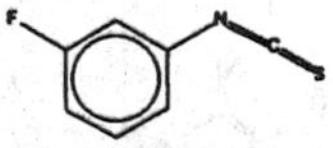

1722

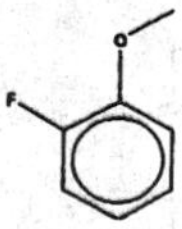

1723

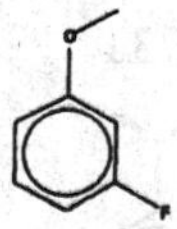

1724

1725

1726

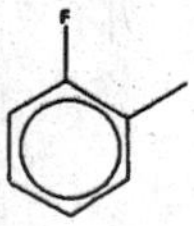

STRUCTURAL FORMULAS OF ORGANIC COMPOUNDS (continued)

In numeric order as they occur in the Table of Physical Constants of Organic Compounds

1727

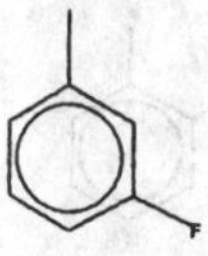

1728

1729

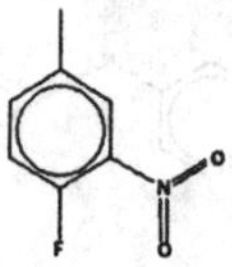

1730

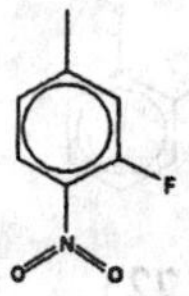

1731

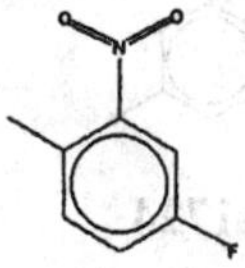

1732

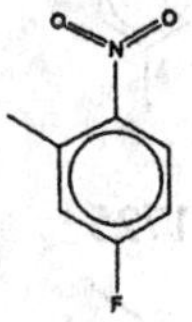

1733

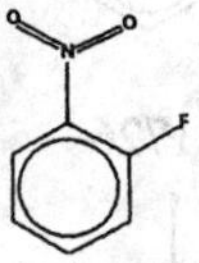

1734

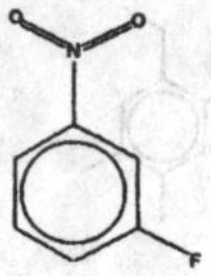

1735

1736

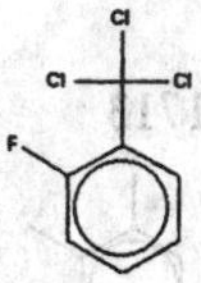

1737

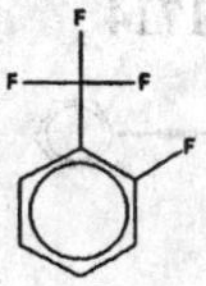

1738

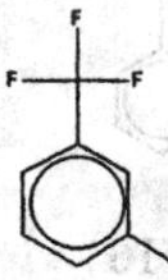

1739

1740

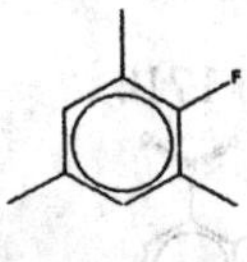

1741

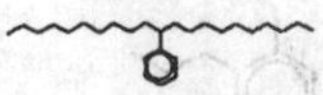

1742

1743

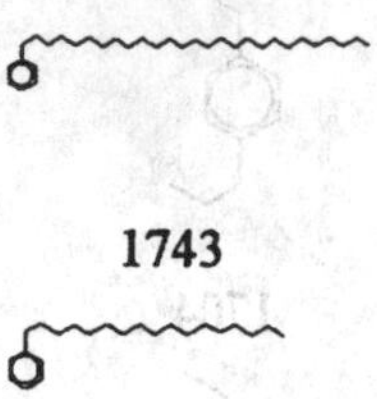

1744

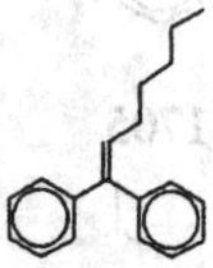

1745

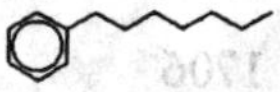

1746

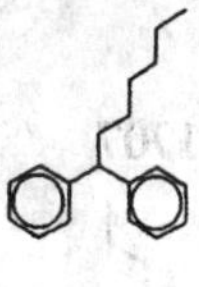

1747

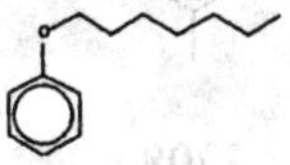

1748

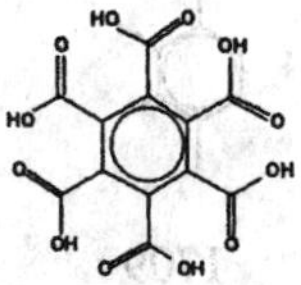

1749

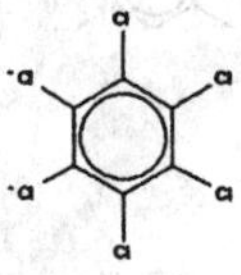

1750

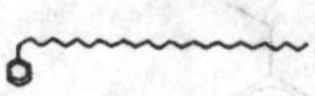

1751

1752

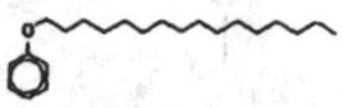

1753

1754

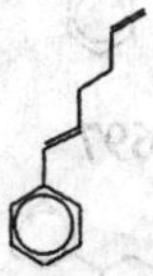

1755

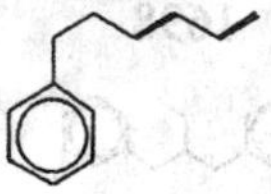

1756

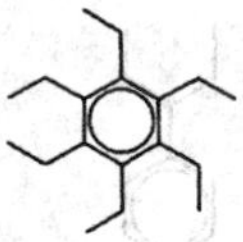

1757

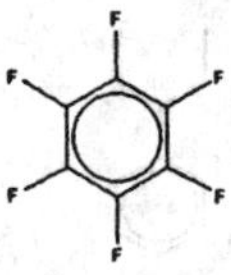

1758

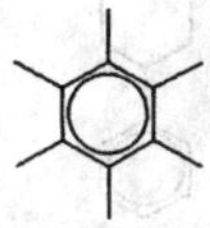

1759

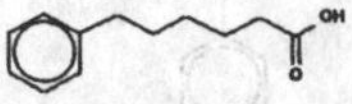

1760

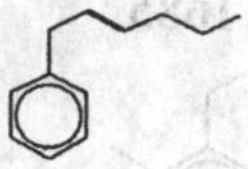

1761

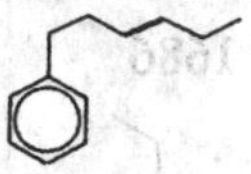

1762

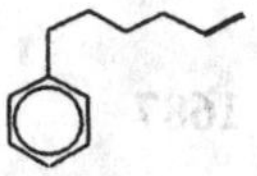

1763

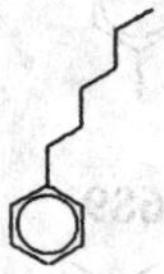

1764

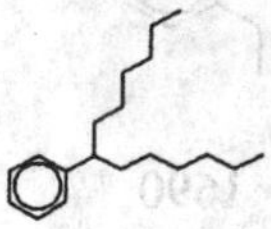

1765

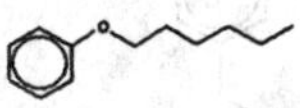

1766

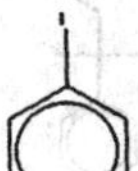

1767

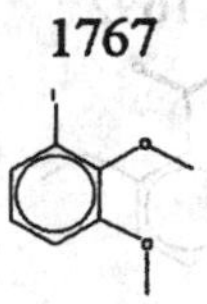

1768

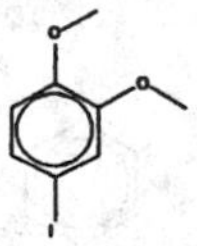

1769

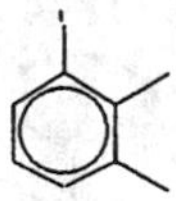

1770

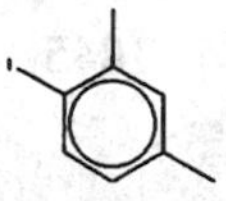

1771

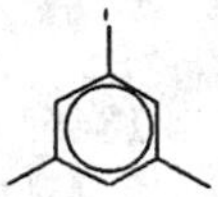

1772

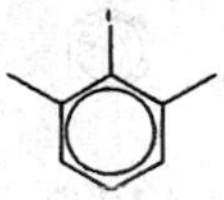

1773

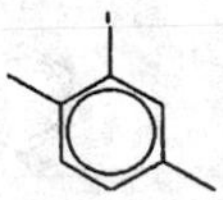

1774

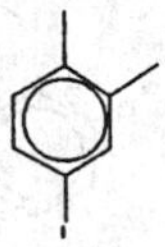

1775

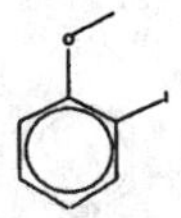

1776

1777

1778

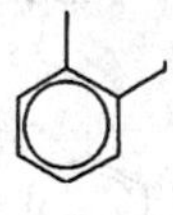

1779

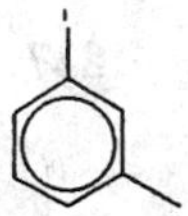

1780

1781

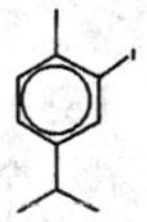

1782

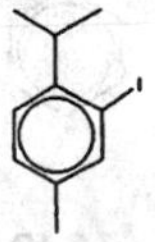

1783

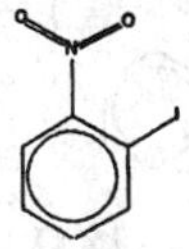

1784

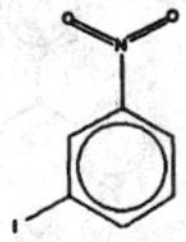

1785

1786

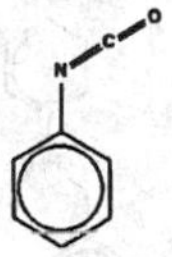

1787

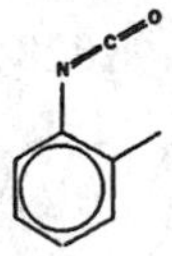

1788

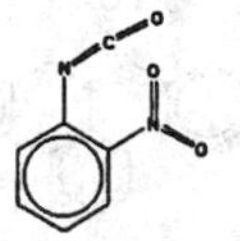

1789

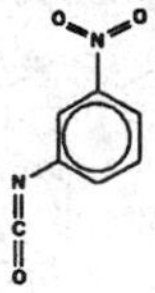

1790

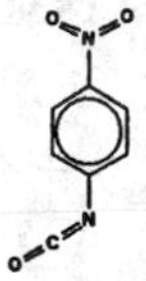

1791

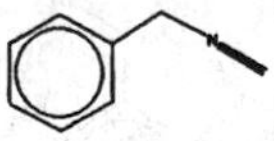

1792

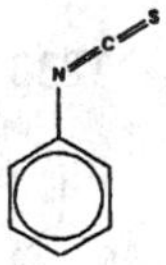

1793

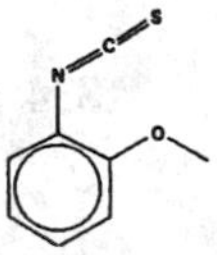

1794

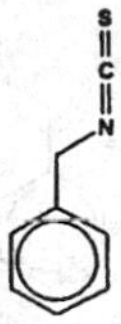

1795

1796

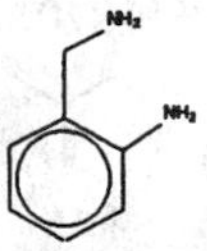

1797

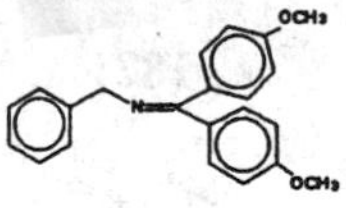

1798

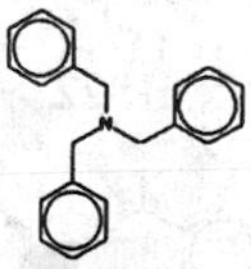

1799

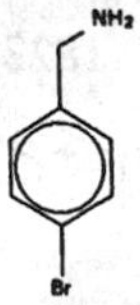

1800

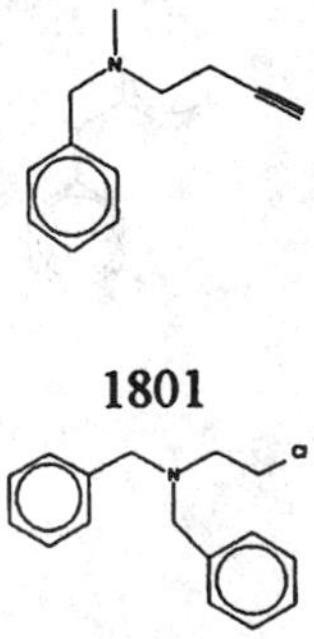

1801

1802

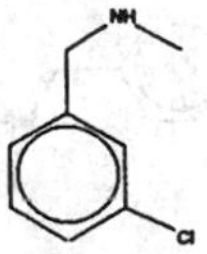

1803

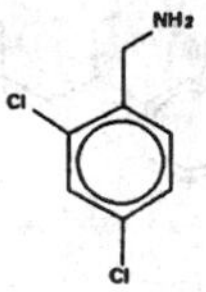

1804

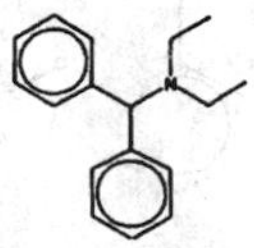

1805

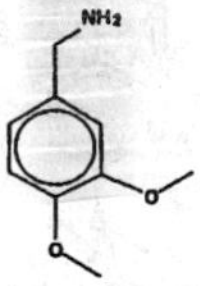

1806

1807

1808

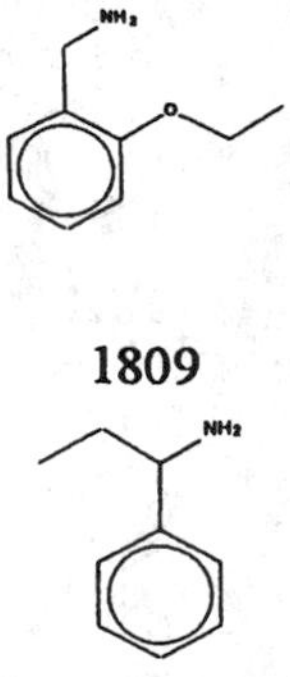

1809

STRUCTURAL FORMULAS OF ORGANIC COMPOUNDS (continued)

In numeric order as they occur in the Table of Physical Constants of Organic Compounds

1810

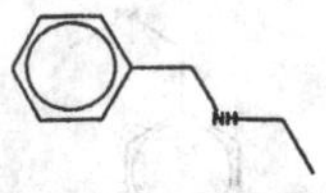

1811

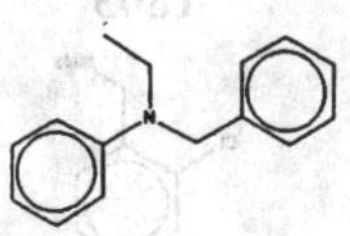

1812

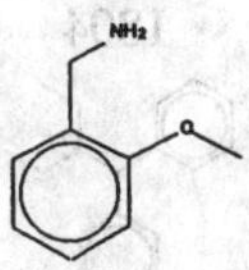

1813

1814

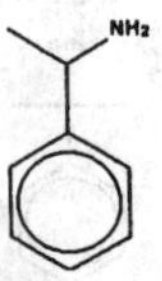

1815

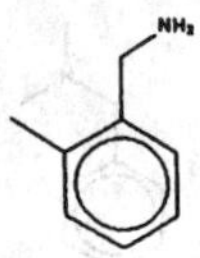

1816

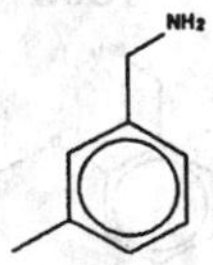

1817

1818

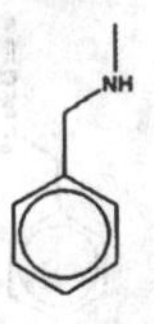

1819

1820

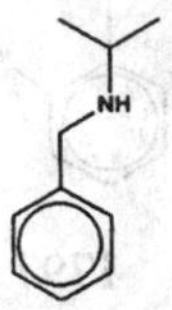

1821

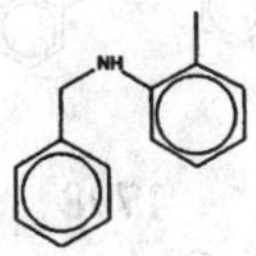

1822

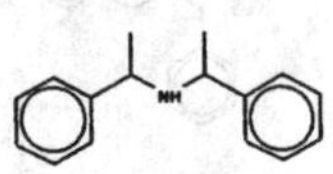

1823

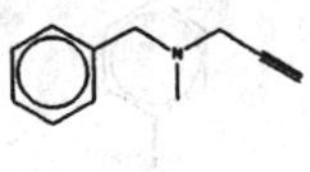

1826

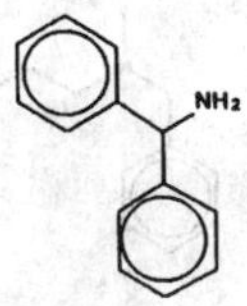

1827

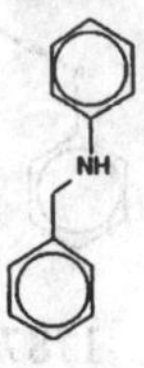

1828

1829

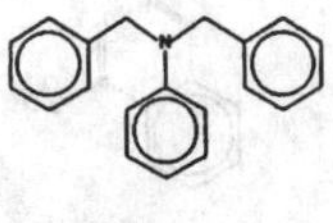

1830

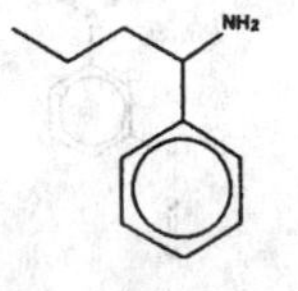

1831

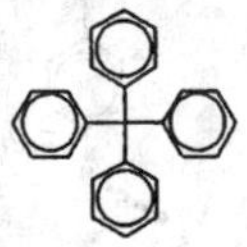

1832

1833

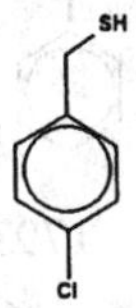

1834

1835

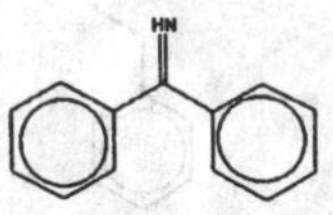

1836

1837

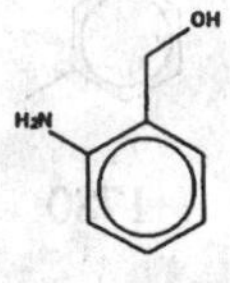

1838

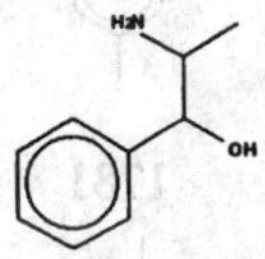

1839

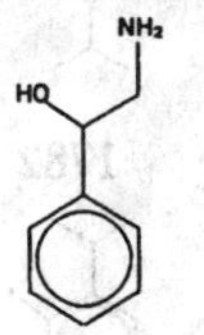

1840

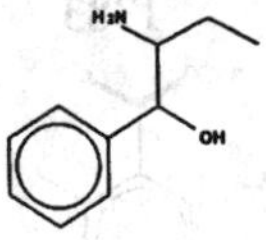

1841

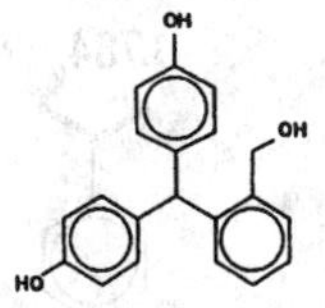

1842

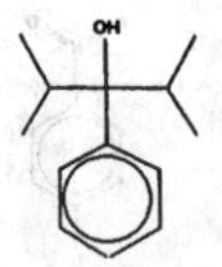

1843

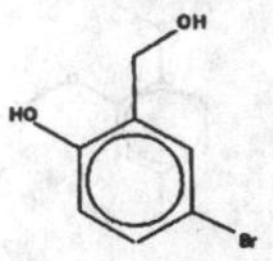

1844

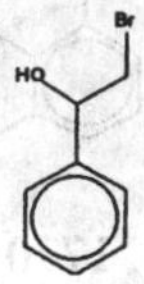

1845

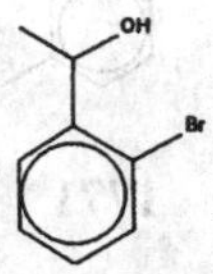

1846

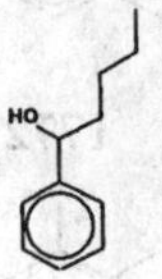

1847

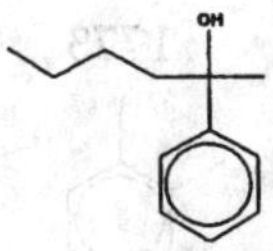

1848

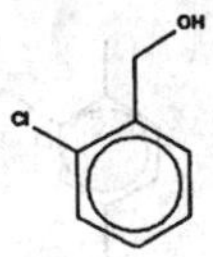

1849

STRUCTURAL FORMULAS OF ORGANIC COMPOUNDS (continued)

In numeric order as they occur in the Table of Physical Constants of Organic Compounds

1850

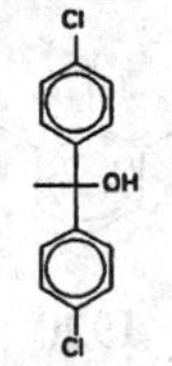

1851

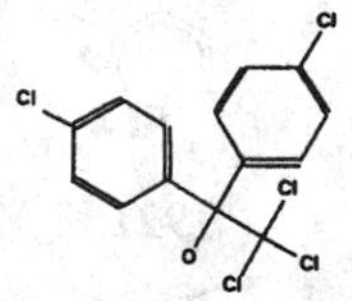

1852

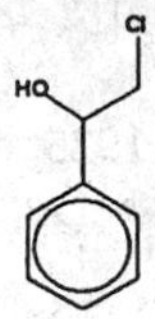

1853

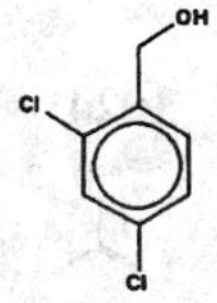

1854

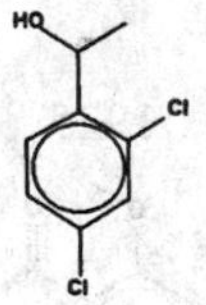

1855

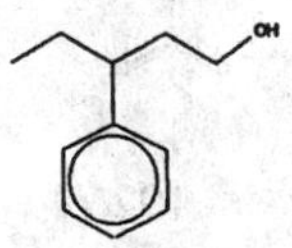

1856

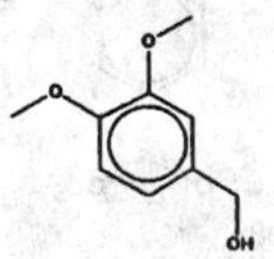

1857

1858

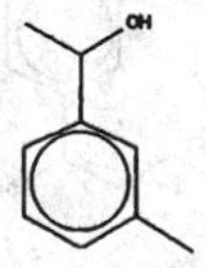

1859

1860

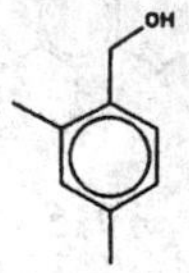

1861

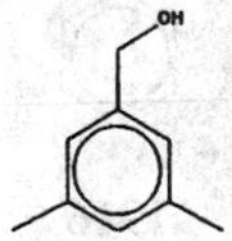

1862

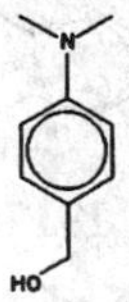

1863

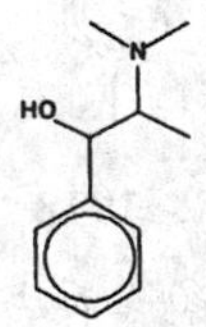

1864

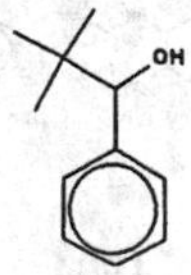

1865

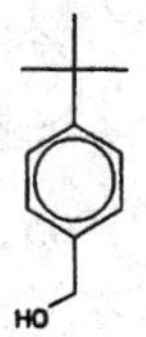

1866

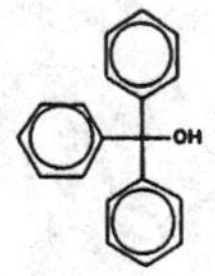

1867

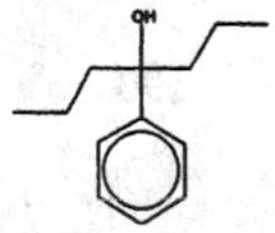

1868

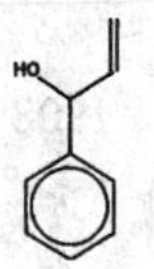

1869

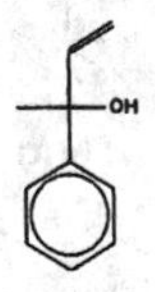

1870

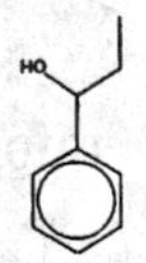

1871

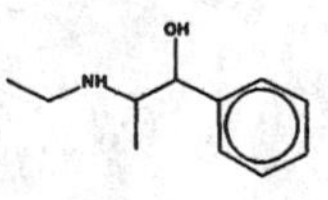

1872

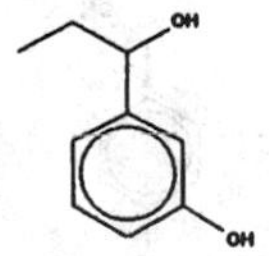

1873

1874

1875

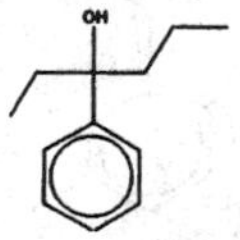

1876

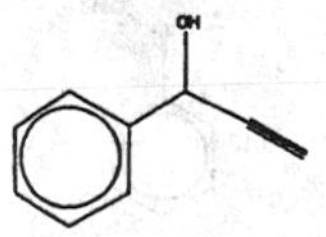

1877

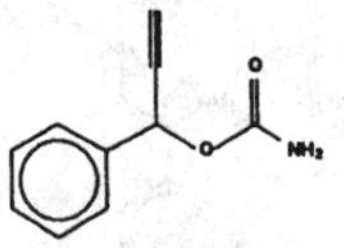

1878

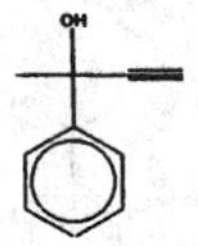

1879

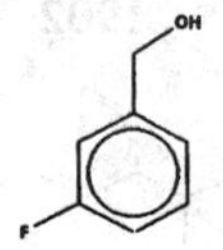

1880

1881

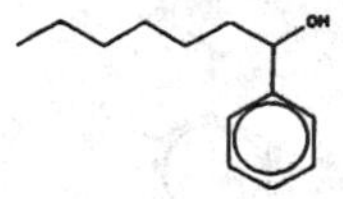

1882

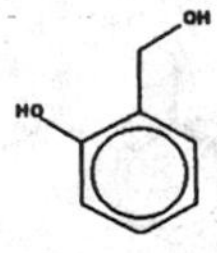

1883

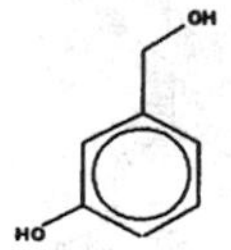

1884

1885

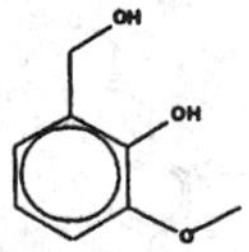

1886

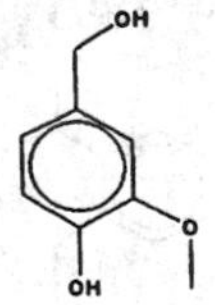

STRUCTURAL FORMULAS OF ORGANIC COMPOUNDS (continued)

In numeric order as they occur in the Table of Physical Constants of Organic Compounds

1887

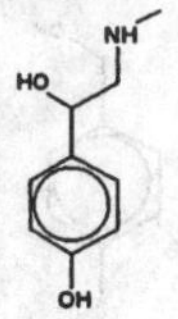

1888

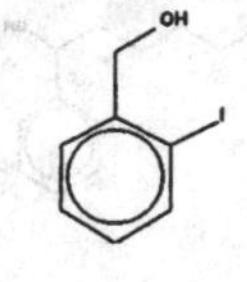

1889

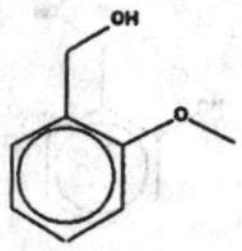

1890

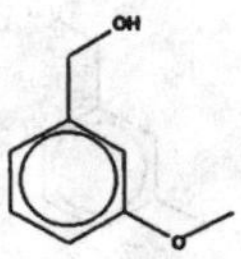

1891

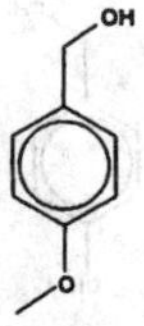

1892

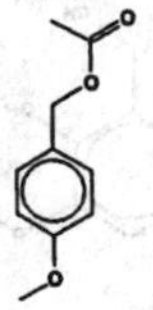

1893

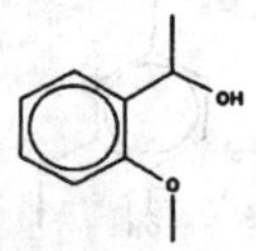

1894

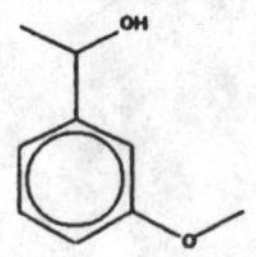

1895

1896

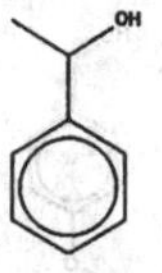

1897

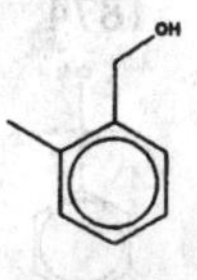

1898

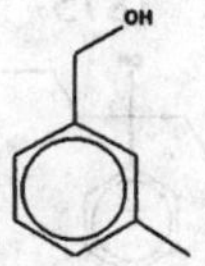

1899

1900

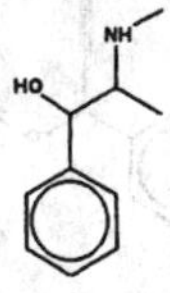

1901

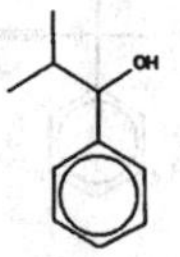

1902

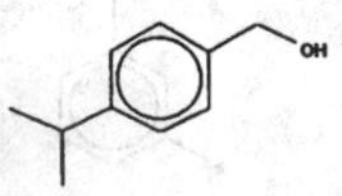

1903

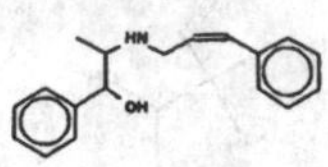

1904

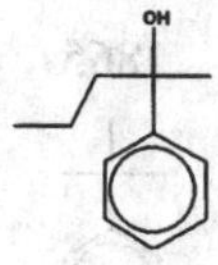

1905

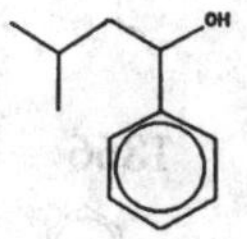

1906

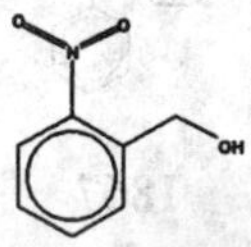

1907

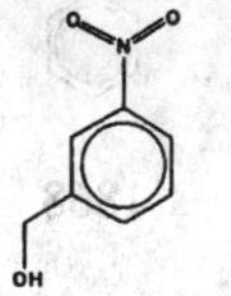

1908

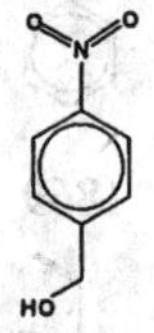

1909

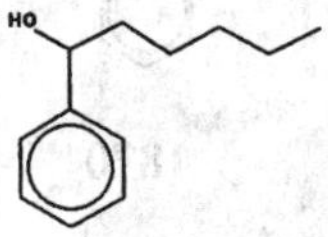

1910

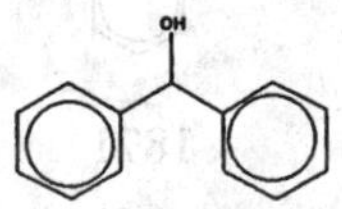

1911

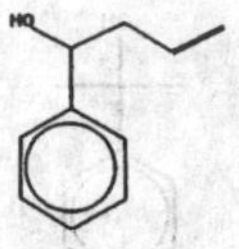

1912

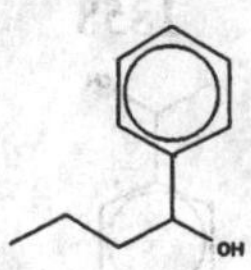

1913

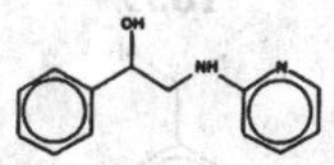

1914

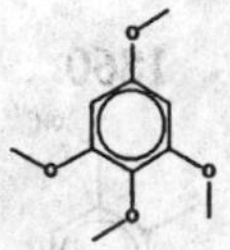

1915

1916

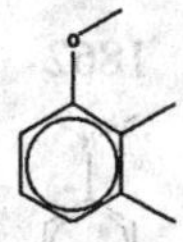

1917

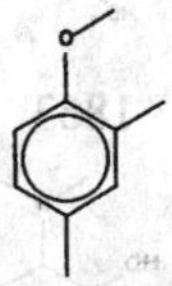

1918

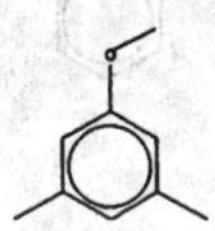

1919

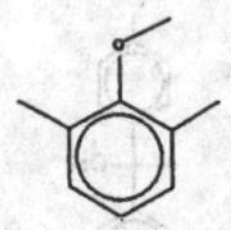

1920

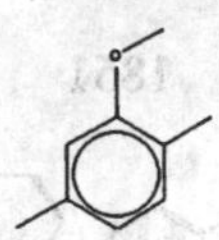

1921

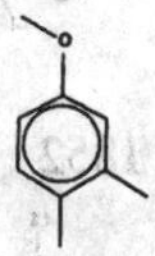

1923

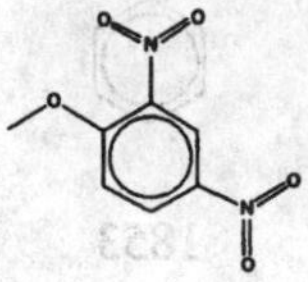

1924

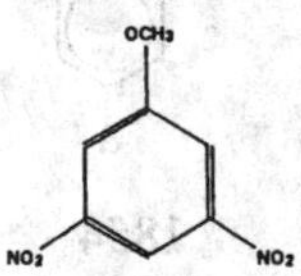

1925

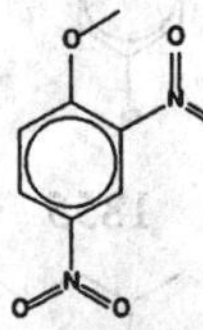

1926

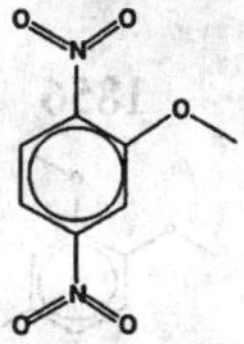

1927

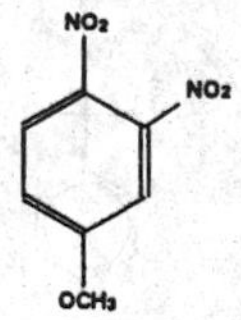

1928

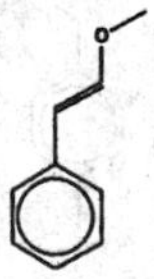

1929

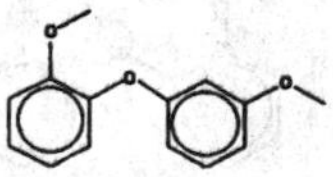

1930

1931

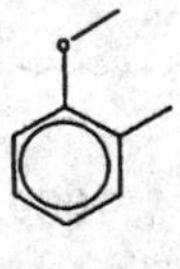

1932

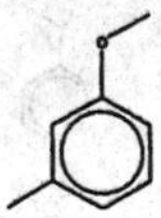

1933

1934

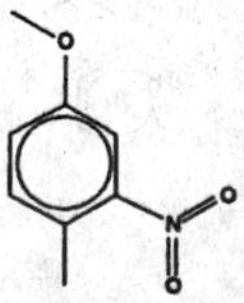

1935

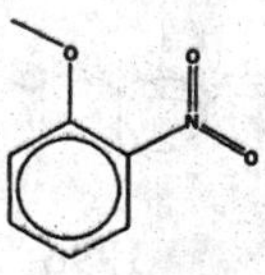

1936

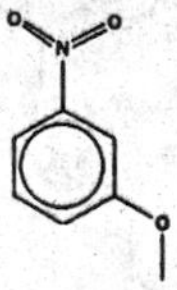

1937

1938

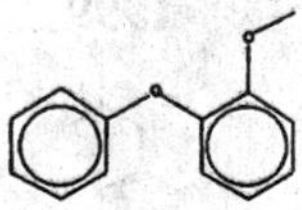

1939

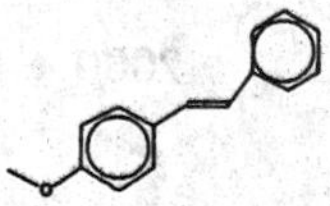

1940

1941

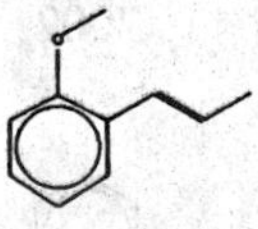

1942

1943

1944

1945

1946

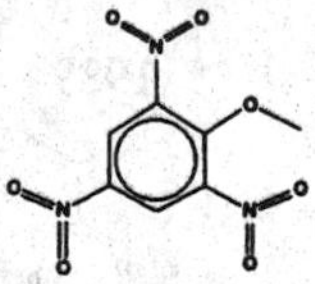

1947

1948

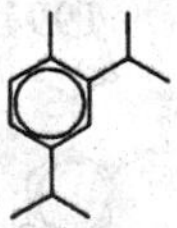

1949

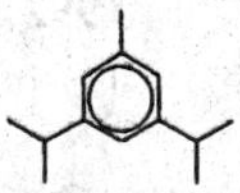

1950

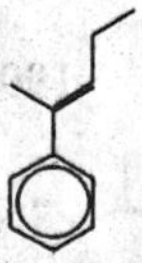

1951

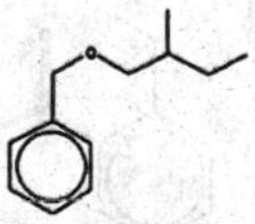

1952

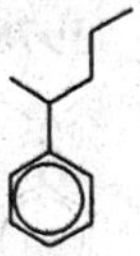

1953

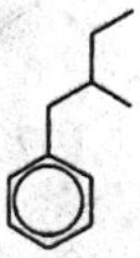

1954

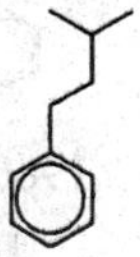

1955

1956

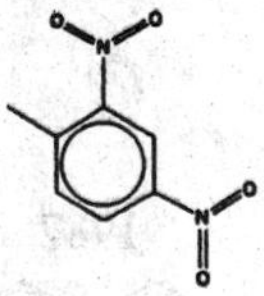

1957

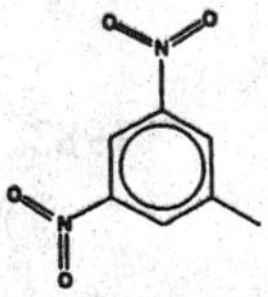

1958

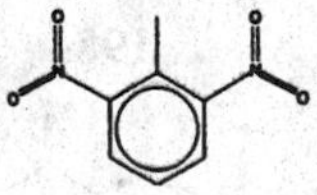

1959

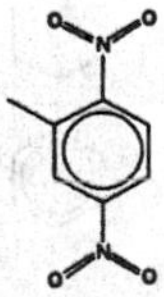

1960

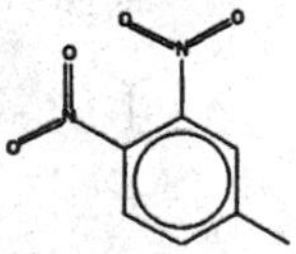

1961

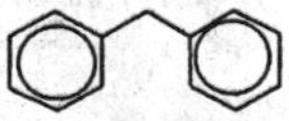

1962

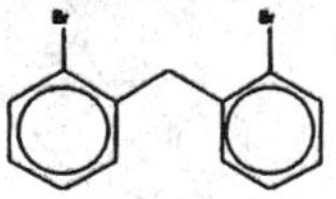

1963

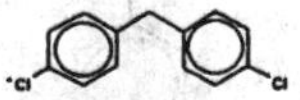

1964

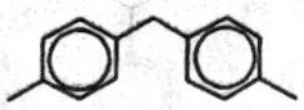

1965

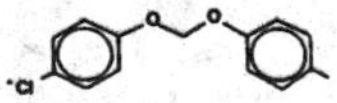

1966

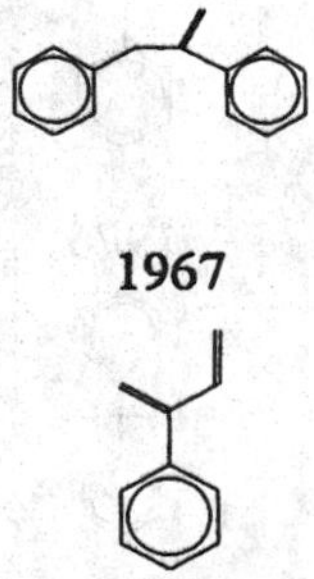

1967

STRUCTURAL FORMULAS OF ORGANIC COMPOUNDS (continued)

In numeric order as they occur in the Table of Physical Constants of Organic Compounds

1968

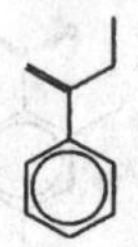

1969

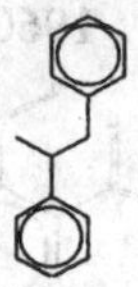

1970

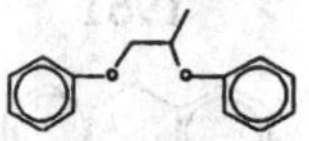

1971

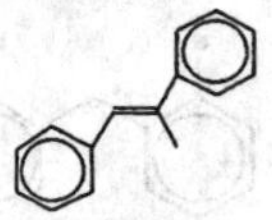

1972

1973

1974

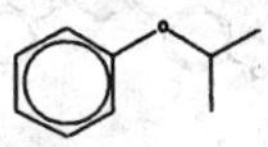

1975

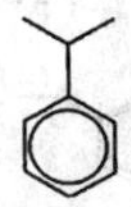

1976

1977

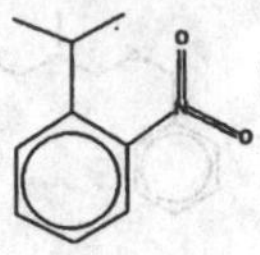

1978

1979

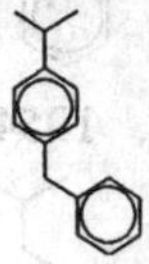

1980

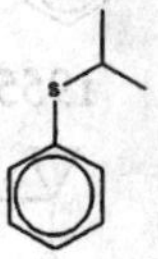

1981

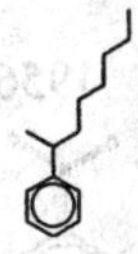

1982

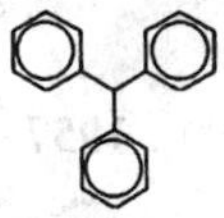

1983

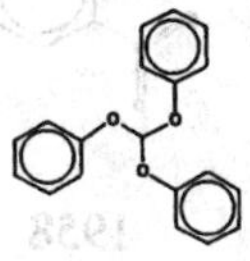

1984

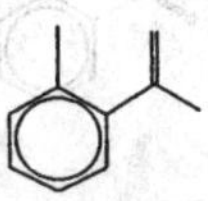

1985

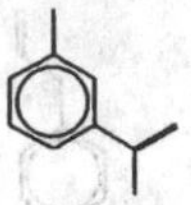

1986

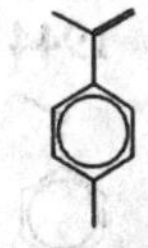

1987

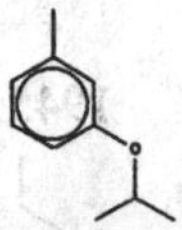

1988

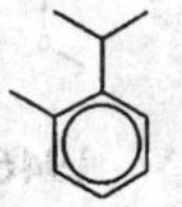

1989

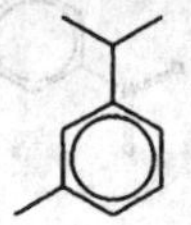

1990

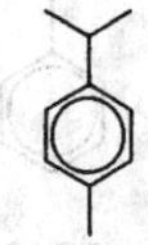

1991

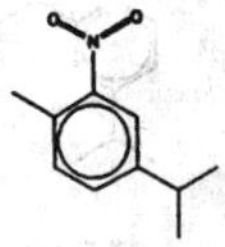

1992

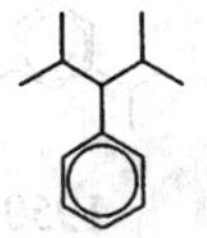

1993

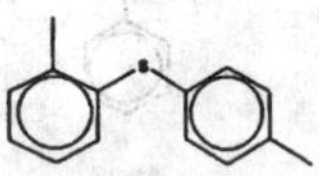

1995

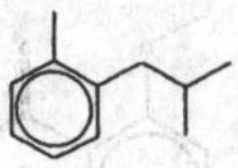

1996

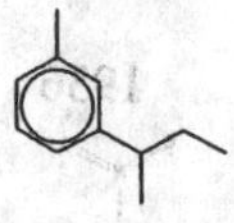

1997

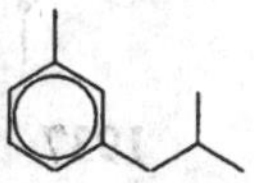

1998

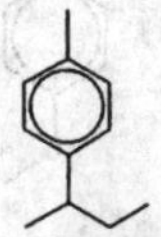

1999

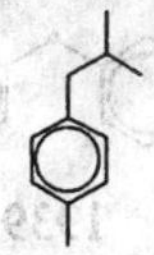

2000

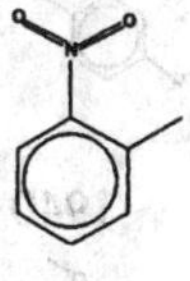

2001

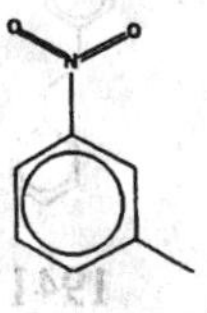

2002

2003

2004

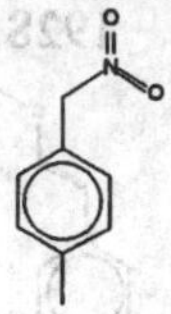

2005

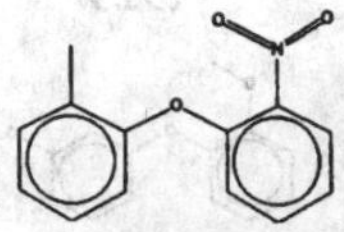

2006

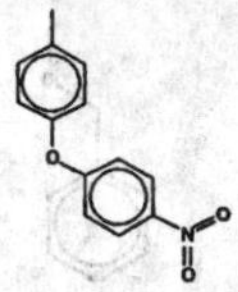

2007

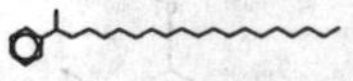

2009

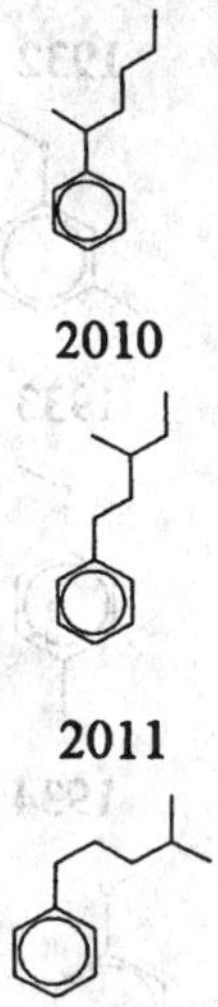

2010

2011

STRUCTURAL FORMULAS OF ORGANIC COMPOUNDS (continued)

In numeric order as they occur in the Table of Physical Constants of Organic Compounds

2012

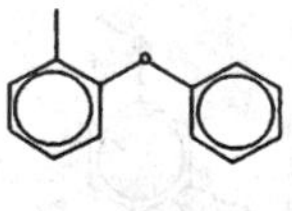

2013

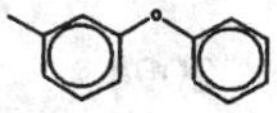

2014

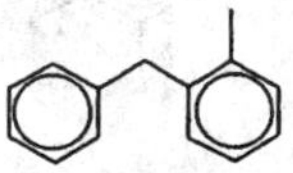

2015

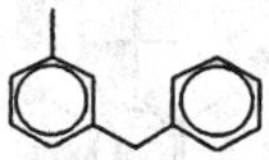

2016

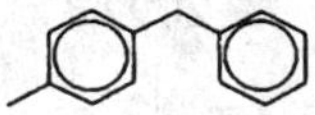

2017

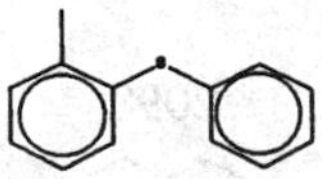

2018

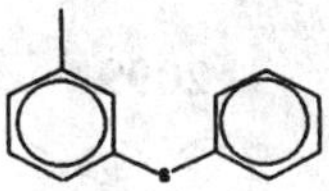

2019

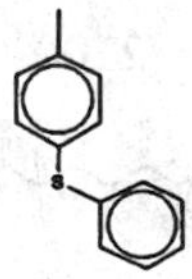

2021

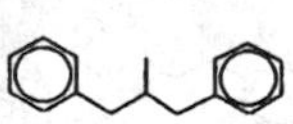

2022

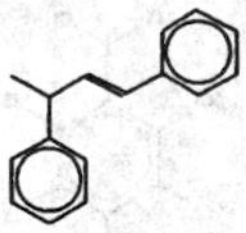

2023

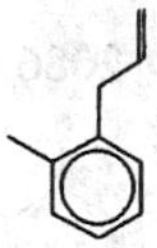

2024

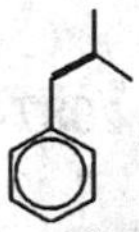

2025

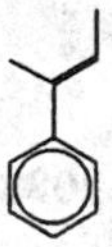

2026

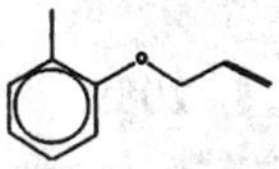

2027

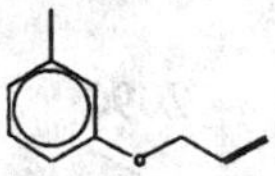

2028

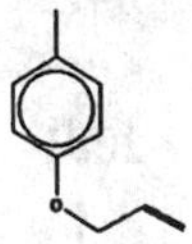

2029

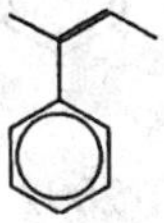

2030

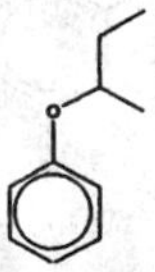

2031

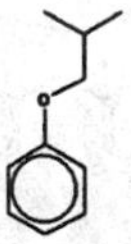

2033

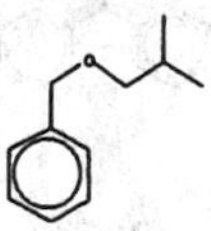

2034

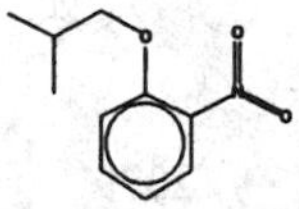

2035

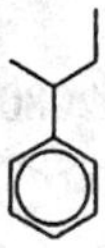

2036

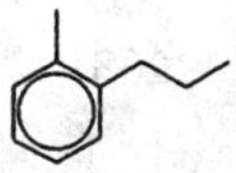

2037

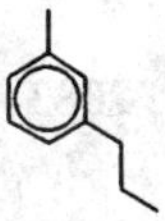

2038

2039

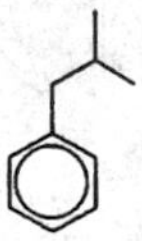

2040

2041

2042

2044

2045

2047

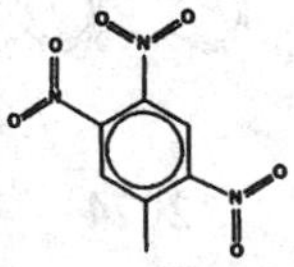

2048

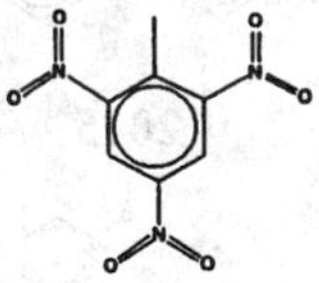

2049

2050

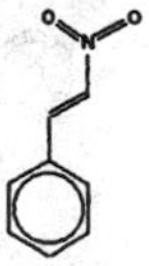

2051

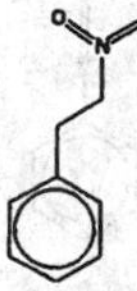

2052

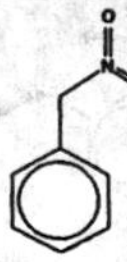

2053

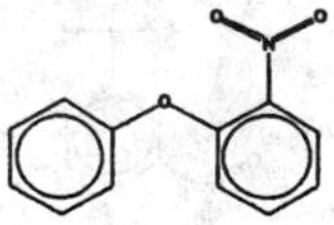

2054

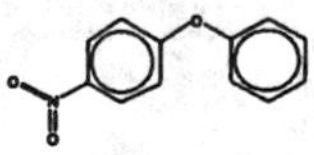

2055

2056

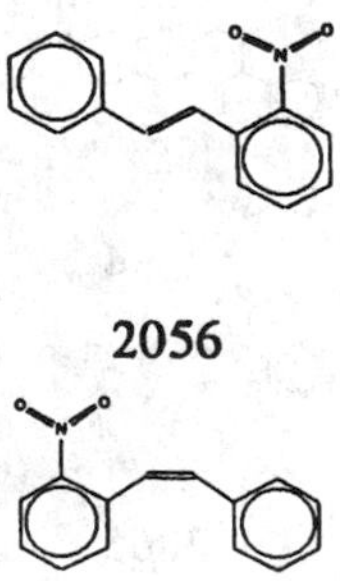

STRUCTURAL FORMULAS OF ORGANIC COMPOUNDS (continued)

In numeric order as they occur in the Table of Physical Constants of Organic Compounds

2057

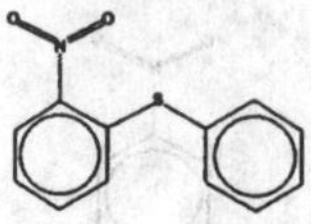

2058

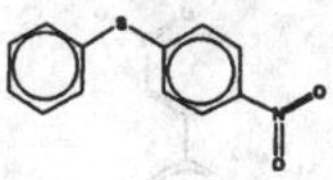

2059

2060

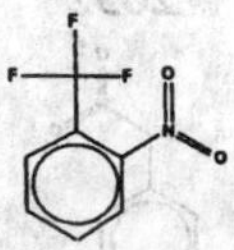

2061

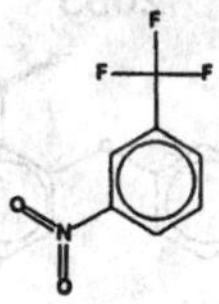

2062

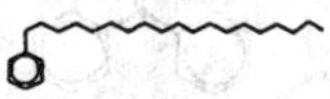

2063

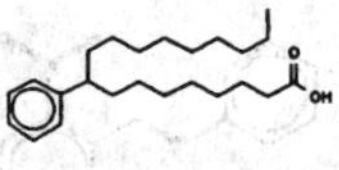

2064

2065

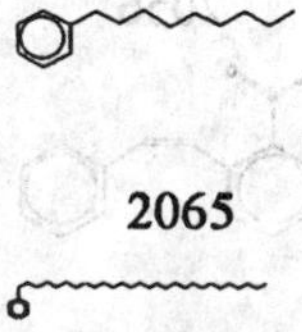

2066

2067

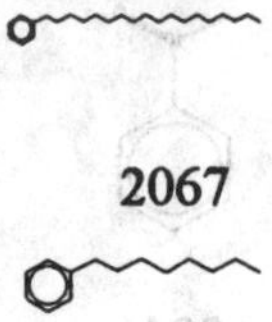

2068

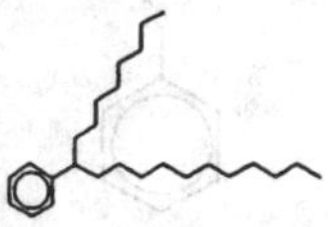

2069

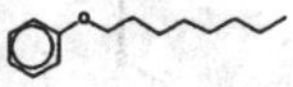

2070

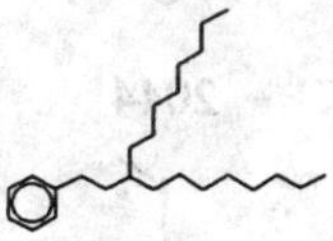

2071

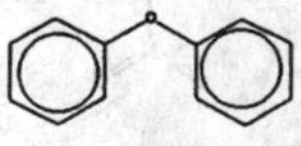

2072

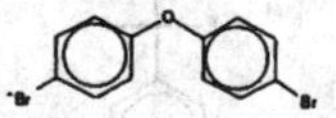

2073

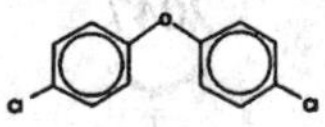

2074

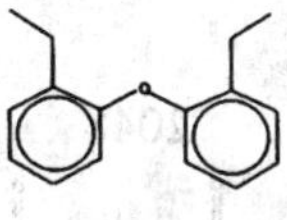

2075

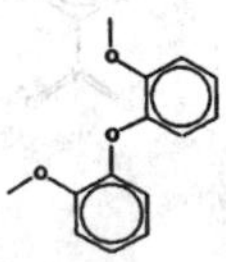

2076

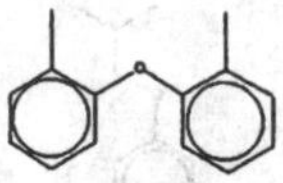

2077

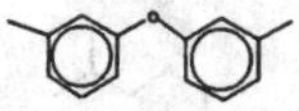

2078

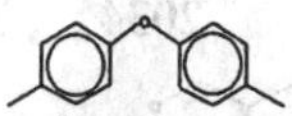

2079

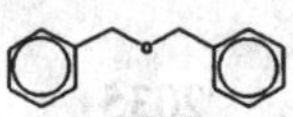

2080

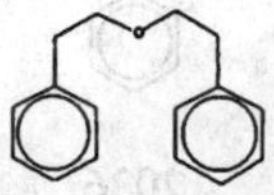

2081

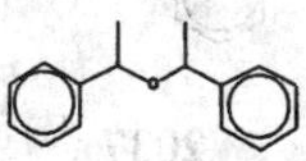

2083

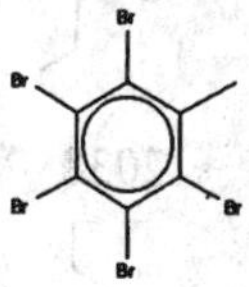

2084

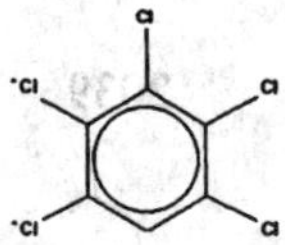

2085

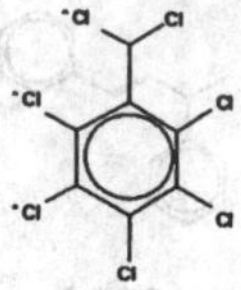

2086

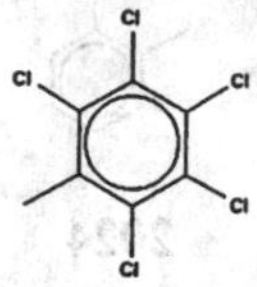

2087

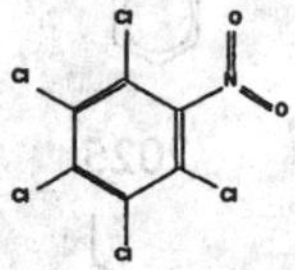

2088

2089

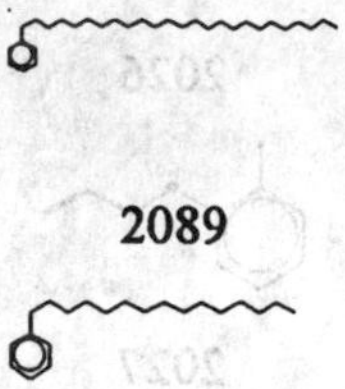

2090

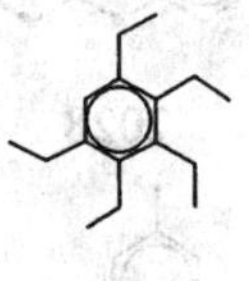

2091

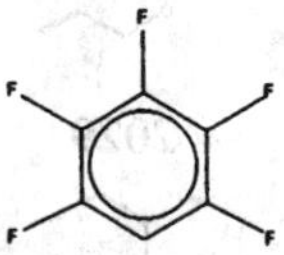

2092

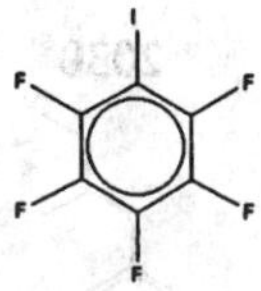

2093

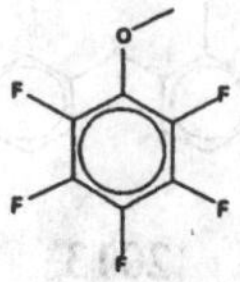

2094

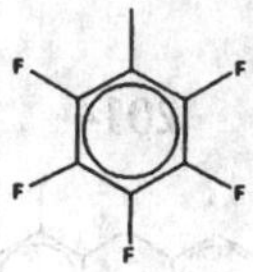

2095

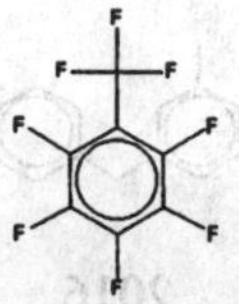

2097

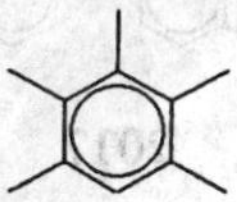

2098

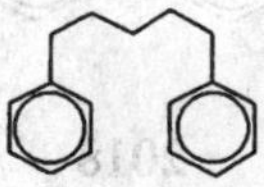

2099

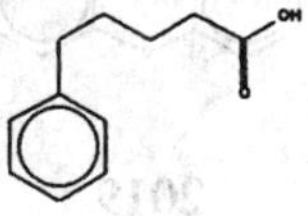

2100

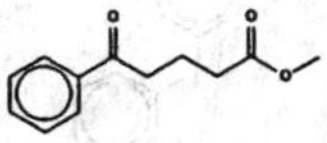

2101

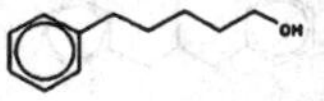

STRUCTURAL FORMULAS OF ORGANIC COMPOUNDS (continued)

In numeric order as they occur in the Table of Physical Constants of Organic Compounds

2102

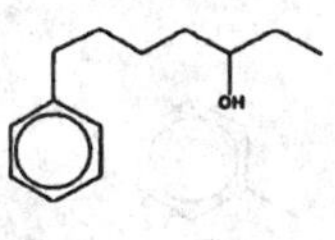

2103

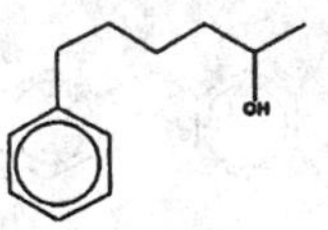

2104

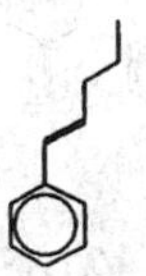

2105

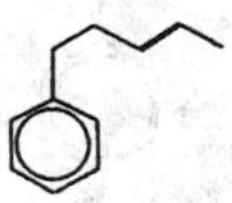

2106

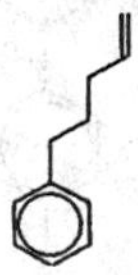

2107

2108

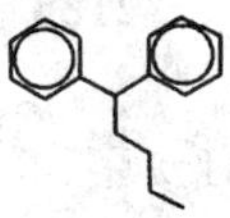

2110

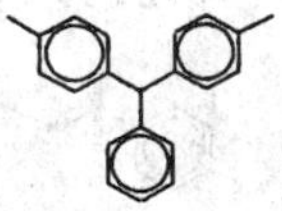

2111

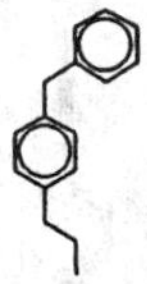

2113

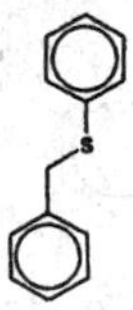

2114

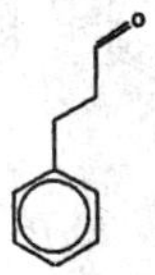

2116

2117

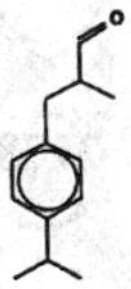

2118

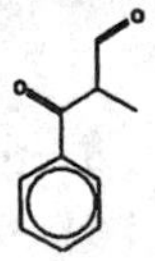

2119

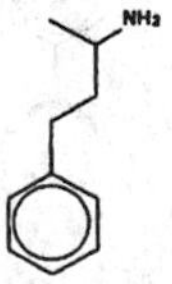

2121

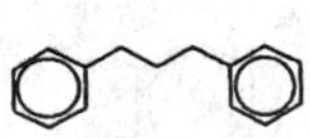

2122

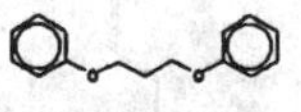

2123

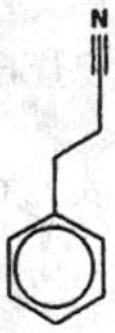

2124

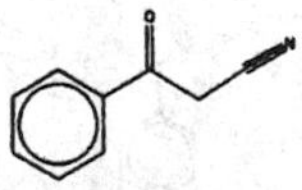

2125

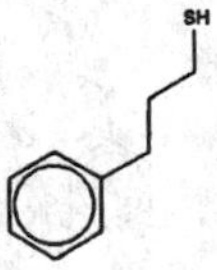

2126

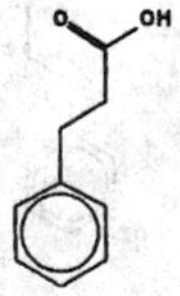

2127

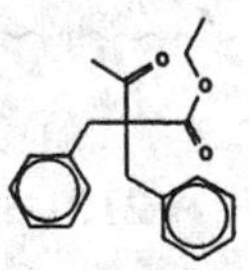

2129

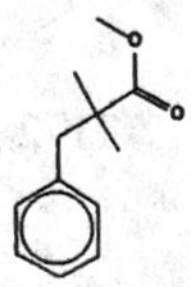

2130

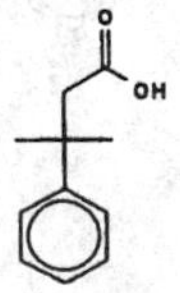

2131

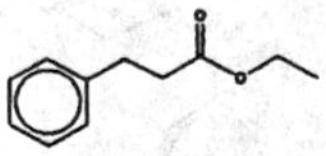

2132

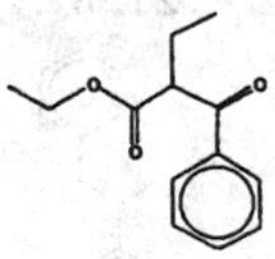

2133

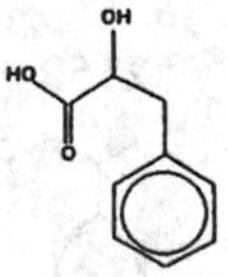

2134

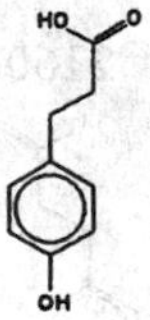

2137

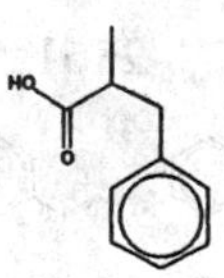

2138

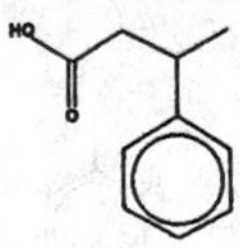

2139

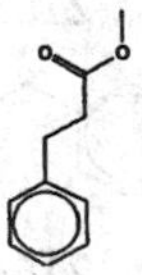

2140

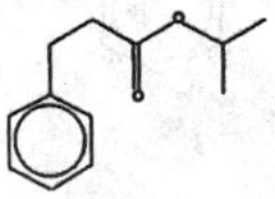

2141

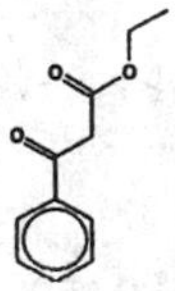

2142

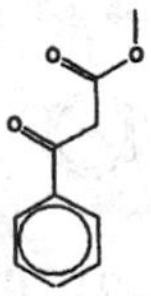

2143

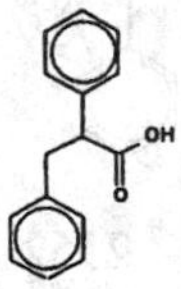

2144

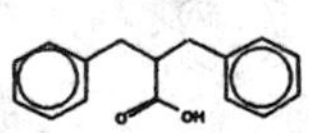

2145

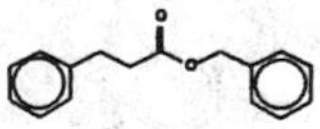

2146

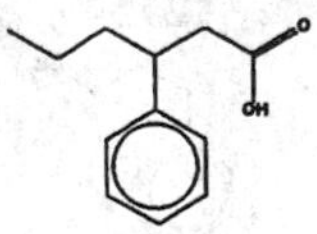

2147

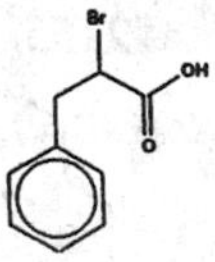

2148

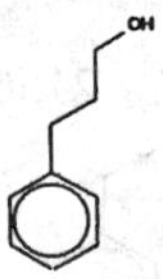

STRUCTURAL FORMULAS OF ORGANIC COMPOUNDS (continued)

In numeric order as they occur in the Table of Physical Constants of Organic Compounds

2149

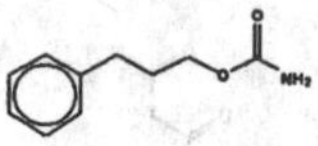

2150

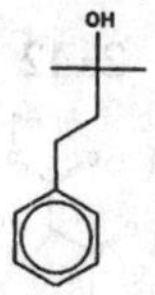

2151

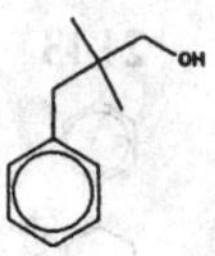

2152

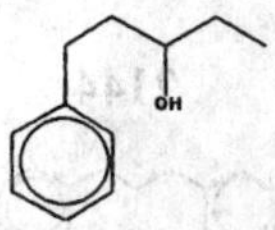

2153

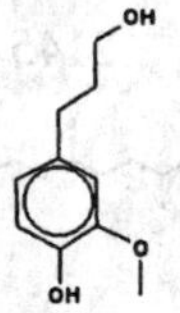

2154

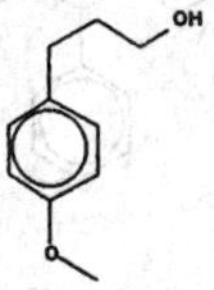

2155

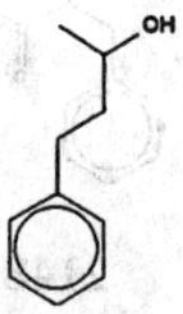

2156

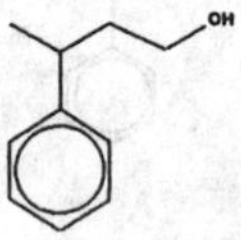

2157

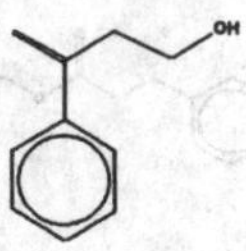

2158

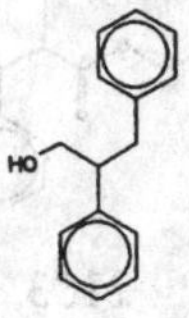

2159

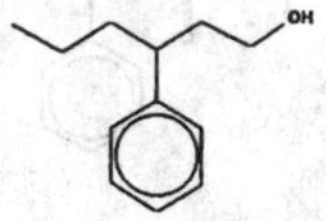

2160

2161

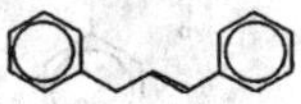

2162

2163

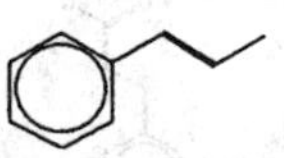

2164

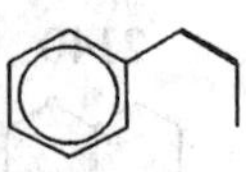

2165

2166

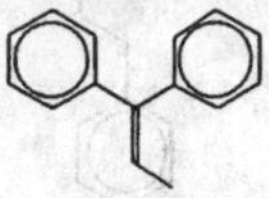

2167

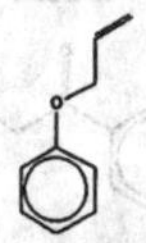

2168

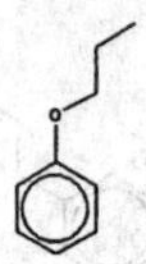

2169

2170

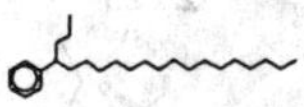

2171

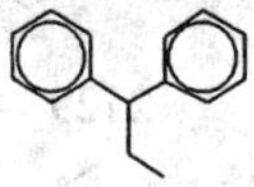

2172

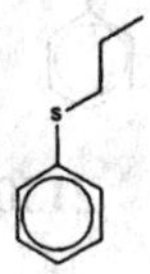

2173

2174

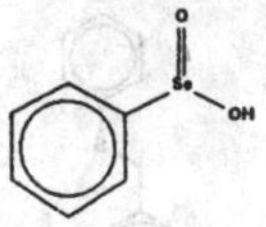

2175

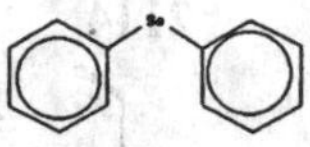

2176

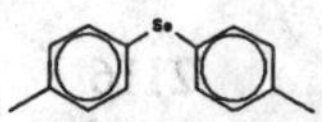

2177

2178

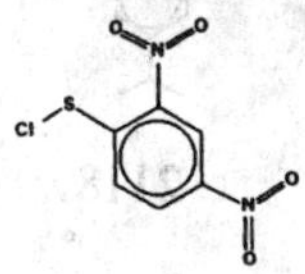

2179

2181

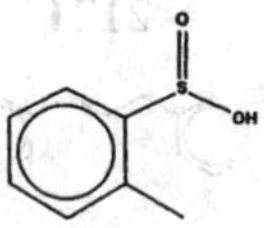

2182

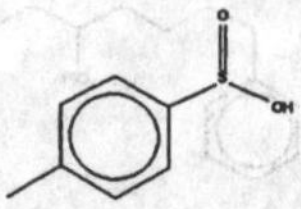

2183

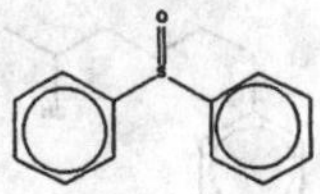

2184

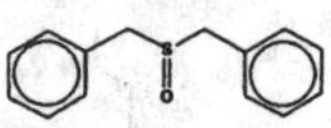

2185

2186

2187

2188

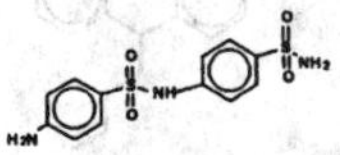

2189

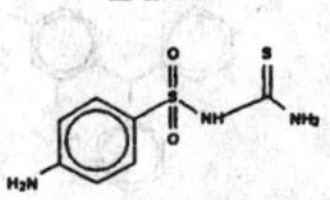

STRUCTURAL FORMULAS OF ORGANIC COMPOUNDS (continued)

In numeric order as they occur in the Table of Physical Constants of Organic Compounds

2190

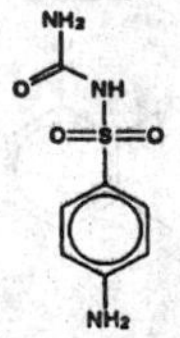

2191

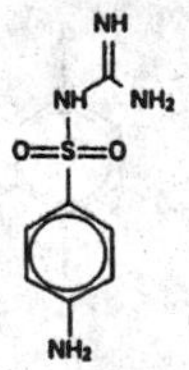

2192

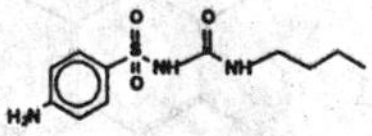

2193

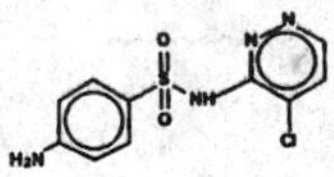

2194

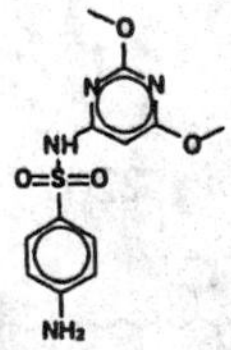

2195

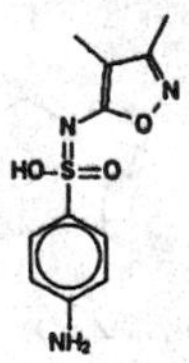

2196

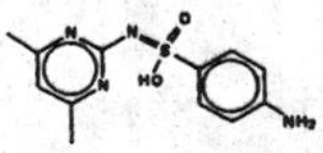

2197

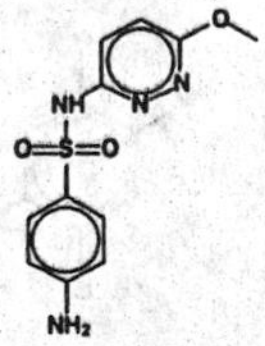

2198

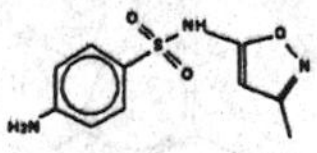

2199

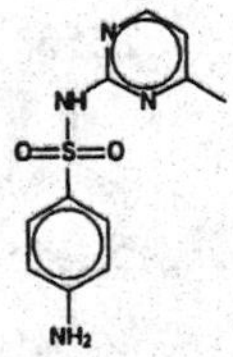

2200

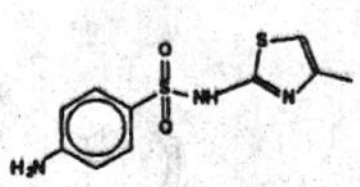

2202

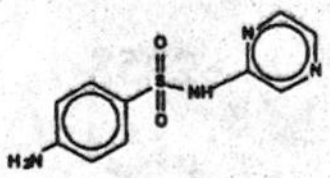

2203

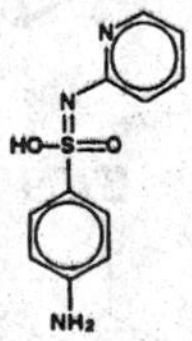

2204

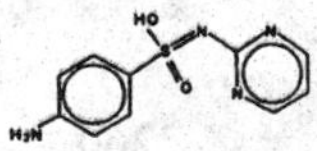

2205

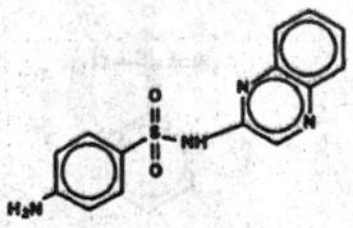

2206

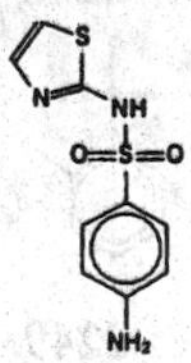

2207

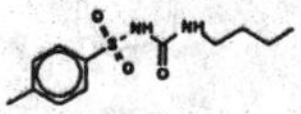

2208

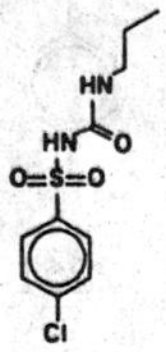

2210

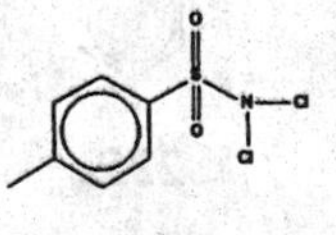

2211

2212

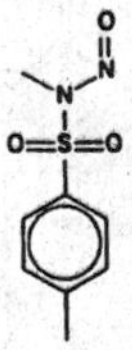

2213

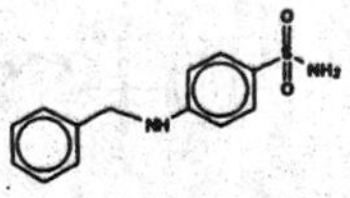

2214

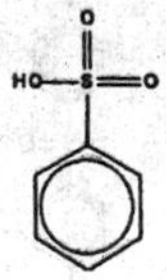

2215

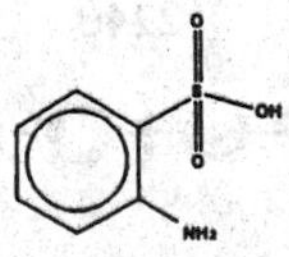

2216

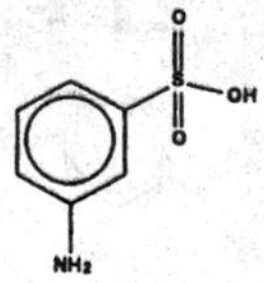

2217

2220

2221

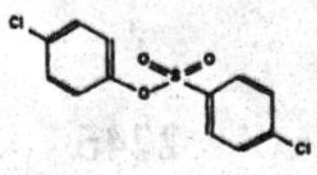

2223

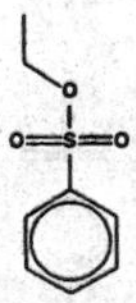

2224

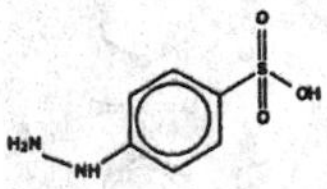

2225

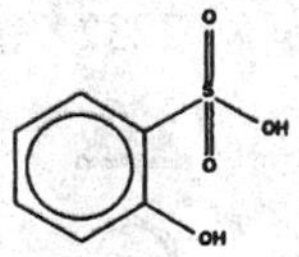

2226

2227

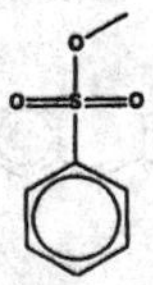

2228

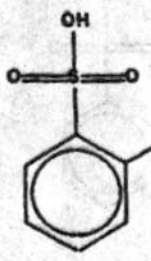

2229

2230

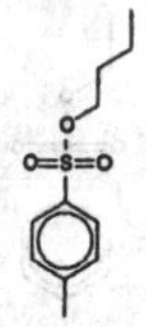

2231

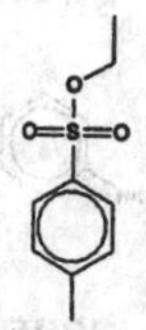

2232

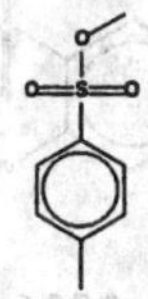

2233

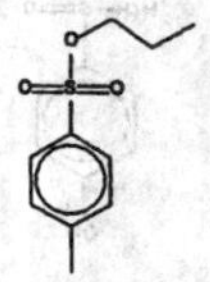

2234

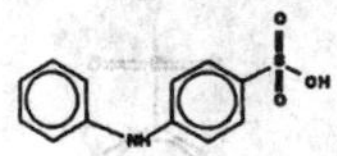

2235

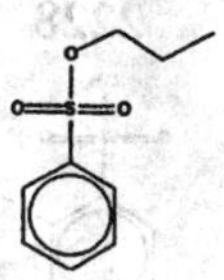

2236

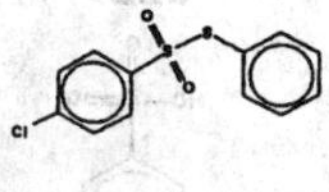

2237

2238

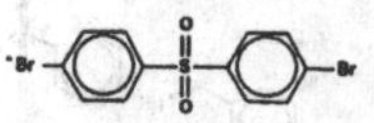

2239

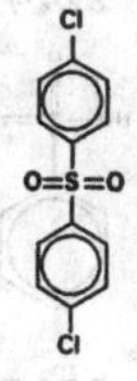

2240

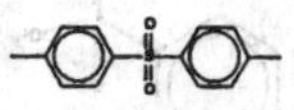

2242

2243

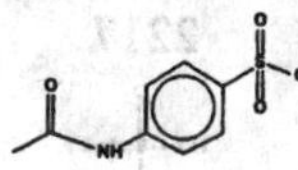

2244

2245

2246

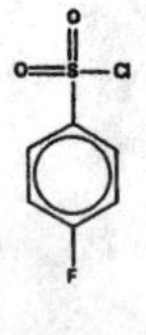

2247

2248

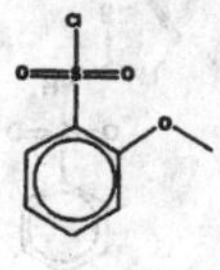

2249

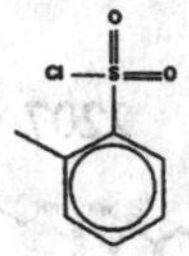

2250

2251

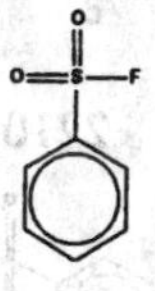

2252

2253

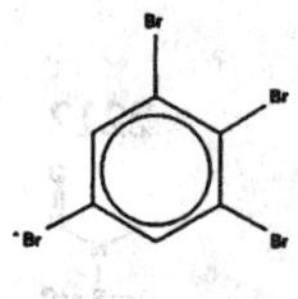

2254

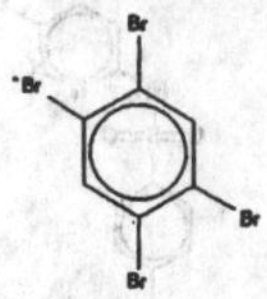

2255

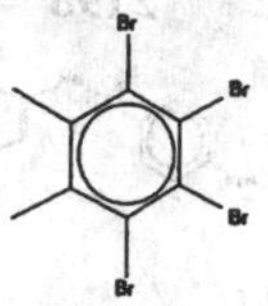

2256

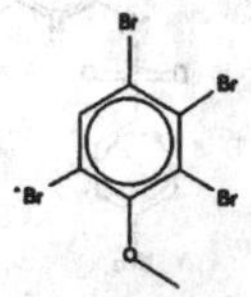

2257

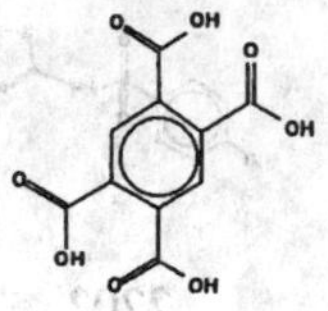

2258

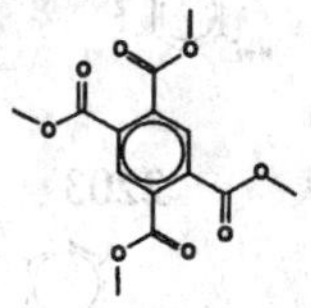

2259

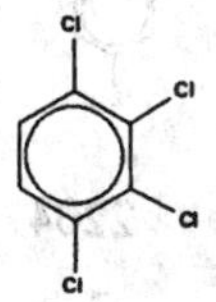

2260

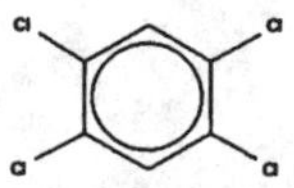

2261

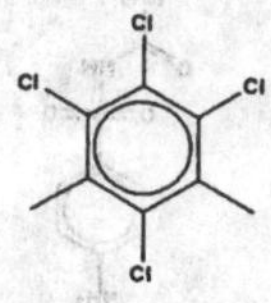

2262

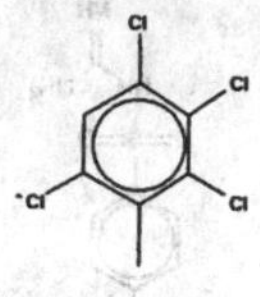

2264

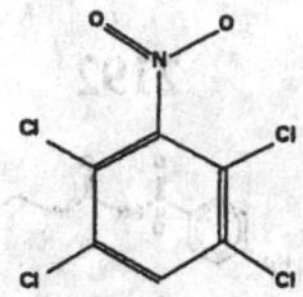

2265

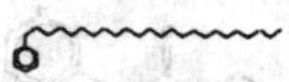

2266

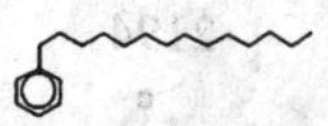

2267

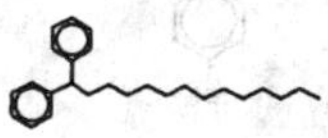

2268

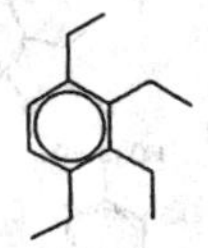

2269

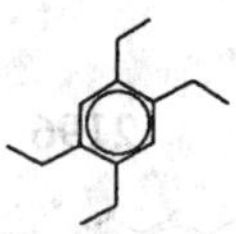

In numeric order as they occur in the Table of Physical Constants of Organic Compounds

2270

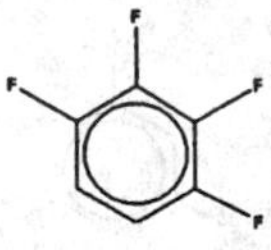

2271

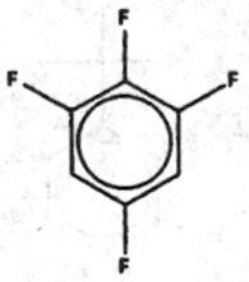

2272

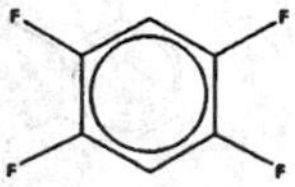

2276

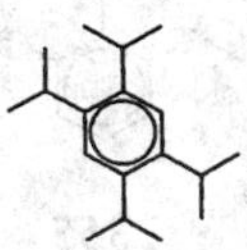

2277

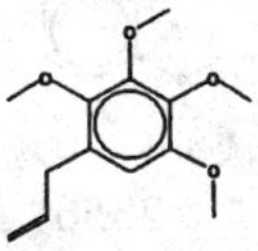

2278

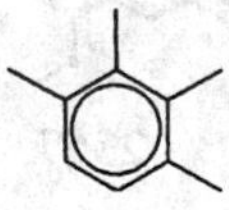

2279

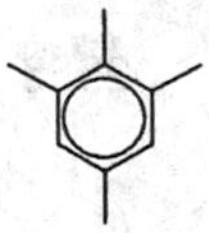

2280

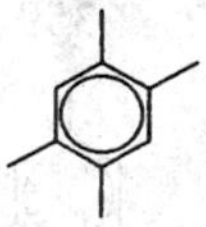

2281

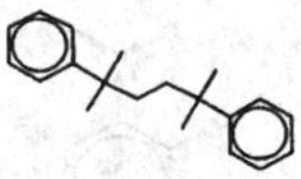

2283

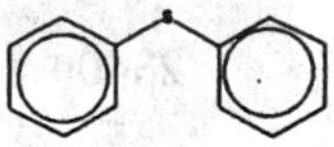

2284

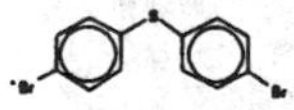

2285

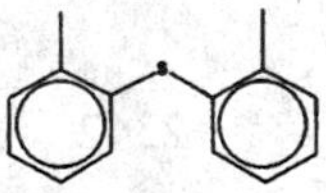

2286

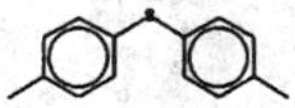

2287

2289

2290

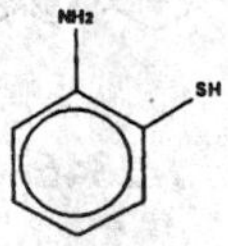

2291

2292

2293

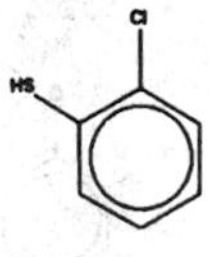

2294

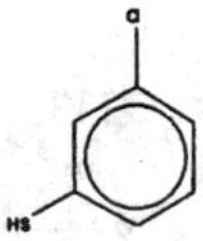

2295

2296

2297

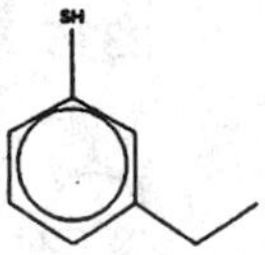

2299

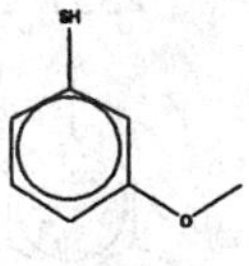

2300

2301

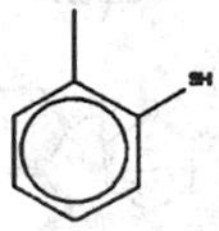

2302

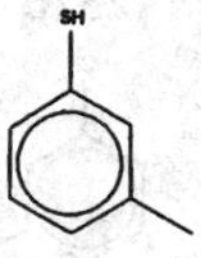

2303

2304

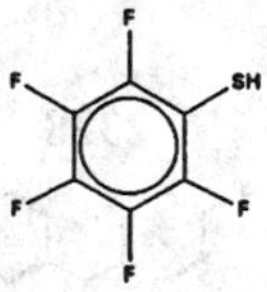

2305

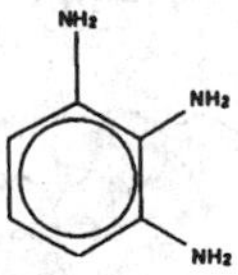

2306

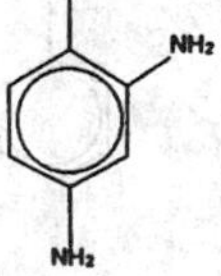

2308

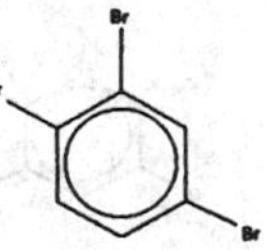

2309

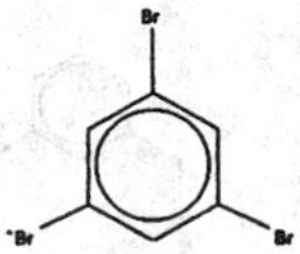

2310

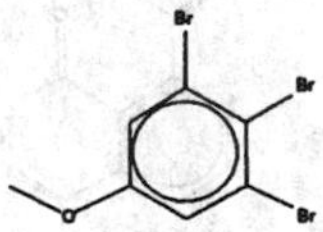

2311

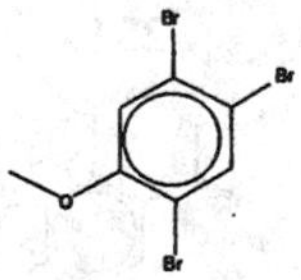

2312

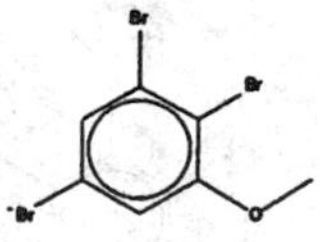

2313

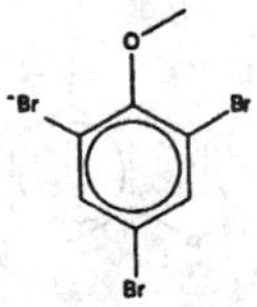

2314

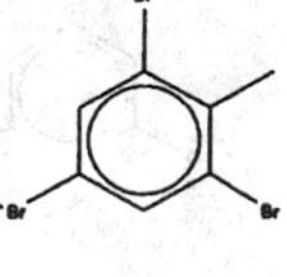

2316

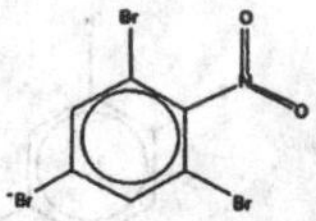

2317

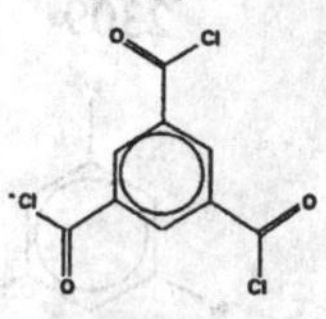

2318

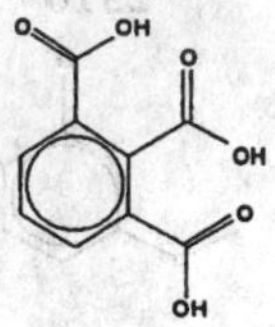

2319

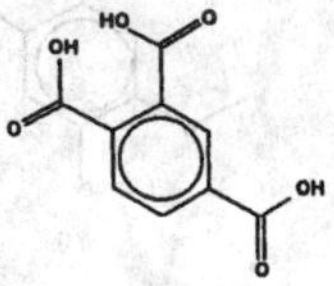

2321

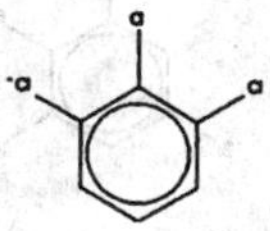

2322

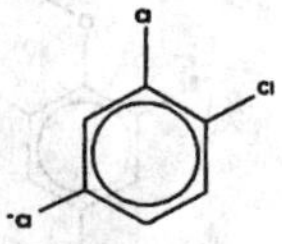

2323

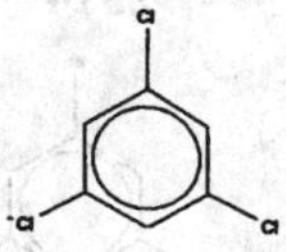

2324

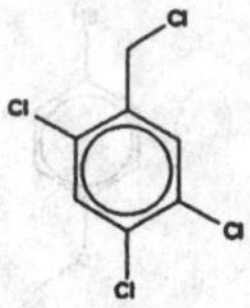

2325

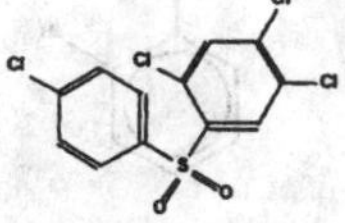

2326

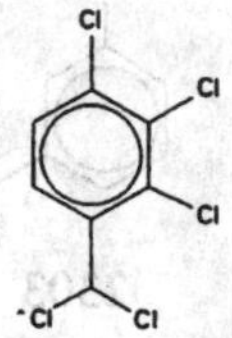

2327

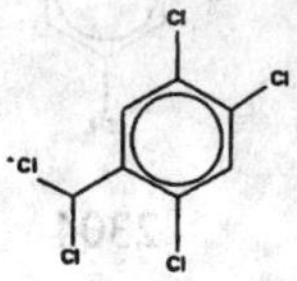

2328

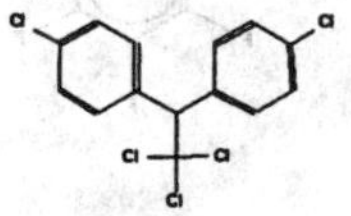

2329

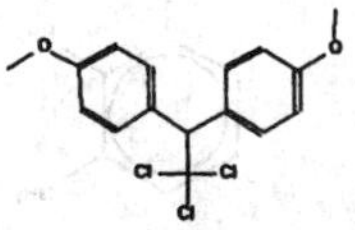

2330

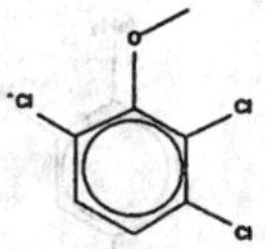

2331

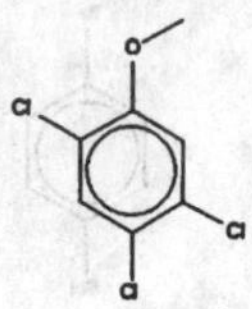

2332

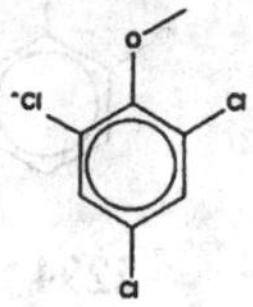

2333

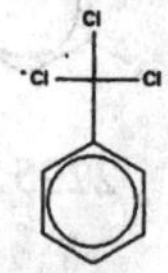

2334

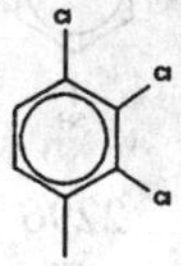

2335

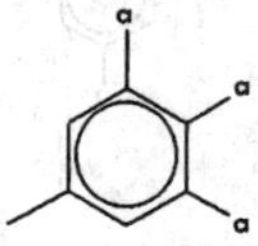

2336

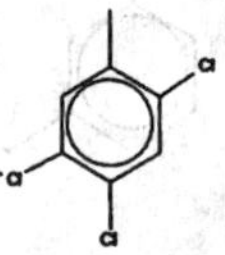

2337

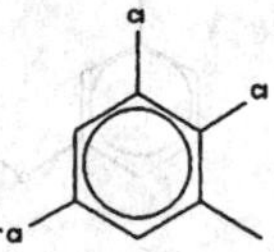

2339

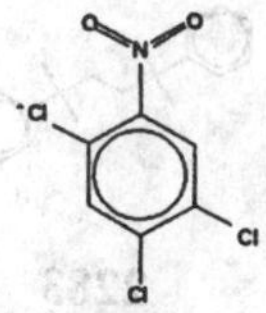

2340

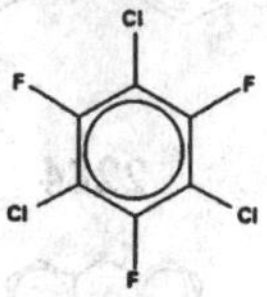

2341

2342

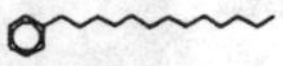

2343

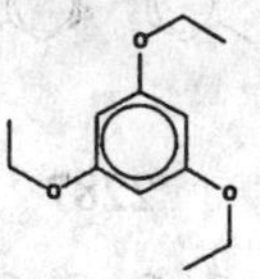

2344

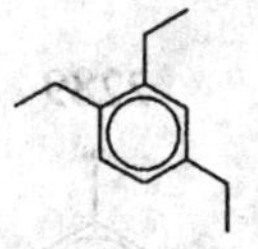

2345

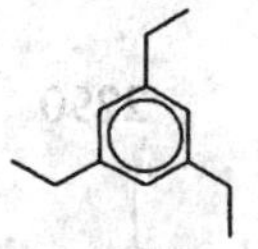

2346

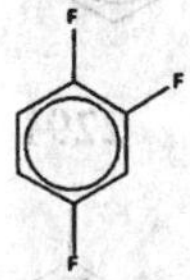

2347

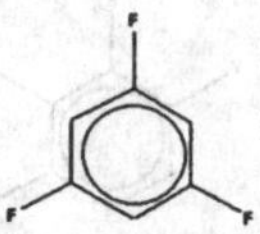

2348

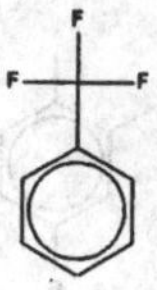

2352

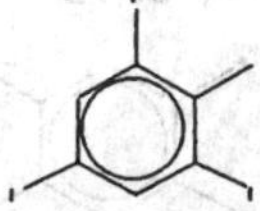

2354

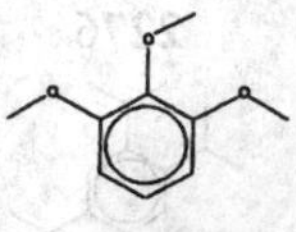

2355

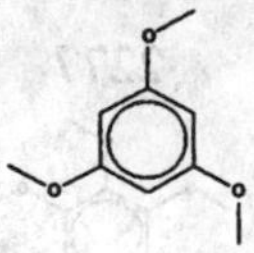

2356

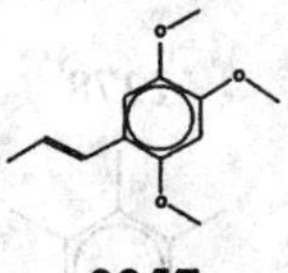

2357

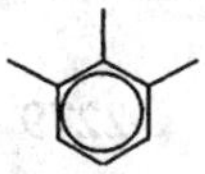

2358

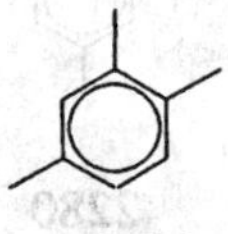

2359

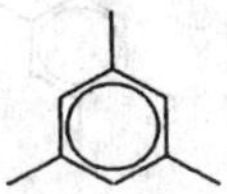

STRUCTURAL FORMULAS OF ORGANIC COMPOUNDS (continued)

In numeric order as they occur in the Table of Physical Constants of Organic Compounds

2360

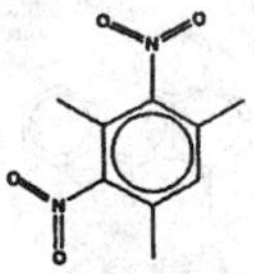

2361

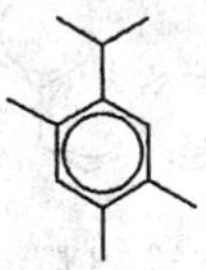

2362

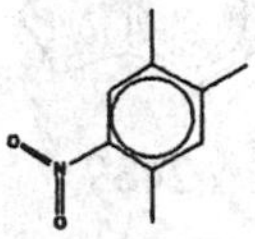

2363

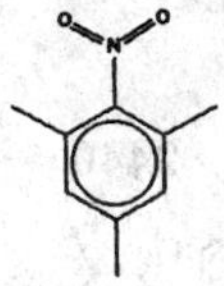

2364

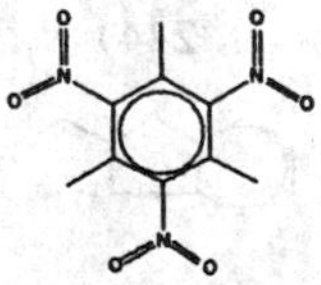

2365

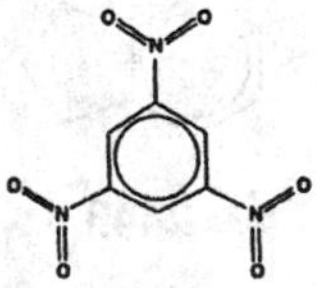

2366

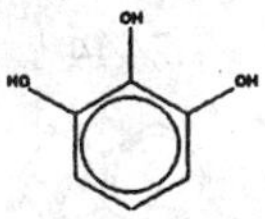

2367

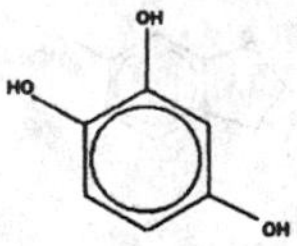

2368

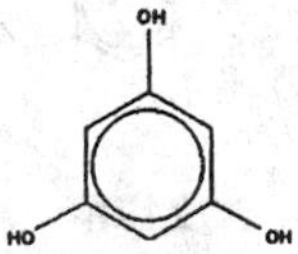

2369

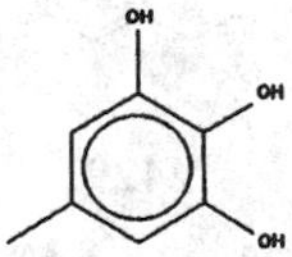

2370

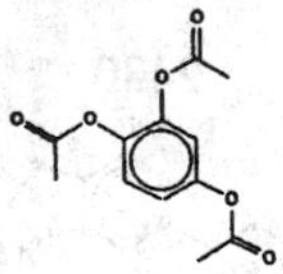

2371

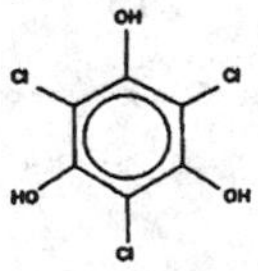

2372

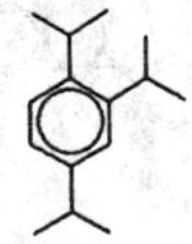

2373

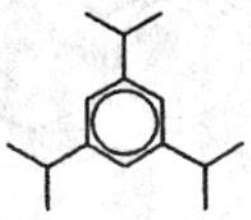

2374

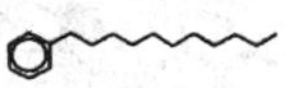

2375

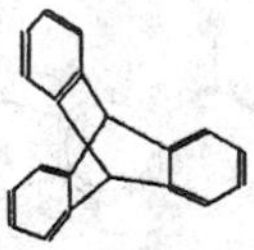

2376

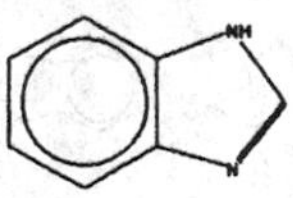

2377

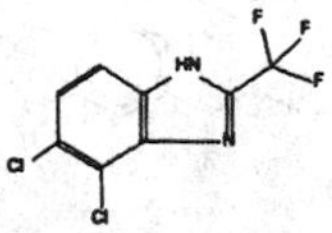

2378

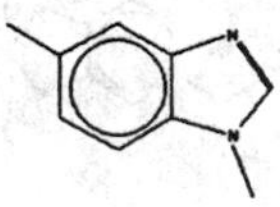

2379

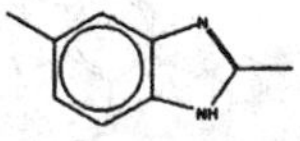

2380

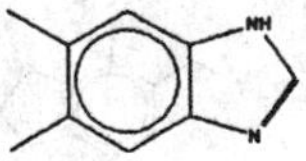

2381

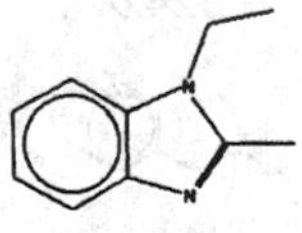

2382

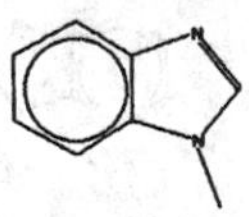

2383

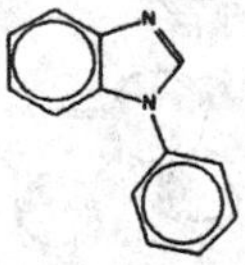

2384

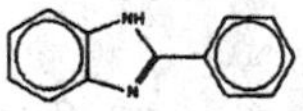

2385

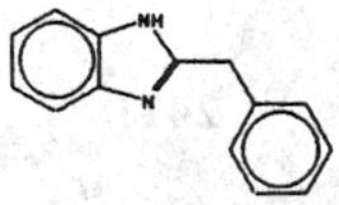

2386

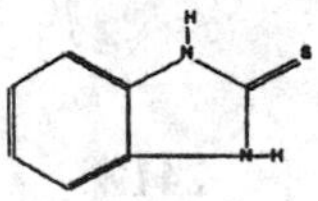

2387

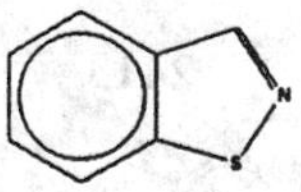

2388

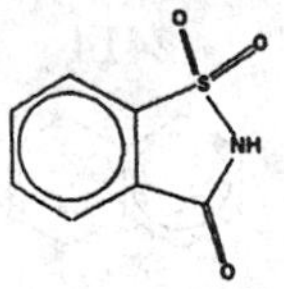

2389

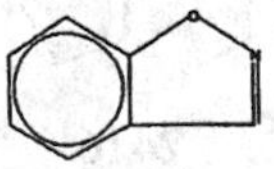

2390

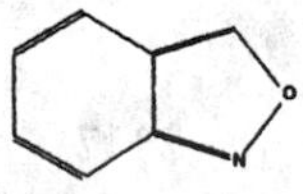

2391

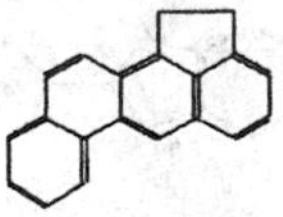

2392

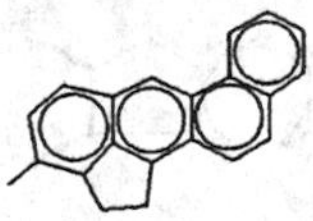

2393

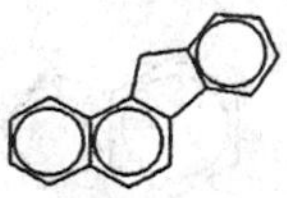

2394

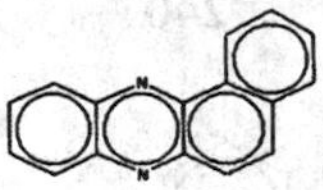

2395

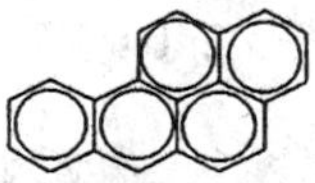

2396

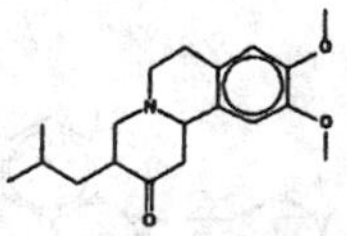

2397

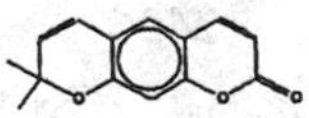

2398

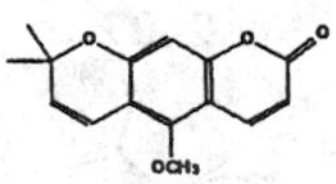

2399

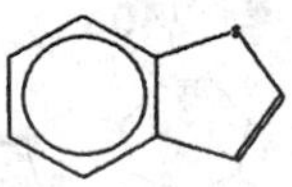

STRUCTURAL FORMULAS OF ORGANIC COMPOUNDS (continued)

In numeric order as they occur in the Table of Physical Constants of Organic Compounds

2400

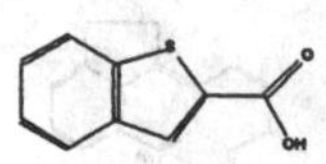

2401

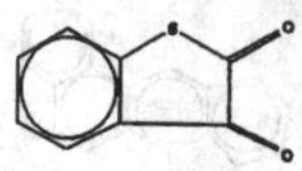

2402

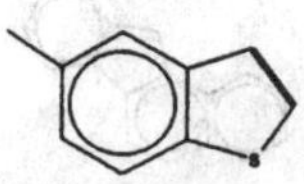

2403

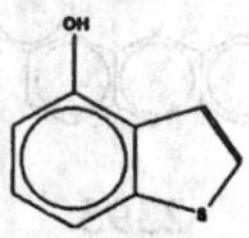

2404

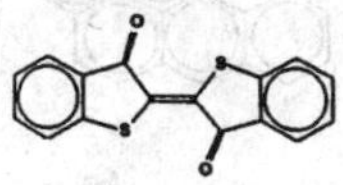

2405

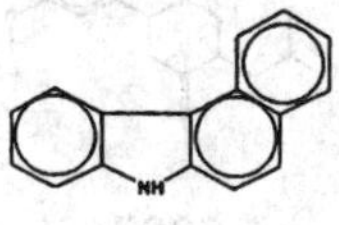

2406

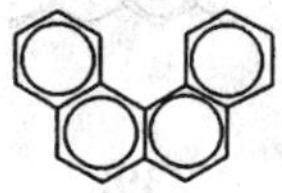

2407

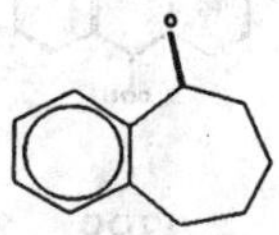

2408

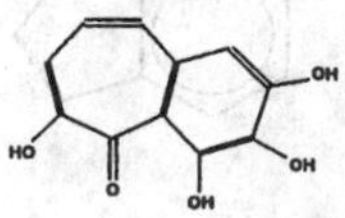

2409

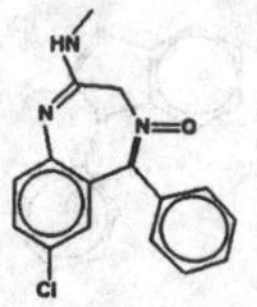

2410

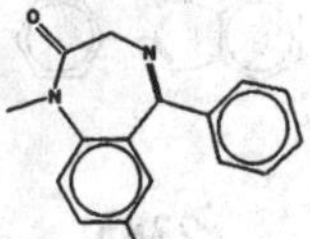

2411

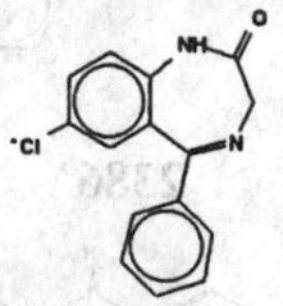

2412

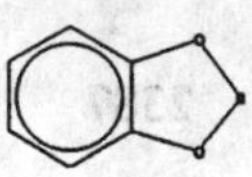

2413

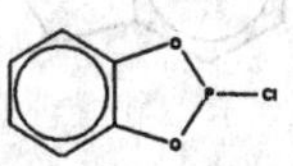

2414

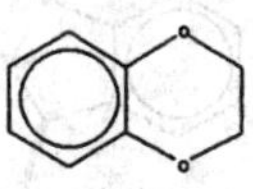

2415

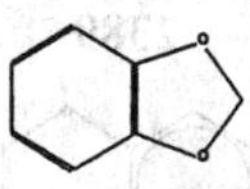

2416

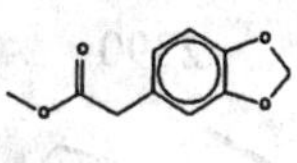

2417

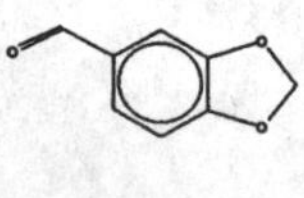

2418

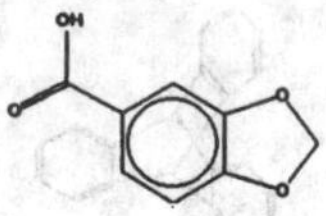

2419

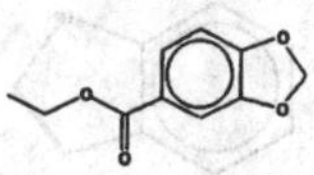

2420

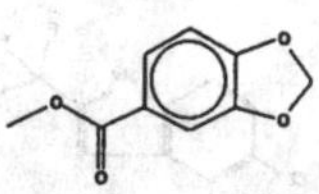

2421

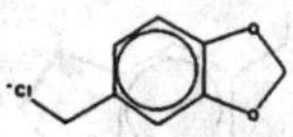

2422

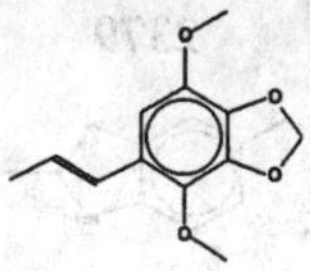

2423

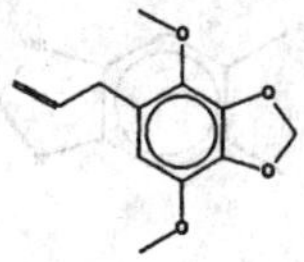

2424

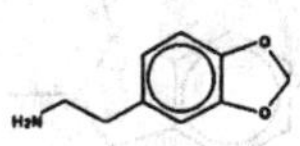

2425

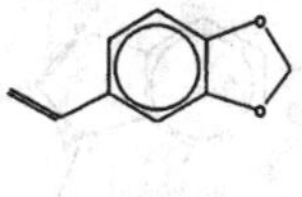

2426

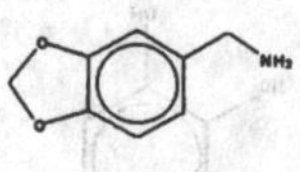

2427

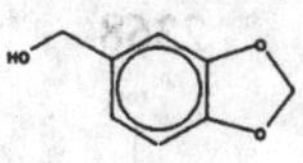

2428

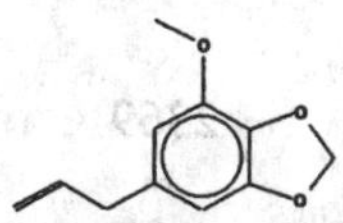

2429

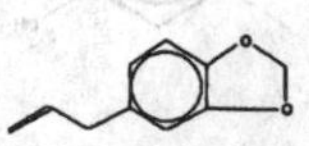

2430

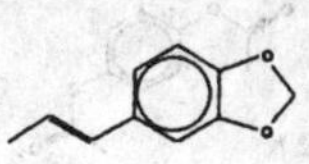

2431

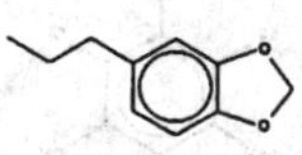

2432

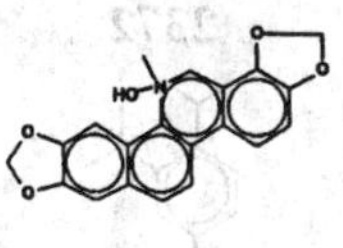

2434

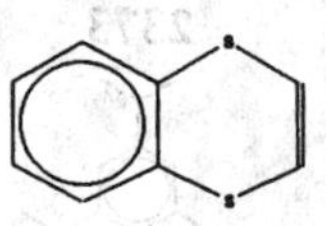

2435

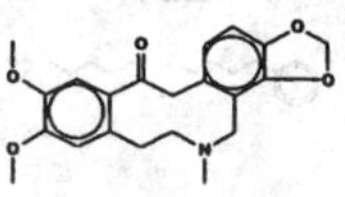

2436

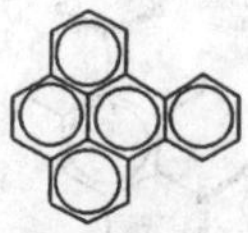

2437

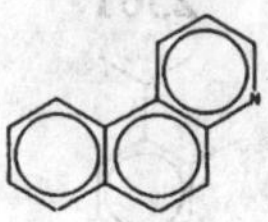

2438

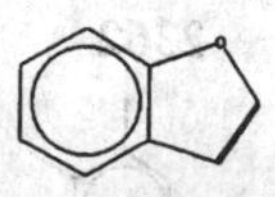

2439

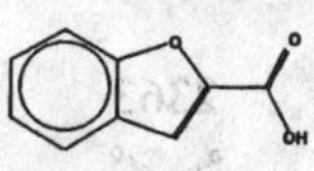

2440

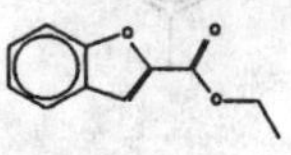

2441

2442

2443

2444

2445

2446

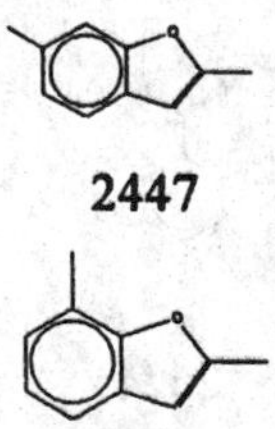

2447

2448

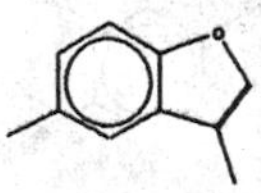

2449

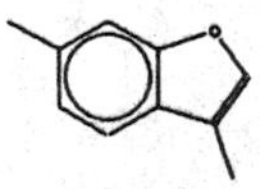

2451

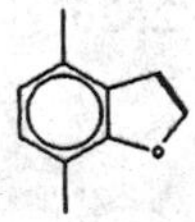

2452

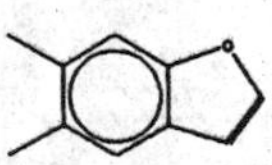

2453

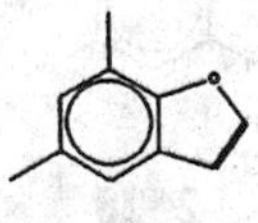

2454

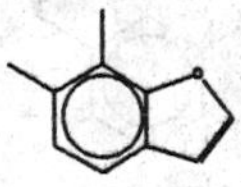

2455

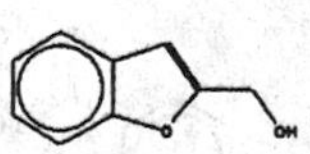

2456

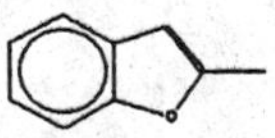

2457

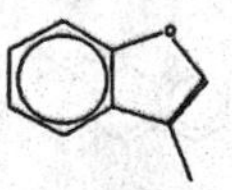

2458

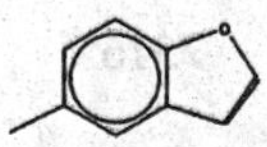

2459

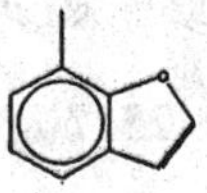

2460

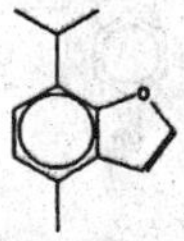

2461

2462

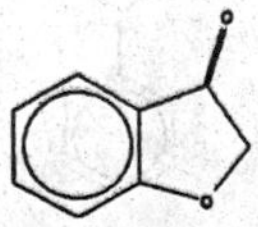

2463

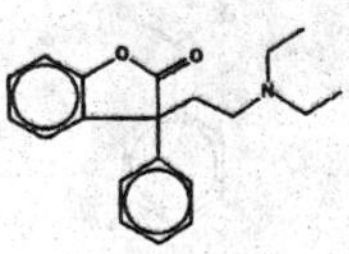

2464

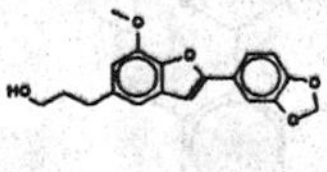

2465

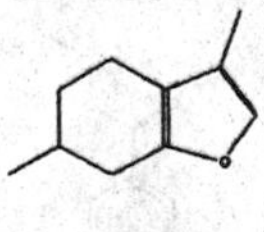

2466

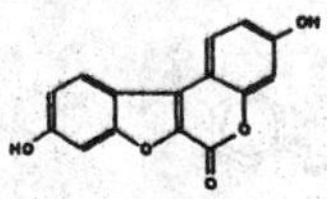

2468

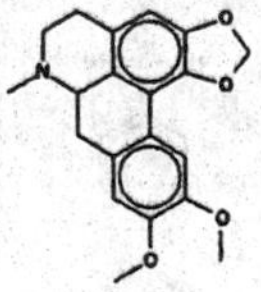

2469

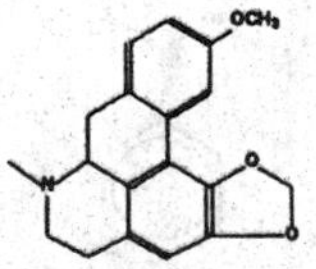

2470

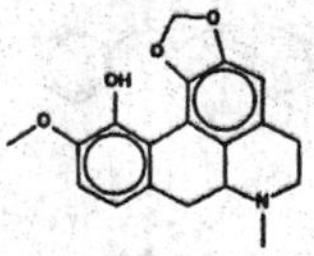

2471

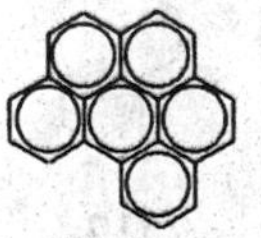

2472

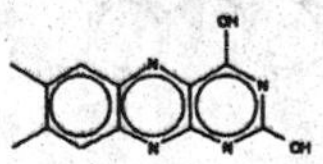

2474

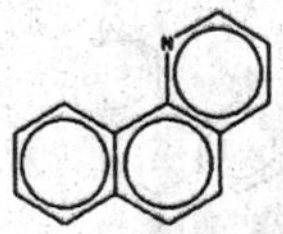

2475

2476

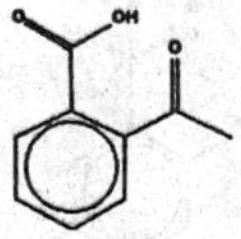

2477

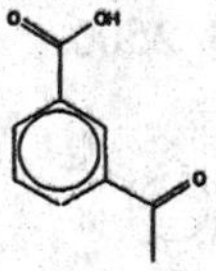

2478

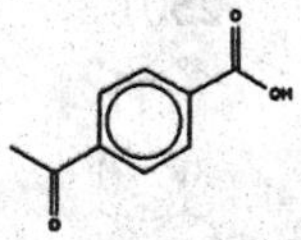

2479

2480

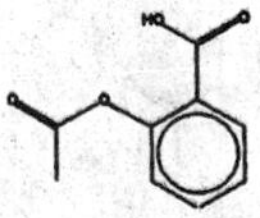

2481

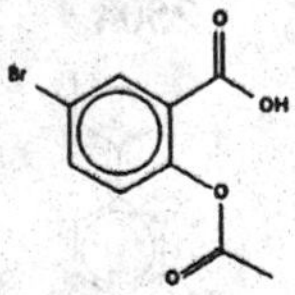

2482

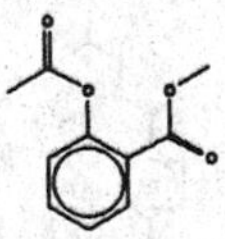

2483

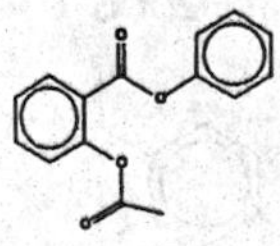

2484

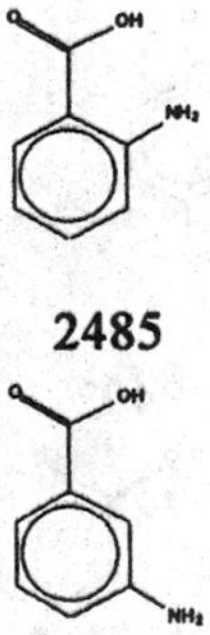

2485

2486

2487

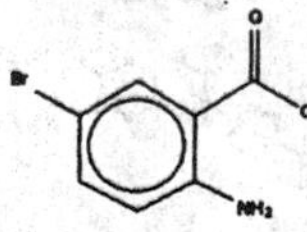

2488

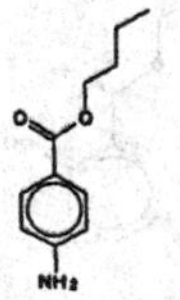

2489

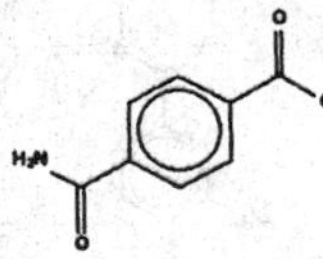

2490

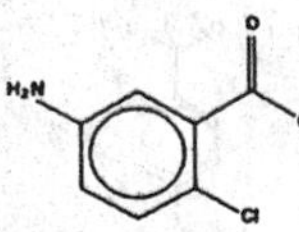

STRUCTURAL FORMULAS OF ORGANIC COMPOUNDS (continued)

In numeric order as they occur in the Table of Physical Constants of Organic Compounds

2491

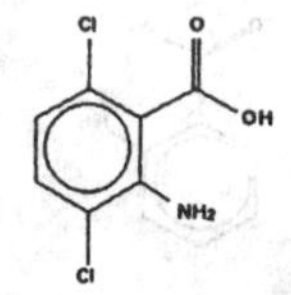

2492

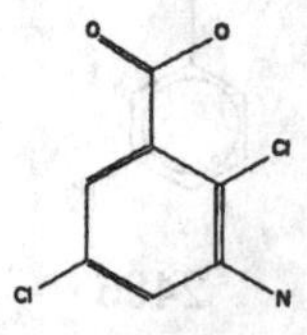

2493

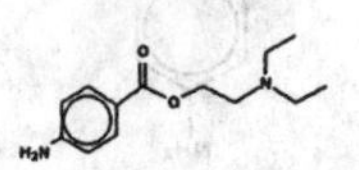

2494

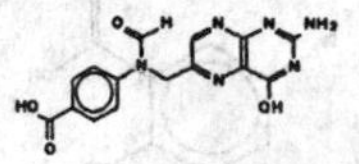

2495

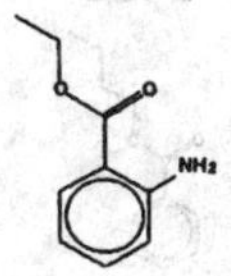

2496

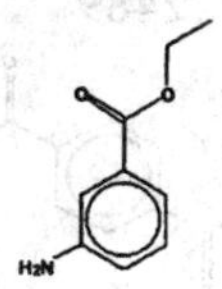

2497

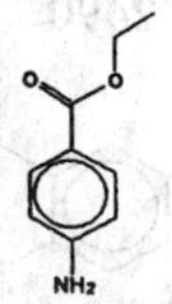

2498

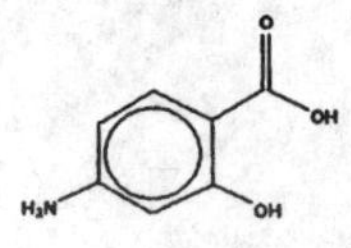

2499

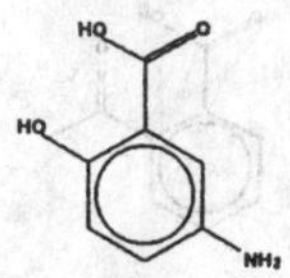

2500

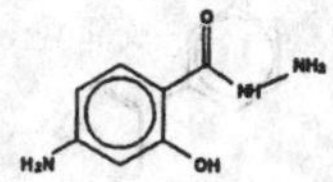

2501

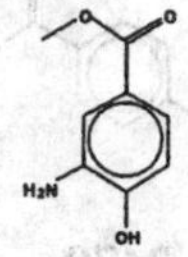

2502

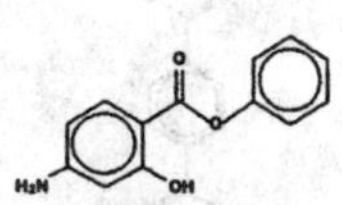

2503

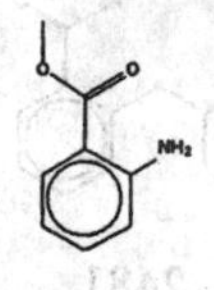

2504

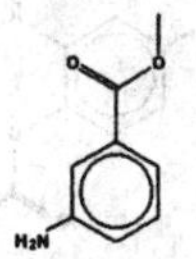

2505

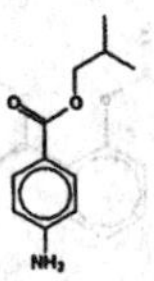

2506

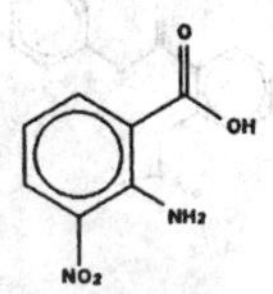

2507

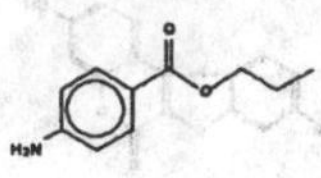

2508

2509

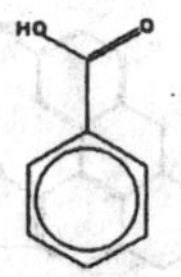

2510

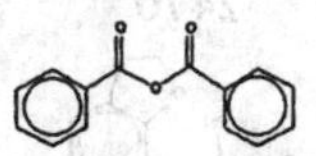

2511

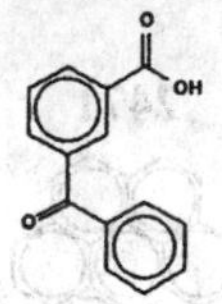

2512

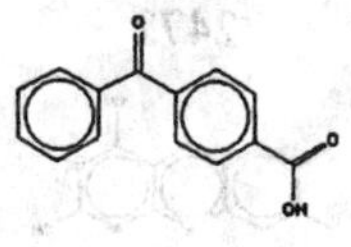

2513

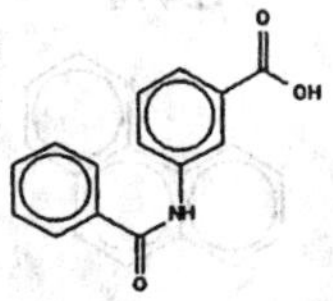

2514

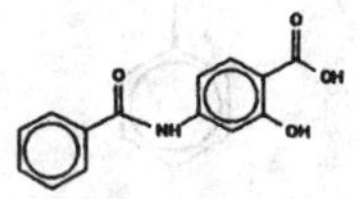

2515

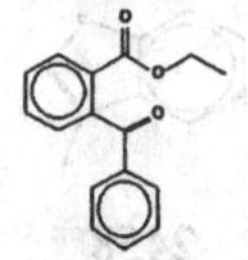

2516

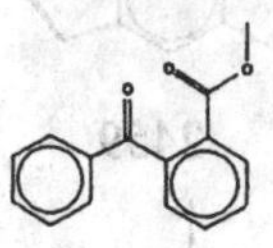

2517

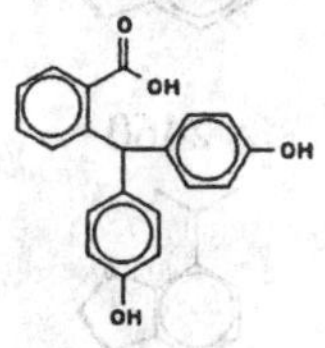

2518

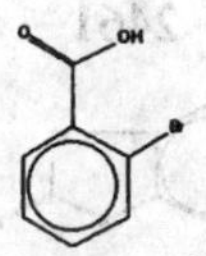

2519

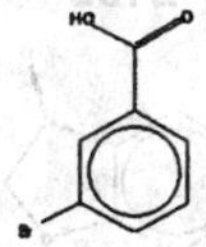

2520

2521

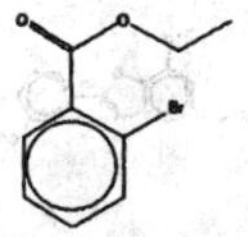

2522

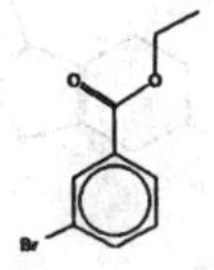

2523

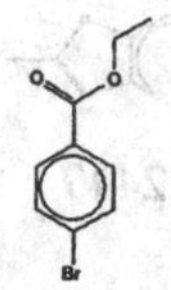

2524

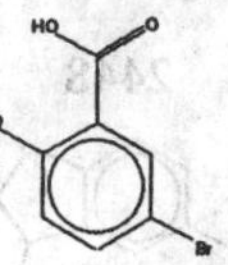

2525

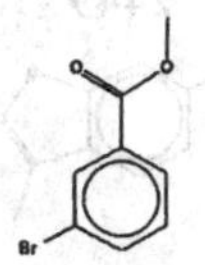

2526

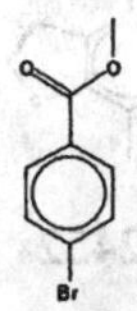

2527

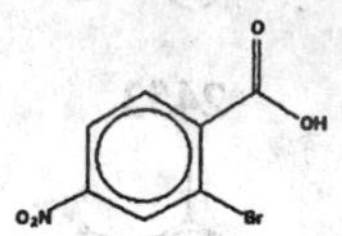

2528

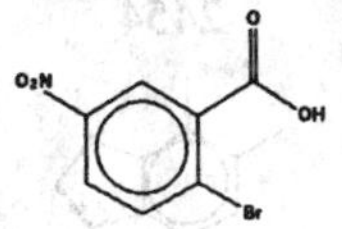

2529

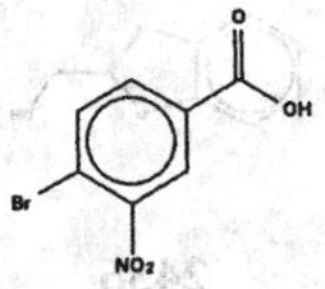

STRUCTURAL FORMULAS OF ORGANIC COMPOUNDS (continued)

In numeric order as they occur in the Table of Physical Constants of Organic Compounds

2530

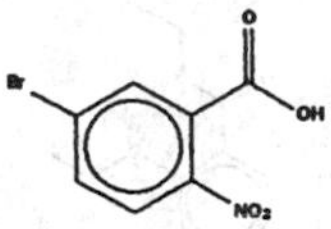

2532

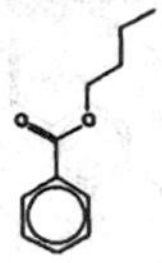

2533

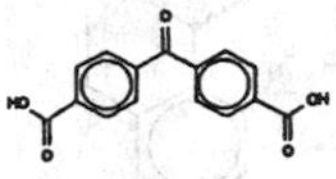

2534

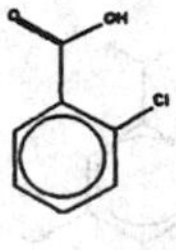

2535

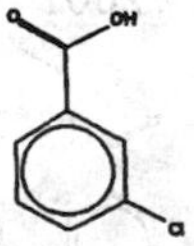

2536

2537

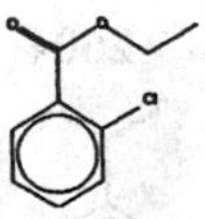

2538

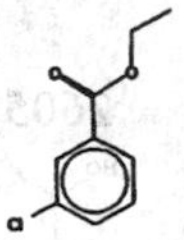

2539

2540

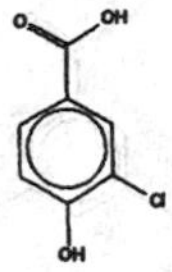

2541

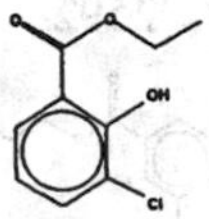

2542

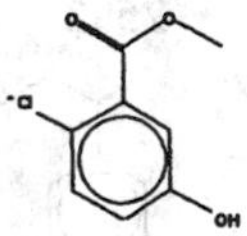

2543

2544

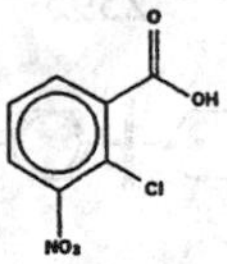

2545

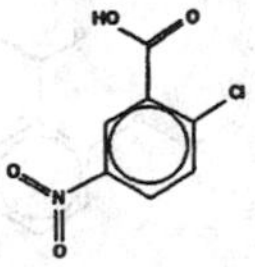

2546

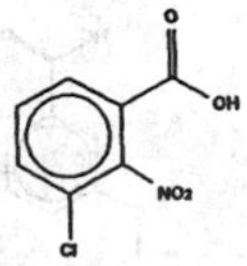

2547

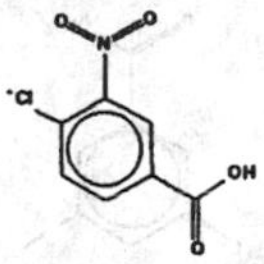

2548

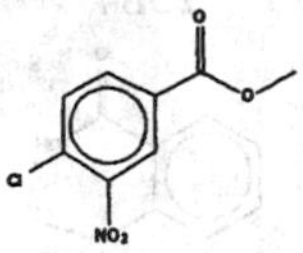

2549

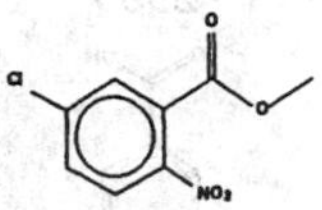

2550

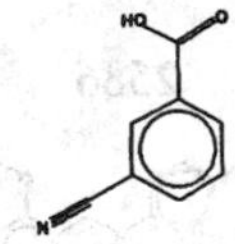

2551

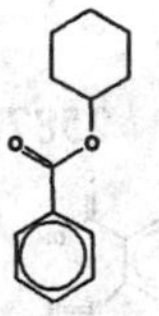

2552

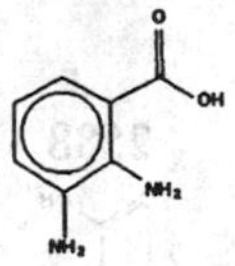

2553

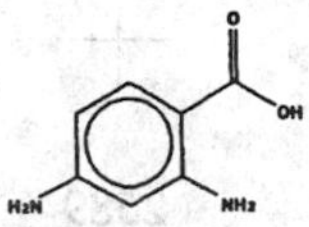

2554

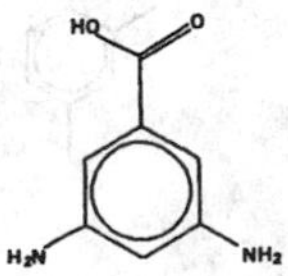

2555

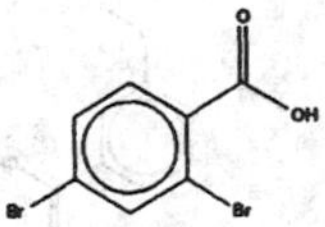

2556

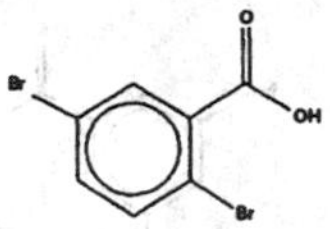

2557

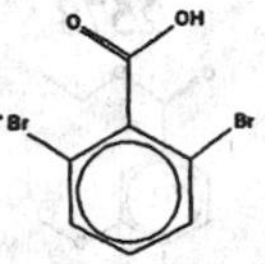

2558

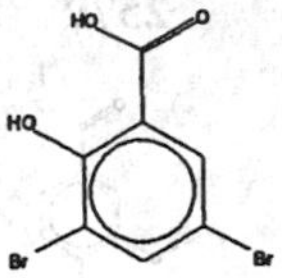

2559

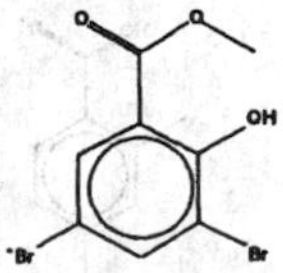

2560

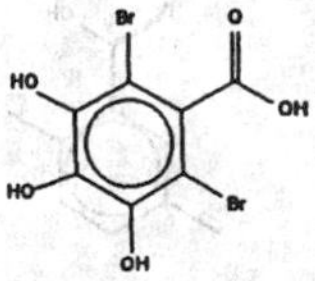

2561

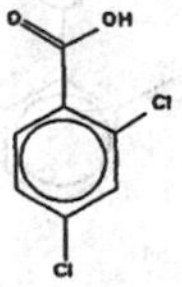

2562

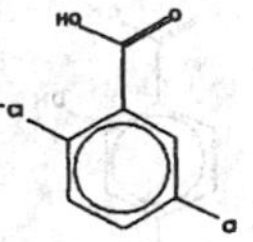

2563

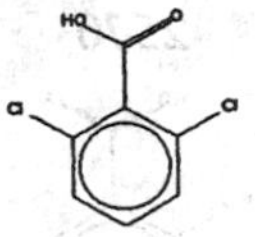

2564

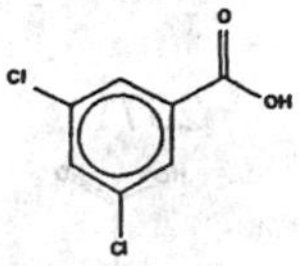

2565

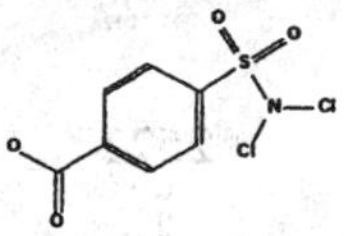

2566

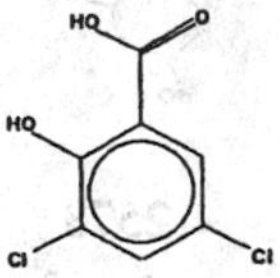

2567

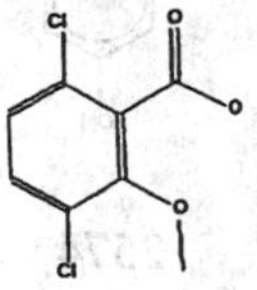

2568

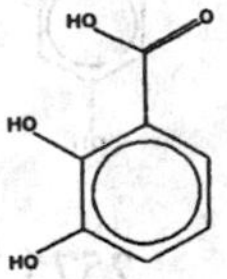

STRUCTURAL FORMULAS OF ORGANIC COMPOUNDS (continued)

In numeric order as they occur in the Table of Physical Constants of Organic Compounds

2569

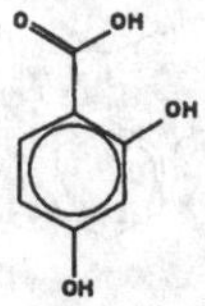

2570

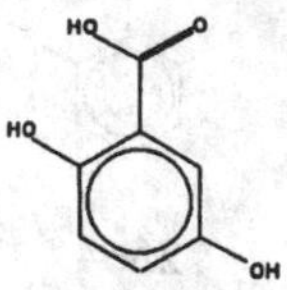

2571

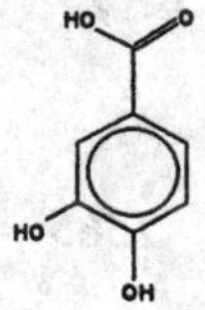

2572

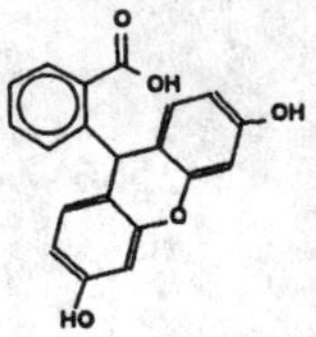

2573

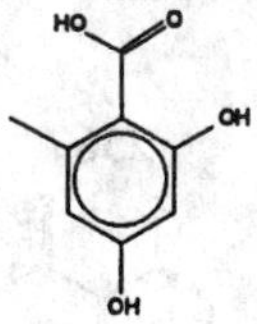

2574

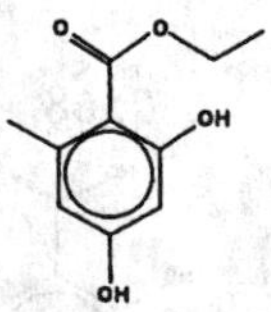

2575

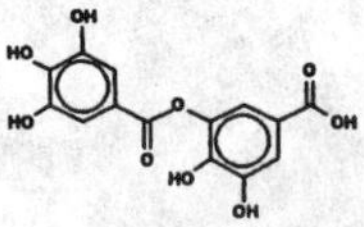

2576

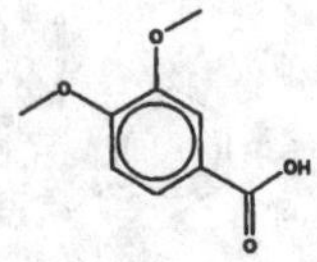

2577

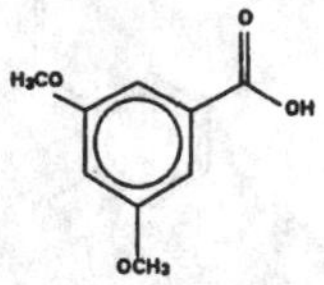

2578

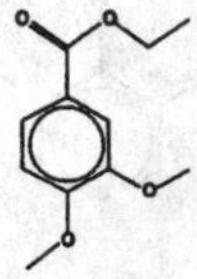

2579

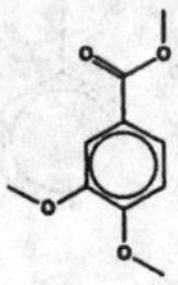

2580

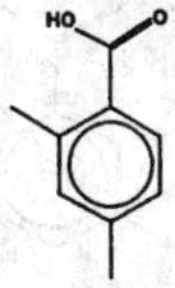

2581

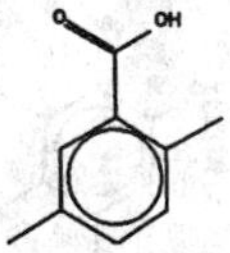

2582

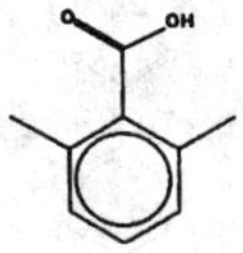

2583

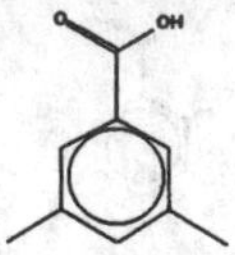

2584

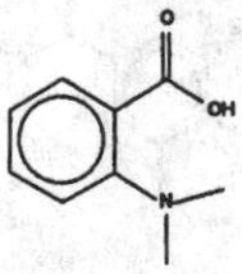

2585

2586

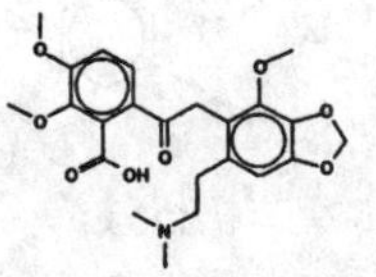

2587

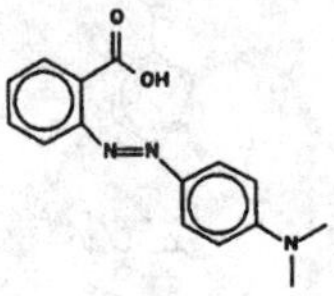

2588

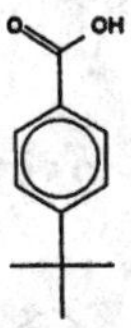

2589

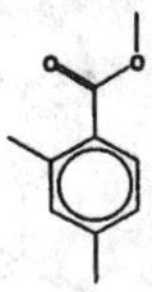

2590

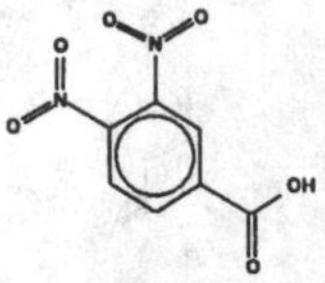

2591

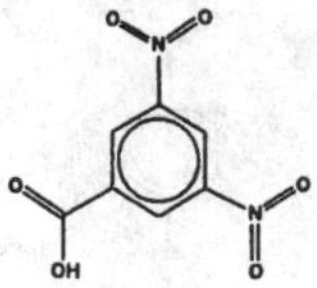

2592

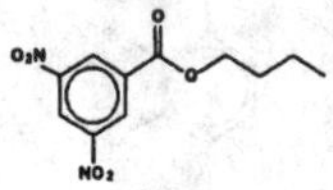

2593

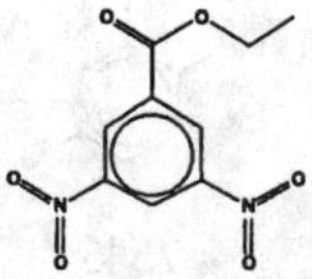

2594

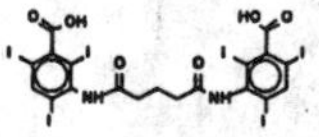

2595

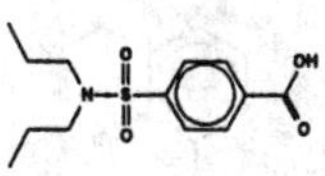

2596

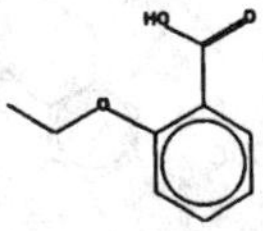

2597

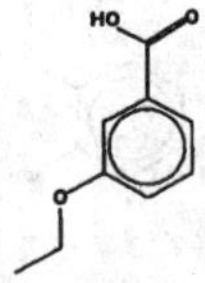

2598

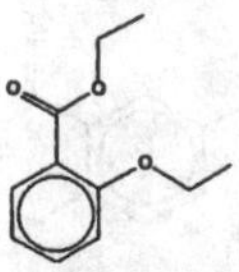

2599

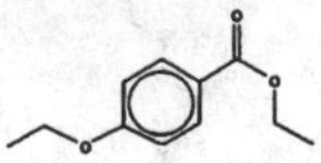

2600

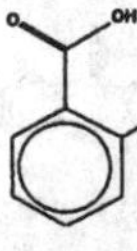

2601

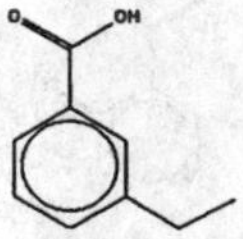

2602

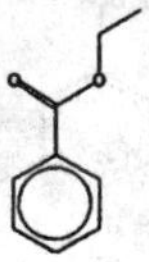

2603

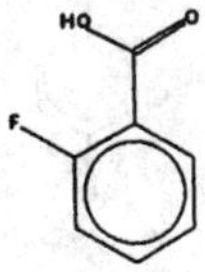

2604

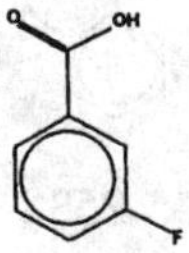

2605

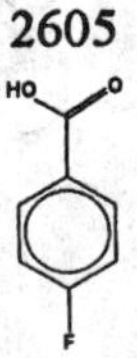

In numeric order as they occur in the Table of Physical Constants of Organic Compounds

2606

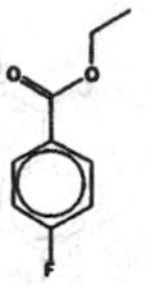

2607

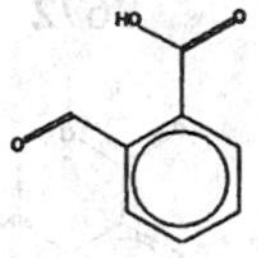

2608

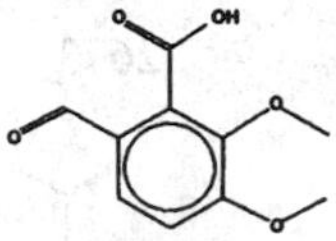

2609

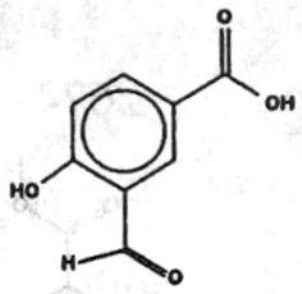

2610

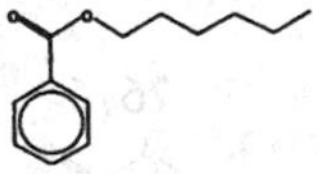

2611

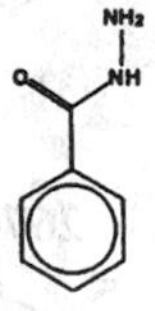

2612

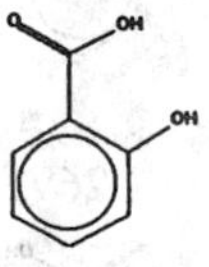

2613

2614

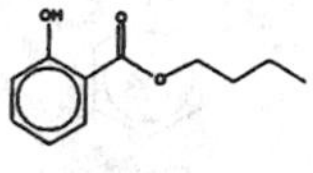

2615

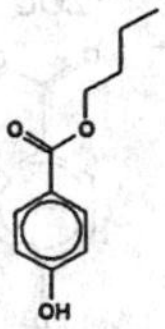

2616

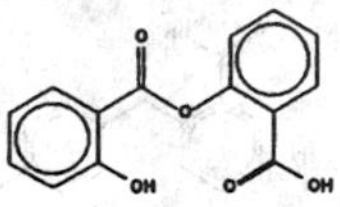

2617

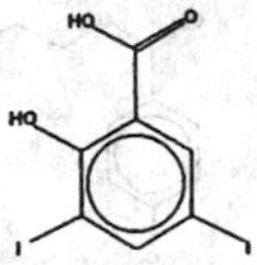

2618

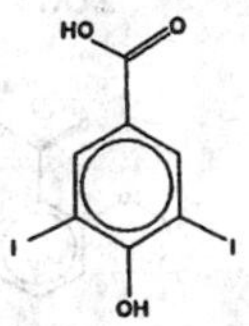

2619

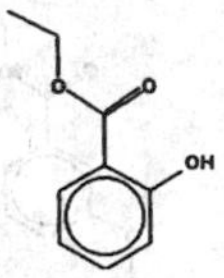

2620

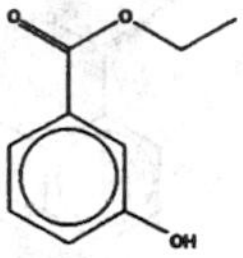

2621

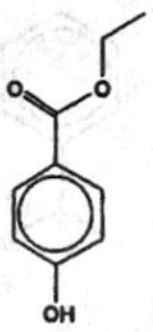

2622

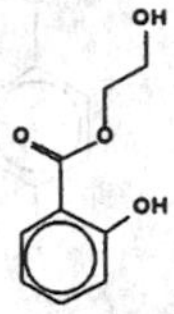

2624

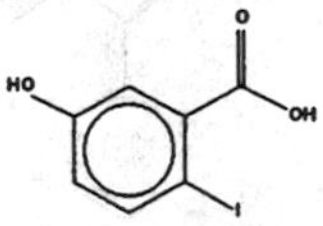

2625

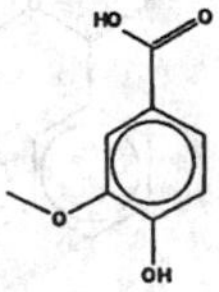

2626

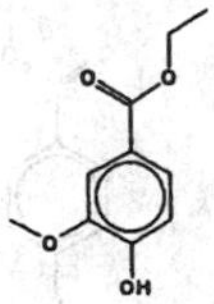

2627

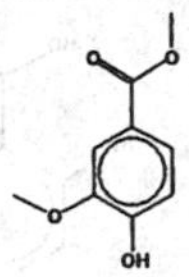

2628

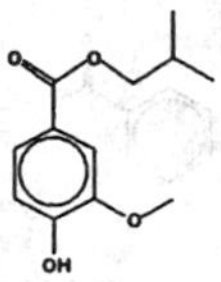

2629

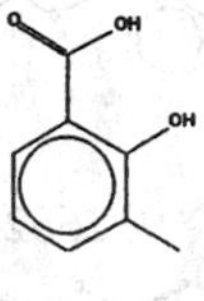

2630

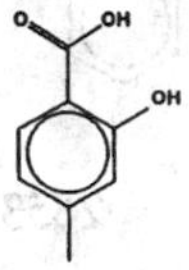

2631

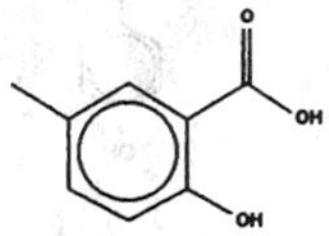

2632

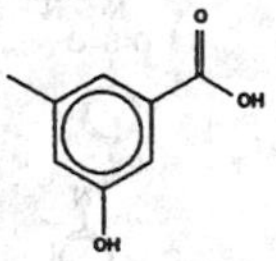

2633

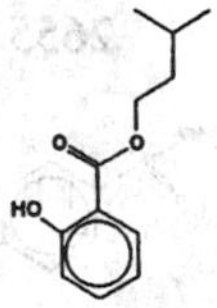

2634

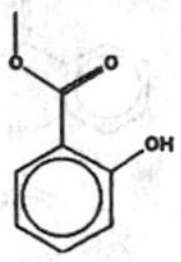

2635

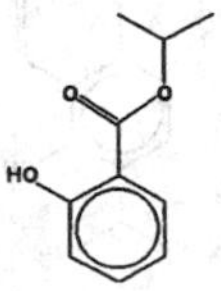

2636

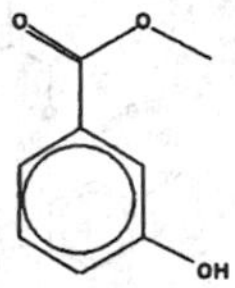

2637

2638

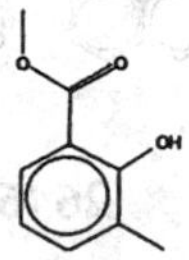

2639

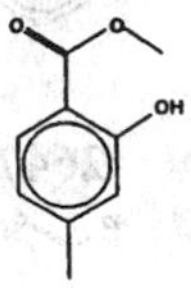

2640

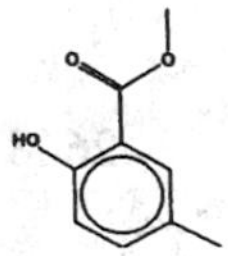

2641

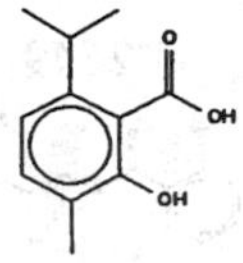

STRUCTURAL FORMULAS OF ORGANIC COMPOUNDS (continued)

In numeric order as they occur in the Table of Physical Constants of Organic Compounds

2642

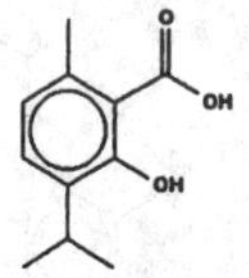

2643

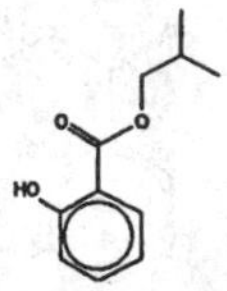

2644

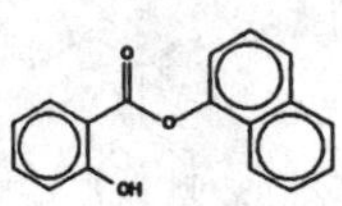

2645

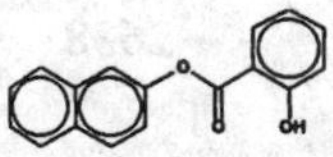

2646

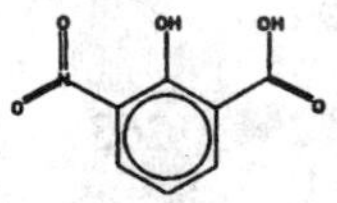

2647

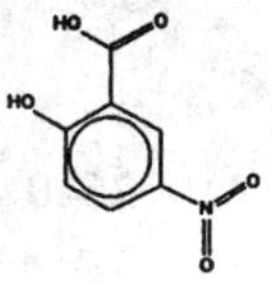

2648

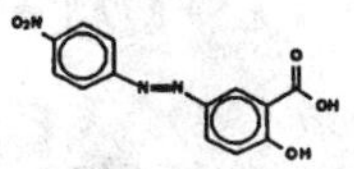

2649

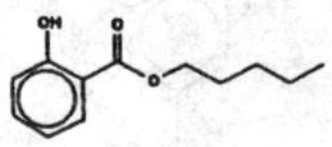

2650

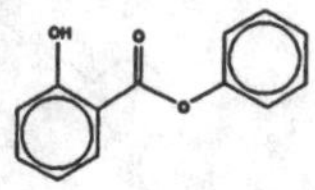

2651

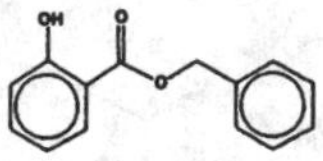

2652

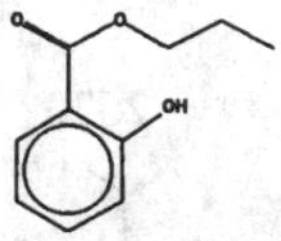

2653

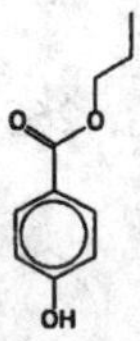

2654

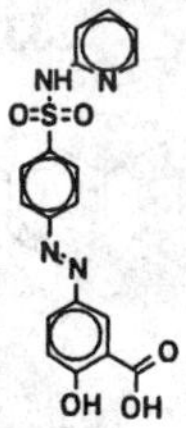

2655

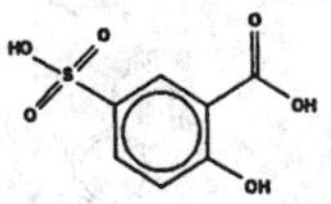

2656

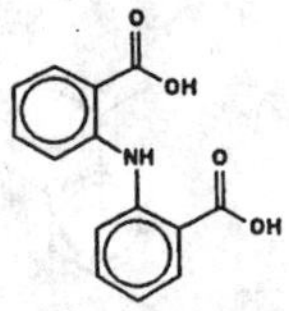

2657

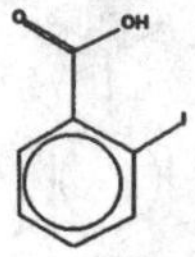

2658

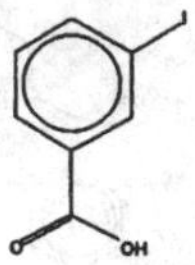

2659

2660

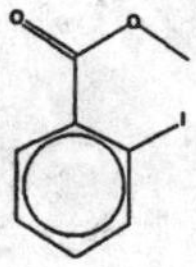

2661

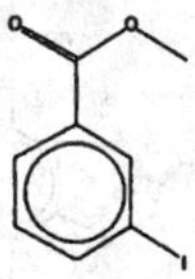

2662

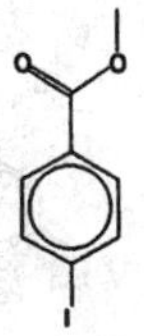

2663

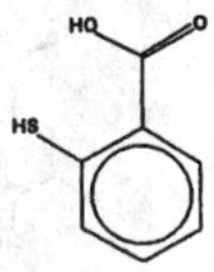

2664

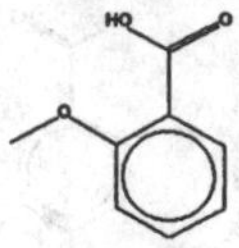

2665

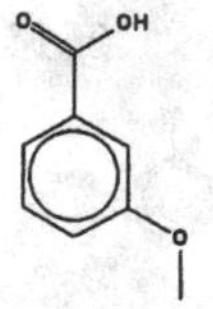

2666

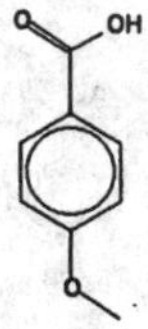

2667

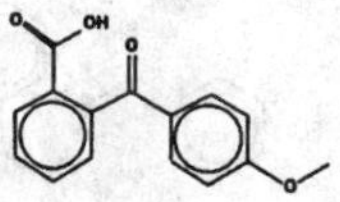

2668

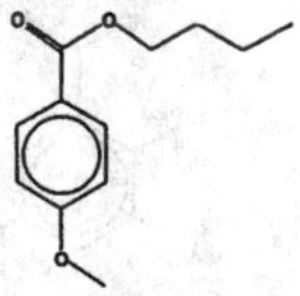

2669

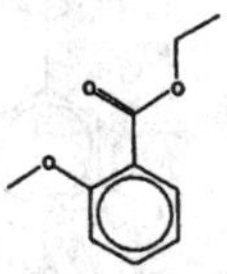

2670

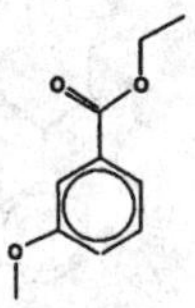

2671

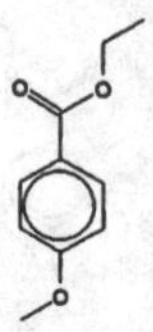

2672

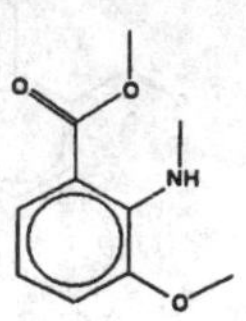

2673

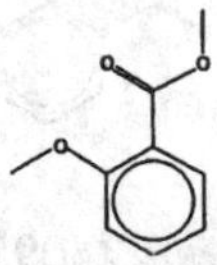

2674

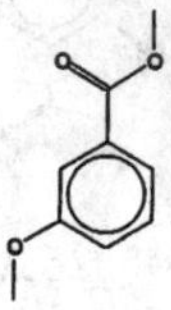

2675

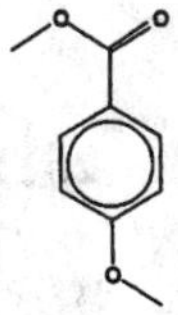

2676

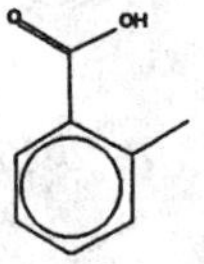

2677

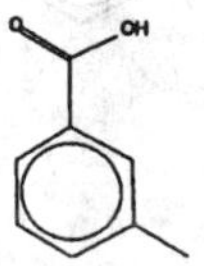

In numeric order as they occur in the Table of Physical Constants of Organic Compounds

2678

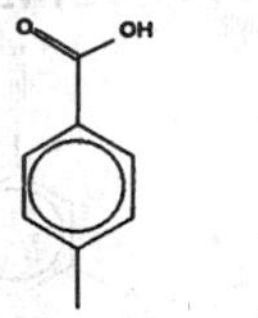

2679

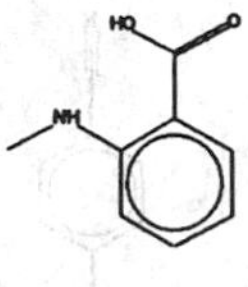

2680

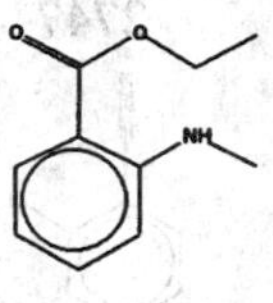

2681

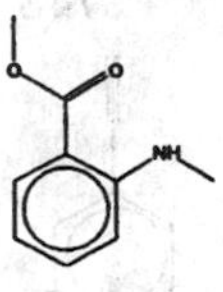

2682

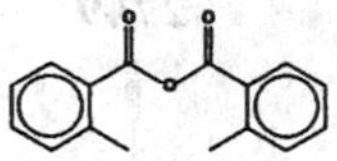

2683

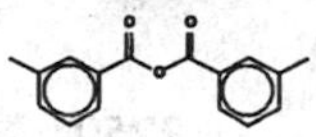

2684

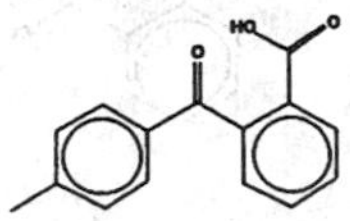

2685

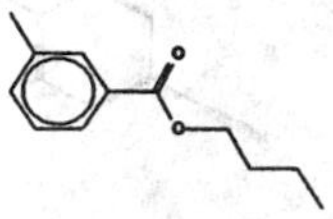

2686

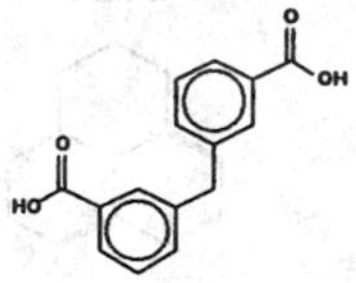

2687

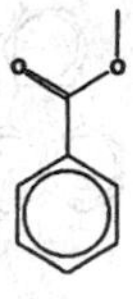

2688

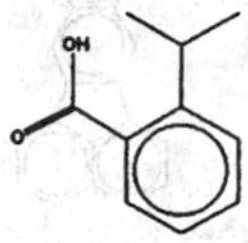

2689

2690

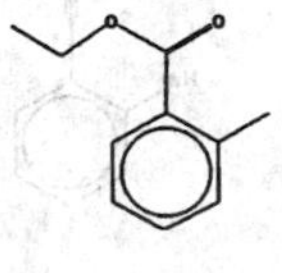

2691

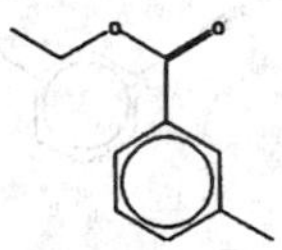

2692

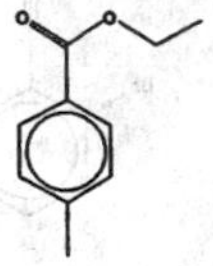

2693

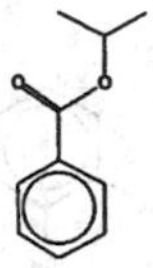

2694

2695

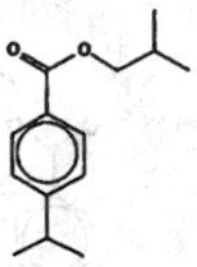

2696

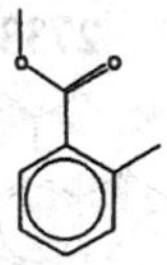

2697

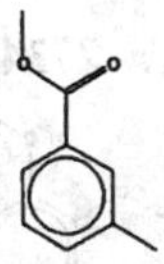

2698

2699

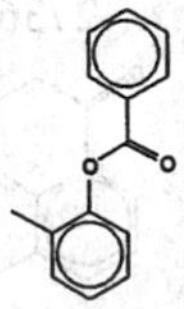

2700

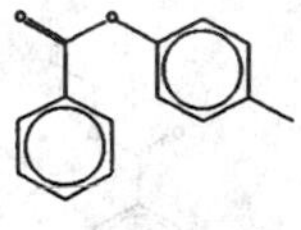

2701

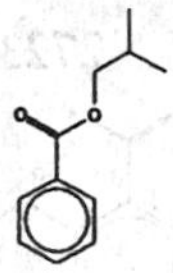

2702

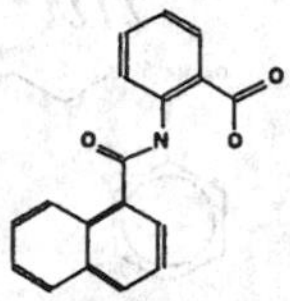

2703

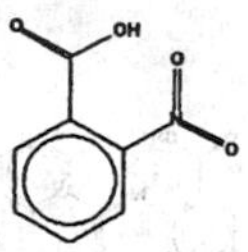

2704

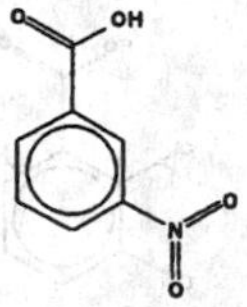

2705

2706

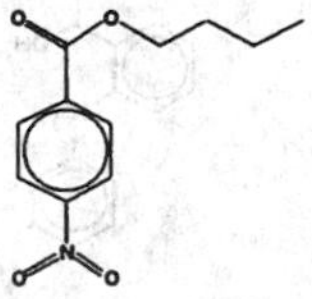

2707

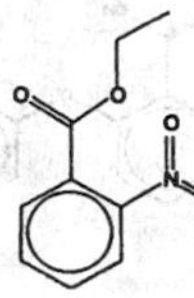

2708

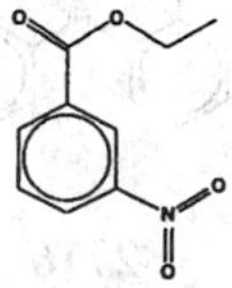

2709

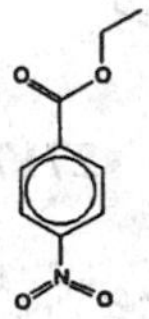

2710

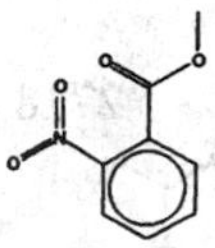

2711

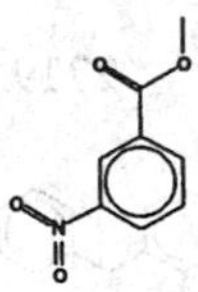

2712

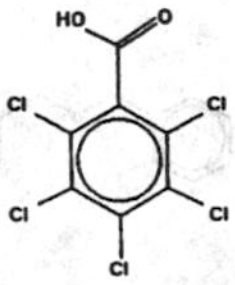

2713

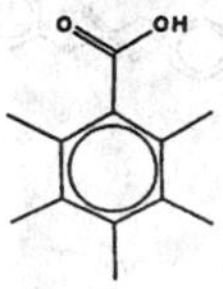

In numeric order as they occur in the Table of Physical Constants of Organic Compounds

2714

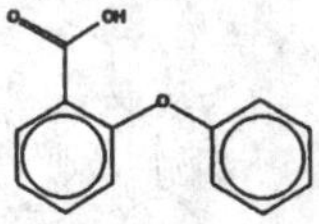

2715

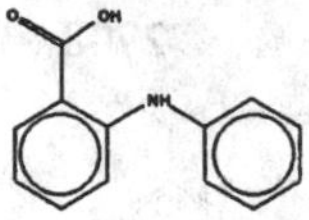

2716

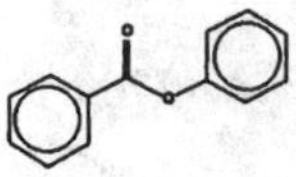

2717

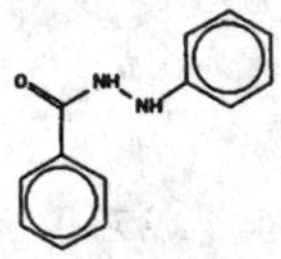

2718

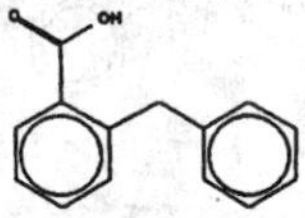

2719

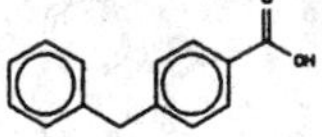

2720

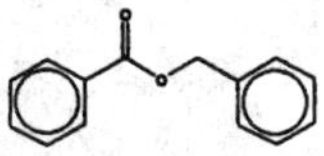

2721

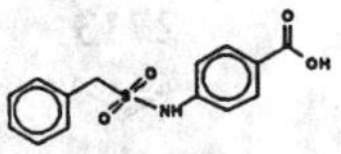

2722

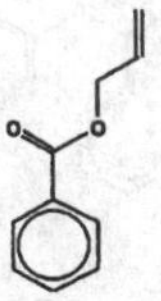

2723

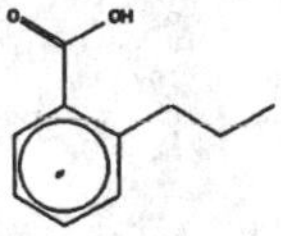

2724

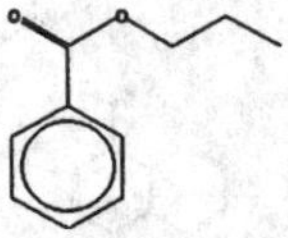

2726

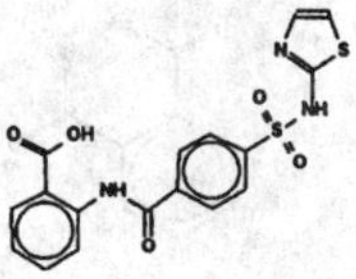

2727

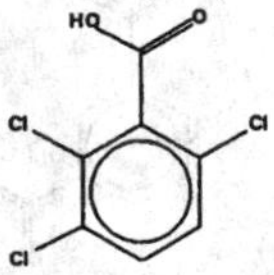

2728

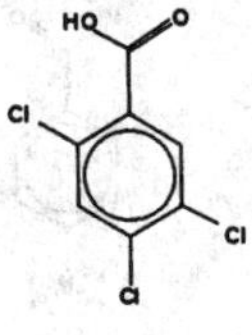

2729

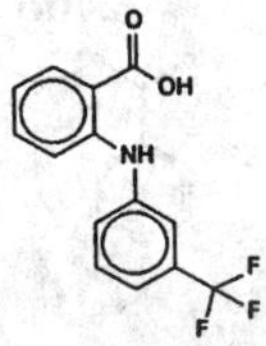

2730

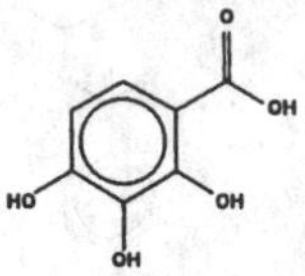

2731

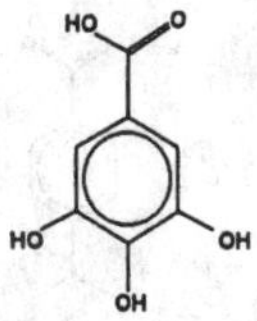

2732

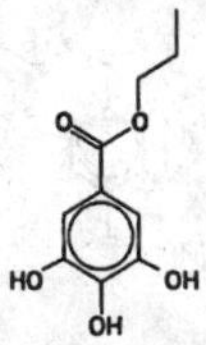

2733

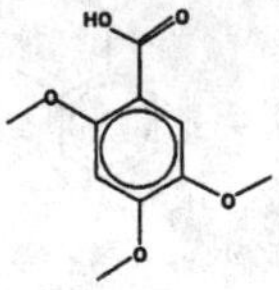

2734

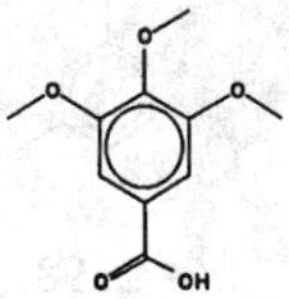

2735

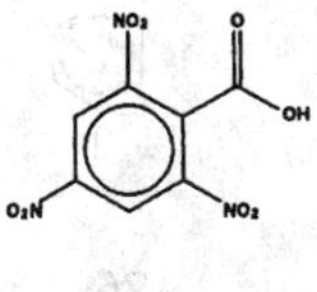

2736

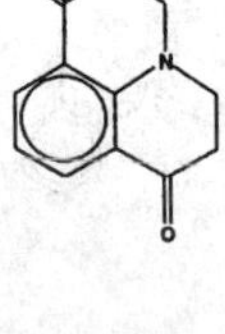

2737

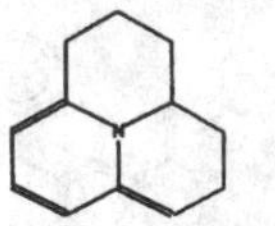

2738

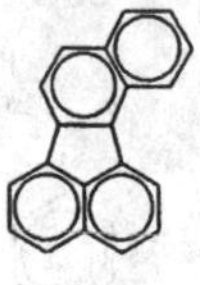

2739

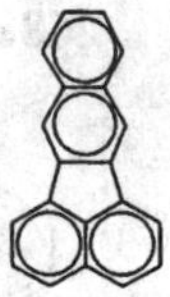

2740

2741

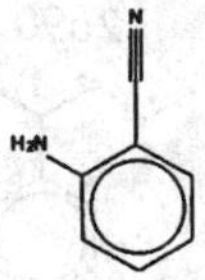

2742

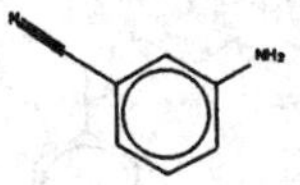

2744

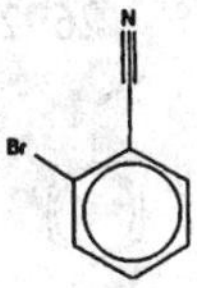

2745

2746

2747

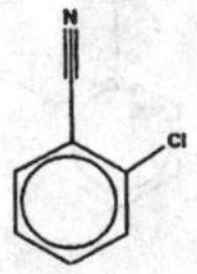

2748

2749

2750

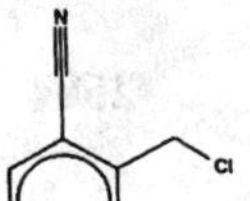

2751

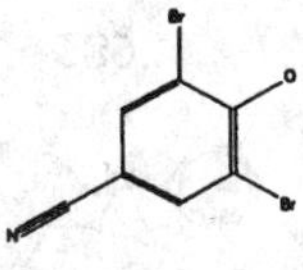

STRUCTURAL FORMULAS OF ORGANIC COMPOUNDS (continued)

In numeric order as they occur in the Table of Physical Constants of Organic Compounds

2752

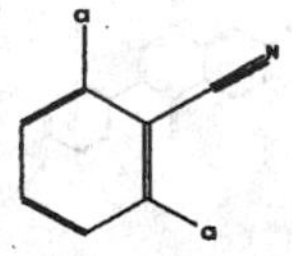

2753

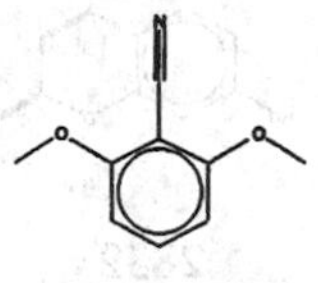

2754

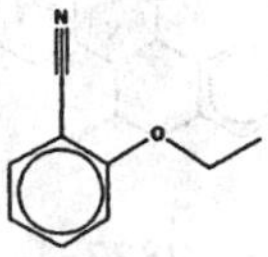

2755

2756

2757

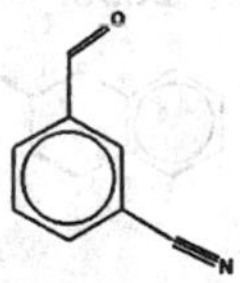

2758

2759

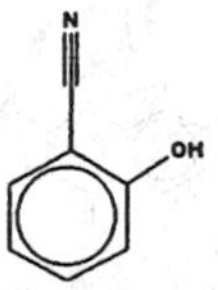

2760

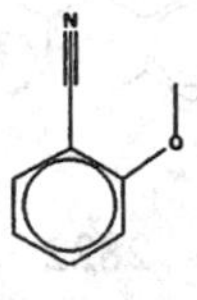

2761

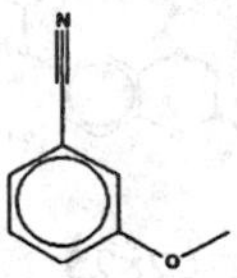

2762

2763

2764

2765

2766

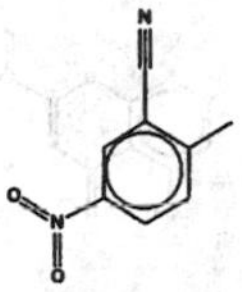

2767

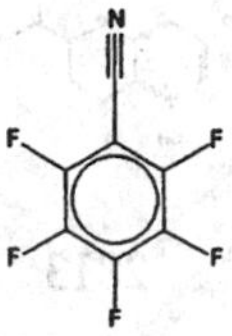

2768

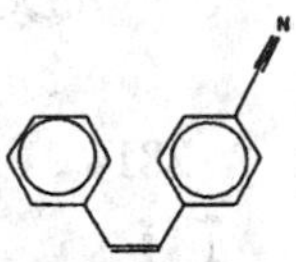

2769

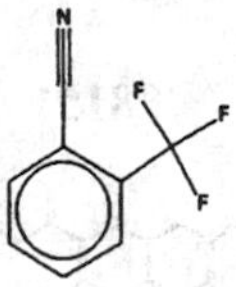

2770

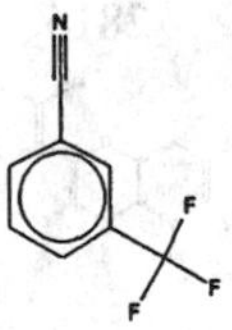

2771

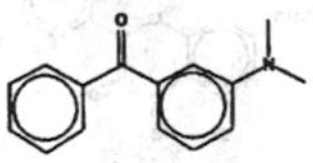

2772

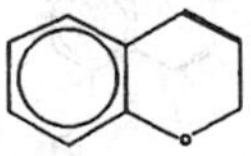

2773

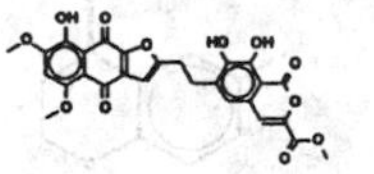

2774

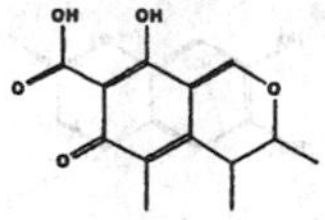

2775

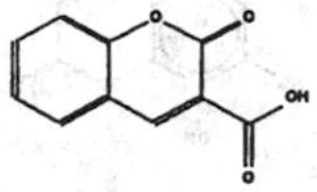

2776

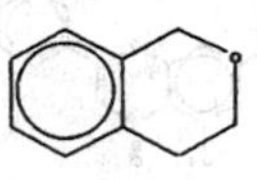

2777

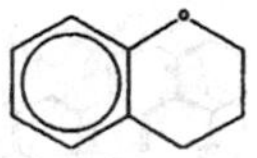

2778

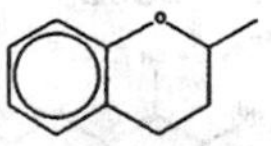

2780

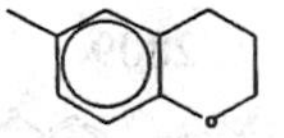

2781

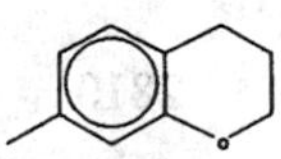

2782

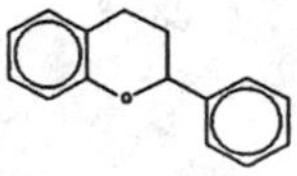

2783

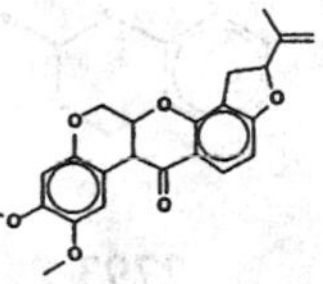

2784

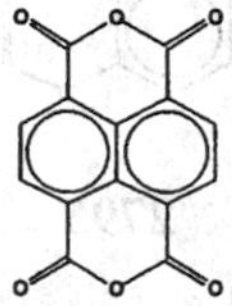

2785

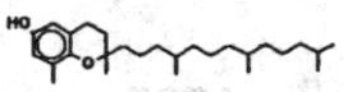

2787

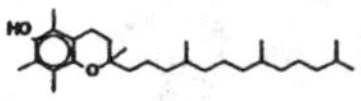

2788

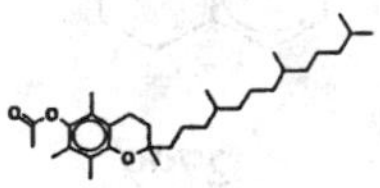

2789

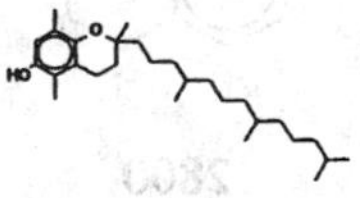

2790

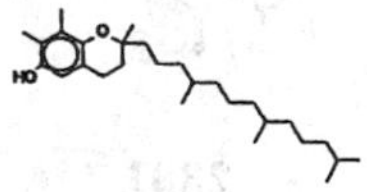

2791

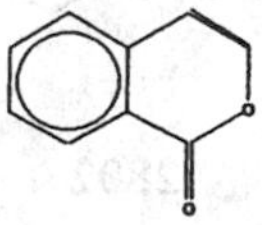

STRUCTURAL FORMULAS OF ORGANIC COMPOUNDS (continued)

In numeric order as they occur in the Table of Physical Constants of Organic Compounds

2792

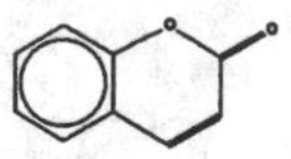

2793

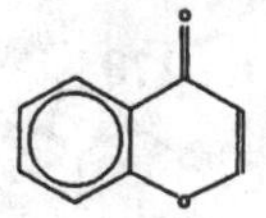

2795

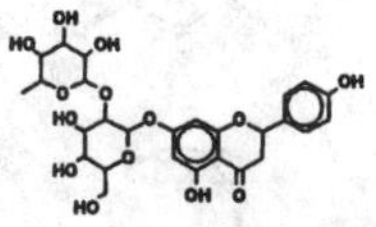

2797

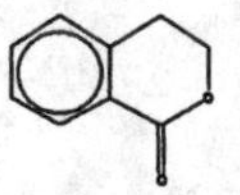

2798

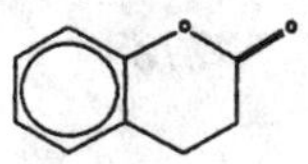

2799

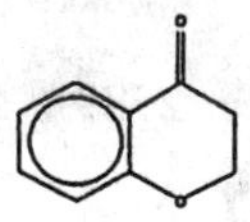

2800

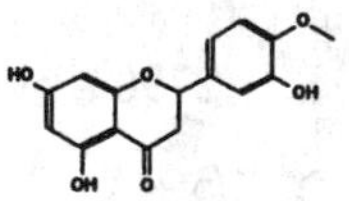

2801

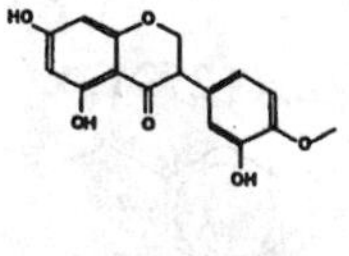

2802

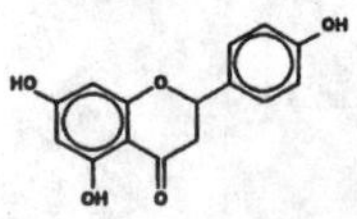

2803

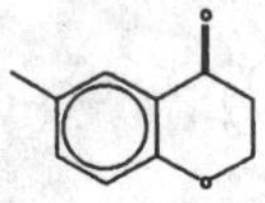

2804

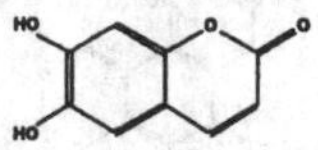

2805

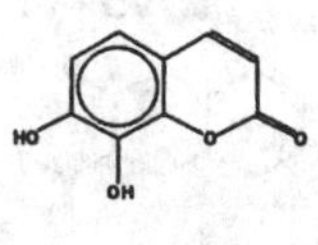

2806

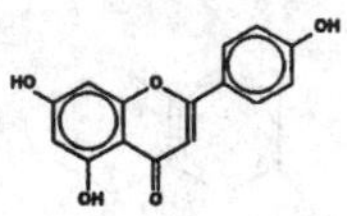

2807

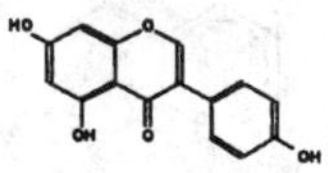

2808

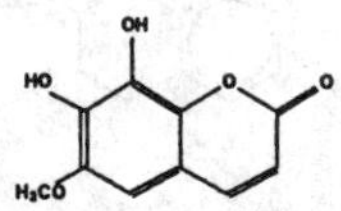

2809

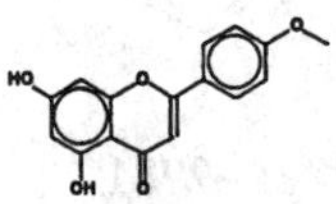

2810

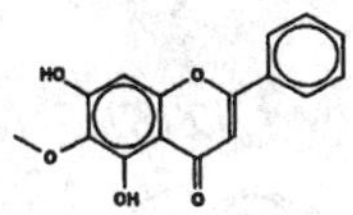

2811

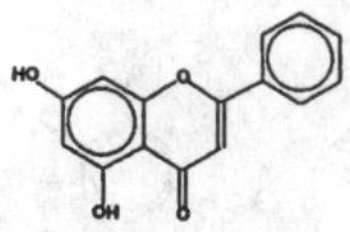

2812

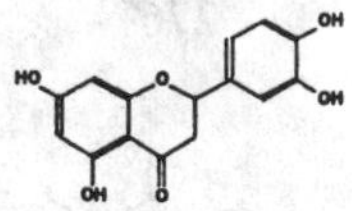

2813

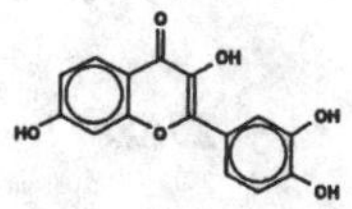

2814

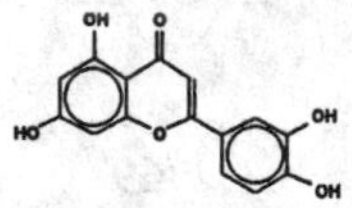

2815

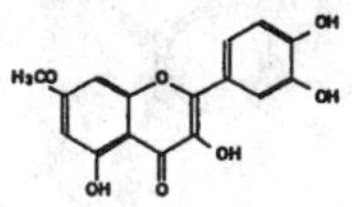

2816

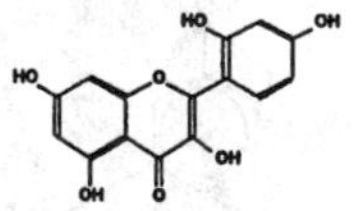

2817

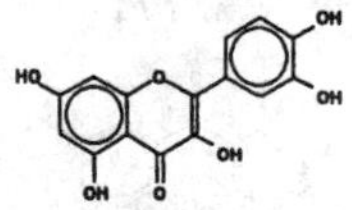

2818

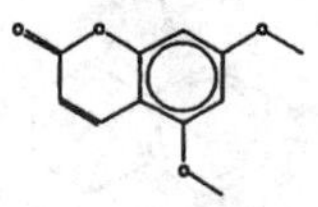

2819

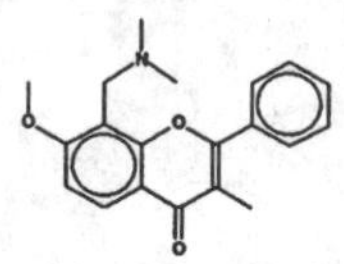

2820

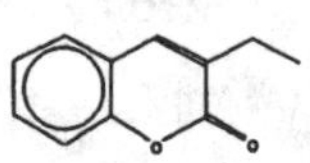

2822

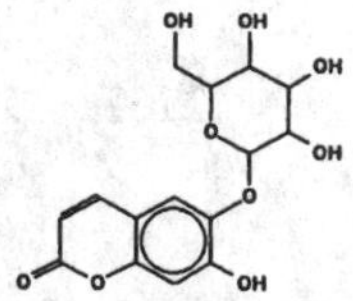

2825

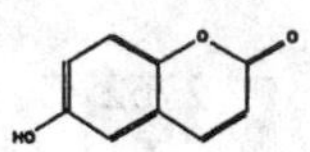

2826

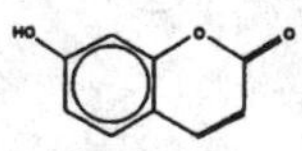

2827

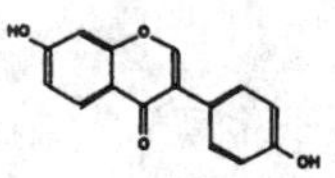

2828

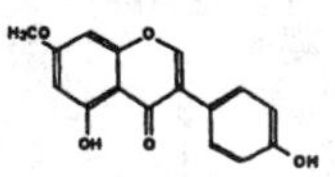

2829

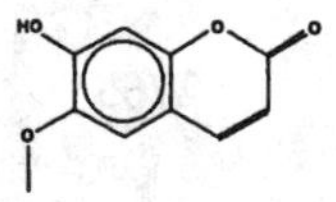

2830

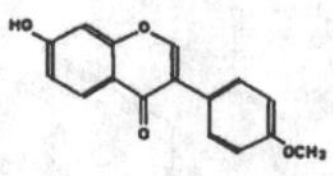

2831

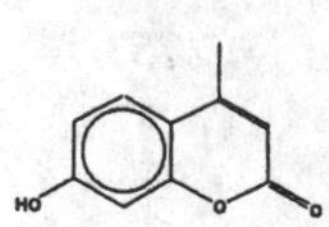

2832

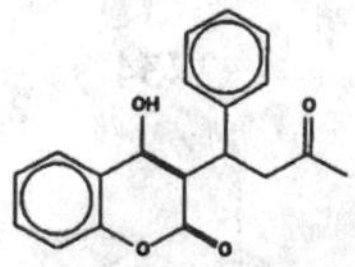

2833

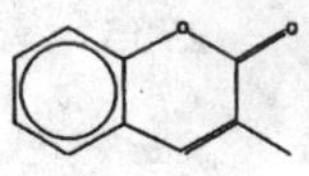

2834

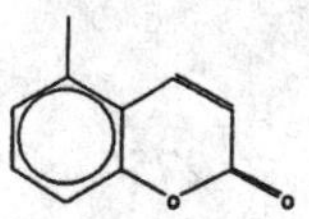

2835

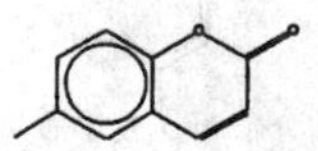

2836

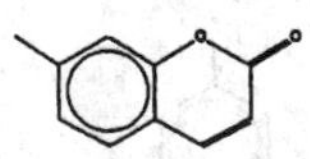

2837

STRUCTURAL FORMULAS OF ORGANIC COMPOUNDS (continued)

In numeric order as they occur in the Table of Physical Constants of Organic Compounds

2838

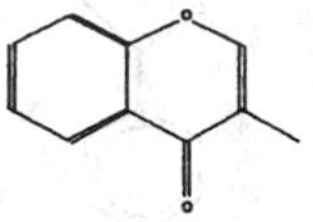

2839

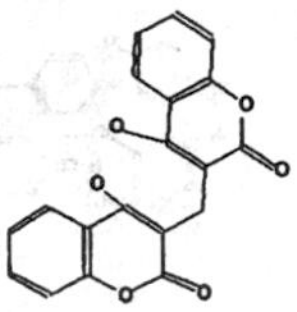

2840

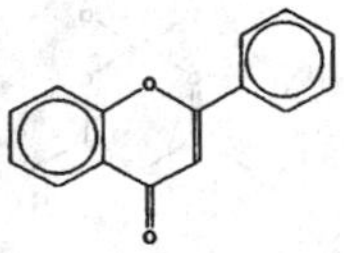

2841

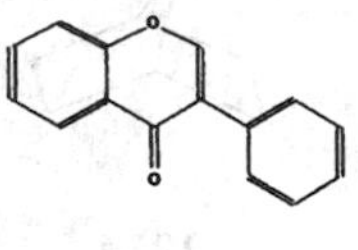

2842

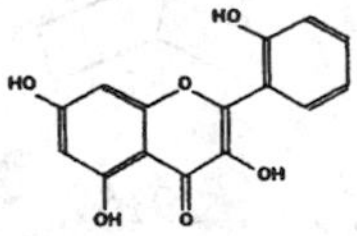

2843

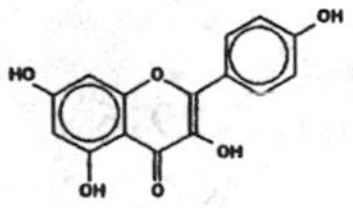

2844

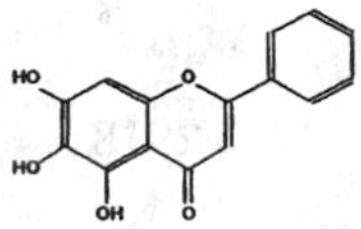

2845

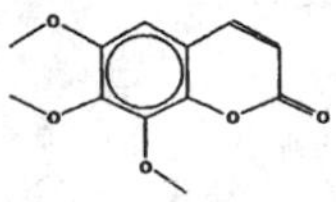

2847

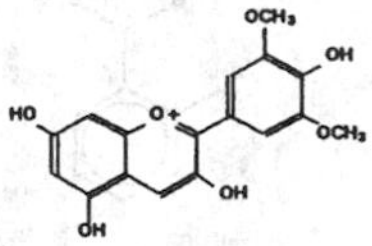

2849

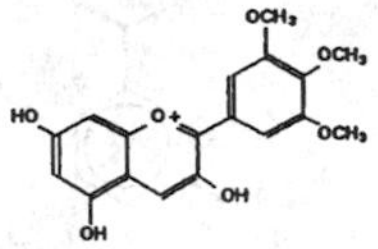

2850

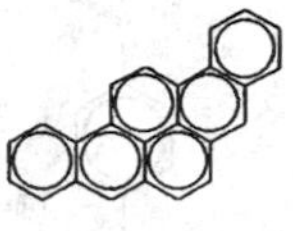

2851

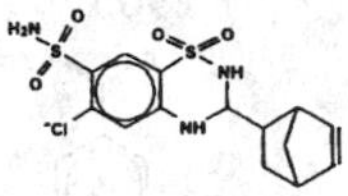

2852

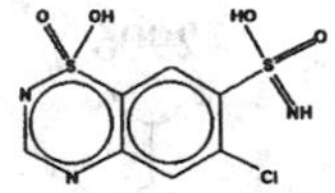

2853

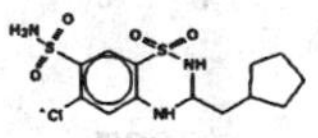

2854

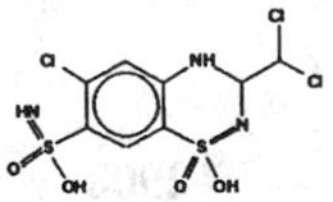

2855

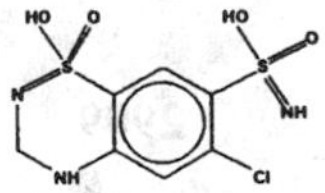

2856

2857

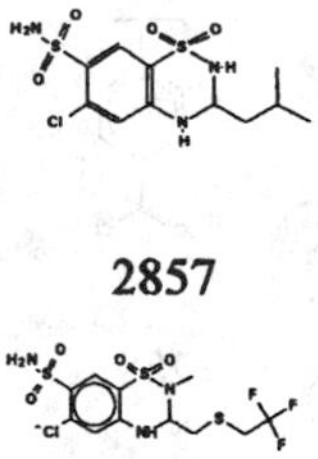

2858

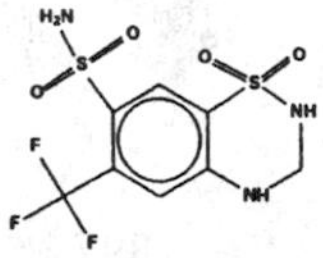

2859

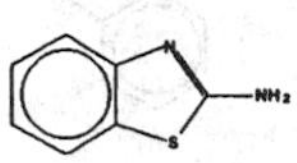

2860

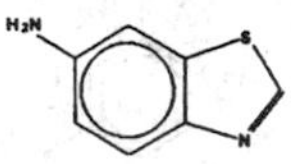

2861

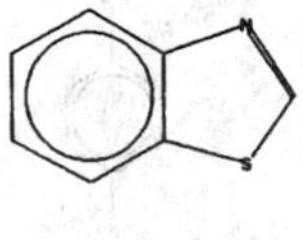

2862

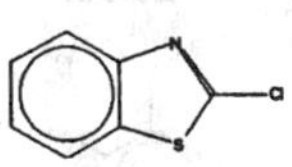

2863

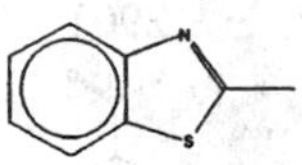

2864

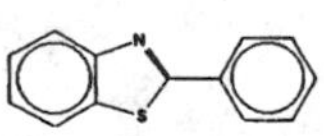

2865

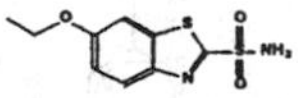

2866

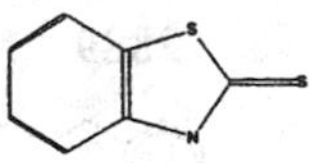

2867

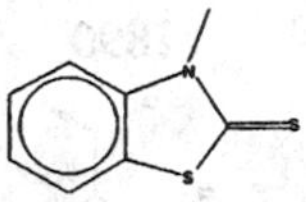

2868

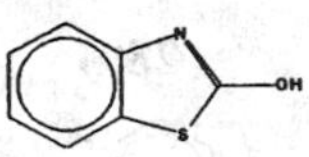

2869

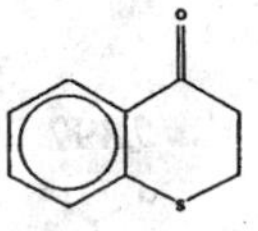

2870

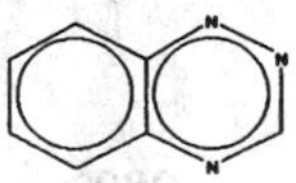

2871

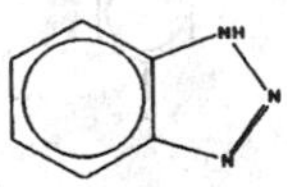

2872

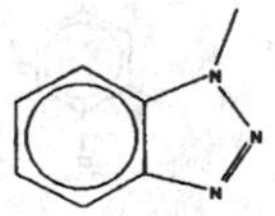

2873

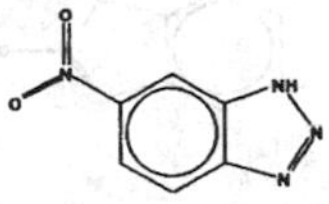

2874

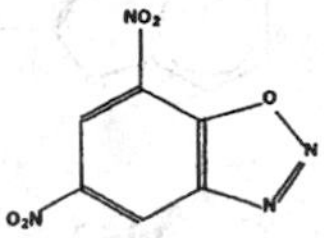

2875

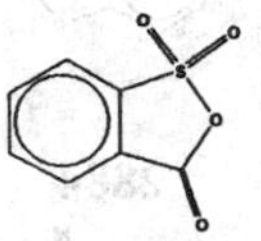

2876

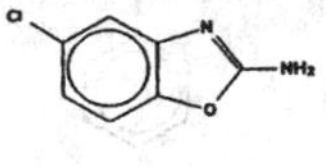

2877

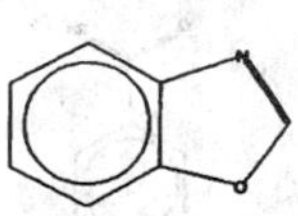

2878

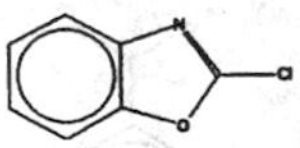

2879

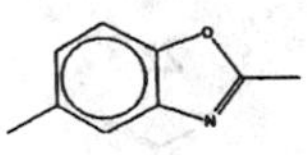

2880

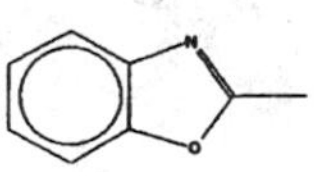

2881

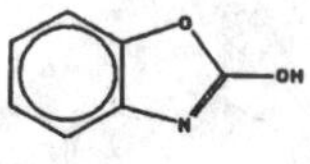

2882

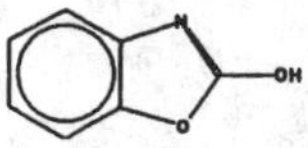

2883

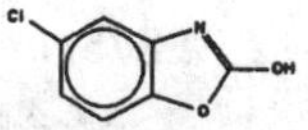

2884

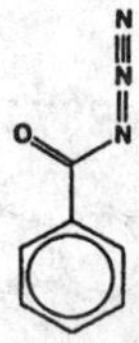

2885

2886

2887

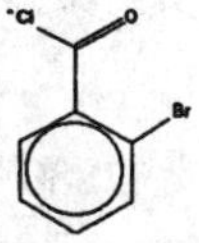

2888

2889

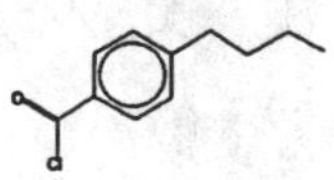

2890

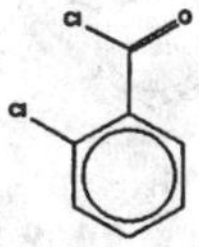

2891

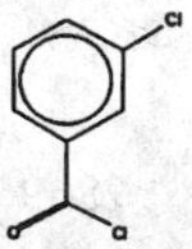

2892

2893

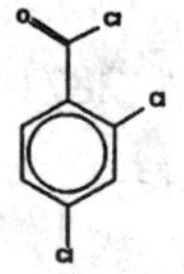

2894

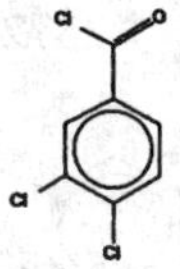

2895

2896

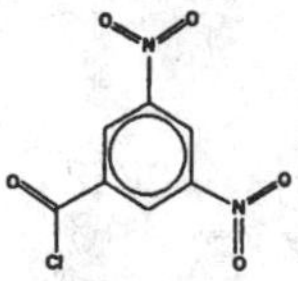

2897

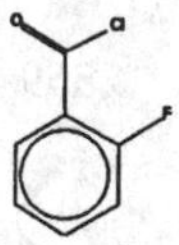

2898

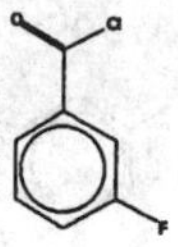

2899

2900

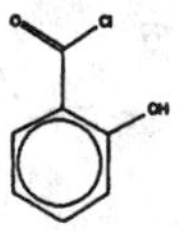

2901

2902

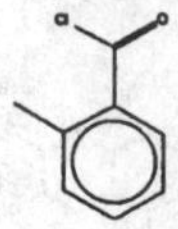

2903

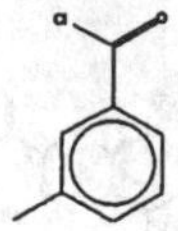

2904

2905

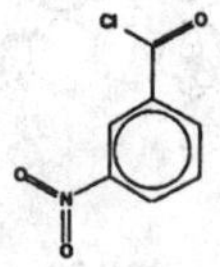

2906

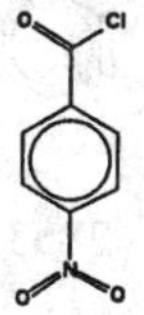

2907

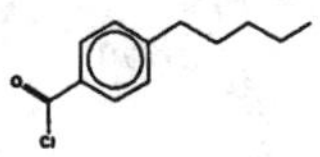

2908

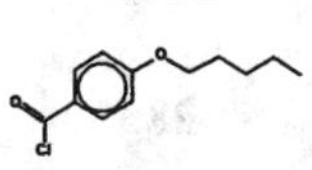

2909

2910

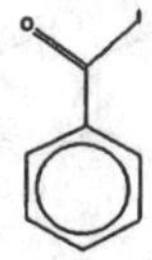

2911

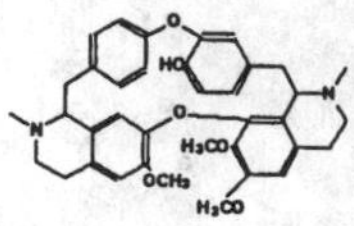

2912

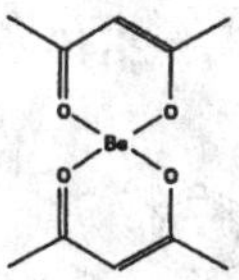

2913

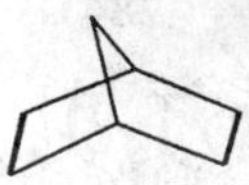

2914

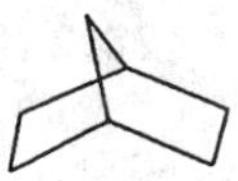

2915

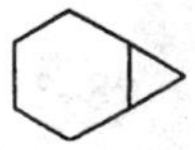

2916

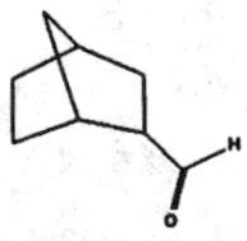

2918

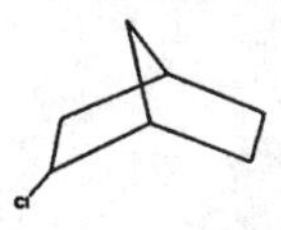

In numeric order as they occur in the Table of Physical Constants of Organic Compounds

2919

2920

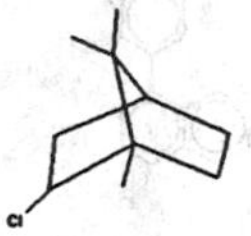

2921

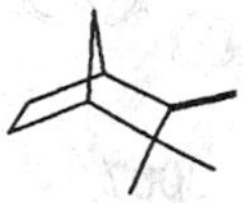

2922

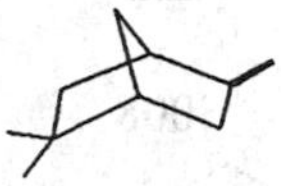

2923

2925

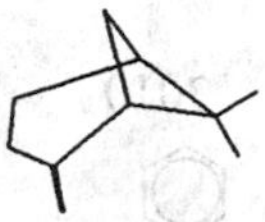

2926

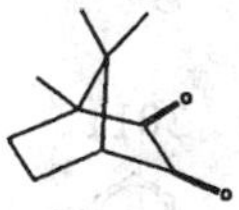

2927

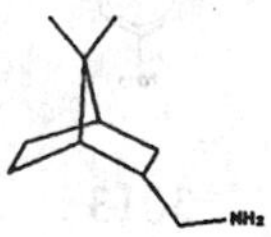

2928

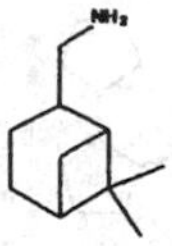

2929

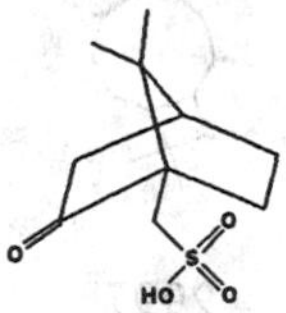

2930

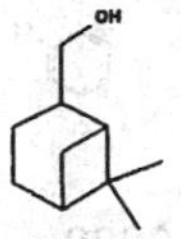

2932

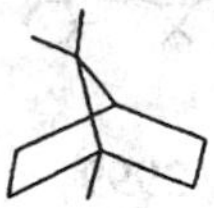

2933

2934

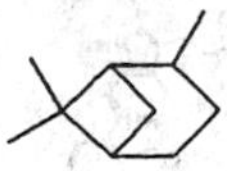

2935

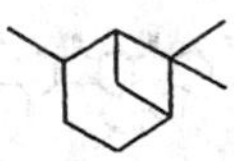

2936

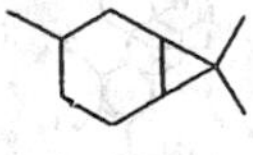

2937

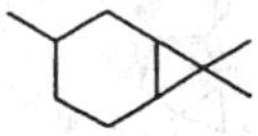

2939

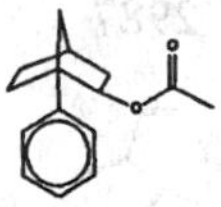

2941

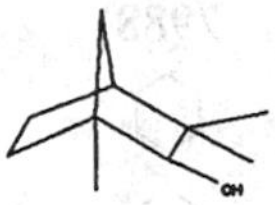

2944

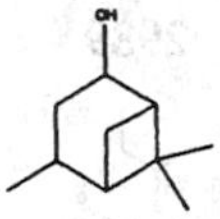

2945

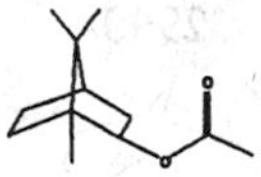

2946

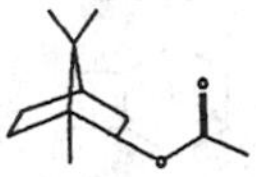

2947

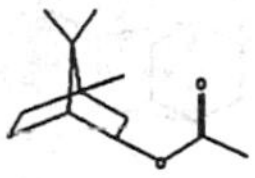

2948

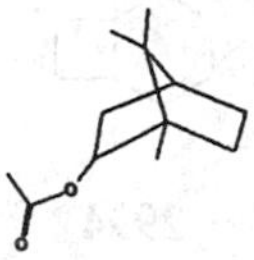

2950

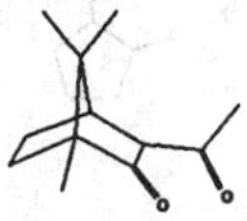

2951

2952

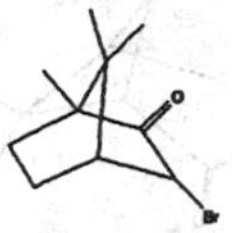

2953

2954

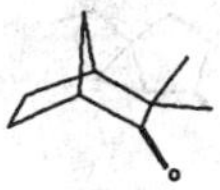

2955

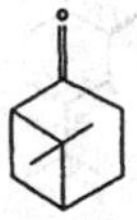

2957

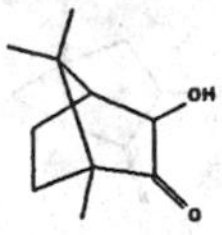

2958

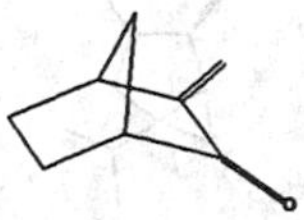

2959

2960

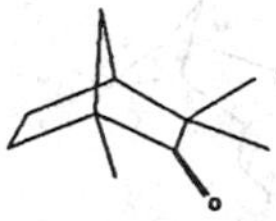

2961

2962

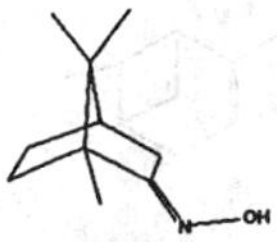

2963

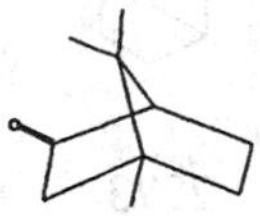

2964

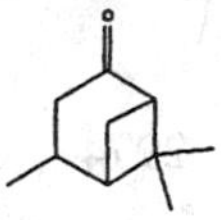

2966

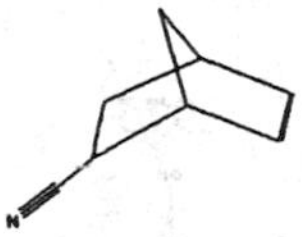

2967

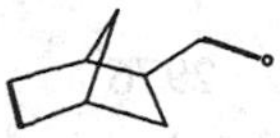

2968

In numeric order as they occur in the Table of Physical Constants of Organic Compounds

2969
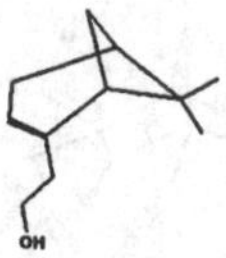
2970
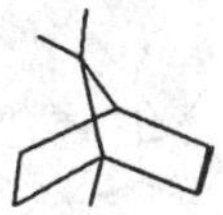
2971
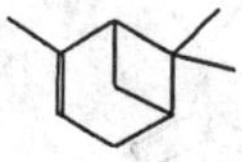
2972
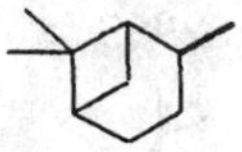
2973
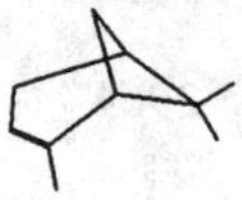
2974
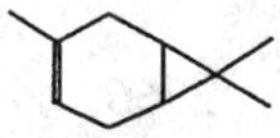
2975
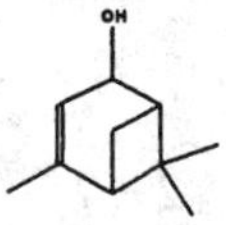
2976
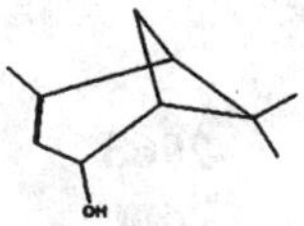
2977

2978
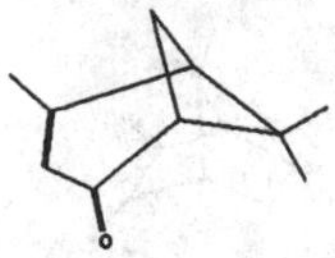
2980
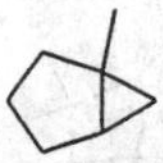
2981
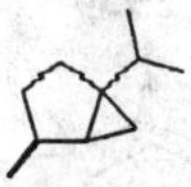
2982
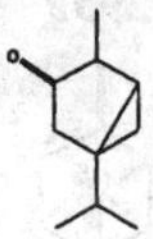
2983
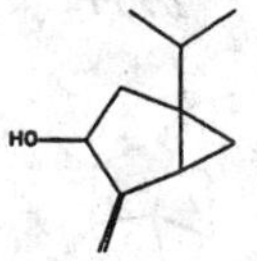
2984
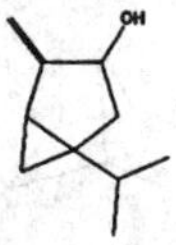
2985
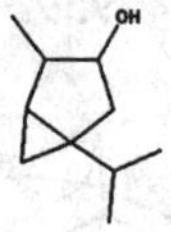
2986
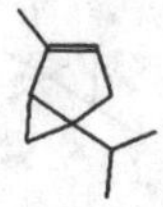
2987
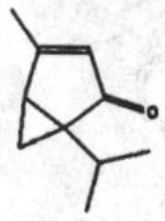
2988
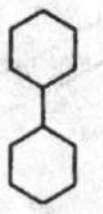
2989
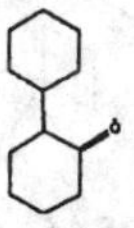
2990
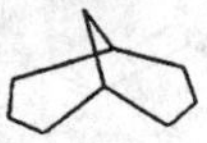
2991
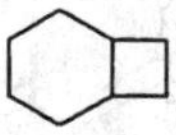
2992
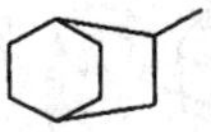
2993
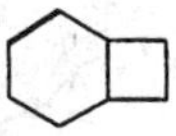
2994
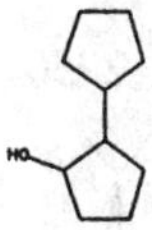
2995
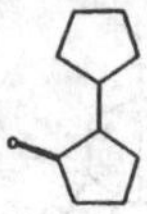
2996
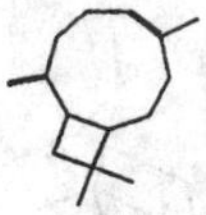
2997
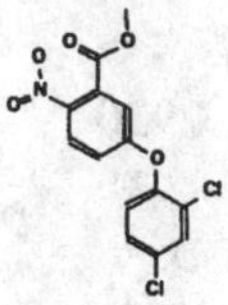
2998
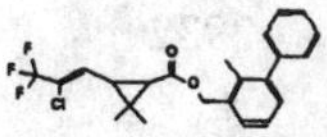
2999
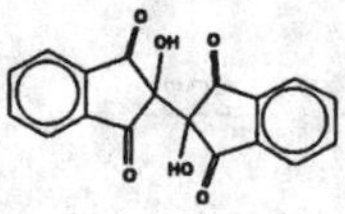
3000
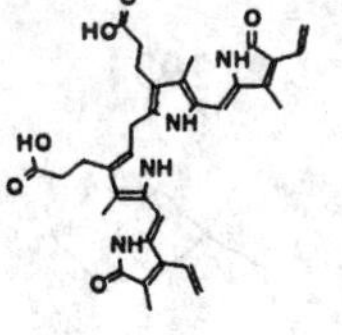
3001
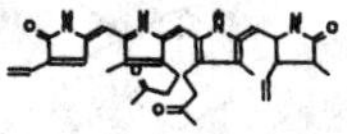
3002
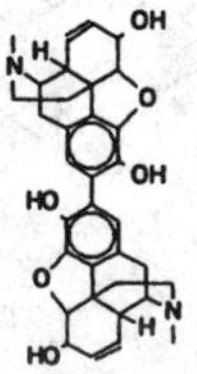
3003
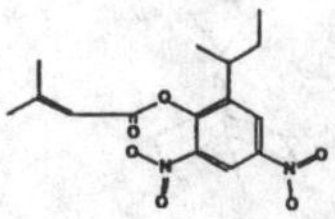
3004
3005
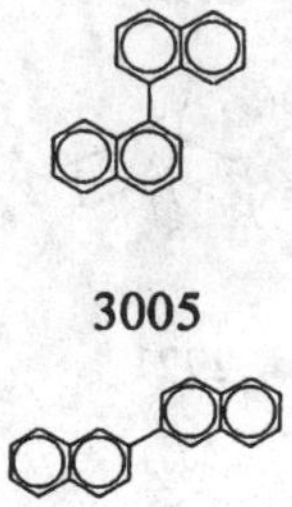
3007
3008
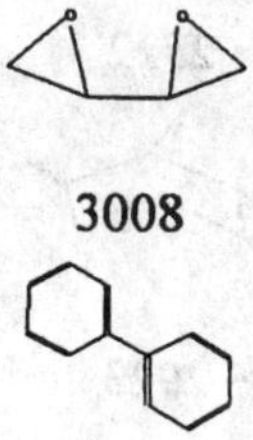
3009
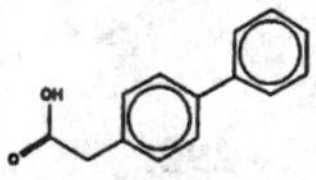
3010
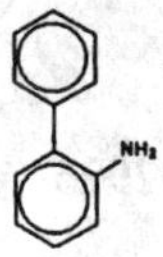
3011
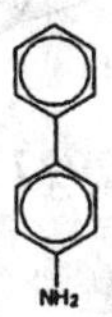
3013
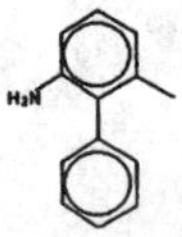

3014

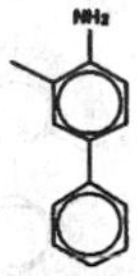

3015

3016

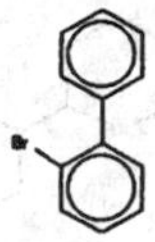

3017

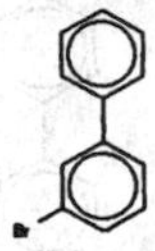

3018

3019

3020

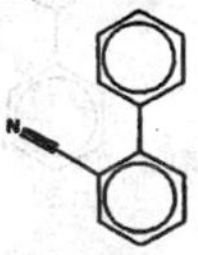

3021

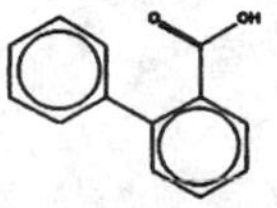

3022

3023

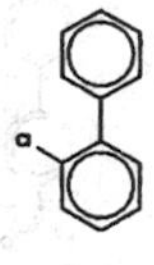

3024

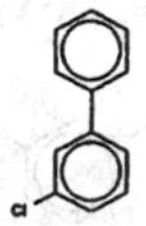

3025

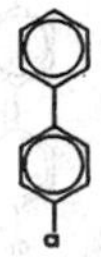

3026

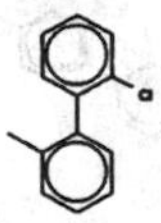

3027

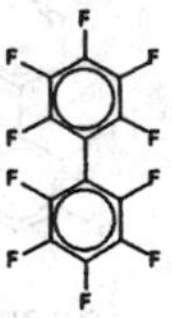

3028

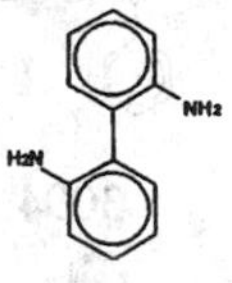

3029

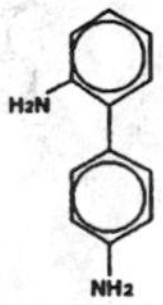

3030

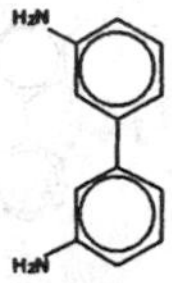

3031

3032

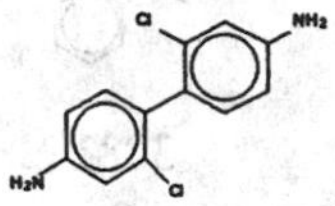

3033

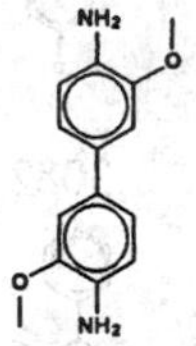

3034

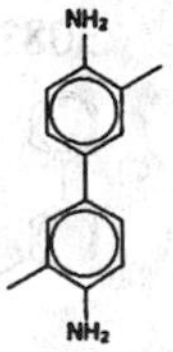

3035

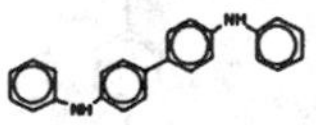

3037

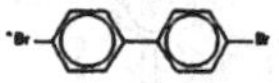

3038

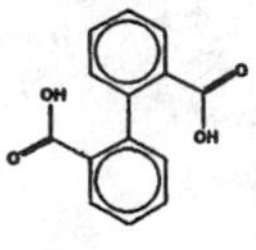

3039

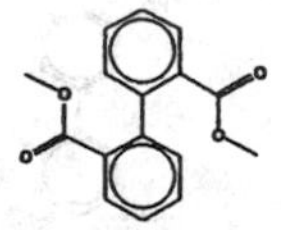

3040

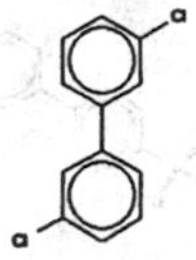

3041

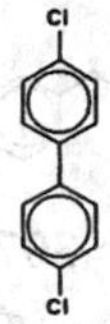

3042

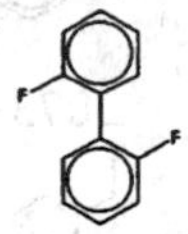

3043

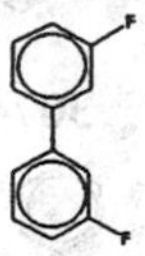

3044

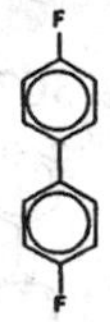

3045

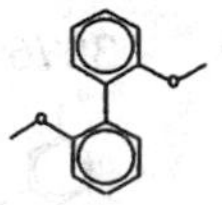

3046

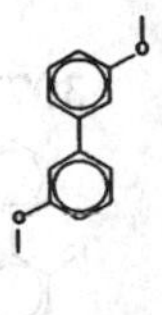

3050

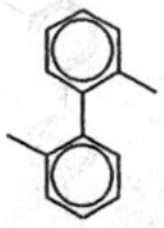

3051

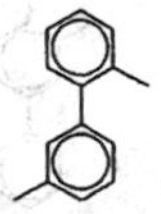

3052

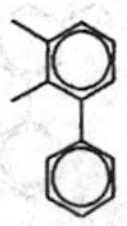

3053

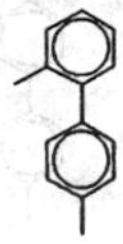

3054

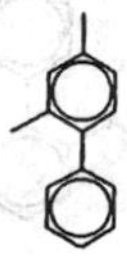

3055

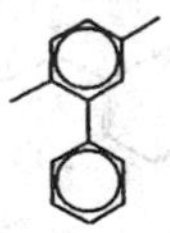

3056

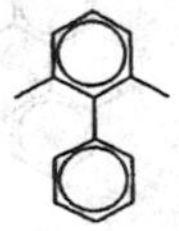

3057

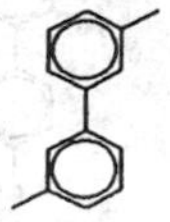

STRUCTURAL FORMULAS OF ORGANIC COMPOUNDS (continued)

In numeric order as they occur in the Table of Physical Constants of Organic Compounds

3058

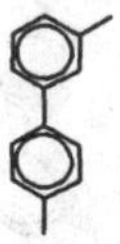

3059

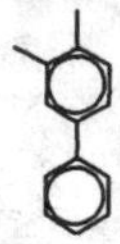

3060

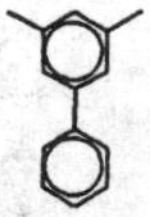

3061

3062

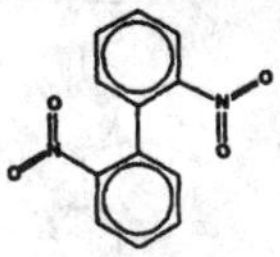

3063

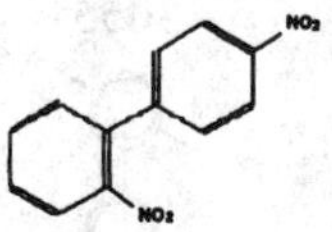

3064

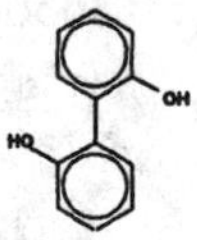

3065

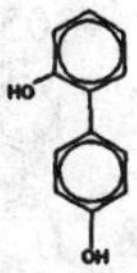

3066

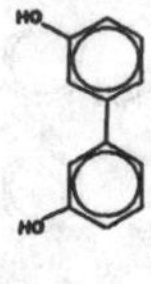

3070

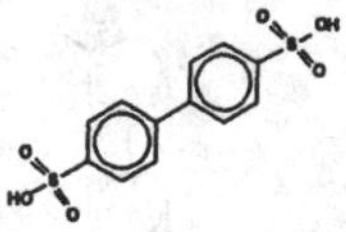

3071

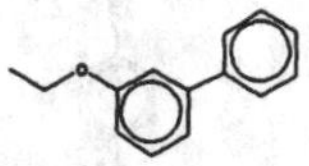

3072

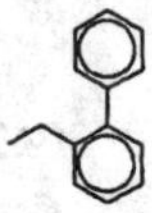

3073

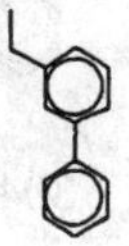

3074

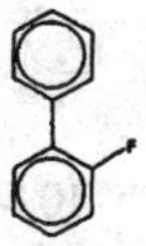

3075

3076

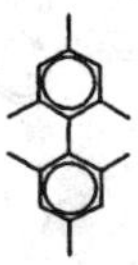

3077

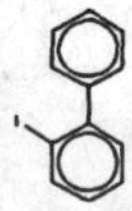

3078

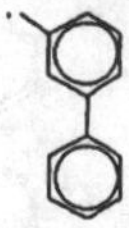

3079

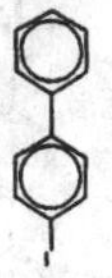

3080

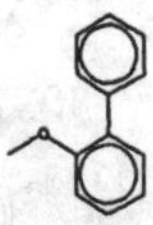

3081

3082

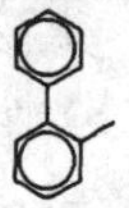

3083

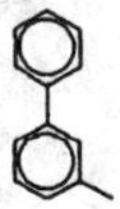

3084

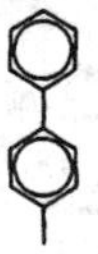

3085

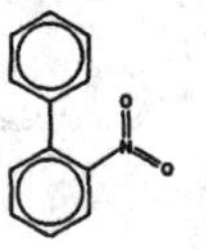

3086

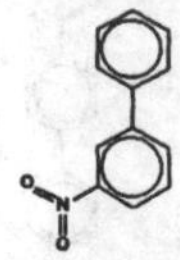

3087

3089

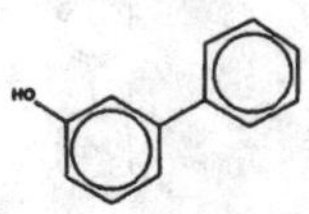

3090

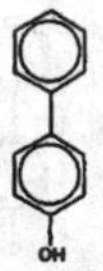

3091

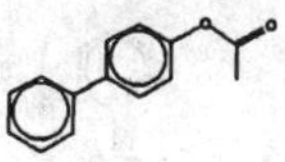

3092

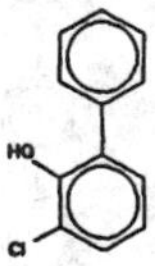

3093

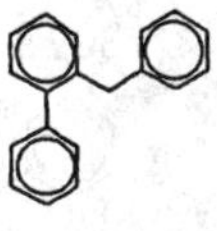

3094

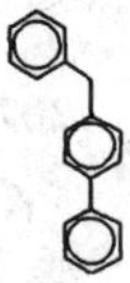

3095

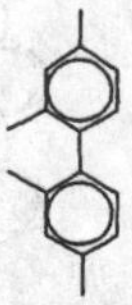

3096

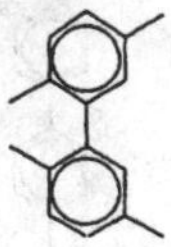

3098

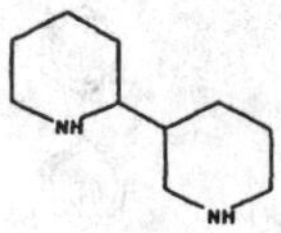

3099

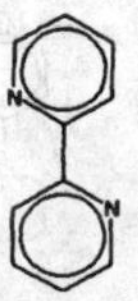

3100

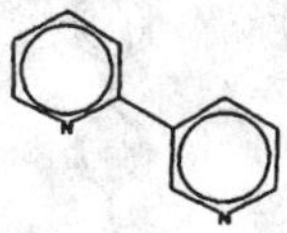

3101

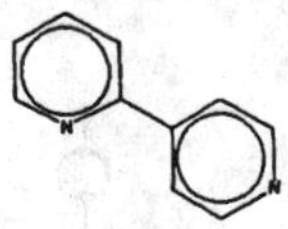

3102

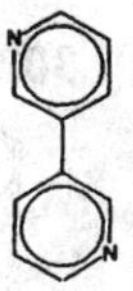

In numeric order as they occur in the Table of Physical Constants of Organic Compounds

3103

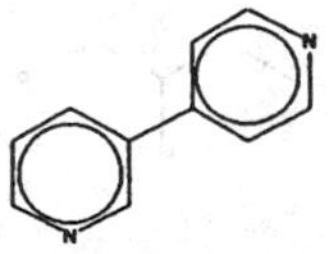

3104

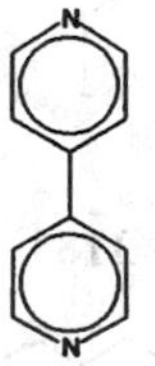

3105

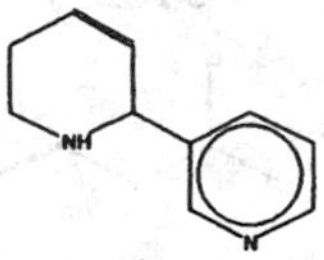

3108

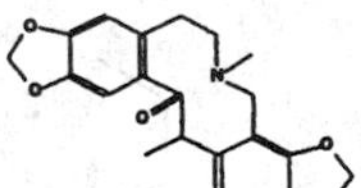

3109

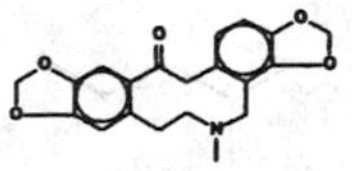

3111

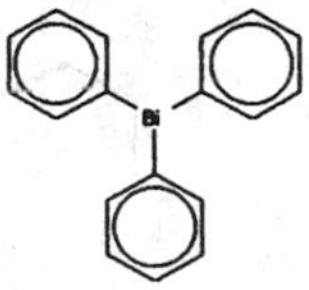

3112

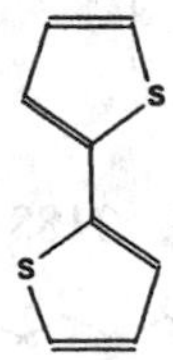

3113

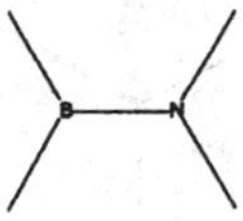

3114

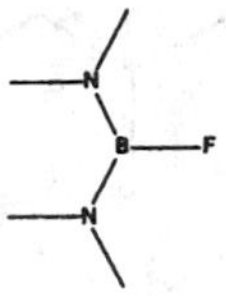

3115

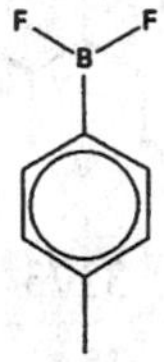

3116

3117

3118

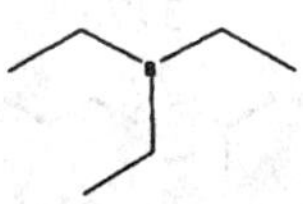

3119

3120

3121

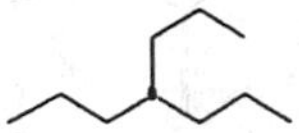

3122

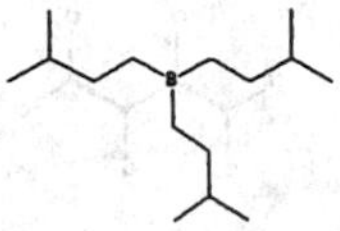

3123

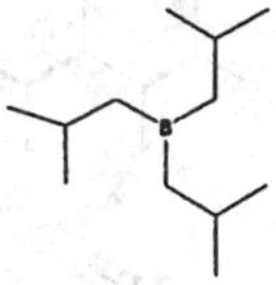

3124

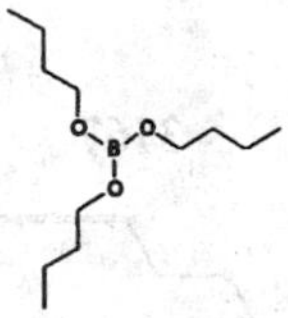

3125

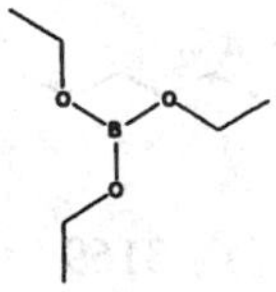

3126

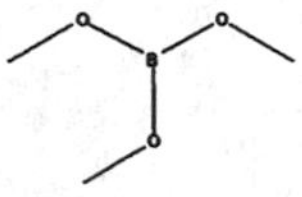

3127

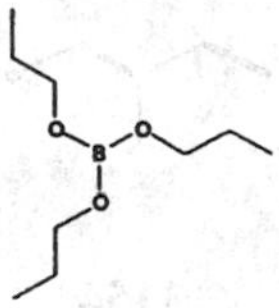

3128

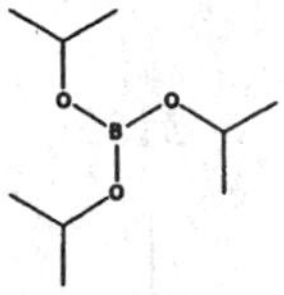

3129

3130

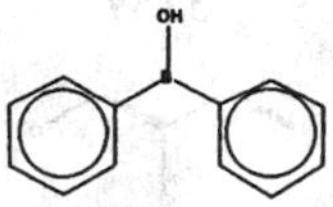

3131

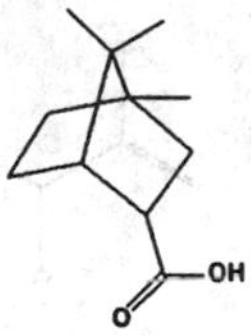

3132

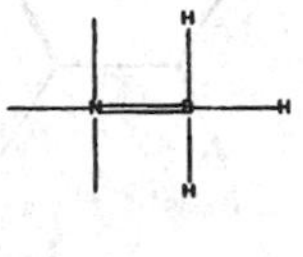

3134

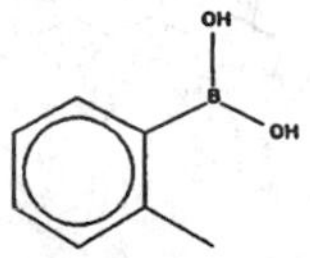

3135

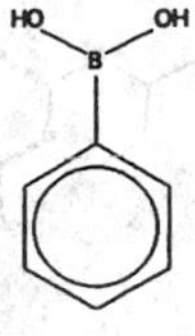

3136

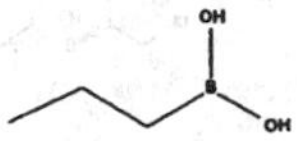

3137

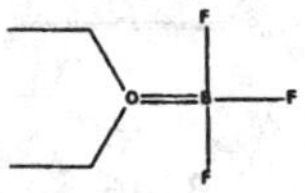

3138

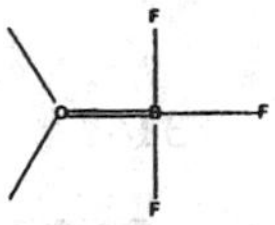

3139

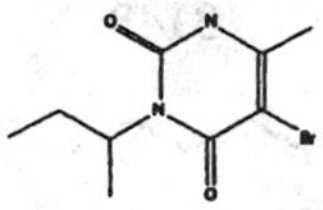

3140

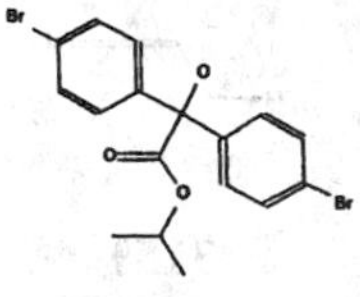

3141

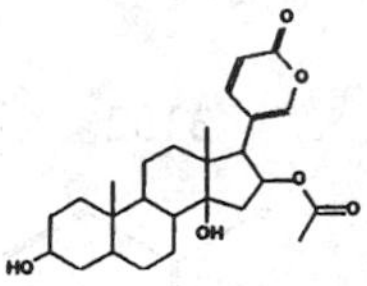

STRUCTURAL FORMULAS OF ORGANIC COMPOUNDS (continued)

In numeric order as they occur in the Table of Physical Constants of Organic Compounds

3143

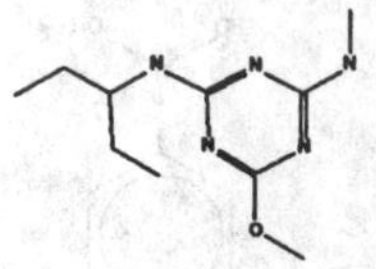

3144

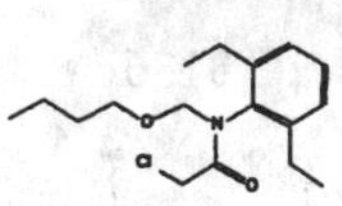

3145

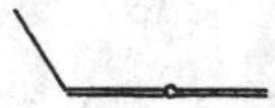

3146

3147

3148

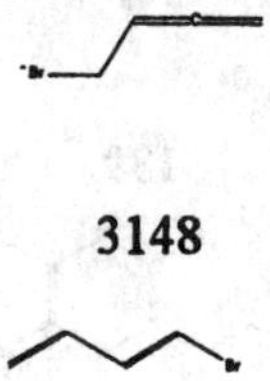

3150

3151

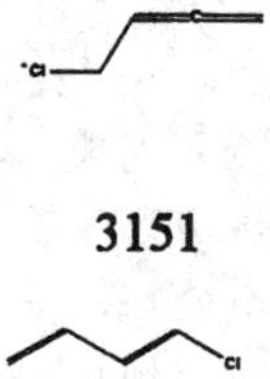

3152

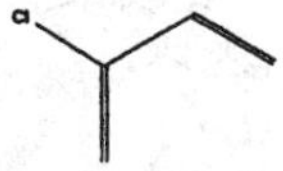

3153

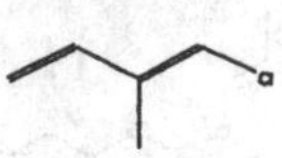

3154

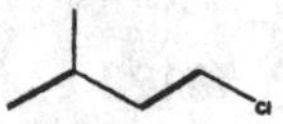

3155

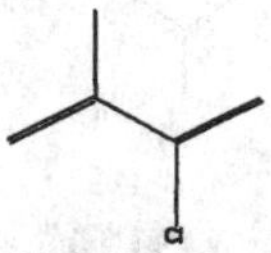

3156

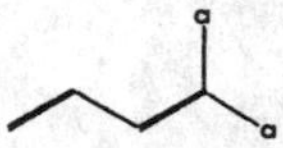

3157

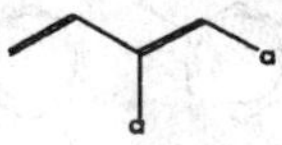

3158

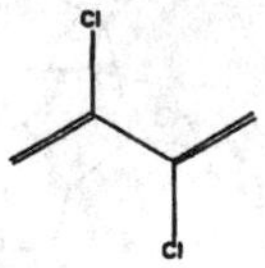

3159

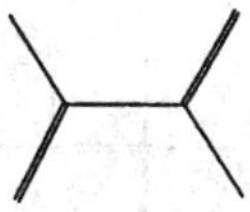

3160

3161

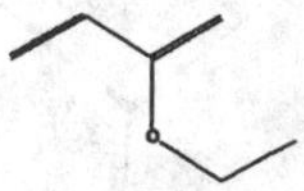

3162

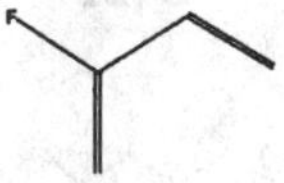

3163

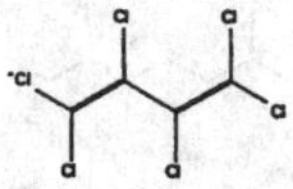

3164

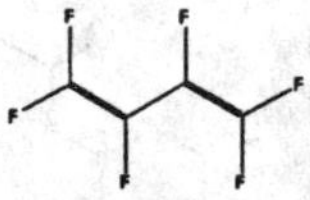

3166

3167

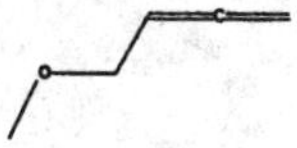

3168

3169

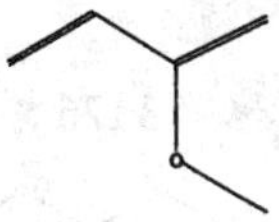

3170

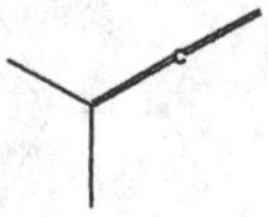

3171

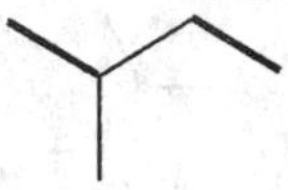

3172

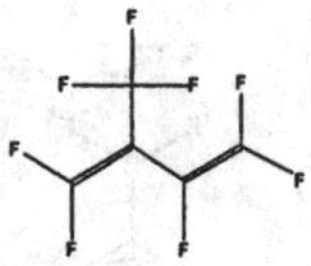

3173

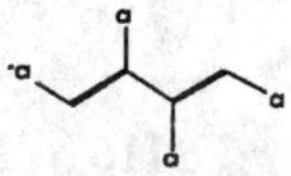

3174

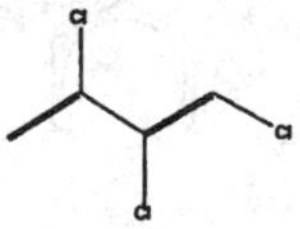

3175

3176

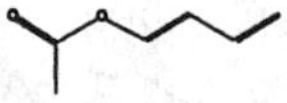

3177

3178

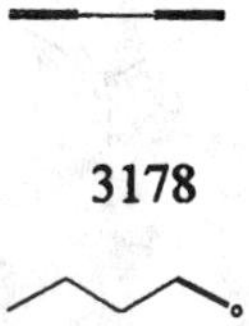

3179

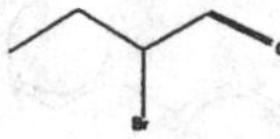

3180

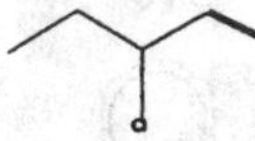

3181

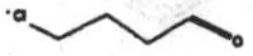

3182

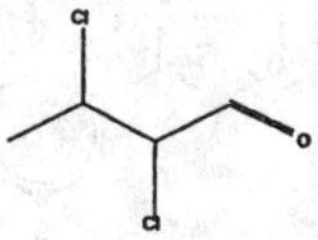

3183

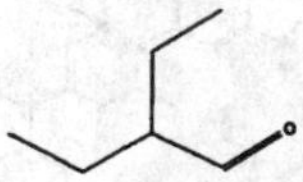

3184

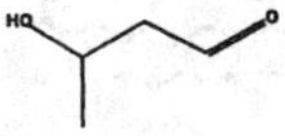

3185

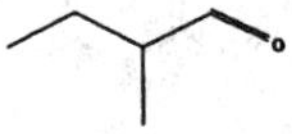

3186

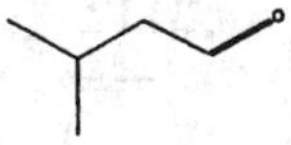

3188

3189

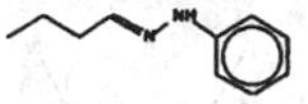

3190

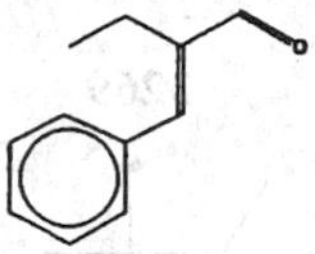

3191

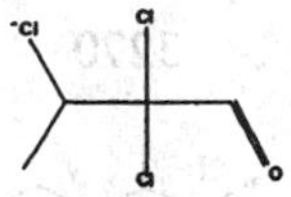

3192

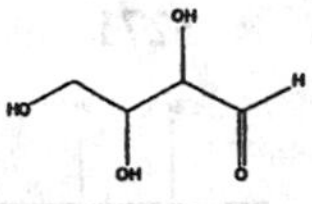

3194

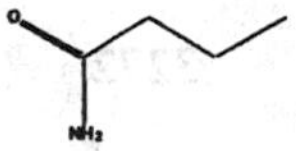

3195

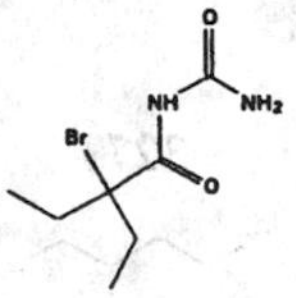

3196

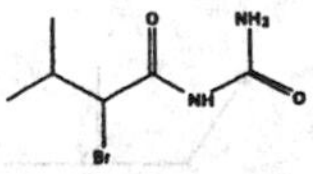

3197

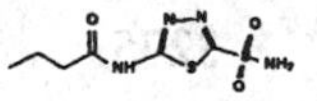

3198

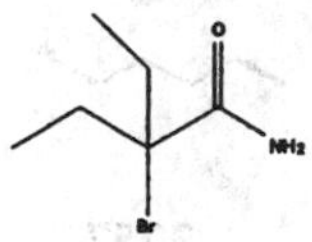

3199

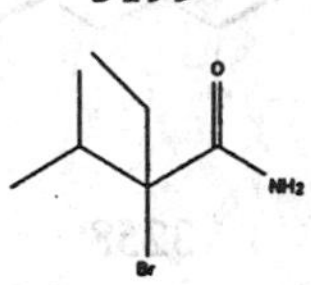

3200

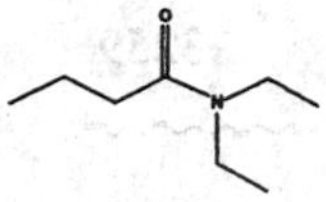

3201

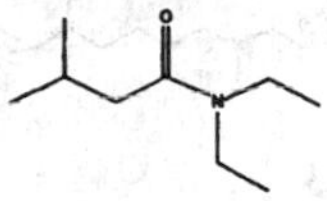

3202

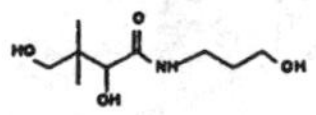

3203

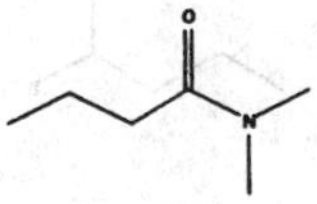

3205

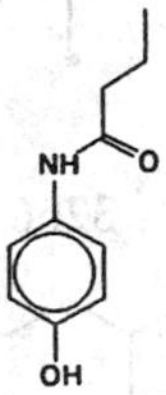

3206

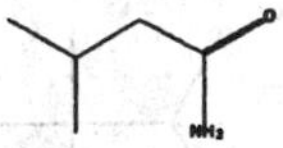

3207

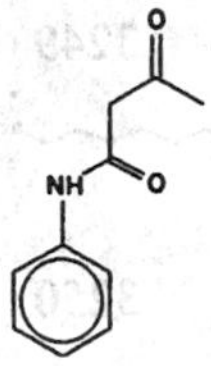

3208

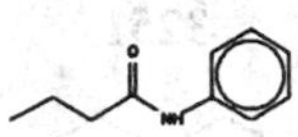

3209

3210

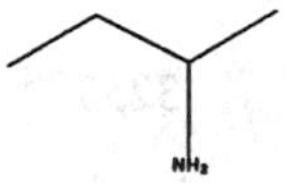

3211

3212

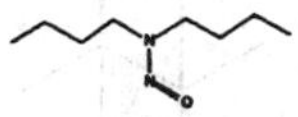

3213

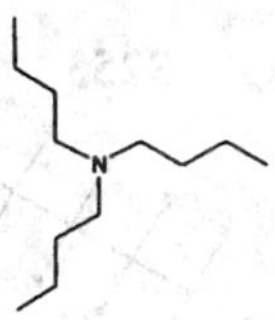

3214

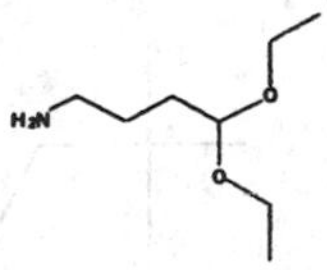

3215

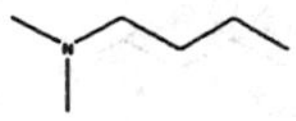

3216

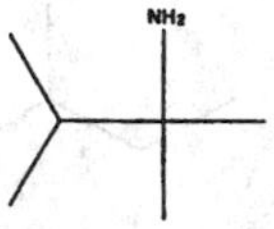

3217

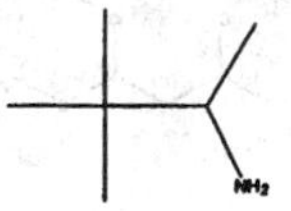

3218

3220

3221

3222

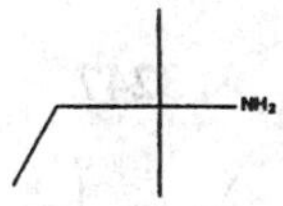

3223

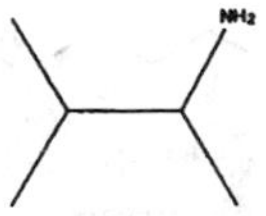

3224

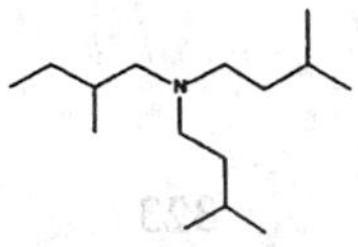

3225

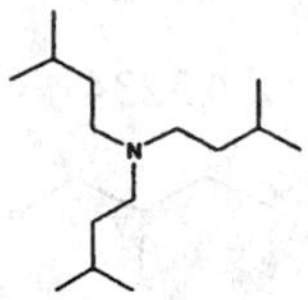

3226

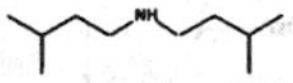

3227

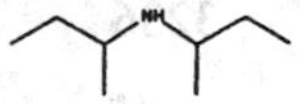

3228

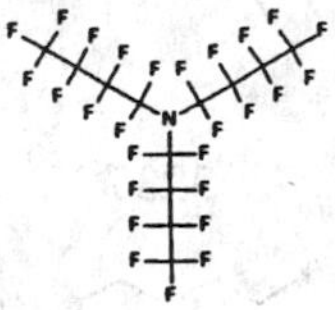

3229

3230

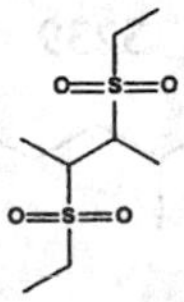

STRUCTURAL FORMULAS OF ORGANIC COMPOUNDS (continued)

In numeric order as they occur in the Table of Physical Constants of Organic Compounds

3231

3232

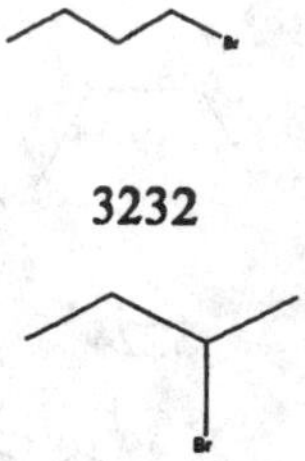

3233

3234

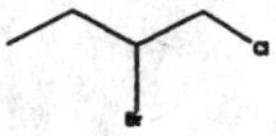

3235

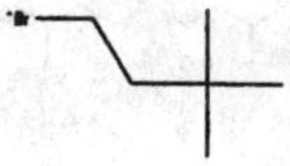

3236

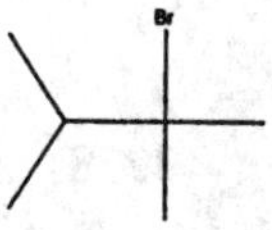

3237

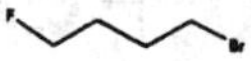

3238

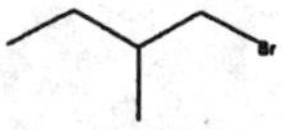

3239

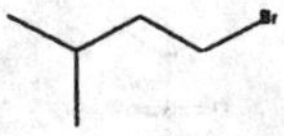

3240

3241

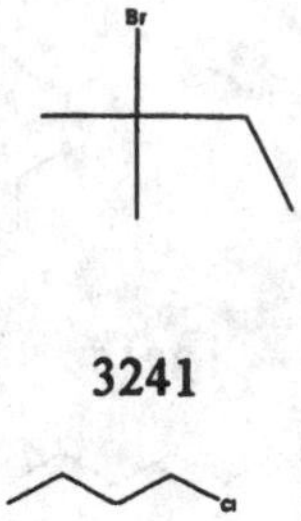

3242

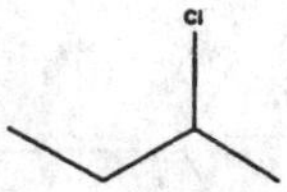

3243

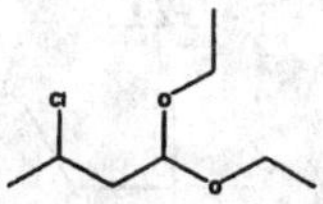

3244

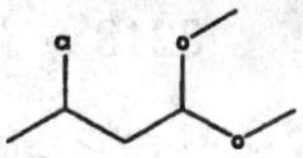

3245

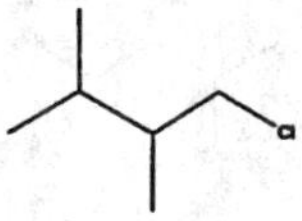

3246

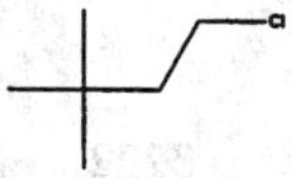

3247

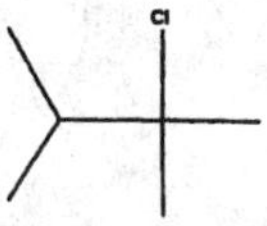

3248

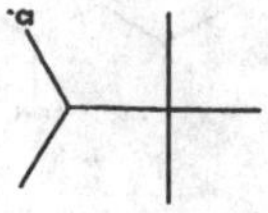

3249

3250

3251

3252

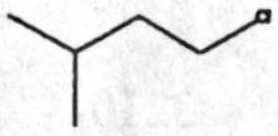

3253

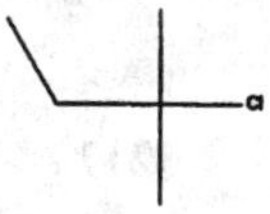

3254

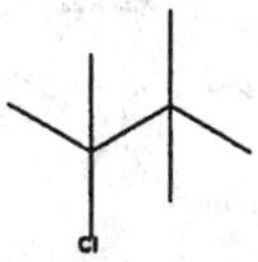

3255

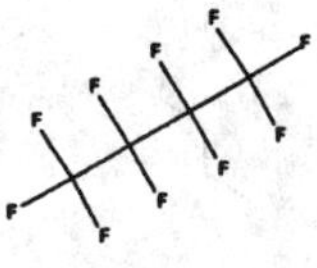

3256

3257

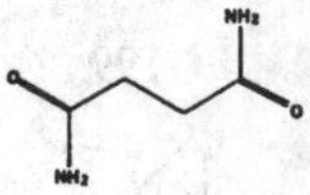

3258

3259

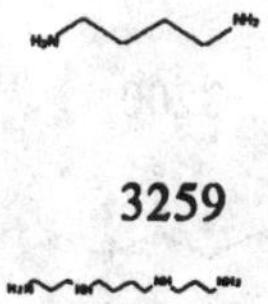

3262

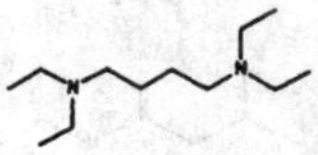

3263

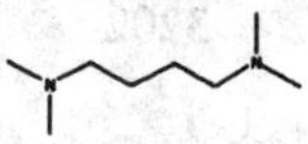

3264

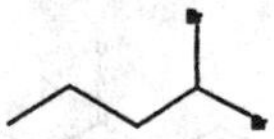

3265

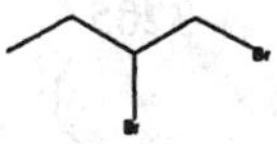

3266

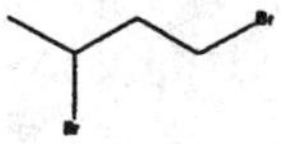

3267

3268

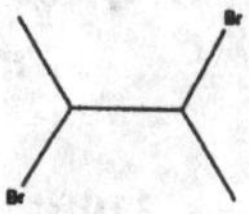

3269

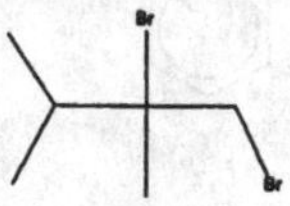

3270

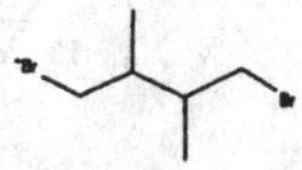

3271

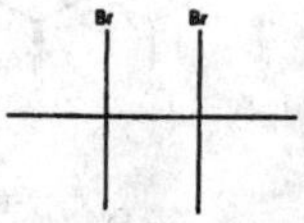

3272

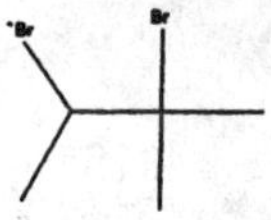

3273

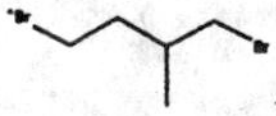

3274

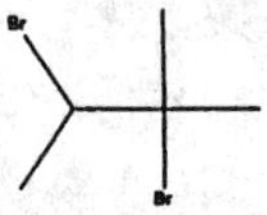

3275

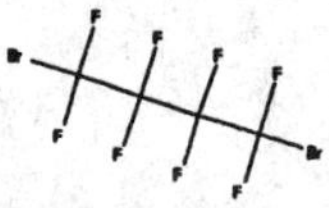

3276

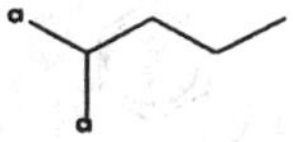

3277

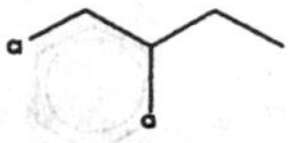

3278

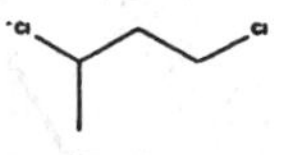

3279

3280

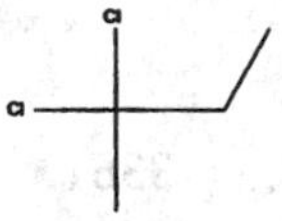

3281

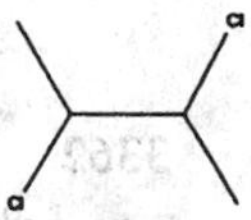

3282

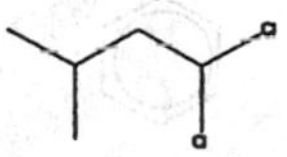

3283

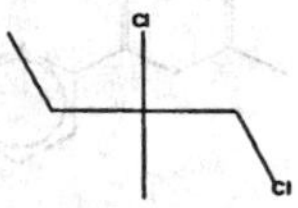

3284

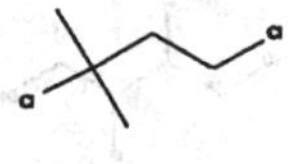

3285

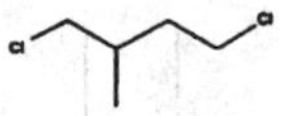

3286

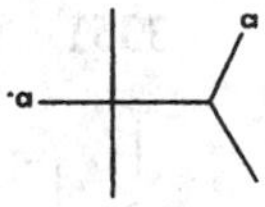

3287

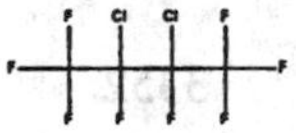

3288

3289

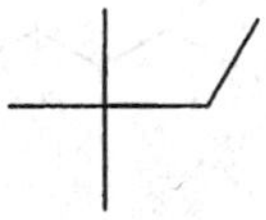

3290

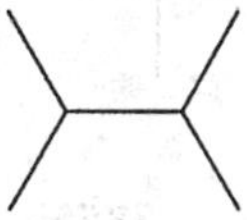

3291

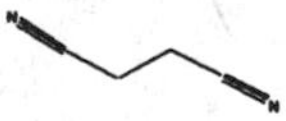

3292

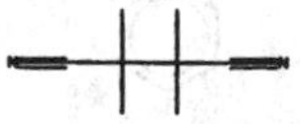

3293

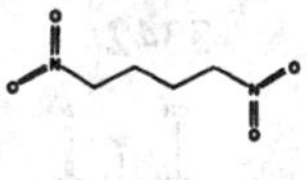

3294

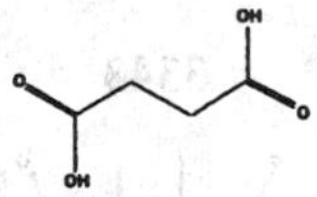

3295

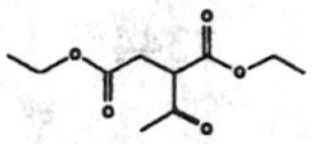

3296

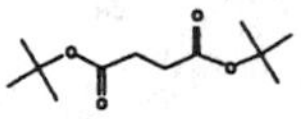

3297

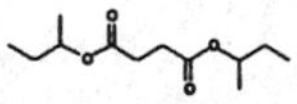

3298

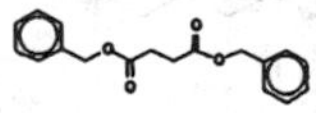

3299

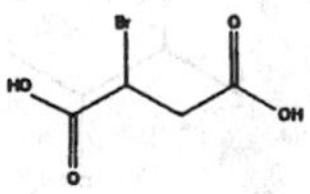

3300

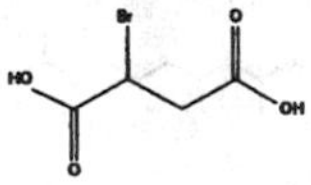

3301

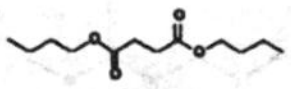

3302

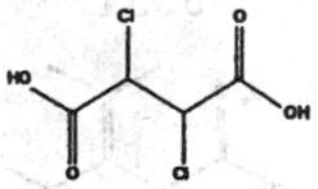

3303

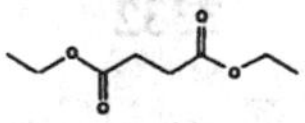

3304

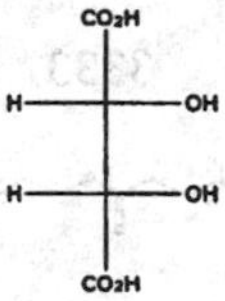

3305

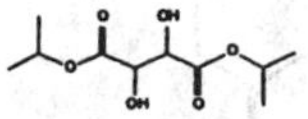

3307

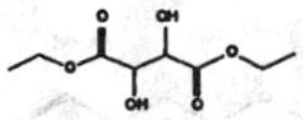

3308

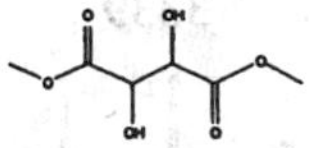

3309

3310

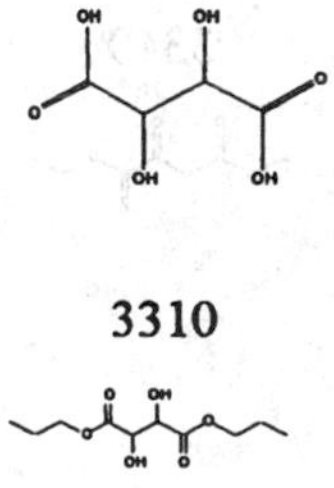

3311

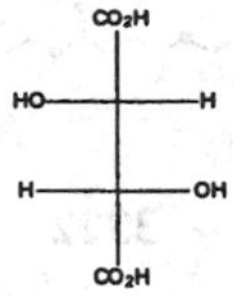

3312

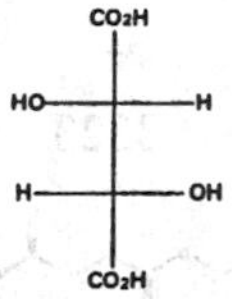

3313

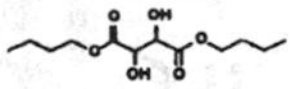

3316

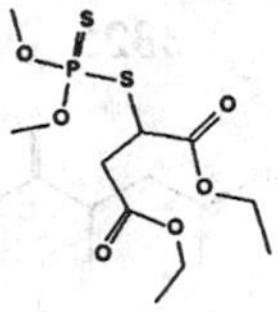

3317

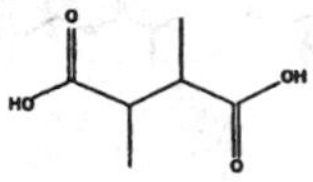

3318

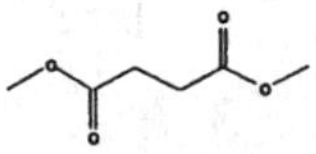

3319

3320

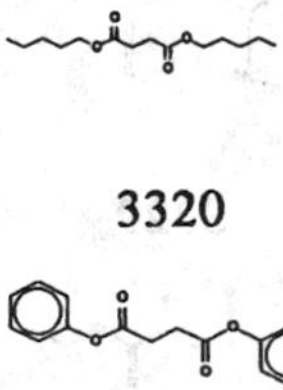

STRUCTURAL FORMULAS OF ORGANIC COMPOUNDS (continued)

In numeric order as they occur in the Table of Physical Constants of Organic Compounds

3321

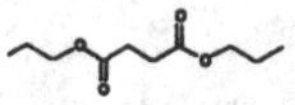

3322

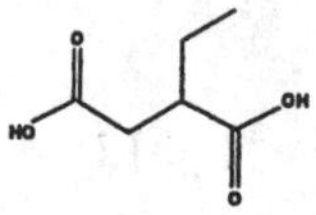

3323

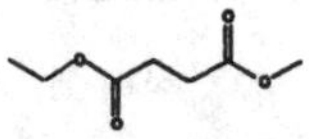

3324

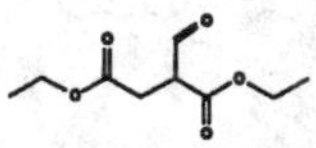

3325

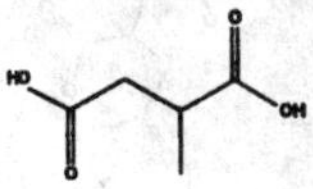

3326

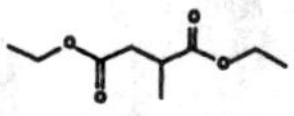

3327

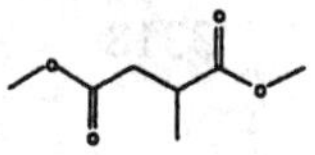

3328

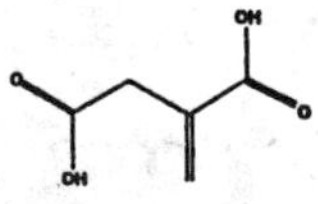

3329

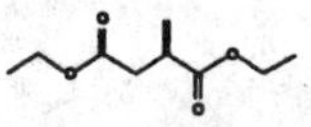

3330

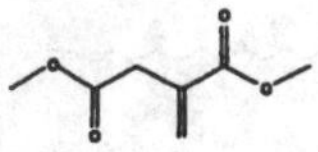

3331

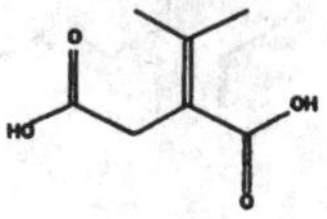

3332

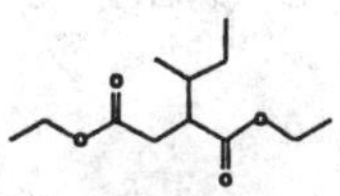

3333

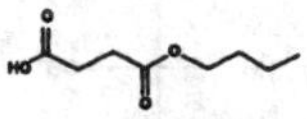

3334

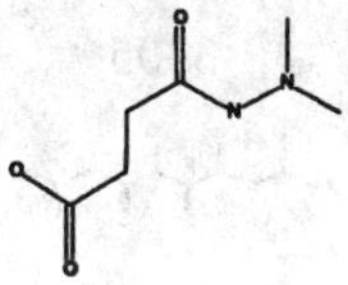

3336

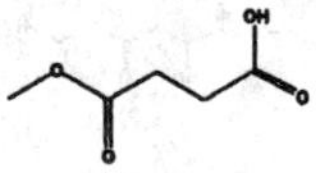

3339

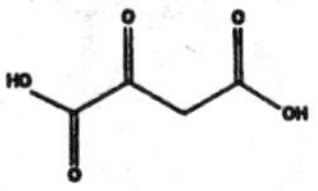

3340

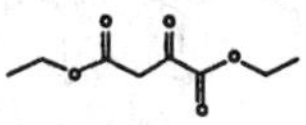

3341

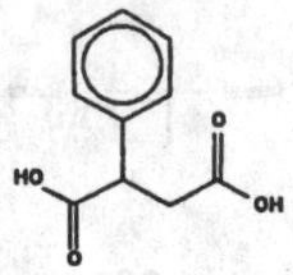

3342

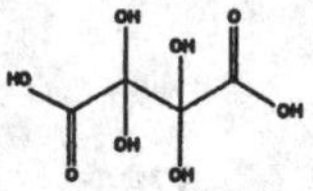

3343

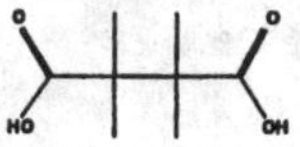

3344

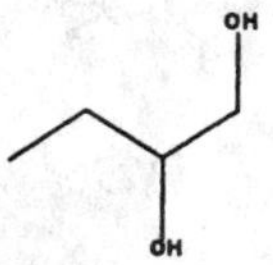

3345

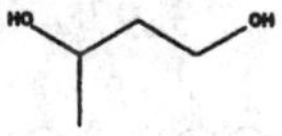

3346

3347

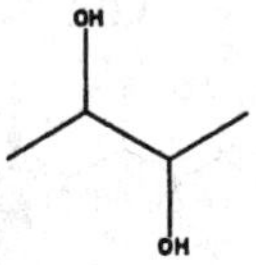

3348

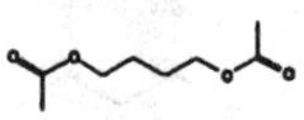

3349

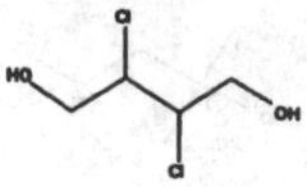

3350

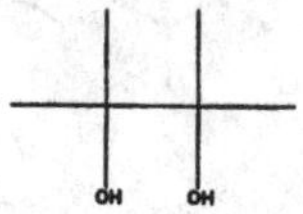

3351

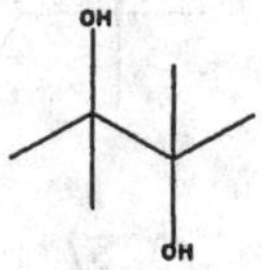

3352

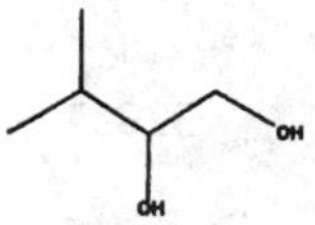

3353

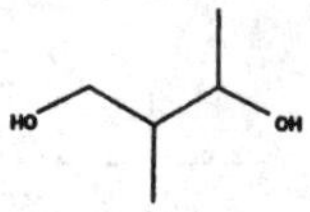

3354

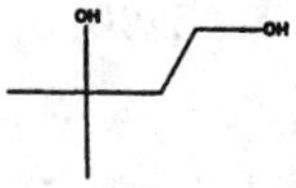

3355

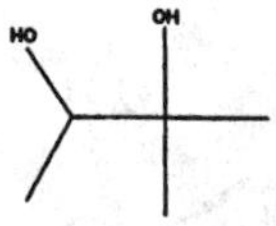

3356

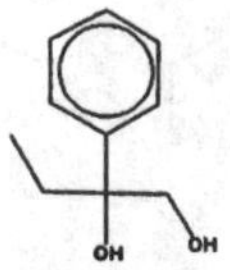

3357

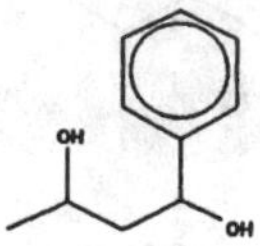

3359

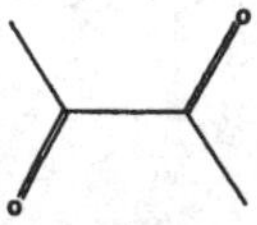

3360

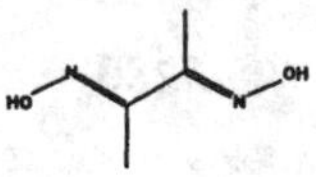

3361

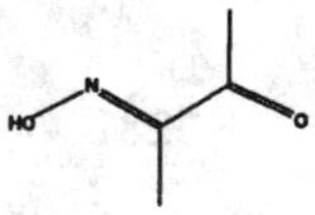

3362

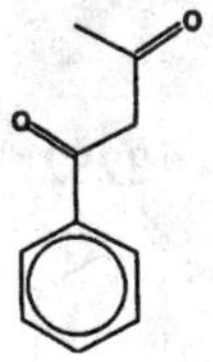

3363

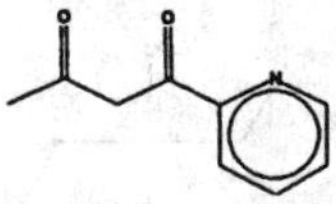

STRUCTURAL FORMULAS OF ORGANIC COMPOUNDS (continued)

In numeric order as they occur in the Table of Physical Constants of Organic Compounds

3364

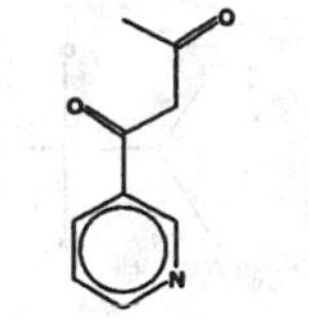

3365

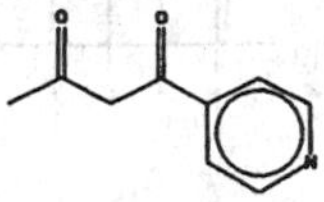

3366

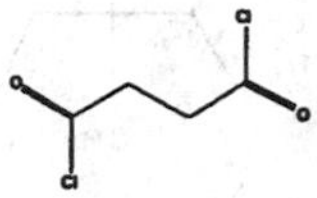

3367

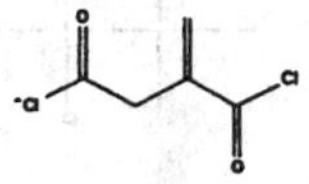

3368

3369

3370

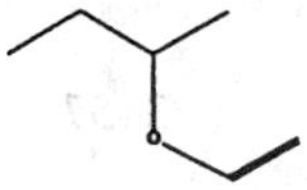

3371

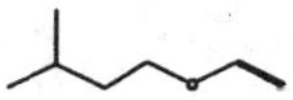

3372

3373

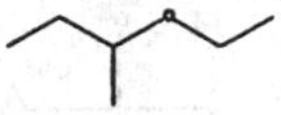

3374

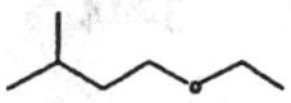

3375

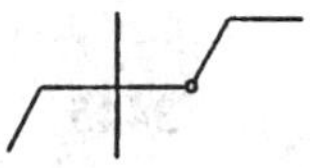

3376

3377

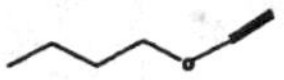

3378

3379

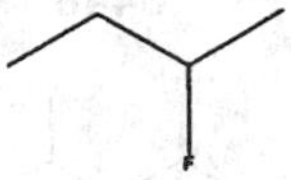

3381

3382

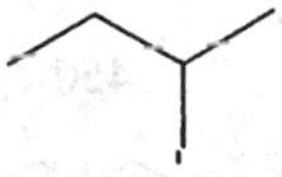

3383

3384

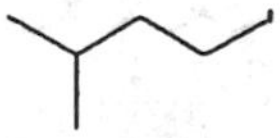

3385

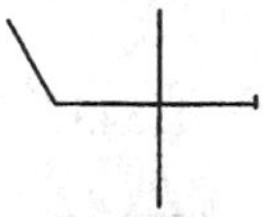

3387

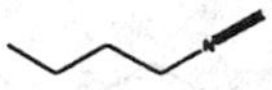

3388

3390

3391

3392

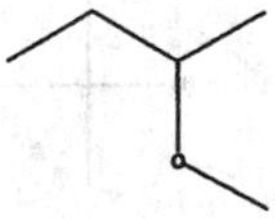

3394

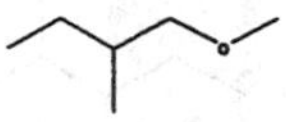

3395

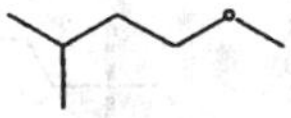

3396

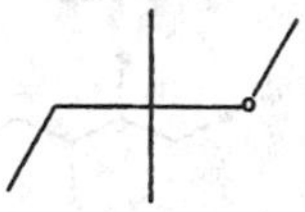

3397

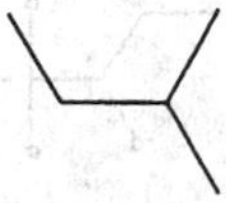

3398

3399

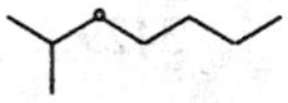

3400

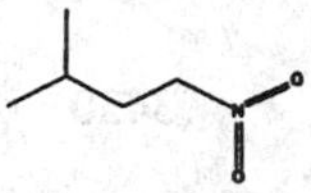

3401

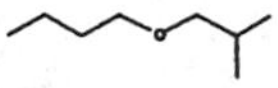

3402

3403

3404

3405

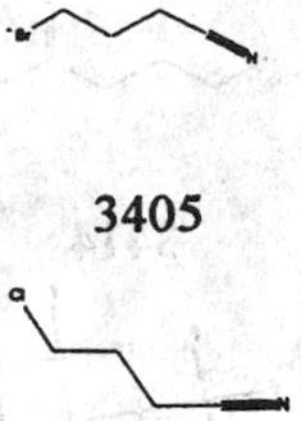

3406

3407

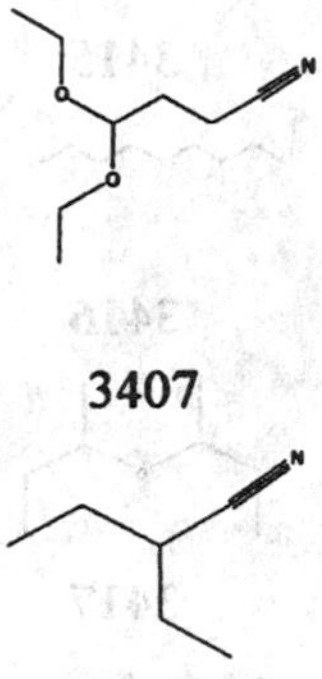

3408

3409

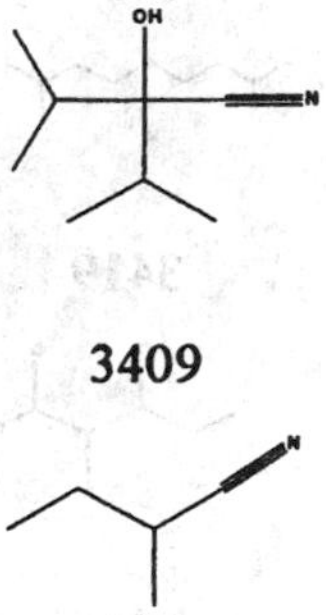

3410

3411

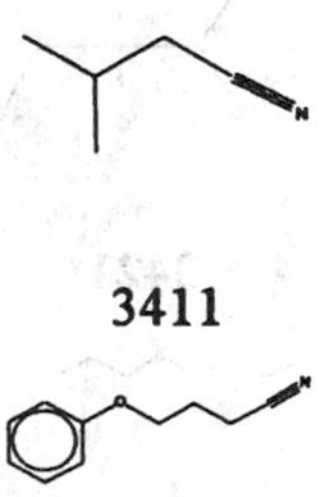

STRUCTURAL FORMULAS OF ORGANIC COMPOUNDS (continued)

In numeric order as they occur in the Table of Physical Constants of Organic Compounds

3412

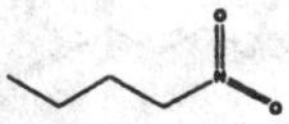

3414

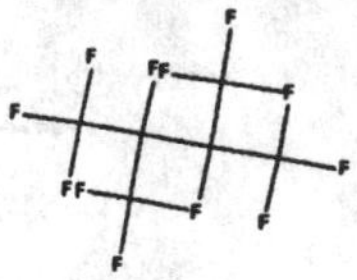

3415

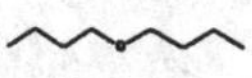

3416

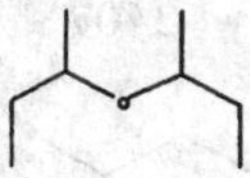

3417

3418

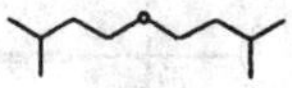

3419

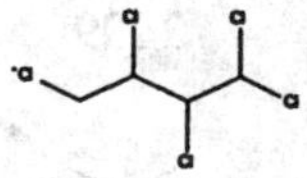

3420

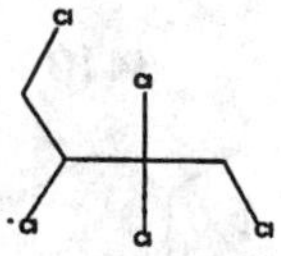

3421

3422

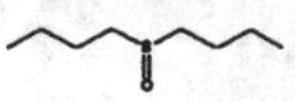

3423

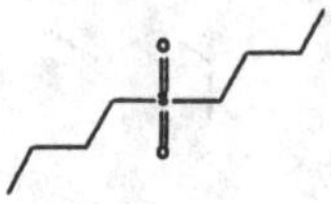

3426

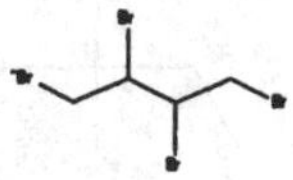

3427

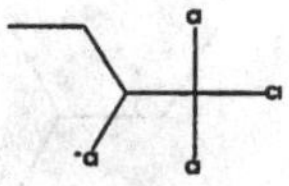

3428

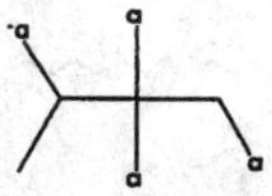

3429

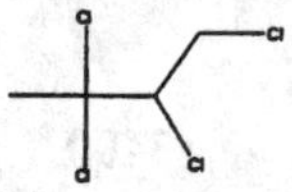

3430

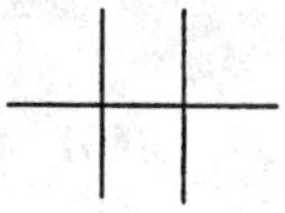

3431

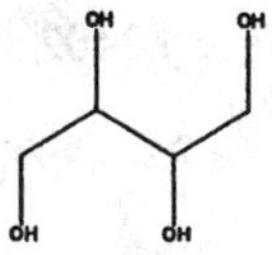

3433

3434

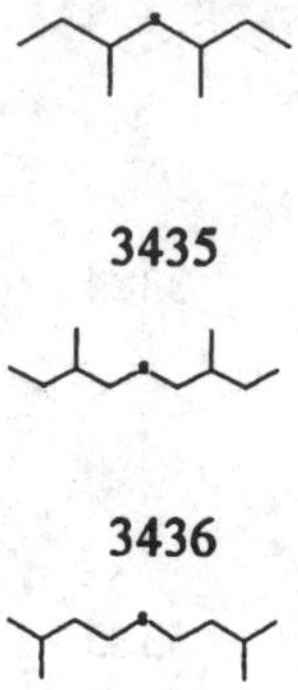

3435

3436

3437

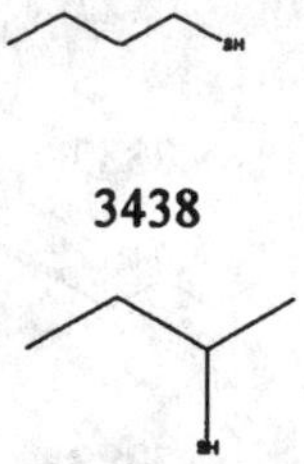

3438

3439

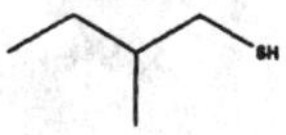

3440

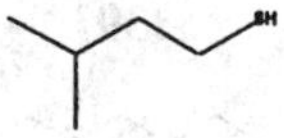

3441

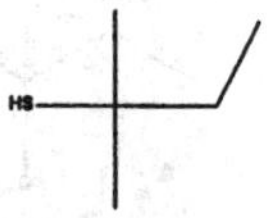

3442

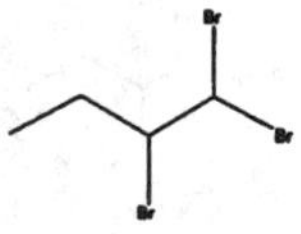

3443

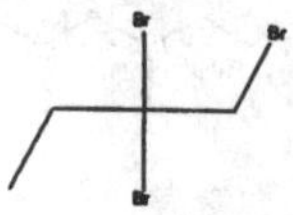

3444

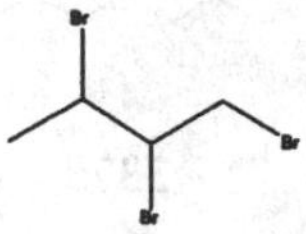

3445

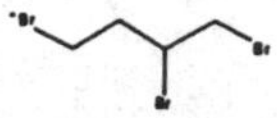

3446

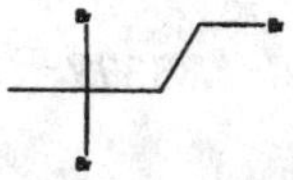

3447

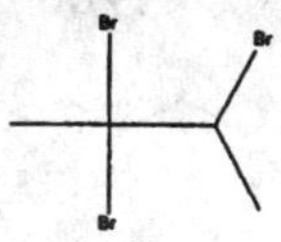

3448

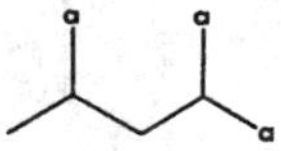

3449

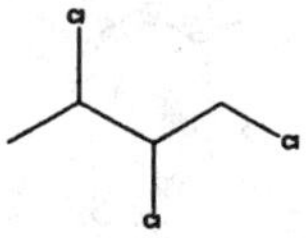

3450

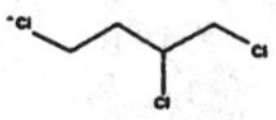

3451

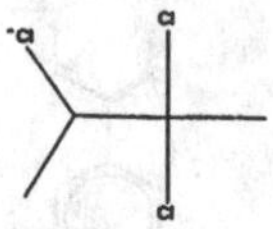

3452

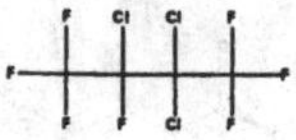

3453

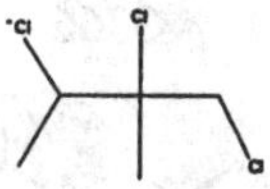

3454

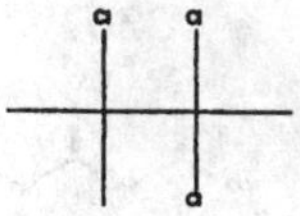

3455

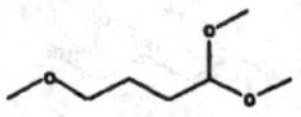

3456

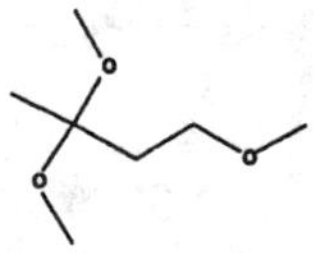

3457

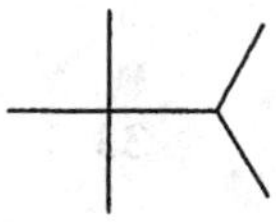

3458

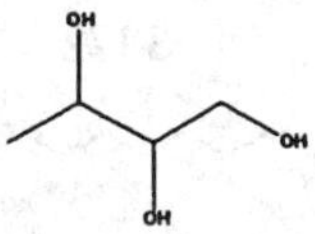

STRUCTURAL FORMULAS OF ORGANIC COMPOUNDS (continued)

In numeric order as they occur in the Table of Physical Constants of Organic Compounds

3459

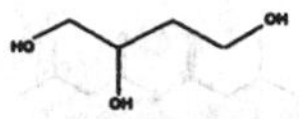

3460

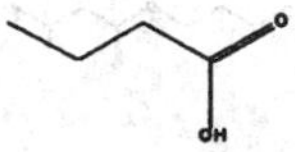

3461

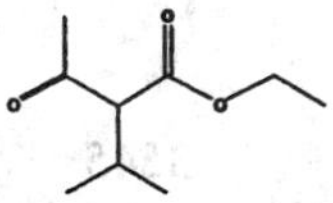

3462

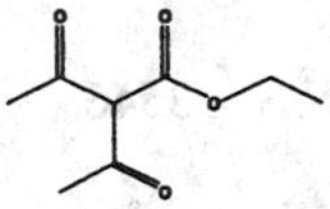

3464

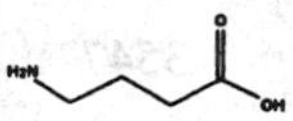

3465

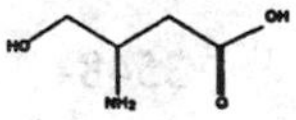

3467

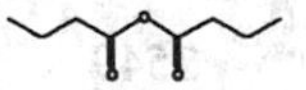

3468

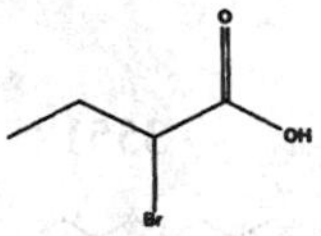

3469

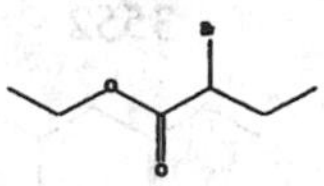

3470

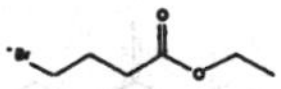

3471

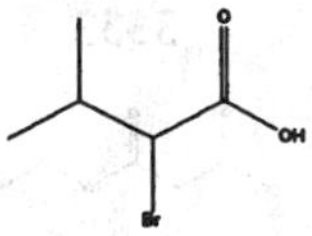

3472

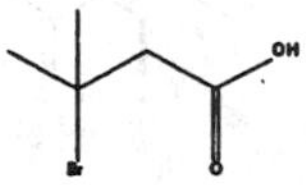

3473

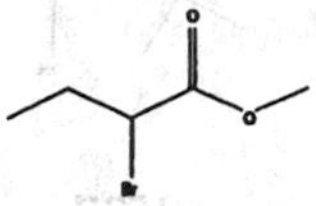

3474

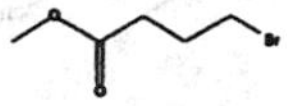

3475

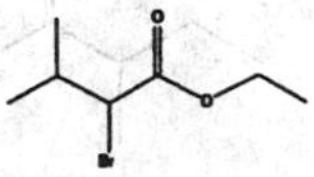

3476

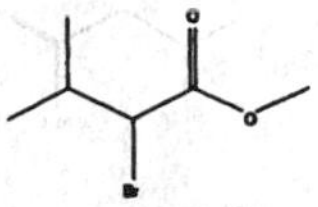

3477

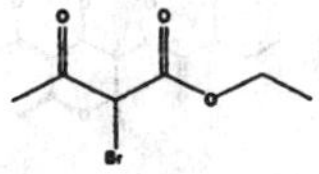

3478

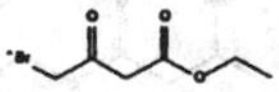

3479

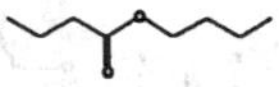

3480

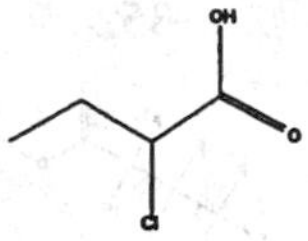

3481

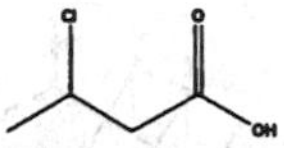

3482

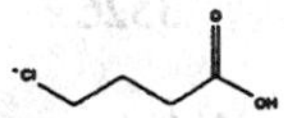

3483

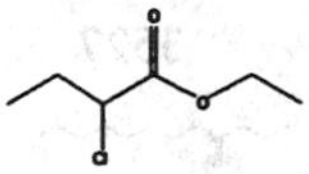

3484

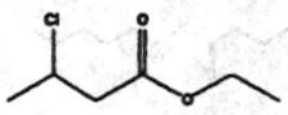

3485

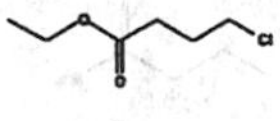

3486

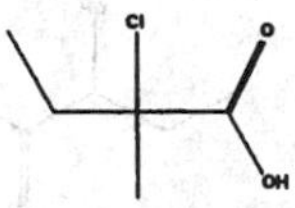

3487

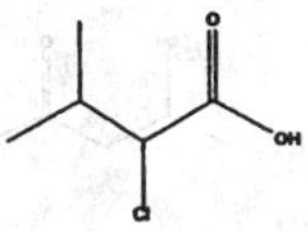

3488

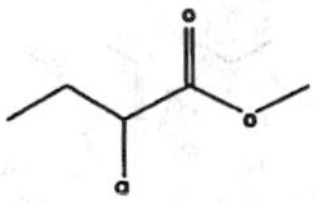

3489

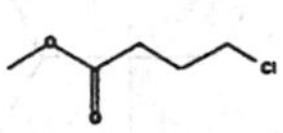

3490

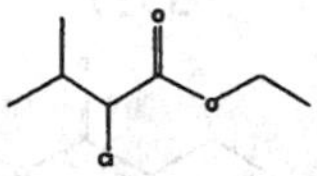

3492

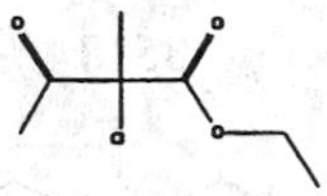

3493

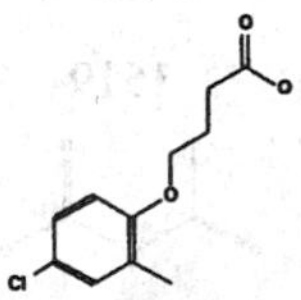

3494

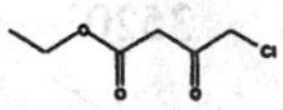

3495

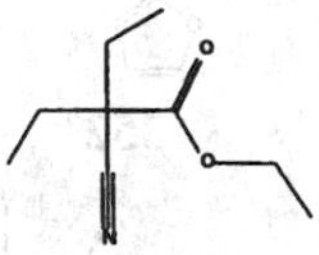

3497

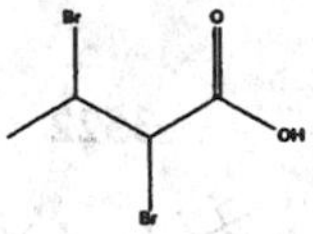

3498

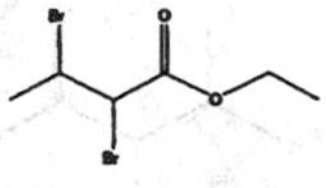

3499

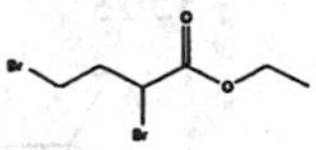

3500

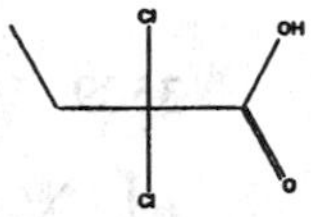

3502

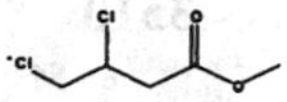

3503

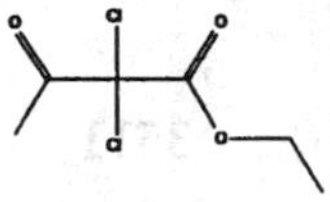

3504

3505

3506

STRUCTURAL FORMULAS OF ORGANIC COMPOUNDS (continued)

In numeric order as they occur in the Table of Physical Constants of Organic Compounds

3507

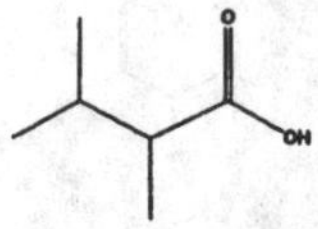

3508

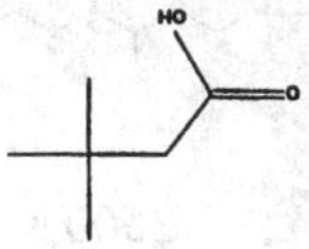

3509

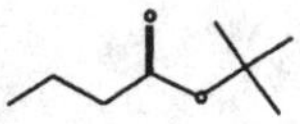

3510

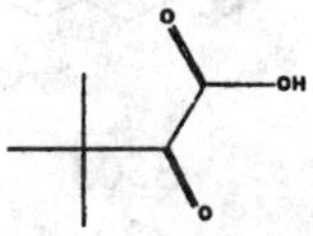

3511

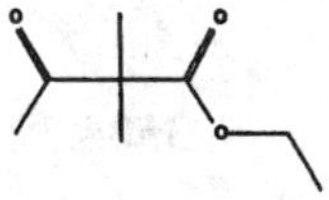

3512

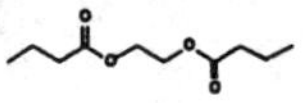

3513

3514

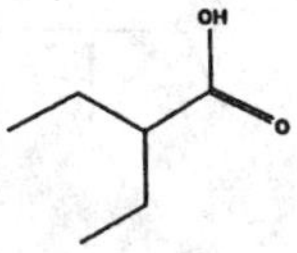

3515

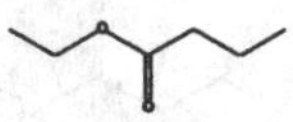

3516

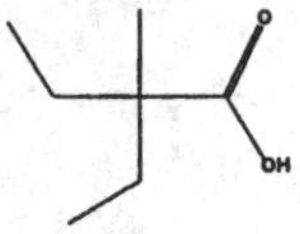

3517

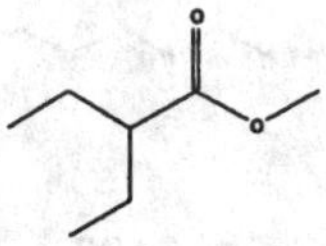

3518

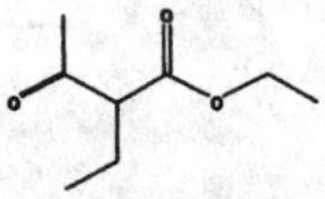

3519

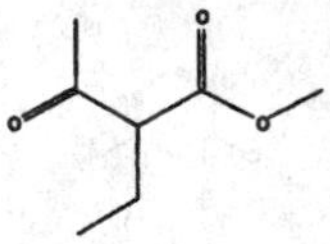

3520

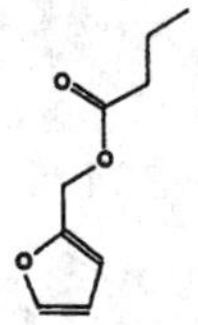

3521

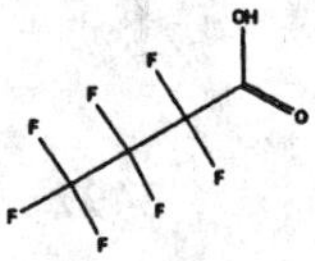

3522

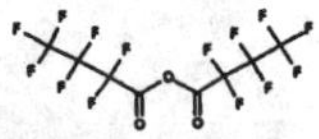

3523

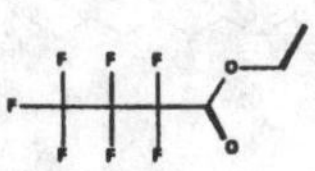

3524

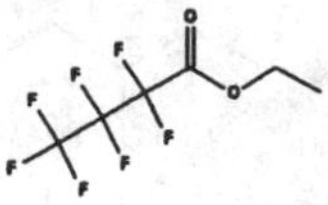

3525

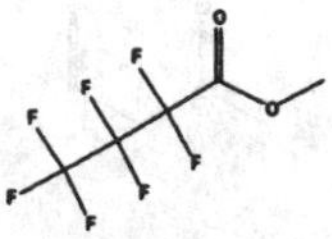

3526

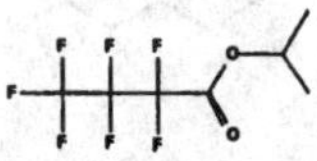

3527

3528

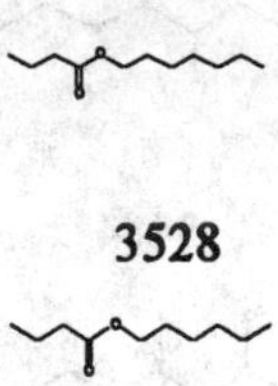

3529

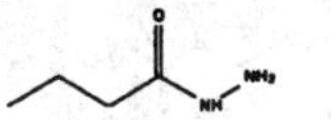

3530

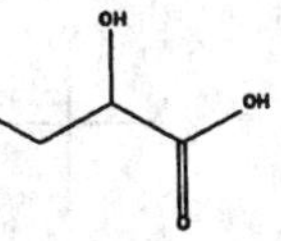

3531

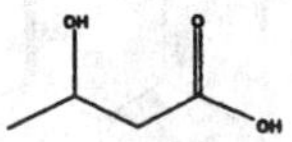

3532

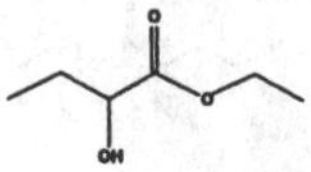

3533

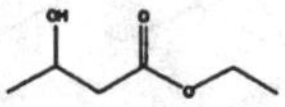

3534

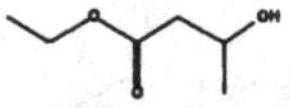

3535

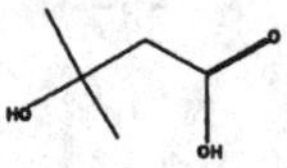

3537

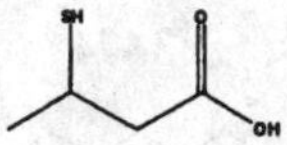

3538

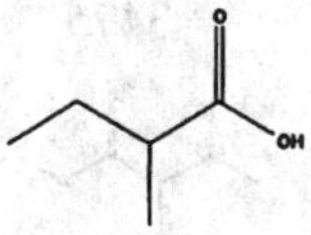

3539

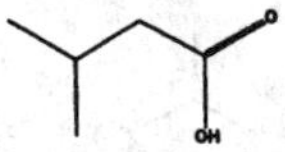

3540

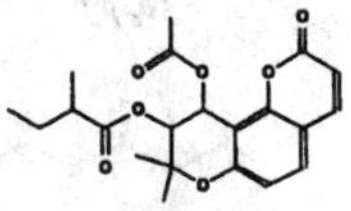

3541

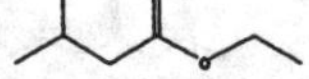

3542

3543

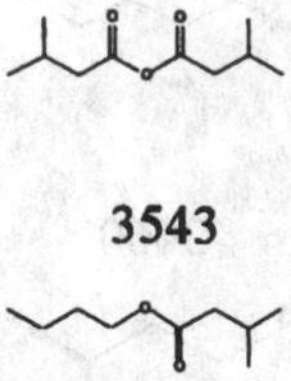

3544

3545

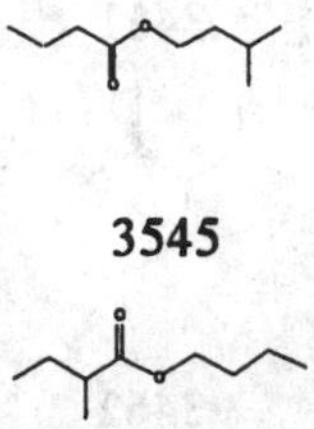

3546

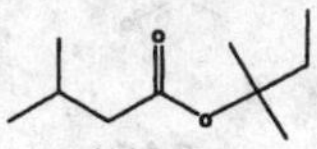

3547

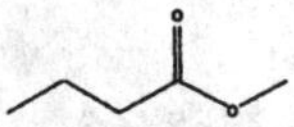

3548

3549

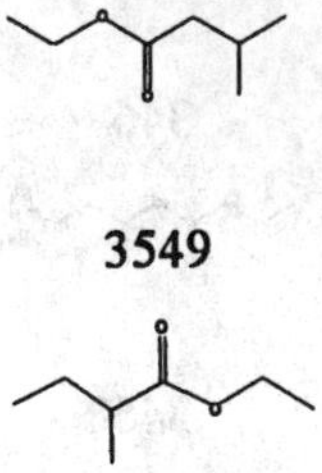

3551

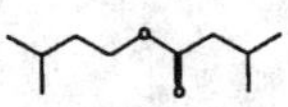

3552

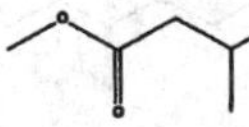

In numeric order as they occur in the Table of Physical Constants of Organic Compounds

3553

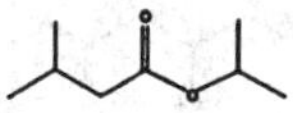

3555

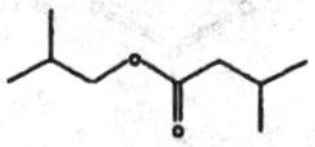

3556

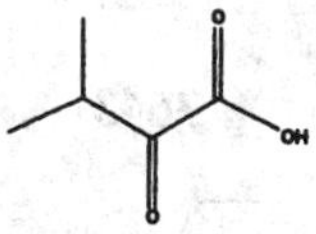

3557

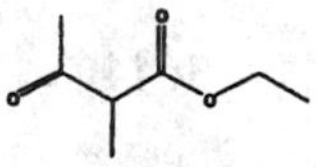

3558

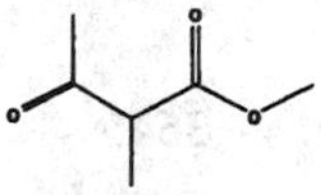

3560

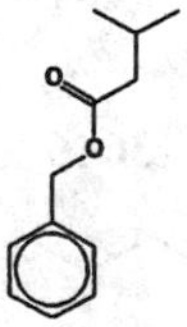

3561

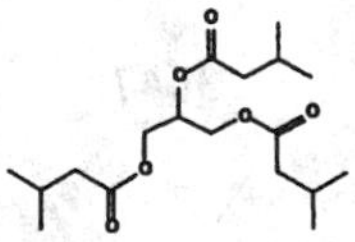

3562

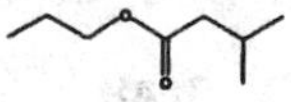

3563

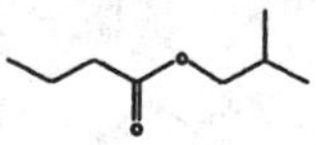

3565

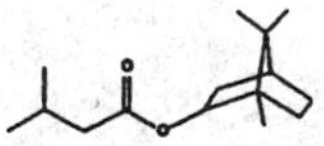

3566

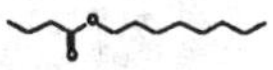

3567

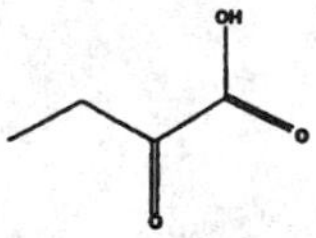

3568

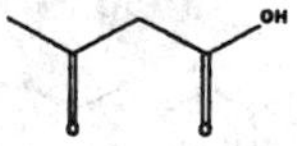

3569

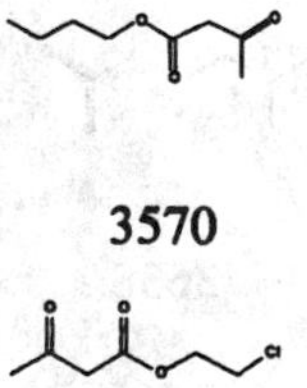

3570

3571

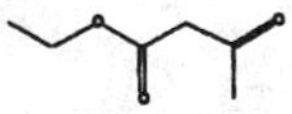

3572

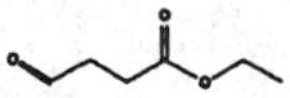

3573

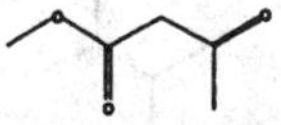

3574

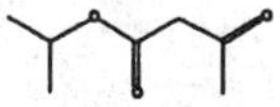

3575

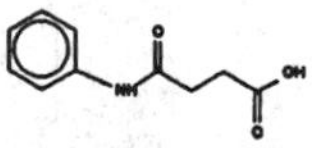

3576

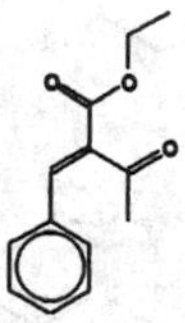

3577

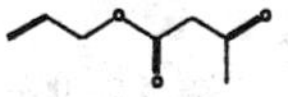

3578

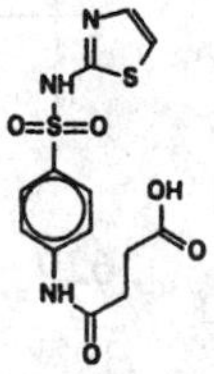

3579

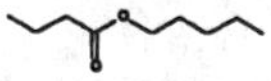

3580

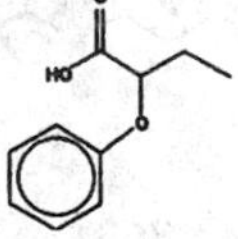

3581

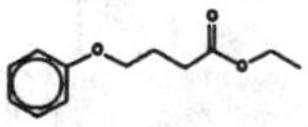

3582

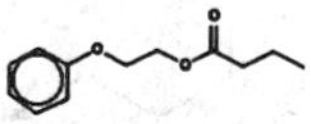

3584

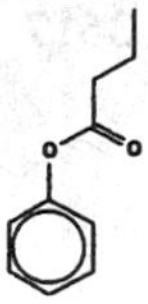

3585

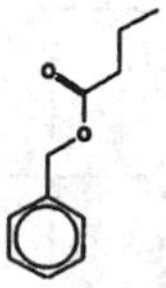

3586

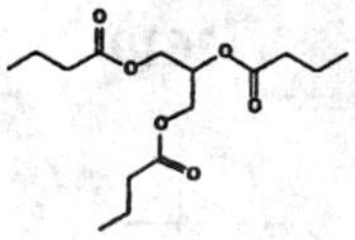

3587

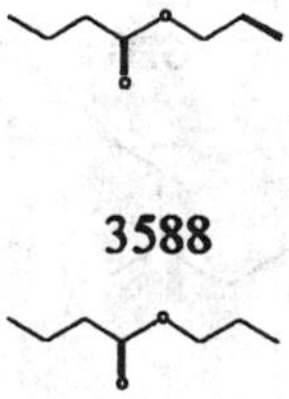

3588

3590

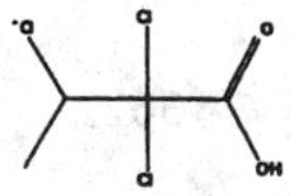

3592

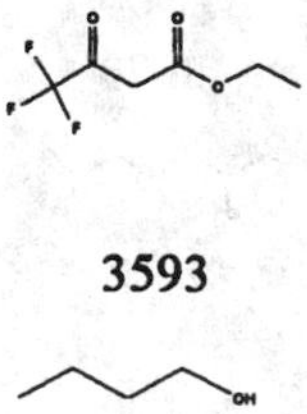

3593

3594

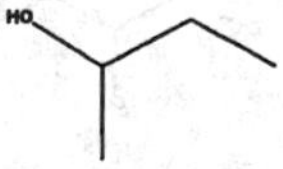

3595

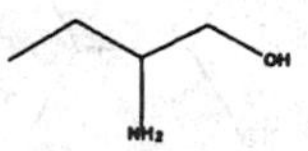

3596

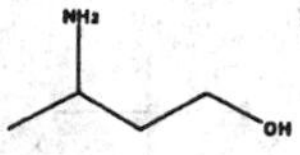

3597

3598

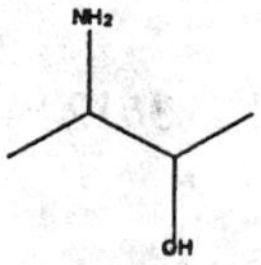

3599

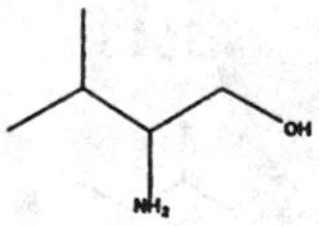

3601

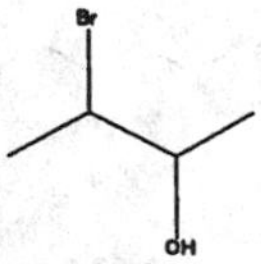

3604

STRUCTURAL FORMULAS OF ORGANIC COMPOUNDS (continued)

In numeric order as they occur in the Table of Physical Constants of Organic Compounds

3605

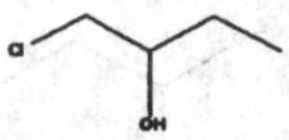

3606

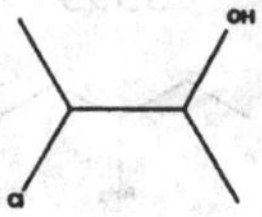

3608

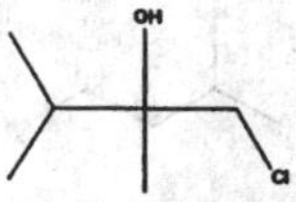

3609

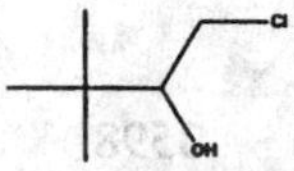

3612

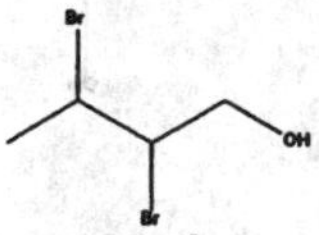

3613

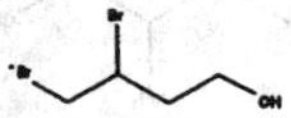

3614

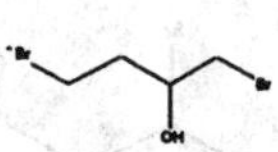

3615

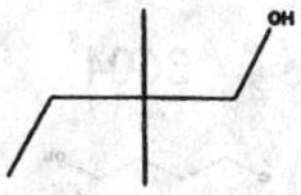

3616

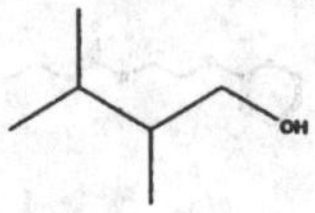

3617

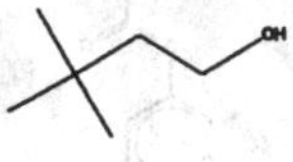

3618

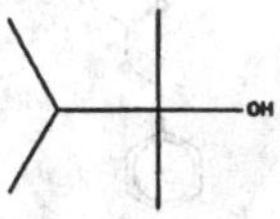

3619

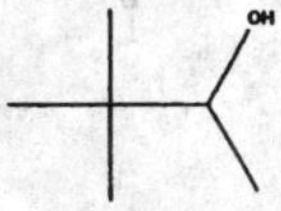

3620

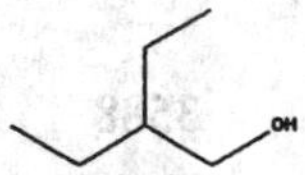

3621

3622

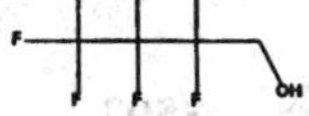

3623

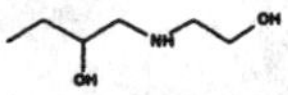

3624

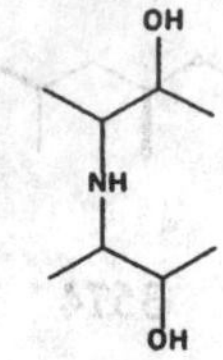

3625

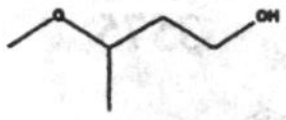

3626

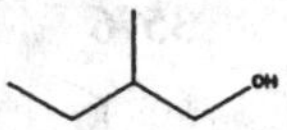

3627

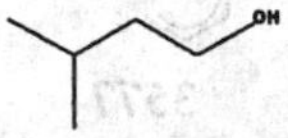

3628

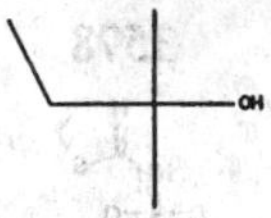

3629

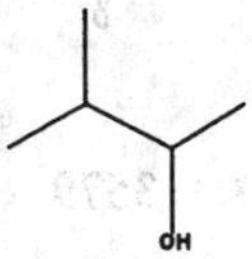

3630

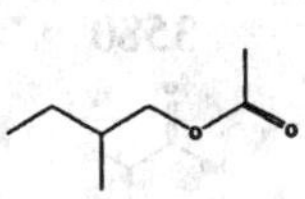

3631

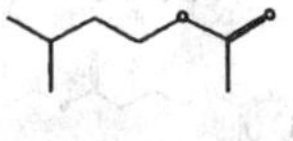

3632

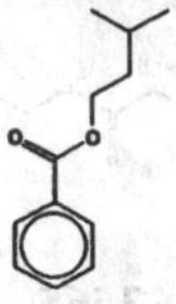

3633

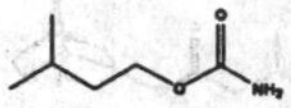

3634

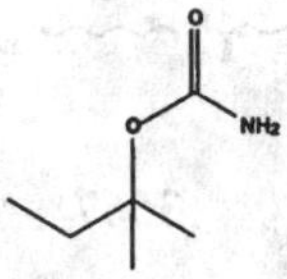

3635

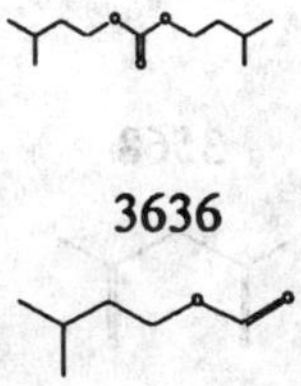

3636

3637

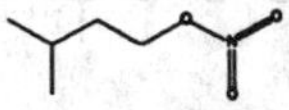

3638

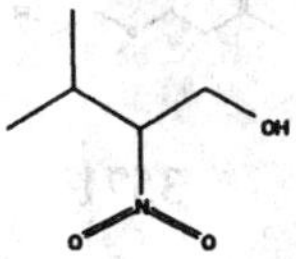

3639

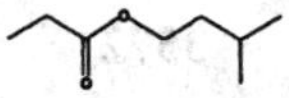

3640

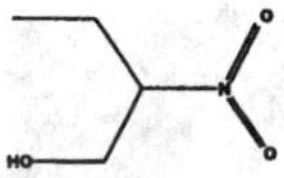

3641

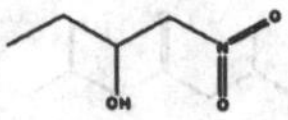

3642

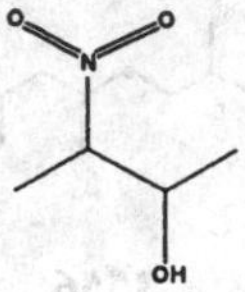

3643

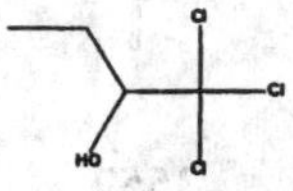

3644

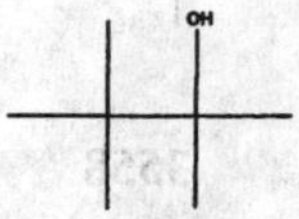

3645

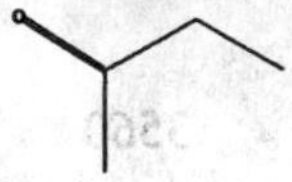

3646

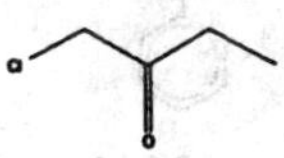

3647

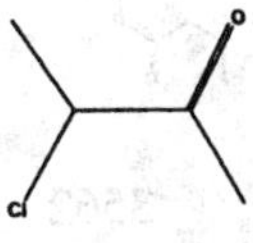

3648

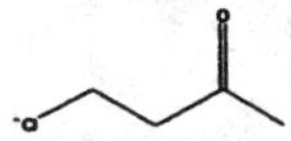

STRUCTURAL FORMULAS OF ORGANIC COMPOUNDS (continued)

In numeric order as they occur in the Table of Physical Constants of Organic Compounds

3649

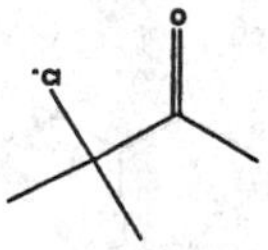

3650

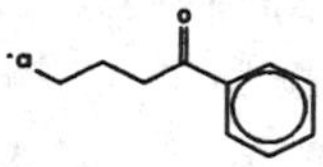

3651

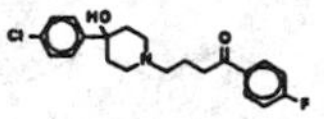

3653

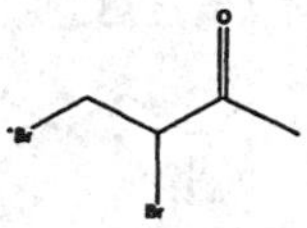

3654

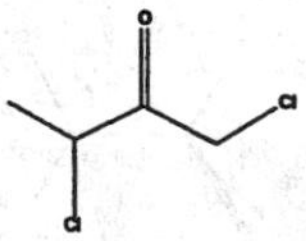

3655

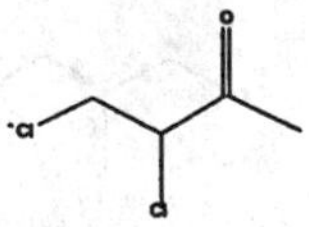

3656

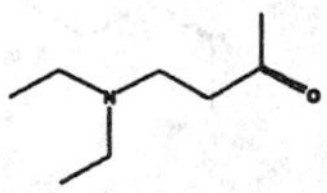

3657

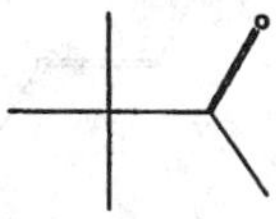

3658

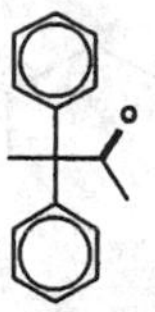

3659

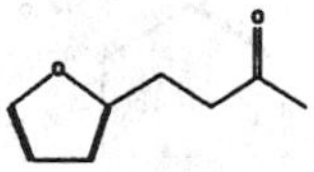

3660

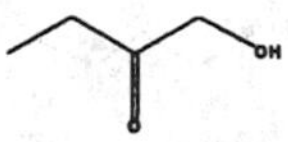

3661

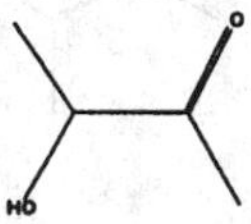

3662

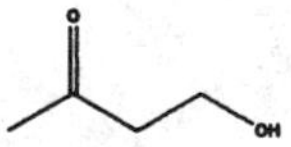

3663

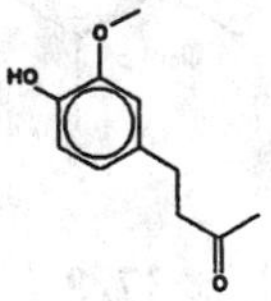

3664

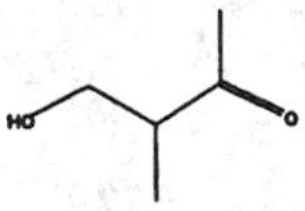

3665

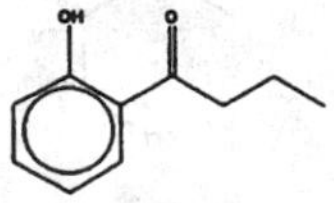

3666

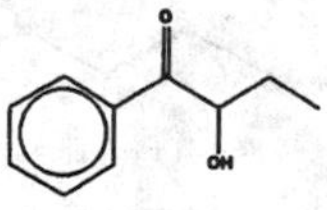

3667

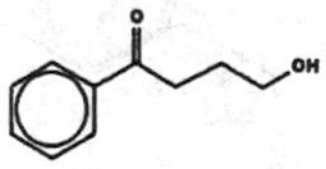

3668

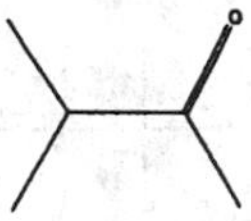

3669

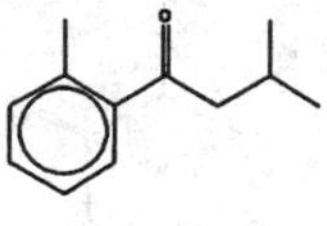

3670

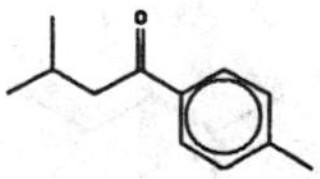

3671

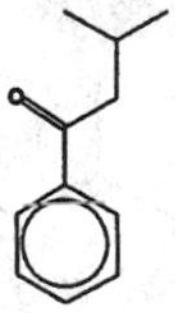

3673

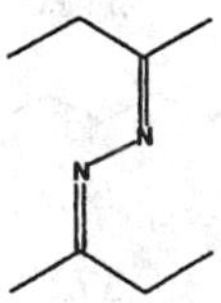

3675

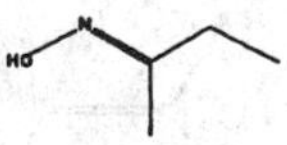

3676

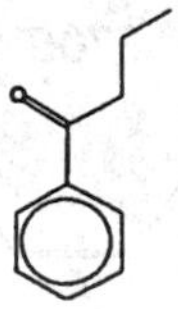

3677

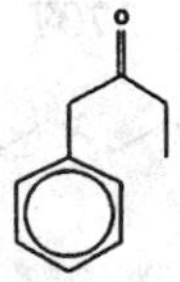

3678

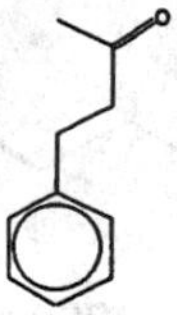

3679

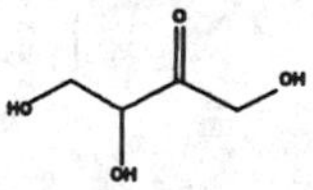

3680

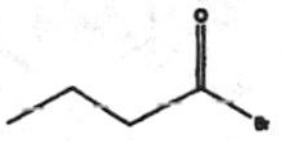

3681

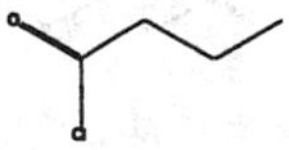

3682

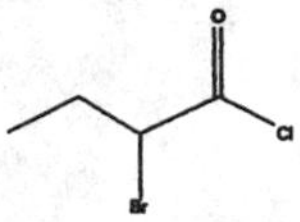

3683

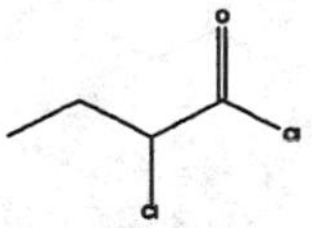

3684

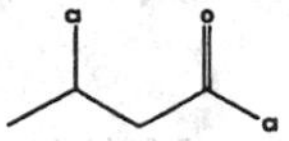

3685

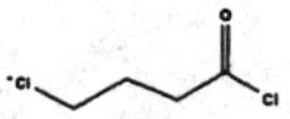

3688

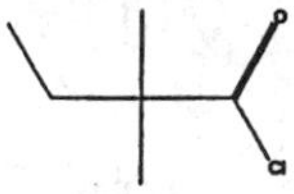

3690

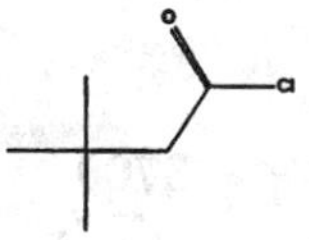

3691

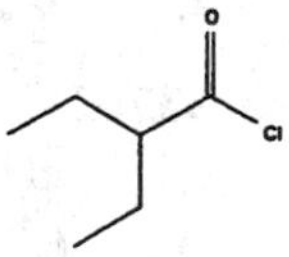

3692

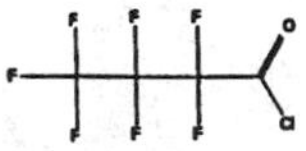

3693

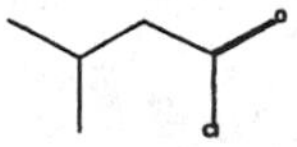

STRUCTURAL FORMULAS OF ORGANIC COMPOUNDS (continued)

In numeric order as they occur in the Table of Physical Constants of Organic Compounds

3695
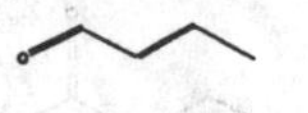

3696
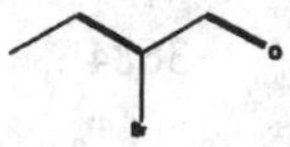

3697
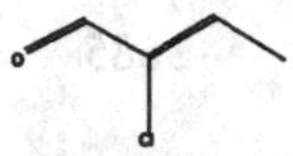

3698
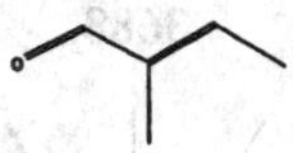

3699
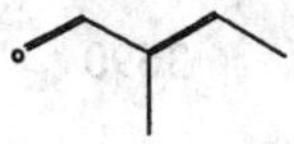

3700
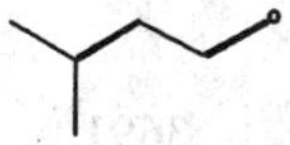

3702
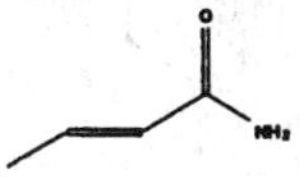

3704

3705
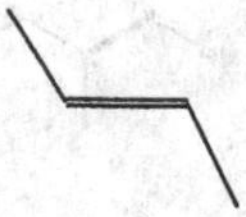

3706

3707

3708
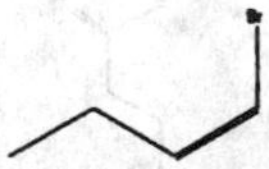

3709
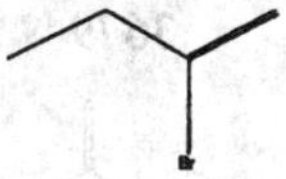

3710
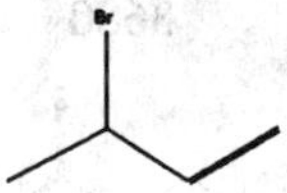

3711

3712

3713
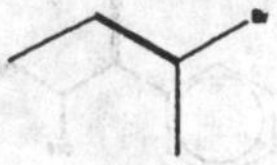

3714
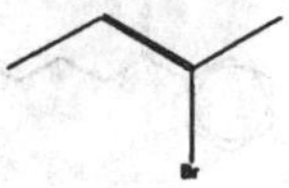

3715
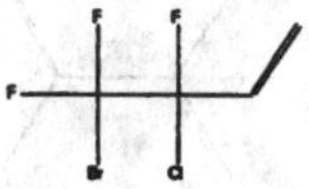

3716
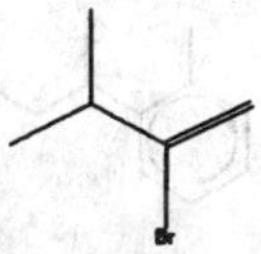

3717
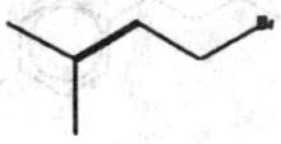

3718
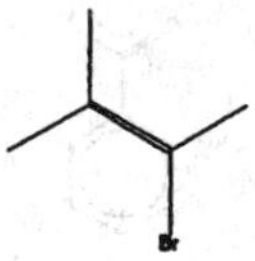

3720

3721
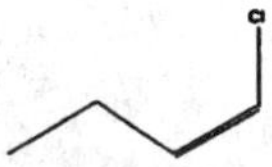

3722
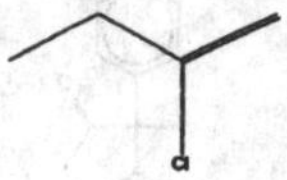

3723
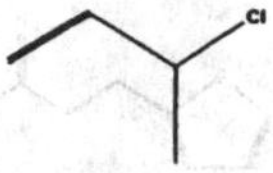

3724

3725
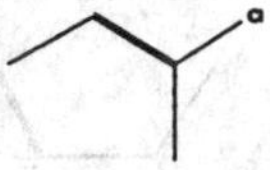

3726
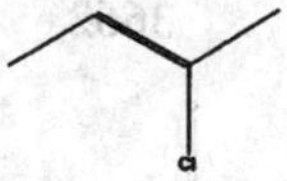

3727
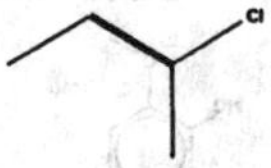

3728
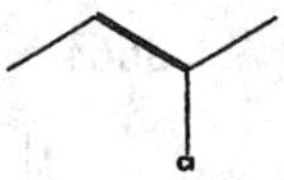

3729
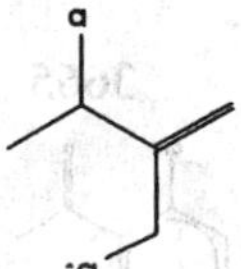

3730
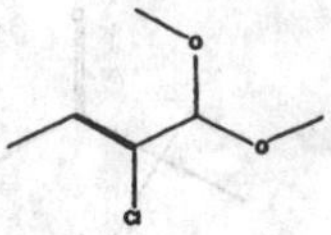

3731
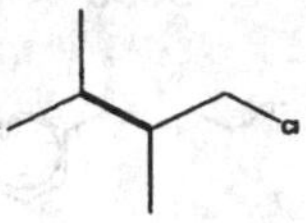

3732

3733
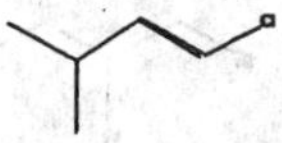

3734
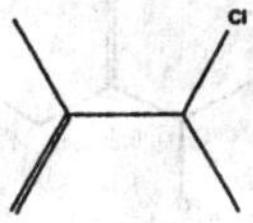

3735
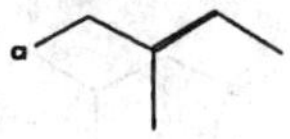

3736
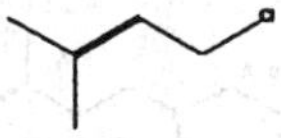

3737
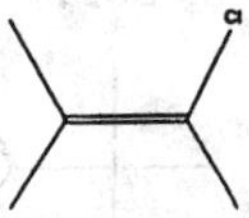

STRUCTURAL FORMULAS OF ORGANIC COMPOUNDS (continued)

In numeric order as they occur in the Table of Physical Constants of Organic Compounds

3738

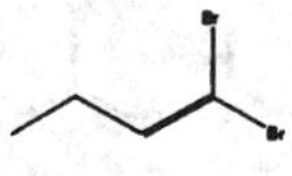

3739

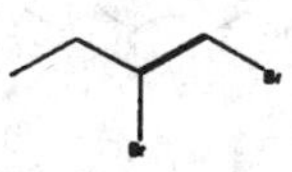

3740

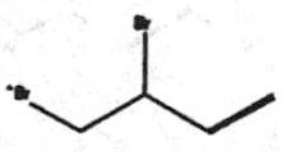

3741

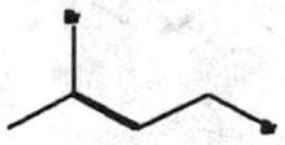

3742

3743

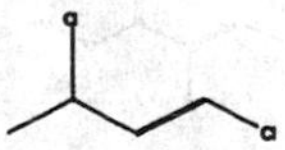

3744

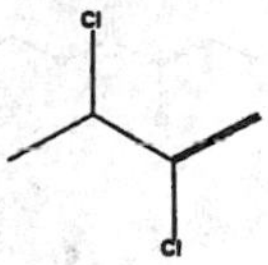

3745

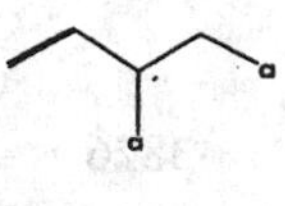

3746

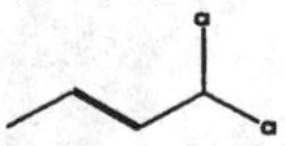

3747

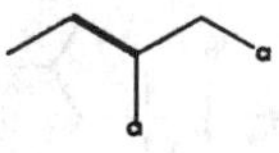

3748

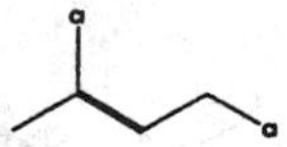

3749

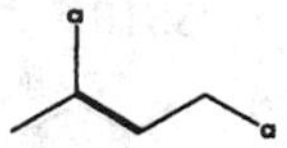

3750

3751

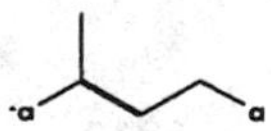

3752

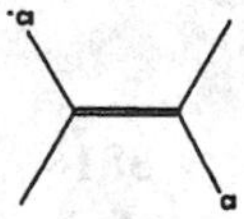

3753

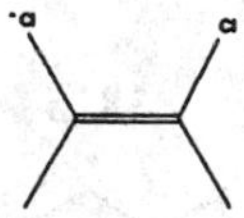

3754

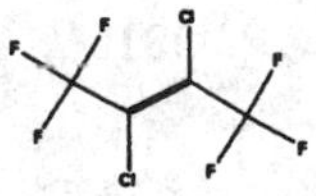

3755

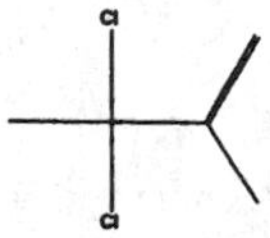

3756

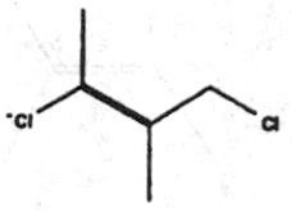

3757

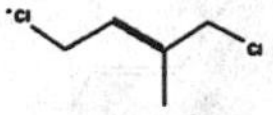

3758

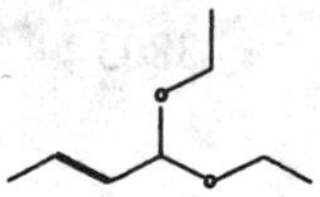

3759

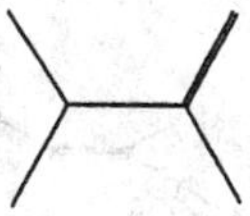

3760

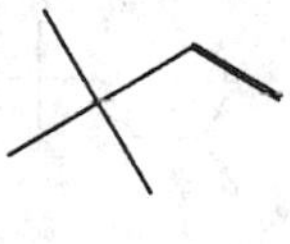

3761

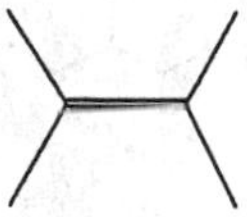

3762

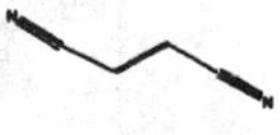

3763

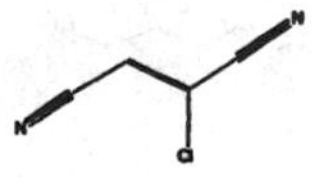

3764

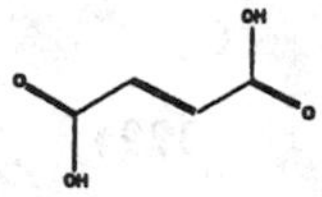

3765

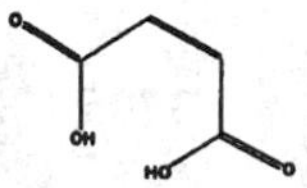

3766

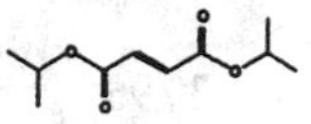

3767

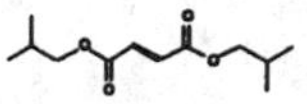

3769

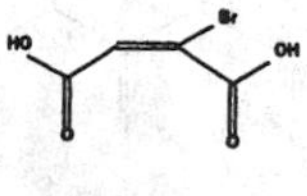

3770

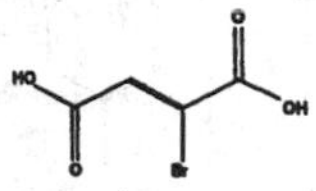

3771

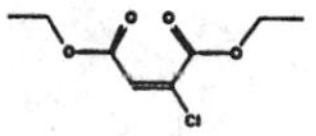

3772

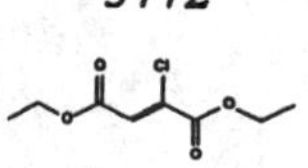

3773

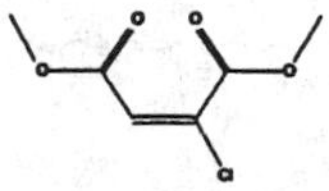

3774

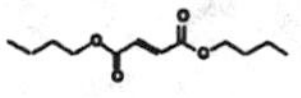

3775

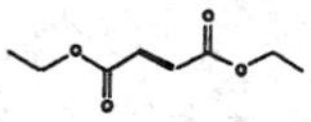

3776

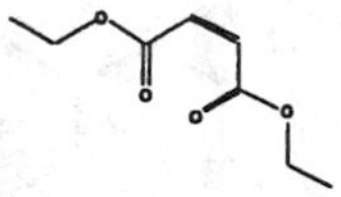

3777

3778

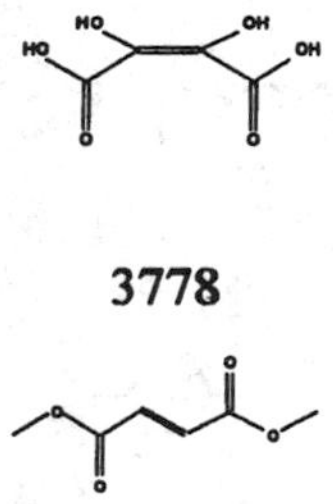

3779

3781

3782

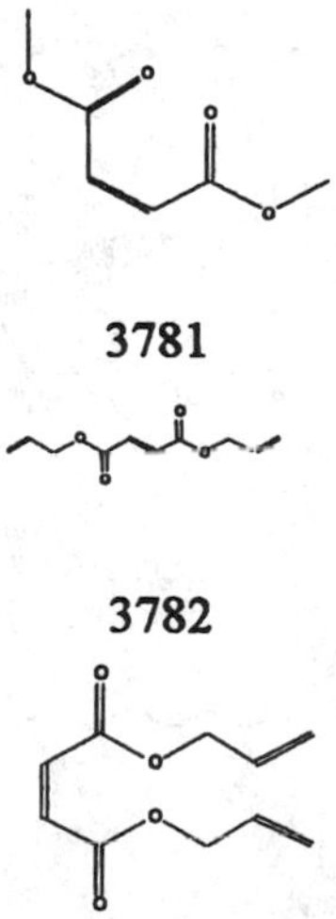

STRUCTURAL FORMULAS OF ORGANIC COMPOUNDS (continued)

In numeric order as they occur in the Table of Physical Constants of Organic Compounds

3783

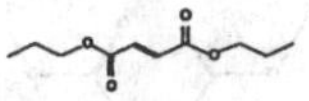

3784

3785

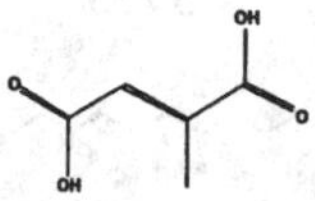

3786

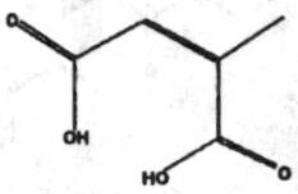

3787

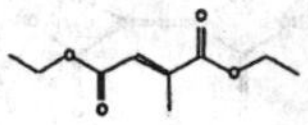

3788

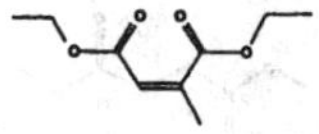

3789

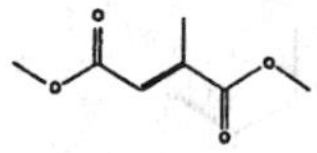

3790

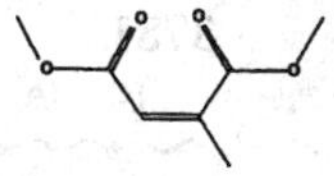

3791

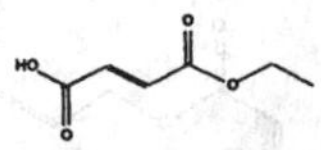

3792

3793

3794

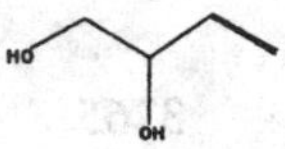

3795

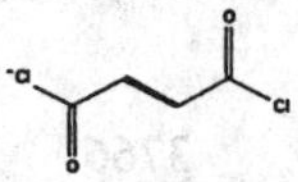

3796

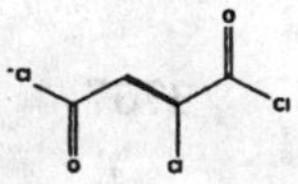

3797

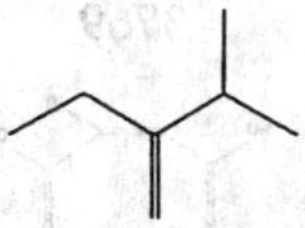

3798

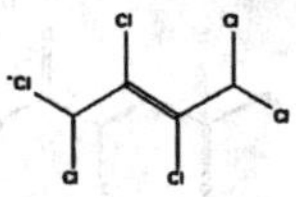

3799

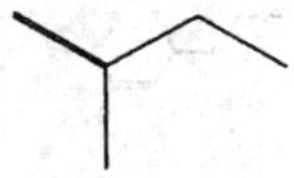

3800

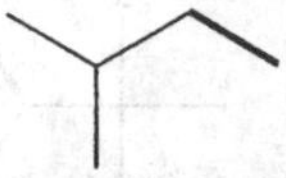

3801

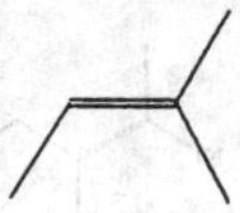

3802

3803

3804

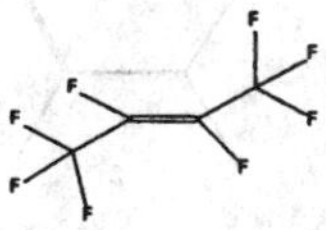

3805

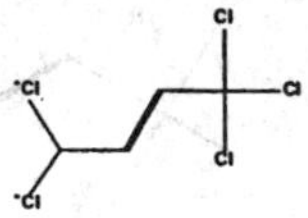

3806

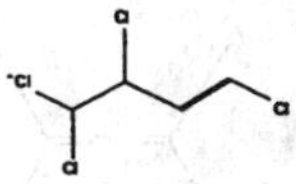

3807

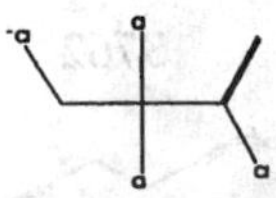

3808

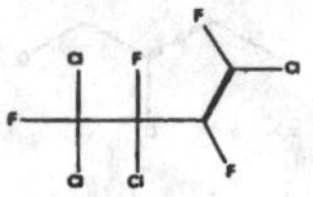

3809

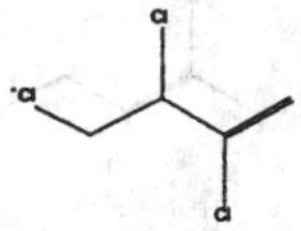

3810

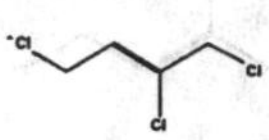

3812

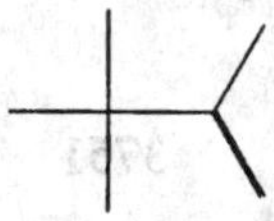

3813

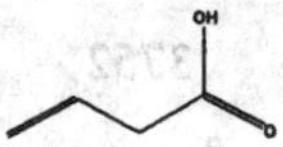

3814

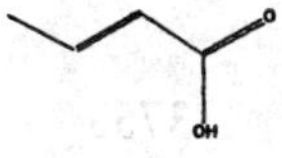

3815

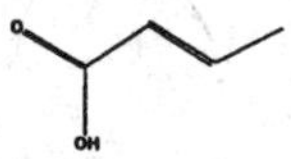

3816

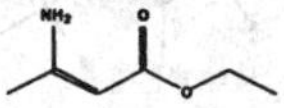

3817

3819

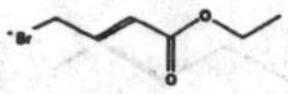

3820

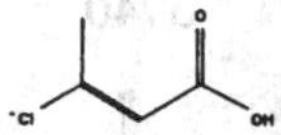

3821

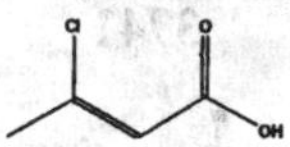

3822

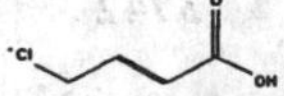

3823

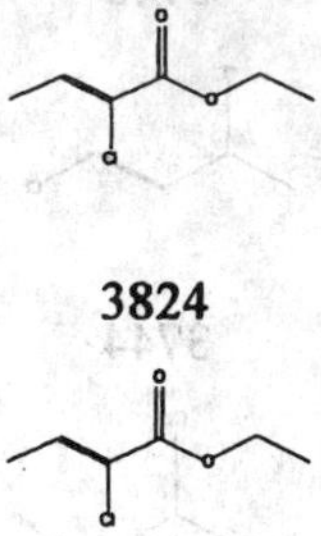

3825

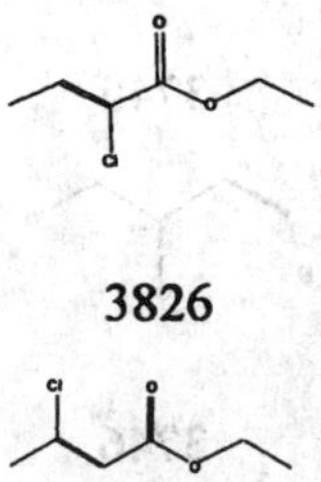

STRUCTURAL FORMULAS OF ORGANIC COMPOUNDS (continued)

In numeric order as they occur in the Table of Physical Constants of Organic Compounds

3827

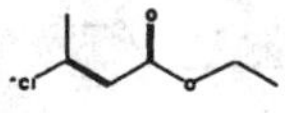

3828

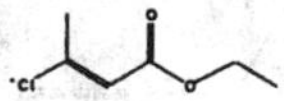

3829

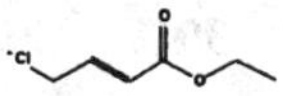

3830

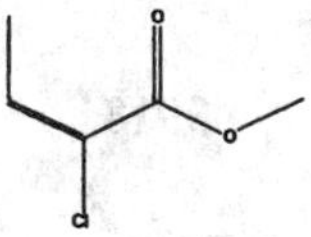

3831

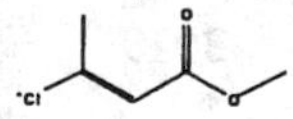

3832

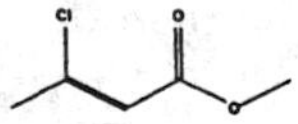

3833

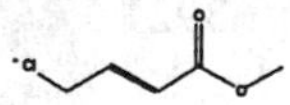

3834

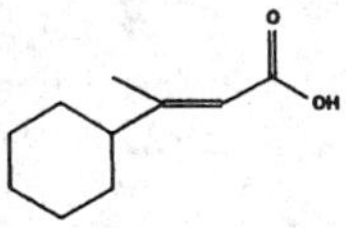

3835

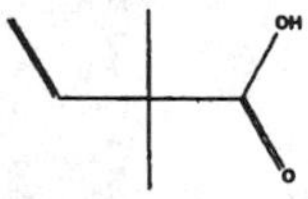

3836

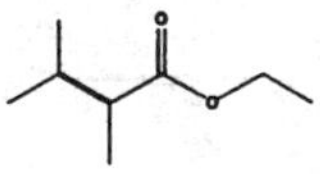

3837

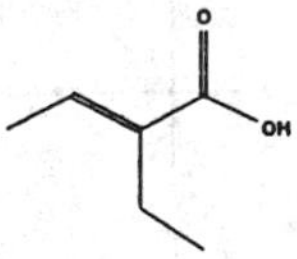

3838

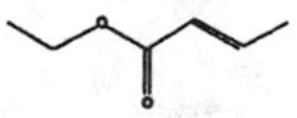

3839

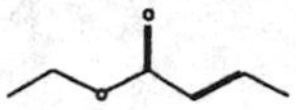

3840

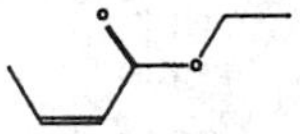

3841

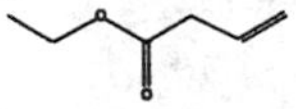

3842

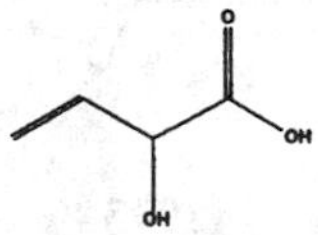

3843

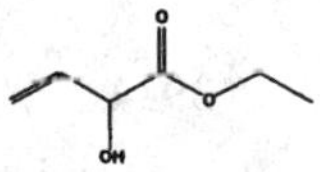

3844

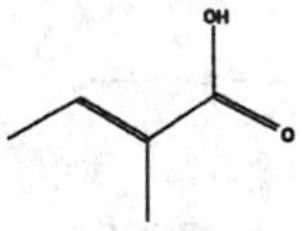

3845

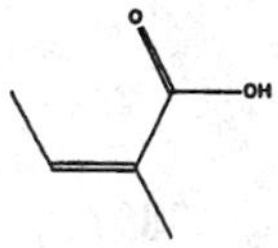

3846

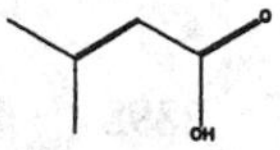

3847

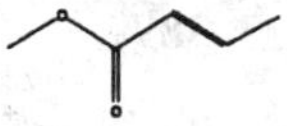

3848

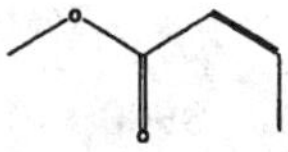

3849

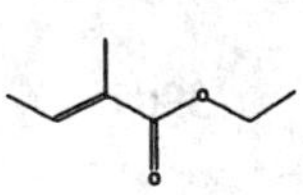

3850

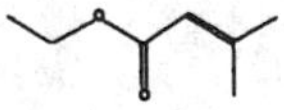

3851

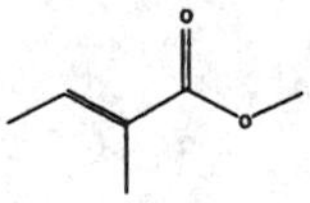

3852

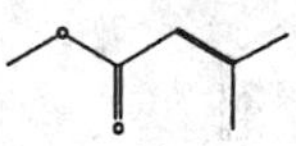

3856

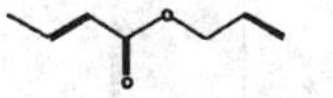

3857

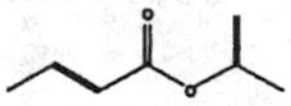

3858

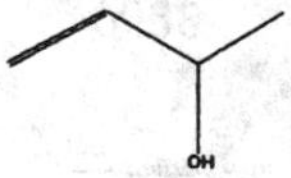

3859

3860

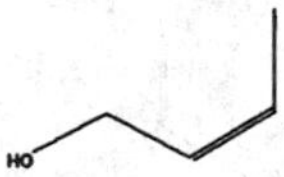

3861

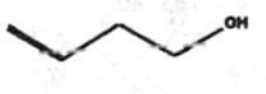

3862

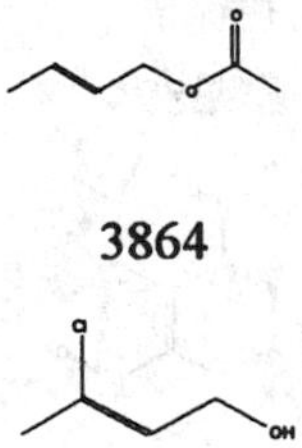

3864

3865

3866

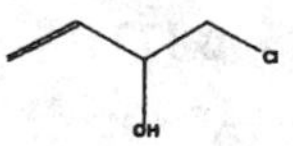

3868

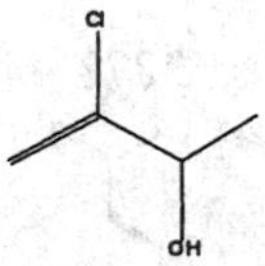

3869

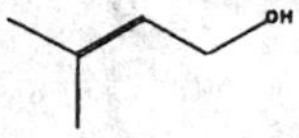

3870

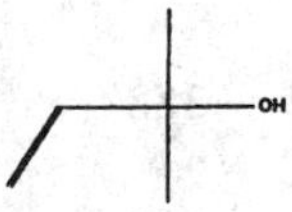

3871

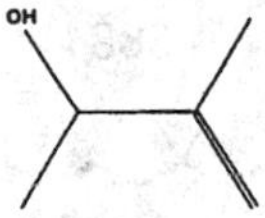

3873

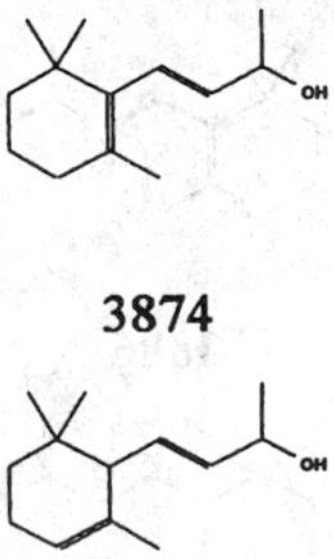

3874

In numeric order as they occur in the Table of Physical Constants of Organic Compounds

3875

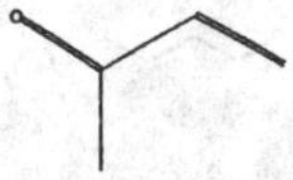

3877

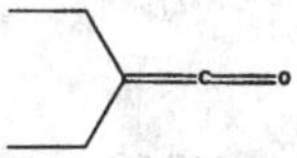

3878

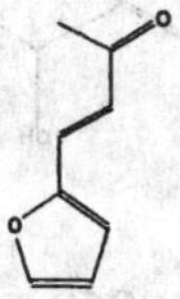

3879

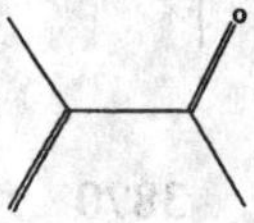

3883

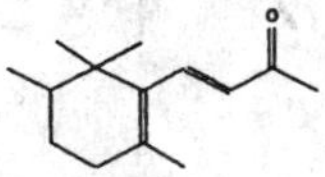

3884

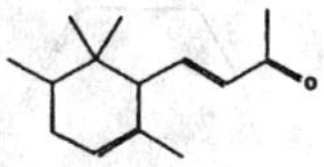

3885

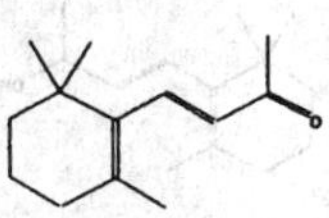

3886

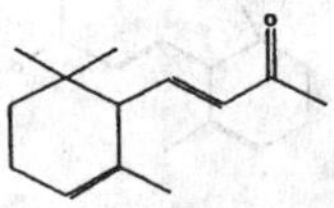

3887

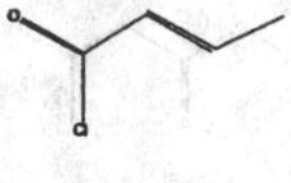

3888

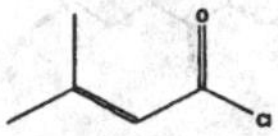

3889

3890

3891

3892

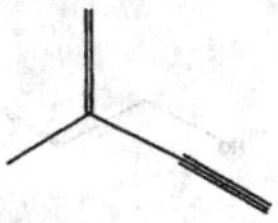

3893

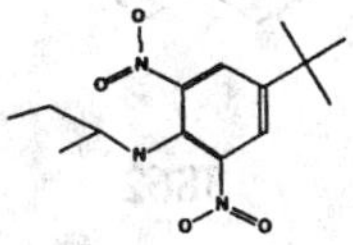

3894

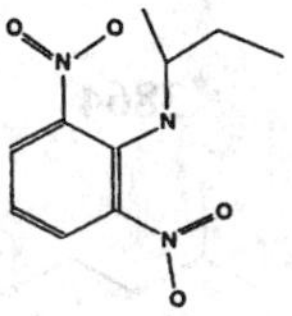

3895

3896

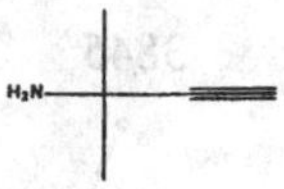

3897

3898

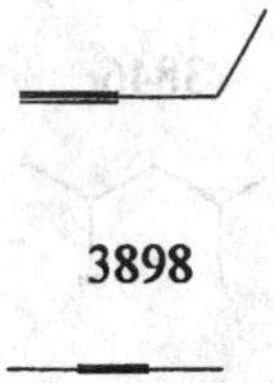

3899

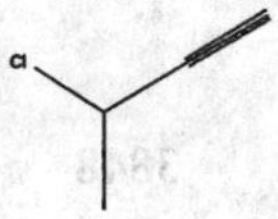

3900

3901

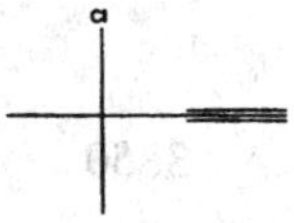

3902

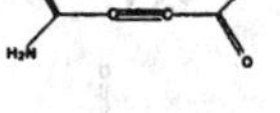

3904

3905

3907

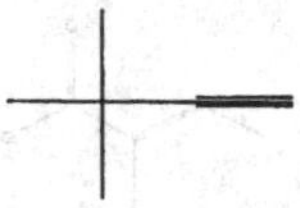

3908

3909

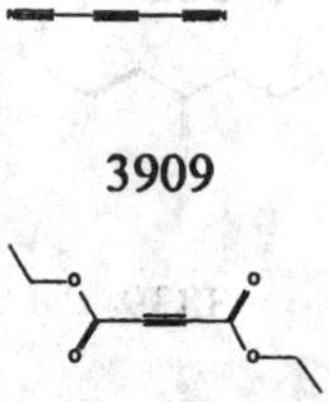

3910

3911

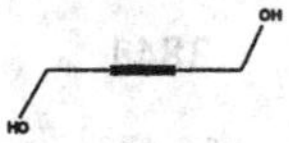

3912

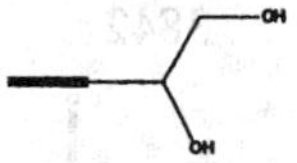

3913

3914

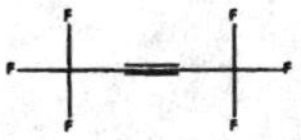

3915

3916

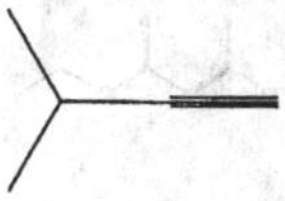

3917

3918

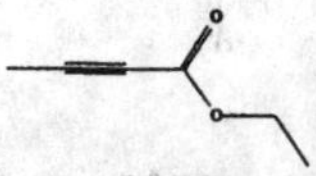

3919

3920

3921

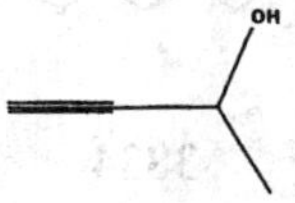

3922

STRUCTURAL FORMULAS OF ORGANIC COMPOUNDS (continued)

In numeric order as they occur in the Table of Physical Constants of Organic Compounds

3923

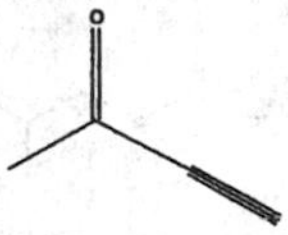

3925

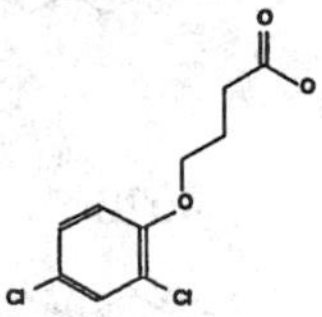

3926

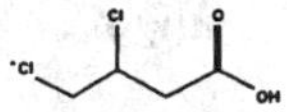

3927

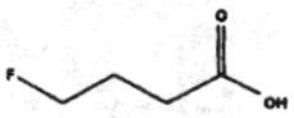

3929

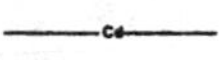

3930

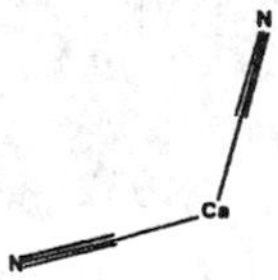

3931

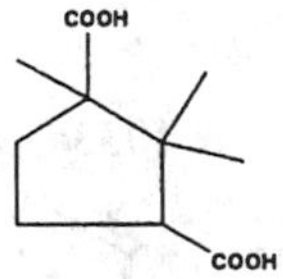

3932

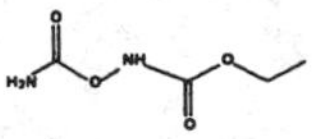

3933

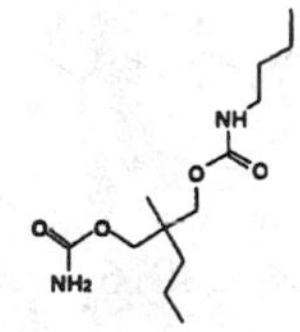

3934

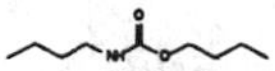

3935

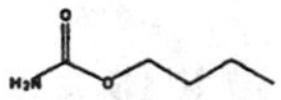

3936

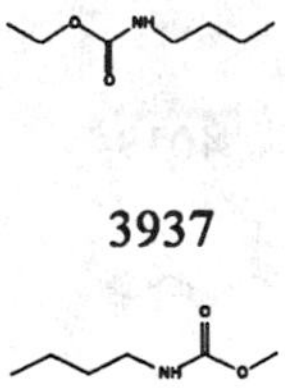

3938

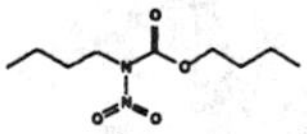

3939

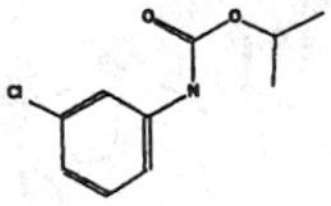

3940

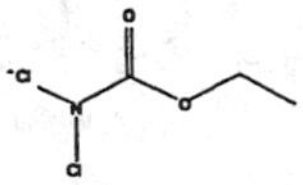

3941

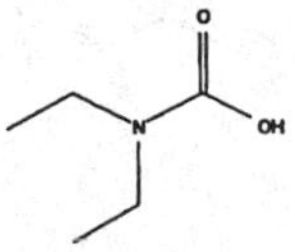

3942

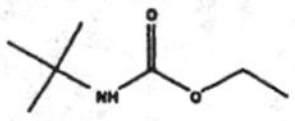

3943

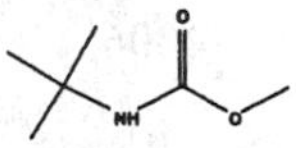

3944

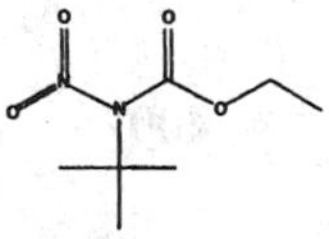

3947

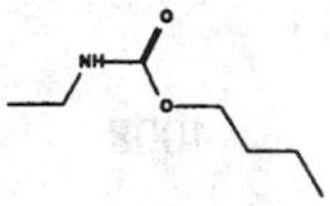

3948

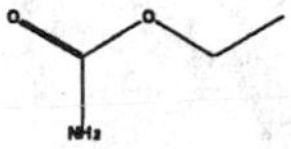

3949

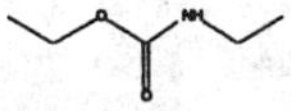

3950

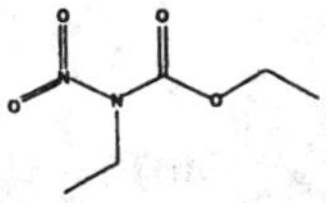

3953

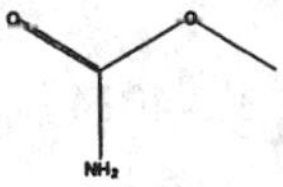

3954

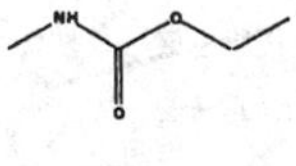

3955

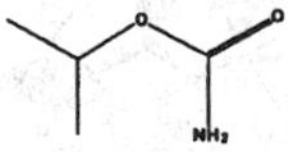

3956

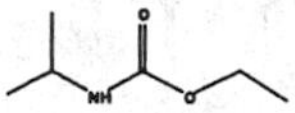

3957

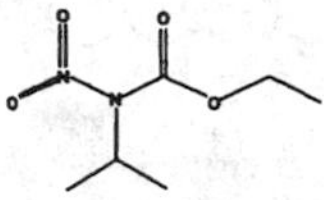

3959

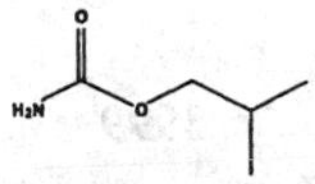

3960

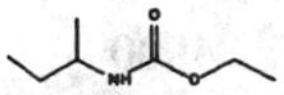

3961

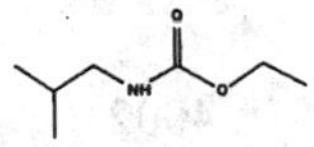

3962

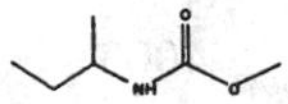

3963

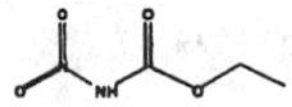

3964

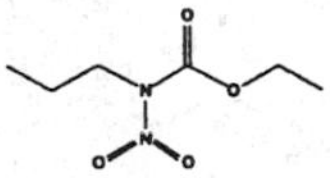

3966

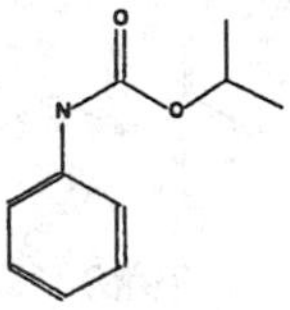

3969

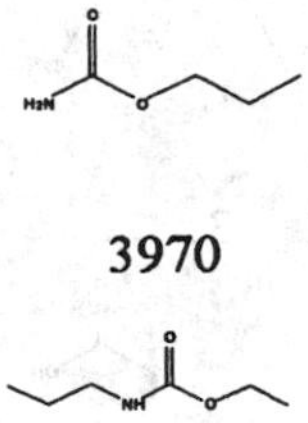

3971

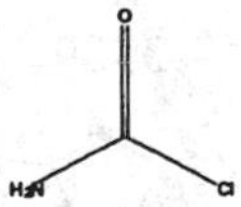

3972

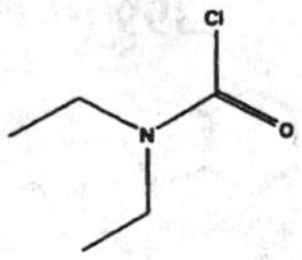

3973

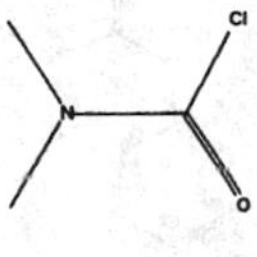

3976

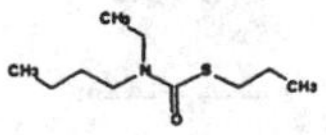

STRUCTURAL FORMULAS OF ORGANIC COMPOUNDS (continued)

In numeric order as they occur in the Table of Physical Constants of Organic Compounds

3978

3979

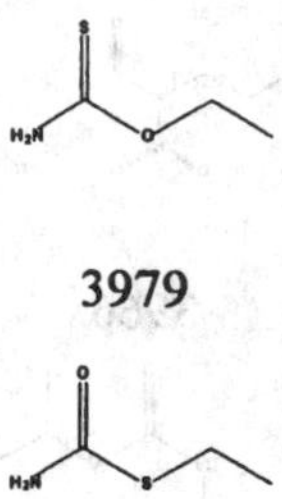

3980

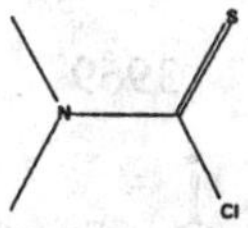

3982

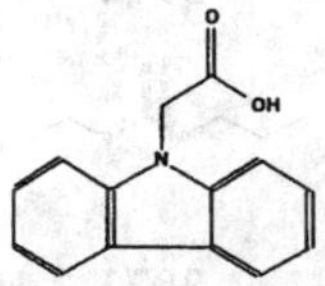

3989

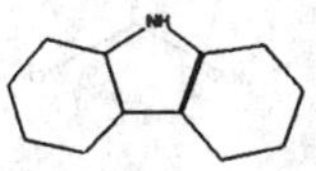

3990

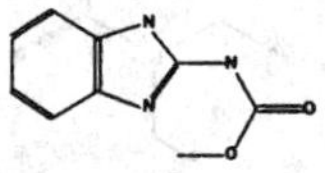

3991

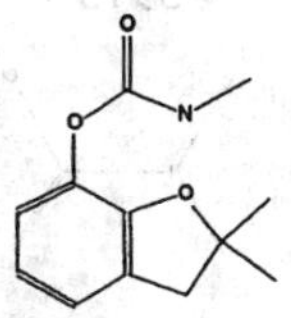

3992

3993

3994

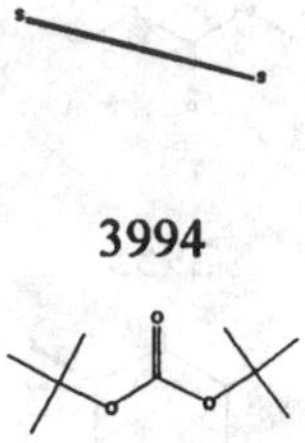

3995

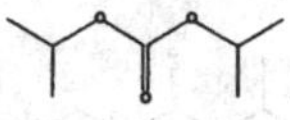

3997

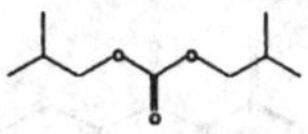

3998

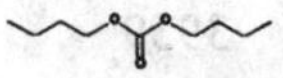

3999

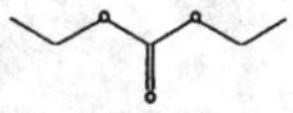

4000

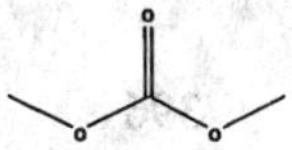

4002

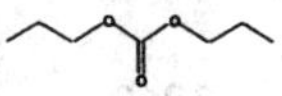

4003

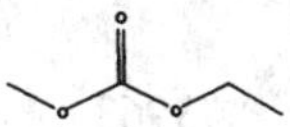

4005

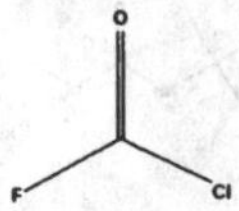

4006

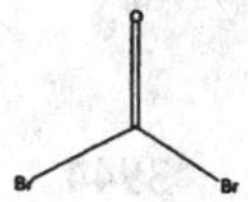

4007

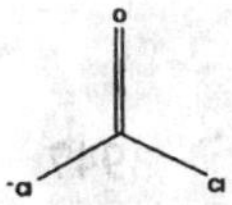

4008

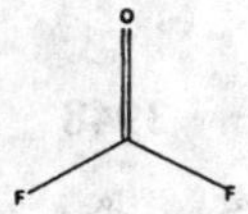

4009

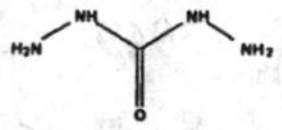

4012

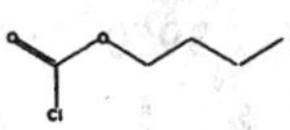

4013

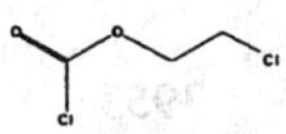

4014

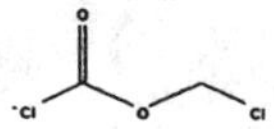

4015

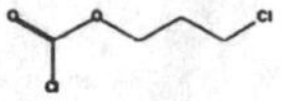

4016

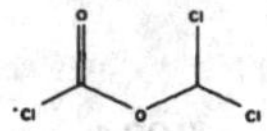

4017

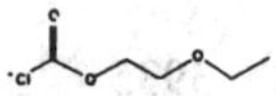

4018

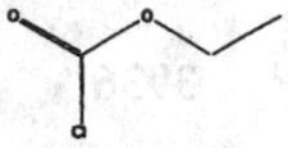

4019

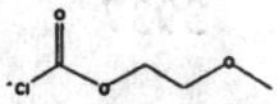

4020

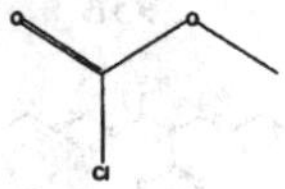

4021

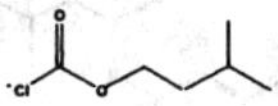

4022

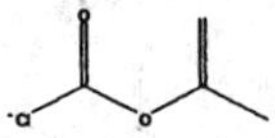

4023

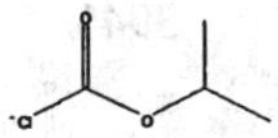

4024

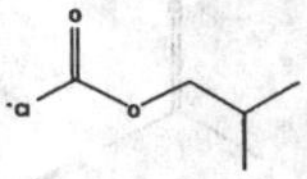

4025

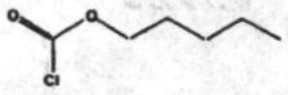

4028

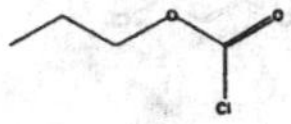

4029

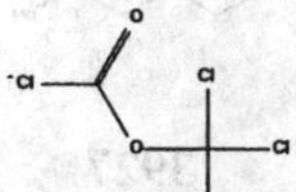

4030

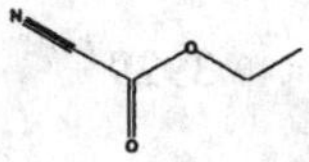

4031

4032

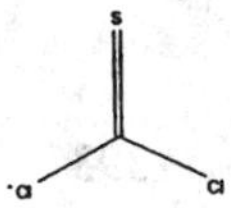

4033

STRUCTURAL FORMULAS OF ORGANIC COMPOUNDS (continued)

In numeric order as they occur in the Table of Physical Constants of Organic Compounds

4034

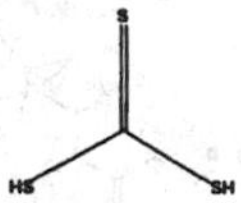

4035

4036

4037

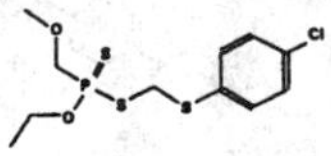

4038

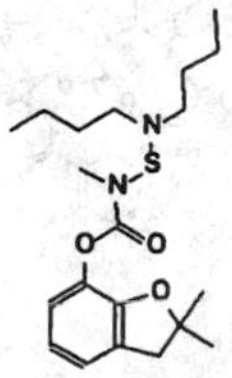

4039

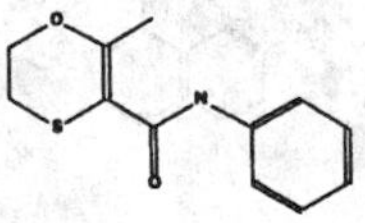

4040

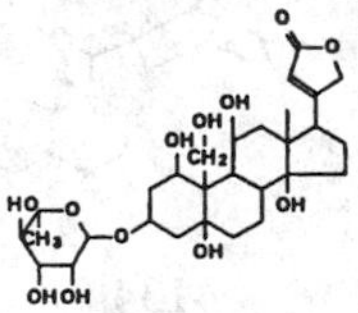

4043

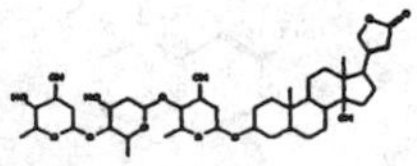

4044

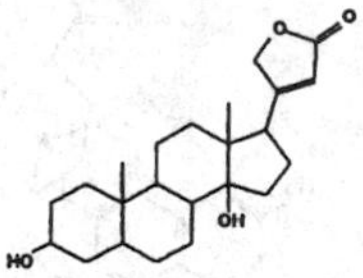

4046

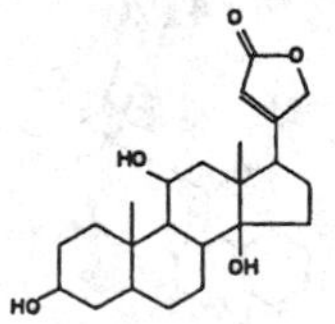

4047

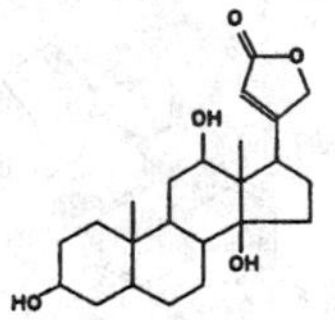

4048

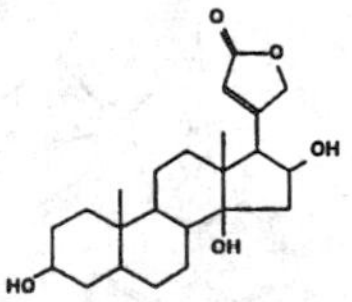

4049

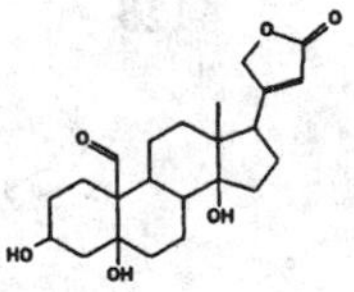

4050

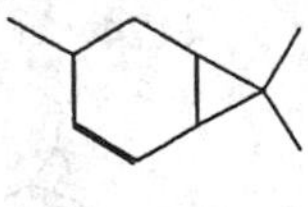

4065

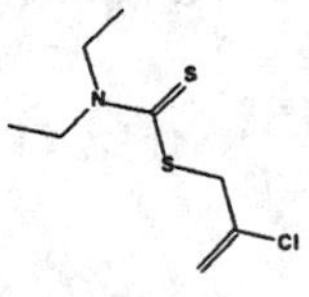

4066

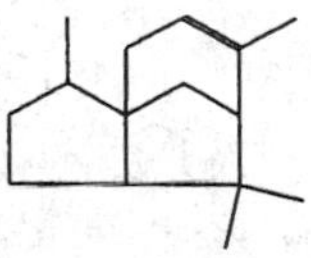

4067

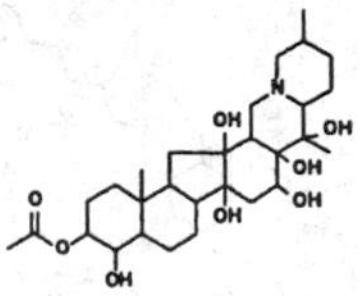

4068

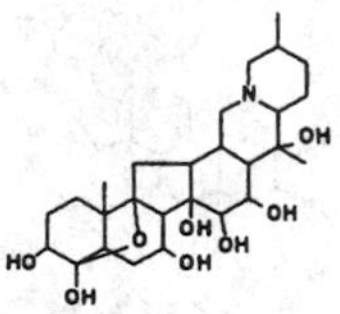

4069

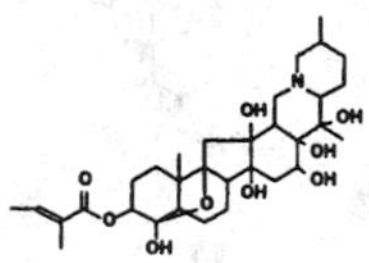

4071

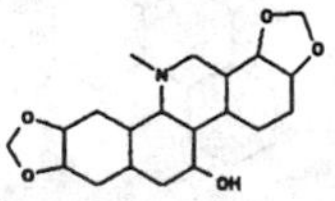

4072

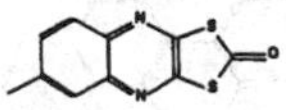

4075

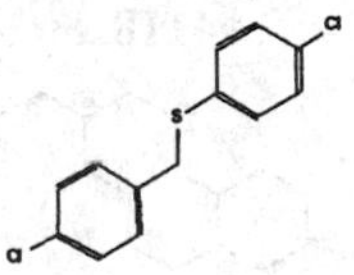

4076

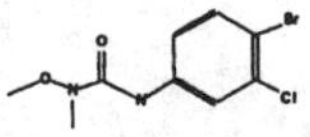

4077

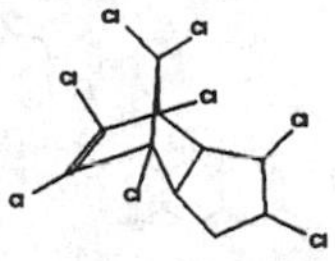

4078

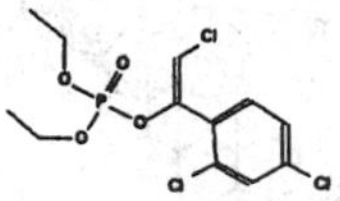

4079

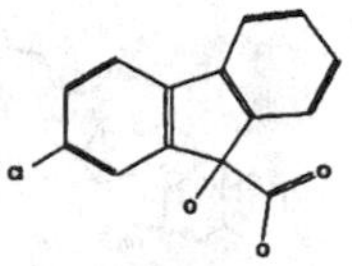

4080

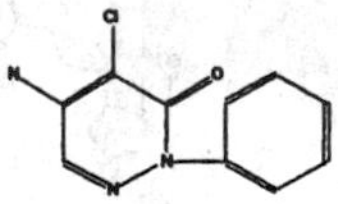

4081

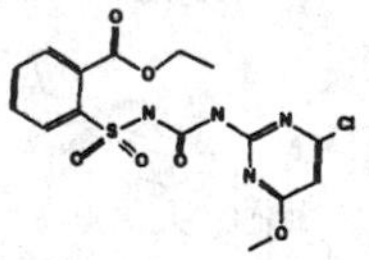

4082

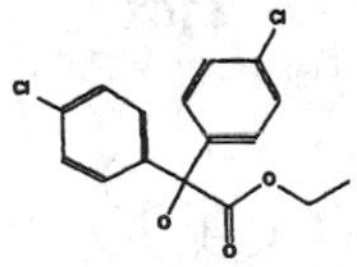

4083

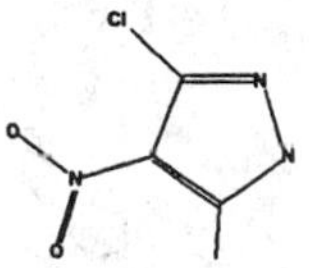

4084

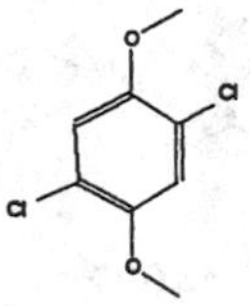

4085

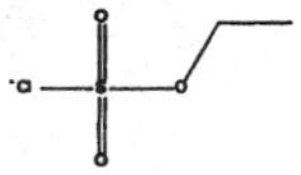

4086

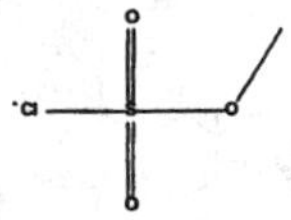

4087

4088

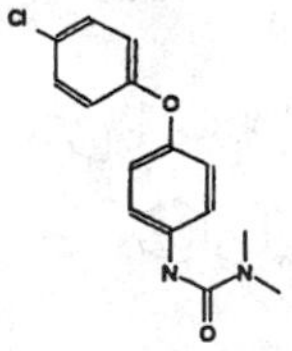

4089

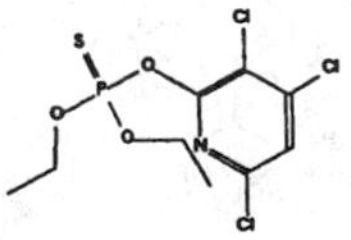

4090

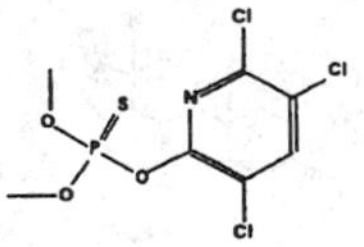

STRUCTURAL FORMULAS OF ORGANIC COMPOUNDS (continued)

In numeric order as they occur in the Table of Physical Constants of Organic Compounds

4091

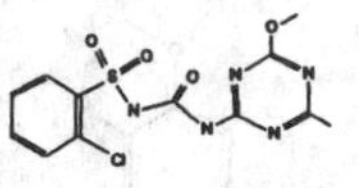

4092

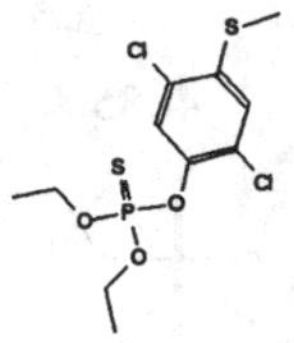

4093

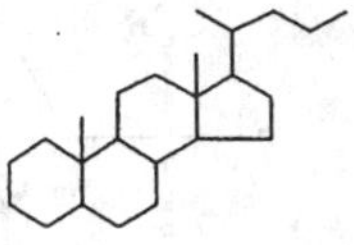

4094

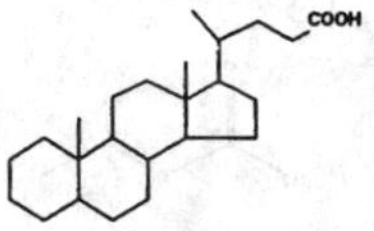

4095

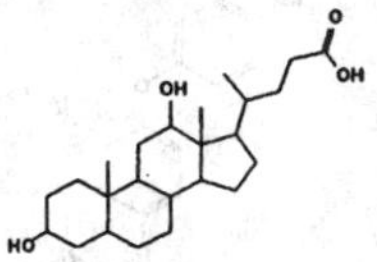

4096

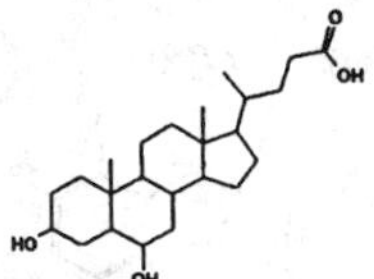

4097

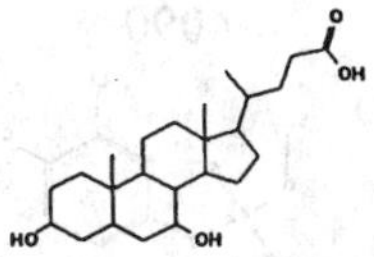

4098

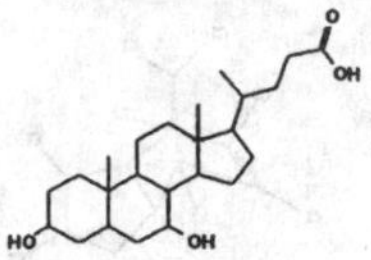

4099

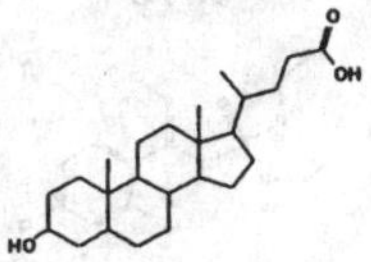

4100

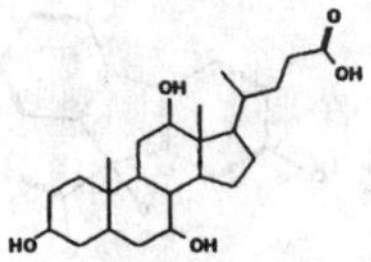

4101

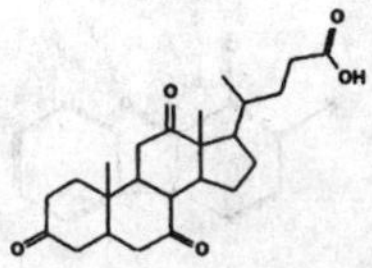

4102

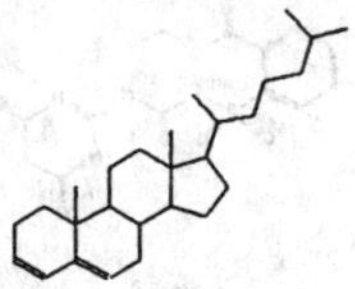

4103

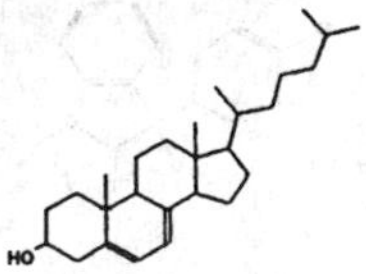

4106

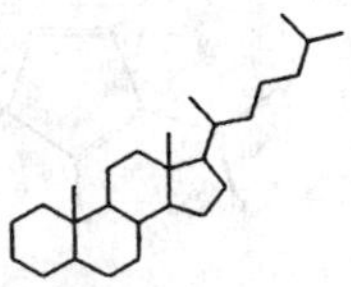

4107

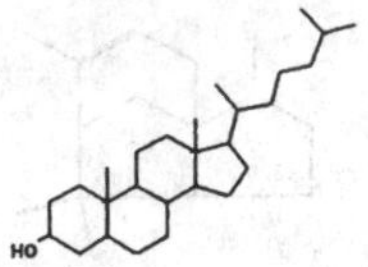

4108

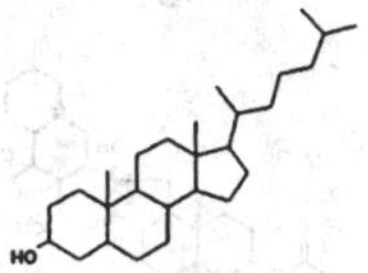

4110

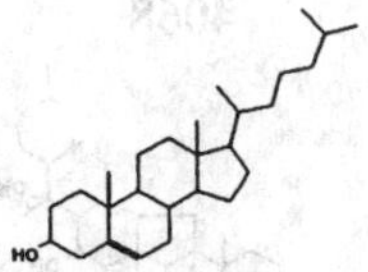

4111

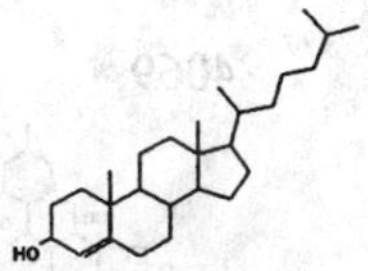

4113

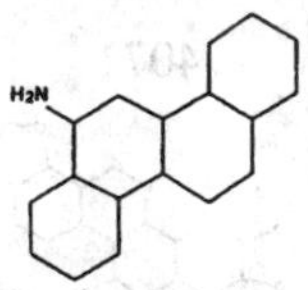

4116

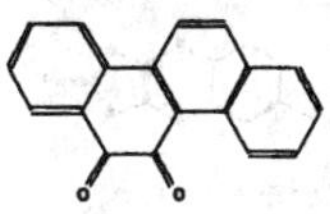

4118

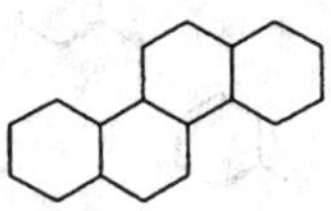

4119

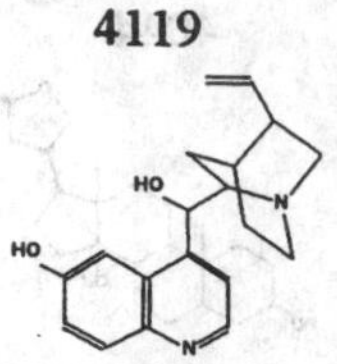

4120

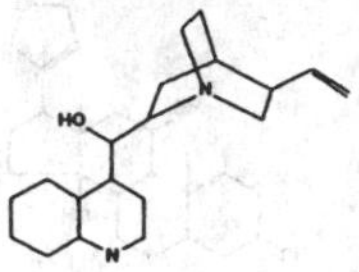

4121

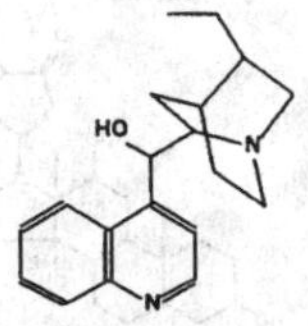

4122

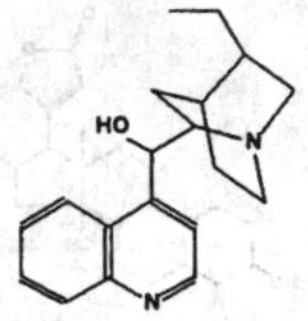

4123

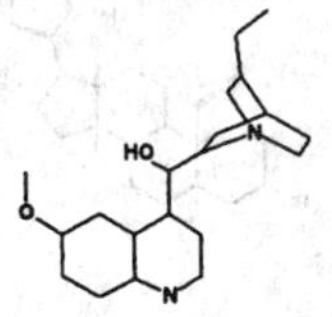

4124

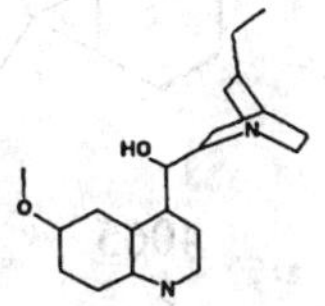

4125

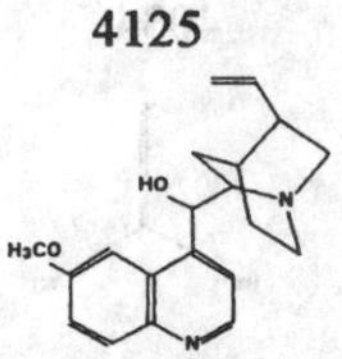

4126

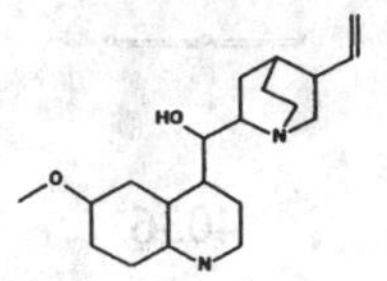

4127

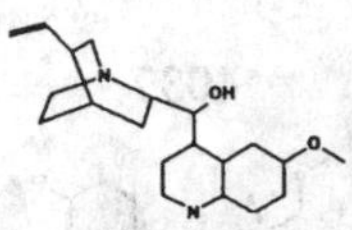

4128

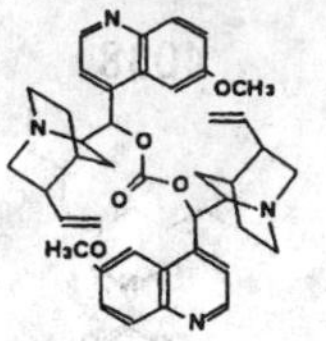

4130

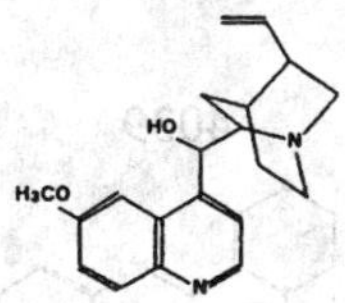

4133

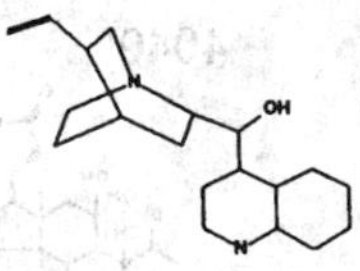

4134

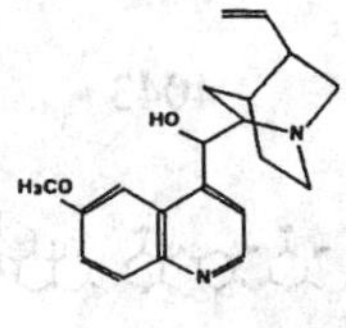

4137

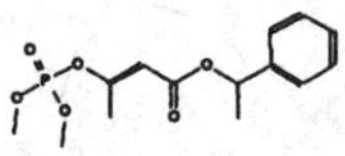

4138

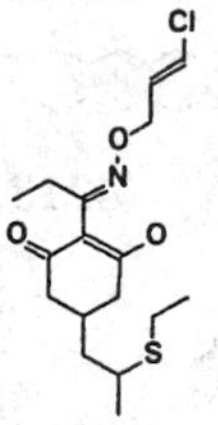

4139

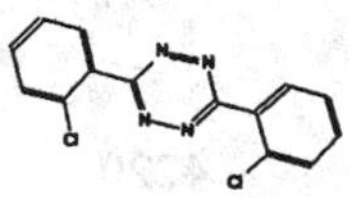

4140

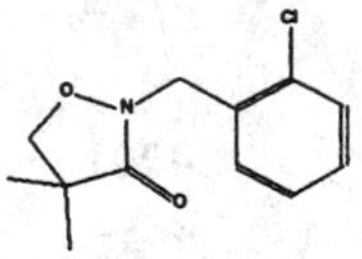

4141

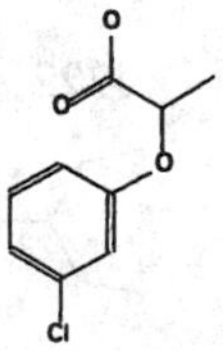

4142

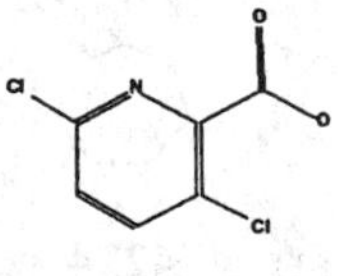

4143

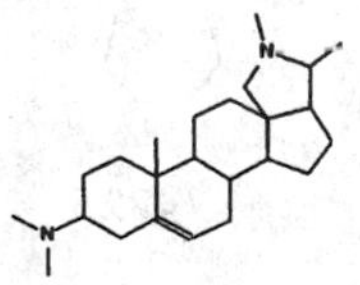

4144

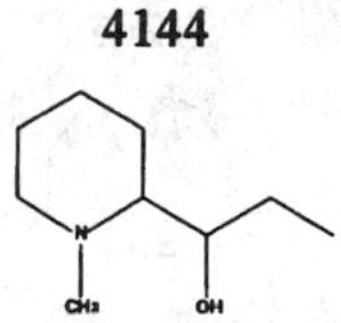

4146

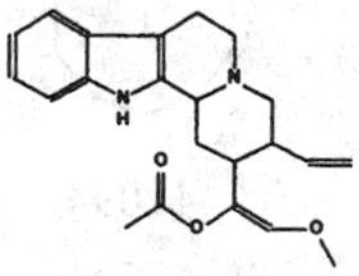

4147

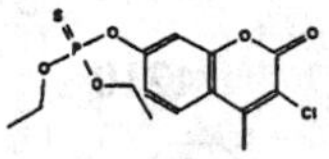

4150

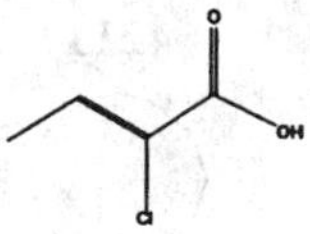

4151

4152

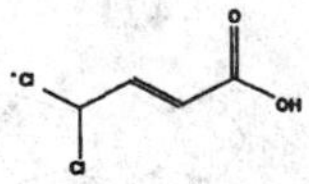

4153

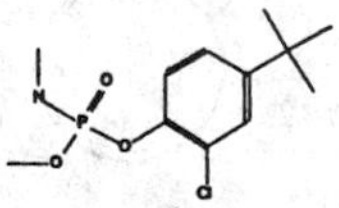

4154

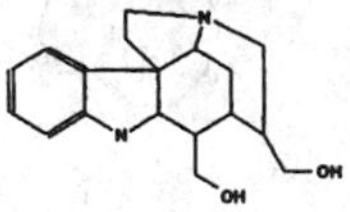

4155

4156

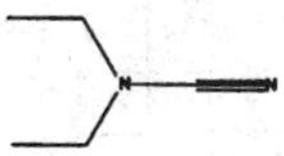

4157

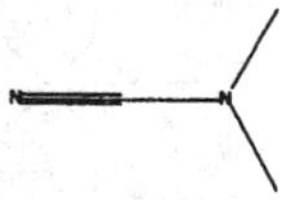

4159

4161

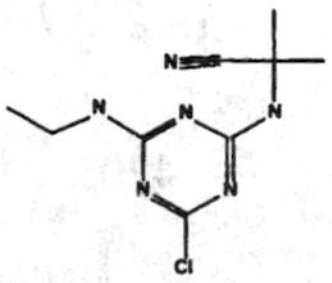

4162

4163

4164

4165

4166

4167

4168

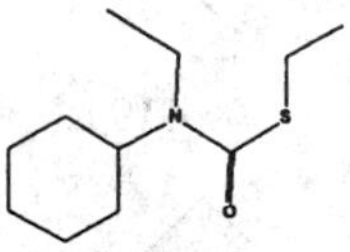

4169

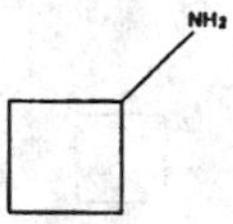

4170

4171

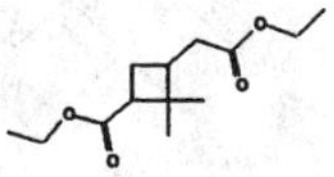

4172

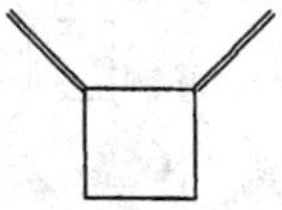

4173

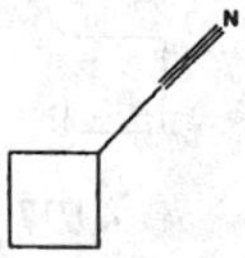

4174

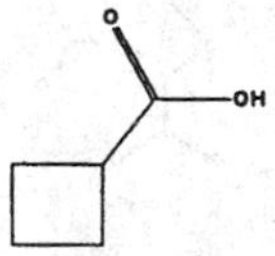

4175

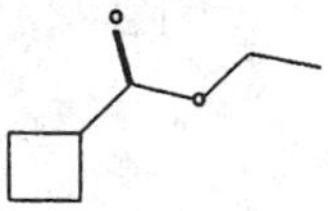

4176

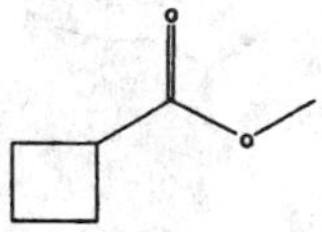

4177

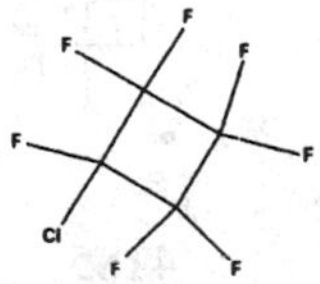

4178

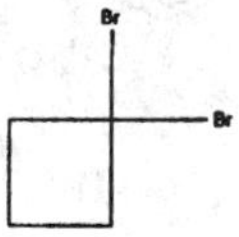

4179

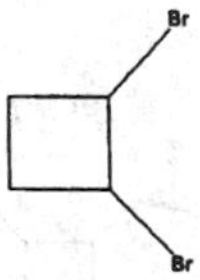

4180

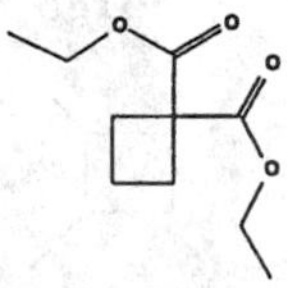

4181

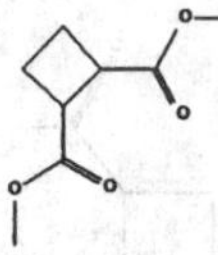

4182

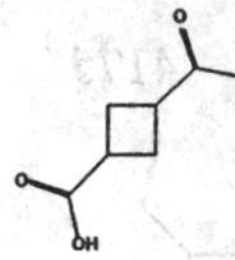

4183

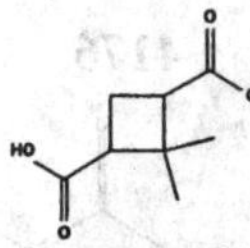

4184

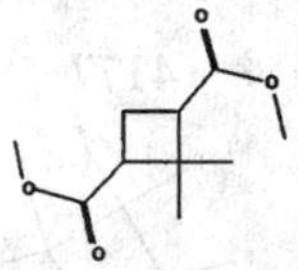

4185

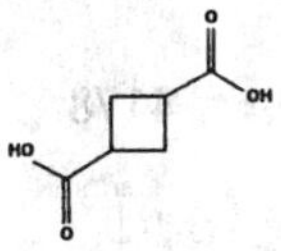

4186

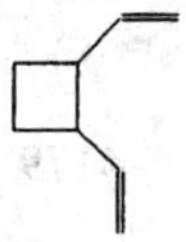

4187

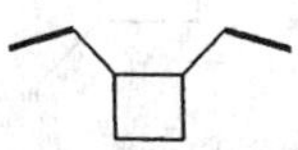

4188

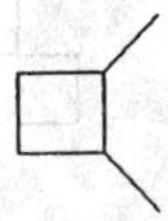

4189

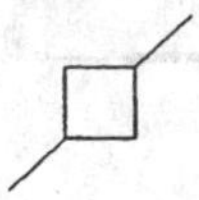

4190

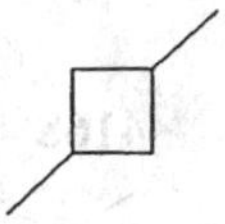

4191

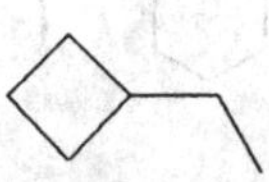

4192

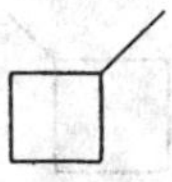

4193

4194

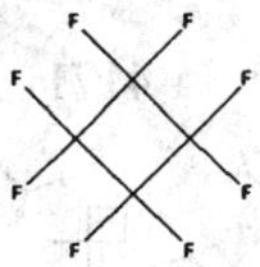

4195

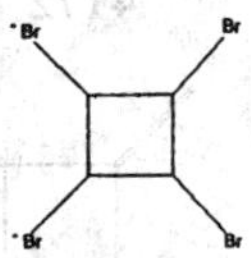

4196

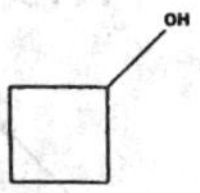

4197

4198

4199

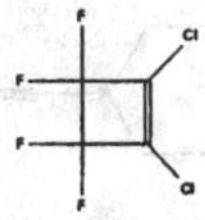

4200

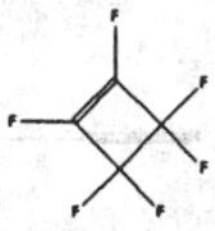

4201

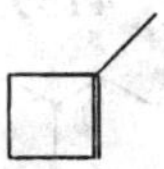

4202

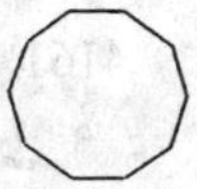

4203

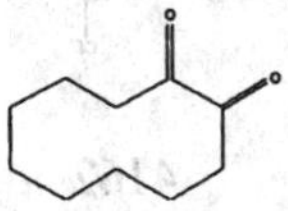

4204

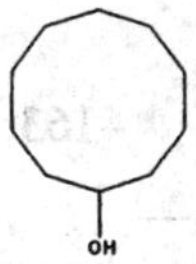

4205

4206

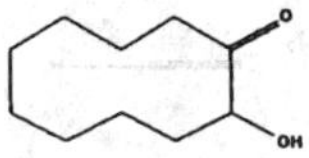

4207

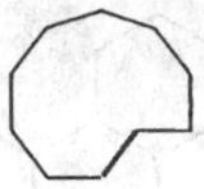

4208

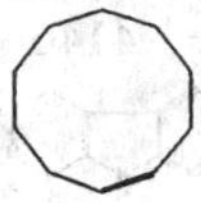

4209

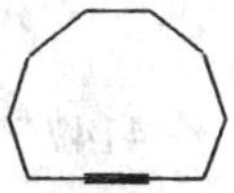

4210

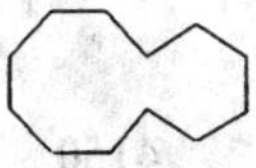

4211

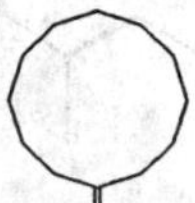

4212

4213

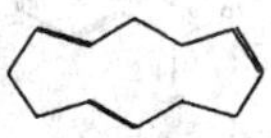

4214

4215

4216

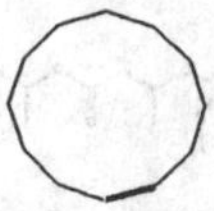

4217

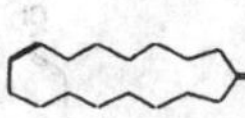

4218

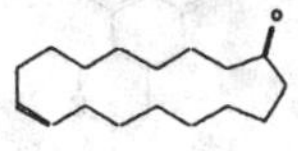

4219

4220

4221

4222

4223

4224

4225

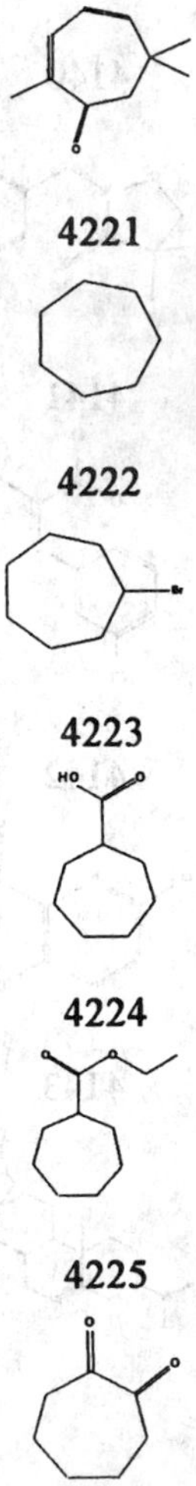

In numeric order as they occur in the Table of Physical Constants of Organic Compounds

4226

4227

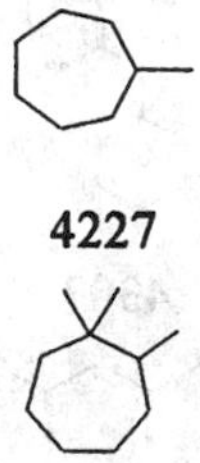

4228

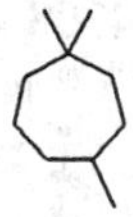

4229

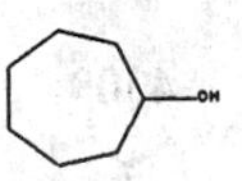

4230

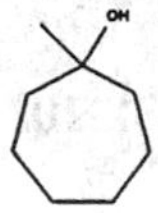

4231

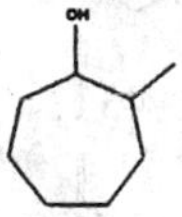

4233

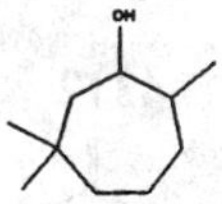

4234

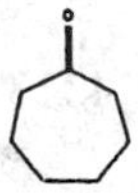

4235

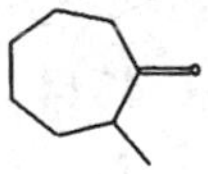

4236

4237

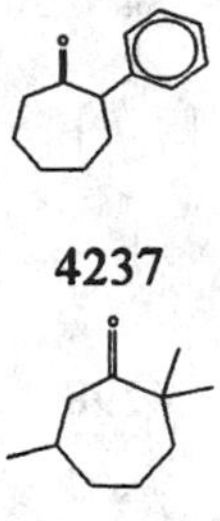

4238

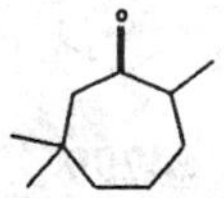

4239

4240

4241

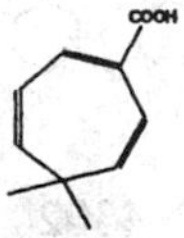

4242

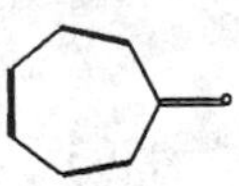

4243

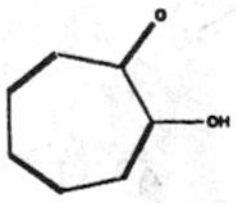

4244

4245

4246

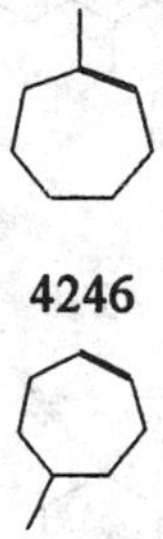

4247

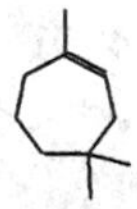

4248

4249

4250

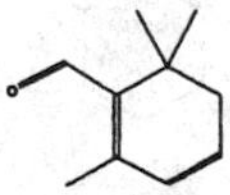

4251

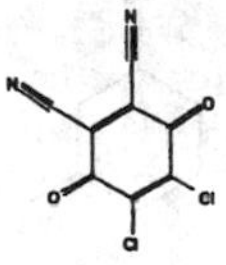

4252

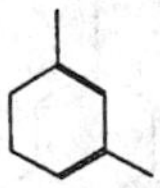

4253

4254

4255

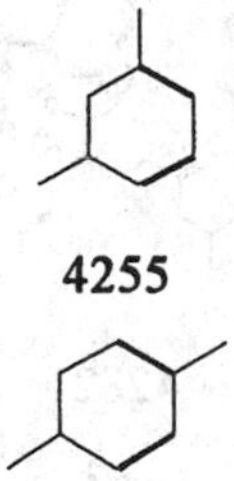

4256

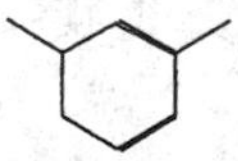

4257

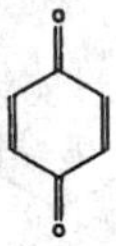

4258

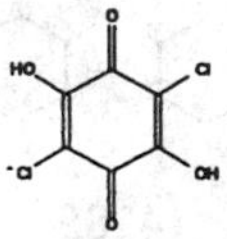

4259

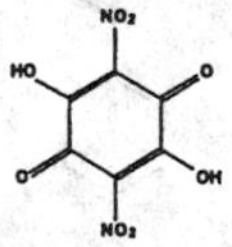

4260

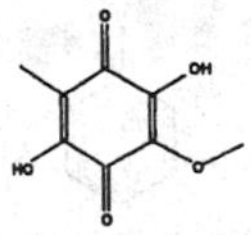

4261

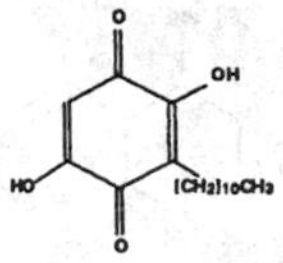

4262

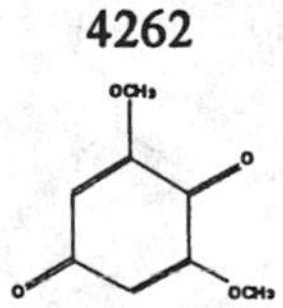

4263

4264

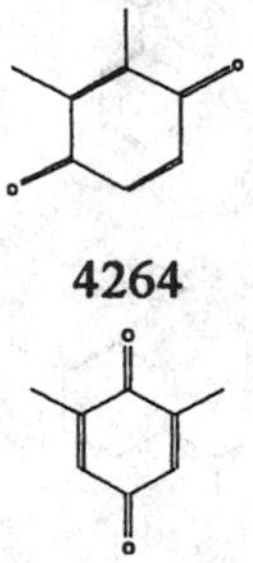

4265

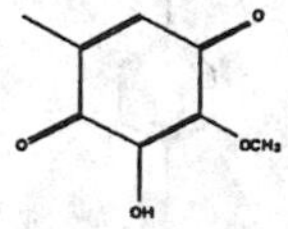

4266

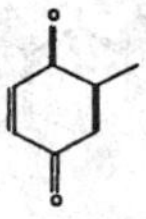

4267

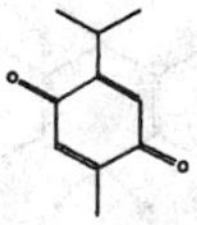

4268

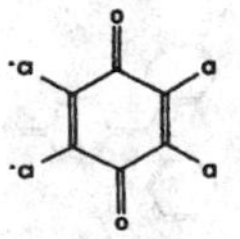

4269

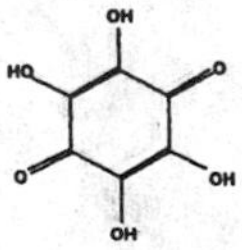

4270

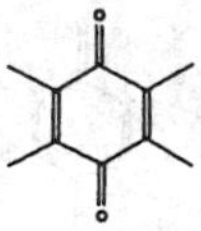

4271

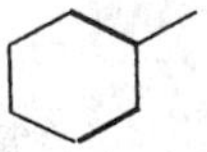

In numeric order as they occur in the Table of Physical Constants of Organic Compounds

4273

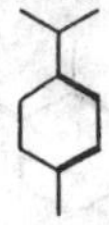

4274

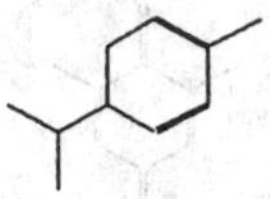

4275

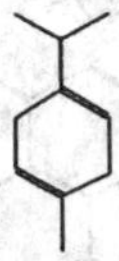

4276

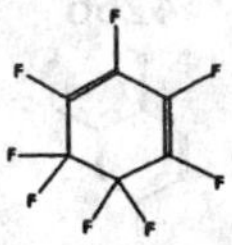

4277

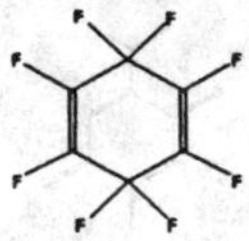

4278

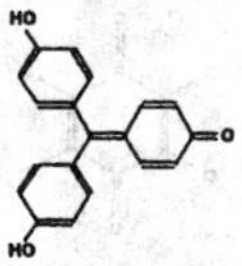

4279

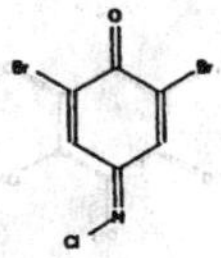

4280

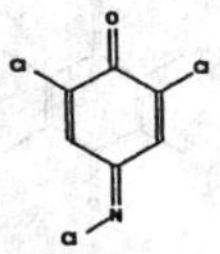

4281

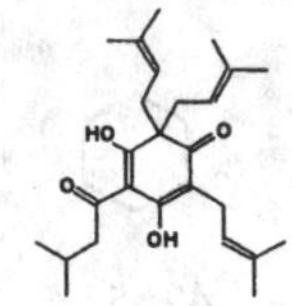

4282

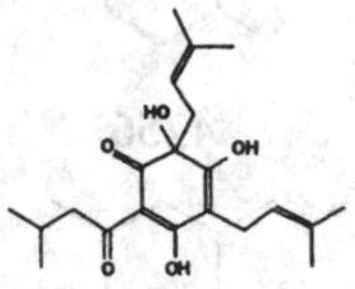

4283

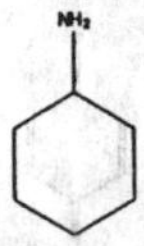

4284

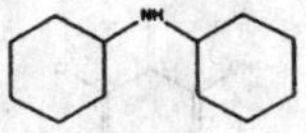

4285

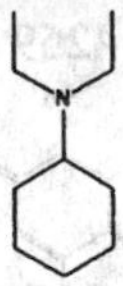

4286

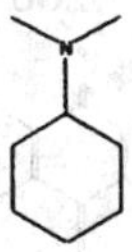

4287

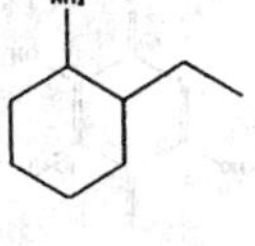

4288

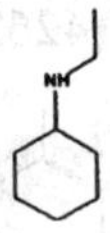

4289

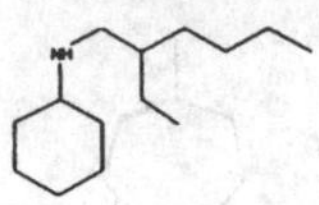

4290

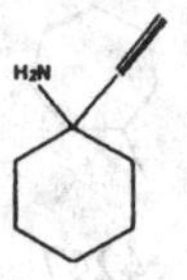

4291

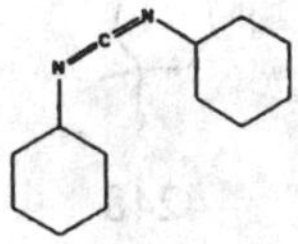

4292

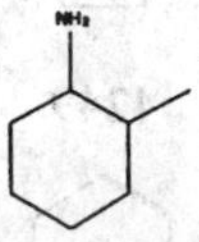

4293

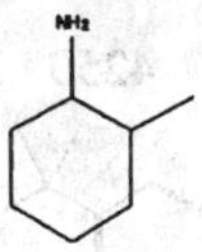

4294

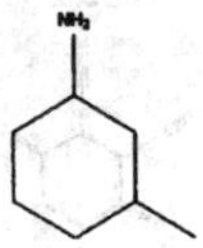

4295

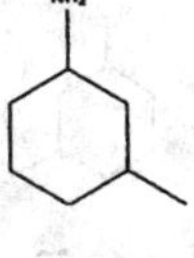

4296

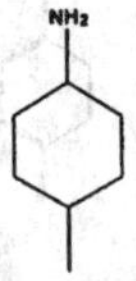

4297

4298

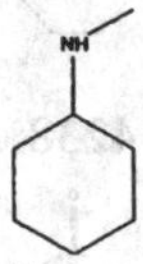

4299

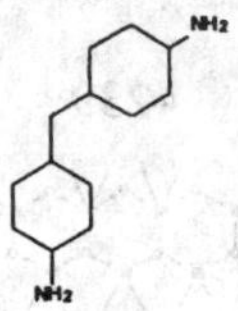

4300

4301

4302

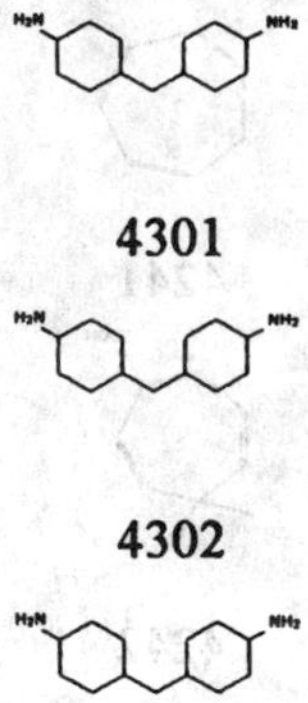

4303

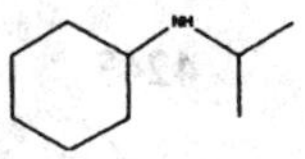

4304

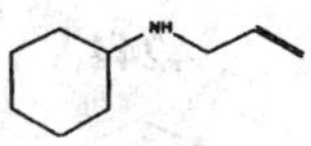

4305

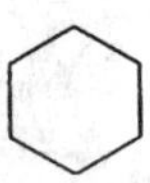

4306

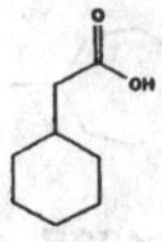

4307

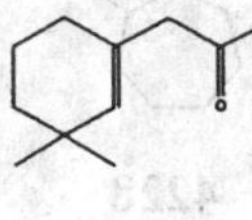

4308

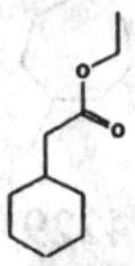

4309

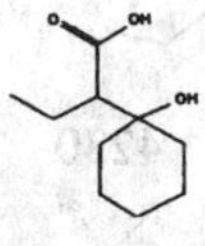

4310

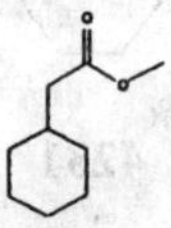

4311

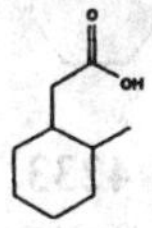

4312

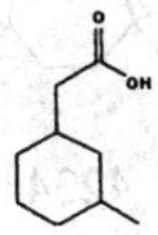

4313

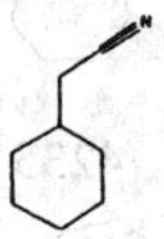

4314

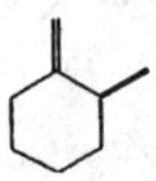

In numeric order as they occur in the Table of Physical Constants of Organic Compounds

4315

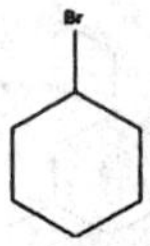

4316

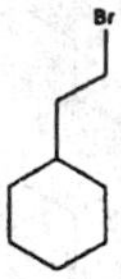

4317

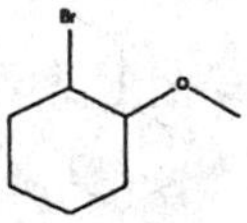

4318

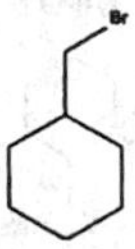

4319

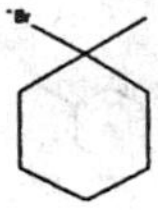

4320

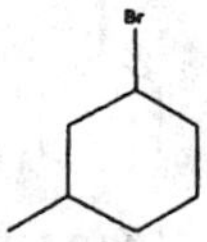

4322

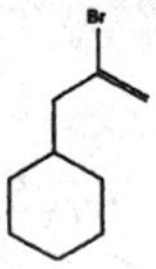

4323

4324

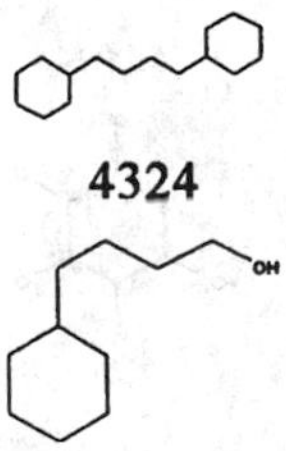

4325

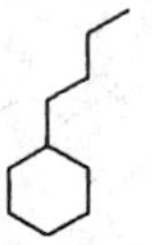

4326

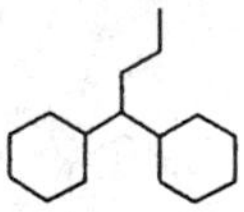

4327

4328

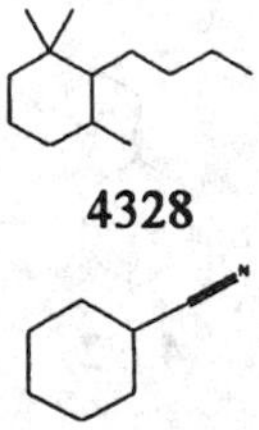

4329

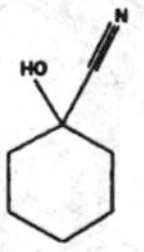

4330

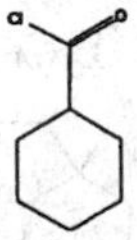

4331

4332

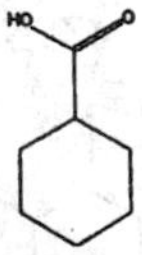

4333

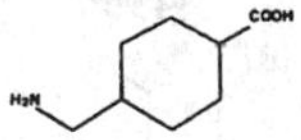

4334

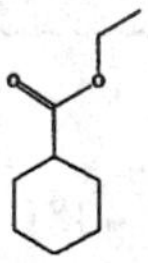

4335

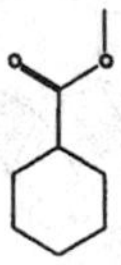

4336

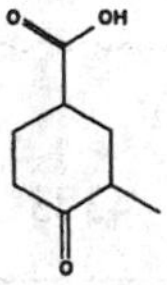

4337

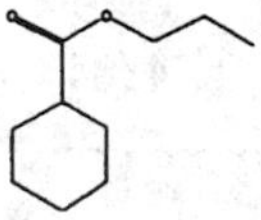

4339

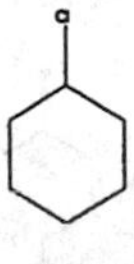

4340

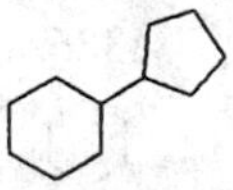

4341

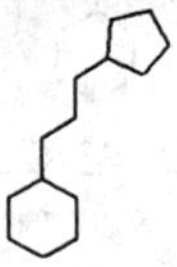

4342

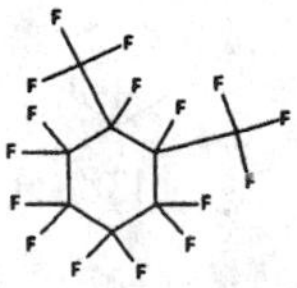

4343

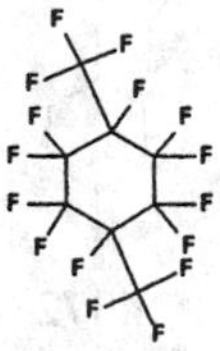

4344

4345

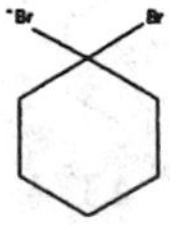

4346

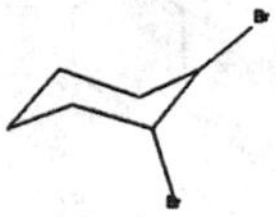

4347

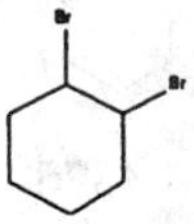

4348

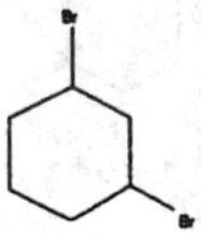

4349

4350

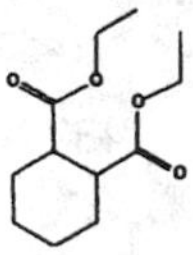

4351

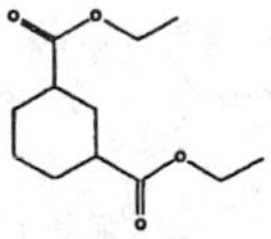

4352

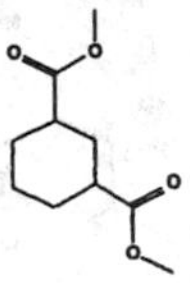

4353

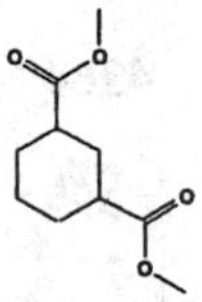

4354

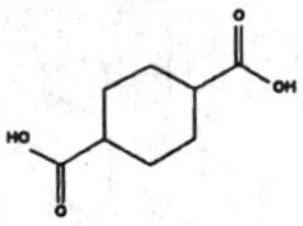

4356

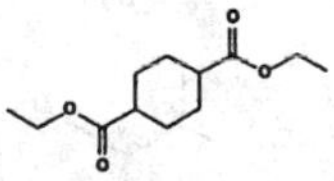

4357

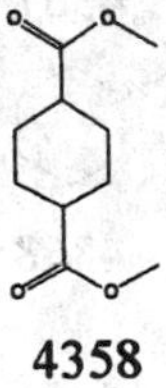

4358

4359

4360

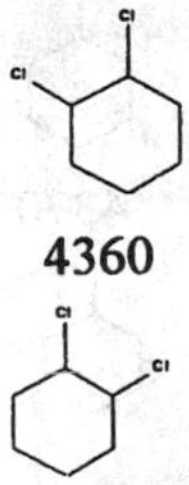

In numeric order as they occur in the Table of Physical Constants of Organic Compounds

4361

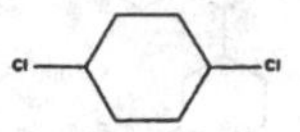

4362

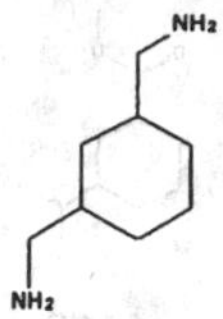

4364

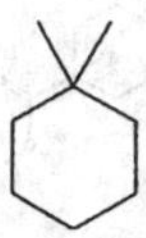

4365

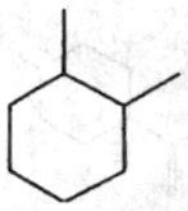

4366

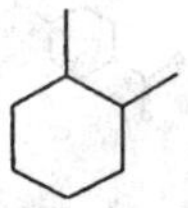

4367

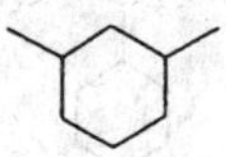

4368

4369

4370

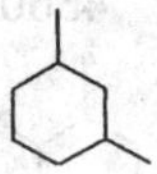

4371

4372

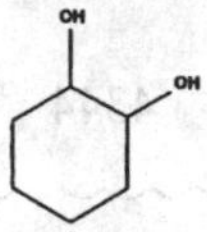

4373

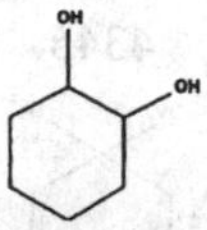

4374

4375

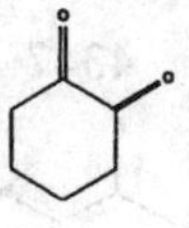

4376

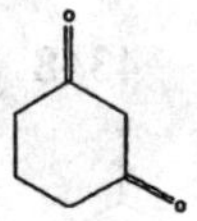

4377

4378

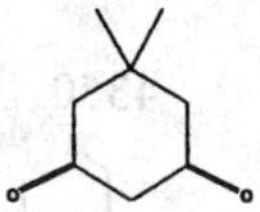

4379

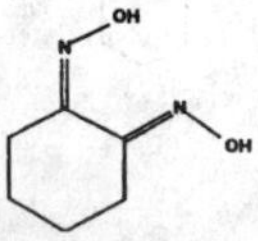

4380

4381

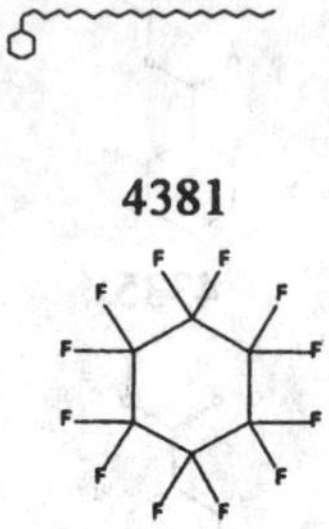

4382

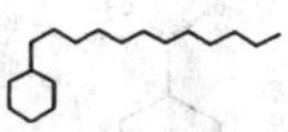

4383

4384

4385

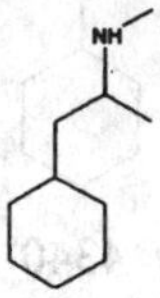

4386

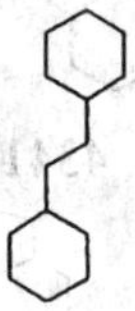

4387

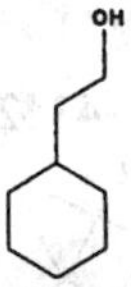

4388

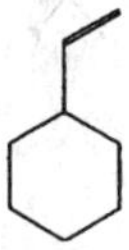

4389

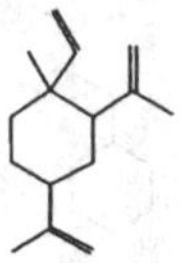

4390

4391

4392

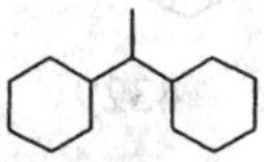

4393

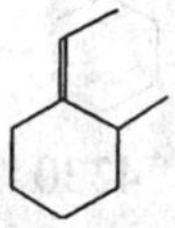

4394

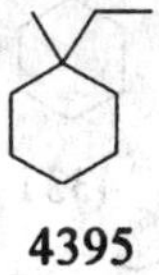

4395

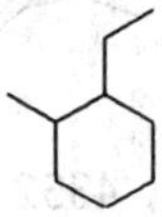

4396

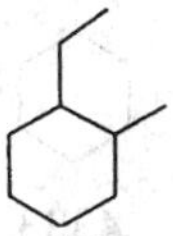

4397

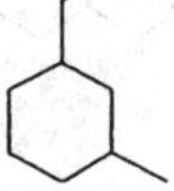

4398

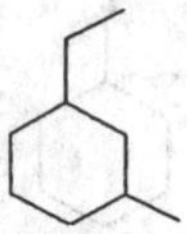

4399

4400

4401

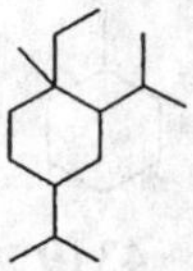

4402

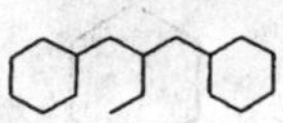

4403

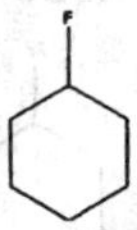

4404

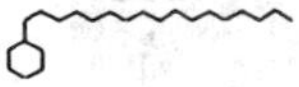

4405

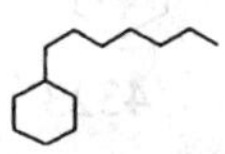

4406

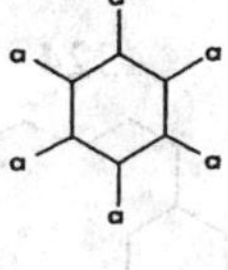

4407

4408

4409

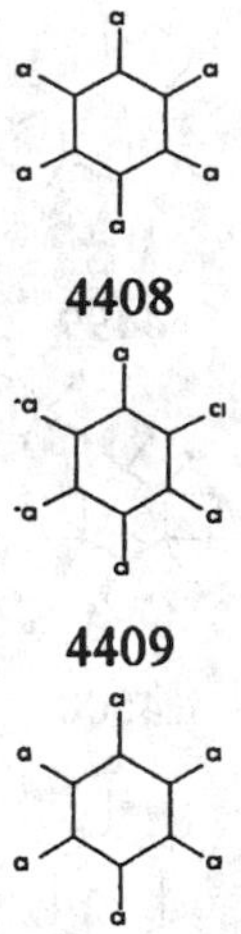

4410

4411

4413

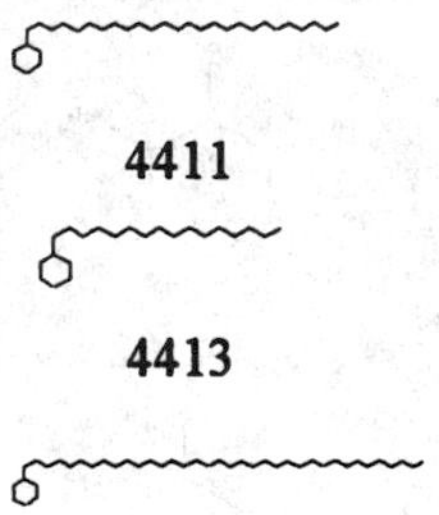

4414

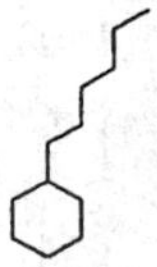

4415

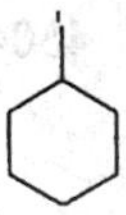

4416

4417

4418

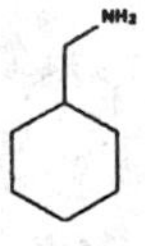

4419

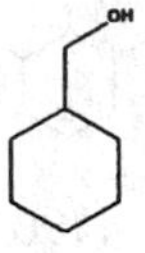

4420

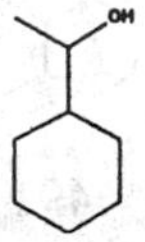

4421

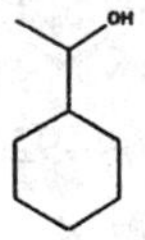

4422

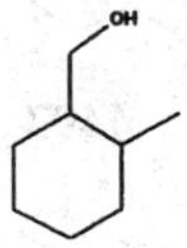

4423

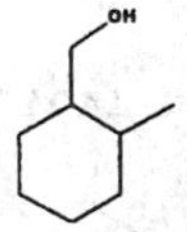

4424

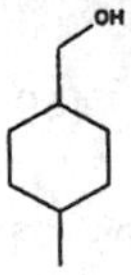

4425

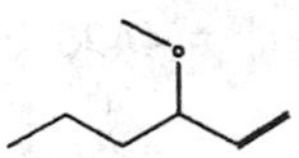

4426

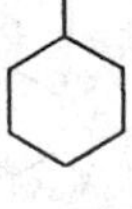

4427

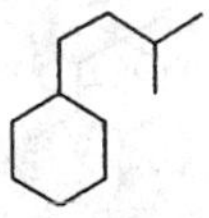

4428

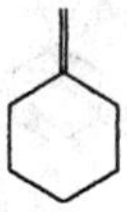

4429

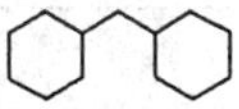

4430

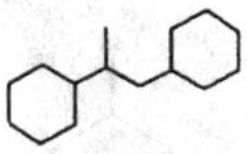

4431

4432

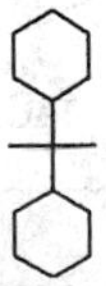

4433

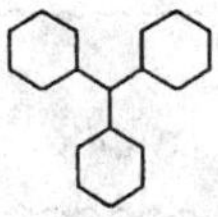

4434

4435

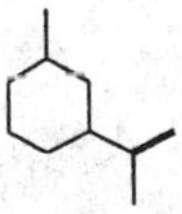

4436

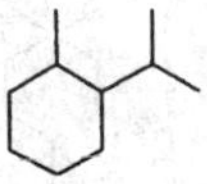

4437

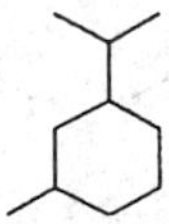

4438

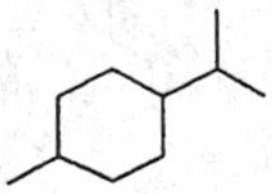

4439

4440

4441

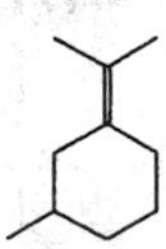

4442

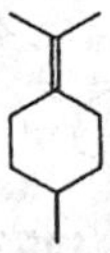

4443

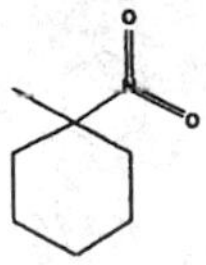

4444

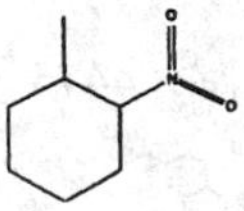

4445

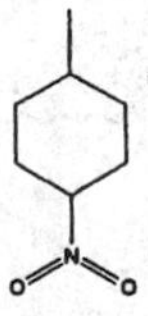

4446

4447

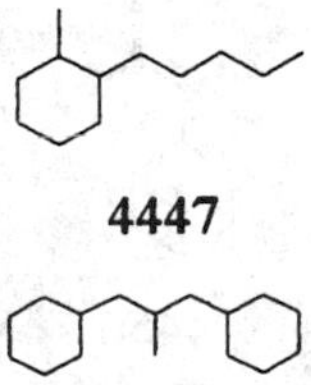

4448

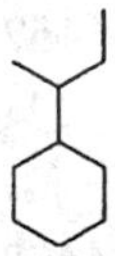

4449

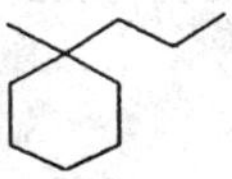

4450

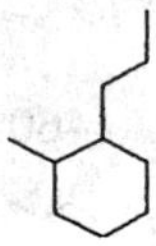

4451

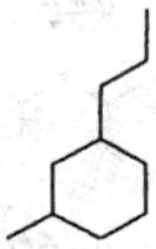

4452

4453

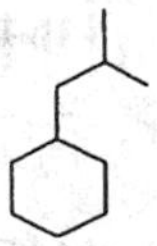

4454

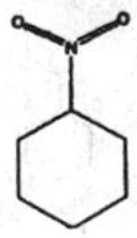

In numeric order as they occur in the Table of Physical Constants of Organic Compounds

4455, 4456, 4457, 4458, 4459, 4460, 4461, 4462, 4463, 4464, 4465, 4466

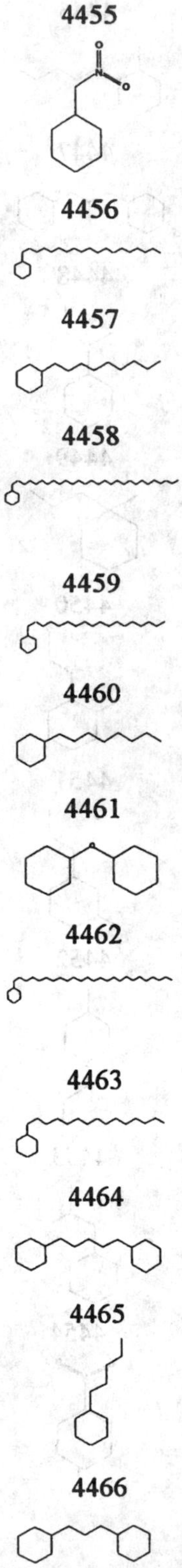

4467, 4468, 4469, 4470, 4471, 4472, 4473, 4474, 4475, 4476

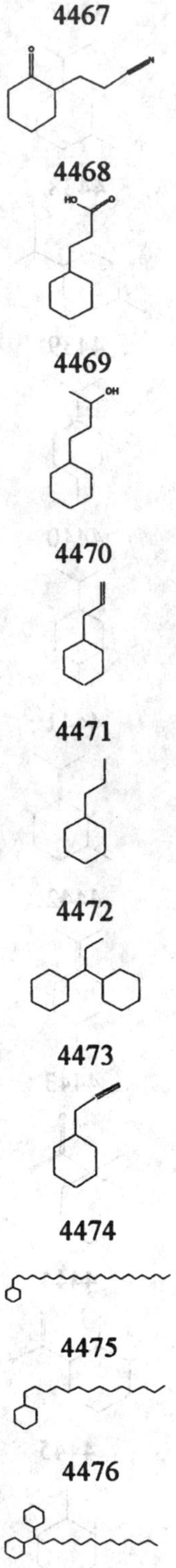

4477, 4478, 4479, 4480, 4481, 4482, 4483, 4484, 4485, 4486, 4487

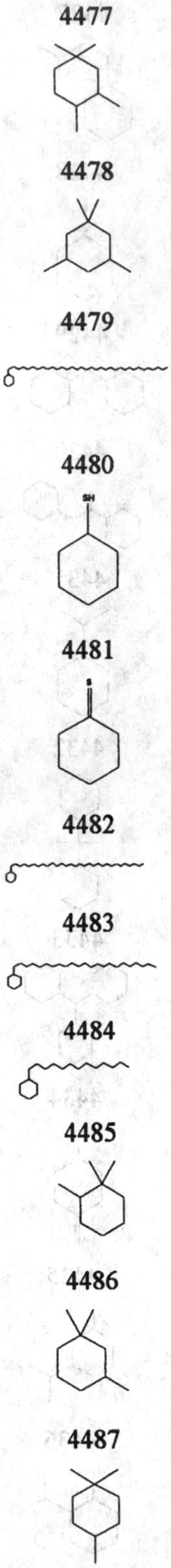

4488, 4489, 4490, 4491, 4492, 4493, 4494, 4495, 4496, 4497

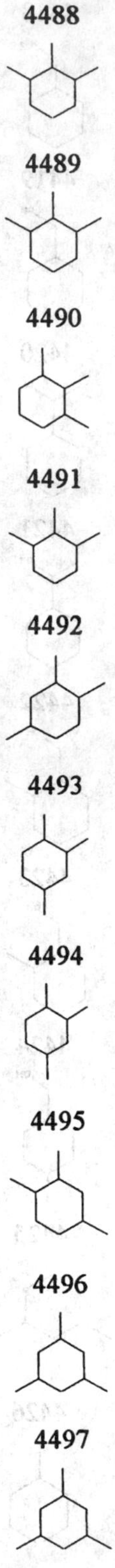

4498, 4499, 4500, 4501, 4502, 4503, 4504, 4505

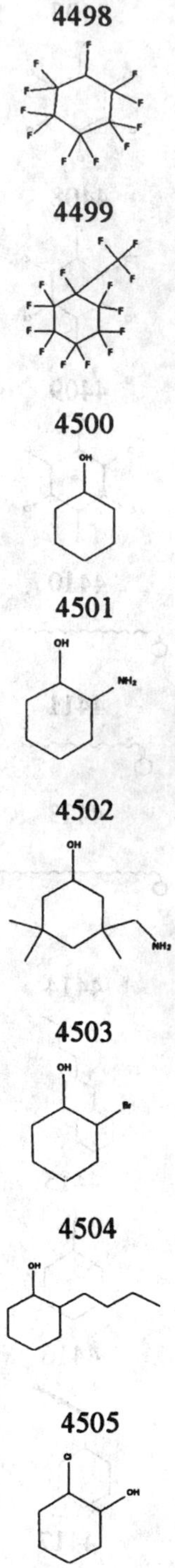

STRUCTURAL FORMULAS OF ORGANIC COMPOUNDS (continued)

In numeric order as they occur in the Table of Physical Constants of Organic Compounds

4507

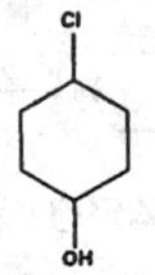

4508

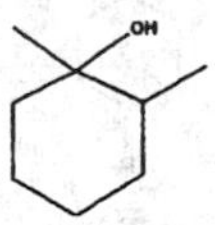

4509

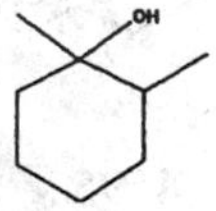

4510

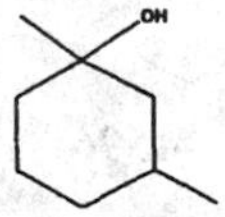

4511

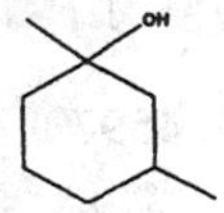

4513

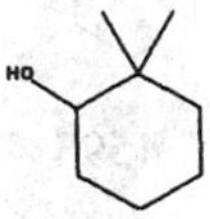

4514

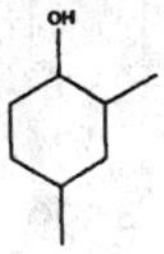

4515

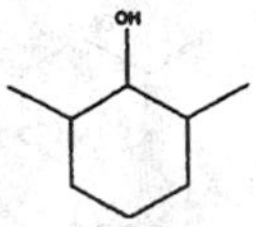

4516

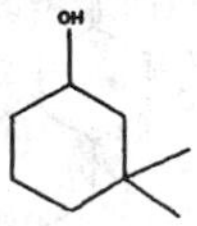

4517

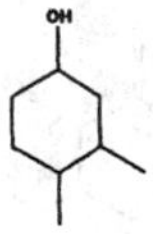

4518

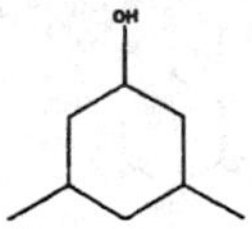

4519

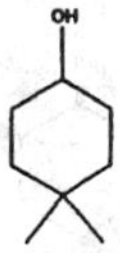

4521

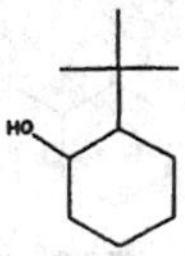

4522

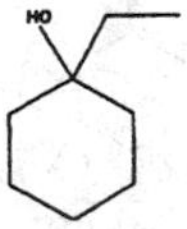

4523

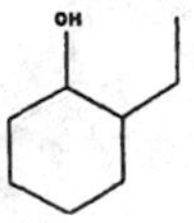

4525

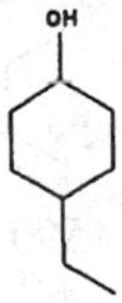

4526

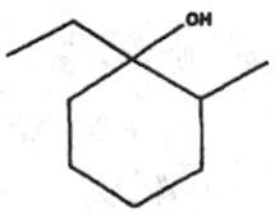

4527

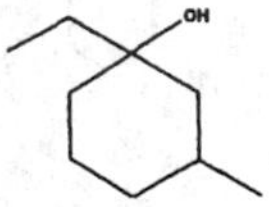

4528

4529

4530

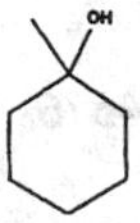

4531

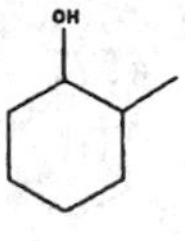

4532

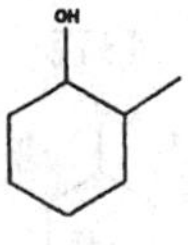

4533

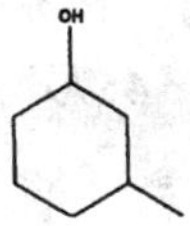

4534

4535

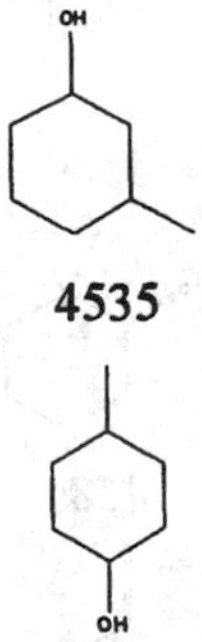

4536

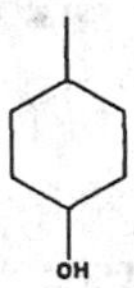

4537

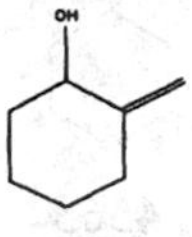

4538

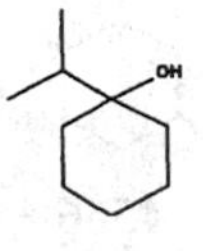

4539

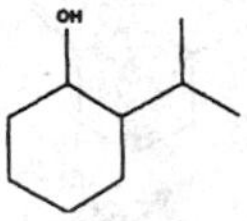

4540

4541

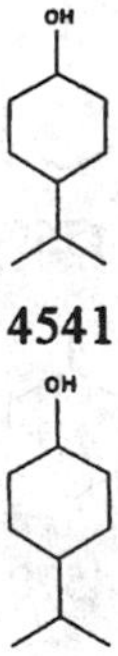

4542

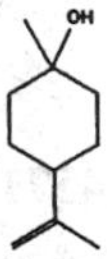

4543

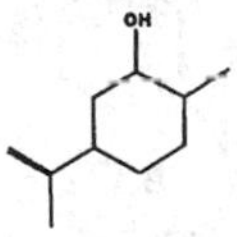

4544

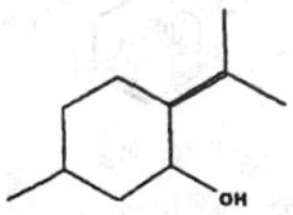

4545

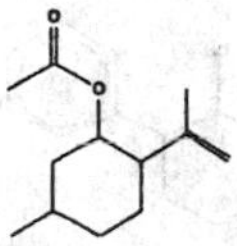

4546

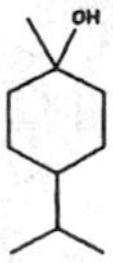

4547

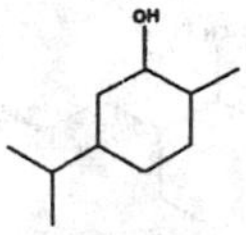

4548

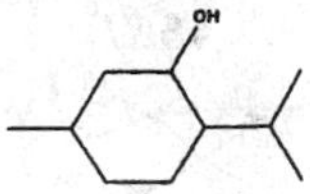

4549

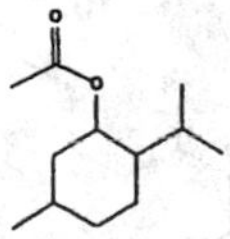

4550

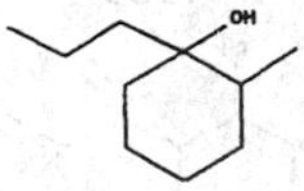

4551

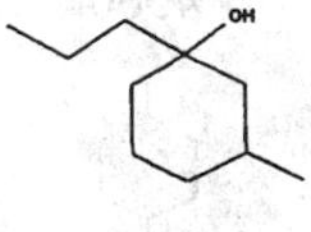

4552

4553

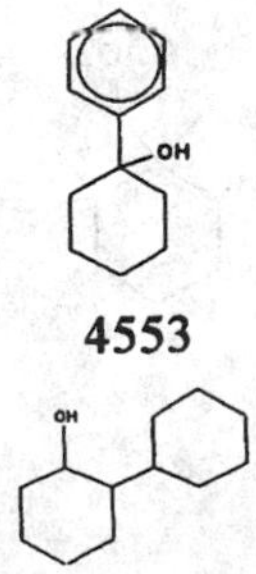

In numeric order as they occur in the Table of Physical Constants of Organic Compounds

4554
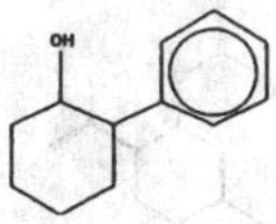
4555
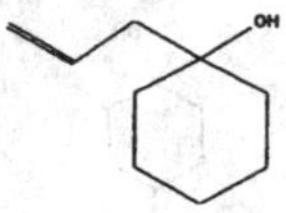
4556
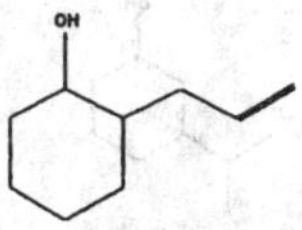
4557
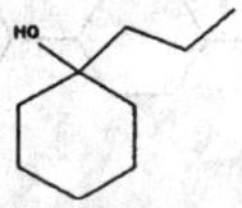
4558
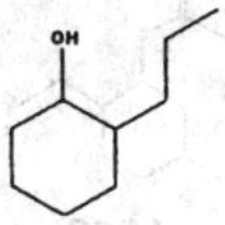
4559
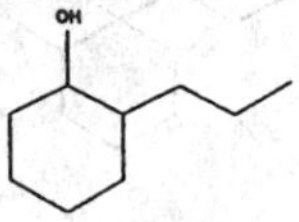
4560
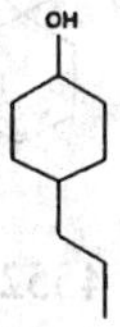
4562
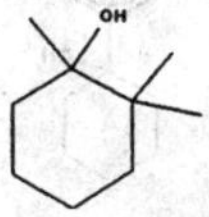
4563
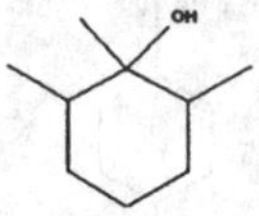
4564
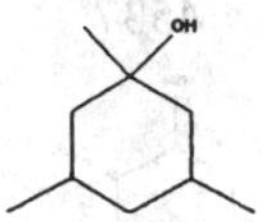
4565
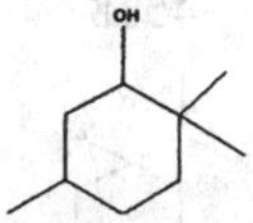
4566
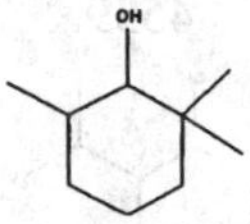
4568
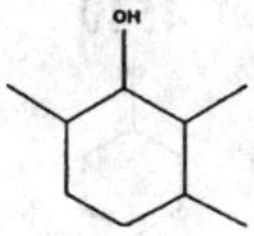
4569
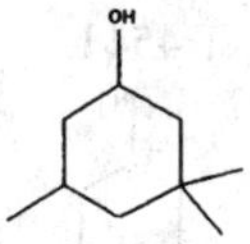
4570
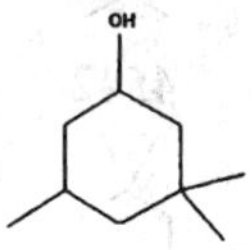
4571
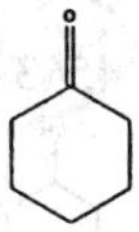
4572
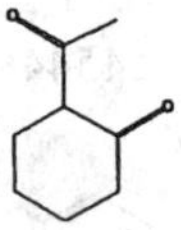
4573
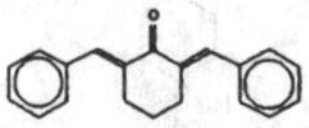
4574
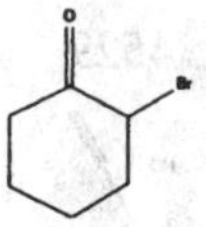
4575
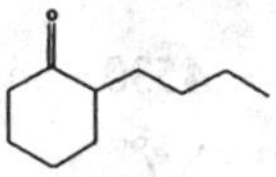
4576
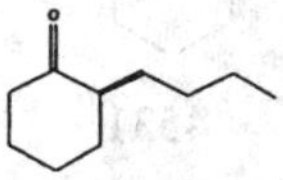
4577
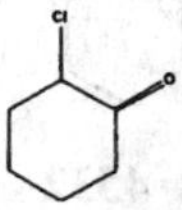
4578

4579
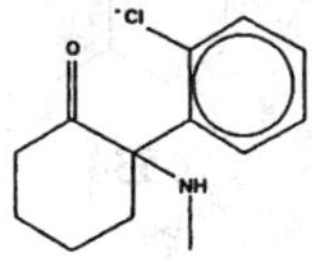
4580
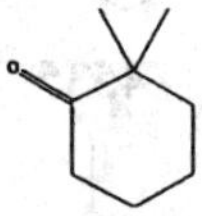
4581
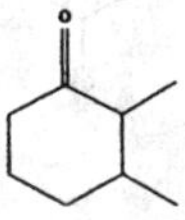
4582
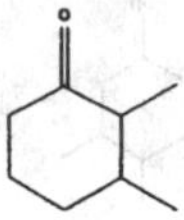
4583
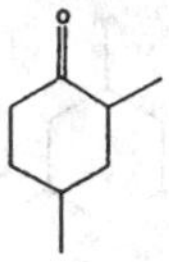
4585
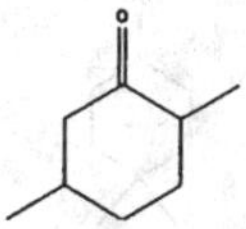
4586
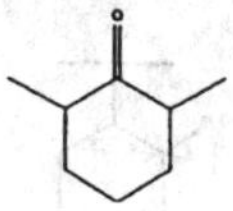
4587
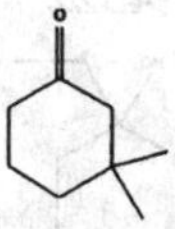
4588
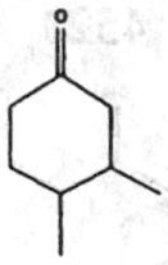
4589
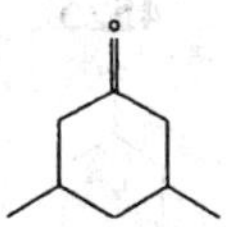
4590
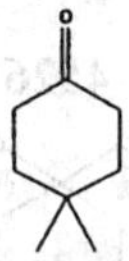
4591
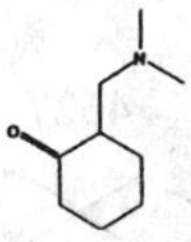
4592
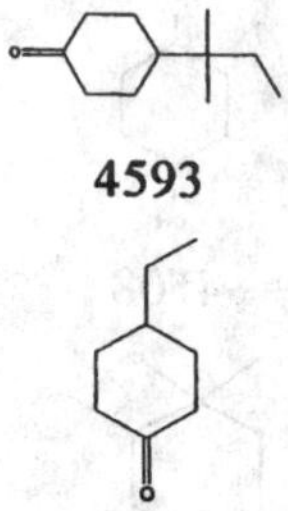
4593
4594
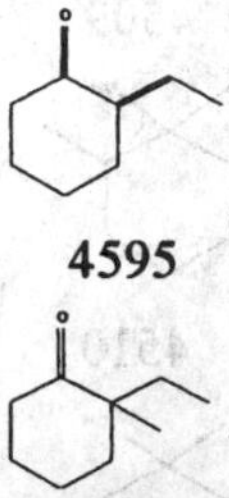
4595
4596
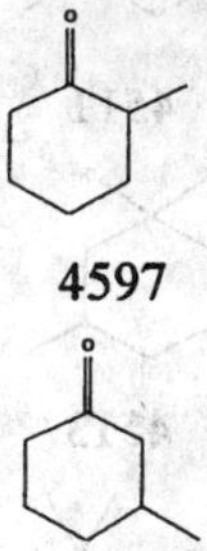
4597
4598
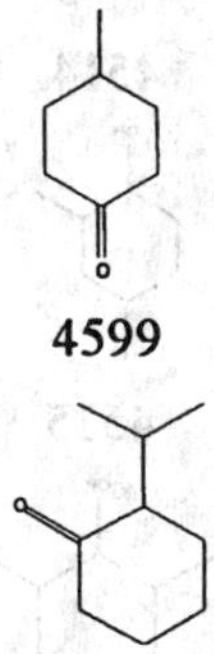
4599
4600
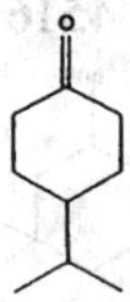

In numeric order as they occur in the Table of Physical Constants of Organic Compounds

4601
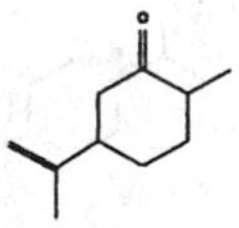
4602
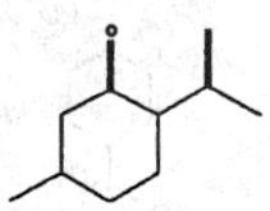
4603
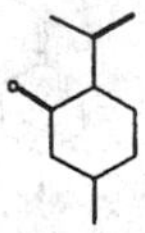
4604
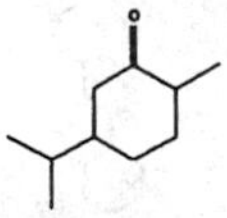
4605
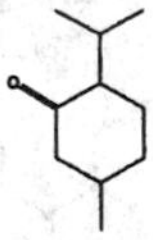
4606
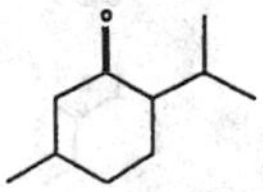
4607
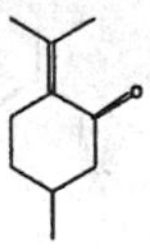
4608
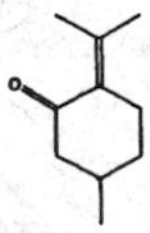
4609
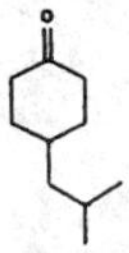
4610
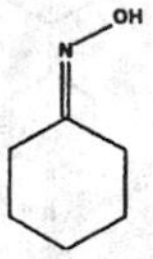

4611
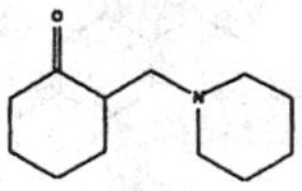

4612
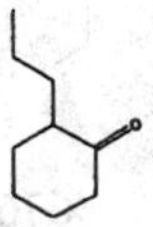
4613
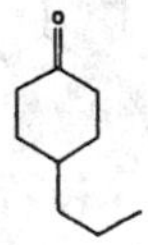
4614
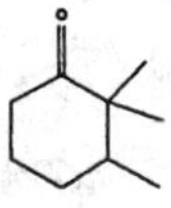
4615
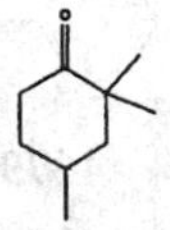
4616
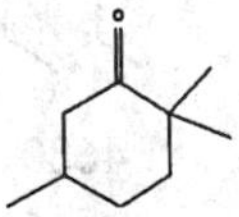
4617
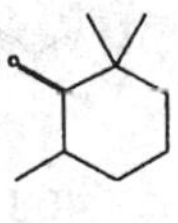
4618
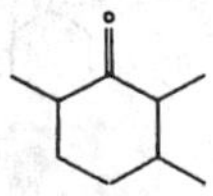
4619
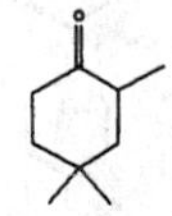
4620
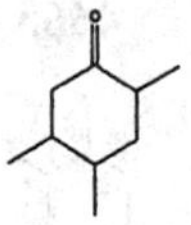
4621
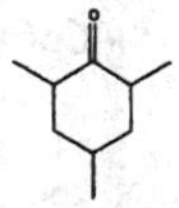
4622
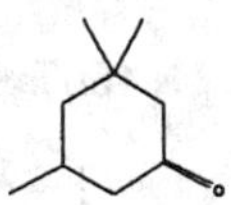
4623
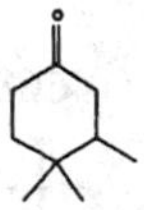
4624
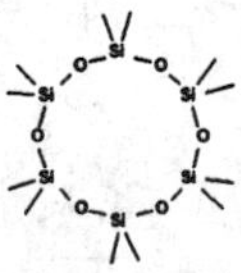

4625
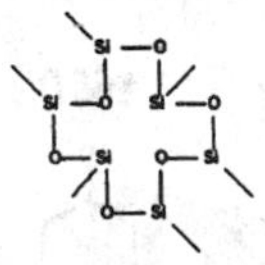

4626
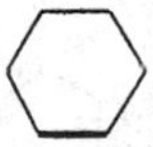
4627
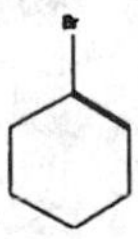

4628
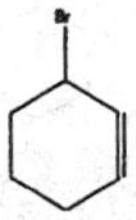

4629
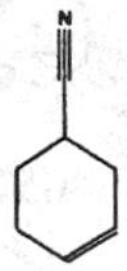

4630
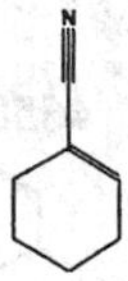

4631
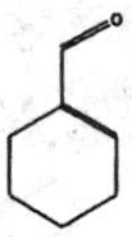
4632
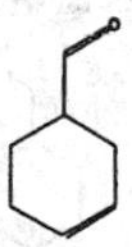
4633
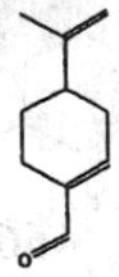
4634
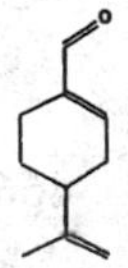
4635
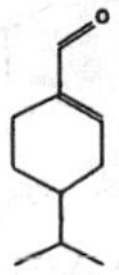
4636
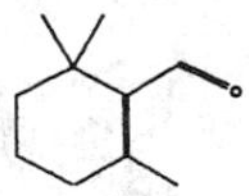
4637
4638

4639
4640
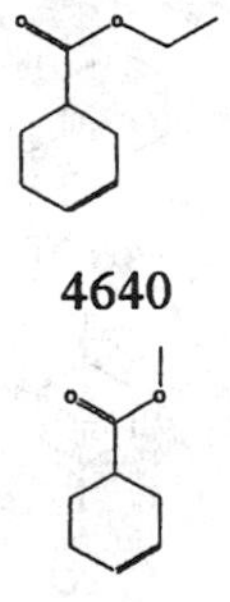
4642
4643
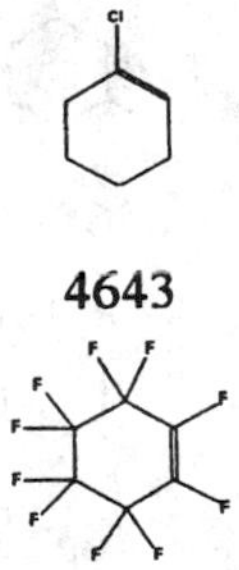

4644
4645
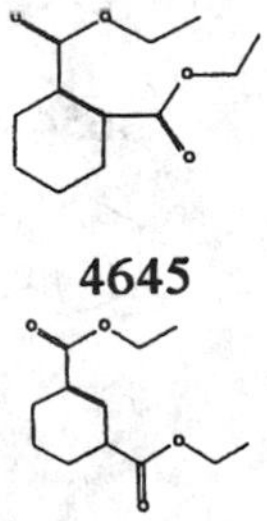

STRUCTURAL FORMULAS OF ORGANIC COMPOUNDS (continued)

In numeric order as they occur in the Table of Physical Constants of Organic Compounds

4646

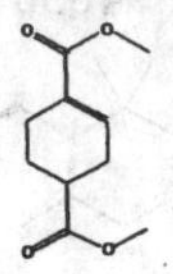

4647

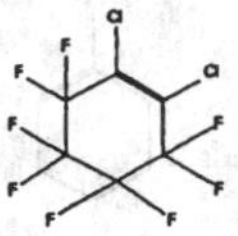

4648

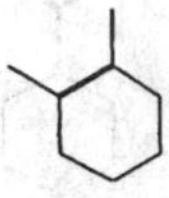

4649

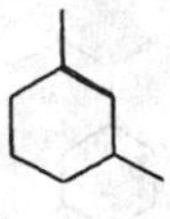

4650

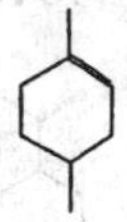

4651

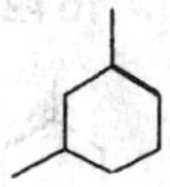

4652

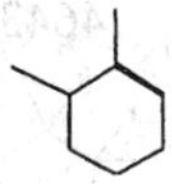

4653

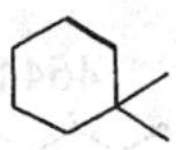

4654

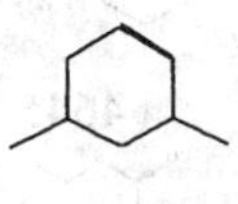

4655

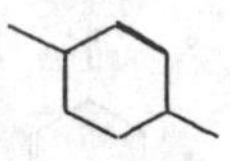

4656

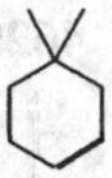

4657

4658

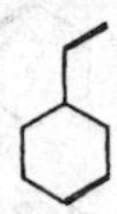

4659

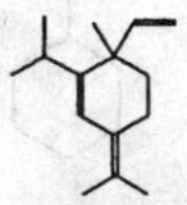

4660

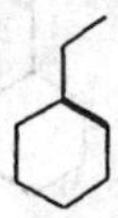

4661

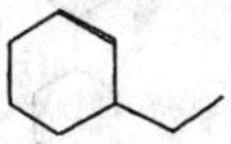

4662

4663

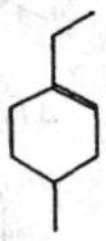

4664

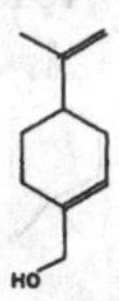

4665

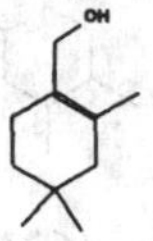

4666

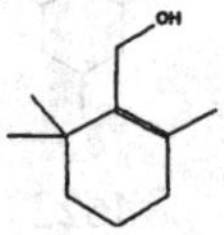

4667

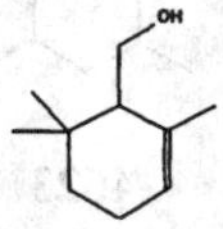

4668

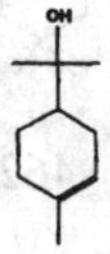

4669

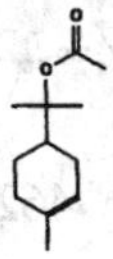

4670

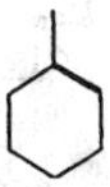

4671

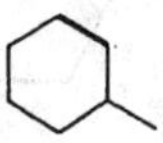

4672

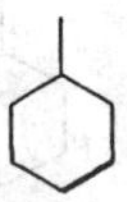

4673

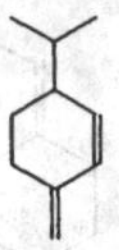

4674

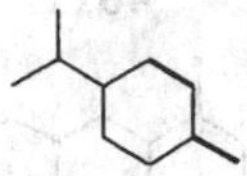

4675

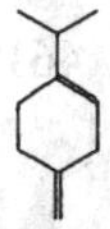

4676

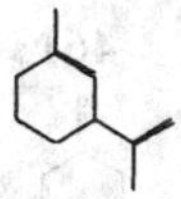

4677

4678

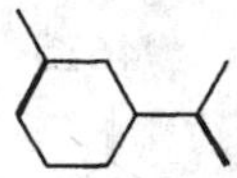

4679

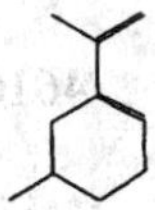

4680

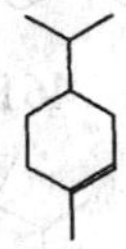

4681

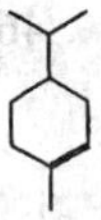

4682

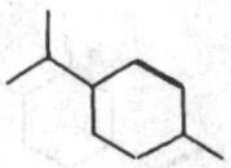

4683

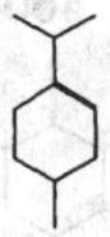

4684

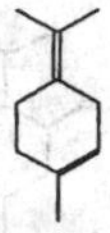

4685, 4686, 4687, 4688, 4689, 4690

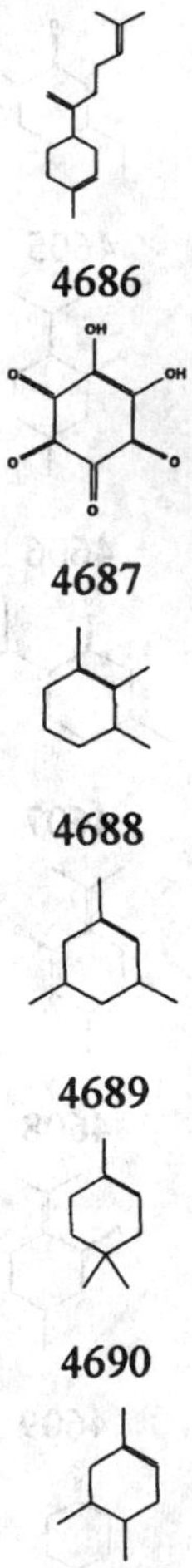

STRUCTURAL FORMULAS OF ORGANIC COMPOUNDS (continued)

In numeric order as they occur in the Table of Physical Constants of Organic Compounds

4691

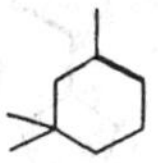

4692

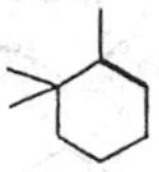

4694

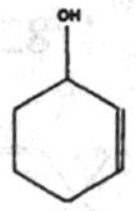

4695

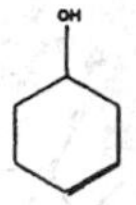

4696

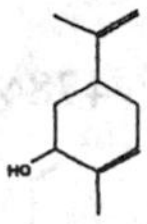

4697

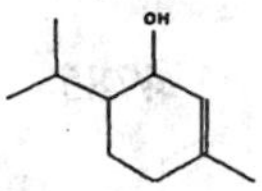

4698

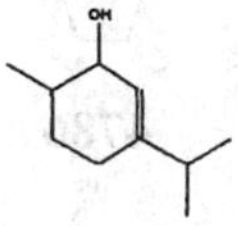

4699

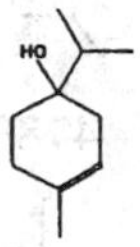

4700

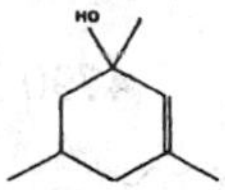

4701

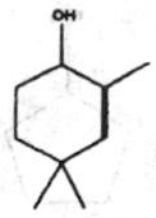

4702

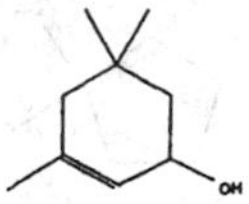

4703

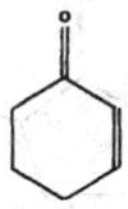

4704

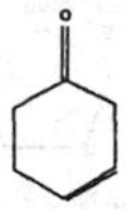

4705

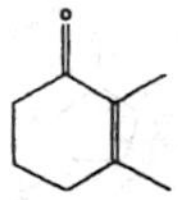

4706

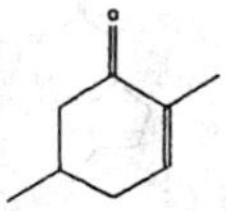

4707

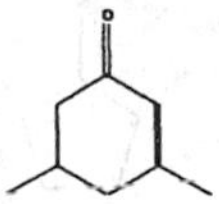

4708

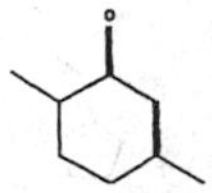

4709

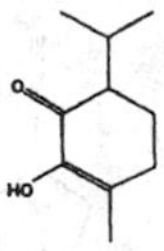

4710

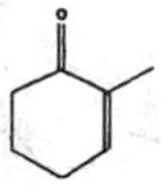

4711

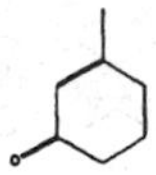

4712

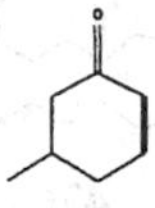

4713

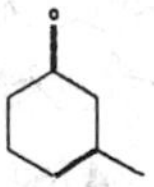

4714

4715

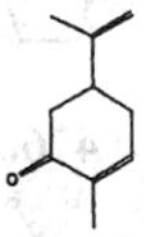

4716

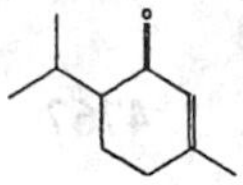

4718

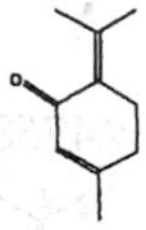

4719

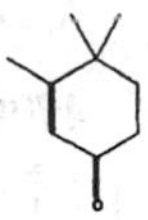

4720

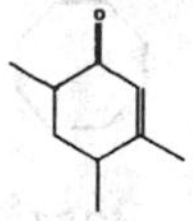

4721

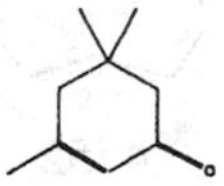

4722

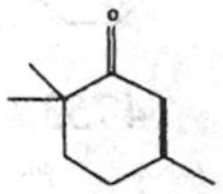

4723

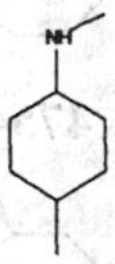

4724

4725

4726

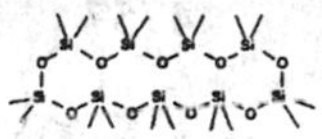

4727

4728

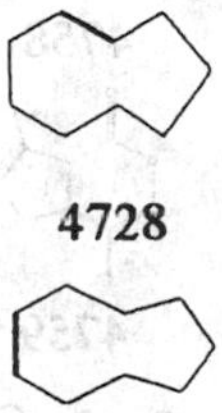

4729

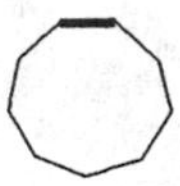

4731

4732

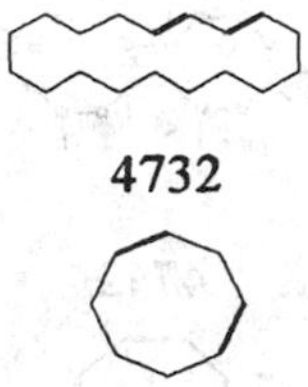

4733

4734

4735

4736

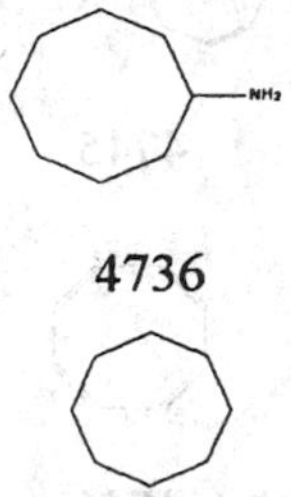

4737

4738

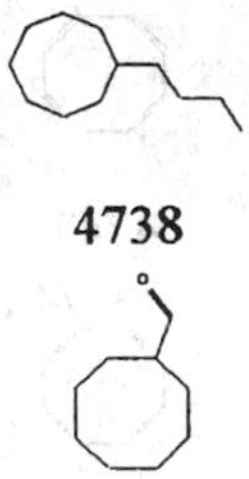

4739

4740

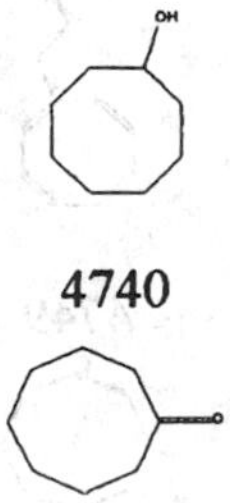

STRUCTURAL FORMULAS OF ORGANIC COMPOUNDS (continued)

In numeric order as they occur in the Table of Physical Constants of Organic Compounds

4741

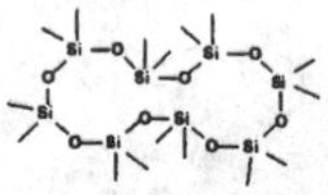

4742

4743

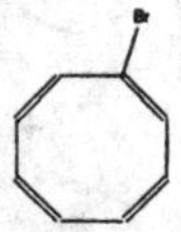

4744

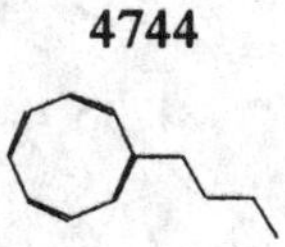

4745

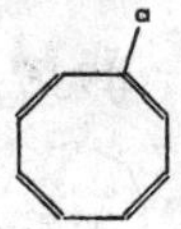

4746

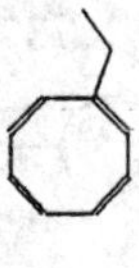

4747

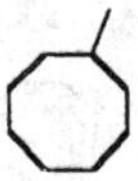

4748

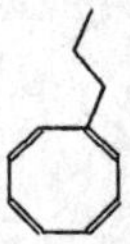

4749

4750

4751

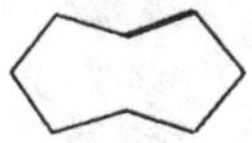

4752

4753

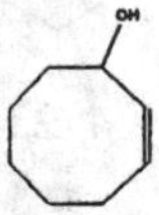

4754

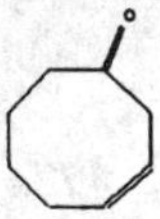

4755

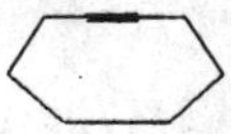

4756

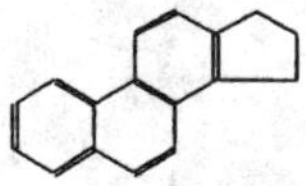

4757

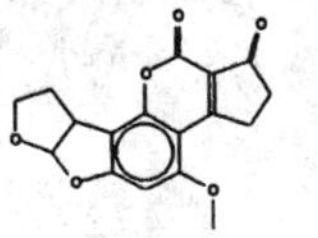

4758

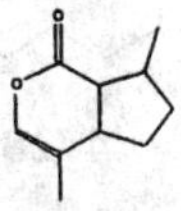

4759

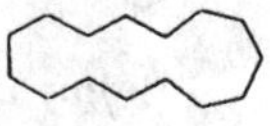

4760

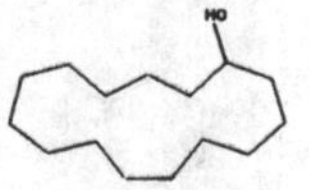

4761

4762

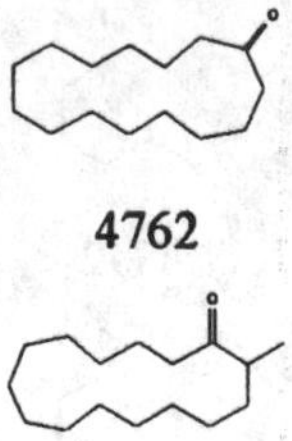

4763

4764

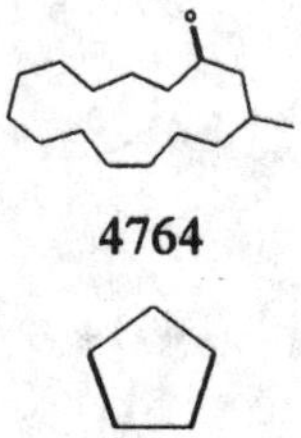

4765

4766

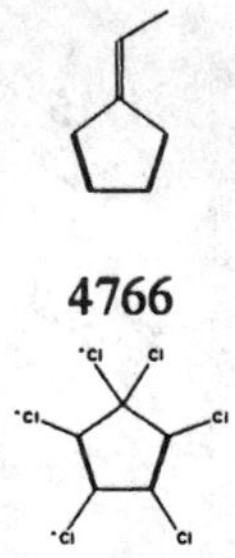

4767

4768

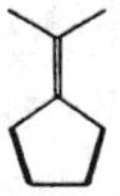

4770

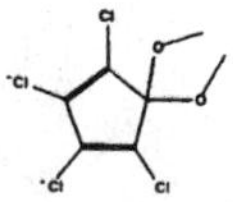

4771

4772

4773

4774

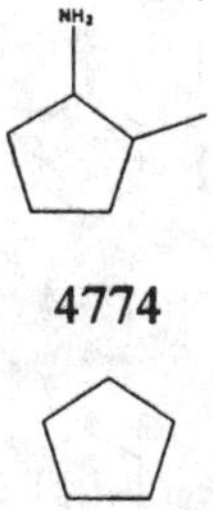

4775

4776

4777

4778

4779

4780

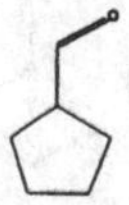

4781

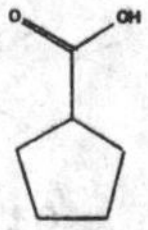

4782

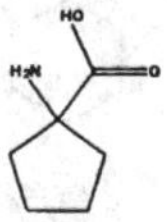

4783

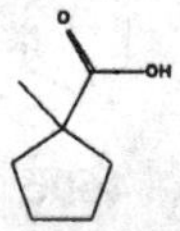

4784

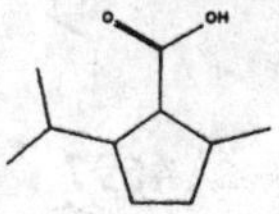

4785

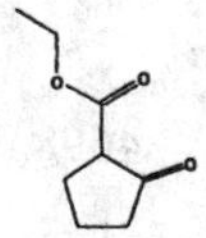

4786

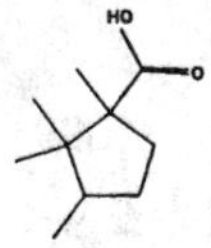

4789

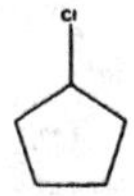

4790

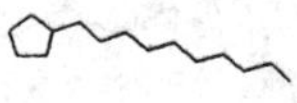

In numeric order as they occur in the Table of Physical Constants of Organic Compounds

4791

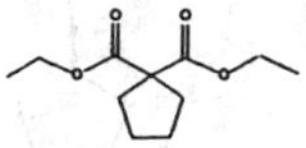

4792

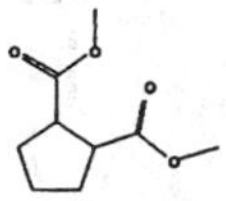

4793

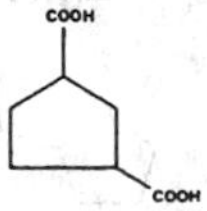

4794

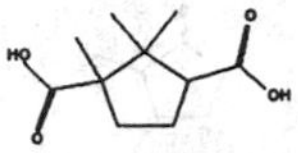

4796

4797

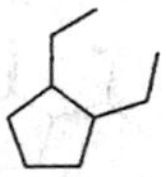

4798

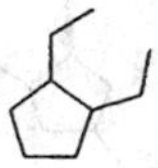

4799

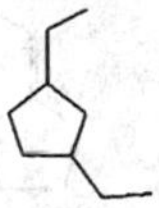

4800

4801

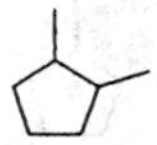

4802

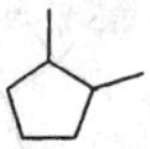

4803

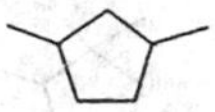

4804

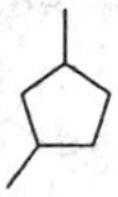

4805

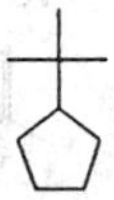

4806

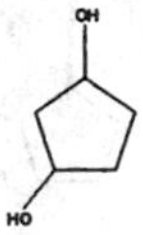

4807

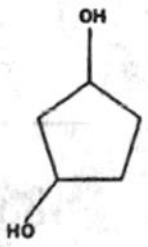

4808

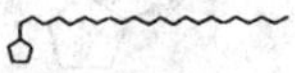

4809

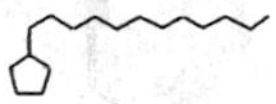

4810

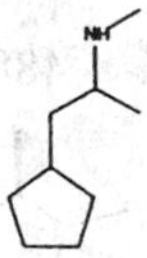

4811

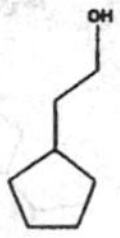

4812

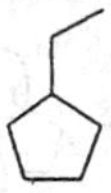

4813

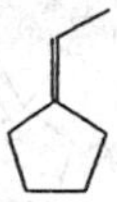

4814

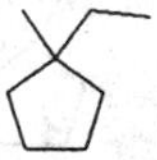

4815

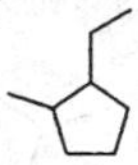

4816

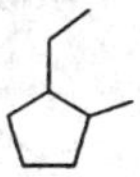

4817

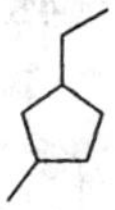

4818

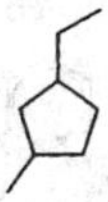

4819

4820

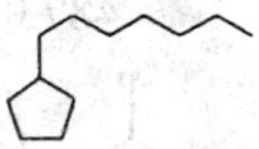

4821

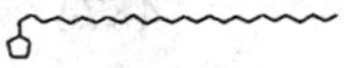

4822

4823

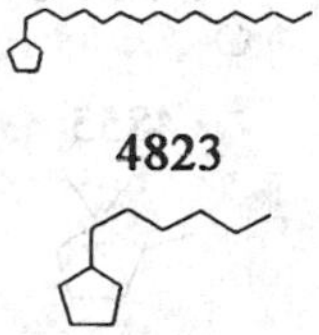

4824

4825

4826

4827

4828

4829

4830

4831

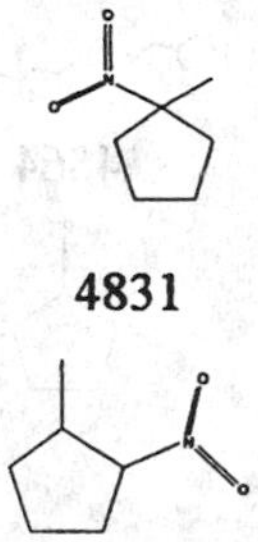

4832

4833

4834

4835

4836

4837

4838

4839

4840

4841

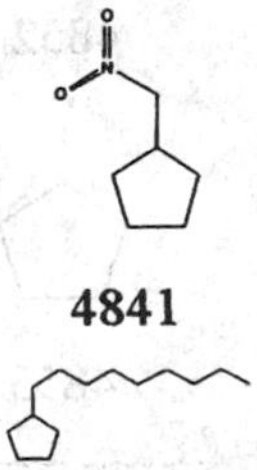

In numeric order as they occur in the Table of Physical Constants of Organic Compounds

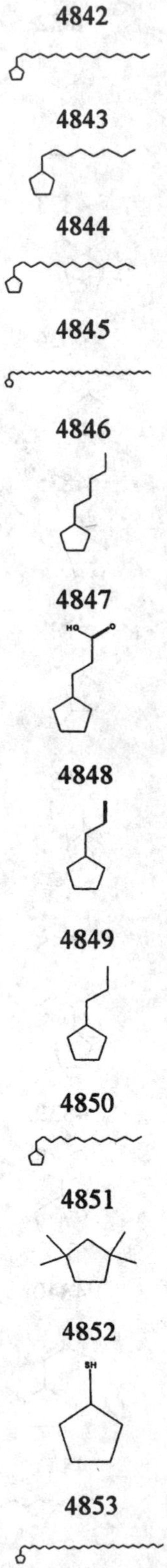

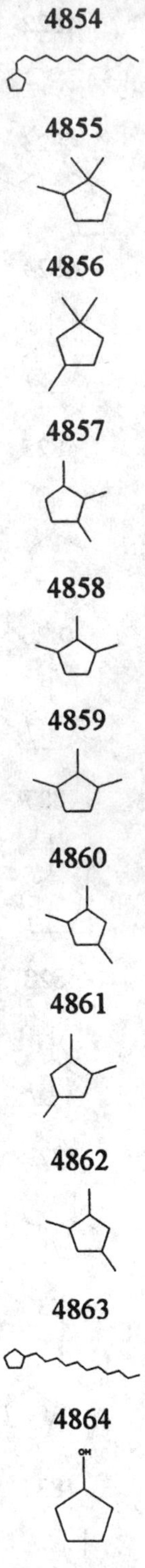

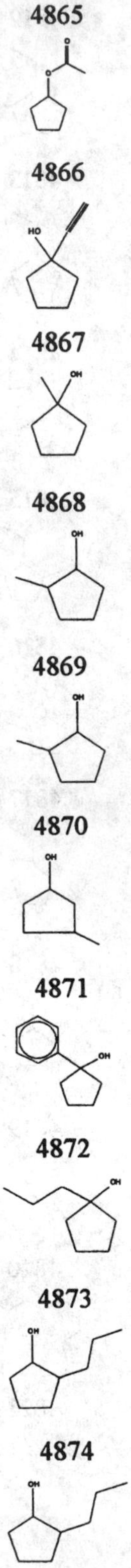

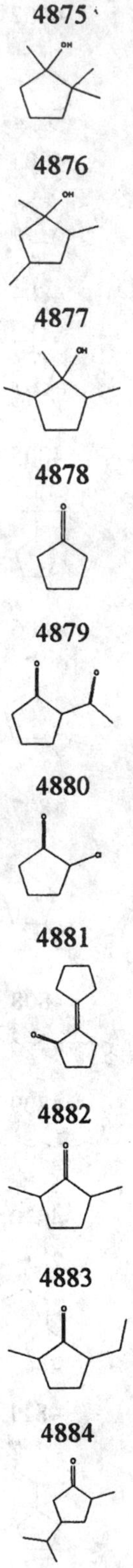

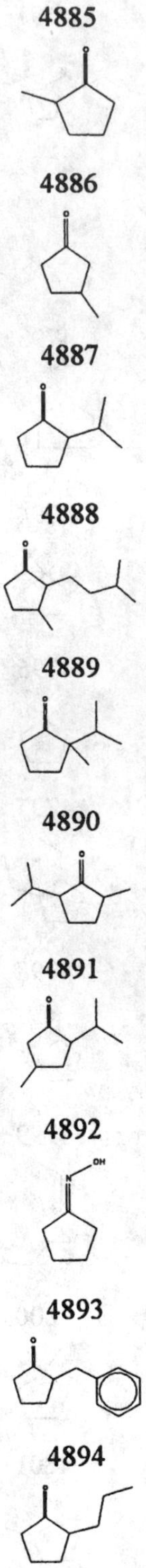

STRUCTURAL FORMULAS OF ORGANIC COMPOUNDS (continued)

In numeric order as they occur in the Table of Physical Constants of Organic Compounds

4895

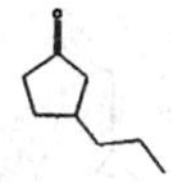

4896

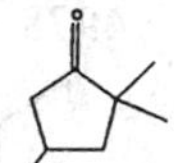

4897

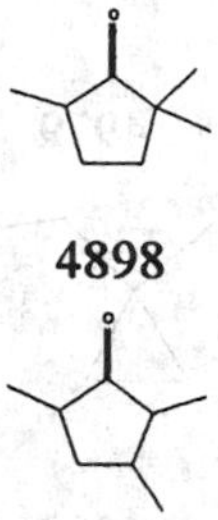

4898

4899

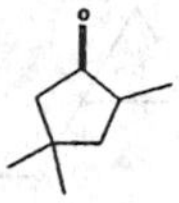

4900

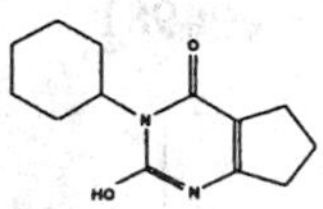

4901

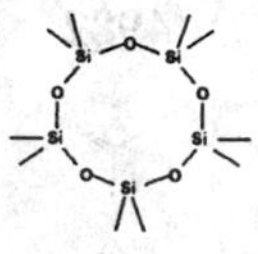

4902

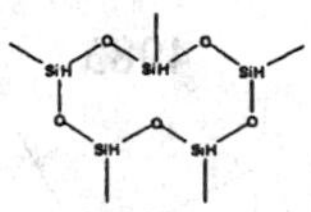

4903

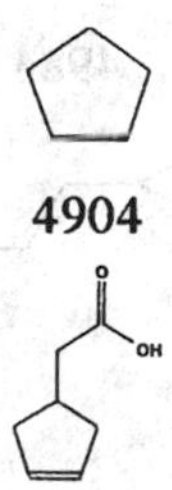

4904

4905

4906

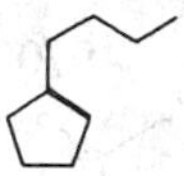

4907

4908

4909

4910

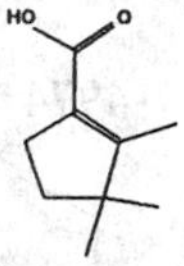

4911

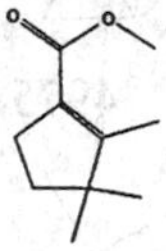

4912

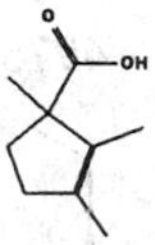

4913

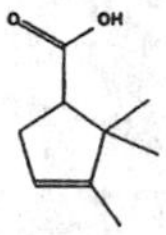

4914

4915

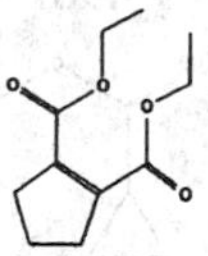

4916

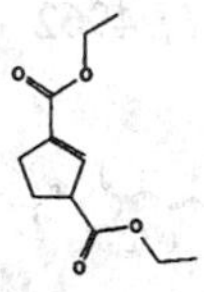

4917

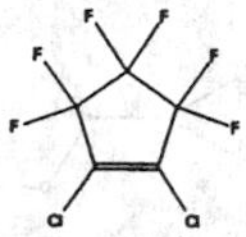

4918

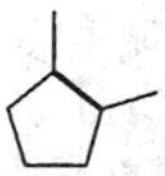

4919

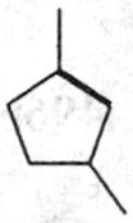

4920

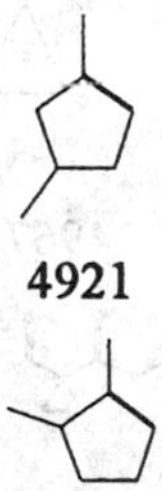

4921

4922

4923

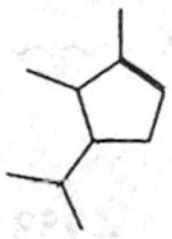

4924

4925

4926

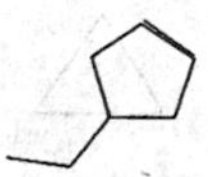

4927

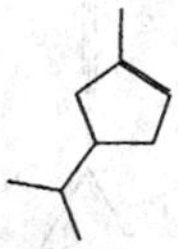

4928

4929

4930

4931

4932

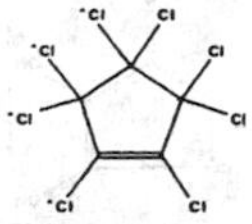

4933

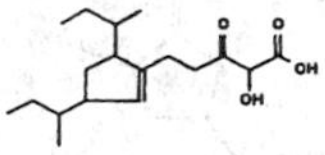

4934

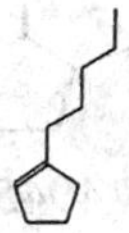

4935

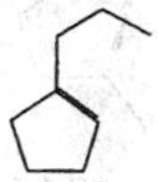

4936

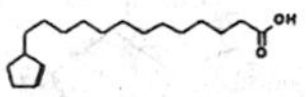

4937

4938

4939

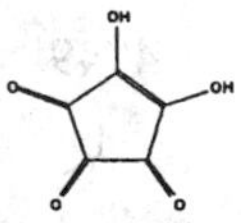

4940

4941

STRUCTURAL FORMULAS OF ORGANIC COMPOUNDS (continued)

In numeric order as they occur in the Table of Physical Constants of Organic Compounds

4942

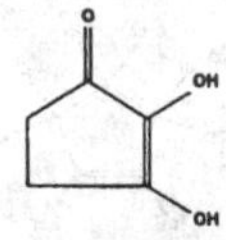

4943

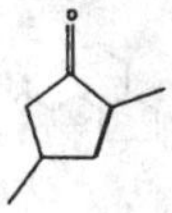

4944

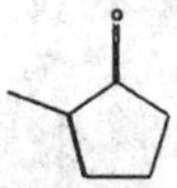

4945

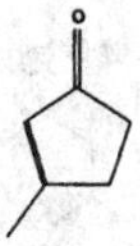

4946

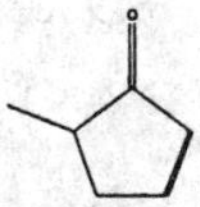

4947

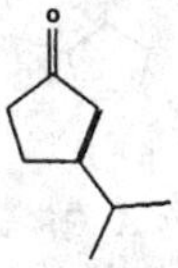

4948

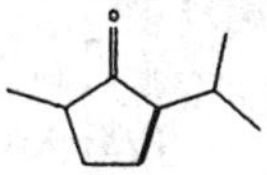

4949

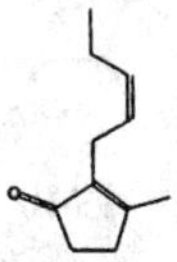

4950

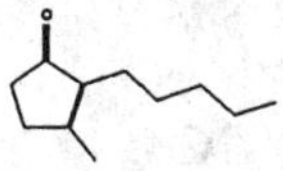

4951

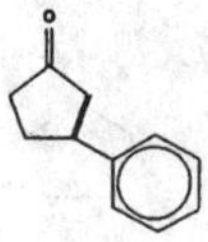

4952

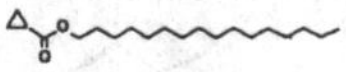

4953

4954

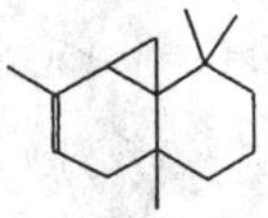

4955

4956

4957

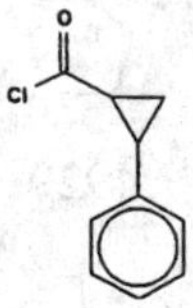

4958

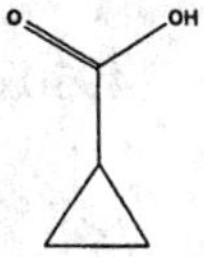

4959

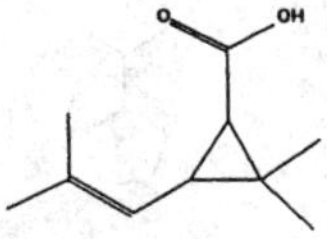

4960

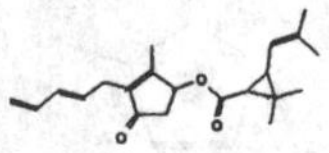

4961

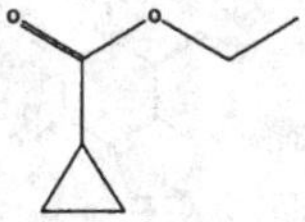

4962

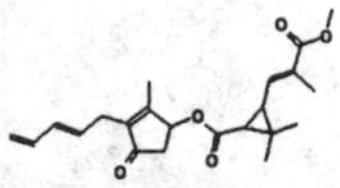

4963

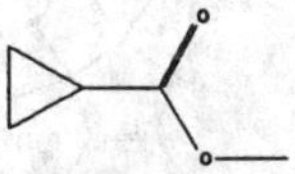

4964

4966

4967

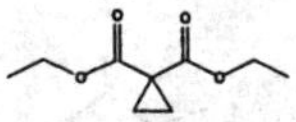

4968

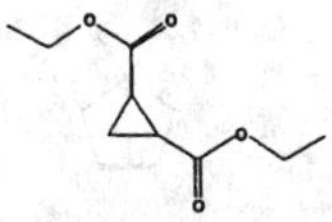

4969

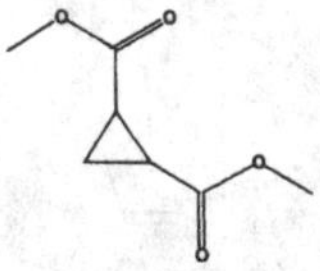

4970

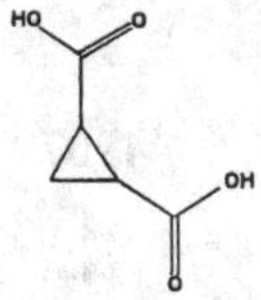

4971

4972

4973

4974

4975

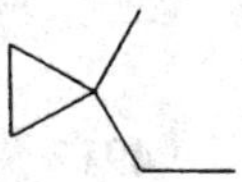

4976

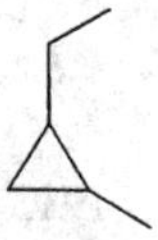

4977

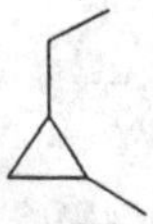

4978

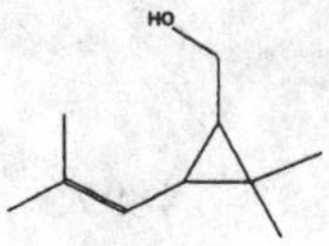

4979

4980

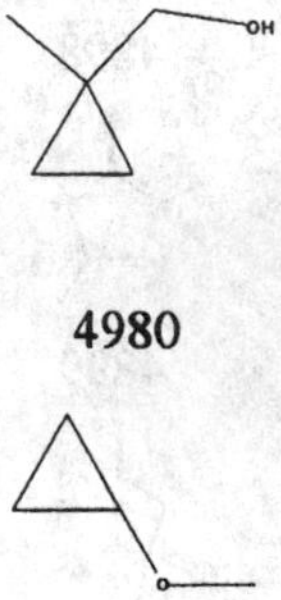

4981

4982

4983

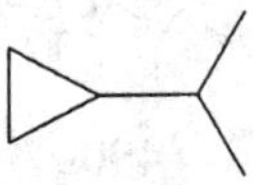

4984

STRUCTURAL FORMULAS OF ORGANIC COMPOUNDS (continued)

In numeric order as they occur in the Table of Physical Constants of Organic Compounds

4985

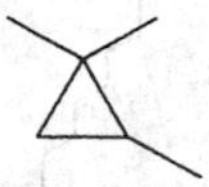

4986

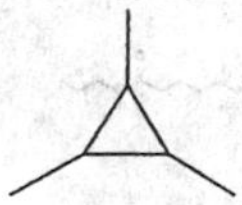

4987

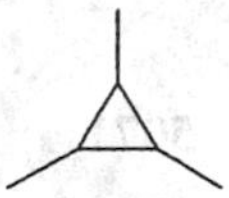

4988

4989

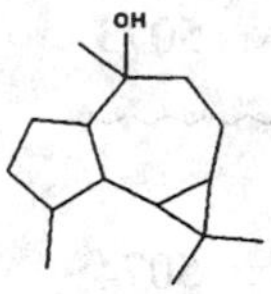

4990

4991

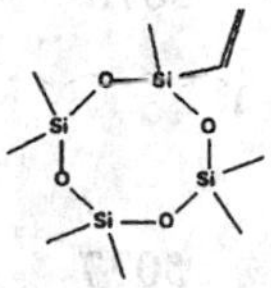

4992

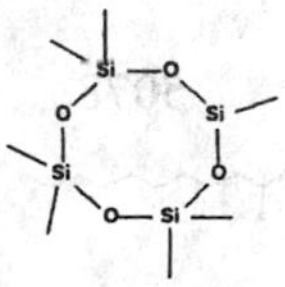

4993

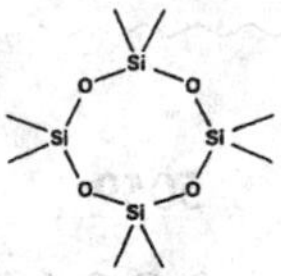

4994

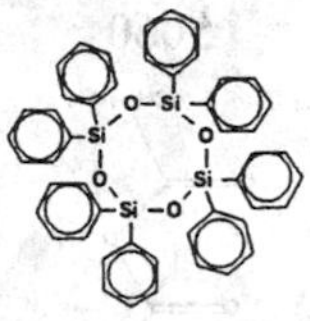

4995

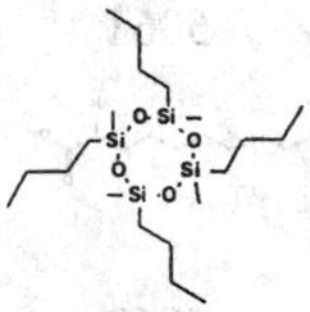

4996

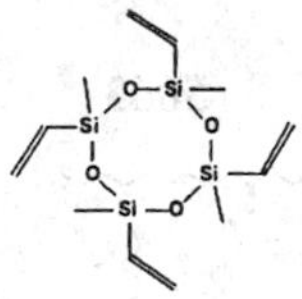

4997

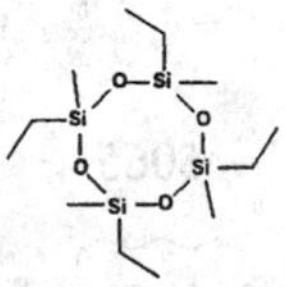

4998

4999

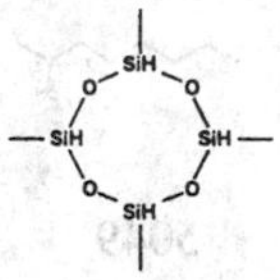

5000

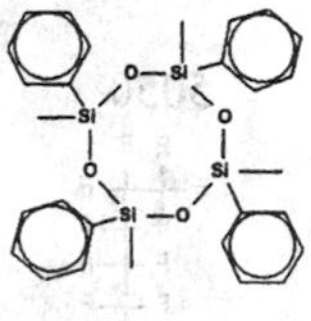

5001

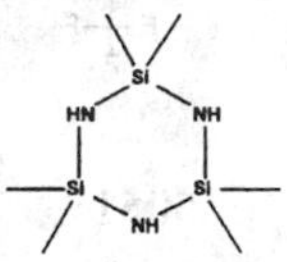

5002

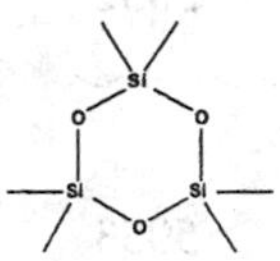

5003

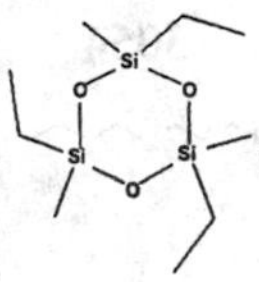

5004

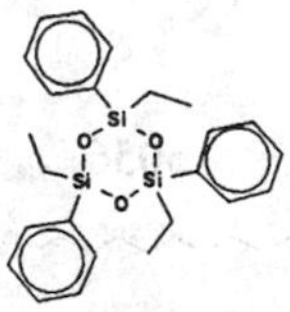

5005

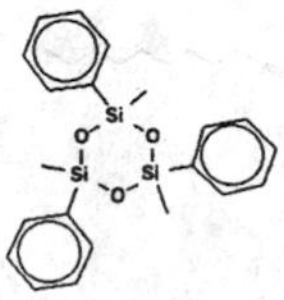

5006

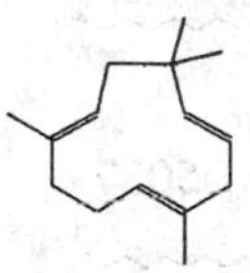

5008

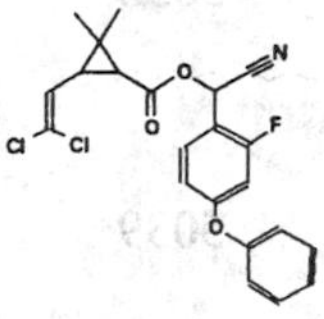

5009

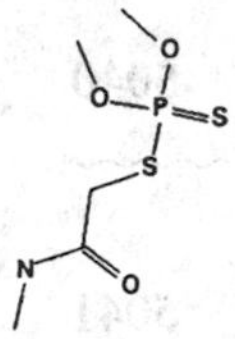

5010

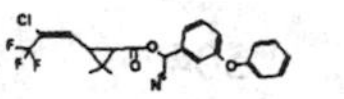

5011

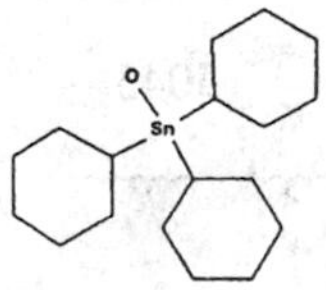

5012

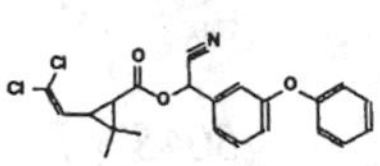

5013

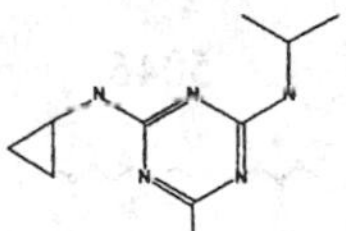

5014

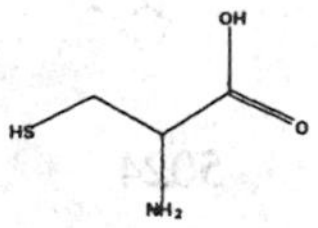

5015

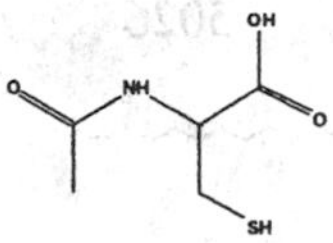

5016

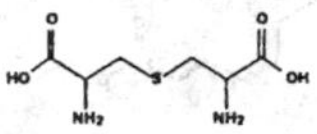

5017

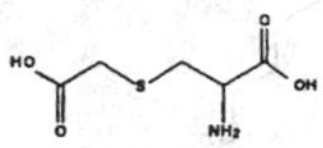

5018

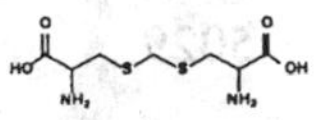

5020

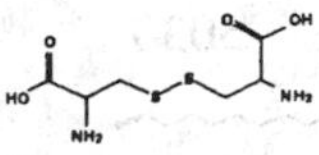

5021

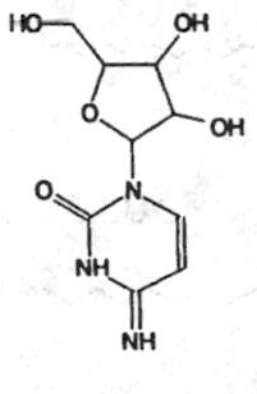

5022

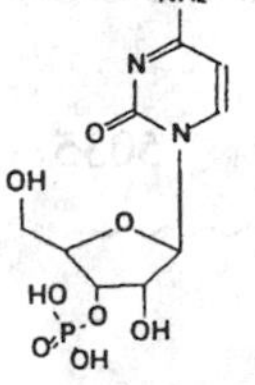

STRUCTURAL FORMULAS OF ORGANIC COMPOUNDS (continued)

In numeric order as they occur in the Table of Physical Constants of Organic Compounds

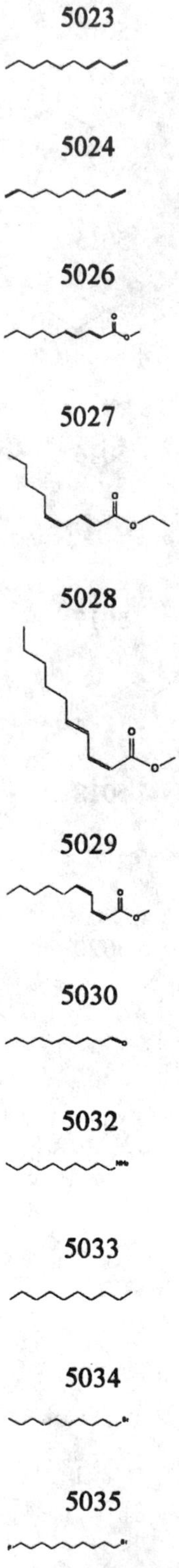

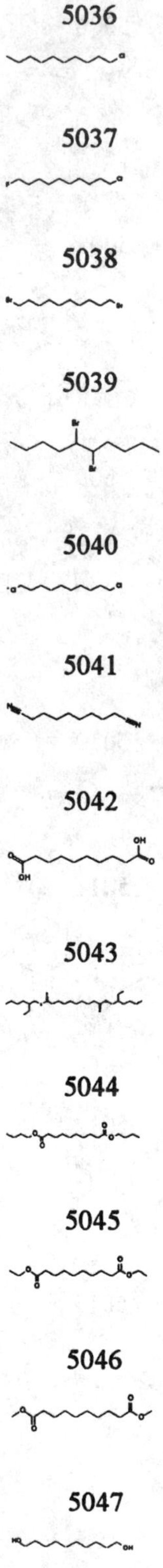

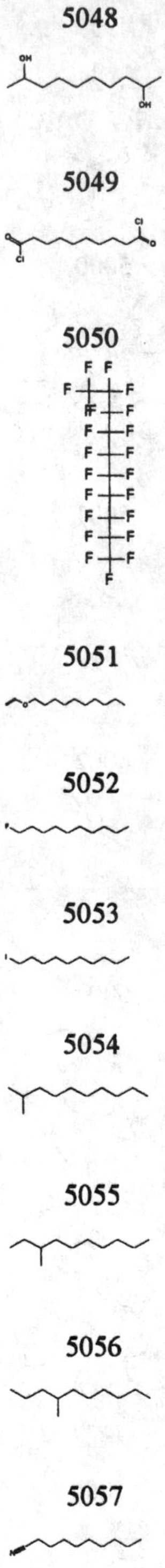

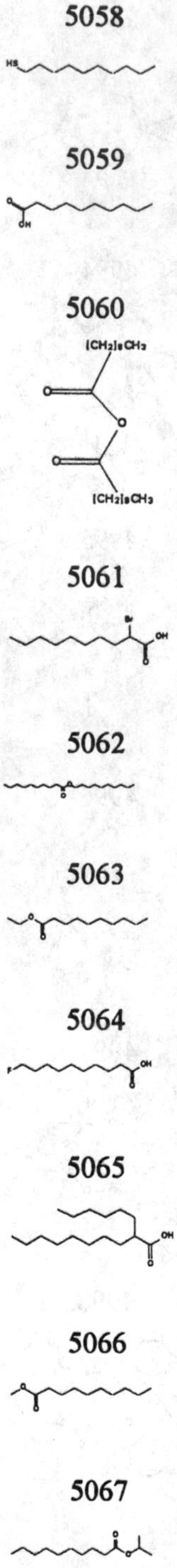

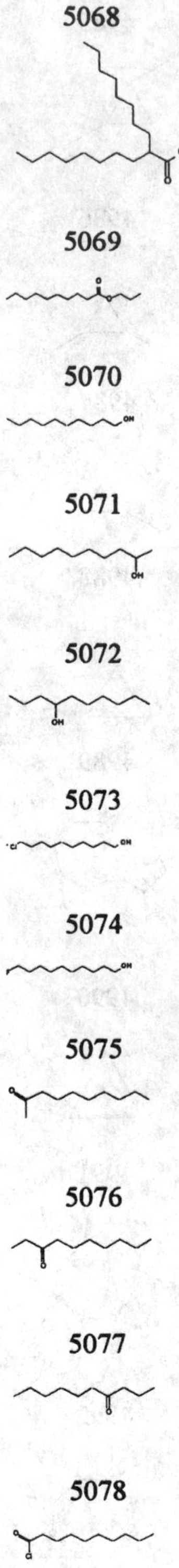

5080

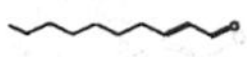

5081

5082

5083

5084

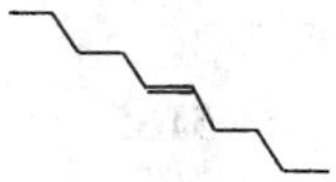

5085

5086

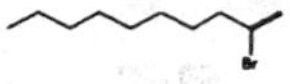

5087

5088

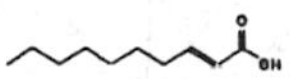

5089

5090

5091

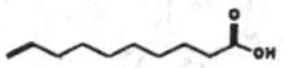

5092

5093

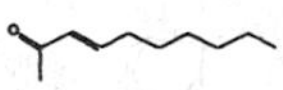

5094

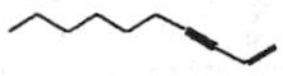

5095

5097

5098

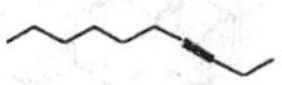

5099

5100

5101

5102

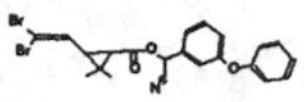

5103

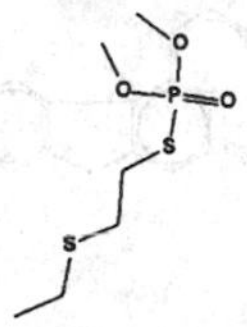

5104

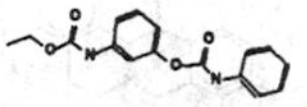

5105

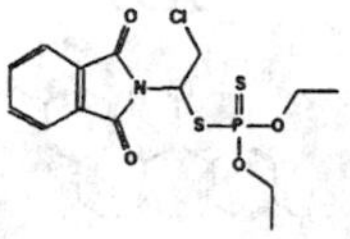

5106

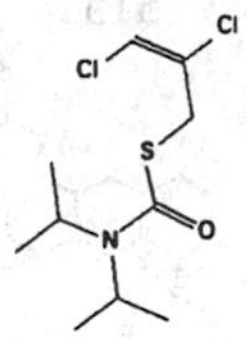

5109

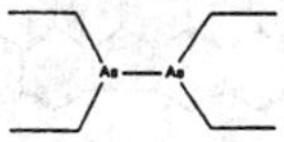

5110

5111

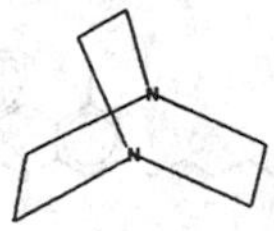

5112

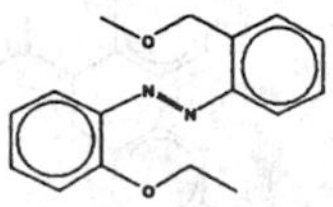

5113

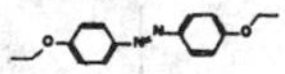

5114

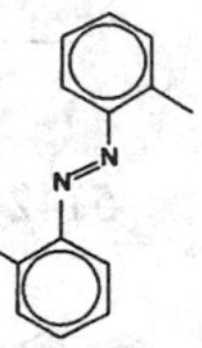

5115

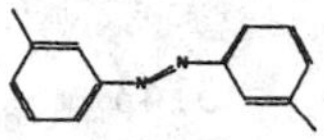

5116

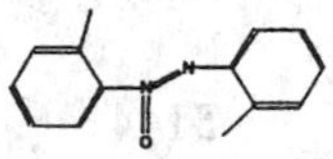

5118

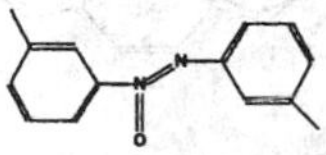

5120

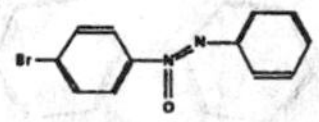

5121

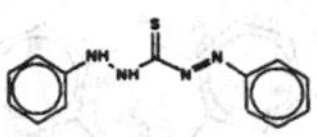

5122

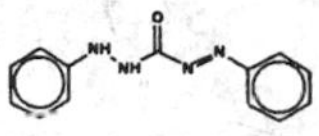

5123

5124, 5125, 5126, 5127, 5128, 5129, 5130, 5131

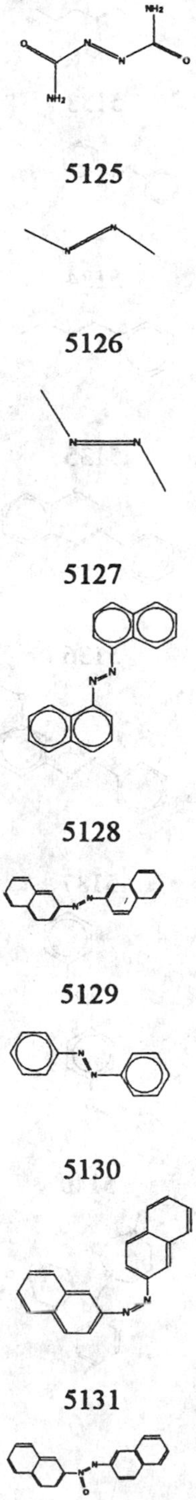

5132

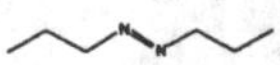

5133

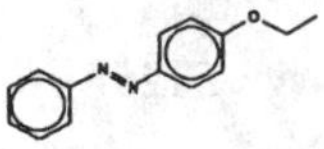

5134

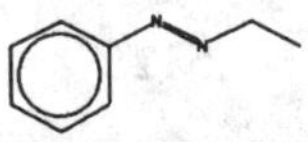

5135

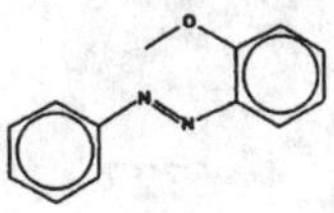

5136

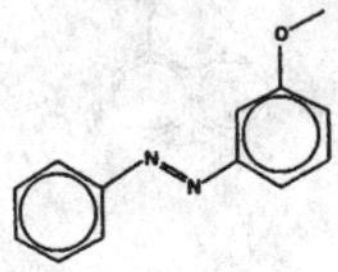

5137

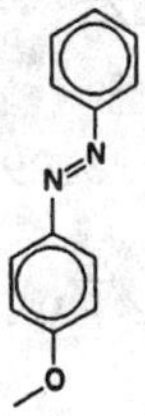

5140

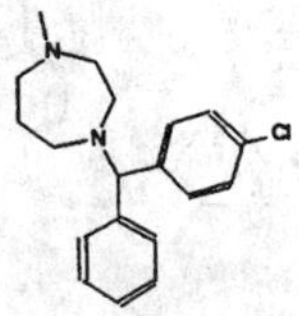

5141

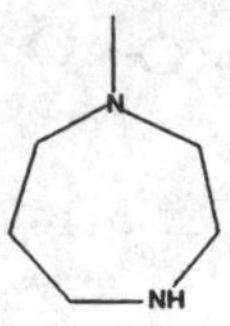

5142

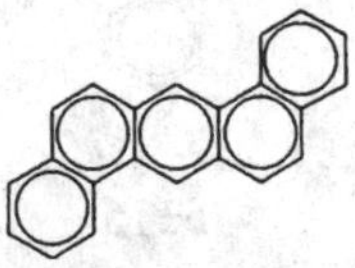

5143

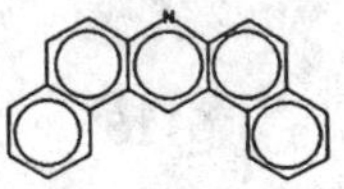

5144

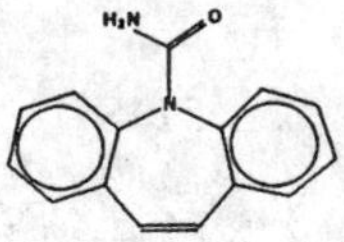

5145

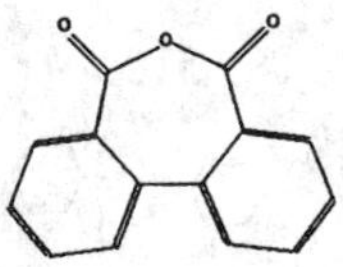

5146

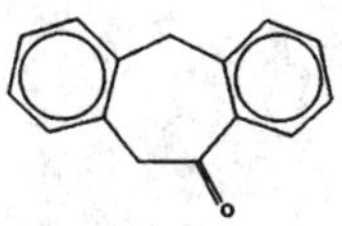

5147

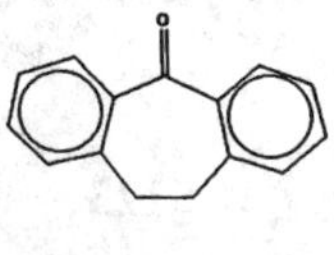

5148

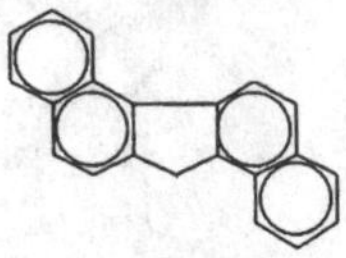

5149

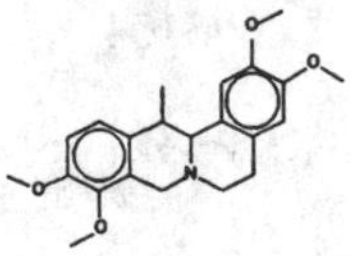

5150

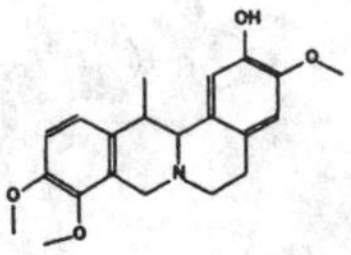

5151

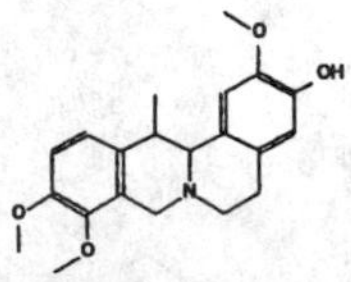

5152

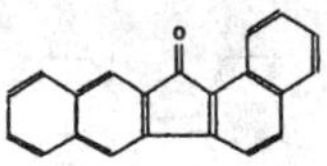

5153

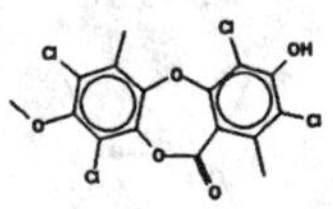

5154

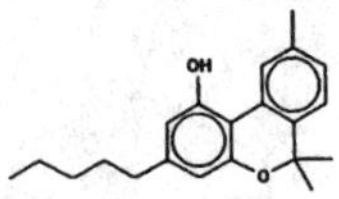

5155

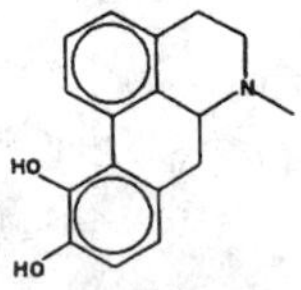

5156

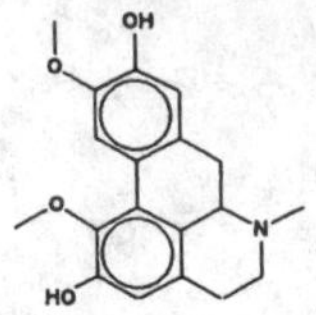

5157

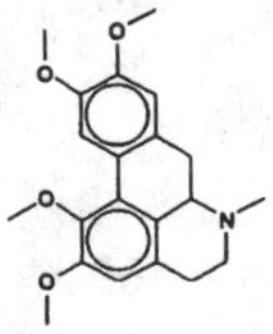

5158

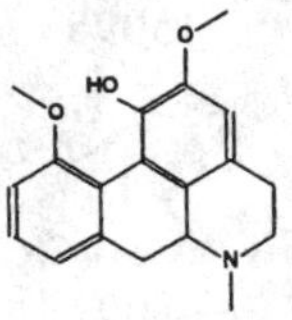

5159

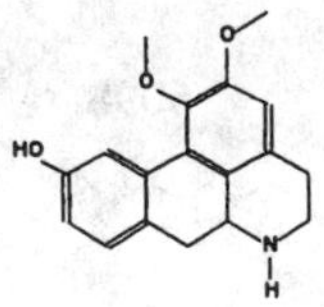

5160

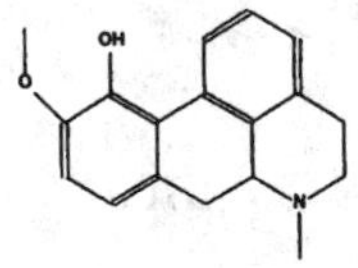

5161

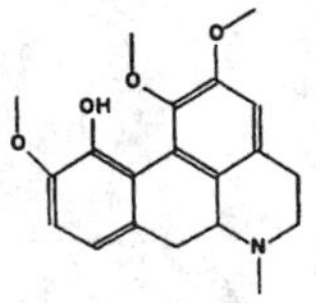

5162

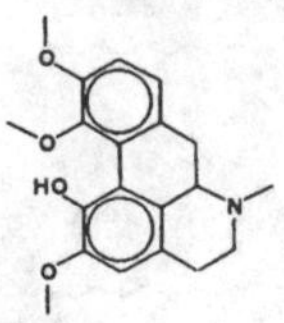

5163

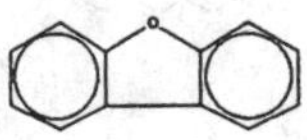

5164

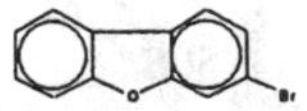

5165

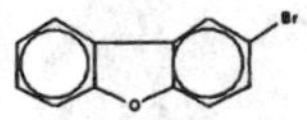

5166

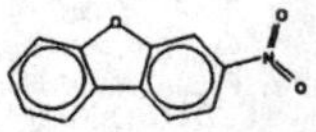

5167

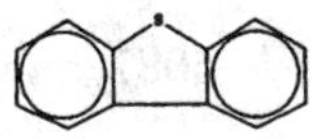

5168

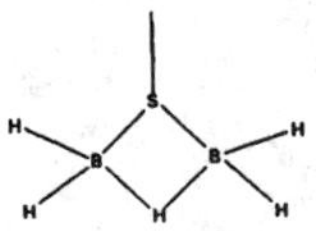

5169

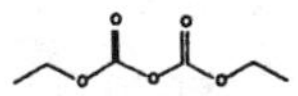

STRUCTURAL FORMULAS OF ORGANIC COMPOUNDS (continued)

In numeric order as they occur in the Table of Physical Constants of Organic Compounds

5170

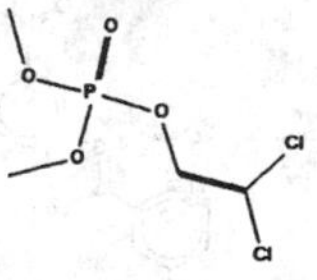

5171

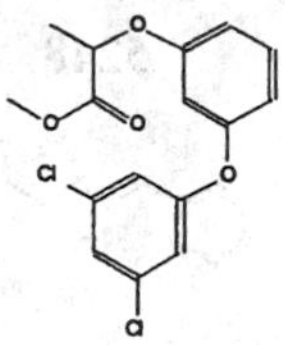

5172

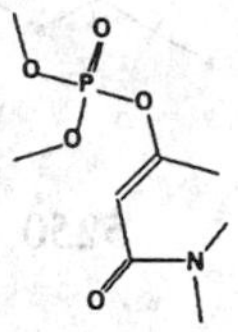

5173

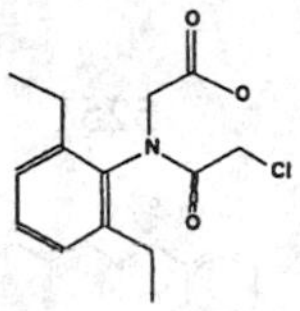

5174

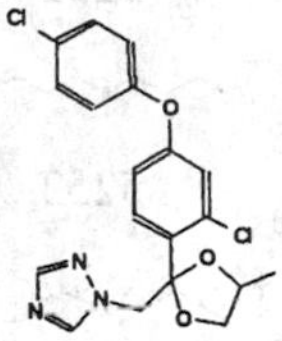

5175

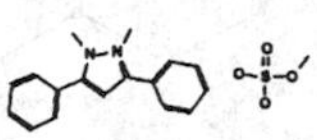

5176

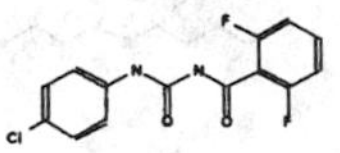

5177

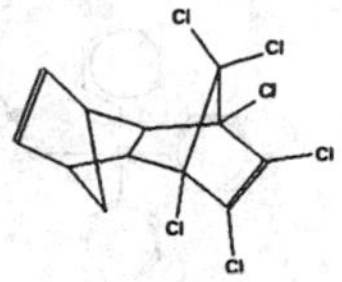

5178

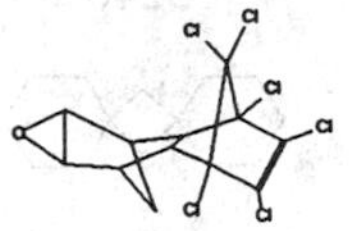

5179

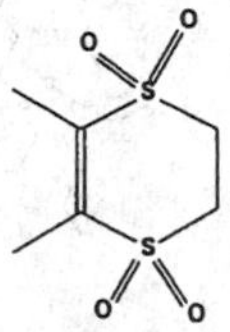

5180

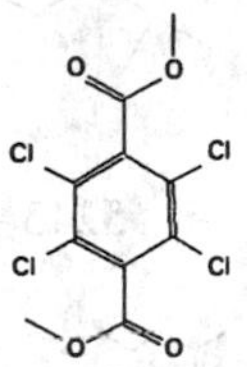

5181

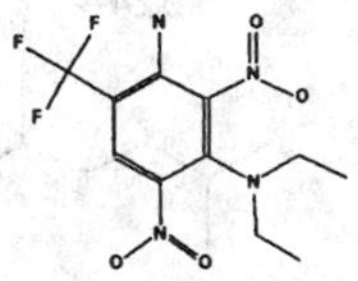

5182

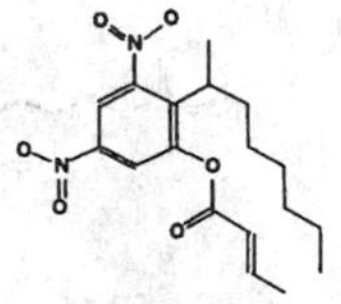

5183

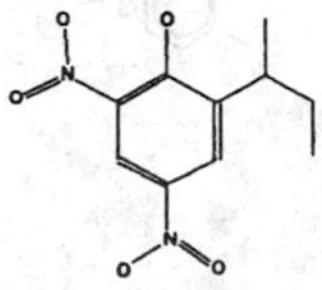

5184

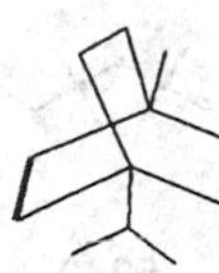

5186

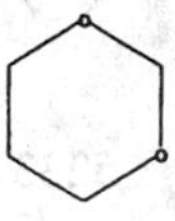

5187

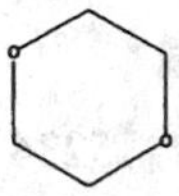

5188

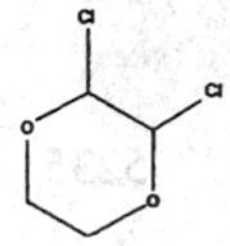

5189

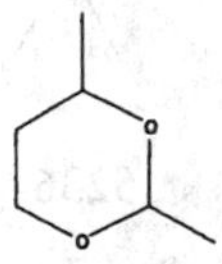

5190

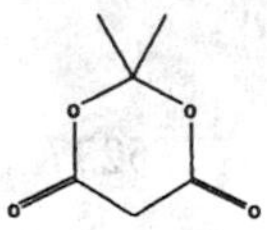

5192

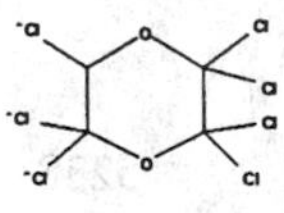

5193

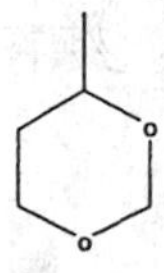

5194

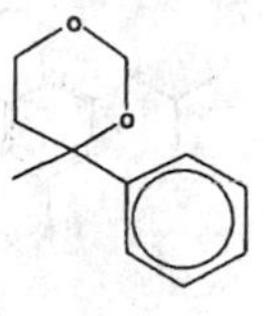

5195

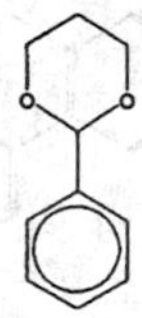

5196

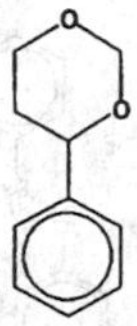

5197

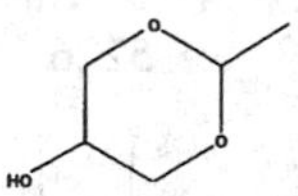

5200

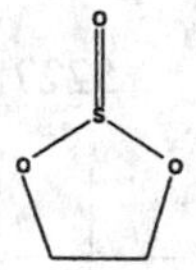

5201

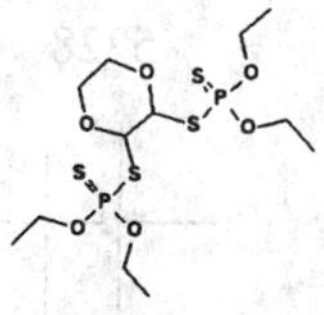

5202

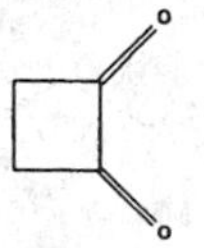

5203

5204

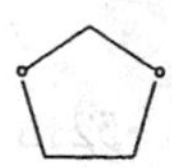

5205

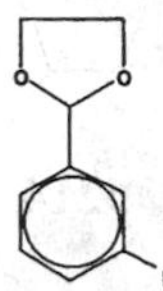

5206

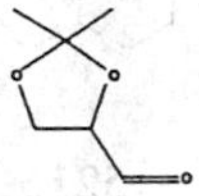

5207

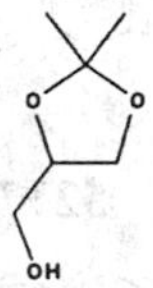

5208

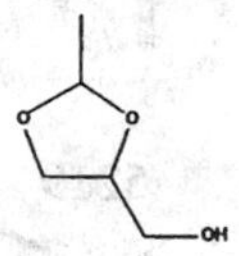

5209

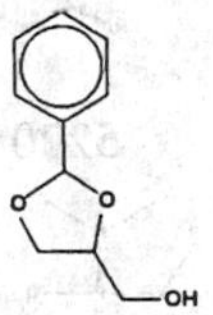

5210

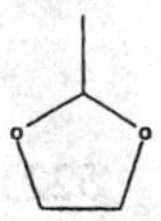

5211

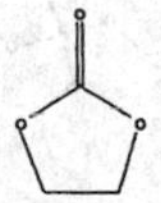

5212

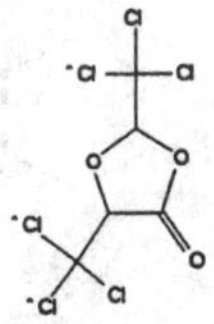

5213

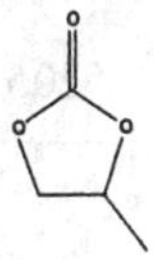

5214

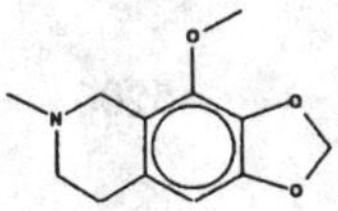

5215

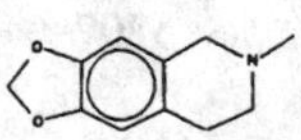

5217

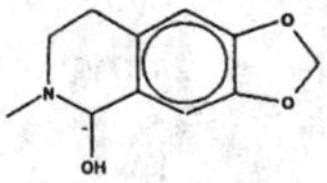

5219

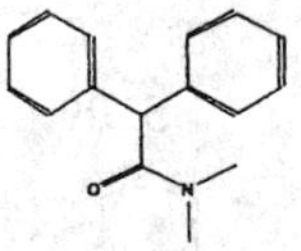

5220

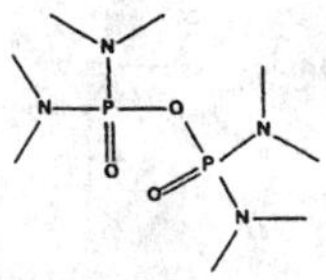

5221

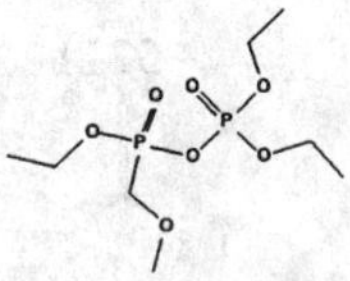

5222

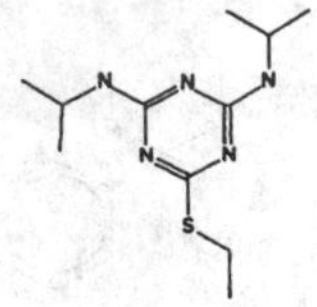

5224

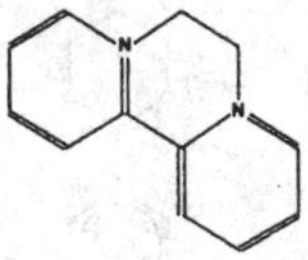

5225

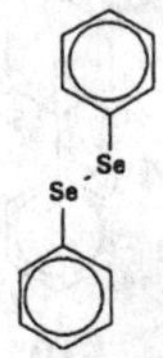

5226

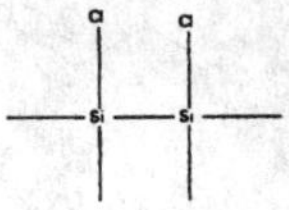

5227

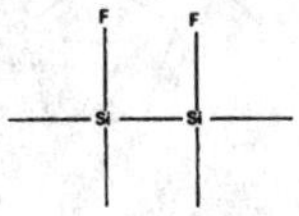

5228

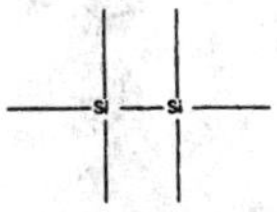

5229

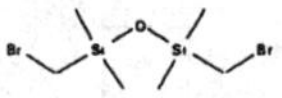

5230

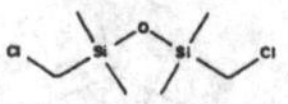

5232

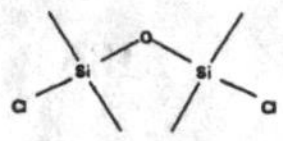

5233

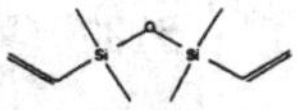

5234

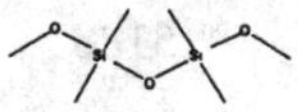

5235

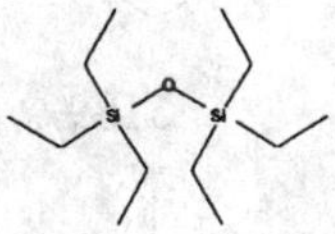

5236

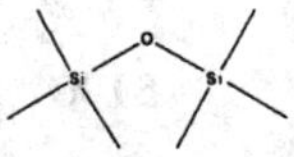

5237

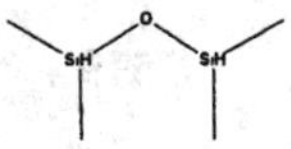

5238

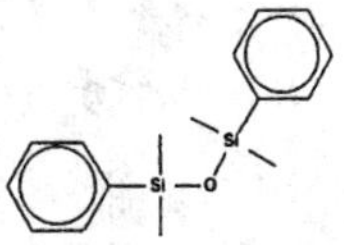

5239

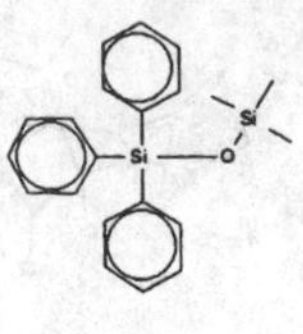

5240

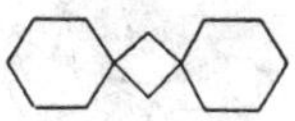

5241

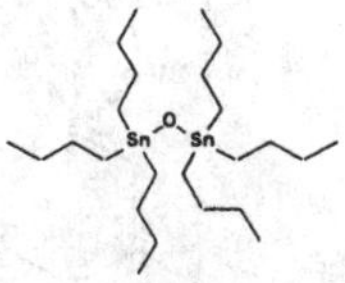

5242

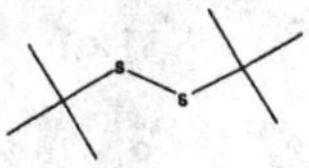

5243

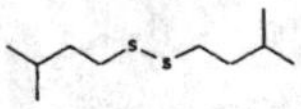

5244

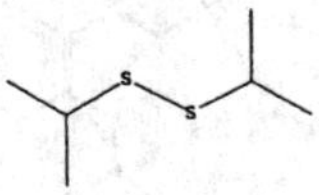

5245

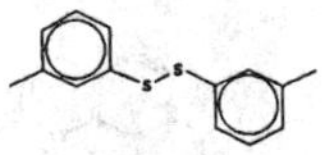

5246

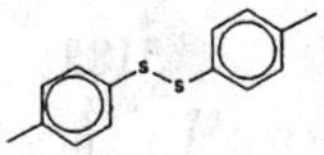

5247

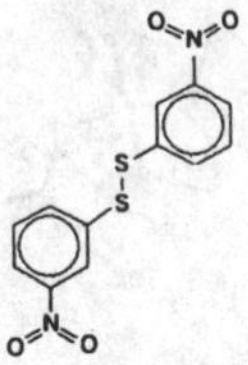

5248

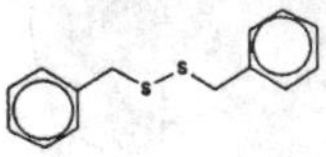

5249

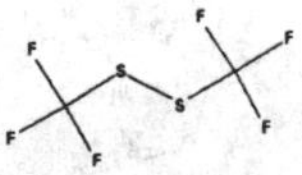

5250

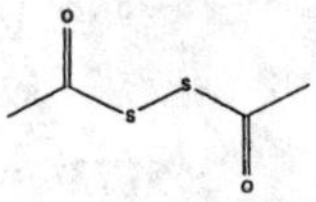

5251

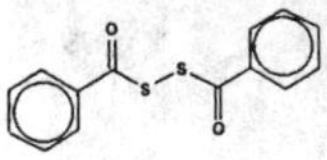

5252

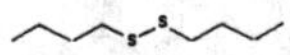

5253

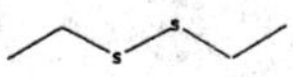

5254

5257

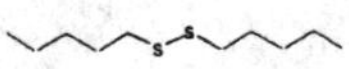

STRUCTURAL FORMULAS OF ORGANIC COMPOUNDS (continued)

In numeric order as they occur in the Table of Physical Constants of Organic Compounds

5258

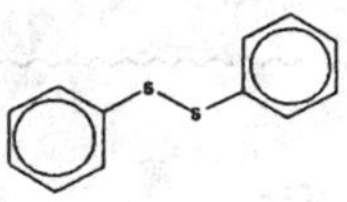

5259

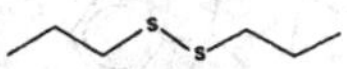

5260

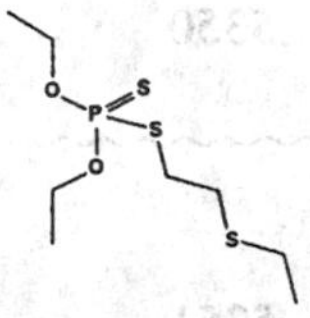

5261

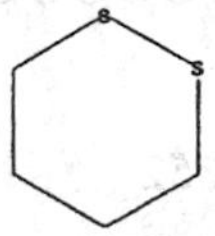

5262

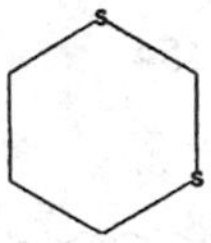

5263

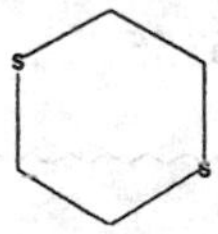

5264

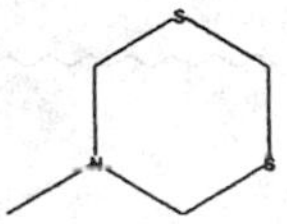

5265

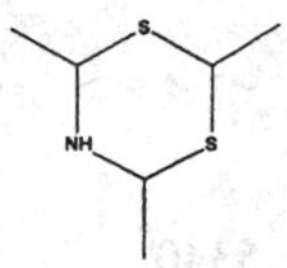

5266

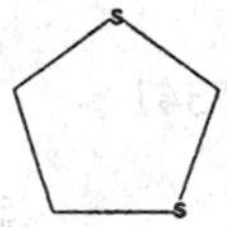

5267

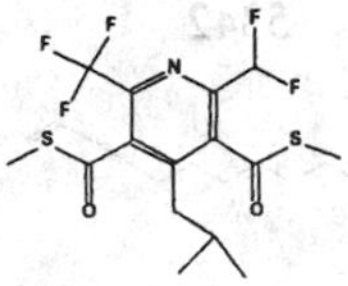

5268

5269

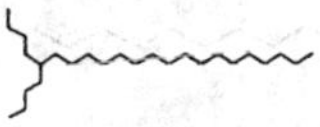

5270

5271

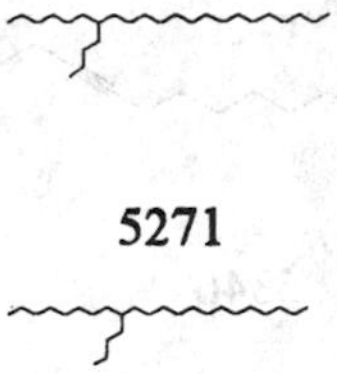

5272

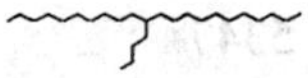

5274

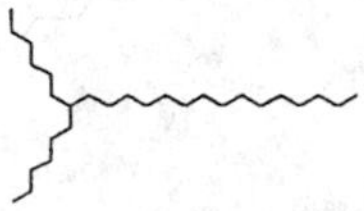

5275

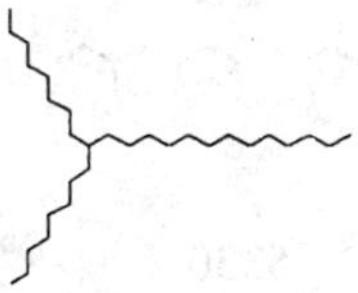

5276

5277

5278

5279

5280

5281

5282

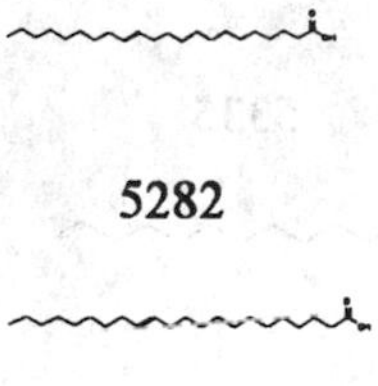

5283

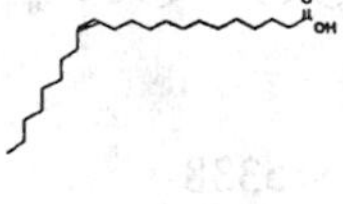

5284

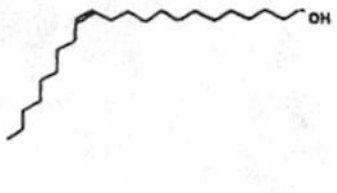

5285

5286

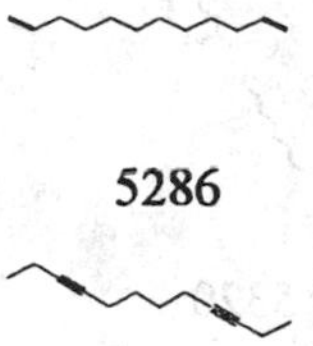

5287

5288

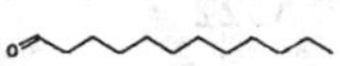

5289

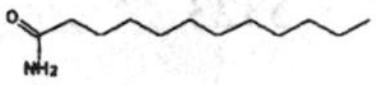

5290

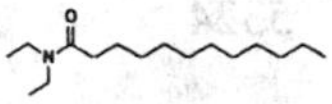

5291

5292

5293

5294

5295

5296

5297

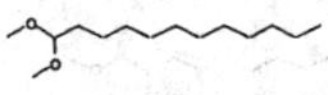

5298

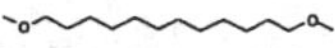

5299

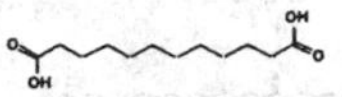

5300

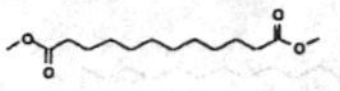

5301

5302

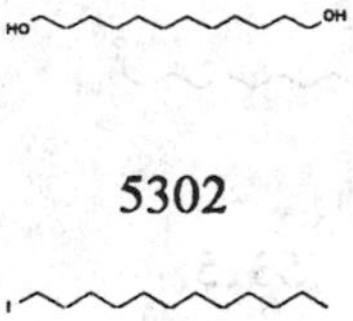

5303

5304

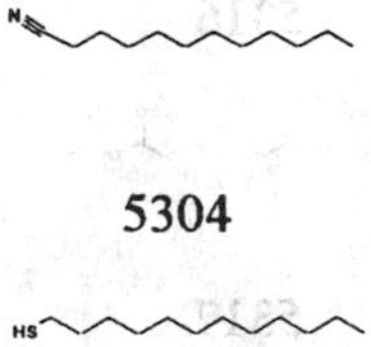

5305

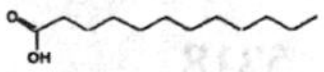

5307

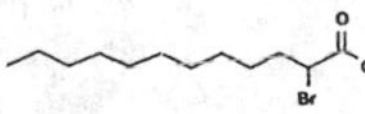

STRUCTURAL FORMULAS OF ORGANIC COMPOUNDS (continued)

In numeric order as they occur in the Table of Physical Constants of Organic Compounds

5308

5309

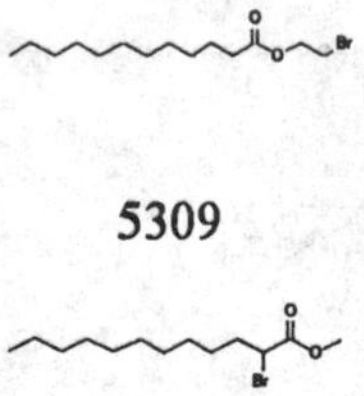

5310

5311

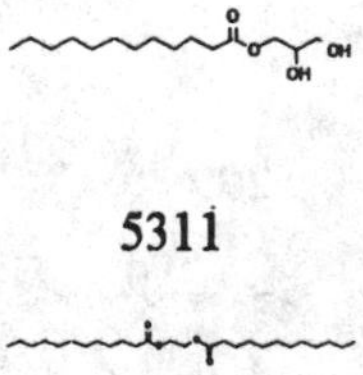

5312

5314

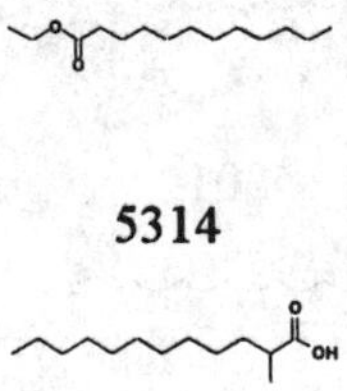

5315

5316

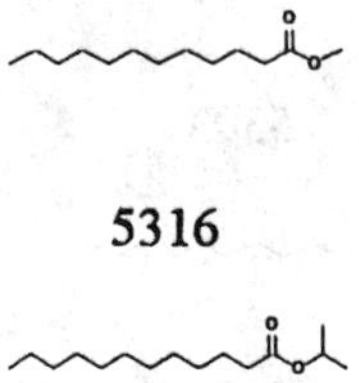

5317

5318

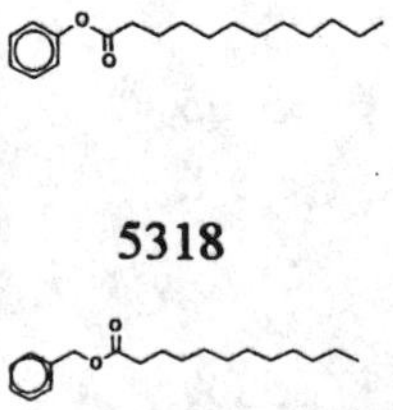

5319

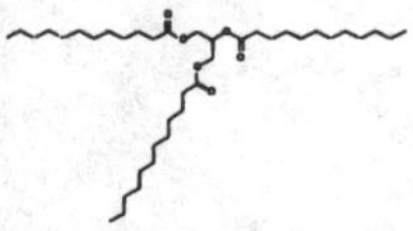

5320

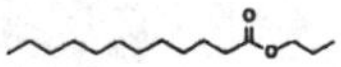

5321

5322

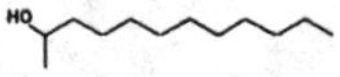

5323

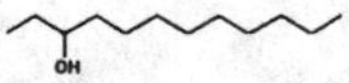

5324

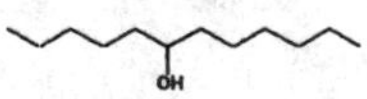

5325

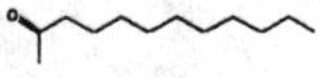

5326

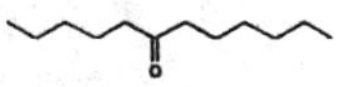

5327

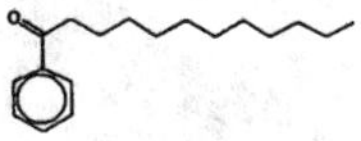

5328

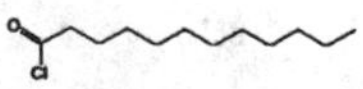

5329

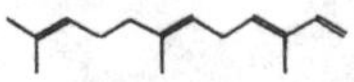

5330

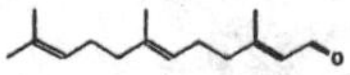

5331

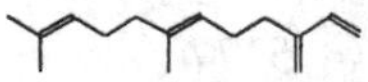

5332

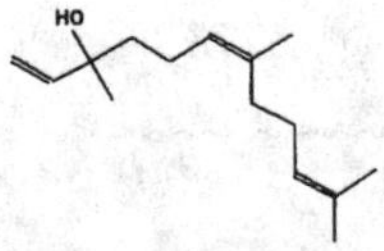

5333

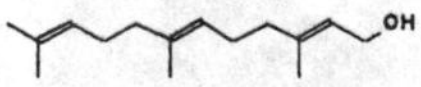

5334

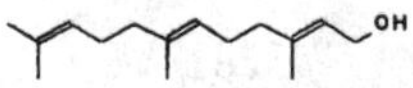

5335

5337

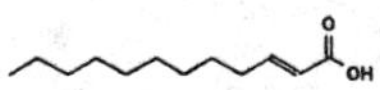

5338

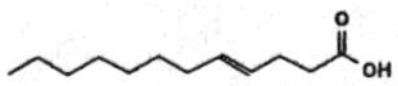

5339

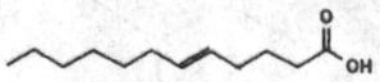

5340

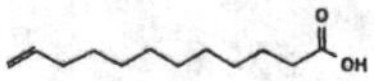

5341

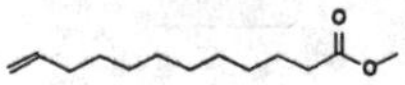

5342

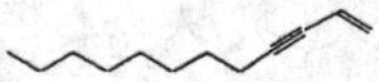

5343

5344

5345

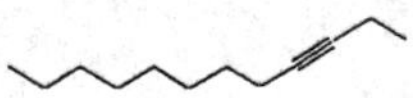

5346

5347

5348

5349

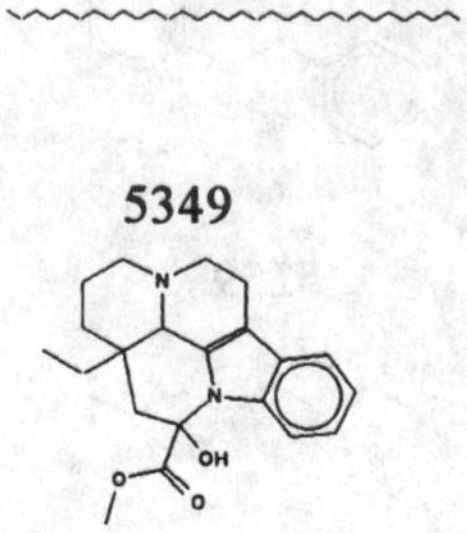

5350

5351

5352

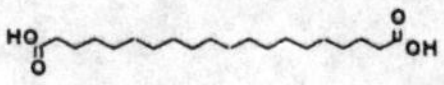

5354

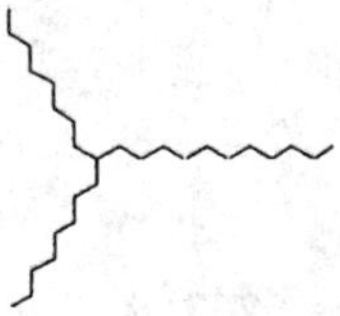

5355

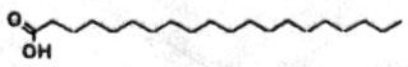

5356

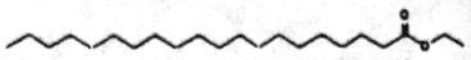

5357

STRUCTURAL FORMULAS OF ORGANIC COMPOUNDS (continued)

In numeric order as they occur in the Table of Physical Constants of Organic Compounds

5358

5359

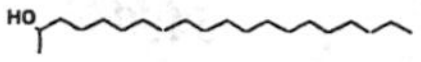

5361

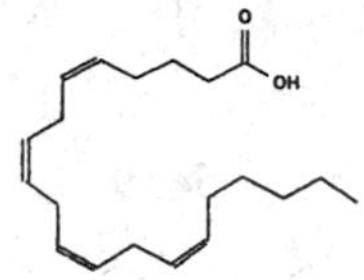

5362

5363

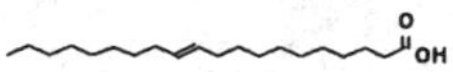

5364

5365

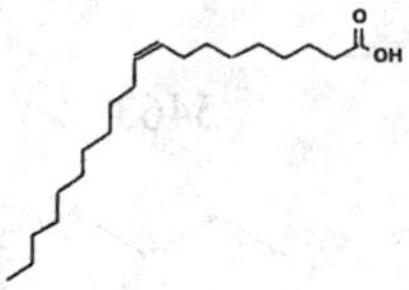

5366

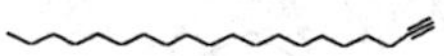

5367

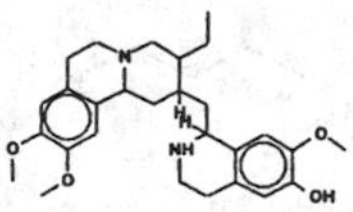

5368

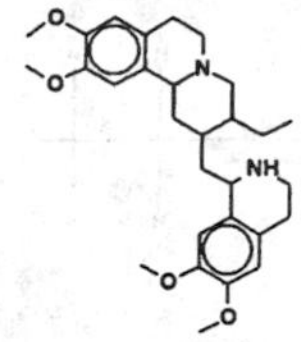

5369

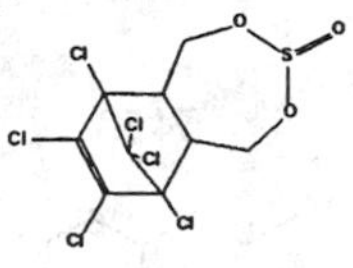

5370

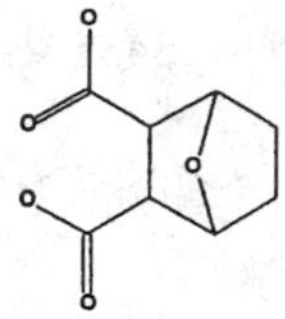

5371

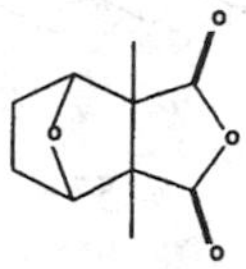

5372

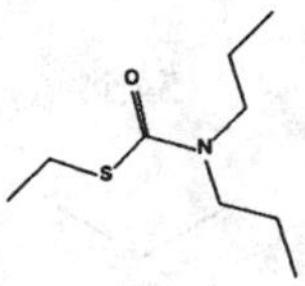

5373

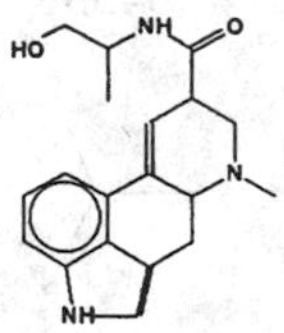

5374

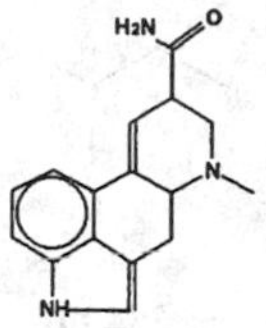

5376

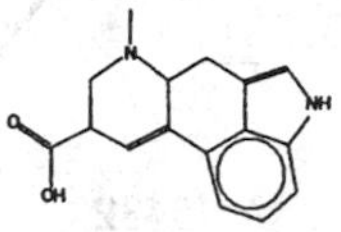

5379

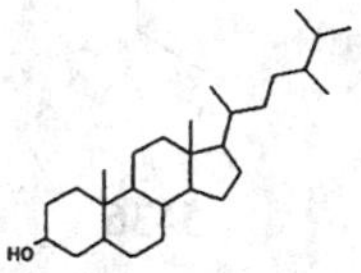

5383

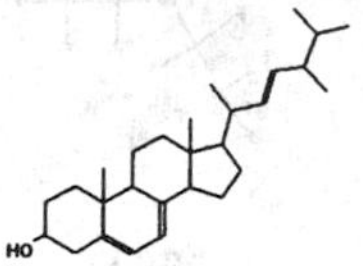

5384

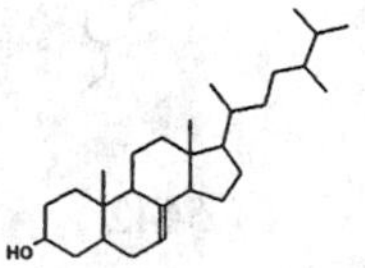

5385

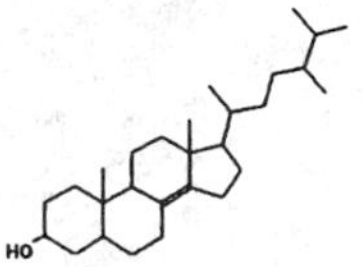

5386

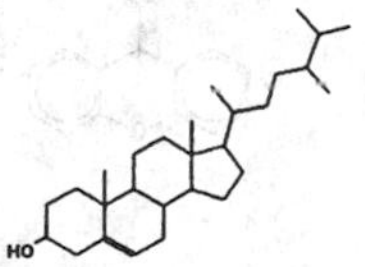

5387

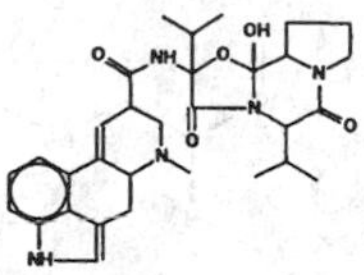

5391

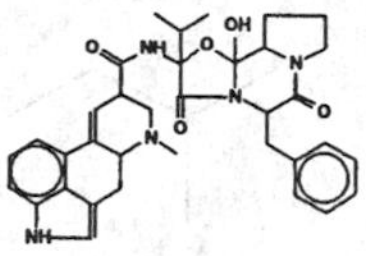

5393

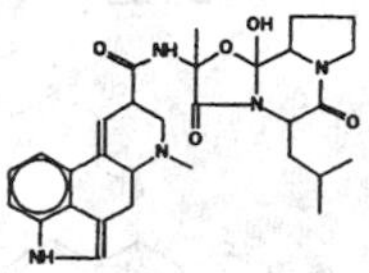

5394

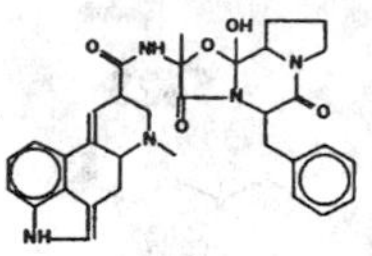

5396

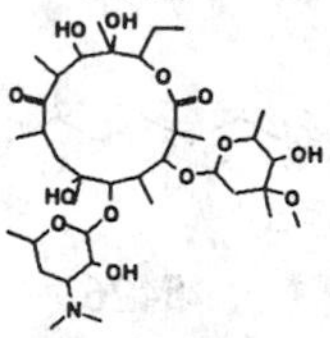

5397

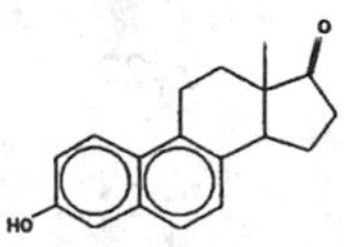

5398

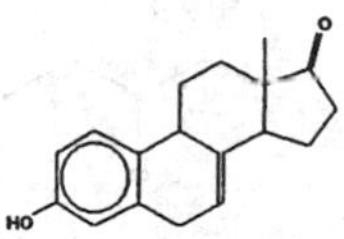

5399

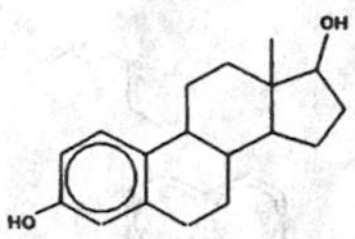

5400

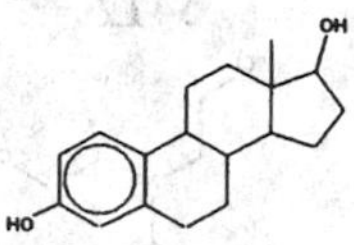

5403

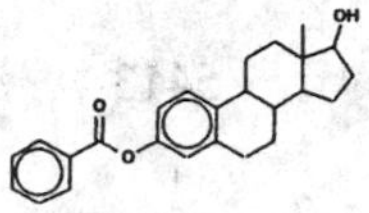

5404

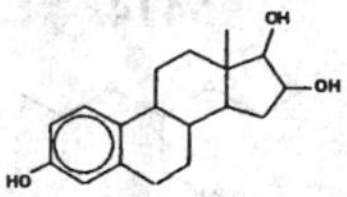

5405

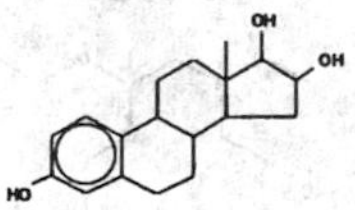

5406

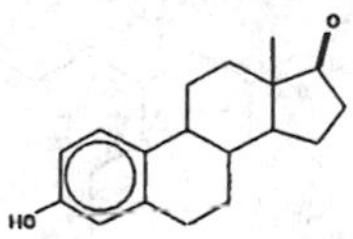

5409

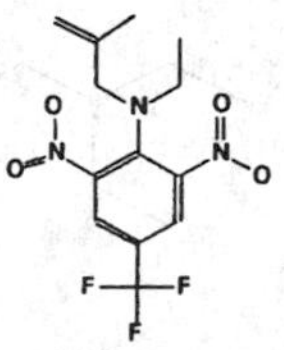

5410

STRUCTURAL FORMULAS OF ORGANIC COMPOUNDS (continued)

In numeric order as they occur in the Table of Physical Constants of Organic Compounds

5411

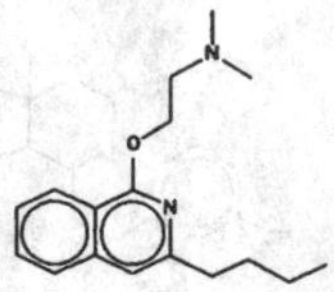

5412

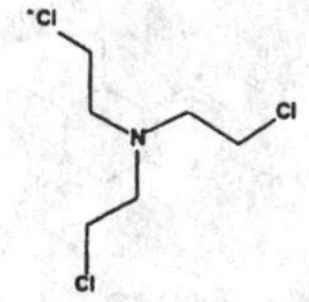

5413

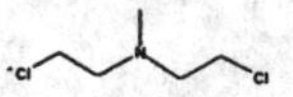

5414

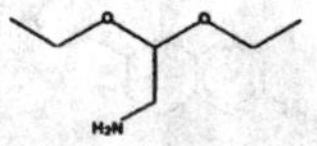

5415

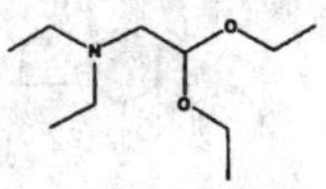

5416

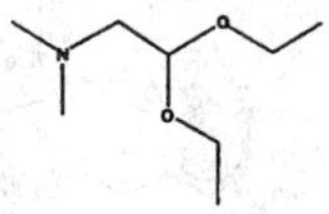

5417

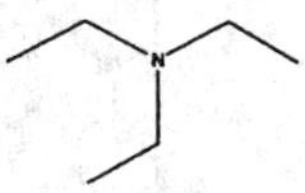

5419

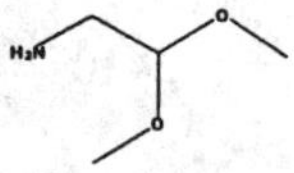

5420

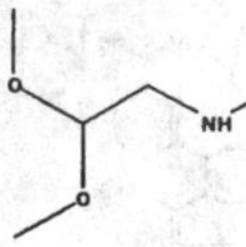

5421

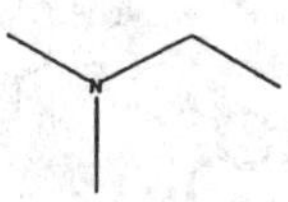

5422

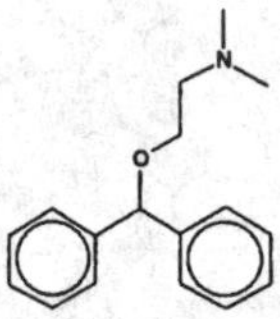

5423

5424

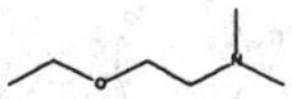

5426

5427

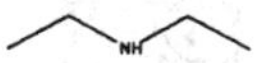

5429

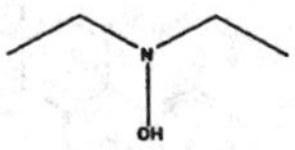

5430

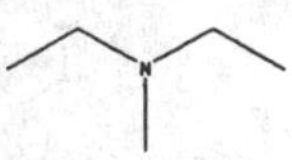

5431

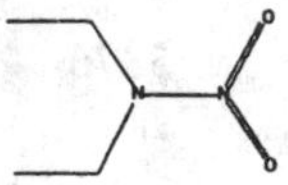

5432

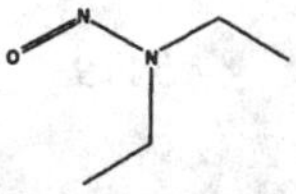

5436

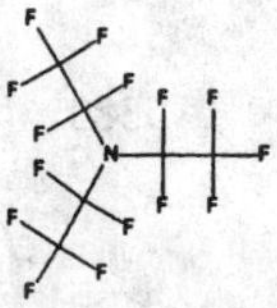

5441

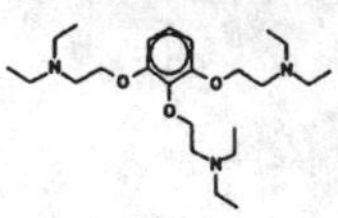

5442

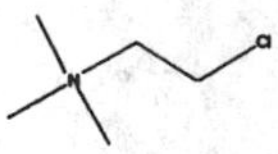

5445

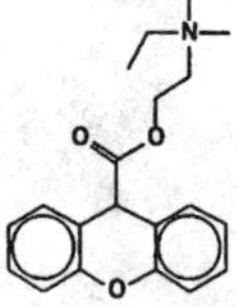

5448

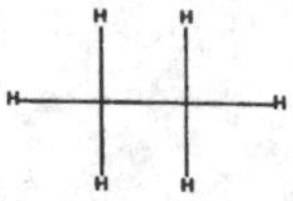

5449

5450

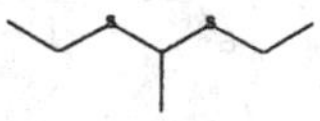

5451

5452

5453

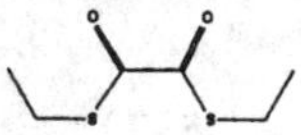

5454

5455

5456

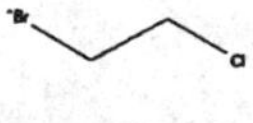

5457

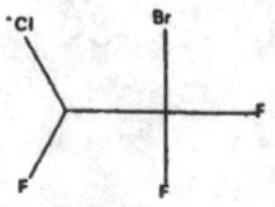

5458

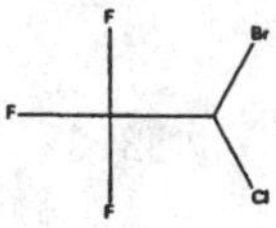

5459

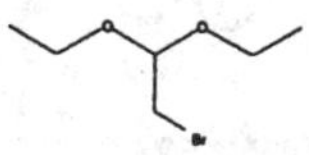

5460

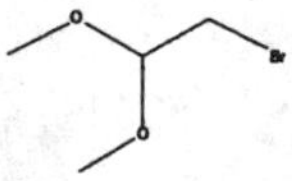

5461

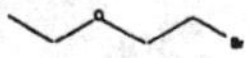

5462

5463

5464

5465

STRUCTURAL FORMULAS OF ORGANIC COMPOUNDS (continued)

In numeric order as they occur in the Table of Physical Constants of Organic Compounds

5466

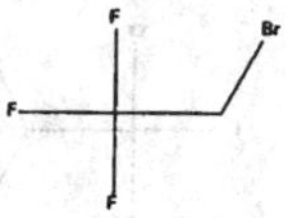

5467

5468

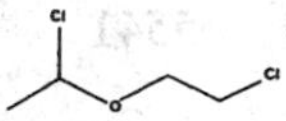

5469

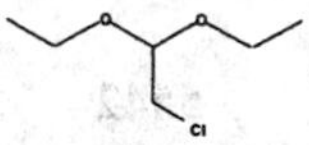

5470

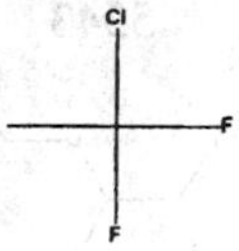

5471

5472

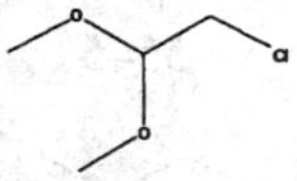

5473

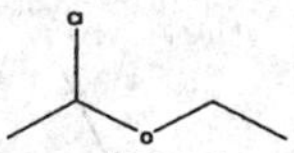

5474

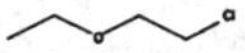

5475

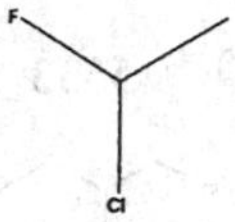

5476

5477

5478

5479

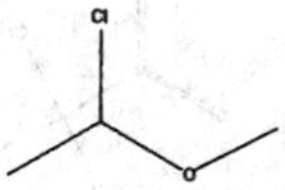

5480

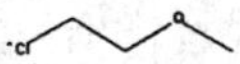

5481

5482

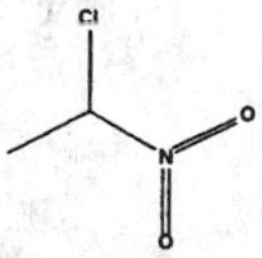

5483

5484

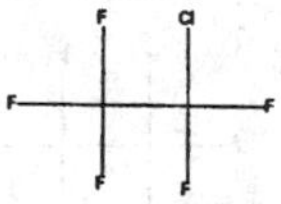

5485

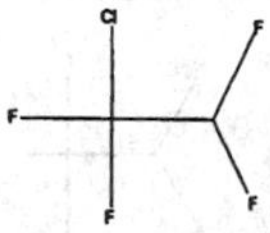

5486

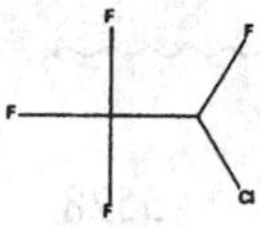

5487

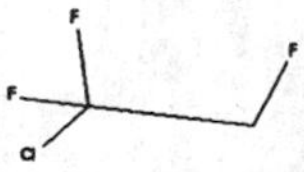

5488

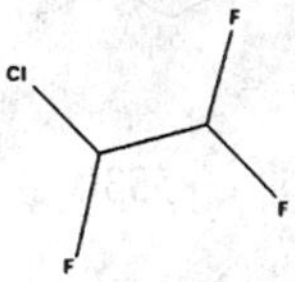

5489

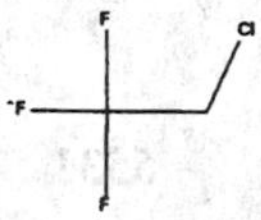

5490

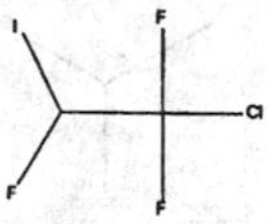

5491

5492

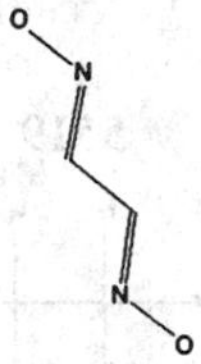

5493

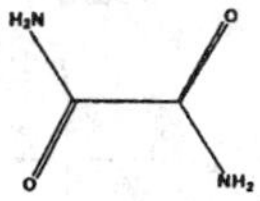

5495

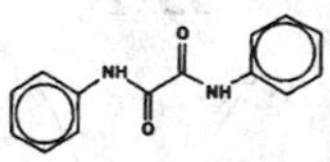

5496

5497

5498

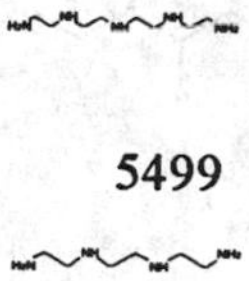

5499

5500

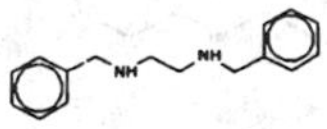

5501

5502

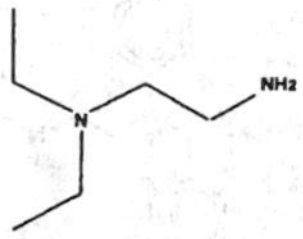

5503

5505

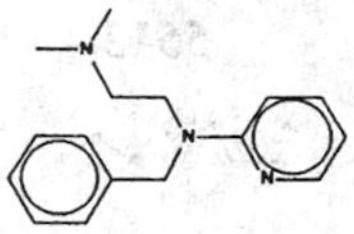

5506

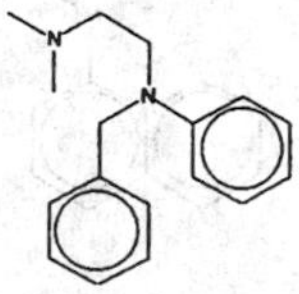

5507

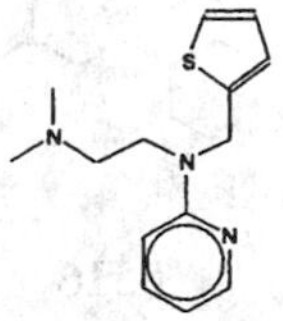

5508

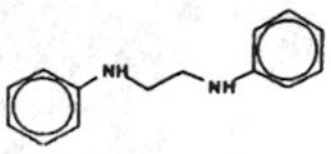

5509

5510

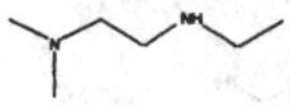

5511

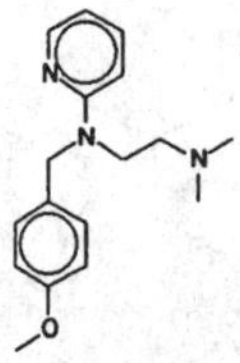

5512

5513

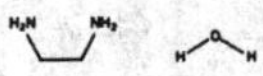

5514

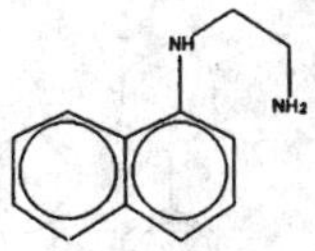

5515

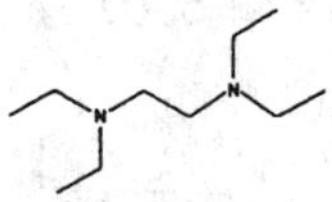

5516

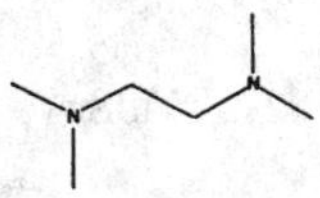

5517

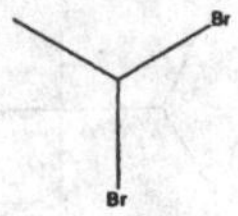

5518

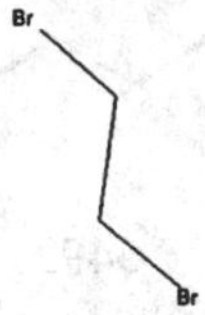

5519

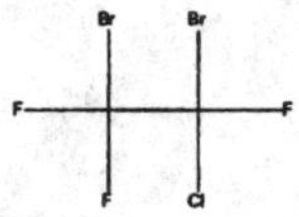

5520

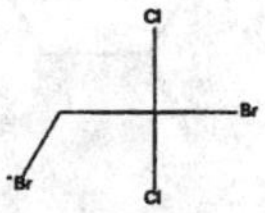

5521

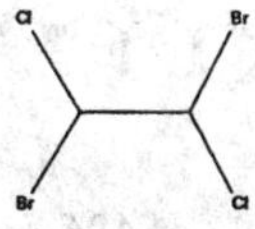

5522

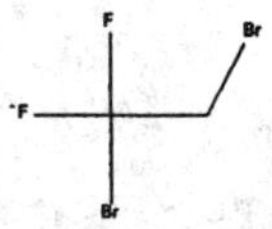

5523

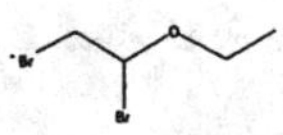

5524

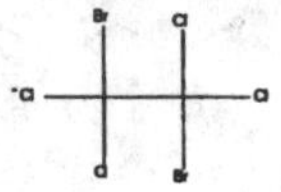

5525

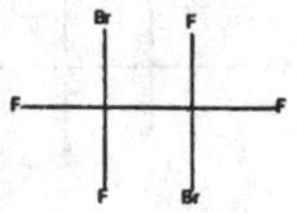

5526

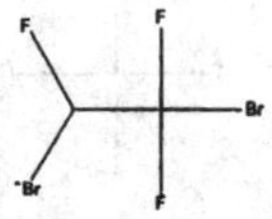

5527

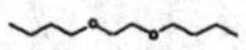

5528

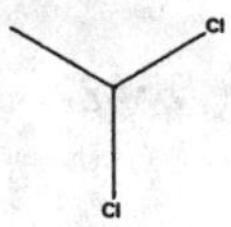

5529

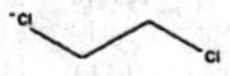

5530

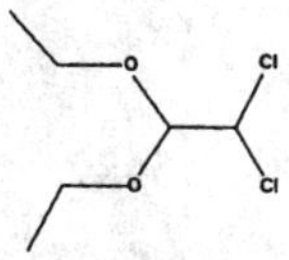

5531

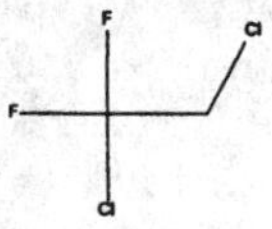

5532

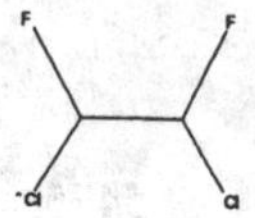

5533

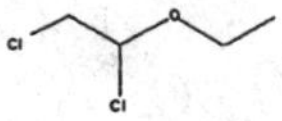

5534

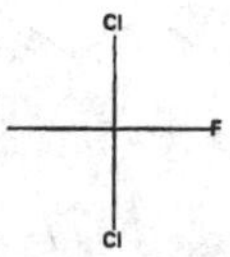

5535

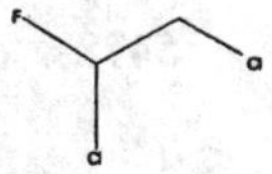

5536

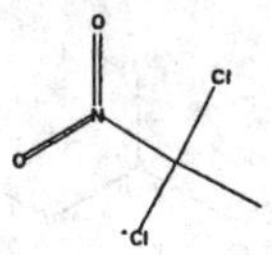

5537

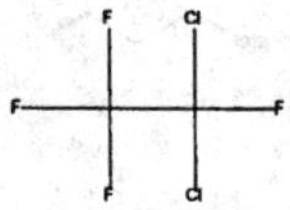

5538

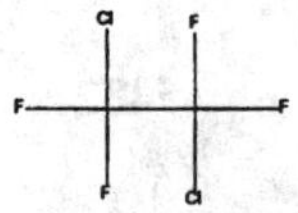

5539

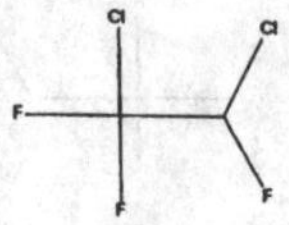

5540

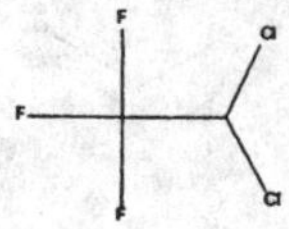

5541

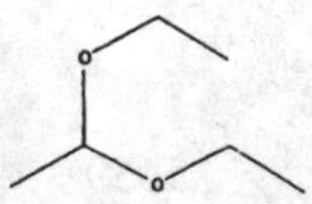

5542

5543

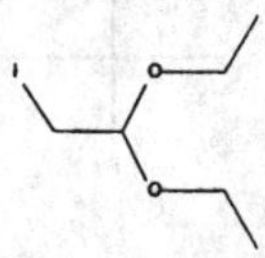

5544

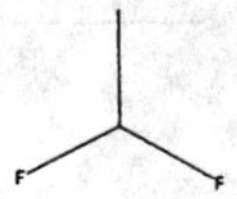

5545

5546

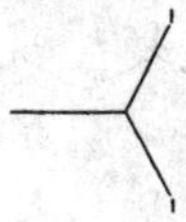

STRUCTURAL FORMULAS OF ORGANIC COMPOUNDS (continued)

In numeric order as they occur in the Table of Physical Constants of Organic Compounds

5547

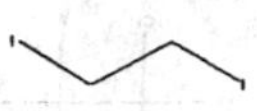

5548

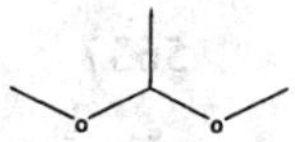

5549

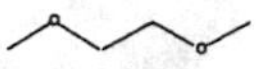

5550

5551

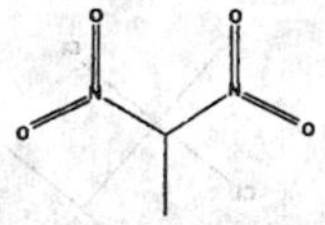

5552

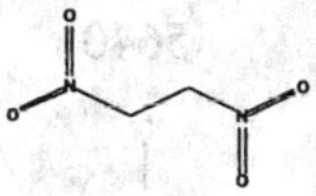

5553

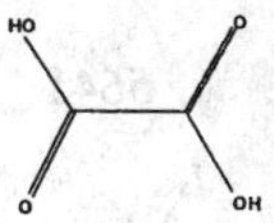

5554

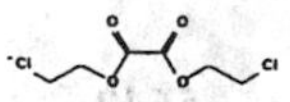

5555

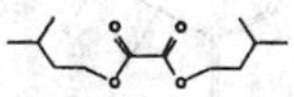

5556

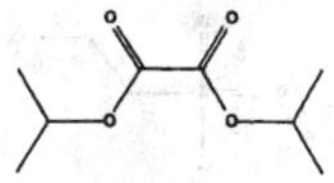

5557

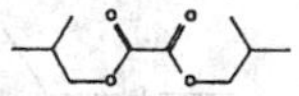

5558

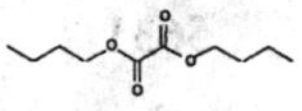

5559

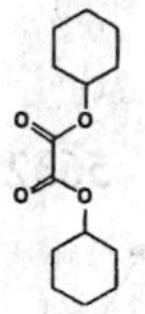

5560

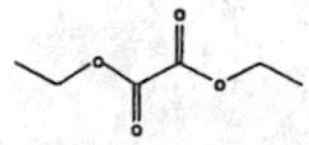

5562

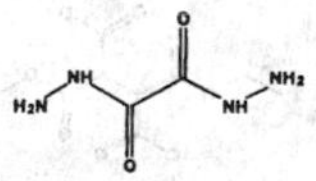

5563

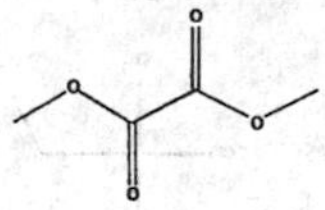

5564

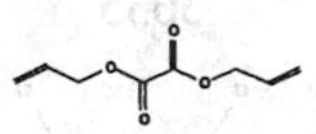

5565

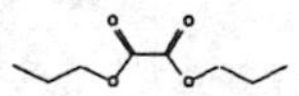

5567

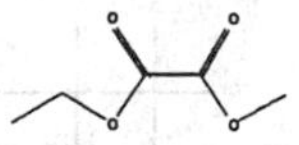

5569

5570

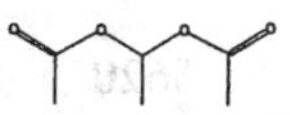

5571

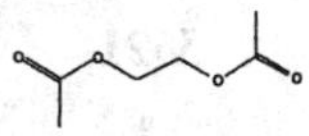

5572

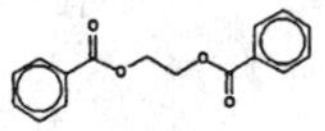

5573

5574

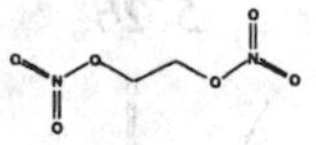

5576

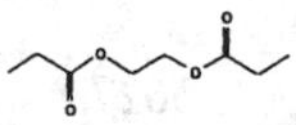

5577

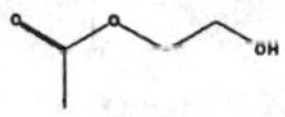

5578

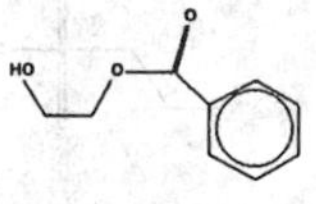

5579

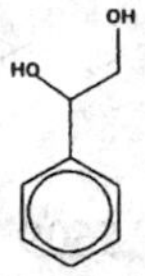

5580

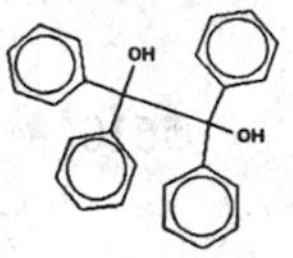

5582

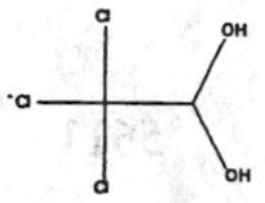

5583

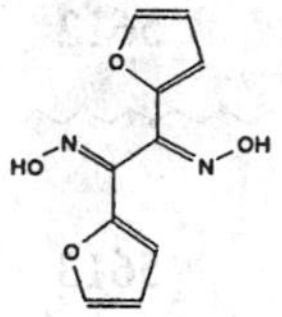

5584

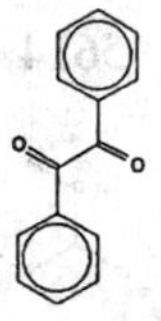

5585

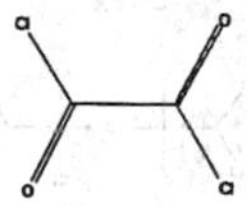

5586

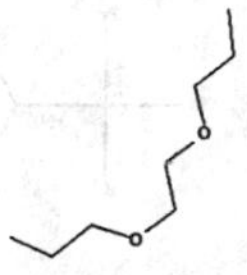

5588

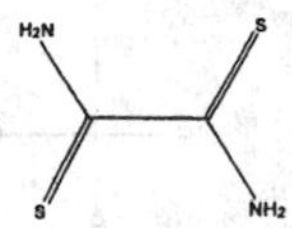

5589

5590

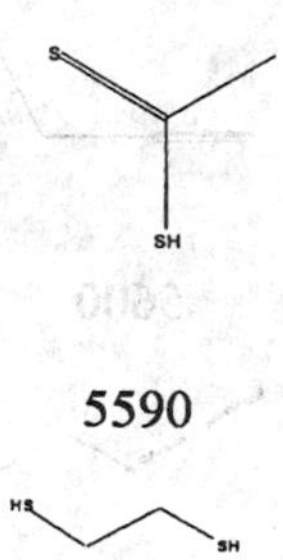

5591

5592

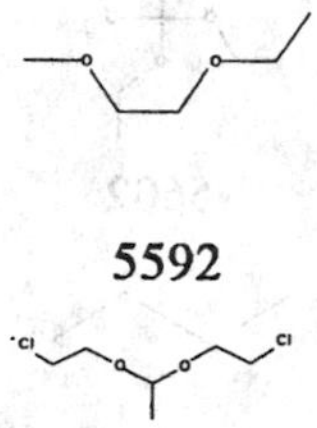

5593

5595

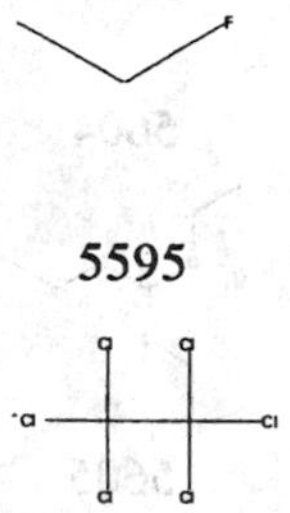

5596

5597

5598

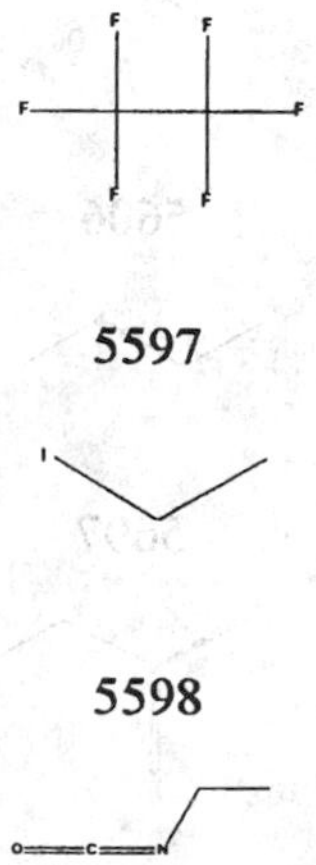

STRUCTURAL FORMULAS OF ORGANIC COMPOUNDS (continued)

In numeric order as they occur in the Table of Physical Constants of Organic Compounds

5599

5600

5601

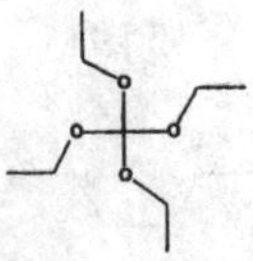

5602

5603

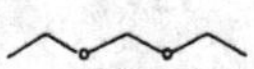

5604

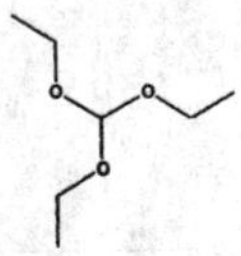

5605

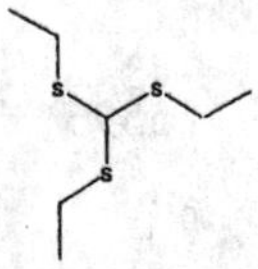

5606

5607

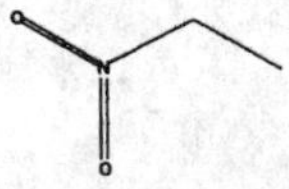

5608

5609

5610

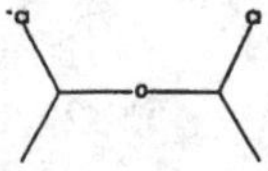

5611

5612

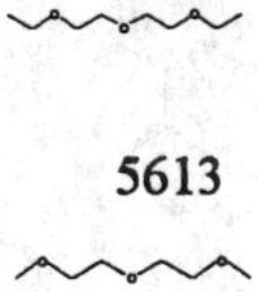

5613

5614

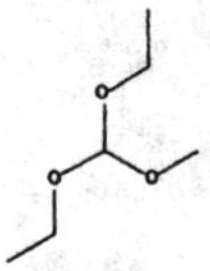

5615

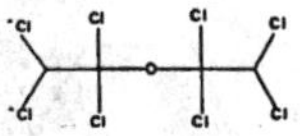

5616

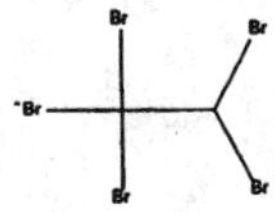

5617

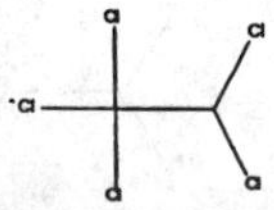

5618

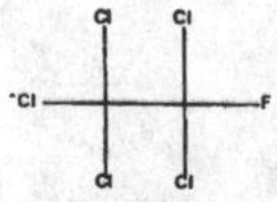

5619

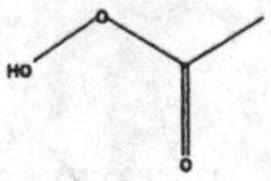

5620

5621

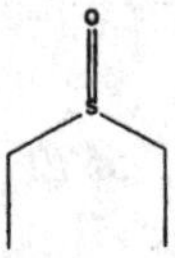

5622

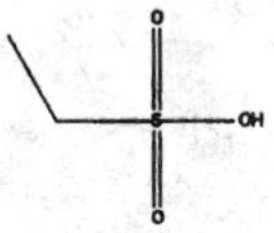

5626

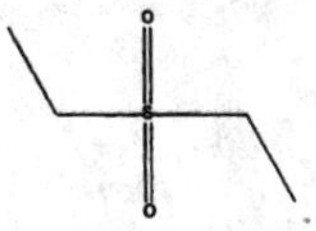

5627

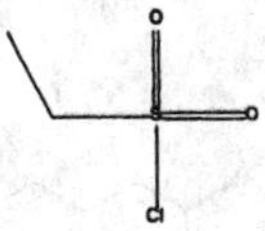

5628

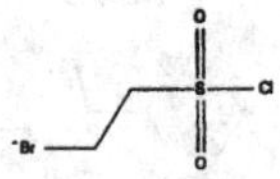

5629

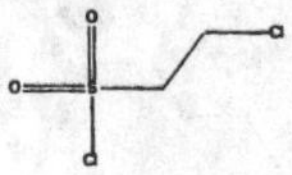

5630

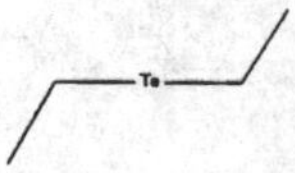

5631

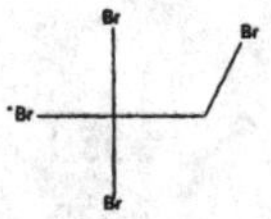

5632

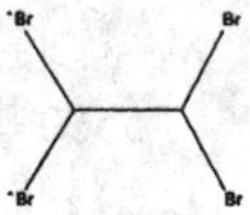

5633

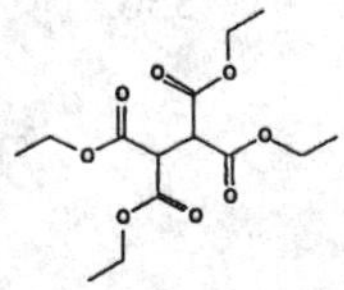

5634

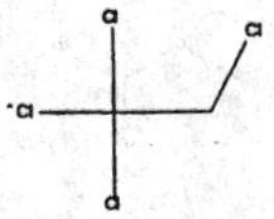

5635

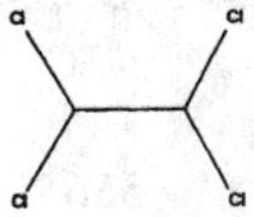

5636

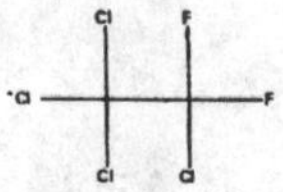

5637

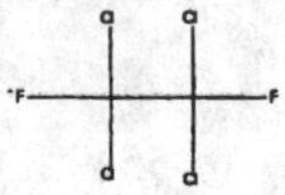

5638

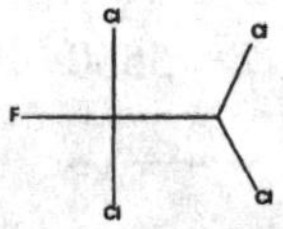

5639

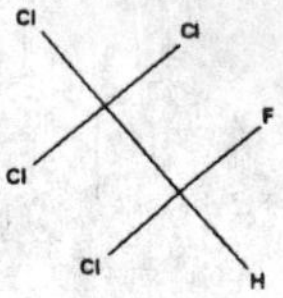

5640

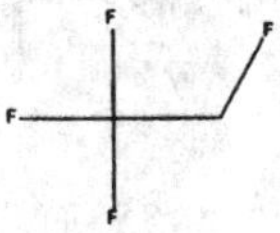

5641

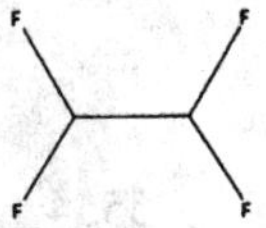

5642

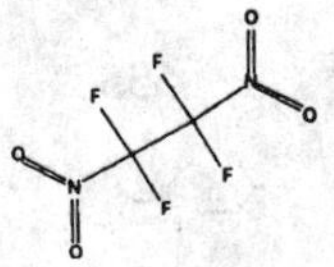

STRUCTURAL FORMULAS OF ORGANIC COMPOUNDS (continued)

In numeric order as they occur in the Table of Physical Constants of Organic Compounds

5643

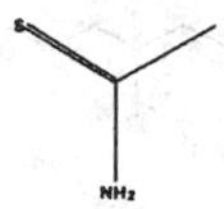

5644

5645

5646

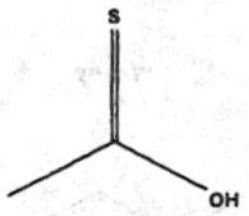

5647

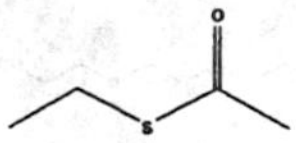

5648

5650

5651

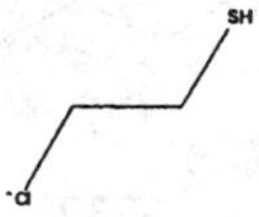

5652

5653

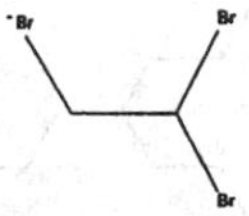

5654

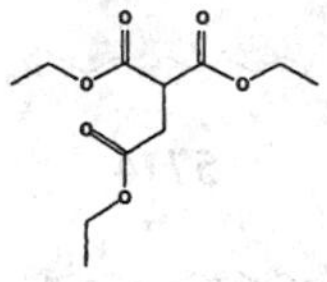

5655

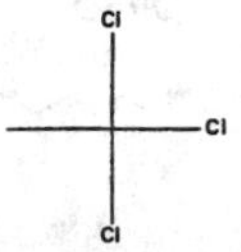

5656

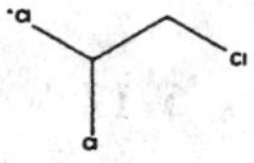

5657

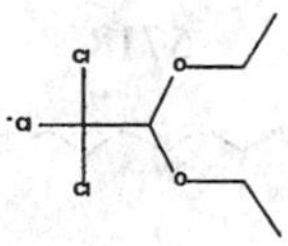

5658

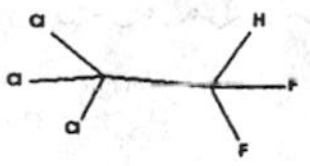

5659

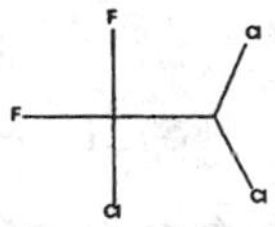

5660

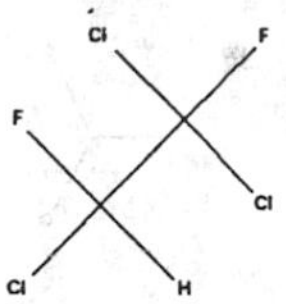

5661

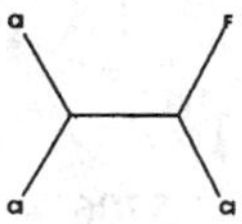

5662

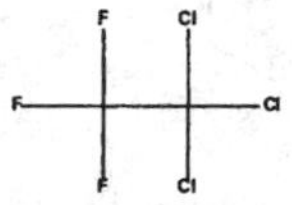

5663

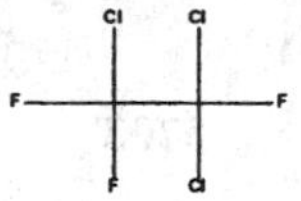

5664

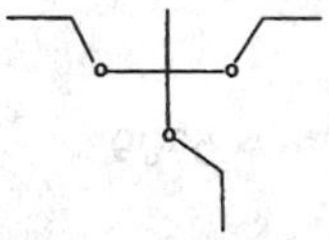

5665

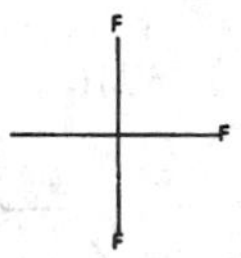

5666

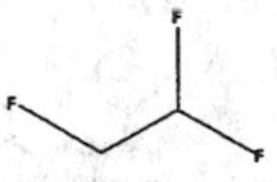

5667

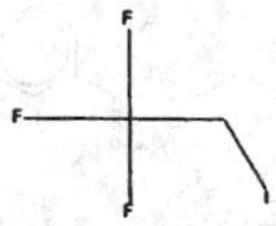

5668

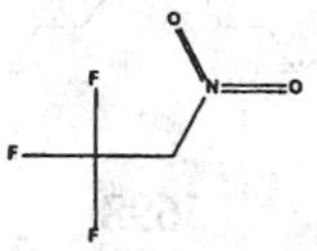

5669

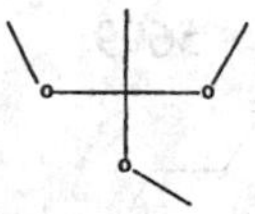

5672

5673

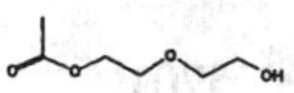

5674

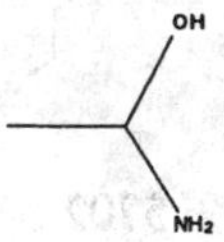

5675

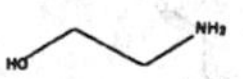

5676

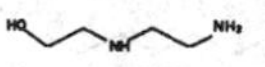

5677

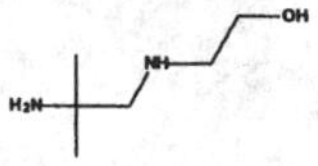

5678

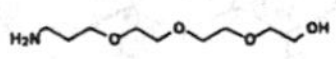

5679

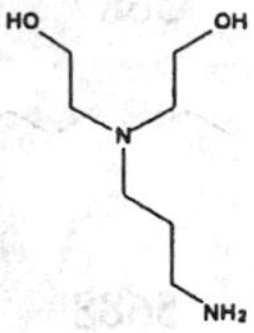

5680

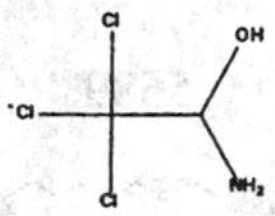

5681

5682

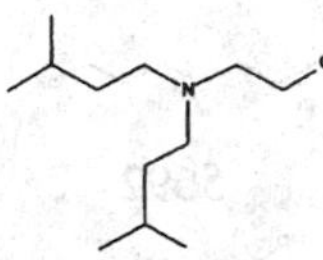

5683

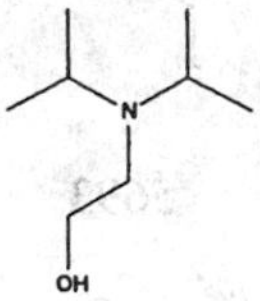

5684

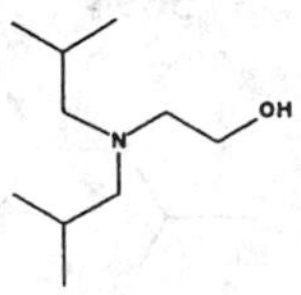

5685

STRUCTURAL FORMULAS OF ORGANIC COMPOUNDS (continued)

In numeric order as they occur in the Table of Physical Constants of Organic Compounds

5686

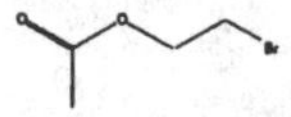

5687

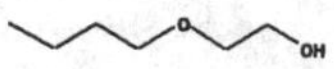

5688

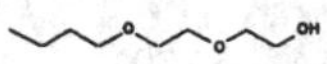

5689

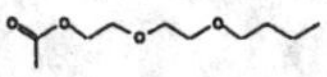

5690

5691

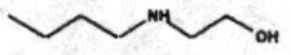

5692

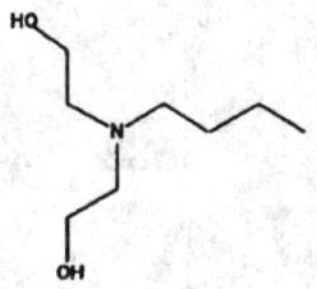

5693

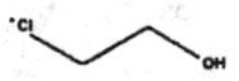

5694

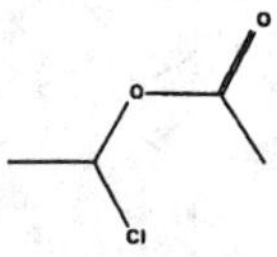

5695

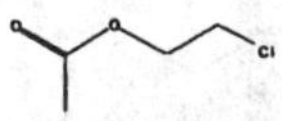

5696

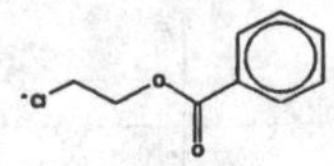

5697

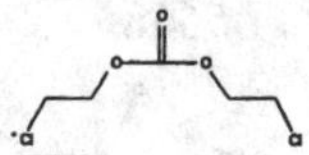

5698

5699

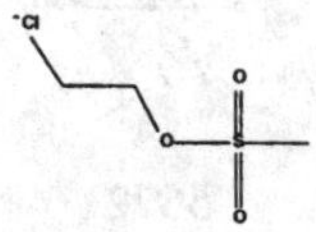

5700

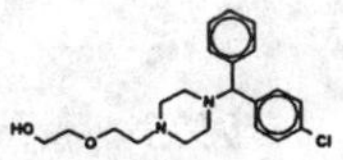

5701

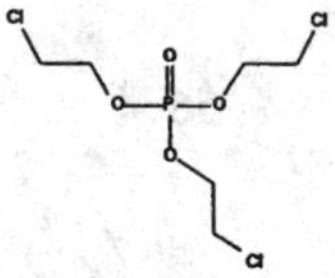

5702

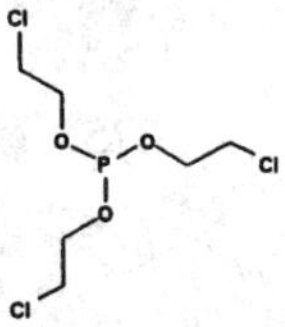

5703

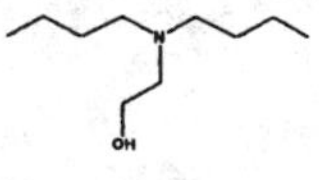

5704

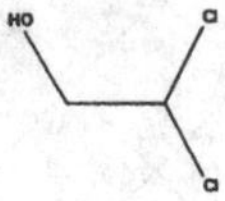

5705

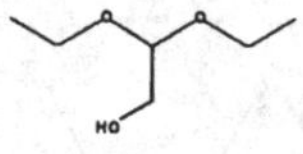

5706

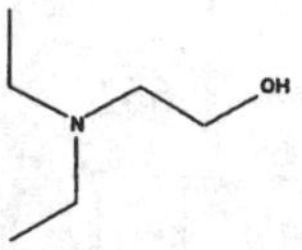

5707

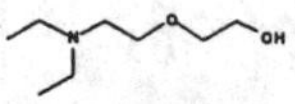

5708

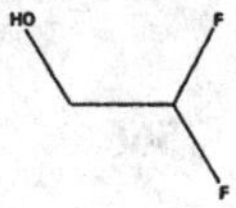

5709

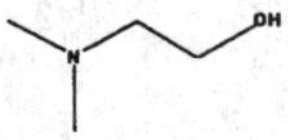

5710

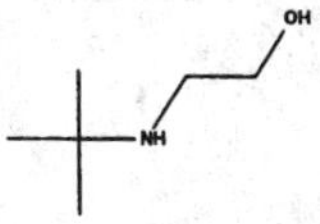

5711

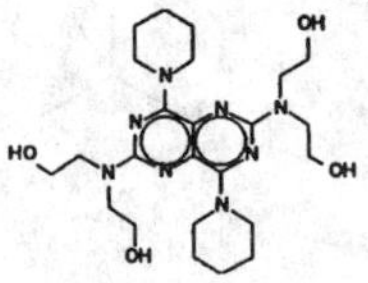

5712

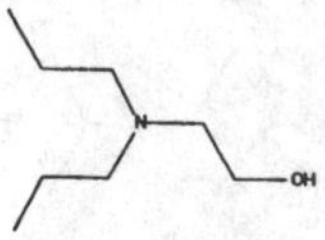

5713

5714

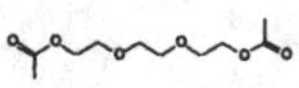

5715

5716

5717

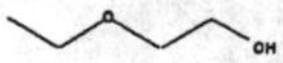

5718

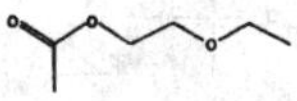

5719

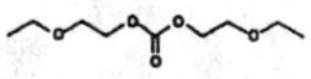

5720

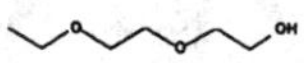

5721

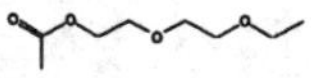

5722

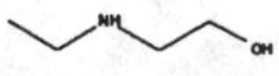

5723

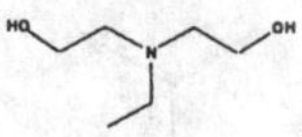

5724

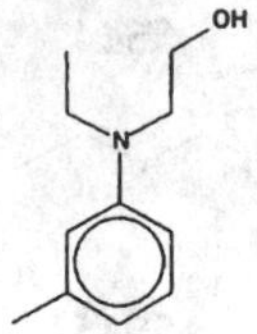

5726

5727

5728

5729

5730

5731

5732

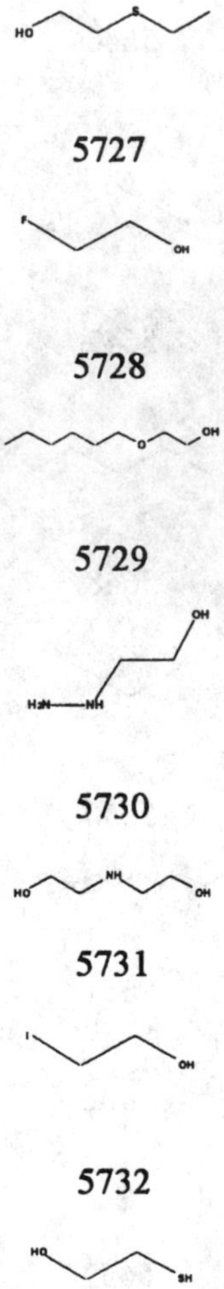

In numeric order as they occur in the Table of Physical Constants of Organic Compounds

5733

5734

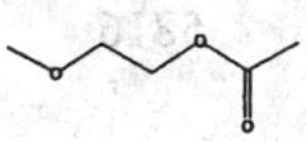

5735

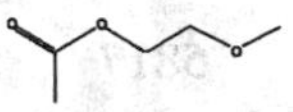

5736

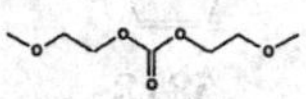

5737

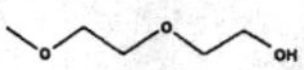

5738

5739

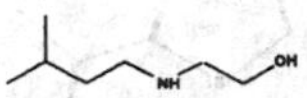

5740

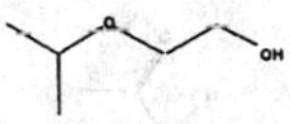

5741

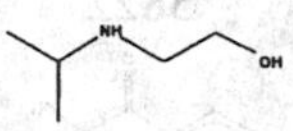

5742

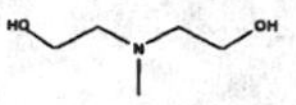

5743

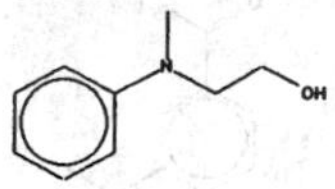

5744

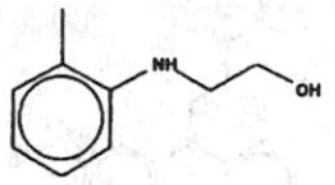

5745

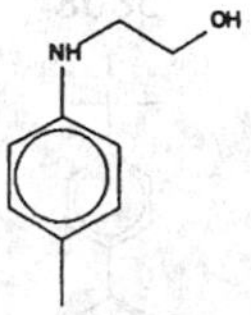

5746

5747

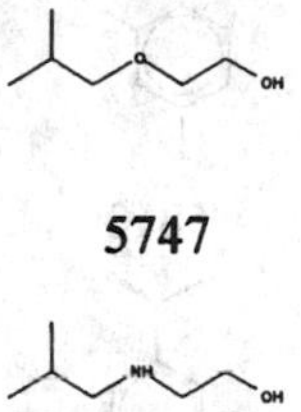

5748

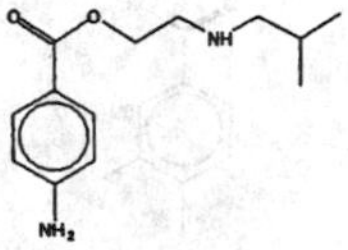

5749

5750

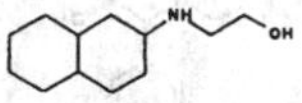

5751

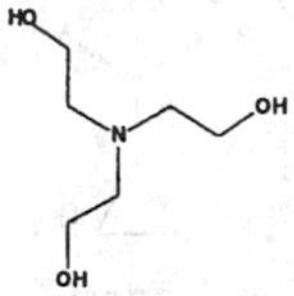

5752

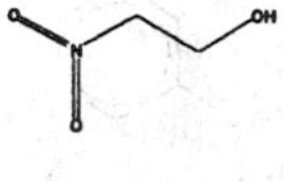

5753

5754

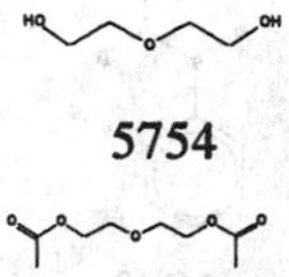

5755

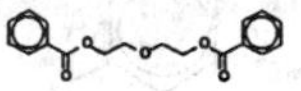

5756

5757

5758

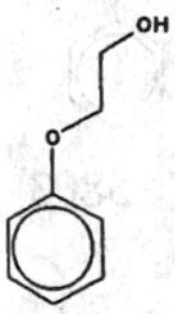

5759

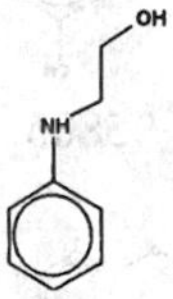

5760

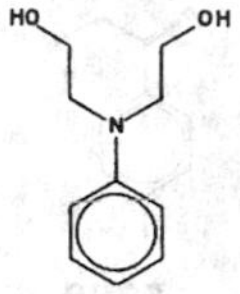

5761

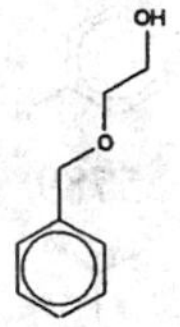

5762

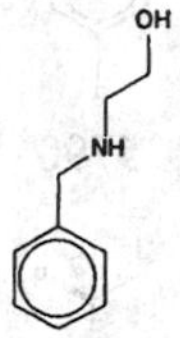

5763

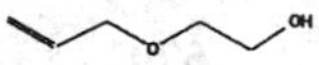

5764

5765

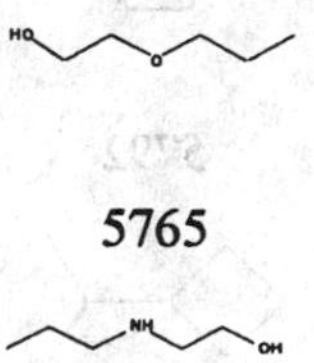

5766

5767

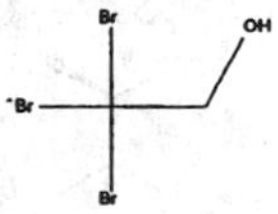

5768

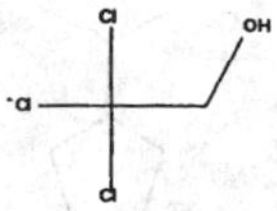

5769

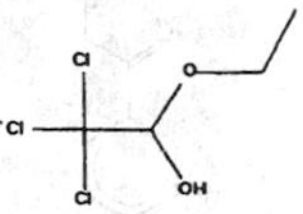

5770

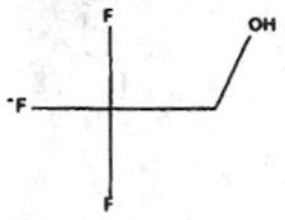

5771

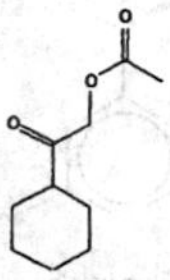

5772

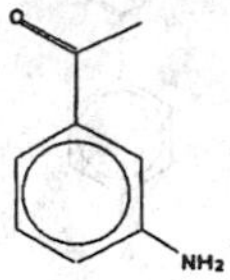

5773

5774

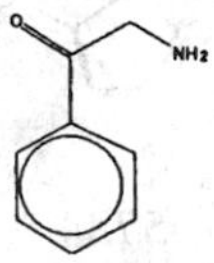

5775

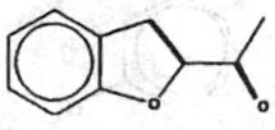

5776

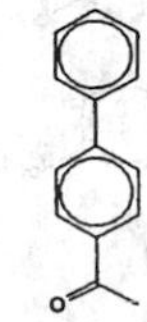

In numeric order as they occur in the Table of Physical Constants of Organic Compounds

5777

5779
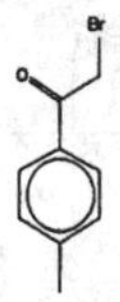
5780
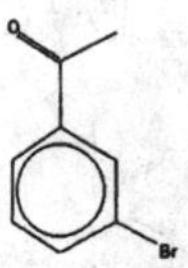
5781

5782
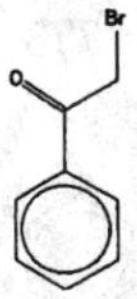
5783
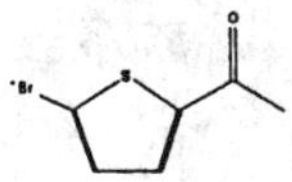
5784
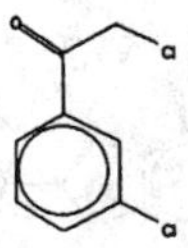
5786
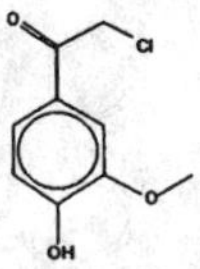

5787
5788
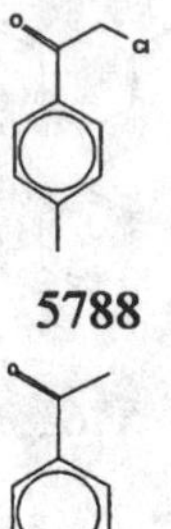
5789

5790
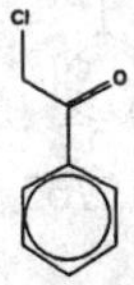
5791
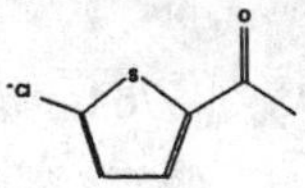
5792
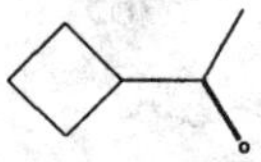
5793
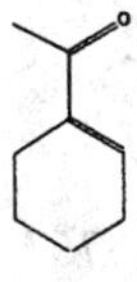
5794
5795

5796
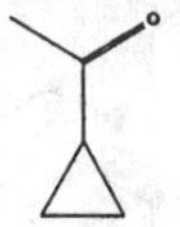
5797
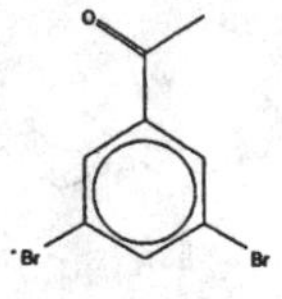
5798
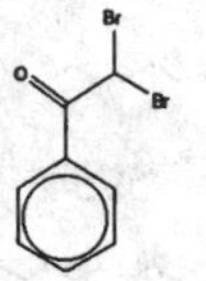
5800
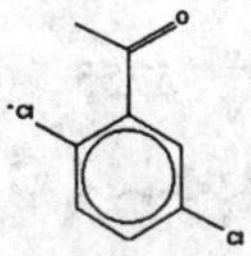
5801
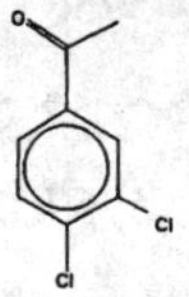
5802
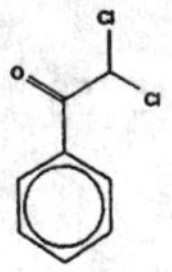
5803
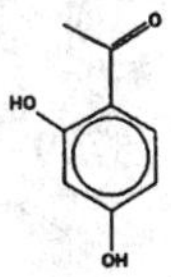
5805
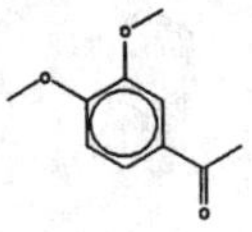

5806
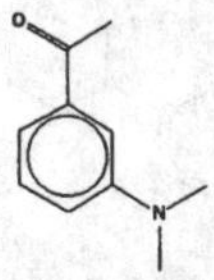
5807
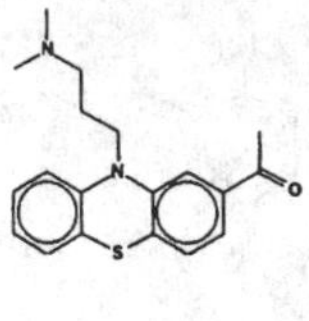
5808

5809
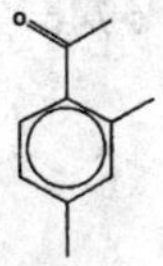
5810
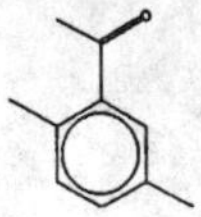
5811
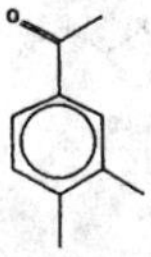
5812
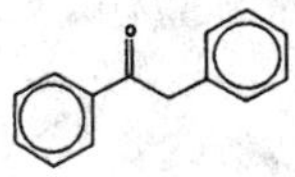
5813
5814
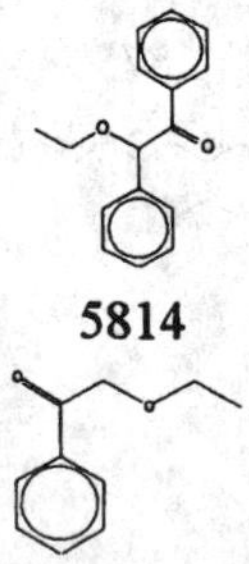

5815

5816
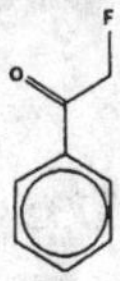
5817
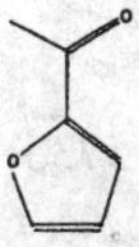
5818
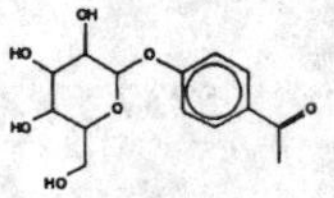
5819
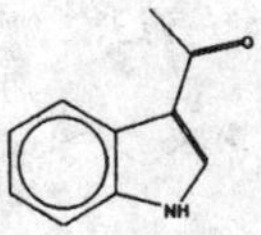
5820
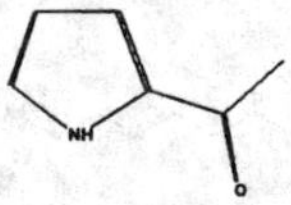
5821
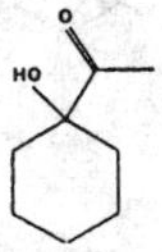
5822
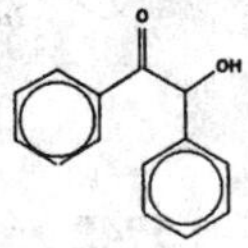

In numeric order as they occur in the Table of Physical Constants of Organic Compounds

5823

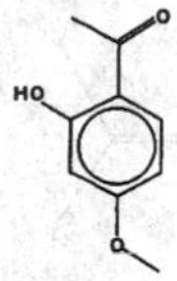

5824

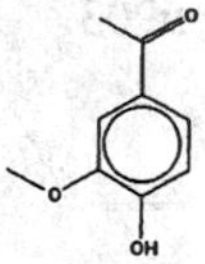

5826

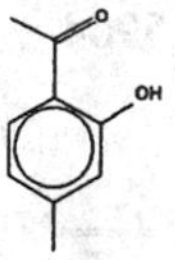

5827

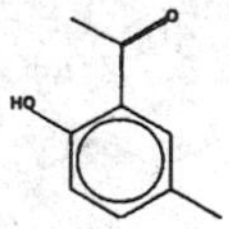

5828

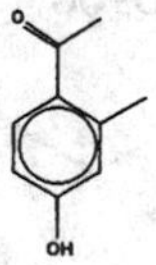

5829

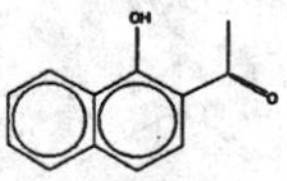

5830

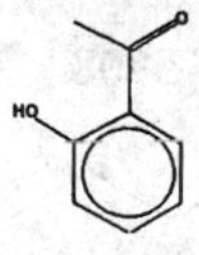

5831

5832

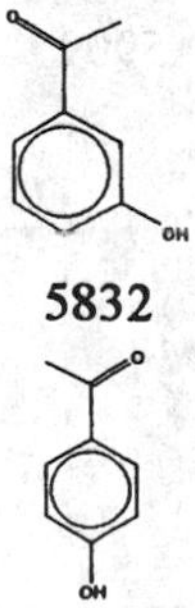

5833

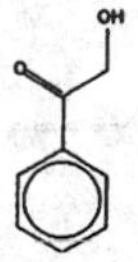

5834

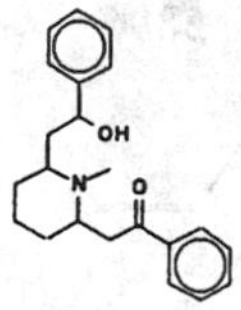

5835

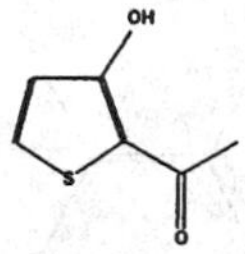

5836

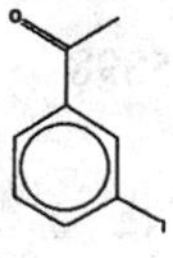

5837

5838

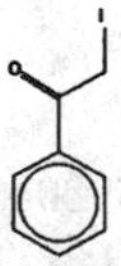

5839

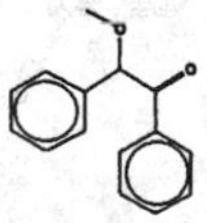

5840

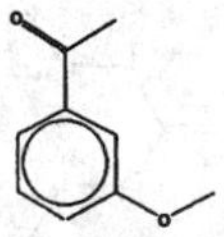

5841

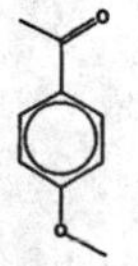

5842

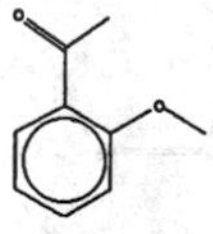

5843

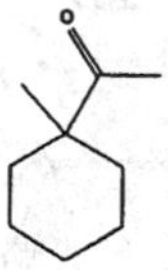

5844

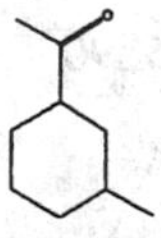

5845

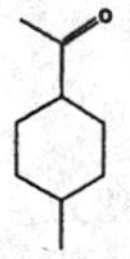

5846

5847

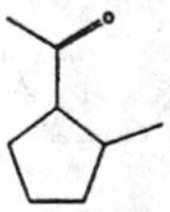

5848

5849

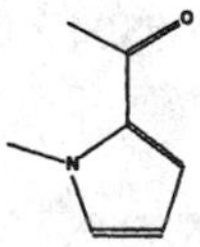

5850

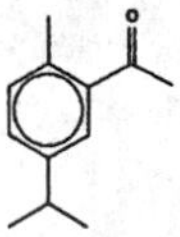

5851

5852

5853

5855

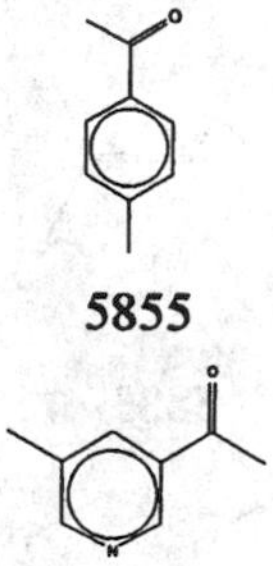

5856

5857

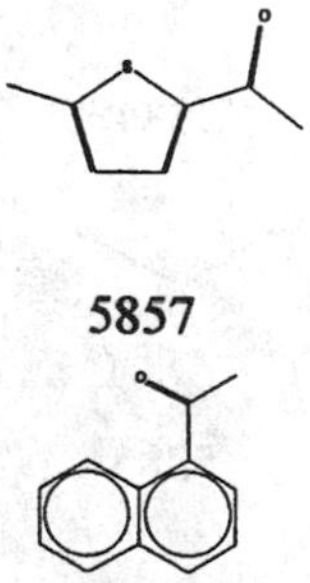

5858

5859

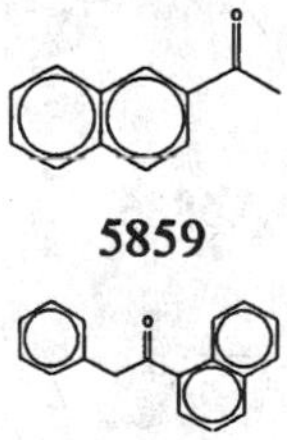

5860

5861

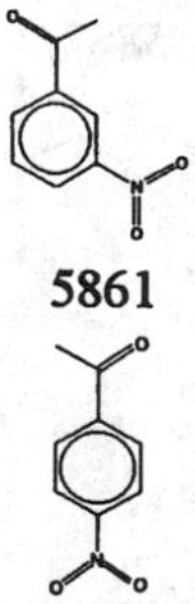

5862

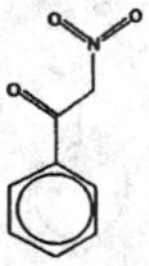

5863

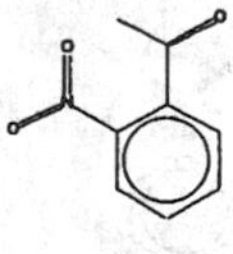

5864

5865

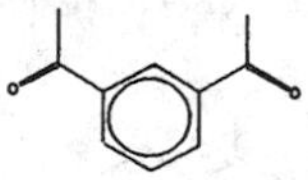

5866

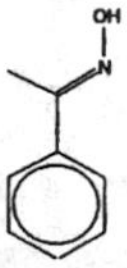

5867

5868

5869

5870

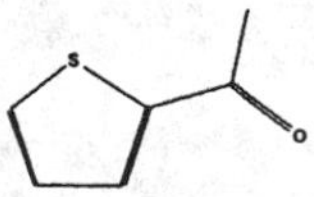

In numeric order as they occur in the Table of Physical Constants of Organic Compounds

5871
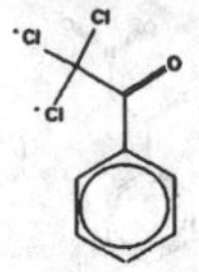
5872
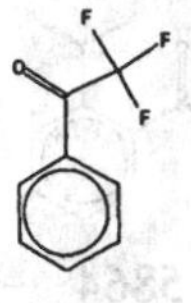
5873
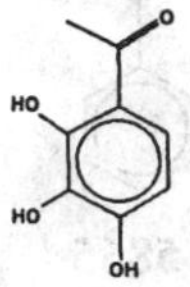
5874
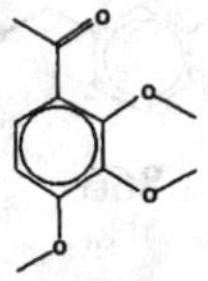
5875
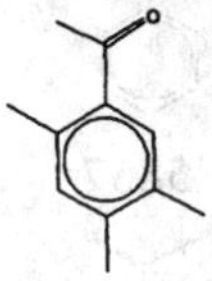
5876
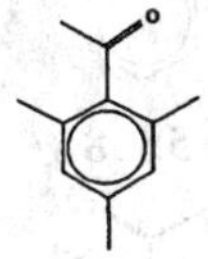
5877
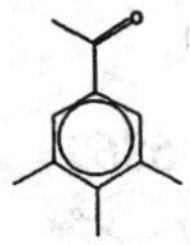
5879
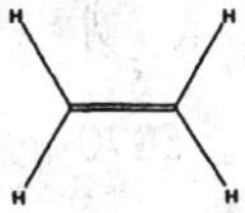
5880

5881

5882
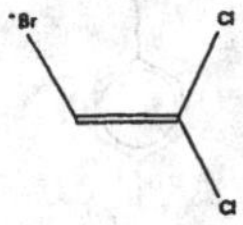
5883
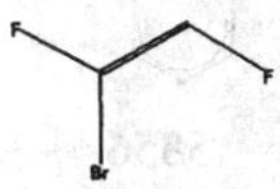
5884

5885
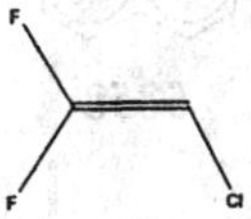
5886
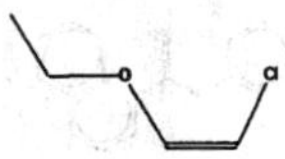
5887

5888

5889
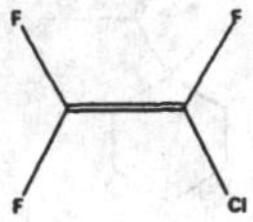
5890
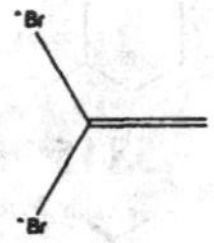
5891
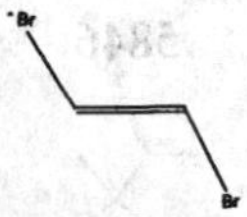
5892

5893
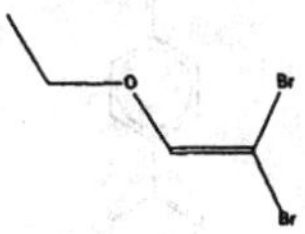
5894
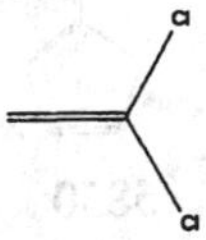
5895
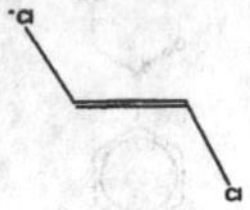
5896

5897
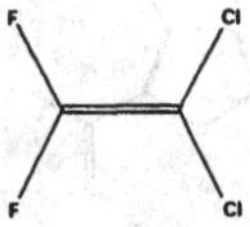
5898
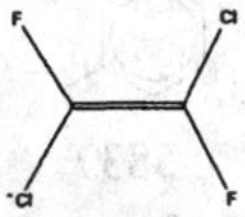
5899
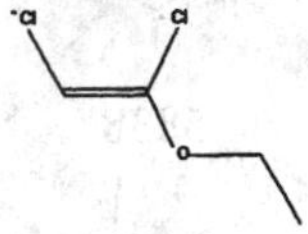
5900
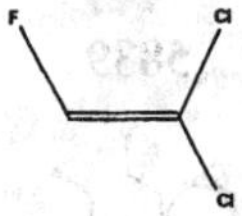
5901
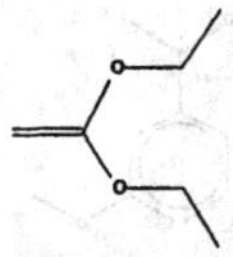
5902
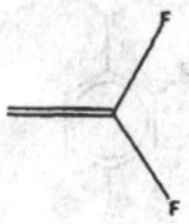
5903

5904

5905

5906

5907

5908

5909
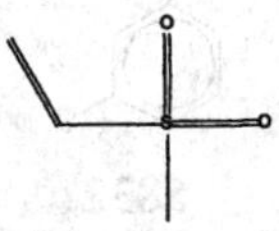

STRUCTURAL FORMULAS OF ORGANIC COMPOUNDS (continued)

In numeric order as they occur in the Table of Physical Constants of Organic Compounds

5910

5911

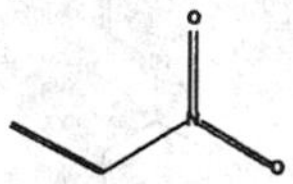

5912

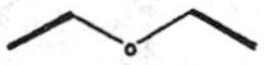

5913

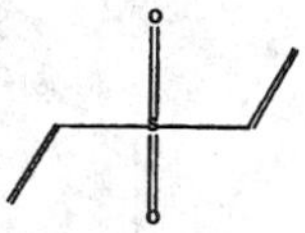

5914

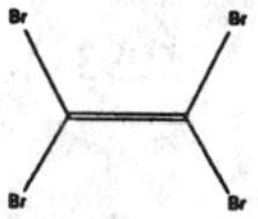

5915

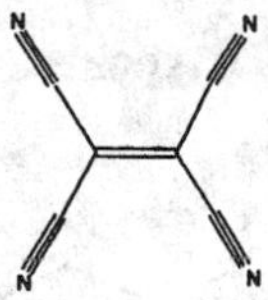

5916

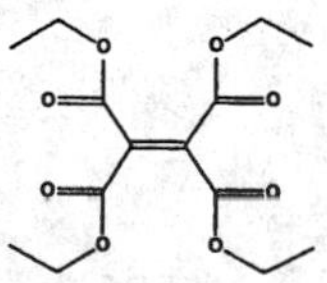

5917

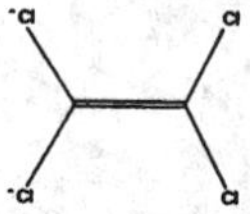

5918

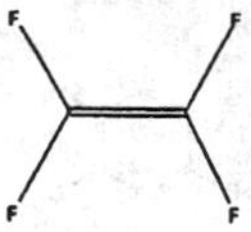

5919

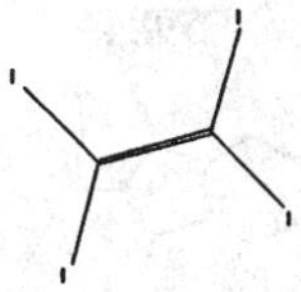

5920

5921

5922

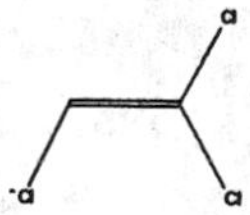

5923

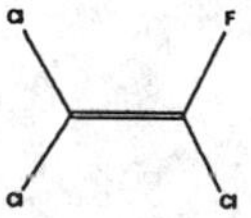

5924

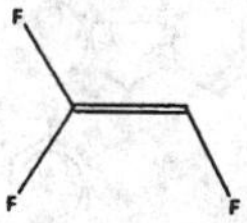

5925

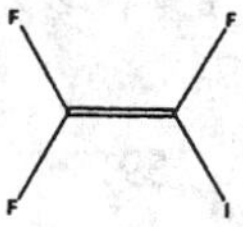

5926

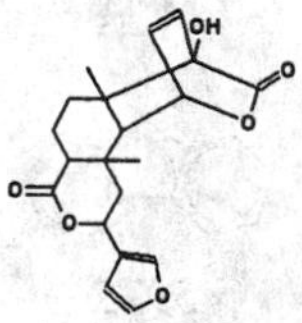

5927

5929

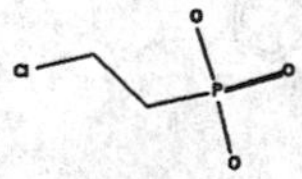

5930

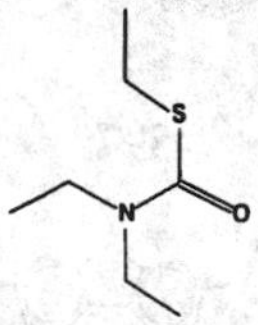

5931

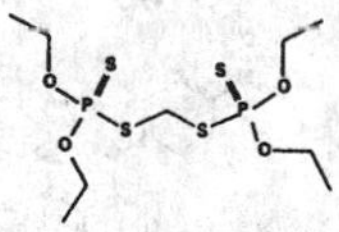

5932

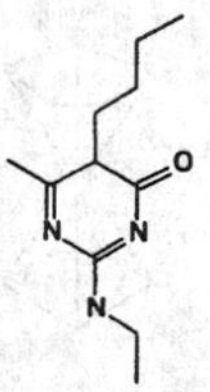

5933

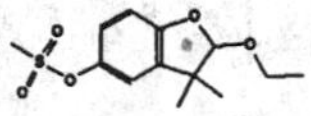

5934

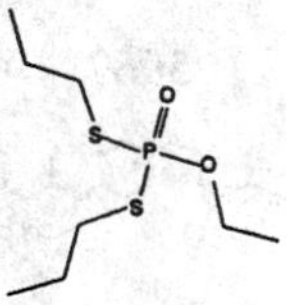

5935

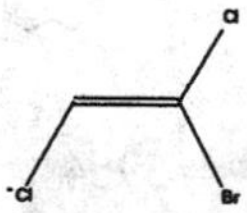

5936

5937

5938

5939

5940

5941

5942

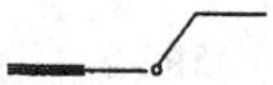

5943

5944

5945

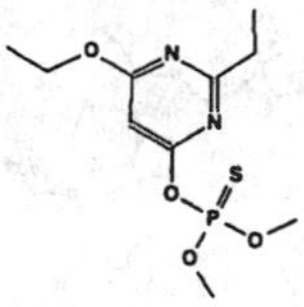

5946

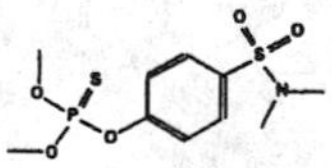

5947

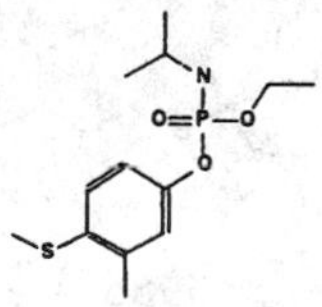

5948

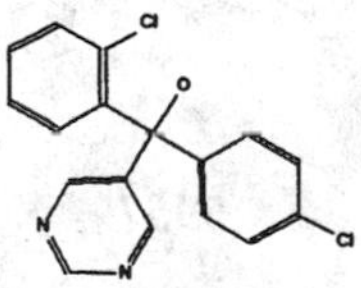

5950

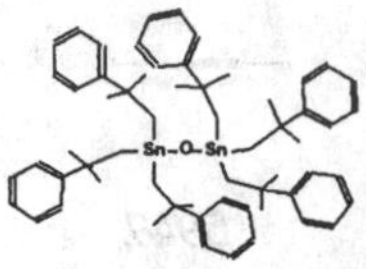

5951

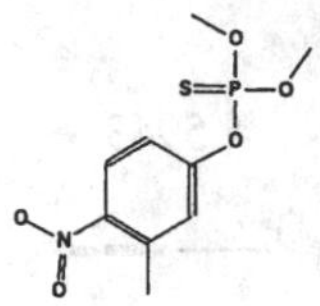

5952

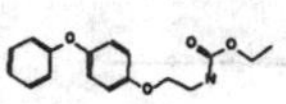

5953

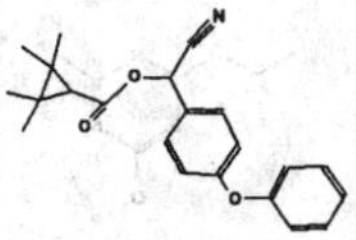

5954

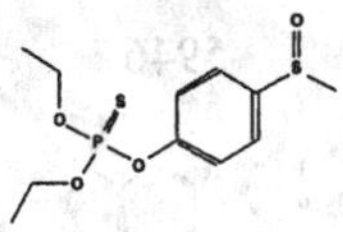

5955

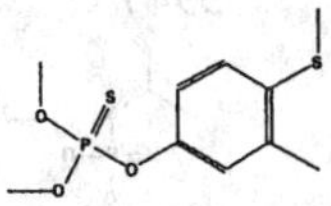

5956

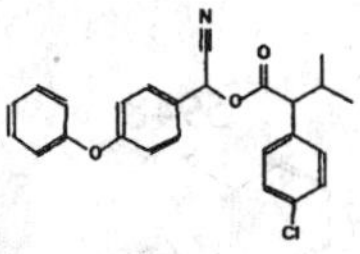

5957

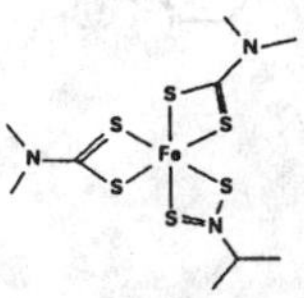

5960

5961

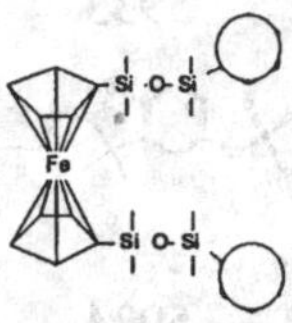

5962

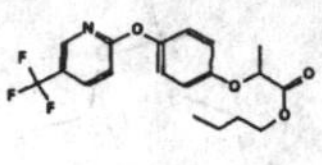

5963

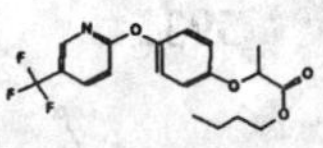

5964

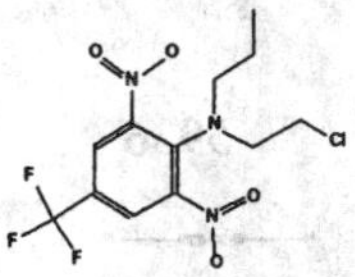

5965

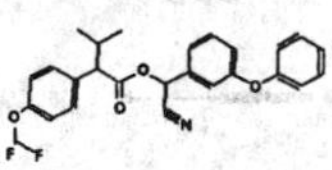

5967

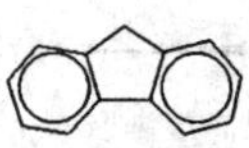

5968

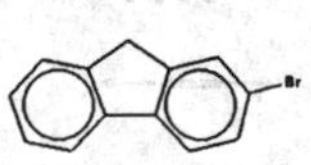

5970

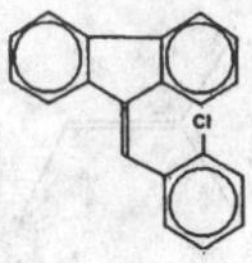

5971

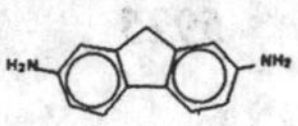

5973

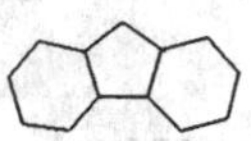

5974

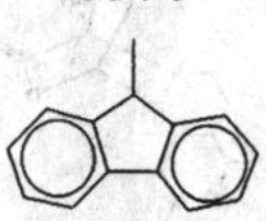

5975

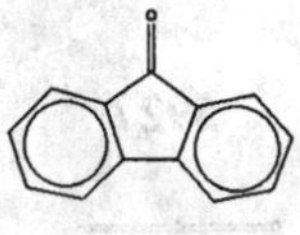

5977

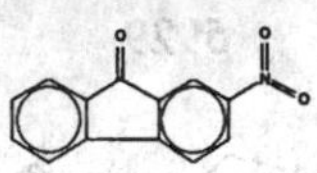

5978

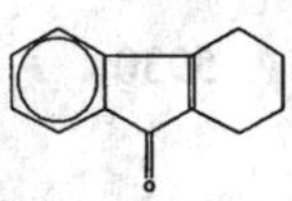

5979

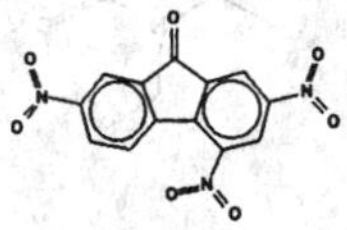

5980

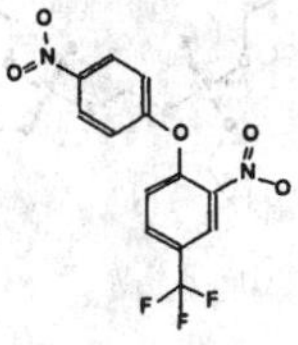

5981

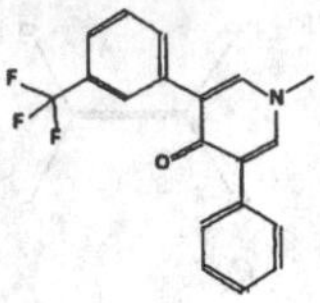

5982

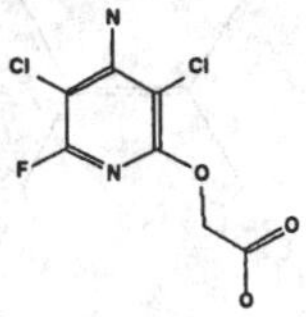

5984

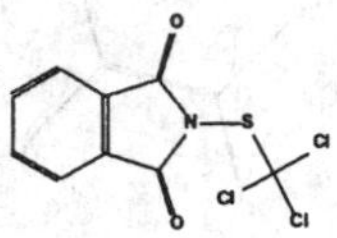

5985

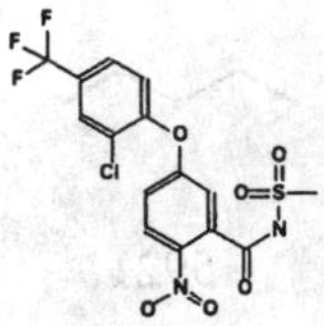

5986

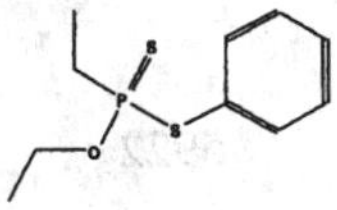

5987

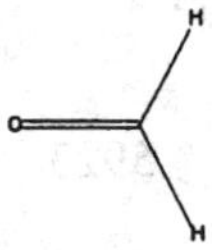

5988

5989

5990

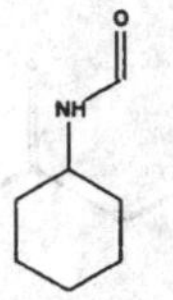

5991

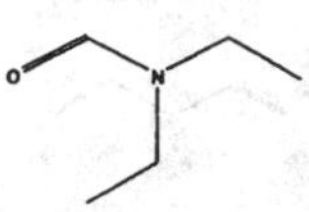

5992

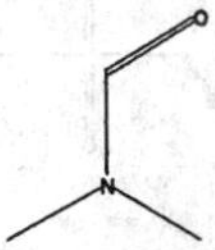

5993

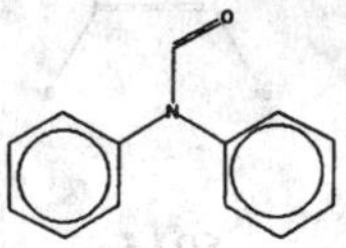

5994

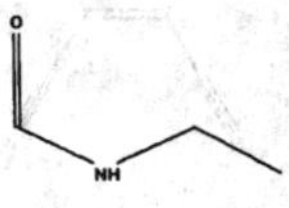

5995

5996

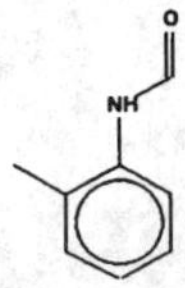

In numeric order as they occur in the Table of Physical Constants of Organic Compounds

5997

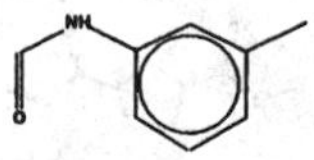

5999

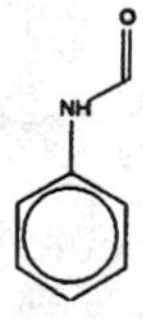

6000

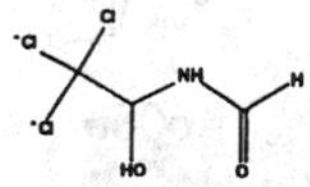

6001

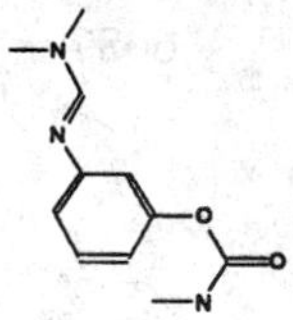

6002

6004

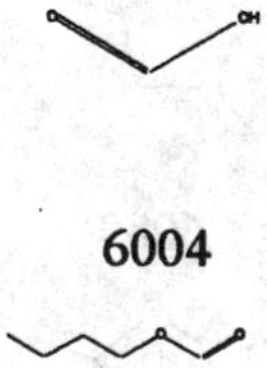

6005

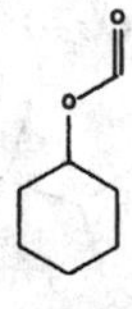

6006

6007

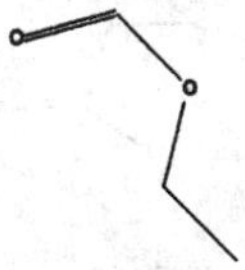

6008

6009

6010

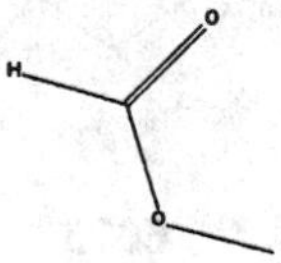

6011

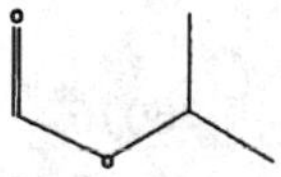

6012

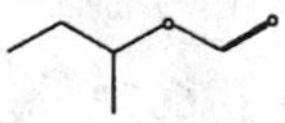

6013

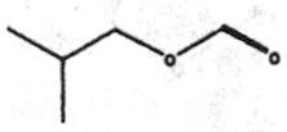

6014

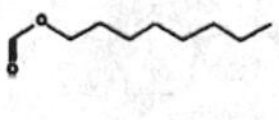

6015

6016

6017

6018

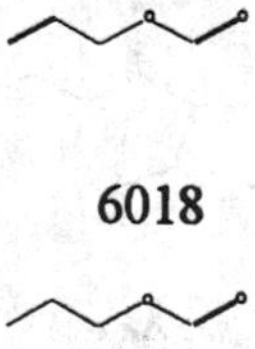

6019

6020

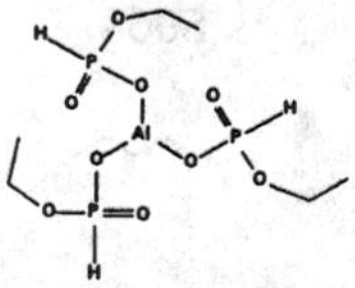

6021

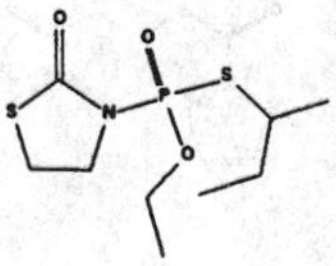

6024

6025

6026

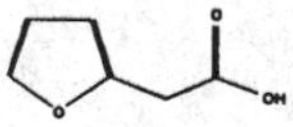

6027

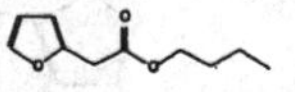

6028

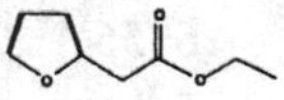

6029

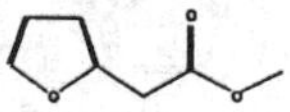

6031

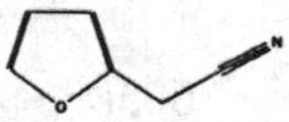

6032

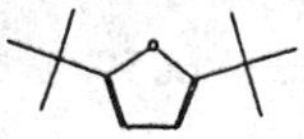

6033

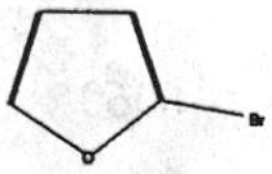

6034

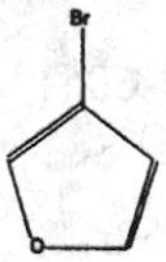

6035

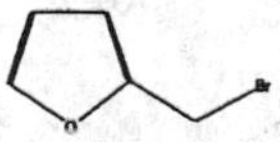

6036

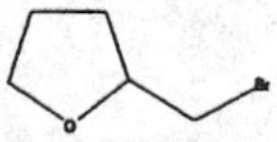

6037

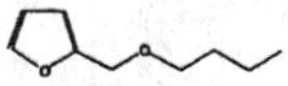

6039

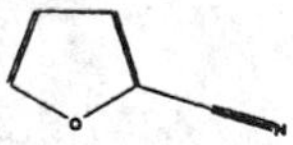

6040

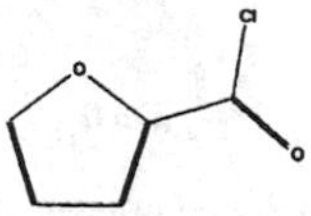

6041

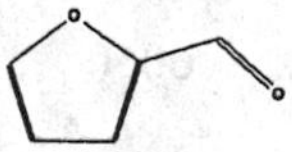

6042

6043

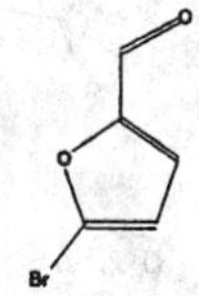

6044

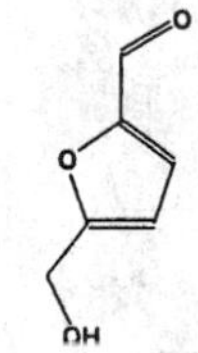

6045

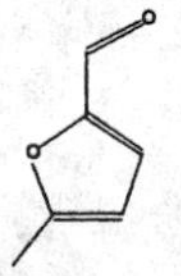

STRUCTURAL FORMULAS OF ORGANIC COMPOUNDS (continued)

In numeric order as they occur in the Table of Physical Constants of Organic Compounds

6046

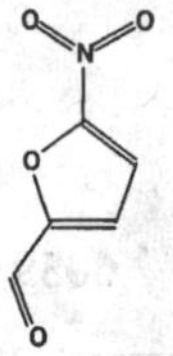

6047

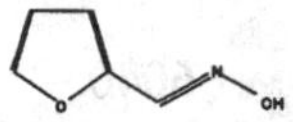

6049

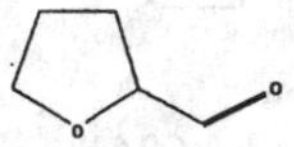

6050

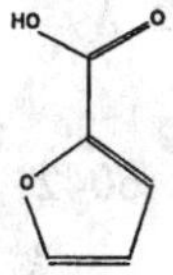

6051

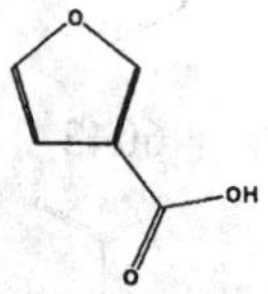

6052

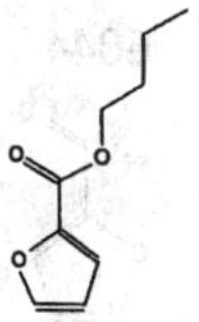

6053

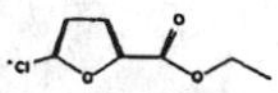

6055

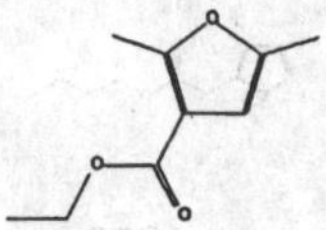

6056

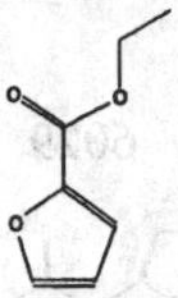

6057

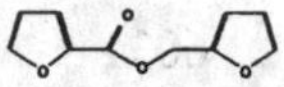

6058

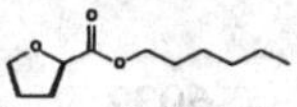

6059

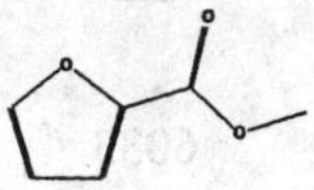

6060

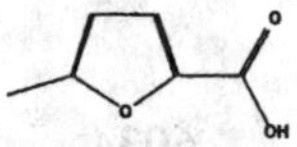

6061

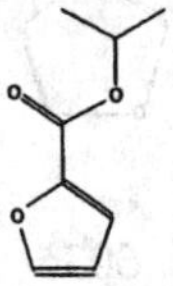

6062

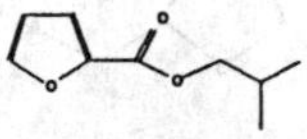

6063

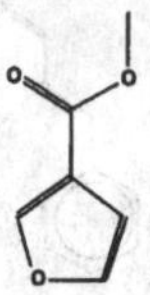

6064

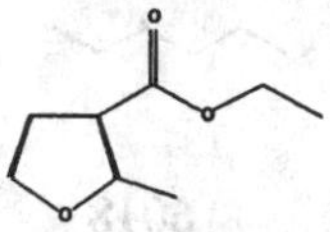

6066

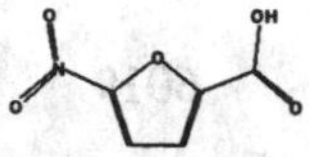

6067

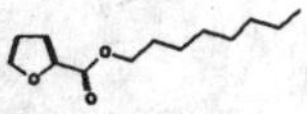

6068

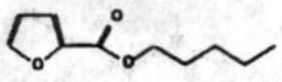

6069

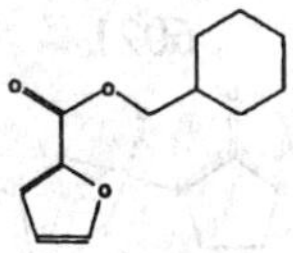

6070

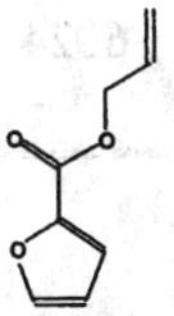

6071

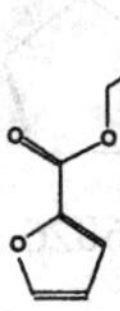

6072

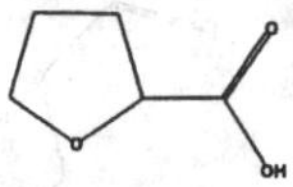

6073

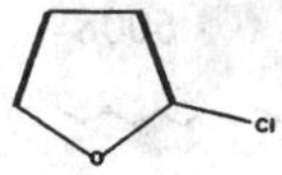

6074

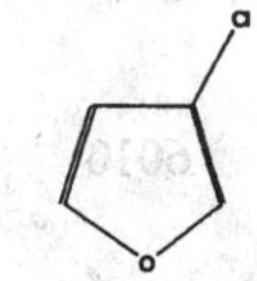

6075

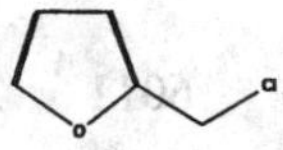

6076

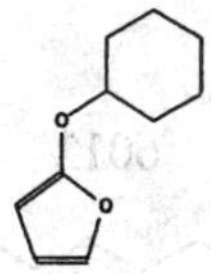

6077

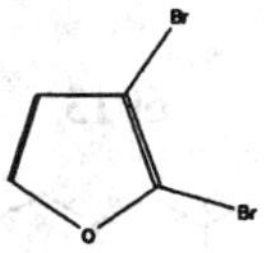

6081

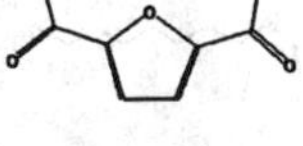

6082

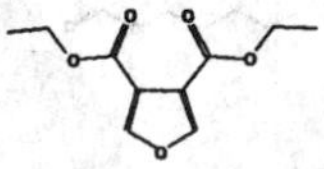

6083

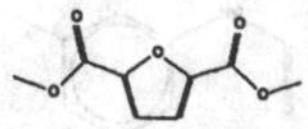

6084

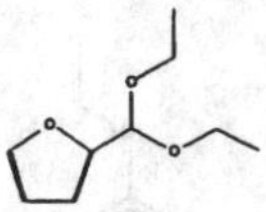

6085

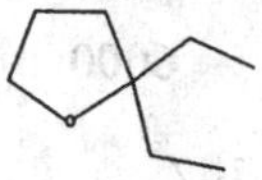

6086

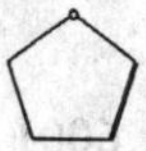

6087

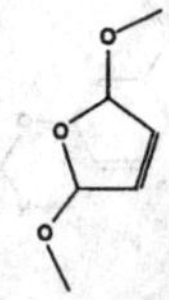

6088

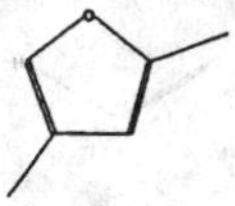

6089

6090

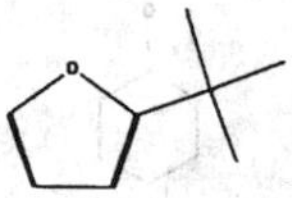

6091

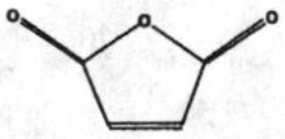

STRUCTURAL FORMULAS OF ORGANIC COMPOUNDS (continued)

In numeric order as they occur in the Table of Physical Constants of Organic Compounds

6092

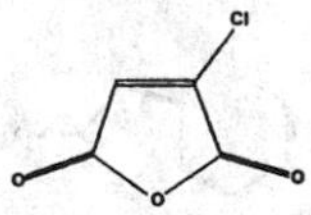

6093

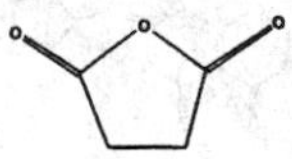

6094

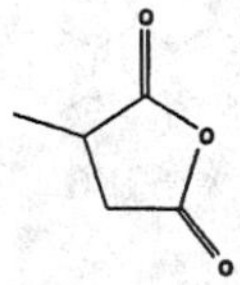

6095

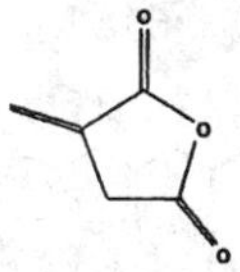

6096

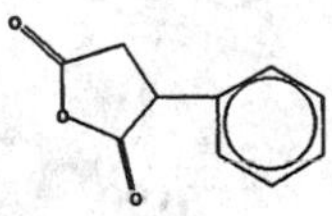

6097

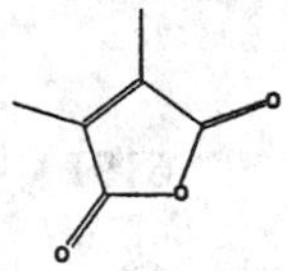

6098

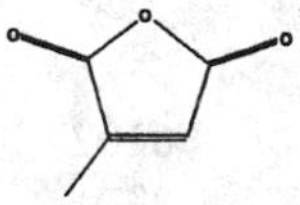

6099

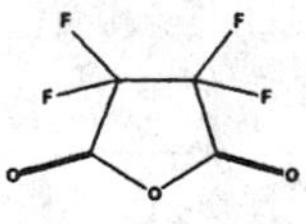

6101

6102

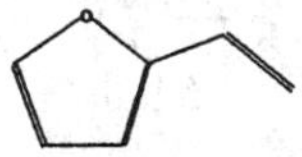

6103

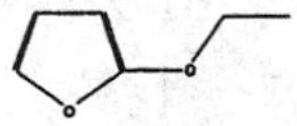

6104

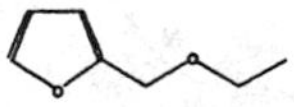

6105

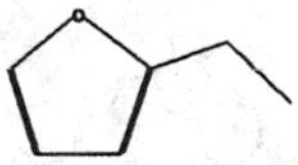

6106

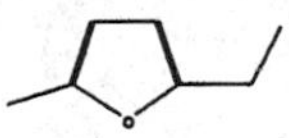

6107

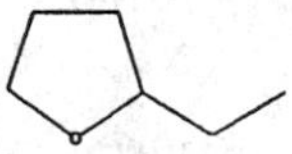

6108

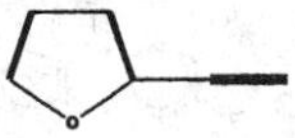

6109

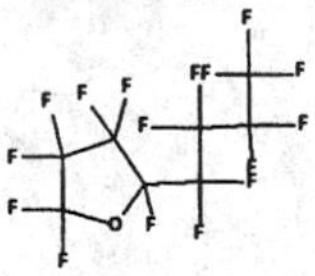

6110

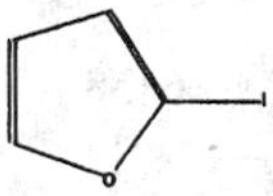

6111

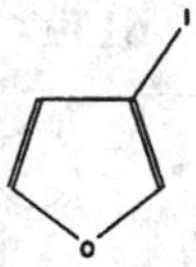

6112

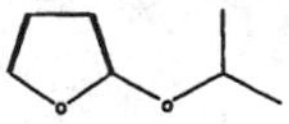

6113

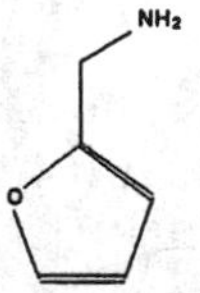

6114

6115

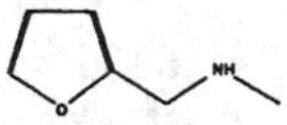

6116

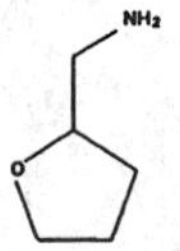

6117

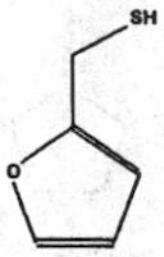

6118

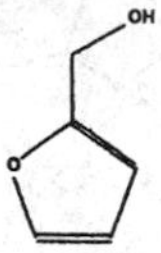

6119

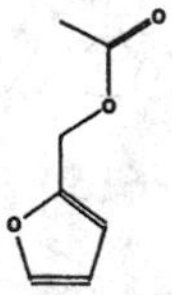

6120

6121

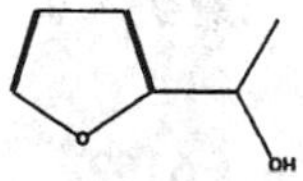

6122

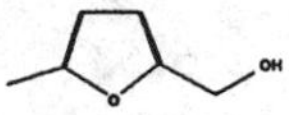

6123

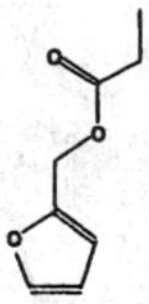

6124

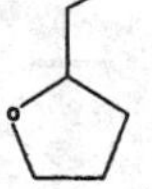

6125

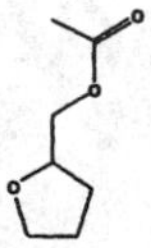

6126

6127

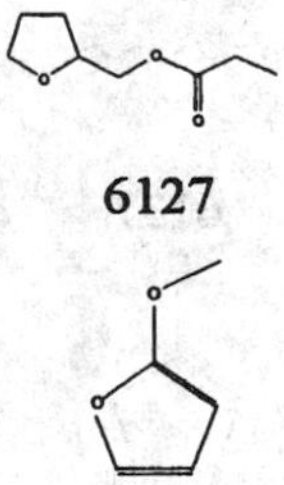

6128

6129

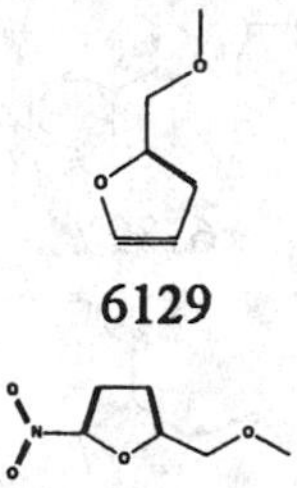

6130

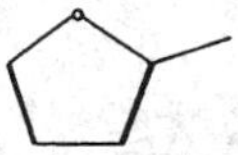

6131

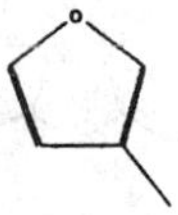

6132

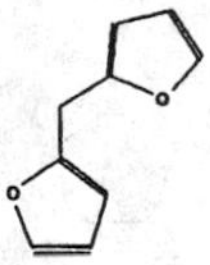

6133

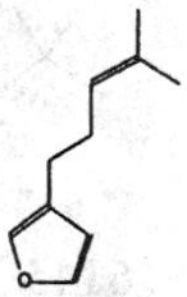

6134

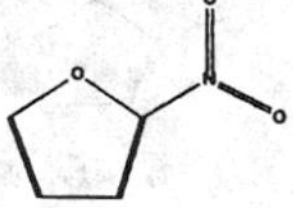

6135

6137

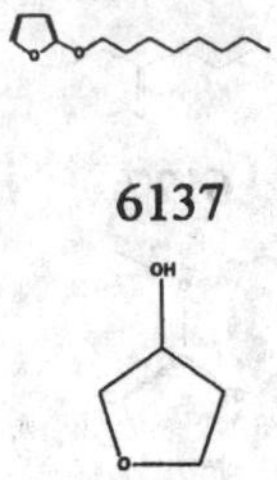

6138

6139

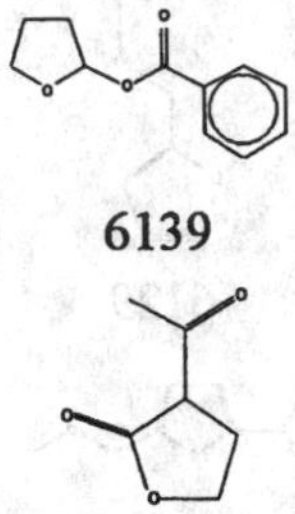

6140

6141

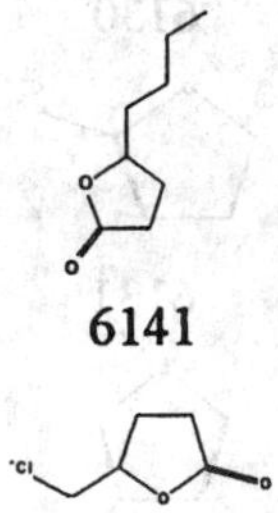

6142

6143

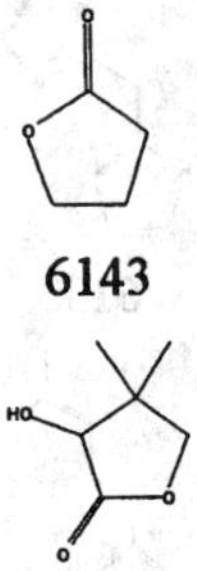

6145

6146

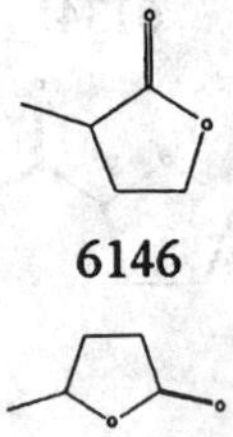

6147

6148

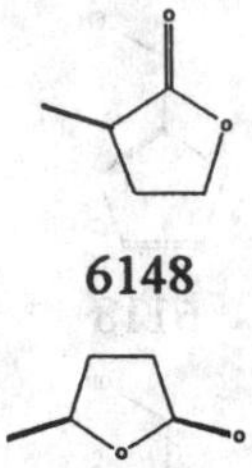

6149

6150

6151

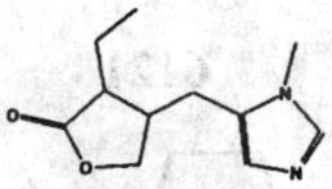

6152

6153

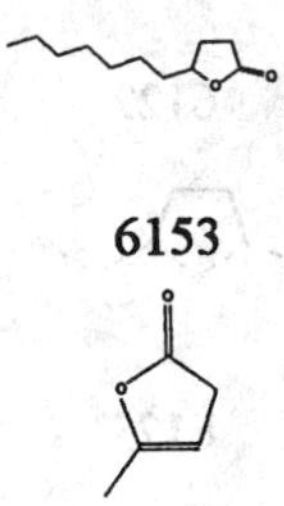

6154

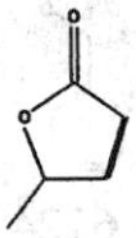

6155

6156

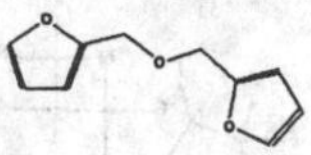

6157

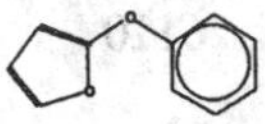

6158

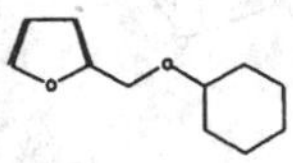

6159

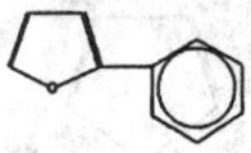

6160

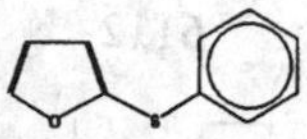

6163

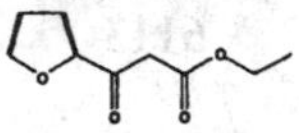

6164

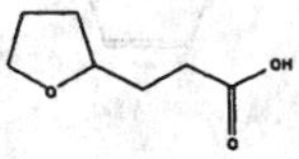

6165

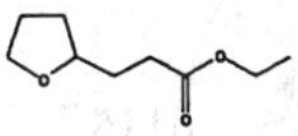

6166

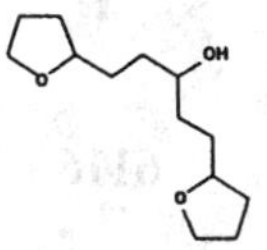

6167

6168

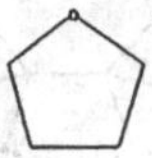

6169

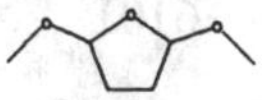

6170

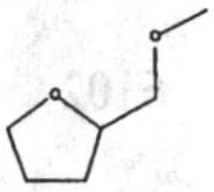

6171

6172

6174

6175

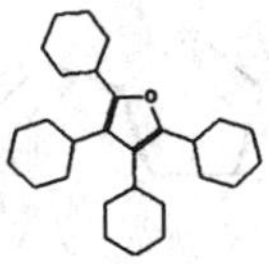

6176

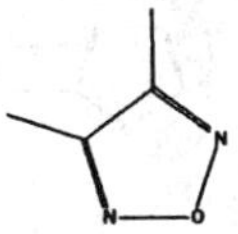

6180

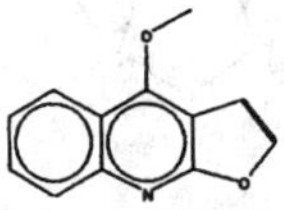

6181

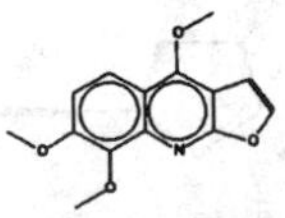

6183

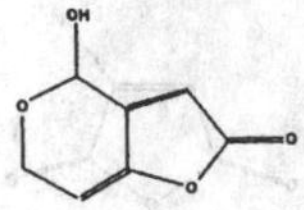

6184

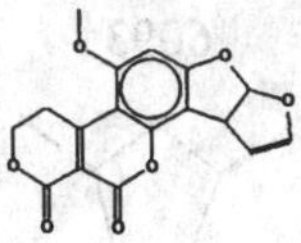

6185

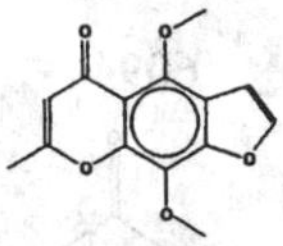

6186

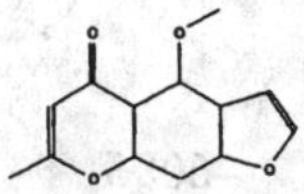

6187

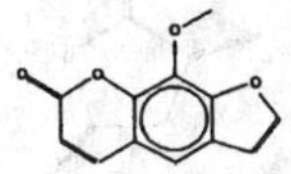

6188

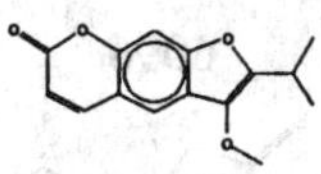

6190

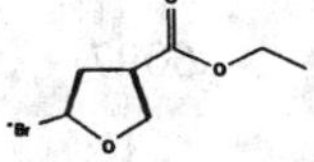

6191

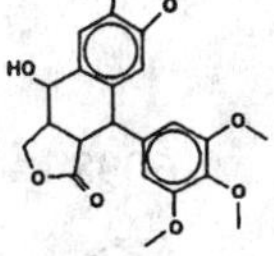

6192

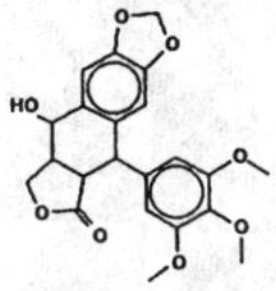

STRUCTURAL FORMULAS OF ORGANIC COMPOUNDS (continued)

In numeric order as they occur in the Table of Physical Constants of Organic Compounds

6194

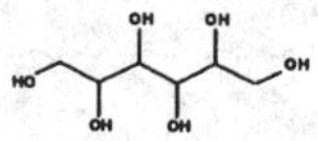

6197

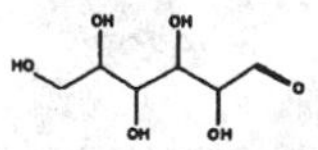

6199

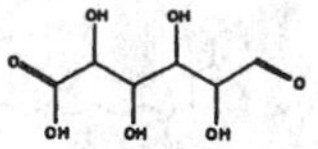

6200

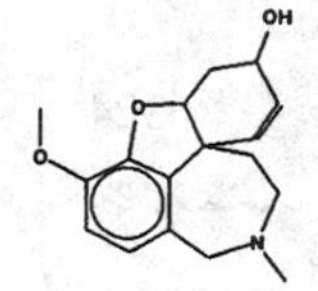

6201

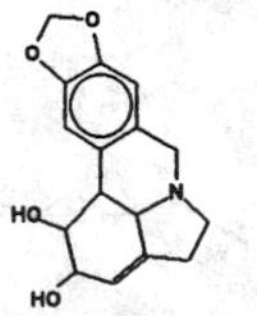

6202

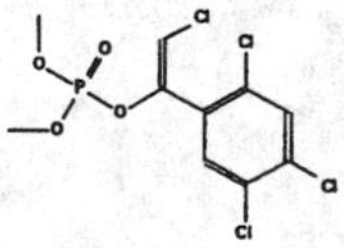

6203

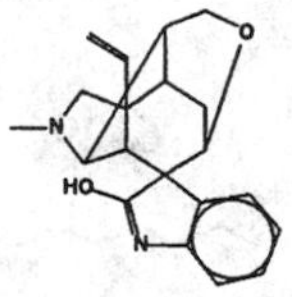

6204

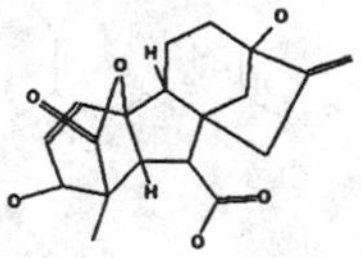

6206

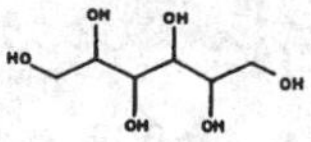

6208

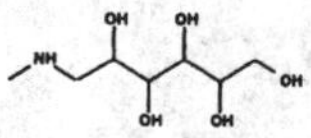

6212

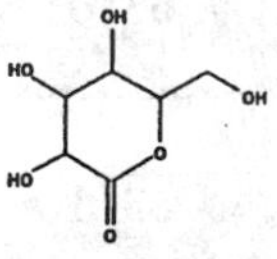

6214

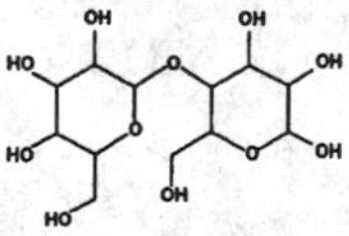

6216

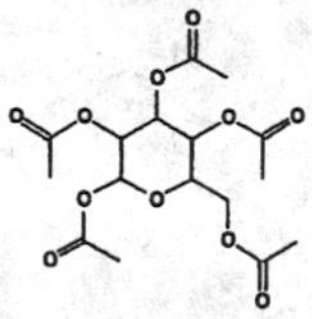

6217

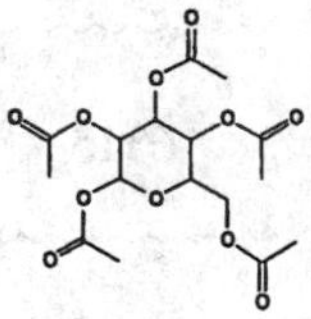

6218

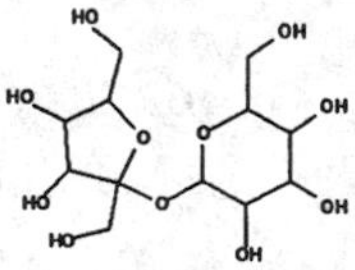

6220

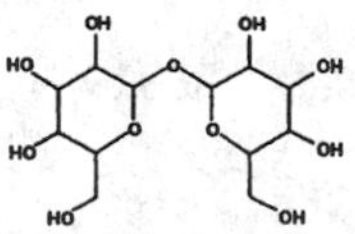

6221

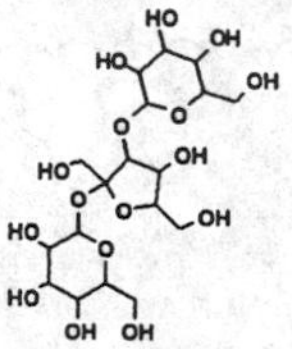

6222

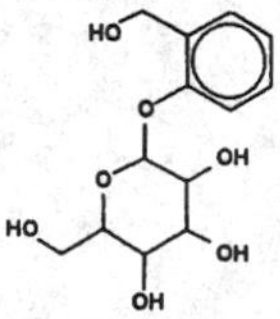

6227

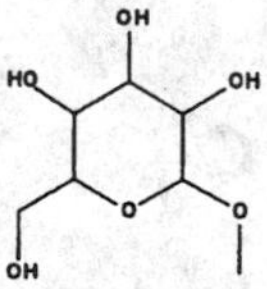

6228

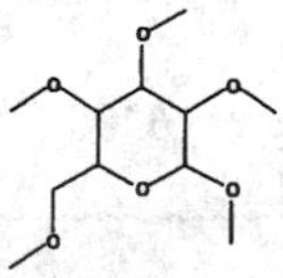

6229

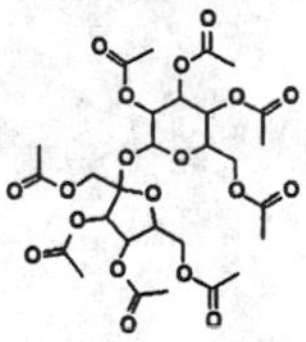

6238

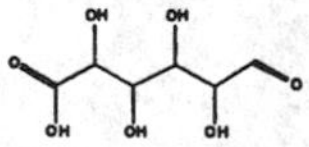

6239

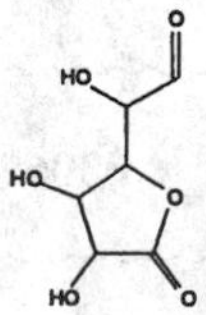

6240

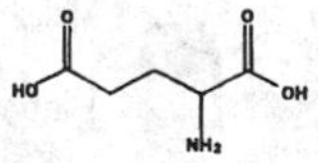

6241

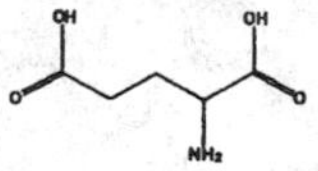

6243

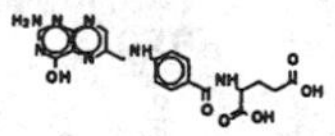

6249

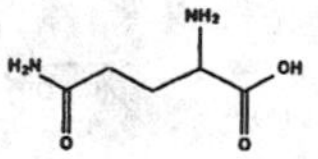

6250

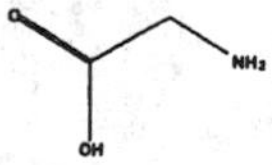

6251

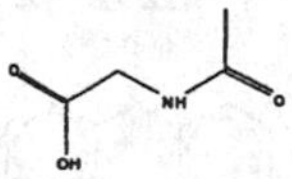

6252

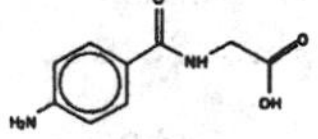

6254

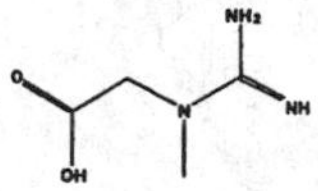

6255

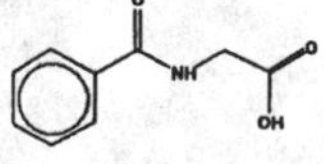

6260

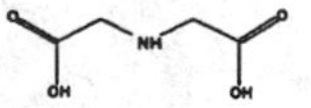

6262

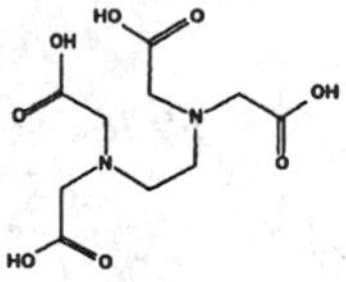

6263

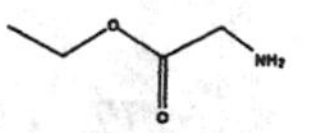

6265

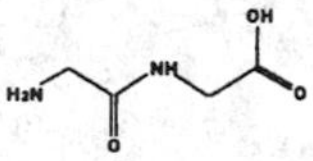

6266

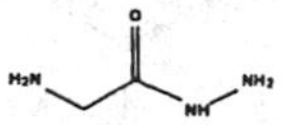

6267

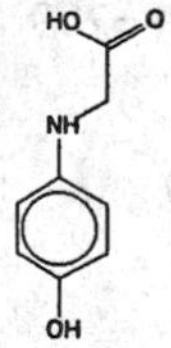

6268

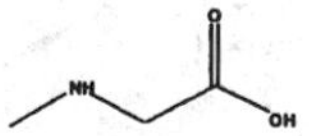

6272

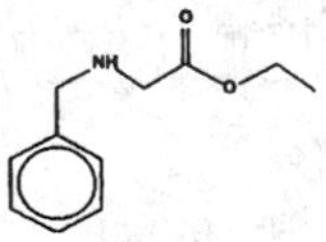

6275

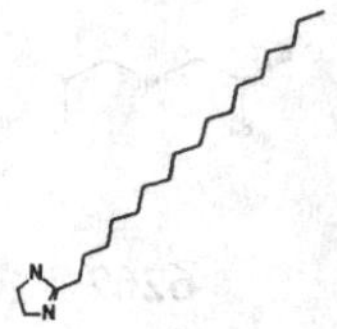

6276

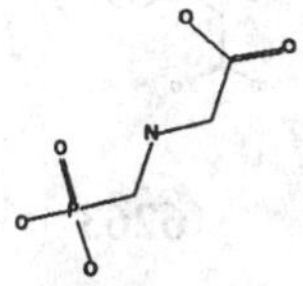

6277

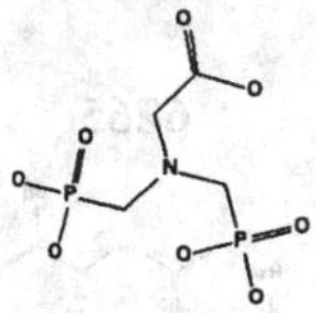

6278

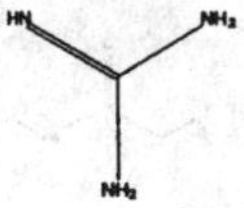

6279

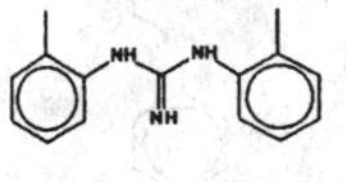

6280

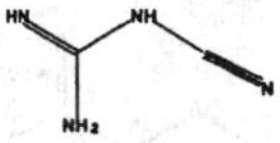

6282

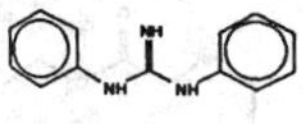

6284

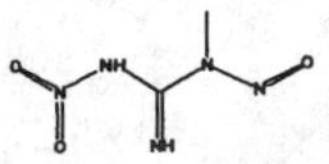

6287

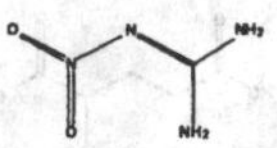

6288

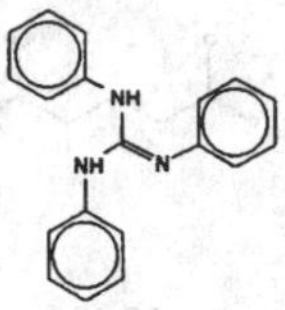

6289

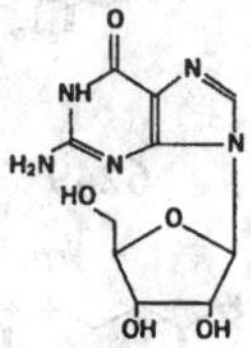

6291

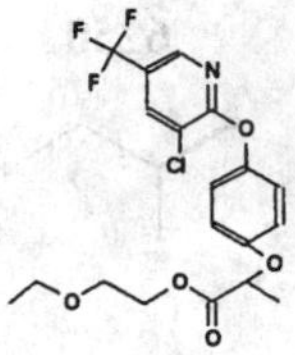

6292

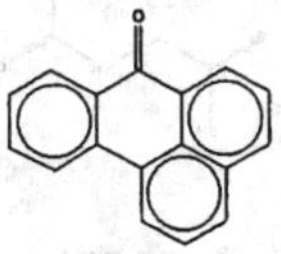

6294

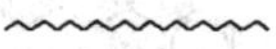

6295

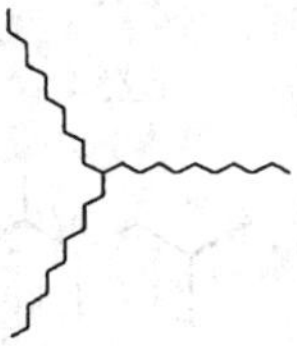

6296

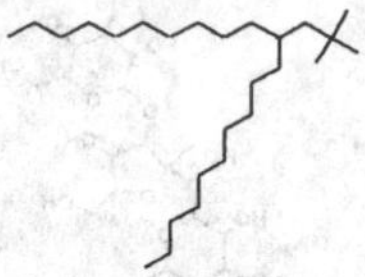

6297

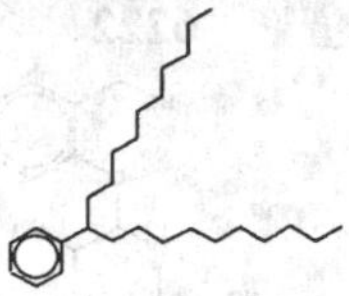

6298

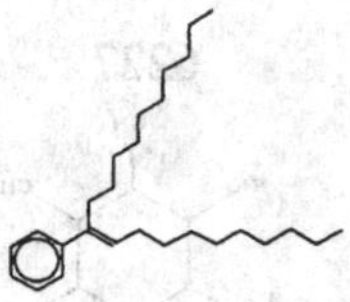

6299

6300

6301

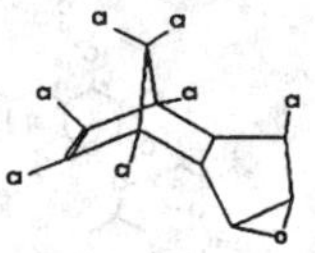

6302

6304

6305

6306

6307

6308

6309

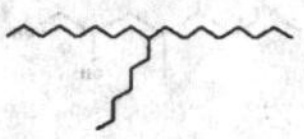

6310

6311

6312

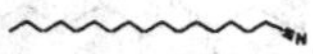

6313

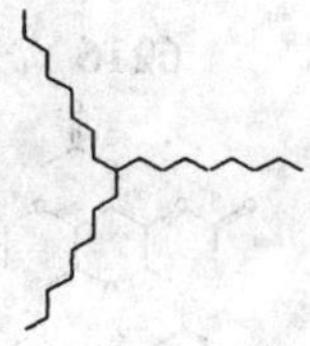

6314

6315

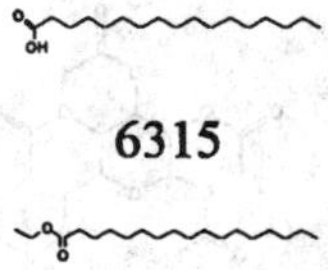

6316

6317

6318

6319

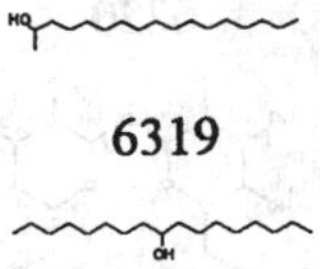

6320

6321

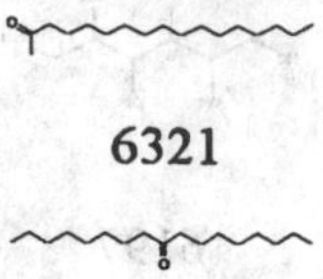

6322

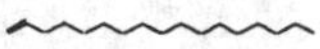

6323

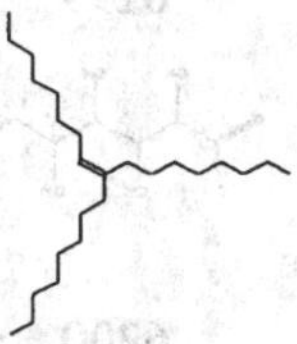

6324

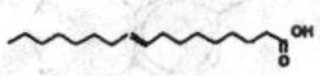

6325

6326

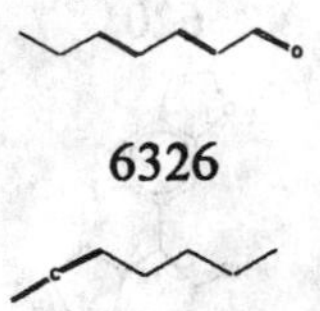

6327

6328

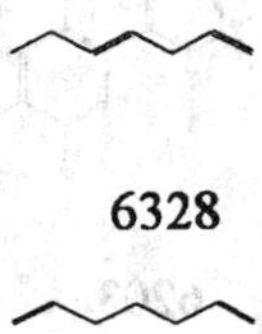

6329

6330

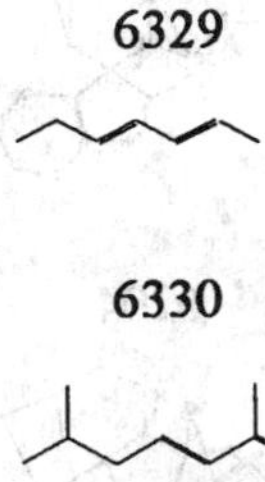

STRUCTURAL FORMULAS OF ORGANIC COMPOUNDS (continued)

In numeric order as they occur in the Table of Physical Constants of Organic Compounds

6331, 6332, 6333, 6334, 6335, 6336, 6337, 6338, 6339, 6340

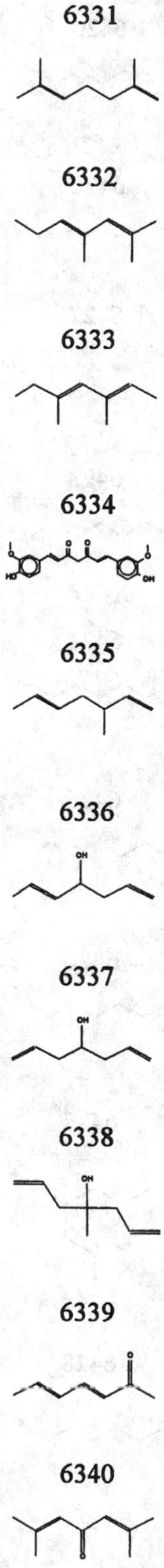

6341, 6342, 6343, 6344, 6345, 6346, 6347, 6348, 6349, 6350, 6351

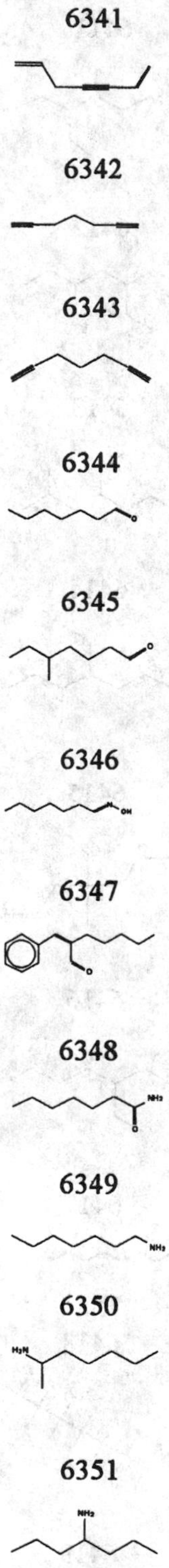

6352, 6353, 6354, 6355, 6356, 6357, 6358, 6359, 6360, 6361, 6362

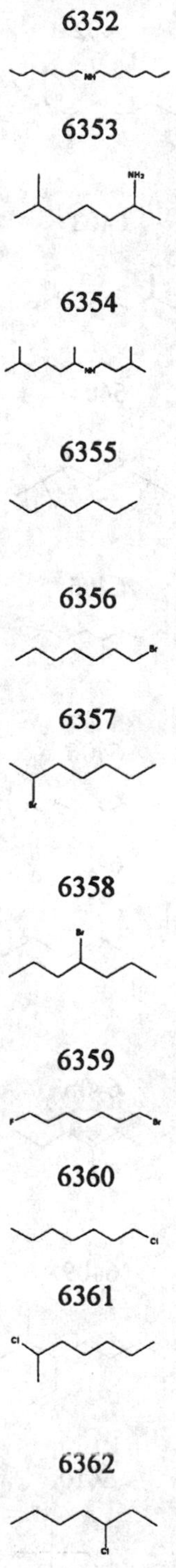

6363, 6364, 6365, 6366, 6367, 6368, 6369, 6370, 6371

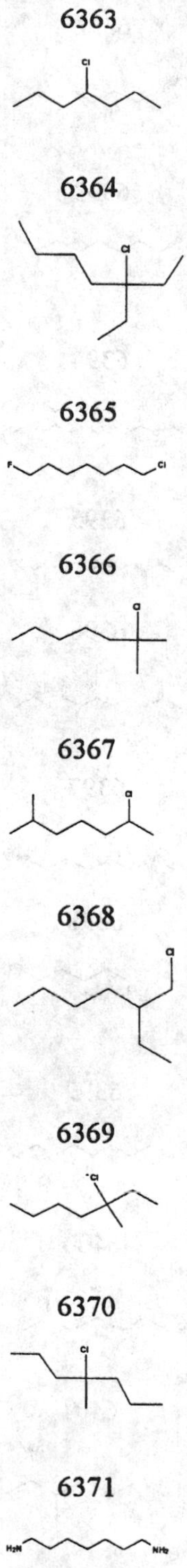

6372, 6373, 6374, 6375, 6376, 6377, 6378, 6379, 6380, 6381

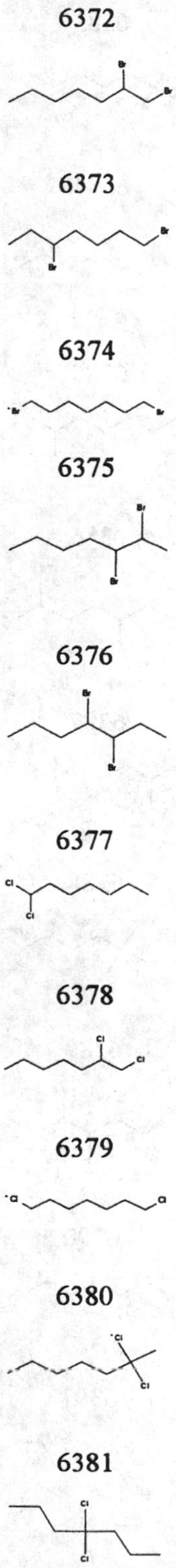

STRUCTURAL FORMULAS OF ORGANIC COMPOUNDS (continued)

In numeric order as they occur in the Table of Physical Constants of Organic Compounds

6383

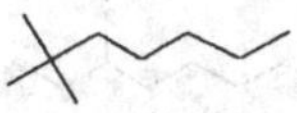

6384

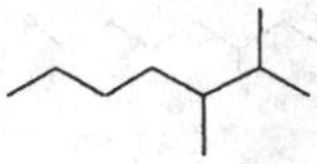

6385

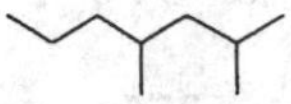

6386

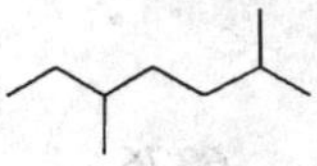

6387

6388

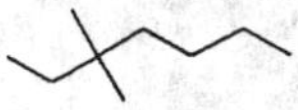

6389

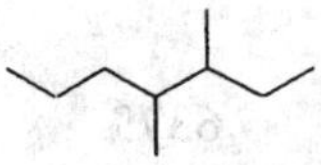

6390

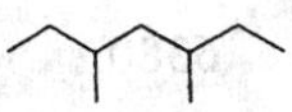

6391

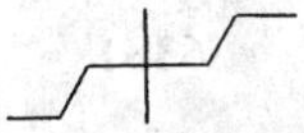

6392

6393

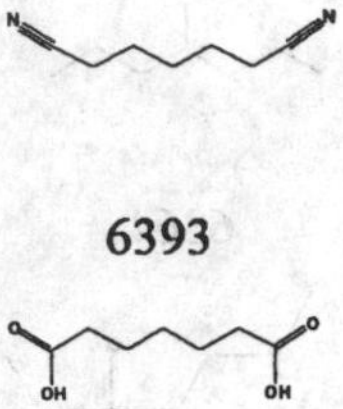

6394

6395

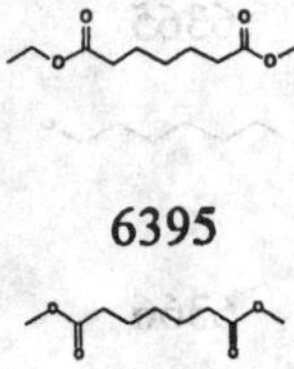

6396

6397

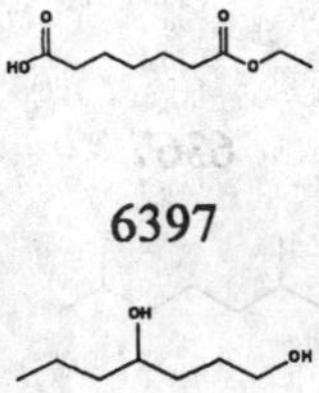

6398

6399

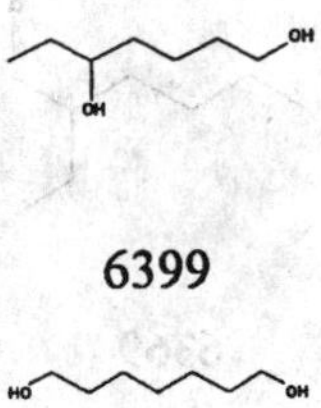

6400

6401

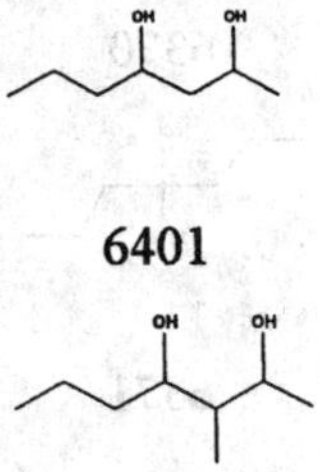

6402

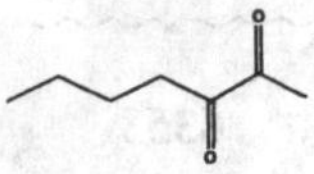

6403

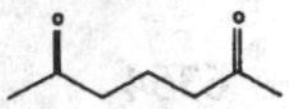

6404

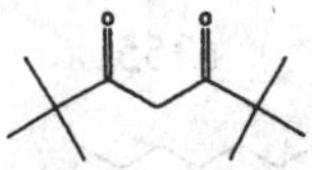

6405

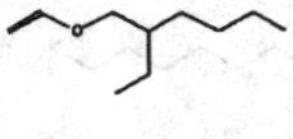

6406

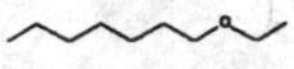

6407

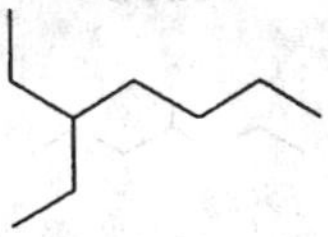

6408

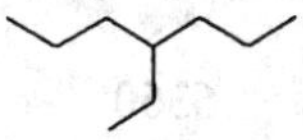

6409

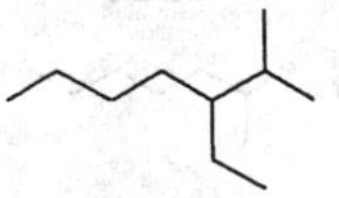

6410

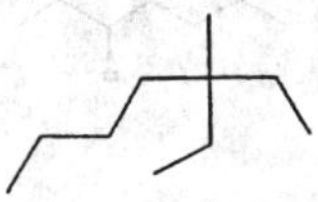

6411

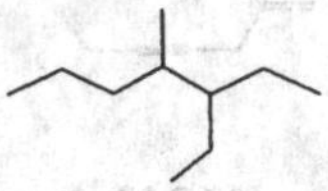

6412

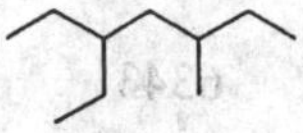

6413

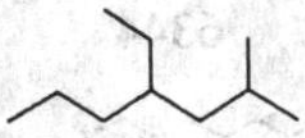

6414

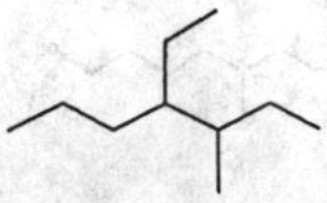

6415

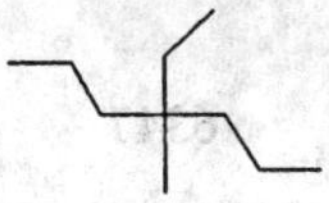

6416

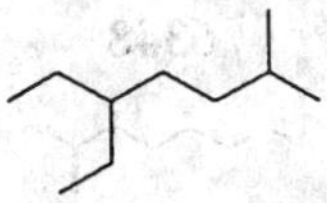

6417

6418

6419

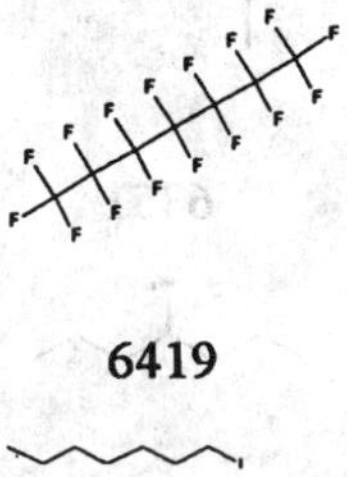

6420

6421

6422

6423

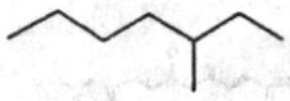

6424

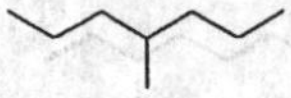

6425

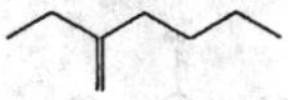

6426

6427

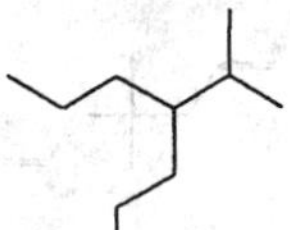

6428

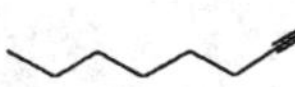

In numeric order as they occur in the Table of Physical Constants of Organic Compounds

6429

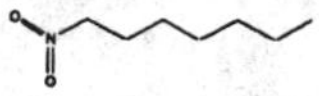

6430

6431

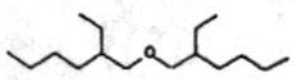

6432

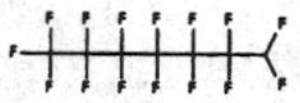

6433

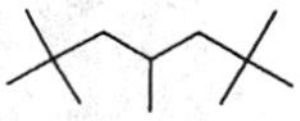

6434

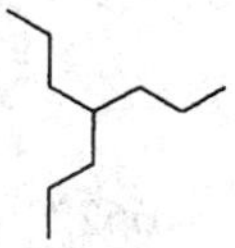

6435

6436

6437

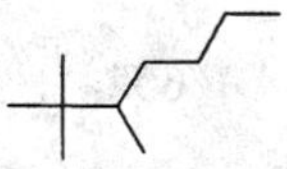

6438

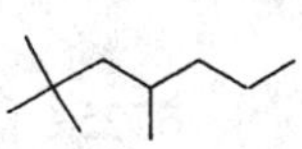

6439

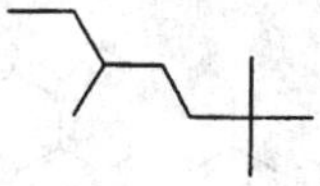

6440

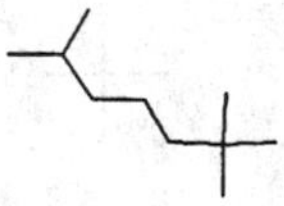

6441

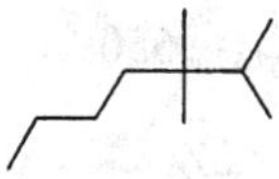

6442

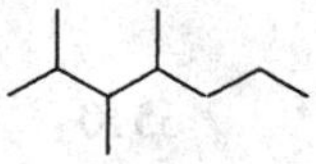

6443

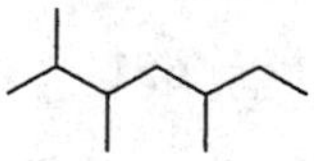

6444

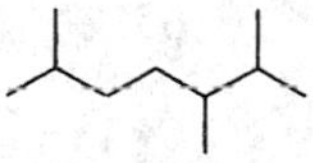

6445

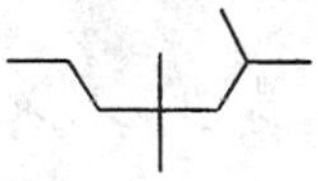

6446

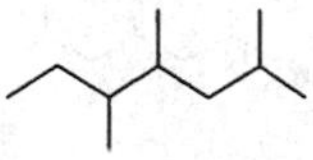

6447

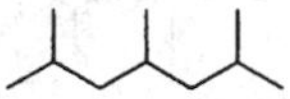

6448

6449

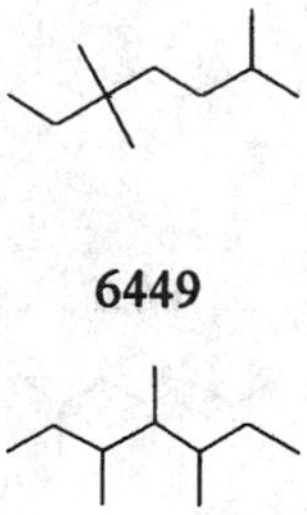

6450

6451

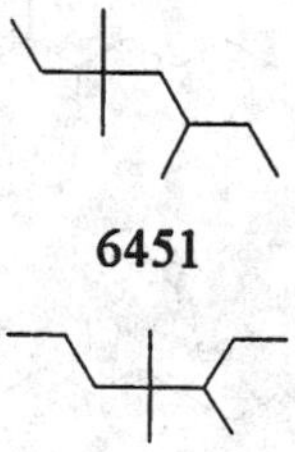

6452

6453

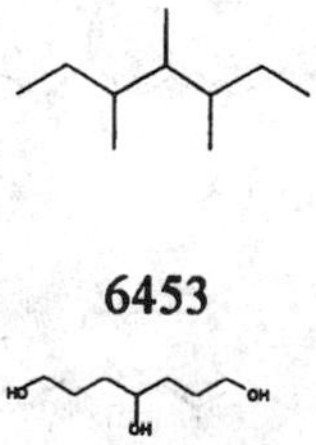

6454

6455

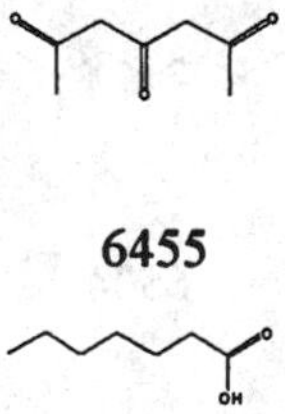

6456

6457

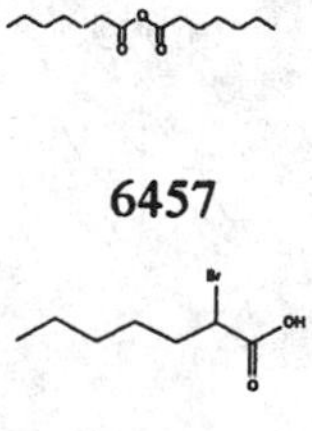

6459

6461

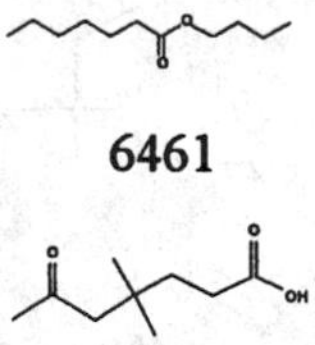

6462

6463

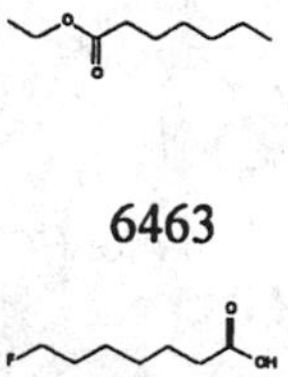

6464

6465

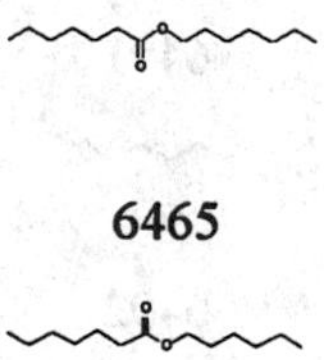

6466

6467

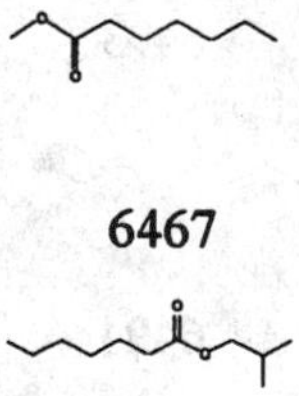

6468

6469

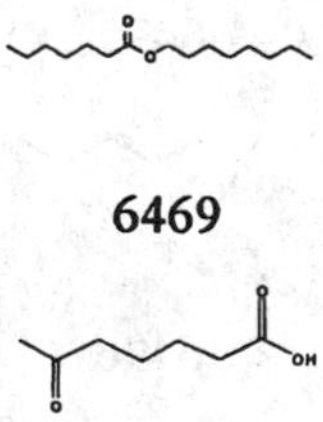

6470

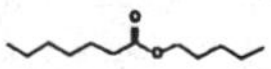

6471

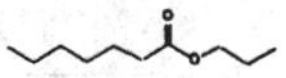

6472

6473

6474

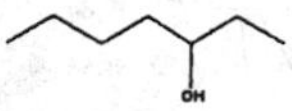

6475

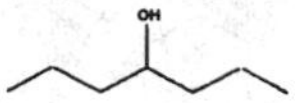

6476

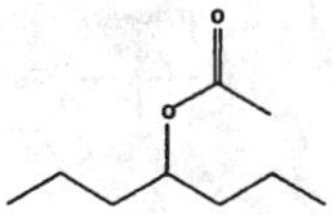

6477

6478

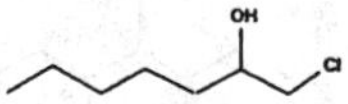

6479

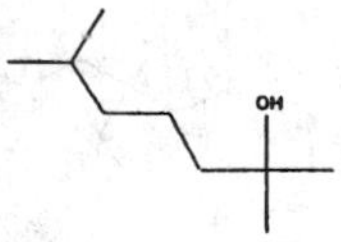

STRUCTURAL FORMULAS OF ORGANIC COMPOUNDS (continued)

In numeric order as they occur in the Table of Physical Constants of Organic Compounds

6480

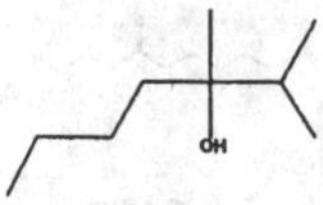

6481

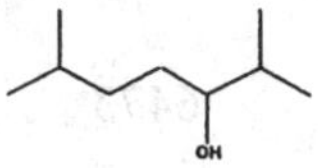

6482

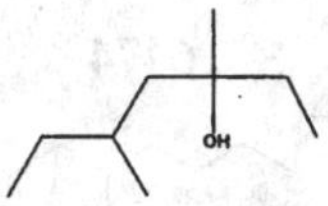

6483

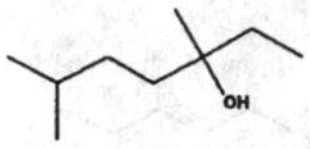

6484

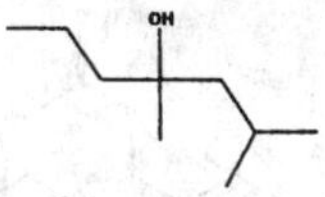

6485

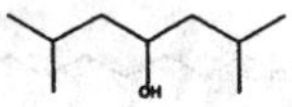

6486

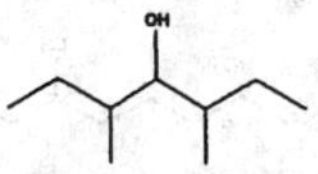

6487

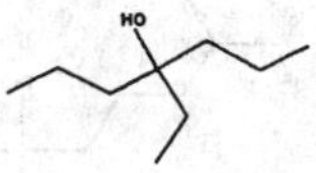

6488

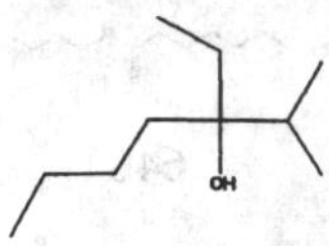

6489

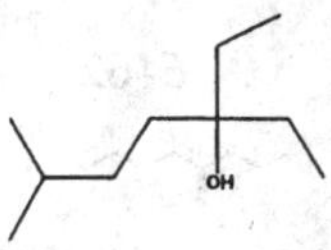

6490

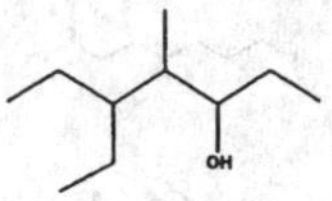

6491

6492

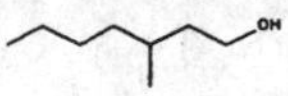

6493

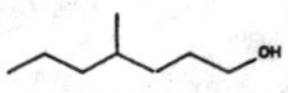

6494

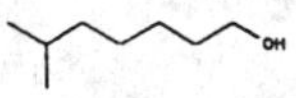

6495

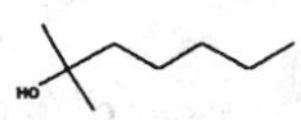

6496

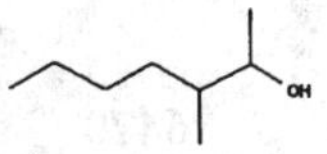

6497

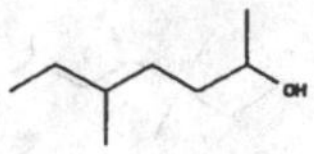

6498

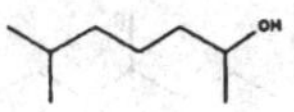

6499

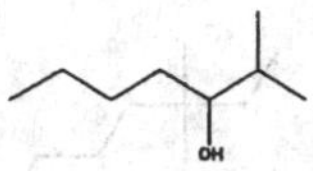

6500

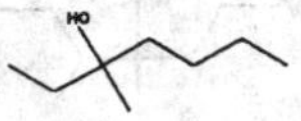

6501

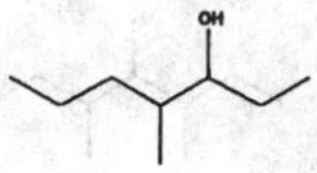

6502

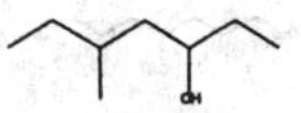

6503

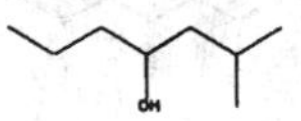

6504

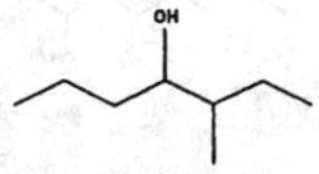

6505

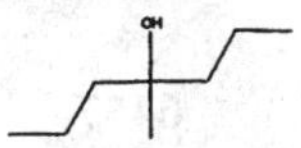

6506

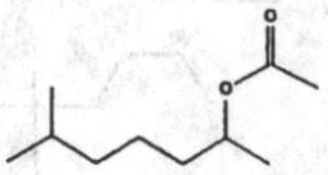

6508

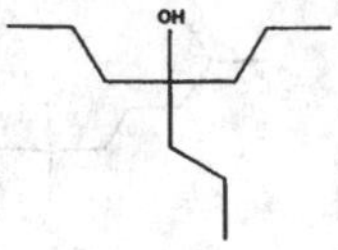

6509

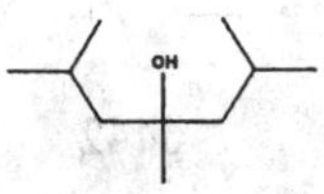

6510

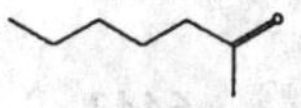

6511

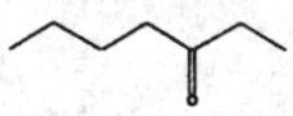

6512

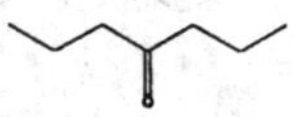

6513

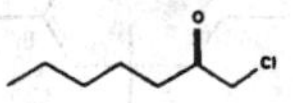

6514

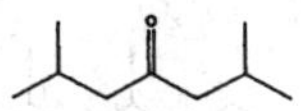

6515

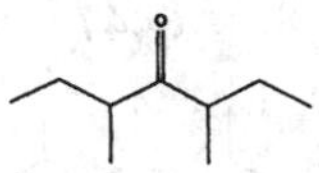

6517

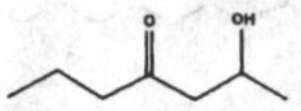

6518

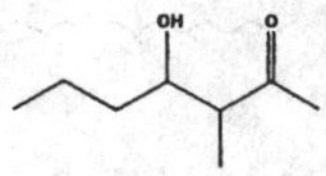

6519

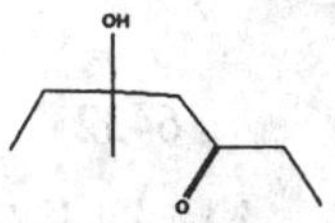

6520

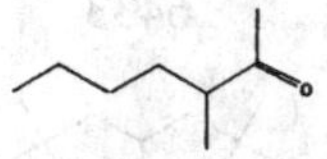

6521

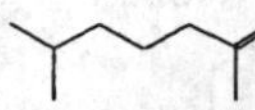

6522

6523

6524

6525

In numeric order as they occur in the Table of Physical Constants of Organic Compounds

6526

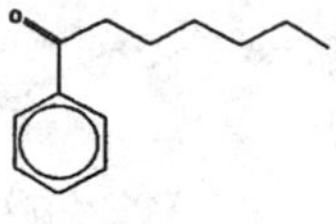

6527

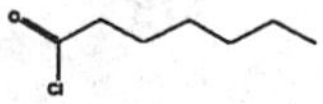

6528

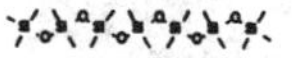

6529

6530

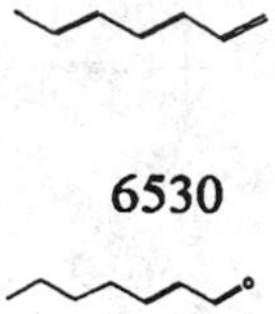

6531

6532

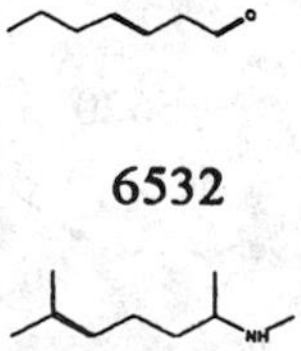

6533

6534

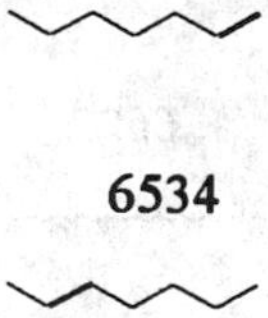

6535

6536

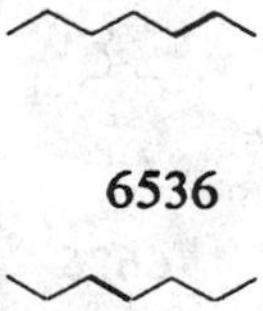

6537

6538

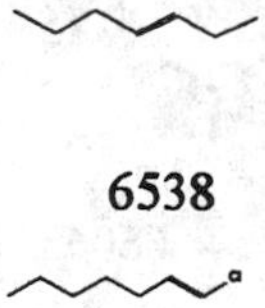

6539

6540

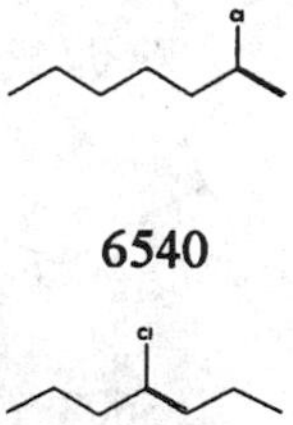

6541

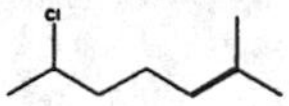

6542

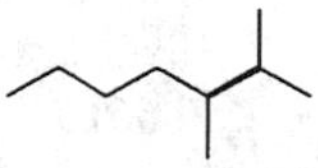

6543

6544

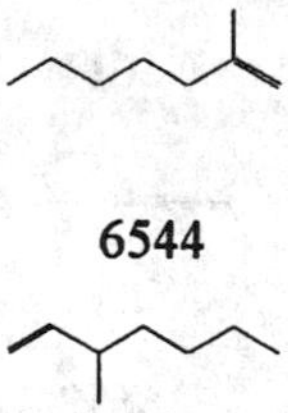

6545

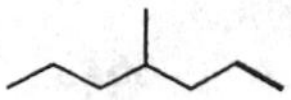

6546

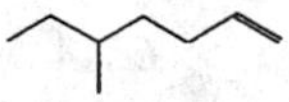

6547

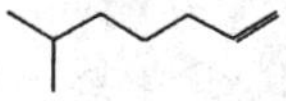

6548

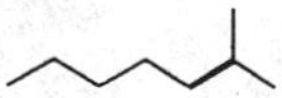

6549

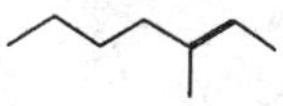

6550

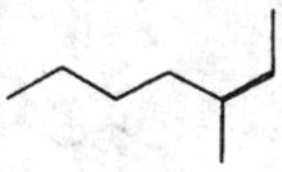

6551

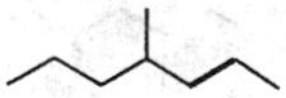

6552

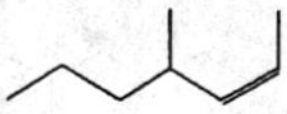

6553

6554

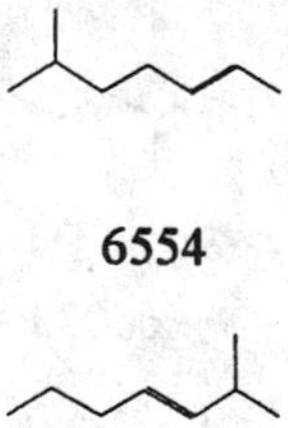

6555

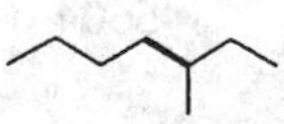

6556

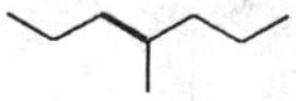

6557

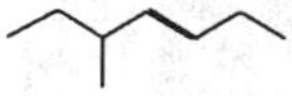

6558

6559

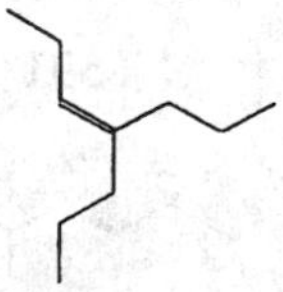

6560

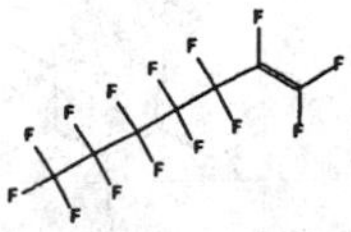

6561

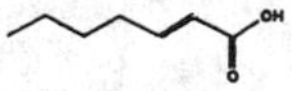

6562

6563

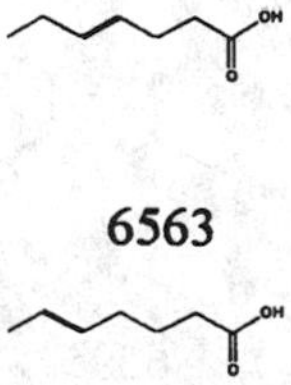

6564

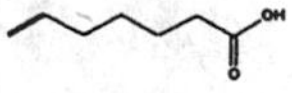

6565

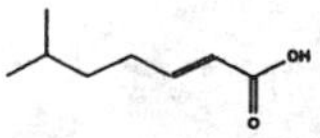

6566

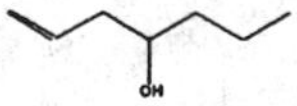

6567

6568

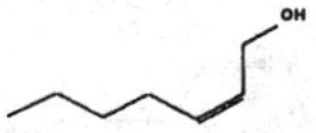

6570

6572

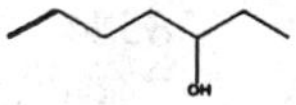

6573

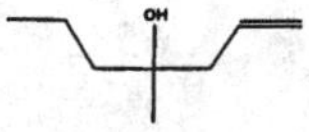

6574

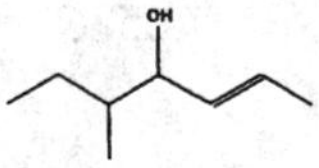

6575

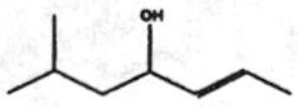

6578

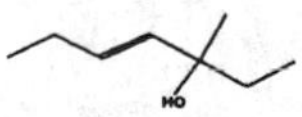

STRUCTURAL FORMULAS OF ORGANIC COMPOUNDS (continued)

In numeric order as they occur in the Table of Physical Constants of Organic Compounds

6579

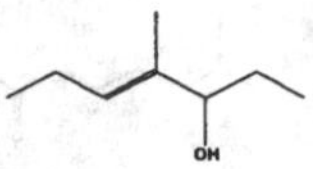

6580

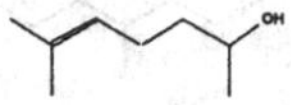

6581

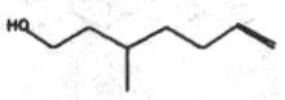

6582

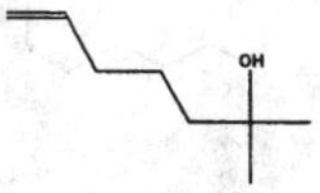

6583

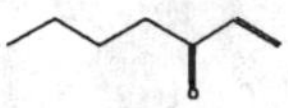

6584

6585

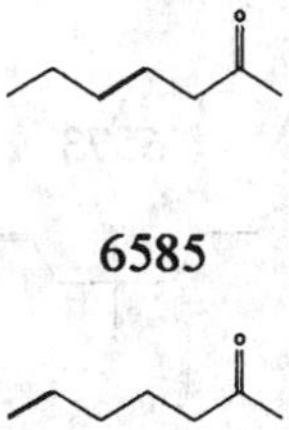

6586

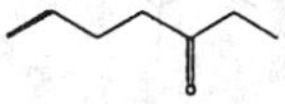

6587

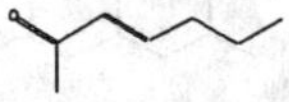

6588

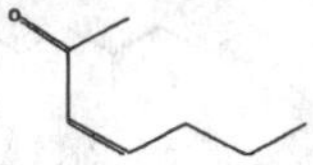

6589

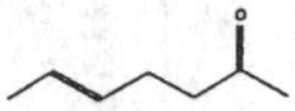

6590

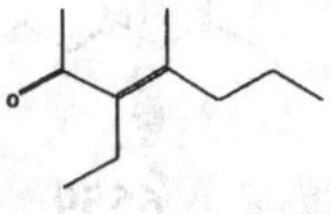

6591

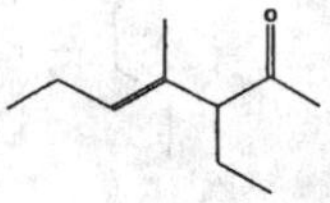

6592

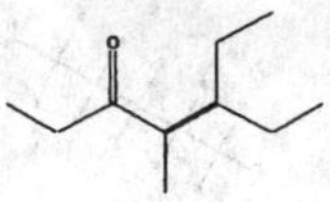

6593

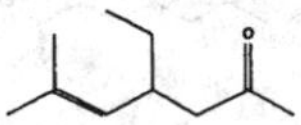

6594

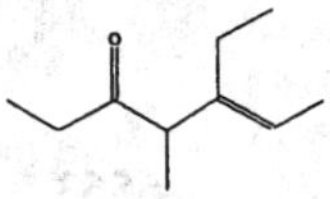

6595

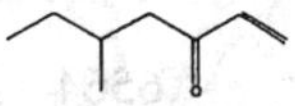

6596

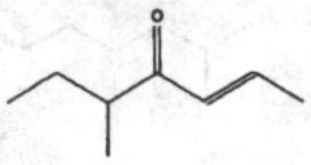

6597

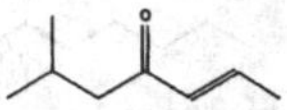

6598

6599

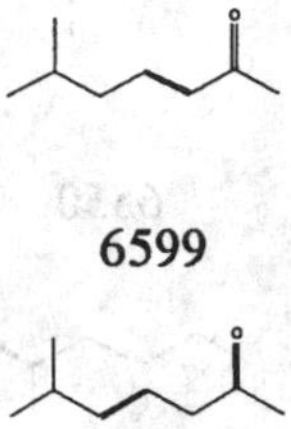

6600

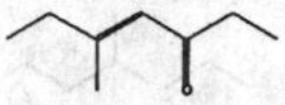

6601

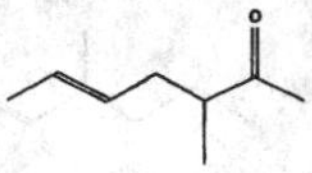

6602

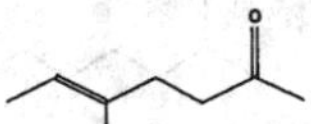

6603

6604

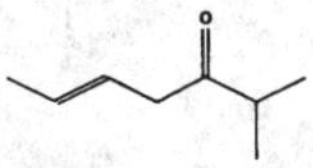

6605

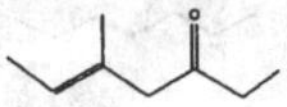

6606

6607

6608

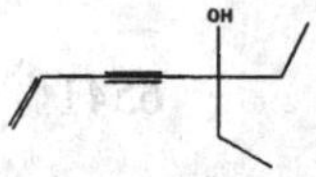

6609

6610

6611

6612

6613

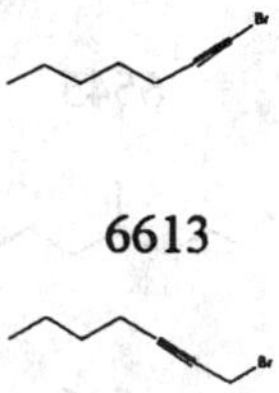

6614

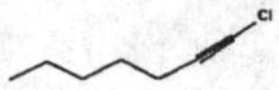

6615

6616

6617

6618

6619

6620

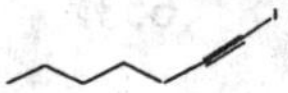

6621

6622

6623

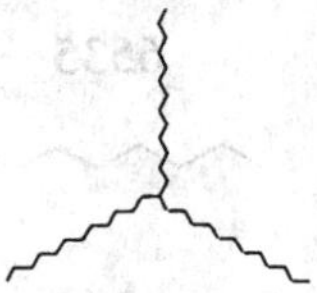

6625

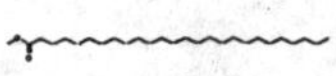

In numeric order as they occur in the Table of Physical Constants of Organic Compounds

6626

6627

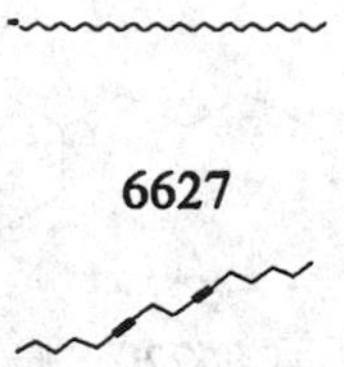

6628

6629

6630

6631

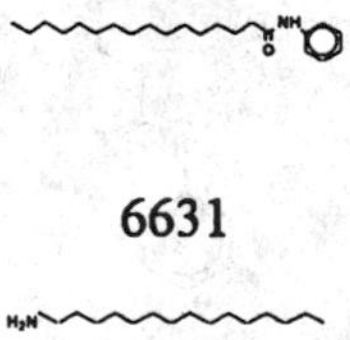

6632

6633

6634

6635

6636

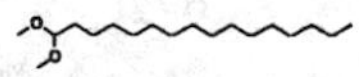

6637

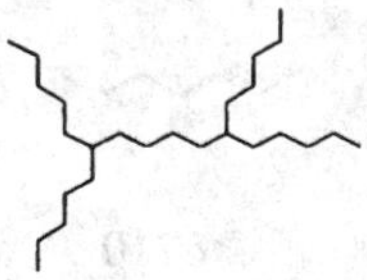

6638

6639

6640

6641

6642

6643

6644

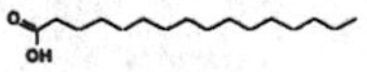

6647

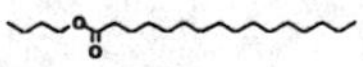

6649

6650

6651

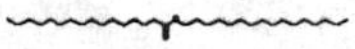

6652

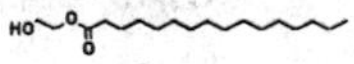

6653

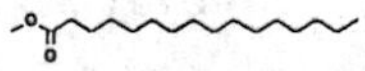

6654

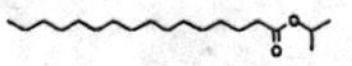

6655

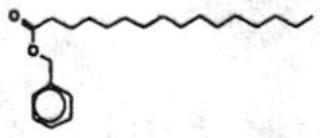

6656

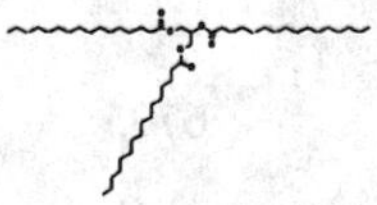

6657

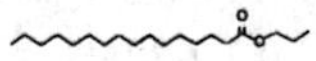

6658

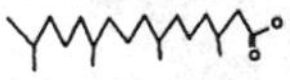

6659

6660

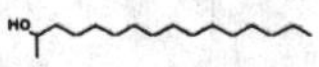

6661

6662

6663

6664

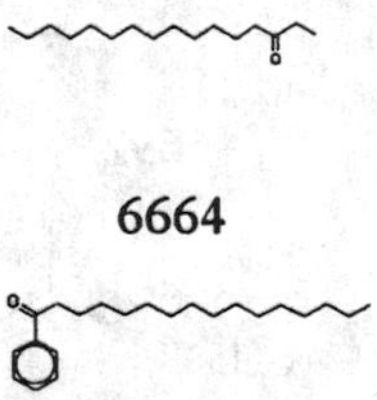

6665

6666

6667

6668

6669

6670

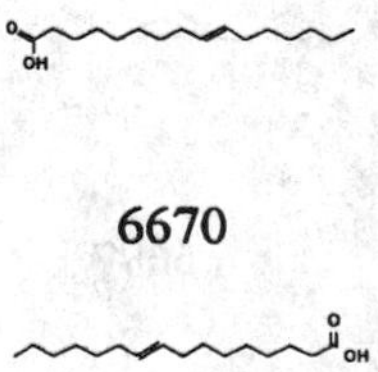

6671

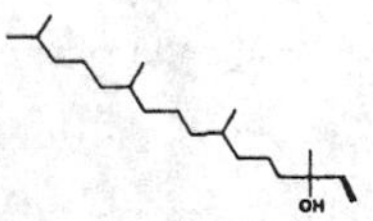

6672

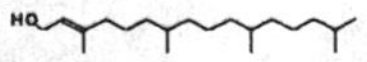

6673

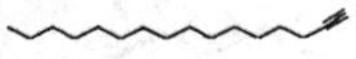

6674

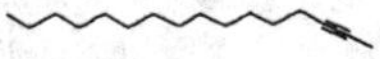

6675

6676

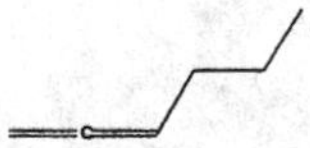

6677

6678

6679

6680

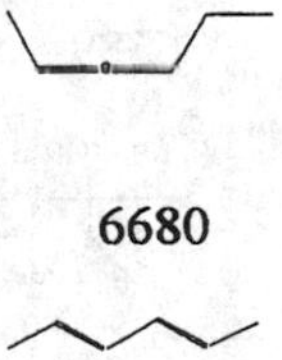

In numeric order as they occur in the Table of Physical Constants of Organic Compounds

6681

6682

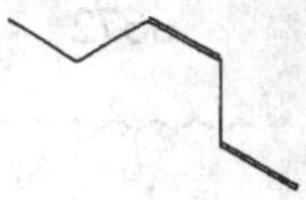

6683

6684

6685

6686

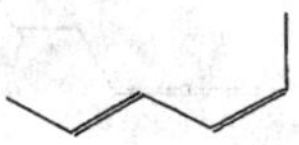

6687

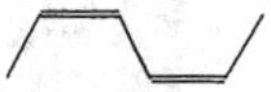

6688

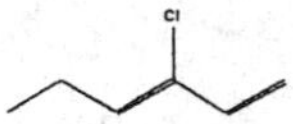

6689

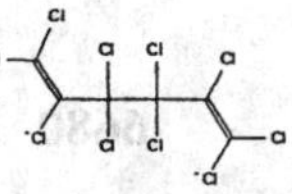

6690

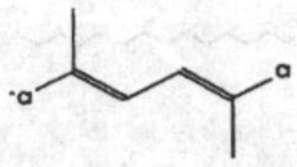

6691

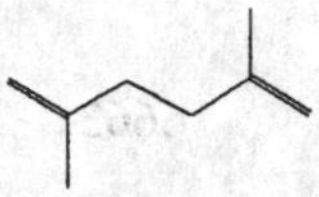

6692

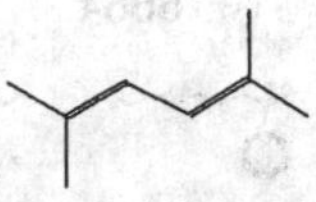

6693

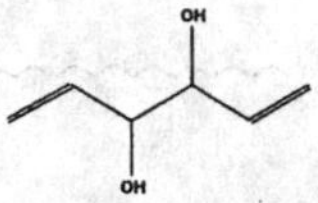

6694

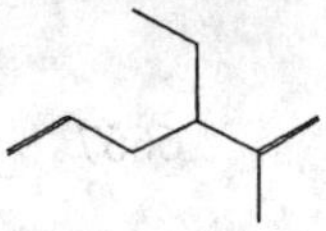

6695

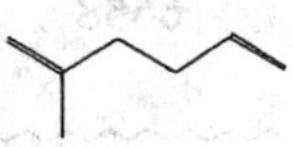

6696

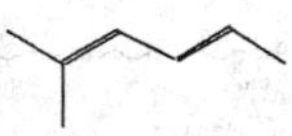

6697

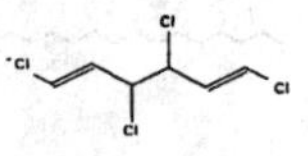

6698

6699

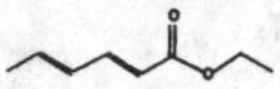

6700

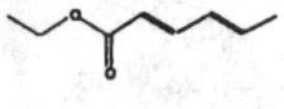

6701

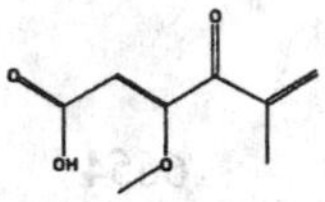

6702

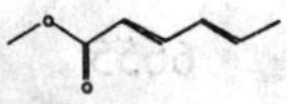

6703

6705

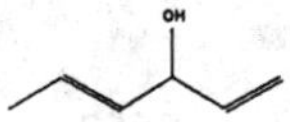

6706

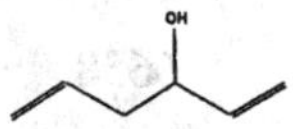

6707

6708

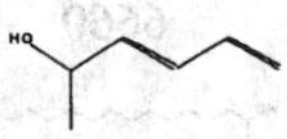

6709

6710

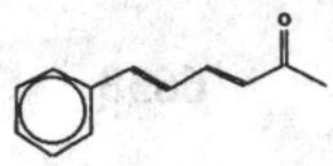

6711

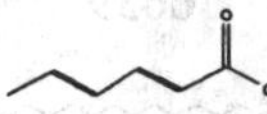

6712

6713

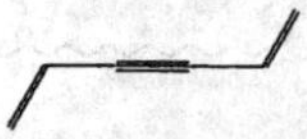

6714

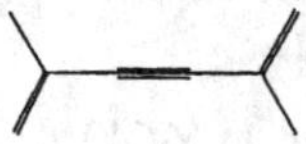

6715

6716

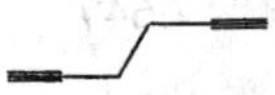

6717

6718

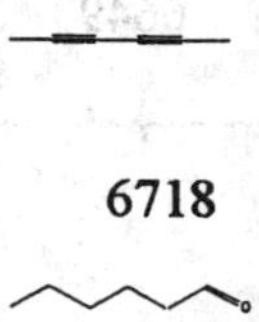

6719

6720

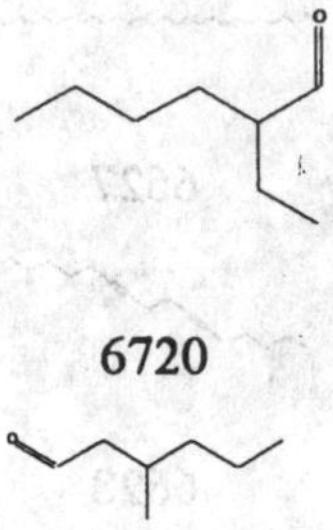

6721

6722

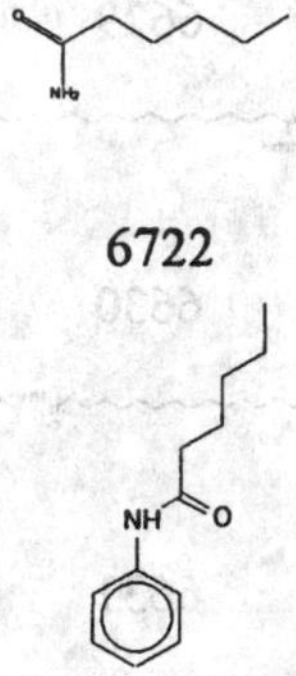

6723

6724

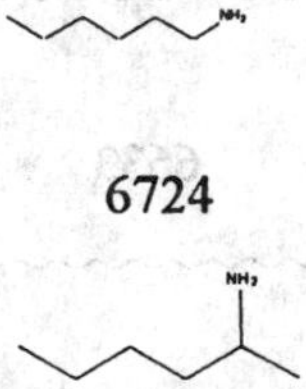

6725

6726

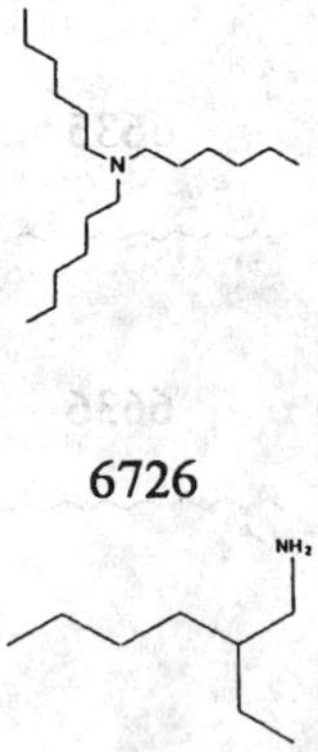

STRUCTURAL FORMULAS OF ORGANIC COMPOUNDS (continued)

In numeric order as they occur in the Table of Physical Constants of Organic Compounds

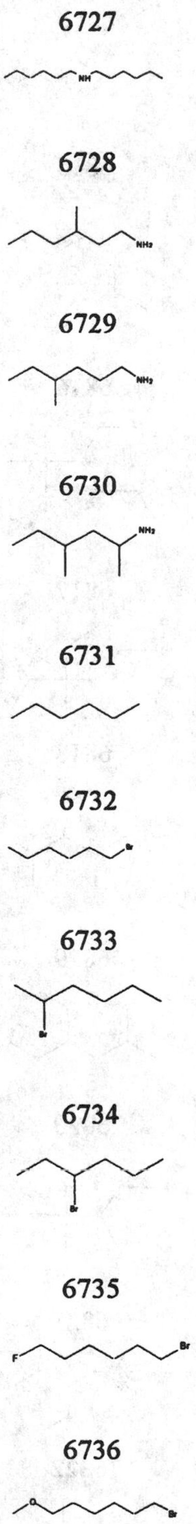

6727, 6728, 6729, 6730, 6731, 6732, 6733, 6734, 6735, 6736

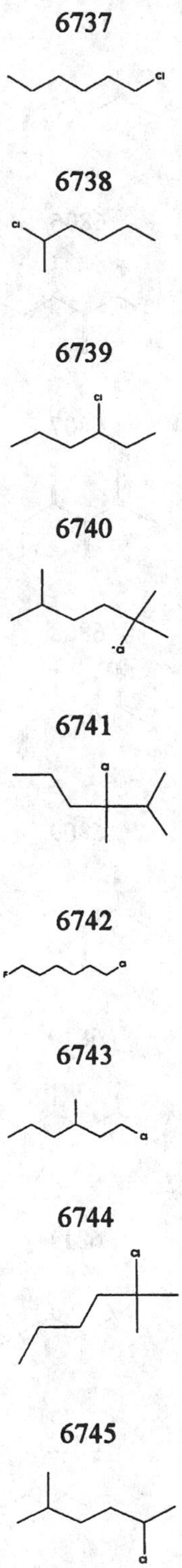

6737, 6738, 6739, 6740, 6741, 6742, 6743, 6744, 6745

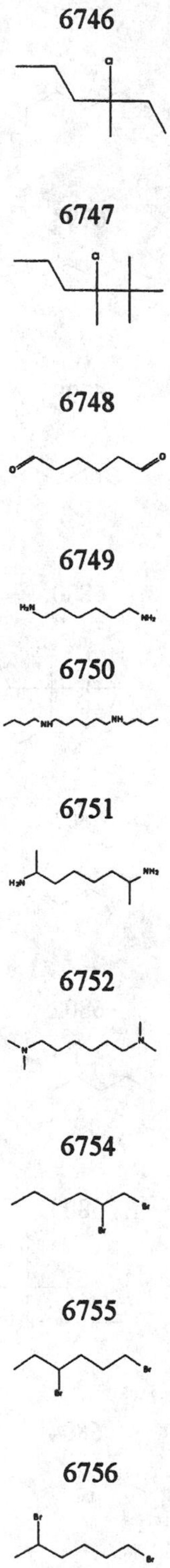

6746, 6747, 6748, 6749, 6750, 6751, 6752, 6754, 6755, 6756

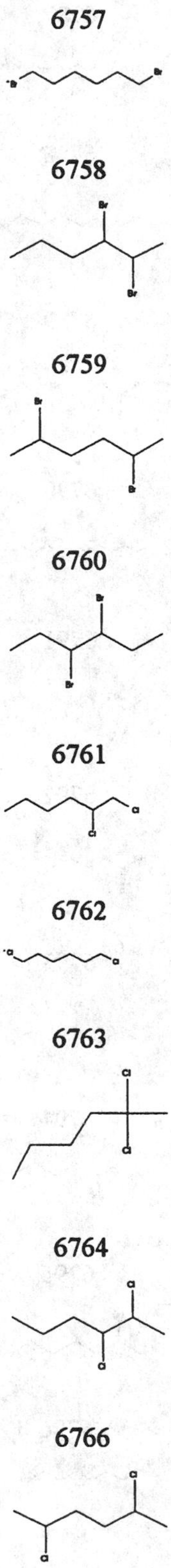

6757, 6758, 6759, 6760, 6761, 6762, 6763, 6764, 6766

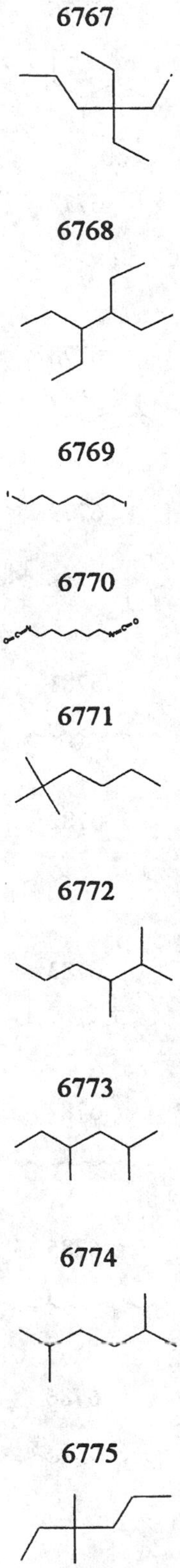

6767, 6768, 6769, 6770, 6771, 6772, 6773, 6774, 6775

In numeric order as they occur in the Table of Physical Constants of Organic Compounds

6776
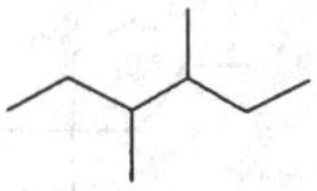

6777
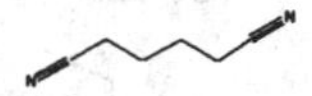

6778
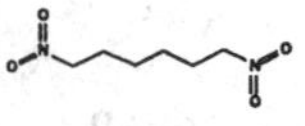

6779
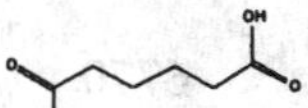

6781, 6782
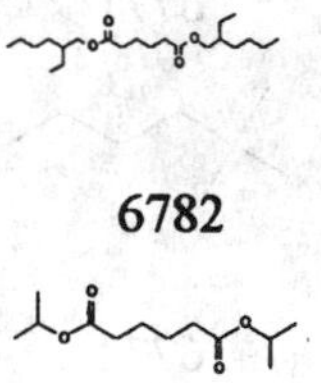

6783, 6784
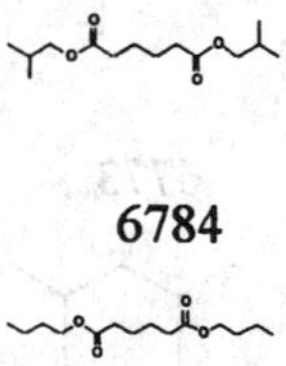

6785, 6786
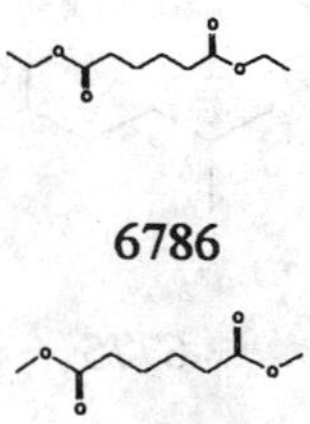

6787
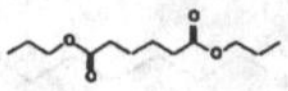

6788
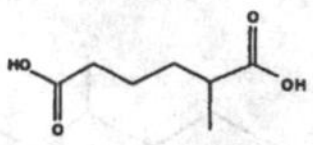

6789
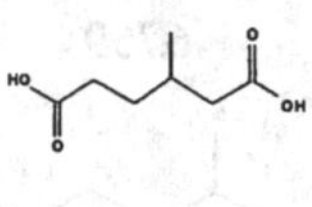

6790
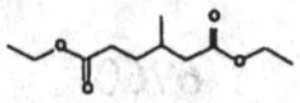

6791
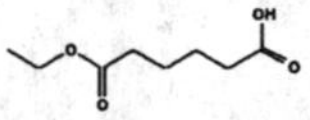

6792
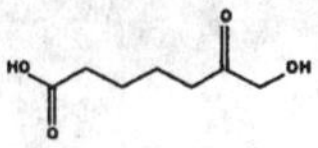

6793, 6794
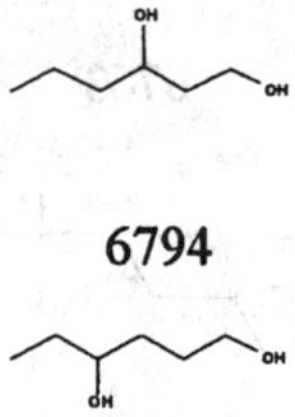

6795, 6796
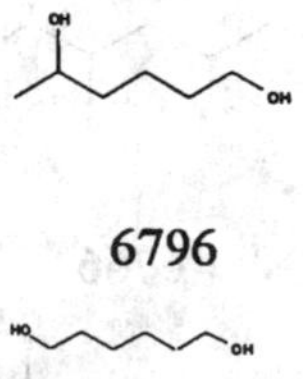

6797
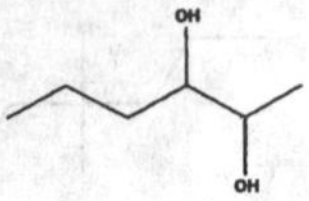

6798
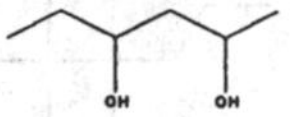

6799
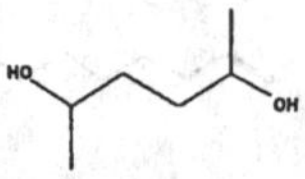

6800
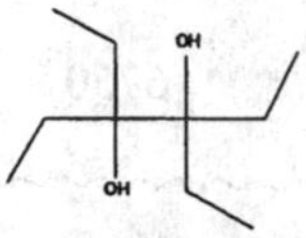

6801
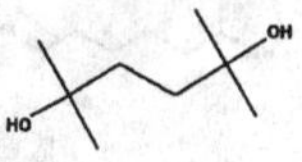

6802
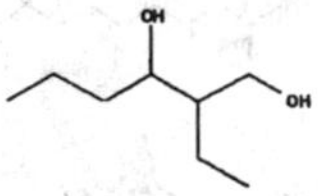

6803
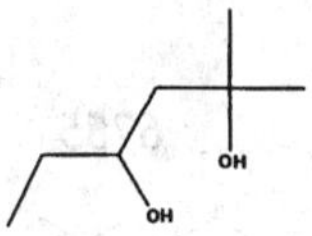

6804
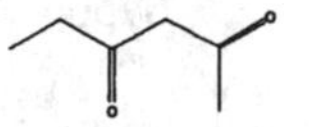

6805
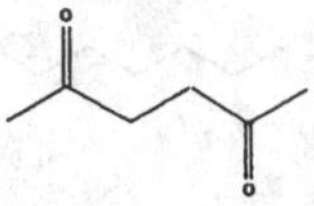

6806
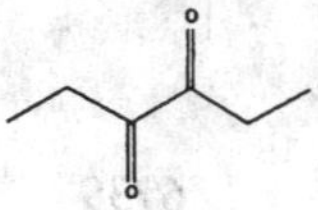

6807
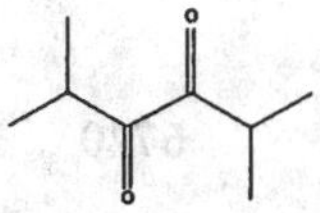

6808
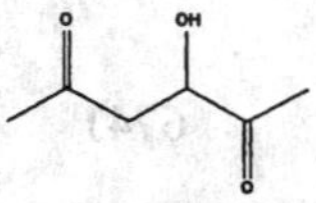

6809
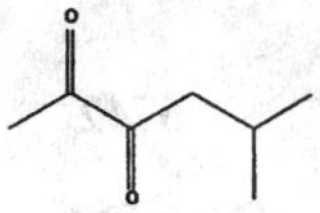

6810
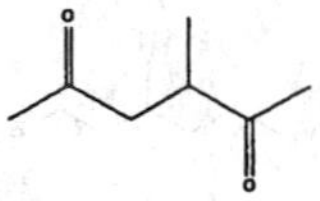

6811
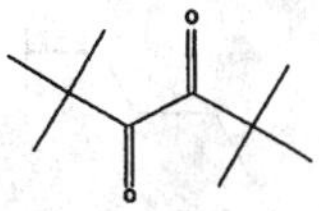

6812, 6813
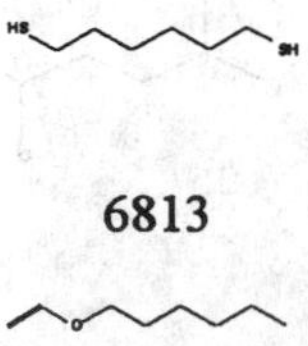

6814

6815
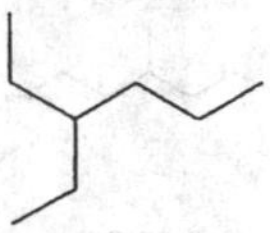

6816
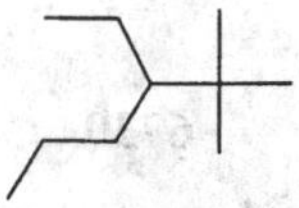

6817
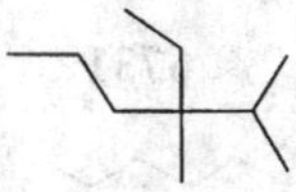

6818
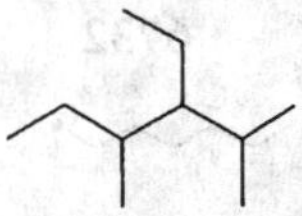

6819
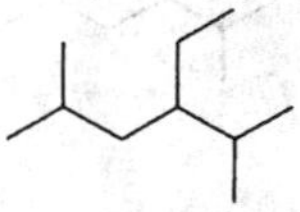

6820
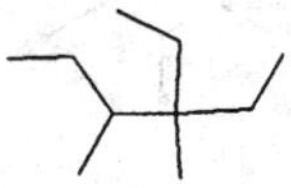

6821
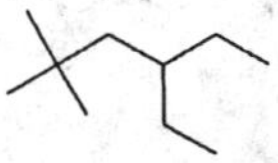

6822
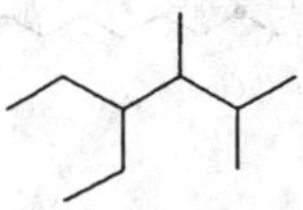

6823

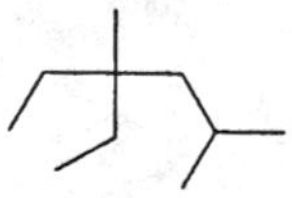

6824

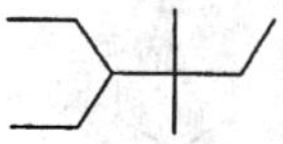

6825

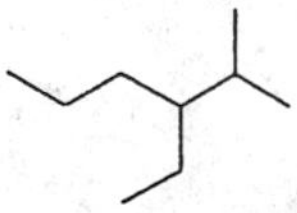

6826

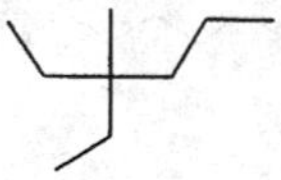

6827

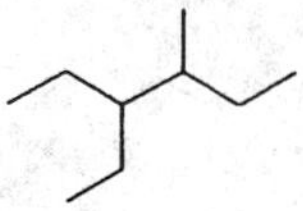

6828

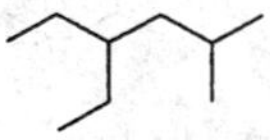

6829

6830

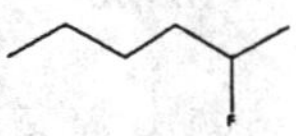

6831

6832

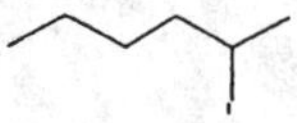

6833

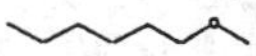

6834

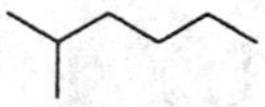

6835

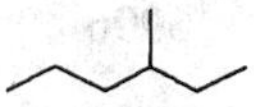

6836

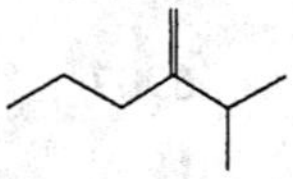

6837

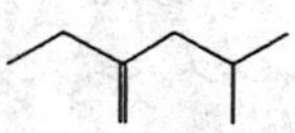

6838

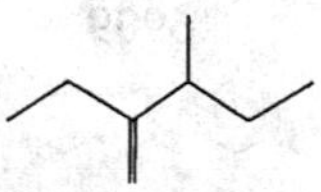

6839

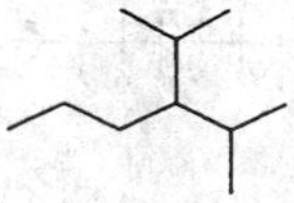

6840

6841

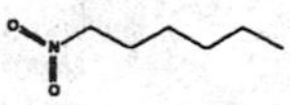

6842

6843

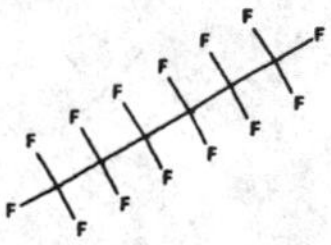

6844

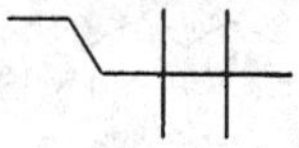

6845

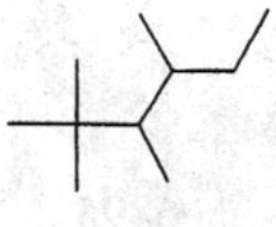

6846

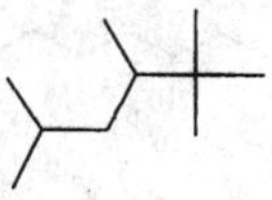

6847

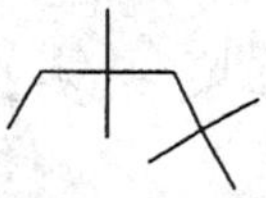

6848

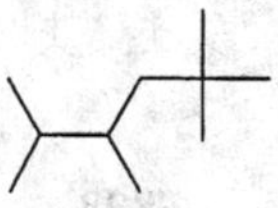

6849

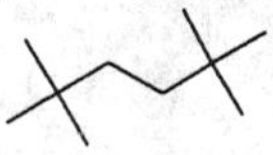

6850

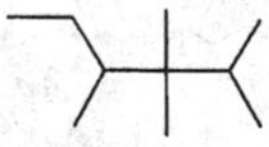

6851

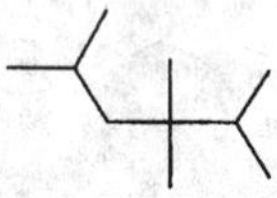

6852

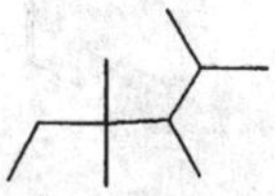

6853

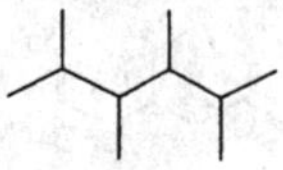

6854

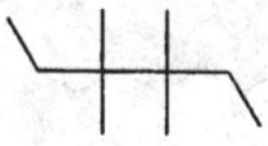

6855

6856

6857

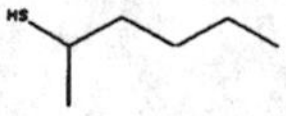

6858

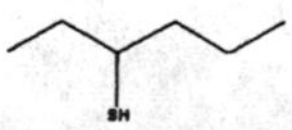

6859

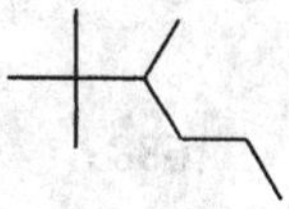

6860

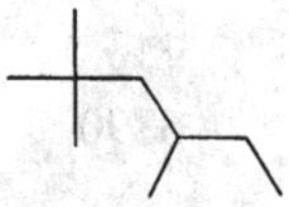

6861

6862

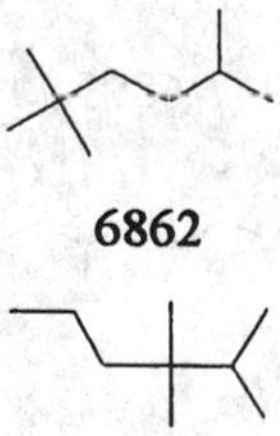

6863

6864

6865

6866

6868

6869

6870

6872

6873

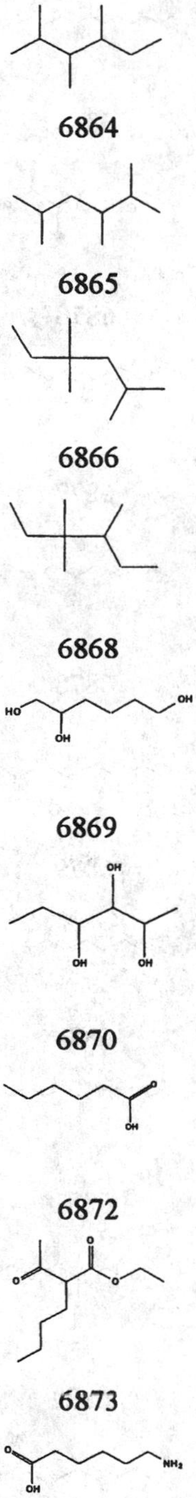

STRUCTURAL FORMULAS OF ORGANIC COMPOUNDS (continued)

In numeric order as they occur in the Table of Physical Constants of Organic Compounds

6874

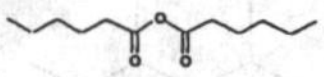

6875

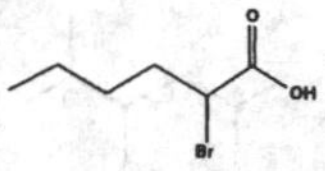

6876

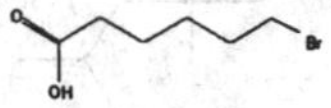

6877

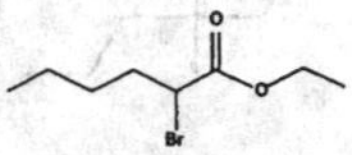

6878

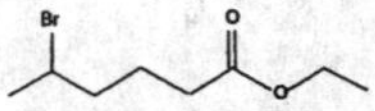

6879

6880

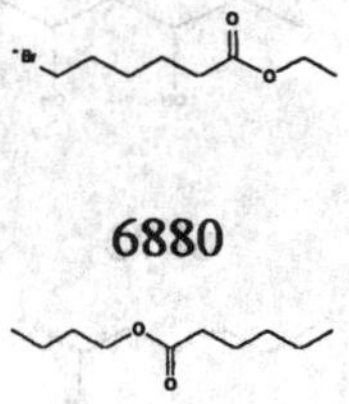

6881

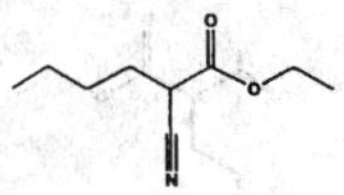

6882

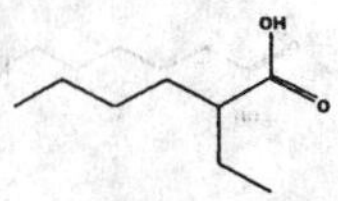

6883

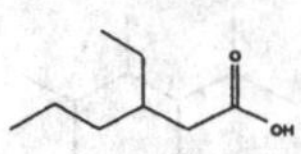

6884

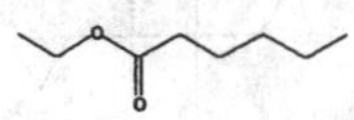

6885

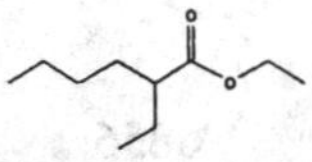

6886

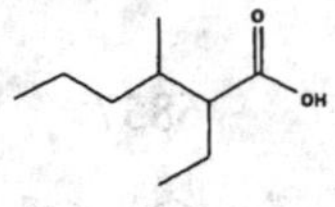

6887

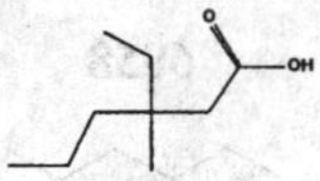

6888

6889

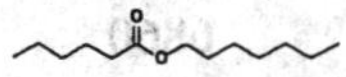

6890

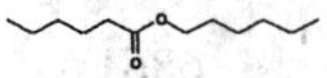

6891

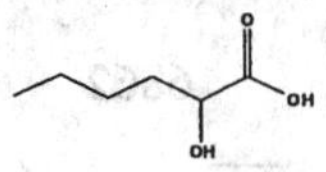

6892

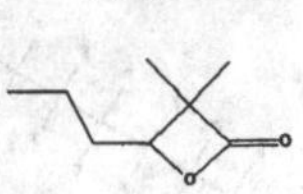

6893

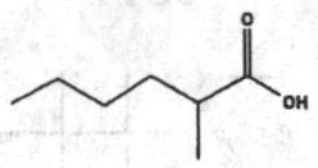

6894

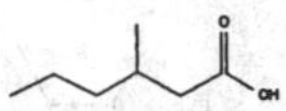

6895

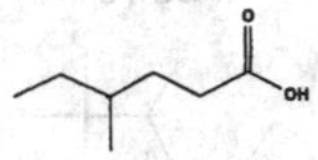

6896

6897

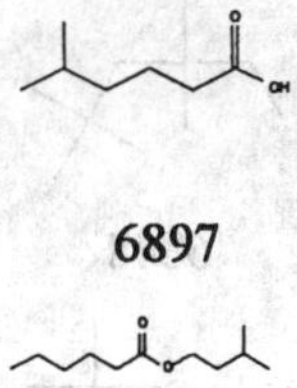

6898

6899

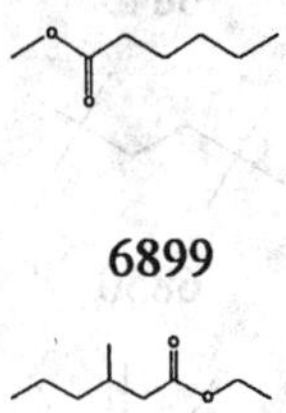

6900

6901

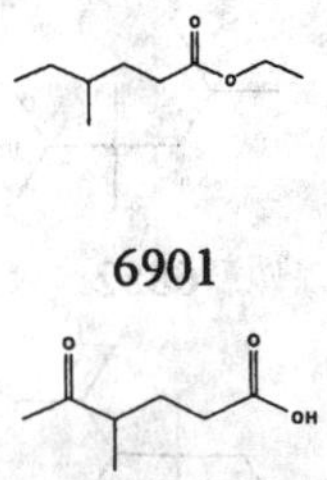

6904

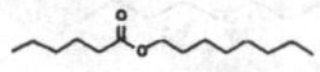

6905

6906

6907

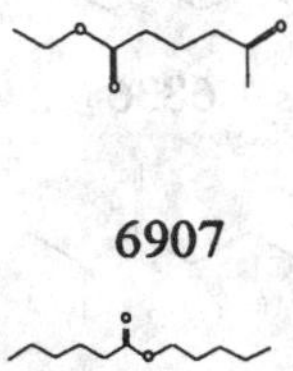

6908

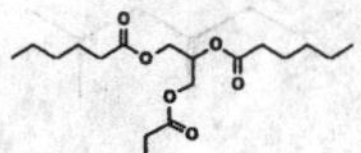

6909

6910

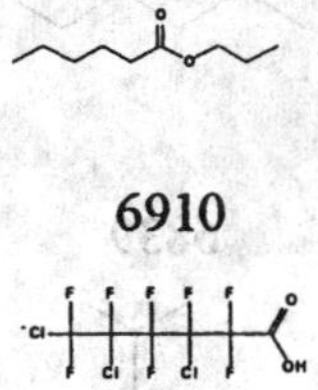

6911

6912

6913

6914

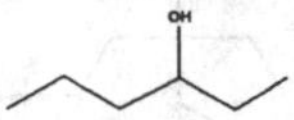

6915

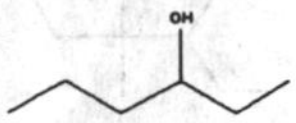

6916

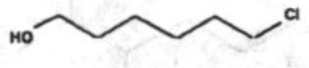

6917

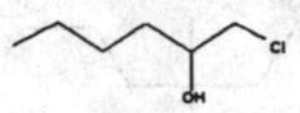

6918

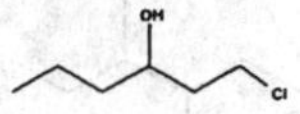

6919

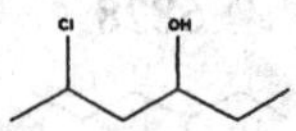

6921

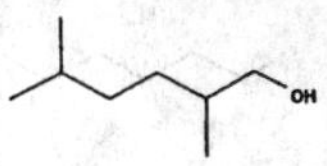

6922

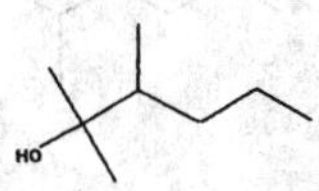

6923

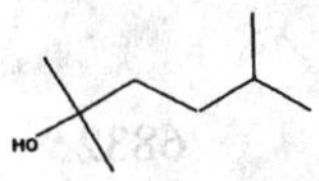

STRUCTURAL FORMULAS OF ORGANIC COMPOUNDS (continued)

In numeric order as they occur in the Table of Physical Constants of Organic Compounds

6924

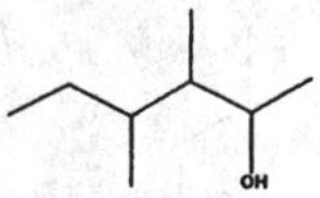

6925

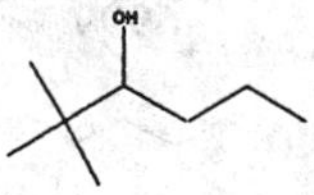

6926

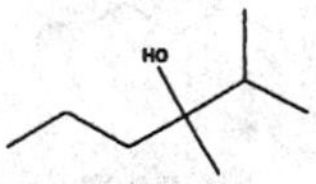

6927

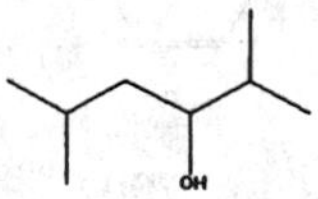

6928

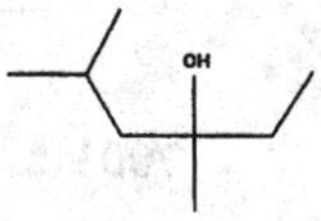

6929

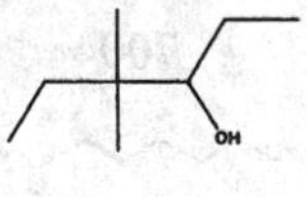

6930

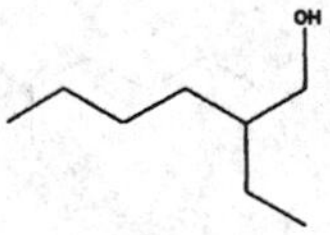

6931

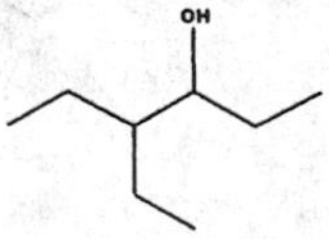

6932

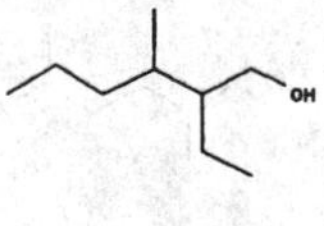

6933

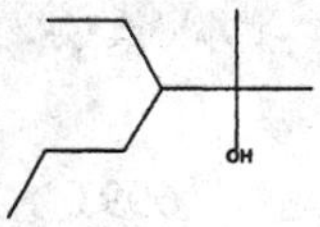

6934

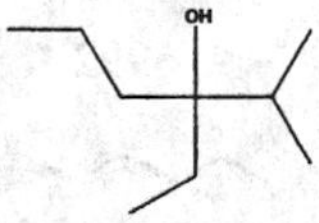

6935

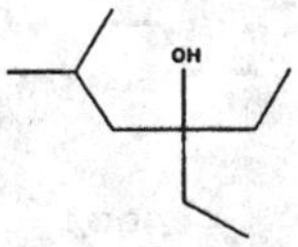

6936

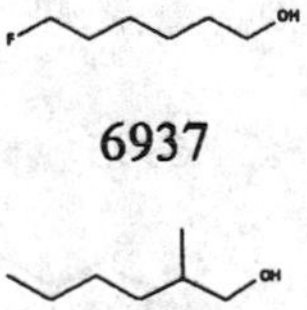

6937

6938

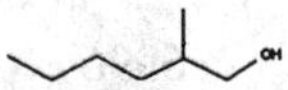

6939

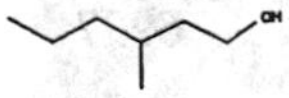

6940

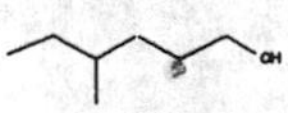

6941

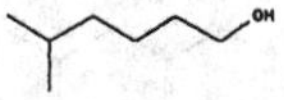

6942

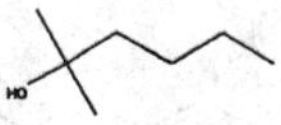

6943

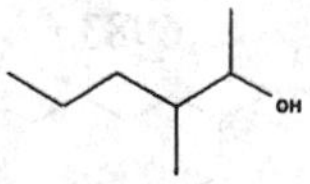

6944

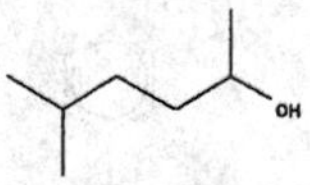

6945

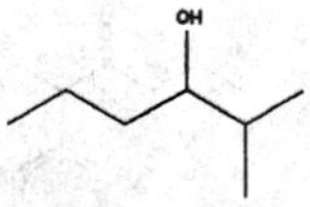

6946

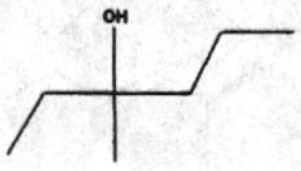

6947

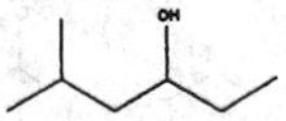

6948

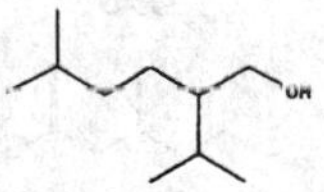

6949

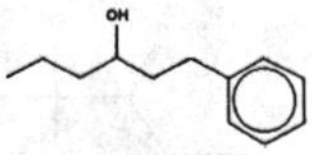

6950

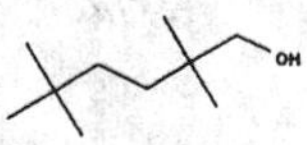

6951

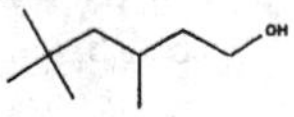

6952

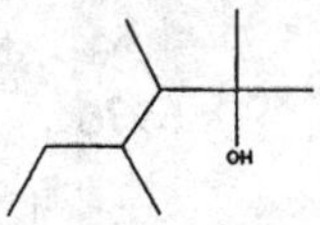

6953

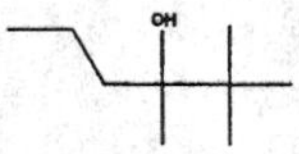

6954

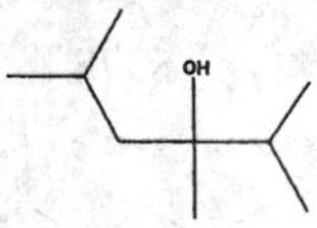

6955

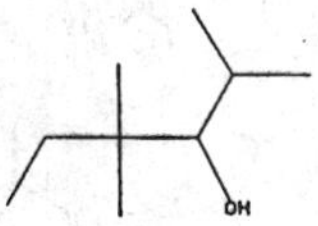

6956

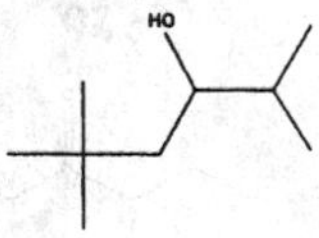

6957

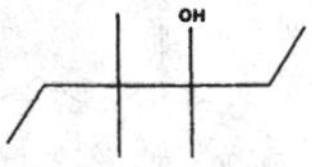

6958

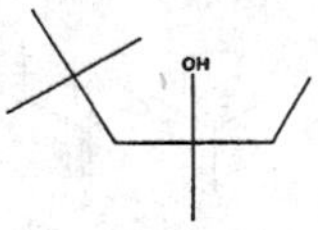

6959

6960

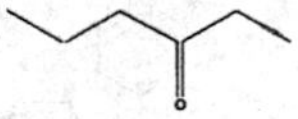

6961

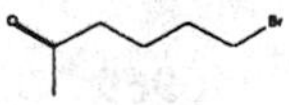

6962

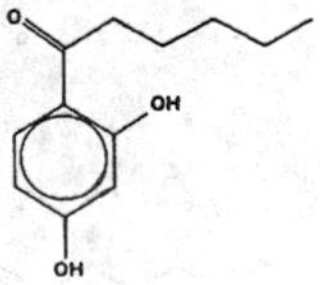

6963

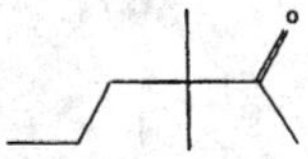

6964

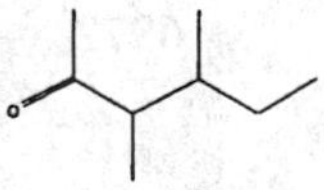

STRUCTURAL FORMULAS OF ORGANIC COMPOUNDS (continued)

In numeric order as they occur in the Table of Physical Constants of Organic Compounds

6965

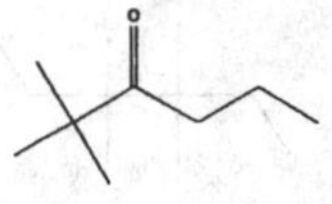

6966

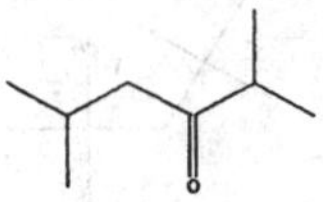

6967

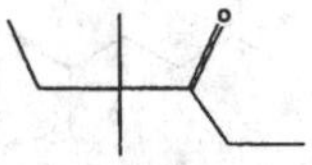

6968

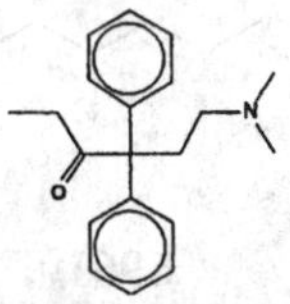

6969

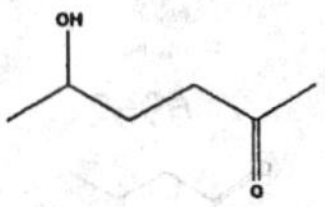

6970

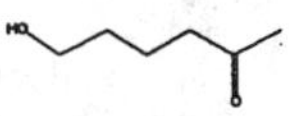

6971

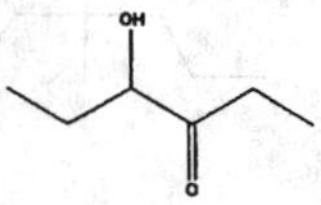

6972

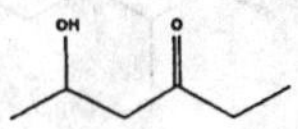

6973

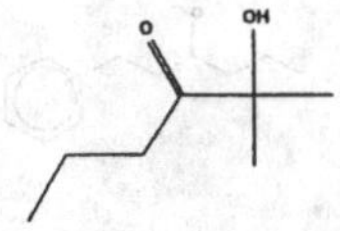

6974

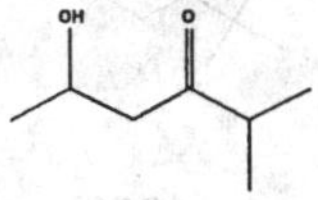

6975

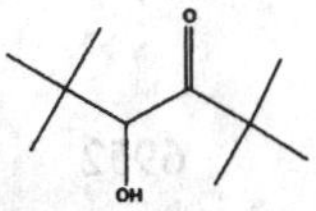

6976

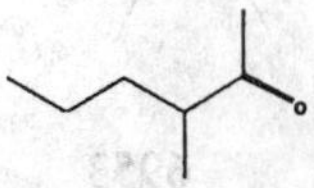

6977

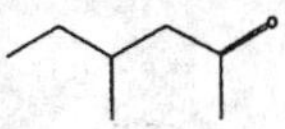

6978

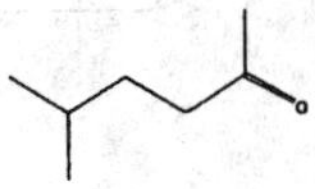

6979

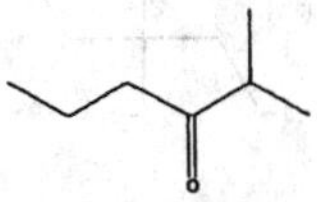

6980

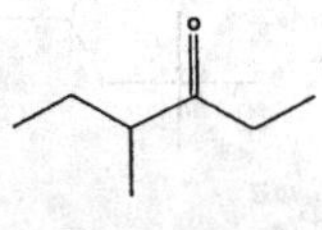

6981

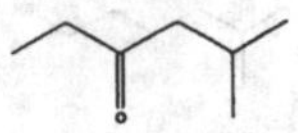

6982

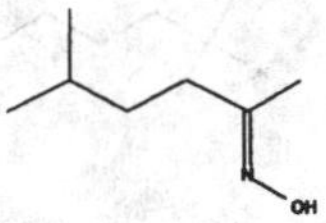

6983

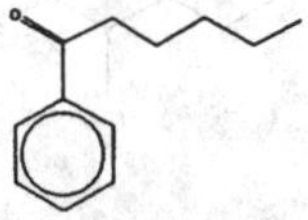

6984

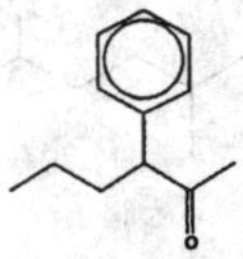

6985

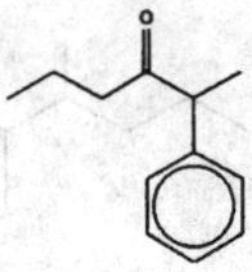

6986

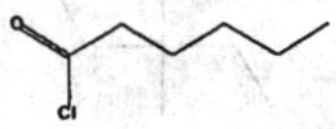

6987

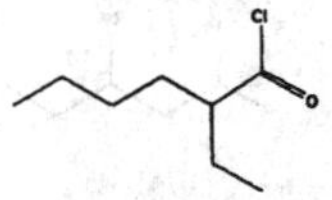

6988

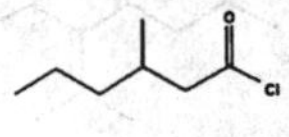

6989

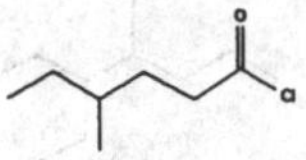

6990

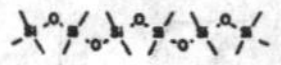

6991

6992

6993

6994

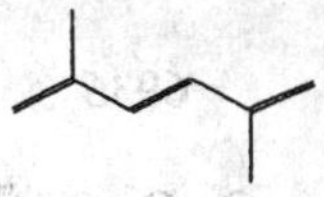

6995

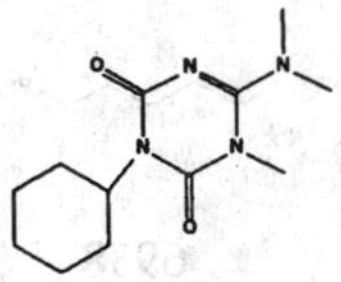

6996

6997

6998

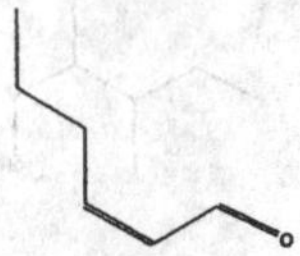

6999

7000

7001

7002

7003

7004

7005

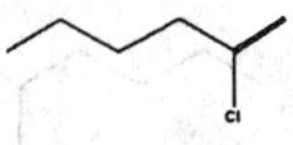

7007

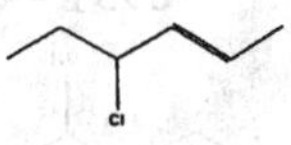

STRUCTURAL FORMULAS OF ORGANIC COMPOUNDS (continued)

In numeric order as they occur in the Table of Physical Constants of Organic Compounds

7008

7009

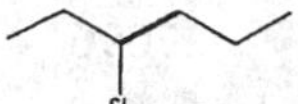

7010

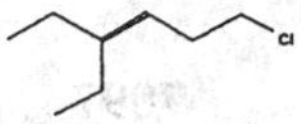

7011

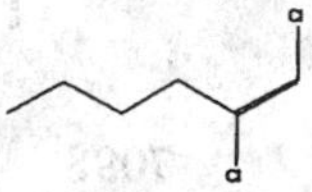

7012

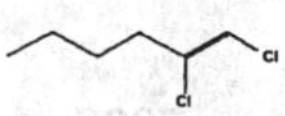

7013

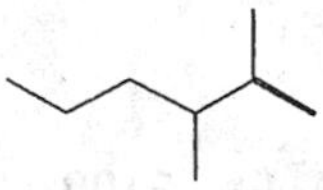

7014

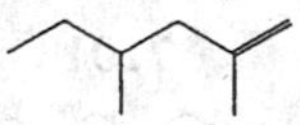

7015

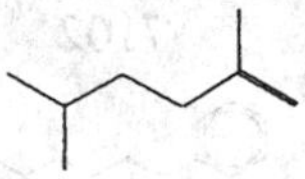

7016

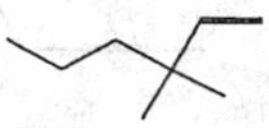

7017

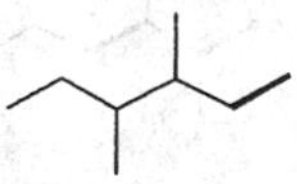

7018

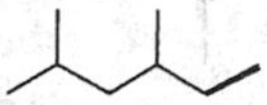

7019

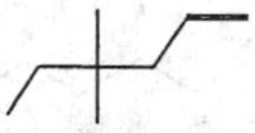

7020

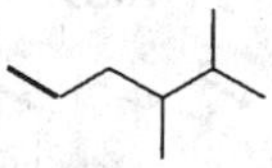

7021

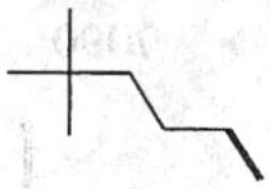

7022

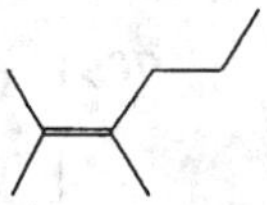

7023

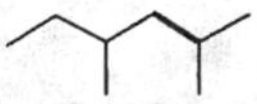

7024

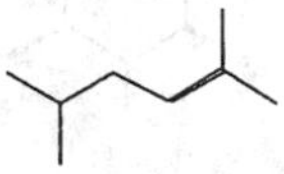

7025

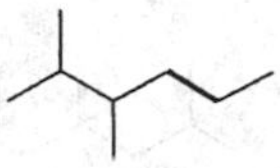

7026

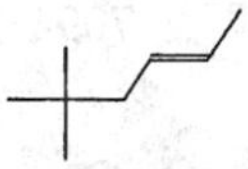

7027

7028

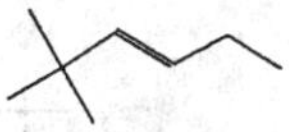

7029

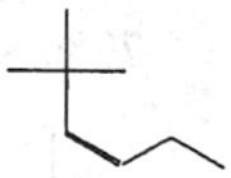

7030

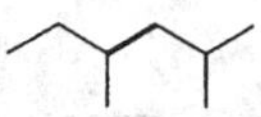

7031

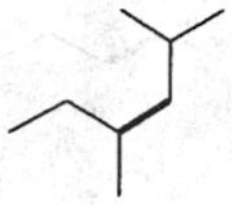

7032

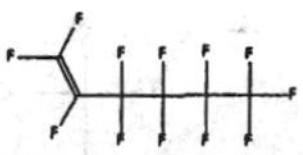

7033

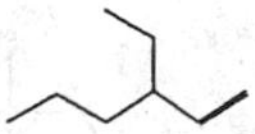

7034

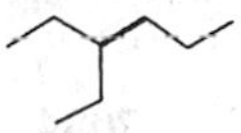

7035

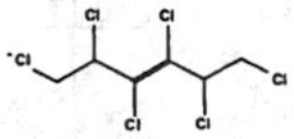

7036

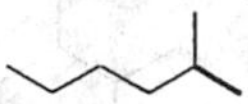

7037

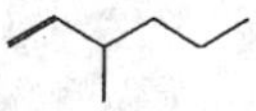

7038

7039

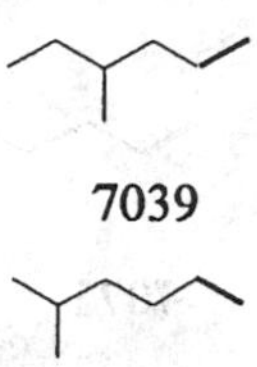

7040

7041

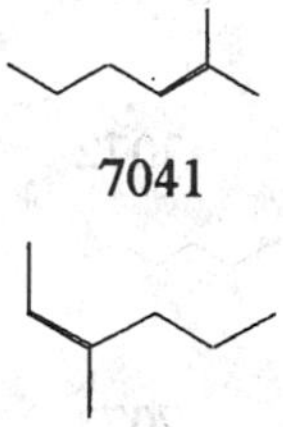

7042

7043

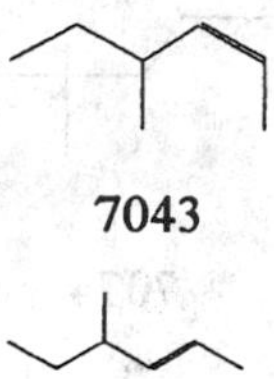

7044

7045

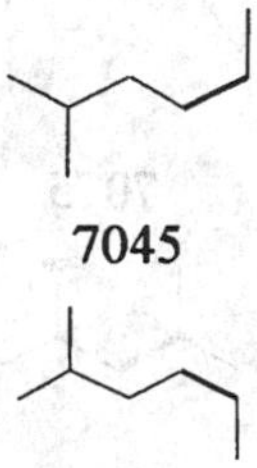

7046

7047

7048

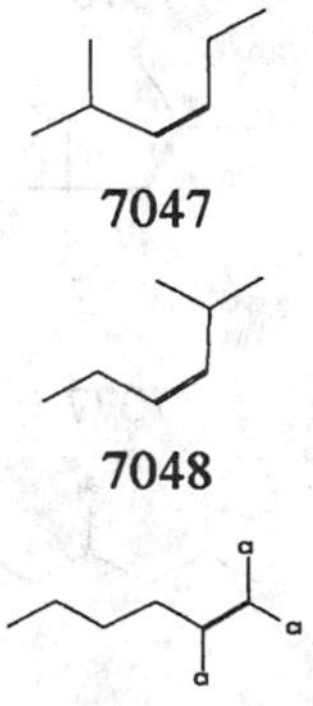

7049

7050

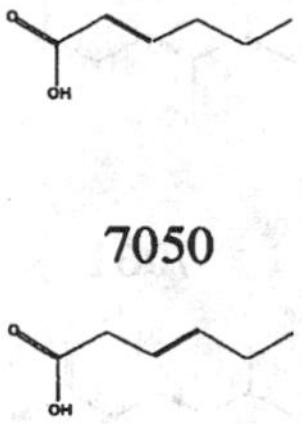

7051

7053

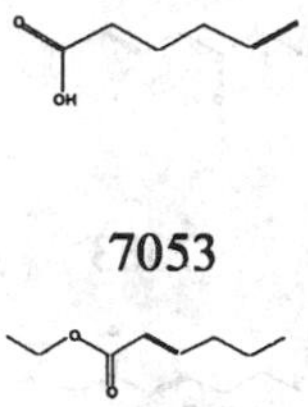

7054

7055

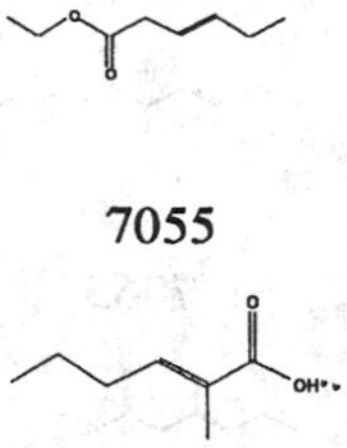

7056

7057

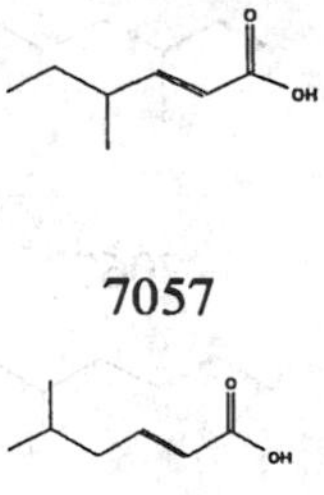

7058

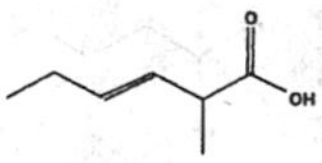

In numeric order as they occur in the Table of Physical Constants of Organic Compounds

7060 7070 7078 7086 7095

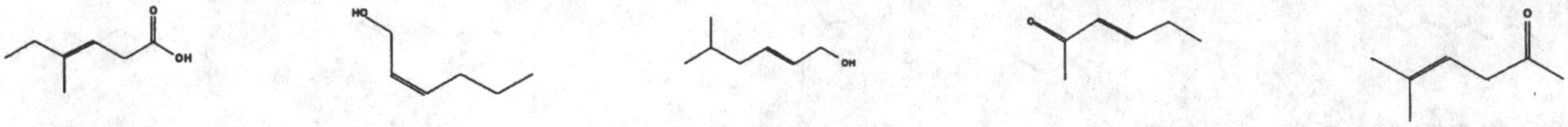

7061 7071 7079

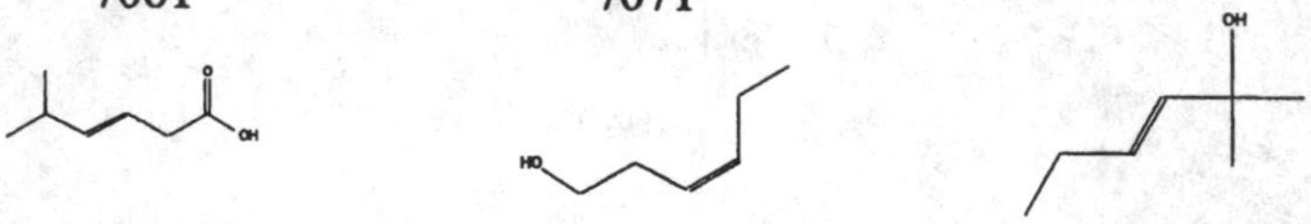

7062

7063

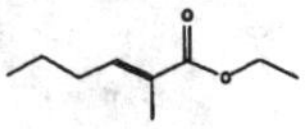

7064

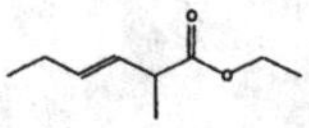

7065

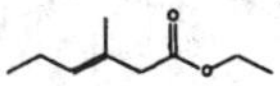

7066

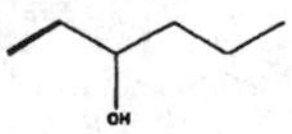

7067

7068

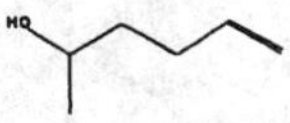

7069

7072

7073

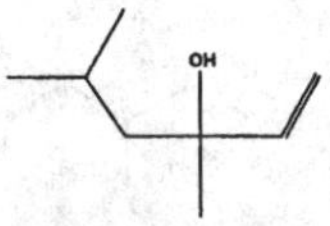

7074

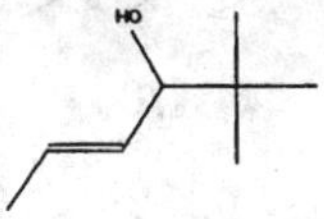

7075

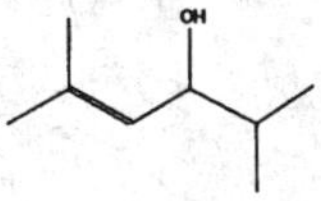

7076

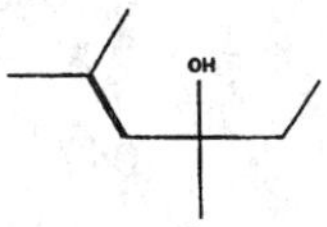

7077

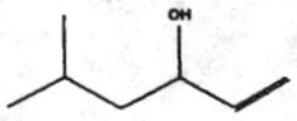

7080

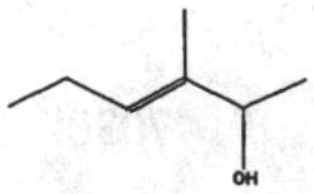

7081

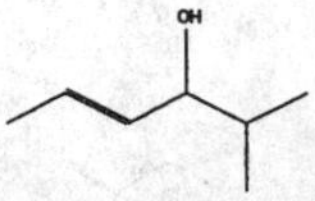

7082

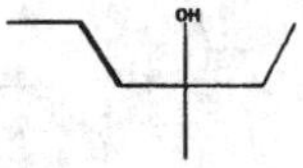

7083

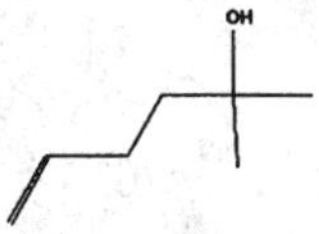

7084

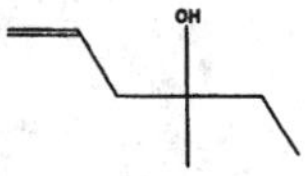

7085

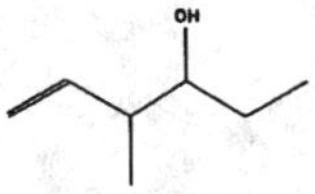

7087

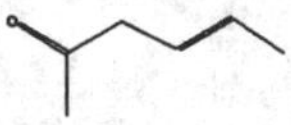

7088

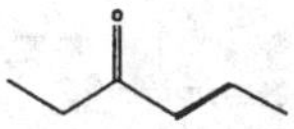

7089

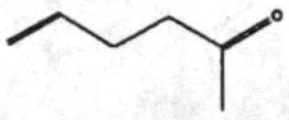

7090

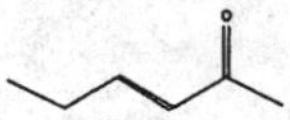

7091

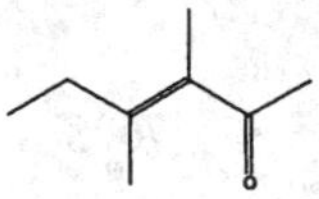

7092

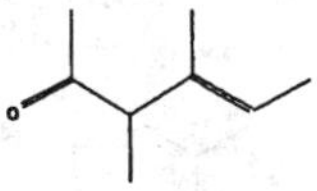

7093

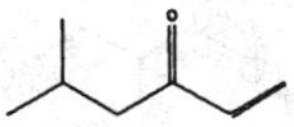

7094

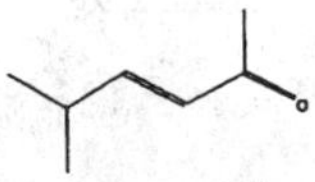

7096

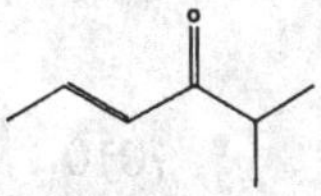

7097

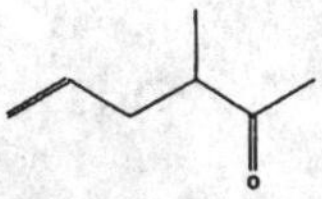

7098

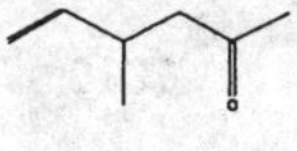

7099

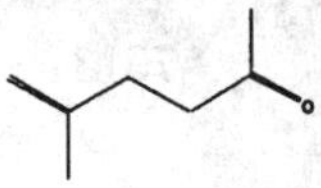

7100

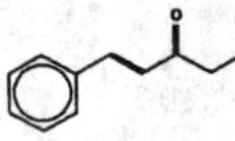

7101

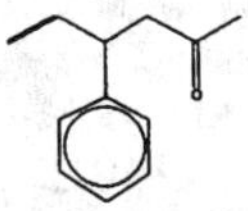

7102

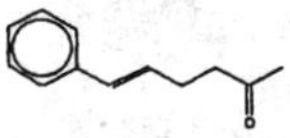

7103

7104

In numeric order as they occur in the Table of Physical Constants of Organic Compounds

7105

7106

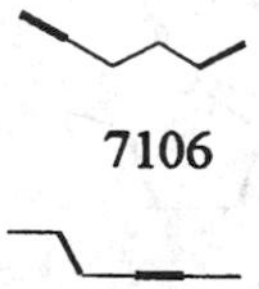

7107

7108

7109

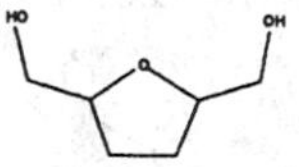

7113

7114

7115

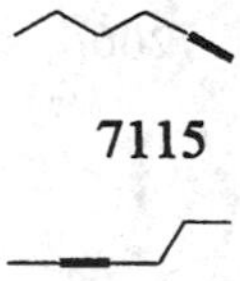

7116

7117

7118

7119

7120

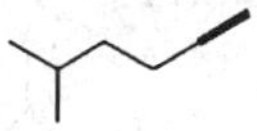

7121

7122

7123

7124

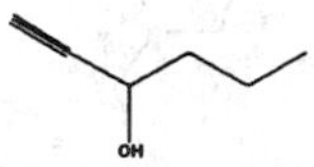

7125

7126

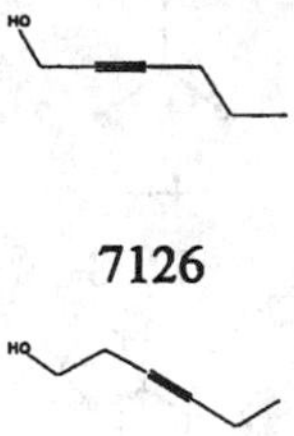

7127

7128

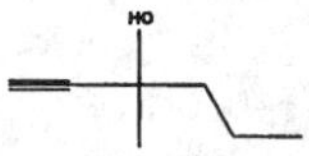

7129

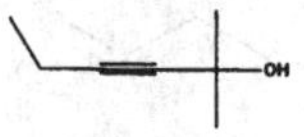

7130

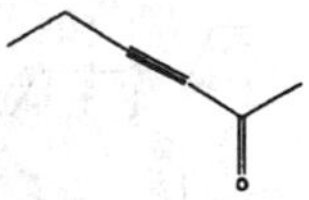

7131

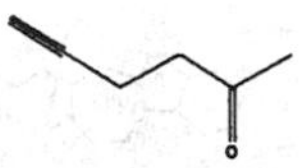

7132

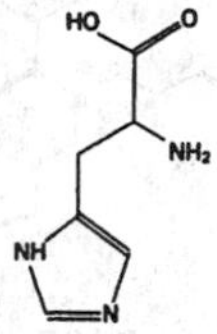

7136

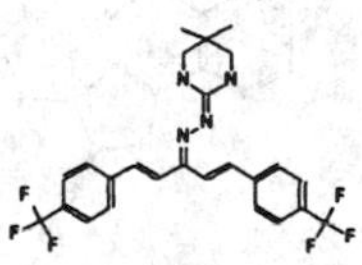

7137

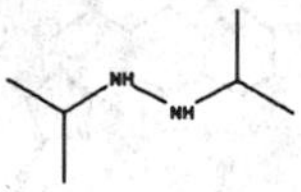

7138

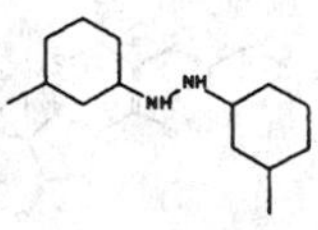

7140

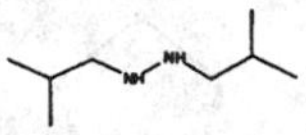

7142

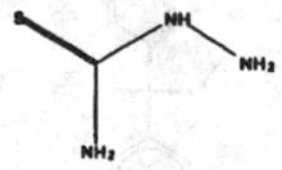

7144

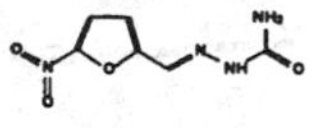

7145

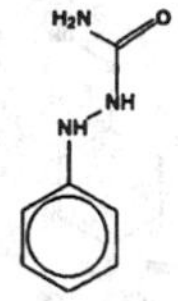

7146

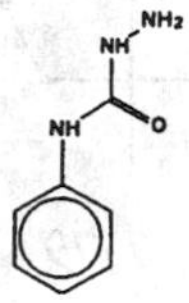

7147

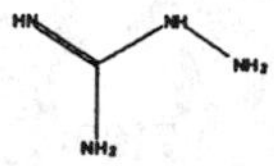

7148

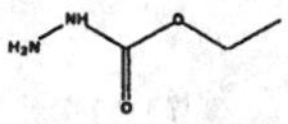

7149

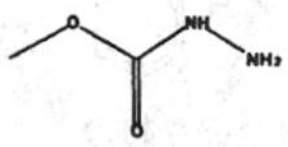

7150

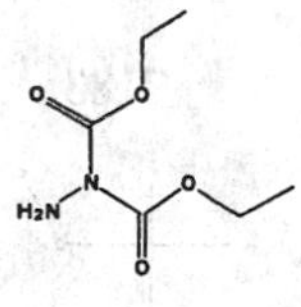

7151

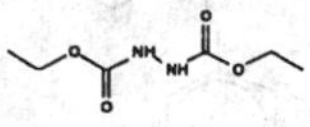

7152

7153

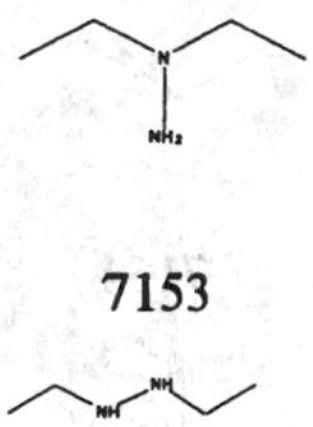

7154

7155

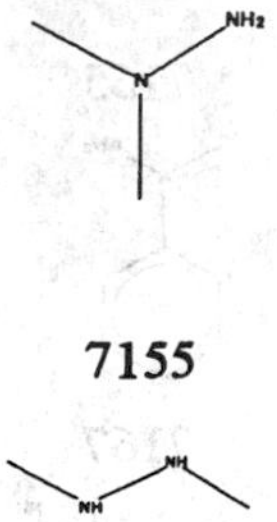

7156

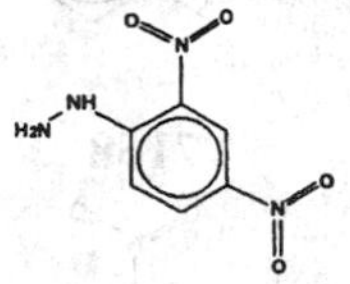

7158

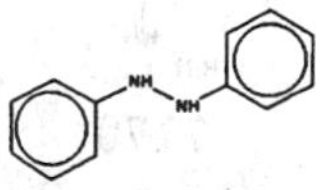

7159

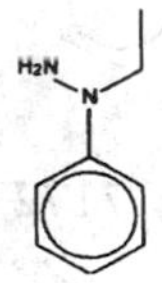

7160
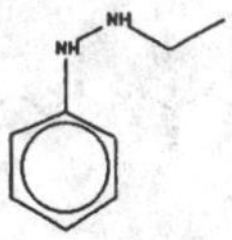

7161

7164
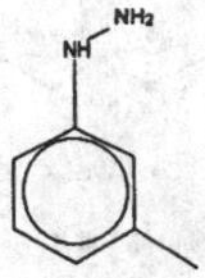

7165

7167
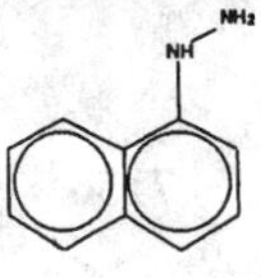

7168
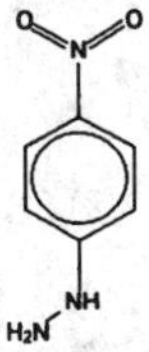

7170
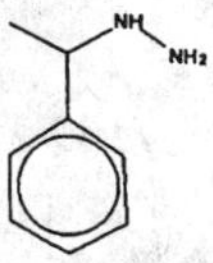

7172
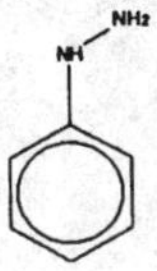

7175
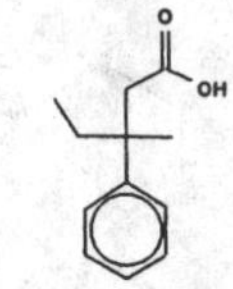

7176

7177
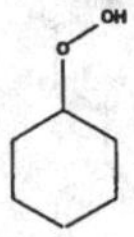

7178
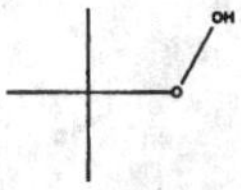

7179

7180

7181
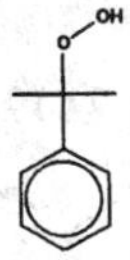

7182
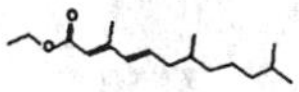

7183
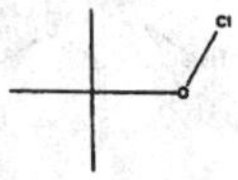

7184
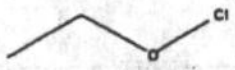

7185
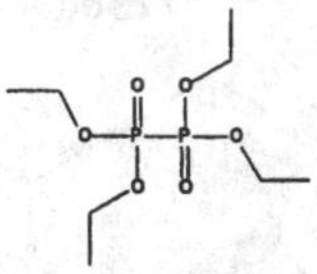

7186
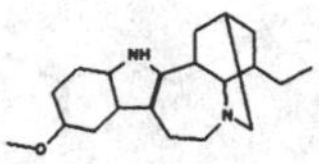

7187
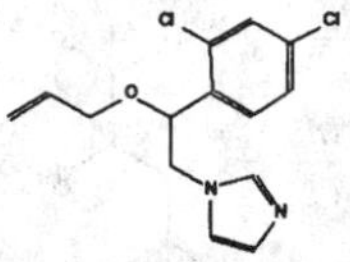

7188
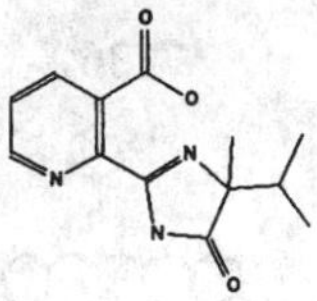

7189
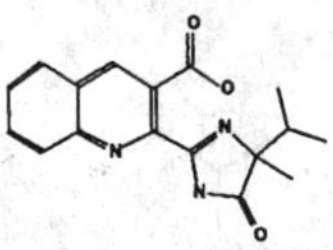

7190
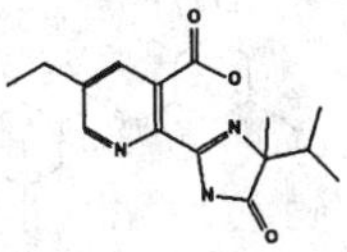

7192
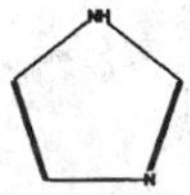

7193
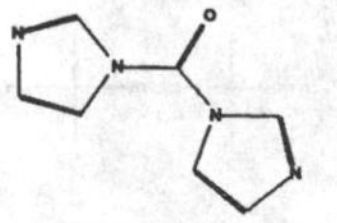

7194
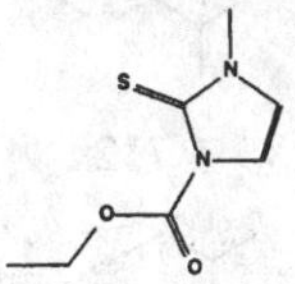

7195
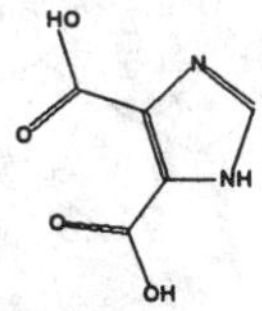

7196
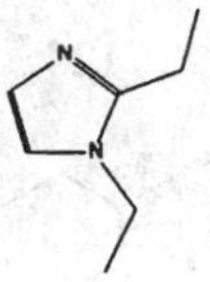

7197
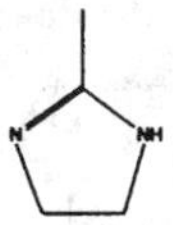

7198
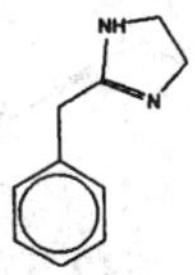

7199
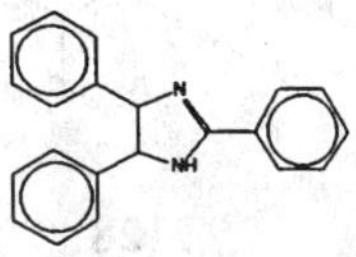

7200
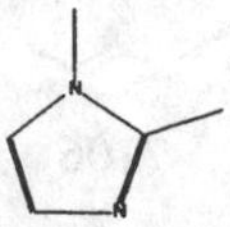

7201
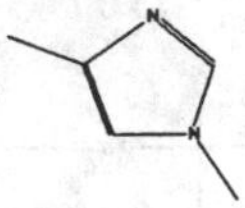

7202
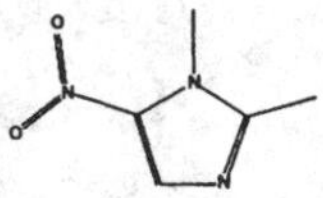

7203
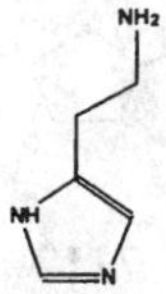

7205

7206
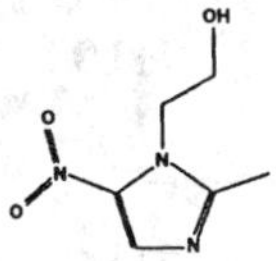

7207

7208

7209
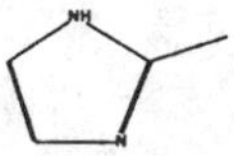

In numeric order as they occur in the Table of Physical Constants of Organic Compounds

7210

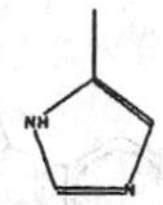

7211

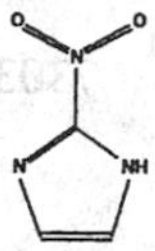

7212

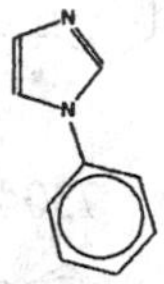

7213

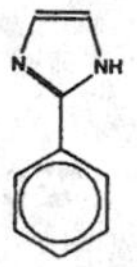

7214

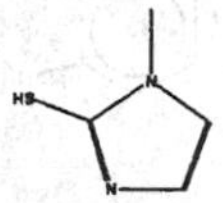

7215

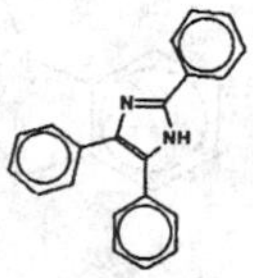

7216

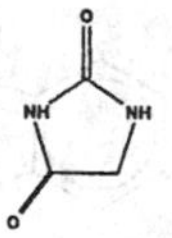

7217

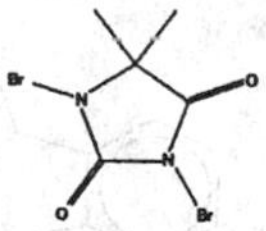

7218

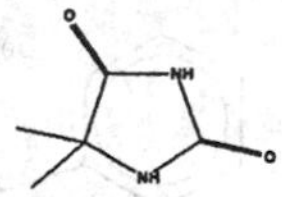

7219

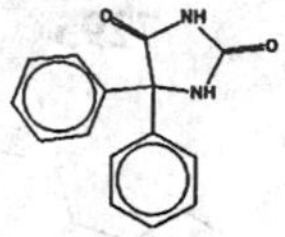

7222

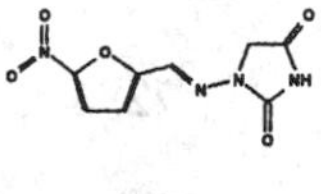

7224

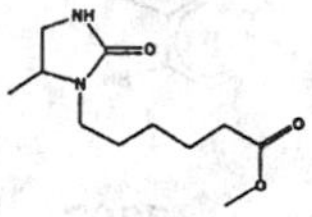

7225

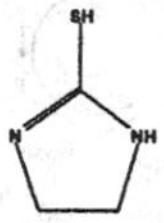

7226

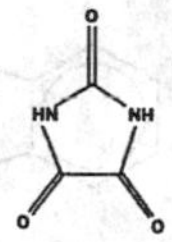

7227

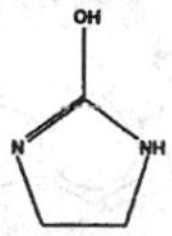

7228

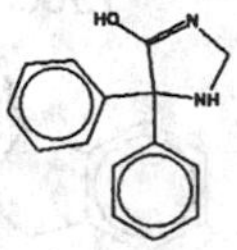

7230

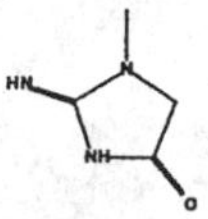

7231

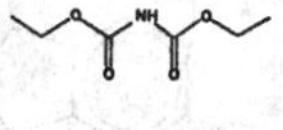

7232

7233

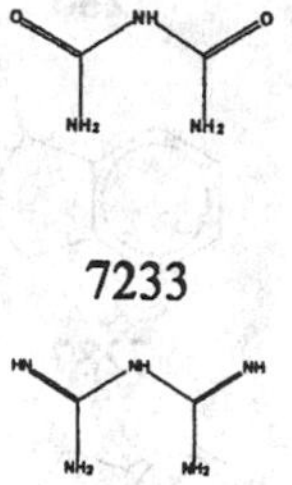

7235

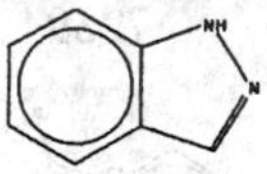

7236

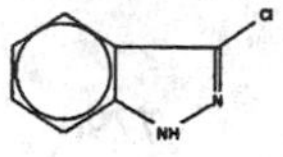

7237

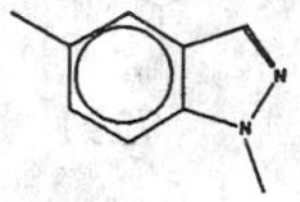

7238

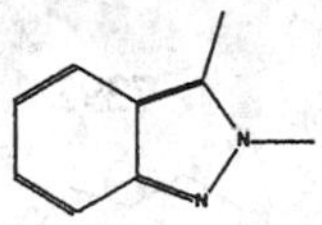

7239

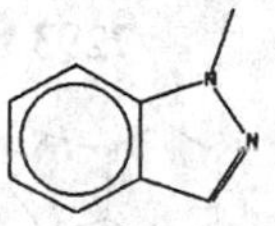

7240

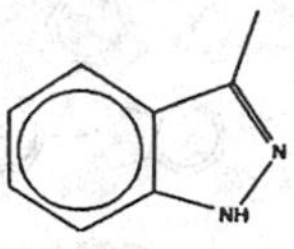

7241

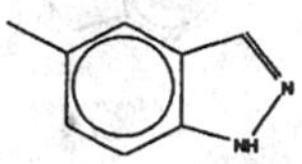

7242

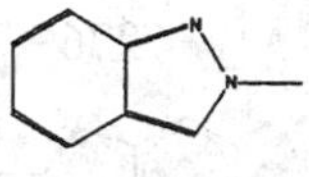

7243

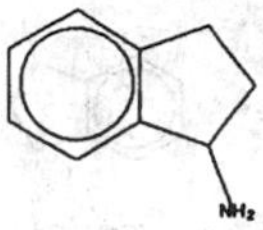

7244

7245

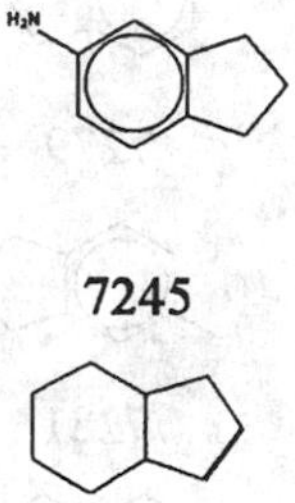

7246

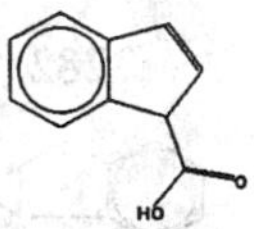

7247

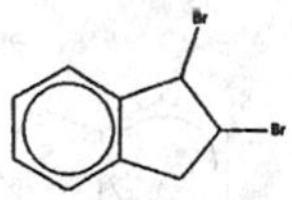

7249

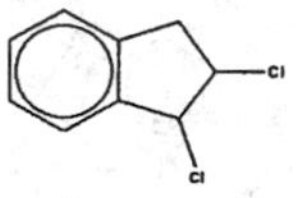

7250

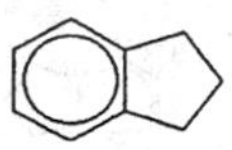

7251

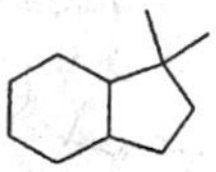

7252

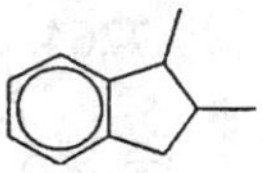

7253

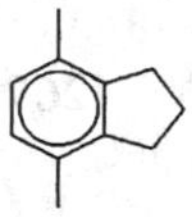

7254

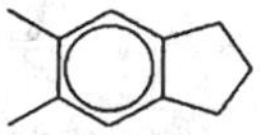

7255

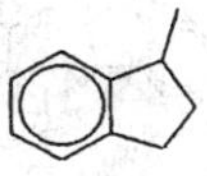

7257

7258

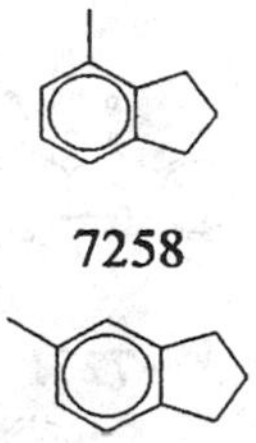

7259

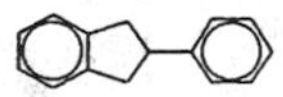

7260

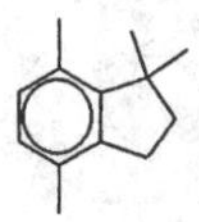

7262

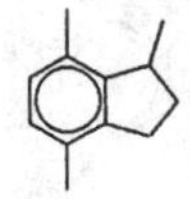

7263

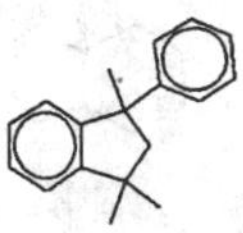

7264

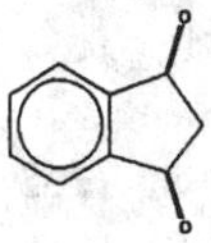

7267

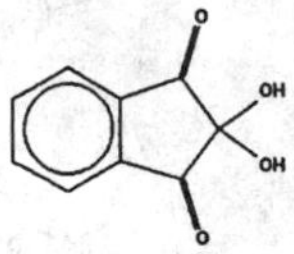

7268

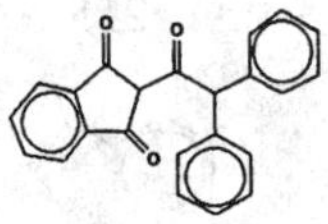

7269

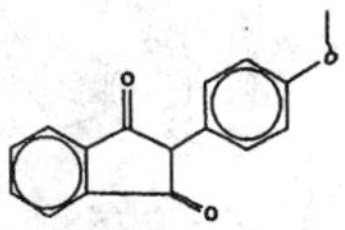

7271

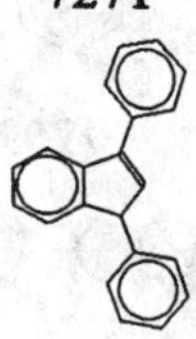

7272

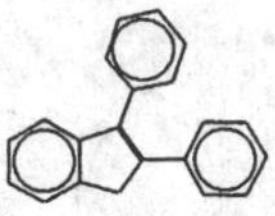

7273

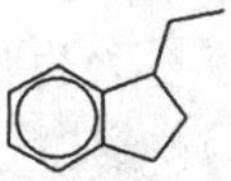

7274

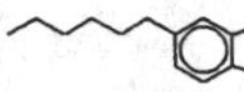

7276

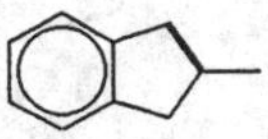

7277

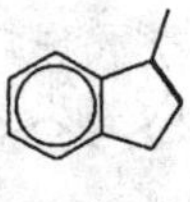

7278

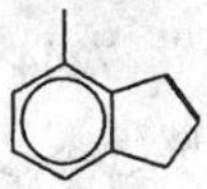

7279

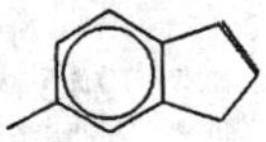

7281

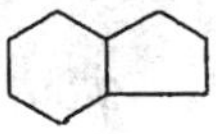

7282

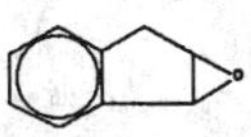

7284

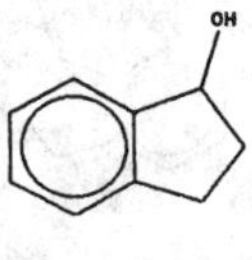

7285

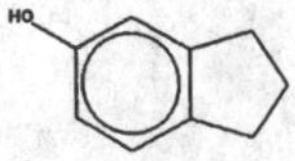

7286

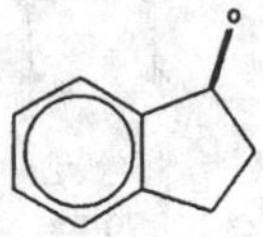

7287

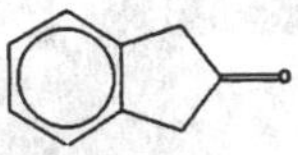

7289

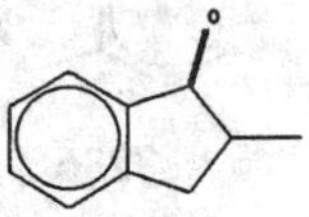

7290

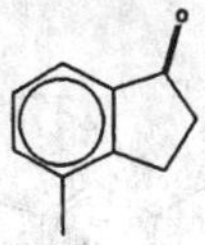

7291

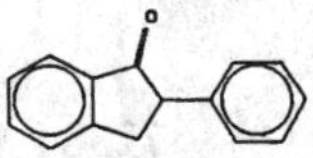

7292

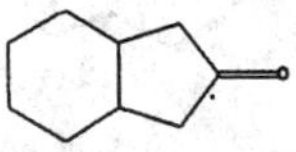

7293

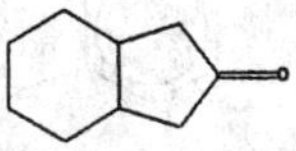

7294

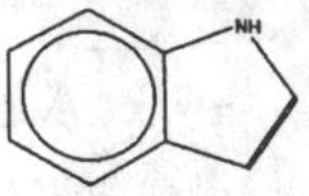

7295

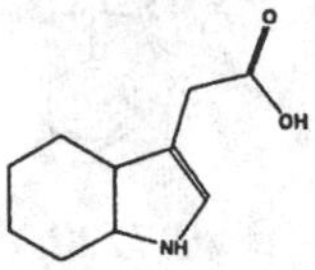

7296

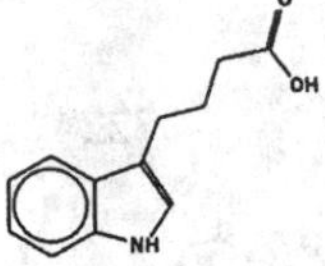

7297

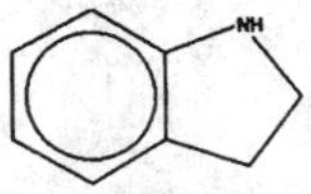

7298

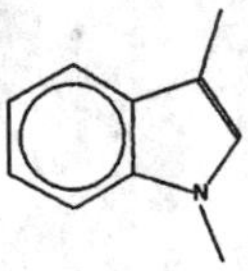

7299

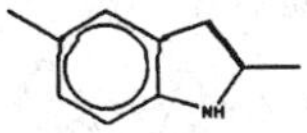

7300

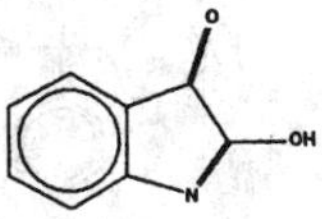

7302

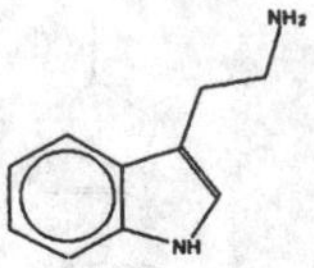

7303

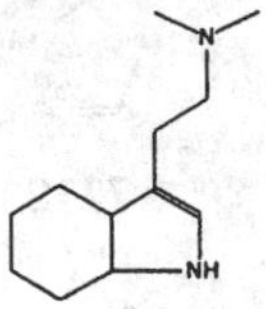

7304

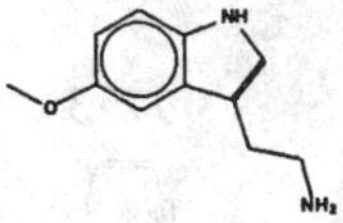

7305

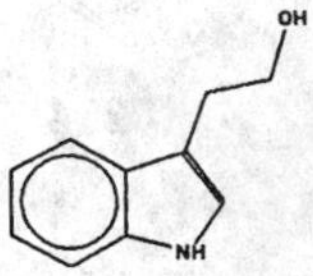

7307

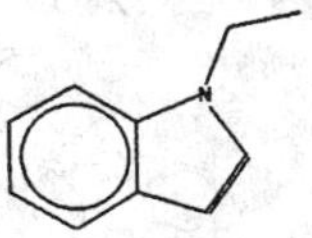

7308

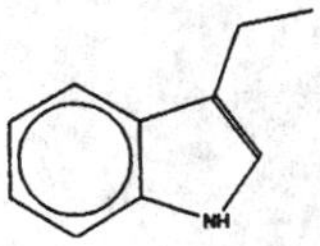

7309

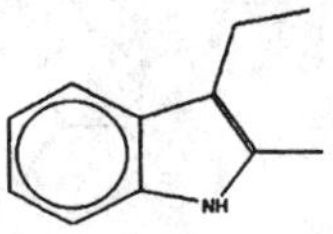

STRUCTURAL FORMULAS OF ORGANIC COMPOUNDS (continued)

In numeric order as they occur in the Table of Physical Constants of Organic Compounds

7310

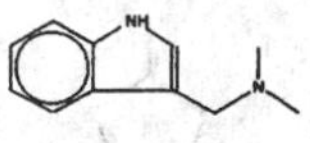

7312

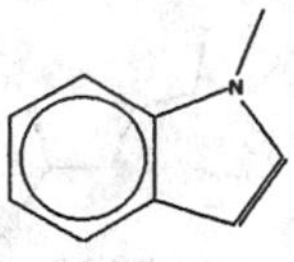

7313

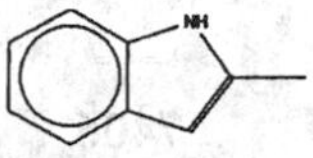

7314

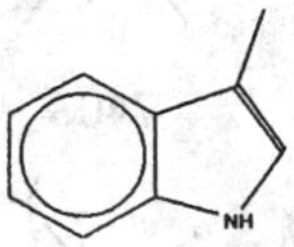

7315

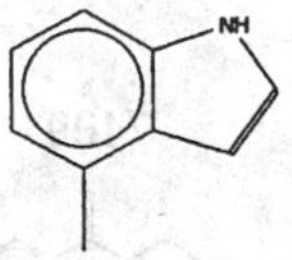

7316

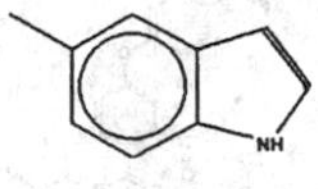

7317

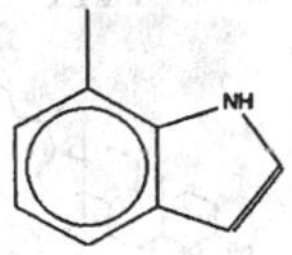

7319

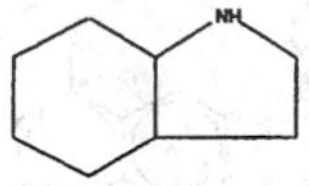

7320

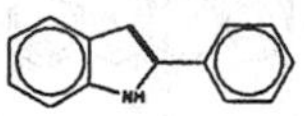

7321

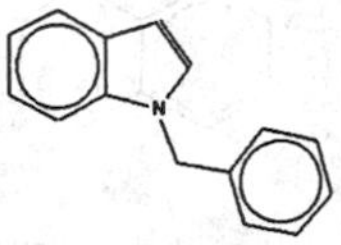

7322

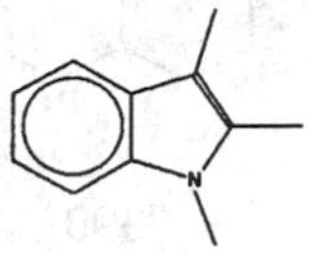

7323

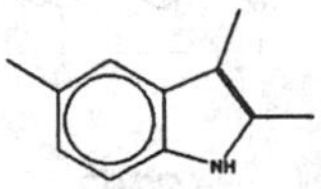

7324

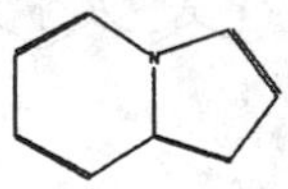

7325

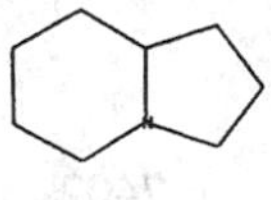

7326

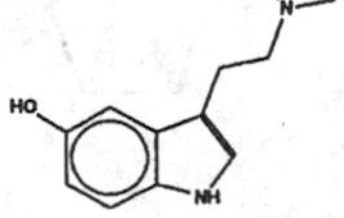

7327

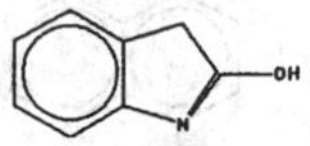

7328

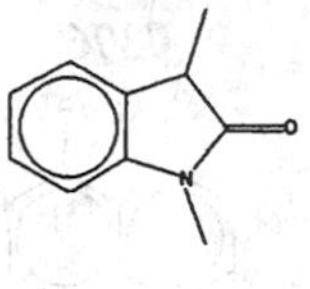

7329

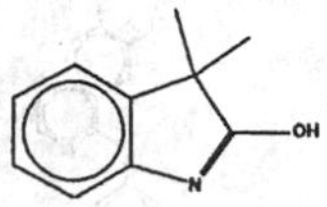

7330

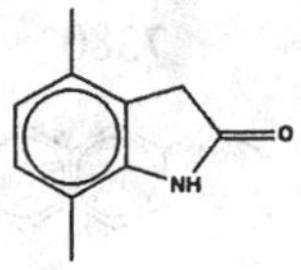

7331

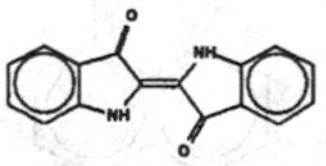

7335

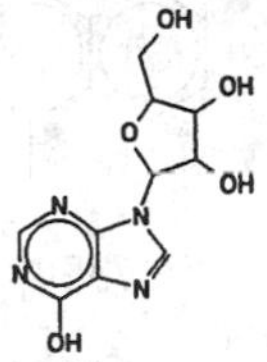

7336

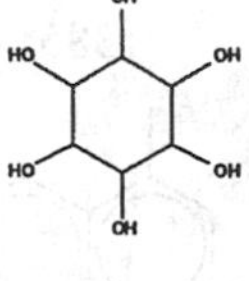

7337

7338

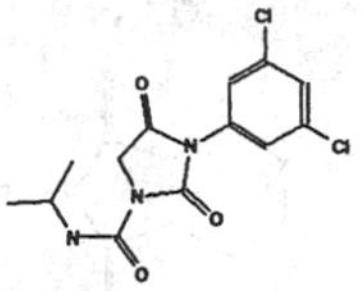

7339

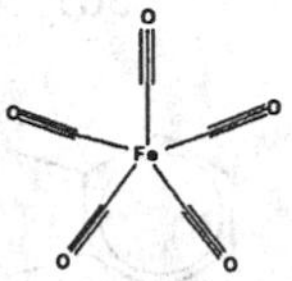

7340

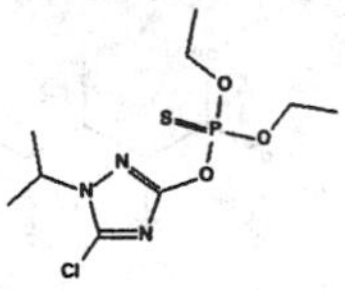

7342

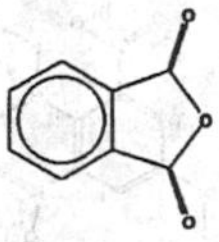

7343

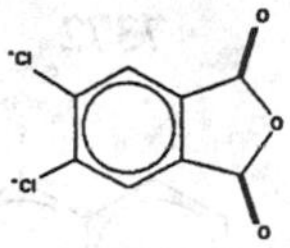

7344

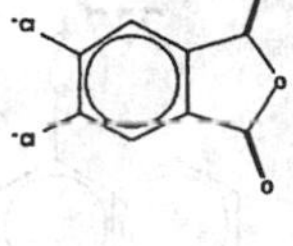

7345

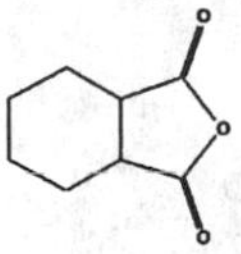

7346

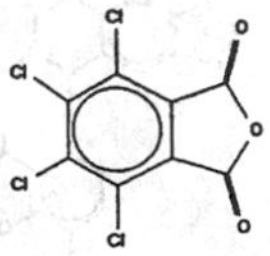

7347

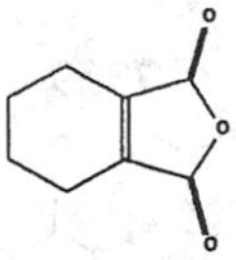

7350

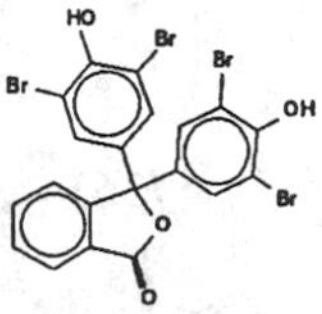

7351

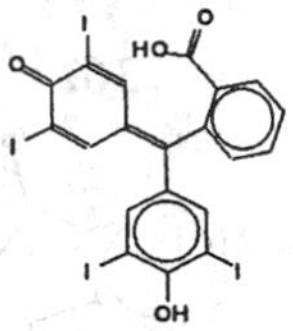

7354

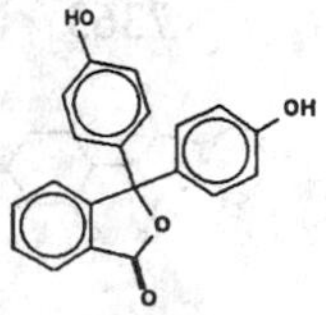

7355

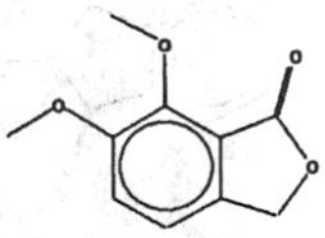

7356

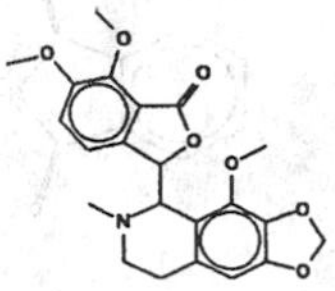

7357

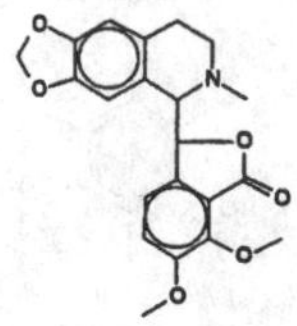

7359

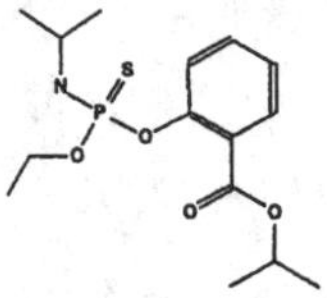

7361

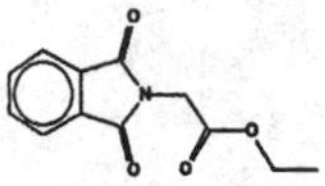

7362

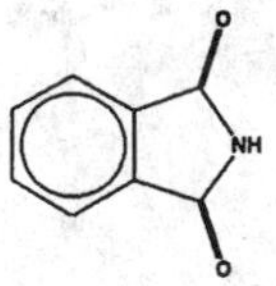

7363

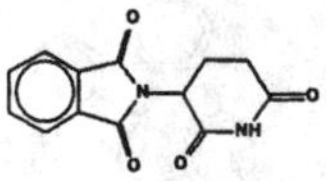

7364

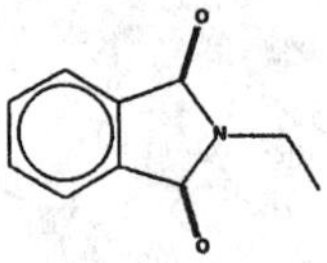

7365

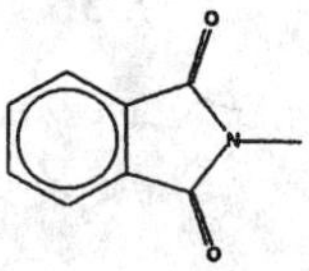

7367

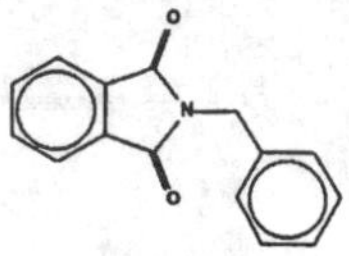

7368

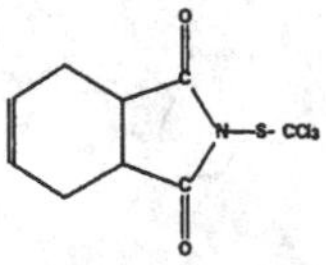

7369

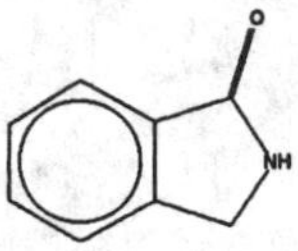

7370

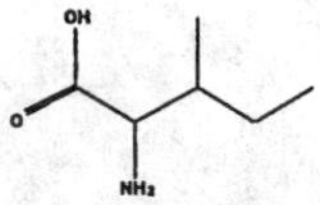

7371

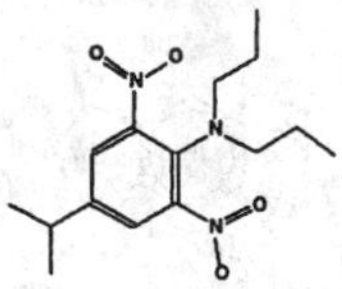

7372

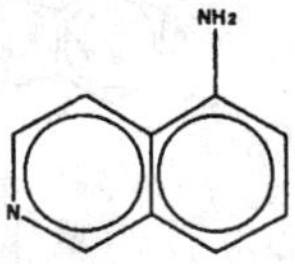

7374

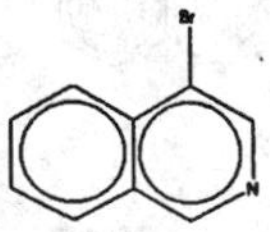

7375

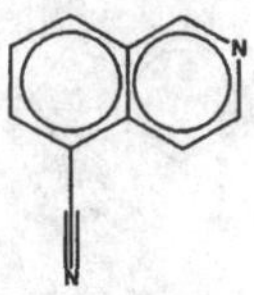

7376

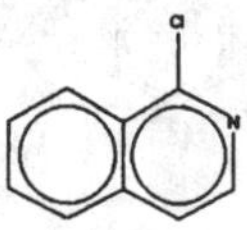

7378

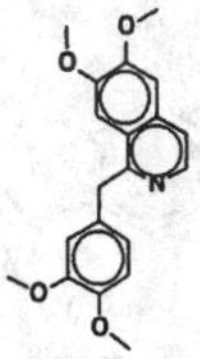

7380

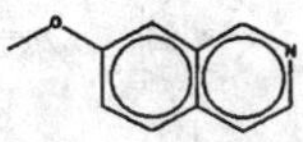

7381

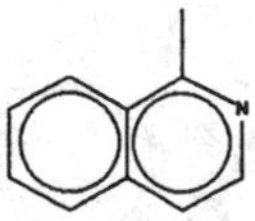

7383

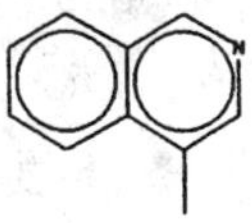

7384

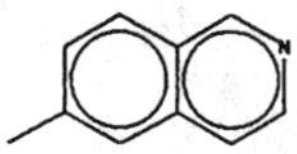

7385

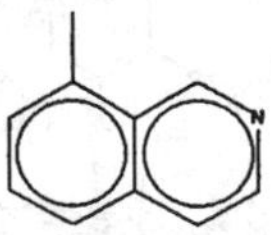

7386

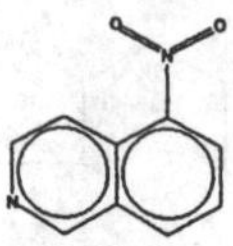

7387

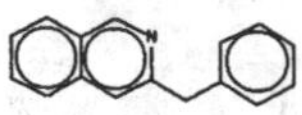

7388

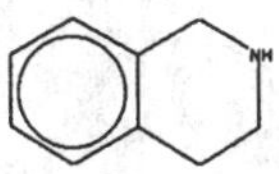

7389

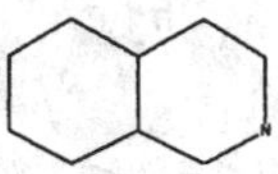

7390

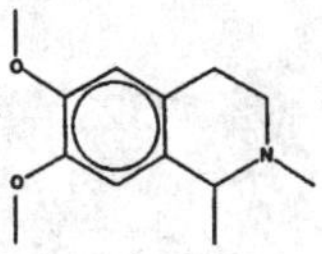

7398

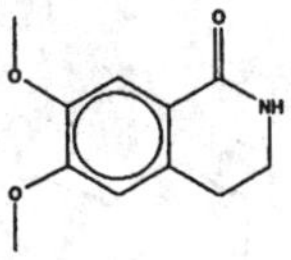

7399

7402

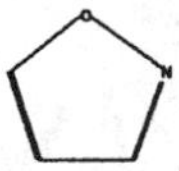

7403

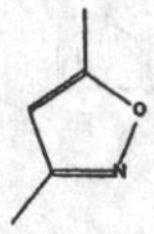

7404

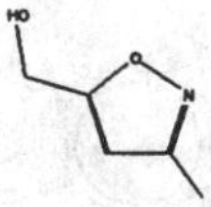

7405

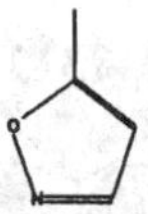

7406

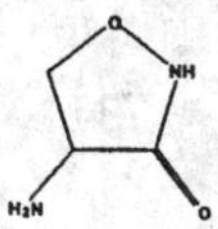

7408

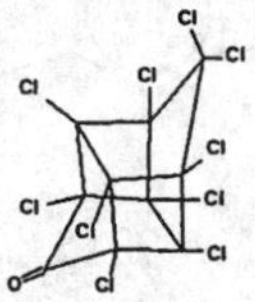

7409

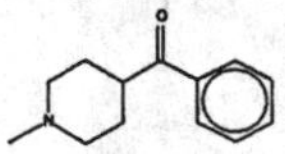

7410

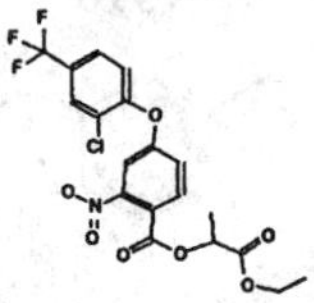

7411

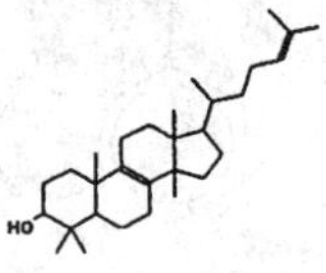

7412

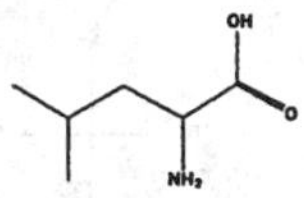

7413

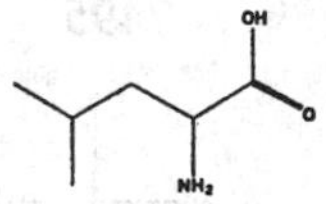

7415

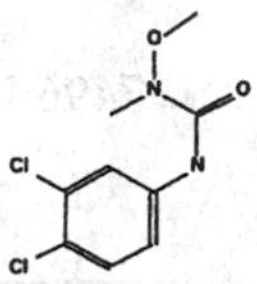

7416

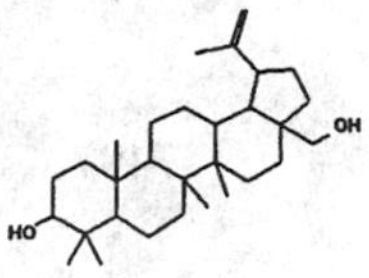

7417

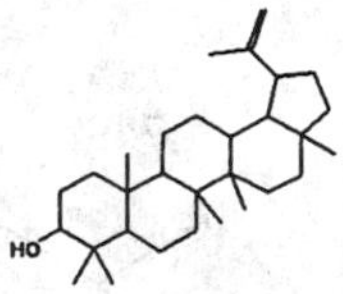

7418

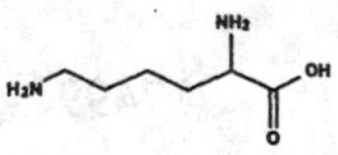

7419

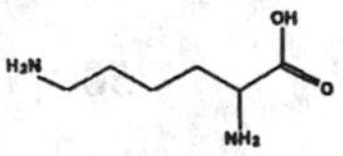

7420

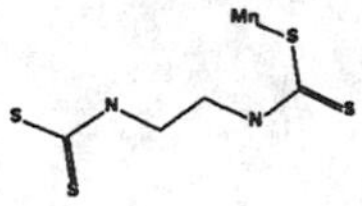

7421

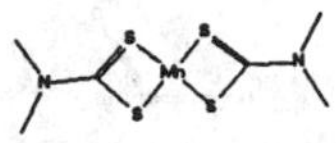

7422

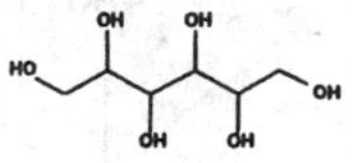

7423

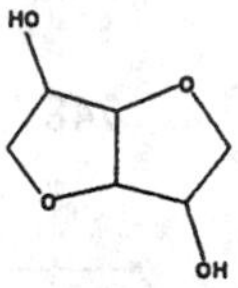

7425

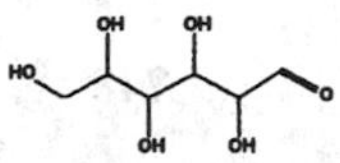

7426

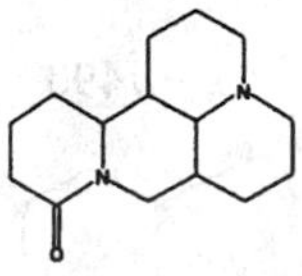

7427

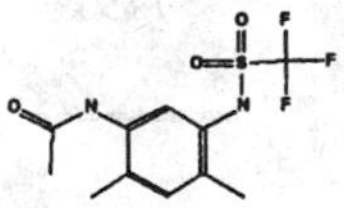

7430

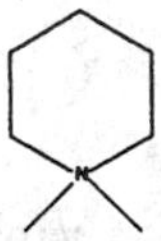

7431

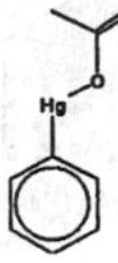

7432

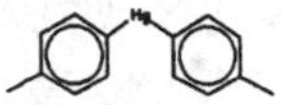

7433

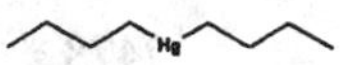

7434

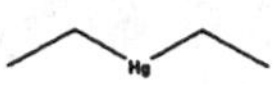

7435

7436

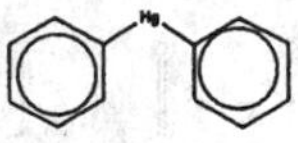

7438

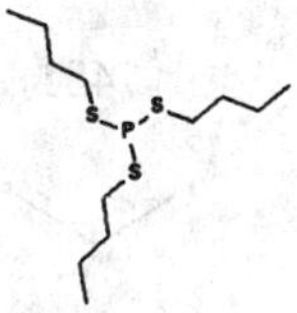

7439

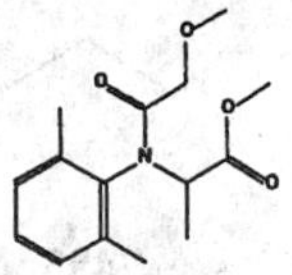

7441

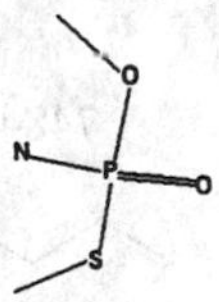

7442

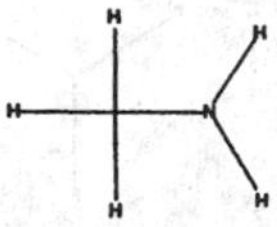

7443

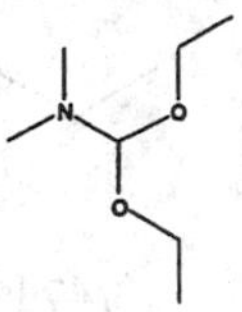

7444

7446

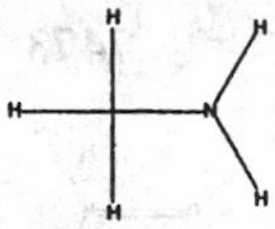

7447

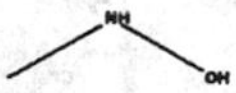

7448

7449

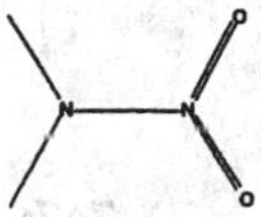

7450

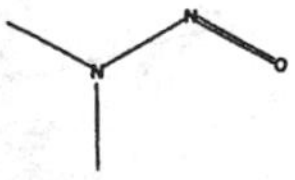

7451

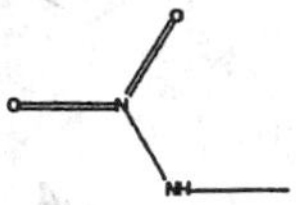

7452

7454

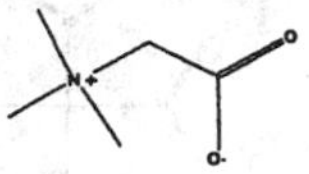

7455

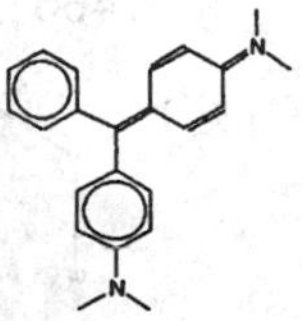

7459

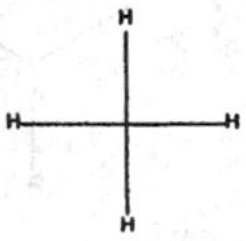

7460

7461

STRUCTURAL FORMULAS OF ORGANIC COMPOUNDS (continued)

In numeric order as they occur in the Table of Physical Constants of Organic Compounds

7462

7463

7464

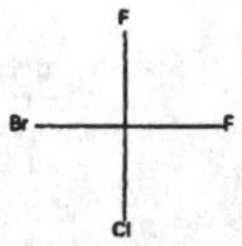

7465

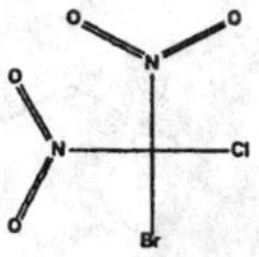

7466

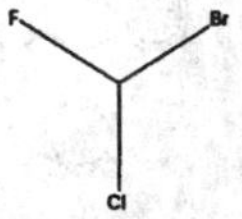

7467

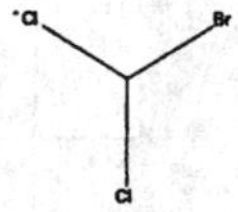

7468

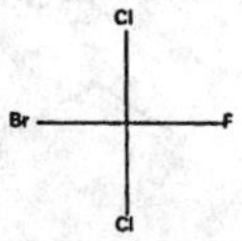

7469

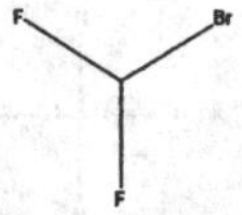

7470

7471

7472

7473

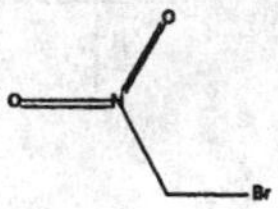

7474

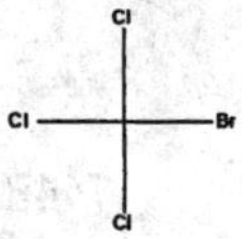

7475

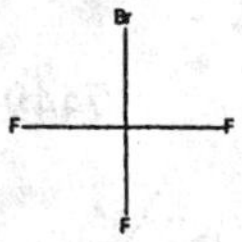

7476

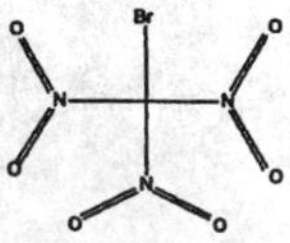

7477

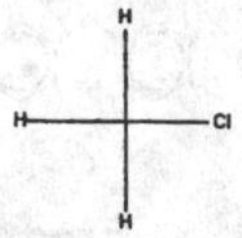

7478

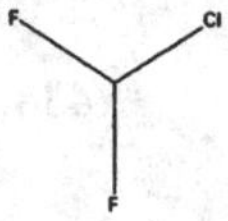

7479

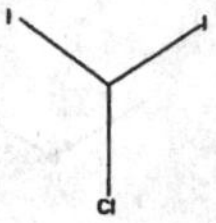

7480

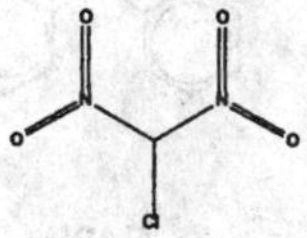

7481

7482

7483

7484

7485

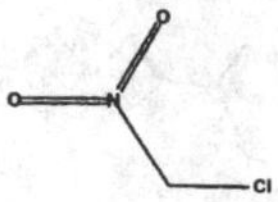

7486

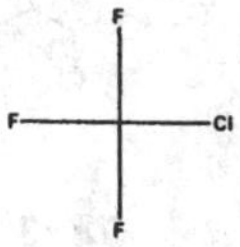

7487

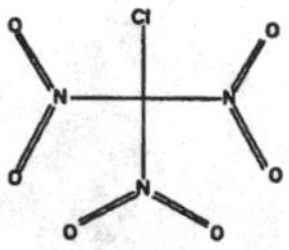

7489

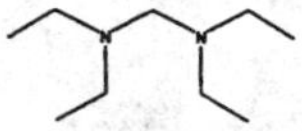

7490

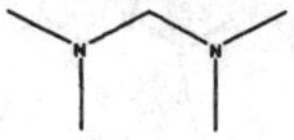

7491

7492

7493

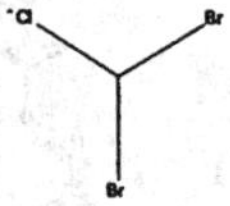

7494

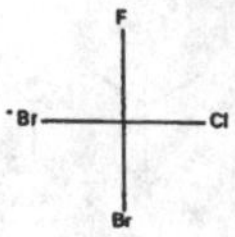

7495

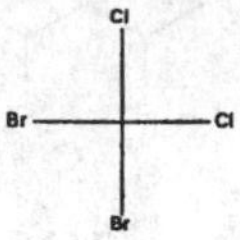

7496

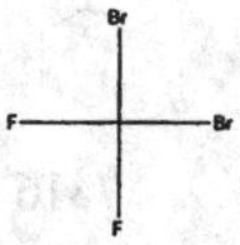

7497

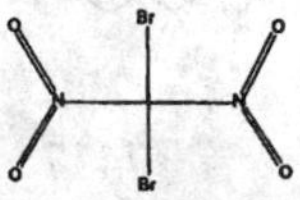

7498

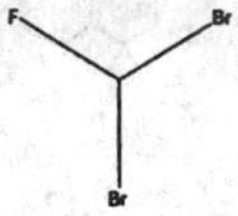

7499

7500

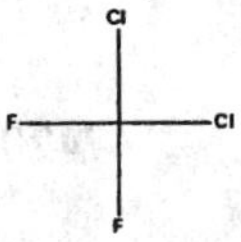

7501

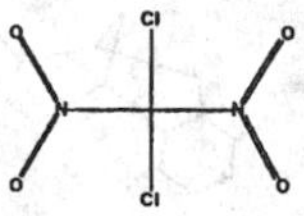

7502

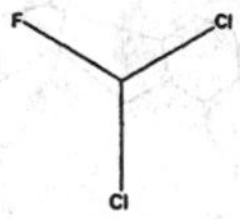

7503

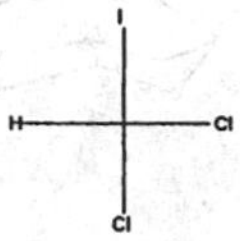

7504

7505

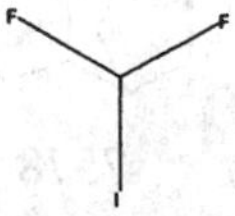

7506

7507

7508

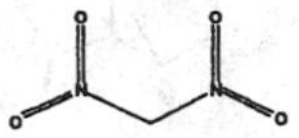

7509

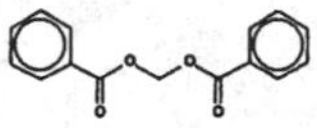

7510

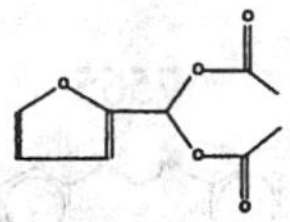

7512

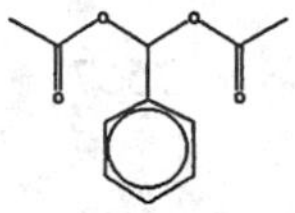

7513

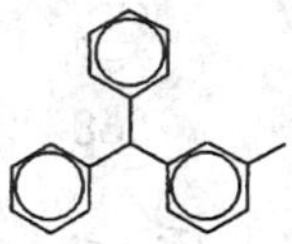

7515

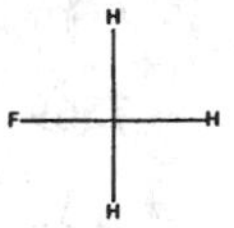

7516

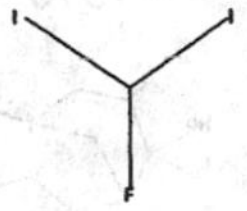

7517

7518

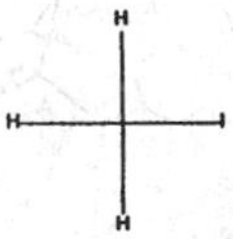

7519

7520

7521

7522

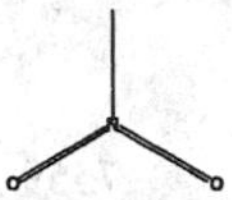

7523

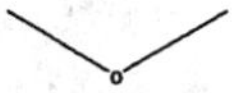

7524

7525

7526

7527

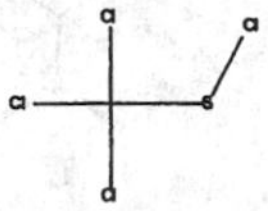

7528

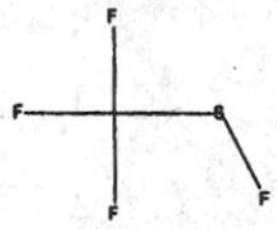

7529

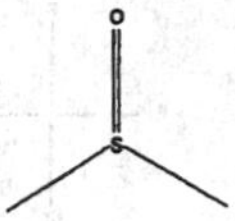

7530

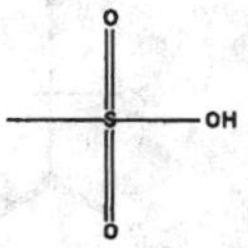

7531

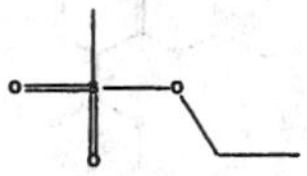

7532

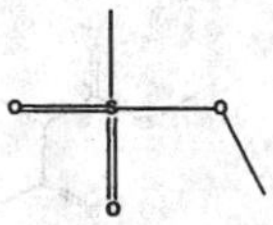

7533

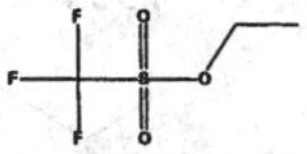

7534

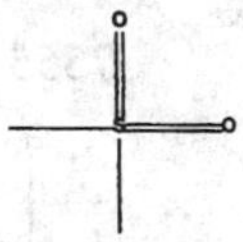

7535

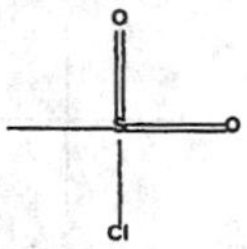

7536

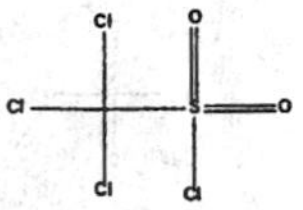

7537

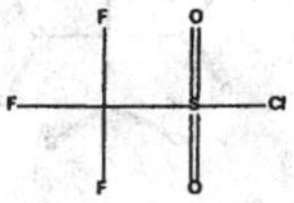

7538

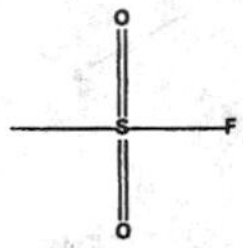

7539

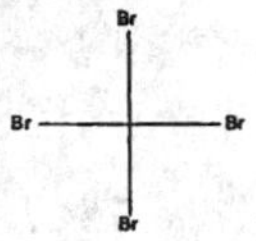

7540

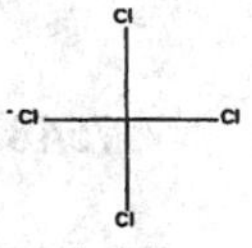

7541

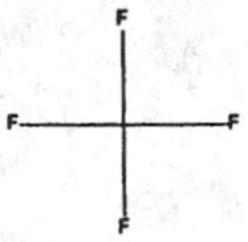

In numeric order as they occur in the Table of Physical Constants of Organic Compounds

7542

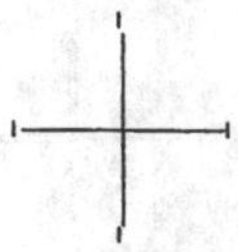

7543

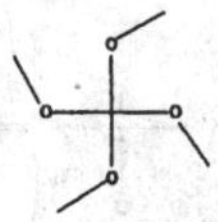

7544

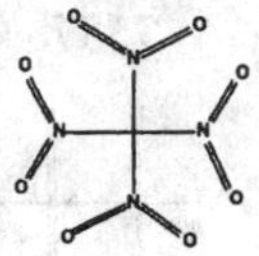

7545

7546

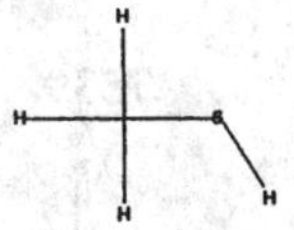

7547

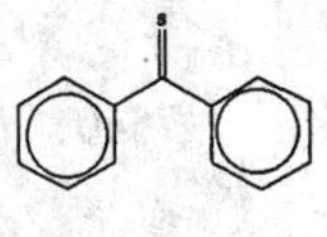

7548

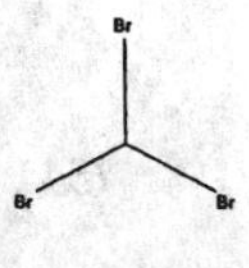

7549

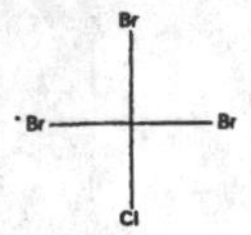

7550

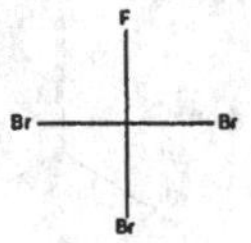

7551

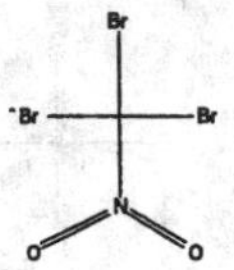

7552

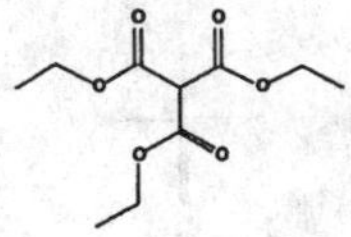

7553

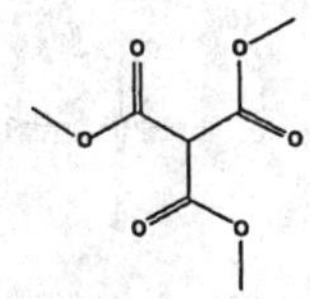

7554

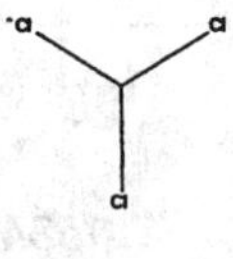

7555

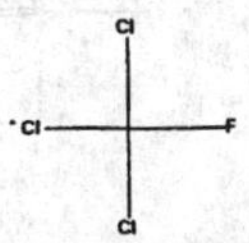

7556

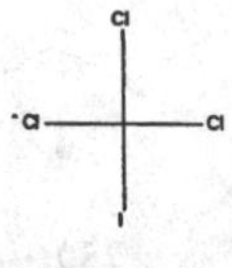

7557

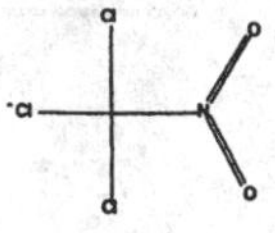

7558

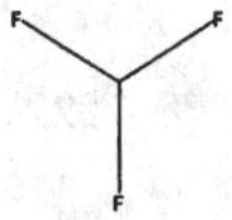

7559

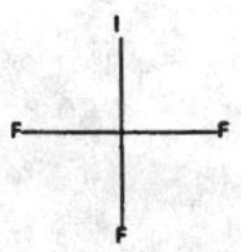

7560

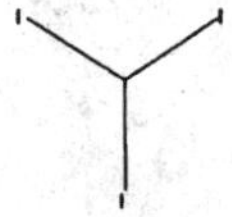

7561

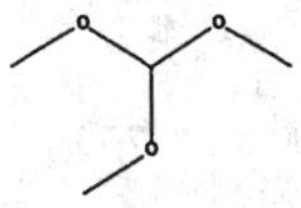

7562

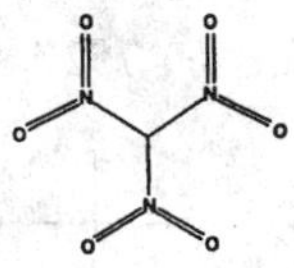

7564

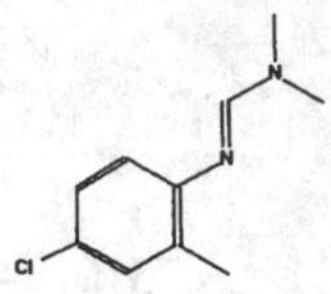

7565

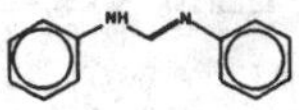

7567

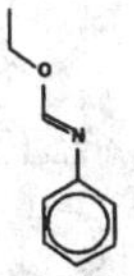

7568

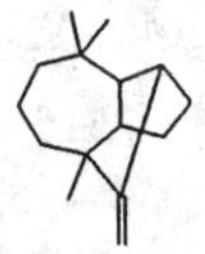

7569

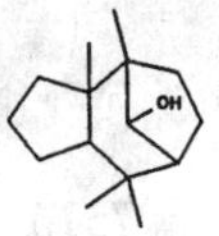

7570

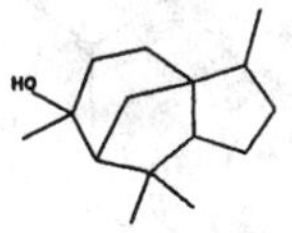

7571

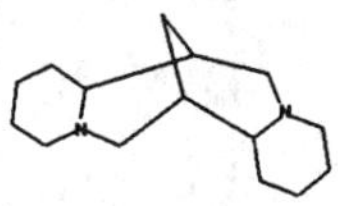

7572

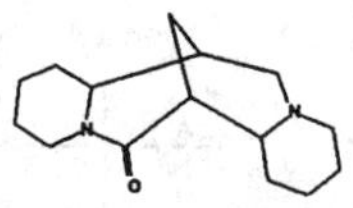

7573

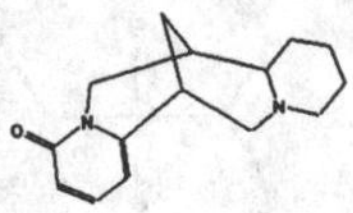

7574

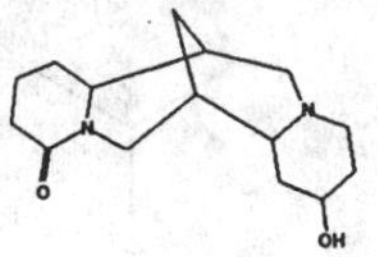

7575

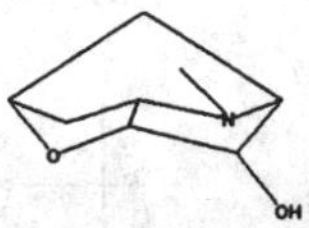

7576

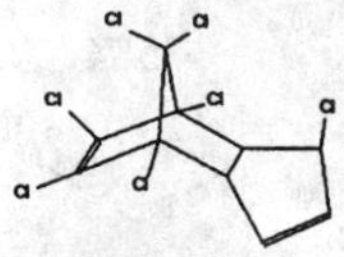

7577

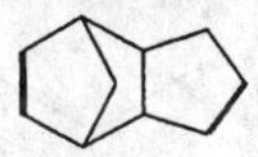

7578

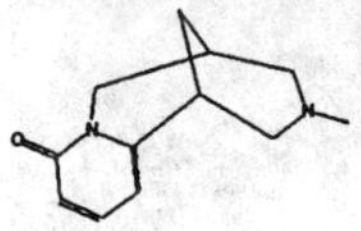

7579

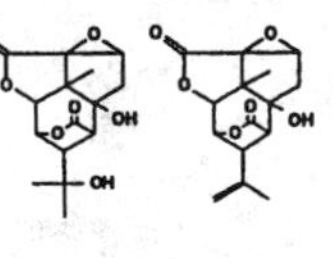

7580

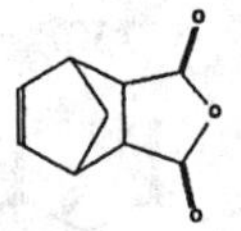

7581

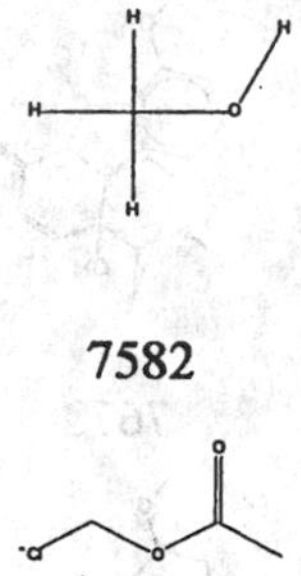

7582

7585

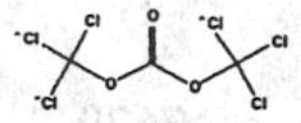

7586

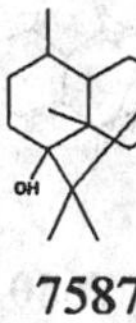

7587

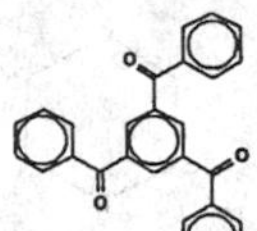

7588

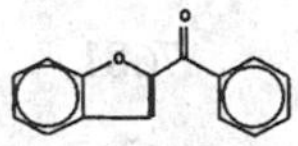

7590

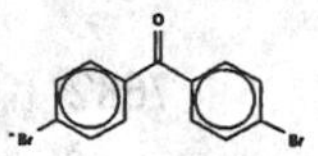

7591

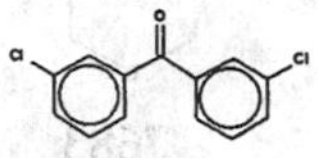

7592

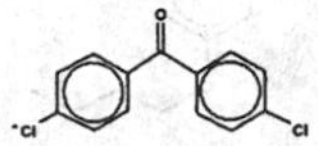

7593

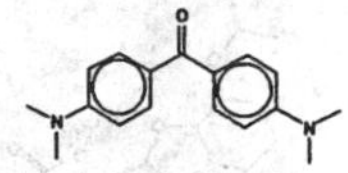

7594

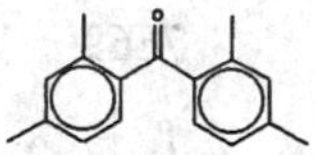

7595

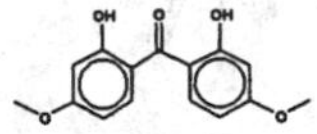

7597

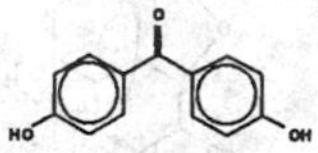

7598

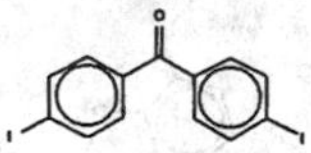

7599

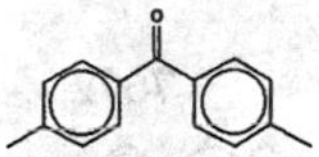

7600

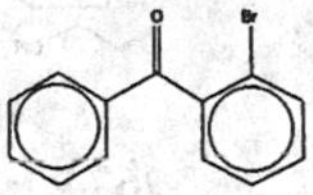

7601

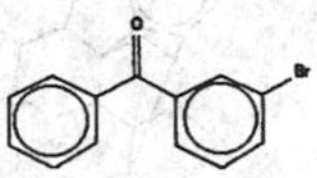

7602

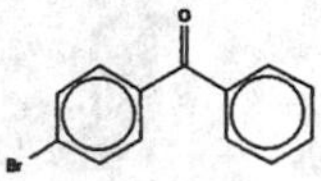

7604

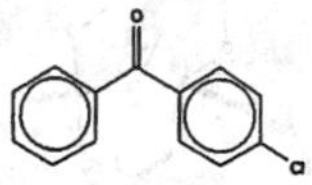

7605

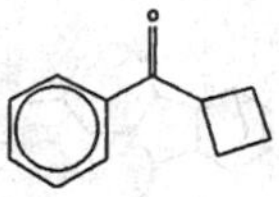

7606

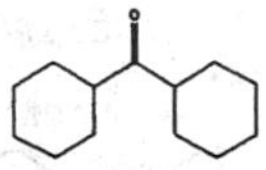

7607

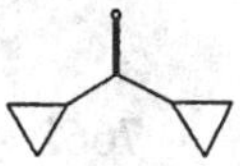

7609

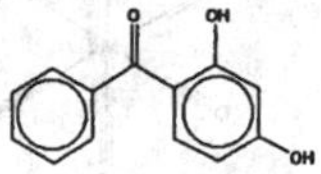

7612

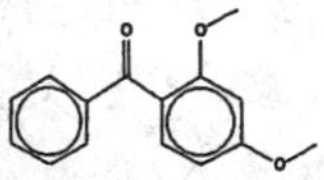

7614

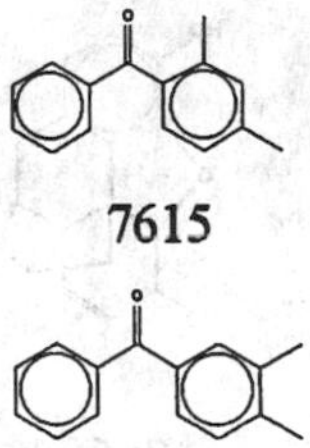

7615

7616

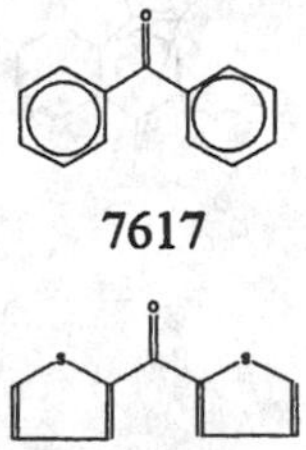

7617

7618

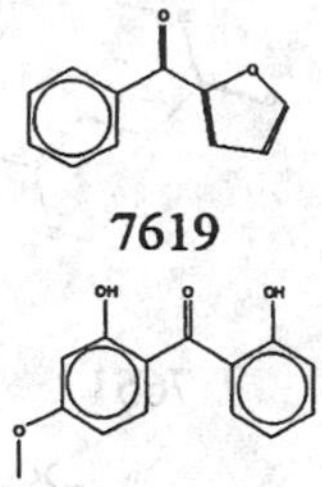

7619

7621

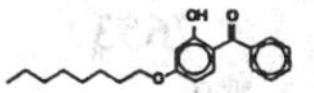

7622

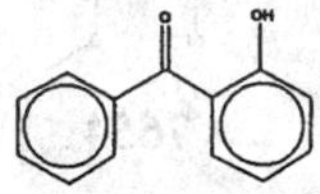

7623

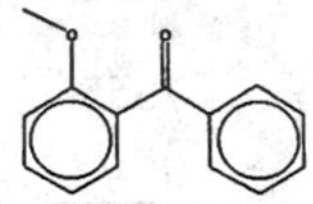

7624

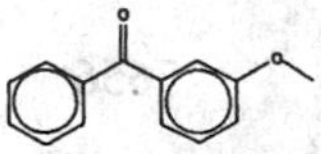

7625

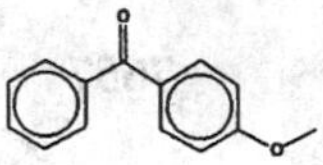

7626

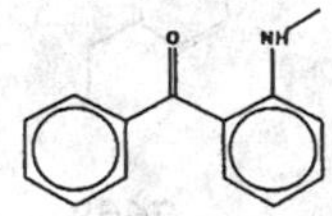

7629

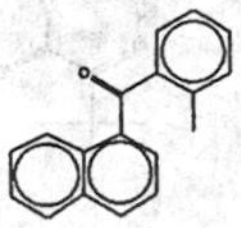

7631

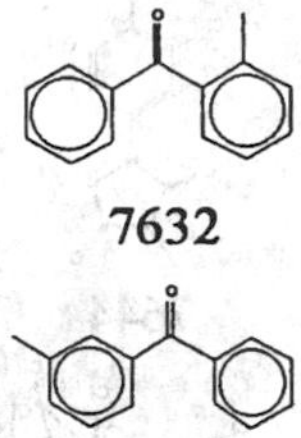

7632

7633

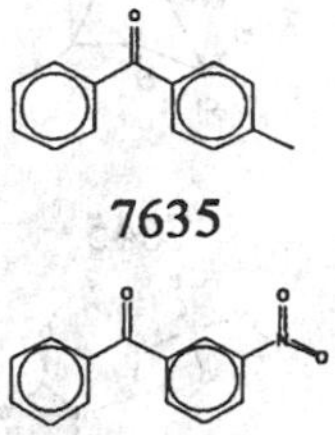

7635

7636

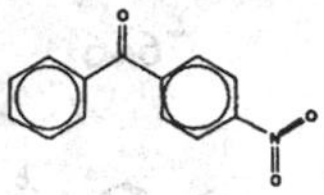

7639

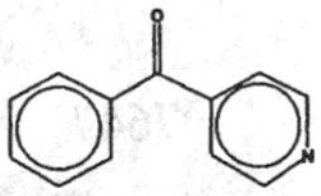

7640

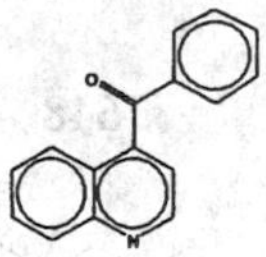

7641

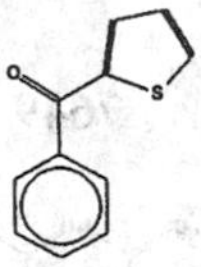

7642

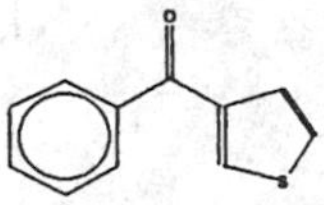

STRUCTURAL FORMULAS OF ORGANIC COMPOUNDS (continued)

In numeric order as they occur in the Table of Physical Constants of Organic Compounds

7643

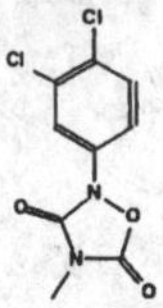

7644

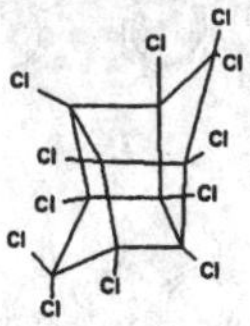

7645

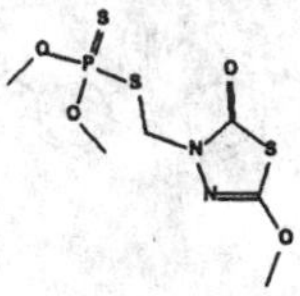

7646

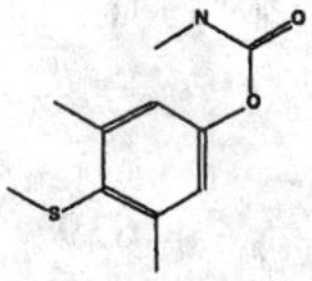

7647

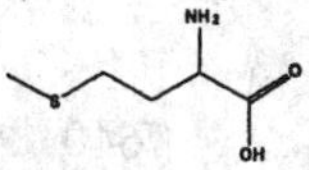

7648

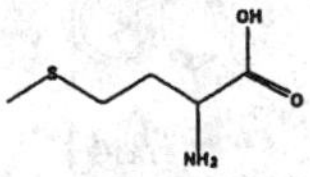

7649

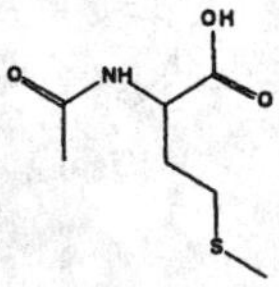

7650

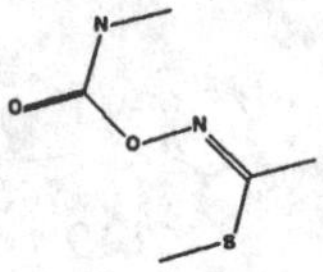

7651

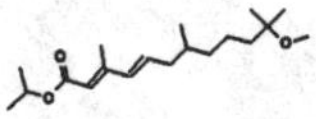

7653

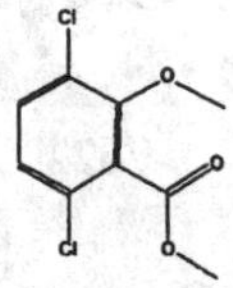

7654

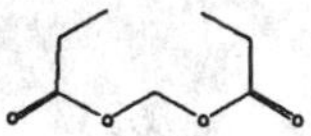

7655

7656

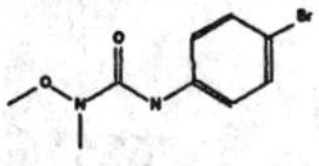

7657

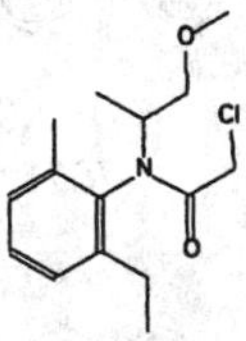

7658

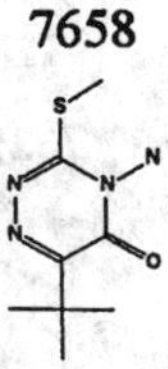

7659

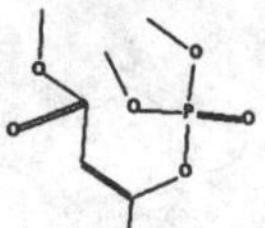

7660

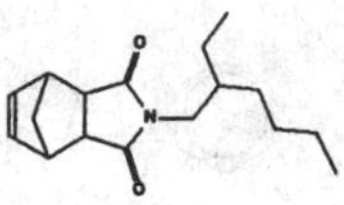

7661

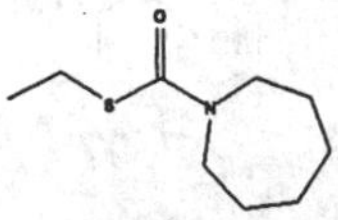

7662

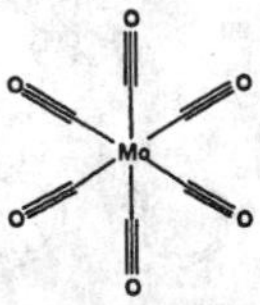

7663

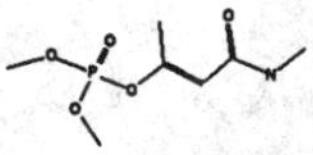

7665

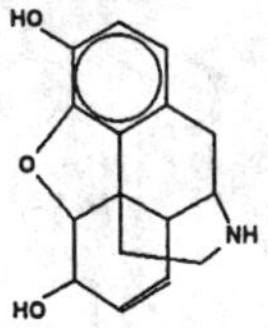

7666

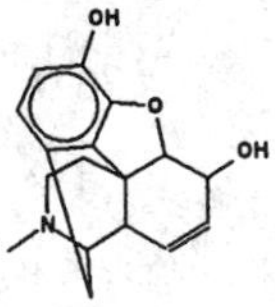

7667

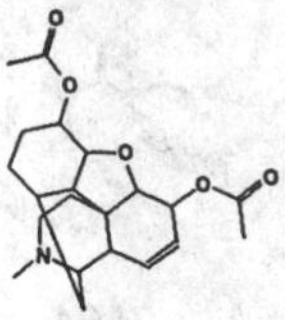

7669

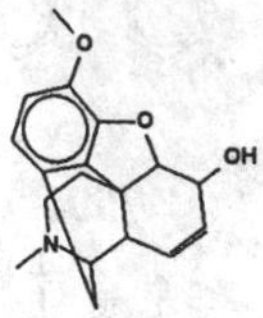

7670

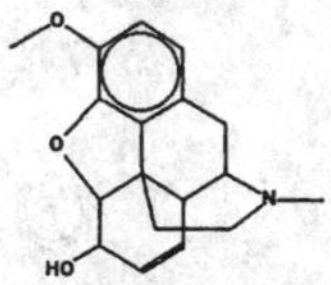

7671

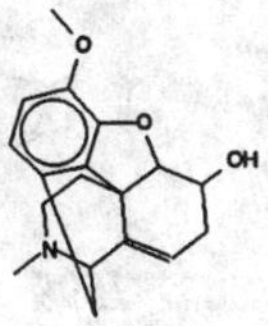

7672

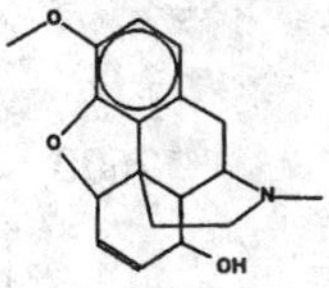

7674

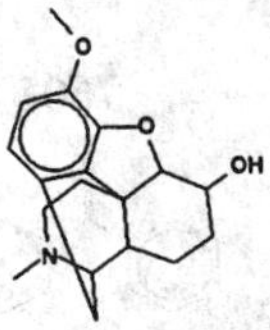

7675

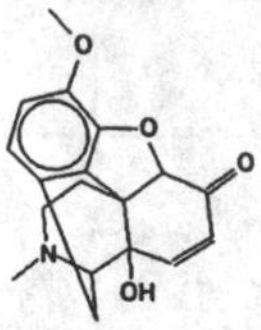

7678

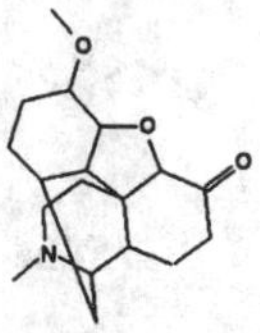

7679

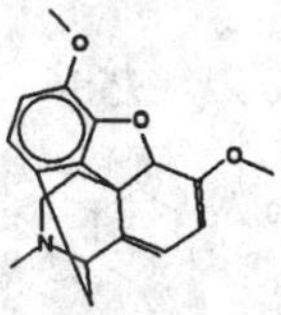

7680

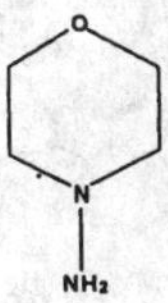

7681

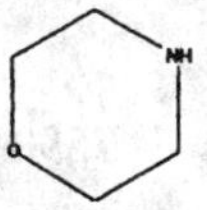

7682

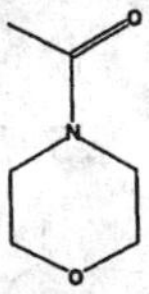

7683

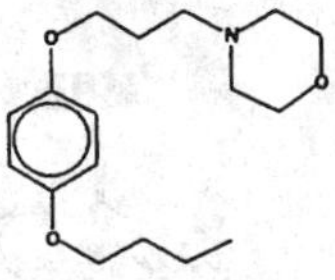

In numeric order as they occur in the Table of Physical Constants of Organic Compounds

7684

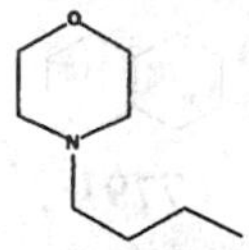

7685

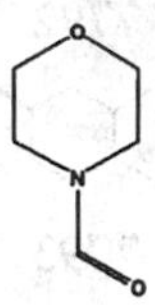

7686

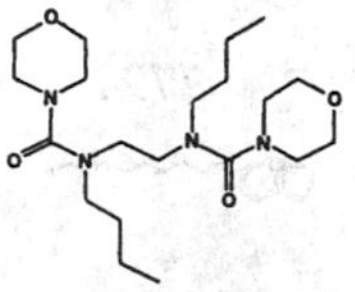

7687

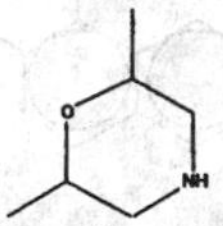

7689

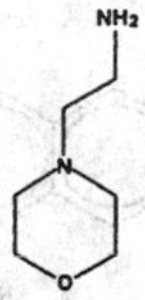

7690

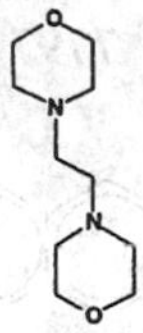

7691

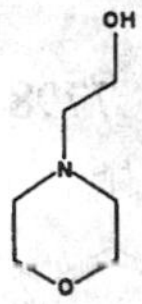

7692

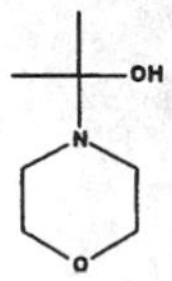

7693

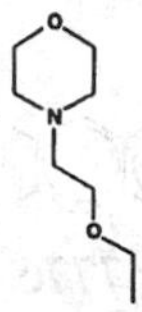

7694

7696

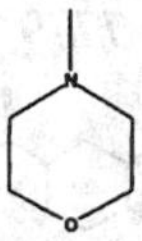

7697

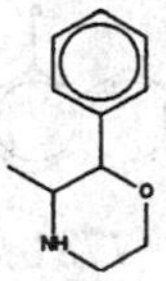

7698

7699

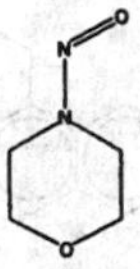

7700

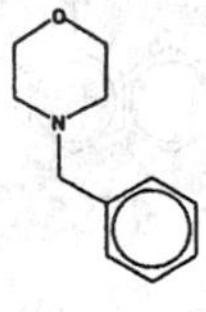

7701

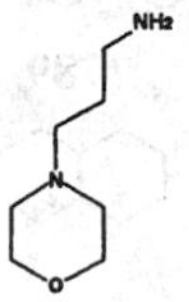

7702

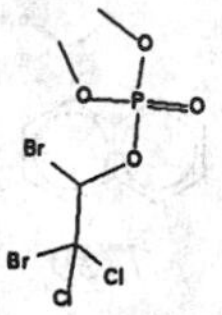

7703

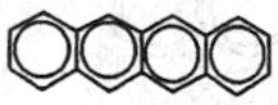

7704

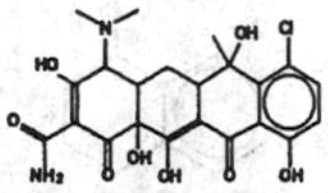

7705

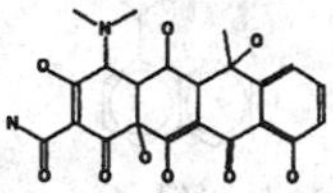

7706

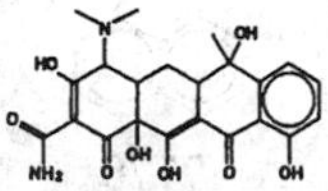

7708

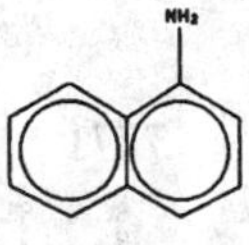

7709

7710

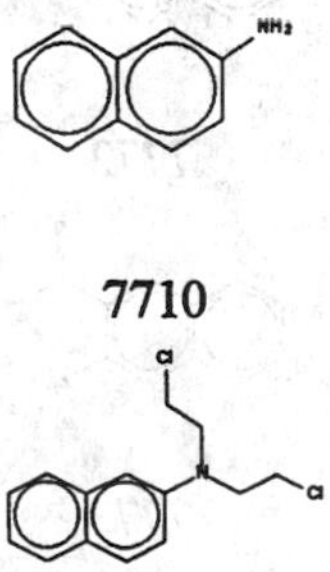

7712

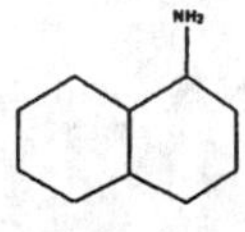

7713

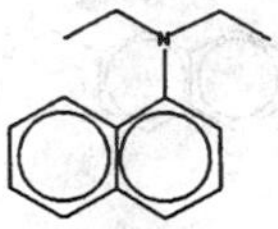

7714

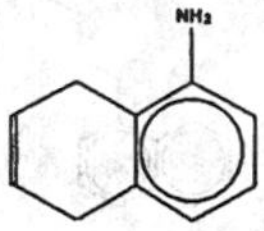

7715

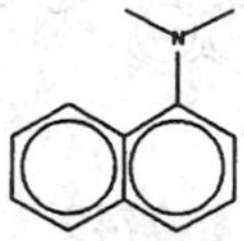

7716

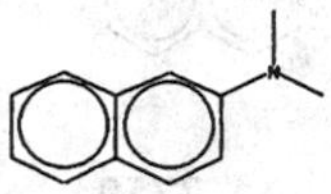

7717

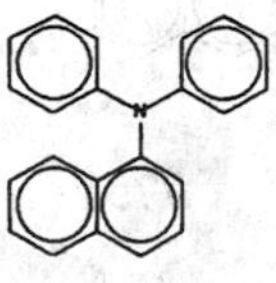

7718

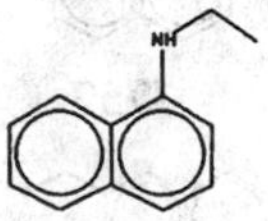

7720

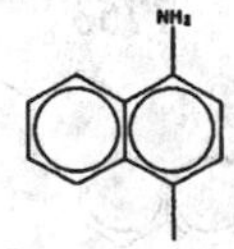

7721

7722

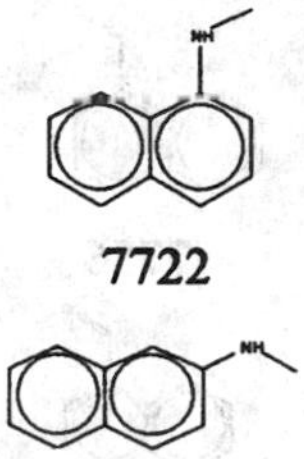

7723

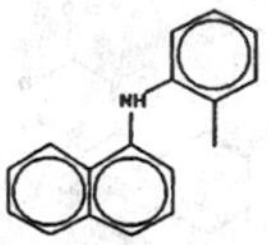

7724

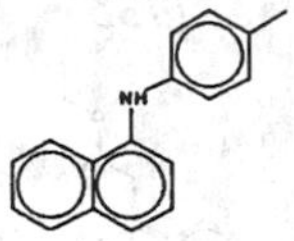

7725

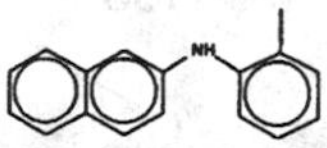

7727

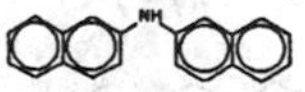

7728

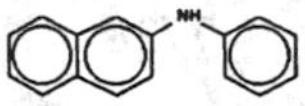

7730

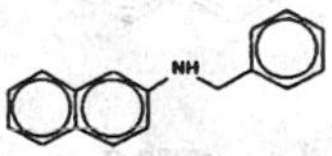

7731

7732

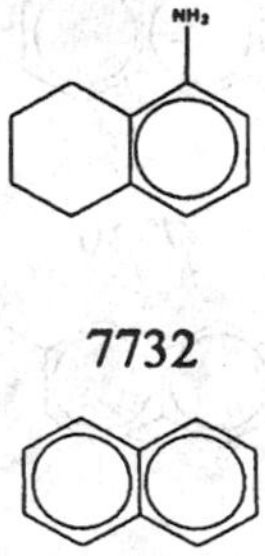

7733

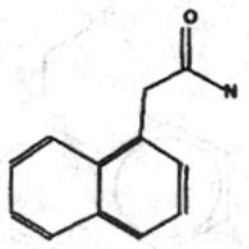

In numeric order as they occur in the Table of Physical Constants of Organic Compounds

7734

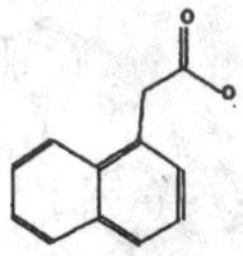

7735

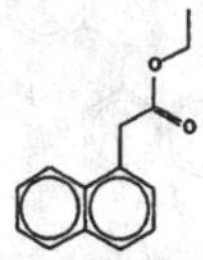

7736

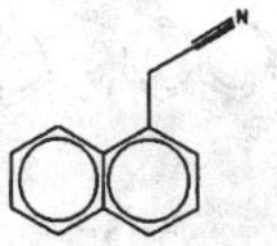

7737

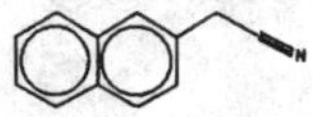

7738

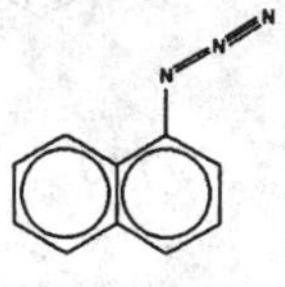

7739

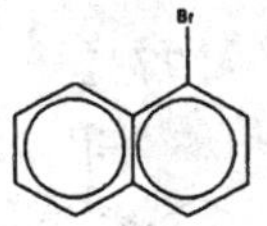

7740

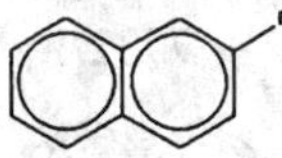

7742

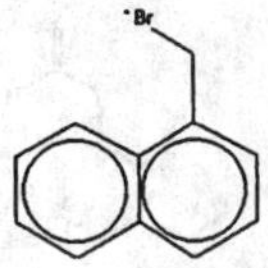

7743

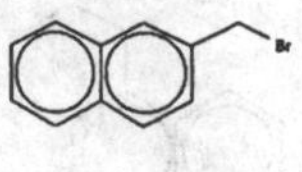

7746

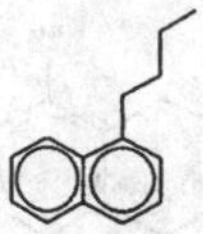

7747

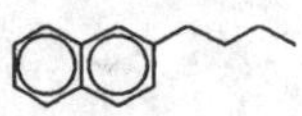

7749

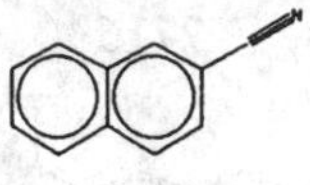

7750

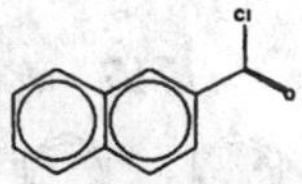

7751

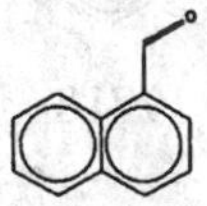

7752

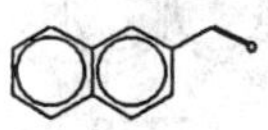

7753

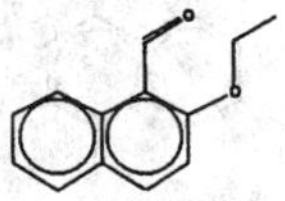

7754

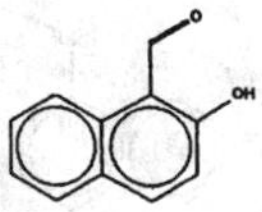

7756

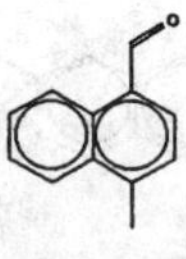

7758

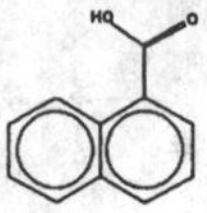

7759

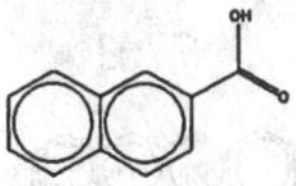

7764

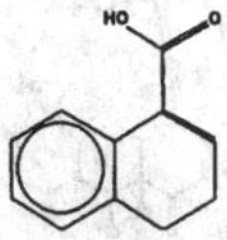

7765

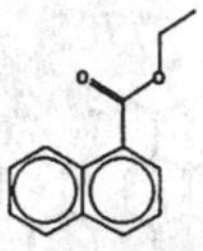

7766

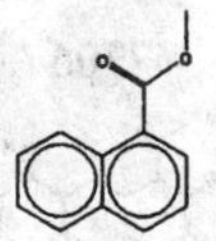

7770

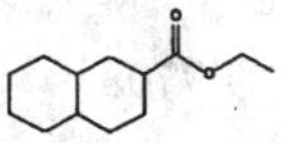

7771

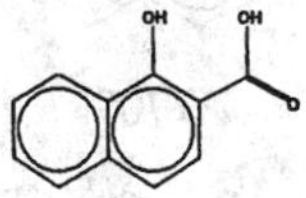

7772

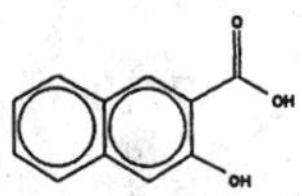

7773

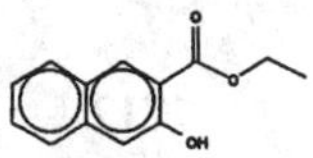

7775

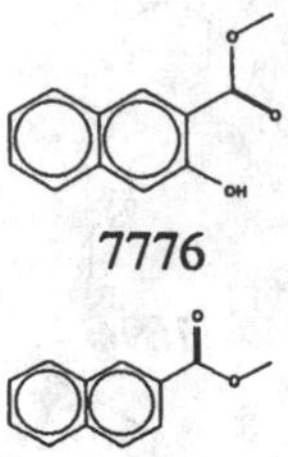

7776

7778

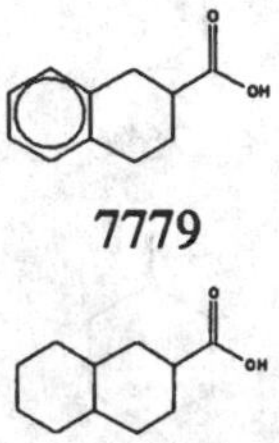

7779

7780

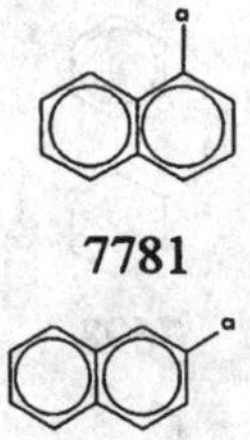

7781

7782

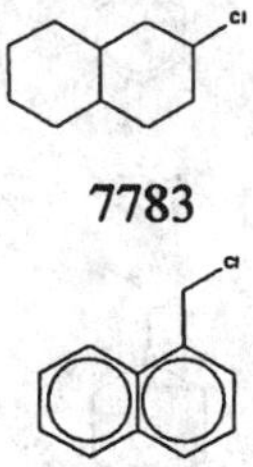

7783

7784

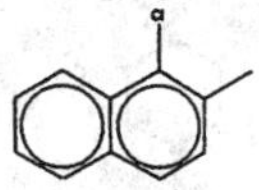

7785

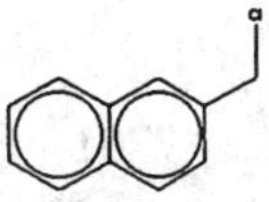

7789

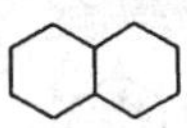

7790

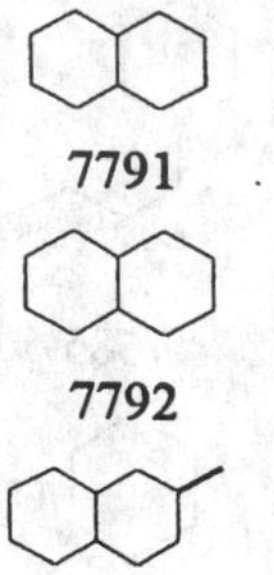

7791

7792

7793

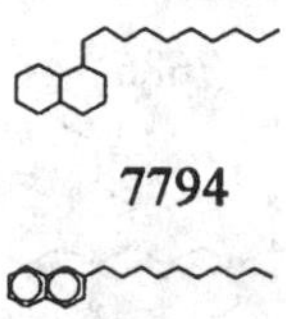

7794

7795

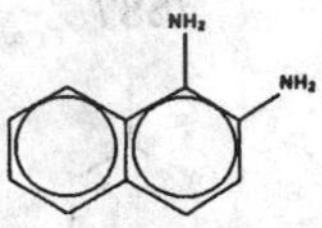

7796

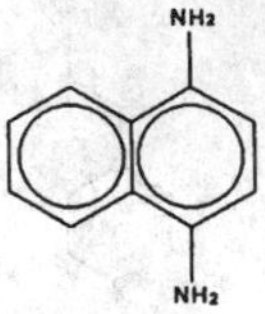

7797

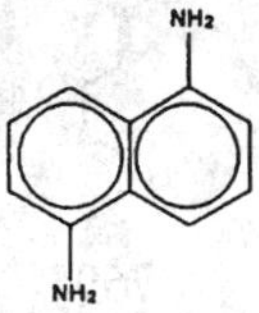

7798

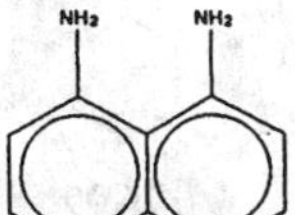

STRUCTURAL FORMULAS OF ORGANIC COMPOUNDS (continued)

In numeric order as they occur in the Table of Physical Constants of Organic Compounds

7799

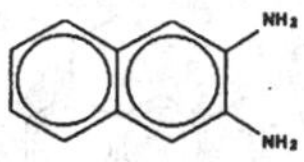

7801

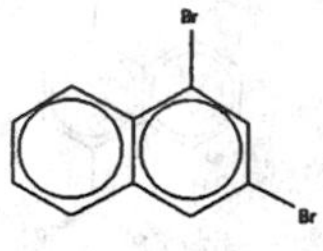

7802

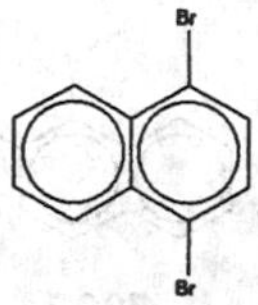

7803

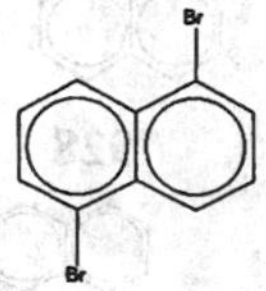

7804

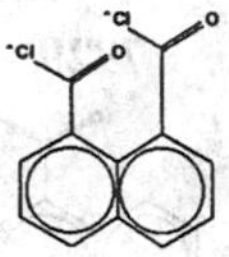

7806

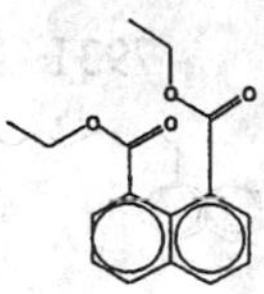

7807

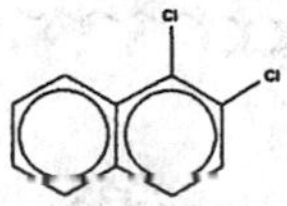

7808

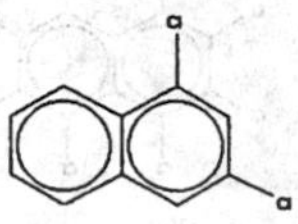

7809

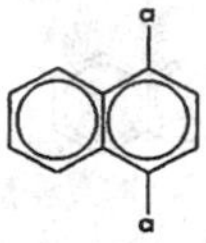

7810

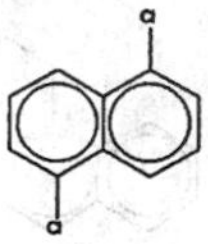

7811

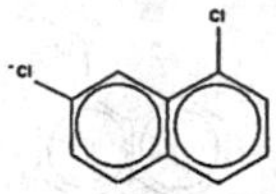

7812

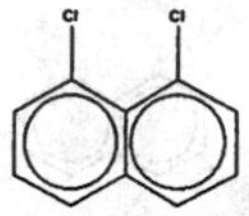

7813

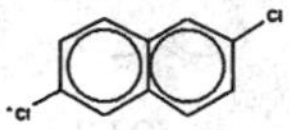

7814

7815

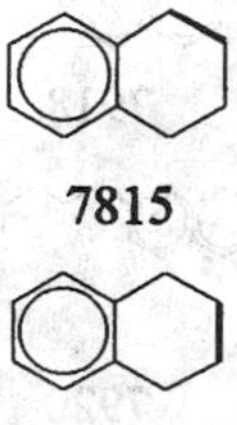

7816

7817

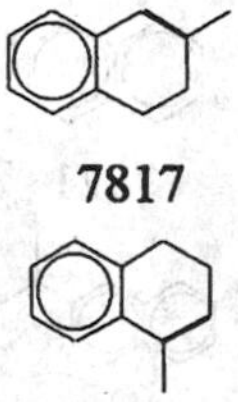

7818

7819

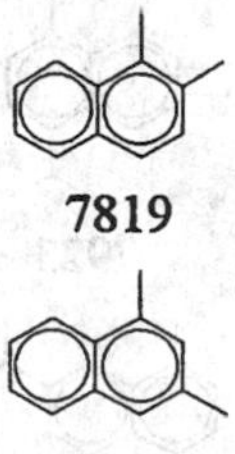

7820

7821

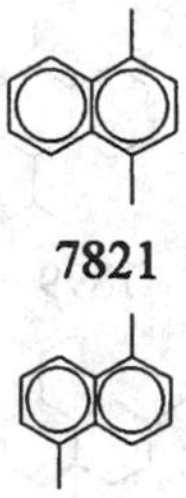

7822

7823

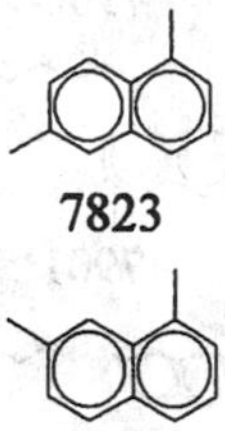

7824

7825

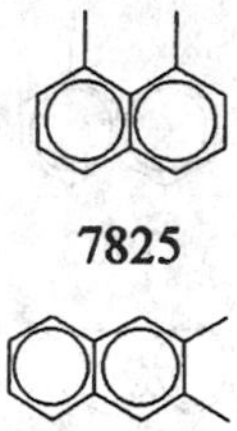

7826

7827

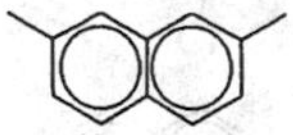

7828

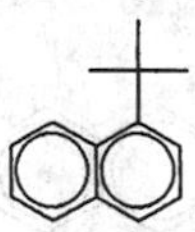

7829

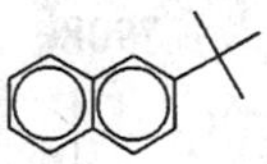

7830

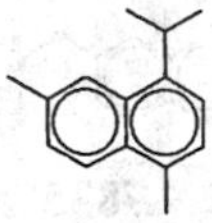

7831

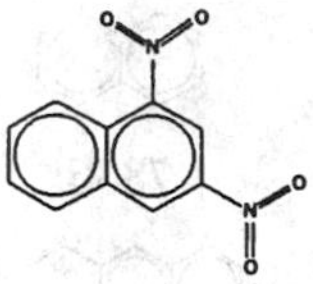

7832

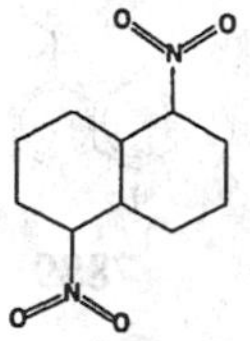

7833

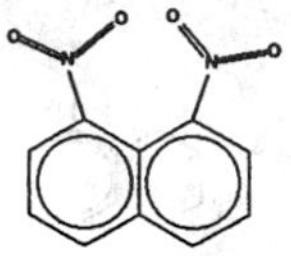

7834

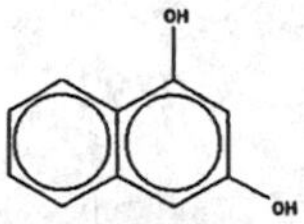

7835

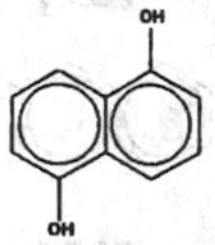

7836

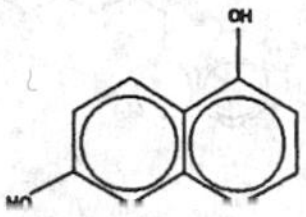

7838

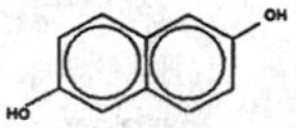

7839

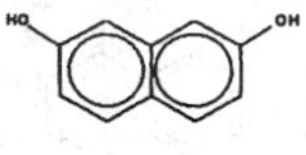

7842

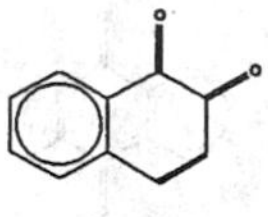

7843

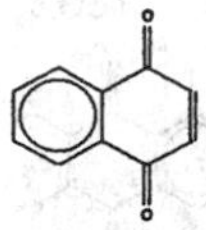

7846

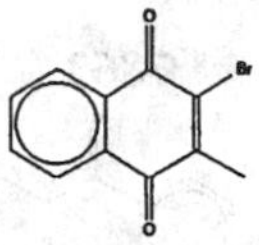

7848

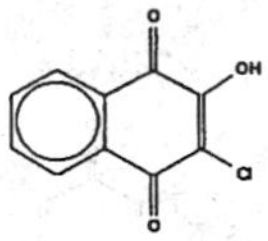

7850

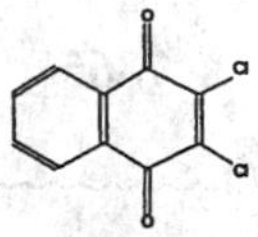

7852

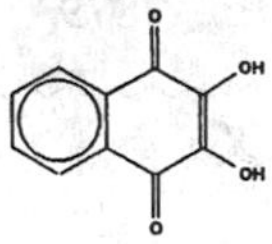

7853

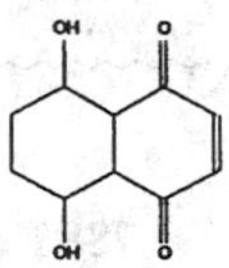

7859

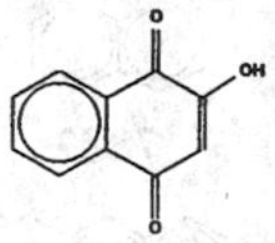

7860
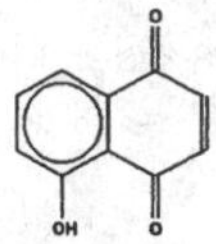
7861
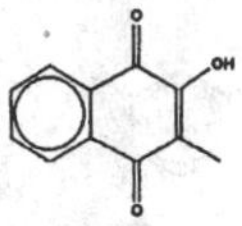
7862
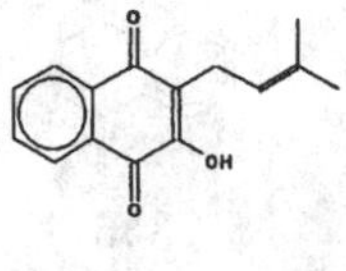
7863
7864
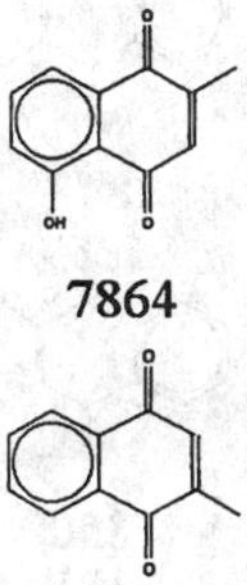
7865
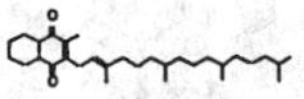
7879
7880
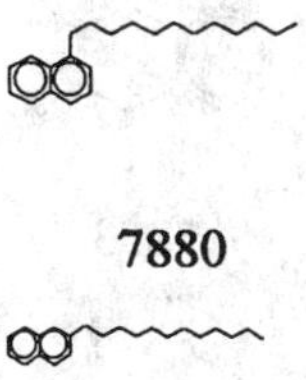
7881
7882
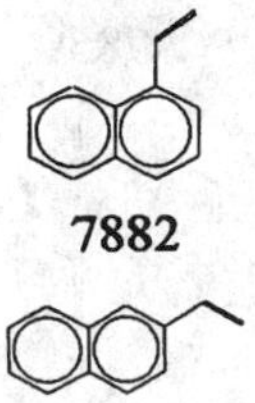
7884
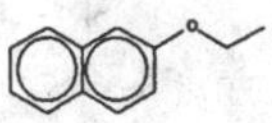
7885
7886
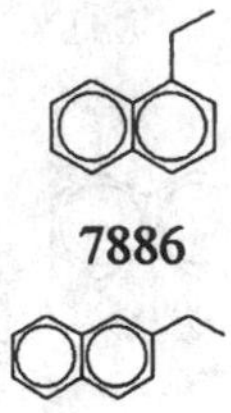
7887
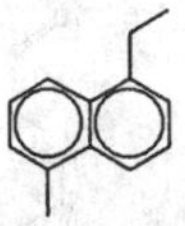
7889
7890
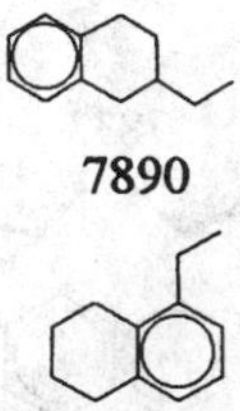
7891
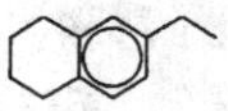
7892

7894
7895
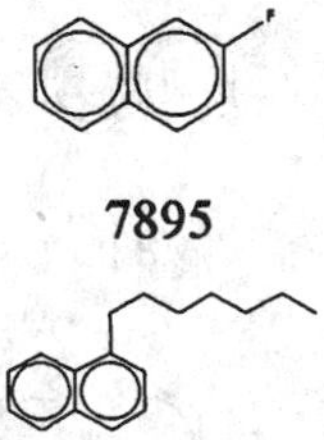
7896
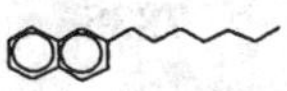
7897
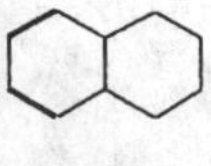
7898
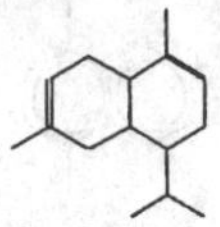
7899
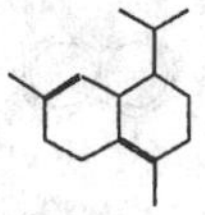
7900
7901
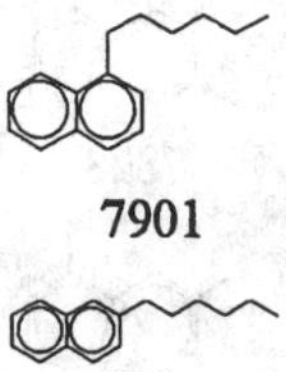
7902
7903
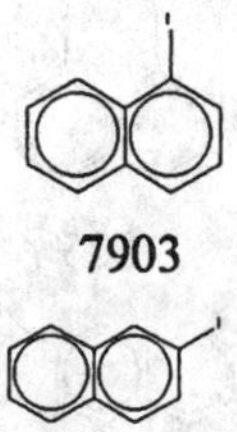
7904
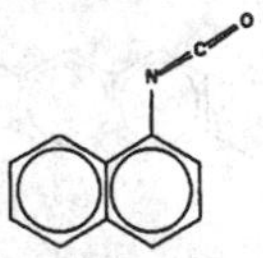
7906
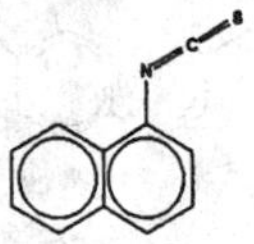
7907
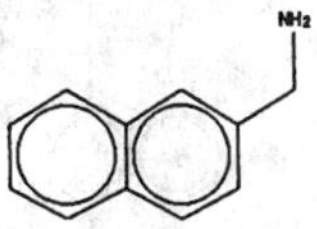
7908
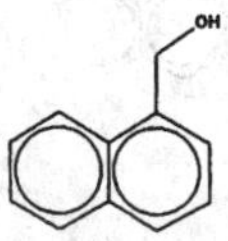
7909
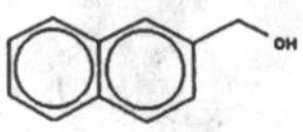
7911
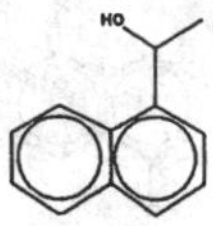
7913
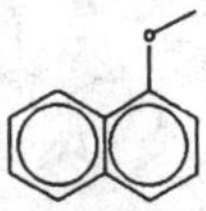
7915
7916

7917
7918
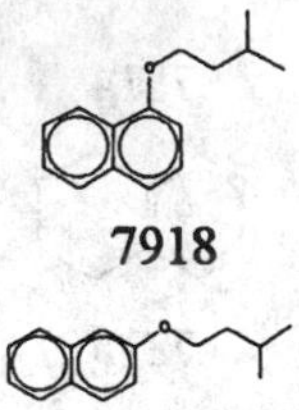
7920
7921
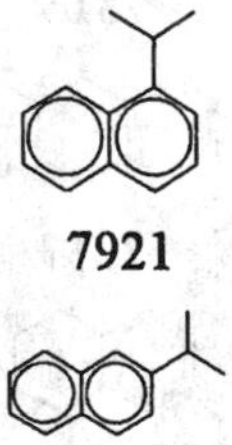
7922
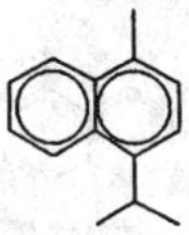
7923
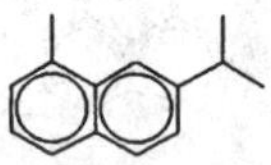
7924
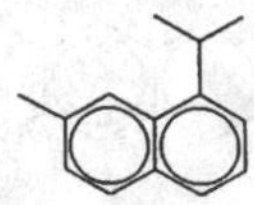
7925
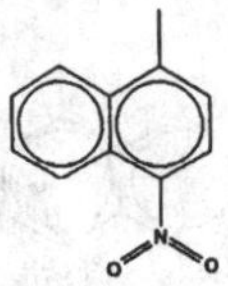
7926
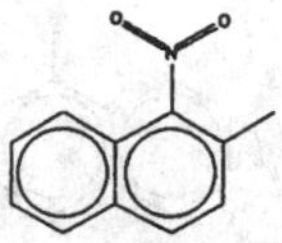
7927
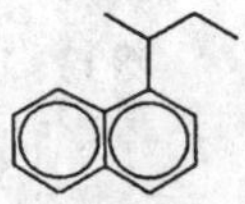
7928
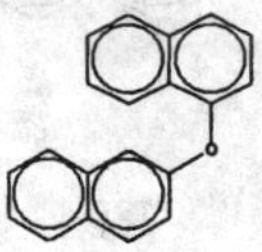
7929
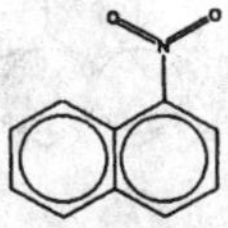
7931
7932
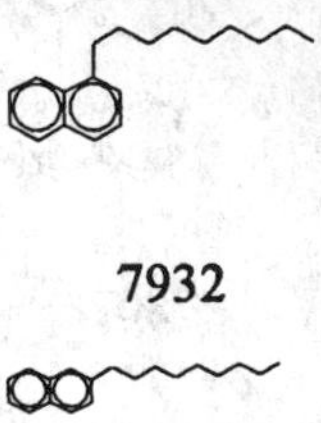
7933
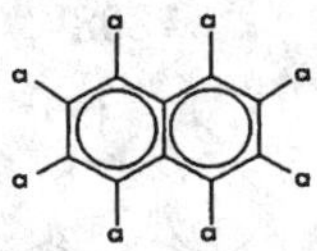

In numeric order as they occur in the Table of Physical Constants of Organic Compounds

7934

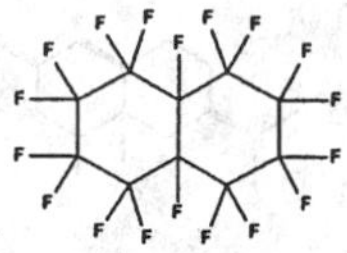

7935

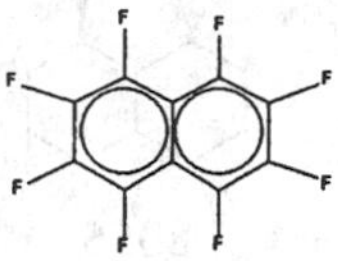

7936

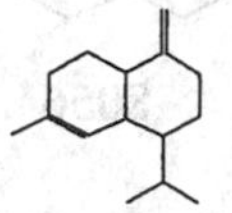

7937

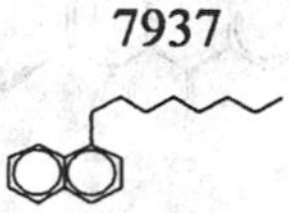

7938

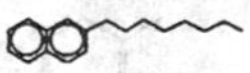

7939

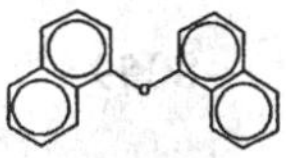

7941

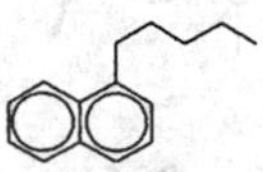

7942

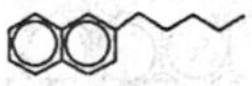

7943

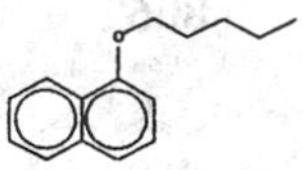

7944

7945

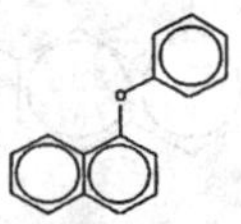

7946

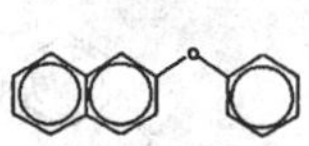

7948

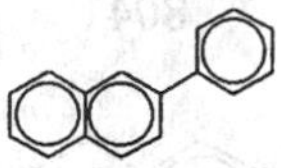

7949

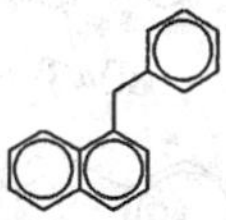

7950

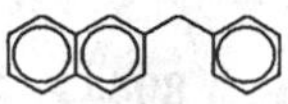

7951

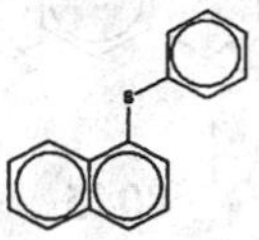

7952

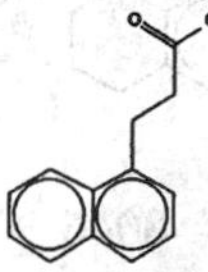

7954

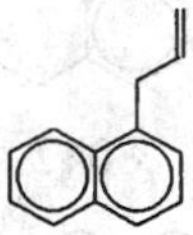

7955

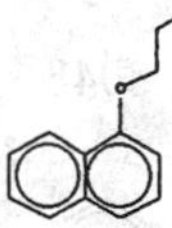

7956

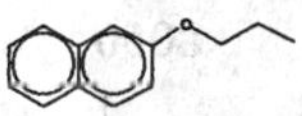

7957

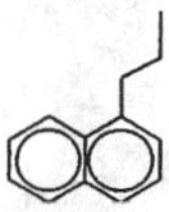

7958

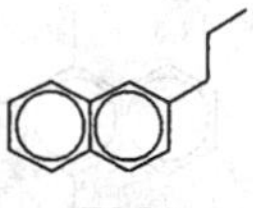

7959

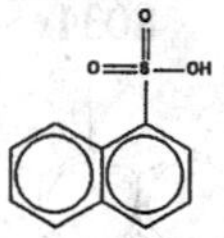

7966

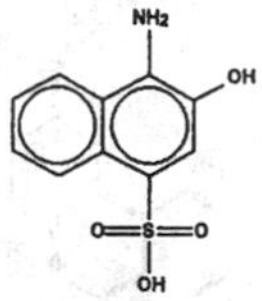

7975

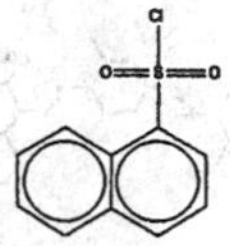

7976

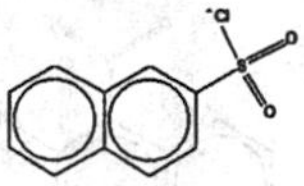

7977

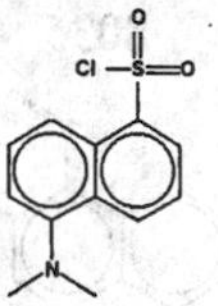

7978

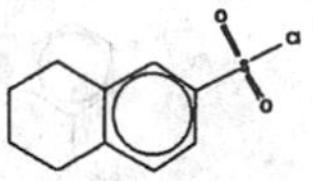

7979

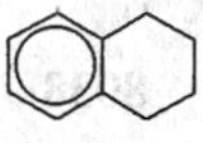

7980

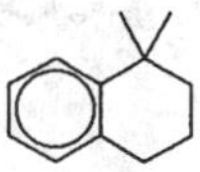

7981, 7982, 7983

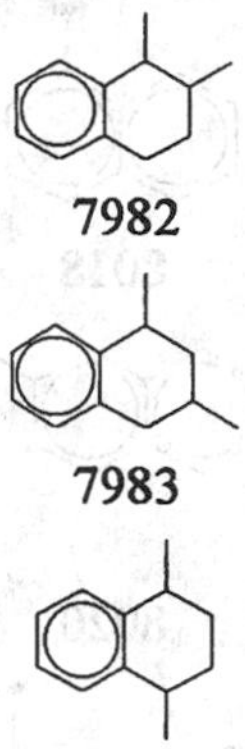

7984, 7985, 7986

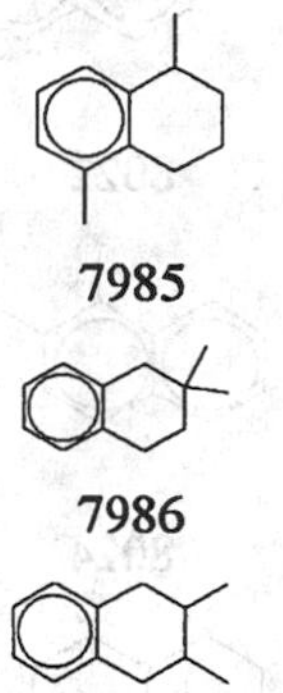

7987, 7988, 7989

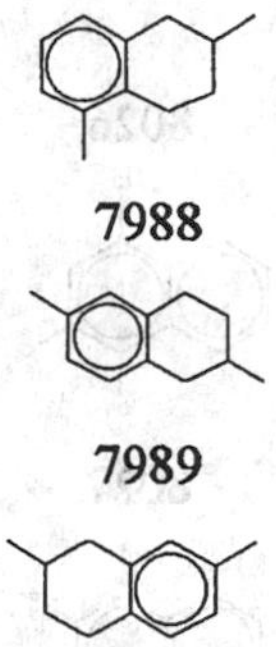

7990, 7991, 7992

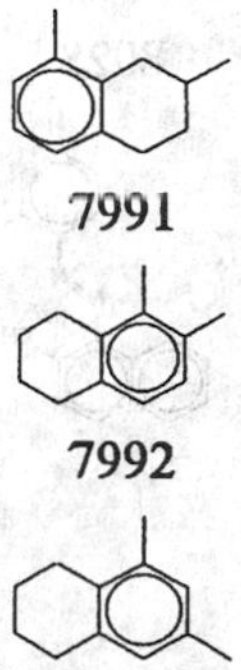

7993, 7994, 7995

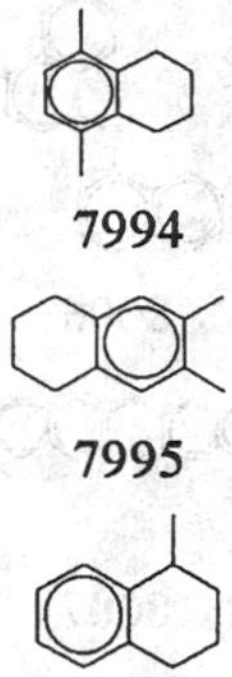

7996, 7997, 7998

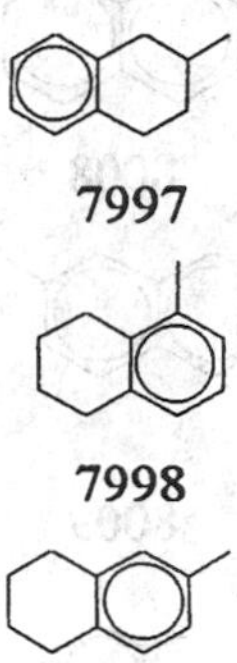

7999, 8000, 8001

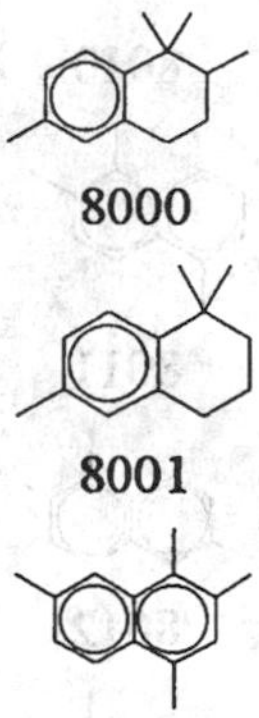

8002, 8003

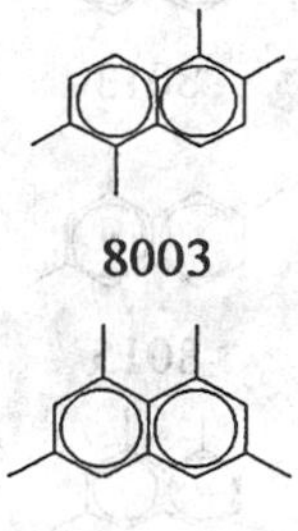

In numeric order as they occur in the Table of Physical Constants of Organic Compounds

8005
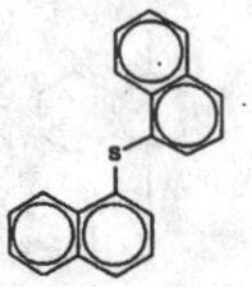
8006
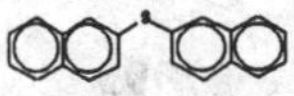
8007
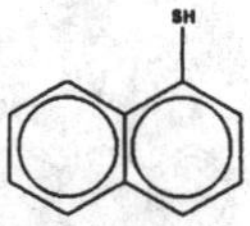
8008
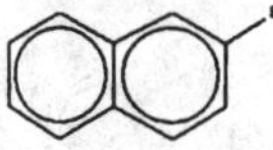
8009
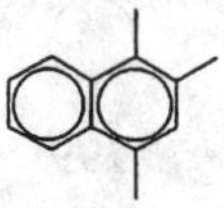
8010
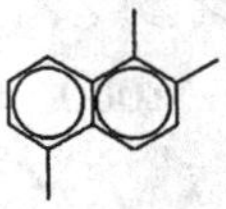
8011
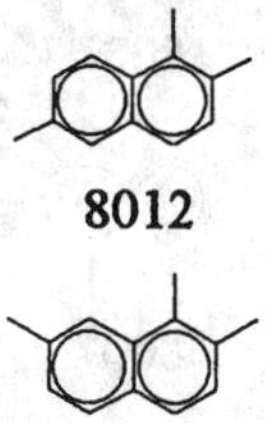
8012
8013
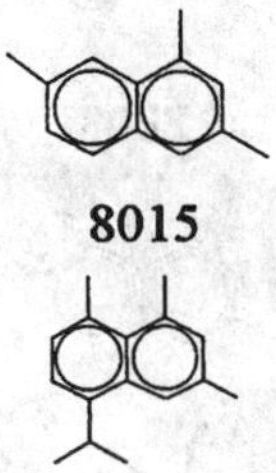
8015

8017
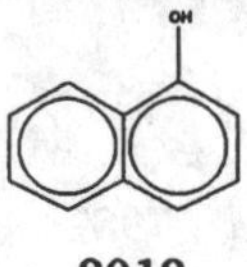
8018
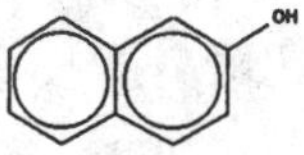
8020
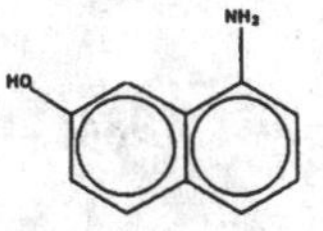
8022
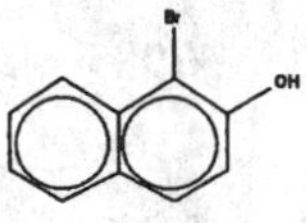
8024
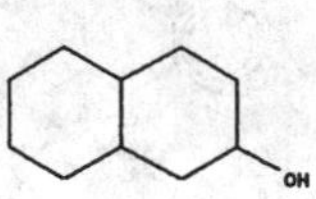
8026
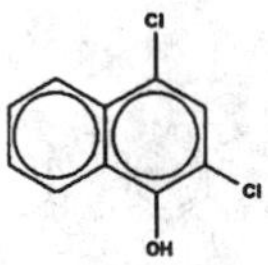
8027
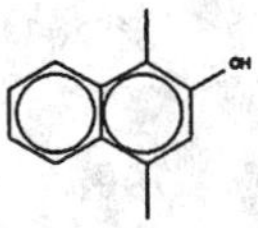
8028
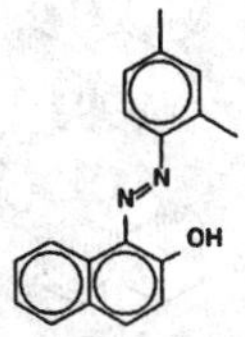

8030
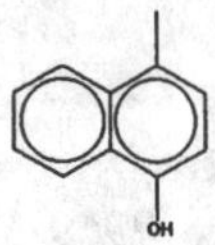
8031
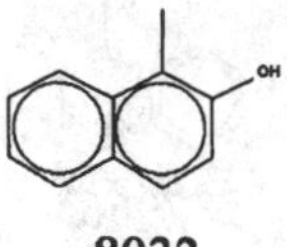
8032
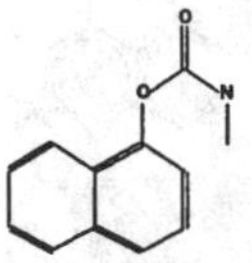
8033
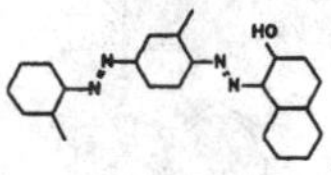
8034
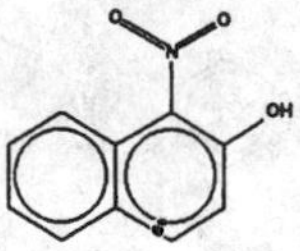
8035
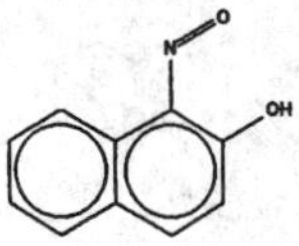
8037
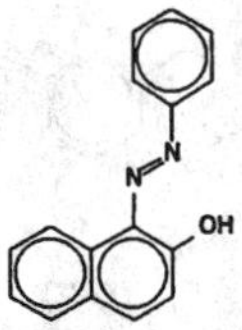
8038
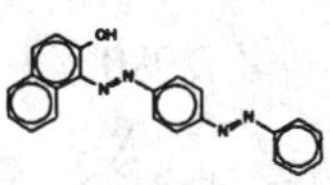

8039
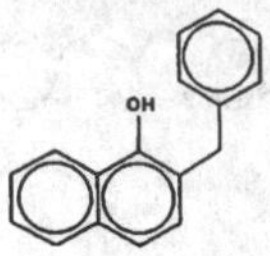
8041
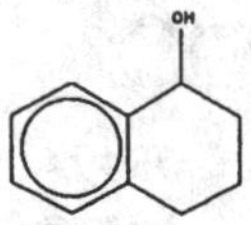
8043
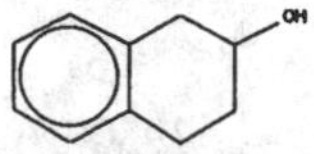
8044
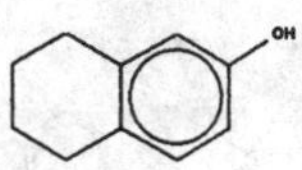
8045
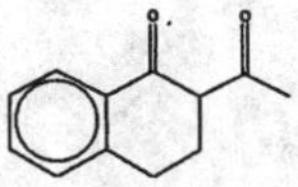
8046
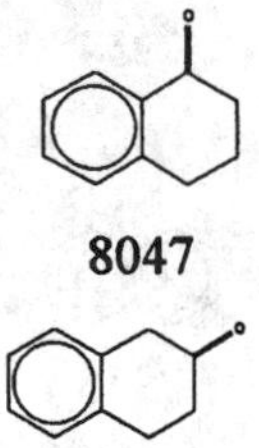
8047
8048
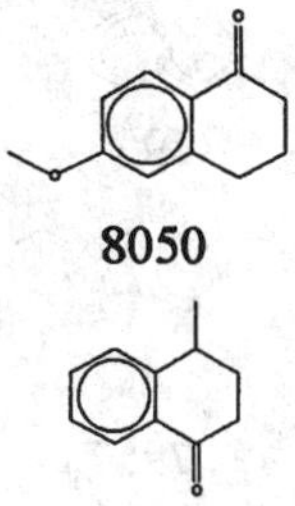
8050

8052
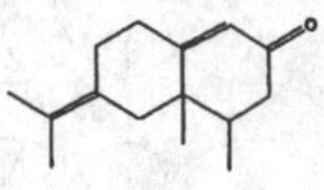
8054
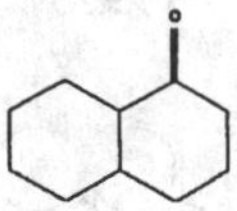
8055
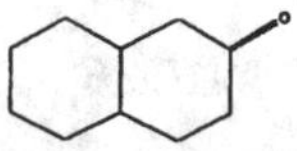
8056
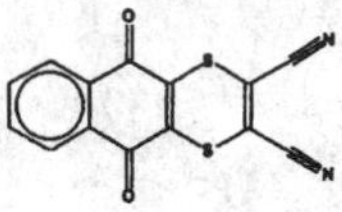
8057
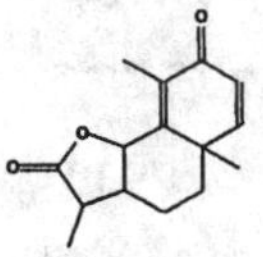
8059
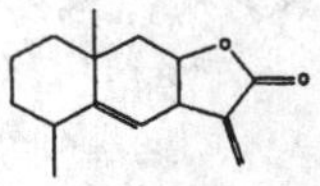
8061
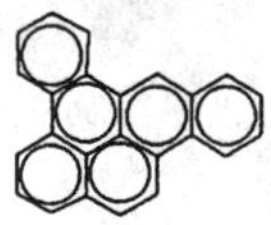
8062
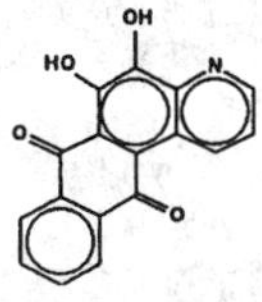
8063
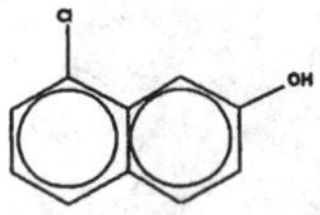

STRUCTURAL FORMULAS OF ORGANIC COMPOUNDS (continued)

In numeric order as they occur in the Table of Physical Constants of Organic Compounds

8064

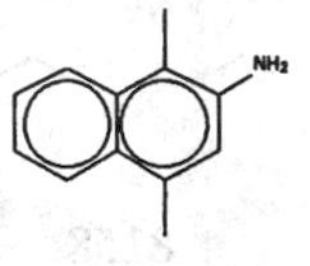

8065

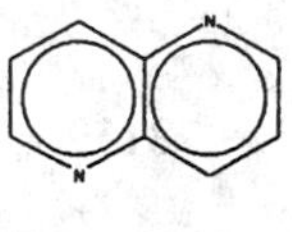

8066

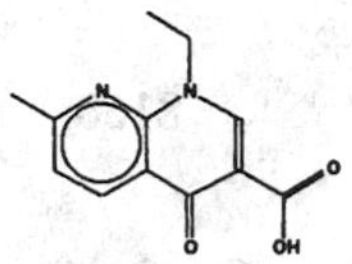

8067

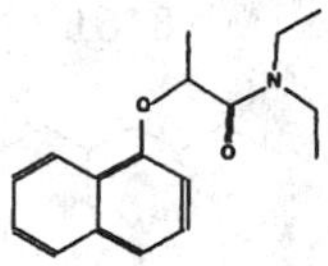

8068

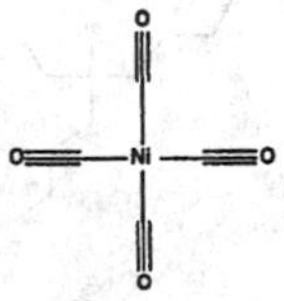

8069

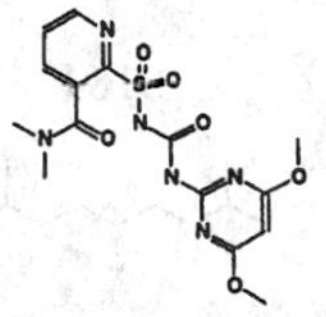

8070

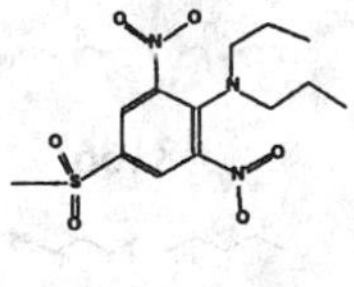

8071

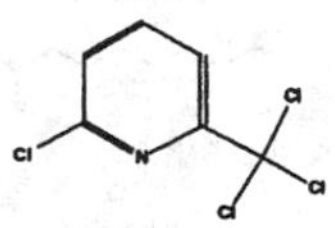

8072

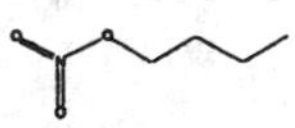

8073

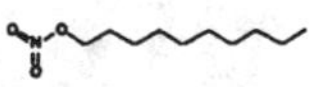

8074

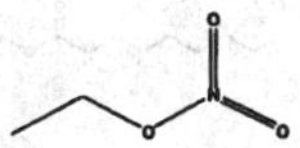

8075

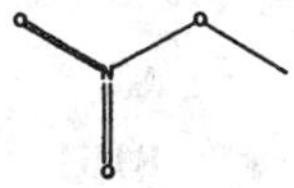

8076

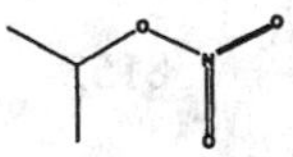

8077

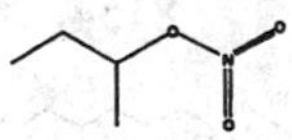

8078

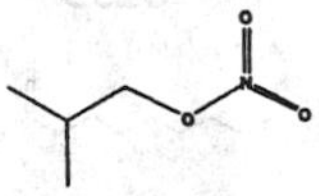

8079

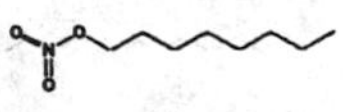

8080

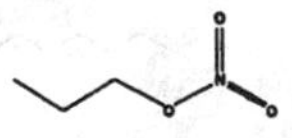

8081

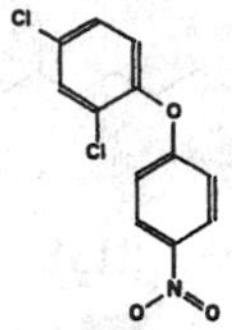

8082

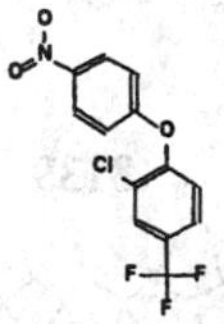

8083

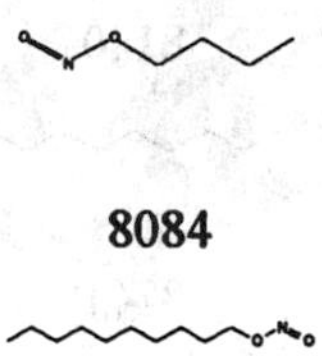

8084

8085

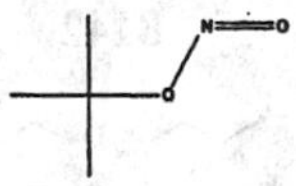

8086

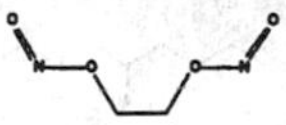

8087

8088

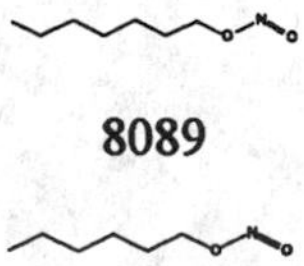

8089

8090

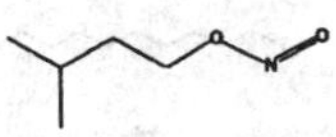

8091

8092

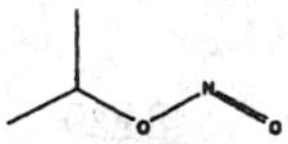

8093

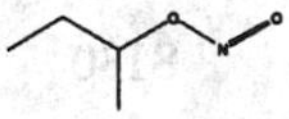

8094

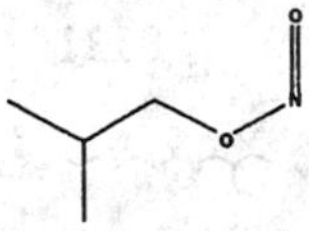

8095

8096

8097

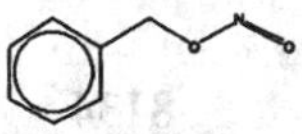

8098

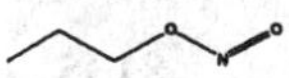

8099

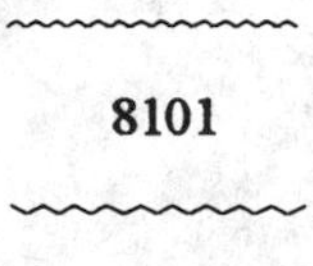

8101

8103

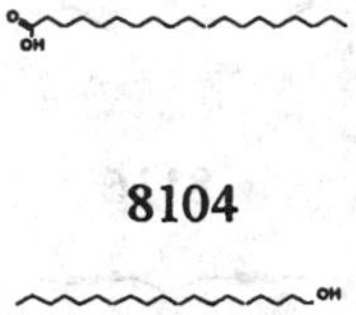

8104

8105

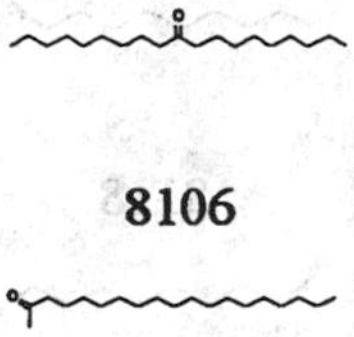

8106

8107

8108

8109

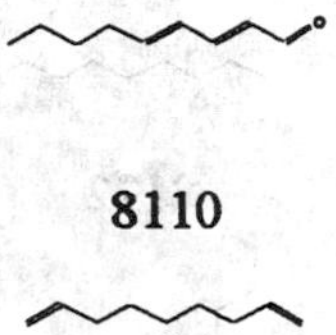

8110

8111

8112

8113

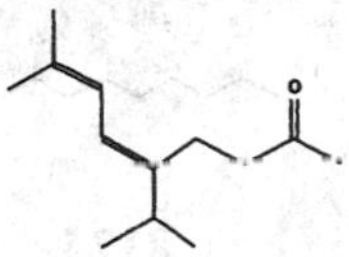

8114

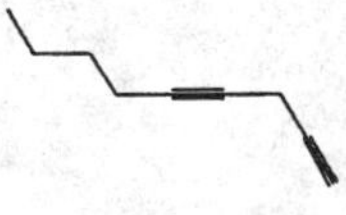

In numeric order as they occur in the Table of Physical Constants of Organic Compounds

8115

8116

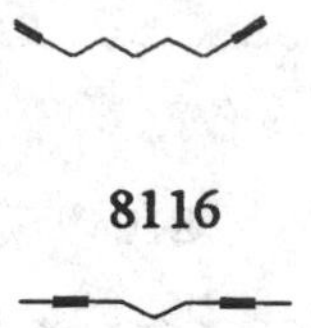

8117

8118

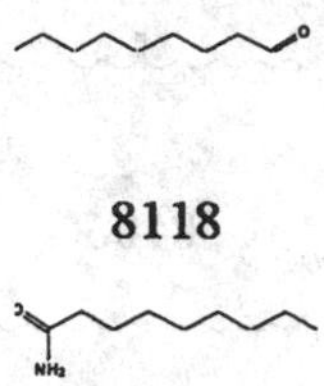

8119

8120

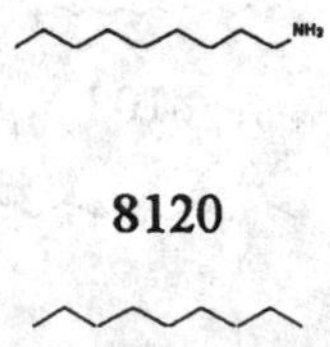

8121

8122

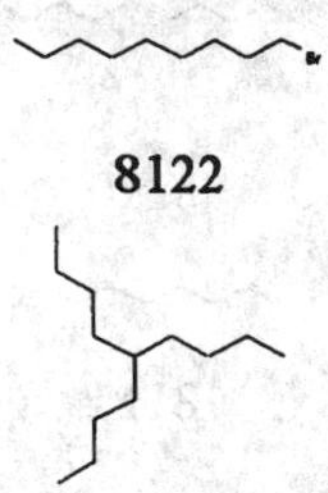

8123

8124

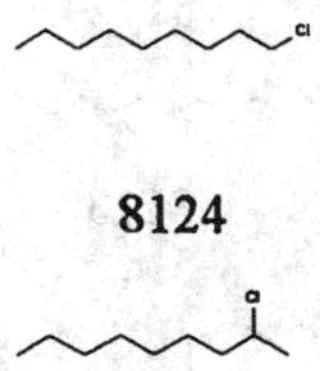

8125

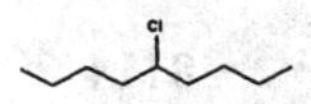

8126

8127

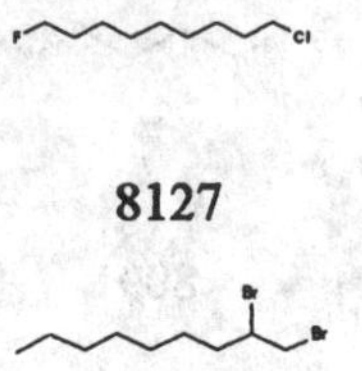

8128

8129

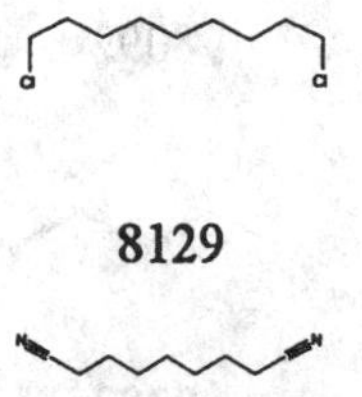

8130

8131

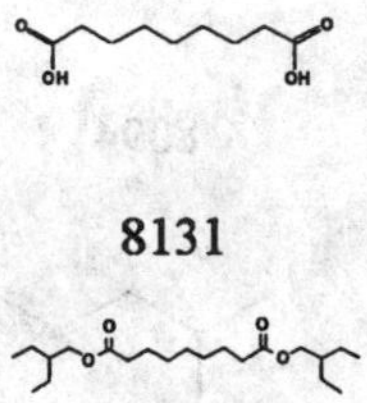

8132

8133

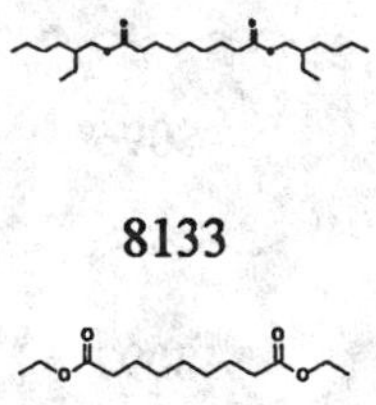

8134

8135

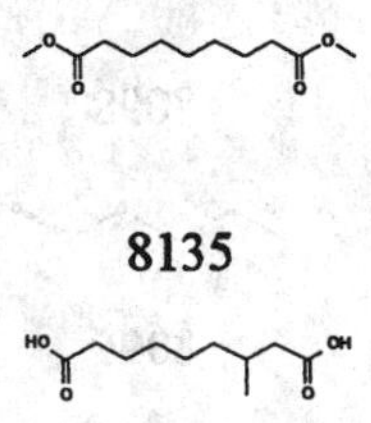

8136

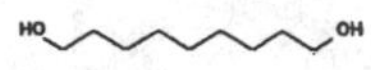

8137

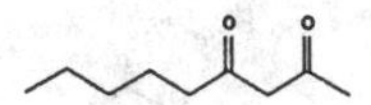

8138

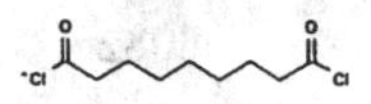

8139

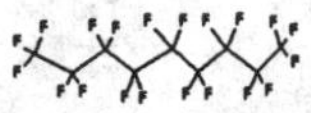

8140

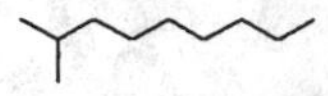

8141

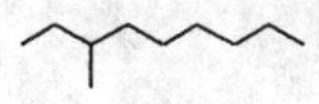

8142

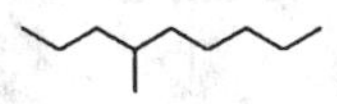

8143

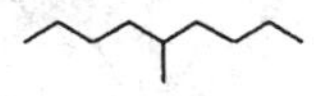

8144

8145

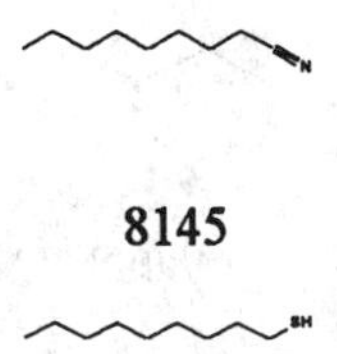

8146

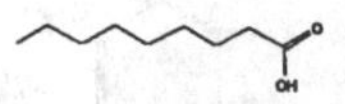

8147

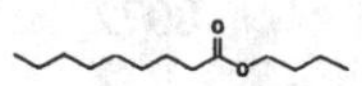

8148

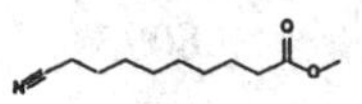

8149

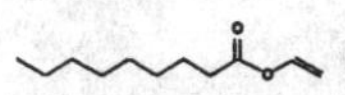

8150

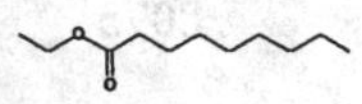

8151

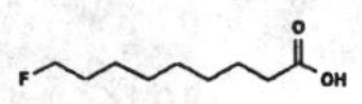

8152

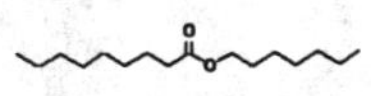

8154

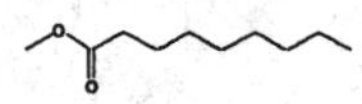

8155

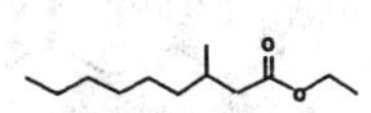

8156

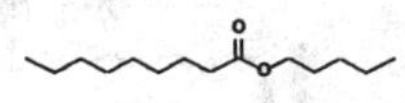

8157

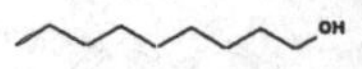

8158

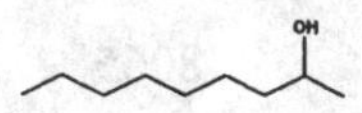

8159

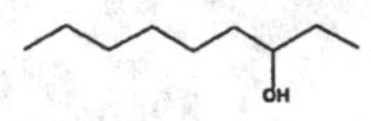

8160

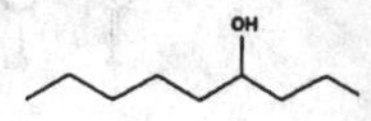

8161

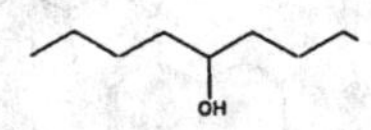

8162

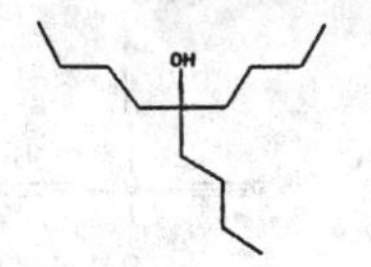

8163

8164

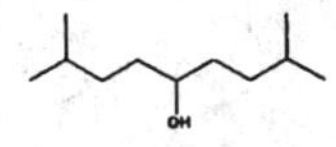

8165

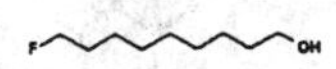

8166

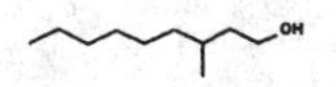

STRUCTURAL FORMULAS OF ORGANIC COMPOUNDS (continued)

In numeric order as they occur in the Table of Physical Constants of Organic Compounds

8167

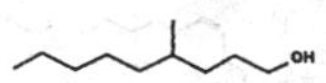

8168

8169

8170

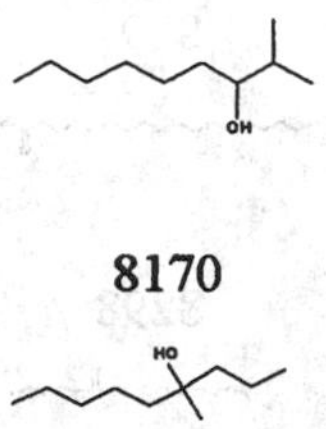

8171

8172

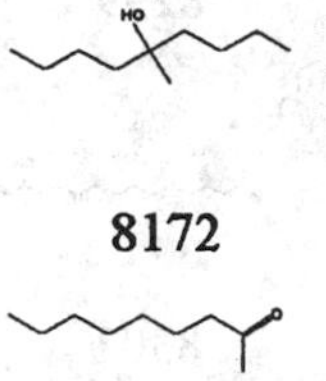

8173

8174

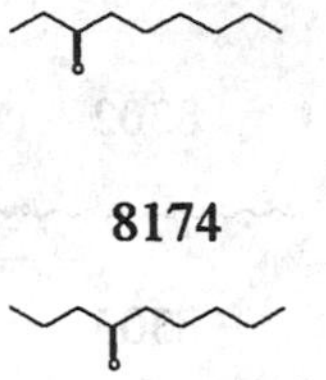

8175

8176

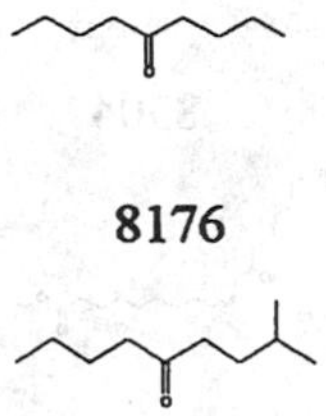

8177

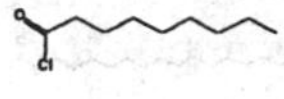

8178

8179

8180

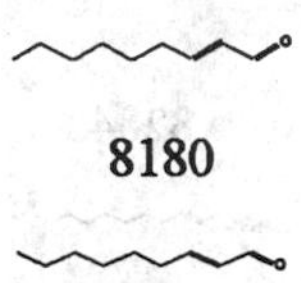

8181

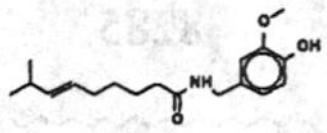

8182

8183

8184

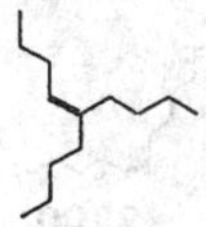

8185

8186

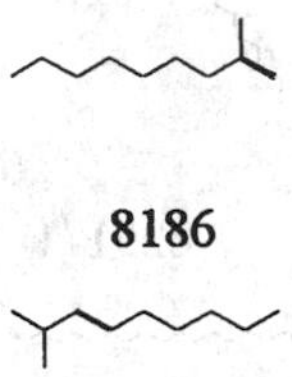

8187

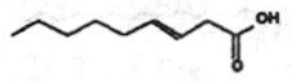

8188

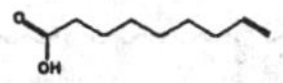

8189

8190

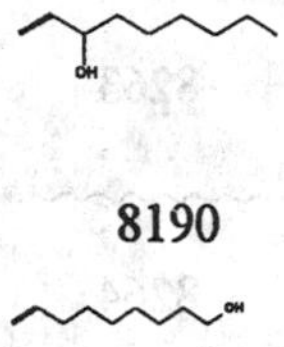

8191

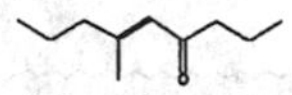

8192

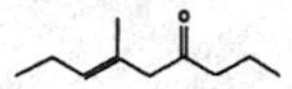

8193

8194

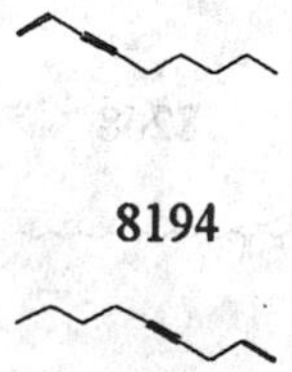

8195

8196

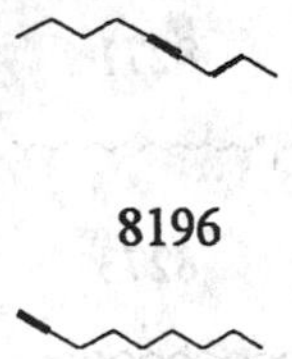

8197

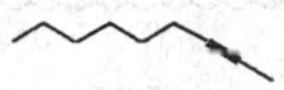

8198

8199

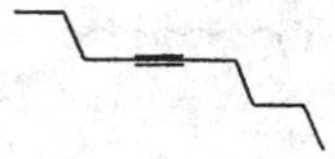

8200

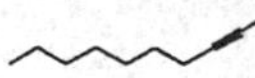

8201

8202

8203

8204

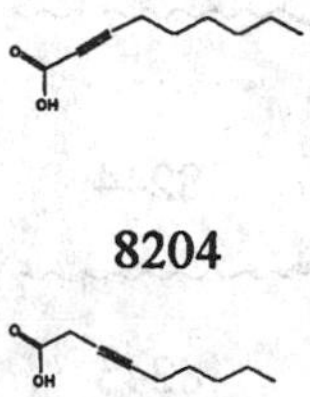

8205

8206

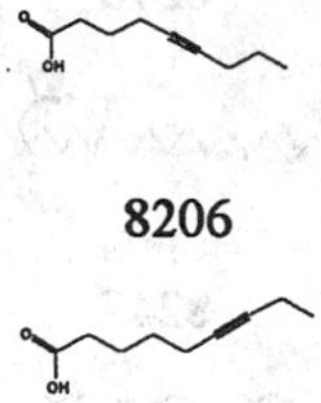

8207

8208

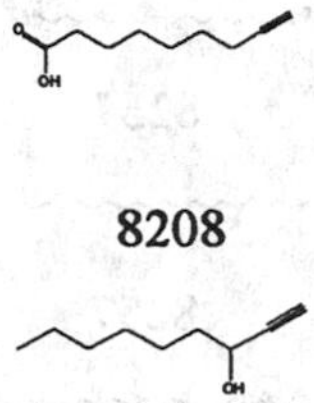

8209

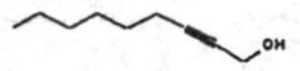

8212

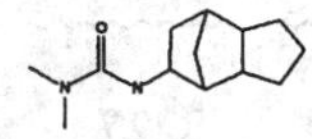

8213

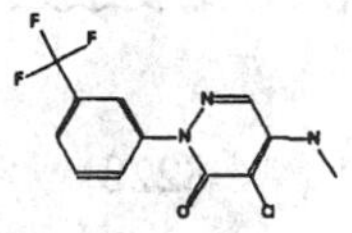

8214

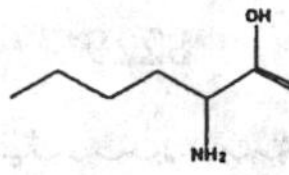

8215

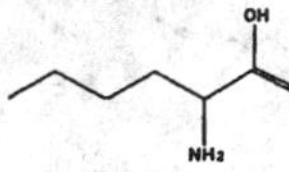

8216

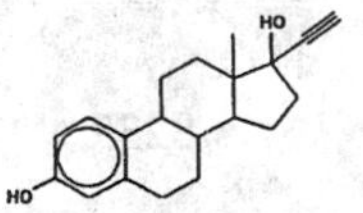

8218

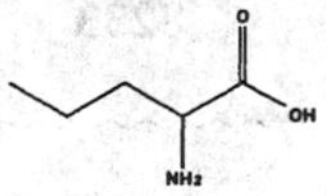

8219

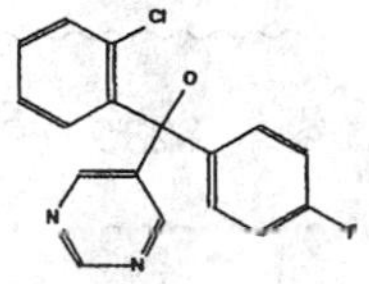

8220

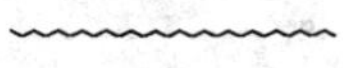

In numeric order as they occur in the Table of Physical Constants of Organic Compounds

8223
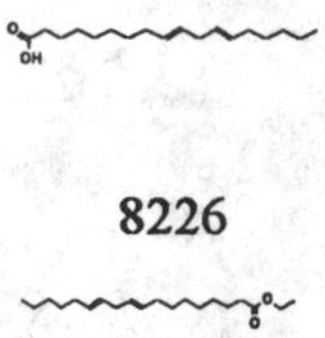
8226

8227
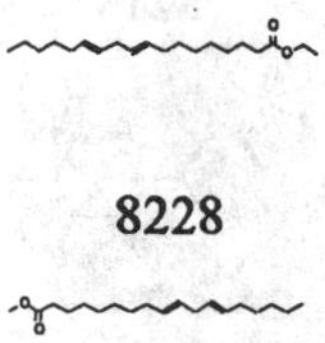
8228

8229
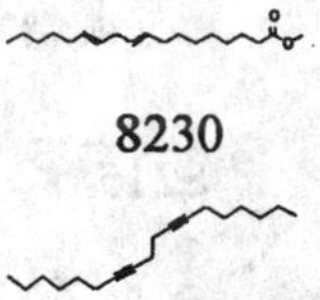
8230

8231
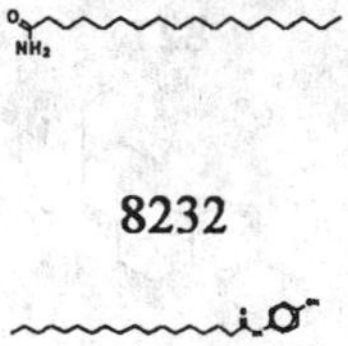
8232

8233
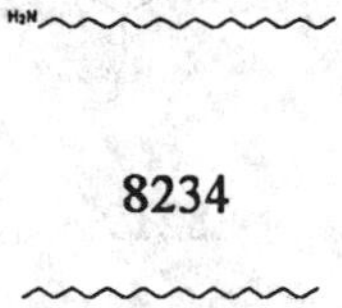
8234

8235
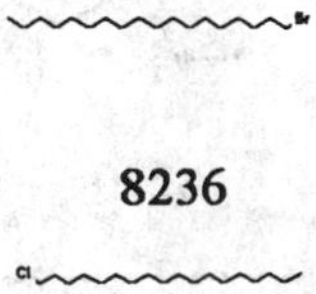
8236

8237
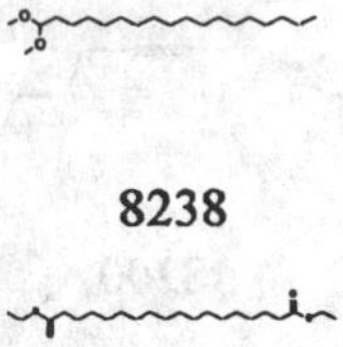
8238

8240
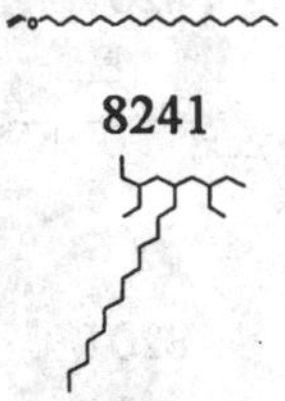
8241

8242

8243

8244
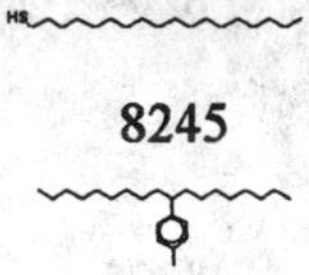
8245

8246
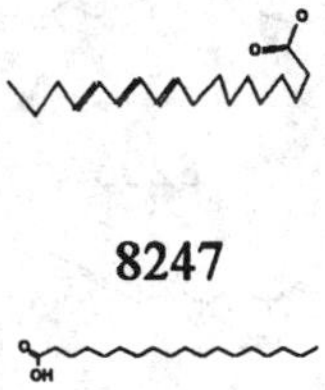
8247

8250
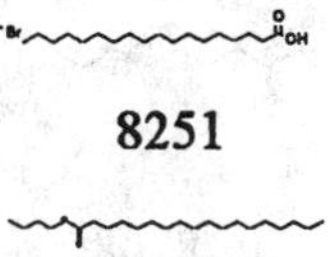
8251

8257
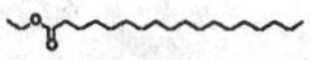

8260
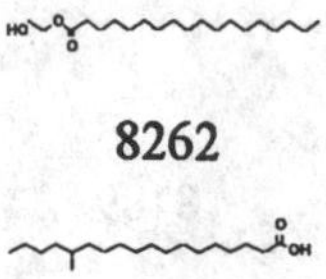
8262

8263
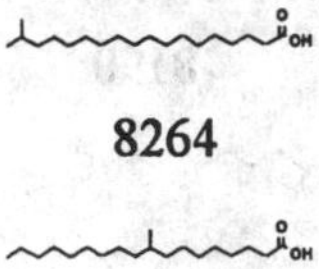
8264

8265
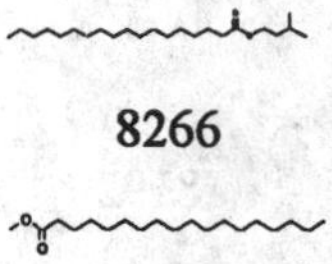
8266

8267
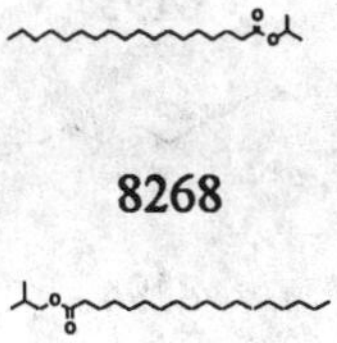
8268

8269
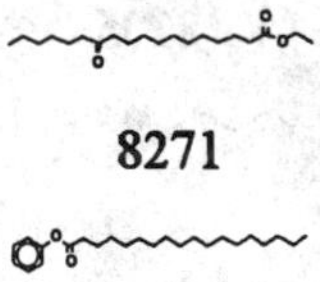
8271

8273
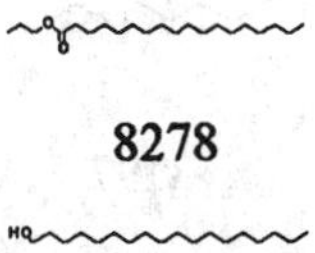
8278

8279

8280

8281

8282

8283
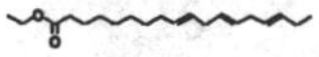

8284
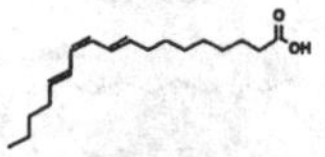

8285

8287

8288

8289
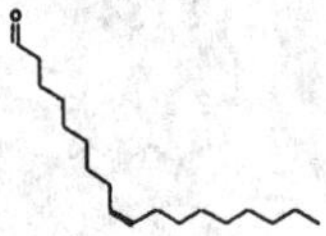

8290
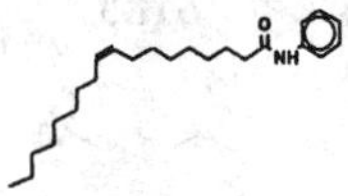

8291

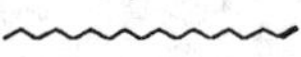

8292

8293
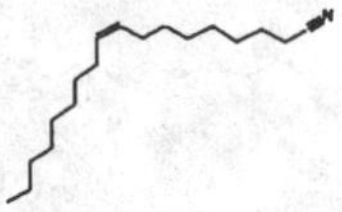

8294
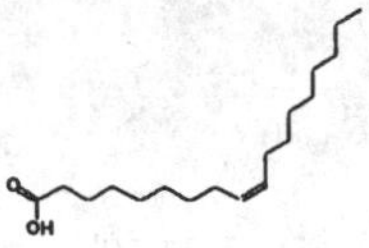

8297
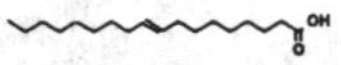

8298
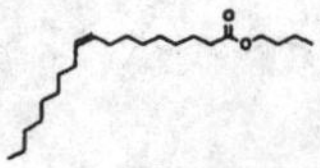

8299
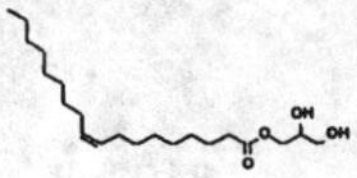

8301
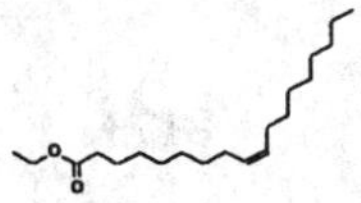

8302
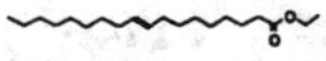

8303
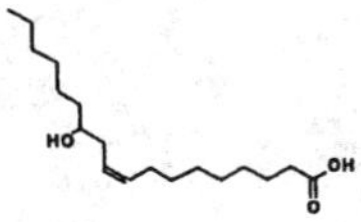

8304
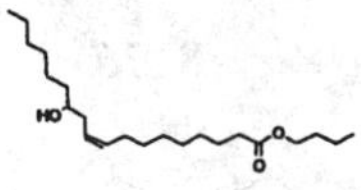

In numeric order as they occur in the Table of Physical Constants of Organic Compounds

8305

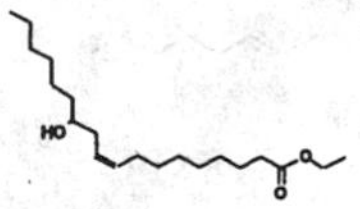

8306

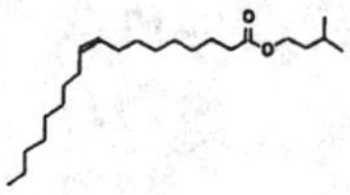

8307

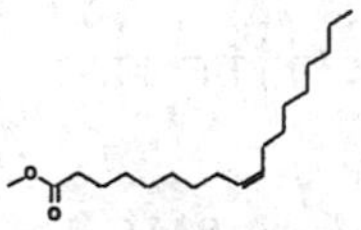

8308

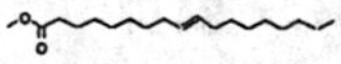

8309

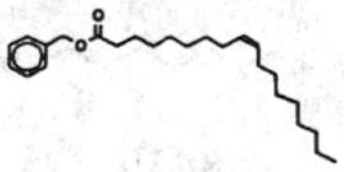

8310

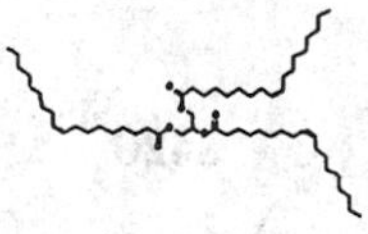

8311

8312

8313

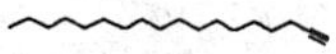

8314

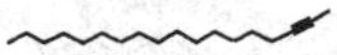

8315

8316

8317

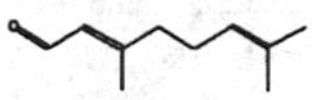

8318

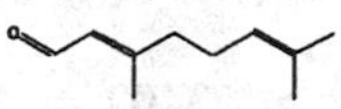

8319

8320

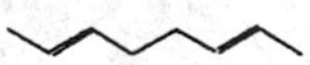

8321

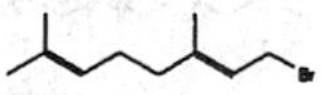

8322

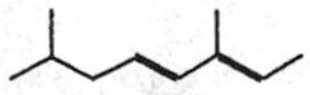

8324

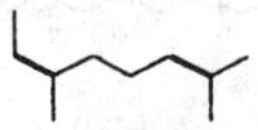

8325

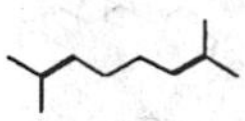

8327

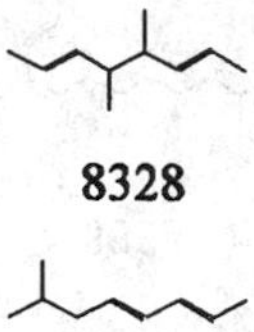

8328

8329

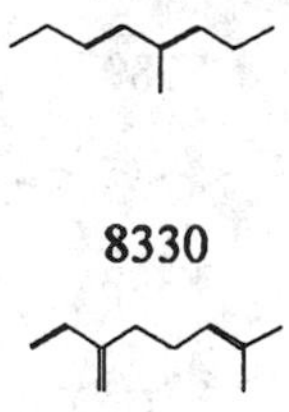

8330

8331

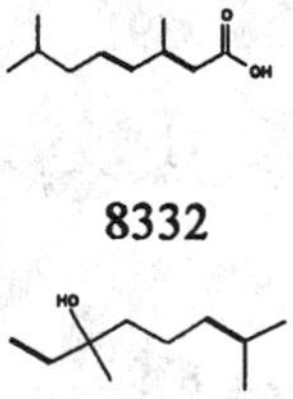

8332

8333

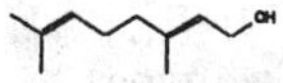

8334

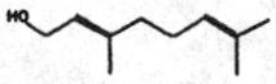

8335

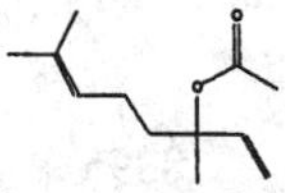

8336

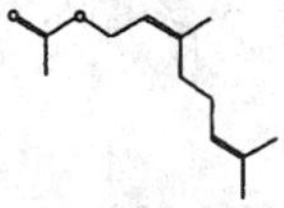

8337

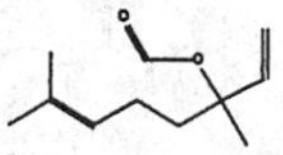

8338

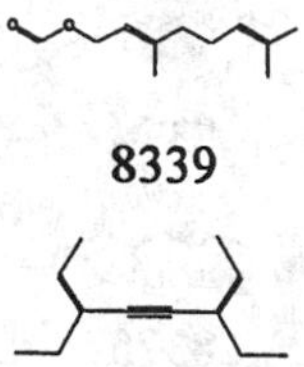

8339

8340

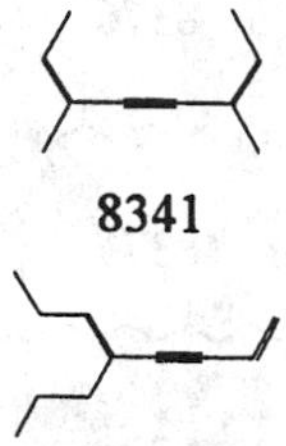

8341

8342

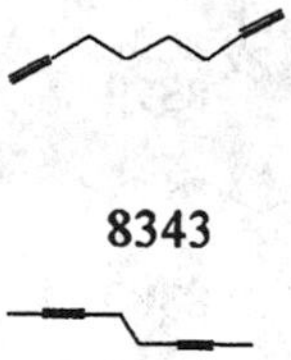

8343

8344

8345

8346

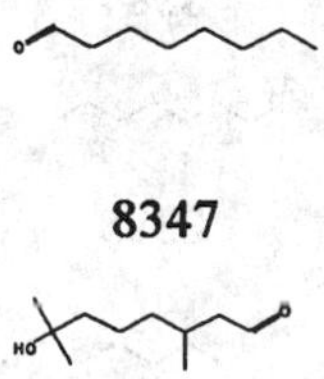

8347

8348

8349

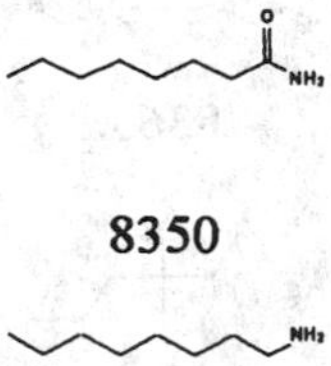

8350

8351

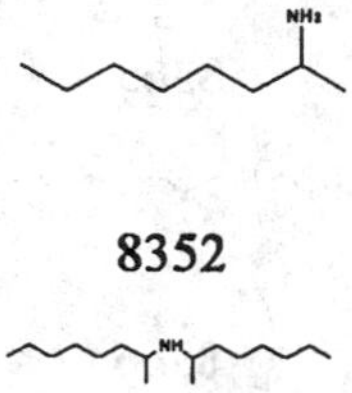

8352

8353

8354

8355

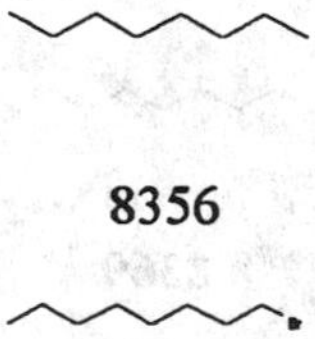

8356

8357

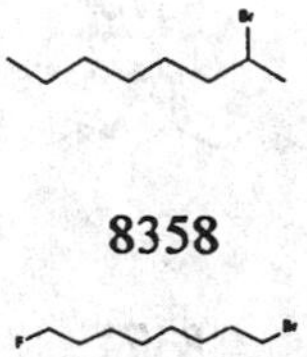

8358

8359

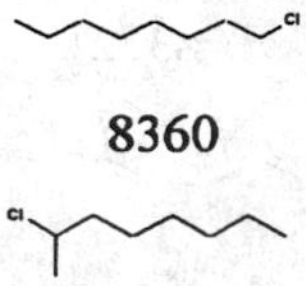

8360

STRUCTURAL FORMULAS OF ORGANIC COMPOUNDS (continued)

In numeric order as they occur in the Table of Physical Constants of Organic Compounds

8361

8363

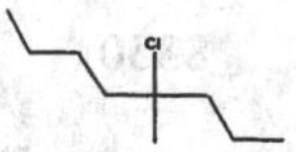

8364

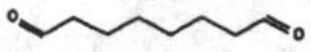

8365

8366

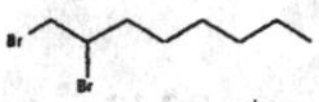

8367

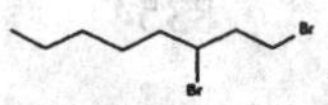

8368

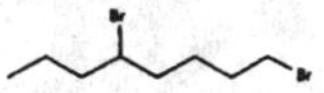

8369

8370

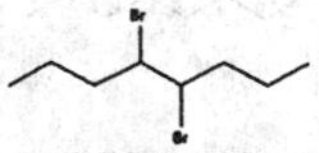

8371

8372

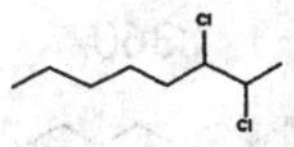

8373

8374

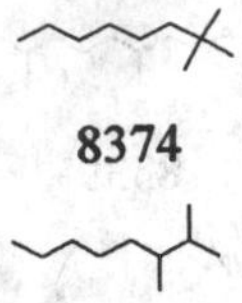

8375

8376

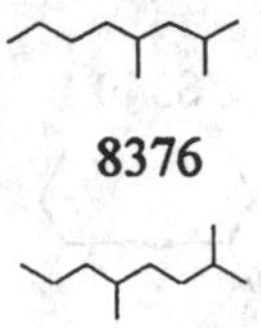

8377

8378

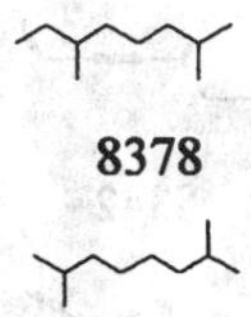

8379

8380

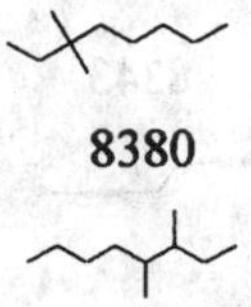

8381

8382

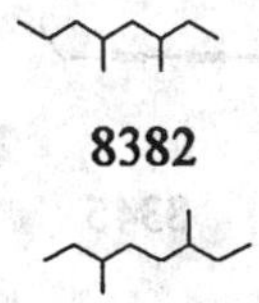

8383

8384

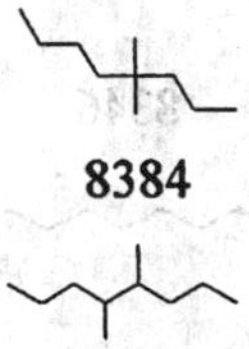

8385

8386

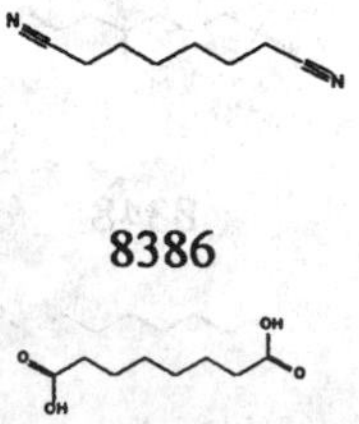

8387

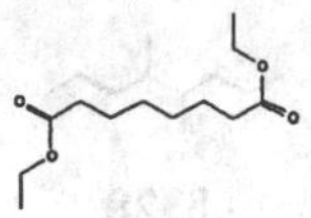

8388

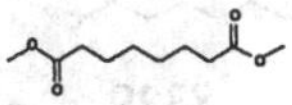

8389

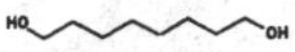

8390

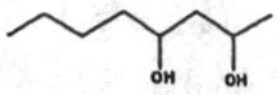

8391

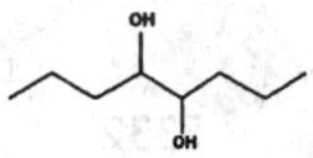

8392

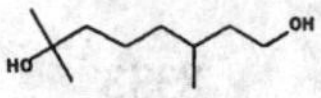

8393

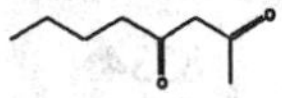

8394

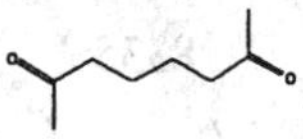

8395

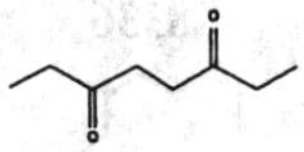

8396

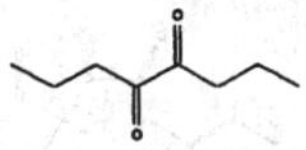

8398

8399

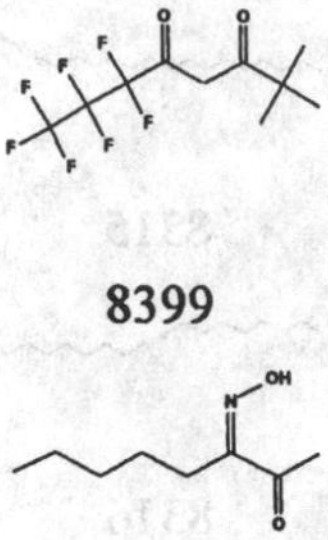

8400

8401

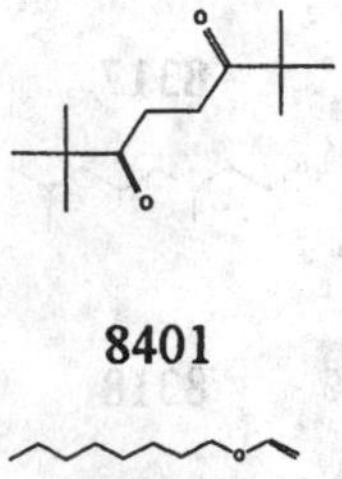

8402

8403

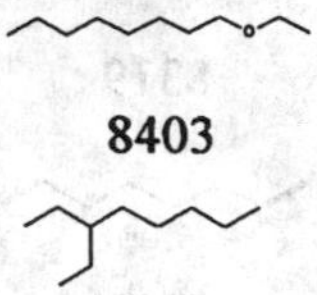

8404

8405

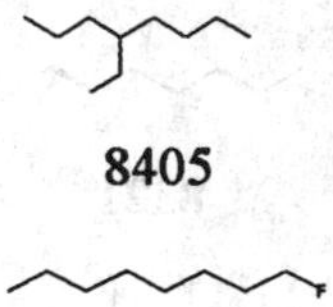

8406

8407

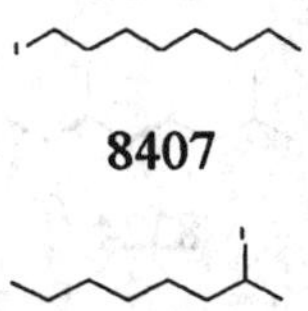

8408

8409

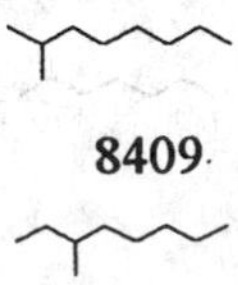

8410

8411

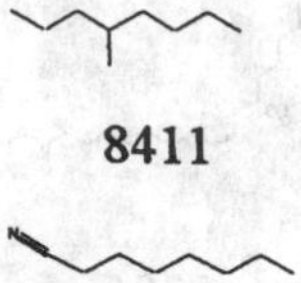

8412

8415

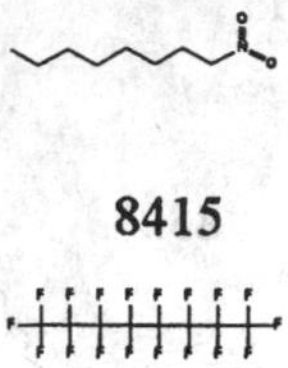

8416

8417

8418

8419

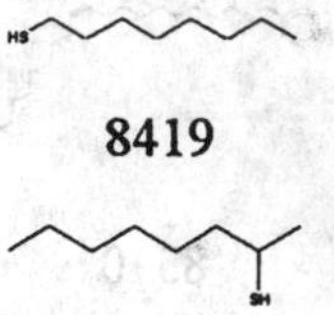

8420

8421

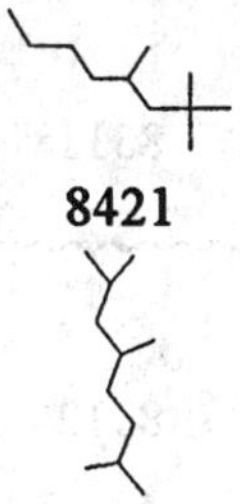

8422

8424

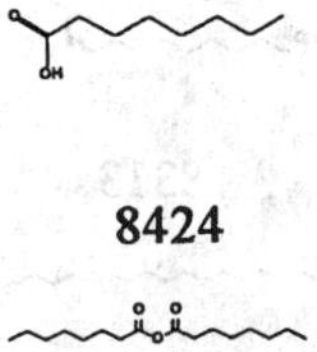

STRUCTURAL FORMULAS OF ORGANIC COMPOUNDS (continued)

In numeric order as they occur in the Table of Physical Constants of Organic Compounds

8425

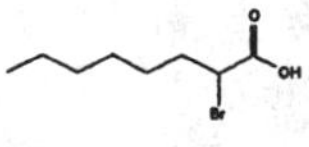

8426

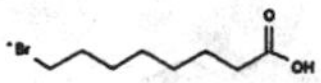

8427

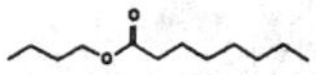

8428

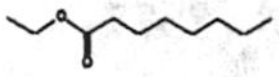

8429

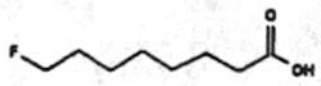

8430

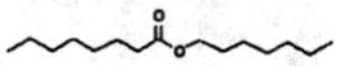

8431

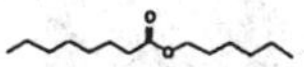

8432

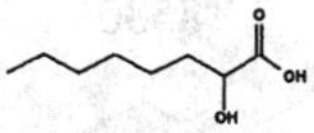

8433

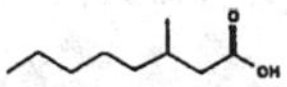

8434

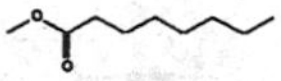

8435

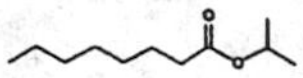

8436

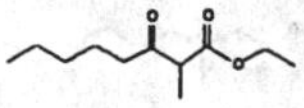

8437

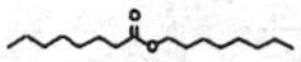

8438

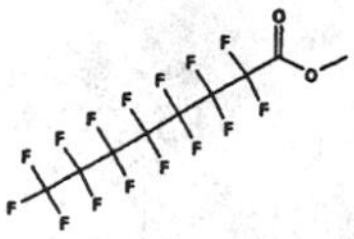

8439

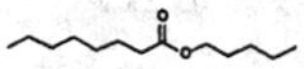

8440

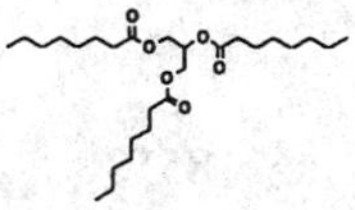

8441

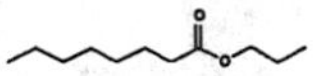

8442

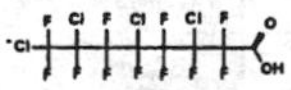

8443

8444

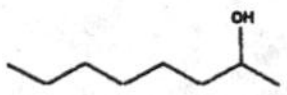

8445

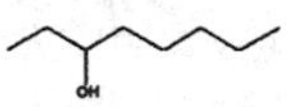

8446

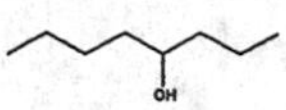

8447

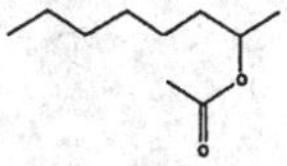

8449

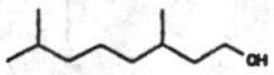

8450

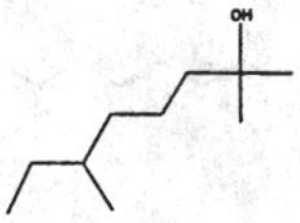

8451

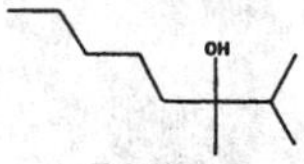

8452

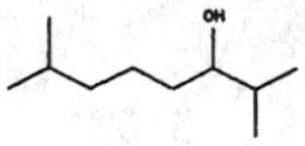

8453

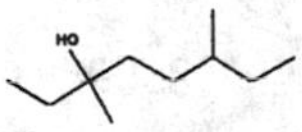

8454

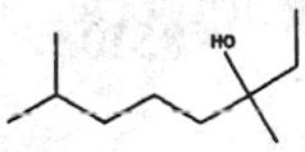

8455

8456

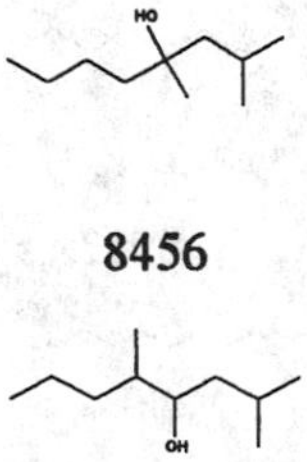

8457

8458

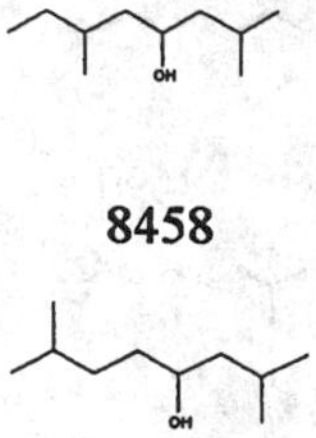

8459

8460

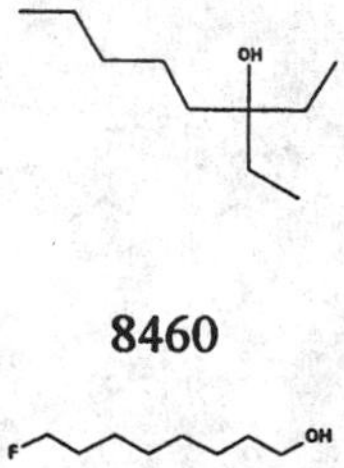

8461

8462

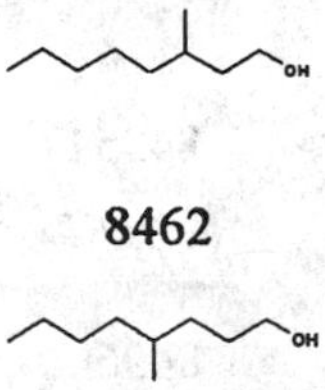

8464

8465

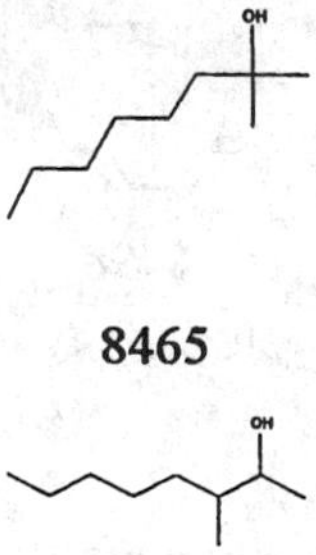

8466

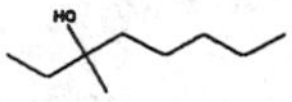

8468

8469

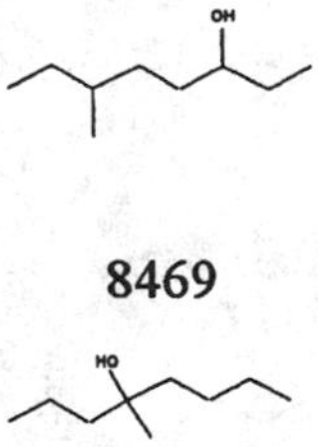

8470

8471

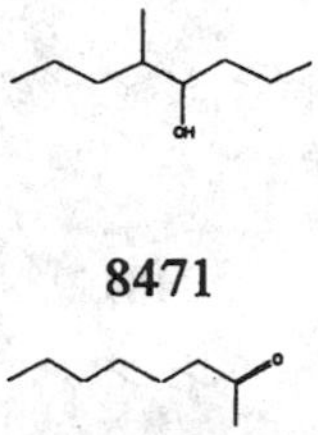

8472

8473

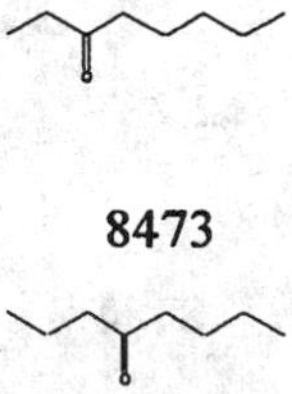

8474

8475

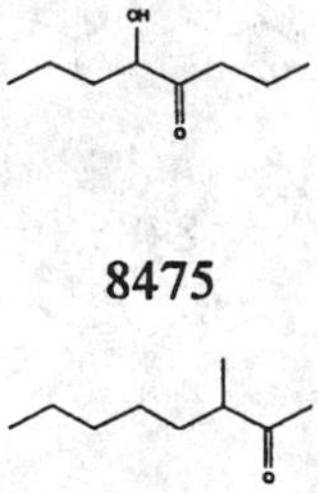

8476

8477

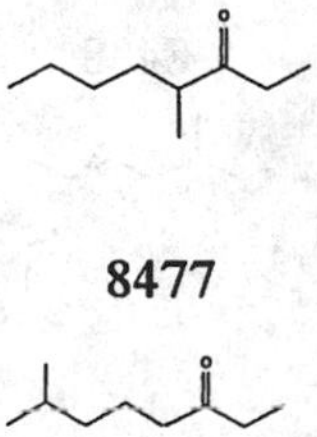

8478

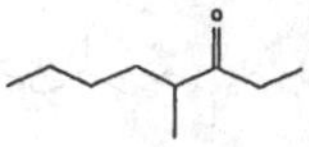

8479

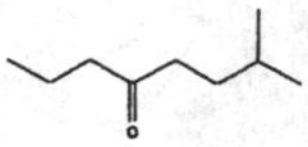

8480

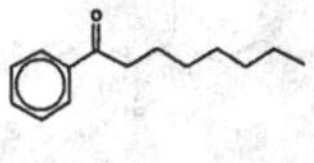

8481

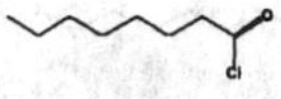

8482

8484

8485

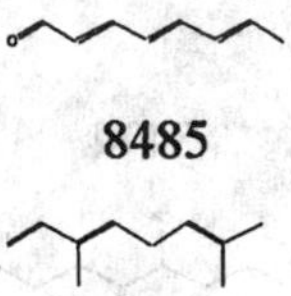

8486

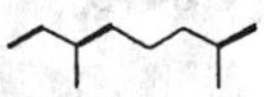

8487

8488

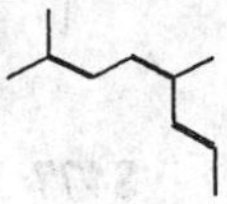

8489

8490

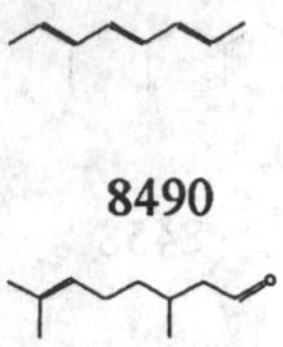

8491

8492

8493

8494

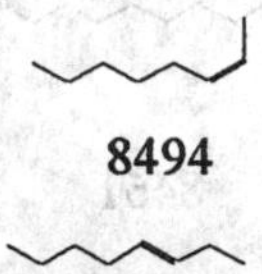

8495

8496

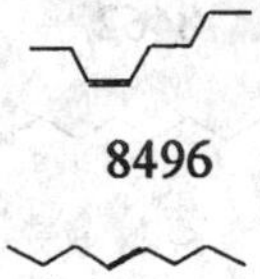

8497

8498

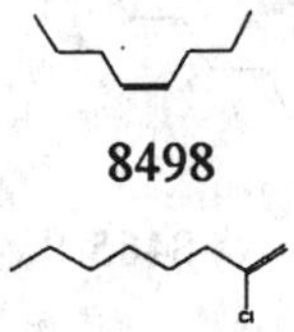

8499

8500

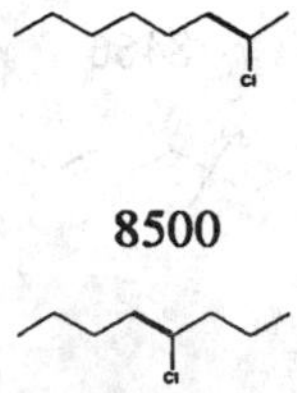

8502

8503

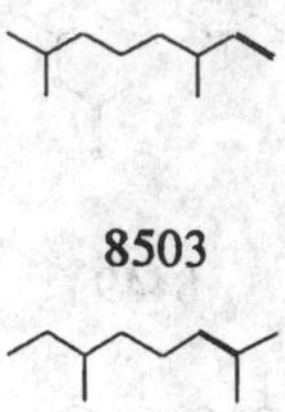

8504

8505

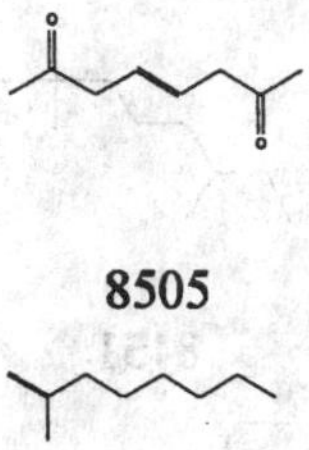

8507

8508

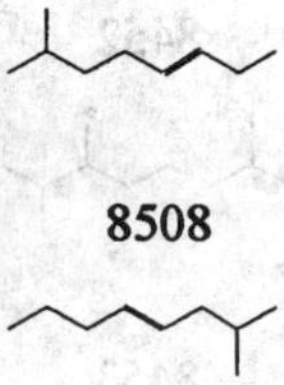

8509

8510

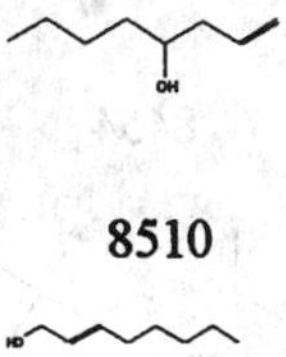

8511

8512

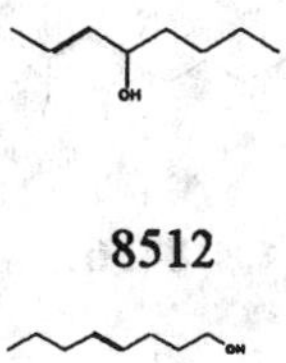

8513

8514

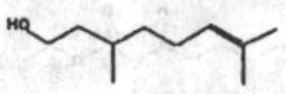

8515

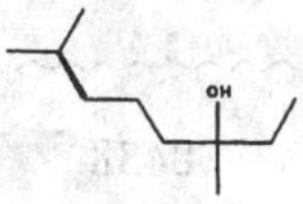

8516

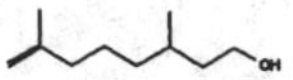

8517

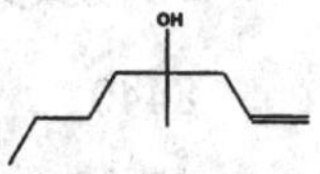

8518

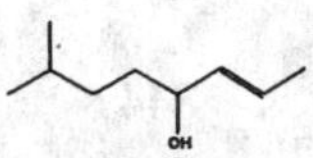

8519

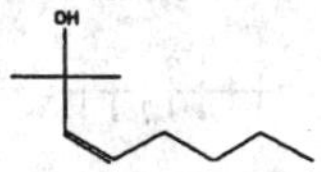

8520

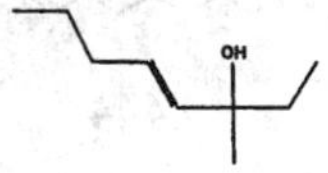

8521

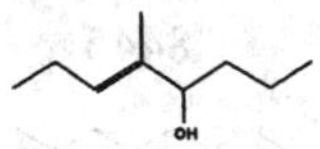

8523

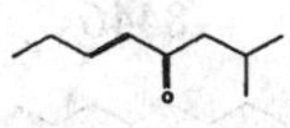

8524

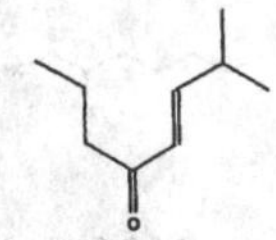

8525

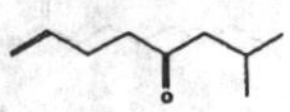

8526

8527

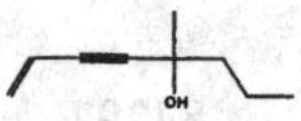

8528

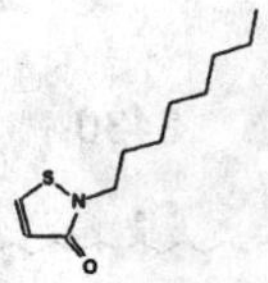

8529

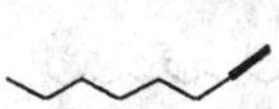

8530

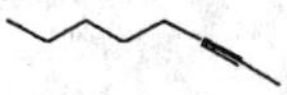

8531

8532

8533

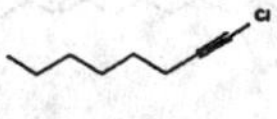

8534

8535

8536

8537

8538

8539

8540

8541

8542

8543

8544

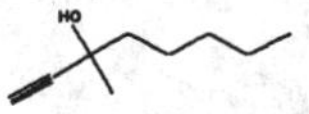

8545

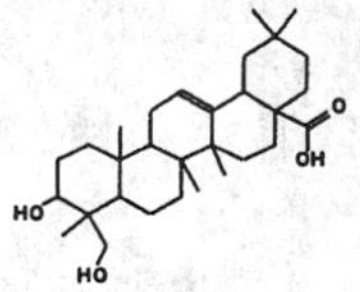

8547

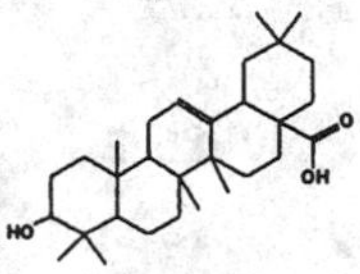

8548

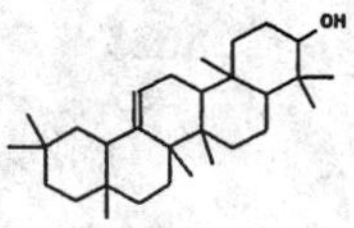

8552

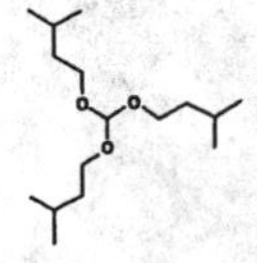

8553

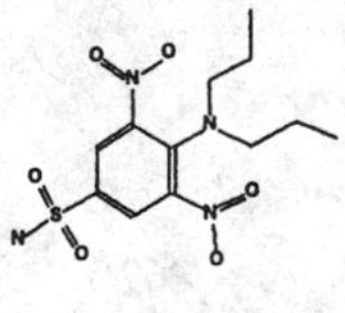

8554

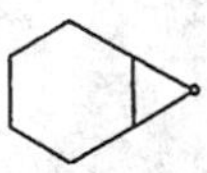

8555

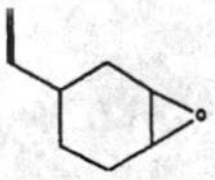

8556

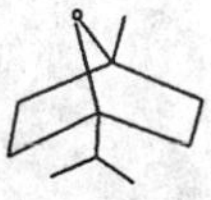

8557

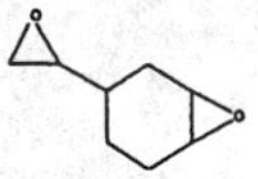

8558

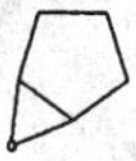

8559

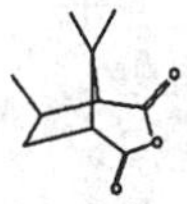

8560

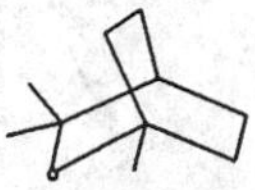

8562

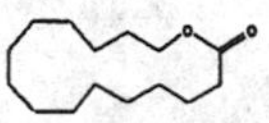

8563

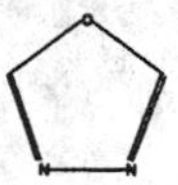

8564

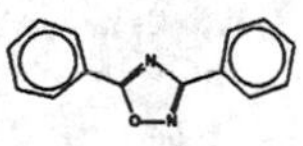

8565

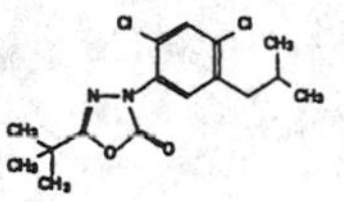

8566

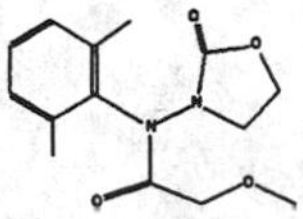

8568

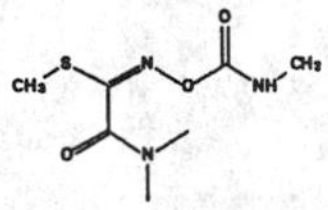

8569

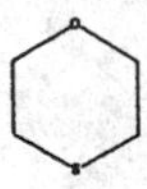

8570

8571

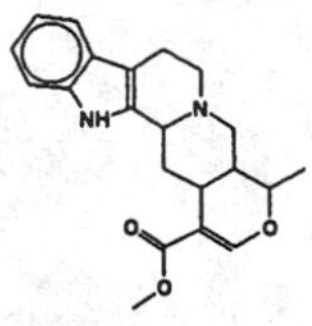

8574

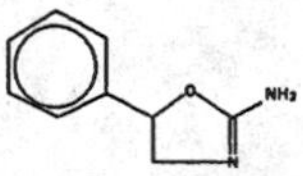

8575

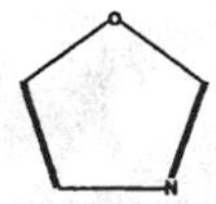

8576

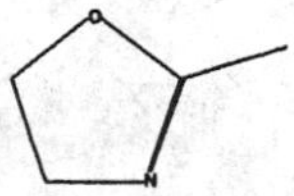

8577

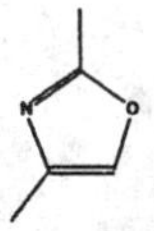

8578

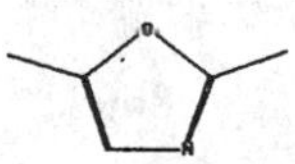

8579

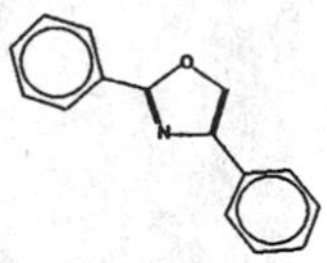

8580

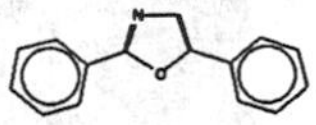

8581, 8582, 8583, 8584, 8585, 8586, 8587

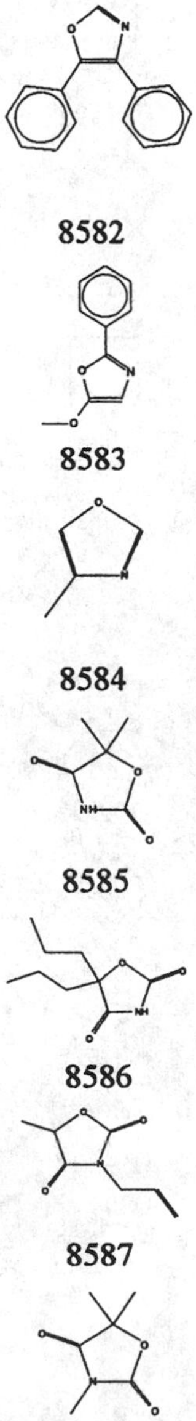

STRUCTURAL FORMULAS OF ORGANIC COMPOUNDS (continued)

In numeric order as they occur in the Table of Physical Constants of Organic Compounds

8588

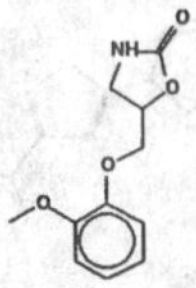

8589

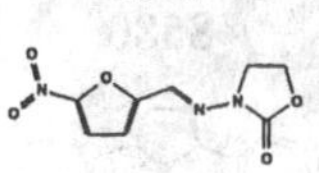

8590

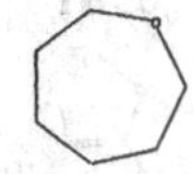

8591

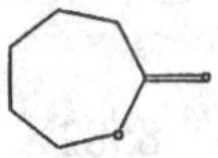

8592

8593

8594

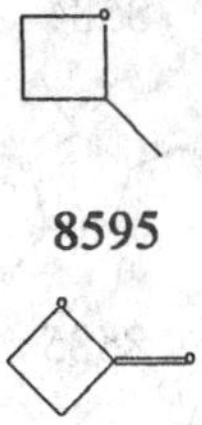

8595

8596

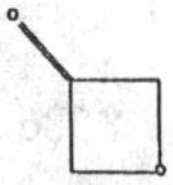

8597

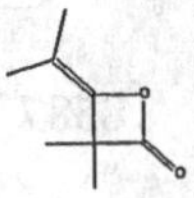

8598

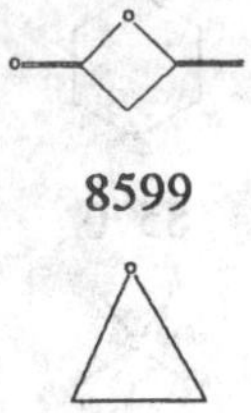

8599

8600

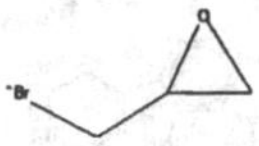

8601

8602

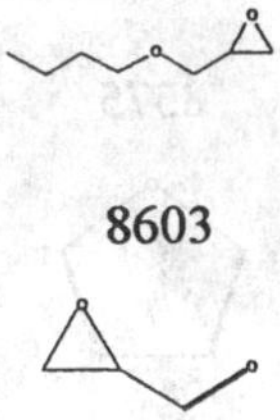

8603

8604

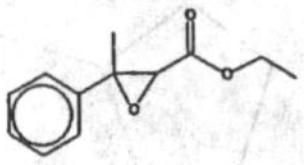

8605

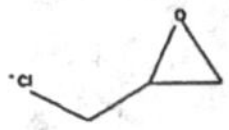

8607

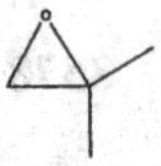

8608

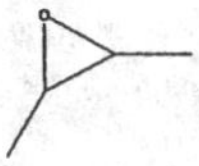

8609

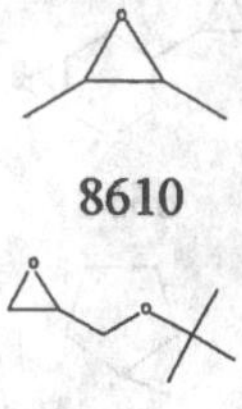

8610

8611

8612

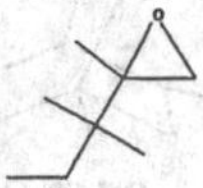

8613

8614

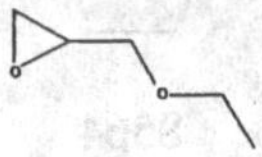

8615

8616

8617

8618

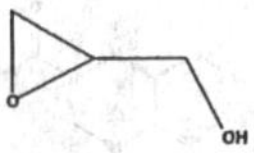

8619

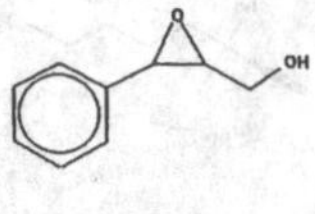

8620

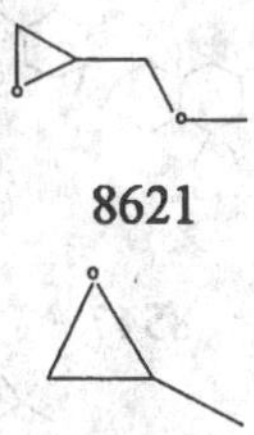

8621

8622

8623

8624

8625

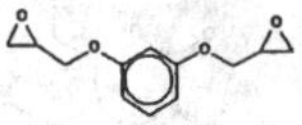

8626

8627

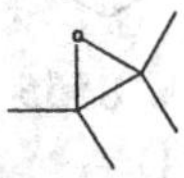

8628

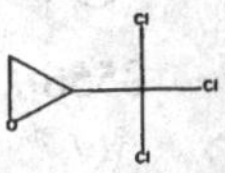

8629

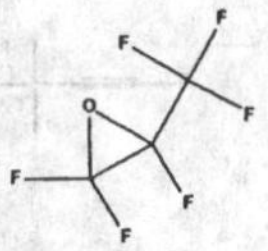

8632

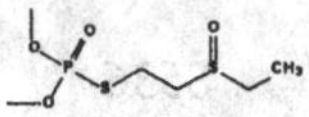

8633

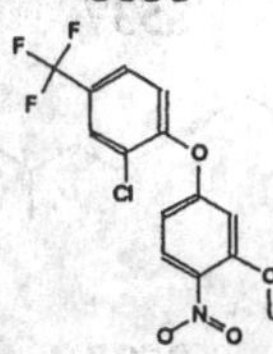

8635

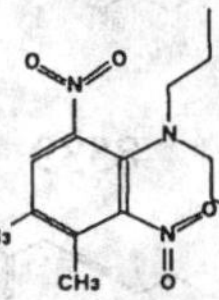

8636

8637

8638

8639

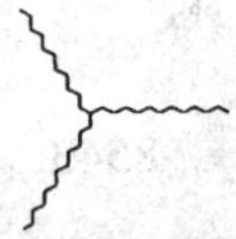

8640

8641

In numeric order as they occur in the Table of Physical Constants of Organic Compounds

8642, 8643

8644, 8645

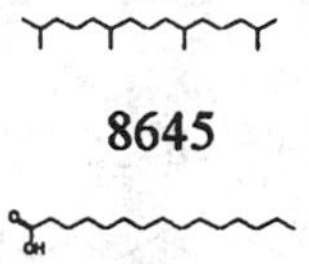

8646, 8647

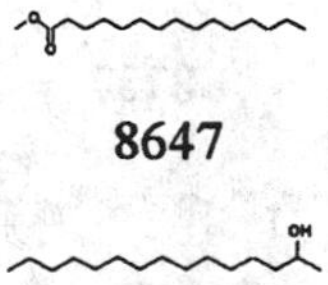

8648, 8649

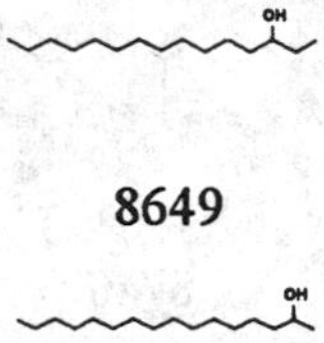

8650, 8651

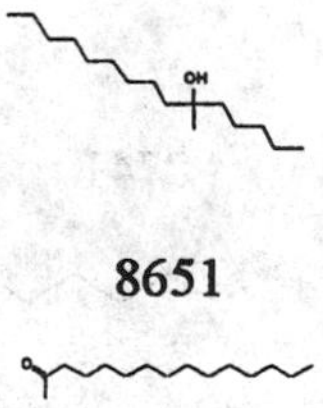

8652, 8653

8654, 8655

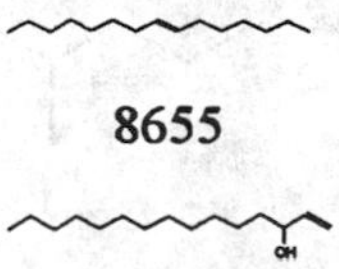

8656, 8657

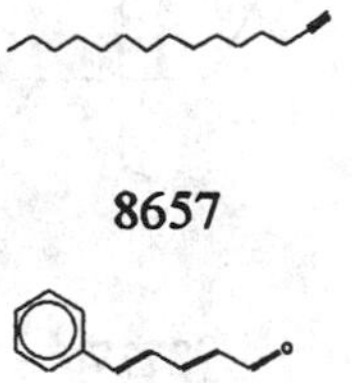

8658, 8659

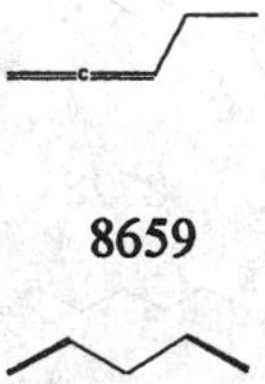

8660, 8661

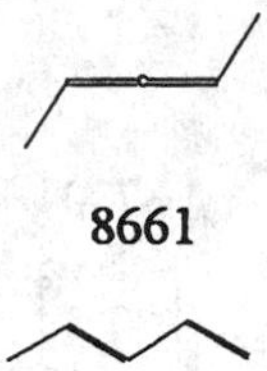

8662, 8663

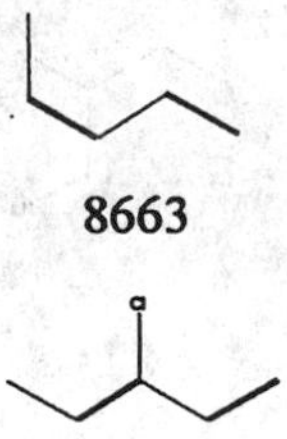

8664, 8665

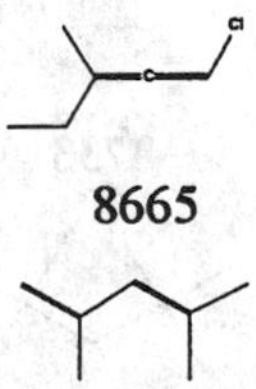

8666

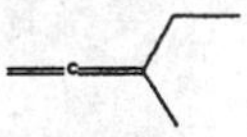

8667, 8668

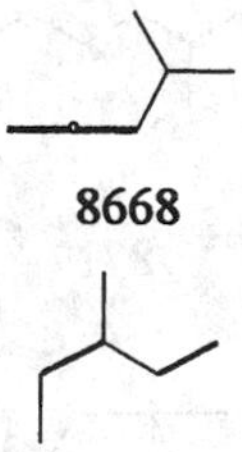

8669

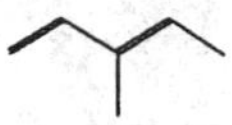

8670

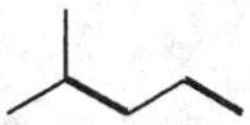

8671

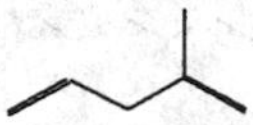

8672

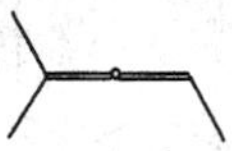

8673

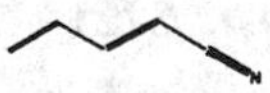

8674

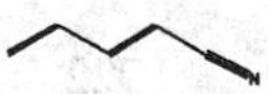

8675

8676

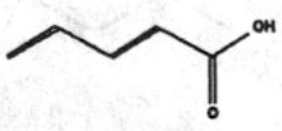

8679, 8680

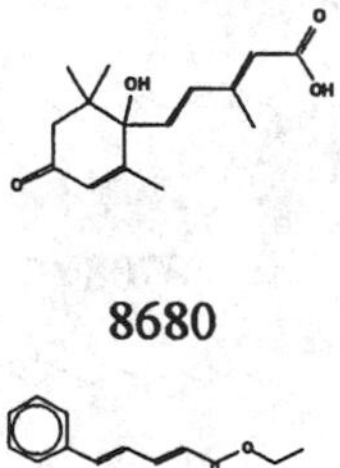

8681

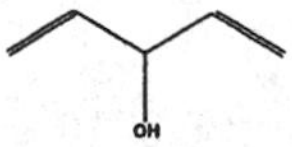

8682

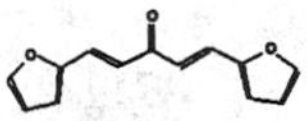

8683

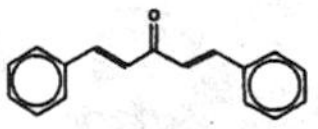

8684

8685

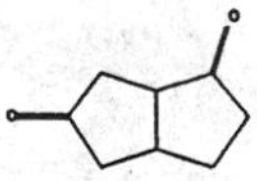

8686

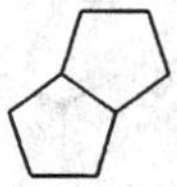

8687

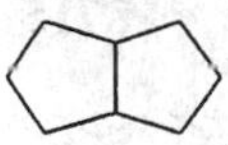

8688

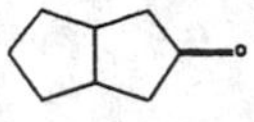

8689

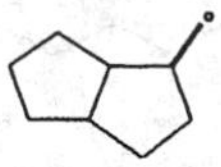

8690

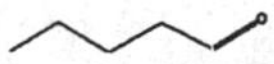

8691

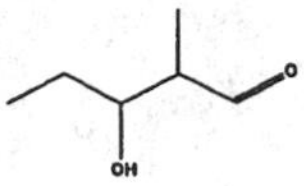

8692

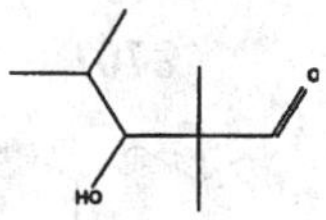

8693

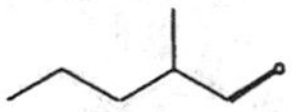

8694

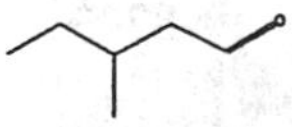

8695

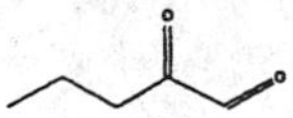

8696

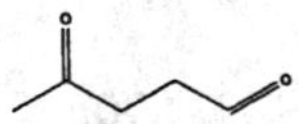

8697

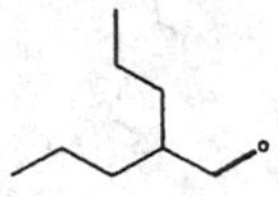

In numeric order as they occur in the Table of Physical Constants of Organic Compounds

8698

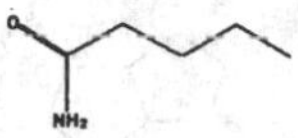

8699

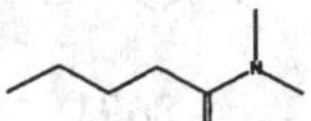

8700

8701

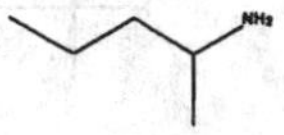

8702

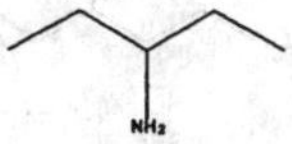

8703

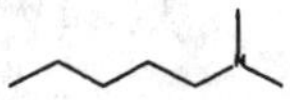

8704

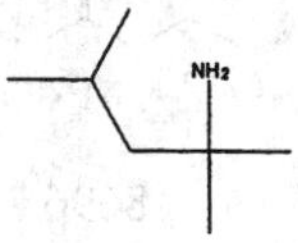

8705

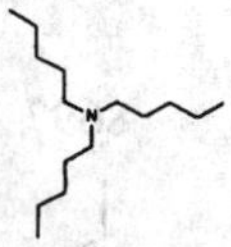

8706

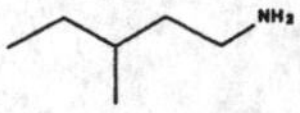

8707

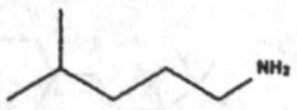

8708

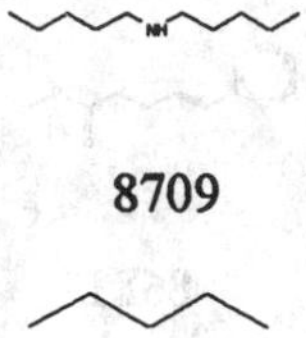

8709

8710

8711

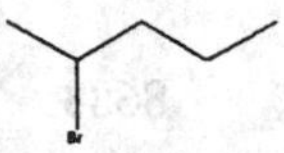

8712

8713

8714

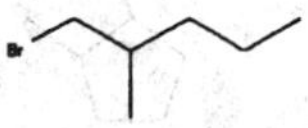

8715

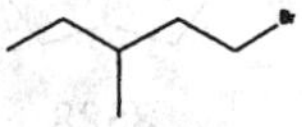

8716

8717

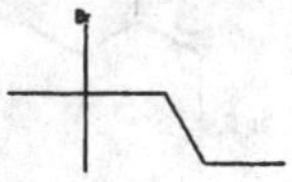

8718

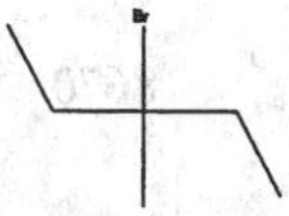

8719

8720

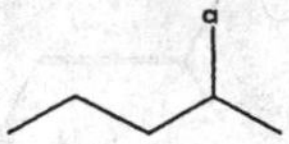

8721

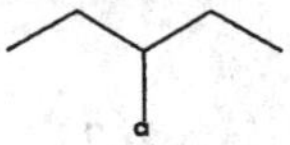

8722

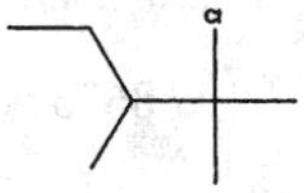

8723

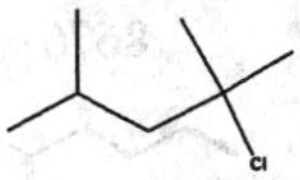

8726

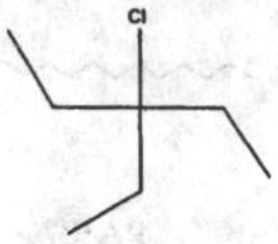

8727

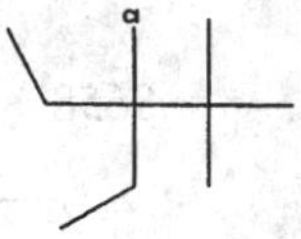

8728

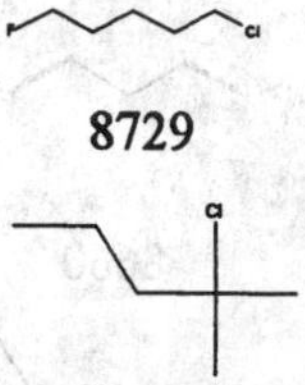

8729

8730

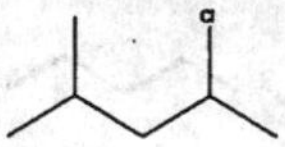

8731

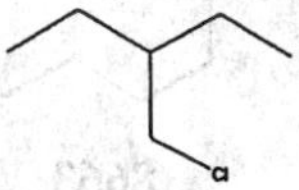

8732

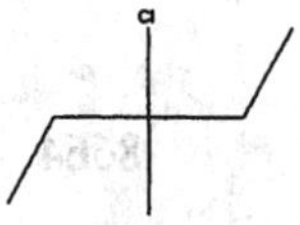

8733

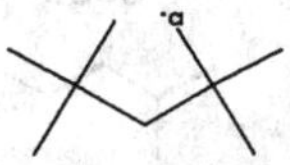

8734

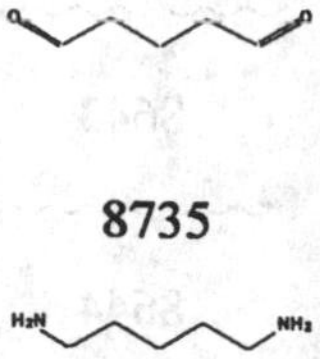

8735

8736

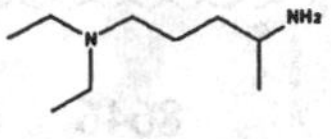

8737

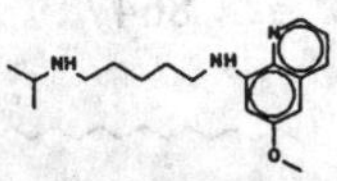

8739

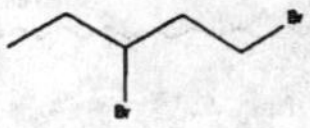

8740

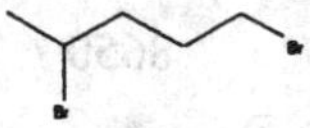

8741

8742

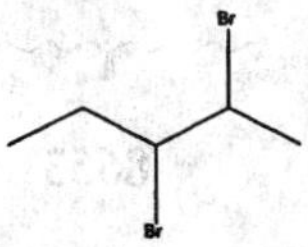

8743

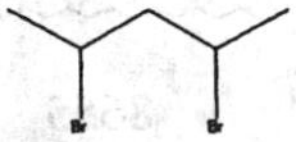

8744

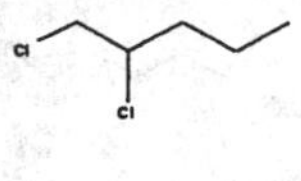

8745

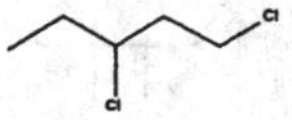

8746

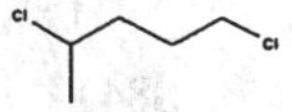

8747

8748

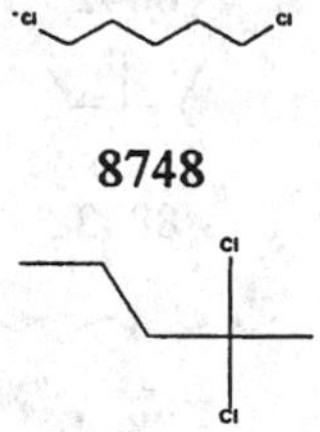

8749

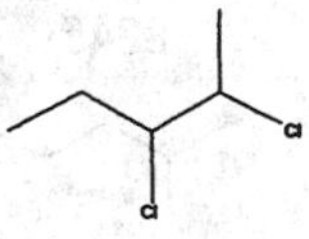

8750

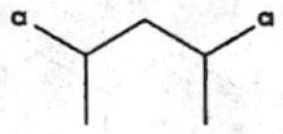

8751

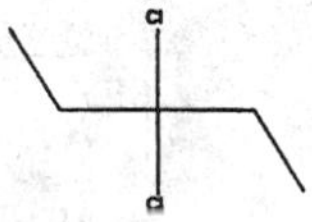

8752

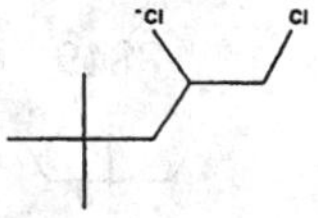

8753

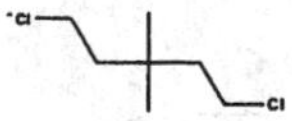

8754

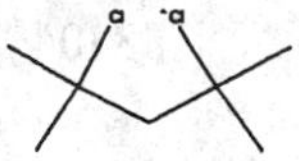

8755

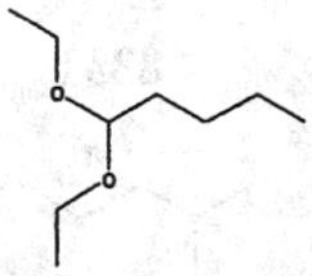

8756

8757

8758

8760

8761

8762

8763

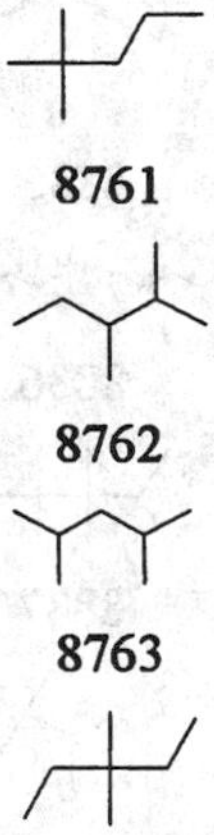

8764

8765

8766

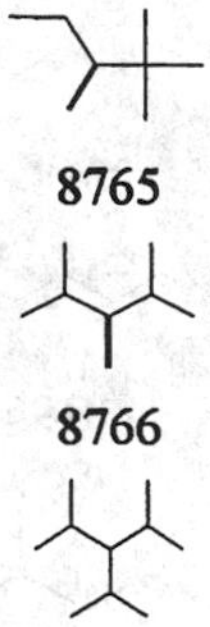

8767

8768

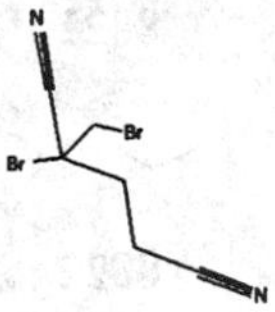

8769

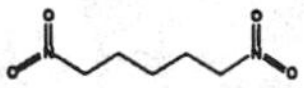

8770

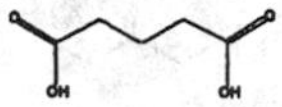

8771

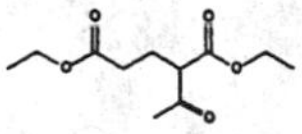

8772

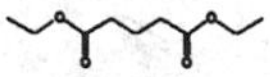

8773

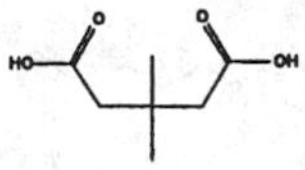

8774

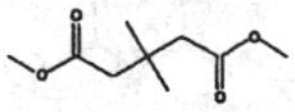

8775

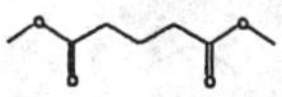

8776

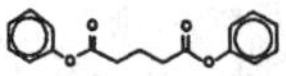

8777

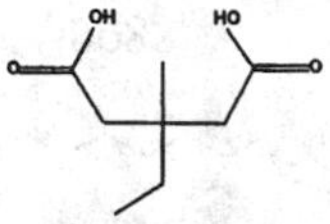

8778

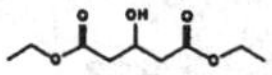

8780

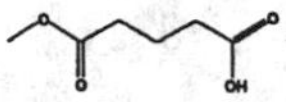

8781

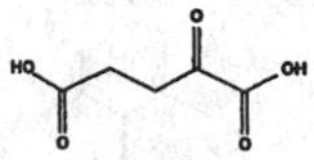

8783

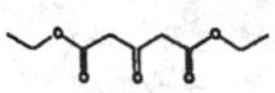

8784

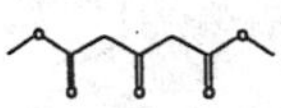

8785

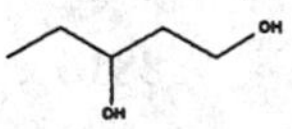

8786

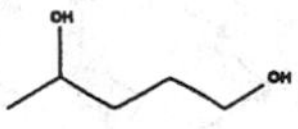

8787

8788

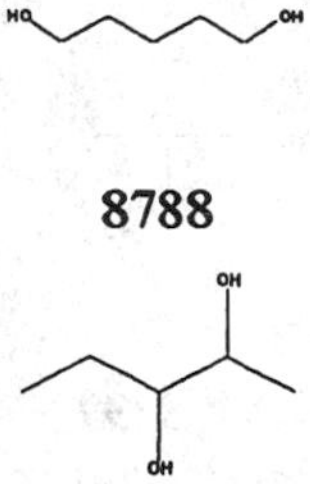

8789

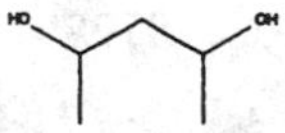

8790

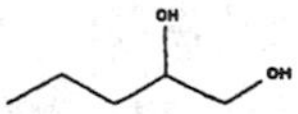

8791

8792

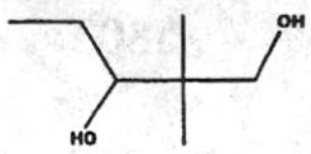

8793

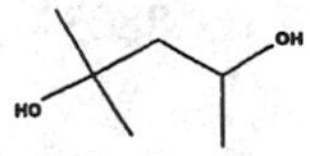

8794

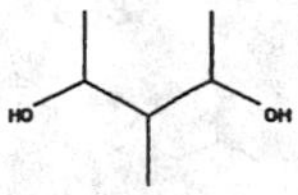

In numeric order as they occur in the Table of Physical Constants of Organic Compounds

8795

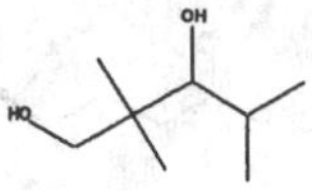

8796

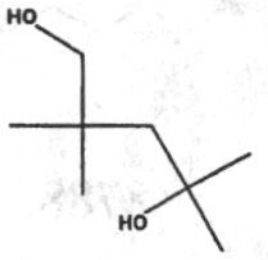

8797

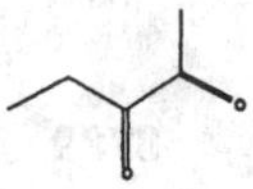

8798

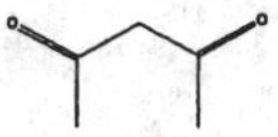

8799

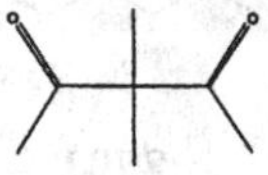

8801

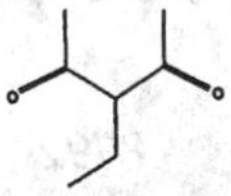

8802

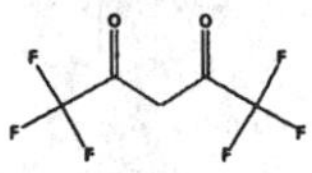

8803

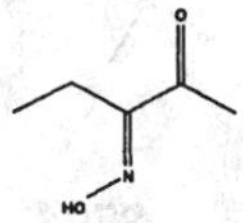

8804

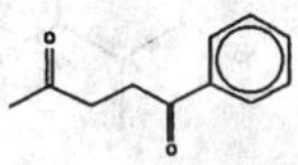

8805

8806

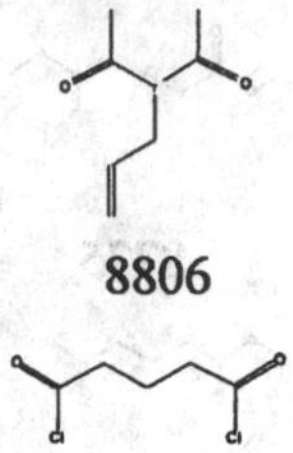

8807

8808

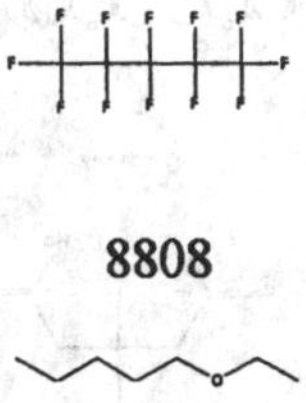

8809

8810

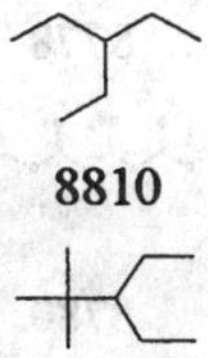

8811

8812

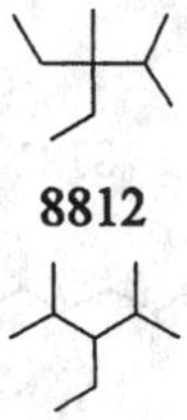

8813

8814

8815

8816

8817

8818

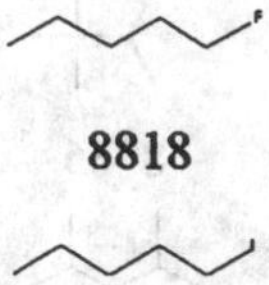

8819

8820

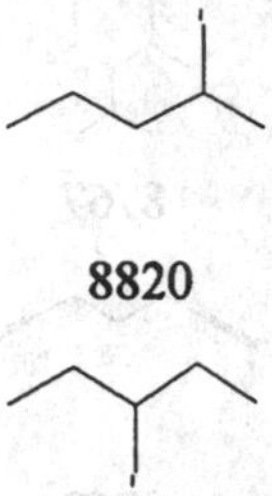

8821

8822

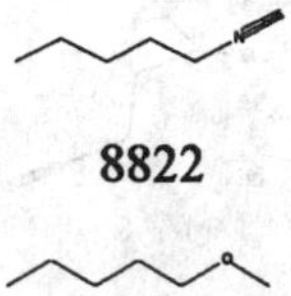

8823

8824

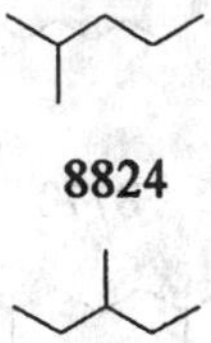

8825

8826

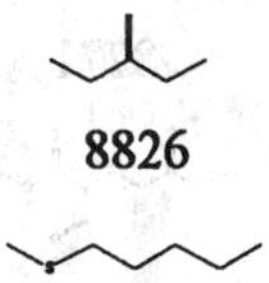

8827

8828

8829

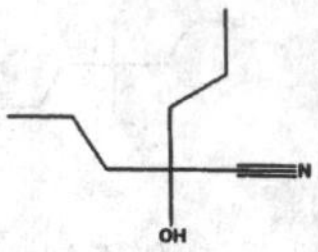

8830

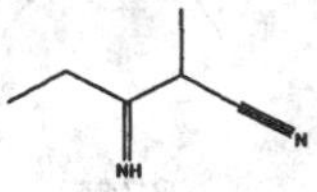

8831

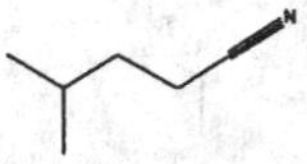

8832

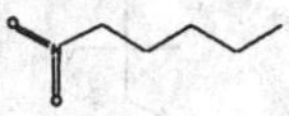

8833

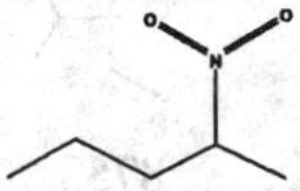

8834

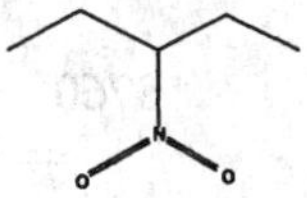

8835

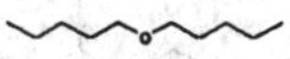

8836

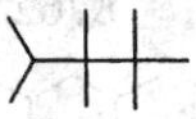

8837

8838

8839

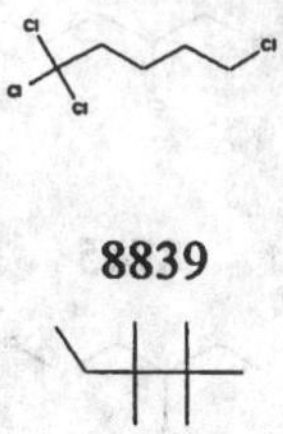

8840

8841

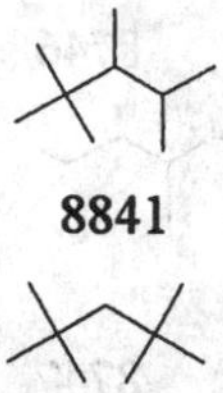

8842

8843

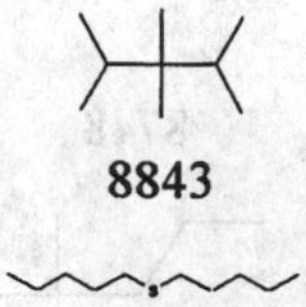

8844

8845

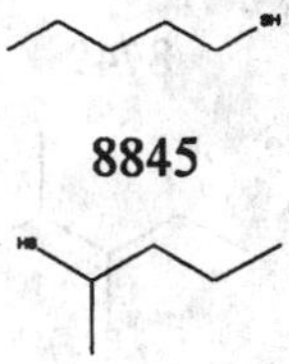

8846

8847

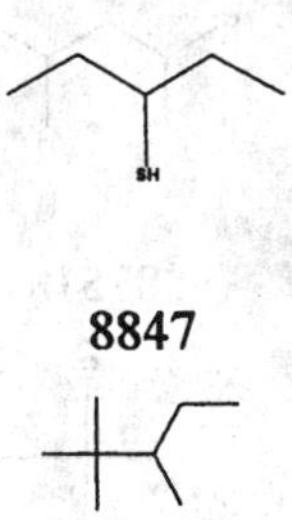

8848

8849

8850

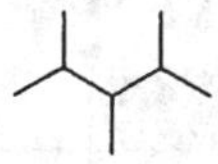

8851

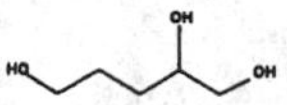

8852

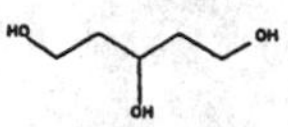

8853

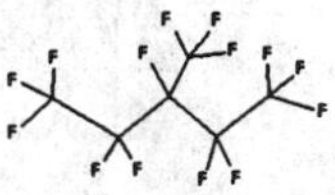

8854

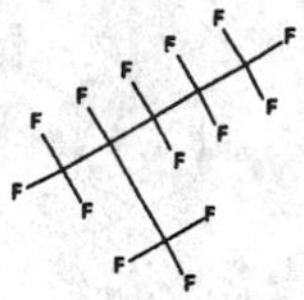

8855

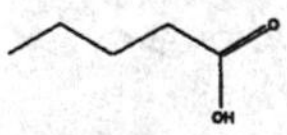

8856

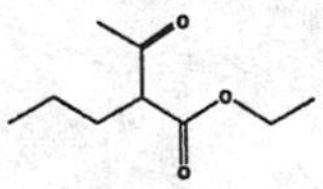

8857

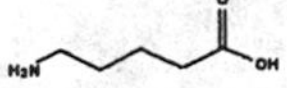

8858

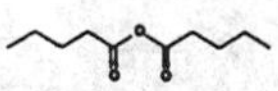

8859

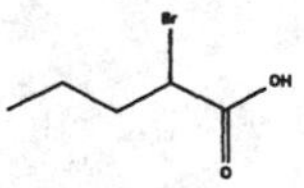

8860

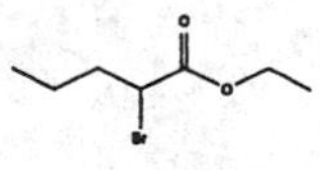

8861

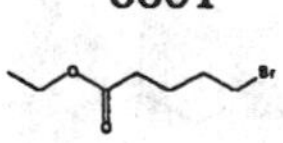

8862

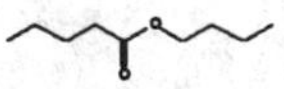

8865

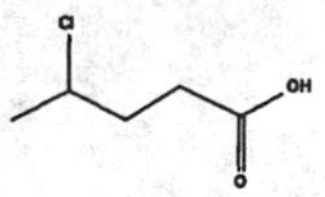

8866

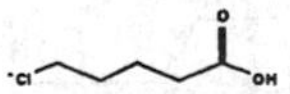

8867

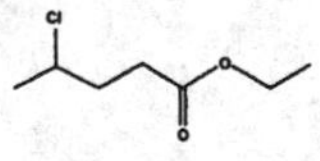

8868

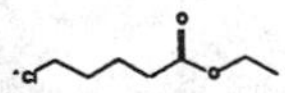

8869

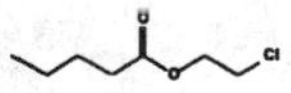

8870

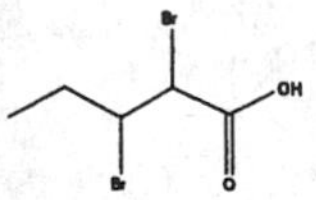

8871

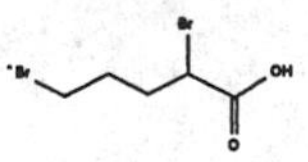

8872

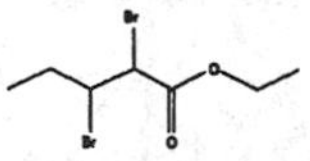

8873

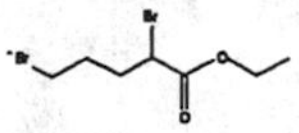

8874

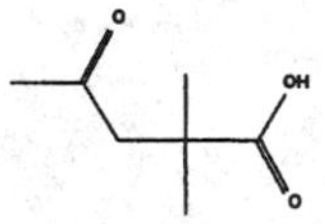

8875

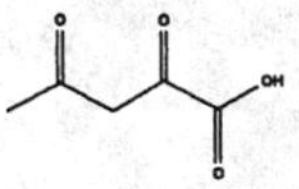

8876

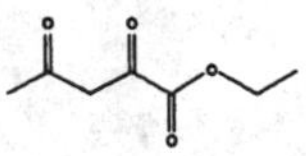

8878

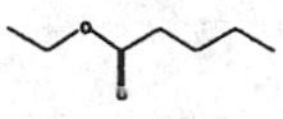

8879

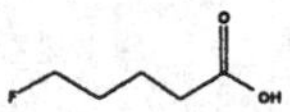

8881

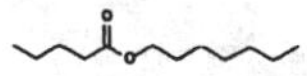

8882

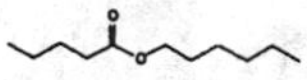

8884

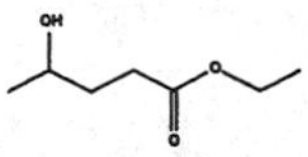

8885

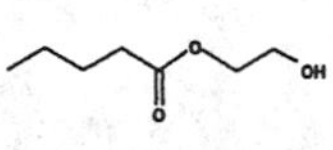

8886

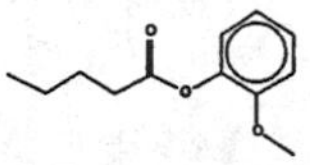

8887

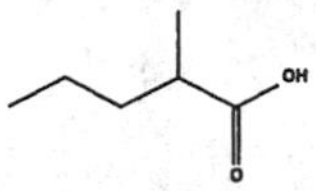

8888

8889

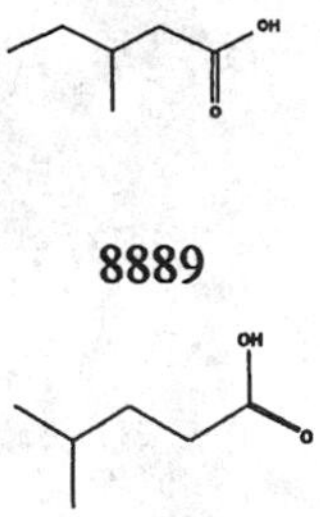

8891

8892

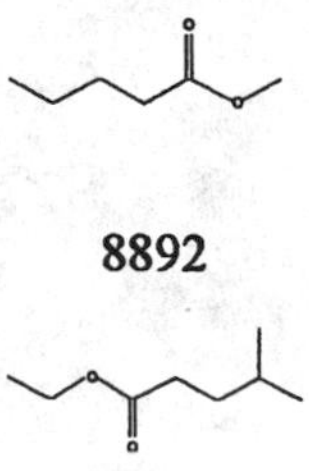

8893

8894

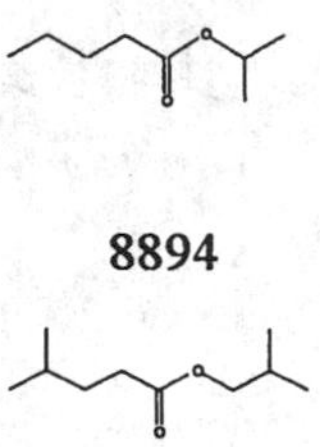

8895

8896

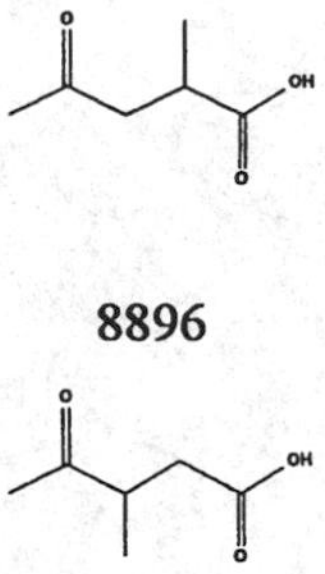

8897

8898

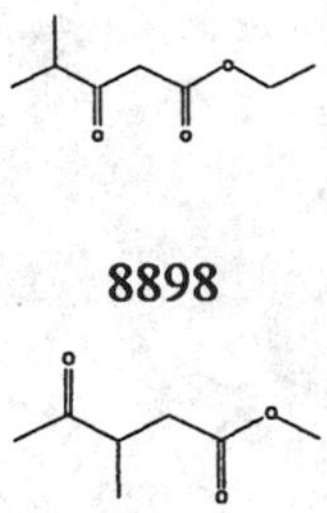

8900

8903

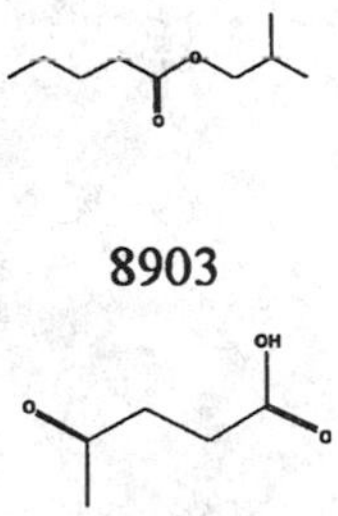

8904

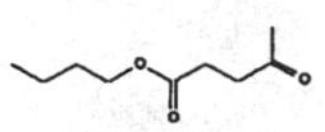

STRUCTURAL FORMULAS OF ORGANIC COMPOUNDS (continued)

In numeric order as they occur in the Table of Physical Constants of Organic Compounds

8905

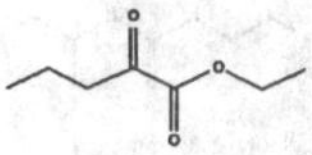

8906

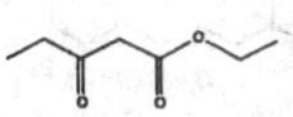

8907

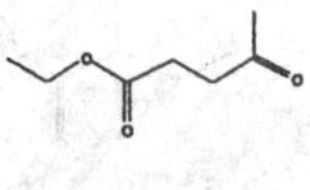

8908

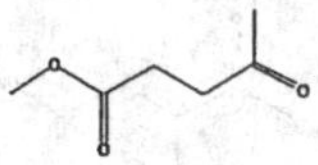

8909

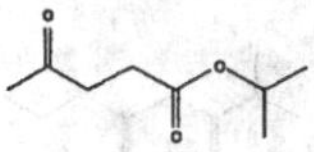

8910

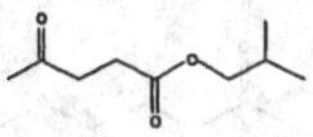

8911

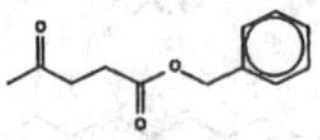

8912

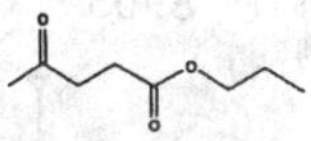

8913

8914

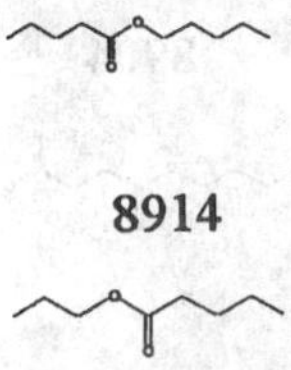

8915

8916

8917

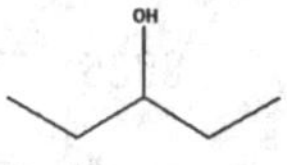

8918

8919

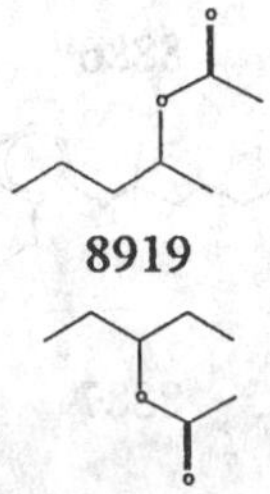

8920

8922

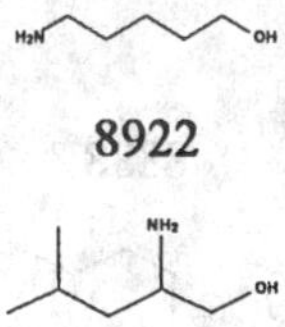

8923

8924

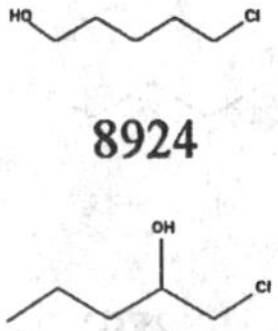

8925

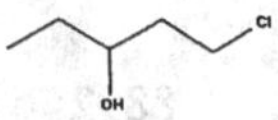

8927

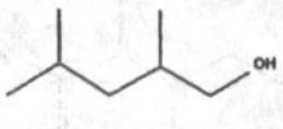

8928

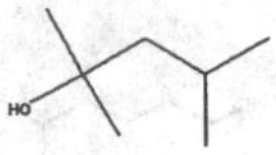

8929

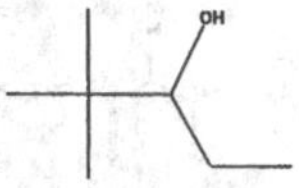

8930

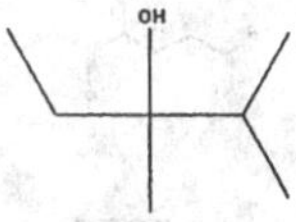

8931

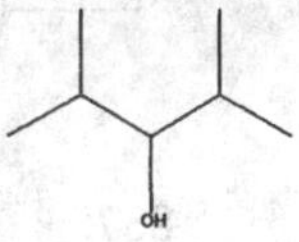

8932

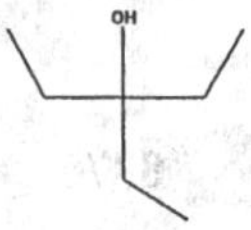

8933

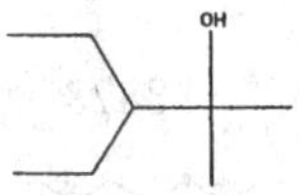

8934

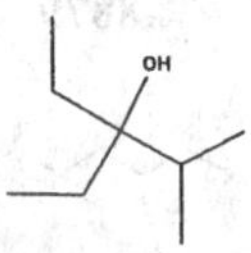

8935

8936

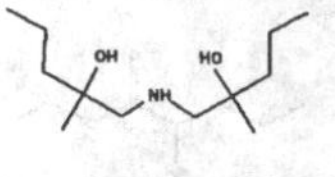

8937

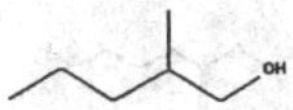

8938

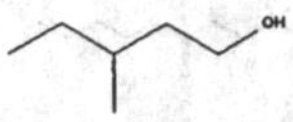

8939

8940

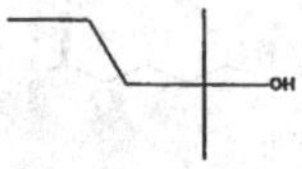

8941

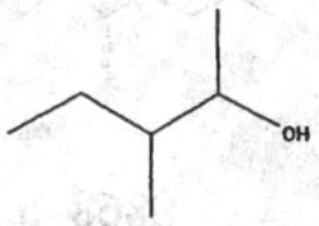

8942

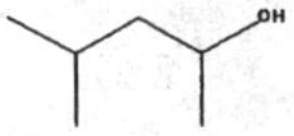

8943

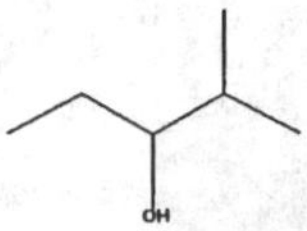

8944

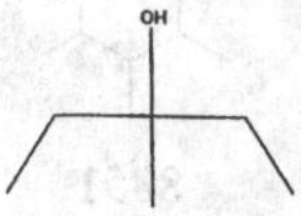

8945

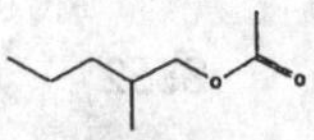

8946

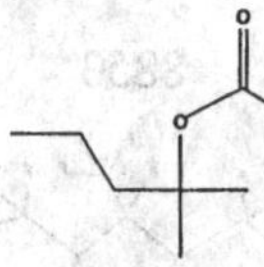

8947

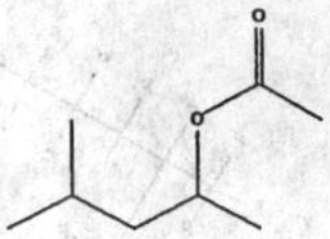

8950

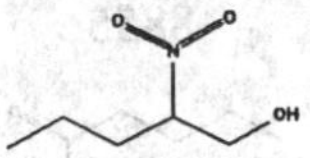

8952

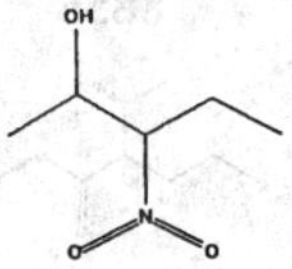

8953

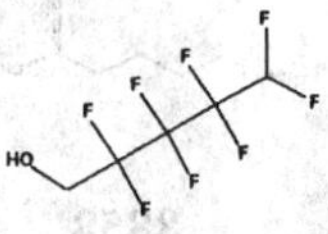

8954

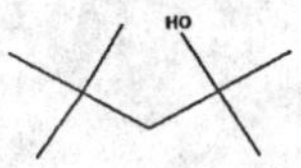

STRUCTURAL FORMULAS OF ORGANIC COMPOUNDS (continued)

In numeric order as they occur in the Table of Physical Constants of Organic Compounds

8955

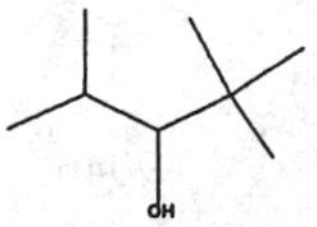

8956

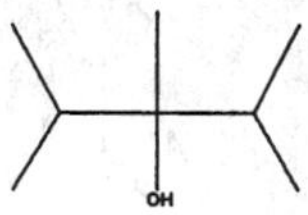

8957

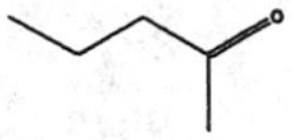

8958

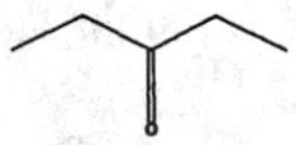

8959

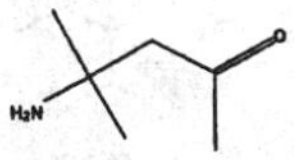

8961

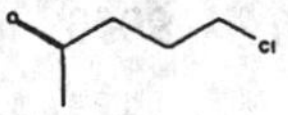

8962

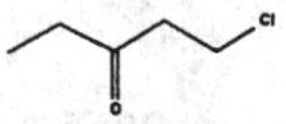

8963

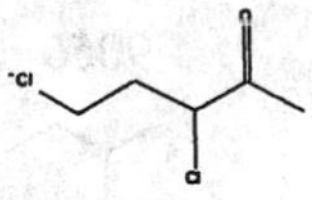

8964

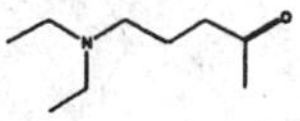

8965

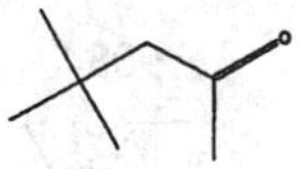

8966

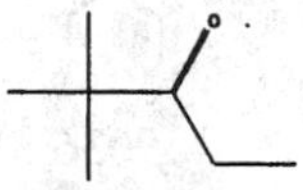

8967

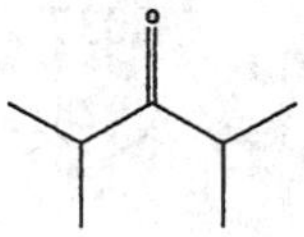

8968

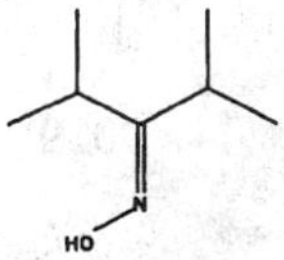

8969

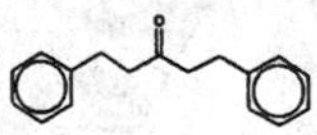

8970

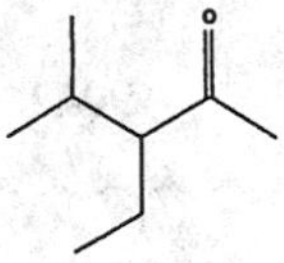

8971

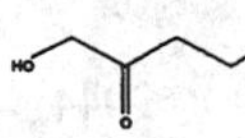

8972

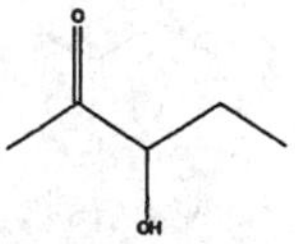

8973

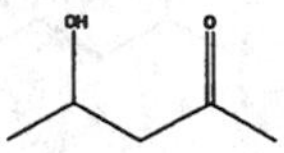

8974

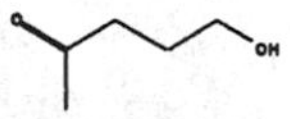

8975

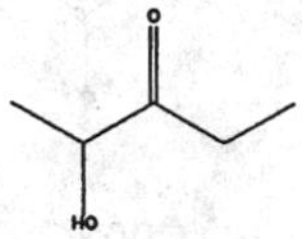

8976

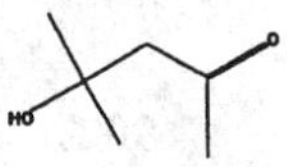

8977

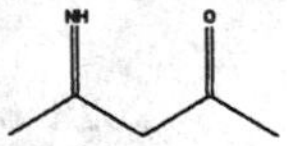

8978

8979

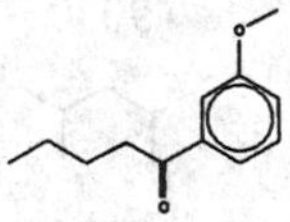

8980

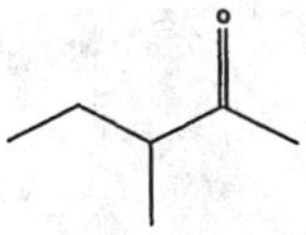

8981

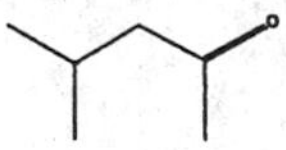

8982

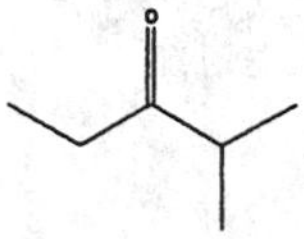

8983

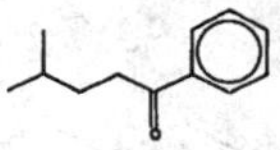

8984

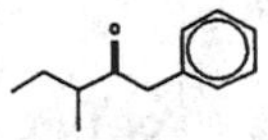

8985

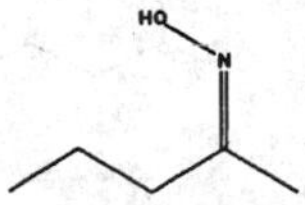

8986

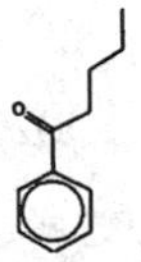

8987

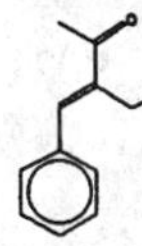

8988

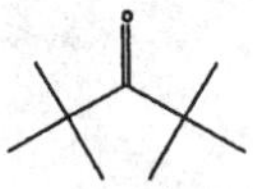

8989

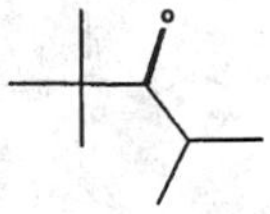

8990

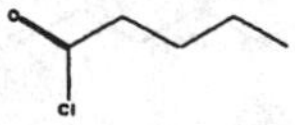

8991

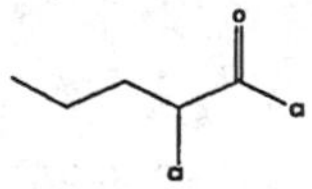

8992

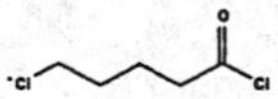

8995

8996

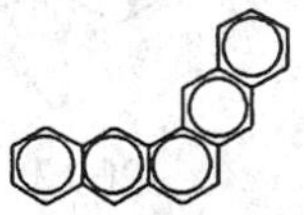

8997

8998

8999

9000

9001

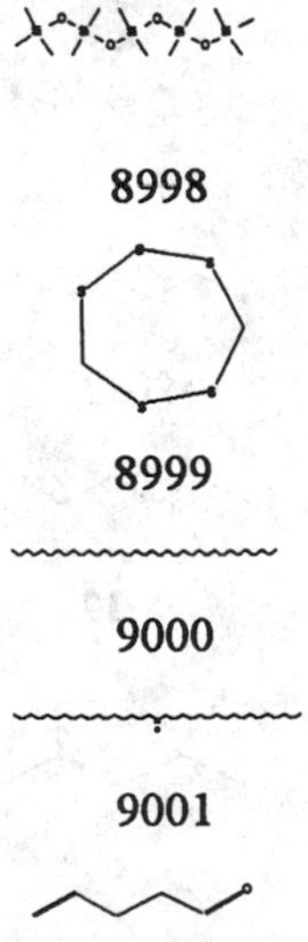

STRUCTURAL FORMULAS OF ORGANIC COMPOUNDS (continued)

In numeric order as they occur in the Table of Physical Constants of Organic Compounds

9002

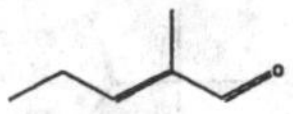

9004

9005

9006

9007

9008

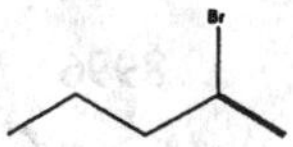

9009

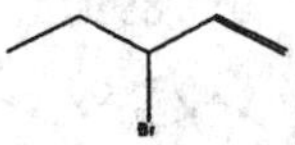

9010

9011

9012

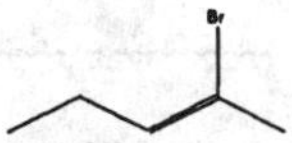

9013

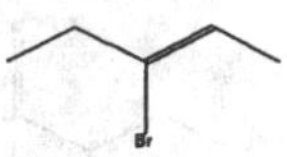

9014

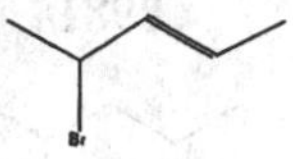

9015

9016

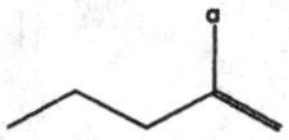

9017

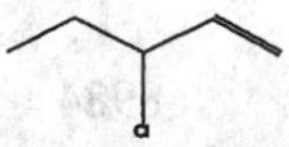

9018

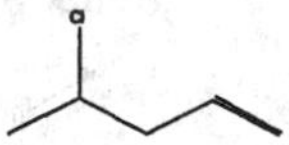

9019

9020

9021

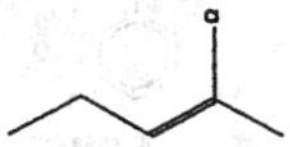

9022

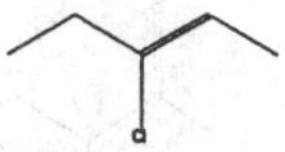

9023

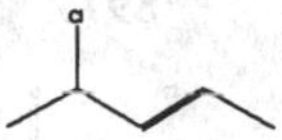

9024

9025

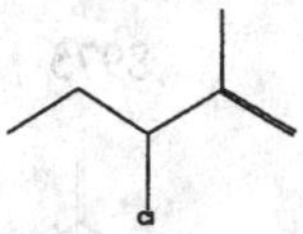

9026

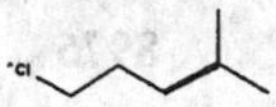

9027

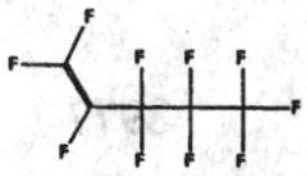

9028

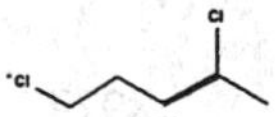

9029

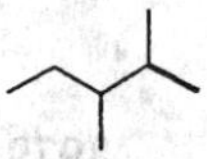

9030

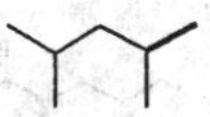

9031

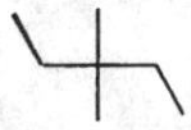

9032

9033

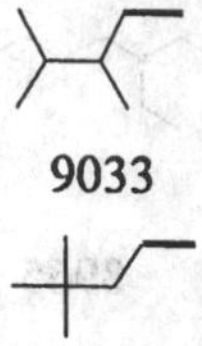

9034

9035

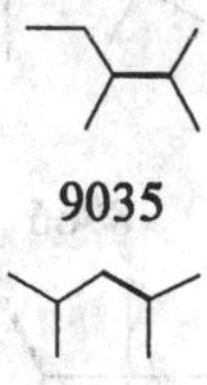

9036

9037

9038

9039

9040

9041

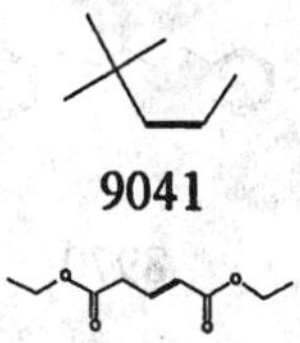

9042

9043

9044

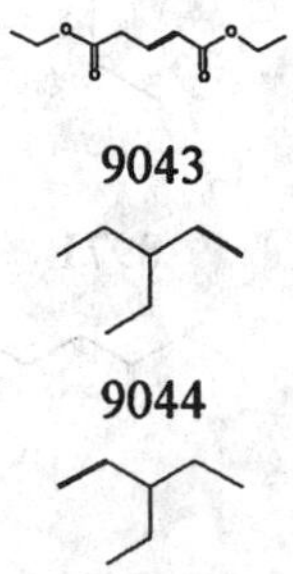

9045

9046

9047

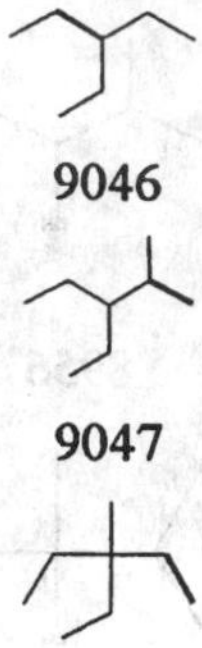

9048

9049

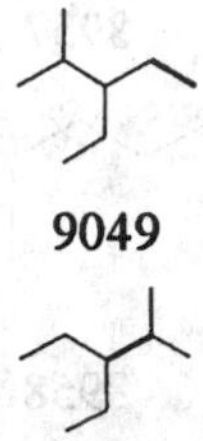

9050

9051

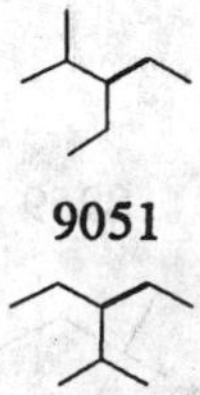

9052

9053

9054

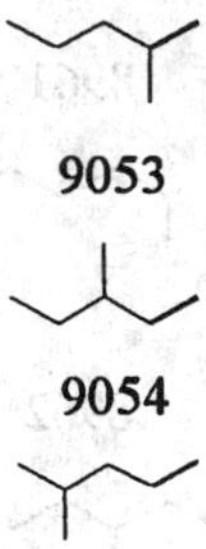

9055

9056

9057

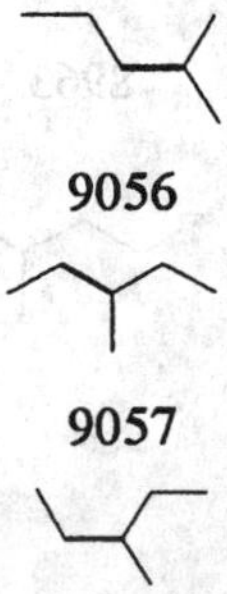

STRUCTURAL FORMULAS OF ORGANIC COMPOUNDS (continued)

In numeric order as they occur in the Table of Physical Constants of Organic Compounds

9058

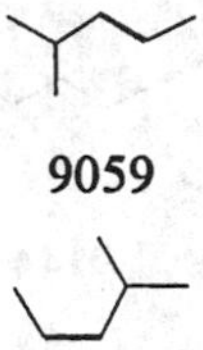

9059

9060

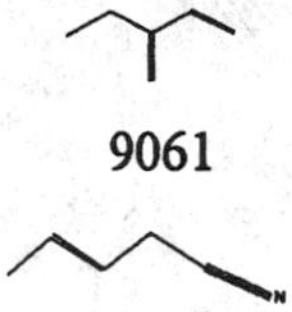

9061

9062

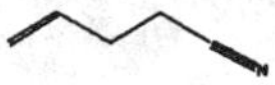

9065

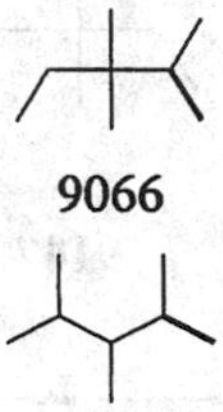

9066

9067

9068

9069

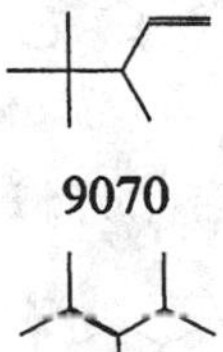

9070

9071

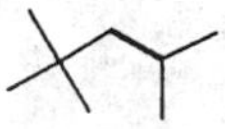

9072

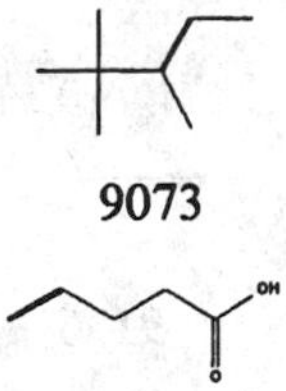

9073

9074

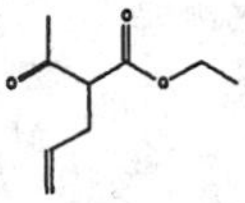

9075

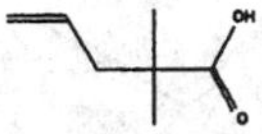

9076

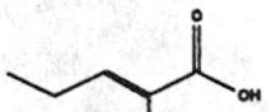

9077

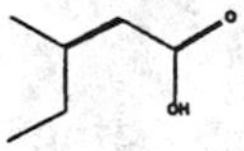

9078

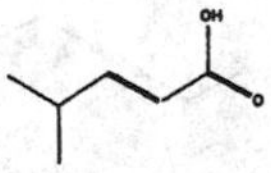

9079

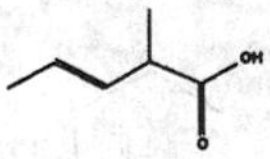

9080

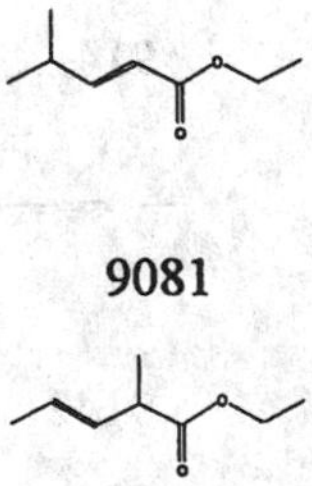

9081

9082

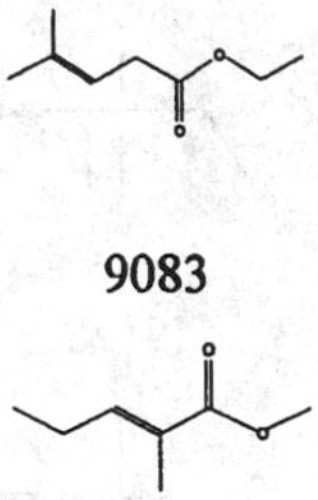

9083

9084

9085

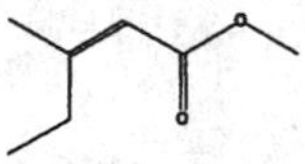

9086

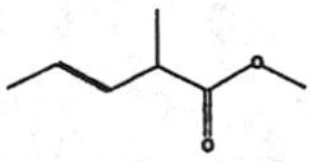

9087

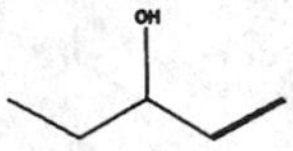

9088

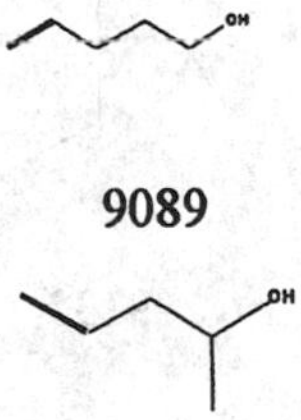

9089

9090

9091

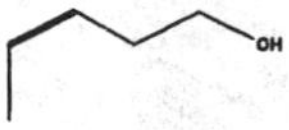

9092

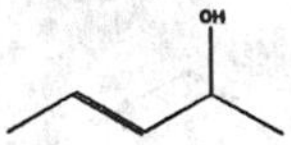

9093

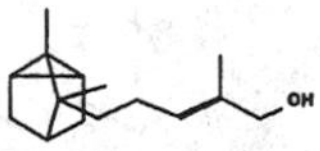

9094

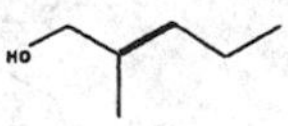

9095

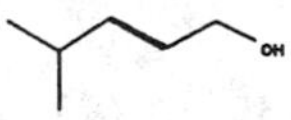

9096

9097

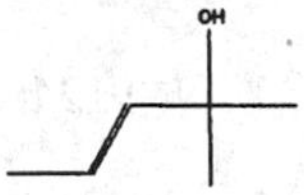

9098

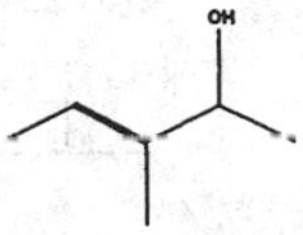

9099

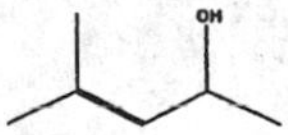

9100

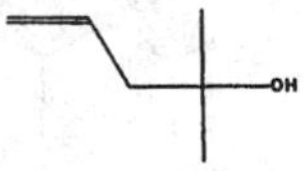

9101

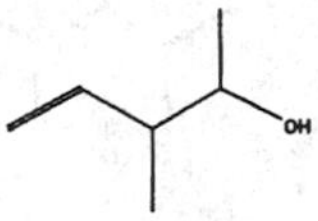

9102

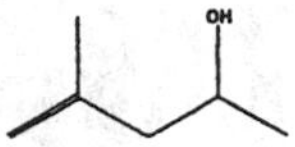

9103

9104

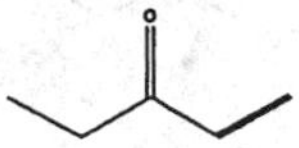

9105

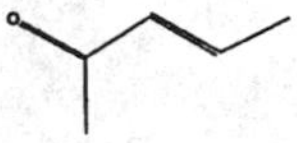

9106

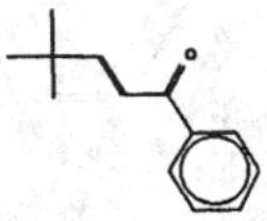

9107

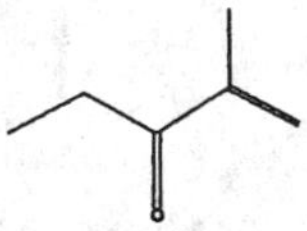

9108

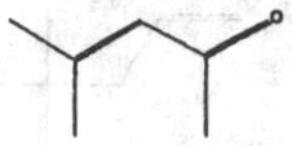

9109

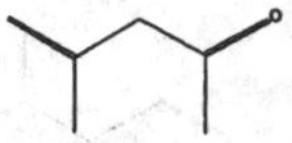

9110

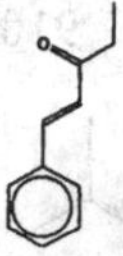

9111

9112

9113

9114

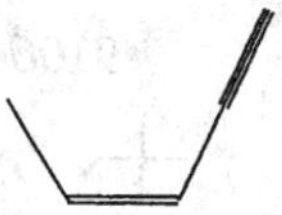

9115

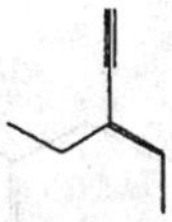

9116

9117

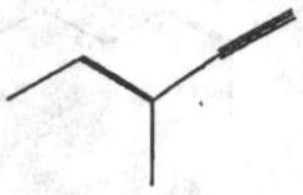

9118

9119

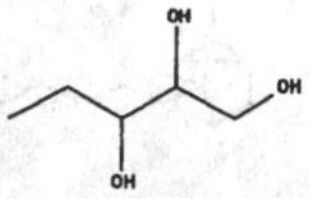

9121

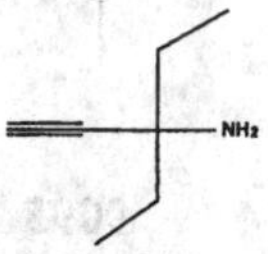

9122

9123

9124

9125

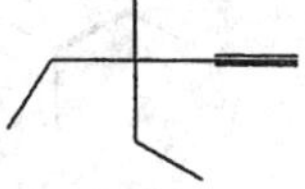

9126

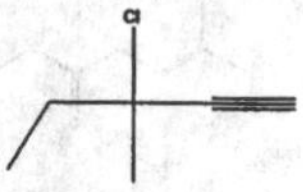

9128

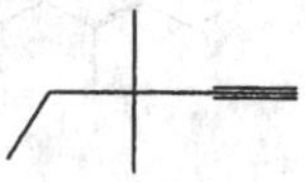

9129

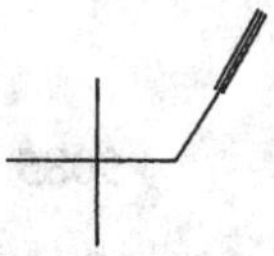

9130

9131

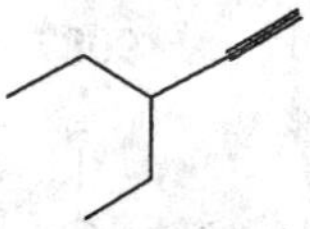

9132

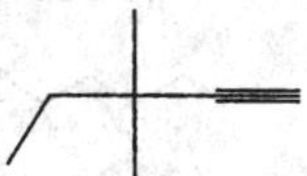

9133

9134

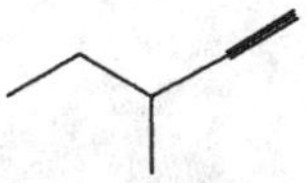

9135

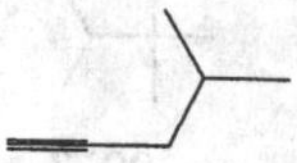

9136

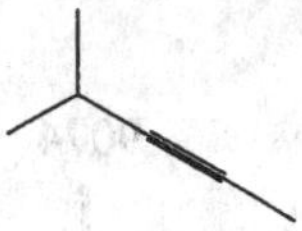

9137

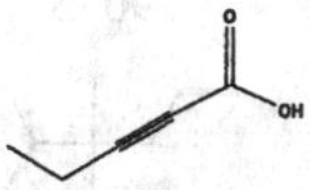

9138

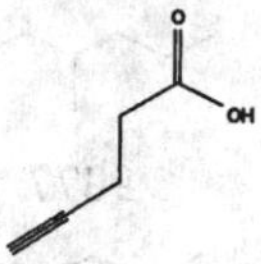

9139

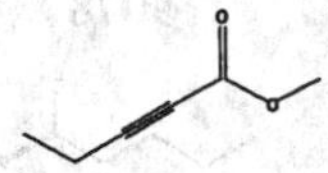

9140

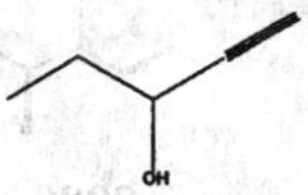

9141

9142

9143

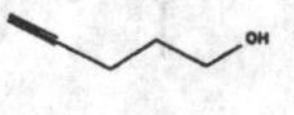

9144

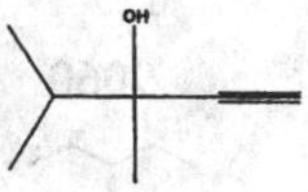

9145

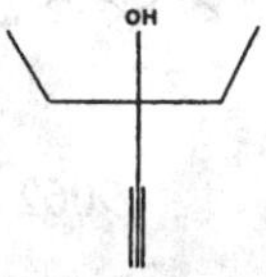

9146

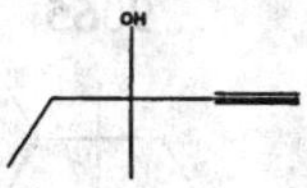

9147

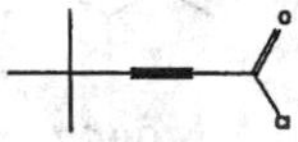

9149

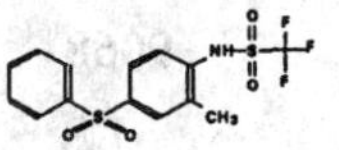

9150

9151

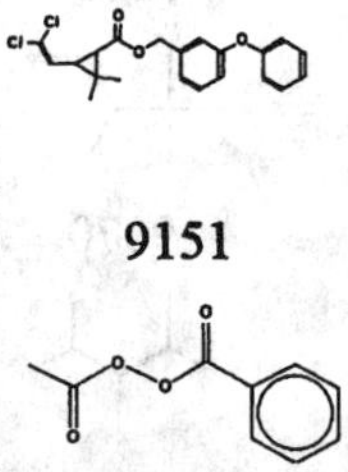

In numeric order as they occur in the Table of Physical Constants of Organic Compounds

9153

9154

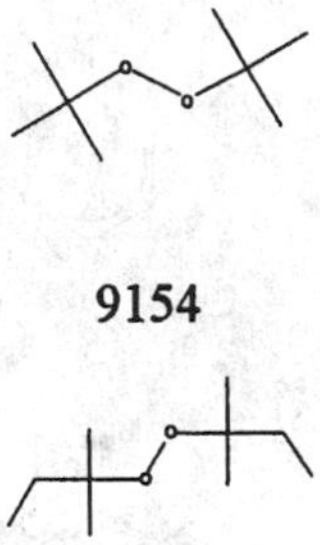

9156

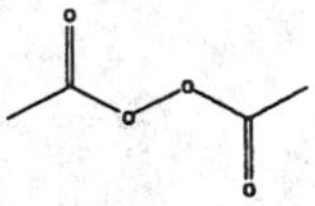

9158

9159

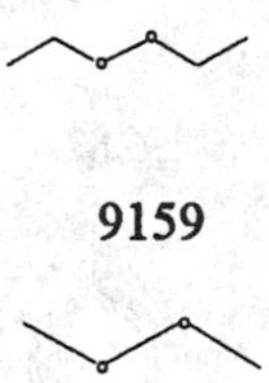

9160

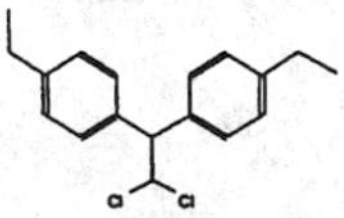

9161

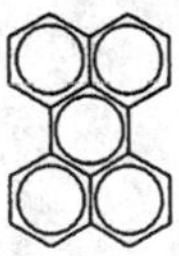

9162

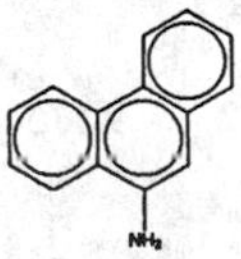

9164

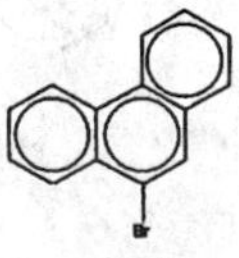

9167

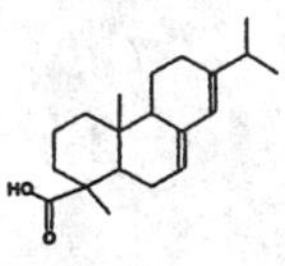

9168

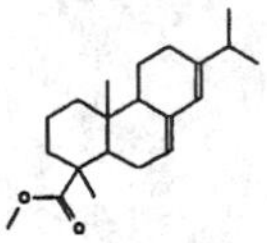

9170

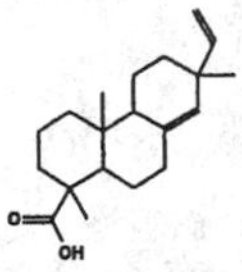

9171

9172

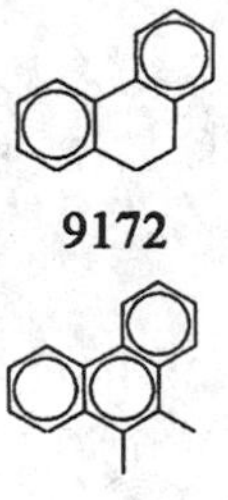

9173

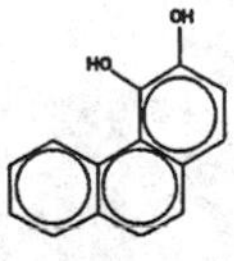

9174

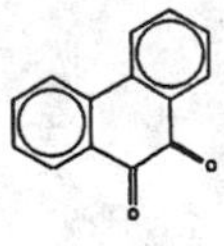

9179

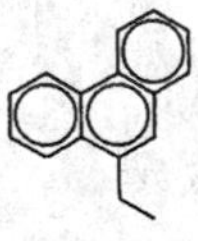

9180

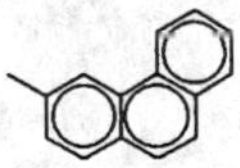

9181

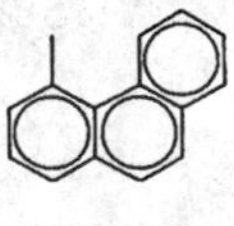

9182

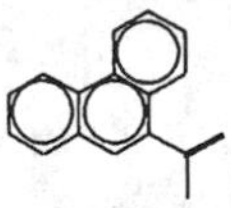

9183

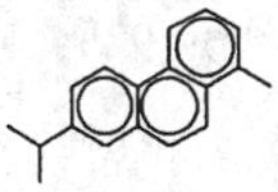

9184

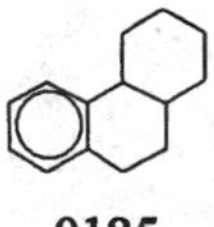

9185

9186

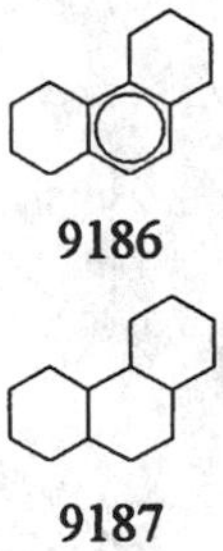

9187

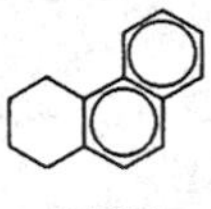

9188

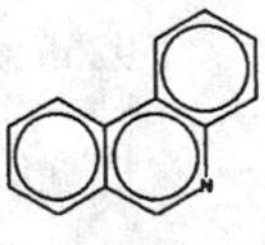

9189

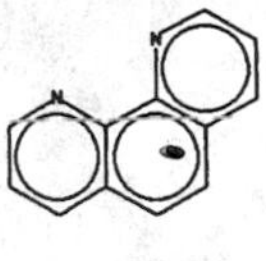

9190

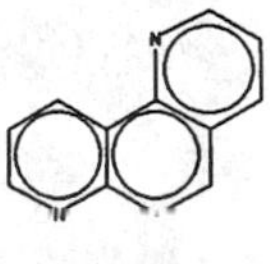

9191

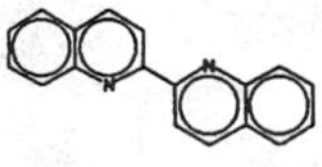

9192

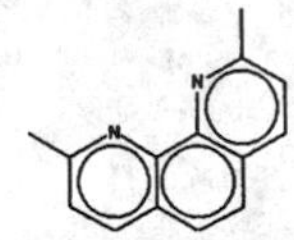

9193

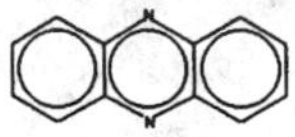

9195

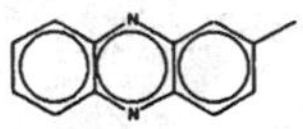

9198

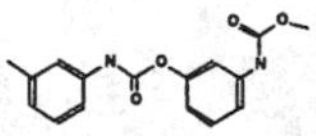

9199

9201

9202

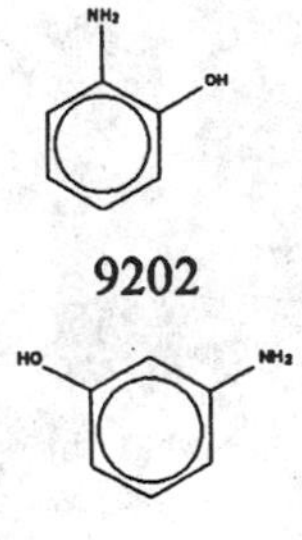

9204

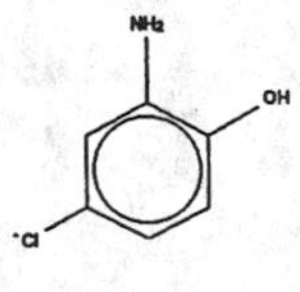

9205

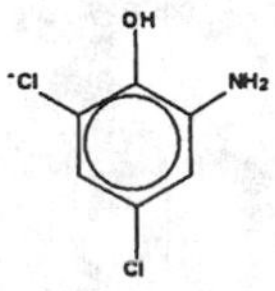

9206

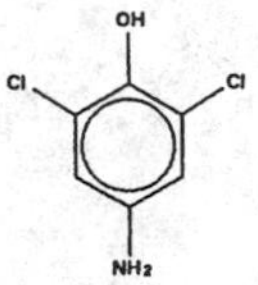

9208

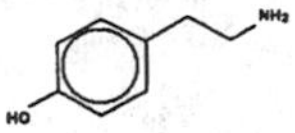

9210

9211

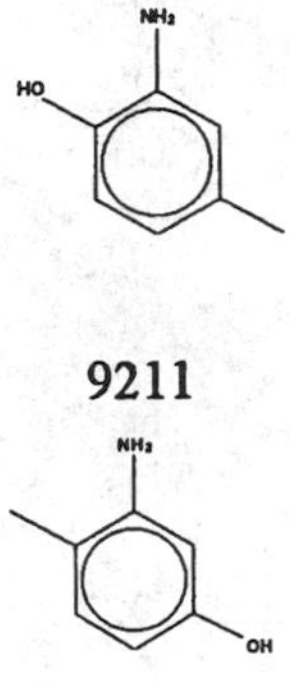

9212

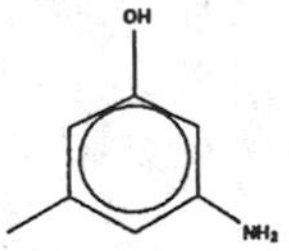

9213

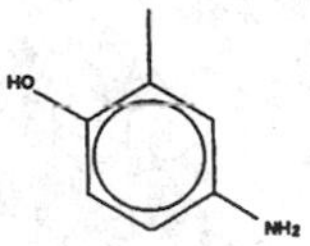

9216

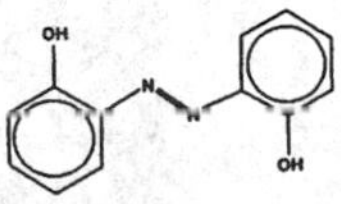

9217

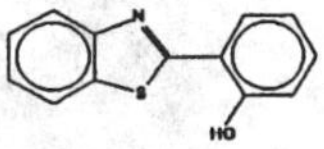

STRUCTURAL FORMULAS OF ORGANIC COMPOUNDS (continued)

In numeric order as they occur in the Table of Physical Constants of Organic Compounds

9218

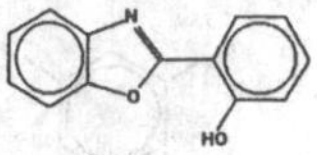

9219

9220

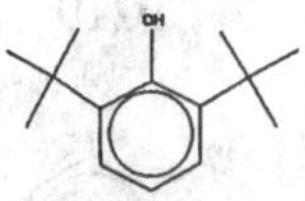

9221

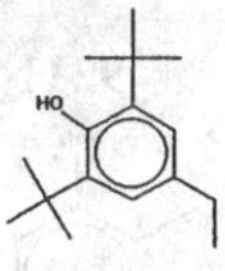

9223

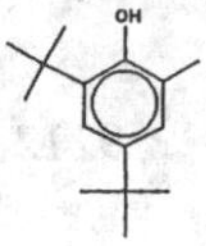

9224

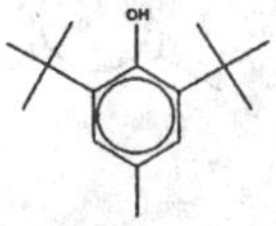

9225

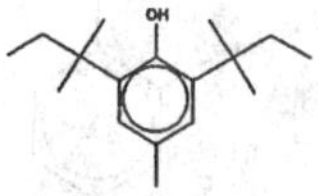

9226

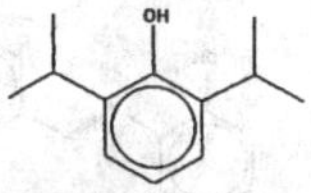

9227

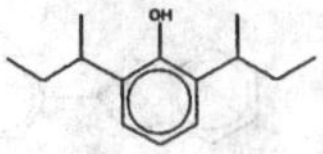

9228

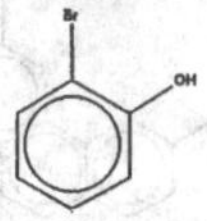

9230

9231

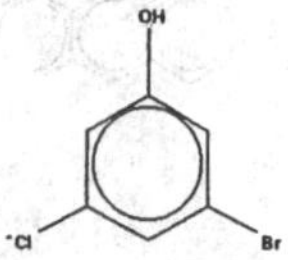

9232

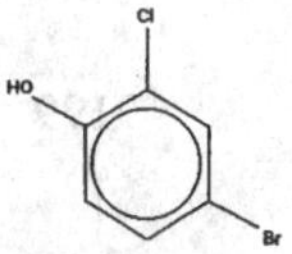

9233

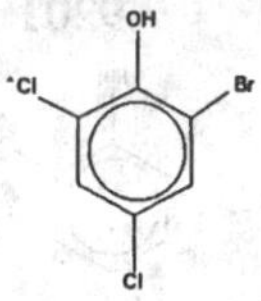

9249

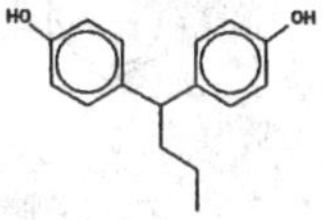

9250

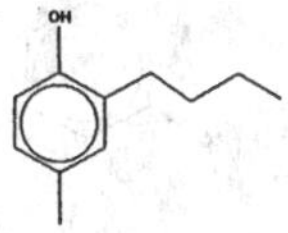

9251

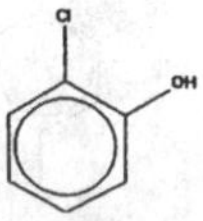

9253

9254

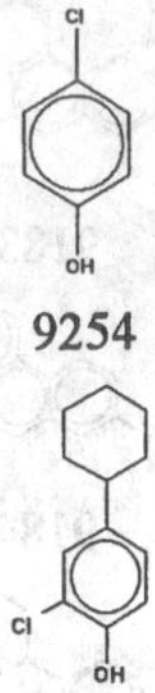

9255

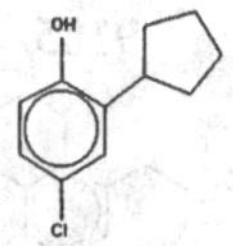

9256

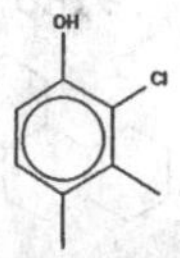

9257

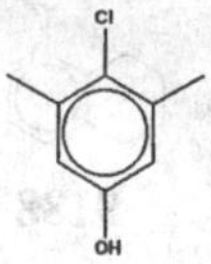

9260

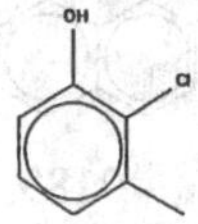

9262

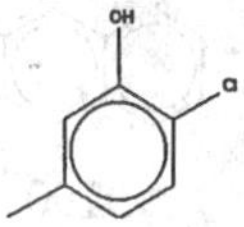

9263

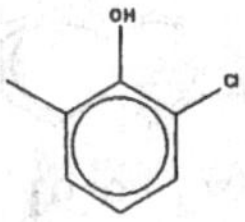

9264

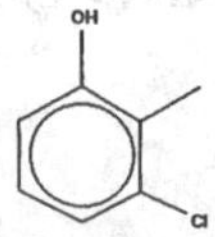

9265

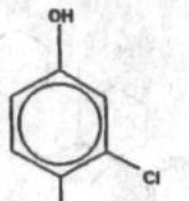

9266

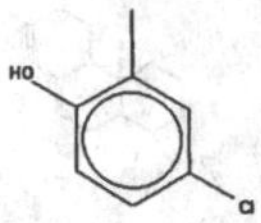

9267

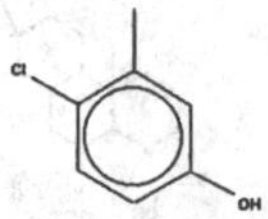

9270

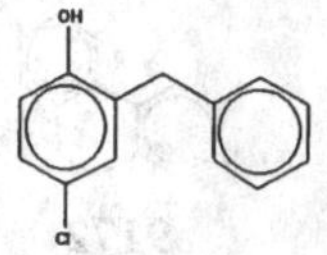

9271

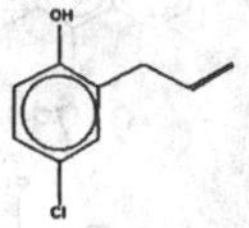

9272

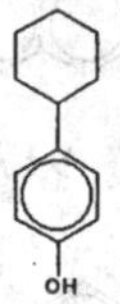

9273

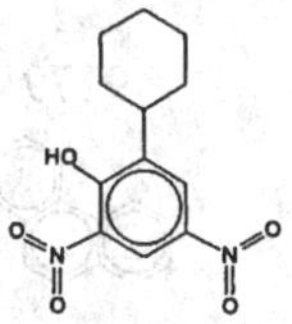

9276

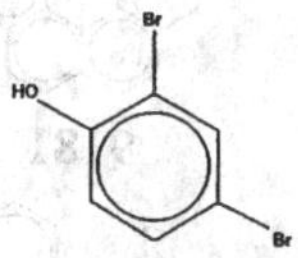

9277

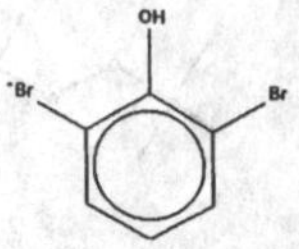

9279

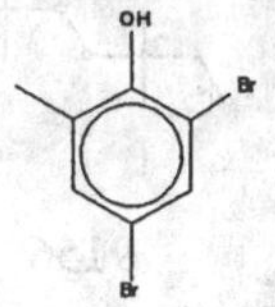

9280

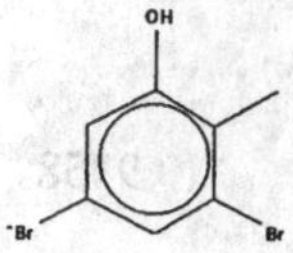

9282

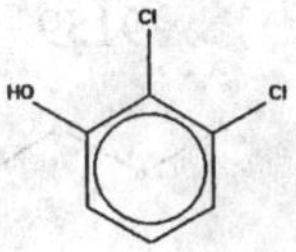

9283

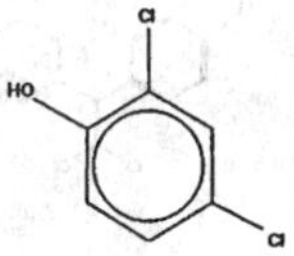

9284

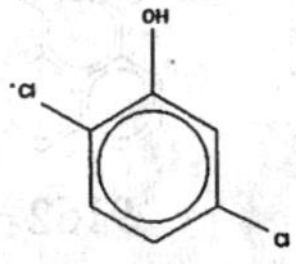

9285

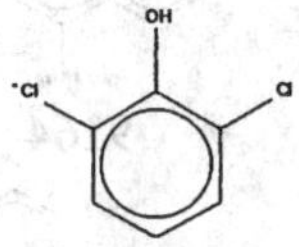

9287

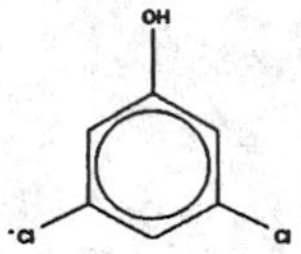

9288

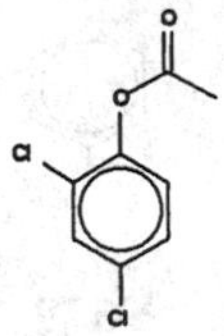

9289

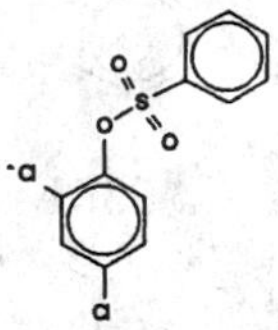

9290

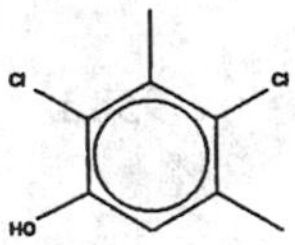

9291

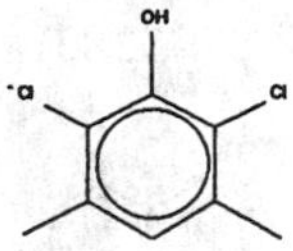

9292

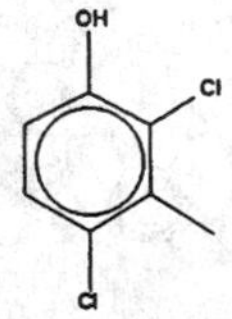

9294

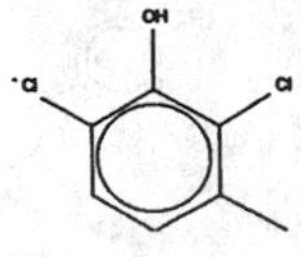

9295

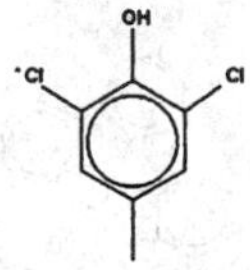

9296

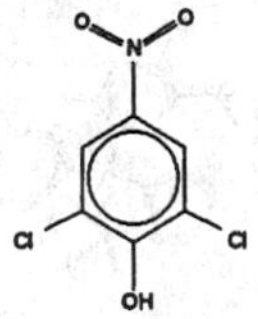

9297

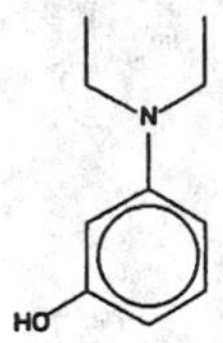

9298

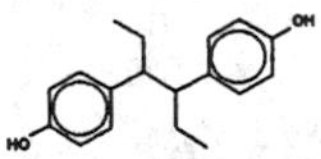

9300

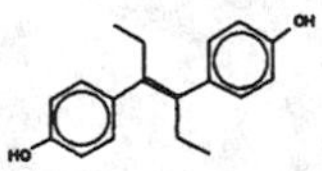

9305

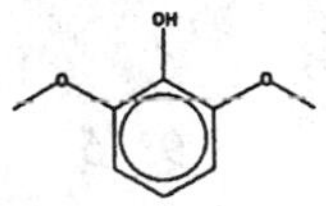

9306

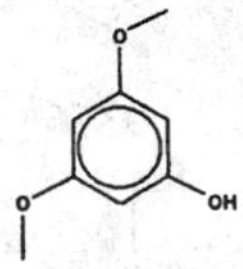

9307

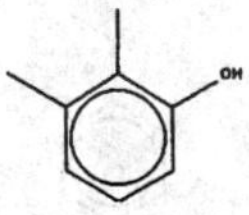

9308

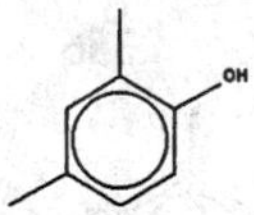

9309

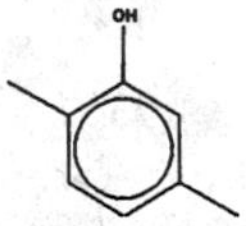

9310

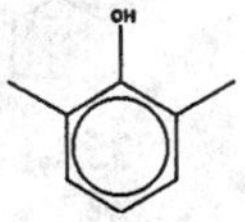

9312

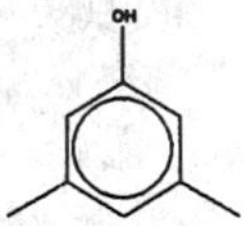

9313

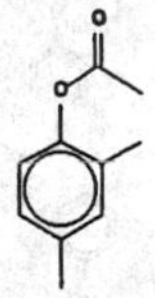

9314

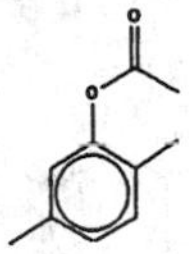

9315

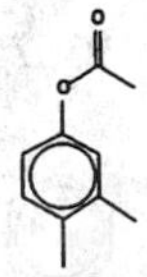

9316

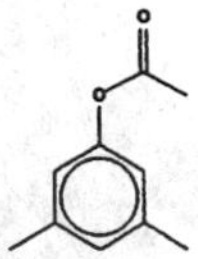

9317

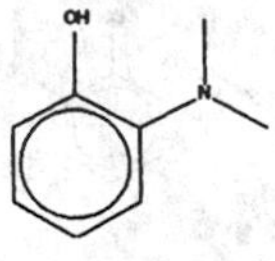

9319

9320

9321

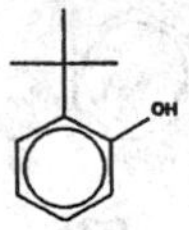

9322

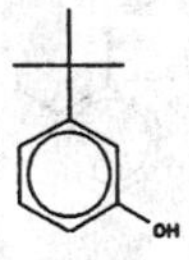

9323

9324

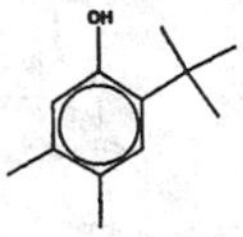

9326

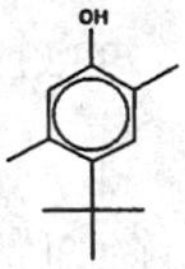

9327

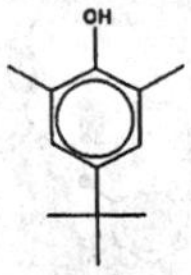

9328

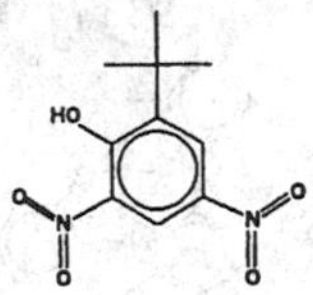

9330

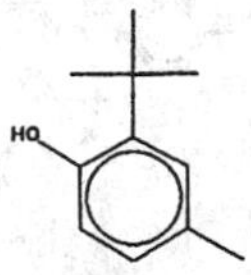

9331

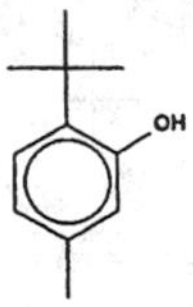

9333

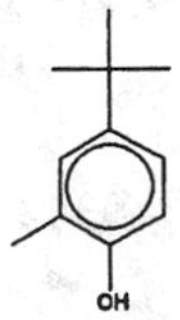

9334

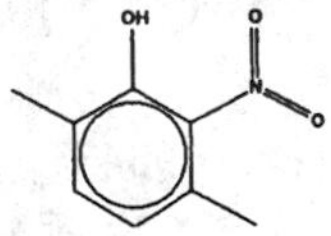

9335

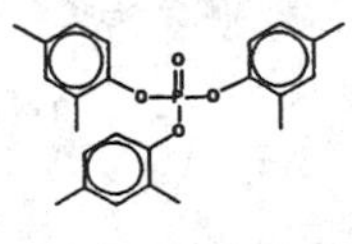

In numeric order as they occur in the Table of Physical Constants of Organic Compounds

9336
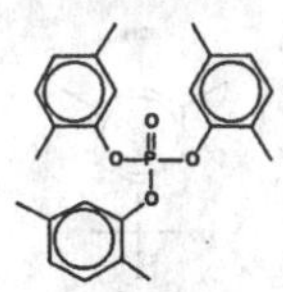
9337
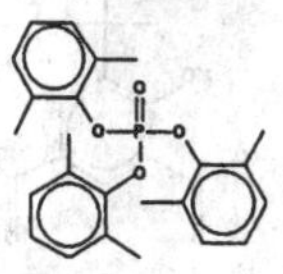
9338
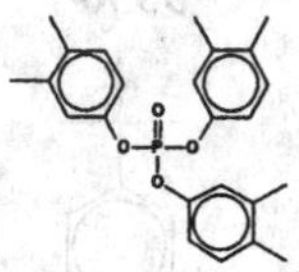
9340

9341
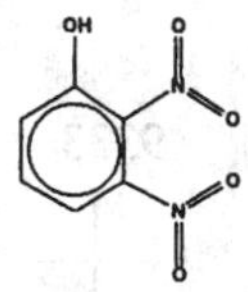
9342
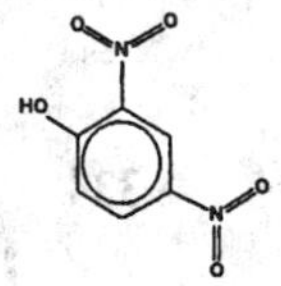
9343
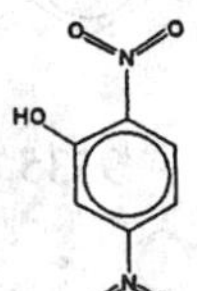
9344
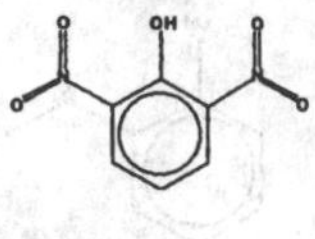
9345
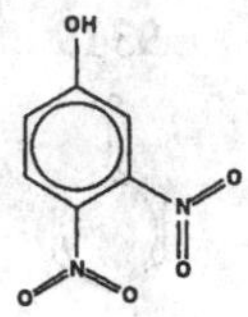
9348
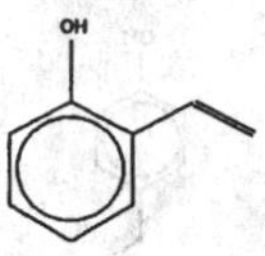
9349
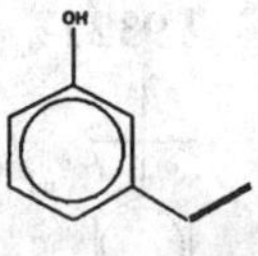
9350
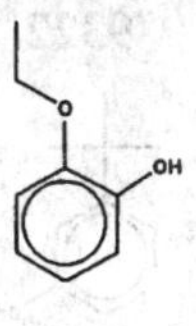
9351
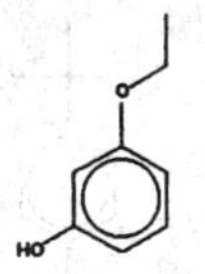
9352
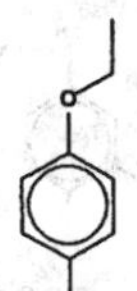
9353
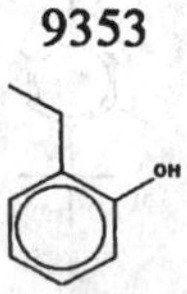
9355

9356
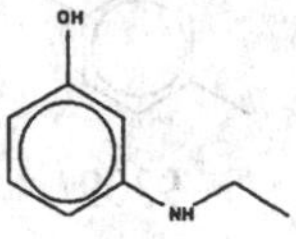
9357
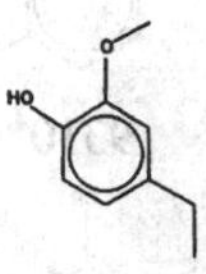
9358
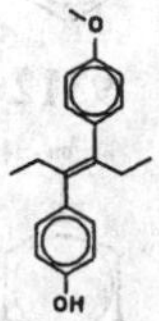
9359
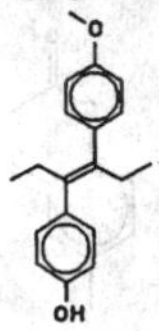
9360
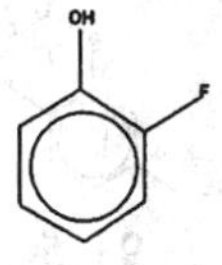
9362

9364
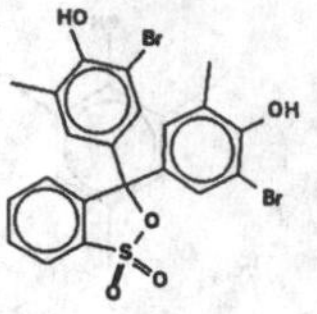
9365
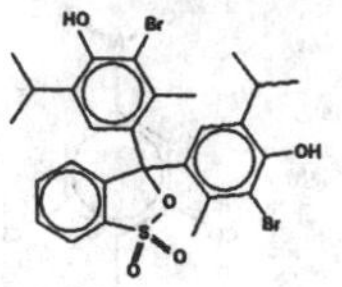
9370
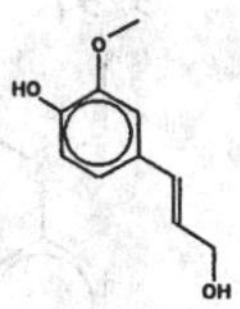
9371
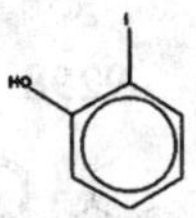
9372
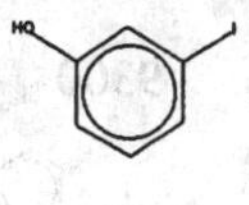
9374
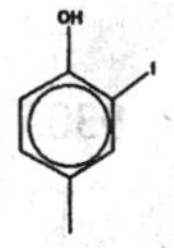
9375
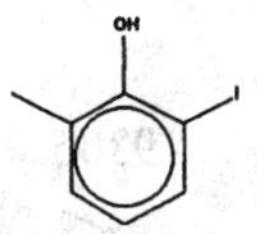
9376
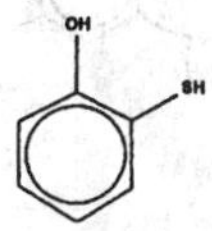
9377
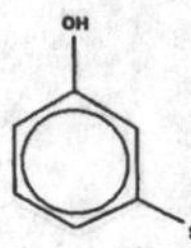
9378

9379
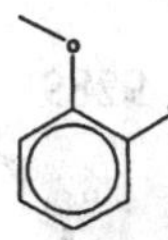
9381

9382
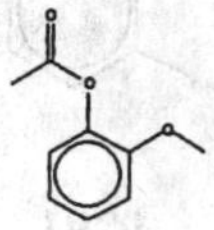
9385
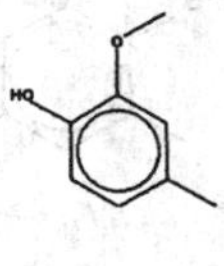
9387
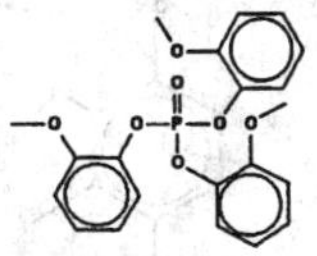
9388
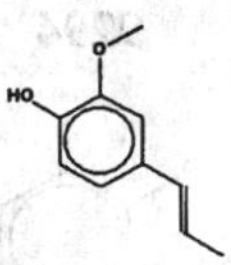

In numeric order as they occur in the Table of Physical Constants of Organic Compounds

9389

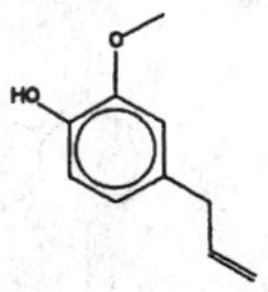

9391

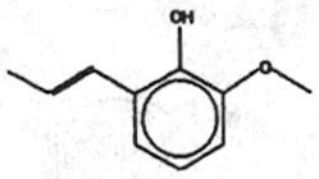

9392

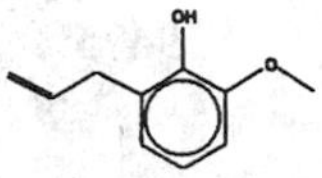

9394

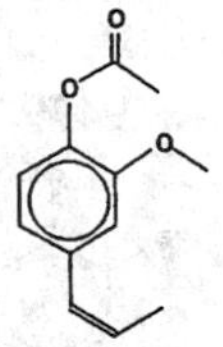

9395

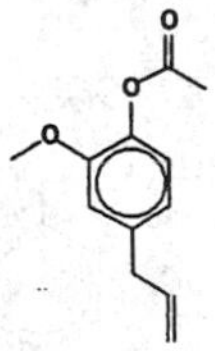

9396

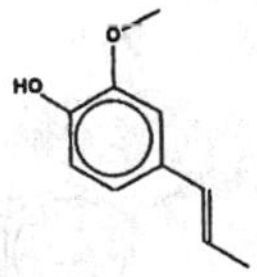

9397

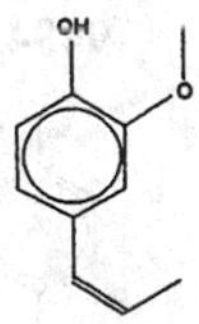

9401

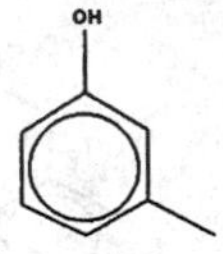

9402

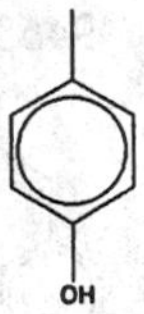

9403

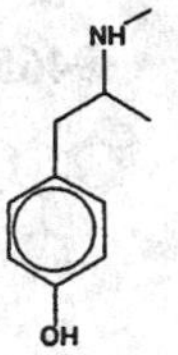

9404

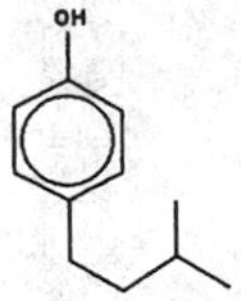

9405

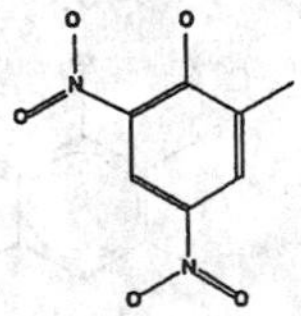

9407

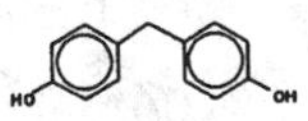

9409

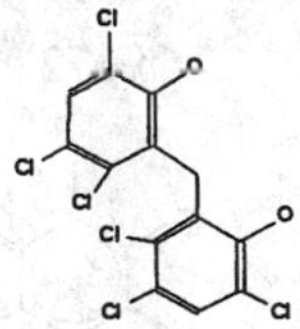

9410

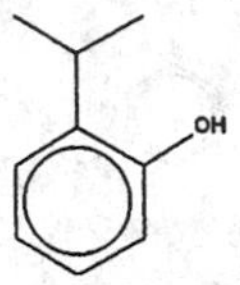

9411

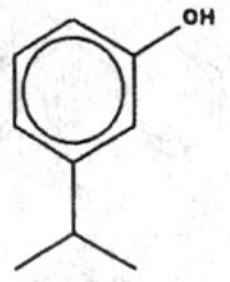

9412

9413

9414

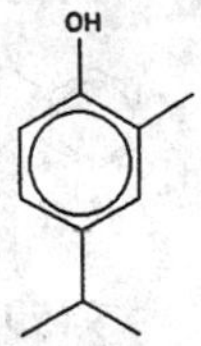

9416

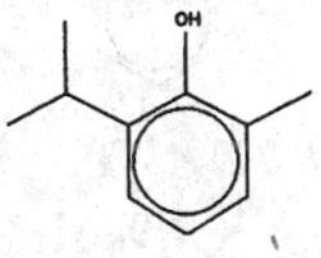

9417

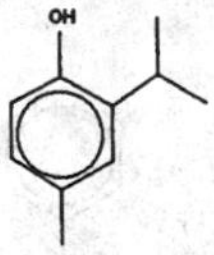

9418

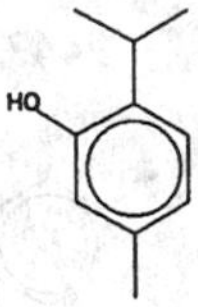

9419

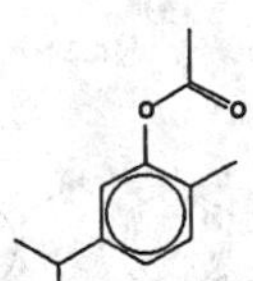

9420

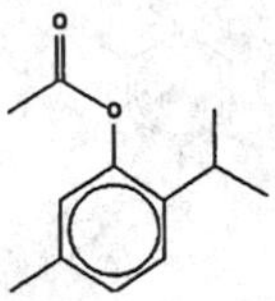

9421

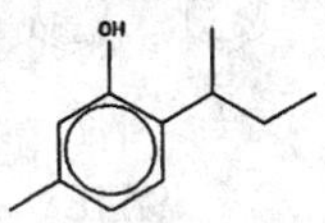

9423

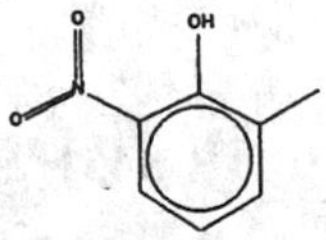

9424

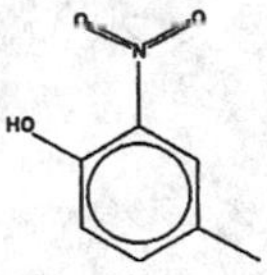

9425

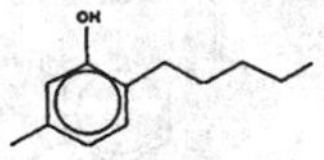

9426

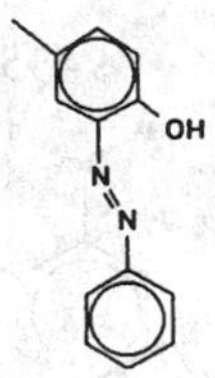

9427

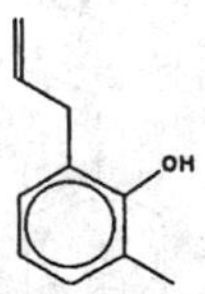

9428

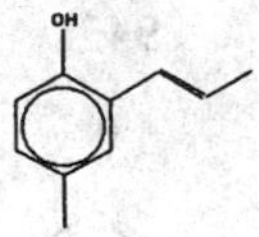

9430

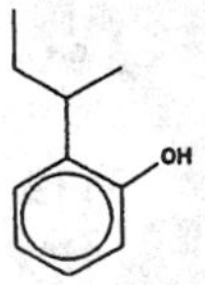

9431

9432

STRUCTURAL FORMULAS OF ORGANIC COMPOUNDS (continued)

In numeric order as they occur in the Table of Physical Constants of Organic Compounds

9434

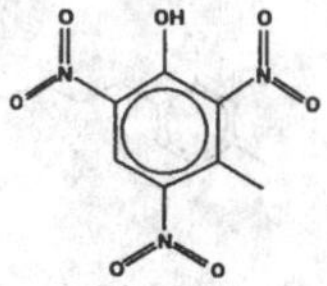

9435

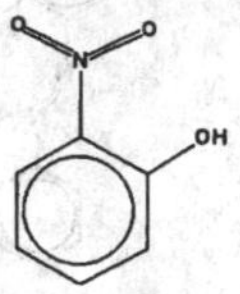

9436

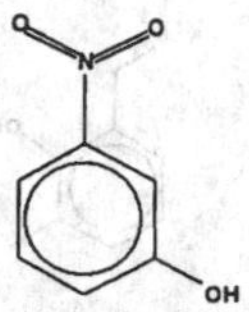

9438

9439

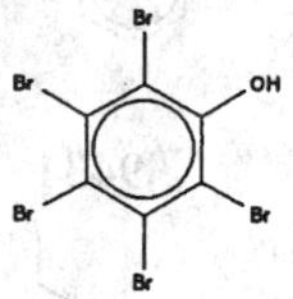

9440

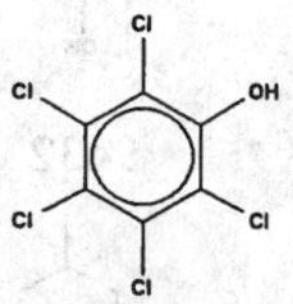

9441

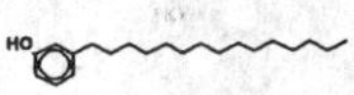

9442

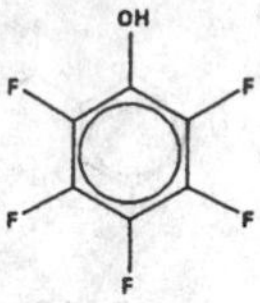

9443

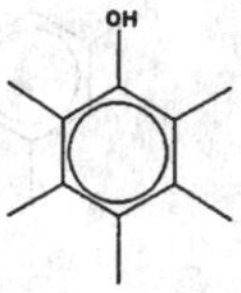

9445

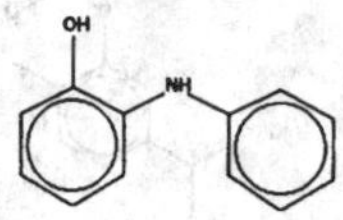

9446

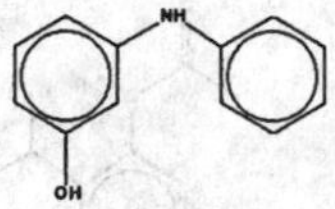

9447

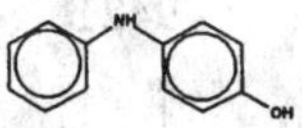

9448

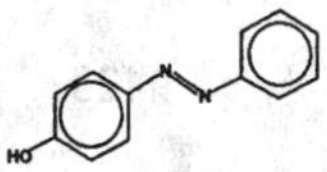

9451

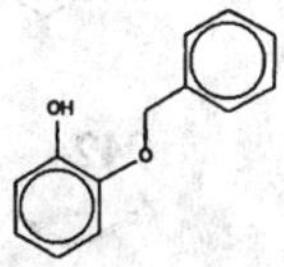

9452

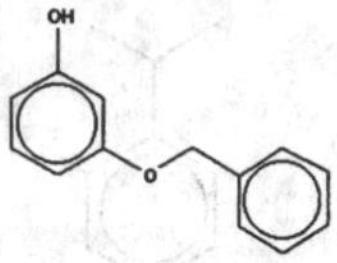

9454

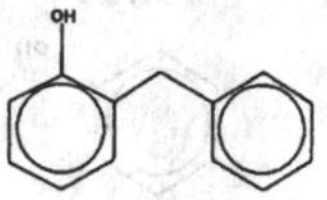

9455

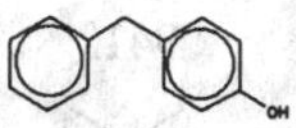

9456

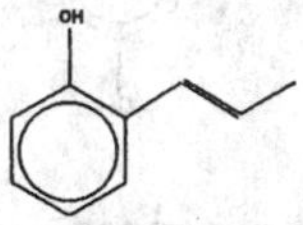

9457

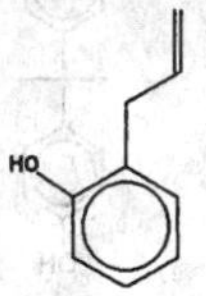

9458

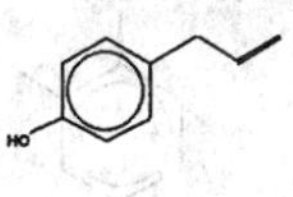

9459

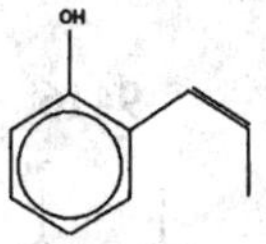

9461

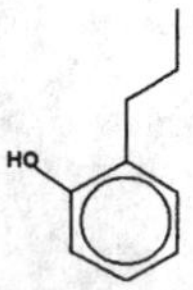

9462

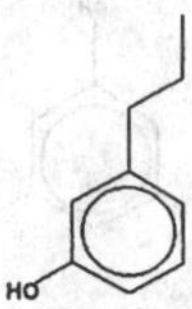

9463

9465

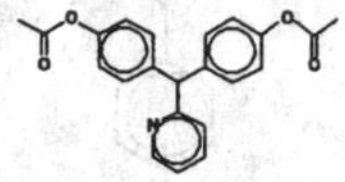

9466

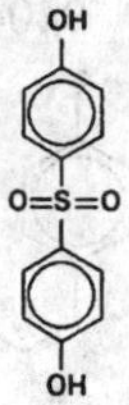

9469

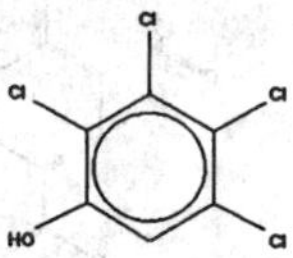

9471

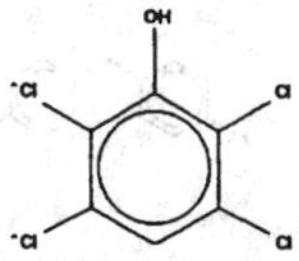

9472

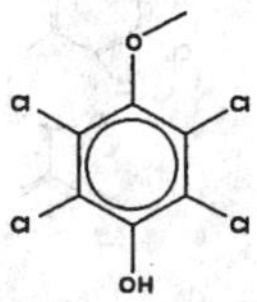

9473

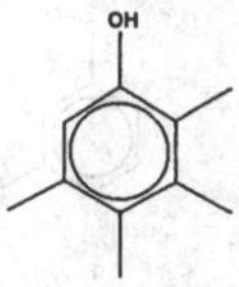

9474

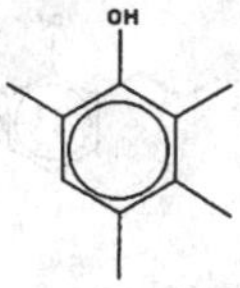

9475

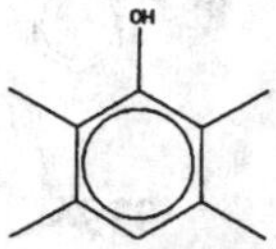

9479

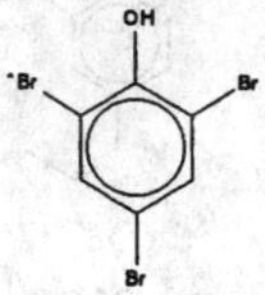

9482

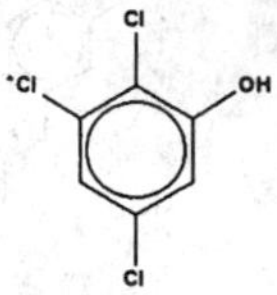

9483

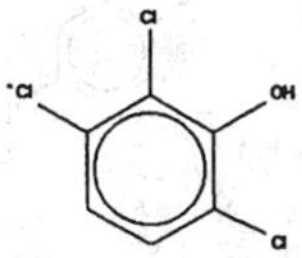

9484

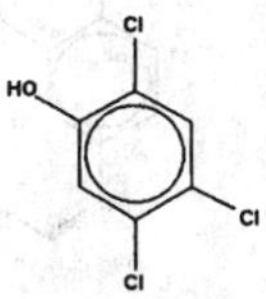

STRUCTURAL FORMULAS OF ORGANIC COMPOUNDS (continued)

In numeric order as they occur in the Table of Physical Constants of Organic Compounds

9485

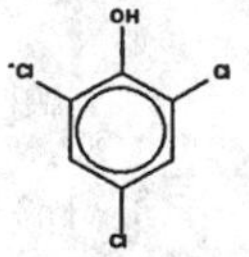

9486

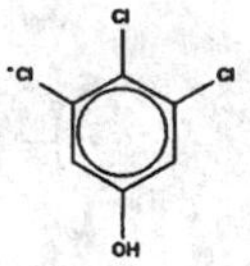

9489

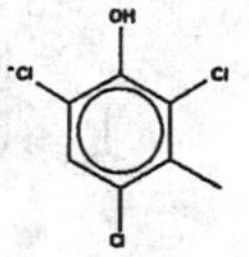

9492

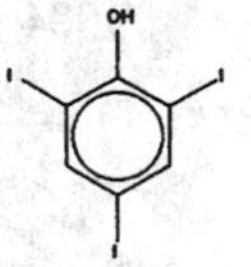

9493

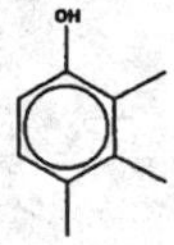

9494

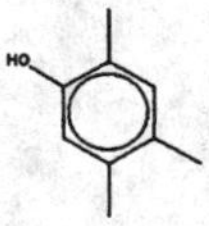

9495

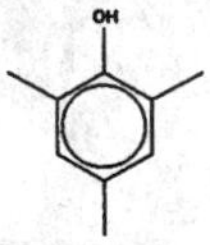

9496

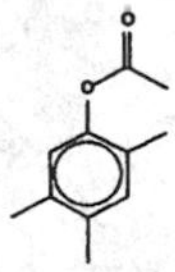

9497

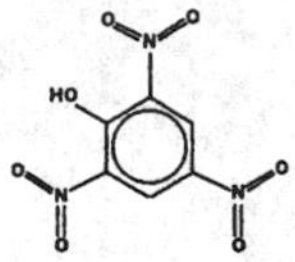

9499

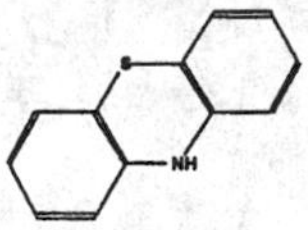

9501

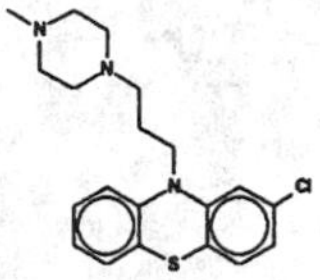

9503

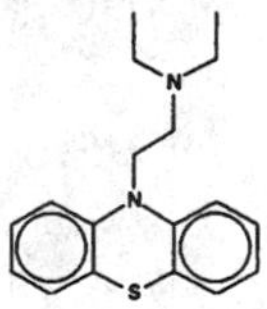

9504

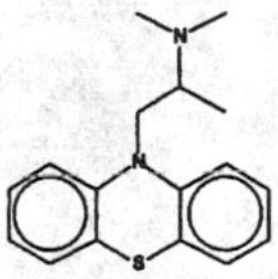

9505

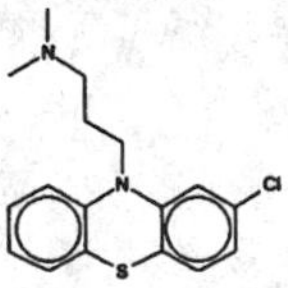

9507

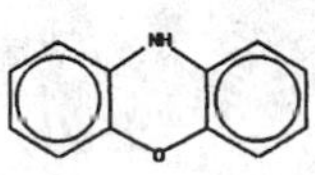

9510

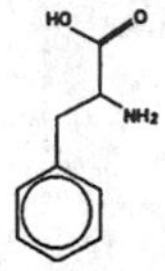

9512

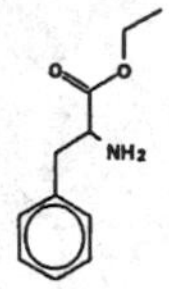

9513

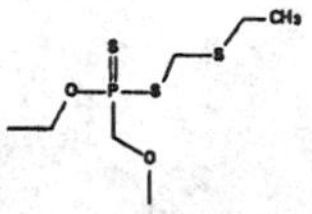

9514

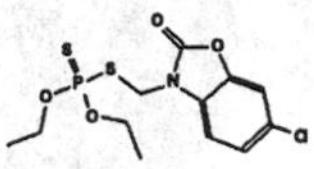

9515

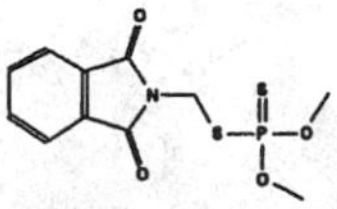

9517

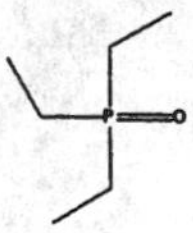

9518

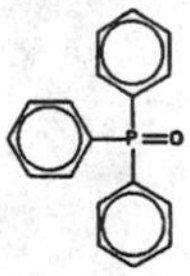

9519

9520

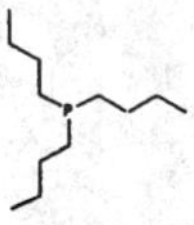

9521

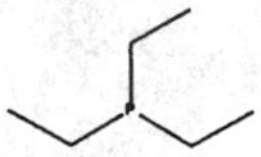

9522

9523

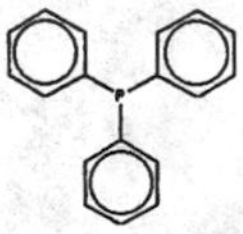

9524

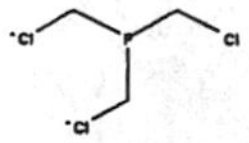

9525

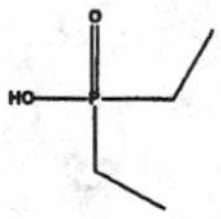

9527

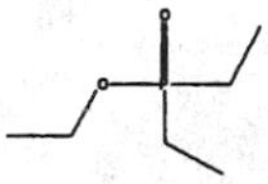

9528

9529

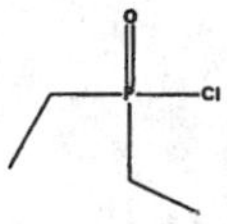

9530

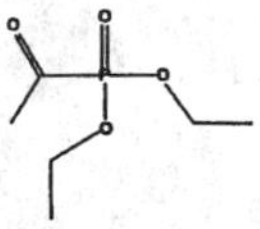

9532

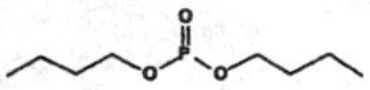

9535

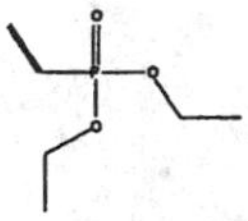

9536

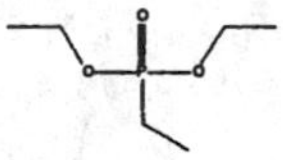

9537

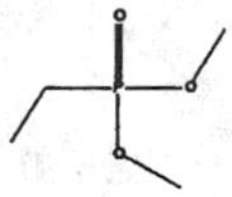

9539

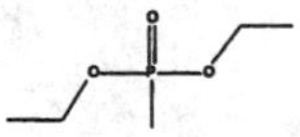

9540

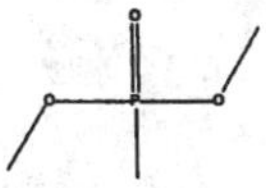

9541

9543

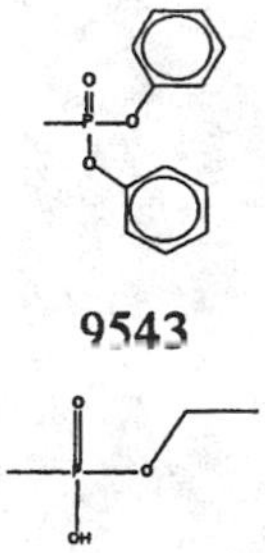

9545
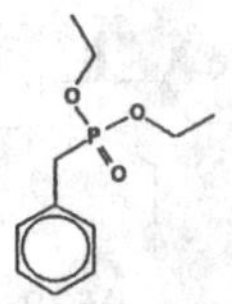

9546
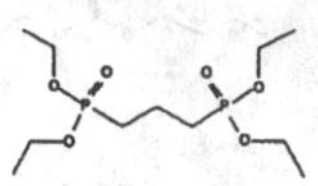

9548
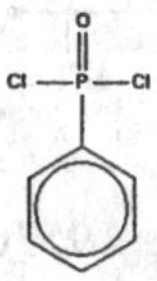

9549
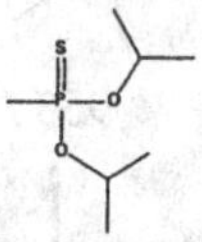

9550
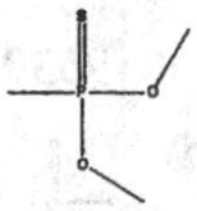

9551
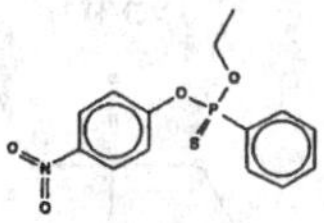

9553
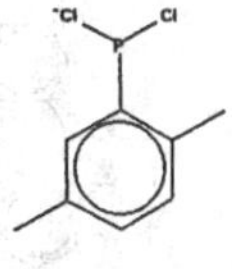

9554
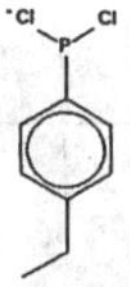

9555
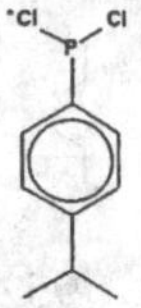

9556

9557
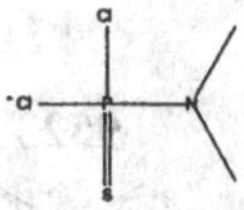

9558
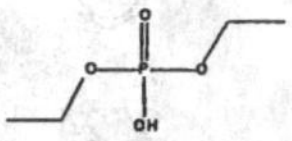

9559
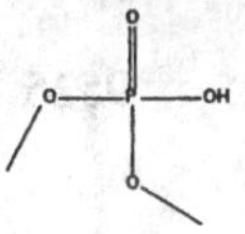

9560
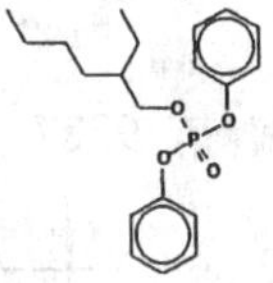

9562
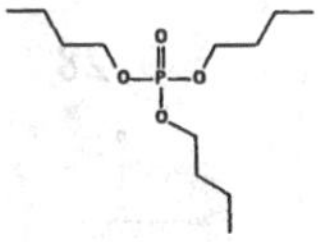

9563
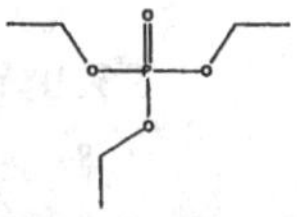

9564
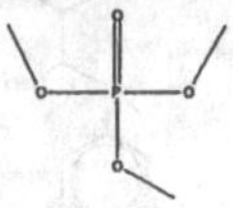

9565
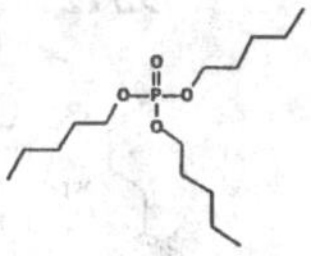

9566
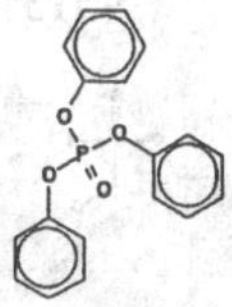

9567
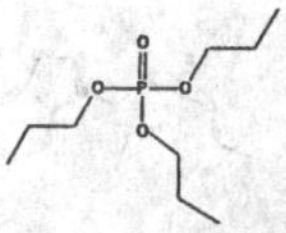

9568
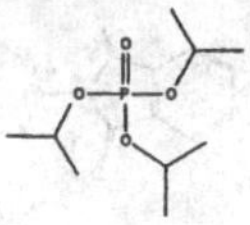

9569
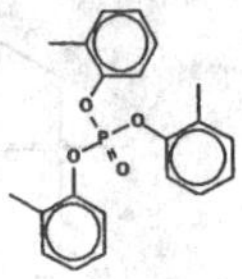

9570
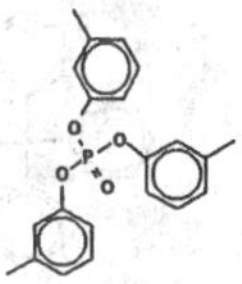

9571
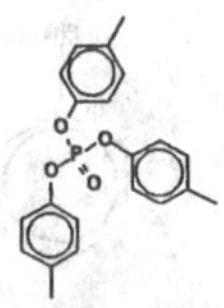

9572
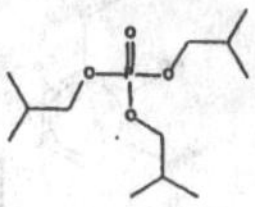

9573
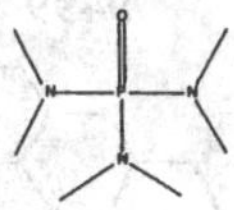

9574
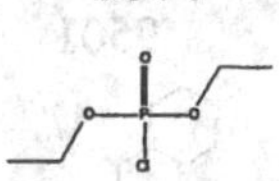

9575
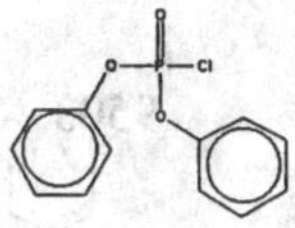

9576
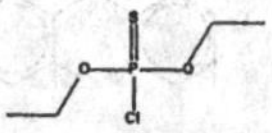

9577
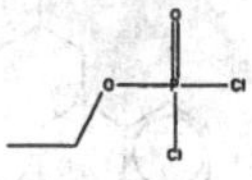

9578
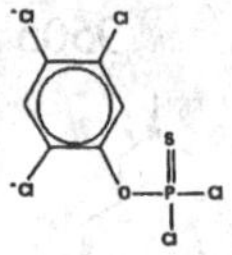

9579
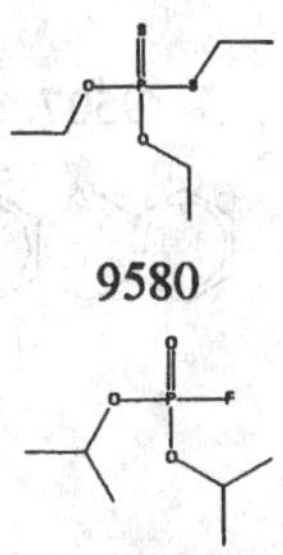

9580

9581
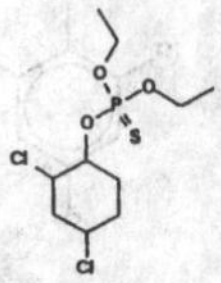

9582
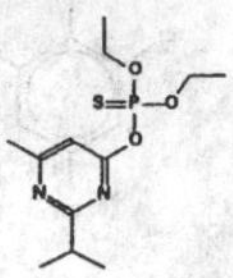

9583
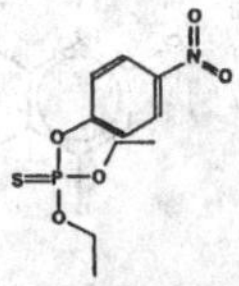

9584
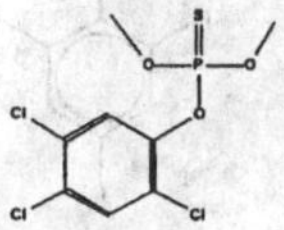

9585
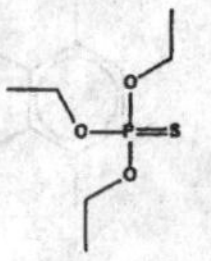

9586
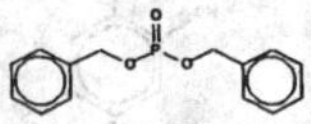

9587
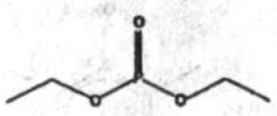

9588
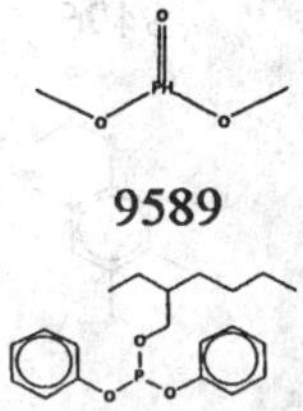

9589

STRUCTURAL FORMULAS OF ORGANIC COMPOUNDS (continued)

In numeric order as they occur in the Table of Physical Constants of Organic Compounds

9590

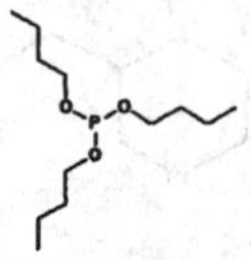

9591

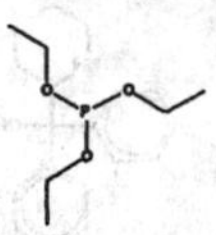

9592

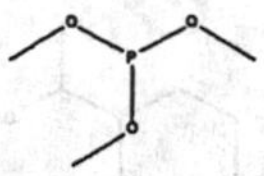

9593

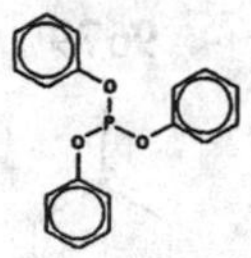

9594

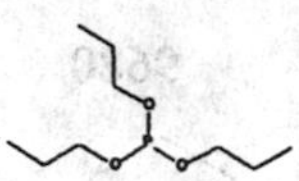

9595

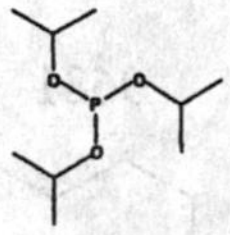

9596

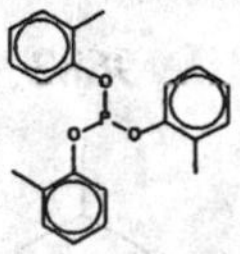

9597

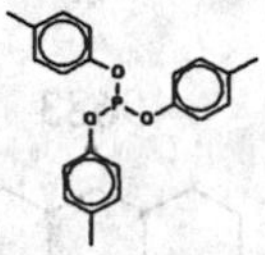

9599

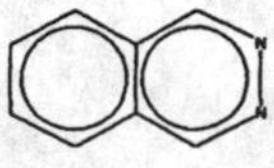

9600

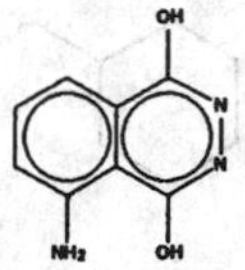

9601

9602

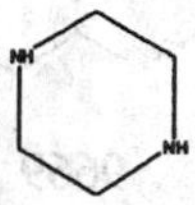

9603

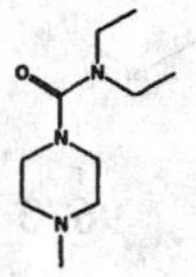

9604

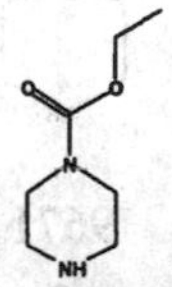

9605

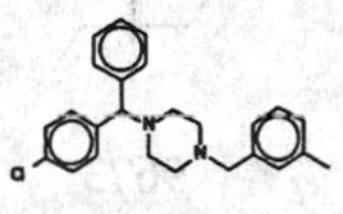

9606

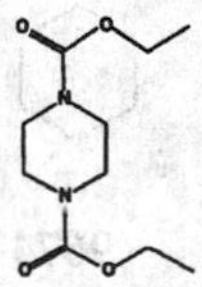

9608

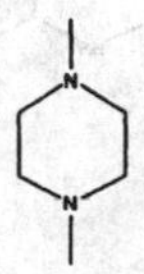

9609

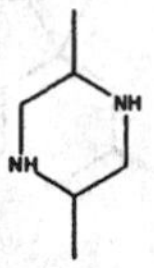

9610

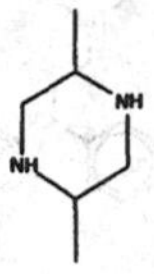

9611

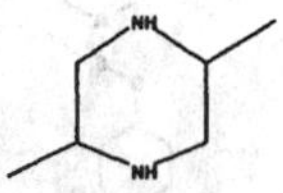

9612

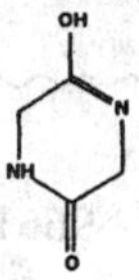

9613

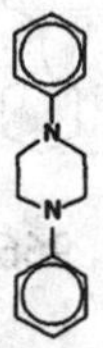

9614

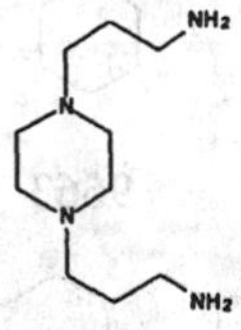

9615

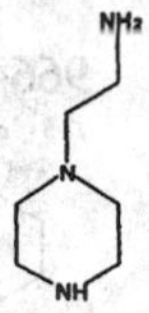

9616

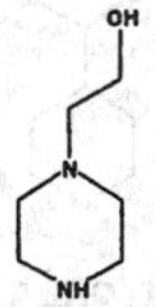

9617

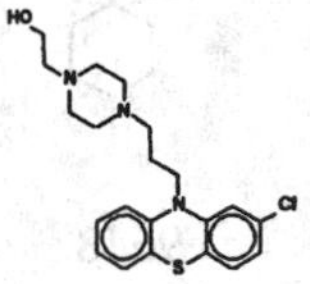

9618

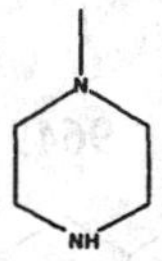

9619

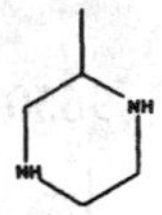

9620

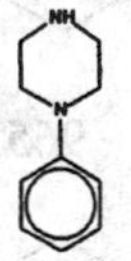

9621

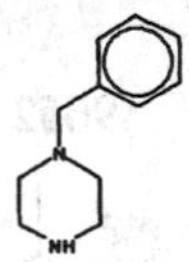

9622

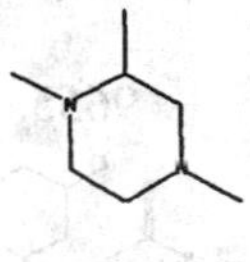

9624

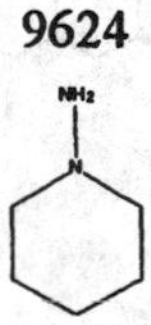

9625

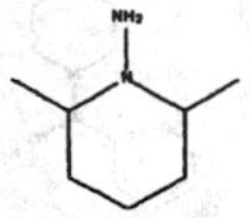

9626

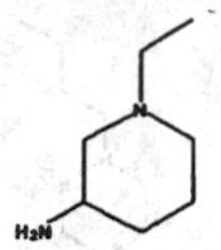

9627

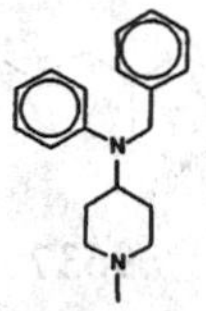

9628

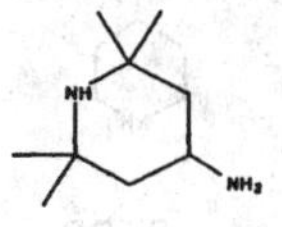

9629

9630

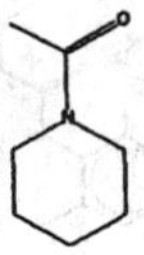

9631

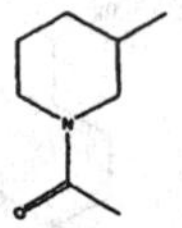

9632

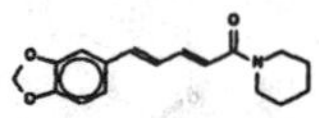

9634

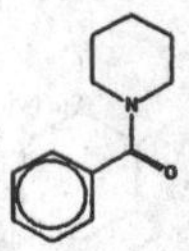

9635

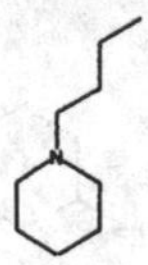

9636

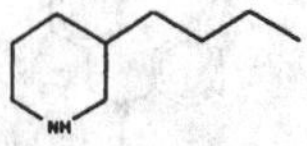

9637

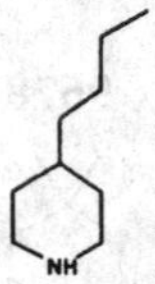

9638

9639

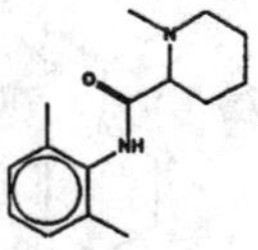

9642

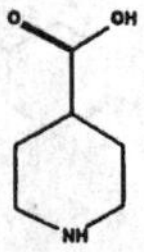

9643

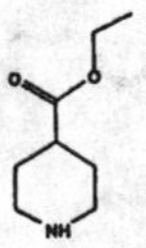

9644

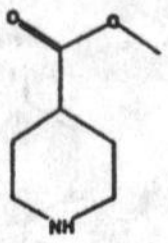

9645

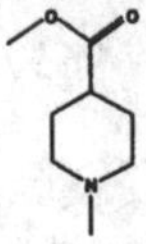

9648

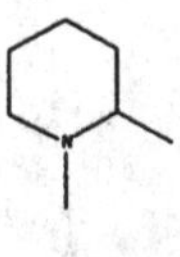

9649

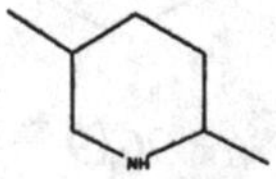

9650

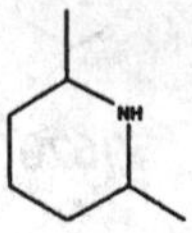

9651

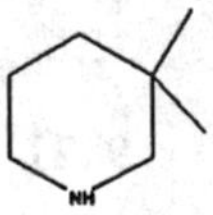

9652

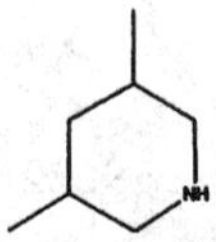

9655

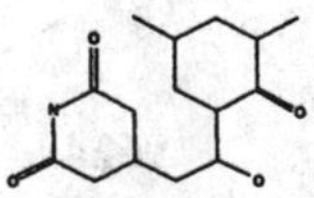

9656

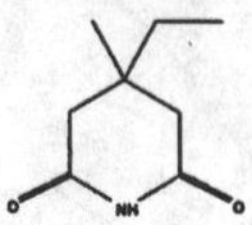

9658

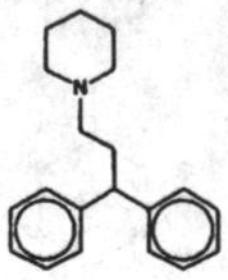

9659

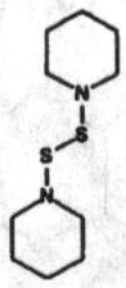

9660

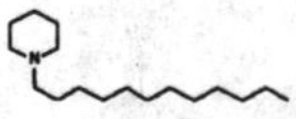

9661

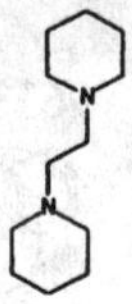

9662

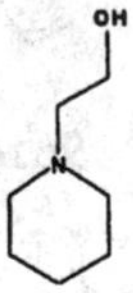

9663

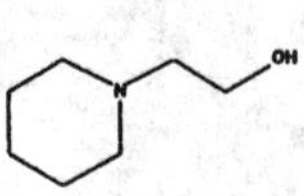

9664

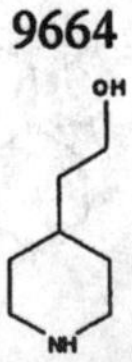

9666

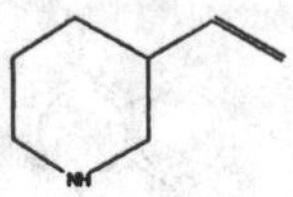

9667

9668

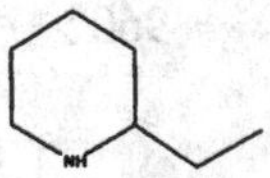

9669

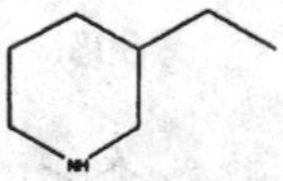

9670

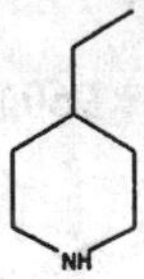

9671

9673

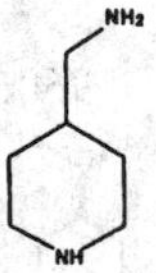

9674

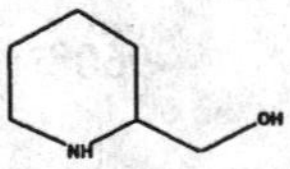

9675

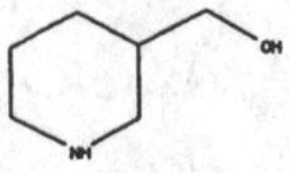

9676

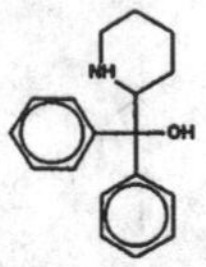

9678

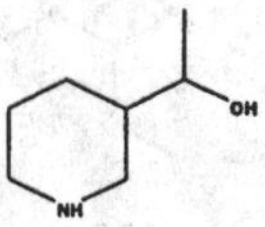

9679

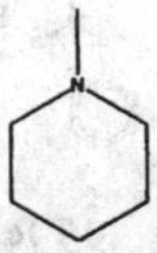

9680

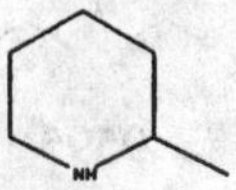

9681

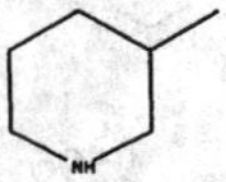

9682

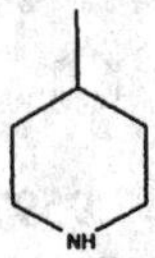

9683

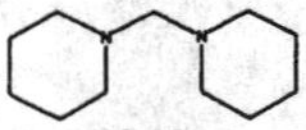

9684

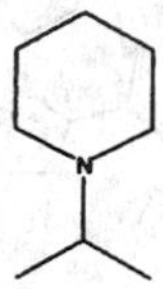

9685

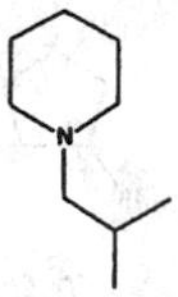

9686

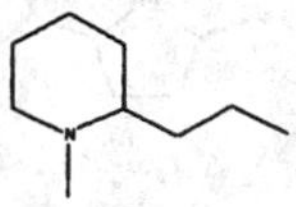

9687

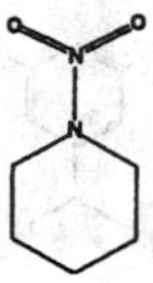

9688

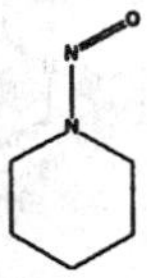

9689

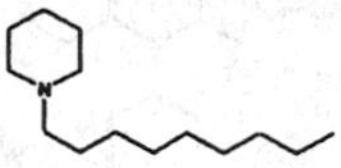

9690

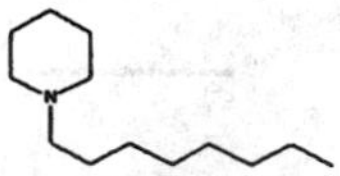

9691

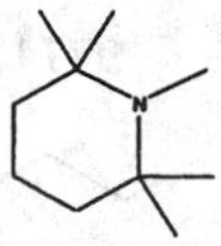

9692

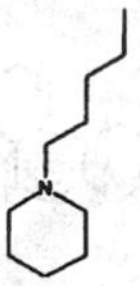

9693

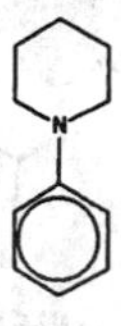

9695

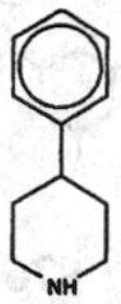

9696

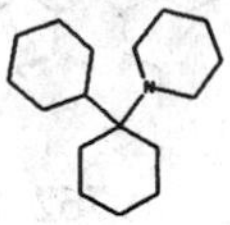

9697

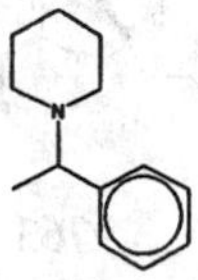

9698

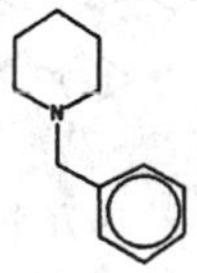

9699

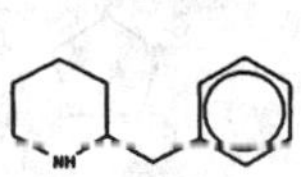

9700

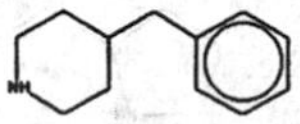

9701

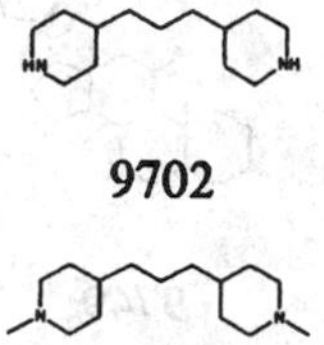

9702

9703

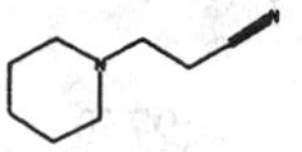

9704

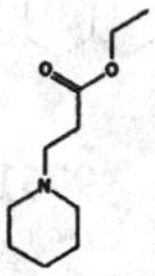

9705

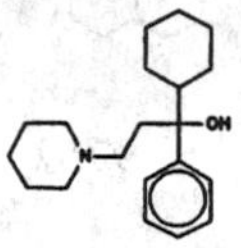

9706

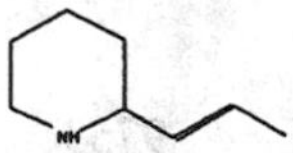

9707

9708

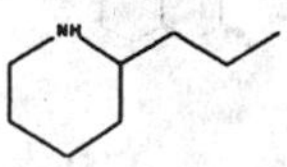

9709

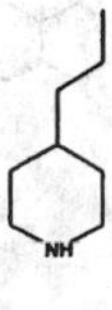

9710

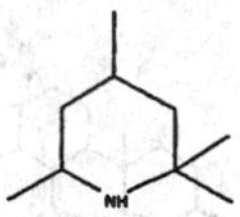

9711

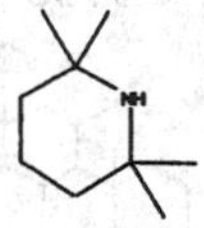

9712

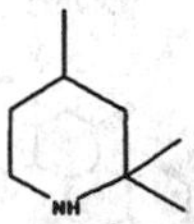

9713

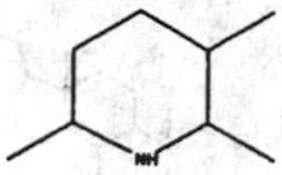

9714

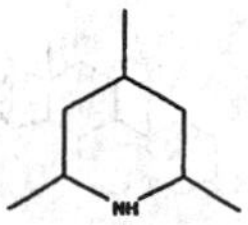

9715

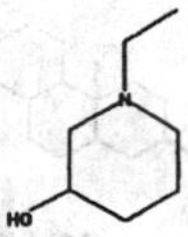

9716

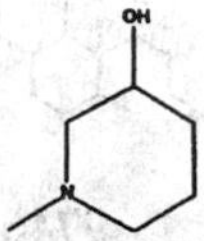

9717

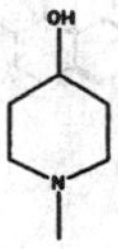

9718

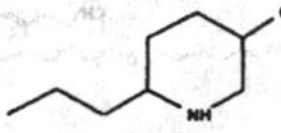

9719

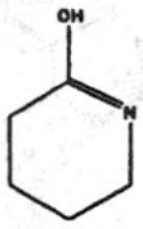

9720

9721

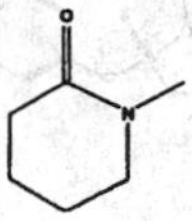

9722

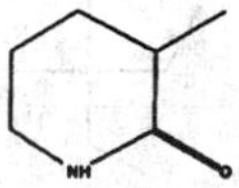

9723

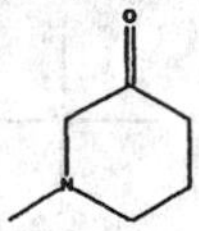

9724

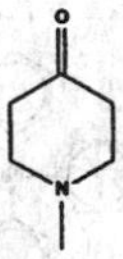

9726

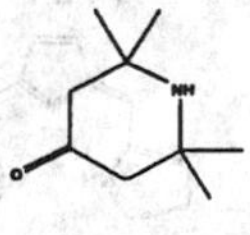

In numeric order as they occur in the Table of Physical Constants of Organic Compounds

9727

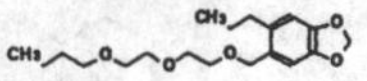

9728

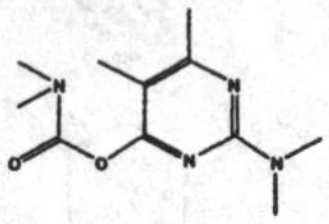

9729

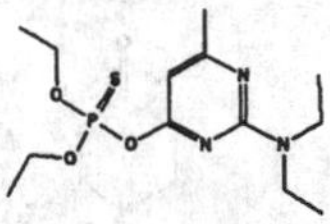

9730

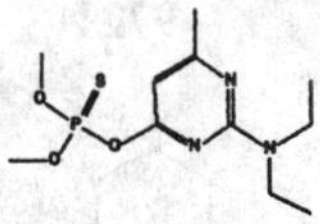

9731

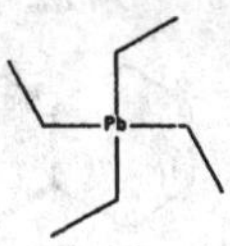

9732

9733

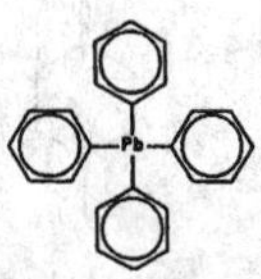

9734

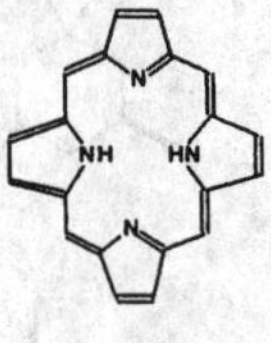

9736

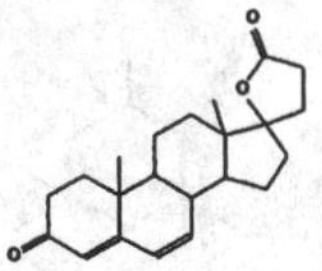

9738

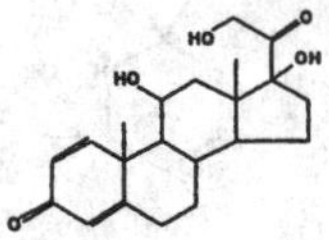

9739

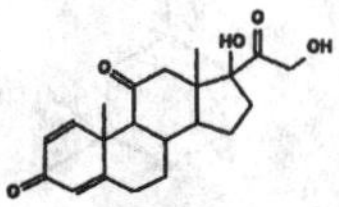

9740

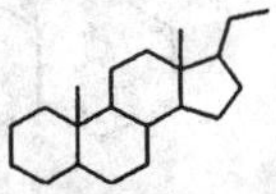

9741

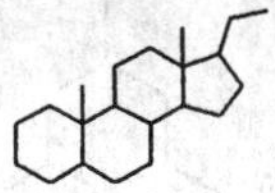

9743

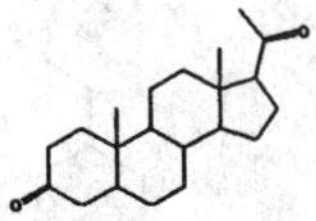

9745

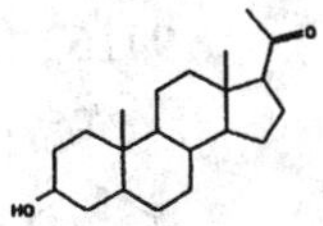

9746

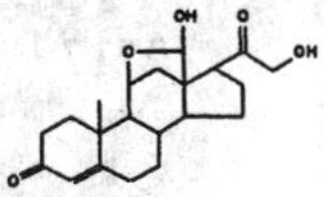

9748

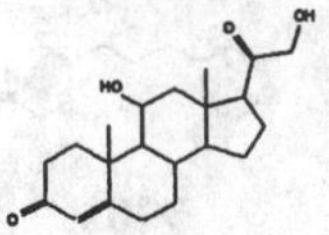

9749

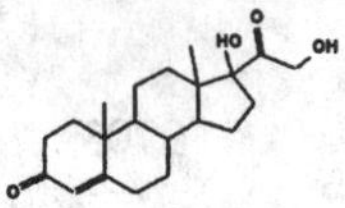

9750

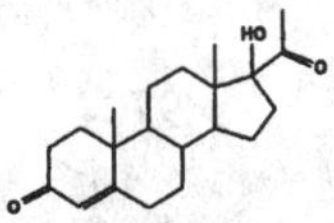

9751

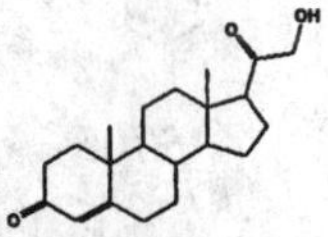

9752

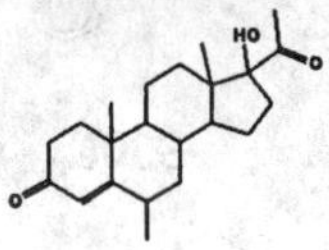

9753

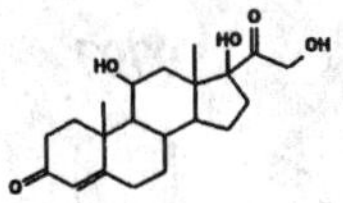

9754

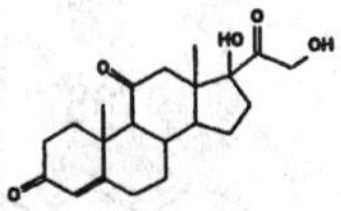

9756

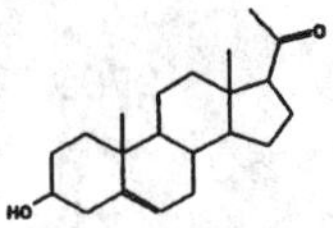

9758

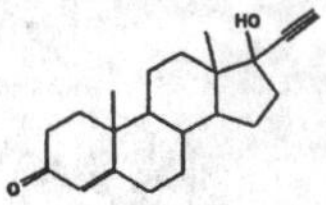

9759

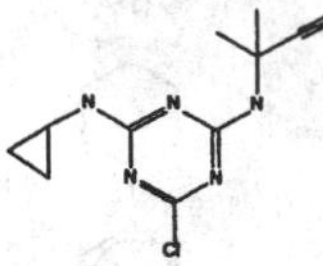

9760

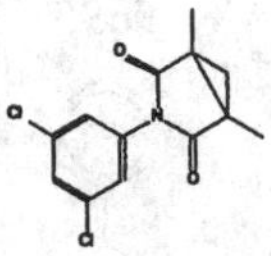

9761

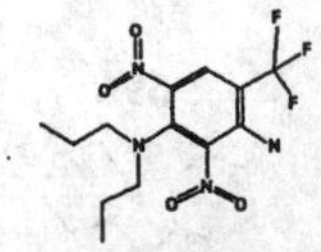

9762

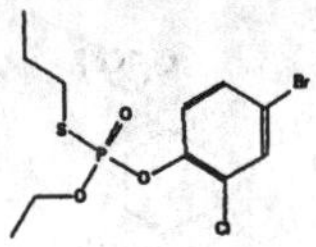

9763

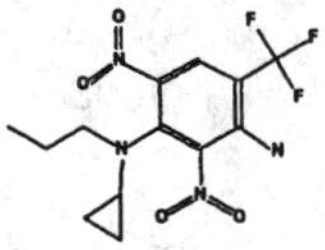

9764

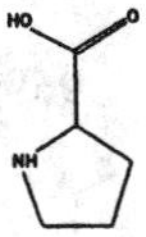

9765

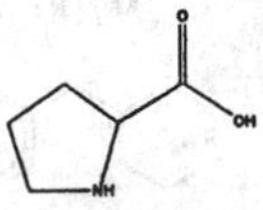

9766

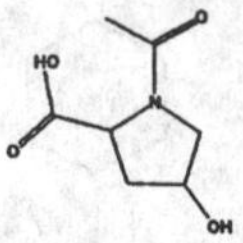

9767

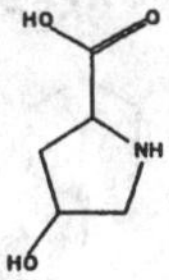

9768

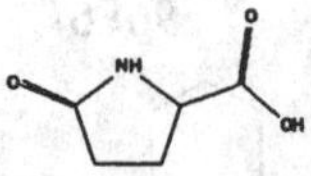

9769

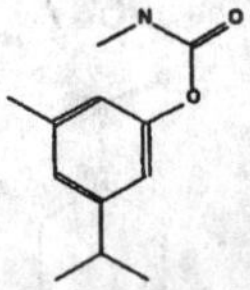

9770

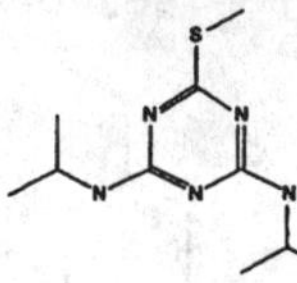

9771

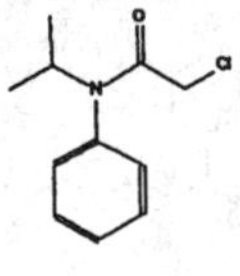

9772

9773

O=C=C=C=O

9774

STRUCTURAL FORMULAS OF ORGANIC COMPOUNDS (continued)

In numeric order as they occur in the Table of Physical Constants of Organic Compounds

9775

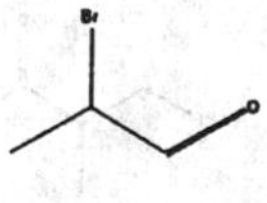

9776

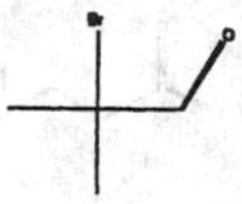

9777

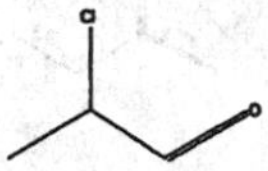

9778

9779

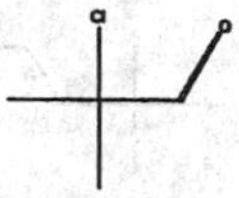

9780

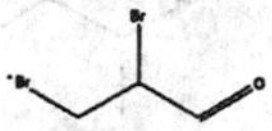

9781

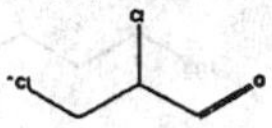

9782

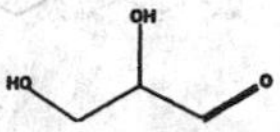

9783

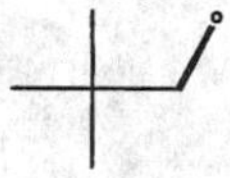

9784

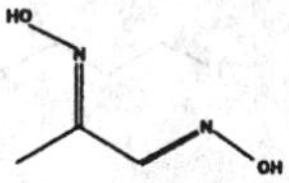

9785

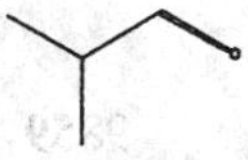

9786

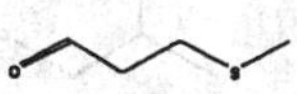

9787

9788

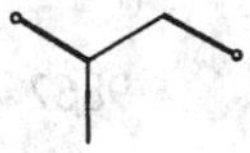

9790

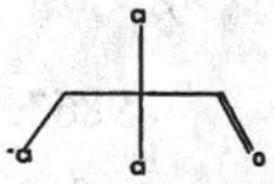

9791

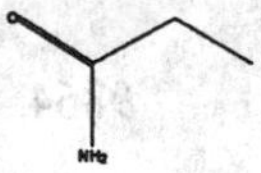

9792

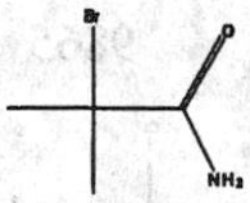

9793

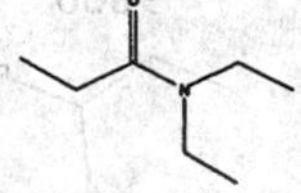

9798

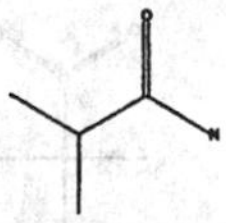

9799

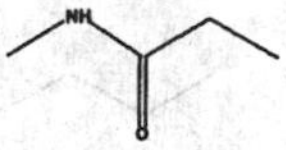

9800

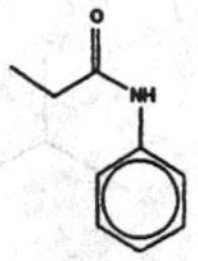

9801

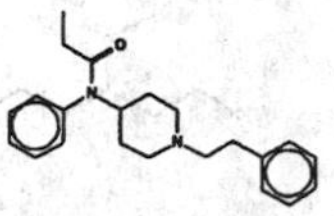

9803

9804

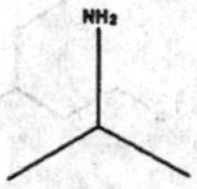

9805

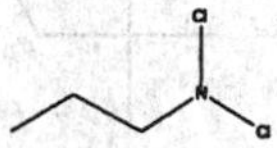

9806

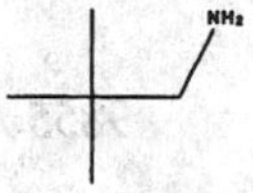

9807

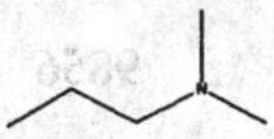

9808

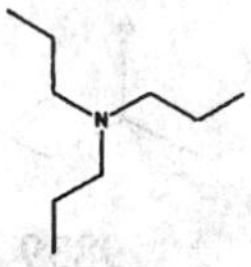

9809

9810

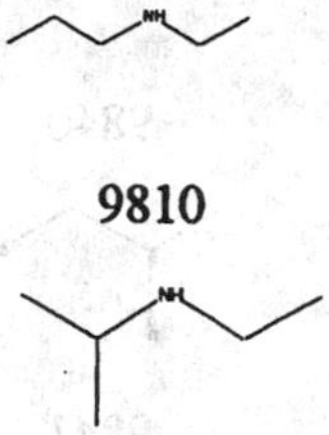

9811

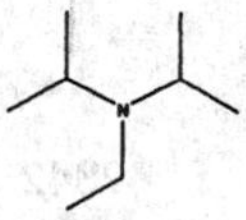

9812

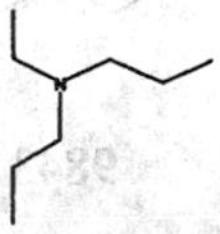

9813

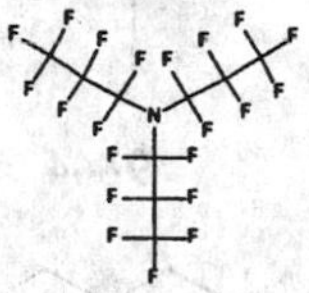

9814

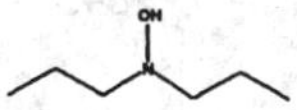

9815

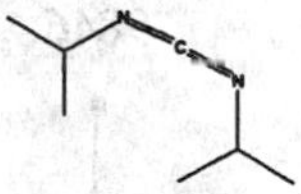

9816

9817

9818

9819

9820

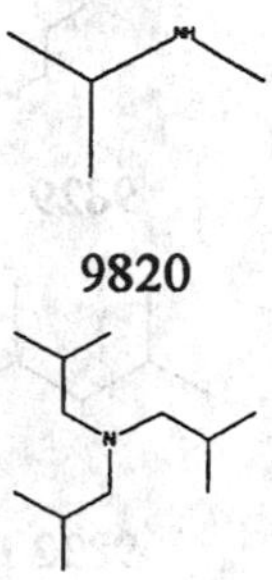

9821

9822

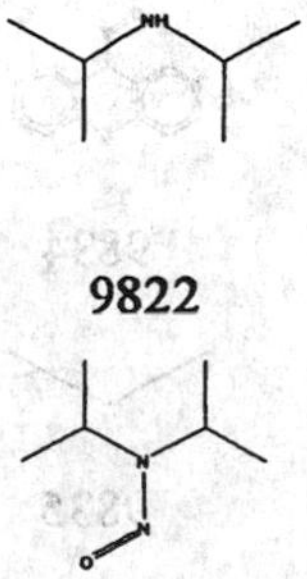

9823

9824

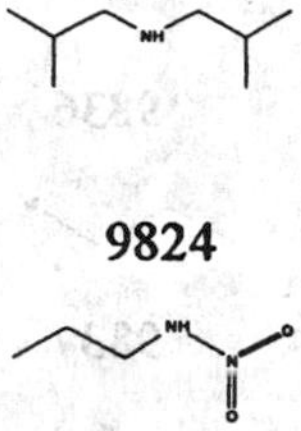

9825

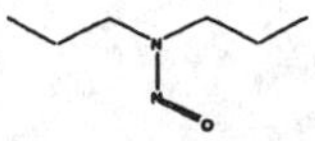

In numeric order as they occur in the Table of Physical Constants of Organic Compounds

9826

9827
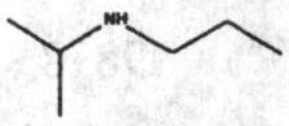
9828
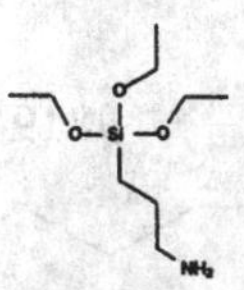
9829
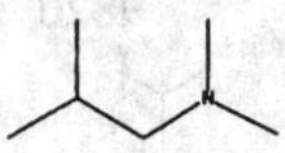
9832
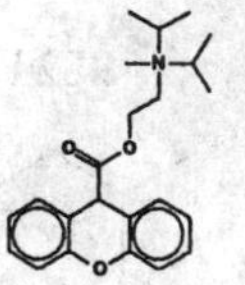
9834

9835
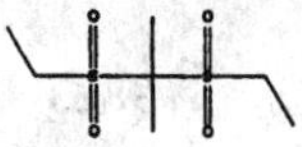
9836
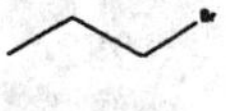
9837
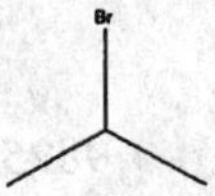
9838
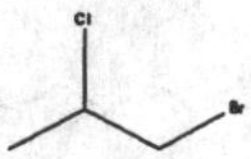
9839

9840
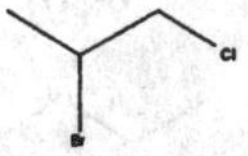
9841
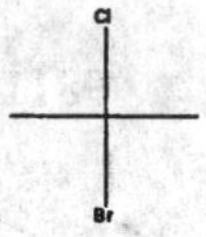
9842
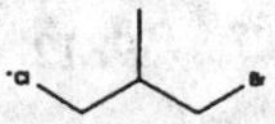
9843
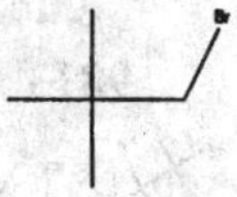
9844

9845
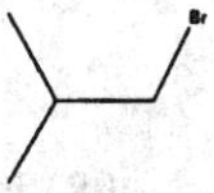
9846

9847
9848
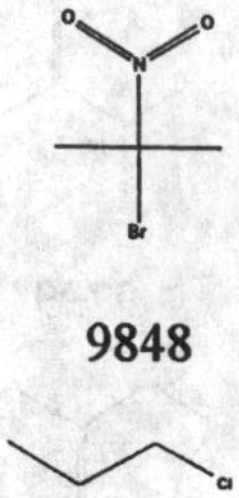
9849
9850
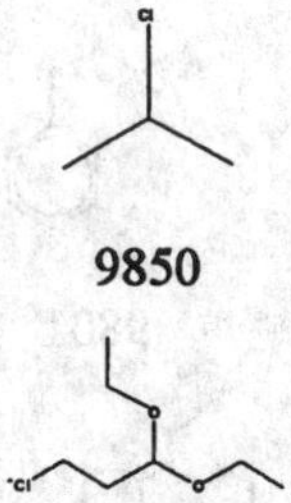
9851
9852
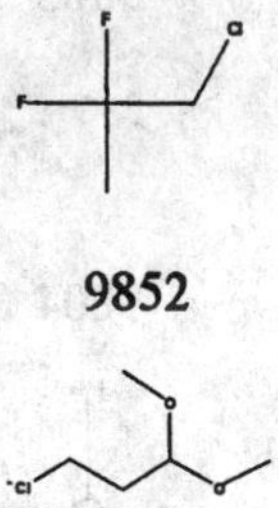
9853
9854
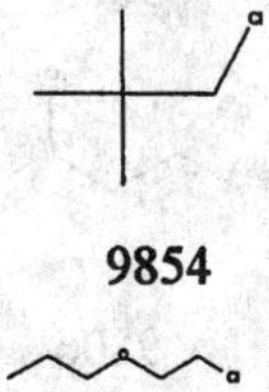
9855
9856
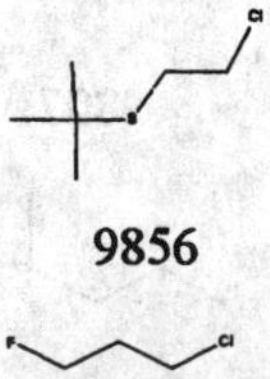
9857
9858
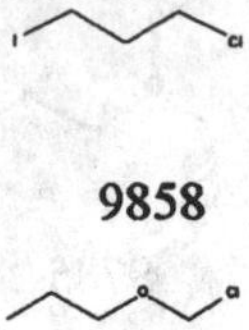
9859
9860
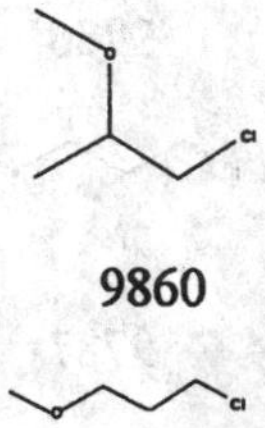
9861
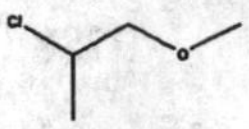
9862
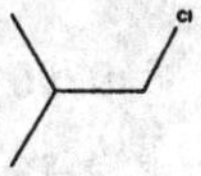
9863
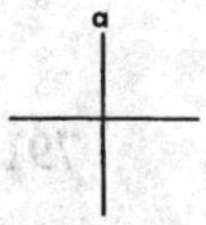
9864

9865
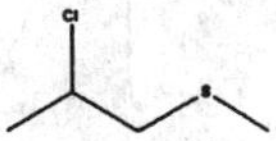
9866
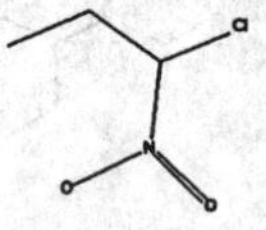
9867
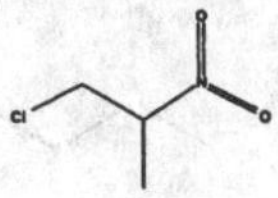
9868
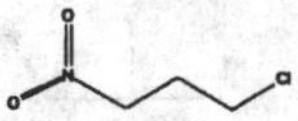
9869
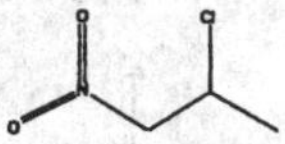
9870
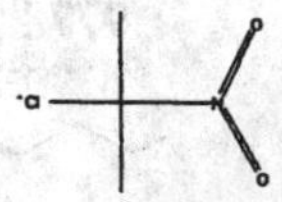
9871
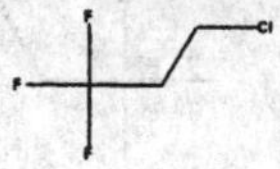
9872
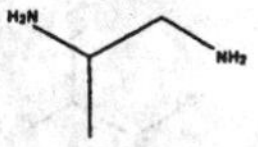
9873

9874

9875
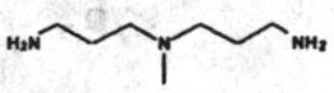

In numeric order as they occur in the Table of Physical Constants of Organic Compounds

9876

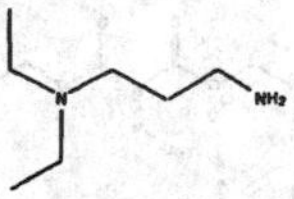

9877

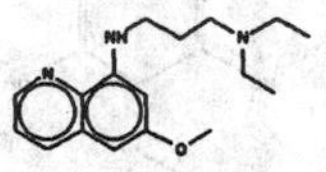

9878

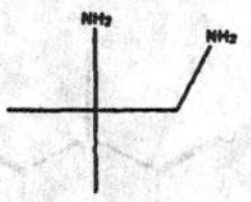

9879

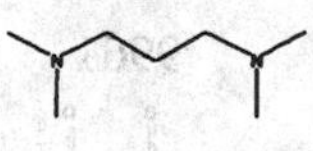

9880

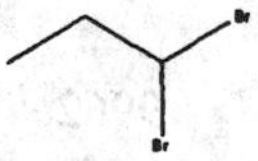

9881

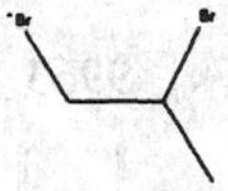

9882

9883

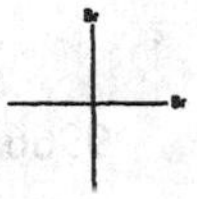

9884

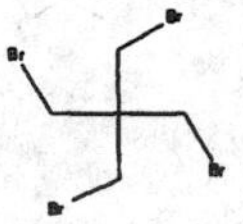

9885

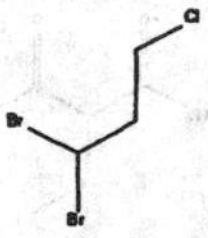

9886

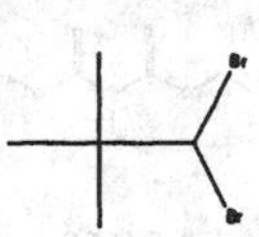

9887

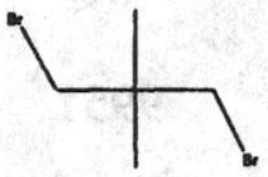

9888

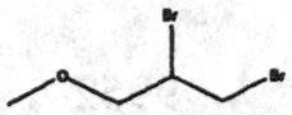

9889

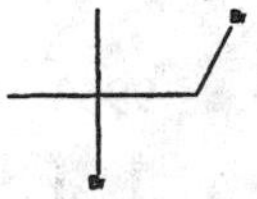

9890

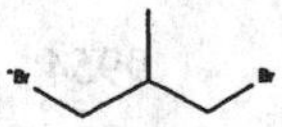

9891

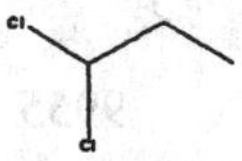

9892

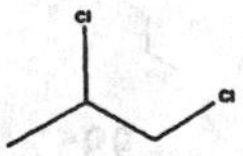

9893

9894

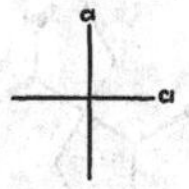

9895

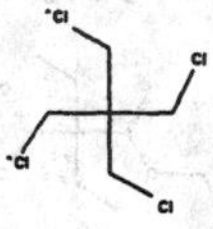

9896

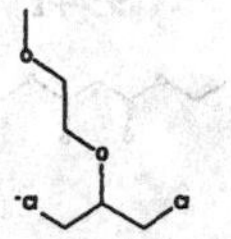

9897

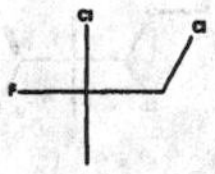

9898

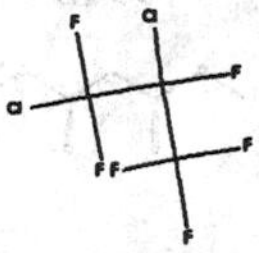

9899

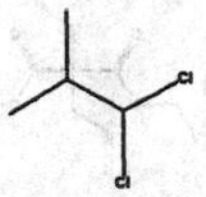

9900

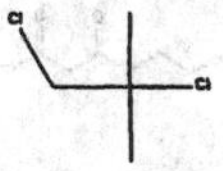

9901

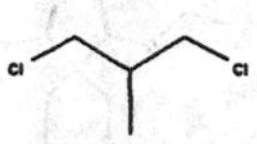

9902

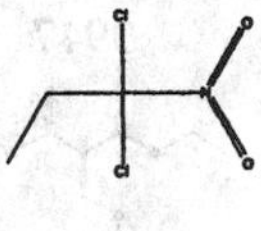

9903

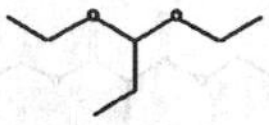

9904

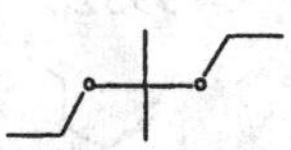

9905

9906

9907

9908

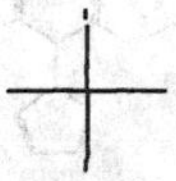

9909

9910

9911

9912

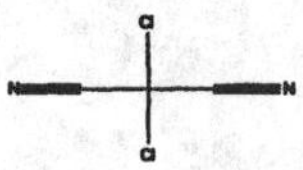

9913

9914

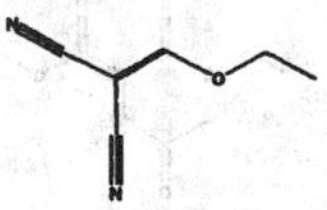

9915

9916

9917

9918

9919

9920

STRUCTURAL FORMULAS OF ORGANIC COMPOUNDS (continued)

In numeric order as they occur in the Table of Physical Constants of Organic Compounds

9921

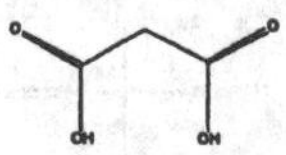

9922

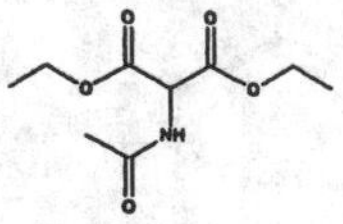

9923

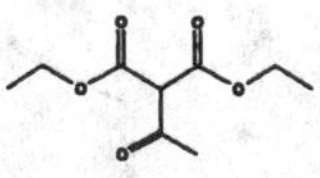

9924

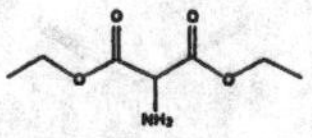

9925

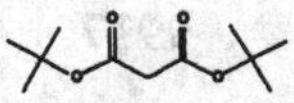

9926

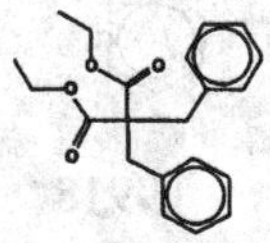

9927

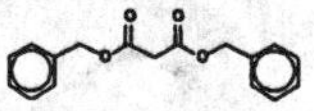

9928

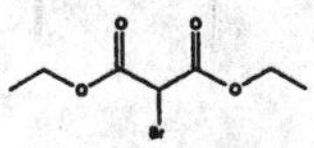

9930

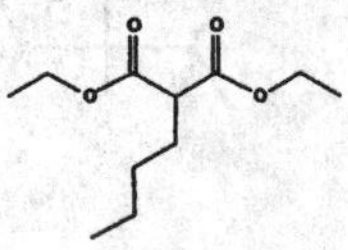

9931

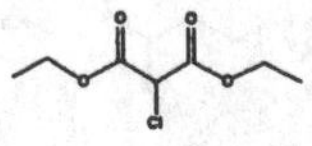

9932

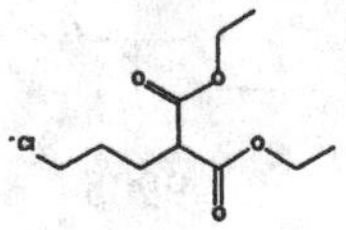

9933

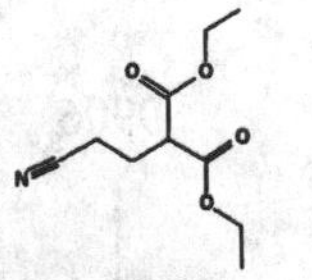

9934

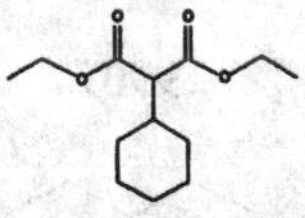

9935

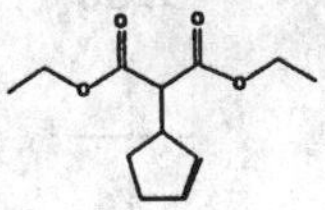

9936

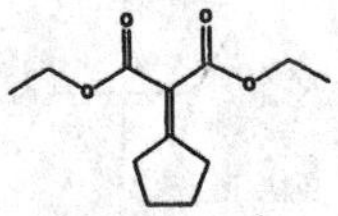

9937

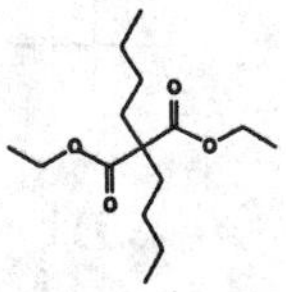

9938

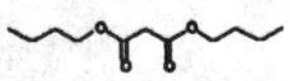

9939

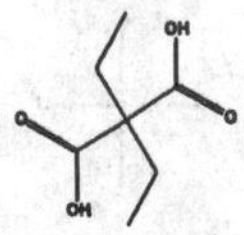

9940

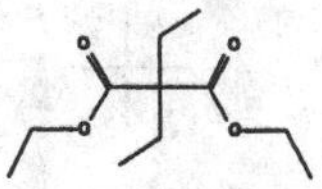

9941

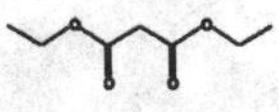

9942

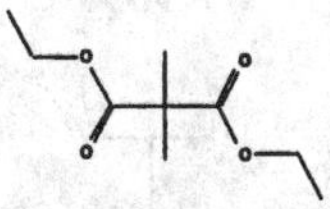

9943

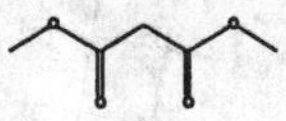

9944

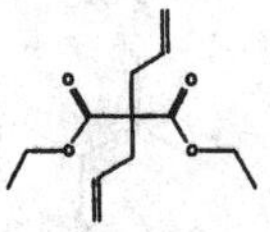

9945

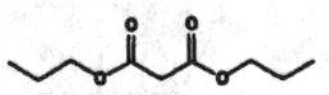

9946

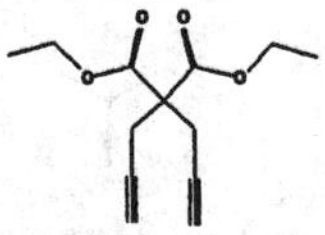

9947

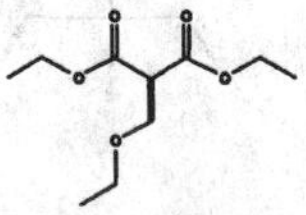

9948

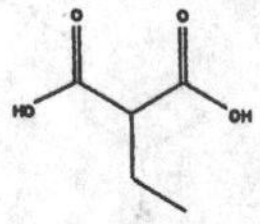

9949

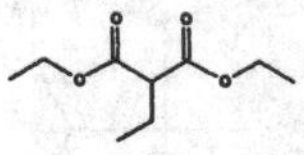

9950

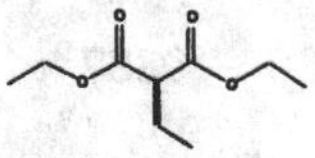

9951

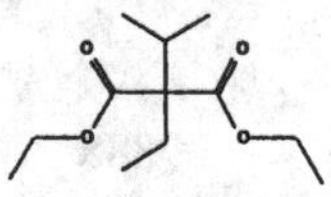

9952

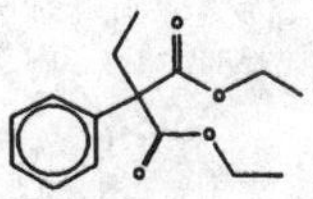

9953

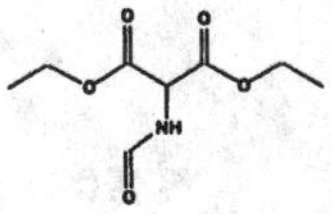

9954

9955

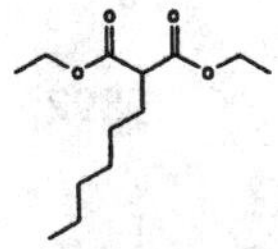

9957

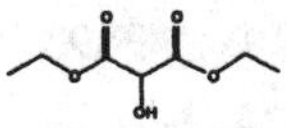

9958

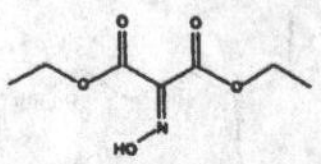

9959

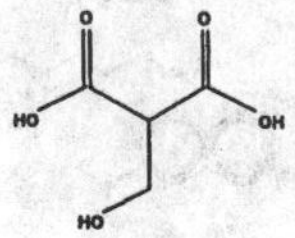

9960

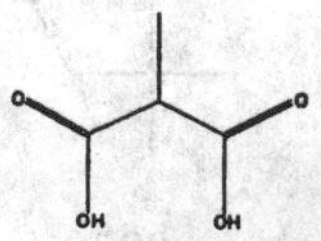

9961

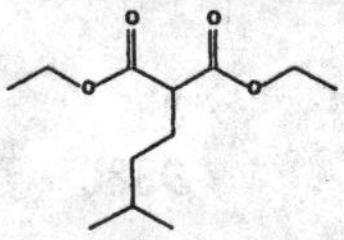

9962

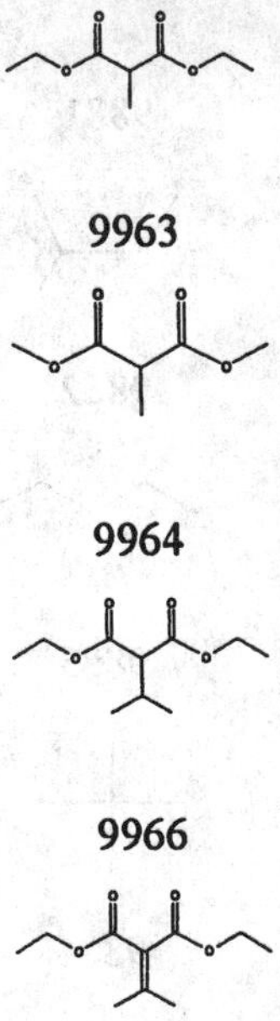

9963

9964

9966

STRUCTURAL FORMULAS OF ORGANIC COMPOUNDS (continued)

In numeric order as they occur in the Table of Physical Constants of Organic Compounds

9967

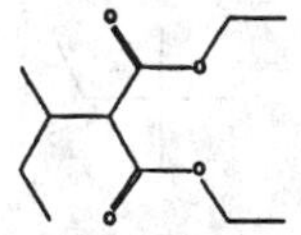

9969

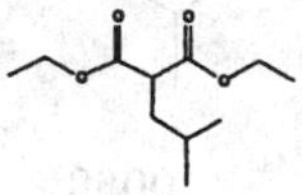

9970

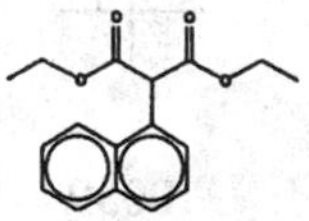

9972

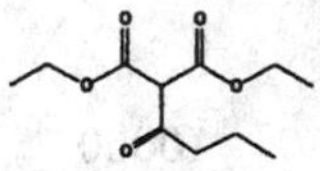

9973

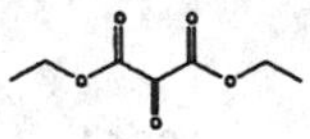

9974

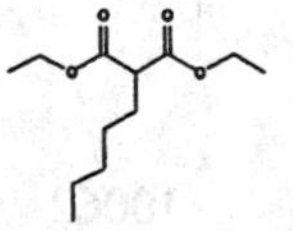

9975

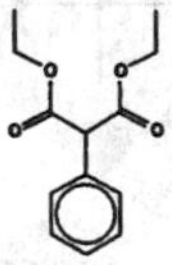

9976

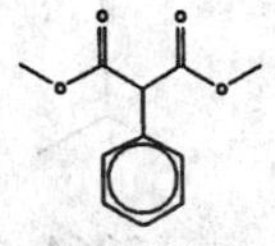

9977

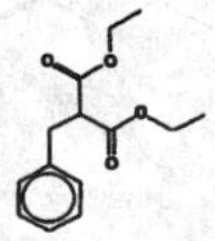

9978

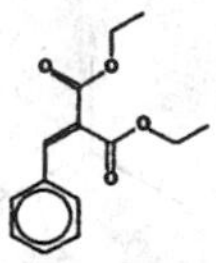

9980

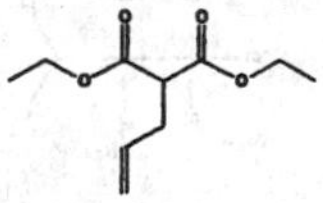

9981

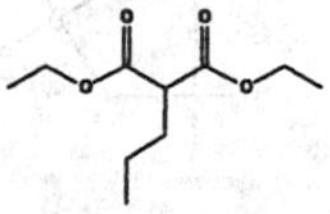

9982

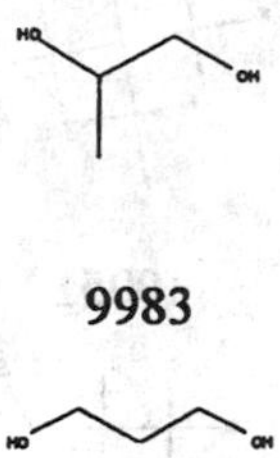

9983

9985

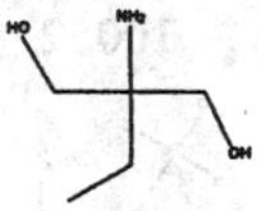

9986

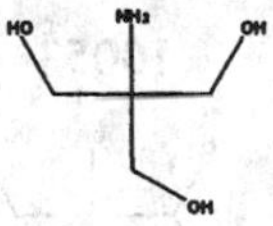

9987

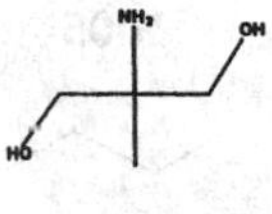

9989

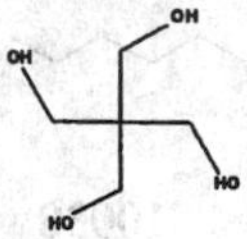

9990

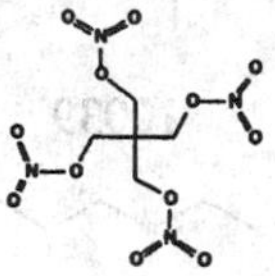

9991

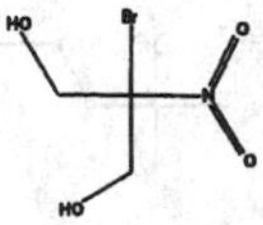

9992

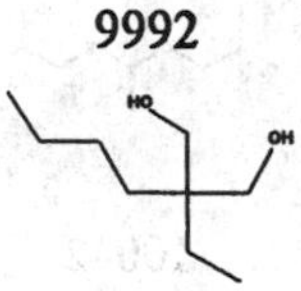

9994

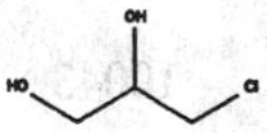

9995

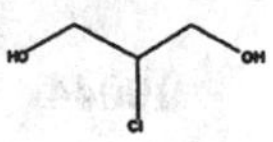

9997

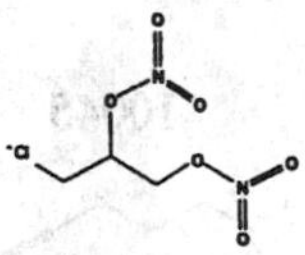

9998

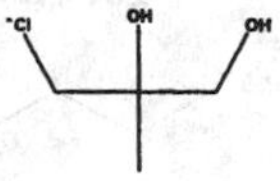

9999

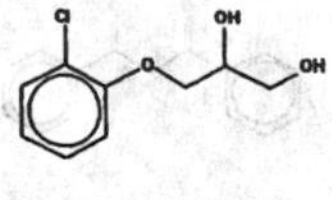

10000

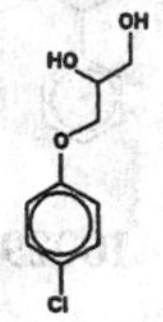

10001

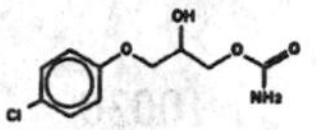

10002

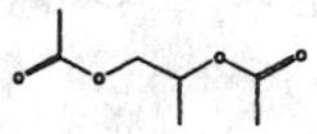

10003

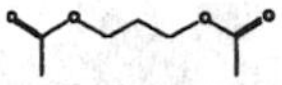

10004

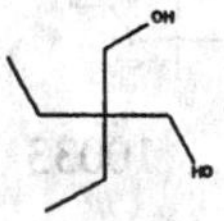

10005

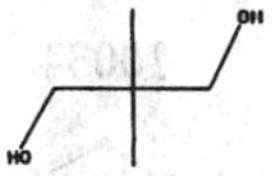

10006

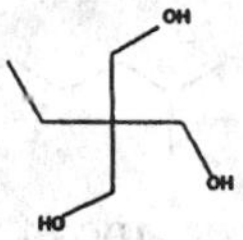

10007

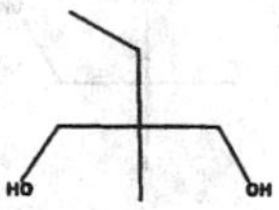

10008

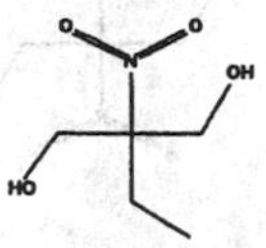

10009

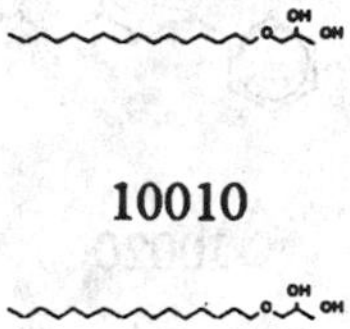

10010

10011

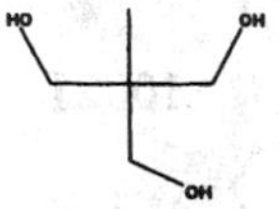

10012

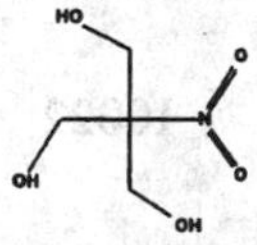

10013

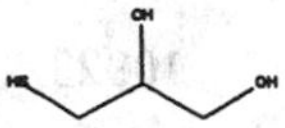

10014

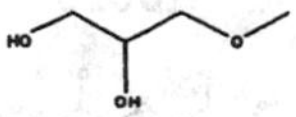

10016

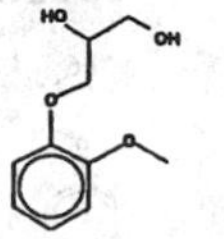

In numeric order as they occur in the Table of Physical Constants of Organic Compounds

10017

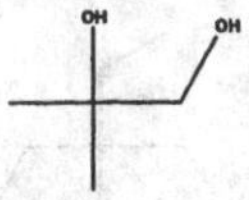

10018

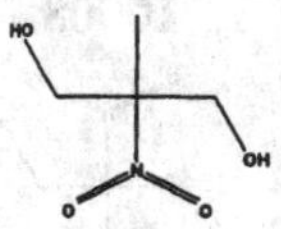

10019

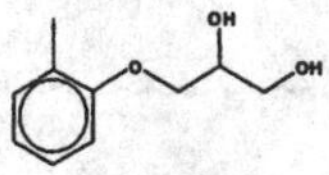

10020

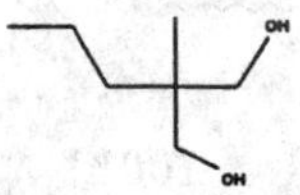

10021

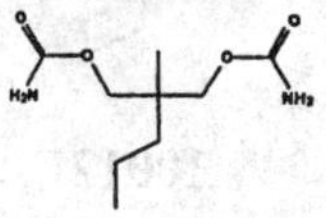

10022

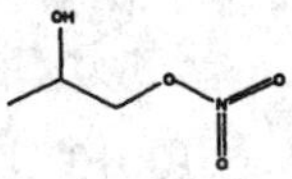

10023

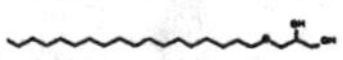

10025

10026

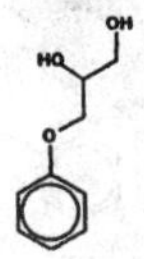

10027

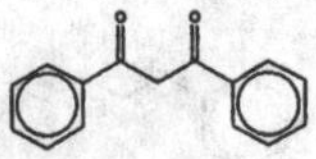

10028

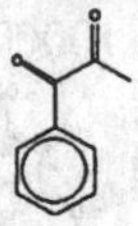

10029

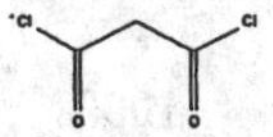

10030

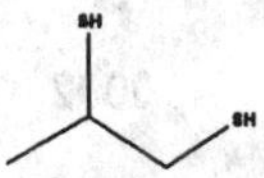

10031

10032

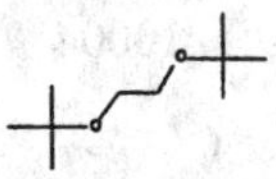

10033

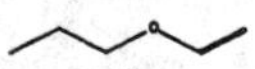

10034

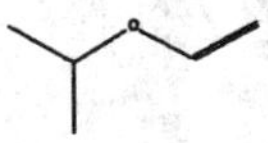

10035

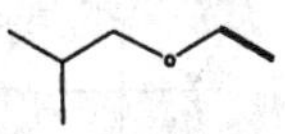

10036

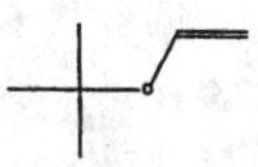

10037

10038

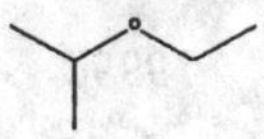

10039

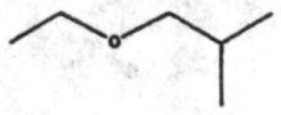

10040

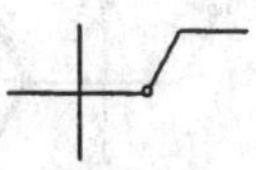

10041

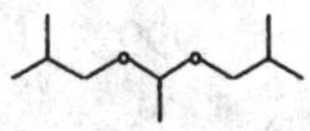

10042

10043

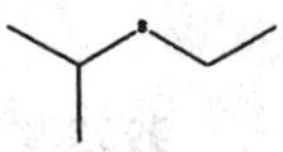

10044

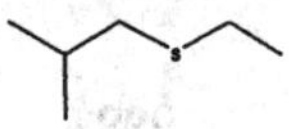

10045

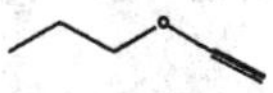

10046

10047

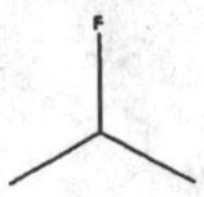

10048

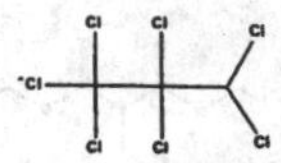

10049

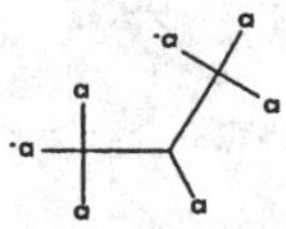

10050

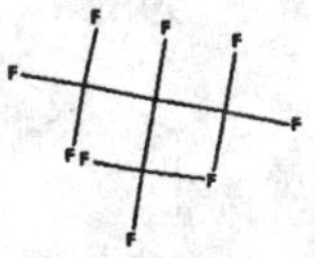

10051

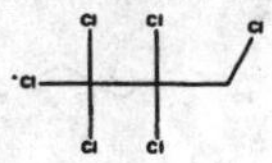

10052

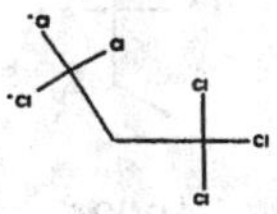

10053

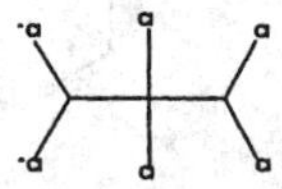

10054

10055

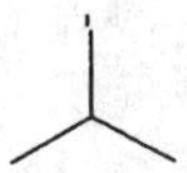

10056

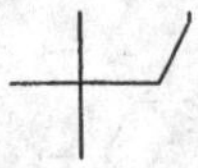

10057

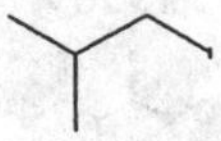

10058

10059

10060

10061

10062

10063

10064

STRUCTURAL FORMULAS OF ORGANIC COMPOUNDS (continued)

In numeric order as they occur in the Table of Physical Constants of Organic Compounds

10065

10066

10067

10068

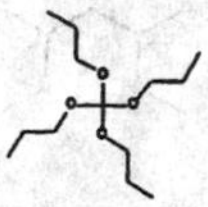

10069

10070

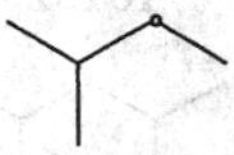

10071

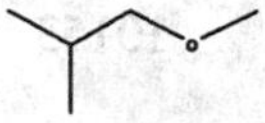

10072

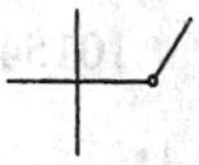

10073

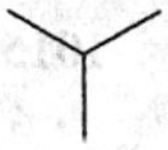

10074

10075

10076

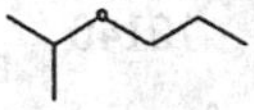

10077

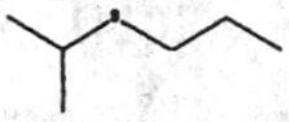

10078

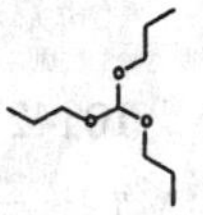

10079

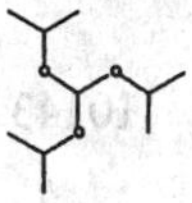

10080

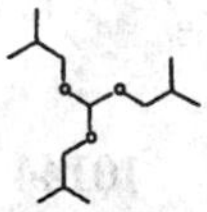

10081

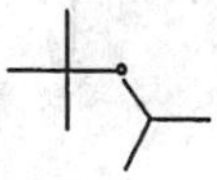

10082

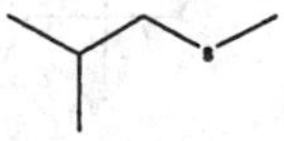

10083

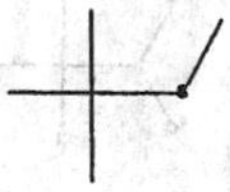

10084

10085

10086

10087

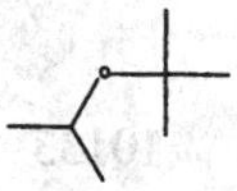

10088

10089

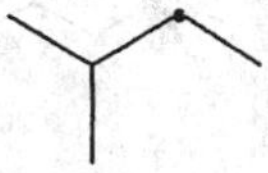

10090

10091

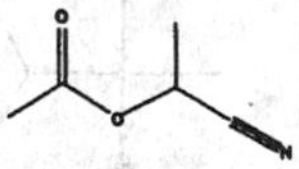

10092

10093

10094

10095

10096

10097

10098

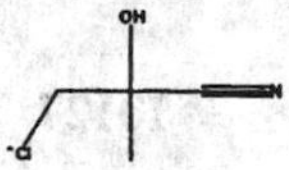

10099

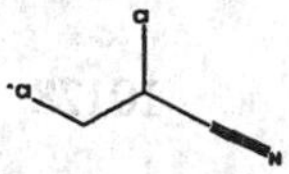

10100

10101

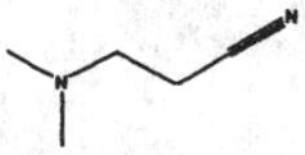

10102

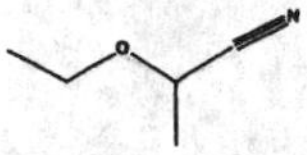

10103

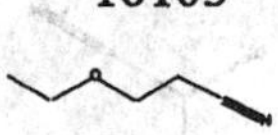

10105

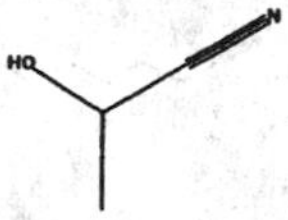

10106

10107

10108

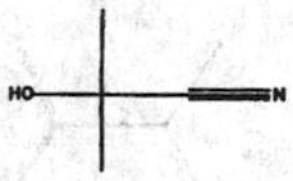

10109

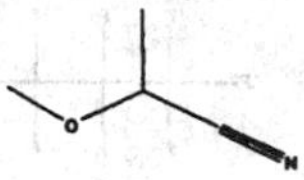

10110

10111

In numeric order as they occur in the Table of Physical Constants of Organic Compounds

10112

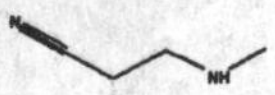

10113

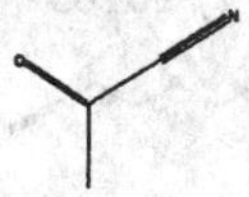

10114

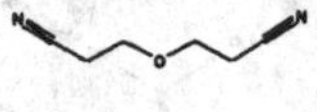

10115

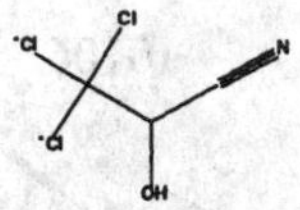

10116

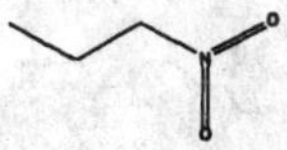

10117

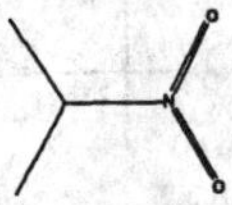

10118

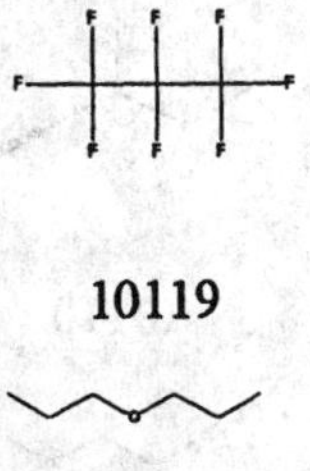

10119

10120

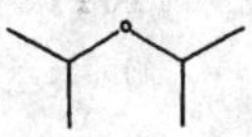

10121

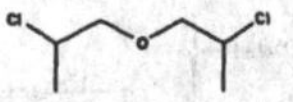

10122

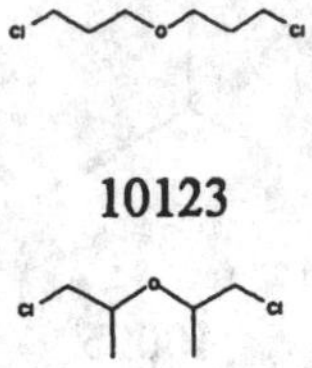

10123

10124

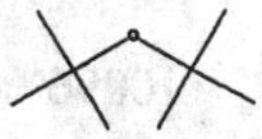

10125

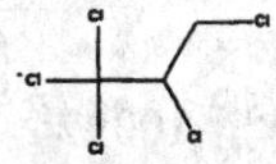

10126

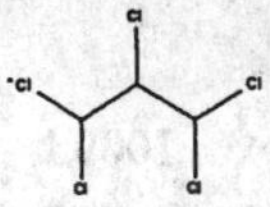

10127

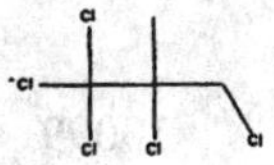

10128

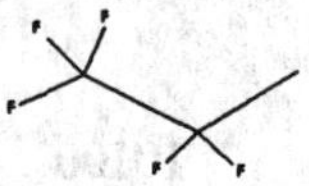

10129

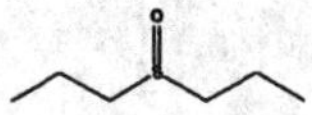

10130

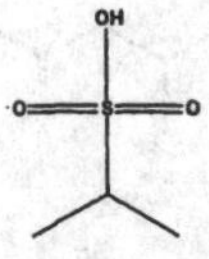

10131

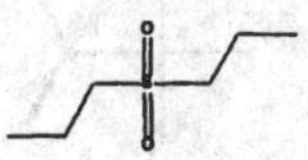

10132

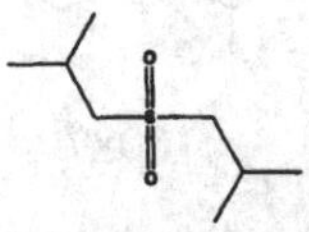

10133

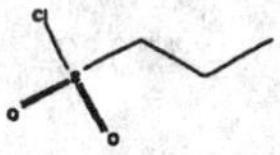

10134

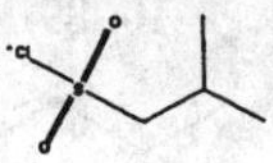

10135

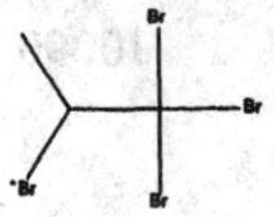

10136

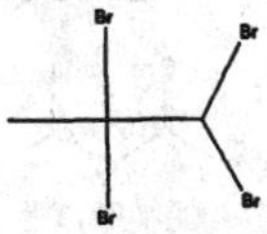

10138

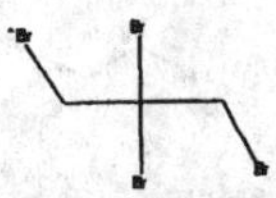

10139

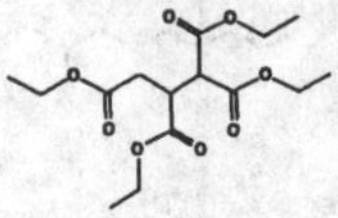

10140

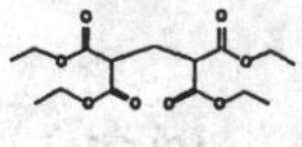

10141

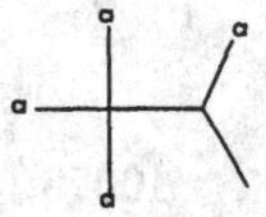

10142

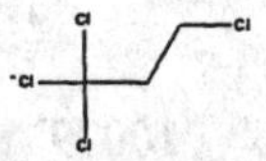

10143

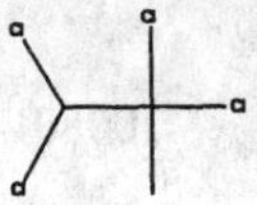

10144

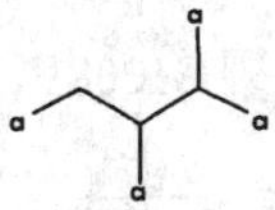

10145

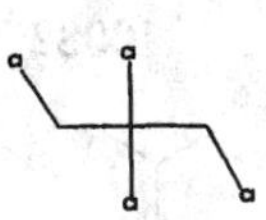

10146

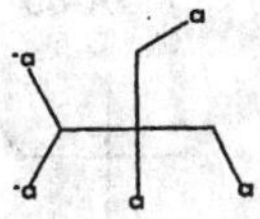

10147

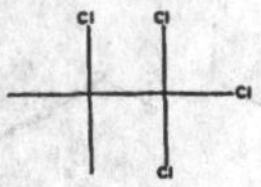

10148

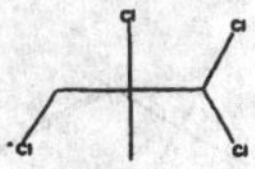

10149

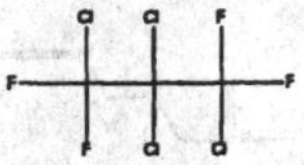

10150

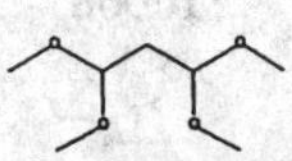

10151

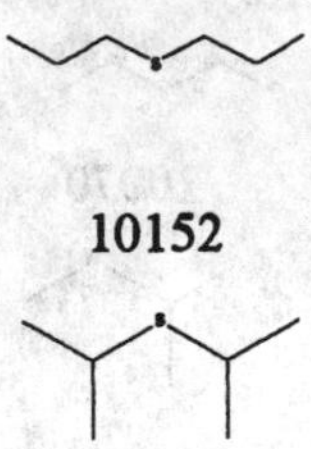

10152

10153

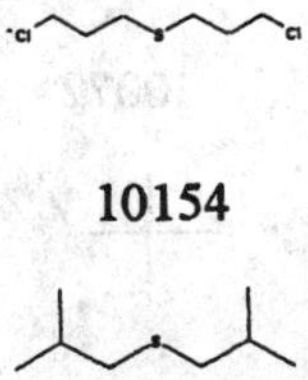

10154

10155

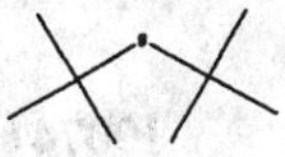

10156

10157

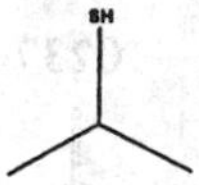

10158

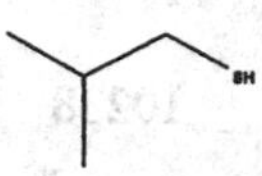

10159

10160

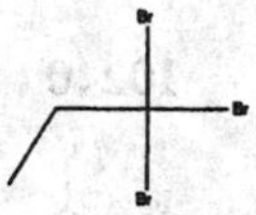

10161

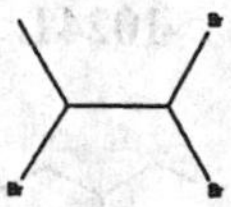

10162

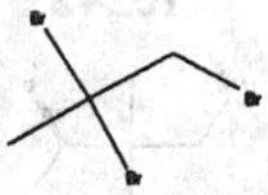

10163

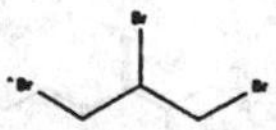

10164

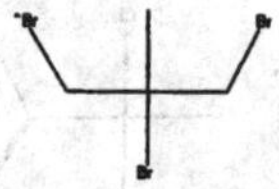

10166

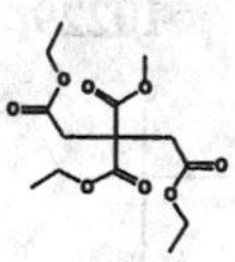

10167

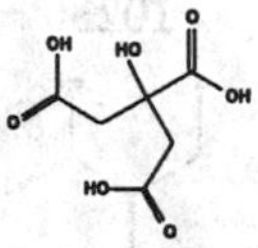

10168

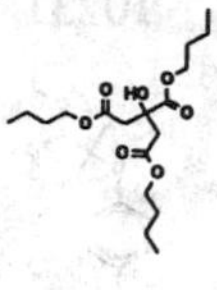

10169

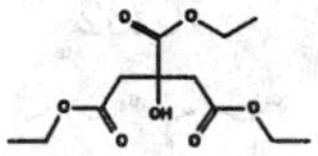

10170

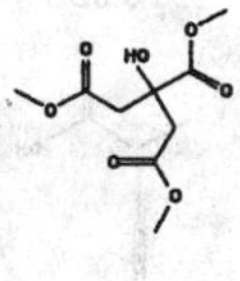

10171

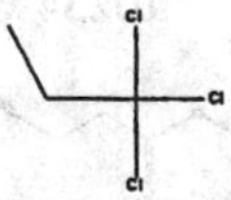

10172

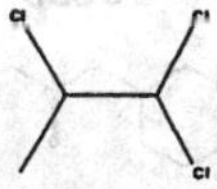

10173

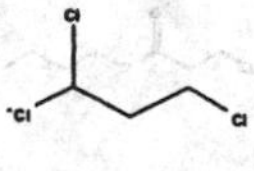

10174

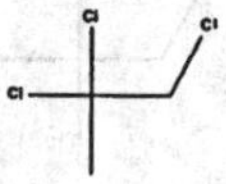

10175

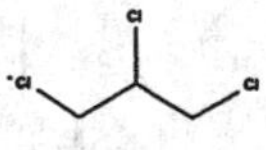

10176

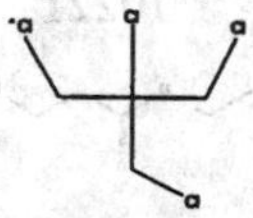

10177

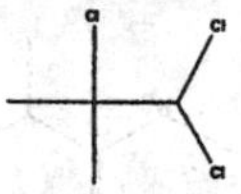

10178

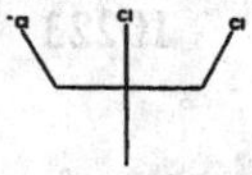

10179

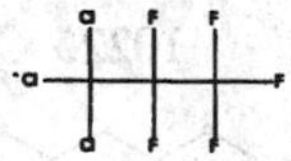

10180

10181

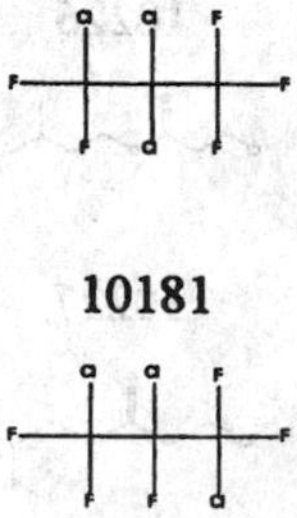

10182

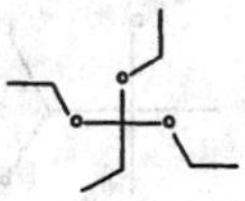

10183

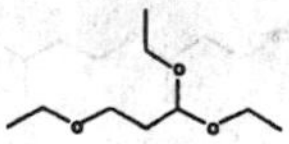

10184

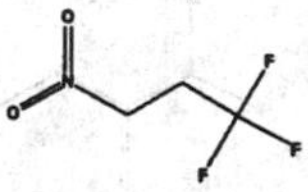

10185

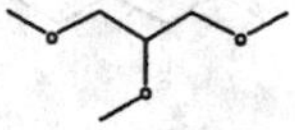

10186

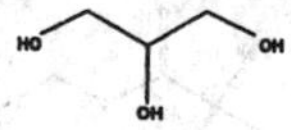

10187

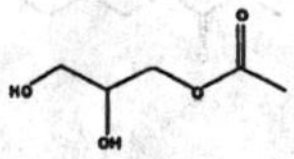

10188

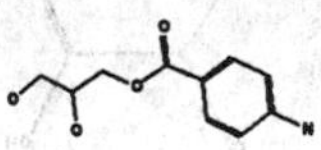

10189

10190

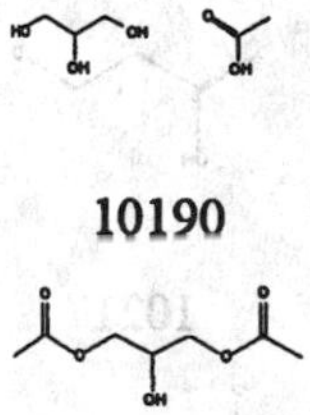

10191

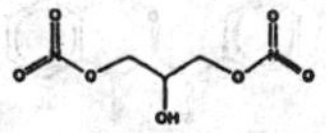

10192

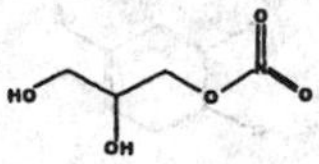

10193

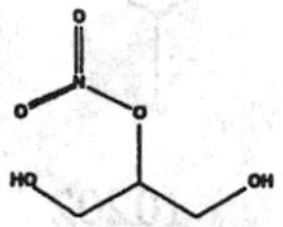

10194

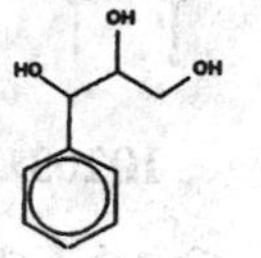

10195

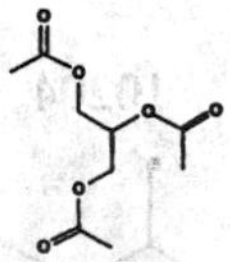

10196

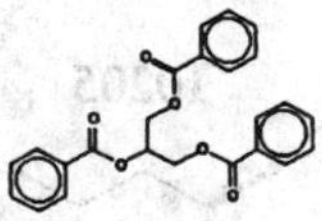

10197

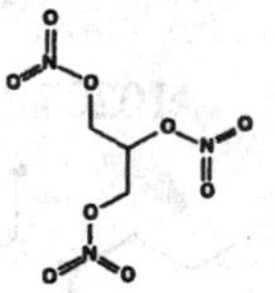

10198

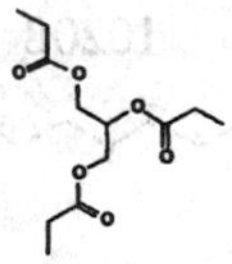

STRUCTURAL FORMULAS OF ORGANIC COMPOUNDS (continued)

In numeric order as they occur in the Table of Physical Constants of Organic Compounds

10199

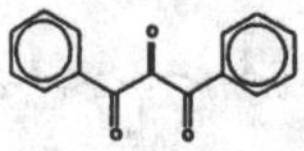

10200

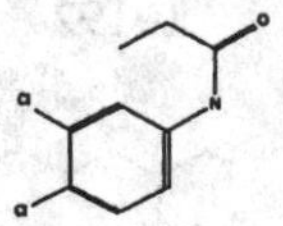

10201

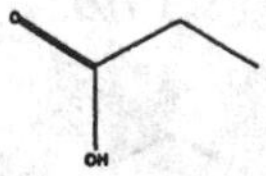

10202

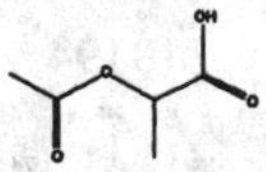

10203

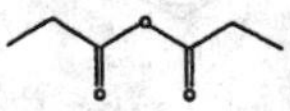

10204

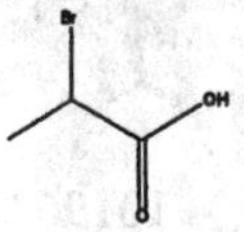

10205

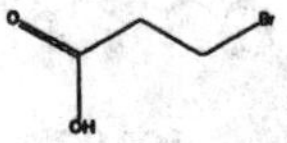

10207

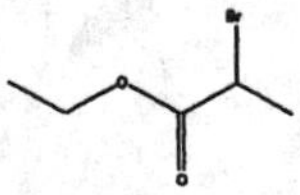

10208

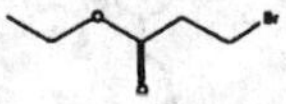

10209

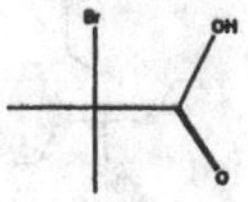

10210

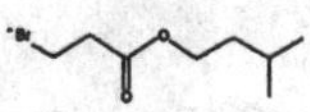

10211

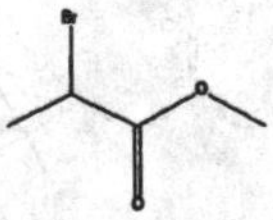

10212

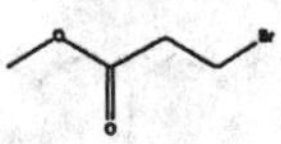

10213

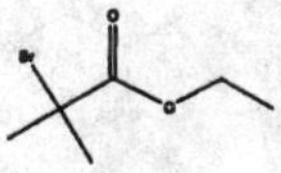

10214

10215

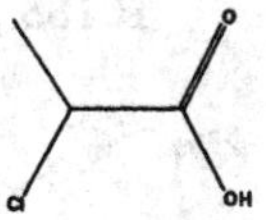

10216

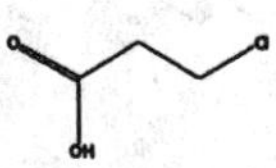

10217

10218

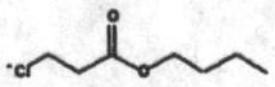

10219

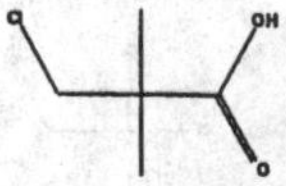

10220

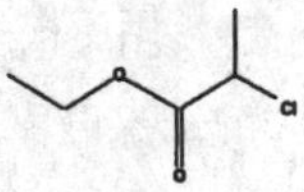

10221

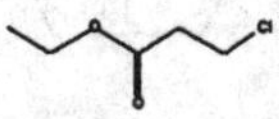

10222

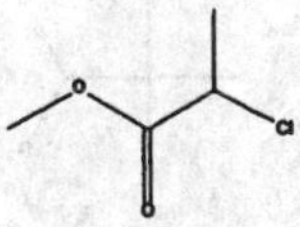

10223

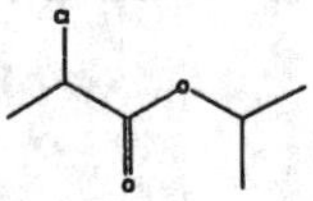

10225

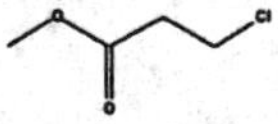

10226

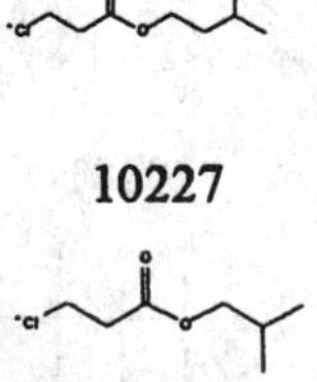

10227

10228

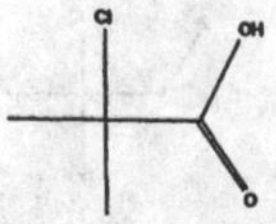

10229

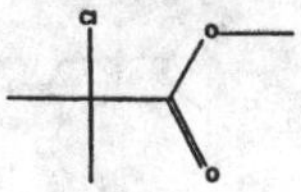

10230

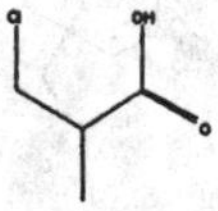

10231

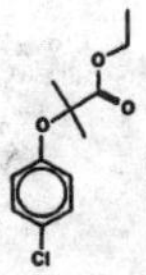

10232

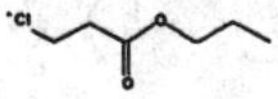

10233

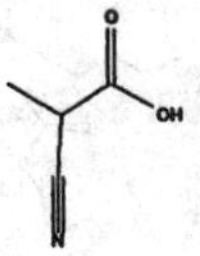

10234

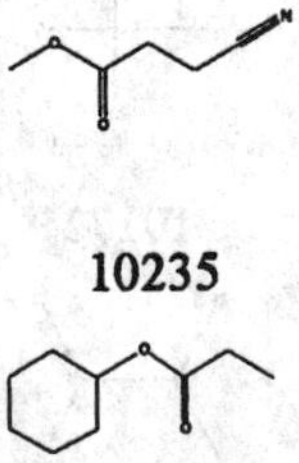

10235

10236

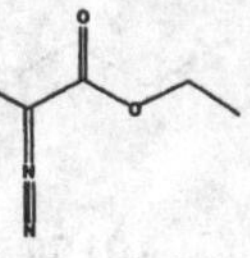

10237

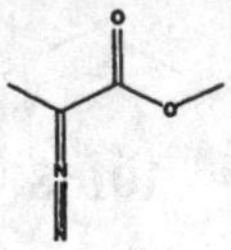

10238

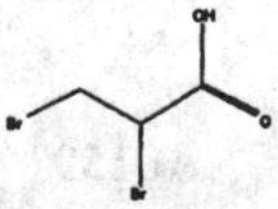

10239

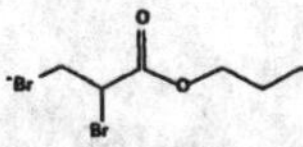

10240

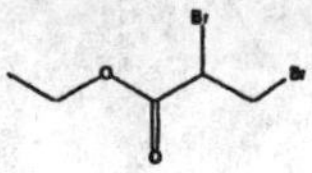

10241

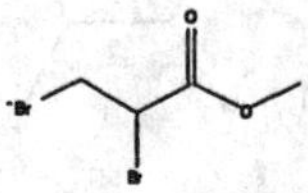

10242

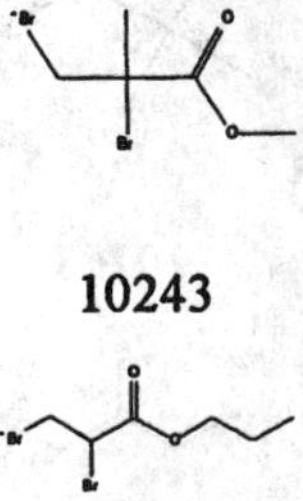

10243

In numeric order as they occur in the Table of Physical Constants of Organic Compounds

10244

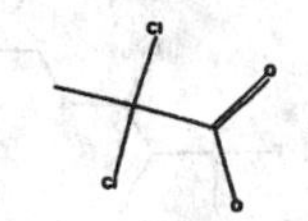

10245

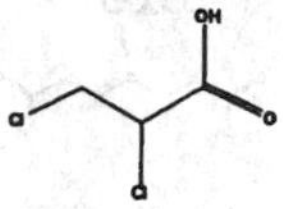

10246

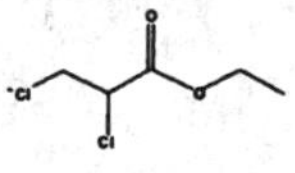

10248

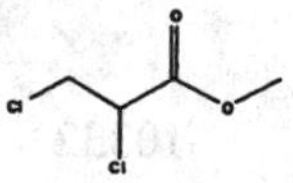

10249

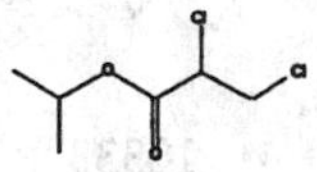

10250

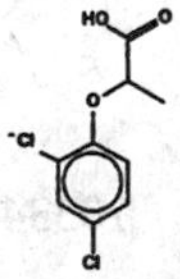

10251

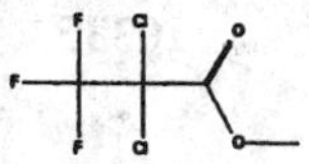

10253

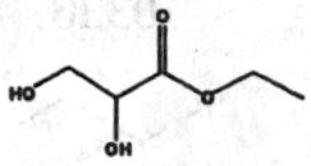

10254

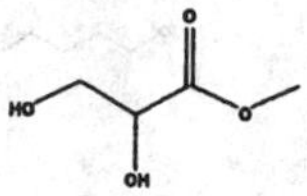

10255

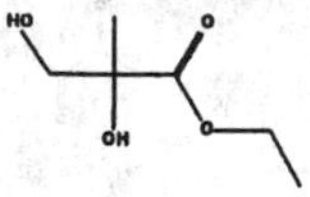

10256

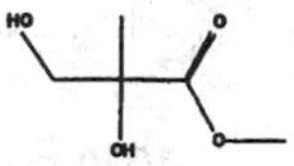

10258

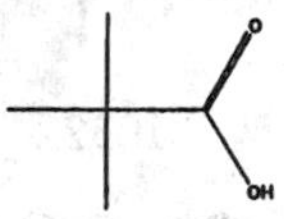

10259

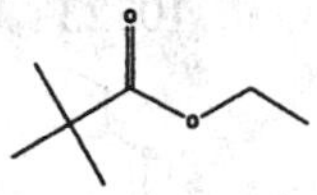

10260

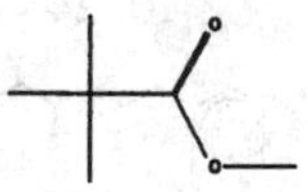

10261

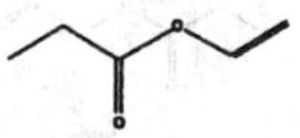

10262

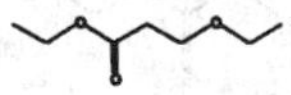

10263

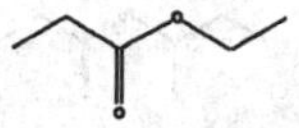

10264

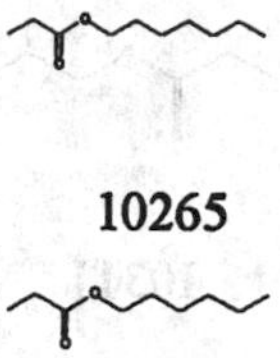

10265

10268

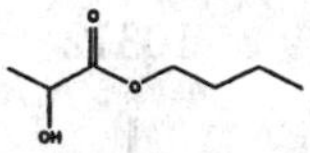

10269

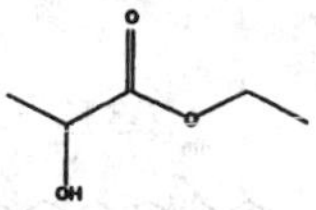

10270

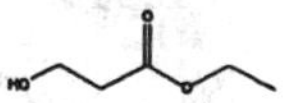

10271

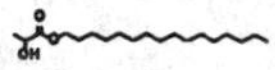

10273

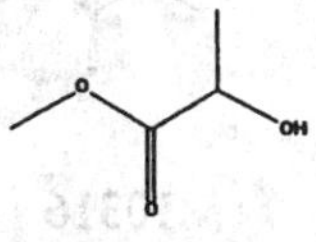

10274

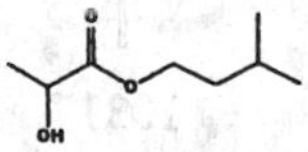

10275

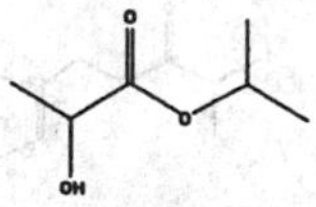

10276

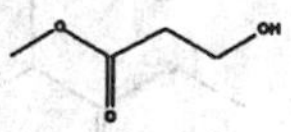

10277

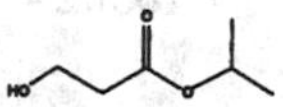

10278

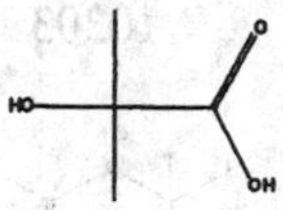

10279

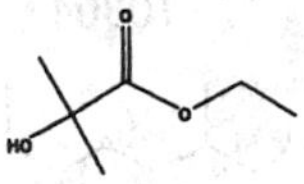

10280

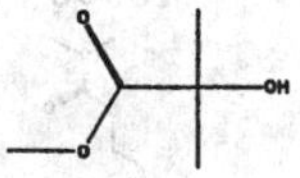

10281

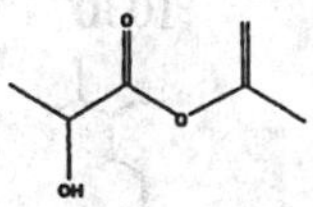

10282

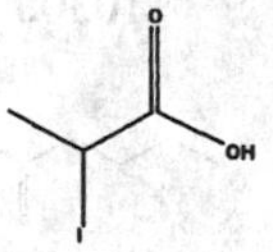

10283

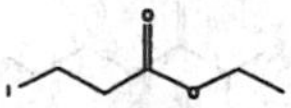

10285

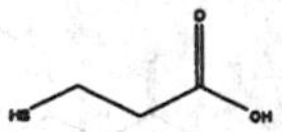

10286

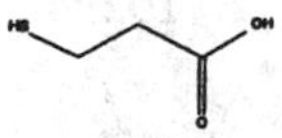

10287

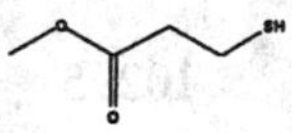

10288

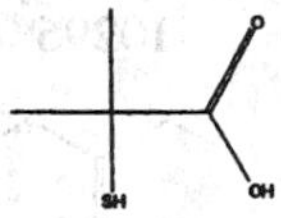

10289

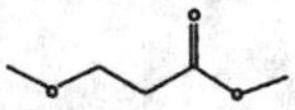

10290

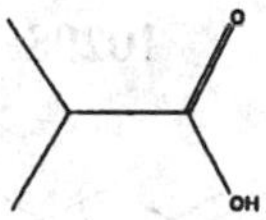

10291

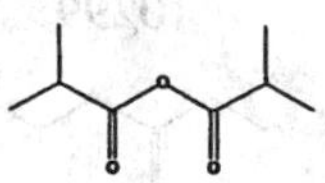

STRUCTURAL FORMULAS OF ORGANIC COMPOUNDS (continued)

In numeric order as they occur in the Table of Physical Constants of Organic Compounds

10292

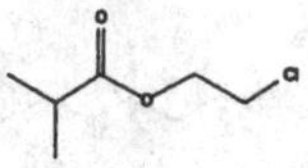

10293

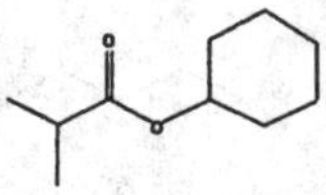

10294

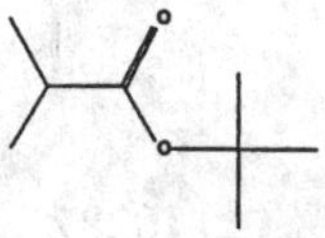

10295

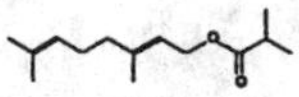

10296

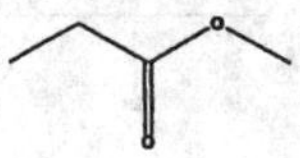

10297

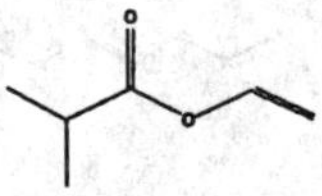

10298

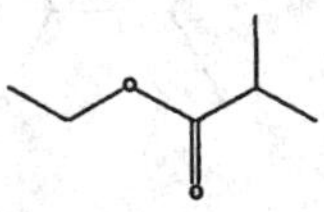

10299

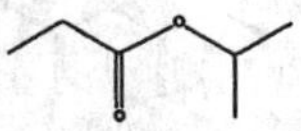

10300

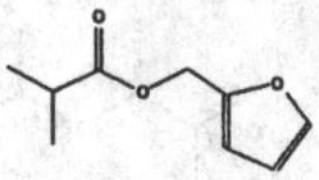

10301

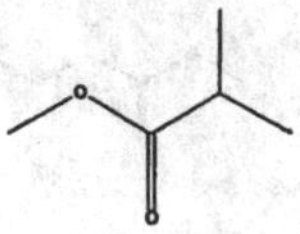

10302

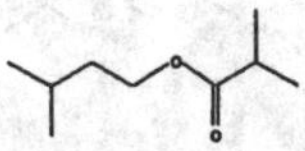

10303

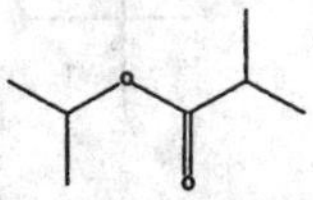

10304

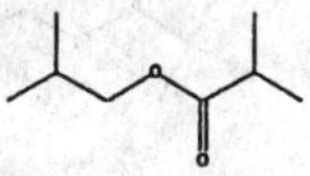

10306

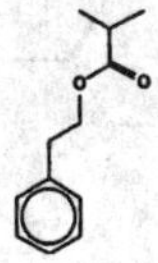

10307

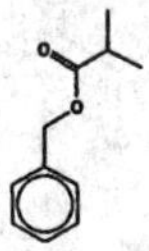

10308

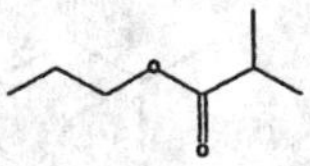

10309

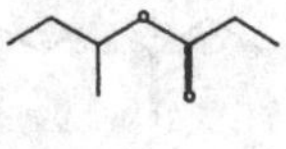

10310

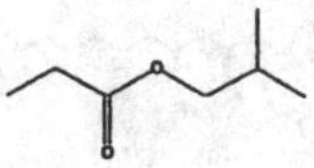

10311

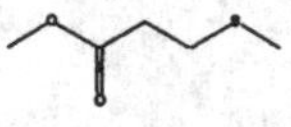

10312

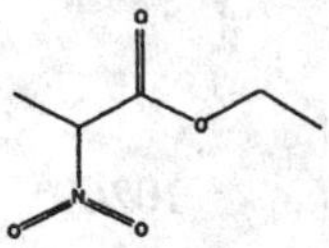

10313

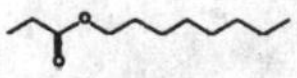

10314

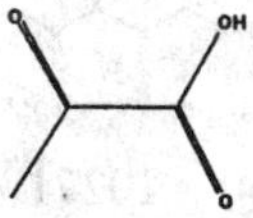

10315

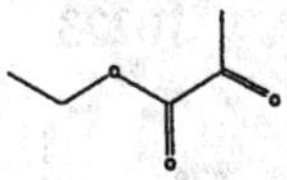

10316

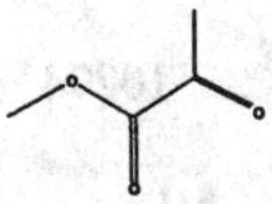

10317

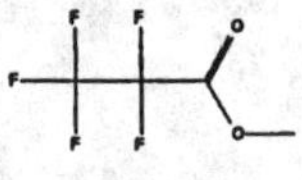

10318

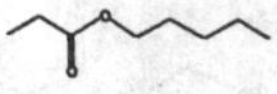

10319

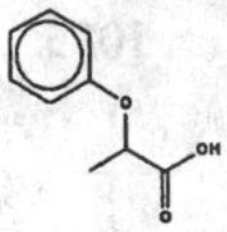

10320

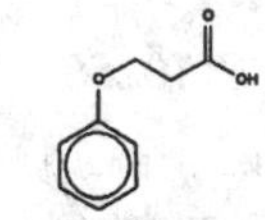

10321

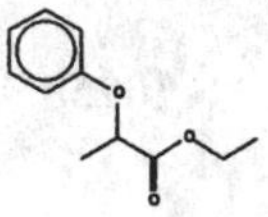

10322

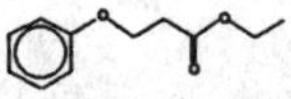

10323

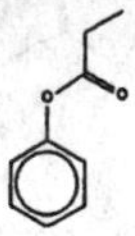

10324

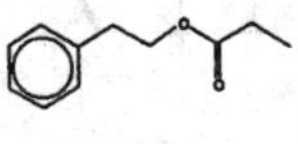

10325

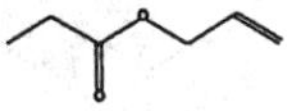

10326

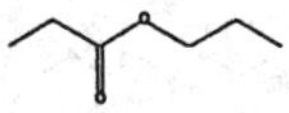

10328

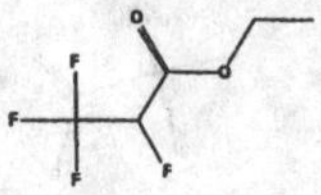

10329

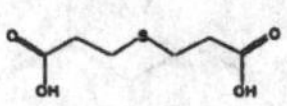

10330

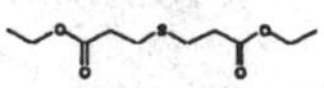

10331

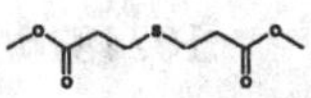

10332

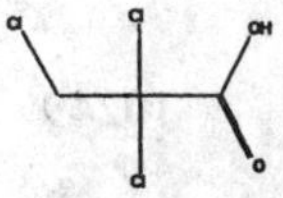

10333, 10334, 10335, 10336

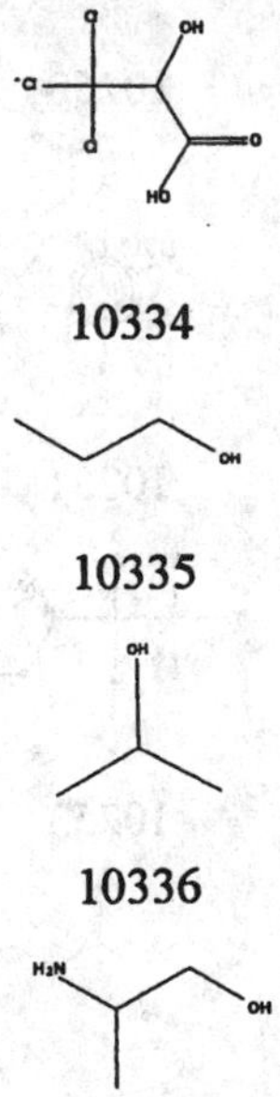

In numeric order as they occur in the Table of Physical Constants of Organic Compounds

10337

10338

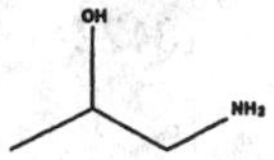

10339

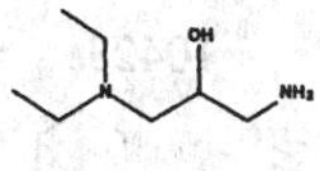

10340

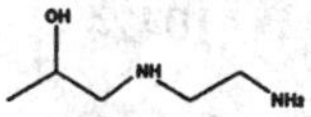

10341

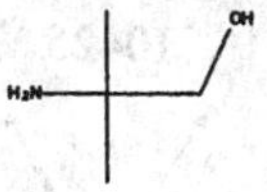

10342

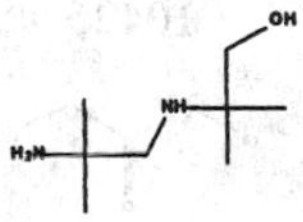

10343

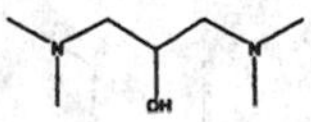

10344

10345

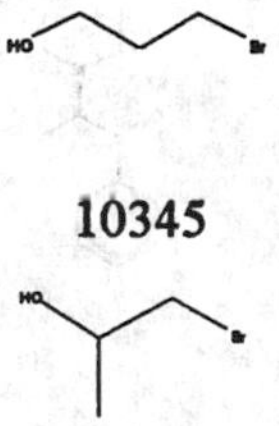

10346

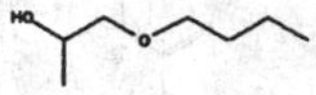

10347

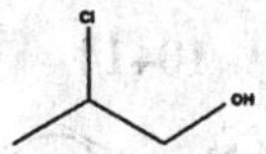

10348

10349

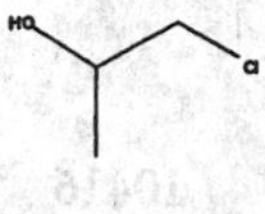

10350

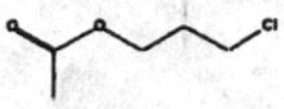

10351

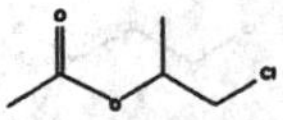

10352

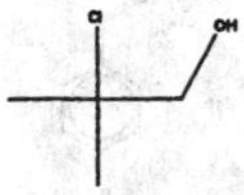

10353

10354

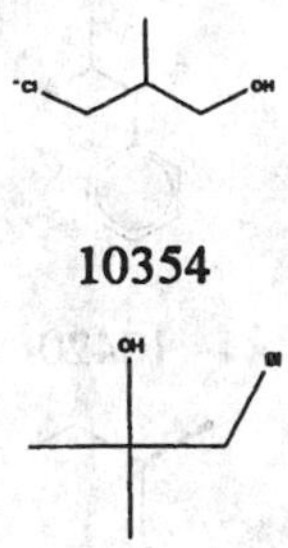

10355

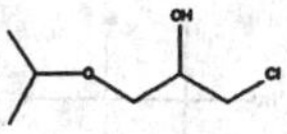

10356

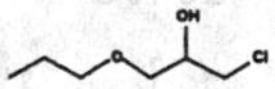

10357

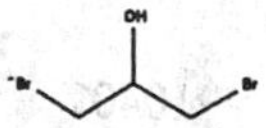

10358

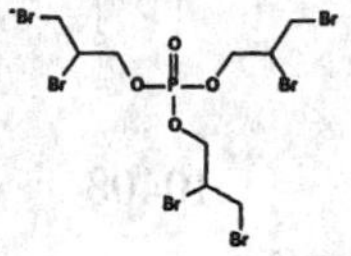

10359

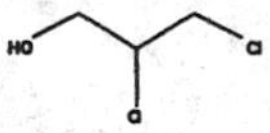

10360

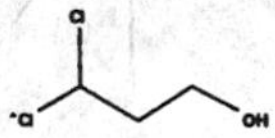

10361

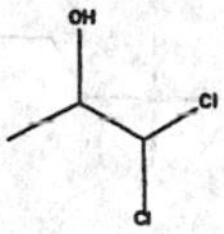

10362

10363

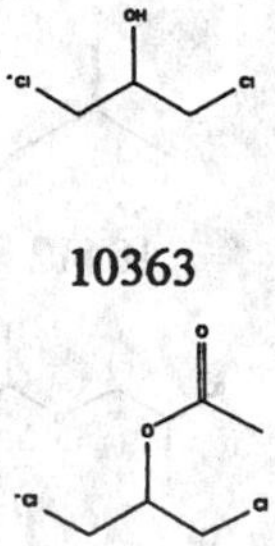

10364

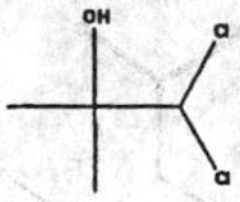

10365

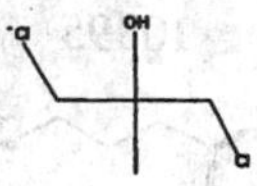

10366

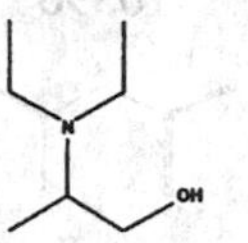

10367

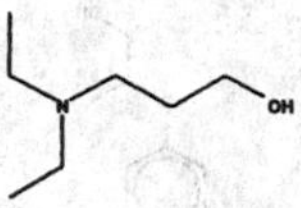

10368

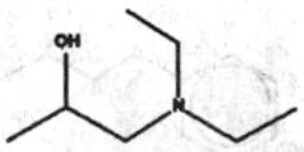

10369

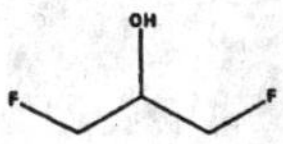

10370

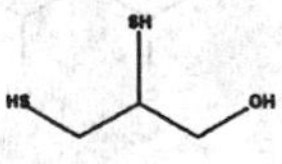

10371

10372

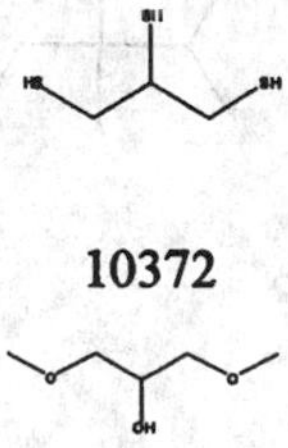

10373

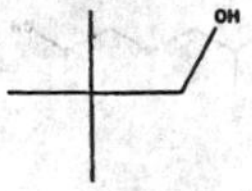

10374

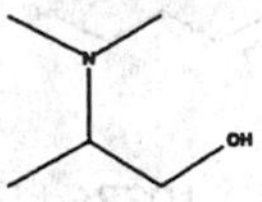

10375

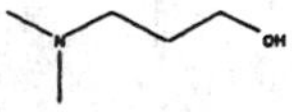

10376

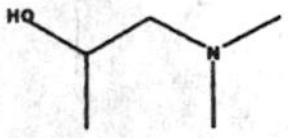

10378

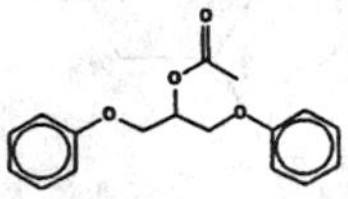

10380

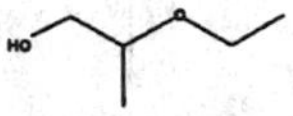

10381

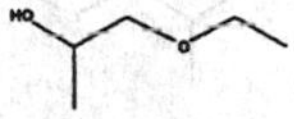

10382

10383

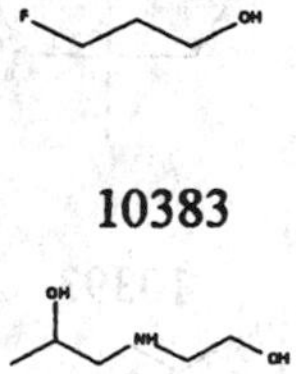

STRUCTURAL FORMULAS OF ORGANIC COMPOUNDS (continued)

In numeric order as they occur in the Table of Physical Constants of Organic Compounds

10384

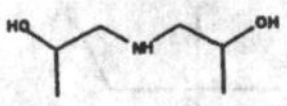

10385

10386

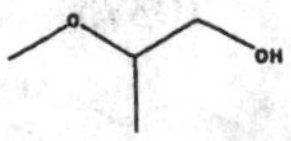

10387

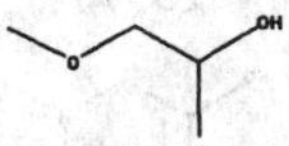

10388

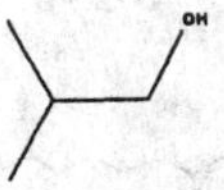

10389

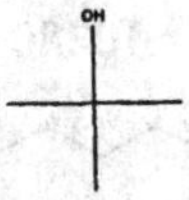

10390

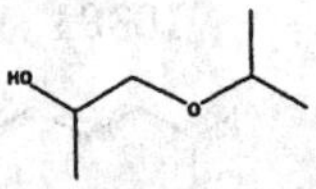

10391

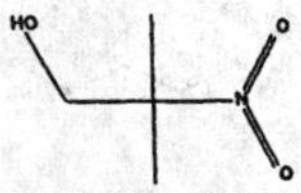

10393

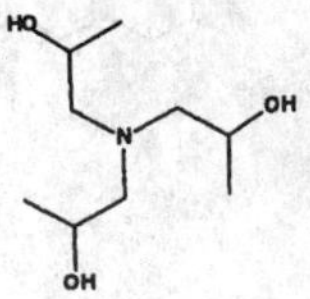

10394

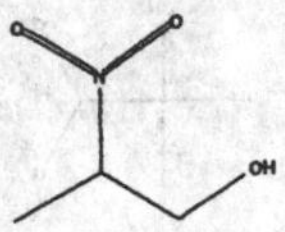

10395

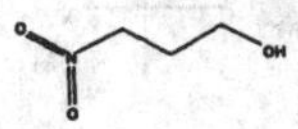

10396

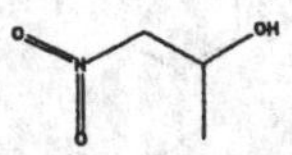

10398

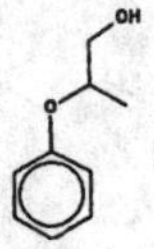

10399

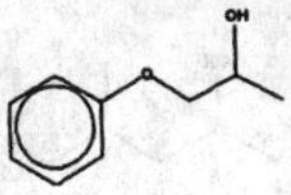

10401

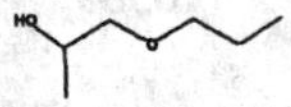

10402

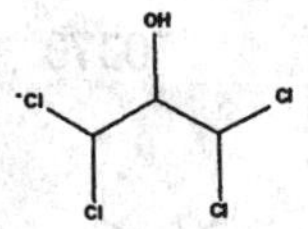

10404

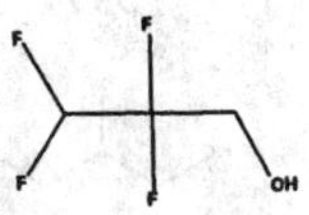

10405

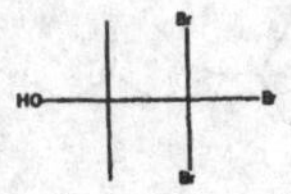

10406

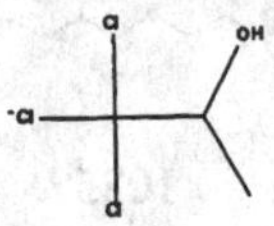

10407

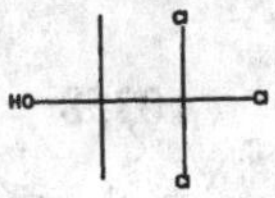

10408

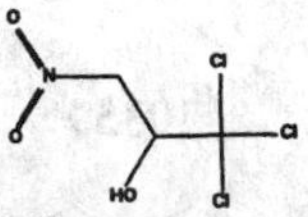

10409

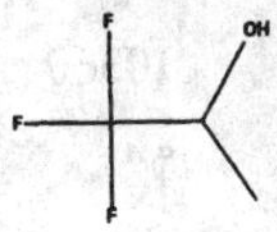

10410

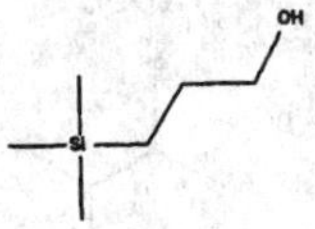

10411

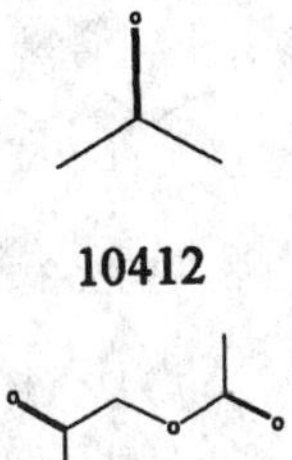

10412

10413

10414

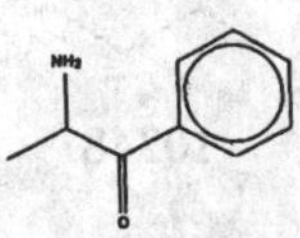

10415

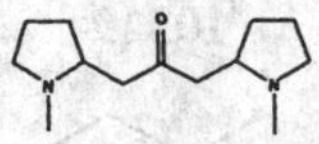

10416

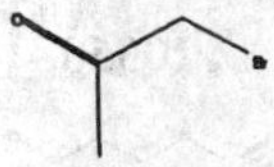

10417

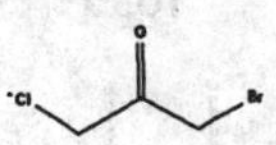

10418

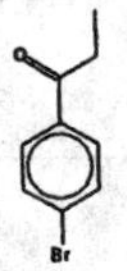

10419

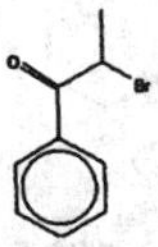

10420

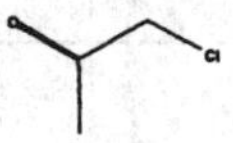

10421

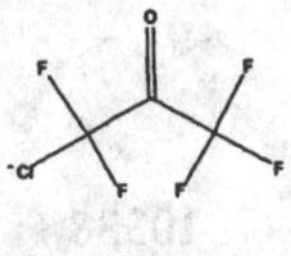

10422

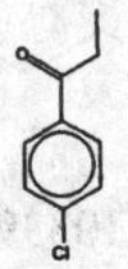

10423

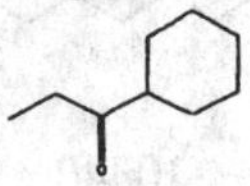

10424

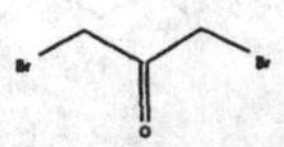

10425

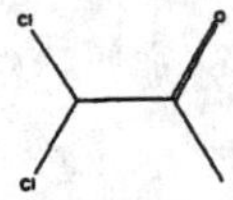

10426

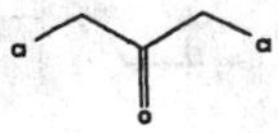

10427

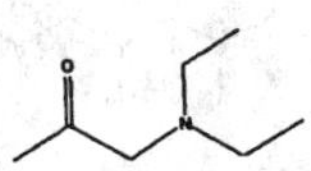

10428

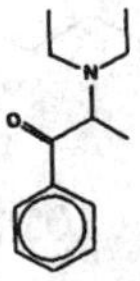

In numeric order as they occur in the Table of Physical Constants of Organic Compounds

10429

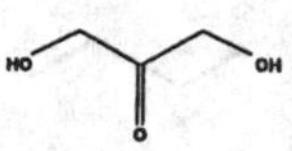

10430

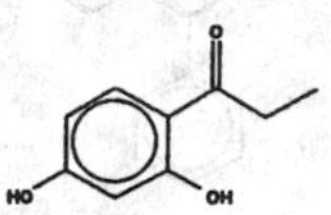

10431

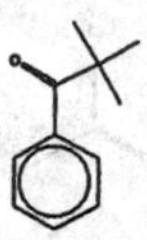

10432

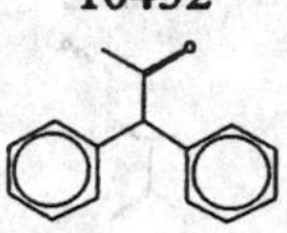

10433

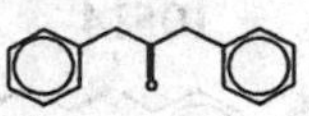

10436

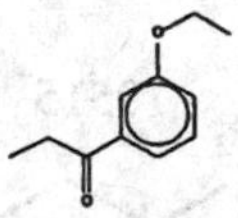

10437

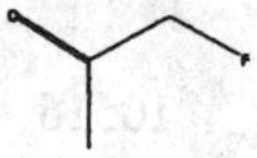

10438

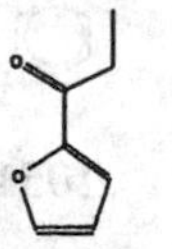

10439

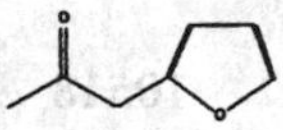

10440

10441

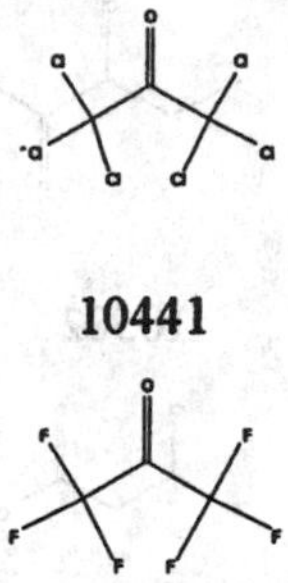

10442

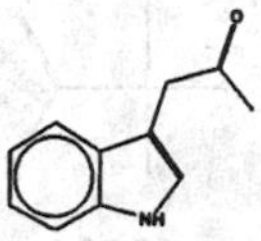

10443

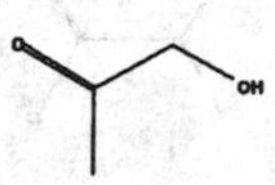

10444

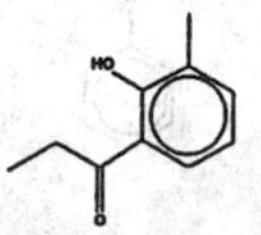

10445

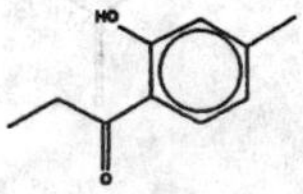

10446

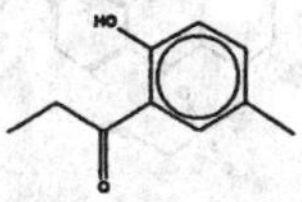

10447

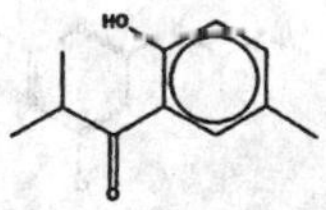

10448

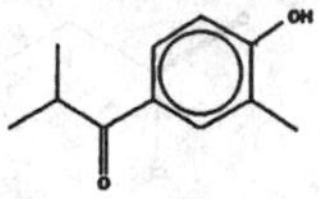

10449

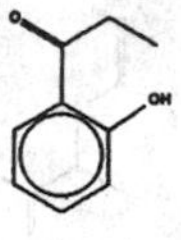

10450

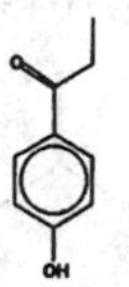

10451

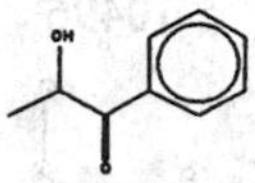

10452

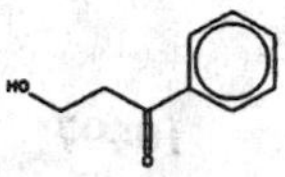

10454

10455

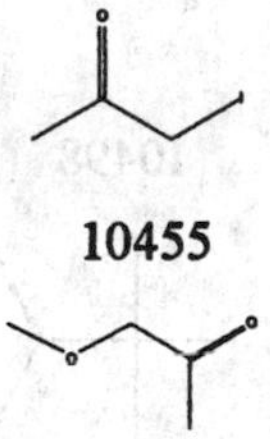

10457

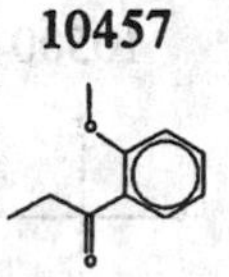

10458

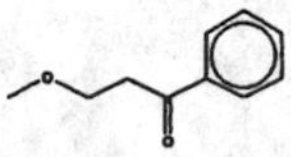

10460

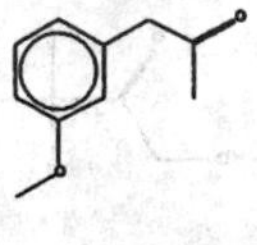

10461

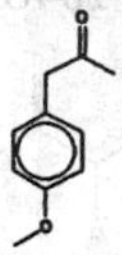

10462

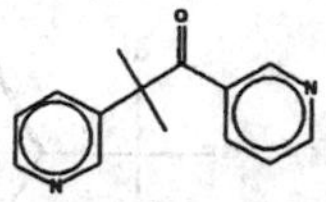

10463

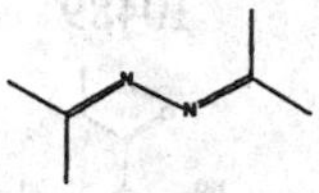

10464

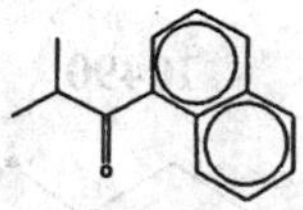

10465

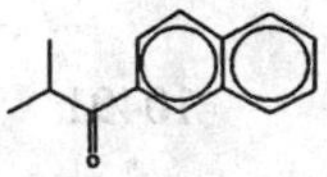

10466

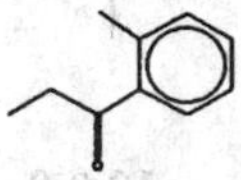

10467

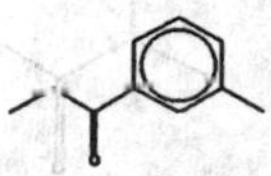

10468

10469

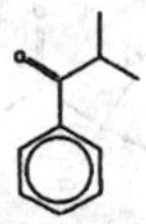

10470

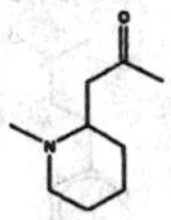

10471

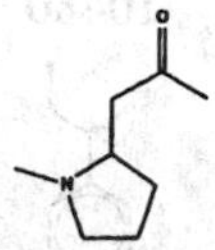

10472

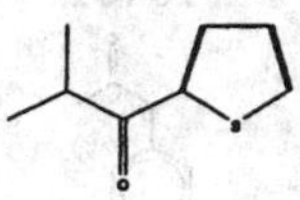

10473

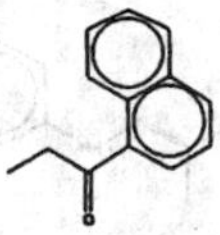

10474

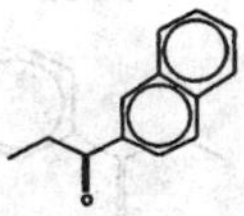

10475

10476

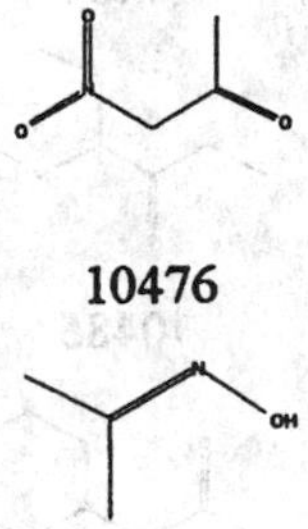

STRUCTURAL FORMULAS OF ORGANIC COMPOUNDS (continued)

In numeric order as they occur in the Table of Physical Constants of Organic Compounds

10478

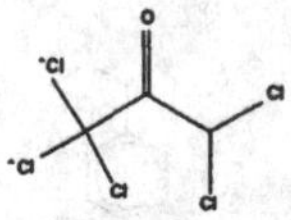

10479

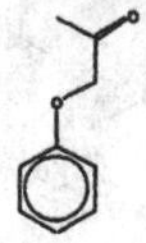

10480

10481

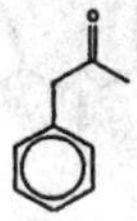

10482

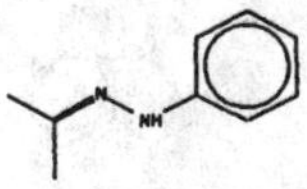

10483

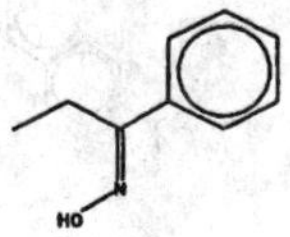

10484

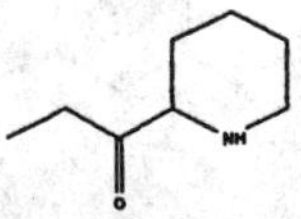

10485

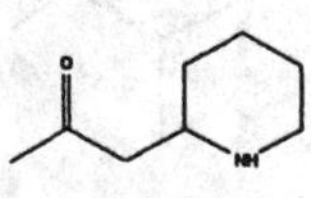

10486

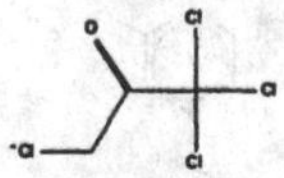

10487

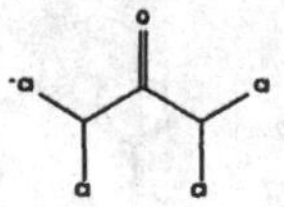

10488

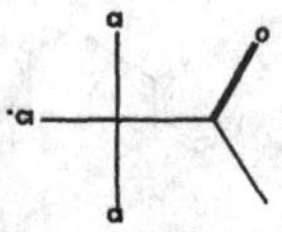

10489

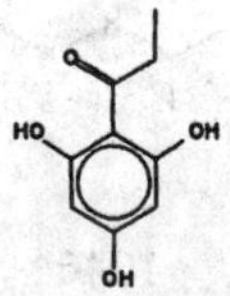

10490

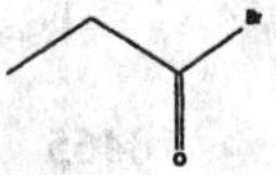

10491

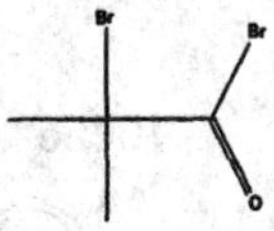

10492

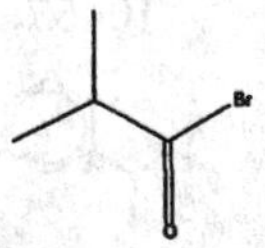

10493

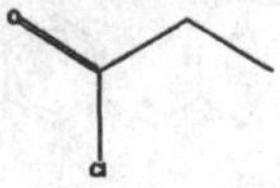

10494

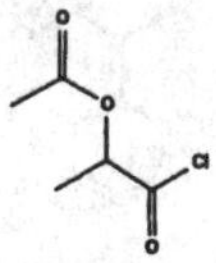

10495

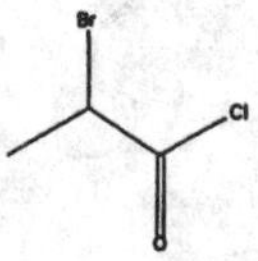

10496

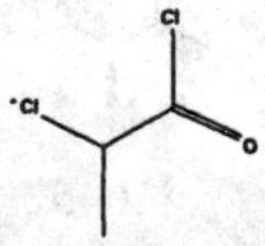

10497

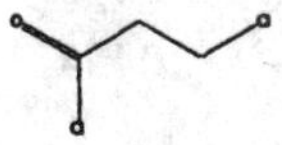

10498

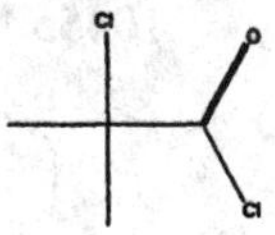

10500

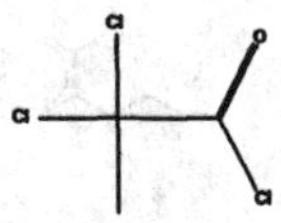

10501

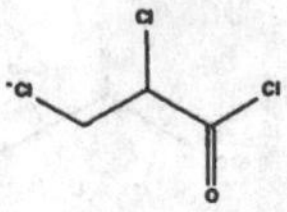

10502

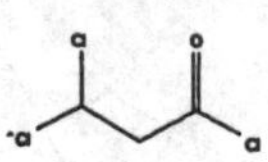

10503

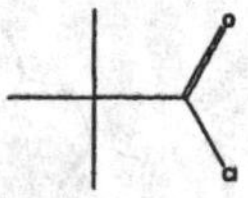

10504

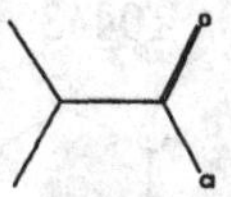

10505

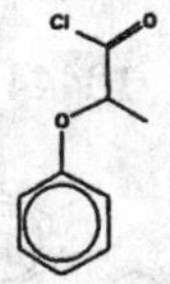

10506

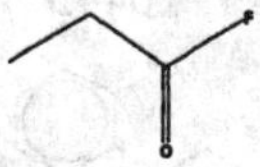

10507

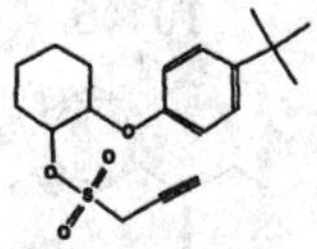

10508

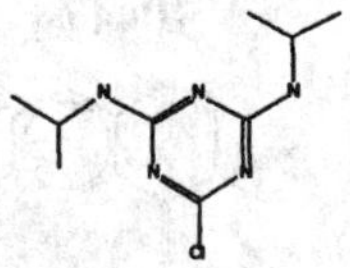

10509

10510

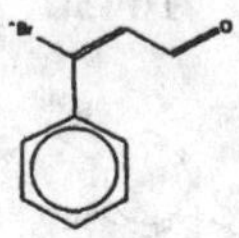

10511

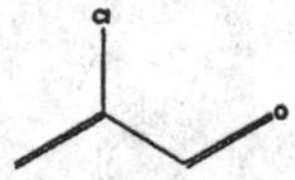

10512

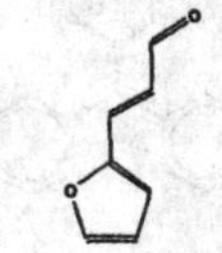

10514

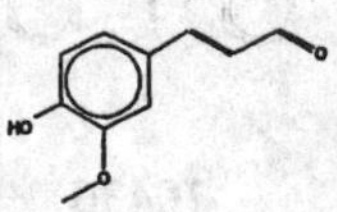

10515

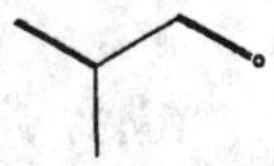

10516

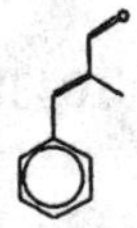

10517

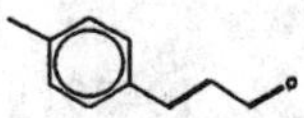

10518

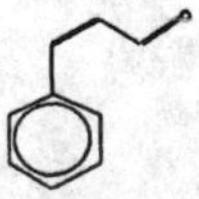

In numeric order as they occur in the Table of Physical Constants of Organic Compounds

10519

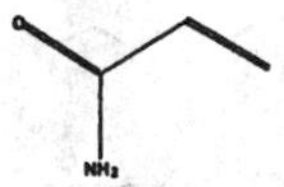

10520

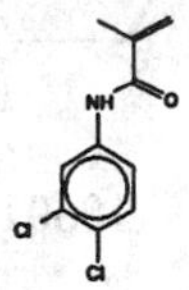

10521

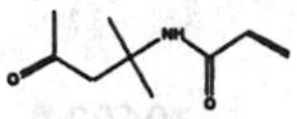

10522

10523

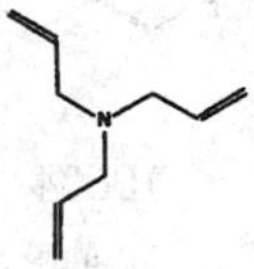

10524

10525

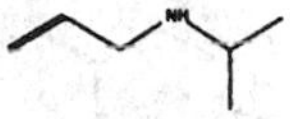

10526

10527

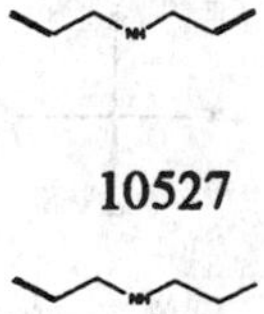

10528

10529

10530

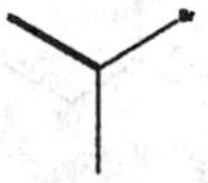

10531

10532

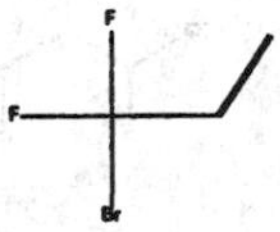

10533

10534

10535

10536

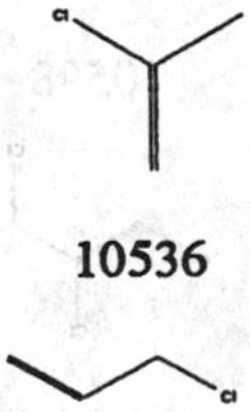

10537

10538

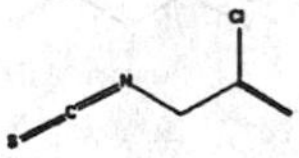

10539

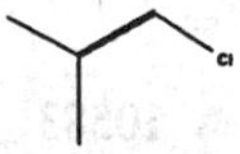

10540

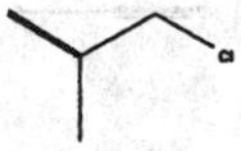

10541

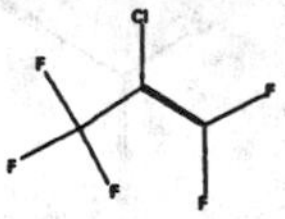

10542

10543

10544

10545

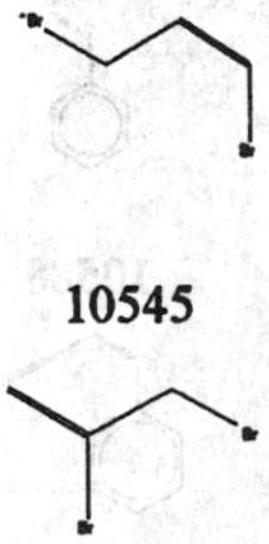

10546

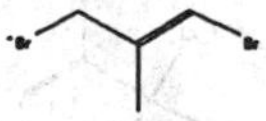

10547

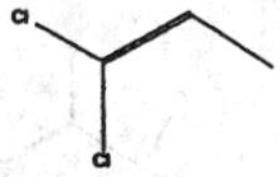

10548

10549

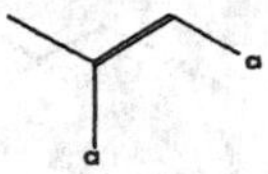

10550

10551

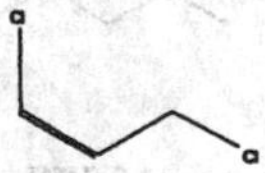

10552

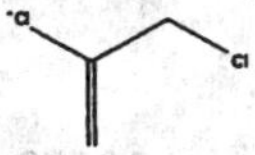

10553

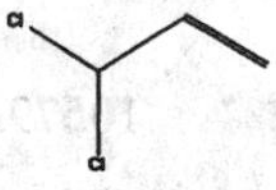

10554

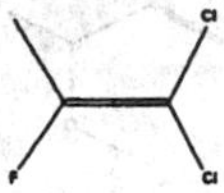

10555

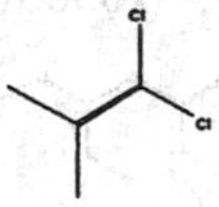

10556

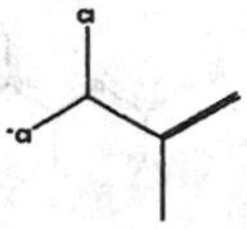

10557

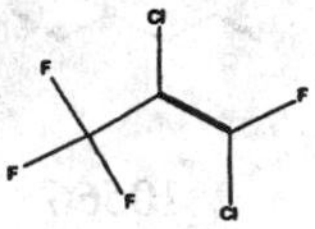

10558

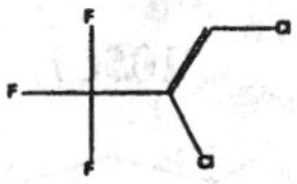

10559

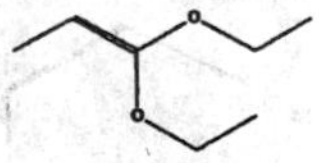

10560

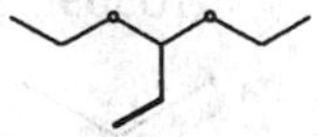

10561

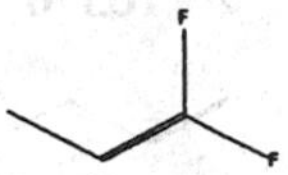

STRUCTURAL FORMULAS OF ORGANIC COMPOUNDS (continued)

In numeric order as they occur in the Table of Physical Constants of Organic Compounds

10562

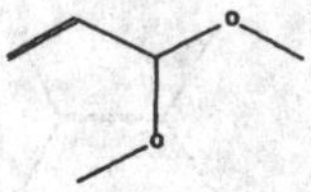

10563

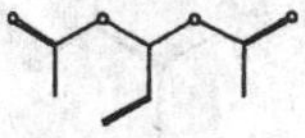

10564

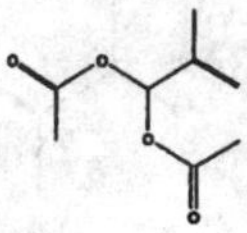

10565

10566

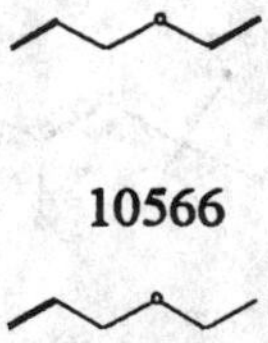

10567

10568

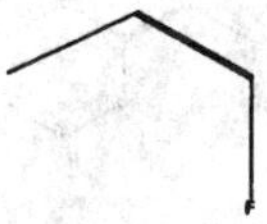

10569

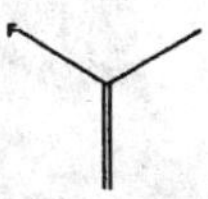

10570

10571

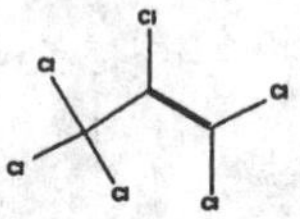

10572

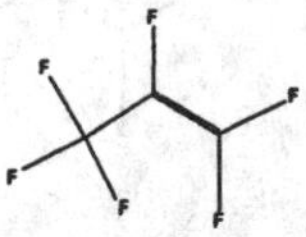

10573

10574

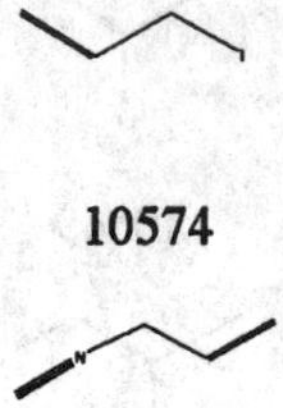

10575

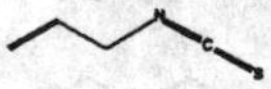

10576

10577

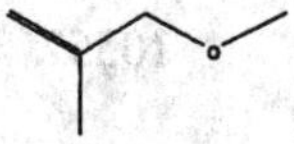

10578

10579

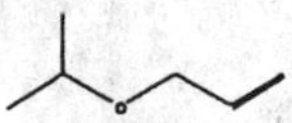

10580

10581

10582

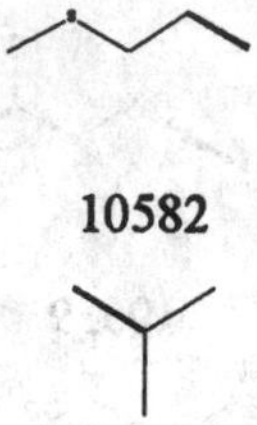

10583

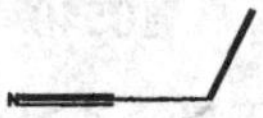

10584

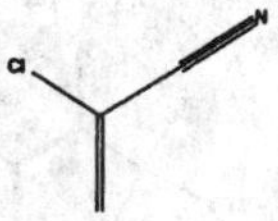

10585

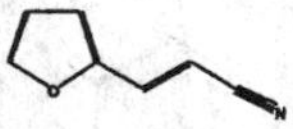

10586

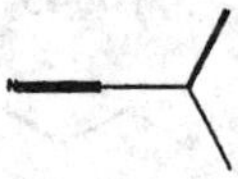

10587

10588

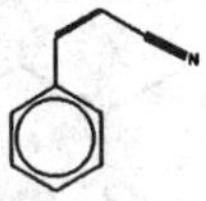

10589

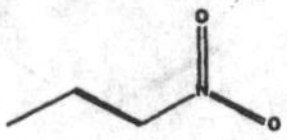

10590

10591

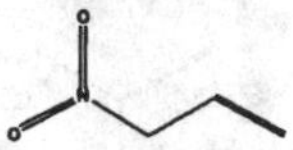

10592

10593

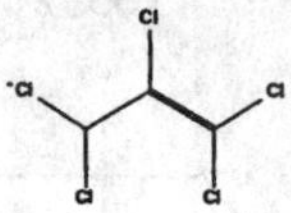

10594

10596

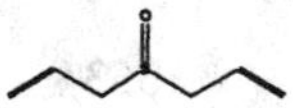

10597

10598

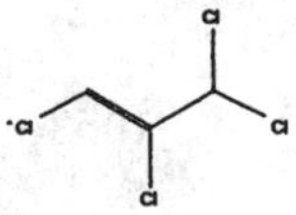

10599

10600

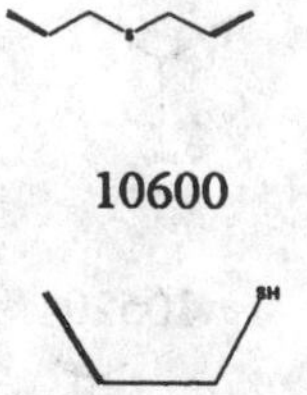

10601

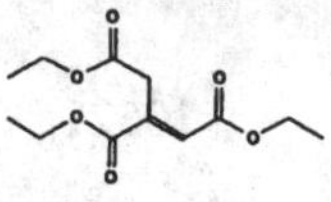

10602

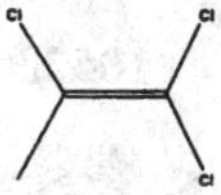

10603

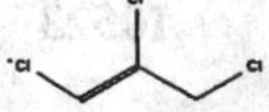

10604

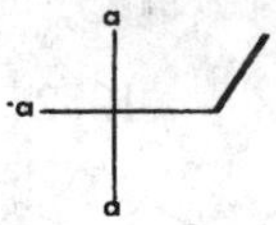

10605

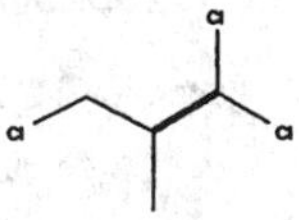

10606

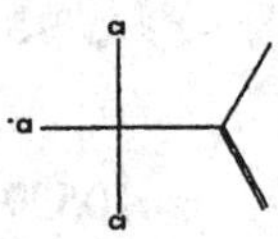

STRUCTURAL FORMULAS OF ORGANIC COMPOUNDS (continued)

In numeric order as they occur in the Table of Physical Constants of Organic Compounds

10607

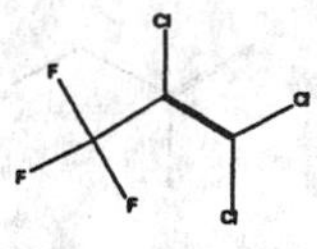

10608

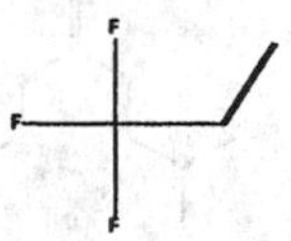

10609

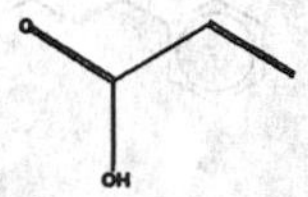

10611

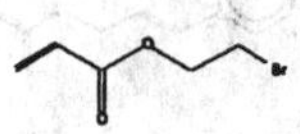

10613

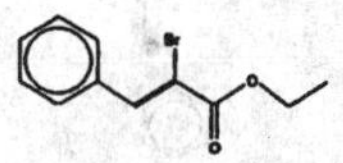

10614

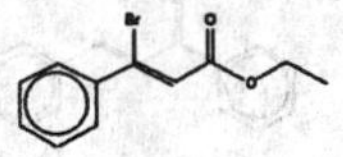

10616

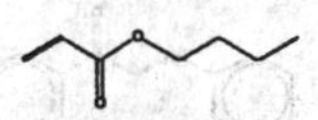

10617

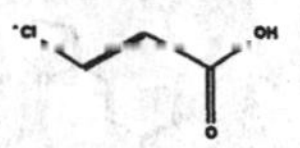

10618

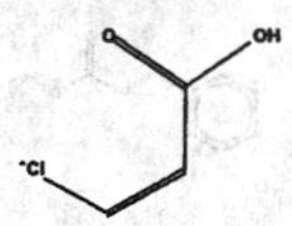

10620

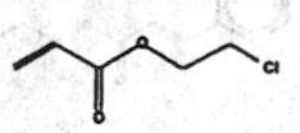

10621

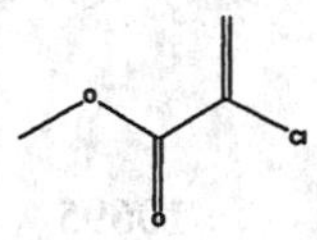

10622

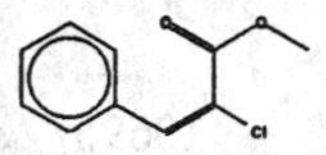

10624

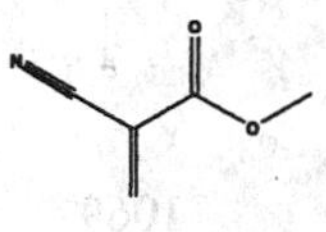

10625

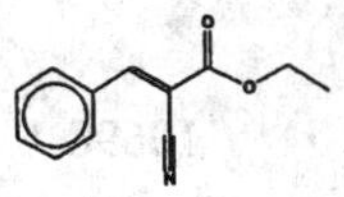

10626

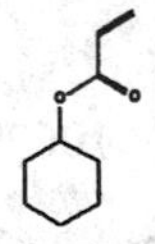

10627

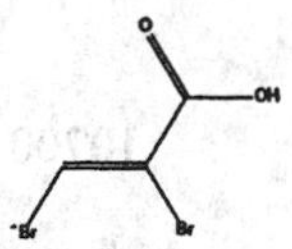

10628

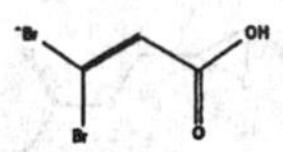

10629

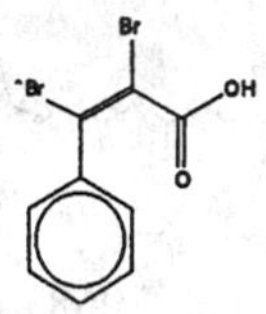

10631

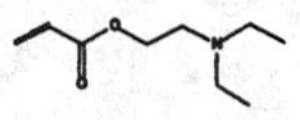

10632

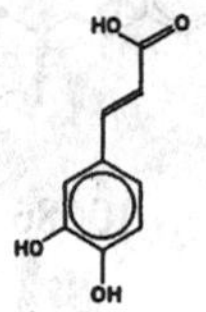

10633

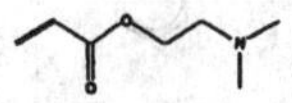

10634

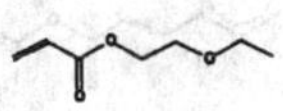

10635

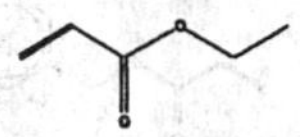

10636

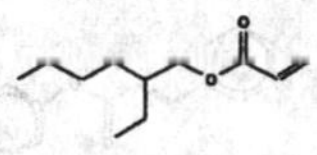

10637

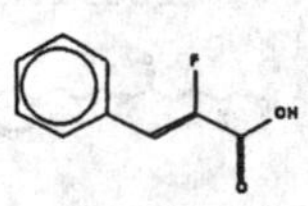

10638

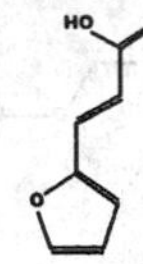

10639

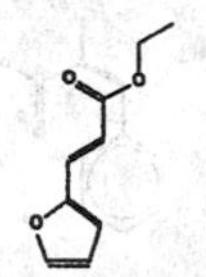

10640

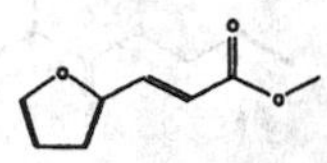

10643

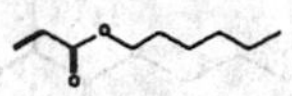

10644

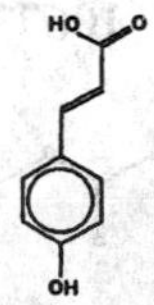

10645

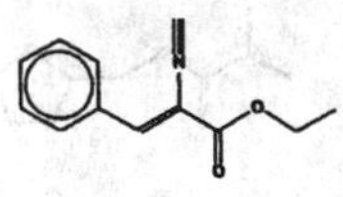

10647

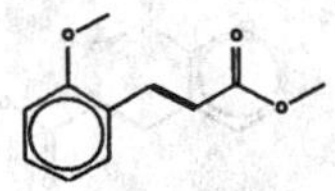

10648

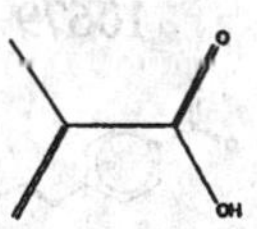

10649

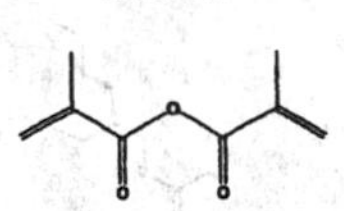

10650

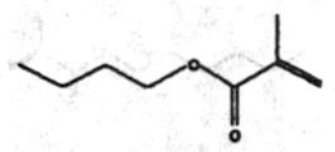

10652

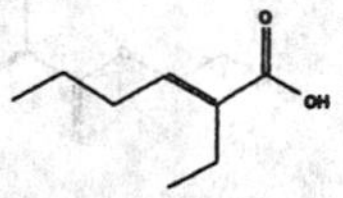

10653

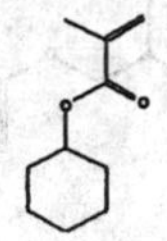

10654

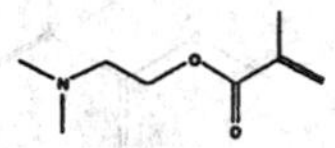

10655

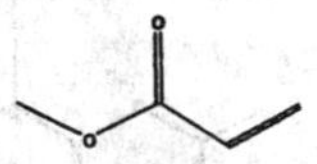

10656

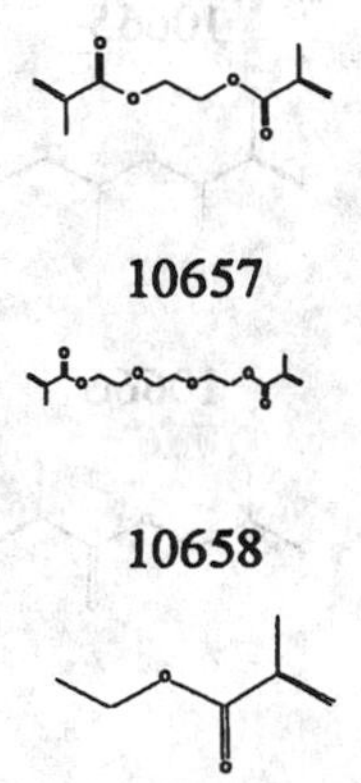

10657

10658

STRUCTURAL FORMULAS OF ORGANIC COMPOUNDS (continued)

In numeric order as they occur in the Table of Physical Constants of Organic Compounds

10659

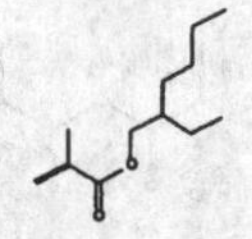

10660

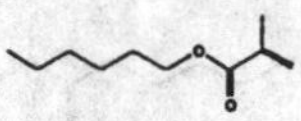

10661

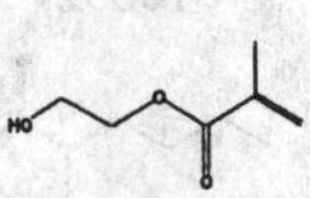

10662

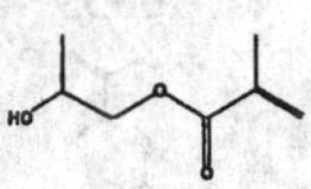

10663

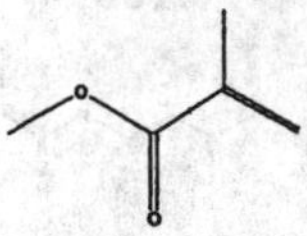

10664

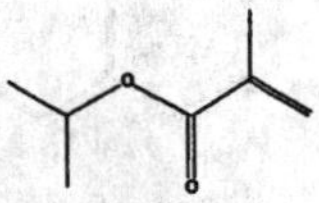

10665

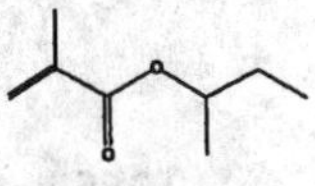

10666

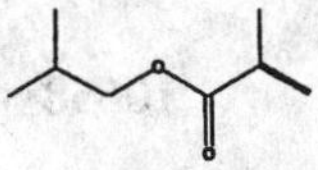

10668

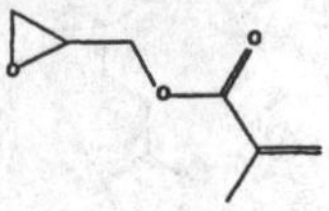

10670

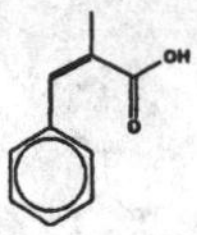

10671

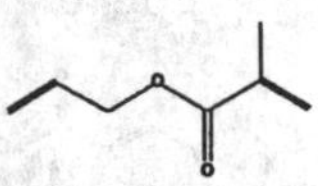

10672

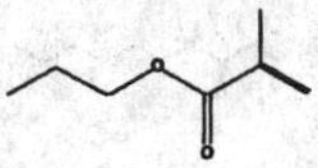

10673

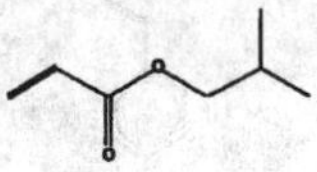

10674

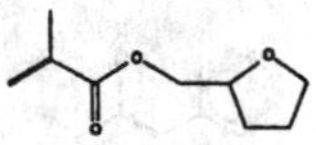

10677

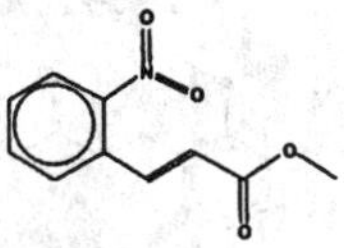

10679

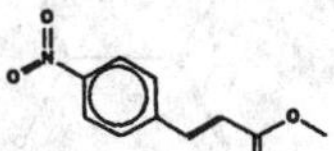

10680

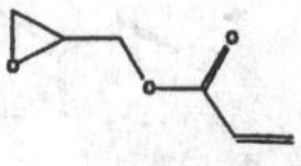

10681

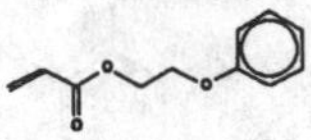

10683

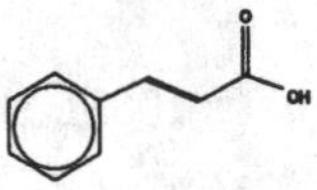

10685

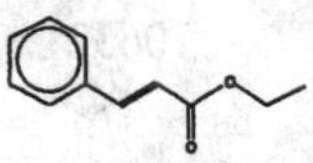

10686

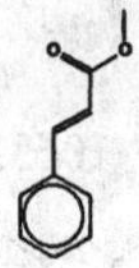

10687

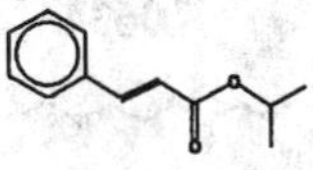

10688

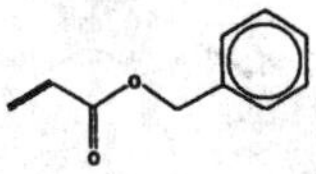

10689

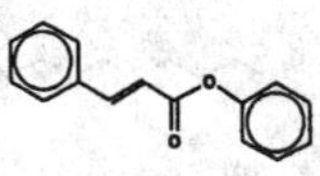

10691

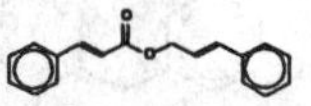

10692

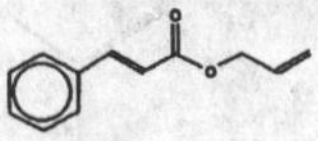

10693

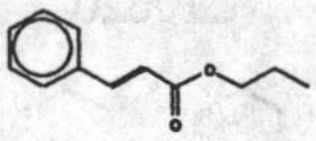

10694

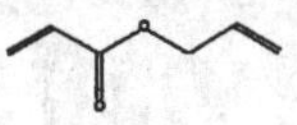

10695

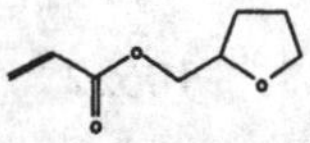

10696

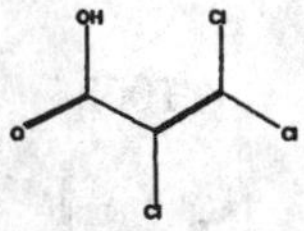

10697

10698

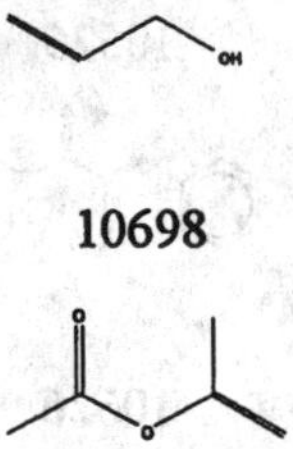

10699

10700

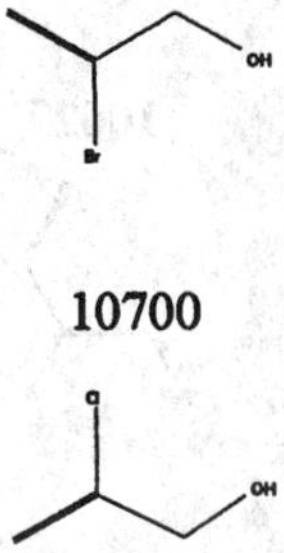

10702

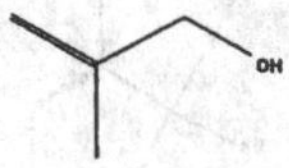

10703

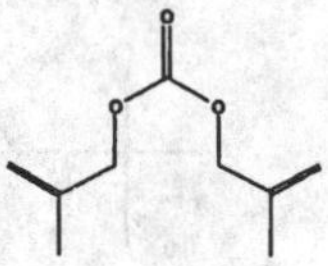

10704

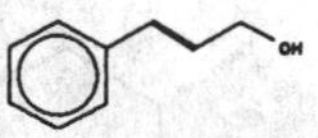

10706

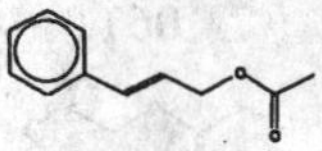

10707

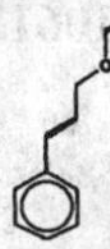

10708

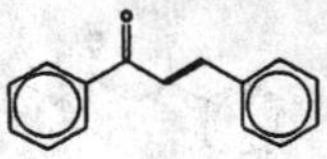

10709

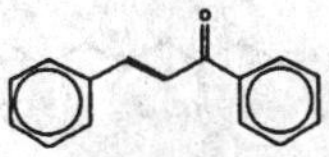

10710

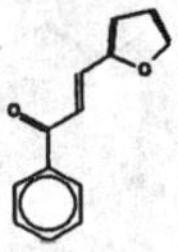

STRUCTURAL FORMULAS OF ORGANIC COMPOUNDS (continued)

In numeric order as they occur in the Table of Physical Constants of Organic Compounds

10711

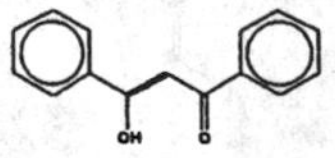

10713

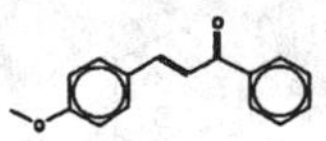

10714

10715

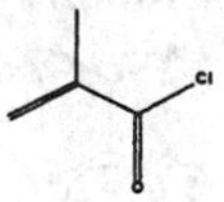

10716

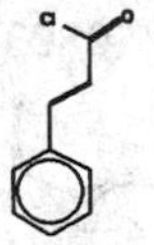

10717

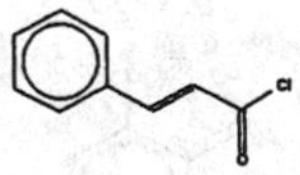

10718

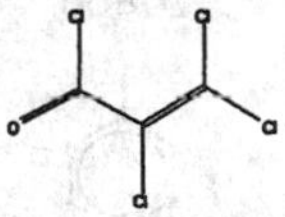

10719

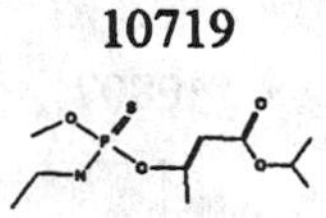

10720

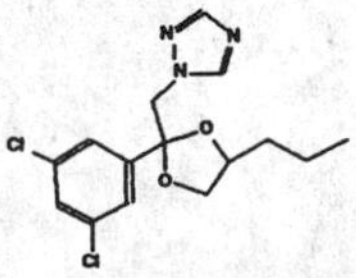

10721

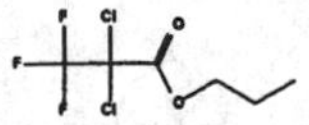

10722

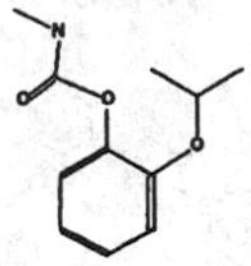

10723

10724

10725

10726

10727

10728

10729

10730

10731

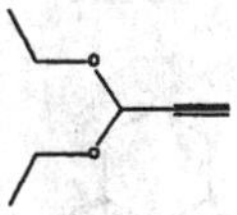

10732

10733

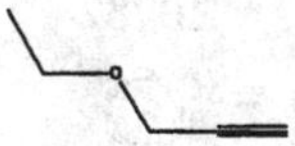

10734

10735

10736

10737

10738

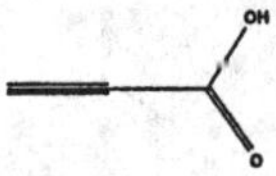

10739

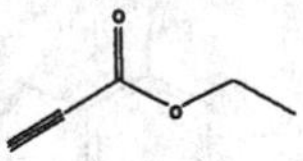

10743

10744

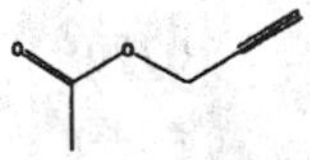

10745

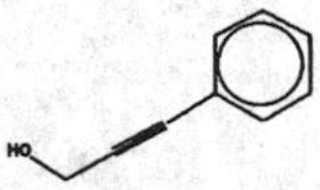

10746

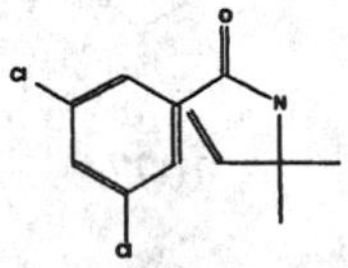

10747

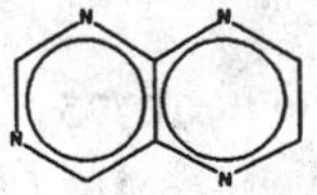

10748

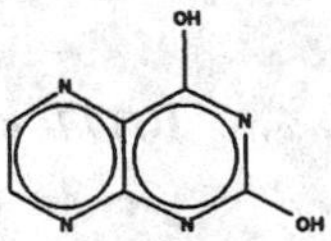

10749

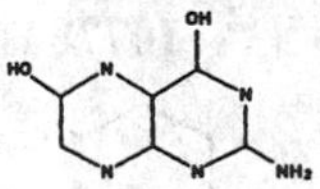

10750

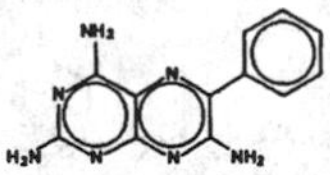

10751

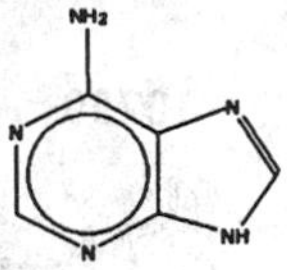

10752

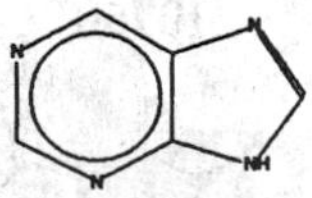

10753

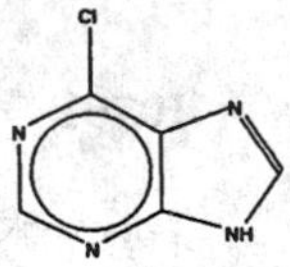

10755

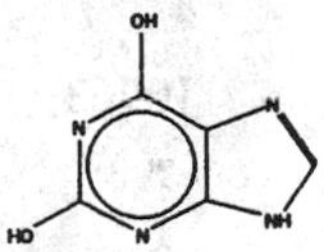

10756

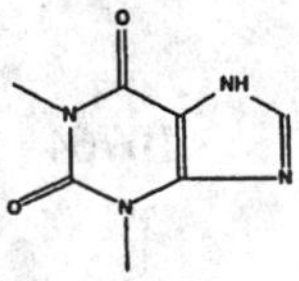

10757

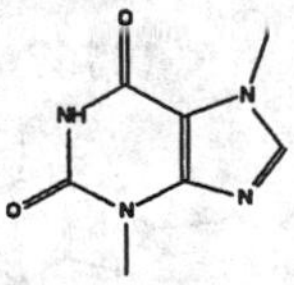

In numeric order as they occur in the Table of Physical Constants of Organic Compounds

10758

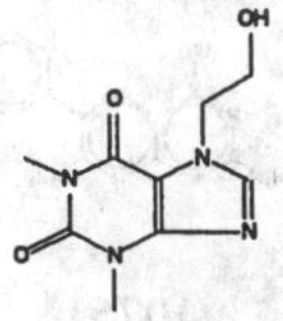

10759

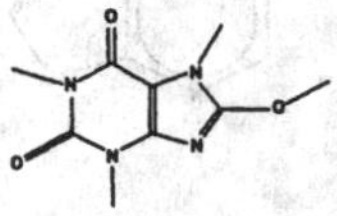

10760

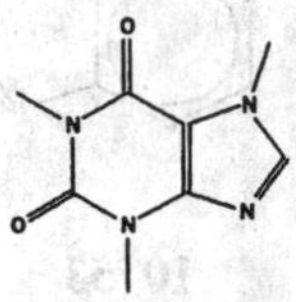

10761

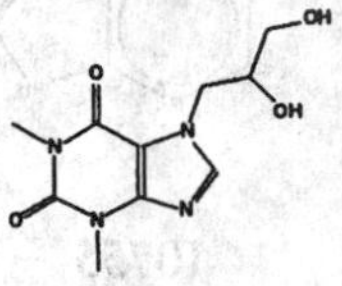

10763

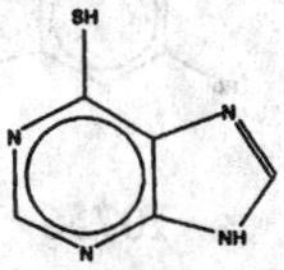

10764

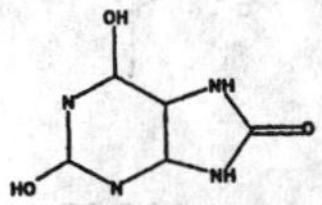

10767

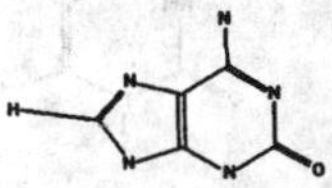

10768

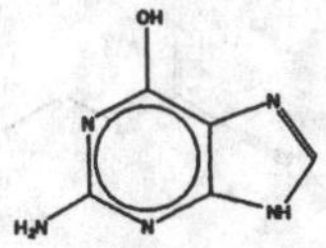

10769

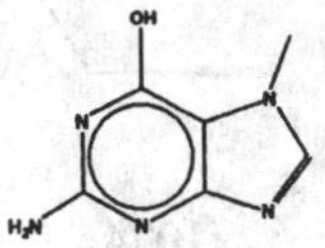

10770

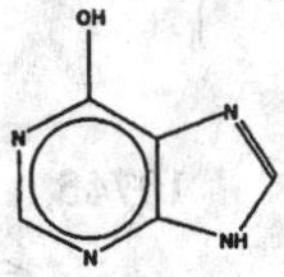

10773

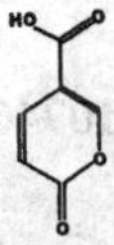

10774

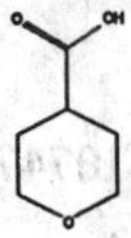

10775

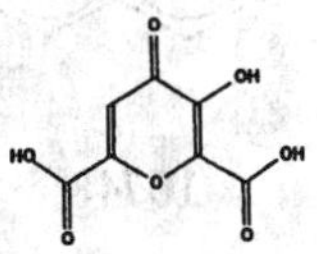

10777

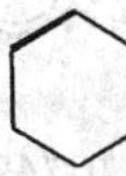

10778

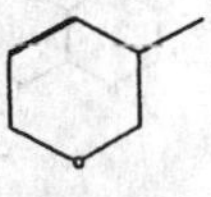

10779

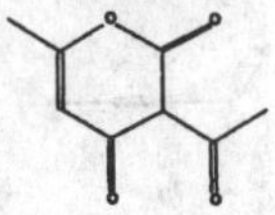

10780

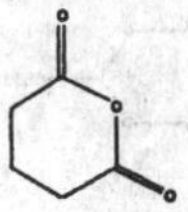

10781

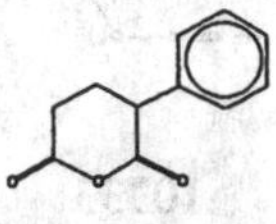

10782

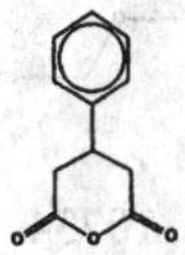

10783

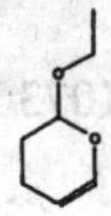

10784

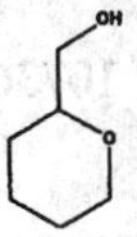

10787

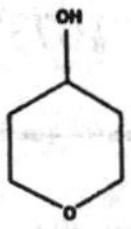

10788

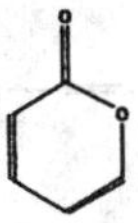

10789

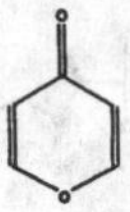

10790

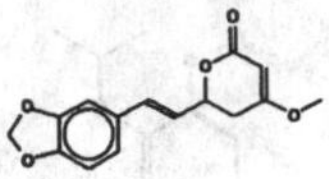

10791

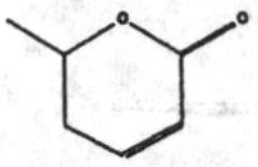

10792

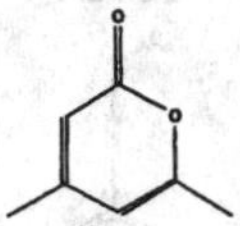

10793

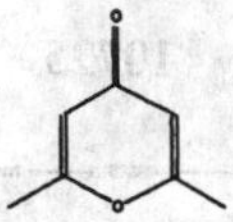

10794

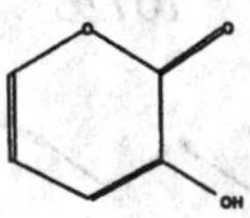

10795

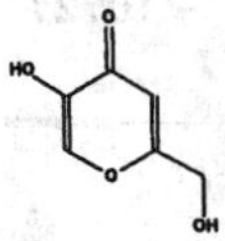

10796

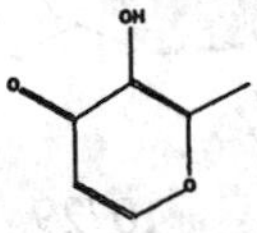

10798

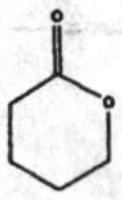

10799

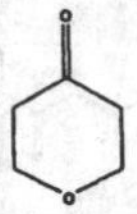

10800

10801

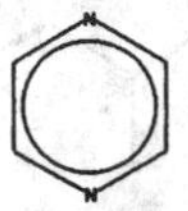

10802

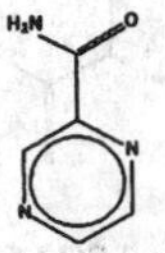

10803

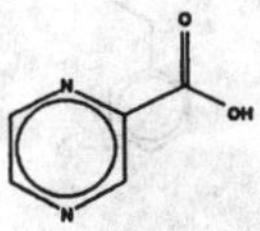

10805

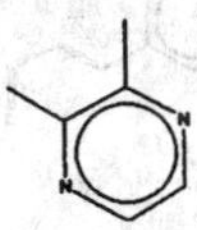

10806

10807

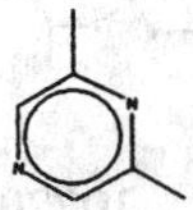

10809

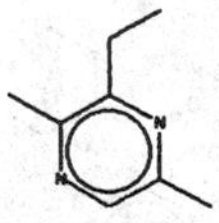

10810

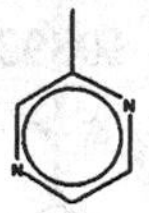

10811

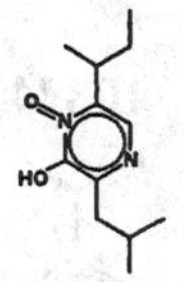

10813

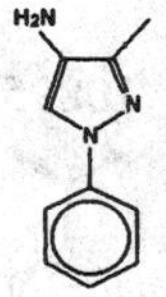

10814

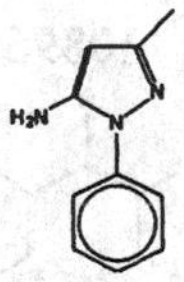

10815

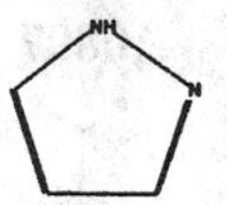

10816

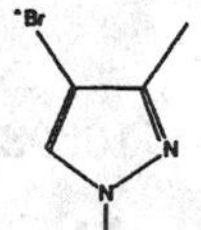

10818

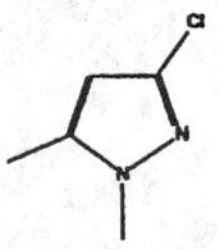

10819

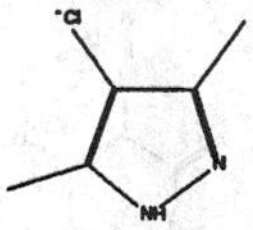

10820

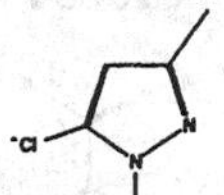

10821

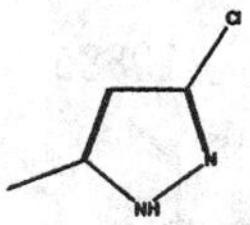

10822

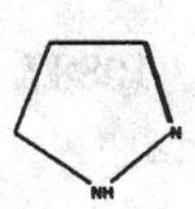

10823

10824

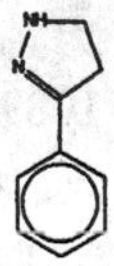

10825

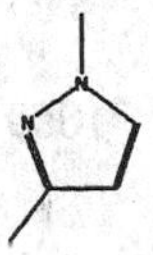

10826

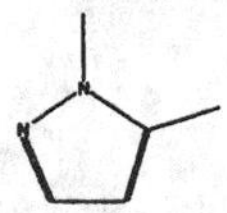

10827

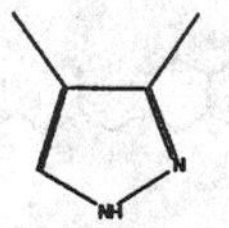

10828

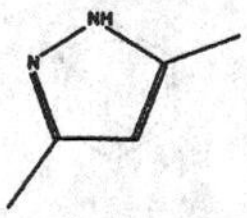

10829

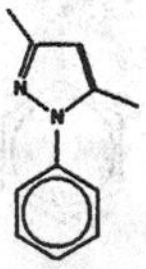

10830

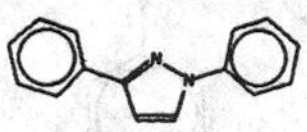

10831

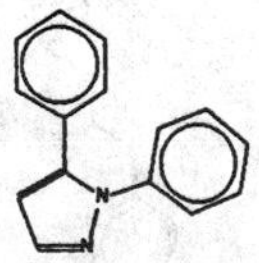

10832

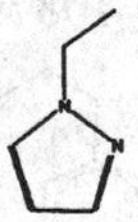

10833

10834

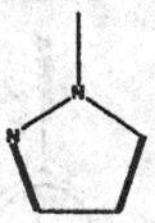

10835

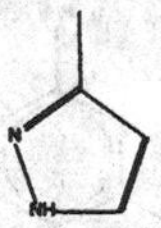

10836

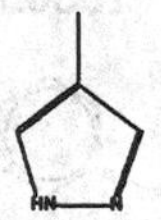

10837

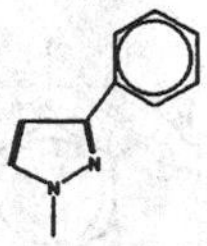

10838

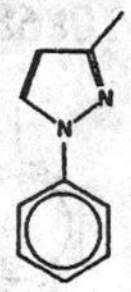

10839

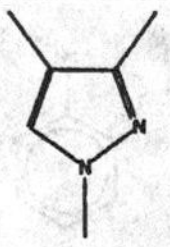

10840

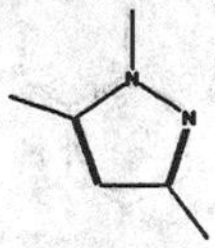

10841

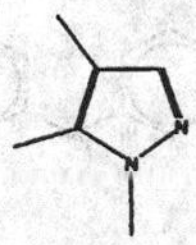

10842

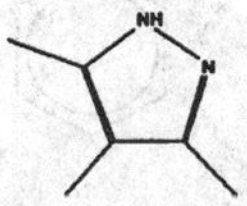

10844

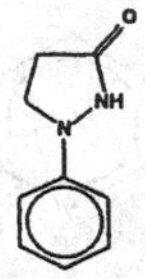

10845

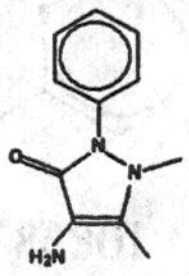

10846

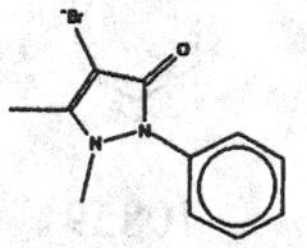

10847

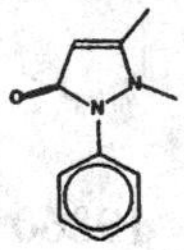

10849

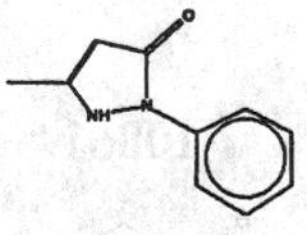

10851

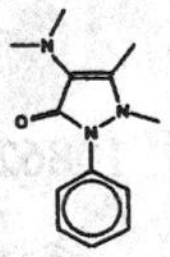

10852

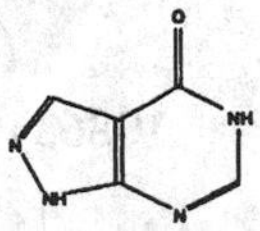

10853

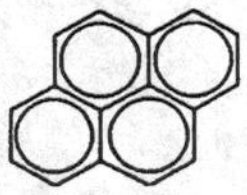

In numeric order as they occur in the Table of Physical Constants of Organic Compounds

10855

10856

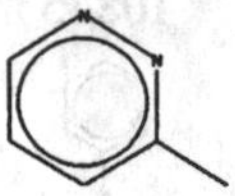

10858

10859

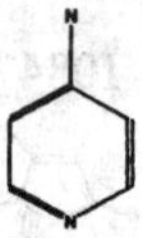

10860

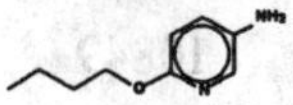

10861

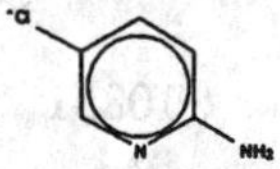

10862

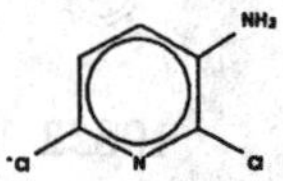

10863

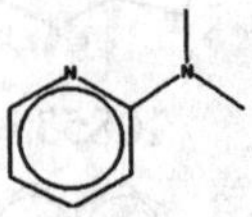

10864

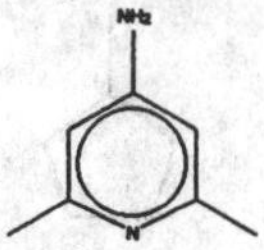

10865

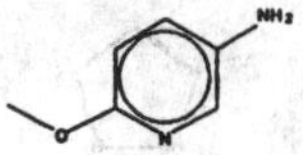

10866

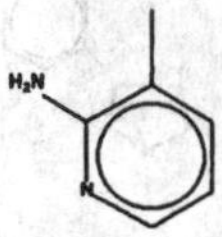

10868

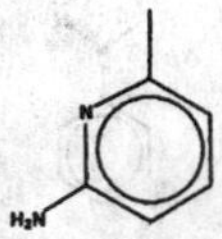

10869

10870

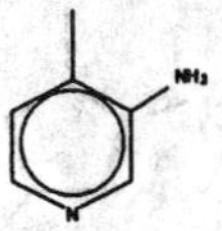

10871

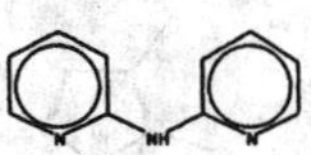

10872

10873

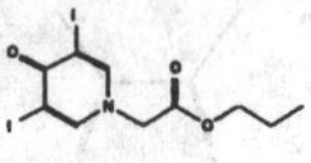

10875

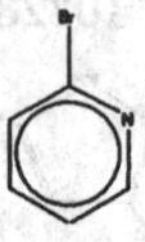

10876

10877

10878

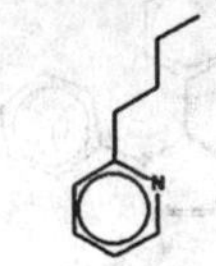

10879

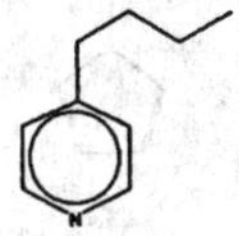

10880

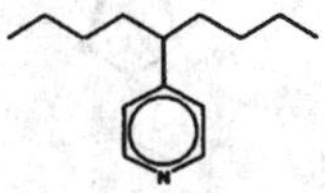

10882

10883

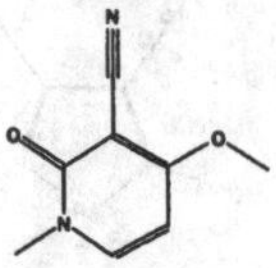

10884

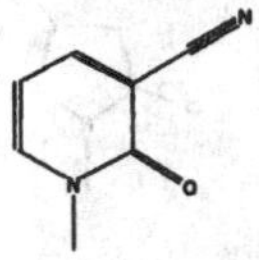

10886

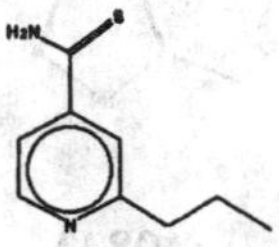

10887

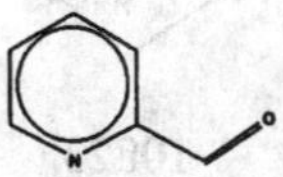

10888

10890

10891

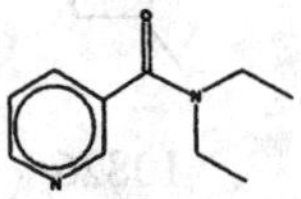

10892

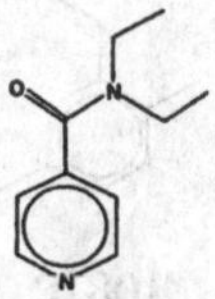

10893

10894

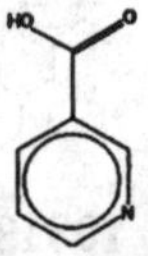

10895

10896

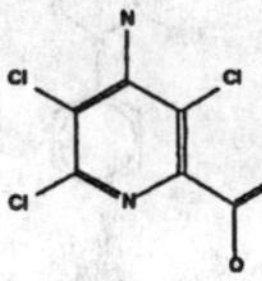

10897

10898

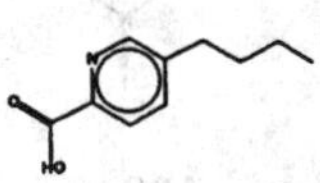

STRUCTURAL FORMULAS OF ORGANIC COMPOUNDS (continued)

In numeric order as they occur in the Table of Physical Constants of Organic Compounds

10899

10901

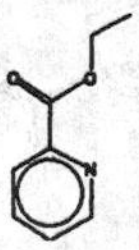

10903

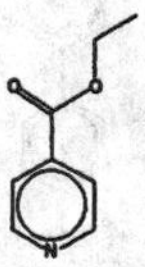

10905

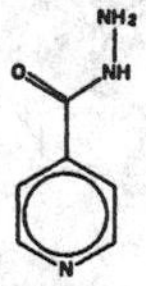

10906

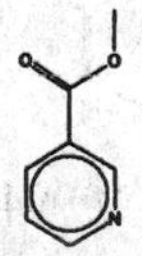

10907

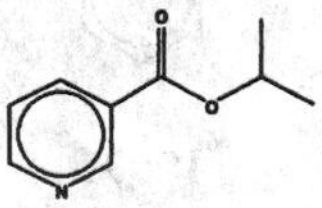

10908

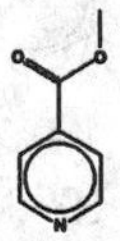

10909

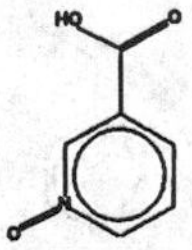

10911

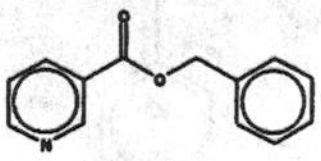

10914

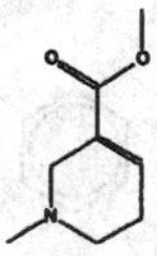

10915

10916

10917

10918

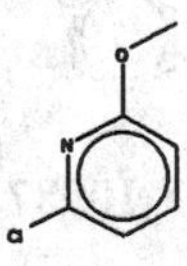

10920

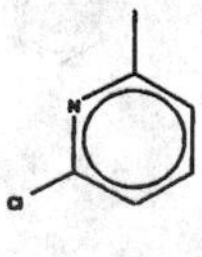

10921

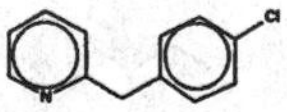

10922

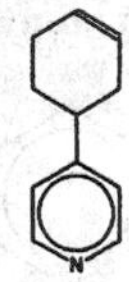

10923

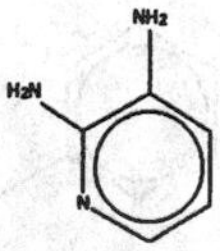

10924

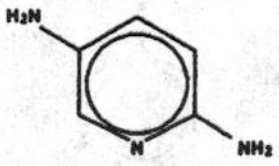

10925

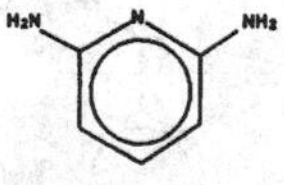

10926

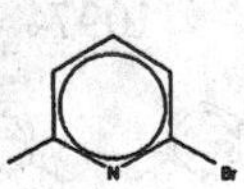

10927

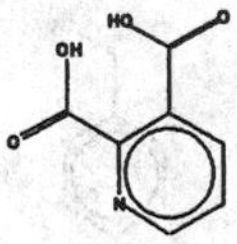

10929

10932

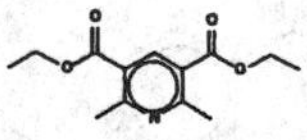

10933

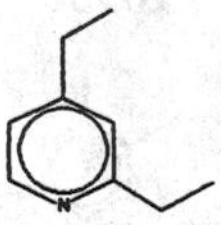

10935

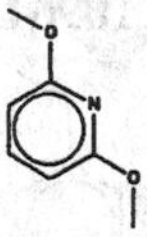

10936

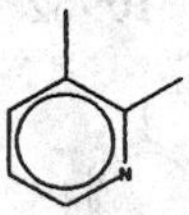

10937

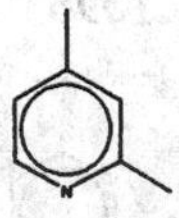

10938

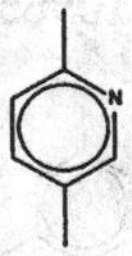

10939

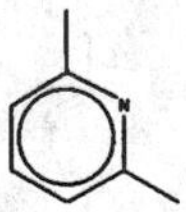

10940

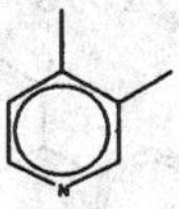

10941

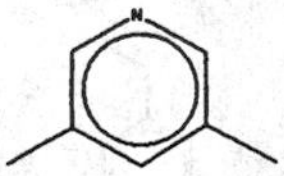

10942

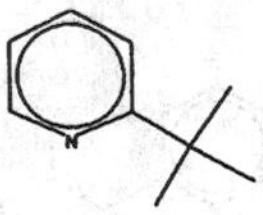

10943

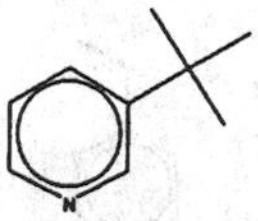

10944

10945

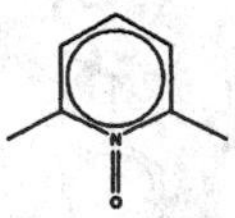

10946

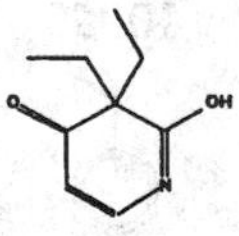

10947

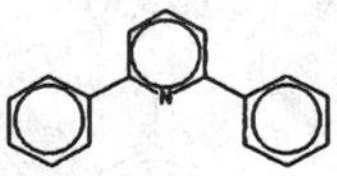

10948

In numeric order as they occur in the Table of Physical Constants of Organic Compounds

10949

10951

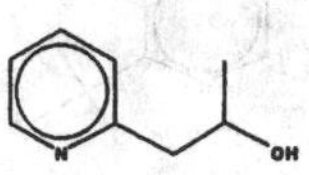

10952

10953

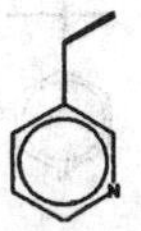

10954

10955

10956

10958

10959

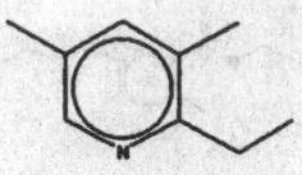

10960

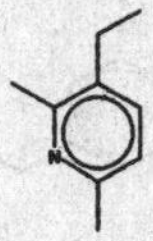

10961

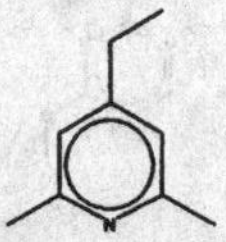

10962

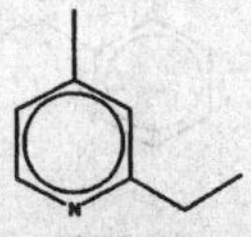

10963

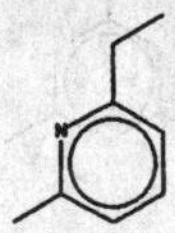

10965

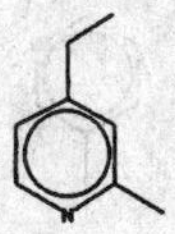

10966

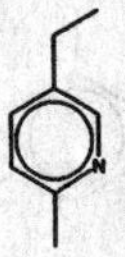

10967

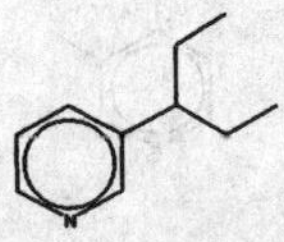

10968

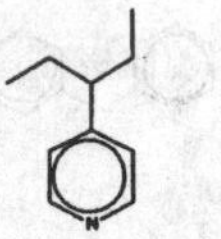

10969

10970

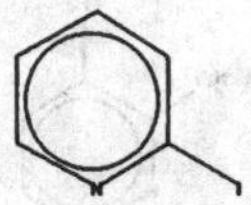

10972

10973

10974

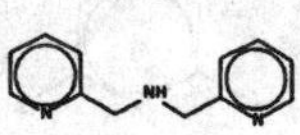

10975

10976

10977

10980

10981

10982

10983

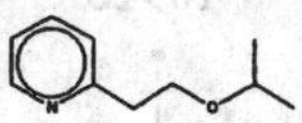

10984

10985

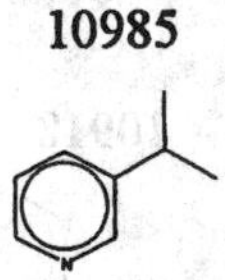

10987

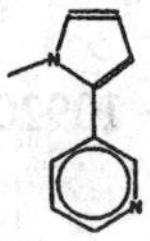

10988

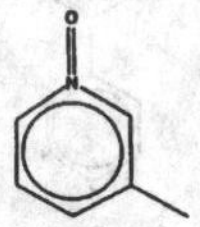

10989

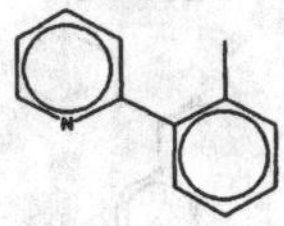

10990

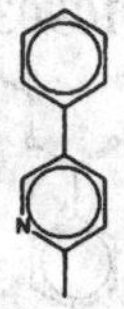

10991

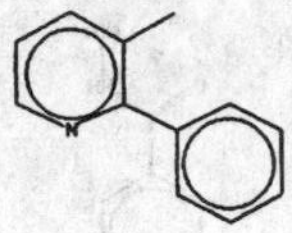

10992

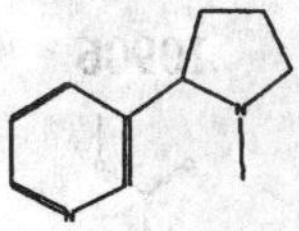

10994

10995

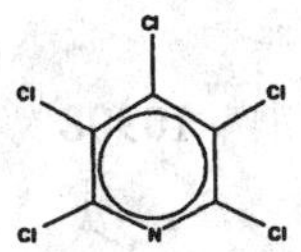

10996

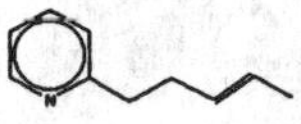

10997

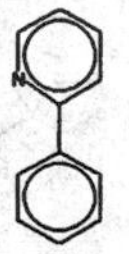

10998

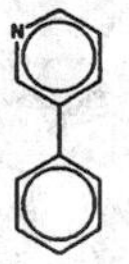

10999

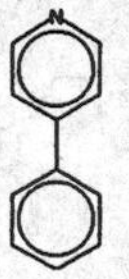

11001

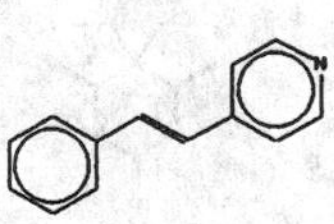

11002

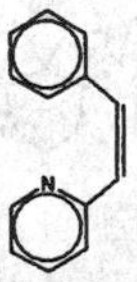

11003

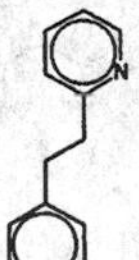

11004

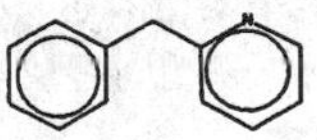

11005

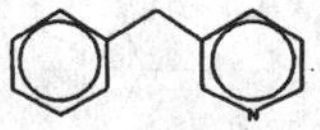

11006

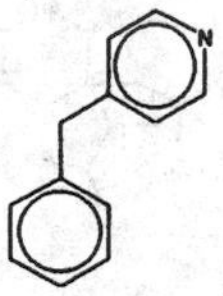

11007

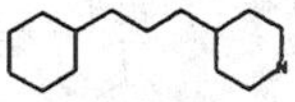

11008

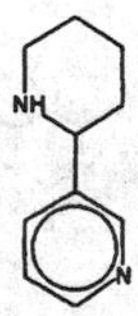

11009

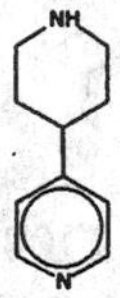

11010

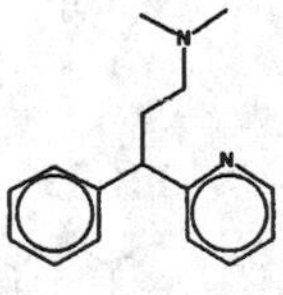

11012

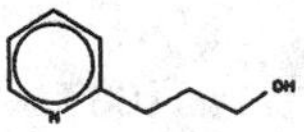

11013

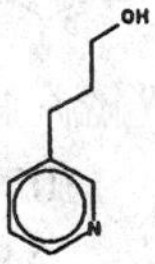

11014

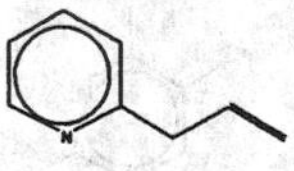

11016

11017

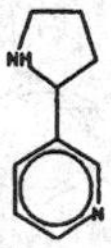

11018

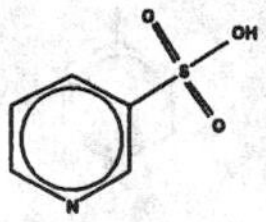

11019

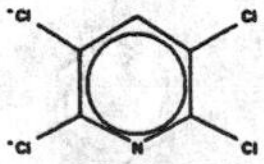

11020

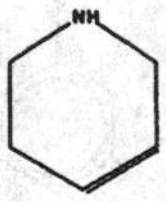

11021

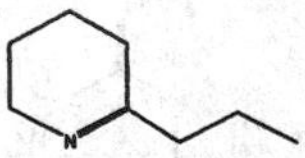

11023

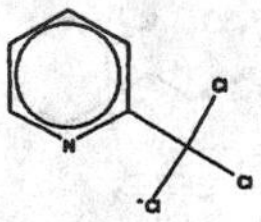

11024

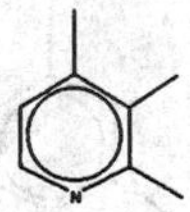

11025

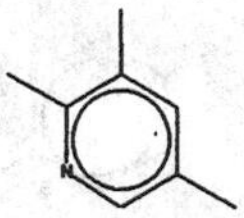

11026

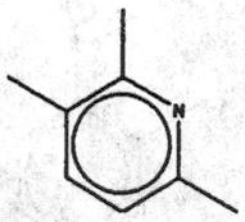

11027

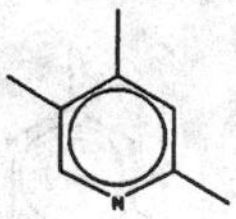

11028

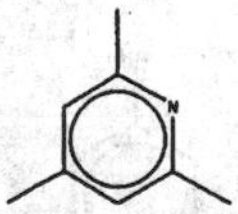

11029

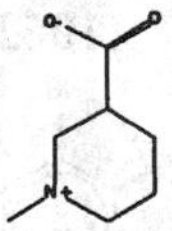

11031

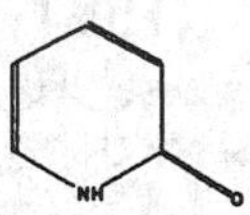

11032

11033

11034

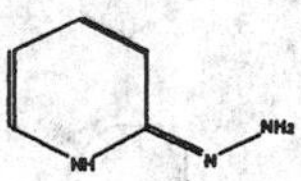

11035

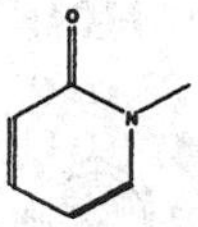

11036

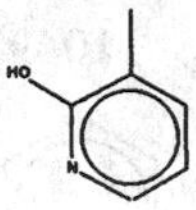

11038

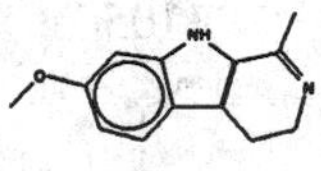

11039

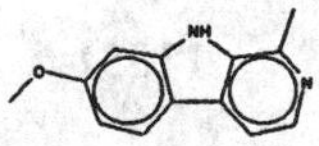

11040

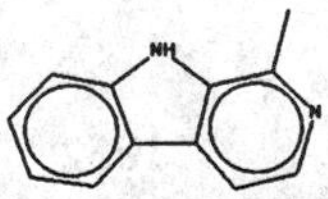

11041

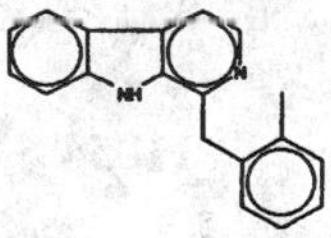

STRUCTURAL FORMULAS OF ORGANIC COMPOUNDS (continued)

In numeric order as they occur in the Table of Physical Constants of Organic Compounds

11042

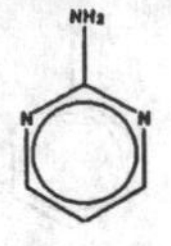

11043

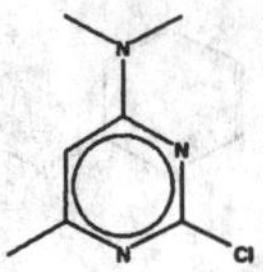

11045

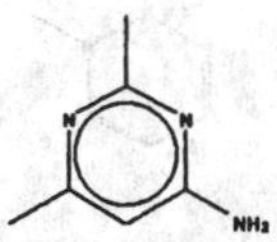

11049

11050

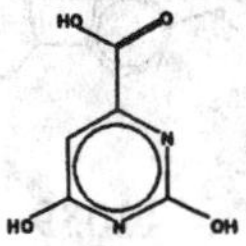

11051

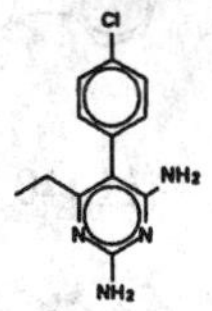

11052

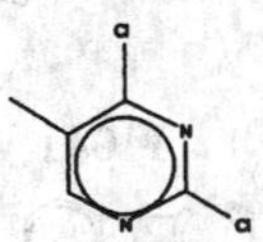

11054

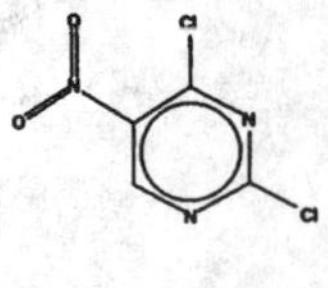

11055

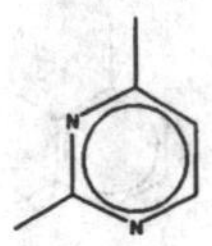

11056

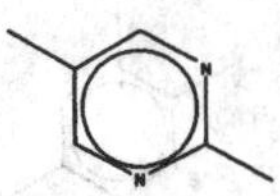

11057

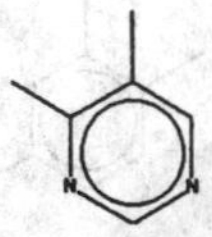

11058

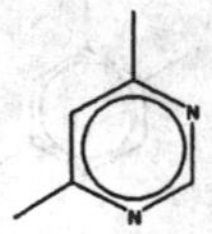

11059

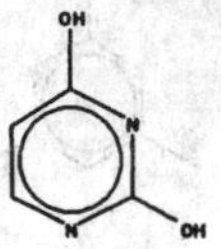

11062

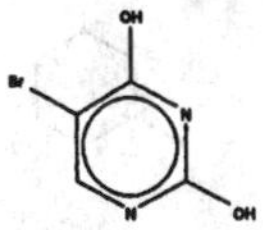

11064

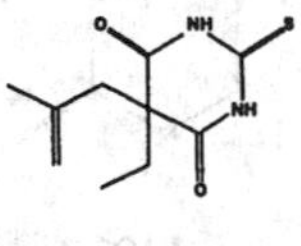

11066

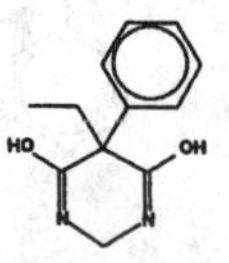

11067

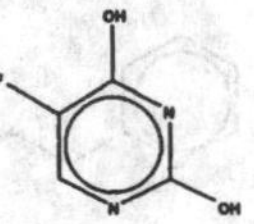

11068

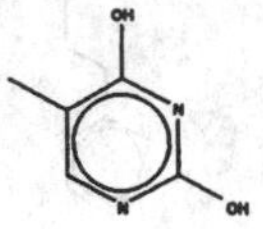

11069

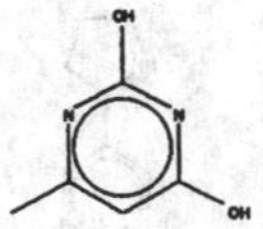

11071

11072

11075

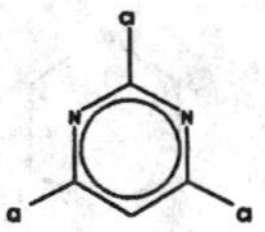

11076

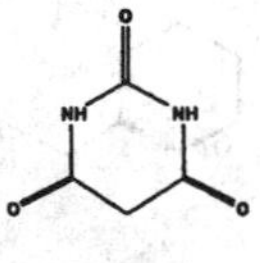

11078

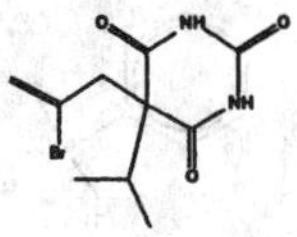

11080

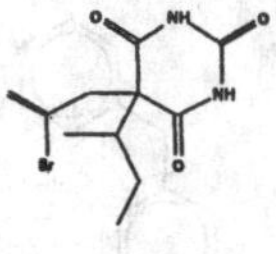

11081

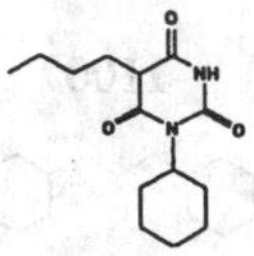

11082

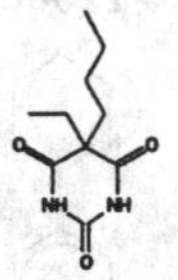

11083

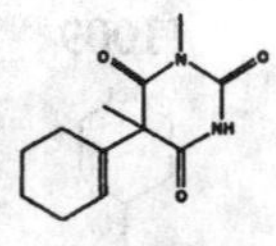

11084

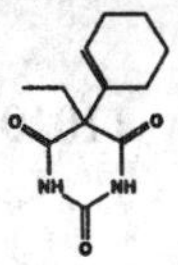

11085

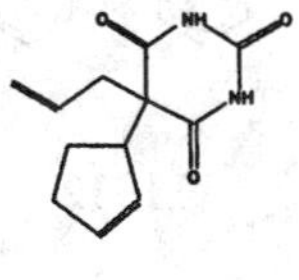

11087

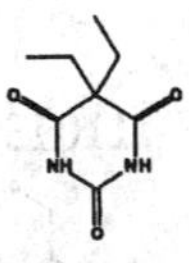

11088

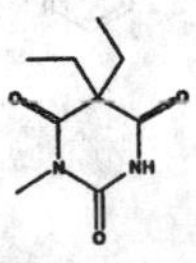

11090

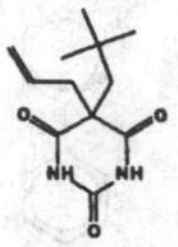

11091

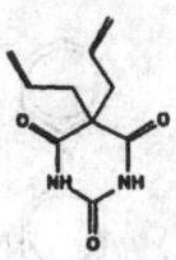

11092

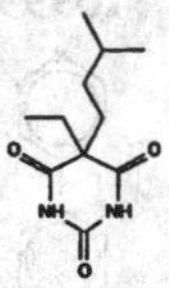

11093

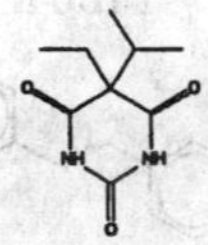

11094

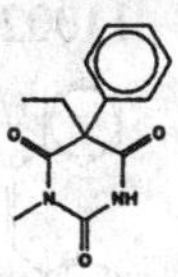

11095

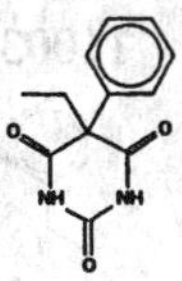

11098

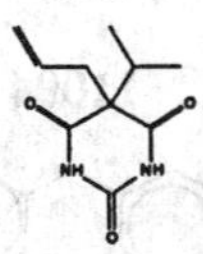

11099

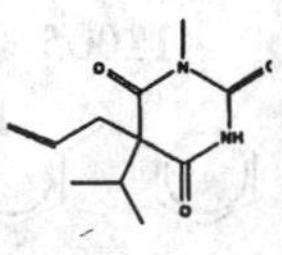

11100

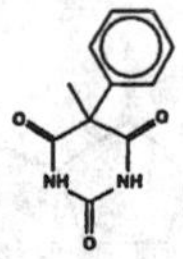

11101

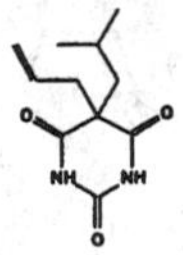

11103

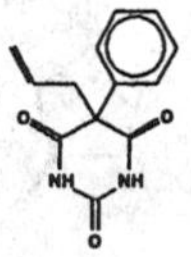

11104

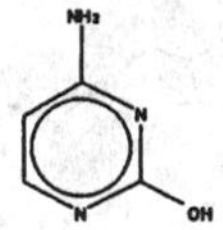

11105

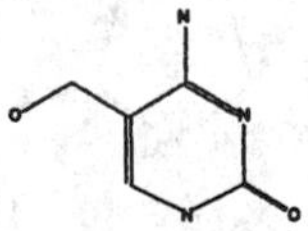

11106

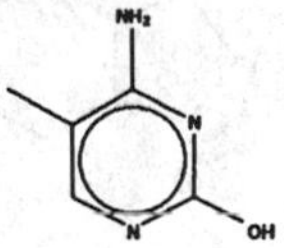

11109

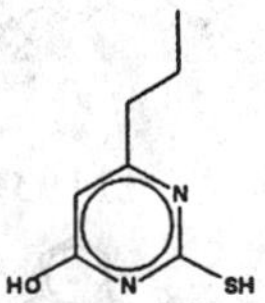

11110

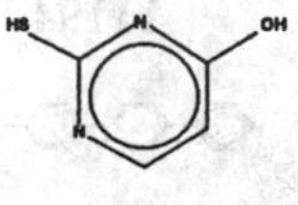

11112

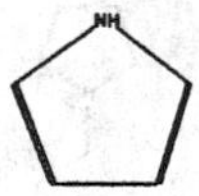

11113

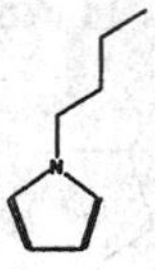

11114

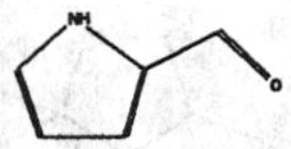

11115

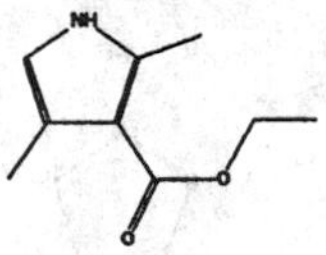

11116

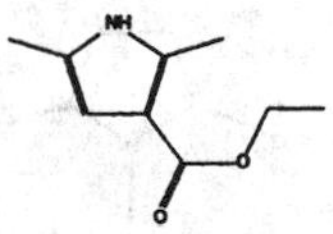

11117

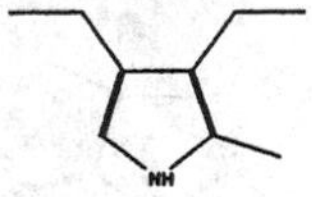

11118

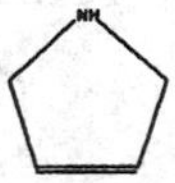

11119

11120

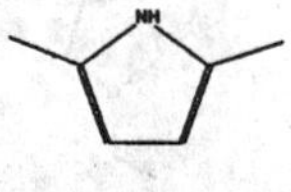

11121

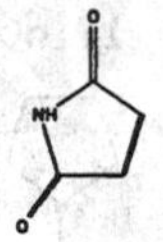

11122

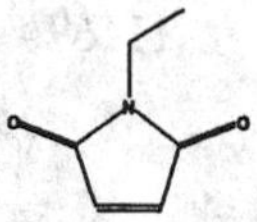

11124

11125

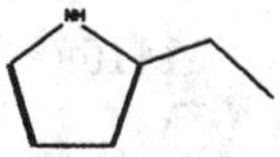

11126

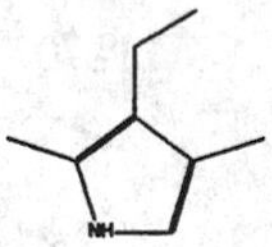

11127

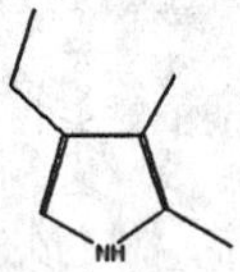

11128

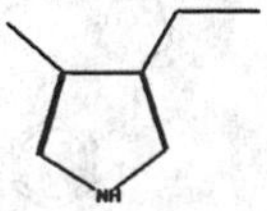

11129

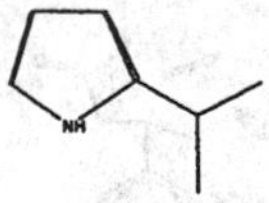

11130

11131

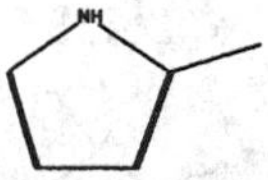

11132

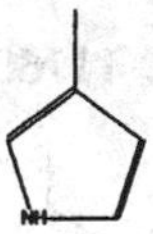

11133

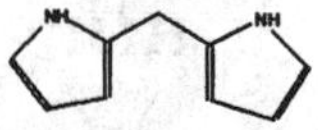

11134

11135

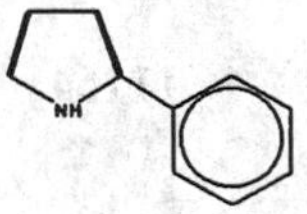

11136

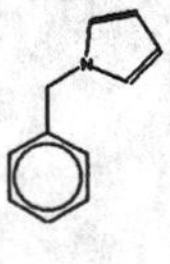

11137

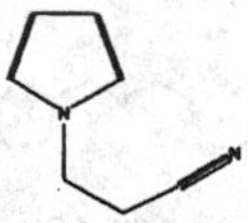

11138

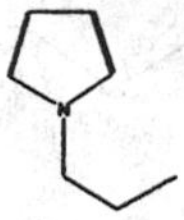

11139

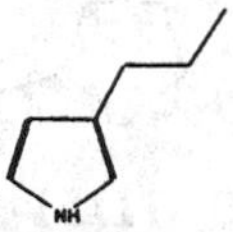

11141

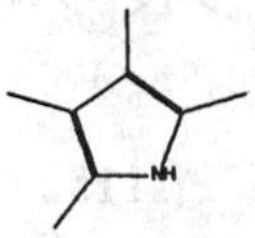

11142

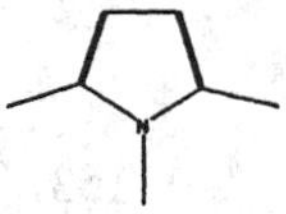

11143

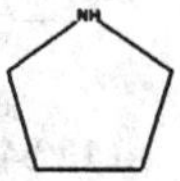

11146

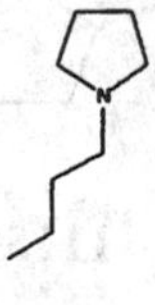

11147

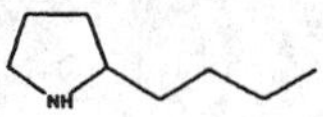

11148

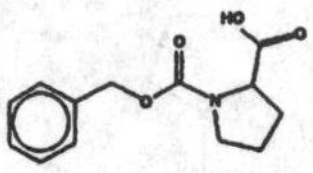

11149

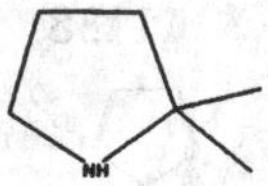

11150

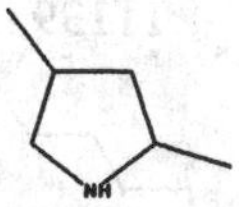

11151

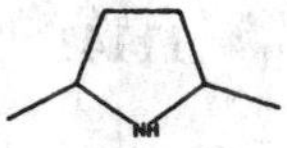

11152

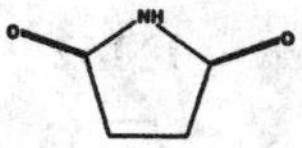

11153

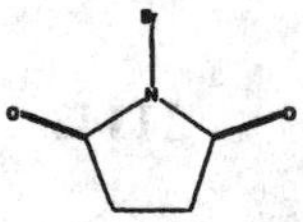

11154

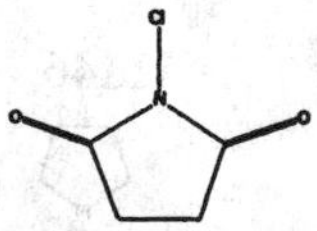

11155

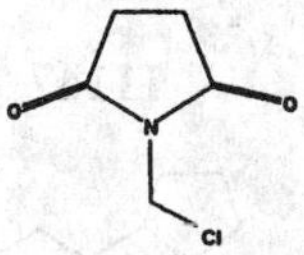

11156

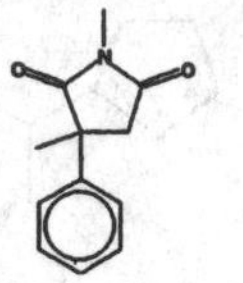

11157

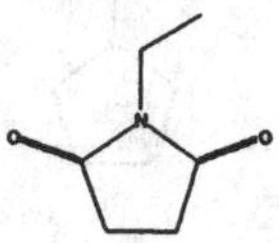

11158

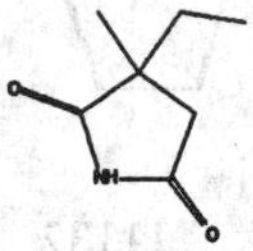

11160

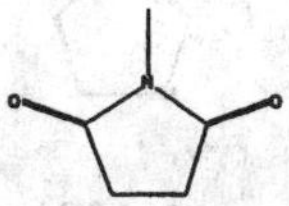

11161

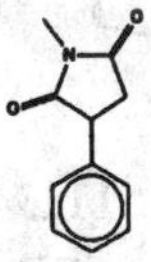

11163

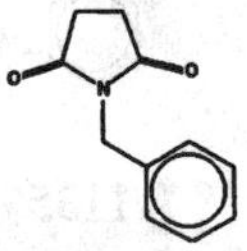

11164

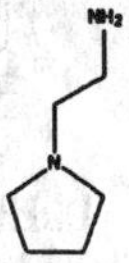

11165

11166

11167

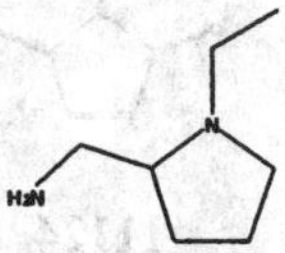

11168

11169

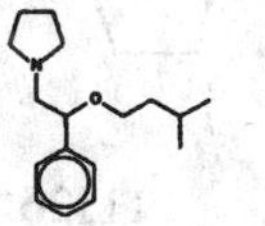

11170

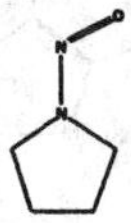

11171

11172

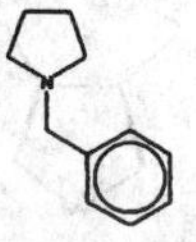

11173

11175

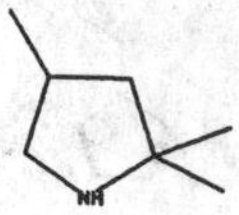

11176

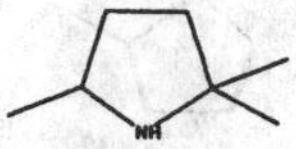

11180

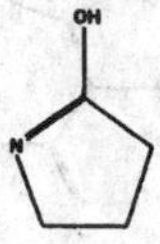

11181

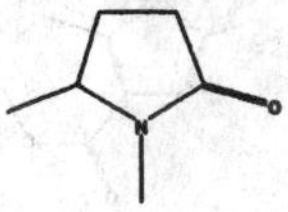

11182

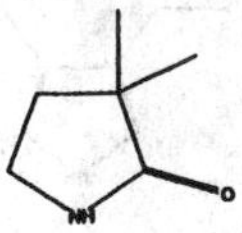

11183

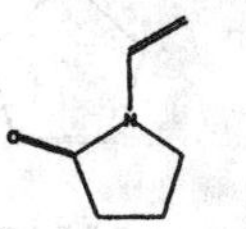

11184

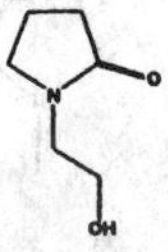

11185

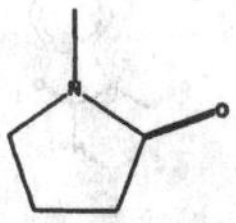

11186

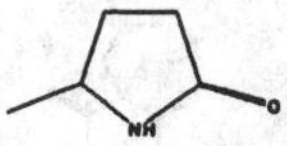

11187

11188

11189

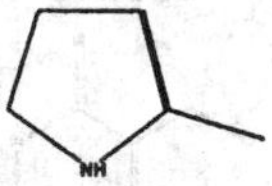

11190

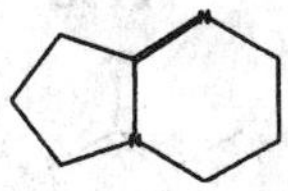

11191

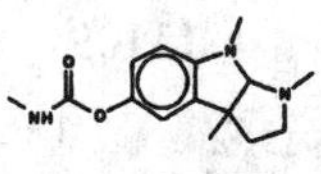

11193

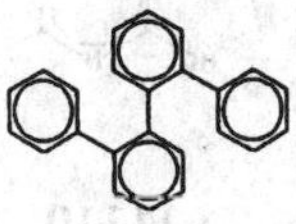

11194

11195

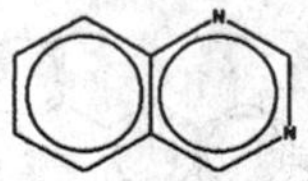

11199

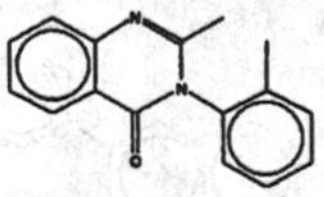

11201

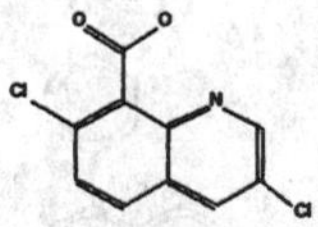

11202

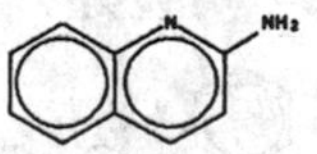

11203

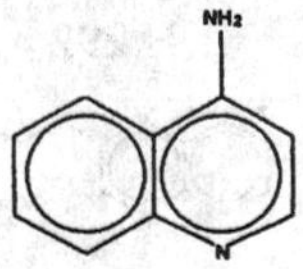

11204

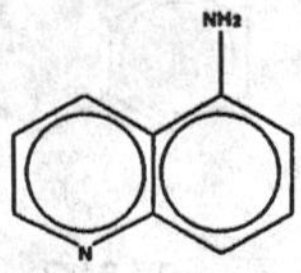

11206

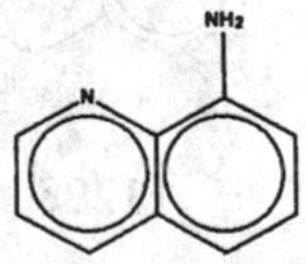

11208

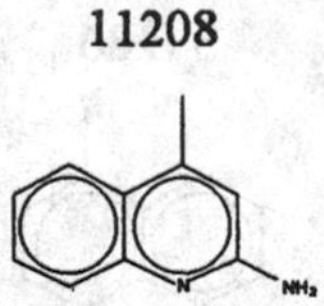

•11209

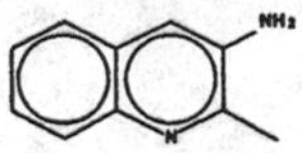

11210

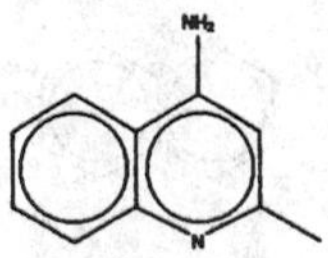

11211

11213

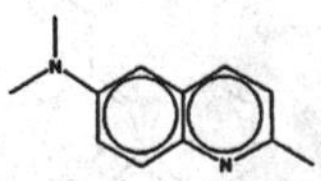

11214

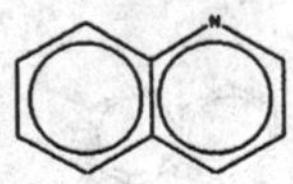

11217

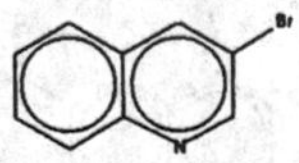

11218

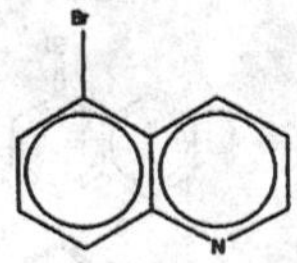

11219

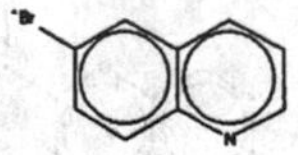

11220

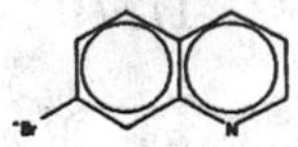

11222

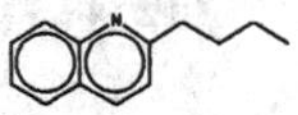

11223

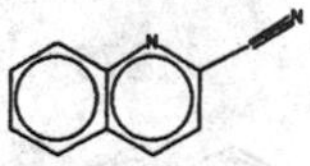

11224

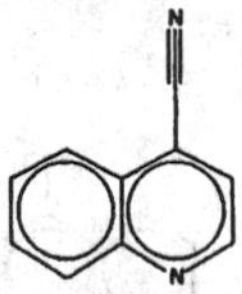

11225

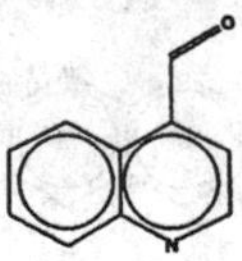

11227

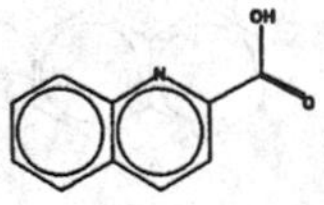

11230

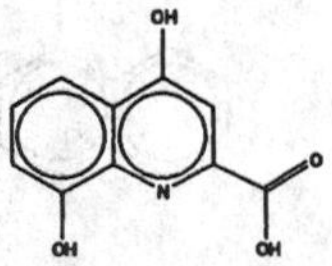

11231

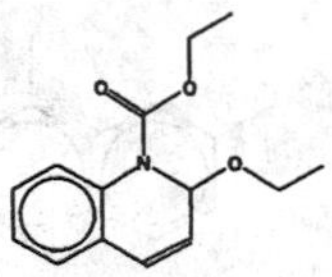

11232

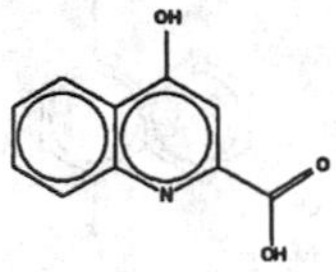

11236

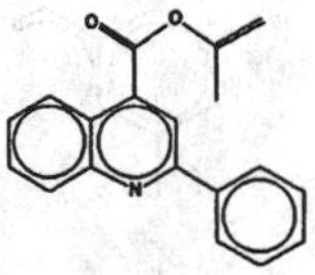

11237

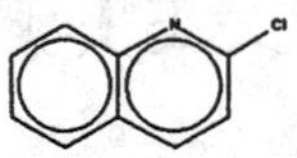

11238

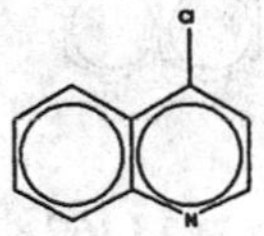

11239

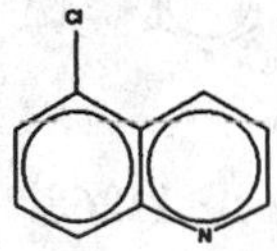

11240

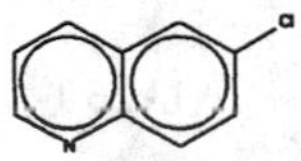

11241

11242

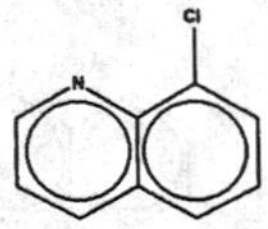

11243

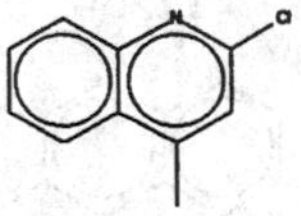

11244

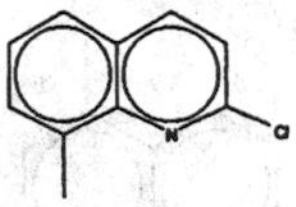

11245

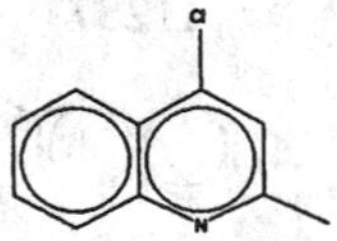

11246

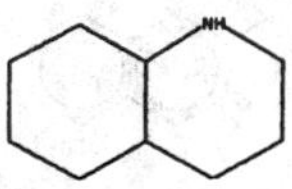

11247

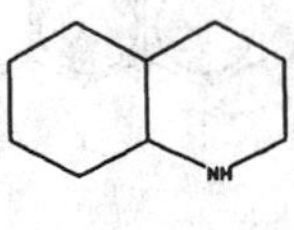

11248

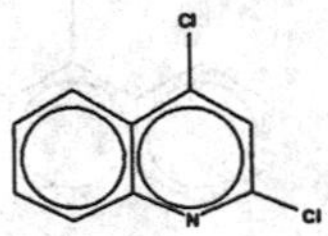

11250

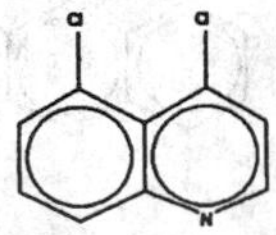

11251

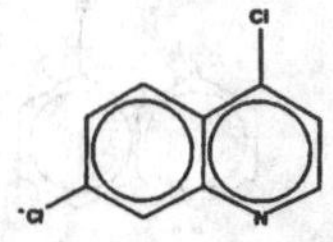

11254

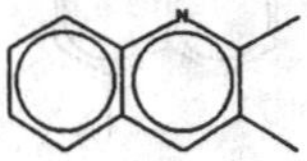

11255

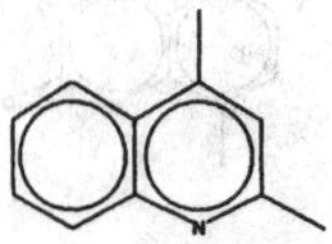

11257

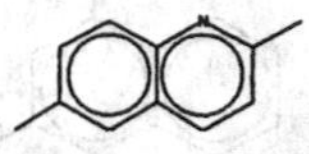

11258

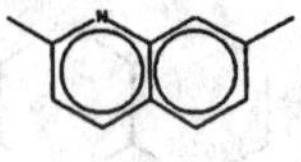

11259

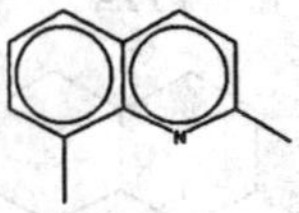

11260

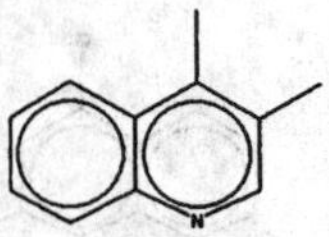

11261

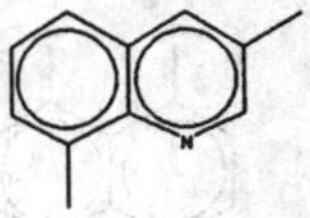

11262

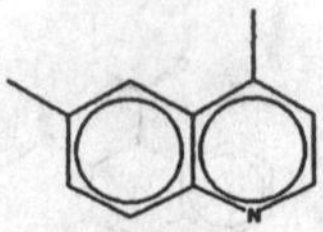

11263

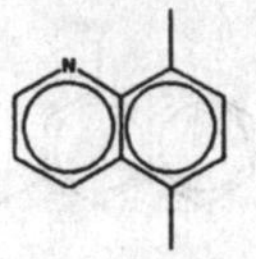

11264

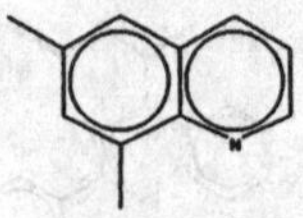

11265

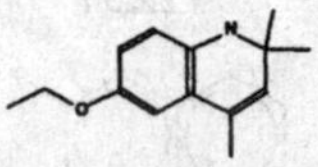

11266

11267

11268

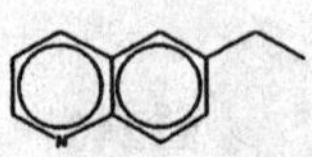

11269

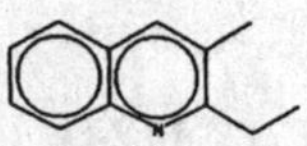

11270

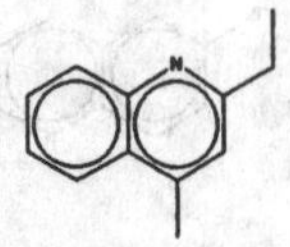

11271

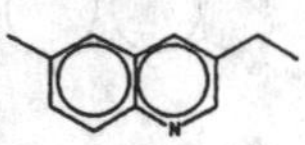

11274

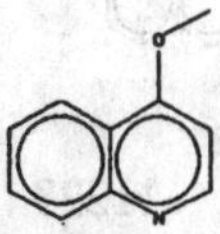

11275

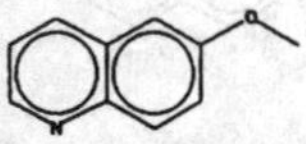

11276

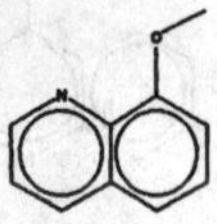

11277

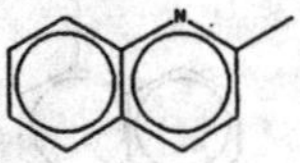

11278

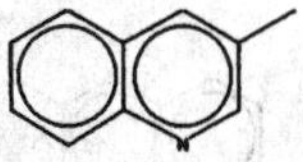

11279

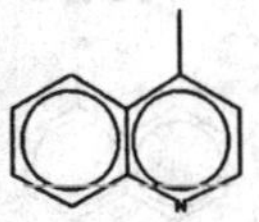

11280

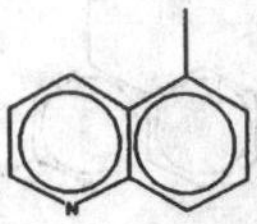

11281

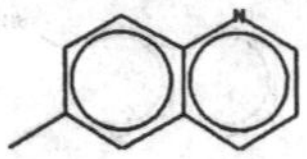

11282

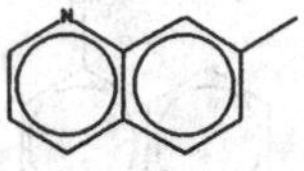

11283

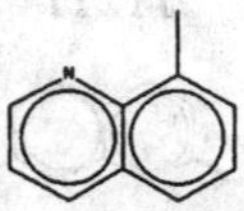

11284

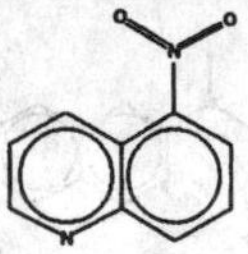

11285

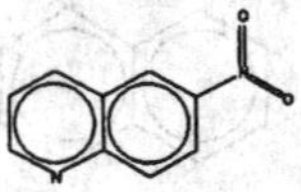

11286

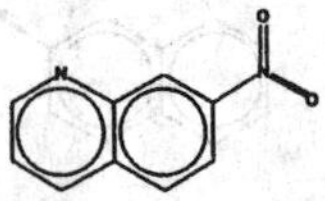

11287

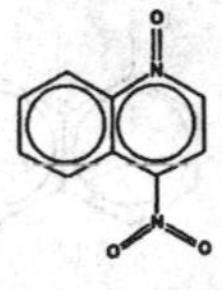

11288

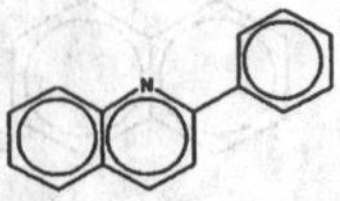

11289

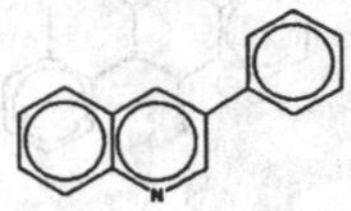

11290

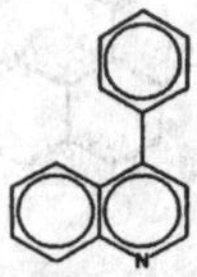

11291

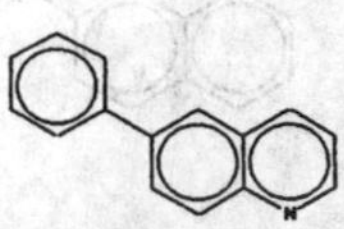

11292

11296

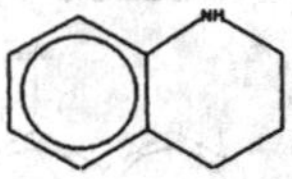

11297

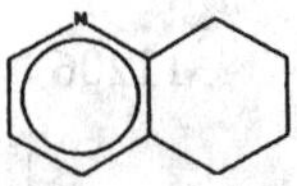

11298

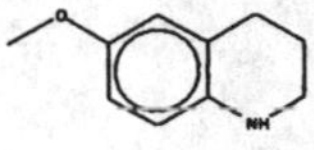

In numeric order as they occur in the Table of Physical Constants of Organic Compounds

11299

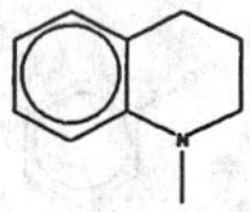

11300

11301

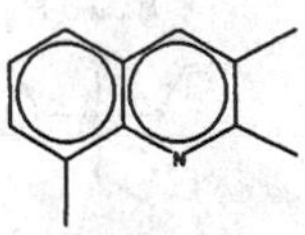

11302

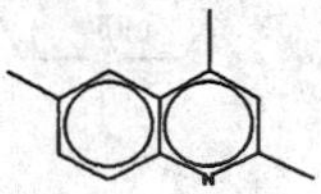

11303

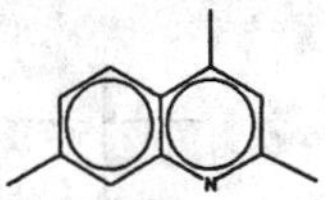

11304

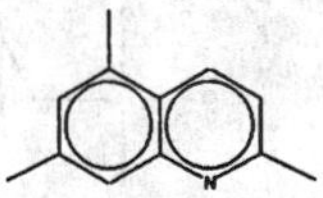

11305

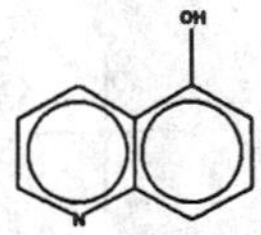

11306

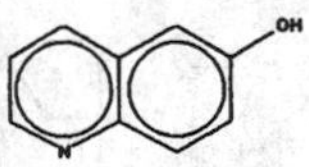

11307

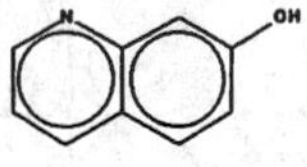

11308

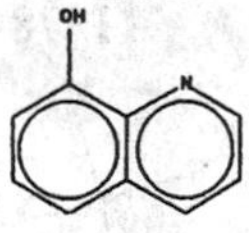

11311

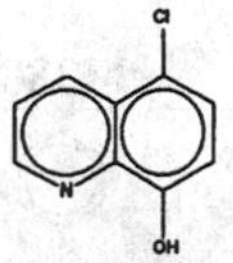

11312

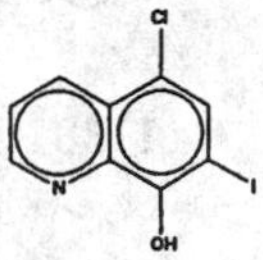

11313

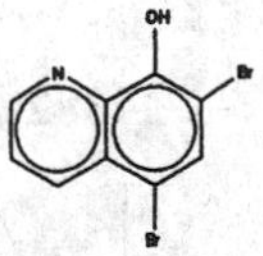

11314

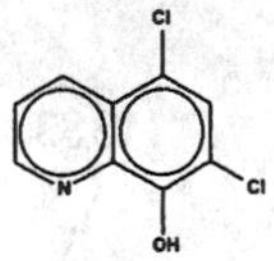

11315

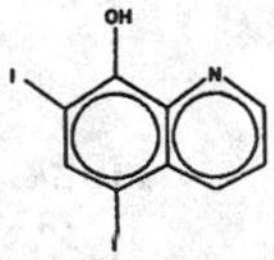

11317

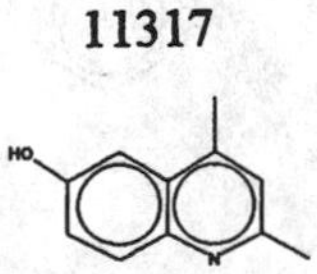

11318

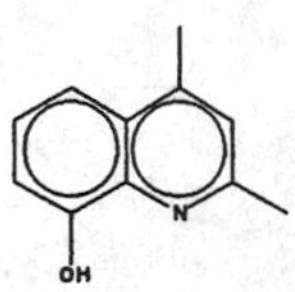

11319

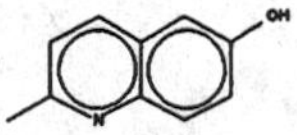

11320

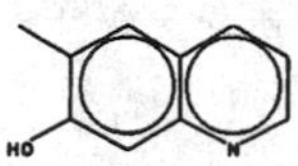

11321

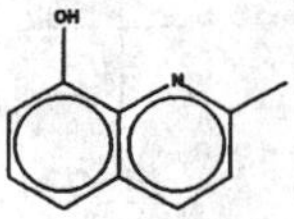

11325

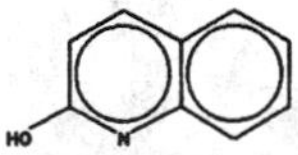

11327

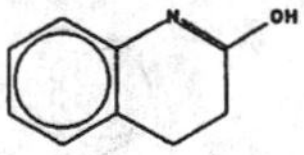

11328

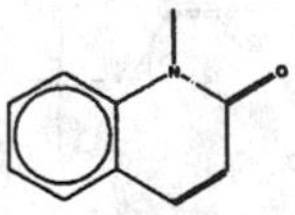

11329

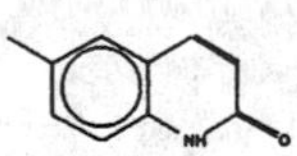

11330

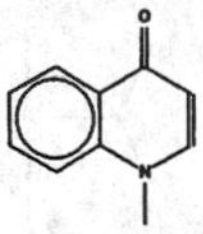

11331

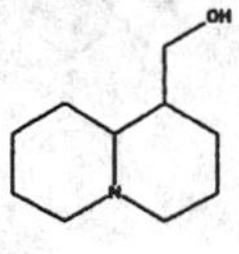

11333

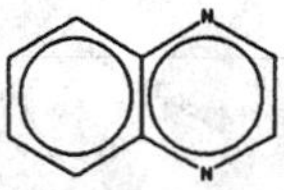

11334

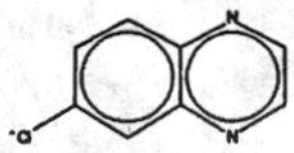

11335

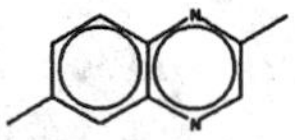

11337

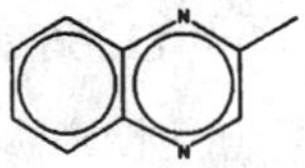

11338

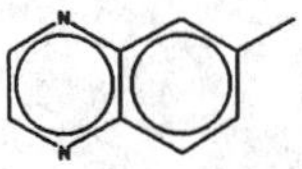

11339

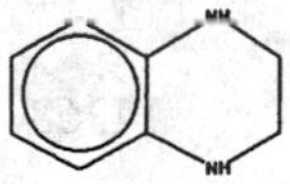

11340

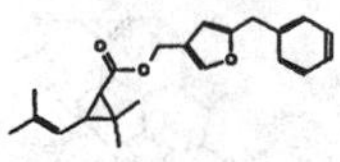

11341

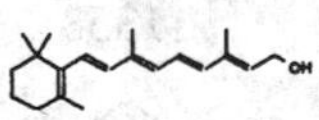

11342

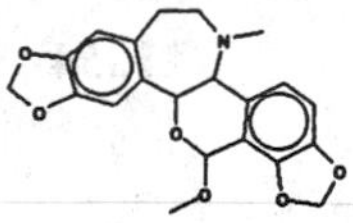

11343

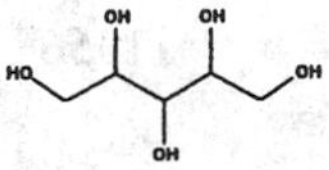

11344

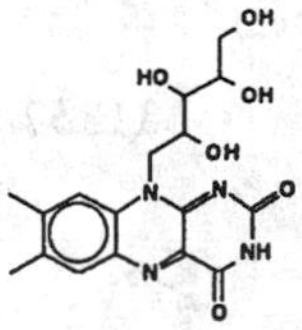

11345

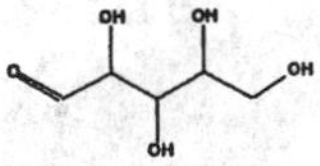

11348

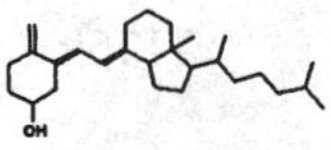

11350

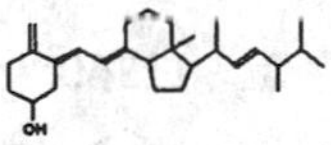

In numeric order as they occur in the Table of Physical Constants of Organic Compounds

11351

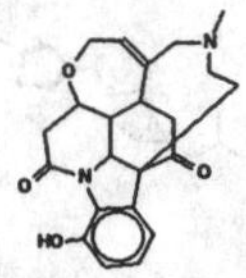

11353

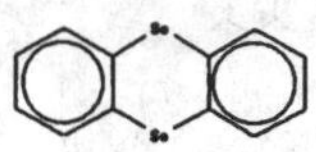

11354

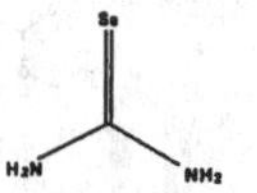

11356

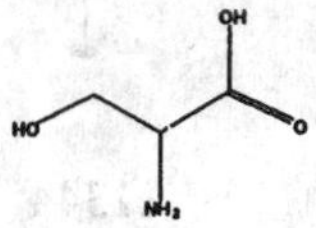

11357

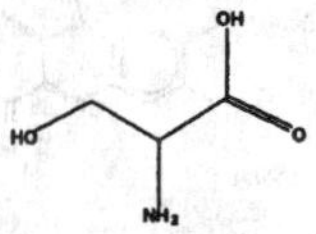

11358

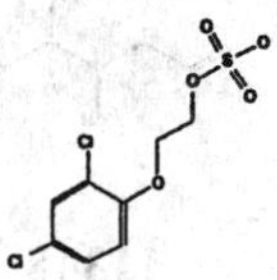

11359

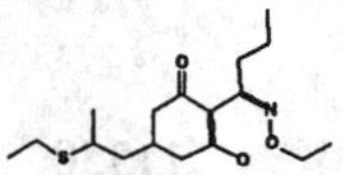

11360

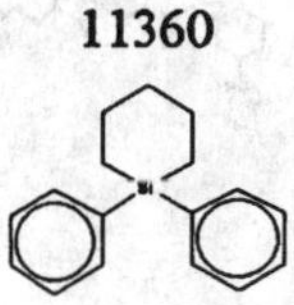

11361

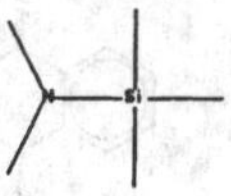

11362

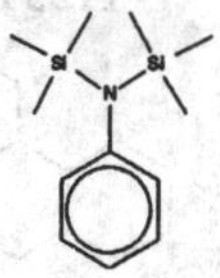

11363

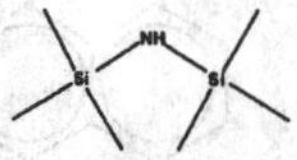

11364

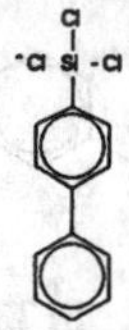

11365

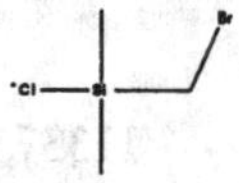

11366

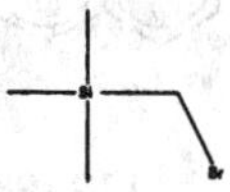

11367

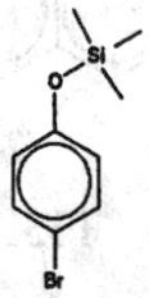

11368

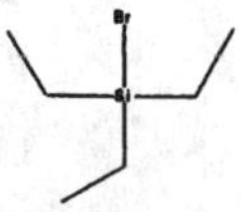

11369

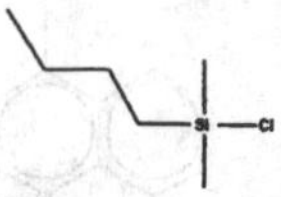

11370

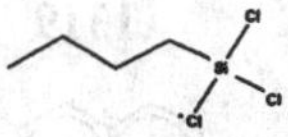

11371

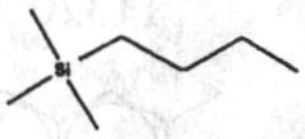

11372

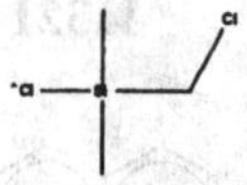

11373

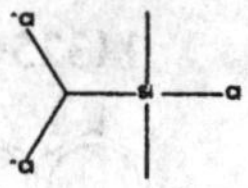

11374

11375

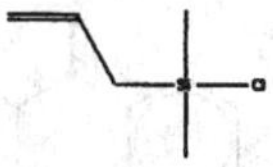

11376

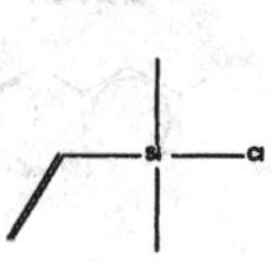

11377

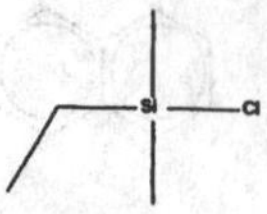

11378

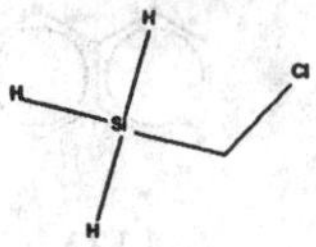

11380

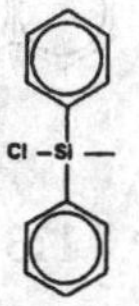

11381

11382

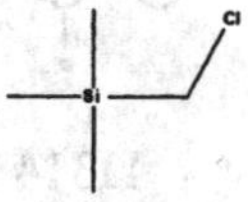

11383

11384

11385

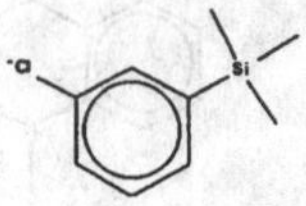

11386

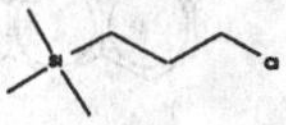

11387

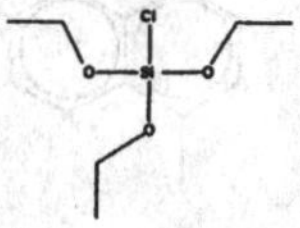

11388

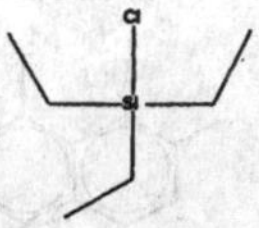

11389

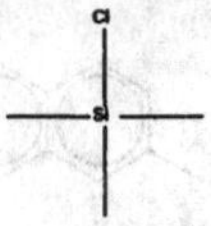

11390

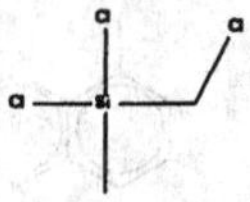

11391

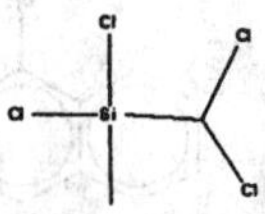

11392

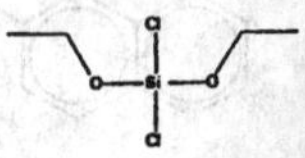

STRUCTURAL FORMULAS OF ORGANIC COMPOUNDS (continued)

In numeric order as they occur in the Table of Physical Constants of Organic Compounds

11393

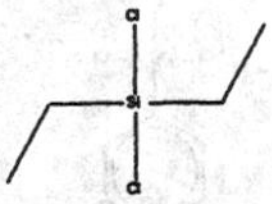

11394

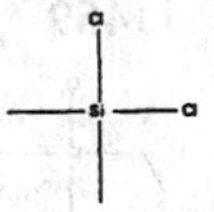

11395

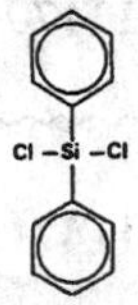

11396

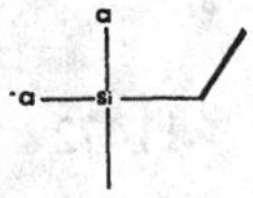

11397

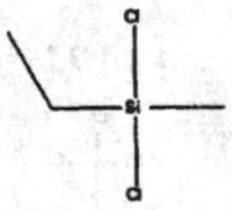

11398

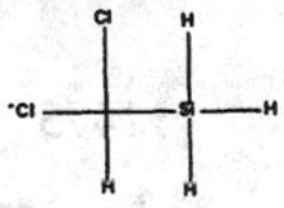

11399

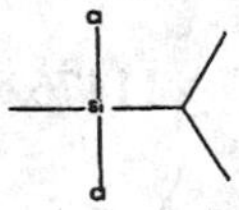

11400

11401

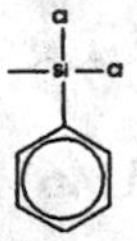

11402

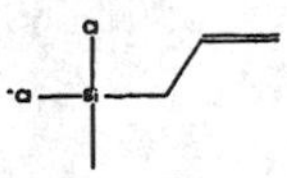

11403

11404

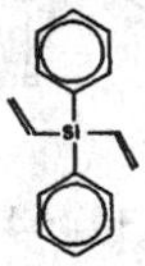

11405

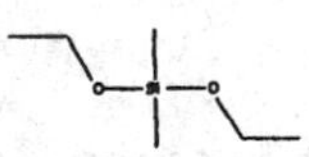

11406

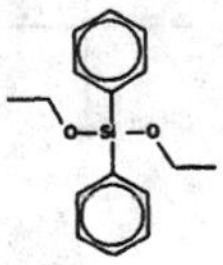

11407

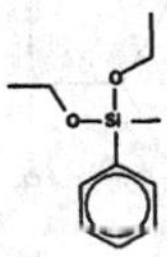

11408

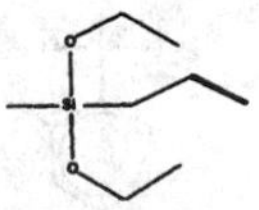

11409

11410

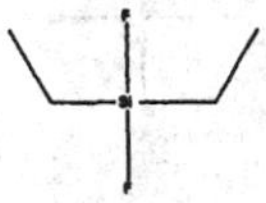

11411

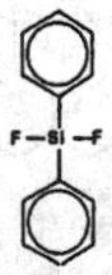

11412

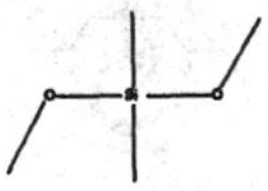

11413

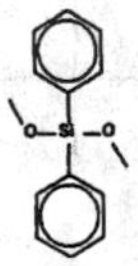

11414

11415

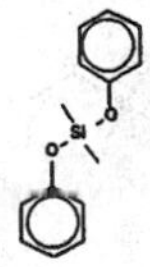

11417

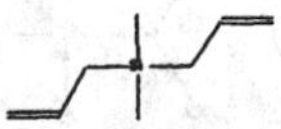

11418

11419

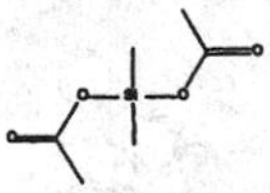

11420

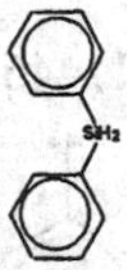

11421

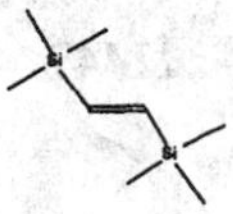

11422

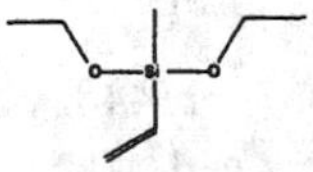

11423

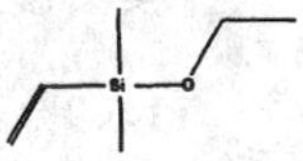

11424

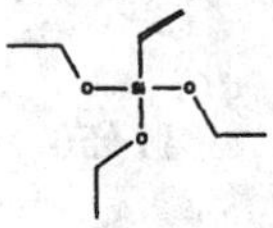

11425

11426

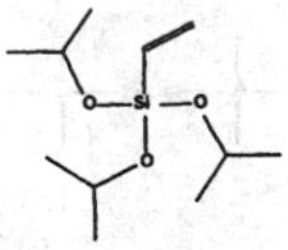

11427

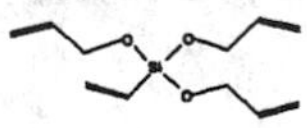

11428

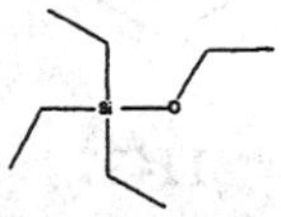

11429

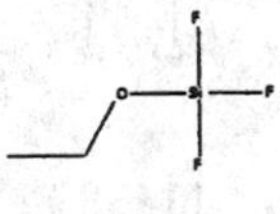

11430

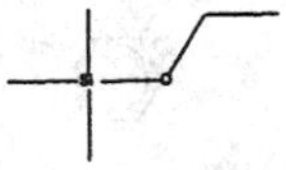

11431

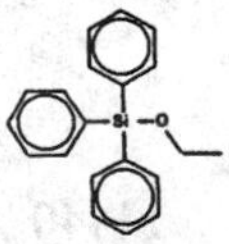

11432

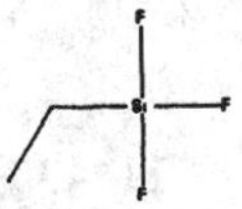

11433

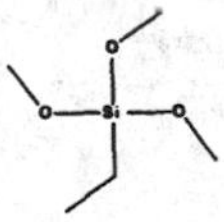

STRUCTURAL FORMULAS OF ORGANIC COMPOUNDS (continued)

In numeric order as they occur in the Table of Physical Constants of Organic Compounds

11434

11435

11436

11437

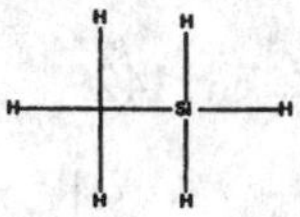

11438

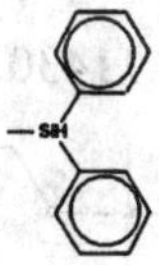

11439

11440

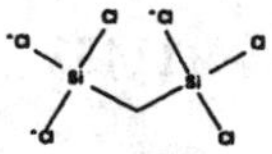

11443

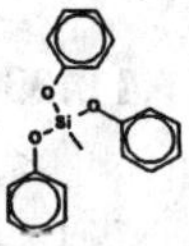

11444

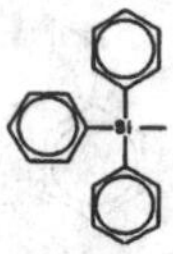

11445

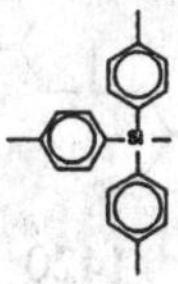

11446

11447

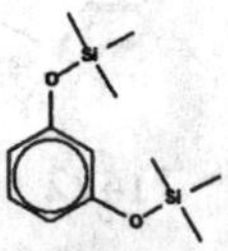

11448

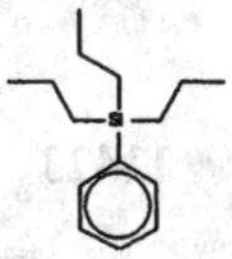

11449

11450

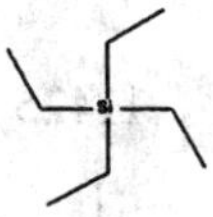

11451

11452

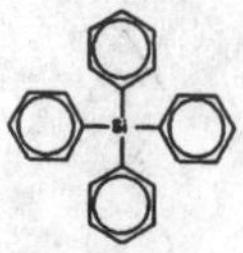

11453

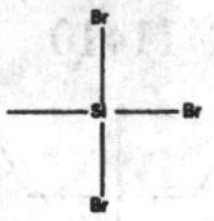

11454

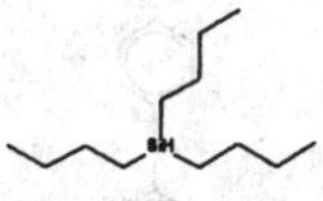

11455

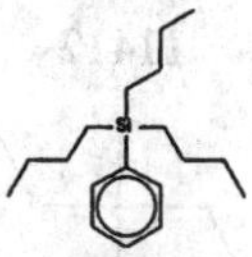

11456

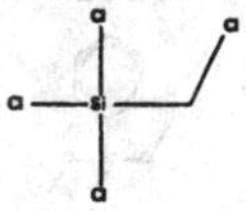

11457

11458

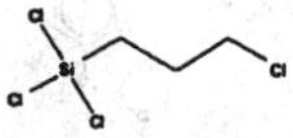

11459

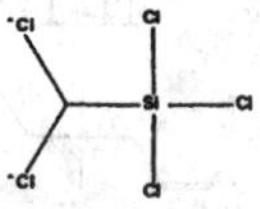

11460

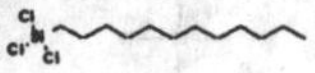

11461

11462

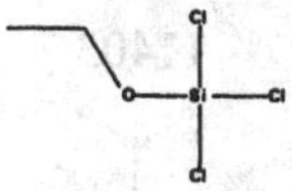

11463

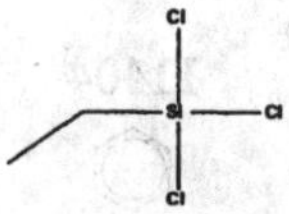

11464

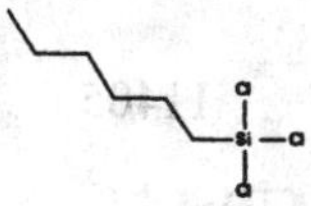

11465

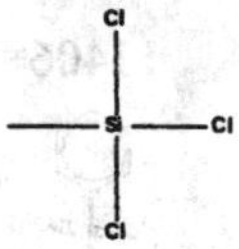

11466

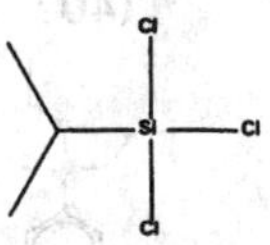

11467

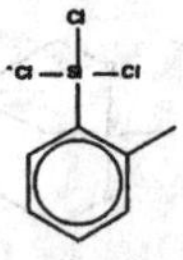

11468

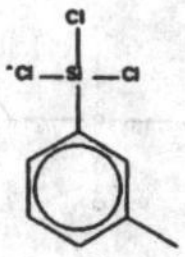

11469

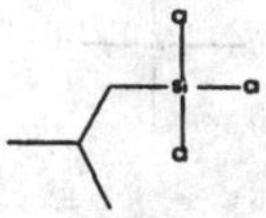

11470

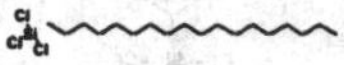

11471

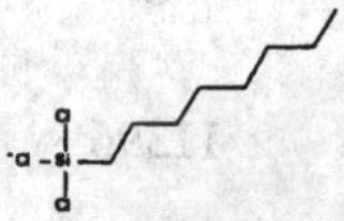

11472

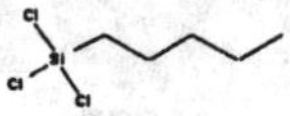

11474

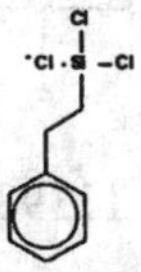

11475

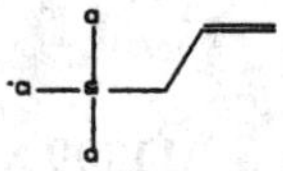

11476

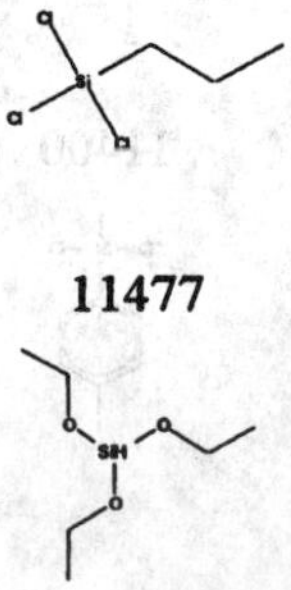

In numeric order as they occur in the Table of Physical Constants of Organic Compounds

11478

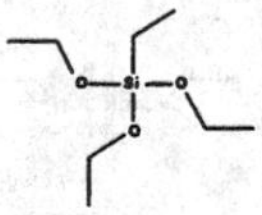

11479

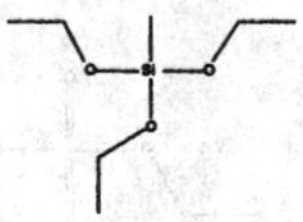

11480

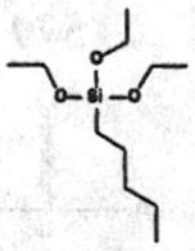

11481

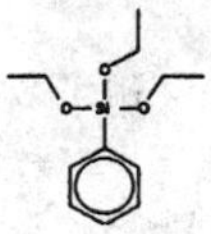

11482

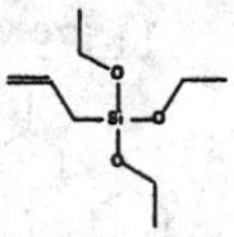

11483

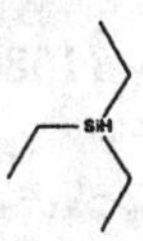

11484

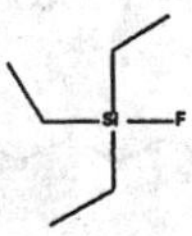

11485

11486

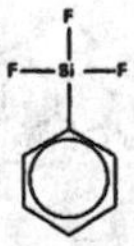

11487

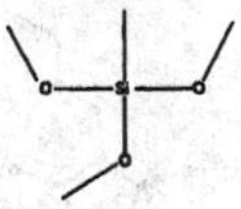

11488

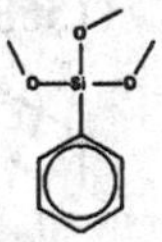

11489

11490

11491

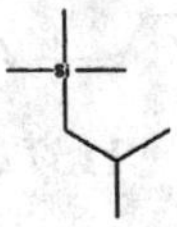

11492

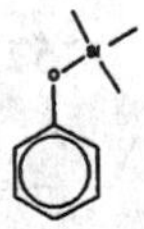

11493

11494

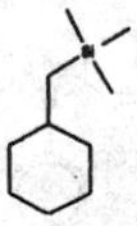

11495

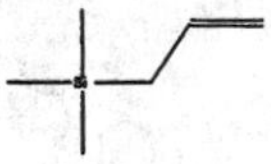

11496

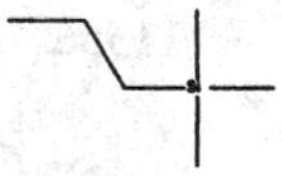

11497

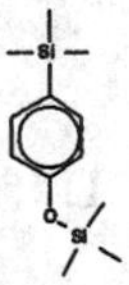

11498

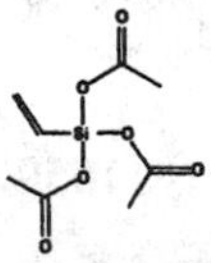

11499

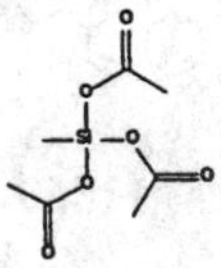

11500

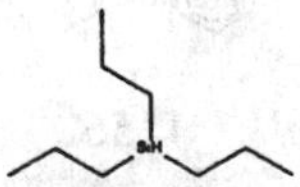

11501

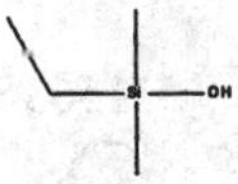

11502

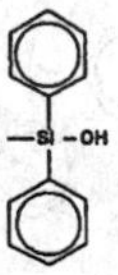

11503

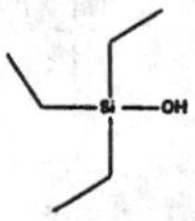

11504

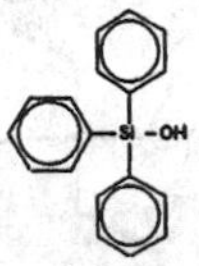

11505

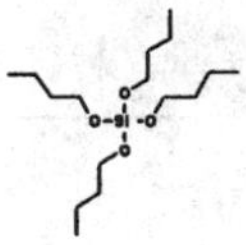

11506

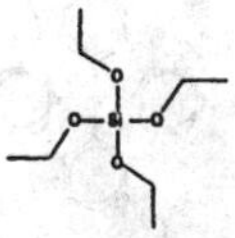

11507

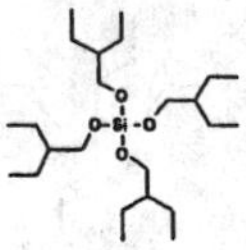

11508

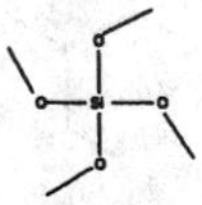

11509

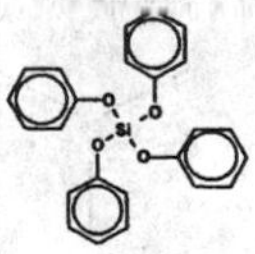

11510

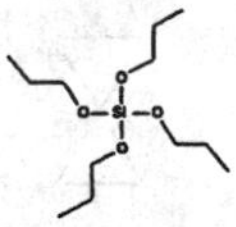

11511

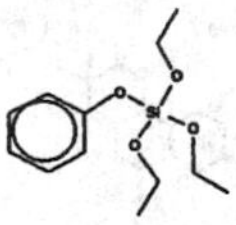

11512

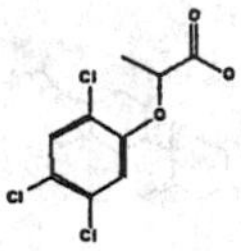

11513

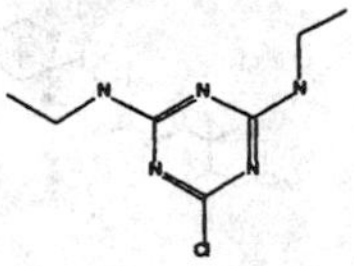

11514

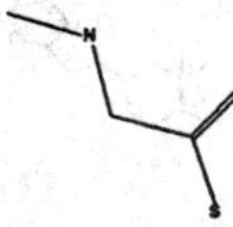

11517

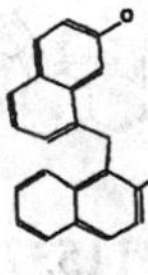

11519

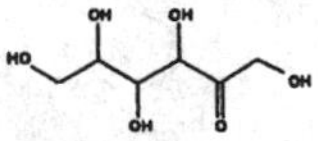

11520

11521

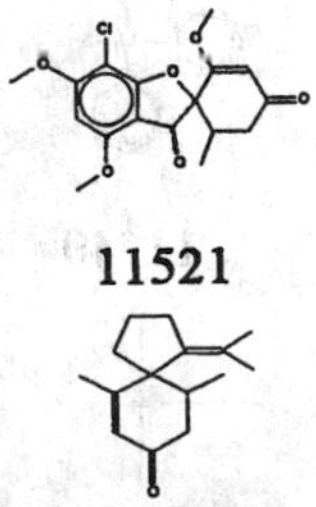

STRUCTURAL FORMULAS OF ORGANIC COMPOUNDS (continued)

In numeric order as they occur in the Table of Physical Constants of Organic Compounds

11527

11528

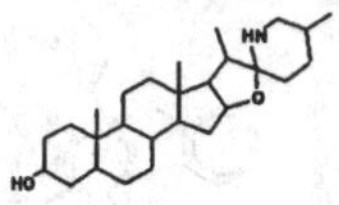

11529

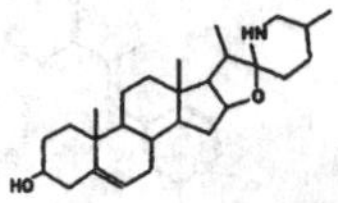

11531

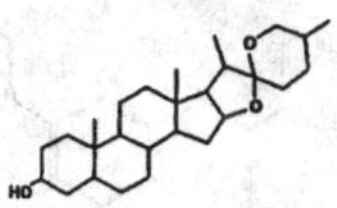

11532

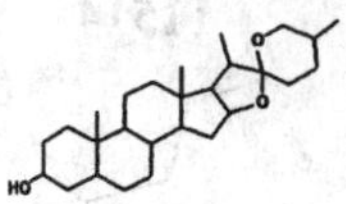

11536

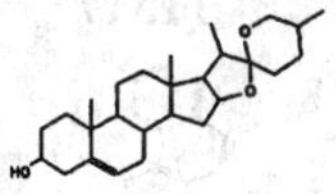

11537

11539

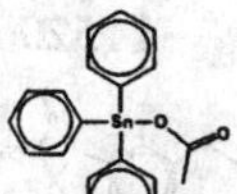

11540

11541

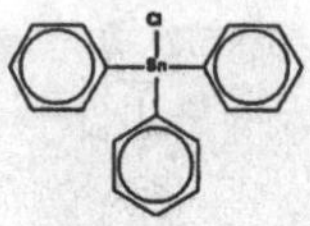

11542

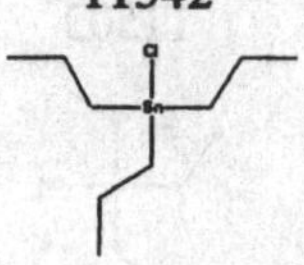

11543

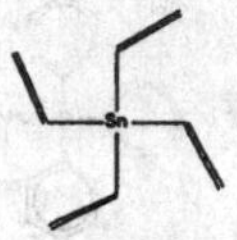

11544

11545

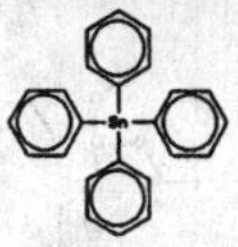

11546

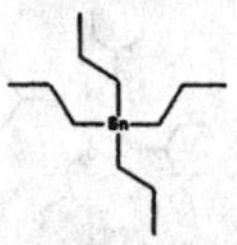

11549

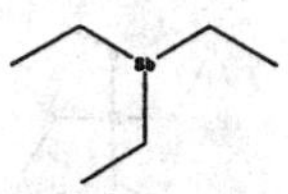

11550

11551

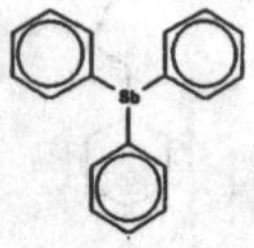

11555

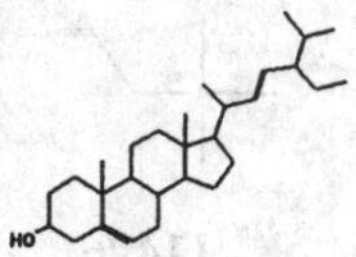

11556

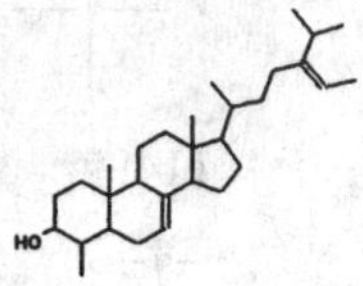

11558

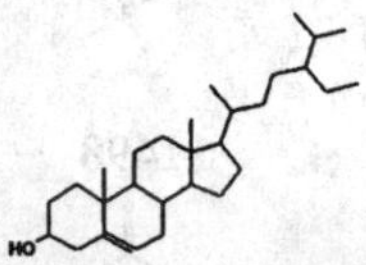

11559

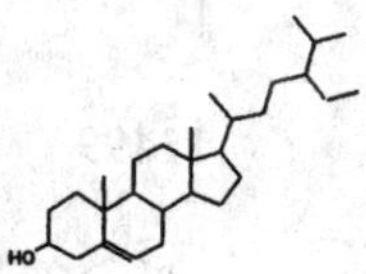

11560

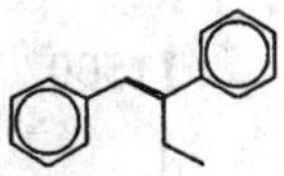

11562

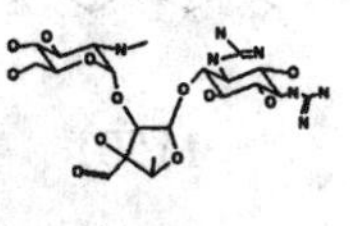

11563

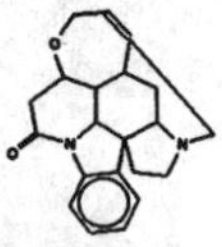

11564

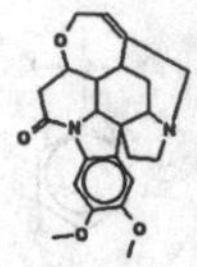

11567

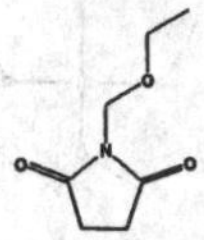

11568

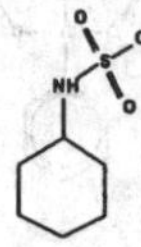

11569

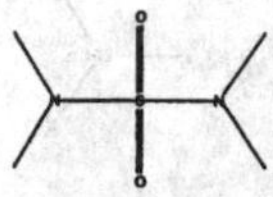

11570

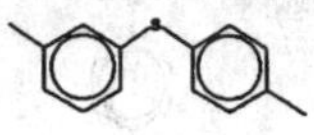

11571

11572

11573

11574

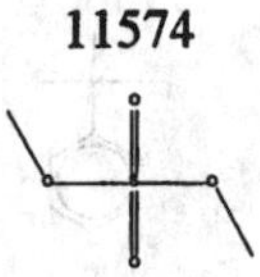

11575

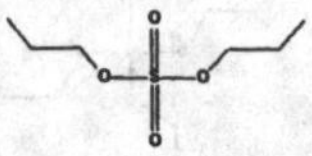

11576

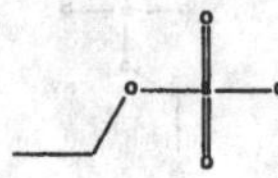

11577

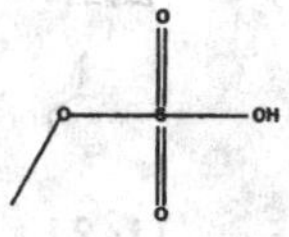

11578

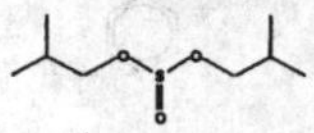

11579

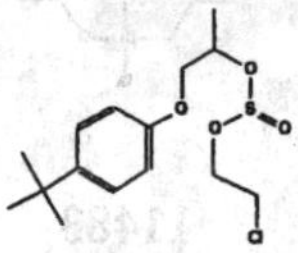

11580

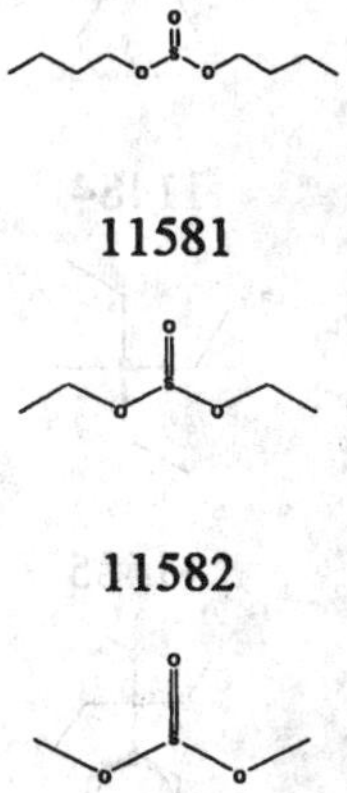

11581

11582

In numeric order as they occur in the Table of Physical Constants of Organic Compounds

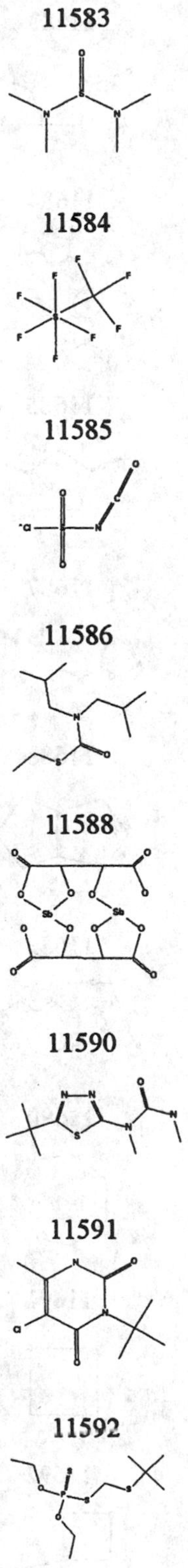

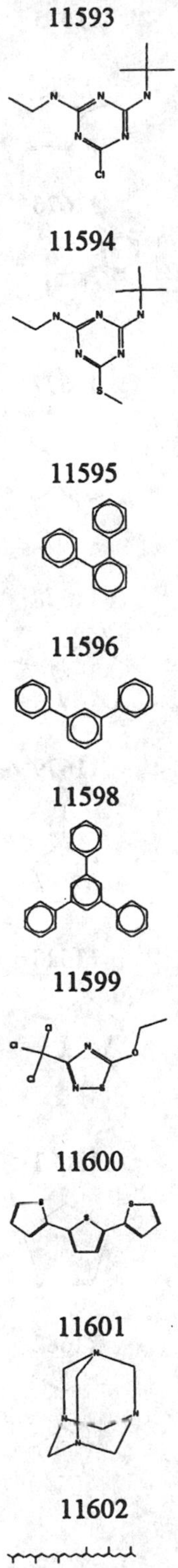

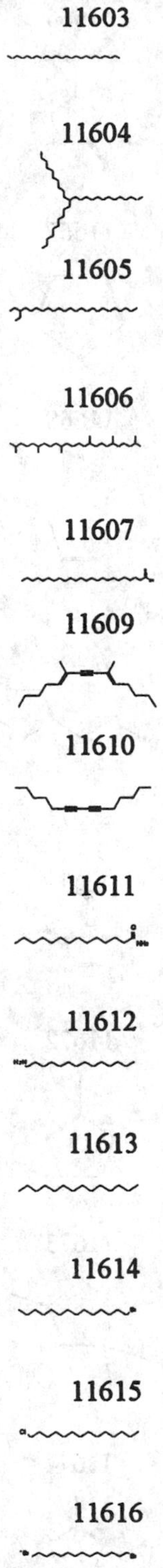

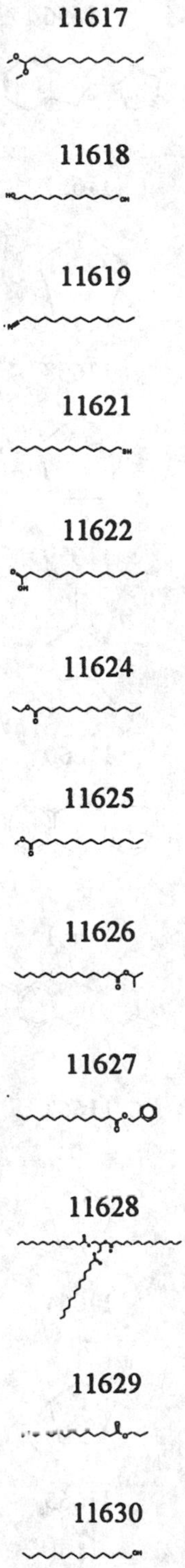

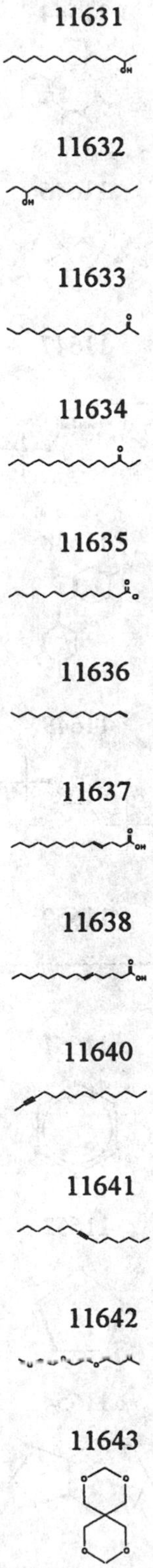

In numeric order as they occur in the Table of Physical Constants of Organic Compounds

11644

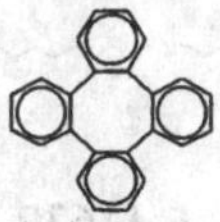

11646

11647

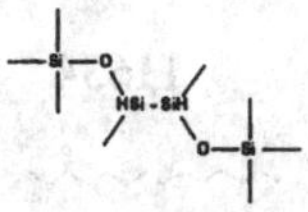

11648

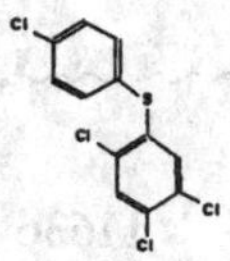

11649

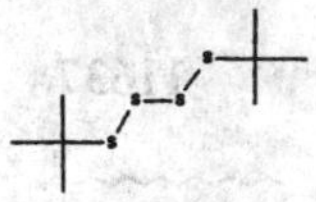

11650

11651

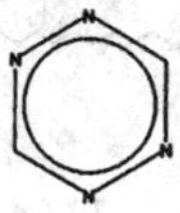

11652

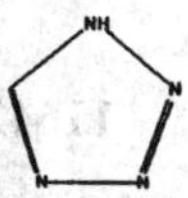

11654

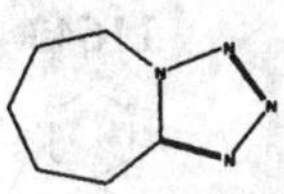

11656

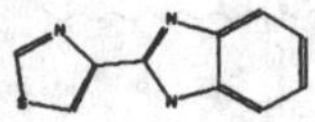

11657

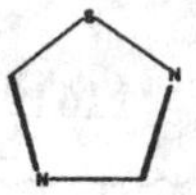

11658

11659

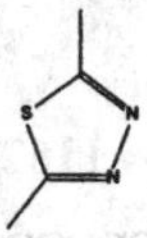

11660

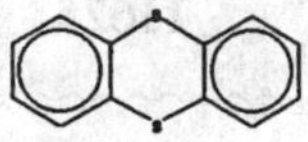

11661

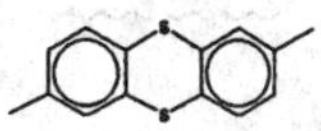

11662

11664

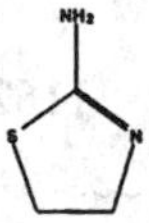

11665

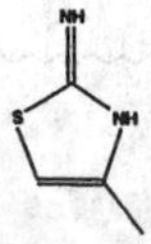

11666

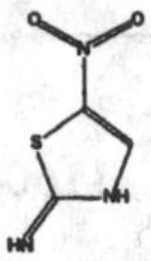

11667

11668

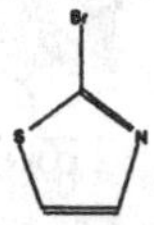

11670

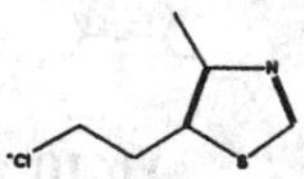

11671

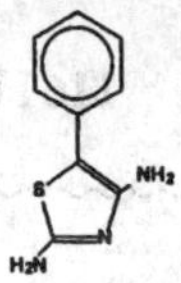

11672

11673

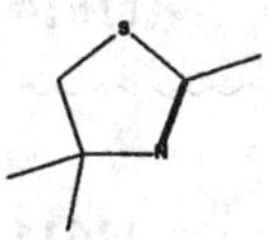

11674

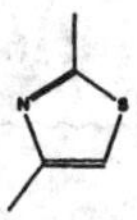

11675

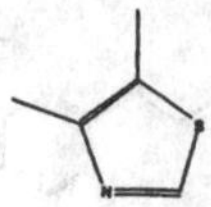

11676

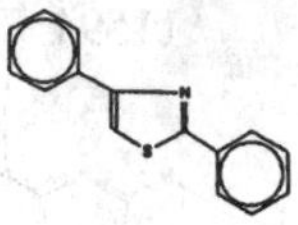

11677

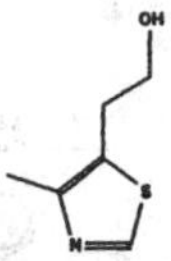

11678

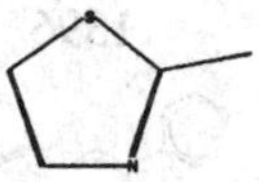

11679

11680

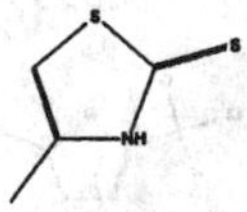

11681

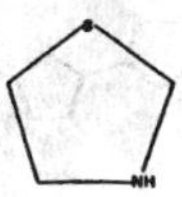

11682

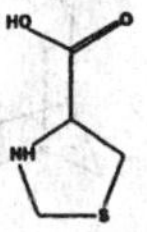

11683

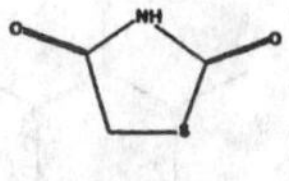

11684

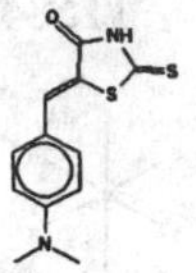

11685

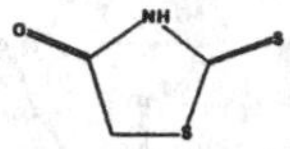

11687

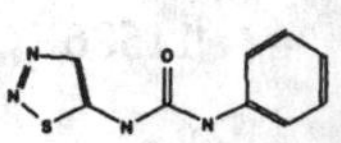

11688

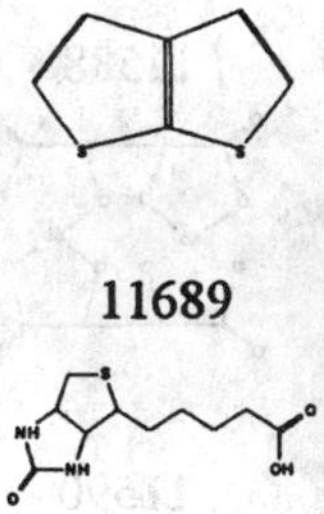

11689

11690

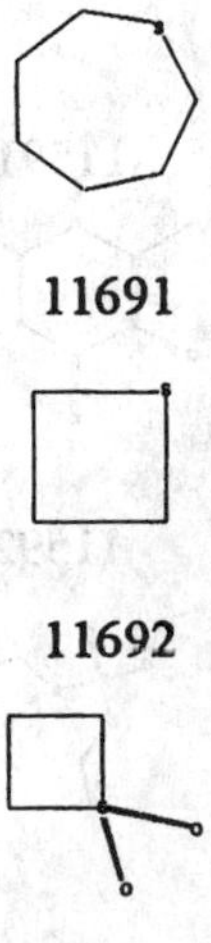

11691

11692

In numeric order as they occur in the Table of Physical Constants of Organic Compounds

11693

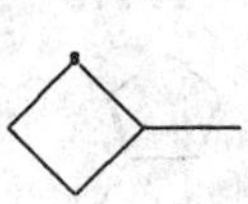

11694

11695

11696

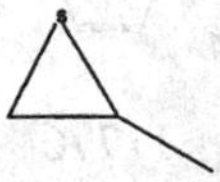

11697

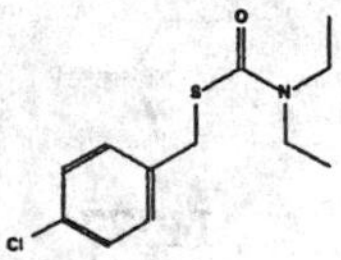

11698

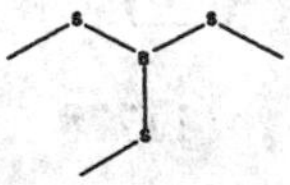

11699

11700

11701

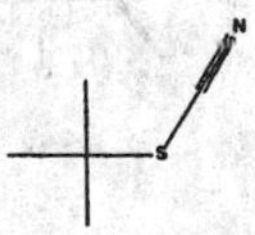

11703

11704

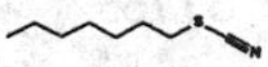

11705

11706

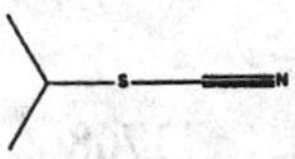

11707

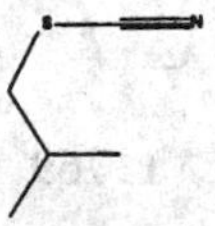

11708

11709

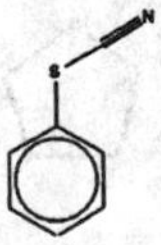

11710

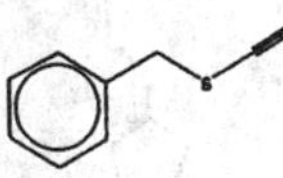

11712

11713

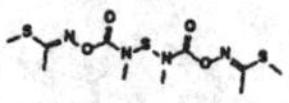

11714

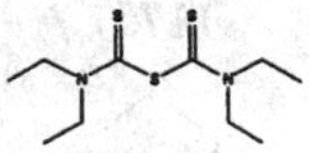

11715

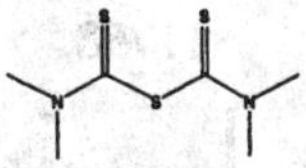

11716

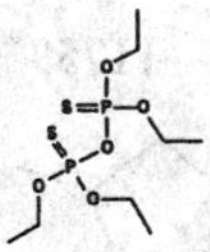

11717

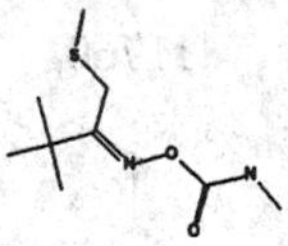

11719

11720

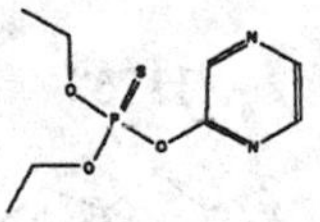

11721

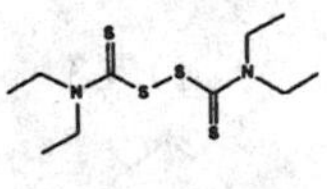

11722

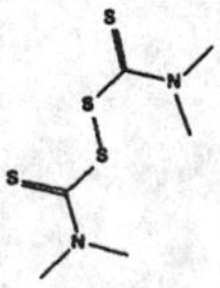

11723

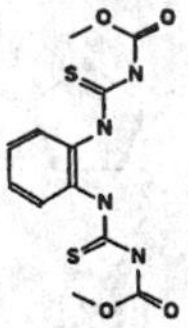

11724

11725

11726

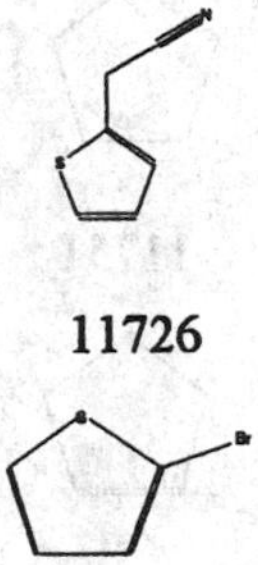

11727

11728

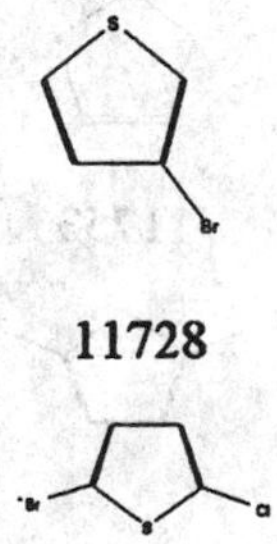

11729

11730

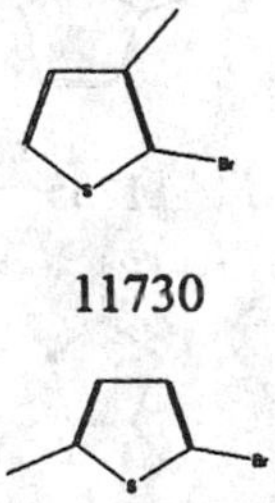

11731

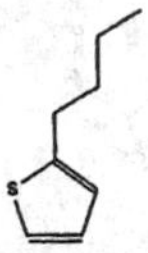

11732

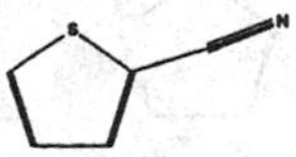

11733

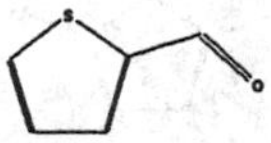

11734

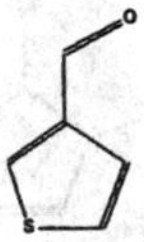

11735

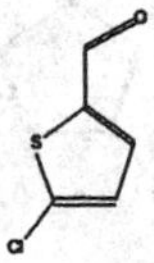

11736

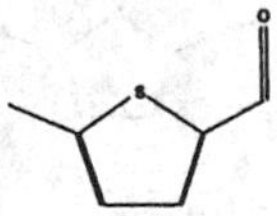

11737

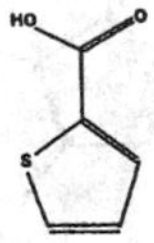

11738

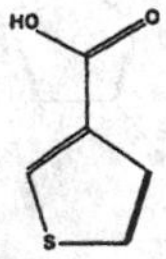

11739

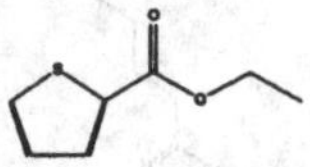

11740

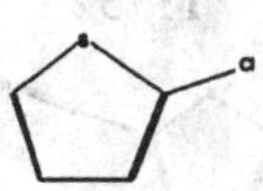

11741

11742

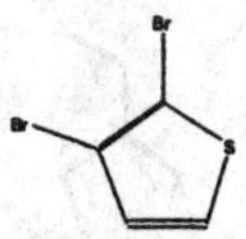

11743

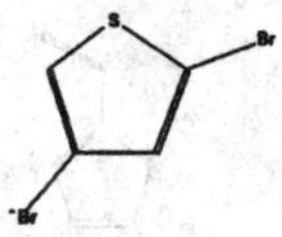

11744

11746

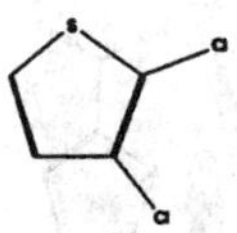

11747

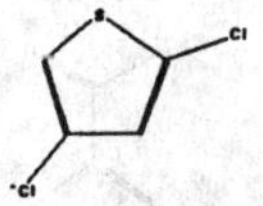

11748

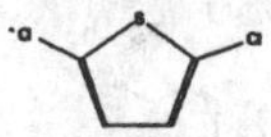

11749

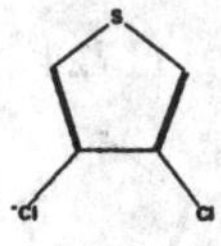

11750

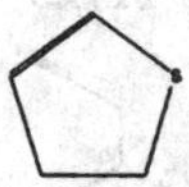

11751

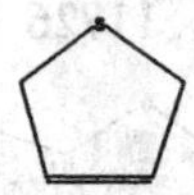

11752

11753

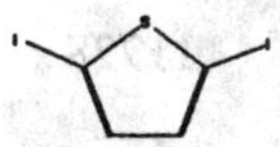

11754

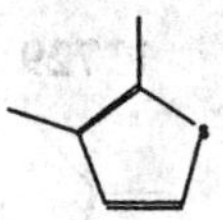

11755

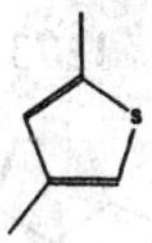

11756

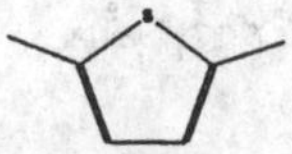

11757

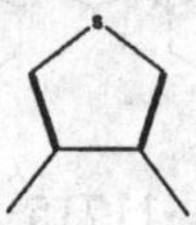

11758

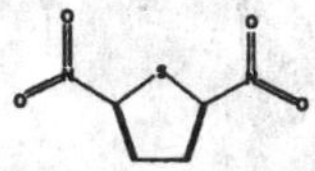

11759

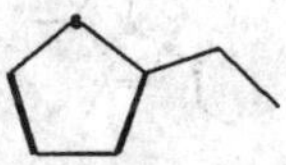

11760

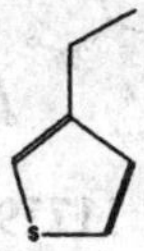

11761

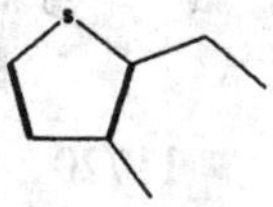

11762

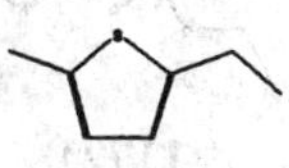

11763

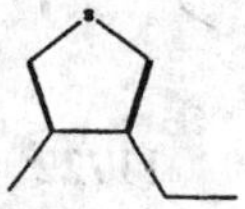

11764

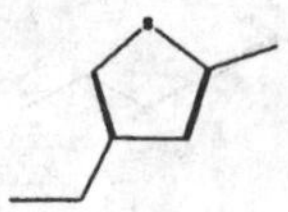

11765

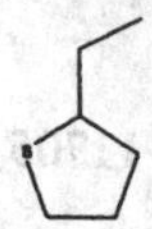

11767

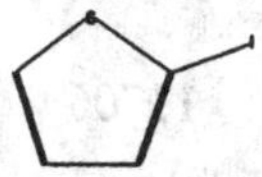

11768

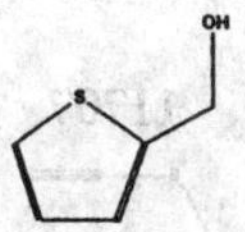

11769

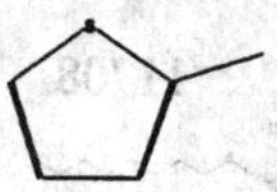

11770

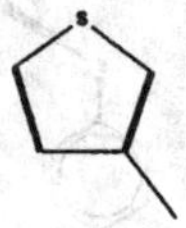

11771

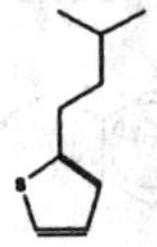

11772

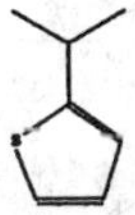

11773

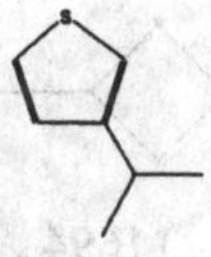

11774

11775

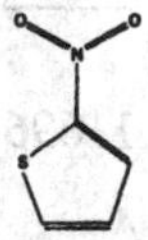

11776

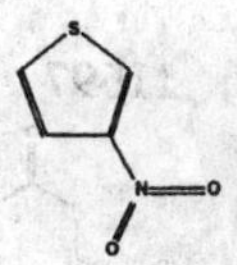

11777

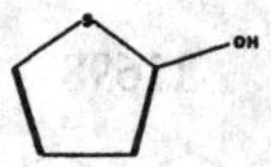

11778

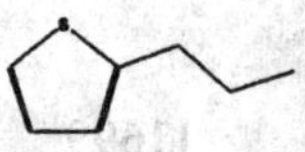

11779

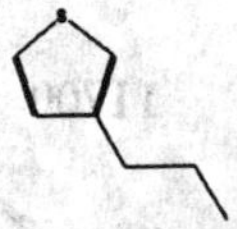

11780

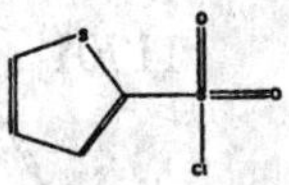

11781
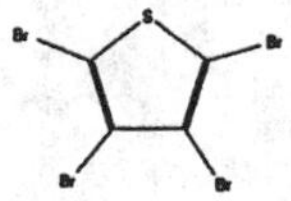

11782
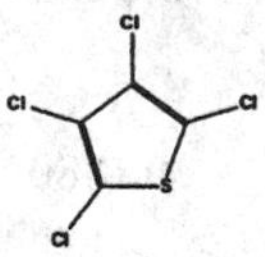

11783
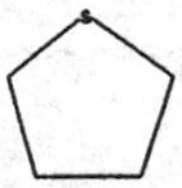

11784
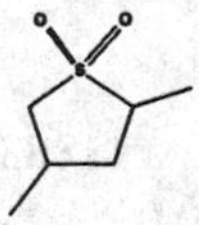

11785
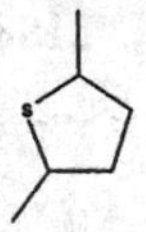

11786
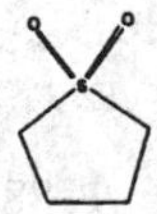

11787
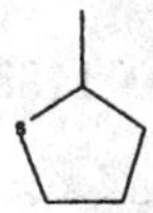

11788

11789
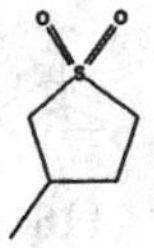

11790
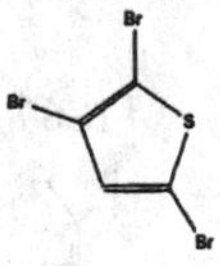

11791
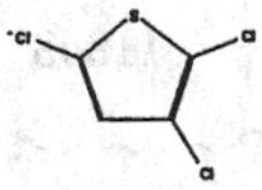

11792
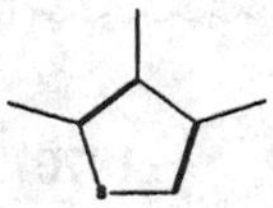

11793
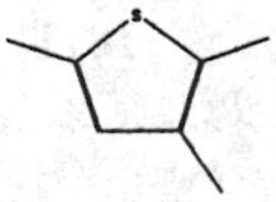

11794
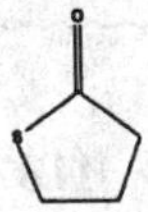

11795
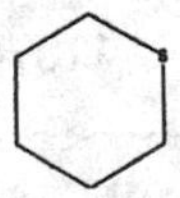

11796
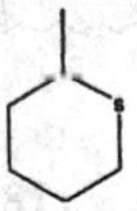

11797
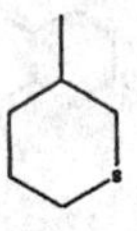

11798
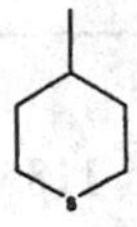

11799
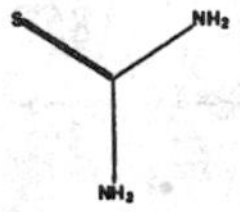

11800

11801
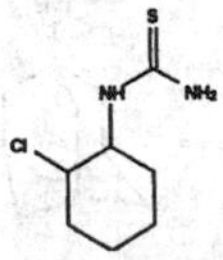

11803
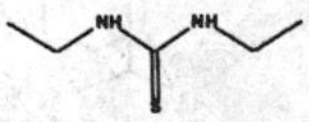

11804
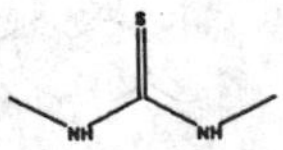

11805
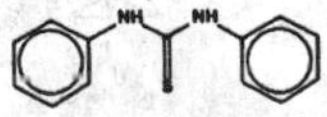

11806
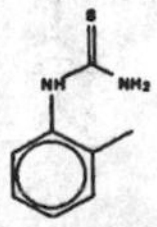

11807
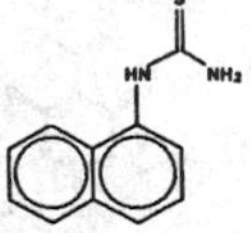

11808
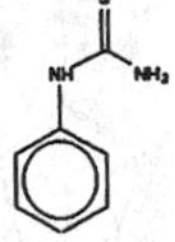

11809
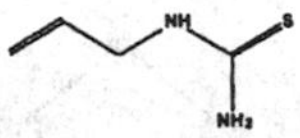

11810
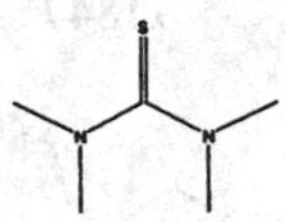

11812
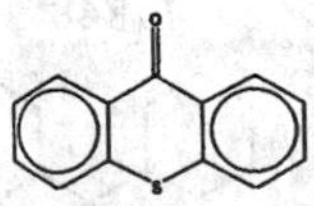

11813
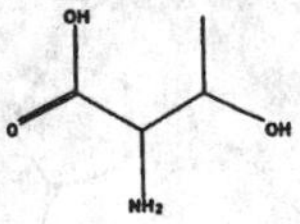

11816
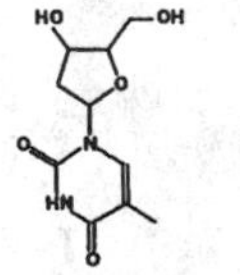

11817
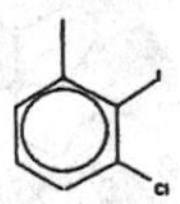

11819

11822

11823
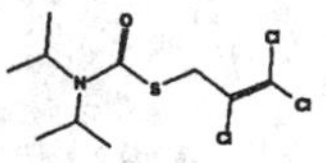

11824
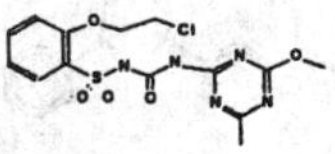

11826
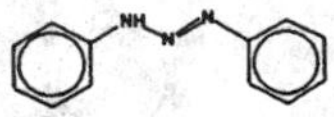

11827
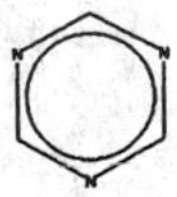

11828
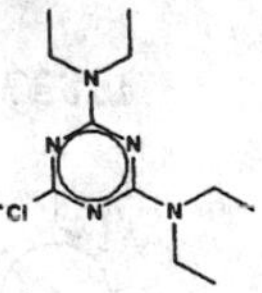

11829
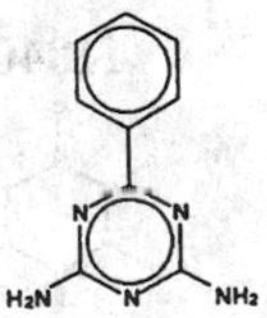

STRUCTURAL FORMULAS OF ORGANIC COMPOUNDS (continued)

In numeric order as they occur in the Table of Physical Constants of Organic Compounds

11831

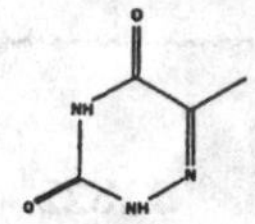

11833

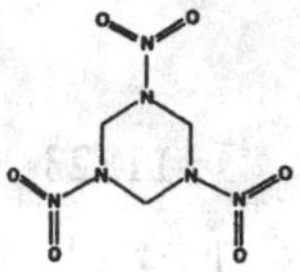

11834

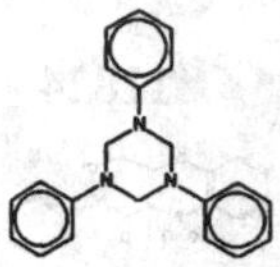

11835

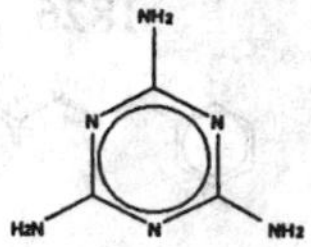

11836

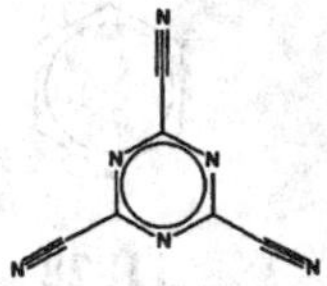

11837

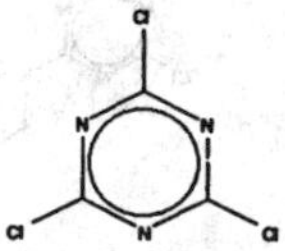

11840

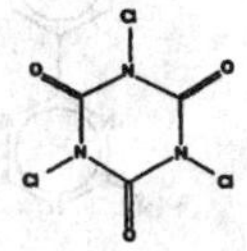

11841

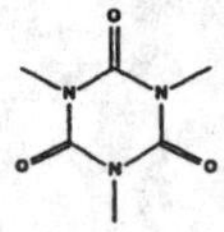

11842

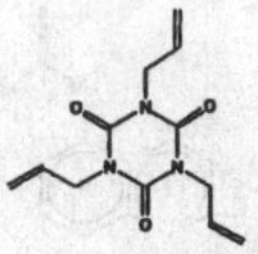

11845

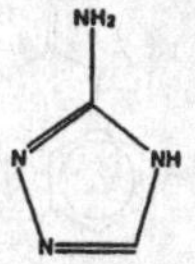

11846

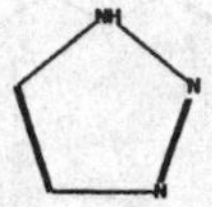

11847

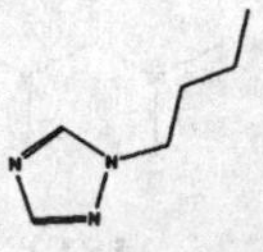

11848

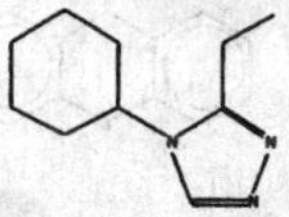

11849

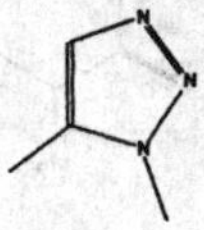

11852

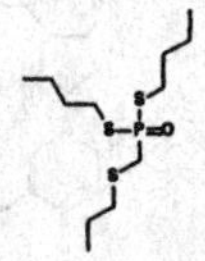

11853

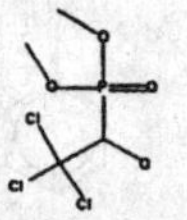

11854

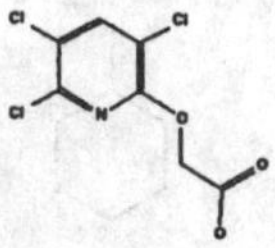

11855

11856

11857

11858

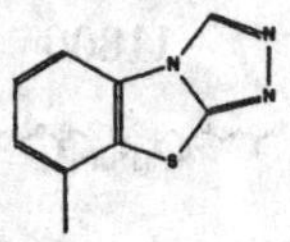

11859

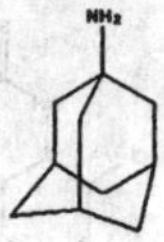

11860

11861

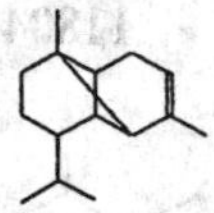

11862

11863

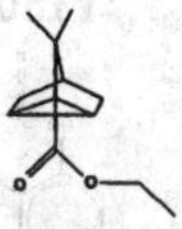

11865

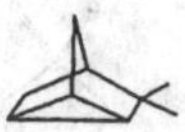

11866

11867

11868

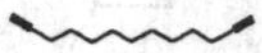

11869

11870

11871

11872

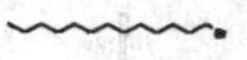

11873

11874

11875

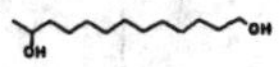

11877

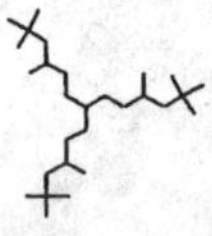

11878

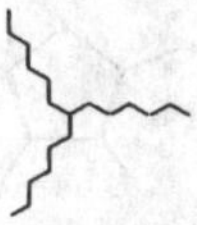

11879

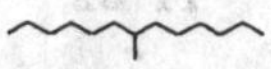

11880

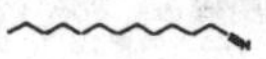

11881

11882

11883

11884

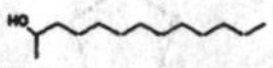

11885

11886

11887

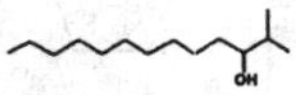

11888

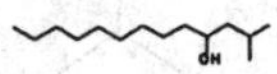

11889

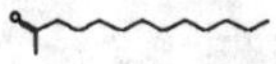

STRUCTURAL FORMULAS OF ORGANIC COMPOUNDS (continued)

In numeric order as they occur in the Table of Physical Constants of Organic Compounds

11890

11891

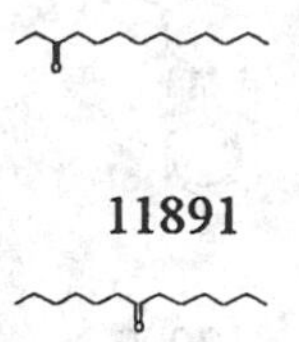

11892

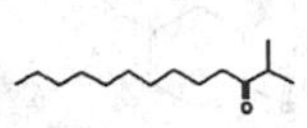

11893

11894

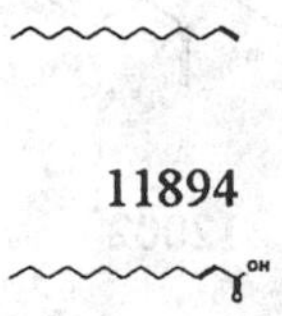

11895

11897

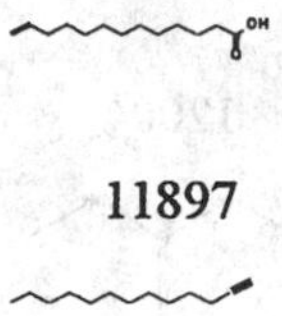

11898

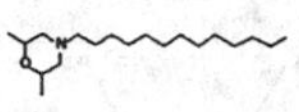

11899

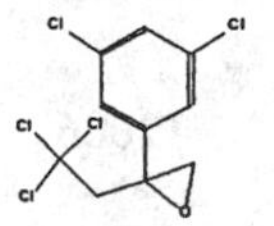

11900

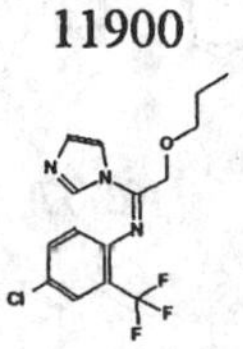

11901

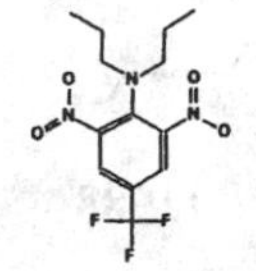

11902

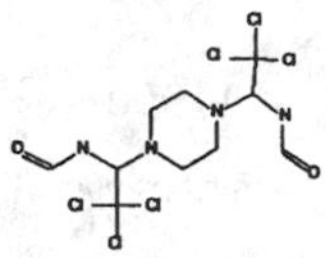

11903

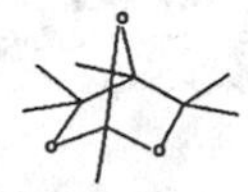

11904

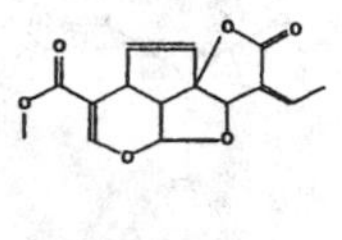

11905

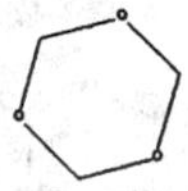

11906

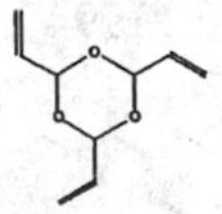

11908

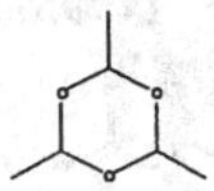

11909

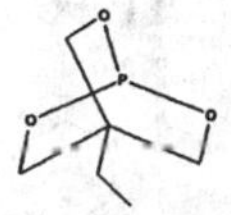

11910

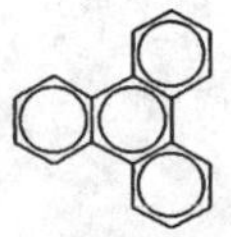

11911

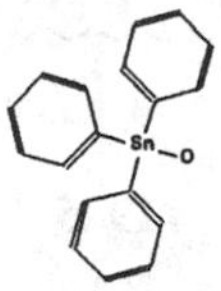

11912

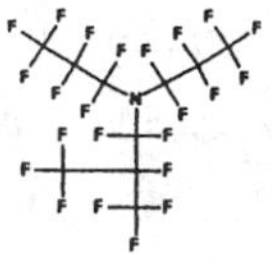

11913

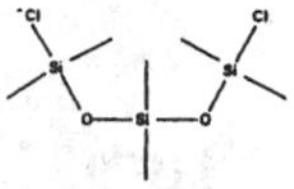

11914

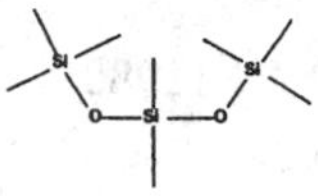

11915

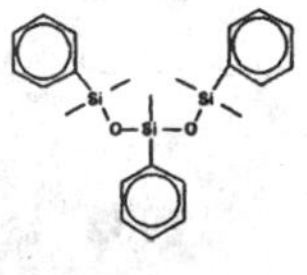

11916

11917

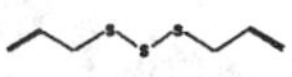

11919

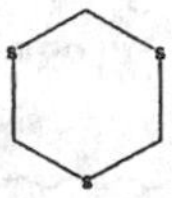

11920

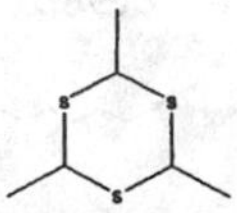

11921

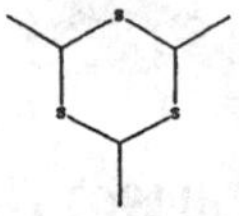

11922

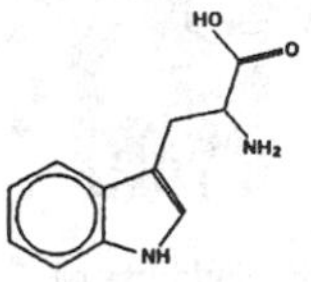

11926

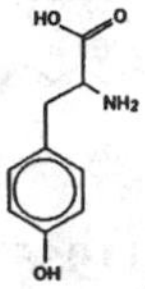

11929

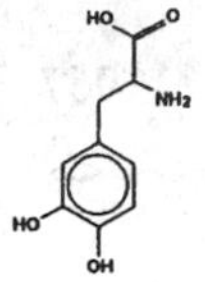

11931

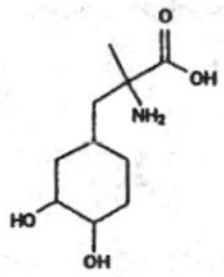

11934

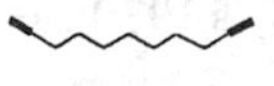

11935

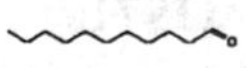

11936

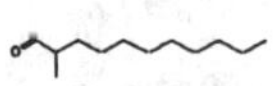

11937

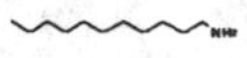

11938

11939

11940

11941

11942

11943

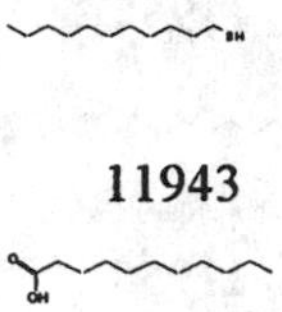

11944

11945

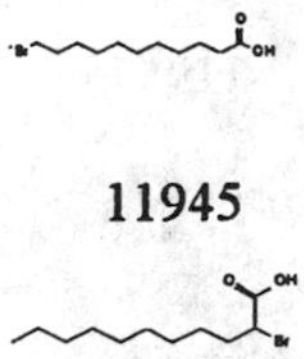

11946

11947

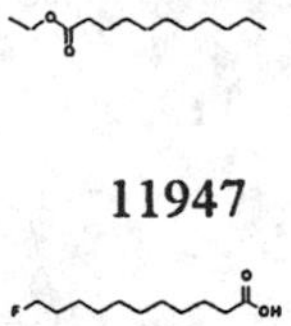

11948

11949

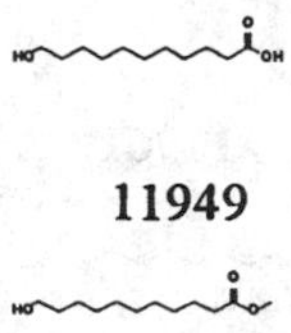

In numeric order as they occur in the Table of Physical Constants of Organic Compounds

11950

11951

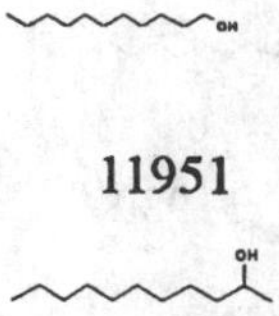

11953

11954

11955

11956

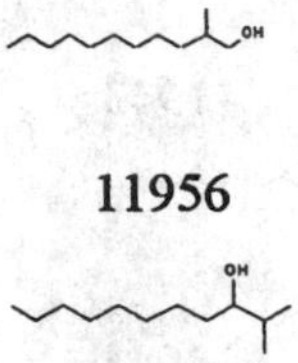

11957

11958

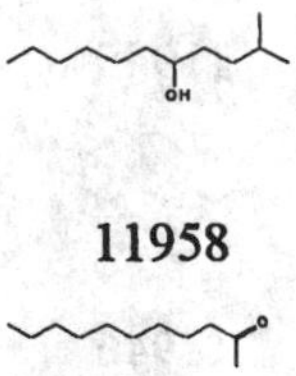

11959

11960

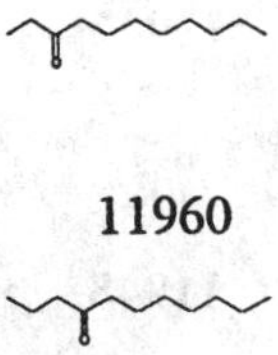

11961

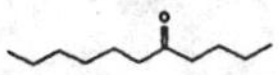

11962

11963

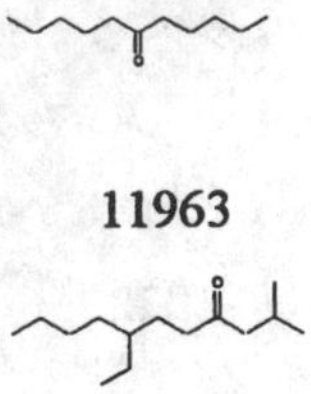

11964

11965

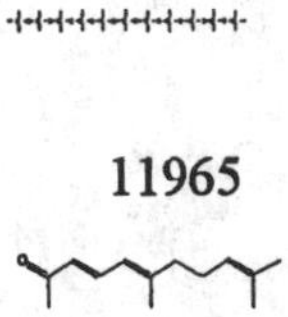

11966

11967

11968

11969

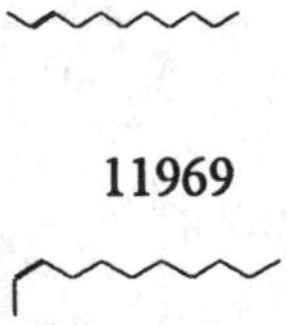

11970

11971

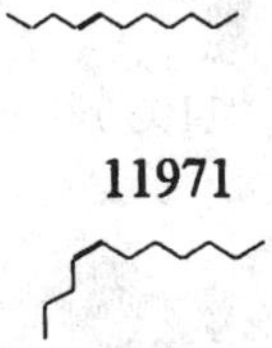

11972

11973

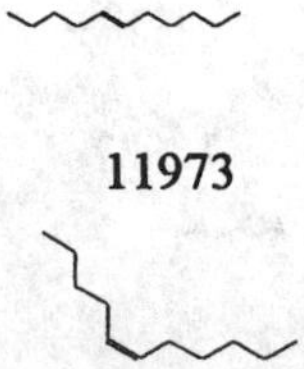

11974

11975

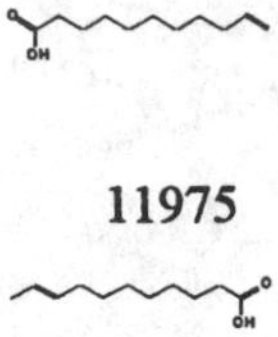

11977

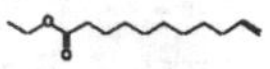

11978

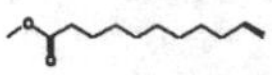

11979

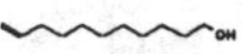

11980

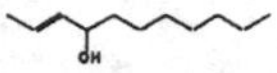

11981

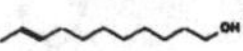

11982

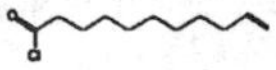

11983

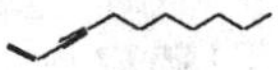

11984

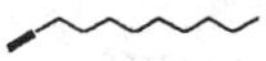

11985

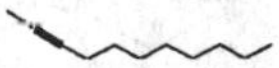

11986

11987

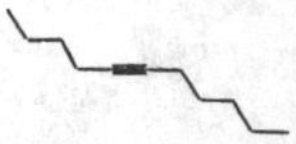

11988

11990

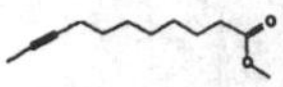

11991

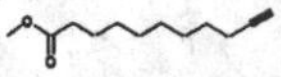

11992

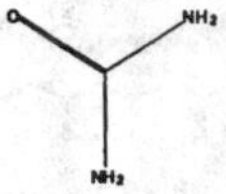

11993

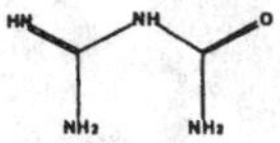

11994

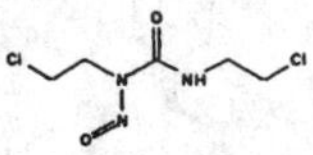

11995

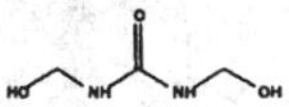

11996

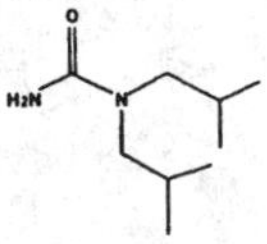

11999

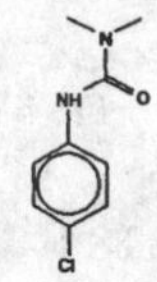

12000

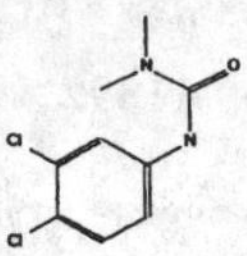

12001

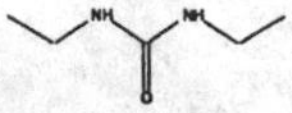

12002

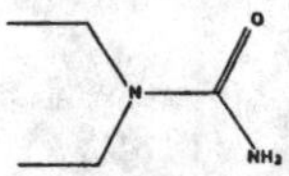

12004

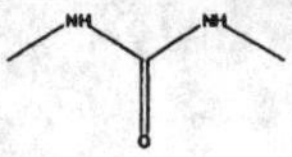

12005

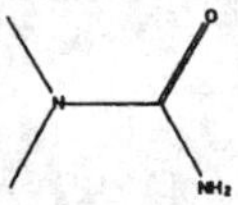

12006

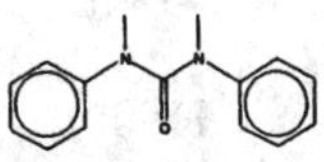

12007

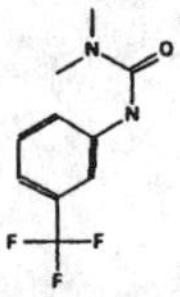

STRUCTURAL FORMULAS OF ORGANIC COMPOUNDS (continued)

In numeric order as they occur in the Table of Physical Constants of Organic Compounds

12009

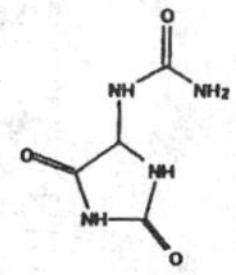

12010

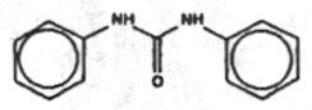

12012

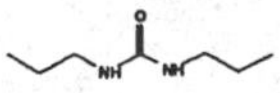

12013

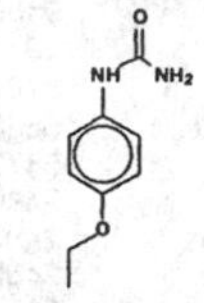

12014

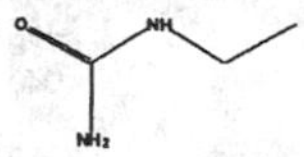

12015

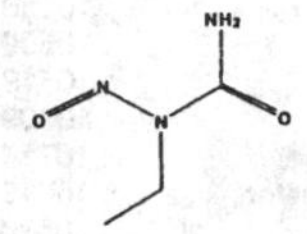

12016

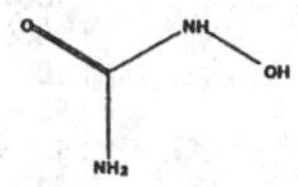

12017

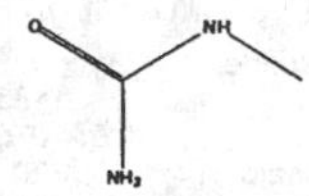

12018

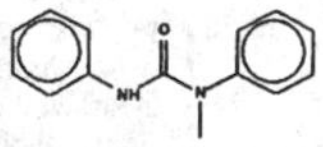

12019

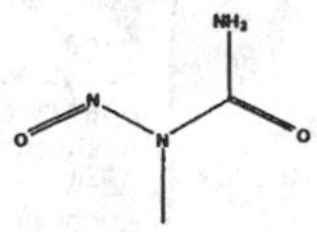

12026

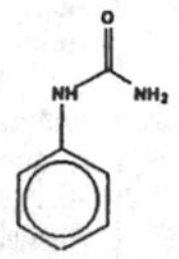

12027

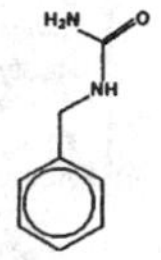

12028

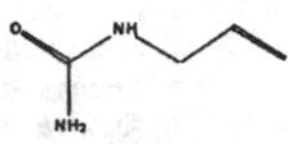

12029

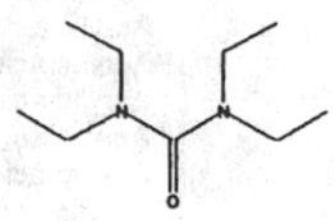

12030

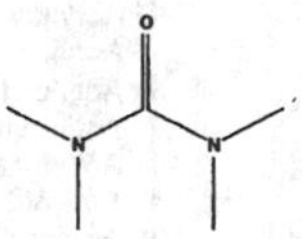

12031

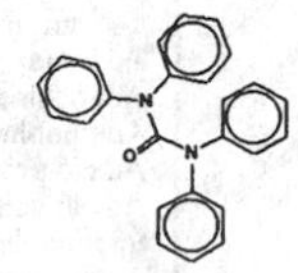

12032

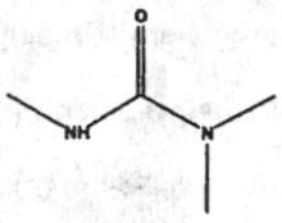

12033

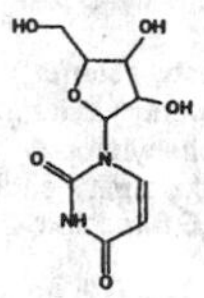

12034

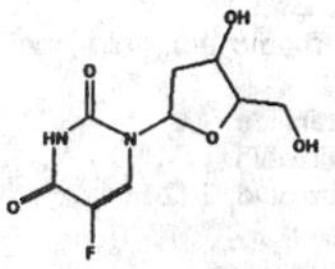

12037

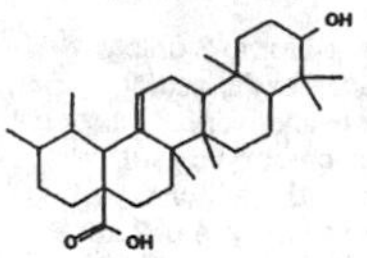

12038

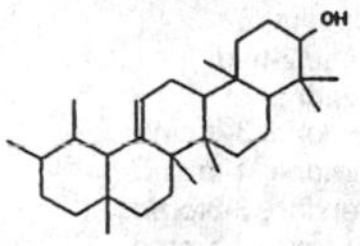

12039

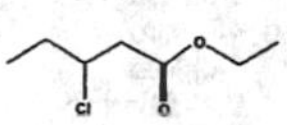

12040

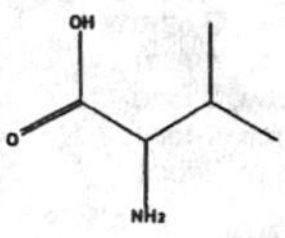

12041

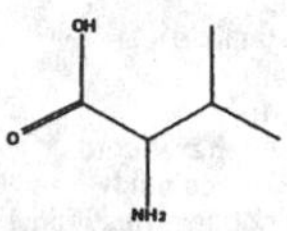

12043

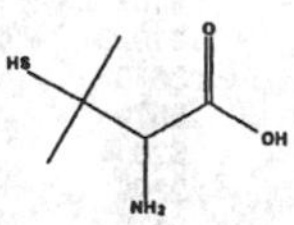

12044

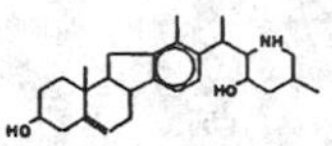

12047

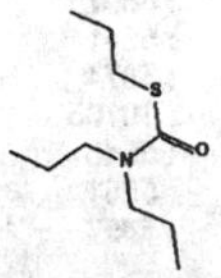

12048

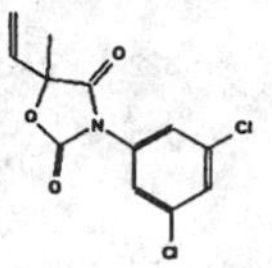

12049

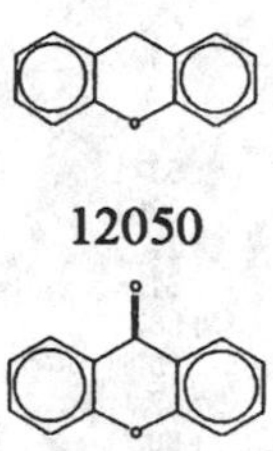

12050

12055

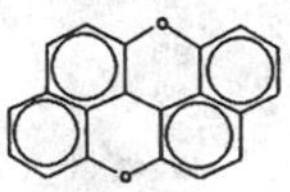

12056

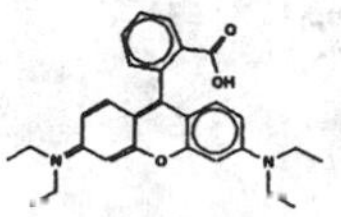

12057

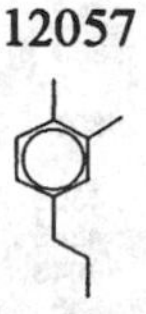

12058

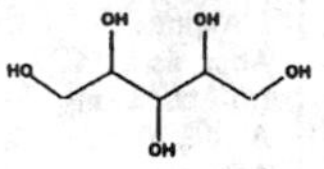

12059

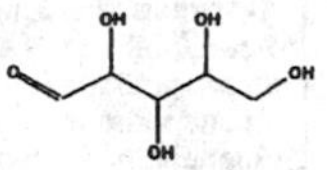

12061

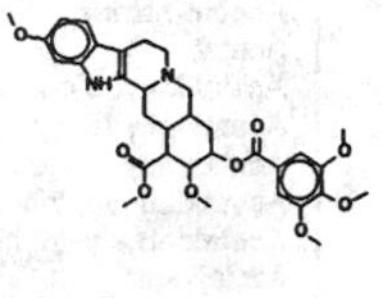

12062

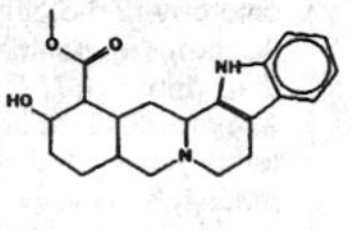

12065

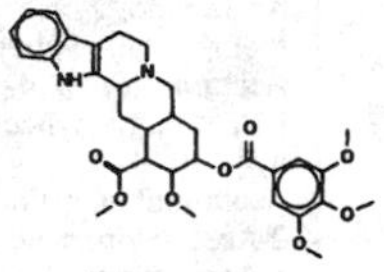

12066

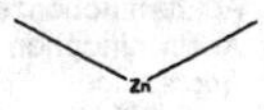

12067

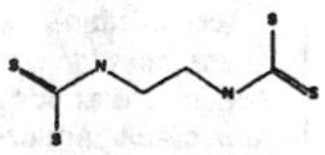

12068

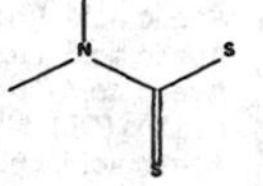

Abadol : 11662
Abamectin : 477
7,13-Abietadien-18-oic acid : 9167
Abietic acid : 9167
L-Abrine : 11923
Abscisic acid : 8679
(+)-Abscisin II : 8679
Abufene : 323
Acacetin : 2809
Acedapsone : 91
Acemin : 6871
Acenaphthalene : 3
Acenaphthaquinone : 8
1-Acenaphthenamine : 2
Acenaphthene : 6
Acenaphthene, 1-amino : 2
Acenaphthene, 5-bromo- : 4
Acenaphthene, 5-chloro- : 5
Acenaphthenedione : 8
Acenaphthene, 5-iodo- : 7
Acenaphthenequinone : 8
Acepramin : 6873
Acepromazine : 5807
Acetal : 5541
Acetaldehyde ammonia : 5674
Acetaldehyde, cyanohydrin : 10105
Acetaldehyde methyl acetal : 5548
Acetaldehyde, tribromo, hydrate : 5581
Acetaldehyde, trichloro-, diethyl acetal : 5657
Acetaldol : 3184
Acetaldoxime : 17
Acetamide, *N*-(butoxymethyl)-2-chloro-*N*-(2,6-diethylphenyl)- : 3144
Acetamide, 2-chloro-*N*-(2,6-diethylphenyl)-*N*-(methoxymethyl)- : 322
Acetamide, 2-chloro-*N*-(ethoxymethyl)-*N*-(2-ethyl-6-methylphenyl)- : 270
Acetamide, 2-chloro-*N*-(2-ethyl-6-methylphenyl)-*N*-(2-methoxy-1-methylethyl)- : 7657
Acetamide, 2-chloro-*N*-(1-methylethyl)-*N*-phenyl- : 9771
Acetamide, *N*-[2,4-dimethyl-5-[[(trifluoromethyl)sulfonyl]amino]phenyl]- : 7427
Acetamidine, hydrochloride : 5671
2-Acetamidoacetic acid : 6251
p-Acetamidoaniline : 29
4-Acetamidobenzenesulfonamide : 31
ε -Acetamidocaproic acid : 6871
2-Acetamido-5-nitrothiazole : 85
p-Acetamidophenol : 67
Acetaminobenzene : 86
Acetaminophen : 67
Acetanilide : 86
Acetanilide, 4'-hydroxy- : 67
Acetanilide, *N*-isopropyl : 74
Acetanilide, 2-nitro- : 80
Acetanilide, *N*-propyl- : 88
Acetanilide, 4'-sulfamoyl- : 31
Acetanilide-*p*-sulfonyl chloride : 2243
p-Acetanisidine : 71
Acetazolamide : 32
Acetic acid amide : 22
Acetic acid, amide,(4(carboxy(propyl ester)methyl)-2-methoxyphenoxy)-*N*,*N*-diethyl : 894
Acetic acid, amide,*N*,*N*-dimethyl : 55
Acetic acid, amide,*N*(4-methoxyphenyl) : 71
Acetic acid, amide, 1-naphthyl : 7733
Acetic acid, [(benzoylamino)oxy]- : 624
Acetic acid, bromide : 291
Acetic acid, (4-bromophenyl)-hydroxy(*DL*) : 889
Acetic acid, chlorophenyl-, ethyl ester, *L*-(+)- : 891
Acetic acid, cyclohexyl : 4306
Acetic acid, cyclopentyl ester : 4865
Acetic acid, (1,1-dimethylbutyl) ester : 8946
Acetic acid, diphenyl,2(diethylamino) ethyl ester,hydrochloride : 938
Acetic acid, fluoride : 301
Acetic acid, (*p*-fluorophenyl)- : 901
Acetic acid, 2-methoxyethyl ester : 5734
Acetic acid, (*N*-methyldodecanamido), sodium salt : 6269
Acetic acid, [3,4-(methylenedioxy)phenyl]-, methyl ester : 2416
Acetic acid, (6-methyl-2-heptyl) ester : 6506
Acetic acid, (6-methyl-3-heptyl) ester : 6507
Acetic acid, (2-methylpentyl) ester : 8945
Acetic acid, (2-methyl-3-pentyl) ester : 8948
Acetic acid, (3-methyl-3-pentyl) ester : 8949
Acetic acid, 2-octyl ester(*DL*) : 8447
Acetic acid, (2-pentyl) ester (*l*) : 8918
Acetic acid, 3-phenylpropyl : 2099
Acetic acid, sulfo, diethyl ester : 184
Acetic acid, [(3,5,6-trichloro-2-pyridinyl)oxy]- : 11854
Acetic acid, trifluoro,(1,2-dimethyl-1-ethylbutyl) ester : 11888
Acetic anhydride : 107
Acetimidic acid : 22
Acetimidic acid, *N*-[(methylcarbamoyl)oxy]thio-, methyl ester : 7650
Acetoacetanilide : 3207
Acetoacetic acid : 3568
Acetoacetic acid, 2-(2-hydroxyethyl)-, γ-lactone : 6139
Acetoacetylaniline : 3207
Acetocinnamone : 3881
Acetoin : 3661
1'-Acetonaphthone, 2-phenyl- : 5859
Acetone : 10411
Acetone, 1-bromo-3-chloro : 10417
Acetonecarboxylic acid : 3568
Acetone cyanohydrin : 10108
Acetonedicarboxylic acid : 8782
Acetone diethyl sulfone : 9835
Acetone monoglycerol ketal : 5207
Acetone, oxime : 10476
Acetone, phenylhydrazone : 10482
Acetonylacetone : 6805
3-(α-Acetonylbenzyl)-4-hydroxycoumarin : 2832
Acetophene,2-nitro : 5863
Acetophenone : 5864
Acetophenone, 3-amino : 5772
Acetophenone, 4-amino : 5773
Acetophenone, 3-bromo : 5780
Acetophenone, 4-bromo : 5781
Acetophenone, 2-bromo- : 5782
Acetophenone, 2-bromo-4'-chloro- : 5778
Acetophenone, 2-bromo-4'-methyl- : 5779
Acetophenone, 3-chloro : 5788
Acetophenone, 4-chloro : 5789
Acetophenone, 2-chloro- : 5790
Acetophenone, 2,4'-dibromo- : 5777
Acetophenone, 3,5-dibromo : 5797
Acetophenone, 2,5-dichloro : 5800
Acetophenone, 3,4-dichloro : 5801
Acetophenone, 2,2-dichloro- : 5802
Acetophenone, 2,4-dihydroxy : 5803
Acetophenone, 2-ethoxy- : 5814
Acetophenone, 4-fluoro : 5815
Acetophenone, 3-iodo : 5836
Acetophenone, 4-iodo : 5837
Acetophenone, 2-iodo- : 5838
Acetoxime : 10476
2-Acetoxyethanol : 5577
Acetoxytributylstannane : 11538
Acetoxytriphenylstannane : 11539
Acetozone : 9151
Aceturic acid : 6251
α-Acetylacetanilide : 3207
Acetyl acetate : 107
Acetylacetone : 8798
Acetylacetone, hexafluoro : 8802
Acetylaminoacetic acid : 6251
p-Acetylaminobenzenesulfonyl chloride : 2243
2-(Acetylamino)fluorene : 64
N-Acetylaniline : 86
3-Acetylaniline : 5772
4-Acetylaniline : 5773
Acetylbenzene : 5864
2-Acetylbenzofuran : 5775
4-Acetylbiphenyl : 5776
p-Acetylbromobenzene : 5781
α-Acetylbutyrolactone : 6139
2-Acetyl-5-chlorothiophene : 5791
Acetyl choline bromide : 5438
Acetyl choline chloride : 5439
1-Acetylcyclohexene : 5793
2-Acetylcyclopentanone : 4879
Acetylcysteine : 5015
Acetyl disulfide : 5250
Acetylene : 5936
Acetylene, bromo- : 5937
Acetylene carbamide : 7191
Acetylene, chloro- : 5938
Acetylene, dibromo- : 5939
Acetylene, dichloro- : 5940
Acetylene, diiodo- : 5941
Acetylene, fluoro- : 5943
Acetylene, phenyl : 1715
Acetylene tetrabromide : 5632
Acetylene tetrachloride : 5635
Acetyleneurea : 7191
2-Acetyl-4-hydroxybutyric acid γ-lactone : 6139
N-Acetylmethionine : 7649
Acetyl methyl carbinol : 3661
Acetyl nitrate : 108
Acetyl peroxide : 9156
Acetyl-*p*-phenylenediamine : 29
Acetylpropionyl : 8797
3-Acetylpyridine : 5868
4-Acetylpyrogallol : 5873
Acetylsalicylic acid : 2480
Acetylsalicylic acid, methyl ester : 2482
Acetylsalicylic acid, phenyl ester : 2483
Acetylsulfanilamide : 31
Acetylsulfanilyl chloride : 2243
Acetylthiourea : 33
Acetyl valeryl : 6402
Acexamic acid : 6871
Aciflоctin : 6779
Acocantherin : 4040
Aconine : 305
Aconitic acid, triethyl ester (*trans*) : 10601
Aconitine : 306
Acorn Sugar : 4074
Acroleic acid : 10609
Acrolein : 10509
Acrolein, diethyl acetal : 10560
Acrylaldehyde : 10509
Acrylamide : 10519
Acrylic acid : 10609
Acrylic acid chloride : 10714
Acrylic acid, chloro-, ethyl ester : 10619
Acrylic acid, 3,3-dibromo- : 10628
Acrylic acid, 3,3-dichloro- : 10630
Acrylic acid, 3-(2-furyl), pentyl ester : 10641
Acrylic acid, 3-(2-furyl), propyl ester : 10642
Acrylic acid, (2-methylbutyl) ester : 10652
Acrylic acid, 3-(1-naphtyl) (*trans*) : 10675
Acrylic aldehyde : 10509
Acrylic amide : 10519
Acrylon : 10583
Acrylonitrile : 10583
Actinoquinol : 11293
Active amyl alcohol : 3626
Activol : 9203
Actriol : 5405
1-Adamantamine : 11859
Adamantane : 11860
Adamantane, 1(pyrid-3-ylmethylamino),dihydrochloride : 7175
1-Adamantylamine : 11859
Adenine : 10751
Adenosine 5'-monophosphate : 320
Adiphenine hydrochloride : 938
Adipic acid : 6779
Adipic acid bis(3-carboxy-2,4,6-triiodoanilide) : 2594
Adipic acid, di-2-ethylbutyl ester : 6780
Adipic acid, 3-methyl (*DL*) : 6789
Adipic acid, 3-methyl, diethyl ester : 6790
Adipic ketone : 4878
Adipiodone : 2594

SYNONYM INDEX OF ORGANIC COMPOUNDS (continued)

Adiponitrile : 6777
Adol : 4900
Adonitol : 11343
Adrenalone : 5804
Adrenochrome : 7301
Adrenosterone : 353
Aflatoxin B2 : 4757
Aflatoxin G1 : 6184
Agaricic acid : 8102
Agaricin : 8102
Agaritine : 6247
Agifutol S : 6264
Agroxone : 141
Ahistan : 9502
Ajmaline : 321
Akalbir : 2842
Alanine, 3-[(carboxymethyl)thio]-, *L-* : 5017
Alant camphor : 8059
Alantolactone : 8059
Alar : 3334
Albutoin : 7229
Alcanfor : 2961
Aldocortene : 9746
Aldocortin : 9746
Aldol : 3184
Aldosterone : 9746
Aldrin : 5177
Aldrin epoxide : 5178
Alizarin : 391
Alizarine Blue : 8062
Alizarine Bordeaux B : 413
Alizarine Brown R : 415
Alizarine Maroon : 382
Alizarine Orange : 399
Alizarine Yellow C : 5873
Alizarine Yellow R : 2648
Alkannin : 7854
Allacil : 11060
Allantoin : 12009
Allene : 9772
Allenolic acid : 7953
Allergosil : 5587
Allicin : 10595
Allidochlor : 44
Alliin : 333
Allobarbital : 11091
Allocholesterol : 4111
Allocyclogeranic acid : 4307
Allomaleic acid : 3764
Allomethadione : 8586
Allopregnane : 9740
Allopurinol : 10852
Alloxan : 11073
Alloxantin : 3106
Allyl acetate : 235
Allylacetic acid : 9073
Allyl alcohol : 10697
Allyl alcohol, 3-chloro : 10701
Allyl alcohol oxide : 8618
Allyl aldehyde : 10509
Allylamine : 10522
Allyl bromide : 10531
Allylcarbamide : 12028
Allylcatechol methylene ether : 2429
Allyl chloride : 10536
Allyl cinnamate : 10692
Allyl crotonate : 3856
Allyl cyanide : 3803
5-Allyl-4,7-dimethoxy-1,3-benzodioxole : 2423
Allyl ether : 10592
Allyl ethyl ether : 10566
Allyl fluoride : 10570
Allyl formate : 6017
Allyl glycidyl ether : 8626
Allyl iodide : 10573
Allyl isocyanide : 10574
Allyl isothiocyanate : 10575
4-Allyl-1,2-(methylenedioxy)benzene : 2429
m-Allylpyrocatechin methylene ether : 2429
Allyl sulfide : 10599
Allyltrichlorosilane : 11475
Allylurea : 12028
Aloe-emodol : 397
Aloperidin : 3651
Aloxidone : 8586
Alstonidine : 10772
Aluminum lactate : 340
Amanozine : 11830
Amantadine : 11859
Amediol : 9987
Amethocaine monohydrochloride : 2531
Amidazophen : 10851
N-Amidinoglycine : 6253
Amidocyanogen : 4155
Amido-G-Acid : 7870
Amikapron : 4333
Aminacrine : 313
Aminate base : 7147
Aminetrimethylboron : 3120
Aminic acid : 6002
Aminitrozole : 85
2-(Aminoacetamido)acetic acid : 6265
p-Aminoacetanilide : 29
2-Aminoacetic acid : 6250
Aminoacetic acid : 6250
Aminoacetic acid, *N,N*-(dicarboethoxy), ethyl ester : 6258
Aminoacetic acid hydrazide : 6266
Aminoacetic acid, *N*-succinyl, ethyl ester : 11145
Aminoacetonitrile : 272
4-Amino-2-anilino-1,3,5-triazine : 11830
4-Aminoanthranilic acid : 2553
1-Aminoanthraquinone : 380
2-Aminoanthraquinone : 381
p-Aminoazobenzene : 837
1,2-Aminoazophenylene : 2871
(*S*)-α-Aminobenzenepropanoic acid : 9510
3-(*p*-Aminobenzenesulfamido)-6-methoxypyridazine : 2197
p-Aminobenzenesulfonamide : 2187
4-(*p*-Aminobenzenesulfonamido)-2,6-dimethoxypyrimidine : 2194
2-(4-Aminobenzenesulfonamido)-4,6-dimethylpyrimidine : 2196
2-(4-Aminobenzenesulfonamido)-4-methylpyrimidine : 2199
2-(*p*-Aminobenzenesulfonamido)-4-methylthiazole : 2200
N'-(4-Aminobenzenesulfonyl)-*N*-butylurea : 2192
3-Aminobenzoic acid, 6-chloro : 2490
4-Aminobenzoic acid, isobutyl ester : 2505
2-Aminobenzothiazole : 2859
6-Aminobenzothiazole : 2860
N-(*p*-Aminobenzoyl)-L(+)-glutamic acid : 6242
N-(*p*-Aminobenzoyl)glycine : 6252
p-Aminobiphenyl : 3011
1-Amino-5-bromonaphthalene : 7711
4-Amino-3-bromoquinoline : 11207
1-Aminobutane : 3209
γ-Aminobutyric acid : 3464
ε-Aminocaproic acid : 6873
(*L*)-2-Amino-3-(carboxymethylthio)propionic acid : 5017
5-Amino-2-chlorobenzoic acid : 2490
2-Amino-5-chlorobenzoxazole : 2876
4-Amino-4'-chlorodiphenyl : 3012
6-Aminochrysene : 4113
Aminocyclobutane : 4169
Aminocyclooctane : 4735
1-Aminocyclopentanecarboxylic acid : 4782
1-Amino-1-deoxysorbitol : 6207
[(4-Amino-3,5-dichloro-6-fluoro-2-pyridyl)oxy]acetic acid : 5982
2-Amino-4,6-dichlorophenol : 9205
1-Amino-3-diethylamino-2-propanol : 10339
2-Amino-1,9-dihydro-9-β-*D*-ribofuranosyl-6H-purin-6-one : 6289
2-Amino-2,2-dimethylethanol : 10341
2-Amino-4,5-dimethylpyrimidine : 11044
p-Aminodiphenylamine : 1283
2-Amino-6,8-disulfonaphthalene : 7870
Aminoethanoic acid : 6250
1-Aminoethanol : 5674
2-Aminoethanol : 5675
N-(Aminoethyl)ethanolamine : 5676
1-(2-Aminoethyl)piperazine : 9615
2-Amino-2-ethyl-1,3-propanediol : 9985
2-Aminoethylsulfonic acid : 5623
Aminoformamidine : 6278
2-Aminoglutaramic acid : 6249
2-Aminoglutaric acid : 6241
Aminoglycol : 9987
Aminoguanidine : 7147
2-Aminohexanoic acid : 8214
p-Aminohippuric acid : 6252
1-Amino-2-hydroxybenzene : 9201
1-Amino-3-hydroxybenzene : 9202
4-Amino-1-hydroxybenzene : 9203
(*S*)-2-Amino-4-hydroxybutanoic acid : 7134
[*R*-(*R**,*S**)]-2-Amino-3-hydroxybutanoic acid : 11813
(*S*)-2-Amino-3-hydroxypropanoic acid : 11357
2-Amino-6-hydroxypurine : 10768
2-Aminohypoxanthine : 10768
(*RS*)-1-Aminoindan : 7243
l-α-Aminoindole-3-propionic acid : 11922
2-Aminoisobutanol : 10341
α-Aminoisobutyric acid : 331
5-Aminoisoquinoline : 7372
2-Aminoisovaleric acid : 12041
(*R*)-2-Amino-3-mercaptopropanoic acid : 5014
Aminomethanamidine : 6278
1-Amino-3-methylbutane : 3220
2-Amino-2-methylbutyric acid : 7401
trans-4-(Aminomethyl)-1-cyclohexanecarboxylic acid : 4333
2-Amino-3-methylpentanoic acid : 7370
2-Amino-4-methylpentanoic acid : 7413
4-Amino-4-methyl-2-pentanone : 8959
3-Amino-5-methylphenol : 9212
4-(Aminomethyl)piperidine : 9673
2-Amino-2-methyl-1,3-propanediol : 9987
7-Amino-2-methylquinoline : 11211
8-Amino-6-methylquinoline : 11212
2-Amino-4-methylthiazole : 11665
2-Amino-4-methylthiobutanoic acid : 7648
3-(Aminomethyl)-3,5,5-trimethylcyclohexanol : 4502
Aminometradine : 11060
2-Aminonaphthalene-6,8-disulfonic acid : 7870
3-Amino-2-naphthoic acid : 7768
1-Amino-8-naphthol-3,6-disulfonic acid : 7874
1-Amino-2-naphthol-4-sulfonic acid : 7966
6-Aminonicotinic acid : 10897
2-Amino-5-nitrothiazole : 11666
(*S*)-2-Aminopentanedioic acid : 6241
(*S*)-2-Aminopentanoic acid : 8218
o-Aminophenol : 9201
m-Aminophenol : 9202
p-Aminophenol : 9203
p-Aminophenylacetic acid : 886
3-Amino-1-phenylbutane : 2119
4-Amino-3-phenylbutyric acid : 2128
5-(4-Aminophenylsulfonamido)-3,4-dimethylisoxazole : 2195
2-Aminophenylsulfonic acid : 2215
3-Aminophenylsulfonic acid : 2216
4-Aminophenylsulfonic acid : 2217
p-Aminophenylsulfonylthiourea : 2189
2-Amino-3-picoline : 10866
2-Amino-4-picoline : 10867
2-Amino-6-picoline : 10868
3-Amino-4-picoline : 10870
1-Aminopropane-1,3-dicarboxylic acid : 6241
3-Aminopropanoic acid : 323
DL-2-Aminopropanoic acid : 324
(*S*)-2-Aminopropanoic acid : 325
DL-1-Amino-2-propanol : 10338
3-Aminopropionitrile : 10092
β-Aminopropionitrile : 10092
4-Aminopropiophenone : 10413
p-Aminopropiophenone : 10413
α-(α-Aminopropyl)benzyl alcohol : 1840
N-(3-Aminopropyl)diethanolamine : 5679
Aminopropylin : 9795
4-(3-Aminopropyl)morpholine : 7701
Aminopropylon : 9795
6-Aminopurine : 10751
4-Aminopyrazole : 10812
2-Aminopyridine : 10857
α-Aminopyridine : 10857

β-Aminopyridine : 10858
4-Aminopyridine : 10859
Aminopyrine : 10851
2-Aminoquinoline : 11202
4-Aminoquinoline : 11203
5-Aminoquinoline : 11204
6-Aminoquinoline : 11205
8-Aminoquinoline : 11206
6-Aminoquinoxaline : 11332
Aminorex : 8574
4-Amino-1-β-*D*-ribofuranosyl-2(1H)-pyrimidinone : 5021
p-Aminosalicylic acid : 2498
5-Aminosalicylic acid : 2499
p-Aminosalicylic acid hydrazide : 2500
α-Aminosuccinamic acid : 468
L-Aminosuccinic acid : 471
2-Amino-1,3-thiazole : 11662
2-Aminothiazole : 11662
Aminoxafen : 8574
Aminozide : 3334
Amiphenazole : 11671
Amitrole : 11845
Amixetrine : 11169
Amizepin : 5144
Ammonium, acetate : 106
Ammonium, (3-carbamoyl-3,3-diphenylpropyl) diisopropyl methyl,iodide : 2120
Ammonium, diethyl(2-hydroxyethyl)methyl-, bromide, xanthene-9-carboxylate : 5445
Ammonium, dimethyl ethyl 3-hydroxyphenyl,chloride : 865
Ammonium formate : 6003
Ammonium, pentamethylenebis trimethyl, bromide : 8738
Ammonium, phenyl trimethyl,iodide : 866
Ammonium, tetraethyl,bromide : 5447
Ammonium, tetramethyl,bromide : 7456
Ammonium, tetramethyl,chloride : 7457
Ammonium, tetramethyl,iodide : 7458
Ammonium, tetrapropyl,iodide : 9833
Amobarbital : 11092
Amolanone : 2463
Amphecloral : 1583
Amphetamine : 1578
Ampyrone : 10845
Amudane : 11520
Amygdalin : 954
Amyl acetate : 227
sec-Amyl acetate (*R*) : 8918
Amylacetylene : 6609
tert-Amyl alcohol : 3628
Amyl alcohol : 8915
sec-Amyl alcohol : 8916
iso-Amylamine : 3223
Amylamine : 8700
Amylbenzene : 2107
d-Amyl bromide : 3238
tert-Amyl bromide : 3240
Amyl bromide : 8710
Amyl butyrate : 3579
Amyl caproate : 6907
tert-Amyl carbamate : 3634
Amyl chloride : 8719
6-*n*-Amyl-*m*-cresol : 9425
Amylene : 3801
α-Amylene : 9004
Amylene dichloride : 3286
2,4-Amylene glycol : 8789
Amyl ether : 8835
Amyl ethyl ketone : 8472
Amylethylmethylcarbinol : 8466
Amyl formate : 6015
Amyl heptanoate : 6470
Amyl hexanoate : 6907
Amyl hexyl ketone : 5326
Amyl iodide : 8818
Amylisopropylmethylcarbinol : 8451
tert-Amyl isovalerate : 3546
Amyl ketone : 11962
Amyl mercaptan : 8844
sec-Amylmercaptan : 8845
Amyl nitrite : 8090
: 8096
Amylol : 8915
4-Amylresorcinol : 1544
Amyl salicylate : 2649
Amyl stearate : 8270
Amyl sulfide : 8843
Amyltrichlorosilane : 11472
Amyl valerate : 8913
β-Amyrin : 8548
α-Amyrin : 12038
Anabasine : 11008
Anagyrine : 7573
Analeptin : 1887
Anastil : 9379
Anatabine : 3105
Anchovyxanthin, all-*trans*- : 4057
Androfluorone : 355
4-Androstene-3,17-dione : 352
Androsterone : 348
Anemonin : 5185
Anesthesin : 2497
Anethole : 1942
Angelic acid : 3845
Anhalamine : 7393
Anhalonidine : 7395
Anhydroecgonine : 491
Anhydron : 2851
Aniline : 633
Aniline, 2-bromo : 638
Aniline, 3-bromo : 639
Aniline, 4-bromo : 640
Aniline, 2-bromo-6-nitro- : 653
Aniline, 4-butyl : 655
Aniline, *N*-butyl- : 656
Aniline, 2-*tert*-butyl : 714
Aniline, 4-*tert*-butyl : 715
Aniline, *N*-*tert*-butyl- : 716
Aniline-4-*sec*-butyl : 813
Aniline, *N*-*sec*-butyl- : 816
Aniline-3-carboxylic acid : 2485
Aniline-4-carboxylic acid : 2486
Aniline, 2-chloro : 658
Aniline, 3-chloro : 659
Aniline, 4-chloro : 660
Aniline, 2-chloro, hydrochloride : 665
Aniline, 2,4-dibromo- : 681
Aniline, 2,6-dibromo- : 682
Aniline, 3,4-dibromo- : 683
Aniline, 2,3-dichloro- : 685
Aniline, 2,4-dichloro- : 686
Aniline, 2,5-dichloro- : 687
Aniline, 2,6-dichloro- : 688
Aniline, 3,4-dichloro- : 689
Aniline, 3,5-dichloro- : 690
Aniline, 2,6-dichloro-4-ethoxy : 691
Aniline, 3,4-diethoxy- : 693
Aniline, *N*,*N*-diethyl-4-nitroso : 699
Aniline, 2,4-difluoro- : 700
Aniline, 2,4-diiodo- : 701
Aniline, 2,6-diisopropyl- : 836
Aniline, 2,6-dimethoxy- : 703
Aniline, *N*,*N*-dimethyl-*p*-[(α,α,α-trifluoro-*m*-tolyl)azo]- : 723
Aniline, 2,3-dinitro- : 724
Aniline, 2,6-dinitro- : 726
Aniline, 3,5-dinitro- : 727
Aniline, hydrochloride : 756
Aniline, 2-methylthio : 819
Aniline nitrate : 821
Aniline, 5-nitro-2-propoxy : 827
p-Anilinesulfonamide : 2187
2-Anilinesulfonic acid : 2215
3-Anilinesulfonic acid : 2216
Aniline-4-sulfonic acid : 2217
Aniline, 2,3,4,5-tetramethyl : 847
Aniline, 2,3,4-trichloro- : 852
p-Anilinobenzenesulfonic acid : 2234
o-Anilinobenzoic acid : 2715
Anilotic acid : 2647
p-Anisaldehyde : 585
Anise alcohol : 1891
p-Anisic acid : 2666
m-Anisic acid, 2-(methylamino)-, methyl ester : 2672
o-Anisidine : 766
m-Anisidine : 767
p-Anisidine : 768
m-Anisidine, 4-nitro- : 776
Anisindione : 7269
Anisole : 1915
Anisole, 3,5-dinitro- : 1924
Anisole, 2,5-dinitro- : 1926
Anisole, 2,3-dinitro : 1922
Anisole, 4-nitro : 1937
Anisole, 2,3,4,6-tetrabromo : 2256
Anisole, 3,4,5-tribromo : 2310
Anisole, 2,4,5-tribromo : 2311
Anisole, 2,3,5-tribromo : 2312
o-(*p*-Anisoyl)benzoic acid : 2667
p-Anisoyl chloride : 2901
p-Anisyl alcohol : 1891
o-Anisyl benzoate : 9383
1-(*p*-Anisyl)ethanol : 1895
[6]Annulene : 867
[8]Annulene : 4742
Anthorine : 475
Anthracene, 9,10-dibenzyl- : 367
Anthracene, 9,3-dimethyl : 378
Anthragallol : 415
Anthralin : 430
Anthranil : 2390
o-Anthranilic acid : 2484
Anthranilic acid, *N*-methyl-, ethyl ester : 2680
Anthranilic acid, 3-nitro- : 2506
Anthranilic acid, *N*-(α,α,α-trifluoro-*m*-tolyl)- : 2729
Anthranol : 432
Anthraquinone : 379
Anthraquinone, 2-amino-1-hydroxy : 383
Anthraquinone, 2,7-dibromo- : 390
9,10-Anthraquinone, 1,7-dihydroxy : 394
Anthrarobin : 429
Anthrarufin : 393
1-Anthroic acid : 368
2-Anthroic acid : 369
9-Anthroic acid : 370
Anthrone : 433
Anthropodeoxycholic acid : 4097
Anticanitic vitamin : 2486
Antifebrin : 86
Antimonate(2-), bis[μ-[2,3-dihydroxybutanedioato(4-)-O1,O2:O3,O4]]di-, dipotassium, trihydrate, stereoisomer : 11588
Antipyrine : 10847
Antipyrine, 4-bromo- : 10846
ANTU : 11807
Apacil : 2498
Apacizin : 2500
Aphylline : 7572
Apigenin : 2806
Apigenin 4'-methyl ether : 2809
Apiole : 2423
Apoatropine : 928
Apocodeine : 5160
Apocynin : 5824
Apomorphine : 5155
6aα-Aporphine-2,9-diol, 1,10-dimethoxy- : 5156
6aα-Aporphine, 1,2,9,10-tetramethoxy- : 5157
6aα-Aporphin-11-ol, 10-methoxy-1,2-(methylenedioxy)- : 2470
6aα-Aporphin-11-ol, 1,2,10-trimethoxy- : 5161
Aprobarbital : 11098
Arabinose (*D*)(β) : 436
Arachic alcohol : 5358
Arachidic acid : 5355
Arachidonic acid : 5361
Aramite : 11579
Arborine : 11200
Arbutin : 6224
Arecaidine : 10913
Arecoline : 10914
Argobyl : 10489
Armstrong's acid : 7867
Arsanilic acid : 456
Arsine, difluoromethyl- : 463
Arsine, oxybis[dimethyl- : 452
Ascaridole : 5184
Ascensil : 10867
Asculetine : 2804
Asmacoril : 2840
Aspartame : 9511

SYNONYM INDEX OF ORGANIC COMPOUNDS (continued)

Aspergillic acid : 10811
Aspidospermine : 473
Aspirin : 2480
Atrolactic acid : 914
Atropic acid : 927
Atropine : 902
Aureothin : 10797
Aurin : 4278
12-Azabenz[a]anthracene : 630
Azacycloheptane : 492
Azacyclohexane : 9629
Azacyclopentane : 11143
7-Azadibenz[a,j]anthracene : 5143
8-Azaguanine : 11851
7-Aza-7H-benzo[c]fluorene : 2405
2-Azanaphthalene : 7373
1-Azanaphthalene : 11214
6-Azathymine : 11831
6-Azauridine : 11832
Azelaic acid : 8130
Azelaic acid, 3-methyl : 8135
1H-Azepine-1-carbothioic acid, hexahydro-, *S*-ethyl ester : 7661
Aziminobenzene : 2871
Azindole : 2376
Azine : 10872
trans-Azobenzene : 5129
cis-Azobenzene : 5130
Azobenzene, 3,3'-bis(dimethylamino) : 361
3-Azobenzenecarboxylic acid, 4-hydroxy-4'-nitro : 2648
Azobenzene, 2,2'-diethoxy- : 5112
Azobenzene, 4-dimethylamino-3'-trifluoromethyl : 723
Azobenzene, 4-methyl- : 5139
Azobenzene, 3-methyl : 5138
Azo, benzene 1'-naphthalene,2'-hydroxy : 8037
2,2'-Azobis(isobutyronitrile) : 10093
1,1'-Azobisnaphthalene : 5127
Azobutane : 5123
Azodicarbonamide : 5124
Azodicarboxylic acid diamide : 5124
Azoformic acid, diamide : 5124
Azole : 11112
Azomethane : 5125
Azomycin : 7211
1,1'-Azonaphthalene : 5127
2,2'-Azonaphthalene : 5128
Azopropane : 5132
Azopyrin : 2654
cis-Azoxybenzene : 5131
Azoxybenzene, 4-bromo- : 5120
Azoxybenzene, 2,2'-dimethyl (*trans*) : 5116
Azoxybenzene, 3,3'-dimethyl (*trans*) : 5118
Azoxybenzene, 2,2'-dimethyl(*trans*) : 5117
Azoxybenzene, 3,3'-dimethyl(*trans*) : 5119
Badische acid : 7964
Baicalein : 2844
Bamipine : 9627
Barbital : 11087
Barbituric acid : 11076
Barbituric acid, 5-allyl-5-isobutyl-2-thio, sodium salt : 11063
Barbituric acid, 5-*sec*-butyl-5-ethyl-2-thio- : 11065
Basedol : 11662
Batyl alcohol : 10023
BCPE : 1850
Bebeerine : 11924
Behenic acid : 5276
Bemegride : 9656
Bendazol : 2385
Benemid : 2595
Benovocylin : 5403
Benzalacetone : 3881
β-Benzalbutyramide : 3703
Benzal chloride : 1382
Benzaldehyde, 3-nitro, diacetate : 7511
Benzalhydrazone : 563
Benzamide, *N*-[[(4-chlorophenyl)amino]carbonyl]-2,6-difluoro- : 5176
Benzamide, 3,5-dichloro-*N*-(1,1-dimethyl-2-propynyl)- : 10746
Benzamide, *N*-(3-tolyl) : 622
Benzanilide : 626
1,2-Benzanthracene : 519
2,3-Benzanthracene : 7703
1,2-Benzanthracene, 3-methyl : 522
Benzanthrone : 6292
Benzathine : 5500
Benzazide : 2884
2-Benzazine : 7373
Benz[b]anthracene : 7703
Benzcurine iodide : 5441
Benzenamine, *N*-butyl-*N*-ethyl-2,6-dinitro-4-(trifluoromethyl)- : 509
Benzenamine, *N*-(2-chloroethyl)-2,6-dinitro-*N*-propyl-4-(trifluoromethyl)- : 5964
Benzenamine, *N*-(cyclopropylmethyl)-2,6-dinitro-*N*-propyl-4-(trifluoromethyl)- : 9763
Benzenamine, 4-(1,1-dimethylethyl)-*N*-(1-methylpropyl)-2,6-dinitro- : 3893
Benzenamine, 2,6-dinitro-*N*,*N*-dipropyl-4-(trifluoromethyl)- : 11901
Benzenamine, *N*,*N*'-1,2-ethanediylbis- : 5508
Benzenamine, *N*-ethyl-*N*-(2-methyl-2-propenyl)-2,6-dinitro-4-(trifluoromethyl)- : 5409
Benzenamine, *N*-(1-ethylpropyl)-3,4-dimethyl-2,6-dinitro- : 8635
Benzenamine, 4-(1-methylethyl)-2,6-dinitro-*N*,*N*-dipropyl- : 7371
Benzenamine, 4-(methylsulfonyl)-2,6-dinitro-*N*,*N*-dipropyl- : 8070
Benzeneacetamide, *N*,*N*-dimethyl-α-phenyl- : 5219
Benzeneacetic acid, 4-bromo-α-(4-bromophenyl)-α-hydroxy-, 1-methylethyl ester : 3140
Benzeneacetic acid, 4-chloro-α-(4-chlorophenyl)-α-hydroxy-, ethyl ester : 4082
Benzeneacetic acid, 4-chloro-α-(1-methylethyl)-, cyano(3-phenoxyphenyl)methyl ester : 5956
Benzeneamine, *N*-(1-methylpropyl)-2,6-dinitro : 3894
Benzene, 3-aminobutyl : 2119
Benzenearsonic acid : 460
Benzene, 1,3-bis(diethylamino) : 1285
Benzeneboronic acid : 3135
Benzene, (1-bromopropyl) (*DL*) : 1086
Benzenecarbonyl chloride : 2886
Benzenecarboxaldehyde : 525
Benzenecarboxylic acid : 2475
Benzene, 1(3-chloroallyl)-4-methoxy : 1231
Benzene, 1-chloro-4-[[(4-chlorophenyl)methyl]thio]- : 4075
Benzene, 2-chloro-1-(3-ethoxy-4-nitrophenoxy)-4-(trifluoromethyl)- : 8633
Benzene, 2-chloro-1-(4-nitrophenoxy)-4-(trifluoromethyl)- : 8082
Benzene, 1-chlorophosphino-4-isopropyl : 9555
Benzene, (1-decylundecyl)- : 6297
1,3-Benzenediamine, *N*3,*N*3-diethyl-2,4-dinitro-6-(trifluoromethyl)- : 5181
1,3-Benzenediamine, 2,4-dinitro-*N*3,*N*3-dipropyl-6-(trifluoromethyl)- : 9761
Benzene, 2,3-dibromo-1-chloro : 1292
Benzene, 3,4-dibromo-1-chloro : 1293
Benzene, 2,6-dibromo-1-chloro : 1294
Benzene, 2,5-dibromo-1-chloro : 1295
Benzene, 1,4-di-*tert*-butyl : 990
1,4 Benzenedicarboxylic acid, 2,3,5,6-tetrachloro-, dimethyl ester : 5180
Benzene, 1,4-dichloro-2,5-dimethoxy- : 4084
Benzene, 2,5-dichloro-1-ethyl : 1377
Benzene, 2,4-dichloro-1-(4-nitrophenoxy)- : 8081
Benzene, 1,3-diisopropyl-4-methyl- : 1948
Benzene, 1,4-dinitro-2-methoxy : 1926
1,3-Benzenedisulfonic acid, diamide,4,5-dichloro : 1559
Benzene, (1-dodecyltridecyl)- : 8638
Benzene, 1-ethyl-2-isopropyl : 1688
Benzene, 1-ethyl-3-isopropyl : 1689
Benzene, 1-ethyl-4-isopropyl : 1690
Benzene, 4-ethyl-1,2,3-trimethyl : 1708
Benzene, 6-ethyl-1,2,4-trimethyl : 1709
Benzene, 5-ethyl-1,2,4-trimethyl : 1710
Benzene, 3-ethyl-1,2,4-trimethyl : 1711
Benzene hexachloride : 4409
Benzene, 1-hydroxy-2-methoxy-4-propenyl,acetate(*cis*) : 9394
Benzene, 1-isobutoxy-2-nitro : 2034
Benzene, mercuriodi- : 7436
Benzenenitrile : 2740
Benzene, 2-nitro-1-(4-nitrophenoxy)-4-(trifluoromethyl)- : 5980
Benzene, 1-nitro-3,4,5-tribromo : 2315
Benzeneselenic acid : 2174
Benzenestibonic acid : 11548
Benzenesulfonamide, 4-(acetylamino)- : 31
Benzenesulfonamide, 4-(dipropylamino)-3,5-dinitro- : 8553
Benzenesulfonic acid, amide,4(4-aminophenylsulfonamido) : 2188
Benzenesulfonic acid, amide,4(benzylamino) : 2213
Benzenesulfonic acid, chloride : 2219
Benzenesulfonic acid, *p*-chlorothio-, *S*-phenyl ester : 2236
Benzenesulfonic acid, (2,4-dichlorophenyl) ester : 9289
Benzenesulfonic acid, fluoride : 2251
Benzenesulfonic acid, guanidine,4-amino : 2191
Benzenesulfonic acid, thiolo, 4-chloro, *S*-phenyl ester : 2236
Benzenesulfonic acid, thiolo,2-methyl,phenyl ester : 11818
Benzene, 1,2,4,5-tetraisopropyl- : 2276
Benzene, 1,2,3-triamino : 2305
Benzene, 1,3,5-tribenzoyl- : 7587
Benzene, 1,3,5-trichloro-2,4,6-trihydroxy : 2371
Benzene, 1,2,4-triisopropyl- : 2372
Benzene, 1,2,3-tris(2(diethylamino) ethoxy),tri(ethyl iodide) : 5441
Benzene, uneicosyl : 1741
Benzestrol : 9303
Benzhydramine : 5422
Benzhydrol, α-(1-naphthyl) : 7910
Benzhydrylamine : 1826
Benzhydryl bromide : 1058
2-(Benzhydryloxy)-*N*,*N*-dimethylethylamine : 5422
p-Benzidine : 3031
Benzidine : 3031
Benzidine, 2,2'-dichloro- : 3032
Benzil : 5584
Benzilic acid : 918
Benzimidazole, 1,5-dimethyl- : 2378
Benzimidazole, 1-ethyl-2-methyl- : 2381
1H-Benzimidazole, 2-(4-thiazolyl)- : 11656
2-Benzimidazolethiol : 2386
Benzo[a]chrysene : 9601
1H-Benzo[a]fluorene, veratraman-3,23-diol deriv. : 12044
2H,8H-Benzo[1,2-b:3,4-b']dipyran, butanoic acid deriv. : 3540
Benzo[b]fluoranthene : 631
Benzo[b]naphthacene : 8636
Benzo[b]quinoline : 314
3,4-Benzocarbazole : 2405
Benzo[def]phenanthrene : 10853
1,2-Benzodiazine : 4136
2,3-Benzodiazine : 9600
1,3-Benzodiazine : 11195
1,4-Benzodiazine : 11333
1,3-Benzodiazole : 2376
Benzodihydrofuran : 2442
1,3-Benzodioxole, 5-[[2-(2-butoxyethoxy)ethoxy]methyl]-6-propyl- : 9727
1,3-Benzodioxol-4-ol, 2,2-dimethyl-, methylcarbamate : 514
Benzo[d,e,f]chrysene : 2395
Benzo[7,8]fluoreno[2,1-b]quinolizine, cevane-3,4,12,14,16,17,20-heptol deriv. : 4067
Benzofuran, 2-hydroxymethyl : 2455
Benzofuran, 7-isopropyl-4-methyl : 2460

7-Benzofuranol, 2,3-dihydro-2,2-dimethyl-, methylcarbamate : 3991
5-Benzofuranpropanol, 7-methoxy-2-[3,4-(methylenedioxy)phenyl]- : 2464
6H-Benzofuro[3a,3,2-ef][2]benzazepin-6-ol, 4a,5,9,10,11,12-hexahydro-3-methoxy-11-methyl-, [4aS-(4aα,6β : 6200
Benzoglyoxaline : 2376
Benzoguanamine : 11829
Benzohydrol : 1910
Benzoic acid amide : 600
Benzoic acid azide : 2884
Benzoic acid, bromide : 2885
Benzoic acid, chloride,2,4-dichloro : 2893
Benzoic acid, chloride,3,4-dichloro : 2894
Benzoic acid, chloride,4-methoxy : 2901
Benzoic acid, 5-[2-chloro-4-(trifluoromethyl)phenoxy]-2-nitro- : 304
Benzoic acid, 3,6-dichloro-2-methoxy-, methyl ester : 7653
Benzoic acid, 5-(2,4-dichlorophenoxy)-2-nitro-, methyl ester : 2997
Benzoic acid, *o*-ethoxy-, ethyl ester : 2598
Benzoic acid, *p*-ethoxy-, ethyl ester : 2599
Benzoic acid, 2-[[ethoxy[(1-methylethyl)amino]phosphinothioyl]oxy]-, 1-methylethyl ester : 7359
Benzoic acid, fluoride : 2909
Benzoic acid, 2-(2-hydroxyethenyl)-, δ-lactone : 2791
Benzoic acid, 2-hydroxy,2-hydroxybenzoate : 2616
Benzoic acid, 4-isopropyl, isobutyl ester : 2695
Benzoic acid, 2-propyl : 2723
Benzoic anhydride : 2510
(±)-Benzoin : 5822
1H,12H-Benzo[i]pyrano[3,4-g]indolizine, 16(15H)-oxaerythrinan-15-one deriv. : 8567
Benzo[d]isothiazole : 2387
Benzonitrile, *o*-ethoxy- : 2754
3,4-Benzophenanthrene : 2406
1,2-Benzophenanthrene : 4114
9,10-Benzophenanthrene : 11910
1,2-Benzophenazine : 2394
Benzophenone 6 : 7595
Benzophenone : 7616
Benzophenone, 2-bromo- : 7600
Benzophenone, 2-carboxy-4'-methoxy- : 2667
Benzophenone, 4,4'-diiodo- : 7598
Benzophenone, 2,4-dimethoxy- : 7612
Benzophenone, 3,4-dimethyl- : 7615
Benzophenone, 2,4'-dimethyl- : 7628
Benzophenone, 4-methyl-2-nitro : 7627
Benzophenone, 2,2',4,4'-tetramethyl- : 7594
Benzopinacol : 5580
1,2-Benzopyran : 2772
1H-Benzopyrazole : 7235
2,3-Benzopyrene : 2395
6,7-Benzopyrene : 2395
3,4-Benzopyrene : 2395
Benzo[d]pyridazine : 9599
Benzopyridine : 7373
1,2-Benzopyrone : 2792
2,3-Benzopyrrole : 7294
6,7-Benzoquinoline : 2473
1,2-Benzoquinolizine-2-one, 9,10-dimethoxy-1,2,3,4,6,7,11b-heptahydro-3-isobutyl : 2396
1,4-Benzoquinone : 4257
p-Benzoquinone : 4257
1,4-Benzoquinone, 2,5-dichloro-3,6-dihydroxy : 4258
p-Benzoquinone, 2,5-dihydroxy-3-methoxy-6-methyl- : 4260
1,4-Benzoquinone, 2,6-dimethoxy : 4262
p-Benzoquinone imine, 2,6-dibromo-*N*-chloro- : 4279
p-Benzoquinone, 2,3,5,6-tetrachloro- : 4268
p-Benzoquinone, tetrahydroxy- : 4269
p-Benzoquinone, tetramethyl- : 4270
4-Benzoquinonimine, *N*,2,6-trichloro : 4280
Benzoresorcinol : 7609
Benzo[rst]pentacene : 2850
Benzosol : 9383
Benzosulfimide : 2388
Benzosulfonazole : 2861
1,2,4-Benzothiadiazine, 6-chloro-3(5-norbornen-2-yl)-7-sulfamyl-2,3,4-trihydro : 2851
1,2,4-Benzothiadiazine-7-sulfonic acid,amide,6-chloro,1,1-dioxide : 2852
1,2,4-Benzothiadiazine-7-sulfonic acid,amide,6-chloro-3-cyclopentylmethyl-2,3,4-trihydro,1,1-dioxide : 2853
1,2,4-Benzothiadiazine-7-sulfonic acid,amide,6-chloro-3(dichloromethyl)-3,4-dihydro,1,1-dioxide : 2854
1,2,4-Benzothiadiazine-7-sulfonic acid,amide,6-chloro-2,3,4-trihydro,1,1-dioxide : 2855
1,2,4-Benzothiadiazine-7-sulfonic acid,amide,6-chloro-3-isobutyl-2,3,4-trihydro : 2856
1,2,4-Benzothiadiazine-7-sulfonic acid,amide,6-chloro-2-methyl-3(2,2,2-trifluoroethylthiomethyl) : 2857
1,2,4-Benzothiadiazine-7-sulfonic acid,amide,6(trifluoromethyl)-2,3,4-trihydro,1,1-dioxide : 2858
1H-2,1,3-Benzothiadiazin-4(3H)-one, 3-(1-methylethyl)-, 2,2-dioxide : 518
Benzothiazole, 6-amino- : 2860
2-Benzothiazolesulfonic acid, amide,6-ethoxy : 2865
2-Benzothiazolethiol : 2866
Benzothiazole-2-thione : 2866
2-Benzothiazolylamine : 2859
Benzothiofuran : 2399
2,3-Benzothiophene : 2399
m-Benzotoluidide : 622
Benzotrifluoride, 3-chloro : 1241
2,3-Benzoxadiazole, 5,7-dinitro : 2874
Benzoxale : 5505
Benzoxazole, 2-hydroxy : 2881
Benzoxiquine : 11309
Benzoylacetic acid ethyl ester : 2141
Benzoylacetonitrile : 2124
Benzoyl *tert*-butyl peroxide : 1128
Benzoylecgonine : 481
Benzoylguaiacol : 9383
Benzoyl-PAS : 2514
Benzoyl peroxide : 9157
1,12-Benzperylene : 2471
Benzphetamine : 1575
Benzpiperylon : 10850
1,2-Benzpyrene : 2436
Benzylacetaldehyde : 2114
Benzyl acetate : 233
Benzylacetic acid : 2126
Benzyl alcohol : 1836
Benzyl alcohol, α-butyl-α-methyl- : 1847
Benzyl alcohol, α,α-dipropyl- : 1867
Benzyl alcohol, α-ethyl-α-propyl- : 1875
Benzyl alcohol, *o*-methyl- : 1897
Benzyl alcohol, *m*-methyl- : 1898
Benzyl alcohol, *p*-methyl- : 1899
Benzyl alcohol, 3,4,5-trimethoxy- : 1914
Benzyl alcohol, α-vinyl- : 1868
Benzylamine : 1795
Benzyl amine, 4-bromo : 1799
Benzyl amine, 2-ethoxy : 1808
Benzylaniline : 1827
Benzyl benzenecarboxylate : 2720
2-Benzylbenzimidazole : 2385
Benzyl benzoate : 2720
Benzyl bromide : 1053
Benzyl bromide, 3,5-dibromo : 1291
Benzyl-*tert*-butanol : 2150
Benzyl *sec*-butyl ketone : 8984
Benzylcarbonochloridate : 4027
Benzyl chloride : 1182
Benzyl chlorocarbonate : 4027
Benzyl chloroformate : 4027
Benzyl *trans*-cinnamate : 10690
Benzyl cyanide : 945
2-Benzylcyclopentanone : 4893
2-[Benzyl(2-dimethylaminoethyl)amino]pyridine : 5505
Benzyl disulfide : 5248
N-Benzylethanolamine : 5762
Benzyl ether : 2079
Benzyl ethyl ether : 1658
Benzyl formate : 6016
Benzyl fumarate : 3768
Benzylideneacetone : 3881
Benzylideneaniline : 840
Benzylidene chloride, 3,4-dichloro : 1359
Benzylidene chloride, 2,3,4-trichloro : 2326
Benzylidene chloride, 2,4,5-trichloro : 2327
Benzylidene hydrazine : 563
Benzylimidobis(*p*-methoxyphenyl)methane : 1797
Benzyl isobutyl ether : 2033
Benzyl levulinate : 8911
Benzyl methyl ether : 1930
2-Benzylnaphthalene : 7950
Benzyl 1-naphthyl ketone : 5859
Benzyl nicotinate : 10911
Benzyl nitrite : 8097
Benzyl oleate : 8309
Benzyloxycarbonyl chloride : 4027
Benzyl palmitate : 6655
o-Benzylphenol : 9454
p-Benzylphenol : 9455
2-Benzylpiperidine : 9699
4-Benzylpiperidine : 9700
2-Benzylpropionic acid : 2137
2-Benzylpyridine : 11004
3-Benzylpyridine : 11005
Benzyl salicylate : 2651
N-Benzylsuccinimide : 11163
N4-Benzylsulfanilamide : 2213
Benzyl sulfide : 2287
p-(Benzylsulfonamido)benzoic acid : 2721
Benzylurea : 12027
Berbamine : 2911
Berberine : 2467
13aβ-Berbin-3-ol, 2,9,10-trimethoxy-13α-methyl- : 5151
Bergenin : 10786
Beryllium acetylacetonate : 2912
Beryllium bis(acetylacetonate) : 2912
Besylic acid : 2214
Betaine : 7454
Bethanidine : 6281
Betonicine : 11178
Betulin : 7416
Biacetyl : 3359
Biallyl : 6678
4,4'-Bianiline : 3031
N,N'-Bianiline : 7158
4,4'-Bi-*o*-anisidine : 3033
Biarsine, tetraethyl : 5109
Bibenzene : 3008
α,α'-Bibenzhydrol : 5580
2,2'-Bibenzoic acid : 3038
2,2'-Bi-*p*-benzoquinone, 3,3'-dihydroxy-5,5'-dimethyl- : 2979
Bicetyl : 5348
Bicine : 6259
Bicolorin : 2822
Bicyclo[5.3.0]decapentaene : 504
cis-Bicyclo[4.1.0]heptane : 2915
Bicyclo[4.1.0]-heptane, 7-aza : 478
Bicyclo[3.3.0]octane-2,6-dione : 8685
Bicyclo[3.3.0]octane-2-one : 8688
cis-Bicyclo[3.3.0]octan-3-one : 8689
Bicyclopentyl, 2-hydroxy : 2994
Bicylo[2.2.1]heptane, 1,3,3,4,6,6-hexamethyl-2,5,7-trioxa : 11903
Biflorine : 3109
Biformyl : 5491
Biguanide : 7233
[2,2'-Biindan]-1,1',3,3'-tetrone, 2,2'-dihydroxy- : 2999
Biisobutyl : 6774
Biisobutyryl : 6807
Biisopropenyl : 3159
Bikhaconitine : 310
Bilirubin : 3000
(*Z,Z*)-Bilirubin : 3000
Biliverdine : 3001
1,1'-Binaphthyl : 3004

2,3,1',8'-Binaphthylene : 2739
Bineopentyl : 6849
Bineophyl : 2281
Biochanin B : 2830
Biotin : 11689
Biphenyl : 3008
p-Biphenylacetic acid : 3009
p-Biphenylamine : 3011
4-Biphenylamine, 4'-methyl- : 3015
Biphenyl, 2-amino-6-methyl : 3013
Biphenyl, 4-amino-4'-methyl : 3015
2,4'-Biphenyldiamine : 3029
2,2'-Biphenyldicarboxylic acid : 3038
Biphenyl, 2,2'-dimethoxy- : 3045
Biphenyl, 3,3'-dimethoxy- : 3046
Biphenyl, 2,2'-dimethoxy-5,5'-dimethyl : 3048
Biphenyl, 2,5'-dimethoxy-2',5-dimethyl : 3049
Biphenyl, 2,3-dimethyl- : 3052
Biphenyl, 2,4-dimethyl- : 3054
Biphenyl, 3,4-dimethyl- : 3059
Biphenyl, 3,5-dimethyl- : 3060
4,4'-Biphenyldisulfonic acid : 3070
o-Biphenylenemethane : 5967
2,2'-Biphenylene oxide : 5163
Biphenyl, 3-ethoxy : 3071
2-Biphenylol : 3088
4-Biphenylol : 3090
2-Biphenylol, 3-chloro- : 3092
2,3'-Bipiperidyl : 3098
Bipropargyl : 6716
Bipropionyl : 6806
3,3'-Bipyridyl : 3102
3,4'-Bipyridyl : 3103
4,4'-Bipyridyl : 3104
2,3'-Bipyridyl : 3100
2,3'-Biquinolyl : 3107
5,5'-Biresorcinol : 3097
Birodan : 2136
Bis(4-acetamidophenyl) sulfone : 91
N,N'-Bis(acetoacetyl)-3,3'-dimethylbenzidine : 3204
Bisacodyl : 9465
Bis(β-amino-β-carboxyethyl) disulfide : 5020
Bis(*p*-aminophenyl) ether : 832
Bis(4-aminophenyl) sulfide : 850
Bis(3-aminopropyl)amine : 9874
1,4-Bis(3-aminopropyl)piperazine : 9614
N,N'-Bis(3-aminopropyl)-1,4-tetramethylenediamine : 3259
1,3-Bis(bromomethyl)tetramethyl-disiloxane : 5229
1,2-Bis(chlorocarbonyl)ethane : 3366
Bis(2-chloroethyl) carbonate : 5697
Bis(2-chloroethyl) ether : 5611
Bis(2-chloroethyl) sulfide : 5645
Bis(chloromethyl) ether : 7525
Bis(*p*-chlorophenoxy)methane : 1965
1,1-Bis(4-chlorophenyl)ethane : 1678
1,1-Bis(4-chlorophenyl)ethanol : 1850
3,6-Bis(2-chlorophenyl)-1,2,4,5-tetrazine : 4139
1,1-Bis(4-chlorophenyl)-2,2,2-trichloroethanol : 1851
1,1-Bis(*p*-chlorophenyl)-2,2,2-trichloroethanol : 1851
Bis(3-chloropropyl) sulfide : 10153
Bis(cyclohexyl)carbodiimide : 4291
Bis(cyclopentadienyl)iron : 5960
1,3-Bis-(dichloromethyl)tetramethyl disiloxane : 5231
Bis(2,2-diethoxyethyl) ether : 5612
Bis-*O,O*-diethylphosphoric anhydride : 5221
Bis(diethylthiocarbamoyl) sulfide : 11714
Bis(dimethylamino)phosphonous anhydride : 5220
2,3-Bis(ethoxycarbonyl)furan : 6082
Bis-(2-ethoxyethyl) amine : 5425
Bis(ethoxymethyl)ether : 5614
Bis(2-ethylhexyl) adipate : 6781
Bis(2-ethylhexyl) decanedioate : 5043
Bis(2-ethylhexyl) phthalate : 1334
Bis(2-ethylhexyl) sebacate : 5043
2,2-Bis(ethylsulfonyl)propane : 9835
Bis(2-furyl)glyoxime : 5583
Bis-(3-hydroxy-2-butyl)amine : 3624
Bishydroxycoumarin : 2839
Bis(hydroxymethyl)acetylene : 3911
Bis-(2-methallyl)carbonate : 10703
Bis(methoxydimethylsilyl) oxide : 5234
Bis(2-methoxyethyl) carbonate : 5736
Bis(2-methylbutyl) sulfide : 3435
Bis[methylmercapto]methane : 7461
Bis(methylthio)methane : 7461
1,2'-Bis-naphthyl ether : 7928
Bis(*m*-nitrophenyl) disulfide : 5247
Bisphenol A : 9413
Bisphenol B : 9433
1,2-Bis(phenylamino)ethane : 5508
Bis-(1-phenylethyl)ether (*DL*) : 2081
Bis(1,2,2,2-tetrachloroethyl) ether : 5615
1,5-Bis(2-tetrahydrofuryl)-3-pentanol : 6166
Bis(*p*-tolyl)mercury : 7432
Bis(tributyltin) oxide : 5241
Bithionol : 9478
Bitoscanate : 1432
Δ 1,1'-Biurea : 5124
Biuret : 7232
Bivinyl : 3146
Bixin : 5108
Boldenone : 343
Boldine : 5156
Boldine dimethyl ether : 5157
Bolusanthin : 2801
Borane, triisopentyl- : 3122
Borane-trimethylamine (1:1) : 3132
Borine, dimethoxy- : 3117
Borine, triisopentyl- : 3122
Bornane, 2-chloro- : 2919
10-Bornanesulfonic acid, 2-oxo-, (1*S*,4*R*)-(+)- : 2929
3-Bornanol, acetate, (1*S*,3*R*,4*S*)-(-)- : 2947
(+)-2-Bornanone : 2961
2-Bornanone, 3-acetyl-17O- : 2950
Borneol, acetate : 2948
Bornyl chloride : 2920
Bornyl formate (*d*) : 2949
Bornyl isovalerate (*d*) : 3565
d-Bornyl isovalerate : 3565
Boron fluoride-diethyl ether compound : 3137
Boron trifluoride compd. with methyl ether : 3138
Boron trifluoride etherate : 3137
Boron trimethoxide : 3126
Brassidic acid : 5282
Brilliant green : 5444
Bristurin : 2858
Bromacetocarbamide : 3195
ω-Bromacetophenone : 5782
Bromal : 20
Bromcarbamide : 3196
Bromcholitin : 5157
Bromcresol green : 9367
Bromcresol purple : 9364
Bromdiethylacetylurea : 3195
Bromindione : 7265
Bromine cyanide : 4165
Bromisovalum : 3196
5-Bromoacenaphthene : 4
N-Bromoacetamide : 36
p-Bromoacetanilide : 37
Bromoacetic acid : 111
Bromoacetic acid, propyl ester : 120
Bromoacetone : 10416
p-Bromoacetophenone : 5701
Bromoacetyl bromide : 292
Bromoacetylene : 5937
5-Bromoacetylsalicylic acid : 2481
o-Bromoaniline : 638
m-Bromoaniline : 639
p-Bromoaniline : 640
5-Bromoanthranilic acid : 2487
4-Bromoantipyrine : 10846
Bromoaspirin : 2481
Bromobenzene : 998
p-Bromobenzenesulfonyl chloride : 2244
p-Bromobenzoic acid : 2520
2-Bromobenzophenone : 7600
p-Bromobenzyl bromide : 1000
p-Bromobenzyl chloride : 1008
α-Bromobenzyl cyanide : 947
4-Bromo-1,2-butadiene : 3147
1-Bromo-1,3-butadiene : 3148
2-Bromo-1-butene : 3709
3-Bromo-1-butene : 3710
(±)-2-Bromobutyric acid : 3468
DL-α-Bromobutyric acid : 3468
(*RS*)-α-Bromobutyric acid : 3468
Bromochloroacetic acid : 112
Bromochloroacetic acid, ethyl ester : 113
1-Bromo-3-chloroacetone : 10417
2-Bromo-1-chlorobutane : 3234
Bromochlorodifluoromethane : 7464
Bromochlorodinitromethane : 7465
1-Bromo-1-chloroethane : 5455
Bromochlorofluoromethane : 7466
Bromochloromethane : 7463
1-Bromo-4-chloronaphthalene : 7741
4-Bromo-2-chlorophenol : 9232
1-Bromo-2-chloropropane : 9838
1-Bromo-3-chloropropane : 9839
2-Bromo-1-chloropropane : 9840
2-Bromo-2-chloropropane : 9841
2-Bromo-5-chlorothiophene : 11728
4-Bromo-3-chloro-3,4,4-trifluoro-1-butene : 3715
1-Bromo-2-chloro-1,1,2-trifluoroethane : 5457
6-Bromo-*m*-cresol : 9237
Bromocycloheptane : 4222
2-Bromo-1-cyclohexanol : 4503
Bromocyclopentane : 4777
2-Bromo-1-decene : 5086
1-Bromo-2-decene : 5087
2-Bromodibenzofuran : 5164
3-Bromodibenzofuran : 5165
Bromodichlorofluoromethane : 7468
Bromodichloromethane : 7467
2-Bromo-2,2-diethylacetamide : 3198
1-Bromo-1,2-difluoroethylene : 5883
Bromodifluoromethane : 7469
2-Bromo-2,3-dimethylbutane : 3236
1-Bromo-1,1-dimethylbutane : 8717
4-Bromo-1,3-dimethylpyrazole : 10816
1-Bromo-2,3-dinitrobenzene : 1024
Bromodiphenylmethane : 1058
2-Bromoethanesulfonyl chloride : 5628
2-Bromoethanol : 5685
2-Bromoethyl ethyl ether : 5461
2-Bromo-2-ethylisovaleramide : 3199
2-Bromofluorene : 5968
Bromofluoroacetic acid : 116
1-Bromo-4-fluorobutane : 3237
1-Bromo-10-fluorodecane : 5035
1-Bromo-12-fluorododecane : 5294
1-Bromo-6-fluorohexane : 6735
Bromofluoromethane : 7470
1-Bromo-5-fluoropentane : 8713
1-Bromo-11-fluoroundecane : 11939
Bromoform : 7548
2-Bromofumaric acid : 3770
3-Bromofuran : 6034
1-Bromoheptadecane : 6307
2-Bromoheptane : 6357
4-Bromoheptane : 6358
2-Bromoheptanoic acid : 6457
1-Bromo-1-heptyne : 6612
1-Bromo-2-heptyne : 6613
2-Bromohexane : 6733
3-Bromohexane : 6734
1-Bromo-3-iodobenzene : 1047
m-Bromoiodobenzene : 1047
1-Bromo-2-iodoethane : 5463
Bromoiodomethane : 7471
α-Bromoisobutyric acid : 10209
4-Bromoisoquinoline : 7374
β-Bromoisovaleric acid : 3472
2-Bromoisovalerylurea : 3196
Bromomethane : 7462
Bromomethoxymethane : 7472
2-Bromo-*p*-methylacetophenone : 5779
2-Bromo-5-methylaniline : 648
(+)-1-Bromo-2-methylbutane : 3238
(*S*)-1-Bromo-2-methylbutane : 3238
2-Bromo-3-methyl-1-butene : 3716
(Bromomethyl)chlorodimethylsilane : 11365
(Bromomethyl)dimethylchlorosilane : 11365

2-Bromo-4-methyl-1-nitrobenzene : 1073
1-Bromo-2-methylpentane : 8714
1-Bromo-3-methylpentane : 8715
2-Bromo-2-methylpentane : 8717
3-Bromo-3-methylpentane : 8718
4-Bromo-3-methylpyrazole : 10817
2-Bromo-3-methylthiophene : 11729
Bromomethyltrimethylsilane : 11366
2-Bromonaphthalene : 7740
5-Bromo-2-naphthoic acid : 7769
3-Bromo-1,2-naphthoquinone : 7844
2-Bromo-4-nitrobenzoic acid : 2527
2-Bromo-5-nitrobenzoic acid : 2528
4-Bromo-3-nitrobenzoic acid : 2529
Bromonitromethane : 7473
2-Bromo-1-nitronaphthalene : 7745
4-Bromo-2-nitrophenol : 9244
6-Bromo-5-nitroquinoline : 11221
3-Bromo-2-nitrotoluene : 1068
4-Bromo-2-nitrotoluene : 1074
18-Bromooctadecanoic acid : 8250
Bromopentafluorobenzene : 1078
2-Bromopentane : 8711
3-Bromopentane : 8712
1-Bromo-1-pentene : 9007
2-Bromo-1-pentene : 9008
3-Bromo-1-pentene : 9009
2-Bromo-2-pentene : 9012
3-Bromo-2-pentene : 9013
5-Bromo-2-pentene : 9015
1-Bromo-1-pentyne : 9124
p-Bromophenacyl bromide : 5777
9-Bromophenanthrene : 9164
o-Bromophenylacetylene : 1040
m-Bromophenylacetylene : 1041
2-Bromo-3-phenyl-2-butene : 3719
(*p*-Bromophenyl)hydrazine : 7141
p-Bromophenyl isocyanate : 1049
Bromophenylmethane : 1053
p-Bromophenyl methyl ketone : 5781
4-Bromophenyl phenyl ether : 1079
3-Bromo-1-propene : 10531
β-Bromopropionic acid : 10205
3-Bromopropylene : 10531
3-Bromopyridine : 10876
4-Bromopyridine : 10877
3-Bromoquinoline : 11217
5-Bromoquinoline : 11218
6-Bromoquinoline : 11219
7-Bromoquinoline : 11220
5-Bromosalicylhydroxamic acid : 606
Bromosaligenin : 1843
trans-2-Bromostilbene : 1080
Bromosuccinic acid : 3300
N-Bromosuccinimide : 11153
2-Bromo-1,3-thiazole : 11668
2-Bromothiophene : 11726
o-Bromotoluene : 1054
m-Bromotoluene : 1055
p-Bromotoluene : 1056
Bromotrichloromethane : 7474
Bromotrifluoromethane : 7475
1-Bromo-2,3,5-trimethylbenzene : 1092
Bromotrimethylmethane : 9846
Bromotrinitromethane : 7476
Bromotriphenylmethane : 1061
11-Bromoundecanoic acid : 11944
2-Bromoundecanoic acid : 11945
5-Bromouracil : 11062
Bromoxynil : 2751
Bromphenol blue : 9366
Bromthymol blue : 9365
Broncholysin : 5015
Bronopol : 9991
p-Brosyl chloride : 2244
Broxyquinoline : 11313
Brucine : 11564
Buclosamide : 607
Bucolome : 11081
5β-Bufa-20,22-dienolide, 3β,14,16β-trihydroxy-, 16-acetate : 3141
Bufexamac : 880
Bufotalin : 3141
Bufotenine : 7326
Bulbocapnine : 2470
1,3-Butadiene, hexafluoro : 3164
1,3-Butadiene, pentafluoro-2-(trifluoromethyl)- : 3172
Butalbital : 11101
Butaldehyde : 3178
Butallylonal : 11080
Butamben : 2488
Butane, 2-amino (*DL*) : 3210
Butane, 2-benzyl-1-phenyl- : 1698
Butane, 1,4-bis(diethylamino) : 3262
Butane, 1-chloro-2-methyl (*DL*) : 3251
Butane, 1,4-diamino, dihydrochloride : 3260
Butane, 1,4-dibromo-2-methyl (*d*) : 3273
1,4-Butanedicarboxylic acid : 6779
Butane, 1,4-dicyclohexyl- : 4323
Butane, 1,1-dicyclohexyl- : 4326
Butane, 1,2:3,4-diepoxy-, *meso-* : 3007
Butane, 2,2-di(2-furyl) : 6038
1,3-Butanediol, sulfite : 5198
Butane, 1,4-diphenyl- : 1099
Butane, 1,1-diphenyl- : 1120
Butane, 1,3-diphenyl (*DL*) : 2020
Butane, 2-iodo (*DL*) : 3382
Butane, 1-iodo-2-methyl (*d*) : 3383
Butane, 2-methyl-1,2,3-trichloro : 3453
Butane, 2-methyl-2,3,3-trichloro : 3454
Butane, 1,1,2,3,4-pentachloro (liquid) : 3419
Butane, 1,2,3,4-tetrabromo (*DL*) : 3426
2-Butanethiol (*DL*) : 3438
Butanimidic acid : 3194
Butanoic acid, amide : 3194
Butanoic acid, amide,3-methyl : 3206
Butanoic acid, *sec*-butyl ester(*d*) : 3564
Butanoic acid, chloride : 3681
Butanoic acid, 4-(2,4-dichlorophenoxy)- : 3925
Butanoic anhydride : 3467
3-Butanolal : 3184
2-Butanol, 3-bromo-, *erythro*-(±)- : 3601
2-Butanol, 3-chloro (*erythro, DL*) : 1605
1-Butanol, 4-cyclohexyl : 4324
2-Butanol, 4-cyclohexyl : 4469
1-Butanol, 3,5-dinitrobenzoate : 2592
1-Butanol, 2,4-diphenyl-, (-)- : 1108
2-Butanol, 3-methyl (*DL*) : 3629
1-Butanol, 2-methyl-4-phenyl (*DL*) : 1107
1-Butanol, 3-methyl-1-phenyl- : 1905
1-Butanol, 3-methyl-2-phenyl : 1611
2-Butanol, 2-methyl-3-phenyl : 1616
2-Butanol, 2-methyl-4-phenyl- : 2150
2-Butanol, 3-methyl-3-phenyl : 1617
2,3-Butanolone : 3661
2-Butanone, 1-(4-chlorophenoxy)-3,3-dimethyl-1-(1H-1,2,4-triazol-1-yl)- : 512
2-Butanone, 3,3-dimethyl-1-(methylthio)-, *O*-[(methylamino)carbonyl]oxime : 11717
Butazolamide : 3197
2-Butenal, 2-bromo, diethyl acetal : 4149
2-Butenal, diethyl acetal : 3758
2-Butenal, 3-ethoxy, diethyl acetal : 3811
trans-2-Butene : 3705
cis-2-Butene : 3706
1-Butene, 2-bromo-4-phenyl : 1001
2-Butene, 1-bromo-4-phenyl- : 1002
2-Butene, 2,3-dichlorohexafluoro- : 3754
trans-Butenedioic acid : 3764
trans-2-Butene-1,4-diol : 3792
cis-2-Butene-1,4-diol : 3793
1-Butene, 1,3-diphenyl- : 2022
1-Butene, 1,2-diphenyl : 11560
2-Butene, 1,1,2,3,4,4-hexachloro (liquid) : 3798
1-Butene, 2-isopropyl-3-methyl- : 8765
1-Butene, 3-methyl-2-isopropyl : 8765
1-Butene-3-ol, 2-methyl : 3871
2-Butene-1-ol, 3-phenyl : 3872
1-Butene, 1,3,4,4-tetrachloro-tetrafluoro : 3808
2-Butenoic acid, 4-chloro (*trans*) : 4151
2-Butenoic acid, 2-chloro, ethyl ester (*trans*) : 3824
2-Butenoic acid, 2-chloro, ethyl ester (*cis*) : 3825
2-Butenoic acid, 3-[(dimethoxyphosphinyl)oxy]-, methyl ester, (*E*)- : 7659
2-Butenoic acid, 3-[(dimethoxyphosphinyl)oxy]-, 1-phenylethyl ester, (*E*)- : 4137
2-Butenoic acid, 2-ethyl (*trans*) : 3837
2-Butenoic acid, 3-[[(ethylamino)methoxyphosphinothioyl]oxy]-, 1-methylethyl ester, (*E*)- : 10719
2-Butenoic acid, 2-(1-methylheptyl)-4,6-dinitrophenyl ester : 5182
2-Butenoic acid, 3-methyl-, 2-(1-methylpropyl)-4,6-dinitrophenyl ester : 3003
3-Buten-2-ol (*DL*) : 3858
3-Buten-2-ol, 2-phenyl- : 1869
3-Buten-1-ol, 3-phenyl- : 2157
trans-1-Butenyl bromide : 3707
1-Butenyl methyl ketone : 7086
Butethal : 11082
Butethamate : 898
Butethamine : 5748
Buthiazide : 2856
Butoform : 2488
2,4-D 2-Butoxyethyl ester : 170
Butyl acetate : 121
sec-Butyl acetate : 214
tert-Butyl acetate : 177
tert-Butylacetic acid : 3508
tert-Butylacetylene : 3907
Butylacetylene : 7114
Butylacrolein : 6530
Butyl acrylate : 10616
Butyl alcohol : 3593
sec-Butyl alcohol : 3594
tert-Butyl alcohol : 10389
Isobutyl alcohol, 1-phenyl : 1901
Butyl aldehyde : 3178
Butylallene : 6326
Butylamine : 3209
sec-Butylamine, (±)- : 3210
tert-Butylamine : 9818
Butyl 4-aminobenzoate : 2488
2-(Butylamino)ethanethiol : 5650
N-Butylaniline : 656
N-tert-Butylaniline : 716
N-sec-Butylaniline : 816
Butylated hydroxyanisole : 9329
Butylbenzene : 1118
tert-Butylbenzene : 1459
sec-Butylbenzene : 2035
Butyl benzoate : 2532
p-tert-Butylbenzoic acid : 2588
α-Butylbenzyl alcohol : 1846
Butyl bromide : 3231
(±)-*sec*-Butyl bromide : 3232
tert-Butyl bromide : 9846
2-*tert*-Butylbutane : 8847
Butyl butanoate : 3479
Butyl butyrate : 3479
Butyl carbamate : 3935
sec-Butylcarbinol : 3626
Butyl Carbitol : 5688
Butyl carbonate : 3998
Butyl Cellosolve : 5687
Butylchloral : 3191
Butyl chloride : 3241
sec-Butyl chloride (*DL*) : 3242
tert-Butyl chloride : 9863
tert-Butyl chloroacetate : 128
Butyl chloroformate : 4012
Butyl citrate : 10168
6-*sec*-Butyl-*m*-cresol : 9421
tert-Butyl cyanide : 10100
Butylcyclohexane : 4325
sec-Butylcyclohexane : 4448
trans-2-Butylcyclohexanol : 4504
2-Butylcyclohexanone : 4575
Butylcyclooctane : 4737
Butylcyclopentane : 4778
1-Butyl-1-cyclopentene : 4906
Butyl 2,3-dibromopropionate : 10239
Butyl dichloroacetate : 163
Butyl 3,5-dinitrobenzoate : 2592
2-*tert*-Butyl-4,6-dinitrophenol : 9328
5-*n*-Butyldocosane : 5269

11-Butyldocosane	: 5272
7-Butyldocosane	: 5270
9-Butyldocosane	: 5271
Butyl enanthate	: 6459
1-Butylene	: 3704
α-Butylene chlorohydrin	: 3605
β-Butylene chlorohydrin	: 3606
1,4-Butylenediamine	: 3258
α-Butylene dibromide	: 3265
1,3-Butylene glycol	: 3345
1,4-Butylene glycol	: 3346
Butylene glycol	: 3346
2,4-D Butyl ester	: 171
Butyl ethanoate	: 121
sec-Butylethene	: 9053
Butyl ether	: 3415
sec-Butyl ether, (±)-	: 3416
Butylethylamine	: 3218
sec-Butyl ethyl amine (*DL*)	: 3219
Butyl ethyl ether	: 3372
sec-Butyl ethyl ether	: 3373
(*RS*)-*S*-*sec*-Butyl *O*-ethyl 2-oxo-1,3-thiazolidin-3-ylphosphonothioate	: 6021
Butyl ethyl sulfide	: 3376
Butyl fluoride	: 3378
sec-Butyl fluoride	: 3379
Butyl formate	: 6004
2-*tert*-Butylfuran	: 6090
Butyl glycidyl ether	: 8602
Butyl heptanoate	: 6459
1-Butylheptyl alcohol	: 11953
2-Butyl-2-hexanol	: 8171
Butyl hexyl ketone	: 11961
tert-Butyl hydroperoxide	: 7178
Butyl 4-hydroxybenzoate	: 2615
Butyl 2-hydroxypropionate	: 10268
tert-Butyl hypochlorite	: 7183
Butylidene chloride	: 3276
2-Butylidenecyclohexanone	: 4576
Butyl iodide	: 3381
tert-Butyl iodide	: 10058
Butyl isobutyl ether	: 3401
tert-Butyl isocyanate	: 10062
Butyl isocyanide	: 3387
Butyl isopropyl ether	: 3399
Butyl isopropyl ketone	: 6522
tert-Butyl isopropyl ketone	: 8989
Butylisopropylmethylcarbinol	: 6480
tert-Butyl isothiocyanate	: 10066
sec-Butyl isothiocyanate (*DL*)	: 3389
Butyl isovalerate	: 3543
Butyl ketone	: 8175
n-Butylmalonic acid	: 9929
Butyl mercaptan	: 3437
tert-Butyl mercaptan	: 10159
Butyl methacrylate	: 10650
Butylmethylamine	: 3221
Butyl *m*-methylbenzoate	: 2685
Butyl methyl ether	: 3391
sec-Butyl methyl ether (*d*)	: 3392
tert-Butyl methyl ether	: 10072
sec-Butyl methyl ether (*DL*)	: 3393
tert-Butyl methyl ketone	: 3657
Butyl methyl ketone	: 6959
1-Butyl-3-(*p*-methylphenylsulfonyl)urea	: 2207
Butyl methyl sulfide	: 3402
tert-Butyl methyl sulfide	: 10083
2-Butylnaphthalene	: 7747
Butyl nitrate	: 8072
Butyl nitrite	: 8083
tert-Butyl nitrite	: 8085
5-Butylnonane	: 8122
Butyl nonanoate	: 8147
5-Butyl-4-nonene	: 8184
Butyl oxide	: 3415
Butylparaben	: 2615
Butyl parasept	: 2615
p-*tert*-Butylphenol	: 9323
o-*sec*-Butylphenol	: 9430
5-Butylpicolinic acid	: 10898
3-Butylpiperidine	: 9636
4-*n*-Butylpiperidine	: 9637
Butyl propionate	: 10214
Butyl propyl ether	: 3421
tert-Butyl propyl ketone	: 6965

Butyl propyl ketone	: 8473
2-Butylpyridine	: 10878
2-*tert*-Butylpyridine	: 10942
3-*tert*-Butylpyridine	: 10943
5-Butyl-2-pyridinecarboxylic acid	: 10898
2-Butylpyrrolidine	: 11147
2-Butylquinoline	: 11222
Butyl stearate	: 8251
1-Butyl-3-sulfanilylurea	: 2192
Butyl sulfide, 2,2'-dimethyl	: 3435
Butyl sulfite	: 11580
tert-Butyl tetrasulfide	: 11649
Butyl thiocyanate	: 11699
tert-Butyl thiocyanate	: 11701
Butylthiol	: 3437
2-Butylthiophene	: 11731
Butyl *m*-toluate	: 2685
p-*tert*-Butyltoluene	: 1466
1-Butyl-*s*-triazole	: 11847
Butyltrichlorosilane	: 11370
Butyl trifluoroacetate	: 265
2-Butyl-1,1,3-trimethylcyclohexane	: 4327
Butyltrimethylsilane	: 11371
Butyl vinyl ether	: 3369
Butyl vinyl ketone	: 6583
2-Butyne-1-al	: 3895
1-Butyne-3,4-diol	: 3912
Butyraldehyde	: 3178
Butyraldehyde, 3-chloro-, diethyl acetal	: 3243
Butyraldehyde, 3-chloro-, dimethyl acetal	: 3244
Butyraldehyde, 3-methyl, oxime	: 3187
Butyraldehyde, oxime	: 3188
Butyraldehyde, phenylhydrazone	: 3189
Butyramide	: 3194
p-Butyramidophenol	: 3205
Butyrhydrazide	: 3529
Butyric acid	: 3460
Butyric acid, 4-acetyl, ethyl ester	: 6906
Butyric acid, 3-bromo-3-methyl-	: 3472
Butyric acid, *sec*-butyl ester (*d*)	: 3564
Butyric acid chloride	: 3681
Butyric acid, 2-chloro-2-methyl-	: 3486
Butyric acid, 2-chloro-3-methyl, ethyl ester	: 3490
Butyric acid, 2-chloro-2-methyl, ethyl ester	: 3491
Butyric acid, 2-chloro-2-methyl-3-oxo, ethyl ester	: 3492
Butyric acid, 2-cyano-2-ethyl-, ethyl ester	: 3495
Butyric acid, 2,2-dichloro-	: 3500
Butyric acid, 3,4-dichloro, butyl ester	: 3501
Butyric acid, 2,3-dimethyl	: 3507
Butyric acid, heptafluoro-, isopropyl ester	: 3526
Butyric acid, 2-hydroxy, ethyl ester (*DL*)	: 3532
Butyric acid, 2-hydroxy, isobutyl ester	: 3536
Butyric acid, 3-methylamino-, ethyl ester	: 3541
Butyric acid, (2-methylbutyl) ester	: 3545
Butyric acid, 3-methyl, isopropyl ester	: 3553
Butyric acid (1-methylpentyl) ester (*d*)	: 3559
Butyric acid, 4-phenoxy-, ethyl ester	: 3581
Butyric acid, phenyl ester	: 3584
Butyric acid, 3-thioxo, ethyl ester	: 3589
Butyric acid, 2,2,3-trichloro-	: 3590
Butyric acid, 2,2,3-trichloro, ethyl ester	: 3591
Butyric aldehyde	: 3178
Butyric anhydride	: 3467
Butyroin	: 8474
γ-Butyrolactone	: 6142
1-Butyronaphthone	: 3674
Butyronitrile	: 3403
Butyrophenone, 4-[4-(*p*-chlorophenyl)-4-hydroxypiperidino]-4'-fluoro-	: 3651
Butyrophenone, 2-hydroxy-	: 3666
Butyrophenone, 4-methyl	: 3672
p-(Butyrylamino)phenol	: 3205
N-Butyryl-*p*-aminophenol	: 3205

Butyryl bromide	: 3680
Butyryl chloride	: 3681
Butyryl chloride, 3-chloro-	: 3684
Butyryl chloride, 2-chloro-3-methyl	: 3686
Butyryl chloride, 4-chloro-3-oxo	: 3687
Butyryl chloride, 2,3-dimethyl	: 3689
Butyryl chloride, 3-methyl	: 3693
Butyryl chloride, 2-methyl (*DL*)	: 3694
Butyryl oxide	: 3467
Butyryl triglyceride	: 3586
C.I. Direct blue 53	: 7877
C.I. Solvent Orange 7	: 8028
C.I. Solvent Orange 35	: 384
Cacodyl	: 5110
Cacodylic acid	: 451
Cacodyl oxide	: 452
Cacotheline	: 11352
Cadalene	: 7830
Cadalene, 3-methyl	: 7905
Cadalene, 8-methyl	: 8015
Cadalene, 5-methyl	: 8014
Cadaverine	: 8735
γ-Cadinene	: 7936
Caffeic acid	: 10632
Caffeine	: 10760
Calcyan	: 3930
Calusterone	: 357
Campesterol	: 5386
Camphane-3-carboxylic acid	: 3131
d-Camphocarboxylic acid	: 2917
Camphocean	: 4855
Campholic acid	: 4786
α-Campholytic acid	: 4913
(+)-Camphor	: 2961
Camphor, 3-acetyl	: 2950
d-Camphor, 10-bromo-	: 2951
Camphor, 3,3-dibromo(*d*)	: 2953
Camphoric acid, dimethyl ester, (+)-	: 4795
(-)-Camphor oxime	: 2962
D-Camphorsulfonic acid	: 2929
α-Camphylamine	: 2927
Canavanine	: 7135
Cannabidiol	: 1537
Cannabinol	: 5154
Canrenone	: 9736
Cantharidin	: 5371
Capraldehyde	: 5030
Capramide	: 5031
Capric acid	: 5059
Capric acid, 10-fluoro	: 5064
Capric alcohol	: 5070
Caprinitrile	: 5057
Caprinoyl chloride	: 5078
Caproaldehyde	: 6718
Caproaldehyde, 3-methyl	: 6720
Caproic acid	: 6870
Caproic acid, 5-bromo,ethyl ester	: 6878
Caproic acid, 6-cyclohexyl	: 4412
Caproic acid, 2,2-dimethyl-3-hydroxy, β-lactone	: 6892
Caproic acid, 6-fluoro	: 6888
Caproic acid, 2-methyl (*DL*)	: 6893
Caprolactam	: 493
Caprolactone	: 8591
Caproleic acid	: 5091
Capronitrile	: 6840
Caproyl alcohol	: 6911
Caproyl chloride	: 6986
Caproyl chloride, 3-methyl	: 6988
Caproyl chloride, 4-methyl	: 6989
Caprylene	: 8491
Caprylic acid	: 8422
Caprylic acid, 8-fluoro	: 8429
Caprylic acid, 3-methyl	: 8433
Caprylic acid, 2-methyl-3-oxo, ethyl ester	: 8436
Caprylic alcohol	: 8443
Caprylic aldehyde	: 8346
Caprylnitrile	: 8411
Capsaicin	: 8181
Capsanthin	: 4063
Captan	: 7368
Carbamamidine	: 6278
Carbamazepine	: 5144
Carbamic acid, 1H-benzimidazol-2-yl-, methyl ester	: 3990

Carbamic acid, [1-[(butylamino)carbonyl]-1H-benzimidazol-2-yl]-, methyl ester : 515
Carbamic acid, *sec*-butyl-, ethyl ester : 3960
Carbamic acid, butyl-, 2-(hydroxymethyl)-2-methylpentyl ester, carbamate : 3933
Carbamic acid, *N*-*sec*-butyl, methyl ester : 3962
Carbamic acid, (3-chlorophenyl)-, 4-chloro-2-butynyl ester : 510
Carbamic acid, *N*,*N*-diethyl : 3941
Carbamic acid, dimethyl-, 2-(dimethylamino)-5,6-dimethyl-4-pyrimidinyl ester : 9728
Carbamic acid, dimethyldithio-, manganese(2+) salt : 7421
Carbamic acid, *N*-ethyl, butyl ester : 3947
Carbamic acid, *N*-ethyl-*N*-nitro, butyl ester : 3951
Carbamic acid, *N*-ethyl-*N*-nitro, ethyl ester : 3950
Carbamic acid, isobutyl ester : 3959
Carbamic acid, isopentyl-, ethyl ester : 3952
Carbamic acid, isopropyl ester : 3955
Carbamic acid, *N*-isopropyl-*N*-nitro, ethyl ester : 3957
Carbamic acid, (3-methylphenyl)-, 3-[(methoxycarbonyl)amino]phenyl ester : 9198
Carbamic acid, [3-[[(phenylamino)-carbonyl]oxy]phenyl]-, ethyl ester : 5104
Carbamic acid, [1,2-phenylenebis-(iminocarbonothioyl)]bis-, dimethyl ester : 11723
Carbamic acid, 3-phenylpropyl ester : 2149
Carbamic acid, 1-phenyl-2-propynyl ester : 1877
Carbamic acid, *N*-propyl-*N*-nitro, ethyl ester : 3964
Carbamide : 11992
Carbamimidoselenoic acid : 11354
Carbamodithioic acid, diethyl-, 2-chloro-2-propenyl ester : 4065
Carbamodithioic acid, dimethyl-, sodium salt : 11514
Carbamonitrile : 4155
Carbamothioic acid, bis(2-methylpropyl)-, *S*-ethyl ester : 11586
Carbamothioic acid, bis(1-methylethyl)-, *S*-(2,3,3-trichloro-2-propenyl) ester : 11823
Carbamothioic acid, bis(1-methylethyl)-, *S*-(2,3-dichloro-2-propenyl) ester : 5106
Carbamothioic acid, cyclohexyl-ethyl-, *S*-ethyl ester : 4168
Carbamothioic acid, diethyl-, *S*-[(4-chlorophenyl)methyl] ester : 11697
Carbamothioic acid, diethyl-, *S*-ethyl ester : 5930
Carbamothioic acid, dipropyl-, *S*-ethyl ester : 5372
Carbamothioic acid, dipropyl-, *S*-propyl ester : 12047
Carbamoyl chloride, *N*,*N*-diethyl : 3972
Carbamoylcholine chloride : 5440
5-Carbamoyl-5H-dibenz[b,f]azepine : 5144
Carbamyl chloride : 3971
Carbanilaldehyde : 5999
Carbanilide : 12010
Carbarsone : 455
Carbaryl : 8032
Carbazide : 4009
(*N*-Carbazolyl)acetic acid : 3982
Carbazone, 1,5-diphenyl : 5122
Carbic anhydride : 7580
Carbimazole : 7194
Carbitol : 5720
Carbitol acetate : 5721
Carbobenzoxy chloride : 4027
Carbocysteine : 5017
Carbodiimide, dicyclohexyl- : 4291
Carbohydrazide : 4009
Carbomethene : 5927
o-Carbomethoxyaniline : 2503
2-Carbomethoxyphenol : 2634
Carbon bisulfide : 3993
Carbon dibromide dichloride : 7495
Carbon diselenide : 4036
Carbonic acid, bis(2-chloroethyl) ester : 5697
Carbonic acid, bis(2-ethoxyethyl) ester : 5719
Carbonic acid, bis(2-methoxyethyl) ester : 5736
Carbonic acid, bis(2-methoxyphenyl) ester : 9384
Carbonic acid, bis(2-methylallyl) ester : 10703
Carbonic acid, dithio-, cyclic *S*,*S*-(6-methyl-2,3-quinoxalinediyl) ester : 4072
Carbonic acid, monochloride,monoethyl ester : 4018
Carbonic anhydride : 3992
Carbon oxide : 4011
Carbon oxysulfide : 4035
Carbon suboxide : 9773
Carbon tetrabromide : 7539
Carbon tetrachloride : 7540
Carbon tetrafluoride : 7541
Carbon tetraiodide : 7542
Carbonyl bromide : 4006
Carbonyl chloride : 4007
Carbonyl chloride fluoride : 4005
Carbonyl cyanide : 9915
N,*N*'-Carbonyldiimidazole : 7193
Carbonyl fluoride : 4008
Carbonyl sulfide : 4035
Carbostyril : 11325
3-Carboxyaniline : 2485
4-Carboxyaniline : 2486
9-Carboxyanthracene : 370
4-Carboxybenzenesulfonamide : 2508
3-Carboxycoumarin : 2775
4-Carboxy-1,3-dimethylbenzene : 2580
2-Carboxyethylamine : 323
2-Carboxyfuran : 6050
4-Carboxymethylbiphenyl : 3009
S-(Carboxymethyl)-*L*-cysteine : 5017
3-[(Carboxymethyl)thio]-*L*-alanine : 5017
m-Carboxynitrobenzene : 2704
p-Carboxynitrobenzene : 2705
4-Carboxyphenol : 2613
4'-Carboxyphenylmethane-sulfonanilide : 2721
4-Carboxyphthalic acid : 2319
3-Carboxypropylamine : 3464
8-Carboxyquinoline : 11229
2-Carboxythiophene : 11737
Carbromal : 3195
Carbromide : 3198
Carbutamide : 2192
Carbyloxime : 6024
Cardilate : 3432
Cardine : 3540
Carditoxin : 4043
Cardrase : 2865
Δ4-Carene (*l*) : 4050
Carfimate : 1877
Carinamide : 2721
Carminic acid : 373
Carmustine : 11994
Carnegine : 7390
Carnosine : 7133
γ-Carotene : 4052
α-Carotene : 4054
β-Carotene, all-*trans*- : 4051
α-Carotene, 6',7'-didehydro-5,6-epoxy-4',5,5',6,7,8-hexahydro-3,3' : 4055
Carvacrol : 9415
Carvacrol, acetate : 9419
Carvacryl acetate : 9419
Carvenol : 4698
Carvenone (*DL*) : 4717
(+)-Carvone : 4715
(-)-β-Caryophyllene : 2996
β-Caryophyllene : 2996
trans-Caryophyllene : 2996
Caryophyllene : 2996
Carzenide : 2508
Casimiroin : 5218
Catapyrin : 11060
Catechol, 5-iodo : 1523
Caulophylline : 7578
CDT : 4212
Cedrin : 8058
Cedrol : 7570
Celidoniol, deoxy- : 8099
Cellocidin : 3902
Cellosolve : 5717
Cephaeline : 5367
Cepharanthine : 8630
Cerotic acid : 6624
Cetane : 6632
1-Cetene : 6667
Cetyl alcohol : 6659
Cetylamine : 6631
Cetyl bromide : 6633
Cetyl chloride : 6634
α-Cetylcitric acid : 8102
Cetyl iodide : 6640
Cetyl lactate : 10271
Cetyl mercaptan : 6643
Cetyl palmitate : 6651
Cetyl stearate : 8258
Cevadic acid : 3844
Cevadine : 4069
CFC 11 : 7555
CFC 12 : 7500
CFC 13 : 7486
CFC 13B1 : 7475
CFC 21 : 7502
CFC 22 : 7478
CFC 113 : 5663
CFC 114 : 5538
CFC 115 : 5484
CFC 123 : 5540
CFC 123a : 5539
C16 Guerbet fatty acid : 5065
Chalcone : 10708
Champacol : 507
Chaulmoogric acid : 4936
Chavicine : 9633
Chavicol : 9458
Cheirolin : 10067
Chelerythrine : 2433
Chelidonic acid : 10776
Chenic acid : 4097
Chenodeoxycholic acid : 4097
Chenodiol : 4097
Chimyl alcohol : 10010
Chitosamine : 6233
Chloral : 21
Chloral alcoholate : 5769
Chloral ammonia : 5680
Chloral, diethyl acetal : 5657
Chloral formamide : 6000
Chloral hydrate : 5582
Chloralide : 5212
Chloramben : 2492
Chlorambucil : 1101
Chloraminophenamide : 1558
Chloramphenicol : 52
Chloramphenicol palmitate : 6648
Chloranil : 4268
Chloranilic acid : 4258
Chlorocyanohydrin : 10115
Chlordantoin : 7220
Chlordimeform : 7564
Chlorfenac : 944
Chlorfenethol : 1850
Chlorfenson : 2221
Chlorindanol : 7283
Chlorine cyanide : 4166
Chlornaphazine : 7710
5-Chloroacenaphthene : 5
2-Chloroacetaldehyde : 12
Chloroacetaldehyde, trimer : 13
Chloroacetamide : 41
Chloroacetamide, *N*-3-chloroallyl : 43
Chloroacetic acid (β) : 123
Chloroacetic acid, methyl ester : 138
Chloroacetic acid, phenyl ester : 147
Chloroacetic acid, propyl ester : 149
Chloroacetic anhydride : 124
Chloroacetone : 10420
m-Chloroacetophenone : 5788
p-Chloroacetophenone : 5789

SYNONYM INDEX OF ORGANIC COMPOUNDS (continued)

ω-Chloroacetophenone : 5790
Chloroacetyl chloride : 295
Chloroacetylene : 5938
9-Chloroacridine : 315
o-Chloroaniline : 658
m-Chloroaniline : 659
p-Chloroaniline : 660
4-Chloro-*o*-anisidine : 667
1-Chloroanthracene : 374
Chlorobenzene : 1131
p-Chlorobenzenesulfonic acid : 2220
1-(*p*-Chlorobenzenesulfonyl)-3-propylurea : 2208
2-Chloro-1,3,5-benzenetricarboxylic acid : 2320
N-(*p*-Chlorobenzhydryl)-*N*'-methylhomopiperazine : 5140
o-Chlorobenzoic acid : 2534
m-Chlorobenzoic acid : 2535
p-Chlorobenzoic acid : 2536
2-Chlorobenzothiazole : 2862
2-Chlorobenzoxazole : 2878
m-Chlorobenzoyl chloride : 2891
4-Chlorobenzyl chloride : 1134
2-(2-Chlorobenzyl)-4,4-dimethyl-1,2-oxazolidin-3-one : 4140
4'-Chloro-4-biphenylamine : 3012
Chlorobromoacetic acid : 112
Chlorobromodifluoromethane : 7464
Chlorobromomethane : 7463
2-Chloro-4-bromophenol : 9232
2-Chloro-1-bromopropane : 9838
1-Chloro-2-bromopropane : 9840
2-Chloro-1,3-butadiene : 3152
1-Chlorobutane : 3241
1-Chloro-2-butanol : 3605
3-Chloro-2-butanol : 3606
Chlorobutanol : 10407
1-Chloro-2-butanone : 3646
1-Chloro-3-butanone : 3648
trans-1-Chloro-1-butene : 3720
cis-1-Chloro-1-butene : 3721
2-Chloro-1-butene : 3722
3-Chloro-1-butene : 3723
3-Chloro-2-buten-1-ol : 3864
4-Chloro-2-buten-1-ol : 3865
1-Chloro-3-buten-2-ol : 3866
3-Chloro-3-buten-2-ol : 3868
Chloro-*tert*-butyl alcohol : 10354
3-Chlorobutyryl chloride : 3684
m-Chlorocarbanilic acid isopropyl ester : 3939
3-Chloro-2-(chloromethyl)-1-butene : 3729
4-Chloro-*o*-cresol : 9266
4-Chloro-*m*-cresol : 9267
6-Chloro-*m*-cresol : 9262
α-Chlorocrotonaldehyde : 3697
γ-Chlorocrotonic acid : 3822
2-Chlorocrotonic acid : 4150
3-Chlorocrotonic acid ethyl ester : 3826
γ-Chlorocrotyl alcohol : 3864
trans-2-Chlorocyclohexanol : 4505
trans-4-Chlorocyclohexanol : 4507
2-Chlorocyclohexanone : 4577
4-Chlorocyclohexanone : 4578
1-Chlorocyclohexene : 4642
2-Chloro-4-cyclohexylphenol : 9254
Chlorocyclooctatetraene : 4745
Chlorocyclopentane : 4789
Chlorocyclopropylmethane : 4966
2-Chloro-*N*,*N*-diallylacetamide : 44
Chlorodiazepoxide : 2409
Chlorodibromomethane : 7493
N-Chloro-2,6-dibromoquinoneimine : 4279
Chloro(dichloromethyl)dimethylsilane : 11373
Chlorodifluoroacetic acid : 127
1-Chloro-1,1-difluoroethane : 5470
1-Chloro-2,2-difluoroethane : 5471
2-Chloro-1,1-difluoroethylene : 5885
Chlorodifluoromethane : 7478
Chlorodiiodomethane : 7479
1-Chloro-2,3-dimethylbutane : 3245
1-Chloro-2,3-dimethyl-2-butene : 3731
1-Chloro-2,3-dimethyl-2-butanol : 3608
1-Chloro-3,3-dimethyl-2-butanol : 3609
2-Chloro-2,5-dimethylhexane : 6740
1-Chloro-3,3-dimethylpropane : 3252
3-Chloro-1,5-dimethylpyrazole : 10818
5-Chloro-1,3-dimethylpyrazole : 10820
Chlorodimethyl(vinylethynyl)methane : 7107
1-Chloro-2,4-dinitrobenzene : 1144
2-Chloro-1,3-dinitrobenzene : 1145
Chlorodinitromethane : 7480
4-Chloro-2,6-dinitrophenol : 9258
3-Chloro-4,6-dinitrophenol : 9259
Chlorodiphenylarsine : 454
2-Chloroethanal : 12
2-Chloroethanethiol : 5651
2-(2-Chloroethoxy)-*N*-[[(4-methoxy-6-methyl-1,3,5-triazi*N*-2-yl)amino]carbonyl]benzene sulfonamide : 11824
β-Chloroethyl acetate : 5695
1-Chloroethyl 2-chloroethyl ether : 5468
2-Chloroethyl chloroformate : 4013
N-(2-Chloroethyl)dibenzylamine hydrochloride : 1801
2-Chloroethyl isobutyrate : 10292
6-Chloro-*N*-ethyl-*N*'-(1-methylethyl)-1,3,5-triazine-2,4-diamine : 476
3-Chloro-3-ethyl-1-pentyne : 9125
1-Chloroethyl propyl ether : 9854
2-Chloroethyl vinyl ether : 5887
Chlorofluoroacetic acid : 131
Chlorofluoroacetic acid, butyl ester : 132
Chlorofluoroacetic acid, methyl ester : 134
Chlorofluoroacetic acid, propyl ester : 135
Chlorofluoroacetyl chloride : 296
1-Chloro-4-fluorobutane : 3250
1-Chloro-10-fluorodecane : 5037
1-Chloro-1-fluoroethane : 5475
1-Chlorofluoroethane : 5475
1-Chloro-6-fluorohexane : 6742
Chlorofluoromethane : 7481
1-Chloro-3-fluoropropane : 9856
Chloroflurazole : 2377
Chloroform : 7554
Chloroformic acid, isopropenyl ester : 4022
Chloroformic digitalin : 4041
2-Chlorofuran : 6073
3-Chlorofuran : 6074
Chloroheptafluorocyclobutane : 4177
2-Chloroheptane : 6361
3-Chloroheptane : 6362
4-Chloroheptane : 6363
7-Chloro-1-heptanol : 6477
1-Chloro-2-heptanol : 6478
1-Chloro-2-heptanone : 6513
1-Chloro-1-heptene : 6538
1-Chloro-1-heptyne : 6614
1-Chloro-5-heptyne : 6615
1-Chloro-4-heptyne : 6616
2-Chlorohexane : 6738
3-Chlorohexane : 6739
1-Chloro-2-hexanol : 6917
1-Chloro-3-hexanol : 6918
1-Chloro-1-hexene : 7004
2-Chloro-1-hexene : 7005
5-Chloro-1-hexene : 7006
4-Chloro-2-hexene : 7007
α-Chloro-*N*-hexylacetamide : 45
α-Chlorohydrin : 9994
1-Chloro-3-hydroxypentane : 8925
3-Chloroindazole : 7236
Chloroiodomethane : 7482
4-Chloroisoprene : 3154
2-Chloroisopropyl ether : 10123
1-Chloroisoquinoline : 7376
Chloromethane : 7477
Chloromethoxymethane : 7483
2-Chloro-*N*-[[(4-methoxy-6-methyl-1,3,5-triazi*N*-2-yl)amino]carbonyl]benzene-sulfonamide : 4091
α-Chloro-*N*-(3-methoxypropyl)acetamide : 46
2-(4-Chloro-6-methoxypyrimidin-2-ylcarbamoylsulfamoyl) benzoic acid, ethyl ester : 4081
2-Chloro-2'-methylbiphenyl : 3026
Chloromethylbutadiene : 3153
3-Chloro-3-methyl-2-butanone : 3649
1-Chloro-3-methyl-1-butene : 3733
1-Chloro-2-methyl-2-butene : 3735
1-Chloro-2-methyl-1-butene : 3732
3-Chloro-2-methyl-1-butene : 3734
2-Chloro-3-methyl-2-butene : 3737
Chloromethyl chloroformate : 4014
Chloromethyl cyanide : 274
2-(Chloromethyl)-2,3-dihydrobenzofuran : 2441
Chloromethyl ethyl ether : 5478
Chloromethyl ethyl ketone : 3646
2-Chloro-2-methylheptane : 6366
2-Chloro-6-methylheptane : 6367
3-Chloro-3-methylheptane : 6369
4-Chloro-4-methylheptane : 6370
2-Chloro-5-methylhexane : 6745
Chloromethyl iodide : 7482
Chloromethyl methyl dichloro silane : 11390
Chloromethyl methyl ether : 7483
1-Chloro-2-methylnaphthalene : 7784
4-Chloro-4-methyloctane : 8363
2-Chloro-2-methyloct-3-yne : 8534
2-Chloro-4-methylpentane : 8730
5-Chloro-2-methyl-2-pentene : 9026
3-Chloro-3-methyl-1-pentyne : 9126
4-(4-Chloro-2-methylphenoxy)butyric acid : 3493
2-Chloromethylpropanol : 10353
1-Chloro-2-methylpropene : 10539
3-Chloro-2-methylpropene : 10540
2-Chloro-8-methylquinoline : 11244
Chloromethylsilane : 11378
2-Chloro-5-methylthiophene : 11741
cis,trans-3-Chloro-4-[4-methyl-2-(1H-1,2,4-triazol-1-ylmethyl)-1,3-dioxolan-2-yl]phenyl 4-chlorophenyl ether : 5174
(Chloromethyl)trichlorosilane : 11456
2-Chloronaphthalene : 7781
5-Chloro-1-naphthalenecarboxylic acid : 7762
5-Chloro-1,4-naphthoquinone : 7847
o-Chloronitrobenzene : 1214
m-Chloronitrobenzene : 1215
p-Chloronitrobenzene : 1216
2-Chloro-3-nitrobenzoic acid : 2544
3-Chloro-2-nitrobenzoic acid : 2546
1-Chloro-1-nitroethane : 5482
2-Chloro-1-nitroethane : 5483
Chloronitromethane : 7485
1-Chloro-5-nitronaphthalene : 7786
1-Chloro-8-nitronaphthalene : 7787
2-Chloro-1-nitronaphthalene : 7788
5-Chloro-2-nitrophenol : 9269
1-Chloro-1-nitropropane : 9866
3-Chloro-1-nitropropane : 9868
1-Chloro-3-nitropropane : 9868
2-Chloro-2-nitropropane : 9870
3-Chloro-5-nitrotoluene : 1207
2-Chlorononane : 8124
5-Chlorononane : 8125
9-Chlorononan-1-ol : 8163
(*Z*)-4-Chloro-4-octene : 8500
1-Chloro-1-octyne : 8533
3-Chloro-1,3-pentadiene : 8663
Chloropentafluoroacetone : 10421
Chloropentafluorobenzene : 1220
Chloropentafluoroethane : 5484
3-Chloropentane : 8721
3-Chloropentanoic acid : 8864
1-Chloro-3-pentanol : 8925
2-Chloro-1-pentene : 9016
3-Chloro-1-pentene : 9017
4-Chloro-1-pentene : 9018
1-Chloro-2-pentene : 9020
2-Chloro-2-pentene : 9021
3-Chloro-2-pentene : 9022
3-Chloroperoxybenzoic acid : 1127
p-Chlorophenacyl bromide : 5778
o-Chlorophenol : 9251
m-Chlorophenol : 9252
p-Chlorophenol : 9253
(*p*-Chlorophenoxy)acetic acid : 146
p'-Chloro-*p*-phenylaniline : 3012

1-(*p*-Chloro-α-phenylbenzyl)-4-methylhomopiperazine : 5140
p-Chlorophenyl *p*-chlorobenzenesulphonate : 2221
(*RS*)-1-*p*-Chlorophenyl-4,4-dimethyl-3-(1H-1,2,4-triazol-1-ylmethyl)pentan-3-ol : 11589
m-Chlorophenyl ethyl ether : 1157
Chlorophenylmethane : 1182
4-Chlorophenyl phenyl ether : 1221
4-(4-Chlorophenyl)-2-phenyl-2-(1H-1,2,4-triazol-1-ylmethyl)butyronitrile : 5949
R-(-)-2-Chloro-1-phenylpropane : 1234
Chlorophenylsilane : 11384
1-(2-Chlorophenyl)thiourea : 11801
p-Chlorophenyl 2,4,5-trichlorophenyl sulfide : 11648
Chloropicrin : 7557
Chloropropamide : 2208
2-Chloro-1,3-propanediol : 9995
3-Chloro-1-propanol : 10348
trans-1-Chloropropene : 10533
cis-1-Chloropropene : 10534
3-Chloro-1-propene : 10536
Chloropropham : 3939
β-Chloropropionic acid : 10216
β-Chloropropionitrile : 10097
(*S*)-2-Chloropropionyl chloride : 10496
3-Chloropropylene : 10536
3-Chloropropyl methyl sulfide : 9864
2-Chloropropyl methyl sulfide : 9865
2-Chloro-2-propylpentane : 6370
6-Chloropurine : 10753
*N*1-(6-Chloro-3-pyridazinyl)sulfanilamide : 2193
2-Chloropyridine : 10915
3-Chloropyridine : 10916
4-Chloropyridine : 10917
2-Chloroquinoline : 11237
4-Chloroquinoline : 11238
5-Chloroquinoline : 11239
6-Chloroquinoline : 11240
7-Chloroquinoline : 11241
8-Chloroquinoline : 11242
6-Chloroquinoxaline : 11334
o-Chlorostyrene : 1149
N-Chlorosuccinimide : 11154
3-Chloro-6-sulfanilamidopyridazine : 2193
1-Chloro-1,1,2,2-tetrafluoroethane : 5485
1-Chloro-1,2,2,2-tetrafluoroethane : 5486
Chlorothalonil : 1317
Chlorothiazide : 2852
5-Chloro-2-thienyl chloride : 11748
2-Chlorothiophene : 11740
Chlorothymol : 9268
o-Chlorotoluene : 1183
m-Chlorotoluene : 1184
p-Chlorotoluene : 1185
3-Chloro-*o*-toluidine : 671
4-Chloro-*o*-toluidine : 673
5-Chloro-*o*-toluidine : 676
4-Chloro-*m*-toluidine : 674
6-Chloro-*m*-toluidine : 669
2-Chloro-*p*-toluidine : 668
3-Chloro-*p*-toluidine : 672
Chlorotrianisene : 1154
Chlorotribromomethane : 7549
1-Chloro-2-(trichloromethyl)benzene : 1237
1-Chloro-3-(trichloromethyl)benzene : 1238
1-Chloro-4-(trichloromethyl)benzene : 1239
2-Chloro-1,1,1-trifluoroethane : 5489
1-Chloro-1,1,2-trifluoroethane : 5487
1-Chloro-1,2,2-trifluoroethane : 5488
Chlorotrifluoroethylene : 5889
Chlorotrifluoromethane : 7486
4-Chloro-3-(trifluoromethyl)nitrobenzene : 1218
4-Chloro-2-(trifluoromethyl)nitrobenzene : 1219
5-[2-Chloro-4-(trifluoromethyl)phenoxy]-*N*-(methylsulfonyl)-2-nitrobenzamide : 5985
(*E*)-1-[1-[[4-Chloro-2-(trifluoromethyl)phenyl]imino]-2-propoxyethyl]-1H-imidazole : 11900
2-[[2-Chloro-4-(trifluoromethyl)phenyl]amino]-3-methylbutanoic acid, (±)-cyano(3-phenoxy phenyl)methyl ester : 5983
(*RS*)-2-[4-(3-Chloro-5-trifluoromethyl-2-pyridyloxy)phenoxy]propionic acid, ethyl ester : 6291
Chlorotriphenylstannane : 11541
4-Chlorovaleric acid : 8865
2-Chlorovinyl ethyl ether : 5886
Chloroxine : 11314
Chloroxylenol : 9257
Chlorphenesin : 10000
Chlorphenesin carbamate : 10001
Chlorpromazine : 9505
Chlortetracycline : 7704
Chlorzoxazone : 2883
Cholagon : 4101
5β-Cholan-24-amide, *N*-(carboxymethyl)-3α,7α,12α-trihydroxy- : 6273
Cholanic acid : 4094
Cholanthrene : 2391
Cholerythrin : 3000
5α-Cholestan-3β-ol : 4107
Δ 5-Cholesten-3β-ol : 4110
Δ 5,7-Cholesterol : 4103
Cholesterol : 4110
Cholesteryl alcohol : 4110
Cholic acid : 4100
Choline, *o*-acetyl,chloride : 5439
Choline, carbamate,chloride : 5440
Choline chloride : 5446
Cholografin : 2594
N-Cholylglycine : 6273
Chroman, 2-methyl- : 2778
Chroman, 6-methyl- : 2780
Chroman, 7-methyl- : 2781
Chroman, 3-methyl : 2779
6-Chromanol, 2,8-dimethyl-2-(4,8,12-trimethyltridecyl)- : 2785
6-Chromanol, 2,5,7,8-tetramethyl-2-(4,8,12-trimethyltridecyl)-, acetate, (+)- : 2788
6-Chromanol, 2,5,8-trimethyl-2-(4,8,12-trimethyltridecyl)- : 2789
6-Chromanol, 2,7,8-trimethyl-2-(4,8,12-trimethyltridecyl)- : 2790
4-Chromanone : 2799
3-Chromene : 2772
1,2-Chromene : 2772
Chromotropic acid : 7875
Chrysamminic acid : 401
Chrysanthemum monocarboxylic acid, methyl ester (*DL*, *trans*) : 4965
Chrysanthenone : 2977
Chrysatropic acid : 2829
Chrysin : 2811
Chrysoidine hydrochloride : 1284
Chrysophanic acid : 398
Chugaev's reagent : 3360
Cicrotoic acid : 3834
Cinchol : 11558
Cinchomeronic acid : 10930
Cinchonamine : 7306
Cinchonidine : 4133
Cinchoninaldehyde : 11225
(+)-Cinchonine : 4120
Cinchonine : 4120
Cinchophen : 11235
Cinchotin : 4122
Cinchotoxine : 10435
trans-Cinnamaldehyde : 10518
Cinnamedrine : 1903
trans-Cinnamic acid : 10682
Cinnamic acid, allyl ester : 10692
Cinnamic acid, benzyl ester (*trans*) : 10690
Cinnamic acid, β-bromo-, ethyl ester : 10614
Cinnamic acid, α-bromo-, methyl ester : 10615
Cinnamic acid, α-chloro-, methyl ester, (*E*)- : 10622
Cinnamic acid, β-chloro-, methyl ester : 10623
Cinnamic acid, α-fluoro- : 10637
Cinnamic acid, isopropyl ester (*trans*) : 10687
Cinnamic acid, 2-nitro, methyl ester (*trans*) : 10677
Cinnamic acid, 3-nitro, methyl ester (*trans*) : 10678
Cinnamic acid, phenyl ester (*trans*) : 10689
Cinnamic acid, propyl ester (*trans*) : 10693
Cinnamic anhydride : 10684
Cinnamoyl chloride : 10716
Cinnamyl cinnamate : 10691
Cirantin : 2794
Citraconic acid : 3786
β-Citraurin : 435
Citrazinic acid : 10900
Citric acid : 10167
Citric acid, trimethyl ester : 10170
Citrinin : 2774
(±)-Citronellol : 8514
(-)-α-Citronellol : 8516
Citropten : 2818
Citrulline : 8551
cis-Civetone : 4218
Civetone : 4218
1,7-Cleve's acid : 7967
Clitizina : 10904
Clofibrate : 10231
Clomethiazole : 11670
Clonitrate : 9997
Clopoxide : 2409
Clorindione : 7266
Clorophene : 9270
Clostebol : 354
Cloxyquin : 11311
Cluytyl alcohol : 8222
Cocaine : 482
Coclaurine : 7396
Codamine : 7392
Codeine : 7669
Codeine, phosphate : 7673
Colchiceine : 92
Colchicine : 93
2,3,4-Collidine : 11024
2,3,5-Collidine : 11025
2,3,6-Collidine : 11026
2,4,5-Collidine : 11027
2,4,6-Collidine : 11028
Collinomycin : 2773
Columbin : 5926
Con-5-enine, 3β-(dimethylamino)- : 4143
Conessine : 4143
Congo red : 7969
Conhydrine (*DL*) (lower m.p.) : 9677
γ-Coniceine : 11021
Coniferin : 6226
Coniferyl alcohol : 9370
Coniine : 9708
Conquinamine : 6177
Copaene : 11861
Copper, [μ-[carbonato(2-)-O:O']]dihydroxydi- : 511
Coproergostane : 5378
Coprostane : 4106
4:5-Coprosten-3-ol : 4111
Coprostenol : 4111
Coprosterol : 4109
Corallin : 4278
Corchorin : 4049
Cordianine : 12009
Cordycepic acid : 7422
Corticosterone : 9748
Cortisone : 9754
Corybulbine : 5151
Corycavamine : 3108
(+)-Corycavine : 3108
Corydaldine : 7398
Corydaline (*d*) : 5149
Corydine : 5162
Corynantheine : 4146
Cotarnine : 5216
Cotoin : 7608
Cotoran : 12007
Coumalic acid : 10773
Coumaran : 2442
p-Coumaric acid : 10644
Coumarilic acid : 2439
Coumarin : 2792

Coumarin-3-carboxylic acid : 2775
Coumarin, 3-chloro-7-hydroxy-4-methyl, *O*-ester with *O,O*-diethylphosphorothioate : 4147
Coumarin, 3-ethyl- : 2820
Coumarin, 7-hydroxy- : 2826
Coumarin, 7-hydroxy-6-methoxy- : 2829
cis-o-Coumarinic acid lactone : 2792
Coumarin, 3-methyl- : 2833
Coumarin, 5-methyl- : 2834
Coumarin, 7-methyl- : 2836
Coumarin, 8-methyl- : 2837
Coumarone : 2438
Coumarone-2-carboxylic acid : 2439
Coumestrol : 2466
Creatine : 6254
Creatinine : 7230
Creosol : 9385
o-Cresol : 9400
m-Cresol : 9401
p-Cresol : 9402
m-Cresol, 5-amino- : 9212
m-Cresol, 6-bromo- : 9237
m-Cresol, 6-*sec*-butyl- : 9421
m-Cresol, 2,4-dichloro : 9292
m-Cresol, 2,6-dichloro- : 9294
p-Cresol, 2,6-di-*tert*-pentyl- : 9225
p-Cresol, 2-iodo- : 9374
o-Cresol, 4-isobutyryl : 10448
o-Cresol, 3-isopropyl, methyl ether : 363
m-Cresol, 6-pentyl- : 9425
o-Cresol, α-phenyl- : 9454
p-Cresol, α-phenyl- : 9455
o-Cresolphthalein : 7353
Cresol red : 9368
o-Cresol, 4,5,6-trichloro- : 9487
p-Cresol, 2,3,6-trichloro- : 9488
m-Cresol, 2,4,6-trichloro- : 9489
o-Cresotic acid : 2629
m-Cresotic acid : 2630
p-Cresotic acid : 2631
o-Cresyl acetate : 211
m-Cresyl acetate : 212
Crimidine : 11043
Crithminic acid : 2678
Croceic acid : 7972
Crocetin (*trans*) : 5107
Croconic acid : 4939
Crodimyl : 2838
Cromaril : 2840
trans-Crotonaldehyde : 3695
Crotonaldehyde, 2-chloro- : 3697
Crotonaldehyde-2-chloro, dimethylacetal : 3730
Crotonaldehyde, diethyl acetal : 3758
Crotonchloral : 3191
trans-Crotonic acid : 3814
cis-Crotonic acid : 3815
Crotonic acid, allyl ester : 3856
Crotonic acid anhydride : 3818
Crotonic acid, 4-chloro- : 3822
Crotonic acid, 2-chloro-, ethyl ester : 3823
Crotonic acid, 3-chloro-, ethyl ester : 3826
Crotonic acid, 4-chloro-, ethyl ester : 3829
Crotonic acid, 4-chloro-, methyl ester : 3833
Crotonic acid, 2,3-dimethyl-, ethyl ester : 3836
Crotonic acid, 2-ethyl-, stereoisomer : 3837
trans-Crotyl alcohol : 3859
cis-Crotyl alcohol : 3860
Cryptocavine : 2436
Cryptopine : 2435
all-*trans*-β-Cryptoxanthin : 4061
Cryptoxanthin : 4061
Cryptoxanthol, all-*trans*- : 4061
Crystal violet : 7453
Cubebin : 6136
Cumene : 1975
Cumene hydroperoxide : 7181
3-*p*-Cumenyl-2-methylpropionaldehyde : 2117
Cumic acid : 2689
Cumic alcohol : 1902
Cumidine : 789
Cuminaldehyde : 590
Cuminic acid : 2689
Cuminol : 1902
Cumotocopherol : 2789
α-Cumyl alcohol : 1857
Cupferron : 759
Cupreine : 4119
Cupreol : 11558
ar-Curcumene : 1468
Curcumin : 6334
Curine : 11925
Curling factor : 11520
Cuscohygrine : 10415
Cusparine : 11216
Cyacetacide : 153
Cyamelide : 11907
Cyanidelonon 1522 : 2817
Cyanoacetamide : 49
Cyanoacetic acid : 150
Cyanoacetic acid, diethyl, ethyl ester : 3495
2-Cyanoacetophenone : 2124
Cyanoacetylene : 10737
N-Cyanoamine : 4155
Cyanobenzene : 2740
2-Cyanocaproic acid, ethyl ester : 6881
Cyanocyclobutane : 4173
1-Cyanocyclohexene : 4630
Cyanocyclopentane : 4779
1-Cyanocyclopentene : 4907
4-Cyano-2,6-dibromophenol : 2751
2-Cyanoethylamine : 10092
(2-Cyanoethyl)benzene : 2123
Cyanoethylene : 10583
Cyano(4-fluoro-3-phenoxyphenyl)methyl 3-(2,2-dichloroethenyl)-2,2-dimethylcyclopropanecarboxylate : 5008
1-Cyano-2-(2-furyl)ethene : 10585
Cyanogen : 5550
Cyanogenamide : 4155
Cyanogen iodide : 7337
Cyanogen nitride : 4155
Cyanoiodomethane : 283
Cyanomethane : 271
Cyanomethyl methyl sulfide : 286
Cyanomethyl phenyl ketone : 2124
Cyano(3-phenoxyphenyl)methyl-3-(2,2-dibromoethenyl)-2,2-dimethylcyclopropanecarboxylate : 5102
Cyano(3-phenoxyphenyl)methyl 2,2,3,3-tetramethylcyclopropanecarboxylate : 5953
1-Cyano-2-phenylethane : 2123
1-Cyanopropane : 3403
2-Cyano-2-propanol : 10108
3-Cyano-1-propene : 3803
2-Cyanoquinoline : 11223
2-Cyanothiophene : 11732
2-Cyanotoluene : 2763
m-Cyanotoluene : 2764
p-Cyanotoluene : 2765
1-Cyanoundecane : 5303
Cyanuric acid, dihydrate : 11839
Cyanuric acid trichloride : 11837
Cyclamal : 2117
Cyclamic acid : 11568
Cyclobarbital : 11084
Cyclobutaneacetic acid, 3-carboxy-2,2-dimethyl-, diethyl ester : 4171
Cyclobutane-1-carboxylic acid : 4174
Cyclobutane-1,1-dicarboxylic acid, diethyl ester : 4180
Cyclobutane-1,2-dicarboxylic acid, dimethyl ester (*trans*) : 4181
Cyclobutane-1,3-dicarboxylic acid (*cis*), 2,2-dimethyl : 4183
Cyclobutane-1,3-dicarboxylic acid, 2,2-dimethyl, dimethyl ester : 4184
Cyclobutane, 1,2-divinyl-, *trans*- : 4187
Cyclobutylamine : 4169
Cyclobutylcarboxylic acid : 4174
Cyclobutyl hydroxide : 4196
Cyclobutyl methyl ketone : 5792
Cyclobutyrol : 4309
Cyclocapron : 4333
β-Cyclocitral : 4636
1,5,9-Cyclododecatriene, (*cis, trans, trans*) : 4213
cis,cis,cis-Cyclododeca-1,5,9-triene : 4214
Cyclodoecanone : 4211
Cyclofenchene : 11866
Cycloform : 2505
β-Cyclogeraniol : 4666
α-Cyclogeraniol : 4667
Cyclogeraniolane : 4486
α-Cyclogeraniolene : 4691
Cyclogeranyl alcohol : 4666
Cycloglycylglycine : 9612
9-Cycloheptadecene-1-one : 4217
Cycloheptyl bromide : 4222
Cycloheptylcarboxylic acid : 4223
1,3-Cyclohexadiene, 3,5-dimethyl : 4256
1,3-Cyclohexadiene, 5-methyl (*DL*) : 4272
Cyclohexanacetic acid, 2-methyl : 4311
Cyclohexanacetic acid, 3-methyl : 4312
Cyclohexane, 1,3-bis(aminomethyl) : 4362
Cyclohexane, 1,3-bis-(hydroxymethyl) : 4363
Cyclohexane, *sec*-butyl- : 4448
Cyclohexane, decafluoro-1,2-bis(trifluoromethyl)- : 4342
Cyclohexane-1,2-dicarboxylic acid, diethyl ester (*trans, DL*) : 4350
Cyclohexane-1,3-dicarboxylic acid, diethyl ester (*cis*) : 4351
Cyclohexane-1,3-dicarboxylic acid, dimethyl ester (*cis*) : 4352
Cyclohexane-1,3-dicarboxylic acid, dimethyl ester (*trans*) : 4353
Cyclohexane-1,4-dicarboxylic acid, diethyl ester (*trans*) : 4356
Cyclohexane-1,4-dicarboxylic acid, dimethyl ester (*cis*) : 4357
Cyclohexane-1,4-dicarboxylic acid, diethyl ester (*cis*) : 4355
Cyclohexane, 2,4-diisopropenyl-1-methyl-1-vinyl-, stereoisomer : 4389
Cyclohexane, 1,2-dimethyl, perfluoro : 4342
Cyclohexane, 1,2-epoxy-4-vinyl : 8555
Cyclohexane, 1,2,3,4,5,6-hexachloro, (α, *DL*) : 4408
Cyclohexane hydroperoxide : 7177
Cyclohexane, 1-hydroxymethyl-2-methyl (*cis*) : 4422
Cyclohexane, 1-hydroxymethyl-2-methyl (*trans*) : 4423
Cyclohexane, 1-isopropenyl-3-methyl : 4435
Cyclohexane, isopropyl- : 4431
Cyclohexane, 1-methylamino-4-methyl : 4723
Cyclohexane, 1-methyl-2-ethyl (*trans*) : 4396
Cyclohexane, 2-methyl-1-ethylidene : 4393
Cyclohexane, 2-nitrophenyl : 1251
1-*L*-1,3,4/2,5-Cyclohexanepentol : 4074
Cyclohexanol, acetate : 155
Cyclohexanol, 2-allyl-, *trans*- : 4556
Cyclohexanol, 2-amino (*trans, DL*) : 4501
Cyclohexanol, 1,2-dimethyl-, stereoisomer : 4508
: 4509
Cyclohexanol, 2-ethyl (*trans, DL*) : 4523
Cyclohexanol, 2-isopropyl-, *cis*- : 4539
Cyclohexanol, 4-isopropyl-, *cis*- : 4540
Cyclohexanol, 4-isopropyl-, *trans*- : 4541
Cyclohexanol-4-isopropyl-2-methyl : 7428
Cyclohexanol, 2-methyl (*cis, DL*) : 4531
Cyclohexanol, 2-phenyl (*cis, DL*) : 4553
Cyclohexanol, 2-phenyl (*trans, DL*) : 4554
Cyclohexanone-4-carboxylic acid, 2-methyl : 4336
Cyclohexanone, 2,4-dimethyl (*trans*, d) : 4583
Cyclohexanone, 2-isopropyl- : 4599
Cyclohexanone, 4-isopropyl- : 4600
Cyclohexanone, 2-isopropyl-5-methyl (*l*) : 4606
1-Cyclohexene-1-carboxaldehyde, 4-isopropenyl- : 4633
1-Cyclohexene-1-carboxaldehyde, 4-isopropyl- : 4635

Cyclohexene-1-carboxylic acid : 4637
1-Cyclohexene, 1-chloro : 4642
Cyclohexene-1,2-dicarboxylic acid, diethyl ester : 4644
Cyclohexene-1,4-dicarboxylic acid, dimethyl ester : 4646
Cyclohexene, 2,4-dimethyl- : 4651
Cyclohexene, 1-isopropenyl-5-methyl (*DL*) : 4679
Cyclohexene, 1-isopropyl-4-methyl (*l*) : 7429
Cyclohexene, 1,1'-(3,7,12,16-tetramethyl-1,3,5,7,9,11,13,15,17-octadecanonaene-1,18-diyl)bis[2,6,6-trimethyl-, : 4051
Cyclohexen-3-ol, 1,3,5-trimethyl : 4700
Cyclohexen-3-ol, 2,6,6-trimethyl : 4701
Cyclohexen-3-ol, 1,5,5-trimethyl : 4702
Cyclohexen-3-one, 1,6,6-trimethyl : 4719
Cyclohexen-3-one, 1,4,6-trimethyl : 4720
2,3-Cyclohexenopyridine : 11297
4-(Cyclohex-3-enyl)pyridine : 10922
Cycloheximide : 9655
Cyclohexyl acetate : 155
Cyclohexylacetic acid : 4306
Cyclohexylacetic acid, methyl ester : 4310
Cyclohexyl alcohol : 4500
Cyclohexylamine : 4283
Cyclohexylamine, 2-ethyl- : 4287
Cyclohexylamine, 2-methyl-, *trans-* : 4293
Cyclohexylbenzene : 1248
Cyclohexyl bromide : 4315
Cyclohexylcarbinol : 4419
Cyclohexyl chloride : 4339
Cyclohexylcyclohexane : 2988
5-Cyclohexyl-5,7-diaza-2,3,4,5,6,7-hexahydroindene-4,6-dione : 4900
Cyclohexyldimethylamine : 4286
1-Cyclohexyleicosane : 4384
Cyclohexyl ethyl ketone : 10423
Cyclohexyl fluoride : 4403
Cyclohexyl 2-furyl ether : 6076
Cyclohexyl hydroperoxide : 7177
Cyclohexylideneacetonitrile : 276
Cyclohexyl iodide : 4415
Cyclohexyl isocyanate : 4416
Cyclohexyl mercaptan : 4480
β-Cyclohexyl-β-methacrylic acid : 3834
Cyclohexylmethyl alcohol : 4419
Cyclohexyl methyl ketone : 5794
3-Cyclohexyl-1-methylpropanol : 4469
3-Cyclohexyl-5,6-trimethyleneuracil : 4900
β-Cyclolavandulol : 4665
Cycloleucine : 4782
Cyclomethiazide : 2853
Cyclomethone : 4378
Cyclonite : 11833
Cyclooctatetraene : 4742
Cyclooctatetraene, chloro : 4745
Cyclooctatetraene, propyl : 4748
Cyclooctylamine : 4735
1H-Cyclopenta[a]phenanthrene, hexadecahydro-10,13-dimethyl-17-(1-methylbutyl)-, [8*R*-[8α,9β,10α,1 : 4093
Cyclopentacycloheptene : 504
Cyclopenta[de]naphthalene : 3
Cyclopentadrin : 4810
Cyclopentamine : 4810
Cyclopentanecarboxylic acid, 2-isopropyl-5-methyl- : 4784
Cyclopentanecarboxylic acid, 2-isopropyl-5-methyl, ethyl ester : 4788
Cyclopentane-1,2-dicarboxylic acid, dimethyl ester (*DL*, *trans*) : 4792
Cyclopentane-1,3-diol (*trans*) : 4807
Cyclopentane, eicosyl : 5351
Cyclopentaneethylamine, *N*,α-dimethyl- : 4810
Cyclopentane, isobutyl- : 4838
Cyclopentane, 1-isopropyl-2-methyl- : 4829
Cyclopentane, 1,2,3-trimethyl-, *trans*-1,2,*cis*-1,3- : 4857
Cyclopentane, 1,2,3-trimethyl-, *cis*-1,2,*cis*-1,3- : 4858
Cyclopentane, 1,2,4-trimethyl-, *cis*-1,2,*cis*-1,4- : 4861
Cyclopentanoic acid : 4781
Cyclopentanone, 2-benzyl- : 4893
Cyclopentanone, 2-isopentyl-3-methyl : 4888
Cyclopentanone, 3-isopropyl-5-methyl : 4884
Cyclopentanone, 2-isopropyl-2-methyl- : 4889
Cyclopentanone, 2-isopropyl-4-methyl : 4891
Cyclopentene-4-carboxylic acid, 2,3,3-trimethyl : 4913
Cyclopentene-1,3-dicarboxylic acid, diethyl ester : 4916
Cyclopentene, 2,3-dimethyl-4-isopropyl : 4923
Cyclopentene, 1-isopropyl-3-methyl- : 4931
Cyclopenten-3-one : 4941
Cyclopenten-3-one, 4-methyl : 4946
1,2-Cyclopentenophenanthrene : 4756
Cyclopenthiazide : 2853
Cyclopentobarbital : 11085
Cyclopentyl acetate : 4865
2-Cyclopentylacetic acid : 4775
Cyclopentyl alcohol : 4864
Cyclopentylamine : 4771
Cyclopentyl bromide : 4777
Cyclopentylcarboxylic acid : 4781
Cyclopentyl chloride : 4789
Cyclopentyl iodide : 4824
Cyclopentyl mercaptan : 4852
Cyclopropanecarboxylic acid, 3-(2,2-dichloroethenyl)-2,2-dimethyl-, cyano(3-phenoxy-phenyl)methyl ester : 5012
Cyclopropanecarboxylic acid, 3-(2,2-dichloroethenyl)-2,2-dimethyl-, (3-phenoxyphenyl), methyl ester : 9150
Cyclopropanecarboxylic acid, 2,2-dimethyl-3-(2-methyl-1-propenyl)-, [5-(phenylmethyl)- 3-furanyl] methyl ester : 11340
Cyclopropanecarboxylic acid, hexadecyl ester : 4952
Cyclopropane-1,2-dicarboxylic acid, diethyl ester (*cis*) : 4968
Cyclopropane-1,2-dicarboxylic acid, dimethyl ester : 4969
Cyclopropane-1,2-dicarboxylic acid (*trans*, *DL*) : 4970
Cyclopropane, 1-methyl-1-ethyl : 4975
Cyclopropylamine : 4953
Cyclopropylbenzene : 1256
Cyclopropyl cyanide : 4956
Cyclopropylmethane : 4981
Cyclopropyl methyl ether : 4980
Cyclopropyl methyl ketone : 5796
Cycloserine : 7406
Cyclotetrasiloxane, 1,3,5,7-tetraethyl-1,3,5,7-tetraphenyl : 4998
Cyclotetrasiloxane, 2,4,6,8-tetramethyl-2,4,6,8-tetraethyl : 4997
Cyclothiazide : 2851
Cymarose : 7112
o-Cymene : 1988
m-Cymene : 1989
p-Cymene : 1990
p-Cymene, 2-iodo : 1781
p-Cymene, 3-iodo- : 1782
p-Cymen-7-ol : 1902
Cymonic acid : 189
Cysteamine : 5649
Cysteine disulfide : 5020
Cythrin : 5965
Cytidine 3'-monophosphate : 5022
Cytofol : 6243
Cytosine : 11104
Cytosine, 5-methyl- : 11106
Cytosine riboside : 5021
2,4-D : 169
2,4-D Isopropyl ester : 173
2,4-D Methyl ester : 172
Dagenan chloride : 2243
Dahl's acid II : 7873
Daidzein : 2827
Daidzein 4'-methyl ether : 2830
Daidzeol : 2827
Dalapon : 10244
Damascenine : 2672
Daminozide : 3334
Dansyl chloride : 7977
Danthron : 395
Daphnetin : 2805
Dapsone : 846
Daptazole : 11671
Datiscetin : 2842
DBMC : 9222
DDT : 2328
Deanol : 5709
Decacil : 2409
α,ω-Decadiene : 5024
2,4-Decadienoic acid (*trans*-2, *trans*-4) : 5025
2,4-Decadienoic acid, methyl ester (*cis*-2, *trans*-4) : 5028
2,4-Decadienoic acid, methyl ester (*cis*-2, *cis*-4) : 5029
Decafluorobutane : 3255
cis-Decahydroquinoline : 11246
Decalin : 7789
cis-Decalin : 7790
trans-Decalin : 7791
Decalin, 2-chloro : 7782
Decalin, 2-methylene (*trans*) : 7792
1-Decalone (*cis*) : 8053
Decamethylcyclopentasiloxane : 4901
Decamethylene dibromide : 5038
Decamethylene dichloride : 5040
Decamethylene glycol : 5047
Decamethyltetrasiloxane : 11646
Decane, 1-cyclopentyl- : 4790
Decane, 1-(1-naphthyl)- : 7793
2-Decanol (*DL*) : 5071
Decasiloxane, dicosamethyl : 5079
Δ 2-Decenoic acid : 5088
Decholin : 4101
Decicaine : 2531
Deconic acid telomer : 5090
Decyl acetate : 157
Decylacetylene : 5343
Decylbenzene : 1258
Decyl cyanide : 11941
Decylenic alcohol : 5092
Decyl fluoride : 5052
Decyl isopropyl ketone : 11892
Decyl mercaptan : 5058
Decyl methyl ketone : 5325
1-Decylnaphthalene : 7793
Decyl nitrate : 8073
Decyl nitrite : 8084
Decyl vinyl ether : 5051
DEET : 612
Dehydroacetic acid : 10779
Dehydrobilirubin : 3001
7-Dehydrocholesterol : 4103
Dehydrocholic acid : 4101
11-Dehydrocorticosterone : 9755
2,2'-Dehydrodimorphine : 3002
Dehydroergosterol : 5380
7-Dehydroestrone : 5398
Dehydromorphine : 3002
Dehydromucic acid : 6081
7-Dehydrositosterol : 11554
Delfinidol chloride : 2849
Delphinic acid : 3539
Delphinidin : 2849
Delphinine : 311
Deltathione : 6264
N-Demethylanhalidine : 7393
Deoxycholic acid : 4095
Deoxycorticosterone : 9751
Deoxyepinephrine : 1533
2-Deoxy-*D*-glucose : 7110
11-Deoxy-17-hydroxycorticosterone : 9749
7-Deoxyhyocholic acid : 4096
1-(2-Deoxy-β-*D*-ribofuranosyl)-5-fluorouracil : 12034
Derris : 2783
Deserpidine : 12065
Desmethyldiazepam : 2411
6-Desoxy-*L*-ascorbic acid : 467
5-Desoxyquercetin : 2813
Desthiobiotin : 7223
Desthiobiotin, methyl ester : 7224
Dexamethasone : 9737

Dexpanthenol	: 3202
Dextroamphetamine sulfate	: 1582
Dextrotartaric acid	: 3309
4,4'-Diacetamidodiphenyl sulfone	: 91
Diacetic acid	: 3568
Diacetin	: 10189
1,3-Diacetin	: 10190
Diacetonamine	: 8959
Diacetone acrylamide	: 10521
Diacetone alcohol	: 8976
Diacetonyl	: 6805
1,4-Diacetoxy-2-butyne	: 3913
1,2-Diacetoxyethane	: 5571
Diacetyl	: 3359
Diacetyldioxime	: 3360
Diacetyl disulfide	: 5250
1,2-Diacetylethane	: 6805
Diacetylmethane	: 8798
Diacetylmorphine	: 7667
Diacetylperoxide	: 9156
Diadol	: 11091
Δ 2-Dialin	: 7815
Diallylamine	: 10526
5,5-Diallylbarbituric acid	: 11091
N,N-Diallyl-2-chloroacetamide	: 44
Diallylcyanamide	: 4159
Diallyl fumarate	: 3781
Diallyl maleate	: 3782
Diallyl oxalate	: 5564
Diallyl sulfide	: 10599
p-Diaminoazobenzene	: 635
2,3-Diaminobenzoic acid	: 2552
2,4-Diaminobenzoic acid	: 2553
3,5-Diaminobenzoic acid	: 2554
o,p'-Diaminobiphenyl	: 3029
4,4'-Diamino-1,1'-biphenyl	: 3031
1,4-Diaminobutane	: 3258
1,8-Diamino-3,6-diazaoctane	: 5499
β,β'-Diamino-β,β'-dicarboxydiethyl disulfide	: 5020
4,4'-Diamino-3,3'-dimethylbiphenyl	: 3034
p-Diaminodiphenyl	: 3031
4,4'-Diaminodiphenylamine	: 1263
p,p'-Diaminodiphenylmethane	: 784
3,3'-Diaminodipropylamine	: 9874
2,7-Diaminofluorene	: 5971
2,6-Diaminohexanoic acid	: 7418
	: 7419
2,6-Diamino-1-methylbenzene	: 1277
1,2-Diaminonaphthalene	: 7795
1,4-Diaminonaphthalene	: 7796
2,3-Diaminonaphthalene	: 7799
(S)-2,5-Diamino-5-oxopentanoic acid	: 6249
(S)-2,5-Diaminopentanoic acid	: 8550
2,3-Diaminophenazine	: 9194
2,4-Diaminophenol	: 9274
2,4-Diamino-5-phenylthiazole	: 11671
1,3-Diaminopropane	: 9873
2,3-Diaminopropionic acid	: 326
2,6-Diaminopurine	: 10754
2,5-Diaminopyridine	: 10924
1,3-Diamino-2-thiourea	: 4033
2,6-Diaminotoluene	: 1277
Diamylamine	: 8708
Diamyl ketone	: 11962
Dianat	: 2567
1,2,3,4-Dianhydroerythritol	: 3007
o,p'-Dianiline	: 3029
1,2-Dianilinoethane	: 5508
o-Dianisidine	: 3033
7,12-Diazabenz[a]anthracene	: 2394
1,2-Diazabenzene	: 10855
1,4-Diazabicyclooctane	: 5111
4,9-Diazadodecane-1,12-diamine	: 3259
2,3-Diazaindole	: 2871
Diazald	: 2212
1,5-Diazanaphthalene	: 8065
3,6-Diazaoctane-1,8-diamine	: 5499
Diazepam	: 2410
1,4-Diazine	: 10801
1,3-Diazine	: 11049
Diazinon	: 9582
Diazoacetic ester	: 158
Diazoaminobenzene	: 11826
1,3-Diazole	: 7192
1,2-Diazole	: 10815
Diazomethane	: 7491
Dibenzalacetone	: 8683
1,2:5,6-Dibenzanthracene	: 5142
2,3:6,7-Dibenzazepine-1-carboxylic acid,amide	: 5144
Dibenz[de,kl]anthracene	: 9161
Dibenzo[a,e]pyrene	: 8061
Dibenzo[a,i]phenanthrene	: 9601
Dibenzo[a,i]pyrene	: 2850
Dibenzo[a,jk]fluorene	: 2738
2,3:6,7-Dibenzoanthracene	: 8636
Dibenzo[b,e]pyridine	: 314
Dibenzo[b,h]pyrene	: 2850
Dibenzo[b,d]pyrrole	: 3981
1,2:6,7-Dibenzo-9-fluorenone	: 5152
Dibenzopyrazine	: 9193
1,2:7,8-Dibenzopyrene	: 2850
1,2:4,5-Dibenzopyrene	: 8061
Dibenzothiapyran	: 11811
Dibenzothiazine	: 9499
2,3,:5,6-Dibenzo-1,4-thiazine, 10(dimethyaminacetyl)	: 9502
Dibenzoylmethane	: 10027
2,3:6,7-Dibenzphenanthrene	: 8996
Dibenzyl	: 1587
Dibenzylamine	: 1828
9,10-Dibenzylanthracene	: 367
Dibenzyl disulfide	: 5248
Dibenzyl phosphite	: 9586
Diborane, methylthio	: 5168
Dibromantine	: 7217
Dibromoacetic acid	: 159
Dibromoacetic acid ethyl ester	: 160
Dibromoacetonitrile	: 277
β,β-Dibromoacrylic acid	: 10628
3,4-Dibromoaniline	: 683
9,10-Dibromoanthracene	: 375
2,7-Dibromoanthraquinone	: 390
o-Dibromobenzene	: 1288
m-Dibromobenzene	: 1289
p-Dibromobenzene	: 1290
2,4-Dibromobenzoic acid	: 2555
2,5-Dibromobenzoic acid	: 2556
2,6-Dibromobenzoic acid	: 2557
3,5-Dibromobenzyl bromide	: 1291
d-3,3-Dibromo-2-bornanone	: 2953
1,1-Dibromobutane	: 3264
1,2-Dibromobutane	: 3265
1,3-Dibromobutane	: 3266
1,4-Dibromobutane	: 3267
2,3-Dibromobutane	: 3268
2,3-Dibromo-1-butanol	: 3612
3,4-Dibromobutanol	: 3613
1,4-Dibromo-2-butanol	: 3614
3,4-Dibromo-2-butanone	: 3653
1,1-Dibromo-1-butene	: 3738
3,4-Dibromo-1-butene	: 3740
1,2-Dibromo-3-butene	: 3740
2,4-Dibromo-2-butene	: 3741
1,3-Dibromo-2-butene	: 3741
1,2-dibromo-1-butene	: 3739
1,4-Dibromo-2-butyne	: 3904
(+)-3,3-Dibromocamphor	: 2953
α,α'-Dibromo-*d*-camphor	: 2953
1,2-Dibromo-3-chlorobenzene	: 1292
1,2-Dibromo-4-chlorobenzene	: 1293
1,3-Dibromo-2-chlorobenzene	: 1294
1,4-Dibromo-2-chlorobenzene	: 1295
Dibromochlorofluoromethane	: 7494
Dibromochloromethane	: 7493
Dibromochloropropane	: 9885
1,2-Dibromo-1-chloro-1,2,2-trifluoroethane	: 5519
1,1-Dibromocyclobutane	: 4178
1,2-Dibromocyclobutane	: 4179
1,1-Dibromocyclohexane	: 4345
cis-1,2-Dibromocyclohexane	: 4346
trans-1,3-Dibromocyclohexane	: 4348
cis-1,4-Dibromocyclohexane	: 4349
1,10-Dibromodecane	: 5038
5,6-Dibromodecane	: 5039
Dibromodichloromethane	: 7495
1,2-Dibromo-2,4-dicyanobutane	: 8768
1,2-Dibromo-1,1-difluoroethane	: 5522
Dibromodifluoromethane	: 7496
1,4-Dibromo-2,3-dimethylbutane	: 3270
2,3-Dibromo-2,3-dimethylbutane	: 3271
1,2-Dibromo-2,3-dimethylbutane	: 3269
1,1-Dibromo-2,2-dimethylpropane	: 9886
Dibromodinitromethane	: 7497
1,1-Dibromoethane	: 5517
1,1-Dibromoethylene	: 5890
trans-1,2-Dibromoethylene	: 5891
cis-1,2-Dibromoethylene	: 5892
1,2-Dibromoethyl methyl ketone	: 3653
Dibromofluoroacetic acid, ethyl ester	: 161
Dibromofluoromethane	: 7498
2,3-Dibromofuran	: 6077
2,5-Dibromofuran	: 6078
Dibromogallic acid	: 2560
1,2-Dibromoheptane	: 6372
1,5-Dibromoheptane	: 6373
1,7-Dibromoheptane	: 6374
2,3-Dibromoheptane	: 6375
3,4-Dibromoheptane	: 6376
1,16-Dibromohexadecane	: 6635
1,2-Dibromohexane	: 6754
1,4-Dibromohexane	: 6755
1,5-Dibromohexane	: 6756
2,3-Dibromohexane	: 6758
1,2-Dibromoindan	: 7247
Dibromomethane	: 7492
1,2-Dibromo-2-methylbutane	: 3272
2,3-Dibromo-2-methylbutane	: 3274
1,3-Dibromo-2-methylpropene	: 10546
Dibromomonochloromethane	: 7493
1,3-Dibromonaphthalene	: 7801
1,4-Dibromonaphthalene	: 7802
1,5-Dibromonaphthalene	: 7803
3,4-Dibromonitrobenzene	: 1307
1,2-Dibromo-4-nitrobenzene	: 1307
2,6-Dibromonitrobenzene	: 1308
1,3-Dibromo-2-nitrobenzene	: 1308
1,2-Dibromononane	: 8127
1,4-Dibromooctafluorobutane	: 3275
1,2-Dibromooctane	: 8366
1,5-Dibromooctane	: 8368
1,8-Dibromooctane	: 8369
1,3-Dibromopentane	: 8739
1,4-Dibromopentane	: 8740
2,3-Dibromopentane	: 8742
2,4-Dibromopentane	: 8743
2,3-Dibromopentanoic acid	: 8870
2,4-Dibromophenol	: 9276
2,6-Dibromophenol	: 9277
3,5-Dibromophenol	: 9278
1,1-Dibromopropane	: 9880
1,2-Dibromopropane	: 9881
1,3-Dibromopropane	: 9882
2,2-Dibromopropane	: 9883
2,3-Dibromopropene	: 10545
2,3-Dibromopropylene	: 10545
3,5-Dibromopyridine	: 10926
2,6-Dibromoquinone-4-chlorimide	: 4279
3,5-Dibromosalicylaldehyde	: 534
3,5-Dibromosalicylic acid	: 2558
1,14-Dibromotetradecane	: 11616
1,2-Dibromotetrafluoroethane	: 5525
1,2-Dibromo-1,1,2,2-tetramethylethane	: 3271
2,3-Dibromothiophene	: 11742
2,4-Dibromothiophene	: 11743
2,5-Dibromothiophene	: 11744
2,3-Dibromotoluene	: 1302
3,4-Dibromotoluene	: 1303
2,6-Dibromotoluene	: 1304
2,5-Dibromotoluene	: 1305
1,2-Dibromo-1,1,2-trifluoroethane	: 5526
1,2-Dibromo-1,1,2 trimethylethane	: 3274
1,11-Dibromoundecane	: 11940
3,4-Dibromo-*o*-xylene	: 1297
Dibucaine hydrochloride	: 11226
Dibutylacetylene	: 5100
Di-*tert*-Butylacyloin	: 6975
Dibutylamine	: 3211
Dibutylbutadiyne	: 5287
Dibutylcarbinol	: 8161
Dibutyl carbonate	: 3998
Dibutyl ether	: 3415
Di-*tert*-Butyl ether	: 10124
N,N'-Dibutylhexamethylenediamine	: 6750
Di-*sec*-butyl ketone	: 6515
Dibutyl ketone	: 8175

Di-*tert*-butyl malonate : 9925
Dibutylmercury : 7433
Di-*tert*-butylmethane : 8841
2,6-Di-*tert*-butyl-4-methylphenol : 9224
Dibutylnitrosamine : 3212
Dibutyl oxide : 3415
2,6-Di-*sec*-butylphenol : 9227
Dibutyl phthalate : 1339
2,6-Di-*tert*-butylpyridine : 10874
Dibutyl sebacate : 5044
Di-*tert*-butyl succinate : 3296
Di-*sec*-butyl succinate : 3297
Dibutyl succinate : 3301
Di-*tert*-butyl sulfide : 10155
Dibutyl sulfide : 3433
Dibutyl sulfite : 11580
Di-*tert*-Butyl tetrasulfide : 11649
Dicamba : 2567
1,3-Dicarboxyacetone : 8782
2,2'-Dicarboxybiphenyl : 3038
2,2'-Dicarboxydiphenylamine : 2656
2,3-Dicarboxypyrazine : 10804
2,5-Dicarboxythiophene : 11745
Dicentrine : 2468
Dichlobenil : 2752
Dichlofenthion : 9581
Dichlone : 7850
Dichloramine-T : 2210
Dichloroacetic acid : 162
Dichloroacetic acid butyl ester : 163
1,1-Dichloroacetone : 10425
1,3-Dichloroacetone : 10426
Dichloroacetonitrile : 278
Dichloroacetyl chloride : 297
Dichloroacetylene : 5940
β,β-Dichloroacrylic acid : 10630
2,4-Dichloro-6-aminophenol : 9205
2,3-Dichloroaniline : 685
2,4-Dichloroaniline : 686
2,5-Dichloroaniline : 687
2,6-Dichloroaniline : 688
3,4-Dichloroaniline : 689
3,5-Dichloroaniline : 690
3,6-Dichloro-*o*-anisic acid : 2567
2,5-Dichlorobenzaldehyde : 535
3,4-Dichlorobenzaldehyde : 536
3,5-Dichlorobenzaldehyde : 537
o-Dichlorobenzene : 1354
m-Dichlorobenzene : 1355
p-Dichlorobenzene : 1356
3,4-Dichlorobenzenecarboxaldehyde : 536
o-Dichlorobenzidine : 3032
2,4-Dichlorobenzoic acid : 2561
2,5-Dichlorobenzoic acid : 2562
2,6-Dichlorobenzoic acid : 2563
3,5-Dichlorobenzoic acid : 2564
2,6-Dichlorobenzonitrile : 2752
2,6-Dichloro-*p*-benzoquinone-4-chlorimide : 4280
2,4-Dichlorobenzoyl chloride : 2893
3,4-Dichlorobenzoyl chloride : 2894
2,4-Dichlorobenzyl alcohol : 1853
2,2-Dichloro-1,1-bis(4-chlorophenyl)ethylene : 1374
1,1-Dichloro-2,2-bis(*p*-chlorophenyl)ethane : 1378
Dichlorobromomethane : 7467
1,1-Dichloro-1,3-butadiene : 3156
1,2-Dichloro-1,3-butadiene : 3157
1,1-Dichlorobutane : 3276
3,4-Dichlorobutane : 3277
1,2-Dichlorobutane : 3277
1,3-Dichlorobutane : 3278
1,4-Dichlorobutane : 3279
2,2-Dichlorobutane : 3280
1,4-Dichloro-2,3-butanediol : 3349
1,3-Dichloro-2-butanone : 3654
1,3-Dichloro-1-butene : 3743
2,3-Dichloro-1-butene : 3744
3,4-Dichloro-1-butene : 3745
1,2-Dichloro-3-butene : 3745
1,1-Dichloro-2-butene : 3746
1,2-Dichloro-2-butene : 3747
1,4-Dichloro-*trans*-2-butene : 3750
3,4-Dichlorobutyric acid : 3926
Dichloro(chloromethyl)methylsilane : 11390
2,6-Dichloro-*m*-cresol : 9294
4,4-Dichlorocrotonic acid : 4152
2,6-Dichlorocyanobenzene : 2752
Dichlorocyanuric acid : 11838
1,1-Dichlorocyclohexane : 4358
cis-1,2-Dichlorocyclohexane : 4359
cis-1,4-Dichlorocyclohexane : 4361
1,10-Dichlorodecane : 5040
2,2'-Dichloro-4,4'-diaminobiphenyl : 3032
Dichloro(dichloromethyl)methylsilane : 11391
2,3-Dichloro-5,6-dicyano-1,4-benzoquinone : 4251
2,3-Dichloro-5,6-dicyanobenzoquinone : 4251
Dichlorodicyano-*p*-benzoquinone : 4251
Dichlorodicyanoquinone : 4251
Dichlorodiethylsilane : 11393
1,2-Dichloro-1,1-difluoroethane : 5531
1,2-Dichloro-1,2-difluoroethane : 5532
1,1-Dichloro-2,2-difluoroethylene : 5897
1,2-Dichloro-1,2-difluoroethylene : 5898
Dichlorodifluoromethane : 7500
2,5-Dichloro-3,6-dihydroxy-*p*-benzoquinone : 4258
2,2'-Dichlorodiisopropyl ether : 10123
2,2-Dichloro-1,1-dimethylethanol : 10364
2,5-Dichloro-2,5-dimethylhexane : 6765
1,2-Dichloro-4,4-dimethylpentane : 8752
1,5-Dichloro-3,3-dimethylpentane : 8753
2,4-Dichloro-2,4-dimethylpentane : 8754
Dichlorodimethylsilane : 11394
Dichlorodinitromethane : 7501
Dichlorodiphenylsilane : 11395
2,2-Dichloroethanol : 5704
2,5-Dichloroethylbenzene : 1377
1,1-Dichloroethylene : 5894
trans-1,2-Dichloroethylene : 5895
cis-1,2-Dichloroethylene : 5896
Dichloroethylmethylsilane : 11397
2,7-Dichlorofluorene : 5972
Dichlorofluoroacetonitrile : 279
1,1-Dichloro-1-fluoroethane : 5534
1,1-Dichloro-2-fluoroethylene : 5900
Dichlorofluoromethane : 7502
1,1-Dichloroheptane : 6377
1,2-Dichloroheptane : 6378
1,7-Dichloroheptane : 6379
2,2-Dichloroheptane : 6380
2,3-Dichlorohexafluoro-2-butene : 3754
1,2-Dichlorohexafluoropropane : 9898
1,2-Dichlorohexane : 6761
1,6-Dichlorohexane : 6762
2,2-Dichlorohexane : 6763
2,3-Dichlorohexane : 6764
(±)-2,5-Dichlorohexane : 6766
Di(5-Chloro-2-hydroxyphenyl)methane : 9408
1,2-Dichloroindan : 7249
Dichloroiodomethane : 7503
2,4-Dichloromesitylene : 1402
Dichloromethane : 7499
1,1-Dichloro-3-methylbutane : 3282
1,2-Dichloro-2-methylbutane : 3283
2,4-Dichloro-2-methylbutane : 3284
2,3-Dichloro-2-methylbutane : 3286
1,4-Dichloro-2-methylbutane : 3285
3,3-Dichloro-2-methyl-1-butene : 3755
1,3-Dichloro-2-methyl-2-butene : 3756
1,4-Dichloro-2-methyl-2-butene : 3757
Dichloromethyl cyanide : 278
(Dichloromethyl)dimethylchlorosilane : 11373
Dichloromethyl methyl dichlorosilane : 11391
Dichloromethylphenylsilane : 11401
1,1-Dichloro-2-methylpropane : 9899
1,2-Dichloro-2-methylpropane : 9900
1,3-Dichloro-2-methylpropane : 9901
2,4-Dichloro-5-methylpyrimidine : 11052
2,6-Dichloro-4-methylpyrimidine : 11053
Dichloromethylsilane : 11398
(Dichloromethyl)trichlorosilane : 11459
1,2-Dichloronaphthalene : 7807
1,3-Dichloronaphthalene : 7808
1,4-Dichloronaphthalene : 7809
1,5-Dichloronaphthalene : 7810
1,7-Dichloronaphthalene : 7811
1,8-Dichloronaphthalene : 7812
2,6-Dichloronaphthalene : 7813
2,3-Dichloro-1-naphthol : 8025
2,6-Dichloro-4-nitroaniline : 692
2,3-Dichloronitrobenzene : 1391
2,4-Dichloronitrobenzene : 1396
2,5-Dichloronitrobenzene : 1395
2,6-Dichloronitrobenzene : 1393
3,4-Dichloronitrobenzene : 1392
3,5-Dichloronitrobenzene : 1394
1,1-Dichloro-1-nitroethane : 5536
1,1-Dichloro-1-nitropropane : 9902
2,4-Dichloro-5-nitropyrimidine : 11054
1,9-Dichlorononane : 8128
2,4-Dichloro-6-(o-chloroanilino)-s-triazine : 360
1,8-Dichlorooctane : 8371
1,2-Dichloropentane : 8744
1,3-Dichloropentane : 8745
1,4-Dichloropentane : 8746
1,5-Dichloropentane : 8747
2,2-Dichloropentane : 8748
2,3-Dichloropentane : 8749
2,4-Dichloropentane : 8750
3,3-Dichloropentane : 8751
3,5-Dichloro-2-pentanone : 8963
2,5-Dichloro-2-pentene : 9028
1,4-Dichloro-3-pentene : 9028
Dichlorophene : 9408
2,3-Dichlorophenol : 9282
2,4-Dichlorophenol : 9283
2,5-Dichlorophenol : 9284
2,6-Dichlorophenol : 9285
3,4-Dichlorophenol : 9286
3,5-Dichlorophenol : 9287
2,4-Dichlorophenyl acetate : 9288
Dichlorophenylarsine : 462
N-(3,5-Dichlorophenyl)-1,2-dimethylcyclopropane-1,2-dicarboximide : 9760
Dichlorophenylphosphine : 9556
Dichlorophenylphosphine sulfide : 9552
1-[[2-(2,4-Dichlorophenyl)-4-propyl-1,3-dioxolan-2-yl]methyl]-1H-1,2,4-triazole : 10720
Dichlorophenylsilane : 11403
(±)-2-(3,5-Dichlorophenyl)-2-(2,2,2-trichloroethyl)oxirane : 11899
4,5-Dichlorophthalic anhydride : 7344
1,1-Dichloropropane : 9891
1,3-Dichloropropane : 9893
1,3-Dichloro-2-propanol : 10362
2,3-Dichloropropene : 10552
3,3-Dichloropropionyl chloride : 10502
N,N-Dichloropropylamine : 9805
1,1-Dichloropropylene : 10547
trans-1,2-Dichloropropylene : 10548
cis-1,2-Dichloropropylene : 10549
trans-1,3-Dichloropropylene : 10550
cis-1,3-Dichloropropylene : 10551
2,3-Dichloropropylene : 10552
3,3-Dichloropropylene : 10553
3,6-Dichloro-2-pyridinecarboxylic acid : 4142
2,4-Dichloroquinoline : 11248
2,7-Dichloroquinoline : 11249
4,7-Dichloroquinoline : 11251
3,7-Dichloroquinoline-8-carboxylic acid : 11201
2,6-Dichloroquinone-4-chloroimide : 4280
2,3-Dichlorostyrene : 1367
2,6-Dichlorostyrene : 1369
3,5-Dichlorostyrene : 1370
2,5-Dichlorostyrene : 1371
2,4-Dichlorostyrene : 1373
p-Dichlorosulfamoylbenzoic Acid : 2565
p-(*N,N*-Dichlorosulfamyl)benzoic acid : 2565
2,5-Dichloroterephthalic acid : 1349
1,2-Dichlorotetrafluorobenzene : 1398
1,3-Dichlorotetrafluorobenzene : 1399
1,4-Dichlorotetrafluorobenzene : 1400
1,1-Dichlorotetrafluoroethane : 5537
1,2-Dichloro-1,1,2,2-tetrafluoroethane : 5538
2,3-Dichlorothiophene : 11746
2,4-Dichlorothiophene : 11747

SYNONYM INDEX OF ORGANIC COMPOUNDS (continued)

2,5-Dichlorothiophene : 11748
3,4-Dichlorothiophene : 11749
2,3-Dichlorotoluene : 1383
3,4-Dichlorotoluene : 1384
2,6-Dichlorotoluene : 1385
3,5-Dichlorotoluene : 1386
2,5-Dichlorotoluene : 1387
2,4-Dichlorotoluene : 1388
N,N-Dichloro-*p*-toluenesulfonamide : 2210
2,2-Dichloro-1,1,1-trifluoroethane : 5540
Dichloroxylenol : 9290
Dichlorprop : 10250
Dicopur M : 141
Dicryl : 10520
Dictamnine : 6180
Dicumarol : 2839
Dicurone : 6239
1,3-Dicyanobenzene : 1315
Dicyanodiamide : 6280
1,2-Dicyanoethane : 3291
1,6-Dicyanohexane : 8385
Dicyanomethane : 9911
1,3-Dicyanopropane : 8767
Dicyclohexylamine : 4284
1,4-Dicyclohexylbutane : 4323
Dicyclohexylcarbodiimide : 4291
1,2-Dicyclohexylethane : 4386
1,1-Dicyclohexylethane : 4392
Dicyclohexylmethane : 4429
1,3-Dicyclohexyl-2-methylpropane : 4447
Dicyclohexyl oxalate : 5559
Dicyclohexylphosphine : 9516
1,2-Dicyclohexylpropane : 4430
1,3-Dicyclohexylpropane : 4466
1,1-Dicyclohexyltetradecane : 4476
Dicyclopropyl ketone : 7607
Dicysteine : 5020
Dieldrin : 5178
Dienestrol : 9302
meso-Diepoxybutane : 3007
Diethadione : 8573
Diethanolamine : 5730
Diethazine : 9503
3,4-Diethoxyaniline : 693
3,4-Diethoxybenzaldehyde : 538
Diethoxydimethylsilane : 11405
Diethoxydiphenylsilane : 11406
1,1-Diethoxyethane : 5541
Diethoxymethane : 5603
3,3-Diethoxy-1-propene : 10560
2,5-Diethoxytoluene : 1410
Diethylacetaldehyde : 3183
Diethylacetic acid : 3514
Diethylacetylene : 7116
O,O-Diethyl acetylphosphonate : 9530
Diethylamine : 5427
2-(Diethylamino)ethanol : 5706
2-(Diethylamino)ethyl *p*-aminobenzoate : 2493
S-[2-(Diethylamino)ethyl] diphenylthiocarbamate : 3977
3-(Diethylamino)-2-hydroxypropylamine : 10339
Diethyl aminomalonate : 9924
2-Diethylamino-1-propanol : 10366
3-(Diethylamino)propylamine : 9876
N,N-Diethylaniline : 695
Diethylarsine : 441
o-Diethylbenzene : 1413
m-Diethylbenzene : 1414
p-Diethylbenzene : 1415
N,N-Diethylbenzhydrylamine : 1804
Diethylbromoacetamide : 3198
Diethyl butanedioate : 3303
Diethylcarbamazine : 9603
N,N'-Diethylcarbanilide : 12003
Diethyl carbinol : 8917
Diethyl Carbitol : 5612
Diethyl carbonate : 3999
Diethyl chloromaleate : 3771
Diethyl chlorophosphate : 9574
Diethyl chlorothiophosphate : 9576
Diethyl 1,1-cyclobutanedicarboxylate : 4180
Diethyl 1-cyclohexene-1,3-dicarboxylate : 4645
1,1-Diethylcyclopentane : 4796
cis-1,2-Diethylcyclopentane : 4797
trans-1,2-Diethylcyclopentane : 4798
1,3-Diethylcyclopentane : 4799
Diethyl 1-cyclopentene-1,2-dicarboxylate : 4915
Diethyl dibenzylmalonate : 9926
Diethyl 2,5-dibromovalerate : 8873
Diethyl dibutylmalonate : 9937
Diethyl dicarbonate : 5169
Diethyl 2,2-diethylmalonate : 9940
Diethyldifluorosilane : 11410
Diethyl dimethylmalonate : 9942
Diethyl dipropargylmalonate : 9946
Diethyl disulfide : 5253
Diethylenediamine : 9602
Diethylenediamine dihydrochloride : 9607
Diethylene glycol : 5753
Diethylene glycol, diacetate : 5754
Diethylene glycol, dibenzoate : 5755
Diethylene glycol diethyl ether : 5612
Diethylene glycol dimethyl ether : 5613
Diethylene glycol, dodecanoate : 5313
Diethylene glycol, monoacetate : 5673
Diethylene glycol monobutyl ether : 5688
Diethylene glycol monoethyl ether : 5720
Diethylene glycol monoethyl ether acetate : 5721
Diethylene glycol monolaurate : 5313
Diethylenetriamine : 5497
Diethyl ether : 5608
Diethyl ether-trifluoroborane (1:1) : 3137
N,N-Diethylethylenediamine : 5502
Diethyl ethylphosphonate : 9536
Diethyl 3,4-furandicarboxylate : 6082
Diethyl glutaconate : 9041
3,4-Diethylhexane : 6768
3,4-Diethyl-3,4-hexanediol : 6800
Diethylhexyl adipate : 6781
1,1-Diethylhydrazine : 7152
1,2-Diethylhydrazine : 7153
Diethyl(2-hydroxyethyl)methylammonium bromide, xanthene-9-carboxylate : 5445
Diethyl 3-hydroxyglutarate : 8778
1,2-Diethylimidazole : 7196
Diethyl isobutylmalonate : 9969
Diethyl isophthalate : 1346
Diethylisopropylcarbinol : 8934
Diethyl isopropylidenemalonate : 9966
N,N-Diethylisovaleramide : 3201
Diethylketene : 3877
Diethyl ketone : 8958
Diethyl maleate : 3776
Diethyl malonate : 9941
Diethylmalonic acid : 9939
Diethyl mercury : 7434
Diethyl mesaconate : 3787
Diethyl 3-methyladipate : 6790
N,N-Diethyl-3-methylbenzamide : 612
3,3-Diethyl-2-methylpentane : 8757
3,4-Diethyl-2-methylpyrrole : 11117
Diethyl oxalacetate : 3340
Diethyl oxalate : 5560
Diethyl oxydiformate : 5169
3,3-Diethylpentane : 8756
Diethylphosphinic acid : 9525
Diethyl phthalate : 1342
Diethyl pimelate : 6394
N,N-Diethyl-*p*-phenylenediamine : 1266
Diethylpropion : 10428
2,4-Diethylpyridine : 10933
Diethyl pyrocarbonate : 5169
Diethylsilane : 11409
Diethylstilbestrol : 9300
Diethylstilbestrol dipropionate : 9301
Diethyl succinate : 3303
Diethyl sulfate : 11573
Diethyl sulfide : 5644
Diethyl sulfite : 11581
Diethyl *DL*-tartrate : 3307
Diethyl terephthalate : 1350
2,2-Diethyltetrahydrofuran : 6085
Diethylvinylethynylcarbinol : 6608
Difluoroacetic acid : 175
Difluoroacetic acid, ethyl ester : 176
2,4-Difluoroaniline : 700
o-Difluorobenzene : 1422
m-Difluorobenzene : 1423
p-Difluorobenzene : 1424
Difluorochlorobromomethane : 7464
1,1-Difluoro-2-chloroethylene : 5885
Difluorochloromethane : 7478
Difluorodibromomethane : 7496
1,1-Difluoro-1,2-dichloroethane : 5531
1,2-Difluoro-1,2-dichloroethane : 5532
1,1-Difluoro-2,2-dichloroethene : 5897
1,1-Difluoro-2,2-dichloroethylene : 5897
Difluorodichloromethane : 7500
4,4'-Difluorodiphenyl : 3044
1,1-Difluoroethane : 5544
1,2-Difluoroethane : 5545
2,2-Difluoroethanol : 5708
1,1-Difluoroethylene : 5902
cis-1,2-Difluoroethylene : 5903
Difluoroiodomethane : 7505
Difluoromethane : 7504
Difluoromethylarsine : 463
1,1-Difluoroperchloroethane : 5636
1,3-Difluoropropane : 9905
2,2-Difluoropropane : 9906
1,3-Difluoro-2-propanol : 10369
1,1-Difluoro-1-propene : 10561
2,2-Difluoropropyl chloride : 9851
2,2-Difluoro-1,1,1,2-tetrachloroethane : 5636
1,2-Difluoro-1,1,2,2-tetrachloroethane : 5637
1,2-Difluorotetrachloroethane : 5637
1,2-Difluoro-1,1,2,2-tetramethyldisilane : 5227
1,2-Difluorotetramethyldisilane : 5227
2,2-Difluoro-1,1,2-trichloroethane : 5659
1,1-Difluoro-1,2,2-trichloroethane : 5659
Difurfurylamine : 6114
Digallic acid : 2575
Digin : 11530
Digitalose : 6198
Digitoflavone : 2814
Digitogenin : 11535
Digitonin : 6195
Digitoxigenin : 4044
Digitoxin : 4043
Digitoxose : 7111
Diglycidyl resorcinol ether : 8625
Diglycine : 6265
Diglycol laurate : 5313
Digoxigenin : 4047
Digoxin : 4041
Diheptylamine : 6352
Dihexyl ketone : 11891
1,2-Dihydroacenaphthylene : 6
9,10-Dihydroanthracene : 376
1,4-Dihydrobenzene : 4249
2,3-Dihydrobenzofuran : 2442
2,3-Dihydro-2-chloromethylbenzofuran : 2441
Dihydrocholesterol : 4107
(+)-Dihydrocinchonine : 4122
Dihydrocinnamaldehyde : 2114
Dihydrocodeine : 7674
Dihydro-*p*-coumaric acid : 2134
2,5-Dihydro-2,5-dimethoxyfuran : 6087
2,3-Dihydro-2,2-dimethyl-7-benzofuranyl (dibutylamino)thio methylcarbamate : 4038
2,3-Dihydro-5,6-dimethyl-1,4-dithiin, 1,1,4,4-tetraoxide : 5179
α-Dihydrofucosterol : 11558
Dihydrofumaric acid : 3294
2,3-Dihydrofuran : 6086
3,4-Dihydro-2H-pyran : 10777
4,6-Dihydro-8-hydroxy-3,4,5-trimethyl-6-oxo-3H-2-benzopyran-7-carboxylic acid : 2774
3,6-Dihydro-6-iminopurine : 10751
1,2-Dihydro-2-iminopyridine : 10857
Dihydroisophorone : 4622
2,3-Dihydro-5-methylbenzofuran : 2444
1,4-Dihydronaphthalene : 7815
3,4-Dihydro-1-naphthoic acid : 7764
2,3-Dihydro-3-oxobenzisosulfonazole : 2388
9,10-Dihydrophenanthrene : 9171
Dihydrosafrole : 2431
Dihydrothebaine : 7664

2,3-Dihydrothiophene : 11750
2,5-Dihydrothiophene : 11751
4,5-Dihydrotoluene : 4271
Dihydroxyacetone : 10429
1,3-Dihydroxy-2-amino-2-methylpropane : 9987
2,5-Dihydroxybenzyl alcohol : 1520
1,3-Dihydroxybutane : 3345
1,4-Dihydroxybutane : 3346
2,3-Dihydroxybutanedioic acid : 3309
1,4-Dihydroxy-2-butyne : 3911
3α,12α-Dihydroxycholanic acid : 4095
3α,6α-Dihydroxy-5β-cholanic acid : 4096
3α,7α-Dihydroxy-5β-cholan-24-oic acid : 4097
3α,7β-Dihydroxy-5β-cholan-24-oic acid : 4098
6,7-Dihydroxycoumarin : 2804
6,7-Dihydroxycoumarin 6-glucoside : 2822
trans-1,4-Dihydroxycyclohexane : 4374
(+)-2,9-Dihydroxy-1,10-dimethoxyaporphine : 5156
3,3'-Dihydroxy-5,5'-dimethylbiquinone : 2979
2,3-Dihydroxy-2,3-dimethylbutane : 3350
4,4'-Dihydroxydiphenylmethane-3,3'-dicarboxylic acid : 2686
1,12-Dihydroxydodecane : 5301
3,17-Dihydroxyestratriene : 5400
Dihydroxyestrin : 5399
1,2-Dihydroxyethylbenzene : 5579
2,2'-Dihydroxyethyl ether monododecanoate : 5313
4,4'-Dihydroxyfuchsone : 4278
5',7-Dihydroxy-4'-glucosyloxyisoflavone : 2824
2,5-Dihydroxyhexane : 6799
1,3-Dihydroxy-5-iodobenzene : 1524
4',7-Dihydroxyisoflavone : 2827
Dihydroxymaleic acid : 3777
5,7-Dihydroxy-4'-methoxyflavone : 2809
5,7-Dihydroxy-6-methoxyflavone : 2810
3,6-Dihydroxy-4-methoxy-2,5-toluquinone : 4260
1,7-Dihydroxy-3-methoxyxanthone : 12054
2,3-Dihydroxy-2-methylbutane : 3355
1,4-Dihydroxy-1-methylbutane : 8786
1,1-Di(hydroxymethyl)ethylamine : 9987
2,5-Dihydroxy-3-methyl-6-methoxy-1,4-benzoquinone : 4260
1,9-Dihydroxynonane : 8136
9,10-Dihydroxyoctadecanoic acid : 8254
2,4-Dihydroxypentane : 8789
2,3-Dihydroxyphenol : 2366
2,5-Dihydroxyphenol : 2367
Dihydroxyphenylborane : 3135
2,3-Dihydroxypropionic acid : 10252
(±)-2,3-Dihydroxypropyl acetate : 10187
3,6-Dihydroxyspiro[xanthene-9,3'-phthalide] : 11522
9,10-Dihydroxystearic acid : 8254
Dihydroxytartaric acid : 3342
4,6-Dihydroxy-*o*-toluic acid : 2573
Diiodoacetylene : 5941
2,4-Diiodoaniline : 701
4,4'-Diiodobenzophenone : 7598
Diiodochloromethane : 7479
1,1-Diiodoethane : 5546
Diiodofluoromethane : 7516
1,6-Diiodohexane : 6769
Diiodomethane : 7506
1,5-Diiodopentane : 8758
3,5-Diiodo-α-phenylphloretic acid : 2135
1,3-Diiodopropane : 9907
2,2-Diiodopropane : 9908
3,5-Diiodosalicylic acid : 2617
Diiodotariric acid : 8300
2,5-Diiodothiophene : 11753
3,5-Diiodotyrosine : 11928
Diisoamylamine : 3226
Diisoamyl ether : 3418
Diisoamyl phthalate : 1335
Diisobutylamine : 9823
Diisobutylcarbinol : 6485
Diisobutylene : 9064
Diisobutyl fumarate : 3767
Diisobutyl ketone : 6514
Diisobutyl oxalate : 5557
Diisobutyl sulfide : 10154
1,1-Diisobutylurea : 11996
2,3-Diisonitrosobutane : 3360
Diisopentyl ether : 3418
Diisopentyl 2-hydroxyethylamine : 5682
Diisopentyl phthalate : 1335
Diisopentyl sulfide : 3436
Diisopropanol : 6799
Diisopropanolamine : 10384
Diisopropenyl : 3159
Diisopropenylacetylene : 6714
Diisopropylamine : 9821
N,*N*-Diisopropyl-2-aminoethanol : 5683
Diisopropylbutadiyne : 8345
Diisopropyl ether : 10120
Diisopropyl fumarate : 3766
Diisopropyl ketone : 8967
Diisopropyl ketone, cyanohydrin : 3408
Diisopropyl ketone oxime : 8968
Diisopropyl sulfide : 10152
Diketene : 8598
Dilauroyl peroxide : 9155
Diluran : 32
Dimedazole : 2380
Dimedone : 4378
Dimefline : 2819
Dimercaprol : 10370
1,4-Dimercaptobutane : 3368
1,2-Dimercaptoethane : 5590
Dimethadione : 8584
1,4:5,8-Dimethanonaphthalene, 1,2,3,4,10,10-hexachloro-6,7-epoxy-1,4,4a,5,6,7,8,8a-octahydro-, endo,*exo*- : 5178
Dimethisoquin : 5411
2,6-Dimethoxyaniline : 703
2,4-Dimethoxybenzaldehyde : 543
2,5-Dimethoxybenzaldehyde : 544
3,5-Dimethoxybenzaldehyde : 546
3,5-Dimethoxybenzoic acid : 2577
2,6-Dimethoxybenzonitrile : 2753
2,4-Dimethoxybenzophenone : 7612
2,6-Dimethoxy-*p*-benzoquinone : 4262
2,2'-Dimethoxybiphenyl : 3045
3,3'-Dimethoxybiphenyl : 3046
Dimethoxyborane : 3117
5,7-Dimethoxycoumarin : 2818
3,3'-Dimethoxy-4,4'-diaminodiphenyl : 3033
2,5-Dimethoxy-2,5-dihydrofuran : 6087
2,4-Dimethoxy-1-nitrobenzene : 1448
1,3-Dimethoxy-4-nitrobenzene : 1448
[(Dimethoxyphosphinothioyl)thio]butanedioic acid diethyl ester : 3316
5,6-Dimethoxyphthalaldehydic acid : 2608
1,10-Dimethoxy-6aα-aporphine-2,9-diol : 5156
1,3-Dimethoxy-2-propanol : 10372
2,6-Dimethoxypyridine : 10935
2-[[[[[(4,6-Dimethoxypyrimidin-2-yl)amino]carbonyl]amino]sulfonyl]methyl]benzoic acid, methyl ester : 516
*N*1-(2,6-Dimethoxy-4-pyrimidinyl)sulfanilamide : 2194
2-(4,6-Dimethoxypyrimidi*N*-2-ylcarbamoylsulfamoyl)-*N*,*N*-dimethylnicotinamide : 8069
2,6-Dimethoxy-*p*-quinone : 4262
Dimethyl acetal : 5548
N,*N*-Dimethylacetamide : 55
3,3-Dimethylacetylacetone : 8799
Dimethylacetylene : 3898
Dimethylacetylenecarbinol : 3922
3,3-Dimethylacrolein : 3700
trans-2,3-Dimethylacrylic acid : 3844
1,3-Dimethylallene : 8660
7,8-Dimethylalloxazine : 2472
Dimethylamide acetate : 55
Dimethylamine : 7448
4-(Dimethylamino)azobenzene : 721
2',3-Dimethyl-4-aminoazobenzene : 798
p-(Dimethylamino)azobenzene-*o*-carboxylic acid : 2587
p-(Dimethylamino)benzal-5-rhodanine : 11684
p-(Dimethylamino)benzoic acid : 2585
4-(Dimethylamino)benzophenone : 7613
3-(Dimethylamino)benzophenone : 2771
4,4'-Dimethylaminobenzophenonimide : 657
2-(Dimethylamino)ethanol : 5709
β-Dimethylaminoethanol diphenylmethyl ether : 5422
β-Dimethylaminoethyl methacrylate : 10654
2-[(2-Dimethylaminoethyl)(*p*-methoxybenzyl)amino]pyridine : 5511
2-[(Dimethylamino)methyl]-cyclohexanone : 4591
3-[(Dimethylamino)methyl]indole : 7310
4-Dimethylaminophenazone : 10851
2-Dimethylamino-1-propanol : 10374
4-[2-(Dimethylamino)propionamido]-antipyrine : 9795
3-(Dimethylamino)propionitrile : 10101
N,*N*-Dimethylaniline : 713
1,3-Dimethylanthracene : 377
Dimethylarsine : 442
2,4-Dimethylbenzaldehyde : 547
2,5-Dimethylbenzaldehyde : 548
9,10-Dimethyl-1,2-benzanthracene : 520
2,4-Dimethylbenzene-carboxaldehyde : 547
3,3'-Dimethylbenzidine : 3034
1,5-Dimethylbenzimidazole : 2378
2,5-Dimethylbenzofuran : 2445
2,6-Dimethylbenzofuran : 2446
3,5-Dimethylbenzofuran : 2448
4,7-Dimethylbenzofuran : 2451
5,6-Dimethylbenzofuran : 2452
5,7-Dimethylbenzofuran : 2453
2,4-Dimethylbenzoic acid : 2580
2,6-Dimethylbenzoic acid : 2582
3,5-Dimethylbenzoic acid : 2583
3,4-Dimethylbenzophenone : 7615
2,4'-Dimethylbenzophenone : 7628
2,5-Dimethylbenzoxazole : 2879
α,α-Dimethylbenzyl alcohol : 1857
m-α-Dimethylbenzyl alcohol : 1858
p-α-Dimethylbenzyl alcohol : 1859
Dimethylbenzylamine : 1807
3,5-Dimethylbenzyl bromide : 1057
2,3-Dimethylbiphenyl : 3052
2,4-Dimethylbiphenyl : 3054
3,4-Dimethylbiphenyl : 3059
3,5-Dimethylbiphenyl : 3060
2,3-Dimethylbutadiene : 3159
1,1-Dimethylbutadiene : 8670
Dimethylbutadiyne : 6717
2,2-Dimethylbutane : 3289
2,3-Dimethylbutane : 3290
2,2-Dimethyl-1-butanol : 3615
3,3-Dimethylbutan-2-ol (*DL*-) : 3619
2,3-Dimethyl-1-butene : 3759
2,3-Dimethyl-2-butene : 3761
2,2-Dimethyl-3-butenoic acid : 3835
4,4-Dimethyl-2-butenoic acid : 9078
1,1-Dimethylbutyl acetate : 8946
1,3-Dimethylbutyl acetate : 8947
3,4'-Dimethylbutyrophenone : 3670
Dimethyl cadmium : 3929
Dimethylcarbamoyl chloride : 3973
Dimethylchloroarsine : 453
Dimethyl chloromaleate : 3773
5,6-Dimethylchrysene : 4115
trans-1,2-Dimethylcyclobutane : 4188
cis-1,3-Dimethylcyclobutane : 4189
trans-1,3-Dimethylcyclobutane : 4190
Dimethyl *trans*-1,2-cyclobutanedicarboxylate : 4181
1,3-Dimethyl-1,3-cyclohexadiene : 4252
1,4-Dimethyl-1,3-cyclohexadiene : 4253
1,5-Dimethyl-1,3-cyclohexadiene : 4254
2,5-Dimethyl-1,3-cyclohexadiene : 4255
1,1-Dimethylcyclohexane : 4364
gem-Dimethylcyclohexane : 4364
cis-1,2-Dimethylcyclohexane : 4365
trans-1,2-Dimethylcyclohexane : 4366
cis-1,3-Dimethylcyclohexane : 4367
cis-1,4-Dimethylcyclohexane : 4368
trans-1,3-Dimethylcyclohexane : 4370
trans-1,4-Dimethylcyclohexane : 4371

Dimethyl *cis*-1,3-cyclohexanedicarboxylate : 4352
cis-Dimethyl 1,4-cyclohexanedicarboxylate : 4357
2,2-Dimethylcyclohexanol : 4513
2,4-Dimethylcyclohexanol : 4514
2,6-Dimethyl-1-cyclohexanol : 4515
3,3-Dimethylcyclohexanol : 4516
3,5-Dimethylcyclohexanol : 4518
4,4-Dimethylcyclohexanol : 4519
cis-1,3-Dimethylcyclohexanol : 4510
cis-2,3-Dimethylcyclohexanone : 4581
2,2-Dimethyl-1-cyclohexanone : 4580
trans-2,3-Dimethylcyclohexanone : 4582
trans-(±)-2,5-Dimethylcyclohexanone : 4585
2,6-Dimethylcyclohexanone : 4586
3,3-Dimethylcyclohexanone : 4587
3,4-Dimethylcyclohexanone : 4588
3,5-Dimethylcyclohexanone : 4589
4,4-Dimethylcyclohexanone : 4590
1,3-Dimethyl-1-cyclohexene : 4649
1,3-Dimethylcyclohexene : 4649
1,4-Dimethyl-1-cyclohexene : 4650
3,3-Dimethylcyclohexene : 4653
3,6-Dimethylcyclohexene : 4655
3,5-Dimethylcyclohexene : 4654
2,5-Dimethyl-2-cyclohexenone : 4706
3,4-Dimethylcyclohexyl alcohol : 4517
1,1-Dimethylcyclopentane : 4800
gem-Dimethylcyclopentane : 4800
cis-1,2-Dimethylcyclopentane : 4801
trans-1,2-Dimethylcyclopentane : 4802
cis-1,3-Dimethylcyclopentane : 4803
trans-1,3-Dimethylcyclopentane : 4804
2,5-Dimethylcyclopentanone : 4882
1,4-Dimethylcyclopentene : 4920
1,3-Dimethylcyclopentene : 4919
1,1-Dimethylcyclopropane : 4971
gem-Dimethylcyclopropane : 4971
cis-1,2-Dimethylcyclopropane : 4972
trans-1,2-Dimethylcyclopropane : 4973
3,3-Dimethyl-4-decyne : 5101
Dimethyldiacetylene : 6717
1,6-Dimethyldiallyl : 8320
2,7-Dimethyl-2,7-diazaoctane : 3263
Dimethyldiazene : 5125
O,O-Dimethyl *S*-(1,2-dicarbethoxyethyl) dithiophosphate : 3316
2,5-Dimethyl-2,5-dichlorohexane : 6765
3,3-Dimethyl-1,5-dichloropentane : 8753
Dimethyldiethoxysilane : 11405
Dimethyldifluoromethane : 9906
S,S'-Dimethyl-2-difluoromethyl-4-isobutyl-6-trifluoromethylpyridine-3,5-dicarbothioate : 5267
5,5-Dimethyldihydroresorcinol : 4378
2,2-Dimethyl-1,3-dihydroxypentane : 8792
Dimethyl diketone : 3359
2,2-Dimethyl-1,3-dioxolane-4-methanol : 5207
Dimethyldiphenoxysilane : 11415
Dimethyldiphenylsilane : 11416
Dimethyl disulfide : 5254
7,8-Dimethylenebenz[a]anthracene : 2391
3,4-Dimethylenedioxybenzaldehyde : 2417
N,N-Dimethylethanamide : 55
Dimethyl ether : 7523
1,1-Dimethylethyl methyl ketone : 3657
Dimethylethynylcarbinol : 3922
Dimethylformaldehyde : 10411
N,N-Dimethylformamide : 5992
3,5-Dimethylfuran : 6088
2,4-Dimethylfuran : 6088
2,5-Dimethylfuran : 6089
2,7-Dimethylfuran : 2447
6,7-Dimethylfuran : 2454
3,6-DimethylfuranBenzofuran : 2449
Dimethyl 2,5-furandicarboxylate : 6083
N,N-Dimethylglycine : 6261
Dimethylglyoxal : 3359
Dimethylglyoxime : 3360
2,2-Dimethylheptane : 6383
2,3-Dimethylheptane : 6384
2,4-Dimethylheptane : 6385
3,6-Dimethylheptane : 6386
2,5-Dimethylheptane : 6386
2,6-Dimethylheptane : 6387
3,3-Dimethylheptane : 6388
3,4-Dimethylheptane : 6389
3,5-Dimethylheptane : 6390
4,4-Dimethylheptane : 6391
2,6-Dimethyl-2-heptanol : 6479
2,3-Dimethyl-3-heptanol : 6480
2,6-Dimethyl-3-heptanol : 6481
3,6-Dimethylheptan-3-ol : 6483
2,4-Dimethyl-4-heptanol : 6484
2,6-Dimethyl-4-heptanol : 6485
3,5-Dimethyl-4-heptanol : 6486
3,5-Dimethyl-4-heptanone : 6515
2,3-Dimethyl-2-heptene : 6542
5,5-Dimethyl-3-heptyne : 6618
2,5-Dimethyl-1,5-hexadien-3-yne : 6714
2,2-Dimethylhexane : 6771
2,3-Dimethylhexane : 6772
2,5-Dimethylhexane : 6774
3,3-Dimethylhexane : 6775
3,4-Dimethylhexane : 6776
2,4-Dimethylhexane : 6773
2,5-Dimethyl-2,5-hexanediol : 6801
2,5-Dimethyl-3,4-hexanedione : 6807
2,3-Dimethyl-2-hexanol : 6922
2,5-Dimethyl-2-hexanol : 6923
2,2-Dimethyl-3-hexanol : 6925
2,5-Dimethyl-3-hexanol : 6927
3,3-Dimethyl-2-hexanone : 6963
3,4-Dimethyl-2-hexanone : 6964
2,2-Dimethyl-3-hexanone : 6965
2,3-Dimethyl-1-hexene : 7013
2,4-Dimethyl-1-hexene : 7014
2,5-Dimethyl-1-hexene : 7015
3,3-Dimethyl-1-hexene : 7016
3,4-Dimethyl-1-hexene : 7017
4,4-Dimethyl-1-hexene : 7019
4,5-Dimethyl-1-hexene : 7020
5,5-Dimethyl-1-hexene : 7021
2,3-Dimethyl-2-hexene : 7022
2,4-Dimethyl-2-hexene : 7023
2,5-Dimethyl-2-hexene : 7024
2,3-Dimethyl-4-hexene : 7025
cis-5,5-Dimethyl-2-hexene : 7027
2,4-Dimethyl-*trans*-3-hexene : 7030
1,1-Dimethyl-5-hexen-1-ol : 6582
2,5-Dimethyl-4-hexen-3-ol : 7075
2,3-Dimethyl-2H-indazole : 7238
1,1-Dimethylhydrazine : 7154
1,2-Dimethylhydrazine : 7155
1,1-Dimethyl-2-hydroxyethylamine : 10341
2,2-Dimethyl-4-hydroxymethyldioxolane : 5207
1,1-Dimethyl-3-hydroxypyrrolidinium bromide α-cyclopentylmandelate : 11179
2,4-Dimethyl-6-hydroxyquinoline : 11317
1,4-Dimethylimidazole : 7201
1,1-Dimethylindan : 7251
1,2-Dimethylindan : 7252
4,7-Dimethylindan : 7253
5,6-Dimethylindan : 7254
1,5-Dimethylindazole : 7237
2,5-Dimethylindole : 7299
4,7-Dimethyl-2-indolinone : 7330
Dimethyl isophthalate : 1347
3,5-Dimethylisoxazole : 7403
Dimethylketol : 3661
Dimethyl ketone : 10411
Dimethyl ketoxime : 10476
Dimethyl mercury : 7435
(*E*)-Dimethyl 1-methyl-3-(methylamino)-3-oxo-1-propenyl phosphate : 7663
2,6-Dimethylmorpholine : 7687
1,2-Dimethylnaphthalene : 7818
1,3-Dimethylnaphthalene : 7819
1,4-Dimethylnaphthalene : 7820
1,5-Dimethylnaphthalene : 7821
1,6-Dimethylnaphthalene : 7822
1,7-Dimethylnaphthalene : 7823
1,8-Dimethylnaphthalene : 7824
2,3-Dimethylnaphthalene : 7825
2,6-Dimethylnaphthalene : 7826
2,7-Dimethylnaphthalene : 7827
N,N-Dimethyl-1-naphthylamine : 7715
2,8-Dimethylnonan-5-ol : 8164
3,3-Dimethyl-4-nonyne : 8201
3,7-Dimethylocta-2,4-diene : 8322
2,7-Dimethyl-2,6-octadiene : 8325
4,5-Dimethyl-2,6-octadiene : 8327
cis-2,6-Dimethyl-2,6-octadiene : 8324
3,6-Dimethyl-2,6-octadiene-4-yne : 8340
2,2-Dimethyloctane : 8373
2,3-Dimethyloctane : 8374
2,4-Dimethyloctane : 8375
2,5-Dimethyloctane : 8376
2,6-Dimethyloctane : 8377
3,3-Dimethyloctane : 8379
3,4-Dimethyloctane : 8380
3,5-Dimethyloctane : 8381
3,6-Dimethyloctane : 8382
4,4-Dimethyloctane : 8383
4,5-Dimethyloctane : 8384
2,6-Dimethyl-2-octanol : 8450
2,7-Dimethyloctan-4-ol : 8458
3,7-Dimethyl-1,3,6-octatriene : 8485
2,6-Dimethyl-7-octene : 8502
2,6-Dimethyl-2-octene : 8503
2,2-Dimethyl-3-octyne : 8535
Dimethylolpropionic acid : 10272
Dimethyl oxalate : 5563
2,4-Dimethyloxazole : 8577
2,5-Dimethyloxazole : 8578
3,3-Dimethyloxetane : 8593
O,O-Dimethyl *S*-[(4-oxo-1,2,3-benzotriazin-3(4H)-yl)methyl]phosphorodithioate : 497
2,2-Dimethylpentane : 8760
2,3-Dimethylpentane : 8761
2,4-Dimethylpentane : 8762
3,3-Dimethylpentane : 8763
3,3-Dimethyl-2,4-pentanedione : 8799
2,4-Dimethyl-2-pentanol : 8928
2,2-Dimethyl-3-pentanol : 8929
2,3-Dimethyl-3-pentanol : 8930
2,2-Dimethyl-3-pentanone : 8966
2,4-Dimethyl-3-pentanone oxime : 8968
2,3-Dimethyl-1-pentene : 9029
2,4-Dimethyl-1-pentene : 9030
3,3-Dimethyl-1-pentene : 9031
3,4-Dimethyl-1-pentene : 9032
2,3-Dimethyl-2-pentene : 9034
2,4-Dimethyl-2-pentene : 9035
3,4-Dimethyl-2-pentene : 9036
2,2-Dimethylpent-4-enoic acid : 9075
3,3-Dimethyl-1-pentyne : 9128
4,4-Dimethyl-1-pentyne : 9129
4,4-Dimethyl-2-pentynoyl chloride : 9147
9,10-Dimethylphenanthrene : 9172
Dimethylphenylcarbinol : 1857
Dimethyl-*p*-phenylenediamine : 1270
Dimethylphenylmethanol : 1857
N-(2,6-Dimethylphenyl)-*N*-(methoxyacetyl)alanine, methyl ester : 7439
N-(2,6-Dimethylphenyl)-2-methoxy-*N*-(2-oxo-3-oxazolidinyl)acetamide : 8566
1,1-Dimethyl-3-phenylpropanol : 2150
Dimethylphosphinic acid : 9528
Dimethylphosphoramidothioic dichloride : 9557
Dimethyl phthalate : 1343
cis-2,5-Dimethylpiperazine : 9609
trans-2,5-Dimethylpiperazine : 9610
cis-2,6-Dimethylpiperazine : 9611
2,5-Dimethylpiperidine : 9649
3,3-Dimethylpiperidine : 9651
3,5-Dimethylpiperidine : 9652
1,1-Dimethylpropargyl alcohol : 3922
Dimethylpropylamine : 9807
1,1-Dimethylpropyl bromide : 3240
1,1-Dimethyl-2-propyn-1-ol : 3922
2,3-Dimethylpyrazine : 10805
2,5-Dimethylpyrazine : 10806
2,6-Dimethylpyrazine : 10807
2,3-Dimethylpyridine : 10936
2,5-Dimethylpyridine : 10938
3,4-Dimethylpyridine : 10940
3,5-Dimethylpyridine : 10941
2,4-Dimethylpyrimidine : 11055
2,5-Dimethylpyrimidine : 11056
4,5-Dimethylpyrimidine : 11057
4,6-Dimethylpyrimidine : 11058

*N*1-(4,6-Dimethyl-2-pyrimidyl)sulfanilamide : 2196
2,4-Dimethylpyrrole : 11119
2,2-Dimethylpyrrolidine : 11149
cis-2,5-Dimethylpyrrolidine : 11151
3,3-Dimethyl-2-pyrrolidone : 11182
4,6-Dimethylquinaldine : 11302
2,3-Dimethylquinoline : 11254
2,4-Dimethylquinoline : 11255
2,7-Dimethylquinoline : 11258
3,4-Dimethylquinoline : 11260
3,8-Dimethylquinoline : 11261
4,6-Dimethylquinoline : 11262
5,8-Dimethylquinoline : 11263
6,8-Dimethylquinoline : 11264
Dimethylsilane : 11414
Dimethylsiloxane cyclic trimer : 5002
2,2'-Dimethylstilbene : 1622
Dimethyl sulfate : 11574
Dimethyl sulfide : 7545
2,4-Dimethylsulfolane : 11784
Dimethyl sulfone : 7534
Dimethyl sulfoxide : 7529
Dimethyl terephthalate : 1351
1,3-Dimethyl-1,2,3,4-tetrahydronaphthalene : 7982
1,1-Dimethyltetralin : 7980
1,4-Dimethyltetralin : 7983
2,3-Dimethyltetralin : 7986
2,5-Dimethyltetralin : 7987
2,7-Dimethyltetralin : 7989
2,8-Dimethyltetralin : 7990
5,6-Dimethyltetralin : 7991
5,7-Dimethyltetralin : 7992
2,5-Dimethyl-1,3,4-thiadiazole : 11659
2,4-Dimethylthiazole : 11674
4,5-Dimethylthiazole : 11675
Dimethyl *N,N'*-[thiobis[[(methylimino)carbonyl]oxyl]]bis[ethanimidothioate] : 11713
2,3-Dimethylthiophene : 11754
2,4-Dimethylthiophene : 11755
2,5-Dimethylthiophene : 11756
3,4-Dimethylthiophene : 11757
N,N'-Dimethylthiourea : 11804
7,8-Dimethyltocol : 2790
N,N-Dimethyl-*o*-toluidine : 860
1,5-Dimethyl-1,2,3-triazole : 11849
N,N-Dimethyltryptamine : 7303
Dimethyl(vinylethyl)carbinol : 7083
Dimethylvinylethynylchloromethane : 7107
Dimethyl zinc : 12066
Dimetridazole : 7202
Dimorpholamine : 7686
Dinaphthalene dioxide : 12055
β,β'-Dinaphthylamine : 7727
Dinaphthylene dioxide : 12055
1,2'-Dinaphthyl ketone : 7634
2,3-Dinitroaniline : 724
2,4-Dinitroaniline : 725
2,6-Dinitroaniline : 726
3,5-Dinitroaniline : 727
3,5-Dinitroanisole : 1924
2,4-Dinitrobenzaldehyde : 551
o-Dinitrobenzene : 1491
m-Dinitrobenzene : 1492
p-Dinitrobenzene : 1493
2,4-Dinitrobenzenesulfenyl chloride : 2178
3,4-Dinitrobenzoic acid : 2590
3,5-Dinitrobenzoic acid : 2591
3,5-Dinitrobenzoyl chloride : 2896
1,4-Dinitrobutane : 3293
4,4'-Dinitrocarbanilide : 11997
Dinitrochloromethane : 7480
2,6-Dinitro-4-chlorophenol : 9258
Dinitrocresol : 9405
4,6-Dinitro-*o*-cyclohexylphenol : 9273
3,3'-Dinitrodiphenyl disulfide : 5247
1,1-Dinitroethane : 5551
1,2-Dinitroethane : 5552
2,4-Dinitro-1-fluorobenzene : 1718
1,6-Dinitrohexane : 6778
Dinitromethane : 7508
1,3-Dinitronaphthalene : 7831
1,5-Dinitronaphthalene : 7832
1,8-Dinitronaphthalene : 7833
1,5-Dinitropentane : 8769
2,3-Dinitrophenol : 9341
2,4-Dinitrophenol : 9342
2,5-Dinitrophenol : 9343
2,6-Dinitrophenol : 9344
3,5-Dinitrophenol : 9346
2,4-Dinitrophenyl fluoride : 1718
(2,4-Dinitrophenyl)hydrazine : 7156
1,1-Dinitropropane : 9918
1,3-Dinitropropane : 9919
2,2-Dinitropropane : 9920
2,4-Dinitroresorcinol : 1517
2,5-Dinitrothiophene : 11758
2,4-Dinitrotoluene : 1956
2,6-Dinitrotoluene : 1958
3,4-Dinitrotoluene : 1960
17,18-Dinor-8βH,9βH,10α-labda-2,13(16),14-triene-19,20-dioic acid, 15,16-epoxy-1β,4,12 : 5926
Dinoroleic acid : 6668
Dioctyl adipate : 6781
Dioctyl phthalate : 1344
Dioctyl sebacate : 5043
Diosgenin : 11536
Diosphenol : 4709
1,3-Dioxacyclohexane : 5186
1,4-Dioxacyclohexane : 5187
1,3-Dioxacyclopentane : 5204
2,15-Dioxa-12,25-diazatricyclo[22.2.2.211,14]-triacontane, carpaine deriv. : 4064
1,4-Dioxane, heptachloro : 5192
m-Dioxane, 4-methyl-4-phenyl- : 5194
1,2-Dioxocyclohexane : 4375
1,4-Dioxocyclohexane : 4377
2,5-Dioxohexane : 6805
1,3-Dioxolane, 2,2-dimethyl-4-hydroxymethyl : 5207
2,4-Dioxopentane : 8798
Dioxybenzone : 7619
3,4-Dioxymethylenebenzoic acid : 2418
Dipentene : 4677
Dipentylamine : 8708
Dipentylamine, 2,2'-dimethyl,2,2'-dihydroxy : 8936
2,5-Di-*tert*-pentylhydroquinone : 1501
Dipentyl ketone : 11962
Diphemethoxidine : 9665
Diphenadione : 7268
Diphenethyl ketone : 8969
Diphenhydramine : 5422
o,o'-Diphenic acid : 3038
Diphenolic acid : 1103
Diphenoxydimethylsilane : 11415
Diphenyl : 3008
Diphenylacetamide : 57
Diphenylacetic acid : 936
Diphenylacetylene : 1714
Diphenylamine : 836
Diphenylamine-2-carboxylic acid : 2715
Diphenylamine-2,2'-dicarboxylic acid : 2656
Diphenylamine-4-sulfonic acid : 2234
Diphenylarsine : 443
N,N'-Diphenylbenzidine : 3035
1,4-Diphenylbutane : 1099
1,1-Diphenylbutane : 1120
2,2-Diphenyl-3-butanone : 3658
1,3-Diphenyl-1-butene : 2022
2,2'-Diphenylcarbazide : 4010
sym-Diphenylcarbazide : 4010
N,N'-Diphenylcarbazide : 4010
Diphenylcarbazone : 5122
Diphenylcarbinol : 1910
1,5-Diphenylcarbohydrazide : 4010
Diphenylcyanamide : 4158
Diphenyldiethoxysilane : 11406
Diphenyl(dimethyl)silane : 11416
Diphenyl diselenide : 5225
1,4-Diphenyl-2,3-dithiabutane : 5248
1,1-Diphenylethane : 1677
2,2-Diphenylethanol : 1615
1,1-Diphenylethene : 1636
Diphenyl ether : 2071
N,N'-Diphenylethylenediamine : 5508
Diphenyl glutarate : 8776
1,3-Diphenylguanidine : 6282
1,1-Diphenylheptane : 1746
1,1-Diphenyl-1-heptene : 1744
1,1-Diphenylhydrazine : 7157
N,N'-Diphenylhydrazine : 7158
1,3-Diphenylindene : 7271
2,3-Diphenylindene : 7272
Diphenyline : 3029
Diphenylketene : 5928
8,10-Diphenyllobelidiol : 9646
8,10-Diphenyllobelidione : 5854
Diphenylmercury : 7436
Diphenylmethane : 1961
Diphenylmethane, 2,2'-dibromo : 1962
Diphenylmethane, 3,3'-dihydroxy : 9406
2-(Diphenylmethoxy)-*N,N*-dimethylethylamine : 5422
Diphenylmethyl alcohol : 1910
Di(phenylmethyl)disulfide : 5248
Diphenyl methylphosphonate : 9541
3,5-Diphenyl-1,2,4-oxadiazole : 8564
2,4-Diphenyloxazole : 8579
4,5-Diphenyloxazole : 8581
1,5-Diphenylpentane : 2098
1,5-Diphenyl-3-pentanone : 8969
N,N'-Diphenyl-*p*-phenylenediamine : 1271
1,3-Diphenylpropane : 2121
1,1-Diphenylpropane : 2171
1,3-Diphenylpyrazole : 10830
1,5-Diphenylpyrazole : 10831
2,6-Diphenylpyridine : 10947
1,1-Diphenylsilacyclohexane : 11360
Diphenylsilane : 11420
Diphenyl succinate : 3320
Diphenyl sulfone : 2237
1,1-Diphenyltetradecane : 2267
2,4-Diphenylthiazole : 11676
Diphenylthiocarbamic acid *S*-[2-(diethylamino)ethyl] ester : 3977
Diphenylthiocarbazone : 5121
sym-Diphenylthiourea : 11805
1,1-Diphenylurea : 12011
Diphosgene : 4029
Dipicrylamine : 864
Dipicryl sulfide : 2288
Dipivaloylmethane : 6404
Diploicin : 5153
Diploicine : 5153
Dipropargyl : 6716
1,3-Dipropoxybenzene : 1556
1,2-Dipropoxyethane : 5586
Dipropylacetylene : 8532
Dipropylamine : 9826
Dipropylcarbinol : 6475
Dipropyl carbonate : 4002
Dipropylene glycol : 10397
Dipropyl ether : 10119
Dipropyl fumarate : 3783
Dipropylglyoxime : 8397
N,N-Dipropylhydroxylamine : 9814
Dipropylisobutylamine, perfluoro : 11912
Dipropyl ketone : 6512
Dipropyl malonate : 9945
Dipropylnitrosamine : 9825
5,5-Dipropyl-2,4-oxazolidinedione : 8585
2,6-Dipropylphenol : 9347
p-(Dipropylsulfamoyl)benzoic acid : 2595
Dipropyl sulfide : 10151
Dipyridamole : 5711
2,2'-Dipyridine : 3099
Dipyrido[1,2-a:2',1'-c]pyrazinediium, 6,7-dihydro-, dibromide : 5224
2,2'-Dipyridyl : 3099
α,α'-Dipyridyl : 3099
3,3'-Dipyridyl : 3102
4,4-Dipyridyl : 3104
γ,γ'-Dipyridyl : 3104
Dipyrine : 10851
Diresorcinol : 3097
Disalcid : 2616
Disalicylic acid : 2616
Disiloxane, 1-(cyclopentadienyl)-1,1,3,3-tetramethyl-3-phenyl-, Fe deriv. : 5961
Disophenol : 9304
Dispiro[5,1,5,1]tetradecane : 5240
Distannoxane, hexakis(2-methyl-2-phenylpropyl)- : 5950
Disulfan : 2188
Disulfiram : 11721

SYNONYM INDEX OF ORGANIC COMPOUNDS (continued)

1,3-Dithiacyclopentane : 5266
Dithianone : 8056
1,3,5-Dithiazine, 4,5-dihydro-5-methyl : 5264
Di-(2-thienyl) ketone : 7617
3,3'-Dithiobis(2-aminopropanoic acid) : 5020
2,4-Dithiobiuret : 11718
4,4'-Dithiodimorpholine : 7688
Dithioglycol : 5590
Dithiooxamide : 5588
Dithiosalicylic acid : 1125
Dithizone : 5121
1,2-Di-*p*-tolylethane : 1590
2,4'-Ditolyl sulfide : 1993
3,4'-Ditolyl sulfide : 11570
Diundecyl ketone : 11857
Diurea glyoxalate : 7191
Diurite : 2852
Diuron : 12000
Diuroral : 2854
Divinyl : 3146
Divinylacetylene : 6713
o-Divinylbenzene : 1403
m-Divinylbenzene : 1404
p-Divinylbenzene : 1405
trans-1,2-Divinylcyclobutane : 4187
Divinyl ether : 5912
Djenkolic acid : 5018
n-Docosane : 5268
Docosane, 1-cyclohexyl- : 4380
Docosane, 11-decyl : 5273
Docosane, 1-phenyl- : 1565
cis-13-Docosenol : 5284
Docosylcyclopentane : 4808
2,4-Dodecadienoic acid, 11-methoxy-3,7,11-trimethyl-, 1-methylethyl ester, (*E*,*E*)- : 7651
2,4-Dodecadienoic acid, 3,7,11-trimethyl-, ethyl ester, (*E*,*E*)- : 7182
Dodecahydrophenanthrene : 9178
Dodecamethylcyclohexasiloxane : 4624
Dodecamethylpentasiloxane : 8997
Dodecanal dimethyl acetal : 5297
1,12-Dodecanediol, dimethyl ether : 5298
Dodecane, 1,1-di-(4-tolyl) : 1567
Dodecane, 1-(1-naphthyl)- : 7879
Dodecane, 1-(2-naphthyl)- : 7880
N-Dodecanoyl-*N*-methylglycine sodium salt : 6269
Dodecanoyl peroxide : 9155
Dodecylmethylketone : 11633
2-Dodecylnaphthalene : 7880
Dodecyltrichlorosilane : 11460
Donaxine : 7310
Dormin : 8679
2,6,8,10,12,14,16,18,20,22,24,26,30-Dotriacontatridecaene, 2,6,10,14,19,23,27,31-octamethyl-, (all-E)- : 4053
Dotriacontylcyclohexane : 4383
Dowicide 9 : 9255
Doxenitoin : 7228
Draconic acid : 2666
Dromostanolone propionate : 351
Drosophilin A : 9472
DTBP : 9153
Dulcin : 12013
Dulcitol : 6194
Dulcose : 6194
Duotal : 9384
Duracaine : 2493
Durene : 2280
Durene aldehyde : 598
Durohydroquinone : 1552
Duroquinone : 4270
1,2-Dyclodecanedione : 4203
Dyphylline : 10761
Dypnone : 3876
Ecgonidine : 491
Ecgonine : 483
Echinochrome A : 7857
Echinopsine : 11330
Echitamine : 4730
Edetic acid : 6262
EEDQ : 11231
Efloxate : 225
Egonol : 2464
Ehrlich's reagent : 550
Eicosamethylnonasiloxane : 8178
Eicosane, 1-cyclohexyl- : 4384
Eicosane, 9-ocytl : 5354
Eicosane, 5-phenyl- : 1119
Eicosane, 3-phenyl- : 1695
Eicosane, 2-phenyl- : 2007
Eicosane, 9-phenyl- : 2068
Eicosane, 4-phenyl- : 2170
cis-5,8,11,14-Eicosatetraenoic acid : 5361
Δ 11-Eicosenoic acid : 5363
5,8,11,14-Eicosotetraenoic acid : 5360
Eiver-pick acid : 7868
Elaidic acid : 8297
Electrocortin : 9746
β-Elemene : 4389
Eleostearic acid : 8246
Embelin : 4261
Emetine : 5368
Emodin : 419
Enallylpropymal : 11099
Enanthaldoxime : 6346
Enanthylic acid : 6455
Enheptin : 11666
Enidin : 2847
ENTPROL : 10379
Eosin : 11525
Ephdine : 2849
(±)-Ephedrine : 1900
Epiandrosterone : 347
Epiborneol, acetate : 2947
DL-α-Epichlorohydrin : 8605
Epicholestanol : 4108
Epicholesterol : 4112
3,17-Epidihydroxyestratriene : 5399
Epiestradiol : 5400
16-Epiestriol : 5405
Epihydrinaldehyde : 8603
Epinephrine : 1521
Epiquinidine : 4125
EPN : 9551
1,2-Epoxybutane : 8615
2,3-Epoxypropanal : 8603
1,2-Epoxypropane : 8621
2,3-Epoxy-1-propanol : 8618
2,3-Epoxypropyl ethyl ether : 8614
Epsilan : 2787
Epsilcapramin : 6873
Equilenin : 5397
Equilin : 5398
Equol : 2786
Ergobasinine : 5373
Ergocornine : 5387
Ergocorninine : 5388
Ergocristine : 5391
Ergocristinine : 5392
Ergocryptine : 5389
Ergocryptinine : 5390
Ergoline-8α-carboxamide, 9,10-didehydro-*N*-((*S*)-2-hydroxy-1-methylethyl)-6-methyl- : 5373
Ergometrinine : 5373
Ergosine : 5393
Ergostanol : 5379
γ-Ergostenol : 5384
α-Ergostenol : 5385
Ergosterol : 5383
Ergotamine : 5394
Ergotaminine : 5395
Ergothioneine : 7204
Eriodictyol : 2812
Eriodictyol 4'-monomethyl ether : 2800
Erucic acid : 5283
Erucyl alcohol : 5284
Erysimupicrone : 4049
Erythraric acid : 3304
meso-Erythritol : 3431
Erythritol : 3431
Erythritol anhydride : 3007
Erythrityl tetranitrate : 3432
β-Erythroidine : 8567
D-Erythrose : 3192
Erythrosin : 11524
L-Erythrulose : 3679
Erytrin : 31
Esculetin : 2804
Esculin : 2822
Esculoside : 2822
d-3,17β-Estradiol : 5399
Estradiol : 5399
α-Estradiol : 5400
17β-Estradiol 3-benzoate : 5403
Estradiol benzoate : 5403
Estragole : 1943
Estra-1,3,5(10)-triene-3,17β-diol : 5399
Estra-1,3,5(10)-triene-3,17α-diol : 5400
Estra-1,3,5(10)-triene-3,16α,17β-triol : 5404
Estratriol : 5404
Estriol : 5404
Estrone : 5406
Ethacridine : 317
Ethacrynic acid : 168
Ethamivan : 611
Ethanal : 11
Ethanamide : 22
Ethanamidine hydrochloride : 5671
Ethanearsonic acid : 457
Ethane, 1,2-bis(di-2-hydroxypropylamino) : 10379
Ethane, 1-chloro-2-iodo-1,1,2-trifluoro : 5490
1,2-Ethanedicarboxylic acid : 3294
Ethane, 1,1-dichloro-2,2-bis(*p*-ethylphenyl)- : 9160
Ethane, 1,2-dicyclohexyl- : 4386
Ethane, 1,1-dicyclohexyl- : 4392
Ethane, 1,2-dinitro-1,1,2,2-tetrafluoro : 5642
Ethanedione : 5491
1,2-Ethanediylbis(chloroformate) : 3366
Ethane, 1-ethoxy-2-methylamino : 5426
Ethane, 1-ethoxy-2-phenoxy- : 1652
Ethanenitrile : 271
Ethane, perbromo : 5594
Ethanephosphonic acid, diethyl ester : 9536
Ethane, 1,1,1,2-tetraphenyl- : 1618
Ethanethiol, 1-phenyl (1) : 1834
Ethanimidothioic acid, 2-(dimethylamino)-*N*-[[(methylamino)carbonyl]oxy]-2-oxo-, methyl ester : 8568
3,5-Ethano-3H-pyrrolo[2,3-d]carbazole, curan-17-ol deriv. : 4154
Ethanoic acid : 100
Ethanoic anhydride : 107
Ethanolamine : 5675
Ethanolamine, *N*,*N*-diisobutyl : 5684
Ethanolamine, *N*-isobutyl : 5747
Ethanolamine, *N*-(β-naphthyl) : 5750
Ethanol, 2-(2,4-dichlorophenoxy)-, hydrogen sulfate, sodium salt : 11358
Ethanol, 1(3,4-dihydroxyphenyl)-2(isopropylamino) : 1522
Ethanol, 2-(diisobutylamino)- : 5684
Ethanol, 2-(diisopentylamino)- : 5682
Ethanol, 2,2-diphenyl- : 1615
Ethanol, 2-(isobutylamino)- : 5747
Ethanol, 2,2'-oxybis-, diethyl ether : 5612
Ethanol, 1(3-tolyl) : 1858
Ethanoylaminoethanoic acid : 6251
Ethanoyl chloride : 294
11aH-12,3,6a-Ethanylylidene-7,9-methanonaphth[2,3-b]azocine, aconitane-3,8,13,14,15-pentol deriv. : 305
Ethenylmethylphenylcarbinol : 1869
Ethenzamide : 615
Ether, allyl *p*-chlorophenyl : 1232
Ether, allyl ethenyl : 10565
Ether, allyl ethyl : 10566
Ether, allyl isopropyl : 10579
Ether, allyl methyl : 10576
Ether, allyl phenyl : 2167
Ether, *p*-allylphenyl methyl : 1943
Ether, allyl propyl : 10594
Ether, allyl *o*-tolyl : 2026
Ether, allyl *m*-tolyl : 2027
Ether, allyl *p*-tolyl : 2028
Ether, allyl vinyl : 10565
Ether, benzyl butyl : 1115
Ether, benzyl chloromethyl : 1181
Ether, benzyl ethyl : 1658
Ether, benzyl isobutyl : 2033

Ether, benzyl isopentyl : 1951
Ether, benzyl methyl : 1930
Ether, benzyl (3-methylbutyl) : 1951
Ether, bis(2-bromoethyl) : 5609
Ether, bis(bromomethyl) : 7524
Ether, bis(*p*-bromophenyl) : 2072
Ether, bis(2-butoxyethyl) : 3417
Ether, bis(1-chloroethyl) : 5610
Ether, bis(chloroethyl) : 5611
Ether, bis(2-chloroethyl) : 5611
Ether, bis(chloromethyl) : 7525
Ether, bis(2-chloro-1-methylethyl) : 10123
Ether, bis(*p*-chlorophenyl) : 2073
Ether, bis(2-chloropropyl) : 10121
Ether, bis(3-chloropropyl) : 10122
Ether, bis(2-ethoxyethyl) : 5612
Ether, bis[2-(2-methoxyethoxy)ethyl] : 8995
Ether, bis(2-methoxyethyl) : 5613
Ether, bis(*o*-methoxyphenyl) : 2075
Ether, 2-bromethyl phenyl : 1033
Ether, 4-bromobutyl phenyl : 1003
Ether, 2-bromocyclohexyl methyl : 4317
Ether, 2-bromoethyl ethyl : 5461
Ether, 2-bromoethyl methyl : 5465
Ether, 6-bromohexyl methyl : 6736
Ether, bromomethyl ethyl : 5464
Ether, bromomethyl methyl : 7472
Ether, 4-bromophenyl hexyl : 1045
Ether, *p*-bromophenyl hexyl : 1045
Ether, *p*-bromophenyl phenyl : 1079
Ether, 3-bromopropyl phenyl : 1085
Ether, 1,3-butadienyl ethyl : 3160
Ether, 2,3-butadienyl methyl : 3167
Ether, 1,3-butadienyl methyl : 3168
Ether, 1-buten-3-ynyl methyl : 3891
Ether, butyl 2-chloroethyl : 3249
Ether, butyl ethyl : 3372
Ether, *sec*-butyl ethyl : 3373
Ether, *tert*-butyl ethyl : 10040
Ether, butyl ethynyl : 3377
Ether, butyl isobutyl : 3401
Ether, butyl isopropyl : 3399
Ether, *tert*-butyl isopropyl : 10081
Ether, butyl methyl : 3391
Ether, *tert*-butyl methyl : 10072
Ether, butyl phenyl : 1114
Ether, *tert*-butyl phenyl : 1458
Ether, *sec*-butyl phenyl : 2030
Ether, butyl propyl : 3421
Ether, *tert*-butyl propyl : 10087
Ether, butyl *m*-tolyl : 1116
Ether, butyl *p*-tolyl : 1117
Ether, butyl vinyl : 3369
Ether, *sec*-butyl vinyl : 3370
Ether, *tert*-butyl vinyl : 10036
Ether, 2-butynyl methyl : 3915
Ether chloratus : 5467
Ether, 1-chloroethyl 2-chloroethyl : 5468
Ether, 1-chloroethyl ethyl : 5473
Ether, 2-chloroethyl ethyl : 5474
Ether, 1-chloroethyl methyl : 5479
Ether, 2-chloroethyl methyl : 5480
Ether, 1-chloroethyl propyl : 9854
Ether, 2-chloroethyl vinyl : 5887
Ether, chloromethyl ethyl : 5478
Ether, 2-chloro-1-methylethyl methyl : 9859
Ether, chloromethyl methyl : 7483
Ether, chloromethyl propyl : 9858
Ether, *p*-chlorophenyl phenyl : 1221
Ether, 3-chloropropyl methyl : 9860
Ether, 2-chloropropyl methyl : 9861
Ether, 2-chloro-α,α,α-trifluoro-*p*-tolyl 3-ethoxy-4-nitrophenyl : 8633
Ether, 2-chloro-α,α,α-trifluoro-*p*-tolyl *p*-nitrophenyl : 8082
Ether, 2-chlorovinyl ethyl : 5886
Ether, cinnamyl ethyl : 1666
Ether cyanatus : 10090
Ether, cyclohexyl methyl : 4425
Ether, cyclohexyl phenyl : 1253
Ether, cyclopropyl methyl : 4980
Ether, decyl vinyl : 5051
Ether, diallyl : 10592
Ether, dibenzyl : 2079
Ether, 1,2-dibromoethyl ethyl : 5523
Ether, 2,3-dibromopropyl methyl : 9888
Ether, dibutyl : 3415
Ether, di-*tert*-butyl : 10124
Ether, 1,2-dichloroethyl ethyl : 5533
Ether, 2,4-dichlorophenyl *p*-nitrophenyl : 8081
Ether, 1,2-dichlorovinyl ethyl : 5899
Ether, dicyclohexyl : 4461
Ether, diethyl : 5608
Ether, diethyl,2-amino : 5423
Ether, diethyl,2-bromo : 5461
Ether, diethyl,1-chloro : 5473
Ether, diethyl,2-chloro : 5474
Ether, diethyl,2,2'-dibromo : 5609
Ether, diethyl,1,2,-dichloro : 5533
Ether, diethyl,2,2,'-dichloro : 5611
Ether, diethyl,2,2'-dicyano : 10114
Ether, diethyl,2,2'-diphenyl : 2074
Ether, diethyl,1,1'-diphenyl(*DL*) : 2081
Ether, difurfuryl : 6156
Ether, dihexdecyl : 6642
Ether, dihexyl : 6842
Ether, dihexyl,2,2'-diethyl : 6431
Ether, diisopropyl : 10120
Ether, diisopropyl,β,β'-dichloro : 10123
Ether, dimethyl : 7523
Ether, dimethyl,chloro : 7483
Ether, 2,2'-dinaphthyl : 7940
Ether, 2,4-dinitrophenyl phenyl : 1495
Ether, 2,4'-dinitro-4-trifluoromethyldiphenyl : 5980
Ether, dioctyl : 8416
Ether, dipentyl : 8835
Ether, diphenyl : 2071
Ether, diphenyl,4-bromo : 1079
Ether, diphenyl,4-chloro : 1221
Ether, diphenyl,4,4'-diamino : 832
Ether, diphenyl,4,4'-dibromo : 2072
Ether, diphenyl,4,4'-dichloro : 2073
Ether, diphenyl,2,2'-dimethoxy : 2075
Ether, diphenyl,2,2'-dimethyl : 2076
Ether, diphenyl,3,3'-dimethyl : 2077
Ether, diphenyl,4,4'-dimethyl : 2078
Ether, diphenyl,2-methoxy : 1938
Ether, diphenyl,3-methyl : 2013
Ether, diphenyl,2-nitro : 2053
Ether, diphenyl,4-nitro : 2054
Ether, dipropyl : 10119
Ether, dipropyl,2,2'-dichloro : 10121
Ether, dipropyl,3,3'-dichloro : 10122
Ether, divinyl : 5912
Ether, 2,3-epoxypropyl ethyl : 8614
Ether, 2,3-epoxypropyl phenyl : 8623
Ether, ethyl ethynyl : 5942
Ether, ethyl furfuryl : 6104
Ether, ethyl heptyl : 6406
Ether, ethyl hexyl : 6814
Ether, 2-ethylhexyl vinyl : 6405
Ether, ethyl isobutyl : 10039
Ether, ethyl isopentyl : 3374
Ether, ethyl isopropyl : 10038
Ether, ethyl methyl : 5602
Ether, ethyl 1-methyleneallyl : 3161
Ether, ethyl octyl : 8402
Ether, ethyl *tert*-pentyl : 3375
Ether, ethyl pentyl : 8808
Ether, ethyl propyl : 10037
Ether, ethyl 1-propynyl : 10732
Ether, ethyl 2-propynyl : 10733
Ether, ethyl styryl : 1651
Ether, ethyl vinyl : 5905
Ether, ethynyl methyl : 5944
Ether, ethynyl phenyl : 1716
Ether, ethynyl propyl : 10045
Ether, furfuryl methyl : 6128
Ether, 2-furyl phenyl : 6157
Ether, heptyl methyl : 6421
Ether, heptyl phenyl : 1747
Ether, hexadecyl phenyl : 1752
Ether, hexadecyl vinyl : 6638
Ether, hexyl phenyl : 1765
Ether hydrochloric : 5467
Ether, iodomethyl methyl : 7519
Ether, isobutyl methyl : 10071
Ether, isobutyl phenyl : 2031
Ether, isobutyl propyl : 10086
Ether, isobutyl vinyl : 10035
Ether, isopentyl methyl : 3395
Ether, isopentyl vinyl : 3371
Ether, isopropyl methyl : 10070
Ether, isopropyl phenyl : 1974
Ether, isopropyl propyl : 10076
Ether, isopropyl *m*-tolyl : 1987
Ether, isopropyl vinyl : 10034
Ether, *m*-methoxyphenyl *o*-methoxyphenyl : 1929
Ether, methyl 2-methylallyl : 10577
Ether, methyl 2-methylbutyl : 3394
Ether, methyl 1-methyleneallyl : 3169
Ether, methyl 2-octynyl : 8536
Ether, methyl *tert*-pentyl : 3396
Ether, methyl pentyl : 8822
Ether, methyl propyl : 10069
Ether, methyl 2-propynyl : 10736
Ether, methyl styryl : 1928
Ether, methyl vinyl : 5908
Ether muriatic : 5467
Ether, 1-naphthyl 2-naphthyl : 7928
Ether, 1-naphthyl phenyl : 7945
Ether, 2-naphthyl phenyl : 7946
Ether, *o*-nitrophenyl phenyl : 2053
Ether, *p*-nitrophenyl phenyl : 2054
Ether, *p*-nitrophenyl *p*-tolyl : 2006
Ether, *p*-nitrophenyl α,α,α-trifluoro-2-nitro-*p*-tolyl : 5980
Ether, octadecyl vinyl : 8240
Ether, octyl phenyl : 2069
Ether, octyl vinyl : 8401
Ether, phenyl *m*-tolyl : 2013
Ether, phenyl propyl : 2168
Ether, phenyl *o*-tolyl : 2012
Ether, phenyl vinyl : 1646
Ether, propyl vinyl : 10033
Ethide : 5536
17α-Ethinyl-17β-estradiol : 8216
Ethionamide : 10885
Ethionine : 991
Ethisterone : 9758
Ethohexadiol : 6802
Ethosuximide : 11158
α-Ethoxyacetaldehyde : 14
2-Ethoxyacetic acid : 180
Ethoxyacetic acid methyl ester : 182
2-Ethoxyacetophenone : 5814
o-Ethoxybenzonitrile : 2754
6-Ethoxy-2-benzothiazolesulfonamide : 2865
2-Ethoxybenzylamine : 1808
3-Ethoxybiphenyl : 3071
N-(Ethoxycarbonyl)aniline : 3965
Ethoxycarbonyl chloride : 4018
Ethoxydiethylphosphine oxide : 9527
2-Ethoxy-2,3-dihydro-3,3-dimethyl-5-benzofuranyl methanesulfonate : 5933
2-Ethoxy-1,3-dinitrobenzene : 1650
2-Ethoxyethyl acetate : 5718
(±)-2-Ethoxy-1-ethyl-2-oxoethyl 5-[2-chloro-4-(trifluoromethyl)phenoxy]-2-nitrobenzoate : 7410
5-Ethoxy-2-hydroxybenzaldehyde : 556
(±)-2-[1-(Ethoxyimino)butyl]-5-[2-(ethylthio)propyl]-3-hydroxy-2-cyclohexen-1-one : 11359
1-Ethoxy-2-methoxyethane : 5591
o-Ethoxy-β-methylstyrene : 1665
2-Ethoxynaphthalene : 7884
2-Ethoxy-1-propanol : 10380
1-Ethoxy-2-propanol : 10381
2-Ethoxypropionitrile : 10102
Ethoxyquin : 11265
8-Ethoxy-5-quinolinesulfonic acid : 11293
Ethoxytrifluorosilicon : 11429
Ethoxzolamide : 2865
N-Ethylacetamide : 61
Ethyl acetate : 186
Ethylacetic acid : 3460
Ethyl acetoacetate : 3571
Ethyl acetylacetate : 3571
Ethylacetylene : 3897
Ethyl acrylate : 10635
Ethyl adipate : 6785
Ethyl alcohol : 5672
Ethylallene : 8658
3-Ethyl-1-allyl-6-aminouracil : 11060
1-Ethylallyl bromide : 9009
1-Ethylallyl chloride : 9017

γ-Ethylallyl chloride : 9020
Ethyl allyl ether : 10566
Ethylamine : 5410
Ethyl aminoacetate : 6263
Ethyl aminobenzoate : 2497
Ethylammonium chloride : 5433
N-Ethylamphetamine : 1576
Ethylamylcarbinol : 8445
Ethyl *tert*-amyl ether : 3375
Ethyl amyl ether : 8808
Ethyl amyl ketone : 8472
2-Ethylaniline : 740
p-Ethylaniline : 742
N-Ethylaniline : 743
Ethyl anisate : 2671
9-Ethylanthracene : 420
Ethyl anthranilate : 2495
Ethyl arachate : 5356
Ethyl arachidate : 5356
Ethyl arsenate : 439
Ethyl arsenite : 440
Ethylarsine : 444
Ethylarsonic acid : 457
Ethyl behenate : 5277
Ethylbenzene : 1668
Ethyl benzeneacetate : 899
Ethyl benzenesulfonate : 2223
Ethyl benzoate : 2602
Ethyl benzoin ether : 5813
Ethyl benzoylacetate : 2141
α-Ethylbenzyl alcohol : 1870
Ethylbenzylamine : 1810
Ethylbenzylaniline : 1811
Ethyl benzyl ether : 1658
Ethyl benzyl ketone : 3677
Ethyl bromide : 5454
Ethyl bromoacetate : 115
2-Ethyl-2-bromobutyramide : 3198
Ethyl bromochloroacetate : 113
2-Ethyl-2-bromo-3-methylbutyramide : 3199
Ethyl α-bromopropionate : 10207
Ethyl butanoate : 3515
2-Ethyl-1-butanol : 3620
2-Ethyl-1-butene : 8825
Ethyl 3-butenoate : 3841
2-Ethylbutyl acetate : 185
Ethylbutylamine : 3218
Ethyl butyl ether : 3372
Ethyl *sec*-butyl ether : 3373
Ethyl *tert*-butyl ether : 10040
Ethyl butyl ketone : 6511
Ethyl *sec*-butyl ketone : 6980
Ethyl *tert*-butyl ketone : 8966
Ethyl butyl sulfide : 3376
Ethyl butyrate : 3515
2-Ethylbutyric acid : 3514
α-Ethylbutyric acid : 3514
Ethyl caprate : 5063
Ethyl caproate : 6884
Ethyl carbanilate : 3965
Ethyl carbitol : 5720
Ethyl carbonate : 3999
Ethyl Cellosolve : 5717
Ethyl cellosolve acetate : 5718
Ethyl chloride : 5467
Ethyl chloroacetate : 130
Ethyl chlorocarbonate : 4018
Ethyl β-chlorocrotonate : 3823
Ethyl chloroethanoate : 130
Ethyl chloroformate : 4018
Ethyl 5-chloro-2-furancarboxylate : 6053
Ethyl 2-chloro-3-methylbutanoate : 3490
Ethyl chloromethyl ether : 5478
Ethyl α-chloropropionate : 10220
Ethyl β-chloropropionate : 10221
Ethyl chlorosilicate : 11462
Ethyl chlorosulfate : 4085
Ethyl 4-chlorovalerate : 8867
Ethyl 5-chlorovalerate : 8868
Ethyl *trans*-cinnamate : 10685
Ethyl cinnamylideneacetate : 8680
Ethyl crotonate : 3838
Ethyl *trans*-crotonate : 3839
cis-Ethyl crotonate : 3840
Ethyl cyanate : 4163
Ethyl cyanide : 10090
Ethyl cyanoacetate : 152
Ethylcyclobutane : 4191
Ethyl cyclobutanecarboxylate : 4175
Ethyl cycloheptanecarboxylate : 4224
Ethylcyclohexane : 4390
1-Ethylcyclohexanol : 4522
4-Ethylcyclohexanol : 4525
4-Ethylcyclohexanone : 4593
3-Ethyl-1-cyclohexene : 4661
4-Ethyl-1-cyclohexene : 4662
2-Ethylcyclohexylamine : 4287
Ethyl cyclohexyl ketone : 10423
Ethylcyclopentane : 4812
4-Ethylcyclopentene : 4926
Ethylcyclopropane : 4974
Ethyl cyclopropanecarboxylate : 4961
Ethyl cyclopropylcarboxylate : 4961
Ethyl 2-diazopropionate : 10236
Ethyl dibromoacetate : 160
Ethyl dibromofluoroacetate : 161
Ethyl 2,3-dibromopentanoate : 8872
Ethyl dibunate : 7968
Ethyl dichloroacetate : 164
Ethyl dichlorophosphate : 9577
Ethyl 3-(diethylamino)propionate : 328
Ethyl diethylphosphinate : 9527
Ethyl difluoroacetate : 176
9-Ethyl-9,10-dihydroanthracene : 421
Ethyl 2,3-dimethyl-2-butenoate : 3836
Ethyl 4,4-dimethylcrotonate : 9080
3-Ethyl-2,2-dimethylhexane : 6816
3-Ethyl-2,3-dimethylhexane : 6817
3-Ethyl-2,4-dimethylhexane : 6818
3-Ethyl-2,5-dimethylhexane : 6819
3-Ethyl-3,4-dimethylhexane : 6820
4-Ethyl-2,3-dimethylhexane : 6822
4-Ethyl-2,4-dimethylhexane : 6823
4-Ethyl-3,3-dimethylhexane : 6824
Ethyl 2,2-dimethylpropionate : 10259
Ethyl 3,5-dinitrobenzoate : 2593
Ethylene : 5879
Ethylene, 1-bromo-2-chloro- : 5881
Ethylene, 1-bromo-1,2-dichloro (dis) : 5935
Ethylene, 1-bromo-1,2-difluoro- : 5883
Ethylene bromohydrin : 5685
Ethylenecarboxamide : 10519
Ethylenecarboxylic acid : 10609
Ethylene chlorohydrin : 5693
Ethylenediamine : 5496
Ethylenediamine, *N*-benzyl-*N'*,*N'*-dimethyl-*N*-phenyl- : 5506
Ethylenediamine, *N*,*N*-dimethyl-*N'*-ethyl : 5510
Ethylene dibromide : 5518
Ethylene, 1,1-dibromo-2-ethoxy : 5893
Ethylene dichloride : 5529
Ethylene dicyanide : 3291
Ethylene diethanoate : 5571
Ethylenedimercaptan : 5590
Ethylenedisulfonic acid : 5587
Ethylene fluorohydrin : 5727
Ethylene glycol : 5569
Ethylene glycol carbonate : 5211
Ethylene glycol diacetate : 5571
Ethylene glycol dibutyl ether : 5527
Ethylene glycol didodecanoate : 5311
Ethylene glycol diethyl ether : 5542
Ethylene glycol diisobutyl ether : 10032
Ethylene glycol dilaurate : 5311
Ethylene glycol dimethyl ether : 5549
Ethylene glycol dinitrate : 5574
Ethylene glycol, dinitrite : 8086
Ethylene glycol, methyl ethyl ether : 5591
Ethylene glycol, monoacetate : 5577
Ethylene glycol monoallyl ether : 5763
Ethylene glycol monobenzyl ether : 5761
Ethylene glycol monobutyl ether : 5687
Ethylene glycol monoethyl ether : 5717
Ethylene glycol monoethyl ether acetate : 5718
Ethylene glycol monomethyl ether : 5733
Ethylene glycol monopentyl ether : 5757
Ethylene glycol monosalicylate : 2622
Ethylene glycol, monovalerate : 8885
Ethyleneimine : 498
Ethylene iodohydrin : 5731
1,8-Ethylenenaphthalene : 6
Ethylene nitrite : 8086
Ethylene oxide : 8599
1,4-Ethylenepiperazine : 5111
Ethylenethiourea : 7225
Ethyleneurea : 7227
Ethyl 2-ethoxybenzoate : 2598
Ethyl 4-ethoxybenzoate : 2599
Ethylethynylcarbinol : 9140
17β-Ethyletiocholane : 9741
Ethyl fluoride : 5593
N-Ethylformamide : 5994
Ethyl formate : 6007
2-Ethylfuran : 6105
Ethyl 2-furyl ether : 6103
γ-Ethyl *L*-glutamate : 6245
Ethyl glycidyl ether : 8614
3-Ethylheptane : 6407
4-Ethylheptane : 6408
Ethyl heptyl ketone : 5076
2-Ethylhexanal : 6719
3-Ethylhexane : 6815
Ethyl hexanoate : 6884
2-Ethylhexanoic acid : 6882
3-Ethylhexanoic acid : 6883
2-Ethyl-2-hexanol : 6500
2-Ethyl-1-hexanol : 6930
2-Ethyl-2-hexene : 6555
3-Ethyl-1-hexene : 7033
Ethyl 2-hexenoate : 7053
Ethyl 3-hexenoate : 7054
2-Ethylhexyl acetate : 187
2-Ethylhexyl acrylate : 10636
2-Ethyl-1-hexylamine : 6726
Ethyl hexyl ketone : 8173
2-Ethylhexyl methacrylate : 10659
Ethyl hydrogen peroxide : 7179
Ethyl hydrogen phosphonate, aluminum salt : 6020
Ethyl hydrogen sulfate : 11576
Ethyl 4-hydroxybenzoate : 2621
Ethyl 2-hydroxy-3-butenoate : 3843
N-Ethylhydroxylamine : 5434
Ethyl 4-hydroxypentanoate : 8884
Ethyl hypochlorite : 7184
Ethylidenecyclohexane : 4391
2-Ethylidenecyclohexanone : 4594
Ethylidenecyclopentane : 4813
Ethylidene diacetate : 5570
Ethylidene dibromide : 5517
Ethylidene dichloride : 5528
Ethylidene diethyl ether : 5541
1-Ethylindan : 7273
Ethyl iodide : 5597
o-Ethyliodobenzene : 1680
Ethyl isobutanoate : 10298
Ethyl isobutyl sulfide : 10044
Ethyl isocyanate : 5598
Ethyl isopentyl ether : 3374
1-Ethyl-2-isopropylacetylene : 7122
Ethylisopropylamine : 9810
Ethyl isopropyl ether : 10038
Ethyl isopropyl ketone : 8982
(*RS*)-5-Ethyl-2-(4-isopropyl-4-methyl-5-oxo-2-imidazolin-2-yl)nicotinic acid : 7190
Ethyl isopropyl sulfide : 10043
Ethyl isothiocyanate : 5600
Ethyl isovalerate : 3548
Ethyl laurate : 5312
Ethyl levulinate : 8907
Ethyl linoleate : 8226
N-Ethylmaleimide : 11122
Ethyl mercaptan : 5648
Ethyl methacrylate : 10658
Ethyl methanesulfonate : 7531
Ethyl methoxyacetate : 206
1-Ethyl-1-methylallene : 8666
Ethyl methyl amine, hydrochloride : 5435
Ethyl 2-(methylamino)benzoate : 2680
1-Ethyl-2-methylbenzimidazole : 2381
2-Ethyl-3-methylbut-1-ene : 3797
Ethyl methyl carbonate : 4003
cis-1-Ethyl-3-methylcyclohexane : 4397
cis-1-Ethyl-4-methylcyclohexane : 4399
trans-1-Ethyl-4-methylcyclohexane : 4400
1-Ethyl-2-methylcyclohexanol : 4526
1-Ethyl-3-methylcyclohexanol : 4527

2-Ethyl-2-methylcyclohexanone	: 4595
1-Ethyl-4-methylcyclohexene	: 4663
1-Ethyl-1-methylcyclopentane	: 4814
cis-1-Ethyl-2-methylcyclopentane	: 4815
cis-1-Ethyl-3-methylcyclopentane	: 4817
trans-1-Ethyl-3-methylcyclopentane	: 4818
2-Ethyl-5-methylcyclopentanone	: 4883
cis-1-Ethyl-2-methylcyclopropane	: 4976
Ethyl methyl ether	: 5602
N-Ethyl-*N'*-(1-methylethyl)-6-(methylthio)1,3,5-triazine-2,4-diamine	: 341
2-Ethyl-5-methylfuran	: 6106
3-Ethyl-3-methylglutaric acid	: 8777
3-Ethyl-4-methylheptane	: 6411
3-Ethyl-5-methylheptane	: 6412
4-Ethyl-2-methylheptane	: 6413
4-Ethyl-3-methylheptane	: 6414
4-Ethyl-4-methylheptane	: 6415
5-Ethyl-5-methyl-3-heptyne	: 6619
3-Ethyl-2-methylhexane	: 6825
3-Ethyl-3-methylhexane	: 6826
4-Ethyl-3-methylhexane	: 6827
3-Ethyl-4-methylhexane	: 6827
4-Ethyl-2-methylhexane	: 6828
Ethyl 3-methylhexanoate	: 6899
Ethyl 4-methylhexanoate	: 6900
2-Ethyl-3-methylhexanoic acid	: 6886
3-Ethyl-3-methylhexanoic acid	: 6887
Ethyl 4-methyl-5-oxohexanoate	: 6902
3-Ethyl-2-methylpentane	: 8813
3-Ethyl-3-methylpentane	: 8814
3-Ethyl-3-methyl-1-pentene	: 9047
3-Ethyl-2-methyl-2-pentene	: 9049
Ethyl 4-methyl-2-pentenoate	: 9080
Ethyl 3-methyl-3-pentenoate	: 9081
Ethyl 4-methyl-3-pentenoate	: 9082
3-Ethyl-3-methyl-1-pentyne	: 9132
1-Ethyl-1-methylpropylbenzene	: 1691
1-Ethyl-3-methylpyrazole	: 10833
2-Ethyl-3-methylquinoline	: 11269
2-Ethyl-4-methylquinoline	: 11270
Ethyl methyl sulfide	: 5606
2-Ethyl-3-methylthiophene	: 11761
2-Ethyl-5-methylthiophene	: 11762
N-Ethylmorpholine	: 7694
1-Ethylnaphthalene	: 7885
2-Ethylnaphthalene	: 7886
Ethyl 1-naphthylacetate	: 7735
Ethyl nitrate	: 8074
Ethyl nitrite	: 8087
Ethyl nitroacetate	: 218
O-Ethyl *O*-*p*-nitrophenyl benzenethiophosphonate	: 9551
2-Ethyl-2-nitro-1,3-propanediol	: 10008
N-Ethyl-*N*-nitrosoethanamine	: 5432
N-Ethyl-*N*-nitrosourea	: 12015
Ethyl nonanoate	: 8150
N-Ethylnorephedrine	: 1871
3-Ethyloctane	: 8403
4-Ethyloctane	: 8404
Ethyl octanoate	: 8428
Ethyl octyl ketone	: 11959
Ethyl oenanthate	: 6462
Ethyl orthoformate	: 5604
Ethyl 3-oxo-3-phenylpropanoate	: 2141
Ethylparaben	: 2621
3-Ethylpentane	: 8809
2-Ethyl-2-pentanol	: 6946
2-Ethyl-1-pentene	: 9043
3-Ethyl-1-pentene	: 9044
3-Ethyl-2-pentene	: 9045
4-Ethyl-3-pentenoic acid	: 7060
Ethyl pentyl ether	: 8808
3-Ethyl-1-pentyne	: 9131
9-Ethylphenanthrene	: 9179
o-Ethylphenol	: 9353
m-Ethylphenol	: 9354
p-Ethylphenol	: 9355
Ethyl phenoxyacetate	: 229
Ethyl 4-phenoxybutyrate	: 3581
Ethyl 2-(4-phenoxyphenoxy)ethylcarbamate	: 5952
Ethyl phenylacetate	: 899
Ethyl *N*-phenylcarbamate	: 3965
1-Ethyl-1-phenylhydrazine	: 7159
1-Ethyl-2-phenylhydrazine	: 7160
Ethyl 5-phenyl-2,4-pentadienoate	: 8680
Ethyl *N*-phenylurethane	: 3965
Ethyl phosphate	: 9563
3-Ethyl-4-picoline	: 10964
4-Ethyl-2-picoline	: 10965
5-Ethyl-2-picoline	: 10966
N-Ethylpiperidine	: 9667
4-Ethylpiperidine	: 9670
1-Ethyl-3-piperidinol	: 9715
5-Ethyl-5-(1-piperidyl)barbituric acid	: 11096
Ethyl 2-propenyl ether	: 10566
Ethylpropiolic acid	: 9137
Ethyl propionate	: 10263
1-Ethyl-2-propylacetylene	: 6611
1-Ethyl-4-propylbenzene	: 1703
1-Ethylpropyl chloride	: 8721
Ethyl propyl ether	: 10037
Ethyl propyl ketone	: 6960
Ethyl propyl sulfide	: 10042
2-Ethylpyridine	: 10956
3-Ethylpyridine	: 10957
4-Ethylpyridine	: 10958
2-Ethylquinoline	: 11266
3-Ethylquinoline	: 11267
6-Ethylquinoline	: 11268
Ethyl salicylate	: 2619
Ethyl silicate	: 11506
Ethyl succinate	: 3303
Ethyl sulfate	: 11576
Ethyl sulfite	: 11581
Ethyl sulfocyanate	: 11703
Ethylsulfonal	: 3230
Ethylsulfonic acid	: 5622
2-(Ethylsulfonyl)ethanol	: 5725
Ethylsulfonylethyl alcohol	: 5725
Ethyl 2,3,3,3-tetrafluoropropionate	: 10328
2-Ethyltetrahydrofuran	: 6107
3-Ethyltetrahydrothiophene	: 11766
1-Ethyltetralin	: 7888
2-Ethyltetralin	: 7889
6-Ethyltetralin	: 7891
Ethyl tetraphosphate	: 11645
Ethyl thiocyanate	: 11703
2-(Ethylthio)ethanol	: 5726
2-Ethylthiophene	: 11759
3-Ethylthiophene	: 11760
3-Ethylthiophenol	: 2297
o-Ethyltoluene	: 1685
m-Ethyltoluene	: 1686
p-Ethyltoluene	: 1687
Ethyl *p*-toluenesulfonate	: 2231
N-Ethyl-4-toluidine	: 748
Ethyl *m*-tolyl sulfide	: 1706
Ethyl tosylate	: 2231
Ethyl tribromoacetate	: 242
Ethyl trichloroacetate	: 248
Ethyl(trichloromethyl)carbinol	: 3643
Ethyltrichlorosilane	: 11463
Ethyltrifluorosilane	: 11432
Ethyltrimethoxysilane	: 11433
1-Ethyl-2,3,5-trimethylbenzene	: 1709
1-Ethyl-2,4,5-trimethylbenzene	: 1710
5-Ethyl-1,2,3-trimethylbenzene	: 1713
Ethyltrimethylethylene	: 9034
3-Ethyl-2,2,3-trimethylpentane	: 8815
3-Ethyl-2,3,4-trimethylpentane	: 8816
Ethyl undecanoate	: 11946
Ethyl undecylate	: 11946
Ethyl urethane	: 3948
O-Ethylurethane	: 3948
Ethyl valerate	: 8878
Ethylvanillin	: 555
Ethylvinylacetylene	: 7103
Ethyl vinyl ether	: 5905
Ethyl vinyl ketone	: 9104
3-Ethyl-*o*-xylene	: 1670
Ethynylbenzene	: 1715
α-Ethynylbenzyl alcohol	: 1876
α-Ethynylbenzyl carbamate	: 1877
1-Ethynyl-1-cyclopentanol	: 4866
Ethynyldimethylcarbinol	: 3922
Ethynylestradiol	: 8216
Ethynyl methyl ketone	: 3923
Ethynylmethylpropylcarbinol	: 7128
Ethynylsilane	: 11435
Etiocholanic acid	: 346
Etofylline	: 10758
Eucalyptol	: 8560
Eucarvone	: 4220
Eucazulen	: 505
4αH-Eudesma-5,11(13)-dien-12-oic acid, 8β-hydroxy-, γ-lactone	: 8059
Eugenol	: 9389
Euonymit	: 6194
Euparin	: 5825
Euprax	: 7229
Evasprin	: 1913
Evodiamine	: 7333
Exaltol	: 4760
Exaltolide	: 8562
Fagarine	: 6179
(*E*,*E*)-α-Farnesene	: 5329
α-Farnesene	: 5329
β-Farnesene	: 5331
Felbinac	: 3009
Fenamic acid	: 2715
Fenamisal	: 2502
Fenchane(*d*)	: 2931
α-Fenchene (*DL*)	: 2923
l-α-Fenchyl alcohol	: 2941
Fenibut	: 2128
Fenicol	: 1870
Fenpiprane	: 9658
Fentanyl	: 9801
Fenticlor	: 9477
Fenyramidol	: 1913
Ferrocene, 1,1'-bis(3-phenyl-1,1,3,3-tetramethyldisiloxanyl)	: 5961
Ferron	: 11295
Fisetin	: 2813
Flacitran	: 2814
Flavan	: 2782
Flavanone, 4'-methoxy-3',5,7-trihydroxy	: 2801
Flavanone, 3',5,7-trihydroxy-4'-methoxy-, 7-(6-*O*-α-*L*-rhamnosyl-*D*-glucoside)	: 2794
Flavone	: 2840
Flexin	: 2876
Flopropione	: 10489
Floxuridine	: 12034
Flufenamic acid	: 2729
Fluimucetin	: 5015
Fluometuron	: 12007
Fluonilid	: 11802
Fluorakil 100	: 63
3',6'-Fluorandiol	: 11522
Fluorene	: 5967
Fluorene, 2-bromo-	: 5968
9H-Fluorene-9-carboxylic acid, 2-chloro-9-hydroxy-	: 4079
Fluorene, 9-(2-chlorobenzylidene)	: 5970
Fluorene, 9-methyl-	: 5974
N-2-Fluorenylacetamide	: 64
Fluorescein	: 11522
Fluorescein, 4',5'-diiodo	: 11523
Fluorescin	: 2572
Fluorine cyanide	: 4167
Fluoroacetic acid amide	: 63
Fluoroacetic acid, methyl ester	: 192
Fluoroacetylene	: 5943
o-Fluoroaniline	: 752
m-Fluoroaniline	: 753
p-Fluoroaniline	: 754
Fluorobenzene	: 1717
4-Fluorobenzeneacetic acid	: 901
p-Fluorobenzoic acid	: 2605
Fluorobromomethane	: 7470
2-Fluoro-1,3-butadiene	: 3162
1-Fluorobutane	: 3378
2-Fluorobutane	: 3379
4-Fluoro-1-butanol	: 3621
4-Fluorobutyric acid	: 3927
Fluorochlorobromomethane	: 7466
α-Fluorocinnamic acid	: 10637
Fluorocyclohexane	: 4403
1-Fluorodecane	: 5052
10-Fluorodecanoic acid	: 5064
10-Fluoro-1-decanol	: 5074
Fluorodichloroacetonitrile	: 279
1-Fluoro-1,1-dichloroethane	: 5534
Fluorodichloromethane	: 7502
Fluorodiiodomethane	: 7516
Fluoroethanoic acid	: 189

Fluoroform : 7558
Fluoroformaldehyde : 6019
1-Fluorohexadecane : 6639
1-Fluorohexane : 6829
2-Fluorohexane : 6830
9α-Fluoro-11β-hydroxy-17-methyltestosterone : 355
Fluoroiodomethane : 7517
Fluoromethane : 7515
1-Fluoronaphthalene : 7893
α-Fluoronaphthalene : 7893
2-Fluoronaphthalene : 7894
β-Fluoronaphthalene : 7894
o-Fluoronitrobenzene : 1733
m-Fluoronitrobenzene : 1734
p-Fluoronitrobenzene : 1735
3-Fluoro-4-nitrotoluene : 1730
8-Fluorooctanoic acid : 8429
1-Fluoropentane : 8817
5-Fluoropentanoic acid : 8879
(*p*-Fluorophenyl)acetic acid : 901
p-Fluorophenylamine : 754
Fluoroplex : 11067
Fluoroprene : 3162
2-Fluoropropane : 10047
3-Fluoro-1-propanol : 10382
2-Fluoropropene : 10569
3-Fluoropropene : 10570
cis-1-Fluoropropene : 10567
trans-1-Fluoropropene : 10568
4-(Fluorosulfonyl)aniline : 2252
o-Fluorotoluene : 1726
m-Fluorotoluene : 1727
p-Fluorotoluene : 1728
Fluorotrichloromethane : 7555
5-Fluorouracil : 11067
5-Fluorouracil 2'-deoxyriboside : 12034
Fluoxymesterone : 355
Fluracil : 11067
Folic acid : 6243
Foligan : 10852
Folinic acid : 6244
Follestrol : 5406
Follicyclin : 5399
Formaldehyde, (2,2'-dichloroisopropyl) ethyl acetal : 9896
Formaldehyde dimethyl mercaptal : 7461
Formaldehyde, trimer : 11905
Formalin : 5987
Formamide, 1,1'-azobis- : 5124
Formamide, *N,N'*-[1,4-piperazinediylbis(2,2,2-trichloroethylidene)]bis- : 11902
Formamidine : 7563
Formanilide : 5999
Formic acid, isopropyl ester : 6011
Formononetin : 2830
N-Formylaniline : 5999
Formylcyclooctane : 4738
Formylformic acid : 223
2-Formylfuran : 6041
Formylpteroic acid : 2494
4-Formylquinoline : 11225
*N*5-Formyl-5,6,7,8-tetrahydrofolic acid : 6244
5-Formyl-5,6,7,8-tetrahydrofolic acid : 6244
3-Formylthiophene : 11734
1-Formyl-2,6,6-trimethyl-1-cyclohexene : 4636
Fraxetin : 2808
Fraxin : 2823
Freon 11 : 7555
Freon 12 : 7500
Freon 12B1 : 7464
Freon 12B2 : 7496
Freon 13 : 7486
Freon 13B1 : 7475
Freon 13I1 : 7559
Freon 14 : 7541
Freon 21 : 7502
Freon 22 : 7478
Freon 23 : 7558
Freon 31 : 7481
Freon 32 : 7504
Freon 41 : 7515
Freon 111 : 5618
Freon 112 : 5637
Freon 112a : 5618
Freon 113 : 5663
Freon 114 : 5538
Freon 114a : 5537
Freon 114B2 : 5525
Freon 115 : 5484
Freon 116 : 5596
Freon 121 : 5638
Freon 122 : 5659
Freon 123 : 5540
Freon 132 : 5532
Freon 133a : 5489
Freon 152 : 5545
Freon 215 : 10179
Freon 218 : 10118
Freon C 318 : 4194
Freon 1110 : 5917
Freon 1111 : 5923
Frigen 11 : 7555
Frigen 11A : 7555
Frigen 12 : 7500
Frigen 13 : 7486
Frigen 22 : 7478
Frigen 113 : 5663
Frigen 113A : 5663
Frigen 114 : 5538
Frigen 114A : 5537
Frigen 122 : 5659
Fuchsin, acid : 2218
Fucoxanthin : 4055
Fulvene : 4767
Fulvene, 6-vinyl(*trans*) : 4769
Fumaric acid : 3764
Fumaric acid, bromo- : 3770
Fumaric acid, diallyl ester : 3781
Fumaric acid dichloride : 3795
Fumaric acid, diisobutyl ester : 3767
Fumaric acid, diisopropyl ester : 3766
Fumaric acid, dipropyl ester : 3783
Fumarine : 3109
Fumaronitrile, chloro- : 3763
Fumaryl chloride : 3795
Fumigacin : 8211
Fumigatin : 4265
Furacrylic acid : 10638
2-Furalacetic acid : 10638
Furaldehyde : 6041
2-Furanacrylic acid : 10638
2-Furanacrylonitrile : 10585
Furan, 2-*tert*-butyl- : 6090
2-Furancarbinol : 6118
Furan, 2(2-cyanovinyl) : 10585
2-Furanmethanimine, 5-nitro-*N*-(2-oxo-3-oxazolidinyl)- : 8589
2-Furanpropionic acid, tetrahydro- : 6164
Furan, tetrahydro-2-ethyl : 6107
2-Furanylmethanol : 6118
Furazolidone : 8589
Furfural : 6041
Furfuran : 6025
Furfuryl alcohol : 6118
Furfuryl alcohol, tetrahydro-, stearate : 8275
Furfurylamine : 6113
Furfurylideneacetic acid : 10638
Furfuryl phenyl ether : 6158
Furfuryl valerate : 8880
Furilazone : 10904
α-Furildioxime : 5583
2-Furoic acid : 6050
β-Furoic acid, 5-bromo, ethyl ester : 6190
2-Furoic acid, 5-chloro-, ethyl ester : 6053
α-Furoic acid, hexyl ester : 6058
3-Furoic acid, 5-methyl- : 6065
α-Furoic acid, octyl ester : 6067
2-Furoic acid, pentyl ester : 6068
Furole : 6041
Furonazide : 10904
2-Furylacetic acid, butyl ester : 6027
2-Furylacetonitrile : 6031
3-(2-Furyl)acrylonitrile : 10585
2-Furylcarbinol : 6118
2-Furyl iodide : 6110
2-Furyl isopropyl ether : 6112
2-Furyl methyl ketone isonicotinoylhydrazone : 10904
2-Furyl octyl ether : 6135
2-Furyl phenyl ether : 6157
2-Furyl phenyl sulfide : 6160
Fusaric acid : 10898
Fussol : 63
Gainex : 2384
Galegine : 6283
Galipine : 11253
Gallacetophenone : 5873
Gallamine triethiodide : 5441
Gallein : 11526
Gallic acid : 2731
Gallic acid, 3-gallate : 2575
Gallodesoxycholic acid : 4097
Gallomide blue : 8062
m-Galloylgallic acid : 2575
Gamarex : 3464
Gardol : 6269
Gastrodigenin : 1884
Geissoschizoline : 4154
Gelseminic acid : 2829
Genetron 132b-B2 : 5522
Genetron 11 : 7555
Genetron 23 : 7558
Genetron 218 : 10118
Genistein : 2807
Genistein, 4'-β-*D*-glucopyranoside : 2824
Genisteol : 2807
Genite : 9289
Gentianic acid : 12054
Gentisic acid : 2570
Gentisin : 12054
Gentisyl alcohol : 1520
trans-Geraniol : 8333
Geraniol, tetrahydro (*DL*) : 8449
Geranyl alcohol : 8333
trans-Geranyl bromide : 8321
Germine : 4068
Geronic acid : 6460
Gibberellic acid : 6204
Gibberellin A3 : 6204
Gibberellin X : 6204
Gibbs' reagent : 4280
Gingerone : 3663
Gitogenin : 11530
Gitoxigenin : 4048
Gitoxin : 4042
d-Glaucine : 5157
Glucamine : 6207
Gluconic acid : 6209
D-Gluconic acid-1,5-lactone : 6212
Gluconic acid, sodium salt : 6213
δ-*D*-Gluconolactone : 6212
1,5-Gluconolactone : 6212
Glucopyranoside, genistein-4', β-*D*- : 2824
D-Glucosaccharic acid : 6205
D-Glucosamine : 6233
L-Glucose, methylamino : 6215
α-*D*-Glucose, pentamethyl ether : 6228
Glucurone : 6239
D-Glucuronolactone : 6239
Glutaconic acid, diethyl ester : 9041
α-Glutamic acid : 6241
Glutamic acid 5-amide : 6249
Glutamic acid, *N*-[*p*-[[(2-amino-5-formyl-5,6,7,8-tetrahydro-4-hydroxy-6-pteridinyl)methyl]amino]benzoyl]-, *L*- : 6244
Glutamic acid, 5-[2-(α-hydroxy-*p*-tolyl)hydrazide], *L*- : 6247
Glutamic acid, monosodium salt : 6248
γ-*L*-Glutamyl-*L*-cysteinylglycine : 6264
Glutaral : 8734
Glutaraldehyde : 8734
Glutaric acid : 8770
Glutaric acid, 3(3-benzoxy-4-methoxyphenyl) : 6892
Glutaric acid dialdehyde : 8734
Glutaric acid dinitrile : 8767
Glutaric acid, diphenyl ester : 8776
Glutaric acid, 3-ethyl-3-methyl- : 8777
Glutaric acid, 3-hydroxy-, diethyl ester : 8778
Glutaric acid, 3-methyl : 8779
Glutaronitrile : 8767
Glutatione : 6264
Glycamine : 6207
Glyceric acid : 10252

Glyceric acid, ethyl ester (*DL*) : 10253
Glyceric acid, methyl ester, *DL*- : 10254
Glyceric acid, 2-methyl, ethyl ester : 10255
Glyceric acid, 2-methyl-, methyl ester : 10256
Glycerin : 10186
Glycerin tributyrate : 3586
Glycerol : 10186
Glycerol acetonide : 5207
Glycerol β-chlorohydrin : 9995
Glycerol diacetate : 10189
Glycerol 1,3-dimethyl ether : 10372
Glycerol, 1,3-diphenyl ether-2-acetate : 10378
Glycerol guaiacolate : 10016
Glycerol, 1-laurate (*DL*) : 5310
Glycerol, 1-(2-methoxyphenyl)ether : 10016
Glycerol, 2-methyl ether : 10015
Glycerol, 1-monoacetate(*DL*) : 10187
Glycerol, 1-monododecanoate(*DL*) : 5310
Glycerol, 1-phenyl : 10194
Glycerol tributanoate : 3586
Glycerol trimethyl ether : 10185
Glyceryl *p*-aminobenzoate : 10188
Glycidaldehyde : 8603
Glycidol : 8618
Glycidyl allyl ether : 8626
Glycine anhydride : 9612
Glycine, *N*,*N*-bis(phosphonomethyl)- : 6277
Glycine, *N*-(chloroacetyl)-*N*-(2,6-diethylphenyl)-, ethyl ester : 5173
Glycine cholate : 6273
Glycine, *N*-(phosphonomethyl)- : 6276
Glycinol : 5675
Glycinonitrile, *N*,*N*-dimethyl- : 281
Glycocholic acid : 6273
Glycocyamine : 6253
Glycolaldehyde : 15
Glycol diacetate : 5571
Glycolic acid : 196
Glycolic aldehyde : 15
Glycol salicylate : 2622
Glyconitrile : 282
Glycopyrrolate : 11179
Glycopyrronium bromide : 11179
Glycosine : 11200
Glycosthene : 6250
Glycurone : 6239
Glycyl alcohol : 10186
α-Glycylglycine : 6265
Glycyrrhetinic acid glycoside : 6230
Glycyrrhizic acid : 6230
Glycyrrhizin : 6230
Glyoxal : 5491
Glyoxaldiureine : 7191
Glyoxaline-5-alanine : 7132
Glyoxylaldehyde : 5491
Glyoxylic acid : 223
Glyoxylic diureide : 12009
Gramine : 7310
Grasex : 21
Griseofulvin, (+)- : 11520
Guaiacol : 9379
Guaiacol benzoate : 9383
Guaiacol carbonate : 9384
Guaiacol glycerol ether : 10016
Guaiacol phosphate : 9387
Guaiacol valerate : 8886
Guaiazulene : 505
Guai-1(5)-en-11-ol : 507
Guaifenesin : 10016
Guaiol : 507
Guajen : 7825
Guanidine, dodecyl-, monoacetate : 5347
Guanidinoacetic acid : 6253
Guanine : 10768
Guanyl glycine : 6253
Guanyl hydrazine : 7147
Gulose, *D*- : 6290
Guvacine : 10912
Gynecormone : 5403
Halazone : 2565
Halon 2302 : 5526
Haloperidol : 3651
Halotestin : 355
Halothane : 5458
Harmaline : 11038
Harman : 11040
Harmine : 11039
Hecogenin : 11534
Hederagenin : 8545
Helenin : 8059
Helicin : 562
Heliotropic acid : 2418
Heliotropin : 2417
Helvolic acid : 8211
Hematein : 628
Hematin : 5959
Hematoidin : 3000
Hematoporphyrin : 9735
Hematoxylin : 629
Hemimellitic acid : 2318
Hemin : 5958
Hemipyocyanine : 9197
Hendecane : 11938
Heneicosylbenzene : 1741
Hepin : 6873
Heptachlor : 7576
1,1,1,2,2,3,3-Heptachloropropane : 10048
1,1,1,2,3,3,3-Heptachloropropane : 10049
Heptacosylbenzene : 1742
Heptadecane, 1-cyclohexyl- : 4404
Heptadecane, 9-phenethyl- : 2070
Heptadecane, 1-phenyl- : 1743
Heptadecyl alcohol : 6317
Heptadecylbenzene : 1743
Heptadecylcyclohexane : 4404
1,6-Heptadiene-4-ol, 4-methyl : 6338
1,6-Heptadiene-3-yne : 6341
Heptaldehyde : 6344
Heptamethylcyclotetrasiloxane : 4992
Heptamethylene chlorohydrin : 6477
Heptamethylene dibromide : 6374
Heptane, 1-cyclopentyl- : 4820
Heptane, 1,1-diphenyl- : 1746
Heptane, 4-isopropyl- : 6427
Heptane, 2-methyl-3-ethyl : 6409
4-Heptanol, 3-methyl(*DL*) : 6504
4-Heptanol-2-one, 3-methyl : 6518
3-Heptanol-5-one, 3-methyl : 6519
4-Heptanol, 4-phenyl- : 1867
1-Heptanol, 1-phenyl- : 1881
3-Heptanol, 7-phenyl : 2102
cis-2-Heptene : 6535
trans-3-Heptene : 6536
cis-3-Heptene : 6537
1-Heptene, 1,1-diphenyl- : 1744
2-Heptene, 3-methyl (*trans*) : 6549
2-Heptene, 3-methyl (*cis*) : 6550
2-Heptene, 4-methyl (*trans*) : 6551
2-Heptene, 4-methyl (*cis*) : 6552
1-Heptene-4-ol : 6572
1-Heptene-4-ol, 4-methyl : 6573
cis-2-Hepten-1-ol : 6568
1-Hepten-5-one : 6586
3-Hepten-2-one (*cis*) : 6588
2-Hepten-6-one, 4-ethyl-2-methyl : 6593
6-Hepten-4-yne-3-ol, 3-ethyl : 6608
Heptyl acetate : 193
Heptylacetylene : 8196
Heptyl alcohol : 6472
Heptylamine : 6349
Heptyl bromide : 6356
2-Heptyl bromide : 6357
4-Heptyl bromide : 6358
Heptyl chloride : 6360
Heptylcyclopentane : 4820
Heptyl formate : 6008
Heptyl heptanoate : 6464
Heptylidene acetone : 5093
Heptyl mercaptan : 6436
Heptyl methyl ketone : 8172
1-Heptylnaphthalene : 7895
2-Heptylnaphthalene : 7896
Heptyl nitrite : 8088
Heptyl propyl ketone : 11960
Heptyl thiocyanate : 11704
2-Heptynyl bromide : 6613
4-Heptynyl chloride : 6616
Herbicide 634 : 4900
Hesperetin : 2800
Hesperetin 7-rutinoside : 2794
Hesperidin : 2794
Hexabromoethane : 5594
Hexachlorane : 4409
Hexachloroacetone : 10440
Hexachlorobenzene : 1749
Hexachloro-1,3-butadiene : 3163
γ-Hexachlorocyclohexane : 4409
Hexachloroethane : 5595
1,2,3,4,5,6-Hexachloro-3-hexene : 7035
Hexachlorophene : 9409
1,1,1,2,2,3-Hexachloropropane : 10051
1,1,1,3,3,3-Hexachloropropane : 10052
1,1,2,2,3,3-Hexachloropropane : 10053
Hexachloropropylene : 10571
Hexacosane, 1-phenyl- : 1750
Hexacose : 6707
Hexacosylcyclohexane : 4410
Hexacosylcyclopentane : 4821
Hexadecamethylcyclooctasiloxane : 4741
Hexadecamethylene dibromide : 6635
Hexadecamethylheptasiloxane : 6528
Hexadecanal dimethyl acetal : 6636
Hexadecane, 1-cyclohexyl- : 4411
Hexadecane, 1-cyclopentyl- : 4822
Hexadecylcyclopentane : 4822
Hexadecyl 3-hydroxy-2-naphthoate : 7774
Hexadecyl mercaptan : 6643
Hexadecyl pyridinium chloride : 11030
1,5-Hexadiene, 2-methyl-3-ethyl : 6694
1,3-Hexadiene, 1-phenyl : 1753
1,5-Hexadiene, 1-phenyl- : 1754
1,3-Hexadiene, 6-phenyl : 1755
1,5-Hexadiene, 3-phenyl : 1625
1,5-Hexadiene-3-yne : 6713
1,3-Hexadien-5-ol : 6708
1,5-Hexadienylbenzene : 1754
Hexaethylbenzene : 1756
1,1,1,3,3,3-Hexaethyldisiloxane : 5235
Hexafluorenium bromide : 6753
Hexafluoro-2,4-pentanedione : 8802
Hexafluoropropylene oxide : 8629
1,2,3,4,5,6-Hexahydroanthracene : 422
Hexahydroaplotaxene : 6322
Hexahydroazepine : 492
Hexahydrobenzene : 4305
Hexahydrobenzoic acid : 4332
Hexahydrobenzyl alcohol : 4419
Hexahydroequilenin : 5402
Hexahydroindan : 7281
Hexahydronicotinic acid : 9641
Hexahydrophthalic anhydride : 7345
Hexahydropyrazine : 9602
3,3',4',5,5',7-Hexahydroxyflavylium chloride : 2849
3,3',4',5,5',7-Hexahydroxy-2-phenylbenzopyrylium chloride : 2849
Hexamethylbenzene : 1758
Hexamethylcyclohexasiloxane : 4625
1,1,3,3,5,5-Hexamethylcyclotrisilazane : 5001
Hexamethylcyclotrisiloxane : 5002
Hexamethyldisilane : 5228
Hexamethyldisilazane : 11363
Hexamethyldisiloxane : 5236
Hexamethylenediamine : 6749
Hexamethylene diiodide : 6769
Hexamethylene diisocyanate : 6770
Hexamethylene glycol : 6796
Hexamethylenetetramine, allyl iodide : 11825
Hexamethylenimine : 492
Hexamethylolmelamine : 7584
Hexane, 1-cyclohexyl- : 4414
Hexane, 1-cyclopentyl- : 4823
Hexane, 2,5-dibromo(*DL*) : 6759
3,4-Dibromohexane : 6760
Hexane, 2,5-dimethyl-2,5-diphenyl- : 2281
Hexane, 2,2-dimethyl-3-ethyl : 6816
Hexane, 2,2-dimethyl-4-ethyl : 6821
Hexane, 2,3-dimethyl-3-ethyl : 6817
Hexane, 2,3-dimethyl-4-ethyl : 6822
Hexane, 2,4-dimethyl-3-ethyl : 6818
Hexane, 2,4-dimethyl-4-ethyl : 6823
Hexane, 2,5-dimethyl-3-ethyl : 6819
Hexane, 3,3-dimethyl-4-ethyl : 6824
Hexane, 3,4-dimethyl-3-ethyl : 6820
Hexane, 2-iodo (*l*) : 6832
Hexane, 3-isopropyl-2-methyl- : 6839
Hexane, 2-methyl-3-isopropyl : 6839

Hexane, 3-phenyl- : 1669
Hexane, 1-phenyl- : 1763
Hexane, 2-phenyl- : 2009
5-Hexanoic acid : 7051
Hexanoic acid, 2-amino(*DL*) : 8215
Hexanoic acid, 6-cyclohexyl : 4412
Hexanoic acid, 2-methyl(*DL*) : 6893
Hexanoic acid, 2,2,3,4,4,5,6,6-octafluoro-3,5,6-trichloro : 6910
3-Hexanol (*l*) : 6915
3-Hexanol, 3,5-dimethyl (*DL*) : 6928
1-Hexanol, 2-isopropyl-5-methyl- : 6948
1-Hexanol, 2-methyl (*DL*) : 6937
1-Hexanol, 4-methyl (*DL*) : 6940
3-Hexanol, 5-methyl (*DL*) : 6947
2-Hexanol-3-one, 2-methyl : 6973
2-Hexanol-4-one, 5-methyl : 6974
2-Hexanol, 2-phenyl- : 1847
3-Hexanol, 3-phenyl : 1875
2-Hexanol, 6-phenyl : 2103
1-Hexanol, 3-phenyl : 2159
Hexatriacontylcyclohexane : 4413
Hexazole : 11848
trans-2-Hexene : 7000
cis-2-Hexene : 7001
3-Hexene, 3-chloro (*cis*) : 7009
1-Hexene, 1,2-dichloro (*trans*) : 7011
1-Hexene 1,2-dichloro (*cis*) : 7012
3-Hexene, 2,4-dimethyl (*cis*) : 7031
3-Hexene, 3-(4-hydroxyphenyl)-4-(4-methoxyphenyl) : 9358
5-Hexene-2-one,5-methyl : 7099
2-Hexene, 1-phenyl : 1760
2-Hexene, 2-phenyl : 2008
2-Hexene-4-yne : 7106
4-Hexenoic acid, 6-(4-hydroxy-6-methoxy-7-methyl-3-oxo-5-phthalanyl)-4-methyl-, (*E*)- : 7052
2-Hexenoic acid, 3-phenyl- : 4135
2-Hexen-4-ol, 5,5-dimethyl : 7074
2-Hexen-4-ol, 2,4-dimethyl : 7076
2-Hexen-5-one : 7087
2-Hexen-5-one, 3,4-dimethyl : 7092
3-Hexen-1-yl chloride : 7008
Hex-5-en-1-yne : 7105
Hexestrol : 9298
Hexobarbital : 11083
Hexocyclium methyl sulfate : 9623
Hexylacetylene : 8529
Hexylamine : 6723
Hexylamine, 3-methyl- : 6728
Hexylamine, 4-methyl- : 6729
Hexyl bromide : 6732
1-Hexyl-1,3-butadiene : 5023
3-Hexylbutanol-1 : 8166
Hexyl chloride : 6737
2-Hexyl chloride : 6738
2-Hexyldecanoic acid : 5065
7-Hexyldocosane : 5274
Hexylene glycol : 8793
Hexyl fluoride : 6829
Hexyl formate : 6009
9-Hexylheptadecane : 6309
Hexyl heptanoate : 6465
Hexyl iodide : 6831
Hexyl mercaptan : 6856
Hexyl methyl ether : 6833
Hexyl methyl ketone : 8471
1-Hexylnaphthalene : 7900
Hexyl nitrite : 8089
1-Hexylpiperidine : 9672
Hexyl propyl ketone : 5077
4-Hexylresorcinol : 1519
Hexyltrichlorosilane : 11404
Hexylvinylcarbinol : 8189
Hexyl vinyl ether : 6813
3-Hexyne, 2,5-dimethyl-2,5-dichloro : 7117
3-Hexyne-2-one : 7130
3-Hexynol : 7126
Hippuric acid : 6255
Hiragonic acid : 6666
Histamine : 7203
Histantin Richter : 9502
Holocaine, hydrochloride : 5670
Homatropine : 911
Homochlorcyclizine : 5140
Homochlorocyclizine : 5140
Homogentisic acid : 895
Homomesityl oxide : 6600
o-Homosalicylic acid : 2629
m-Homosalicylic acid : 2630
p-Homosalicylic acid : 2631
Homovanillic acid : 913
Hordenine : 9320
Humulene : 5006
Humulon : 4282
Hydantoin : 7216
Hydantoin, 3-allyl-5-isobutyl-2-thio- : 7229
Hydnocarpic acid : 4940
Hydracrylic acid : 10267
Hydracrylic acid, isopropyl ester : 10277
Hydracrylonitrile : 10106
Hydrangin : 2826
Hydrastine : 7357
Hydrastinine : 5217
Hydrazinecarboxamide, *N*-amino- : 4009
Hydrazinecarboxamide, 2-phenyl-*N*-(phenylimino)- : 5122
Hydrazinecarboxylic acid, hydrazide : 4009
Hydrazine, 1-isopentyl-1-phenyl : 7162
Hydrazine, 1-methyl-2-(3-tolyl) : 7163
Hydrazine, 1-phenyl-2-(3-tolyl) : 7174
2-Hydrazinoethanol : 5729
Hydrindantin : 2999
Hydrocarbostyril : 11327
Hydrochlorothiazide : 2855
Hydrochol : 4101
Hydrocinchonidine : 4121
Hydrocinchonine : 4122
Hydrocinnamaldehyde, *p*-isopropyl-α-methyl- : 2117
Hydrocinnamaldehyde, *o*-methyl- : 2115
Hydrocinnamic acid : 2126
Hydrocinnamic acid, α-bromo- : 2147
Hydrocinnamic acid, α,α-dimethyl-, methyl ester : 2129
Hydrocinnamic acid, ester with 17β-hydroxyestr-4-en-3-one : 5408
Hydrocinnamic acid, 4-(4-hydroxy-3-iodophenoxy)-3,5-diiodo- : 2136
Hydrocinnamic aldehyde : 2114
Hydrocinnamonitrile : 2123
Hydrocinnamyl alcohol : 2148
Hydrocodone : 7676
Hydrocortisone : 9753
Hydrocortisone acetate : 9747
Hydrocotarnine : 5214
Hydroflumethiazide : 2858
Hydrofuramide : 7488
Hydrogen carboxylic acid : 6002
Hydrogen cyanate (HOCN) : 4162
Hydrogen cyanide : 7176
Hydrohydrastinine : 5215
Hydroquinidine : 4124
Hydroquinine : 4123
p-Hydroquinone : 1499
Hydroquinonecarboxylic acid : 2570
Hydrosorbic acid : 7050
Hydrosorbic acid, ethyl ester : 7054
Hydrosorbic acid, 2-methyl- : 7058
Hydrotrichlorothiazide : 2854
2-Hydroxyacetaldehyde : 15
Hydroxyacetic acid, propyl ester : 199
4-Hydroxy-3-aminobutyric acid : 3465
γ-Hydroxy-β-aminobutyric acid : 3465
1-Hydroxy-3-aminomethyl-3,5,5-trimethylcyclohexane : 4502
Hydroxyamphetamine : 9215
2-Hydroxyaniline : 9201
3-Hydroxyaniline : 9202
4-Hydroxyaniline : 9203
o-Hydroxyanisole : 9379
p-Hydroxybenzaldehyde : 566
Hydroxybenzene : 9199
3-Hydroxybenzoic acid, 4-formyl : 9598
3-Hydroxybenzoic acid, 6-iodo : 2624
4-Hydroxybenzoic acid, 3-iodo : 2623
2-Hydroxybenzoxazole : 2881
o-Hydroxybenzyl alcohol : 1882
p-Hydroxybenzyl alcohol : 1884
2-Hydroxybiphenyl : 3088
2-Hydroxy-3-butenoic acid : 3842
Hydroxy-*tert*-butylamine : 10341
N-(2-Hydroxybutyl)-*N*-(2-hydroxyethyl)amine : 3623
3-Hydroxybutyraldehyde : 3184
4'-Hydroxybutyranilide : 3205
3-Hydroxycamphor : 2957
3-Hydroxy-4-carboxyaniline : 2498
3-Hydroxy-β-carotene : 4061
2-Hydroxy-5-chloroaniline : 9204
2-Hydroxy-3-chlorobiphenyl : 3092
3α-Hydroxy-5α-cholestane : 4108
3β-Hydroxy-5β-cholestanol : 4109
3β-Hydroxycholest-5-ene : 4110
3β-Hydroxycholest-4-ene : 4111
DL-α-Hydroxycinnamic acid : 2133
o-Hydroxycinnamic acid lactone : 2792
Hydroxycodeinone : 7675
7-Hydroxycoumarin 7-glucoside : 2821
1-Hydroxycumene : 1857
Hydroxycyclobutane : 4196
Hydroxycyclopentane : 4864
3-Hydroxy-2-*p*-cymenecarboxylic acid : 2642
2-Hydroxy-1,4-dibromobutane : 3614
12β-Hydroxydigitoxin : 4041
2-Hydroxy-3,5-diiodobenzoic acid : 2617
1-Hydroxy-2,11-dimethoxy-6aα-aporphine : 5158
Hydroxydimethylphosphine oxide : 9528
3-Hydroxy-1,1-dimethylpyrrolidinium bromide α-cyclopentylmandelate : 11179
4-Hydroxydiphenyl : 3090
3-Hydroxy-1,5-diphenylformazan : 5122
16β-Hydroxyestradiol : 5405
2-Hydroxyethanamine : 5675
Hydroxyethanoic acid : 196
2-Hydroxyethyl acetate : 5577
2-Hydroxyethyl 2-hydroxybutyl amine : 3623
1-Hydroxyethyl-4-methoxybenzene : 1895
2-(2-Hydroxyethyl)piperidine : 9663
4-(2-Hydroxyethyl)piperidine : 9664
α-(2-Hydroxyethyl)styrene : 2157
4-Hydroxy-1,6-heptadiene : 6337
3-Hydroxy-1,5-hexadiene : 6706
1-Hydroxy-2,4-hexadiene : 6707
3-Hydroxy-2,5-hexanedione : 6808
5-Hydroxy-2-hexanone : 6969
5-Hydroxy-3-hexanone : 6972
p-Hydroxyhydrocinnamic acid : 2134
Hydroxyhydroquinone : 2367
(Hydroxyimino)cyclohexane : 4610
2-Hydroxyisobutyronitrile : 10108
4-Hydroxyisophthalic acid : 1352
β-Hydroxylactic acid : 10252
Hydroxylamine, *N*,*N*-dipropyl- : 9814
Hydroxylamine, *N*-ethyl- : 5434
Hydroxylamine, *N*-methyl : 7447
6α-Hydroxylithocholic acid : 4096
Hydroxylupanine : 7574
3-Hydroxy-5-methoxybenzaldehyde : 573
1-Hydroxy-2-methoxybenzene : 9379
4-Hydroxy-3-methoxybenzylacetone : 3663
7-Hydroxy-4'-methoxyisoflavone : 2830
3-Hydroxy-2-methoxy-5-methyl-*p*-benzoquinone : 4265
(4-Hydroxy-3-methoxyphenyl)ethyl methyl ketone : 3663
3-Hydroxy-4-methoxy-2,5-toluquinone : 4265
p-Hydroxy-α-[(methylamino)-methyl]benzyl alcohol : 1887
2-Hydroxymethylbenzofuran : 2455
3-Hydroxy-3-methyl-1-butyne : 3922
7-Hydroxy-4-methylcoumarin : 2831
5-Hydroxymethylcytosine : 11105
5-(Hydroxymethyl)-2-furaldehyde : 6044
4-Hydroxy-4-methylheptane : 6505
p-(Hydroxymethyl)nitrobenzene : 1908
o-(Hydroxymethyl)phenol : 1882
p-(Hydroxymethyl)phenol : 1884
2'-Hydroxy-4'-methylpropiophenone : 10445
3-Hydroxy-2-methylundecane : 11956
1-Hydroxy-2-naphthoic acid : 7771
3-Hydroxy-2-naphthoic acid : 7772
2-Hydroxy-1-nitrobutane : 3641
4-Hydroxynonanoic acid lactone : 6149
2'-Hydroxypelargidenolon 1522 : 2816

4-Hydroxy-1-pentanol : 8786
1-Hydroxy-2-pentanone : 8971
2-Hydroxy-3-pentanone : 8975
3-Hydroxy-1-pentyne : 9140
1-Hydroxypent-2-yn-4-ene : 9118
3-Hydroxy-9,10-phenanthraquinone : 9176
2-(β-Hydroxyphenethyl-amino)pyridine : 1913
Hydroxyphenylamine : 9200
o-Hydroxyphenylamine : 9201
m-Hydroxyphenylamine : 9202
N-(4-Hydroxyphenyl)glycine : 6267
1-Hydroxy-3-phenylpropane : 2148
1-(*m*-Hydroxyphenyl)-1-propanol : 1872
DL-2-Hydroxy-3-phenylpropionic acid : 2133
4-Hydroxyphenylpropionic acid : 2134
3-Hydroxy-3-phenylprop-1-yne : 1876
4-Hydroxyphenylsulfonic acid : 2226
17α-Hydroxyprogesterone : 9750
4-Hydroxy-*L*-proline : 9767
3-Hydroxy-1-propene : 10697
2-Hydroxypropionitrile : 10105
(3-Hydroxypropyl)benzene : 2148
1-Hydroxy-1-propylcyclohexane : 4557
3-Hydroxypropylene oxide : 8618
N-(2-Hydroxypropyl)ethylenediamine : 10340
3-Hydroxypropyl fluoride : 10382
2-(2-Hydroxypropyl)pyridine : 10951
2-(3-Hydroxypropyl)pyridine : 11012
(2-Hydroxypropyl)trimethyl ammonium chloride : 9831
4-Hydroxypyrazolo[3,4-d]pyrimidine : 10852
5-Hydroxyquinoline : 11305
7-Hydroxyquinoline : 11307
8-Hydroxyquinoline : 11308
8-Hydroxyquinoline sulfate : 11324
8-Hydroxy-5-quinolinesulfonic acid : 11294
p-Hydroxysalicylic acid : 2569
5-Hydroxysalicylic acid : 2570
2α-Hydroxytigogenin : 11530
2-Hydroxy-*m*-toluic acid : 2629
2-Hydroxy-*p*-toluic acid : 2630
6-Hydroxy-*m*-toluic acid : 2631
2-Hydroxytridecane : 11884
11-Hydroxyundecanoic acid : 11948
4-Hydroxyundecanoic acid lactone : 6152
Hydroxyurea : 12016
Hydroxyzine : 5700
Hygrine : 10471
Hymecromone : 2831
Hyodeoxycholic acid : 4096
Hyoscyamine : 917
Hypothiazide : 2855
Hypoxanthine : 10770
Hypoxanthine riboside : 7335
Ibogaine : 7186
Ibrotamide : 3199
Imecromone : 2831
1H-Imidazole, 1-[2-(2,4-dichlorophenyl)-2-(2-propenyloxy)ethyl]- : 7187
1H-Imidazole, 2-heptadecyl-4,5-dihydro-, monoacetate : 6275
2-Imidazolidine, 2-benzyl : 7198
1-Imidazolidinecarboxamide, 3-(3,5-dichlorophenyl)-*N*-(1-methylethyl)-2,4-dioxo- : 7338
7H-Imidazo[4,5-d]pyrimidine : 10752
2-Iminobenzothiazoline : 2859
Iminodiacetic acid : 6260
2,2'-Iminodibenzoic acid : 2656
Iminourea : 6278
Imperatorin : 6189
Indaconitine : 307
Indan : 7250
Indan, 1-amino (*DL*) : 7243
Indan, 1,2-dibromo- : 7247
Indan, 1,2-dichloro : 7249
Indan, 1,1-diemthyl : 7251
Indan, 1,2-dimethyl- : 7252
Indan, 4,7-dimethyl- : 7253
Indan, 5,6-dimethyl- : 7254
Indan, 1-ethyl- : 7273
Indan, 5-hexyl- : 7274
Indan, 1-methyl- : 7255
Indan, 2-methyl- : 7256
Indan, 4-methyl- : 7257
Indan, 5-methyl- : 7258
2-Indanone : 7287
1-Indanone, 4-methyl- : 7290
1-Indanone, 2-phenyl- : 7291
Indan, 2-phenyl- : 7259
Indan, 1-phenyl-1,3,3-trimethyl : 7263
Indan, 1,1,4,7-tetramethyl- : 7260
Indanthrene : 434
Indan, 1,1,5-trimethyl- : 7261
Indan, 1,4,7-trimethyl : 7262
1-Indenecarboxylic acid : 7246
Indene, 1,3-diphenyl- : 7271
Indene, 2,3-diphenyl- : 7272
Indene, 4-methyl- : 7278
Indene, 6-methyl- : 7279
Indene, 7-methyl- : 7280
Indigo : 7331
Indigo Yellow : 2843
Indoleacetic acid : 7295
Indolebutyric acid : 7296
Indole, 7-methyl- : 7317
Indole, 1,2,3-trimethyl- : 7322
Indole, 2,3,5-trimethyl- : 7323
2-Indolinone, 1-ethyl-3-methyl- : 7332
1H-Indolo[3,2,1-de]pyrido[3,2,1-ij][1,5]naphthyridine, eburnamenine-14-carboxylic acid deriv. : 5349
Indolo[4,3-fg]quinoline, ergoline-8-carboxamide deriv. : 5373
3-Indolylacetone : 10442
Indonaphthene : 7245
Indoxazene : 2389
Inosine, 2-amino- : 6289
Iodipamide : 2594
5-Iodoacenaphthene : 7
Iodoacetic acid : 200
Iodoacetonitrile : 283
Iodoalphionic acid : 2135
p-Iodoaniline : 761
o-Iodoanisole : 1775
Iodobenzene : 1766
p-Iodobenzenesulfonyl chloride : 2247
o-Iodobenzoic acid : 2657
4-Iodo-1,2-butadiene : 3166
4-Iodocatechol : 1523
Iodochlorhydroxyquin : 11312
Iodocyanide (ICN) : 7337
Iodocyclohexane : 4415
Iodocyclopentane : 4824
2-Iodo-*p*-cymene : 1781
1-Iododecane : 5053
Iododifluoromethane : 7505
1-Iodo-3,5-dimethylbenzene : 1771
4-Iododiphenyl : 3079
1-Iodododecane : 5302
Iodoethane : 5597
2-Iodoethanol : 5731
Iodoethene : 5907
Iodoethylene : 5907
Iodoform : 7560
2-Iodofuran : 6110
3-Iodofuran : 6111
1-Iodoheptane : 6419
2-Iodoheptane : 6420
1-Iodo-1-heptyne : 6620
1-Iodohexadecane : 6640
1-Iodohexane : 6831
Iodomethane : 7518
2-Iodo-4-methylaniline : 762
4-Iodo-2-methylbutane : 3384
(Iodomethyl) methyl ether : 7519
1-Iodonaphthalene : 7902
1-Iodooctadecane : 8242
1-Iodooctane : 8406
1-Iodopentane : 8818
3-Iodopentane : 8820
1-Iodo-1-pentyne : 9133
o-Iodophenol : 9371
p-Iodophenol : 9373
1-Iodopropane : 10054
2-Iodopropane : 10055
3-Iodopropene : 10573
3-Iodopropylene : 10573
1-Iodopropyne : 10734
3-Iodopropyne : 10735
2-Iodopyridine : 10970
Iodopyrrole : 11140
Iodoquinol : 11315
4-Iodoquinoline : 11272
Iodostarin : 8300
N-Iodosuccinimide : 11159
Iodotrifluoroethylene : 5925
Iodotrifluoromethane : 7559
3-Iodotyrosine : 11932
3-Iodoveratrole : 1767
4-Iodo-*o*-xylene : 1774
β-Ionol : 3873
α-Ionol : 3874
Irene : 7999
Iron dicyclopentadienyl : 5960
β-Irone : 3883
α-Irone : 3884
Iron pentacarbonyl : 7339
Iron, tris(dimethylcarbamodithioato-*S,S'*)-, (OC-6-11)- : 5957
Isatin : 7300
Isoamylaldehyde : 3186
Isoamylamine : 3220
Isoamyl benzoate : 3632
Isoamyl bromide : 3239
Isoamyl chloride : 3252
Isoamyl cyanide : 8831
Isoamyl ether : 3418
Isoamyl iodide : 3384
Isoamyl isovalerate : 3551
Isoamyl 3-methylbutyrate : 3551
Isoamyl nitrate : 3637
Isoamyl nitrite : 8090
Isoamyl salicylate : 2633
Isoamyl sulfide : 3436
Isobutamben : 2505
Isobutanal : 9785
Isobutane : 10073
Isobutane, 1,1-dichloro : 9899
Isobutane, 1,2-dichloro : 9900
Isobutane, 1,1,1,2,3-pentachloro : 10127
Isobutane, 1,1,1,2-tetrachloro : 10147
Isobutane, 1,1,2-trichloro : 10177
Isobutene : 10578
2-Isobutoxyethanol : 5746
1-Isobutoxy-2-nitrobenzene : 2034
Isobutryic anhydride : 10291
Isobutyl acetate : 215
Isobutyl acrylate : 10673
Isobutyl alcohol : 10388
Isobutyl alcohol, 2-chloro : 10352
Isobutyl alcohol, 3-chloro : 10353
Isobutylamine : 9817
Isobutyl *p*-aminobenzoate : 2505
Isobutylbenzene : 2039
Isobutyl bromide : 9845
Isobutyl butanoate : 3563
Isobutyl carbamate : 3959
Isobutyl chloride : 9862
Isobutyl chlorocarbonate : 4024
Isobutyl chloroformate : 4024
Isobutyl cyanide : 3410
Isobutylcyclopentane : 4838
Isobutyldimethylcarbinol : 8928
Isobutylene chlorohydrin : 10354
Isobutylene, 1,3-dibromo : 10546
Isobutylene glycol, 3-chloro : 9998
Isobutylene (trimer) : 10582
Isobutylethene : 9054
Isobutyl formate : 6013
Isobutylhydrochlorothiazide : 2856
Isobutyl iodide : 10057
Isobutyl isobutanoate : 10304
Isobutyl isocyanate : 10061
Isobutyl isovalerate : 3555
Isobutyl ketone : 6514
Isobutyl mercaptan : 10158
Isobutyl methacrylate : 10666
Isobutyl methyl ether : 10071
Isobutyl methyl ketone : 8981
Isobutyl nitrate : 8078
Isobutyl nitrite : 8094
Isobutyloxycarbonyl chloride : 4024
Isobutyl propionate : 10310
Isobutyl propyl ketone : 6524
4-Isobutylresorcinol : 1539
Isobutyl salicylate : 2643
Isobutyl stearate : 8268

Isobutyl thiocyanate : 11707
Isobutyl 2-tolyl ketone : 3669
Isobutyl 4-tolyl ketone : 3670
Isobutyl urethane : 3961
Isobutyl vinyl ether : 10035
Isobutyl vinyl ketone : 7093
Isobutyramide, 2-bromo : 9792
Isobutyric acid : 10290
Isobutyric acid chloride : 10504
Isobutyric acid, 2-chloroethyl ester : 10292
Isobutyric acid, cyclohexyl ester : 10293
Isobutyric acid, (2-nitropentyl) ester : 10305
Isobutyric acid, 3-phenyl : 2137
1-Isobutyronaphthone : 10464
2-Isobutyronaphthone : 10465
Isobutyronitrile : 10111
Isobutyronitrile, 2-hydroxy-3-chloro : 10098
Isobutyryl chloride, 3-chloro : 10499
Isocaine : 2505
Isocamphoric acid, (*DL*) : 3931
Isochroman : 2776
Isochrysene : 11910
Isocinchomeronic acid : 10929
Isocitrate lactone : 6079
Isocorybulbine : 5150
L-(+)-Isocorydine : 5161
Isocorydine : 5161
Isocoumarin : 2791
Isodurene : 2279
Isoestradiol : 5401
8-Isoestrone : 5407
Isoeugenol : 9388
Isoeugenol, acetate (*cis*) : 9394
Isoflavone : 2841
Isoflavone, 4',7-dihydroxy- : 2827
Isoflavone, 4',5-dihydroxy-7-methoxy- : 2828
Isoflurophate : 9580
Isogeronic acid : 6461
Isoguanine : 10767
Isohexane : 8823
Isohexyl alcohol : 8939
1H-Isoindole-1,3(2H)-dione, 2-[(trichloromethyl)thio]- : 5984
Isolysergic acid : 5375
β-Isomalic acid : 9959
Isometheptene : 6532
Isoniazid : 10905
Isonicotinic acid : 10895
Isonicotinic acid diethylamide : 10892
Isonicotinic acid, (α-methylfurfurylidene)hydrazide : 10904
Isonipecotic acid : 9642
Isonipecotic acid, methyl ester : 9644
Isonipecotic acid, 1-methyl-, methyl ester : 9645
Isonitrosoacetone : 9789
Isonitrosoacetophenone : 876
β-Isonitrosopropane : 10476
Isooctane : 8848
Isopentadiene : 3171
Isopentanal : 3186
Isopentane : 3397
Isopentanoic acid : 3539
Isopentanoyl chloride : 3693
Isopentyl acetate : 3631
Isopentyl alcohol : 3627
Isopentyl alcohol, orthoformate (3:1) : 8552
Isopentyl alcohol, stearate : 8265
Isopentylamine : 3220
Isopentyl bromide : 3239
Isopentyl 3-bromopropionate : 10210
Isopentyl butanoate : 3544
Isopentyl chloride : 3252
N-Isopentylethanolamine : 5739
Isopentyl ether : 3418
Isopentyl formate : 3636
Isopentyl iodide : 3384
Isopentyl isovalerate : 3551
Isopentyl mercaptan : 3440
Isopentyl 3-methylbutanoate : 3551
Isopentyl β-Naphthyl ether : 7918
Isopentyl oleate : 8306
Isopentyl propanoate : 3639
Isopentyl salicylate : 2633
2-Isopentylthiophene : 11771
Isopentyl vinyl ether : 3371
Isophorol : 4702
Isophorone : 4721
Isophthalamide, *N,N,N',N'*-tetraethyl : 1326
Isophthalic acid : 1328
Isophthalic acid, diethyl ester : 1346
Isophytol : 6671
Isopilosine : 6144
Isoprene : 3171
Isopropanesulfonic acid : 10130
Isopropenyl acetate : 10698
Isopropenylacetylene : 3892
Isopropenyl carbinol : 10702
Isopropenyl chloride : 10535
Isopropenyl chloroformate : 4022
4-Isopropenyl-1-cyclohexenecarboxaldehyde : 4633
4-Isopropenyl-1-methylcyclohexene : 4677
Isopropenyl methyl ketone : 3879
9-Isopropenylphenanthrene : 9182
2-Isopropoxyethanol : 5740
1-Isopropoxy-2-propanol : 10390
N-Isopropylacetanilide : 74
Isopropyl acetate : 209
Isopropylacetic acid : 3539
Isopropyl acetoacetate : 3574
Isopropyl alcohol : 10335
Isopropylamine : 9804
2-Isopropylaniline : 788
N-Isopropylaniline : 790
Isopropylbenzene hydroperoxide : 7181
p-Isopropylbenzoic acid : 2689
α-Isopropylbenzyl alcohol : 1901
4-Isopropylbenzyl alcohol : 1902
Isopropyl borate : 3128
Isopropyl bromide : 9837
α-Isopropyl-α-bromobutyramide : 3199
Isopropyl-*tert*-butylcarbinol : 8955
Isopropyl butyrate : 3549
Isopropyl carbamate : 3955
Isopropyl carbanilate : 3966
Isopropyl chloride : 9849
Isopropyl chloroacetate : 139
Isopropyl 3-chlorocarbanilate : 3939
Isopropyl chloroformate : 4023
Isopropyl *N*-(3-chlorophenyl)carbamate : 3939
Isopropyl 2-chloropropionate : 10223
Isopropyl *trans*-cinnamate : 10687
Isopropylcyclohexane : 4431
cis-2-Isopropylcyclohexanol : 4539
cis-4-Isopropylcyclohexanol : 4540
trans-4-Isopropylcyclohexanol : 4541
2-Isopropylcyclohexanone : 4599
4-Isopropylcyclohexanone : 4600
Isopropylcyclopentane : 4828
2-Isopropylcyclopentanone : 4887
7-Isopropyl-1,4-dimethylazulene : 505
N-Isopropylethanolamine : 5741
Isopropyl fluoride : 10047
Isopropyl formate : 6011
Isopropyl heptafluorobutyrate : 3526
4-Isopropylheptane : 6427
Isopropylidene glycerol : 5207
Isopropyl iodide : 10055
Isopropyl isobutyrate : 10303
Isopropyl isocyanate : 10060
Isopropyl isocyanide : 10063
Isopropyl isonitrile : 10063
Isopropyl lactate : 10275
Isopropyl mercaptan : 10157
Isopropylmethylamine : 9819
4-Isopropyl-1-methylcyclopentene : 4927
Isopropyl methyl ether : 10070
2-Isopropyl-5-methyl-1-hexanol : 6948
2-(4-Isopropyl-4-methyl-5-oxo-2-imidazolin-2-yl)nicotinic acid : 7188
(*RS*)-2-(4-Isopropyl-4-methyl-5-oxo-2-imidazolin-2-yl)quinoline-3-carboxylic acid : 7189
Isopropyl methyl sulfide : 10089
Isopropyl myristate : 11626
Isopropyl 1-naphthyl ketone : 10464
Isopropyl nitrate : 8076
Isopropyl nitrite : 8092
2-Isopropyl-1-pentene : 6836
Isopropyl phenylcarbamate : 3966
Isopropylphenylcarbinol : 1901
3-(*p*-Isopropylphenyl)-2-methylpropionaldehyde : 2117
Isopropyl phenyl urethane : 3966
Isopropyl propionate : 10299
Isopropylpropylamine : 9827
Isopropyl propyl sulfide : 10077
3-Isopropylpyridine : 10985
Isopropyl salicylate : 2635
Isopropyl 2-thienyl ketone : 10472
Isopropyl thiocyanate : 11706
2-Isopropylthiophene : 11772
3-Isopropylthiophene : 11773
Isoquinaldine : 7381
Isoquinoline, 5-amino- : 7372
Isoquinoline, 3-benzyl : 7387
Isoquinoline, 3-butyl-1-[2-(dimethylamino)ethoxy]- : 5411
Isoquinoline, 6,7-diethoxy-1(3,4-diethoxybenzyl) : 7377
Isoquinoline, 6,7-dimethoxy-1,2-dimethyl-1,2,3,4-tetrahydro : 7391
Isoquinoline, 8-methoxy-2-methyl-6,7-methylenedioxy-1,2,3,4-tetrahydro : 5214
Isorubijervine : 11516
(+)-Isothebaine : 5158
Isothebaine : 5158
3(2H)-Isothiazolone, 2-octyl- : 8528
Isothiocyanic acid, acetyl : 303
Isothiocyanic acid, *tert*-butyl ester : 10066
Isovaleraldehyde : 3186
Isovaleramide : 3206
Isovaleric acid : 3539
Isovaleric acid chloride : 3693
Isovaleric aldehyde : 3186
Isovaleric amide : 3206
Isovalerylamine : 3220
Isovaleryl chloride : 3693
Isovaleryl diethylamide : 3201
Isoxylaldehyde : 548
Itaconic acid : 3328
Jasmone : 4949
Javanicin : 7855
Jervine : 12046
Jilkon : 6200
Juglone : 7860
Julolidine : 2737
Julolidine, 1,6-dioxo : 2736
Kaempferol : 2843
Kainic acid : 11144
Kelthane : 1851
Ketamine : 4579
Ketene : 5927
Ketene, diethyl- : 3877
β-Ketobutyranilide : 3207
α-Ketoglutaric acid : 8781
3-Ketoglutaric acid : 8782
Ketohydroxyestrin : 5406
Ketone, 2-benzofuranyl methyl : 5775
Ketone, 2-benzofuranyl phenyl : 7588
Ketone, benzyl phenyl : 5812
Ketone, 5-bromo-2-thienyl methyl : 5783
Ketone, α-chlorobenzyl phenyl : 5785
Ketone, 5-chloro-2-thienyl methyl : 5791
Ketone, cyano phenyl : 968
Ketone, cyclobutyl methyl : 5792
Ketone, cyclobutyl phenyl : 7605
Ketone, 1-cyclohexen-1-yl methyl : 5793
Ketone, cyclohexyl methyl : 5794
Ketone, cyclopentyl methyl : 5795
Ketone, cyclopropyl methyl : 5796
Ketone, dicyclohexyl : 7606
Ketone, dicyclopropyl : 7607
Ketone, 6,7-dimethoxy-1-isoquinolyl 3,4-dimethoxyphenyl : 7611
Ketone, 10-[3-(dimethylamino)-propyl]phenothiazin-2-yl methyl : 5807
Ketone, 2-furyl methyl : 5817
Ketone, 2-furyl phenyl : 7618
Ketone, 1-hydroxycyclohexyl methyl : 5821
Ketone, 6-hydroxy-2-isopropenyl-5-benzofuranyl methyl : 5825
Ketone, 3-hydroxy-2-thienyl methyl : 5835
Ketone, indol-3-yl methyl : 5819
Ketone, methyl 1-methylcyclohexyl : 5843
Ketone, methyl 3-methylcyclohexyl : 5844

Ketone, methyl 4-methylcyclohexyl : 5845
Ketone, methyl 1-methylcyclopentyl : 5846
Ketone, methyl 2-methylcyclopentyl : 5847
Ketone, methyl 6-methyl-3-pyridyl : 5855
Ketone, methyl 1-methylpyrrol-2-yl : 5849
Ketone, methyl 5-methyl-2-thienyl : 5856
Ketone, methyl 2-pyridyl : 5867
Ketone, methyl 3-pyridyl : 5868
Ketone, methyl 4-pyridyl : 5869
Ketone, methyl pyrrol-2-yl : 5820
Ketone, methyl styryl : 3881
Ketone, methyl 2-thienyl : 5870
Ketone, 1-naphthyl *o*-tolyl : 7629
Ketone, phenyl 2-pyridyl : 7638
Ketone, phenyl 4-pyridyl : 7639
Ketone, phenyl 2-thienyl : 7641
Ketone, phenyl 3-thienyl : 7642
2-Ketosuccinic acid : 3339
Khellin : 6185
Kojic acid : 10795
KSCN : 11711
Kyanmethin : 11045
Kynurenic acid : 11232
Kynurenine : 1102
Lactic acid (*DL*) : 10266
Lactic acid, butyl ester (*d*) : 10268
Lactic acid, isopropyl ester : 10275
Lactic acid, methyl ester (*DL*) : 10273
Lactic acid, 3,3,3-trichloro- : 10333
Lactobionic acid : 6210
Lactonic acid : 6210
Lactonitrile : 10105
Lactonitrile, 3-chloro-2-methyl- : 10098
Lactonitrile, 3,3,3-trichloro- : 10115
p-Lactophenetide : 9796
Lactyl chloride, acetate (*DL*) : 10494
Lamoryl : 11520
Lanex : 12007
Lanosterol : 7411
Lanoxin : 4041
L-Lanthionine : 5016
Lapachol : 7862
Lappaconitine : 309
Laricic acid : 8102
Laudanidine : 9399
Laudanine : 9398
Laudanosine : 7379
Lauraldehyde : 5288
Lauraldehyde, dimethyl acetal : 5297
Laureline : 2469
Lauric acid : 5305
Lauric acid, phenyl ester : 5317
Lauric acid, propyl ester : 5320
Laurolenic acid : 4912
Laurone : 11857
Lauronitrile : 5303
Lauroyl peroxide : 9155
Lauryl alcohol : 5321
Lauryl bromide : 5293
Lawsone : 7859
Leaf aldehyde violet : 8108
Ledol : 4989
Lemonol : 8333
Lenacil : 4900
Lenthionine : 8998
Lepidine : 11279
Lepidine, 2-amino- : 11208
Lepidine, 2-ethyl- : 11270
Lergitin : 5506
Leucethane : 3948
Leucoline : 7373
Levodopa : 11929
Levopimaric acid : 9169
Levulinic acid : 8903
Levulinic acid, benzyl ester : 8911
Levulinic acid, isopropyl ester : 8909
Levulinic acid, 3-methyl-, methyl ester : 8898
Levulinic acid, propyl ester : 8912
β-Levulose : 6023
Lichenic acid : 3764
Lidocaine : 54
Lignoceric acid : 11607
Limettin : 2818
α-Limonene : 4677
Limonene : 4677
Linalool formate : 8337
Linalool, tetrahydro (*DL*) : 8454
Linalyl formate : 8337
Linamarin : 10104
Linarigenin : 2809
β-Lindane : 4406
δ-Lindane : 4407
Lindane : 4409
Linderic acid : 5338
Linoleic acid : 8223
Linolenic acid : 8287
Lithocholic acid : 4099
Lobelanidine : 9646
Lobelanine : 5854
Lobeline : 5834
Loflucarban : 11802
Longifolene : 7568
Luciculine : 5007
Lumazine : 10748
Lumichrome : 2472
Luminol : 9600
Lumisterol : 5382
Lunacrine : 6182
Lupeol : 7417
3,5-Lupetidine : 9652
Lupinine : 11331
Lupulon : 4281
trans-Lutein : 4058
Luteolin : 2814
2,3-Lutidine : 10936
2,4-Lutidine : 10937
2,5-Lutidine : 10938
2,6-Lutidine : 10939
3,4-Lutidine : 10940
3,5-Lutidine : 10941
Lycomarasmine : 472
trans-Lycopene : 4053
Lycoremine : 6200
Lycorine : 6201
Lycoxanthin : 4060
Lysergamide : 5374
Lysergic acid : 5376
D-Lysergic acid *D*-propanolamide : 5373
Lysidine : 7197
Lysilan : 5140
Maclurin : 7610
Magenta I : 634
Magneson : 1542
Malachite green : 7455
Malathion : 3316
Maleamic acid : 3817
Maleanilic acid : 3853
Maleic acid : 3765
Maleic acid, chloro-, diethyl ester : 3771
Maleic acid, diallyl ester : 3782
Maleic acid, dihydroxy- : 3777
Maleic anhydride : 6091
Maleimide, *N*-phenyl- : 11123
Malidone : 8586
Malilum : 11091
Malonic acid : 9921
Malonic acid, amino-, diethyl ester : 9924
Malonic acid, 2-cyclopentenyl, diethyl ester : 9935
Malonic acid, cyclopentylidene, diethyl ester : 9936
Malonic acid, dibenzyl-, diethyl ester : 9926
Malonic acid, dibutyl-, diethyl ester : 9937
Malonic acid, diethyl-, diethyl ester : 9940
Malonic acid, dimethyl-, diethyl ester : 9942
Malonic acid, dipropyl ester : 9945
Malonic acid, di-2-propynyl-, diethyl ester : 9946
Malonic acid, ethylisopropyl-, diethyl ester : 9951
Malonic acid, (hydroxymethyl)- : 9959
Malonic acid, isobutyl-, diethyl ester : 9969
Malonic acid, isopropylidene-, diethyl ester : 9966
Malonomononitrile, benzylidene,monoethyl ester : 10645
Malononitrile : 9911
Malononitrile, phenyl- : 9916
Maltol : 10796
Malvidin chloride : 2847
Mandelic acid isoamyl ester : 912
Mandelonitrile glucoside : 955
Manganese, [[1,2-ethanediylbis-[carbamodithioato]](2-)]- : 7420
Mannite : 7422
Mannitol, hexanitrate : 7424
Margaric acid : 6314
Margaric aldehyde : 6304
Matrine : 7426
MCPA : 141
Mechlorethamine : 5413
Meclizine : 9605
Meconic acid : 10775
Meconin : 7355
Mecrylate : 10624
Medroxyprogesterone : 9752
Melamine : 11835
Melampyrin : 6194
Melatonin : 69
Meldrum's acid : 5190
Melezitose : 6221
Mellitene : 1758
Mellitic acid : 1748
Menadiol diacetate : 7841
Menadione : 7864
p-Mentha-1,8-dien-7-al : 4633
p-Mentha-1(7),2-diene : 4673
p-Menthane, 4-bromo : 4321
trans-(-)-*p*-Menthan-3-one : 4606
m-Menth-8-ene : 4435
p-Menth-2-ene, 1,4-epidioxy- : 5184
p-Menth-3-en-2-ol : 4698
(±)-Menthol : 4548
Menthol(*DL*) : 4548
Menthone(*d*) : 4605
l-Menthone : 4606
(-)-Menthone : 4606
Meparfynol : 9146
Mephenesin : 10019
Mephenoxalone : 8588
Mephobarbital : 11094
Mepivacaine : 9639
Meprobamate : 10021
β-Mercaptoalanine : 5014
Mercaptobenzene : 2289
Mercaptobenzimidazole : 2386
2-Mercaptobenzoic acid : 2663
2-Mercaptobenzothiazole : 2866
1-Mercaptobutane : 3437
Mercaptocyclopentane : 4852
2-Mercaptoethanol : 5732
3-Mercaptohexane : 6858
Mercaptomethane : 7546
6-Mercaptopurine : 10763
Mercapturic acid : 5015
Mercury, chloro(2-methoxyethyl)- : 11822
Mesaconic acid : 3785
Mesaconic acid, diethyl ester : 3787
Mesalamine : 2499
Mescaline : 1586
Mesitylamine : 859
Mesitylene : 2359
Mesitylene, α-bromo- : 1057
Mesitylene, 2,4-dibromo- : 1313
Mesitylene, 2,4-dichloro- : 1402
Mesitylenic acid : 2583
Mesityl oxide : 9108
Mesoxalylcarbamide : 11073
Mesoxalylurea : 11073
Mestanolone : 350
Mestilbol : 9358
Mesulphen : 11661
Metacrolein : 11906
Metaldehyde II (tetramer) : 19
Metanilic acid : 2216
Methacetin : 71
Methacholine chloride : 9830
Methacrolein : 10515
Methacrolein diacetate : 10564
Methacrylic acid : 10648
Methacrylic acid anhydride : 10649
Methacrylic acid, 3-bromo : 10612
Methacrylic acid chloride : 10715
Methacrylic acid, 2-ethylhexyl ester : 10659
Methadone hydrochloride : 6516
Methallatal : 11064
Methallyl alcohol : 10702
α-Methallyl bromide : 3710
Methaminodiazepoxide : 2409

Methamphetamine : 1573
Methanal : 5987
Methanamide : 5989
Methandrostenolone : 344
Methanearsonic acid : 458
Methanecarbonitrile : 271
Methanecarboxamide : 22
Methanecarboxylic acid : 100
Methane, dicyclohexyl- : 4429
Methane, diphenyl-(3-tolyl) : 7513
Methane, iodo-trichloro : 7556
Methanesulfonic acid chloride : 7535
Methanesulfonic acid fluoride : 7538
Methane, tetraphenyl- : 1831
Methane, tricyclohexyl- : 4433
Methanimidamide, *N,N*-dimethyl-*N'*-[3-[[(methylamino)carbonyl]-oxy]phenyl]-, monohydrochloride : 6001
1H-3a,7-Methanoazulene, cedrene deriv. : 4066
5,9-Methano-2,4,3-benzodioxathiepin, 6,7,8,9,10,10-hexachloro-1,5,5a,6,9,9a-hexahydro-, 3-oxide : 5369
7,9-Methano-7H-oxepino[3,4-a]pyrrolo[2,3-d]carbazole, 9,10-secostrychnidin-10-oic acid deriv. : 11352
Methanoic acid : 6002
Methanoic acid amidine : 7563
4,7-Methano-1H-indene, 1,2,4,5,6,7,8,8-octachloro-2,3,3a,4,7,7a-hexahydro- : 4077
2,5-Methano-2H-indeno[1,2-b]oxirene, 2,3,4,5,6,7,7-heptachloro-1a,1b,5,5a,6,6a-hexahydro-, (1aα,1bβ,2α,5α,5aβ,6β,6aα)- : 6301
4,7-Methano-1H-isoindole-1,3(2H)-dione, 2-(2-ethylhexyl)-3a,4,7,7a-tetrahydro- : 7660
Methanol, 2-tolyl : 1897
Methanol, 3-tolyl : 1898
Methanol, 4-tolyl : 1899
Methantheline bromide : 5445
Methapyrilene : 5507
Methaqualone : 11199
Metharbital : 11088
Methenamine : 11601
1,3,4-Metheno-2H-cyclobuta[cd]pentalen-2-one, 1,1a,3,3a,4,5,5,5a,5b,6-decachloro-octahydro- : 7408
N,N'-Methenyl-*o*-phenylenediamine : 2376
Methestrol : 9299
Methimazole : 7214
Methionamine : 7649
Methionic acid : 7514
Methone : 4378
Methoxone : 141
Methoxsalen : 6187
Methoxyacetaldehyde : 16
p-Methoxyacetanilide : 71
l-Methoxyacetic acid : 210
Methoxyacetic acid, ethyl ester : 206
6-Methoxybaicalein : 2810
4-Methoxybenzoic acid chloride : 2901
4-Methoxybenzoyl chloride : 2901
4-Methoxybenzyl alcohol : 1891
3-Methoxy-1-butanol : 3625
2-Methoxy-4-carboxyphenol : 2625
Methoxychlor : 2329
2-Methoxy-3,6-dichlorobenzoic acid : 2567
2-(2-Methoxyethoxy)ethanol : 5737
2-Methoxyethyl acetate : 5735
2-Methoxyethyl trichloroacetate : 250
2-Methoxyfuran : 6127
7-Methoxyisoquinoline : 7380
2-(Methoxymethyl)-5-nitrofuran : 6129
2-Methoxynaphthalene : 7914
3-Methoxy-4-nitroaniline : 776
4-Methoxyphenol : 9381
4-Methoxy-*m*-phenylenediamine : 1274
1-(4-Methoxyphenyl)ethanol : 1895
5-Methoxy-2-phenyloxazole : 8582
2-Methoxy-1-propanol : 10386
1-Methoxy-2-propanol : 10387
2-Methoxypropionitrile : 10109
3-Methoxypropiophenone : 10456
*N*1-(6-Methoxy-3-pyridazinyl)sulfanilamide : 2197
7-Methoxyquercetin : 2815
4-Methoxyquinoline : 11274
6-Methoxyquinoline : 11275
8-Methoxyquinoline : 11276
6-Methoxyquinoxaline : 11336
6-Methoxy-α-tetralone : 8048
5-Methoxytryptamine : 7304
Methsuximide : 11156
Methyl abietate : 9168
N-Methylacetanilide : 79
Methyl acetate : 208
Methyl acetoacetate : 3573
Methylacetylene : 10727
Methyl acetylsalicylate : 2482
Methyl acrylate : 10655
Methylacrylonitrile : 10586
Methylal : 7507
Methyl alcohol : 7581
Methylallene : 3145
1-Methyl-1-allyl acetone : 7097
Methylamine : 7442
Methyl 2-aminobenzoate : 2503
Methyl *m*-amino-*p*-hydroxybenzoate : 2501
Methyl [(4-aminophenyl)sulfonyl]-carbamate : 474
4-Methyl-2-aminopyridine : 10867
4-Methyl-2-aminothiazole : 11665
N-Methylaniline : 781
2-(*N*-Methylanilino)ethanol : 5743
5-Methyl-*o*-anisidine : 770
1-Methylanthracene : 423
2-Methylanthracene : 424
9-Methylanthracene : 425
Methyl anthranilate : 2503
2-Methylanthraquinone : 409
Methylarsine : 446
3-Methylazelaic acid : 8135
Methyl azide : 7460
p-Methylazobenzene : 5139
3-Methylbenz[a]anthracene : 522
2-Methylbenzenemethanol : 1897
4-Methylbenzenemethanol : 1899
p-Methylbenzenesulfinic acid : 2182
Methyl benzenesulfonate : 2227
N-(4-Methylbenzenesulfonyl)-*N'*-butylurea : 2207
Methyl benzoate : 2687
3-Methylbenzo[b]furan : 2457
5-Methylbenzo[b]furan : 2458
7-Methylbenzo[b]furan : 2459
4'-Methylbenzophenone-2-carboxylic acid : 2684
2-Methylbenzothiazole : 2863
2-Methylbenzoxazole : 2880
α-Methylbenzyl alcohol : 1896
o-Methylbenzyl alcohol : 1897
m-Methylbenzyl alcohol : 1898
p-Methylbenzyl alcohol : 1899
1-Methylbicyclo[3,1,0]hexane : 2980
2-Methylbicyclo[2.2.2]octane : 2992
2-Methylbiphenyl-3-ylmethyl (*Z*)-(1RS,3RS)-3-(2-chloro-3,3,3-trifluoroprop-1-enyl)- 2,2-dimethylcyclopropanecarboxylate : 2998
N-Methylbis(2,4-xylyliminomethyl)amine : 342
Methyl borate : 3126
Methyl bromide : 7462
Methyl 2-bromoacetate : 117
3-Methyl-1-bromobutane : 3239
Methyl α-bromocinnamate : 10615
Methyl *p*-bromophenyl ketone : 5781
2-Methylbutadiene : 3171
1-Methyl-2,3-butadiene : 8658
Methylbutadiyne : 8684
3-Methyl-1,2-butanediol : 3352
2-Methyl-1,3-butanediol : 3353
2-Methyl-2,3-butanediol : 3355
2-Methylbutanenitrile : 3409
(+)-2-Methyl-1-butanethiol : 3439
2-Methyl-2-butanethiol : 3441
1-Methylbutanethiol : 8845
Methyl butanoate : 3547
2-Methyl-1-butanol : 3626
(±)-3-Methyl-2-butanol : 3629
3-Methyl-2-butenal : 3700
2-Methyl-1-butene : 3799
3-Methyl-1-butene : 3800
2-Methyl-2-butene : 3801
trans-2-Methyl-2-butenoic acid : 3844
Methyl 1-butenyl ketone : 7086
1-Methyl-2-butylacetylene : 6610
2-Methylbutyl alcohol : 3626
3-Methylbutyl chloride : 3252
3-Methylbutyl iodide : 3384
Methyl *tert*-butyl ketone : 3657
3-Methylbutyl 3-methylbutanoate : 3551
3-Methylbutyramide : 3206
(±)-2-Methylbutyric acid : 3538
3-Methylbutyryl chloride : 3693
Methyl caproate : 6898
3-Methylcaproic acid, ethyl ester : 6899
4-Methylcaproic acid, ethyl ester : 6900
Methyl caprylate : 8434
Methyl carbamate : 3953
Methylcarbinol : 5672
Methyl carbonate : 4000
Methylcarbonyl fluoride : 301
Methyl chemosept : 2637
Methyl chloride : 7477
Methyl 2-chloroacrylate : 10621
Methyl chlorocarbonate : 4020
Methyl γ-chlorocrotonate : 3833
Methyl chlorofluoroacetate : 134
Methylchloroform : 5655
3-Methyl-3-chloroheptane : 6369
Methyl 4-chloro-3-nitrobenzoate : 2548
Methyl 5-chloro-2-nitrobenzoate : 2549
2-Methyl-4-chlorophenoxyacetic acid : 141
Methyl 2-chloropropionate : 10222
2-Methyl-5-chlorothiophene : 11741
3-Methylcholanthrene : 2392
(24*R*)-Methylcholest-5-en-3β-ol : 5386
2-Methylchroman : 2778
3-Methylchromone : 2838
1-Methylchrysene : 4117
N-Methylconhydrine : 4144
Methylconiine : 9686
3-Methylcoumarin : 2833
5-Methylcoumarin : 2834
7-Methylcoumarin : 2836
8-Methylcoumarin : 2837
3-Methylcrotonaldehyde : 3700
Methyl *trans*-crotonate : 3847
Methyl *cis*-crotonate : 3848
trans-2-Methylcrotonic acid : 3844
Methyl cyanide : 271
Methyl cyanoacetate : 154
4-Methylcyanobenzene : 2765
Methylcyclobutane : 4192
Methyl cyclobutanecarboxylate : 4176
1-Methylcyclobutene : 4201
Methyl cyclobutyl ketone : 5792
Methylcycloheptane : 4226
1-Methyl-1-cycloheptanol : 4230
2-Methylcycloheptanol : 4231
2-Methylcycloheptanone : 4235
1-Methylcycloheptene : 4245
5-Methylcycloheptene : 4246
2-Methyl-1,3-cyclohexadiene : 4271
Methylcyclohexane : 4426
Methyl cyclohexaneacetate : 4310
β-Methylcyclohexaneacrylic acid : 3834
cis-2-Methylcyclohexanemethanol : 4422
trans-2-Methylcyclohexanemethanol : 4423
β-Methylcyclohexanepropenoic acid : 3834
1-Methyl-1-cyclohexanol : 4530
1-Methylcyclohexanol : 4530
(±)-*cis*-2-Methylcyclohexanol : 4531
(±)-*trans*-2-Methylcyclohexanol : 4532
cis-4-Methyl-1-cyclohexanol : 4535
cis-4-Methylcyclohexanol : 4535
trans-4-Methylcyclohexan-1-ol : 4536
trans-4-Methylcyclohexanol : 4536
cis-3-Methylcyclohexanol : 4533
trans-3-Methylcyclohexanol : 4534
(±)-2-Methylcyclohexanone : 4596
(±)-3-Methylcyclohexanone : 4597
4-Methyl-1-cyclohexanone : 4598
4-Methylcyclohexanone : 4598
(±)-3-Methylcyclohexene : 4671
5-Methyl-2-cyclohexenone : 4712

trans-2-Methylcyclohexylamine : 4293
3-Methylcyclohexyl bromide : 4320
Methyl cyclohexylideneacetate : 156
Methyl cyclohexyl ketone : 5794
2-Methylcyclopentadecanone : 4762
Methylcyclopentane : 4826
3-Methylcyclopentaneacetic acid : 4776
1-Methylcyclopentanecarboxylic acid : 4783
cis-2-Methylcyclopentanol : 4868
trans-2-Methylcyclopentanol : 4869
3-Methylcyclopentanol : 4870
(±)-3-Methylcyclopentanone : 4886
3-Methylcyclopentene : 4929
3-Methyl-2-cyclopentenone : 4945
Methylcyclopropane : 4981
Methyl cyclopropanecarboxylate : 4963
2-Methylcyclopropanecarboxylic acid : 4964
Methyl cyclopropylcarboxylate : 4963
Methylcysteine hydrochloride : 5019
5-Methylcytosine : 11106
2-Methyldecane : 5054
3-Methyldecane : 5055
4-Methyldecane : 5056
Methyl decyl ketone : 5325
Methyldiacetylene : 8684
1-Methyldiallyl : 6328
Methyldiazepinone : 2410
3-Methyl-1,2-diazine : 10856
2-Methyl-1,3-diazine : 11070
4-Methyl-1,3-diazine : 11071
5-Methyl-1,3-diazine : 11072
Methyl 2-diazopropionate : 10237
Methyl 3,5-dibromosalicylate : 2559
Methyl dichloroacetate : 166
5-Methyl-2,4-dichloropyrimidine : 11052
N-Methyldiethanolamine : 5742
3-Methyldigitoxose : 7112
Methyl dihydrocinnamate : 2139
α-Methyldihydrocinnamic acid : 2137
Methyl 2,4-dimethylbenzoate : 2589
Methyl α,α-dimethylhydrocinnamate : 2129
Methyl *N*-(2,6-dimethylphenyl)-*N*-(phenylacetyl)-*DL*-alaninate : 513
Methyl 2,2-dimethylpropionate : 10260
Methyldiphenylamine : 807
Methyl 11-dodecenoate : 5341
Methyldopa : 11931
Methyleneaminoacetonitrile : 285
2,2'-Methylenebiphenyl : 5967
4,4'-Methylenebis[2-chloroaniline] : 785
3,3'-Methylenebis[6-hydroxybenzoic acid] : 2686
3,3'-Methylenebis[4-hydroxycoumarin] : 2839
Methylene blue : 9506
Methylene bromide : 7492
α-Methylene butyrolactone : 6147
Methylene chloride : 7499
exo-Methylenecyclobutane : 4193
Methylenecyclohexane : 4428
Methylenecyclopentane : 4827
Methylene difluoride : 7504
1,2-Methylenedioxy-9,10-dimethoxy-6aα-aporphine : 2468
1,2-(Methylenedioxy)-4-propylbenzene : 2431
5,5'-Methylenedisalicylic acid : 2686
Methylenedisulfonic acid : 7514
Methylene iodide : 7506
3-Methylene-2-norbornanone : 2958
Methylene oxide : 5987
Methylenesuccinic acid : 3328
(1*R*,2*S*)-*N*-Methylephedrine : 1863
1-Methylethanesulfonic acid : 10130
N-Methyl-2-ethanolamine : 5738
Methyl ethoxyacetate : 182
Methyl(2-ethoxyethyl)amine : 5426
Methylethylallene : 6679
Methylethylallylcarbinol : 7084
cis-1-Methyl-3-ethylcyclohexane : 4397
cis-1-Methyl-4-ethylcyclohexane : 4399
trans-1-Methyl-4-ethylcyclohexane : 4400
1-Methyl-1-ethylcyclopentane : 4814
cis-1-Methyl-2-ethylcyclopentane : 4815
1-Methyl-*cis*-3-ethylcyclopentane : 4817
1-Methyl-*trans*-3-ethylcyclopentane : 4818
cis-1-Methyl-2-ethylcyclopropane : 4976
trans-1-Methyl-2-ethylcyclopropane : 4977
Methylethyldichlorosilane : 11397
2-Methyl-5-ethylfuran : 6106
3-Methyl-3-ethylhexane : 6826
Methyl ethyl ketone : 3645
1-Methyl-5-ethylnaphthalene : 7887
2-Methyl-3-ethylpentane : 8813
3-Methyl-3-ethylpentane : 8814
Methyl α-ethylstyryl ketone : 8987
5-Methyl-3-ethylthiophene : 11764
Methyl ethynyl ketone : 3923
9-Methylfluorene : 5974
Methyl fluoride : 7515
Methyl fluoroacetate : 192
Methylfluoroform : 5665
N-Methylformamide : 5995
Methyl formate : 6010
2-Methylfuran : 6130
3-Methylfuran : 6131
5-Methyl-3-furancarboxylic acid : 6065
N-Methylglucamine : 6208
N-Methyl-*L*-glucosamidinostreptosidostreptidine : 11562
N-Methyl-α-*L*-glucosamine : 6215
α-Methylglucoside : 6227
Methyl *DL*-glycerate : 10254
1-Methylglycerol : 3458
β-Methylglycerol α,γ-dichlorohydrin : 10365
7-Methylguanine : 10769
2-Methylheptadecane : 6310
9-Methylheptadecane : 6311
3-Methyl-1,5-heptadiene : 6335
3-Methylheptan-2,4-diol : 6401
2-Methylheptane : 6422
4-Methylheptane : 6424
Methyl heptanoate : 6466
3-Methyl-1-heptanol : 6492
4-Methyl-1-heptanol : 6493
6-Methyl-1-heptanol : 6494
2-Methyl-2-heptanol : 6495
3-Methyl-2-heptanol : 6496
5-Methyl-2-heptanol : 6497
3-Methyl-3-heptanol : 6500
4-Methyl-3-heptanol : 6501
6-Methyl-4-heptanol : 6503
2-Methyl-4-heptanol : 6503
4-Methyl-4-heptanol : 6505
3-Methyl-2-heptanone : 6520
2-Methyl-3-heptanone : 6522
2-Methyl-4-heptanone : 6524
3-Methyl-4-heptanone : 6525
2-Methyl-1-heptene : 6543
3-Methyl-1-heptene : 6544
4-Methyl-1-heptene : 6545
5-Methyl-1-heptene : 6546
6-Methyl-1-heptene : 6547
2-Methyl-2-heptene : 6548
3-Methyl-*trans*-2-heptene : 6549
4-Methyl-*trans*-2-heptene : 6551
2-Methyl-5-heptene : 6553
2-Methyl-3-heptene : 6554
3-Methyl-3-heptene : 6555
4-Methyl-3-heptene : 6556
2-Methyl-4-heptene : 6558
6-Methyl-2-hepten-4-ol : 6575
5-Methyl-2-hepten-4-one : 6596
6-Methyl-4-hepten-2-one : 6599
2-Methyl-3-hepten-6-one : 6599
5-Methylhept-4-en-3-one : 6600
3-Methyl-5-hepten-2-one : 6601
5-Methylhept-5-en-3-one : 6605
Methyl hexacosanoate : 6625
Methyl 2,4-hexadienoate : 6702
2-Methylhexa-4,5-dione : 6809
3-Methylhexanal : 6720
2-Methylhexane : 6834
(+)-3-Methylhexane : 6835
Methylhexaneamine : 6730
3-Methyl-2,5-hexanedione : 6810
(*R*)-2-Methyl-1-hexanol : 6938
5-Methyl-1-hexanol : 6941
3-Methyl-2-hexanol : 6943
5-Methyl-2-hexanol : 6944
(±)-2-Methyl-3-hexanol : 6945
3-Methyl-3-hexanol : 6946
(±)-5-Methyl-3-hexanol : 6947
3-Methyl-2-hexanone : 6976
4-Methyl-2-hexanone : 6977
2-Methyl-1-hexene : 7036
3-Methyl-1-hexene : 7037
4-Methyl-1-hexene : 7038
5-Methyl-1-hexene : 7039
2-Methyl-2-hexene : 7040
Methyl 3-hexenoate : 7062
4-Methyl-2-hexenoic acid : 7056
5-Methyl-2-hexenoic acid : 7057
5-Methyl-2-hexenol : 7078
4-Methyl-5-hexen-3-ol : 7085
4-Methyl-5-hexen-2-one : 7098
5-Methyl-1-hexyne : 7120
2-Methyl-3-hexyne : 7122
2-Methyl-3-hexyn-2-ol : 7129
2-Methyl-2H-indazole : 7242
Methylhydrazine : 7161
Methyl hydrocinnamate : 2139
α-Methylhydrocinnamic acid : 2137
Methyl hydrogen sulfate : 11577
Methyl *p*-hydroxybenzoate : 2637
Methyl 11-hydroxyhendecanoate : 11949
Methyl 11-hydroxyundecanoate : 11949
1-Methylindan : 7255
2-Methylindan : 7256
4-Methylindan : 7257
5-Methylindan : 7258
4-Methyl-1-indanone : 7290
1-Methylindazole : 7239
3-Methylindazole : 7240
5-Methylindazole : 7241
1-Methylindene : 7275
2-Methylindene : 7276
3-Methylindene : 7277
4-Methylindene : 7278
6-Methylindene : 7279
7-Methylindene : 7280
7-Methylindole : 7317
Methyl iodide : 7518
6-Methyl-α-ionone : 3884
Methylisobutylcarbinol acetate : 8947
Methyl isobutyrate : 10301
Methyl isocrotonate : 3848
Methyl isocyanate : 7520
: 7655
Methyl isonipecotate : 9644
Methyl isopentyl ketone : 6978
Methylisopropenylacetylene : 9116
Methyl isopropenyl carbinol : 3871
α-Methyl-*p*-isopropylhydrocinnamaldehyde : 2117
Methyl isopropyl ketone : 3668
6-Methyl-3-isopropylsalicylic acid : 2642
1-Methylisoquinoline : 7381
3-Methylisoquinoline : 7382
4-Methylisoquinoline : 7383
6-Methylisoquinoline : 7384
8-Methylisoquinoline : 7385
Methyl isothiocyanate : 7521
Methyl isovalerate : 3552
5-Methylisoxazole : 7405
Methyl ketone : 10411
2-Methyllactonitrile : 10108
Methyl laurate : 5315
6-Methyllepidine : 11262
Methyl linoleate : 8228
Methyl malonate : 9943
Methyl mercaptan : 7546
Methyl mercaptan, perfluoro : 7528
Methyl methacrylate : 10663
Methyl methanesulfonate : 7532
3-Methyl-6-methoxy-2,5-dihydroxy-1,4-benzoquinone : 4260
4-Methyl-4-methoxy-2-pentanone : 8978
Methyl (2-methylbutyl) ether : 3394
Methyl 2-methylbutyl ketone : 6977
Methyl *N*-methylisonipecotinate : 9645
Methyl 4-methyl-5-oxohexanoate : 6903
Methyl 3-methyl-4-oxopentanoate : 8898
Methyl 3-methyl-3-pentenoate : 9086
N-Methylmorpholine : 7696
1-Methylnaphthalene : 7915
α-Methylnaphthalene : 7915
2-Methylnaphthalene : 7916
β-Methylnaphthalene : 7916

4-Methylnaphthalene-1-carboxaldehyde : 7756
Methyl nicotinate : 10906
Methyl nitrate : 8075
Methyl nitrite : 8091
4-Methyl-2-nitrobenzophenone : 7627
3-Methyl-1-nitrobutane : 3400
trans-Methyl *o*-nitrocinnamate : 10677
2-Methylnitrocyclohexane : 4444
1-Methyl-1-nitrocyclohexane : 4443
1-Methyl-4-nitrocyclohexane : 4445
1-Methyl-1-nitrocyclopentane : 4830
1-Methyl-2-nitrocyclopentane : 4831
2-Methyl-2'-nitrodiphenyl ether : 2005
4-Methyl-1-nitronaphthalene : 7925
2-Methyl-1-nitronaphthalene : 7926
N-Methyl-*N'*-nitro-*N*-nitrosoguanidine : 6284
N-Methyl-*N*-nitroso-*p*-toluenesulfonamide : 2212
Methylnitrosourea : 12019
1-Methyl-1-nonadecanol : 5359
2-Methylnonane : 8140
3-Methylnonane : 8141
4-Methylnonane : 8142
5-Methylnonane : 8143
3-Methyl-1-nonanol : 8166
4-Methyl-4-nonanol : 8170
5-Methyl-5-nonanol : 8171
8-Methyl-1-nonanol : 8168
2-Methyl-5-nonanone : 8176
2-Methylnon-1-ene : 8185
Methyl nonyl ketone : 11958
8-Methyl-4-nonyne : 8202
Methyl 9,12-octadecadienoate : 8229
9-Methyloctadecanoic acid : 8264
4-Methyl-3,5-octadiene : 8329
2-Methyloctane : 8408
3-Methyloctane : 8409
4-Methyloctane : 8410
3-Methyloctanoic acid : 8433
3-Methyl-1-octanol : 8461
4-Methyl-1-octanol : 8462
2-Methyl-2-octanol : 8464
3-Methyl-3-octanol : 8466
6-Methyl-3-octanol : 8468
4-Methyl-4-octanol : 8469
3-Methyl-2-octanone : 8475
4-Methyl-3-octanone : 8476
7-Methyl-3-octanone : 8477
7-Methyl-4-octanone : 8479
2-Methyl-1-octene : 8505
2-Methyl-5-octen-4-one : 8523
Methyl octyl ketone : 5075
7-Methyl-3-octyne : 8537
Methyl (2-octyn-1-yl) ether : 8536
Methyl oleate : 8307
4-Methylolphenol : 1884
Methyl orange : 2222
4-Methyloxazole : 8583
Methyloxirane : 8621
Methyl β-oxoheptyl ketone : 8137
2-Methyl-4-oxo-3-(2-propenyl)-2-cyclopenten-1-yl 2,2-dimethyl-3-(2-methyl-1-propenyl) cyclopropane carboxylate : 336
Methylparaben : 2637
Methyl pentadecanoate : 8646
2-Methyl-1-pentadecanol : 8649
3-Methyl-1,2-pentadiene : 8666
cis-2-Methyl-1,3-pentadiene : 8668
4-Methyl-1,3-pentadiene : 8670
2-Methyl-1,4-pentadiene : 8671
3-Methylpentane : 8824
3-Methyl-2,4-pentanediol : 8794
2-Methyl-1-pentanol : 8937
2-Methyl-2-pentanol : 8940
3-Methyl-2-pentanol : 8941
4-Methyl-2-pentanol : 8942
2-Methyl-3-pentanol : 8943
2-Methyl-1-pentene : 9052
3-Methyl-1-pentene : 9053
4-Methyl-1-pentene : 9054
2-Methyl-2-pentene : 9055
(*cis*)-3-Methyl-2-pentene : 9057
4-Methyl-*trans*-2-pentene : 9058
cis-4-Methyl-2-pentene : 9059
4-Methyl-2-pentenoic acid : 9078
cis-3-Methyl-2-pentenoic acid methyl ester : 9085
4-Methyl-3-pentenol : 9096
2-Methyl-3-penten-2-ol : 9097
3-Methyl-3-penten-2-ol : 9098
Methyl 4-pentenyl ketone : 6585
2-Methyl-1-penten-3-yne : 9116
3-Methyl-3-penten-1-yne : 9117
Methylpentylacetylene : 8530
3-Methylpentylamine : 8706
(3-Methylpentyl)benzene : 2010
2-Methylpentyl bromide : 8714
3-Methylpentyl bromide : 8715
1-Methyl-2-pentylcyclohexane : 4446
Methyl pentyl ether : 8822
Methyl pentyl ketone : 6510
Methyl pentyl sulfide : 8826
2-Methylpentyl trifluoroacetate : 268
3-Methyl-1-pentyne : 9134
Methyl 2-pentynoate : 9139
3-Methylphenanthrene : 9180
4-Methylphenanthrene : 9181
2-Methylphenazine : 9195
Methyl phenoxyacetate : 230
2-Methyl-4-phenyl-2-butanol : 2150
1-(3-Methylphenyl)ethanol : 1858
1-(4-Methylphenyl)ethanol : 1859
1-(3-Methylphenyl)ethyl alcohol : 1858
3-Methyl-5-phenylhydantoin : 7221
1-Methyl-2-phenylhydrazine : 7166
Methyl phenyl ketone : 5864
Methyl β-phenylpropionate : 2139
2-Methyl-3-phenylpropionic acid : 2137
1-Methyl-3-phenylpropylamine : 2119
3-Methyl-2-phenylpyridine : 10991
2-Methyl-5-phenylthiophene : 11774
Methyl 2-phenylvinyl ketone : 3881
Methyl picrate : 1946
3-Methylpicric acid : 9434
2-Methylpiperazine : 9619
N-Methylpiperidine : 9679
1-Methyl-1,3-propanediol : 3345
Methyl propionate : 10296
2-Methyl-6-propionylphenol : 10444
1-Methyl-2-propylacetylene : 7115
Methylpropylcarbinol : 8916
2-Methylpropyl chloroformate : 4024
1-Methyl-1-propylcyclohexane : 4449
1-Methyl-2-propylcyclohexane : 4450
1-Methyl-3-propylcyclohexane : 4451
1-Methyl-4-propylcyclohexane : 4452
3-Methyl-1-propylcyclohexanol : 4551
2-Methyl-1-propylcyclohexanol : 4550
1-Methyl-1-propylcyclopentane : 4833
cis-1-Methyl-2-propylcyclopentane : 4834
trans-1-Methyl-2-propylcyclopentane : 4835
cis-1-Methyl-3-propylcyclopentane : 4836
trans-1-Methyl-3-propylcyclopentane : 4837
Methyl propyl ether : 10069
1-Methyl-1-propylhexanol : 8170
Methyl propyl ketone : 8957
Methyl propyl ketone oxime : 8985
Methyl propyl sulfide : 10088
2-Methyl-2-propyltrimethylene butylcarbamate carbamate : 3933
2-Methylpyrazine : 10810
1-Methylpyrazole : 10834
4-Methylpyrazole : 10836
3-Methylpyridazine : 10856
Methyl 3-pyridyl ketone : 5868
2-Methylpyrimidine : 11070
4-Methylpyrimidine : 11071
5-Methylpyrimidine : 11072
N-(4-Methyl-2-pyrimidyl)sulfanilamide : 2199
3-Methylpyrrole : 11132
N-Methylpyrrolidine : 11168
N-Methyl-2-pyrrolidone : 11185
4-Methylquinaldine : 11255
3-Methylquinoline : 11278
5-Methylquinoline : 11280
6-Methylquinoline : 11281
7-Methylquinoline : 11282
8-Methylquinoline : 11283
6-Methyl-7-quinolinol : 11320
2-Methylquinoxaline : 11337
6-Methylquinoxaline : 11338
Methyl red : 2587
6-Methyl-β-resorcylic acid : 2573
Methyl salicylate : 2634
3-Methylsalicylic acid : 2629
4-Methylsalicylic acid : 2630
5-Methylsalicylic acid : 2631
Methylsilane : 11437
Methyl silicate : 11508
Methyl silyl ether : 11436
o-Methylstyrene : 1640
m-Methylstyrene : 1641
p-Methylstyrene : 1642
α-Methylstyrene : 1972
Methyl styryl ketone : 3881
5-Methyl-3-sulfanilamidoisoxazole : 2198
Methyl sulfate : 11577
Methyl sulfocyanate : 11705
3-Methylsulfolane : 11789
Methylsulfonic acid : 7530
N-Methyltaurine : 5624
17-Methyltestosterone : 358
2-Methyltetrahydrofuran : 6171
3-Methyltetrahydrofuran : 6172
2-Methylthiazole : 11678
4-Methylthiazole : 11679
4-Methyl-5-thiazolecarboxylic acid : 11669
4-Methyl-5-thiazoleethanol : 11677
*N*1-(4-Methyl-2-thiazolyl)sulfanilamide : 2200
(Methylthio)acetonitrile : 286
Methyl thiocyanate : 11705
2-(Methylthio)ethanol : 5749
2-Methylthiophene : 11769
3-Methylthiophene : 11770
3-(Methylthio)propionaldehyde : 9786
Methylthiouracil : 11108
8-Methyltocol : 2785
Methyl trichloroacetate : 253
Methyltrichlorosilane : 11465
2-Methyltricosane : 11856
7-Methyltridecane : 11879
Methyl tridecanoate : 11882
12-Methyl-1-tridecanol : 11886
Methyltrimethylene glycol : 3345
3-Methyl-2,4,6-trinitrophenol : 9434
Methyltriphenoxysilane : 11443
Methyltriphenylsilane : 11444
N-Methyltryptophan : 11923
β-Methylumbelliferone : 2831
2-Methyl-1-undecanol : 11955
Methyl undecyl ketone : 11889
Methyl 10-undecynoate : 11991
Methyl 9-undecynoate : 11990
6-Methyluracil : 11069
3-Methyluric acid : 10765
7-Methyluric acid : 10766
2-Methylvaleraldehyde : 8693
Methyl valerate : 8891
Methyl vincaminate : 5349
Methylvinylacetylene : 9111
Methyl vinyl ether : 5908
Methyl vinyl ketone : 3875
Methysticin : 10790
Metiazinic acid : 9500
Metoxon : 3939
Metronidazole : 7206
Metyrapone : 10462
Metyridine : 10979
Michler's Base : 786
Michler's ketone : 7593
Mimosine : 11011
Mirex : 7644
Molybdenum hexacarbonyl : 7662
Monobenzone : 9453
Monofluoroacetic acid : 189
Monomethylarsine : 446
Monophenylphosphine : 9519
Montanic acid : 8221
Montanyl alcohol : 8222
Monuron : 11999
Moquizone : 11198
Morin : 2816
Morphine : 7666
Morphine, *N*-oxide : 7668
Morpholine, 2,6-dimethyl-4-tridecyl- : 11898
Morpholine, *N*-(2-ethoxyethyl) : 7693

Mucic acid : 6193
Murexide : 11097
Muscone : 4763
Mustard gas : 5645
Mustard oil : 10575
Myacide : 1853
Mycophenolic acid : 7052
β-Myrcene : 8330
Myrcene : 8330
Myricyl alcohol : 11821
Myristaldehyde, dimethyl acetal : 11617
Myristic acid : 11622
Myristicin : 2428
Myristonitrile : 11619
Myristoyl chloride : 11635
Nandrolone phenpropionate : 5408
9,10-Naphthacenequinone : 7707
1-Naphthaldehyde, 4-methyl- : 7756
Naphthalene, 2-amino-1,4-dimethyl : 8064
Naphthalene, 2-benzyl- : 7950
Naphthalene, 1,2-diamino : 7795
Naphthalene, 1,4-diamino : 7796
Naphthalene, 1,8-diamino : 7798
Naphthalene, 2,3-diamino : 7799
2,5-Naphthalenedisulfonic acid : 7868
Naphthalene-2,7-disulfonic acid : 7869
Naphthalene, 1,2,3,4,9,10-hexahydro : 7897
Naphthalene, 1-(1-hydroxyethyl) (*DL*) : 7911
Naphthalene, 2-(isopentyloxy)- : 7918
Naphthalene, 1-isopropyl-7-methyl : 7924
Naphthalene, 5-isopropyl-1,3,8-trimethyl- : 8015
Naphthalene, 8-isopropyl-1,2,5-trimethyl- : 8014
1-Naphthalenesulfonic acid, 8-hydroxy, sulfone : 7840
Naphthalene, 2-vinyl- : 7882
2-Naphthalenol, 1,1'-methylenebis-, monosodium salt : 11517
Naphthalic acid : 7805
Naphthalic acid, diethyl ester : 7806
Naphthalol : 2645
Naphthaloyl chloride : 7804
Naphthanil Blue B Base : 3033
1H-Naphth[2',1':4,5]indeno[1,7a-c]pyrrole, con-5-enin-3-amine deriv. : 4143
Naphtho[1,8-de]-1,3,2-dioxathiin, 2-oxide : 7840
1-Naphthoic acid : 7758
2-Naphthoic acid : 7759
2-Naphthoic acid, 5-bromo- : 7769
1-Naphthoic acid, 5-chloro- : 7762
1-Naphthoic acid, 3,4-dihydro- : 7764
2-Naphthoic acid, 1,2,3,4-tetrahydro- : 7778
1-Naphthol : 8017
2-Naphthol : 8018
2-Naphthol, 1-benzyl : 8040
1-Naphthol, 2,3-dichloro : 8025
2-Naphthol, 1,4-dimethyl- : 8027
2-Naphthol-6,8-disulfonic acid : 7871
2-Naphthol-3,6-disulfonic acid : 7876
1-Naphthol-4-sulfonic acid : 7971
1-Naphthol-2-sulfonic acid : 7973
2-Naphthol-6-sulfonic acid : 7974
β-Naphthoquinoline : 2437
1,2-Naphthoquinone : 7842
1,4-Naphthoquinone : 7843
1,4-Naphthoquinone, 3-acetonyl-5,8-dihydroxy-6-methoxy-2-methyl- : 7855
1,2-Naphthoquinone, 3-bromo- : 7844
1,4-Naphthoquinone, 5-chloro- : 7847
1,4-Naphthoquinone, 5,6-dichloro : 7851
1,4-Naphthoquinone, 5-hydroxy-2-methyl : 7863
1,2-Naphthoquinone-4-sulfonic acid,sodium salt : 7970
Naphthoresorcinol : 7834
2-Naphthoxyacetic acid : 217
1-Naphthylacetic acid : 7734
1-Naphthylamine : 7708
2-Naphthylamine : 7709
1-Naphthylamine, 5-bromo- : 7711
β-Naphthylamine-3-carboxylic acid : 7768
1-Naphthylamine, 5,8-dihydro- : 7714
1-Naphthylamine, *N*,*N*-diphenyl- : 7717
2-Naphthylamine-6,8-disulfonic acid : 7870
1-Naphthylamine-4,7-disulfonic acid : 7872
1-Naphthylamine-4,6-disulfonic acid : 7873
2-Naphthylamine-1-sulfonic acid : 7961
1-Naphthylamine-4-sulfonic acid : 7962
1-Naphthylamine-5-sulfonic acid : 7963
1-Naphthylamine-8-sulfonic acid : 7965
1-Naphthylamine, *N*-*o*-tolyl- : 7723
2-Naphthylamine, *N*-*o*-tolyl- : 7725
2-Naphthyl benzoate : 8021
1-Naphthyl bromide : 7739
1-Naphthyl chloride : 7780
1,2-(1,8-Naphthylene)benzene : 5966
Naphthyleneethylene : 6
(±)-1-(α-Naphthyl)ethanol : 7911
1-Naphthyl ether : 7939
N-(1-Naphthyl)ethylenediamine : 5514
1-Naphthyl isocyanate : 7904
1-Naphthyl isothiocyanate : 7906
1-Naphthyl mercaptan : 8007
2-Naphthyl mercaptan : 8008
1-Naphthyl pentyl ether : 7943
2-Naphthyl pentyl ether : 7944
2-Naphthyl 1-pentyl ether : 7944
1-Naphthyl salicylate : 2644
α-Naphthyl salicylate : 2644
2-Naphthyl salicylate : 2645
β-Naphthyl salicylate : 2645
β-Naphthylsulfonic acid : 7960
1-Naphthyl(2-tolyl)amine : 7723
2-Naphthyl(2-tolyl)amine : 7725
1-Naphthylurea : 12022
2-Naphthylurea : 12023
Naptalam : 2702
Narceine : 2586
Narcobarbital : 11079
Naringenin : 2802
Naringin : 2795
Navidrex : 2853
Nealbarbital : 11090
Neocuproine : 9192
Neohexane : 3289
Neopentane : 9910
Neopentane, 1,1-dibromo : 9886
Neopentanoyl chloride : 10503
Neopentyl alcohol : 10373
Neopentyl glycol : 10005
Neophyl chloride : 1143
Neopine : 7671
Nepetalactone : 4758
Nephramid : 32
Nephril : 2857
Nerol : 8334
Neurine : 5878
Niacinamide : 10890
Nialamide : 10910
Nickel tetracarbonyl : 8068
Nicotine : 10992
Nicotinic acid : 10894
Nicotinic acid benzyl ester : 10911
Nicotinyl alcohol : 10976
Nigelline : 2672
Nikethamide : 10891
Nimbecetin : 2843
Ninhydrin : 7267
Nioxime : 4379
Nipagin A : 2621
Nipecotic acid : 9641
Nirvotin : 1877
Nitarsone : 459
Nitramine : 818
Nitranilic acid : 4259
Nitric acid, anhydride with acetic acid : 108
Nitrilotriacetic acid : 6256
5-Nitro-2-acetamidothiazole : 85
α-Nitroacetanilide : 80
Nitroacetic acid, ethyl ester : 218
5-Nitro-2-aminothiazole : 11666
o-Nitroaniline : 822
m-Nitroaniline : 823
p-Nitroaniline : 824
5-Nitro-*o*-anisidine : 775
2-Nitroanisole : 1935
9-Nitroanthracene : 426
5-Nitroanthracene : 426
3-Nitroanthranilic acid : 2506
5-Nitrobarbituric acid : 11102
Nitrobenzene : 2049
p-Nitrobenzeneazosalicylic acid : 2648
m-Nitrobenzenecarboxylic acid : 2704
p-Nitrobenzenecarboxylic acid : 2705
o-Nitrobenzoic acid : 2703
m-Nitrobenzoic acid : 2704
p-Nitrobenzoic acid : 2705
4-Nitrobenzoic acid chloride : 2906
5-Nitrobenzotriazole : 2873
p-Nitrobenzoyl chloride : 2906
p-Nitrobenzyl alcohol : 1908
4-Nitrobenzyl chloride : 1208
p-Nitrobenzyl cyanide : 967
o-Nitrobiphenyl : 3085
p-Nitrobiphenyl : 3087
p-Nitrobromobenzene : 1077
3-Nitro-4-bromobenzoic acid : 2529
2-Nitro-5-bromobenzoic acid : 2530
Nitrobromomethane : 7473
1-Nitrobutane : 3412
2-Nitro-1-butanol : 3640
3-Nitro-2-butanol : 3642
Nitrocarbol : 7522
2-Nitro-3-chlorobenzoic acid : 2546
2-Nitro-5-chlorothiophenol : 9269
Nitrocyclohexane : 4454
Nitrocyclopentane : 4839
3-Nitrodibenzofuran : 5166
p-Nitrodiphenyl : 3087
Nitroerythrol : 3432
Nitroethane : 5607
2-Nitroethanol : 5752
2-Nitro-2-ethyl-1,3-propanediol : 10008
Nitroform : 7562
Nitrofurantoin : 7222
Nitrofurazone : 7144
3-[(5-Nitrofurfurylidene)amino]-2-oxazolidinone : 8589
Nitroglycerin : 10197
Nitroguanidine : 6287
1-Nitroheptane : 6429
1-Nitrohexane : 6841
5-Nitroisoquinoline : 7386
Nitromersol : 7437
Nitromethane : 7522
1-Nitro-3-methylbutane : 3400
(Nitromethyl)cyclohexane : 4455
(Nitromethyl)cyclopentane : 4840
Nitromide : 614
1-Nitronaphthalene : 7929
2-Nitronaphthalene : 7930
1-Nitro-2-naphthol : 8034
1-Nitrooctane : 8412
Nitropentaerythritol : 9990
1-Nitropentane : 8832
2-Nitropentane : 8833
3-Nitropentane : 8834
2-Nitro-1-pentanol : 8950
3-Nitro-2-pentanol : 8952
Nitrophenide : 5247
o-Nitrophenol : 9435
m-Nitrophenol : 9436
p-Nitrophenol : 9437
p-Nitrophenylacetic acid : 933
1-Nitro-4-phenylbenzene : 3087
4-Nitro-*o*-phenylenediamine : 1281
(4-Nitrophenyl)hydrazine : 7168
(4-Nitrophenyl)methanol : 1908
o-Nitrophenylpropiolic acid : 10740
2-Nitrophenyl 2-tolyl ether : 2005
4-Nitrophenylurea : 12025
1-Nitropropane : 10116
2-Nitropropane : 10117
1-Nitro-2-propanol : 10396
5'-Nitro-2'-propoxyacetanilide : 84
4-Nitropyridine *N*-oxide : 10993
5-Nitroquinoline : 11284
6-Nitroquinoline : 11285
7-Nitroquinoline : 11286
4-Nitroquinoline oxide : 11287
3-Nitrosalicylic acid : 2646
5-Nitrosalicylic acid : 2647
Nitrosobenzene : 2059
N-Nitrosodiethylamine : 5432
N-Nitrosodimethylamine : 7450

SYNONYM INDEX OF ORGANIC COMPOUNDS (continued)

p-Nitroso-*N*,*N*-dimethylaniline : 720
p-Nitrosodiphenylamine : 829
N-Nitroso-*N*-diphenylamine : 830
N-Nitrosomorpholine : 7699
1-Nitroso-2-naphthol : 8035
4-Nitrosophenol : 9438
N-Nitrosopiperidine : 9688
N-Nitrosopyrrolidine : 11170
5-Nitrothiazolylamine-2 : 11666
2-Nitrothiophene : 11775
3-Nitrothiophene : 11776
o-Nitrotoluene : 2000
m-Nitrotoluene : 2001
p-Nitrotoluene : 2002
5-Nitro-*o*-toluidine : 801
2-Nitro-4-(trifluoromethyl)-1-chlorobenzene : 1217
Nitrourea : 12024
Nitrous acid, benzyl ester : 8097
Nitrous acid, isopropyl ester : 8092
Nitroxoline : 11323
4-Nitro-*o*-xylene : 1475
Nonacosanyl alcohol : 8100
6,8-Nonadien-2-one, 5-isopropyl-8-methyl-, (*E*)-(*S*)-(+)- : 8113
Nonaldehyde : 8117
2-Nonen-1-al : 8179
Nonyl acetate : 220
Nonyl alcohol : 8157
Nonyl mercaptan : 8145
Nonyl methyl ketone : 11958
19-Norandrostenolone phenylpropionate : 5408
Norboral : 2192
2-Norbornanecarboxaldehyde : 2916
Norbornane, 7,7-dimethyl-2-methylene-, (±)- : 2923
2-Norbornanone, 3-methylene- : 2958
2-Norbornene, 2,3-dimethyl- : 2968
Norcarane : 2915
Norcholanic acid : 8210
Nordazepam : 2411
Nordihydroguaiaretic acid : 1515
Norephedrine, *N*-ethyl : 1871
Norepinephrine : 1500
Norhyoscyamine : 916
Normethadone : 6968
Normorphine : 7665
Normotiroide : 11665
Normuscol : 4760
Nornicotine : 11017
Norpempidine : 9711
cis-Norpinic acid : 4183
Norpinic acid, dimethyl ester : 4184
Norsulfazole : 2206
Nortestosterone phenylpropionate : 5408
Norvalamine : 3209
Noscapine : 7356
Novoldiamine : 8736
Novonal : 9003
Novoscabin : 2720
Ocimene : 8485
Octaacetyl sucrose : 6229
Octabenzone : 7621
Octachlorocyclopentene : 4932
Octacosyl alcohol : 8222
Octacosylbenzene : 2065
Octacosylcyclohexane : 4458
Octadecamethylcyclononasiloxane : 4726
Octadecamethyloctasiloxane : 8482
Octadecanal, dimethyl acetal : 8237
Octadecane, 1-cyclohexyl- : 4459
Octadecane, 1-cyclopentyl- : 4842
1,18-Octadecanedicarboxylic acid : 5352
Octadecane, 1-phenyl- : 2066
3-Octadecanol (*DL*) : 8279
Octadecenyl aldehyde : 8288
Octadecylbenzene : 2066
Octadecyl mercaptan : 8244
Octadecyltrichlorosilane : 11470
α,ω-Octadiene : 8319
2,2,3,3,4,4,5,5-Octafluoro-1-pentanol : 8953
1,2,3,4,4a,9,10,10-Octahydrophenanthrene : 9184
1,2,3,4,5,6,7,8-Octahydrophenanthrene : 9185
Octalene : 5177
Octamethylcyclotetrasiloxane : 4993
(all-E)-2,6,10,14,19,23,27,31-Octamethyl-2,6,8,10,12,14,16,18,20,22,24,26,30-dotriacontatridecaene : 4053
Octamethylene dibromide : 8369
Octamethylpyrophosphoramide : 5220
1,1,1,3,5,7,7,7-Octamethyltetrasiloxane : 11647
Octamethyltrisiloxane : 11914
Octamylamine : 6354
Octane, 2-bromo (*DL*) : 8357
Octane, 1-cyclohexyl- : 4460
Octane, 1-cyclopentyl- : 4843
Octane, 2-phenyl- : 1981
Octane, 1-phenyl- : 2067
2-Octanethiol (*DL*) : 8419
Octane, 2,2,4-trimethyl : 8420
Octanoic acid, 3,5,7,8-tetrachloro-2,2,3,4,4,5,6,6.7,8,8-undecafluoro : 8442
Octatetraene : 8483
Octatrienal : 8484
4-Octene, 4-chloro (*cis*) : 8500
3-Octen-2-ol, 2-methyl (*cis*) : 8519
3-Octen-5-ol, 4-methyl : 8521
1-Octen-5-one, 7-methyl : 8525
1-Octen-3-yne-5-ol, 5-methyl : 8527
Octodecylcyclohexane : 4459
Octylacetylene : 5097
Octyl alcohol : 8443
Octyl bromide : 8356
Octyl chloride : 8359
9-Octyldocosane : 5275
Octyl enanthate : 6468
Octyl fluoride : 8405
Octyl heptanoate : 6468
(±)-2-Octyl iodide : 8407
Octyl mercaptan : 8418
Octyl methyl ketone : 5075
1-Octylnaphthalene : 7937
Octyl phthalate : 1344
Octyltrichlorosilane : 11471
1-Octyne-3-ol, 3-methyl : 8544
2-Octynol : 8543
Oenanthylidene acetone : 5093
Oestra-1,3,5(10)-triene-3,17β-diol : 5399
Oleanolic acid : 8547
Oleic acid : 8294
Oleic acid, benzyl ester : 8309
Oleic acid, isopentyl ester : 8306
Oleyl alcohol : 8312
Opianic acid : 2608
Orcinol : 1531
Orcinolcarboxylic acid : 2573
Oretic : 2855
Orexine : 11196
Orotic acid : 11050
Oroxylin A : 2810
o-Orsellinic acid : 2573
Orthanilic acid : 2215
Orthesin : 2497
Orthocaine : 2501
Orthoform : 2501
Orthoformic acid, triisobutyl ester : 10080
Ouabain : 4040
Ovahormon benzoate : 5403
Ovex : 2221
10H-9-Oxaanthracene : 12049
1-Oxa-2-azacyclopentadiene : 7402
1-Oxa-2-aza-1H-indene : 2389
1-Oxa-3-azaindene : 2877
7-Oxabicyclo[4.1.0]heptane, β,β-carotene-3,3'-diol deriv. : 4056
7-Oxabicyclo[2.2.1]heptane-2,3-dicarboxylic acid, disodium salt : 5370
Oxaceprol : 9766
Oxacyclohexane : 10800
Oxacyclopentadiene : 6025
Oxacyclotetradecane-2,10-dione, 4-[(2,6-dideoxy-3-C-methyl-3-O-methyl-α-*L*-*ribo*-hexopyranosyl)oxy]-14-eth : 5396
1-Oxa-3,4-diazacyclopentadiene : 8563
1,2,4-Oxadiazolidin-3,5-dione, 2-*tert*-butyl-4-(2,4-dichloro-5-isopropoxyphenyl)- : 8565
1,2,4-Oxadiazolidine-3,5-dione, 2-(3,4-dichlorophenyl)-4-methyl- : 7643
Oxalacetic acid : 3339
Oxalacetic ester : 3340
Oxalic acid : 5553
Oxalic acid, diallyl ester : 5564
Oxalic acid dichloride : 5585
Oxalic acid, dicyclohexyl ester : 5559
Oxalic acid dihydrate : 5561
Oxalic acid, diisobutyl ester : 5557
Oxalic acid, disodium salt : 5566
Oxalic acid, lead salt : 5568
Oxalyl chloride : 5585
Oxamic acid : 104
Oxamide : 5493
Oxane : 10800
1,4-OxathiiN-3-carboxamide, 5,6-dihydro-2-methyl-*N*-phenyl- : 4039
1,3-Oxazole : 8575
2,4-Oxazolidinedione, 3-allyl-5-methyl- : 8586
2,4-Oxazolidinedione, 3-(3,5-dichlorophenyl)-5-ethenyl-5-methyl- : 12048
2-Oxazoline, 2-amino-5-phenyl- : 8574
Oxetane, gluconic acid deriv. : 6211
3-Oximino-2-octanone : 8399
3-Oximino-2-pentanone : 8803
1-Oxindene : 2438
Oxindole, 4,7-dimethyl : 7330
Oxindole, 1-methyl-3-ethyl : 7332
Oxiniacic acid : 10909
2-Oxo-3-acetyltetrahydrofuran : 6139
(+)-2-Oxo-10-bornanesulfonic acid : 2929
2-Oxobutanedioic acid diethyl ester : 3340
18-Oxocorticosterone : 9746
Oxoethanoic acid : 223
β-Oxoglutaric acid : 6782
2-Oxo-2H-1-benzopyran-3-carboxylic acid : 2775
Oxole : 6025
2-Oxo-5-methylhex-3-ene : 7094
Oxophenylarsine : 447
2-Oxosuccinic acid : 3339
Oxotremorine : 11188
Oxyacanthine : 8631
Oxybenzone : 7620
Oxycholin : 4101
Oxycinchophen : 11233
Oxydiethylene acetate : 5754
Oxydimorphine : 3002
Oxymethurea : 11995
Oxyphenonium bromide : 5443
Oxytetracycline : 7705
Palmitaldehyde, dimethyl acetal : 6636
Palmitic acid : 6644
Palmitic acid, benzyl ester : 6655
Palmitoleic acid : 6670
Palustric acid : 9166
Pamacyl : 2498
Pamoic acid : 7777
Panthesin : 8926
Pantocaine : 2531
Pantocid : 2565
Pantolactone : 6143
Pantothenic acid : 329
D(+)-Pantothenyl alcohol : 3202
d-Pantothenyl alcohol : 3202
Papaveraldine : 7611
Papaverine : 7378
Parabanic acid : 7226
Paracetamin : 29
Paraldehyde : 11908
Paraldol : 5191
Parasorbic acid : 10791
Parasympatol : 1887
Parathion : 9583
Paraxenol : 3090
Pargyline : 1823
Paroxypropione : 10450
Parsley camphor : 2423
Patchouli alcohol : 7586
Patulin : 6183
Pebulate : 3976
Pelargonic acid : 8146
Pelargonic acid, butyl ester : 8147
Pelargonic acid, 9-fluoro : 8151
Pelargonic acid, heptyl ester : 8152

Pelargonic acid, 3-methyl : 8153
Pelargonic acid 3-methyl, ethyl ester : 8155
Pelargonic acid, pentyl ester : 8156
Pelargonidin chloride : 2848
(-)-Pelletierine : 10485
Pellotine : 7394
Pempidine : 9691
Penicillamine : 12043
Penicillamine cysteine disulfide : 12042
Penicillic acid : 6701
Penicillin G, sodium salt : 11655
Pentabromoacetone : 10477
1,2,3,4,5-Pentabromobenzene : 2082
Pentabromobenzene : 2082
1,1,1,2,2-Pentabromoethane : 5616
Pentabromophenol : 9439
1,2,3,4,5-Pentachlorobenzene : 2084
Pentachlorobenzene : 2084
Pentachlorobenzoic acid : 2712
1,2,2,3,4-Pentachlorobutane : 3420
1,1,1,4,4-Pentachloro-2-butene : 3805
Pentachloroethane : 5617
Pentachlorophenol : 9440
1,1,1,2,3-Pentachloropropane : 10125
1,1,2,3,3-Pentachloropropane : 10126
Pentacosylbenzene : 2088
Pentacosylcyclohexane : 4462
1H-Pentadecafluoroheptane : 6432
Pentadecaldehyde : 8640
Pentadecane, 1-cyclopentyl- : 4844
Pentadecanoic acid, 15-hydroxy,ω-lactone : 8562
sec-Pentadecyl alcohol : 8647
1-Pentadecylamine : 8641
3-Pentadecylcatechol : 1543
Pentadecylic acid : 8645
Pentadecyl methyl ketone : 6320
1,3-Pentadiene-1-carboxylic acid : 6698
2,4-Pentadienoic acid, 5(3,4-metylenedioxyphenyl) : 8677
Pentaerythritol : 9989
Pentaerythrityl tetraacetate : 9988
Pentaerythrityl tetranitrate : 9990
Pentaethylbenzene : 2090
1,2,3,4,5-Pentafluorobenzene : 2091
Pentafluorobenzene : 2091
Pentafluorophenol : 9442
1,1,1,2,2-Pentafluoropropane : 10128
Pentafluoropropionic acid, methyl ester : 10317
2,3,4,5,6-Pentafluorotoluene : 2094
3,5,7,2',4'-Pentahydroxyflavone : 2816
3,5,7,3',4'-Pentahydroxyflavone : 2817
Pentaiodobenzene : 2096
Pentalin : 5617
1,2,3,4,5-Pentamethylbenzene : 2097
Pentamethylbenzene : 2097
Pentamethylbenzoic acid : 2713
2,4,6,8,10-Pentamethylcyclopentasiloxane : 4902
Pentamethylene : 4774
Pentamethylene acetate : 8791
Pentamethylene chloride : 8747
Pentamethylene diiodide : 8758
Pentamethylene glycol : 8787
Pentamethylene oxide : 10800
2,2,4,6,6-Pentamethylheptane : 6433
2,2,3,3,4-Pentamethylpentane : 8836
2,2,3,4,4-Pentamethylpentane : 8837
Pentamethylphenol : 9443
1,1,3,5,5-Pentamethyl-1,3,5-triphenyltrisiloxane : 11915
Pentane, 1-amino-3-methyl : 8706
3-Pentanecarboxylic acid : 3514
Pentane, 2-chloro (*d*) : 8720
Pentane 3-chloro-2,2-dimethyl-3-ethyl : 8727
Pentane, 1,5-dicyclohexyl- : 4464
1,2-Pentanediol (*DL*) : 8790
1,5-Pentanedione : 8734
2,4-Pentanedione, bis,beryllium(II)complex : 2912
Pentane, 1,3-diphenyl- : 1697
Pentane, 1,5-diphenyl- : 2098
Pentane, 1,1-diphenyl- : 2108
Pentane, 2-iodo (*DL*) : 8819
Pentane, 3-isopropyl-2,4-dimethyl- : 8766
Pentane, 2-methyl-3,3-diethyl : 8757
Pentane, 3-methyl-3-phenyl : 1691
Pentane, 3-methyl-1-phenyl : 2010
Pentane, 2-methyl-2-phenyl : 1455
Pentane, perfluoro : 8807
Pentane, perfluoro,2-trifluoro methyl : 8854
Pentane, 3-phenyl : 1701
Pentane, 1-phenyl : 2107
Pentane, 2,2,3-trimethyl-3-ethyl : 8815
Pentane, 2,3,4-trimethyl-3-ethyl : 8816
1,4,5-Pentanetriol : 8851
1,2,3-Pentanetriol : 9119
Pentanoic acid, 4-[(17β)-androstan-17-yl]- : 4094
Pentanoic acid, 5-chloro-4-hydroxy,γ-lactone : 6141
3-Pentanol, 1,5-bis(tetrahydro-2-furyl)- : 6166
1-Pentanol, 2,4-dimethyl (*DL*) : 8927
1-Pentanol, 3-methyl (*DL*) : 8938
1-Pentanol-2-one : 8971
1-Pentanol-4-one : 8974
2-Pentanol-3-one : 8975
3-Pentanol-2-one : 8972
1-Pentanol, 1-phenyl- : 1846
1-Pentanol, 5-phenyl- : 2101
2-Pentanol, 2-phenyl- : 1904
2-Pentanol, 5-phenyl- : 1106
3-Pentanol, 1-phenyl (*d*) : 2152
3-Pentanol, 3-phenyl- : 1855
2-Pentanone, 3-benzylidene- : 8987
2-Pentanone, 3-methyl (*DL*) : 8980
Pentasa : 2499
Pentatriacontylcyclopentane : 4845
Pent-4-en-1-al : 9001
trans-2-Pentene : 9005
cis-2-Pentene : 9006
trans-2-Pentenedioic acid diethyl ester : 9042
1-Pentene, 2-ethyl-4-methyl- : 6837
1-Pentene, 2-isopropyl- : 6836
2-Pentene, 3-methyl (*cis*) : 9057
1-Pentene, 3-methyl-3-ethyl : 9047
1-Pentene, 4-methyl-3-ethyl : 9048
2-Pentene, 2-methyl-3-ethyl : 9049
2-Pentene, 4-methyl-3-ethyl (*trans*) : 9050
2-Pentene, 4-methyl-3-ethyl (*cis*) : 9051
trans-3-Pentenenitrile : 9061
1-Pentene-3-one : 9104
2-Pentene, 3-phenyl- : 1699
2-Pentene, 2-phenyl- : 1950
1-Pentene, 5-phenyl : 2106
1-Pentene, 2-propyl- : 6426
2-Pentenoic acid, 2-methyl (*trans*) : 9076
2-Pentenoic acid, 3-methyl (*cis*) : 9077
3-Pentenyl bromide : 9015
trans-3-Penten-1-yne : 9113
cis-3-Penten-1-yne : 9114
Pentetic acid : 6257
Pentoxone : 8978
Pentyl acetate : 227
sec-Pentyl acetate (*R*) : 8918
sec-Pentyl alcohol : 8916 : 8917
tert-Pentyl bromide : 3240
2-Pentyl bromide : 8711
3-Pentyl bromide : 8712
Pentyl butanoate : 3579
Pentyl butyrate : 3579
Pentylcyclohexane : 4465
Pentylenetetrazole : 11654
Pentyl fluoride : 8817
Pentyl formate : 6015
Pentyl heptanoate : 6470
Pentyl isocyanide : 8821
Pentyl ketone : 11962
Pentyl mercaptan : 8844
3-Pentyl mercaptan : 8846
2-Pentylnaphthalene : 7942
Pentyl nitrite : 8096
Pentyl nonanoate : 8156
p-*tert*-Pentylphenol : 9339
Pentyl phenyl ether : 2109
Pentylpropiolic acid : 8538
4-(3-Pentyl)pyridine : 10968
1-Pentynyl bromide : 9124
Perbenzoic acid : 1126
Perchloroacetone : 10440
Perchlorobenzoic acid : 2712
Perchlorocyclopentadiene : 4766
Perchlorocyclopentene : 4932
Perchloromethyl mercaptan : 7527
Perchloronaphthalene : 7933
Perchloropropylene : 10571
Pereirine : 4154
Perfluoroacetone : 10441
Perfluorobenzene : 1757
Perfluorobutane : 3255
Perfluoro-2-butene : 3804
Perfluoro-2-butyltetrahydrofuran : 6109
Perfluorocyclobutane : 4194
Perfluorocyclohexane : 4381
Perfluorocyclohexene : 4643
Perfluorodecalin : 7934
Perfluorodecane : 5050
Perfluoro-2,3-dimethylbutane : 3414
Perfluoroethane : 5596
Perfluoroheptane : 6418
Perfluoro-1-heptene : 6560
Perfluorohexane : 6843
Perfluoro-1-hexene : 7032
Perfluoroisobutane : 10050
Perfluoro(methylcyclohexane) : 4499
Perfluoro(2-methylpentane) : 8854
Perfluoro-3-methylpentane : 8853
Perfluoronaphthalene : 7935
Perfluorononane : 8139
Perfluorooctane : 8415
Perfluoropentane : 8807
Perfluoropropane : 10118
Perfluoropropene : 10572
Perfluoropropylene oxide : 8629
Perfluorotoluene : 2095
Perillaldehyde : 4633
Perillyl aldehyde : 4633
Permethylbenzoic acid : 2713
Permethyldisilane : 5228
1,4-Peroxido-*p*-menthene-2 : 5184
Peroxyacetic acid : 5619
Perphenazine : 9617
Peucedanin : 6188
Phellandral : 4635
β-Phellandrene : 4673
Phenacemide : 879
Phenacetin : 60
Phenacyl acetate : 5771
Phenacylamine : 5774
Phenacyl bromide : 5782
Phenacyl cyanide : 2124
Phenallymal : 11103
Phenamizole : 11671
9,10-Phenanthraquinodimethane : 9172
9,10-Phenanthraquinone, 3-hydroxy : 9176
Phenanthrene, 9-isopropenyl- : 9182
Phenanthrenequinone : 9174
9-Phenanthroic acid : 9165
o-Phenanthroline : 9189
9-Phenanthryl bromide : 9164
Phenazine, 2,3-diamino- : 9194
Phenbenzamine : 5506
Phencarbamide : 3977
Phencyclidine : 9696
[v-Phenenyltris(oxyethylene)]-tris[triethylammonium iodide] : 5441
Phenetharbital : 11089
Phenethyl alcohol : 1592
Phenethyl alcohol, 2-ethyl : 1603
Phenethyl alcohol, 4-ethyl : 1604
Phenethyl alcohol, α,α,β-trimethyl- : 1616
Phenethylamine, *N*-benzyl-*N*,α-dimethyl-, (+)- : 1575
Phenethylamine, 4-bromo : 1570
Phenethylamine, 2-methoxy : 1577
2-Phenethyl propionate : 10324
o-Phenetidine : 736
p-Phenetidine : 737
o-Phenetidine, 5-nitro : 739
Phenetole : 1649
Phenetole, *m*-chloro- : 1157
Phenetole, 2,6-dinitro- : 1650
Phenetole, 2-propenyl : 1665
Phenetole, 2-propyl : 1667
Phenicarbazide : 7145
Phenicin : 2979

Phenigam : 2128
Phenindione : 7270
Pheniramine : 11010
Phenmetrazine : 7697
Phenobarbital : 11095
Phenol, benzoate : 2716
Phenol, 6-*tert*-butyl-3,4-dimethyl : 9324
Phenol-2-carboxylic acid : 2612
Phenol, 2,6-di-*sec*-butyl- : 9227
Phenol, 3,5-dimethyl-4-(methylthio)-, methylcarbamate : 7646
Phenoldisulfonic acid : 1560
Phenol, 3-(1-hydroxypropyl) : 1872
Phenol, *o*-methoxy-, phosphate (3:1) : 9387
Phenol, 2-(1-methylethoxy)-, methylcarbamate : 10722
Phenol, 3-methyl-5-(1-methylethyl)-, methylcarbamate : 9769
Phenol, 2-methyl-4-(3-methyl pentyl) : 4148
Phenol, 2-(1-methylpropyl)-4,6-dinitro- : 5183
Phenolphthalein : 7354
Phenolphthalein, 5',5''-diisopropyl-2',2''-dimethyl- : 7352
Phenolphthalin : 2517
Phenolphthalol : 1841
o-Phenolsulfonic acid : 2225
p-Phenolsulfonic acid : 2226
Phenolsulfonphthalein : 9363
Phenol, 2,4,5-trichloro-, *O*-phosphorodichloridothioate : 9578
Phenopyrazone : 10843
Phenothiazine, 2-chloro-10(3(4-methyl-1-piperazinyl)propyl) : 9501
Phenoxazine : 9507
Phenoxyacetic acid : 228
Phenoxyacetic acid methyl ester : 230
Phenoxyacetic acid, 2,4,5-trichloro : 256
Phenoxyacetonitrile : 287
2-Phenoxybutyric acid, phenyl ester : 3583
2-Phenoxycarbonylphenol : 2650
2-Phenoxyethanol : 5758
Phenprobamate : 2149
Phensuximide : 11161
Phentermine : 1572
Phentydrone : 5978
Phenylacetaldehyde : 868
α-Phenylacetamide : 878
Phenyl acetate : 231
Phenylacetic acid : 883
Phenylacetic acid, α-chloro, ethyl ester (*d*) : 891
Phenylacetic acid, 3,4-methylenedioxy, methyl ester : 2416
N-Phenylacetoacetamide : 3207
Phenylacetone : 10481
Phenylacetyl chloride : 971
3-(α-Phenyl-β-acetylethyl)-4-hydroxycoumarin : 2832
Phenyl acetylsalicylate : 2483
9-Phenylacridine : 318
L-Phenylalanine, 4-hydroxy- : 11926
1-Phenylallyl alcohol : 1868
Phenylaminoacetic acid : 6270
4-Phenyl-2-aminobutane : 2119
Phenyl *p*-aminosalicylate : 2502
p-Phenylaniline : 3011
9-Phenylanthracene : 428
N-Phenylanthranilic acid : 2715
Phenylarsonic acid : 460
(Phenylazo)thioformic acid 2-phenylhydrazide : 5121
Phenylbenzene : 3008
Phenyl benzoate : 2716
2-Phenylbenzothiazole : 2864
Phenyl biguanide : 7234
Phenylboric acid : 3135
N-Phenylbutanedioic acid monoamide : 3575
1-Phenylbutane-1,3-diol : 3357
2-Phenyl-1,2-butanediol : 3356
4-Phenyl-3-butene-2-one : 3881
4-Phenyl-3-butenoic acid : 3855
Phenylbutylcarbinol : 1846
α-Phenylbutyramide : 881
Phenyl butyrate : 3584
Phenyl carbonate : 4001
Phenyl chloroacetate : 147
Phenyl chloroformate : 4026
2-Phenyl-6-chlorophenol : 3092
Phenylchlorosilane : 11384
2-Phenylchroman : 2782
2-Phenyl-4-chromone : 2840
3-Phenylchromone : 2841
α-Phenylcinnamic acid : 941
Phenyl cyanide : 2740
1-Phenyl-2-cyanoethane : 2123
2-Phenylcycloheptanone : 4236
1-Phenylcyclohexan-1-ol : 4552
trans-2-Phenylcyclopropanecarbonyl chloride : 4957
Phenyldichlorosilane : 11403
N-Phenyldiethanolamine : 5760
Phenyldihydroxyborane : 3135
Phenyl diselenide : 5225
1-Phenyldodecane : 1566
3-Phenyleicosane : 1695
o-Phenylenediamine : 1260
m-Phenylenediamine : 1261
p-Phenylenediamine : 1262
o-Phenylenediamine, *N,N'*-dimethyl : 1269
o-Phenylenethiourea : 2386
1-Phenyl-1,2-ethanediol : 5579
1-Phenylethanol : 1896
Phenylethanolamine : 1839
N-Phenylethanolamine : 5759
Phenylethylamine : 1569
Phenyl ethyl carbinol : 1870
Phenylethyl dimethyl carbinol : 2150
1-Phenylethylene glycol : 5579
Phenylethylene oxide : 8624
9-(2-Phenylethyl)heptadecane : 2070
1-Phenyl-1-ethylhydrazine : 7159
1-Phenylethyl propyl ketone : 6985
Phenylethynylcarbinol : 1876
N-Phenylformamide : 5999
N-Phenylformoguanamine : 11830
2-Phenylfuran : 6159
Phenylglyceryl ether : 10026
Phenyl glycidyl ether : 8623
α-Phenylglycine : 885
N-Phenylglycine : 6270
1-Phenyl-1-heptanol : 1881
7-Phenylheptan-3-ol : 2102
1-Phenyl-1,3-hexadiene : 1753
6-Phenyl-1,3-hexadiene : 1755
6-Phenyl-2-hexanol : 2103
6-Phenyl-5-hexen-2-one : 7102
Phenylhydrazine : 7169
Phenylhydrazine-4-sulfonic acid : 2224
Phenylhydroxylamine : 757
2-Phenylindan : 7259
2-Phenyl-1-indanone : 7291
Phenyl isocyanate : 1786
2-Phenylisopropanol : 1857
Phenyl isothiocyanate : 1792
3-Phenyl-3-ketopropionitrile : 2124
(±)-3-Phenyllactic acid : 2133
DL-β-Phenyllactic acid : 2133
Phenyl laurate : 5317
N-Phenylmaleimide : 11123
Phenylmalononitrile : 9916
Phenyl mercaptan : 2289
Phenylmercuric acetate : 7431
Phenylmethylbarbituric acid : 11100
Phenyl methyl ketone : 5864
2-(Phenylmethyl)phenol : 9454
4-(Phenylmethyl)phenol : 9455
1-Phenyl-2-methylpropyl alcohol : 1901
1-Phenylnaphthalene : 7947
N-Phenyl-β-naphthylamine : 7728
2-Phenyloctane : 1981
5-Phenyl-2,4-pentadienal : 8657
5-Phenylpentanoic acid : 2099
1-Phenyl-1-pentanol : 1846
3-Phenyl-2-pentene : 1699
2-Phenyl-2-pentene : 1950
o-Phenylphenol : 3088
m-Phenylphenol : 3089
p-Phenylphenol : 3090
Phenylphosphine : 9519
Phenyl phthalate : 1345
4-Phenylpiperidine : 9695
3-Phenyl-1-propanal : 2114
1-Phenyl-1,2,3-propanetriol : 10194
3-Phenylpropanoic acid : 2126
2-Phenyl-2-propanol : 1857
1-Phenyl-1-propanol : 1870
3-Phenyl-1-propanol : 2148
Phenylpropanolamine hydrochloride : 1838
1-Phenylpropargyl alcohol : 1876
3-Phenylpropionaldehyde : 2114
β-Phenylpropionic acid methyl ester : 2139
3-Phenylpropionitrile : 2123
3-Phenylpropyl alcohol : 2148
Phenylpropylmethylamine : 1574
1-Phenyl-2-propyn-1-ol : 1876
1-Phenyl-3-pyrazolidinone : 10844
1-Phenyl-2-pyrazoline : 10823
2-Phenylpyridine : 10997
3-Phenylpyridine : 10998
4-Phenylpyridine : 10999
2-Phenylpyrrole : 11135
2-Phenylquinoline : 11288
3-Phenylquinoline : 11289
4-Phenylquinoline : 11290
6-Phenylquinoline : 11291
Phenyl salicylate : 2650
Phenylseleninic acid : 2174
4-Phenylsemicarbazide : 7146
Phenyl stearate : 8271
N-Phenylsuccinamic acid : 3575
2-Phenyl-3-sulfanilamidopyrazole : 2201
N-Phenylsulfanilic acid : 2234
Phenyl sulfide : 2283
Phenylsulfonyl chloride : 2242
Phenylsulfonyl fluoride : 2251
Phenyl thiocyanate : 11709
2-(Phenylthio)furan : 6160
Phenylthiourea : 11808
Phenyltrichlorosilane : 11473
7-Phenyltridecane : 1764
1-Phenyltridecane : 2342
Phenyl trimethyl ammonium iodide : 866
Phenyltripropylsilane : 11448
Phenylurea : 12026
Phenylurethane : 3965
5-Phenylvaleric acid : 2099
δ-Phenylvaleric acid : 2099
Phenylvinylcarbinol : 1868
Phenyramidol : 1913
Phenytoin : 7219
Phenzidole : 2384
Phloionic acid : 8239
Phloionolic acid : 8276
Phloretic acid : 2134
Phloretin : 10453
Phloroglucinol : 2368
Phloropropiophenone : 10489
Pholedrine : 9403
Phorone : 6340
Phosgene : 4007
Phosphine, dichloro-(2,5-dimethylphenyl) : 9553
Phosphonic acid, (2-chloroethyl)- : 5929
Phosphonic acid, dibenzyl ester : 9586
Phosphonic acid, isopropyl- : 9542
Phosphonic acid, methyl diphenyl ester : 9541
Phosphonic acid, thiono, methyl, diisopropyl ester : 9549
Phosphonic acid, (2,2,2-trichloro-1-hydroxyethyl)-, dimethyl ester : 11853
Phosphonodithioic acid, ethyl-, *O*-ethyl *S*-phenyl ester : 5986
Phosphonous dichloride, 2,5-xylyl- : 9553
Phosphoramidic acid, methyl-, 2-chloro-4-(1,1-dimethylethyl)phenyl methyl ester : 4153
Phosphoramidic acid, (1-methylethyl)-, ethyl 3-methyl-4-(methylthio)phenyl ester : 5947
Phosphoramidothioic acid, acetyl-, *O,S*-dimethyl ester : 10
Phosphoramidothioic acid, *O,S*-dimethyl ester : 7441
Phosphoric acid, amide,thiono,dichloride,*N,N*-dimethyl : 9557
Phosphoric acid, 2-chloro-1-(2,4-dichlorophenyl)ethenyl diethyl ester : 4078

Phosphoric acid, 2-chloro-1-(2,4,5-trichlorophenyl)ethenyl dimethyl ester : 6202
Phosphoric acid, 1,2-dibromo-2,2-dichloroethyl dimethyl ester : 7702
Phosphoric acid dichloride,thiono,2,4,5-trichlorophenyl ester : 9578
Phosphoric acid, 2,2-dichloroethenyl dimethyl ester : 5170
Phosphoric acid, 3-(dimethylamino)-1-methyl-3-oxo-1-propenyl dimethyl ester, (*E*)- : 5172
Phosphoric acid, triisobutyl ester : 9572
Phosphorodithioic acid, *O,O*-bis(1-methylethyl) *S*-[2-[(phenylsulfonyl)amino]ethyl] ester : 517
Phosphorodithioic acid, *S*-[[(4-chlorophenyl)thio]methyl] *O,O*-diethyl ester : 4037
Phosphorodithioic acid, *S*-[(6-chloro-2-oxo-3(2H)-benzoxazolyl)methyl]*O,O*-diethyl ester : 9514
Phosphorodithioic acid, *S*-[2-chloro-1-(1,3-dihydro-1,3-dioxo-2H-isoindol-2-yl)ethyl] : 5105
Phosphorodithioic acid, *O,O*-diethyl *S*-[2-(ethylthio)ethyl] ester : 5260
Phosphorodithioic acid, *O,O*-diethyl *S*-[(ethylthio)methyl] ester : 9513
Phosphorodithioic acid, *S*-[(1,3-dihydro-1,3-dioxo-2H-isoindol-2-yl)methyl] *O,O*-dimethyl ester : 9515
Phosphorodithioic acid, *S*-[[(1,1-dimethylethyl)thio]methyl] *O,O*-diethyl ester : 11592
Phosphorodithioic acid, *O,O*-dimethyl *S*-[2-(methylamino)-2-oxoethyl] ester : 5009
Phosphorodithioic acid, *S,S'*-1,4-dioxane-2,3-diyl *O,O,O',O'*-tetraethyl ester : 5201
Phosphorodithioic acid, *O*-ethyl *S,S*-dipropyl ester : 5934
Phosphorodithioic acid, *S,S'*-methylene *O,O,O',O'*-tetraethyl ester : 5931
Phosphorodithioic acid, *S*-[(5-methoxy-2-oxo-1,3,4-thiadiazol-3(2H)-yl)methyl], *O,O*-dimethyl ester : 7645
Phosphorothioic acid, *O*-(4-bromo-2-chlorophenyl) *O*-ethyl *S*-propyl ester : 9762
Phosphorothioic acid, *O*-[2-chloro-l-(l-methylethyl)-lH-imidazol-4-yl] *O,O*-diethyl ester : 7340
Phosphorothioic acid, *O*-[2,5-dichloro-4-(methylthio)phenyl] *O,O*-diethyl ester : 4092
Phosphorothioic acid, *O,O*-diethyl *O*-(3,5,6-trichloro-2-pyridinyl) ester : 4089
Phosphorothioic acid, *O*-[2-(diethylamino)-6-methyl-4-pyrimidinyl] *O,O*-dimethyl ester : 9730
Phosphorothioic acid, *O*-[2-(diethylamino)-6-methyl-4-pyrimidinyl] *O,O*-diethyl ester : 9729
Phosphorothioic acid, *O,O*-diethyl *O*-[4-(methylsulfinyl)phenyl] ester : 5954
Phosphorothioic acid, *O,O*-diethyl *O*-pyrazinyl ester : 11720
Phosphorothioic acid, *O,O*-dimethyl *O*-[3-methyl-4-(methylthio)phenyl] ester : 5955
Phosphorothioic acid, *O,O*-dimethyl *O*-(3,5,6-trichloro-2-pyridinyl) ester : 4090
Phosphorothioic acid, *O*-[4-[(dimethylamino)sulfonyl]phenyl] *O,O*-dimethyl ester : 5946
Phosphorothioic acid, *O,O*-dimethyl *O*-(3-methyl-4-nitrophenyl) ester : 5951
Phosphorothioic acid, *O*-(6-ethoxy-2-ethyl-4-pyrimidinyl) *O,O*-dimethyl ester : 5945
Phosphorothioic acid, *S*-[2-(ethylsulfinyl)ethyl] *O,O*-dimethyl ester : 8632
Phosphorothioic acid, *S*-[2-(ethylthio)ethyl] *O,O*-dimethyl ester : 5103
Phosphorothioic acid, *O,O'*-(thiodi-4,1-phenylene) *O,O,O',O'*-tetramethyl ester : 1
Phosphorotrithious acid, *S,S,S*-tributyl ester : 7438
Phthalamide : 1324
Phthalic acid : 1327
Phthalic acid, 3,5-dichloro- : 1340
Phthalic acid, dioctyl ester : 1344
Phthalic acid, monoamide,*N*(1-naphthyl) : 2702
Phthalic acid, tetrachloro-, ethyl ester : 1353
Phthalic acid, tetrachloro-, monoethyl ester : 1353
Phthalic anhydride : 7342
Phthalic anhydride, 3,4-dichloro : 7343
Phthalic anhydride, 4,5-dichloro- : 7344
Phthalimide : 7362
Phthaloyl chloride : 1318
Phthaloylmalonic acid, diethyl ester : 7248
2-(*N*4-Phthalylaminobenzene-sulfonamide)thiazole : 2726
Phthalylsulfonazole : 2726
Phthalylsulphathiazole : 2726
Phthiocol : 7861
Phycitol : 3431
Phylloquinone : 7865
Physeteric acid : 11638
Physostigmine : 11191
Physoteric acid : 11638
Phytanic acid : 6658
Phytol : 6672
Phytomenadione : 7865
Piazthiole : 11658
Picein : 5818
Picloram : 10896
2-Picoline : 10980
3-Picoline : 10981
4-Picoline : 10982
γ-Picoline, 2-amino : 10867
3-Picoline, 2-phenyl- : 10991
Picolinic acid : 10893
4-Picolyl alcohol : 10977
Picramic acid : 9207
Picric acid : 9497
Picrolonic acid : 10848
Picropodophyllin : 6191
Picrotoxin : 7579
Picryl chloride : 1244
Picryl sulfide : 2288
Pilocarpine : 6151
Pimaric acid : 9170
Pimeclone : 4611
Pimelic acid : 6393
Pimelic acid, diethyl ester : 6394
Pimelic ketone : 4571
Pinacol : 3350
Pinacolone : 3657
Pinacone : 3350
Pinane (*DL*) : 2934
(-)-2-Pinene : 2971
(-)-α-Pinene : 2971
(1*S*,5*S*)-(-)-α-Pinene : 2971
(±)-α-Pinene : 2972
β-Pinene : 2925
2-Pinene : 2973
2-Pinen-7-one : 2977
2-Pinen-4-one, (1*R*,5*R*)-(+)- : 2978
Pinic acid diethyl ester : 4171
(±)-Pinol : 8561
Pinosylvin : 1545
Pipecolic acid : 9640
N,N'-Piperazinedicarboxylic acid, diethyl ester : 9606
Piperidic acid : 3464
Δ 3-Piperidine : 11020
Piperidine, 4-(aminomethyl)- : 9673
Piperidine, 4-benzoyl-*N*-methyl : 7409
Piperidine, 2-benzyl- : 9699
Piperidine, 4-benzyl- : 9700
Piperidine, *N*-*tert*-butyl : 9653
Piperidine, *N*-carboxyethyl, ethyl ester : 9704
Piperidine, 4(4-chlorophenyl)-1(3(4-fluorobenzoyl)propyl)-4-hydroxy : 3651
Piperidine, *N*,2-dimethyl (*DL*) : 9648
Piperidine, 2-ethyl (*DL*) : 9668
Piperidine, 3-ethyl (*DL*) : 9669
Piperidine, 2-(2-hydroxyethyl) : 9663
Piperidine, 4-(2-hydroxyethyl) : 9664
Piperidine, 3-methyl (*DL*) : 9681
Piperidine, *N*-(3-methyl-3-pentyl) : 9671
1-Piperidinepropionic acid, ethyl ester : 9704
Piperidine, 2(1-propionyl) : 10484
Piperidine, 3-vinyl- : 9666
Piperidinium, 1,1-dimethyl-, chloride : 7430
Piperidione : 9654
4-Piperidone, 1,2,3,6-tetramethyl : 9725
Piperine : 9632
Piperitenone : 4718
(±)-Piperitone : 4716
Piperonal : 2417
Piperonaldehyde : 2417
Piperonylic acid : 2418
trans-Piperylene : 8661
cis-Piperylene : 8662
Pipradrol : 9676
Pipsyl chloride : 2247
Pivalamide, *N,N*-diethyl : 9794
Pivaldehyde : 9783
Pivalic acid : 10258
Pivalic acid chloride : 10503
Pivalil : 6811
Pivaloin : 6975
Plasmocid : 9877
Plumericin : 11904
Podocarpa-7,13-dien-15-oic acid, 13-isopropyl- : 9167
Podophyllotoxin : 6192
Polythiazide : 2857
Populin : 6223
Pramoxine : 7683
Prednisolone : 9738
Prednisone : 9739
Pregnanediol : 9742
5β-Pregnane-3,20-dione : 9743
Pregnan-3α-ol-20-one : 9745
4-Pregnene-11β,17α,20β,21-tetrol-3-one : 9757
Pregnenolone : 9756
Prehnitenol : 9473
Prenoxdiazine hydrochloride : 9657
Primeverose : 6237
Primidone : 11066
Pristane : 8644
Probarbital : 11093
Probenecid : 2595
Procainamide hydrochloride : 602
Procaine : 2493
Proflavine : 316
Proformiphen : 2149
Progallin P : 2732
Promethazine : 9504
Prontosil : 2209
Propallylonal : 11078
Propanal, 2-methyl-2-(methylsulfonyl)-, *O*-[(methylamino)carbonyl]oxime : 335
Propanal, 2-methyl-2-(methylthio)-, *O*-[(methylamino)carbonyl]oxime : 334
Propanamide, *N*-(3,4-dichlorophenyl)- : 10200
Propanamide, *N,N*-diethyl-2-(1-naphthalenyloxy)- : 8067
1-Propanearsonic acid : 461
Propane, 1,3-bis(1-methyl-4-piperidyl) : 9702
Propane, 3-bromo-1,2-epoxy (*DL*) : 8600
1-Propanecarboxylic acid : 3460
Propane, 2-chloromethyl-1,1,2,3-tetrachloro : 10146
Propane, 2-chloro-1-phenyl (*l*) : 1234
Propane, 1-cyclohexyl-3-cyclopentyl- : 4341
Propane, 1-cyclopropyl- : 4984
Propane, 1,1-dibenzyl : 1698
Propane, 1,2-dichloro (*DL*) : 9892
Propane, 1,2-dicyclohexyl- : 4430
Propane, 2,2-dicyclohexyl- : 4432
Propane, 1,3-dicyclohexyl- : 4466

Propane, 1,1-dicyclohexyl-	: 4472
Propane, 1,3-dicyclohexyl-2-ethyl	: 4402
Propane, 1,3-dicyclohexyl-2-methyl-	: 4447
1,2-Propanediol, 3-amino(*DL*)	: 9984
1,3-Propanediol, 2-methyl-2-propyl,*N*-butyl carbamate,carbamate	: 3933
Propane, 1,2-diphenoxy	: 1970
Propane, 1,3-dipheny-1-ethyl	: 1697
Propane, 1,3-diphenyl-	: 2121
Propane, 1,1-diphenyl-	: 2171
Propanenitrile, 2-[[4-chloro-6-(cyclopropylamino)-1,3,5-triazin-2-yl]amino]-2-methyl-	: 9759
Propanenitrile, 2-[[4-chloro-6-(ethylamino)-1,3,5-triazin-2-yl]amino]-2-methyl-	: 4161
Propane, 1-nitro-3,3,3-trifluoro	: 10184
2-Propanephosphonic acid	: 9542
1,3-Propanesulfone	: 11694
Propane sultone	: 8570
Propane, 1,2,2,3-tetrachloro-tetrafluoro	: 10149
Propanoic acid, 3-amino-	: 323
Propanoic acid, 2-amino-3-mercapto-, (*R*)-	: 5014
Propanoic acid, chloride, 2-chloro(*d*)	: 10496
Propanoic acid, 2-(3-chlorophenoxy)-	: 4141
Propanoic acid, 2-[4-(2,4-dichlorophenoxy)phenoxy]-, methyl ester	: 5171
Propanoic acid, 2,2-dichloro-3,3,3-trifluoro, propyl ester	: 10721
Propanoic acid, 3,3'-dithiobis[2-amino-, [*R*-(*R**,*R**)]-	: 5020
Propanoic acid, methylene ester	: 7654
Propanoic acid, 2-(2,4,5-trichlorophenoxy)-	: 11512
Propanolamine	: 10337
Propanolamine, *N*-pantoyl-	: 3202
1-Propanol, 2-[2-(2-amino-2-methylpropyl)amino]-2-methyl-	: 10342
2-Propanol, 1-chloro-3-isopropoxy-	: 10355
2-Propanol, 1-diethylamino (*DL*)	: 10368
2-Propanol, 1,1',1'',1'''-(ethylenedinitrilo)tetra-	: 10379
2-Propanol, 1-isopropoxy-	: 10390
2-Propanol, 1-nitro-3,3,3-trichloro	: 10408
2-Propanol, 1,1,1-trifluoro(*DL*)	: 10409
2-Propanone, acetyl-	: 8798
2-Propanone, (methoxyphenyl)-	: 10459
Propantheline bromide	: 9832
Propargylacetic acid	: 9138
Propargyl alcohol	: 10743
Propargyl aldehyde	: 10723
Propargyl bromide	: 10728
Propargyl chloride	: 10729
Propargylic acid, (2-nitrophenyl)	: 10740
2-Propene-1,2-dicarboxylic acid	: 3328
Propene-3,3-diol, 2-methyl,diacetate	: 10564
Propene, 2-methyl(tetramer)	: 10580
2-Propene-1-ol, 1-phenyl	: 1868
Propenoic acid, 2-chloro,ethyl ester	: 10619
Propenoic acid, 2-methyl,butyl ester	: 10651
Propenoic acid, 3(1-naphthyl)(*trans*)	: 10675
2-Propenyl alcohol	: 10697
2-Propenyl bromide	: 10531
2-Propenyl chloride	: 10536
2-Propenyl isothiocyanate	: 10575
N-2-Propenylurea	: 12028
Propesin	: 2507
Propham	: 3966
Propiocine	: 5396
β-Propiolactone	: 8595
Propiolic acid	: 10738
Propiolic acid, propyl-	: 7123
Propionaldehyde	: 9774
Propionaldehyde, 3-(2-tolyl)	: 2115
Propionamide	: 9791
Propionamide, *N*-antipyrinyl-2-(dimethylamino)-	: 9795
Propionic acid	: 10201
Propionic acid bromide	: 10490
Propionic acid, 3-bromo, isopentyl ester	: 10210
Propionic acid, 2-chloro, isobutyl ester (*d*)	: 10224
Propionic acid, 2-diazo-, ethyl ester	: 10236
Propionic acid, 2-diazo-, methyl ester	: 10237
Propionic acid, 2,3-dibromo-, butyl ester	: 10239
Propionic acid, 2,3-dibromo-, propyl ester	: 10243
Propionic acid, 2,2-dichloro-3,3,3-trifluoro-, methyl ester	: 10251
Propionic acid, 3-(diethylamino), ethyl ester	: 328
Propionic acid, 2,3-dihydroxy-2-methyl-, methyl ester	: 10256
Propionic acid, 2,3-diphenyl (*DL*)	: 2143
Propionic acid, 3,3'-iminodi-*N*-methyl, diethyl ester	: 330
Propionic acid, 3,3'-(methylimino)-di-, diethyl ester	: 330
Propionic acid, 2,3,3,3-tetrafluoro-, ethyl ester	: 10328
Propionic acid, 3-(2-tetrahydrofuryl)	: 6164
Propionic acid, 3,3,3-trichloro-2-(2,2,2-trichloro-1-hydroxyethoxy)-, γ-lactone	: 5212
Propionic anhydride	: 10203
Propionitrile, 2-ethoxy-	: 10102
Propionitrile, 2-methoxy-	: 10109
Propionylacetone	: 6804
Propionyl bromide	: 10490
Propionyl chloride	: 10493
Propionyl chloride, 3,3-dichloro	: 10502
Propionyl fluoride	: 10506
Propiophenone	: 10480
Propiophenone, 2'-hydroxy-3'-methyl-	: 10444
Propiophenone, 2'-hydroxy-4'-methyl-	: 10445
Propiophenone, 3-methoxy	: 10456
Propiophenone, 2-methoxy-	: 10457
Propofol	: 9226
2-Propoxyethanol	: 5764
1-Propoxy-2-propanol	: 10401
N-Propylacetanilide	: 88
Propyl acetate	: 236
Propylacetylene	: 9122
Propyl alcohol	: 10334
Propylallene	: 6676
Propylamine	: 9803
Propylamine, *N*,*N*-dichloro	: 9805
Propyl *p*-aminobenzoate	: 2507
Propyl amyl ketone	: 8174
Propylbenzene	: 2169
Propyl benzenesulfonate	: 2235
Propyl benzoate	: 2724
Propyl bromide	: 9836
Propyl bromoacetate	: 120
Propyl butanoate	: 3588
Propyl-*tert*-butylcarbinol	: 6925
Propyl carbamate	: 3969
Propyl carbonate	: 4002
Propyl chloride	: 9848
Propyl chloroacetate	: 149
1-Propyl-3-(*p*-chlorobenzenesulfonyl)urea	: 2208
Propyl chlorocarbonate	: 4028
trans-Propyl cinnamate	: 10693
Propyl cyanide	: 3403
Propylcyclohexane	: 4471
Propyl cyclohexanecarboxylate	: 4337
trans-4-Propylcyclohexanol	: 4560
cis-2-Propylcyclohexanol	: 4558
trans-2-Propylcyclohexanol	: 4559
4-Propylcyclohexanone	: 4613
2-Propylcyclohexanone	: 4612
trans-2-Propylcyclopentanol	: 4874
Propylcyclopentane	: 4849
cis-2-Propylcyclopentanol	: 4873
2-Propylcyclopentanone	: 4894
3-Propylcyclopentanone	: 4895
1-Propyl-1-cyclopentene	: 4935
Propylcyclopropane	: 4984
Propyl 2,3-dibromopropionate	: 10243
Propyldimethylamine	: 9807
Propylene	: 10528
Propylene, 2-bromo-3-cyclohexyl	: 4322
Propylene, 3-bromo-3,3-difluoro	: 10532
Propylene chlorohydrin	: 10347
sec-Propylene chlorohydrin	: 10349
Propylenediamine	: 9872
Propylene dibromide	: 9881
Propylene, 1,3-dibromo (*trans*)	: 10543
Propylene, 1,3-dibromo (*cis*)	: 10544
Propylenedicarboxylic acid	: 3328
Propylene, 1,1-dichloro-2-fluoro	: 10554
Propylene, 1,2-dichloro-3,3,3-trifluoro	: 10558
1,2-Propylene glycol	: 9982
1,3-Propylene glycol	: 9983
Propylene glycol β-monoethyl ether	: 10380
1,2-Propylene glycol 1-propyl ether	: 10401
Propylene oxide	: 8621
1,2-Propylenimine	: 501
N-Propylethanolamine	: 5765
Propylethylene	: 9004
Propyl fluoride	: 10046
Propyl formate	: 6018
Propylformic acid	: 3460
2-Propylfuran	: 6167
Propyl gallate	: 2732
4-Propylheptane	: 6434
Propyl heptanoate	: 6471
4-Propyl-3-heptene	: 6559
1-Propylheptyl alcohol	: 5072
Propyl hexahydrobenzoate	: 4337
Propylhexedrine	: 4385
Propyl hexyl ketone	: 5077
Propyl *p*-hydroxybenzoate	: 2653
Propylidene chloride	: 9891
Propyl iodide	: 10054
Propyliodone	: 10873
Propyl isobutyrate	: 10308
Propyl isocyanate	: 10059
Propyl isopropyl sulfide	: 10077
Propyl isovalerate	: 3562
Propyl laurate	: 5320
Propyl levulinate	: 8912
Propyl mercaptan	: 10156
Propyl methacrylate	: 10672
Propyl nitrate	: 8080
Propyl nitrite	: 8098
5-Propyl-1,5-octadien-3-yne	: 8341
Propylparaben	: 2653
2-Propylpentanal	: 8697
2-Propyl-1-pentene	: 6426
3-Propyl-1-pentene	: 7033
Propyl phosphite	: 9594
4-Propylpiperidine	: 9709
Propyl propionate	: 10326
3-Propylpyrrole	: 11139
2-Propylquinoline	: 11292
Propyl sulfide, 3,3'-dichloro	: 10153
2-Propylthiophene	: 11778
3-Propylthiophene	: 11779
Propylthiouracil	: 11109
N-Propyl-*p*-toluidine	: 815
Propyltrichlorosilane	: 11476
Propyl trifluoroacetate	: 269
Propyl 3,4,5-trihydroxybenzoate	: 2732
Propyne, 1,3-dibromo-	: 10730
Protionamide	: 10886
Protoanemonin	: 6155
Protocatechualdehyde	: 542
Protocatechuic acid	: 2571
Protocatechuic acid methylene ether	: 2418
Protopine	: 3109
Protoverine	: 4070
Provitamin D3	: 4103
Prunetin	: 2828
Prunetol	: 2807
Pseudoaconitine	: 308
Pseudocinchonine	: 4122
Pseudocodeine	: 7672
Pseudoconhydrine	: 9718
Pseudocumene	: 2358
Pseudoionone	: 11965
Pseudojervine	: 12045
Pseudomorphine	: 3002
Pseudopelletierine	: 479
Pseudotropine	: 486
Pteroylglutamic acid	: 6243
Puleganic acid	: 4784
Pulegene	: 4931
Pulegone	: 4608

Purodigin : 4043
Purostrophan : 4040
Purpurin : 416
Purpurogallin : 2408
Putrescine : 3258
Pyocyanine : 9196
Pyran-4-carboxylic acid, tetrahydro- : 10774
Pyrazinamide : 10802
Pyrazinoic acid : 10803
Pyrazino[2,3-d]pyrimidine : 10747
*N*1-2-Pyrazinylsulfanilamide : 2202
Pyrazole, 4-amino-3-methyl-1-phenyl- : 10813
1H-Pyrazole, 3-chloro-5-methyl-4-nitro- : 4083
2-Pyrazoline : 10822
2-Pyrazoline, 1-phenyl- : 10823
2-Pyrazoline, 3-phenyl : 10824
1H-Pyrazolium, 1,2-dimethyl-3,5-diphenyl-, methyl sulfate : 5175
3-Pyrenecarboxylic acid : 10854
Pyrethrin 1 : 4960
Pyrethrin 2 : 4962
Pyribenzamine : 5505
3(2H)-Pyridazinone, 5-amino-4-chloro-2-phenyl- : 4080
3(2H)-Pyridazinone, 4-chloro-5-(methylamino)-2-[3-(trifluoromethyl)phenyl]- : 8213
Pyridine, 3-amino-2,6-dichloro : 10862
Pyridine, 2-benzyl- : 11004
Pyridine, 3-benzyl- : 11005
Pyridine, 2-*tert*-butyl- : 10942
Pyridine, 3-*tert*-butyl- : 10943
Pyridine-3-carbinol : 10976
4-Pyridinecarbinol : 10977
Pyridine, 2-chloro-6-(trichloromethyl)- : 8071
Pyridine, 2,6-di-*tert*-butyl- : 10874
Pyridine, 2-[[2-(dimethylamino)ethyl]-2-thenylamino]- : 5507
Pyridine, 2(2-dimethylaminoethyl-4-methoxybenzylamino) : 5511
Pyridine, 2-(2-hydroxypropyl) : 10951
Pyridine, 2-(2-isopropoxyethyl) : 10983
Pyridine, 3-isopropyl- : 10985
Pyridine *N*-oxide : 10994
Pyridine, 4-(3-pentyl) : 10968
Pyridinium chloride : 10969
4(1H)-Pyridinone, 1-methyl-3-phenyl-5-[3-(trifluoromethyl)phenyl]- : 5981
1,5-Pyridopyridine : 8065
Pyridoxamine dihydrochloride : 10978
Pyridoxine hydrochloride : 10934
α-Pyridylamine : 10857
3-Pyridylamine : 10858
2-(2-Pyridyl)pyridine : 3099
4-(4-Pyridyl)pyridine : 3104
5-[*p*-(2-Pyridylsulfamyl)phenylazo]salicylic acid : 2654
*N*1-2-Pyridylsulfanilamide : 2203
3-Pyridylsulfonic acid : 11018
Pyrilamine : 5511
Pyrimethamine : 11051
Pyrimidine, 2-amino-4,5-dimethyl- : 11044
2,4(1H,3H)-Pyrimidinedione, 5-bromo-6-methyl-3-(1-methylpropyl)- : 3139
2,4(1H,3H)-Pyrimidinedione, 5-chloro-3-(1,1-dimethylethyl)-6-methyl- : 11591
5-Pyrimidinemethanol, α-(2-chlorophenyl)-α-(4-chlorophenyl)- : 5948
5-Pyrimidinemethanol, α-(2-chlorophenyl)-α-(4-fluorophenyl)- : 8219
4(1H)-Pyrimidinone, 5-butyl-2-(ethylamino)-6-methyl- : 5932
*N*1-2-Pyrimidinylsulfanilamide : 2204
Pyrithyldione : 10946
Pyrocalciferol : 5381
Pyrocarbonic acid diethyl ester : 5169
Pyrocatechol : 1497
Pyrochol : 4095
Pyrocoll : 5223
Pyrogallic acid : 2366
Pyrogallol : 2366
L-Pyroglutamic acid : 9768
Pyrolan : 3945
Pyromellitic acid : 2257
Pyromucic acid : 6050
Pyropentylene : 4764
Pyrrole, 2-benzoyl : 7637
Pyrrole-3-propionic acid, 2-[[3-(2-carboxyethyl)-4-methyl-5-[(3-methyl-5-oxo-4-vinyl-3-pyrrolin-2-ylidene)meth : 3001
2-Pyrrolidinecarboxylic acid : 9764
2-Pyrrolidone : 11180
2-Pyrrolidone, 3,3-dimethyl : 11182
3-Pyrroline : 11118
Pyruvaldehyde : 9788
Pyruvic acid : 10314
Pyrylene : 9111
Quercetin : 2817
Quercetin 7-methyl ether : 2815
Quercetin-3-*L*-rhamnoside : 2796
D-Quercitol : 4074
Quercitrin : 2796
Quillaic acid : 8546
Quinaldic acid : 11227
Quinaldine : 11277
Quinaldinic acid : 11227
Quinaldonitrile : 11223
Quinalizarin : 413
Quinamine : 6178
4-Quinazolinone, 2-benzyl-1-methyl : 11200
Quinethazone : 11197
Quinic acid : 4338
Quinidine : 4126
Quinine : 4127
Quinine carbonate : 4128
Quinine, formate : 4129
Quinine hydrochloride : 4130
Quinine salicylate : 4131
Quinine sulfate : 4132
Quininic acid : 11234
Quininone : 4134
Quinizarin : 392
Quinoline, 2-amino- : 11202
Quinoline, 6-amino- : 11205
Quinoline, 4-amino-3-bromo : 11207
Quinoline, 2-amino-4-methyl : 11208
Quinoline, 7-amino-2-methyl : 11211
Quinoline, 8-amino-6-methyl- : 11212
Quinoline, decahydro (*trans*, *DL*) : 11247
Quinoline, 2,4-dimethyl-6-hydroxy : 11317
Quinoline, 2,4-dimethyl-8-hydroxy : 11318
Quinoline, 5-hydroxy : 11305
Quinoline, 7-hydroxy : 11307
Quinoline, 7-hydroxy-6-methyl : 11320
Quinoline, 4-methoxy-2-[3,4-(methylenedioxy)phenethyl]- : 11216
Quinolinic acid : 10927
p-Quinone : 4257
Quinone : 4257
Quinone tetrachloride : 4268
Quinovic acid : 12036
Quinovose : 6235
Quinoxaline, 6-amino- : 11332
*N*1-(2-Quinoxalinyl)sulfanilamide : 2205
Quintozene : 2087
Quinuclidine : 480
3-Quinuclidinol : 485
Racemic tartaric acid : 3311
Radicinin : 10785
Raffinose : 6219
Randox : 44
Raubasine : 8571
Raunescine : 12064
Reductic acid : 4942
Reductinic acid : 4942
Refrigerant 10 : 7540
Refrigerant 11 : 7555
Refrigerant 12 : 7500
Refrigerant 13 : 7486
Refrigerant 13B1 : 7475
Refrigerant 14 : 7541
Refrigerant 20 : 7554
Refrigerant 21 : 7502
Refrigerant 22 : 7478
Refrigerant 31 : 7481
Refrigerant 32 : 7504
Refrigerant 50 : 7459
Refrigerant 113 : 5663
Refrigerant 114 : 5538
Refrigerant 121a : 5639
Refrigerant 122a : 5660
Refrigerant 122b : 5658
Refrigerant 124 : 5486
Refrigerant 133 : 5488
Refrigerant 133b : 5487
Refrigerant 134 : 5641
Refrigerant 142 : 5471
Refrigerant 142b : 5470
Refrigerant 245 : 10128
Refrigerant C317 : 4177
Renazide : 2851
Renese : 2857
Resacetophenone : 5803
Resazurin : 9508
Rescinnamine : 12060
Reserpic acid : 12063
Reserpine : 12061
Resorcinol : 1498
Resorcinol, 4-acetyl : 8490
Resorcinol diglycidyl ether : 8625
Resorcinol dipropyl ether : 1556
Resorcinol, 5-iodo : 1524
Resorcinol, 4-isobutyl- : 1539
Resorcinolphthalein : 11522
β-Resorcylaldehyde : 541
β-Resorcylic acid : 2569
Retene : 9183
Reychler's acid : 2929
Rhamnetin : 2815
Rhamnol : 11558
Rheadine : 11342
Rhein : 371
Rhizopterin : 2494
Rhodamine, B : 12056
Rhodanine : 11685
Rhodanine, 5-[*p*-(dimethylamino)benzylidene]- : 11684
(-)-Rhodinol : 8516
Rhodizonic acid : 4686
Riboflavine : 11344
Riboflavin lumichrome : 2472
β-*D*-Ribofuranoside, adenine-9 : 319
β-*D*-Ribofuranoside, guanine-9 : 6289
1-β-*D*-Ribofuranosyluracil : 12033
Ricinine : 10883
Ricinoleic acid : 8303
Risocaine : 2507
Ronnel : 9584
Roquessine : 4143
p-Rosolic acid : 4278
Rotenone : 2783
Rubeanic acid : 5588
Rubijervine : 11515
Rubixanthin : 4062
Rufigallol : 406
Rutecarpine : 7334
Rutinose : 6236
Sabadine : 4067
Sabinane (*d*) : 11814
Sabinene (*DL*) : 2981
D-Saccharic acid : 6205
Saccharin : 2388
Safranal : 4250
Safrole : 2429
Salicin : 6222
Salicyl alcohol : 1882
Salicylaldehyde : 564
Salicylaldoxime : 579
Salicylamide : 616
Salicylamide *O*-acetic acid : 103
Salicylanilide : 617
Salicylhydroxamic acid : 613
Salicylic acid : 2612
p-Salicylic acid : 2613
Salicylic acid, 4-benzamido- : 2514
Salicylic acid, 3-chloro, ethyl ester : 2541
Salicylic acid, 3,5-dibromo-, methyl ester : 2559
Salicylic ether : 2619
Salicylsulfonic acid : 2655
Saligenol : 1882
Salirepol : 1520
Salsalate : 2616
Salsoline : 7397

Salvarsan : 0275
Sanguinarine : 2432
α-Santalol : 9093
β-Santalol : 9103
Santene : 2968
β-Santenol : 2938
α-Santonin : 8057
Sarcosine : 6268
Sarmentogenin : 4046
Sarpagine : 11347
Sarsasapogenin : 11531
Scarlet red : 8033
Schradan : 5220
Scilliroside : 3142
Scopolamine : 908
Scopoletin : 2829
Scopoline : 7575
Sebacic acid : 5042
Sebacic acid di(2-ethylhexyl) ester : 5043
Sebacic acid, monomitrile, methyl ester : 8148
Sebacil : 4203
Sebacoin : 4206
7,13a-Secoberbin-13a-one, 7-methyl-2,3:9,10-bis(methylenedioxy)- : 3109
Seedrin : 5177
Selanthrene : 11353
Semicarbazide hydrochloride : 7143
Seminose : 7425
Semioxamazide : 105
Senecialdehyde : 3700
Senfoel : 10575
Septochol : 4095
Serenase : 3651
Serpentine : 8572
Silane, allyldiethoxymethyl- : 11408
Silane, methyl tri-(4-tolyl) : 11445
Silane, (oxy-*p*-phenylene)bis[trimethyl- : 11497
Silane, phenyl-tributyl : 11455
Silane, propyl trimethyl : 11496
Silane, tetravinyl- : 11449
Silane, 4-trimethylsilylphenoxy-trimethyl : 11497
Silane, tris(allyloxy)vinyl- : 11427
Silanol, dimethyl ethyl : 11501
3-Silapentane : 11409
2-Silapropane : 11414
Silicon tetraethoxide : 11506
Silicon tetramethoxide : 11508
Silin, 1,1-diphenyl perhydro : 11360
Silmurin : 3142
Siloxane, butylmethyl(cyclic tetramer) : 4995
Sinomenine : 7676
α1-Sitosterol : 11556
β-Sitosterol : 11558
γ-Sitosterol : 11559
Skatole : 7314
Skimmetin : 2826
Skimmianine : 6181
Skimmin : 2821
Smilagenin : 11533
Sodium cyanate : 4164
Sodium *N*-lauroylsarcosinate : 6269
Solanone : 8113
Solasodine : 11529
Solbrol A : 2621
Solestro : 5403
Sophoricoside : 2824
Sorbic acid : 6698
L-Sorbinose : 11519
Sorbitol : 6206
Spantol : 2149
Sparteine : 7571
Spasmocromona : 2838
Spermine : 3259
Spinulosin : 4260
Spiro[3,5,8-ethanylylidene-1H-pyrano[3,4-c]pyridine-10,3'-[3H]indol]-2'(1'H)-one, 5-ethenyl : 6203
Spiro[2.2]pentane : 11527
Spirosal : 2622
Spir[5.]undecane : 11537
Squalane : 11606
Squalene : 11602
Stachydrine : 11177
Stannane, hydroxytriphenyl- : 11911
Stannane, tricyclohexylhydroxy- : 5011
Stanolone : 349
Staticin : 2721
Stearic acid : 8247
Stearic acid, 18-bromo : 8250
Stearic acid, cyclohexyl ester : 8253
Stearic acid, isopentyl ester : 8265
Stearic acid, 14-methyl : 8262
Stearic acid, 9-methyl : 8264
Stearic acid, 12-oxo, ethyl ester : 8269
Stearic acid, pentyl ester : 8270
Stearic acid, phenyl ester : 8271
Stearic acid, 9,10,12,13-tetrabromo, methyl ester : 8274
Stearic acid, tetrahydrofurfuryl ester : 8275
9-Stearolic acid : 8316
Stearyl alcohol : 8278
Stearyl mercaptan : 8244
Stercorin : 4109
Stigmastanol : 11557
Stigmasterol : 11555
trans-Stilbene : 1619
cis-Stilbene : 1620
Stilbene, 2-bromo (*trans*) : 1080
Stilbene, 2-bromo (*cis*) : 1081
4-Stilbenecarbonitrile (*cis*) : 2768
Stilbene, 2,2'-dimethyl- : 1622
Stilbene, 2,4-dinitro (*trans*) : 1496
Stimulina : 6249
Streptomycin, sulfate : 11561
Strodival : 4040
Strophanthidin : 4049
Strychnine : 11563
Strychnine nitrate : 11565
Strychnine, sulfate : 11566
Stylophorine : 4071
Styphnic acid : 1554
Styrallyl alcohol : 1896
Styrene : 1624
Styrene, α-chloro-2,4-dimethyl : 1152
Styrene, β-chloro, α-4-dimethyl : 1188
Styrene, 2,3-dichloro- : 1367
Styrene, 2,6-dichloro- : 1369
Styrene, 3,5-dichloro- : 1370
Styrene, 2,5-dichloro- : 1371
Styrene, 2,4-dichloro- : 1373
Styrene epoxide : 8624
Styrene glycol : 5579
Styrylacetic acid : 3855
Styryl methyl ketone : 3881
Suberic acid : 8386
Suberone : 4234
Suberonitrile : 8385
Succinamide : 3257
Succinanil : 11162
Succinanilic acid : 3575
Succinanilic acid, 4'-(2-thiazolylsulfamoyl)- : 3578
Succinic acid : 3294
Succinic acid, amide,*N*-phenyl : 3575
Succinic acid, *sec*-butyl-, diethyl ester : 3332
Succinic acid, di-*sec*-butyl ester : 3297
Succinic acid, dibutyl ester : 3301
Succinic acid dichloride : 3366
Succinic acid, 2,3-dichloro-, (±)- : 3302
Succinic acid, diethyl ester : 3303
Succinic acid *N,N*-dimethylhydrazide : 3334
Succinic acid dinitrile : 3291
Succinic acid, diphenyl ester : 3320
Succinic acid, ethyl ester : 3335
Succinic acid, mercapto-, diethyl ester, *S*-ester with *O,O*-dimethyl phosphorodithioate : 3316
Succinic acid monoanilide : 3575
Succinic acid, monopropyl ester : 3338
Succinic acid, oxo,diethyl ester : 3340
Succinic acid, pentyl ester : 3337
Succinic acid, propyl ester : 3338
Succinic anhydride : 6093
Succinic anhydride, phenyl (*DL*) : 6096
Succinimide : 11152
Succinimide, *N*-benzyl- : 11163
Succinonitrile : 3291
Succinyl chloride : 3366
Succinylsulphathiazole : 3578
Sucrose : 6218
Sucrose, octaacetate : 6229
Sudan III : 8038
Sulfabenzamide : 603
Sulfabenzpyrazine : 2205
Sulfacetamide : 30
Sulfachlorpyridazine : 2193
Sulfadiazine : 2204
Sulfadigesin : 3578
Sulfadimethoxine : 2194
Sulfafurazole : 2195
Sulfaguanidine : 2191
Sulfamerazine : 2199
Sulfamethazine : 2196
Sulfamethoxazole : 2198
Sulfamethoxypyridazine : 2197
Sulfamethyldiazine : 2199
Sulfamethylthiazole : 2200
p-Sulfamidoaniline : 2187
p-Sulfamoylbenzoic acid : 2508
p-Sulfamylacetanilide : 31
Sulfanilamide : 2187
Sulfanilamide-*N*4-acetate : 31
Sulfanilamide, *N*1-(5-methyl-3-isoxazolyl)- : 2198
Sulfanilamide, *N*1-(thiocarbamoyl)- : 2189
5-Sulfanilamido-1-phenylpyrazole : 2201
2-Sulfanilamidopyrazine : 2202
2-Sulfanilamidopyridine : 2203
o-Sulfanilic acid : 2215
m-Sulfanilic acid : 2216
Sulfanilic acid : 2217
2-(Sulfanilylamino)thiazole : 2206
p-Sulfanilyl fluoride : 2252
*N*4-Sulfanilylsulfanilamide : 2188
Sulfanilylurea : 2190
Sulfaphenazole : 2201
Sulfapyrazine : 2202
Sulfapyridine : 2203
Sulfapyrimidine : 2204
Sulfaquinoxaline : 2205
Sulfasalazine : 2654
Sulfasuccidine : 3578
Sulfathalidine : 2726
Sulfathiazole : 2206
Sulfathiourea : 2189
Sulfazole : 2200
o-Sulfhydrylbenzoic acid : 2663
Sulfide, 3-chloropropyl methyl : 9864
Sulfide, 2-chloropropyl methyl : 9865
Sulfide, ethyl isobutyl : 10044
Sulfide, *o*-tolyl *p*-tolyl : 1993
4,4'-Sulfinyldianiline : 845
Sulfiram : 11714
Sulfisoxazole : 2195
Sulfoacetic acid : 238
o-Sulfobenzoic acid imide : 2388
Sulfodiamine : 91
2-Sulfoethylamine : 5623
Sulfolane : 11786
Sulfolane, 2,4-dimethyl : 11784
3-Sulfolene : 11752
Sulfonethylmethane : 3230
Sulfonmethane : 9835
Sulfonyldiacetic acid : 239
p-Sulfophenol : 2226
p-Sulfophenylhydrazine : 2224
Sulfosalicylic acid : 2655
Sulfotep : 11716
Sulfovinic acid : 11576
Sulfurous acid, 2-[4-(1,1-dimethylethyl)phenoxy]cyclohexyl 2-propynyl ester : 10507
Sulphamidine : 2196
Sulphenone : 1225
Surinamine : 11933
Sylvic acid : 9167
Symclosene : 11840
Synephrine : 1887
Syringaldehyde : 569
Syringidin chloride : 2847
Syringin : 6225
2,4,5-T : 256

2,4,5-T Butoxyethyl ester : 257
2,4,5-T Butyl ester : 258
2,4,5-T *sec*-Butyl ester : 260
2,4,5-T Isopropyl ester : 259
Tabutrex : 3301
Tachysterol : 11349
Tagatose, *D*- : 11587
L-Tartaric acid : 3309
(+)-Tartaric acid : 3309
DL-Tartaric acid : 3311
(±)-Tartaric acid : 3311
(*R*,*S*)-Tartaric acid : 3304
meso-Tartaric acid : 3304
(-)-*D*-Tartaric acid : 3312
D-Tartaric acid : 3312
(*S*,*S*)-Tartaric acid : 3312
Tartaric acid, diethyl ester (*DL*) : 3307
Tartaric acid, diisopropyl ester (*DL*) : 3305
Tartaric acid, dipropyl ester, (+)- : 3310
Tartaric acid, disodium salt : 3314
Tartaric acid, ethyl ester : 3315
Tartronic acid : 9956
Taurine : 5623
Taurocholic acid : 5625
Tebanyl : 2502
Tegosept P : 2653
Tephrosin : 3110
Terephthalic acid : 1329
Terephthalic acid, 2,5-dichloro- : 1349
Terephthalic acid, diethyl ester : 1350
Terpenolic acid : 6030
Terpenylic acid : 6030
o-Terphenyl : 11595
m-Terphenyl : 11596
p-Terphenyl : 11597
β-Terpineol : 4542
Terpinolene : 4684
α-Terthienyl : 11600
Testosterone : 356
Testosterone-17-propionate : 359
1,4,7,10-Tetraazadecane : 5499
1,5,10,14-Tetraazatetradecane : 3259
Tetrabenzocyclooctatetraene : 11644
2,3,4,6-Tetrabromoanisole : 2256
1,2,3,5-Tetrabromobenzene : 2253
2,3,5,6-Tetrabromobenzene : 2254
1,2,4,5-Tetrabromobenzene : 2254
3,4,5,6-Tetrabromo-*o*-cresol : 9468
1,2,3,4-Tetrabromocyclobutane : 4195
1,2,2,2-Tetrabromoethane : 5631
Tetrabromoethylene : 5914
2,3,4,6-Tetrabromophenol : 9467
3',3'',5',5''-Tetrabromophenolphthalein : 7350
1,1,2,2-Tetrabromopropane : 10136
1,2,2,3-Tetrabromopropane : 10138
2,3,4,5-Tetrabromothiophene : 11781
Tetracaine hydrochloride : 2531
1,1,3,3-Tetracarbethoxypropane : 10140
1,2,4,5-Tetracarboxybenzene : 2257
Tetracene : 7703
1,2,3,4-Tetrachlorobenzene : 2259
1,2,4,5-Tetrachlorobenzene : 2260
s-Tetrachlorobenzene : 2260
Tetrachloro-*p*-benzoquinone : 4268
1,1,1,2-Tetrachlorobutane : 3427
1,2,2,3-Tetrachlorobutane : 3428
1,2,3,3-Tetrachlorobutane : 3429
1,3,4,4-Tetrachloro-1-butene : 3806
2,3,3,4-Tetrachloro-1-butene : 3807
1,1,2,3-Tetrachloro-2-(chloromethyl)propane : 10146
Tetrachloro-1,2-difluoroethane : 5637
1,1,1,2-Tetrachloroethane : 5634
Tetrachloroethylene : 5917
1,1,1,2-Tetrachloro-2-fluoroethane : 5639
1,1,1,2-Tetrachloro-2-methylpropane : 10147
1,1,1,5-Tetrachloropentane : 8838
2,3,4,6-Tetrachlorophenate : 9470
2,3,4,5-Tetrachlorophenol : 9469
2,3,5,6-Tetrachlorophenol : 9471
Tetrachlorophthalic anhydride : 7346
1,1,1,2-Tetrachloropropane : 10141
1,1,1,3-Tetrachloropropane : 10142
1,1,2,2-Tetrachloropropane : 10143
1,1,2,3-Tetrachloropropane : 10144
1,2,2,3-Tetrachloropropane : 10145
1,1,3,3-Tetrachloro-2-propanol : 10402
2,3,5,6-Tetrachloropyridine : 11019
Tetrachloro-*p*-quinone : 4268
3,3',4',5-Tetrachlorosalicylanilide : 610
2,3,4,6-Tetrachlorotoluene : 2262
2,3,5,6-Tetrachlorotoluene : 2263
Tetracosamethylhendecasiloxane : 11964
Tetracosamethylundecasiloxane : 11964
Tetracosylbenzene : 2265
Tetracosylcyclohexane : 4474
Tetracyanoethylene : 5915
Tetracycline : 7706
Tetradecamethylcycloheptasiloxane : 4239
Tetradecamethylene dibromide : 11616
Tetradecamethylhexasiloxane : 6990
Tetradecane, 1-cyclopentyl- : 4850
Tetradecane, 1,1-dicyclohexyl : 4476
Tetradecane, 1,1-diphenyl- : 2267
Tetradecanoic acid, chloride : 11635
Tetradecyl alcohol : 11630
Tetradecylsulfonic acid : 11620
Tetradifon : 2325
Tetraethylammonium bromide : 5447
1,2,3,4-Tetraethylbenzene : 2268
1,2,4,5-Tetraethylbenzene : 2269
N,*N*,*N'*,*N'*-Tetraethyl-1,4-butanediamine : 3262
Tetraethyldiarsenic : 5109
Tetraethyl diphosphate : 5221
Tetraethylene glycol : 5756
Tetraethylene glycol dimethyl ether : 8995
Tetraethylenepentamine : 5498
Tetraethyl lead : 9731
Tetraethylmethane : 8756
N,*N*,*N'*,*N'*-Tetraethyl-*m*-phenylenediamine : 1285
N,*N*,*N'*,*N'*-Tetraethylphthalamide : 1325
Tetraethyl 1,1,2,3-propanetetracarboxylate : 10139
Tetraethyl 1,1,3,3-propanetetracarboxylate : 10140
Tetraethyl pyrophosphate : 5221
Tetraethylsilane : 11450
Tetraethyl silicate : 11506
2,3,4,5-Tetrafluorobenzene : 2270
1,2,3,4-Tetrafluorobenzene : 2270
1,2,3,5-Tetrafluorobenzene : 2271
2,3,5,6-Tetrafluorobenzene : 2272
1,2,4,5-Tetrafluorobenzene : 2272
1,1,1,2-Tetrafluoroethane : 5640
1,1,2,2-Tetrafluoroethane : 5641
Tetrafluoroethylene : 5918
2,2,3,3-Tetrafluoro-1-propanol : 10404
Tetrahelicene : 2406
(1-4)-Tetrahydroacenaphthene : 9
Tetrahydrobenzene : 4626
2,3,6,7-Tetrahydro-1H,5H-benzo[ij]quinolizine : 2737
Tetrahydrocortisone : 9744
1,2,3,4-Tetrahydro-1,2-dimethylnaphthalene : 7981
Tetrahydro-5,5-dimethyl-2-(1H)-pyrimidinone[3[4-(trifluoromethyl-styryl)-cinnamylidene- hydrazone : 7136
Tetrahydroeucarvone : 4238
Tetrahydrofuran : 6168
Tetrahydro-2,5-furandimethanol : 7109
Tetrahydrofurfuryl alcohol : 6124
Tetrahydrofurfurylamine : 6116
Tetrahydro-2H-pyran : 10800
5,6,7,8-Tetrahydroisoquinoline : 7389
Tetrahydro-3-methylfuran : 6172
Tetrahydromyrcenol : 8450
5,6,7,8-Tetrahydronaphthalene-2-sulfonyl chloride : 7978
1,2,3,4-Tetrahydro-2-naphthoic acid : 7778
Tetrahydro-1,4-oxazine : 7681
1,2,3,4-Tetrahydrophenanthrene : 9187
3,4,5,6-Tetrahydrophthalic anhydride : 7347
1,2,3,6-Tetrahydropyridine : 11020
5,6,7,8-Tetrahydroquinoline : 11297
Tetrahydroquinone : 4377
1,2,3,4-Tetrahydroquinoxaline : 11339
D-Tetrahydroxyadipic acid : 6205
Tetrahydroxy-1,4-benzoquinone : 4269
3,3',5,5'-Tetrahydroxybiphenyl : 3097
3,4',5,7-Tetrahydroxy-3',5'-dimethoxyflavylium chloride : 2847
3,3',4',7-Tetrahydroxyflavone : 2813
3',4',5,7-Tetrahydroxyflavone : 2814
2',3,5,7-Tetrahydroxyflavone : 2842
3,4',5,7-Tetrahydroxyflavone : 2843
3,5,3',4'-Tetrahydroxy-7-methoxyflavone : 2815
N,*N*,*N'*,*N'*-Tetra(2-hydroxypropyl)ethylenediamine : 10379
1,2,4,5-Tetraiodobenzene : 2275
1,2,3,4-Tetraiodobenzene : 2273
1,2,3,5-Tetraiodobenzene : 2274
Tetraiodoethylene : 5919
Tetraiodomethane : 7542
1,2,4,5-Tetraisopropylbenzene : 2276
Tetrakisdimethylaminophosphonous anhydride : 5220
Tetrakis(2-hydroxypropyl)-ethylenediamine : 10379
Tetralin : 7979
Tetralin, 1,1-dimethyl : 7980
Tetralin, 1,2-dimethyl : 7981
Tetralin, 1,3-dimethyl : 7982
Tetralin, 1,4-dimethyl : 7983
Tetralin, 2,2-dimethyl : 7985
Tetralin, 2,3-dimethyl : 7986
Tetralin, 2,5-dimethyl : 7987
Tetralin, 2,7-dimethyl : 7989
Tetralin, 2,8-dimethyl : 7990
Tetralin, 5,6-dimethyl : 7991
Tetralin, 5,7-dimethyl : 7992
Tetralin, 1-ethyl : 7888
Tetralin, 2-ethyl : 7889
Tetralin, 6-ethyl : 7891
6-Tetralinsulfonyl chloride : 7978
2-Tetralol : 8043
1-Tetralone, 7-ethyl : 8051
Tetramethylammonium bromide : 7456
Tetramethylammonium chloride : 7457
Tetramethylammonium iodide : 7458
2,3,5,6-Tetramethylbenzaldehyde : 598
2,2',4,4'-Tetramethylbenzophenone : 7594
Tetramethyl-*p*-benzoquinone : 4270
2,2,3,3-Tetramethylbutane : 3430
1,1,4,4-Tetramethyl-1,4-butanediol : 6801
1,1,3,4-Tetramethylcyclohexane : 4477
cis-1,1,3,5-Tetramethylcyclohexane : 4478
4-(2,5,6,6-Tetramethyl-2-cyclohexen-1-yl)-3-butene-2-one : 3884
1,1,3,3-Tetramethylcyclopentane : 4851
N,*N*,*N'*,*N'*-Tetramethyl-1,4-diaminobutane : 3263
N,*N*,*N'*,*N'*-Tetramethyldiaminoethane : 5516
Tetramethylene : 4170
2,3-Tetramethyleneaziridine : 478
1,4-Tetramethylenediamine : 3258
Tetramethylenedithiol : 3368
Tetramethylene glycol : 3346
Tetramethylethylene glycol : 3350
2,2,6,6-Tetramethyl-3,5-heptanedione : 6404
2,2,3,3-Tetramethylhexane : 6844
2,2,3,4-Tetramethylhexane : 6845
2,2,3,5-Tetramethylhexane : 6846
2,2,4,4-Tetramethylhexane : 6847
2,2,4,5-Tetramethylhexane : 6848
2,2,5,5-Tetramethylhexane : 6849
2,3,3,4-Tetramethylhexane : 6850
2,3,3,5-Tetramethylhexane : 6851
2,3,4,4-Tetramethylhexane : 6852
2,3,4,5-Tetramethylhexane : 6853
3,3,4,4-Tetramethylhexane : 6854
2,2,5,5-Tetramethyl-3,4-hexanedione : 6811
1,1,4,7-Tetramethylindan : 7260
Tetramethyl lead : 9732
Tetramethylmethanediamine : 7490
1,3,6,8-Tetramethylnaphthalene : 8003
2,2,7,7-Tetramethyl-3,6-octanedione : 8400
2,2,3,3-Tetramethylpentane : 8839
2,2,3,4-Tetramethylpentane : 8840
2,2,4,4-Tetramethylpentane : 8841
2,3,3,4-Tetramethylpentane : 8842
2,3,4,5-Tetramethylphenol : 9473
Tetramethyl-*p*-phenylenediamine : 1287

2,2,4,6-Tetramethylpiperidine : 9710
2,2,6,6-Tetramethylpiperidine : 9711
1,2,5,6-Tetramethyl-4-piperidinone : 9725
Tetramethylputrescine : 3263
2,3,4,6-Tetramethylpyridine : 11022
Tetramethylsilane : 11451
Tetramethyl silicate : 11508
Tetramethylstannane : 11544
Tetramethylsuccinonitrile : 3292
2,2,4,4-Tetramethyltetrahydrofuran : 6174
N,N,N',N'-Tetramethyltetra-methylenediamine : 3263
Tetramethylurea : 12030
Tetranitrin : 3432
Tetranitromethane : 7544
Tetranitropentaerythritol : 9990
2,3,4,6-Tetranitrophenol : 9476
2,4,5,6-Tetraoxohexahydropyrimidine : 11073
1,1,1,2-Tetraphenylethane : 1618
1,1,2,2-Tetraphenyl-1,2-ethanediol : 5580
1,1,2,2-Tetraphenylethylene glycol : 5580
2,3,4,5-Tetraphenylfuran : 6175
Tetraphenylfuran : 6175
Tetraphenylmethane : 1831
Tetraphenylsilane : 11452
Tetraphenylurea : 12031
Tetraphthene : 9
Tetrapropylammonium iodide : 9833
Tetratriacontylcyclohexane : 4479
Tetravinylsilane : 11449
sym-Tetrazine : 11651
Tetrolaldehyde : 3895
Tetrole : 6025
Tetroquinone : 4269
Thalidomide : 7363
Thebaine : 7679
Thebainone : 7677
3-Thenoic acid : 11738
Thenylpyramine : 5507
Theobromine : 10757
Theophylline : 10756
THF glycol : 7109
1-Thia-2-azaindene : 2387
Thiabutazide : 2856
Thiacetazone : 34
Thiachromine-8-ethanol, 2,7-dimethyl- : 11111
Thiacyclohexane : 11795
Thiacyclopentane : 11783
1-Thia-2,4-diazacyclopentadiene : 11657
1,2,4-Thiadiazole, 5-ethoxy-3-(trichloromethyl)- : 11599
1,3,4-Thiadiazole-2-sulfonic acid,amide,5-acetamido : 32
1-Thiaindene : 2399
Thiamine hydrochloride : 11686
Thiamorpholine : 11719
Thianaphthene : 2399
Thiazolsulfone : 11663
2-Thiazolylamine : 11662
4'-(2-Thiazolylsulfamoyl)succinanilic acid : 3578
2-Thienonitrile : 11732
2-Thienyl bromide : 11726
2-Thienyl chloride : 11740
Thioacetaldehyde trimer : 11920
Thioacetamide : 5643
Thioacetic acid : 5646
Thiobenzyl alcohol : 1832
Thiobutyl alcohol : 3437
Thiocarbamide : 11799
Thiochrome : 11111
Thiocyanatobenzene : 11709
Thiocyanic acid, *tert*-butyl ester : 11701
Thiocyanic acid, isopropyl ester : 11706
1-Thiocyanobutane : 11699
Thiocyclohexanone : 4481
2,2'-Thiodiethanol : 5766
Thiodiglycolic acid : 240
Thiodiphenylamine : 9499
3,3'-Thiodipropionic acid : 10329
Thiofuran : 11724
Thioglycerol : 10013
Thioglycolic acid : 202
Thioguanine : 10762
Thiolactic acid : 10285
Thiomedon : 7649
Thionalide : 68
Thionaphthene-2-carboxylic acid : 2400
Thiophene, 2-isobutyryl : 10472
Thiophene, 2-isopentyl- : 11771
Thiophene, 2-isopropyl- : 11772
Thiophene, 3-isopropyl- : 11773
Thiophenol : 2289
Thiophenol, 3-ethyl : 2297
Thiophosgene : 4032
o-Thiosalicylic acid : 2663
Thiosemicarbazide : 7142
Thioserine : 5014
Thiosinamine : 11809
Thiotax : 2866
3-Thiotolene : 11770
2-Thiouracil : 11110
Thioxanthone : 11812
Thiram : 11722
THP : 10800
Threose (*D*-) : 3193
4(10)-Thujene, (±)- : 2981
4(10)-Thujene (*d*) : 11815
4(10)-Thujene-3-ol : 2983
Thujic acid : 4241
Thujopsene : 4954
Thymine : 11068
Thymine 2-desoxyriboside : 11816
Thymol : 9418
Thymol, acetate : 9420
Thymol blue : 9369
2-Thymolcarboxylic acid : 2642
Thymol, 2,6-dibromo : 9281
Thymolphthalein : 7352
o-Thymotic acid : 2642
Thymyl iodide : 1782
Thyroxine : 11930
Tiglic acid : 3844
Tigogenin : 11532
Timonacic : 11682
Tiocarlide : 11800
δ-Tocopherol : 2785
α-Tocopherol : 2787
β-Tocopherol : 2789
γ-Tocopherol : 2790
α-Tocopheryl acetate : 2788
Tokoferol acetate : 2788
Tolazine : 9195
Tolbutamide : 2207
3,3'-Tolidine : 3034
o-Tolidine : 3034
o-Tolualdehyde : 587
p-Tolualdehyde : 589
α-Tolualdehyde : 868
m-Tolualdehyde, 2-hydroxy : 575
o-Toluamide : 619
Toluene : 1947
Toluene, 3-bromo-2-nitro- : 1068
Toluene, 3-bromo-5-nitro- : 1069
Toluene, 3-bromo-4-nitro- : 1073
Toluene, 4-bromo-2-nitro- : 1074
Toluene, 2-bromo-6-nitro : 1070
o-Toluenecarbonitrile : 2763
Toluene, 4-chloro-3-iodo- : 1173
Toluene, 5-chloro-2-iodo : 1174
Toluene, 2-chloro-α-methylimino : 7652
Toluene, 3-chloro-5-nitro- : 1207
Toluene-3,4-diamine : 1276
Toluene-2,6-diamine : 1277
Toluene-2,4-diamine : 1278
Toluene, 2,3-dibromo : 1302
Toluene, 3,4-dibromo : 1303
Toluene, 2,6-dibromo : 1304
Toluene, 2,5-dibromo- : 1305
Toluene, 2,3-dichloro- : 1383
Toluene, 2,5-dichloro- : 1387
Toluene, 2,4-dichloro-α-hydroxy : 1853
Toluene, 2,5-diethoxy : 1410
Toluene, α,4-dihydroxy : 1884
Toluene 2,4-diisocyanate : 1431
Toluene-2,6-diisocyanate : 1430
Toluene, 2,4-diisopropyl- : 1948
Toluene-α,α-diol, *m*-nitro-, diacetate (ester) : 7511
Toluene-3,4-dithiol : 1564
Toluene, 3-fluoro-4-nitro- : 1730
Toluene, 4-fluoro-3-nitro : 1729
Toluene, 3-(1-hydroxyethyl) : 1858
Toluene, α-hydroxy-4-methoxy-α-methyl : 1895
Toluene, α-hydroxy-4-nitro : 1908
m-Toluenenitrile : 2764
p-Toluenenitrile : 2765
Toluene, *p*-(1-octyldecyl)- : 8245
Toluene, α,α,2,3,4-pentachloro- : 2326
Toluene, α,α,2,4,5-pentachloro- : 2327
Toluene, 4-propoxy : 2032
o-Toluenesulfinic acid : 2181
p-Toluenesulfinic acid : 2182
p-Toluenesulfonamide, *N,N*-dichloro- : 2210
p-Toluenesulfonic acid : 2229
p-Toluenesulfonic acid chloride : 2250
p-Toluenesulfonic acid, ethyl ester : 2231
o-Toluenesulfonyl chloride : 2249
p-Toluenesulfonyl chloride : 2250
Toluene, 2,3,4,6-tetrachloro- : 2262
Toluene, 2,3,5,6-tetrachloro- : 2263
α-Toluenethiol, α-methyl-, (*S*)-(-)- : 1834
Toluene, 2,4,6-tribromo- : 2314
Toluene, 2,3,4-trichloro- : 2334
Toluene, 3,4,5-trichloro- : 2335
Toluene, 2,4,5-trichloro- : 2336
Toluene, 2,3,5-trichloro : 2337
Toluene, 2,4,6-triiodo- : 2352
Toluene, 2,3,4-trinitro- : 2046
Toluene, 2,4,5-trinitro- : 2047
o-Toluic acid : 2676
m-Toluic acid : 2677
p-Toluic acid : 2678
o-Toluidine : 778
m-Toluidine : 779
p-Toluidine : 780
m-Toluidine, *N*-benzyl : 1824
p-Toluidine, *N*-benzyl : 1825
m-Toluidine, 6-bromo- : 648
o-Toluidine hydrochloride : 791
p-Toluidine, 2-iodo- : 762
o-Toluidine, 5-iodo : 763
o-Toluidine, 4-(phenylsulfonyl)-*N*-[(trifluoromethyl)sulfono]- : 9149
p-Toluidine, *N*-propyl- : 815
o-Tolunitrile : 2763
m-Tolunitrile : 2764
p-Tolunitrile : 2765
m-Toluquinaldine : 11258
m-Toluquinoline : 11282
2-(*p*-Toluyl)benzoic acid : 2684
o-Tolyl alcohol : 1897
p-Tolyl alcohol : 1899
2-*o*-Tolylaminoethanol : 5744
2-*p*-Tolylaminoethanol : 5745
p-Tolylcarbinol : 1899
1-(4-Tolyl)ethanol : 1859
m-Tolylmethanol : 1898
p-Tolylsulfonylmethylnitrosamide : 2212
o-Tolylthiourea : 11806
Tomatidine : 11528
Tomatine : 6196
Topazone : 8589
Torularhodin : 4059
Tosic acid : 2229
p-Tosyl chloride : 2250
Tranexamic acid : 4333
trans-1,3-Dimethylcyclohexanol : 4511
Traumatic acid : 5336
Trehalose : 6220
Triacetamide : 50
Triacetin : 10195
N,N,N-Triacetylammonia : 50
Triacontyl alcohol : 11821
Triacontylcyclohexane : 4482
Triacontylcyclopentane : 4853
Triallylamine : 10523
1,2,3-Triaminobenzene : 2305
Triamterene : 10750
1,2,3-Triaza-1H-indene : 2871
1,2,4-Triazanaphthalene : 2870
1,3,5-Triazine-2,4-diamine, *N,N'*-bis(1-methylethyl)-6-(methylthio)- : 9770

1,3,5-Triazine-2,4-diamine, 6-chloro-*N,N'*-bis(1-methylethyl)- : 10508
1,3,5-Triazine-2,4-diamine, 6-chloro-*N*-cyclopropyl-*N'*-(1-methylethyl)- : 5013
1,3,5-Triazine-2,4-diamine, 6-chloro-*N,N'*-diethyl- : 11513
1,3,5-Triazine-2,4-diamine, 6-chloro-*N*-(1,1-dimethylethyl)-*N'*-ethyl- : 11593
1,3,5-Triazine-2,4-diamine, *N*-(1,1-dimethylethyl)-*N'*-ethyl-6-(methylthio)- : 11594
1,3,5-Triazine-2,4-diamine, *N*-ethyl-6-methoxy-*N'*-(1-methylpropyl)- : 3143
1,3,5-Triazine-2,4-diamine, 6-(ethylthio)-*N,N'*-bis(1-methylethyl)- : 5222
1,3,5-Triazine-2,4(1H,3H)-dione, 3-cyclohexyl-6-(dimethylamino)-1-methyl- : 6995
1,2,4-Triazin-5(4H)-one, 4-amino-6-(1,1-dimethylethyl)-3-(methylthio)- : 7658
1,2,4-Triazolo[3,4-b]benzothiazole, 5-methyl- : 11858
Tribenzobicyclo[2.2.2]octatriene : 2375
1,3,5-Tribenzoylbenzene : 7587
Tribromoacetamide : 94
Tribromoacetic acid : 241
Tribromoacetic acid, ethyl ester : 242
2,4,6-Tribromoaniline : 851
2,4,5-Tribromoanisole : 2311
1,2,3-Tribromobenzene : 2307
1,2,4-Tribromobenzene : 2308
1,3,5-Tribromobenzene : 2309
1,2,4-Tribromobutane : 3445
1,2,2-Tribromobutane : 3443
1,2,3-Tribromobutane : 3444
1,3,3-Tribromobutane : 3446
2,2,3-Tribromobutane : 3447
Tribromo-*tert*-butyl alcohol : 10405
Tribromochloromethane : 7549
2,4,6-Tribromo-*m*-cresol : 9480
1,1,2-Tribromocyclobutane : 3442
1,1,2-Tribromoethane : 5653
Tribromoethanol : 5767
Tribromofluoromethane : 7550
Tribromomethane : 7548
3,4,5-Tribromonitrobenzene : 2315
Tribromonitromethane : 7551
2,4,6-Tribromophenol : 9479
1,2,3-Tribromopropane : 10163
2,3,5-Tribromothiophene : 11790
2,4,6-Tribromotoluene : 2314
Tribromsalan : 609
Tributylamine : 3213
Tributyl phosphate : 9562
Tributylphosphine : 9520
Tributylsilane : 11454
Tributyltin acetate : 11538
S,S,S-Tributyl trithiophosphate : 11852
Tributyrin : 3586
Tributyryl glyceride : 3586
Tricarballylic acid : 10165
Tricarbethoxymethane : 7552
1,2,3-Tricarboxybenzene : 2318
1,2,4-Tricarboxybenzene : 2319
2,2,2-Trichloroacetaldehyde : 21
Trichloroacetaldehyde hydrate : 5582
Trichloroacetic acid : 243
Trichloroacetic acid, ethyl ester : 248
Trichloroacetic acid, (2-methoxyethyl) ester : 250
Trichloroacetic acid, methyl ester : 253
Trichloroacetic anhydride : 244
Trichloroacetonitrile : 288
Trichloroacetyl bromide : 293
Trichloroacetyl chloride : 300
2,3,4-Trichloroaniline : 852
2,4,6-Trichloroanisole : 2332
1,2,3-Trichlorobenzene : 2321
1,2,4-Trichlorobenzene : 2322
1,3,5-Trichlorobenzene : 2323
2,3,6-Trichlorobenzoic acid : 2727
2,4,5-Trichlorobenzoic acid : 2728
2,2,2-Trichloro-1,1-bis(4-chlorophenyl)ethanol : 1851
Trichlorobromomethane : 7474
1,2,3-Trichloro-1,3-butadiene : 3174
1,2,3-Trichlorobutane : 3449
1,2,4-Trichlorobutane : 3450
2,2,3-Trichlorobutane : 3451
1,1,3-Trichlorobutane : 3448
1,1,1-Trichloro-2-butanol : 3643
2,3,4-Trichloro-1-butene : 3809
1,2,4-Trichloro-2-butene : 3810
Trichlorobutylsilane : 11370
2,2,3-Trichlorobutyraldehyde : 3191
α,α,β-Trichlorobutyraldehyde : 3191
2,2,3-Trichlorobutyric acid : 3590
Trichloro(chloromethyl)silane : 11456
4,5,6-Trichloro-*o*-cresol : 9487
2,3,6-Trichloro-*p*-cresol : 9488
2,4,6-Trichloro-*m*-cresol : 9489
Trichloro(dichloromethyl)silane : 11459
1,1,1-Trichloro-2,2-difluoroethane : 5658
1,2,2-Trichloro-1,2-difluoroethane : 5660
1,1,1-Trichloro-2,2-dihydroxyethane : 5582
Trichlorododecylsilane : 11460
2,2,2-Trichloroethanal : 21
1,1,1-Trichloroethane : 5655
1,1,2-Trichloroethane : 5656
2,2,2-Trichloro-1,1-ethanediol : 5582
2,2,2-Trichloroethanol : 5768
Trichloroethylene : 5922
Trichlorofluoromethane : 7555
Trichlorohexylsilane : 11464
Trichloroiodomethane : 7556
Trichloromethane : 7554
Trichloromethiazide : 2854
(Trichloromethyl)benzene : 2333
1,1,2-Trichloro-2-methylpropane : 10177
2-(Trichloromethyl)pyridine : 11023
3,4,6-Trichloro-2-nitrophenol : 9490
Trichloro perchlorate : 9148
2,3,4-Trichlorophenol : 9481
2,3,5-Trichlorophenol : 9482
2,3,6-Trichlorophenol : 9483
2,4,5-Trichlorophenol : 9484
2,4,6-Trichlorophenol : 9485
3,4,5-Trichlorophenol : 9486
1,1,1-Trichloropropane : 10171
1,1,2-Trichloropropane : 10172
1,1,3-Trichloropropane : 10173
1,2,2-Trichloropropane : 10174
1,2,3-Trichloropropane : 10175
1,1,1-Trichloro-2-propanol : 10406
2,4,6-Trichloropyrimidine : 11075
3',4',5-Trichlorosalicylanilide : 608
2,3,5-Trichlorothiophene : 11791
2,3,4-Trichlorotoluene : 2334
2,4,5-Trichlorotoluene : 2336
2,3,5-Trichlorotoluene : 2337
2,2',2''-Trichlorotriethylamine : 5412
1,3,5-Trichloro-2,4,6-trifluorobenzene : 2340
1,1,1-Trichlorotrifluoroethane : 5662
1,1,2-Trichlorotrifluoroethane : 5663
Trichloro-1,3,5-trihydroxybenzene : 2371
Trichlorovinylsilane : 11461
Trichocid : 85
Tricosylbenzene : 2341
Tricosylcyclohexane : 4483
Tricromyl : 2838
Tricyclenic acid : 11862
Tricyclenic acid, ethyl ester : 11863
Tricyclenic acid, methyl ester : 11864
Tridecamethylene dibromide : 11874
Tridecane, 1-amino : 11870
Tridecane, 1-cyclohexyl- : 4484
Tridecane, 1-cyclopentyl- : 4854
Tridecane, 2-methyl-13-bromo : 11873
Tridecane, 7-phenyl- : 1764
Tridecyl alcohol : 11883
Tridecylcyclohexane : 4484
Trientine : 5499
Triethanolamine : 5751
1,3,5-Triethoxybenzene : 2343
Triethoxysilane : 11477
Triethylamine : 5417
Triethylantimony : 11549
Triethyl arsenate : 439
Triethyl arsenite : 440
Triethylarsine : 448
1,2,4-Triethylbenzene : 2344
1,3,5-Triethylbenzene : 2345
Triethylborane : 3118
Triethylenediamine : 5111
Triethylene glycol : 5713
Triethylene glycol, (3-aminopropyl) ether : 5678
Triethylene glycol, diacetate : 5714
Triethylenemelamine : 11844
Triethylenetetramine : 5499
Triethyl methanetricarboxylate : 7552
Triethyl phosphate : 9563
Triethyl phosphine : 9521
Triethyl phosphite : 9591
Triethylphosphorus : 9521
Triethylsilane : 11483
Triethylstibine : 11549
2,4,6-Triethyl-2,4,6-trimethylcyclotrisiloxane : 5003
2,4,6-Triethyl-2,4,6-triphenylcyclotrisiloxane : 5004
Trifluoroacetic acid : 263
Trifluoroacetic acid, propyl ester : 269
Trifluoroacetonitrile : 289
1,3,5-Trifluorobenzene : 2347
sym-Trifluorobenzene : 2347
Trifluoroboron etherate : 3137
Trifluorobromomethane : 7475
Trifluorochloromethane : 7486
1,1,2-Trifluoroethane : 5666
2,2,2-Trifluoroethanol : 5770
Trifluoroethylene : 5924
Trifluoroethylsilane : 11432
Trifluoroiodomethane : 7559
Trifluoromethane : 7558
o-(Trifluoromethyl)aniline : 855
m-(Trifluoromethyl)aniline : 856
(Trifluoromethyl)benzene : 2348
3'-Trifluoromethyldiphenylamine-2-carboxylic acid : 2729
Trifluoromethylhydrothiazide : 2858
3-(Trifluoromethyl)-1-nitrobenzene : 2061
N-(*m*-Trifluoromethylphenyl)-2-aminobenzoic acid : 2729
N-(3-Trifluoromethylphenyl)-*N',N'*-dimethylurea : 12007
(±)-2-[4-[[5-(Trifluoromethyl)-2-pyridinyl]oxy]phenoxy]propanoic acid, butyl ester : 5962
(*R*)-2-[4-(5-Trifluoromethyl-2-pyridyloxy)phenoxy]propionic acid, butyl ester : 5963
3,3,3-Trifluoropropene : 10608
Triglycol diacetate : 5714
Triglyme : 11642
Trigonelline : 11029
Trihexyphenidyl hydrochloride : 9705
2',3',4'-Trihydroxyacetophenone : 5873
1,3,5-Trihydroxybenzene : 2368
3,4,5-Trihydroxybenzoic acid : 2731
1,2,3-Trihydroxybutane : 3458
Trihydroxyestrin : 5404
4',5,7-Trihydroxyflavone : 2806
1,2,6-Trihydroxyhexane : 6868
4',5,7-Trihydroxyisoflavone : 2807
3',5,7-Trihydroxy-4'-methoxyflavanone : 2800
1,2,3-Trihydroxypropane : 10186
2,4,6-Trihydroxypropiophenone : 10489
1,2,3-Triiodobenzene : 2349
1,2,4-Triiodobenzene : 2350
1,3,5-Triiodobenzene : 2351
Triiodomethane : 7560
2,4,6-Triiodophenol : 9492
3,3',5-Triiodothyropropionic acid : 2136
2,4,6-Triiodotoluene : 2352
Triisobutylamine : 9820
Triisobutyl phosphate : 9572
1,2,4-Triisopropylbenzene : 2372
3,7,12-Triketocholanic acid : 4101
Trimellitic acid : 2319
Trimellitic anhydride : 7341
Trimethadione : 8587
3,4,5-Trimethoxybenzyl alcohol : 1914
Trimethoxyborane : 3126
Trimethoxyphosphine : 9592
1,1,1-Trimethylacetone : 3657
Trimethyl aluminum : 339
Trimethylamine : 7444
Trimethylamineborane : 3132

Trimethylantimony : 11550
Trimethylarsine : 449
1,2,3-Trimethylbenzene : 2357
Trimethylborane : 3119
Trimethyl borate : 3126
2,2,3-Trimethylbutane : 3457
Trimethylbutylsilane : 11371
Trimethylchlorosilane : 11389
Trimethyl citrate : 10170
1,1,2-Trimethylcycloheptane : 4227
1,1,4-Trimethylcycloheptane : 4228
3,3,7-Trimethylcycloheptanol : 4233
2,6,6-Trimethylcycloheptanone : 4238
1,4,4-Trimethyl-1-cycloheptene : 4247
1,1,2-Trimethylcyclohexane : 4485
1,1,3-Trimethylcyclohexane : 4486
1,1,4-Trimethylcyclohexane : 4487
1,2,3-Trimethylcyclohexane : 4488
trans-1,3,5-Trimethylcyclohexane : 4497
1,1,3-Trimethyl-2-cyclohexanol : 4566
2,3,6-Trimethylcyclohexanol : 4568
cis-3,3,5-Trimethylcyclohexanol : 4569
trans-3,5,5-Trimethylcyclohexanol : 4570
trans-3,3,5-Trimethylcyclohexanol : 4570
1,2,2-Trimethylcyclohexanol : 4562
1,2,6-Trimethylcyclohexanol : 4563
1,3,5-Trimethylcyclohexanol : 4564
2,2,5-Trimethylcyclohexanol : 4565
2,2,3-Trimethylcyclohexanone : 4614
2,2,4-Trimethylcyclohexanone : 4615
2,3,6-Trimethylcyclohexanone : 4618
2,4,4-Trimethylcyclohexanone : 4619
2,4,5-Trimethylcyclohexanone : 4620
3,3,5-Trimethylcyclohexanone : 4622
3,4,4-Trimethylcyclohexanone : 4623
2,2,5-Trimethylcyclohexanone : 4616
2,4,6-Trimethylcyclohexanone : 4621
1,4,4-Trimethylcyclohexene : 4689
1,5,5-Trimethylcyclohexene : 4691
1,2,3-Trimethylcyclohexene : 4687
1,3,5-Trimethylcyclohexene : 4688
1,4,5-Trimethylcyclohexene : 4690
1,6,6-Trimethylcyclohexene : 4692
2,4,4-Trimethyl-2-cyclohexen-1-ol : 4701
3,6,6-Trimethyl-2-cyclohexenone : 4722
4-(2,6,6-Trimethyl-1-cyclohexen-1-yl)-3-buten-2-ol : 3873
1,1,2-Trimethylcyclopentane : 4855
1,1,3-Trimethylcyclopentane : 4856
1,2,2-Trimethylcyclopentanol : 4875
1,2,4-Trimethylcyclopentanol : 4876
1,2,5-Trimethylcyclopentanol : 4877
2,3,5-Trimethylcyclopentanol : 4898
2,2,4-Trimethylcyclopentanone : 4896
2,2,5-Trimethylcyclopentanone : 4897
2,4,4-Trimethylcyclopentanone : 4899
1,5,5-Trimethylcyclopentene : 4938
1,1,2-Trimethylcyclopropane : 4985
trans-3,7,11-Trimethyl-1,3,6,10-dodecatetraene : 5329
trans-2,6,10-Trimethyl-2,6,9,11-dodecatetraene : 5329
Trimethylene : 4955
Trimethylene chlorohydrin : 10348
Trimethylene dichloride : 9893
Trimethylene diiodide : 9907
Trimethylene dimercaptan : 10031
Trimethylene glycol : 9983
Trimethylene oxide : 8592
Trimethylene sulfide : 11691
Trimethylene sulfone : 11692
2,2,3-Trimethylheptane : 6437
2,2,4-Trimethylheptane : 6438
2,2,5-Trimethylheptane : 6439
2,2,6-Trimethylheptane : 6440
2,3,3-Trimethylheptane : 6441
2,3,4-Trimethylheptane : 6442
2,3,5-Trimethylheptane : 6443
2,3,6-Trimethylheptane : 6444
2,4,4-Trimethylheptane : 6445
2,4,5-Trimethylheptane : 6446
2,4,6-Trimethylheptane : 6447
2,5,5-Trimethylheptane : 6448
3,3,4-Trimethylheptane : 6449
3,3,5-Trimethylheptane : 6450
3,4,4-Trimethylheptane : 6451
3,4,5-Trimethylheptane : 6452
2,2,3-Trimethylhexane : 6859
2,2,4-Trimethylhexane : 6860
2,2,5-Trimethylhexane : 6861
2,3,3-Trimethylhexane : 6862
2,3,4-Trimethylhexane : 6863
2,3,5-Trimethylhexane : 6864
2,4,4-Trimethylhexane : 6865
3,3,4-Trimethylhexane : 6866
3,5,5-Trimethylhexan-3-ol : 6958
1,1,5-Trimethylindan : 7261
1,4,7-Trimethylindan : 7262
1,2,3-Trimethylindole : 7322
2,3,5-Trimethylindole : 7323
Trimethyl methanetricarboxylate : 7553
1,2,4-Trimethylnaphthalene : 8009
1,2,5-Trimethylnaphthalene : 8010
1,2,6-Trimethylnaphthalene : 8011
2,4,6-Trimethylnaphthalene : 8013
2,4,7-Trimethyloctane : 8421
Trimethylolmelamine : 7583
Trimethylolpropane : 10006
Trimethylolpropane phosphite : 11909
2,2,3-Trimethylpentane : 8847
2,3,3-Trimethylpentane : 8849
2,3,4-Trimethylpentane : 8850
2,2,4-Trimethyl-1,3-pentanediol : 8795
2,4,4-Trimethyl-2-pentanol : 8954
2,2,4-Trimethyl-3-pentanol : 8955
2,4,4-Trimethyl-1-pentene : 9064
2,3,3-Trimethyl-1-pentene : 9065
2,3,4-Trimethyl-1-pentene : 9066
2,2,4-Trimethyl-4-pentene : 9067
3,4,4-Trimethyl-1-pentene : 9069
2,3,4-Trimethyl-2-pentene : 9070
2,4,4-Trimethyl-2-pentene : 9071
3,4,4-Trimethyl-2-pentene : 9072
2,4,5-Trimethylphenyl acetate : 9496
Trimethylphosphine : 9522
Trimethyl phosphite : 9592
1,2,4-Trimethylpiperazine : 9622
2,3,6-Trimethylpiperidine : 9713
2,4,6-Trimethylpiperidine : 9714
1,1,2-Trimethylpropyl bromide : 3236
Trimethylpropylsilane : 11496
1,3,4-Trimethylpyrazole : 10839
1,3,5-Trimethylpyrazole : 10840
1,4,5-Trimethylpyrazole : 10841
2,3,4-Trimethylpyridine : 11024
2,3,5-Trimethylpyridine : 11025
2,3,6-Trimethylpyridine : 11026
2,4,5-Trimethylpyridine : 11027
1,2,5-Trimethylpyrrole : 11142
2,2,4-Trimethylpyrrolidine : 11175
2,4,6-Trimethylquinoline : 11302
2,4,7-Trimethylquinoline : 11303
Trimethylsilane : 11489
(Trimethylsilyl)methyl bromide : 11366
Trimethylstibine : 11550
2,3,4-Trimethylthiophene : 11792
2,3,5-Trimethylthiophene : 11793
5,7,8-Trimethyltocol : 2787
1,1,1-Trimethyl-3,3,3-triphenyldisiloxane : 5239
Trimethylurea : 12032
Trimyristin : 11628
Trinitroacetonitrile : 290
2,3,5-Trinitroanisole : 364
sym-Trinitrobenzene : 2365
2,4,6-Trinitrobenzoic acid : 2735
Trinitro-*m*-cresol : 9434
2,4,6-Trinitro-3,5-dimethyl-*tert*-butylbenzene : 1461
2,4,7-Trinitrofluorenone : 5979
Trinitromethane : 7562
2,3,4-Trinitrotoluene : 2046
2,4,5-Trinitrotoluene : 2047
2,4,6-Trinitrotoluene : 2048
28,29,30-Trinorlanostane : 4105
Triolein : 8310
Trional : 3230
Triopron : 2136
s-Trioxane, 2,4,6-trivinyl- : 11906
3,6,9-Trioxaundecane : 5612
3,7,12-Trioxocholanic acid : 4101
Tripalmitin : 6656
Tripelennamine : 5505
Triphenylamine : 728
Triphenylarsane : 450
Triphenylarsine : 450
Triphenylbismuth : 3111
Triphenylcarbinol : 1866
Triphenylhydrazine : 7173
Triphenylmethane : 1982
Triphenylmethane, 3-methyl : 7513
Triphenylmethanol : 1866
Triphenylmethylsilane : 11444
Triphenyl phosphate : 9566
Triphenylphosphine : 9523
Triphenyltetrazolium chloride : 11653
Triphenyltin acetate : 11539
Triphenyltin chloride : 11541
Tripropylamine : 9808
Tripropyl phosphate : 9567
Tripropyl phosphite : 9594
Tripropylsilane : 11500
Triptycene : 2375
2,4,6-Tripyridyl-s-triazine : 11843
Tris(allyloxy)vinylsilane : 11427
Tris-BP : 10358
Tris-(2-chloroethyl) amine : 5412
Tris(chloromethyl)phosphine : 9524
Tris(2,3-dibromopropyl) phosphate : 10358
Tris(dimethylamino)phosphine oxide : 9573
Tris(hydroxymethyl)nitromethane : 10012
Tristearin : 8272
Trithioacetaldehyde : 11920
Trithiocarbonic acid : 4034
Tri-*o*-tolyl phosphate : 9569
Trityl alcohol : 1866
Tromasedan : 2385
Tromethamine : 9986
Tropacocaine : 488
Tropane : 484
Tropic acid : 915
Tropilidene : 4240
Tropine : 489
Tryptamine : 7302
Tryptophol : 7305
Tsuduranine : 5159
Tsuzuic acid : 11637
Tuaminoheptane : 6350
Tubatoxin : 2783
Tuduic acid : 11637
D-Turanose : 6022
Tybamate : 3933
Tylenol : 67
Tyramine : 9208
Tyrosine, 3,5-dibromo(*L*) : 11927
Ugurol : 4333
Umbelliferone : 2826
Umbelliferone glucoside : 2821
Undecafluorocyclohexane : 4498
trans-9-Undecenoic acid : 11975
(10-Undecenyl)acetic acid : 11895
Undecylenic acid : 11974
Undecyl ketone : 11857
Undecyl mercaptan : 11942
Uneicosane, 11-phenyl : 6297
Unidigin : 4043
Untriacontane : 6299
Uracil : 11059
Uracil mustard : 11061
Uramil : 11077
Urazole : 11850
Urea, *N*'-(4-bromo-3-chlorophenyl)-*N*-methoxy-*N*-methyl- : 4076
Urea, 1-(2-bromo-2-ethylbutanoyl) : 3195
Urea, 1-(2-bromo-3-methylbutanoyl) : 3196
Urea, 3-(*p*-bromophenyl)-1-methoxy-1-methyl- : 7656
Urea, *N*'-[4-(4-chlorophenoxy)phenyl]-*N,N*-dimethyl- : 4088
Urea, *N,N*'-diamino- : 4009
Urea, *N*'-(3,4-dichlorophenyl)-*N*-methoxy-*N*-methyl- : 7415
Urea, 1,1-diisobutyl : 11996
Urea, *N*-[5-(1,1-dimethylethyl)-1,3,4-thiadiazol-2-yl]-*N,N*'-dimethyl- : 11590
Urea, *N,N*-dimethyl-*N*'-(octahydro-4,7-methano-1H-inde*N*-5-yl)-, (3aα,4α, 5α,7α,7aα)- : 8212
Urea hydrochloride : 12020
Urea nitrate : 12021

Urea, *N*-phenyl-*N'*-1,2,3-thiadiazol-5-yl- : 11687
Urea, 1-sulfanilyl-2-thio- : 2189
Urea, 1,1,3,3-tetraethyl- : 12029
5-Ureidohydantoin : 12009
Urethan : 3948
Urethane, *sec*-butyl : 3960
Urethane, isopentyl : 3952
Uric acid : 10764
Uric acid, 3-methyl- : 10765
Uric acid, 7-methyl- : 10766
Uridine 5'-phosphoric acid : 12035
Urinex : 2852
Uriprim : 10852
Urochloralic acid : 6231
Urofort : 11830
Ursodeoxycholic acid : 4098
Ursodiol : 4098
Ursolic acid : 12037
Uteroverdine : 3001
Uvic acid : 3311
Uzarin : 4045
Vaccenic acid : 8295
Valeraldehyde : 8690
Valeraldehyde, 2-propyl- : 8697
Valeric acid : 8855
Valeric acid, 4-acetyl, ethyl ester : 6902
Valeric acid, 4-acetyl, methyl ester : 6903
Valeric acid, *sec*-butyl ester (*d*) : 8899
Valeric acid, 4-chloro- : 8865
Valeric acid, 3-chloro : 8864
Valeric acid, 4-chloro-, ethyl ester : 8867
Valeric acid, 5-chloro-4-hydroxy-, γ-lactone : 6141
Valeric acid, 2,3-dibromo- : 8870
Valeric acid, 2,5-dibromo- : 8871
Valeric acid, 2,3-dibromo, ethyl ester : 8872
Valeric acid, 2,5-dibromo-, ethyl ester : 8873
Valeric acid, 4,4-dimethyl-5-acetyl : 6461
Valeric acid, 2,2-dimethyl-5-acetyl : 6460
Valeric acid, 2-ethyl (*DL*) : 8877
Valeric acid, 5-fluoro- : 8879
Valeric acid, furfuryl ester : 8880
Valeric acid, 4-hydroxy-, ethyl ester : 8884
Valeric acid, 2-hydroxyethyl ester : 8885
Valeric acid, 3-methyl (*DL*) : 8888
Valeric acid, 2-methylene : 8890
Valeric acid, 3-methyl-3-phenyl : 7175
Valeric acid, octyl ester : 8901
Valeric acid, 2-oxo-, ethyl ester : 8905
Valeric acid, 2-phenyl-, (±)- : 943
Valeronitrile : 8827
Valeronitrile, 3-imino-2-methyl : 8830
Valeronitrile, 4-methyl-2-phenyl- : 964
Valeroyl chloride : 8990
Valerylacetone : 8393
Valerylacetyl : 6402
Valeryl chloride, 2-methyl (*DL*) : 8993
Valeryl chloride, 3-methyl (*DL*) : 8994
Valimyl : 3201
Valium : 2410
Vanillic acid : 2625
Vanillin : 574
Vanillylacetone : 3663
Vanilmandelic acid : 896
Vasicine : 11192
Vellosimine : 11346
Veratraldehyde : 545
Veratramine : 12044
Veratric acid : 2576
Veratrole : 1436
Veratrole, 3-iodo : 1767
(*R*)-(+)-Verbenone : 2978
d-Verbenone : 2978
Vernine : 6289
Vesipyrin : 2483
Vibeline : 3540
Vicianose : 6234
Vicine : 11107
(+)-*cis*-Vincamine : 5349
Vincamine : 5349
Vinyl acetate : 179
Vinylacetonitrile : 3803
Vinylacetylene : 3889
Vinyl amide : 10519
α-Vinylbenzyl alcohol : 1868
Vinyl bromide : 5880
Vinyl butyl ketone : 6583
Vinyl butyrate : 3513
Vinylcarbinol : 10697
Vinyl chloride : 5884
Vinyl cyanide : 10583
4-Vinyl-1-cyclohexene dioxide : 8557
Vinylethylacetylene : 7103
Vinylethylene : 3146
Vinyl fluoride : 5906
Vinyl formate : 6006
Vinylformic acid : 10609
Vinylglycolic acid : 3842
1-Vinylheptanol : 8189
Vinylidene chloride : 5894
Vinylidene fluoride : 5902
Vinyl isobutyl ketone : 7093
2-Vinylnaphthalene : 7882
5-Vinyl-2-picoline : 10955
Vinyl propionate : 10261
Vinyltrichlorosilane : 11461
Violaxanthin : 4056
Violaxanthin, all-*trans*- : 4056
Violet-leaf aldehyde : 8108
Violuric acid : 11074
Viquidil : 10434
Visnadine : 3540
Visnagin : 6186
Vitamin A : 11341
Vitamin B1, hydrochloride : 11686
Vitamin B4 : 10751
Vitamin Bc : 6243
Vitamin BX : 2486
Vitamin C : 466
Vitamin D2 : 11350
Vitamin D3 : 11348
Vitamin E : 2787
Vitamin E acetate : 2788
Vitamin K1 : 7865
Vitamin K6 : 7800
Vitamin L1 : 2484
Vomicine : 11351
Warfarin : 2832
Wintergreen oil : 2634
Woodward's Reagent K : 7407
Xanthaurine : 2817
9-Xanthenecarboxylic acid, 2(*N*,*N*-diethylamino)ethyl ester,methylbromide : 5445
Xanthine : 10755
Xanthone : 12050
Xanthone, 2,7-dihydroxy : 12052
Xanthophyll 3 : 4057
all-*trans*-(+)-Xanthophyll : 4058
Xanthophyll : 4058
Xanthopterin : 10749
Xanthoxyletin : 2398
Xanthurenic acid : 11230
Xanthyletin : 2397
o-Xenol : 3088
p-Xenylamine : 3011
o-Xylene : 1452
m-Xylene : 1453
p-Xylene : 1454
m-Xylene, 4-*sec*-butyl- : 1473
m-Xylene, 5-*tert*-butyl- : 1460
m-Xylenediamine : 1433
o-Xylene, 3,4-dibromo- : 1297
o-Xylene, 3,4-dichloro- : 1360
p-Xylene, α,α'-dimethyl- : 1415
p-Xylene, 2,5-dimethyl- : 2280
p-Xylene, α,α'-diphenyl- : 995
o-Xylene, 3-ethyl- : 1670
o-Xylene, 4-ethyl- : 1675
m-Xylene, 2-ethyl- : 1673
m-Xylene, 4-ethyl- : 1671
m-Xylene, 5-ethyl- : 1672
p-Xylene, 2-ethyl- : 1674
o-Xylene, 4-iodo- : 1774
m-Xylene, 5-iodo- : 1771
o-Xylene, 3-isopropyl- : 1469
m-Xylene, 2-isopropyl- : 1471
o-Xylene, 3-propyl- : 1485
m-Xylene, 2-propyl- : 1482
m-Xylene, 4-propyl- : 1487
m-Xylene, 5-propyl- : 1483
p-Xylene, 2-propyl- : 1486
p-Xylene, 2,3,5-trinitro- : 1489
2,3-Xylenol : 9307
2,4-Xylenol : 9308
2,5-Xylenol : 9309
2,6-Xylenol : 9310
3,4-Xylenol : 9311
3,5-Xylenol : 9312
3,4-Xylenol, 6-*tert*-butyl- : 9324
2,3-Xylidine : 704
2,4-Xylidine : 705
2,5-Xylidine : 706
2,6-Xylidine : 707
3,4-Xylidine : 708
3,5-Xylidine : 709
Xylite : 12058
p-Xylotocopherol : 2789
o-Xylotocopherol : 2790
1-(2,5-Xylylazo)-2-naphthol : 8029
Yellow AB : 7729
Yohimbine : 12062
Zeaxanthin : 4057
Zinc, bis(dimethylcarbamodithioato-*S*,*S'*)-, (T-4)- : 12068
Zineb : 12067
Zingerone : 3663
Zoxazolamine : 2876
Zymostanol : 4107

$CBrClF_2$
7464 - Methane, bromochlorodifluoro-
$CBrClN_2O_4$
7465 - Methane, bromochlorodinitro-
$CBrCl_2F$
7468 - Methane, bromodichlorofluoro-
$CBrCl_3$
7474 - Methane, bromotrichloro-
$CBrF_3$
7475 - Methane, bromotrifluoro-
$CBrN$
4165 - Cyanogen bromide
$CBrN_3O_6$
7476 - Methane, bromotrinitro-
CBr_2ClF
7494 - Methane, dibromochlorofluoro-
CBr_2Cl_2
7495 - Methane, dibromodichloro-
CBr_2F_2
7496 - Methane, dibromodifluoro-
$CBr_2N_2O_4$
7497 - Methane, dibromodinitro-
CBr_2O
4006 - Carbonic dibromide
CBr_3Cl
7549 - Methane, tribromochloro-
CBr_3F
7550 - Methane, tribromofluoro-
CBr_3NO_2
7551 - Methane, tribromonitro-
CBr_4
7539 - Methane, tetrabromo-
$CClFO$
4005 - Carbonic chloride fluoride
$CClF_3$
7486 - Methane, chlorotrifluoro-
$CClF_3O_2S$
7537 - Methanesulfonyl chloride, trifluoro-
$CClN$
4166 - Cyanogen chloride
$CClNO_3S$
11585 - Sulfuryl chloride isocyanate
$CClN_3O_6$
7487 - Methane, chlorotrinitro-
CCl_2F_2
7500 - Methane, dichlorodifluoro-
$CCl_2N_2O_4$
7501 - Methane, dichlorodinitro-
CCl_2O
4007 - Carbonic dichloride
CCl_2S
4032 - Carbonothioic dichloride
CCl_3F
7555 - Methane, trichlorofluoro-
CCl_3I
7556 - Methane, trichloroiodo-
CCl_3NO_2
7557 - Methane, trichloronitro-
CCl_4
7540 - Methane, tetrachloro-
CCl_4O_2S
7536 - Methanesulfonyl chloride, trichloro-
CCl_4O_4
9148 - Perchloric acid, trichloromethyl ester
CCl_4S
7527 - Methanesulfenyl chloride, trichloro-
CFN
4167 - Cyanogen fluoride
CF_2O
4008 - Carbonic difluoride
CF_3I
7559 - Methane, trifluoroiodo-
CF_4
7541 - Methane, tetrafluoro-
CF_4S
7528 - Methanesulfenyl fluoride, trifluoro-
CF_8S
11584 - Sulfur, pentafluoro(trifluoromethyl)-, (OC-6-21)-
$CHBrClF$
7466 - Methane, bromochlorofluoro-
$CHBrCl_2$
7467 - Methane, bromodichloro-
$CHBrF_2$
7469 - Methane, bromodifluoro-
$CHBr_2Cl$
7493 - Methane, dibromochloro-
$CHBr_2F$
7498 - Methane, dibromofluoro-
$CHBr_3$
7548 - Methane, tribromo-
$CHClF_2$
7478 - Methane, chlorodifluoro-
$CHClI_2$
7479 - Methane, chlorodiiodo-
$CHClN_2O_4$
7480 - Methane, chlorodinitro-
$CHCl_2F$
7502 - Methane, dichlorofluoro-
$CHCl_2I$
7503 - Methane, dichloroiodo-
$CHCl_3$
7554 - Methane, trichloro-
$CHCl_5Si$
11459 - Silane, trichloro(dichloromethyl)-
$CHFI_2$
7516 - Methane, fluorodiiodo-
$CHFO$
6019 - Formyl fluoride
CHF_2I
7505 - Methane, difluoroiodo-
CHF_3
7558 - Methane, trifluoro-
CHI_3
7560 - Methane, triiodo-
CHN
7176 - Hydrocyanic acid
$CHNO$
4162 - Cyanic acid
6024 - Fulminic acid
7360 - Isofulminic acid
$CHNS$
7399 - Isothiocyanic acid
CHN_3O_6
7562 - Methane, trinitro-
$CHNaO_3$
4004 - Carbonic acid monosodium salt
CH_2BrCl
7463 - Methane, bromochloro-
CH_2BrF
7470 - Methane, bromofluoro-
CH_2BrI
7471 - Methane, bromoiodo-
CH_2BrNO_2
7473 - Methane, bromonitro-
CH_2Br_2
7492 - Methane, dibromo-
CH_2ClF
7481 - Methane, chlorofluoro-
CH_2ClI
7482 - Methane, chloroiodo-
CH_2ClNO
3971 - Carbamic chloride
CH_2ClNO_2
7485 - Methane, chloronitro-
CH_2Cl_2
7499 - Methane, dichloro-
CH_2Cl_4Si
11456 - Silane, trichloro(chloromethyl)-
$CH_2Cl_6Si_2$
11440 - Silane, methylenebis[trichloro-
$CH_2Cu_2O_5$
511 - Basic copper carbonate
CH_2FI
7517 - Methane, fluoroiodo-
CH_2F_2
7504 - Methane, difluoro-
CH_2I_2
7506 - Methane, diiodo-
CH_2N_2
4155 - Cyanamide
7491 - Methane, diazo-
$CH_2N_2O_4$
7508 - Methane, dinitro-
CH_2N_4
11652 - 1H-Tetrazole
CH_2O
5987 - Formaldehyde
$(CH_2O)_x$
8634 - Paraformaldehyde
CH_2O_2
6002 - Formic acid
CH_2S_3
4034 - Carbonotrithioic acid
CH_3AsF_2
463 - Arsonous difluoride, methyl-
CH_3Br
7462 - Methane, bromo-
CH_3Br_3Si
11453 - Silane, tribromomethyl-
CH_3Cl
7477 - Methane, chloro-
CH_3ClO_2S
7535 - Methanesulfonyl chloride
CH_3ClO_3S
4086 - Chlorosulfuric acid, methyl ester
CH_3Cl_3Si
11465 - Silane, trichloromethyl-
CH_3F
7515 - Methane, fluoro-
CH_3FO_2S
7538 - Methanesulfonyl fluoride
CH_3I
7518 - Methane, iodo-
CH_3NO
5988 - Formaldehyde, oxime
5989 - Formamide
CH_3NO_2
7522 - Methane, nitro-
8091 - Nitrous acid, methyl ester
CH_3NO_3
8075 - Nitric acid, methyl ester
CH_3N_3
7460 - Methane, azido
$CH_3N_3O_3$
12024 - Urea, nitro-
CH_4
7459 - Methane
CH_4Cl_2Si
11398 - Silane, dichloromethyl-
CH_4N_2
7563 - Methanimidamide
CH_4N_2O
7566 - Methanimidamide, *N*-hydroxy-
11992 - Urea
$CH_4N_2O_2$
7451 - Methanamine, *N*-nitro-
12016 - Urea, hydroxy-
CH_4N_2S
11799 - Thiourea
CH_4N_2Se
11354 - Selenourea
$CH_4N_4O_2$
6287 - Guanidine, nitro-
CH_4O
7581 - Methanol
CH_4O_2
7180 - Hydroperoxide, methyl
CH_4O_3S
7530 - Methanesulfonic acid

CH_4O_4S
11577 - Sulfuric acid, monomethyl ester
$CH_4O_6S_2$
7514 - Methanedisulfonic acid
CH_4S
7546 - Methanethiol
CH_5As
446 - Arsine, methyl-
CH_5AsO_3
458 - Arsonic acid, methyl-
CH_5ClN_2O
12020 - Urea, monohydrochloride
CH_5ClSi
11378 - Silane, chloromethyl
CH_5N
7442 - Methanamine
CH_5NO
7447 - Methanamine, *N*-hydroxy-
CH_5NO_2
6003 - Formic acid, ammonium salt
CH_5N_3
6278 - Guanidine
$CH_5N_3O_4$
12021 - Urea, mononitrate
CH_5N_3S
7142 - Hydrazinecarbothioamide
CH_5O_3P
9538 - Phosphonic acid, methyl-
CH_6ClN
7446 - Methanamine, hydrochloride
CH_6ClN_3
6285 - Guanidine, monohydrochloride
CH_6ClN_3O
7143 - Hydrazinecarboxamide, monohydrochloride
CH_6N_2
7161 - Hydrazine, methyl-
CH_6N_4
7147 - Hydrazinecarboximidamide
CH_6N_4O
4009 - Carbonic dihydrazide
$CH_6N_4O_3$
6286 - Guanidine, mononitrate
CH_6N_4S
4033 - Carbonothioic dihydrazide
CH_6OSi
11436 - Silane, methoxy
CH_6Si
11437 - Silane, methyl-
CH_8B_2S
5168 - Diborane(6), mu-(methylthio)-
CH_8Si_2
11439 - Silane, methylenebis-
ClN
7337 - Iodine cyanide
CI_4
7542 - Methane, tetraiodo-
$CKNS$
11711 - Thiocyanic acid, potassium salt
$CNNaO$
4164 - Cyanic acid, sodium salt
CN_4O_8
7544 - Methane, tetranitro-
CO
4011 - Carbon monoxide
COS
4035 - Carbon oxide sulfide
CO_2
3992 - Carbon dioxide
CS_2
3993 - Carbon disulfide
CSe_2
4036 - Carbon selenide
C_2BrCl_3O
293 - Acetyl bromide, trichloro-
C_2Br_2
5939 - Ethyne, dibromo-
$C_2Br_2ClF_3$
5519 - Ethane, 1,2-dibromo-1-chloro-1,2,2-trifluoro-
$C_2Br_2Cl_4$
5524 - Ethane, 1,2-dibromo-1,1,2,2-tetrachloro-
$C_2Br_2F_4$
5525 - Ethane, 1,2-dibromo-1,1,2,2-tetrafluoro-
C_2Br_4
5914 - Ethene, tetrabromo-
C_2Br_6
5594 - Ethane, hexabromo-
C_2CaN_2
3930 - Calcium cyanide
C_2ClF_3
5889 - Ethene, chlorotrifluoro-
C_2ClF_5
5484 - Ethane, chloropentafluoro-
C_2Cl_2
5940 - Ethyne, dichloro-
C_2Cl_2FN
279 - Acetonitrile, dichlorofluoro-
$C_2Cl_2F_2$
5897 - Ethene, 1,1-dichloro-2,2-difluoro-
5898 - Ethene, 1,2-dichloro-1,2-difluoro-
$C_2Cl_2F_4$
5537 - Ethane, 1,1-dichloro-1,2,2,2-tetrafluoro-
5538 - Ethane, 1,2-dichloro-1,1,2,2-tetrafluoro-
$C_2Cl_2O_2$
5585 - Ethanedioyl dichloride
C_2Cl_3F
5923 - Ethene, trichlorofluoro-
$C_2Cl_3F_3$
5662 - Ethane, 1,1,1-trichloro-2,2,2-trifluoro-
5663 - Ethane, 1,1,2-trichloro-1,2,2-trifluoro-
C_2Cl_3N
288 - Acetonitrile, trichloro-
C_2Cl_4
5917 - Ethene, tetrachloro-
$C_2Cl_4F_2$
5636 - Ethane, 1,1,1,2-tetrachloro-2,2-difluoro-
5637 - Ethane, 1,1,2,2-tetrachloro-1,2-difluoro-
C_2Cl_4O
300 - Acetyl chloride, trichloro-
$C_2Cl_4O_2$
4029 - Carbonochloridic acid, trichloromethyl ester
C_2Cl_5F
5618 - Ethane, pentachlorofluoro-
C_2Cl_6
5595 - Ethane, hexachloro-
C_2F_3I
5925 - Ethene, trifluoroiodo-
C_2F_3N
289 - Acetonitrile, trifluoro-
C_2F_4
5918 - Ethene, tetrafluoro-
$C_2F_4N_2O_4$
5642 - Ethane, 1,1,2,2-tetrafluoro-1,2-dinitro-
C_2F_6
5596 - Ethane, hexafluoro-
$C_2F_6S_2$
5249 - Disulfide, bis(trifluoromethyl)
C_2HBr
5937 - Ethyne, bromo-
$C_2HBrClF_3$
5457 - Ethane, 1-bromo-2-chloro-1,1,2-trifluoro-
5458 - Ethane, 2-bromo-2-chloro-1,1,1-trifluoro-
C_2HBrCl_2
5882 - Ethene, 2-bromo-1,1-dichloro-
5935 - Ethylene, 1-bromo-1,2-dichloro-, *trans-*
C_2HBrF_2
5883 - Ethene, 1-bromo-1,2-difluoro-
$C_2HBr_2F_3$
5526 - Ethane, 1,2-dibromo-1,1,2-trifluoro-
C_2HBr_2N
277 - Acetonitrile, dibromo-
C_2HBr_3
5921 - Ethene, tribromo-
C_2HBr_3O
20 - Acetaldehyde, tribromo-
$C_2HBr_3O_2$
241 - Acetic acid, tribromo-
C_2HBr_5
5616 - Ethane, pentabromo-
C_2HCl
5938 - Ethyne, chloro-
C_2HClF_2
5885 - Ethene, 2-chloro-1,1-difluoro-
$C_2HClF_2O_2$
127 - Acetic acid, chlorodifluoro-
C_2HClF_3I
5490 - Ethane, 1-chloro-1,1,2-trifluoro-2-iodo-
C_2HClF_4
5485 - Ethane, 1-chloro-1,1,2,2-tetrafluoro-
5486 - Ethane, 1-chloro-1,2,2,2-tetrafluoro-
C_2HCl_2F
5900 - Ethene, 1,1-dichloro-2-fluoro-
C_2HCl_2FO
296 - Acetyl chloride, chlorofluoro-
$C_2HCl_2F_3$
5539 - Ethane, 1,2-dichloro-1,1,2-trifluoro-
5540 - Ethane, 2,2-dichloro-1,1,1-trifluoro-
C_2HCl_2N
278 - Acetonitrile, dichloro-
C_2HCl_3
5922 - Ethene, trichloro-
$C_2HCl_3F_2$
5658 - Ethane, 1,1,1-trichloro-2,2-difluoro
5659 - Ethane, 1,2,2-trichloro-1,1-difluoro-
5660 - Ethane, 1,2,2-trichloro-1,2-difluoro-
C_2HCl_3O
21 - Acetaldehyde, trichloro-
297 - Acetyl chloride, dichloro-
$C_2HCl_3O_2$
243 - Acetic acid, trichloro-
4016 - Carbonochloridic acid, dichloromethyl ester
C_2HCl_4F
5638 - Ethane, 1,1,2,2-tetrachloro-1-fluoro-
5639 - Ethane, 1,1,1,2-tetrachloro-2-fluoro-
C_2HCl_5
5617 - Ethane, pentachloro-
C_2HF
5943 - Ethyne, fluoro-
C_2HF_3
5924 - Ethene, trifluoro-
$C_2HF_3O_2$
263 - Acetic acid, trifluoro-
C_2H_2
5936 - Ethyne
C_2H_2BrCl
5881 - Ethene, 1-bromo-2-chloro-
$C_2H_2BrClO_2$
112 - Acetic acid, bromochloro-
$C_2H_2BrFO_2$
116 - Acetic acid, bromofluoro-
$C_2H_2BrF_3$
5466 - Ethane, 2-bromo-1,1,1-trifluoro-
$C_2H_2Br_2$
5890 - Ethene, 1,1-dibromo-
5891 - Ethene, 1,2-dibromo-, (*E*)-
5892 - Ethene, 1,2-dibromo-, (*Z*)-
$C_2H_2Br_2Cl_2$
5520 - Ethane, 1,2-dibromo-1,1-dichloro-
5521 - Ethane, 1,2-dibromo-1,2-dichloro-
$C_2H_2Br_2F_2$
5522 - Ethane, 1,2-dibromo-1,1-difluoro-
$C_2H_2Br_2O$
292 - Acetyl bromide, bromo-
$C_2H_2Br_2O_2$
159 - Acetic acid, dibromo-
$C_2H_2Br_3NO$
94 - Acetamide, 2,2,2-tribromo-
$C_2H_2Br_4$
5631 - Ethane, 1,1,1,2-tetrabromo-

5632 - Ethane, 1,1,2,2-tetrabromo-

$C_2H_2ClFO_2$
131 - Acetic acid, chlorofluoro-

$C_2H_2ClF_3$
5487 - Ethane, 1-chloro-1,1,2-trifluoro
5488 - Ethane, 1-chloro-1,2,2-trifluoro
5489 - Ethane, 2-chloro-1,1,1-trifluoro-

C_2H_2ClI
5888 - Ethene, 1-chloro-2-iodo-

C_2H_2ClN
274 - Acetonitrile, chloro-

C_2H_2ClNS
11700 - Thiocyanic acid, chloromethyl ester

$C_2H_2Cl_2$
5894 - Ethene, 1,1-dichloro-
5895 - Ethene, 1,2-dichloro-, (*E*)-
5896 - Ethene, 1,2-dichloro-, (*Z*)-

$C_2H_2Cl_2F_2$
5531 - Ethane, 1,2-dichloro-1,1-difluoro-
5532 - Ethane, 1,2-dichloro-1,2-difluoro-

$C_2H_2Cl_2O$
295 - Acetyl chloride, chloro-

$C_2H_2Cl_2O_2$
162 - Acetic acid, dichloro-
4014 - Carbonochloridic acid, chloromethyl ester

$C_2H_2Cl_3F$
5661 - Ethane, 1,1,2-trichloro-2-fluoro-

$C_2H_2Cl_3NO$
95 - Acetamide, 2,2,2-trichloro-

$C_2H_2Cl_4$
5634 - Ethane, 1,1,1,2-tetrachloro-
5635 - Ethane, 1,1,2,2-tetrachloro-

$C_2H_2F_2$
5902 - Ethene, 1,1-difluoro-
5903 - Ethene, 1,2-difluoro-, (*Z*)-

$C_2H_2F_2O_2$
175 - Acetic acid, difluoro-

$C_2H_2F_3I$
5667 - Ethane, 1,1,1-trifluoro-2-iodo-

$C_2H_2F_3NO_2$
5668 - Ethane, 1,1,1-trifluoro-2-nitro-

$C_2H_2F_4$
5640 - Ethane, 1,1,1,2-tetrafluoro-
5641 - Ethane, 1,1,2,2-tetrafluoro-

C_2H_2IN
283 - Acetonitrile, iodo-

$C_2H_2I_2$
5904 - Ethene, 1,2-diiodo-, (*Z*)-

$C_2H_2N_2O$
8563 - 1,3,4-Oxadiazole

$C_2H_2N_2S$
11657 - 1,2,4-Thiadiazole
11658 - 1,2,5-Thiadiazole

$C_2H_2N_4$
11651 - 1,2,4,5-Tetrazine

C_2H_2O
5927 - Ethenone

$C_2H_2O_2$
5491 - Ethanedial

$C_2H_2O_3$
223 - Acetic acid, oxo-

$C_2H_2O_4$
5553 - Ethanedioic acid

C_2H_3Br
5880 - Ethene, bromo-

C_2H_3BrO
291 - Acetyl bromide

$C_2H_3BrO_2$
111 - Acetic acid, bromo-

$C_2H_3Br_3$
5653 - Ethane, 1,1,2-tribromo-

$C_2H_3Br_3O$
5767 - Ethanol, 2,2,2-tribromo-

$C_2H_3Br_3O_2$
5581 - 1,1-Ethanediol, 2,2,2-tribromo-

C_2H_3Cl
5884 - Ethene, chloro-

$C_2H_3ClF_2$
5470 - Ethane, 1-chloro-1,1-difluoro
5471 - Ethane, 2-chloro-1,1-difluoro

C_2H_3ClO
12 - Acetaldehyde, chloro-
294 - Acetyl chloride

$C_2H_3ClO_2$
123 - Acetic acid, chloro-
4020 - Carbonochloridic acid, methyl ester

$C_2H_3Cl_2F$
5534 - Ethane, 1,1-dichloro-1-fluoro-
5535 - Ethane, 1,2-dichloro-1-fluoro-

$C_2H_3Cl_2NO$
51 - Acetamide, 2,2-dichloro-

$C_2H_3Cl_2NO_2$
5536 - Ethane, 1,1-dichloro-1-nitro-

$C_2H_3Cl_3$
5655 - Ethane, 1,1,1-trichloro-
5656 - Ethane, 1,1,2-trichloro-

$C_2H_3Cl_3O$
5768 - Ethanol, 2,2,2-trichloro-

$C_2H_3Cl_3O_2$
5582 - 1,1-Ethanediol, 2,2,2-trichloro-

$C_2H_3Cl_3Si$
11461 - Silane, trichloroethenyl-

C_2H_3F
5906 - Ethene, fluoro-

C_2H_3FO
301 - Acetyl fluoride

$C_2H_3FO_2$
189 - Acetic acid, fluoro-

$C_2H_3F_3$
5665 - Ethane, 1,1,1-trifluoro-
5666 - Ethane, 1,1,2-trifluoro-

$C_2H_3F_3O$
5770 - Ethanol, 2,2,2-trifluoro-

C_2H_3I
5907 - Ethene, iodo-

C_2H_3IO
302 - Acetyl iodide

$C_2H_3IO_2$
200 - Acetic acid, iodo-

$C_2H_3KO_2$
234 - Acetic acid, potassium salt

C_2H_3N
271 - Acetonitrile
7655 - Methylisonitrile

C_2H_3NO
282 - Acetonitrile, hydroxy-
7520 - Methane, isocyanato-

$C_2H_3NO_2$
5911 - Ethene, nitro-

$C_2H_3NO_3$
104 - Acetic acid, aminooxo-

$C_2H_3NO_4$
108 - Acetic acid, anhydride with nitric acid

C_2H_3NS
7521 - Methane, isothiocyanato-
11705 - Thiocyanic acid, methyl ester

$C_2H_3N_3$
11846 - 1H-1,2,3-Triazole

$C_2H_3N_3O_2$
109 - Acetic acid, azido-
11850 - 1,2,4-Triazolidine-3,5-dione

$C_2H_3NaO_2$
237 - Acetic acid, sodium salt

C_2H_4
5879 - Ethene

C_2H_4BrCl
5455 - Ethane, 1-bromo-1-chloro-
5456 - Ethane, 1-bromo-2-chloro-

$C_2H_4BrClO_2S$
5628 - Ethanesulfonyl chloride, 2-bromo-

C_2H_4BrF
5462 - Ethane, 1-bromo-2-fluoro-

C_2H_4BrI
5463 - Ethane, 1-bromo-2-iodo-

C_2H_4BrNO
36 - Acetamide, *N*-bromo-

$C_2H_4Br_2$
5517 - Ethane, 1,1-dibromo-
5518 - Ethane, 1,2-dibromo-

$C_2H_4Br_2O$
7524 - Methane, oxybis[bromo-

C_2H_4ClF
5475 - Ethane, 1-chloro-1-fluoro-
5476 - Ethane, 1-chloro-2-fluoro-

C_2H_4ClI
5477 - Ethane, 1-chloro-2-iodo-

C_2H_4ClNO
41 - Acetamide, 2-chloro-

$C_2H_4ClNO_2$
5482 - Ethane, 1-chloro-1-nitro-
5483 - Ethane, 1-chloro-2-nitro-

$C_2H_4Cl_2$
5528 - Ethane, 1,1-dichloro-
5529 - Ethane, 1,2-dichloro-

$C_2H_4Cl_2O$
5704 - Ethanol, 2,2-dichloro-
7525 - Methane, oxybis[chloro-

$C_2H_4Cl_2O_2S$
5629 - Ethanesulfonyl chloride, 2-chloro-

$C_2H_4Cl_3NO$
5680 - Ethanol, 1-amino-2,2,2-trichloro-

$C_2H_4Cl_4Si$
11391 - Silane, dichloro(dichloromethyl)methyl-

C_2H_4FNO
63 - Acetamide, 2-fluoro-

$C_2H_4F_2$
5544 - Ethane, 1,1-difluoro-
5545 - Ethane, 1,2-difluoro-

$C_2H_4F_2O$
5708 - Ethanol, 2,2-difluoro-

$C_2H_4I_2$
5546 - Ethane, 1,1-diiodo-
5547 - Ethane, 1,2-diiodo-

$C_2H_4N_2$
272 - Acetonitrile, amino-

$C_2H_4N_2O_2$
5492 - Ethanedial, dioxime
5493 - Ethanediamide

$C_2H_4N_2O_4$
5551 - Ethane, 1,1-dinitro-
5552 - Ethane, 1,2-dinitro-
8086 - Nitrous acid, 1,2-ethanediyl ester

$C_2H_4N_2O_6$
5574 - 1,2-Ethanediol, dinitrate

$C_2H_4N_2S_2$
5588 - Ethanedithioamide

$C_2H_4N_4$
6280 - Guanidine, cyano-
11845 - 1H-1,2,4-Triazol-3-amine

$C_2H_4N_4O_2$
5124 - Diazenedicarboxamide

C_2H_4O
8599 - Oxirane
11 - Acetaldehyde

$(C_2H_4O)_x$
7440 - Metaldehyde

C_2H_4OS
5646 - Ethanethioic acid

$C_2H_4O_2$
15 - Acetaldehyde, hydroxy-
100 - Acetic acid
6010 - Formic acid, methyl ester

$C_2H_4O_2S$
202 - Acetic acid, mercapto-

$C_2H_4O_3$
196 - Acetic acid, hydroxy-
5619 - Ethaneperoxoic acid

$C_2H_4O_3S$
5200 - 1,3,2-Dioxathiolane, 2-oxide

$C_2H_4O_4S$
5199 - 1,3,2-Dioxathiolane, 2,2-dioxide

$C_2H_4O_5S$
238 - Acetic acid, sulfo-
C_2H_4S
11695 - Thiirane
$C_2H_4S_2$
5589 - Ethane(dithioic) acid
$C_2H_4S_5$
8998 - 1,2,3,5,6-Pentathiepane
C_2H_4Si
11435 - Silane, ethynyl
$C_2H_5AsF_2$
445 - Arsine, ethyldifluoro-
C_2H_5Br
5454 - Ethane, bromo-
C_2H_5BrO
5685 - Ethanol, 2-bromo-
7472 - Methane, bromomethoxy-
C_2H_5Cl
5467 - Ethane, chloro-
C_2H_5ClO
5693 - Ethanol, 2-chloro-
7184 - Hypochlorous acid, ethyl ester
7483 - Methane, chloromethoxy-
$C_2H_5ClO_2S$
4087 - Chlorosulfurous acid, ethyl ester
5627 - Ethanesulfonyl chloride
$C_2H_5ClO_3S$
4085 - Chlorosulfuric acid, ethyl ester
C_2H_5ClS
5651 - Ethanethiol, 2-chloro-
7484 - Methane, chloro(methylthio)-
$C_2H_5Cl_2O_2P$
9577 - Phosphorodichloridic acid, ethyl ester
$C_2H_5Cl_3OSi$
11462 - Silane, trichloroethoxy-
$C_2H_5Cl_3Si$
11390 - Silane, dichloro(chloromethyl)methyl-
11463 - Silane, trichloroethyl-
C_2H_5F
5593 - Ethane, fluoro-
C_2H_5FO
5727 - Ethanol, 2-fluoro-
$C_2H_5F_3OSi$
11429 - Silane, ethoxytrifluoro-
$C_2H_5F_3Si$
11432 - Silane, ethyltrifluoro-
C_2H_5I
5597 - Ethane, iodo-
C_2H_5IO
5731 - Ethanol, 2-iodo-
7519 - Methane, iodomethoxy-
C_2H_5N
498 - Aziridine
C_2H_5NO
17 - Acetaldehyde, oxime
22 - Acetamide
5995 - Formamide, *N*-methyl-
$C_2H_5NO_2$
3953 - Carbamic acid, methyl ester
5607 - Ethane, nitro-
6250 - Glycine
8087 - Nitrous acid, ethyl ester
$C_2H_5NO_3$
5752 - Ethanol, 2-nitro-
8074 - Nitric acid, ethyl ester
C_2H_5NS
5643 - Ethanethioamide
$C_2H_5N_3O$
5681 - Ethanol, 2-azido-
$C_2H_5N_3O_2$
105 - Acetic acid, aminooxo-, hydrazide
7232 - Imidodicarbonic diamide
12019 - Urea, *N*-methyl-*N*-nitroso-
$C_2H_5N_3S_2$
11718 - Thioimidodicarbonic diamide
$C_2H_5N_5O_3$
6284 - Guanidine, *N*-methyl-*N'*-nitro-*N*-nitroso-

C_2H_6
5448 - Ethane
C_2H_6AsCl
453 - Arsinous chloride, dimethyl-
$C_2H_6BF_3O$
3138 - Boron, trifluoro[oxybis[methane]]-, (T-4)-
C_2H_6Cd
3929 - Cadmium, dimethyl-
$C_2H_6ClO_3P$
5929 - Ethephon
$C_2H_6Cl_2NPS$
9557 - Phosphoramidothioic dichloride, dimethyl-
$C_2H_6Cl_2Si$
11394 - Silane, dichlorodimethyl-
C_2H_6Hg
7435 - Mercury, dimethyl-
$C_2H_6N_2$
5125 - Diazene, dimethyl-
5126 - Diazene, dimethyl-, (*E*)-
$C_2H_6N_2O$
195 - Acetic acid, hydrazide
7450 - Methanamine, *N*-methyl-*N*-nitroso-
12017 - Urea, methyl-
$C_2H_6N_2O_2$
7149 - Hydrazinecarboxylic acid, methyl ester
7449 - Methanamine, *N*-methyl-*N*-nitro-
$C_2H_6N_4O$
11993 - Urea, (aminoiminomethyl)-
$C_2H_6N_4O_2$
5562 - Ethanedioic acid, dihydrazide
C_2H_6O
5672 - Ethanol
7523 - Methane, oxybis-
C_2H_6OS
5732 - Ethanol, 2-mercapto-
7529 - Methane, sulfinylbis-
$C_2H_6O_2$
5569 - 1,2-Ethanediol
7179 - Hydroperoxide, ethyl
9159 - Peroxide, dimethyl
$C_2H_6O_2S$
7534 - Methane, sulfonylbis-
$C_2H_6O_3S$
5622 - Ethanesulfonic acid
7532 - Methanesulfonic acid, methyl ester
11582 - Sulfurous acid, dimethyl ester
$C_2H_6O_4S$
11574 - Sulfuric acid, dimethyl ester
11576 - Sulfuric acid, monoethyl ester
$C_2H_6O_6$
5561 - Ethanedioic acid, dihydrate
$C_2H_6O_6S_2$
5587 - 1,2-Ethanedisulfonic acid
C_2H_6S
5648 - Ethanethiol
7545 - Methane, thiobis-
$C_2H_6S_2$
5254 - Disulfide, dimethyl
5590 - 1,2-Ethanedithiol
C_2H_6Se
7526 - Methane, selenobis-
C_2H_6Zn
12066 - Zinc, dimethyl
C_2H_7As
442 - Arsine, dimethyl-
444 - Arsine, ethyl-
$C_2H_7AsO_2$
451 - Arsinic acid, dimethyl-
$C_2H_7AsO_3$
457 - Arsonic acid, ethyl-
$C_2H_7BO_2$
3117 - Borane, dimethoxy-
3133 - Boronic acid, ethyl-
$C_2H_7ClN_2$
5671 - Ethanimidamide, monohydrochloride
C_2H_7N
5410 - Ethanamine
7448 - Methanamine, *N*-methyl-
C_2H_7NO
5434 - Ethanamine, *N*-hydroxy-
5674 - Ethanol, 1-amino-
5675 - Ethanol, 2-amino-
$C_2H_7NO_2$
106 - Acetic acid, ammonium salt
$C_2H_7NO_3S$
5623 - Ethanesulfonic acid, 2-amino-
C_2H_7NS
5649 - Ethanethiol, 2-amino-
$C_2H_7N_3O$
6266 - Glycine, hydrazide
$C_2H_7N_5$
7233 - Imidodicarbonimidic diamide
$C_2H_7O_2P$
9528 - Phosphinic acid, dimethyl-
$C_2H_7O_3P$
9588 - Phosphorous acid, dimethyl ester
$C_2H_7O_4P$
9559 - Phosphoric acid, dimethyl ester
9561 - Phosphoric acid, monoethyl ester
C_2H_8ClN
5433 - Ethanamine, hydrochloride
$C_2H_8NO_2PS$
7441 - Methamidophos
$C_2H_8N_2$
5496 - 1,2-Ethanediamine
7154 - Hydrazine, 1,1-dimethyl-
7155 - Hydrazine, 1,2-dimethyl-
$C_2H_8N_2O$
5729 - Ethanol, 2-hydrazino-
C_2H_8Si
11414 - Silane, dimethyl-
$C_2H_{10}N_2O$
5513 - 1,2-Ethanediamine, monohydrate
C_2I_2
5941 - Ethyne, diiodo-
C_2I_4
5919 - Ethene, tetraiodo-
C_2N_2
5550 - Ethanedinitrile
C_2N_2S
11572 - Sulfur cyanide
$C_2N_4O_6$
290 - Acetonitrile, trinitro-
$C_2Na_2O_4$
5566 - Ethanedioic acid, disodium salt
C_2O_4Pb
5568 - Ethanedioic acid, lead salt
C_3ClF_5
10541 - 1-Propene, 2-chloro-1,1,3,3,3-pentafluoro-
C_3ClF_5O
10421 - 2-Propanone, 1-chloro-1,1,3,3,3-pentafluoro-
$C_3Cl_2F_4$
10557 - 1-Propene, 1,2-dichloro-1,3,3,3-tetrafluoro-
$C_3Cl_2F_6$
9898 - Propane, 1,2-dichloro-1,1,2,3,3,3-hexafluoro-
$C_3Cl_2N_2$
9912 - Propanedinitrile, dichloro-
$C_3Cl_3F_3$
10607 - 1-Propene, 1,1,2-trichloro-3,3,3-trifluoro-
$C_3Cl_3F_5$
10179 - Propane, 1,1,1-trichloro-2,2,3,3,3-pentafluoro-
10180 - Propane, 1,2,2-trichloro-1,1,3,3,3-pentafluoro-
10181 - Propane, 1,2,3-trichloro-1,1,2,3,3-pentafluoro-
$C_3Cl_3N_3$
11837 - 1,3,5-Triazine, 2,4,6-trichloro-
$C_3Cl_3N_3O_3$
11840 - 1,3,5-Triazine-2,4,6(1H,3H,5H)-trione, 1,3,5-trichloro-
$C_3Cl_4F_4$
10149 - Propane, 1,2,2,3-tetrachloro-1,1,3,3-tetrafluoro-

C_3Cl_4O
10718 - 2-Propenoyl chloride, 2,3,3-trichloro-
C_3Cl_6
10571 - 1-Propene, 1,1,2,3,3,3-hexachloro-
C_3Cl_6O
10440 - 2-Propanone, 1,1,1,3,3,3-hexachloro-
$C_3Cl_6O_2$
262 - Acetic acid, trichloro-, trichloromethyl ester
$C_3Cl_6O_3$
7585 - Methanol, trichloro-, carbonate (2:1)
C_3F_6
10572 - 1-Propene, 1,1,2,3,3,3-hexafluoro-
C_3F_6O
8629 - Oxirane, trifluoro(trifluoromethyl)-
10441 - 2-Propanone, 1,1,1,3,3,3-hexafluoro-
C_3F_8
10118 - Propane, octafluoro-
C_3HBr_5O
10477 - 2-Propanone, 1,1,1,3,3-pentabromo-
$C_3HCl_2F_3$
10558 - Propene, 1,2-dichloro-3,3,3-trifluoro-
$C_3HCl_2N_3O_3$
11838 - 1,3,5-Triazine-2,4,6(1H,3H,5H)-trione, 1,3-dichloro-
$C_3HCl_3O_2$
10696 - 2-Propenoic acid, 2,3,3-trichloro-
C_3HCl_5
10593 - 1-Propene, 1,1,2,3,3-pentachloro-
C_3HCl_5O
10478 - 2-Propanone, 1,1,1,3,3-pentachloro-
C_3HCl_7
10048 - Propane, 1,1,1,2,2,3,3-heptachloro-
10049 - Propane, 1,1,1,2,3,3,3-heptachloro-
C_3HN
10737 - 2-Propynenitrile
C_3H_2BrNS
11668 - Thiazole, 2-bromo-
$C_3H_2Br_2$
10730 - 1-Propyne, 1,3-dibromo-
$C_3H_2Br_2O_2$
10627 - 2-Propenoic acid, 2,3-dibromo-, (*Z*)-
10628 - 2-Propenoic acid, 3,3-dibromo-
C_3H_2ClN
10584 - 2-Propenenitrile, 2-chloro-
$C_3H_2Cl_2O_2$
10029 - Propanedioyl dichloride
10630 - 2-Propenoic acid, 3,3-dichloro-
$C_3H_2Cl_3NO$
10115 - Propanenitrile, 3,3,3-trichloro-2-hydroxy-
$C_3H_2Cl_4$
10598 - 1-Propene, 1,2,3,3-tetrachloro-
$C_3H_2Cl_4O$
10486 - 2-Propanone, 1,1,1,3-tetrachloro-
10487 - 2-Propanone, 1,1,3,3-tetrachloro-
$C_3H_2Cl_6$
10051 - Propane, 1,1,1,2,2,3-hexachloro-
10052 - Propane, 1,1,1,3,3,3-hexachloro-
10053 - Propane, 1,1,2,2,3,3-hexachloro-
$C_3H_2N_2$
9911 - Propanedinitrile
$C_3H_2N_2O_3$
7226 - Imidazolidinetrione
C_3H_2O
10723 - 2-Propynal
$C_3H_2O_2$
10738 - 2-Propynoic acid
C_3H_3Br
10728 - 1-Propyne, 3-bromo-
$C_3H_3BrF_2$
10532 - 1-Propene, 3-bromo-3,3-difluoro-
$C_3H_3BrO_2$
10610 - 2-Propenoic acid, 2-bromo-
C_3H_3Cl
10729 - 1-Propyne, 3-chloro-
C_3H_3ClO
10511 - 2-Propenal, 2-chloro-
10714 - 2-Propenoyl chloride
$C_3H_3ClO_2$
10617 - 2-Propenoic acid, 3-chloro-, (*E*)-
10618 - 2-Propenoic acid, 3-chloro-, (*Z*)-
$C_3H_3ClO_3$
145 - Acetic acid, chlorooxo-, methyl ester
$C_3H_3Cl_2F$
10554 - 1-Propene, 1,1-dichloro-2-fluoro-
$C_3H_3Cl_2N$
10099 - Propanenitrile, 2,3-dichloro-
$C_3H_3Cl_3$
10602 - 1-Propene, 1,1,2-trichloro-
10603 - 1-Propene, 1,2,3-trichloro-
10604 - 1-Propene, 3,3,3-trichloro-
$C_3H_3Cl_3O$
8628 - Oxirane, (trichloromethyl)-
9790 - Propanal, 2,2,3-trichloro-
10488 - 2-Propanone, 1,1,1-trichloro-
10500 - Propanoyl chloride, 2,2-dichloro-
10501 - Propanoyl chloride, 2,3-dichloro-
10502 - Propanoyl chloride, 3,3-dichloro-
$C_3H_3Cl_3O_2$
253 - Acetic acid, trichloro-, methyl ester
10332 - Propanoic acid, 2,2,3-trichloro-
$C_3H_3Cl_3O_3$
10333 - Propanoic acid, 3,3,3-trichloro-2-hydroxy-
$C_3H_3Cl_5$
10125 - Propane, 1,1,1,2,3-pentachloro-
10126 - Propane, 1,1,2,3,3-pentachloro-
$C_3H_3F_3$
10608 - 1-Propene, 3,3,3-trifluoro-
$C_3H_3F_5$
10128 - Propane, 1,1,1,2,2-pentafluoro
C_3H_3I
10734 - 1-Propyne, 1-iodo-
10735 - 1-Propyne, 3-iodo-
C_3H_3N
10583 - 2-Propenenitrile
C_3H_3NO
7402 - Isoxazole
8575 - Oxazole
10113 - Propanenitrile, 2-oxo-
C_3H_3NOS
303 - Acetyl isothiocyanate
$C_3H_3NOS_2$
11685 - 4-Thiazolidinone, 2-thioxo-
$C_3H_3NO_2$
150 - Acetic acid, cyano-
$C_3H_3NO_2S$
11683 - 2,4-Thiazolidinedione
C_3H_3NS
11667 - Thiazole
$C_3H_3N_3$
11827 - 1,3,5-Triazine
$C_3H_3N_3O_2$
7211 - 1H-Imidazole, 2-nitro-
$C_3H_3N_3O_2S$
11666 - 2-Thiazolamine, 5-nitro-
$C_3H_3N_3O_3$
11907 - 1,3,5-Trioxane-2,4,6-triimine
C_3H_4
4990 - Cyclopropene
9772 - 1,2-Propadiene
10727 - 1-Propyne
C_3H_4BrClO
10417 - 2-Propanone, 1-bromo-3-chloro-
10495 - Propanoyl chloride, 2-bromo-
C_3H_4BrN
10094 - Propanenitrile, 2-bromo-
10095 - Propanenitrile, 3-bromo-
$C_3H_4Br_2$
10542 - 1-Propene, 1,1-dibromo-
10543 - 1-Propene, 1,3-dibromo-, (*E*)-
10544 - 1-Propene, 1,3-dibromo-, (*Z*)-
10545 - 1-Propene, 2,3-dibromo-
$C_3H_4Br_2O$
9780 - Propanal, 2,3-dibromo-
10424 - 2-Propanone, 1,3-dibromo-
$C_3H_4Br_2O_2$
10238 - Propanoic acid, 2,3-dibromo-
$C_3H_4Br_4$
10135 - Propane, 1,1,1,2-tetrabromo-
10136 - Propane, 1,1,2,2-tetrabromo-
10137 - Propane, 1,1,3,3-tetrabromo-
10138 - Propane, 1,2,2,3-tetrabromo-
$C_3H_4ClFO_2$
134 - Acetic acid, chlorofluoro-, methyl ester
$C_3H_4ClF_3$
9871 - Propane, 3-chloro-1,1,1-trifluoro-
C_3H_4ClN
10097 - Propanenitrile, 3-chloro-
$C_3H_4Cl_2$
10547 - 1-Propene, 1,1-dichloro-
10548 - 1-Propene, 1,2-dichloro-, (*E*)-
10549 - 1-Propene, 1,2-dichloro-, (*Z*)-
10550 - 1-Propene, 1,3-dichloro-, (*E*)-
10551 - 1-Propene, 1,3-dichloro-, (*Z*)-
10552 - 1-Propene, 2,3-dichloro-
10553 - 1-Propene, 3,3-dichloro-
$C_3H_4Cl_2O$
9781 - Propanal, 2,3-dichloro-
10425 - 2-Propanone, 1,1-dichloro-
10426 - 2-Propanone, 1,3-dichloro-
10496 - Propanoyl chloride, 2-chloro-, (*S*)-
10497 - Propanoyl chloride, 3-chloro-
$C_3H_4Cl_2O_2$
166 - Acetic acid, dichloro-, methyl ester
4013 - Carbonochloridic acid, 2-chloroethyl ester
10244 - Propanoic acid, 2,2-dichloro-
10245 - Propanoic acid, 2,3-dichloro-
$C_3H_4Cl_3NO_2$
6000 - Formamide, *N*-(2,2,2-trichloro-1-hydroxyethyl)-
$C_3H_4Cl_3NO_3$
10408 - 2-Propanol, 1,1,1-trichloro-3-nitro-
$C_3H_4Cl_4$
10141 - Propane, 1,1,1,2-tetrachloro-
10142 - Propane, 1,1,1,3-tetrachloro-
10143 - Propane, 1,1,2,2-tetrachloro-
10144 - Propane, 1,1,2,3-tetrachloro-
10145 - Propane, 1,2,2,3-tetrachloro-
$C_3H_4Cl_4O$
10402 - 2-Propanol, 1,1,3,3-tetrachloro-
10403 - 2-Propanol, 1,1,1,3-tetrachloro-
$C_3H_4F_2$
10561 - 1-Propene, 1,1-difluoro
$C_3H_4F_3NO_2$
10184 - Propane, 1,1,1-trifluoro-3-nitro-
$C_3H_4F_4O$
10404 - 1-Propanol, 2,2,3,3-tetrafluoro-
$C_3H_4N_2$
285 - Acetonitrile, (methyleneamino)-
7192 - 1H-Imidazole
10815 - 1H-Pyrazole
$C_3H_4N_2O$
49 - Acetamide, 2-cyano-
$C_3H_4N_2O_2$
7216 - 2,4-Imidazolidinedione
$C_3H_4N_2S$
11662 - 2-Thiazolamine
C_3H_4O
4988 - Cyclopropanone
5944 - Ethyne, methoxy-
10509 - 2-Propenal
10743 - 2-Propyn-1-ol
$C_3H_4O_2$
6006 - Formic acid, ethenyl ester
8595 - 2-Oxetanone
8596 - 3-Oxetanone
8603 - Oxiranecarboxaldehyde
9788 - Propanal, 2-oxo-
10609 - 2-Propenoic acid
$C_3H_4O_3$
5211 - 1,3-Dioxolan-2-one
10314 - Propanoic acid, 2-oxo-

$C_3H_4O_4$
9921 - Propanedioic acid
$C_3H_4O_5$
9956 - Propanedioic acid, hydroxy-
C_3H_5Br
10529 - 1-Propene, 1-bromo-, (*Z*)-
10530 - 1-Propene, 2-bromo-
10531 - 1-Propene, 3-bromo-
C_3H_5BrO
8600 - Oxirane, (bromomethyl)-, (±)-
9775 - Propanal, 2-bromo-
10416 - 2-Propanone, 1-bromo-
10490 - Propanoyl bromide
10699 - 2-Propen-1-ol, 2-bromo-
$C_3H_5BrO_2$
117 - Acetic acid, bromo-, methyl ester
10204 - Propanoic acid, 2-bromo-, (±)-
10205 - Propanoic acid, 3-bromo-
$C_3H_5Br_2Cl$
9885 - Propane, 1,2-dibromo-3-chloro-
$C_3H_5Br_3$
10160 - Propane, 1,1,1-tribromo-
10161 - Propane, 1,1,2-tribromo-
10162 - Propane, 1,2,2-tribromo-
10163 - Propane, 1,2,3-tribromo-
C_3H_5Cl
10533 - 1-Propene, 1-chloro-, (*E*)-
10534 - 1-Propene, 1-chloro-, (*Z*)-
10535 - 1-Propene, 2-chloro-
10536 - 1-Propene, 3-chloro-
$C_3H_5ClF_2$
9851 - Propane, 1-chloro-2,2-difluoro-
$C_3H_5ClN_2O_6$
9997 - 1,2-Propanediol, 3-chloro-, dinitrate
C_3H_5ClO
8605 - Oxirane, (chloromethyl)-, (±)-
9777 - Propanal, 2-chloro-
9778 - Propanal, 3-chloro-
10420 - 2-Propanone, 1-chloro-
10493 - Propanoyl chloride
10700 - 2-Propen-1-ol, 2-chloro-
10701 - 2-Propen-1-ol, 3-chloro-
$C_3H_5ClO_2$
138 - Acetic acid, chloro-, methyl ester
299 - Acetyl chloride, methoxy-
4018 - Carbonochloridic acid, ethyl ester
7582 - Methanol, chloro-, acetate
10215 - Propanoic acid, 2-chloro-
10216 - Propanoic acid, 3-chloro-
$C_3H_5Cl_2F$
9897 - Propane, 1,2-dichloro-2-fluoro-
$C_3H_5Cl_2NO_2$
3940 - Carbamic acid, dichloro-, ethyl ester
9902 - Propane, 1,1-dichloro-1-nitro-
$C_3H_5Cl_3$
10171 - Propane, 1,1,1-trichloro-
10172 - Propane, 1,1,2-trichloro-
10173 - Propane, 1,1,3-trichloro-
10174 - Propane, 1,2,2-trichloro-
10175 - Propane, 1,2,3-trichloro-
$C_3H_5Cl_3O$
10406 - 2-Propanol, 1,1,1-trichloro-
$C_3H_5Cl_3Si$
11475 - Silane, trichloro-2-propenyl-
C_3H_5F
10567 - 1-Propene, 1-fluoro- (*Z*)
10568 - 1-Propene, 1-fluoro- (*E*)
10569 - 1-Propene, 2-fluoro-
10570 - 1-Propene, 3-fluoro-
C_3H_5FO
8616 - Oxirane, (fluoromethyl)-
10437 - 2-Propanone, 1-fluoro-
10506 - Propanoyl fluoride
$C_3H_5FO_2$
192 - Acetic acid, fluoro-, methyl ester
$C_3H_5F_3O$
10409 - 2-Propanol, 1,1,1-trifluoro-, (±)-
$C_3H_5F_3O_3S$
7533 - Methanesulfonic acid, trifluoro-, ethyl ester
C_3H_5I
10573 - 1-Propene, 3-iodo-
C_3H_5IO
8617 - Oxirane, (iodomethyl)-
10454 - 2-Propanone, 1-iodo-
$C_3H_5IO_2$
10282 - Propanoic acid, 2-iodo-, (±)-
C_3H_5N
5599 - Ethane, isocyano-
10090 - Propanenitrile
10725 - 2-Propyn-1-amine
C_3H_5NO
284 - Acetonitrile, methoxy-
496 - 2-Azetidinone
4163 - Cyanic acid, ethyl ester
5598 - Ethane, isocyanato-
10105 - Propanenitrile, 2-hydroxy-
10106 - Propanenitrile, 3-hydroxy-
10519 - 2-Propenamide
$C_3H_5NO_2$
9789 - Propanal, 2-oxo-, 1-oxime
10589 - 1-Propene, 1-nitro-
10590 - 1-Propene, 2-nitro-
10591 - 1-Propene, 3-nitro-
$C_3H_5NO_3$
10475 - 2-Propanone, 1-nitro-
C_3H_5NS
286 - Acetonitrile, (methylthio)-
5600 - Ethane, isothiocyanato-
11703 - Thiocyanic acid, ethyl ester
$C_3H_5N_3$
10812 - 1H-Pyrazol-4-amine
$C_3H_5N_3O$
153 - Acetic acid, cyano-, hydrazide
$C_3H_5N_3O_9$
10197 - 1,2,3-Propanetriol, trinitrate
$C_3H_5NaO_2$
10327 - Propanoic acid, sodium salt
C_3H_6
4955 - Cyclopropane
10528 - 1-Propene
C_3H_6BrCl
9838 - Propane, 1-bromo-2-chloro-
9839 - Propane, 1-bromo-3-chloro-
9840 - Propane, 2-bromo-1-chloro-
9841 - Propane, 2-bromo-2-chloro-
C_3H_6BrF
9844 - Propane, 1-bromo-3-fluoro-
$C_3H_6BrNO_2$
9847 - Propane, 2-bromo-2-nitro-
$C_3H_6BrNO_4$
9991 - 1,3-Propanediol, 2-bromo-2-nitro-
$C_3H_6Br_2$
9880 - Propane, 1,1-dibromo-
9881 - Propane, 1,2-dibromo-
9882 - Propane, 1,3-dibromo-
9883 - Propane, 2,2-dibromo-
$C_3H_6Br_2O$
10357 - 2-Propanol, 1,3-dibromo-
C_3H_6ClF
9856 - Propane, 1-chloro-3-fluoro-
C_3H_6ClI
9857 - Propane, 1-chloro-3-iodo-
C_3H_6ClNO
3973 - Carbamic chloride, dimethyl-
$C_3H_6ClNO_2$
9866 - Propane, 1-chloro-1-nitro-
9867 - Propane, 1-chloro-2-nitro-
9868 - Propane, 1-chloro-3-nitro-
9869 - Propane, 2-chloro-1-nitro-
9870 - Propane, 2-chloro-2-nitro-
C_3H_6ClNS
3980 - Carbamothioic chloride, dimethyl-
$C_3H_6Cl_2$
9891 - Propane, 1,1-dichloro-
9892 - Propane, 1,2-dichloro-, (±)-
9893 - Propane, 1,3-dichloro-
9894 - Propane, 2,2-dichloro-
$C_3H_6Cl_2O$
10359 - 1-Propanol, 2,3-dichloro-
10360 - 1-Propanol, 3,3-dichloro-
10361 - 2-Propanol, 1,1-dichloro-
10362 - 2-Propanol, 1,3-dichloro-
$C_3H_6Cl_2Si$
11396 - Silane, dichloroethenylmethyl-
$C_3H_6Cl_3P$
9524 - Phosphine, tris(chloromethyl)-
$C_3H_6Cl_4Si$
11458 - Silane, trichloro(3-chloropropyl)-
$C_3H_6F_2$
9905 - Propane, 1,3-difluoro-
9906 - Propane, 2,2-difluoro-
$C_3H_6F_2O$
10369 - 2-Propanol, 1,3-difluoro-
$C_3H_6I_2$
9907 - Propane, 1,3-diiodo-
9908 - Propane, 2,2-diiodo-
$C_3H_6N_2$
4157 - Cyanamide, dimethyl-
10092 - Propanenitrile, 3-amino-
10822 - 1H-Pyrazole, 4,5-dihydro-
$C_3H_6N_2O$
7227 - 2-Imidazolidinone
$C_3H_6N_2OS$
33 - Acetamide, *N*-(aminothioxomethyl)-
$C_3H_6N_2O_2$
27 - Acetamide, *N*-(aminocarbonyl)-
7406 - 3-Isoxazolidinone, 4-amino-, (*R*)-
9784 - Propanal, 2-(hydroxyimino)-, oxime
$C_3H_6N_2O_4$
3963 - Carbamic acid, nitro-, ethyl ester
9918 - Propane, 1,1-dinitro-
9919 - Propane, 1,3-dinitro-
9920 - Propane, 2,2-dinitro-
$C_3H_6N_2O_7$
10191 - 1,2,3-Propanetriol, 1,3-dinitrate
$C_3H_6N_2S$
7225 - 2-Imidazolidinethione
11664 - 2-Thiazolamine, 4,5-dihydro-
$C_3H_6N_6$
11835 - 1,3,5-Triazine-2,4,6-triamine
$C_3H_6N_6O_6$
11833 - 1,3,5-Triazine, hexahydro-1,3,5-trinitro-
C_3H_6O
5908 - Ethene, methoxy-
8592 - Oxetane
8621 - Oxirane, methyl-
9774 - Propanal
10411 - 2-Propanone
10697 - 2-Propen-1-ol
$C_3H_6O_2$
16 - Acetaldehyde, methoxy-
208 - Acetic acid, methyl ester
5204 - 1,3-Dioxolane
6007 - Formic acid, ethyl ester
8618 - Oxiranemethanol
10201 - Propanoic acid
10443 - 2-Propanone, 1-hydroxy-
$C_3H_6O_2S$
204 - Acetic acid, mercapto-, methyl ester
216 - Acetic acid, (methylthio)-
5909 - Ethene, (methylsulfonyl)-
10285 - Propanoic acid, 2-mercapto-
10286 - Propanoic acid, 3-mercapto-
11692 - Thietane, 1,1-dioxide
$C_3H_6O_3$
198 - Acetic acid, hydroxy-, methyl ester
205 - Acetic acid, methoxy-
4000 - Carbonic acid, dimethyl ester
9782 - Propanal, 2,3-dihydroxy-, (±)-
$C_3H_6O_2$
10266 - Propanoic acid, 2-hydroxy-, (±)-

MOLECULAR FORMULA INDEX OF ORGANIC COMPOUNDS (continued)

$C_3H_6O_3$
10267 - Propanoic acid, 3-hydroxy-
10429 - 2-Propanone, 1,3-dihydroxy-
11905 - 1,3,5-Trioxane
$C_3H_6O_3S$
8570 - 1,2-Oxathiolane, 2,2-dioxide
11694 - Thietane, 1-oxide
$C_3H_6O_4$
10252 - Propanoic acid, 2,3-dihydroxy-
C_3H_6S
5910 - Ethene, (methylthio)-
10600 - 2-Propene-1-thiol
11691 - Thietane
11696 - Thiirane, methyl-
$C_3H_6S_2$
5266 - 1,3-Dithiolane
$C_3H_6S_3$
11919 - 1,3,5-Trithiane
C_3H_7Br
9836 - Propane, 1-bromo-
9837 - Propane, 2-bromo-
C_3H_7BrO
5464 - Ethane, (bromomethoxy)-
5465 - Ethane, 1-bromo-2-methoxy-
10344 - 1-Propanol, 3-bromo-
10345 - 2-Propanol, 1-bromo-
C_3H_7Cl
9848 - Propane, 1-chloro-
9849 - Propane, 2-chloro-
C_3H_7ClHgO
11822 - Triadimenol
C_3H_7ClO
5478 - Ethane, (chloromethoxy)-
5479 - Ethane, 1-chloro-1-methoxy-
5480 - Ethane, 1-chloro-2-methoxy-
10347 - 1-Propanol, 2-chloro-
10348 - 1-Propanol, 3-chloro-
10349 - 2-Propanol, 1-chloro-
$C_3H_7ClO_2$
9994 - 1,2-Propanediol, 3-chloro-
9995 - 1,3-Propanediol, 2-chloro-
$C_3H_7ClO_2S$
10133 - 1-Propanesulfonyl chloride
$C_3H_7ClO_3S$
5699 - Ethanol, 2-chloro-, methanesulfonate
C_3H_7ClS
5481 - Ethane, 1-chloro-2-(methylthio)-
$C_3H_7Cl_2N$
9805 - 1-Propanamine, *N,N*-dichloro-
$C_3H_7Cl_3Si$
11373 - Silane, chloro(dichloromethyl)dimethyl-
11466 - Silane, trichloro(1-methylethyl)-
11476 - Silane, trichloropropyl-
C_3H_7F
10046 - Propane, 1-fluoro-
10047 - Propane, 2-fluoro-
C_3H_7FO
10382 - 1-Propanol, 3-fluoro-
C_3H_7I
10054 - Propane, 1-iodo-
10055 - Propane, 2-iodo-
C_3H_7IO
10385 - 1-Propanol, 3-iodo-
C_3H_7N
494 - Azetidine
500 - Aziridine, 1-methyl-
501 - Aziridine, 2-methyl-
4953 - Cyclopropanamine
10522 - 2-Propen-1-amine
$C_3H_7NNaS_2$
11514 - Sodium dimethyldithiocarbamate
C_3H_7NO
72 - Acetamide, *N*-methyl-
5992 - Formamide, *N,N*-dimethyl-
5994 - Formamide, *N*-ethyl-
9787 - Propanal, oxime
9791 - Propanamide
10476 - 2-Propanone, oxime
C_3H_7NOS
3978 - Carbamothioic acid, *O*-ethyl ester
3979 - Carbamothioic acid, *S*-ethyl ester
$C_3H_7NO_2$
323 - β-Alanine
324 - *DL*-Alanine
325 - *L*-Alanine
3948 - Carbamic acid, ethyl ester
6268 - Glycine, *N*-methyl-
8092 - Nitrous acid, 1-methylethyl ester
8098 - Nitrous acid, propyl ester
9797 - Propanamide, 2-hydroxy-, (±)-
10116 - Propane, 1-nitro-
10117 - Propane, 2-nitro-
$C_3H_7NO_2S$
5014 - *L*-Cysteine
$C_3H_7NO_3$
8076 - Nitric acid, 1-methylethyl ester
8080 - Nitric acid, propyl ester
10394 - 1-Propanol, 2-nitro-
10395 - 1-Propanol, 3-nitro-
10396 - 2-Propanol, 1-nitro-
11355 - *D*-Serine
11356 - *DL*-Serine
11357 - *L*-Serine
$C_3H_7NO_4$
10022 - 1,2-Propanediol, 1-nitrate
$C_3H_7NO_5$
10192 - 1,2,3-Propanetriol, 1-nitrate
10193 - 1,2,3-Propanetriol, 2-nitrate
C_3H_7NS
11681 - Thiazolidine
$C_3H_7N_3O_2$
6253 - Glycine, *N*-(aminoiminomethyl)-
12015 - Urea, *N*-ethyl-*N*-nitroso-
$C_3H_7N_3O_5$
11839 - 1,3,5-Triazine-2,4,6(1H,3H,5H)-trione, dihydrate
C_3H_8
9834 - Propane
$C_3H_8BrClSi$
11365 - Silane, (bromomethyl)chlorodimethyl-
$C_3H_8Cl_2Si$
11372 - Silane, chloro(chloromethyl)dimethyl-
11397 - Silane, dichloroethylmethyl-
$C_3H_8NO_5P$
6276 - Glyphosate
$C_3H_8N_2O$
12004 - Urea, *N,N'*-dimethyl-
12005 - Urea, *N,N*-dimethyl-
12014 - Urea, ethyl-
$C_3H_8N_2O_2$
326 - Alanine, 3-amino-
7148 - Hydrazinecarboxylic acid, ethyl ester
9824 - 1-Propanamine, *N*-nitro-
$C_3H_8N_2O_3$
11995 - Urea, *N,N'*-bis(hydroxymethyl)-
$C_3H_8N_2S$
11804 - Thiourea, *N,N'*-dimethyl-
C_3H_8O
5602 - Ethane, methoxy-
10334 - 1-Propanol
10335 - 2-Propanol
C_3H_8OS
5749 - Ethanol, 2-(methylthio)-
$C_3H_8OS_2$
10370 - 1-Propanol, 2,3-dimercapto-
10371 - 2-Propanol, 1,3-dimercapto-
$C_3H_8O_2$
5733 - Ethanol, 2-methoxy-
7507 - Methane, dimethoxy-
9982 - 1,2-Propanediol
9983 - 1,3-Propanediol
$C_3H_8O_2S$
10013 - 1,2-Propanediol, 3-mercapto-
$C_3H_8O_3$
10186 - 1,2,3-Propanetriol
$C_3H_8O_3S$
7531 - Methanesulfonic acid, ethyl ester
10130 - 2-Propanesulfonic acid
C_3H_8S
5606 - Ethane, (methylthio)-
10156 - 1-Propanethiol
10157 - 2-Propanethiol
$C_3H_8S_2$
7461 - Methane, bis(methylthio)-
10030 - 1,2-Propanedithiol
10031 - 1,3-Propanedithiol
C_3H_9Al
339 - Aluminum, trimethyl
C_3H_9As
449 - Arsine, trimethyl-
$C_3H_9AsO_3$
461 - Arsonic acid, propyl-
C_3H_9B
3119 - Borane, trimethyl-
$C_3H_9BO_2$
3136 - Boronic acid, propyl-
$C_3H_9BO_3$
3126 - Boric acid, trimethyl ester
C_3H_9BS
3129 - Borinic acid, dimethylthio-, methyl ester
$C_3H_9BS_3$
11698 - Thioboric acid, trimethyl ester
C_3H_9ClSi
11389 - Silane, chlorotrimethyl-
C_3H_9ClSn
11540 - Stannane, chlorotrimethyl-
C_3H_9N
7444 - Methanamine, *N,N*-dimethyl-
9803 - 1-Propanamine
9804 - 2-Propanamine
C_3H_9NO
5738 - Ethanol, 2-(methylamino)-
10336 - 1-Propanol, 2-amino-, (±)-
10337 - 1-Propanol, 3-amino-
10338 - 2-Propanol, 1-amino-, (±)-
$C_3H_9NO_2$
9984 - 1,2-Propanediol, 3-amino-, (±)-
$C_3H_9NO_3S$
5624 - Ethanesulfonic acid, 2-(methylamino)-
$C_3H_9O_2PS$
9550 - Phosphonothioic acid, methyl-, *O,O*-dimethyl ester
$C_3H_9O_3P$
9540 - Phosphonic acid, methyl-, dimethyl ester
9542 - Phosphonic acid, (1-methylethyl)-
9543 - Phosphonic acid, methyl-, monoethyl ester
9547 - Phosphonic acid, propyl-
9592 - Phosphorous acid, trimethyl ester
$C_3H_9O_4P$
9564 - Phosphoric acid, trimethyl ester
C_3H_9P
9522 - Phosphine, trimethyl-
C_3H_9Sb
11550 - Stibine, trimethyl-
$C_3H_{10}ClN$
5435 - Ethanamine, *N*-methyl-, hydrochloride
7445 - Methanamine, *N,N*-dimethyl-, hydrochloride
$C_3H_{10}N_2$
5512 - 1,2-Ethanediamine, *N*-methyl-
9872 - 1,2-Propanediamine
9873 - 1,3-Propanediamine
$C_3H_{10}Si$
11489 - Silane, trimethyl-
$C_3H_{12}BN$
3120 - Borane, trimethyl-, monoammoniate
3132 - Boron, (*N,N*-dimethylmethanamine)trihydro-, (T-4)-
C_3N_2O
9915 - Propanedinitrile, oxo-

C_3O_2
9773 - 1,2-Propadiene-1,3-dione
$C_4Br_2F_8$
3275 - Butane, 1,4-dibromo-1,1,2,2,3,3,4,4-octafluoro-
C_4Br_4S
11781 - Thiophene, tetrabromo-
C_4ClF_7
4177 - Cyclobutane, 1-chloro-1,2,2,3,3,4,4-heptafluoro-
C_4ClF_7O
3692 - Butanoyl chloride, heptafluoro-
$C_4Cl_2F_4$
4199 - Cyclobutene, 1,2-dichloro-3,3,4,4-tetrafluoro-
$C_4Cl_2F_6$
3754 - 2-Butene, 2,3-dichloro-1,1,1,4,4,4-hexafluoro-
$C_4Cl_2F_8$
3287 - Butane, 2,3-dichloro-1,1,1,2,3,4,4,4-octafluoro-
$C_4Cl_3F_7$
3452 - Butane, 2,2,3-trichloro-1,1,1,3,4,4,4-heptafluoro-
$C_4Cl_4F_4$
3808 - 1-Butene, 1,3,4,4-tetrachloro-1,2,3,4-tetrafluoro-
C_4Cl_4S
11782 - Thiophene, tetrachloro-
C_4Cl_6
3163 - 1,3-Butadiene, 1,1,2,3,4,4-hexachloro-
$C_4Cl_6O_3$
244 - Acetic acid, trichloro-, anhydride
$C_4F_4O_3$
6099 - 2,5-Furandione, 3,3,4,4-tetrafluorodihydro-
C_4F_6
3164 - 1,3-Butadiene, 1,1,2,3,4,4-hexafluoro-
3914 - 2-Butyne, 1,1,1,4,4,4-hexafluoro-
4200 - Cyclobutene, hexafluoro-
$C_4F_6O_3$
264 - Acetic acid, trifluoro-, anhydride
C_4F_8
3804 - 2-Butene, 1,1,1,2,3,4,4,4-octafluoro-
4194 - Cyclobutane, octafluoro-
C_4F_{10}
3255 - Butane, decafluoro-
10050 - Propane, 1,1,1,2,3,3,3-heptafluoro-2-(trifluoromethyl)-
C_4HBr_3S
11790 - Thiophene, 2,3,5-tribromo-
C_4HClN_2
3763 - 2-Butenedinitrile, 2-chloro-, (*Z*)-
C_4HClO_3
6092 - 2,5-Furandione, 3-chloro-
$C_4HCl_2N_3O_2$
11054 - Pyrimidine, 2,4-dichloro-5-nitro-
$C_4HCl_3N_2$
11075 - Pyrimidine, 2,4,6-trichloro-
$C_4HCl_3O_2$
3796 - 2-Butenedioyl dichloride, 2-chloro-, (*Z*)-
C_4HCl_3S
11791 - Thiophene, 2,3,5-trichloro-
$C_4HCl_7O_2$
5192 - 1,4-Dioxane, 2,2,3,3,5,5,6-heptachloro-
$C_4HF_7O_2$
3521 - Butanoic acid, heptafluoro-
C_4HI_4N
11140 - 1H-Pyrrole, 2,3,4,5-tetraiodo-
C_4H_2
3177 - 1,3-Butadiyne
C_4H_2BrClS
11728 - Thiophene, 2-bromo-5-chloro-
$C_4H_2Br_2N_2O_3$
11086 - 2,4,6(1H,3H,5H)-Pyrimidinetrione, 5,5-dibromo-
$C_4H_2Br_2O$
6077 - Furan, 2,3-dibromo-
6078 - Furan, 2,5-dibromo-
$C_4H_2Br_2S$
11742 - Thiophene, 2,3-dibromo-
11743 - Thiophene, 2,4-dibromo-
11744 - Thiophene, 2,5-dibromo-
$C_4H_2Cl_2O_2$
3795 - 2-Butenedioyl dichloride, (*E*)-
$C_4H_2Cl_2S$
11746 - Thiophene, 2,3-dichloro-
11747 - Thiophene, 2,4-dichloro-
11748 - Thiophene, 2,5-dichloro-
11749 - Thiophene, 3,4-dichloro-
$C_4H_2Cl_4$
3173 - 1,3-Butadiene, 1,2,3,4-tetrachloro-
$C_4H_2Cl_6$
3798 - 2-Butene, 1,1,2,3,4,4-hexachloro-
$C_4H_2Cl_8O$
5615 - Ethane, 1,1′-oxybis[1,2,2,2-tetrachloro-
$C_4H_2I_2S$
11753 - Thiophene, 2,5-diiodo-
$C_4H_2N_2$
3762 - 2-Butenedinitrile, (*E*)-
$C_4H_2N_2O_4$
11073 - 2,4,5,6(1H,3H)-Pyrimidinetetrone
$C_4H_2N_2O_4S$
11758 - Thiophene, 2,5-dinitro-
$C_4H_2O_3$
6091 - 2,5-Furandione
$C_4H_3BrClF_3$
3715 - 1-Butene, 4-bromo-3-chloro-3,4,4-trifluoro-
$C_4H_3BrN_2O_2$
11062 - 2,4(1H,3H)-Pyrimidinedione, 5-bromo-
C_4H_3BrO
6033 - Furan, 2-bromo-
6034 - Furan, 3-bromo-
$C_4H_3BrO_4$
3769 - 2-Butenedioic acid, 2-bromo-, (*E*)-
3770 - 2-Butenedioic acid, 2-bromo-, (*Z*)-
C_4H_3BrS
11726 - Thiophene, 2-bromo-
11727 - Thiophene, 3-bromo-
C_4H_3Cl
3890 - 1-Buten-3-yne, 4-chloro-
C_4H_3ClO
6073 - Furan, 2-chloro-
6074 - Furan, 3-chloro-
$C_4H_3ClO_2S_2$
11780 - 2-Thiophenesulfonyl chloride
C_4H_3ClS
11740 - Thiophene, 2-chloro-
$C_4H_3Cl_2F_3O_2$
10251 - Propanoic acid, 2,2-dichloro-3,3,3-trifluoro-, methyl ester
$C_4H_3Cl_3$
3174 - 1,3-Butadiene, 1,2,3-trichloro-
$C_4H_3Cl_5$
3805 - 2-Butene, 1,1,1,4,4-pentachloro-
$C_4H_3Cl_7$
3380 - Butane, 1,1,2,2,3,4,4-heptachloro-
$C_4H_3FN_2O_2$
11067 - 2,4(1H,3H)-Pyrimidinedione, 5-fluoro-
$C_4H_3F_5O_2$
10317 - Propanoic acid, pentafluoro-, methyl ester
$C_4H_3F_7O$
3622 - 1-Butanol, 2,2,3,3,4,4,4-heptafluoro-
C_4H_3IO
6110 - Furan, 2-iodo-
6111 - Furan, 3-iodo-
C_4H_3IS
11767 - Thiophene, 2-iodo-
$C_4H_3NO_2$
11121 - 1H-Pyrrole-2,5-dione
$C_4H_3NO_2S$
11775 - Thiophene, 2-nitro-
11776 - Thiophene, 3-nitro-
$C_4H_3NO_3$
6134 - Furan, 2-nitro-
$C_4H_3N_3O_4$
11074 - 2,4,5,6(1H,3H)-Pyrimidinetetrone, 5-oxime
$C_4H_3N_3O_5$
11102 - 2,4,6(1H,3H,5H)-Pyrimidinetrione, 5-nitro-
C_4H_4
3889 - 1-Buten-3-yne
$C_4H_4BrNO_2$
11153 - 2,5-Pyrrolidinedione, 1-bromo-
$C_4H_4Br_2$
3904 - 2-Butyne, 1,4-dibromo-
$C_4H_4Br_4$
4195 - Cyclobutane, 1,2,3,4-tetrabromo-
$C_4H_4ClNO_2$
11154 - 2,5-Pyrrolidinedione, 1-chloro-
C_4H_4ClNS
10538 - 1-Propene, 2-chloro-3-isothiocyanato-
$C_4H_4ClN_3O_2$
4083 - 5-Chloro-3-methyl-4-nitro-1H-pyrazole
$C_4H_4Cl_2$
3156 - 1,3-Butadiene, 1,1-dichloro-
3157 - 1,3-Butadiene, 1,2-dichloro-
3158 - 1,3-Butadiene, 2,3-dichloro-
3905 - 2-Butyne, 1,4-dichloro-
$C_4H_4Cl_2O_2$
3366 - Butanedioyl dichloride
3687 - Butanoyl chloride, 4-chloro-3-oxo-
4152 - Crotonic acid, 4,4-dichloro-
$C_4H_4Cl_2O_3$
124 - Acetic acid, chloro-, anhydride
$C_4H_4Cl_2O_4$
3302 - Butanedioic acid, 2,3-dichloro-, (±)-
$C_4H_4Cl_4$
3806 - 1-Butene, 1,3,4,4-tetrachloro-
3807 - 1-Butene, 2,3,3,4-tetrachloro-
$C_4H_4Cl_4O_2$
246 - Acetic acid, trichloro-, 2-chloroethyl ester
$C_4H_4INO_2$
11159 - 2,5-Pyrrolidinedione, 1-iodo-
$C_4H_4I_2$
3906 - 2-Butyne, 1,4-diiodo-
$C_4H_4N_2$
3291 - Butanedinitrile
10801 - Pyrazine
10855 - Pyridazine
11049 - Pyrimidine
$C_4H_4N_2OS$
11110 - 4(1H)-Pyrimidinone, 2,3-dihydro-2-thioxo-
$C_4H_4N_2O_2$
3902 - 2-Butynediamide
11059 - 2,4(1H,3H)-Pyrimidinedione
$C_4H_4N_2O_3$
11076 - 2,4,6(1H,3H,5H)-Pyrimidinetrione
$C_4H_4N_2S_2$
11702 - Thiocyanic acid, 1,2-ethanediyl ester
$C_4H_4N_6O$
11851 - 7H-1,2,3-Triazolo[4,5-d]pyrimidin-7-one, 5-amino-1,4-dihydro-
$C_4H_4Na_2O_6$
3314 - Butanedioic acid, 2,3-dihydroxy- [*R*-(*R**,*R**)]-, disodium salt
C_4H_4O
3895 - 2-Butynal
3923 - 3-Butyn-2-one
6025 - Furan
C_4H_4OS
11777 - Thiophene-2-ol
$C_4H_4O_2$
3917 - 2-Butynoic acid
5202 - 1,4-Dioxin
8598 - 2-Oxetanone, 4-methylene-
$C_4H_4O_3$
6093 - 2,5-Furandione, dihydro-
$C_4H_4O_4$
3764 - 2-Butenedioic acid (*E*)-
3765 - 2-Butenedioic acid (*Z*)-
$C_4H_4O_5$
3339 - Butanedioic acid, oxo-

MOLECULAR FORMULA INDEX OF ORGANIC COMPOUNDS (continued)

$C_4H_4O_6$
3777 - 2-Butenedioic acid, 2,3-dihydroxy-, (*Z*)-
C_4H_4S
11724 - Thiophene
C_4H_5Br
3147 - 1,2-Butadiene, 4-bromo-
3148 - 1,3-Butadiene, 1-bromo-
3149 - 1,3-Butadiene, 2-bromo-
$C_4H_5BrN_2$
10817 - 1H-Pyrazole, 4-bromo-3-methyl-
C_4H_5BrO
3696 - 2-Butenal, 2-bromo-
$C_4H_5BrO_2$
10612 - 2-Propenoic acid, 3-bromo-2-methyl-
$C_4H_5BrO_4$
3299 - Butanedioic acid, bromo-, (±)-
3300 - Butanedioic acid, bromo-
$C_4H_5Br_2FO_2$
161 - Acetic acid, dibromofluoro-, ethyl ester
$C_4H_5Br_3O_2$
242 - Acetic acid, tribromo-, ethyl ester
C_4H_5Cl
3150 - 1,2-Butadiene, 4-chloro-
3151 - 1,3-Butadiene, 1-chloro-
3152 - 1,3-Butadiene, 2-chloro-
3899 - 1-Butyne, 3-chloro-
3900 - 2-Butyne, 1-chloro-
$C_4H_5ClN_2$
10821 - 1H-Pyrazole, 3-chloro-5-methyl-
C_4H_5ClO
3697 - 2-Butenal, 2-chloro-
3887 - 2-Butenoyl chloride
10715 - 2-Propenoyl chloride, 2-methyl-
$C_4H_5ClO_2$
3820 - 2-Butenoic acid, 3-chloro-, (*E*)-
3821 - 2-Butenoic acid, 3-chloro-, (*Z*)-
3822 - 2-Butenoic acid, 4-chloro-
4022 - Carbonochloridic acid, 1-methylethenyl ester
4150 - Crotonic acid, 2-chloro-
4151 - Crotonic acid, 4-chloro-, (*E*)-
10621 - 2-Propenoic acid, 2-chloro-, methyl ester
$C_4H_5ClO_3$
144 - Acetic acid, chlorooxo-, ethyl ester
$C_4H_5Cl_3$
3809 - 1-Butene, 2,3,4-trichloro-
3810 - 2-Butene, 1,2,4-trichloro-
10605 - 1-Propene, 1,1,3-trichloro-2-methyl-
10606 - 1-Propene, 3,3,3-trichloro-2-methyl-
$C_4H_5Cl_3O$
3191 - Butanal, 2,2,3-trichloro-
$C_4H_5Cl_3O_2$
248 - Acetic acid, trichloro-, ethyl ester
3590 - Butanoic acid, 2,2,3-trichloro-
$C_4H_5Cl_3O_3$
249 - Acetic acid, trichloro-, 2-hydroxyethyl ester
$C_4H_5Cl_5$
3419 - Butane, 1,1,2,3,4-pentachloro-
3420 - Butane, 1,2,2,3,4-pentachloro-
10127 - Propane, 1,1,1,2,3-pentachloro-2-methyl-
10146 - Propane, 1,1,2,3-tetrachloro-
2-(chloromethyl)-
C_4H_5F
3162 - 1,3-Butadiene, 2-fluoro-
$C_4H_5F_3O_2$
267 - Acetic acid, trifluoro-, ethyl ester
C_4H_5I
3165 - 1,3-Butadiene, 2-iodo-
3166 - 1,2-Butadiene, 4-iodo-
C_4H_5N
3802 - 2-Butenenitrile, (*E*)-
3803 - 3-Butenenitrile
4956 - Cyclopropanecarbonitrile
10574 - 1-Propene, 3-isocyano-
10586 - 2-Propenenitrile, 2-methyl-
11112 - 1H-Pyrrole
C_4H_5NO
7405 - Isoxazole, 5-methyl-
8583 - Oxazole, 4-methyl-
$C_4H_5NO_2$
154 - Acetic acid, cyano-, methyl ester
4030 - Carbonocyanidic acid, ethyl ester
10233 - Propanoic acid, 2-cyano-
11152 - 2,5-Pyrrolidinedione
$C_4H_5NO_3$
3817 - 2-Butenoic acid, 4-amino-4-oxo-, (*Z*)-
C_4H_5NS
10575 - 1-Propene, 3-isothiocyanato-
11678 - Thiazole, 2-methyl-
11679 - Thiazole, 4-methyl-
11712 - Thiocyanic acid, 2-propenyl ester
$C_4H_5NS_2$
11680 - 2(3H)-Thiazolethione, 4-methyl-
$C_4H_5N_3$
11042 - 2-Pyrimidinamine
$C_4H_5N_3O$
11104 - 2(1H)-Pyrimidinone, 4-amino-
$C_4H_5N_3O_2$
11831 - 1,2,4-Triazine-3,5(2H,4H)-dione, 6-methyl-
$C_4H_5N_3O_3$
11077 - 2,4,6(1H,3H,5H)-Pyrimidinetrione, 5-amino-
C_4H_6
3145 - 1,2-Butadiene
3146 - 1,3-Butadiene
3897 - 1-Butyne
3898 - 2-Butyne
4198 - Cyclobutene
$C_4H_6BaO_4$
110 - Acetic acid, barium salt
C_4H_6BrClO
3682 - Butanoyl chloride, 2-bromo-
$C_4H_6BrClO_2$
113 - Acetic acid, bromochloro-, ethyl ester
C_4H_6BrN
3404 - Butanenitrile, 4-bromo-
10096 - Propanenitrile, 2-bromo-2-methyl-
$C_4H_6Br_2$
3738 - 1-Butene, 1,1-dibromo-
3739 - 1-Butene, 1,2-dibromo-
3740 - 1-Butene, 3,4-dibromo-
3741 - 2-Butene, 1,3-dibromo-
3742 - 2-Butene, 1,4-dibromo-, (*E*)-
4178 - Cyclobutane, 1,1-dibromo-
4179 - Cyclobutane, 1,2-dibromo-
10546 - 1-Propene, 1,3-dibromo-2-methyl-
$C_4H_6Br_2O$
3653 - 2-Butanone, 3,4-dibromo-
5893 - Ethene, 1,1-dibromo-2-ethoxy-
10491 - Propanoyl bromide, 2-bromo-2-methyl-
$C_4H_6Br_2O_2$
160 - Acetic acid, dibromo-, ethyl ester
3497 - Butanoic acid, 2,3-dibromo-
10241 - Propanoic acid, 2,3-dibromo-, methyl ester
$C_4H_6Br_4$
3424 - Butane, 1,1,4,4-tetrabromo-
3425 - Butane, 1,2,2,3-tetrabromo-
3426 - Butane, 1,2,3,4-tetrabromo-, *DL*-
$C_4H_6CaO_4$
122 - Acetic acid, calcium salt
$C_4H_6ClFO_2$
133 - Acetic acid, chlorofluoro-, ethyl ester
C_4H_6ClN
3405 - Butanenitrile, 4-chloro-
C_4H_6ClNO
10098 - Propanenitrile, 3-chloro-2-hydroxy-2-methyl-
$C_4H_6Cl_2$
3743 - 1-Butene, 1,3-dichloro-
3744 - 1-Butene, 2,3-dichloro-
3745 - 1-Butene, 3,4-dichloro-
3746 - 2-Butene, 1,1-dichloro-
3747 - 2-Butene, 1,2-dichloro-
3748 - 2-Butene, 1,3-dichloro-, (*E*)-
3749 - 2-Butene, 1,3-dichloro-, (*Z*)-
3750 - 2-Butene, 1,4-dichloro-, (*E*)-
3751 - 2-Butene, 1,4-dichloro-, (*Z*)-
3752 - 2-Butene, 2,3-dichloro-, (*E*)-
3753 - 2-Butene, 2,3-dichloro-, (*Z*)-
10537 - 1-Propene, 3-chloro-2-(chloromethyl)-
10555 - 1-Propene, 1,1-dichloro-2-methyl-
10556 - 1-Propene, 3,3-dichloro-2-methyl-
$C_4H_6Cl_2O$
3182 - Butanal, 2,3-dichloro-
3654 - 2-Butanone, 1,3-dichloro-
3655 - 2-Butanone, 3,4-dichloro-
3683 - Butanoyl chloride, 2-chloro-
3684 - Butanoyl chloride, 3-chloro-
3685 - Butanoyl chloride, 4-chloro-
5899 - Ethene, 1,2-dichloro-1-ethoxy-
10498 - Propanoyl chloride, 2-chloro-2-methyl-
10499 - Propanoyl chloride, 3-chloro-2-methyl-
$C_4H_6Cl_2O_2$
126 - Acetic acid, chloro-, 2-chloroethyl ester
164 - Acetic acid, dichloro-, ethyl ester
3500 - Butanoic acid, 2,2-dichloro-
3926 - Butyric acid, 3,4-dichloro-
4015 - Carbonochloridic acid, 3-chloropropyl ester
5188 - 1,4-Dioxane, 2,3-dichloro-
10248 - Propanoic acid, 2,3-dichloro-, methyl ester
$C_4H_6Cl_2O_3$
165 - Acetic acid, dichloro-, 2-hydroxyethyl ester
10247 - Propanoic acid, 3,3-dichloro-2-hydroxy-2-methyl-
$C_4H_6Cl_3NO$
97 - Acetamide, 2,2,2-trichloro-*N*,*N*-dimethyl-
$C_4H_6Cl_4$
3427 - Butane, 1,1,1,2-tetrachloro-
3428 - Butane, 1,2,2,3-tetrachloro-
3429 - Butane, 1,2,3,3-tetrachloro-
10147 - Propane, 1,1,1,2-tetrachloro-2-methyl-
10148 - Propane, 1,1,2,3-tetrachloro-2-methyl-
10176 - Propane, 1,2,3-trichloro-2-(chloromethyl)-
$C_4H_6F_2O_2$
176 - Acetic acid, difluoro-, ethyl ester
$C_4H_6F_3NO$
99 - Acetamide, 2,2,2-trifluoro-*N*,*N*-dimethyl-
$C_4H_6MnN_2S_4$
7420 - Maneb
$C_4H_6N_2$
7208 - 1H-Imidazole, 1-methyl-
7209 - 1H-Imidazole, 2-methyl-
7210 - 1H-Imidazole, 4-methyl-
10834 - 1H-Pyrazole, 1-methyl-
10835 - 1H-Pyrazole, 3-methyl-
10836 - 1H-Pyrazole, 4-methyl-
$C_4H_6N_2O$
6176 - Furazan, dimethyl-
$C_4H_6N_2O_2$
158 - Acetic acid, diazo-, ethyl ester
9612 - 2,5-Piperazinedione
10237 - Propanoic acid, 2-diazo-, methyl ester
$C_4H_6N_2S$
7214 - 2H-Imidazole-2-thione, 1,3-dihydro-1-methyl-
11659 - 1,3,4-Thiadiazole, 2,5-dimethyl-
11665 - 2-Thiazolamine, 4-methyl-
$C_4H_6N_2S_4Zn$
12067 - Zinc, [[1,2-ethanediylbis
[carbamodithioato]](2-)]-
$C_4H_6N_4O_2$
7191 - Imidazo[4,5-d]imidazole-2,5(1H,3H)-dione,
tetrahydro-
$C_4H_6N_4O_3$
12009 - Urea, (2,5-dioxo-4-imidazolidinyl)-
$C_4H_6N_4O_3S_2$
32 - Acetamide, *N*-[5-(aminosulfonyl)-1,3,4-
thiadiazol-2-yl]-
$C_4H_6N_4O_{12}$
3432 - 1,2,3,4-Butanetetrol, tetranitrate, (*R**,*S**)-
C_4H_6O
3175 - 2,3-Butadien-1-ol
3695 - 2-Butenal, (*E*)-
3875 - 3-Buten-2-one
3919 - 2-Butyn-1-ol

3920 - 3-Butyn-1-ol
3921 - 3-Butyn-2-ol
4197 - Cyclobutanone
5912 - Ethene, 1,1′-oxybis-
5942 - Ethyne, ethoxy-
6086 - Furan, 2,3-dihydro-
8613 - Oxirane, ethenyl-
10515 - 2-Propenal, 2-methyl-
10736 - 1-Propyne, 3-methoxy-

C_4H_6OS
11794 - 2(3H)-Thiophenone, dihydro-

$C_4H_6O_2$
179 - Acetic acid ethenyl ester
3007 - 2,2′-Bioxirane, (*R**,*S**)-
3256 - Butanedial
3359 - 2,3-Butanedione
3813 - 3-Butenoic acid
3814 - 2-Butenoic acid, (*E*)-
3815 - 2-Butenoic acid, (*Z*)-
3911 - 2-Butyne-1,4-diol
3912 - 3-Butyne-1,2-diol
4958 - Cyclopropanecarboxylic acid
5203 - 1,4-Dioxin, 2,3-dihydro-
6017 - Formic acid, 2-propenyl ester
6142 - 2(3H)-Furanone, dihydro-
10648 - 2-Propenoic acid, 2-methyl-
10655 - 2-Propenoic acid, methyl ester

$C_4H_6O_2S$
5913 - Ethene, 1,1′-sulfonylbis-
11752 - Thiophene, 2,5-dihydro-, 1,1-dioxide

$C_4H_6O_2S_2$
5250 - Disulfide, diacetyl

$C_4H_6O_3$
107 - Acetic acid, anhydride
3567 - Butanoic acid, 2-oxo-
3568 - Butanoic acid, 3-oxo-
3842 - 3-Butenoic acid, 2-hydroxy-
5213 - 1,3-Dioxolan-2-one, 4-methyl-
10316 - Propanoic acid, 2-oxo-, methyl ester

$C_4H_6O_4$
101 - Acetic acid, (acetyloxy)-
3294 - Butanedioic acid
5563 - Ethanedioic acid, dimethyl ester
5573 - 1,2-Ethanediol, diformate
9156 - Peroxide, diacetyl
9960 - Propanedioic acid, methyl-

$C_4H_6O_4S$
240 - Acetic acid, 2,2′-thiobis-

$C_4H_6O_5$
226 - Acetic acid, 2,2′-oxybis-
9959 - Propanedioic acid, (hydroxymethyl)-

$C_4H_6O_6$
3304 - Butanedioic acid, 2,3-dihydroxy-, (*R**,*S**)-
3309 - Butanedioic acid, 2,3-dihydroxy- [*R*-(*R**,*R**)]-
3311 - Butanedioic acid, 2,3-dihydroxy-, (*R**,*R**)-(±)-
3312 - Butanedioic acid, 2,3-dihydroxy-, [*S*-(*R**,*R**)]-

$C_4H_6O_6S$
239 - Acetic acid, 2,2′-sulfonylbis-

$C_4H_6O_8$
3342 - Butanedioic acid, tetrahydroxy-

C_4H_6S
5920 - Ethene, 1,1′-thiobis-
11750 - Thiophene, 2,3-dihydro-
11751 - Thiophene, 2,5-dihydro-

C_4H_7Br
3707 - 1-Butene, 1-bromo-, (*E*)-
3708 - 1-Butene, 1-bromo-, (*Z*)-
3709 - 1-Butene, 2-bromo-
3710 - 1-Butene, 3-bromo-
3711 - 1-Butene, 4-bromo-
3712 - 2-Butene, 1-bromo-
3713 - 2-Butene, 2-bromo-, (*E*)-
3714 - 2-Butene, 2-bromo-, (*Z*)-

C_4H_7BrO
3179 - Butanal, 2-bromo-
3680 - Butanoyl bromide
9776 - Propanal, 2-bromo-2-methyl-
10492 - Propanoyl bromide, 2-methyl-

$C_4H_7BrO_2$
115 - Acetic acid, bromo-, ethyl ester
3468 - Butanoic acid, 2-bromo-, (±)-
5686 - Ethanol, 2-bromo-, acetate
10209 - Propanoic acid, 2-bromo-2-methyl-
10211 - Propanoic acid, 2-bromo-, methyl ester, (±)-
10212 - Propanoic acid, 3-bromo-, methyl ester

$C_4H_7Br_2Cl_2O_4P$
7702 - Naled

$C_4H_7Br_3$
3442 - Butane, 1,1,2-tribromo-
3443 - Butane, 1,2,2-tribromo-
3444 - Butane, 1,2,3-tribromo-
3445 - Butane, 1,2,4-tribromo-
3446 - Butane, 1,3,3-tribromo-
3447 - Butane, 2,2,3-tribromo-
10164 - Propane, 1,2,3-tribromo-2-methyl-

$C_4H_7Br_3O$
10405 - 2-Propanol, 1,1,1-tribromo-2-methyl-

C_4H_7Cl
3720 - 1-Butene, 1-chloro-, (*E*)-
3721 - 1-Butene, 1-chloro-, (*Z*)-
3722 - 1-Butene, 2-chloro-
3723 - 1-Butene, 3-chloro-
3724 - 1-Butene, 4-chloro-
3725 - 2-Butene, 1-chloro-, (*E*)-
3726 - 2-Butene, 1-chloro-, (*Z*)-
3727 - 2-Butene, 2-chloro-, (*E*)-
3728 - 2-Butene, 2-chloro-, (*Z*)-
4966 - Cyclopropane, (chloromethyl)-
10539 - 1-Propene, 1-chloro-2-methyl-
10540 - 1-Propene, 3-chloro-2-methyl-

C_4H_7ClO
3180 - Butanal, 2-chloro-
3181 - Butanal, 4-chloro-
3646 - 2-Butanone, 1-chloro-
3647 - 2-Butanone, 3-chloro-
3648 - 2-Butanone, 4-chloro-
3681 - Butanoyl chloride
3863 - 2-Buten-1-ol, 2-chloro-
3864 - 2-Buten-1-ol, 3-chloro-
3865 - 2-Buten-1-ol, 4-chloro-
3866 - 3-Buten-2-ol, 1-chloro-
3867 - 3-Buten-1-ol, 2-chloro-
3868 - 3-Buten-2-ol, 3-chloro-
5886 - Ethene, 1-chloro-2-ethoxy-
5887 - Ethene, (2-chloroethoxy)-
8606 - Oxirane, 2-(chloromethyl)-2-methyl-
9779 - Propanal, 2-chloro-2-methyl-
10504 - Propanoyl chloride, 2-methyl-

$C_4H_7ClO_2$
130 - Acetic acid, chloro-, ethyl ester
298 - Acetyl chloride, ethoxy-
3480 - Butanoic acid, 2-chloro-
3481 - Butanoic acid, 3-chloro-, (±)-
3482 - Butanoic acid, 4-chloro-
4023 - Carbonochloridic acid, 1-methylethyl ester
4028 - Carbonochloridic acid, propyl ester
5694 - Ethanol, 1-chloro-, acetate
5695 - Ethanol, 2-chloro-, acetate
10222 - Propanoic acid, 2-chloro-, methyl ester
10225 - Propanoic acid, 3-chloro-, methyl ester
10228 - Propanoic acid, 2-chloro-2-methyl-
10230 - Propanoic acid, 3-chloro-2-methyl-

$C_4H_7ClO_3$
136 - Acetic acid, chloro-, 2-hydroxyethyl ester
4019 - Carbonochloridic acid, 2-methoxyethyl ester

$C_4H_7Cl_2O_4P$
5170 - Dichlorvos

$C_4H_7Cl_3$
3448 - Butane, 1,1,3-trichloro-
3449 - Butane, 1,2,3-trichloro-
3450 - Butane, 1,2,4-trichloro-
3451 - Butane, 2,2,3-trichloro-
10177 - Propane, 1,1,2-trichloro-2-methyl-
10178 - Propane, 1,2,3-trichloro-2-methyl-

$C_4H_7Cl_3O$
3643 - 2-Butanol, 1,1,1-trichloro-
10407 - 2-Propanol, 1,1,1-trichloro-2-methyl-

$C_4H_7Cl_3O_2$
5769 - Ethanol, 2,2,2-trichloro-1-ethoxy-

$C_4H_7FO_2$
191 - Acetic acid, fluoro-, ethyl ester
3927 - Butyric acid, 4-fluoro-

$C_4H_7IO_2$
201 - Acetic acid, iodo-, ethyl ester
10284 - Propanoic acid, 3-iodo-, methyl ester

C_4H_7N
3403 - Butanenitrile
10063 - Propane, 2-isocyano-
10111 - Propanenitrile, 2-methyl-
10726 - 2-Propyn-1-amine, *N*-methyl-
11118 - 1H-Pyrrole, 2,5-dihydro-

C_4H_7NO
3702 - 2-Butenamide, (*E*)-
8576 - Oxazole, 4,5-dihydro-2-methyl-
10059 - Propane, 1-isocyanato-
10060 - Propane, 2-isocyanato-
10108 - Propanenitrile, 2-hydroxy-2-methyl-
10109 - Propanenitrile, 2-methoxy-
10110 - Propanenitrile, 3-methoxy-
11180 - 2-Pyrrolidinone

$C_4H_7NO_2$
23 - Acetamide, *N*-acetyl-
495 - 2-Azetidinecarboxylic acid
3361 - 2,3-Butanedione, monooxime

$C_4H_7NO_2S$
11682 - 4-Thiazolidinecarboxylic acid

$C_4H_7NO_3$
6251 - Glycine, *N*-acetyl-

$C_4H_7NO_4$
218 - Acetic acid, nitro-, ethyl ester
470 - *DL*-Aspartic acid
471 - *L*-Aspartic acid
3466 - Butanoic acid, 4-amino-2-hydroxy-4-oxo-
6260 - Glycine, *N*-(carboxymethyl)-

C_4H_7NS
10064 - Propane, 1-isothiocyanato-
11672 - Thiazole, 4,5-dihydro-2-methyl-
11706 - Thiocyanic acid, 1-methylethyl ester

$C_4H_7N_3$
11849 - 1H-1,2,3-Triazole, 1,5-dimethyl-

$C_4H_7N_3O$
7230 - 4H-Imidazol-4-one, 2-amino-1,5-dihydro-1-methyl-

C_4H_8
3704 - 1-Butene
3705 - 2-Butene, (*E*)-
3706 - 2-Butene, (*Z*)-
4170 - Cyclobutane
4981 - Cyclopropane, methyl-
10578 - 1-Propene, 2-methyl-

C_4H_8BrCl
3233 - Butane, 1-bromo-4-chloro-
3234 - Butane, 2-bromo-1-chloro-
9842 - Propane, 1-bromo-3-chloro-2-methyl-

C_4H_8BrF
3237 - Butane, 1-bromo-4-fluoro-

C_4H_8BrNO
9792 - Propanamide, 2-bromo-2-methyl-

$C_4H_8Br_2$
3264 - Butane, 1,1-dibromo-
3265 - Butane, 1,2-dibromo-
3266 - Butane, 1,3-dibromo-
3267 - Butane, 1,4-dibromo-
3268 - Butane, 2,3-dibromo-
9889 - Propane, 1,2-dibromo-2-methyl-
9890 - Propane, 1,3-dibromo-2-methyl-

$C_4H_8Br_2O$
3612 - 1-Butanol, 2,3-dibromo-
3613 - 1-Butanol, 3,4-dibromo-
3614 - 2-Butanol, 1,4-dibromo-
5523 - Ethane, 1,2-dibromo-1-ethoxy-

5609 - Ethane, 1,1′-oxybis[2-bromo-
9888 - Propane, 1,2-dibromo-3-methoxy-
C_4H_8ClF
3250 - Butane, 1-chloro-4-fluoro-
$C_4H_8Cl_2$
3276 - Butane, 1,1-dichloro-
3277 - Butane, 1,2-dichloro-
3278 - Butane, 1,3-dichloro-
3279 - Butane, 1,4-dichloro-
3280 - Butane, 2,2-dichloro-
3281 - Butane, 2,3-dichloro-, (*R**,*R**)-(±)-
9899 - Propane, 1,1-dichloro-2-methyl
9900 - Propane, 1,2-dichloro-2-methyl-
9901 - Propane, 1,3-dichloro-2-methyl-
$C_4H_8Cl_2O$
5468 - Ethane, 1-chloro-1-(2-chloroethoxy)-
5533 - Ethane, 1,2-dichloro-1-ethoxy-
5610 - Ethane, 1,1′-oxybis[1-chloro-
5611 - Ethane, 1,1′-oxybis[2-chloro-
10364 - 2-Propanol, 1,1-dichloro-2-methyl-
10365 - 2-Propanol, 1,3-dichloro-2-methyl-
$C_4H_8Cl_2O_2$
3349 - 2,3-Butanediol, 1,4-dichloro-
$C_4H_8Cl_2S$
5645 - Ethane, 1,1′-thiobis[2-chloro-
$C_4H_8Cl_2Si$
11402 - Silane, dichloromethyl-2-propenyl-
$C_4H_8Cl_3O_4P$
11853 - Trichlorfon
$C_4H_8I_2$
3288 - Butane, 1,4-diiodo-
$C_4H_8N_2$
281 - Acetonitrile, (dimethylamino)-
3903 - 2-Butyne-1,4-diamine
7197 - 1H-Imidazole, 4,5-dihydro-2-methyl-
10112 - Propanenitrile, 3-(methylamino)-
$C_4H_8N_2O$
11170 - Pyrrolidine, 1-nitroso-
12028 - Urea, 2-propenyl-
$C_4H_8N_2O_2$
73 - Acetamide, *N*-[(methylamino)carbonyl]-
3257 - Butanediamide
3360 - 2,3-Butanedione, dioxime
7699 - Morpholine, 4-nitroso-
$C_4H_8N_2O_3$
468 - *L*-Asparagine
3932 - Carbamic acid, (aminocarbonyl)-, ethyl ester
6265 - Glycine, *N*-glycyl-
$C_4H_8N_2O_4$
3293 - Butane, 1,4-dinitro-
$C_4H_8N_2S$
11809 - Thiourea, 2-propenyl-
C_4H_8O
3178 - Butanal
3645 - 2-Butanone
3858 - 3-Buten-2-ol, (±)-
3859 - 2-Buten-1-ol, (*E*)-
3860 - 2-Buten-1-ol, (*Z*)-
3861 - 3-Buten-1-ol
4196 - Cyclobutanol
4980 - Cyclopropane, methoxy-
5905 - Ethene, ethoxy-
6168 - Furan, tetrahydro-
8594 - Oxetane, 2-methyl-
8607 - Oxirane, 2,2-dimethyl-
8608 - Oxirane, 2,3-dimethyl-, *cis*-
8609 - Oxirane, 2,3-dimethyl-, *trans*-
9785 - Propanal, 2-methyl-
10576 - 1-Propene, 3-methoxy-
10702 - 2-Propen-1-ol, 2-methyl-
C_4H_8OS
5647 - Ethanethioic acid, *S*-ethyl ester
8569 - 1,4-Oxathiane
9786 - Propanal, 3-(methylthio)-
$C_4H_8O_2$
14 - Acetaldehyde, ethoxy-
186 - Acetic acid ethyl ester
3184 - Butanal, 3-hydroxy-
3460 - Butanoic acid
3660 - 2-Butanone, 1-hydroxy-
3661 - 2-Butanone, 3-hydroxy-
3662 - 2-Butanone, 4-hydroxy-
3792 - 2-Butene-1,4-diol, (*E*)-
3793 - 2-Butene-1,4-diol, (*Z*)-
3794 - 3-Butene-1,2-diol
5186 - 1,3-Dioxane
5187 - 1,4-Dioxane
5210 - 1,3-Dioxolane, 2-methyl-
5716 - Ethanol, 2-(ethenyloxy)-
6011 - Formic acid, 1-methylethyl ester
6018 - Formic acid, propyl ester
6137 - 3-Furanol, tetrahydro-
8615 - Oxirane, ethyl-
8620 - Oxirane, (methoxymethyl)-
10290 - Propanoic acid, 2-methyl-
10296 - Propanoic acid, methyl ester
10455 - 2-Propanone, 1-methoxy-
$C_4H_8O_2S$
188 - Acetic acid, (ethylthio)-
203 - Acetic acid, mercapto-, ethyl ester
3537 - Butanoic acid, 3-mercapto-
10287 - Propanoic acid, 3-mercapto-, methyl ester
10288 - Propanoic acid, 2-mercapto-2-methyl-
11786 - Thiophene, tetrahydro-, 1,1-dioxide
$C_4H_8O_3$
180 - Acetic acid, ethoxy-
197 - Acetic acid, hydroxy-, ethyl ester
207 - Acetic acid, methoxy-, methyl ester
3530 - Butanoic acid, 2-hydroxy-, (±)-
3531 - Butanoic acid, 3-hydroxy-, (±)-
4003 - Carbonic acid, ethyl methyl ester
5577 - 1,2-Ethanediol, monoacetate
10273 - Propanoic acid, 2-hydroxy-, methyl ester, (±)-
10276 - Propanoic acid, 3-hydroxy-, methyl ester
10278 - Propanoic acid, 2-hydroxy-2-methyl-
$C_4H_8O_3S$
5198 - 1,3,2-Dioxathiane, 4-methyl-, 2-oxide
$C_4H_8O_4$
3192 - Butanal, 2,3,4-trihydroxy-, [*R*-(*R**,*R**)]-
3193 - Butanal, 2,3,4-trihydroxy-, [*S*-(*R**,*S**)]-
3679 - 2-Butanone, 1,3,4-trihydroxy-, (*S*)-
10254 - Propanoic acid, 2,3-dihydroxy-, methyl ester, (±)-
C_4H_8S
10581 - 1-Propene, 3-(methylthio)-
11693 - Thietane, 2-methyl-
11783 - Thiophene, tetrahydro-
$C_4H_8S_2$
5261 - 1,2-Dithiane
5262 - 1,3-Dithiane
5263 - 1,4-Dithiane
C_4H_9Br
3231 - Butane, 1-bromo-
3232 - Butane, 2-bromo-, (±)-
9845 - Propane, 1-bromo-2-methyl-
9846 - Propane, 2-bromo-2-methyl-
C_4H_9BrO
3600 - 1-Butanol, 3-bromo-
3601 - 2-Butanol, 3-bromo-, (*R**,*S**)-(±)-
5461 - Ethane, 1-bromo-2-ethoxy-
$C_4H_9BrO_2$
5460 - Ethane, 2-bromo-1,1-dimethoxy-
C_4H_9Cl
3241 - Butane, 1-chloro-
3242 - Butane, 2-chloro-, (±)-
9862 - Propane, 1-chloro-2-methyl-
9863 - Propane, 2-chloro-2-methyl-
C_4H_9ClO
1605 - Benzeneethanol, α-ethyl-α-methyl-, (±)-
3602 - 1-Butanol, 2-chloro-
3604 - 1-Butanol, 4-chloro-
3605 - 2-Butanol, 1-chloro-
3606 - 2-Butanol, 3-chloro-
3607 - 2-Butanol, 4-chloro-
5473 - Ethane, 1-chloro-1-ethoxy-
5474 - Ethane, 1-chloro-2-ethoxy-
7183 - Hypochlorous acid, 1,1-dimethylethyl ester
9858 - Propane, 1-(chloromethoxy)-
9859 - Propane, 1-chloro-2-methoxy-
9860 - Propane, 1-chloro-3-methoxy-
9861 - Propane, 2-chloro-1-methoxy-
10352 - 1-Propanol, 2-chloro-2-methyl-
10353 - 1-Propanol, 3-chloro-2-methyl-
10354 - 2-Propanol, 1-chloro-2-methyl-
$C_4H_9ClO_2$
5472 - Ethane, 2-chloro-1,1-dimethoxy-
5698 - Ethanol, 2-(2-chloroethoxy)-
9998 - 1,2-Propanediol, 3-chloro-2-methyl-
$C_4H_9ClO_2S$
10134 - 1-Propanesulfonyl chloride, 2-methyl-
C_4H_9ClS
9864 - Propane, 1-chloro-3-(methylthio)-
9865 - Propane, 2-chloro-1-(methylthio)-
C_4H_9ClSi
11376 - Silane, chloroethenyldimethyl-
$C_4H_9Cl_3Si$
11370 - Silane, butyltrichloro-
11469 - Silane, trichloro(2-methylpropyl)-
C_4H_9F
3378 - Butane, 1-fluoro-
3379 - Butane, 2-fluoro-
C_4H_9FO
3621 - 1-Butanol, 4-fluoro-
C_4H_9I
3381 - Butane, 1-iodo-
3382 - Butane, 2-iodo-, (±)-
10057 - Propane, 1-iodo-2-methyl-
10058 - Propane, 2-iodo-2-methyl-
C_4H_9N
4169 - Cyclobutanamine
10524 - 2-Propen-1-amine, *N*-methyl-
11143 - Pyrrolidine
C_4H_9NO
55 - Acetamide, *N,N*-dimethyl-
61 - Acetamide, *N*-ethyl-
499 - 1-Aziridineethanol
3188 - Butanal, oxime
3194 - Butanamide
3675 - 2-Butanone, oxime
7681 - Morpholine
9798 - Propanamide, 2-methyl-
9799 - Propanamide, *N*-methyl-
$C_4H_9NO_2$
65 - Acetamide, *N*-(2-hydroxyethyl)-
331 - Alanine, 2-methyl-
3412 - Butane, 1-nitro-
3413 - Butane, 2-nitro-, (±)-
3463 - Butanoic acid, 2-amino-, (±)-
3464 - Butanoic acid, 4-amino-
3954 - Carbamic acid, methyl-, ethyl ester
3955 - Carbamic acid, 1-methylethyl ester
3969 - Carbamic acid, propyl ester
6261 - Glycine, *N,N*-dimethyl-
6263 - Glycine, ethyl ester
8083 - Nitrous acid, butyl ester
8085 - Nitrous acid, 1,1-dimethylethyl ester
8093 - Nitrous acid, 1-methylpropyl ester
8094 - Nitrous acid, 2-methylpropyl ester
10084 - Propane, 2-methyl-1-nitro-
10085 - Propane, 2-methyl-2-nitro-
$C_4H_9NO_3$
3465 - Butanoic acid, 3-amino-4-hydroxy-
3640 - 1-Butanol, 2-nitro-
3641 - 2-Butanol, 1-nitro-
3642 - 2-Butanol, 3-nitro-
7134 - Homoserine
8072 - Nitric acid, butyl ester
8077 - Nitric acid, 1-methylpropyl ester
8078 - Nitric acid, 2-methylpropyl ester
10391 - 1-Propanol, 2-methyl-2-nitro-
11813 - *L*-Threonine

$C_4H_9NO_4$
10018 - 1,3-Propanediol, 2-methyl-2-nitro-
$C_4H_9NO_5$
10012 - 1,3-Propanediol, 2-(hydroxymethyl)-2-nitro-
C_4H_9NS
11719 - Thiomorpholine
$C_4H_9NS_2$
5264 - 4H-1,3,5-Dithiazine, dihydro-5-methyl-
$C_4H_9N_3O_2$
6254 - Glycine, *N*-(aminoiminomethyl)-*N*-methyl-
C_4H_{10}
3229 - Butane
10073 - Propane, 2-methyl-
$C_4H_{10}BF_3O$
3137 - Boron, trifluoro[1,1′-oxybis[ethane]]-, (T-4)-
$C_4H_{10}ClNO_2S$
5019 - *L*-Cysteine, methyl ester, hydrochloride
$C_4H_{10}ClOP$
9529 - Phosphinic chloride, diethyl-
$C_4H_{10}ClO_2PS$
9576 - Phosphorochloridothioic acid, *O,O*-diethyl ester
$C_4H_{10}ClO_3P$
9574 - Phosphorochloridic acid, diethyl ester
$C_4H_{10}Cl_2O_2Si$
11392 - Silane, dichlorodiethoxy-
$C_4H_{10}Cl_2Si$
11393 - Silane, dichlorodiethyl-
11399 - Silane, dichloromethyl(1-methylethyl)-
$C_4H_{10}F_2Si$
11410 - Silane, diethyldifluoro-
$C_4H_{10}Hg$
7434 - Mercury, diethyl-
$C_4H_{10}NO_3PS$
10 - Acephate
$C_4H_{10}N_2$
9602 - Piperazine
$C_4H_{10}N_2O$
28 - Acetamide, *N*-(2-aminoethyl)-
3529 - Butanoic acid, hydrazide
5432 - Ethanamine, *N*-ethyl-*N*-nitroso-
7680 - 4-Morpholinamine
12032 - Urea, trimethyl-
$C_4H_{10}N_2O_2$
5431 - Ethanamine, *N*-ethyl-*N*-nitro-
$C_4H_{10}N_2O_4$
469 - *DL*-Asparagine, monohydrate
$C_4H_{10}O$
3593 - 1-Butanol
3594 - 2-Butanol
5608 - Ethane, 1,1′-oxybis-
10069 - Propane, 1-methoxy-
10070 - Propane, 2-methoxy-
10388 - 1-Propanol, 2-methyl-
10389 - 2-Propanol, 2-methyl-
$C_4H_{10}OS$
5621 - Ethane, 1,1′-sulfinylbis-
5726 - Ethanol, 2-(ethylthio)-
$C_4H_{10}O_2$
3344 - 1,2-Butanediol, (±)-
3345 - 1,3-Butanediol
3346 - 1,4-Butanediol
3347 - 2,3-Butanediol, (*R**,*R**)-(±)-
5548 - Ethane, 1,1-dimethoxy-
5549 - Ethane, 1,2-dimethoxy-
5717 - Ethanol, 2-ethoxy-
7178 - Hydroperoxide, 1,1-dimethylethyl
9158 - Peroxide, diethyl
10017 - 1,2-Propanediol, 2-methyl-
10386 - 1-Propanol, 2-methoxy-
10387 - 2-Propanol, 1-methoxy-
$C_4H_{10}O_2S$
5626 - Ethane, 1,1′-sulfonylbis-
5766 - Ethanol, 2,2′-thiobis-
11571 - Sulfoxylic acid, diethyl ester
$C_4H_{10}O_3$
3458 - 1,2,3-Butanetriol
3459 - 1,2,4-Butanetriol
5753 - Ethanol, 2,2′-oxybis-
7561 - Methane, trimethoxy-
10014 - 1,2-Propanediol, 3-methoxy-
10015 - 1,3-Propanediol, 2-methoxy-
$C_4H_{10}O_3S$
5725 - Ethanol, 2-(ethylsulfonyl)-
11581 - Sulfurous acid, diethyl ester
$C_4H_{10}O_4$
3431 - 1,2,3,4-Butanetetrol, (*R**,*S**)-
$C_4H_{10}O_4S$
11573 - Sulfuric acid, diethyl ester
$C_4H_{10}S$
3437 - 1-Butanethiol
3438 - 2-Butanethiol, (±)-
5644 - Ethane, 1,1′-thiobis-
10088 - Propane, 1-(methylthio)-
10089 - Propane, 2-(methylthio)-
10158 - 1-Propanethiol, 2-methyl-
10159 - 2-Propanethiol, 2-methyl-
$C_4H_{10}S_2$
3368 - 1,4-Butanedithiol
5253 - Disulfide, diethyl
5452 - Ethane, 1,2-bis(methylthio)-
$C_4H_{10}S_3$
5652 - Ethanethiol, 2,2′-thiobis-
11916 - Trisulfide, diethyl
$C_4H_{10}Se$
5620 - Ethane, 1,1′-selenobis-
$C_4H_{10}Te$
5630 - Ethane, 1,1′-tellurobis-
$C_4H_{11}As$
441 - Arsine, diethyl-
$C_4H_{11}BrSi$
11366 - Silane, (bromomethyl)trimethyl-
$C_4H_{11}ClSi$
11377 - Silane, chloroethyldimethyl-
11382 - Silane, (chloromethyl)trimethyl-
$C_4H_{11}N$
3209 - 1-Butanamine
3210 - 2-Butanamine, (±)-
5421 - Ethanamine, *N,N*-dimethyl-
5427 - Ethanamine, *N*-ethyl-
9817 - 1-Propanamine, 2-methyl-
9818 - 2-Propanamine, 2-methyl-
9819 - 2-Propanamine, *N*-methyl-
$C_4H_{11}NO$
3595 - 1-Butanol, 2-amino-, (±)-
3596 - 1-Butanol, 3-amino-
3597 - 1-Butanol, 4-amino-
3598 - 2-Butanol, 3-amino-
5423 - Ethanamine, 2-ethoxy-
5429 - Ethanamine, *N*-ethyl-*N*-hydroxy-
5709 - Ethanol, 2-(dimethylamino)-
5722 - Ethanol, 2-(ethylamino)-
9816 - 1-Propanamine, 3-methoxy-
10341 - 1-Propanol, 2-amino-2-methyl-
$C_4H_{11}NO_2$
5419 - Ethanamine, 2,2-dimethoxy-
5730 - Ethanol, 2,2′-iminobis-
9987 - 1,3-Propanediol, 2-amino-2-methyl-
$C_4H_{11}NO_3$
9986 - 1,3-Propanediol, 2-amino-2-(hydroxymethyl)-
$C_4H_{11}NO_8P_2$
6277 - Glyphosine
$C_4H_{11}O_2P$
9525 - Phosphinic acid, diethyl-
$C_4H_{11}O_3P$
9533 - Phosphonic acid, (1,1-dimethylethyl)-
9537 - Phosphonic acid, ethyl-, dimethyl ester
9544 - Phosphonic acid, (2-methylpropyl)-
9587 - Phosphorous acid, diethyl ester
$C_4H_{11}O_4P$
9558 - Phosphoric acid, diethyl ester
$C_4H_{12}As_2$
5110 - Diarsine, tetramethyl-
$C_4H_{12}As_2O$
452 - Arsinous acid, dimethyl-, anhydride
$C_4H_{12}BFN_2$
3114 - Boranediamine, 1-fluoro-*N,N,N′,N′*-tetramethyl-
$C_4H_{12}BN$
3113 - Boranamine, *N,N*,1,1-tetramethyl-
$C_4H_{12}BrN$
7456 - Methanaminium, *N,N,N*-trimethyl-, bromide
$C_4H_{12}ClN$
5428 - Ethanamine, *N*-ethyl-, hydrochloride
7457 - Methanaminium, *N,N,N*-trimethyl-, chloride
$C_4H_{12}Cl_2N_2$
9607 - Piperazine, dihydrochloride
$C_4H_{12}Cl_2OSi_2$
5232 - Disiloxane, 1,3-dichloro-1,1,3,3-tetramethyl-
$C_4H_{12}Cl_2Si_2$
5226 - Disilane, 1,2-dichloro-1,1,2,2-tetramethyl-
$C_4H_{12}F_2Si_2$
5227 - Disilane, 1,2-difluoro-1,1,2,2-tetramethyl-
$C_4H_{12}IN$
7458 - Methanaminium, *N,N,N*-trimethyl-, iodide
$C_4H_{12}N_2$
3258 - 1,4-Butanediamine
3261 - 2,3-Butanediamine, (*R**,*R**)-(±)-
5503 - 1,2-Ethanediamine, *N,N′*-dimethyl-
5509 - 1,2-Ethanediamine, *N*-ethyl-
7152 - Hydrazine, 1,1-diethyl-
7153 - Hydrazine, 1,2-diethyl-
9878 - 1,2-Propanediamine, 2-methyl-
$C_4H_{12}N_2O$
5676 - Ethanol, 2-[(2-aminoethyl)amino]-
$C_4H_{12}N_2OS$
11583 - Sulfurous diamide, tetramethyl-
$C_4H_{12}N_2O_2S$
11569 - Sulfamide, tetramethyl-
$C_4H_{12}OSi$
11501 - Silanol, ethyldimethyl-
$C_4H_{12}O_2Si$
11412 - Silane, dimethoxydimethyl-
$C_4H_{12}O_3Si$
11487 - Silane, trimethoxymethyl-
$C_4H_{12}O_4$
19 - Acetaldehyde, tetramer
$C_4H_{12}O_4Si$
11508 - Silicic acid (H_4SiO_4), tetramethyl ester
$C_4H_{12}Pb$
9732 - Plumbane, tetramethyl
$C_4H_{12}Si$
11409 - Silane, diethyl-
11451 - Silane, tetramethyl-
$C_4H_{12}Sn$
11544 - Stannane, tetramethyl-
$C_4H_{13}N_3$
5497 - 1,2-Ethanediamine, *N*-(2-aminoethyl)-
$C_4H_{14}Cl_2N_2$
3260 - 1,4-Butanediamine, dihydrochloride
$C_4H_{14}OSi_2$
5237 - Disiloxane, 1,1,3,3-tetramethyl-
$C_4H_{16}O_4Si_4$
4999 - Cyclotetrasiloxane, 2,4,6,8-tetramethyl-
C_4N_2
3908 - 2-Butynedinitrile
C_4NiO_4
8068 - Nickel carbonyl
$C_5Cl_2F_6$
4917 - Cyclopentene, 1,2-dichloro-3,3,4,4,5,5-hexafluoro-
C_5Cl_5N
10995 - Pyridine, pentachloro-
C_5Cl_6
4766 - 1,3-Cyclopentadiene, 1,2,3,4,5,5-hexachloro-
C_5Cl_8
4932 - Cyclopentene, octachloro-
C_5F_8
3172 - 1,3-Butadiene, 1,1,2,4,4-pentafluoro-3-(trifluoromethyl)-

C_5F_{10}
9027 - 1-Pentene, 1,1,2,3,3,4,4,5,5,5-decafluoro-
C_5F_{12}
8807 - Pentane, dodecafluoro-
C_5FeO_5
7339 - Iron carbonyl
C_5HCl_4N
11019 - Pyridine, 2,3,5,6-tetrachloro-
$C_5H_2Cl_6O_3$
5212 - 1,3-Dioxolan-4-one, 2,5-bis(trichloromethyl)-
$C_5H_2F_6O_2$
8802 - 2,4-Pentanedione, 1,1,1,5,5,5-hexafluoro-
$C_5H_2O_5$
4939 - 4-Cyclopentene-1,2,3-trione, 4,5-dihydroxy-
$C_5H_3BrO_2$
6043 - 2-Furancarboxaldehyde, 5-bromo-
$C_5H_3Br_2N$
10926 - Pyridine, 3,5-dibromo-
$C_5H_3ClN_4$
10753 - 1H-Purine, 6-chloro-
$C_5H_3ClO_2$
6040 - 2-Furancarbonyl chloride
$C_5H_3F_7O_2$
3525 - Butanoic acid, heptafluoro-, methyl ester
C_5H_3NO
6039 - 2-Furancarbonitrile
$C_5H_3NO_4$
6046 - 2-Furancarboxaldehyde, 5-nitro-
$C_5H_3NO_5$
6066 - 2-Furancarboxylic acid, 5-nitro-
C_5H_3NS
11732 - 2-Thiophenecarbonitrile
C_5H_4
8684 - 1,3-Pentadiyne
C_5H_4BrN
10875 - Pyridine, 2-bromo-
10876 - Pyridine, 3-bromo-
10877 - Pyridine, 4-bromo-
C_5H_4ClN
10915 - Pyridine, 2-chloro-
10916 - Pyridine, 3-chloro-
10917 - Pyridine, 4-chloro-
C_5H_4ClOS
11735 - 2-Thiophenecarboxaldehyde, 5-chloro-
$C_5H_4Cl_2N_2$
10862 - 3-Pyridinamine, 2,6-dichloro-
11052 - Pyrimidine, 2,4-dichloro-5-methyl-
11053 - Pyrimidine, 2,4-dichloro-6-methyl-
$C_5H_4Cl_2O_2$
3367 - Butanedioyl dichloride, methylene-
$C_5H_4F_8O$
8953 - 1-Pentanol, 2,2,3,3,4,4,5,5-octafluoro-
C_5H_4IN
10970 - Pyridine, 2-iodo-
$C_5H_4N_2O_2$
10803 - Pyrazinecarboxylic acid
$C_5H_4N_2O_3$
10993 - Pyridine, 4-nitro-, 1-oxide
$C_5H_4N_2O_4$
7195 - 1H-Imidazole-4,5-dicarboxylic acid
11050 - 4-Pyrimidinecarboxylic acid, 1,2,3,6-tetrahydro-2,6-dioxo-
$C_5H_4N_4$
10752 - 1H-Purine
$C_5H_4N_4O$
10770 - 6H-Purin-6-one, 1,7-dihydro-
10852 - 4H-Pyrazolo[3,4-d]pyrimidin-4-one, 1,5-dihydro-
$C_5H_4N_4O_2$
10755 - 1H-Purine-2,6-dione, 3,7-dihydro-
$C_5H_4N_4O_3$
10764 - 1H-Purine-2,6,8(3H)-trione, 7,9-dihydro-
$C_5H_4N_4S$
10763 - 6H-Purine-6-thione, 1,7-dihydro-
C_5H_4OS
11733 - 2-Thiophenecarboxaldehyde
11734 - 3-Thiophenecarboxaldehyde
$C_5H_4O_2$
6041 - 2-Furancarboxaldehyde
6042 - 3-Furancarboxaldehyde
6155 - 2(5H)-Furanone, 5-methylene-
10788 - 2H-Pyran-2-one
10789 - 4H-Pyran-4-one
$C_5H_4O_2S$
11737 - 2-Thiophenecarboxylic acid
11738 - 3-Thiophenecarboxylic acid
$C_5H_4O_3$
6050 - 2-Furancarboxylic acid
6051 - 3-Furancarboxylic acid
6095 - 2,5-Furandione, dihydro-3-methylene-
6098 - 2,5-Furandione, 3-methyl-
10794 - 2H-Pyran-2-one, 3-hydroxy-
$C_5H_4O_4$
6054 - 3-Furancarboxylic acid, 4,5-dihydro-5-oxo-
C_5H_5BrO
6035 - Furan, 2-(bromomethyl)-
C_5H_5BrS
11729 - Thiophene, 2-bromo-3-methyl-
11730 - Thiophene, 2-bromo-5-methyl-
$C_5H_5ClN_2$
10861 - 2-Pyridinamine, 5-chloro-
C_5H_5ClO
6075 - Furan, 2-(chloromethyl)-
C_5H_5ClS
11741 - Thiophene, 2-chloro-5-methyl-
$C_5H_5Cl_3N_2OS$
11599 - Terrazole
C_5H_5N
8673 - 2,4-Pentadienenitrile
8674 - 2,4-Pentadienenitrile, (*E*)-
8675 - 2,4-Pentadienenitrile, (*Z*)-
10872 - Pyridine
C_5H_5NO
10994 - Pyridine, 1-oxide
11031 - 2-Pyridinol
11032 - 4-Pyridinol
11033 - 2(1H)-Pyridinone
11114 - 1H-Pyrrole-2-carboxaldehyde
$C_5H_5NO_2$
6047 - 2-Furancarboxaldehyde, oxime, (*E*)-
6048 - 2-Furancarboxaldehyde, oxime, (*Z*)-
10624 - 2-Propenoic acid, 2-cyano-, methyl ester
$C_5H_5NO_2S$
11669 - 5-Thiazolecarboxylic acid, 4-methyl-
$C_5H_5NO_3S$
11018 - 3-Pyridinesulfonic acid
$C_5H_5N_3O$
10802 - Pyrazinecarboxamide
$C_5H_5N_3O_3S$
85 - Acetamide, *N*-(5-nitro-2-thiazolyl)-
$C_5H_5N_5$
10751 - 1H-Purin-6-amine
$C_5H_5N_5O$
10767 - 2H-Purin-2-one, 6-amino-1,3-dihydro-
10768 - 6H-Purin-6-one, 2-amino-1,7-dihydro-
$C_5H_5N_5S$
10762 - 6H-Purine-6-thione, 2-amino-1,7-dihydro-
C_5H_6
3892 - 1-Buten-3-yne, 2-methyl-
4764 - 1,3-Cyclopentadiene
9111 - 1-Penten-3-yne
9112 - 1-Penten-4-yne
9113 - 3-Penten-1-yne, (*E*)-
9114 - 3-Penten-1-yne, (*Z*)-
$C_5H_6Br_2N_2O_2$
7217 - 2,4-Imidazolidinedione, 1,3-dibromo-5,5-dimethyl-
C_5H_6ClN
10969 - Pyridine, hydrochloride
$C_5H_6ClNO_2$
11155 - 2,5-Pyrrolidinedione, 1-(chloromethyl)-
$C_5H_6Cl_2O_2$
8806 - Pentanedioyl dichloride
$C_5H_6F_4O_2$
10328 - Propanoic acid, 2,3,3,3-tetrafluoro-, ethyl ester
$C_5H_6N_2$
8767 - Pentanedinitrile
9913 - Propanedinitrile, dimethyl-
10810 - Pyrazine, methyl-
10856 - Pyridazine, 3-methyl-
10857 - 2-Pyridinamine
10858 - 3-Pyridinamine
10859 - 4-Pyridinamine
11070 - Pyrimidine, 2-methyl-
11071 - Pyrimidine, 4-methyl-
11072 - Pyrimidine, 5-methyl-
$C_5H_6N_2OS$
11108 - 4(1H)-Pyrimidinone, 2,3-dihydro-6-methyl-2-thioxo-
$C_5H_6N_2O_2$
11068 - 2,4(1H,3H)-Pyrimidinedione, 5-methyl-
11069 - 2,4(1H,3H)-Pyrimidinedione, 6-methyl-
$C_5H_6N_6$
10754 - 1H-Purine-2,6-diamine
C_5H_6O
3891 - 1-Buten-3-yne, 1-methoxy-
4941 - 2-Cyclopenten-1-one
6130 - Furan, 2-methyl-
6131 - Furan, 3-methyl-
9118 - 4-Penten-2-yn-1-ol
10771 - Pyran
C_5H_6OS
6117 - 2-Furanmethanethiol
11768 - 2-Thiophenemethanol
$C_5H_6O_2$
6118 - 2-Furanmethanol
6127 - Furan, 2-methoxy-
6147 - 2(3H)-Furanone, dihydro-3-methylene-
6148 - 2(3H)-Furanone, dihydro-5-methylene-
6153 - 2(3H)-Furanone, 5-methyl-
6154 - 2(5H)-Furanone, 5-methyl-
8676 - 2,4-Pentadienoic acid
9137 - 2-Pentynoic acid
9138 - 4-Pentynoic acid
10739 - 2-Propynoic acid, ethyl ester
10744 - 2-Propyn-1-ol, acetate
$C_5H_6O_3$
4942 - 2-Cyclopenten-1-one, 2,3-dihydroxy-
6094 - 2,5-Furandione, dihydro-3-methyl-
10780 - 2H-Pyran-2,6(3H)-dione, dihydro-
$C_5H_6O_4$
3328 - Butanedioic acid, methylene-
3785 - 2-Butenedioic acid, 2-methyl-, (*E*)-
3786 - 2-Butenedioic acid, 2-methyl-, (*Z*)-
4970 - 1,2-Cyclopropanedicarboxylic acid, *trans*-(±)-
8875 - Pentanoic acid, 2,4-dioxo-
$C_5H_6O_5$
8781 - Pentanedioic acid, 2-oxo-
8782 - Pentanedioic acid, 3-oxo-
C_5H_6S
11769 - Thiophene, 2-methyl-
11770 - Thiophene, 3-methyl-
C_5H_7Br
9124 - 1-Pentyne, 1-bromo-
$C_5H_7BrN_2$
10816 - 1H-Pyrazole, 4-bromo-1,3-dimethyl-
$C_5H_7BrO_2$
10611 - 2-Propenoic acid, 2-bromoethyl ester
C_5H_7Cl
3153 - 1,3-Butadiene, 1-chloro-2-methyl-
3154 - 1,3-Butadiene, 1-chloro-3-methyl-
3155 - 1,3-Butadiene, 2-chloro-3-methyl-
3901 - 1-Butyne, 3-chloro-3-methyl-
4914 - Cyclopentene, 3-chloro-
8663 - 1,3-Pentadiene, 3-chloro-
$C_5H_7ClN_2$
10818 - 1H-Pyrazole, 3-chloro-1,5-dimethyl-
10819 - 1H-Pyrazole, 4-chloro-3,5-dimethyl-
10820 - 1H-Pyrazole, 5-chloro-1,3-dimethyl-

C_5H_7ClO
3888 - 2-Butenoyl chloride, 3-methyl-
4880 - Cyclopentanone, 2-chloro-
$C_5H_7ClO_2$
3830 - 2-Butenoic acid, 2-chloro-, methyl ester, (*E*)-
3831 - 2-Butenoic acid, 3-chloro-, methyl ester, (*E*)-
3832 - 2-Butenoic acid, 3-chloro-, methyl ester, (*Z*)-
3833 - 2-Butenoic acid, 4-chloro-, methyl ester
6141 - 2(3H)-Furanone, 5-(chloromethyl)dihydro-
10619 - 2-Propenoic acid, chloro-, ethyl ester
10620 - 2-Propenoic acid, 2-chloroethyl ester
$C_5H_7ClO_3$
10494 - Propanoyl chloride, 2-(acetyloxy)-, (±)-
$C_5H_7Cl_2NO$
43 - Acetamide, 2-chloro-*N*-3-chloroallyl-
$C_5H_7Cl_3O_2$
254 - Acetic acid, trichloro-, 1-methylethyl ester
261 - Acetic acid, trichloro-, propyl ester
$C_5H_7Cl_3O_3$
250 - Acetic acid, trichloro-, 2-methoxyethyl ester
$C_5H_7F_3O_2$
269 - Acetic acid, trifluoro-, propyl ester
C_5H_7I
9133 - 1-Pentyne, 1-iodo-
C_5H_7N
4173 - Cyclobutanecarbonitrile
9061 - 3-Pentenenitrile, (*E*)
9062 - 4-Pentenenitrile
11130 - 1H-Pyrrole, 1-methyl-
11131 - 1H-Pyrrole, 2-methyl-
11132 - 1H-Pyrrole, 3-methyl-
C_5H_7NO
6113 - 2-Furanmethanamine
7403 - Isoxazole, 3,5-dimethyl-
8577 - Oxazole, 2,4-dimethyl-
8578 - Oxazole, 2,5-dimethyl-
9063 - 3-Pentenenitrile, 2-hydroxy-, (*Z*)-
$C_5H_7NO_2$
152 - Acetic acid, cyano-, ethyl ester
7404 - 5-Isoxazolemethanol, 3-methyl-
10091 - Propanenitrile, 2-(acetyloxy)-
10234 - Propanoic acid, 3-cyano-, methyl ester
11160 - 2,5-Pyrrolidinedione, 1-methyl-
$C_5H_7NO_3$
8584 - 2,4-Oxazolidinedione, 5,5-dimethyl-
9768 - *L*-Proline, 5-oxo-
C_5H_7NS
11674 - Thiazole, 2,4-dimethyl-
11675 - Thiazole, 4,5-dimethyl-
$C_5H_7N_3$
10923 - 2,3-Pyridinediamine
10924 - 2,5-Pyridinediamine
10925 - 2,6-Pyridinediamine
11034 - 2(1H)-Pyridinone, hydrazone
11046 - 2-Pyrimidinamine, 4-methyl-
11047 - 2-Pyrimidinamine, 5-methyl-
11048 - 4-Pyrimidinamine, 6-methyl-
$C_5H_7N_3O$
11106 - 2(1H)-Pyrimidinone, 4-amino-5-methyl-
$C_5H_7N_3O_2$
7202 - 1H-Imidazole, 1,2-dimethyl-5-nitro-
11105 - 2(1H)-Pyrimidinone, 4-amino-5-(hydroxymethyl)-
C_5H_8
3170 - 1,2-Butadiene, 3-methyl-
3171 - 1,3-Butadiene, 2-methyl-
3916 - 1-Butyne, 3-methyl-
4193 - Cyclobutane, methylene-
4201 - Cyclobutene, 1-methyl-
4903 - Cyclopentene
8658 - 1,2-Pentadiene
8659 - 1,4-Pentadiene
8660 - 2,3-Pentadiene
8661 - 1,3-Pentadiene, (*E*)-
8662 - 1,3-Pentadiene, (*Z*)-
9122 - 1-Pentyne
9123 - 2-Pentyne
11527 - Spiropentane
C_5H_8BrN
8828 - Pentanenitrile, 5-bromo-
$C_5H_8Br_2O_2$
8870 - Pentanoic acid, 2,3-dibromo-
8871 - Pentanoic acid, 2,5-dibromo-
10240 - Propanoic acid, 2,3-dibromo-, ethyl ester
10242 - Propanoic acid, 2,3-dibromo-2-methyl-, methyl ester
$C_5H_8Br_4$
9884 - Propane, 1,3-dibromo-2,2-bis(bromomethyl)-
$C_5H_8ClFO_2$
135 - Acetic acid, chlorofluoro-, propyl ester
$C_5H_8Cl_2$
3729 - 1-Butene, 3-chloro-2-(chloromethyl)-
3755 - 1-Butene, 3,3-dichloro-2-methyl-
3756 - 2-Butene, 1,3-dichloro-2-methyl-
3757 - 2-Butene, 1,4-dichloro-2-methyl-
9028 - 2-Pentene, 2,5-dichloro-
$C_5H_8Cl_2O$
3686 - Butanoyl chloride, 2-chloro-3-methyl-
8963 - 2-Pentanone, 3,5-dichloro-
8991 - Pentanoyl chloride, 2-chloro-
8992 - Pentanoyl chloride, 5-chloro-
$C_5H_8Cl_2O_2$
167 - Acetic acid, dichloro-, 1-methylethyl ester
3502 - Butanoic acid, 3,4-dichloro-, methyl ester
10246 - Propanoic acid, 2,3-dichloro-, ethyl ester
10363 - 2-Propanol, 1,3-dichloro-, acetate
$C_5H_8Cl_2O_3$
5697 - Ethanol, 2-chloro-, carbonate (2:1)
$C_5H_8Cl_4$
8838 - Pentane, 1,1,1,5-tetrachloro-
9895 - Propane, 1,3-dichloro-2,2-bis(chloromethyl)-
$C_5H_8NNaO_4$
6248 - *L*-Glutamic acid, monosodium salt
$C_5H_8N_2$
7200 - 1H-Imidazole, 1,2-dimethyl-
7201 - 1H-Imidazole, 1,4-dimethyl-
7207 - 1H-Imidazole, 1-ethyl-
10825 - 1H-Pyrazole, 1,3-dimethyl-
10826 - 1H-Pyrazole, 1,5-dimethyl-
10827 - 1H-Pyrazole, 3,4-dimethyl-
10828 - 1H-Pyrazole, 3,5-dimethyl-
10832 - 1H-Pyrazole, 1-ethyl-
$C_5H_8N_2O$
7205 - 1H-Imidazole-1-ethanol
$C_5H_8N_2O_2$
7218 - 2,4-Imidazolidinedione, 5,5-dimethyl-
10236 - Propanoic acid, 2-diazo-, ethyl ester
$C_5H_8N_2O_3$
40 - Acetamide, *N,N'*-carbonylbis-
$C_5H_8N_4O_{12}$
9990 - 1,3-Propanediol, 2,2-bis[(nitrooxy)methyl]-, dinitrate (ester)
C_5H_8O
3167 - 1,2-Butadiene, 4-methoxy-
3168 - 1,3-Butadiene, 1-methoxy-
3169 - 1,3-Butadiene, 2-methoxy-
3698 - 2-Butenal, 2-methyl-
3699 - 2-Butenal, 2-methyl-, (*E*)-
3700 - 2-Butenal, 3-methyl-
3879 - 3-Buten-2-one, 3-methyl-
3915 - 2-Butyne, 1-methoxy-
3922 - 3-Butyn-2-ol, 2-methyl-
4878 - Cyclopentanone
5796 - Ethanone, 1-cyclopropyl-
8558 - 6-Oxabicyclo[3.1.0]hexane
8681 - 1,4-Pentadien-3-ol
9001 - 4-Pentenal
9104 - 1-Penten-3-one
9105 - 3-Penten-2-one, (*E*)-
9140 - 1-Pentyn-3-ol
9141 - 2-Pentyn-1-ol
9142 - 3-Pentyn-1-ol
9143 - 4-Pentyn-1-ol
10045 - Propane, 1-(ethynyloxy)-
10565 - 1-Propene, 3-(ethenyloxy)-
10732 - 1-Propyne, 1-ethoxy-
10733 - 1-Propyne, 3-ethoxy-
10777 - 2H-Pyran, 3,4-dihydro-
$C_5H_8O_2$
235 - Acetic acid, 2-propenyl ester
3844 - 2-Butenoic acid, 2-methyl-, (*E*)-
3845 - 2-Butenoic acid, 2-methyl-, (*Z*)-
3846 - 2-Butenoic acid, 3-methyl-
3847 - 2-Butenoic acid, methyl ester, (*E*)-
3848 - 2-Butenoic acid, methyl ester, (*Z*)-
4174 - Cyclobutanecarboxylic acid
4963 - Cyclopropanecarboxylic acid, methyl ester
4964 - Cyclopropanecarboxylic acid, 2-methyl-
6049 - 2-Furancarboxaldehyde, tetrahydro-
6145 - 2(3H)-Furanone, dihydro-3-methyl-
6146 - 2(3H)-Furanone, dihydro-5-methyl-, (±)-
8695 - Pentanal, 2-oxo-
8696 - Pentanal, 4-oxo-
8734 - Pentanedial
8797 - 2,3-Pentanedione
8798 - 2,4-Pentanedione
9073 - 4-Pentenoic acid
10261 - Propanoic acid, ethenyl ester
10635 - 2-Propenoic acid, ethyl ester
10663 - 2-Propenoic acid, 2-methyl-, methyl ester
10698 - 1-Propen-2-ol, acetate
10798 - 2H-Pyran-2-one, tetrahydro-
10799 - 4H-Pyran-4-one, tetrahydro-
$C_5H_8O_3$
3556 - Butanoic acid, 3-methyl-2-oxo-
3573 - Butanoic acid, 3-oxo-, methyl ester
6072 - 2-Furancarboxylic acid, tetrahydro-
8902 - Pentanoic acid, 2-oxo-
8903 - Pentanoic acid, 4-oxo-
10315 - Propanoic acid, 2-oxo-, ethyl ester
10412 - 2-Propanone, 1-(acetyloxy)-
$C_5H_8O_4$
3325 - Butanedioic acid, methyl-, (±)-
3336 - Butanedioic acid, monomethyl ester
5567 - Ethanedioic acid, ethyl methyl ester
8770 - Pentanedioic acid
9943 - Propanedioic acid, dimethyl ester
9948 - Propanedioic acid, ethyl-
10202 - Propanoic acid, 2-(acetyloxy)-, (±)-
C_5H_9Br
3716 - 1-Butene, 2-bromo-3-methyl-
3717 - 2-Butene, 1-bromo-3-methyl-
3718 - 2-Butene, 2-bromo-3-methyl-
4777 - Cyclopentane, bromo-
9007 - 1-Pentene, 1-bromo-
9008 - 1-Pentene, 2-bromo-
9009 - 1-Pentene, 3-bromo-
9010 - 1-Pentene, 5-bromo-
9011 - 2-Pentene, 1-bromo-
9012 - 2-Pentene, 2-bromo-
9013 - 2-Pentene, 3-bromo-
9014 - 2-Pentene, 4-bromo-
9015 - 2-Pentene, 5-bromo-
C_5H_9BrO
6036 - Furan, 2-(bromomethyl)tetrahydro-
$C_5H_9BrO_2$
120 - Acetic acid, bromo-, propyl ester
3471 - Butanoic acid, 2-bromo-3-methyl-, (±)-
3472 - Butanoic acid, 3-bromo-3-methyl-
3473 - Butanoic acid, 2-bromo-, methyl ester
3474 - Butanoic acid, 4-bromo-, methyl ester
8859 - Pentanoic acid, 2-bromo-, (±)-
10207 - Propanoic acid, 2-bromo-, ethyl ester
10208 - Propanoic acid, 3-bromo-, ethyl ester
C_5H_9Cl
3732 - 1-Butene, 1-chloro-2-methyl-
3733 - 1-Butene, 1-chloro-3-methyl-
3734 - 1-Butene, 3-chloro-2-methyl-
3735 - 2-Butene, 1-chloro-2-methyl-
3736 - 2-Butene, 1-chloro-3-methyl-
3737 - 2-Butene, 2-chloro-3-methyl-

4789 - Cyclopentane, chloro-
9016 - 1-Pentene, 2-chloro-
9017 - 1-Pentene, 3-chloro-
9018 - 1-Pentene, 4-chloro-
9019 - 1-Pentene, 5-chloro-
9020 - 2-Pentene, 1-chloro-
9021 - 2-Pentene, 2-chloro-
9022 - 2-Pentene, 3-chloro-
9023 - 2-Pentene, 4-chloro-
9024 - 2-Pentene, 5-chloro-

C_5H_9ClO
3603 - 1-Butanol, 3-chloro-
3649 - 2-Butanone, 3-chloro-3-methyl-
3693 - Butanoyl chloride, 3-methyl-
3694 - Butanoyl chloride, 2-methyl-, (±)-
8961 - 2-Pentanone, 5-chloro-
8962 - 3-Pentanone, 1-chloro-
8990 - Pentanoyl chloride
10503 - Propanoyl chloride, 2,2-dimethyl-

$C_5H_9ClO_2$
139 - Acetic acid, chloro-, 1-methylethyl ester
149 - Acetic acid, chloro-, propyl ester
3486 - Butanoic acid, 2-chloro-2-methyl-
3487 - Butanoic acid, 2-chloro-3-methyl-
3488 - Butanoic acid, 2-chloro-, methyl ester
3489 - Butanoic acid, 4-chloro-, methyl ester
4012 - Carbonochloridic acid, butyl ester
4024 - Carbonochloridic acid, 2-methylpropyl ester
8863 - Pentanoic acid, 2-chloro-, (±)-
8864 - Pentanoic acid, 3-chloro-
8865 - Pentanoic acid, 4-chloro-
8866 - Pentanoic acid, 5-chloro-
10219 - Propanoic acid, 3-chloro-2,2-dimethyl-
10220 - Propanoic acid, 2-chloro-, ethyl ester
10221 - Propanoic acid, 3-chloro-, ethyl ester
10229 - Propanoic acid, 2-chloro-2-methyl-, methyl ester
10350 - 1-Propanol, 3-chloro-, acetate
10351 - 2-Propanol, 1-chloro-, acetate

$C_5H_9ClO_3$
137 - Acetic acid, chloro-, 2-methoxyethyl ester
4017 - Carbonochloridic acid, 2-ethoxyethyl ester

$C_5H_9Cl_2N_3O_2$
11994 - Urea, *N,N'*-bis(2-chloroethyl)-*N*-nitroso-

$C_5H_9Cl_3$
3453 - Butane, 1,2,3-trichloro-2-methyl-
3454 - Butane, 2,2,3-trichloro-3-methyl-

$C_5H_9FO_2$
8879 - Pentanoic acid, 5-fluoro-

C_5H_9I
4824 - Cyclopentane, iodo-

$C_5H_9IO_2$
10283 - Propanoic acid, 3-iodo-, ethyl ester

C_5H_9N
3387 - Butane, 1-isocyano-
3409 - Butanenitrile, 2-methyl-
3410 - Butanenitrile, 3-methyl-
3896 - 3-Butyn-2-amine, 2-methyl-
8827 - Pentanenitrile
10100 - Propanenitrile, 2,2-dimethyl-
11020 - Pyridine, 1,2,3,6-tetrahydro-
11189 - 2-Pyrroline, 2-methyl-

C_5H_9NO
4892 - Cyclopentanone, oxime
8977 - 2-Pentanone, 4-imino-
9719 - 2-Piperidinone
10061 - Propane, 1-isocyanato-2-methyl-
10062 - Propane, 2-isocyanato-2-methyl-
10102 - Propanenitrile, 2-ethoxy-
10103 - Propanenitrile, 3-ethoxy-
11185 - 2-Pyrrolidinone, 1-methyl-
11186 - 2-Pyrrolidinone, 5-methyl-
11187 - 3-Pyrrolidinone, 1-methyl-

$C_5H_9NO_2$
25 - Acetamide, *N*-acetyl-*N*-methyl-
4839 - Cyclopentane, nitro-
7685 - 4-Morpholinecarboxaldehyde
8803 - 2,3-Pentanedione, 3-oxime
9764 - *L*-Proline
9765 - *DL*-Proline

$C_5H_9NO_2S_2$
10067 - Propane, 1-isothiocyanato-3-(methylsulfonyl)-

$C_5H_9NO_3$
9767 - *L*-Proline, 4-hydroxy-, *trans*-

$C_5H_9NO_3S$
5015 - *L*-Cysteine, *N*-acetyl-

$C_5H_9NO_4$
6240 - *DL*-Glutamic acid
6241 - *L*-Glutamic acid
10312 - Propanoic acid, 2-nitro-, ethyl ester

$C_5H_9NO_4S$
5017 - *L*-Cysteine, *S*-(carboxymethyl)-

$C_5H_9NO_5$
6246 - *DL*-Glutamic acid, 3-hydroxy-

C_5H_9NS
3388 - Butane, 1-isothiocyanato-
3389 - Butane, 2-isothiocyanato-, (±)-
10065 - Propane, 1-isothiocyanato-2-methyl-
10066 - Propane, 2-isothiocyanato-2-methyl-
11699 - Thiocyanic acid, butyl ester
11701 - Thiocyanic acid, 1,1-dimethylethyl ester
11707 - Thiocyanic acid, 2-methylpropyl ester

$C_5H_9N_3$
7203 - 1H-Imidazole-4-ethanamine

C_5H_{10}
3799 - 1-Butene, 2-methyl-
3800 - 1-Butene, 3-methyl-
3801 - 2-Butene, 2-methyl-
4192 - Cyclobutane, methyl-
4774 - Cyclopentane
4971 - Cyclopropane, 1,1-dimethyl-
4972 - Cyclopropane, 1,2-dimethyl-, *cis*-
4973 - Cyclopropane, 1,2-dimethyl-, *trans*-
4974 - Cyclopropane, ethyl-
9004 - 1-Pentene
9005 - 2-Pentene, (*E*)-
9006 - 2-Pentene, (*Z*)-

$C_5H_{10}BrF$
8713 - Pentane, 1-bromo-5-fluoro-

$C_5H_{10}Br_2$
3272 - Butane, 1,2-dibromo-2-methyl-
3273 - Butane, 1,4-dibromo-2-methyl-, (*R*)-
3274 - Butane, 2,3-dibromo-2-methyl-
8739 - Pentane, 1,3-dibromo-
8740 - Pentane, 1,4-dibromo-
8741 - Pentane, 1,5-dibromo-
8742 - Pentane, 2,3-dibromo-
8743 - Pentane, 2,4-dibromo-
9886 - Propane, 1,1-dibromo-2,2-dimethyl-
9887 - Propane, 1,3-dibromo-2,2-dimethyl-

$C_5H_{10}ClF$
8728 - Pentane, 1-chloro-5-fluoro-

$C_5H_{10}ClNO$
3972 - Carbamic chloride, diethyl-

$C_5H_{10}Cl_2$
3282 - Butane, 1,1-dichloro-3-methyl-
3283 - Butane, 1,2-dichloro-2-methyl-
3284 - Butane, 1,3-dichloro-3-methyl-
3285 - Butane, 1,4-dichloro-2-methyl-
3286 - Butane, 2,3-dichloro-2-methyl-
8744 - Pentane, 1,2-dichloro-
8745 - Pentane, 1,3-dichloro-
8746 - Pentane, 1,4-dichloro-
8747 - Pentane, 1,5-dichloro-
8748 - Pentane, 2,2-dichloro-
8749 - Pentane, 2,3-dichloro-
8750 - Pentane, 2,4-dichloro-
8751 - Pentane, 3,3-dichloro-

$C_5H_{10}I_2$
8758 - Pentane, 1,5-diiodo-
8759 - Pentane, 2,4-diiodo-

$C_5H_{10}N_2$
4156 - Cyanamide, diethyl-
10101 - Propanenitrile, 3-(dimethylamino)-

$C_5H_{10}N_2O$
9688 - Piperidine, 1-nitroso-

$C_5H_{10}N_2O_2$
8800 - 2,3-Pentanedione, dioxime
9687 - Piperidine, 1-nitro-

$C_5H_{10}N_2O_2S$
7650 - Methomyl

$C_5H_{10}N_2O_3$
6249 - *L*-Glutamine

$C_5H_{10}N_2O_4$
3950 - Carbamic acid, ethylnitro-, ethyl ester
8769 - Pentane, 1,5-dinitro-

$C_5H_{10}O$
3185 - Butanal, 2-methyl-, (±)-
3186 - Butanal, 3-methyl-
3668 - 2-Butanone, 3-methyl-
3869 - 2-Buten-1-ol, 3-methyl-
3870 - 3-Buten-2-ol, 2-methyl-
3871 - 3-Buten-2-ol, 3-methyl-
4864 - Cyclopentanol
4979 - Cyclopropanemethanol, α-methyl-
6171 - Furan, tetrahydro-2-methyl-
6172 - Furan, tetrahydro-3-methyl-
8593 - Oxetane, 3,3-dimethyl-
8690 - Pentanal
8957 - 2-Pentanone
8958 - 3-Pentanone
9087 - 1-Penten-3-ol
9088 - 4-Penten-1-ol
9089 - 4-Penten-2-ol
9090 - 2-Penten-1-ol, (*E*)-
9091 - 2-Penten-1-ol, (*Z*)-
9092 - 3-Penten-2-ol, (±)-
9783 - Propanal, 2,2-dimethyl-
10033 - Propane, 1-(ethenyloxy)-
10034 - Propane, 2-(ethenyloxy)-
10566 - 1-Propene, 3-ethoxy-
10577 - 1-Propene, 3-methoxy-2-methyl-
10800 - 2H-Pyran, tetrahydro-

$C_5H_{10}OS_2$
4031 - Carbonodithioic acid, *S,S*-diethyl ester

$C_5H_{10}O_2$
209 - Acetic acid, 1-methylethyl ester
236 - Acetic acid, propyl ester
3538 - Butanoic acid, 2-methyl-, (±)-
3539 - Butanoic acid, 3-methyl-
3547 - Butanoic acid, methyl ester
3664 - 2-Butanone, 4-hydroxy-3-methyl-
4806 - 1,3-Cyclopentanediol, *cis*-
4807 - 1,3-Cyclopentanediol, *trans*-
5193 - 1,3-Dioxane, 4-methyl-
5763 - Ethanol, 2-(2-propenyloxy)-
6004 - Formic acid, butyl ester
6012 - Formic acid, 1-methylpropyl ester
6013 - Formic acid, 2-methylpropyl ester
6124 - 2-Furanmethanol, tetrahydro-
8614 - Oxirane, (ethoxymethyl)-
8855 - Pentanoic acid
8971 - 2-Pentanone, 1-hydroxy-
8972 - 2-Pentanone, 3-hydroxy-
8973 - 2-Pentanone, 4-hydroxy-
8974 - 2-Pentanone, 5-hydroxy-
8975 - 3-Pentanone, 2-hydroxy-
10258 - Propanoic acid, 2,2-dimethyl-
10263 - Propanoic acid, ethyl ester
10301 - Propanoic acid, 2-methyl-, methyl ester
10562 - 1-Propene, 3,3-dimethoxy-
10787 - 2H-Pyran-4-ol, tetrahydro-

$C_5H_{10}O_2S$
10311 - Propanoic acid, 3-(methylthio)-, methyl ester
11789 - Thiophene, tetrahydro-3-methyl-, 1,1-dioxide

$C_5H_{10}O_3$
182 - Acetic acid, ethoxy-, methyl ester
199 - Acetic acid, hydroxy-, propyl ester
206 - Acetic acid, methoxy-, ethyl ester
3535 - Butanoic acid, 3-hydroxy-3-methyl-

3999 - Carbonic acid, diethyl ester
5197 - 1,3-Dioxan-5-ol, 2-methyl-
5208 - 1,3-Dioxolane-4-methanol, 2-methyl-
5734 - Ethanol, methoxy-, acetate
5735 - Ethanol, 2-methoxy-, acetate
8883 - Pentanoic acid, 2-hydroxy-
10269 - Propanoic acid, 2-hydroxy-, ethyl ester, (±)-
10270 - Propanoic acid, 3-hydroxy-, ethyl ester
10280 - Propanoic acid, 2-hydroxy-2-methyl-, methyl ester
10289 - Propanoic acid, 3-methoxy-, methyl ester

$C_5H_{10}O_4$
10187 - 1,2,3-Propanetriol, 1-acetate, (±)-
10253 - Propanoic acid, 2,3-dihydroxy-, ethyl ester, (±)-
10256 - Propanoic acid, 2,3-dihydroxy-2-methyl-, methyl ester
10272 - Propanoic acid, 3-hydroxy-2-(hydroxymethyl)-2-methyl-

$C_5H_{10}O_5$
436 - β-*D*-Arabinopyranose
11345 - *D*-Ribose
12059 - *D*-Xylose

$C_5H_{10}S$
4852 - Cyclopentanethiol
11787 - Thiophene, tetrahydro-2-methyl-
11788 - Thiophene, tetrahydro-3-methyl-
11795 - 2H-Thiopyran, tetrahydro-

$C_5H_{11}Br$
3238 - Butane, 1-bromo-2-methyl-, (*S*)-
3239 - Butane, 1-bromo-3-methyl-
3240 - Butane, 2-bromo-2-methyl-
8710 - Pentane, 1-bromo-
8711 - Pentane, 2-bromo-
8712 - Pentane, 3-bromo-
9843 - Propane, 1-bromo-2,2-dimethyl-

$C_5H_{11}Cl$
3251 - Butane, 1-chloro-2-methyl-, (±)-
3252 - Butane, 1-chloro-3-methyl-
3253 - Butane, 2-chloro-2-methyl-
8719 - Pentane, 1-chloro-
8720 - Pentane, 2-chloro-, (+)-
8721 - Pentane, 3-chloro-
9853 - Propane, 1-chloro-2,2-dimethyl-

$C_5H_{11}ClO$
3610 - 2-Butanol, 1-chloro-2-methyl-
3611 - 2-Butanol, 3-chloro-2-methyl-
8923 - 1-Pentanol, 5-chloro-
8924 - 2-Pentanol, 1-chloro-
8925 - 3-Pentanol, 1-chloro-
9854 - Propane, 1-(1-chloroethoxy)-

$C_5H_{11}ClO_2$
9852 - Propane, 3-chloro-1,1-dimethoxy-

$C_5H_{11}ClSi$
11375 - Silane, chlorodimethyl-2-propenyl-

$C_5H_{11}Cl_2N$
5413 - Ethanamine, 2-chloro-*N*-(2-chloroethyl)-*N*-methyl-

$C_5H_{11}Cl_3Si$
11472 - Silane, trichloropentyl-

$C_5H_{11}F$
8817 - Pentane, 1-fluoro-

$C_5H_{11}FO$
8935 - 1-Pentanol, 5-fluoro-

$C_5H_{11}I$
3383 - Butane, 1-iodo-2-methyl-, (*S*)-
3384 - Butane, 1-iodo-3-methyl-
3385 - Butane, 2-iodo-2-methyl-
8818 - Pentane, 1-iodo-
8819 - Pentane, 2-iodo-, (±)-
8820 - Pentane, 3-iodo-
10056 - Propane, 1-iodo-2,2-dimethyl-

$C_5H_{11}N$
4771 - Cyclopentanamine
9629 - Piperidine
11168 - Pyrrolidine, 1-methyl-

$C_5H_{11}NO$
3187 - Butanal, 3-methyl-, oxime
3206 - Butanamide, 3-methyl-
5991 - Formamide, *N,N*-diethyl-
6116 - 2-Furanmethanamine, tetrahydro-
7696 - Morpholine, 4-methyl-
8698 - Pentanamide
8985 - 2-Pentanone, oxime

$C_5H_{11}NO_2$
3400 - Butane, 3-methyl-1-nitro-
3935 - Carbamic acid, butyl ester
3941 - Carbamic acid, diethyl-
3949 - Carbamic acid, ethyl-, ethyl ester
3959 - Carbamic acid, 2-methylpropyl ester
7400 - *DL*-Isovaline
7401 - *L*-Isovaline
7454 - Methanaminium, 1-carboxy-*N,N,N*-trimethyl-, hydroxide, inner salt
8090 - Nitrous acid, 3-methylbutyl ester
8096 - Nitrous acid, pentyl ester
8217 - *DL*-Norvaline
8218 - *L*-Norvaline
8832 - Pentane, 1-nitro-
8833 - Pentane, 2-nitro-
8834 - Pentane, 3-nitro-
8857 - Pentanoic acid, 5-amino-
12040 - *DL*-Valine
12041 - *L*-Valine

$C_5H_{11}NO_2S$
7647 - *DL*-Methionine
7648 - *L*-Methionine
12043 - *D*-Valine, 3-mercapto-

$C_5H_{11}NO_3$
3637 - 1-Butanol, 3-methyl-, nitrate
3638 - 1-Butanol, 3-methyl-2-nitro-
8950 - 1-Pentanol, 2-nitro-
8951 - 2-Pentanol, 1-nitro-
8952 - 2-Pentanol, 3-nitro-

$C_5H_{11}NO_4$
10008 - 1,3-Propanediol, 2-ethyl-2-nitro-

C_5H_{12}
3397 - Butane, 2-methyl-
8709 - Pentane
9910 - Propane, 2,2-dimethyl-

$C_5H_{12}NO_3PS_2$
5009 - Cygon

$C_5H_{12}N_2$
9618 - Piperazine, 1-methyl-
9619 - Piperazine, 2-methyl-
9624 - 1-Piperidinamine

$C_5H_{12}N_2O$
12001 - Urea, *N,N'*-diethyl-
12002 - Urea, *N,N*-diethyl-
12030 - Urea, tetramethyl-

$C_5H_{12}N_2O_2$
8550 - *L*-Ornithine

$C_5H_{12}N_2S$
11803 - Thiourea, *N,N'*-diethyl-
11810 - Thiourea, tetramethyl-

$C_5H_{12}N_4O_3$
7135 - *L*-Homoserine, *O*-[(aminoiminomethyl)amino]-

$C_5H_{12}O$
3391 - Butane, 1-methoxy-
3392 - Butane, 2-methoxy-, (*S*)-
3393 - Butane, 2-methoxy-, (±)-
3626 - 1-Butanol, 2-methyl-
3627 - 1-Butanol, 3-methyl-
3628 - 2-Butanol, 2-methyl-
3629 - 2-Butanol, 3-methyl-, (±)-
8915 - 1-Pentanol
8916 - 2-Pentanol
8917 - 3-Pentanol
10037 - Propane, 1-ethoxy-
10038 - Propane, 2-ethoxy-
10071 - Propane, 1-methoxy-2-methyl-
10072 - Propane, 2-methoxy-2-methyl-
10373 - 1-Propanol, 2,2-dimethyl-

$C_5H_{12}O_2$
3352 - 1,2-Butanediol, 3-methyl-
3353 - 1,3-Butanediol, 2-methyl-
3354 - 1,3-Butanediol, 3-methyl-
3355 - 2,3-Butanediol, 2-methyl-
3625 - 1-Butanol, 3-methoxy-
5591 - Ethane, 1-ethoxy-2-methoxy-
5603 - Ethane, 1,1'-[methylenebis(oxy)]bis-
5740 - Ethanol, 2-(1-methylethoxy)-
5764 - Ethanol, 2-propoxy-
8785 - 1,3-Pentanediol
8786 - 1,4-Pentanediol
8787 - 1,5-Pentanediol
8788 - 2,3-Pentanediol
8789 - 2,4-Pentanediol
8790 - 1,2-Pentanediol, (±)-
9909 - Propane, 2,2-dimethoxy-
10005 - 1,3-Propanediol, 2,2-dimethyl-
10380 - 1-Propanol, 2-ethoxy-
10381 - 2-Propanol, 1-ethoxy-

$C_5H_{12}O_3$
5669 - Ethane, 1,1,1-trimethoxy-
5737 - Ethanol, 2-(2-methoxyethoxy)-
8851 - 1,2,5-Pentanetriol
8852 - 1,3,5-Pentanetriol
9119 - Pentitol, 1,2-dideoxy-
10011 - 1,3-Propanediol, 2-(hydroxymethyl)-2-methyl-
10372 - 2-Propanol, 1,3-dimethoxy-

$C_5H_{12}O_4$
7543 - Methane, tetramethoxy-
9989 - 1,3-Propanediol, 2,2-bis(hydroxymethyl)-

$C_5H_{12}O_5$
11343 - Ribitol
12058 - Xylitol

$C_5H_{12}S$
3402 - Butane, 1-(methylthio)-
3439 - 1-Butanethiol, 2-methyl-, (*S*)-
3440 - 1-Butanethiol, 3-methyl-
3441 - 2-Butanethiol, 2-methyl-
8844 - 1-Pentanethiol
8845 - 2-Pentanethiol
8846 - 3-Pentanethiol
10042 - Propane, 1-(ethylthio)-
10043 - Propane, 2-(ethylthio)-
10082 - Propane, 2-methyl-1-(methylthio)-
10083 - Propane, 2-methyl-2-(methylthio)-

$C_5H_{12}Si$
11425 - Silane, ethenyltrimethyl-

$C_5H_{13}Cl_2N$
5442 - Ethanaminium, 2-chloro-*N,N,N*-trimethyl-, chloride

$C_5H_{13}N$
3220 - 1-Butanamine, 3-methyl-
3221 - 1-Butanamine, *N*-methyl-
3222 - 2-Butanamine, 2-methyl-
3223 - 2-Butanamine, 3-methyl-
5430 - Ethanamine, *N*-ethyl-*N*-methyl-
8700 - 1-Pentanamine
8701 - 2-Pentanamine
8702 - 3-Pentanamine
9806 - 1-Propanamine, 2,2-dimethyl-
9807 - 1-Propanamine, *N,N*-dimethyl-
9809 - 1-Propanamine, *N*-ethyl-
9810 - 2-Propanamine, *N*-ethyl-

$C_5H_{13}NO$
3599 - 1-Butanol, 2-amino-3-methyl-, (±)-
5426 - Ethanamine, 2-ethoxy-*N*-methyl-
5741 - Ethanol, 2-[(1-methylethyl)amino]-
5765 - Ethanol, 2-(propylamino)-
5878 - Ethenaminium, *N,N,N*-trimethyl-, hydroxide
8920 - 1-Pentanol, 5-amino-
8921 - 3-Pentanol, 2-amino-
10374 - 1-Propanol, 2-(dimethylamino)-
10375 - 1-Propanol, 3-(dimethylamino)-
10376 - 2-Propanol, 1-(dimethylamino)-

$C_5H_{13}NO_2$
5420 - Ethanamine, 2,2-dimethoxy-*N*-methyl-
5742 - Ethanol, 2,2′-(methylimino)bis-
9985 - 1,3-Propanediol, 2-amino-2-ethyl-
10383 - 2-Propanol, 1-[(2-hydroxyethyl)amino]-
$C_5H_{13}O_3P$
9539 - Phosphonic acid, methyl-, diethyl ester
$C_5H_{14}ClNO$
5446 - Ethanaminium, 2-hydroxy-*N,N,N*-trimethyl-, chloride
$C_5H_{14}N_2$
7490 - Methanediamine, *N,N,N′,N′*-tetramethyl-
8735 - 1,5-Pentanediamine
$C_5H_{14}N_2O$
10340 - 2-Propanol, 1-[(2-aminoethyl)amino]-
$C_5H_{14}OSi$
11430 - Silane, ethoxytrimethyl-
$C_5H_{14}O_3Si$
11433 - Silane, ethyltrimethoxy-
$C_5H_{15}NSi$
11361 - Silanamine, pentamethyl-
$C_5H_{20}O_5Si_5$
4902 - Cyclopentasiloxane, 2,4,6,8,10-pentamethyl-
C_6BrF_5
1078 - Benzene, bromopentafluoro-
$C_6Br_2F_4$
1312 - Benzene, 1,2-dibromo-3,4,5,6-tetrafluoro-
C_6ClF_5
1220 - Benzene, chloropentafluoro-
$C_6Cl_2F_4$
1398 - Benzene, 1,2-dichloro-3,4,5,6-tetrafluoro
1399 - Benzene, 1,3-dichloro-2,4,5,6-tetrafluoro-
1400 - Benzene, 1,4-dichloro-2,3,5,6-tetrafluoro-
$C_6Cl_2F_8$
4647 - Cyclohexene, 1,2-dichloro-3,3,4,4,5,5,6,6-octafluoro-
$C_6Cl_3F_3$
2340 - Benzene, 1,3,5-trichloro-2,4,6-trifluoro-
$C_6Cl_4O_2$
4268 - 2,5-Cyclohexadiene-1,4-dione, 2,3,5,6-tetrachloro-
$C_6Cl_5NO_2$
2087 - Benzene, pentachloronitro-
C_6Cl_6
1749 - Benzene, hexachloro-
C_6Cl_{10}
6689 - 1,5-Hexadiene, 1,1,2,3,3,4,4,5,6,6-decachloro-
C_6F_5I
2092 - Benzene, pentafluoroiodo-
C_6F_6
1757 - Benzene, hexafluoro-
C_6F_8
4276 - 1,3-Cyclohexadiene, 1,2,3,4,5,5,6,6-octafluoro-
4277 - 1,4-Cyclohexadiene, 1,2,3,3,4,5,6,6-octafluoro-
C_6F_{10}
4643 - Cyclohexene, decafluoro-
C_6F_{12}
4381 - Cyclohexane, dodecafluoro-
7032 - 1-Hexene, 1,1,2,3,3,4,4,5,5,6,6,6-dodecafluoro-
C_6F_{14}
3414 - Butane, 1,1,1,2,3,4,4,4-octafluoro-2,3-bis(trifluoromethyl)-
6843 - Hexane, tetradecafluoro-
8853 - Pentane, 1,1,1,2,2,3,4,4,5,5,5-undecafluoro-3-(trifluoromethyl)
8854 - Pentane, 1,1,1,2,2,3,3,4,5,5,5-undecafluoro-4-(trifluoromethyl)-
$C_6F_{15}N$
5436 - Ethanamine, 1,1,2,2,2-pentafluoro-*N,N*-bis(pentafluoroethyl)-
C_6HBr_5
2082 - Benzene, pentabromo-
C_6HBr_5O
9439 - Phenol, pentabromo-
$C_6HCl_3F_8O_2$
6910 - Hexanoic acid, 3,5,6-trichloro-2,2,3,4,4,5,6,6-octafluoro-
$C_6HCl_4NO_2$
2264 - Benzene, 1,2,4,5-tetrachloro-3-nitro-
C_6HCl_5
2084 - Benzene, pentachloro-
C_6HCl_5O
9440 - Phenol, pentachloro-
C_6HF_5
2091 - Benzene, pentafluoro-
C_6HF_5O
9442 - Phenol, pentafluoro-
C_6HF_5S
2304 - Benzenethiol, pentafluoro-
C_6HF_{11}
4498 - Cyclohexane, undecafluoro-
C_6HI_5
2096 - Benzene, pentaiodo-
$C_6H_2Br_2ClNO$
4279 - 2,5-Cyclohexadien-1-one, 2,6-dibromo-4-(chloroimino)-
$C_6H_2Br_3NO_2$
2315 - Benzene, 1,2,3-tribromo-5-nitro-
2316 - Benzene, 1,3,5-tribromo-2-nitro-
$C_6H_2Br_4$
2253 - Benzene, 1,2,3,5-tetrabromo-
2254 - Benzene, 1,2,4,5-tetrabromo-
$C_6H_2Br_4O$
9467 - Phenol, 2,3,4,6-tetrabromo-
$C_6H_2Br_4O_2$
1551 - 1,4-Benzenediol, 2,3,5,6-tetrabromo-
$C_6H_2ClN_3O_6$
1244 - Benzene, 2-chloro-1,3,5-trinitro-
$C_6H_2Cl_2O_4$
4258 - 2,5-Cyclohexadiene-1,4-dione, 2,5-dichloro-3,6-dihydroxy-
$C_6H_2Cl_3NO$
4280 - 2,5-Cyclohexadien-1-one, 2,6-dichloro-4-(chloroimino)-
$C_6H_2Cl_3NO_2$
2338 - Benzene, 1,2,3-trichloro-5-nitro-
2339 - Benzene, 1,2,4-trichloro-5-nitro-
$C_6H_2Cl_3NO_3$
9490 - Phenol, 3,4,6-trichloro-2-nitro-
$C_6H_2Cl_4$
2259 - Benzene, 1,2,3,4-tetrachloro-
2260 - Benzene, 1,2,4,5-tetrachloro-
$C_6H_2Cl_4O$
9469 - Phenol, 2,3,4,5-tetrachloro-
9470 - Phenol, 2,3,4,6-tetrachloro-
9471 - Phenol, 2,3,5,6-tetrachloro-
$C_6H_2Cl_5OPS$
9578 - Phosphorodichloridothioic acid, *O*-(2,4,5-trichlorophenyl) ester
$C_6H_2F_4$
2270 - Benzene, 1,2,3,4-tetrafluoro-
2271 - Benzene, 1,2,3,5-tetrafluoro-
2272 - Benzene, 1,2,4,5-tetrafluoro-
$C_6H_2I_3NO_2$
2353 - Benzene, 1,2,3-triiodo-5-nitro-
$C_6H_2I_4$
2273 - Benzene, 1,2,3,4-tetraiodo-
2274 - Benzene, 1,2,3,5-tetraiodo-
2275 - Benzene, 1,2,4,5-tetraiodo-
$C_6H_2N_2O_8$
4259 - 2,5-Cyclohexadiene-1,4-dione, 2,5-dihydroxy-3,6-dinitro-
$C_6H_2N_4O_5$
2874 - 1,2,3-Benzoxadiazole, 5,7-dinitro-
$C_6H_2N_4O_9$
9476 - Phenol, 2,3,4,6-tetranitro-
$C_6H_2O_6$
4686 - 5-Cyclohexene-1,2,3,4-tetrone, 5,6-dihydroxy-
$C_6H_3BrCl_2$
1010 - Benzene, 1-bromo-2,3-dichloro-
1011 - Benzene, 1-bromo-2,4-dichloro-
1012 - Benzene, 1-bromo-3,5-dichloro-
1013 - Benzene, 2-bromo-1,3-dichloro-
1014 - Benzene, 2-bromo-1,4-dichloro-
1015 - Benzene, 4-bromo-1,2-dichloro-
$C_6H_3BrCl_2O$
9233 - Phenol, 2-bromo-4,6-dichloro-
$C_6H_3BrN_2O_4$
1024 - Benzene, 1-bromo-2,3-dinitro-
$C_6H_3BrN_2O_5$
9234 - Phenol, 2-bromo-4,6-dinitro-
9235 - Phenol, 4-bromo-2,6-dinitro-
$C_6H_3Br_2Cl$
1292 - Benzene, 1,2-dibromo-3-chloro-
1293 - Benzene, 1,2-dibromo-4-chloro-
1294 - Benzene, 1,3-dibromo-2-chloro-
1295 - Benzene, 1,4-dibromo-2-chloro-
1296 - Benzene, 2,4-dibromo-1-chloro-
$C_6H_3Br_2F$
1300 - Benzene, 2,4-dibromo-1-fluoro-
$C_6H_3Br_2NO_2$
1307 - Benzene, 1,2-dibromo-4-nitro-
1308 - Benzene, 1,3-dibromo-2-nitro-
1309 - Benzene, 1,3-dibromo-5-nitro-
1310 - Benzene, 1,4-dibromo-2-nitro-
1311 - Benzene, 2,4-dibromo-1-nitro-
$C_6H_3Br_3$
2307 - Benzene, 1,2,3-tribromo-
2308 - Benzene, 1,2,4-tribromo-
2309 - Benzene, 1,3,5-tribromo-
$C_6H_3Br_3O$
9479 - Phenol, 2,4,6-tribromo-
$C_6H_3ClNO_3$
9296 - Phenol, 2,6-dichloro-4-nitro-
$C_6H_3ClN_2O_4$
1144 - Benzene, 1-chloro-2,4-dinitro-
1145 - Benzene, 2-chloro-1,3-dinitro-
$C_6H_3ClN_2O_4S$
2178 - Benzenesulfenyl chloride, 2,4-dinitro-
$C_6H_3ClN_2O_5$
9258 - Phenol, 4-chloro-2,6-dinitro-
9259 - Phenol, 5-chloro-2,4-dinitro-
$C_6H_3Cl_2I$
1380 - Benzene, 1,4-dichloro-2-iodo-
$C_6H_3Cl_2NO_2$
1391 - Benzene, 1,2-dichloro-3-nitro-
1392 - Benzene, 1,2-dichloro-4-nitro-
1393 - Benzene, 1,3-dichloro-2-nitro-
1394 - Benzene, 1,3-dichloro-5-nitro-
1395 - Benzene, 1,4-dichloro-2-nitro-
1396 - Benzene, 2,4-dichloro-1-nitro-
4142 - Clopyralid
$C_6H_3Cl_3$
2321 - Benzene, 1,2,3-trichloro-
2322 - Benzene, 1,2,4-trichloro-
2323 - Benzene, 1,3,5-trichloro-
$C_6H_3Cl_3N_2O_2$
10896 - 2-Pyridinecarboxylic acid, 4-amino-3,5,6-trichloro-
$C_6H_3Cl_3O$
9481 - Phenol, 2,3,4-trichloro-
9482 - Phenol, 2,3,5-trichloro-
9483 - Phenol, 2,3,6-trichloro-
9484 - Phenol, 2,4,5-trichloro-
9485 - Phenol, 2,4,6-trichloro-
9486 - Phenol, 3,4,5-trichloro-
$C_6H_3Cl_3O_3$
2371 - 1,3,5-Benzenetriol, 2,4,6-trichloro-
$C_6H_3Cl_4N$
8071 - Nitrapyrin
$C_6H_3FN_2O_4$
1718 - Benzene, 1-fluoro-2,4-dinitro-
$C_6H_3F_2NO_2$
1426 - Benzene, 2,4-difluoro-1-nitro-

$C_6H_3F_3$
2346 - Benzene, 1,2,4-trifluoro-
2347 - Benzene, 1,3,5-trifluoro-
$C_6H_3F_7O_2$
3523 - Butanoic acid, heptafluoro-, ethenyl ester
$C_6H_3I_2NO_3$
9304 - Phenol, 2,6-diiodo-4-nitro-
$C_6H_3I_3$
2349 - Benzene, 1,2,3-triiodo-
2350 - Benzene, 1,2,4-triiodo-
2351 - Benzene, 1,3,5-triiodo-
$C_6H_3I_3O$
9492 - Phenol, 2,4,6-triiodo-
$C_6H_3N_3O_6$
2365 - Benzene, 1,3,5-trinitro-
$C_6H_3N_3O_7$
9497 - Phenol, 2,4,6-trinitro-
$C_6H_3N_3O_8$
1554 - 1,3-Benzenediol, 2,4,6-trinitro-
C_6H_4BrCl
1004 - Benzene, 1-bromo-2-chloro-
1005 - Benzene, 1-bromo-3-chloro-
1006 - Benzene, 1-bromo-4-chloro-
C_6H_4BrClO
9231 - Phenol, 3-bromo-5-chloro-
9232 - Phenol, 4-bromo-2-chloro-
$C_6H_4BrClO_2S$
2244 - Benzenesulfonyl chloride, 4-bromo-
$C_6H_4BrCl_2N$
641 - Benzenamine, 2-bromo-4,6-dichloro-
C_6H_4BrF
1043 - Benzene, 1-bromo-3-fluoro-
1044 - Benzene, 1-bromo-4-fluoro-
C_6H_4BrI
1046 - Benzene, 1-bromo-2-iodo-
1047 - Benzene, 1-bromo-3-iodo-
1048 - Benzene, 1-bromo-4-iodo-
$C_6H_4BrNO_2$
1075 - Benzene, 1-bromo-2-nitro-
1076 - Benzene, 1-bromo-3-nitro-
1077 - Benzene, 1-bromo-4-nitro-
$C_6H_4BrNO_3$
9243 - Phenol, 2-bromo-3-nitro-
9244 - Phenol, 4-bromo-2-nitro-
$C_6H_4BrN_3$
977 - Benzene, 1-azido-4-bromo-
$C_6H_4BrN_3O_4$
646 - Benzenamine, 2-bromo-4,6-dinitro-
$C_6H_4Br_2$
1288 - Benzene, 1,2-dibromo-
1289 - Benzene, 1,3-dibromo-
1290 - Benzene, 1,4-dibromo-
$C_6H_4Br_2O$
9276 - Phenol, 2,4-dibromo-
9277 - Phenol, 2,6-dibromo-
9278 - Phenol, 3,5-dibromo-
$C_6H_4Br_3N$
851 - Benzenamine, 2,4,6-tribromo-
C_6H_4ClF
1166 - Benzene, 1-chloro-2-fluoro-
1167 - Benzene, 1-chloro-3-fluoro-
1168 - Benzene, 1-chloro-4-fluoro-
$C_6H_4ClFO_2S$
2246 - Benzenesulfonyl chloride, 4-fluoro-
C_6H_4ClI
1171 - Benzene, 1-chloro-2-iodo-
1172 - Benzene, 1-chloro-4-iodo-
$C_6H_4ClIO_2S$
2247 - Benzenesulfonyl chloride, 4-iodo-
$C_6H_4ClNO_2$
1214 - Benzene, 1-chloro-2-nitro-
1215 - Benzene, 1-chloro-3-nitro-
1216 - Benzene, 1-chloro-4-nitro-
$C_6H_4ClNO_2S$
2179 - Benzenesulfenyl chloride, 4-nitro-
$C_6H_4ClNO_3$
9269 - Phenol, 5-chloro-2-nitro-
$C_6H_4ClN_3$
978 - Benzene, 1-azido-4-chloro-
$C_6H_4ClO_2P$
2413 - 1,3,2-Benzodioxaphosphole, 2-chloro-
$C_6H_4Cl_2$
1354 - Benzene, 1,2-dichloro-
1355 - Benzene, 1,3-dichloro-
1356 - Benzene, 1,4-dichloro-
$C_6H_4Cl_2N_2O_2$
692 - Benzenamine, 2,6-dichloro-4-nitro-
$C_6H_4Cl_2O$
9282 - Phenol, 2,3-dichloro-
9283 - Phenol, 2,4-dichloro-
9284 - Phenol, 2,5-dichloro-
9285 - Phenol, 2,6-dichloro-
9286 - Phenol, 3,4-dichloro-
9287 - Phenol, 3,5-dichloro-
$C_6H_4Cl_2O_2$
1510 - 1,3-Benzenediol, 4,6-dichloro-
$C_6H_4Cl_2O_2S$
2245 - Benzenesulfonyl chloride, 4-chloro-
$C_6H_4Cl_3N$
852 - Benzenamine, 2,3,4-trichloro-
853 - Benzenamine, 2,4,5-trichloro-
854 - Benzenamine, 2,4,6-trichloro-
11023 - Pyridine, 2-(trichloromethyl)-
$C_6H_4Cl_4Si$
11457 - Silane, trichloro(4-chlorophenyl)-
C_6H_4FI
1719 - Benzene, 1-fluoro-2-iodo-
1720 - Benzene, 1-fluoro-4-iodo-
$C_6H_4FNO_2$
1733 - Benzene, 1-fluoro-2-nitro-
1734 - Benzene, 1-fluoro-3-nitro-
1735 - Benzene, 1-fluoro-4-nitro-
$C_6H_4F_2$
1422 - Benzene, 1,2-difluoro-
1423 - Benzene, 1,3-difluoro-
1424 - Benzene, 1,4-difluoro-
$C_6H_4INO_2$
1783 - Benzene, 1-iodo-2-nitro-
1784 - Benzene, 1-iodo-3-nitro-
1785 - Benzene, 1-iodo-4-nitro-
$C_6H_4IN_3$
979 - Benzene, 1-azido-2-iodo-
$C_6H_4I_2$
1427 - Benzene, 1,2-diiodo-
1428 - Benzene, 1,3-diiodo-
1429 - Benzene, 1,4-diiodo-
$C_6H_4N_2$
10881 - 2-Pyridinecarbonitrile
10882 - 3-Pyridinecarbonitrile
$C_6H_4N_2O_4$
1491 - Benzene, 1,2-dinitro-
1492 - Benzene, 1,3-dinitro-
1493 - Benzene, 1,4-dinitro-
10804 - 2,3-Pyrazinedicarboxylic acid
$C_6H_4N_2O_5$
9341 - Phenol, 2,3-dinitro-
9342 - Phenol, 2,4-dinitro-
9343 - Phenol, 2,5-dinitro-
9344 - Phenol, 2,6-dinitro-
9345 - Phenol, 3,4-dinitro-
9346 - Phenol, 3,5-dinitro-
$C_6H_4N_2O_6$
1517 - 1,3-Benzenediol, 2,4-dinitro-
$C_6H_4N_4$
10747 - Pteridine
$C_6H_4N_4O_2$
2873 - 1H-Benzotriazole, 5-nitro-
10748 - 2,4(1H,3H)-Pteridinedione
$C_6H_4N_4O_6$
863 - Benzenamine, 2,4,6-trinitro-
C_6H_4O
6108 - Furan, 2-ethynyl-
$C_6H_4O_2$
4257 - 2,5-Cyclohexadiene-1,4-dione
$C_6H_4O_4$
10773 - 2H-Pyran-5-carboxylic acid, 2-oxo-
$C_6H_4O_4S$
11745 - 2,5-Thiophenedicarboxylic acid
$C_6H_4O_5$
6079 - 2,3-Furandicarboxylic acid
6080 - 2,4-Furandicarboxylic acid
6081 - 2,5-Furandicarboxylic acid
$C_6H_4O_6$
4269 - 2,5-Cyclohexadiene-1,4-dione, 2,3,5,6-tetrahydroxy-
$C_6H_4S_2$
11688 - Thieno[2,3-b]thiophene
$C_6H_5AsCl_2$
462 - Arsonous dichloride, phenyl-
$C_6H_5AsI_2$
464 - Arsonous diiodide, phenyl-
C_6H_5AsO
447 - Arsine, oxophenyl-
$C_6H_5BF_2$
3116 - Borane, difluorophenyl-
$C_6H_5BO_2$
2412 - 1,3,2-Benzodioxaborole
C_6H_5Br
998 - Benzene, bromo-
$C_6H_5BrN_2O_2$
653 - Benzenamine, 2-bromo-6-nitro-
654 - Benzenamine, 4-bromo-2-nitro-
C_6H_5BrO
9228 - Phenol, 2-bromo-
9229 - Phenol, 3-bromo-
9230 - Phenol, 4-bromo-
C_6H_5BrOS
5783 - Ethanone, 1-(5-bromo-2-thienyl)-
$C_6H_5BrO_2$
1502 - 1,3-Benzenediol, 4-bromo-
1503 - 1,4-Benzenediol, 2-bromo-
C_6H_5BrS
2292 - Benzenethiol, 4-bromo-
$C_6H_5Br_2N$
681 - Benzenamine, 2,4-dibromo-
682 - Benzenamine, 2,6-dibromo-
683 - Benzenamine, 3,4-dibromo-
C_6H_5Cl
1131 - Benzene, chloro-
C_6H_5ClFN
664 - Benzenamine, 2-chloro-6-fluoro-
$C_6H_5ClN_2O_2$
677 - Benzenamine, 5-chloro-2-nitro-
C_6H_5ClO
9251 - Phenol, 2-chloro-
9252 - Phenol, 3-chloro-
9253 - Phenol, 4-chloro-
C_6H_5ClOS
2185 - Benzenesulfinyl chloride
5791 - Ethanone, 1-(5-chloro-2-thienyl)-
$C_6H_5ClO_2$
1504 - 1,2-Benzenediol, 3-chloro-
1505 - 1,2-Benzenediol, 4-chloro-
1506 - 1,3-Benzenediol, 5-chloro-
1507 - 1,4-Benzenediol, 2-chloro-
$C_6H_5ClO_2S$
2242 - Benzenesulfonyl chloride
$C_6H_5ClO_3S$
2220 - Benzenesulfonic acid, 4-chloro-
C_6H_5ClS
2293 - Benzenethiol, 2-chloro-
2294 - Benzenethiol, 3-chloro-
2295 - Benzenethiol, 4-chloro-
$C_6H_5ClS_2$
1563 - 1,3-Benzenedithiol, 4-chloro-
$C_6H_5Cl_2N$
685 - Benzenamine, 2,3-dichloro-
686 - Benzenamine, 2,4-dichloro-
687 - Benzenamine, 2,5-dichloro-
688 - Benzenamine, 2,6-dichloro-
689 - Benzenamine, 3,4-dichloro-

690 - Benzenamine, 3,5-dichloro-
$C_6H_5Cl_2NO$
9205 - Phenol, 2-amino-4,6-dichloro-
9206 - Phenol, 4-amino-2,6-dichloro-
$C_6H_5Cl_2OP$
9548 - Phosphonic dichloride, phenyl-
$C_6H_5Cl_2P$
9556 - Phosphonous dichloride, phenyl-
$C_6H_5Cl_2PS$
9552 - Phosphonothioic dichloride, phenyl-
$C_6H_5Cl_2Sb$
11547 - Stibine, dichlorophenyl-
$C_6H_5Cl_3Si$
11473 - Silane, trichlorophenyl-
C_6H_5F
1717 - Benzene, fluoro-
C_6H_5FO
9360 - Phenol, 2-fluoro-
9361 - Phenol, 3-fluoro-
9362 - Phenol, 4-fluoro-
$C_6H_5FO_2S$
2251 - Benzenesulfonyl fluoride
$C_6H_5F_2N$
700 - Benzenamine, 2,4-difluoro-
$C_6H_5F_3Si$
11486 - Silane, trifluorophenyl-
$C_6H_5F_7O_2$
3524 - Butanoic acid, heptafluoro-, ethyl ester
C_6H_5I
1766 - Benzene, iodo-
C_6H_5IO
9371 - Phenol, 2-iodo-
9372 - Phenol, 3-iodo-
9373 - Phenol, 4-iodo-
$C_6H_5IO_2$
1523 - 1,2-Benzenediol, 4-iodo-
1524 - 1,3-Benzenediol, 5-iodo-
$C_6H_5I_2N$
701 - Benzenamine, 2,4-diiodo-
C_6H_5NO
2059 - Benzene, nitroso-
6031 - 2-Furanacetonitrile
10887 - 2-Pyridinecarboxaldehyde
10888 - 3-Pyridinecarboxaldehyde
10889 - 4-Pyridinecarboxaldehyde
C_6H_5NOS
844 - Benzenamine, *N*-sulfinyl-
$C_6H_5NO_2$
2049 - Benzene, nitro-
9438 - Phenol, 4-nitroso-
10893 - 2-Pyridinecarboxylic acid
10894 - 3-Pyridinecarboxylic acid
10895 - 4-Pyridinecarboxylic acid
$C_6H_5NO_3$
9435 - Phenol, 2-nitro-
9436 - Phenol, 3-nitro-
9437 - Phenol, 4-nitro-
10899 - 3-Pyridinecarboxylic acid, 1,6-dihydro-6-oxo-
10909 - 3-Pyridinecarboxylic acid, 1-oxide
$C_6H_5NO_4$
1541 - 1,3-Benzenediol, 4-nitro-
10900 - 4-Pyridinecarboxylic acid, 1,2-dihydro-6-hydroxy-2-oxo-
C_6H_5NS
11725 - 2-Thiopheneacetonitrile
$C_6H_5N_3$
976 - Benzene, azido-
2871 - 1H-Benzotriazole
$C_6H_5N_3O_4$
724 - Benzenamine, 2,3-dinitro-
725 - Benzenamine, 2,4-dinitro-
726 - Benzenamine, 2,6-dinitro-
727 - Benzenamine, 3,5-dinitro-
$C_6H_5N_3O_5$
9207 - Phenol, 2-amino-4,6-dinitro-
$C_6H_5N_5O_2$
10749 - 4,6-Pteridinedione, 2-amino-1,5-dihydro-
C_6H_6
867 - Benzene
4767 - 1,3-Cyclopentadiene, 5-methylene-
6712 - 1,3-Hexadien-5-yne
6713 - 1,5-Hexadien-3-yne
6715 - 1,4-Hexadiyne
6716 - 1,5-Hexadiyne
6717 - 2,4-Hexadiyne
$C_6H_6AsNO_5$
459 - Arsonic acid, (4-nitrophenyl)-
C_6H_6BrN
638 - Benzenamine, 2-bromo-
639 - Benzenamine, 3-bromo-
640 - Benzenamine, 4-bromo-
$C_6H_6Br_2N_2$
8768 - Pentanedinitrile, 2-bromo-2-(bromomethyl)-
C_6H_6ClN
658 - Benzenamine, 2-chloro-
659 - Benzenamine, 3-chloro-
660 - Benzenamine, 4-chloro-
10919 - Pyridine, 2-chloro-4-methyl-
10920 - Pyridine, 2-chloro-6-methyl-
C_6H_6ClNO
9204 - Phenol, 2-amino-4-chloro-
10918 - Pyridine, 2-chloro-6-methoxy-
$C_6H_6ClN_2O_4$
1559 - 1,3-Benzenedisulfonamide, 4,5-dichloro-
$C_6H_6Cl_2Si$
11403 - Silane, dichlorophenyl-
$C_6H_6Cl_4$
6697 - 2,4-Hexadiene, 1,3,4,6-tetrachloro-
$C_6H_6Cl_6$
4406 - Cyclohexane, 1,2,3,4,5,6-hexachloro-, (1α,2β,3α,4β,5α,6β)-
4407 - Cyclohexane, 1,2,3,4,5,6-hexachloro-, (1α,2α,3α,4β,5α,6β)-
4408 - Cyclohexane, 1,2,3,4,5,6-hexachloro-, (1α,2α,3β,4α,5β,6β)-(±)-
4409 - Cyclohexane, 1,2,3,4,5,6-hexachloro-, (1α,2α,3β,4α,5α,6β)-
7035 - 3-Hexene, 1,2,3,4,5,6-hexachloro-
C_6H_6FN
752 - Benzenamine, 2-fluoro-
753 - Benzenamine, 3-fluoro-
754 - Benzenamine, 4-fluoro-
$C_6H_6FNO_2S$
2252 - Benzenesulfonyl fluoride, 4-amino-
C_6H_6IN
760 - Benzenamine, 3-iodo-
761 - Benzenamine, 4-iodo-
$C_6H_6N_2O$
828 - Benzenamine, 4-nitroso-
9914 - Propanedinitrile, (ethoxymethylene)-
10890 - 3-Pyridinecarboxamide
$C_6H_6N_2O_2$
822 - Benzenamine, 2-nitro-
823 - Benzenamine, 3-nitro-
824 - Benzenamine, 4-nitro-
825 - Benzenamine, *N*-nitro-
10897 - 3-Pyridinecarboxylic acid, 6-amino-
$C_6H_6N_2O_3$
9214 - Phenol, 2-amino-3-nitro-
$C_6H_6N_4O_3$
10765 - 1H-Purine-2,6,8(3H)-trione, 4,9-dihydro-3-methyl-
10766 - 1H-Purine-2,6,8(3H)-trione, 7,9-dihydro-7-methyl-
$C_6H_6N_4O_4$
7144 - Hydrazinecarboxamide, 2-[(5-nitro-2-furanyl)methylene]-
7156 - Hydrazine, (2,4-dinitrophenyl)-
C_6H_6O
6102 - Furan, 2-ethenyl-
9199 - Phenol
C_6H_6OS
5870 - Ethanone, 1-(2-thienyl)-
9376 - Phenol, 2-mercapto-
9377 - Phenol, 3-mercapto-
9378 - Phenol, 4-mercapto-
11736 - 2-Thiophenecarboxaldehyde, 5-methyl-
$C_6H_6O_2$
1497 - 1,2-Benzenediol
1498 - 1,3-Benzenediol
1499 - 1,4-Benzenediol
5817 - Ethanone, 1-(2-furanyl)-
6045 - 2-Furancarboxaldehyde, 5-methyl-
$C_6H_6O_2S$
2180 - Benzenesulfinic acid
5835 - Ethanone, 1-(3-hydroxy-2-thienyl)-
$C_6H_6O_2Se$
2174 - Benzeneseleninic acid
$C_6H_6O_3$
2366 - 1,2,3-Benzenetriol
2367 - 1,2,4-Benzenetriol
2368 - 1,3,5-Benzenetriol
6026 - 2-Furanacetic acid
6044 - 2-Furancarboxaldehyde, 5-(hydroxymethyl)-
6059 - 2-Furancarboxylic acid, methyl ester
6060 - 2-Furancarboxylic acid, 5-methyl-
6063 - 3-Furancarboxylic acid, methyl ester
6065 - 3-Furancarboxylic acid, 5-methyl-
6097 - 2,5-Furandione, 3,4-dimethyl-
10796 - 4H-Pyran-4-one, 3-hydroxy-2-methyl-
$C_6H_6O_3S$
2214 - Benzenesulfonic acid
$C_6H_6O_4$
3910 - 2-Butynedioic acid, dimethyl ester
10795 - 4H-Pyran-4-one, 5-hydroxy-2-(hydroxymethyl)-
$C_6H_6O_4S$
2225 - Benzenesulfonic acid, 2-hydroxy-
2226 - Benzenesulfonic acid, 4-hydroxy-
$C_6H_6O_7S_2$
1560 - 1,3-Benzenedisulfonic acid, 4-hydroxy-
C_6H_6S
2289 - Benzenethiol
$C_6H_6S_2$
1561 - 1,2-Benzenedithiol
1562 - 1,3-Benzenedithiol
C_6H_6Se
2177 - Benzeneselenol
$C_6H_7AsO_3$
460 - Arsonic acid, phenyl-
$C_6H_7BO_2$
3135 - Boronic acid, phenyl-
$C_6H_7BrN_2$
7141 - Hydrazine, (4-bromophenyl)-
$C_6H_7ClN_2$
1264 - 1,2-Benzenediamine, 4-chloro-
C_6H_7ClO
6711 - 2,4-Hexadienoyl chloride, (*E,E*)-
$C_6H_7ClO_4$
3773 - 2-Butenedioic acid, 2-chloro-, dimethyl ester, (*E*)-
C_6H_7ClSi
11384 - Silane, chlorophenyl-
$C_6H_7Cl_2F_3O_2$
10721 - Propionic acid, 2,2-dichloro-3,3,3-trifluoro-, propyl ester
$C_6H_7Cl_2N$
665 - Benzenamine, 2-chloro-, hydrochloride
$C_6H_7F_3O_3$
3592 - Butanoic acid, 4,4,4-trifluoro-3-oxo-, ethyl ester
$C_6H_7KO_2$
6704 - 2,4-Hexadienoic acid, potassium salt, (*E,E*)-
C_6H_7N
633 - Benzenamine
4907 - 1-Cyclopentenecarbonitrile
10980 - Pyridine, 2-methyl-
10981 - Pyridine, 3-methyl-
10982 - Pyridine, 4-methyl-
C_6H_7NO
757 - Benzenamine, *N*-hydroxy-

5820 - Ethanone, 1-(1H-pyrrol-2-yl)-
9200 - Phenol, amino-
9201 - Phenol, 2-amino-
9202 - Phenol, 3-amino-
9203 - Phenol, 4-amino-
10975 - 2-Pyridinemethanol
10976 - 3-Pyridinemethanol
10977 - 4-Pyridinemethanol
10988 - Pyridine, 3-methyl-, 1-oxide
11035 - 2(1H)-Pyridinone, 1-methyl-
11036 - 2(1H)-Pyridinone, 3-methyl-
11037 - 2(1H)-Pyridinone, 4-methyl-

$C_6H_7NO_2$
11122 - 1H-Pyrrole-2,5-dione, 1-ethyl-

$C_6H_7NO_3S$
2215 - Benzenesulfonic acid, 2-amino-
2216 - Benzenesulfonic acid, 3-amino-
2217 - Benzenesulfonic acid, 4-amino-

$C_6H_7NO_4$
6129 - Furan, 2-(methoxymethyl)-5-nitro-

C_6H_7NS
2290 - Benzenethiol, 2-amino-
2291 - Benzenethiol, 4-amino-

$C_6H_7N_3O$
10905 - 4-Pyridinecarboxylic acid, hydrazide

$C_6H_7N_3O_2$
1281 - 1,2-Benzenediamine, 4-nitro-
7168 - Hydrazine, (4-nitrophenyl)-

$C_6H_7N_5O$
10769 - 6H-Purin-6-one, 2-amino-1,7-dihydro-7-methyl

$C_6H_7O_3Sb$
11548 - Stibine oxide, dihydroxyphenyl-

C_6H_7P
9519 - Phosphine, phenyl-

C_6H_8
4172 - Cyclobutane, 1,2-bis(methylene)-
4248 - 1,3-Cyclohexadiene
4249 - 1,4-Cyclohexadiene
6992 - 1,3,5-Hexatriene, (*E*)-
6993 - 1,3,5-Hexatriene, (*Z*)-
7103 - 1-Hexen-3-yne
7104 - 1-Hexen-4-yne
7105 - 1-Hexen-5-yne
7106 - 2-Hexen-4-yne
9116 - 1-Penten-3-yne, 2-methyl-
9117 - 3-Penten-1-yne, 3-methyl-

$C_6H_8AsNO_3$
456 - Arsonic acid, (4-aminophenyl)-

C_6H_8ClN
756 - Benzenamine, hydrochloride

C_6H_8ClNS
11670 - Thiazole, 5-(2-chloroethyl)-4-methyl-

$C_6H_8ClN_3O_4S_2$
1558 - 1,3-Benzenedisulfonamide, 4-amino-6-chloro-

$C_6H_8Cl_2$
6690 - 2,4-Hexadiene, 1,3-dichloro-

$C_6H_8Cl_2O_3$
3503 - Butanoic acid, 2,2-dichloro-3-oxo-, ethyl ester

$C_6H_8Cl_2O_4$
129 - Acetic acid, chloro-, 1,2-ethanediyl ester
5554 - Ethanedioic acid, bis(2-chloroethyl) ester

$C_6H_8N_2$
1260 - 1,2-Benzenediamine
1261 - 1,3-Benzenediamine
1262 - 1,4-Benzenediamine
6777 - Hexanedinitrile
7169 - Hydrazine, phenyl-
10805 - Pyrazine, 2,3-dimethyl-
10806 - Pyrazine, 2,5-dimethyl-
10807 - Pyrazine, 2,6-dimethyl-
10866 - 2-Pyridinamine, 3-methyl-
10867 - 2-Pyridinamine, 4-methyl-
10868 - 2-Pyridinamine, 6-methyl-
10869 - 2-Pyridinamine, *N*-methyl-
10870 - 3-Pyridinamine, 4-methyl-
10971 - 2-Pyridinemethanamine
10972 - 3-Pyridinemethanamine
10973 - 4-Pyridinemethanamine
11055 - Pyrimidine, 2,4-dimethyl-
11056 - Pyrimidine, 2,5-dimethyl-
11057 - Pyrimidine, 4,5-dimethyl-
11058 - Pyrimidine, 4,6-dimethyl-

$C_6H_8N_2O$
9274 - Phenol, 2,4-diamino-
10114 - Propanenitrile, 3,3′-oxybis-
10808 - Pyrazineethanol
10865 - 3-Pyridinamine, 6-methoxy-

$C_6H_8N_2O_2S$
2187 - Benzenesulfonamide, 4-amino-

$C_6H_8N_2O_3$
821 - Benzenamine, nitrate

$C_6H_8N_2O_3S$
2224 - Benzenesulfonic acid, 4-hydrazino-

$C_6H_8N_6O_{18}$
7424 - *D*-Mannitol, hexanitrate

C_6H_8O
4703 - 2-Cyclohexen-1-one
4704 - 3-Cyclohexen-1-one
4908 - 1-Cyclopentene-1-carboxaldehyde
4944 - 2-Cyclopenten-1-one, 2-methyl-
4945 - 2-Cyclopenten-1-one, 3-methyl-
4946 - 2-Cyclopenten-1-one, 5-methyl-
6088 - Furan, 2,4-dimethyl-
6089 - Furan, 2,5-dimethyl-
6105 - Furan, 2-ethyl-
6675 - 2,4-Hexadienal, (*E*,*E*)-
7113 - 3-Hexynal
7130 - 3-Hexyn-2-one
7131 - 5-Hexyn-2-one

$C_6H_8O_2$
3176 - 1,3-Butadien-1-ol, acetate
3918 - 2-Butynoic acid, ethyl ester
4375 - 1,2-Cyclohexanedione
4376 - 1,3-Cyclohexanedione
4377 - 1,4-Cyclohexanedione
4909 - 1-Cyclopentene-1-carboxylic acid
6103 - Furan, 2-ethoxy-
6121 - 2-Furanmethanol, α-methyl-
6122 - 2-Furanmethanol, 5-methyl-
6128 - Furan, 2-(methoxymethyl)-
6698 - 2,4-Hexadienoic acid
7123 - 2-Hexynoic acid
9139 - 2-Pentynoic acid, methyl ester
10694 - 2-Propenoic acid, 2-propenyl ester
10791 - 2H-Pyran-2-one, 5,6-dihydro-6-methyl-, (*S*)-

$C_6H_8O_3$
6139 - 2(3H)-Furanone, 3-acetyldihydro-
10680 - 2-Propenoic acid, oxiranylmethyl ester

$C_6H_8O_4$
3778 - 2-Butenedioic acid (*E*)-, dimethyl ester
3779 - 2-Butenedioic acid (*Z*)-, dimethyl ester
3791 - 2-Butenedioic acid (*E*)-, monoethyl ester
4182 - 1,3-Cyclobutanedicarboxylic acid, *cis*-
4185 - 1,3-Cyclobutanedicarboxylic acid, *trans*-
5190 - 1,3-Dioxane-4,6-dione, 2,2-dimethyl-
9965 - Propanedioic acid, (1-methylethylidene)-
9979 - Propanedioic acid, 2-propenyl-

$C_6H_8O_5$
467 - *L*-Ascorbic acid, 6-deoxy-

$C_6H_8O_6$
465 - *DL*-Ascorbic acid
466 - *L*-Ascorbic acid
6239 - *D*-Glucuronic acid, γ-lactone
10165 - 1,2,3-Propanetricarboxylic acid

$C_6H_8O_7$
10167 - 1,2,3-Propanetricarboxylic acid, 2-hydroxy-

C_6H_8S
11754 - Thiophene, 2,3-dimethyl-
11755 - Thiophene, 2,4-dimethyl-
11756 - Thiophene, 2,5-dimethyl-
11757 - Thiophene, 3,4-dimethyl-
11759 - Thiophene, 2-ethyl-
11760 - Thiophene, 3-ethyl-

C_6H_8Si
11446 - Silane, phenyl-

C_6H_9Br
4627 - Cyclohexene, 1-bromo-
4628 - Cyclohexene, 3-bromo-

C_6H_9BrO
4574 - Cyclohexanone, 2-bromo-

$C_6H_9BrO_2$
3819 - 2-Butenoic acid, 4-bromo-, ethyl ester, (*E*)-

$C_6H_9BrO_3$
3477 - Butanoic acid, 2-bromo-3-oxo-, ethyl ester
3478 - Butanoic acid, 4-bromo-3-oxo-, ethyl ester

C_6H_9Cl
4642 - Cyclohexene, 1-chloro-
6688 - 1,3-Hexadiene, 3-chloro-
8664 - 1,2-Pentadiene, 1-chloro-3-methyl-
9126 - 1-Pentyne, 3-chloro-3-methyl-
9127 - 2-Pentyne, 4-chloro-4-methyl-

$C_6H_9ClN_2$
7172 - Hydrazine, phenyl-, monohydrochloride

C_6H_9ClO
4577 - Cyclohexanone, 2-chloro-
4578 - Cyclohexanone, 4-chloro-

$C_6H_9ClO_2$
3823 - 2-Butenoic acid, 2-chloro-, ethyl ester
3824 - 2-Butenoic acid, 2-chloro-, ethyl ester, (*E*)-
3825 - 2-Butenoic acid, 2-chloro-, ethyl ester, (*Z*)-
3826 - 2-Butenoic acid, 3-chloro-, ethyl ester
3827 - 2-Butenoic acid, 3-chloro-, ethyl ester, (*E*)-
3828 - 2-Butenoic acid, 3-chloro-, ethyl ester, (*Z*)-
3829 - 2-Butenoic acid, 4-chloro-, ethyl ester

$C_6H_9ClO_3$
3494 - Butanoic acid, 4-chloro-3-oxo-, ethyl ester
3570 - Butanoic acid, 3-oxo-, 2-chloroethyl ester

$C_6H_9Cl_3$
7048 - 1-Hexene, 1,1,2-trichloro-

$C_6H_9Cl_3O_2$
245 - Acetic acid, trichloro-, butyl ester
247 - Acetic acid, trichloro-, 1,1-dimethylethyl ester
251 - Acetic acid, trichloro-, 1-methylpropyl ester
252 - Acetic acid, trichloro-, 2-methylpropyl ester
3591 - Butanoic acid, 2,2,3-trichloro-, ethyl ester

$C_6H_9Cl_3O_3$
13 - Acetaldehyde, chloro-, trimer

$C_6H_9F_3O_2$
265 - Acetic acid, trifluoro-, butyl ester
266 - Acetic acid, trifluoro-, 1,1-dimethylethyl ester

C_6H_9N
4779 - Cyclopentanecarbonitrile
11119 - 1H-Pyrrole, 2,4-dimethyl-
11120 - 1H-Pyrrole, 2,5-dimethyl-
11124 - 1H-Pyrrole, 1-ethyl-
11125 - 1H-Pyrrole, 2-ethyl-

C_6H_9NO
6115 - 2-Furanmethanamine, *N*-methyl-
11183 - 2-Pyrrolidinone, 1-ethenyl-

C_6H_9NOS
11677 - 5-Thiazoleethanol, 4-methyl-

$C_6H_9NO_2$
10912 - 3-Pyridinecarboxylic acid, 1,2,5,6-tetrahydro-
11157 - 2,5-Pyrrolidinedione, 1-ethyl-

$C_6H_9NO_3$
50 - Acetamide, *N*,*N*-diacetyl-
8587 - 2,4-Oxazolidinedione, 3,5,5-trimethyl-

$C_6H_9NO_6$
6256 - Glycine, *N*,*N*-bis(carboxymethyl)-

$C_6H_9N_3$
2305 - 1,2,3-Benzenetriamine
2306 - 1,2,4-Benzenetriamine
11044 - 2-Pyrimidinamine, 4,5-dimethyl-
11045 - 4-Pyrimidinamine, 2,6-dimethyl-

$C_6H_9N_3O_2$
759 - Benzenamine, *N*-hydroxy-*N*-nitroso-, ammonium salt
7132 - *L*-Histidine

$C_6H_9N_3O_3$
7206 - 1H-Imidazole-1-ethanol, 2-methyl-5-nitro-

MOLECULAR FORMULA INDEX OF ORGANIC COMPOUNDS (continued)

11841 - 1,3,5-Triazine-2,4,6(1H,3H,5H)-trione, 1,3,5-trimethyl-

C_6H_{10}
3159 - 1,3-Butadiene, 2,3-dimethyl-
3907 - 1-Butyne, 3,3-dimethyl-
4626 - Cyclohexene
4827 - Cyclopentane, methylene-
4928 - Cyclopentene, 1-methyl-
4929 - Cyclopentene, 3-methyl-
4930 - Cyclopentene, 4-methyl-
4982 - Cyclopropane, (1-methylethenyl)-
6676 - 1,2-Hexadiene
6677 - 1,4-Hexadiene
6678 - 1,5-Hexadiene
6679 - 2,3-Hexadiene
6680 - 2,4-Hexadiene
6681 - 1,3-Hexadiene, (*E*)-
6682 - 1,3-Hexadiene, (*Z*)-
6683 - 1,4-Hexadiene, (*E*)-
6684 - 1,4-Hexadiene, (*Z*)-
6685 - 2,4-Hexadiene, (*E,E*)-
6686 - 2,4-Hexadiene, (*E,Z*)-
6687 - 2,4-Hexadiene, (*Z,Z*)-
7114 - 1-Hexyne
7115 - 2-Hexyne
7116 - 3-Hexyne
8666 - 1,2-Pentadiene, 3-methyl-
8667 - 1,2-Pentadiene, 4-methyl-
8668 - 1,3-Pentadiene, 2-methyl-, (*Z*)-
8669 - 1,3-Pentadiene, 3-methyl-
8670 - 1,3-Pentadiene, 4-methyl-
8671 - 1,4-Pentadiene, 2-methyl-
8672 - 2,3-Pentadiene, 2-methyl-
9060 - 1-Pentene, 3-methylene-
9134 - 1-Pentyne, 3-methyl-
9135 - 1-Pentyne, 4-methyl-
9136 - 2-Pentyne, 4-methyl-

$C_6H_{10}Br_2$
4345 - Cyclohexane, 1,1-dibromo-
4346 - Cyclohexane, 1,2-dibromo-, *cis-*
4347 - Cyclohexane, 1,2-dibromo-, *trans-*(±)-
4348 - Cyclohexane, 1,3-dibromo-, *trans-*
4349 - Cyclohexane, 1,4-dibromo-, *cis-*

$C_6H_{10}Br_2O_2$
3498 - Butanoic acid, 2,3-dibromo-, ethyl ester
3499 - Butanoic acid, 2,4-dibromo-, ethyl ester
10243 - Propanoic acid, 2,3-dibromo-, propyl ester

$C_6H_{10}ClFO_2$
132 - Acetic acid, chlorofluoro-, butyl ester

$C_6H_{10}Cl_2$
4358 - Cyclohexane, 1,1-dichloro-
4359 - Cyclohexane, 1,2-dichloro-, *cis-*
4360 - Cyclohexane, 1,2-dichloro-, *trans-*(±)-
4361 - Cyclohexane, 1,4-dichloro-, *cis-*
7011 - 1-Hexene, 1,2-dichloro-, (*E*)-
7012 - 1-Hexene, 1,2-dichloro-, (*Z*)-

$C_6H_{10}Cl_2O_2$
163 - Acetic acid, dichloro-, butyl ester
10249 - Propanoic acid, 2,3-dichloro-, 1-methylethyl ester

$C_6H_{10}Cl_3NO$
96 - Acetamide, 2,2,2-trichloro-*N,N*-diethyl-

$C_6H_{10}N_2$
8830 - Pentanenitrile, 3-imino-2-methyl-
10833 - 1H-Pyrazole, 1-ethyl-3-methyl-
10839 - 1H-Pyrazole, 1,3,4-trimethyl-
10840 - 1H-Pyrazole, 1,3,5-trimethyl-
10841 - 1H-Pyrazole, 1,4,5-trimethyl-
10842 - 1H-Pyrazole, 3,4,5-trimethyl-

$C_6H_{10}N_2O_2$
4379 - 1,2-Cyclohexanedione, dioxime

$C_6H_{10}N_4$
11654 - 5H-Tetrazolo[1,5-a]azepine, 6,7,8,9-tetrahydro-

$C_6H_{10}N_4O_3S_2$
3197 - Butanamide, *N*-[5-(aminosulfonyl)-1,3,4-thiadiazol-2-yl]-

$C_6H_{10}O$
3160 - 1,3-Butadiene, 1-ethoxy-
3161 - 1,3-Butadiene, 2-ethoxy-
3377 - Butane, 1-(ethynyloxy)-
3877 - 1-Buten-1-one, 2-ethyl-
4571 - Cyclohexanone
4694 - 2-Cyclohexen-1-ol
4695 - 3-Cyclohexen-1-ol
4780 - Cyclopentanecarboxaldehyde
4885 - Cyclopentanone, 2-methyl-
4886 - Cyclopentanone, 3-methyl-, (±)-
5792 - Ethanone, 1-cyclobutyl-
6705 - 1,4-Hexadien-3-ol
6706 - 1,5-Hexadien-3-ol
6707 - 2,4-Hexadien-1-ol
6708 - 3,5-Hexadien-2-ol
6709 - 2,4-Hexadien-1-ol, (*E,E*)-
6996 - 2-Hexenal, (*E*)-
6997 - 3-Hexenal, (*E*)-
6998 - 3-Hexenal, (*Z*)-
7086 - 3-Hexen-2-one
7087 - 4-Hexen-2-one
7088 - 4-Hexen-3-one
7089 - 5-Hexen-2-one
7090 - 3-Hexen-2-one, (*E*)-
7124 - 1-Hexyn-3-ol
7125 - 2-Hexyn-1-ol
7126 - 3-Hexyn-1-ol
7127 - 5-Hexyn-2-ol
8554 - 7-Oxabicyclo[4.1.0]heptane
9002 - 2-Pentenal, 2-methyl-
9107 - 1-Penten-3-one, 2-methyl-
9108 - 3-Penten-2-one, 4-methyl-
9109 - 4-Penten-2-one, 4-methyl-
9146 - 1-Pentyn-3-ol, 3-methyl-
10592 - 1-Propene, 3,3′-oxybis-
10778 - 2H-Pyran, 3,6-dihydro-4-methyl-

$C_6H_{10}OS$
10596 - 1-Propene, 3,3′-sulfinylbis-

$C_6H_{10}OS_2$
10595 - 2-Propene-1-sulfinothioic acid, *S*-2-propenyl ester

$C_6H_{10}O_2$
3513 - Butanoic acid, ethenyl ester
3835 - 3-Butenoic acid, 2,2-dimethyl-
3837 - 2-Butenoic acid, 2-ethyl-, (*E*)-
3838 - 2-Butenoic acid, ethyl ester
3839 - 2-Butenoic acid, ethyl ester, (*E*)-
3840 - 2-Butenoic acid, ethyl ester, (*Z*)-
3841 - 3-Butenoic acid, ethyl ester
3851 - 2-Butenoic acid, 2-methyl-, methyl ester, (*E*)-
3852 - 2-Butenoic acid, 3-methyl-, methyl ester
3862 - 2-Buten-1-ol, acetate
4176 - Cyclobutanecarboxylic acid, methyl ester
4781 - Cyclopentanecarboxylic acid
4961 - Cyclopropanecarboxylic acid, ethyl ester
6150 - 2(3H)-Furanone, 5-ethyldihydro-
6693 - 1,5-Hexadiene-3,4-diol, (*R**,*R**)-(±)-
6748 - Hexanedial
6804 - 2,4-Hexanedione
6805 - 2,5-Hexanedione
6806 - 3,4-Hexanedione
7049 - 2-Hexenoic acid
7050 - 3-Hexenoic acid
7051 - 5-Hexenoic acid
7118 - 3-Hexyne-2,5-diol
8591 - 2-Oxepanone
8626 - Oxirane, [(2-propenyloxy)methyl]-
8890 - Pentanoic acid, 2-methylene-
9076 - 2-Pentenoic acid, 2-methyl-, (*E*)-
9077 - 2-Pentenoic acid, 3-methyl-, (*Z*)-
9078 - 2-Pentenoic acid, 4-methyl-
9079 - 3-Pentenoic acid, 2-methyl-
10297 - Propanoic acid, 2-methyl-, ethenyl ester
10325 - Propanoic acid, 2-propenyl ester
10658 - 2-Propenoic acid, 2-methyl-, ethyl ester

$C_6H_{10}O_2S$
3589 - Butanoic acid, 3-thioxo-, ethyl ester
10597 - 1-Propene, 3,3′-sulfonylbis-

$C_6H_{10}O_2S_2$
5453 - Ethanebis(thioic) acid, *S,S*-diethyl ester

$C_6H_{10}O_3$
3510 - Butanoic acid, 3,3-dimethyl-2-oxo-
3558 - Butanoic acid, 2-methyl-3-oxo-, methyl ester
3571 - Butanoic acid, 3-oxo-, ethyl ester
3572 - Butanoic acid, 4-oxo-, ethyl ester
3843 - 3-Butenoic acid, 2-hydroxy-, ethyl ester
5206 - 1,3-Dioxolane-4-carboxaldehyde, 2,2-dimethyl-
6087 - Furan, 2,5-dihydro-2,5-dimethoxy-
6143 - 2(3H)-Furanone, dihydro-3-hydroxy-4,4-dimethyl-, (*R*)-
6808 - 2,5-Hexanedione, 3-hydroxy-
6905 - Hexanoic acid, 5-oxo-
8895 - Pentanoic acid, 2-methyl-4-oxo-
8896 - Pentanoic acid, 3-methyl-4-oxo-
8908 - Pentanoic acid, 4-oxo-, methyl ester
10203 - Propanoic acid, anhydride
10281 - Propanoic acid, 2-hydroxy-, 2-propenyl ester
10661 - 2-Propenoic acid, 2-methyl-, 2-hydroxyethyl ester
10774 - 2H-Pyran-4-carboxylic acid, tetrahydro-

$C_6H_{10}O_4$
102 - Acetic acid, (acetyloxy)-, ethyl ester
3317 - Butanedioic acid, 2,3-dimethyl-
3318 - Butanedioic acid, dimethyl ester
3322 - Butanedioic acid, ethyl-, (*R*)-
3335 - Butanedioic acid, monoethyl ester
5560 - Ethanedioic acid, diethyl ester
5570 - 1,1-Ethanediol, diacetate
5571 - 1,2-Ethanediol, diacetate
6779 - Hexanedioic acid
7423 - *D*-Mannitol, 1,4:3,6-dianhydro-
8779 - Pentanedioic acid, 3-methyl-
8780 - Pentanedioic acid, monomethyl ester
9963 - Propanedioic acid, methyl-, dimethyl ester

$C_6H_{10}O_4S$
10329 - Propanoic acid, 3,3′-thiobis-

$C_6H_{10}O_4S_2$
5179 - Dimethipin

$C_6H_{10}O_5$
5169 - Dicarbonic acid, diethyl ester

$(C_6H_{10}O_5)_x$
6274 - Glycogen

$C_6H_{10}O_6$
3308 - Butanedioic acid, 2,3-dihydroxy-, dimethyl ester, (*R**,*R**)-(±)-
3315 - Butanedioic acid, 2,3-dihydroxy- [*R*-(*R**,*R**)]-, monoethyl ester
6211 - Gluconic acid, β-lactone
6212 - *D*-Gluconic acid, δ-lactone

$C_6H_{10}O_7$
6199 - *D*-Galacturonic acid
6238 - *D*-Glucuronic acid

$C_6H_{10}O_8$
6193 - Galactaric acid
6205 - *D*-Glucaric acid

$C_6H_{10}S$
4481 - Cyclohexanethione
10599 - 1-Propene, 3,3′-thiobis-

$C_6H_{10}S_3$
11917 - Trisulfide, di-2-propenyl

$C_6H_{11}Br$
4315 - Cyclohexane, bromo-

$C_6H_{11}BrN_2O_2$
3196 - Butanamide, *N*-(aminocarbonyl)-2-bromo-3-methyl-

$C_6H_{11}BrO$
4503 - Cyclohexanol, 2-bromo-
6961 - 2-Hexanone, 6-bromo-

$C_6H_{11}BrO_2$
114 - Acetic acid, bromo-, 1,1-dimethylethyl ester
118 - Acetic acid, bromo-, 2-methylpropyl ester

3469 - Butanoic acid, 2-bromo-, ethyl ester
3470 - Butanoic acid, 4-bromo-, ethyl ester
3476 - Butanoic acid, 2-bromo-3-methyl-, methyl ester
6875 - Hexanoic acid, 2-bromo-, (±)-
6876 - Hexanoic acid, 6-bromo-
10213 - Propanoic acid, 2-bromo-2-methyl-, ethyl ester

$C_6H_{11}Cl$
3731 - 2-Butene, 1-chloro-2,3-dimethyl-
4339 - Cyclohexane, chloro-
7004 - 1-Hexene, 1-chloro-
7005 - 1-Hexene, 2-chloro-
7006 - 1-Hexene, 5-chloro-
7007 - 2-Hexene, 4-chloro-
7008 - 3-Hexene, 1-chloro-
7009 - 3-Hexene, 3-chloro-, (*Z*)-
9025 - 1-Pentene, 3-chloro-2-methyl-
9026 - 2-Pentene, 5-chloro-2-methyl-

$C_6H_{11}ClO$
3688 - Butanoyl chloride, 2,2-dimethyl-
3689 - Butanoyl chloride, 2,3-dimethyl-
3690 - Butanoyl chloride, 3,3-dimethyl-
3691 - Butanoyl chloride, 2-ethyl-
4505 - Cyclohexanol, 2-chloro-, *trans*-
4506 - Cyclohexanol, 2-chloro-, *cis*-(±)-
4507 - Cyclohexanol, 4-chloro-, *trans*-
6986 - Hexanoyl chloride
8993 - Pentanoyl chloride, 2-methyl-, (±)-
8994 - Pentanoyl chloride, 3-methyl-, (±)-

$C_6H_{11}ClO_2$
125 - Acetic acid, chloro-, butyl ester
128 - Acetic acid, chloro-, 1,1-dimethylethyl ester
142 - Acetic acid, chloro-, 1-methylpropyl ester
143 - Acetic acid, chloro-, 2-methylpropyl ester
3483 - Butanoic acid, 2-chloro-, ethyl ester
3484 - Butanoic acid, 3-chloro-, ethyl ester
3485 - Butanoic acid, 4-chloro-, ethyl ester
3730 - 2-Butene, 2-chloro-1,1-dimethoxy-
4021 - Carbonochloridic acid, 3-methylbutyl ester
4025 - Carbonochloridic acid, pentyl ester
10223 - Propanoic acid, 2-chloro-, 1-methylethyl ester
10232 - Propanoic acid, 3-chloro-, propyl ester
10292 - Propanoic acid, 2-methyl-, 2-chloroethyl ester

$C_6H_{11}Cl_3O_2$
5657 - Ethane, 1,1,1-trichloro-2,2-diethoxy-

$C_6H_{11}F$
4403 - Cyclohexane, fluoro-

$C_6H_{11}FO_2$
190 - Acetic acid, fluoro-, 1,1-dimethylethyl ester
6888 - Hexanoic acid, 6-fluoro-

$C_6H_{11}I$
4415 - Cyclohexane, iodo-

$C_6H_{11}IO_2$
3928 - Butyric acid, 2-iodo-, ethyl ester

$C_6H_{11}N$
478 - 7-Azabicyclo[4.1.0]heptane
3407 - Butanenitrile, 2-ethyl-
6840 - Hexanenitrile
8821 - Pentane, 1-isocyano-
8831 - Pentanenitrile, 4-methyl-
10526 - 2-Propen-1-amine, *N*-2-propenyl-

$C_6H_{11}NO$
493 - 2H-Azepin-2-one, hexahydro-
3386 - Butane, 1-isocyanato-3-methyl-
4610 - Cyclohexanone, oxime
9638 - 1-Piperidinecarboxaldehyde
9721 - 2-Piperidinone, 1-methyl-
9722 - 2-Piperidinone, 3-methyl-
9723 - 3-Piperidinone, 1-methyl-
9724 - 4-Piperidinone, 1-methyl-
11181 - 2-Pyrrolidinone, 1,5-dimethyl-
11182 - 2-Pyrrolidinone, 3,3-dimethyl-

$C_6H_{11}NO_2$
24 - Acetamide, *N*-acetyl-*N*-ethyl-
3816 - 2-Butenoic acid, 3-amino-, ethyl ester, (*E*)-
4454 - Cyclohexane, nitro-
4782 - Cyclopentanecarboxylic acid, 1-amino-
4830 - Cyclopentane, 1-methyl-1-nitro-
4831 - Cyclopentane, 1-methyl-2-nitro-
4840 - Cyclopentane, (nitromethyl)-
7682 - Morpholine, 4-acetyl-
9640 - 2-Piperidinecarboxylic acid
9641 - 3-Piperidinecarboxylic acid
9642 - 4-Piperidinecarboxylic acid
11184 - 2-Pyrrolidinone, 1-(2-hydroxyethyl)-

$C_6H_{11}NO_3S$
333 - *L*-Alanine, 3-(2-propenylsulfinyl)-, (*S*)-

$C_6H_{11}NO_4$
7231 - Imidodicarbonic acid, diethyl ester

$C_6H_{11}NS$
3390 - Butane, 1-isothiocyanato-3-methyl-
11673 - Thiazole, 4,5-dihydro-2,4,4-trimethyl-

$C_6H_{11}N_2O_4PS_3$
7645 - Methidathion

$C_6H_{11}N_3$
11847 - 1H-1,2,4-Triazole, 1-butyl

$C_6H_{11}NaO_7$
6213 - *D*-Gluconic acid, monosodium salt

$C_6H_{11}O_3P$
11909 - 2,6,7-Trioxa-1-phosphabicyclo[2.2.2]octane, 4-ethyl-

C_6H_{12}
3759 - 1-Butene, 2,3-dimethyl-
3760 - 1-Butene, 3,3-dimethyl-
3761 - 2-Butene, 2,3-dimethyl-
4188 - Cyclobutane, 1,2-dimethyl-, *trans*-
4189 - Cyclobutane, 1,3-dimethyl-, *cis*-
4190 - Cyclobutane, 1,3-dimethyl-, *trans*-
4191 - Cyclobutane, ethyl-
4305 - Cyclohexane
4826 - Cyclopentane, methyl-
4975 - Cyclopropane, 1-ethyl-1-methyl-
4976 - Cyclopropane, 1-ethyl-2-methyl-, *cis*-
4977 - Cyclopropane, 1-ethyl-2-methyl-, *trans*-
4983 - Cyclopropane, (1-methylethyl)-
4984 - Cyclopropane, propyl-
4985 - Cyclopropane, 1,1,2-trimethyl-
4986 - Cyclopropane, 1,2,3-trimethyl-, (1α,2α,3α)-
4987 - Cyclopropane, 1,2,3-trimethyl-, (1α,2α,3β)-
6999 - 1-Hexene
7000 - 2-Hexene, (*E*)-
7001 - 2-Hexene, (*Z*)-
7002 - 3-Hexene, (*E*)-
7003 - 3-Hexene, (*Z*)-
8825 - Pentane, 3-methylene-
9052 - 1-Pentene, 2-methyl-
9053 - 1-Pentene, 3-methyl-
9054 - 1-Pentene, 4-methyl-
9055 - 2-Pentene, 2-methyl-
9056 - 2-Pentene, 3-methyl-, (*E*)-
9057 - 2-Pentene, 3-methyl-, (*Z*)-
9058 - 2-Pentene, 4-methyl-, (*E*)-
9059 - 2-Pentene, 4-methyl-, (*Z*)-

$C_6H_{12}BrF$
6735 - Hexane, 1-bromo-6-fluoro-

$C_6H_{12}BrNO$
3198 - Butanamide, 2-bromo-2-ethyl-

$C_6H_{12}Br_2$
3269 - Butane, 1,2-dibromo-2,3-dimethyl-
3270 - Butane, 1,4-dibromo-2,3-dimethyl-
3271 - Butane, 2,3-dibromo-2,3-dimethyl-
6754 - Hexane, 1,2-dibromo-
6755 - Hexane, 1,4-dibromo-
6756 - Hexane, 1,5-dibromo-
6757 - Hexane, 1,6-dibromo-
6758 - Hexane, 2,3-dibromo-
6759 - Hexane, 2,5-dibromo-, (*R**,*R**)-(±)-
6760 - Hexane, 3,4-dibromo-

$C_6H_{12}ClF$
6742 - Hexane, 1-chloro-6-fluoro-

$C_6H_{12}ClNO$
47 - Acetamide, 2-chloro-*N*-(1-methylpropyl)-

$C_6H_{12}ClNO_2$
46 - Acetamide, 2-chloro-*N*-(3-methoxypropyl)-

$C_6H_{12}Cl_2$
6761 - Hexane, 1,2-dichloro-
6762 - Hexane, 1,6-dichloro-
6763 - Hexane, 2,2-dichloro-
6764 - Hexane, 2,3-dichloro-
6766 - Hexane, 2,5-dichloro-, (*R**,*R**)-(±)-

$C_6H_{12}Cl_2O$
10121 - Propane, 1,1′-oxybis[2-chloro-
10122 - Propane, 1,1′-oxybis[3-chloro-
10123 - Propane, 2,2′-oxybis[1-chloro-

$C_6H_{12}Cl_2O_2$
5449 - Ethane, 1,2-bis(2-chloroethoxy)-
5530 - Ethane, 1,1-dichloro-2,2-diethoxy-
5592 - Ethane, 1,1′-[ethylidenebis(oxy)]bis[2-chloro-
9896 - Propane, 1,3-dichloro-2-(ethoxymethoxy)-

$C_6H_{12}Cl_2S$
10153 - Propane, 1,1′-thiobis[3-chloro-

$C_6H_{12}Cl_3N$
5412 - Ethanamine, 2-chloro-*N*,*N*-bis(2-chloroethyl)-

$C_6H_{12}Cl_3O_3P$
5702 - Ethanol, 2-chloro-, phosphite (3:1)

$C_6H_{12}Cl_3O_4P$
5701 - Ethanol, 2-chloro-, phosphate (3:1)

$C_6H_{12}I_2$
6769 - Hexane, 1,6-diiodo-

$C_6H_{12}MnN_2S_4$
7421 - Manganous dimethyldithiocarbamate

$C_6H_{12}N_2$
273 - Acetonitrile, (butylamino)-
280 - Acetonitrile, (diethylamino)-
5111 - 1,4-Diazabicyclo[2.2.2]octane
10463 - 2-Propanone, (1-methylethylidene)hydrazone

$C_6H_{12}N_2O_2$
5494 - Ethanediamide, *N*,*N*′-diethyl-

$C_6H_{12}N_2O_3$
3334 - Butanedioic acid, mono (2,2-dimethylhydrazide)

$C_6H_{12}N_2O_4$
3957 - Carbamic acid, (1-methylethyl)nitro-, ethyl ester
3964 - Carbamic acid, nitropropyl-, ethyl ester
6778 - Hexane, 1,6-dinitro-
7150 - 1,1-Hydrazinedicarboxylic acid, diethyl ester
7151 - 1,2-Hydrazinedicarboxylic acid, diethyl ester

$C_6H_{12}N_2O_4S$
5016 - *L*-Cysteine, *S*-(2-amino-2-carboxyethyl)-, (*R*)-

$C_6H_{12}N_2O_4S_2$
5020 - *L*-Cystine

$C_6H_{12}N_2S_3$
11715 - Thiodicarbonic diamide ([(H2N)C(*S*)]2*S*), tetramethyl-

$C_6H_{12}N_2S_4$
11722 - Thioperoxydicarbonic diamide ([(H2N)C(*S*)]2S2), tetramethyl-

$C_6H_{12}N_2S_4Zn$
12068 - Ziram

$C_6H_{12}N_4$
11601 - 1,3,5,7-Tetraazatricyclo[3.3.1.13,7]decane

$C_6H_{12}N_6O_3$
7583 - Methanol, (1,3,5-triazine-2,4,6-triyltriimino)tris-

$C_6H_{12}O$
3183 - Butanal, 2-ethyl-
3369 - Butane, 1-(ethenyloxy)-
3370 - Butane, 2-(ethenyloxy)-
3657 - 2-Butanone, 3,3-dimethyl-
4500 - Cyclohexanol
4825 - Cyclopentanemethanol
4867 - Cyclopentanol, 1-methyl-
4868 - Cyclopentanol, 2-methyl-, *cis*-
4869 - Cyclopentanol, 2-methyl-, *trans*-
4870 - Cyclopentanol, 3-methyl-
6107 - Furan, 2-ethyltetrahydro-

6718 - Hexanal
6959 - 2-Hexanone
6960 - 3-Hexanone
7066 - 1-Hexen-3-ol
7067 - 4-Hexen-2-ol
7068 - 5-Hexen-2-ol
7069 - 2-Hexen-1-ol, (*E*)-
7070 - 2-Hexen-1-ol, (*Z*)-
7071 - 3-Hexen-1-ol, (*Z*)-
7072 - 4-Hexen-1-ol, (*E*)-
8590 - Oxepane
8627 - Oxirane, tetramethyl-
8693 - Pentanal, 2-methyl-
8694 - Pentanal, 3-methyl-
8980 - 2-Pentanone, 3-methyl-, (±)-
8981 - 2-Pentanone, 4-methyl-
8982 - 3-Pentanone, 2-methyl-
9094 - 2-Penten-1-ol, 2-methyl-
9095 - 2-Penten-1-ol, 4-methyl-
9096 - 3-Penten-1-ol, 4-methyl-
9097 - 3-Penten-2-ol, 2-methyl-
9098 - 3-Penten-2-ol, 3-methyl-
9099 - 3-Penten-2-ol, 4-methyl-
9100 - 4-Penten-2-ol, 2-methyl-
9101 - 4-Penten-2-ol, 3-methyl-
9102 - 4-Penten-2-ol, 4-methyl-
10035 - Propane, 1-(ethenyloxy)-2-methyl-
10036 - Propane, 2-(ethenyloxy)-2-methyl-
10579 - 1-Propene, 3-(1-methylethoxy)-
10594 - 1-Propene, 3-propoxy-

$C_6H_{12}O_2$
121 - Acetic acid, butyl ester
177 - Acetic acid, 1,1-dimethylethyl ester
214 - Acetic acid, 1-methylpropyl ester
215 - Acetic acid, 2-methylpropyl ester
3506 - Butanoic acid, 2,2-dimethyl-
3507 - Butanoic acid, 2,3-dimethyl-
3508 - Butanoic acid, 3,3-dimethyl-
3514 - Butanoic acid, 2-ethyl-
3515 - Butanoic acid, ethyl ester
3552 - Butanoic acid, 3-methyl-, methyl ester
3636 - 1-Butanol, 3-methyl-, formate
4372 - 1,2-Cyclohexanediol, *cis*-
4373 - 1,2-Cyclohexanediol, *trans*-(±)-
4374 - 1,4-Cyclohexanediol, *trans*-
5189 - 1,3-Dioxane, 2,4-dimethyl-
5901 - Ethene, 1,1-diethoxy-
6015 - Formic acid, pentyl ester
6170 - Furan, tetrahydro-2-(methoxymethyl)-
6870 - Hexanoic acid
6969 - 2-Hexanone, 5-hydroxy-
6970 - 2-Hexanone, 6-hydroxy-
6971 - 3-Hexanone, 4-hydroxy-
6972 - 3-Hexanone, 5-hydroxy-
7177 - Hydroperoxide, cyclohexyl
8691 - Pentanal, 3-hydroxy-2-methyl-
8887 - Pentanoic acid, 2-methyl-, (±)-
8888 - Pentanoic acid, 3-methyl-, (±)-
8889 - Pentanoic acid, 4-methyl-
8891 - Pentanoic acid, methyl ester
8976 - 2-Pentanone, 4-hydroxy-4-methyl-
10260 - Propanoic acid, 2,2-dimethyl-, methyl ester
10298 - Propanoic acid, 2-methyl-, ethyl ester
10299 - Propanoic acid, 1-methylethyl ester
10326 - Propanoic acid, propyl ester
10784 - 2H-Pyran-2-methanol, tetrahydro-

$C_6H_{12}O_2S$
11784 - Thiophene, tetrahydro-2,4-dimethyl-, 1,1-dioxide

$C_6H_{12}O_3$
181 - Acetic acid, ethoxy-, ethyl ester
3532 - Butanoic acid, 2-hydroxy-, ethyl ester, (±)-
3533 - Butanoic acid, 3-hydroxy-, ethyl ester, (±)-
3534 - Butanoic acid, 3-hydroxy-, ethyl ester
5207 - 1,3-Dioxolane-4-methanol, 2,2-dimethyl-
5718 - Ethanol, 2-ethoxy-, acetate
6169 - Furan, tetrahydro-2,5-dimethoxy-
6891 - Hexanoic acid, 2-hydroxy-, (±)-
7109 - Hexitol, 2,5-anhydro-3,4-dideoxy-
10275 - Propanoic acid, 2-hydroxy-, 1-methylethyl ester
10277 - Propanoic acid, 3-hydroxy-, 1-methylethyl ester
10279 - Propanoic acid, 2-hydroxy-2-methyl-, ethyl ester
11908 - 1,3,5-Trioxane, 2,4,6-trimethyl-

$C_6H_{12}O_4$
5673 - Ethanol, 2-[2-(acetyloxy)ethoxy]-
7111 - *D-ribo*-Hexose, 2,6-dideoxy-
10255 - Propanoic acid, 2,3-dihydroxy-2-methyl-, ethyl ester
10257 - Propanoic acid, 2,3-dihydroxy-, propyl ester

$C_6H_{12}O_4Si$
11419 - Silanediol, dimethyl-, diacetate

$C_6H_{12}O_5$
4074 - *D*-chiro-Inositol, 2-deoxy-
6235 - *D*-Glucose, 6-deoxy-
7110 - *D-arabino*-Hexose, 2-deoxy-

$C_6H_{12}O_5S$
184 - Acetic acid, (ethoxysulfonyl)-, ethyl ester

$C_6H_{12}O_6$
337 - *D*-Allose
338 - *D*-Altrose
4073 - *DL*-chiro-Inositol
6023 - β-*D*-Fructose
6197 - *D*-Galactose
6232 - β-*D*-Glucose
6290 - *D*-Gulose
7336 - *scyllo*-Inositol
7425 - *D*-Mannose
11518 - *DL*-Sorbose
11519 - *L*-Sorbose
11587 - *D*-Tagatose

$C_6H_{12}O_7$
6209 - *D*-Gluconic acid

$C_6H_{12}S$
4480 - Cyclohexanethiol
11690 - Thiepane
11765 - Thiophene, 2-ethyltetrahydro-
11766 - Thiophene, 3-ethyltetrahydro-
11785 - Thiophene, tetrahydro-2,5-dimethyl-, *cis*-
11796 - 2H-Thiopyran, tetrahydro-2-methyl-
11797 - 2H-Thiopyran, tetrahydro-3-methyl-
11798 - 2H-Thiopyran, tetrahydro-4-methyl-

$C_6H_{12}S_3$
11920 - 1,3,5-Trithiane, 2,4,6-trimethyl-
11921 - 1,3,5-Trithiane, 2,4,6-trimethyl-, (2α,4α,6β)-

$C_6H_{13}Br$
3235 - Butane, 1-bromo-3,3-dimethyl-
3236 - Butane, 2-bromo-2,3-dimethyl-
6732 - Hexane, 1-bromo-
6733 - Hexane, 2-bromo-
6734 - Hexane, 3-bromo-
8714 - Pentane, 1-bromo-2-methyl-
8715 - Pentane, 1-bromo-3-methyl-
8716 - Pentane, 1-bromo-4-methyl-
8717 - Pentane, 2-bromo-2-methyl-
8718 - Pentane, 3-bromo-3-methyl-

$C_6H_{13}BrO_2$
5459 - Ethane, 2-bromo-1,1-diethoxy-

$C_6H_{13}Cl$
3245 - Butane, 1-chloro-2,3-dimethyl-
3246 - Butane, 1-chloro-3,3-dimethyl-
3247 - Butane, 2-chloro-2,3-dimethyl-
3248 - Butane, 3-chloro-2,2-dimethyl-
6737 - Hexane, 1-chloro-
6738 - Hexane, 2-chloro-
6739 - Hexane, 3-chloro-
8729 - Pentane, 2-chloro-2-methyl-
8730 - Pentane, 2-chloro-4-methyl-
8731 - Pentane, 3-(chloromethyl)-
8732 - Pentane, 3-chloro-3-methyl-

$C_6H_{13}ClO$
3249 - Butane, 1-(2-chloroethoxy)-
3608 - 2-Butanol, 1-chloro-2,3-dimethyl-
3609 - 2-Butanol, 1-chloro-3,3-dimethyl-
6916 - 1-Hexanol, 6-chloro-
6917 - 2-Hexanol, 1-chloro-
6918 - 3-Hexanol, 1-chloro-
6919 - 3-Hexanol, 5-chloro-

$C_6H_{13}ClO_2$
3244 - Butane, 3-chloro-1,1-dimethoxy-
5469 - Ethane, 2-chloro-1,1-diethoxy-
10355 - 2-Propanol, 1-chloro-3-(1-methylethoxy)-
10356 - 2-Propanol, 1-chloro-3-propoxy-

$C_6H_{13}ClS$
9855 - Propane, 2-[(2-chloroethyl)thio]-2-methyl-

$C_6H_{13}Cl_3Si$
11464 - Silane, trichlorohexyl-

$C_6H_{13}F$
6829 - Hexane, 1-fluoro-
6830 - Hexane, 2-fluoro-

$C_6H_{13}FO$
6936 - 1-Hexanol, 6-fluoro-

$C_6H_{13}I$
6831 - Hexane, 1-iodo-
6832 - Hexane, 2-iodo-, (*R*)-

$C_6H_{13}IO_2$
5543 - Ethane, 1,1-diethoxy-2-iodo-

$C_6H_{13}N$
492 - 1H-Azepine, hexahydro-
4283 - Cyclohexanamine
4772 - Cyclopentanamine, 1-methyl-
4773 - Cyclopentanamine, 2-methyl-
9679 - Piperidine, 1-methyl-
9680 - Piperidine, 2-methyl-, (±)-
9681 - Piperidine, 3-methyl-, (±)-
9682 - Piperidine, 4-methyl-
10525 - 2-Propen-1-amine, *N*-(1-methylethyl)-
10527 - 2-Propen-1-amine, *N*-propyl-
11149 - Pyrrolidine, 2,2-dimethyl-
11150 - Pyrrolidine, 2,4-dimethyl-
11151 - Pyrrolidine, 2,5-dimethyl-, *cis*-
11166 - Pyrrolidine, 1-ethyl-

$C_6H_{13}NO$
38 - Acetamide, *N*-butyl-
53 - Acetamide, *N,N*-diethyl-
3203 - Butanamide, *N,N*-dimethyl-
4501 - Cyclohexanol, 2-amino-, *trans*-(±)-
6721 - Hexanamide
7687 - Morpholine, 2,6-dimethyl-
7694 - Morpholine, 4-ethyl-
8959 - 2-Pentanone, 4-amino-4-methyl-
9674 - 2-Piperidinemethanol
9675 - 3-Piperidinemethanol
9716 - 3-Piperidinol, 1-methyl-
9717 - 4-Piperidinol, 1-methyl-
9802 - Propanamide, *N*-propyl-
11165 - 1-Pyrrolidineethanol

$C_6H_{13}NO_2$
332 - β-Alanine, *N*-methyl-, ethyl ester
3633 - 1-Butanol, 3-methyl-, carbamate
3634 - 2-Butanol, 2-methyl-, carbamate
3937 - Carbamic acid, butyl-, methyl ester
3943 - Carbamic acid, (1,1-dimethylethyl)-, methyl ester
3956 - Carbamic acid, (1-methylethyl)-, ethyl ester
3962 - Carbamic acid, (1-methylpropyl)-, methyl ester
3970 - Carbamic acid, propyl-, ethyl ester
6120 - 2-Furanmethanol, 5-(aminomethyl)tetrahydro-
6841 - Hexane, 1-nitro-
6873 - Hexanoic acid, 6-amino-
7370 - *L*-Isoleucine
7412 - *DL*-Leucine
7413 - *L*-Leucine
7691 - 4-Morpholineethanol
8089 - Nitrous acid, hexyl ester
8214 - Norleucine
8215 - *DL*-Norleucine

$C_6H_{13}NO_3S$
11568 - Sulfamic acid, cyclohexyl-

$C_6H_{13}NO_4$
6259 - Glycine, *N,N*-bis(2-hydroxyethyl)-
$C_6H_{13}NO_5$
6233 - *D*-Glucose, 2-amino-2-deoxy-
$C_6H_{13}NS_2$
5265 - 4H-1,3,5-Dithiazine, dihydro-2,4,6-trimethyl-, (2α,4α,6α)-
$C_6H_{13}N_3$
6283 - Guanidine, (3-methyl-2-butenyl)-
$C_6H_{13}N_3O_3$
8551 - *L*-Ornithine, *N*5-(aminocarbonyl)-
$C_6H_{13}O_3P$
9535 - Phosphonic acid, ethenyl-, diethyl ester
$C_6H_{13}O_4P$
9530 - Phosphonic acid, acetyl-, diethyl ester
C_6H_{14}
3289 - Butane, 2,2-dimethyl-
3290 - Butane, 2,3-dimethyl-
6731 - Hexane
8823 - Pentane, 2-methyl-
8824 - Pentane, 3-methyl-
$C_6H_{14}Cl_4OSi_2$
5231 - Disiloxane, 1,3-bis(dichloromethyl)-1,1,3,3-tetramethyl-
$C_6H_{14}FO_3P$
9580 - Phosphorofluoridic acid, bis(1-methylethyl) ester
$C_6H_{14}N_2$
5132 - Diazene, dipropyl
5141 - 1H-1,4-Diazepine, hexahydro-1-methyl-
9608 - Piperazine, 1,4-dimethyl-
9609 - Piperazine, 2,5-dimethyl-, *cis*-
9610 - Piperazine, 2,5-dimethyl-, *trans*-
9611 - Piperazine, 2,6-dimethyl-, *cis*-
9673 - 4-Piperidinemethanamine
11164 - 1-Pyrrolidineethanamine
$C_6H_{14}N_2O$
7689 - 4-Morpholineethanamine
9616 - 1-Piperazineethanol
9822 - 2-Propanamine, *N*-(1-methylethyl)-*N*-nitroso-
9825 - 1-Propanamine, *N*-nitroso-*N*-propyl-
$C_6H_{14}N_2O_2$
7418 - *DL*-Lysine
7419 - *L*-Lysine
$C_6H_{14}N_4O_2$
437 - *DL*-Arginine
438 - *L*-Arginine
$C_6H_{14}O$
3372 - Butane, 1-ethoxy-
3373 - Butane, 2-ethoxy-
3394 - Butane, 1-methoxy-2-methyl-
3395 - Butane, 1-methoxy-3-methyl-
3396 - Butane, 2-methoxy-2-methyl-
3615 - 1-Butanol, 2,2-dimethyl-
3616 - 1-Butanol, 2,3-dimethyl-, (±)-
3617 - 1-Butanol, 3,3-dimethyl-
3618 - 2-Butanol, 2,3-dimethyl-
3619 - 2-Butanol, 3,3-dimethyl-, (±)-
3620 - 1-Butanol, 2-ethyl-
6911 - 1-Hexanol
6912 - 2-Hexanol, (±)-
6913 - 2-Hexanol, (*R*)-
6914 - 3-Hexanol, (±)-
6915 - 3-Hexanol, (*R*)-
8822 - Pentane, 1-methoxy-
8937 - 1-Pentanol, 2-methyl-
8938 - 1-Pentanol, 3-methyl-, (±)-
8939 - 1-Pentanol, 4-methyl-
8940 - 2-Pentanol, 2-methyl-
8941 - 2-Pentanol, 3-methyl-
8942 - 2-Pentanol, 4-methyl-
8943 - 3-Pentanol, 2-methyl-
8944 - 3-Pentanol, 3-methyl-
10039 - Propane, 1-ethoxy-2-methyl-
10040 - Propane, 2-ethoxy-2-methyl-
10076 - Propane, 1-(1-methylethoxy)-
10119 - Propane, 1,1′-oxybis-
10120 - Propane, 2,2′-oxybis-
$C_6H_{14}OS$
10129 - Propane, 1,1′-sulfinylbis-
$C_6H_{14}OSi$
11423 - Silane, ethenylethoxydimethyl-
$C_6H_{14}O_2$
3350 - 2,3-Butanediol, 2,3-dimethyl-
5541 - Ethane, 1,1-diethoxy-
5542 - Ethane, 1,2-diethoxy-
5687 - Ethanol, 2-butoxy-
5746 - Ethanol, 2-(2-methylpropoxy)-
6793 - 1,3-Hexanediol
6794 - 1,4-Hexanediol
6795 - 1,5-Hexanediol
6796 - 1,6-Hexanediol
6797 - 2,3-Hexanediol
6798 - 2,4-Hexanediol
6799 - 2,5-Hexanediol
8793 - 2,4-Pentanediol, 2-methyl-
8794 - 2,4-Pentanediol, 3-methyl-
10007 - 1,3-Propanediol, 2-ethyl-2-methyl-
10390 - 2-Propanol, 1-(1-methylethoxy)-
10401 - 2-Propanol, 1-propoxy-
$C_6H_{14}O_2S$
10131 - Propane, 1,1′-sulfonylbis-
$C_6H_{14}O_3$
5613 - Ethane, 1,1′-oxybis[2-methoxy-
5614 - Ethane, 1,1′-[oxybis(methyleneoxy)]bis-
5705 - Ethanol, 2,2-diethoxy-
5720 - Ethanol, 2-(2-ethoxyethoxy)-
6867 - 1,2,5-Hexanetriol
6868 - 1,2,6-Hexanetriol
6869 - 2,3,4-Hexanetriol
10006 - 1,3-Propanediol, 2-ethyl-2-(hydroxymethyl)-
10185 - Propane, 1,2,3-trimethoxy-
10397 - Propanol, oxybis-
$C_6H_{14}O_4$
5713 - Ethanol, 2,2′-[1,2-ethanediylbis(oxy)]bis-
$C_6H_{14}O_4S$
11575 - Sulfuric acid, dipropyl ester
$C_6H_{14}O_5$
10024 - Propanediol, oxybis-
10025 - 1,2-Propanediol, 3,3′-oxybis-
$C_6H_{14}O_6$
6194 - Galactitol
6206 - *D*-Glucitol
7422 - *D*-Mannitol
$C_6H_{14}S$
3376 - Butane, 1-(ethylthio)-
6856 - 1-Hexanethiol
6857 - 2-Hexanethiol
6858 - 3-Hexanethiol
8826 - Pentane, 1-(methylthio)-
10044 - Propane, 1-(ethylthio)-2-methyl-
10077 - Propane, 1-[(1-methylethyl)thio]-
10151 - Propane, 1,1′-thiobis-
10152 - Propane, 2,2′-thiobis-
$C_6H_{14}S_2$
5244 - Disulfide, bis(1-methylethyl)
5259 - Disulfide, dipropyl
5450 - Ethane, 1,1-bis(ethylthio)-
5451 - Ethane, 1,2-bis(ethylthio)-
6812 - 1,6-Hexanedithiol
$C_6H_{14}Si$
11495 - Silane, trimethyl-2-propenyl-
$C_6H_{15}As$
448 - Arsine, triethyl-
$C_6H_{15}AsO_3$
440 - Arsenous acid, triethyl ester
$C_6H_{15}AsO_4$
439 - Arsenic acid, triethyl ester
$C_6H_{15}B$
3118 - Borane, triethyl
$C_6H_{15}BO_3$
3125 - Boric acid, triethyl ester
$C_6H_{15}BrSi$
11368 - Silane, bromotriethyl-
$C_6H_{15}ClN_2O_2$
5440 - Ethanaminium, 2-[(aminocarbonyl)oxy]-*N,N,N*-trimethyl-, chloride
$C_6H_{15}ClO_3Si$
11387 - Silane, chlorotriethoxy-
$C_6H_{15}ClSi$
11369 - Silane, butylchlorodimethyl-
11386 - Silane, (3-chloropropyl)trimethyl-
11388 - Silane, chlorotriethyl-
$C_6H_{15}FSi$
11484 - Silane, triethylfluoro-
$C_6H_{15}N$
3215 - 1-Butanamine, *N,N*-dimethyl-
3216 - 2-Butanamine, 2,3-dimethyl-
3217 - 2-Butanamine, 3,3-dimethyl-
3218 - 1-Butanamine, *N*-ethyl-
3219 - 2-Butanamine, *N*-ethyl-, (±)-
5417 - Ethanamine, *N,N*-diethyl-
6723 - 1-Hexanamine
6724 - 2-Hexanamine, (±)-
8706 - 1-Pentanamine, 3-methyl-
8707 - 1-Pentanamine, 4-methyl-
9821 - 2-Propanamine, *N*-(1-methylethyl)-
9826 - 1-Propanamine, *N*-propyl-
9827 - 2-Propanamine, *N*-propyl
9829 - 1-Propanamine, *N,N*,2-trimethyl-
$C_6H_{15}NO$
5424 - Ethanamine, 2-ethoxy-*N,N*-dimethyl-
5691 - Ethanol, 2-(butylamino)-
5706 - Ethanol, 2-(diethylamino)-
5710 - Ethanol, 2-[(1,1-dimethylethyl)amino]-
5747 - Ethanol, 2-[(2-methylpropyl)amino]-
8922 - 1-Pentanol, 2-amino-4-methyl-, (±)-
9814 - 1-Propanamine, *N*-hydroxy-*N*-propyl-
$C_6H_{15}NO_2$
3623 - 2-Butanol, 1-[(2-hydroxyethyl)amino]-
5414 - Ethanamine, 2,2-diethoxy-
5723 - Ethanol, 2,2′-(ethylimino)bis-
10384 - 2-Propanol, 1,1′-iminobis-
$C_6H_{15}NO_3$
5751 - Ethanol, 2,2′,2″-nitrilotris-
$C_6H_{15}NO_5$
6207 - *D*-Glucitol, 1-amino-1-deoxy-
$C_6H_{15}NS$
5650 - Ethanethiol, 2-(butylamino)-
$C_6H_{15}N_2$
5510 - 1,2-Ethanediamine, *N′*-ethyl-*N,N*-dimethyl-
$C_6H_{15}N_3$
9615 - 1-Piperazineethanamine
$C_6H_{15}OP$
9517 - Phosphine oxide, triethyl-
$C_6H_{15}O_2P$
9527 - Phosphinic acid, diethyl-, ethyl ester
$C_6H_{15}O_2PS_2$
9579 - Phosphorodithioic acid, *O,O,S*-triethyl ester
$C_6H_{15}O_3P$
9536 - Phosphonic acid, ethyl-, diethyl ester
9591 - Phosphorous acid, triethyl ester
$C_6H_{15}O_3PS$
9585 - Phosphorothioic acid, *O,O,O*-triethyl ester
$C_6H_{15}O_3PS_2$
5103 - Demeton S methyl
$C_6H_{15}O_4P$
9563 - Phosphoric acid, triethyl ester
$C_6H_{15}O_4PS_2$
8632 - Oxydemeton-methyl
$C_6H_{15}P$
9521 - Phosphine, triethyl-
$C_6H_{15}Sb$
11549 - Stibine, triethyl-
$C_6H_{16}Br_2OSi_2$
5229 - Disiloxane, 1,3-bis(bromomethyl)-1,1,3,3-tetramethyl-
$C_6H_{16}ClN$
5418 - Ethanamine, *N,N*-diethyl-, hydrochloride

$C_6H_{16}ClNO$
9831 - 1-Propanaminium, 2-hydroxy-*N,N,N*-trimethyl-, chloride
$C_6H_{16}Cl_2OSi_2$
5230 - Disiloxane, 1,3-bis(chloromethyl)-1,1,3,3-tetramethyl-
$C_6H_{16}N_2$
5501 - 1,2-Ethanediamine, *N,N'*-diethyl-
5502 - 1,2-Ethanediamine, *N,N*-diethyl-
5516 - 1,2-Ethanediamine, *N,N,N',N'*-tetramethyl-
6749 - 1,6-Hexanediamine
7137 - Hydrazine, 1,2-bis(1-methylethyl)-
$C_6H_{16}N_2O$
5677 - Ethanol, 2-[(2-amino-2-methylpropyl)amino]-
$C_6H_{16}N_2O_2$
5715 - Ethanol, 2,2'-(1,2-ethanediyldiimino)bis-
$C_6H_{16}OSi$
10410 - 1-Propanol, 3-(trimethylsilyl)-
11503 - Silanol, triethyl-
$C_6H_{16}O_2Si$
11405 - Silane, diethoxydimethyl-
$C_6H_{16}O_3Si$
11477 - Silane, triethoxy-
$C_6H_{16}Si$
11483 - Silane, triethyl-
11496 - Silane, trimethylpropyl-
$C_6H_{17}N_3$
9874 - 1,3-Propanediamine, *N*-(3-aminopropyl)-
$C_6H_{18}AlO_9P_3$
6020 - Fosetyl-Al
$C_6H_{18}Cl_2O_2Si_3$
11913 - Trisiloxane, 1,5-dichloro-1,1,3,3,5,5-hexamethyl-
$C_6H_{18}N_3OP$
9573 - Phosphoric triamide, hexamethyl-
$C_6H_{18}N_4$
5499 - 1,2-Ethanediamine, *N,N'*-bis(2-aminoethyl)-
$C_6H_{18}OSi_2$
5236 - Disiloxane, hexamethyl-
$C_6H_{18}O_3Si_2$
5234 - Disiloxane, 1,3-dimethoxy-1,1,3,3-tetramethyl-
$C_6H_{18}O_3Si_3$
5002 - Cyclotrisiloxane, hexamethyl-
$C_6H_{18}Si_2$
5228 - Disilane, hexamethyl-
$C_6H_{19}NSi_2$
11363 - Silanamine, 1,1,1-trimethyl-*N*-(trimethylsilyl)-
$C_6H_{21}N_3Si_3$
5001 - Cyclotrisilazane, 2,2,4,4,6,6-hexamethyl-
$C_6H_{24}O_6Si_6$
4625 - Cyclohexasiloxane, 2,4,6,8,10,12-hexamethyl-
$C_6H_{26}O_8$
3351 - 2,3-Butanediol, 2,3-dimethyl-, hexahydrate
C_6MoO_6
7662 - Molybdenum carbonyl
C_6N_4
5915 - Ethenetetracarbonitrile
C_6N_6
11836 - 1,3,5-Triazine-2,4,6-tricarbonitrile
C_7F_5N
2767 - Benzonitrile, pentafluoro-
C_7F_8
2095 - Benzene, pentafluoro(trifluoromethyl)-
C_7F_{14}
4499 - Cyclohexane, undecafluoro(trifluoromethyl)-
6560 - 1-Heptene, 1,1,2,3,3,4,4,5,5,6,6,7,7,7-tetradecafluoro-
C_7F_{16}
6418 - Heptane, hexadecafluoro-
$C_7HCl_5O_2$
2712 - Benzoic acid, pentachloro-
C_7HCl_7
2085 - Benzene, pentachloro(dichloromethyl)-
C_7HF_5O
596 - Benzaldehyde, pentafluoro-
C_7HF_{15}
6432 - Heptane, 1,1,1,2,2,3,3,4,4,5,5,6,6,7,7-pentadecafluoro-
$C_7H_3Br_2NO$
2751 - Benzonitrile, 3,5-dibromo-4-hydroxy-
$C_7H_3Br_5$
2083 - Benzene, pentabromomethyl-
$C_7H_3ClF_3NO_2$
1217 - Benzene, 1-chloro-2-nitro-4-(trifluoromethyl)-
1218 - Benzene, 1-chloro-4-nitro-2-(trifluoromethyl)-
1219 - Benzene, 4-chloro-1-nitro-2-(trifluoromethyl)-
$C_7H_3ClN_2O_5$
2896 - Benzoyl chloride, 3,5-dinitro-
$C_7H_3Cl_2N$
2752 - Benzonitrile, 2,6-dichloro-
$C_7H_3Cl_3O$
2893 - Benzoyl chloride, 2,4-dichloro-
2894 - Benzoyl chloride, 3,4-dichloro-
$C_7H_3Cl_3O_2$
2727 - Benzoic acid, 2,3,6-trichloro-
2728 - Benzoic acid, 2,4,5-trichloro-
$C_7H_3Cl_5$
1401 - Benzene, 1,2-dichloro-4-(trichloromethyl)-
2086 - Benzene, pentachloromethyl-
2326 - Benzene, 1,2,3-trichloro-4-(dichloromethyl)-
2327 - Benzene, 1,2,4-trichloro-5-(dichloromethyl)-
$C_7H_3F_5$
2094 - Benzene, pentafluoromethyl-
$C_7H_3F_5O$
2093 - Benzene, pentafluoromethoxy-
$C_7H_3N_3O_8$
2735 - Benzoic acid, 2,4,6-trinitro-
C_7H_4BrClO
2887 - Benzoyl chloride, 2-bromo-
2888 - Benzoyl chloride, 4-bromo-
$C_7H_4BrF_3$
1089 - Benzene, 1-bromo-2-(trifluoromethyl)-
1090 - Benzene, 1-bromo-3-(trifluoromethyl)-
1091 - Benzene, 1-bromo-4-(trifluoromethyl)-
C_7H_4BrN
2744 - Benzonitrile, 2-bromo-
2745 - Benzonitrile, 3-bromo-
2746 - Benzonitrile, 4-bromo-
C_7H_4BrNO
1049 - Benzene, 1-bromo-4-isocyanato-
$C_7H_4BrNO_4$
2527 - Benzoic acid, 2-bromo-4-nitro-
2528 - Benzoic acid, 2-bromo-5-nitro-
2529 - Benzoic acid, 4-bromo-3-nitro-
2530 - Benzoic acid, 5-bromo-2-nitro-
$C_7H_4Br_2O_2$
534 - Benzaldehyde, 3,5-dibromo-2-hydroxy-
2555 - Benzoic acid, 2,4-dibromo-
2556 - Benzoic acid, 2,5-dibromo-
2557 - Benzoic acid, 2,6-dibromo-
$C_7H_4Br_2O_3$
2558 - Benzoic acid, 3,5-dibromo-2-hydroxy-
$C_7H_4Br_2O_5$
2560 - Benzoic acid, 2,6-dibromo-3,4,5-trihydroxy-
$C_7H_4Br_4O$
2256 - Benzene, 1,2,3,5-tetrabromo-4-methoxy-
9468 - Phenol, 2,3,4,5-tetrabromo-6-methyl-
C_7H_4ClFO
2897 - Benzoyl chloride, 2-fluoro-
2898 - Benzoyl chloride, 3-fluoro-
2899 - Benzoyl chloride, 4-fluoro-
$C_7H_4ClF_3$
1240 - Benzene, 1-chloro-2-(trifluoromethyl)-
1241 - Benzene, 1-chloro-3-(trifluoromethyl)-
1242 - Benzene, 1-chloro-4-(trifluoromethyl)-
C_7H_4ClN
2747 - Benzonitrile, 2-chloro-
2748 - Benzonitrile, 3-chloro-
2749 - Benzonitrile, 4-chloro-
C_7H_4ClNO
1176 - Benzene, 1-chloro-2-isocyanato-
2878 - Benzoxazole, 2-chloro-
$C_7H_4ClNO_2$
2883 - 2(3H)-Benzoxazolone, 5-chloro-
$C_7H_4ClNO_3$
2905 - Benzoyl chloride, 3-nitro-
2906 - Benzoyl chloride, 4-nitro-
$C_7H_4ClNO_4$
2544 - Benzoic acid, 2-chloro-3-nitro-
2545 - Benzoic acid, 2-chloro-5-nitro-
2546 - Benzoic acid, 3-chloro-2-nitro-
2547 - Benzoic acid, 4-chloro-3-nitro-
C_7H_4ClNS
1177 - Benzene, 1-chloro-4-isothiocyanato-
2862 - Benzothiazole, 2-chloro-
$C_7H_4Cl_2O$
535 - Benzaldehyde, 2,5-dichloro-
536 - Benzaldehyde, 3,4-dichloro-
537 - Benzaldehyde, 3,5-dichloro-
2890 - Benzoyl chloride, 2-chloro-
2891 - Benzoyl chloride, 3-chloro-
2892 - Benzoyl chloride, 4-chloro-
$C_7H_4Cl_2O_2$
2561 - Benzoic acid, 2,4-dichloro-
2562 - Benzoic acid, 2,5-dichloro-
2563 - Benzoic acid, 2,6-dichloro-
2564 - Benzoic acid, 3,5-dichloro-
$C_7H_4Cl_2O_3$
2566 - Benzoic acid, 3,5-dichloro-2-hydroxy-
$C_7H_4Cl_3F$
1736 - Benzene, 1-fluoro-2-(trichloromethyl)-
$C_7H_4Cl_3NO_3$
11854 - Triclopyr
$C_7H_4Cl_4$
1237 - Benzene, 1-chloro-2-(trichloromethyl)-
1238 - Benzene, 1-chloro-3-(trichloromethyl)-
1239 - Benzene, 1-chloro-4-(trichloromethyl)-
1359 - Benzene, 1,2-dichloro-4-(dichloromethyl)-
2262 - Benzene, 1,2,3,5-tetrachloro-4-methyl-
2263 - Benzene, 1,2,4,5-tetrachloro-3-methyl-
2324 - Benzene, 1,2,4-trichloro-5-(chloromethyl)-
$C_7H_4Cl_4O_2$
9472 - Phenol, 2,3,5,6-tetrachloro-4-methoxy-
C_7H_4FN
2756 - Benzonitrile, 4-fluoro-
C_7H_4FNS
1721 - Benzene, 1-fluoro-3-isothiocyanato-
$C_7H_4F_3NO_2$
2060 - Benzene, 1-nitro-2-(trifluoromethyl)-
2061 - Benzene, 1-nitro-3-(trifluoromethyl)-
$C_7H_4F_4$
1737 - Benzene, 1-fluoro-2-(trifluoromethyl)-
1738 - Benzene, 1-fluoro-3-(trifluoromethyl)-
1739 - Benzene, 1-fluoro-4-(trifluoromethyl)-
$C_7H_4I_2O_3$
2617 - Benzoic acid, 2-hydroxy-3,5-diiodo-
2618 - Benzoic acid, 4-hydroxy-3,5-diiodo-
$C_7H_4N_2O_3$
1788 - Benzene, 1-isocyanato-2-nitro-
1789 - Benzene, 1-isocyanato-3-nitro-
1790 - Benzene, 1-isocyanato-4-nitro-
$C_7H_4N_2O_5$
551 - Benzaldehyde, 2,4-dinitro-
$C_7H_4N_2O_6$
2590 - Benzoic acid, 3,4-dinitro-
2591 - Benzoic acid, 3,5-dinitro-
$C_7H_4O_4S$
2875 - 3H-2,1-Benzoxathiol-3-one, 1,1-dioxide
$C_7H_4O_6$
10776 - 4H-Pyran-2,6-dicarboxylic acid, 4-oxo-
$C_7H_4O_7$
10775 - 4H-Pyran-2,6-dicarboxylic acid, 3-hydroxy-4-oxo-
C_7H_5BrO
526 - Benzaldehyde, 2-bromo-
527 - Benzaldehyde, 3-bromo-
528 - Benzaldehyde, 4-bromo-
2885 - Benzoyl bromide

$C_7H_5BrO_2$
2518 - Benzoic acid, 2-bromo-
2519 - Benzoic acid, 3-bromo-
2520 - Benzoic acid, 4-bromo-
$C_7H_5BrO_3$
2524 - Benzoic acid, 5-bromo-2-hydroxy-
$C_7H_5Br_3$
1291 - Benzene, 1,3-dibromo-5-(bromomethyl)-
2314 - Benzene, 1,3,5-tribromo-2-methyl-
$C_7H_5Br_3O$
2310 - Benzene, 1,2,3-tribromo-5-methoxy-
2311 - Benzene, 1,2,4-tribromo-5-methoxy-
2312 - Benzene, 1,2,5-tribromo-3-methoxy-
2313 - Benzene, 1,3,5-tribromo-2-methoxy-
9480 - Phenol, 2,4,6-tribromo-3-methyl-
$C_7H_5ClN_2$
7236 - 1H-Indazole, 3-chloro-
$C_7H_5ClN_2O$
2876 - 2-Benzoxazolamine, 5-chloro-
C_7H_5ClO
529 - Benzaldehyde, 2-chloro-
530 - Benzaldehyde, 3-chloro-
531 - Benzaldehyde, 4-chloro-
2886 - Benzoyl chloride
$C_7H_5ClO_2$
532 - Benzaldehyde, 3-chloro-4-hydroxy-
533 - Benzaldehyde, 5-chloro-2-hydroxy-
2534 - Benzoic acid, 2-chloro-
2535 - Benzoic acid, 3-chloro-
2536 - Benzoic acid, 4-chloro-
2900 - Benzoyl chloride, 2-hydroxy-
4026 - Carbonochloridic acid, phenyl ester
$C_7H_5ClO_3$
1127 - Benzenecarboperoxoic acid, 3-chloro-
2540 - Benzoic acid, 3-chloro-4-hydroxy-
$C_7H_5Cl_2F$
1379 - Benzene, (dichlorofluoromethyl)-
$C_7H_5Cl_2FN_2O_3$
5982 - Fluroxypyr
$C_7H_5Cl_2NO_2$
1135 - Benzene, 1-chloro-4-(chloromethyl)-2-nitro-
2491 - Benzoic acid, 2-amino-3,6-dichloro-
2492 - Benzoic acid, 3-amino-2,5-dichloro-
$C_7H_5Cl_2NO_4S$
2565 - Benzoic acid, 4-[(dichloroamino)sulfonyl]-
$C_7H_5Cl_3$
1357 - Benzene, 1,2-dichloro-4-(chloromethyl)-
1358 - Benzene, 1,3-dichloro-5-(chloromethyl)-
2333 - Benzene, (trichloromethyl)-
2334 - Benzene, 1,2,3-trichloro-4-methyl-
2335 - Benzene, 1,2,3-trichloro-5-methyl-
2336 - Benzene, 1,2,4-trichloro-5-methyl-
2337 - Benzene, 1,2,5-trichloro-3-methyl-
$C_7H_5Cl_3O$
2330 - Benzene, 1,2,4-trichloro-3-methoxy-
2331 - Benzene, 1,2,4-trichloro-5-methoxy-
2332 - Benzene, 1,3,5-trichloro-2-methoxy-
9487 - Phenol, 2,3,4-trichloro-6-methyl-
9488 - Phenol, 2,3,6-trichloro-4-methyl-
9489 - Phenol, 2,4,6-trichloro-3-methyl-
C_7H_5FO
559 - Benzaldehyde, 2-fluoro-
560 - Benzaldehyde, 3-fluoro-
561 - Benzaldehyde, 4-fluoro-
2909 - Benzoyl fluoride
$C_7H_5FO_2$
2603 - Benzoic acid, 2-fluoro-
2604 - Benzoic acid, 3-fluoro-
2605 - Benzoic acid, 4-fluoro-
$C_7H_5F_3$
2348 - Benzene, (trifluoromethyl)-
$C_7H_5F_3O$
9491 - Phenol, 3-(trifluoromethyl)-
$C_7H_5HgNO_3$
7437 - Mercury, [2-methyl-5-nitrophenolato(2-)-C6,*O*1]-
C_7H_5IO
580 - Benzaldehyde, 2-iodo-
581 - Benzaldehyde, 3-iodo-
582 - Benzaldehyde, 4-iodo-
2910 - Benzoyl iodide
$C_7H_5IO_2$
2657 - Benzoic acid, 2-iodo-
2658 - Benzoic acid, 3-iodo-
2659 - Benzoic acid, 4-iodo-
$C_7H_5IO_3$
2623 - Benzoic acid, 4-hydroxy-3-iodo-
2624 - Benzoic acid, 5-hydroxy-2-iodo-
$C_7H_5I_3$
2352 - Benzene, 1,3,5-triiodo-2-methyl-
C_7H_5N
2740 - Benzonitrile
C_7H_5NO
1786 - Benzene, isocyanato-
2389 - 1,2-Benzisoxazole
2390 - 2,1-Benzisoxazole
2759 - Benzonitrile, 2-hydroxy-
2877 - Benzoxazole
10585 - 2-Propenenitrile, 3-(2-furanyl)-
C_7H_5NOS
2868 - 2(3H)-Benzothiazolone
$C_7H_5NO_2$
2881 - 2-Benzoxazolol
2882 - 2(3H)-Benzoxazolone
$C_7H_5NO_3$
591 - Benzaldehyde, 2-nitro-
592 - Benzaldehyde, 3-nitro-
593 - Benzaldehyde, 4-nitro-
$C_7H_5NO_3S$
2388 - 1,2-Benzisothiazol-3(2H)-one, 1,1-dioxide
$C_7H_5NO_4$
2703 - Benzoic acid, 2-nitro-
2704 - Benzoic acid, 3-nitro-
2705 - Benzoic acid, 4-nitro-
10927 - 2,3-Pyridinedicarboxylic acid
10928 - 2,4-Pyridinedicarboxylic acid
10929 - 2,5-Pyridinedicarboxylic acid
10930 - 3,4-Pyridinedicarboxylic acid
10931 - 3,5-Pyridinedicarboxylic acid
$C_7H_5NO_5$
2646 - Benzoic acid, 2-hydroxy-3-nitro-
2647 - Benzoic acid, 2-hydroxy-5-nitro-
C_7H_5NS
1792 - Benzene, isothiocyanato-
2387 - 1,2-Benzisothiazole
2861 - Benzothiazole
11709 - Thiocyanic acid, phenyl ester
$C_7H_5NS_2$
2866 - 2(3H)-Benzothiazolethione
$C_7H_5N_3$
2870 - 1,2,4-Benzotriazine
$C_7H_5N_3O$
2884 - Benzoyl azide
$C_7H_5N_3O_5$
614 - Benzamide, 3,5-dinitro-
$C_7H_5N_3O_6$
2046 - Benzene, 1-methyl-2,3,4-trinitro-
2047 - Benzene, 1-methyl-2,4,5-trinitro-
2048 - Benzene, 2-methyl-1,3,5-trinitro-
$C_7H_5N_3O_7$
364 - Anisole, 2,3,5-trinitro-
1946 - Benzene, 2-methoxy-1,3,5-trinitro-
9434 - Phenol, 3-methyl-2,4,6-trinitro-
$C_7H_5N_5O_8$
818 - Benzenamine, *N*-methyl-*N*,2,4,6-tetranitro-
$C_7H_5NaO_2$
2725 - Benzoic acid, sodium salt
C_7H_6BrCl
1007 - Benzene, 1-bromo-3-(chloromethyl)-
1008 - Benzene, 1-bromo-4-(chloromethyl)-
C_7H_6BrNO
604 - Benzamide, 2-bromo-
605 - Benzamide, 3-bromo-
$C_7H_6BrNO_2$
1066 - Benzene, 1-bromo-2-methyl-4-nitro-
1067 - Benzene, 1-(bromomethyl)-3-nitro-
1068 - Benzene, 1-bromo-3-methyl-2-nitro-
1069 - Benzene, 1-bromo-3-methyl-5-nitro-
1070 - Benzene, 1-bromo-2-methyl-3-nitro-
1071 - Benzene, 2-bromo-1-methyl-3-nitro-
1072 - Benzene, 2-bromo-1-methyl-4-nitro-
1073 - Benzene, 2-bromo-4-methyl-1-nitro-
1074 - Benzene, 4-bromo-1-methyl-2-nitro-
2487 - Benzoic acid, 2-amino-5-bromo-
$C_7H_6BrNO_3$
606 - Benzamide, 5-bromo-*N*,2-dihydroxy-
$C_7H_6Br_2$
999 - Benzene, 1-bromo-2-(bromomethyl)-
1000 - Benzene, 1-bromo-4-(bromomethyl)-
1301 - Benzene, (dibromomethyl)-
1302 - Benzene, 1,2-dibromo-3-methyl-
1303 - Benzene, 1,2-dibromo-4-methyl-
1304 - Benzene, 1,3-dibromo-2-methyl-
1305 - Benzene, 1,4-dibromo-2-methyl-
1306 - Benzene, 2,4-dibromo-1-methyl-
$C_7H_6Br_2O$
9279 - Phenol, 2,4-dibromo-6-methyl-
9280 - Phenol, 3,5-dibromo-2-methyl-
C_7H_6ClF
1169 - Benzene, 1-chloro-3-fluoro-2-methyl-
1170 - Benzene, 2-chloro-4-fluoro-1-methyl-
1194 - Benzene, 1-(chloromethyl)-2-fluoro-
1195 - Benzene, 1-(chloromethyl)-4-fluoro-
C_7H_6ClI
1173 - Benzene, 1-chloro-2-iodo-4-methyl-
1174 - Benzene, 4-chloro-1-iodo-2-methyl-
1175 - Benzene, 4-chloro-2-iodo-1-methyl-
11817 - Toluene, 3-chloro-2-iodo-
$C_7H_6ClNO_2$
1203 - Benzene, 1-(chloromethyl)-2-nitro-
1204 - Benzene, 1-chloro-2-methyl-3-nitro-
1205 - Benzene, 1-chloro-2-methyl-4-nitro-
1206 - Benzene, 1-(chloromethyl)-3-nitro-
1207 - Benzene, 1-chloro-3-methyl-5-nitro-
1208 - Benzene, 1-(chloromethyl)-4-nitro-
1209 - Benzene, 1-chloro-4-methyl-2-nitro-
1210 - Benzene, 2-chloro-1-methyl-4-nitro-
1211 - Benzene, 2-chloro-4-methyl-1-nitro-
1212 - Benzene, 4-chloro-1-methyl-2-nitro-
1213 - Benzene, 4-chloro-2-methyl-1-nitro-
2490 - Benzoic acid, 5-amino-2-chloro-
$C_7H_6ClN_3O_4S_2$
2852 - 2H-1,2,4-Benzothiadiazine-7-sulfonamide, 6-chloro-, 1,1-dioxide
$C_7H_6Cl_2$
1132 - Benzene, 1-chloro-2-(chloromethyl)-
1133 - Benzene, 1-chloro-3-(chloromethyl)-
1134 - Benzene, 1-chloro-4-(chloromethyl)-
1382 - Benzene, (dichloromethyl)-
1383 - Benzene, 1,2-dichloro-3-methyl-
1384 - Benzene, 1,2-dichloro-4-methyl-
1385 - Benzene, 1,3-dichloro-2-methyl-
1386 - Benzene, 1,3-dichloro-5-methyl-
1387 - Benzene, 1,4-dichloro-2-methyl-
1388 - Benzene, 2,4-dichloro-1-methyl-
$C_7H_6Cl_2O$
1381 - Benzene, 2,4-dichloro-1-methoxy-
1853 - Benzenemethanol, 2,4-dichloro-
9292 - Phenol, 2,4-dichloro-3-methyl-
9293 - Phenol, 2,4-dichloro-5-methyl-
9294 - Phenol, 2,6-dichloro-3-methyl-
9295 - Phenol, 2,6-dichloro-4-methyl-
$C_7H_6Cl_2S$
1136 - Benzene, 1-chloro-4-[(chloromethyl)thio]-
$C_7H_6Cl_4O_2$
4770 - 1,3-Cyclopentadiene, 1,2,3,4-tetrachloro-5,5-dimethoxy-
$C_7H_6FNO_2$
1729 - Benzene, 1-fluoro-4-methyl-2-nitro-
1730 - Benzene, 2-fluoro-4-methyl-1-nitro-

1731 - Benzene, 4-fluoro-1-methyl-2-nitro-
1732 - Benzene, 4-fluoro-2-methyl-1-nitro-
$C_7H_6F_2$
1425 - Benzene, (difluoromethyl)-
$C_7H_6F_3N$
855 - Benzenamine, 2-(trifluoromethyl)-
856 - Benzenamine, 3-(trifluoromethyl)-
857 - Benzenamine, 4-(trifluoromethyl)-
$C_7H_6N_2$
2376 - 1H-Benzimidazole
2741 - Benzonitrile, 2-amino-
2742 - Benzonitrile, 3-amino-
7235 - 1H-Indazole
$C_7H_6N_2O$
10884 - 3-Pyridinecarbonitrile, 1,2-dihydro-1-methyl-2-oxo-
$C_7H_6N_2O_3$
624 - Benzamide, 2-nitro-
625 - Benzamide, 3-nitro-
$C_7H_6N_2O_4$
1956 - Benzene, 1-methyl-2,4-dinitro-
1957 - Benzene, 1-methyl-3,5-dinitro-
1958 - Benzene, 2-methyl-1,3-dinitro-
1959 - Benzene, 2-methyl-1,4-dinitro-
1960 - Benzene, 4-methyl-1,2-dinitro-
2506 - Benzoic acid, 2-amino-3-nitro-
$C_7H_6N_2O_5$
1922 - Benzene, 1-methoxy-2,3-dinitro-
1923 - Benzene, 1-methoxy-2,4-dinitro-
1924 - Benzene, 1-methoxy-3,5-dinitro-
1925 - Benzene, 2-methoxy-1,3-dinitro-
1926 - Benzene, 2-methoxy-1,4-dinitro-
1927 - Benzene, 4-methoxy-1,2-dinitro-
9405 - Phenol, 2-methyl-4,6-dinitro-
$C_7H_6N_2S$
2386 - 2H-Benzimidazole-2-thione, 1,3-dihydro-
2859 - 2-Benzothiazolamine
2860 - 6-Benzothiazolamine
$C_7H_6N_4O$
7193 - 1H-Imidazole, 1,1′-carbonylbis-
C_7H_6O
525 - Benzaldehyde
4242 - 2,4,6-Cycloheptatrien-1-one
C_7H_6OS
1130 - Benzenecarbothioic acid
$C_7H_6OS_2$
1125 - Benzenecarbodithioic acid, 2-hydroxy-
$C_7H_6O_2$
564 - Benzaldehyde, 2-hydroxy-
565 - Benzaldehyde, 3-hydroxy-
566 - Benzaldehyde, 4-hydroxy-
2415 - 1,3-Benzodioxole
2475 - Benzoic acid
4243 - 2,4,6-Cycloheptatrien-1-one, 2-hydroxy-
4266 - 2,5-Cyclohexadiene-1,4-dione, 2-methyl-
10512 - 2-Propenal, 3-(2-furanyl)-
10513 - 2-Propenal, 3-(2-furanyl)-, (*E*)-
$C_7H_6O_2S$
2663 - Benzoic acid, 2-mercapto-
$C_7H_6O_3$
540 - Benzaldehyde, 2,3-dihydroxy-
541 - Benzaldehyde, 2,4-dihydroxy-
542 - Benzaldehyde, 3,4-dihydroxy-
1126 - Benzenecarboperoxoic acid
2612 - Benzoic acid, 2-hydroxy-
2613 - Benzoic acid, 4-hydroxy-
10638 - 2-Propenoic acid, 3-(2-furanyl)-
$C_7H_6O_4$
2568 - Benzoic acid, 2,3-dihydroxy-
2569 - Benzoic acid, 2,4-dihydroxy-
2570 - Benzoic acid, 2,5-dihydroxy-
2571 - Benzoic acid, 3,4-dihydroxy-
6183 - 4H-Furo[3,2-c]pyran-2(6H)-one, 4-hydroxy-
$C_7H_6O_5$
2730 - Benzoic acid, 2,3,4-trihydroxy-
2731 - Benzoic acid, 3,4,5-trihydroxy-
$C_7H_6O_6S$
2655 - Benzoic acid, 2-hydroxy-5-sulfo-
$C_7H_7BF_2$
3115 - Borane, difluoro(4-methylphenyl)-
C_7H_7Br
1053 - Benzene, (bromomethyl)-
1054 - Benzene, 1-bromo-2-methyl-
1055 - Benzene, 1-bromo-3-methyl-
1056 - Benzene, 1-bromo-4-methyl-
$C_7H_7BrN_2O$
11998 - Urea, (4-bromophenyl)-
C_7H_7BrO
1050 - Benzene, 1-bromo-2-methoxy-
1051 - Benzene, 1-bromo-3-methoxy-
1052 - Benzene, 1-bromo-4-methoxy-
9236 - Phenol, 2-bromo-4-methyl-
9237 - Phenol, 2-bromo-5-methyl-
9238 - Phenol, 3-bromo-2-methyl-
9239 - Phenol, 3-bromo-4-methyl-
9240 - Phenol, 3-bromo-5-methyl-
9241 - Phenol, 4-bromo-2-methyl-
9242 - Phenol, 4-bromo-3-methyl-
$C_7H_7BrO_2$
1843 - Benzenemethanol, 5-bromo-2-hydroxy-
$C_7H_7BrO_3$
6190 - 3-Furoic acid, 5-bromo-, ethyl ester
C_7H_7Cl
1182 - Benzene, (chloromethyl)-
1183 - Benzene, 1-chloro-2-methyl-
1184 - Benzene, 1-chloro-3-methyl-
1185 - Benzene, 1-chloro-4-methyl-
$C_7H_7ClN_2S$
11801 - Thiourea, (2-chlorophenyl)-
C_7H_7ClO
1178 - Benzene, 1-chloro-2-methoxy-
1179 - Benzene, 1-chloro-3-methoxy-
1180 - Benzene, 1-chloro-4-methoxy-
1848 - Benzenemethanol, 2-chloro-
1849 - Benzenemethanol, 4-chloro-
9260 - Phenol, 2-chloro-3-methyl-
9261 - Phenol, 2-chloro-4-methyl-
9262 - Phenol, 2-chloro-5-methyl-
9263 - Phenol, 2-chloro-6-methyl-
9264 - Phenol, 3-chloro-2-methyl-
9265 - Phenol, 3-chloro-4-methyl-
9266 - Phenol, 4-chloro-2-methyl-
9267 - Phenol, 4-chloro-3-methyl-
C_7H_7ClOS
2186 - Benzenesulfinyl chloride, 4-methyl-
$C_7H_7ClO_2S$
2249 - Benzenesulfonyl chloride, 2-methyl-
2250 - Benzenesulfonyl chloride, 4-methyl-
$C_7H_7ClO_3$
6053 - 2-Furancarboxylic acid, 5-chloro-, ethyl ester
$C_7H_7ClO_3S$
2248 - Benzenesulfonyl chloride, 2-methoxy-
C_7H_7ClS
1833 - Benzenemethanethiol, 4-chloro-
$C_7H_7Cl_2N$
1803 - Benzenemethanamine, 2,4-dichloro-
$C_7H_7Cl_2NO_2S$
2210 - Benzenesulfonamide, *N,N*-dichloro-4-methyl-
$C_7H_7Cl_3NO_3PS$
4090 - Chlorpyrifos-methyl
$C_7H_7Cl_3Si$
11467 - Silane, trichloro(2-methylphenyl)-
11468 - Silane, trichloro(3-methylphenyl)-
C_7H_7F
1725 - Benzene, (fluoromethyl)-
1726 - Benzene, 1-fluoro-2-methyl-
1727 - Benzene, 1-fluoro-3-methyl-
1728 - Benzene, 1-fluoro-4-methyl-
C_7H_7FO
1722 - Benzene, 1-fluoro-2-methoxy-
1723 - Benzene, 1-fluoro-3-methoxy-
1724 - Benzene, 1-fluoro-4-methoxy-
1879 - Benzenemethanol, 3-fluoro-
1880 - Benzenemethanol, 4-fluoro-
$C_7H_7F_7O_2$
3526 - Butanoic acid, heptafluoro-, 1-methylethyl ester
C_7H_7I
1777 - Benzene, (iodomethyl)-
1778 - Benzene, 1-iodo-2-methyl-
1779 - Benzene, 1-iodo-3-methyl-
1780 - Benzene, 1-iodo-4-methyl-
C_7H_7IO
1775 - Benzene, 1-iodo-2-methoxy-
1776 - Benzene, 1-iodo-4-methoxy-
1888 - Benzenemethanol, 2-iodo-
9374 - Phenol, 2-iodo-4-methyl-
9375 - Phenol, 2-iodo-6-methyl-
C_7H_7N
10952 - Pyridine, 2-ethenyl-
10953 - Pyridine, 3-ethenyl-
10954 - Pyridine, 4-ethenyl-
C_7H_7NO
594 - Benzaldehyde, oxime, (*E*)-
595 - Benzaldehyde, oxime, (*Z*)-
600 - Benzamide
5867 - Ethanone, 1-(2-pyridinyl)-
5868 - Ethanone, 1-(3-pyridinyl)-
5869 - Ethanone, 1-(4-pyridinyl)-
5999 - Formamide, *N*-phenyl-
$C_7H_7NO_2$
579 - Benzaldehyde, 2-hydroxy-, oxime
616 - Benzamide, 2-hydroxy-
2000 - Benzene, 1-methyl-2-nitro-
2001 - Benzene, 1-methyl-3-nitro-
2002 - Benzene, 1-methyl-4-nitro-
2052 - Benzene, (nitromethyl)-
2484 - Benzoic acid, 2-amino-
2485 - Benzoic acid, 3-amino-
2486 - Benzoic acid, 4-amino-
8097 - Nitrous acid, phenylmethyl ester
10906 - 3-Pyridinecarboxylic acid, methyl ester
10908 - 4-Pyridinecarboxylic acid, methyl ester
11029 - Pyridinium, 3-carboxy-1-methyl-, hydroxide, inner salt
$C_7H_7NO_2S$
2043 - Benzene, 1-(methylthio)-2-nitro-
2044 - Benzene, 1-(methylthio)-4-nitro-
$C_7H_7NO_3$
613 - Benzamide, *N*,2-dihydroxy-
1906 - Benzenemethanol, 2-nitro-
1907 - Benzenemethanol, 3-nitro-
1908 - Benzenemethanol, 4-nitro-
1935 - Benzene, 1-methoxy-2-nitro-
1936 - Benzene, 1-methoxy-3-nitro-
1937 - Benzene, 1-methoxy-4-nitro-
2498 - Benzoic acid, 4-amino-2-hydroxy-
2499 - Benzoic acid, 5-amino-2-hydroxy-
9422 - Phenol, 2-methyl-4-nitro-
9423 - Phenol, 2-methyl-6-nitro-
9424 - Phenol, 4-methyl-2-nitro-
$C_7H_7NO_4$
9386 - Phenol, 2-methoxy-3-nitro-
$C_7H_7NO_4S$
2508 - Benzoic acid, 4-(aminosulfonyl)-
$C_7H_7N_3$
980 - Benzene, (azidomethyl)-
981 - Benzene, 1-azido-2-methyl-
982 - Benzene, 1-azido-4-methyl-
2872 - 1H-Benzotriazole, 1-methyl-
$C_7H_7N_3O_3$
12025 - Urea, (4-nitrophenyl)-
C_7H_8
1947 - Benzene, methyl-
2913 - Bicyclo[2.2.1]hepta-2,5-diene
4240 - 1,3,5-Cycloheptatriene
4765 - 1,3-Cyclopentadiene, 5-ethylidene-
6341 - 1,6-Heptadien-3-yne
6342 - 1,5-Heptadiyne
6343 - 1,6-Heptadiyne

C_7H_8BrN
647 - Benzenamine, 2-bromo-4-methyl-
648 - Benzenamine, 2-bromo-5-methyl-
649 - Benzenamine, 3-bromo-4-methyl-
650 - Benzenamine, 4-bromo-2-methyl-
651 - Benzenamine, 4-bromo-3-methyl-
652 - Benzenamine, 5-bromo-2-methyl-
1799 - Benzenemethanamine, 4-bromo-
C_7H_8ClN
668 - Benzenamine, 2-chloro-4-methyl-
669 - Benzenamine, 2-chloro-5-methyl-
670 - Benzenamine, 2-chloro-*N*-methyl-
671 - Benzenamine, 3-chloro-2-methyl-
672 - Benzenamine, 3-chloro-4-methyl-
673 - Benzenamine, 4-chloro-2-methyl-
674 - Benzenamine, 4-chloro-3-methyl-
675 - Benzenamine, 4-chloro-*N*-methyl-
676 - Benzenamine, 5-chloro-2-methyl-
C_7H_8ClNO
666 - Benzenamine, 2-chloro-5-methoxy-
667 - Benzenamine, 4-chloro-2-methoxy-
$C_7H_8ClN_3O_4S_2$
2855 - 2H-1,2,4-Benzothiadiazine-7-sulfonamide, 6-chloro-3,4-dihydro-, 1,1-dioxide
$C_7H_8Cl_2Si$
11401 - Silane, dichloromethylphenyl-
C_7H_8FN
755 - Benzenamine, 4-fluoro-2-methyl-
C_7H_8IN
762 - Benzenamine, 2-iodo-4-methyl-
763 - Benzenamine, 5-iodo-2-methyl-
$C_7H_8N_2$
563 - Benzaldehyde, hydrazone
11137 - 1H-Pyrrole-1-propanenitrile
$C_7H_8N_2O$
89 - Acetamide, *N*-3-pyridinyl-
601 - Benzamide, 2-amino-
806 - Benzenamine, *N*-methyl-*N*-nitroso-
2611 - Benzoic acid, hydrazide
12026 - Urea, phenyl-
$C_7H_8N_2O_2$
799 - Benzenamine, 2-methyl-3-nitro-
800 - Benzenamine, 2-methyl-4-nitro-
801 - Benzenamine, 2-methyl-5-nitro-
802 - Benzenamine, 2-methyl-6-nitro-
803 - Benzenamine, 4-methyl-2-nitro-
804 - Benzenamine, *N*-methyl-2-nitro-
805 - Benzenamine, *N*-methyl-4-nitro-
2552 - Benzoic acid, 2,3-diamino-
2553 - Benzoic acid, 2,4-diamino-
2554 - Benzoic acid, 3,5-diamino-
$C_7H_8N_2O_3$
775 - Benzenamine, 2-methoxy-5-nitro-
776 - Benzenamine, 3-methoxy-4-nitro-
777 - Benzenamine, 5-methoxy-2-nitro-
$C_7H_8N_2S$
11808 - Thiourea, phenyl-
$C_7H_8N_4O_2$
10756 - 1H-Purine-2,6-dione, 3,7-dihydro-1,3-dimethyl-
10757 - 1H-Purine-2,6-dione, 3,7-dihydro-3,7-dimethyl-
C_7H_8O
1836 - Benzenemethanol
1915 - Benzene, methoxy-
9400 - Phenol, 2-methyl-
9401 - Phenol, 3-methyl-
9402 - Phenol, 4-methyl-
C_7H_8OS
2040 - Benzene, (methylsulfinyl)-
2299 - Benzenethiol, 3-methoxy-
2300 - Benzenethiol, 4-methoxy-
5856 - Ethanone, 1-(5-methyl-2-thienyl)-
$C_7H_8O_2$
1527 - 1,2-Benzenediol, 3-methyl-
1528 - 1,2-Benzenediol, 4-methyl-
1529 - 1,3-Benzenediol, 2-methyl-
1530 - 1,3-Benzenediol, 4-methyl-
1531 - 1,3-Benzenediol, 5-methyl-
1532 - 1,4-Benzenediol, 2-methyl-
1882 - Benzenemethanol, 2-hydroxy-
1883 - Benzenemethanol, 3-hydroxy-
1884 - Benzenemethanol, 4-hydroxy-
9379 - Phenol, 2-methoxy-
9380 - Phenol, 3-methoxy-
9381 - Phenol, 4-methoxy-
10438 - 1-Propanone, 1-(2-furanyl)-
10439 - 2-Propanone, 1-(2-furanyl)-
10792 - 2H-Pyran-2-one, 4,6-dimethyl-
10793 - 4H-Pyran-4-one, 2,6-dimethyl-
$C_7H_8O_2S$
2181 - Benzenesulfinic acid, 2-methyl-
2182 - Benzenesulfinic acid, 4-methyl-
11739 - 2-Thiophenecarboxylic acid, ethyl ester
$C_7H_8O_3$
1520 - 1,4-Benzenediol, 2-(hydroxymethyl)-
1525 - 1,2-Benzenediol, 3-methoxy-
1526 - 1,3-Benzenediol, 5-methoxy-
2369 - 1,2,3-Benzenetriol, 5-methyl-
6029 - 2-Furanacetic acid, methyl ester
6056 - 2-Furancarboxylic acid, ethyl ester
6119 - 2-Furanmethanol, acetate
6161 - 2-Furanpropanoic acid
$C_7H_8O_3S$
2227 - Benzenesulfonic acid, methyl ester
2228 - Benzenesulfonic acid, 2-methyl-
2229 - Benzenesulfonic acid, 4-methyl-
C_7H_8S
1832 - Benzenemethanethiol
2041 - Benzene, (methylthio)-
2301 - Benzenethiol, 2-methyl-
2302 - Benzenethiol, 3-methyl-
2303 - Benzenethiol, 4-methyl-
$C_7H_8S_2$
1564 - 1,2-Benzenedithiol, 4-methyl-
$C_7H_9AsN_2O_4$
455 - Arsonic acid, [4-[(aminocarbonyl)amino]phenyl]-
$C_7H_9BO_2$
3134 - Boronic acid, (2-methylphenyl)-
C_7H_9Cl
7107 - 1-Hexen-3-yne, 5-chloro-5-methyl-
C_7H_9ClO
9147 - 2-Pentynoyl chloride, 4,4-dimethyl-
C_7H_9ClSi
11381 - Silane, chloromethylphenyl-
C_7H_9N
778 - Benzenamine, 2-methyl-
779 - Benzenamine, 3-methyl-
780 - Benzenamine, 4-methyl-
781 - Benzenamine, *N*-methyl-
1795 - Benzenemethanamine
4629 - 3-Cyclohexene-1-carbonitrile
4630 - 1-Cyclohexenecarbonitrile
4905 - 1-Cyclopentene-1-acetonitrile
10936 - Pyridine, 2,3-dimethyl-
10937 - Pyridine, 2,4-dimethyl-
10938 - Pyridine, 2,5-dimethyl-
10939 - Pyridine, 2,6-dimethyl-
10940 - Pyridine, 3,4-dimethyl-
10941 - Pyridine, 3,5-dimethyl-
10956 - Pyridine, 2-ethyl-
10957 - Pyridine, 3-ethyl-
10958 - Pyridine, 4-ethyl-
C_7H_9NO
758 - Benzenamine, *N*-hydroxy-4-methyl-
766 - Benzenamine, 2-methoxy-
767 - Benzenamine, 3-methoxy-
768 - Benzenamine, 4-methoxy-
1837 - Benzenemethanol, 2-amino-
5849 - Ethanone, 1-(1-methyl-1H-pyrrol-2-yl)-
9209 - Phenol, 2-amino-3-methyl-
9210 - Phenol, 2-amino-4-methyl-
9211 - Phenol, 3-amino-4-methyl-
9212 - Phenol, 3-amino-5-methyl-
9213 - Phenol, 4-amino-2-methyl-
10945 - Pyridine, 2,6-dimethyl-, 1-oxide
10950 - 2-Pyridineethanol
$C_7H_9NO_2$
2509 - Benzoic acid, ammonium salt
10935 - Pyridine, 2,6-dimethoxy-
$C_7H_9NO_3$
8586 - 2,4-Oxazolidinedione, 5-methyl-3-(2-propenyl)-
C_7H_9NS
819 - Benzenamine, 2-(methylthio)-
820 - Benzenamine, 4-(methylthio)-
$C_7H_9N_3O$
7145 - Hydrazinecarboxamide, 2-phenyl-
7146 - Hydrazinecarboxamide, *N*-phenyl-
$C_7H_9N_3O_2$
2500 - Benzoic acid, 4-amino-2-hydroxy-, hydrazide
$C_7H_9N_3O_2S_2$
2189 - Benzenesulfonamide, 4-amino-*N*-(aminothioxomethyl)-
$C_7H_9N_3O_3S$
2190 - Benzenesulfonamide, 4-amino-*N*-(aminocarbonyl)-
C_7H_{10}
4219 - 1,3-Cycloheptadiene
4271 - 1,3-Cyclohexadiene, 2-methyl-
4272 - 1,3-Cyclohexadiene, 5-methyl-, (±)-
6529 - 1,3,5-Heptatriene
6606 - 1-Hepten-3-yne
9115 - 3-Penten-1-yne, 3-ethyl-
$C_7H_{10}ClN$
791 - Benzenamine, 2-methyl-, hydrochloride
792 - Benzenamine, *N*-methyl-, hydrochloride
$C_7H_{10}ClN_3$
11043 - 4-Pyrimidinamine, 2-chloro-*N*,*N*,6-trimethyl-
$C_7H_{10}N_2$
1275 - 1,2-Benzenediamine, 3-methyl-
1276 - 1,2-Benzenediamine, 4-methyl-
1277 - 1,3-Benzenediamine, 2-methyl-
1278 - 1,3-Benzenediamine, 4-methyl-
1279 - 1,4-Benzenediamine, 2-methyl-
1280 - 1,4-Benzenediamine, *N*-methyl-
1796 - Benzenemethanamine, 2-amino-
4159 - Cyanamide, di-2-propenyl-
6392 - Heptanedinitrile
7164 - Hydrazine, (3-methylphenyl)-
7165 - Hydrazine, 1-methyl-1-phenyl-
7166 - Hydrazine, 1-methyl-2-phenyl-
10863 - 2-Pyridinamine, *N*,*N*-dimethyl-
10864 - 4-Pyridinamine, 2,6-dimethyl-
10948 - 2-Pyridineethanamine
10949 - 4-Pyridineethanamine
$C_7H_{10}N_2O$
1273 - 1,2-Benzenediamine, 4-methoxy-
1274 - 1,3-Benzenediamine, 4-methoxy-
$C_7H_{10}N_2OS$
11109 - 4(1H)-Pyrimidinone, 2,3-dihydro-6-propyl-2-thioxo-
$C_7H_{10}N_2O_2S$
7194 - 1H-Imidazole-1-carboxylic acid, 2,3-dihydro-3-methyl-2-thioxo-, ethyl ester
$C_7H_{10}N_4O_2S$
2191 - Benzenesulfonamide, 4-amino-*N*-(aminoiminomethyl)-
$C_7H_{10}O$
4631 - 1-Cyclohexene-1-carboxaldehyde
4632 - 3-Cyclohexene-1-carboxaldehyde
4710 - 2-Cyclohexen-1-one, 2-methyl-
4711 - 2-Cyclohexen-1-one, 3-methyl-
4712 - 2-Cyclohexen-1-one, 5-methyl-
4713 - 3-Cyclohexen-1-one, 3-methyl-
4714 - 3-Cyclohexen-1-one, 4-methyl-
4866 - Cyclopentanol, 1-ethynyl-
4943 - 2-Cyclopenten-1-one, 2,4-dimethyl-
6106 - Furan, 2-ethyl-5-methyl-
6167 - Furan, 2-propyl-

6325 - 2,4-Heptadienal, (*E,E*)-
6339 - 3,5-Heptadien-2-one
7607 - Methanone, dicyclopropyl-

$C_7H_{10}O_2$
3856 - 2-Butenoic acid, 2-propenyl ester
3857 - 2-Butenoic acid, 2-propenyl ester, (*E*)-
4225 - 1,2-Cycloheptanedione
4637 - 1-Cyclohexene-1-carboxylic acid
4638 - 3-Cyclohexene-1-carboxylic acid
4879 - Cyclopentanone, 2-acetyl-
4904 - 3-Cyclopentene-1-acetic acid
6104 - Furan, 2-(ethoxymethyl)-
6112 - Furan, 2-isopropoxy-
6702 - 2,4-Hexadienoic acid, methyl ester
6703 - 2,4-Hexadienoic acid, methyl ester, (*E,E*)-
10671 - 2-Propenoic acid, 2-methyl-, 2-propenyl ester

$C_7H_{10}O_3$
3577 - Butanoic acid, 3-oxo-, 2-propenyl ester
6454 - 2,4,6-Heptanetrione
10668 - 2-Propenoic acid, 2-methyl-, oxiranylmethyl ester

$C_7H_{10}O_4$
3330 - Butanedioic acid, methylene-, dimethyl ester
3331 - Butanedioic acid, (1-methylethylidene)-
3789 - 2-Butenedioic acid, 2-methyl-, dimethyl ester, (*E*)-
3790 - 2-Butenedioic acid, 2-methyl-, dimethyl ester, (*Z*)-
4793 - 1,3-Cyclopentanedicarboxylic acid, *cis*-
4969 - 1,2-Cyclopropanedicarboxylic acid, dimethyl ester
8876 - Pentanoic acid, 2,4-dioxo-, ethyl ester
10563 - 2-Propene-1,1-diol, diacetate

$C_7H_{10}O_5$
8784 - Pentanedioic acid, 3-oxo-, dimethyl ester
9973 - Propanedioic acid, oxo-, diethyl ester

$C_7H_{10}O_6$
7553 - Methanetricarboxylic acid, trimethyl ester

$C_7H_{10}S$
11761 - Thiophene, 2-ethyl-3-methyl-
11762 - Thiophene, 2-ethyl-5-methyl-
11763 - Thiophene, 3-ethyl-4-methyl-
11764 - Thiophene, 4-ethyl-2-methyl-
11772 - Thiophene, 2-(1-methylethyl)-
11773 - Thiophene, 3-(1-methylethyl)-
11778 - Thiophene, 2-propyl-
11779 - Thiophene, 3-propyl-
11792 - Thiophene, 2,3,4-trimethyl-
11793 - Thiophene, 2,3,5-trimethyl-

$C_7H_{10}Si$
11442 - Silane, methylphenyl-

$C_7H_{11}Br$
6612 - 1-Heptyne, 1-bromo-
6613 - 2-Heptyne, 1-bromo-

$C_7H_{11}BrO_4$
9928 - Propanedioic acid, bromo-, diethyl ester

$C_7H_{11}Cl$
2918 - Bicyclo[2.2.1]heptane, 2-chloro-
6614 - 1-Heptyne, 1-chloro-
6615 - 2-Heptyne, 7-chloro-
6616 - 3-Heptyne, 7-chloro-
9125 - 1-Pentyne, 3-chloro-3-ethyl-

$C_7H_{11}ClO$
4330 - Cyclohexanecarbonyl chloride

$C_7H_{11}ClO_3$
3492 - Butanoic acid, 2-chloro-2-methyl-3-oxo-, ethyl ester

$C_7H_{11}ClO_4$
9931 - Propanedioic acid, chloro-, diethyl ester
9996 - 1,2-Propanediol, 3-chloro-, diacetate

$C_7H_{11}Cl_3O_2$
255 - Acetic acid, trichloro-, 3-methylbutyl ester

$C_7H_{11}I$
6620 - 1-Heptyne, 1-iodo-

$C_7H_{11}N$
4328 - Cyclohexanecarbonitrile
11128 - 1H-Pyrrole, 3-ethyl-4-methyl-
11129 - Pyrrole, 2-isopropyl-
11138 - 1H-Pyrrole, 1-propyl-
11139 - 1H-Pyrrole, 3-propyl-
11142 - 1H-Pyrrole, 1,2,5-trimethyl-

$C_7H_{11}NO$
4329 - Cyclohexanecarbonitrile, 1-hydroxy-
4416 - Cyclohexane, isocyanato-

$C_7H_{11}NO_2$
151 - Acetic acid, cyano-, butyl ester
9720 - 4-Piperidinone, 1-acetyl-
10913 - 3-Pyridinecarboxylic acid, 1,2,5,6-tetrahydro-1-methyl-
11158 - 2,5-Pyrrolidinedione, 3-ethyl-3-methyl-

$C_7H_{11}NO_3$
11567 - Succinimide, *N*-(ethoxymethyl)-

$C_7H_{11}NO_4$
9766 - *L*-Proline, 1-acetyl-4-hydroxy-, *trans*-

$C_7H_{11}NO_5$
9958 - Propanedioic acid, (hydroxyimino)-, diethyl ester

$C_7H_{11}NS$
4417 - Cyclohexane, isothiocyanato-

C_7H_{12}
2914 - Bicyclo[2.2.1]heptane
2915 - Bicyclo[4.1.0]heptane
2980 - Bicyclo[3,1,0]hexane, 1-methyl-
4244 - Cycloheptene
4428 - Cyclohexane, methylene-
4670 - Cyclohexene, 1-methyl-
4671 - Cyclohexene, 3-methyl-, (±)-
4672 - Cyclohexene, 4-methyl-
4813 - Cyclopentane, ethylidene-
4918 - Cyclopentene, 1,2-dimethyl-
4919 - Cyclopentene, 1,3-dimethyl-
4920 - Cyclopentene, 1,4-dimethyl-
4921 - Cyclopentene, 1,5-dimethyl-
4922 - Cyclopentene, 4,4-dimethyl-
4924 - Cyclopentene, 1-ethyl-
4925 - Cyclopentene, 3-ethyl-
4926 - Cyclopentene, 4-ethyl-
6326 - 1,2-Heptadiene
6327 - 1,4-Heptadiene
6328 - 1,5-Heptadiene
6329 - 2,4-Heptadiene
6609 - 1-Heptyne
6610 - 2-Heptyne
6611 - 3-Heptyne
6695 - 1,5-Hexadiene, 2-methyl-
6696 - 2,4-Hexadiene, 2-methyl-
7120 - 1-Hexyne, 5-methyl-
7121 - 2-Hexyne, 5-methyl-
7122 - 3-Hexyne, 2-methyl-
8665 - 1,3-Pentadiene, 2,4-dimethyl-
9128 - 1-Pentyne, 3,3-dimethyl-
9129 - 1-Pentyne, 4,4-dimethyl-
9130 - 2-Pentyne, 4,4-dimethyl-
9131 - 1-Pentyne, 3-ethyl-

$C_7H_{12}Br_2O_2$
8872 - Pentanoic acid, 2,3-dibromo-, ethyl ester
8873 - Pentanoic acid, 2,5-dibromo-, ethyl ester
10239 - Propanoic acid, 2,3-dibromo-, butyl ester

$C_7H_{12}ClN_5$
11513 - Simazine

$C_7H_{12}N_2$
7196 - 1H-Imidazole, 1,2-diethyl-
11190 - Pyrrolo[1,2-a]pyrimidine, 2,3,4,6,7,8-hexahydro-

$C_7H_{12}O$
4234 - Cycloheptanone
4331 - Cyclohexanecarboxaldehyde
4537 - Cyclohexanol, 2-methylene-
4596 - Cyclohexanone, 2-methyl-, (±)-
4597 - Cyclohexanone, 3-methyl-, (±)-
4598 - Cyclohexanone, 4-methyl-
4882 - Cyclopentanone, 2,5-dimethyl-
5795 - Ethanone, 1-cyclopentyl-
6336 - 1,5-Heptadien-4-ol
6337 - 1,6-Heptadien-4-ol
6530 - 2-Heptenal
6531 - 3-Heptenal
6583 - 1-Hepten-3-one
6584 - 4-Hepten-2-one
6585 - 6-Hepten-2-one
6586 - 6-Hepten-3-one
6587 - 3-Hepten-2-one, (*E*)-
6588 - 3-Hepten-2-one, (*Z*)-
6589 - 5-Hepten-2-one, (*E*)-
7093 - 1-Hexen-3-one, 5-methyl-
7094 - 3-Hexen-2-one, 5-methyl-
7095 - 4-Hexen-2-one, 5-methyl-
7096 - 4-Hexen-3-one, 2-methyl-
7097 - 5-Hexen-2-one, 3-methyl-
7098 - 5-Hexen-2-one, 4-methyl-
7099 - 5-Hexen-2-one, 5-methyl-
7128 - 1-Hexyn-3-ol, 3-methyl-
7129 - 3-Hexyn-2-ol, 2-methyl-
9144 - 1-Pentyn-3-ol, 3,4-dimethyl-
9145 - 1-Pentyn-3-ol, 3-ethyl-

$C_7H_{12}O_2$
3587 - Butanoic acid, 2-propenyl ester
3849 - 2-Butenoic acid, 2-methyl-, ethyl ester, (*E*)-
3850 - 2-Butenoic acid, 3-methyl-, ethyl ester
4175 - Cyclobutanecarboxylic acid, ethyl ester
4332 - Cyclohexanecarboxylic acid
4775 - Cyclopentaneacetic acid
4783 - Cyclopentanecarboxylic acid, 1-methyl-
4865 - Cyclopentanol, acetate
6005 - Formic acid, cyclohexyl ester
6402 - 2,3-Heptanedione
6403 - 2,6-Heptanedione
6561 - 2-Heptenoic acid
6562 - 4-Heptenoic acid
6563 - 5-Heptenoic acid
6564 - 6-Heptenoic acid
6809 - 2,3-Hexanedione, 5-methyl-
6810 - 2,5-Hexanedione, 3-methyl-
7055 - 2-Hexenoic acid, 2-methyl-, (*E*)-
7056 - 2-Hexenoic acid, 4-methyl-
7057 - 2-Hexenoic acid, 5-methyl-
7058 - 3-Hexenoic acid, 2-methyl-
7059 - 3-Hexenoic acid, 3-methyl-
7060 - 3-Hexenoic acid, 4-methyl-
7061 - 4-Hexenoic acid, 5-methyl-
7062 - 3-Hexenoic acid, methyl ester
8799 - 2,4-Pentanedione, 3,3-dimethyl-
8801 - 2,4-Pentanedione, 3-ethyl-
9075 - 4-Pentenoic acid, 2,2-dimethyl-
9083 - 2-Pentenoic acid, 2-methyl-, methyl ester
9084 - 2-Pentenoic acid, 3-methyl-, methyl ester, (*E*)-
9085 - 2-Pentenoic acid, 3-methyl-, methyl ester, (*Z*)-
9086 - 3-Pentenoic acid, 3-methyl-, methyl ester
10616 - 2-Propenoic acid, butyl ester
10664 - 2-Propenoic acid, 2-methyl-, 1-methylethyl ester
10672 - 2-Propenoic acid, 2-methyl-, propyl ester
10673 - 2-Propenoic acid, 2-methylpropyl ester
10731 - 1-Propyne, 3,3-diethoxy-
10783 - 2H-Pyran, 2-ethoxy-3,4-dihydro-

$C_7H_{12}O_3$
3519 - Butanoic acid, 2-ethyl-3-oxo-, methyl ester
3557 - Butanoic acid, 2-methyl-3-oxo-, ethyl ester
3574 - Butanoic acid, 3-oxo-, 1-methylethyl ester
6125 - 2-Furanmethanol, tetrahydro-, acetate
6164 - 2-Furanpropanoic acid, tetrahydro-
6469 - Heptanoic acid, 6-oxo-
6901 - Hexanoic acid, 4-methyl-5-oxo-
8874 - Pentanoic acid, 2,2-dimethyl-4-oxo-
8898 - Pentanoic acid, 3-methyl-4-oxo-, methyl ester
8905 - Pentanoic acid, 2-oxo-, ethyl ester
8906 - Pentanoic acid, 3-oxo-, ethyl ester
8907 - Pentanoic acid, 4-oxo-, ethyl ester
10634 - 2-Propenoic acid, 2-ethoxyethyl ester
10662 - 2-Propenoic acid, 2-methyl-, 2-hydroxypropyl ester

$C_7H_{12}O_4$
3323 - Butanedioic acid, ethyl methyl ester
3327 - Butanedioic acid, methyl-, dimethyl ester
3338 - Butanedioic acid, monopropyl ester
6393 - Heptanedioic acid
6788 - Hexanedioic acid, 2-methyl-
6789 - Hexanedioic acid, 3-methyl-, (±)-
6792 - Hexanedioic acid, monomethyl ester
7654 - Methylene bispropionate
8773 - Pentanedioic acid, 3,3-dimethyl-
8775 - Pentanedioic acid, dimethyl ester
9929 - Propanedioic acid, butyl-
9939 - Propanedioic acid, diethyl-
9941 - Propanedioic acid, diethyl ester
9968 - Propanedioic acid, (2-methylpropyl)-
10002 - 1,2-Propanediol, diacetate
10003 - 1,3-Propanediol, diacetate
11643 - 2,4,8,10-Tetraoxaspiro[5.5]undecane

$C_7H_{12}O_5$
9957 - Propanedioic acid, hydroxy-, diethyl ester
10189 - 1,2,3-Propanetriol, diacetate
10190 - 1,2,3-Propanetriol, 1,3-diacetate

$C_7H_{12}O_6$
4338 - Cyclohexanecarboxylic acid, 1,3,4,5-tetrahydroxy-, [1*R*-(1α,3α,4α,5β)]-

$C_7H_{12}O_6Si$
11499 - Silanetriol, methyl-, triacetate

$C_7H_{13}Br$
4222 - Cycloheptane, bromo-
4318 - Cyclohexane, (bromomethyl)-
4319 - Cyclohexane, 1-bromo-1-methyl-
4320 - Cyclohexane, 1-bromo-3-methyl-

$C_7H_{13}BrN_2O_2$
3195 - Butanamide, *N*-(aminocarbonyl)-2-bromo-2-ethyl-

$C_7H_{13}BrO$
4317 - Cyclohexane, 1-bromo-2-methoxy-

$C_7H_{13}BrO_2$
3475 - Butanoic acid, 2-bromo-3-methyl-, ethyl ester
6457 - Heptanoic acid, 2-bromo-
6458 - Heptanoic acid, 7-bromo-
8860 - Pentanoic acid, 2-bromo-, ethyl ester
8861 - Pentanoic acid, 5-bromo-, ethyl ester
10206 - Propanoic acid, 3-bromo-, butyl ester

$C_7H_{13}Cl$
6538 - 1-Heptene, 1-chloro-
6539 - 1-Heptene, 2-chloro-
6540 - 3-Heptene, 4-chloro-

$C_7H_{13}ClO$
6513 - 2-Heptanone, 1-chloro-
6527 - Heptanoyl chloride
6988 - Hexanoyl chloride, 3-methyl-
6989 - Hexanoyl chloride, 4-methyl-

$C_7H_{13}ClO_2$
3490 - Butanoic acid, 2-chloro-3-methyl-, ethyl ester
3491 - Butanoic acid, 2-chloro-2-methyl-, ethyl ester
8867 - Pentanoic acid, 4-chloro-, ethyl ester
8868 - Pentanoic acid, 5-chloro-, ethyl ester
8869 - Pentanoic acid, 2-chloroethyl ester
10217 - Propanoic acid, 2-chloro-, butyl ester
10218 - Propanoic acid, 3-chloro-, butyl ester
10224 - Propanoic acid, 2-chloro-, 2-methylpropyl ester, (*R*)-
10227 - Propanoic acid, 3-chloro-, 2-methylpropyl ester
12039 - Valeric acid, 3-chloro-, ethyl ester

$C_7H_{13}FO_2$
6463 - Heptanoic acid, 7-fluoro-

$C_7H_{13}N$
480 - 1-Azabicyclo[2.2.2]octane
6428 - Heptanenitrile
9121 - 1-Pentyn-3-amine, 3-ethyl-
9666 - Piperidine, 3-ethenyl-

$C_7H_{13}NO$
485 - 1-Azabicyclo[2.2.2]octan-3-ol
5990 - Formamide, *N*-cyclohexyl-
9630 - Piperidine, 1-acetyl-

$C_7H_{13}NO_2$
4443 - Cyclohexane, 1-methyl-1-nitro-
4444 - Cyclohexane, 1-methyl-2-nitro-
4445 - Cyclohexane, 1-methyl-4-nitro-
4455 - Cyclohexane, (nitromethyl)-
9644 - 4-Piperidinecarboxylic acid, methyl ester
10633 - 2-Propenoic acid, 2-(dimethylamino)ethyl ester
11177 - Pyrrolidinium, 2-carboxy-1,1-dimethyl-, hydroxide, inner salt, (*S*)-

$C_7H_{13}NO_3$
10107 - Propanenitrile, 3-[2-(2-hydroxyethoxy)ethoxy]-
11178 - Pyrrolidinium, 2-carboxy-4-hydroxy-1,1-dimethyl-, hydroxide, inner salt, (2*S-trans*)-

$C_7H_{13}NO_3S$
7649 - *L*-Methionine, *N*-acetyl-

$C_7H_{13}NO_4$
6245 - *L*-Glutamic acid, 5-ethyl ester
9924 - Propanedioic acid, amino-, diethyl ester

$C_7H_{13}N_3O_3S$
8568 - Oxamyl

$C_7H_{13}O_6P$
7659 - Mevinphos

C_7H_{14}
3797 - 1-Butene, 2-ethyl-3-methyl-
3812 - 1-Butene, 2,3,3-trimethyl-
4221 - Cycloheptane
4426 - Cyclohexane, methyl-
4800 - Cyclopentane, 1,1-dimethyl-
4801 - Cyclopentane, 1,2-dimethyl-, *cis*-
4802 - Cyclopentane, 1,2-dimethyl-, *trans*-
4803 - Cyclopentane, 1,3-dimethyl-, *cis*-
4804 - Cyclopentane, 1,3-dimethyl-, *trans*-
4812 - Cyclopentane, ethyl-
6533 - 1-Heptene
6534 - 2-Heptene, (*E*)-
6535 - 2-Heptene, (*Z*)-
6536 - 3-Heptene, (*E*)-
6537 - 3-Heptene, (*Z*)-
7036 - 1-Hexene, 2-methyl-
7037 - 1-Hexene, 3-methyl-
7038 - 1-Hexene, 4-methyl-
7039 - 1-Hexene, 5-methyl-
7040 - 2-Hexene, 2-methyl-
7041 - 2-Hexene, 3-methyl-, (*Z*)-
7042 - 2-Hexene, 4-methyl-, (*E*)-
7043 - 2-Hexene, 4-methyl-, (*Z*)-
7044 - 2-Hexene, 5-methyl-, (*E*)-
7045 - 2-Hexene, 5-methyl-, (*Z*)-
7046 - 3-Hexene, 2-methyl-, (*E*)-
7047 - 3-Hexene, 2-methyl-, (*Z*)-
9029 - 1-Pentene, 2,3-dimethyl-
9030 - 1-Pentene, 2,4-dimethyl-
9031 - 1-Pentene, 3,3-dimethyl-
9032 - 1-Pentene, 3,4-dimethyl-
9033 - 1-Pentene, 4,4-dimethyl-
9034 - 2-Pentene, 2,3-dimethyl-
9035 - 2-Pentene, 2,4-dimethyl-
9036 - 2-Pentene, 3,4-dimethyl-
9037 - 2-Pentene, 3,4-dimethyl-, (*E*)-
9038 - 2-Pentene, 3,4-dimethyl-, (*Z*)-
9039 - 2-Pentene, 4,4-dimethyl-, (*E*)-
9040 - 2-Pentene, 4,4-dimethyl-, (*Z*)-
9043 - 1-Pentene, 2-ethyl-
9044 - 1-Pentene, 3-ethyl-
9045 - 2-Pentene, 3-ethyl-

$C_7H_{14}BrF$
6359 - Heptane, 1-bromo-7-fluoro-

$C_7H_{14}BrNO$
3199 - Butanamide, 2-bromo-2-ethyl-3-methyl-

$C_7H_{14}Br_2$
6372 - Heptane, 1,2-dibromo-
6373 - Heptane, 1,5-dibromo-
6374 - Heptane, 1,7-dibromo-
6375 - Heptane, 2,3-dibromo-
6376 - Heptane, 3,4-dibromo-

$C_7H_{14}ClF$
6365 - Heptane, 1-chloro-7-fluoro-

$C_7H_{14}Cl_2$
6377 - Heptane, 1,1-dichloro-
6378 - Heptane, 1,2-dichloro-
6379 - Heptane, 1,7-dichloro-
6380 - Heptane, 2,2-dichloro-
6381 - Heptane, 4,4-dichloro-
8752 - Pentane, 1,2-dichloro-4,4-dimethyl-
8753 - Pentane, 1,5-dichloro-3,3-dimethyl-
8754 - Pentane, 2,4-dichloro-2,4-dimethyl-

$C_7H_{14}NO_5P$
7663 - Monocrotophos

$C_7H_{14}N_2$
9815 - 2-Propanamine, *N,N'*-methanetetraylbis-

$C_7H_{14}N_2O_2$
9604 - 1-Piperazinecarboxylic acid, ethyl ester

$C_7H_{14}N_2O_2S$
334 - Aldicarb

$C_7H_{14}N_2O_4$
3944 - Carbamic acid, (1,1-dimethylethyl)nitro-, ethyl ester
3951 - Carbamic acid, ethylnitro-, methyl ester

$C_7H_{14}N_2O_4S$
335 - Aldoxycarb

$C_7H_{14}N_2O_4S_2$
5018 - *L*-Cysteine, *S,S'*-methylenebis-

$C_7H_{14}O$
3371 - Butane, 1-(ethenyloxy)-3-methyl-
4229 - Cycloheptanol
4419 - Cyclohexanemethanol
4425 - Cyclohexane, methoxy-
4530 - Cyclohexanol, 1-methyl-
4531 - Cyclohexanol, 2-methyl-, *cis*-(±)-
4532 - Cyclohexanol, 2-methyl-, *trans*-(±)-
4533 - Cyclohexanol, 3-methyl-, *cis*-(±)-
4534 - Cyclohexanol, 3-methyl-, *trans*-(±)-
4535 - Cyclohexanol, 4-methyl-, *cis*-
4536 - Cyclohexanol, 4-methyl-, *trans*-
4811 - Cyclopentaneethanol
6173 - Furan, tetrahydro-2-propyl-
6344 - Heptanal
6510 - 2-Heptanone
6511 - 3-Heptanone
6512 - 4-Heptanone
6566 - 1-Hepten-4-ol
6567 - 2-Hepten-1-ol, (*E*)-
6568 - 2-Hepten-1-ol, (*Z*)-
6569 - 2-Hepten-4-ol, (±)-
6570 - 3-Hepten-2-ol
6571 - 6-Hepten-2-ol
6572 - 6-Hepten-3-ol
6720 - Hexanal, 3-methyl-
6976 - 2-Hexanone, 3-methyl-
6977 - 2-Hexanone, 4-methyl-
6978 - 2-Hexanone, 5-methyl-
6979 - 3-Hexanone, 2-methyl-
6980 - 3-Hexanone, 4-methyl-
6981 - 3-Hexanone, 5-methyl-
7077 - 1-Hexen-3-ol, 5-methyl-
7078 - 2-Hexen-1-ol, 5-methyl-
7079 - 3-Hexen-2-ol, 2-methyl-
7080 - 3-Hexen-2-ol, 3-methyl-
7081 - 4-Hexen-3-ol, 2-methyl-
7082 - 4-Hexen-3-ol, 3-methyl-
7083 - 5-Hexen-2-ol, 2-methyl-
7084 - 5-Hexen-3-ol, 3-methyl-
7085 - 5-Hexen-3-ol, 4-methyl-
8965 - 2-Pentanone, 4,4-dimethyl-
8966 - 3-Pentanone, 2,2-dimethyl-
8967 - 3-Pentanone, 2,4-dimethyl-

$C_7H_{14}O_2$
227 - Acetic acid, pentyl ester
3516 - Butanoic acid, 2-ethyl-2-methyl-
3517 - Butanoic acid, 2-ethyl-, methyl ester
3548 - Butanoic acid, 3-methyl-, ethyl ester
3549 - Butanoic acid, 1-methylethyl ester

3550 - Butanoic acid, 2-methyl-, ethyl ester, (*S*)-
3588 - Butanoic acid, propyl ester
3630 - 1-Butanol, 2-methyl-, acetate
3631 - 1-Butanol, 3-methyl-, acetate
6009 - Formic acid, hexyl ester
6455 - Heptanoic acid
6517 - 4-Heptanone, 2-hydroxy-
6893 - Hexanoic acid, 2-methyl-, (±)-
6894 - Hexanoic acid, 3-methyl-
6895 - Hexanoic acid, 4-methyl-, (±)-
6896 - Hexanoic acid, 5-methyl-
6898 - Hexanoic acid, methyl ester
6973 - 3-Hexanone, 2-hydroxy-2-methyl-
6974 - 3-Hexanone, 5-hydroxy-2-methyl-
8602 - Oxirane, (butoxymethyl)-
8610 - Oxirane, [(1,1-dimethylethoxy)methyl]-
8877 - Pentanoic acid, 2-ethyl-, (±)-
8878 - Pentanoic acid, ethyl ester
8918 - 2-Pentanol, acetate, (*R*)-
8919 - 3-Pentanol, acetate
8978 - 2-Pentanone, 4-methoxy-4-methyl-
10214 - Propanoic acid, butyl ester
10259 - Propanoic acid, 2,2-dimethyl-, ethyl ester
10303 - Propanoic acid, 2-methyl-, 1-methylethyl ester
10308 - Propanoic acid, 2-methyl-, propyl ester
10309 - Propanoic acid, 1-methylpropyl ester
10310 - Propanoic acid, 2-methylpropyl ester
10559 - 1-Propene, 1,1-diethoxy-
10560 - 1-Propene, 3,3-diethoxy-

$C_7H_{14}O_3$
3995 - Carbonic acid, bis(1-methylethyl) ester
4002 - Carbonic acid, dipropyl ester
8884 - Pentanoic acid, 4-hydroxy-, ethyl ester
8885 - Pentanoic acid, 2-hydroxyethyl ester
10262 - Propanoic acid, 3-ethoxy-, ethyl ester
10268 - Propanoic acid, 2-hydroxy-, butyl ester, (*R*)-

$C_7H_{14}O_4$
3505 - Butanoic acid, 2,3-dihydroxypropyl ester
7112 - *ribo*-Hexose, 2,6-dideoxy-3-*O*-methyl-

$C_7H_{14}O_5$
5736 - Ethanol, 2-methoxy-, carbonate (2:1)
6198 - Galactose, 6-deoxy-3-*O*-methyl-

$C_7H_{14}O_6$
6227 - α-*D*-Glucopyranoside, methyl

$C_7H_{15}Br$
6356 - Heptane, 1-bromo-
6357 - Heptane, 2-bromo-
6358 - Heptane, 4-bromo-

$C_7H_{15}BrO$
6736 - Hexane, 1-bromo-6-methoxy-

$C_7H_{15}Cl$
3254 - Butane, 2-chloro-2,3,3-trimethyl-
6360 - Heptane, 1-chloro-
6361 - Heptane, 2-chloro-
6362 - Heptane, 3-chloro-
6363 - Heptane, 4-chloro-
6743 - Hexane, 1-chloro-3-methyl-
6744 - Hexane, 2-chloro-2-methyl-
6745 - Hexane, 2-chloro-5-methyl-
6746 - Hexane, 3-chloro-3-methyl-
8722 - Pentane, 2-chloro-2,3-dimethyl-
8723 - Pentane, 2-chloro-2,4-dimethyl-
8724 - Pentane, 3-chloro-2,3-dimethyl-
8725 - Pentane, 4-chloro-2,2-dimethyl-
8726 - Pentane, 3-chloro-3-ethyl-

$C_7H_{15}ClO$
6477 - 1-Heptanol, 7-chloro-
6478 - 2-Heptanol, 1-chloro-

$C_7H_{15}ClO_2$
9850 - Propane, 3-chloro-1,1-diethoxy-

$C_7H_{15}F$
6417 - Heptane, 1-fluoro-

$C_7H_{15}FO$
6491 - 1-Heptanol, 7-fluoro-

$C_7H_{15}I$
6419 - Heptane, 1-iodo-
6420 - Heptane, 2-iodo-

$C_7H_{15}N$
502 - Azocine, octahydro-
4292 - Cyclohexanamine, 2-methyl-, *cis*-
4293 - Cyclohexanamine, 2-methyl-, *trans*-
4294 - Cyclohexanamine, 3-methyl-, *cis*-
4295 - Cyclohexanamine, 3-methyl-, *trans*-
4296 - Cyclohexanamine, 4-methyl-, *cis*-
4297 - Cyclohexanamine, 4-methyl-, *trans*-
4298 - Cyclohexanamine, *N*-methyl-
4418 - Cyclohexanemethanamine
9648 - Piperidine, 1,2-dimethyl-, (±)-
9649 - Piperidine, 2,5-dimethyl-
9650 - Piperidine, 2,6-dimethyl-
9651 - Piperidine, 3,3-dimethyl-
9652 - Piperidine, 3,5-dimethyl-
9667 - Piperidine, 1-ethyl-
9668 - Piperidine, 2-ethyl-, (±)-
9669 - Piperidine, 3-ethyl-, (±)-
9670 - Piperidine, 4-ethyl-
11173 - Pyrrolidine, 1-propyl-
11174 - Pyrrolidine, 3-propyl-
11175 - Pyrrolidine, 2,2,4-trimethyl-
11176 - Pyrrolidine, 2,2,5-trimethyl-

$C_7H_{15}NO$
6346 - Heptanal, oxime
6348 - Heptanamide
6982 - 2-Hexanone, 5-methyl-, oxime
8699 - Pentanamide, *N*,*N*-dimethyl-
8968 - 3-Pentanone, 2,4-dimethyl-, oxime
9662 - 1-Piperidineethanol
9663 - 2-Piperidineethanol
9664 - 4-Piperidineethanol
9678 - 3-Piperidinemethanol, 1-methyl-
9715 - 3-Piperidinol, 1-ethyl-
9793 - Propanamide, *N*,*N*-diethyl-
10427 - 2-Propanone, 1-(diethylamino)-

$C_7H_{15}NOS$
5930 - Ethiolate

$C_7H_{15}NO_2$
3541 - Butanoic acid, 3-(methylamino)-, ethyl ester
3936 - Carbamic acid, butyl-, ethyl ester
3942 - Carbamic acid, (1,1-dimethylethyl)-, ethyl ester
3947 - Carbamic acid, ethyl-, butyl ester
3960 - Carbamic acid, (1-methylpropyl)-, ethyl ester
3961 - Carbamic acid, (2-methylpropyl)-, ethyl ester
6429 - Heptane, 1-nitro-
7692 - 4-Morpholineethanol, α-methyl-
8088 - Nitrous acid, heptyl ester

$C_7H_{15}NO_5$
6215 - α-*L*-Glucopyranose, 2-(methylamino)-2-deoxy-

C_7H_{16}
3457 - Butane, 2,2,3-trimethyl-
6355 - Heptane
6834 - Hexane, 2-methyl-
6835 - Hexane, 3-methyl-, (*S*)-
8760 - Pentane, 2,2-dimethyl-
8761 - Pentane, 2,3-dimethyl-
8762 - Pentane, 2,4-dimethyl-
8763 - Pentane, 3,3-dimethyl-
8809 - Pentane, 3-ethyl-

$C_7H_{16}BrNO_2$
5438 - Ethanaminium, 2-(acetyloxy)-*N*,*N*,*N*-trimethyl-, bromide

$C_7H_{16}NCl$
7430 - Mepiquat chloride

$C_7H_{16}ClNO_2$
5439 - Ethanaminium, 2-(acetyloxy)-*N*,*N*,*N*-trimethyl-, chloride

$C_7H_{16}N_2$
9622 - Piperazine, 1,2,4-trimethyl-
9625 - 1-Piperidinamine, 2,6-dimethyl-
9626 - 3-Piperidinamine, 1-ethyl-
11167 - 2-Pyrrolidinemethanamine, 1-ethyl-

$C_7H_{16}N_2O$
7701 - 4-Morpholinepropanamine
12012 - Urea, *N*,*N′*-dipropyl-

$C_7H_{16}O$
3374 - Butane, 1-ethoxy-3-methyl-
3375 - Butane, 2-ethoxy-2-methyl-
3399 - Butane, 1-(1-methylethoxy)-
3421 - Butane, 1-propoxy-
3644 - 2-Butanol, 2,3,3-trimethyl-
6472 - 1-Heptanol
6473 - 2-Heptanol, (±)-
6474 - 3-Heptanol, (*S*)-
6475 - 4-Heptanol
6833 - Hexane, 1-methoxy-
6937 - 1-Hexanol, 2-methyl-, (±)-
6938 - 1-Hexanol, 2-methyl-, (*R*)-
6939 - 1-Hexanol, 3-methyl-, (±)-
6940 - 1-Hexanol, 4-methyl-, (±)-
6941 - 1-Hexanol, 5-methyl-
6942 - 2-Hexanol, 2-methyl-
6943 - 2-Hexanol, 3-methyl-
6944 - 2-Hexanol, 5-methyl-
6945 - 3-Hexanol, 2-methyl-, (±)-
6946 - 3-Hexanol, 3-methyl-
6947 - 3-Hexanol, 5-methyl-, (±)-
8808 - Pentane, 1-ethoxy-
8927 - 1-Pentanol, 2,4-dimethyl-, (±)-
8928 - 2-Pentanol, 2,4-dimethyl-
8929 - 3-Pentanol, 2,2-dimethyl-
8930 - 3-Pentanol, 2,3-dimethyl-
8931 - 3-Pentanol, 2,4-dimethyl-
8932 - 3-Pentanol, 3-ethyl-
10081 - Propane, 2-methyl-2-(1-methylethoxy)-
10086 - Propane, 2-methyl-1-propoxy-
10087 - Propane, 2-methyl-2-propoxy-

$C_7H_{16}O_2$
5757 - Ethanol, 2-(pentyloxy)-
6397 - 1,4-Heptanediol
6398 - 1,5-Heptanediol
6399 - 1,7-Heptanediol
6400 - 2,4-Heptanediol
6803 - 2,4-Hexanediol, 2-methyl-
8792 - 1,3-Pentanediol, 2,2-dimethyl-
9903 - Propane, 1,1-diethoxy-
9904 - Propane, 2,2-diethoxy-
10004 - 1,3-Propanediol, 2,2-diethyl-
10020 - 1,3-Propanediol, 2-methyl-2-propyl-
10074 - Propane, 1,1′-[methylenebis(oxy)]bis-
10346 - 2-Propanol, 1-butoxy-

$C_7H_{16}O_2Si$
11422 - Silane, ethenyldiethoxymethyl-

$C_7H_{16}O_3$
3455 - Butane, 1,1,3-trimethoxy-
3456 - Butane, 1,3,3-trimethoxy-
5604 - Ethane, 1,1′,1″-[methylidynetris(oxy)]tris-
6453 - 1,4,7-Heptanetriol

$C_7H_{16}O_4$
10150 - Propane, 1,1,3,3-tetramethoxy-

$C_7H_{16}O_4S_2$
9835 - Propane, 2,2-bis(ethylsulfonyl)-

$C_7H_{16}S$
6436 - 1-Heptanethiol

$C_7H_{16}S_3$
5605 - Ethane, 1,1′,1″-[methylidynetris(thio)]tris-

$C_7H_{17}N$
6349 - 1-Heptanamine
6350 - 2-Heptanamine
6351 - 4-Heptanamine
6728 - 1-Hexanamine, 3-methyl-
6729 - 1-Hexanamine, 4-methyl-
6730 - 2-Hexanamine, 4-methyl-
8703 - 1-Pentanamine, *N*,*N*-dimethyl-
8704 - 2-Pentanamine, 2,4-dimethyl-

$C_7H_{17}NO$
5739 - Ethanol, 2-[(3-methylbutyl)amino]-
10366 - 1-Propanol, 2-(diethylamino)-
10367 - 1-Propanol, 3-(diethylamino)-

10368 - 2-Propanol, 1-(diethylamino)-, (±)-

$C_7H_{17}NO_2$

7443 - Methanamine, 1,1-diethoxy-*N,N*-dimethyl-

$C_7H_{17}NO_5$

6208 - *D*-Glucitol, 1-deoxy-1-(methylamino)-

$C_7H_{17}O_2PS$

9549 - Phosphonothioic acid, methyl-, *O,O*-bis(1-methylethyl) ester

$C_7H_{17}O_2PS_3$

9513 - Phorate

$C_7H_{18}N_2$

6371 - 1,7-Heptanediamine

9876 - 1,3-Propanediamine, *N,N*-diethyl-

9879 - 1,3-Propanediamine, *N,N,N',N'*-tetramethyl-

$C_7H_{18}N_2O$

10339 - 2-Propanol, 1-amino-3-(diethylamino)-

10343 - 2-Propanol, 1,3-bis(dimethylamino)-

$C_7H_{18}N_2O_2$

5679 - Ethanol, 2,2'-[(3-aminopropyl)imino]bis-

$C_7H_{18}O_3Si$

11479 - Silane, triethoxymethyl-

$C_7H_{18}Si$

11371 - Silane, butyltrimethyl-

11491 - Silane, trimethyl(2-methylpropyl)-

$C_7H_{19}N_3$

9875 - 1,3-Propanediamine, *N*-(3-aminopropyl)-*N*-methyl-

$C_7H_{22}O_4Si_4$

4992 - Cyclotetrasiloxane, heptamethyl-

$C_8Cl_2N_2O_2$

4251 - 1,4-Cyclohexadiene-1,2-dicarbonitrile, 4,5-dichloro-3,6-dioxo-

$C_8Cl_4N_2$

1317 - 1,3-Benzenedicarbonitrile, 2,4,5,6-tetrachloro-

$C_8Cl_4O_3$

7346 - 1,3-Isobenzofurandione, 4,5,6,7-tetrachloro-

$C_8F_{14}O_3$

3522 - Butanoic acid, heptafluoro-, anhydride

C_8F_{16}

4342 - Cyclohexane, 1,1,2,2,3,3,4,4,5,6-decafluoro-5,6-bis(trifluoromethyl)-

4343 - Cyclohexane, 1,1,2,2,3,4,4,5,5,6-decafluoro-3,6-bis(trifluoromethyl)-

$C_8F_{16}O$

6109 - Furan, 2,2,3,3,4,4,5-heptafluorotetrahydro-5-(nonafluorobutyl)-

C_8F_{18}

8415 - Octane, octadecafluoro-

$C_8HCl_4F_{11}O_2$

8442 - Octanoic acid, 3,5,7,8-tetrachloro-2,2,3,4,4,5,6,6,7,8,8-undecafluoro-

$C_8H_2Cl_2O_3$

7343 - 1,3-Isobenzofurandione, 4,5-dichloro-

7344 - 1,3-Isobenzofurandione, 5,6-dichloro-

$C_8H_3Cl_2F_3N_2$

2377 - 1H-Benzimidazole, 4,5-dichloro-2-(trifluoromethyl)-

$C_8H_4Cl_2O_2$

1318 - 1,2-Benzenedicarbonyl dichloride

1319 - 1,3-Benzenedicarbonyl dichloride

1320 - 1,4-Benzenedicarbonyl dichloride

$C_8H_4Cl_2O_4$

1340 - 1,2-Benzenedicarboxylic acid, 3,5-dichloro-

1349 - 1,4-Benzenedicarboxylic acid, 2,5-dichloro-

$C_8H_4Cl_6$

996 - Benzene, 1,4-bis(trichloromethyl)-

$C_8H_4F_2O_2$

1321 - 1,2-Benzenedicarbonyl difluoride

$C_8H_4F_3N$

2769 - Benzonitrile, 2-(trifluoromethyl)-

2770 - Benzonitrile, 3-(trifluoromethyl)-

$C_8H_4F_6$

997 - Benzene, 1,3-bis(trifluoromethyl)-

$C_8H_4N_2$

1315 - 1,3-Benzenedicarbonitrile

1316 - 1,4-Benzenedicarbonitrile

$C_8H_4N_2S_2$

1432 - Benzene, 1,4-diisothiocyanato-

$C_8H_4O_2S$

2401 - Benzo[b]thiophene-2,3-dione

$C_8H_4O_3$

7342 - 1,3-Isobenzofurandione

C_8H_5Br

1040 - Benzene, 1-bromo-2-ethynyl-

1041 - Benzene, 1-bromo-3-ethynyl-

1042 - Benzene, 1-bromo-4-ethynyl-

$C_8H_5ClN_2$

11334 - Quinoxaline, 6-chloro-

$C_8H_5Cl_3O$

5871 - Ethanone, 2,2,2-trichloro-1-phenyl-

$C_8H_5Cl_3O_2$

944 - Benzeneacetic acid, 2,3,6-trichloro-

$C_8H_5Cl_3O_3$

256 - Acetic acid, (2,4,5-trichlorophenoxy)-

$C_8H_5F_3O$

5872 - Ethanone, 2,2,2-trifluoro-1-phenyl-

$C_8H_5F_6N$

637 - Benzenamine, 3,5-bis(trifluoromethyl)-

C_8H_5NO

968 - Benzeneacetonitrile, α-oxo-

2757 - Benzonitrile, 3-formyl-

2758 - Benzonitrile, 4-formyl-

$C_8H_5NO_2$

2550 - Benzoic acid, 3-cyano-

7300 - 1H-Indole-2,3-dione

7362 - 1H-Isoindole-1,3(2H)-dione

C_8H_6

1715 - Benzene, ethynyl-

C_8H_6BrClO

5778 - Ethanone, 2-bromo-1-(4-chlorophenyl)-

C_8H_6BrN

947 - Benzeneacetonitrile, α-bromo-

948 - Benzeneacetonitrile, 2-bromo-

$C_8H_6Br_2O$

5777 - Ethanone, 2-bromo-1-(4-bromophenyl)-

5797 - Ethanone, 1-(3,5-dibromophenyl)-

5798 - Ethanone, 2,2-dibromo-1-phenyl-

$C_8H_6Br_2O_3$

2559 - Benzoic acid, 3,5-dibromo-2-hydroxy-, methyl ester

$C_8H_6Br_4$

2255 - Benzene, 1,2,3,4-tetrabromo-5,6-dimethyl-

C_8H_6ClN

949 - Benzeneacetonitrile, 2-chloro-

950 - Benzeneacetonitrile, 3-chloro-

2750 - Benzonitrile, 2-(chloromethyl)-

$C_8H_6ClNO_4$

2548 - Benzoic acid, 4-chloro-3-nitro-, methyl ester

2549 - Benzoic acid, 5-chloro-2-nitro-, methyl ester

$C_8H_6Cl_2$

1367 - Benzene, 1,2-dichloro-3-ethenyl-

1368 - Benzene, 1,2-dichloro-4-ethenyl-

1369 - Benzene, 1,3-dichloro-2-ethenyl-

1370 - Benzene, 1,3-dichloro-5-ethenyl-

1371 - Benzene, 1,4-dichloro-2-ethenyl-

1372 - Benzene, (2,2-dichloroethenyl)-

1373 - Benzene, 2,4-dichloro-1-ethenyl-

$C_8H_6Cl_2O$

973 - Benzeneacetyl chloride, α-chloro-

5784 - Ethanone, 2-chloro-1-(4-chlorophenyl)-

5799 - Ethanone, 1-(2,4-dichlorophenyl)-

5800 - Ethanone, 1-(2,5-dichlorophenyl)-

5801 - Ethanone, 1-(3,4-dichlorophenyl)-

5802 - Ethanone, 2,2-dichloro-1-phenyl-

$C_8H_6Cl_2O_2$

9288 - Phenol, 2,4-dichloro-, acetate

$C_8H_6Cl_2O_3$

169 - Acetic acid, (2,4-dichlorophenoxy)-

2567 - Benzoic acid, 3,6-dichloro-2-methoxy-

$C_8H_6Cl_3NO$

98 - Acetamide, 2,2,2-trichloro-*N*-phenyl-

$C_8H_6Cl_4$

989 - Benzene, 1,4-bis(dichloromethyl)-

2261 - Benzene, 1,2,3,5-tetrachloro-4,6-dimethyl-

C_8H_6FN

952 - Benzeneacetonitrile, 2-fluoro-

953 - Benzeneacetonitrile, 4-fluoro-

$C_8H_6N_2$

4136 - Cinnoline

8065 - 1,5-Naphthyridine

9599 - Phthalazine

11195 - Quinazoline

11333 - Quinoxaline

$C_8H_6N_2O_2$

965 - Benzeneacetonitrile, 2-nitro-

966 - Benzeneacetonitrile, 3-nitro-

967 - Benzeneacetonitrile, 4-nitro-

2766 - Benzonitrile, 2-methyl-5-nitro-

$C_8H_6N_4O_5$

7222 - 2,4-Imidazolidinedione, 1-[[(5-nitro-2-furanyl)methylene]amino]-

$C_8H_6N_4O_8$

3106 -[5,5'Bipyrimidine]2,2',4,4',6,6'(1H,1'H,3H,3'H,5H,5'H)-hexone-5,5'-dihydroxy

C_8H_6O

1716 - Benzene, (ethynyloxy)-

2438 - Benzofuran

C_8H_6OS

2403 - Benzo[b]thiophene-4-ol

$C_8H_6O_2$

1322 - 1,3-Benzenedicarboxaldehyde

1323 - 1,4-Benzenedicarboxaldehyde

2461 - 2(3H)-Benzofuranone

2462 - 3(2H)-Benzofuranone

7349 - 1(3H)-Isobenzofuranone

$C_8H_6O_3$

934 - Benzeneacetic acid, α-oxo-

2417 - 1,3-Benzodioxole-5-carboxaldehyde

2607 - Benzoic acid, 2-formyl-

$C_8H_6O_4$

1327 - 1,2-Benzenedicarboxylic acid

1328 - 1,3-Benzenedicarboxylic acid

1329 - 1,4-Benzenedicarboxylic acid

2418 - 1,3-Benzodioxole-5-carboxylic acid

2609 - Benzoic acid, 3-formyl-4-hydroxy-

9598 - *p*-Phthalaldehydic acid, 3-hydroxy-

$C_8H_6O_5$

1352 - 1,3-Benzenedicarboxylic acid, 4-hydroxy-

C_8H_6S

2399 - Benzo[b]thiophene

$C_8H_6S_2$

2434 - 1,4-Benzodithiin

3112 - 2,2'-Bithiophene

C_8H_7Br

1025 - Benzene, (1-bromoethenyl)-

1026 - Benzene, 1-bromo-2-ethenyl-

1027 - Benzene, 1-bromo-3-ethenyl-

1028 - Benzene, 1-bromo-4-ethenyl-

1029 - Benzene, (2-bromoethenyl)-, (*E*)-

1031 - Benzene, (2-bromoethenyl)-, (*Z*)-

4743 - 1,3,5,7-Cyclooctatetraene, 1-bromo-

C_8H_7BrO

5780 - Ethanone, 1-(3-bromophenyl)-

5781 - Ethanone, 1-(4-bromophenyl)-

5782 - Ethanone, 2-bromo-1-phenyl-

$C_8H_7BrO_2$

119 - Acetic acid, bromo-, phenyl ester

888 - Benzeneacetic acid, 4-bromo-

2525 - Benzoic acid, 3-bromo-, methyl ester

2526 - Benzoic acid, 4-bromo-, methyl ester

$C_8H_7BrO_3$

889 - Benzeneacetic acid, 4-bromo-α-hydroxy-, (±)-

C_8H_7Cl

1148 - Benzene, (1-chloroethenyl)-

1149 - Benzene, 1-chloro-2-ethenyl-

1150 - Benzene, 1-chloro-3-ethenyl-

1151 - Benzene, 1-chloro-4-ethenyl-

1153 - Benzene, (2-chloroethenyl)-, (*E*)-

1155 - Benzene, (2-chloroethenyl)-, (*Z*)-

4745 - 1,3,5,7-Cyclooctatetraene, 1-chloro-

C_8H_7ClO
- 971 - Benzeneacetyl chloride
- 2902 - Benzoyl chloride, 2-methyl-
- 2903 - Benzoyl chloride, 3-methyl-
- 2904 - Benzoyl chloride, 4-methyl-
- 5788 - Ethanone, 1-(3-chlorophenyl)-
- 5789 - Ethanone, 1-(4-chlorophenyl)-
- 5790 - Ethanone, 2-chloro-1-phenyl-

$C_8H_7ClO_2$
- 147 - Acetic acid, chloro-, phenyl ester
- 2421 - 1,3-Benzodioxole, 5-(chloromethyl)-
- 2543 - Benzoic acid, 4-chloro-, methyl ester
- 2901 - Benzoyl chloride, 4-methoxy-
- 4027 - Carbonochloridic acid, phenylmethyl ester

$C_8H_7ClO_3$
- 146 - Acetic acid, (4-chlorophenoxy)-
- 2542 - Benzoic acid, 5-chloro-2-hydroxy-, methyl ester

C_8H_7F
- 1633 - Benzene, 1-ethenyl-2-fluoro-
- 1634 - Benzene, 1-ethenyl-3-fluoro-
- 1635 - Benzene, 1-ethenyl-4-fluoro-

C_8H_7FO
- 5815 - Ethanone, 1-(4-fluorophenyl)-
- 5816 - Ethanone, 2-fluoro-1-phenyl-

$C_8H_7FO_2$
- 901 - Benzeneacetic acid, 4-fluoro-

$C_8H_7F_3$
- 2045 - Benzene, 1-methyl-4-(trifluoromethyl)-

C_8H_7IO
- 5836 - Ethanone, 1-(3-iodophenyl)-
- 5837 - Ethanone, 1-(4-iodophenyl)-
- 5838 - Ethanone, 2-iodo-1-phenyl-

$C_8H_7IO_2$
- 2660 - Benzoic acid, 2-iodo-, methyl ester
- 2661 - Benzoic acid, 3-iodo-, methyl ester
- 2662 - Benzoic acid, 4-iodo-, methyl ester

C_8H_7N
- 945 - Benzeneacetonitrile
- 1791 - Benzene, (isocyanomethyl)-
- 2763 - Benzonitrile, 2-methyl-
- 2764 - Benzonitrile, 3-methyl-
- 2765 - Benzonitrile, 4-methyl-
- 7294 - 1H-Indole
- 7324 - Indolizine

C_8H_7NO
- 287 - Acetonitrile, phenoxy-
- 956 - Benzeneacetonitrile, α-hydroxy-, (±)-
- 1787 - Benzene, 1-isocyanato-2-methyl-
- 2760 - Benzonitrile, 2-methoxy-
- 2761 - Benzonitrile, 3-methoxy-
- 2762 - Benzonitrile, 4-methoxy-
- 2880 - Benzoxazole, 2-methyl-
- 7327 - 2H-Indol-2-one, 1,3-dihydro-
- 7369 - 1H-Isoindol-1-one, 2,3-dihydro-

C_8H_7NOS
- 1793 - Benzene, 1-isothiocyanato-2-methoxy-

$C_8H_7NO_2$
- 876 - Benzeneacetaldehyde, α-oxo-, aldoxime
- 1644 - Benzene, 1-ethenyl-3-nitro-
- 1645 - Benzene, 1-ethenyl-4-nitro-
- 2050 - Benzene, (2-nitroethenyl)-, (*E*)-
- 8414 - Octane, 3-nitro-

$C_8H_7NO_3$
- 875 - Benzeneacetaldehyde, 2-nitro-
- 2489 - Benzoic acid, 4-(aminocarbonyl)-
- 5860 - Ethanone, 1-(3-nitrophenyl)-
- 5861 - Ethanone, 1-(4-nitrophenyl)-
- 5862 - Ethanone, 2-nitro-1-phenyl-

$C_8H_7NO_4$
- 219 - Acetic acid, 2-nitrophenyl ester
- 933 - Benzeneacetic acid, 4-nitro-
- 1330 - 1,3-Benzenedicarboxylic acid, 2-amino-
- 1331 - 1,3-Benzenedicarboxylic acid, 5-amino-
- 2710 - Benzoic acid, 2-nitro-, methyl ester
- 2711 - Benzoic acid, 3-nitro-, methyl ester

C_8H_7NS
- 1794 - Benzene, (isothiocyanatomethyl)-
- 2863 - Benzothiazole, 2-methyl-
- 11710 - Thiocyanic acid, phenylmethyl ester

$C_8H_7NS_2$
- 2867 - 2(3H)-Benzothiazolethione, 3-methyl-

$C_8H_7N_3$
- 11332 - 6-Quinoxalinamine

$C_8H_7N_3O$
- 5863 - Ethanone, 1-(2-nitrophenyl)-

$C_8H_7N_3O_2$
- 9600 - 1,4-Phthalazinedione, 5-amino-2,3-dihydro-

$C_8H_7N_3O_5$
- 8589 - 2-Oxazolidinone, 3-[[(5-nitro-2-furanyl)methylene]amino]-

$C_8H_7N_3O_6$
- 1489 - Benzene, 1,4-dimethyl-2,3,5-trinitro-
- 1490 - Benzene, 2,4-dimethyl-1,3,5-trinitro-

C_8H_8
- 1624 - Benzene, ethenyl-
- 4742 - 1,3,5,7-Cyclooctatetraene
- 4769 - 1,3-Cyclopentadiene, 5-(2-propenylidene)-, (*E*)-

C_8H_8BrNO
- 37 - Acetamide, *N*-(4-bromophenyl)-

$C_8H_8Br_2$
- 983 - Benzene, 1,2-bis(bromomethyl)-
- 984 - Benzene, 1,3-bis(bromomethyl)-
- 985 - Benzene, 1,4-bis(bromomethyl)-
- 1297 - Benzene, 1,2-dibromo-3,4-dimethyl-
- 1298 - Benzene, 1,5-dibromo-2,4-dimethyl-
- 1299 - Benzene, (1,2-dibromoethyl)-

C_8H_8ClN
- 7652 - Methylamine, *N*-(*o*-chlorobenzylidene)-

C_8H_8ClNO
- 48 - Acetamide, *N*-(4-chlorophenyl)-
- 3974 - Carbamic chloride, methylphenyl-

$C_8H_8ClNO_3S$
- 2243 - Benzenesulfonyl chloride, 4-(acetylamino)-

$C_8H_8Cl_2$
- 986 - Benzene, 1,2-bis(chloromethyl)-
- 987 - Benzene, 1,3-bis(chloromethyl)-
- 988 - Benzene, 1,4-bis(chloromethyl)-
- 1360 - Benzene, 1,2-dichloro-3,4-dimethyl-
- 1361 - Benzene, 1,3-dichloro-2,5-dimethyl-
- 1362 - Benzene, 1,4-dichloro-2,5-dimethyl-
- 1363 - Benzene, 1,5-dichloro-2,4-dimethyl-
- 1364 - Benzene, 2,3-dichloro-1,4-dimethyl-
- 1376 - Benzene, (1,2-dichloroethyl)-
- 1377 - Benzene, 1,4-dichloro-2-ethyl-
- 1390 - Benzene, 1-(dichloromethyl)-4-methyl-

$C_8H_8Cl_2NaO_5S$
- 11358 - Sesone

$C_8H_8Cl_2O$
- 1375 - Benzene, 2,4-dichloro-1-ethoxy-
- 1854 - Benzenemethanol, 2,4-dichloro-α-methyl-
- 9290 - Phenol, 2,4-dichloro-3,5-dimethyl-
- 9291 - Phenol, 2,6-dichloro-3,5-dimethyl-

$C_8H_8Cl_2O_2$
- 4084 - Chloroneb

$C_8H_8Cl_3N_3O_4S_2$
- 2854 - 2H-1,2,4-Benzothiadiazine-7-sulfonamide, 6-chloro-3-(dichloromethyl)-3,4-dihydro-, 1,1-dioxide

$C_8H_8Cl_3O_3PS$
- 9584 - Phosphorothioic acid, *O*,*O*-dimethyl *O*-(2,4,5-trichlorophenyl) ester

$C_8H_8F_3N_3O_4S_2$
- 2858 - 2H-1,2,4-Benzothiadiazine-7-sulfonamide, 3,4-dihydro-6-(trifluoromethyl)-, 1,1-dioxide

$C_8H_8HgO_2$
- 7431 - Mercury, (acetato-*O*)phenyl-

$C_8H_8N_2$
- 946 - Benzeneacetonitrile, 4-amino-
- 2382 - 1H-Benzimidazole, 1-methyl-
- 2743 - Benzonitrile, 5-amino-2-methyl-
- 4160 - Cyanamide, methylphenyl-
- 7239 - 1H-Indazole, 1-methyl-
- 7240 - 1H-Indazole, 3-methyl-
- 7241 - 1H-Indazole, 5-methyl-
- 7242 - 2H-Indazole, 2-methyl-

$C_8H_8N_2O_2$
- 1324 - 1,2-Benzenedicarboxamide
- 10883 - 3-Pyridinecarbonitrile, 1,2-dihydro-4-methoxy-1-methyl-2-oxo-

$C_8H_8N_2O_3$
- 80 - Acetamide, 2-nitro-*N*-phenyl-
- 81 - Acetamide, *N*-(2-nitrophenyl)-
- 82 - Acetamide, *N*-(3-nitrophenyl)-
- 83 - Acetamide, *N*-(4-nitrophenyl)-

$C_8H_8N_2O_5$
- 1650 - Benzene, 2-ethoxy-1,3-dinitro-

C_8H_8O
- 587 - Benzaldehyde, 2-methyl-
- 588 - Benzaldehyde, 3-methyl-
- 589 - Benzaldehyde, 4-methyl-
- 868 - Benzeneacetaldehyde
- 1646 - Benzene, (ethenyloxy)-
- 2442 - Benzofuran, 2,3-dihydro-
- 5864 - Ethanone, 1-phenyl-
- 8624 - Oxirane, phenyl-
- 9348 - Phenol, 2-ethenyl-
- 9349 - Phenol, 3-ethenyl-

$C_8H_8O_2$
- 231 - Acetic acid, phenyl ester
- 575 - Benzaldehyde, 2-hydroxy-3-methyl-
- 576 - Benzaldehyde, 2-hydroxy-4-methyl-
- 577 - Benzaldehyde, 2-hydroxy-5-methyl-
- 578 - Benzaldehyde, 2-hydroxy-6-methyl-
- 583 - Benzaldehyde, 2-methoxy-
- 584 - Benzaldehyde, 3-methoxy-
- 585 - Benzaldehyde, 4-methoxy-
- 883 - Benzeneacetic acid
- 2414 - 1,4-Benzodioxin, 2,3-dihydro-
- 2676 - Benzoic acid, 2-methyl-
- 2677 - Benzoic acid, 3-methyl-
- 2678 - Benzoic acid, 4-methyl-
- 2687 - Benzoic acid, methyl ester
- 3878 - 3-Buten-2-one, 4-(2-furanyl)-
- 4263 - 2,5-Cyclohexadiene-1,4-dione, 2,3-dimethyl-
- 4264 - 2,5-Cyclohexadiene-1,4-dione, 2,6-dimethyl-
- 5830 - Ethanone, 1-(2-hydroxyphenyl)-
- 5831 - Ethanone, 1-(3-hydroxyphenyl)-
- 5832 - Ethanone, 1-(4-hydroxyphenyl)-
- 5833 - Ethanone, 2-hydroxy-1-phenyl-
- 6016 - Formic acid, phenylmethyl ester

$C_8H_8O_3$
- 228 - Acetic acid, phenoxy-
- 570 - Benzaldehyde, 2-hydroxy-3-methoxy-
- 571 - Benzaldehyde, 2-hydroxy-5-methoxy-
- 572 - Benzaldehyde, 3-hydroxy-4-methoxy-
- 573 - Benzaldehyde, 3-hydroxy-5-methoxy-
- 574 - Benzaldehyde, 4-hydroxy-3-methoxy-
- 903 - Benzeneacetic acid, α-hydroxy-, (*S*)-
- 904 - Benzeneacetic acid, 2-hydroxy-
- 905 - Benzeneacetic acid, 3-hydroxy-
- 906 - Benzeneacetic acid, 4-hydroxy-
- 1540 - 1,2-Benzenediol, monoacetate
- 2427 - 1,3-Benzodioxole-5-methanol
- 2629 - Benzoic acid, 2-hydroxy-3-methyl-
- 2630 - Benzoic acid, 2-hydroxy-4-methyl-
- 2631 - Benzoic acid, 2-hydroxy-5-methyl-
- 2632 - Benzoic acid, 3-hydroxy-5-methyl-
- 2634 - Benzoic acid, 2-hydroxy-, methyl ester
- 2636 - Benzoic acid, 3-hydroxy-, methyl ester
- 2637 - Benzoic acid, 4-hydroxy-, methyl ester
- 2664 - Benzoic acid, 2-methoxy-
- 2665 - Benzoic acid, 3-methoxy-
- 2666 - Benzoic acid, 4-methoxy-
- 5803 - Ethanone, 1-(2,4-dihydroxyphenyl)-
- 6070 - 2-Furancarboxylic acid, 2-propenyl ester
- 7347 - 1,3-Isobenzofurandione, 4,5,6,7-tetrahydro-
- 10640 - 2-Propenoic acid, 3-(2-furanyl)-, methyl ester

$C_8H_8O_4$
895 - Benzeneacetic acid, 2,5-dihydroxy-
2573 - Benzoic acid, 2,4-dihydroxy-6-methyl-
2625 - Benzoic acid, 4-hydroxy-3-methoxy-
4262 - 2,5-Cyclohexadiene-1,4-dione, 2,6-dimethoxy-
4265 - 2,5-Cyclohexadiene-1,4-dione, 3-hydroxy-2-methoxy-5-methyl-
5873 - Ethanone, 1-(2,3,4-trihydroxyphenyl)-
10779 - 2H-Pyran-2,4(3H)-dione, 3-acetyl-6-methyl-
$C_8H_8O_5$
4260 - 2,5-Cyclohexadiene-1,4-dione, 2,5-dihydroxy-3-methoxy-6-methyl-
6083 - 2,5-Furandicarboxylic acid, dimethyl ester
C_8H_9Br
1017 - Benzene, 1-bromo-2,3-dimethyl-
1018 - Benzene, 1-bromo-2,4-dimethyl-
1019 - Benzene, 1-bromo-3,5-dimethyl-
1020 - Benzene, 2-bromo-1,3-dimethyl-
1021 - Benzene, 2-bromo-1,4-dimethyl-
1022 - Benzene, 4-bromo-1,2-dimethyl-
1034 - Benzene, (1-bromoethyl)-, (±)-
1035 - Benzene, 1-bromo-2-ethyl-
1036 - Benzene, 1-bromo-3-ethyl-
1037 - Benzene, 1-bromo-4-ethyl-
1038 - Benzene, (2-bromoethyl)-
1039 - Benzene, (1-bromoethyl)-, (*R*)-
1062 - Benzene, 1-(bromomethyl)-2-methyl-
1063 - Benzene, 1-(bromomethyl)-3-methyl-
1064 - Benzene, 1-(bromomethyl)-4-methyl-
C_8H_9BrO
1032 - Benzene, 1-bromo-4-ethoxy-
1033 - Benzene, (2-bromoethoxy)-
1595 - Benzeneethanol, 4-bromo-
1844 - Benzenemethanol, α-(bromomethyl)-
1845 - Benzenemethanol, 2-bromo-α-methyl-
$C_8H_9BrO_2$
1016 - Benzene, 4-bromo-1,2-dimethoxy-
C_8H_9Cl
1137 - Benzene, 1-chloro-2,3-dimethyl-
1138 - Benzene, 1-chloro-2,4-dimethyl-
1139 - Benzene, 2-chloro-1,3-dimethyl-
1140 - Benzene, 2-chloro-1,4-dimethyl-
1141 - Benzene, 4-chloro-1,2-dimethyl-
1160 - Benzene, (1-chloroethyl)-, (±)-
1161 - Benzene, 1-chloro-2-ethyl-
1162 - Benzene, 1-chloro-3-ethyl-
1163 - Benzene, 1-chloro-4-ethyl-
1164 - Benzene, (2-chloroethyl)-
1198 - Benzene, 1-(chloromethyl)-2-methyl-
1199 - Benzene, 1-(chloromethyl)-3-methyl-
1200 - Benzene, 1-(chloromethyl)-4-methyl-
C_8H_9ClO
1156 - Benzene, 1-chloro-2-ethoxy-
1157 - Benzene, 1-chloro-3-ethoxy-
1158 - Benzene, 1-chloro-4-ethoxy-
1159 - Benzene, (2-chloroethoxy)-
1181 - Benzene, [(chloromethoxy)methyl]-
1197 - Benzene, 1-(chloromethyl)-4-methoxy-
1596 - Benzeneethanol, 3-chloro-
1597 - Benzeneethanol, 4-chloro-
1852 - Benzenemethanol, α-(chloromethyl)-
9256 - Phenol, 2-chloro-3,4-dimethyl-
9257 - Phenol, 4-chloro-3,5-dimethyl-
C_8H_9ClS
1165 - Benzene, [(2-chloroethyl)thio]-
$C_8H_9Cl_2NO$
691 - Benzenamine, 2,6-dichloro-4-ethoxy-
$C_8H_9Cl_2P$
9553 - Phosphonous dichloride, (2,5-dimethylphenyl)-
9554 - Phosphonous dichloride, (4-ethylphenyl)-
$C_8H_9Cl_3Si$
11474 - Silane, trichloro(2-phenylethyl)-
C_8H_9FO
1654 - Benzene, 1-ethoxy-2-fluoro-
1655 - Benzene, 1-ethoxy-3-fluoro-
1656 - Benzene, 1-ethoxy-4-fluoro-
C_8H_9I
1680 - Benzene, 1-ethyl-2-iodo-
1681 - Benzene, 1-ethyl-4-iodo-
1769 - Benzene, 1-iodo-2,3-dimethyl-
1770 - Benzene, 1-iodo-2,4-dimethyl-
1771 - Benzene, 1-iodo-3,5-dimethyl-
1772 - Benzene, 2-iodo-1,3-dimethyl-
1773 - Benzene, 2-iodo-1,4-dimethyl-
1774 - Benzene, 4-iodo-1,2-dimethyl-
C_8H_9IO
1657 - Benzene, 1-ethoxy-4-iodo-
$C_8H_9IO_2$
1767 - Benzene, 1-iodo-2,3-dimethoxy-
1768 - Benzene, 4-iodo-1,2-dimethoxy-
C_8H_9N
733 - Benzenamine, 2-ethenyl-
734 - Benzenamine, 3-ethenyl-
735 - Benzenamine, 4-ethenyl-
2966 - Bicyclo[2.2.1]hept-5-ene-2-carbonitrile
7297 - 1H-Indole, 2,3-dihydro-
7452 - Methanamine, *N*-(phenylmethylene)-
10955 - Pyridine, 5-ethenyl-2-methyl-
11014 - Pyridine, 2-(2-propenyl)-
C_8H_9NO
86 - Acetamide, *N*-phenyl-
619 - Benzamide, 2-methyl-
620 - Benzamide, *N*-methyl-
878 - Benzeneacetamide
5772 - Ethanone, 1-(3-aminophenyl)-
5773 - Ethanone, 1-(4-aminophenyl)-
5774 - Ethanone, 2-amino-1-phenyl-
5855 - Ethanone, 1-(6-methyl-3-pyridinyl)-
5866 - Ethanone, 1-phenyl-, oxime
5996 - Formamide, *N*-(2-methylphenyl)-
5997 - Formamide, *N*-(3-methylphenyl)-
5998 - Formamide, *N*-methyl-*N*-phenyl-
$C_8H_9NO_2$
67 - Acetamide, *N*-(4-hydroxyphenyl)-
618 - Benzamide, 4-methoxy-
885 - Benzeneacetic acid, α-amino-
886 - Benzeneacetic acid, 4-amino-
1474 - Benzene, 1,2-dimethyl-3-nitro-
1475 - Benzene, 1,2-dimethyl-4-nitro-
1476 - Benzene, 1,3-dimethyl-2-nitro-
1477 - Benzene, 1,3-dimethyl-5-nitro-
1478 - Benzene, 1,4-dimethyl-2-nitro-
1479 - Benzene, 2,4-dimethyl-1-nitro-
1692 - Benzene, 1-ethyl-2-nitro-
1693 - Benzene, 1-ethyl-3-nitro-
1694 - Benzene, 1-ethyl-4-nitro-
2004 - Benzene, 1-methyl-4-(nitromethyl)-
2051 - Benzene, (2-nitroethyl)-
2426 - 1,3-Benzodioxole-5-methanamine
2503 - Benzoic acid, 2-amino-, methyl ester
2504 - Benzoic acid, 3-amino-, methyl ester
2679 - Benzoic acid, 2-(methylamino)-
6270 - Glycine, *N*-phenyl-
10901 - 2-Pyridinecarboxylic acid, ethyl ester
10902 - 3-Pyridinecarboxylic acid, ethyl ester
10903 - 4-Pyridinecarboxylic acid, ethyl ester
$C_8H_9NO_3$
1612 - Benzeneethanol, 2-nitro-
1662 - Benzene, 1-ethoxy-2-nitro-
1663 - Benzene, 1-ethoxy-3-nitro-
1664 - Benzene, 1-ethoxy-4-nitro-
1934 - Benzene, 4-methoxy-1-methyl-2-nitro-
2501 - Benzoic acid, 3-amino-4-hydroxy-, methyl ester
6267 - Glycine, *N*-(4-hydroxyphenyl)-
9334 - Phenol, 3,6-dimethyl-2-nitro-
$C_8H_9NO_4$
1445 - Benzene, 1,2-dimethoxy-4-nitro-
1446 - Benzene, 1,3-dimethoxy-5-nitro-
1447 - Benzene, 1,4-dimethoxy-2-nitro-
1448 - Benzene, 2,4-dimethoxy-1-nitro-
C_8H_{10}
1452 - Benzene, 1,2-dimethyl-
1453 - Benzene, 1,3-dimethyl-
1454 - Benzene, 1,4-dimethyl-
1668 - Benzene, ethyl-
4749 - 1,3,5-Cyclooctatriene
4750 - 1,3,6-Cyclooctatriene
4768 - 1,3-Cyclopentadiene, 5-(1-methylethylidene)-
6714 - 1,5-Hexadien-3-yne, 2,5-dimethyl-
8342 - 1,7-Octadiyne
8343 - 2,6-Octadiyne
8344 - 3,5-Octadiyne
8483 - 1,3,5,7-Octatetraene
$C_8H_{10}BrN$
643 - Benzenamine, 2-bromo-*N*,*N*-dimethyl-
644 - Benzenamine, 3-bromo-*N*,*N*-dimethyl-
645 - Benzenamine, 4-bromo-*N*,*N*-dimethyl-
1570 - Benzeneethanamine, 4-bromo-
$C_8H_{10}ClN$
662 - Benzenamine, 2-chloro-*N*,*N*-dimethyl-
663 - Benzenamine, 4-chloro-*N*,*N*-dimethyl-
1802 - Benzenemethanamine, 3-chloro-*N*-methyl-
$C_8H_{10}Cl_2Si$
11400 - Silane, dichloromethyl(4-methylphenyl)-
$C_8H_{10}K_2O_{15}Sb_2$
11588 - Tartar emetic
$C_8H_{10}N_2$
18 - Acetaldehyde, phenylhydrazone
5134 - Diazene, ethylphenyl-
11339 - Quinoxaline, 1,2,3,4-tetrahydro-
$C_8H_{10}N_2O$
29 - Acetamide, *N*-(4-aminophenyl)-
720 - Benzenamine, *N*,*N*-dimethyl-4-nitroso-
750 - Benzenamine, *N*-ethyl-*N*-nitroso-
12027 - Urea, (phenylmethyl)-
$C_8H_{10}N_2O_2$
718 - Benzenamine, *N*,*N*-dimethyl-2-nitro-
719 - Benzenamine, *N*,*N*-dimethyl-3-nitro-
$C_8H_{10}N_2O_3$
739 - Benzenamine, 2-ethoxy-5-nitro-
$C_8H_{10}N_2O_3S$
30 - Acetamide, *N*-[(4-aminophenyl)sulfonyl]-
31 - Acetamide, *N*-[4-(aminosulfonyl)phenyl]-
2212 - Benzenesulfonamide, *N*,4-dimethyl-*N*-nitroso-
$C_8H_{10}N_2O_4$
11011 - 1(4H)-Pyridinepropanoic acid, α-amino-3-hydroxy-4-oxo-, (*S*)-
$C_8H_{10}N_2O_4S$
474 - Asulam
$C_8H_{10}N_2S$
10885 - 4-Pyridinecarbothioamide, 2-ethyl-
11806 - Thiourea, (2-methylphenyl)-
$C_8H_{10}N_4O_2$
10760 - 1H-Purine-2,6-dione, 3,7-dihydro-1,3,7-trimethyl-
$C_8H_{10}N_6O_7$
11097 - 2,4,6(1H,3H,5H)-Pyrimidinetrione, 5-[(hexahydro-2,4,6-trioxo-5-pyrimidinyl)imino]-, monoammonium salt
$C_8H_{10}Na_2O_5$
5370 - Endothall, disodium salt
$C_8H_{10}O$
1592 - Benzeneethanol
1649 - Benzene, ethoxy-
1896 - Benzenemethanol, α-methyl-
1897 - Benzenemethanol, 2-methyl-
1898 - Benzenemethanol, 3-methyl-
1899 - Benzenemethanol, 4-methyl-
1930 - Benzene, (methoxymethyl)-
1931 - Benzene, 1-methoxy-2-methyl-
1932 - Benzene, 1-methoxy-3-methyl-
1933 - Benzene, 1-methoxy-4-methyl-
2916 - Bicyclo[2.2.1]heptane-2-carboxaldehyde
2958 - Bicyclo[2.2.1]heptan-2-one, 3-methylene-
2967 - Bicyclo[2.2.1]hept-5-ene-2-carboxaldehyde
8484 - 2,4,6-Octatrienal
9307 - Phenol, 2,3-dimethyl-
9308 - Phenol, 2,4-dimethyl-
9309 - Phenol, 2,5-dimethyl-

9310 - Phenol, 2,6-dimethyl-
9311 - Phenol, 3,4-dimethyl-
9312 - Phenol, 3,5-dimethyl-
9353 - Phenol, 2-ethyl-
9354 - Phenol, 3-ethyl-
9355 - Phenol, 4-ethyl-

$C_8H_{10}OS$
2296 - Benzenethiol, 4-ethoxy-
10472 - 1-Propanone, 2-methyl-1-(2-thienyl)-

$C_8H_{10}O_2$
1434 - 1,3-Benzenedimethanol
1435 - 1,4-Benzenedimethanol
1436 - Benzene, 1,2-dimethoxy-
1437 - Benzene, 1,3-dimethoxy-
1438 - Benzene, 1,4-dimethoxy-
1511 - 1,2-Benzenediol, 4,5-dimethyl-
1512 - 1,3-Benzenediol, 2,5-dimethyl-
1513 - 1,3-Benzenediol, 4,5-dimethyl-
1514 - 1,3-Benzenediol, 4,6-dimethyl-
1518 - 1,3-Benzenediol, 4-ethyl-
1606 - Benzeneethanol, 3-hydroxy-
1889 - Benzenemethanol, 2-methoxy-
1890 - Benzenemethanol, 3-methoxy-
1891 - Benzenemethanol, 4-methoxy-
3659 - 2-Butanone, 4-(2-furanyl)-
5579 - 1,2-Ethanediol, 1-phenyl-
5758 - Ethanol, 2-phenoxy-
8685 - 1,5-Pentalenedione, hexahydro-
9350 - Phenol, 2-ethoxy-
9351 - Phenol, 3-ethoxy-
9352 - Phenol, 4-ethoxy-
9385 - Phenol, 2-methoxy-4-methyl-

$C_8H_{10}O_2S$
1704 - Benzene, (ethylsulfonyl)-

$C_8H_{10}O_3$
1885 - Benzenemethanol, 2-hydroxy-3-methoxy-
1886 - Benzenemethanol, 4-hydroxy-3-methoxy-
3818 - 2-Butenoic acid, anhydride
6028 - 2-Furanacetic acid, ethyl ester
6061 - 2-Furancarboxylic acid, 1-methylethyl ester
6064 - 3-Furancarboxylic acid, 2-methyl-, ethyl ester
6071 - 2-Furancarboxylic acid, propyl ester
6123 - 2-Furanmethanol, propanoate
6162 - 2-Furanpropanoic acid, methyl ester
7345 - 1,3-Isobenzofurandione, hexahydro-
9305 - Phenol, 2,6-dimethoxy-
9306 - Phenol, 3,5-dimethoxy-
10649 - 2-Propenoic acid, 2-methyl-, anhydride

$C_8H_{10}O_3S$
2223 - Benzenesulfonic acid, ethyl ester
2232 - Benzenesulfonic acid, 4-methyl-, methyl ester

$C_8H_{10}O_4$
3909 - 2-Butynedioic acid, diethyl ester
3913 - 2-Butyne-1,4-diol, diacetate
5564 - Ethanedioic acid, di-2-propenyl ester
6701 - 2,5-Hexadienoic acid, 3-methoxy-5-methyl-4-oxo-

$C_8H_{10}S$
1705 - Benzene, (ethylthio)-
1834 - Benzenemethanethiol, α-methyl-, (*S*)-
2042 - Benzene, [(methylthio)methyl]-
2297 - Benzenethiol, 3-ethyl-
2298 - Benzenethiol, 4-ethyl-

$C_8H_{11}ClO_4$
3771 - 2-Butenedioic acid, 2-chloro-, diethyl ester, (*E*)-
3772 - 2-Butenedioic acid, 2-chloro-, diethyl ester, (*Z*)-

$C_8H_{11}ClSi$
11374 - Silane, chlorodimethylphenyl-

$C_8H_{11}Cl_2N_3O_2$
11061 - 2,4(1H,3H)-Pyrimidinedione, 5-[bis(2-chloroethyl)amino]-

$C_8H_{11}Cl_3O_7$
6231 - β-*D*-Glucopyranosiduronic acid, 2,2,2-trichloroethyl

$C_8H_{11}N$
276 - Acetonitrile, cyclohexylidene-
704 - Benzenamine, 2,3-dimethyl-
705 - Benzenamine, 2,4-dimethyl-
706 - Benzenamine, 2,5-dimethyl-
707 - Benzenamine, 2,6-dimethyl-
708 - Benzenamine, 3,4-dimethyl-
709 - Benzenamine, 3,5-dimethyl-
710 - Benzenamine, *N*,2-dimethyl-
711 - Benzenamine, *N*,3-dimethyl-
712 - Benzenamine, *N*,4-dimethyl-
713 - Benzenamine, *N*,*N*-dimethyl-
740 - Benzenamine, 2-ethyl-
741 - Benzenamine, 3-ethyl-
742 - Benzenamine, 4-ethyl-
743 - Benzenamine, *N*-ethyl-
1569 - Benzeneethanamine
1814 - Benzenemethanamine, α-methyl-, (±)-
1815 - Benzenemethanamine, 2-methyl-
1816 - Benzenemethanamine, 3-methyl-
1817 - Benzenemethanamine, 4-methyl-
1818 - Benzenemethanamine, *N*-methyl-
10962 - Pyridine, 2-ethyl-4-methyl-
10963 - Pyridine, 2-ethyl-6-methyl-
10964 - Pyridine, 3-ethyl-4-methyl-
10965 - Pyridine, 4-ethyl-2-methyl-
10966 - Pyridine, 5-ethyl-2-methyl-
10984 - Pyridine, 2-(1-methylethyl)-
10985 - Pyridine, 3-(1-methylethyl)-
10986 - Pyridine, 4-(1-methylethyl)-
11015 - Pyridine, 2-propyl-
11016 - Pyridine, 4-propyl-
11024 - Pyridine, 2,3,4-trimethyl-
11025 - Pyridine, 2,3,5-trimethyl-
11026 - Pyridine, 2,3,6-trimethyl-
11027 - Pyridine, 2,4,5-trimethyl-
11028 - Pyridine, 2,4,6-trimethyl-

$C_8H_{11}NO$
736 - Benzenamine, 2-ethoxy-
737 - Benzenamine, 4-ethoxy-
770 - Benzenamine, 2-methoxy-5-methyl-
771 - Benzenamine, 3-methoxy-4-methyl-
772 - Benzenamine, 4-methoxy-2-methyl-
773 - Benzenamine, 5-methoxy-2-methyl-
774 - Benzenamine, 2-methoxy-6-methyl-
1593 - Benzeneethanol, β-amino-
1594 - Benzeneethanol, 2-amino-
1812 - Benzenemethanamine, 2-methoxy-
1813 - Benzenemethanamine, 4-methoxy-
1839 - Benzenemethanol, α-(aminomethyl)-
5759 - Ethanol, 2-(phenylamino)-
9208 - Phenol, 4-(2-aminoethyl)-
9317 - Phenol, 2-(dimethylamino)-
9318 - Phenol, 3-(dimethylamino)-
9319 - Phenol, 4-(dimethylamino)-
9356 - Phenol, 3-(ethylamino)-
10951 - 2-Pyridineethanol, α-methyl-
10979 - Pyridine, 2-(2-methoxyethyl)-
11012 - 2-Pyridinepropanol
11013 - 3-Pyridinepropanol

$C_8H_{11}NO_2$
702 - Benzenamine, 2,5-dimethoxy-
703 - Benzenamine, 2,6-dimethoxy-

$C_8H_{11}NO_2S$
2211 - Benzenesulfonamide, *N*,4-dimethyl-

$C_8H_{11}NO_3$
1500 - 1,2-Benzenediol, 4-(2-amino-1-hydroxyethyl)-, (*R*)-

$C_8H_{11}NO_4$
11145 - 1-Pyrrolidineacetic acid, 2,5-dioxo-, ethyl ester

$C_8H_{11}N_3O_6$
11832 - 1,2,4-Triazine-3,5(2H,4H)-dione, 2-β-*D*-ribofuranosyl-

$C_8H_{11}N_5$
7234 - Imidodicarbonimidic diamide, *N*-phenyl-

$C_8H_{11}O_4$
5191 - 1,3-Dioxane-2-ethanol, 4-hydroxy-α,6-dimethyl-

C_8H_{12}
2993 - Bicyclo[4.2.0]oct-2-ene
4186 - Cyclobutane, 1,2-diethenyl-, *cis*-
4187 - Cyclobutane, 1,2-diethenyl-, *trans*-
4252 - 1,3-Cyclohexadiene, 1,3-dimethyl-
4253 - 1,3-Cyclohexadiene, 1,4-dimethyl-
4254 - 1,3-Cyclohexadiene, 1,5-dimethyl-
4255 - 1,3-Cyclohexadiene, 2,5-dimethyl-
4256 - 1,3-Cyclohexadiene, 2,6-dimethyl-
4314 - Cyclohexane, 1,2-bis(methylene)-
4657 - Cyclohexene, 1-ethenyl-
4658 - Cyclohexene, 4-ethenyl-
4732 - 1,4-Cyclooctadiene
4733 - 1,5-Cyclooctadiene
4734 - 1,5-Cyclooctadiene, (*Z*,*Z*)-
4755 - Cyclooctyne
6994 - 1,3,5-Hexatriene, 2,5-dimethyl-
8489 - 2,4,6-Octatriene, (*E*,*E*,*E*)-
8526 - 1-Octen-3-yne

$C_8H_{12}ClN$
717 - Benzenamine, *N*,*N*-dimethyl-, hydrochloride

$C_8H_{12}ClNO$
44 - Acetamide, 2-chloro-*N*,*N*-di-2-propenyl-

$C_8H_{12}ClNO_3$
10934 - 3,4-Pyridinedimethanol, 5-hydroxy-6-methyl-, hydrochloride

$C_8H_{12}Cl_2$
7117 - 3-Hexyne, 2,5-dichloro-2,5-dimethyl-

$C_8H_{12}N_2$
1267 - 1,2-Benzenediamine, *N*,*N*-dimethyl-
1268 - 1,3-Benzenediamine, *N*,*N*-dimethyl-
1269 - 1,2-Benzenediamine, *N*,*N*′-dimethyl-
1270 - 1,4-Benzenediamine, *N*,*N*-dimethyl-
1433 - 1,3-Benzenedimethanamine
3292 - Butanedinitrile, tetramethyl-
7159 - Hydrazine, 1-ethyl-1-phenyl-
7160 - Hydrazine, 1-ethyl-2-phenyl-
7163 - Hydrazine, 1-methyl-2-(3-methylphenyl)-
7170 - Hydrazine, (1-phenylethyl)-
8385 - Octanedinitrile
10809 - Pyrazine, 3-ethyl-2,5-dimethyl-

$C_8H_{12}N_2O$
1272 - 1,2-Benzenediamine, 4-ethoxy-

$C_8H_{12}N_2O_2$
6770 - Hexane, 1,6-diisocyanato-

$C_8H_{12}N_2O_3$
11087 - 2,4,6(1H,3H,5H)-Pyrimidinetrione, 5,5-diethyl-

$C_8H_{12}N_4$
10093 - Propanenitrile, 2,2′-azobis[2-methyl-

$C_8H_{12}O$
4529 - Cyclohexanol, 1-ethynyl-
4594 - Cyclohexanone, 2-ethylidene-
4705 - 2-Cyclohexen-1-one, 2,3-dimethyl-
4706 - 2-Cyclohexen-1-one, 2,5-dimethyl-
4707 - 2-Cyclohexen-1-one, 3,5-dimethyl-
4708 - 2-Cyclohexen-1-one, 3,6-dimethyl-
4754 - 3-Cycloocten-1-one
4947 - 2-Cyclopenten-1-one, 3-(1-methylethyl)-
5793 - Ethanone, 1-(1-cyclohexen-1-yl)-
6090 - Furan, 2-(1,1-dimethylethyl)-
8509 - 1-Octen-4-ol
8555 - 7-Oxabicyclo[4.1.0]heptane, 3-ethenyl-
8688 - 2(1H)-Pentalenone, hexahydro-
8689 - 2(1H)-Pentalenone, hexahydro-, *cis*-

$C_8H_{12}O_2$
4378 - 1,3-Cyclohexanedione, 5,5-dimethyl-
4572 - Cyclohexanone, 2-acetyl-
4640 - 3-Cyclohexene-1-carboxylic acid, methyl ester
6699 - 2,4-Hexadienoic acid, ethyl ester
6700 - 2,4-Hexadienoic acid, ethyl ester, (*E*,*E*)-
8504 - 4-Octene-2,7-dione
8538 - 2-Octynoic acid
8539 - 3-Octynoic acid

8540 - 5-Octynoic acid
8541 - 7-Octynoic acid
8557 - 7-Oxabicyclo[4.1.0]heptane, 3-oxiranyl-
8597 - 2-Oxetanone, 3,3-dimethyl-4-(1-methylethylidene)-
8805 - 2,4-Pentanedione, 3-(2-propenyl)-

$C_8H_{12}O_3$
4336 - Cyclohexanecarboxylic acid, 3-methyl-4-oxo-
4785 - Cyclopentanecarboxylic acid, 2-oxo-, ethyl ester
10695 - 2-Propenoic acid, (tetrahydro-2-furanyl)methyl ester

$C_8H_{12}O_4$
3462 - Butanoic acid, 2-acetyl-3-oxo-, ethyl ester
3775 - 2-Butenedioic acid (*E*)-, diethyl ester
3776 - 2-Butenedioic acid (*Z*)-, diethyl ester
4181 - 1,2-Cyclobutanedicarboxylic acid, dimethyl ester, *trans*-
4183 - 1,3-Cyclobutanedicarboxylic acid, 2,2-dimethyl-, *cis*-
4354 - 1,4-Cyclohexanedicarboxylic acid, *trans*-
6030 - 3-Furanacetic acid, tetrahydro-2,2-dimethyl-5-oxo-
10564 - 2-Propene-1,1-diol, 2-methyl-, diacetate

$C_8H_{12}O_5$
3340 - Butanedioic acid, oxo-, diethyl ester

$C_8H_{12}O_6Si$
11498 - Silanetriol, ethenyl-, triacetate

$C_8H_{12}S$
11731 - Thiophene, 2-butyl-

$C_8H_{12}Si$
11418 - Silane, dimethylphenyl-
11449 - Silane, tetraethenyl-

$C_8H_{12}Sn$
11543 - Stannane, tetraethenyl-

$C_8H_{13}Cl$
8533 - 1-Octyne, 1-chloro-

$C_8H_{13}F_3O_2$
268 - Acetic acid, trifluoro-, 2-methylpentyl ester

$C_8H_{13}N$
4290 - Cyclohexanamine, 1-ethynyl-
4313 - Cyclohexaneacetonitrile
11113 - 1H-Pyrrole, 1-butyl-
11126 - 1H-Pyrrole, 3-ethyl-2,4-dimethyl-
11127 - 1H-Pyrrole, 4-ethyl-2,3-dimethyl-
11141 - 1H-Pyrrole, 2,3,4,5-tetramethyl-

$C_8H_{13}NO$
490 - 8-Azabicyclo[3.2.1]octan-3-one, 8-methyl-

$C_8H_{13}NO_2$
7575 - 2,5-Methano-2H-furo[3,2-b]pyrrol-6-ol, hexahydro-4-methyl-, (2α,3aβ,5α,6β,6aβ
9656 - 2,6-Piperidinedione, 4-ethyl-4-methyl-
10914 - 3-Pyridinecarboxylic acid, 1,2,5,6-tetrahydro-1-methyl-, methyl ester

$C_8H_{13}NO_3$
8573 - 2H-1,3-Oxazine-2,4(3H)-dione, 5,5-diethyldihydro-

$C_8H_{13}NO_5$
9953 - Propanedioic acid, (formylamino)-, diethyl ester

$C_8H_{13}N_2O_3PS$
11720 - Thionazin

C_8H_{14}
2991 - Bicyclo[4.2.0]octane
4245 - Cycloheptene, 1-methyl-
4246 - Cycloheptene, 5-methyl-
4388 - Cyclohexane, ethenyl-
4391 - Cyclohexane, ethylidene-
4434 - Cyclohexane, 1-methyl-4-methylene-
4648 - Cyclohexene, 1,2-dimethyl-
4649 - Cyclohexene, 1,3-dimethyl-
4650 - Cyclohexene, 1,4-dimethyl-
4651 - Cyclohexene, 1,5-dimethyl-
4652 - Cyclohexene, 1,6-dimethyl-
4653 - Cyclohexene, 3,3-dimethyl-
4654 - Cyclohexene, 3,5-dimethyl-
4655 - Cyclohexene, 3,6-dimethyl-
4656 - Cyclohexene, 4,4-dimethyl-
4660 - Cyclohexene, 1-ethyl-
4661 - Cyclohexene, 3-ethyl-
4662 - Cyclohexene, 4-ethyl-
4751 - Cyclooctene, (*E*)-
4752 - Cyclooctene, (*Z*)-
4848 - Cyclopentane, 2-propenyl-
4935 - Cyclopentene, 1-propyl-
4937 - Cyclopentene, 1,2,3-trimethyl-
4938 - Cyclopentene, 1,5,5-trimethyl-
6335 - 1,5-Heptadiene, 3-methyl-
6691 - 1,5-Hexadiene, 2,5-dimethyl-
6692 - 2,4-Hexadiene, 2,5-dimethyl-
8319 - 1,7-Octadiene
8320 - 2,6-Octadiene
8529 - 1-Octyne
8530 - 2-Octyne
8531 - 3-Octyne
8532 - 4-Octyne
8686 - Pentalene, octahydro-, *cis*-
8687 - Pentalene, octahydro-, *trans*-
9132 - 1-Pentyne, 3-ethyl-3-methyl-

$C_8H_{14}ClNS_2$
4065 - CDEC

$C_8H_{14}ClN_5$
476 - Atrazine

$C_8H_{14}Cl_2N_2O_2$
10978 - 3-Pyridinemethanol, 4-(aminomethyl)-5-hydroxy-6-methyl-, dihydrochloride

$C_8H_{14}Cl_2O_2$
3501 - Butanoic acid, 3,4-dichloro-, butyl ester

$C_8H_{14}N_2$
275 - Acetonitrile, (cyclohexylamino)-
9703 - 1-Piperidinepropanenitrile

$C_8H_{14}N_4OS$
7658 - Metribuzin

$C_8H_{14}O$
4235 - Cycloheptanone, 2-methyl-
4580 - Cyclohexanone, 2,2-dimethyl-
4581 - Cyclohexanone, 2,3-dimethyl-, *cis*-
4582 - Cyclohexanone, 2,3-dimethyl-, *trans*-
4583 - Cyclohexanone, 2,4-dimethyl-, (2*S*-*trans*)-
4584 - Cyclohexanone, 2,4-dimethyl-, *cis*-
4585 - Cyclohexanone, 2,5-dimethyl-, *trans*-(±)-
4586 - Cyclohexanone, 2,6-dimethyl-
4587 - Cyclohexanone, 3,3-dimethyl-
4588 - Cyclohexanone, 3,4-dimethyl-
4589 - Cyclohexanone, 3,5-dimethyl-
4590 - Cyclohexanone, 4,4-dimethyl-
4593 - Cyclohexanone, 4-ethyl-
4740 - Cyclooctanone
4753 - 2-Cycloocten-1-ol
4883 - Cyclopentanone, 2-ethyl-5-methyl-
4887 - Cyclopentanone, 2-(1-methylethyl)-
4894 - Cyclopentanone, 2-propyl-
4895 - Cyclopentanone, 3-propyl-
4896 - Cyclopentanone, 2,2,4-trimethyl-
4897 - Cyclopentanone, 2,2,5-trimethyl-
4898 - Cyclopentanone, 2,3,5-trimethyl-
4899 - Cyclopentanone, 2,4,4-trimethyl-
5794 - Ethanone, 1-cyclohexyl-
5846 - Ethanone, 1-(1-methylcyclopentyl)-
5847 - Ethanone, 1-(2-methylcyclopentyl)-
6338 - 1,6-Heptadien-4-ol, 4-methyl-
6595 - 1-Hepten-3-one, 5-methyl-
6596 - 2-Hepten-4-one, 5-methyl-
6597 - 2-Hepten-4-one, 6-methyl-
6598 - 3-Hepten-2-one, 6-methyl-
6599 - 4-Hepten-2-one, 6-methyl-
6600 - 4-Hepten-3-one, 5-methyl-
6601 - 5-Hepten-2-one, 3-methyl-
6602 - 5-Hepten-2-one, 5-methyl-
6603 - 5-Hepten-2-one, 6-methyl-
6604 - 5-Hepten-3-one, 2-methyl-
6605 - 5-Hepten-3-one, 5-methyl-
7091 - 3-Hexen-2-one, 3,4-dimethyl-
7092 - 4-Hexen-2-one, 3,4-dimethyl-
8543 - 2-Octyn-1-ol

$C_8H_{14}O_2$
155 - Acetic acid, cyclohexyl ester
3836 - 2-Butenoic acid, 2,3-dimethyl-, ethyl ester
4223 - Cycloheptanecarboxylic acid
4306 - Cyclohexaneacetic acid
4335 - Cyclohexanecarboxylic acid, methyl ester
4776 - Cyclopentaneacetic acid, 3-methyl-
4847 - Cyclopentanepropanoic acid
5821 - Ethanone, 1-(1-hydroxycyclohexyl)-
6140 - 2(3H)-Furanone, 5-butyldihydro-
6565 - 2-Heptenoic acid, 6-methyl-
6807 - 3,4-Hexanedione, 2,5-dimethyl-
6892 - Hexanoic acid, 3-hydroxy-2,2-dimethyl-, β-lactone
7053 - 2-Hexenoic acid, ethyl ester
7054 - 3-Hexenoic acid, ethyl ester
7119 - 3-Hexyne-2,5-diol, 2,5-dimethyl-
8364 - Octanedial
8393 - 2,4-Octanedione
8394 - 2,7-Octanedione
8395 - 3,6-Octanedione
8396 - 4,5-Octanedione
9080 - 2-Pentenoic acid, 4-methyl-, ethyl ester
9081 - 3-Pentenoic acid, 3-methyl-, ethyl ester
9082 - 3-Pentenoic acid, 4-methyl-, ethyl ester
10650 - 2-Propenoic acid, 2-methyl-, butyl ester
10651 - 2-Propenoic acid, methylbutyl ester
10652 - 2-Propenoic acid, 2-methylbutyl ester
10665 - 2-Propenoic acid, 2-methyl-, 1-methylpropyl ester
10666 - 2-Propenoic acid, 2-methyl-, 2-methylpropyl ester

$C_8H_{14}O_3$
3467 - Butanoic acid, anhydride
3511 - Butanoic acid, 2,2-dimethyl-3-oxo-, ethyl ester
3518 - Butanoic acid, 2-ethyl-3-oxo-, ethyl ester
3569 - Butanoic acid, 3-oxo-, butyl ester
6126 - 2-Furanmethanol, tetrahydro-, propanoate
6903 - Hexanoic acid, 4-methyl-5-oxo-, methyl ester
6906 - Hexanoic acid, 5-oxo-, ethyl ester
8897 - Pentanoic acid, 4-methyl-3-oxo-, ethyl ester
8909 - Pentanoic acid, 4-oxo-, 1-methylethyl ester
8912 - Pentanoic acid, 4-oxo-, propyl ester
10291 - Propanoic acid, 2-methyl-, anhydride

$C_8H_{14}O_4$
3303 - Butanedioic acid, diethyl ester
3333 - Butanedioic acid, monobutyl ester
3343 - Butanedioic acid, tetramethyl-
3348 - 1,4-Butanediol, diacetate
5556 - Ethanedioic acid, bis(1-methylethyl) ester
5565 - Ethanedioic acid, dipropyl ester
5576 - 1,2-Ethanediol, dipropanoate
6786 - Hexanedioic acid, dimethyl ester
6791 - Hexanedioic acid, monoethyl ester
8386 - Octanedioic acid
8777 - Pentanedioic acid, 3-ethyl-3-methyl-
9962 - Propanedioic acid, methyl-, diethyl ester

$C_8H_{14}O_4S$
10331 - Propanoic acid, 3,3′-thiobis-, dimethyl ester

$C_8H_{14}O_5$
5754 - Ethanol, 2,2′-oxybis-, diacetate

$C_8H_{14}O_6$
3307 - Butanedioic acid, 2,3-dihydroxy-, diethyl ester, (*R**,*R**)-(±)-

$C_8H_{15}Br$
4316 - Cyclohexane, (2-bromoethyl)-

$C_8H_{15}BrO_2$
4149 - Crotonaldehyde, 2-bromo-, diethyl acetal
6877 - Hexanoic acid, 2-bromo-, ethyl ester, (±)-
6878 - Hexanoic acid, 5-bromo-, ethyl ester
6879 - Hexanoic acid, 6-bromo-, ethyl ester
8425 - Octanoic acid, 2-bromo-
8426 - Octanoic acid, 8-bromo-
10210 - Propanoic acid, 3-bromo-, 3-methylbutyl ester

$C_8H_{15}Cl$
6541 - 2-Heptene, 6-chloro-2-methyl-
7010 - 3-Hexene, 1-chloro-4-ethyl-
8498 - 1-Octene, 2-chloro-
8499 - 2-Octene, 2-chloro-
8500 - 4-Octene, 4-chloro-, (*Z*)-
8501 - 2-Octene, 4-chloro-
$C_8H_{15}ClO$
6987 - Hexanoyl chloride, 2-ethyl-
8481 - Octanoyl chloride
$C_8H_{15}ClO_2$
10226 - Propanoic acid, 3-chloro-, 3-methylbutyl ester
$C_8H_{15}FO_2$
8429 - Octanoic acid, 8-fluoro-
$C_8H_{15}N$
484 - 8-Azabicyclo[3.2.1]octane, 8-methyl-
7319 - 1H-Indole, octahydro-
7325 - Indolizine, octahydro-
8411 - Octanenitrile
9706 - Piperidine, 2-(1-propenyl)-, (±)-
11021 - Pyridine, 2,3,4,5-tetrahydro-6-propyl-
$C_8H_{15}NO$
486 - 8-Azabicyclo[3.2.1]octan-3-ol, 8-methyl-, *exo*
489 - 8-Azabicyclo[3.2.1]octan-3-ol, 8-methyl-, *endo*
3408 - Butanenitrile, 2-hydroxy-3-methyl-2-(1-methylethyl)-
8829 - Pentanenitrile, 2-hydroxy-2-propyl-
9631 - Piperidine, 1-acetyl-3-methyl-
10471 - 2-Propanone, 1-(1-methyl-2-pyrrolidinyl)-, (*R*)-
10484 - 1-Propanone, 1-(2-piperidinyl)-
10485 - 2-Propanone, 1-(2-piperidinyl)-, (*R*)-
$C_8H_{15}NO_2$
3406 - Butanenitrile, 4,4-diethoxy-
4333 - Cyclohexanecarboxylic acid, 4-(aminomethyl)-, *trans*-
8399 - 2,3-Octanedione, 3-oxime
9643 - 4-Piperidinecarboxylic acid, ethyl ester
9645 - 4-Piperidinecarboxylic acid, 1-methyl-, methyl ester
10654 - 2-Propenoic acid, 2-methyl-, 2-(dimethylamino)ethyl ester
$C_8H_{15}NO_3$
6871 - Hexanoic acid, 6-(acetylamino)-
$C_8H_{15}NS$
11704 - Thiocyanic acid, heptyl ester
C_8H_{16}
4226 - Cycloheptane, methyl-
4364 - Cyclohexane, 1,1-dimethyl-
4365 - Cyclohexane, 1,2-dimethyl-, *cis*-
4366 - Cyclohexane, 1,2-dimethyl-, *trans*-
4367 - Cyclohexane, 1,3-dimethyl-, *cis*-
4368 - Cyclohexane, 1,4-dimethyl-, *cis*-
4370 - Cyclohexane, 1,3-dimethyl-, *trans*-
4371 - Cyclohexane, 1,4-dimethyl-, *trans*-
4390 - Cyclohexane, ethyl-
4736 - Cyclooctane
4814 - Cyclopentane, 1-ethyl-1-methyl-
4815 - Cyclopentane, 1-ethyl-2-methyl-, *cis*-
4816 - Cyclopentane, 1-ethyl-2-methyl-, *trans*-
4817 - Cyclopentane, 1-ethyl-3-methyl-, *cis*-
4818 - Cyclopentane, 1-ethyl-3-methyl-, *trans*-
4828 - Cyclopentane, (1-methylethyl)-
4849 - Cyclopentane, propyl-
4855 - Cyclopentane, 1,1,2-trimethyl-
4856 - Cyclopentane, 1,1,3-trimethyl-
4857 - Cyclopentane, 1,2,3-trimethyl-, (1α,2β,3α)-
4858 - Cyclopentane, 1,2,3-trimethyl-, (1α,2α,3α)-
4859 - Cyclopentane, 1,2,3-trimethyl-, (1α,2α,3β)-
4860 - Cyclopentane, 1,2,4-trimethyl-, (1α,2β,4α)-
4861 - Cyclopentane, 1,2,4-trimethyl-, (1α,2α,4α)-
4862 - Cyclopentane, 1,2,4-trimethyl-, (1α,2α,4β)-
6425 - Heptane, 3-methylene-
6426 - Heptane, 4-methylene-
6543 - 1-Heptene, 2-methyl-
6544 - 1-Heptene, 3-methyl-
6545 - 1-Heptene, 4-methyl-
6546 - 1-Heptene, 5-methyl-
6547 - 1-Heptene, 6-methyl-
6548 - 2-Heptene, 2-methyl-
6549 - 2-Heptene, 3-methyl-, (*E*)-
6550 - 2-Heptene, 3-methyl-, (*Z*)-
6551 - 2-Heptene, 4-methyl-, (*E*)-
6552 - 2-Heptene, 4-methyl-, (*Z*)-
6553 - 2-Heptene, 6-methyl-
6554 - 3-Heptene, 2-methyl-
6555 - 3-Heptene, 3-methyl-
6556 - 3-Heptene, 4-methyl-
6557 - 3-Heptene, 5-methyl-
6558 - 3-Heptene, 6-methyl-
6836 - Hexane, 2-methyl-3-methylene-
6837 - Hexane, 2-methyl-4-methylene-
6838 - Hexane, 3-methyl-4-methylene-
7013 - 1-Hexene, 2,3-dimethyl-
7014 - 1-Hexene, 2,4-dimethyl-
7015 - 1-Hexene, 2,5-dimethyl-
7016 - 1-Hexene, 3,3-dimethyl-
7017 - 1-Hexene, 3,4-dimethyl-
7018 - 1-Hexene, 3,5-dimethyl-
7019 - 1-Hexene, 4,4-dimethyl-
7020 - 1-Hexene, 4,5-dimethyl-
7021 - 1-Hexene, 5,5-dimethyl-
7022 - 2-Hexene, 2,3-dimethyl-
7023 - 2-Hexene, 2,4-dimethyl-
7024 - 2-Hexene, 2,5-dimethyl-
7025 - 2-Hexene, 4,5-dimethyl-
7026 - 2-Hexene, 5,5-dimethyl-, (*E*)-
7027 - 2-Hexene, 5,5-dimethyl-, (*Z*)-
7028 - 3-Hexene, 2,2-dimethyl-, (*E*)-
7029 - 3-Hexene, 2,2-dimethyl-, (*Z*)-
7030 - 3-Hexene, 2,4-dimethyl-, (*E*)-
7031 - 3-Hexene, 2,4-dimethyl-, (*Z*)-
7033 - 1-Hexene, 3-ethyl-
7034 - 3-Hexene, 3-ethyl-
8491 - 1-Octene
8492 - 2-Octene, (*E*)-
8493 - 2-Octene, (*Z*)-
8494 - 3-Octene, (*E*)-
8495 - 3-Octene, (*Z*)-
8496 - 4-Octene, (*E*)-
8497 - 4-Octene, (*Z*)-
8764 - Pentane, 2,2-dimethyl-3-methylene-
8765 - Pentane, 2,4-dimethyl-3-methylene-
9046 - 1-Pentene, 3-ethyl-2-methyl-
9047 - 1-Pentene, 3-ethyl-3-methyl-
9048 - 1-Pentene, 3-ethyl-4-methyl-
9049 - 2-Pentene, 3-ethyl-2-methyl-
9050 - 2-Pentene, 3-ethyl-4-methyl-, (*E*)-
9051 - 2-Pentene, 3-ethyl-4-methyl-, (*Z*)-
9064 - Pentene, 2,4,4-trimethyl-
9065 - 1-Pentene, 2,3,3-trimethyl-
9066 - 1-Pentene, 2,3,4-trimethyl-
9067 - 1-Pentene, 2,4,4-trimethyl-
9068 - 1-Pentene, 3,3,4-trimethyl-
9069 - 1-Pentene, 3,4,4-trimethyl-
9070 - 2-Pentene, 2,3,4-trimethyl-
9071 - 2-Pentene, 2,4,4-trimethyl-
9072 - 2-Pentene, 3,4,4-trimethyl-
$C_8H_{16}BrF$
8358 - Octane, 1-bromo-8-fluoro-
$C_8H_{16}Br_2$
8366 - Octane, 1,2-dibromo-
8367 - Octane, 1,4-dibromo-
8368 - Octane, 1,5-dibromo-
8369 - Octane, 1,8-dibromo-
8370 - Octane, 4,5-dibromo-
$C_8H_{16}ClF$
8361 - Octane, 1-chloro-8-fluoro-
$C_8H_{16}ClNO$
42 - Acetamide, 2-chloro-*N*,*N*-bis(1-methylethyl)-
45 - Acetamide, 2-chloro-*N*-hexyl-
$C_8H_{16}Cl_2$
6765 - Hexane, 2,5-dichloro-2,5-dimethyl-
8371 - Octane, 1,8-dichloro-
8372 - Octane, 2,3-dichloro-
$C_8H_{16}NO_5P$
5172 - Dicrotophos
$C_8H_{16}N_2$
3673 - 2-Butanone, (1-methylpropylidene)hydrazone
$C_8H_{16}N_2O_2$
8397 - 4,5-Octanedione, dioxime
$C_8H_{16}N_2O_2S_2$
7688 - Morpholine, 4,4′-dithiobis-
$C_8H_{16}N_2O_3$
7414 - *DL*-Leucine, *N*-glycyl-
$C_8H_{16}N_2O_4S_2$
12042 - *D*-Valine, 3-[(2-amino-2-carboxyethyl)dithio]-, (*R*)-
$C_8H_{16}O$
4230 - Cycloheptanol, 1-methyl-
4231 - Cycloheptanol, 2-methyl-
4232 - Cycloheptanol, 4-methyl-
4387 - Cyclohexaneethanol
4420 - Cyclohexanemethanol, α-methyl-
4421 - Cyclohexanemethanol, α-methyl-, (*S*)-
4422 - Cyclohexanemethanol, 2-methyl-, *cis*-
4423 - Cyclohexanemethanol, 2-methyl-, *trans*-
4424 - Cyclohexanemethanol, 4-methyl-
4508 - Cyclohexanol, 1,2-dimethyl-, *cis*-
4509 - Cyclohexanol, 1,2-dimethyl-, *trans*-
4510 - Cyclohexanol, 1,3-dimethyl-, *cis*-
4511 - Cyclohexanol, 1,3-dimethyl-, *trans*-
4512 - Cyclohexanol, 1,4-dimethyl-, *cis*-
4513 - Cyclohexanol, 2,2-dimethyl-
4514 - Cyclohexanol, 2,4-dimethyl-
4515 - Cyclohexanol, 2,6-dimethyl-
4516 - Cyclohexanol, 3,3-dimethyl-
4517 - Cyclohexanol, 3,4-dimethyl-
4518 - Cyclohexanol, 3,5-dimethyl-
4519 - Cyclohexanol, 4,4-dimethyl-
4522 - Cyclohexanol, 1-ethyl-
4523 - Cyclohexanol, 2-ethyl-, *trans*-(±)-
4524 - Cyclohexanol, 2-ethyl-, *cis*-(±)-
4525 - Cyclohexanol, 4-ethyl-
4739 - Cyclooctanol
4872 - Cyclopentanol, 1-propyl-
4873 - Cyclopentanol, 2-propyl-, *cis*-
4874 - Cyclopentanol, 2-propyl-, *trans*-
4875 - Cyclopentanol, 1,2,2-trimethyl-
4876 - Cyclopentanol, 1,2,4-trimethyl-
4877 - Cyclopentanol, 1,2,5-trimethyl-
6085 - Furan, 2,2-diethyltetrahydro-
6174 - Furan, tetrahydro-2,2,4,4-tetramethyl-
6345 - Heptanal, 5-methyl-
6520 - 2-Heptanone, 3-methyl-
6521 - 2-Heptanone, 6-methyl-
6522 - 3-Heptanone, 2-methyl-
6523 - 3-Heptanone, 6-methyl-
6524 - 4-Heptanone, 2-methyl-
6525 - 4-Heptanone, 3-methyl-
6573 - 1-Hepten-4-ol, 4-methyl-
6574 - 2-Hepten-4-ol, 5-methyl-
6575 - 2-Hepten-4-ol, 6-methyl-
6576 - 3-Hepten-2-ol, 2-methyl-
6577 - 4-Hepten-2-ol, 2-methyl-
6578 - 4-Hepten-3-ol, 3-methyl-
6579 - 4-Hepten-3-ol, 4-methyl-
6580 - 5-Hepten-2-ol, 6-methyl-
6581 - 6-Hepten-1-ol, 3-methyl-
6582 - 6-Hepten-2-ol, 2-methyl-
6719 - Hexanal, 2-ethyl-
6813 - Hexane, 1-(ethenyloxy)-
6963 - 2-Hexanone, 3,3-dimethyl-
6964 - 2-Hexanone, 3,4-dimethyl-
6965 - 3-Hexanone, 2,2-dimethyl-
6966 - 3-Hexanone, 2,5-dimethyl-
6967 - 3-Hexanone, 4,4-dimethyl-
7073 - 1-Hexen-3-ol, 3,5-dimethyl-
7074 - 4-Hexen-3-ol, 2,2-dimethyl-
7075 - 4-Hexen-3-ol, 2,5-dimethyl-

7076 - 4-Hexen-3-ol, 3,5-dimethyl-
8346 - Octanal
8471 - 2-Octanone
8472 - 3-Octanone
8473 - 4-Octanone
8510 - 2-Octen-1-ol
8511 - 2-Octen-4-ol
8512 - 4-Octen-1-ol
8513 - 5-Octen-1-ol
8612 - Oxirane, 2-(2,2-dimethylpropyl)-2-methyl-
8697 - Pentanal, 2-propyl-
8970 - 2-Pentanone, 3-ethyl-4-methyl-
8989 - 3-Pentanone, 2,2,4-trimethyl-

$C_8H_{16}O_2$
185 - Acetic acid, 2-ethylbutyl ester
194 - Acetic acid, hexyl ester
3479 - Butanoic acid, butyl ester
3509 - Butanoic acid, 1,1-dimethylethyl ester
3553 - Butanoic acid, 3-methyl-, 1-methylethyl ester
3562 - Butanoic acid, 3-methyl-, propyl ester
3563 - Butanoic acid, 2-methylpropyl ester
3564 - Butanoic acid, 1-methylpropyl ester, (*S*)-
3639 - 1-Butanol, 3-methyl-, propanoate
3758 - 2-Butene, 1,1-diethoxy-
4363 - 1,3-Cyclohexanedimethanol
6008 - Formic acid, heptyl ester
6466 - Heptanoic acid, methyl ester
6518 - 2-Heptanone, 4-hydroxy-3-methyl-
6519 - 3-Heptanone, 5-hydroxy-5-methyl-
6882 - Hexanoic acid, 2-ethyl-
6883 - Hexanoic acid, 3-ethyl-
6884 - Hexanoic acid, ethyl ester
8422 - Octanoic acid
8474 - 4-Octanone, 5-hydroxy-
8692 - Pentanal, 3-hydroxy-2,2,4-trimethyl-
8892 - Pentanoic acid, 4-methyl-, ethyl ester
8893 - Pentanoic acid, 1-methylethyl ester
8914 - Pentanoic acid, propyl ester
8945 - 1-Pentanol, 2-methyl-, acetate
8946 - 2-Pentanol, 2-methyl-, acetate
8947 - 2-Pentanol, 4-methyl-, acetate
8948 - 3-Pentanol, 2-methyl-, acetate
8949 - 3-Pentanol, 3-methyl-, acetate
10294 - Propanoic acid, 2-methyl-, 1,1-dimethylethyl ester
10304 - Propanoic acid, 2-methyl-, 2-methylpropyl ester
10318 - Propanoic acid, pentyl ester

$C_8H_{16}O_3$
3536 - Butanoic acid, 2-hydroxy-, 2-methylpropyl ester
8432 - Octanoic acid, 2-hydroxy-
10274 - Propanoic acid, 2-hydroxy-, 3-methylbutyl ester

$C_8H_{16}O_4$
174 - Acetic acid, diethoxy-, ethyl ester
5721 - Ethanol, 2-(2-ethoxyethoxy)-, acetate

$C_8H_{16}Si$
11417 - Silane, dimethyldi-2-propenyl-

$C_8H_{17}Br$
8356 - Octane, 1-bromo-
8357 - Octane, 2-bromo-, (±)-

$C_8H_{17}Cl$
6366 - Heptane, 2-chloro-2-methyl-
6367 - Heptane, 2-chloro-6-methyl-
6368 - Heptane, 3-(chloromethyl)-
6369 - Heptane, 3-chloro-3-methyl-
6370 - Heptane, 4-chloro-4-methyl-
6740 - Hexane, 2-chloro-2,5-dimethyl-
6741 - Hexane, 3-chloro-2,3-dimethyl-
8359 - Octane, 1-chloro-
8360 - Octane, 2-chloro-
8733 - Pentane, 2-chloro-2,4,4-trimethyl-

$C_8H_{17}ClO$
8448 - 1-Octanol, 8-chloro-

$C_8H_{17}ClO_2$
3243 - Butane, 3-chloro-1,1-diethoxy-

$C_8H_{17}Cl_3Si$
11471 - Silane, trichlorooctyl-

$C_8H_{17}F$
8405 - Octane, 1-fluoro-

$C_8H_{17}FO$
8460 - 1-Octanol, 8-fluoro-

$C_8H_{17}I$
8406 - Octane, 1-iodo-
8407 - Octane, 2-iodo-, (±)-

$C_8H_{17}N$
503 - 1H-Azonine, octahydro-
4286 - Cyclohexanamine, *N*,*N*-dimethyl-
4287 - Cyclohexanamine, 2-ethyl-
4288 - Cyclohexanamine, *N*-ethyl-
4723 - Cyclohexylamine, *N*,4-dimethyl-
4735 - Cyclooctanamine
9684 - Piperidine, 1-(1-methylethyl)-
9707 - Piperidine, 1-propyl-
9708 - Piperidine, 2-propyl-, (*S*)-
9709 - Piperidine, 4-propyl-
9712 - Piperidine, 2,2,4-trimethyl-
9713 - Piperidine, 2,3,6-trimethyl-
9714 - Piperidine, 2,4,6-trimethyl-
11146 - Pyrrolidine, 1-butyl-
11147 - Pyrrolidine, 2-butyl-

$C_8H_{17}NO$
58 - Acetamide, *N*,*N*-dipropyl-
3200 - Butanamide, *N*,*N*-diethyl-
3656 - 2-Butanone, 4-(diethylamino)-
4520 - Cyclohexanol, 3-(dimethylamino)-
7684 - Morpholine, 4-butyl-
8348 - Octanal, oxime
8349 - Octanamide
9677 - 2-Piperidinemethanol, α-ethyl-, (*R**,*S**)-(±)-
9718 - 3-Piperidinol, 6-propyl-, (3*S*-*trans*)-

$C_8H_{17}NO_2$
3952 - Carbamic acid, (3-methylbutyl)-, ethyl ester
7693 - Morpholine, 4-(2-ethoxyethyl)-
8095 - Nitrous acid, octyl ester
8412 - Octane, 1-nitro-
8413 - Octane, 2-nitro-, (+)-
8423 - Octanoic acid, 2-amino-, (±)-

$C_8H_{17}NO_3$
8079 - Nitric acid, octyl ester

C_8H_{18}
3430 - Butane, 2,2,3,3-tetramethyl-
6422 - Heptane, 2-methyl-
6423 - Heptane, 3-methyl-, (*S*)-
6424 - Heptane, 4-methyl-
6771 - Hexane, 2,2-dimethyl-
6772 - Hexane, 2,3-dimethyl-
6773 - Hexane, 2,4-dimethyl-
6774 - Hexane, 2,5-dimethyl-
6775 - Hexane, 3,3-dimethyl-
6776 - Hexane, 3,4-dimethyl-
6815 - Hexane, 3-ethyl-
8355 - Octane
8813 - Pentane, 3-ethyl-2-methyl-
8814 - Pentane, 3-ethyl-3-methyl-
8847 - Pentane, 2,2,3-trimethyl-
8848 - Pentane, 2,2,4-trimethyl-
8849 - Pentane, 2,3,3-trimethyl-
8850 - Pentane, 2,3,4-trimethyl-

$C_8H_{18}ClNO_2$
9830 - 1-Propanaminium, 2-(acetyloxy)-*N*,*N*,*N*-trimethyl-, chloride

$C_8H_{18}Hg$
7433 - Mercury, dibutyl-

$C_8H_{18}N_2$
4362 - 1,3-Cyclohexanedimethanamine
5123 - Diazene, dibutyl

$C_8H_{18}N_2O$
3212 - 1-Butanamine, *N*-butyl-*N*-nitroso-

$C_8H_{18}O$
3401 - Butane, 1-(2-methylpropoxy)-
3415 - Butane, 1,1′-oxybis-
3416 - Butane, 2,2′-oxybis-, (±)-
6421 - Heptane, 1-methoxy-
6492 - 1-Heptanol, 3-methyl-
6493 - 1-Heptanol, 4-methyl-
6494 - 1-Heptanol, 6-methyl-
6495 - 2-Heptanol, 2-methyl-
6496 - 2-Heptanol, 3-methyl-
6497 - 2-Heptanol, 5-methyl-
6498 - 2-Heptanol, 6-methyl-
6499 - 3-Heptanol, 2-methyl-, (±)-
6500 - 3-Heptanol, 3-methyl-
6501 - 3-Heptanol, 4-methyl-
6502 - 3-Heptanol, 5-methyl-
6503 - 4-Heptanol, 2-methyl-
6504 - 4-Heptanol, 3-methyl-, (*R**,*S**)-(±)-
6505 - 4-Heptanol, 4-methyl-
6814 - Hexane, 1-ethoxy-
6921 - 1-Hexanol, 2,5-dimethyl-
6922 - 2-Hexanol, 2,3-dimethyl-
6923 - 2-Hexanol, 2,5-dimethyl-
6924 - 2-Hexanol, 3,4-dimethyl-
6925 - 3-Hexanol, 2,2-dimethyl-
6926 - 3-Hexanol, 2,3-dimethyl-
6927 - 3-Hexanol, 2,5-dimethyl-
6928 - 3-Hexanol, 3,5-dimethyl-, (±)-
6929 - 3-Hexanol, 4,4-dimethyl-
6930 - 1-Hexanol, 2-ethyl-
6931 - 3-Hexanol, 3-ethyl-
8443 - 1-Octanol
8444 - 2-Octanol, (±)-
8445 - 3-Octanol
8446 - 4-Octanol, (±)-
8933 - 2-Pentanol, 3-ethyl-2-methyl-
8934 - 3-Pentanol, 3-ethyl-2-methyl-
8954 - 2-Pentanol, 2,4,4-trimethyl-
8955 - 3-Pentanol, 2,2,4-trimethyl-
8956 - 3-Pentanol, 2,3,4-trimethyl-
10124 - Propane, 2,2′-oxybis[2-methyl-

$C_8H_{18}OS$
3422 - Butane, 1,1′-sulfinylbis-

$C_8H_{18}OSi_2$
5233 - Disiloxane, 1,3-diethenyl-1,1,3,3-tetramethyl-

$C_8H_{18}O_2$
5586 - Ethane, 1,1-dipropoxy-
5728 - Ethanol, 2-(hexyloxy)-
6401 - 2,4-Heptanediol, 3-methyl-
6801 - 2,5-Hexanediol, 2,5-dimethyl-
6802 - 1,3-Hexanediol, 2-ethyl-
8389 - 1,8-Octanediol
8390 - 2,4-Octanediol
8391 - 4,5-Octanediol, (±)-
8795 - 1,3-Pentanediol, 2,2,4-trimethyl-
8796 - 1,4-Pentanediol, 2,2,4-trimethyl-
9153 - Peroxide, bis(1,1-dimethylethyl)
9993 - 1,3-Propanediol, 2-butyl-2-methyl-

$C_8H_{18}O_2S$
3423 - Butane, 1,1′-sulfonylbis-
10132 - Propane, 1,1′-sulfonylbis[2-methyl-

$C_8H_{18}O_2Si$
11408 - Silane, diethoxymethyl-2-propenyl-

$C_8H_{18}O_3$
5612 - Ethane, 1,1′-oxybis[2-ethoxy-
5664 - Ethane, 1,1,1-triethoxy-
5688 - Ethanol, 2-(2-butoxyethoxy)-
6920 - 1-Hexanol, 3,5-dimethoxy-

$C_8H_{18}O_3S$
11578 - Sulfurous acid, bis(2-methylpropyl) ester
11580 - Sulfurous acid, dibutyl ester

$C_8H_{18}O_3Si$
11424 - Silane, ethenyltriethoxy-

$C_8H_{18}O_4$
11642 - 2,5,8,11-Tetraoxadodecane

$C_8H_{18}O_4S_2$
3230 - Butane, 2,2-bis(ethylsulfonyl)-

$C_8H_{18}O_5$
5756 - Ethanol, 2,2′-[oxybis(2,1-ethanediyloxy)]bis-

$C_8H_{18}S$
3433 - Butane, 1,1′-thiobis-

3434 - Butane, 2,2′-thiobis-
8418 - 1-Octanethiol
8419 - 2-Octanethiol, (±)-
10154 - Propane, 1,1′-thiobis[2-methyl-
10155 - Propane, 2,2′-thiobis[2-methyl-
$C_8H_{18}S_2$
5242 - Disulfide, bis(1,1-dimethylethyl)
5252 - Disulfide, dibutyl
$C_8H_{18}S_4$
11649 - Tetrasulfide, bis(1,1-dimethylethyl)
$C_8H_{18}Si_2$
11434 - Silane, 1,2-ethynediylbis[trimethyl-
$C_8H_{19}N$
3211 - 1-Butanamine, *N*-butyl-
3227 - 2-Butanamine, *N*-(1-methylpropyl)-
6353 - 2-Heptanamine, 6-methyl-, (±)-
6726 - 1-Hexanamine, 2-ethyl-
8350 - 1-Octanamine
8351 - 2-Octanamine, (±)-
9811 - 2-Propanamine, *N*-ethyl-*N*-(1-methylethyl)-
9812 - 1-Propanamine, *N*-ethyl-*N*-propyl-
9823 - 1-Propanamine, 2-methyl-*N*-(2-methylpropyl)-
$C_8H_{19}NO$
5683 - Ethanol, 2-[bis(1-methylethyl)amino]-
5712 - Ethanol, 2-(dipropylamino)-
$C_8H_{19}NO_2$
3214 - 1-Butanamine, 4,4-diethoxy-
3624 - 2-Butanol, 3,3′-iminobis-
5416 - Ethanamine, 2,2-diethoxy-*N*,*N*-dimethyl-
5425 - Ethanamine, 2-ethoxy-*N*-(2-ethoxyethyl)-
5692 - Ethanol, 2,2′-(butylimino)bis-
5707 - Ethanol, 2-[2-(diethylamino)ethoxy]-
$C_8H_{19}O_2PS_2$
5934 - Ethoprop
$C_8H_{19}O_2PS_3$
5260 - Disulfoton
$C_8H_{19}O_3P$
9531 - Phosphonic acid, bis(1,1-dimethylethyl) ester
9532 - Phosphonic acid, dibutyl ester
$C_8H_{20}As_2$
5109 - Diarsine, tetraethyl-
$C_8H_{20}BrN$
5447 - Ethanaminium, *N*,*N*,*N*-triethyl-, bromide
$C_8H_{20}N_2$
3263 - 1,4-Butanediamine, *N*,*N*,*N*′,*N*′-tetramethyl-
6751 - 2,5-Hexanediamine, 2,5-dimethyl-
7140 - Hydrazine, 1,2-bis(2-methylpropyl)-
8365 - 1,8-Octanediamine
$C_8H_{20}N_2O$
10342 - 1-Propanol, 2-[(2-amino-2-methylpropyl)amino]-2-methyl-
$C_8H_{20}OSi$
11428 - Silane, ethoxytriethyl-
$C_8H_{20}O_3P_2$
9526 - Phosphinic acid, diethyl-, anhydride
$C_8H_{20}O_3Si$
11478 - Silane, triethoxyethyl-
$C_8H_{20}O_4Si$
11506 - Silicic acid (H4SiO4), tetraethyl ester
$C_8H_{20}O_5P_2S_2$
11716 - Thiodiphosphoric acid ([(HO)2P(*S*)]2O), tetraethyl ester
$C_8H_{20}O_6P_2$
7185 - Hypophosphoric acid, tetraethyl ester
$C_8H_{20}O_7P_2$
5221 - Diphosphoric acid, tetraethyl ester
$C_8H_{20}Pb$
9731 - Plumbane, tetraethyl
$C_8H_{20}Si$
11450 - Silane, tetraethyl-
$C_8H_{20}Si_2$
11421 - Silane, 1,2-ethenediylbis[trimethyl-, (*E*)-
$C_8H_{23}N_5$
5498 - 1,2-Ethanediamine, *N*-(2-aminoethyl)-*N*′-[2-[(2-aminoethyl)amino]ethyl]-
$C_8H_{24}N_4O_3P_2$
5220 - Diphosphoramide, octamethyl-
$C_8H_{24}O_2Si_3$
11914 - Trisiloxane, octamethyl-
$C_8H_{24}O_4Si_4$
4993 - Cyclotetrasiloxane, octamethyl-
$C_8H_{26}O_3Si_4$
11647 - Tetrasiloxane, 1,1,1,3,5,7,7,7-octamethyl-
$C_8I_4O_3$
7348 - 1,3-Isobenzofurandione, 4,5,6,7-tetraiodo-
C_9F_{20}
8139 - Nonane, icosofluoro
$C_9F_{21}N$
9813 - 1-Propanamine, 1,1,2,2,3,3,3-heptafluoro-*N*,*N*-bis(heptafluoropropyl)-
$C_9H_3Cl_3O_3$
2317 - 1,3,5-Benzenetricarbonyl trichloride
$C_9H_3F_{15}O_2$
8438 - Octanoic acid, pentadecafluoro-, methyl ester
$C_9H_4Cl_3NO_2S$
5984 - Folpet
$C_9H_4O_5$
7341 - 5-Isobenzofurancarboxylic acid, 1,3-dihydro-1,3-dioxo-
$C_9H_5BrN_2O_2$
11221 - Quinoline, 6-bromo-5-nitro-
$C_9H_5Br_2NO$
11313 - 8-Quinolinol, 5,7-dibromo-
C_9H_5ClINO
11312 - 8-Quinolinol, 5-chloro-7-iodo-
$C_9H_5ClO_6$
2320 - 1,3,5-Benzenetricarboxylic acid, 2-chloro-
$C_9H_5Cl_2N$
11248 - Quinoline, 2,4-dichloro-
11249 - Quinoline, 2,7-dichloro-
11250 - Quinoline, 4,5-dichloro-
11251 - Quinoline, 4,7-dichloro-
11252 - Quinoline, 5,8-dichloro-
$C_9H_5Cl_2NO$
11314 - 8-Quinolinol, 5,7-dichloro-
$C_9H_5Cl_3N_4$
360 - Anilazine
$C_9H_5I_2NO$
11315 - 8-Quinolinol, 5,7-diiodo-
$C_9H_5NO_4$
10740 - 2-Propynoic acid, 3-(2-nitrophenyl)-
C_9H_6BrN
7374 - Isoquinoline, 4-bromo-
11217 - Quinoline, 3-bromo-
11218 - Quinoline, 5-bromo-
11219 - Quinoline, 6-bromo-
11220 - Quinoline, 7-bromo-
C_9H_6BrNO
11310 - 8-Quinolinol, 5-bromo-
11326 - 2(1H)-Quinolinone, 4-bromo-
$C_9H_6Br_2O_2$
10629 - 2-Propenoic acid, 2,3-dibromo-3-phenyl-, (*Z*)-
C_9H_6ClN
7376 - Isoquinoline, 1-chloro-
11237 - Quinoline, 2-chloro-
11238 - Quinoline, 4-chloro-
11239 - Quinoline, 5-chloro-
11240 - Quinoline, 6-chloro-
11241 - Quinoline, 7-chloro-
11242 - Quinoline, 8-chloro-
C_9H_6ClNO
11311 - 8-Quinolinol, 5-chloro-
$C_9H_6Cl_2N_2O_3$
7643 - Methazole
$C_9H_6Cl_6O_3S$
5369 - Endosulfan
C_9H_6IN
11272 - Quinoline, 4-iodo-
11273 - Quinoline, 6-iodo-
$C_9H_6INO_4S$
11295 - 5-Quinolinesulfonic acid, 8-hydroxy-7-iodo-
$C_9H_6N_2$
9916 - Propanedinitrile, phenyl-
$C_9H_6N_2O_2$
1430 - Benzene, 1,3-diisocyanato-2-methyl-
1431 - Benzene, 2,4-diisocyanato-1-methyl-
7386 - Isoquinoline, 5-nitro-
11284 - Quinoline, 5-nitro-
11285 - Quinoline, 6-nitro-
11286 - Quinoline, 7-nitro-
$C_9H_6N_2O_3$
11287 - Quinoline, 4-nitro-, 1-oxide
11323 - 8-Quinolinol, 5-nitro-
C_9H_6O
10724 - 2-Propynal, 3-phenyl-
$C_9H_6OS_2$
7617 - Methanone, di-2-thienyl-
$C_9H_6O_2$
2791 - 1H-2-Benzopyran-1-one
2792 - 2H-1-Benzopyran-2-one
2793 - 4H-1-Benzopyran-4-one
7264 - 1H-Indene-1,3(2H)-dione
$C_9H_6O_2S$
2400 - Benzo[b]thiophene-2-carboxylic acid
$C_9H_6O_3$
2439 - 2-Benzofurancarboxylic acid
2775 - 2H-1-Benzopyran-3-carboxylic acid, 2-oxo-
2825 - 2H-1-Benzopyran-2-one, 6-hydroxy-
2826 - 2H-1-Benzopyran-2-one, 7-hydroxy-
$C_9H_6O_4$
2804 - 2H-1-Benzopyran-2-one, 6,7-dihydroxy-
2805 - 2H-1-Benzopyran-2-one, 7,8-dihydroxy-
7267 - 1H-Indene-1,3(2H)-dione, 2,2-dihydroxy-
$C_9H_6O_6$
2318 - 1,2,3-Benzenetricarboxylic acid
2319 - 1,2,4-Benzenetricarboxylic acid
$C_9H_7BrN_2$
11207 - 4-Quinolinamine, 3-bromo-
11215 - Quinoline, 6-amino-3-bromo-
C_9H_7BrO
10510 - 2-Propenal, 3-bromo-3-phenyl-
$C_9H_7BrO_4$
2481 - Benzoic acid, 2-(acetyloxy)-5-bromo-
C_9H_7ClO
10716 - 2-Propenoyl chloride, 3-phenyl-
10717 - 2-Propenoyl chloride, 3-phenyl-, (*E*)-
$C_9H_7Cl_3O_3$
11512 - Silvex
$C_9H_7FO_2$
10637 - 2-Propenoic acid, 2-fluoro-3-phenyl-
C_9H_7N
7373 - Isoquinoline
10587 - 2-Propenenitrile, 3-phenyl-, (*E*)-
10588 - 2-Propenenitrile, 3-phenyl-, (*Z*)-
11214 - Quinoline
C_9H_7NO
2124 - Benzenepropanenitrile, β-oxo-
11305 - 5-Quinolinol
11306 - 6-Quinolinol
11307 - 7-Quinolinol
11308 - 8-Quinolinol
11325 - 2(1H)-Quinolinone
$C_9H_7NO_2$
7365 - 1H-Isoindole-1,3(2H)-dione, 2-methyl-
$C_9H_7NO_4$
10676 - 2-Propenoic acid, 3-(2-nitrophenyl)-, (*E*)-
$C_9H_7NO_4S$
11294 - 5-Quinolinesulfonic acid, 8-hydroxy-
$C_9H_7N_3S$
11858 - Tricyclazole
C_9H_8
2173 - Benzene, 1-propynyl-
7245 - 1H-Indene
$C_9H_8Br_2$
7247 - 1H-Indene, 1,2-dibromo-2,3-dihydro-
$C_9H_8Cl_2$
1397 - Benzene, (3,3-dichloro-2-propenyl)-
7249 - 1H-Indene, 1,2-dichloro-2,3-dihydro-

$C_9H_8Cl_2O_3$
172 - Acetic acid, (2,4-dichlorophenoxy)-, methyl ester
7653 - Methyl 3,6-dichloro-*o*-anisate
10250 - Propanoic acid, 2-(2,4-dichlorophenoxy)-
$C_9H_8Cl_3NO_2S$
7368 - 1H-Isoindole-1,3(2H)-dione, 3a,4,7,7a-tetrahydro-2-[(trichloromethyl)thio]-
$C_9H_8NO_3S$
11324 - 8-Quinolinol, sulfate (2:1) (salt)
$C_9H_8N_2$
7212 - 1H-Imidazole, 1-phenyl-
7213 - 1H-Imidazole, 2-phenyl-
7372 - 5-Isoquinolinamine
11202 - 2-Quinolinamine
11203 - 4-Quinolinamine
11204 - 5-Quinolinamine
11205 - 6-Quinolinamine
11206 - 8-Quinolinamine
11337 - Quinoxaline, 2-methyl-
11338 - Quinoxaline, 6-methyl-
$C_9H_8N_2O$
11336 - Quinoxaline, 6-methoxy-
$C_9H_8N_2O_6$
2593 - Benzoic acid, 3,5-dinitro-, ethyl ester
$C_9H_8N_4OS$
11687 - Thidiazuron
C_9H_8O
1876 - Benzenemethanol, α-ethynyl-
2456 - Benzofuran, 2-methyl-
2457 - Benzofuran, 3-methyl-
2458 - Benzofuran, 5-methyl-
2459 - Benzofuran, 7-methyl-
2772 - 2H-1-Benzopyran
7282 - 6H-Indeno[1,2-b]oxirene, 1a,6a-dihydro-
7286 - 1H-Inden-1-one, 2,3-dihydro-
7287 - 2H-Inden-2-one, 1,3-dihydro-
10518 - 2-Propenal, 3-phenyl-, (*E*)-
10745 - 2-Propyn-1-ol, 3-phenyl-
C_9H_8OS
2869 - 4H-1-Benzothiopyran-4-one, 2,3-dihydro-
$C_9H_8O_2$
927 - Benzeneacetic acid, α-methylene-
2425 - 1,3-Benzodioxole, 5-ethenyl-
2455 - 2-Benzofuranmethanol
2797 - 1H-2-Benzopyran-1-one, 3,4-dihydro-
2798 - 2H-1-Benzopyran-2-one, 3,4-dihydro-
2799 - 4H-1-Benzopyran-4-one, 2,3-dihydro-
6132 - Furan, 2,2′-methylenebis-
10028 - 1,2-Propanedione, 1-phenyl-
10682 - 2-Propenoic acid, 3-phenyl-, (*E*)-
10683 - 2-Propenoic acid, 3-phenyl-, (*Z*)-
$C_9H_8O_3$
2476 - Benzoic acid, 2-acetyl-
2477 - Benzoic acid, 3-acetyl-
2478 - Benzoic acid, 4-acetyl-
7580 - 4,7-Methanoisobenzofuran-1,3-dione, 3a,4,7,7a-tetrahydro-, (3aα,4α,7α,7aα)-
10644 - 2-Propenoic acid, 3-(4-hydroxyphenyl)-
$C_9H_8O_4$
2420 - 1,3-Benzodioxole-5-carboxylic acid, methyl ester
2480 - Benzoic acid, 2-(acetyloxy)-
9151 - Peroxide, acetyl benzoyl
10632 - 2-Propenoic acid, 3-(3,4-dihydroxyphenyl)-
C_9H_8S
2402 - Benzo[b]thiophene, 5-methyl-
C_9H_9Br
1082 - Benzene, 1-bromo-4-(1-propenyl)-
1083 - Benzene, 1-bromo-4-(2-propenyl)-
1084 - Benzene, (3-bromo-1-propenyl)-
C_9H_9BrO
5779 - Ethanone, 2-bromo-1-(4-methylphenyl)-
10418 - 1-Propanone, 1-(4-bromophenyl)-
10419 - 1-Propanone, 2-bromo-1-phenyl-
$C_9H_9BrO_2$
2147 - Benzenepropanoic-β-14C acid, α-bromo-
2521 - Benzoic acid, 2-bromo-, ethyl ester
2522 - Benzoic acid, 3-bromo-, ethyl ester
2523 - Benzoic acid, 4-bromo-, ethyl ester
5205 - 1,3-Dioxolane, 2-(3-bromophenyl)-
$C_9H_9Br_2NO_3$
11927 - *L*-Tyrosine, 3,5-dibromo-
C_9H_9Cl
1226 - Benzene, (1-chloro-1-propenyl)-
1227 - Benzene, (2-chloro-1-propenyl)-
1228 - Benzene, (3-chloro-2-propenyl)-
1229 - Benzene, 1-chloro-3-(1-propenyl)-, (*E*)-
1230 - Benzene, (3-chloro-1-propenyl)-, (*E*)-
C_9H_9ClO
1232 - Benzene, 1-chloro-4-(2-propenyloxy)-
2160 - Benzenepropanoyl chloride
2441 - Benzofuran, 2-(chloromethyl)-2,3-dihydro-
5787 - Ethanone, 2-chloro-1-(4-methylphenyl)-
7283 - 1H-Inden-4-ol, 7-chloro-2,3-dihydro-
9271 - Phenol, 4-chloro-2-(2-propenyl)-
10422 - 1-Propanone, 1-(4-chlorophenyl)-
$C_9H_9ClO_2$
140 - Acetic acid, chloro-, 4-methylphenyl ester
148 - Acetic acid, chloro-, phenylmethyl ester
2537 - Benzoic acid, 2-chloro-, ethyl ester
2538 - Benzoic acid, 3-chloro-, ethyl ester
2539 - Benzoic acid, 4-chloro-, ethyl ester
5696 - Ethanol, 2-chloro-, benzoate
10505 - Propanoyl chloride, 2-phenoxy-
$C_9H_9ClO_3$
141 - Acetic acid, (4-chloro-2-methylphenoxy)-
2541 - Benzoic acid, 3-chloro-2-hydroxy-, ethyl ester
4141 - Cloprop
5786 - Ethanone, 2-chloro-1-(4-hydroxy-3-methoxyphenyl)-
$C_9H_9Cl_2NO$
10200 - Propanil
$C_9H_9FO_2$
2606 - Benzoic acid, 4-fluoro-, ethyl ester
$C_9H_9I_2NO_3$
11928 - Tyrosine, 3,5-diiodo-
C_9H_9N
959 - Benzeneacetonitrile, α-methyl-
960 - Benzeneacetonitrile, 2-methyl-
961 - Benzeneacetonitrile, 3-methyl-
962 - Benzeneacetonitrile, 4-methyl-
2123 - Benzenepropanenitrile
7312 - 1H-Indole, 1-methyl-
7313 - 1H-Indole, 2-methyl-
7314 - 1H-Indole, 3-methyl-
7315 - 1H-Indole, 4-methyl-
7316 - 1H-Indole, 5-methyl-
7317 - 1H-Indole, 7-methyl-
C_9H_9NO
957 - Benzeneacetonitrile, 2-methoxy-
958 - Benzeneacetonitrile, 4-methoxy-
2754 - Benzonitrile, 2-ethoxy-
2755 - Benzonitrile, 4-ethoxy-
2879 - Benzoxazole, 2,5-dimethyl-
7311 - 1H-Indole, 7-methoxy-
11327 - 2(1H)-Quinolinone, 3,4-dihydro-
$C_9H_9NO_2$
2753 - Benzonitrile, 2,6-dimethoxy-
3363 - 1,3-Butanedione, 1-(2-pyridinyl)-
3364 - 1,3-Butanedione, 1-(3-pyridinyl)-
3365 - 1,3-Butanedione, 1-(4-pyridinyl)-
$C_9H_9NO_3$
6255 - Glycine, *N*-benzoyl-
7301 - 1H-Indole-5,6-dione, 2,3-dihydro-3-hydroxy-1-methyl-
$C_9H_9NO_4$
103 - Acetic acid, [2-(aminocarbonyl)phenoxy]-
524 - Benzadox
2707 - Benzoic acid, 2-nitro-, ethyl ester
2708 - Benzoic acid, 3-nitro-, ethyl ester
2709 - Benzoic acid, 4-nitro-, ethyl ester
$C_9H_9N_3O_2$
3990 - Carbendazim
$C_9H_9N_3O_2S_2$
2206 - Benzenesulfonamide, 4-amino-*N*-2-thiazolyl-
11663 - 2-Thiazolamine, 5-[(4-aminophenyl)sulfonyl]-
$C_9H_9N_3O_4S_2$
3975 - Carbamodithioic acid, dimethyl-, 2,4-dinitrophenyl ester
$C_9H_9N_3O_6$
2364 - Benzene, 1,3,5-trimethyl-2,4,6-trinitro-
$C_9H_9N_3S$
11671 - 2,4-Thiazolediamine, 5-phenyl-
$C_9H_9N_5$
11829 - 1,3,5-Triazine-2,4-diamine, 6-phenyl-
11830 - 1,3,5-Triazine-2,4-diamine, *N*-phenyl-
C_9H_{10}
1256 - Benzene, cyclopropyl-
1640 - Benzene, 1-ethenyl-2-methyl-
1641 - Benzene, 1-ethenyl-3-methyl-
1642 - Benzene, 1-ethenyl-4-methyl-
1972 - Benzene, (1-methylethenyl)-
2162 - Benzene, 1-propenyl-
2163 - Benzene, 1-propenyl-, (*E*)-
2164 - Benzene, 1-propenyl-, (*Z*)-
2165 - Benzene, 2-propenyl-
4747 - 1,3,5,7-Cyclooctatetraene, 1-methyl-
7250 - 1H-Indene, 2,3-dihydro-
$C_9H_{10}BrClN_2O_2$
4076 - Chlorbromuron
$C_9H_{10}Br_2$
1313 - Benzene, 2,4-dibromo-1,3,5-trimethyl-
$C_9H_{10}Cl_2$
1402 - Benzene, 2,4-dichloro-1,3,5-trimethyl-
$C_9H_{10}Cl_2N_2O$
12000 - Urea, *N′*-(3,4-dichlorophenyl)-*N,N*-dimethyl-
$C_9H_{10}Cl_2N_2O_2$
7415 - Linuron
$C_9H_{10}INO_3$
11932 - *L*-Tyrosine, 3-iodo-
$C_9H_{10}N_2$
2378 - 1H-Benzimidazole, 1,5-dimethyl-
2379 - 1H-Benzimidazole, 2,5-dimethyl-
2380 - 1H-Benzimidazole, 5,6-dimethyl-
7237 - 1H-Indazole, 1,5-dimethyl-
7238 - 2H-Indazole, 2,3-dimethyl-
10823 - 1H-Pyrazole, 4,5-dihydro-1-phenyl-
10824 - 1H-Pyrazole, 4,5-dihydro-3-phenyl-
11133 - 1H-Pyrrole, 2,2′-methylenebis-
$C_9H_{10}N_2O$
8574 - 2-Oxazolamine, 4,5-dihydro-5-phenyl-
10844 - 3-Pyrazolidinone, 1-phenyl-
$C_9H_{10}N_2O_2$
879 - Benzeneacetamide, *N*-(aminocarbonyl)-
$C_9H_{10}N_2O_3$
6252 - Glycine, *N*-(4-aminobenzoyl)-
$C_9H_{10}N_2O_3S_2$
2865 - 2-Benzothiazolesulfonamide, 6-ethoxy-
$C_9H_{10}N_2O_4$
2360 - Benzene, 1,3,5-trimethyl-2,4-dinitro-
$C_9H_{10}O$
547 - Benzaldehyde, 2,4-dimethyl-
548 - Benzaldehyde, 2,5-dimethyl-
549 - Benzaldehyde, 3,5-dimethyl-
558 - Benzaldehyde, 2-ethyl-
872 - Benzeneacetaldehyde, α-methyl-
873 - Benzeneacetaldehyde, 2-methyl-
874 - Benzeneacetaldehyde, 4-methyl-
1637 - Benzene, 1-ethenyl-2-methoxy-
1638 - Benzene, 1-ethenyl-3-methoxy-
1639 - Benzene, 1-ethenyl-4-methoxy-
1868 - Benzenemethanol, α-ethenyl-
1928 - Benzene, (2-methoxyethenyl)-
2114 - Benzenepropanal
2167 - Benzene, (2-propenyloxy)-
2443 - Benzofuran, 2,3-dihydro-2-methyl-
2444 - Benzofuran, 2,3-dihydro-5-methyl-
2776 - 1H-2-Benzopyran, 3,4-dihydro-
2777 - 2H-1-Benzopyran, 3,4-dihydro-
5851 - Ethanone, 1-(2-methylphenyl)-

5852 - Ethanone, 1-(3-methylphenyl)-
5853 - Ethanone, 1-(4-methylphenyl)-
7284 - 1H-Inden-1-ol, 2,3-dihydro-
7285 - 1H-Inden-5-ol, 2,3-dihydro-
8622 - Oxirane, 2-methyl-2-phenyl-
9456 - Phenol, 2-(1-propenyl)-
9457 - Phenol, 2-(2-propenyl)-
9458 - Phenol, 4-(2-propenyl)-
9459 - Phenol, 2-(1-propenyl)-, (*E*)-
10480 - 1-Propanone, 1-phenyl-
10481 - 2-Propanone, 1-phenyl-
10704 - 2-Propen-1-ol, 3-phenyl-, (*E*)-
10705 - 2-Propen-1-ol, 3-phenyl-, (*Z*)-

$C_9H_{10}O_2$
211 - Acetic acid, 2-methylphenyl ester
212 - Acetic acid, 3-methylphenyl ester
213 - Acetic acid, 4-methylphenyl ester
233 - Acetic acid, phenylmethyl ester
552 - Benzaldehyde, 2-ethoxy-
553 - Benzaldehyde, 3-ethoxy-
554 - Benzaldehyde, 4-ethoxy-
586 - Benzaldehyde, 4-methoxy-3-methyl-
870 - Benzeneacetaldehyde, 2-methoxy-
871 - Benzeneacetaldehyde, 4-methoxy-
924 - Benzeneacetic acid, α-methyl-, (±)-
925 - Benzeneacetic acid, 3-methyl-
926 - Benzeneacetic acid, 4-methyl-
929 - Benzeneacetic acid, methyl ester
1547 - 1,2-Benzenediol, 3-(2-propenyl)-
2126 - Benzenepropanoic acid
2580 - Benzoic acid, 2,4-dimethyl-
2581 - Benzoic acid, 2,5-dimethyl-
2582 - Benzoic acid, 2,6-dimethyl-
2583 - Benzoic acid, 3,5-dimethyl-
2600 - Benzoic acid, 2-ethyl-
2601 - Benzoic acid, 3-ethyl-
2602 - Benzoic acid, ethyl ester
2696 - Benzoic acid, 2-methyl-, methyl ester
2697 - Benzoic acid, 3-methyl-, methyl ester
2698 - Benzoic acid, 4-methyl-, methyl ester
5826 - Ethanone, 1-(2-hydroxy-4-methylphenyl)-
5827 - Ethanone, 1-(2-hydroxy-5-methylphenyl)-
5828 - Ethanone, 1-(4-hydroxy-2-methylphenyl)-
5840 - Ethanone, 1-(3-methoxyphenyl)-
5841 - Ethanone, 1-(4-methoxyphenyl)-
5842 - Ethanone, 2-methoxy-1-phenyl-
8619 - Oxiranemethanol, 3-phenyl-
8623 - Oxirane, (phenoxymethyl)-
10323 - Propanoic acid, phenyl ester
10449 - 1-Propanone, 1-(2-hydroxyphenyl)-
10450 - 1-Propanone, 1-(4-hydroxyphenyl)-
10451 - 1-Propanone, 2-hydroxy-1-phenyl-
10452 - 1-Propanone, 3-hydroxy-1-phenyl-
10479 - 2-Propanone, 1-phenoxy-

$C_9H_{10}O_3$
230 - Acetic acid, phenoxy-, methyl ester
543 - Benzaldehyde, 2,4-dimethoxy-
544 - Benzaldehyde, 2,5-dimethoxy-
545 - Benzaldehyde, 3,4-dimethoxy-
546 - Benzaldehyde, 3,5-dimethoxy-
555 - Benzaldehyde, 3-ethoxy-4-hydroxy-
556 - Benzaldehyde, 5-ethoxy-2-hydroxy-
910 - Benzeneacetic acid, α-hydroxy-, methyl ester, (±)-
914 - Benzeneacetic acid, α-hydroxy-α-methyl-
915 - Benzeneacetic acid, α-(hydroxymethyl)-
921 - Benzeneacetic acid, 2-methoxy-
922 - Benzeneacetic acid, 4-methoxy-
2133 - Benzenepropanoic acid, α-hydroxy-, (±)-
2134 - Benzenepropanoic acid, 4-hydroxy-
2596 - Benzoic acid, 2-ethoxy-
2597 - Benzoic acid, 3-ethoxy-
2619 - Benzoic acid, 2-hydroxy-, ethyl ester
2620 - Benzoic acid, 3-hydroxy-, ethyl ester
2621 - Benzoic acid, 4-hydroxy-, ethyl ester
2638 - Benzoic acid, 2-hydroxy-3-methyl-, methyl ester
2639 - Benzoic acid, 2-hydroxy-4-methyl-, methyl ester
2640 - Benzoic acid, 2-hydroxy-5-methyl-, methyl ester
2673 - Benzoic acid, 2-methoxy-, methyl ester
2674 - Benzoic acid, 3-methoxy-, methyl ester
2675 - Benzoic acid, 4-methoxy-, methyl ester
5578 - 1,2-Ethanediol, monobenzoate
5823 - Ethanone, 1-(2-hydroxy-4-methoxyphenyl)-
5824 - Ethanone, 1-(4-hydroxy-3-methoxyphenyl)-
9382 - Phenol, 2-methoxy-, acetate
10319 - Propanoic acid, 2-phenoxy-, (±)-
10320 - Propanoic acid, 3-phenoxy-
10430 - 1-Propanone, 1-(2,4-dihydroxyphenyl)-
10639 - 2-Propenoic acid, 3-(2-furanyl)-, ethyl ester

$C_9H_{10}O_4$
567 - Benzaldehyde, 2-hydroxy-4,6-dimethoxy-
568 - Benzaldehyde, 3-hydroxy-4,5-dimethoxy-
569 - Benzaldehyde, 4-hydroxy-3,5-dimethoxy-
913 - Benzeneacetic acid, 4-hydroxy-3-methoxy-
2576 - Benzoic acid, 3,4-dimethoxy-
2577 - Benzoic acid, 3,5-dimethoxy-
2622 - Benzoic acid, 2-hydroxy-, 2-hydroxyethyl ester
2627 - Benzoic acid, 4-hydroxy-3-methoxy-, methyl ester
6163 - 2-Furanpropanoic acid, β-oxo-, ethyl ester
10489 - 1-Propanone, 1-(2,4,6-trihydroxyphenyl)-

$C_9H_{10}O_5$
896 - Benzeneacetic acid, α,4-dihydroxy-3-methoxy-
7510 - Methanediol, 2-furanyl-, diacetate

$C_9H_{11}Br$
1057 - Benzene, 1-(bromomethyl)-3,5-dimethyl-
1059 - Benzene, 1-bromo-4-(1-methylethyl)-
1060 - Benzene, (2-bromo-1-methylethyl)-
1086 - Benzene, (1-bromopropyl)-, (±)-
1087 - Benzene, (2-bromopropyl)-
1088 - Benzene, (3-bromopropyl)-
1092 - Benzene, 1-bromo-2,3,5-trimethyl-
1093 - Benzene, 1-bromo-2,4,5-trimethyl-
1094 - Benzene, 2-bromo-1,3,5-trimethyl-

$C_9H_{11}BrN_2O_2$
7656 - Metobromuron

$C_9H_{11}BrO$
1085 - Benzene, (3-bromopropoxy)-

$C_9H_{11}Cl$
1186 - Benzene, 1-(chloromethyl)-2,4-dimethyl-
1189 - Benzene, (1-chloro-1-methylethyl)-
1190 - Benzene, 1-chloro-2-(1-methylethyl)-
1191 - Benzene, 1-(chloromethyl)-4-ethyl-
1192 - Benzene, 1-chloro-4-(1-methylethyl)-
1193 - Benzene, (2-chloro-1-methylethyl)-
1234 - Benzene, (2-chloropropyl)-, (*R*)-
1235 - Benzene, (3-chloropropyl)-
1243 - Benzene, 2-chloro-1,3,5-trimethyl-

$C_9H_{11}ClN_2O$
11999 - Urea, *N'*-(4-chlorophenyl)-*N,N*-dimethyl-

$C_9H_{11}ClO$
1233 - Benzene, (3-chloropropoxy)-

$C_9H_{11}ClO_3$
9999 - 1,2-Propanediol, 3-(2-chlorophenoxy)-
10000 - 1,2-Propanediol, 3-(4-chlorophenoxy)-

$C_9H_{11}ClS$
1236 - Benzene, [(3-chloropropyl)thio]-

$C_9H_{11}Cl_2P$
9555 - Phosphonous dichloride, [4-(1-methylethyl)phenyl]-

$C_9H_{11}Cl_3NO_3PS$
4089 - Chlorpyrifos

$C_9H_{11}F$
1740 - Benzene, 2-fluoro-1,3,5-trimethyl-

$C_9H_{11}FN_2O_5$
12034 - Uridine, 2'-deoxy-5-fluoro-

$C_9H_{11}N$
787 - Benzenamine, 2-(1-methylethenyl)-
841 - Benzenamine, *N*-2-propenyl-
5437 - Ethanamine, *N*-(phenylmethylene)-
7243 - 1H-Inden-1-amine, 2,3-dihydro-, (±)-
7244 - 1H-Inden-5-amine, 2,3-dihydro-
7388 - Isoquinoline, 1,2,3,4-tetrahydro-
7389 - Isoquinoline, 5,6,7,8-tetrahydro-
11296 - Quinoline, 1,2,3,4-tetrahydro-
11297 - Quinoline, 5,6,7,8-tetrahydro-

$C_9H_{11}NO$
76 - Acetamide, *N*-(2-methylphenyl)-
77 - Acetamide, *N*-(3-methylphenyl)-
78 - Acetamide, *N*-(4-methylphenyl)-
79 - Acetamide, *N*-methyl-*N*-phenyl-
87 - Acetamide, *N*-(phenylmethyl)-
550 - Benzaldehyde, 4-(dimethylamino)-
7567 - Methanimidic acid, *N*-phenyl-, ethyl ester
9800 - Propanamide, *N*-phenyl-
10413 - 1-Propanone, 1-(4-aminophenyl)-
10414 - 1-Propanone, 2-amino-1-phenyl-
10483 - 1-Propanone, 1-phenyl-, oxime

$C_9H_{11}NO_2$
70 - Acetamide, *N*-(2-methoxyphenyl)-
71 - Acetamide, *N*-(4-methoxyphenyl)-
615 - Benzamide, 2-ethoxy-
1977 - Benzene, 1-(1-methylethyl)-2-nitro-
1978 - Benzene, 1-(1-methylethyl)-4-nitro-
2003 - Benzene, (1-methyl-1-nitroethyl)-
2362 - Benzene, 1,2,4-trimethyl-5-nitro-
2363 - Benzene, 1,3,5-trimethyl-2-nitro-
2424 - 1,3-Benzodioxole-5-ethanamine
2495 - Benzoic acid, 2-amino-, ethyl ester
2496 - Benzoic acid, 3-amino-, ethyl ester
2497 - Benzoic acid, 4-amino-, ethyl ester
2584 - Benzoic acid, 2-(dimethylamino)-
2585 - Benzoic acid, 4-(dimethylamino)-
2681 - Benzoic acid, 2-(methylamino)-, methyl ester
3958 - Carbamic acid, methylphenyl-, methyl ester
3965 - Carbamic acid, phenyl-, ethyl ester
9509 - *DL*-Phenylalanine
9510 - *L*-Phenylalanine
10907 - 3-Pyridinecarboxylic acid, 1-methylethyl ester

$C_9H_{11}NO_3$
5804 - Ethanone, 1-(3,4-dihydroxyphenyl)-2-(methylamino)-
11926 - *L*-Tyrosine

$C_9H_{11}NO_4$
1443 - Benzene, 1-(dimethoxymethyl)-3-nitro-
1444 - Benzene, 1-(dimethoxymethyl)-4-nitro-
11929 - *L*-Tyrosine, 3-hydroxy-

C_9H_{12}
1685 - Benzene, 1-ethyl-2-methyl-
1686 - Benzene, 1-ethyl-3-methyl-
1687 - Benzene, 1-ethyl-4-methyl-
1975 - Benzene, (1-methylethyl)-
2169 - Benzene, propyl-
2357 - Benzene, 1,2,3-trimethyl-
2358 - Benzene, 1,2,4-trimethyl-
2359 - Benzene, 1,3,5-trimethyl-
8114 - 1,4-Nonadiyne
8115 - 1,8-Nonadiyne
8116 - 2,7-Nonadiyne

$C_9H_{12}NO_5PS$
5951 - Fenitrothion

$C_9H_{12}N_2$
10482 - 2-Propanone, phenylhydrazone
11017 - Pyridine, 3-(2-pyrrolidinyl)-, (*S*)-

$C_9H_{12}N_2O_2$
12013 - Urea, (4-ethoxyphenyl)-

$C_9H_{12}N_2O_3$
827 - Benzenamine, 5-nitro-2-propoxy-

$C_9H_{12}N_2O_6$
12033 - Uridine

$C_9H_{12}N_2S$
10886 - 4-Pyridinecarbothioamide, 2-propyl-

$C_9H_{12}N_4O_3$
10758 - 1H-Purine-2,6-dione, 3,7-dihydro-7-(2-hydroxyethyl)-1,3-dimethyl-
10759 - 1H-Purine-2,6-dione, 3,7-dihydro-8-methoxy-1,3,7-trimethyl-

$C_9H_{12}N_6$
11844 - 1,3,5-Triazine, 2,4,6-tris(1-aziridinyl)-
$C_9H_{12}O$
1607 - Benzeneethanol, α-methyl-
1608 - Benzeneethanol, β-methyl-
1609 - Benzeneethanol, 2-methyl-
1610 - Benzeneethanol, 4-methyl-
1658 - Benzene, (ethoxymethyl)-
1659 - Benzene, 1-ethoxy-2-methyl-
1660 - Benzene, 1-ethoxy-3-methyl-
1661 - Benzene, 1-ethoxy-4-methyl-
1682 - Benzene, 1-ethyl-2-methoxy-
1683 - Benzene, 1-ethyl-3-methoxy-
1684 - Benzene, 1-ethyl-4-methoxy-
1857 - Benzenemethanol, α,α-dimethyl-
1858 - Benzenemethanol, α,3-dimethyl-
1859 - Benzenemethanol, α,4-dimethyl-
1860 - Benzenemethanol, 2,4-dimethyl-
1861 - Benzenemethanol, 3,5-dimethyl-
1870 - Benzenemethanol, α-ethyl-
1916 - Benzene, 1-methoxy-2,3-dimethyl-
1917 - Benzene, 1-methoxy-2,4-dimethyl-
1918 - Benzene, 1-methoxy-3,5-dimethyl-
1919 - Benzene, 2-methoxy-1,3-dimethyl-
1920 - Benzene, 2-methoxy-1,4-dimethyl-
1921 - Benzene, 4-methoxy-1,2-dimethyl-
1974 - Benzene, (1-methylethoxy)-
2148 - Benzenepropanol
2168 - Benzene, propoxy-
9410 - Phenol, 2-(1-methylethyl)-
9411 - Phenol, 3-(1-methylethyl)-
9412 - Phenol, 4-(1-methylethyl)-
9461 - Phenol, 2-propyl-
9462 - Phenol, 3-propyl-
9463 - Phenol, 4-propyl-
9493 - Phenol, 2,3,4-trimethyl-
9494 - Phenol, 2,4,5-trimethyl-
9495 - Phenol, 2,4,6-trimethyl-
$C_9H_{12}O_2$
1439 - Benzene, 1,2-dimethoxy-3-methyl-
1440 - Benzene, 1,2-dimethoxy-4-methyl-
1441 - Benzene, 1,3-dimethoxy-5-methyl-
1535 - 1,2-Benzenediol, 4-(1-methylethyl)-
1548 - 1,2-Benzenediol, 4-propyl-
1549 - 1,3-Benzenediol, 4-propyl-
1550 - 1,3-Benzenediol, 5-propyl-
1553 - 1,3-Benzenediol, 2,4,6-trimethyl-
1872 - Benzenemethanol, α-ethyl-3-hydroxy-
1893 - Benzenemethanol, 2-methoxy-α-methyl-
1894 - Benzenemethanol, 3-methoxy-α-methyl-
1895 - Benzenemethanol, 4-methoxy-α-methyl-
5761 - Ethanol, 2-(phenylmethoxy)-
7181 - Hydroperoxide, 1-methyl-1-phenylethyl
9357 - Phenol, 4-ethyl-2-methoxy-
9460 - Phenol, 2-propoxy-
10398 - 1-Propanol, 2-phenoxy-
10399 - 2-Propanol, 1-phenoxy-
$C_9H_{12}O_3$
1856 - Benzenemethanol, 3,4-dimethoxy-
2354 - Benzene, 1,2,3-trimethoxy-
2355 - Benzene, 1,3,5-trimethoxy-
3520 - Butanoic acid, 2-furanylmethyl ester
6052 - 2-Furancarboxylic acid, butyl ester
6055 - 3-Furancarboxylic acid, 2,5-dimethyl-, ethyl ester
6062 - 2-Furancarboxylic acid, 2-methylpropyl ester
10026 - 1,2-Propanediol, 3-phenoxy-
10194 - 1,2,3-Propanetriol, 1-phenyl-
10300 - Propanoic acid, 2-methyl-, 2-furanylmethyl ester
11906 - 1,3,5-Trioxane, 2,4,6-triethenyl-
$C_9H_{12}O_3S$
2231 - Benzenesulfonic acid, 4-methyl-, ethyl ester
2235 - Benzenesulfonic acid, propyl ester
$C_9H_{12}S$
1706 - Benzene, 1-(ethylthio)-3-methyl-
1707 - Benzene, 1-(ethylthio)-4-methyl-
1980 - Benzene, [(1-methylethyl)thio]-
2125 - Benzenepropanethiol
2172 - Benzene, (propylthio)-
$C_9H_{13}BrN_2O_2$
3139 - Bromacil
$C_9H_{13}BrOSi$
11367 - Silane, (4-bromophenoxy)trimethyl-
$C_9H_{13}ClN_2O_2$
11591 - Terbacil
$C_9H_{13}ClN_6$
4161 - Cyanazine
$C_9H_{13}ClOSi$
11383 - Silane, (4-chlorophenoxy)trimethyl-
$C_9H_{13}ClSi$
11379 - Silane, (chloromethyl)dimethylphenyl-
11385 - Silane, (3-chlorophenyl)trimethyl-
$C_9H_{13}N$
745 - Benzenamine, 2-ethyl-6-methyl-
746 - Benzenamine, *N*-ethyl-2-methyl-
747 - Benzenamine, *N*-ethyl-3-methyl-
748 - Benzenamine, *N*-ethyl-4-methyl-
749 - Benzenamine, *N*-ethyl-*N*-methyl-
788 - Benzenamine, 2-(1-methylethyl)-
789 - Benzenamine, 4-(1-methylethyl)-
790 - Benzenamine, *N*-(1-methylethyl)-
842 - Benzenamine, 2-propyl-
843 - Benzenamine, *N*-propyl-
858 - Benzenamine, 2,4,5-trimethyl-
859 - Benzenamine, 2,4,6-trimethyl-
860 - Benzenamine, *N*,*N*,2-trimethyl-
861 - Benzenamine, *N*,*N*,3-trimethyl-
862 - Benzenamine, *N*,*N*,4-trimethyl-
1578 - Benzeneethanamine, α-methyl-, (±)-
1579 - Benzeneethanamine, β-methyl-
1580 - Benzeneethanamine, 4-methyl-
1581 - Benzeneethanamine, *N*-methyl-
1806 - Benzenemethanamine, α,α-dimethyl-
1807 - Benzenemethanamine, *N*,*N*-dimethyl-
1809 - Benzenemethanamine, α-ethyl-
1810 - Benzenemethanamine, *N*-ethyl-
10878 - Pyridine, 2-butyl-
10879 - Pyridine, 4-butyl-
10933 - Pyridine, 2,4-diethyl-
10942 - Pyridine, 2-(1,1-dimethylethyl)-
10943 - Pyridine, 3-(1,1-dimethylethyl)-
10944 - Pyridine, 4-(1,1-dimethylethyl)-
10959 - Pyridine, 2-ethyl-3,5-dimethyl-
10960 - Pyridine, 3-ethyl-2,6-dimethyl-
10961 - Pyridine, 4-ethyl-2,6-dimethyl-
11022 - Pyridine, 2,3,4,6-tetramethyl-
$C_9H_{13}NO$
769 - Benzenamine, 2-methoxy-*N*,*N*-dimethyl-
1577 - Benzeneethanamine, 2-methoxy-
1808 - Benzenemethanamine, 2-ethoxy-
1862 - Benzenemethanol, 4-(dimethylamino)-
4467 - Cyclohexanepropanenitrile, 2-oxo-
5743 - Ethanol, 2-(methylphenylamino)-
5744 - Ethanol, 2-[(2-methylphenyl)amino]-
5745 - Ethanol, 2-[(4-methylphenyl)amino]-
5762 - Ethanol, 2-[(phenylmethyl)amino]-
9215 - Phenol, 4-(2-aminopropyl)-, (±)-
$C_9H_{13}NO_2$
491 - 8-Azabicyclo[3.2.1]oct-2-ene-2-carboxylic acid, 8-methyl-, (1*R*)-
1533 - 1,2-Benzenediol, 4-[2-(methylamino)ethyl]-
1805 - Benzenemethanamine, 3,4-dimethoxy-
1887 - Benzenemethanol, 4-hydroxy-α-[(methylamino)methyl]-
10946 - 2,4(1H,3H)-Pyridinedione, 3,3-diethyl-
11115 - 1H-Pyrrole-3-carboxylic acid, 2,4-dimethyl-, ethyl ester
11116 - 1H-Pyrrole-3-carboxylic acid, 2,5-dimethyl-, ethyl ester
$C_9H_{13}NO_3$
1521 - 1,2-Benzenediol, 4-[1-hydroxy-2-(methylamino)ethyl]-, (*R*)-
$C_9H_{13}N_2O_9P$
12035 - 5′-Uridylic acid
$C_9H_{13}N_3O_2$
11060 - 2,4(1H,3H)-Pyrimidinedione, 6-amino-3-ethyl-1-(2-propenyl)-
$C_9H_{13}N_3O_5$
5021 - Cytidine
C_9H_{14}
2968 - Bicyclo[2.2.1]hept-2-ene, 2,3-dimethyl-
4473 - Cyclohexane, 2-propynyl-
4729 - Cyclononyne
6607 - 1-Hepten-4-yne, 6,6-dimethyl-
7108 - 3-Hexen-1-yne, 3-propyl-
8193 - 1-Nonen-3-yne
8194 - 1-Nonen-4-yne
8195 - 2-Nonen-4-yne
11865 - Tricyclo[2.2.1.02,6]heptane, 3,3-dimethyl-
$C_9H_{14}ClNO$
1838 - Benzenemethanol, α-(1-aminoethyl)-, hydrochloride, (*R**,*S**)-(±)-
$C_9H_{14}ClN_5$
5013 - Cyprazine
$C_9H_{14}Cl_2O_2$
8138 - Nonanedioyl dichloride
$C_9H_{14}IN$
866 - Benzenaminium, *N*,*N*,*N*-trimethyl-, iodide
$C_9H_{14}N_2$
8129 - Nonanedinitrile
$C_9H_{14}N_2O$
10860 - 3-Pyridinamine, 6-butoxy-
$C_9H_{14}N_2O_3$
11088 - 2,4,6(1H,3H,5H)-Pyrimidinetrione, 5,5-diethyl-1-methyl-
11093 - 2,4,6(1H,3H,5H)-Pyrimidinetrione, 5-ethyl-5-(1-methylethyl)-
$C_9H_{14}N_3O_8P$
5022 - 3′-Cytidylic acid
$C_9H_{14}N_4O_3$
7133 - *L*-Histidine, *N*-β-alanyl-
$C_9H_{14}O$
2954 - Bicyclo[2.2.1]heptan-2-one, 3,3-dimethyl-, (1*S*)-
2955 - Bicyclo[3.1.1]heptan-2-one, 6,6-dimethyl-, (1*R*)-
4719 - 2-Cyclohexen-1-one, 3,4,4-trimethyl-
4720 - 2-Cyclohexen-1-one, 3,4,6-trimethyl-
4721 - 2-Cyclohexen-1-one, 3,5,5-trimethyl-
4722 - 2-Cyclohexen-1-one, 3,6,6-trimethyl-
4948 - 2-Cyclopenten-1-one, 5-methyl-2-(1-methylethyl)-
6340 - 2,5-Heptadien-4-one, 2,6-dimethyl-
6608 - 6-Hepten-4-yn-3-ol, 3-ethyl-
7292 - 2H-Inden-2-one, octahydro-, *cis*-
7293 - 2H-Inden-2-one, octahydro-, *trans*-
8108 - 2,6-Nonadienal
8109 - 2,4-Nonadienal, (*E*,*E*)-
8527 - 7-Octen-5-yn-4-ol, 4-methyl-
$C_9H_{14}OSi$
11492 - Silane, trimethylphenoxy-
$C_9H_{14}O_2$
156 - Acetic acid, cyclohexylidene-, methyl ester
4639 - 3-Cyclohexene-1-carboxylic acid, ethyl ester
4910 - 1-Cyclopentene-1-carboxylic acid, 2,3,3-trimethyl-
4912 - 2-Cyclopentene-1-carboxylic acid, 1,2,3-trimethyl-
4913 - 3-Cyclopentene-1-carboxylic acid, 2,2,3-trimethyl-
6037 - Furan, 2-(butoxymethyl)-
8203 - 2-Nonynoic acid
8204 - 3-Nonynoic acid
8205 - 5-Nonynoic acid
8206 - 6-Nonynoic acid
8207 - 8-Nonynoic acid
8542 - 2-Octynoic acid, methyl ester
10626 - 2-Propenoic acid, cyclohexyl ester

$C_9H_{14}O_3$
6084 - Furan, 2-(diethoxymethyl)-
9074 - 4-Pentenoic acid, 2-acetyl-, ethyl ester
10674 - 2-Propenoic acid, 2-methyl-, (tetrahydro-2-furanyl)methyl ester
10703 - 2-Propen-1-ol, 2-methyl-, carbonate (2:1)
$C_9H_{14}O_3Si$
11488 - Silane, trimethoxyphenyl-
$C_9H_{14}O_4$
3329 - Butanedioic acid, methylene-, diethyl ester
3787 - 2-Butenedioic acid, 2-methyl-, diethyl ester, (*E*)-
3788 - 2-Butenedioic acid, 2-methyl-, diethyl ester, (*Z*)-
4792 - 1,2-Cyclopentanedicarboxylic acid, dimethyl ester, *trans*-(±)-
4967 - 1,1-Cyclopropanedicarboxylic acid, diethyl ester
4968 - 1,2-Cyclopropanedicarboxylic acid, diethyl ester, *cis*-
9041 - 2-Pentenedioic acid, diethyl ester
9042 - 2-Pentenedioic acid, diethyl ester, (*E*)-
9950 - Propanedioic acid, ethylidene-, diethyl ester
$C_9H_{14}O_5$
3324 - Butanedioic acid, formyl-, diethyl ester
8783 - Pentanedioic acid, 3-oxo-, diethyl ester
9923 - Propanedioic acid, acetyl-, diethyl ester
$C_9H_{14}O_6$
10195 - 1,2,3-Propanetriol, triacetate
$C_9H_{14}O_7$
10170 - 1,2,3-Propanetricarboxylic acid, 2-hydroxy-, trimethyl ester
$C_9H_{14}S$
11771 - Thiophene, 2-(3-methylbutyl)-
$C_9H_{14}Si$
11493 - Silane, trimethylphenyl-
$C_9H_{15}AlO_9$
340 - Aluminum, tris(2-hydroxypropanoato-*O*1,*O*2)-
$C_9H_{15}Br$
4322 - Cyclohexane, (2-bromo-2-propenyl)-
$C_9H_{15}Br_6O_4P$
10358 - 1-Propanol, 2,3-dibromo-, phosphate (3:1)
$C_9H_{15}Cl$
8200 - 1-Nonyne, 1-chloro-
8534 - 3-Octyne, 2-chloro-2-methyl-
$C_9H_{15}N$
10523 - 2-Propen-1-amine, *N*,*N*-di-2-propenyl-
11117 - 1H-Pyrrole, 3,4-diethyl-2-methyl-
$C_9H_{15}NO$
479 - 9-Azabicyclo[3.3.1]nonan-3-one, 9-methyl-
$C_9H_{15}NO_2$
3495 - Butanoic acid, 2-cyano-2-ethyl-, ethyl ester
6881 - Hexanoic acid, 2-cyano-, ethyl ester
9654 - 2,4-Piperidinedione, 3,3-diethyl-
10521 - 2-Propenamide, *N*-(1,1-dimethyl-3-oxobutyl)-
$C_9H_{15}NO_3$
483 - 8-Azabicyclo[3.2.1]octane-2-carboxylic acid, 3-hydroxy-8-methyl-, [1*R*-(*exo,exo*)]-
8585 - 2,4-Oxazolidinedione, 5,5-dipropyl-
$C_9H_{15}NO_5$
9922 - Propanedioic acid, (acetylamino)-, diethyl ester
$C_9H_{15}N_3O_2S$
7204 - 1H-Imidazole-4-ethanaminium, α-carboxy-2,3-dihydro-*N*,*N*,*N*-trimethyl-2-thioxo-, hydroxide, in
$C_9H_{15}N_3O_7$
472 - *L*-Aspartic acid, *N*-[2-[(2-amino-2-oxoethyl)amino]-2-carboxyethyl]-
C_9H_{16}
2990 - Bicyclo[3.3.1]nonane
2992 - Bicyclo[2.2.2]octane, 2-methyl-
4393 - Cyclohexane, 1-ethylidene-2-methyl-
4470 - Cyclohexane, 2-propenyl-
4663 - Cyclohexene, 1-ethyl-4-methyl-
4687 - Cyclohexene, 1,2,3-trimethyl-
4688 - Cyclohexene, 1,3,5-trimethyl-
4689 - Cyclohexene, 1,4,4-trimethyl-
4690 - Cyclohexene, 1,4,5-trimethyl-
4691 - Cyclohexene, 1,5,5-trimethyl-
4692 - Cyclohexene, 1,6,6-trimethyl-
4693 - Cyclohexene, 1,5,6-trimethyl-
4727 - Cyclononene, (*E*)-
4728 - Cyclononene, (*Z*)-
4906 - Cyclopentene, 1-butyl-
4927 - Cyclopentene, 4-isopropyl-1-methyl-
4931 - Cyclopentene, 3-methyl-1-(1-methylethyl)-
6330 - 1,3-Heptadiene, 2,6-dimethyl-
6331 - 1,5-Heptadiene, 2,6-dimethyl-
6332 - 2,4-Heptadiene, 2,4-dimethyl-
6333 - 2,4-Heptadiene, 3,5-dimethyl-
6617 - 3-Heptyne, 2,6-dimethyl-
6618 - 3-Heptyne, 5,5-dimethyl-
6694 - 1,5-Hexadiene, 3-ethyl-2-methyl-
7281 - 1H-Indene, octahydro-
8110 - 1,8-Nonadiene
8111 - 2,7-Nonadiene
8196 - 1-Nonyne
8197 - 2-Nonyne
8198 - 3-Nonyne
8199 - 4-Nonyne
8328 - 2,4-Octadiene, 7-methyl-
8329 - 3,5-Octadiene, 4-methyl-
8537 - 3-Octyne, 7-methyl-
$C_9H_{16}ClN_5$
10508 - Propazine
11593 - Terbuthylazine
$C_9H_{16}N_4OS$
11590 - Tebuthiuron
$C_9H_{16}O$
2938 - Bicyclo[2.2.1]heptan-2-ol, 1,7-dimethyl-, (*exo,syn*)-
4555 - Cyclohexanol, 1-(2-propenyl)-
4556 - Cyclohexanol, 2-(2-propenyl)-, *trans*-
4595 - Cyclohexanone, 2-ethyl-2-methyl-
4599 - Cyclohexanone, 2-(1-methylethyl)-
4600 - Cyclohexanone, 4-(1-methylethyl)-
4612 - Cyclohexanone, 2-propyl-
4613 - Cyclohexanone, 4-propyl-
4614 - Cyclohexanone, 2,2,3-trimethyl-
4615 - Cyclohexanone, 2,2,4-trimethyl-
4616 - Cyclohexanone, 2,2,5-trimethyl-
4617 - Cyclohexanone, 2,2,6-trimethyl-
4618 - Cyclohexanone, 2,3,6-trimethyl-
4619 - Cyclohexanone, 2,4,4-trimethyl-
4620 - Cyclohexanone, 2,4,5-trimethyl-
4621 - Cyclohexanone, 2,4,6-trimethyl-
4622 - Cyclohexanone, 3,3,5-trimethyl-
4623 - Cyclohexanone, 3,4,4-trimethyl-
4700 - 2-Cyclohexen-1-ol, 1,3,5-trimethyl-
4701 - 2-Cyclohexen-1-ol, 2,4,4-trimethyl-
4702 - 2-Cyclohexen-1-ol, 3,5,5-trimethyl-
4725 - Cyclononanone
4738 - Cyclooctanecarboxaldehyde
4884 - Cyclopentanone, 4-isopropyl-2-methyl-
4889 - Cyclopentanone, 2-methyl-2-(1-methylethyl)-
4890 - Cyclopentanone, 2-methyl-5-(1-methylethyl)-
4891 - Cyclopentanone, 4-methyl-2-(1-methylethyl)-
5843 - Ethanone, 1-(1-methylcyclohexyl)-
5844 - Ethanone, 1-(3-methylcyclohexyl)-
5845 - Ethanone, 1-(4-methylcyclohexyl)-
8112 - 2,6-Nonadien-1-ol
8179 - 2-Nonenal
8180 - 2-Nonenal, (*E*)-
8208 - 1-Nonyn-3-ol
8209 - 2-Nonyn-1-ol
8522 - 1-Octen-3-one, 5-methyl-
8523 - 5-Octen-4-one, 2-methyl-
8524 - 5-Octen-4-one, 7-methyl-
8525 - 7-Octen-4-one, 2-methyl-
8536 - 2-Octyne, 1-methoxy-
8544 - 1-Octyn-3-ol, 3-methyl-
10423 - 1-Propanone, 1-cyclohexyl-
$C_9H_{16}O_2$
4310 - Cyclohexaneacetic acid, methyl ester
4311 - Cyclohexaneacetic acid, 2-methyl-
4312 - Cyclohexaneacetic acid, 3-methyl-
4334 - Cyclohexanecarboxylic acid, ethyl ester
4468 - Cyclohexanepropanoic acid
6149 - 2(3H)-Furanone, dihydro-5-pentyl-
7063 - 2-Hexenoic acid, 2-methyl-, ethyl ester
7064 - 3-Hexenoic acid, 2-methyl-, ethyl ester
7065 - 3-Hexenoic acid, 3-methyl-, ethyl ester
8137 - 2,4-Nonanedione
8187 - 3-Nonenoic acid
8188 - 8-Nonenoic acid
10235 - Propanoic acid, cyclohexyl ester
10643 - 2-Propenoic acid, hexyl ester
$C_9H_{16}O_3$
3461 - Butanoic acid, 2-acetyl-3-methyl-, ethyl ester
6165 - 2-Furanpropanoic acid, tetrahydro-, ethyl ester
6460 - Heptanoic acid, 2,2-dimethyl-6-oxo-
6461 - Heptanoic acid, 4,4-dimethyl-6-oxo-
6902 - Hexanoic acid, 4-methyl-5-oxo-, ethyl ester
8856 - Pentanoic acid, 2-acetyl-, ethyl ester
8904 - Pentanoic acid, 4-oxo-, butyl ester
8910 - Pentanoic acid, 4-oxo-, 2-methylpropyl ester
$C_9H_{16}O_4$
3326 - Butanedioic acid, methyl-, diethyl ester
3337 - Butanedioic acid, monopentyl ester
6395 - Heptanedioic acid, dimethyl ester
6396 - Heptanedioic acid, monoethyl ester
8130 - Nonanedioic acid
8772 - Pentanedioic acid, diethyl ester
8774 - Pentanedioic acid, 3,3-dimethyl-, dimethyl ester
8791 - 1,5-Pentanediol, diacetate
9942 - Propanedioic acid, dimethyl-, diethyl ester
9945 - Propanedioic acid, dipropyl ester
9949 - Propanedioic acid, ethyl-, diethyl ester
$C_9H_{16}O_5$
8778 - Pentanedioic acid, 3-hydroxy-, diethyl ester
$C_9H_{17}ClN_3O_3PS$
7340 - Isazophos
$C_9H_{17}ClO$
8177 - Nonanoyl chloride
$C_9H_{17}FO_2$
8151 - Nonanoic acid, 9-fluoro-
$C_9H_{17}IN_4$
11825 - 3,5,7-Triaza-1-azoniatricyclo[3.3.1.13,7]decane, 1-(2-propenyl)-, iodide
$C_9H_{17}N$
4304 - Cyclohexanamine, *N*-2-propenyl-
8144 - Nonanenitrile
11246 - Quinoline, decahydro-, *cis*-
11247 - Quinoline, decahydro-, *trans*-(±)-
$C_9H_{17}NO$
4591 - Cyclohexanone, 2-[(dimethylamino)methyl]-
9003 - 4-Pentenamide, 2,2-diethyl-
9725 - 4-Piperidinone, 1,2,3,6-tetramethyl-
9726 - 4-Piperidinone, 2,2,6,6-tetramethyl-
10470 - 2-Propanone, 1-(1-methyl-2-piperidinyl)-, (±)-
$C_9H_{17}NOS$
7661 - Molinate
$C_9H_{17}NO_2$
10631 - 2-Propenoic acid, 2-(diethylamino)ethyl ester
$C_9H_{17}NO_4$
10305 - Propanoic acid, 2-methyl-, 2-nitropentyl ester
$C_9H_{17}NO_5$
329 - β-Alanine, *N*-(2,4-dihydroxy-3,3-dimethyl-1-oxobutyl)-, (*R*)-
$C_9H_{17}NS$
11708 - Thiocyanic acid, octyl ester
$C_9H_{17}N_5S$
341 - Ametryn
C_9H_{18}
4394 - Cyclohexane, 1-ethyl-1-methyl-
4395 - Cyclohexane, 1-ethyl-2-methyl-, *cis*-
4396 - Cyclohexane, 1-ethyl-2-methyl-, *trans*-

4397 - Cyclohexane, 1-ethyl-3-methyl-, *cis*-
4398 - Cyclohexane, 1-ethyl-3-methyl-, *trans*-
4399 - Cyclohexane, 1-ethyl-4-methyl-, *cis*-
4400 - Cyclohexane, 1-ethyl-4-methyl-, *trans*-
4431 - Cyclohexane, (1-methylethyl)-
4471 - Cyclohexane, propyl-
4485 - Cyclohexane, 1,1,2-trimethyl-
4486 - Cyclohexane, 1,1,3-trimethyl-
4487 - Cyclohexane, 1,1,4-trimethyl-
4488 - Cyclohexane, 1,2,3-trimethyl-
4489 - Cyclohexane, 1,2,3-trimethyl-, (1α,2β,3α)-
4490 - Cyclohexane, 1,2,3-trimethyl-, (1α,2α,3α)-
4491 - Cyclohexane, 1,2,3-trimethyl-, (1α,2α,3β)-
4492 - Cyclohexane, 1,2,4-trimethyl-, (1α,2α,4α)-
4493 - Cyclohexane, 1,2,4-trimethyl-, (1α,2β,4α)-
4494 - Cyclohexane, 1,2,4-trimethyl-, (1α,2α,4β)-
4495 - Cyclohexane, 1,2,4-trimethyl-, (1α,2β,4β)-
4496 - Cyclohexane, 1,3,5-trimethyl-, (1α,3α,5α)-
4497 - Cyclohexane, 1,3,5-trimethyl-, (1α,3α,5β)-
4724 - Cyclononane
4778 - Cyclopentane, butyl-
4796 - Cyclopentane, 1,1-diethyl-
4797 - Cyclopentane, 1,2-diethyl-, *cis*-
4798 - Cyclopentane, 1,2-diethyl-, *trans*-
4799 - Cyclopentane, 1,3-diethyl-
4805 - Cyclopentane, (1,1-dimethylethyl)-
4829 - Cyclopentane, 1-methyl-2-(1-methylethyl)-
4832 - Cyclopentane, (1-methylpropyl)-
4833 - Cyclopentane, 1-methyl-1-propyl-
4834 - Cyclopentane, 1-methyl-2-propyl-, *cis*-
4835 - Cyclopentane, 1-methyl-2-propyl-, *trans*-
4836 - Cyclopentane, 1-methyl-3-propyl-, *cis*-
4837 - Cyclopentane, 1-methyl-3-propyl-, *trans*-
4838 - Cyclopentane, (2-methylpropyl)-
4851 - Cyclopentane, 1,1,3,3-tetramethyl-
6542 - 2-Heptene, 2,3-dimethyl-
8182 - 1-Nonene
8183 - 3-Nonene, (*E*)-
8505 - 1-Octene, 2-methyl-
8506 - 2-Octene, 3-methyl-
8507 - 3-Octene, 7-methyl-
8508 - 4-Octene, 2-methyl-

$C_9H_{18}Br_2$

8127 - Nonane, 1,2-dibromo-

$C_9H_{18}ClF$

8126 - Nonane, 1-chloro-9-fluoro-

$C_9H_{18}Cl_2$

6382 - Heptane, 2,6-dichloro-2,6-dimethyl-
8128 - Nonane, 1,9-dichloro-

$C_9H_{18}FeN_3S_6$

5957 - Ferbam

$C_9H_{18}NO_3PS_2$

6021 - Fosthiazate

$C_9H_{18}N_2O_2S$

11717 - Thiofanox

$C_9H_{18}N_2O_4$

3938 - Carbamic acid, *N*-butyl-*N*-nitro,butyl ester
10021 - 1,3-Propanediol, 2-methyl-2-propyl-, dicarbamate

$C_9H_{18}N_6O_6$

7584 - Methanol, (1,3,5-triazine-2,4,6-triyltrinitrilo)hexakis-

$C_9H_{18}O$

4526 - Cyclohexanol, 1-ethyl-2-methyl-
4527 - Cyclohexanol, 1-ethyl-3-methyl-
4538 - Cyclohexanol, 1-(1-methylethyl)-
4539 - Cyclohexanol, 2-(1-methylethyl)-, *cis*-
4540 - Cyclohexanol, 4-(1-methylethyl)-, *cis*-
4541 - Cyclohexanol, 4-(1-methylethyl)-, *trans*-
4557 - Cyclohexanol, 1-propyl-
4558 - Cyclohexanol, 2-propyl-, *cis*-
4559 - Cyclohexanol, 2-propyl-, *trans*-
4560 - Cyclohexanol, 4-propyl-, *trans*-
4561 - Cyclohexanol, 4-propyl-, *cis*-
4562 - Cyclohexanol, 1,2,2-trimethyl-
4563 - Cyclohexanol, 1,2,6-trimethyl-
4564 - Cyclohexanol, 1,3,5-trimethyl-
4565 - Cyclohexanol, 2,2,5-trimethyl-
4566 - Cyclohexanol, 2,2,6-trimethyl-
4567 - Cyclohexanol, 2,3,3-trimethyl-
4568 - Cyclohexanol, 2,3,6-trimethyl-
4569 - Cyclohexanol, 3,3,5-trimethyl-, *cis*-
4570 - Cyclohexanol, 3,3,5-trimethyl-, *trans*-
6514 - 4-Heptanone, 2,6-dimethyl-
6515 - 4-Heptanone, 3,5-dimethyl-
8117 - Nonanal
8172 - 2-Nonanone
8173 - 3-Nonanone
8174 - 4-Nonanone
8175 - 5-Nonanone
8189 - 1-Nonen-3-ol
8190 - 8-Nonen-1-ol
8475 - 2-Octanone, 3-methyl-
8476 - 3-Octanone, 4-methyl-
8477 - 3-Octanone, 7-methyl-
8478 - 4-Octanone, 3-methyl-
8479 - 4-Octanone, 7-methyl-
8517 - 1-Octen-4-ol, 4-methyl-
8518 - 2-Octen-4-ol, 7-methyl-
8519 - 3-Octen-2-ol, 2-methyl-, (*Z*)-
8520 - 4-Octen-3-ol, 3-methyl-
8521 - 5-Octen-4-ol, 5-methyl-
8988 - 3-Pentanone, 2,2,4,4-tetramethyl-

$C_9H_{18}O_2$

193 - Acetic acid, heptyl ester
3543 - Butanoic acid, 3-methyl-, butyl ester
3544 - Butanoic acid, 3-methylbutyl ester
3545 - Butanoic acid, 2-methyl-, butyl ester
3555 - Butanoic acid, 3-methyl-, 2-methylpropyl ester
3579 - Butanoic acid, pentyl ester
6014 - Formic acid, octyl ester
6462 - Heptanoic acid, ethyl ester
6476 - 4-Heptanol, acetate
6886 - Hexanoic acid, 2-ethyl-3-methyl-
6887 - Hexanoic acid, 3-ethyl-3-methyl-
6899 - Hexanoic acid, 3-methyl-, ethyl ester
6900 - Hexanoic acid, 4-methyl-, ethyl ester
6909 - Hexanoic acid, propyl ester
8146 - Nonanoic acid
8433 - Octanoic acid, 3-methyl-
8434 - Octanoic acid, methyl ester
8862 - Pentanoic acid, butyl ester
8899 - Pentanoic acid, 1-methylpropyl ester, (+)-
8900 - Pentanoic acid, 2-methylpropyl ester
10265 - Propanoic acid, hexyl ester
10302 - Propanoic acid, 2-methyl-, 3-methylbutyl ester

$C_9H_{18}O_3$

3994 - Carbonic acid, bis(1,1-dimethylethyl) ester
3997 - Carbonic acid, bis(2-methylpropyl) ester
3998 - Carbonic acid, dibutyl ester

$C_9H_{18}O_5$

5719 - Ethanol, 2-ethoxy-, carbonate (2:1)

$C_9H_{19}Br$

8121 - Nonane, 1-bromo-

$C_9H_{19}Cl$

6364 - Heptane, 3-chloro-3-ethyl-
6747 - Hexane, 3-chloro-2,2,3-trimethyl-
8123 - Nonane, 1-chloro-
8124 - Nonane, 2-chloro-
8125 - Nonane, 5-chloro-
8362 - Octane, 3-chloro-3-methyl-
8363 - Octane, 4-chloro-4-methyl-
8727 - Pentane, 3-chloro-3-ethyl-2,2-dimethyl-

$C_9H_{19}ClO$

8163 - 1-Nonanol, 9-chloro-

$C_9H_{19}FO$

8165 - 1-Nonanol, 9-fluoro-

$C_9H_{19}N$

4303 - Cyclohexanamine, *N*-(1-methylethyl)-
4810 - Cyclopentaneethanamine, *N*,α-dimethyl-
6532 - 5-Hepten-2-amine, *N*,6-dimethyl-
9635 - Piperidine, 1-butyl-
9636 - Piperidine, 3-butyl-
9637 - Piperidine, 4-butyl-
9647 - Piperidine, 2,5-diethyl-
9653 - Piperidine, 1-(1,1-dimethylethyl)-
9685 - Piperidine, 1-(2-methylpropyl)-
9686 - Piperidine, 1-methyl-2-propyl-, (*S*)-
9710 - Piperidine, 2,2,4,6-tetramethyl-
9711 - Piperidine, 2,2,6,6-tetramethyl-

$C_9H_{19}NO$

3201 - Butanamide, *N*,*N*-diethyl-3-methyl-
4144 - Conhydrine, *N*-methyl-
8118 - Nonanamide
8964 - 2-Pentanone, 5-(diethylamino)-
9794 - Propanamide, *N*,*N*-diethyl-2,2-dimethyl-

$C_9H_{19}NOS$

5372 - EPTC

$C_9H_{19}NO_2$

328 - β-Alanine, *N*,*N*-diethyl-, ethyl ester
3934 - Carbamic acid, butyl-, butyl ester

$C_9H_{19}NO_4$

3202 - Butanamide, 2,4-dihydroxy-*N*-(3-hydroxypropyl)-3,3-dimethyl-, (*R*)-

C_9H_{20}

6383 - Heptane, 2,2-dimethyl-
6384 - Heptane, 2,3-dimethyl-
6385 - Heptane, 2,4-dimethyl-
6386 - Heptane, 2,5-dimethyl-
6387 - Heptane, 2,6-dimethyl-
6388 - Heptane, 3,3-dimethyl-
6389 - Heptane, 3,4-dimethyl-
6390 - Heptane, 3,5-dimethyl-
6391 - Heptane, 4,4-dimethyl-
6407 - Heptane, 3-ethyl-
6408 - Heptane, 4-ethyl-
6825 - Hexane, 3-ethyl-2-methyl-
6826 - Hexane, 3-ethyl-3-methyl-
6827 - Hexane, 3-ethyl-4-methyl-
6828 - Hexane, 4-ethyl-2-methyl-
6859 - Hexane, 2,2,3-trimethyl-
6860 - Hexane, 2,2,4-trimethyl-
6861 - Hexane, 2,2,5-trimethyl-
6862 - Hexane, 2,3,3-trimethyl-
6863 - Hexane, 2,3,4-trimethyl-
6864 - Hexane, 2,3,5-trimethyl-
6865 - Hexane, 2,4,4-trimethyl-
6866 - Hexane, 3,3,4-trimethyl-
8120 - Nonane
8408 - Octane, 2-methyl-
8409 - Octane, 3-methyl-
8410 - Octane, 4-methyl-
8756 - Pentane, 3,3-diethyl-
8810 - Pentane, 3-ethyl-2,2-dimethyl-
8811 - Pentane, 3-ethyl-2,3-dimethyl-
8812 - Pentane, 3-ethyl-2,4-dimethyl-
8839 - Pentane, 2,2,3,3-tetramethyl-
8840 - Pentane, 2,2,3,4-tetramethyl-
8841 - Pentane, 2,2,4,4-tetramethyl-
8842 - Pentane, 2,3,3,4-tetramethyl-

$C_9H_{20}N_2$

9628 - 4-Piperidinamine, 2,2,6,6-tetramethyl-

$C_9H_{20}N_2O$

11996 - Urea, *N*,*N*-bis(2-methylpropyl)-
12029 - Urea, tetraethyl-

$C_9H_{20}O$

6406 - Heptane, 1-ethoxy-
6479 - 2-Heptanol, 2,6-dimethyl-
6480 - 3-Heptanol, 2,3-dimethyl-
6481 - 3-Heptanol, 2,6-dimethyl-
6482 - 3-Heptanol, 3,5-dimethyl-
6483 - 3-Heptanol, 3,6-dimethyl-
6484 - 4-Heptanol, 2,4-dimethyl-
6485 - 4-Heptanol, 2,6-dimethyl-
6486 - 4-Heptanol, 3,5-dimethyl-
6487 - 4-Heptanol, 4-ethyl-
6932 - 1-Hexanol, 2-ethyl-3-methyl-
6933 - 2-Hexanol, 3-ethyl-2-methyl-
6934 - 3-Hexanol, 3-ethyl-2-methyl-
6935 - 3-Hexanol, 3-ethyl-5-methyl-

6951 - 1-Hexanol, 3,5,5-trimethyl-
6952 - 2-Hexanol, 2,3,4-trimethyl-
6953 - 3-Hexanol, 2,2,3-trimethyl-
6954 - 3-Hexanol, 2,3,5-trimethyl-
6955 - 3-Hexanol, 2,4,4-trimethyl-
6956 - 3-Hexanol, 2,5,5-trimethyl-
6957 - 3-Hexanol, 3,4,4-trimethyl-
6958 - 3-Hexanol, 3,5,5-trimethyl-
8157 - 1-Nonanol
8158 - 2-Nonanol, (±)-
8159 - 3-Nonanol, (±)-
8160 - 4-Nonanol, (*S*)-
8161 - 5-Nonanol
8461 - 1-Octanol, 3-methyl-
8462 - 1-Octanol, 4-methyl-
8463 - 1-Octanol, 7-methyl-
8464 - 2-Octanol, 2-methyl-
8465 - 2-Octanol, 3-methyl-
8466 - 3-Octanol, 3-methyl-
8467 - 3-Octanol, 4-methyl-
8468 - 3-Octanol, 6-methyl-
8469 - 4-Octanol, 4-methyl-
8470 - 4-Octanol, 5-methyl-

$C_9H_{20}O_2$
3398 - Butane, 1,1′-[methylenebis(oxy)]bis-
8136 - 1,9-Nonanediol
8755 - Pentane, 1,1-diethoxy-
9992 - 1,3-Propanediol, 2-butyl-2-ethyl-
10075 - Propane, 1,1′-[methylenebis(oxy)]bis[2-methyl-

$C_9H_{20}O_3$
10182 - Propane, 1,1,1-triethoxy-
10183 - Propane, 1,1,3-triethoxy-

$C_9H_{20}O_3Si$
11482 - Silane, triethoxy-2-propenyl-

$C_9H_{20}O_4$
5601 - Ethane, 1,1′,1″1‴-[methanetetrayl tetrakis(oxy)]tetrakis-

$C_9H_{20}S$
8145 - 1-Nonanethiol

$C_9H_{21}B$
3121 - Borane, tripropyl-

$C_9H_{21}BO_3$
3127 - Boric acid, tripropyl ester
3128 - Boric acid, tris(1-methylethyl) ester

$C_9H_{21}ClSn$
11542 - Stannane, chlorotripropyl-

$C_9H_{21}N$
8119 - 1-Nonanamine
9808 - 1-Propanamine, *N,N*-dipropyl-

$C_9H_{21}NO_3$
10393 - 2-Propanol, 1,1′,1″-nitrilotris-

$C_9H_{21}NO_4$
5678 - Ethanol, 2-[2-[2-(3-aminopropoxy) ethoxy]ethoxy]-

$C_9H_{21}O_2PS_3$
11592 - Terbufos

$C_9H_{21}O_3P$
9594 - Phosphorous acid, tripropyl ester
9595 - Phosphorous acid, tris(1-methylethyl) ester

$C_9H_{21}O_4P$
9567 - Phosphoric acid, tripropyl ester
9568 - Phosphoric acid, tris(1-methylethyl) ester

$C_9H_{22}N_2$
7489 - Methanediamine, *N,N,N′,N′*-tetraethyl-
8736 - 1,4-Pentanediamine, *N*1,*N*1-diethyl-

$C_9H_{22}O_4P_2S_4$
5931 - Ethion

$C_9H_{22}Si$
11500 - Silane, tripropyl-

$C_9H_{23}NO_3Si$
9828 - 1-Propanamine, 3-(triethoxysilyl)-

$C_9H_{24}O_3Si_3$
5003 - Cyclotrisiloxane, 2,4,6-triethyl-2,4,6-trimethyl-

$C_9H_{24}O_4Si_4$
4991 - Cyclotetrasiloxane, ethenylheptamethyl-

$C_{10}Cl_8$
7933 - Naphthalene, octachloro-

$C_{10}Cl_{10}O$
7408 - Kepone

$C_{10}Cl_{12}$
7644 - 1,3,4-Metheno-1H-cyclobuta[cd]pentalene, 1,1a,2,2,3,3a,4,5,5,5a,5b, 6-dodecachlorooctahydro-

$C_{10}F_8$
7935 - Naphthalene, octafluoro-

$C_{10}F_{18}$
7934 - Naphthalene, octadecafluorodecahydro-

$C_{10}F_{22}$
5050 - Decane, docosafluoro-

$C_{10}F_{23}N$
11912 - Tripropylamine, eicosafluoro-2-(trifluoromethyl)-

$C_{10}H_4Cl_2O_2$
7849 - 1,2-Naphthalenedione, 3,4-dichloro-
7850 - 1,4-Naphthalenedione, 2,3-dichloro-
7851 - 1,4-Naphthalenedione, 5,6-dichloro-

$C_{10}H_4N_4O_8$
8004 - Naphthalene, 1,3,6,8-tetranitro-

$C_{10}H_5BrO_2$
7844 - 1,2-Naphthalenedione, 3-bromo-

$C_{10}H_5BrO_3$
7845 - 1,4-Naphthalenedione, 2-bromo-3-hydroxy-

$C_{10}H_5ClO_2$
7847 - 1,4-Naphthalenedione, 5-chloro-

$C_{10}H_5ClO_3$
7848 - 1,4-Naphthalenedione, 2-chloro-3-hydroxy-

$C_{10}H_5Cl_2NO_2$
11201 - Quinclorac

$C_{10}H_5Cl_7$
7576 - 4,7-Methano-1H-indene, 1,4,5,6,7,8,8-heptachloro-3a,4,7,7a-tetrahydro-

$C_{10}H_5Cl_7O$
6301 - Heptachlor epoxide

$C_{10}H_5N_3O_6$
8016 - Naphthalene, 1,3,5-trinitro-

$C_{10}H_5NaO_5S$
7970 - 1-Naphthalenesulfonic acid, 3,4-dihydro-3,4-dioxo-, sodium salt

$C_{10}H_6BrCl$
7741 - Naphthalene, 1-bromo-4-chloro-

$C_{10}H_6BrNO_2$
7744 - Naphthalene, 1-bromo-3-nitro-
7745 - Naphthalene, 2-bromo-1-nitro-

$C_{10}H_6Br_2$
7801 - Naphthalene, 1,3-dibromo-
7802 - Naphthalene, 1,4-dibromo-
7803 - Naphthalene, 1,5-dibromo-

$C_{10}H_6ClNO_2$
7786 - Naphthalene, 1-chloro-5-nitro-
7787 - Naphthalene, 1-chloro-8-nitro-
7788 - Naphthalene, 2-chloro-1-nitro-

$C_{10}H_6Cl_2$
7807 - Naphthalene, 1,2-dichloro-
7808 - Naphthalene, 1,3-dichloro-
7809 - Naphthalene, 1,4-dichloro-
7810 - Naphthalene, 1,5-dichloro-
7811 - Naphthalene, 1,7-dichloro-
7812 - Naphthalene, 1,8-dichloro-
7813 - Naphthalene, 2,6-dichloro-

$C_{10}H_6Cl_2O$
8025 - 1-Naphthalenol, 2,3-dichloro-
8026 - 1-Naphthalenol, 2,4-dichloro-

$C_{10}H_6Cl_4O_4$
1353 - 1,2-Benzenedicarboxylic acid, 3,4,5,6-tetrachloro-, monoethyl ester
5180 - Dimethyl tetrachloroterephthalate

$C_{10}H_6Cl_8$
4077 - Chlordane

$C_{10}H_6N_2$
7375 - 5-Isoquinolinecarbonitrile
11223 - 2-Quinolinecarbonitrile
11224 - 4-Quinolinecarbonitrile

$C_{10}H_6N_2OS_2$
4072 - Chinomethionat

$C_{10}H_6N_2O_2$
5223 - 5H,10H-Dipyrrolo[1,2-a:1′,2′-d]pyrazine-5,10-dione

$C_{10}H_6N_2O_4$
7831 - Naphthalene, 1,3-dinitro-
7832 - Naphthalene, 1,5-dinitro-
7833 - Naphthalene, 1,8-dinitro-

$C_{10}H_6O_2$
7842 - 1,2-Naphthalenedione
7843 - 1,4-Naphthalenedione

$C_{10}H_6O_3$
7859 - 1,4-Naphthalenedione, 2-hydroxy-
7860 - 1,4-Naphthalenedione, 5-hydroxy-

$C_{10}H_6O_3S$
7840 - 1,8-Naphthalenediol, cyclic sulfite

$C_{10}H_6O_4$
7852 - 1,4-Naphthalenedione, 2,3-dihydroxy-
7853 - 1,4-Naphthalenedione, 5,8-dihydroxy-

$C_{10}H_6O_8$
2257 - 1,2,4,5-Benzenetetracarboxylic acid

$C_{10}H_7Br$
7739 - Naphthalene, 1-bromo-
7740 - Naphthalene, 2-bromo-

$C_{10}H_7BrO$
8022 - 2-Naphthalenol, 1-bromo-

$C_{10}H_7Cl$
7780 - Naphthalene, 1-chloro-
7781 - Naphthalene, 2-chloro-

$C_{10}H_7ClO$
8023 - 2-Naphthalenol, 6-chloro-
8063 - 2-Naphthol, 8-chloro-

$C_{10}H_7ClO_2S$
7975 - 1-Naphthalenesulfonyl chloride
7976 - 2-Naphthalenesulfonyl chloride

$C_{10}H_7Cl_5O$
11899 - Tridiphane

$C_{10}H_7F$
7893 - Naphthalene, 1-fluoro-
7894 - Naphthalene, 2-fluoro-

$C_{10}H_7I$
7902 - Naphthalene, 1-iodo-
7903 - Naphthalene, 2-iodo-

$C_{10}H_7NO$
11225 - 4-Quinolinecarboxaldehyde

$C_{10}H_7NO_2$
7929 - Naphthalene, 1-nitro-
7930 - Naphthalene, 2-nitro-
8035 - 2-Naphthalenol, 1-nitroso-
11123 - 1H-Pyrrole-2,5-dione, 1-phenyl-
11227 - 2-Quinolinecarboxylic acid
11228 - 7-Quinolinecarboxylic acid
11229 - 8-Quinolinecarboxylic acid

$C_{10}H_7NO_3$
8034 - 2-Naphthalenol, 1-nitro-
11232 - 2-Quinolinecarboxylic acid, 4-hydroxy-

$C_{10}H_7NO_4$
11230 - 2-Quinolinecarboxylic acid, 4,8-dihydroxy-

$C_{10}H_7N_3$
7738 - Naphthalene, 1-azido-

$C_{10}H_7N_3S$
11656 - Thiabendazole

$C_{10}H_8$
504 - Azulene
7732 - Naphthalene

$C_{10}H_8BrN$
7711 - 1-Naphthalenamine, 5-bromo-

$C_{10}H_8ClN$
11243 - Quinoline, 2-chloro-4-methyl-
11244 - Quinoline, 2-chloro-8-methyl-
11245 - Quinoline, 4-chloro-2-methyl-

$C_{10}H_8ClN_3O$
4080 - Chloridazon

$C_{10}H_8N_2$
3099 - 2,2′-Bipyridine
3100 - 2,3′-Bipyridine

3101 - 2,4′-Bipyridine
3102 - 3,3′-Bipyridine
3103 - 3,4′-Bipyridine
3104 - 4,4′-Bipyridine
9917 - Propanedinitrile, (phenylmethyl)-

$C_{10}H_8N_2O_4$
5583 - Ethanedione, di-2-furanyl-, dioxime

$C_{10}H_8N_4O_5$
10848 - 3H-Pyrazol-3-one, 2,4-dihydro-5-methyl-4-nitro-2-(4-nitrophenyl)-

$C_{10}H_8O$
3924 - 3-Butyn-2-one, 4-phenyl-
6159 - Furan, 2-phenyl-
8017 - 1-Naphthalenol
8018 - 2-Naphthalenol

$C_{10}H_8OS$
6160 - Furan, 2-(phenylthio)-

$C_{10}H_8O_2$
2833 - 2H-1-Benzopyran-2-one, 3-methyl-
2834 - 2H-1-Benzopyran-2-one, 5-methyl-
2835 - 2H-1-Benzopyran-2-one, 6-methyl-
2836 - 2H-1-Benzopyran-2-one, 7-methyl-
2837 - 2H-1-Benzopyran-2-one, 8-methyl-
2838 - 4H-1-Benzopyran-4-one, 3-methyl-
5775 - Ethanone, 1-(2-benzofuranyl)-
6157 - Furan, 2-phenoxy-
7246 - 1H-Indene-1-carboxylic acid
7834 - 1,3-Naphthalenediol
7835 - 1,5-Naphthalenediol
7836 - 1,6-Naphthalenediol
7837 - 1,7-Naphthalenediol
7838 - 2,6-Naphthalenediol
7839 - 2,7-Naphthalenediol
10742 - 2-Propynoic acid, 3-phenyl-, methyl ester

$C_{10}H_8O_3$
2831 - 2H-1-Benzopyran-2-one, 7-hydroxy-4-methyl-
6096 - 2,5-Furandione, dihydro-3-phenyl-, (±)-

$C_{10}H_8O_3S$
7959 - 1-Naphthalenesulfonic acid
7960 - 2-Naphthalenesulfonic acid

$C_{10}H_8O_4$
2829 - 2H-1-Benzopyran-2-one, 7-hydroxy-6-methoxy-
5185 - 1,7-Dioxadispiro[4.0.4.2]dodeca-3,9-diene-2,8-dione, *trans*-
6057 - 2-Furancarboxylic acid, 2-furanylmethyl ester

$C_{10}H_8O_4S$
7971 - 1-Naphthalenesulfonic acid, 4-hydroxy-
7972 - 1-Naphthalenesulfonic acid, 7-hydroxy-
7973 - 2-Naphthalenesulfonic acid, 1-hydroxy-
7974 - 2-Naphthalenesulfonic acid, 6-hydroxy-

$C_{10}H_8O_5$
2808 - 2H-1-Benzopyran-2-one, 7,8-dihydroxy-6-methoxy-

$C_{10}H_8O_6S_2$
7867 - 1,5-Naphthalenedisulfonic acid
7868 - 1,6-Naphthalenedisulfonic acid
7869 - 2,7-Naphthalenedisulfonic acid

$C_{10}H_8O_7S_2$
7871 - 1,3-Naphthalenedisulfonic acid, 7-hydroxy-
7876 - 2,7-Naphthalenedisulfonic acid, 3-hydroxy-

$C_{10}H_8O_8S_2$
7875 - 2,7-Naphthalenedisulfonic acid, 4,5-dihydroxy-

$C_{10}H_8S$
8007 - 1-Naphthalenethiol
8008 - 2-Naphthalenethiol

$C_{10}H_9BrO_2$
10615 - 2-Propenoic acid, 2-bromo-3-phenyl-, methyl ester

$C_{10}H_9ClN_4O_2S$
2193 - Benzenesulfonamide, 4-amino-*N*-(6-chloro-3-pyridazinyl)-

$C_{10}H_9ClO$
4957 - Cyclopropanecarbonyl chloride, 2-phenyl-, *trans*-

$C_{10}H_9ClO_2$
10622 - 2-Propenoic acid, 2-chloro-3-phenyl-, methyl ester, (*E*)-
10623 - 2-Propenoic acid, 3-chloro-3-phenyl-, methyl ester

$C_{10}H_9ClO_3$
972 - Benzeneacetyl chloride, α-(acetyloxy)-

$C_{10}H_9Cl_2NO$
10520 - 2-Propenamide, *N*-(3,4-dichlorophenyl)-2-methyl-

$C_{10}H_9Cl_4O_4P$
6202 - Gardona

$C_{10}H_9N$
7381 - Isoquinoline, 1-methyl-
7382 - Isoquinoline, 3-methyl-
7383 - Isoquinoline, 4-methyl-
7384 - Isoquinoline, 6-methyl-
7385 - Isoquinoline, 8-methyl-
7708 - 1-Naphthalenamine
7709 - 2-Naphthalenamine
11134 - 1H-Pyrrole, 1-phenyl-
11135 - 1H-Pyrrole, 2-phenyl-
11277 - Quinoline, 2-methyl-
11278 - Quinoline, 3-methyl-
11279 - Quinoline, 4-methyl-
11280 - Quinoline, 5-methyl-
11281 - Quinoline, 6-methyl-
11282 - Quinoline, 7-methyl-
11283 - Quinoline, 8-methyl-

$C_{10}H_9NO$
5819 - Ethanone, 1-(1H-indol-3-yl)-
7380 - Isoquinoline, 7-methoxy-
8019 - 2-Naphthalenol, 6-amino-
8020 - 2-Naphthalenol, 8-amino-
11274 - Quinoline, 4-methoxy-
11275 - Quinoline, 6-methoxy-
11276 - Quinoline, 8-methoxy-
11319 - 6-Quinolinol, 2-methyl-
11320 - 7-Quinolinol, 6-methyl-
11321 - 8-Quinolinol, 2-methyl-
11322 - 8-Quinolinol, 6-methyl-
11328 - 2(1H)-Quinolinone, 1-methyl-
11329 - 2(1H)-Quinolinone, 6-methyl-
11330 - 4(1H)-Quinolinone, 1-methyl-

$C_{10}H_9NO_2$
1877 - Benzenemethanol, α-ethynyl-, carbamate
7295 - 1H-Indole-3-acetic acid
7364 - 1H-Isoindole-1,3(2H)-dione, *N*-ethyl-
8582 - Oxazole, 5-methoxy-2-phenyl-
11162 - 2,5-Pyrrolidinedione, 1-phenyl-

$C_{10}H_9NO_3$
3853 - 2-Butenoic acid, 4-oxo-4-(phenylamino)-, (*Z*)-

$C_{10}H_9NO_3S$
7961 - 1-Naphthalenesulfonic acid, 2-amino-
7962 - 1-Naphthalenesulfonic acid, 4-amino-
7963 - 1-Naphthalenesulfonic acid, 5-amino-
7964 - 1-Naphthalenesulfonic acid, 7-amino-
7965 - 1-Naphthalenesulfonic acid, 8-amino-
7967 - 2-Naphthalenesulfonic acid, 8-amino-

$C_{10}H_9NO_4$
10677 - 2-Propenoic acid, 3-(2-nitrophenyl)-, methyl ester, (*E*)-
10678 - 2-Propenoic acid, 3-(3-nitrophenyl)-, methyl ester, (*E*)-
10679 - 2-Propenoic acid, 3-(4-nitrophenyl)-, methyl ester, (*E*)-

$C_{10}H_9NO_4S$
7966 - 1-Naphthalenesulfonic acid, 4-amino-3-hydroxy-

$C_{10}H_9NO_6S_2$
7870 - 1,3-Naphthalenedisulfonic acid, 7-amino-
7872 - 1,6-Naphthalenedisulfonic acid, 4-amino-
7873 - 1,7-Naphthalenedisulfonic acid, 4-amino-

$C_{10}H_9NO_7S_2$
7874 - 2,7-Naphthalenedisulfonic acid, 4-amino-5-hydroxy-

$C_{10}H_9N_3$
10871 - 2-Pyridinamine, *N*-2-pyridinyl-

$C_{10}H_{10}$
1097 - Benzene, 1,3-butadienyl-, (*E*)-
1098 - Benzene, 1,3-butadienyl-, (*Z*)-
1124 - Benzene, 3-butynyl-
1403 - Benzene, 1,2-diethenyl-
1404 - Benzene, 1,3-diethenyl-
1405 - Benzene, 1,4-diethenyl-
1967 - Benzene, (1-methylene-2-propenyl)-
7275 - 1H-Indene, 1-methyl-
7276 - 1H-Indene, 2-methyl-
7277 - 1H-Indene, 3-methyl-
7278 - 1H-Indene, 4-methyl-
7279 - 1H-Indene, 6-methyl-
7280 - 1H-Indene, 7-methyl-
7814 - Naphthalene, 1,2-dihydro-
7815 - Naphthalene, 1,4-dihydro-

$C_{10}H_{10}Cl_2O_3$
3925 - Butyrac 118

$C_{10}H_{10}Fe$
5960 - Ferrocene

$C_{10}H_{10}N_2$
7167 - Hydrazine, 1-naphthalenyl-
7795 - 1,2-Naphthalenediamine
7796 - 1,4-Naphthalenediamine
7797 - 1,5-Naphthalenediamine
7798 - 1,8-Naphthalenediamine
7799 - 2,3-Naphthalenediamine
10837 - 1H-Pyrazole, 1-methyl-3-phenyl-
10838 - 1H-Pyrazole, 3-methyl-1-phenyl-
10987 - Pyridine, 3-(1-methyl-1H-pyrrol-2-yl)-
11208 - 2-Quinolinamine, 4-methyl-
11209 - 3-Quinolinamine, 2-methyl-
11210 - 4-Quinolinamine, 2-methyl-
11211 - 7-Quinolinamine, 2-methyl-
11212 - 8-Quinolinamine, 6-methyl-
11335 - Quinoxaline, 2,6-dimethyl-

$C_{10}H_{10}N_2O$
10849 - 3H-Pyrazol-3-one, 1,2-dihydro-5-methyl-2-phenyl-

$C_{10}H_{10}N_2O_2$
7221 - 2,4-Imidazolidinedione, 3-methyl-5-phenyl-

$C_{10}H_{10}N_4O_2S$
2202 - Benzenesulfonamide, 4-amino-*N*-pyrazinyl-
2204 - Benzenesulfonamide, 4-amino-*N*-2-pyrimidinyl-

$C_{10}H_{10}O$
1878 - Benzenemethanol, α-ethynyl-α-methyl-
2445 - Benzofuran, 2,5-dimethyl-
2446 - Benzofuran, 2,6-dimethyl-
2447 - Benzofuran, 2,7-dimethyl-
2448 - Benzofuran, 3,5-dimethyl-
2449 - Benzofuran, 3,6-dimethyl-
2450 - Benzofuran, 4,6-dimethyl-
2451 - Benzofuran, 4,7-dimethyl-
2452 - Benzofuran, 5,6-dimethyl-
2453 - Benzofuran, 5,7-dimethyl-
2454 - Benzofuran, 6,7-dimethyl-
3701 - 2-Butenal, 3-phenyl-
3880 - 2-Buten-1-one, 1-phenyl-
3881 - 3-Buten-2-one, 4-phenyl-
3882 - 3-Buten-2-one, 4-phenyl-, (*E*)-
7289 - 1H-Inden-1-one, 2,3-dihydro-2-methyl-
7290 - 1H-Inden-1-one, 2,3-dihydro-4-methyl-
8046 - 1(2H)-Naphthalenone, 3,4-dihydro-
8047 - 2(1H)-Naphthalenone, 3,4-dihydro-
10516 - 2-Propenal, 2-methyl-3-phenyl-
10517 - 2-Propenal, 3-(4-methylphenyl)-

$C_{10}H_{10}O_2$
2118 - Benzenepropanal, α-methyl-β-oxo-
2429 - 1,3-Benzodioxole, 5-(2-propenyl)-
2430 - 1,3-Benzodioxole, 5-(1-propenyl)-, (*E*)-
2722 - Benzoic acid, 2-propenyl ester
2803 - 4H-1-Benzopyran-4-one, 2,3-dihydro-6-methyl-
3362 - 1,3-Butanedione, 1-phenyl-

3854 - 2-Butenoic acid, 3-phenyl-, (*Z*)-
3855 - 3-Butenoic acid, 4-phenyl-
5865 - Ethanone, 1,1′-(1,3-phenylene)bis-
10670 - 2-Propenoic acid, 2-methyl-3-phenyl-, (*Z*)-
10686 - 2-Propenoic acid, 3-phenyl-, methyl ester, (*E*)-
10688 - 2-Propenoic acid, phenylmethyl ester
10707 - 2-Propen-1-ol, 3-phenyl-, formate

$C_{10}H_{10}O_2S_2$
6101 - Furan, 2,2′-[dithiobis(methylene)]bis-

$C_{10}H_{10}O_3$
935 - Benzeneacetic acid, α-oxo-, ethyl ester
2142 - Benzenepropanoic acid, β-oxo-, methyl ester
2479 - Benzoic acid, 4-acetyl-, methyl ester
5771 - Ethanone, 2-(acetyloxy)-1-phenyl-
6156 - Furan, 2,2′-[oxybis(methylene)]bis-
10514 - 2-Propenal, 3-(4-hydroxy-3-methoxyphenyl)-

$C_{10}H_{10}O_4$
1343 - 1,2-Benzenedicarboxylic acid, dimethyl ester
1347 - 1,3-Benzenedicarboxylic acid, dimethyl ester
1348 - 1,3-Benzenedicarboxylic acid, 4,6-dimethyl-
1351 - 1,4-Benzenedicarboxylic acid, dimethyl ester
1508 - 1,2-Benzenediol, diacetate
1509 - 1,4-Benzenediol, diacetate
2416 - 1,3-Benzodioxole-5-acetic acid, methyl ester
2419 - 1,3-Benzodioxole-5-carboxylic acid, ethyl ester
2482 - Benzoic acid, 2-(acetyloxy)-, methyl ester
3341 - Butanedioic acid, phenyl-, (±)-
7355 - 1(3H)-Isobenzofuranone, 6,7-dimethoxy-

$C_{10}H_{10}O_5$
2608 - Benzoic acid, 6-formyl-2,3-dimethoxy-

$C_{10}H_{11}Br$
1001 - Benzene, (3-bromo-3-butenyl)-
1002 - Benzene, (4-bromo-2-butenyl)-
3719 - 2-Butene, 2-bromo-3-phenyl-

$C_{10}H_{11}Cl$
1152 - Benzene, 1-(1-chloroethenyl)-2,4-dimethyl-
1188 - Benzene, 1-(2-chloro-1-methylethenyl)-4-methyl-

$C_{10}H_{11}ClO$
974 - Benzeneacetyl chloride, α-ethyl-
1231 - Benzene, 1-(3-chloro-2-propenyl)-4-methoxy-
3650 - 1-Butanone, 4-chloro-1-phenyl-

$C_{10}H_{11}ClO_2$
891 - Benzeneacetic acid, α-chloro-, ethyl ester, (*S*)-

$C_{10}H_{11}ClO_2S$
7978 - 2-Naphthalenesulfonyl chloride, 5,6,7,8-tetrahydro-

$C_{10}H_{11}F_3N_2O$
12007 - Urea, *N,N*-dimethyl-*N′*-[3-(trifluoromethyl)phenyl]-

$C_{10}H_{11}F_7O_2$
8398 - 3,5-Octanedione, 6,6,7,7,8,8,8-heptafluoro-2,2-dimethyl-

$C_{10}H_{11}I_2NO_3$
10873 - 1(4H)-Pyridineacetic acid, 3,5-diiodo-4-oxo-, propyl ester

$C_{10}H_{11}N$
951 - Benzeneacetonitrile, α-ethyl-
7298 - 1H-Indole, 1,3-dimethyl-
7299 - 1H-Indole, 2,5-dimethyl-
7307 - 1H-Indole, 1-ethyl-
7308 - 1H-Indole, 3-ethyl-
7714 - 1-Naphthalenamine, 5,8-dihydro-

$C_{10}H_{11}NO$
3411 - Butanenitrile, 4-phenoxy-
7305 - 1H-Indole-3-ethanol
7328 - 2H-Indol-2-one, 1,3-dihydro-1,3-dimethyl-
7329 - 2H-Indol-2-one, 1,3-dihydro-3,3-dimethyl-
7330 - 2H-Indol-2-one, 1,3-dihydro-4,7-dimethyl-

$C_{10}H_{11}NO_2$
26 - Acetamide, *N*-acetyl-*N*-phenyl-
3207 - Butanamide, 3-oxo-*N*-phenyl-
6114 - 2-Furanmethanamine, *N*-(2-furanylmethyl)-

$C_{10}H_{11}NO_3$
224 - Acetic acid, oxo(phenylamino)-, ethyl ester
327 - *DL*-Alanine, *N*-benzoyl-
3575 - Butanoic acid, 4-oxo-4-(phenylamino)-

$C_{10}H_{11}N_3$
10813 - 1H-Pyrazol-4-amine, 3-methyl-1-phenyl-
10814 - 1H-Pyrazol-5-amine, 3-methyl-1-phenyl-

$C_{10}H_{11}N_3O_2S_2$
2200 - Benzenesulfonamide, 4-amino-*N*-(4-methyl-2-thiazolyl)-

$C_{10}H_{11}N_3O_3S$
2198 - Benzenesulfonamide, 4-amino-*N*-(5-methyl-3-isoxazolyl)-

$C_{10}H_{12}$
1109 - Benzene, 1-butenyl-, (*E*)-
1110 - Benzene, 1-butenyl-, (*Z*)-
1111 - Benzene, 2-butenyl-
1112 - Benzene, 3-butenyl-
1245 - Benzene, cyclobutyl-
1626 - Benzene, 1-ethenyl-2,4-dimethyl-
1627 - Benzene, 1-ethenyl-3,5-dimethyl-
1628 - Benzene, 2-ethenyl-1,4-dimethyl-
1629 - Benzene, 4-ethenyl-1,2-dimethyl-
1630 - Benzene, 1-ethenyl-2-ethyl-
1631 - Benzene, 1-ethenyl-3-ethyl-
1632 - Benzene, 1-ethenyl-4-ethyl-
1955 - Benzene, (1-methylcyclopropyl)-
1968 - Benzene, (1-methylenepropyl)-
1984 - Benzene, 1-methyl-2-(1-methylethenyl)-
1985 - Benzene, 1-methyl-3-(1-methylethenyl)-
1986 - Benzene, 1-methyl-4-(1-methylethenyl)-
2023 - Benzene, 1-methyl-2-(2-propenyl)-
2024 - Benzene, (2-methyl-1-propenyl)-
2025 - Benzene, (1-methyl-1-propenyl)-, (*E*)-
2029 - Benzene, (1-methyl-1-propenyl)-, (*Z*)-
4746 - 1,3,5,7-Cyclooctatetraene, 1-ethyl-
7255 - 1H-Indene, 2,3-dihydro-1-methyl-
7256 - 1H-Indene, 2,3-dihydro-2-methyl-
7257 - 1H-Indene, 2,3-dihydro-4-methyl-
7258 - 1H-Indene, 2,3-dihydro-5-methyl-
7577 - 4,7-Methano-1H-indene, 3a,4,7,7a-tetrahydro-, (3aα,4α,7α,7aα)-
7979 - Naphthalene, 1,2,3,4-tetrahydro-

$C_{10}H_{12}Br_2O$
9281 - Phenol, 2,4-dibromo-3-methyl-6-(1-methylethyl)-

$C_{10}H_{12}ClNO_2$
3939 - Carbamic acid, (3-chlorophenyl)-, 1-methylethyl ester

$C_{10}H_{12}ClNO_4$
10001 - 1,2-Propanediol, 3-(4-chlorophenoxy)-, 1-carbamate

$C_{10}H_{12}ClN_3O_3S$
11197 - 6-Quinazolinesulfonamide, 7-chloro-2-ethyl-1,2,3,4-tetrahydro-4-oxo-

$C_{10}H_{12}N_2$
2381 - 1H-Benzimidazole, 1-ethyl-2-methyl-
3105 - 2,3′-Bipyridine, 1,2,3,6-tetrahydro-, (*S*)-
7198 - 1H-Imidazole, 4,5-dihydro-2-(phenylmethyl)-
7302 - 1H-Indole-3-ethanamine

$C_{10}H_{12}N_2O_3$
1102 - Benzenebutanoic acid, α,2-diamino-γ-oxo-
11091 - 2,4,6(1H,3H,5H)-Pyrimidinetrione, 5,5-di-2-propenyl-

$C_{10}H_{12}N_2O_3S$
518 - Bentazon

$C_{10}H_{12}N_2O_4$
2282 - Benzene, 1,2,4,5-tetramethyl-3,6-dinitro-

$C_{10}H_{12}N_2O_5$
5183 - Dinoseb
9328 - Phenol, 2-(1,1-dimethylethyl)-4,6-dinitro-

$C_{10}H_{12}N_3O_3PS_2$
497 - Azinphos-methyl

$C_{10}H_{12}N_4OS$
34 - Acetamide, *N*-[4-[[(aminothioxomethyl)hydrazono]methyl]phenyl]-

$C_{10}H_{12}N_4O_5$
7335 - Inosine

$C_{10}H_{12}O$
590 - Benzaldehyde, 4-(1-methylethyl)-
599 - Benzaldehyde, 2,4,6-trimethyl-
1651 - Benzene, (2-ethoxyethenyl)-
1869 - Benzenemethanol, α-ethenyl-α-methyl-
1911 - Benzenemethanol, α-2-propenyl-
1941 - Benzene, 1-methoxy-2-(1-propenyl)-
1942 - Benzene, 1-methoxy-4-(1-propenyl)-
1943 - Benzene, 1-methoxy-4-(2-propenyl)-
1944 - Benzene, 1-methoxy-4-(1-propenyl)-, (*E*)-
2026 - Benzene, 1-methyl-2-(2-propenyloxy)-
2027 - Benzene, 1-methyl-3-(2-propenyloxy)-
2028 - Benzene, 1-methyl-4-(2-propenyloxy)-
2115 - Benzenepropanal, 2-methyl-
2116 - Benzenepropanal, 4-methyl-
2157 - Benzenepropanol, γ-methylene-
2778 - 2H-1-Benzopyran, 3,4-dihydro-2-methyl-
2779 - 2H-1-Benzopyran, 3,4-dihydro-3-methyl-
2780 - 2H-1-Benzopyran, 3,4-dihydro-6-methyl-
2781 - 2H-1-Benzopyran, 3,4-dihydro-7-methyl-
3676 - 1-Butanone, 1-phenyl-
3677 - 2-Butanone, 1-phenyl-
3678 - 2-Butanone, 4-phenyl-
3872 - 2-Buten-1-ol, 3-phenyl-
5809 - Ethanone, 1-(2,4-dimethylphenyl)-
5810 - Ethanone, 1-(2,5-dimethylphenyl)-
5811 - Ethanone, 1-(3,4-dimethylphenyl)-
8041 - 1-Naphthalenol, 1,2,3,4-tetrahydro-
8042 - 1-Naphthalenol, 5,6,7,8-tetrahydro-
8043 - 2-Naphthalenol, 1,2,3,4-tetrahydro-
8044 - 2-Naphthalenol, 5,6,7,8-tetrahydro-
9427 - Phenol, 2-methyl-6-(2-propenyl)-
9428 - Phenol, 4-methyl-2-(1-propenyl)-
9429 - Phenol, 4-methyl-2-(2-propenyl)-
10466 - 1-Propanone, 1-(2-methylphenyl)-
10467 - 1-Propanone, 1-(3-methylphenyl)-
10468 - 1-Propanone, 1-(4-methylphenyl)-
10469 - 1-Propanone, 2-methyl-1-phenyl-

$C_{10}H_{12}O_2$
232 - Acetic acid, 2-phenylethyl ester
897 - Benzeneacetic acid, α-ethyl-
899 - Benzeneacetic acid, ethyl ester
1100 - Benzenebutanoic acid
2137 - Benzenepropanoic acid, α-methyl-
2138 - Benzenepropanoic acid, β-methyl-, (±)-
2139 - Benzenepropanoic acid, methyl ester
2431 - 1,3-Benzodioxole, 5-propyl-
2589 - Benzoic acid, 2,4-dimethyl-, methyl ester
2688 - Benzoic acid, 2-(1-methylethyl)-
2689 - Benzoic acid, 4-(1-methylethyl)-
2690 - Benzoic acid, 2-methyl-, ethyl ester
2691 - Benzoic acid, 3-methyl-, ethyl ester
2692 - Benzoic acid, 4-methyl-, ethyl ester
2693 - Benzoic acid, 1-methylethyl ester
2723 - Benzoic acid, *o*-propyl-
2724 - Benzoic acid, propyl ester
3584 - Butanoic acid, phenyl ester
3665 - 1-Butanone, 1-(2-hydroxyphenyl)-
3666 - 1-Butanone, 2-hydroxy-1-phenyl-
3667 - 1-Butanone, 4-hydroxy-1-phenyl-
4241 - 1,3,6-Cycloheptatriene-1-carboxylic acid, 5,5-dimethyl-
4267 - 2,5-Cyclohexadiene-1,4-dione, 2-methyl-5-(1-methylethyl)-
4270 - 2,5-Cyclohexadiene-1,4-dione, 2,3,5,6-tetramethyl-
5195 - 1,3-Dioxane, 2-phenyl-
5196 - 1,3-Dioxane, 4-phenyl-
5814 - Ethanone, 2-ethoxy-1-phenyl-
9313 - Phenol, 2,4-dimethyl-, acetate
9314 - Phenol, 2,5-dimethyl-, acetate
9315 - Phenol, 3,4-dimethyl-, acetate
9316 - Phenol, 3,5-dimethyl-, acetate
9388 - Phenol, 2-methoxy-4-(1-propenyl)-
9389 - Phenol, 2-methoxy-4-(2-propenyl)-
9390 - Phenol, 2-methoxy-5-(2-propenyl)-
9391 - Phenol, 2-methoxy-6-(1-propenyl)-

9392 - Phenol, 2-methoxy-6-(2-propenyl)-
9396 - Phenol, 2-methoxy-4-(1-propenyl)-, (*E*)-
9397 - Phenol, 2-methoxy-4-(1-propenyl)-, (*Z*)-
10444 - 1-Propanone, 1-(2-hydroxy-3-methylphenyl)-
10445 - 1-Propanone, 1-(2-hydroxy-4-methylphenyl)-
10446 - 1-Propanone, 1-(2-hydroxy-5-methylphenyl)-
10456 - 1-Propanone, 1-(3-methoxyphenyl)-
10457 - 1-Propanone, 2-methoxy-1-phenyl-
10458 - 1-Propanone, 3-methoxy-1-phenyl-
10459 - 2-Propanone, 1-(methoxyphenyl)-
10460 - 2-Propanone, 1-(3-methoxyphenyl)-
10461 - 2-Propanone, 1-(4-methoxyphenyl)-

$C_{10}H_{12}O_3$
229 - Acetic acid, phenoxy-, ethyl ester
557 - Benzaldehyde, 4-ethoxy-3-methoxy-
869 - Benzeneacetaldehyde, 3,4-dimethoxy-
907 - Benzeneacetic acid, α-hydroxy-, ethyl ester, (±)-
909 - Benzeneacetic acid, α-hydroxy-, ethyl ester
1892 - Benzenemethanol, 4-methoxy-, acetate
2635 - Benzoic acid, 2-hydroxy-, 1-methylethyl ester
2652 - Benzoic acid, 2-hydroxy-, propyl ester
2653 - Benzoic acid, 4-hydroxy-, propyl ester
2669 - Benzoic acid, 2-methoxy-, ethyl ester
2670 - Benzoic acid, 3-methoxy-, ethyl ester
2671 - Benzoic acid, 4-methoxy-, ethyl ester
3580 - Butanoic acid, 2-phenoxy-
5209 - 1,3-Dioxolane-4-methanol, 2-phenyl-
5805 - Ethanone, 1-(3,4-dimethoxyphenyl)-
9370 - Phenol, 4-(3-hydroxy-1-propenyl)-2-methoxy-
10642 - 2-Propenoic acid, 3-(2-furanyl)-, propyl ester

$C_{10}H_{12}O_4$
2574 - Benzoic acid, 2,4-dihydroxy-6-methyl-, ethyl ester
2579 - Benzoic acid, 3,4-dimethoxy-, methyl ester
2626 - Benzoic acid, 4-hydroxy-3-methoxy-, ethyl ester
3781 - 2-Butenedioic acid (*E*)-, di-2-propenyl ester
3782 - 2-Butenedioic acid (*Z*)-, di-2-propenyl ester
5371 - 4,7-Epoxyisobenzofuran-1,3-dione, hexahydro-3a,7a-dimethyl-, (3aα,4β,7β,7aα)-

$C_{10}H_{12}O_5$
2732 - Benzoic acid, 3,4,5-trihydroxy-, propyl ester
2733 - Benzoic acid, 2,4,5-trimethoxy-
2734 - Benzoic acid, 3,4,5-trimethoxy-
6082 - 3,4-Furandicarboxylic acid, diethyl ester

$C_{10}H_{13}Br$
1023 - Benzene, 1-bromo-4-(1,1-dimethylethyl)-
1065 - Benzene, 2-bromo-1-methyl-4-(1-methylethyl)-

$C_{10}H_{13}BrN_2O_3$
11078 - 2,4,6(1H,3H,5H)-Pyrimidinetrione, 5-(2-bromo-2-propenyl)-5-(1-methylethyl)-

$C_{10}H_{13}BrO$
1003 - Benzene, (4-bromobutoxy)-

$C_{10}H_{13}Cl$
1142 - Benzene, 1-chloro-4-(1,1-dimethylethyl)-
1143 - Benzene, (2-chloro-1,1-dimethylethyl)-
1201 - Benzene, 2-chloro-1-methyl-4-(1-methylethyl)-
1202 - Benzene, 2-chloro-4-methyl-1-(1-methylethyl)-

$C_{10}H_{13}ClN_2$
7564 - Methanimidamide, *N'*-(4-chloro-2-methylphenyl)-*N,N*-dimethyl-

$C_{10}H_{13}ClN_2O_3S$
2208 - Benzenesulfonamide, 4-chloro-*N*-[(propylamino)carbonyl]-

$C_{10}H_{13}ClN_6$
9759 - Procyazine

$C_{10}H_{13}ClO$
1598 - Benzeneethanol, 4-chloro-α,α-dimethyl-
9268 - Phenol, 4-chloro-5-methyl-2-(1-methylethyl)-

$C_{10}H_{13}Cl_2O_3PS$
9581 - Phosphorothioic acid, *O*-(2,4-dichlorophenyl) *O,O*-diethyl ester

$C_{10}H_{13}I$
1781 - Benzene, 2-iodo-1-methyl-4-(1-methylethyl)-
1782 - Benzene, 2-iodo-4-methyl-1-(1-methylethyl)-

$C_{10}H_{13}N$
7731 - 1-Naphthalenamine, 5,6,7,8-tetrahydro-
10996 - Pyridine, 2-(3-pentenyl)-
11171 - Pyrrolidine, 1-phenyl-
11299 - Quinoline, 1,2,3,4-tetrahydro-1-methyl-

$C_{10}H_{13}NO$
56 - Acetamide, *N*-(2,4-dimethylphenyl)-
62 - Acetamide, *N*-ethyl-*N*-phenyl-
75 - Acetamide, *N*-methyl-*N*-(2-methylphenyl)-
627 - Benzamide, *N,N*,4-trimethyl-
881 - Benzeneacetamide, α-ethyl-
3208 - Butanamide, *N*-phenyl-
5806 - Ethanone, 1-[3-(dimethylamino)phenyl]-
11298 - Quinoline, 1,2,3,4-tetrahydro-6-methoxy-

$C_{10}H_{13}NO_2$
59 - Acetamide, *N*-(2-ethoxyphenyl)-
60 - Acetamide, *N*-(4-ethoxyphenyl)-
1467 - Benzene, 1-(1,1-dimethylethyl)-3-nitro-
1991 - Benzene, 1-methyl-4-(1-methylethyl)-2-nitro-
2128 - Benzenepropanoic acid, β-(aminomethyl)-
2149 - Benzenepropanol, carbamate
2507 - Benzoic acid, 4-amino-, propyl ester
2680 - Benzoic acid, 2-(methylamino)-, ethyl ester
3205 - Butanamide, *N*-(4-hydroxyphenyl)-
3966 - Carbamic acid, phenyl-, 1-methylethyl ester
3967 - Carbamic acid, (phenylmethyl)-, ethyl ester
6271 - Glycine, *N*-phenyl-, ethyl ester
10898 - 2-Pyridinecarboxylic acid, 5-butyl-

$C_{10}H_{13}NO_3$
2034 - Benzene, 1-(2-methylpropoxy)-2-nitro-
2672 - Benzoic acid, 3-methoxy-2-(methylamino)-, methyl ester
11933 - *L*-Tyrosine, *N*-methyl-

$C_{10}H_{13}NO_4$
10188 - 1,2,3-Propanetriol, 1-(4-aminobenzoate)
11931 - *L*-Tyrosine, 3-hydroxy-α-methyl-

$C_{10}H_{13}N_3O_4$
3894 - *N*-*sec*-Butyl-2,6-dinitrobenzene

$C_{10}H_{13}N_5O_4$
319 - Adenosine

$C_{10}H_{13}N_5O_5$
6289 - Guanosine

$C_{10}H_{14}$
1118 - Benzene, butyl-
1413 - Benzene, 1,2-diethyl-
1414 - Benzene, 1,3-diethyl-
1415 - Benzene, 1,4-diethyl-
1459 - Benzene, (1,1-dimethylethyl)-
1670 - Benzene, 1-ethyl-2,3-dimethyl-
1671 - Benzene, 1-ethyl-2,4-dimethyl-
1672 - Benzene, 1-ethyl-3,5-dimethyl-
1673 - Benzene, 2-ethyl-1,3-dimethyl-
1674 - Benzene, 2-ethyl-1,4-dimethyl-
1675 - Benzene, 4-ethyl-1,2-dimethyl-
1988 - Benzene, 1-methyl-2-(1-methylethyl)-
1989 - Benzene, 1-methyl-3-(1-methylethyl)-
1990 - Benzene, 1-methyl-4-(1-methylethyl)-
2035 - Benzene, 1-methylpropyl
2036 - Benzene, 1-methyl-2-propyl-
2037 - Benzene, 1-methyl-3-propyl-
2038 - Benzene, 1-methyl-4-propyl-
2039 - Benzene, (2-methylpropyl)-
2278 - Benzene, 1,2,3,4-tetramethyl-
2279 - Benzene, 1,2,3,5-tetramethyl-
2280 - Benzene, 1,2,4,5-tetramethyl-
7897 - Naphthalene, 1,2,3,4,4a,8a-hexahydro-
8340 - 2,6-Octadien-4-yne, 3,6-dimethyl-
8345 - 3,5-Octadiyne, 2,7-dimethyl-

$C_{10}H_{14}BeO_4$
2912 - Beryllium, bis(2,4-pentanedionato-*O,O'*)-, (T-4)-

$C_{10}H_{14}BrN$
642 - Benzenamine, 4-bromo-*N,N*-diethyl-

$C_{10}H_{14}Br_2O$
2953 - Bicyclo[2.2.1]heptan-2-one, 3,3-dibromo-1,7,7-trimethyl-, (1*R*)-

$C_{10}H_{14}ClN$
661 - Benzenamine, 4-chloro-*N,N*-diethyl-

$C_{10}H_{14}Cl_6N_4O_2$
11902 - Triforine

$C_{10}H_{14}NO_5PS$
9583 - Phosphorothioic acid, *O,O*-diethyl *O*-(4-nitrophenyl) ester

$C_{10}H_{14}N_2$
3189 - Butanal, phenylhydrazone
9620 - Piperazine, 1-phenyl-
10992 - Pyridine, 3-(1-methyl-2-pyrrolidinyl)-, (*S*)-
11008 - Pyridine, 3-(2-piperidinyl)-, (*S*)-
11009 - Pyridine, 4-(4-piperidinyl)-

$C_{10}H_{14}N_2O$
699 - Benzenamine, *N,N*-diethyl-4-nitroso-
10891 - 3-Pyridinecarboxamide, *N,N*-diethyl-
10892 - 4-Pyridinecarboxamide, *N,N*-diethyl-

$C_{10}H_{14}N_2O_2$
698 - Benzenamine, *N,N*-diethyl-4-nitro-

$C_{10}H_{14}N_2O_2S$
11064 - 4,6(1*H*,5*H*)-Pyrimidinedione, 5-ethyldihydro-5-(2-methyl-2-propenyl)-2-thioxo-

$C_{10}H_{14}N_2O_3$
11098 - 2,4,6(1H,3H,5H)-Pyrimidinetrione, 5-(1-methylethyl)-5-(2-propenyl)-

$C_{10}H_{14}N_2O_5$
11816 - Thymidine

$C_{10}H_{14}N_4O_4$
10761 - 1H-Purine-2,6-dione, 7-(2,3-dihydroxypropyl)-3,7-dihydro-1,3-dimethyl-

$C_{10}H_{14}N_5O_7P$
320 - 5'-Adenylic acid

$C_{10}H_{14}O$
1114 - Benzene, butoxy-
1458 - Benzene, (1,1-dimethylethoxy)-
1600 - Benzeneethanol, α,α-dimethyl-
1602 - Benzeneethanol, β-ethyl-
1603 - Benzeneethanol, 2-ethyl-
1604 - Benzeneethanol, 4-ethyl-
1653 - Benzene, 1-ethoxy-4-ethyl-
1874 - Benzenemethanol, α-ethyl-α-methyl-, (*R*)-
1901 - Benzenemethanol, α-(1-methylethyl)-
1902 - Benzenemethanol, 4-(1-methylethyl)-
1912 - Benzenemethanol, α-propyl-, (*R*)-
1945 - Benzene, 1-methoxy-4-propyl-
1987 - Benzene, 1-methyl-3-(1-methylethoxy)-
2030 - Benzene, (1-methylpropoxy)-
2031 - Benzene, (2-methylpropoxy)-
2032 - Benzene, 1-methyl-4-propoxy-
2155 - Benzenepropanol, α-methyl-
2156 - Benzenepropanol, γ-methyl-
2465 - Benzofuran, 4,5,6,7-tetrahydro-3,6-dimethyl-
2956 - Bicyclo[3.1.1]heptan-3-one, 6,6-dimethyl-2-methylene-, (1*S*)-
2977 - Bicyclo[3.1.1]hept-2-en-6-one, 2,7,7-trimethyl-
2978 - Bicyclo[3.1.1]hept-3-en-2-one, 4,6,6-trimethyl-, (1*R*)-
2987 - Bicyclo[3.1.0]hex-3-en-2-one, 4-methyl-1-(1-methylethyl)-
4220 - 2,4-Cycloheptadien-1-one, 2,6,6-trimethyl-
4250 - 1,3-Cyclohexadiene-1-carboxaldehyde, 2,6,6-trimethyl-
4633 - 1-Cyclohexene-1-carboxaldehyde, 4-(1-methylethenyl)-
4634 - 1-Cyclohexene-1-carboxaldehyde, 4-(1-methylethenyl)-, (*R*)-
4715 - 2-Cyclohexen-1-one, 2-methyl-5-(1-methylethenyl)-, (*S*)-
4718 - 2-Cyclohexen-1-one, 3-methyl-6-(1-methylethylidene)-
4881 - Cyclopentanone, 2-cyclopentylidene-
6133 - Furan, 3-(4-methyl-3-pentenyl)-
9246 - Phenol, 2-butyl-
9247 - Phenol, 3-butyl-
9248 - Phenol, 4-butyl-
9321 - Phenol, 2-(1,1-dimethylethyl)-

9322 - Phenol, 3-(1,1-dimethylethyl)-
9323 - Phenol, 4-(1,1-dimethylethyl)-
9414 - Phenol, 2-methyl-4-(1-methylethyl)-
9415 - Phenol, 2-methyl-5-(1-methylethyl)-
9416 - Phenol, 2-methyl-6-(1-methylethyl)-
9417 - Phenol, 4-methyl-2-(1-methylethyl)-
9418 - Phenol, 5-methyl-2-(1-methylethyl)-
9430 - Phenol, 2-(1-methylpropyl)-
9431 - Phenol, 4-(1-methylpropyl)-
9432 - Phenol, 4-(2-methylpropyl)-
9473 - Phenol, 2,3,4,5-tetramethyl-
9474 - Phenol, 2,3,4,6-tetramethyl-
9475 - Phenol, 2,3,5,6-tetramethyl-

$C_{10}H_{14}O_2$
1406 - Benzene, 1,2-diethoxy-
1407 - Benzene, 1,3-diethoxy-
1408 - Benzene, 1,4-diethoxy-
1516 - 1,2-Benzenediol, 4-(1,1-dimethylethyl)-
1536 - 1,2-Benzenediol, 3-methyl-6-(1-methylethyl)-
1538 - 1,4-Benzenediol, 2-methyl-5-(1-methylethyl)-
1539 - 1,3-Benzenediol, 4-(2-methylpropyl)-
1552 - 1,4-Benzenediol, 2,3,5,6-tetramethyl-
1652 - Benzene, (2-ethoxyethoxy)-
1873 - Benzenemethanol, α-ethyl-4-methoxy-
2154 - Benzenepropanol, 4-methoxy-
2926 - Bicyclo[2.2.1]heptane-2,3-dione, 1,7,7-trimethyl-, (1*R*)-
3356 - 1,2-Butanediol, 2-phenyl-
3357 - 1,3-Butanediol, 1-phenyl-
3358 - 1,3-Butanediol, 3-phenyl-
4758 - Cyclopenta[c]pyran-1(4aH)-one, 5,6,7,7a-tetrahydro-4,7-dimethyl-
6076 - Furan, 2-(cyclohexyloxy)-
9245 - Phenol, 2-butoxy-
10400 - 1-Propanol, 3-(phenylmethoxy)-
11862 - Tricyclo[2.2.1.02,6]heptane-1-carboxylic acid, 7,7-dimethyl-

$C_{10}H_{14}O_3$
1599 - Benzeneethanol, 2,5-dimethoxy-
2153 - Benzenepropanol, 4-hydroxy-3-methoxy-
6027 - 2-Furanacetic acid, butyl ester
6068 - 2-Furancarboxylic acid, pentyl ester
8559 - 3-Oxabicyclo[3.2.1]octane-2,4-dione, 1,8,8-trimethyl-, (±)-
8880 - Pentanoic acid, 2-furanylmethyl ester
10019 - 1,2-Propanediol, 3-(2-methylphenoxy)-

$C_{10}H_{14}O_3S$
2219 - Benzenesulfonic acid, *sec*-butyl ester
2233 - Benzenesulfonic acid, 4-methyl-, propyl ester

$C_{10}H_{14}O_4$
1914 - Benzenemethanol, 3,4,5-trimethoxy-
4646 - 1-Cyclohexene-1,4-dicarboxylic acid, dimethyl ester
10016 - 1,2-Propanediol, 3-(2-methoxyphenoxy)-
10656 - 2-Propenoic acid, 2-methyl-, 1,2-ethanediyl ester

$C_{10}H_{15}BrO$
2951 - Bicyclo[2.2.1]heptan-2-one, 1-(bromomethyl)-7,7-dimethyl-, (1*S*)-
2952 - Bicyclo[2.2.1]heptan-2-one, 3-bromo-1,7,7-trimethyl-, (1*R-endo*)-

$C_{10}H_{15}N$
655 - Benzenamine, 4-butyl-
656 - Benzenamine, *N*-butyl-
694 - Benzenamine, 2,6-diethyl-
695 - Benzenamine, *N,N*-diethyl-
714 - Benzenamine, 2-(1,1-dimethylethyl)-
715 - Benzenamine, 4-(1,1-dimethylethyl)-
716 - Benzenamine, *N*-(1,1-dimethylethyl)-
793 - Benzenamine, 2-methyl-5-(1-methylethyl)-
794 - Benzenamine, 4-methyl-*N*-(1-methylethyl)-
812 - Benzenamine, 2-(1-methylpropyl)-
813 - Benzenamine, 4-(1-methylpropyl)-
814 - Benzenamine, 4-(2-methylpropyl)-
815 - Benzenamine, 4-methyl-*N*-propyl-
816 - Benzenamine, *N*-(1-methylpropyl)-
817 - Benzenamine, *N*-(2-methylpropyl)-
847 - Benzenamine, 2,3,4,5-tetramethyl-
848 - Benzenamine, 2,3,4,6-tetramethyl-
849 - Benzenamine, *N,N*,2,6-tetramethyl-
1572 - Benzeneethanamine, α,α-dimethyl-
1573 - Benzeneethanamine, *N*,α-dimethyl-, (*S*)-
1574 - Benzeneethanamine, *N*,β-dimethyl-
1819 - Benzenemethanamine, 4-(1-methylethyl)-
1820 - Benzenemethanamine, *N*-(1-methylethyl)-
1830 - Benzenemethanamine, α-propyl-
2119 - Benzenepropanamine, α-methyl-
10967 - Pyridine, 2-(1-ethylpropyl)-
10968 - Pyridine, 4-(1-ethylpropyl)-

$C_{10}H_{15}NO$
1840 - Benzenemethanol, α-(1-aminopropyl)-
1900 - Benzenemethanol, α-[1-(methylamino)ethyl]-, (*R**,*S**)-(±)-
9297 - Phenol, 3-(diethylamino)-
9320 - Phenol, 4-[2-(dimethylamino)ethyl]-
9403 - Phenol, 4-[2-(methylamino)propyl]-
10983 - Pyridine, 2-[2-(1-methylethoxy)ethyl]-

$C_{10}H_{15}NO_2$
693 - Benzenamine, 3,4-diethoxy-
1571 - Benzeneethanamine, 3,4-dimethoxy-
5760 - Ethanol, 2,2′-(phenylimino)bis-

$C_{10}H_{15}NO_4$
9933 - Propanedioic acid, (2-cyanoethyl)-, diethyl ester
11144 - 3-Pyrrolidineacetic acid, 2-carboxy-4-(1-methylethenyl)-, [2*S*-(2α,3β,4β)]-

$C_{10}H_{15}N_3$
6281 - Guanidine, *N,N′*-dimethyl-*N″*-(phenylmethyl)-

$C_{10}H_{15}OPS_2$
5986 - Fonofos

$C_{10}H_{15}O_3PS_2$
5955 - Fenthion

$C_{10}H_{16}$
2921 - Bicyclo[2.2.1]heptane, 2,2-dimethyl-3-methylene-, (±)-
2922 - Bicyclo[2.2.1]heptane, 2,2-dimethyl-5-methylene-
2923 - Bicyclo[2.2.1]heptane, 7,7-dimethyl-2-methylene-, (±)-
2924 - Bicyclo[3.1.1]heptane, 6,6-dimethyl-2-methylene-, (1*R*)-
2925 - Bicyclo[3.1.1]heptane, 6,6-dimethyl-2-methylene-
2970 - Bicyclo[2.2.1]hept-2-ene, 1,7,7-trimethyl-
2971 - Bicyclo[3.1.1]hept-2-ene, 2,6,6-trimethyl-, (1*S*)-
2972 - Bicyclo[3.1.1]hept-2-ene, 2,6,6-trimethyl-, (±)-
2973 - Bicyclo[3.1.1]hept-2-ene, 2,6,6-trimethyl
2974 - Bicyclo[4.1.0]hept-3-ene, 3,7,7-trimethyl-, (1*S*)-
2981 - Bicyclo[3.1.0]hexane, 4-methylene-1-(1-methylethyl)-, (±)-
2986 - Bicyclo[3.1.0]hex-2-ene, 2-methyl-5-(1-methylethyl)-
4050 - 4-Carene, (1*S*,3*R*,6*R*)-(-)-
4209 - Cyclodecyne
4273 - 1,3-Cyclohexadiene, 1-methyl-4-(1-methylethyl)-
4274 - 1,3-Cyclohexadiene, 2-methyl-5-(1-methylethyl)-, (±)-
4275 - 1,4-Cyclohexadiene, 1-methyl-4-(1-methylethyl)-
4673 - Cyclohexene, 3-methylene-6-(1-methylethyl)-
4674 - Cyclohexene, 3-methylene-6-(1-methylethyl)-, (+)-
4675 - Cyclohexene, 4-methylene-1-(1-methylethyl)-
4676 - Cyclohexene, 1-methyl-3-(1-methylethenyl)-
4677 - Cyclohexene, 1-methyl-4-(1-methylethenyl)-
4678 - Cyclohexene, 1-methyl-5-(1-methylethenyl)-
4679 - Cyclohexene, 5-methyl-1-(1-methylethenyl)-, (±)-
4684 - Cyclohexene, 1-methyl-4-(1-methylethylidene)-
5094 - 1-Decen-3-yne
5095 - 1-Decen-4-yne
5096 - 2-Decen-4-yne
8330 - 1,6-Octadiene, 7-methyl-3-methylene-
8485 - 1,3,6-Octatriene, 3,7-dimethyl-
8486 - 1,3,7-Octatriene, 3,7-dimethyl-
8487 - 2,4,6-Octatriene, 2,6-dimethyl-, (*E,E*)-
8488 - 2,4,6-Octatriene, 2,6-dimethyl-, (*E,Z*)-
11815 - 4(10)-Thujene, (+)-
11860 - Tricyclo[3.3.1.13,7]decane
11866 - Tricyclo[2.2.1.02,6]heptane, 1,3,3-trimethyl-
11867 - Tricyclo[2.2.1.02,6]heptane, 1,7,7-trimethyl-

$C_{10}H_{16}ClNO$
865 - Benzenaminium, *N*-ethyl-3-hydroxy-*N,N*-dimethyl-, chloride

$C_{10}H_{16}Cl_2O_2$
5049 - Decanedioyl dichloride

$C_{10}H_{16}Cl_3NOS$
11823 - Triallate

$C_{10}H_{16}NO_5PS_2$
5946 - Famphur

$C_{10}H_{16}N_2$
1265 - 1,3-Benzenediamine, *N,N*-diethyl-
1266 - 1,4-Benzenediamine, *N,N*-diethyl-
1286 - 1,2-Benzenediamine, *N,N,N′,N′*-tetramethyl-
1287 - 1,4-Benzenediamine, *N,N,N′,N′*-tetramethyl-
5041 - Decanedinitrile

$C_{10}H_{16}N_2OS$
7229 - 4-Imidazolidinone, 5-(2-methylpropyl)-3-(2-propenyl)-2-thioxo-

$C_{10}H_{16}N_2O_2S$
11065 - 4,6(1H,5H)-Pyrimidinedione, 5-ethyldihydro-5-(1-methylpropyl)-2-thioxo-

$C_{10}H_{16}N_2O_3$
11082 - 2,4,6(1H,3H,5H)-Pyrimidinetrione, 5-butyl-5-ethyl-

$C_{10}H_{16}N_2O_3S$
11689 - 1H-Thieno[3,4-d]imidazole-4-pentanoic acid, hexahydro-2-oxo-, [3a*S*-(3aα,4β,6aα)]-

$C_{10}H_{16}N_2O_8$
6262 - Glycine, *N,N′*-1,2-ethanediylbis[*N*-(carboxymethyl)-

$C_{10}H_{16}N_4O_7$
11107 - 4(1H)-Pyrimidinone, 2,6-diamino-5-(β-*D*-glucopyranosyloxy)-

$C_{10}H_{16}O$
2960 - Bicyclo[2.2.1]heptan-2-one, 1,3,3-trimethyl-, (±)-
2961 - Bicyclo[2.2.1]heptan-2-one, 1,7,7-trimethyl-, (1*R*)-
2963 - Bicyclo[2.2.1]heptan-2-one, 4,7,7-trimethyl-, (1*S*)-
2964 - Bicyclo[3.1.1]heptan-2-one, 4,6,6-trimethyl-, [1*R*-(1α,4α,5α)]-
2965 - Bicyclo[3.1.1]heptan-3-one, 2,6,6-trimethyl-, [1*S*-(1α,2β,5α)]-
2975 - Bicyclo[3.1.1]hept-3-en-2-ol, 4,6,6-trimethyl-, (1α,2β,5α)-
2976 - Bicyclo[3.1.1]hept-3-en-2-ol, 4,6,6-trimethyl-, (1α,2α,5α)-
2983 - Bicyclo[3.1.0]hexan-3-ol, 4-methylene-1-(1-methylethyl)-
2984 - Bicyclo[3.1.0]hexan-3-ol, 4-methylene-1-(1-methylethyl)-, [1*S*-(1α,3β,5α)]-
2995 - [1,1′-Bicyclopentyl]-2-one
4576 - Cyclohexanone, 2-butylidene-
4601 - Cyclohexanone, 2-methyl-5-(1-methylethenyl)-, (2*S-trans*)-
4602 - Cyclohexanone, 5-methyl-2-(1-methylethenyl)-
4603 - Cyclohexanone, 5-methyl-2-(1-methylethenyl)-, *trans*-
4607 - Cyclohexanone, 5-methyl-2-(1-methylethylidene)-
4608 - Cyclohexanone, 5-methyl-2-(1-methylethylidene)-, (*R*)-
4635 - 1-Cyclohexene-1-carboxaldehyde, 4-(1-methylethyl)-

4636 - 1-Cyclohexene-1-carboxaldehyde, 2,6,6-trimethyl-
4664 - 1-Cyclohexene-1-methanol, 4-(1-methylethenyl)-
4696 - 2-Cyclohexen-1-ol, 2-methyl-5-(1-methylethenyl)-
4716 - 2-Cyclohexen-1-one, 3-methyl-6-(1-methylethyl)-, (±)-
4717 - 2-Cyclohexen-1-one, 6-methyl-3-(1-methylethyl)-, (±)-
8053 - 1(2H)-Naphthalenone, octahydro-, *cis-*
8054 - 1(2H)-Naphthalenone, octahydro-, *trans-*
8055 - 2(1H)-Naphthalenone, octahydro-
8317 - 2,6-Octadienal, 3,7-dimethyl-, (*E*)-
8318 - 2,6-Octadienal, 3,7-dimethyl-, (*Z*)-
8561 - 6-Oxabicyclo[3.2.1]oct-3-ene, 4,7,7-trimethyl-, (±)-

$C_{10}H_{16}O_2$
2957 - Bicyclo[2.2.1]heptan-2-one, 3-hydroxy-1,7,7-trimethyl-
3834 - 2-Butenoic acid, 3-cyclohexyl-
4203 - 1,2-Cyclodecanedione
4307 - Δ1,α-Cyclohexaneacetic acid, 3,3-dimethyl-
4709 - 2-Cyclohexen-1-one, 2-hydroxy-3-methyl-6-(1-methylethyl)-
4911 - 1-Cyclopentene-1-carboxylic acid, 2,3,3-trimethyl-, methyl ester
4959 - Cyclopropanecarboxylic acid, 2,2-dimethyl-3-(2-methyl-1-propenyl)-, (1*R-cis*)-
5025 - 2,4-Decadienoic acid, (*E,E*)-
5184 - 2,3-Dioxabicyclo[2.2.2]oct-5-ene, 1-methyl-4-(1-methylethyl)-
8331 - 2,4-Octadienoic acid, 3,7-dimethyl-
10653 - 2-Propenoic acid, 2-methyl-, cyclohexyl ester

$C_{10}H_{16}O_4$
3766 - 2-Butenedioic acid (*E*)-, bis(1-methylethyl) ester
3783 - 2-Butenedioic acid (*E*)-, dipropyl ester
3784 - 2-Butenedioic acid (*Z*)-, dipropyl ester
3931 - Camphoric acid, *trans*-(±)-
4180 - 1,1-Cyclobutanedicarboxylic acid, diethyl ester
4184 - 1,3-Cyclobutanedicarboxylic acid, 2,2-dimethyl-, dimethyl ester
4352 - 1,3-Cyclohexanedicarboxylic acid, dimethyl ester, *cis-*
4353 - 1,3-Cyclohexanedicarboxylic acid, dimethyl ester, *trans-*
4357 - 1,4-Cyclohexanedicarboxylic acid, dimethyl ester, *cis-*
4794 - 1,3-Cyclopentanedicarboxylic acid, 1,2,2-trimethyl-, (1*R-cis*)-
9966 - Propanedioic acid, (1-methylethylidene)-, diethyl ester
9980 - Propanedioic acid, 2-propenyl-, diethyl ester

$C_{10}H_{16}O_4S$
2929 - Bicyclo[2.2.1]heptane-1-methanesulfonic acid, 7,7-dimethyl-2-oxo-, (1*S*)-

$C_{10}H_{16}O_5$
3295 - Butanedioic acid, acetyl-, diethyl ester
9947 - Propanedioic acid, (ethoxymethylene)-, diethyl ester

$C_{10}H_{16}O_6$
7552 - Methanetricarboxylic acid, triethyl ester

$C_{10}H_{16}Si$
11490 - Silane, trimethyl(4-methylphenyl)-
11494 - Silane, trimethyl(phenylmethyl)-

$C_{10}H_{17}Br$
8321 - 2,6-Octadiene, 1-bromo-3,7-dimethyl-, (*E*)-

$C_{10}H_{17}Cl$
2919 - Bicyclo[2.2.1]heptane, 2-chloro-1,7,7-trimethyl-
2920 - Bicyclo[2.2.1]heptane, 2-chloro-1,7,7-trimethyl-, *endo-*
7782 - Naphthalene, 2-chlorodecahydro-

$C_{10}H_{17}ClO_4$
9932 - Propanedioic acid, (3-chloropropyl)-, diethyl ester

$C_{10}H_{17}Cl_2NOS$
5106 - Diallate

$C_{10}H_{17}N$
11859 - Tricyclo[3.3.1.13,7]decan-1-amine

$C_{10}H_{17}NO$
2962 - Bicyclo[2.2.1]heptan-2-one, 1,7,7-trimethyl-, oxime, (1*S*)-

$C_{10}H_{17}NO_6$
6258 - Glycine, *N,N*-bis(ethoxycarbonyl)-, ethyl ester
10104 - Propanenitrile, 2-(β-*D*-glucopyranosyloxy)-2-methyl-

$C_{10}H_{17}N_2O_4PS$
5945 - Etrimfos

$C_{10}H_{17}N_3$
11848 - 4H-1,2,4-Triazole, 4-cyclohexyl-3-ethyl-

$C_{10}H_{17}N_3O_6S$
6264 - Glycine, *N*-(*N-L*-γ-glutamyl-*L*-cysteinyl)-

$C_{10}H_{18}$
2931 - Bicyclo[2.2.1]heptane, 1,3,3-trimethyl-, (+)-
2932 - Bicyclo[2.2.1]heptane, 1,7,7-trimethyl-
2933 - Bicyclo[2.2.1]heptane, 2,2,3-trimethyl-
2934 - Bicyclo[3.1.1]heptane, 2,6,6-trimethyl-, (1α,2α,5α)-(±)-
2935 - Bicyclo[3.1.1]heptane, 2,6,6-trimethyl-, [1*R*-(1α,2β,5α)]-
2936 - Bicyclo[4.1.0]heptane, 3,7,7-trimethyl-, [1*S*-(1α,3β,6α)]-
2937 - Bicyclo[4.1.0]heptane, 3,7,7-trimethyl-, [1*S*-(1α,3α,6α)]-
2982 - Bicyclo[3.1.0]hexane, 4-methyl-1-(1-methylethyl)-
4207 - Cyclodecene, (*E*)-
4208 - Cyclodecene, (*Z*)-
4247 - Cycloheptene, 1,4,4-trimethyl-
4435 - Cyclohexane, 1-methyl-3-(1-methylethenyl)-
4441 - Cyclohexane, 1-methyl-3-(1-methylethylidene)-
4442 - Cyclohexane, 1-methyl-4-(1-methylethylidene)-
4680 - Cyclohexene, 1-methyl-4-(1-methylethyl)-
4681 - Cyclohexene, 1-methyl-4-(1-methylethyl)-, (*R*)-
4682 - Cyclohexene, 3-methyl-6-(1-methylethyl)-, (3*R-trans*)-
4683 - Cyclohexene, 4-methyl-1-(1-methylethyl)-, (*R*)-
4923 - Cyclopentene, 1,5-dimethyl-4-(1-methylethyl)-
4934 - Cyclopentene, 1-pentyl-
5023 - 1,3-Decadiene
5024 - 1,9-Decadiene
5097 - 1-Decyne
5098 - 3-Decyne
5099 - 4-Decyne
5100 - 5-Decyne
6619 - 3-Heptyne, 5-ethyl-5-methyl-
7429 - *p*-Menth-3-ene, (*S*)-(-)-
7789 - Naphthalene, decahydro-
7790 - Naphthalene, decahydro-, *cis-*
7791 - Naphthalene, decahydro-, *trans-*
8202 - 4-Nonyne, 8-methyl-
8322 - 2,4-Octadiene, 3,7-dimethyl-
8323 - 2,5-Octadiene, 2,6-dimethyl-
8324 - 2,6-Octadiene, 2,6-dimethyl-, (*Z*)-
8325 - 2,6-Octadiene, 2,7-dimethyl-
8326 - 2,6-Octadiene, 3,6-dimethyl-
8327 - 2,6-Octadiene, 4,5-dimethyl-
8535 - 3-Octyne, 2,2-dimethyl-
11814 - Thujane, (1*S*,4*R*,5*S*)-(+)-

$C_{10}H_{18}N_2O_3$
7223 - 4-Imidazolidinehexanoic acid, 5-methyl-2-oxo-, (4*R-cis*)-

$C_{10}H_{18}N_2O_4$
9606 - 1,4-Piperazinedicarboxylic acid, diethyl ester

$C_{10}H_{18}N_4O_4S_3$
11713 - Thiodicarb

$C_{10}H_{18}O$
2930 - Bicyclo[3.1.1]heptane-2-methanol, 6,6-dimethyl-, [1*S*-(1α,2β,5α)]-
2940 - Bicyclo[2.2.1]heptan-2-ol, 1,3,3-trimethyl-, *endo*-(±)-
2941 - Bicyclo[2.2.1]heptan-2-ol, 1,3,3-trimethyl-, (1*S-endo*)-
2942 - Bicyclo[2.2.1]heptan-2-ol, 1,7,7-trimethyl-, *endo*-(±)-
2943 - Bicyclo[2.2.1]heptan-2-ol, 1,7,7-trimethyl-, *exo*-(±)-
2944 - Bicyclo[3.1.1]heptan-2-ol, 4,6,6-trimethyl-, [1*R*-(1α,2α,4α,5α)]-
2985 - Bicyclo[3.1.0]hexan-3-ol, 4-methyl-1-(1-methylethyl)-
2994 - [1,1′-Bicyclopentyl]-2-ol
3652 - 1-Butanone, 1-cyclohexyl-
4205 - Cyclodecanone
4237 - Cycloheptanone, 2,2,6-trimethyl-
4238 - Cycloheptanone, 2,6,6-trimethyl-
4542 - Cyclohexanol, 1-methyl-4-(1-methylethenyl)-
4543 - Cyclohexanol, 2-methyl-5-(1-methylethenyl)-, [1*S*-(1α,2β,5α)]-
4544 - Cyclohexanol, 5-methyl-2-(1-methylethenyl)-, [1*R*-(1α,2β,5α)]-
4575 - Cyclohexanone, 2-butyl-
4604 - Cyclohexanone, 2-methyl-5-(1-methylethyl)-, (2*S-trans*)-
4605 - Cyclohexanone, 5-methyl-2-(1-methylethyl)-, (2*R-cis*)-
4606 - Cyclohexanone, 5-methyl-2-(1-methylethyl)-, (2*S-trans*)-
4609 - Cyclohexanone, 4-(2-methylpropyl)-
4665 - 1-Cyclohexene-1-methanol, 2,4,4-trimethyl-
4666 - 1-Cyclohexene-1-methanol, 2,6,6-trimethyl-
4667 - 2-Cyclohexene-1-methanol, 2,6,6-trimethyl-
4668 - 3-Cyclohexene-1-methanol, α,α,4-trimethyl-, (±)-
4697 - 2-Cyclohexen-1-ol, 3-methyl-6-(1-methylethyl)-
4698 - 2-Cyclohexen-1-ol, 6-methyl-3-(1-methylethyl)-
4699 - 3-Cyclohexen-1-ol, 4-methyl-1-(1-methylethyl)-
4978 - Cyclopropanemethanol, 2,2-dimethyl-3-(2-methyl-1-propenyl)-
5080 - 2-Decenal
5081 - 3-Decenal
5093 - 3-Decen-2-one
6590 - 3-Hepten-2-one, 3-ethyl-4-methyl-
6591 - 4-Hepten-2-one, 3-ethyl-4-methyl-
6592 - 4-Hepten-3-one, 5-ethyl-4-methyl-
6593 - 5-Hepten-2-one, 4-ethyl-6-methyl-
6594 - 5-Hepten-3-one, 5-ethyl-4-methyl-
8024 - 2-Naphthalenol, decahydro-
8191 - 5-Nonen-4-one, 6-methyl-
8192 - 6-Nonen-4-one, 6-methyl-
8332 - 1,6-Octadien-3-ol, 3,7-dimethyl-, (±)-
8333 - 2,6-Octadien-1-ol, 3,7-dimethyl-, (*E*)-
8334 - 2,6-Octadien-1-ol, 3,7-dimethyl-, (*Z*)-
8490 - 6-Octenal, 3,7-dimethyl-
8556 - 7-Oxabicyclo[2.2.1]heptane, 1-methyl-4-(1-methylethyl)-
8560 - 2-Oxabicyclo[2.2.2]octane, 1,3,3-trimethyl-

$C_{10}H_{18}O_2$
3496 - Butanoic acid, cyclohexyl ester
4206 - Cyclodecanone, 2-hydroxy-
4224 - Cycloheptanecarboxylic acid, ethyl ester
4308 - Cyclohexaneacetic acid, ethyl ester
4337 - Cyclohexanecarboxylic acid, propyl ester
4784 - Cyclopentanecarboxylic acid, 2-methyl-5-(1-methylethyl)-
4786 - Cyclopentanecarboxylic acid, 1,2,2,3-tetramethyl-
5088 - 2-Decenoic acid
5089 - 3-Decenoic acid

5090 - 4-Decenoic acid
5091 - 9-Decenoic acid
6811 - 3,4-Hexanedione, 2,2,5,5-tetramethyl-
10293 - Propanoic acid, 2-methyl-, cyclohexyl ester
10660 - 2-Propenoic acid, 2-methyl-, hexyl ester

$C_{10}H_{18}O_3$
3504 - Butanoic acid, 2,2-diethyl-3-oxo-, ethyl ester
3542 - Butanoic acid, 3-methyl-, anhydride
4309 - Cyclohexaneacetic acid, α-ethyl-1-hydroxy-
6872 - Hexanoic acid, 2-acetyl-, ethyl ester
8858 - Pentanoic acid, anhydride
11903 - 2,5,7-Trioxabicyclo[2.2.1]heptane, 1,3,3,4,6,6-hexamethyl-

$C_{10}H_{18}O_4$
3321 - Butanedioic acid, dipropyl ester
3512 - Butanoic acid, 1,2-ethanediyl ester
5042 - Decanedioic acid
5557 - Ethanedioic acid, bis(2-methylpropyl) ester
5558 - Ethanedioic acid, dibutyl ester
6785 - Hexanedioic acid, diethyl ester
8135 - Nonanedioic acid, 3-methyl-
8388 - Octanedioic acid, dimethyl ester
8601 - Oxirane, 2,2′-[1,4-butanediylbis(oxymethylene)]bis-
9964 - Propanedioic acid, (1-methylethyl)-, diethyl ester
9981 - Propanedioic acid, propyl-, diethyl ester

$C_{10}H_{18}O_4S$
10330 - Propanoic acid, 3,3′-thiobis-, diethyl ester

$C_{10}H_{18}O_6$
3305 - Butanedioic acid, 2,3-dihydroxy-, bis(1-methylethyl) ester, (*R**,*R**)-(±)-
3310 - Butanedioic acid, 2,3-dihydroxy- [*R*-(*R**,*R**)]-, dipropyl ester
5714 - Ethanol, 2,2′-[1,2-ethanediylbis(oxy)]bis-, diacetate

$C_{10}H_{19}Br$
4321 - Cyclohexane, 1-bromo-4-methyl-1-(1-methylethyl)-
5086 - 1-Decene, 2-bromo-
5087 - 2-Decene, 1-bromo-

$C_{10}H_{19}BrO_2$
5061 - Decanoic acid, 2-bromo-

$C_{10}H_{19}ClO$
5078 - Decanoyl chloride

$C_{10}H_{19}FO_2$
5064 - Decanoic acid, 10-fluoro-

$C_{10}H_{19}N$
2927 - Bicyclo[2.2.1]heptane-1-methanamine, 7,7-dimethyl-
2928 - Bicyclo[3.1.1]heptane-2-methanamine, 6,6-dimethyl-, [1*S*-(1α,2β,5α)]-
5057 - Decanenitrile
7712 - 1-Naphthalenamine, decahydro-

$C_{10}H_{19}NO$
11331 - 2H-Quinolizine-1-methanol, octahydro-, (1*R*-*trans*)-

$C_{10}H_{19}NO_2$
9704 - 1-Piperidinepropanoic acid, ethyl ester

$C_{10}H_{19}N_5O$
3143 - *sec*-Bumeton

$C_{10}H_{19}N_5S$
9770 - Prometryn
11594 - Terbutryn

$C_{10}H_{19}O_6PS_2$
3316 - Butanedioic acid, [(dimethoxyphosphinothioyl)thio]-, diethyl ester

$C_{10}H_{20}$
4202 - Cyclodecane
4227 - Cycloheptane, 1,1,2-trimethyl-
4228 - Cycloheptane, 1,1,4-trimethyl-
4325 - Cyclohexane, butyl-
4369 - Cyclohexane, (1,1-dimethylethyl)-
4436 - Cyclohexane, 1-methyl-2-(1-methylethyl)-
4437 - Cyclohexane, 1-methyl-3-(1-methylethyl)-
4438 - Cyclohexane, 1-methyl-4-(1-methylethyl)-
4439 - Cyclohexane, 1-methyl-4-(1-methylethyl)-, *cis*-
4440 - Cyclohexane, 1-methyl-4-(1-methylethyl)-, *trans*-
4448 - Cyclohexane, (1-methylpropyl)-
4449 - Cyclohexane, 1-methyl-1-propyl-
4450 - Cyclohexane, 1-methyl-2-propyl-
4451 - Cyclohexane, 1-methyl-3-propyl-
4452 - Cyclohexane, 1-methyl-4-propyl-
4453 - Cyclohexane, (2-methylpropyl)-
4477 - Cyclohexane, 1,1,3,4-tetramethyl-
4478 - Cyclohexane, 1,1,3,5-tetramethyl-, *cis*-
4846 - Cyclopentane, pentyl-
5082 - 1-Decene
5083 - 4-Decene
5084 - 5-Decene, (*E*)-
5085 - 5-Decene, (*Z*)-
6559 - 3-Heptene, 4-propyl-
8185 - 1-Nonene, 2-methyl-
8186 - 3-Nonene, 2-methyl-
8502 - 1-Octene, 3,7-dimethyl-
8503 - 2-Octene, 2,6-dimethyl-

$C_{10}H_{20}BrF$
5035 - Decane, 1-bromo-10-fluoro-

$C_{10}H_{20}Br_2$
5038 - Decane, 1,10-dibromo-
5039 - Decane, 5,6-dibromo-

$C_{10}H_{20}ClF$
5037 - Decane, 1-chloro-10-fluoro-

$C_{10}H_{20}Cl_2$
5040 - Decane, 1,10-dichloro-

$C_{10}H_{20}NO_4PS$
10719 - Propetamphos

$C_{10}H_{20}N_2$
3098 - 2,3′-Bipiperidine

$C_{10}H_{20}N_2O_2$
7690 - Morpholine, 4,4′-(1,2-ethanediyl)bis-

$C_{10}H_{20}N_2S_2$
9659 - Piperidine, 1,1′-dithiobis-

$C_{10}H_{20}N_2S_3$
11714 - Thiodicarbonic diamide ([(H2N)C(*S*)]2*S*), tetraethyl-

$C_{10}H_{20}N_2S_4$
11721 - Thioperoxydicarbonic diamide ([(H2N)C(*S*)]2S2), tetraethyl-

$C_{10}H_{20}O$
4204 - Cyclodecanol
4233 - Cycloheptanol, 2,6,6-trimethyl-
4324 - Cyclohexanebutanol
4469 - Cyclohexanepropanol, α-methyl-
4504 - Cyclohexanol, 2-butyl-, *trans*-
4521 - Cyclohexanol, 2-(1,1-dimethylethyl)-
4546 - Cyclohexanol, 1-methyl-4-(1-methylethyl)-
4547 - Cyclohexanol, 2-methyl-5-(1-methylethyl)-, [1*R*-(1α,2α,5β)]-
4548 - Cyclohexanol, 5-methyl-2-(1-methylethyl)-, (1α,2β,5α)-(±)-
4550 - Cyclohexanol, 2-methyl-1-propyl-
4551 - Cyclohexanol, 3-methyl-1-propyl-
5030 - Decanal
5075 - 2-Decanone
5076 - 3-Decanone
5077 - 4-Decanone
5092 - 9-Decen-1-ol
6405 - Heptane, 3-[(ethenyloxy)methyl]-
7428 - *m*-Menthan-6-ol
8176 - 5-Nonanone, 2-methyl-
8401 - Octane, 1-(ethenyloxy)-
8514 - 6-Octen-1-ol, 3,7-dimethyl-, (±)-
8515 - 6-Octen-3-ol, 3,7-dimethyl-
8516 - 7-Octen-1-ol, 3,7-dimethyl-, (*S*)-

$C_{10}H_{20}O_2$
187 - Acetic acid, 2-ethylhexyl ester
222 - Acetic acid, octyl ester
3528 - Butanoic acid, hexyl ester
3546 - Butanoic acid, 3-methyl-, 1,1-dimethylpropyl ester
3551 - Butanoic acid, 3-methyl-, 3-methylbutyl ester
3559 - Butanoic acid, 1-methylpentyl ester, (*S*)-
5059 - Decanoic acid
6471 - Heptanoic acid, propyl ester
6506 - 2-Heptanol, 6-methyl-, acetate
6507 - 3-Heptanol, 6-methyl-, acetate
6880 - Hexanoic acid, butyl ester
6885 - Hexanoic acid, 2-ethyl-, ethyl ester
6975 - 3-Hexanone, 4-hydroxy-2,2,5,5-tetramethyl-
8153 - Nonanoic acid, 3-methyl-
8154 - Nonanoic acid, methyl ester
8347 - Octanal, 7-hydroxy-3,7-dimethyl-
8428 - Octanoic acid, ethyl ester
8447 - 2-Octanol, acetate, (±)-
8894 - Pentanoic acid, 4-methyl-, 2-methylpropyl ester
8913 - Pentanoic acid, pentyl ester
10264 - Propanoic acid, heptyl ester

$C_{10}H_{20}O_3$
3811 - 2-Butene, 1,1,3-triethoxy-

$C_{10}H_{20}O_4$
5689 - Ethanol, 2-(2-butoxyethoxy)-, acetate

$C_{10}H_{21}Br$
5034 - Decane, 1-bromo-

$C_{10}H_{21}Cl$
5036 - Decane, 1-chloro-

$C_{10}H_{21}ClO$
5073 - 1-Decanol, 10-chloro-

$C_{10}H_{21}F$
5052 - Decane, 1-fluoro-

$C_{10}H_{21}FO$
5074 - 1-Decanol, 10-fluoro-

$C_{10}H_{21}I$
5053 - Decane, 1-iodo-

$C_{10}H_{21}N$
4285 - Cyclohexanamine, *N*,*N*-diethyl-
4385 - Cyclohexaneethanamine, *N*,α-dimethyl-
9691 - Piperidine, 1,2,2,6,6-pentamethyl-
9692 - Piperidine, 1-pentyl-

$C_{10}H_{21}NO$
4502 - Cyclohexanol, 3-(aminomethyl)-3,5,5-trimethyl-
5031 - Decanamide

$C_{10}H_{21}NOS$
3976 - Carbamothioic acid, butylethyl-, *S*-propyl ester
12047 - Vernolate

$C_{10}H_{21}NO_2$
8084 - Nitrous acid, decyl ester

$C_{10}H_{21}NO_3$
8073 - Nitric acid, decyl ester

$C_{10}H_{21}N_3O$
9603 - 1-Piperazinecarboxamide, *N*,*N*-diethyl-4-methyl-

$C_{10}H_{22}$
5033 - Decane
6409 - Heptane, 3-ethyl-2-methyl-
6410 - Heptane, 3-ethyl-3-methyl-
6411 - Heptane, 3-ethyl-4-methyl-
6412 - Heptane, 3-ethyl-5-methyl-
6413 - Heptane, 4-ethyl-2-methyl-
6414 - Heptane, 4-ethyl-3-methyl-
6415 - Heptane, 4-ethyl-4-methyl-
6416 - Heptane, 5-ethyl-2-methyl-
6427 - Heptane, 4-(1-methylethyl)-
6434 - Heptane, 4-propyl-
6437 - Heptane, 2,2,3-trimethyl-
6438 - Heptane, 2,2,4-trimethyl-
6439 - Heptane, 2,2,5-trimethyl-
6440 - Heptane, 2,2,6-trimethyl-
6441 - Heptane, 2,3,3-trimethyl-
6442 - Heptane, 2,3,4-trimethyl-
6443 - Heptane, 2,3,5-trimethyl-
6444 - Heptane, 2,3,6-trimethyl-
6445 - Heptane, 2,4,4-trimethyl-
6446 - Heptane, 2,4,5-trimethyl-
6447 - Heptane, 2,4,6-trimethyl-
6448 - Heptane, 2,5,5-trimethyl-
6449 - Heptane, 3,3,4-trimethyl-

6450 - Heptane, 3,3,5-trimethyl-
6451 - Heptane, 3,4,4-trimethyl-
6452 - Heptane, 3,4,5-trimethyl-
6767 - Hexane, 3,3-diethyl-
6768 - Hexane, 3,4-diethyl-
6816 - Hexane, 3-ethyl-2,2-dimethyl-
6817 - Hexane, 3-ethyl-2,3-dimethyl-
6818 - Hexane, 3-ethyl-2,4-dimethyl-
6819 - Hexane, 3-ethyl-2,5-dimethyl-
6820 - Hexane, 3-ethyl-3,4-dimethyl-
6821 - Hexane, 4-ethyl-2,2-dimethyl-
6822 - Hexane, 4-ethyl-2,3-dimethyl-
6823 - Hexane, 4-ethyl-2,4-dimethyl-
6824 - Hexane, 4-ethyl-3,3-dimethyl-
6839 - Hexane, 2-methyl-3-(1-methylethyl)-
6844 - Hexane, 2,2,3,3-tetramethyl-
6845 - Hexane, 2,2,3,4-tetramethyl-
6846 - Hexane, 2,2,3,5-tetramethyl-
6847 - Hexane, 2,2,4,4-tetramethyl-
6848 - Hexane, 2,2,4,5-tetramethyl-
6849 - Hexane, 2,2,5,5-tetramethyl-
6850 - Hexane, 2,3,3,4-tetramethyl-
6851 - Hexane, 2,3,3,5-tetramethyl-
6852 - Hexane, 2,3,4,4-tetramethyl-
6853 - Hexane, 2,3,4,5-tetramethyl-
6854 - Hexane, 3,3,4,4-tetramethyl-
8140 - Nonane, 2-methyl-
8141 - Nonane, 3-methyl-
8142 - Nonane, 4-methyl-
8143 - Nonane, 5-methyl-
8373 - Octane, 2,2-dimethyl-
8374 - Octane, 2,3-dimethyl-
8375 - Octane, 2,4-dimethyl-
8376 - Octane, 2,5-dimethyl-
8377 - Octane, 2,6-dimethyl-
8378 - Octane, 2,7-dimethyl-
8379 - Octane, 3,3-dimethyl-
8380 - Octane, 3,4-dimethyl-
8381 - Octane, 3,5-dimethyl-
8382 - Octane, 3,6-dimethyl-
8383 - Octane, 4,4-dimethyl-
8384 - Octane, 4,5-dimethyl-
8403 - Octane, 3-ethyl-
8404 - Octane, 4-ethyl-
8757 - Pentane, 3,3-diethyl-2-methyl-
8766 - Pentane, 2,4-dimethyl-3-(1-methylethyl)-
8815 - Pentane, 3-ethyl-2,2,3-trimethyl-
8816 - Pentane, 3-ethyl-2,3,4-trimethyl-
8836 - Pentane, 2,2,3,3,4-pentamethyl-
8837 - Pentane, 2,2,3,4,4-pentamethyl-

$C_{10}H_{22}O$
3418 - Butane, 1,1′-oxybis[3-methyl-
5070 - 1-Decanol
5071 - 2-Decanol, (±)-
5072 - 4-Decanol
6488 - 3-Heptanol, 3-ethyl-2-methyl-
6489 - 3-Heptanol, 3-ethyl-6-methyl-
6490 - 3-Heptanol, 5-ethyl-4-methyl-
6508 - 4-Heptanol, 4-propyl-
6509 - 4-Heptanol, 2,4,6-trimethyl-
6948 - 1-Hexanol, 5-methyl-2-(1-methylethyl)-
6950 - 3-Hexanol, 2,2,5,5-tetramethyl-
8166 - 1-Nonanol, 3-methyl-
8167 - 1-Nonanol, 4-methyl-
8168 - 1-Nonanol, 8-methyl
8169 - 3-Nonanol, 2-methyl-
8170 - 4-Nonanol, 4-methyl-
8171 - 5-Nonanol, 5-methyl-
8402 - Octane, 1-ethoxy-
8449 - 1-Octanol, 3,7-dimethyl-, (±)-
8450 - 2-Octanol, 2,6-dimethyl-
8451 - 3-Octanol, 2,3-dimethyl-
8452 - 3-Octanol, 2,7-dimethyl-
8453 - 3-Octanol, 3,6-dimethyl-
8454 - 3-Octanol, 3,6-dimethyl-(+-)-
8455 - 4-Octanol, 2,4-dimethyl-
8456 - 4-Octanol, 2,5-dimethyl-
8457 - 4-Octanol, 2,6-dimethyl-
8458 - 4-Octanol, 2,7-dimethyl-
8459 - 3-Octanol, 3-ethyl-
8835 - Pentane, 1,1′-oxybis-

$C_{10}H_{22}O_2$
5047 - 1,10-Decanediol
5048 - 2,9-Decanediol
5527 - Ethane, 1,2-dibutoxy
6800 - 3,4-Hexanediol, 3,4-diethyl-
8392 - 1,7-Octanediol, 3,7-dimethyl-
9154 - Peroxide, bis(1,1-dimethylpropyl)
10032 - Propane, 1,1′-[1,2-ethanediylbis(oxy)]bis[2-methyl-
10041 - Propane, 1,1′-[ethylidenebis(oxy)]bis[2-methyl-

$C_{10}H_{22}O_3$
10078 - Propane, 1,1′,1″-[methylidynetris(oxy)]tris-
10079 - Propane, 2,2′,2″-[methylidynetris(oxy)]tris-

$C_{10}H_{22}O_4$
5690 - Ethanol, 2-[2-(2-butoxyethoxy)ethoxy]-

$C_{10}H_{22}O_5$
8995 - 2,5,8,11,14-Pentaoxapentadecane

$C_{10}H_{22}S$
3435 - Butane, 1,1′-thiobis[2-methyl-
3436 - Butane, 1,1′-thiobis[3-methyl-
5058 - 1-Decanethiol
8843 - Pentane, 1,1′-thiobis-

$C_{10}H_{22}S_2$
5243 - Disulfide, bis(3-methylbutyl)
5257 - Disulfide, dipentyl

$C_{10}H_{23}N$
3226 - 1-Butanamine, 3-methyl-*N*-(3-methylbutyl)-
5032 - 1-Decanamine
8708 - 1-Pentanamine, *N*-pentyl-

$C_{10}H_{23}NO$
5684 - Ethanol, 2-[bis(2-methylpropyl)amino]-
5703 - Ethanol, 2-(dibutylamino)-

$C_{10}H_{23}NO_2$
5415 - Ethanamine, 2,2-diethoxy-*N,N*-diethyl-

$C_{10}H_{24}N_2$
5515 - 1,2-Ethanediamine, *N,N,N′,N′*-tetraethyl-
6752 - 1,6-Hexanediamine, *N,N,N′,N′*-tetramethyl-

$C_{10}H_{24}N_4$
9614 - 1,4-Piperazinedipropanamine

$C_{10}H_{26}N_4$
3259 - 1,4-Butanediamine, *N,N′*-bis(3-aminopropyl)-

$C_{10}H_{30}O_3Si_4$
11646 - Tetrasiloxane, decamethyl-

$C_{10}H_{30}O_5Si_5$
4901 - Cyclopentasiloxane, decamethyl-

$C_{11}H_7BrO_2$
7761 - 1-Naphthalenecarboxylic acid, 5-bromo-
7769 - 2-Naphthalenecarboxylic acid, 5-bromo-
7846 - 1,4-Naphthalenedione, 2-bromo-3-methyl-

$C_{11}H_7ClO$
7750 - 2-Naphthalenecarbonyl chloride

$C_{11}H_7ClO_2$
7762 - 1-Naphthalenecarboxylic acid, 5-chloro-
7763 - 1-Naphthalenecarboxylic acid, 8-chloro-

$C_{11}H_7N$
7748 - 1-Naphthalenecarbonitrile
7749 - 2-Naphthalenecarbonitrile

$C_{11}H_7NO$
7904 - Naphthalene, 1-isocyanato-

$C_{11}H_7NO_4$
7767 - 1-Naphthalenecarboxylic acid, 5-nitro-

$C_{11}H_7NS$
7906 - Naphthalene, 1-isothiocyanato-

$C_{11}H_8O$
7751 - 1-Naphthalenecarboxaldehyde
7752 - 2-Naphthalenecarboxaldehyde

$C_{11}H_8OS$
7641 - Methanone, phenyl-2-thienyl-
7642 - Methanone, phenyl-3-thienyl-

$C_{11}H_8O_2$
7618 - Methanone, 2-furanylphenyl-
7754 - 1-Naphthalenecarboxaldehyde, 2-hydroxy-
7758 - 1-Naphthalenecarboxylic acid
7759 - 2-Naphthalenecarboxylic acid
7864 - 1,4-Naphthalenedione, 2-methyl-

$C_{11}H_8O_3$
7771 - 2-Naphthalenecarboxylic acid, 1-hydroxy-
7772 - 2-Naphthalenecarboxylic acid, 3-hydroxy-
7861 - 1,4-Naphthalenedione, 2-hydroxy-3-methyl-
7863 - 1,4-Naphthalenedione, 5-hydroxy-2-methyl-

$C_{11}H_8O_4$
7856 - 1,4-Naphthalenedione, 3,5-dihydroxy-2-methyl-

$C_{11}H_8O_5$
2408 - 5H-Benzocyclohepten-5-one, 2,3,4,6-tetrahydroxy-

$C_{11}H_9Br$
7742 - Naphthalene, 1-(bromomethyl)-
7743 - Naphthalene, 2-(bromomethyl)-

$C_{11}H_9Cl$
7783 - Naphthalene, 1-(chloromethyl)-
7784 - Naphthalene, 1-chloro-2-methyl-
7785 - Naphthalene, 2-(chloromethyl)-

$C_{11}H_9Cl_2NO_2$
510 - Barban

$C_{11}H_9N$
10997 - Pyridine, 2-phenyl-
10998 - Pyridine, 3-phenyl-
10999 - Pyridine, 4-phenyl-

$C_{11}H_9NO$
7637 - Methanone, phenyl-1H-pyrrol-2-yl-
7757 - 1-Naphthalenecarboxamide

$C_{11}H_9NO_2$
7760 - 1-Naphthalenecarboxylic acid, 5-amino-
7768 - 2-Naphthalenecarboxylic acid, 3-amino-
7925 - Naphthalene, 1-methyl-4-nitro-
7926 - Naphthalene, 2-methyl-1-nitro-

$C_{11}H_9NO_3$
11234 - 4-Quinolinecarboxylic acid, 6-methoxy-

$C_{11}H_{10}$
7915 - Naphthalene, 1-methyl-
7916 - Naphthalene, 2-methyl-

$C_{11}H_{10}NO_3$
7695 - Morpholine, 4-(2-hydroxybenzoyl)-

$C_{11}H_{10}N_2O$
90 - Acetamide, *N*-4-quinolinyl-
12022 - Urea, 1-naphthalenyl-
12023 - Urea, 2-naphthalenyl-

$C_{11}H_{10}N_2O_3$
11100 - 2,4,6(1H,3H,5H)-Pyrimidinetrione, 5-methyl-5-phenyl-

$C_{11}H_{10}N_2S$
11807 - Thiourea, 1-naphthalenyl-

$C_{11}H_{10}O$
4951 - 2-Cyclopenten-1-one, 3-phenyl-
7908 - 1-Naphthalenemethanol
7909 - 2-Naphthalenemethanol
7913 - Naphthalene, 1-methoxy-
7914 - Naphthalene, 2-methoxy-
8030 - 1-Naphthalenol, 4-methyl-
8031 - 2-Naphthalenol, 1-methyl-
8657 - 2,4-Pentadienal, 5-phenyl-

$C_{11}H_{10}O_2$
2820 - 2H-1-Benzopyran-2-one, 3-ethyl-
6158 - Furan, 2-(phenoxymethyl)-
7764 - 1-Naphthalenecarboxylic acid, 3,4-dihydro-
10741 - 2-Propynoic acid, 3-phenyl-, ethyl ester

$C_{11}H_{10}O_3$
2440 - 2-Benzofurancarboxylic acid, ethyl ester
10781 - 2H-Pyran-2,6(3H)-dione, dihydro-3-phenyl-
10782 - 2H-Pyran-2,6(3H)-dione, dihydro-4-phenyl-

$C_{11}H_{10}O_4$
2818 - 2H-1-Benzopyran-2-one, 5,7-dimethoxy-

$C_{11}H_{10}S$
11774 - Thiophene, 2-methyl-5-phenyl-

$C_{11}H_{11}BrN_2O$
10846 - 3H-Pyrazol-3-one, 4-bromo-1,2-dihydro-1,5-dimethyl-2-phenyl-

MOLECULAR FORMULA INDEX OF ORGANIC COMPOUNDS (continued)

$C_{11}H_{11}BrO_2$
10613 - 2-Propenoic acid, 2-bromo-3-phenyl-, ethyl ester
10614 - 2-Propenoic acid, 3-bromo-3-phenyl-, ethyl ester, (*Z*)-

$C_{11}H_{11}Cl_3O_3$
259 - Acetic acid, (2,4,5-trichlorophenoxy)-, 1-methylethyl ester

$C_{11}H_{11}N$
7720 - 1-Naphthalenamine, 4-methyl-
7721 - 1-Naphthalenamine, *N*-methyl-
7722 - 2-Naphthalenamine, *N*-methyl-
7907 - 2-Naphthalenemethanamine
11136 - 1H-Pyrrole, 1-(phenylmethyl)-
11254 - Quinoline, 2,3-dimethyl-
11255 - Quinoline, 2,4-dimethyl-
11256 - Quinoline, 2,5-dimethyl-
11257 - Quinoline, 2,6-dimethyl-
11258 - Quinoline, 2,7-dimethyl-
11259 - Quinoline, 2,8-dimethyl-
11260 - Quinoline, 3,4-dimethyl-
11261 - Quinoline, 3,8-dimethyl-
11262 - Quinoline, 4,6-dimethyl-
11263 - Quinoline, 5,8-dimethyl-
11264 - Quinoline, 6,8-dimethyl-
11266 - Quinoline, 2-ethyl-
11267 - Quinoline, 3-ethyl-
11268 - Quinoline, 6-ethyl-

$C_{11}H_{11}NO$
10442 - 2-Propanone, 1-(1H-indol-3-yl)-
11316 - 4-Quinolinol, 2,8-dimethyl-
11317 - 6-Quinolinol, 2,4-dimethyl-
11318 - 8-Quinolinol, 2,4-dimethyl-

$C_{11}H_{11}NO_2$
893 - Benzeneacetic acid, α-cyano-, ethyl ester
11161 - 2,5-Pyrrolidinedione, 1-methyl-3-phenyl-
11163 - 2,5-Pyrrolidinedione, 1-(phenylmethyl)-

$C_{11}H_{11}NO_4S$
7407 - Isoxazolium, 2-ethyl-5-(3-sulfophenyl)-, hydroxide, inner salt
11293 - 5-Quinolinesulfonic acid, 8-ethoxy-

$C_{11}H_{11}NO_6$
7511 - Methanediol, (3-nitrophenyl)-, diacetate (ester)

$C_{11}H_{11}N_3O_2S$
2203 - Benzenesulfonamide, 4-amino-*N*-2-pyridinyl-

$C_{11}H_{12}$
7816 - Naphthalene, 1,2-dihydro-3-methyl-
7817 - Naphthalene, 1,2-dihydro-4-methyl-

$C_{11}H_{12}Cl_2N_2O_5$
52 - Acetamide, 2,2-dichloro-*N*-[2-hydroxy-1-(hydroxymethyl)-2-(4-nitrophenyl)ethyl]-, [*R*-(*R**,*R**)]-

$C_{11}H_{12}Cl_2O_3$
173 - Acetic acid, (2,4-dichlorophenoxy)-, 1-methylethyl ester

$C_{11}H_{12}Cl_3N$
1583 - Benzeneethanamine, α-methyl-*N*-(2,2,2-trichloroethylidene)-

$C_{11}H_{12}NO_4PS_2$
9515 - Phosmet

$C_{11}H_{12}N_2$
7800 - 1,4-Naphthalenediamine, 2-methyl-
10829 - 1H-Pyrazole, 3,5-dimethyl-1-phenyl-

$C_{11}H_{12}N_2O$
10847 - 3H-Pyrazol-3-one, 1,2-dihydro-1,5-dimethyl-2-phenyl-
11192 - Pyrrolo[2,1-b]quinazolin-3-ol, 1,2,3,9-tetrahydro-, (*R*)-

$C_{11}H_{12}N_2O_2$
11922 - *L*-Tryptophan

$C_{11}H_{12}N_2O_6$
2592 - Benzoic acid, 3,5-dinitro-, butyl ester

$C_{11}H_{12}N_4O_2S$
2199 - Benzenesulfonamide, 4-amino-*N*-(4-methyl-2-pyrimidinyl)-

$C_{11}H_{12}N_4O_3S$
2197 - Benzenesulfonamide, 4-amino-*N*-(6-methoxy-3-pyridazinyl)-

$C_{11}H_{12}O$
2407 - 5H-Benzocyclohepten-5-one, 6,7,8,9-tetrahydro-
3190 - Butanal, 2-(phenylmethylene)-
7288 - 1H-Inden-1-one, 2,3-dihydro-3,3-dimethyl-
7605 - Methanone, cyclobutylphenyl-
8049 - 1(2H)-Naphthalenone, 3,4-dihydro-2-methyl-
8050 - 1(2H)-Naphthalenone, 3,4-dihydro-4-methyl-
9110 - 1-Penten-3-one, 1-phenyl-

$C_{11}H_{12}O_2$
7778 - 2-Naphthalenecarboxylic acid, 1,2,3,4-tetrahydro-
7779 - 2-Naphthalenecarboxylic acid, 5,6,7,8-tetrahydro-
8048 - 1(2H)-Naphthalenone, 3,4-dihydro-6-methoxy-
8804 - 1,4-Pentanedione, 1-phenyl-
10685 - 2-Propenoic acid, 3-phenyl-, ethyl ester, (*E*)-
10706 - 2-Propen-1-ol, 3-phenyl-, acetate, (*E*)-

$C_{11}H_{12}O_3$
2141 - Benzenepropanoic acid, β-oxo-, ethyl ester
2428 - 1,3-Benzodioxole, 4-methoxy-6-(2-propenyl)-
6138 - 2-Furanol, tetrahydro-, benzoate
10647 - 2-Propenoic acid, 3-(2-methoxyphenyl)-, methyl ester
10681 - 2-Propenoic acid, 2-phenoxyethyl ester

$C_{11}H_{12}O_4$
7512 - Methanediol, phenyl-, diacetate
9976 - Propanedioic acid, phenyl-, dimethyl ester

$C_{11}H_{13}ClF_3N_3O_4S_3$
2857 - 2H-1,2,4-Benzothiadiazine-7-sulfonamide, 6-chloro-3,4-dihydro-2-methyl-3-[[(2,2,2-trifluoroethyl)thio]methyl]-

$C_{11}H_{13}ClO$
2889 - Benzoyl chloride, 4-butyl-
2895 - Benzoyl chloride, 4-(1,1-dimethylethyl)-
9255 - Phenol, 4-chloro-2-cyclopentyl-

$C_{11}H_{13}ClO_3$
3493 - Butanoic acid, 4-(4-chloro-2-methylphenoxy)-

$C_{11}H_{13}F_3N_2O_3S$
7427 - Mefluidide

$C_{11}H_{13}F_3N_4O_4$
5181 - Dinitramine

$C_{11}H_{13}N$
963 - Benzeneacetonitrile, α-(1-methylethyl)-
970 - Benzeneacetonitrile, α-propyl-
1823 - Benzenemethanamine, *N*-methyl-*N*-2-propynyl-
7309 - 1H-Indole, 3-ethyl-2-methyl-
7322 - 1H-Indole, 1,2,3-trimethyl-
7323 - 1H-Indole, 2,3,5-trimethyl-
10922 - Pyridine, 4-(3-cyclohexen-1-yl)-

$C_{11}H_{13}NO$
3703 - 3-Butenamide, 3-methyl-4-phenyl-
7332 - 2H-Indol-2-one, 1-ethyl-1,3-dihydro-3-methyl-

$C_{11}H_{13}NO_2$
5215 - 1,3-Dioxolo[4,5-g]isoquinoline, 5,6,7,8-tetrahydro-6-methyl-

$C_{11}H_{13}NO_3$
5217 - 1,3-Dioxolo[4,5-g]isoquinolin-5-ol, 5,6,7,8-tetrahydro-6-methyl-
7398 - 1(2H)-Isoquinolinone, 3,4-dihydro-6,7-dimethoxy-

$C_{11}H_{13}NO_4$
514 - Bendiocarb
2706 - Benzoic acid, 4-nitro-, butyl ester
8588 - 2-Oxazolidinone, 5-[(2-methoxyphenoxy)methyl]-

$C_{11}H_{13}N_3O$
10845 - 3H-Pyrazol-3-one, 4-amino-1,2-dihydro-1,5-dimethyl-2-phenyl-

$C_{11}H_{13}N_3O_3S$
2195 - Benzenesulfonamide, 4-amino-*N*-(3,4-dimethyl-5-isoxazolyl)-

$C_{11}H_{14}$
1254 - Benzene, cyclopentyl-
1643 - Benzene, 1-ethenyl-4-(1-methylethyl)-
1647 - Benzene, 2-ethenyl-1,3,5-trimethyl-
1699 - Benzene, (1-ethyl-1-propenyl)-
1700 - Benzene, (1-ethyl-2-propenyl)-
1950 - Benzene, (1-methyl-1-butenyl)-
2104 - Benzene, 1-pentenyl-
2105 - Benzene, 3-pentenyl-
2106 - Benzene, 4-pentenyl-
4748 - 1,3,5,7-Cyclooctatetraene, propyl-
7251 - 1H-Indene, 2,3-dihydro-1,1-dimethyl-
7252 - 1H-Indene, 2,3-dihydro-1,2-dimethyl-
7253 - 1H-Indene, 2,3-dihydro-4,7-dimethyl-
7254 - 1H-Indene, 2,3-dihydro-5,6-dimethyl-
7273 - 1H-Indene, 1-ethyl-2,3-dihydro-
7995 - Naphthalene, 1,2,3,4-tetrahydro-1-methyl-
7996 - Naphthalene, 1,2,3,4-tetrahydro-2-methyl-
7997 - Naphthalene, 1,2,3,4-tetrahydro-5-methyl-
7998 - Naphthalene, 1,2,3,4-tetrahydro-6-methyl-

$C_{11}H_{14}ClNO$
9771 - Propachlor

$C_{11}H_{14}ClNO_2$
607 - Benzamide, *N*-butyl-4-chloro-2-hydroxy-

$C_{11}H_{14}N_2$
7310 - 1H-Indole-3-methanamine, *N*,*N*-dimethyl-

$C_{11}H_{14}N_2O$
7304 - 1H-Indole-3-ethanamine, 5-methoxy-

$C_{11}H_{14}N_2O_4$
84 - Acetamide, *N*-(5-nitro-2-propoxyphenyl)-

$C_{11}H_{14}O$
598 - Benzaldehyde, 2,3,5,6-tetramethyl-
1665 - Benzene, 1-ethoxy-2-(1-propenyl)-
1666 - Benzene, (3-ethoxy-1-propenyl)-
3671 - 1-Butanone, 3-methyl-1-phenyl-
3672 - 1-Butanone, 1-(4-methylphenyl)-
4871 - Cyclopentanol, 1-phenyl-
5848 - Ethanone, 1-[4-(1-methylethyl)phenyl]-
5875 - Ethanone, 1-(2,4,5-trimethylphenyl)-
5876 - Ethanone, 1-(2,4,6-trimethylphenyl)-
5877 - Ethanone, 1-(3,4,5-trimethylphenyl)-
8986 - 1-Pentanone, 1-phenyl-
10431 - 1-Propanone, 2,2-dimethyl-1-phenyl-

$C_{11}H_{14}O_2$
900 - Benzeneacetic acid, α-ethyl-, methyl ester
930 - Benzeneacetic acid, α-(1-methylethyl)-
943 - Benzeneacetic acid, α-propyl-, (±)-
1104 - Benzenebutanoic acid, γ-methyl-
1449 - Benzene, 1,2-dimethoxy-4-(1-propenyl)-
1450 - Benzene, 1,2-dimethoxy-4-(2-propenyl)-
1451 - Benzene, 1,2-dimethoxy-4-(1-propenyl)-, (*Z*)-
2099 - Benzenepentanoic acid
2130 - Benzenepropanoic acid, β,β-dimethyl-
2131 - Benzenepropanoic acid, ethyl ester
2532 - Benzoic acid, butyl ester
2588 - Benzoic acid, 4-(1,1-dimethylethyl)-
2694 - Benzoic acid, 4-(1-methylethyl)-, methyl ester
2701 - Benzoic acid, 2-methylpropyl ester
3585 - Butanoic acid, phenylmethyl ester
5194 - 1,3-Dioxane, 4-methyl-4-phenyl-
9496 - Phenol, 2,4,5-trimethyl-, acetate
10307 - Propanoic acid, 2-methyl-, phenylmethyl ester
10324 - Propanoic acid, 2-phenylethyl ester
10436 - 1-Propanone, 3-ethoxy-1-phenyl-
10447 - 1-Propanone, 1-(2-hydroxy-5-methylphenyl)-2-methyl-
10448 - 1-Propanone, 1-(4-hydroxy-3-methylphenyl)-2-methyl-

$C_{11}H_{14}O_3$
538 - Benzaldehyde, 3,4-diethoxy-
923 - Benzeneacetic acid, 4-methoxy-, ethyl ester
1128 - Benzenecarboperoxoic acid, 1,1-dimethylethyl ester
2598 - Benzoic acid, 2-ethoxy-, ethyl ester
2599 - Benzoic acid, 4-ethoxy-, ethyl ester
2614 - Benzoic acid, 2-hydroxy-, butyl ester

2615 - Benzoic acid, 4-hydroxy-, butyl ester
2641 - Benzoic acid, 2-hydroxy-3-methyl-6-(1-methylethyl)-
2642 - Benzoic acid, 2-hydroxy-6-methyl-3-(1-methylethyl)-
2643 - Benzoic acid, 2-hydroxy-, 2-methylpropyl ester
3663 - 2-Butanone, 4-(4-hydroxy-3-methoxyphenyl)-
10321 - Propanoic acid, 2-phenoxy-, ethyl ester
10322 - Propanoic acid, 3-phenoxy-, ethyl ester

$C_{11}H_{14}O_4$
2578 - Benzoic acid, 3,4-dimethoxy-, ethyl ester
5874 - Ethanone, 1-(2,3,4-trimethoxyphenyl)-

$C_{11}H_{15}BrClO_3PS$
9762 - Profenofos

$C_{11}H_{15}BrN_2O_3$
11079 - 2,4,6(1H,3H,5H)-Pyrimidinetrione, 5-(2-bromo-2-propenyl)-1-methyl-5-(1-methylethyl)-
11080 - 2,4,6(1H,3H,5H)-Pyrimidinetrione, 5-(2-bromo-2-propenyl)-5-(1-methylpropyl)-

$C_{11}H_{15}Cl_2O_3PS_2$
4092 - Chlorthiophos

$C_{11}H_{15}N$
9693 - Piperidine, 1-phenyl-
9694 - Piperidine, 3-phenyl-
9695 - Piperidine, 4-phenyl-
11172 - Pyrrolidine, 1-(phenylmethyl)-

$C_{11}H_{15}NO$
74 - Acetamide, *N*-(1-methylethyl)-*N*-phenyl-
88 - Acetamide, *N*-phenyl-*N*-propyl-
539 - Benzaldehyde, 4-(diethylamino)-
882 - Benzeneacetamide, α-(1-methylethyl)-
7697 - Morpholine, 3-methyl-2-phenyl-
7698 - Morpholine, 4-(4-methylphenyl)-
7700 - Morpholine, 4-(phenylmethyl)-
8960 - 1-Pentanone, 1-(4-aminophenyl)-

$C_{11}H_{15}NO_2$
2488 - Benzoic acid, 4-amino-, butyl ester
2505 - Benzoic acid, 4-amino-, 2-methylpropyl ester
3968 - Carbamic acid, phenyl-, 2-methylpropyl ester
6272 - Glycine, *N*-(phenylmethyl)-, ethyl ester
7397 - 6-Isoquinolinol, 1,2,3,4-tetrahydro-7-methoxy-1-methyl-, (*S*)-
9512 - *L*-Phenylalanine, ethyl ester

$C_{11}H_{15}NO_2S$
7646 - Methiocarb

$C_{11}H_{15}NO_3$
7393 - 8-Isoquinolinol, 1,2,3,4-tetrahydro-6,7-dimethoxy-
9796 - Propanamide, *N*-(4-ethoxyphenyl)-2-hydroxy-
10722 - Propoxur

$C_{11}H_{15}N_2NaO_2S$
11063 - 4,6(1H,5H)-Pyrimidinedione, dihydro-5-(2-methylpropyl)-5-(2-propenyl)-2-thioxo-, monosodium salt

$C_{11}H_{16}$
1121 - Benzene, 1-butyl-2-methyl-
1122 - Benzene, 1-butyl-3-methyl-
1123 - Benzene, 1-butyl-4-methyl-
1416 - Benzene, 1,2-diethyl-3-methyl-
1417 - Benzene, 1,2-diethyl-4-methyl-
1418 - Benzene, 1,3-diethyl-2-methyl-
1419 - Benzene, 1,3-diethyl-5-methyl-
1420 - Benzene, 1,4-diethyl-2-methyl-
1421 - Benzene, 2,4-diethyl-1-methyl-
1464 - Benzene, 1-(1,1-dimethylethyl)-2-methyl-
1465 - Benzene, 1-(1,1-dimethylethyl)-3-methyl-
1466 - Benzene, 1-(1,1-dimethylethyl)-4-methyl-
1469 - Benzene, 1,2-dimethyl-3-(1-methylethyl)-
1470 - Benzene, 1,2-dimethyl-4-(1-methylethyl)-
1471 - Benzene, 1,3-dimethyl-2-(1-methylethyl)-
1472 - Benzene, 2,4-dimethyl-1-(1-methylethyl)-
1480 - Benzene, (1,1-dimethylpropyl)-
1481 - Benzene, (1,2-dimethylpropyl)-
1482 - Benzene, 1,3-dimethyl-2-propyl-
1483 - Benzene, 1,3-dimethyl-5-propyl-
1484 - Benzene, (2,2-dimethylpropyl)-
1485 - Benzene, 1,2-dimethyl-3-propyl-
1486 - Benzene, 1,4-dimethyl-2-propyl-
1487 - Benzene, 2,4-dimethyl-1-propyl-
1688 - Benzene, 1-ethyl-2-(1-methylethyl)-
1689 - Benzene, 1-ethyl-3-(1-methylethyl)-
1690 - Benzene, 1-ethyl-4-(1-methylethyl)-
1701 - Benzene, (1-ethylpropyl)-
1702 - Benzene, 1-ethyl-2-propyl-
1703 - Benzene, 1-ethyl-4-propyl-
1708 - Benzene, 1-ethyl-2,3,4-trimethyl-
1709 - Benzene, 1-ethyl-2,3,5-trimethyl-
1710 - Benzene, 1-ethyl-2,4,5-trimethyl-
1711 - Benzene, 2-ethyl-1,3,4-trimethyl-
1712 - Benzene, 2-ethyl-1,3,5-trimethyl-
1713 - Benzene, 5-ethyl-1,2,3-trimethyl-
1952 - Benzene, (1-methylbutyl)-
1953 - Benzene, (2-methylbutyl)-
1954 - Benzene, (3-methylbutyl)-
1994 - Benzene, 1-methyl-2-(1-methylpropyl)-
1995 - Benzene, 1-methyl-2-(2-methylpropyl)-
1996 - Benzene, 1-methyl-3-(1-methylpropyl)-
1997 - Benzene, 1-methyl-3-(2-methylpropyl)-
1998 - Benzene, 1-methyl-4-(1-methylpropyl)-
1999 - Benzene, 1-methyl-4-(2-methylpropyl)-
2097 - Benzene, pentamethyl-
2107 - Benzene, pentyl-
8341 - 1,5-Octadien-3-yne, 5-propyl-
11934 - 1,10-Undecadiyne
12057 - *o*-Xylene, 4-propyl-

$C_{11}H_{16}ClN_3O_2$
6001 - Formetanate hydrochloride

$C_{11}H_{16}ClN_3O_4S_2$
2856 - 2H-1,2,4-Benzothiadiazine-7-sulfonamide, 6-chloro-3,4-dihydro-3-(2-methylpropyl)-, 1,1-dioxide

$C_{11}H_{16}ClO_2PS_3$
4037 - Carbophenothion

$C_{11}H_{16}N_2$
9621 - Piperazine, 1-(phenylmethyl)-

$C_{11}H_{16}N_2O_2$
6151 - 2(3H)-Furanone, 3-ethyldihydro-4-[(1-methyl-1H-imidazol-5-yl)methyl]-, (3*S*-*cis*)-

$C_{11}H_{16}N_2O_3$
11099 - 2,4,6(1H,3H,5H)-Pyrimidinetrione, 1-methyl-5-(1-methylethyl)-5-(2-propenyl)-
11101 - 2,4,6(1H,3H,5H)-Pyrimidinetrione, 5-(2-methylpropyl)-5-(2-propenyl)-

$C_{11}H_{16}O$
363 - Anisole, 3-isopropyl-2-methyl-
1106 - Benzenebutanol, α-methyl-
1107 - Benzenebutanol, β-methyl-, (±)-
1115 - Benzene, (butoxymethyl)-
1116 - Benzene, 1-butoxy-3-methyl-
1117 - Benzene, 1-butoxy-4-methyl-
1462 - Benzene, 1-(1,1-dimethylethyl)-4-methoxy-
1611 - Benzeneethanol, β-(1-methylethyl)-
1616 - Benzeneethanol, α,α,β-trimethyl-
1617 - Benzeneethanol, α,β,β-trimethyl-
1667 - Benzene, 1-ethoxy-2-propyl-
1846 - Benzenemethanol, α-butyl-
1855 - Benzenemethanol, α,α-diethyl-
1864 - Benzenemethanol, α-(1,1-dimethylethyl)-
1865 - Benzenemethanol, 4-(1,1-dimethylethyl)-
1904 - Benzenemethanol, α-methyl-α-propyl-
1905 - Benzenemethanol, α-(2-methylpropyl)-
2033 - Benzene, [(2-methylpropoxy)methyl]-
2101 - Benzenepentanol
2109 - Benzene, (pentyloxy)-
2150 - Benzenepropanol, α,α-dimethyl-
2151 - Benzenepropanol, β,β-dimethyl-
2152 - Benzenepropanol, α-ethyl-, (*S*)-
4949 - 2-Cyclopenten-1-one, 3-methyl-2-(2-pentenyl)-, (*Z*)-
9250 - Phenol, 2-butyl-4-methyl-
9330 - Phenol, 2-(1,1-dimethylethyl)-4-methyl-
9331 - Phenol, 2-(1,1-dimethylethyl)-5-methyl-
9332 - Phenol, 2-(1,1-dimethylethyl)-6-methyl-
9333 - Phenol, 4-(1,1-dimethylethyl)-2-methyl-
9339 - Phenol, 4-(1,1-dimethylpropyl)-
9404 - Phenol, 4-(3-methylbutyl)-
9421 - Phenol, 5-methyl-2-(1-methylpropyl)-
9443 - Phenol, pentamethyl-
9444 - Phenol, 4-pentyl-

$C_{11}H_{16}O_2$
1105 - Benzenebutanol, 4-methoxy-
1410 - Benzene, 1,4-diethoxy-2-methyl-
1442 - Benzene, 1,4-dimethoxy-2-(1-methylethyl)-
1534 - 1,3-Benzenediol, 4-(3-methylbutyl)-
1544 - 1,3-Benzenediol, 4-pentyl-
9329 - Phenol, (1,1-dimethylethyl)-4-methoxy-
11864 - Tricyclo[2.2.1.02,6]heptane-1-carboxylic acid, 7,7-dimethyl-, methyl ester

$C_{11}H_{16}O_3$
1409 - Benzene, (diethoxymethoxy)-
2917 - Bicyclo[2.2.1]heptane-2-carboxylic acid, 4,7,7-trimethyl-3-oxo-
4641 - 2-Cyclohexene-1-carboxylic acid, 2,6-dimethyl-4-oxo-, ethyl ester
6058 - 2-Furancarboxylic acid, hexyl ester

$C_{11}H_{16}O_3S$
2230 - Benzenesulfonic acid, 4-methyl-, butyl ester

$C_{11}H_{16}O_4$
4915 - 1-Cyclopentene-1,2-dicarboxylic acid, diethyl ester
4916 - 1-Cyclopentene-1,3-dicarboxylic acid, diethyl ester

$C_{11}H_{17}Cl_3N_2O_2S$
7220 - 2,4-Imidazolidinedione, 5-(1-ethylpentyl)-3-[(trichloromethyl)thio]-

$C_{11}H_{17}N$
362 - Aniline, *p*-isopentyl-
696 - Benzenamine, *N*,*N*-diethyl-2-methyl-
697 - Benzenamine, *N*,*N*-diethyl-4-methyl-
783 - Benzenamine, *N*-(3-methylbutyl)-
833 - Benzenamine, 2,3,4,5,6-pentamethyl-
1576 - Benzeneethanamine, *N*-ethyl-α-methyl-

$C_{11}H_{17}NO$
1863 - Benzenemethanol, α-[1-(dimethylamino)ethyl]-, [*R*-(*R**,*S**)]-
1871 - Benzenemethanol, α-[1-(ethylamino)ethyl]-, (*R**,*S**)-
5724 - Ethanol, 2-[ethyl(3-methylphenyl)amino]-
10392 - 1-Propanol, 3-[methyl(phenylmethyl)amino]-

$C_{11}H_{17}NO_3$
1522 - 1,2-Benzenediol, 4-[1-hydroxy-2-[(1-methylethyl)amino]ethyl]-
1586 - Benzeneethanamine, 3,4,5-trimethoxy-

$C_{11}H_{17}N_3O_3$
11096 - 2,4,6(1H,3H,5H)-Pyrimidinetrione, 5-ethyl-5-(1-piperidinyl)-

$C_{11}H_{17}N_3O_3S$
2192 - Benzenesulfonamide, 4-amino-*N*-[(butylamino)carbonyl]-

$C_{11}H_{17}O_3P$
9545 - Phosphonic acid, (phenylmethyl)-, diethyl ester

$C_{11}H_{17}O_4PS_2$
5954 - Fensulfothion

$C_{11}H_{18}$
7792 - Naphthalene, decahydro-2-methylene-, *trans*
11983 - 1-Undecen-3-yne

$C_{11}H_{18}N_2$
5504 - 1,2-Ethanediamine, *N*,*N*-dimethyl-*N′*-(phenylmethyl)-
7162 - Hydrazine, 1-(3-methylbutyl)-1-phenyl-

$C_{11}H_{18}N_2O_3$
11092 - 2,4,6(1H,3H,5H)-Pyrimidinetrione, 5-ethyl-5-(3-methylbutyl)-

$C_{11}H_{18}N_4O_2$
9728 - Pirimicarb

$C_{11}H_{18}O$
2959 - Bicyclo[2.2.1]heptan-2-one, 1,4,7,7-tetramethyl-
2969 - Bicyclo[3.1.1]hept-2-ene-2-ethanol, 6,6-dimethyl-

4950 - 2-Cyclopenten-1-one, 3-methyl-2-pentyl-
$C_{11}H_{18}O_2$
2949 - Bicyclo[2.2.1]heptan-2-ol, 1,7,7-trimethyl-, formate, (1*R*-*endo*)-
3131 - 3-Bornanecarboxylic acid
4965 - Cyclopropanecarboxylic acid, 2,2-dimethyl-3-(2-methyl-1-propenyl)-, methyl ester, *trans*-(±)-
5026 - 2,4-Decadienoic acid, methyl ester, (*E,E*)-
5027 - 2,4-Decadienoic acid, methyl ester, (*E,Z*)-
5028 - 2,4-Decadienoic acid, methyl ester, (*Z,E*)-
5029 - 2,4-Decadienoic acid, methyl ester, (*Z,Z*)-
8337 - 1,6-Octadien-3-ol, 3,7-dimethyl-, formate
8338 - 2,6-Octadien-1-ol, 3,7-dimethyl-, formate, (*E*)-
11988 - 6-Undecynoic acid
11989 - 9-Undecynoic acid
$C_{11}H_{18}O_2Si$
11407 - Silane, diethoxymethylphenyl-
$C_{11}H_{18}O_3Si$
11427 - Silane, ethenyltris(2-propenyloxy)-
$C_{11}H_{18}O_4$
4791 - 1,1-Cyclopentanedicarboxylic acid, diethyl ester
$C_{11}H_{18}O_5$
8771 - Pentanedioic acid, 2-acetyl-, diethyl ester
9972 - Propanedioic acid, (1-oxobutyl)-, diethyl ester
$C_{11}H_{18}O_6$
5654 - 1,1,2-Ethanetricarboxylic acid, triethyl ester
$C_{11}H_{19}ClO$
11982 - 10-Undecenoyl chloride
$C_{11}H_{19}NOS$
8528 - Octhilinone
$C_{11}H_{19}NO_2$
8148 - Nonanoic acid, 9-cyano-, methyl ester
$C_{11}H_{19}N_3O$
5932 - Ethirimol
$C_{11}H_{20}$
4340 - Cyclohexane, cyclopentyl-
8201 - 4-Nonyne, 3,3-dimethyl-
11537 - Spiro[5.5]undecane
11984 - 1-Undecyne
11985 - 2-Undecyne
11986 - 4-Undecyne
11987 - 5-Undecyne
$C_{11}H_{20}ClN_5$
11828 - 1,3,5-Triazine-2,4-diamine, 6-chloro-*N,N,N′,N′*-tetraethyl-
$C_{11}H_{20}N_2O_3$
7224 - 4-Imidazolidinehexanoic acid, 5-methyl-2-oxo-, methyl ester, (4*R*-*cis*)-
$C_{11}H_{20}N_3O_3PS$
9730 - Pirimiphos-methyl
$C_{11}H_{20}O$
4592 - Cyclohexanone, 4-(1,1-dimethylpropyl)-
4888 - Cyclopentanone, 3-methyl-2-(3-methylbutyl)-
$C_{11}H_{20}O_2$
6152 - 2(3H)-Furanone, 5-heptyldihydro-
6404 - 3,5-Heptanedione, 2,2,6,6-tetramethyl-
8149 - Nonanoic acid, ethenyl ester
10636 - 2-Propenoic acid, 2-ethylhexyl ester
11974 - 10-Undecenoic acid
11975 - 9-Undecenoic acid, (*E*)-
11976 - 9-Undecenoic acid, (*Z*)-
$C_{11}H_{20}O_3$
8436 - Octanoic acid, 2-methyl-3-oxo-, ethyl ester
$C_{11}H_{20}O_4$
6394 - Heptanedioic acid, diethyl ester
6790 - Hexanedioic acid, 3-methyl-, diethyl ester
8134 - Nonanedioic acid, dimethyl ester
9925 - Propanedioic acid, bis(1,1-dimethylethyl) ester
9930 - Propanedioic acid, butyl-, diethyl ester
9938 - Propanedioic acid, dibutyl ester
9940 - Propanedioic acid, diethyl-, diethyl ester
9967 - Propanedioic acid, (1-methylpropyl)-, diethyl ester
9969 - Propanedioic acid, (2-methylpropyl)-, diethyl ester
9971 - Propanedioic acid, octyl-
$C_{11}H_{20}O_{10}$
6234 - *D*-Glucose, 6-*O*-α-*L*-arabinopyranosyl-
6237 - *D*-Glucose, 6-*O*-β-*D*-xylopyranosyl-
$C_{11}H_{21}BrO_2$
11944 - Undecanoic acid, 11-bromo-
11945 - Undecanoic acid, 2-bromo-
$C_{11}H_{21}FO_2$
11947 - Undecanoic acid, 11-fluoro-
$C_{11}H_{21}N$
11941 - Undecanenitrile
$C_{11}H_{21}NOS$
4168 - Cycloate
$C_{11}H_{21}NO_4$
330 - β-Alanine, *N*-(3-ethoxy-3-oxopropyl)-*N*-methyl-, ethyl ester
$C_{11}H_{21}N_5S$
5222 - Dipropetryn
$C_{11}H_{22}$
4427 - Cyclohexane, (3-methylbutyl)-
4465 - Cyclohexane, pentyl-
4823 - Cyclopentane, hexyl-
11966 - 1-Undecene
11967 - 3-Undecene
11968 - 2-Undecene, (*E*)-
11969 - 2-Undecene, (*Z*)-
11970 - 4-Undecene, (*E*)-
11971 - 4-Undecene, (*Z*)-
11972 - 5-Undecene, (*E*)-
11973 - 5-Undecene, (*Z*)-
$C_{11}H_{22}BrF$
11939 - Undecane, 1-bromo-11-fluoro-
$C_{11}H_{22}Br_2$
11940 - Undecane, 1,11-dibromo-
$C_{11}H_{22}N_2$
9683 - Piperidine, 1,1′-methylenebis-
$C_{11}H_{22}O$
11935 - Undecanal
11958 - 2-Undecanone
11959 - 3-Undecanone
11960 - 4-Undecanone
11961 - 5-Undecanone
11962 - 6-Undecanone
11979 - 10-Undecen-1-ol
11980 - 2-Undecen-4-ol
11981 - 9-Undecen-1-ol
$C_{11}H_{22}O_2$
220 - Acetic acid, nonyl ester
3527 - Butanoic acid, heptyl ester
5066 - Decanoic acid, methyl ester
6459 - Heptanoic acid, butyl ester
6467 - Heptanoic acid, 2-methylpropyl ester
6897 - Hexanoic acid, 3-methylbutyl ester
6907 - Hexanoic acid, pentyl ester
8150 - Nonanoic acid, ethyl ester
8435 - Octanoic acid, 1-methylethyl ester
8441 - Octanoic acid, propyl ester
8882 - Pentanoic acid, hexyl ester
10313 - Propanoic acid, octyl ester
11943 - Undecanoic acid
$C_{11}H_{22}O_3$
3635 - 1-Butanol, 3-methyl-, carbonate (2:1)
11948 - Undecanoic acid, 11-hydroxy-
$C_{11}H_{22}O_6$
6228 - α-*D*-Glucopyranoside, methyl 2,3,4,6-tetra-*O*-methyl-
$C_{11}H_{23}N$
9671 - Piperidine, 1-(1-ethyl-1-methylpropyl)-
9672 - Piperidine, 1-hexyl-
$C_{11}H_{23}NOS$
11586 - Sutan
$C_{11}H_{24}$
5054 - Decane, 2-methyl-
5055 - Decane, 3-methyl-
5056 - Decane, 4-methyl-
8420 - Octane, 2,2,4-trimethyl-
8421 - Octane, 2,4,7-trimethyl
11938 - Undecane
$C_{11}H_{24}O$
8164 - 5-Nonanol, 2,8-dimethyl-
11950 - 1-Undecanol
11951 - 2-Undecanol, (±)-
11952 - 3-Undecanol, (*R*)-
11953 - 5-Undecanol
11954 - 6-Undecanol
$C_{11}H_{24}O_3Si$
11426 - Silane, ethenyltris(1-methylethoxy)-
$C_{11}H_{24}S$
11942 - 1-Undecanethiol
$C_{11}H_{25}N$
11937 - 1-Undecanamine
$C_{11}H_{26}O_3Si$
11480 - Silane, triethoxypentyl-
$C_{11}H_{26}O_6P_2$
9546 - Phosphonic acid, 1,3-propanediylbis-, tetraethyl ester
$C_{11}H_{28}Br_2N_2$
8738 - 1,5-Pentanediaminium, *N,N,N,N′,N′,N′*-hexamethyl-, dibromide
$C_{12}F_{10}$
3027 - 1,1′-Biphenyl, 2,2′,3,3′,4,4′,5,5′,6,6′-decafluoro-
$C_{12}F_{27}N$
3228 - 1-Butanamine, 1,1,2,2,3,3,4,4,4-nonafluoro-*N,N*-bis(nonafluorobutyl)-
$C_{12}H_4N_6O_{12}S$
2288 - Benzene, 1,1′-thiobis[2,4,6-trinitro-
$C_{12}H_5N_7O_{12}$
864 - Benzenamine, 2,4,6-trinitro-*N*-(2,4,6-trinitrophenyl)-
$C_{12}H_6Cl_2O_2$
7804 - 1,8-Naphthalenedicarbonyl dichloride
$C_{12}H_6Cl_4O_2S$
2325 - Benzene, 1,2,4-trichloro-5-[(4-chlorophenyl)sulfonyl]-
9478 - Phenol, 2,2′-thiobis[4,6-dichloro-
$C_{12}H_6Cl_4S$
11648 - Tetrasul
$C_{12}H_6O_2$
8 - 1,2-Acenaphthylenedione
$C_{12}H_6O_{12}$
1748 - Benzenehexacarboxylic acid
$C_{12}H_7BrO$
5164 - Dibenzofuran, 2-bromo-
5165 - Dibenzofuran, 3-bromo-
$C_{12}H_7Cl_2NO_3$
8081 - Nitrofen
$C_{12}H_7NO_3$
5166 - Dibenzofuran, 3-nitro-
$C_{12}H_7NO_4$
9508 - 3H-Phenoxazin-3-one, 7-hydroxy-, 10-oxide
$C_{12}H_8$
3 - Acenaphthylene
7892 - Naphthalene, 1-ethynyl-
$C_{12}H_8Br_2$
3037 - 1,1′-Biphenyl, 4,4′-dibromo-
$C_{12}H_8Br_2O$
2072 - Benzene, 1,1′-oxybis[4-bromo-
$C_{12}H_8Br_2O_2S$
2238 - Benzene, 1,1′-sulfonylbis[4-bromo-
$C_{12}H_8Br_2S$
2284 - Benzene, 1,1′-thiobis[4-bromo-
$C_{12}H_8Cl_2$
3040 - 1,1′-Biphenyl, 3,3′-dichloro-
3041 - 1,1′-Biphenyl, 4,4′-dichloro-
$C_{12}H_8Cl_2O$
2073 - Benzene, 1,1′-oxybis[4-chloro-
$C_{12}H_8Cl_2O_2S$
2239 - Benzene, 1,1′-sulfonylbis[4-chloro-
9477 - Phenol, 2,2′-thiobis[4-chloro-
$C_{12}H_8Cl_2O_3S$
2221 - Benzenesulfonic acid, 4-chloro-, 4-chlorophenyl ester
9289 - Phenol, 2,4-dichloro-, benzenesulfonate

$C_{12}H_8Cl_6$
5177 - 1,4 5,8-Dimethanonaphthalene,1,2,3,4,10,10-hexachloro-1,4,4a,5,8,8a-hexahydro-,
$C_{12}H_8Cl_6O$
5178 - 2,7 3,6-Dimethanonaphth[2,3-b]oxirene, 3,4,5,6,9,9-hexachloro-1a,2,2a,3,6,6a,7,7a-octahydro-,(1aα,2β,2aα,3β,6β,6aα,7β,7aα)-
$C_{12}H_8F_2$
3042 - 1,1′-Biphenyl, 2,2′-difluoro-
3043 - 1,1′-Biphenyl, 3,3′-difluoro-
3044 - 1,1′-Biphenyl, 4,4′-difluoro-
$C_{12}H_8N_2$
9189 - 1,10-Phenanthroline
9190 - 1,7-Phenanthroline
9191 - 4,7-Phenanthroline
9193 - Phenazine
$C_{12}H_8N_2O$
9197 - 1-Phenazinol
$C_{12}H_8N_2O_4$
3062 - 1,1′-Biphenyl, 2,2′-dinitro-
3063 - 1,1′-Biphenyl, 2,4′-dinitro-
$C_{12}H_8N_2O_4S_2$
5247 - Disulfide, bis(3-nitrophenyl)
$C_{12}H_8N_2O_5$
1495 - Benzene, 2,4-dinitro-1-phenoxy-
$C_{12}H_8O$
5163 - Dibenzofuran
$C_{12}H_8O_3$
8060 - Naphtho[1,2-c]furan-1,3-dione, 4,5-dihydro-
$C_{12}H_8O_4$
6187 - 7H-Furo[3,2-g][1]benzopyran-7-one, 9-methoxy-
7805 - 1,8-Naphthalenedicarboxylic acid
$C_{12}H_8S$
5167 - Dibenzothiophene
$C_{12}H_8S_2$
11660 - Thianthrene
$C_{12}H_8S_3$
11600 - 2,2′:5′,2″-Terthiophene
$C_{12}H_8Se_2$
11353 - Selenanthrene
$C_{12}H_9Br$
4 - Acenaphthylene, 5-bromo-1,2-dihydro-
3016 - 1,1′-Biphenyl, 2-bromo-
3017 - 1,1′-Biphenyl, 3-bromo-
3018 - 1,1′-Biphenyl, 4-bromo-
$C_{12}H_9BrN_2O$
5120 - Diazene, (4-bromophenyl)phenyl-, 1-oxide
$C_{12}H_9BrO$
1079 - Benzene, 1-bromo-4-phenoxy-
$C_{12}H_9Cl$
5 - Acenaphthylene, 5-chloro-1,2-dihydro-
3023 - 1,1′-Biphenyl, 2-chloro-
3024 - 1,1′-Biphenyl, 3-chloro-
3025 - 1,1′-Biphenyl, 4-chloro-
$C_{12}H_9ClF_3N_3O$
8213 - Norflurazon
$C_{12}H_9ClO$
1221 - Benzene, 1-chloro-4-phenoxy-
3092 - [1,1′-Biphenyl]-2-ol, 3-chloro-
$C_{12}H_9ClO_2S$
1225 - Benzene, 1-chloro-4-(phenylsulfonyl)-
$C_{12}H_9ClO_2S_2$
2236 - Benzenesulfonothioic acid, 4-chloro-, *S*-phenyl ester
$C_{12}H_9Cl_2NO_3$
12048 - Vinclozolin
$C_{12}H_9Cl_3Si$
11364 - Silane, [1,1′-biphenyl]-4-yltrichloro-
$C_{12}H_9F$
3074 - 1,1′-Biphenyl, 2-fluoro-
3075 - 1,1′-Biphenyl, 4-fluoro-
$C_{12}H_9I$
7 - Acenaphthylene, 1,2-dihydro-5-iodo-
3077 - 1,1′-Biphenyl, 2-iodo-
3078 - 1,1′-Biphenyl, 3-iodo-
3079 - 1,1′-Biphenyl, 4-iodo-
$C_{12}H_9N$
3981 - 9H-Carbazole
7736 - 1-Naphthaleneacetonitrile
7737 - 2-Naphthaleneacetonitrile
$C_{12}H_9NO$
7638 - Methanone, phenyl-2-pyridinyl-
7639 - Methanone, phenyl-4-pyridinyl-
9507 - 10H-Phenoxazine
$C_{12}H_9NO_2$
3085 - 1,1′-Biphenyl, 2-nitro-
3086 - 1,1′-Biphenyl, 3-nitro-
3087 - 1,1′-Biphenyl, 4-nitro-
6180 - Furo[2,3-b]quinoline, 4-methoxy-
$C_{12}H_9NO_2S$
2057 - Benzene, 1-nitro-2-(phenylthio)-
2058 - Benzene, 1-nitro-4-(phenylthio)-
$C_{12}H_9NO_3$
2053 - Benzene, 1-nitro-2-phenoxy-
2054 - Benzene, 1-nitro-4-phenoxy-
$C_{12}H_9NS$
9499 - 10H-Phenothiazine
$C_{12}H_9N_3O_4$
1542 - 1,3-Benzenediol, 4-[(4-nitrophenyl)azo]-
$C_{12}H_{10}$
6 - Acenaphthylene, 1,2-dihydro-
3008 - 1,1′-Biphenyl
7881 - Naphthalene, 1-ethenyl-
7882 - Naphthalene, 2-ethenyl-
$C_{12}H_{10}AsCl$
454 - Arsinous chloride, diphenyl-
$C_{12}H_{10}ClN$
678 - Benzenamine, 3-chloro-*N*-phenyl-
679 - Benzenamine, 4-chloro-*N*-phenyl-
3012 - [1,1′-Biphenyl]-4-amine, 4′-chloro-
10921 - Pyridine, 2-[(4-chlorophenyl)methyl]-
$C_{12}H_{10}ClO_3P$
9575 - Phosphorochloridic acid, diphenyl ester
$C_{12}H_{10}Cl_2N_2$
3032 - [1,1′-Biphenyl]-4,4′-diamine, 2,2′-dichloro-
$C_{12}H_{10}Cl_2Si$
11395 - Silane, dichlorodiphenyl-
$C_{12}H_{10}F_2Si$
11411 - Silane, difluorodiphenyl-
$C_{12}H_{10}Hg$
7436 - Mercury, diphenyl-
$C_{12}H_{10}N_2$
5129 - Diazene, diphenyl-, (*E*)-
5130 - Diazene, diphenyl-, (*Z*)-
11040 - 9H-Pyrido[3,4-b]indole, 1-methyl-
$C_{12}H_{10}N_2O$
829 - Benzenamine, 4-nitroso-*N*-phenyl-
830 - Benzenamine, *N*-nitroso-*N*-phenyl-
5131 - Diazene, diphenyl-, 1-oxide, (*Z*)-
9448 - Phenol, 4-(phenylazo)-
$C_{12}H_{10}N_2O_2$
826 - Benzenamine, 4-nitro-*N*-phenyl-
9216 - Phenol, 2,2′-azobis-
$C_{12}H_{10}N_4$
9194 - 2,3-Phenazinediamine
$C_{12}H_{10}N_4O_2$
2472 - Benzo[g]pteridine-2,4(1H,3H)-dione, 7,8-dimethyl-
$C_{12}H_{10}O$
2071 - Benzene, 1,1′-oxybis-
3088 - [1,1′-Biphenyl]-2-ol
3089 - [1,1′-Biphenyl]-3-ol
3090 - [1,1′-Biphenyl]-4-ol
5857 - Ethanone, 1-(1-naphthalenyl)-
5858 - Ethanone, 1-(2-naphthalenyl)-
7756 - 1-Naphthalenecarboxaldehyde, 4-methyl-
$C_{12}H_{10}OS$
2183 - Benzene, 1,1′-sulfinylbis-
$C_{12}H_{10}O_2$
3064 - [1,1′-Biphenyl]-2,2′-diol
3065 - [1,1′-Biphenyl]-2,4′-diol
3066 - [1,1′-Biphenyl]-3,3′-diol
5829 - Ethanone, 1-(1-hydroxy-2-naphthalenyl)-
7734 - 1-Naphthaleneacetic acid
7755 - 1-Naphthalenecarboxaldehyde, 4-methoxy-
7766 - 1-Naphthalenecarboxylic acid, methyl ester
7776 - 2-Naphthalenecarboxylic acid, methyl ester
$C_{12}H_{10}O_2S$
2237 - Benzene, 1,1′-sulfonylbis-
$C_{12}H_{10}O_3$
217 - Acetic acid, (2-naphthalenyloxy)-
6069 - 2-Furancarboxylic acid, phenylmethyl ester
7775 - 2-Naphthalenecarboxylic acid, 3-hydroxy-, methyl ester
$C_{12}H_{10}O_4$
3097 - [1,1′-Biphenyl]-3,3′,5,5′-tetrol
8677 - 2,4-Pentadienoic acid, 5-(1,3-benzodioxol-5-yl)-
8678 - 2,4-Pentadienoic acid, 5-(1,3-benzodioxol-5-yl)-, (*E,E*)-
$C_{12}H_{10}O_4S$
9466 - Phenol, 4,4′-sulfonylbis-
$C_{12}H_{10}O_6S_2$
3070 - [1,1′-Biphenyl]-4,4′-disulfonic acid
$C_{12}H_{10}O_7$
7857 - 1,4-Naphthalenedione, 2-ethyl-3,5,6,7,8-pentahydroxy-
7858 - 1,4-Naphthalenedione, 6-ethyl-2,3,5,7,8-pentahydroxy-
$C_{12}H_{10}S$
2283 - Benzene, 1,1′-thiobis-
$C_{12}H_{10}S_2$
5258 - Disulfide, diphenyl
$C_{12}H_{10}Se$
2175 - Benzene, 1,1′-selenobis-
$C_{12}H_{10}Se_2$
5225 - Diselenide, diphenyl
$C_{12}H_{11}As$
443 - Arsine, diphenyl-
$C_{12}H_{11}BO$
3130 - Borinic acid, diphenyl-
$C_{12}H_{11}Cl_2NO$
10746 - Propyzamide
$C_{12}H_{11}N$
2 - 1-Acenaphthylenamine, 1,2-dihydro-
836 - Benzenamine, *N*-phenyl-
3010 - [1,1′-Biphenyl]-2-amine
3011 - [1,1′-Biphenyl]-4-amine
10989 - Pyridine, 2-(4-methylphenyl)-
10990 - Pyridine, 2-methyl-5-phenyl-
10991 - Pyridine, 3-methyl-2-phenyl-
11004 - Pyridine, 2-(phenylmethyl)-
11005 - Pyridine, 3-(phenylmethyl)-
11006 - Pyridine, 4-(phenylmethyl)-
$C_{12}H_{11}NO$
834 - Benzenamine, 2-phenoxy-
835 - Benzenamine, 3-phenoxy-
7733 - 1-Naphthaleneacetamide
9445 - Phenol, 2-(phenylamino)-
9446 - Phenol, 3-(phenylamino)-
9447 - Phenol, 4-(phenylamino)-
$C_{12}H_{11}NOS$
68 - Acetamide, 2-mercapto-*N*-2-naphthalenyl-
$C_{12}H_{11}NO_2$
66 - Acetamide, *N*-(2-hydroxy-1-naphthalenyl)-
2736 - 1H,5H-Benzo[ij]quinolizine-1,6(7H)-dione, 2,3-dihydro-
8032 - 1-Naphthalenol, methylcarbamate
10625 - 2-Propenoic acid, 2-cyano-3-phenyl-, ethyl ester
10645 - 2-Propenoic acid, 2-isocyano-3-phenyl-, ethyl ester
$C_{12}H_{11}NO_3S$
2234 - Benzenesulfonic acid, 4-(phenylamino)-
$C_{12}H_{11}NO_4$
5218 - 1,3-Dioxolo[4,5-h]quinolin-8(9H)-one, 6-methoxy-9-methyl-
7361 - 2H-Isoindole-2-acetic acid, 1,3-dihydro-1,3-dioxo-, ethyl ester

$C_{12}H_{11}N_3$
837 - Benzenamine, 4-(phenylazo)-
11826 - 1-Triazene, 1,3-diphenyl-
$C_{12}H_{11}N_3O_2$
10904 - 4-Pyridinecarboxylic acid, [1-(2-furanyl)ethylidene]hydrazide
$C_{12}H_{11}N_7$
10750 - 2,4,7-Pteridinetriamine, 6-phenyl-
$C_{12}H_{11}O_3P$
9534 - Phosphonic acid, diphenyl ester
$C_{12}H_{12}$
7818 - Naphthalene, 1,2-dimethyl-
7819 - Naphthalene, 1,3-dimethyl-
7820 - Naphthalene, 1,4-dimethyl-
7821 - Naphthalene, 1,5-dimethyl-
7822 - Naphthalene, 1,6-dimethyl-
7823 - Naphthalene, 1,7-dimethyl-
7824 - Naphthalene, 1,8-dimethyl-
7825 - Naphthalene, 2,3-dimethyl-
7826 - Naphthalene, 2,6-dimethyl-
7827 - Naphthalene, 2,7-dimethyl-
7885 - Naphthalene, 1-ethyl-
7886 - Naphthalene, 2-ethyl-
$C_{12}H_{12}N_22Br$
5224 - Diquat dibromide
$C_{12}H_{12}ClNO_2S$
7977 - 1-Naphthalenesulfonyl chloride, 5-(dimethylamino)-
$C_{12}H_{12}ClN_5O_4S$
4091 - Chlorsulfuron
$C_{12}H_{12}N_2$
1282 - 1,2-Benzenediamine, *N*-phenyl-
1283 - 1,4-Benzenediamine, *N*-phenyl-
3028 - [1,1′-Biphenyl]-2,2′-diamine
3029 - [1,1′-Biphenyl]-2,4′-diamine
3030 - [1,1′-Biphenyl]-3,3′-diamine
3031 - [1,1′-Biphenyl]-4,4′-diamine
7157 - Hydrazine, 1,1-diphenyl-
7158 - Hydrazine, 1,2-diphenyl-
$C_{12}H_{12}N_2O$
832 - Benzenamine, 4,4′-oxybis-
$C_{12}H_{12}N_2OS$
845 - Benzenamine, 4,4′-sulfinylbis-
$C_{12}H_{12}N_2OS_2$
11684 - 4-Thiazolidinone, 5-[[4-(dimethylamino)phenyl]methylene]-2-thioxo-
$C_{12}H_{12}N_2O_2S$
846 - Benzenamine, 4,4′-sulfonylbis-
$C_{12}H_{12}N_2O_3$
8066 - 1,8-Naphthyridine-3-carboxylic acid, 1-ethyl-1,4-dihydro-7-methyl-4-oxo-
11095 - 2,4,6(1H,3H,5H)-Pyrimidinetrione, 5-ethyl-5-phenyl-
$C_{12}H_{12}N_2S$
850 - Benzenamine, 4,4′-thiobis-
$C_{12}H_{12}N_4$
635 - Benzenamine, 4,4′-azobis-
$C_{12}H_{12}O$
6710 - 3,5-Hexadien-2-one, 6-phenyl-
7883 - Naphthalene, 1-ethoxy-
7884 - Naphthalene, 2-ethoxy-
7911 - 1-Naphthalenemethanol, α-methyl-, (±)-
8027 - 2-Naphthalenol, 1,4-dimethyl-
$C_{12}H_{12}O_2$
8045 - 1(2H)-Naphthalenone, 2-acetyl-3,4-dihydro-
10692 - 2-Propenoic acid, 3-phenyl-, 2-propenyl ester
$C_{12}H_{12}O_5$
2845 - 2H-1-Benzopyran-2-one, 6,7,8-trimethoxy-
10785 - 4H,5H-Pyrano[4,3-b]pyran-4,5-dione, 2,3-dihydro-3-hydroxy-2-methyl-7-(1-propenyl)-, [2*S*-[2α,3β,7(*E*)]]-
$C_{12}H_{12}O_6$
2370 - 1,2,4-Benzenetriol, triacetate
$C_{12}H_{12}Si$
11420 - Silane, diphenyl-
$C_{12}H_{13}ClF_3N_3O_4$
5964 - Fluchloralin
$C_{12}H_{13}ClN_4$
1284 - 1,3-Benzenediamine, 4-(phenylazo)-, monohydrochloride
11051 - 2,4-Pyrimidinediamine, 5-(4-chlorophenyl)-6-ethyl-
$C_{12}H_{13}Cl_3O_3$
258 - Acetic acid, (2,4,5-trichlorophenoxy)-, butyl ester
260 - Acetic acid, (2,4,5-trichlorophenoxy)-, 1-methylpropyl ester
$C_{12}H_{13}N$
3989 - 1H-Carbazole, 2,3,4,9-tetrahydro-
7715 - 1-Naphthalenamine, *N,N*-dimethyl-
7716 - 2-Naphthalenamine, *N,N*-dimethyl-
7718 - 1-Naphthalenamine, *N*-ethyl-
7719 - 2-Naphthalenamine, *N*-ethyl-
8064 - 2-Naphthylamine, 1,4-dimethyl-
11269 - Quinoline, 2-ethyl-3-methyl-
11270 - Quinoline, 2-ethyl-4-methyl-
11271 - Quinoline, 3-ethyl-6-methyl-
11292 - Quinoline, 2-propyl-
11300 - Quinoline, 2,3,6-trimethyl-
11301 - Quinoline, 2,3,8-trimethyl-
11302 - Quinoline, 2,4,6-trimethyl-
11303 - Quinoline, 2,4,7-trimethyl-
11304 - Quinoline, 2,5,7-trimethyl-
$C_{12}H_{13}NO$
5750 - Ethanol, 2-(2-naphthalenylamino)-
$C_{12}H_{13}NO_2$
7296 - 1H-Indole-3-butanoic acid
11156 - 2,5-Pyrrolidinedione, 1,3-dimethyl-3-phenyl-
$C_{12}H_{13}NO_2S$
4039 - Carboxin
$C_{12}H_{13}N_3$
1263 - 1,4-Benzenediamine, *N*-(4-aminophenyl)-
10974 - 2-Pyridinemethanamine, *N*-(2-pyridinylmethyl)-
$C_{12}H_{13}N_3O_4S_2$
2188 - Benzenesulfonamide, 4-amino-*N*-[4-(aminosulfonyl)phenyl]-
$C_{12}H_{14}$
9 - Acenaphthylene, 1,2,2a,3,4,5-hexahydro-
1246 - Benzene, 1-cyclohexen-1-yl-
1247 - Benzene, 2-cyclohexen-1-yl-
1625 - Benzene, (1-ethenyl-3-butenyl)-
1753 - Benzene, 1,3-hexadienyl-
1754 - Benzene, 1,5-hexadienyl-
1755 - Benzene, 3,5-hexadienyl-
$C_{12}H_{14}ClNO_2$
4140 - Clomazone
$C_{12}H_{14}ClN_5O_2S$
2209 - Benzenesulfonamide, 4-[(2,4-diaminophenyl)azo]-
$C_{12}H_{14}Cl_2O_3$
171 - Acetic acid, (2,4-dichlorophenoxy)-, butyl ester
$C_{12}H_{14}Cl_3O_4P$
4078 - Chlorfenvinphos
$C_{12}H_{14}N_2$
5514 - 1,2-Ethanediamine, *N*-1-naphthalenyl-
11213 - 6-Quinolinamine, *N,N*,2-trimethyl-
$C_{12}H_{14}N_2O_2$
11066 - 4,6(1H,5H)-Pyrimidinedione, 5-ethyldihydro-5-phenyl-
11923 - *L*-Tryptophan, *N*-methyl-
$C_{12}H_{14}N_2O_3$
11085 - 2,4,6(1H,3H,5H)-Pyrimidinetrione, 5-(2-cyclopenten-1-yl)-5-(2-propenyl)-
$C_{12}H_{14}N_2O_5$
6242 - *L*-Glutamic acid, *N*-(4-aminobenzoyl)-
9273 - Phenol, 2-cyclohexyl-4,6-dinitro-
$C_{12}H_{14}N_4OS$
11111 - 5H-Pyrimido[4,5-d]thiazolo[3,2-a]pyrimidine-8-ethanol, 2,7-dimethyl-
$C_{12}H_{14}N_4O_2S$
2196 - Benzenesulfonamide, 4-amino-*N*-(4,6-dimethyl-2-pyrimidinyl)-
$C_{12}H_{14}N_4O_4S$
2194 - Benzenesulfonamide, 4-amino-*N*-(2,6-dimethoxy-4-pyrimidinyl)-
$C_{12}H_{14}N_4O_4S_2$
11723 - Thiophanate-methyl
$C_{12}H_{14}O$
2460 - Benzofuran, 4-methyl-7-(1-methylethyl)-
4893 - Cyclopentanone, 2-(phenylmethyl)-
7101 - 5-Hexen-2-one, 4-phenyl-
7102 - 5-Hexen-2-one, 6-phenyl-
8051 - 1(2H)-Naphthalenone, 7-ethyl-3,4-dihydro-
8987 - 2-Pentanone, 3-(phenylmethylene)-
$C_{12}H_{14}O_2$
4135 - Cinnamic acid, β-propyl-
6038 - Furan, 2,2′-*sec*-butylidenedi-
10687 - 2-Propenoic acid, 3-phenyl-, 1-methylethyl ester, (*E*)-
10693 - 2-Propenoic acid, 3-phenyl-, propyl ester, (*E*)-
$C_{12}H_{14}O_3$
884 - Benzeneacetic acid, α-acetyl-, ethyl ester
2100 - Benzenepentanoic acid, δ-oxo-, methyl ester
8604 - Oxiranecarboxylic acid, 3-methyl-3-phenyl-, ethyl ester
8911 - Pentanoic acid, 4-oxo-, phenylmethyl ester
9393 - Phenol, 2-methoxy-4-(1-propenyl)-, acetate, (*E*)-
9394 - Phenol, 2-methoxy-4-(1-propenyl)-, acetate, (*Z*)-
9395 - Phenol, 2-methoxy-4-(2-propenyl)-, acetate
$C_{12}H_{14}O_4$
1259 - 1,4-Benzenediacetic acid, dimethyl ester
1342 - 1,2-Benzenedicarboxylic acid, diethyl ester
1346 - 1,3-Benzenedicarboxylic acid, diethyl ester
1350 - 1,4-Benzenedicarboxylic acid, diethyl ester
2422 - 1,3-Benzodioxole, 4,7-dimethoxy-5-(1-propenyl)-, (*E*)-
2423 - 1,3-Benzodioxole, 4,7-dimethoxy-5-(2-propenyl)-
8625 - Oxirane, 2,2′-[1,3-phenylenebis(oxymethylene)]bis-
$C_{12}H_{15}Br$
1009 - Benzene, 1-bromo-4-cyclohexyl-
$C_{12}H_{15}ClNO_4PS_2$
9514 - Phosalone
$C_{12}H_{15}ClO$
2907 - Benzoyl chloride, 4-pentyl-
9254 - Phenol, 2-chloro-4-cyclohexyl-
$C_{12}H_{15}ClO_2$
2908 - Benzoyl chloride, 4-(pentyloxy)-
$C_{12}H_{15}ClO_3$
10231 - Propanoic acid, 2-(4-chlorophenoxy)-2-methyl-, ethyl ester
$C_{12}H_{15}N$
964 - Benzeneacetonitrile, α-(2-methylpropyl)-
1800 - Benzenemethanamine, *N*-3-butynyl-*N*-methyl-
2737 - 1H,5H-Benzo[ij]quinolizine, 2,3,6,7-tetrahydro-
$C_{12}H_{15}NO$
9634 - Piperidine, 1-benzoyl-
$C_{12}H_{15}NO_2$
1251 - Benzene, 1-cyclohexyl-2-nitro-
1252 - Benzene, 1-cyclohexyl-4-nitro-
$C_{12}H_{15}NO_3$
3991 - Carbofuran
5214 - 1,3-Dioxolo[4,5-g]isoquinoline, 5,6,7,8-tetrahydro-4-methoxy-6-methyl-
$C_{12}H_{15}NO_4$
5216 - 1,3-Dioxolo[4,5-g]isoquinolin-5-ol, 5,6,7,8-tetrahydro-4-methoxy-6-methyl-
$C_{12}H_{15}N_3O_3$
11842 - 1,3,5-Triazine-2,4,6(1H,3H,5H)-trione, 1,3,5-tri-2-propenyl-

$C_{12}H_{15}N_3O_6$
1461 - Benzene, 1-(1,1-dimethylethyl)-3,5-dimethyl-2,4,6-trinitro-

$C_{12}H_{16}$
1248 - Benzene, cyclohexyl-
1760 - Benzene, 2-hexenyl-
1761 - Benzene, 3-hexenyl-
1762 - Benzene, 5-hexenyl-
1973 - Benzene, 1-(1-methylethenyl)-4-(1-methylethyl)-
2008 - Benzene, (1-methyl-1-pentenyl)-
4744 - 1,3,5,7-Cyclooctatetraene, 1-butyl-
7261 - 1H-Indene, 2,3-dihydro-1,1,5-trimethyl-
7262 - 1H-Indene, 2,3-dihydro-1,4,7-trimethyl-
7888 - Naphthalene, 1-ethyl-1,2,3,4-tetrahydro-
7889 - Naphthalene, 2-ethyl-1,2,3,4-tetrahydro-
7890 - Naphthalene, 5-ethyl-1,2,3,4-tetrahydro-
7891 - Naphthalene, 6-ethyl-1,2,3,4-tetrahydro-
7980 - Naphthalene, 1,2,3,4-tetrahydro-1,1-dimethyl-
7981 - Naphthalene, 1,2,3,4-tetrahydro-1,2-dimethyl-
7982 - Naphthalene, 1,2,3,4-tetrahydro-1,3-dimethyl-
7983 - Naphthalene, 1,2,3,4-tetrahydro-1,4-dimethyl-
7984 - Naphthalene, 1,2,3,4-tetrahydro-1,5-dimethyl-
7985 - Naphthalene, 1,2,3,4-tetrahydro-2,2-dimethyl-
7986 - Naphthalene, 1,2,3,4-tetrahydro-2,3-dimethyl-
7987 - Naphthalene, 1,2,3,4-tetrahydro-2,5-dimethyl-
7988 - Naphthalene, 1,2,3,4-tetrahydro-2,6-dimethyl-
7989 - Naphthalene, 1,2,3,4-tetrahydro-2,7-dimethyl-
7990 - Naphthalene, 1,2,3,4-tetrahydro-2,8-dimethyl-
7991 - Naphthalene, 1,2,3,4-tetrahydro-5,6-dimethyl-
7992 - Naphthalene, 1,2,3,4-tetrahydro-5,7-dimethyl-
7993 - Naphthalene, 1,2,3,4-tetrahydro-5,8-dimethyl-
7994 - Naphthalene, 1,2,3,4-tetrahydro-6,7-dimethyl-

$C_{12}H_{16}ClNOS$
11697 - Thiobencarb

$C_{12}H_{16}N_2$
7303 - 1H-Indole-3-ethanamine, *N,N*-dimethyl-

$C_{12}H_{16}N_2O$
7326 - 1H-Indol-5-ol, 3-[2-(dimethylamino)ethyl]-
7578 - 1,5-Methano-8H-pyrido[1,2-a][1,5]diazocin-8-one, 1,2,3,4,5,6-hexahydro-3-methyl-, (1*R*)-

$C_{12}H_{16}N_2O_3$
11083 - 2,4,6(1H,3H,5H)-Pyrimidinetrione, 5-(1-cyclohexen-1-yl)-1,5-dimethyl-
11084 - 2,4,6(1H,3H,5H)-Pyrimidinetrione, 5-(1-cyclohexen-1-yl)-5-ethyl-

$C_{12}H_{16}N_2O_5$
1463 - Benzene, 1-(1,1-dimethylethyl)-2-methoxy-4-methyl-3,5-dinitro-

$C_{12}H_{16}O$
1253 - Benzene, (cyclohexyloxy)-
3669 - 1-Butanone, 3-methyl-1-(2-methylphenyl)-
3670 - 1-Butanone, 3-methyl-1-(4-methylphenyl)-
4552 - Cyclohexanol, 1-phenyl-
4553 - Cyclohexanol, 2-phenyl-, *cis*-(±)-
4554 - Cyclohexanol, 2-phenyl-, *trans*-(±)-
5808 - Ethanone, 1-[4-(1,1-dimethylethyl)phenyl]-
5850 - Ethanone, 1-[2-methyl-5-(1-methylethyl)phenyl]-
6983 - 1-Hexanone, 1-phenyl-
6984 - 2-Hexanone, 3-phenyl-
6985 - 3-Hexanone, 2-phenyl-
8983 - 1-Pentanone, 4-methyl-1-phenyl-
8984 - 2-Pentanone, 3-methyl-1-phenyl-
9272 - Phenol, 4-cyclohexyl-

$C_{12}H_{16}O_2$
890 - Benzeneacetic acid, α-butyl-
932 - Benzeneacetic acid, 2-methylpropyl ester
1759 - Benzenehexanoic acid
2129 - Benzenepropanoic acid, α,α-dimethyl-, methyl ester
2140 - Benzenepropanoic acid, 1-methylethyl ester
2146 - Benzenepropanoic acid, β-propyl-
2685 - Benzoic acid, 3-methyl-, butyl ester
2713 - Benzoic acid, pentamethyl-
3560 - Butanoic acid, 3-methyl-, phenylmethyl ester
3632 - 1-Butanol, 3-methyl-, benzoate
7175 - Hydrocinnamic acid, β-ethyl-β-methyl-
8979 - 1-Pentanone, 1-(3-methoxyphenyl)-
9419 - Phenol, 2-methyl-5-(1-methylethyl)-, acetate
9420 - Phenol, 5-methyl-2-(1-methylethyl)-, acetate
10306 - Propanoic acid, 2-methyl-, 2-phenylethyl ester

$C_{12}H_{16}O_3$
2356 - Benzene, 1,2,4-trimethoxy-5-(1-propenyl)-
2633 - Benzoic acid, 2-hydroxy-, 3-methylbutyl ester
2649 - Benzoic acid, 2-hydroxy-, pentyl ester
2668 - Benzoic acid, 4-methoxy-, butyl ester
3581 - Butanoic acid, 4-phenoxy-, ethyl ester
3582 - Butanoic acid, 2-phenoxyethyl ester
6962 - 1-Hexanone, 1-(2,4-dihydroxyphenyl)-
8886 - Pentanoic acid, 2-methoxyphenyl ester
10641 - 2-Propenoic acid, 3-(2-furanyl)-, pentyl ester

$C_{12}H_{16}O_4$
2628 - Benzoic acid, 4-hydroxy-3-methoxy-, 2-methylpropyl ester

$C_{12}H_{16}O_7$
6224 - β-*D*-Glucopyranoside, 4-hydroxyphenyl

$C_{12}H_{17}BrO$
1045 - Benzene, 1-bromo-4-(hexyloxy)-

$C_{12}H_{17}N$
680 - Benzenamine, *N*-cyclohexyl-
9698 - Piperidine, 1-(phenylmethyl)-
9699 - Piperidine, 2-(phenylmethyl)-
9700 - Piperidine, 4-(phenylmethyl)-

$C_{12}H_{17}NO$
39 - Acetamide, *N*-butyl-*N*-phenyl-
612 - Benzamide, *N,N*-diethyl-3-methyl-
6722 - Hexanamide, *N*-phenyl-

$C_{12}H_{17}NO_2$
9769 - Promecarb

$C_{12}H_{17}NO_3$
611 - Benzamide, *N,N*-diethyl-4-hydroxy-3-methoxy-
880 - Benzeneacetamide, 4-butoxy-*N*-hydroxy-
7395 - 8-Isoquinolinol, 1,2,3,4-tetrahydro-6,7-dimethoxy-1-methyl-, (*S*)-

$C_{12}H_{17}N_3O_4$
6247 - *L*-Glutamic acid, 5-[2-[4-(hydroxymethyl)phenyl]hydrazide]

$C_{12}H_{18}$
992 - Benzene, 1,2-bis(1-methylethyl)-
993 - Benzene, 1,3-bis(1-methylethyl)-
994 - Benzene, 1,4-bis(1-methylethyl)-
1455 - Benzene, (1,1-dimethylbutyl)-
1460 - Benzene, 1-(1,1-dimethylethyl)-3,5-dimethyl-
1473 - Benzene, 2,4-dimethyl-1-(1-methylpropyl)-
1557 - Benzene, 1,4-dipropyl-
1669 - Benzene, (1-ethylbutyl)-
1691 - Benzene, (1-ethyl-1-methylpropyl)-
1758 - Benzene, hexamethyl-
1763 - Benzene, hexyl-
2009 - Benzene, (1-methylpentyl)-
2010 - Benzene, (3-methylpentyl)-
2011 - Benzene, (4-methylpentyl)-
2344 - Benzene, 1,2,4-triethyl-
2345 - Benzene, 1,3,5-triethyl-
2361 - Benzene, 1,2,4-trimethyl-5-(1-methylethyl)-
4212 - 1,5,9-Cyclododecatriene
4213 - 1,5,9-Cyclododecatriene, (*Z,E,E*)-
4214 - 1,5,9-Cyclododecatriene, (*Z,Z,Z*)-
5286 - 3,9-Dodecadiyne
5287 - 5,7-Dodecadiyne
8339 - 2,6-Octadien-4-yne, 3,6-diethyl-

$C_{12}H_{18}Cl_2N_4OS$
11686 - Thiazolium, 3-[(4-amino-2-methyl-5-pyrimidinyl)methyl]-5-(2-hydroxyethyl)-4-methyl- chloride, monohydrochlorid

$C_{12}H_{18}N_2O$
11188 - 2-Pyrrolidinone, 1-[4-(1-pyrrolidinyl)-2-butynyl]-

$C_{12}H_{18}N_2O_3$
11090 - 2,4,6(1H,3H,5H)-Pyrimidinetrione, 5-(2,2-dimethylpropyl)-5-(2-propenyl)-

$C_{12}H_{18}N_2O_3S$
2207 - Benzenesulfonamide, *N*-[(butylamino)carbonyl]-4-methyl-

$C_{12}H_{18}N_4O$
7171 - Hydrazine, phenyl-, hemihydrate

$C_{12}H_{18}N_4O_6S$
8553 - Oryzalin

$C_{12}H_{18}O$
1488 - Benzene, 1-(1,1-dimethylpropyl)-4-methoxy-
1765 - Benzene, (hexyloxy)-
1847 - Benzenemethanol, α-butyl-α-methyl-
1875 - Benzenemethanol, α-ethyl-α-propyl-
1909 - Benzenemethanol, α-pentyl-
1951 - Benzene, [(3-methylbutoxy)methyl]-
2103 - Benzenepentanol, α-methyl-
2159 - Benzenepropanol, γ-propyl-
6949 - 3-Hexanol, 1-phenyl-
9226 - Phenol, 2,6-bis(1-methylethyl)-
9324 - Phenol, 2-(1,1-dimethylethyl)-4,5-dimethyl-
9325 - Phenol, 2-(1,1-dimethylethyl)-4,6-dimethyl-
9326 - Phenol, 4-(1,1-dimethylethyl)-2,5-dimethyl-
9327 - Phenol, 4-(1,1-dimethylethyl)-2,6-dimethyl-
9340 - Phenol, 4-(1,1-dimethylpropyl)-2-methyl-
9347 - Phenol, 2,6-dipropyl-
9425 - Phenol, 5-methyl-2-pentyl-

$C_{12}H_{18}O_2$
1519 - 1,3-Benzenediol, 4-hexyl-
1555 - Benzene, 1,2-dipropoxy-
1556 - Benzene, 1,3-dipropoxy-
2950 - Bicyclo[2.2.1]heptan-2-one, 3-(acetyl-17*O*)-1,7,7-trimethyl-
11863 - Tricyclo[2.2.1.02,6]heptane-1-carboxylic acid, 7,7-dimethyl-, ethyl ester

$C_{12}H_{18}O_3$
2343 - Benzene, 1,3,5-triethoxy-

$C_{12}H_{18}O_4$
4644 - 1-Cyclohexene-1,2-dicarboxylic acid, diethyl ester
4645 - 1-Cyclohexene-1,3-dicarboxylic acid, diethyl ester
9935 - Propanedioic acid, 2-cyclopenten-1-yl-, diethyl ester
9936 - Propanedioic acid, cyclopentylidene-, diethyl ester

$C_{12}H_{18}O_5$
10669 - 2-Propenoic acid, 2-methyl-, oxydi-2,1-ethanediyl ester

$C_{12}H_{18}O_6$
10601 - 1-Propene-1,2,3-tricarboxylic acid, triethyl ester, (*E*)-

$C_{12}H_{19}ClNO_3P$
4153 - Crufomate

$C_{12}H_{19}N$
636 - Benzenamine, 2,6-bis(1-methylethyl)-
730 - Benzenamine, *N,N*-dipropyl-

$C_{12}H_{19}NO$
738 - Benzenamine, 3-ethoxy-*N,N*-diethyl-

$C_{12}H_{20}$
5342 - 1-Dodecen-3-yne

$C_{12}H_{20}As_2Cl_2O_5$
9275 - Phenol, 4,4′-(1,2-diarsenediyl)bis[2-amino-, dihydrochloride

$C_{12}H_{20}N_2O_2$
10811 - 2(1H)-Pyrazinone, 1-hydroxy-6-(1-methylpropyl)-3-(2-methylpropyl)-, (+)-

$C_{12}H_{20}N_4O_2$
6995 - Hexazinone

$C_{12}H_{20}O$
2989 - [1,1′-Bicyclohexyl]-2-one
6032 - Furan, 2,5-bis(1,1-dimethylethyl)-

$C_{12}H_{20}O_2$
2945 - Bicyclo[2.2.1]heptan-2-ol, 1,7,7-trimethyl-, acetate, *endo*-(±)-
2946 - Bicyclo[2.2.1]heptan-2-ol, 1,7,7-trimethyl-, acetate, *exo*-(±)-
2947 - Bicyclo[2.2.1]heptan-2-ol, 4,7,7-trimethyl-, acetate, (1*R*-*endo*)-

4545 - Cyclohexanol, 5-methyl-2-(1-methylethenyl)-, acetate, [1*R*-(1α,2β,5α)]-
4669 - 3-Cyclohexene-1-methanol, α,α,4-trimethyl-, acetate
6135 - Furan, 2-(octyloxy)-
8335 - 1,6-Octadien-3-ol, 3,7-dimethyl-, acetate, (*R*)-
8336 - 2,6-Octadien-1-ol, 3,7-dimethyl-, acetate, (*Z*)-
11990 - 9-Undecynoic acid, methyl ester
11991 - 10-Undecynoic acid, methyl ester

$C_{12}H_{20}O_3Si$
11481 - Silane, triethoxyphenyl-

$C_{12}H_{20}O_4$
3767 - 2-Butenedioic acid (*E*)-, bis(2-methylpropyl) ester
3774 - 2-Butenedioic acid (*E*)-, dibutyl ester
4350 - 1,2-Cyclohexanedicarboxylic acid, diethyl ester, *trans*-(±)-
4351 - 1,3-Cyclohexanedicarboxylic acid, diethyl ester, *cis*-
4355 - 1,4-Cyclohexanedicarboxylic acid, diethyl ester, *cis*-
4356 - 1,4-Cyclohexanedicarboxylic acid, diethyl ester, *trans*-
4795 - 1,3-Cyclopentanedicarboxylic acid, 1,2,2-trimethyl-, dimethyl ester, (1*R*-*cis*)-
5336 - 2-Dodecenedioic acid, (*E*)-

$C_{12}H_{20}O_4Si$
11511 - Silicic acid (H_4SiO_4), triethyl phenyl ester

$C_{12}H_{20}O_6$
10198 - 1,2,3-Propanetriol, tripropanoate

$C_{12}H_{20}O_7$
10169 - 1,2,3-Propanetricarboxylic acid, 2-hydroxy-, triethyl ester

$C_{12}H_{20}Si$
11485 - Silane, triethylphenyl-

$C_{12}H_{21}NO$
4611 - Cyclohexanone, 2-(1-piperidinylmethyl)-

$C_{12}H_{21}N_2O_3PS$
9582 - Phosphorothioic acid, *O*,*O*-diethyl *O*-[6-methyl-2-(1-methylethyl)-4-pyrimidinyl] ester

$C_{12}H_{22}$
2988 - 1,1′-Bicyclohexyl
4215 - Cyclododecene, (*E*)-
4216 - Cyclododecene, (*Z*)-
5101 - 4-Decyne, 3,3-dimethyl-
5285 - 1,11-Dodecadiene
5343 - 1-Dodecyne
5344 - 2-Dodecyne
5345 - 3-Dodecyne
5346 - 6-Dodecyne

$C_{12}H_{22}O$
4211 - Cyclododecanone
4461 - Cyclohexane, 1,1′-oxybis-

$C_{12}H_{22}OSi_2$
11497 - Silane, trimethyl[4-[(trimethylsilyl)oxy]phenyl]-

$C_{12}H_{22}O_2$
2948 - Bicyclo[2.2.1]heptan-2-ol, 1,7,7-trimethyl-, acetate, *endo*-
4412 - Cyclohexanehexanoic acid
4549 - Cyclohexanol, 5-methyl-2-(1-methylethyl)-, acetate, [1*R*-(1α,2β,5α)]-
4787 - Cyclopentanecarboxylic acid, 3-isopropyl-1-methyl-
4788 - Cyclopentanecarboxylic acid, 2-methyl-5-(1-methylethyl)-, ethyl ester
5337 - 2-Dodecenoic acid
5338 - 4-Dodecenoic acid
5339 - 5-Dodecenoic acid
5340 - 11-Dodecenoic acid
8400 - 3,6-Octanedione, 2,2,7,7-tetramethyl-
10659 - 2-Propenoic acid, 2-methyl-, 2-ethylhexyl ester
11978 - 10-Undecenoic acid, methyl ester

$C_{12}H_{22}O_2Si_2$
11447 - Silane, [1,3-phenylenebis(oxy)]bis[trimethyl-

$C_{12}H_{22}O_3$
210 - Acetic acid, [[5-methyl-2-(1-methylethyl)cyclohexyl]oxy]-, [1*R*-(1α,2β,5α)]-
6874 - Hexanoic acid, anhydride

$C_{12}H_{22}O_4$
3296 - Butanedioic acid, bis(1,1-dimethylethyl) ester
3297 - Butanedioic acid, bis(1-methylpropyl) ester
3301 - Butanedioic acid, dibutyl ester
3332 - Butanedioic acid, (1-methylpropyl)-, diethyl ester
5046 - Decanedioic acid, dimethyl ester
5299 - Dodecanedioic acid
5555 - Ethanedioic acid, bis(3-methylbutyl) ester
6782 - Hexanedioic acid, bis(1-methylethyl) ester
6787 - Hexanedioic acid, dipropyl ester
8387 - Octanedioic acid, diethyl ester
9951 - Propanedioic acid, ethyl(1-methylethyl)-, diethyl ester
9961 - Propanedioic acid, (3-methylbutyl)-, diethyl ester
9974 - Propanedioic acid, pentyl-, diethyl ester

$C_{12}H_{22}O_6$
3313 - Butanedioic acid, 2,3-dihydroxy- [*R*-(*R**,*R**)]-, dibutyl ester

$C_{12}H_{22}O_{10}$
6236 - *D*-Glucose, 6-*O*-(6-deoxy-α-*L*-mannopyranosyl)-

$C_{12}H_{22}O_{11}$
6022 - *D*-Fructopyranose, 3-*O*-α-*D*-glucopyranosyl-
6214 - β-*D*-Glucopyranose, 4-*O*-β-*D*-galactopyranosyl-
6218 - α-*D*-Glucopyranoside, β-*D*-fructofuranosyl
6220 - α-*D*-Glucopyranoside, α-*D*-glucopyranosyl

$C_{12}H_{22}O_{12}$
6210 - *D*-Gluconic acid, 4-*O*-β-*D*-galactopyranosyl-

$C_{12}H_{23}BrO_2$
5307 - Dodecanoic acid, 2-bromo-

$C_{12}H_{23}ClO$
5328 - Dodecanoyl chloride

$C_{12}H_{23}N$
4284 - Cyclohexanamine, *N*-cyclohexyl-
5303 - Dodecanenitrile

$C_{12}H_{23}NSi_2$
11362 - Silanamine, 1,1,1-trimethyl-*N*-phenyl-*N*-(trimethylsilyl)-

$C_{12}H_{23}P$
9516 - Phosphine, dicyclohexyl-

$C_{12}H_{24}$
4210 - Cyclododecane
4414 - Cyclohexane, hexyl-
4446 - Cyclohexane, 1-methyl-2-pentyl-
4737 - Cyclooctane, butyl-
4820 - Cyclopentane, heptyl-
5335 - 1-Dodecene
10582 - 1-Propene, 2-methyl-, trimer

$C_{12}H_{24}BrF$
5294 - Dodecane, 1-bromo-12-fluoro-

$C_{12}H_{24}Br_2$
5296 - Dodecane, 1,12-dibromo-

$C_{12}H_{24}N_2$
9661 - Piperidine, 1,1′-(1,2-ethanediyl)bis-

$C_{12}H_{24}O$
5051 - Decane, 1-(ethenyloxy)-
5288 - Dodecanal
5325 - 2-Dodecanone
5326 - 6-Dodecanone
11936 - Undecanal, 2-methyl-

$C_{12}H_{24}O_2$
157 - Acetic acid, decyl ester
3566 - Butanoic acid, octyl ester
5063 - Decanoic acid, ethyl ester
5305 - Dodecanoic acid
6470 - Heptanoic acid, pentyl ester
6890 - Hexanoic acid, hexyl ester
8155 - Nonanoic acid, 3-methyl-, ethyl ester
8427 - Octanoic acid, butyl ester
8881 - Pentanoic acid, heptyl ester

$C_{12}H_{24}O_3$
11949 - Undecanoic acid, 11-hydroxy-, methyl ester

$C_{12}H_{24}O_4Si_4$
4996 - Cyclotetrasiloxane, 2,4,6,8-tetraethenyl-2,4,6,8-tetramethyl-

$C_{12}H_{25}Br$
5293 - Dodecane, 1-bromo-

$C_{12}H_{25}Cl$
5295 - Dodecane, 1-chloro-

$C_{12}H_{25}Cl_3Si$
11460 - Silane, trichlorododecyl-

$C_{12}H_{25}I$
5302 - Dodecane, 1-iodo-

$C_{12}H_{25}NO$
5289 - Dodecanamide

$C_{12}H_{26}$
5292 - Dodecane
6433 - Heptane, 2,2,4,6,6-pentamethyl-

$C_{12}H_{26}O$
5321 - 1-Dodecanol
5322 - 2-Dodecanol
5323 - 3-Dodecanol
5324 - 6-Dodecanol
6842 - Hexane, 1,1′-oxybis-
11955 - 1-Undecanol, 2-methyl-
11956 - 3-Undecanol, 2-methyl-
11957 - 5-Undecanol, 2-methyl-

$C_{12}H_{26}O_2$
5301 - 1,12-Dodecanediol

$C_{12}H_{26}O_3$
3417 - Butane, 1,1′-[oxybis(2,1-ethanediyloxy)]bis-

$C_{12}H_{26}O_6P_2S_4$
5201 - Dioxathion

$C_{12}H_{26}S$
5304 - 1-Dodecanethiol
6855 - Hexane, 1,1′-thiobis-

$C_{12}H_{27}B$
3123 - Borane, tris(2-methylpropyl)-

$C_{12}H_{27}BO_3$
3124 - Boric acid, tributyl ester

$C_{12}H_{27}N$
3213 - 1-Butanamine, *N*,*N*-dibutyl-
5291 - 1-Dodecanamine
6727 - 1-Hexanamine, *N*-hexyl-
9820 - 1-Propanamine, 2-methyl-*N*,*N*-bis(2-methylpropyl)-

$C_{12}H_{27}NO$
5682 - Ethanol, 2-[bis(3-methylbutyl)amino]-

$C_{12}H_{27}NO_2$
8936 - 2-Pentanol, 1,1′-iminobis[2-methyl-

$C_{12}H_{27}OPS_3$
11852 - *S*,*S*,*S*-Tributyl phosphorotrothioate

$C_{12}H_{27}O_3P$
9590 - Phosphorous acid, tributyl ester

$C_{12}H_{27}O_4P$
9562 - Phosphoric acid tributyl ester
9572 - Phosphoric acid, tris(2-methylpropyl) ester

$C_{12}H_{27}P$
9520 - Phosphine, tributyl-

$C_{12}H_{27}PS_3$
7438 - Merphos

$C_{12}H_{28}IN$
9833 - 1-Propanaminium, *N*,*N*,*N*-tripropyl-, iodide

$C_{12}H_{28}N_2$
3262 - 1,4-Butanediamine, *N*,*N*,*N*′,*N*′-tetraethyl-

$C_{12}H_{28}O_4Si$
11510 - Silicic acid (H4SiO4), tetrapropyl ester

$C_{12}H_{28}Si$
11454 - Silane, tributyl-

$C_{12}H_{28}Sn$
11546 - Stannane, tetrapropyl-

$C_{12}H_{30}OSi_2$
5235 - Disiloxane, hexaethyl-

$C_{12}H_{30}O_{13}P_4$
11645 - Tetraphosphoric acid, hexaethyl ester

$C_{12}H_{32}O_4Si_4$
4997 - Cyclotetrasiloxane, 2,4,6,8-tetraethyl-2,4,6,8-tetramethyl-
$C_{12}H_{36}O_4Si_5$
8997 - Pentasiloxane, dodecamethyl-
$C_{12}H_{36}O_6Si_6$
4624 - Cyclohexasiloxane, dodecamethyl-
$C_{13}H_5N_3O_7$
5979 - 9H-Fluoren-9-one, 2,4,7-trinitro-
$C_{13}H_6Cl_6O_2$
9409 - Phenol, 2,2′-methylenebis[3,4,6-trichloro-
$C_{13}H_7ClF_3NO_3$
8082 - Nitrofluorfen
$C_{13}H_7ClO$
5976 - 9H-Fluoren-9-one, 2-chloro-
$C_{13}H_7Cl_4NO_2$
610 - Benzamide, 3,5-dichloro-*N*-(3,4-dichlorophenyl)-2-hydroxy-
$C_{13}H_7F_3N_2O_5$
5980 - Fluorodifen
$C_{13}H_7NO_3$
5977 - 9H-Fluoren-9-one, 2-nitro-
$C_{13}H_8Br_2O$
7590 - Methanone, bis(4-bromophenyl)-
$C_{13}H_8Br_3NO_2$
609 - Benzamide, 3,5-dibromo-*N*-(4-bromophenyl)-2-hydroxy-
$C_{13}H_8ClN$
315 - Acridine, 9-chloro-
$C_{13}H_8Cl_2$
5972 - 9H-Fluorene, 2,7-dichloro-
$C_{13}H_8Cl_2O$
7591 - Methanone, bis(3-chlorophenyl)-
7592 - Methanone, bis(4-chlorophenyl)-
7603 - Methanone, (2-chlorophenyl)(4-chlorophenyl)-
$C_{13}H_8Cl_3NO_2$
608 - Benzamide, 5-chloro-*N*-(3,4-dichlorophenyl)-2-hydroxy-
$C_{13}H_8I_2O$
7598 - Methanone, bis(4-iodophenyl)-
$C_{13}H_8O$
5975 - 9H-Fluoren-9-one
$C_{13}H_8OS$
11812 - 9H-Thioxanthen-9-one
$C_{13}H_8O_2$
12050 - 9H-Xanthen-9-one
$C_{13}H_8O_4$
12051 - 9H-Xanthen-9-one, 1,3-dihydroxy-
12052 - 9H-Xanthen-9-one, 2,7-dihydroxy-
12053 - 9H-Xanthen-9-one, 3,6-dihydroxy-
$C_{13}H_9Br$
5968 - 9H-Fluorene, 2-bromo-
$C_{13}H_9BrO$
7600 - Methanone, (2-bromophenyl)phenyl-
7601 - Methanone, (3-bromophenyl)phenyl-
7602 - Methanone, (4-bromophenyl)phenyl-
$C_{13}H_9ClO$
7604 - Methanone, (4-chlorophenyl)phenyl-
$C_{13}H_9Cl_2FN_2S$
11802 - Thiourea, *N*-(3,5-dichlorophenyl)-*N*′-(4-fluorophenyl)-
$C_{13}H_9N$
314 - Acridine
2437 - Benzo[f]quinoline
2473 - Benzo[g]quinoline
2474 - Benzo[h]quinoline
3020 - [1,1′-Biphenyl]-2-carbonitrile
9188 - Phenanthridine
$C_{13}H_9NOS$
9217 - Phenol, 2-(2-benzothiazolyl)-
$C_{13}H_9NO_2$
9218 - Phenol, 2-(2-benzoxazolyl)-
$C_{13}H_9NO_3$
7635 - Methanone, (3-nitrophenyl)phenyl-
7636 - Methanone, (4-nitrophenyl)phenyl-
$C_{13}H_9NS$
2864 - Benzothiazole, 2-phenyl-
$C_{13}H_9N_3O_5$
2648 - Benzoic acid, 2-hydroxy-5-[(4-nitrophenyl)azo]-
$C_{13}H_{10}$
5967 - 9H-Fluorene
$C_{13}H_{10}Br_2$
1962 - Benzene, 1,1′-methylenebis[2-bromo-
$C_{13}H_{10}Cl_2$
1389 - Benzene, 1,1′-(dichloromethylene)bis-
1963 - Benzene, 1,1′-methylenebis[4-chloro-
$C_{13}H_{10}Cl_2O_2$
1965 - Benzene, 1,1′-[methylenebis(oxy)]bis[4-chloro-
9408 - Phenol, 2,2′-methylenebis[4-chloro-
$C_{13}H_{10}Cl_2S$
4075 - Chlorbenside
$C_{13}H_{10}N_2$
312 - 4-Acridinamine
313 - 9-Acridinamine
764 - Benzenamine, *N,N*′-methanetetraylbis-
2383 - 1H-Benzimidazole, 1-phenyl-
2384 - 1H-Benzimidazole, 2-phenyl-
4158 - Cyanamide, diphenyl-
9195 - Phenazine, 2-methyl-
$C_{13}H_{10}N_2O$
9196 - Phenazinium, 1-hydroxy-5-methyl-, hydroxide, inner salt
$C_{13}H_{10}N_2O_4$
7363 - 1H-Isoindole-1,3(2H)-dione, 2-(2,6-dioxo-3-piperidinyl)-
$C_{13}H_{10}N_4O_5$
11997 - Urea, *N,N*′-bis(4-nitrophenyl)-
$C_{13}H_{10}O$
7616 - Methanone, diphenyl-
12049 - 9H-Xanthene
$C_{13}H_{10}O_2$
597 - Benzaldehyde, 3-phenoxy-
2716 - Benzoic acid, phenyl ester
3021 - [1,1′-Biphenyl]-2-carboxylic acid
3022 - [1,1′-Biphenyl]-4-carboxylic acid
7622 - Methanone, (2-hydroxyphenyl)phenyl-
10675 - 2-Propenoic acid, 3-(1-naphthalenyl)-, (*E*)-
10710 - 2-Propen-1-one, 3-(2-furanyl)-1-phenyl-
$C_{13}H_{10}O_3$
2650 - Benzoic acid, 2-hydroxy-, phenyl ester
2714 - Benzoic acid, 2-phenoxy-
4001 - Carbonic acid, diphenyl ester
7596 - Methanone, bis(2-hydroxyphenyl)-
7597 - Methanone, bis(4-hydroxyphenyl)-
7609 - Methanone, (2,4-dihydroxyphenyl)phenyl-
8682 - 1,4-Pentadien-3-one, 1,5-di-2-furanyl-
$C_{13}H_{10}O_4$
6186 - 5H-Furo[3,2-g][1]benzopyran-5-one, 4-methoxy-7-methyl-
$C_{13}H_{10}O_6$
7610 - Methanone, (3,4-dihydroxyphenyl)(2,4,6-trihydroxyphenyl)-
$C_{13}H_{10}S$
7547 - Methanethione, diphenyl-
11811 - 9H-Thioxanthene
$C_{13}H_{11}Br$
1058 - Benzene, 1,1′-(bromomethylene)bis-
$C_{13}H_{11}Cl$
1187 - Benzene, 1,1′-(chloromethylene)bis-
1224 - Benzene, 1-chloro-4-(phenylmethyl)-
3026 - Biphenyl, 2-chloro-2′-methyl-
$C_{13}H_{11}ClO$
9270 - Phenol, 4-chloro-2-(phenylmethyl)-
$C_{13}H_{11}Cl_2NO_2$
9760 - Procymidone
$C_{13}H_{11}N$
840 - Benzenamine, *N*-(phenylmethylene)-
1835 - Benzenemethanimine, α-phenyl-
3986 - 9H-Carbazole, 3-methyl-
3987 - 9H-Carbazole, 9-methyl-
11000 - Pyridine, 2-(2-phenylethenyl)-, (*E*)-
11001 - Pyridine, 4-(2-phenylethenyl)-, (*E*)-
11002 - Pyridine, 2-(2-phenylethenyl)-, (*Z*)-
$C_{13}H_{11}NO$
626 - Benzamide, *N*-phenyl-
5993 - Formamide, *N,N*-diphenyl-
9450 - Phenol, 2-[(phenylimino)methyl]-
$C_{13}H_{11}NO_2$
617 - Benzamide, 2-hydroxy-*N*-phenyl-
2715 - Benzoic acid, 2-(phenylamino)-
10911 - 3-Pyridinecarboxylic acid, phenylmethyl ester
$C_{13}H_{11}NO_3$
2005 - Benzene, 1-methyl-2-(2-nitrophenoxy)-
2006 - Benzene, 1-methyl-4-(4-nitrophenoxy)-
2502 - Benzoic acid, 4-amino-2-hydroxy-, phenyl ester
6179 - Furo[2,3-b]quinoline, 4,8-dimethoxy-
$C_{13}H_{11}NS$
1129 - Benzenecarbothioamide, *N*-phenyl-
$C_{13}H_{11}N_3$
316 - 3,6-Acridinediamine
$C_{13}H_{12}$
632 - 1H-Benz[e]indene, 2,3-dihydro-
1961 - Benzene, 1,1′-methylenebis-
3082 - 1,1′-Biphenyl, 2-methyl-
3083 - 1,1′-Biphenyl, 3-methyl-
3084 - 1,1′-Biphenyl, 4-methyl-
7954 - Naphthalene, 1-(2-propenyl)-
$C_{13}H_{12}Cl_2N_2$
785 - Benzenamine, 4,4′-methylenebis[2-chloro-
$C_{13}H_{12}Cl_2O_4$
168 - Acetic acid, [2,3-dichloro-4-(2-methylene-1-oxobutyl)phenoxy]-
$C_{13}H_{12}N_2$
5138 - Diazene, (3-methylphenyl)phenyl-
5139 - Diazene, (4-methylphenyl)phenyl-
5971 - 9H-Fluorene-2,7-diamine
7565 - Methanimidamide, *N,N*′-diphenyl-
$C_{13}H_{12}N_2O$
2717 - Benzoic acid, 2-phenylhydrazide
5135 - Diazene, (2-methoxyphenyl)phenyl-
5136 - Diazene, (3-methoxyphenyl)phenyl-
5137 - Diazene, (4-methoxyphenyl)phenyl-
7589 - Methanone, bis(3-aminophenyl)-
9426 - Phenol, 4-methyl-2-(phenylazo)-
11039 - 9H-Pyrido[3,4-b]indole, 7-methoxy-1-methyl-
12010 - Urea, *N,N*′-diphenyl-
12011 - Urea, *N,N*-diphenyl-
$C_{13}H_{12}N_2O_3$
11103 - 2,4,6(1H,3H,5H)-Pyrimidinetrione, 5-phenyl-5-(2-propenyl)-
$C_{13}H_{12}N_2O_3S$
603 - Benzamide, *N*-[(4-aminophenyl)sulfonyl]-
$C_{13}H_{12}N_2S$
11805 - Thiourea, *N,N*′-diphenyl-
$C_{13}H_{12}N_4O$
5122 - Diazenecarboxylic acid, phenyl-, 2-phenylhydrazide
$C_{13}H_{12}N_4S$
5121 - Diazenecarbothioic acid, phenyl-, 2-phenylhydrazide
$C_{13}H_{12}O$
1910 - Benzenemethanol, α-phenyl-
2012 - Benzene, 1-methyl-2-phenoxy-
2013 - Benzene, 1-methyl-3-phenoxy-
3080 - 1,1′-Biphenyl, 2-methoxy-
3081 - 1,1′-Biphenyl, 4-methoxy-
5978 - 9H-Fluoren-9-one, 1,2,3,4-tetrahydro-
9454 - Phenol, 2-(phenylmethyl)-
9455 - Phenol, 4-(phenylmethyl)-
10473 - 1-Propanone, 1-(1-naphthalenyl)-
10474 - 1-Propanone, 1-(2-naphthalenyl)-
$C_{13}H_{12}O_2$
1546 - 1,4-Benzenediol, 2-(phenylmethyl)-
1938 - Benzene, 1-methoxy-2-phenoxy-
7753 - 1-Naphthalenecarboxaldehyde, 2-ethoxy-
7765 - 1-Naphthalenecarboxylic acid, ethyl ester
7770 - 2-Naphthalenecarboxylic acid, ethyl ester
7952 - 1-Naphthalenepropanoic acid

9406 - Phenol, 3,3′-methylenebis-
9407 - Phenol, 4,4′-methylenebis-
9451 - Phenol, 2-(phenylmethoxy)-
9452 - Phenol, 3-(phenylmethoxy)-
9453 - Phenol, 4-(phenylmethoxy)-

$C_{13}H_{12}O_2S$
2112 - Benzene, [(phenylmethyl)sulfonyl]-

$C_{13}H_{12}O_2S_2$
11818 - *o*-Toluenesulfonic acid, thio-, *S*-phenyl ester

$C_{13}H_{12}O_3$
5825 - Ethanone, 1-[6-hydroxy-2-(1-methylethenyl)-5-benzofuranyl]-
7773 - 2-Naphthalenecarboxylic acid, 3-hydroxy-, ethyl ester
7953 - 2-Naphthalenepropanoic acid, 6-hydroxy-

$C_{13}H_{12}S$
2017 - Benzene, 1-methyl-2-(phenylthio)-
2018 - Benzene, 1-methyl-3-(phenylthio)-
2019 - Benzene, 1-methyl-4-(phenylthio)-
2113 - Benzene, [(phenylmethyl)thio]-

$C_{13}H_{13}ClSi$
11380 - Silane, chloromethyldiphenyl-

$C_{13}H_{13}Cl_2N_3O_3$
7338 - Iprodione

$C_{13}H_{13}N$
807 - Benzenamine, *N*-methyl-*N*-phenyl-
808 - Benzenamine, 3-methyl-*N*-phenyl-
839 - Benzenamine, 4-(phenylmethyl)-
1826 - Benzenemethanamine, α-phenyl-
1827 - Benzenemethanamine, *N*-phenyl-
3013 - [1,1′-Biphenyl]-2-amine, 6-methyl-
3014 - [1,1′-Biphenyl]-4-amine, 3-methyl-
3015 - [1,1′-Biphenyl]-4-amine, 4′-methyl-
11003 - Pyridine, 2-(2-phenylethyl)-

$C_{13}H_{13}N_3$
6282 - Guanidine, *N,N′*-diphenyl-

$C_{13}H_{13}N_3O_5S_2$
3578 - Butanoic acid, 4-oxo-4-[[4-[(2-thiazolylamino)sulfonyl]phenyl]amino]-

$C_{13}H_{13}O_3P$
9541 - Phosphonic acid, methyl-, diphenyl ester

$C_{13}H_{14}$
7887 - Naphthalene, 1-ethyl-5-methyl-
7920 - Naphthalene, 1-(1-methylethyl)-
7921 - Naphthalene, 2-(1-methylethyl)-
7957 - Naphthalene, 1-propyl-
7958 - Naphthalene, 2-propyl-
8009 - Naphthalene, 1,2,4-trimethyl-
8010 - Naphthalene, 1,2,5-trimethyl-
8011 - Naphthalene, 1,2,6-trimethyl-
8012 - Naphthalene, 1,2,7-trimethyl-
8013 - Naphthalene, 1,3,7-trimethyl-

$C_{13}H_{14}F_3N_3O_4$
5409 - Ethalfluralin

$C_{13}H_{14}N_2$
784 - Benzenamine, 4,4′-methylenebis-
7174 - Hydrazobenzene, 3-methyl-

$C_{13}H_{14}N_2O$
1913 - Benzenemethanol, α-[(2-pyridinylamino)methyl]-

$C_{13}H_{14}N_2O_2S$
2213 - Benzenesulfonamide, 4-[(phenylmethyl)amino]-

$C_{13}H_{14}N_2O_3$
11094 - 2,4,6(1H,3H,5H)-Pyrimidinetrione, 5-ethyl-1-methyl-5-phenyl-

$C_{13}H_{14}N_2O_{13}$
11038 - 3H-Pyrido[3,4-b]indole, 4,9-dihydro-7-methoxy-1-methyl-

$C_{13}H_{14}N_4O$
4010 - Carbonic dihydrazide, 2,2′-diphenyl-

$C_{13}H_{14}O$
7955 - Naphthalene, 1-propoxy-
7956 - Naphthalene, 2-propoxy-

$C_{13}H_{14}OSi$
11502 - Silanol, methyldiphenyl-

$C_{13}H_{14}O_2$
8680 - 2,4-Pentadienoic acid, 5-phenyl-, ethyl ester

$C_{13}H_{14}O_3$
3576 - Butanoic acid, 3-oxo-2-(phenylmethylene)-, ethyl ester

$C_{13}H_{14}O_5$
2774 - 3H-2-Benzopyran-7-carboxylic acid, 4,6-dihydro-8-hydroxy-3,4,5-trimethyl-6-oxo-, (3*R*-*trans*)-

$C_{13}H_{14}Si$
11438 - Silane, methyldiphenyl-

$C_{13}H_{15}N$
11222 - Quinoline, 2-butyl-

$C_{13}H_{15}NO_4$
11148 - 1,2-Pyrrolidinedicarboxylic acid, 1-(phenylmethyl) ester, (*S*)-

$C_{13}H_{15}N_3O_2$
3945 - Carbamic acid, dimethyl-, 3-methyl-1-phenyl-1H-pyrazol-5-yl ester

$C_{13}H_{15}N_3O_3$
7188 - Imazapyr

$C_{13}H_{16}ClNO$
4579 - Cyclohexanone, 2-(2-chlorophenyl)-2-(methylamino)-, (±)-

$C_{13}H_{16}F_3N_3O_4$
509 - Balan
11901 - Trifluralin

$C_{13}H_{16}N_2O_2$
69 - Acetamide, *N*-[2-(5-methoxy-1H-indol-3-yl)ethyl]-

$C_{13}H_{16}O$
4236 - Cycloheptanone, 2-phenyl-
7100 - 1-Hexen-3-one, 5-methyl-1-phenyl-
9106 - 1-Penten-3-one, 4,4-dimethyl-1-phenyl-

$C_{13}H_{16}O_2$
1412 - Benzene, (3,3-diethoxy-1-propynyl)-
2551 - Benzoic acid, cyclohexyl ester

$C_{13}H_{16}O_3$
2132 - Benzenepropanoic acid, α-ethyl-β-oxo-, ethyl ester

$C_{13}H_{16}O_4$
9946 - Propanedioic acid, di-2-propynyl-, diethyl ester
9975 - Propanedioic acid, phenyl-, diethyl ester

$C_{13}H_{16}O_7$
562 - Benzaldehyde, 2-(β-*D*-glucopyranosyloxy)-

$C_{13}H_{17}F_3N_4O_4$
9761 - Prodiamine

$C_{13}H_{17}NO$
7409 - Ketone, 1-methyl-4-piperidyl phenyl

$C_{13}H_{17}NO_4$
10932 - 3,5-Pyridinedicarboxylic acid, 2,6-dimethyl-, diethyl ester

$C_{13}H_{17}N_3O$
10851 - 3H-Pyrazol-3-one, 4-(dimethylamino)-1,2-dihydro-1,5-dimethyl-2-phenyl-

$C_{13}H_{18}$
7260 - 1H-Indene, 2,3-dihydro-1,1,4,7-tetramethyl-
8000 - Naphthalene, 1,2,3,4-tetrahydro-1,1,6-trimethyl-

$C_{13}H_{18}ClN_3O_4S_2$
2853 - 2H-1,2,4-Benzothiadiazine-7-sulfonamide, 6-chloro-3-(cyclopentylmethyl)-3,4-dihydro-, 1,1-dioxide

$C_{13}H_{18}N_2O_2$
4900 - 1H-Cyclopentapyrimidine-2,4(3H,5H)-dione, 3-cyclohexyl-6,7-dihydro-

$C_{13}H_{18}O$
1250 - Benzene, 1-cyclohexyl-2-methoxy-
2117 - Benzenepropanal, α-methyl-4-(1-methylethyl)-
6526 - 1-Heptanone, 1-phenyl-

$C_{13}H_{18}O_2$
2610 - Benzoic acid, hexyl ester
8611 - Oxirane, [[4-(1,1-dimethylethyl)phenoxy]methyl]-

$C_{13}H_{18}O_3$
912 - Benzeneacetic acid, α-hydroxy-, 3-methylbutyl ester

$C_{13}H_{18}O_4$
2277 - Benzene, 1,2,3,4-tetramethoxy-5-(2-propenyl)-

$C_{13}H_{18}O_5S$
5933 - Ethofumesate

$C_{13}H_{18}O_7$
6222 - β-*D*-Glucopyranoside, 2-(hydroxymethyl)phenyl

$C_{13}H_{19}N$
9697 - Piperidine, 1-(1-phenylethyl)-

$C_{13}H_{19}NO$
10428 - 1-Propanone, 2-(diethylamino)-1-phenyl-

$C_{13}H_{19}NO_2$
7390 - Isoquinoline, 1,2,3,4-tetrahydro-6,7-dimethoxy-1,2-dimethyl-, (±)-
7391 - Isoquinoline, 1,2,3,4-tetrahydro-6,7-dimethoxy-1,2-dimethyl-

$C_{13}H_{19}NO_3$
7394 - 8-Isoquinolinol, 1,2,3,4-tetrahydro-6,7-dimethoxy-1,2-dimethyl-, (*S*)-

$C_{13}H_{19}NO_4S$
2595 - Benzoic acid, 4-[(dipropylamino)sulfonyl]-

$C_{13}H_{19}N_3O_4$
8635 - Pendimethalin

$C_{13}H_{19}N_3O_6S$
8070 - Nitralin

$C_{13}H_{20}$
1745 - Benzene, heptyl-
1948 - Benzene, 1-methyl-2,4-bis(1-methylethyl)-
1949 - Benzene, 1-methyl-3,5-bis(1-methylethyl)-
1992 - Benzene, [2-methyl-1-(1-methylethyl)propyl]-
11868 - 1,12-Tridecadiyne

$C_{13}H_{20}N_2O_2$
2493 - Benzoic acid, 4-amino-, 2-(diethylamino)ethyl ester
5748 - Ethanol, 2-[(2-methylpropyl)amino]-, 4-aminobenzoate (ester)

$C_{13}H_{20}O$
1747 - Benzene, (heptyloxy)-
1842 - Benzenemethanol, α,α-bis(1-methylethyl)-
1867 - Benzenemethanol, α,α-dipropyl-
1881 - Benzenemethanol, α-hexyl-
2102 - Benzenepentanol, α-ethyl-
3885 - 3-Buten-2-one, 4-(2,6,6-trimethyl-1-cyclohexen-1-yl)-, (*E*)-
3886 - 3-Buten-2-one, 4-(2,6,6-trimethyl-2-cyclohexen-1-yl)-, (*E*)-(±)-
4148 - *o*-Cresol, 4-(3-methylpentyl)-
11965 - 3,5,9-Undecatrien-2-one, 6,10-dimethyl-

$C_{13}H_{20}O_3$
6067 - 2-Furancarboxylic acid, octyl ester

$C_{13}H_{20}O_4$
9944 - Propanedioic acid, di-2-propenyl-, diethyl ester

$C_{13}H_{20}O_8$
9988 - 1,3-Propanediol, 2,2-bis[(acetyloxy)methyl]-, diacetate

$C_{13}H_{21}N$
10874 - Pyridine, 2,6-bis(1,1-dimethylethyl)-

$C_{13}H_{22}$
5973 - 1H-Fluorene, dodecahydro-

$C_{13}H_{22}ClN_3O$
602 - Benzamide, 4-amino-*N*-[2-(diethylamino)ethyl]-, monohydrochloride

$C_{13}H_{22}NO_3PS$
5947 - Fenamiphos

$C_{13}H_{22}N_2$
4291 - Cyclohexanamine, *N,N′*-methanetetraylbis-

$C_{13}H_{22}N_2O$
8212 - Norea

$C_{13}H_{22}O$
3873 - 3-Buten-2-ol, 4-(2,6,6-trimethyl-1-cyclohexen-1-yl)-
3874 - 3-Buten-2-ol, 4-(2,6,6-trimethyl-2-cyclohexen-1-yl)-
7606 - Methanone, dicyclohexyl-

8113 - 6,8-Nonadien-2-one, 8-methyl-5-(1-methylethyl)-, [*S*-(*E*)]-

$C_{13}H_{22}O_4$
4171 - Cyclobutaneacetic acid, 3-(ethoxycarbonyl)-2,2-dimethyl-, ethyl ester
9934 - Propanedioic acid, cyclohexyl-, diethyl ester

$C_{13}H_{24}$
4429 - Cyclohexane, 1,1′-methylenebis-
11897 - 1-Tridecyne

$C_{13}H_{24}N_2O$
10415 - 2-Propanone, 1,3-bis(1-methyl-2-pyrrolidinyl)-, (*R**,*S**)-

$C_{13}H_{24}N_3O_3PS$
9729 - Pirimiphos-ethyl

$C_{13}H_{24}O_2$
5341 - 11-Dodecenoic acid, methyl ester
10646 - 2-Propenoic acid, isodecyl ester
11894 - 2-Tridecenoic acid
11895 - 12-Tridecenoic acid
11977 - 10-Undecenoic acid, ethyl ester

$C_{13}H_{24}O_3$
6166 - 2-Furanpropanol, tetrahydro-α-[2-(tetrahydro-2-furanyl)ethyl]-

$C_{13}H_{24}O_4$
8133 - Nonanedioic acid, diethyl ester
9955 - Propanedioic acid, hexyl-, diethyl ester

$C_{13}H_{25}BrO_2$
5309 - Dodecanoic acid, 2-bromo-, methyl ester

$C_{13}H_{25}N$
11880 - Tridecanenitrile

$C_{13}H_{26}$
4327 - Cyclohexane, 2-butyl-1,1,3-trimethyl-
4405 - Cyclohexane, heptyl-
4843 - Cyclopentane, octyl-
8184 - 4-Nonene, 5-butyl-
11893 - 1-Tridecene

$C_{13}H_{26}Br_2$
11874 - Tridecane, 1,13-dibromo-

$C_{13}H_{26}N_2$
4299 - Cyclohexanamine, 4,4′-methylenebis-
4300 - Cyclohexanamine, 4,4′-methylenebis-, [cis(*cis*)]-
4301 - Cyclohexanamine, 4,4′-methylenebis-, [*trans*(*cis*)]-
4302 - Cyclohexanamine, 4,4′-methylenebis-, [trans(*trans*)]-
9701 - Piperidine, 4,4′-(1,3-propanediyl)bis-

$C_{13}H_{26}N_2O_4$
3933 - Carbamic acid, butyl-,2-[[(aminocarbonyl)oxy]methyl]-2-methylpentyl ester

$C_{13}H_{26}O$
11869 - Tridecanal
11889 - 2-Tridecanone
11890 - 3-Tridecanone
11891 - 7-Tridecanone

$C_{13}H_{26}O_2$
5067 - Decanoic acid, 1-methylethyl ester
5069 - Decanoic acid, propyl ester
5314 - Dodecanoic acid, 2-methyl-
5315 - Dodecanoic acid, methyl ester
6465 - Heptanoic acid, hexyl ester
6889 - Hexanoic acid, heptyl ester
8147 - Nonanoic acid, butyl ester
8439 - Octanoic acid, pentyl ester
8901 - Pentanoic acid, octyl ester
11881 - Tridecanoic acid
11946 - Undecanoic acid, ethyl ester

$C_{13}H_{27}Br$
11872 - Tridecane, 1-bromo-

$C_{13}H_{27}N$
9690 - Piperidine, 1-octyl-

$C_{13}H_{28}$
8122 - Nonane, 5-butyl-
11871 - Tridecane

$C_{13}H_{28}O$
8162 - 5-Nonanol, 5-butyl-
11883 - 1-Tridecanol
11884 - 2-Tridecanol
11885 - 3-Tridecanol

$C_{13}H_{28}O_2$
11875 - 1,12-Tridecanediol
11876 - 1,13-Tridecanediol

$C_{13}H_{28}O_3$
10080 - Propane, 1,1′,1″-[methylidynetris(oxy)]tris[2-methyl-

$C_{13}H_{28}O_4$
10068 - Propane, 1,1′,1″,1‴-[methanetetrayltetrakis(oxy)]tetrakis-

$C_{13}H_{29}N$
6354 - 2-Heptanamine, 6-methyl-*N*-(3-methylbutyl)-
11870 - 1-Tridecanamine

$C_{14}H_4N_2O_2S_2$
8056 - Naphtho[2,3-b]-1,4-dithiin-2,3-dicarbonitrile, 5,10-dihydro-5,10-dioxo-

$C_{14}H_4N_4O_{12}$
401 - 9,10-Anthracenedione, 1,8-dihydroxy-2,4,5,7-tetranitro-

$C_{14}H_4O_6$
2784 - [2]Benzopyrano[6,5,4-def][2]benzopyran-1,3,6,8-tetrone

$C_{14}H_6Br_2O_2$
390 - 9,10-Anthracenedione, 2,7-dibromo-

$C_{14}H_6Cl_4O_4$
9152 - Peroxide, bis(2,4-dichlorobenzoyl)

$C_{14}H_6N_2O_6$
405 - 9,10-Anthracenedione, 1,5-dinitro-

$C_{14}H_7BrO_2$
385 - 9,10-Anthracenedione, 1-bromo-
386 - 9,10-Anthracenedione, 2-bromo-

$C_{14}H_7ClF_3NO_5$
304 - Acifluorfen

$C_{14}H_7ClO_2$
387 - 9,10-Anthracenedione, 1-chloro-
388 - 9,10-Anthracenedione, 2-chloro-

$C_{14}H_7NO_4$
410 - 9,10-Anthracenedione, 1-nitro-
411 - 9,10-Anthracenedione, 2-nitro-

$C_{14}H_7NO_6$
399 - 9,10-Anthracenedione, 1,2-dihydroxy-3-nitro-
400 - 9,10-Anthracenedione, 1,2-dihydroxy-4-nitro-

$C_{14}H_8Br_2$
375 - Anthracene, 9,10-dibromo-

$C_{14}H_8Cl_2N_4$
4139 - Clofentezine

$C_{14}H_8Cl_4$
1374 - Benzene, 1,1′-(dichloroethenylidene)bis[4-chloro-

$C_{14}H_8O_2$
379 - 9,10-Anthracenedione
9174 - 9,10-Phenanthrenedione

$C_{14}H_8O_3$
407 - 9,10-Anthracenedione, 1-hydroxy-
408 - 9,10-Anthracenedione, 2-hydroxy-
5145 - Dibenz[c,e]oxepin-5,7-dione
5969 - 9H-Fluorene-2-carboxylic acid, 9-oxo-
9175 - 9,10-Phenanthrenedione, 2-hydroxy-
9176 - 9,10-Phenanthrenedione, 3-hydroxy-

$C_{14}H_8O_4$
391 - 9,10-Anthracenedione, 1,2-dihydroxy-
392 - 9,10-Anthracenedione, 1,4-dihydroxy-
393 - 9,10-Anthracenedione, 1,5-dihydroxy-
394 - 9,10-Anthracenedione, 1,7-dihydroxy-
395 - 9,10-Anthracenedione, 1,8-dihydroxy-
396 - 9,10-Anthracenedione, 2,7-dihydroxy-

$C_{14}H_8O_5$
415 - 9,10-Anthracenedione, 1,2,3-trihydroxy-
416 - 9,10-Anthracenedione, 1,2,4-trihydroxy-
417 - 9,10-Anthracenedione, 1,2,6-trihydroxy-
418 - 9,10-Anthracenedione, 1,2,7-trihydroxy-

$C_{14}H_8O_6$
412 - 9,10-Anthracenedione, 1,2,5,6-tetrahydroxy-
413 - 9,10-Anthracenedione, 1,2,5,8-tetrahydroxy-
414 - 9,10-Anthracenedione, 1,3,5,7-tetrahydroxy-

$C_{14}H_8O_8$
406 - 9,10-Anthracenedione, 1,2,3,5,6,7-hexahydroxy-

$C_{14}H_9Br$
9164 - Phenanthrene, 9-bromo-

$C_{14}H_9Cl$
374 - Anthracene, 1-chloro-

$C_{14}H_9ClF_2N_2O_2$
5176 - Diflubenzuron

$C_{14}H_9ClO_3$
4079 - Chlorflurecol

$C_{14}H_9Cl_2NO_5$
2997 - Bifenox

$C_{14}H_9Cl_5$
2328 - Benzene, 1,1′-(2,2,2-trichloroethylidene)bis[4-chloro-

$C_{14}H_9Cl_5O$
1851 - Benzenemethanol, 4-chloro-α-(4-chlorophenyl)-α-(trichloromethyl)-

$C_{14}H_9NO_2$
380 - 9,10-Anthracenedione, 1-amino-
381 - 9,10-Anthracenedione, 2-amino-
426 - Anthracene, 9-nitro-
7366 - 1H-Isoindole-1,3(2H)-dione, 2-phenyl-

$C_{14}H_9NO_3$
383 - 9,10-Anthracenedione, 2-amino-1-hydroxy-

$C_{14}H_9NO_4$
382 - 9,10-Anthracenedione, 3-amino-1,2-dihydroxy-

$C_{14}H_{10}$
366 - Anthracene
1714 - Benzene, 1,1′-(1,2-ethynediyl)bis-
9163 - Phenanthrene

$C_{14}H_{10}Cl_2$
1365 - Benzene, 1,1′-(1,2-dichloro-1,2-ethenediyl)bis-, (*E*)-
1366 - Benzene, 1,1′-(1,2-dichloro-1,2-ethenediyl)bis-, (*Z*)-

$C_{14}H_{10}Cl_4$
1378 - Benzene, 1,1′-(2,2-dichloroethylidene)bis[4-chloro-

$C_{14}H_{10}F_3NO_2$
2729 - Benzoic acid, 2-[[3-(trifluoromethyl)phenyl]amino]-

$C_{14}H_{10}N_2O$
8564 - 1,2,4-Oxadiazole, 3,5-diphenyl-

$C_{14}H_{10}N_2O_2$
389 - 9,10-Anthracenedione, 1,5-diamino-

$C_{14}H_{10}N_2O_4$
1494 - Benzene, 1,1′-(1,2-dinitro-1,2-ethenediyl)bis-, (*Z*)-
1496 - Benzene, 2,4-dinitro-1-(2-phenylethenyl)-, (*E*)-

$C_{14}H_{10}O$
431 - 1-Anthracenol
432 - 9-Anthracenol
433 - 9(10H)-Anthracenone
5928 - Ethenone, diphenyl-

$C_{14}H_{10}O_2$
5584 - Ethanedione, diphenyl-
9173 - 3,4-Phenanthrenediol

$C_{14}H_{10}O_2S_2$
5251 - Disulfide, dibenzoyl

$C_{14}H_{10}O_3$
429 - 1,2,10-Anthracenetriol
430 - 1,8,9-Anthracenetriol
2510 - Benzoic acid, anhydride
2511 - Benzoic acid, 3-benzoyl-
2512 - Benzoic acid, 4-benzoyl-

$C_{14}H_{10}O_4$
3038 - [1,1′-Biphenyl]-2,2′-dicarboxylic acid
9157 - Peroxide, dibenzoyl

$C_{14}H_{10}O_5$
2616 - Benzoic acid, 2-hydroxy-, 2-carboxyphenyl ester
12054 - 9H-Xanthen-9-one, 1,7-dihydroxy-3-methoxy-

$C_{14}H_{10}O_6$
2979 - [Bi-1,4-cyclohexadien-1-yl]-3,3′,6,6′-tetrone, 2,2′-dihydroxy-4,4′-dimethyl-
$C_{14}H_{10}O_9$
2575 - Benzoic acid, 3,4-dihydroxy-5-[(3,4,5-trihydroxybenzoyl)oxy]-
$C_{14}H_{11}Br$
1030 - Benzene, 1,1′-(bromoethenylidene)bis-
1080 - Benzene, 1-bromo-2-(2-phenylethenyl)-, (*E*)-
1081 - Benzene, 1-bromo-2-(2-phenylethenyl)-, (*Z*)-
$C_{14}H_{11}Cl$
1146 - Benzene, 1,1′-(1-chloro-1,2-ethenediyl)bis-, (*E*)-
1147 - Benzene, 1,1′-(1-chloro-1,2-ethenediyl)bis-, (*Z*)-
1222 - Benzene, 1-chloro-2-(2-phenylethenyl)-, (*E*)-
1223 - Benzene, 1-chloro-3-(2-phenylethenyl)-, (*E*)-
$C_{14}H_{11}ClO$
975 - Benzeneacetyl chloride, α-phenyl-
5785 - Ethanone, 2-chloro-1,2-diphenyl-
$C_{14}H_{11}N$
365 - 2-Anthracenamine
969 - Benzeneacetonitrile, α-phenyl-
7320 - 1H-Indole, 2-phenyl-
9162 - 9-Phenanthrenamine
$C_{14}H_{11}NO$
3983 - 9H-Carbazole, 9-acetyl-
$C_{14}H_{11}NO_2$
2055 - Benzene, 1-nitro-2-(2-phenylethenyl)-, (*E*)-
2056 - Benzene, 1-nitro-2-(2-phenylethenyl)-, (*Z*)-
3982 - 9H-Carbazole-9-acetic acid
$C_{14}H_{11}NO_3$
2513 - Benzoic acid, 3-(benzoylamino)-
7627 - Methanone, (4-methyl-2-nitrophenyl)phenyl-
7630 - Methanone, (4-methylphenyl)(4-nitrophenyl)-
$C_{14}H_{11}NO_4$
2514 - Benzoic acid, 4-(benzoylamino)-2-hydroxy-
2656 - Benzoic acid, 2,2′-iminobis-
$C_{14}H_{12}$
376 - Anthracene, 9,10-dihydro-
1619 - Benzene, 1,1′-(1,2-ethenediyl)bis-, (*E*)-
1620 - Benzene, 1,1′-(1,2-ethenediyl)bis-, (*Z*)-
1636 - Benzene, 1,1′-ethenylidenebis-
5974 - 9H-Fluorene, 9-methyl-
9171 - Phenanthrene, 9,10-dihydro-
$C_{14}H_{12}Cl_2$
1678 - Benzene, 1,1′-ethylidenebis[4-chloro-
$C_{14}H_{12}Cl_2O$
1850 - Benzenemethanol, 4-chloro-α-(4-chlorophenyl)-α-methyl-
$C_{14}H_{12}F_3NO_4S_2$
9149 - Perfluidone
$C_{14}H_{12}N_2$
2385 - 1H-Benzimidazole, 2-(phenylmethyl)-
9192 - 1,10-Phenanthroline, 2,9-dimethyl-
11196 - Quinazoline, 3,4-dihydro-3-phenyl-
$C_{14}H_{12}N_2O_2$
5495 - Ethanediamide, *N,N*′-diphenyl-
9449 - Phenol, 4-(phenylazo)-, acetate (ester)
$C_{14}H_{12}N_2S$
782 - Benzenamine, 4-(6-methyl-2-benzothiazolyl)-
$C_{14}H_{12}N_4O_2S$
2205 - Benzenesulfonamide, 4-amino-*N*-2-quinoxalinyl-
$C_{14}H_{12}O$
877 - Benzeneacetaldehyde, α-phenyl-
5776 - Ethanone, 1-[1,1′-biphenyl]-4-yl-
5812 - Ethanone, 1,2-diphenyl-
7631 - Methanone, (2-methylphenyl)phenyl-
7632 - Methanone, (3-methylphenyl)phenyl-
7633 - Methanone, (4-methylphenyl)phenyl-
$C_{14}H_{12}O_2$
936 - Benzeneacetic acid, α-phenyl-
1545 - 1,3-Benzenediol, 5-(2-phenylethenyl)-, (*E*)-
2699 - Benzoic acid, 2-methylphenyl ester
2700 - Benzoic acid, 4-methylphenyl ester
2718 - Benzoic acid, 2-(phenylmethyl)-
2719 - Benzoic acid, 4-(phenylmethyl)-
2720 - Benzoic acid, phenylmethyl ester
3009 - [1,1′-Biphenyl]-4-acetic acid
3091 - [1,1′-Biphenyl]-4-ol, acetate
5822 - Ethanone, 2-hydroxy-1,2-diphenyl-, (±)-
7623 - Methanone, (2-methoxyphenyl)phenyl-
7624 - Methanone, (3-methoxyphenyl)phenyl-
7625 - Methanone, (4-methoxyphenyl)phenyl-
$C_{14}H_{12}O_3$
918 - Benzeneacetic acid, α-hydroxy-α-phenyl-
2397 - 2H,8H-Benzo[1,2-b:5,4-b′]dipyran-2-one, 8,8-dimethyl-
2651 - Benzoic acid, 2-hydroxy-, phenylmethyl ester
7620 - Methanone, (2-hydroxy-4-methoxyphenyl)phenyl-
9383 - Phenol, 2-methoxy-, benzoate
$C_{14}H_{12}O_4$
7608 - Methanone, (2,6-dihydroxy-4-methoxyphenyl)phenyl-
7619 - Methanone, (2-hydroxy-4-methoxyphenyl)(2-hydroxyphenyl)-
$C_{14}H_{12}O_5$
6185 - 5H-Furo[3,2-g][1]benzopyran-5-one, 4,9-dimethoxy-7-methyl-
$C_{14}H_{12}S_2$
11661 - Thianthrene, 2,7-dimethyl-
$C_{14}H_{13}N$
809 - Benzenamine, 2-methyl-*N*-(phenylmethylene)-
810 - Benzenamine, 3-methyl-*N*-(phenylmethylene)-
811 - Benzenamine, 4-methyl-*N*-(phenylmethylene)-
3985 - 9H-Carbazole, 9-ethyl-
$C_{14}H_{13}NO$
35 - Acetamide, *N*-[1,1′-biphenyl]-2-yl-
57 - Acetamide, *N,N*-diphenyl-
621 - Benzamide, *N*-(2-methylphenyl)-
622 - Benzamide, *N*-(3-methylphenyl)-
623 - Benzamide, *N*-(4-methylphenyl)-
7626 - Methanone, [2-(methylamino)phenyl]phenyl-
$C_{14}H_{13}NO_4$
6181 - Furo[2,3-b]quinoline, 4,7,8-trimethoxy-
$C_{14}H_{13}NO_4S$
2721 - Benzoic acid, 4-[[(phenylmethyl)sulfonyl]amino]-
$C_{14}H_{14}$
1587 - Benzene, 1,1′-(1,2-ethanediyl)bis-
1677 - Benzene, 1,1′-ethylidenebis-
2014 - Benzene, 1-methyl-2-(phenylmethyl)-
2015 - Benzene, 1-methyl-3-(phenylmethyl)-
2016 - Benzene, 1-methyl-4-(phenylmethyl)-
3050 - 1,1′-Biphenyl, 2,2′-dimethyl-
3051 - 1,1′-Biphenyl, 2,3′-dimethyl-
3052 - 1,1′-Biphenyl, 2,3-dimethyl-
3053 - 1,1′-Biphenyl, 2,4′-dimethyl-
3054 - 1,1′-Biphenyl, 2,4-dimethyl-
3055 - 1,1′-Biphenyl, 2,5-dimethyl-
3056 - 1,1′-Biphenyl, 2,6-dimethyl-
3057 - 1,1′-Biphenyl, 3,3′-dimethyl-
3058 - 1,1′-Biphenyl, 3,4′-dimethyl-
3059 - 1,1′-Biphenyl, 3,4-dimethyl-
3060 - 1,1′-Biphenyl, 3,5-dimethyl-
3061 - 1,1′-Biphenyl, 4,4′-dimethyl-
3072 - 1,1′-Biphenyl, 2-ethyl-
3073 - 1,1′-Biphenyl, 3-ethyl-
9187 - Phenanthrene, 1,2,3,4-tetrahydro-
$C_{14}H_{14}Cl_2N_2O$
7187 - Imazalil
$C_{14}H_{14}Hg$
7432 - Mercury, bis(4-methylphenyl)-
$C_{14}H_{14}NO_4PS$
9551 - Phosphonothioic acid, phenyl-, *O*-ethyl *O*-(4-nitrophenyl) ester
$C_{14}H_{14}N_2$
732 - Benzenamine, 4,4′-(1,2-ethenediyl)bis-, (*E*)-
5114 - Diazene, bis(2-methylphenyl)-
5115 - Diazene, bis(3-methylphenyl)-, (*E*)-
$C_{14}H_{14}N_2O$
5116 - Diazene, bis(2-methylphenyl)-, 1-oxide, (*Z*)-
5117 - Diazene, bis(2-methylphenyl)-, 1-oxide, (*E*)-
5118 - Diazene, bis(3-methylphenyl)-, 1-oxide, (*Z*)-
5119 - Diazene, bis(3-methylphenyl)-, 1-oxide, (*E*)-
5133 - Diazene, (4-ethoxyphenyl)phenyl-
10462 - 1-Propanone, 2-methyl-1,2-di-3-pyridinyl-
12018 - Urea, *N*-methyl-*N,N*′-diphenyl-
$C_{14}H_{14}N_3NaO_3S$
2222 - Benzenesulfonic acid, 4-[[4-(dimethylamino)phenyl]azo]-, sodium salt
$C_{14}H_{14}O$
1613 - Benzeneethanol, α-phenyl-
1614 - Benzeneethanol, α-phenyl-, (*S*)-
1615 - Benzeneethanol, β-phenyl-
2076 - Benzene, 1,1′-oxybis[2-methyl-
2077 - Benzene, 1,1′-oxybis[3-methyl-
2078 - Benzene, 1,1′-oxybis[4-methyl-
2079 - Benzene, 1,1′-[oxybis(methylene)]bis-
3071 - 1,1′-Biphenyl, 3-ethoxy-
3674 - 1-Butanone, 1-(1-naphthalenyl)-
10464 - 1-Propanone, 2-methyl-1-(1-naphthalenyl)-
10465 - 1-Propanone, 2-methyl-1-(2-naphthalenyl)-
$C_{14}H_{14}OS$
2184 - Benzene, 1,1′-[sulfinylbis(methylene)]bis-
$C_{14}H_{14}O_2$
1588 - Benzene, 1,1′-[1,2-ethanediylbis(oxy)]bis-
3045 - 1,1′-Biphenyl, 2,2′-dimethoxy-
3046 - 1,1′-Biphenyl, 3,3′-dimethoxy-
3047 - 1,1′-Biphenyl, 4,4′-dimethoxy-
3067 - [1,1′-Biphenyl]-2,2′-diol, 3,3′-dimethyl-
3068 - [1,1′-Biphenyl]-2,2′-diol, 5,5′-dimethyl-
5575 - 1,2-Ethanediol, 1,2-diphenyl-, (*R**,*R**)-(±)-
7735 - 1-Naphthaleneacetic acid, ethyl ester
$C_{14}H_{14}O_2S$
2240 - Benzene, 1,1′-sulfonylbis[4-methyl-
2241 - Benzene, 1,1′-[sulfonylbis(methylene)]bis-
$C_{14}H_{14}O_3$
1929 - Benzene, 1-methoxy-2-(3-methoxyphenoxy)-
2075 - Benzene, 1,1′-oxybis[2-methoxy-
2786 - 2H-1-Benzopyran-7-ol, 3,4-dihydro-3-(4-hydroxyphenyl)-, (*S*)-
$C_{14}H_{14}O_8$
2258 - 1,2,4,5-Benzenetetracarboxylic acid, tetramethyl ester
$C_{14}H_{14}S$
1993 - Benzene, 1-methyl-2-[(4-methylphenyl)thio]-
2285 - Benzene, 1,1′-thiobis[2-methyl-
2286 - Benzene, 1,1′-thiobis[4-methyl-
2287 - Benzene, 1,1′-[thiobis(methylene)]bis-
11570 - Sulfide, *m*-tolyl *p*-tolyl
$C_{14}H_{14}S_2$
5245 - Disulfide, bis(3-methylphenyl)
5246 - Disulfide, bis(4-methylphenyl)
5248 - Disulfide, bis(phenylmethyl)
$C_{14}H_{14}Se$
2176 - Benzene, 1,1′-selenobis[4-methyl-
$C_{14}H_{15}Cl_2N$
7710 - 2-Naphthalenamine, *N,N*-bis(2-chloroethyl)-
$C_{14}H_{15}N$
751 - Benzenamine, *N*-ethyl-*N*-phenyl-
795 - Benzenamine, 2-methyl-*N*-(2-methylphenyl)-
796 - Benzenamine, 3-methyl-*N*-(3-methylphenyl)-
797 - Benzenamine, 4-methyl-*N*-(4-methylphenyl)-
838 - Benzenamine, 2-(2-phenylethyl)-
1584 - Benzeneethanamine, α-phenyl-
1821 - Benzenemethanamine, *N*-(2-methylphenyl)
1824 - Benzenemethanamine, *N*-(3-methylphenyl)-
1825 - Benzenemethanamine, *N*-(4-methylphenyl)-
1828 - Benzenemethanamine, *N*-(phenylmethyl)-
11007 - Pyridine, 4-(3-phenylpropyl)-
$C_{14}H_{15}N_3$
721 - Benzenamine, *N,N*-dimethyl-4-(phenylazo)-
798 - Benzenamine, 2-methyl-4-[(2-methylphenyl)azo]-
$C_{14}H_{15}O_3P$
9586 - Phosphorous acid, bis(phenylmethyl) ester
$C_{14}H_{16}$
422 - Anthracene, 1,2,3,4,5,6-hexahydro-

7746 - Naphthalene, 1-butyl-
7828 - Naphthalene, 1-(1,1-dimethylethyl)-
7829 - Naphthalene, 2-(1,1-dimethylethyl)-
7922 - Naphthalene, 1-methyl-4-(1-methylethyl)-
7923 - Naphthalene, 1-methyl-7-(1-methylethyl)-
7924 - Naphthalene, 7-methyl-1-(1-methylethyl)-
7927 - Naphthalene, 1-(1-methylpropyl)-
8001 - Naphthalene, 1,2,4,7-tetramethyl-
8002 - Naphthalene, 1,2,5,6-tetramethyl-
8003 - Naphthalene, 1,3,6,8-tetramethyl-

$C_{14}H_{16}ClN_3O_2$
512 - Bayleton

$C_{14}H_{16}ClN_3O_4S_2$
2851 - 2H-1,2,4-Benzothiadiazine-7-sulfonamide, 3-bicyclo[2.2.1]hept-5-en-2-yl-6-chloro-3,4-dihydro-, 1,1-dioxide

$C_{14}H_{16}ClN_5O_5S$
11824 - Triasulfuron

$C_{14}H_{16}ClO_5PS$
4147 - Coumaphos

$C_{14}H_{16}F_3N_3O_4$
9763 - Profluralin

$C_{14}H_{16}N_2$
731 - Benzenamine, 4,4′-(1,2-ethanediyl)bis-
3034 - [1,1′-Biphenyl]-4,4′-diamine, 3,3′-dimethyl-
5508 - 1,2-Ethanediamine, *N,N′*-diphenyl-
7138 - Hydrazine, 1,2-bis(3-methylphenyl)-
7139 - Hydrazine, 1,2-bis(4-methylphenyl)-

$C_{14}H_{16}N_2O_2$
3033 - [1,1′-Biphenyl]-4,4′-diamine, 3,3′-dimethoxy-

$C_{14}H_{16}N_2O_3$
11089 - 2,4,6(1H,3H,5H)-Pyrimidinetrione, 5,5-diethyl-1-phenyl-

$C_{14}H_{16}O_2Si$
11413 - Silane, dimethoxydiphenyl-
11415 - Silane, dimethyldiphenoxy-

$C_{14}H_{16}O_4$
9978 - Propanedioic acid, (phenylmethylene)-, diethyl ester

$C_{14}H_{16}O_9$
10786 - Pyrano[3,2-c][2]benzopyran-6(2H)-one, 3,4,4a,10b-tetrahydro-3,4,8,10-tetrahydroxy-2-(hydroxymethyl)-9-methoxy-[2R-(2α,3β,4α,4aα,10bβ)]-

$C_{14}H_{16}Si$
11416 - Silane, dimethyldiphenyl-

$C_{14}H_{17}ClNO_4PS_2$
5105 - Dialifor

$C_{14}H_{17}Cl_3O_4$
257 - Acetic acid, (2,4,5-trichlorophenoxy)-, 2-butoxyethyl ester

$C_{14}H_{17}N$
7713 - 1-Naphthalenamine, *N,N*-diethyl-

$C_{14}H_{17}NO_3$
11231 - 1(2H)-Quinolinecarboxylic acid, 2-ethoxy-, ethyl ester

$C_{14}H_{17}NO_6$
955 - Benzeneacetonitrile, α-(β-*D*-glucopyranosyloxy)-

$C_{14}H_{18}$
427 - Anthracene, 1,2,3,4,5,6,7,8-octahydro-
7747 - Naphthalene, 2-butyl-
9184 - Phenanthrene, 1,2,3,4,4a,9,10,10a-octahydro-
9185 - Phenanthrene, 1,2,3,4,5,6,7,8-octahydro-

$C_{14}H_{18}Cl_2O_4$
170 - Acetic acid, (2,4-dichlorophenoxy)-, 2-butoxyethyl ester

$C_{14}H_{18}N_2O_4$
8566 - Oxadixyl

$C_{14}H_{18}N_2O_5$
9511 - *L*-Phenylalanine, *N-L*-α-aspartyl-, 1-methyl ester

$C_{14}H_{18}N_4O_3$
515 - Benomyl

$C_{14}H_{18}O$
6347 - Heptanal, 2-(phenylmethylene)-

$C_{14}H_{18}O_4$
9977 - Propanedioic acid, (phenylmethyl)-, diethyl ester

$C_{14}H_{18}O_7$
5818 - Ethanone, 1-[4-(β-*D*-glucopyranosyloxy)phenyl]-

$C_{14}H_{19}Cl_2NO_2$
1101 - Benzenebutanoic acid, 4-[bis(2-chloroethyl)amino]-

$C_{14}H_{19}NO$
11265 - Quinoline, 6-ethoxy-1,2-dihydro-2,2,4-trimethyl-

$C_{14}H_{19}N_3S$
5507 - 1,2-Ethanediamine, *N,N*-dimethyl-*N′*-2-pyridinyl-*N′*-(2-thienylmethyl)-

$C_{14}H_{19}O_6P$
4137 - Ciodrin

$C_{14}H_{20}$
1249 - Benzene, (1-cyclohexylethyl)-
1255 - Benzene, (3-cyclopentylpropyl)-
7999 - Naphthalene, 1,2,3,4-tetrahydro-1,1,2,6-tetramethyl-

$C_{14}H_{20}ClNO_2$
270 - Acetochlor
322 - Alachlor

$C_{14}H_{20}O$
8480 - 1-Octanone, 1-phenyl-

$C_{14}H_{20}O_2$
2695 - Benzoic acid, 4-(1-methylethyl)-, 2-methylpropyl ester

$C_{14}H_{20}O_8$
5916 - Ethenetetracarboxylic acid, tetraethyl ester

$C_{14}H_{21}N_3O_4$
3893 - Butralin

$C_{14}H_{22}$
990 - Benzene, 1,4-bis(1,1-dimethylethyl)-
1981 - Benzene, (1-methylheptyl)-
2067 - Benzene, octyl-
2268 - Benzene, 1,2,3,4-tetraethyl-
2269 - Benzene, 1,2,4,5-tetraethyl-
9178 - Phenanthrene, dodecahydro-
11610 - 6,8-Tetradecadiyne

$C_{14}H_{22}N_2O$
54 - Acetamide, 2-(diethylamino)-*N*-(2,6-dimethylphenyl)-

$C_{14}H_{22}N_2O_3$
11081 - 2,4,6(1H,3H,5H)-Pyrimidinetrione, 5-butyl-1-cyclohexyl-

$C_{14}H_{22}O$
2069 - Benzene, (octyloxy)-
3883 - 3-Buten-2-one, 4-(2,5,6,6-tetramethyl-1-cyclohexen-1-yl)-
3884 - 3-Buten-2-one, 4-(2,5,6,6-tetramethyl-2-cyclohexen-1-yl)-
9219 - Phenol, 2,4-bis(1,1-dimethylethyl)-
9220 - Phenol, 2,6-bis(1,1-dimethylethyl)-
9227 - Phenol, 2,6-bis(1-methylpropyl)-

$C_{14}H_{22}O_2$
1314 - Benzene, 1,4-dibutoxy-
1411 - Benzene, 1-(diethoxymethyl)-4-(1-methylethyl)-
4528 - Cyclohexanol, 1,1′-(1,2-ethynediyl)bis-

$C_{14}H_{22}O_4$
5559 - Ethanedioic acid, dicyclohexyl ester

$C_{14}H_{22}O_6$
10657 - 2-Propenoic acid, 2-methyl-, 1,2-ethanediylbis(oxy-2,1-ethanediyl) ester

$C_{14}H_{22}O_8$
5633 - 1,1,2,2-Ethanetetracarboxylic acid, tetraethyl ester
10166 - 1,2,3-Propanetricarboxylic acid, 2-(acetyloxy)-, triethyl ester

$C_{14}H_{23}N$
684 - Benzenamine, *N,N*-dibutyl-
831 - Benzenamine, 4-octyl-
10880 - Pyridine, 4-(1-butylpentyl)-

$C_{14}H_{23}N_3O_{10}$
6257 - Glycine, *N,N*-bis[2-[bis(carboxymethyl)amino]ethyl]-

$C_{14}H_{24}$
5240 - Dispiro[5.1.5.1]tetradecane
9186 - Phenanthrene, tetradecahydro-

$C_{14}H_{244}NO_4PS_3$
517 - Bensulide

$C_{14}H_{24}N_2$
1285 - 1,3-Benzenediamine, *N,N,N′,N′*-tetraethyl-

$C_{14}H_{24}O_2$
10295 - Propanoic acid, 2-methyl-, 3,7-dimethyl-2,6-octadienyl ester, (*E*)-

$C_{14}H_{26}$
4341 - Cyclohexane, (3-cyclopentylpropyl)-
4386 - Cyclohexane, 1,1′-(1,2-ethanediyl)bis-
4392 - Cyclohexane, 1,1′-ethylidenebis-
11640 - 2-Tetradecyne
11641 - 7-Tetradecyne

$C_{14}H_{26}O_2$
11637 - 4-Tetradecenoic acid
11638 - 5-Tetradecenoic acid
11639 - 9-Tetradecenoic acid
11896 - 12-Tridecenoic acid, methyl ester

$C_{14}H_{26}O_3$
183 - Acetic acid, ethoxy-, 5-methyl-2-(1-methylethyl)cyclohexyl ester, (1α,2β,5α)-
6456 - Heptanoic acid, anhydride

$C_{14}H_{26}O_4$
3319 - Butanedioic acid, dipentyl ester
5045 - Decanedioic acid, diethyl ester
5300 - Dodecanedioic acid, dimethyl ester
6783 - Hexanedioic acid, bis(2-methylpropyl) ester
6784 - Hexanedioic acid, dibutyl ester

$C_{14}H_{27}BrO_2$
5308 - Dodecanoic acid, 2-bromoethyl ester

$C_{14}H_{27}ClO$
11635 - Tetradecanoyl chloride

$C_{14}H_{27}N$
11619 - Tetradecanenitrile

$C_{14}H_{28}$
4460 - Cyclohexane, octyl-
4841 - Cyclopentane, nonyl-
11636 - 1-Tetradecene

$C_{14}H_{28}Br_2$
11616 - Tetradecane, 1,14-dibromo-

$C_{14}H_{28}O$
11633 - 2-Tetradecanone
11634 - 3-Tetradecanone
11892 - 3-Tridecanone, 2-methyl-
11963 - 4-Undecanone, 7-ethyl-2-methyl-

$C_{14}H_{28}O_2$
178 - Acetic acid, dodecyl ester
5312 - Dodecanoic acid, ethyl ester
6464 - Heptanoic acid, heptyl ester
6904 - Hexanoic acid, octyl ester
8156 - Nonanoic acid, pentyl ester
8431 - Octanoic acid, hexyl ester
11622 - Tetradecanoic acid
11882 - Tridecanoic acid, methyl ester

$C_{14}H_{29}Br$
11614 - Tetradecane, 1-bromo-
11873 - Tridecane, 1-bromo-12-methyl-

$C_{14}H_{29}Cl$
11615 - Tetradecane, 1-chloro-

$C_{14}H_{29}N$
4289 - Cyclohexanamine, *N*-(2-ethylhexyl)-
9689 - Piperidine, 1-nonyl-

$C_{14}H_{29}NO$
11611 - Tetradecanamide

$C_{14}H_{30}$
11613 - Tetradecane
11879 - Tridecane, 7-methyl-

$C_{14}H_{30}O$
6430 - Heptane, 1,1′-oxybis-
11630 - 1-Tetradecanol
11631 - 2-Tetradecanol

11632 - 3-Tetradecanol
11886 - 1-Tridecanol, 12-methyl-
11887 - 3-Tridecanol, 2-methyl-
11888 - 4-Tridecanol, 2-methyl

$C_{14}H_{30}O_2$
5297 - Dodecane, 1,1-dimethoxy-
5298 - Dodecane, 1,12-dimethoxy-
11618 - 1,14-Tetradecanediol

$C_{14}H_{30}O_2Sn$
11538 - Stannane, (acetyloxy)tributyl-

$C_{14}H_{30}O_3S$
11620 - 1-Tetradecanesulfonic acid

$C_{14}H_{30}S$
6435 - Heptane, 1,1′-thiobis-
11621 - 1-Tetradecanethiol

$C_{14}H_{31}N$
6352 - 1-Heptanamine, *N*-heptyl-
11612 - 1-Tetradecanamine

$C_{14}H_{32}N_2$
6750 - 1,6-Hexanediamine, *N,N′*-dibutyl-

$C_{14}H_{32}N_2O_4$
10379 - 2-Propanol, 1,1′,1″,1‴-(1,2-ethanediyldinitrilo)tetrakis-

$C_{14}H_{42}O_5Si_6$
6990 - Hexasiloxane, tetradecamethyl-

$C_{14}H_{42}O_7Si_7$
4239 - Cycloheptasiloxane, tetradecamethyl-

$C_{15}H_8O_4$
372 - 2-Anthracenecarboxylic acid, 9,10-dihydro-9,10-dioxo-

$C_{15}H_8O_5$
2466 - 6H-Benzofuro[3,2-c][1]benzopyran-6-one, 3,9-dihydroxy-

$C_{15}H_8O_6$
371 - 2-Anthracenecarboxylic acid, 9,10-dihydro-4,5-dihydroxy-9,10-dioxo-

$C_{15}H_9BrO_2$
7265 - 1H-Indene-1,3(2H)-dione, 2-(4-bromophenyl)-

$C_{15}H_9ClO_2$
7266 - 1H-Indene-1,3(2H)-dione, 2-(4-chlorophenyl)-

$C_{15}H_{10}ClF_3N_2O_6S$
5985 - Fomesafen

$C_{15}H_{10}O_2$
368 - 1-Anthracenecarboxylic acid
369 - 2-Anthracenecarboxylic acid
370 - 9-Anthracenecarboxylic acid
409 - 9,10-Anthracenedione, 2-methyl-
2840 - 4H-1-Benzopyran-4-one, 2-phenyl-
2841 - 4H-1-Benzopyran-4-one, 3-phenyl-
7270 - 1H-Indene-1,3(2H)-dione, 2-phenyl-
7588 - Methanone, 2-benzofuranylphenyl-
9165 - 9-Phenanthrenecarboxylic acid

$C_{15}H_{10}O_3$
10199 - Propanetrione, diphenyl-

$C_{15}H_{10}O_4$
398 - 9,10-Anthracenedione, 1,8-dihydroxy-3-methyl-
2811 - 4H-1-Benzopyran-4-one, 5,7-dihydroxy-2-phenyl-
2827 - 4H-1-Benzopyran-4-one, 7-hydroxy-3-(4-hydroxyphenyl)-

$C_{15}H_{10}O_5$
397 - 9,10-Anthracenedione, 1,8-dihydroxy-3-(hydroxymethyl)-
419 - 9,10-Anthracenedione, 1,3,8-trihydroxy-6-methyl-
1332 - 1,2-Benzenedicarboxylic acid, 3-benzoyl-
2533 - Benzoic acid, 4,4′-carbonylbis-
2806 - 4H-1-Benzopyran-4-one, 5,7-dihydroxy-2-(4-hydroxyphenyl)-
2807 - 4H-1-Benzopyran-4-one, 5,7-dihydroxy-3-(4-hydroxyphenyl)-
2844 - 4H-1-Benzopyran-4-one, 5,6,7-trihydroxy-2-phenyl-

$C_{15}H_{10}O_6$
2813 - 4H-1-Benzopyran-4-one, 2-(3,4-dihydroxyphenyl)-3,7-dihydroxy-
2814 - 4H-1-Benzopyran-4-one, 2-(3,4-dihydroxyphenyl)-5,7-dihydroxy-
2842 - 4H-1-Benzopyran-4-one, 3,5,7-trihydroxy-2-(2-hydroxyphenyl)-
2843 - 4H-1-Benzopyran-4-one, 3,5,7-trihydroxy-2-(4-hydroxyphenyl)-

$C_{15}H_{10}O_7$
2816 - 4H-1-Benzopyran-4-one, 2-(2,4-dihydroxyphenyl)-3,5,7-trihydroxy-
2817 - 4H-1-Benzopyran-4-one, 2-(3,4-dihydroxyphenyl)-3,5,7-trihydroxy-

$C_{15}H_{11}ClF_3NO_4$
8633 - Oxyfluorfen

$C_{15}H_{11}ClN_2O$
2411 - 2H-1,4-Benzodiazepin-2-one, 7-chloro-1,3-dihydro-5-phenyl-

$C_{15}H_{11}ClO_5$
2848 - 1-Benzopyrylium, 3,5,7-trihydroxy-2-(4-hydroxyphenyl)-, chloride

$C_{15}H_{11}ClO_7$
2849 - 1-Benzopyrylium, 3,5,7-trihydroxy-2-(3,4,5-trihydroxyphenyl)-, chloride

$C_{15}H_{11}I_3O_4$
2136 - Benzenepropanoic acid, 4-(4-hydroxy-3-iodophenoxy)-3,5-diiodo-

$C_{15}H_{11}I_4NO_4$
11930 - *L*-Tyrosine, *O*-(4-hydroxy-3,5-diiodophenyl)-3,5-diiodo-

$C_{15}H_{11}N$
2768 - Benzonitrile, 4-(2-phenylethenyl)-, (*Z*)-
11288 - Quinoline, 2-phenyl-
11289 - Quinoline, 3-phenyl-
11290 - Quinoline, 4-phenyl-
11291 - Quinoline, 6-phenyl-

$C_{15}H_{11}NO$
8579 - Oxazole, 2,4-diphenyl-
8580 - Oxazole, 2,5-diphenyl-
8581 - Oxazole, 4,5-diphenyl-

$C_{15}H_{11}NO_2$
384 - 9,10-Anthracenedione, 1-amino-2-methyl-
7367 - 1H-Isoindole-1,3(2H)-dione, 2-(phenylmethyl)-

$C_{15}H_{11}NS$
11676 - Thiazole, 2,4-diphenyl-

$C_{15}H_{12}$
423 - Anthracene, 1-methyl-
424 - Anthracene, 2-methyl-
425 - Anthracene, 9-methyl-
9180 - Phenanthrene, 3-methyl-
9181 - Phenanthrene, 4-methyl-

$C_{15}H_{12}I_2O_3$
2135 - Benzenepropanoic acid, 4-hydroxy-3,5-diiodo-α-phenyl-

$C_{15}H_{12}N_2$
10830 - 1H-Pyrazole, 1,3-diphenyl-
10831 - 1H-Pyrazole, 1,5-diphenyl-

$C_{15}H_{12}N_2O$
5144 - 5H-Dibenz[b,f]azepine-5-carboxamide

$C_{15}H_{12}N_2O_2$
7219 - 2,4-Imidazolidinedione, 5,5-diphenyl-
10843 - 3,5-Pyrazolidinedione, 1,4-diphenyl-

$C_{15}H_{12}N_2O_3$
7488 - Methanediamine, 1-(2-furanyl)-*N,N′*-bis(2-furanylmethylene)-

$C_{15}H_{12}N_6O_4$
2494 - Benzoic acid, 4-[[(2-amino-1,4-dihydro-4-oxo-6-pteridinyl)methyl]formylamino]-

$C_{15}H_{12}O$
5146 - 10H-Dibenzo[a,d]cyclohepten-10-one, 5,11-dihydro-
5147 - 5H-Dibenzo[a,d]cyclohepten-5-one, 10,11-dihydro-
7291 - 1H-Inden-1-one, 2,3-dihydro-2-phenyl-
10708 - 2-Propen-1-one, 1,3-diphenyl-
10709 - 2-Propen-1-one, 1,3-diphenyl-, (*E*)-

$C_{15}H_{12}O_2$
941 - Benzeneacetic acid, α-(phenylmethylene)-
10027 - 1,3-Propanedione, 1,3-diphenyl-
10689 - 2-Propenoic acid, 3-phenyl-, phenyl ester, (*E*)-
10711 - 2-Propen-1-one, 3-hydroxy-1,3-diphenyl-

$C_{15}H_{12}O_3$
2516 - Benzoic acid, 2-benzoyl-, methyl ester
2684 - Benzoic acid, 2-(4-methylbenzoyl)-

$C_{15}H_{12}O_4$
2483 - Benzoic acid, 2-(acetyloxy)-, phenyl ester
2667 - Benzoic acid, 2-(4-methoxybenzoyl)-
7509 - Methanediol, dibenzoate

$C_{15}H_{12}O_5$
2802 - 4H-1-Benzopyran-4-one, 2,3-dihydro-5,7-dihydroxy-2-(4-hydroxyphenyl)-, (*S*)-

$C_{15}H_{12}O_6$
2686 - Benzoic acid, 3,3′-methylenebis[6-hydroxy-
2812 - 4H-1-Benzopyran-4-one, 2-(3,4-dihydroxyphenyl)-2,3-dihydro-5,7-dihydroxy-, (*S*)-

$C_{15}H_{13}N$
7318 - 1H-Indole, 3-methyl-2-phenyl-
7321 - 1H-Indole, 1-(phenylmethyl)-

$C_{15}H_{13}NO$
64 - Acetamide, *N*-9H-fluoren-2-yl-

$C_{15}H_{13}NO_2S$
9500 - 10H-Phenothiazine-2-acetic acid, 10-methyl-

$C_{15}H_{14}$
1966 - Benzene, 1,1′-(1-methylene-1,2-ethanediyl)bis-
1971 - Benzene, 1,1′-(1-methyl-1,2-ethenediyl)bis-
2161 - Benzene, 1,1′-(1-propene-1,3-diyl)bis-
2166 - Benzene, 1,1′-(1-propenylidene)bis-
7259 - 1H-Indene, 2,3-dihydro-2-phenyl-

$C_{15}H_{14}F_3N_3$
723 - Benzenamine, *N,N*-dimethyl-4-[[3-(trifluoromethyl)phenyl]azo]-

$C_{15}H_{14}N_2$
765 - Benzenamine, *N,N′*-methanetetraylbis[4-methyl-

$C_{15}H_{14}N_2O$
7228 - 4-Imidazolidinone, 5,5-diphenyl-

$C_{15}H_{14}N_4O_2S$
2201 - Benzenesulfonamide, 4-amino-*N*-(1-phenyl-1H-pyrazol-5-yl)-

$C_{15}H_{14}O$
1939 - Benzene, 1-methoxy-4-(2-phenylethenyl)-, (*E*)-
2782 - 2H-1-Benzopyran, 3,4-dihydro-2-phenyl-
7599 - Methanone, bis(4-methylphenyl)-
7614 - Methanone, (2,4-dimethylphenyl)phenyl-
7615 - Methanone, (3,4-dimethylphenyl)phenyl-
7628 - Methanone, (2-methylphenyl)(4-methylphenyl)-
10432 - 2-Propanone, 1,1-diphenyl-
10433 - 2-Propanone, 1,3-diphenyl-

$C_{15}H_{14}O_2$
931 - Benzeneacetic acid, α-methyl-α-phenyl-
2143 - Benzenepropanoic acid, α-phenyl-, (±)-
5839 - Ethanone, 2-methoxy-1,2-diphenyl-

$C_{15}H_{14}O_3$
920 - Benzeneacetic acid, α-hydroxy-α-phenyl-, methyl ester
3996 - Carbonic acid, bis(2-methylphenyl) ester
7612 - Methanone, (2,4-dimethoxyphenyl)phenyl-
7862 - 1,4-Naphthalenedione, 2-hydroxy-3-(3-methyl-2-butenyl)-

$C_{15}H_{14}O_4$
2398 - 2H,8H-Benzo[1,2-b:5,4-b′]dipyran-2-one, 5-methoxy-8,8-dimethyl-
6188 - 7H-Furo[3,2-g][1]benzopyran-7-one, 3-methoxy-2-(1-methylethyl)-
7841 - 1,4-Naphthalenediol, 2-methyl-, diacetate

$C_{15}H_{14}O_5$
7595 - Methanone, bis(2-hydroxy-4-methoxyphenyl)-
9384 - Phenol, 2-methoxy-, carbonate (2:1)
10453 - 1-Propanone, 3-(4-hydroxyphenyl)-1-(2,4,6-trihydroxyphenyl)-

10790 - 2H-Pyran-2-one, 6-[2-(1,3-benzodioxol-5-yl)ethenyl]-5,6-dihydro-4-methoxy-, [*R*-(*E*)]-

$C_{15}H_{14}O_6$
2846 - 2H-1-Benzopyran-3,5,7-triol, 2-(3,4-dihydroxyphenyl)-3,4-dihydro-, (2*S*-*cis*)-
7248 - 2H-Indene-2,2-dicarboxylic acid, 1,3-dihydro-1,3-dioxo-, diethyl ester
7855 - 1,4-Naphthalenedione, 5,8-dihydroxy-6-methoxy-2-methyl-3-(2-oxopropyl)-
11904 - 2H,4aH-1,4,5-Trioxadicyclopent[a,hi]indene-7-carboxylic acid, 3-ethylidene-3,3a,7a,9b-tetrahydro-2-oxo-, methyl ester, [3aS-(3E,3aα,4aβ,7aβ,9aR*,9bβ)]-

$C_{15}H_{15}ClF_3N_3O$
11900 - Triflumizole

$C_{15}H_{15}ClN_2O_2$
4088 - Chloroxuron

$C_{15}H_{15}ClN_4O_6S$
4081 - Chlorimuron

$C_{15}H_{15}NO$
2771 - Benzophenone, 3-(dimethylamino)-
7613 - Methanone, [4-(dimethylamino)phenyl]phenyl-

$C_{15}H_{15}NO_2$
3946 - Carbamic acid, diphenyl-, ethyl ester

$C_{15}H_{15}N_3O$
317 - 3,9-Acridinediamine, 7-ethoxy-

$C_{15}H_{15}N_3O_2$
2587 - Benzoic acid, 2-[[4-(dimethylamino) phenyl]azo]-

$C_{15}H_{16}$
1696 - Benzene, 1-ethyl-4-(phenylmethyl)-
1964 - Benzene, 1,1′-methylenebis[4-methyl-
1969 - Benzene, 1,1′-(1-methyl-1,2-ethanediyl)bis-
1976 - Benzene, 1,1′-(1-methylethylidene)bis-
2121 - Benzene, 1,1′-(1,3-propanediyl)bis-
2171 - Benzene, 1,1′-propylidenebis-

$C_{15}H_{16}F_5NO_2S_2$
5267 - Dithiopyr

$C_{15}H_{16}N_2O$
12006 - Urea, *N,N′*-dimethyl-*N,N′*-diphenyl-

$C_{15}H_{16}O$
1940 - Benzene, 1-methoxy-4-(1-phenylethyl)-
2158 - Benzenepropanol, β-phenyl-

$C_{15}H_{16}O_2$
1970 - Benzene, 1,1′-[(1-methyl-1,2-ethanediyl)bis(oxy)]bis-
2122 - Benzene, 1,1′-[1,3-propanediylbis(oxy)]bis-
9413 - Phenol, 4,4′-(1-methylethylidene)bis-
9464 - Phenol, 4,4′-propylidenebis-

$C_{15}H_{16}O_3$
10377 - 2-Propanol, 1,3-diphenoxy-

$C_{15}H_{16}O_8$
2821 - 2H-1-Benzopyran-2-one, 7-(β-*D*-glucopyranosyloxy)-

$C_{15}H_{16}O_9$
2822 - 2H-1-Benzopyran-2-one, 6-(β-*D*-glucopyranosyloxy)-7-hydroxy-

$C_{15}H_{17}Cl_2N_3O_2$
10720 - Propiconazole

$C_{15}H_{17}N$
1811 - Benzenemethanamine, *N*-ethyl-*N*-phenyl-

$C_{15}H_{17}N_3$
6279 - Guanidine, *N,N′*-bis(2-methylphenyl)-

$C_{15}H_{18}$
505 - Azulene, 1,4-dimethyl-7-(1-methylethyl)-
506 - Azulene, 4,8-dimethyl-2-(1-methylethyl)-
7830 - Naphthalene, 1,6-dimethyl-4-(1-methylethyl)-
7941 - Naphthalene, 1-pentyl-
7942 - Naphthalene, 2-pentyl-

$C_{15}H_{18}Cl_2N_2O_3$
8565 - Oxadiazon

$C_{15}H_{18}N_2O_6$
3003 - Binapacryl

$C_{15}H_{18}N_6O_6S$
8069 - Nicosulfuron

$C_{15}H_{18}O$
7917 - Naphthalene, 1-(3-methylbutoxy)-
7918 - Naphthalene, 2-(3-methylbutoxy)-
7943 - Naphthalene, 1-(pentyloxy)-
7944 - Naphthalene, 2-(pentyloxy)-

$C_{15}H_{18}O_2$
2939 - Bicyclo[2.2.1]heptan-2-ol, 1-phenyl-, acetate

$C_{15}H_{18}O_3$
8057 - Naphtho[1,2-b]furan-2,8(3H,4H)-dione, 3a,5,5a,9b-tetrahydro-3,5a,9-trimethyl-,

$C_{15}H_{18}O_6$
8058 - Naphtho[1,2-b]furan-2,8(3H,4H)-dione, 3a,5,5a,9b-tetrahydro-4,5-dihydroxy-9-methoxy-3,5a-dimethyl-

$C_{15}H_{19}NO_2$
487 - 8-Azabicyclo[3.2.1]octan-3-ol, 8-methyl-, benzoate (ester), *endo*-
488 - 8-Azabicyclo[3.2.1]octan-3-ol, 8-methyl-, benzoate (ester), *exo*-

$C_{15}H_{19}N_3O_3$
7190 - Imazethapyr

$C_{15}H_{20}N_2O$
7573 - 7,14-Methano-4H,6H-dipyrido[1,2-a1′,2′-e][1,5]diazocin-4-one, 7,7a,8,9,10,11,13,14-octahydro-,

$C_{15}H_{20}O_2$
6293 - Helenine
8059 - Naphtho[2,3-b]furan-2(3H)-one, 3a,5,6,7,8,8a,9,9a-octahydro-5,8a-dimethyl-3-methylene-,

$C_{15}H_{20}O_4$
8679 - 2,4-Pentadienoic acid, 5-(1-hydroxy-2,6,6-trimethyl-4-oxo-2-cyclohexen-1-yl)-3-methyl-, [*S*-(*Z,E*)]-
9952 - Propanedioic acid, ethylphenyl-, diethyl ester

$C_{15}H_{21}NO_4$
7439 - Metalaxyl

$C_{15}H_{21}N_3O_2$
11191 - Pyrrolo[2,3-b]indol-5-ol, 1,2,3,3a,8,8a-hexahydro-1,3a,8-trimethyl-, methylcarbamate (ester), (3a*S*-*cis*)-

$C_{15}H_{22}$
1468 - Benzene, 1-(1,5-dimethyl-4-hexenyl)-4-methyl-
7274 - 1H-Indene, 5-hexyl-2,3-dihydro-

$C_{15}H_{22}ClNO_2$
7657 - Metolachlor

$C_{15}H_{22}N_2O$
9639 - 2-Piperidinecarboxamide, *N*-(2,6-dimethylphenyl)-1-methyl-

$C_{15}H_{22}O$
8052 - 2(3H)-Naphthalenone, 4,4a,5,6,7,8-hexahydro-4,4a-dimethyl-6-(1-methylethylidene)-, (4*R*-*cis*)-
11521 - Spiro[4.5]dec-6-en-8-one, 6,10-dimethyl-2-(1-methylethylidene)-, (5*R*-*cis*)-

$C_{15}H_{23}ClO_4S$
11579 - Sulfurous acid, 2-chloroethyl 2-[4-(1,1-dimethylethyl)phenoxy]-1-methylethyl ester

$C_{15}H_{23}NO_4$
9655 - 2,6-Piperidinedione, 4-[2-(3,5-dimethyl-2-oxocyclohexyl)-2-hydroxyethyl]-,

$C_{15}H_{23}N_3O_4$
7371 - Isopropalin

$C_{15}H_{24}$
2064 - Benzene, nonyl-
2372 - Benzene, 1,2,4-tris(1-methylethyl)-
2373 - Benzene, 1,3,5-tris(1-methylethyl)-
2996 - Bicyclo[7.2.0]undec-4-ene, 4,11,11-trimethyl-8-methylene-, [1*R*-(1*R**,4E,9*S**)]-
4066 - Cedrene
4389 - Cyclohexane, 1-ethenyl-1-methyl-2,4-bis(1-methylethenyl)-, (1α,2β,4β)-
4659 - Cyclohexene, 6-ethenyl-6-methyl-1-(1-methylethyl)-3-(1-methylethylidene)-, (*S*)-
4685 - Cyclohexene, 1-methyl-4-(5-methyl-1-methylene-4-hexenyl)-, (*S*)-
4954 - Cyclopropa[d]naphthalene, 1,1a,4,4a,5,6,7,8-octahydro-2,4a,8,8-tetramethyl-, [1a*S*-(1aα,4aβ,8a*R**)]-
5006 - 1,4,8-Cycloundecatriene, 2,6,6,9-tetramethyl-, (*E,E,E*)-
5329 - 1,3,6,10-Dodecatetraene, 3,7,11-trimethyl-, (*E,E*)-
5331 - 1,6,10-Dodecatriene, 7,11-dimethyl-3-methylene-, (*E*)-
7568 - 1,4-Methanoazulene, decahydro-4,8,8-trimethyl-9-methylene-, [1*S*-(1α,3aβ,4α,8aβ)]-
7898 - Naphthalene, 1,2,4a,5,8,8a-hexahydro-4,7-dimethyl-1-(1-methylethyl)-, [1*S*-(1α,4aβ,8aα)]-
7899 - Naphthalene, 1,2,3,5,6,8a-hexahydro-4,7-dimethyl-1-(1-methylethyl)-, (1*S*-*cis*)-
7936 - Naphthalene, 1,2,3,4,4a,5,6,8a-octahydro-7-methyl-4-methylene-1-(1-methylethyl)-, (1α,4aβ,8aα)-
11861 - Tricyclo[4.4.0.02,7]dec-3-ene, 1,3-dimethyl-8-(1-methylethyl)-, stereoisomer

$C_{15}H_{24}NO_4PS$
7359 - Isofenphos

$C_{15}H_{24}N_2O$
7426 - Matridin-15-one
7572 - 7,14-Methano-2H,6H-dipyrido[1,2-a 1′,2′-e][1,5]diazocin-6-one, dodecahydro-, [7*R*-(7α,7aβ,14α,14aalp

$C_{15}H_{24}N_2O_2$
7574 - 7,14-Methano-4H,6H-dipyrido[1,2-a 1′,2′-e][1,5]diazocin-4-one, dodecahydro-9-hydroxy-,

$C_{15}H_{24}O$
5330 - 2,6,10-Dodecatrienal, 3,7,11-trimethyl-
9093 - 2-Penten-1-ol, 5-(2,3-dimethyltricyclo [2.2.1.02,6]hept-3-yl)-2-methyl-, stereoisomer
9103 - 2-Penten-1-ol, 2-methyl-5-(2-methyl-3-methylenebicyclo[2.2.1]hept-2-yl)-, [1*S*-[1α,2α(*Z*),4α]
9222 - Phenol, 2,4-bis(1,1-dimethylethyl)-5-methyl-
9223 - Phenol, 2,4-bis(1,1-dimethylethyl)-6-methyl-
9224 - Phenol, 2,6-bis(1,1-dimethylethyl)-4-methyl-

$C_{15}H_{24}O_8$
10139 - 1,1,2,3-Propanetetracarboxylic acid, tetraethyl ester
10140 - 1,1,3,3-Propanetetracarboxylic acid, tetraethyl ester

$C_{15}H_{25}ClN_2O_2$
2531 - Benzoic acid, 4-(butylamino)-, 2-(dimethylamino)ethyl ester, monohydrochloride

$C_{15}H_{26}N_2$
7571 - 7,14-Methano-2H,6H-dipyrido[1,2-a 1′,2′-e][1,5]diazocine, dodecahydro-,

$C_{15}H_{26}O$
507 - 5-Azulenemethanol, 1,2,3,4,5,6,7,8-octahydro-α,α,3,8-tetramethyl-,
4989 - 1H-Cycloprop[e]azulen-4-ol, decahydro-1,1,4,7tetramethyl,(1a*R*,4*R*,4a*S*,7*R*,7a*S*,7b*S*)-
5332 - 1,6,10-Dodecatrien-3-ol, 3,7,11-trimethyl-, [*S*-(*Z*)]-
5333 - 2,6,10-Dodecatrien-1-ol, 3,7,11-trimethyl-, (*E,E*)-
5334 - 2,6,10-Dodecatrien-1-ol, 3,7,11-trimethyl-, (*Z,E*)-
7569 - 1,4-Methanoazulen-9-ol, decahydro-1,5,5,8a-tetramethyl-, [1*R*-(1α,3aβ,4α,8aβ,9*S**)]
7570 - 1H-3a,7-Methanoazulen-6-ol, octahydro-3,6,8,8-tetramethyl-,
7586 - 1,6-Methanonaphthalen-1(2H)-ol, octahydro-4,8a,9,9-tetramethyl-,

$C_{15}H_{26}O_2$
3565 - Butanoic acid, 3-methyl-, 1,7,7-trimethylbicyclo[2.2.1]hept-2-yl ester, (1*R*-*endo*)-

$C_{15}H_{26}O_6$
3586 - Butanoic acid, 1,2,3-propanetriyl ester

$C_{15}H_{26}Si$
11448 - Silane, phenyltripropyl-
$C_{15}H_{28}$
4430 - Cyclohexane, 1,1′-(1-methyl-1,2-ethanediyl)bis-
4432 - Cyclohexane, 1,1′-(1-methylethylidene)bis-
4466 - Cyclohexane, 1,1′-(1,3-propanediyl)bis-
4472 - Cyclohexane, 1,1′-propylidenebis-
8656 - 1-Pentadecyne
$C_{15}H_{28}NNaO_3$
6269 - Glycine, *N*-methyl-*N*-(1-oxododecyl)-, sodium salt
$C_{15}H_{28}O$
4761 - Cyclopentadecanone
$C_{15}H_{28}O_2$
3554 - Butanoic acid, 3-methyl-, 5-methyl-2-(1-methylethyl)cyclohexyl ester
8562 - Oxacyclohexadecan-2-one
$C_{15}H_{28}O_4$
9937 - Propanedioic acid, dibutyl-, diethyl ester
$C_{15}H_{30}$
4401 - Cyclohexane, 1-ethyl-1-methyl-2,4-bis(1-methylethyl)-, [1*R*-(1α,2β,4β)]-
4457 - Cyclohexane, nonyl-
4759 - Cyclopentadecane
4790 - Cyclopentane, decyl-
8653 - 1-Pentadecene
8654 - 7-Pentadecene
$C_{15}H_{30}N_2$
9702 - Piperidine, 4,4′-(1,3-propanediyl)bis[1-methyl-
$C_{15}H_{30}O$
4760 - Cyclopentadecanol
8640 - Pentadecanal
8651 - 2-Pentadecanone
8652 - 8-Pentadecanone
8655 - 1-Pentadecen-3-ol
$C_{15}H_{30}O_2$
5316 - Dodecanoic acid, 1-methylethyl ester
5320 - Dodecanoic acid, propyl ester
6468 - Heptanoic acid, octyl ester
8430 - Octanoic acid, heptyl ester
8645 - Pentadecanoic acid
11625 - Tetradecanoic acid, methyl ester
$C_{15}H_{30}O_4$
5310 - Dodecanoic acid, 2,3-dihydroxypropyl ester, (±)-
$C_{15}H_{31}Br$
8643 - Pentadecane, 1-bromo-
$C_{15}H_{32}$
8642 - Pentadecane
$C_{15}H_{32}O$
8647 - 2-Pentadecanol
8648 - 3-Pentadecanol
$C_{15}H_{33}B$
3122 - Borane, tris(3-methylbutyl)-
$C_{15}H_{33}N$
3224 - 1-Butanamine, 2-methyl-*N*,*N*-bis(2-methylbutyl)-
3225 - 1-Butanamine, 3-methyl-*N*,*N*-bis(3-methylbutyl)-
8641 - 1-Pentadecanamine
8705 - 1-Pentanamine, *N*,*N*-dipentyl-
$C_{13}H_{29}N_3 \cdot C_2H_4O_2$
5347 - Dodine
$C_{15}H_{33}O_4P$
9565 - Phosphoric acid, tripentyl ester
$C_{16}H_8O_2S_2$
2404 - Benzo[b]thiophen-3(2H)-one, 2-(3-oxobenzo[b]thien-2(3H)-ylidene)-
$C_{16}H_9N_3Na_2O_{10}S_2$
7878 - 2,7-Naphthalenedisulfonic acid, 4,5-dihydroxy-3-[(4-nitrophenyl)azo]-, disodium salt
$C_{16}H_{10}$
5966 - Fluoranthene
10853 - Pyrene
$C_{16}H_{10}Cl_4O_5$
5153 - 11H-Dibenzo[b,e][1,4]dioxepin-11-one, 2,4,7,9-tetrachloro-3-hydroxy-8-methoxy-1,6-dimethyl-
$C_{16}H_{10}N_2$
2394 - Benzo[a]phenazine
$C_{16}H_{10}N_2O_2$
7331 - 3H-Indol-3-one, 2-(1,3-dihydro-3-oxo-2H-indol-2-ylidene)-1,2-dihydro-
$C_{16}H_{11}N$
2405 - 7H-Benzo[c]carbazole
$C_{16}H_{11}NO$
7640 - Methanone, phenyl-4-quinolinyl-
$C_{16}H_{11}NO_2$
7866 - 1,4-Naphthalenedione, 2-(phenylamino)-
11235 - 4-Quinolinecarboxylic acid, 2-phenyl-
11309 - 8-Quinolinol, benzoate (ester)
$C_{16}H_{11}NO_3$
11233 - 4-Quinolinecarboxylic acid, 3-hydroxy-2-phenyl-
$C_{16}H_{12}$
7947 - Naphthalene, 1-phenyl-
7948 - Naphthalene, 2-phenyl-
$C_{16}H_{12}N_2O$
8036 - 1-Naphthalenol, 2-(phenylazo)-
8037 - 2-Naphthalenol, 1-(phenylazo)-
$C_{16}H_{12}O$
6100 - Furan, 2,5-diphenyl-
7945 - Naphthalene, 1-phenoxy-
7946 - Naphthalene, 2-phenoxy-
$C_{16}H_{12}O_2$
402 - 9,10-Anthracenedione, 1,4-dimethyl-
403 - 9,10-Anthracenedione, 2,3-dimethyl-
404 - 9,10-Anthracenedione, 2,6-dimethyl-
$C_{16}H_{12}O_3$
7269 - 1H-Indene-1,3(2H)-dione, 2-(4-methoxyphenyl)-
$C_{16}H_{12}O_4$
2830 - 4H-1-Benzopyran-4-one, 7-hydroxy-3-(4-methoxyphenyl)-
3780 - 2-Butenedioic acid (*Z*)-, diphenyl ester
$C_{16}H_{12}O_5$
2809 - 4H-1-Benzopyran-4-one, 5,7-dihydroxy-2-(4-methoxyphenyl)-
2810 - 4H-1-Benzopyran-4-one, 5,7-dihydroxy-6-methoxy-2-phenyl-
2828 - 4H-1-Benzopyran-4-one, 5-hydroxy-3-(4-hydroxyphenyl)-7-methoxy-
$C_{16}H_{12}O_6$
628 - Benz[b]indeno[1,2-d]pyran-9(6H)-one, 6a,7-dihydro-3,4,6a,10-tetrahydroxy-
$C_{16}H_{12}O_7$
2815 - 4H-1-Benzopyran-4-one, 2-(3,4-dihydroxyphenyl)-3,5-dihydroxy-7-methoxy-
$C_{16}H_{12}S$
7951 - Naphthalene, 1-(phenylthio)-
$C_{16}H_{13}ClN_2O$
2410 - 2H-1,4-Benzodiazepin-2-one, 7-chloro-1,3-dihydro-1-methyl-5-phenyl-
$C_{16}H_{13}N$
7387 - Isoquinoline, 3-(phenylmethyl)-
7728 - 2-Naphthalenamine, *N*-phenyl-
$C_{16}H_{13}N_3$
7729 - 2-Naphthalenamine, 1-(phenylazo)-
$C_{16}H_{14}$
377 - Anthracene, 1,3-dimethyl-
378 - Anthracene, 2,10-dimethyl-
420 - Anthracene, 9-ethyl-
1095 - Benzene, 1,1′-(1,3-butadiene-1,4-diyl)bis-, (*E*,*E*)-
1096 - Benzene, 1,1′-(1,3-butadiene-1,4-diyl)bis-, (*Z*,*Z*)-
9172 - Phenanthrene, 9,10-dimethyl-
9179 - Phenanthrene, 9-ethyl-
$C_{16}H_{14}ClN_3O$
2409 - 3H-1,4-Benzodiazepin-2-amine, 7-chloro-*N*-methyl-5-phenyl-, 4-oxide
$C_{16}H_{14}Cl_2O_3$
4082 - Chlorobenzilate
$C_{16}H_{14}Cl_2O_4$
5171 - Diclofop-methyl
$C_{16}H_{14}N_2O$
11199 - 4(3H)-Quinazolinone, 2-methyl-3-(2-methylphenyl)-
11200 - 4(1H)-Quinazolinone, 1-methyl-2-(phenylmethyl)-
$C_{16}H_{14}O$
3876 - 2-Buten-1-one, 1,3-diphenyl-
$C_{16}H_{14}O_2$
10690 - 2-Propenoic acid, 3-phenyl-, phenylmethyl ester, (*E*)-
10712 - 2-Propen-1-one, 3-(3-methoxyphenyl)-1-phenyl-
10713 - 2-Propen-1-one, 3-(4-methoxyphenyl)-1-phenyl-
$C_{16}H_{14}O_3$
887 - Benzeneacetic acid, anhydride
2515 - Benzoic acid, 2-benzoyl-, ethyl ester
2682 - Benzoic acid, 2-methyl-, anhydride
2683 - Benzoic acid, 3-methyl-, anhydride
$C_{16}H_{14}O_4$
3039 - [1,1′-Biphenyl]-2,2′-dicarboxylic acid, dimethyl ester
3320 - Butanedioic acid, diphenyl ester
5572 - 1,2-Ethanediol, dibenzoate
6189 - 7H-Furo[3,2-g][1]benzopyran-7-one, 9-[(3-methyl-2-butenyl)oxy]-
$C_{16}H_{14}O_6$
629 - Benz[b]indeno[1,2-d]pyran-3,4,6a,9,10(6H)-pentol, 7,11b-dihydro-, *cis*-(+)-
2800 - 4H-1-Benzopyran-4-one, 2,3-dihydro-5,7-dihydroxy-2-(3-hydroxy-4-methoxyphenyl)-, (*S*)-
2801 - 4H-1-Benzopyran-4-one, 2,3-dihydro-5,7-dihydroxy-3-(3-hydroxy-4-methoxyphenyl)-
$C_{16}H_{15}ClO_2$
892 - Benzeneacetic acid, α-chloro-α-phenyl-, ethyl ester
$C_{16}H_{15}Cl_3O_2$
2329 - Benzene, 1,1′-(2,2,2-trichloroethylidene)bis[4-methoxy-
$C_{16}H_{16}$
421 - Anthracene, 9-ethyl-9,10-dihydro-
1113 - Benzene, 1,1′-(1-butenylidene)bis-
1456 - Benzene, 1,1′-(1,2-dimethyl-1,2-ethenediyl)bis-, (*E*)-
1457 - Benzene, 1,1′-(1,2-dimethyl-1,2-ethenediyl)bis-, (*Z*)-
1622 - Benzene, 1,1′-(1,2-ethenediyl)bis[2-methyl-
2022 - Benzene, 1,1′-(3-methyl-1-propene-1,3-diyl)bis-
11560 - Stilbene, α-ethyl-
$C_{16}H_{16}N_2OS$
9502 - 10H-Phenothiazine, 10-[(dimethylamino)acetyl]-
$C_{16}H_{16}N_2O_2$
5375 - Ergoline-8-carboxylic acid, 9,10-didehydro-6-methyl-, (8α)-
5376 - Ergoline-8-carboxylic acid, 9,10-didehydro-6-methyl-, (8β)-
$C_{16}H_{16}N_2O_4$
5104 - Desmedipham
9198 - Phenmedipham
$C_{16}H_{16}N_2O_4S$
91 - Acetamide, *N*,*N*′-(sulfonyldi-4,1-phenylene)bis-
$C_{16}H_{16}O$
3658 - 2-Butanone, 3,3-diphenyl-
$C_{16}H_{16}O_2$
939 - Benzeneacetic acid, α-phenyl-, ethyl ester
940 - Benzeneacetic acid, 2-phenylethyl ester
1621 - Benzene, 1,1′-(1,2-ethenediyl)bis[4-methoxy-

2144 - Benzenepropanoic acid, α-(phenylmethyl)-
2145 - Benzenepropanoic acid, phenylmethyl ester
3019 - [1,1′-Biphenyl]-4-butanoic acid
5813 - Ethanone, 2-ethoxy-1,2-diphenyl-

$C_{16}H_{16}O_3$
919 - Benzeneacetic acid, α-hydroxy-α-phenyl-, ethyl ester
3583 - Butanoic acid, 2-phenoxy-, phenyl ester

$C_{16}H_{16}O_4$
7806 - 1,8-Naphthalenedicarboxylic acid, diethyl ester

$C_{16}H_{16}O_5$
7854 - 1,4-Naphthalenedione, 5,8-dihydroxy-6-(1-hydroxy-4-methyl-3-pentenyl)-, (*S*)-

$C_{16}H_{16}Si$
11404 - Silane, diethenyldiphenyl-

$C_{16}H_{17}N$
722 - Benzenamine, *N*,*N*-dimethyl-4-(2-phenylethenyl)-, (*E*)-
3984 - 9H-Carbazole, 9-butyl-

$C_{16}H_{17}NO$
5219 - Diphenamid

$C_{16}H_{17}NO_3$
7665 - Morphinan-3,6-diol, 7,8-didehydro-4,5-epoxy-, (5α,6α)-

$C_{16}H_{17}NO_4$
6201 - Galanthan-1,2-diol, 3,12-didehydro-9,10-[methylenebis(oxy)]-, (1α,2β)-

$C_{16}H_{17}N_2NaO_4S$
11655 - 4-Thia-1-azabicyclo[3.2.0]heptane-2-carboxylic acid, 3,3-dimethyl-7-oxo-6-[(phenylacetyl)amino]-

$C_{16}H_{17}N_3O$
5374 - Ergoline-8-carboxamide, 9,10-didehydro-6-methyl-, (8β)-

$C_{16}H_{18}$
1099 - Benzene, 1,1′-(1,4-butanediyl)bis-
1120 - Benzene, 1,1′-butylidenebis-
1589 - Benzene, 1,1′-(1,2-ethanediyl)bis[3-methyl-
1590 - Benzene, 1,1′-(1,2-ethanediyl)bis[4-methyl-
1676 - Benzene, 1,1′-(1-ethyl-1,2-ethanediyl)bis-
1979 - Benzene, 1-(1-methylethyl)-4-(phenylmethyl)-
2020 - Benzene, 1,1′-(1-methyl-1,3-propanediyl)bis-, (±)-
2021 - Benzene, 1,1′-(2-methyl-1,3-propanediyl)bis-
2111 - Benzene, 1-(phenylmethyl)-4-propyl-
3095 - 1,1′-Biphenyl, 2,2′,4,4′-tetramethyl-
3096 - 1,1′-Biphenyl, 2,2′,5,5′-tetramethyl-

$C_{16}H_{18}ClN_3S$
9506 - Phenothiazin-5-ium, 3,7-bis(dimethylamino)-, chloride

$C_{16}H_{18}N_2$
9613 - Piperazine, 1,4-diphenyl-

$C_{16}H_{18}N_2O_2$
5112 - Diazene, bis(2-ethoxyphenyl)-
5113 - Diazene, bis(4-ethoxyphenyl)-

$C_{16}H_{18}N_2O_3$
6144 - 2(3H)-Furanone, dihydro-3-(hydroxyphenylmethyl)-4-[(1-methyl-1H-imidazol-5-yl)methyl]-,

$C_{16}H_{18}N_4O_2$
10910 - 4-Pyridinecarboxylic acid, 2-[3-oxo-3-[(phenylmethyl)amino]propyl]hydrazide

$C_{16}H_{18}N_4O_7S$
516 - Bensulfuron-methyl

$C_{16}H_{18}O$
1108 - Benzenebutanol, β-phenyl-, (*R*)-
2074 - Benzene, 1,1′-oxybis[2-ethyl-
2080 - Benzene, 1,1′-(oxydi-2,1-ethanediyl)bis-
2081 - Benzene, 1,1′-(oxydiethylidene)bis-, (*R**,*R**)-(±)-

$C_{16}H_{18}O_2$
3048 - 1,1′-Biphenyl, 2,2′-dimethoxy-5,5′-dimethyl-
3049 - 1,1′-Biphenyl, 2,5′-dimethoxy-2′,5-dimethyl-
3069 - [1,1′-Biphenyl]-4,4′-diol, 3,3′,5,5′-tetramethyl-
9249 - Phenol, 4,4′-butylidenebis-
9433 - Phenol, 4,4′-(1-methylpropylidene)bis-

$C_{16}H_{18}O_{10}$
2823 - 2H-1-Benzopyran-2-one, 8-(β-*D*-glucopyranosyloxy)-7-hydroxy-6-methoxy-

$C_{16}H_{19}Cl_2N$
1801 - Benzenemethanamine, *N*-(2-chloroethyl)-*N*-(phenylmethyl)-, hydrochloride

$C_{16}H_{19}N$
744 - Benzenamine, 2-ethyl-*N*-(2-ethylphenyl)-
1585 - Benzeneethanamine, *N*-(2-phenylethyl)-
1822 - Benzenemethanamine, α-methyl-*N*-(1-phenylethyl)-

$C_{16}H_{19}NO_3$
6182 - Furo[2,3-b]quinolin-4(2H)-one, 3,9-dihydro-8-methoxy-9-methyl-2-(1-methylethyl)-, (*R*)-
8567 - 16(15H)-Oxaerythrinan-15-one, 1,2,6,7-tetradehydro-14,17-dihydro-3-methoxy-, (3β)-

$C_{16}H_{19}NO_4$
481 - 8-Azabicyclo[3.2.1]octane-2-carboxylic acid, 3-(benzoyloxy)-8-methyl-, [1*R*-(*exo*,*exo*)]-

$C_{16}H_{20}$
7900 - Naphthalene, 1-hexyl-
7901 - Naphthalene, 2-hexyl-
7905 - Naphthalene, 1-isopropyl-2,4,7-trimethyl-
8014 - Naphthalene, 1,2,5-trimethyl-8-(1-methylethyl)-
8015 - Naphthalene, 1,3,8-trimethyl-5-(1-methylethyl)-

$C_{16}H_{20}N_2$
3036 - [1,1′-Biphenyl]-4,4′-diamine, tetramethyl-
5500 - 1,2-Ethanediamine, *N*,*N*′-bis(phenylmethyl)-
11010 - 2-Pyridinepropanamine, *N*,*N*-dimethyl-γ-phenyl-

$C_{16}H_{20}N_4$
361 - Aniline, 3,3′-azobis[*N*,*N*-dimethyl-

$C_{16}H_{20}O_2Si$
11406 - Silane, diethoxydiphenyl-

$C_{16}H_{20}O_6P_2S_3$
1 - Abate

$C_{16}H_{21}NO_3$
911 - Benzeneacetic acid, α-hydroxy-, 8-methyl-8-azabicyclo[3.2.1]oct-3-yl ester, *endo*-(±)-
916 - Benzeneacetic acid, α-(hydroxymethyl)-, 8-azabicyclo[3.2.1]oct-3-yl ester, [3(*S*)-*endo*]-

$C_{16}H_{21}N_3$
5505 - 1,2-Ethanediamine, *N*,*N*-dimethyl-*N*′-(phenylmethyl)-*N*′-2-pyridinyl-

$C_{16}H_{22}ClNO_3$
5173 - Diethatyl-ethyl

$C_{16}H_{22}N_4O_2$
9795 - Propanamide, *N*-(2,3-dihydro-1,5-dimethyl-3-oxo-2-phenyl-1H-pyrazol-4-yl)-2-(dimethylamino)-

$C_{16}H_{22}OSi_2$
5238 - Disiloxane, 1,1,3,3-tetramethyl-1,3-diphenyl-

$C_{16}H_{22}O_4$
1336 - 1,2-Benzenedicarboxylic acid, bis(2-methylpropyl) ester
1338 - 1,4-Benzenedicarboxylic acid, bis(2-methylpropyl) ester
1339 - 1,2-Benzenedicarboxylic acid, dibutyl ester

$C_{16}H_{22}O_6$
1333 - 1,2-Benzenedicarboxylic acid, bis(2-ethoxyethyl) ester

$C_{16}H_{22}O_8$
6226 - β-*D*-Glucopyranoside, 4-(3-hydroxy-1-propenyl)-2-methoxyphenyl

$C_{16}H_{22}O_{11}$
6216 - α-*D*-Glucopyranose, pentaacetate
6217 - β-*D*-Glucopyranose, pentaacetate

$C_{16}H_{23}ClN_3O$
11589 - Tebuconazole

$C_{16}H_{24}N_2O_2$
1325 - 1,2-Benzenedicarboxamide, *N*,*N*,*N*′,*N*′-tetraethyl-
1326 - 1,3-Benzenedicarboxamide, *N*,*N*,*N*′,*N*′-tetraethyl-

$C_{16}H_{25}NO_2$
898 - Benzeneacetic acid, α-ethyl-, 2-(diethylamino)ethyl ester

$C_{16}H_{26}$
1258 - Benzene, decyl-
2090 - Benzene, pentaethyl-
6627 - 6,10-Hexadecadiyne
11609 - 5,9-Tetradecadien-7-yne, 6,9-dimethyl-

$C_{16}H_{26}O$
9221 - Phenol, 2,6-bis(1,1-dimethylethyl)-4-ethyl-

$C_{16}H_{26}O_2$
1501 - 1,4-Benzenediol, 2,5-bis(1,1-dimethylpropyl)-
6666 - 6,10,14-Hexadecatrienoic acid

$C_{16}H_{28}O_2$
4940 - 2-Cyclopentene-1-undecanoic acid, (*R*)-

$C_{16}H_{30}$
4323 - Cyclohexane, 1,1′-(1,4-butanediyl)bis-
4326 - Cyclohexane, 1,1′-butylidenebis-
4447 - Cyclohexane, 1,1′-(2-methyl-1,3-propanediyl)bis-
6673 - 1-Hexadecyne
6674 - 2-Hexadecyne

$C_{16}H_{30}O$
4762 - Cyclopentadecanone, 2-methyl-
4763 - Cyclopentadecanone, 3-methyl-

$C_{16}H_{30}O_2$
6668 - 7-Hexadecenoic acid
6669 - 9-Hexadecenoic acid
6670 - 9-Hexadecenoic acid, (*Z*)-

$C_{16}H_{30}O_3$
8424 - Octanoic acid, anhydride

$C_{16}H_{31}ClO$
6665 - Hexadecanoyl chloride

$C_{16}H_{31}N$
6641 - Hexadecanenitrile

$C_{16}H_{32}$
4344 - Cyclohexane, decyl-
4863 - Cyclopentane, undecyl-
6667 - 1-Hexadecene
10580 - 1-Propene, 2-methyl-, tetramer

$C_{16}H_{32}Br_2$
6635 - Hexadecane, 1,16-dibromo-

$C_{16}H_{32}O$
6628 - Hexadecanal
6663 - 3-Hexadecanone

$C_{16}H_{32}O_2$
5065 - Decanoic acid, 2-hexyl-
6644 - Hexadecanoic acid
8152 - Nonanoic acid, heptyl ester
8437 - Octanoic acid, octyl ester
8646 - Pentadecanoic acid, methyl ester
11624 - Tetradecanoic acid, ethyl ester

$C_{16}H_{32}O_4$
5313 - Dodecanoic acid, 2-(2-hydroxyethoxy)ethyl ester

$C_{16}H_{33}Br$
6633 - Hexadecane, 1-bromo-

$C_{16}H_{33}Cl$
6634 - Hexadecane, 1-chloro-

$C_{16}H_{33}F$
6639 - Hexadecane, 1-fluoro-

$C_{16}H_{33}I$
6640 - Hexadecane, 1-iodo-

$C_{16}H_{33}NO$
5290 - Dodecanamide, *N*,*N*-diethyl-
6629 - Hexadecanamide

$C_{16}H_{34}$
6632 - Hexadecane

$C_{16}H_{34}O$
6431 - Heptane, 3,3′-[oxybis(methylene)]bis-
6659 - 1-Hexadecanol
6660 - 2-Hexadecanol
6661 - 3-Hexadecanol
8416 - Octane, 1,1′-oxybis-
8649 - 1-Pentadecanol, 2-methyl-
8650 - 6-Pentadecanol, 6-methyl-

$C_{16}H_{34}O_2$
11617 - Tetradecane, 1,1-dimethoxy-
$C_{16}H_{34}O_3$
8552 - Orthoformic acid, triisopentyl ester
$C_{16}H_{34}S$
6643 - 1-Hexadecanethiol
8417 - Octane, 1,1′-thiobis-
$C_{16}H_{35}N$
6631 - 1-Hexadecanamine
8352 - 2-Octanamine, *N*-(1-methylheptyl)-
8354 - 1-Octanamine, *N*-octyl-
$C_{16}H_{36}O_4Si$
11505 - Silicic acid (H_4SiO_4), tetrabutyl ester
$C_{16}H_{48}O_6Si_7$
6528 - Heptasiloxane, hexadecamethyl-
$C_{16}H_{48}O_8Si_8$
4741 - Cyclooctasiloxane, hexadecamethyl-
$C_{17}H_9NO_4$
8062 - Naphtho[2,3-f]quinoline-7,12-dione, 5,6-dihydroxy-
$C_{17}H_{10}O$
6292 - 7H-Benz[de]anthracen-7-one
$C_{17}H_{10}O_2$
10854 - 1-Pyrenecarboxylic acid
$C_{17}H_{11}N$
630 - Benz[c]acridine
$C_{17}H_{12}$
2393 - 11H-Benzo[a]fluorene
$C_{17}H_{12}ClFN_2O$
8219 - Nuarimol
$C_{17}H_{12}Cl_2N_2O$
5948 - Fenarimol
$C_{17}H_{12}O_2$
8021 - 2-Naphthalenol, benzoate
$C_{17}H_{12}O_3$
2644 - Benzoic acid, 2-hydroxy-, 1-naphthalenyl ester
2645 - Benzoic acid, 2-hydroxy-, 2-naphthalenyl ester
$C_{17}H_{12}O_7$
6184 - 1H,12H-Furo[3′,2′:4,5]furo[2,3-h]pyrano[3,4-c][1]benzopyran-1,12-dione, 3,4,7a,10a-tetrahydro-5-meth
$C_{17}H_{13}N$
10947 - Pyridine, 2,6-diphenyl-
$C_{17}H_{13}N_3O_5S_2$
2726 - Benzoic acid, 2-[[[4-[(2-thiazolylamino) sulfonyl]phenyl]amino]carbonyl]-
$C_{17}H_{14}$
4756 - 15H-Cyclopenta[a]phenanthrene, 16,17-dihydro-
7949 - Naphthalene, 1-(phenylmethyl)-
7950 - Naphthalene, 2-(phenylmethyl)-
9182 - Phenanthrene, 9-(1-methylethenyl)-
$C_{17}H_{14}O$
7912 - 1-Naphthalenemethanol, α-phenyl-
8039 - 1-Naphthalenol, 2-(phenylmethyl)-
8040 - 2-Naphthalenol, 1-(phenylmethyl)-
8683 - 1,4-Pentadien-3-one, 1,5-diphenyl-
$C_{17}H_{14}O_6$
4757 - Cyclopenta[c]furo[3′,2′:4,5]furo[2,3-h][1]benzopyran-1,11-dione, 2,3,6a,8,9,9a-hexahydro-4-methoxy-, (6a*R*-
$C_{17}H_{15}ClF_3NO_4$
6291 - Haloxyfop-(2-ethoxyethyl)
$C_{17}H_{15}ClO_7$
2847 - 1-Benzopyrylium, 3,5,7-trihydroxy-2-(4-hydroxy-3,5-dimethoxyphenyl)-, chloride
$C_{17}H_{15}N$
7723 - 1-Naphthalenamine, *N*-(2-methylphenyl)-
7724 - 1-Naphthalenamine, *N*-(4-methylphenyl)-
7725 - 2-Naphthalenamine, *N*-(2-methylphenyl)-
7730 - 2-Naphthalenamine, *N*-(phenylmethyl)-
$C_{17}H_{16}Br_2O_3$
3140 - Bromopropylate
$C_{17}H_{16}O_2$
942 - Benzeneacetic acid, α-(phenylmethylene)-, ethyl ester
$C_{17}H_{16}O_4$
8776 - Pentanedioic acid, diphenyl ester
9927 - Propanedioic acid, bis(phenylmethyl) ester
$C_{17}H_{17}ClO_6$
11520 - Spiro[benzofuran-2(3H),1′-[2]cyclohexene]-3,4′-dione, 7-chloro-2′,4,6-trimethoxy-
$C_{17}H_{17}NO_2$
5155 - 4H-Dibenzo[de,g]quinoline-10,11-diol, 5,6,6a,7-tetrahydro-6-methyl-, (*R*)-
$C_{17}H_{17}N_3O_3$
7189 - Imazaquin
$C_{17}H_{18}O$
7594 - Methanone, bis(2,4-dimethylphenyl)-
8969 - 3-Pentanone, 1,5-diphenyl-
$C_{17}H_{18}O_4$
1103 - Benzenebutanoic acid, 4-hydroxy-γ-(4-hydroxyphenyl)-γ-methyl-
9970 - Propanedioic acid, 1-naphthalenyl-, diethyl ester
10378 - 2-Propanol, 1,3-diphenoxy-, acetate
$C_{17}H_{19}ClN_2S$
9505 - 10H-Phenothiazine-10-propanamine, 2-chloro-*N,N*-dimethyl-
$C_{17}H_{19}NO_3$
7396 - 7-Isoquinolinol, 1,2,3,4-tetrahydro-1-[(4-hydroxyphenyl)methyl]-6-methoxy-, (*S*)-
7666 - Morphinan-3,6-diol, 7,8-didehydro-4,5-epoxy-17-methyl- (5α,6α)-
9632 - Piperidine, 1-[5-(1,3-benzodioxol-5-yl)-1-oxo-2,4-pentadienyl]-, (*E,E*)-
9633 - Piperidine, 1-[5-(1,3-benzodioxol-5-yl)-1-oxo-2,4-pentadienyl]-, (*Z,Z*)-
$C_{17}H_{19}NO_4$
5952 - Fenoxycarb
7668 - Morphinan-3,6-diol, 7,8-didehydro-4,5-epoxy-17-methyl- (5α,6α)-, 17-oxide
$C_{17}H_{20}$
1697 - Benzene, 1,1′-(1-ethyl-1,3-propanediyl)bis-
1698 - Benzene, 1,1′-(2-ethyl-1,3-propanediyl)bis-
2098 - Benzene, 1,1′-(1,5-pentanediyl)bis-
2108 - Benzene, 1,1′-pentylidenebis-
$C_{17}H_{20}N_2O$
7593 - Methanone, bis[4-(dimethylamino)phenyl]-
12003 - Urea, *N,N*′-diethyl-*N,N*′-diphenyl-
$C_{17}H_{20}N_2S$
9504 - 10H-Phenothiazine-10-ethanamine, *N,N*,α-trimethyl-
$C_{17}H_{20}N_4O_6$
11344 - Riboflavin
$C_{17}H_{20}O_6$
7052 - 4-Hexenoic acid, 6-(1,3-dihydro-4-hydroxy-6-methoxy-7-methyl-3-oxo-5-isobenzofuranyl)-4-methyl-, (*F*)-
$C_{17}H_{20}Si$
11360 - Silacyclohexane, 1,1-diphenyl-
$C_{17}H_{21}N$
1575 - Benzeneethanamine, *N*,α-dimethyl-*N*-(phenylmethyl)-, (+)-
1804 - Benzenemethanamine, *N,N*-diethyl-α-phenyl-
$C_{17}H_{21}NO$
5422 - Ethanamine, 2-(diphenylmethoxy)-*N,N*-dimethyl-
$C_{17}H_{21}NO_2$
928 - Benzeneacetic acid, α-methylene-, 8-methyl-8-azabicyclo[3.2.1]oct-3-yl ester, *endo*-
8067 - Napropamide
$C_{17}H_{21}NO_3$
6200 - Galanthamine
$C_{17}H_{21}NO_4$
482 - 8-Azabicyclo[3.2.1]octane-2-carboxylic acid, 3-(benzoyloxy)-8-methyl-, methyl ester, [1*R*-(*exo,exo*)]-
908 - Benzeneacetic acid, α-(hydroxymethyl)-, 9-methyl-3-oxa-9-azatricyclo[3.3.1.02,4]non-7-yl ester, [(7S)-(1α,2β,4β,5α,7β)]-
$C_{17}H_{21}N_3$
657 - Benzenamine, 4,4′-carbonimidoylbis[*N,N*-dimethyl-
$C_{17}H_{22}$
7895 - Naphthalene, 1-heptyl-
7896 - Naphthalene, 2-heptyl-
$C_{17}H_{22}N_2$
786 - Benzenamine, 4,4′-methylenebis[*N,N*-dimethyl-
5506 - 1,2-Ethanediamine, *N,N*-dimethyl-*N*′-phenyl-*N*′-(phenylmethyl)-
$C_{17}H_{23}NO_3$
902 - Benzeneacetic acid, α-(hydroxymethyl)- 8-methyl-8-azabicyclo[3.2.1]oct-3-yl ester *endo*-(±)-
917 - Benzeneacetic acid, α-(hydroxymethyl)-, 8-methyl-8-azabicyclo[3.2.1]oct-3-yl ester, [3(*S*)-*endo*]-
$C_{17}H_{23}N_3O$
5511 - 1,2-Ethanediamine, *N*-[(4-methoxyphenyl) methyl]-*N*′,*N*′-dimethyl-*N*-2-pyridinyl-
$C_{17}H_{24}N_2O$
5411 - Ethanamine, 2-[(3-butyl-1-isoquinolinyl)oxy]-*N,N*-dimethyl-
$C_{17}H_{24}O_9$
6225 - β-*D*-Glucopyranoside, 4-(3-hydroxy-1-propenyl)-2,6-dimethoxyphenyl
$C_{17}H_{25}N$
9696 - Piperidine, 1-(1-phenylcyclohexyl)-
$C_{17}H_{25}NO_2$
7660 - MGK 264
$C_{17}H_{25}N_3O$
9877 - 1,3-Propanediamine, *N,N*-diethyl-*N*′-(6-methoxy-8-quinolinyl)-
$C_{17}H_{26}ClNO_2$
3144 - Butachlor
$C_{17}H_{26}ClNO_3S$
4138 - Clethodim
$C_{17}H_{26}O_4$
4261 - 2,5-Cyclohexadiene-1,4-dione, 2,5-dihydroxy-3-undecyl-
$C_{17}H_{27}NO$
11169 - Pyrrolidine, 1-[2-(3-methylbutoxy)-2-phenylethyl]-
$C_{17}H_{27}NO_3$
7683 - Morpholine, 4-[3-(4-butoxyphenoxy)propyl]-
$C_{17}H_{28}$
2374 - Benzene, undecyl-
$C_{17}H_{28}O$
9225 - Phenol, 2,6-bis(1,1-dimethylpropyl)-4-methyl-
$C_{17}H_{29}NO_3S$
11359 - Sethoxydim
$C_{17}H_{30}O$
4217 - 9-Cycloheptadecen-1-one
4218 - 9-Cycloheptadecen-1-one, (*Z*)-
$C_{17}H_{30}O_2$
7182 - Hydroprene
$C_{17}H_{32}$
4402 - Cyclohexane, 1,1′-(2-ethyl-1,3-propanediyl)bis-
4464 - Cyclohexane, 1,1′-(1,5-pentanediyl)bis-
$C_{17}H_{32}O_2$
6324 - 9-Heptadecenoic acid
$C_{17}H_{33}N$
6312 - Heptadecanenitrile
$C_{17}H_{34}$
4809 - Cyclopentane, dodecyl-
6322 - 1-Heptadecene
$C_{17}H_{34}Br_2$
6308 - Heptadecane, 1,17-dibromo-
$C_{17}H_{34}O$
6304 - Heptadecanal
6320 - 2-Heptadecanone
6321 - 9-Heptadecanone

$C_{17}H_{34}O_2$
6314 - Heptadecanoic acid
6653 - Hexadecanoic acid, methyl ester
11626 - Tetradecanoic acid, 1-methylethyl ester
11629 - Tetradecanoic acid, propyl ester
$C_{17}H_{35}Br$
6307 - Heptadecane, 1-bromo-
$C_{17}H_{35}N$
9660 - Piperidine, 1-dodecyl-
$C_{17}H_{36}$
6306 - Heptadecane
$C_{17}H_{36}O$
6317 - 1-Heptadecanol
6318 - 2-Heptadecanol
6319 - 9-Heptadecanol
$C_{17}H_{37}N$
6305 - 1-Heptadecanamine
8353 - 1-Octanamine, *N*-methyl-*N*-octyl-
$C_{18}H_{10}O_2$
4116 - 5,6-Chrysenedione
7707 - 1,2-Naphthacenedione
$C_{18}H_{10}O_6$
2999 - [2,2′-Bi-1H-indene]-1,1′,3,3′(2H,2′H)-tetrone, 2,2′-dihydroxy-
$C_{18}H_{12}$
519 - Benz[a]anthracene
2406 - Benzo[c]phenanthrene
4114 - Chrysene
7703 - Naphthacene
11910 - Triphenylene
$C_{18}H_{12}N_2$
3107 - 2,3′-Biquinoline
$C_{18}H_{12}N_6$
11843 - 1,3,5-Triazine, 2,4,6-tri-2-pyridinyl-
$C_{18}H_{13}N$
4113 - 6-Chrysenamine
$C_{18}H_{13}NO_3$
2702 - Benzoic acid, 2-[(1-naphthalenylamino)carbonyl]-
$C_{18}H_{13}N_3O$
7334 - Indolo[2′,3′:3,4]pyrido[2,1-b]quinazolin-5(7H)-one, 8,13-dihydro-
$C_{18}H_{14}$
11595 - 1,1′:2′,1″-Terphenyl
11596 - 1,1′:3′,1″-Terphenyl
11597 - 1,1′:4′,1″-Terphenyl
$C_{18}H_{14}N_4O_5S$
2654 - Benzoic acid, 2-hydroxy-5-[[4-[(2-pyridinylamino)sulfonyl]phenyl]azo]-
$C_{18}H_{14}O$
5859 - Ethanone, 1-(1-naphthalenyl)-2-phenyl-
7629 - Methanone, (2-methylphenyl)-1-naphthalenyl-
$C_{18}H_{14}O_3$
10684 - 2-Propenoic acid, 3-phenyl-, anhydride
$C_{18}H_{15}As$
450 - Arsine, triphenyl-
$C_{18}H_{15}Bi$
3111 - Bismuthine, triphenyl-
$C_{18}H_{15}ClSn$
11541 - Stannane, chlorotriphenyl-
$C_{18}H_{15}N$
728 - Benzenamine, *N*,*N*-diphenyl-
$C_{18}H_{15}OP$
9518 - Phosphine oxide, triphenyl-
$C_{18}H_{15}O_3P$
9593 - Phosphorous acid, triphenyl ester
$C_{18}H_{15}O_4P$
9566 - Phosphoric acid, triphenyl ester
$C_{18}H_{15}P$
9523 - Phosphine, triphenyl-
$C_{18}H_{15}Sb$
11551 - Stibine, triphenyl-
$C_{18}H_{16}N_2$
1271 - 1,4-Benzenediamine, *N*,*N*′-diphenyl-
7173 - Hydrazine, triphenyl-
$C_{18}H_{16}N_2O$
8028 - 2-Naphthalenol, 1-[(2,4-dimethylphenyl)azo]-
8029 - 2-Naphthalenol, 1-[(2,5-dimethylphenyl)azo]-
$C_{18}H_{16}OSi$
11504 - Silanol, triphenyl-
$C_{18}H_{16}OSn$
11911 - Triphenyltin hydroxide
$C_{18}H_{16}O_2$
9177 - 9,10-Phenanthrenedione, 1-methyl-7-(1-methylethyl)-
10691 - 2-Propenoic acid, 3-phenyl-, 3-phenyl-2-propenyl ester
$C_{18}H_{16}O_4$
3768 - 2-Butenedioic acid (*E*)-, bis(phenylmethyl) ester
$C_{18}H_{18}$
9183 - Phenanthrene, 1-methyl-7-(1-methylethyl)-
$C_{18}H_{18}O_2$
5397 - Estra-1,3,5,7,9-pentaen-17-one, 3-hydroxy-
9302 - Phenol, 4,4′-(1,2-diethylidene-1,2-ethanediyl)bis-
$C_{18}H_{18}O_4$
3298 - Butanedioic acid, bis(phenylmethyl) ester
$C_{18}H_{18}O_5$
5755 - Ethanol, 2,2′-oxybis-, dibenzoate
$C_{18}H_{18}O_6$
3306 - Butanedioic acid, 2,3-dihydroxy- [*R*-(*R**,*R**)]-, bis(phenylmethyl) ester
$C_{18}H_{19}NO_2$
5160 - 4H-Dibenzo[de,g]quinolin-11-ol, 5,6,6a,7-tetrahydro-10-methoxy-6-methyl-, (*R*)-
$C_{18}H_{19}NO_3$
5159 - 4H-Dibenzo[de,g]quinolin-10-ol, 5,6,6a,7-tetrahydro-1,2-dimethoxy-, (*R*)-
$C_{18}H_{19}NO_4$
7675 - Morphinan-6-one, 7,8-didehydro-4,5-epoxy-14-hydroxy-3-methoxy-17-methyl-, (5α)-
$C_{18}H_{20}$
7263 - 1H-Indene, 2,3-dihydro-1,1,3-trimethyl-3-phenyl-
$C_{18}H_{20}Cl_2$
9160 - Perthane
$C_{17}H_{17}N_2 \cdot CH_3O_4S$
5175 - Difenzoquat methyl sulfate
$C_{18}H_{20}O_2$
5398 - Estra-1,3,5(10),7-tetraen-17-one, 3-hydroxy-
9300 - Phenol, 4,4′-(1,2-diethyl-1,2-ethenediyl)bis-, (*E*)-
$C_{18}H_{21}NO$
9676 - 2-Piperidinemethanol, α,α-diphenyl-
$C_{18}H_{21}NO_3$
7669 - Morphinan-6-ol, 7,8-didehydro-4,5-epoxy-3-methoxy-17-methyl-, (5α,6α)-
7670 - Morphinan-6-ol, 7,8-didehydro-4,5-epoxy-3-methoxy-17-methyl-, (5α,6β)-
7671 - Morphinan-6-ol, 8,14-didehydro-4,5-epoxy-3-methoxy-17-methyl-, (5α,6α)-
7672 - Morphinan-8-ol, 6,7-didehydro-4,5-epoxy-3-methoxy-17-methyl-, (5α,8β)-
7677 - Morphinan-6-one, 7,8-didehydro-4-hydroxy-3-methoxy-17-methyl-
7678 - Morphinan-6-one, 4,5-epoxy-3-methoxy-17-methyl-, (5α)-
$C_{18}H_{22}$
3076 - 1,1′-Biphenyl, 2,2′,4,4′,6,6′-hexamethyl-
$C_{18}H_{22}N_2S$
9503 - 10H-Phenothiazine-10-ethanamine, *N*,*N*-diethyl-
$C_{18}H_{22}O_2$
5406 - Estra-1,3,5(10)-trien-17-one, 3-hydroxy-
5407 - Estra-1,3,5(10)-trien-17-one, 3-hydroxy-, (8α)-
9298 - Phenol, 4,4′-(1,2-diethyl-1,2-ethanediyl)bis-, (*R**,*S**)-
$C_{18}H_{22}O_4$
1515 - 1,2-Benzenediol, 4,4′-(2,3-dimethyl-1,4-butanediyl)bis-
$C_{18}H_{23}ClN_2O_2$
5670 - Ethanimidamide, *N*,*N*′-bis(4-ethoxyphenyl)-, monohydrochloride
$C_{18}H_{23}NO_3$
7674 - Morphinan-6-ol, 4,5-epoxy-3-methoxy-17-methyl-, (5α,6α)-
$C_{18}H_{24}$
7937 - Naphthalene, 1-octyl-
7938 - Naphthalene, 2-octyl-
$C_{18}H_{24}NO_7P$
7673 - Morphinan-6-ol, 7,8-didehydro-4,5-epoxy-3-methoxy-17-methyl-, (5α,6α)-, phosphate (1 1) (salt)
$C_{18}H_{24}N_2O_6$
5182 - Dinocap
$C_{18}H_{24}O_2$
5399 - Estra-1,3,5(10)-triene-3,17-diol (17β)-
5400 - Estra-1,3,5(10)-triene-3,17-diol, (17α)-
5401 - Estra-1,3,5(10)-triene-3,17-diol, (8α,17β)-
5402 - Estra-5,7,9-triene-3,17-diol, (3β,17β)-
$C_{18}H_{24}O_3$
5404 - Estra-1,3,5(10)-triene-3,16,17-triol, (16α,17β)-
5405 - Estra-1,3,5(10)-triene-3,16,17-triol, (16β,17β)-
$C_{18}H_{26}O_4$
1335 - 1,2-Benzenedicarboxylic acid, bis(3-methylbutyl) ester
$C_{18}H_{27}NO_3$
8181 - 6-Nonenamide, *N*-[(4-hydroxy-3-methoxyphenyl)methyl]-8-methyl-, (*E*)-
$C_{18}H_{27}NO_5$
894 - Benzeneacetic acid, 4-[2-(diethylamino)-2-oxoethoxy]-3-methoxy-, propyl ester
$C_{18}H_{27}N_3O$
8737 - 1,5-Pentanediamine, *N*-(6-methoxy-8-quinolinyl)-*N*′-(1-methylethyl)-
$C_{18}H_{28}N_2O_4S$
1582 - Benzeneethanamine, α-methyl-, (*S*)-, sulfate (2:1)
$C_{18}H_{28}O$
5327 - 1-Dodecanone, 1-phenyl-
$C_{18}H_{28}O_2$
5317 - Dodecanoic acid, phenyl ester
$C_{18}H_{30}$
1566 - Benzene, dodecyl-
1756 - Benzene, hexaethyl-
2276 - Benzene, 1,2,4,5-tetrakis(1-methylethyl)-
4118 - Chrysene, octadecahydro-
8230 - 7,11-Octadecadiyne
$C_{18}H_{30}O$
9498 - Phenol, 2,4,6-tris(1,1-dimethylethyl)-
$C_{18}H_{30}O_2$
8246 - 9,11,13-Octadecanetrienoic acid (*Z*,*Z*,*Z*)
8282 - 9,11,13-Octadecatrienoic acid, (*E*,*E*,*E*)-
8284 - 9,11,13-Octadecatrienoic acid, (*E*,*Z*,*E*)-
8286 - 9,11,13-Octadecatrienoic acid, (*Z*,*Z*,*E*)-
8287 - 9,12,15-Octadecatrienoic acid, (*Z*,*Z*,*Z*)-
$C_{18}H_{30}O_4$
4933 - 1-Cyclopentene-1-pentanoic acid, δ-hydroxy-3,5-bis(1-methylpropyl)-β-oxo-
$C_{18}H_{32}$
4731 - 1,3-Cyclooctadecadiene
$C_{18}H_{32}I_2$
8225 - 9,12-Octadecadienoic acid
$C_{18}H_{32}I_2O_2$
8300 - 6-Octadecenoic acid, 6,7-diiodo-
$C_{18}H_{32}N_2O_5S$
8926 - 1-Pentanol, 2-(diethylamino)-4-methyl-, 4-aminobenzoate (ester), monomethanesulfonate (salt)
$C_{18}H_{32}O_2$
4936 - 2-Cyclopentene-1-tridecanoic acid, (*S*)-
8223 - 9,12-Octadecadienoic acid (*Z*,*Z*)-
8224 - 10,12-Octadecadienoic acid, (*E*,*E*)-
8316 - 9-Octadecynoic acid
$C_{18}H_{32}O_5$
9120 - Pentonic acid, 5-C-[3,5-bis(1-methylpropyl)-1-cyclopenten-1-yl]-4-deoxy-
$C_{18}H_{32}O_6$
3561 - Butanoic acid, 3-methyl-, 1,2,3-propanetriyl ester

$C_{18}H_{32}O_7$
10168 - 1,2,3-Propanetricarboxylic acid, 2-hydroxy-, tributyl ester
$C_{18}H_{32}O_{16}$
6219 - α-*D*-Glucopyranoside, β-*D*-fructofuranosyl *O*-α-*D*-galactopyranosyl-(1
6221 - α-*D*-Glucopyranoside, *O*-α-*D*-glucopyranosyl-(13)-β-*D*-fructofuranosyl
$C_{18}H_{32}Si$
11455 - Silane, tributylphenyl-
$C_{18}H_{33}N$
8293 - 9-Octadecenenitrile, (*Z*)-
$C_{18}H_{34}$
8313 - 1-Octadecyne
8314 - 2-Octadecyne
8315 - 9-Octadecyne
$C_{18}H_{34}O$
8288 - 9-Octadecenal
8289 - 9-Octadecenal, (*Z*)-
$C_{18}H_{34}OSn$
5011 - Cyhexatin
$C_{18}H_{34}O_2$
8294 - 9-Octadecenoic acid (*Z*)-
8295 - 11-Octadecenoic acid, (*E*)-
8296 - 2-Octadecenoic acid, (*E*)-
8297 - 9-Octadecenoic acid, (*E*)-
$C_{18}H_{34}O_3$
8303 - 9-Octadecenoic acid, 12-hydroxy-, [*R*-(*Z*)]-
$C_{18}H_{34}O_4$
5044 - Decanedioic acid, dibutyl ester
6780 - Hexanedioic acid, bis(2-ethylbutyl) ester
$C_{18}H_{34}O_6$
8239 - Octadecanedioic acid, 9,10-dihydroxy-, (*R**,*R**)-(±)-
$C_{18}H_{35}BrO_2$
8250 - Octadecanoic acid, 18-bromo-
$C_{18}H_{35}ClO$
8281 - Octadecanoyl chloride
$C_{18}H_{35}N$
8243 - Octadecanenitrile
$C_{18}H_{36}$
4382 - Cyclohexane, dodecyl-
4854 - Cyclopentane, tridecyl-
8291 - 1-Octadecene
8292 - 9-Octadecene
$C_{18}H_{36}O$
6638 - Hexadecane, 1-(ethenyloxy)-
8280 - 3-Octadecanone
8311 - 9-Octadecen-1-ol, (*E*)-
8312 - 9-Octadecen-1-ol, (*Z*)-
$C_{18}H_{36}O_2$
5068 - Decanoic acid, 2-octyl-
6316 - Heptadecanoic acid, methyl ester
6650 - Hexadecanoic acid, ethyl ester
6662 - 1-Hexadecanol, acetate
8247 - Octadecanoic acid
$C_{18}H_{36}O_3$
6652 - Hexadecanoic acid, 2-hydroxyethyl ester
$C_{18}H_{36}O_4$
8254 - Octadecanoic acid, 9,10-dihydroxy-
$C_{18}H_{36}O_5$
8276 - Octadecanoic acid, 9,10,18-trihydroxy-, (*R**,*R**)-
$C_{18}H_{37}Br$
8235 - Octadecane, 1-bromo-
$C_{18}H_{37}Cl$
8236 - Octadecane, 1-chloro-
$C_{18}H_{37}Cl_3Si$
11470 - Silane, trichlorooctadecyl-
$C_{18}H_{37}I$
8242 - Octadecane, 1-iodo-
$C_{18}H_{37}NO$
8231 - Octadecanamide
$C_{18}H_{38}$
6310 - Heptadecane, 2-methyl-
6311 - Heptadecane, 9-methyl-
8234 - Octadecane
$C_{18}H_{38}O$
8278 - 1-Octadecanol
8279 - 3-Octadecanol, (±)-
$C_{18}H_{38}O_2$
6636 - Hexadecane, 1,1-dimethoxy-
$C_{18}H_{38}S$
8244 - 1-Octadecanethiol
$C_{18}H_{39}N$
6725 - 1-Hexanamine, *N*,*N*-dihexyl-
8233 - 1-Octadecanamine
$C_{18}H_{54}O_7Si_8$
8482 - Octasiloxane, octadecamethyl-
$C_{18}H_{54}O_9Si_9$
4726 - Cyclononasiloxane, octadecamethyl-
$C_{19}H_{10}Br_4O_5S$
9366 - Phenol, 4,4′-(3H-2,1-benzoxathiol-3-ylidene)bis[2,6-dibromo-, *S*,*S*-dioxide
$C_{19}H_{12}O_6$
2839 - 2H-1-Benzopyran-2-one, 3,3′-methylenebis[4-hydroxy-
$C_{19}H_{13}N$
318 - Acridine, 9-phenyl-
$C_{19}H_{14}$
521 - Benz[a]anthracene, 11-methyl-
522 - Benz[a]anthracene, 3-methyl-
523 - Benz[a]anthracene, 8-methyl-
4117 - Chrysene, 1-methyl-
$C_{19}H_{14}F_3NO$
5981 - Fluridone
$C_{19}H_{14}O_3$
4278 - 2,5-Cyclohexadien-1-one, 4-[bis(4-hydroxyphenyl)methylene]-
$C_{19}H_{14}O_5S$
9363 - Phenol, 4,4′-(3H-2,1-benzoxathiol-3-ylidene)bis-, *S*,*S*-dioxide
$C_{19}H_{15}Br$
1061 - Benzene, 1,1′,1″-(bromomethylidyne)tris-
$C_{19}H_{15}Cl$
1196 - Benzene, 1,1′,1″-(chloromethylidyne)tris-
$C_{19}H_{15}ClF_3NO_7$
7410 - Lactofen
$C_{19}H_{15}ClN_4$
11653 - 2H-Tetrazolium, 2,3,5-triphenyl-, chloride
$C_{19}H_{15}N$
3988 - 9H-Carbazole, 9-(phenylmethyl)-
$C_{19}H_{15}NO_2$
11236 - 4-Quinolinecarboxylic acid, 2-phenyl-, 2-propenyl ester
$C_{19}H_{16}$
1982 - Benzene, 1,1′,1″-methylidynetris-
3093 - 1,1′-Biphenyl, 2-(phenylmethyl)-
3094 - 1,1′-Biphenyl, 4-(phenylmethyl)-
$C_{19}H_{16}N_2$
11041 - 9H-Pyrido[3,4-b]indole, 1-[(2-methylphenyl)methyl]-
$C_{19}H_{16}O$
1866 - Benzenemethanol, α,α-diphenyl-
$C_{19}H_{16}O_3$
1983 - Benzene, 1,1′,1″-[methylidynetris(oxy)]tris-
$C_{19}H_{16}O_4$
2832 - 2H-1-Benzopyran-2-one, 4-hydroxy-3-(3-oxo-1-phenylbutyl)-
$C_{19}H_{16}O_5$
225 - Acetic acid, [(4-oxo-2-phenyl-4H-1-benzopyran-7-yl)oxy]-, ethyl ester
$C_{19}H_{17}Cl_2N_3O_3$
5174 - Difenoconazole
$C_{19}H_{17}N$
729 - Benzenamine, 4-(diphenylmethyl)-
$C_{19}H_{17}NO_3$
11216 - Quinoline, 2-[2-(1,3-benzodioxol-5-yl)ethyl]-4-methoxy-
$C_{19}H_{17}N_3$
6288 - Guanidine, *N*,*N*′,*N*″-triphenyl-
$C_{19}H_{17}N_3O$
7333 - Indolo[2′,3′:3,4]pyrido[2,1-b]quinazolin-5(7H)-one, 8,13,13b,14-tetrahydro-14-methyl-, (*S*)-
$C_{19}H_{18}O_3Si$
11443 - Silane, methyltriphenoxy-
$C_{19}H_{18}O_5$
2464 - 5-Benzofuranpropanol, 2-(1,3-benzodioxol-5-yl)-7-methoxy-
$C_{19}H_{18}Si$
11444 - Silane, methyltriphenyl-
$C_{19}H_{19}NO_3$
2469 - 5H-Benzo[g]-1,3-benzodioxolo[6,5,4-de]quinoline, 6,7,7a,8-tetrahydro-11-methoxy-7-methyl-, (*R*)-
$C_{19}H_{19}NO_4$
2470 - 5H-Benzo[g]-1,3-benzodioxolo[6,5,4-de]quinolin-12-ol, 6,7,7a,8-tetrahydro-11-methoxy-7-methyl-, (*S*)-
$C_{19}H_{19}N_7O_6$
6243 - *L*-Glutamic acid, *N*-[4-[[(2-amino-1,4-dihydro-4-oxo-6-pteridinyl)methyl]amino]benzoyl]-
$C_{19}H_{20}F_3NO_4$
5962 - Fluazipop-butyl
5963 - Fluazipop-P-butyl
$C_{19}H_{20}N_2O$
11346 - Sarpagan-17-al
$C_{19}H_{21}NO_3$
5158 - 4H-Dibenzo[de,g]quinolin-1-ol, 5,6,6a,7-tetrahydro-2,11-dimethoxy-6-methyl-, (*S*)-
7679 - Morphinan, 6,7,8,14-tetradehydro-4,5-epoxy-3,6-dimethoxy-17-methyl-, (5α)-
$C_{19}H_{21}NO_4$
5156 - 4H-Dibenzo[de,g]quinoline-2,9-diol, 5,6,6a,7-tetrahydro-1,10-dimethoxy-6-methyl-, (*S*)-
$C_{19}H_{22}$
1744 - Benzene, 1,1′-(1-heptenylidene)bis-
$C_{19}H_{22}N_2O$
4120 - Cinchonan-9-ol, (9*S*)-
4133 - Cinchonan-9-ol, (8α,9*R*)-
10435 - 1-Propanone, 3-(3-ethenyl-4-piperidinyl)-1-(4-quinolinyl)-, (3*R*-*cis*)-
$C_{19}H_{22}N_2OS$
5807 - Ethanone, 1-[10-[3-(dimethylamino)propyl]-10H-phenothiazin-2-yl]-
$C_{19}H_{22}N_2O_2$
4119 - Cinchonan-6′,9-diol, (8α,9*R*)-
11347 - Sarpagan-10,17-diol
$C_{19}H_{22}O_2$
9358 - Phenol, 4-[1-ethyl-2-(4-methoxyphenyl)-1-butenyl]-
9359 - Phenol, 4-[1-ethyl-2-(4-methoxyphenyl)-1-butenyl]-, (*E*)-
$C_{19}H_{22}O_6$
6204 - Gibb-3-ene-1,10-dicarboxylic acid, 2,4a,7-trihydroxy-1-methyl-8-methylene-, 1,4a-lactone,
$C_{19}H_{23}ClN_2$
5140 - 1H-1,4-Diazepine, 1-[(4chlorophenyl) phenylmethyl]hexahydro-4-methyl-
$C_{19}H_{23}NO$
1903 - Benzenemethanol, α-[1-[methyl(3-phenyl-2-propenyl)amino]ethyl]-
$C_{19}H_{23}NO_3$
7664 - Morphinan, 6,7-didehydro-4,5-epoxy-3,6-dimethoxy-17-methyl-, (5α)-
$C_{19}H_{23}NO_4$
7676 - Morphinan-6-one, 7,8-didehydro-4-hydroxy-3,7-dimethoxy-17-methyl-, (9α,13α,14α)-
$C_{19}H_{23}N_3$
342 - Amitraz
$C_{19}H_{23}N_3O_2$
5373 - Ergoline-8-carboxamide, 9,10-didehydro-*N*-(2-hydroxy-1-methylethyl)-6-methyl-, [8α(*S*)]-
$C_{19}H_{24}$
1746 - Benzene, 1,1′-heptylidenebis-

$C_{19}H_{24}N_2$
9627 - 4-Piperidinamine, 1-methyl-*N*-phenyl-*N*-(phenylmethyl)-
$C_{19}H_{24}N_2O$
4121 - Cinchonan-9-ol, 10,11-dihydro-, (8α,9*R*)-
4122 - Cinchonan-9-ol, 10,11-dihydro-, (9*S*)-
7306 - 1H-Indole-3-ethanol, 2-(5-ethenyl-1-azabicyclo[2.2.2]oct-2-yl)-
$C_{19}H_{24}N_2OS$
3977 - Carbamothioic acid, diphenyl-, *S*-[2-(diethylamino)ethyl] ester
$C_{19}H_{24}N_2O_2$
6177 - 3aH-Furo[2,3-b]indol-3a-ol, 8a-(5-ethenyl-1-azabicyclo[2.2.2]oct-2-yl)-2,3,8,8a-tetrahydro-
6178 - 3aH-Furo[2,3-b]indol-3a-ol, 8a-(5-ethenyl-1-azabicyclo[2.2.2]oct-2-yl)-2,3,8,8a-tetrahydro-
$C_{19}H_{24}O_3$
353 - Androst-4-ene-3,11,17-trione
$C_{19}H_{26}$
7931 - Naphthalene, 1-nonyl-
7932 - Naphthalene, 2-nonyl-
$C_{19}H_{26}N_2O$
4154 - Curan-17-ol, (16α)-
$C_{19}H_{26}O_2$
343 - Androsta-1,4-dien-3-one, 17-hydroxy-, (17β)-
352 - Androst-4-ene-3,17-dione
$C_{19}H_{26}O_3$
336 - Allethrin
$C_{19}H_{26}O_4S$
10507 - Propargite
$C_{19}H_{27}ClO_2$
354 - Androst-4-en-3-one, 4-chloro-17-hydroxy-, (17β)-
$C_{19}H_{27}NO_3$
2396 - 2H-Benzo[a]quinolizin-2-one, 1,3,4,6,7,11b-hexahydro-9,10-dimethoxy-3-(2-methylpropyl)-
$C_{19}H_{28}BrNO_3$
11179 - Pyrrolidinium, 3-[(cyclopentylhydroxyphenylacetyl)oxy]-1,1-dimethyl-, bromide
$C_{19}H_{28}O_2$
356 - Androst-4-en-3-one, 17-hydroxy-, (17β)-
$C_{19}H_{30}O_2$
347 - Androstan-17-one, 3-hydroxy-, (3β,5α)-
348 - Androstan-17-one, 3-hydroxy-, (3α,5α)-
349 - Androstan-3-one, 17-hydroxy-, (5α,17β)-
5318 - Dodecanoic acid, phenylmethyl ester
$C_{19}H_{30}O_5$
9727 - Piperonyl butoxide
$C_{19}H_{32}$
345 - Androstane
$C_{20}H_{27}O_4P$
9560 - Phosphoric acid, 2-ethylhexyl diphenyl ester
$C_{20}H_{28}$
7793 - Naphthalene, 1-decyl-
7794 - Naphthalene, 2-decyl-
$C_{20}H_{28}O_2$
344 - Androsta-1,4-dien-3-one, 17-hydroxy-17-methyl-, (17β)-
$C_{20}H_{28}O_3S$
7968 - 1-Naphthalenesulfonic acid, 3,6-bis(1,1-dimethylethyl)-, ethyl ester
$C_{20}H_{29}FO_3$
355 - Androst-4-en-3-one, 9-fluoro-11,17-dihydroxy-17-methyl-, (11β,17β)-
$C_{20}H_{30}ClN_3O_2$
11226 - 4-Quinolinecarboxamide, 2-butoxy-*N*-[2-(diethylamino)ethyl]-, monohydrochloride
$C_{20}H_{30}O$
11341 - Retinol
$C_{20}H_{30}O_2$
358 - Androst-4-en-3-one, 17-hydroxy-17-methyl-, (17β)-
9166 - 1-Phenanthrenecarboxylic acid, 1,2,3,4,4a,5,6,9,10,10a-decahydro-1,4a-dimethyl-7-(1-methylethyl)-,
9167 - 1-Phenanthrenecarboxylic acid, 1,2,3,4,4a,4b,5,6,10,10a-decahydro-1,4a-dimethyl-7-(1-methylethyl)-,
9169 - 1-Phenanthrenecarboxylic acid, 1,2,3,4,4a,4b,5,9,10,10a-decahydro-1,4a-dimethyl-7-(1-methylethyl)-,
9170 - 1-Phenanthrenecarboxylic acid, 7-ethenyl-1,2,3,4,4a,4b,5,6,7,9,10,10a-dodecahydro-1,4a,7-trimethyl-, [1R-(1α,4aβ,4bα,7β,10aα)]-
$C_{20}H_{32}ClNO$
9705 - 1-Piperidinepropanol, α-cyclohexyl-α-phenyl-, hydrochloride
$C_{20}H_{32}N_2O_3S$
4038 - Carbosulfan
$C_{20}H_{32}O_2$
346 - Androstane-17-carboxylic acid, (5β,17β)-
350 - Androstan-3-one, 17-hydroxy-17-methyl-, (5α,17β)-
5360 - 5,8,11,14-Eicosatetraenoic acid
5361 - 5,8,11,14-Eicosatetraenoic acid, (all-Z)-
$C_{20}H_{34}$
2266 - Benzene, tetradecyl-
$C_{20}H_{34}O_2$
8283 - 9,12,15-Octadecatrienoic acid, ethyl ester, (*Z,Z,Z*)-
$C_{20}H_{36}O_2$
8226 - 9,12-Octadecadienoic acid (*Z,Z*)-, ethyl ester
8227 - 9,12-Octadecadienoic acid, ethyl ester
$C_{20}H_{38}$
5366 - 1-Eicosyne
$C_{20}H_{38}N_4O_4$
7686 - 4-Morpholinecarboxamide, *N,N'*-1,2-ethanediylbis[*N*-butyl-
$C_{20}H_{38}O_2$
4952 - Cycloprate
5363 - 11-Eicosenoic acid
5364 - 9-Eicosenoic acid
5365 - 9-Eicosenoic acid, (*Z*)-
8301 - 9-Octadecenoic acid (*Z*)-, ethyl ester
8302 - 9-Octadecenoic acid, ethyl ester, (*E*)-
$C_{20}H_{38}O_3$
5060 - Decanoic acid, anhydride
8269 - Octadecanoic acid, 12-oxo-, ethyl ester
8305 - 9-Octadecenoic acid, 12-hydroxy-, ethyl ester, [*R*-(*Z*)]-
$C_{20}H_{38}O_4$
5352 - Eicosanedioic acid
$C_{20}H_{40}$
4475 - Cyclohexane, tetradecyl-
4844 - Cyclopentane, pentadecyl-
5362 - 1-Eicosene
$C_{20}H_{40}O$
6671 - 1-Hexadecen-3-ol, 3,7,11,15-tetramethyl-
6672 - 2-Hexadecen-1-ol, 3,7,11,15-tetramethyl-, [*R*-[*R**,*R**-(*E*)]]-
8240 - Octadecane, 1-(ethenyloxy)-
$C_{20}H_{40}O_2$
221 - Acetic acid, octadecyl ester
5062 - Decanoic acid, decyl ester
5355 - Eicosanoic acid
6647 - Hexadecanoic acid, butyl ester
6658 - Hexadecanoic acid, 3,7,11,15-tetramethyl-
8257 - Octadecanoic acid, ethyl ester
$C_{20}H_{40}O_3$
8260 - Octadecanoic acid, 2-hydroxyethyl ester
$C_{20}H_{42}$
5350 - Eicosane
$C_{20}H_{42}O$
5358 - 1-Eicosanol
5359 - 2-Eicosanol
$C_{20}H_{42}O_2$
8237 - Octadecane, 1,1-dimethoxy-
$C_{20}H_{48}O_4Si_4$
4995 - Cyclotetrasiloxane, 2,4,6,8-tetrabutyl-2,4,6,8-tetramethyl-
$C_{20}H_{60}O_8Si_9$
8178 - Nonasiloxane, eicosamethyl-
$C_{21}H_{12}O$
5152 - 13H-Dibenzo[a,h]fluoren-13-one
$C_{21}H_{13}N$
5143 - Dibenz[a,j]acridine
$C_{21}H_{14}$
5148 - 13H-Dibenzo[a,g]fluorene
$C_{21}H_{14}Br_4O_5S$
9367 - Phenol, 4,4'-(3H-2,1-benzoxathiol-3-ylidene)bis[2,6-dibromo-3-methyl-, *S,S*-dioxide
$C_{21}H_{14}O$
7634 - Methanone, 1-naphthalenyl-2-naphthalenyl-
$C_{21}H_{16}$
2392 - Benz[j]aceanthrylene, 1,2-dihydro-3-methyl-
7271 - 1H-Indene, 1,3-diphenyl-
7272 - 1H-Indene, 2,3-diphenyl-
7919 - Naphthalene, 1,1'-methylenebis-
$C_{21}H_{16}Br_2O_5S$
9364 - Phenol, 4,4'-(3H-2,1-benzoxathiol-3-ylidene)bis[2-bromo-6-methyl-, *S,S*-dioxide
$C_{21}H_{16}N_2$
7215 - 1H-Imidazole, 2,4,5-triphenyl-
$C_{21}H_{16}N_2O$
12008 - Urea, *N,N'*-di-1-naphthalenyl-
$C_{21}H_{16}NaO_2$
11517 - Sonar
$C_{21}H_{18}N_2$
7199 - 1H-Imidazole, 4,5-dihydro-2,4,5-triphenyl-, *cis*-
$C_{21}H_{18}O_5S$
9368 - Phenol, 4,4'-(3H-2,1-benzoxathiol-3-ylidene)bis[2-methyl-, *S,S*-dioxide
$C_{21}H_{19}NO_5$
2433 - [1,3]Benzodioxolo[5,6-c]phenanthridinium, 1,2-dimethoxy-12-methyl-
$C_{21}H_{20}$
2110 - Benzene, 1,1'-(phenylmethylene)bis[4-methyl-
$C_{21}H_{20}Cl_2O_3$
9150 - Permethrin
$C_{21}H_{20}N_2O_3$
8572 - Oxayohimbanium, 3,4,5,6,16,17 hexadehydro-16-(methoxycarbonyl)-19-methyl-, (19α)-
$C_{21}H_{20}O_6$
6334 - 1,6-Heptadiene-3,5-dione, 1,7-bis(4-hydroxy-3-methoxyphenyl)-, (*E,E*)-
$C_{21}H_{20}O_{10}$
2824 - 4H-1-Benzopyran-4-one, 3-[4-(β-*D*-glucopyranosyloxy)phenyl]-5,7-dihydroxy-
$C_{21}H_{20}O_{11}$
2796 - 4H-1-Benzopyran-4-one, 3-[(6-deoxy-α-*L*-mannopyranosyl)oxy]-2-(3,4-dihydroxyphenyl)-5,7-dihydroxy-
$C_{21}H_{21}N$
1798 - Benzenemethanamine, *N,N*-bis(phenylmethyl)-
$C_{21}H_{21}NO_5$
3108 - Bis[1,3]benzodioxolo[4,5-c:5',6'-g]azecin-13(5H)-one, 4,6,7,14-tetrahydro-5,14-dimethyl-, (*R*)-
$C_{21}H_{21}NO_6$
7357 - 1(3H)-Isobenzofuranone, 6,7-dimethoxy-3-(5,6,7,8-tetrahydro-6-methyl-1,3-dioxolo[4,5-g]isoquinolin-5-yl)-, [*S*-
11342 - Rheadan, 8-methoxy-16-methyl-2,3 10,11-bis[methylenebis(oxy)]-, (8β)-
$C_{21}H_{21}N_3$
11834 - 1,3,5-Triazine, hexahydro-1,3,5-triphenyl-
$C_{21}H_{21}N_3O_7$
11352 - 9,10-Secostrychnidin-10-oic acid, 2,3-dihydro-4-nitro-2,3-dioxo-
$C_{21}H_{21}O_3P$
9596 - Phosphorous acid, tris(2-methylphenyl) ester
9597 - Phosphorous acid, tris(4-methylphenyl) ester
$C_{21}H_{21}O_4P$
9569 - Phosphoric acid, tris(2-methylphenyl) ester
9570 - Phosphoric acid, tris(3-methylphenyl) ester
9571 - Phosphoric acid, tris(4-methylphenyl) ester

$C_{21}H_{21}O_7P$
9387 - Phenol, 2-methoxy-, phosphate (3:1)
$C_{21}H_{21}Sb$
11552 - Stibine, tris(3-methylphenyl)-
11553 - Stibine, tris(4-methylphenyl)-
$C_{21}H_{22}N_2O_2$
11563 - Strychnidin-10-one
$C_{21}H_{23}ClFNO_2$
3651 - 1-Butanone, 4-[4-(4-chlorophenyl)-4-hydroxy-1-piperidinyl]-1-(4-fluorophenyl)-
$C_{21}H_{23}NO_5$
2435 - Benzo[e]-1,3-dioxolo[4,5-l][2]benzazecin-12(5H)-one, 4,6,7,13-tetrahydro-9,10-dimethoxy-5-methyl-
7667 - Morphinan-3,6-diol, 7,8-didehydro-4,5-epoxy-17-methyl- (5α,6α)-, diacetate (ester)
$C_{21}H_{23}NO_6$
92 - Acetamide, *N*-(5,6,7,9-tetrahydro-10-hydroxy-1,2,3-trimethoxy-9-oxobenzo[a]heptalen-7-yl)-, (*S*)-
$C_{21}H_{23}N_3O_5$
11565 - Strychnidin-10-one, mononitrate
$C_{21}H_{24}N_2O_3$
8571 - Oxayohimban-16-carboxylic acid, 16,17-didehydro-19-methyl-, methyl ester, (19α)-
$C_{21}H_{24}OSi_2$
5239 - Disiloxane, 1,1,1-trimethyl-3,3,3-triphenyl-
$C_{21}H_{24}O_3Si_3$
5005 - Cyclotrisiloxane, 2,4,6-trimethyl-2,4,6-triphenyl-
$C_{21}H_{24}O_4$
9926 - Propanedioic acid, bis(phenylmethyl)-, diethyl ester
$C_{21}H_{24}O_7$
3540 - Butanoic acid, 2-methyl-, 10-(acetyloxy)-9,10-dihydro-8,8-dimethyl-2-oxo-2H,8H-benzo[1,2-b 3,4-b′]dipyran-9-yl
$C_{21}H_{25}NO_4$
5150 - 6H-Dibenzo[a,g]quinolizin-2-ol, 5,8,13,13a-tetrahydro-3,9,10-trimethoxy-13-methyl-, (13*S*-*trans*)-
5151 - 6H-Dibenzo[a,g]quinolizin-3-ol, 5,8,13,13a-tetrahydro-2,9,10-trimethoxy-13-methyl-, (13*S*-*trans*)-
5157 - 4H-Dibenzo[de,g]quinoline, 5,6,6a,7-tetrahydro-1,2,9,10-tetramethoxy-6-methyl-, (*S*)-
$C_{21}H_{26}BrNO_3$
5445 - Ethanaminium, *N*,*N*-diethyl-*N*-methyl-2-[(9H-xanthen-9-ylcarbonyl)oxy]-, bromide
$C_{21}H_{26}ClN_3OS$
9617 - 1-Piperazineethanol, 4-[3-(2-chloro-10H-phenothiazin-10-yl)propyl]-
$C_{21}H_{26}N_2O_3$
5349 - Eburnamenine-14-carboxylic acid, 14,15-dihydro-14-hydroxy-, methyl ester, (3α,14β,16α)-
12062 - Yohimban-16-carboxylic acid, 17-hydroxy-, methyl ester, (16α,17α)-
$C_{21}H_{26}N_2O_4$
4129 - Cinchonan-9-ol, 6′-methoxy-, (8α,9*R*)-, monoformate (salt)
$C_{21}H_{26}O_2$
5154 - 6H-Dibenzo[b,d]pyran-1-ol, 6,6,9-trimethyl-3-pentyl-
$C_{21}H_{26}O_3$
7621 - Methanone, [2-hydroxy-4-(octyloxy)phenyl]phenyl-
$C_{21}H_{26}O_5$
9739 - Pregna-1,4-diene-3,11,20-trione, 17,21-dihydroxy-
$C_{21}H_{27}ClN_2O_2$
5700 - Ethanol, 2-[2-[4-[(4-chlorophenyl)phenylmethyl]-1-piperazinyl]ethoxy]-
$C_{21}H_{27}NO_4$
7379 - Isoquinoline, 1-[(3,4-dimethoxyphenyl)methyl]-1,2,3,4-tetrahydro-6,7-dimethoxy-2-methyl-, (*S*)-
$C_{21}H_{28}ClNO$
6516 - 3-Heptanone, 6-(dimethylamino)-4,4-diphenyl-, hydrochloride
$C_{21}H_{28}O_2$
9758 - Pregn-4-en-20-yn-3-one, 17-hydroxy-, (17α)-
$C_{21}H_{28}O_3$
4960 - Cyclopropanecarboxylic acid, 2,2-dimethyl-3-(2-methyl-1-propenyl)-, 2-methyl-4-oxo-3-(2,4-pentadienyl)-2-cyclopenten-1-yl ester, [1R-[1α[S*(Z)],3β]]-
$C_{21}H_{28}O_4$
9755 - Pregn-4-ene-3,11,20-trione, 21-hydroxy-
$C_{21}H_{28}O_5$
9738 - Pregna-1,4-diene-3,20-dione, 11,17,21-trihydroxy-, (11β)-
9746 - Pregn-4-en-18-al, 11,21-dihydroxy-3,20-dioxo-, (11β)-
9754 - Pregn-4-ene-3,11,20-trione, 17,21-dihydroxy-
$C_{21}H_{30}O_2$
1537 - 1,3-Benzenediol, 2-[3-methyl-6-(1-methylethenyl)-2-cyclohexen-1-yl]-5-pentyl-, (1*R*-*trans*)-
$C_{21}H_{30}O_3$
9750 - Pregn-4-ene-3,20-dione, 17-hydroxy-
9751 - Pregn-4-ene-3,20-dione, 21-hydroxy-
$C_{21}H_{30}O_4$
9748 - Pregn-4-ene-3,20-dione, 11,21-dihydroxy-, (11β)-
9749 - Pregn-4-ene-3,20-dione, 17,21-dihydroxy-
$C_{21}H_{30}O_5$
4282 - 2,4-Cyclohexadien-1-one, 3,5,6-trihydroxy-4,6-bis(3-methyl-2-butenyl)-2-(3-methyl-1-oxobutyl)-, (*R*)-
9753 - Pregn-4-ene-3,20-dione, 11,17,21-trihydroxy-, (11β)-
$C_{21}H_{32}O_2$
9168 - 1-Phenanthrenecarboxylic acid, 1,2,3,4,4a,4b,5,6,10,10a-decahydro-1,4a-dimethyl-7-(1-methylethyl)-, methyl ester, [1R-(1α,4aβ,4bα,10aα)]-
9743 - Pregnane-3,20-dione, (5β)-
9756 - Pregn-5-en-20-one, 3-hydroxy-, (3β)-
$C_{21}H_{32}O_5$
9744 - Pregnane-11,20-dione, 3,17,21-trihydroxy-, (3α,5β)-
9757 - Pregn-4-en-3-one, 11,17,20,21-tetrahydroxy-, (11β,20*R*)-
$C_{21}H_{34}BrNO_3$
5443 - Ethanaminium, 2-[(cyclohexylhydroxyphenylacetyl)oxy]-*N*,*N*-diethyl-*N*-methyl-, bromide
$C_{21}H_{34}O_2$
9745 - Pregnan-20-one, 3-hydroxy-, (3α,5β)-
11627 - Tetradecanoic acid, phenylmethyl ester
$C_{21}H_{36}$
2089 - Benzene, pentadecyl-
9740 - Pregnane, (5α)-
9741 - Pregnane, (5β)-
$C_{21}H_{36}N_2O_5S$
9623 - Piperazinium, 4-(2-cyclohexyl-2-hydroxy-2-phenylethyl)-1,1-dimethyl-, methyl sulfate (salt)
$C_{21}H_{36}O$
9441 - Phenol, 3-pentadecyl-
$C_{21}H_{36}O_2$
1543 - 1,2-Benzenediol, 3-pentadecyl-
9742 - Pregnane-3,20-diol, (3α,5β,20*S*)-
$C_{21}H_{38}ClN$
11030 - Pyridinium, 1-hexadecyl-, chloride
$C_{21}H_{38}O_6$
6908 - Hexanoic acid, 1,2,3-propanetriyl ester
$C_{21}H_{39}N_7O_{12}$
11562 - Streptomycin
$C_{21}H_{40}O_4$
8131 - Nonanedioic acid, bis(2-ethylbutyl) ester
8299 - 9-Octadecenoic acid (*Z*)-, 2,3-dihydroxypropyl ester
$C_{21}H_{41}O_2$
5357 - Eicosanoic acid, methyl ester
$C_{21}H_{42}$
4463 - Cyclohexane, pentadecyl-
4822 - Cyclopentane, hexadecyl-
$C_{21}H_{42}O_2$
8267 - Octadecanoic acid, 1-methylethyl ester
8273 - Octadecanoic acid, propyl ester
$C_{21}H_{42}O_4$
8255 - Octadecanoic acid, 2,3-dihydroxypropyl ester, (±)-
$C_{21}H_{44}$
6294 - Heneicosane
$C_{21}H_{44}O_3$
10023 - 1,2-Propanediol, 3-(octadecyloxy)-
$C_{22}H_{12}$
2471 - Benzo[ghi]perylene
$C_{22}H_{14}$
5142 - Dibenz[a,h]anthracene
8636 - Pentacene
8996 - Pentaphene
9601 - Picene
$C_{22}H_{16}N_4O$
8038 - 2-Naphthalenol, 1-[[4-(phenylazo)phenyl]azo]-
$C_{22}H_{17}N$
7717 - 1-Naphthalenamine, *N*,*N*-diphenyl-
$C_{22}H_{18}Cl_2FNO_3$
5008 - Cyfluthrin
$C_{22}H_{18}O_4$
1337 - 1,2-Benzenedicarboxylic acid, bis(phenylmethyl) ester
7353 - 1(3H)-Isobenzofuranone, 3,3-bis(4-hydroxy-3-methylphenyl)-
$C_{22}H_{19}Br_2NO_3$
5102 - Deltamethrin
$C_{22}H_{19}Cl_2NO_3$
5012 - Cypermethrin
$C_{22}H_{19}NO_4$
9465 - Phenol, 4,4′-(2-pyridinylmethylene)bis-, diacetate (ester)
$C_{22}H_{20}O_{13}$
373 - 2-Anthracenecarboxylic acid, 7-β-*D*-glucopyranosyl-9,10-dihydro-3,5,6,8-tetrahydroxy-1-methyl-9,10-dioxo-
$C_{22}H_{21}NO_2$
1797 - Benzenemethanamine, *N*-[bis(4-methoxyphenyl)methylene]-
$C_{22}H_{22}O_8$
6191 - Furo[3′,4′:6,7]naphtho[2,3-d]-1,3-dioxol-6(5aH)-one, 5,8,8a,9-tetrahydro-9-hydroxy-5-(3,4,5-trimethoxyphenyl)-
6192 - Furo[3′,4′:6,7]naphtho[2,3-d]-1,3-dioxol-6(5aH)-one, 5,8,8a,9-tetrahydro-9-hydroxy-5-(3,4,5-trimetho
$C_{22}H_{23}ClN_2O_8$
7704 - 2-Naphthacenecarboxamide, 7-chloro-4-(dimethylamino)-1,4,4a,5,5a,6,11,12a-octahydro-3,6,10,12,12a-pentahydroxy-6-methyl-1,11-dioxo, [4S (4α,4aα,5aα,6β,12aα)]-
$C_{22}H_{23}NO_3$
5953 - Fenpropathrin
$C_{22}H_{23}NO_6$
10797 - 4H-Pyran-4-one, 2-methoxy-3,5-dimethyl-6-[tetrahydro-4-[2-methyl-3-(4-nitrophenyl)-2-propenylidene]-2-furanyl]
$C_{22}H_{23}NO_7$
7356 - 1(3H)-Isobenzofuranone, 6,7-dimethoxy-3-(5,6,7,8-tetrahydro-4-methoxy-6-methyl-1,3-dioxolo[4,5-g]isoquinolin-5
$C_{22}H_{24}N_2O_4$
3204 - Butanamide, *N*,*N*′-(3,3′-dimethyl[1,1′-biphenyl]-4,4′-diyl)bis[3-oxo-

10772 - 2H-Pyran-5-carboxylic acid, 3,4-dihydro-3-(hydroxymethyl)-2-methyl-4-[(9-methyl-9H-pyrido[3,4-b]indol-1-yl)methyl]-, methyl ester, [2S-(2α,3β,4β)]-
11351 - 16,19-Secostrychnidine-10,16-dione, 4-hydroxy-19-methyl-
$C_{22}H_{24}N_2O_8$
7706 - 2-Naphthacenecarboxamide, 4-(dimethylamino)-1,4,4a,5,5a,6,11,12a-octahydro-3,6,10,12,12a-pentahydroxy-6-methyl-1,11-dioxo-, [4S-(4α,4aα,5aα,6β,12aα)]-
$C_{22}H_{24}N_2O_9$
7705 - 2-Naphthacenecarboxamide, 4-(dimethylamino)-1,4,4a,5,5a,6,11,12a-octahydro-3,5,6,10,12,12a-hexahydroxy-6-methyl-1,11-dioxo-,[4S-(4α,4aα,5aα,6β,12aα)]-
$C_{22}H_{24}Si$
11445 - Silane, methyltri-*p*-tolyl-
$C_{22}H_{25}NO_2$
5854 - Ethanone, 2,2′-(1-methyl-2,6-piperidinediyl)bis[1-phenyl-, *cis*-
$C_{22}H_{25}NO_6$
93 - Acetamide, *N*-(5,6,7,9-tetrahydro-1,2,3,10-tetramethoxy-9-oxobenzo[a]heptalen-7-yl)-, (*S*)-
$C_{22}H_{25}N_3O$
10850 - 3H-Pyrazol-3-one, 1,2-dihydro-2-(1-methyl-4-piperidinyl)-5-phenyl-4-(phenylmethyl)-
$C_{22}H_{26}N_2O_3$
4146 - Corynan-16-carboxylic acid, 16,17,18,19-tetradehydro-17-methoxy-, methyl ester, (16E)-
$C_{22}H_{26}O_3$
11340 - (-)-*cis*-Resmethrin
$C_{22}H_{27}NO_2$
5834 - Ethanone, 2-[6-(2-hydroxy-2-phenylethyl)-1-methyl-2-piperidinyl]-1-phenyl-, [2*R*-[2α,6α(*S**)]]-
$C_{22}H_{27}NO_4$
5149 - 6H-Dibenzo[a,g]quinolizine, 5,8,13,13a-tetrahydro-2,3,9,10-tetramethoxy-13-methyl-, (13*S*-*trans*)-
$C_{22}H_{28}N_2O$
9801 - Propanamide, *N*-phenyl-*N*-[1-(2-phenylethyl)-4-piperidinyl]-
$C_{22}H_{28}N_2O_5$
12063 - Yohimban-16-carboxylic acid, 18-hydroxy-11,17-dimethoxy-, (3β,16β,17α,18β,20
$C_{22}H_{28}O_3$
9736 - Pregna-4,6-diene-21-carboxylic acid, 17-hydroxy-3-oxo-, γ-lactone, (17α)-
$C_{22}H_{28}O_5$
4962 - Cyclopropanecarboxylic acid, 3-(3-methoxy-2-methyl-3-oxo-1-propenyl)-2,2-dimethyl-, 2-methyl-4-oxo-3-(2,4-pentadienyl)-2-cyclopenten-1-yl ester [1R-[1α[S*(Z)],3β(E)]]-
$C_{22}H_{29}FO_5$
9737 - Pregna-1,4-diene-3,20-dione, 9-fluoro-11,17,21-trihydroxy-16-methyl-, (11β,16α)-
$C_{22}H_{29}NO_2$
9646 - 2,6-Piperidinediethanol, 1-methyl-α,α′-diphenyl-, [2α(*R**),6α(*S**)]-
$C_{22}H_{30}N_2O_2$
473 - Aspidospermidine, 1-acetyl-17-methoxy-
$C_{22}H_{30}N_2O_5$
4730 - 2,4(1H)-Cyclo-3,4-secoakuammilanium, 3,17-dihydroxy-16-(methoxycarbonyl)-4-methyl-, (3β,16*R*)-
$C_{22}H_{32}$
7879 - Naphthalene, 1-dodecyl-
7880 - Naphthalene, 2-dodecyl-
$C_{22}H_{32}O_3$
359 - Androst-4-en-3-one, 17-(1-oxopropoxy)-, (17β)-
9752 - Pregn-4-ene-3,20-dione, 17-hydroxy-6-methyl-, (6α)-
$C_{22}H_{33}NO_2$
475 - Atisine
$C_{22}H_{35}NO_3$
5007 - 7,20-Cycloveatchane-1,12,15-triol, 21-ethyl-4-methyl-16-methylene-, (1α,12α,15β)-
$C_{22}H_{36}O$
6664 - 1-Hexadecanone, 1-phenyl-
$C_{22}H_{37}NO$
6630 - Hexadecanamide, *N*-phenyl-
$C_{22}H_{38}$
1751 - Benzene, hexadecyl-
$C_{22}H_{38}O$
1752 - Benzene, (hexadecyloxy)-
$C_{22}H_{40}O_7$
8102 - 1,2,3-Nonadecanetricarboxylic acid, 2-hydroxy-
$C_{22}H_{42}O_2$
5281 - 13-Docosenoic acid
5282 - 13-Docosenoic acid, (*E*)-
5283 - 13-Docosenoic acid, (*Z*)-
8298 - 9-Octadecenoic acid (*Z*)-, butyl ester
10667 - 2-Propenoic acid, 2-methyl-, octadecyl ester
$C_{22}H_{42}O_3$
8304 - 9-Octadecenoic acid, 12-hydroxy-, butyl ester, [*R*-(*Z*)]-
$C_{22}H_{42}O_4$
6781 - Hexanedioic acid, bis(2-ethylhexyl) ester
8238 - Octadecanedioic acid, diethyl ester
$C_{22}H_{44}$
4411 - Cyclohexane, hexadecyl-
5280 - 1-Docosene
$C_{20}H_{40}N_2{\bullet}C_2H_4O_2$
6275 - Glyodin
$C_{22}H_{44}O$
5284 - 13-Docosen-1-ol, (*Z*)-
$C_{22}H_{44}O_2$
5276 - Docosanoic acid
5356 - Eicosanoic acid, ethyl ester
8251 - Octadecanoic acid, butyl ester
8268 - Octadecanoic acid, 2-methylpropyl ester
$C_{22}H_{46}$
5268 - Docosane
$C_{22}H_{46}O$
5279 - 1-Docosanol
$C_{22}H_{66}O_9Si_{10}$
5079 - Decasiloxane, docosamethyl-
$C_{23}H_{16}O_3$
7268 - 1H-Indene-1,3(2H)-dione, 2-(diphenylacetyl)-
$C_{23}H_{16}O_6$
7777 - 2-Naphthalenecarboxylic acid, 4,4′-methylenebis[3-hydroxy-
$C_{23}H_{18}O$
7910 - 1-Naphthalenemethanol, α,α-diphenyl-
$C_{23}H_{19}ClF_3NO_3$
5010 - Cyhalothrin
$C_{23}H_{21}ClO_3$
1154 - Benzene, 1,1′,1″-(1-chloro-1-ethenyl-2-ylidene)tris[4-methoxy-
$C_{23}H_{22}ClF_3O_2$
2998 - Bifenthrin
$C_{23}H_{22}O_6$
2783 - [1]Benzopyrano[3,4-b]furo[2,3-h][1]benzopyran-6(6aH)-one, 1,2,12,12a-tetrahydro-8,9-dimethoxy-2-(1-methylethen
$C_{23}H_{22}O_7$
3110 - 3H-Bis[1]benzopyrano[3,4-b:6′,5′-e]pyran-7(7aH)-one, 13,13a-dihydro-7a-hydroxy-9,10-dimethoxy-3,3-dimethyl
$C_{23}H_{23}O_6$
9747 - Pregn-4-ene-3,20-dione, 21-(acetyloxy)-11,17-dihydroxy-, (11β)-
$C_{23}H_{25}ClN_2$
7455 - Methanaminium, *N*-[4-[[4-(dimethylamino)phenyl]phenylmethylene]-2,5-cyclohexadien-1-ylidene]-*N*-methyl-, chlorid
$C_{23}H_{26}N_2O_4$
11564 - Strychnidin-10-one, 2,3-dimethoxy-
$C_{23}H_{27}NO_8$
2586 - Benzoic acid, 6-[[6-[2-(dimethylamino)ethyl]-4-methoxy-1,3-benzodioxol-5-yl]acetyl]-2,3-dimethoxy-
$C_{23}H_{28}ClN_3O$
9657 - Piperidine, 1-[2-[3-(2,2-diphenylethyl)-1,2,4-oxadiazol-5-yl]ethyl]-, monohydrochloride
$C_{23}H_{30}BrNO_3$
9832 - 2-Propanaminium, *N*-methyl-*N*-(1-methylethyl)-*N*-[2-[(9H-xanthen-9-ylcarbonyl)oxy]ethyl]-, bromide
$C_{23}H_{30}O_2Si_3$
11915 - Trisiloxane, 1,1,3,5,5-pentamethyl-1,3,5-triphenyl-
$C_{23}H_{32}N_2O_2S$
11800 - Thiourea, *N,N′*-bis[4-(3-methylbutoxy)phenyl]-
$C_{23}H_{32}O_6$
4049 - Card-20(22)-enolide, 3,5,14-trihydroxy-19-oxo-, (3β,5β)-
$C_{23}H_{33}IN_2O$
2120 - Benzenepropanaminium, γ-(aminocarbonyl)-*N*-methyl-*N,N*-bis(1-methylethyl)-γ-phenyl-, iod
$C_{23}H_{34}O_4$
4044 - Card-20(22)-enolide, 3,14-dihydroxy-, (3β,5β)-
$C_{23}H_{34}O_5$
4046 - Card-20(22)-enolide, 3,11,14-trihydroxy-, (3β,5β,11α)-
4047 - Card-20(22)-enolide, 3,12,14-trihydroxy-, (3β,5β,12β)-
4048 - Card-20(22)-enolide, 3,14,16-trihydroxy-, (3β,5β,16β)-
$C_{23}H_{36}O_3$
351 - Androstan-3-one, 2-methyl-17-(1-oxopropoxy)-, (2α,5α,17β)-
$C_{23}H_{38}O_2$
6655 - Hexadecanoic acid, phenylmethyl ester
8210 - 24-Norcholan-23-oic acid, (5β)-
$C_{23}H_{40}$
1743 - Benzene, heptadecyl-
$C_{23}H_{44}O_2$
8306 - 9-Octadecenoic acid (*Z*)-, 3-methylbutyl ester
$C_{23}H_{44}O_3$
8275 - Octadecanoic acid, (tetrahydro-2-furanyl)methyl ester
$C_{23}H_{44}O_4$
9954 - Propanedioic acid, hexadecyl-, diethyl ester
$C_{23}H_{46}$
4404 - Cyclohexane, heptadecyl-
4842 - Cyclopentane, octadecyl-
$C_{23}H_{46}O$
11857 - 12-Tricosanone
$C_{23}H_{46}O_2$
5278 - Docosanoic acid, methyl ester
8265 - Octadecanoic acid, 3-methylbutyl ester
8270 - Octadecanoic acid, pentyl ester
$C_{23}H_{48}$
6309 - Heptadecane, 9-hexyl-
11855 - Tricosane
$C_{24}H_{12}$
4145 - Coronene
$C_{24}H_{14}$
2850 - Benzo[rst]pentaphene
8061 - Naphtho[1,2,3,4-def]chrysene
$C_{24}H_{16}$
11644 - Tetraphenylene
$C_{24}H_{18}$
11193 - 1,1′:2′,1″:2″,1‴-Quaterphenyl
11194 - 1,1′:4′,1″:4″,1‴-Quaterphenyl
11598 - 1,1′:3′,1″-Terphenyl, 5′-phenyl-
$C_{24}H_{20}N_2$
3035 - [1,1′-Biphenyl]-4,4′-diamine, *N,N′*-diphenyl-
$C_{24}H_{20}N_4O$
8033 - 2-Naphthalenol, 1-[[2-methyl-4-[(2-methylphenyl)azo]phenyl]azo]-

$C_{24}H_{20}O_4Si$
11509 - Silicic acid (H4SiO4), tetraphenyl ester
$C_{24}H_{20}O_6$
10196 - 1,2,3-Propanetriol, tribenzoate
$C_{24}H_{20}Pb$
9733 - Plumbane, tetraphenyl-
$C_{24}H_{20}Si$
11452 - Silane, tetraphenyl-
$C_{24}H_{20}Sn$
11545 - Stannane, tetraphenyl-
$C_{24}H_{27}O_4P$
9335 - Phenol, 2,4-dimethyl-, phosphate (3:1)
9336 - Phenol, 2,5-dimethyl-, phosphate (3:1)
9337 - Phenol, 2,6-dimethyl-, phosphate (3:1)
9338 - Phenol, 3,4-dimethyl-, phosphate (3:1)
$C_{24}H_{28}O_4$
9301 - Phenol, 4,4′-(1,2-diethyl-1,2-ethenediyl)bis-, dipropanoate, (*E*)-
$C_{24}H_{29}NO_4$
7377 - Isoquinoline, 1-[(3,4-diethoxyphenyl)methyl]-6,7-diethoxy-
$C_{24}H_{30}O_3Si_3$
5004 - Cyclotrisiloxane, 2,4,6-triethyl-2,4,6-triphenyl-
$C_{24}H_{34}O_5$
4101 - Cholan-24-oic acid, 3,7,12-trioxo-, (5β)-
$C_{24}H_{38}O_4$
1334 - 1,2-Benzenedicarboxylic acid, bis(2-ethylhexyl) ester
1344 - 1,2-Benzenedicarboxylic acid, dioctyl ester
$C_{24}H_{39}NO$
8290 - 9-Octadecenamide, *N*-phenyl-, (*Z*)-
$C_{24}H_{40}N_2$
4143 - Con-5-enin-3-amine, *N,N*-dimethyl-, (3β)-
$C_{24}H_{40}N_8O_4$
5711 - Ethanol, 2,2′,2″,2‴-[(4,8-di-1-piperidinylpyrimido[5,4-d]pyrimidine-
$C_{24}H_{40}O_2$
1257 - Benzenedecanoic acid, ι-octyl-
2063 - Benzenenonanoic acid, θ-nonyl-
4094 - Cholan-24-oic acid
8271 - Octadecanoic acid, phenyl ester
$C_{24}H_{40}O_3$
4099 - Cholan-24-oic acid, 3-hydroxy-, (3α,5β)-
$C_{24}H_{40}O_4$
4095 - Cholan-24-oic acid, 3,12-dihydroxy-, (3α,5β,12α)-
4096 - Cholan-24-oic acid, 3,6-dihydroxy-, (3α,5β,6α)-
4097 - Cholan-24-oic acid, 3,7-dihydroxy-, (3α,5β,7α)-
4098 - Cholan-24-oic acid, 3,7-dihydroxy-, (3α,5β,7β)-
$C_{24}H_{40}O_5$
4100 - Cholan-24-oic acid, 3,7,12-trihydroxy-, (3α,5β,7α,12α)-
$C_{24}H_{41}NO_2$
8232 - Octadecanamide, *N*-(4-hydroxyphenyl)-
$C_{24}H_{42}$
2066 - Benzene, octadecyl-
4093 - Cholane
$C_{24}H_{46}O_2$
8253 - Octadecanoic acid, cyclohexyl ester
11608 - 15-Tetracosenoic acid
$C_{24}H_{46}O_3$
5306 - Dodecanoic acid, anhydride
$C_{24}H_{46}O_4$
5353 - Eicosanedioic acid, diethyl ester
9155 - Peroxide, bis(1-oxododecyl)
$C_{24}H_{48}$
4459 - Cyclohexane, octadecyl-
$C_{24}H_{48}O_2$
5277 - Docosanoic acid, ethyl ester
11607 - Tetracosanoic acid
$C_{24}H_{50}$
11603 - Tetracosane
11856 - Tricosane, 2-methyl-
$C_{24}H_{52}O_4Si$
11507 - Silicic acid (H4SiO4), tetrakis(2-ethylbutyl) ester
$C_{24}H_{54}OSn_2$
5241 - Distannoxane, hexabutyl-
$C_{24}H_{72}O_{10}Si_{11}$
11964 - Undecasiloxane, tetracosamethyl-
$C_{25}H_{20}$
1831 - Benzene, 1,1′,1″,1‴-methanetetrayltetrakis-
$C_{25}H_{20}N_2O$
12031 - Urea, tetraphenyl-
$C_{25}H_{22}ClNO_3$
5956 - Fenvalerate
$C_{25}H_{22}OSi$
11441 - Silane, (2-methylphenoxy)triphenyl-
$C_{25}H_{24}F_6N_4$
7136 - Hydramethylnon
$C_{25}H_{27}ClN_2$
9605 - Piperazine, 1-[(4-chlorophenyl)phenylmethyl]-4-[(3-methylphenyl)methyl]-
$C_{25}H_{28}O_3$
5403 - Estra-1,3,5(10)-triene-3,17-diol (17β)-, 3-benzoate
$C_{25}H_{30}ClN_3$
7453 - Methanaminium, *N*-[4-[bis[4-(dimethylamino)phenyl]methylene]-2,5-cyclohexadien-1-ylidene]-*N*-methyl-, chloride
$C_{25}H_{30}O_4$
5108 - 6,6′-Diapo-ψ,ψ-carotenedioic acid, monomethyl ester, 9-*cis*-
$C_{25}H_{40}O_2$
8309 - 9-Octadecenoic acid (*Z*)-, phenylmethyl ester
$C_{25}H_{41}NO_9$
305 - Aconitane-3,8,13,14,15-pentol, 20-ethyl-1,6,16-trimethoxy-4-(methoxymethyl)-,
$C_{25}H_{44}$
2062 - Benzene, nonadecyl-
2070 - Benzene, (3-octylundecyl)-
8245 - Octadecane, 9-*p*-tolyl-
$C_{25}H_{48}O_4$
8132 - Nonanedioic acid, bis(2-ethylhexyl) ester
$C_{25}H_{50}$
4456 - Cyclohexane, nonadecyl-
5351 - Eicosane, 1-cyclopentyl-
6323 - 8-Heptadecene, 9-octyl-
$C_{25}H_{52}$
6313 - Heptadecane, 9-octyl-
8637 - Pentacosane
$C_{26}H_{16}$
6621 - Hexacene
$C_{26}H_{20}$
1623 - Benzene, 1,1′,1″,1‴-(1,2-ethenediylidene)tetrakis-
$C_{26}H_{22}$
1591 - Benzene, 1,1′,1″,1‴-(1,2-ethanediylidene)tetrakis-
1618 - Benzene, 1,1′,1″,1‴-(1-ethanyl-2-ylidyne)tetrakis-
$C_{26}H_{22}ClF_3N_2O_3$
5983 - Fluvalinate
$C_{26}H_{22}O_2$
5580 - 1,2-Ethanediol, 1,1,2,2-tetraphenyl-
$C_{26}H_{23}F_2NO_4$
5965 - Flucythrinate
$C_{26}H_{36}O_6$
3141 - Bufa-20,22-dienolide, 16-(acetyloxy)-3,14-dihydroxy-, (3β,5β,16β)-
$C_{26}H_{38}$
1567 - Benzene, 1,1′-dodecylidenebis[4-methyl-
2267 - Benzene, 1,1′-tetradecylidenebis-
$C_{26}H_{38}O_4$
4281 - 2,4-Cyclohexadien-1-one, 3,5-dihydroxy-2,6,6-tris(3-methyl-2-butenyl)-4-(3-methyl-1-oxobutyl)-
$C_{26}H_{43}NO_6$
6273 - Glycine, *N*-[(3α,5β,7α,12α)-3,7,12-trihydroxy-24-oxocholan-24-yl]-
$C_{26}H_{45}NO_7S$
5625 - Ethanesulfonic acid, 2-[[(3α,5β,7α,12α)-3,7,12-trihydroxy-24-oxocholan-24-yl]amino]-
$C_{26}H_{46}$
1119 - Benzene, (1-butylhexadecyl)-
1568 - Benzene, eicosyl-
1695 - Benzene, (1-ethyloctadecyl)-
2007 - Benzene, (1-methylnonadecyl)-
2068 - Benzene, (1-octyldodecyl)-
2170 - Benzene, (1-propylheptadecyl)-
$C_{26}H_{50}$
4476 - Cyclohexane, 1,1′-tetradecylidenebis-
$C_{26}H_{50}O_4$
5043 - Decanedioic acid, bis(2-ethylhexyl) ester
5311 - Dodecanoic acid, 1,2-ethanediyl ester
$C_{26}H_{52}$
4384 - Cyclohexane, eicosyl-
4819 - Cyclopentane, heneicosyl-
$C_{26}H_{52}O_2$
6624 - Hexacosanoic acid
$C_{26}H_{52}O_6$
8259 - Octadecanoic acid, 2-[2-[2-(2-hydroxyethoxy)ethoxy]ethoxy]ethyl ester
$C_{26}H_{54}$
5269 - Docosane, 5-butyl-
5270 - Docosane, 7-butyl-
5271 - Docosane, 9-butyl-
5272 - Docosane, 11-butyl-
6296 - Heneicosane, 11-(2,2-dimethylpropyl)-
6622 - Hexacosane
6637 - Hexadecane, 6,11-dipentyl-
8241 - Octadecane, 3-ethyl-5-(2-ethylbutyl)-
11605 - Tetracosane, 3-ethyl-
$C_{26}H_{54}O$
6626 - 1-Hexacosanol
$C_{27}H_{18}O_3$
7587 - Methanone, 1,3,5-benzenetriyltris[phenyl-
$C_{27}H_{20}O_{12}$
2773 - 1H-2-Benzopyran-3-carboxylic acid, 6-[2-(4,9-dihydro-8-hydroxy-5,7-dimethoxy-4,9-dioxonaphtho[2,3-b]furan-2-yl
$C_{27}H_{28}Br_2O_5S$
9365 - Phenol, 4,4′-(3H-2,1-benzoxathiol-3-ylidene)bis[2-bromo-3-methyl-6-(1-methylethyl)-, *S,S*-dioxide
$C_{27}H_{30}O_5S$
9369 - Phenol, 4,4′-(3H-2,1-benzoxathiol-3-ylidene)bis[5-methyl-2-(1-methylethyl)-, *S,S*-dioxide
$C_{27}H_{32}N_2O_6$
4131 - Cinchonan-9-ol, 6′-methoxy-, (8α,9*R*)-, mono(2-hydroxybenzoate) (salt)
$C_{27}H_{32}O_{14}$
2795 - 4H-1-Benzopyran-4-one, 7-[[2-*O*-(6-deoxy-α-*L*-mannopyranosyl)-β-*D*-glucopyranosyl]oxy]-2,3-dihydro-5-hy
$C_{27}H_{34}N_2O_4S$
5444 - Ethanaminium, *N*-[4-[[4-(diethylamino)phenyl]phenylmethylene]-2,5-cyclohexadien-1-ylidene]-*N*-ethyl-, sulfate (1
$C_{27}H_{34}O_3$
5408 - Estr-4-en-3-one, 17-(1-oxo-3-phenylpropoxy)-, (17β)-
$C_{27}H_{39}NO_2$
12044 - Veratraman-3,23-diol, 14,15,16,17-tetradehydro-, (3β,23β)-
$C_{27}H_{39}NO_3$
12046 - Veratraman-11-one, 17,23-epoxy-3-hydroxy-, (3β,23β)-
$C_{27}H_{40}O_3$
7774 - 2-Naphthalenecarboxylic acid, 3-hydroxy-, hexadecyl ester

$C_{27}H_{42}Cl_2N_2O_6$
6648 - Hexadecanoic acid, 2-[(dichloroacetyl)amino]-3-hydroxy-3-(4-nitrophenyl)propyl ester, [*R*-(*R**,*R**)]-

$C_{27}H_{42}O_3$
11536 - Spirost-5-en-3-ol, (3β,25*R*)-

$C_{27}H_{42}O_4$
11534 - Spirostan-12-one, 3-hydroxy-, (3β,5α,25*R*)-

$C_{27}H_{43}NO_2$
11515 - Solanid-5-ene-3,12-diol, (3β,12α)-
11516 - Solanid-5-ene-3,18-diol, (3β)-
11529 - Spirosol-5-en-3-ol, (3β,22α,25*R*)-

$C_{27}H_{43}NO_8$
4068 - Cevane-3,4,7,14,15,16,20-heptol, 4,9-epoxy-, (3β,4α,7α,15α,16β)-

$C_{27}H_{43}NO_9$
4070 - Cevane-3,4,6,7,14,15,16,20-octol, 4,9-epoxy-, (3β,4α,6α,7α,15α,16β

$C_{27}H_{44}$
4102 - Cholesta-3,5-diene

$C_{27}H_{44}O$
4103 - Cholesta-5,7-dien-3-ol, (3β)-
4104 - Cholesta-8,24-dien-3-ol, (3β,5α)-
11348 - 9,10-Secocholesta-5,7,10(19)-trien-3-ol, (3β,5Z,7E)-

$C_{27}H_{44}O_3$
11531 - Spirostan-3-ol, (3β,5β,25*S*)-
11532 - Spirostan-3-ol, (3β,5α,25*R*)-
11533 - Spirostan-3-ol, (3β,5β,25*R*)-

$C_{27}H_{44}O_4$
11530 - Spirostan-2,3-diol, (2α,3β,5α,25*R*)-

$C_{27}H_{44}O_5$
11535 - Spirostan-2,3,15-triol, (2α,3β,5α,15β,25*R*)-

$C_{27}H_{45}NO_2$
11528 - Spirosolan-3-ol, (3β,5α,22β,25*S*)-

$C_{27}H_{46}$
6298 - 10-Heneicosene, 11-phenyl-

$C_{27}H_{46}O$
4110 - Cholest-5-en-3-ol (3β)-
4111 - Cholest-4-en-3-ol, (3β)-
4112 - Cholest-5-en-3-ol, (3α)-

$C_{27}H_{46}O_2$
2785 - 2H-1-Benzopyran-6-ol, 3,4-dihydro-2,8-dimethyl-2-(4,8,12-trimethyltridecyl)-, [2*R*-[2*R**(4*R**,8*R**)]]-

$C_{27}H_{48}$
1741 - Benzene, heneicosyl-
4105 - Cholestane, (5α)-
4106 - Cholestane, (5β)-
6297 - Heneicosane, 11-phenyl-

$C_{27}H_{48}O$
4107 - Cholestanol
4108 - Cholestan-3-ol, (3α,5α)-
4109 - Cholestan-3-ol, (3β,5β)-

$C_{27}H_{50}O_6$
8440 - Octanoic acid, 1,2,3-propanetriyl ester

$C_{27}H_{54}$
4808 - Cyclopentane, docosyl-

$C_{27}H_{54}O$
6303 - 14-Heptacosanone

$C_{27}H_{54}O_2$
6625 - Hexacosanoic acid, methyl ester

$C_{27}H_{56}$
6302 - Heptacosane

$C_{28}H_{14}N_2O_4$
434 - 5,9,14,18-Anthrazinetetrone, 6,15-dihydro-

$C_{28}H_{20}O$
6175 - Furan, tetraphenyl-

$C_{28}H_{22}$
367 - Anthracene, 9,10-bis(phenylmethyl)-

$C_{28}H_{22}O_3$
937 - Benzeneacetic acid, α-phenyl-, anhydride

$C_{28}H_{24}N_2O_7$
8549 - Orcein

$C_{28}H_{30}O_4$
7352 - 1(3H)-Isobenzofuranone, 3,3-bis[4-hydroxy-2-methyl-5-(1-methylethyl)phenyl]-

$C_{28}H_{32}ClN_2O_3$
12056 - Xanthylium, 9-(2-carboxyphenyl)-3,6-bis(diethylamino)-, chloride

$C_{28}H_{32}O_4Si_4$
5000 - Cyclotetrasiloxane, 2,4,6,8-tetramethyl-2,4,6,8-tetraphenyl-

$C_{28}H_{34}O_{15}$
2794 - 4H-1-Benzopyran-4-one, 7-[[6-*O*-(6-deoxy-α-*L*-mannopyranosyl)-β-*D*-glucopyranosyl]oxy]-2,3-dih

$C_{28}H_{38}N_2O_4$
5367 - Emetan-6′-ol, 7′,10,11-trimethoxy-

$C_{28}H_{38}O_{19}$
6229 - α-*D*-Glucopyranoside, 1,3,4,6-tetra-*O*-acetyl-β-*D*-fructofuranosyl, tetraacetate

$C_{28}H_{42}O$
5380 - Ergosta-5,7,9(11),22-tetraen-3-ol, (3β,22E)-

$C_{28}H_{44}O$
5381 - Ergosta-5,7,22-trien-3-ol, (3β,10α,22E)-
5382 - Ergosta-5,7,22-trien-3-ol, (3β,9β,10α,22E)-
5383 - Ergosta-5,7,22-trien-3-ol, (3β,22E)-
11349 - 9,10-Secoergosta-5(10),6,8,22-tetraen-3-ol, (3β,6E,22E)-
11350 - 9,10-Secoergosta-5,7,10(19),22-tetraen-3-ol, (3β,5Z,7E,22E)-

$C_{28}H_{48}O$
357 - Androst-4-en-3-one, 17-hydroxy-7,17-dimethyl-, (7β,17β)-
5384 - Ergost-7-en-3-ol, (3β,5α)-
5385 - Ergost-8(14)-en-3-ol, (3β,5α)-
5386 - Ergost-5-en-3-ol, (3β,24*R*)-

$C_{28}H_{48}O_2$
2789 - 2H-1-Benzopyran-6-ol, 3,4-dihydro-2,5,8-trimethyl-2-(4,8,12-trimethyltridecyl)-
2790 - 2H-1-Benzopyran-6-ol, 3,4-dihydro-2,7,8-trimethyl-2-(4,8,12-trimethyltridecyl)-

$C_{28}H_{50}$
1565 - Benzene, docosyl-
5377 - Ergostane, (5α)-
5378 - Ergostane, (5β)-

$C_{28}H_{50}N_2O_4$
4064 - Carpaine

$C_{28}H_{50}O$
5379 - Ergostan-3-ol, (3β,5α)-

$C_{28}H_{54}O_3$
11623 - Tetradecanoic acid, anhydride

$C_{28}H_{56}$
4380 - Cyclohexane, docosyl-

$C_{28}H_{56}O_2$
8221 - Octacosanoic acid

$C_{28}H_{58}$
5274 - Docosane, 7-hexyl-
5354 - Eicosane, 9-octyl-
8220 - Octacosane
11877 - Tridecane, 2,2,4,10,12,12-hexamethyl-7-(3,5,5-trimethylhexyl)-

$C_{28}H_{58}O$
8222 - 1-Octacosanol

$C_{29}H_{40}N_2O_4$
5368 - Emetan, 6′,7′,10,11-tetramethoxy-

$C_{29}H_{44}O_{12}$
4040 - Card-20(22)-enolide, 3-[(6-deoxy-α-*L*-mannopyranosyl)oxy]-1,5,11,14,19-pentahydroxy-, (1β,3β

$C_{29}H_{47}NO_8$
4067 - Cevane-3,4,12,14,16,17,20-heptol, 3-acetate, (3β,4α,16β)-

$C_{29}H_{48}O$
11554 - Stigmasta-5,7-dien-3-ol, (3β)-
11555 - Stigmasta-5,22-dien-3-ol, (3β,22E)-

$C_{29}H_{50}O$
11558 - Stigmast-5-en-3-ol, (3β)-
11559 - Stigmast-5-en-3-ol, (3β,24*S*)-

$C_{29}H_{50}O_2$
2787 - 2H-1-Benzopyran-6-ol, 3,4-dihydro-2,5,7,8-tetramethyl-2-(4,8,12-trimethyltridecyl)-, [2*R*-[2*R**(4*R**,8*R**)]]-

$C_{29}H_{52}$
2341 - Benzene, tricosyl-

$C_{29}H_{52}O$
11557 - Stigmastan-3-ol, (3β,5α)-

$C_{29}H_{58}$
4483 - Cyclohexane, tricosyl-

$C_{29}H_{60}$
8099 - Nonacosane

$C_{29}H_{60}O$
8100 - 1-Nonacosanol

$C_{30}H_{34}O_{13}$
7579 - 3,6-Methano-8H-1,5,7-trioxacyclopenta[ij]cycloprop[a]azulene-4,8(3H)-dione, hexahydro-2a-hydroxy

$C_{30}H_{37}N_5O_5$
5393 - Ergotaman-3′,6′,18-trione, 12′-hydroxy-2′-methyl-5′-(2-methylpropyl)-,

$C_{30}H_{40}O_2$
435 - 8′-Apo-β,ψ-carotenal, 3-hydroxy-, (3*R*)-

$C_{30}H_{42}FeO_2Si_4$
5961 - Ferrocene, 1,1′-bis(1,1,3,3-tetramethyl-3-phenyldisiloxanyl)-

$C_{30}H_{46}O_5$
8546 - Olean-12-en-28-oic acid, 3,16-dihydroxy-23-oxo-, (3β,4α,16α)-
12036 - Urs-12-ene-27,28-dioic acid, 3-hydroxy-, (3β)-

$C_{30}H_{48}O_3$
8547 - Olean-12-en-28-oic acid, 3-hydroxy-, (3β)-
12037 - Urs-12-en-28-oic acid, 3-hydroxy-, (3β)-

$C_{30}H_{48}O_4$
8545 - Olean-12-en-28-oic acid, 3,23-dihydroxy-, (3β,4α)-

$C_{30}H_{50}$
11602 - 2,6,10,14,18,22-Tetracosahexaene, 2,6,10,15,19,23-hexamethyl-, (all-E)-

$C_{30}H_{50}O$
7411 - Lanosta-8,24-dien-3-ol, (3β)-
7417 - Lup-20(29)-en-3-ol, (3β)-
8548 - Olean-12-en-3-ol, (3β)-
11556 - Stigmasta-7,24(28)-dien-3-ol, 4-methyl-, (3β,4α,5α,24Z)-
12038 - Urs-12-en-3-ol, (3β)-

$C_{30}H_{50}O_2$
7416 - Lup-20(29)-ene-3,28-diol, (3β)-

$C_{30}H_{54}$
2265 - Benzene, tetracosyl-

$C_{30}H_{60}$
4474 - Cyclohexane, tetracosyl-

$C_{30}H_{60}I_3N_3O_3$
5441 - Ethanaminium, 2,2′,2″-[1,2,3-benzenetriyltris(oxy)]tris[*N*,*N*,*N*-triethyl-, tr

$C_{30}H_{60}O_2$
11820 - Triacontanoic acid

$C_{30}H_{62}$
5275 - Docosane, 9-octyl-
11606 - Tetracosane, 2,6,10,15,19,23-hexamethyl-
11819 - Triacontane

$C_{30}H_{62}O$
11821 - 1-Triacontanol

$C_{31}H_{36}N_2O_8$
12064 - Yohimban-16-carboxylic acid, 17-hydroxy-18-[(3,4,5-trimethoxybenzoyl)oxy]-, methyl ester, (3β,16β

$C_{31}H_{39}N_5O_5$
5387 - Ergotaman-3′,6′,18-trione, 12′-hydroxy-2′,5′-bis(1-methylethyl)-,
5388 - Ergotaman-3′,6′,18-trione, 12′-hydroxy-2′,5′-bis(1-methylethyl)-,

$C_{31}H_{46}O_2$
7865 - 1,4-Naphthalenedione, 2-methyl-3-(3,7,11,15-tetramethyl-2-hexadecenyl)-, [*R*-[*R**,*R**-(*E*)]]-

$C_{31}H_{52}O_3$
2788 - 2H-1-Benzopyran-6-ol, 3,4-dihydro-2,5,7,8-tetramethyl-2-(4,8,12-trimethyltridecyl)-, acetate, [2*R*-[2*R**(4*R**,8*R**

$C_{31}H_{56}$
2088 - Benzene, pentacosyl-
8638 - Pentacosane, 13-phenyl-

$C_{31}H_{62}$
4462 - Cyclohexane, pentacosyl-
4821 - Cyclopentane, hexacosyl-

$C_{31}H_{62}O$
6300 - 16-Hentriacontanone

$C_{31}H_{64}$
6295 - Heneicosane, 11-decyl-
6299 - Hentriacontane

$C_{32}H_{22}N_6Na_2O_6S_2$
7969 - 1-Naphthalenesulfonic acid, 3,3′-[[1,1′-biphenyl]-4,4′-diylbis(azo)]bis[4-amino-, disodium salt

$C_{32}H_{38}N_2O_8$
12065 - Yohimban-16-carboxylic acid, 17-methoxy-18-[(3,4,5-trimethoxybenzoyl)oxy]-, methyl ester, (3β,16β

$C_{32}H_{40}O_4Si_4$
4998 - Cyclotetrasiloxane, 2,4,6,8-tetraethyl-2,4,6,8-tetraphenyl-

$C_{32}H_{41}N_5O_5$
5389 - Ergotaman-3′,6′,18-trione, 12′-hydroxy-2′-(1-methylethyl)-5′-(2-methylpropyl)-

$C_{32}H_{44}N_2O_8$
309 - Aconitane-4,8,9-triol, 20-ethyl-1,14,16-trimethoxy-, 4-[2-(acetylamino)benzoate], (1α,14α,16β)-

$C_{32}H_{44}O_{12}$
3142 - Bufa-4,20,22-trienolide, 6-(acetyloxy)-3-(β-*D*-glucopyranosyloxy)-8,14-dihydroxy-, (3β,6β)-

$C_{32}H_{49}NO_9$
4069 - Cevane-3,4,12,14,16,17,20-heptol, 4,9-epoxy-, 3-(2-methyl-2-butenoate), [3β(*Z*),4α,16β]-

$C_{32}H_{58}$
1750 - Benzene, hexacosyl-

$C_{32}H_{62}O_3$
6646 - Hexadecanoic acid, anhydride

$C_{32}H_{64}$
4410 - Cyclohexane, hexacosyl-

$C_{32}H_{64}O_2$
6651 - Hexadecanoic acid, hexadecyl ester

$C_{32}H_{66}$
5273 - Docosane, decyl-
5348 - Dotriacontane

$C_{32}H_{66}O$
6642 - Hexadecane, 1,1′-oxybis-

$C_{33}H_{34}N_4O_6$
3001 - 21H-Biline-8,12-dipropanoic acid, 3,18-diethenyl-1,19,22,24-tetrahydro-2,7,13,17-tetramethyl-1,19-dioxo-

$C_{33}H_{35}N_5O_5$
5395 - Ergotaman-3′,6′,18-trione, 12′-hydroxy-2′-methyl-5′-(phenylmethyl)-,

$C_{33}H_{36}N_4O_6$
3000 - 21H-Biline-8,12-dipropanoic acid, 2,17-diethenyl-1,10,19,22,23,24-hexahydro-3,7,13,18-tetramethyl-1,19-dioxo-

$C_{33}H_{40}N_2O_9$
12061 - Yohimban-16-carboxylic acid, 11,17-dimethoxy-18-[(3,4,5-trimethoxybenzoyl)oxy]-, methyl ester,

$C_{33}H_{41}N_5O_5$
5390 - Ergotaman-3′,6′,18-trione, 12′-hydroxy-2′-(1-methylethyl)-5′-(2-methylpropyl)-, (5′α,8α)-

$C_{33}H_{44}O_8$
8211 - 29-Nordammara-1,17(20),24-trien-21-oic acid, 6,16-bis(acetyloxy)-3,7-dioxo-,

$C_{33}H_{45}NO_9$
311 - Aconitane-8,13,14-triol, 1,6,16-trimethoxy-4-(methoxymethyl)-20-methyl-, 8-acetate 14-benzoate,

$C_{33}H_{49}NO_8$
12045 - Veratraman-11-one, 17,23-epoxy-3-(β-*D*-glucopyranosyloxy)-, (3β,23β)-

$C_{33}H_{55}N_5O_5$
5394 - Ergotaman-3′,6′,18-trione, 12′-hydroxy-2′-methyl-5′-(phenylmethyl)-, (5′α)-

$C_{33}H_{60}$
1742 - Benzene, heptacosyl-

$C_{34}H_{24}N_6Na_4O_{14}S_4$
7877 - 1,3-Naphthalenedisulfonic acid, 6,6′-[(3,3′-dimethyl[1,1′-biphenyl]-4,4′-diyl)bis(azo)]bis[4-amino-5-hydroxy-,

$C_{34}H_{32}ClFeN_4O_4$
5958 - Ferrate(2-), chloro[7,12-diethenyl-3,8,13,17-tetramethyl-21H,23H-porphine-2,18-dipropanoato(4-)-*N*21,*N*22,*N*23,

$C_{34}H_{33}FeN_4O_5$
5959 - Ferrate(2-), [7,12-diethenyl-3,8,13,17-tetramethyl-21H,23H-porphine-2,18-dipropanoato(4-)-*N*21,*N*22,*N*23,*N*24

$C_{34}H_{36}N_2O_6$
3002 - [2,2′-Bimorphinan]-3,3′,6,6′-tetrol, 7,7′,8,8′-tetradehydro-4,5:4′,5′-diepoxy-17,17′dimethyl-

$C_{34}H_{38}N_4O_6$
9735 - 21H,23H-Porphine-2,18-dipropanoic acid, 7,12-bis(1-hydroxyethyl)-3,8,13,17-tetramethyl-

$C_{34}H_{47}NO_{10}$
307 - Aconitane-3,8,13,14-tetrol, 20-ethyl-1,6,16-trimethoxy-4-(methoxymethyl)-, 8-acetate 14-benzoate, (1α,3alp

$C_{34}H_{47}NO_{11}$
306 - Aconitane-3,8,13,14,15-pentol, 20-ethyl-1,6,16-trimethoxy-4-(methoxymethyl)-, 8-acetate 14-benzoate,

$C_{34}H_{62}$
2065 - Benzene, octacosyl-

$C_{34}H_{66}O_4$
6649 - Hexadecanoic acid, 1,2-ethanediyl ester

$C_{34}H_{68}$
4458 - Cyclohexane, octacosyl-

$C_{34}H_{68}O_2$
8258 - Octadecanoic acid, hexadecyl ester

$C_{34}H_{70}$
11604 - Tetracosane, 11-decyl-
11650 - Tetratriacontane

$C_{35}H_{39}N_5O_5$
5391 - Ergotaman-3′,6′,18-trione, 12′-hydroxy-2′-(1-methylethyl)-5′-(phenylmethyl)-
5392 - Ergotaman-3′,6′,18-trione, 12′-hydroxy-2′-(1-methylethyl)-5′-(phenylmethyl)-

$C_{35}H_{42}N_2O_9$
12060 - Yohimban-16-carboxylic acid, 11,17-dimethoxy-18-[[1-oxo-3-(3,4,5-trimethoxyphenyl)-2-propenyl]oxy]-, methyl ester, (3β,16β,17α,18β,20α)-

$C_{35}H_{54}O_{14}$
4045 - Card-20(22)-enolide, 3-[(6-*O*-β-*D*-glucopyranosyl-β-D-glucopyranosyl)oxy]-14-hydroxy-,

$C_{35}H_{70}$
4853 - Cyclopentane, triacontyl-

$C_{35}H_{70}O$
9000 - 18-Pentatriacontanone

$C_{35}H_{72}$
8999 - Pentatriacontane

$C_{36}H_{38}N_2O_6$
11924 - Tubocuraran-7′,12′-diol, 6,6′-dimethoxy-2,2′-dimethyl-, (1′α)-
11925 - Tubocuraran-7′,12′-diol, 6,6′-dimethoxy-2,2′-dimethyl-, (1β)-

$C_{36}H_{42}Br_2N_2$
6753 - 1,6-Hexanediaminium, *N,N*′-di-9H-fluoren-9-yl-*N,N,N*′,*N*′-tetramethyl-dibromide

$C_{36}H_{51}NO_{11}$
310 - Aconitane-8,13,14-triol, 20-ethyl-1,6,16-trimethoxy-4-(methoxymethyl)-, 8-acetate 14-(3,4-dimethoxybenzoate),

$C_{36}H_{51}NO_{12}$
308 - Aconitane-3,8,13,14-tetrol, 20-ethyl-1,6,16-trimethoxy-4-(methoxymethyl)-, 8-acetate 14-(3,4-dimethoxybenzoate),(1α,3α,6α,14α,16β)-

$C_{36}H_{70}CaO_4$
8252 - Octadecanoic acid, calcium salt

$C_{36}H_{70}O_3$
8249 - Octadecanoic acid, anhydride

$C_{36}H_{70}O_4Pb$
8261 - Octadecanoic acid, lead salt

$C_{36}H_{70}O_4Zn$
8277 - Octadecanoic acid, zinc salt

$C_{36}H_{72}$
4482 - Cyclohexane, triacontyl-

$C_{36}H_{74}$
6991 - Hexatriacontane
8639 - Pentacosane, 13-undecyl-

$C_{37}H_{38}N_2O_6$
8630 - Oxyacanthan, 6′,12′-dimethoxy-2,2′-dimethyl-6,7-[methylenebis(oxy)]-

$C_{37}H_{40}N_2O_6$
2911 - Berbaman-12-ol, 6,6′,7-trimethoxy-2,2′-dimethyl-
8631 - Oxyacanthan-12′-ol, 6,6′,7-trimethoxy-2,2′-dimethyl-

$C_{37}H_{67}NO_{13}$
5396 - Erythromycin

$C_{38}H_{74}O_4$
8256 - Octadecanoic acid, 1,2-ethanediyl ester

$C_{38}H_{76}$
4383 - Cyclohexane, dotriacontyl-

$C_{38}H_{78}$
6623 - Hexacosane, 13-dodecyl-

$C_{39}H_{74}O_6$
5319 - Dodecanoic acid, 1,2,3-propanetriyl ester

$C_{40}H_{36}$
4054 - β,ε-Carotene, (6′*R*)-

$C_{40}H_{50}N_4O_8S$
4132 - Cinchonan-9-ol, 6′-methoxy-, (8α,9*R*)-, sulfate (2:1) (salt)

$C_{40}H_{52}O_2$
4059 - β,ψ-Caroten-16′-oic acid, 3′,4′-didehydro-

$C_{40}H_{56}$
4051 - β,β-Carotene
4052 - β,ψ-Carotene
4053 - ψ,ψ-Carotene

$C_{40}H_{56}O$
4060 - ψ,ψ-Caroten-16-ol
4061 - β,β-Caroten-3-ol, (3*R*)-
4062 - β,ψ-Caroten-3-ol, (3*R*)-

$C_{40}H_{56}O_2$
4057 - β,β-Carotene-3,3′-diol, (3*R*,3′*R*)-
4058 - β,ε-Carotene-3,3′-diol, (3*R*,3′*R*,6′*R*)-

$C_{40}H_{56}O_3$
4063 - β,κ-Caroten-6′-one, 3,3′-dihydroxy-, (3*R*,3′*S*,5′*R*)-

$C_{40}H_{56}O_4$
4056 - β,β-Carotene-3,3′-diol, 5,6:5′,6′-diepoxy-5,5′,6,6′-tetrahydro-

$C_{40}H_{60}O_6$
4055 - β,β-Carotene, 3′-(acetyloxy)-6′,7′-didehydro-5,6-epoxy-5,5′,6,6′,7,8-hexahydro-3,5′-dihydroxy-8-oxo-,

$C_{40}H_{80}$
4479 - Cyclohexane, tetratriacontyl-
4845 - Cyclopentane, pentatriacontyl

$C_{41}H_{46}N_4O_5$
4128 - Cinchonan-9-ol, 6′-methoxy-, carbonate (2:1) (este*R*), (8α,9*R*)-

$C_{41}H_{64}O_{13}$
4043 - Card-20(22)-enolide, 3-[(*O*-2,6-dideoxy-β-*D*-ribo-hexopyranosyl-(1-4)-*O*-2,6-dideoxy-β-*D*-ribo-hexopyranosyl-(1-4)-2,6-dideoxy-β-*D*, 14-dihydroxy

$C_{41}H_{64}O_{14}$
4041 - Card-20(22)-enolide, 3-[(*O*-2,6-dideoxy-β-*D*-ribo-hexopyranosyl-(1-4)-*O*-2,6-dideoxy-β-*D*-ribo-hexopyranosyl-(1-4)-2,6-dideoxy-β-*D*, 12,14-dihydroxy
4042 - Card-20(22)-enolide, 3-[(*O*-2,6-dideoxy-β-*D*-ribo-hexopyranosyl-(1-4)-*O*-2,6-dideoxy-β-*D*-ribo-hexopyranosyl-(1-4)-2,6-dideoxy-β-*D*, 14,16-dihydroxy

$C_{42}H_{46}N_4O_8S$
11566 - Strychnidin-10-one, sulfate (2:1)

$C_{42}H_{62}O_{16}$
6230 - α-*D*-Glucopyranosiduronic acid, (3β,20β)-20-carboxy-11-oxo-30-norolean-12-en-3-yl 2-*O*-

$C_{42}H_{84}$
4413 - Cyclohexane, hexatriacontyl-

$C_{42}H_{84}N_{14}O_{36}S_3$
11561 - *D*-Streptamine, *O*-2-deoxy-2-(methylamino)-α-*L*-glucopyranosyl-(12)-*O*-5-deoxy-3-C-formyl-α-L-lyxofuranosyl-(1-4)-N,N,N′ -55-bis(aminoiminomethyl)-, sulfate (2:3)(salt)

$C_{43}H_{88}$
11918 - Tritetracontane

$C_{45}H_{86}O_6$
11628 - Tetradecanoic acid, 1,2,3-propanetriyl ester

$C_{48}H_{40}O_4Si_4$
4994 - Cyclotetrasiloxane, octaphenyl-

$C_{48}H_{72}O_{14}$
477 - Avermectin B1a

$C_{48}H_{93}AlO_6$
6645 - Hexadecanoic acid, aluminum salt

$C_{50}H_{83}NO_{21}$
6196 - β-*D*-Galactopyranoside, (3β,5α,22β,25*S*)-spirosolan-3-yl *O*-β-

$C_{51}H_{98}O_6$
6656 - Hexadecanoic acid, 1,2,3-propanetriyl ester

$C_{54}H_{105}AlO_6$
8248 - Octadecanoic acid, aluminum salt

$C_{56}H_{92}O_{29}$
6195 - β-*D*-Galactopyranoside, (2α,3β,5α,15β,25*R*)-2,15-dihydroxyspirostan-3-yl

$C_{57}H_{104}O_6$
8310 - 9-Octadecenoic acid (*Z*)-, 1,2,3-propanetriyl ester

$C_{57}H_{110}O_6$
8272 - Octadecanoic acid, 1,2,3-propanetriyl ester

$C_{60}H_{78}OSn_2$
5950 - Fenbutatin oxide

$C_{66}H_{103}N_{17}O_{16}S$
508 - Bacitracin

CAS NUMBER INDEX OF ORGANIC COMPOUNDS

50-00-0 : 5987
50-01-1 : 6285
50-02-2 : 9737
50-03-3 : 9747
50-06-6 : 11095
50-10-2 : 5443
50-11-3 : 11088
50-14-6 : 11350
50-22-6 : 9748
50-23-7 : 9753
50-24-8 : 9738
50-27-1 : 5404
50-28-2 : 5399
50-29-3 : 2328
50-30-6 : 2563
50-31-7 : 2727
50-32-8 : 2395
50-34-0 : 9832
50-35-1 : 7363
50-36-2 : 482
50-42-0 : 938
50-44-2 : 10763
50-50-0 : 5403
50-53-3 : 9505
50-55-5 : 12061
50-69-1 : 11345
50-70-4 : 6206
50-71-5 : 11073
50-78-2 : 2480
50-79-3 : 2562
50-81-7 : 466
50-82-8 : 2728
50-84-0 : 2561
50-85-1 : 2630
50-89-5 : 11816
50-91-9 : 12034
51-03-6 : 9727
51-12-7 : 10910
51-17-2 : 2376
51-18-3 : 11844
51-20-7 : 11062
51-21-8 : 11067
51-26-3 : 2136
51-28-5 : 9342
51-34-3 : 908
51-35-4 : 9767
51-36-5 : 2564
51-41-2 : 1500
51-43-4 : 1521
51-45-6 : 7203
51-48-9 : 11930
51-52-5 : 11109
51-55-8 : 902
51-63-8 : 1582
51-66-1 : 71
51-67-2 : 9208
51-75-2 : 5413
51-79-6 : 3948
51-80-9 : 7490
51-83-2 : 5440
52-28-8 : 7673
52-31-3 : 11084
52-39-1 : 9746
52-43-7 : 11091
52-49-3 : 9705
52-51-7 : 9991
52-52-8 : 4782
52-67-5 : 12043
52-68-6 : 11853
52-85-7 : 5946
52-86-8 : 3651
52-90-4 : 5014
53-03-2 : 9739
53-05-4 : 9744
53-06-5 : 9754
53-16-7 : 5406
53-41-8 : 348
53-46-3 : 5445
53-70-3 : 5142
53-89-4 : 10850
53-96-3 : 64
54-04-6 : 1586
54-06-8 : 7301
54-11-5 : 10992
54-25-1 : 11832
54-36-4 : 10462
54-85-3 : 10905
54-95-5 : 11654
55-10-7 : 896
55-18-5 : 5432
55-21-0 : 600
55-22-1 : 10895
55-38-9 : 5955
55-43-6 : 1801
55-63-0 : 10197
55-73-2 : 6281
55-91-4 : 9580
56-03-1 : 7233
56-04-2 : 11108
56-12-2 : 3464
56-18-8 : 9874
56-23-5 : 7540
56-25-7 : 5371
56-29-1 : 11083
56-33-7 : 5238
56-35-9 : 5241
56-36-0 : 11538
56-38-2 : 9583
56-40-6 : 6250
56-41-7 : 325
56-45-1 : 11357
56-49-5 : 2392
56-53-1 : 9300
56-54-2 : 4126
56-55-3 : 519
56-57-5 : 11287
56-72-4 : 4147
56-75-7 : 52
56-81-5 : 10186
56-82-6 : 9782
56-84-8 : 471
56-85-9 : 6249
56-86-0 : 6241
56-87-1 : 7419
56-89-3 : 5020
57-00-1 : 6254
57-06-7 : 10575
57-08-9 : 6871
57-10-3 : 6644
57-11-4 : 8247
57-13-6 : 11992
57-14-7 : 7154
57-15-8 : 10407
57-24-9 : 11563
57-27-2 : 7666
57-41-0 : 7219
57-43-2 : 11092
57-44-3 : 11087
57-47-6 : 11191
57-50-1 : 6218
57-53-4 : 10021
57-55-6 : 9982
57-57-8 : 8595
57-62-5 : 7704
57-63-6 : 8216
57-66-9 : 2595
57-67-0 : 2191
57-68-1 : 2196
57-71-6 : 3361
57-74-9 : 4077
57-85-2 : 359
57-87-4 : 5383
57-88-5 : 4110
57-91-0 : 5400
57-92-1 : 11562
57-97-6 : 520
58-00-4 : 5155
58-05-9 : 6244
58-08-2 : 10760
58-14-0 : 11051
58-15-1 : 10851
58-18-4 : 358
58-22-0 : 356
58-25-3 : 2409
58-27-5 : 7864
58-32-2 : 5711
58-38-8 : 9501
58-39-9 : 9617
58-46-8 : 2396
58-54-8 : 168
58-55-9 : 10756
58-56-0 : 10934
58-61-7 : 319
58-63-9 : 7335
58-72-0 : 1648
58-73-1 : 5422
58-74-2 : 7378
58-85-5 : 11689
58-86-6 : 12059
58-89-9 : 4409
58-90-2 : 9470
58-93-5 : 2855
58-94-6 : 2852
58-95-7 : 2788
58-96-8 : 12033
58-97-9 : 12035
59-00-7 : 11230
59-02-9 : 2787
59-23-4 : 6197
59-26-7 : 10891
59-30-3 : 6243
59-31-4 : 11325
59-40-5 : 2205
59-46-1 : 2493
59-47-2 : 10019
59-48-3 : 7327
59-49-4 : 2882
59-50-7 : 9267
59-51-8 : 7647
59-52-9 : 10370
59-66-5 : 32
59-67-6 : 10894
59-87-0 : 7144
59-88-1 : 7172
59-89-2 : 7699
59-92-7 : 11929
59-98-3 : 7198
60-00-4 : 6262
60-01-5 : 3586
60-09-3 : 837
60-10-6 : 5121
60-11-7 : 721
60-12-8 : 1592
60-18-4 : 11926
60-23-1 : 5649
60-24-2 : 5732
60-27-5 : 7230
60-29-7 : 5608
60-31-1 : 5439
60-32-2 : 6873
60-33-3 : 8223
60-34-4 : 7161
60-35-5 : 22
60-41-3 : 11566
60-51-5 : 5009
60-54-8 : 7706
60-56-0 : 7214
60-57-1 : 5178
60-70-8 : 12044
60-80-0 : 10847
60-82-2 : 10453
60-87-7 : 9504
60-91-3 : 9503
61-00-7 : 5807
61-12-1 : 11226
61-19-8 : 320
61-50-7 : 7303
61-54-1 : 7302
61-73-4 : 9506
61-78-9 : 6252
61-80-3 : 2876
61-82-5 : 11845
61-90-5 : 7413
62-23-7 : 2705
62-38-4 : 7431
62-44-2 : 60
62-50-0 : 7531
62-51-1 : 9830
62-53-3 : 633
62-54-4 : 122
62-55-5 : 5643
62-56-6 : 11799
62-57-7 : 331
62-59-9 : 4069
62-73-7 : 5170
62-75-9 : 7450
62-76-0 : 5566
62-90-8 : 5408
63-05-8 : 352
63-25-2 : 8032
63-68-3 : 7648
63-74-1 : 2187
63-75-2 : 10914
63-91-2 : 9510
63-98-9 : 879
64-04-0 : 1569
64-10-8 : 12026
64-17-5 : 5672
64-18-6 : 6002
64-19-7 : 100
64-20-0 : 7456
64-65-3 : 9656
64-67-5 : 11573
64-69-7 : 200
64-77-7 : 2207
64-85-7 : 9751
64-86-8 : 93
65-29-2 : 5441
65-45-2 : 616
65-46-3 : 5021
65-49-6 : 2498
65-64-5 : 7170
65-71-4 : 11068
65-82-7 : 7649
65-85-0 : 2475
65-86-1 : 11050
66-02-4 : 11928
66-22-8 : 11059
66-23-9 : 5438
66-25-1 : 6718
66-27-3 : 7532
66-28-4 : 4049
66-32-0 : 11565
66-56-8 : 9341
66-71-7 : 9189
66-75-1 : 11061
66-76-2 : 2839
66-77-3 : 7751
66-81-9 : 9655
66-99-9 : 7752
67-03-8 : 11686
67-20-9 : 7222
67-43-6 : 6257
67-45-8 : 8589
67-47-0 : 6044
67-48-1 : 5446
67-51-6 : 10828
67-52-7 : 11076
67-56-1 : 7581
67-63-0 : 10335
67-64-1 : 10411
67-66-3 : 7554
67-68-5 : 7529
67-71-0 : 7534
67-72-1 : 5595
67-97-0 : 11348
68-11-1 : 202
68-12-2 : 5992
68-26-8 : 11341
68-35-9 : 2204
68-36-0 : 996
68-41-7 : 7406
68-88-2 : 5700
68-94-0 : 10770
68-96-2 : 9750
69-24-9 : 10435
69-57-8 : 11655
69-65-8 : 7422
69-72-7 : 2612
69-89-6 : 10755
69-91-0 : 885
69-93-2 : 10764
70-07-5 : 8588
70-11-1 : 5782
70-18-8 : 6264
70-22-4 : 11188
70-25-7 : 6284
70-26-8 : 8550
70-29-1 : 5821
70-30-4 : 9409
70-34-8 : 1718
70-47-3 : 468
70-54-2 : 7418
70-69-9 : 10413
70-70-2 : 10450
70-78-0 : 11932
71-00-1 : 7132
71-23-8 : 10334
71-30-7 : 11104
71-36-3 : 3593
71-41-0 : 8915
71-43-2 : 867
71-44-3 : 3259
71-55-6 : 5655
71-63-6 : 4043
71-81-8 : 2120
71-91-0 : 5447
72-14-0 : 2206
72-18-4 : 12041
72-19-5 : 11813
72-23-1 : 9755
72-43-5 : 2329
72-44-6 : 11199
72-48-0 : 391
72-54-8 : 1378
72-55-9 : 1374
72-56-0 : 9160
72-63-9 : 344
73-22-3 : 11922
73-24-5 : 10751
73-31-4 : 69
73-32-5 : 7370
73-40-5 : 10768
73-49-4 : 11197
74-11-3 : 2536
74-31-7 : 1271
74-39-5 : 1542
74-79-3 : 438
74-82-8 : 7459
74-83-9 : 7462
74-84-0 : 5448
74-85-1 : 5879
74-86-2 : 5936
74-87-3 : 7477
74-88-4 : 7518
74-89-5 : 7442
74-90-8 : 7176
74-93-1 : 7546
74-95-3 : 7492
74-96-4 : 5454
74-97-5 : 7463
74-98-6 : 9834
74-99-7 : 10727
75-00-3 : 5467
75-01-4 : 5884
75-02-5 : 5906
75-03-6 : 5597
75-04-7 : 5410
75-05-8 : 271
75-07-0 : 11
75-08-1 : 5648
75-09-2 : 7499
75-10-5 : 7504
75-11-6 : 7506
75-12-7 : 5989
75-15-0 : 3993
75-17-2 : 5988
75-18-3 : 7545
75-19-4 : 4955
75-21-8 : 8599
75-22-9 : 3132
75-24-1 : 339
75-25-2 : 7548
75-26-3 : 9837
75-27-4 : 7467
75-28-5 : 10073
75-29-6 : 9849
75-30-9 : 10055
75-31-0 : 9804
75-33-2 : 10157
75-34-3 : 5528
75-35-4 : 5894
75-36-5 : 294
75-37-6 : 5544
75-38-7 : 5902
75-39-8 : 5674
75-43-4 : 7502
75-44-5 : 4007
75-45-6 : 7478
75-46-7 : 7558
75-47-8 : 7560
75-50-3 : 7444
75-52-5 : 7522
75-54-7 : 11398
75-55-8 : 501

75-56-9 : 8621
75-57-0 : 7457
75-58-1 : 7458
75-60-5 : 451
75-61-6 : 7496
75-62-7 : 7474
75-63-8 : 7475
75-64-9 : 9818
75-65-0 : 10389
75-66-1 : 10159
75-68-3 : 5470
75-69-4 : 7555
75-71-8 : 7500
75-72-9 : 7486
75-73-0 : 7541
75-74-1 : 9732
75-75-2 : 7530
75-76-3 : 11451
75-77-4 : 11389
75-78-5 : 11394
75-79-6 : 11465
75-80-9 : 5767
75-81-0 : 5520
75-82-1 : 5522
75-83-2 : 3289
75-84-3 : 10373
75-85-4 : 3628
75-86-5 : 10108
75-87-6 : 21
75-88-7 : 5489
75-89-8 : 5770
75-91-2 : 7178
75-93-4 : 11577
75-94-5 : 11461
75-95-6 : 5616
75-96-7 : 241
75-97-8 : 3657
75-98-9 : 10258
75-99-0 : 10244
76-00-6 : 10406
76-01-7 : 5617
76-02-8 : 300
76-03-9 : 243
76-04-0 : 127
76-05-1 : 263
76-06-2 : 7557
76-08-4 : 10405
76-09-5 : 3350
76-11-9 : 5636
76-12-0 : 5637
76-13-1 : 5663
76-14-2 : 5538
76-15-3 : 5484
76-16-4 : 5596
76-17-5 : 10181
76-19-7 : 10118
76-20-0 : 3230
76-24-4 : 3106
76-28-8 : 4046
76-30-2 : 3342
76-36-8 : 3191
76-37-9 : 10404
76-39-1 : 10391
76-43-7 : 355
76-44-8 : 7576
76-45-9 : 4070
76-49-3 : 2948
76-57-3 : 7669
76-59-5 : 9365
76-60-8 : 9367
76-61-9 : 9369
76-62-0 : 7350
76-65-3 : 2463
76-67-5 : 9952
76-68-6 : 11085
76-76-6 : 11093
76-80-2 : 3110
76-83-5 : 1196
76-84-6 : 1866
76-87-9 : 11911
76-89-1 : 920
76-93-7 : 918
76-94-8 : 11100
77-02-1 : 11098
77-03-2 : 9654
77-04-3 : 10946
77-06-5 : 6204
77-09-8 : 7354
77-10-1 : 9696
77-16-7 : 11904
77-25-8 : 9940
77-26-9 : 11101
77-28-1 : 11082
77-40-7 : 9433
77-41-8 : 11156
77-42-9 : 9103
77-46-3 : 91
77-47-4 : 4766
77-48-5 : 7217
77-49-6 : 10018
77-52-1 : 12037
77-53-2 : 7570
77-59-8 : 11528
77-60-1 : 11532
77-63-4 : 5000
77-65-6 : 3195
77-67-8 : 11158
77-71-4 : 7218
77-74-7 : 8944
77-75-8 : 9146
77-76-9 : 9909
77-77-0 : 5913
77-78-1 : 11574
77-79-2 : 11752
77-83-8 : 8604
77-84-9 : 10007
77-85-0 : 10011
77-86-1 : 9986
77-89-4 : 10166
77-92-9 : 10167
77-93-0 : 10169
77-94-1 : 10168
77-95-2 : 4338
77-99-6 : 10006
78-00-2 : 9731
78-07-9 : 11478
78-08-0 : 11424
78-09-1 : 5601
78-10-4 : 11506
78-11-5 : 9990
78-13-7 : 11507
78-26-2 : 10020
78-27-3 : 4529
78-30-8 : 9569
78-32-0 : 9571
78-34-2 : 5201
78-38-6 : 9536
78-39-7 : 5664
78-40-0 : 9563
78-48-8 : 11852
78-54-6 : 4528
78-59-1 : 4721
78-62-6 : 11405
78-67-1 : 10093
78-74-0 : 5653
78-75-1 : 9881
78-77-3 : 9845
78-78-4 : 3397
78-79-5 : 3171
78-80-8 : 3892
78-81-9 : 9817
78-82-0 : 10111
78-83-1 : 10388
78-84-2 : 9785
78-85-3 : 10515
78-88-6 : 10552
78-89-7 : 10347
78-90-0 : 9872
78-92-2 : 3594
78-93-3 : 3645
78-94-4 : 3875
78-95-5 : 10420
78-97-7 : 10105
78-98-8 : 9788
78-99-9 : 9891
79-00-5 : 5656
79-01-6 : 5922
79-03-8 : 10493
79-04-9 : 295
79-05-0 : 9791
79-06-1 : 10519
79-07-2 : 41
79-08-3 : 111
79-09-4 : 10201
79-10-7 : 10609
79-11-8 : 123
79-14-1 : 196
79-15-2 : 36
79-16-3 : 72
79-17-4 : 7147
79-19-6 : 7142
79-20-9 : 208
79-21-0 : 5619
79-22-1 : 4020
79-24-3 : 5607
79-27-6 : 5632
79-28-7 : 5914
79-29-8 : 3290
79-30-1 : 10504
79-31-2 : 10290
79-34-5 : 5635
79-35-6 : 5897
79-36-7 : 297
79-37-8 : 5585
79-38-9 : 5889
79-40-3 : 5588
79-41-4 : 10648
79-42-5 : 10285
79-43-6 : 162
79-44-7 : 3973
79-46-9 : 10117
79-49-2 : 10477
79-53-8 : 10421
79-54-9 : 9169
79-55-0 : 9691
79-57-2 : 7705
79-58-3 : 11515
79-63-0 : 7411
79-69-6 : 3884
79-70-9 : 3883
79-74-3 : 1501
79-77-6 : 3885
79-83-4 : 329
80-00-2 : 1225
80-05-7 : 9413
80-06-8 : 1850
80-07-9 : 2239
80-08-0 : 846
80-09-1 : 9466
80-10-4 : 11395
80-11-5 : 2212
80-13-7 : 2565
80-14-8 : 11915
80-15-9 : 7181
80-18-2 : 2227
80-26-2 : 4669
80-32-0 : 2193
80-33-1 : 2221
80-35-3 : 2197
80-40-0 : 2231
80-42-2 : 2235
80-46-6 : 9339
80-48-8 : 2232
80-55-7 : 10279
80-56-8 : 2973
80-59-1 : 3844
80-62-6 : 10663
80-63-7 : 10621
80-69-3 : 9956
80-72-8 : 4942
80-92-2 : 9742
80-97-7 : 4107
81-04-9 : 7867
81-07-2 : 2388
81-08-3 : 2875
81-13-0 : 3202
81-15-2 : 1461
81-16-3 : 7961
81-20-9 : 1476
81-23-2 : 4101
81-24-3 : 5625
81-25-4 : 4100
81-30-1 : 2784
81-38-9 : 2469
81-54-9 : 416
81-61-8 : 413
81-64-1 : 392
81-77-6 : 434
81-81-2 : 2832
81-88-9 : 12056
81-90-3 : 2517
81-92-5 : 1841
82-02-0 : 6185
82-05-3 : 6292
82-12-2 : 406
82-28-0 : 384
82-29-1 : 417
82-34-8 : 410
82-35-9 : 405
82-40-6 : 6182
82-44-0 : 387
82-45-1 : 380
82-54-2 : 5216
82-57-5 : 6186
82-58-6 : 5376
82-62-2 : 9490
82-66-6 : 7268
82-68-8 : 2087
82-71-3 : 1554
82-75-7 : 7965
82-86-0 : 8
83-07-8 : 10845
83-12-5 : 7270
83-13-6 : 9975
83-14-7 : 7394
83-25-0 : 11162
83-27-2 : 9967
83-32-9 : 6
83-33-0 : 7286
83-34-1 : 7314
83-40-9 : 2629
83-41-0 : 1474
83-42-1 : 1204
83-44-3 : 4095
83-45-4 : 11557
83-46-5 : 11558
83-47-6 : 11559
83-48-7 : 11555
83-49-8 : 4096
83-53-4 : 7802
83-54-5 : 11330
83-56-7 : 7835
83-60-3 : 12063
83-66-9 : 1463
83-67-0 : 10757
83-68-1 : 7800
83-72-7 : 7859
83-73-8 : 11315
83-74-9 : 7186
83-79-4 : 2783
83-81-8 : 1325
83-88-5 : 11344
83-95-4 : 6181
84-11-7 : 9174
84-15-1 : 11595
84-16-2 : 9298
84-17-3 : 9302
84-26-4 : 7334
84-31-1 : 4134
84-52-6 : 5022
84-54-8 : 409
84-55-9 : 10434
84-58-2 : 4251
84-61-7 : 1341
84-62-8 : 1345
84-65-1 : 379
84-66-2 : 1342
84-68-4 : 3032
84-69-5 : 1336
84-74-2 : 1339
84-79-7 : 7862
84-80-0 : 7865
84-86-6 : 7962
84-87-7 : 7971
84-88-8 : 11294
84-89-9 : 7963
84-95-7 : 7713
84-99-1 : 2398
85-00-7 : 5224
85-01-8 : 9163
85-02-9 : 2437
85-23-4 : 4260
85-29-0 : 7603
85-34-7 : 944
85-38-1 : 2646
85-41-6 : 7362
85-42-7 : 7345
85-44-9 : 7342
85-46-1 : 7975
85-47-2 : 7959
85-55-2 : 2684
85-64-3 : 9398
85-66-5 : 9196
85-73-4 : 2726
85-74-5 : 7873
85-75-6 : 7872
85-82-5 : 8029
85-83-6 : 8033
85-84-7 : 7729
85-86-9 : 8038
85-90-5 : 2838
85-91-6 : 2681
85-95-0 : 9303
85-97-2 : 3092
85-98-3 : 12003
86-00-0 : 3085
86-21-5 : 11010
86-26-0 : 3080
86-28-2 : 3985
86-29-3 : 969
86-30-6 : 830
86-34-0 : 11161
86-48-6 : 7771
86-50-0 : 497
86-52-2 : 7783
86-53-3 : 7748
86-55-5 : 7758
86-56-6 : 7715
86-57-7 : 7929
86-59-9 : 11229
86-60-2 : 7964
86-65-7 : 7870
86-68-0 : 11234
86-73-7 : 5967
86-74-8 : 3981
86-75-9 : 11309
86-76-0 : 5164
86-78-2 : 8737
86-80-6 : 5411
86-84-0 : 7904
86-86-2 : 7733
86-87-3 : 7734
86-88-4 : 11807
86-89-5 : 7941
86-98-6 : 11251
87-00-3 : 911
87-10-5 : 609
87-13-8 : 9947
87-17-2 : 617
87-19-4 : 2643
87-20-7 : 2633
87-24-1 : 2690
87-25-2 : 2495
87-28-5 : 2622
87-31-0 : 2874
87-39-8 : 11074
87-40-1 : 2332
87-41-2 : 7349
87-42-3 : 10753
87-44-5 : 2996
87-47-8 : 3945
87-51-4 : 7295
87-52-5 : 7310
87-58-1 : 11140
87-59-2 : 704
87-60-5 : 671
87-61-6 : 2321
87-62-7 : 707
87-64-9 : 9263
87-65-0 : 9285
87-66-1 : 2366
87-68-3 : 3163
87-69-4 : 3309
87-73-0 : 6205
87-79-6 : 11519
87-81-0 : 11587
87-83-2 : 2083
87-85-4 : 1758
87-86-5 : 9440
87-88-7 : 4258
87-90-1 : 11840
87-92-3 : 3313
87-99-0 : 12058
88-04-0 : 9257
88-05-1 : 859

88-06-2	: 9485
88-09-5	: 3514
88-10-8	: 3972
88-12-0	: 11183
88-13-1	: 11738
88-14-2	: 6050
88-15-3	: 5870
88-16-4	: 1240
88-17-5	: 855
88-18-6	: 9321
88-20-0	: 2228
88-21-1	: 2215
88-60-8	: 9331
88-65-3	: 2518
88-67-5	: 2657
88-68-6	: 601
88-69-7	: 9410
88-72-2	: 2000
88-73-3	: 1214
88-74-4	: 822
88-75-5	: 9435
88-85-7	: 5183
88-87-9	: 9258
88-88-0	: 1244
88-89-1	: 9497
88-95-9	: 1318
88-96-0	: 1324
88-99-3	: 1327
89-00-9	: 10927
89-01-0	: 10804
89-05-4	: 2257
89-31-6	: 7397
89-37-2	: 3975
89-39-4	: 1447
89-54-3	: 2490
89-55-4	: 2524
89-56-5	: 2631
89-57-6	: 2499
89-58-7	: 1478
89-59-8	: 1212
89-60-1	: 1209
89-61-2	: 1395
89-62-3	: 803
89-68-9	: 9268
89-69-0	: 2339
89-71-4	: 2696
89-72-5	: 9430
89-73-6	: 613
89-74-7	: 5809
89-75-8	: 2893
89-79-2	: 4544
89-82-7	: 4608
89-83-8	: 9418
89-84-9	: 5803
89-86-1	: 2569
89-87-2	: 1479
89-92-9	: 1062
89-93-0	: 1815
89-95-2	: 1897
89-96-3	: 1161
89-98-5	: 529
90-00-6	: 9353
90-01-7	: 1882
90-02-8	: 564
90-04-0	: 766
90-05-1	: 9379
90-11-9	: 7739
90-12-0	: 7915
90-13-1	: 7780
90-14-2	: 7902
90-15-3	: 8017
90-19-7	: 2815
90-20-0	: 7874
90-26-6	: 881
90-27-7	: 897
90-33-5	: 2831
90-39-1	: 7571
90-41-5	: 3010
90-42-6	: 2989
90-43-7	: 3088
90-44-8	: 433
90-45-9	: 313
90-47-1	: 12050
90-59-5	: 534
90-65-3	: 6701
90-69-7	: 5834
90-74-4	: 6236
90-80-2	: 6212
90-81-3	: 1900
90-84-6	: 10428
90-86-8	: 1903
90-89-1	: 9603
90-90-4	: 7602
90-94-8	: 7593
90-98-2	: 7592
90-99-3	: 1187
91-00-9	: 1826
91-01-0	: 1910
91-02-1	: 7638
91-08-7	: 1430
91-10-1	: 9305
91-13-4	: 983
91-14-5	: 1403
91-16-7	: 1436
91-17-8	: 7789
91-18-9	: 10747
91-19-0	: 11333
91-20-3	: 7732
91-21-4	: 7388
91-22-5	: 11214
91-23-6	: 1935
91-40-7	: 2715
91-49-6	: 39
91-53-2	: 11265
91-56-5	: 7300
91-57-6	: 7916
91-58-7	: 7781
91-59-8	: 7709
91-60-1	: 8008
91-62-3	: 11281
91-63-4	: 11277
91-64-5	: 2792
91-65-6	: 4285
91-66-7	: 695
91-68-9	: 9297
91-73-6	: 1829
91-76-9	: 11829
91-78-1	: 11834
91-80-5	: 5507
91-81-6	: 5505
91-84-9	: 5511
91-88-3	: 5724
91-96-3	: 3204
92-06-8	: 11596
92-13-7	: 6151
92-24-0	: 7703
92-35-3	: 11111
92-36-4	: 782
92-41-1	: 7869
92-43-3	: 10844
92-48-8	: 2835
92-51-3	: 2988
92-52-4	: 3008
92-54-6	: 9620
92-59-1	: 1811
92-61-5	: 2829
92-62-6	: 316
92-66-0	: 3018
92-67-1	: 3011
92-69-3	: 3090
92-70-6	: 7772
92-71-7	: 8580
92-82-0	: 9193
92-83-1	: 12049
92-84-2	: 9499
92-85-3	: 11660
92-86-4	: 3037
92-87-5	: 3031
92-91-1	: 5776
92-92-2	: 3022
92-93-3	: 3087
92-94-4	: 11597
92-99-9	: 11213
93-01-6	: 7974
93-02-7	: 544
93-03-8	: 1856
93-04-9	: 7914
93-05-0	: 1266
93-07-2	: 2576
93-08-3	: 5858
93-09-4	: 7759
93-10-7	: 11227
93-11-8	: 7976
93-14-1	: 10016
93-15-2	: 1450
93-16-3	: 1449
93-18-5	: 7884
93-22-1	: 7942
93-25-4	: 921
93-26-5	: 70
93-28-7	: 9395
93-35-6	: 2826
93-37-8	: 11258
93-39-0	: 2821
93-42-5	: 68
93-44-7	: 8021
93-50-5	: 667
93-51-6	: 9385
93-52-7	: 1299
93-53-8	: 872
93-54-9	: 1870
93-55-0	: 10480
93-56-1	: 5579
93-58-3	: 2687
93-59-4	: 1126
93-60-7	: 10906
93-61-8	: 5998
93-71-0	: 44
93-72-1	: 11512
93-76-5	: 256
93-78-7	: 259
93-79-8	: 258
93-88-9	: 1574
93-89-0	: 2602
93-90-3	: 5743
93-91-4	: 3362
93-97-0	: 2510
93-98-1	: 626
93-99-2	: 2716
94-02-0	: 2141
94-07-5	: 1887
94-08-6	: 2692
94-09-7	: 2497
94-11-1	: 173
94-12-2	: 2507
94-13-3	: 2653
94-14-4	: 2505
94-20-2	: 2208
94-25-7	: 2488
94-26-8	: 2615
94-30-4	: 2671
94-33-7	: 5578
94-36-0	: 9157
94-41-7	: 10708
94-44-0	: 10911
94-46-2	: 3632
94-49-5	: 5572
94-53-1	: 2418
94-58-6	: 2431
94-59-7	: 2429
94-62-2	: 9632
94-65-5	: 4612
94-67-7	: 579
94-68-8	: 746
94-69-9	: 5996
94-70-2	: 736
94-71-3	: 9350
94-74-6	: 141
94-75-7	: 169
94-80-4	: 171
94-81-5	: 3493
94-82-6	: 3925
94-96-2	: 6802
95-00-1	: 1803
95-01-2	: 541
95-05-6	: 11714
95-06-7	: 4065
95-11-4	: 2066
95-13-6	: 7245
95-14-7	: 2871
95-15-8	: 2399
95-16-9	: 2861
95-20-5	: 7313
95-21-6	: 2880
95-25-0	: 2883
95-43-2	: 3193
95-45-4	: 3360
95-46-5	: 1054
95-47-6	: 1452
95-48-7	: 9400
95-49-8	: 1183
95-50-1	: 1354
95-51-2	: 658
95-52-3	: 1726
95-53-4	: 778
95-54-5	: 1260
95-55-6	: 9201
95-56-7	: 9228
95-57-8	: 9251
95-59-0	: 5188
95-63-6	: 2358
95-64-7	: 708
95-65-8	: 9311
95-66-9	: 1138
95-68-1	: 705
95-69-2	: 673
95-70-5	: 1279
95-71-6	: 1532
95-72-7	: 1140
95-73-8	: 1388
95-74-9	: 672
95-75-0	: 1384
95-76-1	: 689
95-77-2	: 9286
95-78-3	: 706
95-79-4	: 676
95-80-7	: 1278
95-81-8	: 669
95-82-9	: 687
95-83-0	: 1264
95-84-1	: 9210
95-85-2	: 9204
95-86-3	: 9274
95-87-4	: 9309
95-92-1	: 5560
95-93-2	: 2280
95-94-3	: 2260
95-95-4	: 9484
96-00-4	: 4206
96-01-5	: 4203
96-02-6	: 6092
96-04-8	: 6402
96-05-9	: 10671
96-09-3	: 8624
96-11-7	: 10163
96-12-8	: 9885
96-14-0	: 8824
96-18-4	: 10175
96-19-5	: 10603
96-21-9	: 10357
96-22-0	: 8958
96-23-1	: 10362
96-24-2	: 9994
96-26-4	: 10429
96-27-5	: 10013
96-29-7	: 3675
96-31-1	: 12004
96-32-2	: 117
96-33-3	: 10655
96-34-4	: 138
96-35-5	: 198
96-37-7	: 4826
96-40-2	: 4914
96-41-3	: 4864
96-43-5	: 11740
96-45-7	: 7225
96-47-9	: 6171
96-48-0	: 6142
96-49-1	: 5211
96-50-4	: 11662
96-54-8	: 11130
96-76-4	: 9219
96-77-5	: 1560
96-80-0	: 5683
96-82-2	: 9210
96-88-8	: 9639
96-91-3	: 9207
96-97-9	: 2647
96-99-1	: 2547
97-00-7	: 1144
97-02-9	: 725
97-05-2	: 2655
97-16-5	: 9289
97-17-6	: 9581
97-18-7	: 9478
97-23-4	: 9408
97-24-5	: 9477
97-25-6	: 6231
97-30-3	: 6227
97-39-2	: 6279
97-53-0	: 9389
97-54-1	: 9388
97-56-3	: 798
97-59-6	: 12009
97-62-1	: 10298
97-63-2	: 10658
97-65-4	: 3328
97-72-3	: 10291
97-74-5	: 11715
97-77-8	: 11721
97-85-8	: 10304
97-86-9	: 10666
97-88-1	: 10650
97-90-5	: 10656
97-94-9	: 3118
97-95-0	: 3620
97-96-1	: 3183
97-97-2	: 5472
97-99-4	: 6124
98-00-0	: 6118
98-01-1	: 6041
98-02-2	: 6117
98-03-3	: 11733
98-04-4	: 866
98-05-5	: 460
98-06-6	: 1459
98-07-7	: 2333
98-08-8	: 2348
98-09-9	: 2242
98-11-3	: 2214
98-13-5	: 11473
98-15-7	: 1241
98-16-8	: 856
98-17-9	: 9491
98-19-1	: 1460
98-27-1	: 9333
98-29-3	: 1516
98-46-4	: 2061
98-50-0	: 456
98-51-1	: 1466
98-54-4	: 9323
98-56-6	: 1242
98-58-8	: 2244
98-59-9	: 2250
98-60-2	: 2245
98-61-3	: 2247
98-62-4	: 2252
98-66-8	: 2220
98-67-9	: 2226
98-71-5	: 2224
98-72-6	: 459
98-73-7	: 2588
98-79-3	: 9768
98-80-6	: 3135
98-81-7	: 1025
98-82-8	: 1975
98-83-9	: 1972
98-85-1	: 1896
98-86-2	: 5864
98-87-3	: 1382
98-88-4	: 2886
98-89-5	: 4332
98-91-9	: 1130
98-92-0	: 10890
98-94-2	: 4286
98-95-3	: 2049
98-96-4	: 10802
98-97-5	: 10803
98-98-6	: 10893
99-02-5	: 5788
99-03-6	: 5772
99-04-7	: 2677
99-05-8	: 2485
99-07-0	: 9318
99-08-1	: 2001
99-09-2	: 823
99-11-6	: 10900
99-12-7	: 1477
99-14-9	: 10165
99-17-2	: 6223
99-20-7	: 6220
99-30-9	: 692
99-31-0	: 1331
99-32-1	: 10776
99-33-2	: 2896

99-34-3 : 2591
99-35-4 : 2365
99-36-5 : 2697
99-45-6 : 5804
99-48-9 : 4696
99-50-3 : 2571
99-51-4 : 1475
99-52-5 : 800
99-53-6 : 9422
99-54-7 : 1392
99-55-8 : 801
99-56-9 : 1281
99-59-2 : 775
99-61-6 : 592
99-62-7 : 993
99-63-8 : 1319
99-65-0 : 1492
99-71-8 : 9431
99-73-0 : 5777
99-75-2 : 2698
99-76-3 : 2637
99-77-4 : 2709
99-82-1 : 4438
99-84-3 : 4675
99-85-4 : 4275
99-86-5 : 4273
99-87-6 : 1990
99-88-7 : 789
99-89-8 : 9412
99-90-1 : 5781
99-91-2 : 5789
99-92-3 : 5773
99-93-4 : 5832
99-94-5 : 2678
99-96-7 : 2613
99-97-8 : 862
99-98-9 : 1270
99-99-0 : 2002
100-00-5 : 1216
100-01-6 : 824
100-02-7 : 9437
100-06-1 : 5841
100-07-2 : 2901
100-09-4 : 2666
100-10-7 : 550
100-12-9 : 1694
100-13-0 : 1645
100-14-1 : 1208
100-15-2 : 805
100-16-3 : 7168
100-17-4 : 1937
100-18-5 : 994
100-19-6 : 5861
100-20-9 : 1320
100-21-0 : 1329
100-22-1 : 1287
100-25-4 : 1493
100-26-5 : 10929
100-28-7 : 1790
100-29-8 : 1664
100-36-7 : 5502
100-37-8 : 5706
100-39-0 : 1053
100-40-3 : 4658
100-41-4 : 1668
100-42-5 : 1624
100-43-6 : 10954
100-44-7 : 1182
100-45-8 : 4629
100-46-9 : 1795
100-47-0 : 2740
100-49-2 : 4419
100-50-5 : 4632
100-51-6 : 1836
100-52-7 : 525
100-53-8 : 1832
100-54-9 : 10882
100-55-0 : 10976
100-60-7 : 4298
100-61-8 : 781
100-63-0 : 7169
100-64-1 : 4610
100-65-2 : 757
100-66-3 : 1915
100-68-5 : 2041
100-69-6 : 10952
100-70-9 : 10881
100-71-0 : 10956
100-72-1 : 10784
100-74-3 : 7694
100-75-4 : 9688
100-76-5 : 480
100-79-8 : 5207
100-80-1 : 1641
100-81-2 : 1816
100-83-4 : 565
100-84-5 : 1932
100-86-7 : 1600
100-88-9 : 11568
100-97-0 : 11601
101-01-9 : 6288
101-02-0 : 9593
101-05-3 : 360
101-10-0 : 4141
101-14-4 : 785
101-17-7 : 678
101-18-8 : 9446
101-21-3 : 3939
101-27-9 : 510
101-31-5 : 917
101-38-2 : 4280
101-39-3 : 10516
101-40-6 : 4385
101-41-7 : 929
101-43-9 : 10653
101-53-1 : 9455
101-54-2 : 1283
101-55-3 : 1079
101-57-5 : 2234
101-60-0 : 9734
101-61-1 : 786
101-76-8 : 1963
101-77-9 : 784
101-80-4 : 832
101-81-5 : 1961
101-82-6 : 11004
101-83-7 : 4284
101-84-8 : 2071
101-90-6 : 8625
101-91-7 : 3205
101-97-3 : 899
101-99-5 : 3965
102-01-2 : 3207
102-02-3 : 7234
102-04-5 : 10433
102-06-7 : 6282
102-07-8 : 12010
102-08-9 : 11805
102-09-0 : 4001
102-13-6 : 932
102-14-7 : 3575
102-20-5 : 940
102-25-0 : 2345
102-27-2 : 747
102-45-4 : 4810
102-47-6 : 1357
102-50-1 : 772
102-51-2 : 1273
102-52-3 : 10150
102-53-4 : 7489
102-54-5 : 5960
102-56-7 : 702
102-60-3 : 10379
102-69-2 : 9808
102-70-5 : 10523
102-71-6 : 5751
102-76-1 : 10195
102-79-4 : 5692
102-81-8 : 5703
102-82-9 : 3213
102-85-2 : 9590
102-86-3 : 6725
102-92-1 : 10716
102-94-3 : 10683
102-97-6 : 1820
103-01-5 : 6270
103-02-6 : 10482
103-03-7 : 7145
103-05-9 : 2150
103-09-3 : 187
103-11-7 : 10636
103-12-8 : 2209
103-16-2 : 9453
103-17-3 : 4075
103-19-5 : 5246
103-23-1 : 6781
103-24-2 : 8132
103-25-3 : 2139
103-28-6 : 10307
103-29-7 : 1587
103-30-0 : 1619
103-32-2 : 1827
103-34-4 : 7688
103-37-7 : 3585
103-38-8 : 3560
103-43-5 : 3298
103-44-6 : 6405
103-45-7 : 232
103-48-0 : 10306
103-49-1 : 1828
103-50-4 : 2079
103-55-9 : 5504
103-63-9 : 1038
103-65-1 : 2169
103-67-3 : 1818
103-69-5 : 743
103-70-8 : 5999
103-71-9 : 1786
103-72-0 : 1792
103-73-1 : 1649
103-74-2 : 10950
103-75-3 : 10783
103-76-4 : 9616
103-79-7 : 10481
103-80-0 : 971
103-81-1 : 878
103-82-2 : 883
103-83-3 : 1807
103-84-4 : 86
103-85-5 : 11808
103-88-8 : 37
103-89-9 : 78
103-90-2 : 67
103-95-7 : 2117
103-99-1 : 8232
104-01-8 : 922
104-03-0 : 933
104-04-1 : 83
104-06-3 : 34
104-07-4 : 8253
104-09-6 : 874
104-13-2 : 655
104-15-4 : 2229
104-21-2 : 1892
104-22-3 : 2213
104-29-0 : 10000
104-36-9 : 1314
104-45-0 : 1945
104-46-1 : 1942
104-47-2 : 958
104-50-7 : 6140
104-51-8 : 1118
104-52-9 : 1235
104-53-0 : 2114
104-57-4 : 6016
104-61-0 : 6149
104-63-2 : 5762
104-65-4 : 10707
104-66-5 : 1588
104-67-6 : 6152
104-72-3 : 1258
104-75-6 : 6726
104-76-7 : 6930
104-78-9 : 9876
104-80-3 : 7109
104-81-4 : 1064
104-82-5 : 1200
104-83-6 : 1134
104-84-7 : 1817
104-85-8 : 2765
104-87-0 : 589
104-88-1 : 531
104-90-5 : 10966
104-91-6 : 9438
104-92-7 : 1052
104-93-8 : 1933
104-94-9 : 768
104-96-1 : 820
105-03-3 : 8131
105-05-5 : 1415
105-06-6 : 1405
105-07-7 : 2758
105-13-5 : 1891
105-14-6 : 8964
105-30-6 : 8937
105-31-7 : 7124
105-34-0 : 154
105-36-2 : 115
105-37-3 : 10263
105-38-4 : 10261
105-39-5 : 130
105-40-8 : 3954
105-41-9 : 6730
105-42-0 : 6977
105-45-3 : 3573
105-46-4 : 214
105-48-6 : 139
105-50-0 : 8783
105-53-3 : 9941
105-54-4 : 3515
105-55-5 : 11803
105-56-6 : 152
105-57-7 : 5541
105-58-8 : 3999
105-59-9 : 5742
105-60-2 : 493
105-66-8 : 3588
105-67-9 : 9308
105-68-0 : 3639
105-70-4 : 10190
105-72-6 : 3512
105-74-8 : 9155
105-75-9 : 3774
105-83-9 : 9875
105-86-2 : 8338
105-99-7 : 6784
106-02-5 : 8562
106-07-0 : 8259
106-19-4 : 6787
106-23-0 : 8490
106-24-1 : 8333
106-25-2 : 8334
106-26-3 : 8318
106-27-4 : 3544
106-28-5 : 5333
106-30-9 : 6462
106-31-0 : 3467
106-32-1 : 8428
106-33-2 : 5312
106-35-4 : 6511
106-36-5 : 10326
106-37-6 : 1290
106-38-7 : 1056
106-39-8 : 1006
106-40-1 : 640
106-41-2 : 9230
106-42-3 : 1454
106-43-4 : 1185
106-44-5 : 9402
106-45-6 : 2303
106-46-7 : 1356
106-47-8 : 660
106-48-9 : 9253
106-49-0 : 780
106-50-3 : 1262
106-51-4 : 4257
106-52-5 : 9717
106-53-6 : 2292
106-54-7 : 2295
106-57-0 : 9612
106-58-1 : 9608
106-63-8 : 10673
106-65-0 : 3318
106-68-3 : 8472
106-69-4 : 6868
106-70-7 : 6898
106-73-0 : 6466
106-74-1 : 10634
106-79-6 : 5046
106-86-5 : 8555
106-87-6 : 8557
106-88-7 : 8615
106-90-1 : 10680
106-91-2 : 10668
106-92-3 : 8626
106-93-4 : 5518
106-94-5 : 9836
106-95-6 : 10531
106-96-7 : 10728
106-97-8 : 3229
106-98-9 : 3704
106-99-0 : 3146
107-00-6 : 3897
107-02-8 : 10509
107-03-9 : 10156
107-04-0 : 5456
107-05-1 : 10536
107-06-2 : 5529
107-07-3 : 5693
107-08-4 : 10054
107-10-8 : 9803
107-11-9 : 10522
107-12-0 : 10090
107-13-1 : 10583
107-14-2 : 274
107-15-3 : 5496
107-16-4 : 282
107-18-6 : 10697
107-19-7 : 10743
107-20-0 : 12
107-21-1 : 5569
107-22-2 : 5491
107-25-5 : 5908
107-29-9 : 17
107-30-2 : 7483
107-31-3 : 6010
107-34-6 : 461
107-35-7 : 5623
107-37-9 : 11475
107-39-1 : 9067
107-40-4 : 9071
107-41-5 : 8793
107-43-7 : 7454
107-46-0 : 5236
107-47-1 : 10155
107-48-2 : 8612
107-49-3 : 5221
107-50-6 : 4239
107-51-7 : 11914
107-52-8 : 6990
107-53-9 : 11964
107-59-5 : 128
107-68-6 : 5624
107-70-0 : 8978
107-72-2 : 11472
107-74-4 : 8392
107-75-5 : 8347
107-80-2 : 3266
107-81-3 : 8711
107-82-4 : 3239
107-83-5 : 8823
107-84-6 : 3252
107-85-7 : 3220
107-86-8 : 3700
107-87-9 : 8957
107-88-0 : 3345
107-89-1 : 3184
107-91-5 : 49
107-92-6 : 3460
107-93-7 : 3814
107-94-8 : 10216
107-95-9 : 323
107-96-0 : 10286
107-97-1 : 6268
107-98-2 : 10387
108-01-0 : 5709
108-03-2 : 10116
108-05-4 : 179
108-08-7 : 8762
108-10-1 : 8981
108-11-2 : 8942
108-12-3 : 3693
108-16-7 : 10376
108-18-9 : 9821
108-19-0 : 7232
108-20-3 : 10120
108-21-4 : 209
108-22-5 : 10698
108-23-6 : 4023
108-24-7 : 107
108-27-0 : 11186
108-28-1 : 6155
108-30-5 : 6093
108-31-6 : 6091
108-32-7 : 5213

108-36-1	: 1289
108-37-2	: 1005
108-38-3	: 1453
108-39-4	: 9401
108-40-7	: 2302
108-41-8	: 1184
108-42-9	: 659
108-43-0	: 9252
108-44-1	: 779
108-45-2	: 1261
108-46-3	: 1498
108-47-4	: 10937
108-48-5	: 10939
108-50-9	: 10807
108-52-1	: 11046
108-55-4	: 10780
108-56-5	: 3340
108-57-6	: 1404
108-59-8	: 9943
108-60-1	: 10123
108-64-5	: 3548
108-67-8	: 2359
108-68-9	: 9312
108-69-0	: 709
108-70-3	: 2323
108-73-6	: 2368
108-75-8	: 11028
108-77-0	: 11837
108-78-1	: 11835
108-82-7	: 6485
108-83-8	: 6514
108-84-9	: 8947
108-85-0	: 4315
108-86-1	: 998
108-87-2	: 4426
108-88-3	: 1947
108-89-4	: 10982
108-90-7	: 1131
108-91-8	: 4283
108-93-0	: 4500
108-94-1	: 4571
108-95-2	: 9199
108-97-4	: 10789
108-98-5	: 2289
108-99-6	: 10981
109-01-3	: 9618
109-02-4	: 7696
109-04-6	: 10875
109-06-8	: 10980
109-07-9	: 9619
109-08-0	: 10810
109-09-1	: 10915
109-12-6	: 11042
109-16-0	: 10657
109-19-3	: 3543
109-21-7	: 3479
109-43-3	: 5044
109-49-9	: 7089
109-52-4	: 8855
109-53-5	: 10035
109-56-8	: 5741
109-57-9	: 11809
109-59-1	: 5740
109-60-4	: 236
109-61-5	: 4028
109-63-7	: 3137
109-64-8	: 9882
109-65-9	: 3231
109-66-0	: 8709
109-67-1	: 9004
109-69-3	: 3241
109-70-6	: 9839
109-73-9	: 3209
109-74-0	: 3403
109-75-1	: 3803
109-76-2	: 9873
109-77-3	: 9911
109-78-4	: 10106
109-79-5	: 3437
109-80-8	: 10031
109-81-9	: 5512
109-82-0	: 285
109-83-1	: 5738
109-84-2	: 5729
109-86-4	: 5733
109-87-5	: 7507
109-89-7	: 5427
109-90-0	: 5598
109-92-2	: 5905
109-93-3	: 5912
109-94-4	: 6007
109-95-5	: 8087
109-96-6	: 11118
109-97-7	: 11112
109-98-8	: 10822
109-99-9	: 6168
110-00-9	: 6025
110-01-0	: 11783
110-02-1	: 11724
110-03-2	: 6801
110-04-3	: 5587
110-05-4	: 9153
110-06-5	: 5242
110-12-3	: 6978
110-13-4	: 6805
110-14-5	: 3257
110-15-6	: 3294
110-16-7	: 3765
110-17-8	: 3764
110-18-9	: 5516
110-19-0	: 215
110-22-5	: 9156
110-27-0	: 11626
110-38-3	: 5063
110-39-4	: 3566
110-40-7	: 5045
110-41-8	: 11936
110-42-9	: 5066
110-43-0	: 6510
110-44-1	: 6698
110-45-2	: 3636
110-46-3	: 8090
110-49-6	: 5735
110-52-1	: 3267
110-53-2	: 8710
110-54-3	: 6731
110-56-5	: 3279
110-57-6	: 3750
110-58-7	: 8700
110-59-8	: 8827
110-60-1	: 3258
110-61-2	: 3291
110-62-3	: 8690
110-63-4	: 3346
110-65-6	: 3911
110-66-7	: 8844
110-67-8	: 10110
110-68-9	: 3221
110-69-0	: 3188
110-70-3	: 5503
110-71-4	: 5549
110-72-5	: 5509
110-73-6	: 5722
110-74-7	: 6018
110-75-8	: 5887
110-76-9	: 5423
110-77-0	: 5726
110-78-1	: 10059
110-80-5	: 5717
110-81-6	: 5253
110-82-7	: 4305
110-83-8	: 4626
110-85-0	: 9602
110-86-1	: 10872
110-87-2	: 10777
110-88-3	: 11905
110-89-4	: 9629
110-91-8	: 7681
110-93-0	: 6603
110-94-1	: 8770
110-95-2	: 9879
110-96-3	: 9823
110-97-4	: 10384
110-99-6	: 226
111-01-3	: 11606
111-02-4	: 11602
111-03-5	: 8299
111-06-8	: 6647
111-11-5	: 8434
111-12-6	: 8542
111-13-7	: 8471
111-14-8	: 6455
111-15-9	: 5718
111-16-0	: 6393
111-17-1	: 10329
111-18-2	: 6752
111-19-3	: 5049
111-20-6	: 5042
111-21-7	: 5714
111-24-0	: 8741
111-25-1	: 6732
111-26-2	: 6723
111-27-3	: 6911
111-28-4	: 6707
111-29-5	: 8787
111-30-8	: 8734
111-31-9	: 6856
111-34-2	: 3369
111-40-0	: 5497
111-41-1	: 5676
111-42-2	: 5730
111-43-3	: 10119
111-44-4	: 5611
111-45-5	: 5763
111-46-6	: 5753
111-47-7	: 10151
111-48-8	: 5766
111-49-9	: 492
111-51-3	: 3263
111-55-7	: 5571
111-56-8	: 5307
111-60-4	: 8260
111-61-5	: 8257
111-62-6	: 8301
111-64-8	: 8481
111-65-9	: 8355
111-66-0	: 8491
111-68-2	: 6349
111-69-3	: 6777
111-70-6	: 6472
111-71-7	: 6344
111-74-0	: 5501
111-75-1	: 5691
111-76-2	: 5687
111-77-3	: 5737
111-78-4	: 4733
111-81-9	: 11978
111-82-0	: 5315
111-83-1	: 8356
111-84-2	: 8120
111-85-3	: 8359
111-86-4	: 8350
111-87-5	: 8443
111-88-6	: 8418
111-90-0	: 5720
111-92-2	: 3211
111-96-6	: 5613
112-04-9	: 11470
112-05-0	: 8146
112-06-1	: 193
112-10-7	: 8267
112-12-9	: 11958
112-13-0	: 5078
112-14-1	: 222
112-15-2	: 5721
112-16-3	: 5328
112-17-4	: 157
112-20-9	: 8119
112-23-2	: 6008
112-24-3	: 5499
112-25-4	: 5728
112-26-5	: 5449
112-27-6	: 5713
112-29-8	: 5034
112-30-1	: 5070
112-31-2	: 5030
112-32-3	: 6014
112-34-5	: 5688
112-36-7	: 5612
112-37-8	: 11943
112-38-9	: 11974
112-39-0	: 6653
112-40-3	: 5292
112-41-4	: 5335
112-42-5	: 11950
112-43-6	: 11979
112-44-7	: 11935
112-46-9	: 11981
112-47-0	: 5047
112-48-1	: 5527
112-49-2	: 11642
112-51-6	: 5257
112-52-7	: 5295
112-53-8	: 5321
112-54-9	: 5288
112-55-0	: 5304
112-57-2	: 5498
112-58-3	: 6842
112-60-7	: 5756
112-61-8	: 8266
112-62-9	: 8307
112-63-0	: 8228
112-64-1	: 11635
112-66-3	: 178
112-67-4	: 6665
112-70-9	: 11883
112-71-0	: 11614
112-72-1	: 11630
112-73-2	: 3417
112-76-5	: 8281
112-79-8	: 8297
112-80-1	: 8294
112-82-3	: 6633
112-85-6	: 5276
112-86-7	: 5283
112-88-9	: 8291
112-89-0	: 8235
112-91-4	: 8293
112-92-5	: 8278
112-95-8	: 5350
113-00-8	: 6278
113-15-5	: 5394
113-48-4	: 7660
114-07-8	: 5396
114-25-0	: 3001
114-26-1	: 10722
114-91-0	: 10979
115-07-1	: 10528
115-10-6	: 7523
115-11-7	: 10578
115-17-3	: 20
115-18-4	: 3870
115-19-5	: 3922
115-20-8	: 5768
115-21-9	: 11463
115-24-2	: 9835
115-25-3	: 4194
115-29-7	: 5369
115-32-2	: 1851
115-37-7	: 7679
115-38-8	: 11094
115-39-9	: 9366
115-40-2	: 9364
115-43-5	: 11103
115-53-7	: 7676
115-56-0	: 11064
115-61-7	: 11349
115-63-9	: 9623
115-69-5	: 9987
115-70-8	: 9985
115-71-9	: 9093
115-76-4	: 10004
115-77-5	: 9989
115-80-0	: 10182
115-84-4	: 9992
115-86-6	: 9566
115-90-2	: 5954
115-96-8	: 5701
115-99-1	: 8337
116-06-3	: 334
116-09-6	: 10443
116-14-3	: 5918
116-15-4	: 10572
116-16-5	: 10440
116-17-6	: 9595
116-26-7	: 4250
116-29-0	: 2325
116-38-1	: 865
116-43-8	: 3578
116-44-9	: 2202
116-54-1	: 166
116-58-5	: 9757
116-63-2	: 7966
117-08-8	: 7346
117-10-2	: 395
117-12-4	: 393
117-18-0	: 2264
117-34-0	: 936
117-37-3	: 7269
117-39-5	: 2817
117-73-7	: 12064
117-78-2	: 372
117-79-3	: 381
117-80-6	: 7850
117-81-7	: 1334
117-84-0	: 1344
117-93-1	: 66
117-99-7	: 7622
118-00-3	: 6289
118-08-1	: 7357
118-10-5	: 4120
118-32-1	: 7871
118-34-3	: 6225
118-41-2	: 2734
118-44-5	: 7718
118-46-7	: 8020
118-55-8	: 2650
118-58-1	: 2651
118-61-6	: 2619
118-69-4	: 1385
118-71-8	: 10796
118-74-1	: 1749
118-75-2	: 4268
118-76-3	: 4686
118-78-5	: 11077
118-79-6	: 9479
118-83-2	: 1219
118-90-1	: 2676
118-91-2	: 2534
118-92-3	: 2484
118-93-4	: 5830
118-96-7	: 2048
119-13-1	: 2785
119-20-0	: 2494
119-26-6	: 7156
119-27-7	: 1923
119-28-8	: 7967
119-33-5	: 9424
119-36-8	: 2634
119-41-5	: 225
119-44-8	: 10749
119-48-2	: 7686
119-59-5	: 845
119-60-8	: 7606
119-61-9	: 7616
119-64-2	: 7979
119-65-3	: 7373
119-67-5	: 2607
119-68-6	: 2679
119-84-6	: 2798
119-90-4	: 3033
119-93-7	: 3034
120-07-0	: 5760
120-12-7	: 366
120-14-9	: 545
120-15-0	: 11298
120-18-3	: 7960
120-20-7	: 1571
120-21-8	: 539
120-22-9	: 699
120-23-0	: 217
120-25-2	: 557
120-29-6	: 489
120-32-1	: 9270
120-33-2	: 2691
120-36-5	: 10250
120-43-4	: 9604
120-46-7	: 10027
120-47-8	: 2621
120-48-9	: 2706
120-50-3	: 2701
120-51-4	: 2720
120-55-8	: 5755
120-57-0	: 2417
120-61-6	: 1351
120-66-1	: 76
120-71-8	: 770
120-72-9	: 7294
120-73-0	: 10752
120-75-2	: 2863
120-80-9	: 1497
120-82-1	: 2322
120-83-2	: 9283
120-85-4	: 9622
120-87-6	: 8254

120-89-8	: 7226
120-92-3	: 4878
120-93-4	: 7227
120-94-5	: 11168
120-97-8	: 1559
121-06-2	: 9337
121-14-2	: 1956
121-17-5	: 1217
121-21-1	: 4960
121-29-9	: 4962
121-30-2	: 1558
121-32-4	: 555
121-33-5	: 574
121-34-6	: 2625
121-43-7	: 3126
121-44-8	: 5417
121-45-9	: 9592
121-46-0	: 2913
121-47-1	: 2216
121-57-3	: 2217
121-59-5	: 455
121-60-8	: 2243
121-61-9	: 31
121-66-4	: 11666
121-69-7	: 713
121-71-1	: 5831
121-72-2	: 861
121-73-3	: 1215
121-75-5	: 3316
121-79-9	: 2732
121-81-3	: 614
121-82-4	: 11833
121-86-8	: 1210
121-89-1	: 5860
121-90-4	: 2905
121-91-5	: 1328
121-92-6	: 2704
121-98-2	: 2675
122-00-9	: 5853
122-01-0	: 2892
122-03-2	: 590
122-04-3	: 2906
122-07-6	: 5420
122-09-8	: 1572
122-11-2	: 2194
122-14-5	: 5951
122-20-3	: 10393
122-25-8	: 2686
122-28-1	: 82
122-32-7	: 8310
122-34-9	: 11513
122-35-0	: 10505
122-37-2	: 9447
122-39-4	: 836
122-40-7	: 6347
122-42-9	: 3966
122-46-3	: 212
122-48-5	: 3663
122-51-0	: 5604
122-52-1	: 9591
122-57-6	: 3881
122-59-8	: 228
122-60-1	: 8623
122-62-3	: 5043
122-66-7	: 7158
122-69-0	: 10691
122-70-3	: 10324
122-73-6	: 1951
122-78-1	: 868
122-79-2	: 231
122-80-5	: 29
122-84-9	: 10461
122-87-2	: 6267
122-88-3	: 146
122-95-2	: 1408
122-97-4	: 2148
122-98-5	: 5759
122-99-6	: 5758
123-00-2	: 7701
123-01-3	: 1566
123-02-4	: 2342
123-03-5	: 11030
123-04-6	: 6368
123-05-7	: 6719
123-06-8	: 9914
123-07-9	: 9355
123-08-0	: 566
123-11-5	: 585
123-15-9	: 8693
123-19-3	: 6512
123-20-6	: 3513
123-22-8	: 9587
123-25-1	: 3303
123-29-5	: 8150
123-30-8	: 9203
123-31-9	: 1499
123-32-0	: 10806
123-35-3	: 8330
123-38-6	: 9774
123-39-7	: 5995
123-42-2	: 8976
123-43-3	: 238
123-45-5	: 239
123-51-3	: 3627
123-54-6	: 8798
123-56-8	: 11152
123-62-6	: 10203
123-63-7	: 11908
123-66-0	: 6884
123-72-8	: 3178
123-73-9	: 3695
123-75-1	: 11143
123-76-2	: 8903
123-77-3	: 5124
123-80-8	: 5576
123-82-0	: 6350
123-83-1	: 5510
123-84-2	: 10340
123-86-4	: 121
123-88-6	: 11822
123-90-0	: 11719
123-91-1	: 5187
123-92-2	: 3631
123-93-3	: 240
123-95-5	: 8251
123-98-8	: 8138
123-99-9	: 8130
124-02-7	: 10526
124-04-9	: 6779
124-06-1	: 11624
124-07-2	: 8422
124-09-4	: 6749
124-10-7	: 11625
124-11-8	: 8182
124-12-9	: 8411
124-13-0	: 8346
124-17-4	: 5689
124-18-5	: 5033
124-19-6	: 8117
124-21-0	: 5425
124-22-1	: 5291
124-26-5	: 8231
124-30-1	: 8233
124-38-9	: 3992
124-40-3	: 7448
124-42-5	: 5671
124-47-0	: 12021
124-48-1	: 7493
124-58-3	: 458
124-63-0	: 7535
124-68-5	: 10341
124-70-9	: 11396
124-73-2	: 5525
124-80-1	: 4067
124-83-4	: 4794
124-87-8	: 7579
125-15-5	: 11351
125-20-2	: 7352
125-24-6	: 3002
125-28-0	: 7674
125-29-1	: 7678
125-33-7	: 11066
125-55-3	: 11079
126-00-1	: 1103
126-07-8	: 11520
126-11-4	: 10012
126-14-7	: 6229
126-17-0	: 11529
126-18-1	: 11533
126-19-2	: 11531
126-29-4	: 4056
126-30-7	: 10005
126-33-0	: 11786
126-54-5	: 11643
126-68-1	: 9585
126-71-6	: 9572
126-72-7	: 10358
126-73-8	: 9562
126-81-8	: 4378
126-84-1	: 9904
126-98-7	: 10586
126-99-8	: 3152
127-00-4	: 10349
127-06-0	: 10476
127-07-1	: 12016
127-08-2	: 234
127-09-3	: 237
127-17-3	: 10314
127-18-4	: 5917
127-19-5	: 55
127-25-3	: 9168
127-27-5	: 9170
127-29-7	: 308
127-40-2	: 4058
127-48-0	: 8587
127-63-9	: 2237
127-66-2	: 1878
127-69-5	: 2195
127-71-9	: 603
127-79-7	: 2199
127-91-3	: 2925
128-04-1	: 11514
128-08-5	: 11153
128-09-6	: 11154
128-13-2	: 4098
128-20-1	: 9745
128-23-4	: 9743
128-27-8	: 5381
128-33-6	: 4104
128-37-0	: 9224
128-39-2	: 9220
128-50-7	: 2969
128-53-0	: 11122
128-62-1	: 7356
128-68-7	: 2979
129-00-0	: 10853
129-43-1	: 407
129-44-2	: 389
129-64-6	: 7580
129-66-8	: 2735
129-79-3	: 5979
130-15-4	: 7843
130-16-5	: 11311
130-26-7	: 11312
130-73-4	: 9299
130-80-3	: 9301
130-85-8	: 7777
130-86-9	: 3109
130-89-2	: 4130
130-90-5	: 4129
130-95-0	: 4127
131-01-1	: 12065
131-09-9	: 388
131-11-3	: 1343
131-28-2	: 2586
131-53-3	: 7619
131-54-4	: 7595
131-56-6	: 7609
131-57-7	: 7620
131-58-8	: 7631
131-73-7	: 864
131-89-5	: 9273
131-91-9	: 8035
132-57-0	: 7972
132-60-5	: 11235
132-64-9	: 5163
132-65-0	: 5167
132-66-1	: 2702
132-75-2	: 7736
132-86-5	: 7834
133-06-2	: 7368
133-07-3	: 5984
133-08-4	: 9930
133-11-9	: 2502
133-13-1	: 9949
133-14-2	: 9152
133-26-6	: 6188
133-32-4	: 7296
133-37-9	: 3311
133-53-9	: 9290
133-58-4	: 7437
133-59-5	: 2249
133-67-5	: 2854
133-90-4	: 2492
133-91-5	: 2617
134-04-3	: 2848
134-11-2	: 2596
134-20-3	: 2503
134-31-6	: 11324
134-32-7	: 7708
134-49-6	: 7697
134-55-4	: 2483
134-58-7	: 11851
134-62-3	: 612
134-81-6	: 5584
134-84-9	: 7633
134-85-0	: 7604
134-96-3	: 569
135-00-2	: 7641
135-01-3	: 1413
135-02-4	: 583
135-09-1	: 2858
135-19-3	: 8018
135-20-6	: 759
135-44-4	: 8926
135-48-8	: 8636
135-58-0	: 11661
135-67-1	: 9507
135-68-2	: 3012
135-70-6	: 11194
135-88-6	: 7728
135-97-7	: 486
135-98-8	: 2035
136-35-6	: 11826
136-44-7	: 10188
136-47-0	: 2531
136-60-7	: 2532
136-72-1	: 8678
136-77-6	: 1519
136-78-7	: 11358
136-79-8	: 739
136-80-1	: 5744
136-95-8	: 2859
137-00-8	: 11677
137-04-2	: 665
137-05-3	: 10624
137-06-4	: 2301
137-07-5	: 2290
137-16-6	: 6269
137-17-7	: 858
137-19-9	: 1510
137-26-8	: 11722
137-30-4	: 12068
137-32-6	: 3626
137-40-6	: 10327
137-43-9	: 4777
137-58-6	: 54
138-41-0	: 2508
138-52-3	: 6222
138-53-4	: 955
138-86-3	: 4677
138-87-4	: 4542
138-89-6	: 720
139-13-9	: 6256
139-40-2	: 10508
139-45-7	: 10198
139-65-1	: 850
139-66-2	: 2283
139-85-5	: 542
139-87-7	: 5723
139-93-5	: 9275
140-08-9	: 5702
140-10-3	: 10682
140-11-4	: 233
140-18-1	: 148
140-22-7	: 4010
140-25-0	: 5318
140-28-3	: 5500
140-29-4	: 945
140-31-8	: 9615
140-39-6	: 213
140-40-9	: 85
140-55-6	: 9718
140-57-8	: 11579
140-65-8	: 7683
140-67-0	: 1943
140-76-1	: 10955
140-77-2	: 4847
140-80-7	: 8736
140-82-9	: 5707
140-87-4	: 153
140-88-5	: 10635
140-95-4	: 11995
141-03-7	: 3301
141-04-8	: 6783
141-05-9	: 3776
141-06-0	: 8914
141-10-6	: 11965
141-12-8	: 8336
141-20-8	: 5313
141-22-0	: 8303
141-27-5	: 8317
141-28-6	: 6785
141-32-2	: 10616
141-43-5	: 5675
141-46-8	: 15
141-57-1	: 11476
141-62-8	: 11646
141-63-9	: 8997
141-66-2	: 5172
141-75-3	: 3681
141-78-6	: 186
141-79-7	: 9108
141-82-2	: 9921
141-83-3	: 11993
141-84-4	: 11685
141-86-6	: 10925
141-90-2	: 11110
141-91-3	: 7687
141-93-5	: 1414
141-97-9	: 3571
142-04-1	: 756
142-08-5	: 11033
142-09-6	: 10660
142-26-7	: 65
142-28-9	: 9893
142-29-0	: 4903
142-30-3	: 7119
142-47-2	: 6248
142-50-7	: 5332
142-60-9	: 10313
142-61-0	: 6986
142-62-1	: 6870
142-64-3	: 9607
142-68-7	: 10800
142-73-4	: 6260
142-77-8	: 8298
142-82-5	: 6355
142-83-6	: 6675
142-84-7	: 9826
142-91-6	: 6654
142-92-7	: 194
142-96-1	: 3415
143-07-7	: 5305
143-08-8	: 8157
143-10-2	: 5058
143-13-5	: 220
143-15-7	: 5293
143-16-8	: 6727
143-22-6	: 5690
143-24-8	: 8995
143-27-1	: 6631
143-28-2	: 8312
143-50-0	: 7408
143-62-4	: 4044
143-74-8	: 9363
144-19-4	: 8795
144-49-0	: 189
144-55-8	: 4004
144-62-7	: 5553
144-68-3	: 4057
144-79-6	: 11380
144-80-9	: 30
144-83-2	: 2203
145-13-1	: 9756
145-73-3	: 5370
145-94-8	: 7283
146-06-5	: 4128
146-48-5	: 12062
147-71-7	: 3312
147-73-9	: 3304
147-82-0	: 851
147-85-3	: 9764
147-93-3	: 2663
148-03-8	: 2789

148-24-3	: 11308
148-25-4	: 7875
148-53-8	: 570
148-75-4	: 7876
148-79-8	: 11656
148-86-7	: 3091
149-29-1	: 6183
149-30-4	: 2866
149-32-6	: 3431
149-57-5	: 6882
149-73-5	: 7561
149-74-6	: 11401
149-91-7	: 2731
150-13-0	: 2486
150-19-6	: 9380
150-25-4	: 6259
150-30-1	: 9509
150-46-9	: 3125
150-50-5	: 7438
150-60-7	: 5248
150-61-8	: 5508
150-68-5	: 11999
150-69-6	: 12013
150-76-5	: 9381
150-77-6	: 5515
150-78-7	: 1438
150-86-7	: 6672
151-10-0	: 1437
151-13-3	: 8304
151-18-8	: 10092
151-19-9	: 8453
151-56-4	: 498
151-67-7	: 5458
152-16-9	: 5220
152-58-9	: 9749
152-93-2	: 11107
152-95-4	: 2824
154-17-6	: 7110
154-41-6	: 1838
154-42-7	: 10762
154-93-8	: 11994
156-08-1	: 1575
156-10-5	: 829
156-38-7	: 906
156-43-4	: 737
156-59-2	: 5896
156-60-5	: 5895
156-87-6	: 10337
157-40-4	: 11527
180-43-8	: 11537
184-97-4	: 5240
189-55-9	: 2850
191-07-1	: 4145
191-24-2	: 2471
191-28-6	: 12055
192-65-4	: 8061
192-97-2	: 2436
195-19-7	: 2406
198-55-0	: 9161
205-25-4	: 2405
205-82-3	: 2738
205-99-2	: 631
206-44-0	: 5966
207-08-9	: 2739
207-83-0	: 5148
208-96-8	: 3
212-74-8	: 11644
213-46-7	: 9601
217-59-4	: 11910
218-01-9	: 4114
222-93-5	: 8996
224-42-0	: 5143
225-51-4	: 630
225-61-6	: 2394
229-87-8	: 9188
230-07-9	: 9191
230-27-3	: 2474
230-46-6	: 9190
238-84-6	: 2393
250-84-0	: 11688
253-52-1	: 9599
253-66-7	: 4136
253-82-7	: 11195
254-04-6	: 2772
254-79-5	: 8065
254-87-5	: 2870
255-50-5	: 2434
258-31-1	: 6621
260-36-6	: 2473
260-94-6	: 314
261-31-4	: 11811
262-30-6	: 11353
271-44-3	: 7235
271-58-9	: 2390
271-89-6	: 2438
271-95-4	: 2389
272-16-2	: 2387
273-53-0	: 2877
274-07-7	: 2412
274-09-9	: 2415
274-40-8	: 7324
275-51-4	: 504
278-30-8	: 2991
279-23-2	: 2914
280-57-9	: 5111
280-65-9	: 2990
281-23-2	: 11860
285-67-6	: 8558
286-08-8	: 2915
286-18-0	: 478
286-20-4	: 8554
287-23-0	: 4170
287-27-4	: 11691
287-92-3	: 4774
288-13-1	: 10815
288-14-2	: 7402
288-32-4	: 7192
288-36-8	: 11846
288-39-1	: 11658
288-42-6	: 8575
288-47-1	: 11667
288-92-6	: 11657
288-94-8	: 11652
288-99-3	: 8563
289-80-5	: 10855
289-95-2	: 11049
290-37-9	: 10801
290-67-5	: 5202
290-87-9	: 11827
290-96-0	: 11651
291-21-4	: 11919
291-64-5	: 4221
292-46-6	: 8998
292-64-8	: 4736
293-55-0	: 4724
293-96-9	: 4202
294-62-2	: 4210
295-48-7	: 4759
297-97-2	: 11720
298-02-2	: 9513
298-04-4	: 5260
298-12-4	: 223
298-45-3	: 2470
298-46-4	: 5144
298-81-7	: 6187
298-96-4	: 11653
299-84-3	: 9584
299-86-5	: 4153
300-38-9	: 11927
300-57-2	: 2165
300-62-9	: 1578
300-76-5	: 7702
300-87-8	: 7403
301-00-8	: 8285
301-12-2	: 8632
301-21-3	: 9399
302-17-0	: 5582
302-27-2	: 306
302-72-7	: 324
302-84-1	: 11356
303-04-8	: 3754
303-38-8	: 2568
304-21-2	: 11038
304-84-7	: 611
305-01-1	: 2804
305-03-3	: 1101
305-84-0	: 7133
305-85-1	: 9304
306-08-1	: 913
306-44-5	: 9789
306-83-2	: 5540
306-94-5	: 7934
306-98-9	: 4342
307-11-9	: 11912
307-34-6	: 8415
307-45-9	: 5050
308-24-7	: 4498
309-00-2	: 5177
311-89-7	: 3228
312-40-3	: 11411
312-84-5	: 11355
313-72-4	: 7935
314-13-6	: 7877
314-40-9	: 3139
315-30-0	: 10852
317-52-2	: 6753
319-85-7	: 4406
319-86-8	: 4407
319-88-0	: 2340
319-89-1	: 4269
320-72-9	: 2566
321-28-8	: 1722
321-38-0	: 7893
321-60-8	: 3074
323-09-1	: 7894
324-74-3	: 3075
326-56-7	: 2420
326-59-0	: 2416
326-62-5	: 952
327-54-8	: 2272
327-57-1	: 8214
328-39-2	: 7412
328-42-7	: 3339
328-50-7	: 8781
328-74-5	: 637
328-96-1	: 723
329-71-5	: 9343
330-54-1	: 12000
330-55-2	: 7415
331-39-5	: 10632
332-77-4	: 6087
333-20-0	: 11711
333-41-5	: 9582
333-93-7	: 3260
334-28-1	: 6463
334-42-9	: 6359
334-43-0	: 6365
334-48-5	: 5059
334-56-5	: 5052
334-59-8	: 5064
334-61-2	: 5035
334-62-3	: 5037
334-64-5	: 5074
334-88-3	: 7491
335-36-4	: 6109
335-44-4	: 3452
335-48-8	: 3275
335-57-9	: 6418
336-19-6	: 4647
336-59-4	: 3522
338-65-8	: 5471
338-83-0	: 9813
339-43-5	: 2192
343-65-7	: 1102
344-04-7	: 1078
344-07-0	: 1220
345-35-7	: 1194
346-18-9	: 2857
348-51-6	: 1166
348-52-7	: 1719
348-54-9	: 752
348-70-9	: 135
349-88-2	: 2246
350-03-8	: 5868
350-46-9	: 1735
350-50-5	: 1725
350-51-6	: 1634
350-90-3	: 10637
352-11-4	: 1195
352-32-9	: 1728
352-33-0	: 1168
352-34-1	: 1720
352-70-5	: 1727
352-91-0	: 9844
352-93-2	: 5644
352-97-6	: 6253
353-25-3	: 8429
353-29-7	: 5294
353-36-6	: 5593
353-42-4	: 3138
353-49-1	: 4005
353-50-4	: 4008
353-54-8	: 7550
353-55-9	: 7494
353-58-2	: 7468
353-59-3	: 7464
353-82-2	: 279
353-83-3	: 5667
353-85-5	: 289
353-89-9	: 11432
354-04-1	: 5526
354-06-3	: 5457
354-11-0	: 5639
354-12-1	: 5658
354-14-3	: 5638
354-15-4	: 5660
354-21-2	: 5659
354-23-4	: 5539
354-25-6	: 5485
354-26-7	: 5490
354-51-8	: 5519
354-56-3	: 5618
354-58-5	: 5662
354-92-7	: 10050
354-96-1	: 3414
355-02-2	: 4499
355-04-4	: 8854
355-20-4	: 3287
355-25-9	: 3255
355-42-0	: 6843
355-63-5	: 6560
355-68-0	: 4381
355-75-9	: 4643
355-80-6	: 8953
356-16-1	: 5642
356-24-1	: 3525
356-27-4	: 3524
356-28-5	: 3523
357-20-0	: 3808
357-49-3	: 10721
357-57-3	: 11564
357-67-5	: 11089
357-70-0	: 6200
358-06-5	: 11410
358-43-0	: 11484
358-99-6	: 5883
359-01-3	: 3379
359-02-4	: 5900
359-10-4	: 5885
359-11-5	: 5924
359-13-7	: 5708
359-25-1	: 116
359-28-4	: 5661
359-29-5	: 5923
359-32-0	: 296
359-35-3	: 5641
359-37-5	: 5925
359-70-6	: 5436
360-68-9	: 4109
360-89-4	: 3804
363-51-9	: 664
363-72-4	: 2091
366-18-7	: 3099
367-11-3	: 1422
367-12-4	: 9360
367-23-7	: 2346
367-25-9	: 700
367-64-6	: 265
368-34-3	: 132
368-43-4	: 2251
368-47-8	: 11486
368-77-4	: 2770
368-98-9	: 3116
370-14-9	: 9403
371-40-4	: 754
371-41-5	: 9362
371-62-0	: 5727
372-09-8	: 150
372-18-9	: 1423
372-19-0	: 753
372-20-3	: 9361
372-31-6	: 3592
372-38-3	: 2347
372-46-3	: 4403
372-54-3	: 6830
372-64-5	: 5249
372-75-8	: 8551
372-93-0	: 3621
373-05-7	: 6888
373-14-8	: 6829
373-28-4	: 6735
373-32-0	: 6936
373-44-4	: 8365
373-49-9	: 6670
373-52-4	: 7470
373-53-5	: 7517
373-80-8	: 11584
374-07-2	: 5537
374-25-4	: 3715
374-77-6	: 4343
375-01-9	: 3622
375-16-6	: 3692
375-22-4	: 3521
375-83-7	: 6432
375-96-2	: 8139
376-27-2	: 8438
376-87-4	: 9027
377-41-3	: 4177
377-70-8	: 4276
377-93-5	: 4199
378-68-7	: 10251
378-75-6	: 10317
381-61-3	: 3162
381-73-7	: 175
382-45-6	: 353
383-63-1	: 267
383-66-4	: 269
383-90-4	: 3114
384-04-3	: 3172
384-22-5	: 2060
386-17-4	: 7351
388-82-9	: 3042
389-08-2	: 8066
389-40-2	: 2093
392-56-3	: 1757
392-69-8	: 1740
392-83-6	: 1089
392-85-8	: 1737
393-52-2	: 2897
394-46-7	: 1633
396-01-0	: 10750
396-64-5	: 3043
398-23-2	: 3044
399-92-8	: 10328
400-52-2	: 266
401-56-9	: 133
401-78-5	: 1090
401-80-9	: 1738
402-31-3	: 997
402-43-7	: 1091
402-44-8	: 1739
402-67-5	: 1734
403-42-9	: 5815
403-43-0	: 2899
404-72-8	: 1721
404-86-4	: 8181
405-50-5	: 901
405-99-2	: 1635
406-74-6	: 190
407-25-0	: 264
407-75-0	: 8879
407-97-6	: 8713
407-98-7	: 8728
408-16-2	: 6491
408-27-5	: 8460
408-38-8	: 6639
420-04-2	: 4155
420-05-3	: 4162
420-12-2	: 11695
420-24-6	: 463
420-26-8	: 10047
420-45-1	: 9906
420-46-2	: 5665
420-90-6	: 10532
420-97-3	: 9897
420-99-5	: 9851
421-04-5	: 5487
421-06-7	: 5466
421-83-0	: 7537
425-23-0	: 3526
425-75-2	: 7533
428-59-1	: 8629
430-40-0	: 445
430-51-3	: 10437
430-57-9	: 5535

430-63-7 : 10561
430-66-0 : 5666
430-71-7 : 10506
430-95-5 : 10554
431-03-8 : 3359
431-06-1 : 5532
431-07-2 : 5488
431-27-6 : 10558
431-52-7 : 10607
431-53-8 : 10557
432-25-7 : 4636
433-52-3 : 134
434-03-7 : 9758
434-13-9 : 4099
434-16-2 : 4103
434-45-7 : 5872
434-64-0 : 2095
434-90-2 : 3027
436-05-5 : 11925
437-38-7 : 9801
437-50-3 : 12054
438-08-4 : 346
439-14-5 : 2410
442-16-0 : 317
442-51-3 : 11039
443-48-1 : 7206
443-83-4 : 1169
444-27-9 : 11682
445-29-4 : 2603
445-69-2 : 1321
446-10-6 : 1731
446-11-7 : 1729
446-33-3 : 1732
446-34-4 : 1730
446-35-5 : 1426
446-52-6 : 559
446-72-0 : 2807
447-31-4 : 5785
447-53-0 : 7814
447-60-9 : 2769
450-95-3 : 5816
451-13-8 : 895
451-40-1 : 5812
451-46-7 : 2606
451-80-9 : 1654
452-35-7 : 2865
452-58-4 : 10923
452-71-1 : 755
452-73-3 : 1170
452-86-8 : 1528
453-13-4 : 10369
453-18-9 : 192
453-20-3 : 6137
454-14-8 : 10415
454-31-9 : 176
455-14-1 : 857
455-31-2 : 1425
455-32-3 : 2909
455-38-9 : 2604
456-22-4 : 2605
456-47-3 : 1879
456-48-4 : 560
456-49-5 : 1723
457-87-4 : 1576
458-03-7 : 1655
458-35-5 : 9370
458-36-6 : 10514
458-37-7 : 6334
458-88-8 : 9708
459-22-3 : 953
459-26-7 : 1656
459-56-3 : 1880
459-57-4 : 561
459-60-9 : 1724
459-67-6 : 4940
459-72-3 : 191
459-73-4 : 6263
460-00-4 : 1044
460-12-8 : 3177
460-13-9 : 10046
460-19-5 : 5550
460-35-5 : 9871
460-55-9 : 11429
461-11-0 : 6460
461-35-8 : 10184
461-58-5 : 6280
461-72-3 : 7216
461-98-3 : 11045
462-02-2 : 11907
462-06-6 : 1717
462-08-8 : 10858
462-18-0 : 11891
462-23-7 : 3927
462-38-4 : 9856
462-39-5 : 9905
462-43-1 : 10382
462-72-6 : 3237
462-73-7 : 3250
462-94-2 : 8735
462-95-3 : 5603
463-04-7 : 8096
463-11-6 : 8405
463-16-1 : 8151
463-17-2 : 11947
463-23-0 : 8126
463-24-1 : 8165
463-33-2 : 11939
463-40-1 : 8287
463-49-0 : 9772
463-51-4 : 5927
463-52-5 : 7563
463-58-1 : 4035
463-71-8 : 4032
463-72-9 : 3971
463-82-1 : 9910
463-88-7 : 5878
464-06-2 : 3457
464-10-8 : 7551
464-15-3 : 2932
464-17-5 : 2970
464-41-5 : 2920
464-49-3 : 2961
464-72-2 : 5580
464-85-7 : 6178
464-86-8 : 6177
464-88-0 : 4786
465-24-7 : 7569
465-42-9 : 4063
465-74-7 : 12036
465-99-6 : 8545
466-14-8 : 3199
466-43-3 : 475
466-49-9 : 473
466-81-9 : 8567
466-96-6 : 7672
466-97-7 : 7665
467-14-1 : 7671
467-55-0 : 11534
467-60-7 : 9676
467-85-6 : 6968
467-98-1 : 7677
468-28-0 : 4281
468-45-1 : 11516
469-59-0 : 12046
470-40-6 : 4954
470-49-5 : 8874
470-67-7 : 8556
470-82-6 : 8560
470-90-6 : 4078
470-99-5 : 4702
471-04-5 : 6461
471-12-5 : 2982
471-16-9 : 2984
471-25-0 : 10738
471-35-2 : 5110
471-44-3 : 131
471-46-5 : 5493
471-47-6 : 104
471-87-4 : 11177
471-95-4 : 3141
472-20-8 : 4666
472-70-8 : 4061
472-93-5 : 4052
473-02-9 : 11865
473-06-3 : 2977
473-19-8 : 2933
473-30-3 : 11663
473-34-7 : 2210
473-81-4 : 10252
473-91-6 : 4937
473-98-3 : 7416
474-25-9 : 4097
474-40-8 : 11556
474-62-4 : 5386
474-69-1 : 5382
474-77-1 : 4112
474-86-2 : 5398
475-03-6 : 8000
475-20-7 : 7568
475-25-2 : 628
475-31-0 : 6273
475-38-7 : 7853
475-67-2 : 5161
475-81-0 : 5157
476-28-8 : 6201
476-32-4 : 4071
476-45-9 : 7855
476-69-7 : 5162
476-70-0 : 5156
477-27-0 : 92
477-32-7 : 3540
477-47-4 : 6191
477-60-1 : 11924
477-75-8 : 2375
477-89-4 : 5218
477-90-7 : 10786
478-40-0 : 7856
478-43-3 : 371
478-61-5 : 2911
478-71-7 : 9173
478-94-4 : 5374
478-95-5 : 5375
479-00-5 : 5373
479-13-0 : 2466
479-18-5 : 10761
479-21-0 : 7608
479-22-1 : 4259
479-23-2 : 2391
479-27-6 : 7798
479-45-8 : 818
479-59-4 : 2737
480-11-5 : 2810
480-15-9 : 2842
480-16-0 : 2816
480-40-0 : 2811
480-41-1 : 2802
480-44-4 : 2809
480-64-8 : 2573
480-68-2 : 11102
480-72-8 : 9
480-91-1 : 7369
481-06-1 : 8057
481-20-9 : 4106
481-21-0 : 4105
481-26-5 : 9741
481-29-8 : 347
481-37-8 : 483
481-39-0 : 7860
481-42-5 : 7863
481-49-2 : 8630
481-72-1 : 397
481-74-3 : 398
482-05-3 : 3038
482-28-0 : 7306
482-44-0 : 6189
482-66-6 : 4756
482-68-8 : 11347
482-74-6 : 2435
482-89-3 : 7331
483-04-5 : 8571
483-17-0 : 5367
483-18-1 : 5368
483-55-6 : 7861
483-64-7 : 2672
483-65-8 : 9183
483-76-1 : 7899
483-78-3 : 7830
484-11-7 : 9192
484-29-7 : 6180
484-47-9 : 7215
484-67-3 : 9472
484-73-1 : 5223
484-89-9 : 4265
484-93-5 : 491
485-31-4 : 3003
485-47-2 : 7267
485-64-3 : 4121
485-65-4 : 4122
485-71-2 : 4133
485-72-3 : 2830
485-89-2 : 11233
486-25-9 : 5975
486-34-0 : 8012
486-35-1 : 2805
486-39-5 : 7396
486-47-5 : 7377
486-66-8 : 2827
486-70-4 : 11331
486-84-0 : 11040
486-86-2 : 7578
486-89-5 : 7573
487-06-9 : 2818
487-10-5 : 5127
487-19-4 : 10987
487-21-8 : 10748
487-27-4 : 7575
487-68-3 : 599
487-79-6 : 11144
487-93-4 : 7326
488-10-8 : 4949
488-17-5 : 1527
488-23-3 : 2278
488-43-7 : 6207
488-59-5 : 7336
488-70-0 : 9473
488-71-1 : 848
488-73-3 : 4074
488-81-3 : 11343
488-86-8 : 4939
488-87-9 : 1512
488-92-6 : 11128
488-93-7 : 6051
488-97-1 : 11866
488-98-2 : 1736
489-84-9 : 505
489-86-1 : 507
489-98-5 : 863
490-02-8 : 10811
490-03-9 : 4709
490-06-2 : 1536
490-10-8 : 4758
490-11-9 : 10930
490-53-9 : 7390
490-55-1 : 11671
490-64-2 : 2733
490-65-3 : 7923
490-79-9 : 2570
490-91-5 : 4267
491-04-3 : 4697
491-09-8 : 4718
491-14-5 : 9120
491-18-9 : 11127
491-31-6 : 2791
491-34-9 : 11299
491-35-0 : 11279
491-37-2 : 2799
491-38-3 : 2793
491-67-8 : 2844
491-70-3 : 2814
491-88-3 : 6144
492-17-1 : 3029
492-22-8 : 11812
492-27-3 : 11232
492-38-6 : 927
492-80-8 : 657
492-89-7 : 1543
492-97-7 : 3112
492-99-9 : 4379
493-01-6 : 7790
493-02-7 : 7791
493-05-0 : 2776
493-08-3 : 2777
493-09-4 : 2414
493-49-2 : 7398
493-52-7 : 2587
493-57-2 : 2401
494-03-1 : 7710
494-12-2 : 2782
494-40-6 : 2356
494-47-3 : 7488
494-52-0 : 11008
494-55-3 : 5215
494-90-6 : 2465
494-97-3 : 11017
494-99-5 : 1440
495-08-9 : 1520
495-40-9 : 3676
495-41-0 : 3880
495-45-4 : 3876
495-61-4 : 4685
495-69-2 : 6255
495-76-1 : 2427
495-85-2 : 10790
495-91-0 : 9633
496-10-6 : 7281
496-11-7 : 7250
496-15-1 : 7297
496-16-2 : 2442
496-41-3 : 2439
496-46-8 : 7191
496-49-1 : 10471
496-64-0 : 10794
496-67-3 : 3196
496-72-0 : 1276
496-73-1 : 1530
496-74-2 : 1564
496-77-5 : 8474
496-78-6 : 9494
497-03-0 : 3699
497-04-1 : 9995
497-06-3 : 3794
497-18-7 : 4009
497-20-1 : 4767
497-26-7 : 5210
497-30-3 : 7204
497-32-5 : 2922
497-39-2 : 9222
497-59-6 : 10775
497-76-7 : 6224
498-00-0 : 1886
498-02-2 : 5824
498-15-7 : 2974
498-23-7 : 3786
498-24-8 : 3785
498-59-9 : 5018
498-60-2 : 6042
498-62-4 : 11734
498-67-9 : 1379
498-94-2 : 9642
498-95-3 : 9641
498-96-4 : 10912
499-04-7 : 10913
499-06-9 : 2583
499-75-2 : 9415
499-80-9 : 10928
499-81-0 : 10931
499-89-8 : 4241
500-05-0 : 10773
500-22-1 : 10888
500-38-9 : 1515
500-44-7 : 11011
500-49-2 : 1550
500-55-0 : 928
500-99-2 : 9306
501-15-5 : 1533
501-19-9 : 9390
501-24-6 : 9441
501-30-4 : 10795
501-52-0 : 2126
501-53-1 : 4027
501-65-5 : 1714
501-92-8 : 9458
501-97-3 : 2134
502-41-0 : 4229
502-42-1 : 4234
502-44-3 : 8591
502-49-8 : 4740
502-56-7 : 8175
502-59-0 : 6354
502-61-4 : 5329
502-65-8 : 4053
502-72-7 : 4761
502-73-8 : 6300
502-99-8 : 8486
503-01-5 : 6532
503-09-3 : 8616
503-17-3 : 3898
503-28-6 : 5125
503-29-7 : 494
503-30-0 : 8592
503-38-8 : 4029
503-40-2 : 7514
503-46-8 : 4691
503-60-6 : 3736
503-64-0 : 3815

503-66-2 : 10267
503-74-2 : 3539
503-76-4 : 9869
503-80-0 : 452
503-93-5 : 4220
504-02-9 : 4376
504-03-0 : 9650
504-15-4 : 1531
504-20-1 : 6340
504-24-5 : 10859
504-29-0 : 10857
504-31-4 : 10788
504-53-0 : 9000
504-57-4 : 8105
504-61-0 : 3859
504-63-2 : 9983
504-64-3 : 9773
504-78-9 : 11681
505-20-4 : 5261
505-22-6 : 5186
505-23-7 : 5262
505-29-3 : 5263
505-32-8 : 6671
505-34-0 : 10067
505-48-6 : 8386
505-60-2 : 5645
505-84-0 : 10074
505-92-0 : 5338
506-03-6 : 10010
506-12-7 : 6314
506-23-0 : 8284
506-24-1 : 8316
506-30-9 : 5355
506-31-0 : 5364
506-32-1 : 5361
506-33-2 : 5282
506-42-3 : 8311
506-46-7 : 6624
506-48-9 : 8221
506-50-3 : 11820
506-52-5 : 6626
506-68-3 : 4165
506-77-4 : 4166
506-78-5 : 7337
506-80-9 : 4036
506-82-1 : 3929
506-85-4 : 6024
506-89-8 : 12020
506-93-4 : 6286
506-96-7 : 291
507-02-8 : 302
507-09-5 : 5646
507-19-7 : 9846
507-20-0 : 9863
507-25-5 : 7542
507-32-4 : 457
507-36-8 : 3240
507-40-4 : 7183
507-42-6 : 5581
507-45-9 : 3286
507-47-1 : 5680
507-60-8 : 3142
508-02-1 : 8547
508-32-7 : 11867
508-44-1 : 5185
508-54-3 : 7675
508-65-6 : 4068
509-12-6 : 2938
509-14-8 : 7544
509-15-9 : 6203
509-20-6 : 305
509-64-8 : 7670
509-87-5 : 11096
510-15-6 : 4002
510-20-3 : 9939
510-90-7 : 11063
511-07-9 : 5392
511-08-0 : 5391
511-09-1 : 5389
511-10-4 : 5390
511-18-2 : 8210
511-20-6 : 5377
511-21-7 : 5378
511-34-2 : 11535
511-67-1 : 11086
511-70-6 : 3198
511-96-6 : 11530
512-04-9 : 11536
512-12-9 : 8585
512-13-0 : 2941
512-16-3 : 4309
512-48-1 : 9003
512-56-1 : 9564
512-60-7 : 11862
512-69-6 : 6219
512-85-6 : 5184
513-02-0 : 9568
513-08-6 : 9567
513-12-2 : 5725
513-23-5 : 2985
513-31-5 : 10545
513-35-9 : 3801
513-36-0 : 9862
513-37-1 : 10539
513-38-2 : 10057
513-42-8 : 10702
513-44-0 : 10158
513-81-5 : 3159
513-86-0 : 3661
513-88-2 : 10425
513-92-8 : 5919
513-96-2 : 10115
514-10-3 : 9167
514-12-5 : 2953
514-92-1 : 4059
515-12-8 : 4401
515-25-3 : 11178
515-30-0 : 914
515-40-2 : 1143
515-46-8 : 2223
515-49-1 : 2189
515-59-3 : 2200
515-82-2 : 6000
515-83-3 : 5769
515-84-4 : 248
515-88-8 : 2944
515-90-2 : 2964
515-94-6 : 326
515-96-8 : 105
516-05-2 : 9960
516-06-3 : 12040
516-12-1 : 11159
516-78-9 : 5384
516-85-8 : 5380
516-95-0 : 4108
517-04-4 : 5401
517-06-6 : 5407
517-07-7 : 5402
517-09-9 : 5397
517-10-2 : 4111
517-22-6 : 11126
517-23-7 : 6139
517-25-9 : 7562
517-28-2 : 629
517-60-2 : 1748
517-66-8 : 2468
517-82-8 : 7857
517-92-0 : 401
517-97-5 : 5159
518-05-8 : 7805
518-17-2 : 7333
518-28-5 : 6192
518-44-5 : 2572
518-61-6 : 9502
518-69-4 : 5149
518-75-2 : 2774
518-77-4 : 5151
518-82-1 : 419
519-02-8 : 7426
519-05-1 : 2608
519-09-5 : 481
519-34-6 : 7610
519-37-9 : 10758
519-44-8 : 1517
519-72-2 : 1804
519-73-3 : 1982
519-87-9 : 57
520-03-6 : 7366
520-18-3 : 2843
520-26-3 : 2794
520-33-2 : 2800
520-36-5 : 2806
520-45-6 : 10779
520-85-4 : 9752
521-04-0 : 11554
521-11-9 : 350
521-12-0 : 351
521-18-6 : 349
521-24-4 : 7970
521-31-3 : 9600
521-35-7 : 5154
521-74-4 : 11313
521-85-7 : 3108
522-12-3 : 2796
522-27-0 : 5583
522-57-6 : 7611
522-66-7 : 4123
522-75-8 : 2404
523-27-3 : 375
523-31-9 : 1337
523-47-7 : 7898
523-80-8 : 2423
524-15-2 : 6179
524-30-1 : 2823
524-34-5 : 11236
524-36-7 : 10978
524-40-3 : 10883
524-42-5 : 7842
524-63-0 : 4119
524-80-1 : 3982
524-96-9 : 1797
525-06-4 : 5928
525-15-5 : 11041
525-37-1 : 7868
525-64-4 : 5971
525-68-8 : 11253
525-82-6 : 2840
526-08-9 : 2201
526-31-8 : 11923
526-35-2 : 8586
526-55-6 : 7305
526-73-8 : 2357
526-75-0 : 9307
526-84-1 : 3777
526-85-2 : 9493
526-86-3 : 4263
526-95-4 : 6209
526-99-8 : 6193
527-07-1 : 6213
527-17-3 : 4270
527-18-4 : 1552
527-35-5 : 9475
527-52-6 : 7111
527-53-7 : 2279
527-55-9 : 1513
527-60-6 : 9495
527-61-7 : 4264
527-62-8 : 9205
527-69-5 : 6040
527-72-0 : 11737
527-73-1 : 7211
527-84-4 : 1988
527-85-5 : 619
527-89-9 : 1125
527-93-5 : 5153
528-21-2 : 5873
528-23-4 : 4784
528-29-0 : 1491
528-44-9 : 2319
528-45-0 : 2590
528-48-3 : 2813
528-53-0 : 2849
528-71-2 : 9197
528-75-6 : 551
528-76-7 : 2178
528-79-0 : 9420
528-81-4 : 467
520-00-0 : 4602
529-08-8 : 506
529-16-8 : 2968
529-17-9 : 484
529-19-1 : 2763
529-20-4 : 587
529-21-5 : 10964
529-28-2 : 1775
529-33-9 : 8041
529-34-0 : 8046
529-35-1 : 8042
529-36-2 : 8007
529-64-6 : 915
529-65-7 : 62
529-86-2 : 432
529-92-0 : 11216
530-14-3 : 5818
530-22-3 : 2464
530-40-5 : 10892
530-43-8 : 6648
530-44-9 : 7613
530-48-3 : 1636
530-50-7 : 7157
530-55-2 : 4262
530-62-1 : 7193
530-78-9 : 2729
530-85-8 : 10740
530-91-6 : 8043
530-93-8 : 8047
531-02-2 : 3097
531-18-0 : 7584
531-29-3 : 6226
531-37-3 : 9383
531-39-5 : 8886
531-75-9 : 2822
531-81-7 : 2775
531-84-0 : 7774
531-91-9 : 3035
531-95-3 : 2786
532-18-3 : 7727
532-24-1 : 490
532-27-4 : 5790
532-32-1 : 2725
532-48-9 : 5825
532-54-7 : 876
532-82-1 : 1284
532-96-7 : 2717
533-18-6 : 211
533-24-4 : 1544
533-30-2 : 2860
533-32-4 : 3201
533-45-9 : 11670
533-48-2 : 7223
533-50-6 : 3679
533-58-4 : 9371
533-68-6 : 3469
533-70-0 : 701
533-73-3 : 2367
533-75-5 : 4243
533-86-8 : 8300
533-98-2 : 3265
534-00-9 : 3238
534-07-6 : 10426
534-13-4 : 11804
534-15-6 : 5548
534-22-5 : 6130
534-26-9 : 7197
534-52-1 : 9405
534-59-8 : 9929
534-84-9 : 4611
534-85-0 : 1282
535-11-5 : 10207
535-13-7 : 10220
535-15-9 : 164
535-46-6 : 11548
535-75-1 : 9640
535-77-3 : 1989
535-80-8 : 2535
535-83-1 : 11029
535-87-5 : 2554
535-89-7 : 11043
536-08-3 : 2575
536-17-4 : 11684
536-25-4 : 2501
536-33-4 : 10885
536-38-9 : 5778
536-50-5 : 1859
536-57-2 : 2182
536-59-4 : 4664
536-60-7 : 1902
536-66-3 : 2689
536-69-6 : 10898
536-74-3 : 1715
536-75-4 : 10958
536-78-7 : 10957
536-88-9 : 10965
536-89-0 : 7164
536-90-3 : 767
536-95-8 : 2721
537-17-7 : 11830
537-26-8 : 488
537-29-1 : 916
537-45-1 : 4279
537-46-2 : 1573
537-47-3 : 7146
537-49-5 : 11933
537-64-4 : 7432
537-65-5 : 1263
537-91-7 : 5247
537-92-8 : 77
538-08-9 : 4159
538-23-8 : 8440
538-24-9 : 5319
538-32-9 : 12027
538-39-6 : 1590
538-41-0 : 635
538-43-2 : 10026
538-44-3 : 9106
538-49-8 : 11000
538-51-2 : 840
538-56-7 : 10684
538-58-9 : 8683
538-62-5 : 5122
538-64-7 : 3768
538-68-1 : 2107
538-74-9 : 2287
538-75-0 : 4291
538-81-8 : 1095
538-86-3 : 1930
538-93-2 : 2039
539-03-7 : 48
539-08-2 : 9796
539-15-1 : 9320
539-23-1 : 10860
539-30-0 : 1658
539-47-9 : 10638
539-52-6 : 6133
539-74-2 : 10208
539-80-0 : 4242
539-82-2 : 8878
539-86-6 : 10595
539-88-8 : 8907
539-89-9 : 3961
539-90-2 : 3563
539-92-4 : 3997
540-07-8 : 6907
540-08-9 : 6321
540-09-0 : 11857
540-10-3 : 6651
540-18-1 : 3579
540-36-3 : 1424
540-37-4 : 761
540-38-5 : 9373
540-42-1 : 10310
540-47-6 : 4980
540-51-2 : 5685
540-54-5 : 9848
540-61-4 : 272
540-63-6 : 5590
540-67-0 : 5602
540-69-2 : 6003
540-73-8 : 7155
540-80-7 : 8085
540-82-9 : 11576
540-84-1 : 8848
540-88-5 : 177
540-97-6 : 4624
541-01-5 : 6528
541-02-6 : 4901
541-05-9 : 5002
541-16-2 : 9925
541-20-8 : 8738
541-28-6 : 3384
541-31-1 : 3440
541-33-3 : 3270
541-35-5 : 3194
541-41-3 : 4018
541-42-4 : 8092
541-46-8 : 3206
541-47-9 : 3846
541-50-4 : 3568
541-53-7 : 11718
541-58-2 : 11674
541-59-3 : 11121
541-73-1 : 1355
541-88-8 : 124
541-91-3 : 4763
542-05-2 : 8782

CAS NUMBER INDEX OF ORGANIC COMPOUNDS (continued)

542-08-5 : 3574
542-10-9 : 5570
542-15-4 : 821
542-18-7 : 4339
542-28-9 : 10798
542-37-0 : 3546
542-46-1 : 4218
542-50-7 : 6303
542-52-9 : 3998
542-54-1 : 8831
542-55-2 : 6013
542-56-3 : 8094
542-58-5 : 5695
542-59-6 : 5577
542-69-8 : 3381
542-76-7 : 10097
542-81-4 : 5481
542-85-8 : 5600
542-88-1 : 7525
542-90-5 : 11703
542-91-6 : 11409
542-92-7 : 4764
543-20-4 : 3366
543-21-5 : 3902
543-24-8 : 6251
543-27-1 : 4024
543-28-2 : 3959
543-29-3 : 8078
543-38-4 : 7135
543-59-9 : 8719
543-67-9 : 8098
543-75-9 : 5203
543-80-6 : 110
543-83-9 : 6283
543-86-2 : 3633
543-87-3 : 3637
544-00-3 : 3226
544-01-4 : 3418
544-02-5 : 3436
544-10-5 : 6737
544-13-8 : 8767
544-16-1 : 8083
544-25-2 : 4240
544-35-4 : 8226
544-40-1 : 3433
544-62-7 : 10023
544-63-8 : 11622
544-65-0 : 11637
544-66-1 : 11638
544-72-9 : 8286
544-73-0 : 8282
544-76-3 : 6632
544-77-4 : 6640
544-85-4 : 5348
544-97-8 : 12066
545-06-2 : 288
545-26-6 : 4048
545-47-1 : 7417
545-93-7 : 11078
546-06-5 : 4143
546-33-8 : 5004
546-43-0 : 8059
546-45-2 : 5005
546-56-5 : 4994
546-97-4 : 5926
547-25-1 : 6022
547-44-4 : 2190
547-52-4 : 2188
547-58-0 : 2222
547-63-7 : 10301
547-65-9 : 6147
547-81-9 : 5405
547-91-1 : 11295
548-40-3 : 8631
548-51-6 : 2642
548-62-9 : 7453
548-80-1 : 7878
548-98-1 : 4093
550-10-7 : 5214
550-24-3 : 4261
550-44-7 : 7365
550-60-7 : 8034
550-74-3 : 10848
550-82-3 : 9508
550-97-0 : 2644
551-01-9 : 9877
551-06-4 : 7906
551-09-7 : 5514
551-62-2 : 2270
551-76-8 : 9489
551-77-9 : 9488
551-78-0 : 9487
551-88-2 : 8834
551-92-8 : 7202
552-16-9 : 2703
552-30-7 : 7341
552-32-9 : 81
552-41-0 : 5823
552-45-4 : 1198
552-58-9 : 2812
552-59-0 : 2828
552-70-5 : 479
552-72-7 : 9646
552-79-4 : 1863
552-82-9 : 807
552-89-6 : 591
552-94-3 : 2616
553-03-7 : 11327
553-17-3 : 9384
553-19-5 : 2397
553-20-8 : 84
553-26-4 : 3104
553-39-9 : 7953
553-60-6 : 10907
553-69-5 : 1913
553-79-7 : 827
553-82-2 : 1381
553-86-6 : 2461
553-90-2 : 5563
553-94-6 : 1021
553-97-9 : 4266
554-00-7 : 686
554-01-8 : 11106
554-12-1 : 10296
554-14-3 : 11769
554-21-2 : 5212
554-35-8 : 10104
554-68-7 : 5418
554-70-1 : 9521
554-84-7 : 9436
555-03-3 : 1936
555-10-2 : 4673
555-16-8 : 593
555-21-5 : 967
555-30-6 : 11931
555-35-1 : 6645
555-43-1 : 8272
555-44-2 : 6656
555-45-3 : 11628
555-57-7 : 1823
555-59-9 : 3853
555-77-1 : 5412
555-89-5 : 1965
556-10-5 : 12025
556-22-9 : 6275
556-24-1 : 3552
556-27-4 : 333
556-50-3 : 6265
556-52-5 : 8618
556-56-9 : 10573
556-61-6 : 7521
556-64-9 : 11705
556-67-2 : 4993
556-68-3 : 4741
556-69-4 : 8482
556-70-7 : 5079
556-71-8 : 4726
556-82-1 : 3869
556-88-7 : 6287
556-89-8 : 12024
556-96-7 : 1019
557-00-6 : 3562
557-05-1 : 8277
557-11-9 : 12028
557-17-5 : 10069
557-24-4 : 3817
557-25-5 : 3505
557-30-2 : 5492
557-31-3 : 10566
557-40-4 : 10592
557-59-5 : 11607
557-61-9 : 8222
557-66-4 : 5433
557-68-6 : 7471
557-89-1 : 453
557-91-5 : 5517
557-93-7 : 10530
557-98-2 : 10535
557-99-3 : 301
558-13-4 : 7539
558-17-8 : 10058
558-25-8 : 7538
558-30-5 : 8607
558-37-2 : 3760
558-38-3 : 10352
558-42-9 : 10354
558-43-0 : 10017
559-70-6 : 8548
560-08-7 : 3931
560-21-4 : 8849
560-22-5 : 9068
560-23-6 : 9065
560-95-2 : 7476
561-07-9 : 311
561-20-6 : 11352
561-25-1 : 7664
561-27-3 : 7667
561-83-1 : 11090
561-94-4 : 5393
562-49-2 : 8763
562-74-3 : 4699
563-03-1 : 9387
563-04-2 : 9570
563-12-2 : 5931
563-16-6 : 6775
563-41-7 : 7143
563-45-1 : 3800
563-46-2 : 3799
563-47-3 : 10540
563-52-0 : 3723
563-57-5 : 10553
563-58-6 : 10547
563-70-2 : 7473
563-78-0 : 3759
563-79-1 : 3761
563-80-4 : 3668
563-83-7 : 9798
563-84-8 : 3606
564-00-1 : 3007
564-02-3 : 8847
564-03-4 : 9069
564-04-5 : 8966
564-36-3 : 5387
564-37-4 : 5388
565-00-4 : 2921
565-53-7 : 161
565-59-3 : 8761
565-60-6 : 8941
565-63-9 : 3845
565-64-0 : 10245
565-67-3 : 8943
565-69-5 : 8982
565-75-3 : 8850
565-76-4 : 9066
565-77-5 : 9070
565-80-0 : 8967
567-18-0 : 7973
568-02-5 : 8062
568-21-8 : 5158
568-93-4 : 399
568-99-0 : 383
569-08-4 : 394
569-31-3 : 7355
569-34-6 : 10759
569-41-5 : 7824
569-51-7 : 2318
569-57-3 : 1154
569-64-2 : 7455
569-65-3 : 9605
569-77-7 : 2408
570-08-1 : 9923
570-22-9 : 7195
570-24-1 : 802
571-58-4 : 7820
571-61-9 : 7821
572-59-8 : 4125
572-83-8 : 386
572-93-0 : 396
573-17-1 : 9164
573-20-6 : 7841
573-26-2 : 75
573-33-1 : 7199
573-56-8 : 9344
573-58-0 : 7969
573-97-7 : 8022
573-98-8 : 7818
574-09-4 : 5813
574-12-9 : 2841
574-39-0 : 3983
574-84-5 : 2808
575-37-1 : 7823
575-38-2 : 7837
575-41-7 : 7819
575-43-9 : 7822
575-44-0 : 7836
575-74-6 : 607
576-22-7 : 1020
576-23-8 : 1017
576-24-9 : 9282
576-26-1 : 9310
576-55-6 : 9468
576-83-0 : 1094
577-16-2 : 5851
577-19-5 : 1075
577-27-5 : 4989
577-33-3 : 429
577-37-7 : 7572
577-55-9 : 992
577-56-0 : 2476
577-59-3 : 5863
577-71-9 : 9345
577-91-3 : 2135
578-07-4 : 312
578-54-1 : 740
578-57-4 : 1050
578-58-5 : 1931
578-66-5 : 11206
578-67-6 : 11305
578-68-7 : 11203
578-76-7 : 10769
579-04-4 : 7112
579-07-7 : 10028
579-10-2 : 79
579-18-0 : 2511
579-21-5 : 5854
579-44-2 : 5822
579-60-2 : 9392
579-66-8 : 694
579-75-9 : 2664
579-92-0 : 2656
579-94-2 : 183
580-02-9 : 2482
580-13-2 : 7740
580-15-4 : 11205
580-16-5 : 11306
580-20-1 : 11307
580-22-3 : 11202
580-48-3 : 11828
580-51-8 : 3089
581-08-8 : 59
581-40-8 : 7825
581-42-0 : 7826
581-43-1 : 7838
581-45-3 : 11009
581-46-4 : 3102
581-47-5 : 3101
581-49-7 : 3105
581-50-0 : 3100
581-55-5 : 7512
581-89-5 : 7930
582-08-1 : 5128
582-16-1 : 7827
582-17-2 : 7839
582-22-9 : 1579
582-24-1 : 5833
582-33-2 : 2496
582-60-5 : 2380
582-61-6 : 2884
582-62-7 : 3671
582-77-4 : 622
582-78-5 : 623
583-03-9 : 1846
583-04-0 : 2722
583-05-1 : 8804
583-33-5 : 6052
583-39-1 : 2386
583-48-2 : 6776
583-50-6 : 3192
583-53-9 : 1288
583-55-1 : 1046
583-58-4 : 10940
583-61-9 : 10936
583-68-6 : 647
583-69-7 : 1503
583-70-0 : 1018
583-71-1 : 1022
583-75-5 : 650
583-78-8 : 9284
583-86-8 : 8276
584-02-1 : 8917
584-04-3 : 10371
584-12-3 : 6033
584-27-0 : 3331
584-70-3 : 621
584-79-2 : 336
584-84-9 : 1431
584-87-2 : 2609
584-90-7 : 5114
584-92-9 : 8399
584-94-1 : 6772
584-98-5 : 3299
584-99-6 : 3769
585-32-0 : 1806
585-34-2 : 9322
585-48-8 : 10874
585-68-2 : 6054
585-74-0 : 5852
585-76-2 : 2519
585-79-5 : 1076
585-81-9 : 2632
586-11-8 : 9346
586-27-6 : 4698
586-37-8 : 5840
586-38-9 : 2665
586-39-0 : 1644
586-42-5 : 2477
586-61-8 : 1059
586-62-9 : 4684
586-75-4 : 2888
586-76-5 : 2520
586-77-6 : 645
586-78-7 : 1077
586-84-5 : 6129
586-89-0 : 2478
586-95-8 : 10977
586-96-9 : 2059
586-98-1 : 10975
587-02-0 : 741
587-03-1 : 1898
587-04-2 : 530
587-54-2 : 2513
587-61-1 : 10873
587-85-9 : 7436
587-90-6 : 11997
588-01-2 : 11906
588-46-5 : 87
588-47-6 : 817
588-52-3 : 5113
588-63-6 : 1085
588-67-0 : 1115
588-72-7 : 1029
588-73-8 : 1031
588-96-5 : 1032
589-08-2 : 1581
589-09-3 : 841
589-10-6 : 1033
589-14-0 : 6120
589-15-1 : 1000
589-16-2 : 742
589-17-3 : 1008
589-18-4 : 1899
589-21-9 : 7141
589-29-7 : 1435
589-33-3 : 11113
589-38-8 : 6960
589-40-2 : 6012
589-43-5 : 6773
589-44-6 : 3465
589-53-7 : 6424
589-55-9 : 6475
589-59-3 : 3555
589-63-9 : 8473
589-75-3 : 8427
589-87-7 : 1048
589-92-4 : 4598

589-93-5	: 10938
589-98-0	: 8445
590-01-2	: 10214
590-02-3	: 125
590-11-4	: 5892
590-12-5	: 5891
590-13-6	: 10529
590-16-9	: 5463
590-18-1	: 3706
590-19-2	: 3145
590-26-1	: 5904
590-35-2	: 8760
590-36-3	: 8940
590-42-1	: 10066
590-50-1	: 8965
590-60-3	: 3634
590-66-9	: 4364
590-67-0	: 4530
590-73-8	: 6771
590-86-3	: 3186
590-90-9	: 3662
590-92-1	: 10205
590-93-2	: 3917
591-07-1	: 27
591-08-2	: 33
591-09-3	: 108
591-11-7	: 6154
591-12-8	: 6153
591-17-3	: 1055
591-18-4	: 1047
591-19-5	: 639
591-20-8	: 9229
591-22-0	: 10941
591-27-5	: 9202
591-31-1	: 584
591-34-4	: 10309
591-35-5	: 9287
591-47-9	: 4672
591-49-1	: 4670
591-50-4	: 1766
591-60-6	: 3569
591-62-8	: 3936
591-68-4	: 8862
591-76-4	: 6834
591-78-6	: 6959
591-80-0	: 9073
591-82-2	: 10065
591-84-4	: 11707
591-87-7	: 235
591-93-5	: 8659
591-95-7	: 8658
591-96-8	: 8660
592-01-8	: 3930
592-13-2	: 6774
592-20-1	: 10412
592-22-3	: 5250
592-27-8	: 6422
592-34-7	: 4012
592-35-8	: 3935
592-41-6	: 6999
592-42-7	: 6678
592-44-9	: 6676
592-45-0	: 6677
592-46-1	: 6680
592-49-4	: 6679
592-50-7	: 8817
592-51-8	: 9062
592-55-2	: 5461
592-57-4	: 4248
592-65-4	: 10154
592-76-7	: 6533
592-80-3	: 8935
592-82-5	: 3388
592-84-7	: 0004
592-88-1	: 10599
592-90-5	: 8590
593-03-3	: 6661
593-08-8	: 11889
593-12-4	: 8358
593-14-6	: 8361
593-45-3	: 8234
593-49-7	: 6302
593-50-0	: 11821
593-51-1	: 7446
593-52-2	: 446
593-53-3	: 7515
593-57-7	: 442
593-59-9	: 444
593-60-2	: 5880
593-61-3	: 5937
593-63-5	: 5938
593-66-8	: 5907
593-70-4	: 7481
593-71-5	: 7482
593-74-8	: 7435
593-75-9	: 7655
593-77-1	: 7447
593-79-3	: 7526
593-81-7	: 7445
593-88-4	: 449
593-90-8	: 3119
593-92-0	: 5890
593-95-3	: 4006
593-96-4	: 5455
593-98-6	: 7466
594-02-5	: 5546
594-03-6	: 5589
594-04-7	: 7503
594-08-1	: 4034
594-09-2	: 9522
594-10-5	: 11550
594-11-6	: 4981
594-15-0	: 7549
594-16-1	: 9883
594-18-3	: 7495
594-20-7	: 9894
594-22-9	: 7556
594-27-4	: 11544
594-34-3	: 9889
594-36-5	: 3253
594-37-6	: 9900
594-38-7	: 3385
594-39-8	: 3222
594-42-3	: 7527
594-44-5	: 5627
594-45-6	: 5622
594-47-8	: 94
594-51-4	: 3274
594-52-5	: 3236
594-56-9	: 3812
594-57-0	: 3247
594-58-1	: 10228
594-60-5	: 3618
594-61-6	: 10278
594-65-0	: 95
594-70-7	: 10085
594-71-8	: 9870
594-72-9	: 5536
594-73-0	: 5594
594-81-0	: 3271
594-82-1	: 3430
594-83-2	: 3644
594-89-8	: 10048
595-30-2	: 8559
595-37-9	: 3506
595-38-0	: 8724
595-39-1	: 7400
595-40-4	: 7401
595-41-5	: 8930
595-44-8	: 9902
595-49-3	: 9920
595-89-1	: 9733
595-90-4	: 11545
596-27-0	: 7353
596-29-2	: 7358
596-43-0	: 1061
596-51-0	: 11179
596-75-8	: 9937
597-04-6	: 3511
597-05-7	: 8934
597-00-1	: 10008
597-12-6	: 6221
597-32-0	: 10365
597-33-1	: 9998
597-35-3	: 5626
597-49-9	: 8932
597-50-2	: 9517
597-52-4	: 11503
597-55-7	: 9926
597-67-1	: 11428
597-71-7	: 9988
597-72-8	: 10068
597-76-2	: 6931
597-77-3	: 6935
597-90-0	: 6487
597-93-3	: 8162
597-96-6	: 6946
598-01-6	: 6505
598-02-7	: 9558
598-03-8	: 10131
598-04-9	: 3423
598-05-0	: 11575
598-09-4	: 8606
598-16-3	: 5921
598-17-4	: 9880
598-19-6	: 10699
598-21-0	: 292
598-22-1	: 10490
598-23-2	: 3916
598-25-4	: 3170
598-31-2	: 10416
598-38-9	: 5704
598-45-8	: 10063
598-50-5	: 12017
598-53-8	: 10070
598-55-0	: 3953
598-56-1	: 5421
598-57-2	: 7451
598-58-3	: 8075
598-61-8	: 4192
598-74-3	: 3223
598-76-5	: 9899
598-77-6	: 10172
598-78-7	: 10215
598-82-3	: 10266
598-88-9	: 5898
598-92-5	: 5482
598-94-7	: 12005
598-96-9	: 9072
598-98-1	: 10260
598-99-2	: 253
599-01-9	: 10333
599-04-2	: 6143
599-66-6	: 2240
599-70-2	: 1704
599-79-1	: 2654
599-91-7	: 2233
599-97-3	: 5657
599-99-5	: 242
600-00-0	: 10213
600-05-5	: 10238
600-06-6	: 3245
600-07-7	: 3538
600-11-3	: 8749
600-13-5	: 4150
600-14-6	: 8797
600-15-7	: 3530
600-17-9	: 3842
600-18-0	: 3567
600-22-6	: 10316
600-25-9	: 9866
600-30-6	: 3497
600-36-2	: 8931
600-40-8	: 5551
601-75-2	: 9948
601-76-3	: 9918
601-77-4	: 9822
601-84-3	: 2557
601-88-7	: 1393
602-03-9	: 724
602-21-1	: 1353
602-27-7	: 1489
602-29-9	: 2046
602-37-9	: 7787
602-38-0	: 7833
602-55-1	: 428
602-56-2	: 318
602-60-8	: 426
602-64-2	: 415
602-65-3	: 418
602-82-4	: 1332
602-92-6	: 2560
602-96-0	: 2364
602-99-3	: 9434
603-26-9	: 7513
603-32-7	: 450
603-33-8	: 3111
603-34-9	: 728
603-35-0	: 9523
603-36-1	: 11551
603-38-3	: 729
603-39-4	: 2110
603-45-2	: 4278
603-50-9	: 9465
603-52-1	: 3946
603-54-3	: 12011
603-69-0	: 3462
603-71-4	: 2363
603-76-9	: 7312
603-81-6	: 2552
603-83-8	: 799
603-85-0	: 9214
604-53-5	: 3004
604-61-5	: 2515
604-68-2	: 6216
604-69-3	: 6217
604-83-1	: 9172
604-88-6	: 1756
605-01-6	: 2090
605-02-7	: 7947
605-03-8	: 11290
605-27-6	: 411
605-32-3	: 408
605-37-8	: 7852
605-39-0	: 3050
605-42-5	: 390
605-50-5	: 1335
605-54-9	: 1333
605-63-0	: 7786
605-65-2	: 7977
605-71-0	: 7832
605-79-8	: 7634
605-81-2	: 6228
605-82-3	: 421
605-83-4	: 420
605-85-6	: 5859
605-99-2	: 10765
606-17-7	: 2594
606-18-8	: 2506
606-20-2	: 1958
606-21-3	: 1145
606-22-4	: 726
606-23-5	: 7264
606-27-9	: 2710
606-28-0	: 2516
606-35-9	: 1946
606-37-1	: 7831
606-43-9	: 11328
606-45-1	: 2673
606-46-2	: 696
606-81-5	: 3063
606-88-2	: 7173
606-97-3	: 3093
606-99-5	: 751
607-00-1	: 5993
607-31-8	: 11274
607-32-9	: 7386
607-34-1	: 11284
607-42-1	: 368
607-50-1	: 7919
607-52-3	: 7939
607-53-4	: 8005
607-56-7	: 12008
607-81-8	: 9977
607-85-2	: 2635
607-86-3	: 2682
607-90-9	: 2652
607-91-0	: 2428
607-97-6	: 3518
607-99-8	: 2313
608-07-1	: 7304
608-21-9	: 2307
608-23-1	: 1137
608-25-3	: 1529
608-26-4	: 9260
608-27-5	: 685
608-28-6	: 1772
608-29-7	: 2349
608-30-0	: 682
608-31-1	: 688
608-32-2	: 2305
608-33-3	: 9277
608-50-4	: 2360
608-66-2	: 6194
608-69-5	: 3308
608-71-9	: 9439
608-89-9	: 3315
608-90-2	: 2082
608-93-5	: 2084
608-96-8	: 2096
608-98-0	: 1553
609-02-9	: 9963
609-08-5	: 9962
609-09-6	: 9973
609-11-0	: 3498
609-12-1	: 3475
609-13-2	: 3477
609-14-3	: 3557
609-19-8	: 9486
609-22-3	: 9279
609-23-4	: 9492
609-25-6	: 2369
609-26-7	: 8813
609-29-0	: 8803
609-31-4	: 3640
609-36-9	: 9765
609-39-2	: 6134
609-40-5	: 11775
609-46-1	: 2225
609-65-4	: 2890
609-72-3	: 860
609-73-4	: 1783
610-02-6	: 2730
610-15-1	: 624
610-16-2	: 2584
610-17-3	: 718
610-25-3	: 2047
610-34-4	: 2707
610-39-9	: 1960
610-46-8	: 377
610-48-0	: 423
610-50-4	: 431
610-66-2	: 965
610-67-3	: 1662
610-69-5	: 219
610-71-9	: 2556
610-72-0	: 2581
610-89-9	: 9074
610-91-3	: 2362
610-97-9	: 2660
610-99-1	: 10449
611-00-7	: 2555
611-01-8	: 2580
611-03-0	: 2553
611-06-3	: 1396
611-07-4	: 9269
611-10-9	: 4785
611-12-1	: 10366
611-13-2	: 6059
611-14-3	: 1685
611-15-4	: 1640
611-19-8	: 1132
611-20-1	: 2759
611-21-2	: 710
611-32-5	: 11283
611-33-6	: 11242
611-34-7	: 11204
611-35-8	: 11238
611-43-8	: 3051
611-45-0	: 7949
611-49-4	: 7928
611-61-0	: 3053
611-62-1	: 3065
611-69-8	: 1901
611-70-1	: 10469
611-73-4	: 934
611-79-0	: 7589
611-92-7	: 12006
611-94-9	: 7625
611-95-0	: 2512
611-97-2	: 7599
611-99-4	: 7597
612-00-0	: 1677
612-01-1	: 12018
612-08-8	: 5109
612-12-4	: 986
612-13-5	: 2750
612-15-7	: 1637
612-16-8	: 1889
612-17-9	: 7815
612-19-1	: 2600
612-22-6	: 1692
612-23-7	: 1203
612-25-9	: 1906
612-28-2	: 804

612-35-1	: 2718
612-37-3	: 10766
612-55-5	: 7903
612-57-7	: 11240
612-58-8	: 11278
612-60-2	: 11282
612-61-3	: 11241
612-62-4	: 11237
612-64-6	: 750
612-71-5	: 11598
612-75-9	: 3057
612-76-0	: 3066
612-78-2	: 3005
612-81-7	: 3107
612-94-2	: 7948
612-95-3	: 11291
612-96-4	: 11288
612-97-5	: 11196
613-03-6	: 2370
613-08-1	: 369
613-12-7	: 424
613-13-8	: 365
613-21-8	: 11319
613-29-6	: 684
613-31-0	: 376
613-33-2	: 3061
613-37-6	: 3081
613-39-8	: 9613
613-42-3	: 3094
613-43-4	: 5112
613-45-6	: 543
613-46-7	: 7749
613-48-9	: 697
613-50-3	: 11285
613-51-4	: 11286
613-59-2	: 7950
613-69-4	: 552
613-70-7	: 9382
613-75-2	: 7510
613-77-4	: 11249
613-78-5	: 2645
613-80-9	: 7940
613-81-0	: 8006
613-84-3	: 577
613-88-7	: 956
613-89-8	: 5774
613-90-1	: 968
613-91-2	: 5866
613-93-4	: 620
613-94-5	: 2611
613-97-8	: 749
614-00-6	: 806
614-16-4	: 2124
614-18-6	: 10902
614-21-1	: 5862
614-27-7	: 2142
614-29-9	: 1613
614-33-5	: 10196
614-34-6	: 2700
614-39-1	: 602
614-45-9	: 1128
614-47-1	: 10709
614-54-0	: 1881
614-68-6	: 1787
614-71-1	: 1659
614-72-2	: 1156
614-75-5	: 904
614-78-8	: 11806
614-96-0	: 7316
614-99-3	: 6056
615-05-4	: 1274
615-09-8	: 6163
615-10-1	: 6071
615-11-2	: 6057
615-13-4	: 7287
615-18-9	: 2878
615-20-3	: 2862
615-30-5	: 8691
615-36-1	: 638
615-37-2	: 1778
615-38-3	: 4531
615-39-4	: 4532
615-41-8	: 1171
615-42-9	: 1427
615-52-1	: 5567
615-54-3	: 2308
615-55-4	: 683
615-57-6	: 681
615-58-7	: 9276
615-59-8	: 1305
615-60-1	: 1141
615-62-3	: 9265
615-65-6	: 668
615-67-8	: 1507
615-68-9	: 2350
615-71-4	: 2306
615-74-7	: 9262
615-79-2	: 8876
615-81-6	: 5556
615-83-8	: 8860
615-84-9	: 5494
615-85-0	: 5453
615-87-2	: 1298
615-89-4	: 1514
615-98-5	: 5565
615-99-6	: 5564
616-02-4	: 6098
616-06-8	: 8215
616-12-6	: 9056
616-19-3	: 9901
616-20-6	: 8721
616-21-7	: 3277
616-23-9	: 10359
616-24-0	: 8702
616-25-1	: 9087
616-27-3	: 3646
616-28-4	: 3912
616-31-9	: 8846
616-38-6	: 4000
616-39-7	: 5430
616-40-0	: 7152
616-42-2	: 11582
616-43-3	: 11132
616-44-4	: 11770
616-45-5	: 11180
616-47-7	: 7208
616-55-7	: 9223
616-91-1	: 5015
617-00-5	: 795
617-02-7	: 2699
617-05-0	: 2626
617-09-4	: 3996
617-30-1	: 6797
617-31-2	: 8883
617-33-4	: 160
617-35-6	: 10315
617-45-8	: 470
617-50-5	: 10303
617-51-6	: 10275
617-52-7	: 3330
617-53-8	: 3789
617-54-9	: 3790
617-60-7	: 5309
617-65-2	: 6240
617-73-2	: 8432
617-75-4	: 448
617-78-7	: 8809
617-80-1	: 3407
617-83-4	: 4156
617-84-5	: 5991
617-85-6	: 11549
617-86-7	: 11483
617-88-9	: 6075
617-89-0	: 6113
617-90-3	: 6039
617-92-5	: 11124
617-94-7	: 1857
618-31-5	: 1301
618-32-6	: 2885
618-34-8	: 1148
618-36-0	: 1814
618-38-2	: 2910
618-40-6	: 7165
618-41-7	: 2180
618-42-8	: 9630
618-45-1	: 9411
618-46-2	: 2891
618-51-9	: 2658
618-62-2	: 1394
618-65-5	: 562
618-68-8	: 2144
618-71-3	: 2593
618-76-8	: 2618
618-80-4	: 9296
618-85-9	: 1957
618-87-1	: 727
618-89-3	: 2525
618-91-7	: 2661
618-95-1	: 2711
618-98-4	: 2708
619-02-3	: 4601
619-12-5	: 9598
619-15-8	: 1959
619-20-5	: 2601
619-23-8	: 1206
619-25-0	: 1907
619-31-8	: 719
619-33-0	: 5530
619-39-6	: 5068
619-41-0	: 5779
619-42-1	: 2526
619-44-3	: 2662
619-52-3	: 4683
619-58-9	: 2659
619-60-3	: 9319
619-73-8	: 1908
619-82-9	: 4354
619-84-1	: 2585
619-99-8	: 6815
620-02-0	: 6045
620-03-1	: 6047
620-05-3	: 1777
620-11-1	: 8919
620-12-2	: 10193
620-13-3	: 1063
620-14-4	: 1686
620-16-6	: 1162
620-17-7	: 9354
620-18-8	: 9349
620-19-9	: 1199
620-20-2	: 1133
620-22-4	: 2764
620-23-5	: 588
620-24-6	: 1883
620-32-6	: 2241
620-40-6	: 1798
620-42-8	: 9597
620-43-9	: 3224
620-47-3	: 2015
620-63-3	: 3561
620-71-3	: 9800
620-72-4	: 119
620-73-5	: 147
620-80-4	: 3576
620-81-5	: 5495
620-82-6	: 5559
620-83-7	: 2016
620-85-9	: 1696
620-86-0	: 2719
620-88-2	: 2054
620-92-8	: 9407
620-93-9	: 797
620-94-0	: 2286
620-95-1	: 11005
620-99-5	: 5670
621-08-9	: 2184
621-14-7	: 3320
621-15-8	: 6722
621-23-8	: 2355
621-25-0	: 7174
621-26-1	: 7138
621-27-2	: 9462
621-31-8	: 9356
621-32-9	: 1660
621-34-1	: 9351
621-36-3	: 925
621-37-4	: 905
621-50-1	: 966
621-51-2	: 2597
621-52-3	: 1663
621-59-0	: 572
621-62-5	: 5469
621-63-6	: 5705
621-64-7	: 9825
621-70-5	: 6908
621-72-7	: 2385
621-76-1	: 10078
621-77-2	: 8705
621-87-4	: 10479
621-95-4	: 731
622-00-4	: 3306
622-04-8	: 10377
622-08-2	: 5761
622-09-3	: 7693
622-15-1	: 7565
622-16-2	: 764
622-24-2	: 1164
622-26-4	: 9664
622-29-7	: 7452
622-31-1	: 594
622-32-2	: 595
622-36-6	: 7166
622-37-7	: 976
622-38-8	: 1705
622-39-9	: 11015
622-40-2	: 7691
622-42-4	: 2052
622-45-7	: 155
622-47-9	: 926
622-57-1	: 748
622-60-6	: 1661
622-61-7	: 1158
622-62-8	: 9352
622-63-9	: 1707
622-78-6	: 1794
622-79-7	: 980
622-80-0	: 843
622-82-2	: 7160
622-85-5	: 2168
622-86-6	: 1159
622-93-5	: 10367
622-96-8	: 1687
622-97-9	: 1642
622-98-0	: 1163
623-00-7	: 2746
623-03-0	: 2749
623-05-2	: 1884
623-08-5	: 712
623-09-6	: 1280
623-10-9	: 758
623-12-1	: 1180
623-15-4	: 3878
623-17-6	: 6119
623-18-7	: 10640
623-19-8	: 6123
623-20-1	: 10639
623-21-2	: 3520
623-22-3	: 10642
623-24-5	: 985
623-25-6	: 988
623-26-7	: 1316
623-27-8	: 1323
623-30-3	: 10512
623-34-7	: 3285
623-36-9	: 9002
623-39-2	: 10014
623-40-5	: 8985
623-42-7	: 3547
623-43-8	: 3847
623-46-1	: 5533
623-47-2	: 10739
623-48-3	: 201
623-49-4	: 4030
623-50-7	: 197
623-51-8	: 203
623-53-0	: 4003
623-56-3	: 6981
623-59-6	: 73
623-60-9	: 10351
623-65-4	: 6646
623-66-5	: 8424
623-68-7	: 3818
623-69-8	: 10372
623-70-1	: 3839
623-71-2	: 10221
623-72-3	: 10270
623-73-4	: 158
623-76-7	: 12001
623-78-9	: 3949
623-80-3	: 4031
623-81-4	: 11581
623-84-7	: 10002
623-85-8	: 3970
623-86-9	: 102
623-87-0	: 10191
623-91-6	: 3775
623-93-8	: 8161
623-95-0	: 12012
623-96-1	: 4002
623-97-2	: 5697
624-03-3	: 6649
624-04-4	: 5311
624-08-8	: 6319
624-09-9	: 6464
624-13-5	: 8441
624-16-8	: 5077
624-17-9	: 8133
624-20-4	: 6754
624-24-8	: 8891
624-29-3	: 4368
624-31-7	: 1780
624-38-4	: 1429
624-41-9	: 3630
624-42-0	: 6523
624-43-1	: 10192
624-44-2	: 6982
624-45-3	: 8908
624-48-6	: 3779
624-49-7	: 3778
624-54-4	: 10318
624-57-7	: 8617
624-60-2	: 5435
624-61-3	: 5939
624-64-6	: 3705
624-65-7	: 10729
624-66-8	: 10734
624-67-9	: 10723
624-70-4	: 5477
624-72-6	: 5545
624-73-7	: 5547
624-74-8	: 5941
624-75-9	: 283
624-76-0	: 5731
624-79-3	: 5599
624-81-7	: 5434
624-82-8	: 7566
624-83-9	: 7520
624-85-1	: 7184
624-89-5	: 5606
624-90-8	: 7460
624-91-9	: 8091
624-92-0	: 5254
624-95-3	: 3617
624-96-4	: 3284
624-97-5	: 9100
625-01-4	: 4085
625-04-7	: 8959
625-06-9	: 8928
625-08-1	: 3535
625-23-0	: 6942
625-25-2	: 6495
625-27-4	: 9055
625-28-5	: 3410
625-30-9	: 8701
625-31-0	: 9089
625-36-5	: 10497
625-37-6	: 3702
625-38-7	: 3813
625-44-5	: 10071
625-45-6	: 205
625-46-7	: 10591
625-47-8	: 5483
625-48-9	: 5752
625-50-3	: 61
625-52-5	: 12014
625-54-7	: 10038
625-55-8	: 6011
625-56-9	: 7582
625-57-0	: 3978
625-58-1	: 8074
625-59-2	: 11706
625-60-5	: 5647
625-65-0	: 9035
625-66-1	: 3282
625-67-2	: 8750
625-68-3	: 3481
625-69-4	: 8789
625-71-8	: 3531
625-74-1	: 10084
625-76-3	: 7508
625-77-4	: 23
625-80-9	: 10152
625-82-1	: 11119
625-84-3	: 11120
625-86-5	: 6089

625-88-7	: 11753
625-92-3	: 10926
625-95-6	: 1779
625-96-7	: 4597
625-98-9	: 1167
626-00-6	: 1428
626-01-7	: 760
626-02-8	: 9372
626-04-0	: 1562
626-10-8	: 3771
626-13-1	: 796
626-15-3	: 984
626-16-4	: 987
626-17-5	: 1315
626-18-6	: 1434
626-19-7	: 1322
626-20-0	: 1638
626-21-1	: 10933
626-23-3	: 3227
626-26-6	: 3434
626-27-7	: 6456
626-29-9	: 11623
626-31-3	: 3297
626-33-5	: 6524
626-35-7	: 218
626-36-8	: 3932
626-37-9	: 3963
626-39-1	: 2309
626-41-5	: 9278
626-43-7	: 690
626-44-8	: 2351
626-48-2	: 11069
626-51-7	: 8779
626-53-9	: 6454
626-55-1	: 10876
626-58-4	: 9682
626-60-8	: 10916
626-61-9	: 10917
626-62-0	: 4415
626-64-2	: 11032
626-67-5	: 9679
626-70-0	: 6788
626-77-7	: 6909
626-82-4	: 6880
626-84-6	: 5736
626-85-7	: 11580
626-86-8	: 6791
626-87-9	: 8740
626-88-0	: 8716
626-89-1	: 8939
626-90-4	: 3187
626-91-5	: 3395
626-92-6	: 8746
626-94-8	: 7068
626-95-9	: 8786
626-96-0	: 8696
626-97-1	: 8698
626-99-3	: 8676
627-00-9	: 3482
627-02-1	: 10039
627-03-2	: 180
627-04-3	: 188
627-05-4	: 3412
627-07-6	: 9824
627-08-7	: 10076
627-09-8	: 10744
627-11-2	: 4013
627-12-3	: 3969
627-13-4	: 8080
627-16-7	: 10730
627-18-9	: 10344
627-19-0	: 9122
627-20-3	: 9006
627-21-4	: 9123
627-22-5	: 3151
627-26-9	: 3802
627-27-0	: 3861
627-30-5	: 10348
627-31-6	: 9907
627-32-7	: 10385
627-37-2	: 10524
627-39-4	: 9787
627-40-7	: 10576
627-41-8	: 10736
627-42-9	: 5480
627-44-1	: 7434
627-45-2	: 5994
627-48-5	: 4163
627-51-0	: 5920
627-52-1	: 11572
627-53-2	: 5620
627-54-3	: 5630
627-58-7	: 6691
627-59-8	: 6944
627-63-4	: 3795
627-67-8	: 3400
627-70-3	: 10463
627-73-6	: 3323
627-82-7	: 10025
627-83-8	: 8256
627-88-3	: 8265
627-89-4	: 8306
627-90-7	: 11946
627-91-8	: 6792
627-93-0	: 6786
627-96-3	: 6756
627-97-4	: 6548
627-98-5	: 6941
628-02-4	: 6721
628-03-5	: 3390
628-04-6	: 3374
628-05-7	: 8832
628-08-0	: 3862
628-09-1	: 10350
628-11-5	: 4015
628-12-6	: 4019
628-13-7	: 10969
628-16-0	: 6716
628-17-1	: 8818
628-20-6	: 3405
628-21-7	: 3288
628-28-4	: 3391
628-29-5	: 3402
628-30-8	: 10064
628-32-0	: 10037
628-33-1	: 10733
628-34-2	: 5474
628-37-5	: 9158
628-41-1	: 4249
628-44-4	: 8464
628-46-6	: 6896
628-50-2	: 4021
628-61-5	: 8360
628-62-6	: 6348
628-63-7	: 227
628-64-8	: 4017
628-66-0	: 10003
628-67-1	: 3348
628-68-2	: 5754
628-71-7	: 6609
628-72-8	: 6329
628-73-9	: 6840
628-76-2	: 8747
628-77-3	: 8758
628-80-8	: 8822
628-81-9	: 3372
628-83-1	: 11699
628-89-7	: 5698
628-92-2	: 4244
628-96-6	: 5574
628-97-7	: 6650
629-01-6	: 8349
629-03-8	: 6757
629-04-9	: 6356
629-05-0	: 8529
629-06-1	: 6360
629-08-3	: 6428
629-09-4	: 6769
629-11-8	: 6796
629-14-1	: 5542
629-15-2	: 5573
629-16-3	: 8086
629-17-4	: 11702
629-19-6	: 5259
629-20-9	: 4742
629-23-2	: 11634
629-27-6	: 8406
629-30-1	: 6399
629-31-2	: 6346
629-32-3	: 6421
629-33-4	: 6009
629-35-6	: 7433
629-36-7	: 10122
629-37-8	: 8412
629-39-0	: 8079
629-40-3	: 8385
629-41-4	: 8389
629-43-6	: 8088
629-45-8	: 5252
629-46-9	: 8095
629-49-2	: 5344
629-50-5	: 11871
629-54-9	: 6629
629-59-4	: 11613
629-60-7	: 11880
629-62-9	: 8642
629-63-0	: 11619
629-64-1	: 6430
629-65-2	: 6435
629-66-3	: 8106
629-70-9	: 6662
629-72-1	: 8643
629-73-2	: 6667
629-74-3	: 6673
629-75-4	: 6674
629-78-7	: 6306
629-79-8	: 6641
629-80-1	: 6628
629-82-3	: 8416
629-89-0	: 8313
629-90-3	: 6304
629-92-5	: 8101
629-93-6	: 8242
629-94-7	: 6294
629-96-9	: 5358
629-97-0	: 5268
629-98-1	: 5284
629-99-2	: 8637
630-01-3	: 6622
630-02-4	: 8220
630-03-5	: 8099
630-04-6	: 6299
630-06-8	: 6991
630-07-9	: 8999
630-08-0	: 4011
630-10-4	: 11354
630-13-7	: 9908
630-16-0	: 5631
630-17-1	: 9843
630-18-2	: 10100
630-19-3	: 9783
630-20-6	: 5634
630-25-1	: 5524
630-51-3	: 3343
630-60-4	: 4040
630-72-8	: 290
630-76-2	: 1831
630-95-5	: 7910
631-01-6	: 8546
631-28-7	: 10164
631-36-7	: 11450
631-40-3	: 9833
631-57-2	: 10113
631-61-8	: 106
631-64-1	: 159
632-05-3	: 3444
632-07-5	: 10233
632-14-4	: 12032
632-15-5	: 11757
632-16-6	: 11754
632-21-3	: 10487
632-22-4	: 12030
632-32-6	: 5385
632-46-2	: 2582
632-50-8	: 1591
632-51-9	: 1623
632-56-4	: 5633
632-77-9	: 412
632-80-4	: 7348
632-82-6	: 414
632-83-7	: 385
632-89-3	: 12031
632-92-8	: 1490
632-99-5	: 634
633-03-4	: 5444
634-19-5	: 5978
634-36-6	: 2354
634-41-3	: 7723
634-43-5	: 7724
634-47-9	: 11243
634-66-2	: 2259
634-67-3	: 852
634-68-4	: 2273
634-89-9	: 2253
634-92-4	: 2274
634-93-5	: 854
634-95-7	: 12002
635-03-0	: 10139
635-10-9	: 2258
635-11-0	: 2276
635-27-8	: 11239
635-46-1	: 11296
635-65-4	: 3000
635-67-6	: 1508
635-81-4	: 2269
635-88-1	: 7918
635-90-5	: 11134
635-93-8	: 533
636-04-4	: 1129
636-09-9	: 1350
636-10-2	: 7162
636-21-5	: 791
636-28-2	: 2254
636-30-6	: 853
636-31-7	: 2275
636-36-2	: 6891
636-41-9	: 11131
636-46-4	: 1352
636-53-3	: 1346
636-60-2	: 3325
636-72-6	: 11768
636-73-7	: 11018
636-82-8	: 4637
636-98-6	: 1785
637-03-6	: 447
637-07-0	: 10231
637-12-7	: 8248
637-27-4	: 10323
637-47-8	: 7139
637-50-3	: 2162
637-55-8	: 8271
637-57-0	: 10679
637-59-2	: 1088
637-64-9	: 6125
637-65-0	: 6126
637-69-4	: 1639
637-78-5	: 10299
637-87-6	: 1172
637-88-7	: 4377
637-89-8	: 9378
637-92-3	: 10040
637-98-9	: 3979
638-00-6	: 11755
638-02-8	: 11756
638-04-0	: 4367
638-07-3	: 3494
638-08-4	: 8249
638-10-8	: 3850
638-11-9	: 3549
638-17-5	: 5265
638-20-0	: 40
638-21-1	: 9519
638-23-3	: 5017
638-25-5	: 8439
638-28-8	: 6738
638-29-9	: 8990
638-37-9	: 3256
638-41-5	: 4025
638-45-9	: 6831
638-46-0	: 3376
638-49-3	: 6015
638-51-7	: 8089
638-53-9	: 11881
638-54-0	: 8364
638-58-4	: 11011
638-60-8	: 11640
638-65-3	: 8243
638-67-5	: 11855
638-68-6	: 11819
638-73-3	: 7479
638-95-9	: 12038
639-46-3	: 7668
639-58-7	: 11541
639-81-6	: 5395
640-19-7	: 63
640-61-9	: 2211
641-06-5	: 50
641-16-7	: 9476
641-36-1	: 5160
641-74-7	: 7423
641-85-0	: 9740
641-91-8	: 8010
641-96-3	: 11193
642-28-4	: 7912
642-32-0	: 2268
642-44-4	: 11060
642-84-2	: 608
643-28-7	: 788
643-58-3	: 3082
643-60-7	: 7393
643-65-2	: 7632
643-75-4	: 10199
643-84-5	: 2847
643-93-6	: 3083
644-08-6	: 3084
644-15-5	: 7725
644-21-3	: 7159
644-30-4	: 1468
644-31-5	: 9151
644-32-6	: 5251
644-35-9	: 9461
644-49-5	: 10308
644-71-3	: 9445
644-80-4	: 3770
644-90-6	: 8423
644-97-3	: 9556
644-98-4	: 10984
645-00-1	: 1784
645-09-0	: 625
645-12-5	: 6066
645-13-6	: 5848
645-36-3	: 5414
645-41-0	: 3225
645-45-4	: 2160
645-49-8	: 1620
645-55-6	: 825
645-56-7	: 9463
645-59-0	: 2123
645-66-9	: 5306
645-67-0	: 8912
645-69-2	: 3319
645-96-5	: 2177
646-04-8	: 9005
646-05-9	: 9111
646-06-0	: 5204
646-07-1	: 8889
646-13-9	: 8268
646-14-0	: 6841
646-19-5	: 6371
646-20-8	: 6392
646-30-0	: 8103
646-31-1	: 11603
650-69-1	: 435
653-37-2	: 596
655-48-1	: 5575
655-86-7	: 9194
659-04-1	: 10678
659-49-4	: 828
659-70-1	: 3551
659-86-9	: 10735
660-68-4	: 5428
660-88-8	: 8857
661-11-0	: 6417
661-19-8	: 5279
661-68-7	: 5227
661-97-2	: 9898
666-99-9	: 8102
670-54-2	: 5915
670-96-2	: 7213
671-56-7	: 3866
672-13-9	: 571
672-15-1	: 7134
673-31-4	: 2149
673-32-5	: 2173
673-79-0	: 10330
674-76-0	: 9058
674-82-8	: 8598
675-09-2	: 10792
675-20-7	: 9719
677-21-4	: 10608
677-68-9	: 10149
678-26-2	: 8807
679-37-8	: 7185
680-31-9	: 9573
681-06-1	: 9550

681-84-5	: 11508
682-01-9	: 11510
682-30-4	: 9535
683-08-9	: 9539
683-50-1	: 9777
683-51-2	: 10511
683-68-1	: 5521
683-72-7	: 51
684-16-2	: 10441
684-84-4	: 3353
684-93-5	: 12019
685-63-2	: 3164
685-73-4	: 6199
685-87-0	: 9928
685-91-6	: 53
688-14-2	: 7414
688-52-8	: 7108
688-71-1	: 3127
688-74-4	: 3124
688-84-6	: 10659
689-89-4	: 6703
689-97-4	: 3889
690-02-8	: 9159
690-08-4	: 9039
690-37-9	: 8954
690-92-6	: 7029
690-93-7	: 7028
691-37-2	: 9054
691-38-3	: 9059
691-83-8	: 3788
692-24-0	: 7046
692-35-3	: 9854
692-42-2	: 441
692-45-5	: 6006
692-50-2	: 3914
692-86-4	: 11977
693-02-7	: 7114
693-05-0	: 10112
693-13-0	: 9815
693-23-2	: 5299
693-39-0	: 6429
693-54-9	: 5075
693-58-3	: 8121
693-61-8	: 11968
693-62-9	: 11970
693-65-2	: 8835
693-72-1	: 8295
693-89-0	: 4928
693-93-6	: 8583
693-95-8	: 11679
693-98-1	: 7209
694-05-3	: 11020
694-28-0	: 4880
694-31-5	: 10826
694-35-9	: 4925
694-48-4	: 10825
694-53-1	: 11446
694-59-7	: 10994
694-80-4	: 1004
694-81-5	: 11057
694-85-9	: 11035
695-06-7	: 6150
695-12-5	: 4388
695-28-3	: 4653
695-34-1	: 10867
695-53-4	: 8584
695-84-1	: 9348
695-98-7	: 11025
696-28-6	: 462
696-29-7	: 4431
696-30-0	: 10986
696-41-3	: 581
696-44-6	: 711
696-59-3	: 6169
696-62-8	: 1776
696-63-9	: 2300
696-71-9	: 4739
697-11-0	: 4200
697-86-9	: 641
698-00-0	: 643
698-01-1	: 662
698-27-1	: 576
698-63-5	: 6046
698-69-1	: 663
698-87-3	: 1607
698-88-4	: 1372
699-02-5	: 1610
699-08-1	: 1657
699-09-2	: 2296
699-17-2	: 3659
699-30-9	: 6099
700-12-9	: 2097
700-75-4	: 769
700-88-9	: 1254
701-57-5	: 2044
701-97-3	: 4468
702-28-3	: 4969
702-54-5	: 8573
703-32-2	: 11252
703-61-7	: 11248
703-80-0	: 5819
704-01-8	: 1286
704-38-1	: 7617
704-79-0	: 3854
706-78-5	: 4932
706-79-6	: 4917
708-06-5	: 7754
708-76-9	: 567
708-82-7	: 10629
709-09-1	: 1445
709-98-8	: 10200
710-43-0	: 4968
711-79-5	: 5829
712-48-1	: 454
713-36-0	: 2014
716-79-0	: 2384
717-21-5	: 10710
717-74-8	: 2373
719-79-9	: 1120
723-46-6	: 2198
723-62-6	: 370
726-42-1	: 765
726-44-3	: 2122
732-11-6	: 9515
732-26-3	: 9498
737-89-3	: 7726
741-58-2	: 517
742-20-1	: 2853
747-90-0	: 4102
750-90-3	: 4131
753-89-9	: 9853
754-05-2	: 11425
755-25-9	: 7032
756-79-6	: 9540
757-58-4	: 11645
759-05-7	: 3556
759-36-4	: 9964
759-73-9	: 12015
759-94-4	: 5372
760-20-3	: 9053
760-21-4	: 8825
760-23-6	: 3745
760-40-7	: 10408
760-55-4	: 9971
760-67-8	: 6987
760-78-1	: 8217
760-79-2	: 3203
760-93-0	: 10649
761-06-8	: 10015
762-21-0	: 3909
762-42-5	: 3910
762-49-2	: 5462
762-50-5	: 5476
762-62-9	: 9033
762-63-0	: 9040
762-72-1	: 11495
763-29-1	: 9052
763-30-4	: 8671
763-69-9	: 10262
763-89-3	: 9096
763-93-9	: 7086
764-01-2	: 3919
764-13-6	: 6692
764-33-0	: 7123
764-35-2	: 7115
764-42-1	: 3762
764-47-6	: 10033
764-48-7	: 5716
764-49-8	: 11712
764-56-7	: 6342
764-60-3	: 7125
764-73-8	: 8343
764-85-2	: 8177
764-93-2	: 5097
764-96-5	: 11973
764-97-6	: 11972
765-03-7	: 5343
765-05-9	: 5051
765-09-3	: 11872
765-13-9	: 8656
765-27-5	: 5366
765-30-0	: 4953
765-34-4	: 8603
765-42-4	: 4979
765-43-5	: 5796
765-47-9	: 4918
765-58-2	: 11730
765-85-5	: 4176
765-87-7	: 4375
766-00-7	: 4811
766-05-2	: 4328
766-07-4	: 7177
766-08-5	: 11442
766-09-6	: 9667
766-20-1	: 5189
766-39-2	: 6097
766-46-1	: 1040
766-51-8	: 1178
766-53-0	: 2992
766-77-8	: 11418
766-79-0	: 9684
766-81-4	: 1041
766-84-7	: 2748
766-90-5	: 2164
766-92-7	: 2042
766-93-8	: 5990
766-94-9	: 1646
766-96-1	: 1042
767-03-3	: 4904
767-10-2	: 11146
767-12-4	: 4516
767-54-4	: 4570
767-58-8	: 7255
767-59-9	: 7275
767-60-2	: 7277
767-88-4	: 10884
767-99-7	: 2029
768-00-3	: 2025
768-22-9	: 7282
768-32-1	: 11493
768-33-2	: 11374
768-39-8	: 3115
768-49-0	: 2024
768-52-5	: 790
768-56-9	: 1112
768-66-1	: 9711
768-94-5	: 11859
769-06-2	: 849
769-25-5	: 1647
769-68-6	: 951
769-92-6	: 715
770-09-2	: 11494
770-35-4	: 10399
771-56-2	: 2094
771-61-9	: 9442
771-62-0	: 2304
771-97-1	: 7799
771-98-2	: 1246
771-99-3	: 9695
772-00-9	: 5196
772-01-0	: 5195
772-17-8	: 2138
773-76-2	: 11314
773-82-0	: 2767
774-40-3	: 909
775-12-2	: 11420
775-51-9	: 4277
775-56-4	: 11407
776-35-2	: 9171
776-74-9	: 1058
776-75-0	: 9634
776-76-1	: 11438
777-22-0	: 1981
777-37-7	: 1218
778-22-3	: 1976
778-24-5	: 11416
778-25-6	: 11502
778-28-9	: 2230
778-66-5	: 2166
779-02-2	: 425
779-51-1	: 1971
779-84-0	: 9450
780-69-8	: 11481
781-35-1	: 10432
782-05-8	: 1457
782-06-9	: 1456
784-50-9	: 5969
786-19-6	: 4037
790-69-2	: 11802
791-28-6	: 9518
791-29-7	: 11444
791-31-1	: 11504
793-23-7	: 995
799-53-1	: 5239
804-63-7	: 4132
811-93-8	: 9878
811-97-2	: 5640
812-01-1	: 4086
812-03-3	: 10141
813-76-3	: 9525
813-78-5	: 9559
814-49-3	: 9574
814-67-5	: 10030
814-68-6	: 10714
814-78-8	: 3879
815-17-8	: 3510
815-24-7	: 8988
815-58-7	: 10718
815-66-7	: 6975
816-11-5	: 3517
816-39-7	: 10424
816-79-5	: 9045
817-91-4	: 6493
817-95-8	: 181
818-23-5	: 8652
818-38-2	: 8772
818-92-8	: 10570
819-07-8	: 5668
819-44-3	: 7107
819-93-2	: 3502
821-06-7	: 3742
821-07-8	: 6992
821-08-9	: 6713
821-09-0	: 9088
821-10-3	: 3905
821-11-4	: 3792
821-25-0	: 6377
821-55-6	: 8172
821-67-0	: 5132
821-76-1	: 6379
821-95-4	: 11966
821-96-5	: 11969
821-98-7	: 11971
821-99-8	: 8128
822-06-0	: 6770
822-23-1	: 221
822-28-6	: 6638
822-35-5	: 4198
822-36-6	: 7210
822-50-4	: 4802
822-66-2	: 4695
822-67-3	: 4694
822-84-4	: 11776
822-85-5	: 4574
822-87-7	: 4577
823-17-6	: 4654
823-40-5	: 1277
823-76-7	: 5794
824-11-3	: 11909
824-22-6	: 7257
824-42-0	: 575
824-47-5	: 1193
824-55-5	: 1186
824-63-5	: 7256
824-72-6	: 9548
824-94-2	: 1197
825-25-2	: 4881
825-51-4	: 8024
825-94-5	: 11457
826-18-6	: 2104
826-36-8	: 9726
826-73-3	: 2407
826-74-4	: 7881
826-77-7	: 11262
826-81-3	: 11321
827-08-7	: 1312
827-15-6	: 2092
827-16-7	: 11841
827-52-1	: 1248
827-54-3	: 7882
828-01-3	: 2133
829-83-4	: 443
829-84-5	: 9516
830-13-7	: 4211
830-89-7	: 7229
831-81-2	: 1224
831-91-4	: 2113
832-64-4	: 9181
832-71-3	: 9180
834-12-8	: 341
835-11-0	: 7596
835-64-3	: 9218
836-30-6	: 826
837-45-6	: 9165
838-41-5	: 8579
838-45-9	: 1697
838-95-9	: 722
841-73-6	: 11081
842-07-9	: 8037
846-48-0	: 343
848-53-3	: 5140
865-71-4	: 8853
868-18-8	: 3314
868-77-9	: 10661
868-85-9	: 9588
869-29-4	: 10563
869-50-1	: 9996
870-23-5	: 10600
870-63-3	: 3717
870-74-6	: 8977
871-28-3	: 9112
871-83-0	: 8140
871-84-1	: 8342
872-05-9	: 5082
872-10-6	: 8843
872-31-1	: 11727
872-50-4	: 11185
872-53-7	: 4780
872-55-9	: 11759
872-85-5	: 10889
872-93-5	: 11789
873-32-5	: 2747
873-49-4	: 1256
873-66-5	: 2163
873-76-7	: 1849
873-94-9	: 4622
874-23-7	: 4572
874-35-1	: 7258
874-41-9	: 1671
874-60-2	: 2904
874-63-5	: 1918
874-79-3	: 2172
874-90-8	: 2762
875-30-9	: 7298
875-40-1	: 2262
875-51-4	: 654
875-59-2	: 5828
876-05-1	: 4793
877-09-8	: 2261
877-11-2	: 2086
877-43-0	: 11257
877-44-1	: 2344
877-48-5	: 9314
877-53-2	: 9313
877-65-6	: 1865
877-82-7	: 9316
878-93-3	: 8064
879-97-0	: 9327
880-09-1	: 9683
880-93-3	: 7925
881-03-8	: 7926
881-67-4	: 1444
882-25-7	: 4148
882-33-7	: 5258
883-93-2	: 2864
883-99-8	: 7775
886-50-0	: 11594
886-58-8	: 1979
886-74-8	: 10001
886-77-1	: 8682
888-71-1	: 8564
897-78-9	: 4573
900-95-8	: 11539
910-86-1	: 11800
917-61-3	: 4164

917-92-0	: 3907
917-93-1	: 9779
918-00-3	: 10488
918-07-0	: 3254
918-79-6	: 8692
918-82-1	: 9128
918-84-3	: 8732
919-12-0	: 9132
919-19-7	: 9530
919-30-2	: 9828
919-86-8	: 5103
919-94-8	: 3375
920-37-6	: 10584
920-46-7	: 10715
921-08-4	: 3487
921-13-1	: 7480
921-47-1	: 6863
922-28-1	: 6389
922-55-4	: 5016
922-59-8	: 9134
922-62-3	: 9057
922-65-6	: 8681
923-06-8	: 3300
923-26-2	: 10662
923-99-9	: 9594
924-16-3	: 3212
924-41-4	: 6706
924-43-6	: 8093
924-50-5	: 3852
924-52-7	: 8077
925-15-5	: 3321
925-78-0	: 8173
926-02-3	: 10036
926-26-1	: 3296
926-55-6	: 9116
926-56-7	: 8670
926-63-6	: 9807
926-64-7	: 281
926-65-8	: 10034
926-82-9	: 6390
927-49-1	: 11962
927-54-8	: 7006
927-62-8	: 3215
927-68-4	: 5686
927-73-1	: 3724
927-74-2	: 3920
927-80-0	: 5942
928-40-5	: 6795
928-45-0	: 8072
928-49-4	: 7116
928-50-7	: 9019
928-51-8	: 3604
928-56-3	: 5886
928-65-4	: 11464
928-68-7	: 6521
928-80-3	: 5076
928-92-7	: 7072
928-94-9	: 7070
928-95-0	: 7069
928-96-1	: 7071
929-55-5	: 8348
929-61-3	: 8402
929-62-4	: 8401
929-77-1	: 5278
930-02-9	: 8240
930-18-7	: 4972
930-21-2	: 496
930-22-3	: 8613
930-27-8	: 6131
930-28-9	: 4789
930-30-3	: 4941
930-36-9	: 10834
930-37-0	: 8620
930-55-2	: 11170
930-66-5	: 4642
930-68-7	: 4703
930-87-0	: 11142
930-89-2	: 4815
930-90-5	: 4816
931-10-2	: 4293
931-20-4	: 9721
931-56-6	: 4425
931-77-1	: 4319
931-87-3	: 4752
931-89-5	: 4751
931-97-5	: 4329
932-01-4	: 4519
932-16-1	: 5849
932-32-1	: 670
932-39-8	: 4797
932-40-1	: 4798
932-43-4	: 4834
932-44-5	: 4835
932-53-6	: 11831
932-56-9	: 4235
932-66-1	: 5793
932-77-4	: 1007
932-96-7	: 675
933-05-1	: 4865
933-21-1	: 4728
933-36-8	: 4616
933-48-2	: 4569
933-67-5	: 7317
933-75-5	: 9483
933-78-8	: 9482
933-88-0	: 2902
933-98-2	: 1670
934-00-9	: 1525
934-34-9	: 2868
934-53-2	: 1189
934-74-7	: 1672
934-80-5	: 1675
935-05-7	: 8097
935-07-9	: 18
935-08-0	: 5134
935-12-6	: 6164
935-13-7	: 6161
935-31-9	: 4208
935-95-5	: 9471
936-48-1	: 10824
936-53-8	: 10823
936-58-3	: 1911
936-72-1	: 2026
937-14-4	: 1127
937-20-2	: 5784
937-32-6	: 2179
937-33-7	: 716
938-16-9	: 10431
938-25-0	: 7795
938-33-0	: 11276
938-39-6	: 11326
938-45-4	: 10446
938-73-8	: 615
939-23-1	: 10999
939-26-4	: 7743
939-27-5	: 7886
939-48-0	: 2693
939-52-6	: 3650
939-55-9	: 5696
939-83-3	: 2766
939-87-7	: 4957
940-41-0	: 11474
940-49-8	: 2033
940-54-5	: 3189
941-69-5	: 11123
941-98-0	: 5857
942-01-8	: 3989
942-06-3	: 7344
942-92-7	: 6983
943-14-6	: 2528
943-15-7	: 1991
943-27-1	: 5808
944-22-9	: 5986
945-51-7	: 2183
947-73-9	: 9162
947-84-2	: 3021
947-91-1	: 877
948-32-3	: 2372
948-65-2	: 7320
948-97-0	: 1966
948-98-1	: 1146
948-99-2	: 1147
949-87-1	: 5139
950-37-8	: 7645
951-86-0	: 1365
952-47-6	: 9426
952-97-6	: 2058
955-83-9	: 6100
957-51-7	: 5219
959-33-1	: 10713
961-11-5	: 6202
961-71-7	: 5506
964-68-1	: 2533
976-71-6	: 9736
982-43-4	: 9657
992-94-9	: 11437
993-00-0	: 11378
993-07-7	: 11489
993-13-5	: 9538
994-05-8	: 3396
994-25-2	: 8726
994-30-9	: 11388
994-49-0	: 5235
996-98-5	: 5562
997-49-9	: 11698
998-29-8	: 11500
998-30-1	: 11477
998-40-3	: 9520
998-41-4	: 11454
998-93-6	: 6358
998-95-8	: 6363
999-21-3	: 3782
999-52-0	: 6362
999-55-3	: 10694
999-78-0	: 9130
999-79-1	: 9127
999-81-5	: 5442
999-97-3	: 11363
1000-49-3	: 11371
1000-50-6	: 11369
1000-86-8	: 8665
1001-53-2	: 28
1001-89-4	: 6361
1002-28-4	: 7126
1002-69-3	: 5036
1002-84-2	: 8645
1003-03-8	: 4771
1003-09-4	: 11726
1003-10-7	: 11794
1003-29-8	: 11114
1003-30-1	: 6107
1003-31-2	: 11732
1003-56-1	: 11036
1003-64-1	: 4391
1003-73-2	: 10988
1003-78-7	: 11784
1003-90-3	: 11141
1004-36-0	: 10793
1004-66-6	: 1919
1004-77-9	: 4599
1005-64-7	: 1109
1005-67-0	: 7684
1006-06-0	: 1874
1006-31-1	: 2263
1007-26-7	: 1484
1007-32-5	: 3677
1008-88-4	: 10998
1008-89-5	: 10997
1009-14-9	: 8986
1009-67-2	: 2137
1009-81-0	: 1754
1009-93-4	: 5001
1010-48-6	: 2130
1012-72-2	: 990
1012-84-6	: 2712
1013-08-7	: 9187
1013-83-8	: 7769
1013-88-3	: 1835
1013-96-3	: 10676
1016-77-9	: 7601
1016-94-0	: 9195
1017-56-7	: 7583
1024-57-3	: 6301
1025-15-6	: 11842
1048-08-4	: 11452
1056-77-5	: 6175
1066-27-9	: 11435
1066-45-1	: 11540
1067-08-9	: 8814
1067-20-5	: 8756
1068-19-5	: 6391
1068-57-1	: 195
1068-87-7	: 8812
1068-90-2	: 9922
1069-53-0	: 6864
1070-14-0	: 6705
1070-32-2	: 6492
1070-34-4	: 3335
1070-71-9	: 10737
1070-78-6	: 10142
1070-83-3	: 3508
1070-87-7	: 8841
1071-26-7	: 6383
1071-73-4	: 8974
1071-81-4	: 6849
1071-83-6	: 6276
1071-94-9	: 6589
1071-98-3	: 3908
1072-05-5	: 6387
1072-16-8	: 8378
1072-21-5	: 6748
1072-36-2	: 8224
1072-39-5	: 5281
1072-43-1	: 11696
1072-44-2	: 500
1072-52-2	: 499
1072-53-3	: 5199
1072-83-9	: 5820
1072-91-9	: 10840
1072-98-6	: 10861
1073-06-9	: 1043
1073-07-0	: 4732
1073-23-0	: 10945
1073-67-2	: 1151
1074-11-9	: 1376
1074-17-5	: 2036
1074-43-7	: 2037
1074-55-1	: 2038
1074-58-4	: 275
1074-92-6	: 1464
1075-22-5	: 7254
1075-38-3	: 1465
1075-74-7	: 2106
1076-26-2	: 8031
1076-55-7	: 9391
1076-61-5	: 7994
1076-66-0	: 1625
1077-16-3	: 1763
1078-04-2	: 7260
1078-19-9	: 8048
1078-21-3	: 2128
1078-30-4	: 11228
1078-71-3	: 1745
1079-71-6	: 427
1080-16-6	: 5130
1080-32-6	: 9545
1081-34-1	: 11600
1081-75-0	: 2121
1081-77-2	: 2064
1083-56-3	: 1099
1086-80-2	: 2472
1088-11-5	: 2411
1093-58-9	: 354
1095-90-5	: 6516
1111-74-6	: 11414
1111-97-3	: 3901
1112-37-4	: 9529
1112-39-6	: 11412
1112-48-7	: 11368
1112-55-6	: 11449
1112-56-7	: 11543
1113-12-8	: 11417
1113-30-0	: 3113
1113-68-4	: 25
1113-74-2	: 8968
1114-09-6	: 3302
1114-51-8	: 9793
1114-71-2	: 3976
1114-76-7	: 3200
1115-11-3	: 3698
1115-12-4	: 9915
1115-30-6	: 3295
1116-24-1	: 58
1116-39-8	: 3123
1110-40-1	: 9020
1116-61-6	: 3121
1117-19-7	: 9945
1117-55-1	: 8431
1117-59-5	: 8882
1118-09-8	: 11491
1118-68-9	: 6261
1118-71-4	: 6404
1118-84-9	: 3577
1119-06-8	: 6465
1119-19-3	: 3895
1119-33-1	: 6245
1119-40-0	: 8775
1119-46-6	: 8866
1119-49-9	: 38
1119-51-3	: 9010
1119-60-4	: 6564
1119-65-9	: 6610
1120-07-6	: 8118
1120-16-7	: 5289
1120-21-4	: 11938
1120-28-1	: 5357
1120-29-2	: 5287
1120-36-1	: 11636
1120-48-5	: 8354
1120-56-5	: 4193
1120-59-8	: 11750
1120-62-3	: 4929
1120-64-5	: 8576
1120-71-4	: 8570
1120-72-5	: 4885
1120-73-6	: 4944
1120-87-2	: 10877
1120-97-4	: 5193
1121-07-9	: 11160
1121-18-2	: 4710
1121-24-0	: 9376
1121-37-5	: 7607
1121-55-7	: 10953
1121-60-4	: 10887
1121-92-2	: 502
1121-95-5	: 4875
1122-20-9	: 4705
1122-25-4	: 4594
1122-39-0	: 11027
1122-42-5	: 1773
1122-54-9	: 5869
1122-60-7	: 4454
1122-62-9	: 5867
1122-69-6	: 10963
1122-81-2	: 11016
1122-82-3	: 4417
1122-83-4	: 844
1122-91-4	: 528
1123-00-8	: 4775
1123-09-7	: 4707
1123-27-9	: 5821
1123-34-8	: 4555
1123-84-8	: 1371
1123-85-9	: 1608
1123-86-0	: 10423
1123-95-1	: 11105
1123-96-2	: 10959
1124-05-6	: 1362
1124-07-8	: 9293
1124-20-5	: 1985
1124-27-2	: 4442
1124-33-0	: 10993
1125-60-6	: 7372
1125-74-2	: 1547
1125-78-6	: 8044
1125-80-0	: 7382
1126-09-6	: 9643
1126-18-7	: 4575
1126-46-1	: 2543
1126-75-6	: 2031
1126-78-9	: 656
1126-79-0	: 1114
1127-76-0	: 7885
1128-08-1	: 4950
1128-54-7	: 10838
1128-76-3	: 2538
1129-47-1	: 10293
1129-50-6	: 3208
1129-89-1	: 4216
1131-16-4	: 10829
1131-18-6	: 10814
1101-00-0	: 9272
1131-62-0	: 5805
1131-63-1	: 7779
1132-21-4	: 2577
1132-39-4	: 2175
1132-66-7	: 1765
1133-80-8	: 5968
1134-23-2	: 4168
1134-62-9	: 7747
1135-12-2	: 839
1140-14-3	: 7614
1140-16-5	: 7628
1142-70-7	: 11080
1142-97-8	: 2236

1143-38-0	: 430
1144-74-7	: 7636
1146-98-1	: 7265
1146-99-2	: 7266
1148-11-4	: 11148
1149-24-2	: 10932
1151-15-1	: 2667
1154-59-2	: 610
1165-39-5	: 6184
1165-48-6	: 2819
1174-72-7	: 11509
1184-60-7	: 10569
1185-33-7	: 3615
1185-55-3	: 11487
1186-31-8	: 6573
1186-53-4	: 8840
1186-73-8	: 7553
1187-03-7	: 12029
1187-13-9	: 3837
1187-58-2	: 9799
1188-33-6	: 7443
1189-71-5	: 11585
1189-99-7	: 6448
1190-22-3	: 3278
1190-34-7	: 6605
1190-39-2	: 9938
1190-63-2	: 8258
1190-83-6	: 6440
1191-04-4	: 7049
1191-08-8	: 3368
1191-41-9	: 8283
1191-43-1	: 6812
1191-75-9	: 6668
1191-95-3	: 4197
1191-96-4	: 4974
1191-99-7	: 6086
1192-18-3	: 4801
1192-28-5	: 4892
1192-30-9	: 6036
1192-37-6	: 4428
1192-62-7	: 5817
1192-88-7	: 4631
1193-02-8	: 2291
1193-12-0	: 9651
1193-16-4	: 4294
1193-17-5	: 4295
1193-18-6	: 4711
1193-35-7	: 6085
1193-46-0	: 4513
1193-47-1	: 4580
1193-70-0	: 4894
1193-72-2	: 1011
1193-74-4	: 11044
1193-81-3	: 4420
1193-82-4	: 2040
1194-02-1	: 2756
1194-65-6	: 2752
1195-31-9	: 4681
1195-32-0	: 1986
1195-42-2	: 4303
1196-31-2	: 4605
1196-39-0	: 7383
1196-58-3	: 1701
1196-67-4	: 3701
1196-79-8	: 7299
1197-18-8	: 4333
1197-37-1	: 1272
1197-40-6	: 6132
1197-55-3	: 886
1198-14-7	: 11310
1198-37-4	: 11255
1198-59-0	: 1398
1198-61-4	: 1399
1198-62-5	: 1400
1200-03-9	: 1003
1200-73-3	: 5194
1201-26-9	: 10442
1202-08-0	: 5850
1202-34-2	: 10871
1203-39-0	: 7845
1204-78-0	: 3015
1205-02-3	: 327
1205-64-7	: 808
1205-71-6	: 679
1205-91-0	: 1509
1207-69-8	: 315
1210-35-1	: 5147
1214-24-0	: 12053
1241-94-7	: 9560
1260-17-9	: 373
1300-94-3	: 9425
1322-67-4	: 9178
1330-61-6	: 10646
1400-62-0	: 8549
1405-86-3	: 6230
1405-87-4	: 508
1407-14-3	: 6293
1420-07-1	: 9328
1421-14-3	: 894
1423-60-5	: 3923
1435-50-3	: 1014
1435-53-6	: 1300
1435-55-8	: 4124
1436-43-7	: 11223
1441-87-8	: 2900
1445-23-4	: 9324
1445-45-0	: 5669
1445-73-4	: 9724
1446-34-0	: 3006
1447-26-3	: 6600
1450-14-2	: 5228
1450-31-3	: 7547
1450-58-4	: 6048
1450-72-2	: 5827
1450-81-3	: 8871
1453-17-4	: 4254
1453-24-3	: 4660
1453-25-4	: 4245
1453-58-3	: 10835
1454-80-4	: 3028
1454-84-8	: 8104
1454-85-9	: 6317
1455-20-5	: 11731
1455-21-6	: 8145
1457-85-8	: 224
1458-99-7	: 9023
1459-00-3	: 1060
1459-09-2	: 1751
1459-10-5	: 2266
1459-14-9	: 1039
1459-93-4	: 1347
1460-16-8	: 4223
1460-88-4	: 4911
1460-98-6	: 1948
1462-03-9	: 4867
1462-12-0	: 9950
1462-27-7	: 3652
1462-34-6	: 5468
1462-84-6	: 11026
1463-17-8	: 11259
1467-05-6	: 1191
1467-79-4	: 4157
1468-39-9	: 3542
1470-94-6	: 7285
1471-03-0	: 10594
1471-96-1	: 7858
1472-90-8	: 8238
1475-13-4	: 1854
1476-07-9	: 1666
1476-11-5	: 3751
1477-55-0	: 1433
1482-15-1	: 9144
1482-91-3	: 8483
1483-60-9	: 1473
1484-08-8	: 3984
1484-12-4	: 3987
1484-19-1	: 7308
1484-84-0	: 9663
1484-85-1	: 2424
1486-75-5	: 4215
1487-18-9	: 6102
1489-47-0	: 8167
1489-57-2	: 4271
1489-60-7	: 4201
1490-05-7	: 7428
1493-01-2	: 7516
1493-02-3	: 6019
1493-03-4	: 7505
1493-27-2	: 1733
1495-50-7	: 4167
1498-51-7	: 9577
1498-65-3	: 9557
1501-04-8	: 2100
1501-06-0	: 8771
1501-27-5	: 8780
1501-60-6	: 8668
1502-05-2	: 4204
1502-06-3	: 4205
1503-53-3	: 2481
1504-54-7	: 3872
1504-58-1	: 10745
1504-70-7	: 10614
1504-75-2	: 10517
1511-62-2	: 7469
1515-76-0	: 3176
1515-80-6	: 6702
1515-95-3	: 1684
1516-80-9	: 11431
1517-05-1	: 5681
1518-75-8	: 11779
1518-86-1	: 9215
1519-36-4	: 402
1519-59-1	: 11002
1520-21-4	: 735
1520-45-2	: 1698
1520-46-3	: 2021
1521-51-3	: 4628
1522-22-1	: 8802
1522-46-9	: 3461
1526-73-4	: 7848
1527-89-5	: 2761
1528-30-9	: 4827
1529-17-5	: 11492
1529-41-5	: 950
1530-03-6	: 2171
1530-05-8	: 1746
1530-20-7	: 1744
1532-97-4	: 7374
1534-26-5	: 11890
1538-87-0	: 5479
1539-42-0	: 10974
1540-28-9	: 8856
1540-29-0	: 6872
1540-34-7	: 8801
1541-23-7	: 6328
1547-87-1	: 99
1550-09-0	: 6742
1551-06-0	: 11125
1551-09-3	: 11139
1551-21-9	: 10089
1551-27-5	: 11778
1551-32-2	: 11765
1551-44-6	: 3496
1551-88-8	: 4581
1551-89-9	: 4582
1552-12-1	: 4734
1552-67-6	: 7053
1552-95-0	: 8680
1555-80-2	: 887
1556-18-9	: 4824
1558-17-4	: 11058
1558-24-3	: 11459
1558-25-4	: 11456
1558-31-2	: 11391
1558-33-4	: 11390
1559-02-0	: 4967
1559-81-5	: 7995
1560-06-1	: 1111
1560-09-4	: 1110
1560-11-8	: 4909
1560-89-0	: 6310
1561-10-0	: 6900
1561-20-2	: 10630
1563-66-2	: 3991
1563-83-3	: 24
1563-87-7	: 26
1565-71-5	: 1855
1565-86-2	: 1905
1569-01-3	: 10401
1569-02-4	: 10381
1569-43-3	: 7076
1569-44-4	: 7084
1569-59-1	: 9101
1569-60-4	: 6580
1569-69-3	: 4480
1570-45-2	: 10903
1570-64-5	: 9266
1573-17-7	: 3913
1573-28-0	: 6483
1573-58-6	: 3174
1574-33-0	: 9117
1574-40-9	: 9114
1574-41-0	: 8662
1575-61-7	: 8992
1576-13-2	: 9464
1576-95-0	: 9091
1576-96-1	: 9090
1577-22-6	: 7051
1578-21-8	: 10628
1579-40-4	: 2078
1582-09-8	: 11901
1585-06-4	: 1653
1585-07-5	: 1037
1587-04-8	: 2023
1587-20-8	: 10170
1587-26-4	: 3753
1587-29-7	: 3752
1587-41-3	: 7501
1588-44-9	: 1762
1589-47-5	: 10386
1589-60-2	: 4552
1590-08-5	: 8049
1591-31-7	: 3079
1592-23-0	: 8252
1592-38-7	: 7909
1595-04-6	: 1122
1595-05-7	: 1123
1595-06-8	: 1994
1595-11-5	: 1121
1595-16-0	: 1998
1596-84-5	: 3334
1599-41-3	: 10180
1599-67-3	: 5280
1600-37-9	: 10593
1601-00-9	: 5847
1603-40-3	: 10866
1603-79-8	: 935
1603-91-4	: 11665
1604-01-9	: 11021
1604-02-0	: 4872
1604-11-1	: 3327
1606-08-2	: 4340
1609-47-8	: 5169
1609-86-5	: 10062
1609-93-4	: 10618
1610-24-8	: 4433
1610-29-3	: 9094
1611-50-3	: 3942
1611-52-5	: 3952
1611-65-0	: 3386
1613-32-7	: 11292
1613-34-9	: 11266
1613-45-2	: 10044
1613-51-0	: 11795
1615-14-1	: 7205
1615-70-9	: 8673
1615-75-4	: 5475
1615-80-1	: 7153
1617-18-1	: 3841
1617-90-9	: 5349
1618-26-4	: 7461
1619-28-9	: 4947
1619-34-7	: 485
1619-56-3	: 3495
1619-57-4	: 3504
1619-62-1	: 9942
1620-14-0	: 10427
1622-10-2	: 6373
1622-32-8	: 5629
1623-14-9	: 9561
1626-09-1	: 8394
1629-58-9	: 9104
1630-77-9	: 5903
1630-94-0	: 4971
1631-82-9	: 11381
1631-84-1	: 11403
1632-16-2	: 6425
1632-76-4	: 10856
1632-83-3	: 2382
1633-90-5	: 6858
1634-04-4	: 10072
1634-09-9	: 7746
1635-02-5	: 7091
1635-61-6	: 677
1637-31-6	: 3173
1638-22-8	: 9248
1638-26-2	: 4800
1638-63-7	: 972
1639-09-4	: 6436
1640-89-7	: 4812
1641-40-3	: 2413
1643-49-8	: 5970
1646-26-0	: 5775
1646-88-4	: 335
1647-08-1	: 7019
1647-16-1	: 5024
1647-23-0	: 3235
1647-26-3	: 4316
1649-08-7	: 5531
1653-19-6	: 3158
1653-31-2	: 11884
1653-32-3	: 11632
1653-34-5	: 8647
1653-40-3	: 6494
1653-57-2	: 8084
1654-86-0	: 5062
1655-70-5	: 2075
1655-71-6	: 1929
1656-48-0	: 10114
1657-52-9	: 1222
1663-67-8	: 10029
1666-13-3	: 5225
1666-96-2	: 11289
1667-01-2	: 5876
1667-04-5	: 1243
1670-46-8	: 4879
1671-75-6	: 6526
1672-46-4	: 4047
1674-10-8	: 4648
1674-30-2	: 1852
1674-33-5	: 8744
1674-37-9	: 8480
1674-38-0	: 5327
1674-56-2	: 10338
1675-69-0	: 8129
1678-80-4	: 4492
1678-81-5	: 4489
1678-82-6	: 4440
1678-91-7	: 4390
1678-92-8	: 4471
1678-93-9	: 4325
1678-97-3	: 4488
1678-98-4	: 4453
1679-06-7	: 6857
1679-07-8	: 4852
1679-09-0	: 3441
1679-36-3	: 7130
1679-47-6	: 6145
1680-51-9	: 7998
1680-53-1	: 7922
1680-58-6	: 7927
1681-22-7	: 7999
1686-41-5	: 4237
1689-82-3	: 9448
1689-84-5	: 2751
1690-75-1	: 9645
1694-19-5	: 1939
1695-04-1	: 1938
1696-20-4	: 7682
1698-60-8	: 4080
1702-17-6	: 4142
1703-46-4	: 1862
1703-52-2	: 6106
1704-15-0	: 10711
1706-11-2	: 1920
1706-73-6	: 1546
1708-32-3	: 11751
1708-39-0	: 5209
1709-03-1	: 46
1710-98-1	: 2895
1711-06-4	: 2903
1711-07-5	: 2898
1712-64-7	: 8076
1717-00-6	: 5534
1718-50-9	: 2098
1719-53-5	: 11393
1719-57-9	: 11372
1719-58-0	: 11376
1721-89-7	: 11254
1721-93-3	: 7381
1723-94-0	: 7690
1725-74-2	: 7035
1726-12-1	: 2108
1726-14-3	: 1113
1729-67-5	: 10241

1731-79-9 : 5300
1731-84-6 : 8154
1731-88-0 : 11882
1731-92-6 : 6316
1732-08-7 : 6395
1732-09-8 : 8388
1732-10-1 : 8134
1733-12-6 : 9368
1738-25-6 : 10101
1738-36-9 : 284
1739-84-0 : 7200
1740-97-2 : 9414
1741-83-9 : 8826
1745-16-0 : 2105
1745-81-9 : 9457
1746-11-8 : 2443
1746-13-0 : 2167
1746-77-6 : 3955
1754-62-7 : 10686
1755-01-7 : 7577
1755-05-1 : 8686
1758-10-7 : 2027
1758-33-4 : 8608
1758-85-6 : 1421
1758-88-9 : 1674
1759-53-1 : 4958
1759-58-6 : 4804
1759-64-4 : 4652
1759-81-5 : 4930
1759-88-2 : 11439
1760-46-9 : 937
1761-71-3 : 4299
1768-31-6 : 10478
1772-10-7 : 1996
1776-37-0 : 7241
1777-82-8 : 1853
1779-81-3 : 11664
1780-31-0 : 11052
1781-78-8 : 4755
1790-22-3 : 3450
1792-41-2 : 2379
1792-81-0 : 4372
1794-84-9 : 7485
1795-01-3 : 11760
1795-04-6 : 11792
1795-05-7 : 11793
1795-09-1 : 11787
1795-15-9 : 4460
1795-16-0 : 4344
1795-17-1 : 4382
1795-18-2 : 4475
1795-20-6 : 4843
1795-21-7 : 4790
1795-22-8 : 4850
1795-26-2 : 4497
1795-27-3 : 4496
1795-48-8 : 10060
1796-05-0 : 1494
1804-15-5 : 9784
1805-61-4 : 9404
1806-29-7 : 3064
1807-36-9 : 2837
1809-02-5 : 3155
1809-05-8 : 8820
1809-10-5 : 8712
1809-19-4 : 9532
1809-26-3 : 9014
1812-51-7 : 3072
1814-88-6 : 10128
1817-47-6 : 1978
1817-57-8 : 3924
1817-73-8 : 646
1818-07-1 : 2069
1820-09-3 : 2976
1821-02-9 : 8902
1821-12-1 : 1100
1821-36-9 : 680
1821-39-2 : 842
1822-74-8 : 5910
1822-86-2 : 3149
1823-91-2 : 959
1824-81-3 : 10868
1825-30-5 : 7810
1825-31-6 : 7809
1825-62-3 : 11430
1825-82-7 : 11462
1826-14-8 : 11676
1830-54-2 : 8784
1830-95-1 : 3120
1832-53-7 : 9543
1833-51-8 : 11379
1836-75-5 : 8081
1838-59-1 : 6017
1838-77-3 : 7085
1839-88-9 : 4490
1843-05-6 : 7621
1845-30-3 : 2975
1846-70-4 : 8203
1850-14-2 : 7543
1855-63-6 : 4630
1859-76-3 : 7626
1860-21-5 : 247
1860-27-1 : 3399
1861-21-8 : 11099
1861-32-1 : 5180
1861-40-1 : 509
1863-63-4 : 2509
1864-92-2 : 738
1866-31-5 : 10692
1867-37-4 : 9917
1868-53-7 : 7498
1871-52-9 : 4749
1871-57-4 : 10537
1871-58-5 : 10178
1871-76-7 : 975
1871-96-1 : 5041
1873-25-2 : 3605
1873-29-6 : 10061
1873-54-7 : 11267
1873-92-3 : 11402
1875-88-3 : 1597
1877-72-1 : 2550
1878-68-8 : 888
1879-06-7 : 5845
1879-09-0 : 9325
1883-32-5 : 1615
1885-14-9 : 4026
1885-29-6 : 2741
1885-38-7 : 10587
1888-57-9 : 6966
1888-71-7 : 10571
1888-85-3 : 3370
1896-62-4 : 3882
1897-45-6 : 1317
1899-24-7 : 6043
1904-98-9 : 10754
1912-21-6 : 10319
1912-24-9 : 476
1917-15-3 : 6060
1918-00-9 : 2567
1918-02-1 : 10896
1918-16-7 : 9771
1921-70-6 : 8644
1928-30-9 : 11856
1928-38-7 : 172
1929-73-3 : 170
1929-77-7 : 12047
1929-82-4 : 8071
1931-60-8 : 3367
1932-04-3 : 9661
1937-54-8 : 8113
1937-62-8 : 8308
1938-17-6 : 1257
1938-22-3 : 2063
1940-18-7 : 4522
1942-45-6 : 8532
1942-46-7 : 5100
1943-16-4 : 7487
1943-54-0 : 9291
1945-53-5 : 9100
1951-11-7 : 3684
1967-25-5 : 11998
1969-43-3 : 6406
1969-73-9 : 875
1971-46-6 : 7322
1973-22-4 : 1035
1974-04-5 : 6357
1975-44-6 : 7767
1975-78-6 : 5057
1982-47-4 : 4088
1985-57-5 : 1455
1985-59-7 : 7980
1985-97-3 : 1691
1990-29-0 : 338
2004-67-3 : 9102
2004-69-5 : 9113
2004-70-8 : 8661
2006-14-6 : 10675
2008-41-5 : 11586
2009-74-7 : 6598
2009-83-8 : 6916
2016-42-4 : 11612
2016-57-1 : 5032
2018-90-8 : 7907
2021-28-5 : 2131
2025-40-3 : 10625
2027-17-0 : 7921
2027-19-2 : 7958
2028-63-9 : 3921
2029-94-9 : 538
2031-67-6 : 11479
2032-35-1 : 5459
2032-65-7 : 7646
2033-24-1 : 5190
2035-94-1 : 1602
2036-41-1 : 11072
2037-31-2 : 2294
2038-03-1 : 7689
2039-82-9 : 1028
2039-83-0 : 1368
2039-85-2 : 1150
2039-86-3 : 1027
2039-87-4 : 1149
2039-88-5 : 1026
2039-89-6 : 1628
2039-93-2 : 1968
2040-07-5 : 5875
2040-14-4 : 10466
2040-95-1 : 4778
2040-96-2 : 4849
2042-37-7 : 2744
2043-38-1 : 2856
2043-61-0 : 4331
2044-08-8 : 4627
2045-79-6 : 1577
2047-21-4 : 5877
2049-66-3 : 9951
2049-67-4 : 9041
2049-73-2 : 1407
2049-74-3 : 5719
2049-80-1 : 9980
2049-94-7 : 1954
2049-95-8 : 1480
2050-01-3 : 10302
2050-03-5 : 1488
2050-04-6 : 2109
2050-07-9 : 8983
2050-08-0 : 2649
2050-14-8 : 9216
2050-20-6 : 6394
2050-23-9 : 8387
2050-24-0 : 1419
2050-32-0 : 4256
2050-33-1 : 4255
2050-43-3 : 56
2050-46-6 : 1406
2050-47-7 : 2072
2050-48-8 : 2238
2050-60-4 : 5558
2050-61-5 : 5557
2050-67-1 : 3040
2050-68-2 : 3041
2050-69-3 : 7807
2050-73-9 : 7811
2050-74-0 : 7812
2050-76-2 : 8026
2050-77-3 : 5053
2050-78-4 : 8073
2050-87-5 : 11917
2050-89-7 : 3030
2050-92-2 : 8708
2050-95-5 : 3635
2051-00-5 : 5555
2051-04-9 : 5243
2051-10-7 : 4116
2051-25-4 : 5023
2051-30-1 : 8377
2051-31-2 : 5072
2051-32-3 : 8459
2051-33-4 : 6948
2051-49-2 : 6874
2051-53-8 : 793
2051-60-7 : 3023
2051-61-8 : 3024
2051-62-9 : 3025
2051-78-7 : 3587
2051-84-5 : 783
2051-90-3 : 1389
2051-97-0 : 11136
2051-98-1 : 4
2052-01-9 : 10209
2052-06-4 : 642
2052-07-5 : 3016
2052-14-4 : 2614
2052-15-5 : 8904
2055-40-5 : 1643
2057-43-4 : 10996
2057-49-0 : 11007
2065-23-8 : 230
2065-70-5 : 7813
2077-13-6 : 1190
2078-54-8 : 9226
2079-95-0 : 11621
2081-44-9 : 10787
2082-59-9 : 8858
2082-76-0 : 5060
2083-91-2 : 11361
2084-19-7 : 8845
2084-45-9 : 1363
2085-88-3 : 8622
2086-83-1 : 2467
2090-14-4 : 4118
2090-89-3 : 5748
2091-29-4 : 6669
2093-20-1 : 5673
2095-57-0 : 11065
2100-17-6 : 9001
2101-86-2 : 982
2101-88-4 : 977
2103-64-2 : 11526
2104-64-5 : 9551
2106-54-9 : 6910
2108-92-1 : 4358
2109-66-2 : 7692
2110-78-3 : 10280
2111-75-3 : 4633
2113-47-5 : 35
2113-51-1 : 3077
2113-57-7 : 3017
2113-58-8 : 3086
2114-00-3 : 10419
2114-39-8 : 1087
2114-42-3 : 4470
2116-62-3 : 11003
2116-65-6 : 11006
2121-66-6 : 10140
2122-70-5 : 7735
2123-27-5 : 1373
2123-28-6 : 1367
2130-23-6 : 8504
2131-18-2 : 2089
2131-38-6 : 8013
2131-43-3 : 8002
2131-55-7 : 1177
2132-80-1 : 3047
2136-81-4 : 1238
2136-89-2 : 1237
2136-95-0 : 2085
2138-22-9 : 1505
2138-43-4 : 1535
2141-62-0 : 10103
2142-01-0 : 7367
2142-06-5 : 11163
2142-63-4 : 5780
2142-73-6 : 5810
2146-37-4 : 4813
2146-38-5 : 4924
2150-18-7 : 10962
2150-38-1 : 2579
2155-30-8 : 10273
2155-42-2 : 1370
2157-31-5 : 8792
2157-50-8 : 10483
2158-55-6 : 4228
2159-75-3 : 5123
2162-92-7 : 6761
2162-98-3 : 5040
2162-99-4 : 8371
2163-00-0 : 6762
2163-44-2 : 9934
2163-48-6 : 9981
2164-08-1 : 4900
2164-09-2 : 10520
2164-17-2 : 12007
2164-19-4 : 4292
2167-39-7 : 8594
2168-93-6 : 3422
2170-03-8 : 6095
2171-96-2 : 11436
2173-56-0 : 8913
2174-64-3 : 1526
2175-91-9 : 4768
2176-62-7 : 10995
2176-98-9 : 11546
2177-32-4 : 7093
2177-47-1 : 7276
2180-43-0 : 6949
2180-68-9 : 8674
2180-69-0 : 8675
2182-66-3 : 11419
2189-60-8 : 2067
2196-23-8 : 6529
2197-37-7 : 8225
2198-20-1 : 4207
2198-61-0 : 6897
2198-72-3 : 6508
2198-75-6 : 7808
2199-51-1 : 11115
2199-52-2 : 11116
2203-34-1 : 3607
2203-35-2 : 3603
2203-80-7 : 7120
2206-38-4 : 1253
2206-48-6 : 1250
2206-89-5 : 10620
2207-01-4 : 4365
2207-03-6 : 4370
2207-04-7 : 4371
2207-27-4 : 4770
2207-50-3 : 8574
2210-28-8 : 10672
2211-67-8 : 3281
2211-68-9 : 3727
2211-69-0 : 3728
2211-70-3 : 3722
2212-67-1 : 7661
2213-08-3 : 332
2213-23-2 : 6385
2213-32-3 : 9030
2213-43-6 : 9624
2214-14-4 : 1955
2216-12-8 : 2053
2216-15-1 : 698
2216-30-0 : 6386
2216-32-2 : 6408
2216-33-3 : 8409
2216-34-4 : 8410
2216-36-6 : 8124
2216-67-3 : 7722
2216-68-4 : 7721
2216-69-5 : 7913
2216-70-8 : 8328
2216-81-1 : 10264
2216-87-7 : 11959
2216-92-4 : 6271
2216-94-6 : 10741
2217-06-3 : 2288
2217-07-4 : 730
2217-14-3 : 3310
2217-41-6 : 7731
2217-45-0 : 847
2217-60-9 : 1500
2219-66-1 : 3904
2219-82-1 : 9332
2224-37-5 : 8951
2227-13-6 : 11648
2230-70-8 : 4619
2231-57-4 : 4033
2233-00-3 : 10604
2233-29-6 : 11024
2234-13-1 : 7933
2234-16-4 : 5799
2234-20-0 : 1626
2237-30-1 : 2742
2239-78-3 : 6657

2243-27-8	8144
2243-30-3	833
2243-32-5	2713
2243-35-8	5771
2243-42-7	2714
2243-43-8	3411
2243-53-0	3855
2243-55-2	7167
2243-61-0	7796
2243-62-1	7797
2243-71-2	400
2243-76-7	2648
2243-80-3	7635
2243-81-4	7757
2243-83-6	7750
2243-89-2	11302
2243-94-9	8016
2243-98-3	11984
2244-07-7	11941
2244-16-8	4715
2254-94-6	2867
2257-35-4	10696
2258-58-4	9086
2259-96-3	2851
2270-20-4	2099
2272-45-9	811
2279-76-7	11542
2288-18-8	1967
2291-80-7	3968
2294-43-1	1083
2294-71-5	900
2294-72-6	11987
2294-94-2	1618
2295-31-0	11683
2295-58-1	10489
2303-16-4	5106
2303-17-5	11823
2305-13-7	2153
2306-88-9	8437
2308-38-5	3509
2310-17-0	9514
2310-98-7	9841
2311-59-3	5067
2312-35-8	10507
2313-65-7	6943
2314-78-5	11157
2314-97-8	7559
2315-68-6	2724
2316-50-9	9234
2316-64-5	1843
2319-29-1	5031
2319-61-1	4392
2320-30-1	4589
2321-07-5	11522
2323-81-1	8868
2328-24-7	924
2338-12-7	2873
2344-70-9	2155
2344-71-0	1106
2344-80-1	11382
2344-83-4	11386
2345-26-8	10295
2345-27-9	11633
2345-28-0	8651
2345-61-1	10617
2346-00-1	11672
2346-81-8	6739
2350-19-8	6367
2351-13-5	5229
2351-97-5	9080
2358-84-1	10669
2362-10-9	5230
2362-12-1	9241
2364-59-2	3581
2365-48-2	204
2366-52-1	3378
2367-82-0	2271
2370-88-9	4999
2371-19-9	6520
2373-51-5	7484
2381-31-9	523
2382-43-6	9831
2384-73-8	6606
2384-75-0	6538
2384-85-2	5098
2384-86-3	5099
2384-90-9	6326
2385-70-8	3468
2385-85-5	7644
2388-14-9	1973
2393-23-9	1813
2396-53-4	2080
2396-60-3	5137
2396-63-6	6343
2396-65-8	8115
2396-78-3	7062
2396-83-0	7054
2396-84-1	6700
2398-09-6	4189
2398-10-9	4190
2398-16-5	4182
2398-37-0	1051
2398-65-4	2068
2398-66-5	2007
2398-68-7	1568
2398-81-4	10909
2399-48-6	10695
2400-01-3	1764
2400-02-4	1695
2400-03-5	2170
2400-04-6	1119
2401-22-1	1173
2401-24-3	666
2401-73-2	5232
2402-06-4	4973
2402-79-1	11019
2404-35-5	4222
2408-20-0	10325
2408-37-9	4617
2409-52-1	3329
2409-55-4	9330
2412-73-9	2551
2415-72-7	4984
2417-04-1	3069
2417-90-5	10095
2418-31-7	3787
2419-73-0	3349
2420-16-8	532
2423-01-0	4906
2423-10-1	8289
2424-92-2	5352
2424-98-8	10297
2425-28-7	1844
2425-54-9	11615
2425-66-3	9867
2425-79-8	8601
2426-02-0	7347
2426-08-6	8602
2426-54-2	10631
2430-00-4	96
2430-22-0	8463
2431-24-5	6540
2431-30-3	7117
2431-50-7	3809
2431-52-9	3420
2431-54-1	3810
2432-12-4	9295
2432-63-5	3784
2436-84-2	6330
2436-85-3	7716
2436-92-2	11260
2436-93-3	11264
2436-96-6	3062
2437-03-8	7719
2437-25-4	5303
2437-56-1	11893
2437-73-2	11300
2437-76-5	1065
2437-88-9	2343
2437-95-8	2972
2437-98-1	88
2438-03-1	2723
2438-04-2	2688
2438-12-2	4668
2438-19-9	10641
2438-72-4	880
2439-01-2	4072
2439-10-3	5347
2439-35-2	10633
2439-99-8	6277
2443-03-0	4837
2443-04-1	4836
2443-91-6	9886
2444-37-3	216
2444-89-5	2073
2445-76-3	10265
2445-82-1	2833
2445-83-2	2836
2447-54-3	2432
2450-71-7	10725
2455-24-5	10674
2459-05-4	3791
2459-09-8	10908
2459-24-7	7766
2459-25-8	7776
2462-85-3	8229
2462-94-4	5363
2463-53-8	8179
2463-63-0	6530
2464-37-1	4079
2467-09-6	3098
2467-10-9	8838
2470-68-0	6352
2473-01-0	8123
2476-37-1	5800
2486-07-9	1495
2489-86-3	7954
2492-22-0	8324
2493-02-9	1049
2495-35-4	10688
2497-21-4	7088
2498-75-1	522
2499-95-8	10643
2505-06-8	4246
2506-41-4	7785
2508-29-4	8920
2512-81-4	9648
2514-91-2	11815
2516-34-9	4169
2516-96-3	2545
2517-04-6	495
2517-43-3	3625
2523-37-7	5974
2523-55-9	4297
2523-56-0	4296
2524-04-1	9576
2524-09-6	9579
2524-37-0	2574
2524-52-9	10901
2524-64-3	9575
2525-02-2	1548
2528-38-3	9565
2528-61-2	6527
2531-80-8	10312
2532-58-3	4803
2536-38-1	8250
2545-59-7	257
2547-61-7	7536
2550-04-1	11482
2550-06-3	11458
2550-21-2	6976
2550-22-3	7097
2550-26-7	3678
2550-28-9	7131
2550-36-9	4318
2553-04-0	7623
2553-19-7	11406
2554-06-5	4996
2555-49-9	229
2562-38-1	4839
2563-97-5	98
2565-39-1	6586
2568-33-4	3354
2568-90-3	3398
2568-91-4	10075
2570-12-9	4747
2570-26-5	8641
2571-39-3	7615
2575-20-4	3658
2579-20-6	4362
2579-22-8	10724
2583-25-7	9979
2586-89-2	6611
2591-86-8	9638
2593-15-9	11599
2594-20-9	3956
2594-21-0	3937
2595-97-3	337
2601-89-0	10099
2605-18-7	1940
2612-33-1	9997
2612-46-6	6993
2613-61-8	6447
2613-65-2	4818
2613-66-3	4817
2613-69-6	4858
2613-72-1	4861
2614-88-2	6711
2620-50-0	2426
2621-46-7	1192
2621-78-5	3967
2622-08-4	9596
2622-21-1	4657
2622-60-8	2383
2622-89-1	3130
2623-23-6	4549
2623-50-9	11263
2623-54-3	2923
2623-82-7	8425
2623-84-9	11945
2623-95-2	5061
2624-01-3	6457
2625-30-1	4455
2625-31-2	4840
2627-95-4	5233
2631-37-0	9769
2639-63-6	3528
2641-89-6	1551
2642-63-9	5801
2642-98-0	4113
2648-61-5	5802
2652-13-3	8178
2655-83-6	1157
2657-65-0	3426
2675-77-6	4084
2676-33-7	10269
2678-54-8	5901
2679-87-0	3373
2681-83-6	6875
2681-92-7	8859
2687-25-4	1275
2688-77-9	7379
2688-84-8	834
2689-59-0	7618
2689-88-5	9946
2690-08-6	8417
2706-56-1	10948
2713-09-9	5943
2717-42-2	8009
2717-44-4	7816
2718-25-4	11342
2719-27-9	4330
2719-52-0	1952
2720-41-4	4481
2721-38-2	4796
2722-36-3	2156
2724-59-6	8263
2725-82-8	1036
2734-70-5	703
2736-37-0	10492
2736-40-5	3691
2738-19-4	7040
2739-12-0	792
2741-16-4	1974
2745-25-7	6031
2745-26-8	6026
2747-53-7	9098
2757-90-6	6247
2758-18-1	4945
2759-28-6	9621
2761-24-2	11480
2761-84-4	5339
2765-04-0	11920
2765-11-9	8640
2765-18-6	7957
2768-41-4	3915
2769-64-4	3387
2777-66-4	11991
2778-68-9	2936
2781-85-3	4990
2782-57-2	11838
2782-91-4	11810
2785-74-2	1511
2785-89-9	9357
2790-09-2	1348
2791-29-9	6636
2798-73-4	3891
2804-50-4	10541
2807-30-9	5764
2807-54-7	3781
2808-71-1	4661
2808-76-6	4649
2808-77-7	4651
2808-79-9	4650
2808-80-2	4434
2809-64-5	7997
2809-67-8	8530
2809-69-0	6717
2810-04-0	11739
2810-69-7	2255
2815-34-1	9610
2816-57-1	4586
2817-71-2	10832
2819-48-9	4314
2819-86-5	9443
2820-37-3	10827
2825-00-5	10797
2825-79-8	8296
2834-05-1	11944
2835-21-4	10574
2835-33-8	11014
2835-81-6	3463
2835-96-3	9213
2835-97-4	9209
2836-00-2	9211
2836-03-5	1267
2836-04-6	1268
2837-89-0	5486
2845-89-8	1179
2847-72-5	5056
2848-25-1	1540
2855-08-5	3246
2856-63-5	949
2858-66-4	10485
2859-67-8	11013
2859-68-9	11012
2859-78-1	1016
2867-05-2	2986
2867-47-2	10654
2867-59-6	3596
2867-63-2	4893
2868-37-3	4963
2869-34-3	11870
2870-04-4	1673
2873-18-9	11728
2873-74-7	8806
2873-89-4	661
2873-97-4	10521
2874-74-0	5314
2876-35-9	7829
2876-44-0	7938
2876-45-1	7896
2876-46-2	7901
2876-51-9	7937
2876-52-0	7895
2876-53-1	7900
2876-62-2	3674
2876-63-3	10473
2882-98-6	4841
2883-02-5	4457
2883-07-0	4341
2883-08-1	4447
2883-12-7	1255
2883-45-6	6337
2885-00-9	8244
2886-52-4	10445
2887-61-8	3665
2890-62-2	5843
2892-18-4	7100
2896-60-8	1518
2899-90-3	8950
2902-69-4	5871
2902-96-7	10394
2905-56-8	9698
2912-62-1	973
2917-26-2	6643
2917-47-7	10410
2918-13-0	6583
2919-05-3	9118
2919-23-5	4196
2921-88-2	4089
2922-51-2	6320
2923-68-4	8442
2933-74-6	5745
2935-44-6	6799

2935-90-2	: 10287
2941-19-7	: 1830
2941-20-0	: 1809
2941-55-1	: 5930
2943-70-6	: 5231
2944-49-2	: 1916
2947-60-6	: 961
2947-61-7	: 962
2955-65-9	: 8395
2955-88-6	: 11165
2959-96-8	: 10781
2961-47-9	: 10880
2969-81-5	: 3470
2971-79-1	: 9644
2973-00-4	: 7497
2973-27-5	: 11224
2977-31-3	: 1617
2978-58-7	: 3896
2979-19-3	: 4587
2980-71-4	: 8185
2983-26-8	: 5523
2983-37-1	: 6885
2984-40-9	: 3806
2987-53-3	: 819
2987-77-1	: 11485
2996-92-1	: 11488
2998-18-7	: 10665
3000-79-1	: 9680
3001-72-7	: 11190
3002-24-2	: 6804
3007-91-8	: 7770
3007-97-4	: 7765
3008-39-7	: 4225
3010-02-4	: 280
3010-04-6	: 273
3012-37-1	: 11710
3016-19-1	: 8487
3017-67-2	: 5086
3017-68-3	: 3714
3017-70-7	: 3718
3017-71-8	: 3713
3017-95-6	: 9840
3017-96-7	: 9838
3018-09-5	: 5881
3018-12-0	: 278
3019-04-3	: 10454
3019-20-3	: 1980
3019-25-8	: 5792
3022-41-1	: 4209
3024-72-4	: 2894
3027-13-2	: 10460
3031-05-8	: 8011
3031-66-1	: 7118
3031-73-0	: 7180
3031-74-1	: 7179
3032-32-4	: 2491
3034-53-5	: 11668
3035-75-4	: 11877
3036-66-6	: 3168
3040-44-6	: 9662
3041-40-5	: 9916
3042-22-6	: 11135
3042-50-0	: 1486
3043-33-2	: 8672
3047-38-9	: 4907
3051-09-0	: 11097
3054-92-0	: 8956
3054-95-3	: 10560
3055-14-9	: 1949
3058-47-7	: 2043
3060-89-7	: 7656
3062-81-5	: 3122
3066-71-5	: 10626
3068-00-6	: 3459
3073-66-3	: 4486
3073-92-5	: 3421
3074-61-1	: 4935
3074-64-4	: 6542
3074-71-3	: 6384
3074-75-7	: 6828
3074-76-8	: 6826
3074-77-9	: 6827
3075-84-1	: 3096
3077-16-5	: 7698
3081-24-1	: 9512
3083-23-6	: 8628
3085-53-8	: 5997
3087-62-5	: 6211
3088-41-3	: 9703
3096-47-7	: 5976
3096-52-4	: 5977
3101-60-8	: 8611
3102-33-8	: 9105
3112-88-7	: 2112
3113-98-2	: 4421
3118-97-6	: 8028
3121-83-3	: 9993
3128-06-1	: 6905
3128-07-2	: 6469
3129-39-3	: 7846
3129-90-6	: 7399
3132-99-8	: 527
3140-92-9	: 11743
3140-93-0	: 11742
3141-12-6	: 440
3141-24-0	: 11790
3141-27-3	: 11744
3142-58-3	: 8799
3142-66-3	: 8972
3142-84-5	: 6608
3144-16-9	: 2929
3144-54-5	: 6962
3147-55-5	: 2558
3152-68-9	: 9110
3153-36-4	: 3485
3153-37-5	: 3489
3156-70-5	: 10589
3156-73-8	: 10396
3156-74-9	: 3641
3163-07-3	: 1541
3163-27-7	: 7742
3167-49-5	: 10897
3172-52-9	: 11748
3173-53-3	: 4416
3173-79-3	: 8597
3174-67-2	: 8785
3174-83-2	: 2157
3175-23-3	: 10174
3176-62-3	: 7240
3178-22-1	: 4369
3178-23-2	: 4429
3178-24-3	: 4466
3178-29-8	: 6434
3179-63-3	: 10375
3180-09-4	: 9246
3187-94-8	: 6073
3188-13-4	: 5478
3189-22-8	: 7311
3194-15-8	: 10438
3195-24-2	: 9944
3196-15-4	: 3473
3199-61-9	: 2440
3202-84-4	: 7695
3208-16-0	: 6105
3208-22-8	: 6173
3209-22-1	: 1391
3211-48-1	: 4183
3212-75-7	: 4753
3213-79-4	: 1269
3217-86-5	: 9802
3218-02-8	: 4418
3221-61-2	: 8408
3228-04-4	: 9416
3228-99-7	: 9895
3229-00-3	: 9884
3230-23-7	: 9670
3232-84-6	: 11850
3234-64-8	: 9121
3238-38-8	: 9474
3238-40-2	: 6081
3238-62-8	: 9677
3238-75-3	: 5712
3240-09-3	: 7099
3243-42-3	: 7952
3244-88-0	: 2218
3252-43-5	: 277
3254-93-1	: 7228
3256-88-0	: 10990
3260-87-5	: 9264
3261-62-9	: 1580
3268-49-3	: 9786
3268-79-9	: 11700
3277-26-7	: 5237
3278-46-4	: 10332
3280-08-8	: 1616
3280-51-1	: 6708
3282-30-2	: 10503
3283-12-3	: 9528
3288-04-8	: 1793
3289-28-9	: 4334
3290-06-0	: 1358
3296-05-7	: 978
3299-38-5	: 3656
3319-15-1	: 1895
3320-83-0	: 1176
3320-86-3	: 1788
3320-87-4	: 1789
3321-50-4	: 4386
3329-48-4	: 7073
3329-56-4	: 3377
3333-23-1	: 7764
3333-44-6	: 6138
3333-52-6	: 3292
3337-71-1	: 474
3338-15-6	: 10444
3344-70-5	: 5296
3347-22-6	: 8056
3350-30-9	: 4725
3350-78-5	: 3888
3351-28-8	: 4117
3351-86-8	: 4055
3352-87-2	: 5290
3354-58-3	: 9427
3355-17-7	: 3900
3358-28-9	: 6174
3368-16-9	: 941
3373-53-3	: 10767
3375-23-3	: 8036
3377-71-7	: 7321
3377-86-4	: 6733
3377-87-5	: 6734
3383-96-8	: 1
3384-04-1	: 1233
3386-33-2	: 8236
3393-64-4	: 3664
3393-78-0	: 2284
3395-79-7	: 1443
3395-91-3	: 10212
3399-21-1	: 4357
3400-45-1	: 4781
3402-74-2	: 2006
3402-76-4	: 7945
3404-57-7	: 6558
3404-58-8	: 7033
3404-61-3	: 7037
3404-63-5	: 9060
3404-67-9	: 6838
3404-71-5	: 9043
3404-72-6	: 9029
3404-73-7	: 9031
3404-77-1	: 7016
3404-78-2	: 7024
3404-80-6	: 6837
3411-95-8	: 9217
3416-24-8	: 6233
3419-71-4	: 4689
3424-93-9	: 618
3426-01-5	: 10843
3430-27-1	: 10870
3433-37-2	: 9674
3433-80-5	: 999
3435-28-7	: 11048
3437-89-6	: 8987
3437-95-4	: 11767
3438-46-8	: 11071
3439-97-2	: 11443
3440-02-0	: 11415
3445-11-2	: 11184
3446-98-8	: 11147
3452-07-1	: 5362
3452-09-3	: 8196
3452-97-9	: 6951
3454-07-7	: 1632
3457-58-7	: 2003
3458-28-4	: 7425
3460-18-2	: 1310
3460-20-6	: 2315
3460-24-0	: 1295
3460-67-1	: 10904
3463-26-1	: 10837
3463-40-9	: 2316
3463-92-1	: 4064
3468-99-3	: 939
3476-89-9	: 11339
3478-88-4	: 7594
3497-00-5	: 9552
3508-00-7	: 6307
3508-78-9	: 8805
3508-94-9	: 930
3510-70-1	: 11496
3512-80-9	: 10864
3521-91-3	: 6566
3522-94-9	: 6861
3524-62-7	: 5839
3524-73-0	: 7039
3524-75-2	: 4848
3528-17-4	: 2869
3535-67-9	: 1925
3536-29-6	: 2158
3538-65-6	: 3529
3540-95-2	: 9658
3544-25-0	: 946
3547-04-4	: 1678
3552-01-0	: 4538
3554-74-3	: 9716
3555-84-8	: 7612
3558-69-8	: 10947
3567-38-2	: 1877
3570-55-6	: 5652
3570-58-9	: 5699
3574-40-1	: 3157
3574-43-4	: 5682
3581-87-1	: 11678
3581-91-7	: 11675
3582-71-6	: 11913
3586-12-7	: 835
3586-14-9	: 2013
3587-57-3	: 9858
3587-60-8	: 1181
3588-30-5	: 3169
3593-00-8	: 6563
3594-37-4	: 3364
3598-14-9	: 287
3600-24-6	: 11916
3607-78-1	: 10052
3609-53-8	: 2479
3610-02-4	: 2403
3613-42-1	: 367
3615-21-2	: 2377
3616-56-6	: 5416
3616-57-7	: 5415
3634-92-2	: 8273
3637-01-2	: 5811
3637-61-4	: 4825
3638-35-5	: 4983
3638-64-0	: 5911
3643-64-9	: 4688
3657-07-6	: 245
3658-79-5	: 8755
3669-80-5	: 11948
3673-79-8	: 10242
3674-09-7	: 10248
3674-10-0	: 10363
3674-13-3	: 10240
3674-75-7	: 9179
3675-68-1	: 3442
3675-69-2	: 3443
3678-62-4	: 10919
3680-02-2	: 5909
3681-78-5	: 5320
3682-35-7	: 11843
3683-19-0	: 7043
3683-22-5	: 7042
3689-24-5	: 11716
3690-04-8	: 9795
3697-27-6	: 4115
3699-01-2	: 2019
3710-30-3	: 8319
3710-43-8	: 6088
3710-84-7	: 5429
3711-34-0	: 7137
3711-37-3	: 7140
3721-95-7	: 4174
3722-73-4	: 2781
3722-74-5	: 2780
3725-05-1	: 6714
3725-07-3	: 8340
3725-30-2	: 4750
3728-43-6	: 11490
3730-60-7	: 6923
3731-51-9	: 10971
3731-52-0	: 10972
3731-53-1	: 10973
3735-90-8	: 3977
3741-00-2	: 4846
3741-38-6	: 5200
3742-38-9	: 4926
3742-42-5	: 4662
3743-22-4	: 9317
3744-02-3	: 9109
3757-32-2	: 8910
3761-94-2	: 4230
3763-39-1	: 4991
3763-55-1	: 4062
3764-01-0	: 11075
3768-43-2	: 9722
3768-60-3	: 11583
3768-63-6	: 11569
3769-23-1	: 7038
3769-41-3	: 9452
3773-93-1	: 5208
3774-03-6	: 5197
3779-29-1	: 4180
3780-58-3	: 6894
3788-32-7	: 4838
3790-71-4	: 5334
3810-26-2	: 4951
3810-74-0	: 11561
3835-64-1	: 1864
3837-38-5	: 404
3839-50-7	: 4765
3840-31-1	: 1914
3848-12-2	: 126
3849-33-0	: 10049
3850-30-4	: 3217
3852-09-3	: 10289
3853-80-3	: 10202
3856-25-5	: 11861
3857-25-8	: 6122
3862-11-1	: 9338
3862-12-2	: 9335
3864-18-4	: 3052
3867-18-3	: 733
3875-51-2	: 4828
3875-52-3	: 4805
3875-68-1	: 12051
3877-15-4	: 10088
3877-19-8	: 7996
3878-55-5	: 3336
3884-88-6	: 8246
3910-35-8	: 7263
3913-71-1	: 5080
3913-85-7	: 5088
3916-64-1	: 6339
3917-15-5	: 10565
3920-53-4	: 6453
3937-45-9	: 4422
3937-46-0	: 4423
3937-56-2	: 8136
3938-95-2	: 10259
3938-96-3	: 206
3943-74-6	: 2627
3943-77-9	: 2578
3944-36-3	: 10390
3955-26-8	: 2324
3958-03-0	: 11781
3958-38-1	: 4727
3958-57-4	: 1067
3959-07-7	: 1799
3962-77-4	: 1926
3963-78-8	: 382
3964-56-5	: 9232
3964-58-7	: 2540
3964-61-2	: 9254
3968-85-2	: 1953
3970-35-2	: 2544
3970-62-5	: 8929
3971-28-6	: 4363
3972-56-3	: 1142
3972-65-4	: 1023
3973-62-4	: 9694
3974-99-0	: 254
3976-34-9	: 3056
3976-36-1	: 3095
3978-81-2	: 10944

CAS NUMBER INDEX OF ORGANIC COMPOUNDS (continued)

3982-64-7	1483
3982-66-9	12057
3982-67-0	1712
3988-03-2	7590
3991-61-5	2012
4001-73-4	604
4008-48-4	11323
4016-11-9	8614
4018-65-9	1504
4021-75-4	3612
4028-23-3	11375
4032-86-4	6388
4032-92-2	6445
4032-93-3	6444
4032-94-4	8375
4038-04-4	9044
4041-09-2	4882
4043-71-4	2430
4044-65-9	1432
4048-30-0	6578
4048-31-1	6574
4048-32-2	6581
4049-81-4	6695
4050-45-7	7000
4053-34-3	11329
4054-38-0	4219
4057-42-5	8503
4065-80-9	4537
4068-78-4	2542
4074-24-2	3322
4074-43-5	9247
4079-52-1	5842
4088-60-2	3860
4088-81-7	9175
4091-39-8	3647
4095-09-4	11453
4095-22-1	11772
4096-20-2	9693
4096-21-3	11171
4096-34-8	4704
4100-80-5	6094
4101-68-2	5038
4104-01-2	9025
4107-62-4	10234
4110-44-5	8379
4110-50-3	10042
4110-77-4	1153
4113-12-6	6642
4114-28-7	7151
4114-31-2	7148
4117-15-1	11934
4124-31-6	244
4124-88-3	8187
4126-78-7	4226
4127-45-1	4985
4128-31-8	8444
4130-08-9	11498
4130-42-1	9221
4131-74-2	10331
4132-71-2	1442
4132-77-8	1470
4142-85-2	11440
4143-41-3	5126
4145-94-2	11673
4147-51-7	5222
4147-89-1	11362
4156-16-5	7407
4160-52-5	3672
4160-80-9	10782
4161-60-8	8973
4164-28-7	7449
4166-46-5	6926
4167-74-2	9432
4167-77-5	4791
4169-04-4	10398
4170-24-5	3480
4170-46-1	11466
4171-83-9	2057
4173-44-8	6710
4175-54-6	7983
4179-19-5	1441
4180-23-8	1944
4185-62-0	7745
4185-63-1	7788
4187-86-4	9140
4187-87-5	1876
4192-77-2	10685
4200-95-7	6305
4205-23-6	6290
4206-75-1	11384
4208-49-5	6070
4208-64-4	6121
4209-24-9	5787
4209-90-9	6925
4214-28-2	1770
4215-86-5	2011
4218-48-8	1690
4219-24-3	7050
4219-49-2	6652
4224-70-8	6876
4225-85-8	11244
4228-00-6	5317
4229-91-8	6167
4253-34-3	11499
4253-89-8	5244
4253-91-2	10129
4254-02-8	4779
4255-62-3	4590
4258-93-9	4449
4259-00-1	4855
4259-43-2	10179
4264-29-3	2055
4265-25-2	2456
4265-97-8	8430
4268-36-4	3933
4271-30-1	6242
4279-22-5	3280
4279-70-3	1993
4279-76-9	1716
4280-28-8	1927
4282-24-0	6079
4282-28-4	6080
4282-31-9	11745
4282-32-0	6083
4282-40-0	6419
4283-80-1	8717
4285-42-1	3974
4286-47-9	6778
4286-49-1	3293
4288-84-0	10355
4291-79-6	4450
4291-80-9	4451
4291-81-0	4452
4291-98-9	4934
4292-19-7	5302
4292-75-5	4414
4292-92-6	4465
4295-06-1	11245
4301-39-7	6053
4303-44-0	9855
4313-03-5	6325
4318-37-0	5141
4318-76-7	10924
4319-49-7	7680
4325-48-8	8729
4325-49-9	6366
4325-82-0	9099
4325-97-7	5651
4328-94-3	8852
4339-05-3	7128
4340-76-5	5359
4341-76-8	3918
4342-61-4	5226
4346-18-3	3584
4350-41-8	10921
4351-54-6	6005
4354-56-7	4412
4358-59-2	3848
4358-75-2	3216
4358-87-6	910
4358-88-7	907
4360-12-7	321
4360-96-7	9959
4361-06-2	9968
4362-01-0	4135
4363-04-6	8019
4363-93-3	11225
4373-13-1	7817
4375-14-8	7319
4376-23-2	7090
4377-37-1	11023
4383-05-5	1885
4383-18-0	1904
4388-87-8	6807
4388-88-9	6811
4389-53-1	2641
4390-75-4	3408
4392-24-9	1084
4392-30-7	1245
4393-06-0	1868
4394-11-0	3103
4394-85-8	7685
4395-73-7	1819
4395-79-3	1201
4395-80-6	1202
4395-81-7	1782
4396-98-9	1847
4397-05-1	1842
4398-65-6	6744
4403-69-4	1796
4405-42-9	11385
4407-36-7	10704
4412-16-2	5337
4413-16-5	1249
4422-95-1	2317
4426-11-3	4173
4426-51-1	5198
4427-56-9	9417
4428-13-1	1601
4433-10-7	3054
4433-11-8	3059
4433-63-0	3133
4435-14-7	4313
4435-18-1	276
4435-50-1	3458
4436-59-3	4238
4436-96-8	1867
4437-18-7	6035
4437-20-1	6101
4437-22-3	6156
4437-23-4	6158
4437-50-7	6810
4437-51-8	6806
4439-20-7	5715
4439-24-1	5746
4441-57-0	4324
4441-90-1	9965
4442-79-9	4387
4443-55-4	4384
4444-12-6	6666
4445-06-1	4459
4445-07-2	2066
4445-08-3	8245
4447-60-3	10079
4455-26-9	8353
4457-00-5	4823
4463-33-6	1439
4467-06-5	10989
4467-88-3	7271
4468-42-2	1669
4471-05-0	1909
4481-08-7	6198
4481-30-5	1481
4482-03-5	3076
4484-72-4	11460
4484-80-4	251
4485-09-0	8174
4485-13-6	6559
4485-16-9	6556
4489-23-0	1082
4491-19-4	307
4492-01-7	10830
4493-42-9	5027
4497-29-4	7524
4510-34-3	10705
4516-69-2	4856
4518-10-9	2504
4520-29-0	11447
4523-45-9	7720
4524-77-0	9233
4525-36-4	6165
4525-44-4	11397
4534-74-1	4525
4535-66-4	5684
4537-00-2	7763
4537-05-7	2285
4542-61-4	3117
4549-31-9	6374
4549-32-0	8369
4549-74-0	8669
4553-07-5	893
4562-36-1	4042
4573-05-1	4252
4573-09-5	4897
4588-18-5	8505
4593-16-2	9631
4594-78-9	4467
4597-87-9	10869
4599-94-4	5152
4604-28-8	1155
4606-07-9	4961
4606-65-9	9675
4609-89-6	8833
4609-92-1	8414
4619-74-3	9480
4620-70-6	5710
4625-24-5	2980
4628-21-1	3726
4630-20-0	3986
4630-82-4	4335
4635-59-0	3685
4636-16-2	5838
4645-15-2	4461
4654-39-1	1595
4655-34-9	10664
4662-96-8	1589
4663-22-3	4982
4667-38-3	11392
4667-99-6	11387
4669-01-6	4844
4670-56-8	2639
4671-03-8	11848
4672-38-2	9547
4675-18-7	8581
4676-51-1	3326
4685-47-6	1921
4694-12-6	4899
4695-31-2	10288
4701-36-4	1699
4702-34-5	2797
4705-34-4	1621
4705-94-6	8027
4706-81-4	11631
4706-89-2	1472
4712-55-4	9534
4721-34-0	9544
4721-37-3	9542
4726-14-1	8070
4727-17-7	4760
4730-22-7	6498
4731-34-4	2076
4731-84-4	9249
4734-90-1	4754
4736-48-5	4214
4737-41-1	8731
4740-00-5	11788
4744-08-5	9903
4747-05-1	3161
4747-07-3	6833
4747-15-3	1928
4747-21-1	9819
4749-27-3	10606
4749-28-4	10590
4749-31-9	10127
4753-75-7	6115
4753-80-4	11690
4755-36-6	4247
4755-77-5	144
4766-33-0	7711
4766-57-8	11505
4767-03-7	10272
4771-47-5	2546
4771-80-6	4638
4773-53-9	10364
4775-09-1	9527
4775-86-4	8800
4780-79-4	7908
4784-77-4	3712
4789-40-6	6032
4795-29-3	6116
4795-86-2	2935
4798-44-1	7066
4798-46-3	7077
4798-60-1	7081
4798-61-2	8511
4798-62-3	6575
4798-64-5	8518
4799-68-2	10400
4806-58-0	4986
4806-59-1	4987
4806-61-5	4191
4810-09-7	6544
4815-57-0	1557
4823-47-6	10611
4829-04-3	5266
4830-99-3	7273
4831-43-0	11182
4832-17-1	8055
4835-11-4	6750
4838-00-0	7242
4839-46-7	8773
4850-28-6	4862
4850-32-2	4832
4860-03-1	6634
4877-77-4	1081
4877-93-4	3045
4884-24-6	2995
4884-25-7	2994
4887-30-3	6904
4891-38-7	10742
4894-61-5	3725
4894-62-6	6335
4897-84-1	3474
4900-30-5	8110
4901-51-3	9469
4904-61-4	4212
4911-55-1	8684
4911-65-3	1236
4912-92-9	7251
4914-91-4	9038
4914-92-5	9037
4915-21-3	6028
4915-22-4	6029
4915-23-5	6027
4916-63-6	6994
4920-84-7	1448
4920-99-4	1689
4923-77-7	4395
4923-78-8	4396
4923-84-6	9533
4926-76-5	4398
4926-78-7	4399
4926-90-3	4394
4930-98-7	11034
4944-94-9	632
4945-28-2	11301
4945-47-5	9627
4945-48-6	9635
4946-13-8	2298
4949-44-4	8906
4957-14-6	1964
4964-71-0	11218
4965-36-0	11220
4972-29-6	2185
4974-21-4	246
4974-27-0	8320
4975-21-7	6176
4984-01-4	8502
4984-85-4	6971
4985-70-0	374
4985-85-7	5679
4996-48-9	6068
5006-66-6	10899
5008-52-6	5007
5008-69-5	10082
5008-73-1	10077
5009-11-0	7104
5009-27-8	4988
5020-21-3	7246
5022-29-7	7364
5026-76-6	6547
5029-66-3	10284
5029-67-4	10970
5035-52-9	11547
5053-43-0	11070
5058-19-5	10878
5069-26-1	11774
5075-33-2	7129
5075-92-3	11181
5076-20-0	8627
5077-67-8	3660
5090-41-5	8288
5097-93-8	11001
5100-98-1	11817
5103-42-4	2999

5112-21-0 : 9999
5113-93-9 : 4682
5131-66-8 : 10346
5132-75-2 : 6468
5137-45-1 : 5591
5145-64-2 : 11138
5145-99-3 : 10043
5150-80-1 : 6341
5150-93-6 : 3333
5153-67-3 : 2050
5159-41-1 : 1888
5160-99-6 : 1997
5161-04-6 : 1999
5161-13-7 : 11785
5161-16-0 : 11796
5161-17-1 : 11798
5162-44-7 : 3711
5162-48-1 : 8955
5166-35-8 : 3734
5166-53-0 : 7094
5171-84-6 : 6854
5182-44-5 : 1596
5183-77-7 : 4347
5183-79-9 : 4360
5194-50-3 : 6686
5194-51-4 : 6685
5195-37-9 : 7982
5195-40-4 : 7981
5208-49-1 : 4050
5209-18-7 : 2161
5209-33-6 : 5
5216-25-1 : 1239
5216-32-0 : 1366
5217-05-0 : 4783
5221-17-0 : 9780
5223-59-6 : 1676
5234-68-4 : 4039
5251-93-4 : 524
5258-50-4 : 11797
5259-65-4 : 4714
5259-98-3 : 8923
5263-87-6 : 11275
5265-18-9 : 10414
5271-38-5 : 5749
5271-39-6 : 1679
5272-02-6 : 6369
5274-50-0 : 9554
5281-18-5 : 563
5283-66-9 : 11471
5285-18-7 : 8677
5285-87-0 : 11709
5292-21-7 : 4306
5292-43-3 : 114
5292-53-5 : 9978
5307-19-7 : 10951
5311-96-6 : 7150
5314-37-4 : 3070
5314-55-6 : 11433
5324-00-5 : 7272
5325-97-3 : 9185
5326-81-8 : 45
5327-44-6 : 1924
5328-01-8 : 7883
5332-06-9 : 3404
5332-24-1 : 11217
5332-25-2 : 11219
5332-52-5 : 11942
5332-73-0 : 9816
5335-75-1 : 10879
5337-03-1 : 10774
5337-70-2 : 8552
5337-72-4 : 4515
5337-93-9 : 10468
5339-85-5 : 1594
5340-36-3 : 8466
5340-41-0 : 6953
5342-31-4 : 7509
5344-20-7 : 8707
5344-55-8 : 3590
5344-82-1 : 11801
5344-90-1 : 1837
5345-01-7 : 8777
5349-18-8 : 2032
5349-55-3 : 10281
5349-60-0 : 1873
5350-47-0 : 7630
5356-83-2 : 11423
5362-55-0 : 9095
5363-64-4 : 6813
5367-28-2 : 1213
5368-81-0 : 2674
5370-25-2 : 5783
5371-48-2 : 9119
5379-20-4 : 1627
5380-40-5 : 6069
5390-04-5 : 9143
5390-71-6 : 9861
5390-72-7 : 9859
5392-86-9 : 1375
5395-43-7 : 11553
5395-75-5 : 5451
5396-24-7 : 149
5396-38-3 : 1462
5396-58-7 : 3355
5396-91-8 : 8969
5398-08-3 : 9961
5398-10-7 : 9955
5398-25-4 : 8742
5398-75-4 : 9177
5399-02-0 : 6312
5405-13-0 : 1821
5405-15-2 : 1825
5405-17-4 : 1824
5405-41-4 : 3534
5405-79-8 : 6965
5406-12-2 : 2116
5406-18-8 : 2154
5407-98-7 : 7605
5408-57-1 : 8477
5408-86-6 : 3268
5410-97-9 : 5166
5411-50-7 : 1307
5411-56-3 : 1845
5412-92-0 : 8352
5413-05-8 : 884
5414-19-7 : 5609
5414-21-1 : 8828
5416-94-4 : 11704
5419-55-6 : 3128
5421-04-5 : 912
5424-21-5 : 11053
5426-65-3 : 10846
5429-85-6 : 8290
5432-61-1 : 4289
5432-85-9 : 4600
5434-27-5 : 9887
5441-51-0 : 4593
5441-52-1 : 4518
5445-24-9 : 4557
5447-97-2 : 9847
5447-99-4 : 8952
5448-43-1 : 11334
5451-80-9 : 8881
5451-85-4 : 8901
5452-37-9 : 4735
5452-75-5 : 4308
5453-44-1 : 3857
5453-80-5 : 2967
5454-28-4 : 6459
5454-79-5 : 4533
5455-24-3 : 8396
5458-48-0 : 1252
5458-59-3 : 8435
5459-58-5 : 151
5459-93-8 : 4288
5460-32-2 : 1768
5461-51-8 : 74
5465-09-8 : 4588
5465-13-4 : 2282
5468-97-3 : 1598
5469-19-2 : 1093
5470-02-0 : 9707
5470-28-0 : 9606
5470-47-3 : 882
5470-91-7 : 10712
5472-38-8 : 3324
5500-21-0 : 4956
5502-88-5 : 4680
5503-12-8 : 4634
5507-44-8 : 11422
5510-99-6 : 9227
5515-83-3 : 328
5519-42-6 : 10842
5519-50-6 : 9723
5535-49-9 : 1165
5556-07-0 : 5835
5557-31-3 : 8292
5558-29-2 : 963
5558-31-6 : 964
5558-66-7 : 931
5558-78-1 : 970
5560-69-0 : 7968
5563-78-0 : 4547
5581-35-1 : 1583
5581-75-9 : 1759
5582-82-1 : 6500
5586-15-2 : 5256
5587-63-3 : 4910
5587-79-1 : 4948
5588-20-5 : 7220
5589-96-8 : 112
5597-27-3 : 2958
5597-81-9 : 7782
5597-89-7 : 8687
5598-13-0 : 4090
5609-09-6 : 6587
5614-32-4 : 3160
5617-41-4 : 4405
5617-42-5 : 4820
5617-92-5 : 4978
5630-56-8 : 7598
5634-22-0 : 1565
5634-30-0 : 4809
5636-65-7 : 7061
5637-96-7 : 2070
5638-26-6 : 6336
5648-29-3 : 5614
5650-40-8 : 10451
5650-41-9 : 10452
5650-75-9 : 8890
5661-71-2 : 503
5663-96-7 : 8538
5666-21-7 : 10527
5668-93-9 : 3073
5669-09-0 : 10041
5675-22-9 : 6327
5675-51-4 : 5301
5676-58-4 : 2879
5683-33-0 : 10863
5683-44-3 : 8794
5685-06-3 : 11680
5687-92-3 : 11692
5689-04-3 : 7292
5697-85-8 : 838
5699-58-1 : 8875
5699-74-1 : 8829
5703-21-9 : 869
5703-26-4 : 871
5703-52-6 : 2146
5704-20-1 : 8975
5715-23-1 : 4517
5728-52-9 : 3009
5736-03-8 : 5206
5743-97-5 : 9186
5744-03-6 : 5973
5750-00-5 : 3248
5756-43-4 : 6814
5763-61-1 : 1805
5765-44-6 : 7405
5773-56-8 : 1614
5775-82-6 : 10816
5779-94-2 : 548
5779-95-3 : 549
5781-53-3 : 145
5787-31-5 : 3232
5792-36-9 : 10430
5794-88-7 : 2407
5798-75-4 : 2523
5798-79-8 : 947
5798-88-9 : 3472
5798-94-7 : 606
5802-82-4 : 6625
5805-76-5 : 2381
5807-64-7 : 3039
5807-76-1 : 1096
5809-07-4 : 6103
5813-64-9 : 9806
5814-42-6 : 10392
5814-85-7 : 1969
5836-66-8 : 9888
5837-78-5 : 3849
5842-00-2 : 5650
5846-43-5 : 4559
5856-77-9 : 3688
5858-82-6 : 3680
5857-36-3 : 8989
5857-86-3 : 4558
5859-45-0 : 7784
5867-45-8 : 89
5870-61-1 : 5882
5870-93-9 : 3527
5876-87-9 : 5285
5877-55-4 : 809
5877-58-7 : 810
5878-19-3 : 10455
5881-17-4 : 8403
5882-44-0 : 717
5891-21-4 : 8961
5897-76-7 : 1840
5902-51-2 : 11591
5906-73-0 : 11501
5908-87-2 : 5277
5910-87-2 : 8109
5910-89-4 : 10805
5911-04-6 : 8141
5911-08-0 : 4966
5912-58-3 : 5694
5912-86-7 : 9397
5912-87-8 : 9393
5913-85-9 : 9706
5915-41-3 : 11593
5917-47-5 : 9660
5921-54-0 : 3673
5921-73-3 : 8209
5921-84-6 : 6476
5930-28-9 : 9206
5932-68-3 : 9396
5943-35-1 : 11649
5944-41-2 : 10942
5950-19-6 : 3649
5951-67-7 : 4659
5959-52-4 : 7768
5963-77-9 : 9137
5965-66-2 : 6214
5966-51-8 : 10343
5976-47-6 : 10700
5984-58-7 : 6353
5985-23-9 : 6246
5986-55-0 : 7586
6001-87-2 : 10225
6004-60-0 : 5795
6006-06-0 : 11895
6006-33-3 : 4484
6006-34-4 : 4854
6006-90-2 : 8638
6006-95-7 : 4463
6012-97-1 : 11782
6020-51-5 : 7224
6031-02-3 : 2009
6032-29-7 : 8916
6035-49-0 : 2845
6040-62-6 : 8058
6044-68-4 : 10562
6050-13-1 : 5145
6051-43-0 : 2489
6051-52-1 : 1869
6057-60-9 : 3019
6061-06-9 : 3156
6061-10-5 : 8433
6064-27-3 : 5326
6065-59-4 : 9974
6065-82-3 : 174
6065-90-3 : 8752
6066-03-6 : 10565
6068-69-5 : 816
6069-71-2 : 2954
6069-97-2 : 2937
6069-98-3 : 4439
6078-26-8 : 310
6080-79-1 : 9125
6086-22-2 : 11847
6088-94-4 : 8114
6089-09-4 : 9138
6089-12-9 : 3600
6091-52-7 : 4716
6091-58-3 : 3351
6091-64-1 : 2521
6093-68-1 : 2825
6094-02-6 : 7036
6102-15-4 : 4641
6108-61-8 : 6687
6109-22-4 : 422
6111-61-1 : 3643
6111-78-0 : 521
6111-88-2 : 8733
6111-99-5 : 10236
6114-18-7 : 8302
6117-80-2 : 3793
6118-14-5 : 3858
6119-92-2 : 5182
6120-13-4 : 2919
6125-21-9 : 9919
6125-24-2 : 2051
6127-92-0 : 3827
6127-93-1 : 3828
6130-75-2 : 2331
6131-24-4 : 6835
6131-25-5 : 6423
6134-66-3 : 3503
6135-29-1 : 262
6136-67-0 : 7624
6137-08-2 : 8475
6137-15-1 : 8476
6138-90-5 : 8321
6139-84-0 : 3181
6140-17-6 : 2045
6140-65-4 : 4908
6140-80-3 : 10579
6142-95-6 : 1412
6149-41-3 : 10276
6149-46-8 : 8884
6152-31-4 : 7171
6153-16-8 : 4674
6153-56-6 : 5561
6158-45-8 : 7920
6159-55-3 : 11192
6161-50-8 : 3046
6162-79-4 : 3951
6163-64-0 : 10083
6163-66-2 : 10124
6163-75-3 : 9537
6164-98-3 : 7564
6165-44-2 : 4323
6166-86-5 : 4902
6166-87-6 : 4625
6168-72-5 : 10336
6174-95-4 : 5916
6175-23-1 : 8137
6175-49-1 : 5325
6195-92-2 : 4886
6196-58-3 : 5757
6196-60-7 : 9047
6202-04-6 : 11839
6213-90-7 : 3821
6214-25-1 : 3832
6214-28-4 : 3820
6222-35-1 : 10235
6223-78-5 : 6765
6225-06-5 : 8699
6231-18-1 : 10935
6236-88-0 : 4400
6258-66-8 : 1833
6261-22-9 : 9141
6265-26-5 : 6166
6267-24-9 : 5605
6268-37-7 : 1228
6269-92-7 : 8366
6270-16-2 : 3642
6270-34-4 : 6061
6270-55-9 : 10300
6270-56-0 : 6104
6272-38-4 : 9451
6272-40-8 : 7588
6274-16-4 : 3950
6279-86-3 : 7552
6280-03-1 : 8149
6280-96-2 : 9460
6280-98-4 : 1555
6284-40-8 : 6208
6284-84-0 : 9609
6285-05-8 : 10422
6285-06-9 : 9145
6287-38-3 : 536
6290-24-0 : 2598
6290-49-9 : 207

6292-82-6 : 9710
6294-31-1 : 6855
6294-89-9 : 7149
6298-37-9 : 11332
6302-94-9 : 5264
6304-18-3 : 10985
6306-30-5 : 6984
6308-98-1 : 1585
6309-50-8 : 5308
6310-09-4 : 5791
6310-21-0 : 714
6311-60-0 : 1309
6314-28-9 : 2400
6315-03-3 : 2695
6315-60-2 : 330
6315-96-4 : 10474
6319-21-7 : 5135
6319-40-0 : 2529
6320-03-2 : 2293
6320-40-7 : 2314
6322-01-6 : 10339
6322-49-2 : 3648
6326-44-9 : 9953
6338-45-0 : 7201
6341-60-2 : 9970
6341-97-5 : 9288
6344-28-1 : 5786
6344-72-5 : 11338
6346-09-4 : 3214
6351-10-6 : 7284
6361-23-5 : 535
6372-01-6 : 3831
6374-70-5 : 5146
6378-11-6 : 4087
6378-65-0 : 6890
6380-21-8 : 9456
6380-24-1 : 1451
6380-28-5 : 9419
6380-34-3 : 464
6382-13-4 : 8270
6402-36-4 : 5336
6414-69-3 : 10283
6423-02-5 : 6758
6423-29-6 : 7804
6436-90-4 : 6272
6443-91-0 : 5944
6443-92-1 : 6535
6453-99-2 : 7642
6482-24-2 : 5465
6482-34-4 : 3995
6493-77-2 : 4640
6493-83-0 : 10457
6496-48-6 : 11176
6498-47-1 : 3868
6513-13-9 : 12039
6526-72-3 : 1977
6531-35-7 : 403
6538-02-9 : 5379
6553-48-6 : 4187
6556-12-3 : 6238
6568-58-7 : 4731
6573-52-0 : 4729
6579-55-1 : 10383
6592-85-4 : 5217
6597-78-0 : 7653
6600-40-4 : 8218
6607-66-5 : 3456
6609-56-9 : 2760
6609-57-0 : 2754
6617-04-5 : 4311
6622-76-0 : 3851
6624-76-6 : 8100
6626-15-9 : 1502
6626-32-0 : 9347
6627-55-0 : 9236
6627-72-1 : 2942
6627-74-3 : 4667
6628-00-8 : 4304
6628-04-2 : 11210
6628-06-4 : 9429
6628-18-8 : 5452
6628-21-3 : 10246
6628-77-9 : 10865
6628-79-1 : 8896
6628-80-4 : 4505
6628-97-3 : 7866
6629-96-5 : 5192
6630-33-7 : 526
6639-30-1 : 2336
6639-82-3 : 11336
6640-27-3 : 9261
6640-77-3 : 2685
6641-83-4 : 8895
6669-13-2 : 1458
6682-71-9 : 7253
6688-11-5 : 4738
6693-29-4 : 4302
6693-30-7 : 4301
6693-31-8 : 4300
6697-12-7 : 6664
6703-78-2 : 6298
6703-80-6 : 6297
6703-82-8 : 4819
6704-31-0 : 8596
6705-31-3 : 10808
6709-22-4 : 4913
6709-39-3 : 6331
6728-26-3 : 6996
6734-98-1 : 7007
6738-23-4 : 1917
6739-34-0 : 4337
6740-88-1 : 4579
6742-54-7 : 2374
6748-95-4 : 436
6753-98-6 : 5006
6765-39-5 : 6322
6776-19-8 : 3840
6780-13-8 : 5513
6780-49-0 : 7567
6781-42-6 : 5865
6781-98-2 : 1139
6784-18-5 : 4890
6785-23-5 : 4863
6789-80-6 : 6998
6789-88-4 : 2610
6789-94-2 : 9626
6790-27-8 : 5345
6790-47-2 : 4307
6795-75-1 : 5935
6802-75-1 : 9966
6812-38-0 : 4411
6812-39-1 : 4822
6812-78-8 : 8516
6814-58-0 : 4083
6818-07-1 : 6901
6829-40-9 : 9924
6829-41-0 : 9958
6831-89-6 : 10831
6832-98-0 : 6630
6836-38-0 : 5324
6843-66-9 : 11413
6846-11-3 : 7221
6848-84-6 : 8769
6849-18-9 : 9082
6850-36-8 : 4287
6850-57-3 : 1812
6852-54-6 : 5437
6861-64-9 : 7
6871-44-9 : 4730
6873-15-0 : 11200
6874-98-2 : 11346
6876-23-9 : 4366
6886-16-4 : 6921
6890-03-5 : 4520
6891-45-8 : 9250
6894-69-5 : 4912
6897-76-3 : 8014
6897-88-7 : 8015
6912-05-6 : 4923
6917-76-6 : 11377
6921-35-3 : 8593
6921-40-0 : 7738
6921-64-8 : 5826
6923-20-2 : 10549
6923-22-4 : 7663
6931-71-1 : 6800
6933-10-4 : 651
6938-94-9 : 6782
6939-75-9 : 8911
6940-09-6 : 8275
6940-76-7 : 9857
6940-78-9 : 3233
6941-69-1 : 129
6942-99-0 : 1313
6943-58-4 : 10356
6946-29-8 : 2500
6946-35-6 : 2668
6946-88-9 : 3338
6950-43-2 : 2530
6950-84-1 : 12022
6951-08-2 : 2419
6952-59-6 : 2745
6959-06-4 : 3624
6963-44-6 : 8791
6964-04-1 : 6401
6967-43-7 : 3623
6969-16-0 : 11894
6971-51-3 : 1890
6973-79-1 : 10206
6974-10-3 : 7361
6974-77-2 : 9842
6975-60-6 : 10439
6975-92-4 : 7015
6975-98-0 : 5054
6975-99-1 : 5346
6976-00-7 : 11963
6976-72-3 : 6889
6982-25-8 : 3347
6983-79-5 : 5108
6986-48-7 : 5610
6995-30-8 : 7905
6995-79-5 : 4374
6996-92-5 : 2174
6998-82-9 : 4352
7005-72-3 : 1221
7012-16-0 : 5972
7013-11-8 : 3744
7021-04-7 : 889
7035-03-2 : 957
7036-98-8 : 8191
7040-43-9 : 6090
7044-96-4 : 7654
7045-89-8 : 11903
7058-01-7 : 4448
7065-46-5 : 3690
7069-38-7 : 10548
7081-78-9 : 5473
7086-07-9 : 10147
7087-68-5 : 9811
7094-26-0 : 4485
7094-27-1 : 4487
7094-34-0 : 7591
7098-07-9 : 7207
7098-21-7 : 11918
7101-96-4 : 11215
7116-86-1 : 7021
7119-92-8 : 5431
7119-94-0 : 9687
7132-64-1 : 8646
7133-68-8 : 3358
7137-56-6 : 1251
7144-05-0 : 9673
7145-20-2 : 7022
7146-60-3 : 8374
7148-74-5 : 10495
7149-70-4 : 1066
7149-75-9 : 674
7152-15-0 : 8897
7152-88-7 : 2628
7153-14-2 : 4576
7154-66-7 : 2887
7154-73-6 : 11164
7154-75-8 : 9135
7154-79-2 : 8839
7154-80-5 : 6450
7163-25-9 : 7773
7164-98-9 : 7212
7168-55-0 : 3048
7169-34-8 : 2462
7170-38-9 : 10320
7187-01-1 : 10585
7193-78-4 : 5844
7200-25-1 : 437
7205-90-5 : 1136
7208-05-1 : 8577
7208-92-6 : 5554
7209-38-3 : 9614
7214-50-8 : 4712
7216-56-0 : 8488
7220-26-0 : 6818
7220-81-7 : 4757
7225-64-1 : 6313
7225-66-3 : 11878
7235-40-7 : 4051
7236-47-7 : 3703
7239-24-9 : 9829
7242-17-3 : 3780
7250-95-5 : 7712
7251-61-8 : 11337
7252-83-7 : 5460
7283-69-4 : 3767
7283-70-7 : 3766
7283-96-7 : 11735
7287-19-6 : 9770
7287-81-2 : 1858
7291-33-0 : 97
7297-25-8 : 3432
7300-03-0 : 6555
7307-55-3 : 11937
7311-34-4 : 546
7314-06-9 : 732
7314-37-6 : 11620
7315-32-4 : 2425
7318-67-4 : 6684
7319-00-8 : 6683
7321-48-4 : 8500
7321-55-3 : 9913
7328-33-8 : 5026
7332-93-6 : 8695
7335-01-5 : 9672
7335-02-6 : 9690
7335-06-0 : 11166
7335-07-1 : 11173
7335-25-3 : 2537
7335-26-4 : 2669
7335-27-5 : 2539
7344-34-5 : 7278
7351-74-8 : 7803
7357-93-9 : 3797
7359-72-0 : 2334
7367-38-6 : 8184
7369-50-8 : 1693
7371-67-7 : 4181
7372-85-2 : 3055
7372-92-1 : 7280
7379-12-6 : 6979
7383-20-2 : 8208
7383-90-6 : 3058
7385-78-6 : 9032
7385-82-2 : 7044
7391-39-1 : 6881
7398-82-5 : 989
7399-49-7 : 1984
7399-50-0 : 10967
7400-08-0 : 10644
7403-66-9 : 42
7415-31-8 : 3748
7417-19-8 : 1599
7417-48-3 : 8666
7423-69-0 : 7018
7425-45-8 : 3483
7425-47-0 : 3928
7425-48-1 : 3484
7428-48-0 : 8261
7433-56-9 : 5084
7433-78-5 : 5085
7439-33-0 : 4185
7443-55-2 : 4534
7446-43-7 : 9814
7459-46-3 : 5654
7462-74-0 : 9792
7466-38-8 : 5133
7474-10-4 : 4720
7474-83-1 : 7844
7488-99-5 : 4054
7493-82-5 : 6470
7495-97-8 : 9526
7498-57-9 : 7737
7499-65-2 : 7744
7521-80-4 : 11370
7523-44-6 : 3865
7524-63-2 : 7988
7525-62-4 : 1631
7526-26-3 : 9541
7529-63-7 : 9697
7539-25-5 : 364
7554-65-6 : 10836
7564-63-8 : 1630
7567-22-8 : 4743
7568-92-5 : 1593
7568-93-6 : 1839
7570-26-5 : 5552
7570-98-1 : 7951
7572-29-4 : 5940
7583-53-1 : 9678
7611-43-0 : 472
7611-86-1 : 3721
7611-87-2 : 3720
7614-93-9 : 2022
7615-57-8 : 11836
7616-22-0 : 2790
7619-08-1 : 8227
7623-01-0 : 4997
7623-10-1 : 10499
7623-11-2 : 3683
7623-13-4 : 10501
7642-04-8 : 8493
7642-09-3 : 7003
7642-10-6 : 6537
7642-15-1 : 8497
7658-08-4 : 6235
7661-39-4 : 11222
7661-55-4 : 11280
7665-72-7 : 8610
7667-55-2 : 4491
7667-58-5 : 4494
7667-59-6 : 4493
7667-60-9 : 4495
7681-84-7 : 6049
7683-59-2 : 1522
7688-21-3 : 7001
7693-52-9 : 9244
7696-51-7 : 11129
7697-46-3 : 7637
7700-17-6 : 4137
7712-60-9 : 9026
7712-66-5 : 11814
7731-28-4 : 4535
7731-29-5 : 4536
7735-33-3 : 8869
7740-33-2 : 3589
7745-91-7 : 649
7745-93-9 : 1072
7756-94-7 : 10582
7766-23-6 : 9238
7771-44-0 : 5360
7773-60-6 : 9358
7778-87-2 : 6471
7779-80-8 : 6467
7781-98-8 : 2620
7782-21-0 : 943
7785-26-4 : 2971
7786-34-7 : 7659
7786-44-9 : 8112
7787-72-6 : 7792
7789-89-1 : 10171
7789-90-4 : 9790
7789-92-6 : 10183
7789-99-3 : 8945
9005-79-2 : 6274
10008-73-8 : 6148
10021-92-8 : 4353
10022-60-3 : 6780
10024-74-5 : 1822
10031-82-0 : 554
10031-87-5 : 185
10042-29-2 : 268
10048-32-5 : 10791
10061-01-5 : 10551
10061-02-6 : 10550
10071-60-0 : 9020
10075-38-4 : 3749
10088-95-6 : 10785
10124-73-9 : 7005
10130-87-7 : 2248
10130-91-3 : 4566
10136-52-4 : 6324
10138-10-0 : 3572
10138-89-3 : 3455
10140-89-3 : 9781
10143-54-1 : 10107
10143-60-9 : 6431
10147-36-1 : 10133
10151-95-8 : 80
10152-76-8 : 10581

10160-87-9	: 10731
10160-99-3	: 6627
10166-08-2	: 873
10193-50-7	: 9406
10203-08-4	: 537
10203-28-8	: 5322
10203-30-2	: 5323
10203-58-4	: 9969
10210-64-7	: 2912
10212-74-5	: 3960
10220-20-9	: 9659
10222-95-4	: 2361
10226-29-6	: 6961
10229-10-4	: 9142
10230-68-9	: 10475
10233-13-3	: 5316
10236-47-2	: 2795
10240-08-1	: 8030
10250-47-2	: 8949
10257-92-8	: 7318
10259-22-0	: 2670
10265-92-6	: 7441
10273-90-2	: 10991
10276-09-2	: 3835
10283-15-5	: 9256
10289-68-6	: 11885
10292-98-5	: 2963
10293-06-8	: 2952
10297-09-3	: 8541
10297-38-8	: 11571
10299-30-6	: 6867
10302-94-0	: 3772
10307-61-6	: 3550
10309-50-9	: 2959
10311-74-7	: 1622
10311-84-9	: 5105
10312-83-1	: 16
10315-89-6	: 9685
10316-00-4	: 7700
10317-10-9	: 10353
10321-71-8	: 9078
10323-40-7	: 3471
10324-58-0	: 9692
10327-08-9	: 10204
10334-26-6	: 2926
10339-67-0	: 6602
10340-91-7	: 1791
10342-83-3	: 10418
10343-99-4	: 11246
10373-81-6	: 2957
10394-35-1	: 2378
10403-60-8	: 3451
10410-35-2	: 2448
10420-90-3	: 6712
10420-91-4	: 6715
10424-29-0	: 3341
10428-64-5	: 3272
10435-81-1	: 8419
10436-75-6	: 794
10439-23-3	: 2186
10443-65-9	: 10610
10453-86-8	: 11340
10463-48-6	: 3740
10473-14-0	: 3871
10476-95-6	: 10564
10478-02-1	: 2592
10478-12-3	: 9083
10486-19-8	: 11869
10487-71-5	: 3887
10487-96-4	: 4871
10488-25-2	: 4539
10488-94-5	: 8436
10489-97-1	: 4045
10495-45-1	: 10132
10498-35-8	: 4359
10500-57-9	: 11297
10503-96-5	: 3249
10508-09-5	: 9154
10519-06-9	: 1117
10519-33-2	: 5093
10521-91-2	: 2101
10522-26-6	: 11955
10524-08-0	: 9018
10528-67-3	: 4469
10544-63-5	: 3838
10568-38-4	: 1683
10574-17-1	: 2030
10574-36-4	: 7041
10574-37-5	: 9034
10575-87-8	: 6378
10577-44-3	: 1941
10588-10-0	: 8900
10602-34-3	: 3758
10604-59-8	: 7307
10605-21-7	: 3990
10606-73-2	: 891
11024-24-1	: 6195
11028-42-5	: 4066
12069-69-1	: 511
12122-67-7	: 12067
12321-04-9	: 5961
12427-38-2	: 7420
13012-59-4	: 9864
13014-24-9	: 1401
13019-20-0	: 6522
13019-22-2	: 5092
13023-00-2	: 3500
13024-80-1	: 1750
13027-43-5	: 1650
13030-26-7	: 2778
13038-21-6	: 8190
13045-99-3	: 3798
13047-06-8	: 7600
13054-87-0	: 3595
13057-17-5	: 7472
13057-19-7	: 7519
13063-43-9	: 9912
13065-07-1	: 7989
13071-79-9	: 11592
13073-29-5	: 9423
13073-35-3	: 991
13086-84-5	: 9531
13099-50-8	: 1227
13102-31-3	: 9449
13105-52-7	: 3934
13114-62-0	: 12023
13116-53-5	: 10145
13116-60-4	: 10143
13121-70-5	: 5011
13129-23-2	: 6063
13138-51-7	: 3429
13147-06-3	: 11639
13151-04-7	: 6546
13151-05-8	: 6545
13151-17-2	: 7045
13151-34-3	: 5055
13153-11-2	: 11694
13163-73-0	: 4604
13165-77-0	: 2181
13172-91-3	: 6557
13176-46-0	: 3478
13194-48-4	: 5934
13195-80-7	: 10542
13206-46-7	: 9776
13222-26-9	: 10498
13249-58-6	: 1030
13250-46-9	: 303
13254-34-7	: 6479
13258-63-4	: 10949
13269-52-8	: 7002
13279-87-3	: 3448
13290-74-9	: 1205
13313-91-2	: 261
13325-10-5	: 3597
13327-31-6	: 11273
13329-40-3	: 5837
13347-42-7	: 9255
13351-61-6	: 2151
13351-73-0	: 2872
13356-08-6	: 6950
13360-45-7	: 4076
13360-61-7	: 8653
13360-63-9	: 3218
13360-65-1	: 10809
13361-35-8	: 143
13361-36-9	: 137
13361-63-2	: 9129
13362-52-2	: 11876
13367-29-8	: 2993
13388-93-7	: 5846
13389-42-9	: 8492
13398-94-2	: 1606
13402-32-9	: 1308
13402-35-2	: 4746
13402-36-3	: 4748
13402-37-4	: 4744
13403-37-7	: 8605
13417-43-1	: 3735
13423-15-9	: 6172
13436-48-1	: 7239
13444-24-1	: 9715
13463-39-3	: 8068
13463-40-6	: 7339
13466-40-5	: 8657
13466-41-6	: 11037
13471-42-6	: 6915
13475-76-8	: 5272
13475-77-9	: 5354
13475-78-0	: 6416
13475-79-1	: 8766
13475-81-5	: 6844
13475-82-6	: 6433
13481-70-4	: 9294
13482-62-7	: 9524
13491-79-7	: 4521
13505-34-5	: 6403
13511-38-1	: 10219
13513-82-1	: 1893
13529-27-6	: 6084
13532-18-8	: 10311
13545-04-5	: 3317
13552-31-3	: 9984
13556-55-3	: 7985
13556-58-6	: 7888
13602-13-6	: 3747
13603-04-8	: 11150
13603-21-9	: 9636
13618-93-4	: 7325
13632-93-4	: 1416
13632-94-5	: 1420
13632-95-6	: 1418
13643-05-5	: 8667
13665-04-8	: 5798
13672-18-9	: 7730
13678-54-1	: 11764
13679-46-4	: 6128
13679-70-4	: 11736
13679-74-8	: 5856
13684-56-5	: 5104
13684-63-4	: 9198
13688-75-0	: 11468
13698-16-3	: 3940
13698-87-8	: 1326
13706-86-0	: 6809
13721-54-5	: 7103
13732-80-4	: 1417
13794-14-4	: 3580
13799-90-1	: 1349
13808-64-5	: 10817
13811-01-3	: 4274
13828-34-7	: 4441
13831-30-6	: 101
13835-81-9	: 11467
13855-77-1	: 3964
13862-07-2	: 9665
13865-48-0	: 2018
13877-91-3	: 8485
13898-58-3	: 2514
13898-73-2	: 4678
13905-48-1	: 4320
13909-73-4	: 5874
13937-08-1	: 9957
13939-06-5	: 7662
13956-29-1	: 1537
13963-35-4	: 2017
13979-28-7	: 3836
13984-57-1	: 6906
13993-65-2	: 9500
13997-70-1	: 1232
13997-73-4	: 9271
14007-64-8	: 898
14010-23-2	: 6315
14021-92-2	: 5048
14045-26-2	: 9671
14062-18-1	: 923
14062-78-3	: 627
14064-10-9	: 9931
14064-43-8	: 1223
14064-68-7	: 2768
14072-86-7	: 4656
14073-97-3	: 4606
14077-58-8	: 298
14090-87-0	: 8393
14092-20-7	: 7106
14108-88-4	: 7993
14116-69-9	: 6234
14122-00-0	: 9280
14159-48-9	: 10130
14167-59-0	: 11650
14179-94-3	: 9126
14180-63-3	: 10596
14181-05-6	: 11145
14188-79-5	: 7794
14214-31-4	: 10538
14222-60-7	: 10886
14248-22-7	: 2129
14252-42-7	: 5450
14255-23-3	: 7023
14272-54-9	: 9115
14273-06-4	: 10732
14282-76-9	: 11729
14287-61-7	: 3507
14296-80-1	: 4172
14303-70-9	: 11629
14304-30-4	: 5087
14315-14-1	: 2402
14321-27-8	: 1810
14331-54-5	: 11055
14352-61-5	: 4310
14371-10-9	: 10518
14379-80-7	: 6266
14400-94-3	: 9467
14401-73-1	: 5797
14411-75-7	: 1471
14436-32-9	: 5091
14444-77-0	: 1409
14446-69-6	: 9653
14452-30-3	: 5836
14459-29-1	: 9735
14472-14-1	: 9242
14476-30-3	: 10162
14476-37-0	: 11960
14484-64-1	: 5957
14490-79-0	: 11608
14548-31-3	: 7105
14548-46-0	: 7639
14558-14-6	: 8003
14575-93-0	: 2965
14595-35-8	: 3783
14596-92-0	: 6682
14602-62-1	: 10161
14620-52-1	: 5297
14620-53-2	: 11617
14620-55-4	: 8237
14630-40-1	: 11434
14631-45-9	: 10102
14660-52-7	: 8861
14676-29-0	: 6409
14685-29-1	: 4995
14686-13-6	: 6534
14686-14-7	: 6536
14689-97-5	: 5592
14697-46-2	: 8851
14716-89-3	: 7404
14719-83-6	: 2548
14720-74-2	: 6438
14721-66-5	: 6658
14737-94-1	: 10622
14752-60-4	: 9124
14752-61-5	: 9133
14752-75-1	: 1743
14804-32-1	: 1682
14804-59-2	: 10510
14813-68-4	: 8345
14845-35-3	: 4706
14845-55-7	: 4887
14847-51-9	: 9237
14850-22-7	: 8495
14850-23-8	: 8496
14852-31-4	: 6660
14869-39-7	: 5814
14919-01-8	: 8494
14924-53-9	: 4175
14938-35-3	: 9444
14963-40-7	: 4946
14979-39-6	: 6501
14996-78-2	: 4236
15014-25-2	: 9927
15068-90-3	: 2950
15104-61-7	: 10126
15111-56-5	: 4639
15116-17-3	: 1534
15121-84-3	: 1612
15150-39-7	: 140
15164-44-0	: 582
15192-80-0	: 8489
15220-85-6	: 10580
15232-76-5	: 8531
15232-96-9	: 1247
15240-91-2	: 1800
15250-29-0	: 10670
15268-49-2	: 10086
15299-99-7	: 8067
15301-40-3	: 11293
15320-72-6	: 3833
15329-10-9	: 4708
15333-22-9	: 3829
15339-36-3	: 7421
15356-70-4	: 4548
15358-48-2	: 7574
15361-99-6	: 2277
15409-60-6	: 4591
15411-43-5	: 734
15430-98-5	: 8654
15457-05-3	: 5980
15466-93-0	: 4511
15466-94-1	: 4510
15469-77-9	: 5089
15489-90-4	: 5959
15501-33-4	: 10056
15519-73-0	: 3068
15521-18-3	: 10374
15543-70-1	: 4965
15570-12-4	: 2299
15600-01-8	: 10053
15606-95-8	: 439
15644-80-1	: 11316
15647-08-2	: 9589
15647-11-7	: 4502
15657-96-2	: 10091
15679-02-4	: 4188
15706-73-7	: 3545
15721-05-8	: 4992
15726-15-5	: 6525
15727-65-8	: 7892
15764-04-2	: 8052
15764-16-6	: 547
15797-21-4	: 4795
15802-97-8	: 10841
15802-99-0	: 10839
15825-70-4	: 7424
15826-80-9	: 2981
15840-60-5	: 7047
15843-48-8	: 5568
15854-58-7	: 10647
15869-80-4	: 6407
15869-85-9	: 8143
15869-86-0	: 8404
15869-87-1	: 8373
15869-89-3	: 8376
15869-92-8	: 8380
15869-93-9	: 8381
15869-94-0	: 8382
15869-95-1	: 8383
15869-96-2	: 8384
15870-10-7	: 6543
15874-03-0	: 6637
15877-57-3	: 8694
15890-36-5	: 4541
15890-40-1	: 4859
15901-49-2	: 5003
15909-76-9	: 10254
15918-08-8	: 6426
15922-53-9	: 11849
15932-80-6	: 4607
15950-66-0	: 9481
15953-45-4	: 10821
15953-73-8	: 10819
15971-29-6	: 7755
15972-60-8	: 322
15980-15-1	: 8569
16003-14-8	: 6160
16009-13-5	: 5958
16020-17-0	: 8001
16021-20-8	: 1702

16066-09-4	: 11647
16096-32-5	: 7315
16106-59-5	: 7020
16109-68-5	: 5120
16133-49-6	: 777
16136-84-8	: 10534
16136-85-9	: 10533
16147-07-2	: 1446
16177-46-1	: 4186
16179-36-5	: 8885
16183-46-3	: 3666
16188-57-1	: 9374
16197-90-3	: 3822
16245-79-7	: 831
16246-07-4	: 3947
16292-88-9	: 776
16302-35-5	: 10778
16306-39-1	: 9184
16308-92-2	: 1860
16315-07-4	: 1922
16326-97-9	: 4806
16326-98-0	: 4807
16357-59-8	: 11231
16369-05-4	: 3599
16369-17-8	: 8922
16369-21-4	: 5765
16386-93-9	: 9075
16387-70-5	: 8344
16387-72-7	: 11610
16409-46-4	: 3554
16419-60-6	: 3134
16420-13-6	: 3980
16423-68-0	: 11524
16426-64-5	: 2527
16432-53-4	: 6794
16433-43-5	: 1104
16435-50-0	: 9024
16452-01-0	: 771
16484-17-6	: 7293
16491-15-9	: 4921
16503-53-0	: 2147
16509-46-9	: 8335
16518-62-0	: 644
16520-62-0	: 1124
16532-02-8	: 11365
16538-93-5	: 4737
16545-78-1	: 9061
16560-43-3	: 11272
16580-23-7	: 4436
16580-24-8	: 4437
16582-38-0	: 1207
16587-71-6	: 4592
16605-36-0	: 4435
16619-12-8	: 7291
16629-19-9	: 11780
16631-63-3	: 4833
16650-52-5	: 7762
16661-99-7	: 4349
16672-87-0	: 5929
16674-04-7	: 10230
16694-52-3	: 9868
16696-65-4	: 11940
16714-77-5	: 3654
16726-67-3	: 7761
16736-42-8	: 8325
16737-44-3	: 1983
16744-89-1	: 7083
16745-94-1	: 7017
16746-02-4	: 6836
16746-86-4	: 7013
16746-87-5	: 7014
16747-25-4	: 6859
16747-26-5	: 6860
16747-28-7	: 6862
16747-30-1	: 6865
16747-31-2	: 6866
16747-32-3	: 8810
16747-33-4	: 8811
16747-38-9	: 8842
16747-42-5	: 6848
16747-44-7	: 8836
16747-45-8	: 8837
16747-50-5	: 4814
16749-11-4	: 4361
16751-59-0	: 6351
16752-77-5	: 7650
16754-49-7	: 10080
16789-46-1	: 6825
16789-51-8	: 7034
16790-49-1	: 3197
16805-78-0	: 9578
16813-18-6	: 6318
16837-43-7	: 2118
16841-48-8	: 10597
16874-33-2	: 6072
16883-48-0	: 4860
16889-72-8	: 10294
16898-52-5	: 9701
16932-49-3	: 2753
16939-57-4	: 1097
16957-70-3	: 9076
16980-61-3	: 4512
16995-35-0	: 10486
17005-59-3	: 11383
17015-11-1	: 6914
17021-26-0	: 357
17042-16-9	: 6980
17057-82-8	: 7252
17057-88-4	: 3060
17057-92-0	: 7887
17059-44-8	: 1485
17059-45-9	: 1482
17059-52-8	: 2459
17071-47-5	: 3401
17082-09-6	: 10717
17082-12-1	: 5129
17085-91-5	: 7828
17091-40-6	: 5747
17094-21-2	: 3558
17096-37-6	: 3796
17102-64-6	: 6709
17113-33-6	: 6159
17176-77-1	: 9586
17199-29-0	: 903
17206-52-9	: 4595
17216-08-9	: 8045
17216-62-5	: 9933
17226-28-7	: 3416
17226-34-5	: 7009
17228-64-7	: 10918
17236-58-7	: 11777
17249-75-1	: 11747
17249-76-2	: 11749
17249-77-3	: 11791
17249-79-5	: 11746
17249-82-0	: 11741
17283-45-3	: 2946
17297-04-0	: 1108
17299-41-1	: 4719
17301-94-9	: 8142
17302-01-1	: 6410
17302-02-2	: 6767
17302-04-4	: 6415
17312-63-9	: 8122
17348-59-3	: 10081
17356-19-3	: 4866
17372-87-1	: 11525
17406-45-0	: 6196
17432-37-0	: 598
17447-00-6	: 9085
17447-01-7	: 9084
17478-66-9	: 5138
17484-36-5	: 1934
17496-14-9	: 7289
17534-15-5	: 1561
17556-48-8	: 10409
17577-93-4	: 8523
17587-22-3	: 8398
17609-31-3	: 8484
17618-76-7	: 6554
17627-77-9	: 7395
17639-93-9	: 10222
17640-26-5	: 182
17648-05-4	: 2281
17655-74-2	: 1651
17658-32-1	: 8924
17672-88-7	: 2422
17679-92-4	: 8526
17696-11-6	: 8426
17696-37-6	: 9326
17696-64-9	: 142
17704-30-2	: 165
17715-00-3	: 4473
17742-04-0	: 7528
17745-45-8	: 3136
17773-65-8	: 3737
17788-00-0	: 9292
17789-14-9	: 5205
17804-35-2	: 515
17837-41-1	: 11693
17849-38-6	: 1848
17851-27-3	: 1710
17878-44-3	: 11367
17880-86-3	: 10502
17912-17-3	: 8368
17937-68-7	: 11404
17952-11-3	: 8808
17972-08-6	: 7652
17988-31-7	: 11427
18002-79-4	: 11360
18023-33-1	: 11426
18023-36-4	: 11511
18030-61-0	: 11364
18036-81-2	: 11497
18169-57-8	: 11469
18171-59-0	: 11373
18178-59-1	: 11421
18181-80-1	: 3140
18187-24-1	: 5234
18231-53-3	: 8764
18236-57-2	: 11400
18236-89-0	: 11399
18240-50-1	: 6114
18243-41-9	: 11366
18261-92-2	: 8280
18262-85-6	: 1709
18281-05-5	: 5356
18282-40-1	: 1680
18282-59-2	: 1015
18295-59-5	: 8697
18309-32-5	: 2978
18338-40-4	: 3449
18362-36-2	: 578
18362-97-5	: 8893
18368-63-3	: 10920
18381-45-8	: 3406
18388-45-9	: 11408
18397-07-4	: 4154
18398-36-2	: 7849
18423-69-3	: 6136
18441-43-5	: 2458
18444-79-6	: 11521
18476-57-8	: 8327
18479-51-1	: 8515
18479-57-7	: 8450
18492-37-0	: 2960
18495-23-3	: 8536
18495-26-6	: 6613
18495-30-2	: 10144
18510-29-7	: 11455
18521-07-8	: 8519
18523-48-3	: 109
18530-30-8	: 2917
18530-56-8	: 8212
18589-29-2	: 6420
18598-63-5	: 5019
18649-64-4	: 6108
18668-68-3	: 3147
18685-70-6	: 4073
18699-48-4	: 1338
18719-43-2	: 9932
18720-65-5	: 6502
18729-48-1	: 4870
18747-42-7	: 10470
18748-27-1	: 11578
18752-92-6	: 11445
18758-34-4	: 4998
18773-77-8	: 4160
18786-24-8	: 8572
18787-64-9	: 6663
18791-49-6	: 10282
18794-84-8	: 5331
18812-62-9	: 7079
18829-56-6	: 8180
18839-90-2	: 9359
18840-45-4	: 12042
18854-56-3	: 5586
18858-65-6	: 11441
18869-72-2	: 6311
18904-54-6	: 4146
18905-91-4	: 6973
18913-31-0	: 3175
18917-91-4	: 340
18932-14-4	: 8420
18936-17-9	: 3409
18937-76-3	: 11990
18963-00-3	: 10176
18963-01-4	: 10148
18970-44-0	: 1688
18971-59-0	: 8821
18979-60-7	: 1549
18979-62-9	: 1539
18992-39-7	: 10402
18992-80-8	: 5806
18999-28-5	: 6561
19037-72-0	: 4922
19044-88-3	: 8553
19074-59-0	: 9336
19089-47-5	: 10380
19113-78-1	: 6928
19145-60-9	: 487
19145-96-1	: 4356
19155-24-9	: 7329
19163-05-4	: 3129
19177-04-9	: 1987
19184-10-2	: 10567
19184-67-9	: 8774
19221-28-4	: 3165
19230-27-4	: 1294
19246-38-9	: 4346
19246-39-0	: 3601
19269-28-4	: 6720
19317-11-4	: 5330
19329-89-6	: 10274
19354-27-9	: 6170
19374-46-0	: 4857
19393-45-4	: 3773
19393-92-1	: 1013
19395-58-5	: 11198
19396-83-9	: 2916
19398-47-1	: 3614
19398-53-9	: 8743
19398-61-9	: 1387
19398-75-5	: 4799
19398-77-7	: 6768
19402-87-0	: 3988
19404-07-0	: 5191
19420-29-2	: 7946
19434-65-2	: 9778
19438-10-9	: 2636
19447-29-1	: 8197
19472-74-3	: 948
19481-82-4	: 10094
19482-57-6	: 8535
19489-10-2	: 4397
19493-31-3	: 3026
19493-44-8	: 7376
19523-57-0	: 7753
19549-71-4	: 6480
19549-73-6	: 6481
19549-74-7	: 6482
19549-77-0	: 6484
19549-79-2	: 6486
19549-84-9	: 6515
19549-97-4	: 6617
19550-03-9	: 6922
19550-05-1	: 6924
19550-07-3	: 6927
19550-09-5	: 6929
19550-10-8	: 6964
19550-14-2	: 6967
19550-40-4	: 4655
19550-48-2	: 4920
19564-40-0	: 2115
19594-02-6	: 1652
19613-78-4	: 1024
19617-44-8	: 7231
19653-33-9	: 9704
19655-60-8	: 11268
19666-30-9	: 8565
19686-73-8	: 10345
19689-18-0	: 5083
19694-02-1	: 10854
19700-97-1	: 6693
19718-45-7	: 7956
19735-89-8	: 10849
19752-55-7	: 1012
19780-63-3	: 8933
19780-66-6	: 9046
19780-67-7	: 9049
19780-90-6	: 6798
19780-96-2	: 8164
19781-10-3	: 8451
19781-11-4	: 8458
19781-68-1	: 4976
19781-69-2	: 4977
19781-73-8	: 4404
19781-77-2	: 6572
19812-64-7	: 11618
19814-71-2	: 2077
19819-98-8	: 1609
19821-84-2	: 6612
19832-98-5	: 8050
19860-56-1	: 10256
19866-51-4	: 9077
19879-11-9	: 4508
19879-12-0	: 4509
19889-37-3	: 3516
19890-00-7	: 2956
19891-74-8	: 4060
19902-08-0	: 2924
19915-11-8	: 8689
19942-78-0	: 11708
19947-22-9	: 1700
19961-27-4	: 9810
19967-57-8	: 9775
19987-13-4	: 9865
20009-26-1	: 7955
20024-90-2	: 1703
20027-77-4	: 7991
20030-32-4	: 4690
20063-92-7	: 8183
20089-07-0	: 3439
20098-48-0	: 2338
20177-02-0	: 9028
20184-89-8	: 8198
20184-91-2	: 8199
20185-55-1	: 2694
20193-20-8	: 9809
20213-30-3	: 7917
20231-81-6	: 4045
20232-11-5	: 7279
20237-34-7	: 6681
20244-71-7	: 5888
20247-89-6	: 2008
20266-65-3	: 10022
20278-84-6	: 6446
20278-85-7	: 6443
20278-87-9	: 6449
20278-88-0	: 6451
20278-89-1	: 6452
20279-53-2	: 6062
20281-83-8	: 8938
20281-85-0	: 3616
20281-86-1	: 6912
20281-91-8	: 3619
20291-91-2	: 6816
20291-95-6	: 6439
20327-65-5	: 10568
20333-41-9	: 5245
20354-26-1	: 7643
20357-79-3	: 7247
20359-55-1	: 8979
20395-25-9	: 10173
20442-79-9	: 3078
20474-93-5	: 3856
20485-41-0	: 11669
20589-85-9	: 10598
20599-21-7	: 8534
20599-27-3	: 9011
20634-92-8	: 9812
20637-49-4	: 10185
20668-29-5	: 11261
20699-48-3	: 3261
20731-93-5	: 1875
20734-71-8	: 9386
20739-58-6	: 8543
20748-86-1	: 6400
20754-04-5	: 8478
20769-85-1	: 10491
20809-46-5	: 8479
20820-82-0	: 11022
20830-75-5	: 4041
20850-43-5	: 2421
20893-30-5	: 11725

20972-43-4	: 5131
20984-33-2	: 11322
21020-24-6	: 3899
21020-26-8	: 9131
21020-27-9	: 9136
21040-45-9	: 10706
21040-59-5	: 7392
21087-29-6	: 1230
21087-64-9	: 7658
21102-13-6	: 6952
21129-27-1	: 4546
21179-48-6	: 10239
21211-65-4	: 11133
21232-53-1	: 361
21266-88-6	: 6375
21266-90-0	: 6376
21293-29-8	: 8679
21296-92-4	: 7323
21299-26-3	: 4578
21326-62-5	: 3611
21352-22-7	: 11209
21370-71-8	: 8054
21391-98-0	: 4635
21400-25-9	: 10602
21436-44-2	: 2683
21472-86-6	: 2335
21490-63-1	: 8609
21504-43-8	: 10559
21531-91-9	: 6793
21535-97-7	: 2457
21564-91-0	: 7984
21564-92-1	: 7986
21570-35-4	: 6503
21571-91-5	: 8751
21617-18-5	: 11250
21633-79-4	: 9972
21655-48-1	: 9611
21693-54-9	: 7992
21700-31-2	: 10125
21702-79-4	: 2559
21725-46-2	: 4161
21777-84-4	: 1992
21849-97-8	: 7840
21856-89-3	: 6970
21884-26-4	: 8909
21889-88-3	: 6585
21890-35-7	: 3148
21915-53-7	: 8619
21923-23-9	: 4092
21948-47-0	: 8372
21964-23-8	: 9013
21964-44-3	: 8189
21968-17-2	: 9827
21974-48-1	: 9714
21984-93-0	: 6065
21987-21-3	: 11886
21994-77-4	: 7065
21994-78-5	: 7064
22029-76-1	: 3873
22037-28-1	: 6034
22037-73-6	: 3710
22038-57-9	: 3830
22056-82-2	: 14
22077-55-0	: 2176
22104-78-5	: 8510
22118-12-3	: 3682
22128-62-7	: 4014
22128-63-8	: 4016
22139-77-1	: 1545
22144-60-1	: 1912
22160-12-9	: 6893
22160-39-0	: 8887
22160-40-3	: 8888
22100-41-4	: 8895
22202-65-9	: 11989
22224-92-6	: 5947
22227-75-4	: 10556
22232-54-8	: 7194
22253-11-8	: 7259
22287-02-1	: 8176
22319-28-4	: 6592
22331-38-0	: 5351
22349-03-7	: 4456
22364-68-7	: 960
22374-89-6	: 2119
22381-86-8	: 11980
22398-09-0	: 9709
22401-25-8	: 9546
22409-91-2	: 10322
22418-49-1	: 10577
22421-97-2	: 10229
22445-41-6	: 1771
22483-09-6	: 5419
22520-40-7	: 8391
22524-32-9	: 113
22531-06-2	: 8051
22531-20-0	: 7891
22539-65-7	: 1469
22545-12-6	: 1603
22545-13-7	: 1604
22564-83-6	: 7429
22564-99-4	: 8332
22567-21-1	: 4543
22610-63-5	: 8255
22618-23-1	: 9315
22621-74-5	: 2947
22646-79-3	: 4646
22663-64-5	: 8166
22672-74-8	: 5150
22692-70-2	: 11560
22717-57-3	: 2640
22726-00-7	: 605
22734-04-9	: 4905
22767-95-9	: 2140
22767-96-0	: 2145
22768-19-0	: 6550
22768-20-3	: 6549
22781-23-3	: 514
22839-47-0	: 9511
22868-76-4	: 11056
22900-08-9	: 4540
22922-67-4	: 7004
22924-15-8	: 553
22927-13-5	: 558
22936-86-3	: 5013
23010-00-6	: 3733
23010-04-0	: 3283
23012-11-5	: 8578
23048-13-7	: 4943
23063-36-7	: 1390
23074-36-4	: 3709
23079-65-4	: 1116
23097-98-5	: 6618
23103-98-2	: 9728
23132-52-7	: 1467
23135-22-0	: 8568
23144-52-7	: 8448
23184-66-9	: 3144
23204-80-0	: 1229
23287-26-5	: 2638
23308-82-9	: 1894
23378-11-2	: 3732
23399-70-4	: 1174
23418-37-3	: 8469
23418-38-4	: 8170
23422-53-9	: 6001
23431-48-3	: 2028
23438-77-9	: 4722
23444-65-7	: 7854
23470-12-4	: 7127
23505-41-1	: 9729
23511-70-8	: 3582
23564-05-8	: 11723
23578-35-0	: 6751
23580-51-0	: 8544
23580-52-1	: 10960
23617-71-2	: 2589
23619-59-2	: 9459
23676-09-7	: 2599
23708-56-7	: 11954
23733-68-8	: 4717
23769-39-3	: 11921
23843-52-9	: 8239
23947-60-6	: 5932
23950-58-5	: 10746
24127-62-6	: 10615
24152-39-4	: 9637
24247-53-8	: 3696
24264-08-2	: 3877
24280-93-1	: 7052
24306-18-1	: 6323
24307-26-4	: 7430
24331-72-4	: 9794
24332-22-7	: 6584
24342-04-9	: 9139
24346-56-3	: 2132
24356-00-1	: 9017
24393-70-2	: 2943
24395-10-6	: 6571
24398-88-7	: 2522
24410-50-2	: 2449
24410-51-3	: 2446
24410-52-4	: 2452
24425-40-9	: 7244
24425-97-6	: 10051
24438-17-3	: 7328
24446-63-7	: 942
24544-04-5	: 636
24545-81-1	: 2987
24549-06-2	: 745
24557-10-6	: 10627
24579-70-2	: 3941
24580-48-1	: 4550
24580-52-7	: 4551
24602-86-6	: 11898
24612-75-7	: 4477
24618-31-3	: 4317
24622-72-8	: 11169
24634-61-5	: 6704
24644-78-8	: 7290
24677-78-9	: 540
24716-09-4	: 890
24724-07-0	: 11949
24734-68-7	: 2125
24764-97-4	: 3179
24796-87-0	: 4503
24815-24-5	: 12060
24840-05-9	: 10588
24844-28-8	: 4556
24885-45-8	: 9375
24887-75-0	: 345
24910-63-2	: 9036
24932-49-8	: 1297
24948-66-1	: 5095
24964-64-5	: 2757
24965-84-2	: 4596
24973-49-7	: 3020
25006-60-4	: 167
25013-16-5	: 9329
25044-01-3	: 9107
25057-89-0	: 518
25109-28-8	: 1009
25109-57-3	: 3653
25117-74-2	: 2755
25118-28-9	: 6755
25144-04-1	: 4869
25144-05-2	: 4868
25148-87-2	: 3756
25167-70-8	: 9064
25172-42-3	: 4873
25172-43-4	: 4874
25182-84-7	: 10395
25186-47-4	: 1386
25201-40-5	: 6338
25229-42-9	: 3834
25234-83-7	: 8898
25245-33-4	: 1767
25265-71-8	: 10397
25309-64-2	: 1681
25311-71-1	: 7359
25312-34-9	: 3874
25312-65-6	: 4094
25346-31-0	: 8718
25346-32-1	: 8730
25346-33-2	: 8714
25354-97-6	: 5065
25354-98-7	: 8649
25394-75-6	: 10772
25395-22-6	: 103
25395-31-7	: 10189
25414-22-6	: 6127
25415-67-2	: 8892
25415-70-7	: 8894
25419-36-7	: 7990
25419-37-8	: 7987
25542-62-5	: 6879
25611-78-3	: 1584
25641-98-9	: 1340
25659-22-7	: 7087
25695-77-6	: 10689
25790-55-0	: 3150
25871-69-6	: 7587
26073-26-7	: 10500
26106-95-6	: 3602
26116-12-1	: 11167
26118-38-7	: 6963
26120-52-5	: 4253
26153-88-8	: 8703
26171-83-5	: 3344
26186-01-6	: 8107
26186-02-7	: 11897
26190-82-9	: 11256
26198-63-0	: 9892
26225-79-6	: 5933
26259-45-0	: 3143
26260-02-6	: 580
26303-90-2	: 5090
26311-17-1	: 5424
26311-33-1	: 7063
26330-51-8	: 3476
26340-58-9	: 4151
26370-28-5	: 8108
26399-36-0	: 9763
26438-26-6	: 7931
26438-27-7	: 7793
26438-28-8	: 7879
26464-32-4	: 3488
26465-81-6	: 7288
26472-41-3	: 4282
26473-49-4	: 3537
26489-01-0	: 8514
26513-20-2	: 1265
26530-20-1	: 8528
26531-85-1	: 6237
26533-33-5	: 8169
26549-24-6	: 6913
26549-25-7	: 6474
26608-06-0	: 5165
26644-46-2	: 11902
26730-14-3	: 11879
26754-48-3	: 6030
26771-11-9	: 4959
26963-33-7	: 11771
27063-27-0	: 11208
27129-86-8	: 1057
27129-87-9	: 1861
27267-69-2	: 2773
27314-13-2	: 8213
27356-52-1	: 11269
27387-79-7	: 10218
27464-82-0	: 11659
27532-76-9	: 378
27598-85-2	: 9200
27610-88-4	: 8400
27644-49-1	: 8465
27655-41-0	: 7375
27701-01-5	: 3943
27779-01-7	: 4158
27831-13-6	: 1629
27876-94-4	: 5107
27993-42-6	: 8984
28056-54-4	: 4896
28123-70-8	: 8125
28148-04-1	: 9890
28249-77-6	: 11697
28261-97-4	: 5273
28300-74-5	: 11588
28320-88-9	: 8362
28320-89-0	: 6364
28332-44-7	: 7095
28466-26-4	: 10812
28467-92-7	: 3190
28469-92-3	: 1369
28664-03-1	: 4171
28685-60-1	: 3958
28715-26-6	: 2451
28788-62-7	: 2889
28823-41-8	: 6696
28832-55-5	: 3180
28905-12-6	: 6232
28921-35-9	: 6064
28994-41-4	: 9454
28995-89-3	: 8004
29003-73-4	: 163
29022-31-9	: 8201
29030-84-0	: 6689
29040-46-8	: 2445
29072-93-3	: 10087
29091-05-2	: 5181
29091-21-2	: 9761
29106-32-9	: 4936
29110-74-5	: 7236
29113-63-1	: 6055
29136-19-4	: 2062
29172-20-1	: 6111
29204-02-2	: 5365
29232-93-7	: 9730
29276-40-2	: 7707
29289-13-2	: 762
29342-44-7	: 6740
29342-53-8	: 2918
29394-58-9	: 3383
29400-42-8	: 8211
29472-95-5	: 9421
29488-27-5	: 11773
29538-77-0	: 4507
29554-49-2	: 4745
29555-02-0	: 4964
29559-27-1	: 2004
29559-52-2	: 10177
29560-84-7	: 10701
29604-75-9	: 1296
29606-79-9	: 4603
29624-17-7	: 4348
29653-38-1	: 3926
29682-41-5	: 1380
29780-00-5	: 11896
29823-16-3	: 8873
29843-58-1	: 3757
29865-90-5	: 568
29882-57-3	: 8720
29883-15-6	: 954
29897-82-3	: 11172
29916-45-8	: 3269
29921-50-4	: 8063
29943-42-8	: 10799
29949-19-7	: 7511
29972-79-0	: 5341
30090-17-6	: 814
30122-12-4	: 8745
30168-50-4	: 4830
30273-11-1	: 813
30361-33-2	: 5025
30389-18-5	: 4290
30433-57-9	: 10833
30515-28-7	: 6458
30525-89-4	: 8634
30538-80-8	: 9689
30544-34-4	: 6077
30560-19-1	: 10
30600-19-2	: 10619
30614-77-8	: 6082
30673-60-0	: 5069
30685-95-1	: 3886
30689-41-9	: 4916
30752-19-3	: 1045
30926-60-4	: 1188
30964-01-3	: 8207
31053-99-3	: 1092
31059-19-5	: 7944
31116-84-4	: 10459
31161-71-4	: 11627
31202-12-7	: 363
31218-83-4	: 10719
31252-42-3	: 9700
31283-43-9	: 8498
31340-44-0	: 10247
31366-00-4	: 1713
31367-46-1	: 6496
31508-08-4	: 6607
31508-12-0	: 8194
31543-75-6	: 1306
31599-60-7	: 1769
31599-61-8	: 1774
31642-67-8	: 8188
31656-92-5	: 981
31699-35-1	: 8116
31702-33-7	: 10605
31720-69-1	: 4562
31766-07-1	: 2771
31772-05-1	: 11874
31805-48-8	: 11761
31844-95-8	: 9008
31844-96-9	: 3716
31849-78-2	: 3708

31883-98-4	: 4713
31915-94-3	: 1098
32018-88-5	: 7760
32064-78-1	: 8524
32116-65-7	: 4889
32121-06-5	: 10544
32121-07-6	: 10543
32161-06-1	: 9720
32166-40-8	: 8053
32315-10-9	: 7585
32322-73-9	: 47
32328-03-3	: 8778
32337-74-9	: 8664
32360-05-7	: 10667
32367-54-7	: 7889
32395-96-3	: 1747
32433-28-6	: 90
32449-92-6	: 6239
32460-00-7	: 6078
32460-20-1	: 6190
32541-33-6	: 8925
32607-54-8	: 8865
32620-08-9	: 3707
32665-23-9	: 3553
32666-56-1	: 7714
32718-56-2	: 5734
32723-67-4	: 586
32750-14-4	: 3067
32764-34-4	: 6507
32768-54-0	: 1383
32777-26-7	: 4224
32809-16-8	: 9760
32830-97-0	: 8962
32838-55-4	: 9699
32854-75-4	: 309
32889-48-8	: 9759
33018-91-6	: 6396
33083-83-9	: 11961
33089-61-1	: 342
33092-82-9	: 4501
33184-48-4	: 1175
33245-39-5	: 5964
33357-44-7	: 11270
33401-05-7	: 10098
33429-70-8	: 2327
33429-72-0	: 8725
33467-76-4	: 6567
33515-82-1	: 4233
33553-93-4	: 8754
33560-15-5	: 252
33560-17-7	: 249
33567-59-8	: 870
33622-26-3	: 5094
33629-47-9	: 3893
33662-96-3	: 10292
33683-44-2	: 6972
33695-59-9	: 10109
33738-48-6	: 7756
33742-81-3	: 4178
33820-53-0	: 7371
33829-48-0	: 7465
33840-38-9	: 4877
33843-55-9	: 4700
33877-11-1	: 1834
33880-83-0	: 4389
33933-78-7	: 8171
33933-79-8	: 8455
33941-07-0	: 10771
33966-50-6	: 3210
33978-71-1	: 11957
33996-33-7	: 9766
34008-71-4	: 10436
34014-18-1	: 11590
34069-94-8	: 293
34103-97-4	: 4876
34238-52-3	: 9022
34238-81-8	: 5136
34240-76-1	: 5739
34256-82-1	: 270
34263-68-8	: 6729
34314-06-2	: 3036
34316-15-9	: 2433
34451-18-8	: 10268
34570-59-7	: 10136
34619-03-9	: 3994
34757-14-7	: 10237
34786-24-8	: 1706
34840-79-4	: 1364
34859-98-8	: 8946
34874-30-1	: 11117
34885-03-5	: 4424
34887-14-4	: 8748
34893-50-0	: 9649
34973-41-6	: 3380
35000-22-7	: 10525
35008-86-7	: 1761
35018-15-6	: 11149
35018-28-1	: 11175
35021-70-6	: 1752
35099-89-9	: 4227
35120-10-6	: 286
35161-71-8	: 10726
35182-51-5	: 10968
35194-37-7	: 6562
35205-71-1	: 7059
35205-79-9	: 8153
35216-11-6	: 11641
35223-80-4	: 120
35242-05-8	: 4504
35246-18-5	: 7309
35274-05-6	: 10271
35280-53-6	: 136
35305-13-6	: 9686
35323-91-2	: 2846
35355-36-3	: 2454
35364-90-0	: 1411
35365-59-4	: 8315
35367-38-5	: 5176
35383-51-8	: 3153
35383-59-6	: 3686
35413-38-8	: 4615
35432-36-1	: 10134
35449-34-4	: 250
35472-56-1	: 2680
35502-06-8	: 9852
35554-44-0	: 7187
35569-54-1	: 11552
35573-93-4	: 9850
35608-64-1	: 3533
35673-03-1	: 1226
35691-65-7	: 8768
35794-11-7	: 9652
35897-16-6	: 8948
35911-17-2	: 10546
35951-33-8	: 8723
35951-36-1	: 6382
36045-56-4	: 6903
36049-78-2	: 8407
36065-15-3	: 2962
36069-05-3	: 12045
36076-25-2	: 1259
36190-77-9	: 5750
36215-07-3	: 9860
36301-29-8	: 1995
36357-38-7	: 5855
36386-49-9	: 2940
36386-52-4	: 2945
36402-31-0	: 3609
36441-31-3	: 8040
36441-32-4	: 8039
36448-60-9	: 10472
36556-06-6	: 7389
36566-80-0	: 7122
36653-82-4	: 6659
36678-06-5	: 3167
36691-77-7	: 11888
36701-01-6	: 8880
36734-19-7	: 7338
36768-62-4	: 9628
36823-84-4	: 2908
36825-36-2	: 11207
36847-51-5	: 3499
36854-57-6	: 974
36895-62-2	: 11825
36903-89-6	: 8363
36903-92-1	: 8521
36903-95-4	: 8329
36917-36-9	: 10961
36994-79-3	: 2352
37025-57-3	: 1871
37050-06-9	: 8537
37273-91-9	: 7440
37434-59-6	: 9976
37470-46-5	: 2623
37493-31-5	: 6162
37493-70-2	: 11953
37549-83-0	: 7056
37549-89-6	: 7031
37674-63-8	: 9079
37674-67-2	: 7074
37688-96-3	: 11616
37710-49-9	: 8663
37746-78-4	: 3819
37806-29-4	: 1808
37845-14-0	: 8060
37866-06-1	: 3731
37924-13-3	: 9149
37935-39-0	: 3492
37951-49-8	: 10456
37973-84-5	: 11975
37985-18-5	: 11701
38031-78-6	: 10943
38204-89-6	: 1361
38237-74-0	: 8960
38256-94-9	: 5426
38260-54-7	: 5945
38300-67-3	: 3445
38460-95-6	: 11982
38487-94-4	: 2103
38511-07-8	: 4645
38514-02-2	: 8461
38514-03-3	: 8462
38552-72-6	: 6601
38577-97-8	: 11523
38624-62-3	: 3826
38628-39-6	: 11868
38651-65-9	: 2955
38661-81-3	: 1034
38661-82-4	: 1160
38727-55-8	: 5173
38738-60-2	: 4676
38827-66-6	: 11517
38862-78-1	: 1397
38870-89-2	: 299
38939-88-7	: 1211
39029-41-9	: 7936
39052-12-5	: 693
39052-57-8	: 2736
39075-90-6	: 9245
39076-02-3	: 3962
39118-35-9	: 8527
39121-37-4	: 6519
39135-39-2	: 9625
39148-24-8	: 6020
39178-11-5	: 5255
39191-07-6	: 1802
39196-18-4	: 11717
39228-29-0	: 10677
39251-86-0	: 6058
39251-88-2	: 6067
39257-08-4	: 4614
39273-81-9	: 6599
39478-78-9	: 652
39511-08-5	: 10513
39513-75-2	: 2803
39515-51-0	: 597
39622-79-2	: 1330
39669-95-9	: 1755
39713-71-8	: 11151
39755-03-8	: 3667
39761-61-0	: 7027
39782-38-2	: 3371
39782-43-9	: 7026
39928-72-8	: 6141
39966-95-5	: 3427
39989-39-4	: 7380
40058-87-5	: 10223
40203-74-5	: 156
40225-75-0	: 8468
40239-35-8	: 11892
40242-15-7	: 7847
40248-63-3	: 210
40248-84-8	: 9377
40323-88-4	: 11762
40385-54-4	: 1073
40441-35-8	: 4623
40466-95-3	: 9235
40487-42-1	: 8635
40514-70-3	: 4393
40527-16-0	: 8582
40571-45-7	: 4772
40575-42-6	: 8509
40589-38-6	: 3890
40596-69-8	: 7651
40604-49-7	: 8023
40605-42-3	: 3864
40614-52-6	: 3363
40646-07-9	: 6397
40649-36-3	: 4613
40650-41-7	: 7261
40734-75-6	: 1002
40738-26-9	: 5310
40745-44-6	: 2
40775-09-5	: 1741
40960-69-8	: 4554
40960-73-4	: 4553
41055-92-9	: 6513
41065-91-2	: 6883
41085-43-2	: 1071
41096-46-2	: 7182
41198-08-7	: 9762
41223-14-7	: 4773
41284-12-2	: 5615
41295-64-1	: 3687
41433-81-2	: 9954
41589-42-8	: 9936
41635-77-2	: 1753
41653-96-7	: 7057
41658-69-9	: 10096
41692-47-1	: 6899
41755-60-6	: 6655
41761-12-0	: 6766
41814-78-2	: 11858
41851-34-7	: 4430
41867-20-3	: 3816
41901-72-8	: 1410
42027-23-6	: 8788
42061-72-3	: 4609
42067-48-1	: 9051
42067-49-2	: 9050
42101-38-2	: 3755
42131-85-1	: 9016
42131-89-5	: 6763
42185-47-7	: 4618
42235-39-2	: 5353
42245-37-4	: 8706
42286-84-0	: 2834
42345-82-4	: 5899
42398-73-2	: 7384
42412-84-0	: 10321
42474-20-4	: 8739
42474-21-5	: 6372
42551-55-3	: 3598
42569-16-4	: 9092
42576-02-3	: 2997
42597-26-2	: 2127
42775-75-7	: 7890
42842-08-0	: 10045
42852-95-9	: 6215
42874-01-1	: 8082
42874-03-3	: 8633
43036-06-2	: 11137
43121-43-3	: 512
43197-78-0	: 6746
43222-48-6	: 5175
44829-76-7	: 6595
44855-57-4	: 8351
44914-03-6	: 10652
45223-18-5	: 6635
47172-89-4	: 8776
48145-04-6	: 10681
49542-66-7	: 5678
49763-65-7	: 2907
49845-33-2	: 11054
49852-35-9	: 6597
50375-10-5	: 2330
50402-72-7	: 9713
50407-18-6	: 7238
50461-74-0	: 8905
50468-22-9	: 3352
50471-44-8	: 12048
50552-30-2	: 7101
50592-87-5	: 6736
50594-66-6	: 304
50599-73-0	: 6989
50623-57-9	: 8147
50670-64-9	: 2743
50689-17-3	: 6074
50710-40-2	: 3244
50840-23-8	: 11047
50868-72-9	: 773
50868-73-0	: 774
50876-32-9	: 4478
50876-33-0	: 4851
50902-82-4	: 6887
51034-46-9	: 3154
51060-05-0	: 7360
51115-02-7	: 4931
51116-73-5	: 8715
51152-12-6	: 2930
51218-45-2	: 7657
51225-20-8	: 691
51235-04-2	: 6995
51282-49-6	: 2549
51284-68-5	: 5116
51299-55-9	: 4620
51308-99-7	: 8163
51309-10-5	: 5073
51333-70-1	: 8111
51338-27-3	: 5171
51437-67-3	: 5115
51500-32-4	: 10818
51525-97-4	: 10137
51556-10-6	: 6614
51575-85-0	: 6616
51630-58-1	: 5956
51637-47-9	: 8864
51666-60-3	: 4883
51686-78-3	: 1311
51707-55-2	: 11687
51750-65-3	: 6847
51756-08-2	: 3519
51760-90-8	: 3689
51772-30-6	: 10467
51786-70-0	: 3243
51806-20-3	: 5543
51807-53-5	: 7196
51952-42-2	: 9015
52033-97-3	: 4888
52152-71-3	: 3382
52152-72-4	: 8819
52182-15-7	: 919
52208-62-5	: 2056
52244-70-9	: 1105
52262-38-1	: 9281
52315-07-8	: 5012
52324-03-5	: 9147
52358-73-3	: 7801
52387-50-5	: 7067
52390-72-4	: 6473
52414-97-8	: 1068
52418-81-2	: 6918
52460-86-3	: 892
52488-28-5	: 1069
52497-07-1	: 3743
52562-19-3	: 787
52645-53-1	: 9150
52708-03-9	: 8160
52744-85-1	: 10645
52780-23-1	: 1506
52802-07-0	: 6917
52896-87-4	: 6427
52896-88-5	: 6413
52896-89-6	: 6414
52896-90-9	: 6412
52896-91-0	: 6411
52896-92-1	: 6437
52896-93-2	: 6441
52896-95-4	: 6442
52896-99-8	: 6821
52897-00-4	: 6817
52897-01-5	: 6822
52897-03-7	: 6823
52897-04-8	: 6819
52897-05-9	: 6824
52897-06-0	: 6820
52897-08-2	: 6845
52897-09-3	: 6846
52897-10-6	: 6850
52897-11-7	: 6851
52897-12-8	: 6852
52897-15-1	: 6853
52897-16-2	: 8757
52897-17-3	: 8815
52897-19-5	: 8816

52914-66-6	4762
52918-63-5	5102
53022-14-3	3565
53045-71-9	9009
53068-88-5	6902
53078-85-6	648
53109-18-5	4933
53152-98-0	9681
53172-84-2	1950
53175-28-3	3697
53178-20-4	3242
53188-23-1	6023
53220-82-9	7741
53252-08-7	6591
53252-19-0	7096
53252-20-3	8192
53252-21-4	7092
53346-71-7	8648
53381-90-1	9725
53440-12-3	7778
53491-32-0	2441
53535-68-5	10417
53566-37-3	7121
53584-29-5	10009
53588-92-4	5464
53608-85-8	4322
53608-93-8	9935
53660-21-2	6478
53663-23-3	2353
53663-26-6	8148
53776-69-5	2081
53778-43-1	4975
53780-34-0	7427
53859-78-2	4213
53878-96-9	3903
53889-94-4	9428
53894-19-2	10361
53966-53-3	8186
53977-99-4	7048
54105-76-9	4427
54106-40-0	2005
54131-44-1	2927
54244-90-5	6590
54268-02-9	10138
54298-97-4	8525
54305-87-2	6764
54340-87-3	7262
54383-22-1	4373
54410-69-4	2010
54411-01-7	4446
54429-56-0	5628
54454-10-3	10820
54460-46-7	4952
54460-96-7	10121
54462-68-9	6759
54462-70-3	3270
54467-95-7	979
54484-63-8	1377
54527-68-3	3570
54553-14-9	11155
54573-13-6	6620
54630-50-1	6497
54638-10-7	8918
54644-23-4	1231
54676-39-0	4327
54715-57-0	9259
54737-45-0	1080
54774-99-1	10249
54819-86-2	10217
54829-48-0	6110
54833-31-7	4464
54833-34-0	4402
54837-90-0	815
54852-73-2	3071
54862-92-9	6517
54885-00-6	7640
54889-55-3	7274
54890-00-5	4326
54922-86-0	10032
54934-90-6	4432
54934-91-7	4472
55030-56-3	3739
55038-01-2	2455
55057-45-9	10494
55066-53-0	8305
55073-86-4	6950
55124-79-3	6309
55130-16-0	8309
55156-16-6	8980
55182-83-7	11988
55268-62-7	1567
55268-63-8	2267
55282-10-5	6296
55282-12-7	8241
55282-14-9	5271
55282-15-0	5270
55282-16-1	5269
55282-17-2	11605
55283-68-6	5409
55285-14-8	4038
55289-35-5	1070
55290-64-7	5179
55319-83-0	5275
55320-06-4	6295
55334-08-2	4476
55335-06-3	11854
55373-86-9	5274
55429-84-0	11604
55449-46-2	1234
55454-22-3	6568
55505-26-5	8168
55517-73-2	6623
55517-89-0	8639
55563-72-9	10458
55665-79-7	7060
55696-02-1	3944
55702-49-3	3894
55751-54-7	812
55775-41-2	3182
55789-02-1	1872
55877-01-5	6790
55882-21-8	10153
55944-70-2	6477
56072-27-6	6518
56103-67-4	9225
56106-37-7	1556
56180-61-1	8688
56216-12-7	3823
56245-02-4	2034
56456-42-9	1496
56523-26-3	8322
56630-31-0	8206
56630-33-2	8204
56630-34-3	8205
56688-75-6	4671
56694-97-4	11864
56694-98-5	11863
56739-95-8	1781
56745-61-0	6969
56800-09-0	3746
56908-88-4	1291
56911-77-4	2074
56920-82-2	6037
56961-23-0	2371
56961-24-1	2320
56961-32-1	2541
56961-77-4	1010
56961-82-1	2326
56961-84-3	1359
56961-86-5	2337
56961-93-4	7851
56962-04-0	9231
56962-07-3	7343
57074-96-1	8539
57129-69-8	6146
57150-66-0	9553
57179-35-8	573
57223-18-4	8193
57296-00-1	10613
57323-93-0	6988
57380-83-1	1402
57392-55-9	255
57403-35-7	1135
57404-54-3	9176
57456-98-1	3185
57502-49-5	9666
57526-28-0	3694
57576-09-7	4545
57605-95-5	7911
57706-88-4	8454
57772-57-3	2624
57837-19-1	7439
57885-43-5	10211
57933-83-2	4022
57968-71-5	3307
58138-08-2	11899
58138-81-1	3669
58167-01-4	3305
58190-94-6	3491
58210-03-0	4568
58371-98-5	8963
58474-80-9	5081
58588-28-6	6919
58593-78-5	1563
58616-95-8	4970
58618-39-6	7806
58625-77-7	3655
58766-17-9	6745
58873-44-2	9182
59020-74-5	2447
59022-71-8	7330
59113-36-9	10024
59204-02-3	8449
59255-95-7	653
59357-07-2	6974
59368-15-9	4443
59376-54-4	184
59433-08-8	9669
59434-05-8	11758
59654-13-6	6832
59669-26-0	11713
59697-51-7	7011
59697-55-1	7012
59734-23-5	8470
59756-60-4	5981
59777-92-3	4231
59889-45-1	8722
59905-55-4	2983
59956-48-8	118
60012-63-7	2102
60096-09-5	6398
60111-14-0	7082
60168-88-9	5948
60207-90-1	10720
60212-29-5	11985
60212-31-9	11986
60251-57-2	8357
60291-32-9	4408
60468-23-7	9007
60512-85-8	10687
60669-40-1	11967
60671-36-5	11956
60698-31-9	6157
60703-31-3	7075
60710-39-6	9239
60761-04-8	8561
60814-29-1	11335
60836-07-9	6509
60899-39-0	7880
60956-23-2	1303
60956-24-3	1293
60956-26-5	1074
61050-68-8	8397
61228-10-2	6619
61231-45-6	7717
61531-45-1	8156
61539-75-1	8370
61551-49-3	7978
61563-25-5	1302
61589-68-2	8991
61592-89-0	1962
61675-14-7	7098
61764-94-1	6370
61792-07-2	260
61827-85-8	1487
61827-86-9	1708
61827-87-0	1711
61827-89-2	5286
61827-93-8	4853
61827-98-3	4845
61828-04-4	2341
61828-05-5	2265
61828-06-6	2088
61828-07-7	4380
61828-08-8	4483
61828-09-9	4474
61828-10-2	4462
61828-11-3	4410
61828-13-5	4458
61828-15-7	4482
61828-17-9	4383
61828-19-1	4479
61828-21-5	4413
61828-25-9	1742
61828-26-0	2065
61838-78-6	10464
61847-78-7	7030
61847-80-1	9048
61847-97-0	8314
61886-67-7	7932
61892-85-1	6808
61949-83-5	7243
61971-91-3	3670
62016-13-1	6839
62016-38-0	8421
62016-48-2	3394
62016-53-9	4842
62016-55-1	4808
62016-58-4	4821
62059-56-7	4351
62067-44-1	4527
62080-86-8	8274
62088-36-2	4663
62108-06-9	8517
62108-66-1	10232
62108-68-3	10227
62108-70-7	10226
62127-46-2	3446
62127-47-3	3447
62127-49-5	10135
62127-61-1	10160
62154-77-2	2297
62155-41-3	2111
62168-25-6	3264
62184-67-2	11766
62184-82-1	4919
62184-83-2	4938
62261-05-6	3957
62353-69-9	8685
62476-56-6	10862
62496-53-1	8753
62521-69-1	3453
62624-30-0	465
62690-62-4	7897
62692-40-4	1001
62706-16-5	7008
62882-00-2	7385
63019-98-7	3014
63157-81-3	10194
63284-71-9	8219
63370-69-4	4776
63468-05-3	9097
63790-14-7	1086
63927-44-6	6877
64057-79-0	10703
64161-50-8	2951
64165-34-0	11317
64168-11-2	9702
64207-03-0	10447
64257-84-7	5953
64287-26-9	1285
64334-96-9	11211
64339-43-1	1524
64346-45-8	3807
64379-30-2	8704
64501-77-5	8508
64502-89-2	8971
64531-26-6	8533
64632-72-0	12052
64653-59-4	744
64902-72-3	4091
64930-16-1	3741
64965-91-9	2453
65144-02-7	9797
65248-43-3	6985
65423-25-8	5340
65438-35-9	13
65469-88-7	3357
65530-93-0	6728
65732-90-3	11518
65786-09-6	6380
65786-11-0	6539
65927-60-8	6954
66050-98-4	6938
66056-06-2	19
66225-16-9	6552
66225-17-0	6551
66225-35-2	6803
66235-62-9	3608
66295-45-2	9549
66395-23-1	4585
66398-52-5	3466
66577-03-5	11763
66577-17-1	7924
66610-39-7	3392
66719-29-7	8759
66719-37-7	6488
66719-40-2	6489
66719-53-7	8456
66719-54-8	8457
66719-55-9	8452
66731-94-0	6490
66793-72-4	6956
66793-74-6	6957
66793-80-4	8467
66793-92-8	6955
66794-02-3	6933
66794-03-4	6934
66794-04-5	6932
66810-87-5	6958
66898-39-3	2820
66997-45-3	10146
67191-37-1	1665
67329-04-8	7340
67451-76-7	4312
67485-29-4	7136
67535-07-3	10255
67590-77-6	1760
67632-66-0	9148
67700-22-5	2159
67700-26-9	8512
67747-70-0	9021
67885-08-9	3166
67952-57-2	6506
68057-83-0	3532
68077-28-1	10601
68107-05-1	6724
68165-06-0	11187
68266-67-1	1360
68359-37-5	5008
68384-70-3	3541
68420-93-9	11212
68694-11-1	11900
68723-25-1	7629
68750-16-3	5677
69112-21-6	6997
69190-65-4	3811
69248-35-7	3332
69295-21-2	3729
69305-42-6	9496
69321-60-4	1304
69371-59-1	7102
69377-81-7	5982
69498-28-8	3273
69542-91-2	4514
69564-68-7	2881
69668-88-8	6588
69704-44-5	3262
69745-49-9	4692
69770-98-5	4891
69806-50-4	5962
69813-63-4	1970
69833-18-7	469
69947-01-9	9805
70110-24-6	10496
70116-68-6	3629
70124-77-5	5965
70127-93-4	7237
70202-92-5	4915
70396-13-3	6615
70401-56-8	2770
70644-46-1	10922
70690-24-3	8367
70715-19-4	10983
70732-43-3	8202
70732-44-4	8341
70732-45-5	5101
70786-82-2	8867
71159-31-4	10378
71172-57-1	8970
71173-15-4	2939
71200-79-8	3763
71284-96-3	8025
71297-97-7	5118
71605-85-1	8152

71608-10-1 : 9334
71626-11-4 : 513
71633-43-7 : 11303
71745-63-6 : 9340
71747-37-0 : 2152
71751-41-2 : 477
71783-56-7 : 7391
72178-02-0 : 5985
72312-48-2 : 4687
72622-74-3 : 10342
72762-00-6 : 11031
72935-12-7 : 3049
73120-52-2 : 5298
73178-43-5 : 9042
73206-57-2 : 10448
73210-25-0 : 4565
73383-24-1 : 3738
73398-15-9 : 6694
73454-83-8 : 6690
73513-50-5 : 7058
73522-42-6 : 2928
73548-71-7 : 7025
73548-72-8 : 6553
73557-60-5 : 2310
73642-91-8 : 8127
73741-61-4 : 4701
73758-54-0 : 3486
73918-56-6 : 1570
73931-44-9 : 2312
74051-80-2 : 11359
74112-36-0 : 8447
74115-24-5 : 4139
74204-00-5 : 9240
74219-20-8 : 2949
74221-86-6 : 4444
74244-64-7 : 4217
74283-48-0 : 3610
74346-30-8 : 1152
74421-05-9 : 6332
74513-58-9 : 10693
74581-94-5 : 6886
74683-66-2 : 8158
74742-08-8 : 8159
74742-10-2 : 5071
74744-28-8 : 11983
74744-36-8 : 5342
74764-56-0 : 6594
74778-22-6 : 8446
74925-48-7 : 7249
75055-73-1 : 3365
75455-41-3 : 3867
75579-88-3 : 6345
76078-85-8 : 8135
76149-14-9 : 1523
76429-68-0 : 2444
76469-08-4 : 8540
76472-83-8 : 3013
76619-89-1 : 9212
77053-92-0 : 7078
77295-79-5 : 5893
77392-54-2 : 3638
77437-98-0 : 6582
77464-06-3 : 11996
77501-63-4 : 7410
77732-09-3 : 8566
77753-21-0 : 3805
77753-24-3 : 3419
77825-53-7 : 3824
77825-54-8 : 3825
77866-58-1 : 4560
77928-86-0 : 5039
77958-21-5 : 6604
78277-23-3 : 10690
78738-36-0 : 10368
78738-37-1 : 9668
78938-11-1 : 11448
79127-80-3 : 5952
79241-46-6 : 5963
79504-01-1 : 3234
79630-70-9 : 3428
79762-76-8 : 10243
79912-55-3 : 8872
79912-57-5 : 8870
80204-19-9 : 9012
80325-37-7 : 6541
80656-12-8 : 4792
80832-54-8 : 556
80864-10-4 : 8796
81177-02-8 : 6789
81280-12-8 : 6579
81334-34-1 : 7188
81335-37-7 : 7189
81335-77-5 : 7190
81726-82-1 : 6308
81777-89-1 : 4140
81925-81-7 : 6596
82097-50-5 : 11824
82322-93-8 : 4895
82507-04-8 : 7010
82584-73-4 : 8600
82657-04-3 : 2998
82906-78-3 : 9555
83055-99-6 : 516
83682-72-8 : 10360
83863-33-6 : 763
84087-01-4 : 11201
84098-45-3 : 10277
84258-49-1 : 7332
84583-53-9 : 11320
85733-97-7 : 8936
86051-37-8 : 8155
86073-38-3 : 8264
86623-80-5 : 9081
86661-53-2 : 8727
86668-33-9 : 8507
87018-30-2 : 3613
87237-48-7 : 6291
87541-87-5 : 10623
88472-61-1 : 8269
88947-16-4 : 10403
89033-70-5 : 4179
89123-63-7 : 10612
89223-57-4 : 4829
89300-69-6 : 10253
89533-67-5 : 7113
89583-12-0 : 6760
89583-61-9 : 9896
89796-76-9 : 6381
89886-25-9 : 4523
89895-45-4 : 4445
89896-73-1 : 6531
90112-26-8 : 4898
90112-75-7 : 6565
90112-82-6 : 6892
90162-24-6 : 8390
90200-61-6 : 4232
90200-83-2 : 8513
90202-08-7 : 6878
90202-21-4 : 8499
90210-56-3 : 7387
90226-23-6 : 4723
90357-58-7 : 199
90499-41-5 : 1611
90638-48-5 : 10651
90644-60-3 : 8195
90645-54-8 : 4621
90645-61-7 : 4884
90676-55-4 : 8520
90722-14-8 : 8200
90760-75-1 : 4564
90769-70-3 : 4927
90841-14-8 : 3719
90925-48-7 : 3356
90952-10-6 : 6920
90982-32-4 : 4081
91049-43-3 : 8790
91057-79-3 : 4184
91465-08-6 : 5010
91572-15-5 : 5168
91840-99-2 : 3438
91890-87-8 : 3843
91913-99-4 : 3490
91965-23-0 : 3131
92040-00-1 : 7409
92687-41-7 : 4644
92737-91-2 : 6504
93713-40-7 : 10187
93921-42-7 : 4583
93972-93-1 : 6869
94347-45-2 : 8863
94434-64-7 : 8262
94826-08-1 : 4144
94942-89-9 : 2143
95642-49-2 : 8830
95835-77-1 : 2460
95970-07-3 : 2256
95970-10-8 : 2311
96034-00-3 : 3435
96097-69-7 : 11818
96244-13-2 : 4563
96836-97-4 : 4350
97073-23-9 : 10484
97412-23-2 : 9394
97479-78-2 : 3337
97886-45-8 : 5267
97961-66-5 : 7055
98070-91-8 : 3454
98203-04-4 : 11221
98272-34-5 : 6112
98431-97-1 : 11567
98551-14-5 : 4149
98790-22-8 : 4561
98886-44-3 : 6021
98930-89-3 : 11887
98991-54-9 : 6570
99129-21-2 : 4138
99172-57-3 : 6076
99365-26-1 : 2801
99706-68-0 : 11875
99814-65-0 : 8655
99980-00-4 : 4152
100121-73-1 : 6038
100130-25-4 : 43
100224-75-7 : 7627
100295-82-7 : 6945
100295-83-8 : 6947
100296-26-2 : 6499
100314-90-7 : 6135
100319-48-0 : 8339
100367-45-1 : 6697
100791-95-5 : 11189
100983-12-8 : 10210
101144-90-5 : 1667
101257-63-0 : 6743
101257-71-0 : 9712
101257-79-8 : 4195
101427-54-7 : 11609
101567-36-6 : 4336
101654-30-2 : 6741
101715-60-0 : 8331
101870-06-8 : 6688
101935-28-8 : 6333
101935-40-4 : 9243
102153-88-8 : 4831
102370-18-3 : 4526
102449-95-6 : 6747
102851-06-9 : 5983
102871-67-0 : 11304
103095-51-8 : 10813
103275-78-1 : 2219
103697-14-9 : 8230
103985-40-6 : 4665
104035-66-7 : 6593
104068-42-0 : 4787
104177-72-2 : 362
104514-49-0 : 1292
105065-24-5 : 4679
105401-59-0 : 7175
105688-74-2 : 11271
105728-23-2 : 11247
107494-37-1 : 11952
107534-96-3 : 11589
107574-57-2 : 10465
107770-92-3 : 11570
108365-83-9 : 3730
108438-40-0 : 7943
108836-86-8 : 8650
108965-84-0 : 5028
108965-86-2 : 5029
110318-09-7 : 6699
110383-31-8 : 7080
110795-27-2 : 3938
111768-02-6 : 8927
111768-04-8 : 6937
111768-05-9 : 6940
111768-08-2 : 6939
111772-90-8 : 11873
111823-35-9 : 8765
111897-18-8 : 8279
111991-09-4 : 8069
112456-46-9 : 2934
112489-85-7 : 6096
113426-22-5 : 8506
113666-84-1 : 11951
114180-21-1 : 3251
114369-43-6 : 5949
114489-96-2 : 10224
115113-98-9 : 6569
115310-98-0 : 11318
116435-30-4 : 2931
116435-95-1 : 10257
116529-73-8 : 3906
116530-63-3 : 10305
116530-99-5 : 4788
116632-47-4 : 11174
116668-34-9 : 2450
116668-40-7 : 5096
116668-44-1 : 6577
116668-45-2 : 6576
116668-48-5 : 8323
116668-49-6 : 8326
116668-50-9 : 8501
116697-35-9 : 4524
116723-89-8 : 5117
116723-90-1 : 5119
116723-93-4 : 3863
116723-94-5 : 3501
116723-95-6 : 3536
116723-96-7 : 3559
116723-97-8 : 3591
116724-10-8 : 3219
116724-11-9 : 3389
116724-15-3 : 4355
116724-17-5 : 4567
116724-18-6 : 4693
116779-77-2 : 3424
116779-78-3 : 3425
116781-85-2 : 3413
116781-86-3 : 4272
116783-11-0 : 1107
116783-12-1 : 1605
116783-21-2 : 2020
116783-23-4 : 3393
116783-28-9 : 4506
116783-29-0 : 4584
116836-06-7 : 7163
116836-10-3 : 4321
116836-12-5 : 8413
116836-13-6 : 8522
116836-16-9 : 8921
116836-19-2 : 3583
116836-20-5 : 7248
116836-21-6 : 9647
116836-30-7 : 11976
116836-32-9 : 8899
116836-55-6 : 3564
116862-65-8 : 4769
116908-78-2 : 9063
116908-83-9 : 8877
116908-84-0 : 8993
116908-85-1 : 8994
119446-68-3 : 5174
127665-32-1 : 6258

MELTING POINT INDEX OF ORGANIC COMPOUNDS

In this index the temperature in °C is given in boldface. The numbers immediately below a temperature entry are identification numbers (given in the first column of the table *Physical Constants of Organic Compounds*) for compounds whose melting point values begin with the same integer as the temperature entry. Thus, the numbers that appear between temperature entries of 3 and 4, for example, indicate the compounds that have melting points equal to or greater than 3.00°C and less than 4.00°C. The entries following **-8** are compounds melting between -8.00°C and -8.99°C, inclusive. The entries for **0** temperature are compounds with melting points between -0.99 and +0.99°C.

-205
4011

-196
5430

-187
9834

-185
3704
10528

-183
7541

-182
5448
7459

-181
7486

-177
4981

-172
7475

-169
5879
8845

-168
3800

-165
3438
9004

-162
8824

-161
3119
4192

-160
4930
5906

-159
3397
7464
10046
10073

-158
5889
7500

-157
3759
7478

-156
10572
11437
11483

-155
7558

-153
5884
8823
9053
9054

-151
5927
9006

-150
8615
11414

-149
4973
4974

-148
8659

-147
5648
6857
10118

-145
3171
7491

-144
5902

-143
446
4814
5593

-142
4191
4826
4856
6019

-141
7001
7038
7046
7515
7523
10984

-140
4972
5421
6678
8662
9005
9058
10034
10578

-139
5880
6999

-138
3229
3706
4035
4812
4985
5467
5885
6683
9056

-137
3799
7003
7029
8658
8720
10535
10557

-136
442
3145
6536
6537
7504
9033
9772

-135
4903
7502
9040
9052
9055
11378
11462
11489

-134
3378
4193
4804
8763
9029
9031
9057
9059
10534
10536
11409

-133
3709
3801
4803
7000
7481
10421

-132
3164
3413
4861
4862

-131
3242
5918
7114
8825

-130
3117
3152
4860
4975
5470
5898
6611
7040
9862
10157
10541
11392

-129
3255
3804
4813
8709
10697

-128
3290
6695

-127
4007
4955
6864
9035
9044
10037

-126
4426
4928
5938
6775
8495
8712
10530

-125
3897
7042
7120
8660
10441

-124
3372
7044
9030
9036
9037
10334

-123
11
3241
4835
6501
7546
8760

-122
4188
5894
5908
7505
8812
9848
10043
11429

-121
3379
6424
6771
8840

-120
4670
6423
6860
8817

-119
4980
6835
8762
9845
10531

-118
4797
4921
4924
5454
6533
6834
7041
8497
8809
9034

-117
3627
3728
3914
4802
4849
5485
5544
7444
8668
9849
10042

-116
4858
5608
5897
6384
6862
7122

-115
3391
3437
3760
4671
4672
4838
5905
7002
7022
7466
8853
9039

-114
3713
5417
5672
6407
8665
8813
10119
10607

-113
3170
3892
5548
5602
6383
6865
8410
9038
9070
10088
10156
10529

-112
294
3231
3232
3239
3993
4857
4859
4983
5085
8599
8847
9046
10035
10036

-111
3403
3714
4008
4390
4828
5597
5665

-110
279
4219
4848
5525
7496
7555
8494
8846
8990
9136
9836

-109
3812
4660
4971
6422
6534
8850
9123
10558
11546
11695

-108
3146
4658
4778
4818
4902
5900
5923
6168
6542
10072
10116

-107
4497
5540
6554
8409
8848
9773
11527

-106
4428
4815
6494
9071
9871
11973

-105
3222
3705
3727
4816
4917
5489
5606
6440
6498
6861
7507
8721
10154
11432
11463

-104
3219
3252
3382
4834
6830
9135
9906

-103
3381
4626
5534
5640
5644
6829
7116
7483
8531
8940

-102
4982
6387
6681
7036
8842
9679
10151
10727

-101
1631
3410
5168
5253
5531
5532
6866
8491
8532
10054
10089
10388
11672

-100
3707
4935
5541
5596
5726
5912
6405
7179
7693
8493
8849
9159
9892
10646
11787

-99
2169
3548
5233

5484
6010
8142
8719
8810
9893
10533
10573
11451

-98
208
214
215
3289
3515
7545
10580
11549

-97
3402
5204
7477
7499
7581
8592
10074
11971

-96
291
1975
3178
5528
6686
8827
11393
11783

-95
1686
3376
3415
3588
4453
4677
4796
4798
4805
5639
6013
6731
8711
9804
11461

-94
1668
1947
4471
6102
6737
8074
8707
8826
10040
10411
10493

-93
179
236
3118
3636
3668
4774
5466
7442
7462
8496
9067
9808
11398

-92
3113
3369
5055
5538
5987
6018
7121
7448
8862
10090
10698

-91
3479
3564
4494
6004
6130
6502
6774
8690
8878
9897
10117
11696

-90
4170
4370
4918
4966
5230
6355
6492
6543
8814
8942
9122
10055
10499
10504
10636
10958
11465

-89
3681
3916
4248
4431
4439
5607
5641
7115
9837
10214
10259
11760
11785

-88
3593
3594
4366
4851
4906
5882
6713
6843
7687
8710
9521
10081
10298
10522

-87
449
1118
1727
3803
4368
4438
6986
7463
8143
8492
8661
10296
10335
10509
11405
11466
11584

-86
1641
2236
3525
3645
4162
4440
4493
5881
9817

-85
1123
1893
3547
3577
4490
4491
4764
5259
5733
6025
6343
6499
8609
8818
9522
10120
10599
11703

-84
301
1672
3125
3129
4648
5254
5666
5922
8141
8981
9803
10301

-83
186
1414
4495
4749
4846
4934
5935
6500
6527
6732
8075
9938
10583

-82
1161
1472
2037
4167
4771
5638
6505
9130
11506

-81
1738
6503
6609
6712
6856
8182
8766
11788

-80
187
194
3281
3512
4018
4156
4400
5238
5410
5752
5896
5936
6840
6895
7124
8408
8608
8997
9774
10179
10304
10575
11283
11914

-79
1685
3433
4240
4625
6007
6680
8419
8529
10031
10992

-78
121
3528
3631
3907
5126
5414
5419
5539
6528
7498
8132
8913
10152
10195
10729
11410

-77
498
4492
5056
8505
8749
8915
9945

-76
3159
3695
8320
8957
10582
10957
11646

-75
1630
2107
3467
3586
4367
4885
6691
8844
9129
10091
10326

-74
107
1190
1419
3280
3436
3761
4325
4425
4656
5542
5687
6696
6831
8140
8325
11986
11987

-73
209
1995
3253
3579
4597
4823
5084
5100
6015
6417
7550
8916
9100
9823
10263
10318
11691

-72
7180
8111
8747
9109
10181
10571
11916

-71
1988
4443
6898
10111
10310
10635
11740

-70
227
494
1112
3213
3216
4086
4647
4734
4765
5634
5717
5729
5735
5887
6331
6474
6694
6930
7891
8610
8914
9158

-69
1423
4733
4800
5244
5527
5549
6360
8835
8917
8967
9065
9107
11770

-68
1640
3287
4894
5613
5688
5796
6406
6909
11762

-67
1035
1990
3754
5693
6433
6459
6781
6884
8453

-66
1675
2345
4489
5082
5236
5603
5940
6462
7518
8841
9818
10980
11969

-65
264
2095
3265
4486
4999
6658
9785
10584

-64
741
2971
2973
6428
6880
7557
7696
8185
8327
8405
8612
8675
8965
10141
10616
10937

-63
743
993
1149
1989
2038
3920
5517
6471
7554
8481
8482
8882
9826
10956
11769
11970

-62
1163
1626
1671
1687
1726
3211
4750
6009
6089
8603
11550
11756
11967

-61
1763
2164
2925
3745
3901
5522
5718
6497
8530
9821
9943
10673
11972

-60
696
860
1627
1949
2036
3137
3417
3617
4200
5413
5535
5992
6908
8177
8673
9982
10715
11764
11797

-59
1060
1483
4650
4751
5709
6990
9108
10173
11707

-58
3398
4886
9839
11796

-57
21
781
992
1098
1459
3527
4029
4316
4465
6335
7154
7467
7684
8359
10265
11143

-56
1241
1728
3121
3992
4244
4315
4342
5537
6356
6466
6718
8355
8742
8858
9563
9851

11130

-55
1038
1162
1334
1669
2972
3566
5516
5911
6890
6959
6960
8356
8372
8700
9148
9154
9881
11492
11544

-54
3879
4801
6844
8378
9928
10986

-53
1674
2935
3368
4703
4732
4820
5892
7281
8120
10291
11913

-52
1026
1404
1437
1466
1930
3657
7125
7161
7492
10409

-51
166
233
2039
2724
3186
3802
4878
5611
6418
7250
8241
8699
8821
8831
8843
9556
10097
10743
11387

-50
193
1464
2924
3223
4197
5266
5425
5678
5714
5723
6470
6495
8196
8793
9941
10264
10315
11002
11658

-49
1632
1670
3209
3739
4249
4365
4496
5360
5361
5427
5895
8418
9141
9855
10800
11368
11754
11966

-48
1745
2024
3428
3751
4735
5043
5054
5213
6419
6465
11020
11132
11373
11968

-47
1184
1422
1453
1932
2091
3640
4358
6907
6911
9909
10634
10663

-46
2271
8441
8881
9564
10106
10148
10290
11028

-45
636
1131
1182
1724
2172
3571
5186
5612
5728
5815
7520
7655
8131
8285
8340
8406
8411
8445
8966
10203
10643
10985
11642

-44
559
1025
1643
1715
2359
3114
3226
4499
4843
5097
5193
5565
6685
6759
7890
8976
10420
10943
11585

-43
271
740
1037
1109
1166
2358
3522
3999
4036
4199
4339
4414
4419
4996
4997
5770
6142
6344
6436
7789
7996
8428
8674
10555
10917
12066

-42
288
1415
1717
3847
3864
5635
7790
7889
8427
8775
8815
8901
9227
9918
10149
10313

-41
1465
1719
1739
4369
5051
5590
5642
6475
6514
6874
7525
8888
10872
10944

-40
1342
1629
2165
3203
4194
4225
4598
4896
5560
5872
6323
6802
7998
8245
8434
8446
8600
9153
10105
10656
10875
11246
11389
11645
11748
11767

-39
1055
1722
3592
5497
6014
6511
8123
8741
8958

-38
222
445
695
1981
4901
5888
6041
6766
8147
8684
8837
10058
11606
11724
11977

-37
1647
1693
1915
2093
3279
4872
5232
7051
10130
10563
11579
11879

-36
849
1716
1740
2067
3116
3177
4371
4461
5656
5663
8150
8836
9915
10416

-35
192
1139
1183
1339
1723
1725
3569
3813
4769
4953
5052
5335
5529
6453
6510
6620
7979
8158
8228
8229
8487
8624
10586
11131

-34
752
1628
1635
1642
1931
2602
4454
5078
6889
7274
7516
7524
8144
8393
8439
8740
9882
11747

-33
745
1242
1747
1974
2021
3973
4364
6464
6472
6512
8595
8756
8855
8889
9894
10942
11657
11667

-32
138
684
1138
1933
2989
3898
4908
4979
5689
6784
8310

-31
232
779
1078
1256
1413
1766
3514
4617
5036
5571
6146
6392
8444

-30
998
1624
1815
1832
1973
2016
2042
2898
4405
5558
5655
7791
7915
8292
8353
8431
8687
9553
9638
9732
9799
9942
9973
10140
10604
11985

-29
463
899
982
1649
2092
2094
2163
3126
3188
3301
3539
3675
4345
4485
4841
5033
5034
5653
8121
8767
10810
11592

-28
1764
2015
2348
2909
3870
5617
5708
6577
6904
7061
7522
11453
11798
11878

-27
976
1054
1592
1655
1720
1779
2162
2168
2788
3073
3574
4571
4992
5550
8115
8156
8689
9983
10052
10466
10876
11574
11978

-26
220
525
1168
1178
1379
1828
2070
3895
4034
4239
5058
5520
5521
5727
5913
8298
8605
8733
10269
11468
11743
11746
11792

-25
25
452
1175
1452
1478
2283
2357
4899
5260
5721
8988
9863
11467
11938
11984

-24
1355
1458
1696
1992
2064
2304
2885
3268
3457
4341
5939
6394
8772
9681
11573

-23
945
1148
1424
1972
2025
2051
2279
2764
2903
4472
4951
7997
8798
8944
11112
11185
11415
11542

-22
152
154
158
295
1059
2532
3261
3936
4430
4790
5574

5917
6468
6723
7083
7540
7941
9073
11281

-21
130
1005
1174
1792
2442
3163
3303
4711
4713
4855
5245
5742
6296
6395
6812
9820
9824
9870
9919
10972

-20
718
1986
2302
3217
4392
4460
4580
5063
5201
7177
7339
7493
7865
8145
8488
8954
9246
9853
9907
10168
10201
10354
10396
11242
11728
12057

-19
462
503
1114
1765
2913
3007
3444
3779
3798
4716
5343
6309
6561
6724
6785
7746
8068
8117
8307
10108

-18
55
1460
2523
2960
3445
3937
4213
4429
4784
5066
5810
6150
6349
6662
7900
8133
8137
8437
8543
8597
8787
10981
11421
11486

-17
253
443
994
1018
1044
1132
1354
1382
1677
1681
3521
4212
4283
4466
4864
5057
5328
5622
8569
10443
11742
11934

-16
12
202
793
1654
1673
3267
5053
5456
5585
6637
6675
7822
7920
8091
8209
8471
9846
9910
10938
11394
11791

-15
157
263
1384
1712
1836
3414
3941
4723
5477
6780
6787
7701
8152
8863
10404
10481
11946

-14
656
705
778
1258
2289
3138
3506
3618
3960
4003
5904
6118
7469
8443
9416
9874
10175
10537
11214

-13
1388
1710
1874
2710
2740
2763
2995
3678
3774
4464
4596
5686
5931
6313
6727
7176
7823
7885
8955
9630
9631
9697
11893

-12
796
1004
1192
1692
1694
2108
2687
4752
5569
6456
6849
6992
6993
7293
8223
8932
10463
10798
11419
11636

-11
1246
4572
4601
5072
5200
5652
7828
8287
9629
10940

-10
561
659
1173
1303
1644
2000
3247
3319
4326
4457
4863
5001
5044
5312
5753
5766
6806
7917
7934
8430
8474
8807
11023

-9
547
1047
1306
1977
3628
4535
4766
5077
5292
5293
5295
5344
5555
6994
7893
8839
8897
8986
10155

-8
188
1656
2098
2634
3494
3776
4221
4721
6748
7155
7814
7895
7957
8173
8203
10855

-7
564
700
807
1031
1289
2373
4242
4781
5713
6176
6455
7556
7886
8172
8416
9357
9389
10518
10950
10973

-6
633
1141
1240
1733
2018
2278
3072
3835
4166
4360
4606
5110
5304
5756
5891
6564
6716
7129
7208
7819
7992
8598
9321
9456
9457
9703
9925
10177
10592
10939
10941
10976
11744

-5
1567
1620
2347
2374
2614
2918
3321
3460
4533
4809
5600
6805
7474
8157
9687
11871

-4
2333
2890
3929
5030
7479
7681
8187
9354
10467
10588
11070
11639
11849

-3
4792
5995
6870
8175
8205
9186

-2
742
859
1450
2061
2100
2589
3056
3173
3468
4347
4532
4742
5242
5412
7543
7747
7780
7829
7942
7986
7993
8225
8929
9586
9680
9957
10180
11055
11705
11935

-1
175
658
767
1053
1094
1217
1364
2123
2640
2790
2904
2955
3359
3425
3595
3919
3921
4174
4359
4624
5003
5049
5071
5811
5945
6040
6757
6811
7245
7253
7901
7937
7958
8001
8119
8293
8385
8388
8424
8591
8653
8983
9708
10440
11003
11508
11635
11784
11868
11942
11953
11979

0
178
363
754
1022
1113
1140
1171
1238
2886
2956
3016
3082
3248
3447
3507
3775
3909
4000
4214
4284
4344
4447
4534
5271
5302
5338
5339
5354
5619
6669
6670
6782
7691
7818
8134
8560
9491
9608
9661
10159
10338
10469
10877
11277
11690
11720
12030

1
671
694
737
948
1050
1090
1105
1301
1467
1609
1612
1662
2079
2175
2296
2767
2861
2910
3316
3750
3766
3922
4179
4348
4632
4768
5076
5270
5273
5988
6565
6777
7892
7896
8053
8148
8315
8400
8556
9281
9349
9415
9548
10446
10478
10901
11015
11128
11610
11649
11697
11749
11789
11961
11988

2
713
1032
1097
1387
1639
2754
2897
3083
3452
3793
5045
5061
5198
5677
5830
5877
5963
5989
6053
6667
6875
7315
7487
7632

7902
8188
8538
8791
11263
11698
11897
11960
11976

3
1566
1914
2272
2787
5184
10982
11057

4
571
2227
2776
2777
2988
3924
5303
8116
9693
11615

5
173
856
867
1304
1305
1343
1757
2049
2643
3288
3619
4854
5315
6310
7497
7883
8161
8302
8387
9228
9376
9385
10737
11613
11614

6
766
1383
2014
2121
2430
3092
3508
4305
4912
5070
7178
7506
7739
8312
8902
9583
9706
9783
11120
11640
11688
11872
11882
11981

7
668
1028
1029
1185
1209
1224
1248
1288
1297
1604
2671
2878
2946
3274
3347
4229
4531
5041
5780
5955
6098
7652
7820
9461
10468
11621

8
1229
1230
1286
1371
1636
2069
3043
3333
3335
4513
5275
5318
5697
5842
6002
7548
7931
8046
8540
9390
9414
10364
10788
10902

9
549
709
989
1021
1046
1360
1426
1479
2880
2899
2978
3057
4346
4850
5062
5518
6078
6769
6792
7465
8639
8642
8735
8758
9251
10738
11008
11279
11777
11874
11880

10
1085
1935
2249
2269
2342
3665
4202
4541
5240
5326
5429
5513
5675
5870
5875
6101
6295
6396
6463
6636
6786
7292
7381
7551
7994
8440
8656
9889
10066
10138
10552
10685
11604
11701
11886
11945

11
644
707
950
1346
1772
2004
2268
2429
2887
4168
4386
4516
4518
4724
5187
5298
5496
6322
6477
9529
9569
9596
10049
10436
11171

12
9
97
212
529
1233
1312
1407
1817
2412
3348
3672
3676
4323
4382
4714
5073
5088
5499
5800
6324
6665
7050
7738
7932
7938
8013
8146
8402
8649
9077
9401
9534
10337
10609
11004
11006
11624
11908
11951
11959

13
162
216
554
1052
1104
1454
1979
2171
2495
2966
3609
4403
4509
4775
5228
5645
5868
5962
6623
6654
6905
7212
7544
7712
8012
8294
8308
8969
9361
9702
10197
10314
10368
11183
11217

14
250
597
599
755
806
1746
2219
2242
2770
2863
3058
4357
4511
4736
5075
5621
5662
5758
5998
6692
7682
7921
8011
8206
9578
9694
9926
11869
11962

15
339
464
706
1151
1318
1361
1474
1476
2001
2019
2301
2975
3338
3815
4299
4432
5040
5491
5874
6673
6703
7561
7562
7793
8043
8049
8369
8412
9340
9375
9410
9458
9614
9730
10100
10806
10869
11136
11950
11958

16
100
109
563
680
1187
1912
2161
2266
2322
2892
2893
3024
3481
3482
4468
4519
5869
6526
6638
6647
7057
8422
9185
9360
9377
9430
9700
9975
10163
10266
10648
10908
11127
11278
11362

17
530
575
715
1934
2068
2267
3026
3337
3644
4638
4844
4993
5032
5220
5313
5337
5808
5855
6190
6634
7476
7529
8204
8291
9628
9662
10132
11937
11952

18
275
639
962
1079
1430
1449
2419
2769
3190
3307
3896
4361
4484
4787
4958
5089
5138
5570
5754
5961
6153
6632
6633
6639
7322
8047
8151
8539
8646
8866
8876
9226
9353
9525
10186
10286
10480
11637

19
1023
1462
1825
1833
2681
2900
3050
3318
3598
3650
4904
5274
5322
5768
6643
8207
8541
8643
8799
9250
11056
11280
11599
11619
11625
11795
11975

20
596
598
831
977
978
1431
1799
1896
2402
2421
2596
2994
3346
3366
3880
3908
5250
5340
5751
5774
5789
5802
5864
5920
6297
6657
6674
7169
7530
7532
7794
7988
8162
10323
11184
11296
11627
11638
11842
11887

21
478
526
552
1062
1136
1158
1735
1942
1944
2012
2361
2541
2720
3487
3942
4822
5325
5826
6072
6693
7685
8015
9454

22
538
875
956
1218
1380
1436
1860
1876
2089
2111
3060
3313
4236
5074
5314
6306
6399
8159
8275
8313
8480
9248
9315
9325
9463
10129
10444
10587
10922
11049
11075
11262

23
735
848
1759

1880
4508
4577
5321
5364
6529
8281
8754
9444
10596
10615
10903
11846
11884

24

39
685
949
1130
1197
1211
1410
1440
1444
1600
1777
2150
2503
2760
2862
2901
2976
3236
4475
4512
5060
5317
5363
5365
5637
6640
6650
7171
7282
7944
8640
8890
9076
9308
9425
10322
10452
10639
11219
11974

25

127
247
1015
1213
1344
1385
1401
1718
1891
1961
2277
2798
2894
3792
4500
4530
4739
5323
7201
7689
7815
8265
9570
9593
9673
10204
10341
10389
10891
11058
11180
11954

26

647
649
672
676
940
1011
1386
1729
2071
2154
2290
2455
2606
3078
4538
5091
5500
7373
7880
7918
8619
8680
9411
9462
10051
10085
10191
10424
10742
11052
11157
11275
11434

27

96
666
1120
1296
1427
1562
1731
1732
1751
2639
3258
3573
3943
4510
4866
5453
5856
6057
6983
7123
7702
7879
8251
8343
8655
9256
9333
9347
11259
11786
11817
11828
11870
11949

28

72
258
1056
1068
1135
1159
1250
1381
1547
1561
1902
1964
2232
2672
4205
4567
5038
5291
5362
5463
5730
5853
5863
6315
6340
6371
6800
7333
8163
8234
8235
8236
8267
8268
8273
9294
9711
10352
10438
11333
11570
11780
11943

29

502
669
744
772
947
1237
1564
1585
1637
1645
1657
1695
1916
1976
2007
2066
2423
2638
2765
2869
2948
3040
3059
3080
3259
4463
4505
4740
5816
6307
6791
7117
7552
7699
9348
9350
9378
9404
9717
9814
10048
10131
10439
10623
10881
11054
11790
11823

30

1
46
673
808
1119
1176
1302
1475
1890
1907
2073
2246
2413
2440
2707
3667
4842
5146
5147
5188
5324
6134
6316
6653
6707
6709
7150
7571
7943
8240
8244
8270
8314
8391
8504
9146
9156
9395
10865
11072
11605
11782
11889

31

112
505
774
810
999
1134
1405
1563
2170
2877
3556
4332
4529
4741
4806
5059
5300
6044
6352
6458
6641
8454
8638
8896
9332
9382
9400
9913
10228
11035
11236
11241
11567
11583
11632
11694
11883
11890
11994

32

119
140
506
578
638
968
1007
1214
1375
1743
1814
1970
2040
2060
2399
2525
2884
3422
3612
4160
4217
4218
4515
4542
5261
5307
5865
7247
7345
7577
7736
7770
7783
8101
9252
9379
9699
9911
9978
10789
11071
11885

33

652
760
838
868
942
2698
2992
3112
3293
3567
4306
4411
5048
5136
5755
5799
5817
6092
6403
6668
6879
7751
7756
8010
8054
8257
8864
8903
9187
9229
9396
10378
10622
10866
10951
11633
11891

34

221
262
517
839
919
987
1080
1084
1269
1333
1396
1598
1826
2151
2231
2756
3023
3305
3816
4291
4412
4522
4725
5281
5283
5838
5857
6056
6094
6151
6312
6949
7755
7916
8041
8242
8883
8968
9150
9171
9334
9402
9496
9763
10417
10704
10705
11005
11012
11220
11238
11631
11634

35

189
594
663
811
855
907
1014
1064
1082
1293
1586
1768
1811
2706
2803
3071
3490
4148
4329
5284
6046
6628
6876
7564
8299
8354
8395
8429
8647
9078
9374
9541
10233
10258
10433
10945
11475

36

595
972
1280
1325
1358
1663
1780
1857
2137
2317
2799
2905
3046
3296
3568
4301
4506
4867
5211
5350
5366
5798
6258
6304
6655
7049
7409
7521
7807
8311
9417
9424
9551
9726
10640
10686
10716
10835
11894
11947

37

26
580
583
835
1073
1204
1357
1827
2063
2417
2622
3085
4082
4307
4476
4569
4875
5194
6778
7244
7308
7376
7714
7748
7884
8107
8262
8641
9151
9306
9442
9459
9985
10422
10717
10838
10840
11856

38

642
804
857
1212
1595
1784
1897
1936
2152
2185
2387
2682
3330
4136
4206
4404
4540
4548
4637
5046
5068
5280
5351
5841
6308
7138
7451

7495
7731
8269
8398
8426
8648
8920
9110
9182
9237
9276
10585
11237
11490
11540
11630
11895

39
260
688
1033
1222
1844
2005
2504
2680
2745
2940
3878
4590
4646
5118
5119
5552
5990
7205
7645
8266
8651
9220
9295
9719
10690
10988
11282

40
74
219
322
762
874
1057
1257
1295
1428
2062
2737
3133
3663
3912
3994
4203
4204
4375
4668
4807
4913
4932
5171
5183
5615
5636
6294
6469
7622
7887
8259
8264
9199
9254
9787
10007
11499

41
290
487
539
627
1003
1030
1071
1734
1752
1788
2441
2748
2954
3020
3095
3658
3882
3939
3978
4459
4553
4959
5135
5195
5296
5306
6374
6749
6796
7374
7623
7686
7951
7956
8243
8290
9269
9584
9658
10216
10219
10271
10517
10868
11274
11753
11881

42
1070
1126
1205
1321
1525
1704
2510
2888
3052
3980
4089
4819
5745
5964
7286
7600
7638
8585
8625
8657
9253
9322
10482
10683
10906
11245
11298
11865
11899

43
490
545
591
780
892
1075
1319
1392
2113
2303
2334
2343
2463
2543
2578
3013
3014
3350
3842
4090
4356
4666
5305
5559
5806
5814
6382
6663
6799
7100
8652
8977
9106
9371
9992
9993
11186
11240
11710

44
147
454
536
570
1215
1337
1350
1406
1568
2308
2363
2626
2768
2939
3471
3530
3903
3958
5268
5288
5710
6660
7309
7624
8135
8253
8295
8394
8581
8922
9221
9435
10384
10445
10691
11521

45
17
47
129
259
661
834
1058
1251
1314
1767
1864
2330
2335
2337
2619
3351
3411
3423
3529
3837
3845
4155
4267
4456
4502
4521
4808
5241
5554
5578
7321
7947
8136
8297
8685
9262
9283
9289
9310
9317
9355
9946
10393
10408
10426
10824
11122
11239
11665
11864

46
124
546
648
691
946
1177
1206
1616
2138
2291
2747
2782
4700
5265
5974
5999
6631
7148
7303
7441
7512
7553
7946
8587
8830
9324
9331
9489
9514
9696
10282
10342
10414
10432
11034
11053
11114
11775

47
19
531
773
825
897
1074
2022
2114
2259
2601
2708
2771
2791
5246
5327
5953
6357
6869
7615
7616
9606
10067
10288
10710
10807
10818
11855
11892

48
42
693
763
1390
1504
1536
1741
1966
2354
2549
2941
3187
3583
3977
4384
4433
6298
6320
7621
7785
8246
8316
9270
9271
9362
9517
9603
9822
9953
10209
10418
11195
11247
11643

49
88
159
488
827
1125
2287
2675
3084
3298
3419
3531
3926
3948
3967
5010
5025
5064
5771
5839
5947
6305
6454
6659
6689
7380
7708
8284
9155
9418
9450
9634
11276
11509
11901

50
111
116
345
687
1008
1203
2461
2542
3199
3306
3363
3911
3945
4243
5277
5356
5519
5781
5782
5827
6278
6461
6661
7157
7187
7231
8300
8483
9137
9232
9566
10245
10406
10462
10475
10513
11616
11906

51
28
556
708
1108
1259
1273
1287
1565
1789
1871
2002
2078
2158
2245
2321
2381
2482
2741
2921
3010
3096
4349
4566
5239
5779
5805
5835
6652
7698
7720
7750
8052
8279
8280
8795
9223
9266
9329
9330
9432
9976
10625
10645
10792
10801
10882
11225
11774

52
4
43
544
667
690
795
1099
1100
1587
1878
1959
1969
2147
2166
2179
2474
2516
4165
4539
5009
5750
5791
5823
6091
6950
7263
7294
7716
8233
8271
8365
8489
8645
8768
9597
10373
10823
11156
11218
11289

53
770
836
1144
1146
1270
1356
1730
1776
2420
3011
3742
3935
3965
5581
5946
5952
6317
6321
7510
7547
9181
9441
9920
10977
11551
11623
11909

54
210
479
840
1516
1783
1929
1937
2097
2341
2355
2661
2742
3029
3906
3953
4380
5115
5262
5278
5353
5357
5563
6096
6318
6651
7284
7572
7903
8238

8637
8776
10193
10483
10512
11335
11603
11622

55
62
75
645
986
1338
1963
2744
3093
4263
4570
5114
5192
5620
6642
7328
7549
7663
7710
7740
7945
8009
9265
9722
10831

56
577
771
975
1076
1395
1678
1839
2058
2110
2248
2422
2628
3356
3362
4554
4821
5137
5214
5311
5616
5769
5790
5858
5914
6622
6635
7210
7242
7586
7641
7742
7743
8045
8224
8484
8635
9160
9219
9236
9239
9240
9260
9277
9305
10780
10837
11099
11231
11595
11628

57
579
581
768
1172
1434
1540
1790
2099
2127
2186
2265
2286
2339
2709
3130
3291
4127
4483
4615
4892
5409
5582
5760
5787
6346
6962
7606
8044
8106
8258
8278
8870
9138
9381
9383
10008
10857
11269
11400
11560
11717
11944

58
475
535
592
765
910
943
1252
1307
1593
1804
1960
2427
2488
2723
2944
3329
3336
3498
3984
5916
6161
6271
7035
7285
7449
7781
7906
7950
7978
8296
8803
9279
9282
9292
9483
10006
10827
11155
11336

59
243
650
1067
1207
1397
1402
1438
1543
1853
2088
2515
2793
3088
3408
3633
3768
4211
4474
5434
6302
6664
7287
7305
7596
7633
7766
7806
7907
7949
8399
9284
10435
10708
10709
11243
11654
11818

60
420
576
719
939
1010
1548
1650
1664
1821
2050
2072
2130
2212
2481
2579
3576
3590
3763
3893
3969
3996
4153
4210
4940
5008
5116
5812
5866
6291
6797
7163
7239
7316
7640
8260
8348
8682
8998
9504
9695
10434
10474
11257
11875
11905

61
87
450
1391
1796
1885
1899
2054
2281
2295
2332
2493
2750
2755
2762
2953
3101
3432
3820
3821
4300
4759
5282
6314
6319
6891
7174
7313
7625
7894
7929
8104
9431
9675
9970
10004
10115
10192
10476
11244
11256
11258
11290
11608
11989

62
568
925
928
1311
1539
1714
1750
2122
2323
2592
2933
3086
3103
3365
4381
4462
5258
5813
5996
6283
6644
7237
7513
7752
7801
7808
8286
8792
9179
9222
9356
9412
9482
9619
10020
10205
11134
11804

63
65
123
686
930
966
1000
1275
4552
4761
5225
5310
5359
6625
7642
7811
8071
8099
8274
8389
9242
9268
9312
9344
10618
11303
11341
11572
11900

64
7
489
557
1042
1279
1508
1615
1742
2043
2056
2285
2505
2627
2688
2872
3844
3963
4302
4410
5002
5378
6646
6788
7629
7908
8220
8463
9164
9241
9659
10009
10010
10612
10858
11158
11334
11752

65
77
224
537
1013
1101
1313
1394
1528
2153
2214
2834
4853
5264
5267
5358
7620
7824
8105
8181
9180
9311
10332
10486
10712
10994
11318
11431
11620
11659
11819
11948

66
150
509
640
830
934
1210
1261
1283
1341
1958
2065
2382
3049
4280
4282
5215
5859
6656
7741
7749
7798
7882
7911
9230
9352
9488
9914
10238
11182

67
101
195
528
761
1274
1347
1366
1457
1534
1537
1613
1614
2059
2220
2228
2356
3027
3198
3959
5129
5579
6299
6717
6765
7603
7809
8263
9267
9701
10026
11145
11248
11867

68
54
422
858
1006
1363
1519
1527
2252
2406
2600
2615
3008
3098
3102
3985
4458
4501
4936
5105
5785
5797
6026
6710
7382
7730
7975
8082
9233
9285
9286
9287
11254
11302

69
493
847
957
1300
1829
1862
1910
1946
2006
2176
2998
3846
3983
4266
5348
5970
6095
7224
7626
8103
8247
9445
9452
9484
9485
9663
9674
9789
11444
11547

70
5
567
660
1096
1298
1451
1850
1965
2314
3003
3791
5012
6435
7711
7977
8042
8081
8432
9231
9386
9423
9470
9916
10019
10023
10027
10199
10815
11206
11331
11857

71
822
1208
1272
1362
1495
1927
1956
2183
2250
2362
2683
2700
2716
2792
3048
3814
4482

5130
5139
5248
5933
7439
7925
8282
9224
9326
9999
10610
10932
11160
11721

72
543
551
689
1294
1393
1408
1447
1544
1648
2044
2338
2584
2820
3070
3099
3946
4106
4264
4599
5279
6649
7511
7639
7774
8249
9307
9338
9494
9515
11161
11552
11650

73
496
497
852
887
1093
1223
1292
1345
1848
1883
2055
2292
2636
3074
3120
3357
3780
4158
4520
5572
5626
5680
5793
5993
7149
7914
8039
9293
9447
9495
9547
10127
10677
11133
11212
11260
11321
11569
11996

74
550
653
679
969
1906
2620
2896
3039
3075
3472
4383
4845
5047
5368
5508
5929
6594
7246
7347
8255
8580
8960
9309
9370
9542
10689
11284
11328

75
510
920
1299
1477
1849
1903
2507
2870
2906
4075
7324
7349
7690
7775
8064
8065
8067
8250
8999
9003
9797
11308
11340
11692
12002
12032

76
703
883
1264
1448
1900
1983
2244
2757
2835
2952
3230
3361
5174
5355
5775
5801
6991
8059
8584
8874
10196
10696
10817
11876

77
555
582
698
900
984
1851
2331
3111
3140
5154
5633
6303
7291
7604
7776
9319
9487
9571
9771
10630
10999
11853

78
427
854
1066
1072
1535
2211
2711
2951
3025
3089
3917
4377
4479
4544
6293
7650
7863
8048
9190
9235
9274
9297
10000
10016
11115
11148
11776
11803
11809

79
23
50
59
196
513
681
797
990
1282
1840
1938
2075
2280
2403
7238
7364
7585
7637
7724
7930
8256
8676
9336
10170
10477
10711
10713
11009
11810
12003

80
22
683
723
1526
2048
2124
2181
4102
4105
4413
4760
6219
6266
6347
6626
7276
7732
9393
9474
9523
11030
11827

81
425
574
651
2526
3028
4309
5276
5301
5767
5860
5861
5941
5978
6975
7563
7601
7909
7926
7928
7976
8008
9258
9278
9391
9446
9493
9791
10377
10812
11758

82
200
512
620
702
1216
1549
1550
1971
2057
2312
2336
4507
7602
7821
9327
10247
10278
10947

83
374
377
641
674
757
1012
1069
1244
1320
1622
1837
1913
2210
2548
2644
3822
4001
4151
4279
4709
5547
6043
6300
6701
7203
7737
7754
7802
8222
9290
9481
9741
9988
11612
11675

84
378
729
965
1308
1892
2180
2326
5247
5975
8003
8022
8100
8633
9455
9480
9740
10514
10519
11081
11348
11538

85
238
423
1310
1463
1590
2247
2360
2997
3094
3364
3540
3834
4545
5133
5377
6188
7271
7317
7384
7773
7804
8030
9473
10062
10627
10830
11618
11664
11918
12028

86
342
534
881
901
953
1326
1877
2084
2182
2221
2465
3207
3634
3968
5163
5837
6229
7570
8796
9264
9318
9405
10617
10628
11288
11300

87
13
70
682
699
922
995
1290
1511
1863
1882
2307
2329
2860
2914
3143
3497
3855
5131
7447
7612
7912
7935
8014
8777
8779
9291
9769
9801
10722
11043
11596
11607

88
10
341
1145
2313
2743
3974
6624
6801
7423
7735
10774
10813
11345
11821

89
1276
1322
1369
1374
1446
1601
3091
3987
7192
7379
7812
9000
9384
9449
10391
11680
11848

90
717
1492
1505
1797
2144
2580
2867
3081
3131
3308
3315
3510
3966
4093
4610
5191
5820
5833
6030
6648
7617
8221
8254
8565
9599
9728
10001
10429
10871
10946
11019
11123
11702
12059

91
3
507
607
786
1798
2350
2833
3018
5255
6048
6100
6630
7318
7588
7960
9387
9917
11000
11273

92
720
784
799
1048
1127
1523
1524
1888
2497
4579
6143
7283
7539
7613
9244
9259
9490
9528
10200
11445
11676
12014

93
6
926
1957
2263
2310
3030
3063
3786
3955
4147
4281
4336
5120
7415
9373
10794
11251
11662
11820
12058

94
80
81
1225
1353
1923
1982
2262
2437
2593
3044
3132
3189
4039
4135
5190
5381
5584
5783
5980
7723
10523
10670
11121
11226
5715
6789
7314
7778
7827
7840
8351
8573
8770
9617
9895
10320
10407
10878
10898
10961
11252

95
611
701
983
1679
2378
2479
2645
2749
3701
5840
6722
7119
7290
7576
7635
7656
7725
7787
8017
9205
9238
9339
9479
10781
11223
11322

96
98
739
758
802
853
1291
1908
2483
3322
3762
4076
4592
5778
5831
6097
6348
7599
9422
9436
9922
11196

97
15
1926
2306
2383
2652
2653
3208
3400
5674
5715
6789
7314
7778
7827
7840
8351
8573
8770
9617
9895
10320
10407
10878
10898
10961
11252

98
223
228
482
937
1445
1518
1588
2133
2607
2759
5181
5772
7514
7795
9323
10811
11272
11826
12015

99
18
51
334
457
504
911
1278
2178
2253
2264
2370
3015
5000
5102
5167
5199
5649
5854
6995
7509
7788
8118
8567
8875
9163
10430
11339
11651

100
533
988
1520
2758
2840
2871
3430
3493
4152
4372
5005
7377
7438
9280
10194
10629
10867
12049

101
1024
2664
2938
5561
5618
5784
5829
6721
7112
8025
8063
8276
9183
9486
11213
11920

102
84
798
1129
1260
2149
2462
4109
5786
6189
7355
9406
11589

103
36
57
79
338
1502
2305
2676
3076
3778
6023
7729
7948
8579
8773
11163
11224
11541

104
1497
2047
2229
2845
5177
6871
7387
7745
8034
8566
9301
9654
9929
11343
11594
11611
12017

105
801
1530
1546
1924
2311
4373
4376
4989
5222
7307
7649
7825
7940
9157
9800
9979
10021
10782
11173
11191
11708
11993
12012

106
927
1277
1309
2766
4077
4786
4998
5369
5773
5862
6393
8130
8698
9152
9602
9665
9718
10870
11215
12004
12018

107
364
593
1456
1531
1872
2126
2665
3136
6629
7197
7424
7810
7864
8021
8026
9188
10828
11031
11033

108
45
63
312
540
565
850
917
1494
1507
2328
2912
4134
5031
7141
7534
7575
7728
8349
9343
9426
9538
12048

109
486
996
1378
1498
2459
2677
2819
3064
3979
5015
5433
5832
6060
7272
7919
8035
8231
8564
8568
9434
10779
10845
11715

110
76
326
601
1461
2837
4104
5164
5289
5966
7386
7939
8005
9987
10924
11204
11291

111
68
376
654
882
1503
3104
4083
4270
5777
6183
6206
7394
7786
11141

112
2046
2315
4409
4528
5263
6083
7111
7361
7569
7674
7826
8031
9968
10926
10933
12001

113
569
823
938
1196
1510
1843
2256
2714
2970
3067
3079
4125
5968
6216
7240
7709
7800
7841
8683
9437
9467

114
86
106
153
572
2469
2473
2476
2662
2746
3021
3087
3194
3342
3852
5967
7566
8040
9609
9611
9948
10847
11205

115
1154
1886
2284
2567
2611
2864
3325
4118
4257
5367
5643
5824
7299
7726
7753
8781
9257
9342
9471
9627
10319
10833
10919
11079
11253
11822

116
803
888
914
2349
2582
4265
5212
5217
6003
7367
8023
9469
9472
9971
10442
10814
10848
11350

117
69
566
721
967
1323
1435
2621
2689
4573
7167
7241
7488
9189
9320
9358
9359
10250
10819
11116
11781

118
361
521
775
902
915
1284
1491
1506
1925
2276
2352
2464
2718
2962
3042
3195
3426
3925
5382
5696
5948
7302
8002
8248
9195
9372
9475
9610
9796
10821
11193
11250

119
172
889
1922
2085
2118
2753
3019
3988
4097
6065
6093
6198
7193
9234
9544
9655
9770
10862
11836
11911

120
141
359
1529
2655
3031
3989
5104
5157
6000
7358
7587
7646
7796
9433
10775
10923
11199
11249

121
35
41
49
94
315
2401
3320
3431
4064
4793
5165
5776
5825
7304
7323
7343
7560
8018
9677
10925

11539

122
168
520
792
851
955
1541
2309
2365
2475
5575
6051
9453
9497
10448
10846
12006

123
225
1509
4909
5160
6177
7194
7643
7834
9202
9218
9743
11661
12019

124
921
1619
1884
2174
2604
2727
3066
7296
7591
7630
7688
8016
9761
11310

125
622
1532
2316
3061
4143
5352
5625
5949
5976
6205
7301
7764
7868
7974
8261
9215
9346
9757
9835
10333
10678
10968
10995

126
728
1514
2572
2603
3062
3349
6200
7627
7658
8060
8219
8239
9656
10844
11152
11995

127
600
677
837
1077
2642
2917
6207
6270
7392
7671
9296
9939
11042
11553

128
337
351
724
2207
2237
2396
2836
5299
5435
5741
5828
5972
6208
7146
7327
7372
7843
9443
10520
10849
11082
11683
11811

129
56
240
285
876
1756
2369
2875
3193
4115
9798
10005
10329
11135
11737

130
514
573
2134
2650
2732
3765
5042
5834
7342
7608
8277
10890
11037
11221
11311

131
71
777
818
896
1429
2397
2637
3034
3854
5112
5385
6136
6209
7158
7227
7264
7803
9217
9498
9632
9991
11001
11080
11202

132
241
305
630
905
1432
2410
2574
2581
2859
2919
2920
3317
4111
5216
7357
7425
7744
9464
9664
9766
10793
11286
11517

133
89
615
800
2366
2398
2729
3703
6050
6180
8232
9196
9272
9465
10682
11208
11992

134
461
903
2184
2405
3012
4084
5251
6217
6698
7199
7365
9345
10851
11587

135
2
541
826
2480
4071
4154
4756
4910
5219
7139
7151
7734
8027
9921
9960

136
373
617
657
1513
1939
2273
2371
2641
3254
5149
5347
7233
7338
7634
7910
9210
9512
10684
10886
10893
11198

137
60
731
2188
2597
3033
3206
5374
5822
7375
7578
7584
8574
9249
9337
10790
10861

138
518
2586
2882
5950
7202
7265
7310
7636
7836
8036
8782
9162
10842
11101
11738

139
532
1489
2260
2868
3205
5256
6235
7595
7862
9212
10747
11085
11548

140
169
402
524
629
916
1332
2367
2530
2534
2553
5019
7411
7536
7813
7959
8550
9204
9476
9990
10413
10884
11558

141
726
1262
1351
2704
2881
3769
4107
4112
4407
7052
8553
9751
10638
11036
11098
12016

142
95
616
625
2394
4261
6179
6231
7173
7298
7565
7717
8102
9959

143
829
913
1365
2119
2501
4182
4374
6892
7234
7319
9149
9173
9198
11060

144
474
985
1496
1618
2087
2192
2258
2563
2752
5370
5379
6186
7209
7609
8386
8588
9172
9341
9438
9500
9502
11554
11557
11834

145
447
621
2225
2467
2733
2736
2990
7266
8032
9299
9782
10104
12020

146
426
2112
2325
2484
2667
2684
5380
5384
5411
5438
5803
6288
7110
7214
7268
7326
7337
7389
7842
10969
11083
11800
11801

147
435
619
824
936
1517
2190
2239
2531
2616
2703
3304
3466
5441
7378
7592
9243
12026

148
226
904
1538
2274
2646
2841
3575
4110
6187
7186
7235
7831
9792
11211
12027

149
2243
2559
2818
3041
3108
4129
4378
5007
5162
7213
7236
7854
9745
10450
11032
11854

150
52
304
722
1271
1553
2498
2518
2557
2560
2608
4103
4939
5439
6232
6282
7270
8070
8630
9169
9209
9646
9736
10018
10770
10853
11088
11140
11154

151
918
2631
3651
3991
4088
4142
7369
7677
8006
10910
11862

152
78
432
477
805
833
906
1265
2241
3975
6779
7846
12021

153
256
542
646
895
1061
2502
3068
4052
6211
6221
6753
7431
8029
9413
10167
10795
11285

154
40
880
933
1095
2562
3196
3334
4009
6185
7329
9802
11203
11287
11504
11805
11808
11837

155
82, 356, 433, 436, 605, 2491, 2519, 3045, 3777, 5180, 5981, 7406, 7779, 7860, 9448, 10746, 11722, 11902

156
93, 146, 217, 428, 522, 523, 1545, 7269, 7952, 9507, 9518, 11090, 11103, 11162, 11227

157
357, 878, 2556, 5122, 5175, 5386, 7223, 7669, 8857, 9211, 9304, 9784, 9956, 10637, 11092, 11652, 12067

158
431, 460, 480, 623, 1263, 2535, 2719, 2874, 3139, 5185, 7168, 8926, 9197, 10758, 10797, 10859, 11832, 12000, 12024

159
2240, 2612, 4130, 4408, 5111, 7330, 9192, 9492, 9832, 11332, 11660, 11845

160
360, 458, 519, 604, 2082, 3004, 3009, 3331, 4061, 4062, 5932, 6241, 6301, 7206, 7395, 8679, 10993, 11046, 11064, 11209

161
610, 944, 2832, 3299, 3339, 3702, 7758, 9403, 10761, 10796, 11200

162
1315, 3065, 4338, 5113, 7341, 7664, 7676, 9407, 9768, 10679, 11518, 11806

163
358, 387, 626, 727, 759, 1512, 2511, 2657, 4094, 5156, 5711, 7847, 8634, 9884, 10885, 11327, 11590, 11671, 11802

164
310, 1340, 1866, 2135, 2561, 2588, 3037, 6054, 7221, 7580, 9166, 9208, 9303, 10998, 12007

165
33, 327, 333, 343, 1758, 2187, 2629, 3032, 4146, 5121, 5179, 5336, 6238, 9613, 9831, 10012, 11519, 12033, 12056, 12064

166
29, 344, 618, 2527, 2545, 2590, 2728, 2738, 3090, 3134, 5971, 6199, 6273, 8028, 9409, 9746, 9760, 10165, 11556, 11940

167
602, 2353, 5013, 5583, 9398, 11502

168
37, 465, 467, 613, 631, 2663, 2717, 2959, 3341, 4055, 4060, 4124, 4161, 5402, 6022, 6227, 7295, 7422, 7704, 9206

169
764, 776, 2524, 3142, 3309, 7574, 9207, 10405, 11065, 11210, 11568

170
67, 1522, 2376, 2796, 3292, 4010, 4033, 4072, 4259, 5221, 5383, 5588, 6197, 6292, 7706, 7971, 9300, 9759, 9965, 11555, 11685, 11999

171
1103, 2213, 2540, 2583, 3271, 4185, 7188, 7542, 7670, 7763, 9986, 10905

172
941, 1499, 2096, 2238, 2477, 2734, 3312, 3817, 4123, 5960, 7145, 7368, 7727, 8069, 8800, 9735, 9830, 11091, 11723

173
476, 828, 2485, 4049, 5587, 5873, 6242, 6603, 7190, 7589, 7667, 7833, 7861, 9167, 9216, 11084, 11711, 11713, 12013

174
455, 1493, 1785, 2391, 2555, 2623, 4126, 5148, 5445, 7721, 9201, 9440, 9477, 11095, 11153, 12050

175
562, 845, 846, 1554, 3047, 3107, 3302, 3328, 4053, 4183, 4970, 5178, 5391, 5494, 6175, 7143, 7398, 8057, 8572, 10489, 10753

176
90, 624, 931, 2573, 2783, 4063, 4091, 5979, 7356, 9193, 9213, 10759, 11094, 11591, 11598, 11841

177
409, 2630, 4095, 5004, 5671, 5984, 6181, 6239, 7590, 8210, 8212, 9191, 9408, 11324

178
92, 347, 2774, 2961, 5399, 5492, 6203, 7218, 7844, 10147, 11089, 11312, 11564, 11593, 11799

179
48, 430, 2678, 6279, 7593, 8252, 11314

180
73, 725, 1501, 2392, 2528, 2679, 5451, 5957, 6223, 7837, 7953, 11102, 11323, 11337, 11859

181
349, 603, 2395, 2436, 2576, 2866, 5166, 5401, 7672, 7851, 7856, 9748, 9795, 10850, 11078, 11353, 11512, 11718

182
2189, 2197, 2254, 2547, 4139, 5387, 5580, 6285, 10863, 12005

183
30, 2544, 2587, 2715, 4051, 4933, 6192, 6334, 7142, 7228, 9755, 11045, 12031

184
318, 411, 614, 1490, 1887, 2351, 2876, 3770, 7427, 7705, 7849, 8213, 9399, 10856

185
348, 1515, 2577, 2605, 2666, 2702, 4108, 5161, 6178, 6218, 6249, 6261, 7759, 11533

186
385, 2963, 4081, 4099, 5595, 6066, 6226, 7306, 8397, 9298, 9657, 10873, 11816, 11824, 12038

187
239, 516, 2193, 2385, 3005, 3294, 4054, 5150, 5919, 6144, 7393, 9203, 9499, 11229, 11858

188
1533, 2486, 2490, 2564, 2658, 7344, 9478

189
354, 832, 2393, 2786, 2865, 4128, 4794, 5553, 6194, 6236, 10772, 12011

190
821, 2120, 2751, 2775, 5127, 5144, 7136, 7232, 7320, 7797, 9744, 10272, 11087

191
466, 692, 1102, 2191, 2195, 2883, 5396, 5443, 5670, 6245, 6255

192
350, 664

2203
2439
3853
4379
5819
6225
9533
9756
10802
10864
11179

193
64
395
407
863
3578
6259
7679
7839
7866
8019
10804
11047

194
782
1838
2509
2831
7156
7700

195
320
451
481
2500
2565
2595
2929
3931
4131
5155
5373
5818
6264
7771
7850
7859
8038
11306
12042

196
398
3932
4058
5011
5403
6281
9120
11313
11682

197
7217
7933
8548
9177
11048
11910

198
756
2196
2624
2911
3036
3110
4096
4100
5108
6252
7678
11807
12043

199
311
1281
2470
2512
2570
2926
5915
6224
6240
7799
7869
11325

200
323
392
511
634
886
1542
2318
2492
2571
3343
4040
4994
11159
11531
11566

201
1841
2552
9365
10703
10883

202
307
4119
4260
5524
7845
10904
11529
11666
12035

203
2194
2379
2529
3464
4098
5158
6220
7134
7225
7300
7579
11074
11093
11207

204
306
2829
3785
5159
10011
11932

205
384
386
483
2380
2568
2591
2823
2957
4080
6873
7385
7757
9765
10773
11532
11536
11833

206
321
2234
3311
4730
5017
6251
8020
11061
11233
12044

207
6222
6247
8004
10934

208
429
473
2478
2506
2942
3109
3986
4056
5128
5269
7855
9468

209
424
2712
6246
9174
11812

210
104
403
2632
2713
4113
4133
5440
6234
6237
7229
7366
7597
7611
11315
11528
11843

211
388
1521
2282
2625
4059
6280
10644
10675
11192
11687
11831

212
2943
3204
4942
5124
5389
6268
7760
8211

213
2377
4069
5394
10508
11319
11597
11928

214
308
1591
1621
2613
2857
4251
5152
9752
11235
11317

215
366
791
879
3982
4057
6265
7453
8677
8678
9749
11044
11096

216
83
2411
3465
5143
7417
7768
7848
8631
9214
10752

217
309
370
437
495
1500
2739
2873
3902
5145
6286
11041

218
27
2368
5371
5375
7335
9170
9367
10896
11338

219
31
2319
2487
2550
3135
7832
11109
11663
12022
12023

220
2566
2654
2821
4068
5985
6230
7216
7396
7838
7857
7858
8056
9753
11100
11520
11919

221
103
105
485
2730
2856
3069
4070
5400
7189
7397
8559
9369
9764
10785
11924
11925

222
353
1324
4047
7610
7772
8551
9754

223
397
508
2261
2435
3141
7353
7430
9747

224
866
954
1316
2086
5977
7418
7419

225
34
1623
2569
10632
10803

226
317
375
6079
10978
11305
11513
11829
11998

227
472
609
2800
2801
7446
9302

228
346
491
1559
1749
2388
2554
2558
2735
3022
5388
5393
5428
6191
9501
9733
10927
11011
11111
11357
11545

229
2418
2647
2721
4121
8066
9439
11355

230
1327
2517
2826
2979
5021
6276
7456
12065

231
410
732
2808
2810
5349

232
456
606
5153
5982
7853
10913
11689

233
1552
3038
5022
8061
9739
10843
11051
11644

234
2851
4048
6204
9598

235
66
319
468
665
2617
4132
5804
6516
9738
11086
11177
11830
11930

236
2199
2409
4074
7162
7719
10894
11040
11452
12010

237
2200
2618
4101
5151
5392
6195
11329

238
365
2364
2546
2853
7144
7362
7598
10760
10786
12025
12060

239
2828
4116
5176
5398
5442
6287
6289
11107
11307
12009

240
2400
5014
5376
7867
9466

241
313
5624
7267
7767
9364
12062

242
404
2585
2705
2846
6256
7414
11515
11516
12063

243
2536
2686
7172
9742
11653
12046
12055

244
399
438
864
1551
2609
5562
7226
10676
11320

245
3360
5390
6184
6244
6262
6267
7432
7762
11358
11927

246
367
3981
7440
9511
11356
11840

247
608
2205
3035
6260

248
4041
4686
11076
11686

249
390
2209
10928
11228
11850

250
628
635
1317
2128
2825
2999
6243
7416
11038
12068

251
368
2202
2802

252
2513
5395
11178

253
380
2648
2731
3106
4044
4073
7352

254
1558
2275
2520
5407
5495
6214
7346
10909
10929

255
2204
4043
4114
6193
7666
8589

256
2830
4117
4262
10930
11073
11813

257
419
8996
9165
11342

258
383
4067
5397
8571
9705
9989

259
416
7334
12051

260
32
2514
5020
5406
5418
5566
7133
7805
11295

261
8
7761
9641

262
2255
2794
2805
7354
7391
7835
9737
10776
11316

263
2809
7222
10453

264
85
2844
9194
9640
12061

265
4120

266
1348
2432
6080
8058
11326
11534
12054

267
2812

268
2575
3257
4122
5223
11860

269
4045
5142
8062

270
355
471
2659
2854
2858
6196
7363
7769
8423
11106
11684

271
11530

272
9758

273
991
2726
7665
10756
11039

274
2408
2824
2855
7668
7870
9767
10854
11201

275
7215
7675

276
2257
2804

277
470
2842
2843
7445
9161
11929

279
9366

280
393
3260
4046
6201
9833
11344
11669

281
369
2773
2850
7648
11066
11535

282
1831
3002
6253
7647
7852
11232
11351

283
2499
4258
9510
11067

284
7175
7370
9175
9509
12037

285
316
2320
2811
4042
5447
11234

286
379
1748
2224
5107
7219
11293

287
3764
4757
7132
7211
7970
11563

288
2083
2217
5404

289
391
400
11230
11922

290
91
2508
2839
4268
4900
5405
6250
7195
7204

291
372

292
885

293
394
2384
5016
7412
7413
7454
11933

294
7707
8546

295
2815
10912
11525
11565
11923

296
2656
7350
12008

297
325

298
2386
12036
12040

300
324
2489
3006
3990
5568
7191
7230
11851
11931

301
2807
8214
8738

302
10754

303
2816
6254
8217

304
381
3463
12045

305
5446
11346

306
408
1349

307
2594
8218

308
4278
7351

309
234

310
1352
3097
8547
10899
11062

312
4354
9612
10897
11997

313
415
10763

315
7777
10895
11522
12041

316
2817
7400
11068

319
389

320
11194
11347

321
371

322
11104
11294

323
2827

324
10931

327
7348
8215

328
237
5623

329
2814

330
417
2813
4782
9176
9600
11108
12052

333
8545

335
331

336
9642

337
5224

338
11059

340
412
1330

341
5969

342
6081

343
11926

345
11050
11835

347
1328
2806

348
10748

350
2852
5493
7408
10852
12053

353
396
7336

357
7703
10757
11018

359
2404
11745

360
414
1331
9734
10751
10768

365
2533

368
9601

370
10769

374
418

380
6621

385
405
2466

390
7331

420
7457

437
4145

450
2784

485
7644

550
4164

BOILING POINT INDEX OF ORGANIC COMPOUNDS

In this index the temperature in °C is given in boldface. The numbers immediately below a temperature entry are identification numbers (given in the first column of the table *Physical Constants of Organic Compounds*) for compounds whose normal boiling point values begin with the same integer as the temperature entry. Thus, the numbers that appear between temperature entries of **3** and **4**, for example, indicate the compounds that have normal boiling points equal to or greater than 3.00°C and less than 4.00°C. The entries following **-8** are compounds boiling between -8.00°C and -8.99°C, inclusive. The entries for **0** temperature are compounds with normal boiling points between -0.99 and +0.99°C.

-191
4011

-161
7459

-128
7541

-103
5879

-88
5448

-85
5902

-84
4008
5936

-82
7558

-81
7486

-78
3992
5596
7515

-75
5918

-72
5906

-68
289

-57
7475
11437

-51
5924
7504

-50
4035

-49
5927

-47
5665
10528

-46
4167

-42
9834

-40
7478

-37
5484
5593

-36
10118

-34
9772

-32
4955

-30
5938

-29
7500
10572

-27
5889
10441

-26
5640
6019

-24
3914
5544
7477
7523
10569

-23
7491
10727

-22
7559
11435

-21
5550

-20
3119
11414
11584

-19
5641
5987

-18
5885

-17
10128
10608

-13
5884

-12
5486
8091

-11
10073

-10
5485

-9
5470
7481

-7
11429

-6
3704
7442
10578

-5
4194

-4
3146
11432

-3
7464
10570

-2
10046

-1
3255

0
3229
3705
4981
9906

1
3804
5126

2
446
4198
7444

3
3706
5538
7462

4
5537
5937

5
3889
4200
5666
5908
7546

6
3164
5489
7448
9773
10541
11489

7
5602
11378

8
3897
4007
7502
10421

9
9910

10
3145
3177
8599

12
3162
4170
5467
5487

13
4166

14
9159

15
5880
11439

16
5410
5475

17
5488

18
8087

19
5883
5897
7470

20
11
301
3800
4971

21
5898
7505

22
10535

23
4162
7555

25
3117
3379
7496

26
3898
3916
5466
7176
8659
11451

27
500
3397
5540

28
4973
5539
5912

29
8807
9004

30
5545
5925
9027
10070

31
3799
5894
6010
6025
9804

32
3378
3892
5534
10534

33
279
5940

34
3171
5249
5608
9898

35
4974
5471
5648
5905
9849

36
442
444
4192
5421
7466
8709
9005
9006

37
3907
4972
5900
7184
7545
9522
10533

38
3692
3801
4201
5454

39
264
3172
5233
7520
10069
11527

40
3170
7499
8092
9122

41
3760
4764
9905
11398

42
4193
7507
7518
8661
9112
10532
10737

44
4903
4980
8658
8662
9114
9818
10506
10576

45
9871
10536

46
3993
5531
5532
5662
9848
12066

47
5525
5663
8592
9803
10557

48
5895
8098
8660
9774
10530

49
3289
4774

50
294
4953
5458
5491
5942
9819
9863

52
449
4381
4643
5457
8607
9113
10157
10509

53
5168
7517
9054
10522
10558

54
4985
5667
6007
6086
6099
8802
9053
10038

55
3759
5427
8684
9851
10034
10072
11425

56
208
498
725
3890
4975
5907
6843
8609
8671
9059
9123
10411

57
3290
4034
4190
4277
5528
7032
7526
8854
9134
10529
11409

58
3722
4188
4976
4983
5642
8853
9058
10071
10729

59
3152
3393
3414
4977
4987
5476
6678
7483
8594
9111
9837
10317

60
3392
4189
5598
5896
8608
8823
10723
11389

61
267
4165
7554
9135
11410

62
3210
3727
4498
8817

9052

63
494
3287
3721
4276
5585
6999
7154
8085
8615
8824
10037
10736

64
3723
4006
4172
4929
5548
6130
7498
7581
8825
9785
10159
10524

65
3113
4930
6131
6168
6677
6683
9158
9915
10033
10600
11465

66
4986
5430
5606
6684
7003
9117
9807
10565

67
501
3126
4199
7000
7002
8094
9055
9057
9817
10156
10566
10577

68
3151
3159
3242
3720
3754
3899
6011
6731
7001
7463
8093
8613
9862
10120
10515
10539

69
296
4984
5910
8575
9810
9894

70
3391
3728
4020
4191
4982
5436
7105
8666
8667
9056
9128
10531
11394

71
3129
4826
5237
5462
5479
5659
5923
7114
9836
10540

72
179
5597
5660
8672
9033
9788
10040
10180
11696

73
263
3761
4032
5535
5658
6681
6682
9136
9846
10181

74
3178
5655
5770
10060
10179

75
2347
3169
3724
4827
4928
8668
9060
10036
10045
10714

76
291
463
3901
3908
4499
5202
5526
6676
7540
8670
9039
9129
10547
11430

77
186
3209
3222
3812
7183
8669
9031
9783
10437
10548
10583

78
3241
5204
5672
6073
6171
6679
6992
6993
8083
10409
10554
11544

79
3523
3645
3896
5599
8760
10296

80
867
1757
3457
3525
4248
4305
6018
6074
6680
7494
8593
8762
9032
9040
9829
9856
10493
10655
10771
11550
11720

81
271
3370
3373
3875
5210
6560
6715
7116
7155
9030
9809
10039
11168

82
269
1423
4169
4626
5490
6418
6685
9806
10335
10389
10733
11412

83
266
5455
5478
5529
6017
6686
6712
7037
7490
9035
9130
9821
9909
10035
10059
10076
10579
10725
10726
11376

84
1717
2271
3726
3923
5881
5920
6325
7115
9029
9044
9131
9853
10089
10553
10732
11724

85
12
288
2091
3223
3253
3438
3713
3725
4249
5549
6687
6713
7039
7046
7103
7153
8616
11495

86
3396
3708
3710
6172
6716
6830
7038
7043
7047
7180
8763
9777
10777
11143
11361

87
3733
4800
5922
7042
7104
7161
7472
8074
10063
10081

88
209
3150
3167
3359
3709
3877
4922
4966
5603
6695
7044
7106
8583
9891
9897
10158
10562
10584
10607
10800

89
1424
2913
3797
5417
7045
8702
8761
9036
9038
10055
10728
11377
11496

90
2272
2346
3394
4000
4803
4917
6834
7467
8627
8701
9779
10119
10586
11118

91
3168
3221
3232
3395
4802
4804
5305
5636
6829
9022
9037
9845
10261
10594
11461

92
16
3185
3186
3372
4919
5480
5522
5637
5644
5890
6105
6835
7036
7120
10113
10301
10499
10504
10581
11396

93
2980
3148
3155
3714
4920
5473
5519
5646
6089
6327
6533
7435
8665
8809
9017
9574
10549

94
445
1422
2270
3161
3369
3668
3707
3734
5203
5227
6328
9043
10552
10592
10698
11658

95
177
3118
3215
3524
3622
4018
5078
5708
6088
6536
6537
7040
7041
7122
7179
7402
9841
10088
11691

96
3220
3732
3737
5668
6432
9016
9021
9045
9115
9827
9892
10525
10585
10715
11189

97
21
3275
3858
3870
4925
6012
7492
8720
8721
9018
9034
9912
10090
10334
10697

98
3116
3158
3219
3252
3437
3452
3711
3879
4926
5911
6013
6355
6534
6535
8086
10574
11382

99
3251
3441
3594
3715
3915
4197
4801
4921
5236
6102
6609
7152
8090
8822
8848
9001
9861
10083
10263
10635
11423

100
176
196
683
3154
4343
4426
7028
7516
8076
8663
10058
10087
10663
11463

101
236
1738
3231
4272
4342
4403
5187
6002
7522
8958
9064
9067
9132
9844
10260
11397
11462
11486

102
265
2348
3217
3375
3547
3628
3681
3695
3752
4022
4428

4672
5541
5591
5661
6109
7121
8558
8957
9126
10054
10490
11487

103
235
1739
3716
3900
4812
5727
6033
6034
7339
8690
9023
9104
9859
10111
10573

104
192
2095
3157
3216
3280
3377
3712
3922
4671
4856
7016
7018
7021
7026
7561
8096
8700
8765
9019
9069
9071
10297
10551
11166

105
441
2914
4023
4363
4632
4918
6108
6326
7029
7474
7484
8415
8846
9068
9899
10086

106
295
3114
3430
3522
3526
3657
3895
3921
4924
5186
6004
6771
7027
7525
9629
9843
9900
10061
10100
11149
11151
11693

107
3153
3180
3218
4003
4014
4271
5423
5456
5474
5882
6611
7019
7030
9008
9048
9679
10043
10124
10388
10503
11585

108
297
302
3240
3399
4771
5517
5620
5669
5887
5891
6329
7550
8577
8719
8797
9020
9024
9065
9066
10171
10328
10555
10556
11020
11491

109
453
3736
4770
4860
4938
5254
6107
6773
6774
7020
7031
7482
8990
9046
9775
9826
9858
10299
10404
10511
11483

110
19
1947
3147
3160
3735
4016
4670
4857
5464
5465
5619
6127
6341
6606
6837
7013
7023
7025
7033
8080
8764
8847
9012
9732
10298
10496
10734
11484

111
3248
5796
6335
6544
6696
6775
7014
7015
8576
9153
9592
9860
10526
11375

112
214
278
299
3165
3247
3371
3374
3629
3731
3744
4187
5892
6343
6545
6554
6557
6610
6838
7017
7024
7557
8603
8695
8845
9047
9072
9097
10082
10149
10527
10550
11130
11176
11750
11769

113
3276
3861
4647
5228
5651
5656
5935
6529
6546
6547
6836
7005
8620
8730
8850
8982
9776
9883
10373
10743

114
1737
3250
3693
3871
4789
4855
4963
5132
5193
5607
6167
6551
6552
6691
7543
8849
9050
9854
9904
10604
10702
11405
11827
11905

115
17
1726
1727
3246
3647
4028
4030
4244
4813
5426
5512
5769
6558
6772
7533
8319
8681
8813
9049
9087
10735
10801
10872
11371
11372
11770

116
134
215
997
1728
2915
3513
3552
3694
3745
4861
4862
5189
5610
5638
5639
5647
7034
7582
7696
8732
8917
8981
9013
9051
9070
9089
10455
11150
11366

117
100
300
1220
2094
3403
3593
3649
3698
3699
4240
4656
4859
5496
6417
6422
6424
6426
6553
6724
6776
7009
8578
8693
8711
8808
8980
9124
9805
9816
9840
10492
10658
10778
11475

118
2236
3421
3618
3848
4221
4851
5513
6106
6423
6815
8605
8712
8814
9014
9107
9680
9838
10042
10259
10500
10602
11456
11571
11667

119
145
284
3281
3439
3718
3803
3841
4364
4371
4653
5542
5888
6090
6543
7521
8590
8916
9007
9100
10109
10387
10420
11175
11402
11446
11492

120
191
3125
3239
3382
3387
3416
3440
3619
3802
4219
4367
5503
5886
6425
7006
7493
9146
9893
10117
10152
10303
10425
10739
11501

121
3238
3515
3521
3847
3859
4772
4814
4816
4817
4818
4937
5424
5516
5917
6174
6555
6556
6998
7004
7022
7501
8491
8694
8940
9025
9092
10057
10694
10744
11390
11399
11466
11508
11657
11783

122
127
4434
6548
6549
6550
6688
6738
7405
7485
7519
8495
8496
8497
8606
8704
8706
8841
8944
9105
10326
11751

123
3402
3636
3860
4366
4773
4858
4979
5536
5694
6714
6739
6960
7007
7538
8078
8494
8703
9011
9878
9903
10325
10560
11049
11476

124
3245
3277
3385
3887
4196
4368
4370
4654
5482
5733
6861
8077
8320
9109
10174
10376
11433
11908

125
298
1893
3137
3409
3743
3753
4036
4655
4848
5821
6103
6527
6607
7077
7137
8139
8355
8492
8493
8707
8966
8967
9010
9101
9140
9681
10542
10664
11363

126
121
274
3746
3999
4038
4651
4828
6833
6860
7544
8529
8598
8731
8844
8943

8965
9811
10294
10498
11582

127
1167
3410
4255
4314
4649
5461
5472
6959
7087
8891
9648
9650
10085
10349
10369
10382
10591
10834

128
3115
3175
3626
4024
4029
4388
4650
4658
4815
5693
5899
7681
8614
8723
8725
8748
10354
10597
11678
11740

129
133
138
2045
3286
3680
3920
4254
4365
5509
6717
7089
7443
8089
8710
9102
9521
10980
11112
11124

130
339
1168
3166
3282
3381
3549
3683
3690
3747
3749
4878
5634
6015
6617
6806
6865
8022
8718
9108
9667
9682
9778
9813
10386
11449

131
207
1131
3627
3644
4256
4390
4652
4849
4935
5518
5600
6718
6864
8532
8554
8942
9608
9787
10116
10381
11365

132
117
190
303
3550
3592
3688
3748
3862
4470
4852
5205
6128
6383
6385
6723
6730
7503
8687
8751
8919
10077
10172
10222
10495
10673
10783
11173
11453
11705

133
99
175
3236
4425
4662
4780
5429
6111
6706
6828
6859
8072
8531
8810
8840
8928
9026
9144
9880
10309
10347
10463
10559
10606
11422
11477
11679

134
3227
3278
3283
3700
4086
4226
4283
4661
4961
4999
5002
5510
5709
6692
6980
7066
8526
8941
9099
10044
11434
11759
11787

135
3237
3427
3548
3689
4176
4252
4487
4956
5618
5717
6112
6387
6391
6737
6746
6914
6915
6979
6981
8342
8929
8989
10184
10229
10308
10316
10700
11392

136
1668
2991
3517
3615
3838
3840
3852
4245
4253
4391
4486
4867
4941
6173
6386
6390
7108
8673
8812
9002
9901
10102
10476
11756
11760

137
144
281
1078
1166
1241
3386
3646
4660
4777
5630
6388
6849
6862
6913
7067
7097
7098
7128
8530
8600
8915
9649
9651
10280
10310
10390
10810
10825
11417

138
448
492
1242
1454
2093
2993
3007
3235
3839
4496
4648
4752
4931
5232
5792
6332
6809
6825
7088
7694
8798
8931
9090
9091
9145
9618
9812
10416
10537
10599
11070
11865

139
107
1453
3606
3851
4885
5234
6351
6539
6540
6863
7068
7081
7084
7471
8508
8686
8749
8930
9823
9873
10096
10731
10832
11369
11788

140
268
1425
1725
2968
3128
3413
3415
3630
3691
3808
3869
4497
4689
4691
4693
4742
4864
4946
5420
5477
5481
5614
6170
6333
6384
6389
6826
6827
6866
6912
7085
7086
7090
8612
8819
8839
8953
9062
9098
10066
10074
10084
10380
11442
11755

141
3605
3611
3831
4688
5716
6330
6408
6614
8714
8827
8842
8993
8994
9088
9881
10201
10410
10545
10609
10672
11754
11795

142
166
206
3553
3587
3631
3832
4012
4339
4488
4493
4495
4642
4927
6049
6350
6586
6734
6857
7083
7124
7556
7934
8110
8405
8410
8507
8717
8918
8932
9061
9668
9678
9866
10151
10289
10603
11071
11132
11479
11785

143
293
1715
2304
3588
3617
4751
4829
5604
5735
6331
6407
6720
6733
6813
6814
6942
6945
6946
6976
7403
8408
8726
8728
9839
11469

144
130
1452
3376
3451
3616
4489
4805
4886
5050
5502
6362
6363
6402
6512
6807
6977
6978
8409
8505
8811
9652
9879
10211
10273
10497
10822
10939
10981
11388

145
1624
3407
3866
4485
4657
4690
4732
4749
5533
5664
5695
5740
6042
6113
6169
6542
6585
6694
8506
8716
8750
8820
8826
9442
9714
9902
10173
10443
10616
10982
11138
11421
11459
11672
11757
11866

146
182
2970
3234
3284
3488
3509
3888
4492
4494
4692
4833
4835
4882
4908
5501
5590
5635
5704
6085
6965
6994
6996
7073
7129
7687
8182
8756
8878
9602
10214
10345
11674
11703

147
135
3384
3620
3697
3758
3995
4298
4798
6039
6447
6511
6821
6848
6947
6963
6966
7527
7528
8183
8489
8569
8947
8954
8972
9624
9691
9815
10220
10361
10505
10665
10701

10917
11131
11673
11912

148
292
3383
3637
3686
3919
4836
4837
4838
4868
4870
5226
6438
6440
6846
6967
7096
7461
8715
8744
8949
9712
9799
10185
10304
10916
10956
11370
11540

149
198
227
4173
4400
4663
4736
5460
5764
6104
6115
6263
6728
6898
7131
7473
7548
8328
8329
8628
8833
8937
8948
9622
9684
10155
10351
10488
10826
11373
11391

150
128
139
452
478
1043
3739
3756
4293
4398
4687
4733
4796
4799
4869
5685
6439
7099
7495
8120
8196
8563
8955
9889
10279
10374
10401
11055
11726

151
1044
1090
2931
2986
3401
3610
3682
4295
4396
4490
4491
4734
6337
6421
6445
6510
6531
6743
6856
6943
6944
8111
8939
9028
9084
9360
9707
10223
11386

152
1170
1240
1975
2922
3564
3675
3751
4085
4297
4394
4399
4478
4834
4897
5768
6344
6448
6566
6928
7095
7127
7450
8610
8971
8975
8988
9669
10030
10141
10364
10575
10833
11867

153
253
1915
3412
3448
3456
3850
4292
4294
4296
4797
5872
5992
6116
6525
6569
6589
6729
6847
6851
6986
8327
8501
8646
8938
8946
9619
9666
9847
10064
10143
11072
11706
11772
11796
11914

154
1169
1722
2909
3188
3601
4021
4431
4832
5141
5253
6524
6587
6819
6923
7092
7524
8199
8326
8502
8626
8834
8970
9141
9142
9143
9670
9814
10269
10290
10699
11146
11428
11450
11503
11741

155
164
1646
2923
2971
3562
3729
3750
3836
4013
4530
4571
4768
6009
6353
6450
6538
6732
8088
8373
8821
9626
9842
10315
10523
10666
10806
10807
11408

156
998
2165
2972
2973
3563
3849
4395
4397
4471
4778
4875
5718
6176
6336
6349
6444
6446
6475
6495
6816
6853
6925
7071
8375
8831
8933
8956
8991
9710
9711
9808
10225
10605
10611
10805
10938
11387
11418

157
203
1398
1724
2982
3455
3609
3625
4473
4866
4906
4944
4945
5469
6117
6434
6437
6474
6845
6911
6927
7069
7070
7660
8198
8383
8766
8818
9096
9591
10142
10175
10192
10193
10820
11765
11773
11814

158
181
2921
3258
3264
3994
4319
4393
4480
4755
4876
4896
4898
5732
5738
5763
5795
6338
6412
6413
6427
6522
6926
6964
7549
8376
8438
8934
10718
10937
11478
11581
11675
11778
11797
11798

159
1183
1723
2169
2992
3121
3211
3389
3795
3863
4175
4331
5601
5617
6416
6473
6573
7072
7091
7434
8378
8381
8837
8924
8927
9095
10338
10368
10952
10984
11058
11174
11727

160
197
1091
3598
3639
3660
4178
4418
4477
4500
4899
5746
6063
6360
6415
6441
6443
6559
6804
6844
6922
6929
6931
8377
8382
8697
8978
9519
9611
9685
10065
10650
10651
10652
10839
10963
11424
11543
11762

161
149
1184
1686
2932
3187
3268
3279
3828
3830
3864
6041
6087
6358
6442
6449
6451
6505
6604
6823
6852
7607
8186
8197
8379
8617
10936
11549
11712
11761

162
131
185
1185
1687
2767
2918
3572
4286
5613
5686
6005
6121
6452
6504
6515
6818
6820
6822
6824
7075
7126
7535
7537
8115
8384
8595
8713
8746
9609
9610
10178
10221
10598
10648
10660
10953
11384
11748
11792

163
142
167
1786
3400
3460
3483
3608
3918
4157
4825
4923
5414
5463
5563
5706
6409
6410
6500
6599
6719
6764
6768
6840
6988
8344
8380
8404
8473
8503
8892
8945
10110
10177
10213
10375
10406
10491
11368
11764
11779

164
1641
2359
2924
2934
2981
3191
4228
4288
4451
4694
4695
4883
5921
6078
6503
6520
6523
6612
6616
6850
6937
6938
8351
8374
10258
10350
11125
11681
11793
11815

165
55
132
440
1685
1972
3434
3435
3905
4531
4596
4627
4943
4992
5110
6411
6414
6839
6924
6957
7489
8114
8142
8143
8322
8500
9625

10075
10145
10341
10348
10957
11419
11763
11766

166
484
2092
2925
2933
3265
3479
3654
4315
4437
4522
4824
6357
6406
6496
6530
6600
6615
6767
6817
6971
7054
7680
8325
8403
8836
9653
9708
10262
10366
10799
11164
11903

167
248
596
1089
2164
3449
3532
3973
4247
4532
4534
4682
4683
5024
5705
6077
6499
6521
6884
6973
6989
7281
7429
8140
8141
8330
8472
8499
8914
8976
9094
9882
10079
10275
10407
11015
11747

168
115
189
499
3263
3285
3388
3473
3555
4002
4533
4728
5687
6742
8185
8324
8396
8420
8421
8985
9706
9876
10302
10318
10958
11119
11506

169
277
1459
1640
1649
2289
2358
2935
2936
2937
2990
3813
3815
4050
4597
4902
5023
5570
5722
6514
6726
6811
6941
7078
8498
8555
8602
8815
8816
9913
10372
11493

170
143
199
280
700
956
1930
3556
3954
4287
4435
4438
4440
4598
4703
4704
4714
5082
5083
5416
5459
5847
6036
6497
6501
6597
6854
6939
6950
6953
6955
7340
7536
8339
8340
8597
9588
10203
10292
10840
10915
10942
11028
11113
11647
11906

171
194
1654
1740
1931
2974
3573
3643
3673
4358
4369
4436
4453
4673
4674
4810
5084
5085
5675
6118
6476
6484
6596
7140
8346
8445
8520
8921
9857
10032
10041
10103
10154
10182
10346
10412
10941
11026
11120
11142
11668

172
1194
1642
1655
2039
2415
3132
3272
3351
3418
3763
4179
4416
4439
4449
4580
4585
5893
6368
6502
6935
8360
8471
8509
8757
8832
8894
9867
9869
10031
10091
11129
11472
11500

173
155
161
560
1256
1355
1656
3546
3685
4442
4535
4675
4681
4872
5200
5483
5567
5741
6040
6479
6480
6483
6761
6940
7010
8556
8799
8897
8925
9908
10101
10426
10876
11690
11746

174
20
1356
2438
3266
3350
3489
3956
4273
4441
4529
4536
4680
5033
5097
5573
5627
6466
6485
6498
6675
7053
8095
8478
8544
8899
9080
9251
9686
9963
10336
10365
10634
10962
11021
11147

175
254
559
752
1932
1933
1984
1989
2162
3233
3355
3603
4274
4452
4586
4676
4678
4887
4993
5266
5817
6481
6580
8511
8752
10097
11707
11719
11729

176
120
1111
1974
2357
3491
3539
3823
3949
4450
4515
4538
5197
5432
5710
6472
8446
8560
9635
9709
9966
10017
10224
10362
10841
12030

177
487
1112
1990
3469
3476
3538
3558
3824
3953
4015
4513
4514
4584
5098
5100
5134
5244
5991
6433
6436
6532
6766
6899
6934
7250
8519
8973
9890
10985
11057

178
1161
1988
2163
3228
3595
4234
4581
4617
4677
4710
4724
4729
5611
6008
6124
6356
6567
6568
7117
8464
8479
8517
8742
8801
9361
9491
10360
10966
10986
11730

179
102
201
525
1182
1379
3127
3398
3490
3544
3545
4448
4583
4830
4934
5546
6119
6598
6921
8350
8444
8735
8747
8900
8902
10125
10144
10208
10276
10417
10439
10935
10940
10965
11426

180
1354
1818
3571
3826
4325
4330
4587
4839
4846
5698
5794
6703
6900
8026
8116
8469
10232
10563
10582
10809
10887

181
561
1054
1414
1807
3322
4320
4524
4564
6059
6137
6381
6509
6831
8968
9199
9540
9943
10343
11403
11458
11731
11767

182
754
1919
1968
2023
2037
2877
3347
3428
3454
3511
3519
3662
4526
4589
4640
4890
5452
5765
6487
7245
7506
8477
8905
10355
10538
11440

183
116
125
312
1055
1118
1162
1415
1672
1720
2024
2038
2173
2177
3390
3955
4230
4275
4335
4419
4616
4811
4891
4894
4958
5530
6428
8359
8523
8561
8803
10105
10150
10246
10291
10486
10487
10626
10920

184
633
1056
1138
1163
1413
1659
3453
3485
3814
3827
4679
5450
5595
5726
6507
6751
6930
9887

9918
10217
10359
11025
11913

185
53
283
1458
1484
1787
1795
1985
1986
2036
3361
3433
3446
3533
3534
3835
3845
3948
4229
4235
4516
4712
5198
5264
5551
5560
5757
6133
7319
7452
8474
8862
8869
8963
9362
9888
9920
10092
10194
10215
10593
10784
10918
11016
11700
11749
11834

186
1139
1658
1674
2709
3203
3474
3475
3506
3574
3579
3664
3762
3972
4519
4621
4684
5052
5109
5843
6486
6492
8401
8402
8419
8855
11699

187
261
856
1140
1481
1630
1682
1701
1814
3557
4518
4566
4588
5056
5208
6045
6377
6462
6506
6519
6909
7449
7791
8174
8788
9587
9982
10123
10244
10270
10337
10815
10933
10961
11165

188
118
252
753
1060
1137
1149
1661
1671
1719
1766
2442
2756
3123
3173
3226
4156
4422
4565
5001
5055
5577
5612
5615
5653
6493
6494
8175
8357
8487
9073
9256
9628
10121
10284
10960
11027

189
123
1110
1465
1675
2168
2770
2898
3510
4360
4420
4517
4570
4622
4706
4948
5054
7529
7539
8451
9263
10367
10668
12039

190
1124
1466
1631
1673
2459
3344
3508
3551
3928
3997
4174
4555
4618
4735
4895
5556
5571
5683
6037
6590
7255
8124
8173
8835
8885
9647
10002
10134
10148
10162
10265
10312
10787
11014
11464
11494

191
262
1190
1700
2167
2740
3507
4231
4237
4614
4619
4884
4964
6084
6508
7251
8117
8860
8906
9400
9793
9820
10227
10564

192
1151
1480
1632
1660
1689
1917
2938
3405
3470
3811
3970
4285
4423
5515
5867
8160
8434
8739
9997
10519
10878
10919
11024
11441
11490
11732
11837
11966
11968
11971
11972
11973

193
163
193
1179
1688
2041
2954
2960
3366
3778
4593
4620
4701
5259
5737
6125
6488
6841
6932
8158
8161
8189
8452
10235
10875
11967
11970

194
160
162
713
860
1141
1670
1810
1832
1918
1920
1996
1997
2025
2029
2030
3492
3514
3960
4375
4639
5415
5425
5752
6429
6805
6951
8393
8512
8624
9228
9539
9637
9822
10078
10943
11646

195
25
868
1199
1817
1953
1954
1987
2301
2302
2303
2927
3368
3820
4208
4568
4595
5437
5520
5521
5592
6123
6346
6699
6700
6982
7208
7256
7790
8159
8172
8191
8443
8447
8457
8481
8689
8887
9261
10485
10967
11374
11938

196
231
243
251
781
849
1005
1006
1690
1806
1994
1995
1999
2280
3214
3318
3327
3536
3577
3794
3969
4334
4427
4723
4740
4892
5712
5720
5815
5845
6056
6092
6758
6810
7197
8454
8867
8908
9262
9636
9885
10423
10863
10944
11771
11969
11984

197
24
564
1164
1180
1472
1569
1616
1617
1637
1683
1699
1998
2031
2443
2456
2457
2886
3267
3846
4031
4567
4612
5569
5657
5844
6340
7897
8793
8888
9564
9942
11733

198
1109
1178
1192
1198
1312
1385
1684
2279
2340
2458
3518
3570
3844
4207
5574
5994
6014
6061
7277
8332
9536
9692
10959
10964
11791
11986
11987

199
174
187
423
588
1021
1035
1148
1153
1471
1734
1834
1916
1952
2687
2940
4551
5003
5263
5691
5747
5995
7201
7275
8418
8459
8789
8805
8922
9079
9317
10126
10209
11126

200
154
587
595
778
780
844
1176
1387
1464
1470
1820
2664
2880
3353
3467
4313
4754
5547
5754
6145
8169
8356
8889
8898
9673
9717
9941
10161
10869
11127

201
1053
1200
1386
1388
1879
2105
2106
2878
3461
4310
4558
4711
4713
5629
5793
6043
7792
8736
9310
9402
9962
11473

202
1034
1036
1469
1483
1857
1921
1950
2061
3354
3779
3936
4202
4569
5860
5864
6091
7258
7532
8119
8171
8453
8458
8708
8786
9401
9662
9663
10274
10283
11117
11185
11659

203
124
126
205
743

779
790
872
993
1020
1039
1417
1578
1702
1816
2251
3659
3742
3789
3917
4465
4708
4823
4911
5076
5527
6016
7051
7585
8176
8913
8964
10204
10242
10440
10971
10979
11022
12018

204
245
589
749
992
1004
1019
1037
1439
1486
1643
3641
4209
4605
5230
6142
6419
6762
9353
10293
10835
10906
11247
11846
11985

205
61
72
152
662
1018
1122
1243
1382
1419
1421
1455
1572
1639
1691
1703
1705
1735
1836
1896
2107
2278
2293
2763
2769
3597
4385
4454
5739
6126
6749
6797
6870
7055
7119
7257
7324
7689
8123
8411
8868
8907
9379
9726
10363

206
180
711
968
1416
1436
1487
1581
1809
1815
2026
2294
2295
3027
3200
3352
3404
3447
3503
4359
4610
5077
5431
6146
7200
7252
7693
7775
7814
8372
8463
9594
9794
9825
10052
10241
10836
10882
11246
11385
11401

207
710
1123
1157
1383
1420
1426
1460
1482
1574
2433
2919
2920
2961
3345
3516
3959
3998
4606
5497
6070
7279
7979
10788
11768

208
111
211
575
658
1121
1384
1418
1980
2009
2323
2983
2984
3323
3330
3528
4238
4387
4546
4560
4575
4707
4722
5030
5285
5728
6055
6467
6471
6796
6877
6880
7050
7207
7276
8412
8428
9074
9949
10005
10226
10855
10868
10908
11537
11573
12057

209
58
558
696
740
1026
1647
1669
1681
2955
2985
3837
4699
6154
6752
7278
7280
7935
8790
8877
8909
8974
8977
9237
10003
10571
10879
10914
11583
12029

210
220
222
712
769
994
1114
1156
1485
1579
1653
1880
2032
2042
2049
2444
2606
2757
3462
3487
3790
4220
4542
4791
4901
4909
5072
5075
5346
5827
6032
6071
8407
8796
9283
9308
9973
10245
10264
10653
11578
11743
11744

211
529
862
1051
1228
1778
1780
1874
1928
1945
2297
2298
2724
3201
3436
4308
5071
5565
5576
6798
6948
7815
8794
9284
9309
10176
10323
10818

212
212
213
861
1028
1063
1438
1573
1712
2027
2033
2414
2602
2965
3520
3591
4316
4781
5869
7273
7894
8449
10278
10789
11075

213
233
530
531
1142
1158
1377
1667
1709
1710
1731
1779
2322
2764
2902
2959
2963
3443
3496
4613
4720
5335
5870
6098
6894
7684
8154
8157
8792
9236
9410
9791
10234
10948

214
705
706
741
1017
1022
1580
2028
3495
3775
4600
4947
5684
5851
6801
7567
8775
8876
8890
9076
9252
9671
9983
10240
10856
11170
11383

215
112
707
716
1052
1186
1286
1600
1713
1898
1943
2345
2390
2696
2777
3163
3445
3542
4337
4340
4667
4721
4967
5343
6150
6893
7893
8177
8184
8591
9563
9802
9964
10127

216
695
746
929
1050
1133
1201
1202
1711
1762
1904
1949
2447
2451
2693
3194
3213
3442
3504
3968
4233
4548
5292
5645
6896
6961
7049
8167
9307
9435
10051
10122
10481
10996
11642
12058

217
246
255
577
742
748
1062
1132
1437
1732
1878
1899
2011
2060
2104
2446
2765
3303
3326
4446
4547
4719
5451
5564
6053
6895
7061
7732
8122
8542
8806
9078
9350
9355
9376
9585
9704
9798
10480
10968
11181

218
547
670
1059
1159
1267
1289
1290
1592
2321
2344
2454
2879
2885
2944
3952
4609
5721
5830
6799
7293
7389
7560
8912
9354
9720
9911
10053
10828
11114
11407
11698
11739
11742
11777

219
841
1038
1088
1235
1254
1406
1706
1859
1861
1870
2450
2903
4501
4783
4969
5729
7196
9568
9672
9688
10466
10798
10903
11053

220
548
668
709
794
1064
1440
1707
1708
1973
1992
2010
2172
2387
2445
2448
2698
2987
3444
3633
4362
4604
4668
4698
5032
5820
5852
5868
5989
6175
7284
7510
7512
7701
7995
8145
8335

9253
9285
9456
9457
9461
9495
9615
10014
10431
10469

221
549
704
747
788
873
874
1792
1830
2333
2361
2399
2452
2697
2948
4417
4601
4785
5707
6872
6906
7980
7996
8121
8920
9312
9385
9721
9957
9981
10106
10819
11454

222
22
843
1217
1361
1362
1761
1941
2000
2449
2453
2634
2892
2964
4549
6062
6165
6403
6455
8741
8863
9638
9800
9931
9980
10028
10163
10243
10696
10866
10926
10969
11297

223
23
576
1115
1134
1143
1363
1855
1901
2008
2116
2119
2749
2945
3776
4635
5915
6097
6563
7253
7433
9266
9321
10080
10339

224
766
1083
1651
1897
1948
2114
2299
2425
2778
3773
4180
4414
4608
4820
4996
5066
6371
7138
8144
8514
8856
10068
10881
10902
11775

225
41
242
789
816
1094
1288
1372
2745
2891
3527
3584
4181
4543
5036
5324
5762
6897
7292
7699
8334
8365
8406
8698
9264
9416
9556
10007
11379
11569
11688
11776

226
842
1014
1049
1476
1680
1763
2052
2904
3206
5252
5853
5914
6459
6561
6564
6842
6907
7057
7231
7983
8441
8882
9313
9630
9677
10385
10922
10972
11467
11468
11510

227
490
899
1172
1321
1721
1819
2330
2690
2978
6970
7691
7712
8150
8557
8858
9311
9460
9664
10146
10640
11959
11961

228
708
900
953
1812
1911
2300
2431
3329
3676
3677
3788
4696
4775
5809
6372
6882
7165
8880
8910
9250
9265
9411
9417
9430
9462
10307
10313
10688
11546
11951
11954
11962

229
38
638
669
697
1116
1117
1769
1772
1842
2109
3348
3787
4184
5557
6161
7297
7998
8317
9945
10472
10479
10612
11333
11952
11953

230
526
556
578
659
1057
1232
1364
1652
1771
1848
2292
2337
3419
3967
5080
6800
6855
7985
8333
8866
9260
9332
9412
9414
9459
9532
9605
9683
9940
10397
10973
11580

231
151
584
663
745
959
1023
1032
1360
1665
1666
1774
2336
2861
4520
4715
5070
5688
5736
6050
6881
6886
7166
7239
9295
10342
11958

232
147
232
535
660
736
817
859
952
1012
1218
1381
1692
1860
1912
2001
2097
2589
2692
2747
2995
3224
4267
4332
4716
5449
5789
5810
5856
6910
7388
8997
9245
9418
9427
9463
9494
9573
9875
9923
10015
10639
11471
11481
11709
11710
11893
12032

233
489
945
1376
2127
3678
5229
5235
6052
8440
9232
9287
9335
10399
10842
11457
11636
11782

234
51
527
714
819
858
1065
1093
1171
1942
1944
2290
2429
2691
2780
3124
3635
4638
4951
5913
7982
7997
9293
9951
10020
11134
11160
11897

235
540
590
770
815
1011
1215
1247
1407
1555
1605
1849
2148
2354
2638
2969
2994
3179
3225
3346
3793
3958
4717
5453
6190
6984
7889
7981
8795
9241
9246
9267
9315
10253
10467
11052
11704
11871

236
439
1008
1305
1577
1813
1865
1905
1951
2746
3671
5092
5677
6727
7289
7987
7990
8772
9229
9292
9357
9493
9718
10468
11157
11485
11511

237
814
1015
1375
2539
2870
3590
4776
6795
6812
7159
7312
7989
8047
8904
9041
9042
9294
9314
9316
9323
9330
9333
9415
9429
9432
9604
10239
11172
11182
11214

238
599
813
990
1082
1092
1204
1462
1488
1561
1776
2002
2139
2296
2373
2635
2863
2890
2988
3911
4634
7534
9230
9276
9458
9930
12026

239
287
676
986
2155
2652
3585
4766
5676
6094
6565
7685
7984
8349
8422
8787
9631

240
95
537
565
647
650
651
675
904
1174
1248
1303
1474
1478
1745
1765
3512
3624
4212
5034
5058
5840
6875
7988
8000
8427
9322
9474
9520
10004
10320
11447

241
486
674
715
793
951
1357
1729
1775
2332
2460
4637

4913
5063
5558
5697
5876
6985
7160
7888
7916
8366
8371
8791
8896
9431
9961
10254
11195
11818

242
730
962
1013
1175
1212
1216
1397
1693
2701
2894
4461
5213
6397
7553
8705
11224
11474
11937

243
583
656
672
694
768
1402
1609
1794
2537
2538
2639
2887
3190
4843
5057
5632
5814
5998
7169
7286
7373
8623
9381
9517
9944
10321
10901

244
157
673
960
1441
1610
2260
2334
2372
2640
2675
3566
4405
4664
5788
6802
7891
7915
9288
10398
10580
11337

245
230
553
671
686
870
985
1010
1214
1239
1497
1562
1694
3560
4306
4624
4997
5689
5719
5753
5758
5826
5842
5866
6470
6757
6785
6990
7890
8881
8986
9212
9420
9496
9687
9698
9996
10483
10628
10792
11266
11274
11810
11950

246
479
963
1304
1322
1323
1408
2335
2673
2888
5325
5811
5875
6890
9239
9257
9351
9352
9419
9421
9485
9616
10518
10864
11277
11299
11411
11920

247
229
536
571
932
1269
1433
1479
2131
2401
3818
4210
5196
5243
5723
5742
5790
9475
9484
10048
10419
10992
11136
11283

248
552
585
772
961
1410
1527
2674
2862
3074
3669
5211
5626
5682
7244
7381
7575
9247
9248
9327
10516
11186
11861
11978
11981

249
9
554
1173
1177
1205
1603
1889
1902
2461
2475
5802
5850
5960
7382
9325
9722
9972
10049
10305
10588
11338

250
1233
1604
1657
1799
2051
2269
2532
3321
4429
7715
8783
9392
9444
10306
10384
11019
11035
11979

251
639
687
767
771
949
987
1431
1475
2598
3582
3672
5774
6321
9554
9938
10447
10451
10793
11180
11296

252
661
685
1047
1048
1246
1884
1890
2268
2744
2750
4182
5322
7307
7991
7994
9567
10707
10858
11640
11963

253
586
773
1380
1556
2430
3075
5195
7285
7294
7552
7992
9286
9389
9390
9553
11613
11941

254
140
737
783
907
1016
1450
1467
1510
2259
2331
2521
3044
4786
5848
6348
6394
7993
8146

255
462
649
848
871
970
1238
1275
1298
1392
2050
2355
2363
2681
2760
3082
3295
4910
5696
6721
8041
8983
9277
9404
10838
10870
11259
11849
12012

256
79
935
1074
1077
1293
2503
2762
3008
3050
3297
3417
4392
5194
5761
5871
6869
7383
7781
9226
9231
9347
9719
11239
11505

257
362
636
1046
1260
1359
1391
3565
4787
5781
7192
8830
9227
9694
9695
10704
10705
10716
10717
11282
11942

258
1075
1280
1296
1320
1396
1528
1936
2071
2755
4949
5841
6430
6790
7298
7385
7568
7885
7886
9268
9324
9548
9693
10821
11281

259
644
847
1295
1597
1891
2276
2600
2676
3670
4223
5291
7780
10189
10195
10456
10460
11278

260
62
75
667
1210
1287
2328
2670
2754
3112
3426
5321
5973
5990
7162
7237
7605
7700
8128
8439
9340
10656
11012
11790
11998

261
655
690
950
1209
1253
1266
1593
2123
2522
2643
2669
3362
3632
3882
5787
6093
6465
6725
6889
7242
8777
8901
9305
10686
11891

262
854
1197
2685
2901
3056
4066
4841
6399
7826
9339
9992
9993
11238
11279
11280
11567
11836
12010

263
682
924
1270
1507
1758
2012
2040
2417
2523
2741
4352
5053
5295
5808
6164
6374
7210
7819
7823
9219
10587
11240
11254
11862
11889
12001

264
74
645
964
1237
1292
1663
1793
2067
2989
4460
4795
7822
9326
9572
11256
11258
11977

265
178
570
582
883
926
1076
1249
1276
1294
1411
1529
1961
2362
4502
6983
7109
7384
7821
7827
8392
8886
9224
9489
9661
10132
10319
10674
10706
10741
10880
11255

266
1934
2680
2895
3072
3291
5499
7314
7317
7818
7954
8042
8601

9318
9388
9473
10976
11237
11257
11263
11267

267
29
1262
1395
1536
1612
1662
1747
3084
5315
5555
5928
7209
7315
7316
9391
9443
9699
11241
11308
11321
11892

268
88
1250
2495
2580
4084
5059
5730
7820
7825
7920
7921
8388
8653
8656
9329
10459
10461
11264
11335

269
1069
1073
1796
2541
2671
3098
6431
6456
7235
7904
7913
9223
9555
9955
10687
11245
11269
12004

270
493
544
642
702
853
1268
1449
1451
1530
1892
1906
2117
2649
2752
2872
2912
3051
4289
4341
5771
5784
6528
6891
7824
8559
8642
9186
9700
10779
11017
11261
11331

271
897
1535
2076
2309
2614
2672
5312
5999
6352
8369
10685
10997
11774

272
689
820
1255
1677
2013
2137
2610
2798
3054
3083
4386
5212
7313
9221
9697
10829
10998
11135
11270

273
641
1279
1837
2324
2723
3099
5579
6271
7328
7927
7958
8604
10823
10859
11262
11975

274
684
1477
1553
1937
2582
3023
3060
3080
3301
6905
7376
7898
7914
7957
9278
11841
11883
11887

275
691
1881
2308
2599
2707
2710
3305
5692
6293
6904
8044
8059
8995
9486
11217
11870
11974

276
1319
1498
1514
2262
2428
2440
2666
2905
3026
4468
5293
5332
7212
7829
9297
11008
11789

277
833
1265
1297
1636
1935
2034
2084
2090
2102
2661
3053
5303
5304
6460
6464
8431
11004
11416

278
1512
2326
2416
2633
5654
5690
9498
11209

279
6
538
680
2015
2126
4790
5759
7308
7731
7828

280
3
224
943
1784
1969
2014
2064
2081
2166
2327
2660
3057
3308
3505
3766
3974
6458
7240
7883
8442
10837
10995
11033
11218
11302
11303
11943

281
39
545
1318
2171
2636
3101
3307
7739
7740
9395
9516
10987
10999
11219
11248
11318
11784

282
719
1347
1835
1976
3059
4457
5766
7374
7626
7884
7923
7924
8387
8424
9222
9393
11222

283
1343
1401
1532
1664
2579
3073
4472
4932
6526
7322
7332
8352
8416
9225
10679
11276
11301

284
822
1221
1513
1587
2077
3024
4430
6667
6673
11013
11271
11298
11631

285
555
574
1261
1313
1428
1429
1499
1516
1958
1971
2069
2078
2285
2419
2551
3096
3774
5713
5744
6791
7364
7618
7690
7713
7811
7813
8018
8480
9280
10170
10693
10925
11300

286
738
958
1964
2016
2382
2627
2791
3088
4432
5714
6152
6632
7238
7365
8767
9479
9620
10638
11413

287
1120
1427
1531
3095
5163
5745
5805
11005
11244
11304
11630
11786

288
1351
1785
1811
1938
2339
3411
4351
4408
5996
7809
8008
8017
10670
11006
11242

289
1966
2742
3058
5772
6639
7746
9562
11003
11036
11339

290
543
600
1100
1538
1783
2314
4741
6468
7349
7776
7801
8430
8564
10186
10637
11220
11758

291
620
1492
1676
3102
3423
4466
4995
7773
7783
7808
8133
8652
11115
11116
11260
11612

292
852
1278
2232
2626
2833
3025
4989
5238
7274
7309
7747
7751
11009
11872

293
807
2020
4326
5129
6783
7955
11880

294
427
632
1444
2423
2496
5773
7241
7721
7830
8651
9272
10169

295
618
751
1272
1342
2578
3061
3100
4447
5967
6777
7342
9185
10613
11184
11380
11624
11625

296
76
1113
1307
1336
1718
1822
2283
2356
2381
3576
4863
5831
5874
7807
8644
11243
11560
11615

297
1493
1696
2621
2708
3016
3076
3103
5824
5857
7323
8354

298
1258
1589
1756
1856
1910
2079
2313
5302
6435

299
1224
2820
3010
4344
7748
8647

300
839
851
1714
2121
2370
2378
2733
3017
6322
7329
7361
7641
7708
9977
10682

301
2046
2086
2175
2562

2792
9093
10932

302
836
1346
1350
1821
2161
3011
3855
5858
6138
6306
7902
11406

303
77
2021
2310
2422
3310
5215
7737
7741

304
70
86
1697
1698
1826
2264
2835
7908
10647
11211
11319

305
376
799
3062
3071
3090
3104
4323
5045
7616
7637
7716
7718
7750
7956
9884
11395

306
1827
4001
4402
7709
7749
7764
8437
10473
11275

307
78
1619
2017
2311
3045
3586
7941
8063
8178
8641
10432
10464
10871
11614

308
834
2108
2312
7263
7631
7770
7903
11037

309
2018
2366
2699

310
831
840
1196
1979
2158
2497
2753
3018
3658
5258
7765
7802
7942
11204
11621

311
1584
2022
2111
2436
6310
11628
12049

312
625
946
1824
2439
4809
5139
6659
9454

313
1282
2073
5828
7344
8313
10465
10474

314
809
2716
2717
6660
7930
10813

315
810
835
1144
1145
2365
6656
7639
8027
9606

316
795
808
2374
2700
7719
8234
9599

317
624
1099
2019
3041
3674
7628
7632
7638
7722
10710

318
811
1491
2074
2080
7917

319
5
796
1825
10847
11213

320
1024
1678
2054
2651
3040
3064
3079
3085
3313
4299
5812
6320
9525
9634
11208

321
421
7614

322
1146
6631
7900
7943
8643
9455
11007
11964

323
2720
4409

324
6317
7901

325
1749
2145
4433
4464
7571
11000
11328

326
89
5776
6634
7617
7803
11781

327
1929
2098
3989
4854
8107

328
1566
3046
5756
8259

329
2253
4763
7343
8101
9701

330
797
2075
3320
3431
5701
9447

331
764
4382
10433

332
824
2782
3561
5167
7604
11595

333
48
1746
6641
7596
8064
8311
10814
11210

334
1519
2085
7599
7947

335
4
679
2867
2882
5751
7944
7946
8278

336
2305
6305
6633

337
258
454
2143
5993
6779
7594

338
7369
9218
10831

339
366
678
2072
2122
2474
5133
8579

340
1339
2256
2306
3087
5137
5366
7213
7895
9163
9446

341
4850
5362
5498
5975
7615
7896
11811

342
3065
3876

343
3021
3987
4217
4218
5350
6962
7624
8251

344
314
5044
5822
6100
7291
11431

345
1333
7600
7948
8104

346
2342
4484
8233

347
5584

348
9188

349
5239
6307
6312
7945

350
1317
2379
7602
7949
7950
9180
9195
11606
12006

351
2516
6644
12050

352
1095
2437
8236
8969

353
4118
7592

354
1283
3981

355
35
2714
3013
4844
7625

356
5358
6294
7937

357
3037
5359
6640
7513
7938

359
1982
2266

360
450
1585
1591
2510
2868
3946
4475
7588
7912
8294
8580
9190
9593
11306

362
8235
8243

363
3029
11288
11596

365
728
3986
7629
11660

366
7931

367
5280

368
4822
5268

371
2864
9499

372
7932

373
2089
4463
11812

374
2255

375
9583

376
11597

377
379
9528

379
2237
7793

380
1866
11855

383
8242

384
1334
5966

385
1751
1798
4411

387
7794

388
8648

390
9183

391
4842
11603

393
11163

395
7590
7728

397
1743
4404
10947

398
784

400
5172
11162

401
3031
8637

402
7725

404
318
10853

405
2393
7730

406
2240
4476

408
2066

409
4459

410
9569

412
6622

413
5351

414
7880

415
7879

417
428
6653
11509

419
2062

420
1623
4456
4819
11193
11545

425
11910

429
1568

430
4384

431
1831
8220

433
4808

434
782

438
519

439
1741

440
8099

442
6302

443
8266

448
1565
2405
4114

449
4380
11819

452
3005

458
6299

459
417
4483

462
418
11598

467
4474
5348

468
4821

471
7727

476
4462

480
2739

484
4410

490
8999

498
4853

499
4458

514
4482

519
9601

525
4145

527
4383

531
4845

540
4479

551
4413

DIAMAGNETIC SUSCEPTIBILITY OF SELECTED ORGANIC COMPOUNDS

When a material is placed in a magnetic field H, a magnetization M is induced in the material which is related to H by $M = \kappa H$, where κ is called the volume susceptibility. Since H and M have the same dimensions, κ is dimensionless. A more useful parameter is the molar susceptibility χ_m, defined by

$$\chi_m = \kappa V_m = \kappa M/\rho$$

where V_m is the molar volume of the substance, M the molar mass, and ρ the mass density. When the cgs system is used, the customary unit for χ_m is $cm^3\ mol^{-1}$; the corresponding SI unit is $m^3\ mol^{-1}$. Substances with no unpaired electrons are called diamagnetic; they have negative values of χ_m.

This table gives values of the diamagnetic susceptibility for about 400 common organic compounds. All values refer to room temperature and atmospheric pressure and to the physical form that is stable under these conditions. Substances are arranged by molecular formula in Hill order. A more extensive table may be found in Reference 1.

In keeping with customary practice, the molar susceptibility is given here in units appropriate to the cgs system. These values should be multiplied by 4π to obtain values for use in SI equations (where the magnetic field strength H has units of $A\ m^{-1}$).

REFERENCES

1. *Landolt-Börnstein, Numerical Data and Functional Relationships in Science and Technology, New Series,* II/16, *Diamagnetic Susceptibility,* Gupta, R. R., Ed., Springer-Verlag, Heidelberg, 1986.
2. Barter, C., Meisenheimer, R. G., and Stevenson, D. P., *J. Phys. Chem.* 64, 1312, 1960.
3. Broersma, S., *J. Chem. Phys.* 17, 873, 1949.

Molecular Formula	Compound	$-\chi_m/10^{-6}\ cm^3\ mol^{-1}$
$CBrCl_3$	Bromotrichloromethane	73.2
CBr_4	Tetrabromomethane	93.7
$CClF_3$	Chlorotrifluoromethane	45.3
$CClN$	Cyanogen chloride	32.4
CCl_2F_2	Dichlorodifluoromethane	52.2
CCl_2O	Carbonyl chloride	47.9
CCl_3F	Trichlorofluoromethane	58.7
CCl_3NO_2	Trichloronitromethane	75.3
CCl_4	Tetrachloromethane	66.8
$CHBrCl_2$	Bromodichloromethane	66.3
$CHBr_3$	Tribromomethane	82.6
$CHCl_3$	Trichloromethane	58.9
CHI_3	Triiodomethane	117.1
CH_2BrCl	Bromochloromethane	55.1
CH_2Br_2	Dibromomethane	65.1
CH_2Cl_2	Dichloromethane	46.6
CH_2I_2	Diiodomethane	93.1
CH_2N_2	Cyanamide	24.8
CH_2O	Formaldehyde	18.6
CH_2O_2	Formic acid	19.9
CH_3Br	Bromomethane	42.8
CH_3Cl	Chloromethane	32.0
CH_3F	Fluoromethane	17.8
CH_3I	Iodomethane	57.2
CH_3NO	Formamide	23.0
CH_3NO_2	Nitromethane	21.0
CH_4	Methane	17.4
CH_4N_2O	Urea	33.5
CH_4O	Methanol	21.4
CH_5N	Methylamine	27.0
CI_4	Tetraiodomethane	136
CN_4O_8	Tetranitromethane	43.0
C_2ClF_3	Chlorotrifluoroethylene	49.1
C_2Cl_4	Tetrachloroethylene	81.6
C_2Cl_6	Hexachloroethane	112.8
C_2HCl_3	Trichloroethylene	65.8
C_2HCl_3O	Trichloroacetaldehyde	73.0
C_2HCl_3O	Dichloroacetyl chloride	69.0
$C_2HCl_3O_2$	Trichloroacetic acid	73.0
C_2HCl_5	Pentachloroethane	99.1
$C_2HF_3O_2$	Trifluoroacetic acid	43.3
C_2H_2	Acetylene	20.8
$C_2H_2Br_4$	1,1,2,2-Tetrabromoethane	123.4
$C_2H_2Cl_2$	1,1-Dichloroethylene	49.2
$C_2H_2Cl_2$	*cis*-1,2-Dichloroethylene	51.0
$C_2H_2Cl_2$	*trans*-1,2-Dichloroethylene	48.9
$C_2H_2Cl_4$	1,1,2,2-Tetrachloroethane	89.8
C_2H_3Cl	Chloroethylene	35.9
C_2H_3ClO	Acetyl chloride	39.3
C_2H_3N	Acetonitrile	27.8
C_2H_4	Ethylene	18.8
$C_2H_4Br_2$	1,2-Dibromoethane	78.9
$C_2H_4Cl_2$	1,1-Dichloroethane	57.4
$C_2H_4Cl_2$	1,2-Dichloroethane	59.6
C_2H_4O	Acetaldehyde	22.2
C_2H_4O	Ethylene oxide	30.5
$C_2H_4O_2$	Acetic acid	31.8
$C_2H_4O_2$	Methyl formate	31.1
C_2H_5Br	Bromoethane	78.8
C_2H_5Cl	Chloroethane	69.9
C_2H_5I	Iodoethane	69.1
C_2H_5NO	Acetamide	33.9
$C_2H_5NO_2$	Nitroethane	35.4
$C_2H_5NO_2$	Glycine	39.6
C_2H_6	Ethane	26.8
C_2H_6O	Ethanol	33.7
C_2H_6O	Dimethyl ether	26.3
$C_2H_6O_2$	Ethylene glycol	38.9
C_2H_6S	Ethanethiol	47.0
C_2H_6S	Dimethyl sulfide	44.9
$C_2H_8N_2$	1,2-Ethanediamine	46.5
C_2N_2	Cyanogen	21.6

Molecular Formula	Compound	$-\chi_m/10^{-6}$ cm^3 mol^{-1}
C_3H_4	Allene	25.3
$C_3H_4O_2$	Vinyl formate	34.7
C_3H_5Br	3-Bromopropene	58.6
C_3H_5Cl	2-Chloropropene	47.8
C_3H_5Cl	3-Chloropropene	47.8
C_3H_5N	Propanenitrile	38.6
C_3H_6	Propene	30.7
C_3H_6	Cyclopropane	39.2
C_3H_6O	Allyl alcohol	36.7
C_3H_6O	Propanal	34.2
C_3H_6O	Acetone	33.8
C_3H_6O	Methyloxirane	42.5
$C_3H_6O_2$	Propanoic acid	43.2
$C_3H_6O_2$	Ethyl formate	42.4
C_3H_7Br	1-Bromopropane	65.6
C_3H_7Br	2-Bromopropane	65.1
C_3H_7Cl	1-Chloropropane	56.0
C_3H_7I	1-Iodopropane	84.3
C_3H_7N	Allylamine	40.1
$C_3H_7NO_2$	1-Nitropropane	45.0
$C_3H_7NO_2$	2-Nitropropane	45.4
$C_3H_7NO_2$	Ethyl carbamate	57.0
C_3H_8	Propane	38.6
C_3H_8O	1-Propanol	44.8
C_3H_8O	2-Propanol	45.7
$C_3H_8O_2$	1,3-Propylene glycol	50.2
$C_3H_8O_2$	Dimethoxymethane	47.3
$C_3H_8O_3$	Glycerol	57.1
$C_4H_2O_3$	Maleic anhydride	35.8
$C_4H_4N_2$	Pyrazine	37.8
$C_4H_4N_2$	Pyrimidine	43.1
C_4H_4O	Furan	43.1
$C_4H_4O_3$	Succinic anhydride	47.5
$C_4H_4O_4$	Maleic acid	49.6
$C_4H_4O_4$	Fumaric acid	49.1
C_4H_4S	Thiophene	57.3
C_4H_5N	Pyrrole	48.6
C_4H_6	1,2-Butadiene	35.6
C_4H_6	1,3-Butadiene	32.1
$C_4H_6O_2$	Vinyl acetate	46.4
$C_4H_6O_3$	Acetic anhydride	52.8
$C_4H_6O_4$	Succinic acid	58.0
$C_4H_6O_4$	Dimethyl oxalate	55.7
C_4H_7N	Butanenitrile	50.4
C_4H_8	1-Butene	41.0
C_4H_8	*cis*-2-Butene	42.6
C_4H_8	*trans*-2-Butene	43.3
C_4H_8	Isobutene	40.8
C_4H_8	Cyclobutane	40.0
C_4H_8O	Ethyl vinyl ether	47.9
C_4H_8O	1,2-Epoxybutane	54.8
C_4H_8O	Butanal	45.9
C_4H_8O	2-Butanone	45.6
$C_4H_8O_2$	Butanoic acid	55.2
$C_4H_8O_2$	2-Methylpropanoic acid	56.1
$C_4H_8O_2$	Propyl formate	55.0
$C_4H_8O_2$	Ethyl acetate	54.1
$C_4H_8O_2$	Methyl propanoate	54.5
$C_4H_8O_2$	1,4-Dioxane	52.2
C_4H_9Br	1-Bromobutane	77.1
C_4H_9Br	1-Bromo-2-methylpropane	79.9
C_4H_9Cl	1-Chlorobutane	67.1
C_4H_9Cl	2-Chlorobutane	67.4
C_4H_9I	1-Iodobutane	93.6
C_4H_9N	Pyrrolidine	54.8
C_4H_9NO	Morpholine	55.0
C_4H_{10}	Butane	50.3
C_4H_{10}	Isobutane	50.5
$C_4H_{10}O$	1-Butanol	56.4
$C_4H_{10}O$	2-Butanol	57.6
$C_4H_{10}O$	2-Methyl-1-propanol	57.6
$C_4H_{10}O$	2-Methyl-2-propanol	56.6
$C_4H_{10}O$	Diethyl ether	55.5
$C_4H_{10}O_2$	1,3-Butanediol	61.8
$C_4H_{10}O_2$	1,4-Butanediol	61.8
$C_4H_{10}S$	1-Butanethiol	70.2
$C_4H_{11}N$	Butylamine	58.9
$C_4H_{11}N$	Isobutylamine	59.8
$C_4H_{11}N$	Diethylamine	56.8
$C_5H_4O_2$	Furfural	47.2
C_5H_5N	Pyridine	48.7
$C_5H_6O_2$	Furfuryl alcohol	61.0
$C_5H_7NO_2$	Ethyl cyanoacetate	67.3
C_5H_8	2-Methyl-1,3-butadiene	46.0
C_5H_8O	Cyclopentanone	51.6
$C_5H_8O_2$	Methyl methacrylate	57.3
$C_5H_8O_2$	2,4-Pentanedione	54.9
C_5H_{10}	1-Pentene	54.6
C_5H_{10}	2-Methyl-2-butene	54.7
C_5H_{10}	Cyclopentane	56.2
$C_5H_{10}O$	Cyclopentanol	64.0
$C_5H_{10}O$	Pentanal	57.5
$C_5H_{10}O$	2-Pentanone	57.5
$C_5H_{10}O$	3-Pentanone	57.7
$C_5H_{10}O_2$	Pentanoic acid	66.5
$C_5H_{10}O_2$	3-Methylbutanoic acid	67.7
$C_5H_{10}O_2$	Butyl formate	65.8
$C_5H_{10}O_2$	Isobutyl formate	66.8
$C_5H_{10}O_2$	Propyl acetate	65.9
$C_5H_{10}O_2$	Isopropyl acetate	67.0
$C_5H_{10}O_2$	Ethyl propanoate	66.3
$C_5H_{10}O_2$	Tetrahydrofurfuryl alcohol	69.4
$C_5H_{10}O_3$	Diethyl carbonate	75.4
$C_5H_{11}N$	Piperidine	64.2
C_5H_{12}	Pentane	61.5
C_5H_{12}	Isopentane	63.0
C_5H_{12}	Neopentane	63.0
$C_5H_{12}O$	1-Pentanol	67.0
$C_5H_{12}O$	2-Pentanol	69.1
$C_5H_{12}O_2$	1,5-Pentanediol	73.5
$C_5H_{13}N$	Pentylamine	69.3
C_6Cl_6	Hexachlorobenzene	147.0
$C_6H_4ClNO_2$	1-Chloro-2-nitrobenzene	75.5
$C_6H_4ClNO_2$	1-Chloro-3-nitrobenzene	77.2
$C_6H_4ClNO_2$	1-Chloro-4-nitrobenzene	74.7
$C_6H_4Cl_2$	*o*-Dichlorobenzene	84.4

Molecular Formula	Compound	$-\chi_m/10^{-6}$ cm^3 mol^{-1}
$C_6H_4Cl_2$	*m*-Dichlorobenzene	84.1
$C_6H_4Cl_2$	*p*-Dichlorobenzene	81.7
$C_6H_4O_2$	*p*-Benzoquinone	36
C_6H_5Br	Bromobenzene	78.4
C_6H_5Cl	Chlorobenzene	69.5
C_6H_5ClO	*o*-Chlorophenol	77.3
C_6H_5ClO	*m*-Chlorophenol	77.6
C_6H_5ClO	*p*-Chlorophenol	77.7
C_6H_5F	Fluorobenzene	58.4
C_6H_5I	Iodobenzene	92.0
$C_6H_5NO_2$	Nitrobenzene	61.9
$C_6H_5NO_3$	*o*-Nitrophenol	68.9
$C_6H_5NO_3$	*m*-Nitrophenol	65.9
$C_6H_5NO_3$	*p*-Nitrophenol	66.9
C_6H_6	Benzene	54.8
C_6H_6ClN	*o*-Chloroaniline	79.5
C_6H_6ClN	*m*-Chloroaniline	76.6
C_6H_6ClN	*p*-Chloroaniline	76.7
$C_6H_6N_2O_2$	*o*-Nitroaniline	67.4
$C_6H_6N_2O_2$	*m*-Nitroaniline	69.7
$C_6H_6N_2O_2$	*p*-Nitroaniline	68.0
C_6H_6O	Phenol	60.6
$C_6H_6O_2$	*p*-Hydroquinone	64.7
$C_6H_6O_2$	Pyrocatechol	68.2
$C_6H_6O_2$	Resorcinol	67.2
C_6H_7N	Aniline	62.4
C_6H_7N	4-Methylpyridine	59.8
C_6H_8	1,4-Cyclohexadiene	48.7
$C_6H_8N_2$	*o*-Phenylenediamine	72.5
$C_6H_8N_2$	*m*-Phenylenediamine	70.4
$C_6H_8N_2$	*p*-Phenylenediamine	70.7
C_6H_{10}	1,5-Hexadiene	55.1
C_6H_{10}	1-Hexyne	64.5
C_6H_{10}	Cyclohexene	58.0
$C_6H_{10}O$	Cyclohexanone	62.0
$C_6H_{10}O_3$	Ethyl acetoacetate	71.7
$C_6H_{10}O_4$	Diethyl oxalate	81.7
C_6H_{12}	1-Hexene	66.4
C_6H_{12}	2,3-Dimethyl-2-butene	65.9
C_6H_{12}	Cyclohexane	68
C_6H_{12}	Methylcyclopentane	70.2
$C_6H_{12}O$	Hexanal	69.4
$C_6H_{12}O$	2-Hexanone	69.2
$C_6H_{12}O$	3-Hexanone	69.0
$C_6H_{12}O$	4-Methyl-2-pentanone	69.7
$C_6H_{12}O$	Cyclohexanol	73.4
$C_6H_{12}O_2$	Hexanoic acid	78.1
$C_6H_{12}O_2$	Isopentyl formate	78.4
$C_6H_{12}O_2$	Isobutyl acetate	78.7
$C_6H_{12}O_2$	Propyl propanoate	77.7
$C_6H_{12}O_3$	Paraldehyde	86.1
C_6H_{14}	Hexane	74.1
C_6H_{14}	2-Methylpentane	75.3
C_6H_{14}	3-Methylpentane	75.5
C_6H_{14}	2,2-Dimethylbutane	76.2
C_6H_{14}	2,3-Dimethylbutane	76.2
$C_6H_{14}O$	1-Hexanol	79.5
$C_6H_{14}O$	4-Methyl-2-pentanol	80.4

Molecular Formula	Compound	$-\chi_m/10^{-6}$ cm^3 mol^{-1}
$C_6H_{14}O$	Dipropyl ether	79.4
$C_6H_{14}O_2$	1,6-Hexanediol	84.3
$C_6H_{14}O_2$	1,1-Diethoxyethane	81.4
$C_6H_{14}O_6$	*D*-Glucitol	107.8
$C_6H_{15}N$	Triethylamine	83.3
C_7H_5N	Benzonitrile	65.2
C_7H_6O	Benzaldehyde	60.7
$C_7H_6O_2$	Salicylaldehyde	66.8
$C_7H_6O_3$	Salicylic acid	75
C_7H_7Br	*p*-Bromotoluene	88.7
C_7H_7Cl	*o*-Chlorotoluene	82.4
C_7H_7Cl	*m*-Chlorotoluene	79.7
C_7H_7Cl	*p*-Chlorotoluene	80.3
C_7H_7Cl	(Chloromethyl)benzene	81.6
C_7H_7NO	Benzamide	72.0
$C_7H_7NO_2$	*o*-Nitrotoluene	72.2
$C_7H_7NO_2$	*m*-Nitrotoluene	72.7
$C_7H_7NO_2$	*p*-Nitrotoluene	73.3
C_7H_8	Toluene	65.6
C_7H_8O	*o*-Cresol	73.3
C_7H_8O	*m*-Cresol	72.2
C_7H_8O	*p*-Cresol	72.4
C_7H_8O	Benzyl alcohol	71.8
C_7H_8O	Anisole	72.2
C_7H_9N	*o*-Methylaniline	74.9
C_7H_9N	*m*-Methylaniline	74.6
C_7H_9N	*p*-Methylaniline	72.5
C_7H_9N	*N*-Methylaniline	74.1
C_7H_9N	2,4-Dimethylpyridine	71.3
C_7H_9N	2,6-Dimethylpyridine	72.5
C_7H_9NO	*o*-Methoxyaniline [*o*-Anisidine]	79.1
$C_7H_{12}O_4$	Diethyl malonate	92.6
C_7H_{14}	1-Heptene	77.8
C_7H_{14}	Cycloheptane	73.9
C_7H_{14}	Methylcyclohexane	78.9
$C_7H_{14}O$	1-Heptanal	81.0
$C_7H_{14}O$	2-Heptanone	80.5
$C_7H_{14}O$	3-Heptanone	80.7
$C_7H_{14}O$	4-Heptanone	80.5
$C_7H_{14}O$	2,4-Dimethyl-3-pentanone	81.1
$C_7H_{14}O_2$	Heptanoic acid	89.0
$C_7H_{14}O_2$	Pentyl acetate	88.9
$C_7H_{14}O_2$	Isopentyl acetate	89.4
$C_7H_{14}O_2$	Butyl propanoate	89.1
$C_7H_{14}O_2$	Ethyl 3-methylbutanoate	91.1
C_7H_{16}	Heptane	85.2
C_7H_{16}	3-Ethylpentane	86.2
C_7H_{16}	2,2-Dimethylpentane	87.0
C_7H_{16}	2,3-Dimethylpentane	87.5
C_7H_{16}	2,4-Dimethylpentane	87.5
C_7H_{16}	3,3-Dimethylpentane	89.5
$C_7H_{16}O$	1-Heptanol	91.7
$C_7H_{16}O$	4-Heptanol	92.1
$C_8H_4O_3$	Phthalic anhydride	66.7
$C_8H_6O_4$	Phthalic acid	83.6
$C_8H_6O_4$	Isophthalic acid	84.6
$C_8H_6O_4$	Terephthalic acid	83.5
C_8H_7N	Benzeneacetonitrile	76.9

DIAMAGNETIC SUSCEPTIBILITY OF SELECTED ORGANIC COMPOUNDS (continued)

Molecular Formula	Compound	$-\chi_m/10^{-6}$ cm^3 mol^{-1}
C_8H_7N	Indole	85.0
C_8H_8	Styrene	68.2
C_8H_8O	Acetophenone	72.5
$C_8H_8O_2$	*o*-Toluic acid	84.3
$C_8H_8O_2$	*m*-Toluic acid	83.0
$C_8H_8O_2$	*p*-Toluic acid	82.4
$C_8H_8O_2$	Benzeneacetic acid	82.4
$C_8H_8O_2$	Methyl benzoate	81.6
$C_8H_8O_3$	Methyl salicylate	86.6
C_8H_{10}	Ethylbenzene	77.3
C_8H_{10}	*o*-Xylene	77.7
C_8H_{10}	*m*-Xylene	76.4
C_8H_{10}	*p*-Xylene	77.0
$C_8H_{10}O$	Phenetole	84.5
$C_8H_{11}N$	*N*-Ethylaniline	85.6
$C_8H_{11}N$	*N,N*-Dimethylaniline	85.1
$C_8H_{11}N$	2,4,6-Trimethylpyridine	83.1
$C_8H_{14}O_4$	Ethyl succinate	105.0
C_8H_{16}	1-Octene	88.8
C_8H_{16}	Cyclooctane	85.3
$C_8H_{16}O_2$	Octanoic acid	99.5
$C_8H_{16}O_2$	Hexyl acetate	100.9
$C_8H_{17}Cl$	1-Chlorooctane	114.9
C_8H_{18}	Octane	96.6
C_8H_{18}	4-Methylheptane	97.3
C_8H_{18}	3-Ethylhexane	97.8
C_8H_{18}	3,4-Dimethylhexane	99.1
C_8H_{18}	2,2,4-Trimethylpentane	99.1
C_8H_{18}	2,3,4-Trimethylpentane	99.8
$C_8H_{18}O$	1-Octanol	101.6
$C_8H_{19}N$	Dibutylamine	103.7
C_9H_7N	Quinoline	86.1
C_9H_7N	Isoquinoline	83.9
C_9H_8	Indene	83
C_9H_{10}	Isopropenylbenzene	80.0
$C_9H_{10}O_2$	Ethyl benzoate	93.8
$C_9H_{10}O_2$	Benzyl acetate	93.2
C_9H_{12}	Propylbenzene	89.1
C_9H_{12}	Isopropylbenzene [Cumene]	89.5
C_9H_{12}	1,3,5-Trimethylbenzene [Mesitylene]	92.3
C_9H_{18}	1-Nonene	100.1
$C_9H_{18}O$	2,6-Dimethyl-4-heptanone	104.3
C_9H_{20}	Nonane	108.1
$C_{10}H_7Br$	1-Bromonaphthalene	123.6
$C_{10}H_7Cl$	1-Chloronaphthalene	107.6
$C_{10}H_8$	Naphthalene	91.6
$C_{10}H_8$	Azulene	123.7
$C_{10}H_8O$	1-Naphthol	96.2
$C_{10}H_8O$	2-Naphthol	96.8
$C_{10}H_9N$	1-Naphthalenamine	92.5
$C_{10}H_9N$	2-Naphthalenamine	98.0
$C_{10}H_{10}O_2$	Safrole	97.5
$C_{10}H_{10}O_4$	Dimethyl terephthalate	101.6
$C_{10}H_{14}$	Butylbenzene	100.7
$C_{10}H_{14}$	*tert*-Butylbenzene	101.8
$C_{10}H_{14}$	Isobutylbenzene	101.7
$C_{10}H_{14}$	*p*-Cymene	102.8
$C_{10}H_{14}$	1,2,4,5-Tetramethylbenzene	101.2
$C_{10}H_{14}O$	*p-tert*-Butylphenol	108.0
$C_{10}H_{15}N$	*N,N*-Diethylaniline	107.9
$C_{10}H_{16}$	*d*-Limonene	98.0
$C_{10}H_{16}$	α-Pinene	100.7
$C_{10}H_{16}$	β-Pinene	101.9
$C_{10}H_{16}O$	Camphor, (+)	103.0
$C_{10}H_{18}$	*cis*-Decahydronaphthalene	107.0
$C_{10}H_{18}$	*trans*-Decahydronaphthalene	107.6
$C_{10}H_{22}$	Decane	119.5
$C_{11}H_{10}$	1-Methylnaphthalene	102.9
$C_{11}H_{10}$	2-Methylnaphthalene	102.7
$C_{11}H_{24}$	Undecane	131.8
$C_{12}H_8$	Acenaphthylene	111.6
$C_{12}H_9N$	Carbazole	119.9
$C_{12}H_{10}$	Acenaphthene	109.9
$C_{12}H_{10}$	Biphenyl	103.3
$C_{12}H_{10}N_2$	Azobenzene	106.8
$C_{12}H_{11}N$	Diphenylamine	108.4
$C_{12}H_{14}O_4$	Diethyl phthalate	127.5
$C_{12}H_{18}$	Hexamethylbenzene	122.5
$C_{12}H_{24}O_2$	Dodecanoic acid	113.0
$C_{13}H_9N$	Acridine	118.8
$C_{13}H_{10}O$	Benzophenone	109.6
$C_{13}H_{12}$	Diphenylmethane	116.0
$C_{13}H_{28}$	Tridecane	153.7
$C_{14}H_8O_2$	9,10-Anthracenedione	113.0
$C_{14}H_{10}$	Anthracene	129.8
$C_{14}H_{10}$	Phenanthrene	127.6
$C_{14}H_{10}$	Diphenylacetylene	116
$C_{14}H_{10}O_2$	Benzil	106.8
$C_{14}H_{12}O_2$	Benzyl benzoate	132.2
$C_{14}H_{14}$	1,2-Diphenylethane	127.8
$C_{14}H_{28}O_2$	Tetradecanoic acid [Myristic acid]	176.0
$C_{14}H_{30}$	Tetradecane	166.2
$C_{16}H_{10}$	Pyrene	147
$C_{16}H_{32}O_2$	Hexadecanoic acid [Palmitic acid]	198.6
$C_{16}H_{34}$	Hexadecane	187.6
$C_{16}H_{34}O$	1-Hexadecanol	183.5
$C_{18}H_{12}$	Chrysene	148.0
$C_{18}H_{14}$	*o*-Terphenyl	150.4
$C_{18}H_{14}$	*m*-Terphenyl	155.5
$C_{18}H_{14}$	*p*-Terphenyl	156.0
$C_{18}H_{34}O_2$	*cis*-9-Octadecenoic acid [Oleic acid]	208.5
$C_{18}H_{36}O_2$	Octadecanoic acid [Stearic acid]	220.8
$C_{20}H_{12}$	Perylene	167.5

Section 4
Properties of the Elements and Inorganic Compounds

The Elements 4-1
Physical Constants of Inorganic Compounds 4-37
Formula Index of Inorganic Compounds 4-97
CAS Registry Number Index of Inorganic Compounds 4-112
Physical Properties of the Rare Earth Metals 4-122
Melting, Boiling, and Critical Temperatures of the Elements 4-132
Heat Capacity of the Elements at 25°C 4-133
Vapor Pressure of the Metallic Elements 4-134
Density of Molten Elements and Representative Salts 4-137
Magnetic Susceptibility of the Elements and Inorganic Compounds 4-141
Index of Refraction of Inorganic Liquids 4-147
Physical and Optical Properties of Minerals 4-149
Crystallographic Data on Minerals 4-156

THE ELEMENTS

C. R. Hammond

One of the most striking facts about the elements is their unequal distribution and occurrence in nature. Present knowledge of the chemical composition of the universe, obtained from the study of the spectra of stars and nebulae, indicates that hydrogen is by far the most abundant element and may account for more than 90% of the atoms or about 75% of the mass of the universe. Helium atoms make up most of the remainder. All of the other elements together contribute only slightly to the total mass.

The chemical composition of the universe is undergoing continuous change. Hydrogen is being converted into helium, and helium is being changed into heavier elements. As time goes on, the ratio of heavier elements increases relative to hydrogen. Presumably, the process is not reversible.

Burbidge, Burbidge, Fowler, and Hoyle, and more recently, Peebles, Penzias, and others have studied the synthesis of elements in stars. To explain all of the features of the nuclear abundance curve — obtained by studies of the composition of the earth, meteorites, stars, etc. — it is necessary to postulate that the elements were originally formed by at least eight different processes: (1) hydrogen burning, (2) helium burning, (3) χ process, (4) *e* process, (5) *s* process, (6) *r* process, (7) *p* process, and (8) the X process. The X process is thought to account for the existence of light nuclei such as D, Li, Be, and B. Common metals such as Fe, Cr, Ni, Cu, Ti, Zn, etc. were likely produced early in the history of our galaxy. It is also probable that most of the heavy elements on earth and elsewhere in the universe were originally formed in supernovae, or in the hot interior of stars.

Studies of the solar spectrum have led to the identification of 67 elements in the sun's atmosphere; however, all elements cannot be identified with the same degree of certainty. Other elements may be present in the sun, although they have not yet been detected spectroscopically. The element helium was discovered on the sun before it was found on earth. Some elements such as scandium are relatively more plentiful in the sun and stars than here on earth.

Minerals in lunar rocks brought back from the moon on the Apollo missions consist predominantly of *plagioclase* {$(Ca,Na)(Al,Si)O_4O_8$} and *pyroxene* {$(Ca,Mg,Fe)_2Si_2O_6$} — two minerals common in terrestrial volcanic rock. No new elements have been found on the moon that cannot be accounted for on earth; however, three minerals, *armalcolite* {$(Fe,Mg)Ti_2O_5$}, *pyroxferroite* {$CaFe_6(SiO_3)_7$}, and *tranquillityite* {$Fe_8(Zr,Y)Ti_3Si_3O_2$}, are new. The oldest known terrestrial rocks are about 4 billion years old. One rock, known as the "Genesis Rock," brought back from the Apollo 15 Mission, is about 4.15 billion years old. This is only about one-half billion years younger than the supposed age of the moon and solar system. Lunar rocks appear to be relatively enriched in refractory elements such as chromium, titanium, zirconium, and the rare earths, and impoverished in volatile elements such as the alkali metals, in chlorine, and in noble metals such as nickel, platinum, and gold.

Even older than the "Genesis Rock" are *carbonaceous chondrites,* a type of meteorite that has fallen to earth and has been studied. These are some of the most primitive objects of the solar system yet found. The grains making up these objects probably condensed directly out the gaseous nebula from which the sun and planets were born. Most of the condensation of the grains probably was completed within 50,000 years of the time the disk of the nebula was first formed — about 4.6 billion years ago. It is now thought that this type of meteorite may contain a small percentage of presolar dust grains. The relative abundances of the elements of these meteorites are about the same as the abundances found in the solar chromosphere.

The X-ray fluorescent spectrometer sent with the Viking I spacecraft to Mars shows that the Martian soil contains about 12 to 16% iron, 14 to 15% silicon, 3 to 8% calcium, 2 to 7% aluminum, and one-half to 2% titanium. The gas chromatograph — mass spectrometer on Viking II found no trace of organic compounds that should be present if life ever existed there.

F. W. Clarke and others have carefully studied the composition of rocks making up the crust of the earth. Oxygen accounts for about 47% of the crust, by weight, while silicon comprises about 28% and aluminum about 8%. These elements, plus iron, calcium, sodium, potassium, and magnesium, account for about 99% of the composition of the crust.

Many elements such as tin, copper, zinc, lead, mercury, silver, platinum, antimony, arsenic, and gold, which are so essential to our needs and civilization, are among some of the rarest elements in the earth's crust. These are made available to us only by the processes of concentration in ore bodies. Some of the so-called *rare-earth* elements have been found to be much more plentiful than originally thought and are about as abundant as uranium, mercury, lead, or bismuth. The least abundant rare-earth or *lanthanide* element, thulium, is now believed to be more plentiful on earth than silver, cadmium, gold, or iodine, for example. Rubidium, the 16th most abundant element, is more plentiful than chlorine while its compounds are little known in chemistry and commerce.

It is now thought that at least 24 elements are essential to living matter. The four most abundant in the human body are hydrogen, oxygen, carbon, and nitrogen. The seven next most common, in order of abundance, are calcium, phosphorus, chlorine, potassium, sulfur, sodium, and magnesium. Iron, copper, zinc, silicon, iodine, cobalt, manganese, molybdenum, fluorine, tin, chromium, selenium, and vanadium are needed and play a role in living matter. Boron is also thought essential for some plants, and it is possible that aluminum, nickel, and germanium may turn out to be necessary.

Ninety-one elements occur naturally on earth. Minute traces of plutonium-244 have been discovered in rocks mined in Southern California. This discovery supports the theory that heavy elements were produced during creation of the solar system. While technetium and promethium have not yet been found naturally on earth, they have been found to be present in stars. Technetium has been identified in the spectra of certain "late" type stars, and promethium lines have been identified in the spectra of a faintly visible star HR465 in Andromeda. Promethium must have been made very recently near the star's surface for no known isotope of this element has a half-life longer than 17.7 years.

It has been suggested that californium is present in certain stellar explosions known as supernovae; however, this has not been proved. At present no elements are found elsewhere in the universe that cannot be accounted for here on earth.

All atomic mass numbers from 1 to 238 are found naturally on earth except for masses 5 and 8. About 285 relatively stable and 67 naturally radioactive isotopes occur on earth totaling 352. In addition, the neutron, technetium, promethium, and the transuranic elements (lying beyond uranium) have now been produced artificially. Scientists at the Lawrence Berkeley National Laboratory, in June 1999, reported they had found evidence of an isotope of Element 118 and its immediate decay products, Element Nos. 116 and 114. The sequence of events tend to reinforce the theories that have predicted, since the 1970s, an "island of stability" for nuclei with approximately 114 protons and 184 neutrons. This "island" refers to nuclei in which the decay lasts for a period of time instead of a decay that occurs instantly.

Laboratory processes have now extended the radioactive element mass numbers beyond 238 to 293. Each element from atomic numbers 1 to 110 is known to have at least one radioactive isotope. As of December 1999, 3288 isotopes and isomers were thought to be known and recognized. Many stable and radioactive isotopes are now produced and distributed by the Oak Ridge National Laboratory, Oak Ridge, Tenn., U.S.A., to customers licensed by the U.S. Department of Energy.

The nucleus of an atom is characterized by the number of protons it contains, denoted by Z, and by the number of neutrons, *N*. Isotopes of an element have the same value of Z, but different values of *N*. The *mass number A*, is the sum of Z and *N*. For example, Uranium-238 has a mass number of 238, and contains 92 protons and 146 neutrons.

There is evidence that the definition of chemical elements must be broadened to include the electron. Several compounds known as *electrides*, have recently been made of alkaline metal elements and electrons. A relatively stable combination of a positron and electron, known as *positronium*, has also been studied.

In addition to the proton, neutron, and electron, there are considerably more than 100 other fundamental particles which have been discovered or hypothesized. The majority of these fall into one of two classes, *leptons* or *hadrons*. The leptons comprise just four known particles, the *electron*, the *muon* (μ meson), and two kinds of *neutrinos*. The muon is essentially similar to the electron and has a charge of –1, but it is 200 times heavier. The neutrino is either of two stable particles of small (probably zero) rest mass, carrying no charge. Also there are four *antileptons*, identical to the corresponding leptons in some respects, such as mass, but they have properties exactly opposite those of the leptons. The *positron*, for example, is an antilepton, with a charge of +1. Leptons cannot be broken into smaller units and are considered to be elementary. On the other hand, hadrons are complex and thought to have internal structure. Protons and neutrons, which make up atomic nuclei, are hadrons.

Elementary particle physics is not yet clearly understood, but groupings and arrangements of these particles have been made resembling the periodic table of chemical elements. This has led to the speculation that hadrons are composed of simpler components called *quarks*. Quarks are presumed to be elementary particles. Quarks come in a variety of "flavors". Two kinds of quarks — up and down — combine to form the protons and neutrons of ordinary matter. Three other quarks, known as {strange", "charm", and "bottom", are found in particle accelerators, and one of these may exist in the center of very dense, massive stars. Theorists have thought that a sixth "top" quark, much more massive than the others, should exist to complete the quark family of sub-atomic particles. In 1994, scientists using Fermilab's Tevatron accelerator may have "glimpsed" the top quark. There is presently no evidence that quarks exist in isolation. Many physicists now hold that all the matter and energy in the universe is controlled by four fundamental natural forces: the electromagnetic force, gravity, a weak nuclear force, and a strong nuclear force. Each of these natural forces is passed back and forth among the basic particles of matter by unique force-carrying particles. The electromagnetic force is carried by the *photon*, the weak nuclear force by the intermediate vector *boson*, and gravity by the *gravitron*. There is now evidence of the existence of a particle, known as the *gluon*, that binds quarks together by carrying the strong nuclear force. *Hadrons* are thought to be particles composed of quarks and gluons.

The available evidence leads to the conclusion that elements 89 (actinium) through 103 (lawrencium) are chemically similar to the rare-earth or lanthanide elements (elements 57 to 71, inclusive). These elements therefore have been named *actinides* after the first member of this series. Those elements beyond uranium that have been produced artificially have the following names and symbols: neptunium, 93 (Np); plutonium, 94 (Pu); americium, 95 (Am); curium, 96 (Cm); berkelium, 97 (Bk); californium, 98 (Cf); einsteinium, 99 (Es); fermium, 100 (Fm); mendelevium, 101 (Md); nobelium, 102 (No); and lawrencium, 103 (Lr). It is now claimed that Elements 104 through 112 have been produced and identified. More recently, Elements 118, 116, and 114 are now thought to be found. In August 1997, the International Union of Pure and Applied Chemistry (IUPAC) gave final approval to the following names for Elements 104 to 109: Element 104 — rutherfordium (Rf); Element 105 — dubnium (Db); Element 106 — seaborgium (Sg); Element 107 — bohrium (Bh); Element 108 — hassium (Hs); and Element 109 — meitnerium (Mt). The recently discovered elements 110, 111, 112, 114, 116, and 118 have not yet been named, but may carry temporary names as designated by the International Union of Pure and Applied Chemistry. IUPAC recommends that until the existence of a new element is proven to their satisfaction, the elements are to have names and symbols derived according to these precise and simple rules: The name is based on the digits in the element's atomic number. Each digit is replaced with these expressions, with the end using the usual –ium suffix as follows: **0 nil, 1 un, 2 bi, 3 tri, 4 quad, 5 pent, 6 hex, 7 sept, 8 oct, 9 enn**. Double letter i's are not used, as for example Ununbiium, but would be Ununbium. The symbol used would be the first letter of the three main syllables. For example, Element 126 would be Unbihexium, with the symbol Ubh. It is thought there is a good possibility of producing elements beyond Element 118, and discovering, in the near future, Elements 117, 115, and 113 by altering the beams of ions and the targets from those now being used.

There are many claims in the literature of the existence of various allotropic modifications of the elements, some of which are based on doubtful or incomplete evidence. Also, the physical properties of an element may change drastically by the presence of small amounts of impurities. With new methods of purification, which are now able to produce elements with 99.9999% purity, it has been necessary to restudy the properties of the elements. For example, the melting point of thorium changes by several hundred degrees by the presence of a small percentage of ThO_2 as an impurity. Ordinary commercial tungsten is brittle and can be worked only with difficulty. Pure tungsten, however, can be cut with a hacksaw, forged, spun, drawn, or extruded. In general, the value of a physical property given here applies to the pure element, when it is known.

Many of the chemical elements and their compounds are toxic and should be handled with due respect and care. In recent years there has been a greatly increased knowledge and awareness of the health hazards associated with chemicals, radioactive materials, and other agents. Anyone working with the elements and certain of their compounds should become thoroughly familiar with the proper safeguards to be taken. Information on specific hazards and recommended exposure limits may also be found in Section 16. Reference should also be made to publications such as the following:

1. *Code of Federal Regulations, Title 29, Labor.* With additions found in issues of the *Federal Register*.
2. *Code of Federal Regulations, Title 10, Energy.* With additions found in issues of the *Federal Register*. (Published by the U.S. Government Printing Office. Supt. of Documents.)
3. *Occupational Safety and Health Reporter* (latest edition with amendments and corrections), Bureau of National Affairs, Washington, D.C.
4. *Atomic Energy Law Reporter,* Commerce Clearing House, Chicago, IL.
5. *Nuclear Regulation Reporter,* Commerce Clearing House, Chicago, IL.
6. *TLVs® Threshold Limit Values for Chemical Substances and Physical Agents* is issued annually be the American Conference of Governmental Industrial Hygienists, Cincinnati, Ohio.
7. *The Sigma Aldrich Library of Regulatory and Safety Data. Vol. 3,* Robert E. Lenga and Kristine L. Volonpal, Sigma Chemical Co. and Aldrich Chemical Co., Inc. 1993.
8. *Hazardous Chemicals Desk Reference,* Richard J. Lewis, Sr., 4th ed., John Wiley & Sons, New York, Dec. 1997.

The prices of elements as indicated in this article are intended to be only a rough guide. Prices may vary, over time, widely with supplier, quantity, and purity.

Actinium — (Gr. *aktis, aktinos,* beam or ray), Ac; at. wt. (227); at. no. 89; m.p. 1051°C, b.p. 3200 ± 300°C (est.); sp. gr. 10.07 (calc.). Discovered by Andre Debierne in 1899 and independently by F. Giesel in 1902. Occurs naturally in association with uranium minerals. Thirty four isotopes and isomers are now recognized. All are radioactive. Actinium-227, a decay product of uranium-235, is an alpha and beta emitter with a 21.77-year half-life. Its principal decay products are thorium-227 (18.72-day half-life), radium-223 (11.4-day half-life), and a number of short-lived products including radon, bismuth, polonium, and lead isotopes. In equilibrium with its decay products, it is a powerful source of alpha rays. Actinium metal has been prepared by the reduction of actinium fluoride with lithium vapor at about 1100 to 1300°C. The chemical behavior of actinium is similar to that of the rare earths, particularly lanthanum. Purified actinium comes into equilibrium with its decay products at the end of 185 days, and then decays according to its 21.77-year half-life. It is about 150 times as active as radium, making it of value in the production of neutrons. Actinium-225, with a purity of 99%, is available from the Oak Ridge National Laboratory to holders of a permit for about $500/millicurie, plus packing charges.

Aluminum — (L. *alumen, alum*), Al; at. wt. 26.981539(5); at. no. 13; f.p. 660.323°C; b.p. 2519°C; sp. gr. 2.6989 (20°C); valence 3. The ancient Greeks and Romans used *alum* in medicine as an astringent, and as a mordant in dyeing. In 1761 de Morveau proposed the name *alumine* for the base in alum, and Lavoisier, in 1787, thought this to be the oxide of a still undiscovered metal. Wohler is generally credited with having isolated the metal in 1827, although an impure form was prepared by Oersted two years earlier. In 1807, Davy proposed the name *alumium* for the metal, undiscovered at that time, and later agreed to change it to *aluminum*. Shortly thereafter, the name *aluminium* was adopted to conform with the "ium" ending of most elements, and this spelling is now in use elsewhere in the world. *Aluminium* was also the accepted spelling in the U.S. until 1925, at which time the American Chemical Society officially decided to use the name *aluminum* thereafter in their publications. The method of obtaining aluminum metal by the electrolysis of alumina dissolved in *cryolite* was discovered in 1886 by Hall in the U.S. and at about the same time by Heroult in France. Cryolite, a natural ore found in Greenland, is no longer widely used in commercial production, but has been replaced by an artificial mixture of sodium, aluminum, and calcium fluorides. *Bauxite,* an impure hydrated oxide ore, is found in large deposits in Jamaica, Australia, Suriname, Guyana, Russia, Arkansas, and elsewhere. The Bayer process is most commonly used today to refine bauxite so it can be accommodated in the Hall-Heroult refining process, used to make most aluminum. Aluminum can now be produced from clay, but the process is not economically feasible at present. Aluminum is the most abundant metal to be found in the earth's crust (8.1%), but is never found free in nature. In addition to the minerals mentioned above, it is found in feldspars, granite, and in many other common minerals. Twenty-two isotopes and isomers are known. Natural aluminum is made of one isotope, ^{27}Al. Pure aluminum, a silvery-white metal, possesses many desirable characteristics. It is light, nontoxic, has a pleasing appearance, can easily be formed, machined, or cast, has a high thermal conductivity, and has excellent corrosion resistance. It is nonmagnetic and nonsparking, stands second among metals in the scale of malleability, and sixth in ductility. It is extensively used for kitchen utensils, outside building decoration, and in thousands of industrial applications where a strong, light, easily constructed material is needed. Although its electrical conductivity is only about 60% that of copper, it is used in electrical transmission lines because of its light weight. Pure aluminum is soft and lacks strength, but it can be alloyed with small amounts of copper, magnesium, silicon, manganese, and other elements to impart a variety of useful properties. These alloys are of vital importance in the construction of modern aircraft and rockets. Aluminum, evaporated in a vacuum, forms a highly reflective coating for both visible light and radiant heat. These coatings soon form a thin layer of the protective oxide and do not deteriorate as do silver coatings. They have found application in coatings for telescope mirrors, in making decorative paper, packages, toys, and in many other uses. The compounds of greatest importance are aluminum oxide, the sulfate, and the soluble sulfate with potassium (alum). The oxide, alumina, occurs naturally as ruby, sapphire, corundum, and emery, and is used in glassmaking and refractories. Synthetic ruby and sapphire have found application in the construction of lasers for producing coherent light. In 1852, the price of aluminum was about $1200/kg, and just before Hall's discovery in 1886, about $25/kg. The price rapidly dropped to 60¢ and has been as low as 33¢/kg. The price in December 1999 was about $1.50/kg.

Americium — (the Americas), Am; at. wt. 243; at. no. 95; m.p. 1176°C; b.p. 2011°C; sp. gr. 13.67 (20°C); valence 2, 3, 4, 5, or 6. Americium was the fourth transuranium element to be discovered; the isotope ^{241}Am was identified by Seaborg, James, Morgan, and Ghiorso late in 1944 at the wartime Metallurgical Laboratory of the University of Chicago as the result of successive neutron capture reactions by plutonium isotopes in a nuclear reactor:

$$^{239}Pu(n,\gamma)\rightarrow{}^{240}Pu(n,\gamma)\rightarrow{}^{241}Pu\xrightarrow{\beta}{}^{241}Am$$

Since the isotope ^{241}Am can be prepared in relatively pure form by extraction as a decay product over a period of years from strongly neutron-bombarded plutonium, ^{241}Pu, this isotope is used for much of the chemical investigation of this element. Better suited is the isotope ^{243}Am due to its longer half-life (7.37×10^3 years as compared to 432.2 years for ^{241}Am). A mixture of the isotopes ^{241}Am, ^{242}Am, and ^{243}Am can be prepared by intense neutron irradiation of ^{241}Am according to the reactions $^{241}Am\ (n, \gamma) \rightarrow {}^{242}Am\ (n, \gamma) \rightarrow {}^{243}Am$. Nearly isotopically pure ^{243}Am can be prepared by a sequence of neutron bombardments and chemical separations as follows: neutron bombardment of ^{241}Am yields ^{242}Pu by the reactions $^{241}Am\ (n, \gamma) \rightarrow {}^{242}Am \rightarrow {}^{242}Pu$, after chemical separation the ^{242}Pu can be transformed to ^{243}Am via the reactions $^{242}Pu\ (n, \gamma) \rightarrow {}^{243}Pu \rightarrow {}^{243}Am$, and the ^{243}Am can be chemically separated. Fairly pure ^{242}Pu can be prepared more simply by very intense neutron irradiation of ^{239}Pu as the result of successive neutron-capture reactions. Seventeen radioactive isotopes and isomers are now recognized. Americium metal has been prepared by reducing the trifluoride with barium vapor at 1000 to 1200°C or the dioxide by lanthanum metal. The luster of freshly prepared americium metal is white and more silvery than plutonium or neptunium prepared in the same manner. It appears to be more malleable than uranium or neptunium and tarnishes slowly in dry air at room temperature. Americium is thought to exist in two forms: an alpha form which has a double hexagonal close-packed structure and a loose-packed cubic beta form. Americium must be handled with great care to avoid personal contamination. As little as 0.03 μCi of ^{241}Am is the maximum permissible total body burden. The alpha activity from ^{241}Am is about three times that of radium. When gram quantities of ^{241}Am are handled, the intense gamma activity makes exposure a serious problem. Americium dioxide, AmO_2, is the most important oxide. AmF_3, AmF_4, $AmCl_3$, $AmBr_3$, AmI_3, and other compounds have been prepared. The isotope ^{241}Am has been used as a portable source for gamma radiography. It has also been used as a radioactive glass thickness gage for the flat glass industry, and as a source of ionization for smoke detectors. Americium-243 (99%) is available from the Oak Ridge National Laboratory at a cost of about $750/g plus packing charges.

Antimony — (Gr. *anti* plus *monos* — a metal not found alone), Sb; at. wt. 121.760(1); at. no. 51; m.p. 630.63°C; b.p. 1587°C; sp. gr. 6.691 (20°C); valence 0, –3, +3, or +5. Antimony was recognized in compounds by the ancients and was known as a metal at the beginning of the 17th century and possibly much earlier. It is not abundant, but is found in over 100 mineral species. It is sometimes found native, but more frequently as the sulfide, *stibnite* (Sb_2S_3); it is also found as antimonides of the heavy metals, and as oxides. It is extracted from the sulfide by roasting to the oxide, which is reduced by salt and scrap iron; from its oxides it is also prepared by reduction with carbon. Two allotropic forms of antimony exist: the normal stable, metallic form, and the amorphous gray form. The so-called explosive antimony is an ill-defined material always containing an appreciable amount of halogen; therefore, it no longer warrants consideration as a separate allotrope. The yellow form, obtained by oxidation of *stibine*, SbH_3, is probably impure, and is not a distinct form. Natural antimony is made of two stable isotopes, ^{121}Sb and ^{123}Sb. Forty five other radioactive isotopes and isomers are now recognized. Metallic antimony is an extremely brittle metal of a flaky, crystalline texture. It is bluish white and has a metallic luster. It is not acted on by air at room temperature, but burns brilliantly when heated with the formation of white fumes of Sb_2O_3. It is a poor conductor of heat and electricity, and has a hardness of 3 to 3.5. Antimony, available commercially with a purity of 99.999 + %, is finding use in semiconductor technology for making infrared detectors, diodes, and Hall-effect devices. Commercial-grade antimony is widely used in alloys with percentages ranging from 1 to 20. It greatly increases the hardness and mechanical strength of lead. Batteries, antifriction alloys, type metal, small arms and tracer bullets, cable sheathing, and minor products use about half the metal produced. Compounds taking up the other half are oxides, sulfides, sodium antimonate, and antimony trichloride. These are used in manufacturing flame-proofing compounds, paints, ceramic enamels, glass, and pottery. Tartar emetic (hydrated potassium antimonyl tartate) has been used in medicine. Antimony and many of its compounds are toxic. Antimony costs about $140/kg or about $12/g (99.999%).

Argon — (Gr. *argos*, inactive), Ar; at. wt. 39.948(1); at. no. 18; m.p. –189.35°C; b.p. –185.85°C; t_c -122.28; density 1.7837 g/l. Its presence in air was suspected by Cavendish in 1785, discovered by Lord Rayleigh and Sir William Ramsay in 1894. The gas is prepared by fractionation of liquid air, the atmosphere containing 0.94% argon. The atmosphere of Mars contains 1.6% of ^{40}Ar and 5 p.p.m. of ^{36}Ar. Argon is two and one half times as soluble in water as nitrogen, having about the same solubility as oxygen. It is recognized by the characteristic lines in the red end of the spectrum. It is used in electric light bulbs and in fluorescent tubes at a pressure of about 400 Pa, and in filling photo tubes, glow tubes, etc. Argon is also used as an inert gas shield for arc welding and cutting, as a blanket for the production of titanium and other reactive elements, and as a protective atmosphere for growing silicon and germanium crystals. Argon is colorless and odorless, both as a gas and liquid. It is available in high-purity form. Commercial argon is available at a cost of about 3¢ per cubic foot. Argon is considered to be a very inert gas and is not known to form true chemical compounds, as do krypton, xenon, and radon. However, it does form a hydrate having a dissociation pressure of 105 atm at 0°C. Ion molecules such as $(ArKr)^+$, $(ArXe)^+$, $(NeAr)^+$ have been observed spectroscopically. Argon also forms a clathrate with β-hydroquinone. This clathrate is stable and can be stored for a considerable time, but a true chemical bond does not exist. Van der Waals' forces act to hold the argon. Naturally occurring argon is a mixture of three isotopes. Seventeen other radioactive isotopes are now known to exist. Commercial argon is priced at about $70/300 cu. ft. or 8.5 cu. meters.

Arsenic — (L. *arsenicum*, Gr. *arsenikon*, yellow orpiment, identified with *arsenikos*, male, from the belief that metals were different sexes; Arabic, *Az-zernikh*, the orpiment from Persian *zerni-zar*, gold), As; at. wt. 74.92160(2); at. no. 33; valence –3, 0, +3 or +5. Elemental arsenic occurs in two solid modifications: yellow, and gray or metallic, with specific gravities of 1.97, and 5.73, respectively. Gray arsenic, the ordinary stable form, has a triple point of 817°C and sublimes at 614°C and has a critical temperature of 1400°C. Several other allotropic forms of arsenic are reported in the literature. It is believed that Albertus Magnus obtained the element in 1250 A.D. In 1649 Schroeder published two methods of preparing the element. It is found native, in the sulfides *realgar* and *orpiment*, as arsenides and sulfarsenides of heavy metals, as the oxide, and as arsenates. *Mispickel*, arsenopyrite, (FeSAs) is the most common mineral, from which on heating the arsenic sublimes leaving ferrous sulfide. The element is a steel gray, very brittle, crystalline, semimetallic solid; it tarnishes in air, and when heated is rapidly oxidized to arsenous oxide (As_2O_3) with the odor of garlic. Arsenic and its compounds are poisonous. Exposure to arsenic and its compounds should not exceed 0.2 mg/m^3 as elemental As during an 8-h work day. These values, however, are being studied, and may be lowered. Arsenic is also used in bronzing, pyrotechny, and for hardening and improving the sphericity of shot. The most important compounds are white arsenic (As_2O_3), the sulfide, Paris green $3Cu(AsO_2)_2 \cdot Cu(C_2H_3O_2)_2$, calcium arsenate, and lead arsenate; the last three have been used as agricultural insecticides and poisons. Marsh's test makes use of the formation and ready decomposition of arsine (AsH_3). Arsenic is available in high-purity form. It is finding increasing uses as a doping agent in solid-state devices such as transistors. Gallium arsenide is used as a laser material to convert electricity directly into coherent light. Natural arsenic is made of one isotope ^{75}As. Thirty other radioactive isotopes and isomers are known. Arsenic (99%) costs about $175/kg. Purified arsenic (99.9995%) costs about $3.50/g.

Astatine — (Gr. *astatos*, unstable), At; at. wt. (210); at. no. 85; m.p. 300°C (est.); valence probably 1, 3, 5, or 7. Synthesized in 1940 by D. R. Corson, K. R. MacKenzie, and E. Segre at the University of California by bombarding bismuth with alpha particles. The longest-lived isotope, ^{210}At, has a half-life of only 8.1 hours. Thirty-six other isotopes and isomers are now known. Minute quantities of ^{215}At, ^{218}At, and ^{219}At exist in equilibrium in nature with naturally occurring uranium and thorium isotopes, and traces of ^{217}At are equilibrium with ^{233}U and ^{239}Np resulting from interaction of thorium and uranium with naturally produced neutrons. The total amount of astatine present in the earth's crust, however, is probably less than 1 oz. Astatine can be produced by bombarding bismuth with energetic alpha particles to obtain the relatively long-lived $^{209-211}At$, which can be distilled from the target by heating it in air. Only about 0.05 μg of astatine has been prepared to date. The "time of flight" mass spectrometer has been used to confirm that this highly radioactive halogen behaves chemically very much like other halogens, particularly iodine. The interhalogen compounds AtI, AtBr, and AtCl are known to form, but it is not yet known if astatine forms diatomic astatine molecules. HAt and CH_3At (methyl astatide) have been detected. Astatine is said to be more metallic that iodine, and, like iodine, it probably accumulates in the thyroid gland. Workers at the Brookhaven National Laboratory have recently used reactive scattering in crossed molecular beams to identify and measure elementary reactions involving astatine.

Barium — (Gr. *barys*, heavy), Ba; at. wt. 137.327(7), at. no. 56; m.p. 727°C; b.p. 1897°C; sp. gr. 3.5 (20°C); valence 2. Baryta was distinguished from lime by Scheele in 1774; the element was discovered by Sir Humphrey Davy in 1808. It is found only in combination with other elements, chiefly in *barite* or *heavy spar* (sulfate) and *witherite* (carbonate) and is prepared by electrolysis of the chloride. Large deposits of barite are found in China, Germany, India, Morocco, and in the U.S. Barium is a metallic element, soft, and when pure is silvery white like lead; it belongs to the alkaline earth group, resembling calcium chemically. The metal oxidizes very easily and should be kept under petroleum or other suitable oxygen-free liquids to exclude air. It is decomposed by water or alcohol. The metal is used as a "getter" in vacuum tubes. The most important compounds are the peroxide (BaO_2), chloride, sulfate, carbonate, nitrate, and chlorate. Lithopone, a pigment containing barium sulfate and zinc sulfide, has good covering power, and does not darken in the presence of sulfides. The sulfate, as permanent white or *blanc fixe,* is also used in paint, in X-ray diagnostic work, and in glassmaking. *Barite* is extensively used as a weighting agent in oilwell drilling fluids, and also in making rubber. The carbonate has been used as a rat poison, while the nitrate and chlorate give green colors in pyrotechny. The impure sulfide phosphoresces after exposure to the light. The compounds and the metal are not expensive. Barium metal (99.2 + % pure) costs about $3/g. All barium compounds that are water or acid soluble are poisonous. Naturally occurring barium is a mixture of seven stable isotopes. Thirty six other radioactive isotopes and isomers are known to exist.

Berkelium — (*Berkeley,* home of the University of California), Bk; at. wt. (247); at. no. 97; m.p. 1050°C; valence 3 or 4; sp. gr. 14 (est.). Berkelium, the eighth member of the actinide transition series, was discovered in December 1949 by Thompson, Ghiorso, and Seaborg, and was the fifth transuranium element synthesized. It was produced by cyclotron bombardment of milligram amounts of ^{241}Am with helium ions at Berkeley, California. The first isotope produced had a mass number of 243 and decayed with a half-life of 4.5 hours. Thirteen isotopes are now known and have been synthesized. The existence of ^{249}Bk, with a half-life of 320 days, makes it feasible to isolate berkelium in weighable amounts so that its properties can be investigated with macroscopic quantities. One of the first visible amounts of a pure berkelium compound, berkelium chloride, was produced in 1962. It weighed 3 billionth of a gram. Berkelium probably has not yet been prepared in elemental form, but it is expected to be a silvery metal, easily soluble in dilute mineral acids, and readily oxidized by air or oxygen at elevated temperatures to form the oxide. X-ray diffraction methods have been used to identify the following compounds: BkO_2, BkO_3, BkF_3, BkCl, and BkOCl. As with other actinide elements, berkelium tends to accumulate in the skeletal system. The maximum permissible body burden of ^{249}Bk in the human skeleton is about 0.0004 μg. Because of its rarity, berkelium presently has no commercial or technological use. Berkelium most likely resembles terbium with respect to chemical properties. Berkelium-249 is available from O.R.N.L. at a cost of $185/μg plus packing charges.

Beryllium — (Gr. *beryllos, beryl;* also called Glucinium or Glucinum, Gr. *glykys,* sweet), Be; at. wt. 9.012182(3); at no. 4; m.p. 1287°C; b.p. 2471°C; sp. gr. 1.848 (20°C); valence 2. Discovered as the oxide by Vauquelin in beryl and in emeralds in 1798. The metal was isolated in 1828 by Wohler and by Bussy independently by the action of potassium on beryllium chloride. Beryllium is found in some 30 mineral species, the most important of which are *bertrandite, beryl, chrysoberyl,* and *phenacite. Aquamarine* and *emerald* are precious forms of *beryl.* Beryllium minerals are found in the U.S., Brazil, Russia, Kazakhstan, and elsewhere. Colombia is known for its emeralds. *Beryl* ($3BeO \cdot Al_2O_3 \cdot 6SiO_2$) and *bertrandite* ($4BeO \cdot 2SiO_2 \cdot H_2O$) are the most important commercial sources of the element and its compounds. Most of the metal is now prepared by reducing beryllium fluoride with magnesium metal. Beryllium metal did not become readily available to industry until 1957. The metal, steel gray in color, has many desirable properties. It is one of the lightest of all metals, and has one of the highest melting points of the light metals. Its modulus of elasticity is about one third greater than that of steel. It resists attack by concentrated nitric acid, has excellent thermal conductivity, and is nonmagnetic. It has a high permeability to X-rays, and when bombarded by alpha particles, as from radium or polonium, neutrons are produced in the ratio of about 30 neutrons/million alpha particles. At ordinary temperatures beryllium resists oxidation in air, although its ability to scratch glass is probably due to the formation of a thin layer of the oxide. Beryllium is used as an alloying agent in producing beryllium copper which is extensively used for springs, electrical contacts, spot-welding electrodes, and nonsparking tools. It has found application as a structural material for high-speed aircraft, missiles, spacecraft, and communication satellites. It is being used in the windshield frame, brake discs, support beams, and other structural components of the space shuttle. Because beryllium is relatively transparent to X-rays, ultra-thin Be-foil is finding use in X-ray lithography for reproduction of microminiature integrated circuits. Natural beryllium is made of ^{9}Be and is stable. Eight other radioactive isotopes are known.

Beryllium is used in nuclear reactors as a reflector or moderator for it has a low thermal neutron absorption cross section. It is used in gyroscopes, computer parts, and instruments where lightness, stiffness, and dimensional stability are required. The oxide has a very high melting point and is also used in nuclear work and ceramic applications. Beryllium and its salts are toxic and should be handled with the greatest of care. Beryllium and its compounds should not be tasted to verify the sweetish nature of beryllium (as did early experimenters). The metal, its alloys, and its salts can be handled safely if certain work codes are observed, but no attempt should be made to work with beryllium before becoming familiar with proper safeguards. Beryllium metal is available at a cost of about $4.50/g (99.5% pure).

Bismuth — (Ger. *Weisse Masse,* white mass; later *Wisuth* and *Bisemutum*), Bi; at. wt. 208.98038(2); at. no. 83; m.p. 271.4°C; b.p. 1564°C; sp. gr. 9.747 (20°C); valence 3 or 5. In early times bismuth was confused with tin and lead. Claude Geoffroy the Younger showed it to be distinct from lead in 1753. It is a white crystalline, brittle metal with a pinkish tinge. It occurs native. The most important ores are *bismuthinite* or bismuth glance (Bi_2S_3) and *bismite* (Bi_2O_3). Peru, Japan, Mexico, Bolivia, and Canada are major bismuth producers. Much of the bismuth produced in the U.S. is obtained as a by-product in refining lead, copper, tin, silver, and gold ores. Bismuth is the most diamagnetic of all metals, and the thermal conductivity is lower than any metal, except mercury. It has a high electrical resistance, and has the highest Hall effect of any metal (i.e., greatest increase in electrical resistance when placed in a magnetic field). "Bismanol" is a permanent magnet of high coercive force, made of MnBi, by the U.S. Naval Surface Weapons Center. Bismuth expands 3.32% on solidification. This property makes bismuth alloys particularly suited to the making of sharp castings of objects subject to damage by high temperatures. With other metals such as tin, cadmium, etc., bismuth forms low-melting alloys which are extensively used for safety devices in fire detection and extinguishing systems. Bismuth is used in producing malleable irons and is finding use as a catalyst for making acrylic fibers. When bismuth is heated in air it burns with a blue flame, forming yellow fumes of the oxide. The metal is also used as a thermocouple material, and has found application as a carrier for U^{235} or U^{233} fuel in atomic reactors. Its soluble salts are characterized by forming insoluble basic salts on the addition of water, a property sometimes used in detection work. Bismuth oxychloride is used extensively in cosmetics. Bismuth subnitrate and subcarbonate are used in medicine. Natural bismuth contains only one isotope ^{209}Bi. Forty-four isotopes and isomers of bismuth are known. Bismuth metal (99.5%) costs about $250/kg.

Bohrium — (Named after Niels Bohr [1885-1962], Danish atomic and nuclear physicist.) Bh; at.wt. [262]. at.no. 107. Bohrium is expected to have chemical properties similar to rhenium. This element was synthesized and unambiguously identified in 1981 using the Universal Linear Accelerator (UNILAC) at the Gesellschaft für Schwerionenforschung (G.S.I.) in Darmstadt, Germany. The discovery team was led by Armbruster and Münzenberg. The reaction producing the element was proposed and applied earlier by a Dubna Group led by Oganessian in 1976. A target of ^{209}Bi was bombarded by a beam of ^{54}Cr ions. In 1983 experiments at Dubna using the 157-inch cyclotron, produced $^{262}107$ by the reaction $^{209}Bi + ^{54}Cr$. The alpha decay of ^{246}Cf, the sixth member in the decay chain of $^{262}107$, served to establish a 1-neutron reaction channel. The IUPAC adopted the name ***Bohrium*** with the symbol Bh for Element 107 in August 1997. Five isotopes of bohrium are now recognized. One isotope of bohrium appears to have a relatively long life of 15 seconds. Work on this relatively long-lived isotope has been performed with the 88-inch cyclotron at the Lawrence-Berkeley National Laboratory.

Boron — (Ar. *Buraq,* Pers. *Burah*), B; at. wt. 10.811(7); at. no. 5; m.p. 2075°C; b.p. 4000°C; sp. gr. of crystals 2.34, of amorphous variety 2.37; valence 3. Boron compounds have been known for thousands of years, but the element was not discovered until 1808 by Sir Humphry Davy and by Gay-Lussac and Thenard. The element is not found free in nature, but occurs as orthoboric acid usually in certain volcanic spring waters and as borates in *borax* and *colemanite. Ulexite,* another boron mineral, is interesting as it is nature's own version of "fiber optics." Important sources of boron are the ores *rasorite (kernite)* and *tincal (borax ore).* Both of these ores are found in the Mojave Desert. *Tincal* is the most important source of boron from the Mojave. Extensive *borax* deposits are also found in Turkey. Boron exists naturally as 19.9% ^{10}B isotope and 80.1% ^{11}B isotope. Ten other isotopes of boron are known. High-purity crystalline boron may be prepared by the vapor phase reduction of boron trichloride or tribromide with hydrogen on electrically heated filaments. The impure, or amorphous, boron, a brownish-black powder, can be obtained by heating the trioxide with magnesium powder. Boron of 99.9999% purity has been produced and is available commercially. Elemental boron has an energy band gap of 1.50 to 1.56 eV, which is higher than that of either silicon or germanium. It has interesting optical characteristics, transmitting portions of the infrared, and is a poor conductor of electricity at room temperature, but a good conductor at high temperature. Amorphous boron is used in pyrotechnic flares to provide a distinctive green color, and in rockets as an igniter. By far the most commercially important boron compound in terms of dollar sales is $Na_2B_4O_7 \cdot 5H_2O$. This pentahydrate is used in very large quantities in the manufacture of insulation fiberglass and sodium perborate bleach. Boric acid is also an important boron compound with major markets in textile fiberglass and in cellulose insulation as a flame retardant. Next in order of importance is borax ($Na_2B_4O_7 \cdot 10H_2O$) which is used principally in laundry products. Use of borax as a mild antiseptic is minor in terms of dollars and tons. Boron compounds are also extensively used in the manufacture of borosilicate glasses. The isotope boron-10 is used as a control for nuclear reactors, as a shield for nuclear radiation, and in instruments used for detecting neutrons. Boron nitride has remarkable properties and can be used to make a material as hard as diamond. The nitride also behaves like an electrical insulator but conducts heat like a metal. It also has lubricating properties similar to graphite. The hydrides are easily oxidized with considerable energy liberation, and have been studied for use as rocket fuels. Demand is increasing for boron filaments, a high-strength, lightweight material chiefly employed for advanced aerospace structures. Boron is similar to carbon in that it has a capacity to form stable covalently bonded molecular networks. Carboranes, metalloboranes, phosphacarboranes, and other families comprise thousands of compounds. Crystalline boron (99.5%) costs about $30/g. Amorphous boron (94–96%) costs about $1.50/g. Elemental boron and the borates are not considered to be toxic, and they do not require special care in handling. However, some of the more exotic boron hydrogen compounds are definitely toxic and do require care.

Bromine — (Gr. *bromos,* stench), Br; at. wt. 79.904(1); at. no. 35; m.p. –7.2°C; b.p. 58.8°C; t_c 315°C; density of gas 7.59 g/l, liquid 3.12 (20°C); valence 1, 3, 5, or 7. Discovered by Balard in 1826, but not prepared in quantity until 1860. A member of the halogen group of elements, it is obtained from natural brines from wells in Michigan and Arkansas. Little bromine is extracted today from seawater, which contains only about 85 ppm. Bromine is the only liquid nonmetallic element. It is a heavy, mobile, reddish-brown liquid, volatilizing readily at room temperature to a red vapor with a strong disagreeable odor, resembling chlorine, and having a very irritating effect on the eyes and throat; it is readily soluble in water or carbon disulfide, forming a red solution, is less active than chlorine but more so than iodine; it unites readily with many elements and has a bleaching action; when spilled on the skin it produces painful sores. It presents a serious health hazard, and maximum safety precautions should be taken when handling it. Much of the bromine output in the U.S. was used in the production of ethylene dibromide, a lead scavenger used in making gasoline antiknock compounds. Lead in gasoline, however, has been drastically reduced, due to environmental considerations. This will greatly affect future production of bromine. Bromine is also used in making fumigants, flameproofing agents, water purification compounds, dyes, medicinals, sanitizers, inorganic bromides for photography, etc. Organic bromides are also important. Natural bromine is made of two isotopes, ^{79}Br and ^{81}Br. Thirty-four isotopes and isomers are known. Bromine costs about $100/kg.

Cadmium — (L. *cadmia;* Gr. *kadmeia* — ancient name for calamine, zinc carbonate), Cd; at. wt. 112.411(8); at. no. 48; m.p. 321.07°C; b.p. 767°C; sp. gr. 8.65 (20°C); valence 2. Discovered by Stromeyer in 1817 from an impurity in zinc carbonate. Cadmium most often occurs in small quantities associated with zinc ores, such as *sphalerite* (ZnS). *Greenockite* (CdS) is the only mineral of any consequence bearing cadmium. Almost all cadmium is obtained as a by-product in the treatment of zinc, copper, and lead ores. It is a soft, bluish-white metal which is easily cut with a knife. It is similar in many respects to zinc. It is a component of some of the lowest melting alloys; it is used in bearing alloys with low coefficients of friction and great resistance to fatigue; it is used extensively in electroplating, which accounts for about 60% of its use. It is also used in many types of solder, for standard E.M.F. cells, for Ni-Cd batteries, and as a barrier to control atomic fission. The market for Ni-Cd batteries is expected to grow significantly in the next few years. Cadmium compounds are used in black and white television phosphors and in blue and green phosphors for color TV tubes. It forms a number of salts, of which the sulfate is most common; the sulfide is used as a yellow pigment. Cadmium and solutions of its compounds are toxic. Failure to appreciate the toxic properties of cadmium may cause workers to be unwittingly exposed to dangerous fumes. Some silver solders, for example, contain cadmium and should be handled with care. Serious toxicity problems have been found from long-term exposure and work with cadmium plating baths. Cadmium is present in certain phosphate rocks. This has raised concerns that the long-term use of certain phosphate fertilizers might pose a health hazard from levels of cadmium that might enter the food chain. In 1927 the International Conference on Weights and Measures redefined the meter in terms of the wavelength of the red cadmium spectral line (i.e. 1 m = 1,553,164.13 wavelengths). This definition has been changed (see under Krypton). The current price of cadmium is about $2/g (99.5%). It is available in high purity form for about $300/kg. Natural cadmium is made of eight isotopes. Thirty four other isotopes and isomers are now known and recognized.

Calcium — (L. *calx,* lime), Ca; at. wt. 40.078(4); at. no. 20; m.p. 842°C; b.p. 1484°C; sp. gr. 1.55 (20°C); valence 2. Though lime was prepared by the Romans in the first century under the name calx, the metal was not discovered until 1808. After learning that Berzelius and Pontin prepared calcium amalgam by electrolyzing lime in mercury, Davy was able to isolate the impure metal. Calcium is a metallic element, fifth in abundance in the earth's crust, of which it forms more than 3%. It is an essential constituent of leaves, bones, teeth, and shells. Never found in nature uncombined, it occurs abundantly as *limestone* ($CaCO_3$), *gypsum* ($CaSO_4 \cdot 2H_2O$), and *fluorite* (CaF_2); *apatite* is the fluorophosphate or chlorophosphate of calcium. The metal has a silvery color, is rather hard, and is prepared by electrolysis of the fused chloride to which calcium fluoride is added to lower the melting point. Chemically it is one of the alkaline earth elements; it readily forms a white coating of oxide in air, reacts with water, burns with a yellow-red flame, forming largely the oxide. The metal is used as a reducing agent in preparing other metals such as thorium, uranium, zirconium, etc., and is used as a deoxidizer, desulfurizer, and inclusion modifier for various ferrous and nonferrous alloys. It is also used as an alloying agent for aluminum, beryllium, copper, lead, and magnesium alloys, and serves as a "getter" for residual gases in vacuum tubes, etc. Its natural and prepared compounds are widely used. Quicklime (CaO), made by heating limestone and changed into slaked lime by the careful addition of water, is the great cheap base of chemical industry with countless uses. Mixed with sand it hardens as mortar and plaster by taking up carbon dioxide from the air. Calcium from limestone is an important element in Portland cement. The solubility of the carbonate in water containing carbon dioxide causes the formation of caves with stalactites and stalagmites and is responsible for hardness in water. Other important compounds are the carbide (CaC_2), chloride ($CaCl_2$), cyanamide ($CaCN_2$), hypochlorite ($Ca(OCl)_2$), nitrate ($Ca(NO_3)_2$), and sulfide (CaS). Calcium sulfide is phosphorescent after being exposed to light. Natural calcium contains six isotopes. Sixteen other radioactive isotopes are known. Metallic calcium (99.5%) costs about $225/kg.

Californium — (State and University of California), Cf; at. wt. (251); m.p. 900°C; at. no. 98. Californium, the sixth transuranium element to be discovered, was produced by Thompson, Street, Ghioirso, and Seaborg in 1950 by bombarding microgram quantities of ^{242}Cm with 35 MeV helium ions in the Berkeley 60-inch cyclotron. Californium (III) is the only ion stable in aqueous solutions, all attempts to reduce or oxidize californium (III) having failed. The isotope ^{249}Cf results from the beta decay of ^{249}Bk while the heavier isotopes are produced by intense neutron irradiation by the reactions:

$$^{249}Bk(n,\gamma)\rightarrow{}^{250}Bk\xrightarrow{\beta}{}^{250}Cf \text{ and } {}^{249}Cf(n,\gamma)\rightarrow{}^{250}Cf$$

followed by

$$^{250}Cf(n,\gamma)\rightarrow{}^{251}Cf(n,\gamma)\rightarrow{}^{252}Cf$$

The existence of the isotopes ^{249}Cf, ^{250}Cf, ^{251}Cf, and ^{252}Cf makes it feasible to isolate californium in weighable amounts so that its properties can be investigated with macroscopic quantities. Californium-252 is a very strong neutron emitter. One microgram releases 170 million neutrons per minute, which presents biological hazards. Proper safeguards should be used in handling californium. Twenty isotopes of californium are now recognized. ^{249}Cf and ^{252}Cf have half-lives of 351 years and 900 years, respectively. In 1960 a few tenths of a microgram of californium trichloride, $CfCl_3$, californium oxychloride, CfOCl, and californium oxide, Cf_2O_3, were first prepared. Reduction of californium to its metallic state has not yet been accomplished. Because californium is a very efficient source of neutrons, many new uses are expected for it. It has already found use in neutron moisture gages and in well-logging (the determination of water and oil-bearing layers). It is also being used as a portable neutron source for discovery of metals such as gold or silver by on-the-spot activation analysis. ^{252}Cf is now being offered for sale by the Oak Ridge National Laboratory (O.R.N.L.) at a cost of \$60/μg and ^{249}Cf at a cost of \$185/μg plus packing charges. It has been suggested that californium may be produced in certain stellar explosions, called *supernovae,* for the radioactive decay of ^{254}Cf (55-day half-life) agrees with the characteristics of the light curves of such explosions observed through telescopes. This suggestion, however, is questioned. Californium is expected to have chemical properties similar to dysprosium.

Carbon — (L. *carbo,* charcoal), C; at. wt. 12.0107(8); at. no. 6; sublimes at 3642°C; triple point (graphite-liquid-gas), 4492°C at a pressure of 101.325 kPa; sp. gr. amorphous 1.8 to 2.1, graphite 1.9 to 2.3, diamond 3.15 to 3.53 (depending on variety); gem diamond 3.513 (25°C); valence 2, 3, or 4. Carbon, an element of prehistoric discovery, is very widely distributed in nature. It is found in abundance in the sun, stars, comets, and atmospheres of most planets. Carbon in the form of microscopic diamonds is found in some meteorites. Natural diamonds are found in *kimberlite* or *lamporite* of ancient formations called "pipes," such as found in South Africa, Arkansas, and elsewhere. Diamonds are now also being recovered from the ocean floor off the Cape of Good Hope. About 30% of all industrial diamonds used in the U.S. are now made synthetically. The energy of the sun and stars can be attributed at least in part to the well-known carbon-nitrogen cycle. Carbon is found free in nature in three allotropic forms: amorphous, graphite, and diamond. A fourth form, known as "white" carbon, is now thought to exist. Graphite is one of the softest known materials while diamond is one of the hardest. Graphite exists in two forms: alpha and beta. These have identical physical properties, except for their crystal structure. Naturally occurring graphites are reported to contain as much as 30% of the rhombohedral (beta) form, whereas synthetic materials contain only the alpha form. The hexagonal alpha type can be converted to the beta by mechanical treatment, and the beta form reverts to the alpha on heating it above 1000°C. In 1969 a new allotropic form of carbon was produced during the sublimation of pyrolytic graphite at low pressures. Under free-vaporization conditions above ~2550 K, "white" carbon forms as small transparent crystals on the edges of the basal planes of graphite. The interplanar spacings of "white" carbon are identical to those of carbon form noted in the graphitic gneiss from the Ries (meteoritic) Crater of Germany. "White" carbon is a transparent birefringent material. Little information is presently available about this allotrope. Of recent interest is the discovery of all-carbon molecules, known as "buckyballs" or fullerenes, which have a number of unusual properties. These interesting molecules, consisting of 60 or 70 carbon atoms linked together, seem capable of withstanding great pressure and trapping foreign atoms inside their network of carbon. They are said to be capable of magnetism and superconductivity and have potential as a nonlinear optical material. Buckyball films are reported to remain superconductive at temperatures as high as 45 K. In combination, carbon is found as carbon dioxide in the atmosphere of the earth and dissolved in all natural waters. It is a component of great rock masses in the form of carbonates of calcium (limestone), magnesium, and iron. Coal, petroleum, and natural gas are chiefly hydrocarbons. Carbon is unique among the elements in the vast number and variety of compounds it can form. With hydrogen, oxygen, nitrogen, and other elements, it forms a very large number of compounds, carbon atom often being linked to carbon atom. There are close to ten million known carbon compounds, many thousands of which are vital to organic and life processes. Without carbon, the basis for life would be impossible. While it has been thought that silicon might take the place of carbon in forming a host of similar compounds, it is now not possible to form stable compounds with very long chains of silicon atoms. The atmosphere of Mars contains 96.2% CO_2. Some of the most important compounds of carbon are carbon dioxide (CO_2), carbon monoxide (CO), carbon disulfide (CS_2), chloroform ($CHCl_3$), carbon tetrachloride (CCl_4), methane (CH_4), ethylene (C_2H_4), acetylene (C_2H_2), benzene (C_6H_6), ethyl alcohol (C_2H_5OH), acetic acid (CH_3COOH), and their derivatives. Carbon has fifteen isotopes. Natural carbon consists of 98.89% ^{12}C and 1.11% ^{13}C. In 1961 the International Union of Pure and Applied Chemistry adopted the isotope carbon-12 as the basis for atomic weights. Carbon-14, an isotope with a half-life of 5715 years, has been widely used to date such materials as wood, archeological specimens, etc. A new brittle form of carbon, known as "glassy carbon", has been developed. It can be obtained with high purity. It has a high resistance to corrosion, has good thermal stability, and is structurally impermeable to both gases and liquids. It has a randomized structure, making it useful in ultra-high technology applications, such as crystal growing, crucibles for high-temperature use, etc. Both Types 1 and 2 glassy carbon are available at a cost of about $35/10gms.

Cerium — (named for the asteroid *Ceres,* which was discovered in 1801 only 2 years before the element), Ce; at. wt. 140.115(4); at. no. 58; m.p. 798°C; b.p. 3424°C; sp. gr. 6.770 (25°C); valence 3 or 4. Discovered in 1803 by Klaproth and by Berzelius and Hisinger; metal prepared by Hillebrand and Norton in 1875. Cerium is the most abundant of the metals of the so-called rare earths. It is found in a number of minerals including *allanite* (also known as *orthite*), *monazite, bastnasite, cerite,* and *samarskite.* Monazite and bastnasite are presently the two most important sources of cerium. Large deposits of monazite found on the beaches of Travancore, India, in river sands in Brazil, and deposits of *allanite* in the western United States, and *bastnasite* in Southern California will supply cerium, thorium, and the other rare-earth metals for many years to come. Metallic cerium is prepared by metallothermic reduction techniques, such as by reducing cerous fluoride with calcium, or by electrolysis of molten cerous chloride or other cerous halides. The metallothermic technique is used to produce high-purity cerium. Cerium is especially interesting because of its variable electronic structure. The energy of the inner 4f level is nearly the same as that of the outer or valence electrons, and only small amounts of energy are required to change the relative occupancy of these electronic levels. This gives rise to dual valency states. For example, a volume change of about 10% occurs when cerium is subjected to high pressures or low temperatures. It appears that the valence changes from about 3 to 4 when it is cooled or compressed. The low temperature behavior of cerium is complex. Four allotropic modifications are thought to exist: cerium at room temperature and at atmospheric pressure is known as γ cerium. Upon cooling to –16°C, γ cerium changes to β cerium. The remaining γ cerium starts to change to α cerium when cooled to –172°C, and the transformation is complete at –269°C. α Cerium has a density of 8.16; δ cerium exists above 726°C. At atmospheric pressure, liquid cerium is more dense than its solid form at the melting point. Cerium is an iron-gray lustrous metal. It is malleable, and oxidizes very readily at room temperature, especially in moist air. Except for europium, cerium is the most reactive of the "rare-earth" metals. It slowly decomposes in cold water, and rapidly in hot water. Alkali solutions and dilute and concentrated acids attack the metal rapidly. The pure metal is likely to ignite if scratched with a knife. Ceric salts are orange red or yellowish; cerous salts are usually white. Cerium is a component of misch metal, which is extensively used in the manufacture of pyrophoric alloys for cigarette lighters, etc. Natural cerium is stable and contains four isotopes. Thirty-two other radioactive isotopes and isomers are known. While cerium is not radioactive, the impure commercial grade may contain traces of thorium, which is radioactive. The oxide is an important constituent of incandescent gas mantles and it is emerging as a hydrocarbon catalyst in "self-cleaning" ovens. In this application it can be incorporated into oven walls to prevent the collection of cooking residues. As ceric sulfate it finds extensive use as a volumetric oxidizing agent in quantitative analysis. Cerium compounds are used in the manufacture of glass, both as a component and as a decolorizer. The oxide is finding increased use as a glass polishing agent instead of rouge, for it is much faster than rouge in polishing glass surfaces. Cerium compounds are finding use in automobile exhaust catalysts. Cerium is also finding use in making permanent magnets. Cerium, with other rare earths, is used in carbon-arc lighting, especially in the motion picture industry. It is also finding use as an important catalyst in petroleum refining and in metallurgical and nuclear applications. In small lots, cerium costs about $5/g (99.9%).

Cesium — (L. *caesius,* sky blue), Cs; at. wt. 132.90545(2); at. no. 55; m.p. 28.44°C; b.p. 671°C; sp. gr. 1.873 (20°C); valence 1. Cesium was discovered spectroscopically by Bunsen and Kirchhoff in 1860 in mineral water from Durkheim. Cesium, an alkali metal, occurs in *lepidolite, pollucite* (a hydrated silicate of aluminum and cesium), and in other sources. One of the world's richest sources of cesium is located at Bernic Lake, Manitoba. The deposits are estimated to contain 300,000 tons of pollucite, averaging 20% cesium. It can be isolated by electrolysis of the fused cyanide and by a number of other methods. Very pure, gas-free cesium can be prepared by thermal decomposition of cesium azide. The metal is characterized by a spectrum containing two bright lines in the blue along with several others in the red, yellow, and green. It is silvery white, soft, and ductile. It is the most electropositive and most alkaline element. Cesium, gallium, and mercury are the only three metals that are liquid at room temperature. Cesium reacts explosively with cold water, and reacts with ice at temperatures above –116°C. Cesium hydroxide, the strongest base known, attacks glass. Because of its great affinity for oxygen the metal is used as a "getter" in electron tubes. It is also used in photoelectric cells, as well as a catalyst in the hydrogenation of certain organic compounds. The metal has recently found application in ion propulsion systems. Cesium is used in atomic clocks, which are accurate to 5 s in 300 years. A second of time is now defined as being the duration of 9,192,631,770 periods of the radiation corresponding to the transition between the two hyper-fine levels of the ground state of the cesium-133 atom. Its chief compounds are the chloride and the nitrate. Cesium has 52 isotopes and isomers with masses ranging from 112 to 148. The present price of cesium is about $65/g (99.98%) sealed in a glass ampoule.

Chlorine — (Gr. *chloros,* greenish yellow), Cl; at. wt. 35.4527(9); at. no. 17; m.p. –101.5°C; b.p. –34.04°C; t_c 143.8°C; density 3.214 g/l; sp. gr. 1.56 (–33.6°C); valence 1, 3, 5, or 7. Discovered in 1774 by Scheele, who thought it contained oxygen; named in 1810 by Davy, who insisted it was an element. In nature it is found in the combined state only, chiefly with sodium as common salt (NaCl), *carnallite* ($KMgCl_3 \cdot 6H_2O$), and *sylvite* (KCl). It is a member of the halogen (salt-forming) group of elements and is obtained from chlorides by the action of oxidizing agents and more often by electrolysis; it is a greenish-yellow gas, combining directly with nearly all elements. At 10°C one volume of water dissolves 3.10 volumes of chlorine, at 30°C only 1.77 volumes. Chlorine is widely used in making many everyday products. It is used for producing safe drinking water the world over. Even the smallest water supplies are now usually chlorinated. It is also extensively used in the production of paper products, dyestuffs, textiles, petroleum products, medicines, antiseptics, insecticides, foodstuffs, solvents, paints, plastics, and many other consumer products. Most of the chlorine produced is used in the manufacture of chlorinated compounds for sanitation, pulp bleaching, disinfectants, and textile processing. Further use is in the manufacture of chlorates, chloroform, carbon tetrachloride, and in the extraction of bromine. Organic chemistry demands much from chlorine, both as an oxidizing agent and in substitution, since it often brings desired properties in an organic compound when substituted for hydrogen, as in one form of synthetic rubber. Chlorine is a respiratory irritant. The gas irritates the mucous membranes and the liquid burns the skin. As little as 3.5 ppm can be detected as an odor, and 1000 ppm is likely to be fatal after a few deep breaths. It was used as a war gas in 1915. Natural chlorine contains two isotopes. Twenty other isotopes and isomers are known.

Chromium — (Gr. *chroma,* color), Cr; at. wt. 51.9961(6); at. no. 24; m.p. 1907°C; b.p. 2671°C; sp. gr. 7.18 to 7.20 (20°C); valence chiefly 2, 3, or 6. Discovered in 1797 by Vauquelin, who prepared the metal the next year, chromium is a steel-gray, lustrous, hard metal that takes a high polish. The principal ore is *chromite* ($FeCr_2O_4$), which is found in Zimbabwe, Russia, South Africa, Turkey, Iran, Albania, Finland, Democratic Republic of Madagascar, the Philippines, and elsewhere. The U.S. has no appreciable chromite ore reserves. The metal is usually produced by reducing the oxide with aluminum. Chromium is used to harden steel, to manufacture stainless steel, and to form many useful alloys. Much is used in plating to produce a hard, beautiful surface and to prevent corrosion. Chromium is used to give glass an emerald green color. It finds wide use as a catalyst. All compounds of chromium are colored; the most important are the chromates of sodium and potassium (K_2CrO_4) and the dichromates ($K_2Cr_2O_7$) and the potassium and ammonium chrome alums, as $KCr(SO_4)_2 \cdot 12H_2O$. The dichromates are used as oxidizing agents in quantitative analysis, also in tanning leather.

Other compounds are of industrial value; lead chromate is chrome yellow, a valued pigment. Chromium compounds are used in the textile industry as mordants, and by the aircraft and other industries for anodizing aluminum. The refractory industry has found chromite useful for forming bricks and shapes, as it has a high melting point, moderate thermal expansion, and stability of crystalline structure. Chromium is an essential trace element for human health. Many chromium compounds, however, are acutely toxic, chronically toxic, and may be carcinogenic. They should be handled with proper safeguards. Natural chromium contains four isotopes. Twenty other isotopes are known. Chromium metal (99.95%) costs about $600/kg. Commercial grade chromium (99%) costs about $75/kg.

Cobalt — (*Kobald,* from the German, goblin or evil spirit, *cobalos,* Greek, mine), Co; at. wt. 58.93320(1); at. no. 27; m.p. 1495°C; b.p. 2927°C; sp. gr. 8.9 (20°C); valence 2 or 3. Discovered by Brandt about 1735. Cobalt occurs in the mineral *cobaltite, smaltite,* and *erythrite,* and is often associated with nickel, silver, lead, copper, and iron ores, from which it is most frequently obtained as a by-product. It is also present in meteorites. Important ore deposits are found in Congo-Kinshasa, Australia, Zambia, Russia, Canada, and elsewhere. The U.S. Geological Survey has announced that the bottom of the north central Pacific Ocean may have cobalt-rich deposits at relatively shallow depths in waters close to the Hawaiian Islands and other U.S. Pacific territories. Cobalt is a brittle, hard metal, closely resembling iron and nickel in appearance. It has a magnetic permeability of about two thirds that of iron. Cobalt tends to exist as a mixture of two allotropes over a wide temperature range; the β-form predominates below 400°C, and the α above that temperature. The transformation is sluggish and accounts in part for the wide variation in reported data on physical properties of cobalt. It is alloyed with iron, nickel and other metals to make Alnico, an alloy of unusual magnetic strength with many important uses. Stellite alloys, containing cobalt, chromium, and tungsten, are used for high-speed, heavy-duty, high temperature cutting tools, and for dies. Cobalt is also used in other magnet steels and stainless steels, and in alloys used in jet turbines and gas turbine generators. The metal is used in electroplating because of its appearance, hardness, and resistance to oxidation. The salts have been used for centuries for the production of brilliant and permanent blue colors in porcelain, glass, pottery, tiles, and enamels. It is the principal ingredient in Sevre's and Thenard's blue. A solution of the chloride ($CoCl_2 \cdot 6H_2O$) is used as sympathetic ink. The cobalt ammines are of interest; the oxide and the nitrate are important. Cobalt carefully used in the form of the chloride, sulfate, acetate, or nitrate has been found effective in correcting a certain mineral deficiency disease in animals. Soils should contain 0.13 to 0.30 ppm of cobalt for proper animal nutrition.Cobalt is found in Vitamin B-12, which is essential for human nutrition. Cobalt of 99.9+% purity is priced at about$500/kg. Cobalt-60, an artificial isotope, is an important gamma ray source, and is extensively used as a tracer and a radiotherapeutic agent. Single compact sources of Cobalt-60 vary from about $1 to $10/curie, depending on quantity and specific activity. Thirty isotopes and isomers of cobalt are known.

Columbium — See Niobium.

Copper — (L. *cuprum,* from the island of Cyprus), Cu; at. wt. 63.546(3); at. no. 29; f.p. 1084.62 °C; b.p. 2562°C; sp. gr. 8.96 (20°C); valence 1 or 2. The discovery of copper dates from prehistoric times. It is said to have been mined for more than 5000 years. It is one of man's most important metals. Copper is reddish colored, takes on a bright metallic luster, and is malleable, ductile, and a good conductor of heat and electricity (second only to silver in electrical conductivity). The electrical industry is one of the greatest users of copper. Copper occasionally occurs native, and is found in many minerals such as *cuprite, malachite, azurite, chalcopyrite,* and *bornite.* Large copper ore deposits are found in the U.S., Chile, Zambia, Zaire, Peru, and Canada. The most important copper ores are the sulfides, oxides, and carbonates. From these, copper is obtained by smelting, leaching, and by electrolysis. Its alloys, brass and bronze, long used, are still very important; all American coins are now copper alloys; monel and gun metals also contain copper. The most important compounds are the oxide and the sulfate, blue vitriol; the latter has wide use as an agricultural poison and as an algicide in water purification. Copper compounds such as Fehling's solution are widely used in analytical chemistry in tests for sugar. High-purity copper (99.999 + %) is readily available commercially. The price of commercial copper has fluctuated widely. The price of copper in December 1999 was about $1.75/kg. Natural copper contains two isotopes. Twenty-six other radioactive isotopes and isomers are known.

Curium — (Pierre and Marie Curie), Cm; at. wt. (247); at. no. 96; m.p. 1345°C; sp. gr. 13.51 (calc.); valence 3 and 4. Although curium follows americium in the periodic system, it was actually known before americium and was the third transuranium element to be discovered. It was identified by Seaborg, James, and Ghiorso in 1944 at the wartime Metallurgical Laboratory in Chicago as a result of helium-ion bombardment of ^{239}Pu in the Berkeley, California, 60-inch cyclotron. Visible amounts (30 μg) of ^{242}Cm, in the form of the hydroxide, were first isolated by Werner and Perlman of the University of California in 1947. In 1950, Crane, Wallmann, and Cunningham found that the magnetic susceptibility of microgram samples of CmF_3 was of the same magnitude as that of GdF_3. This provided direct experimental evidence for assigning an electronic configuration to Cm^{+3}. In 1951, the same workers prepared curium in its elemental form for the first time. Sixteen isotopes of curium are now known. The most stable, ^{247}Cm, with a half-life of 16 million years, is so short compared to the earth's age that any primordial curium must have disappeared long ago from the natural scene. Minute amounts of curium probably exist in natural deposits of uranium, as a result of a sequence of neutron captures and β decays sustained by the very low flux of neutrons naturally present in uranium ores. The presence of natural curium, however, has never been detected. ^{242}Cm and ^{244}Cm are available in multigram quantities. ^{248}Cm has been produced only in milligram amounts. Curium is similar in some regards to gadolinium, its rare-earth homolog, but it has a more complex crystal structure. Curium is silver in color, is chemically reactive, and is more electropositive than aluminum. CmO_2, Cm_2O_3, CmF_3, CmF_4, $CmCl_3$, $CmBr_3$, and CmI_3 have been prepared. Most compounds of trivalent curium are faintly yellow in color. ^{242}Cm generates about three watts of thermal energy per gram. This compares to one-half watt per gram of ^{238}Pu. This suggests use for curium as a power source. ^{244}Cm is now offered for sale by the O.R.N.L. at $185/mg plus packing charges. ^{248}Cm is available at a cost of $160/μg, plus packing charges, from the O.R.N.L. Curium absorbed into the body accumulates in the bones, and is therefore very toxic as its radiation destroys the red-cell forming mechanism. The maximum permissible total body burden of ^{244}Cm (soluble) in a human being is 0.3 μCi (microcurie).

Deuterium, an isotope of hydrogen — see Hydrogen.

Dubnium — (named after the Joint Institute of Nuclear Research in Dubna, Russia). Db; at.wt. [262]; at.no. 105. In 1967 G. N. Flerov reported that a Soviet team working at the Joint Institute for Nuclear Research at Dubna may have produced a few atoms of $^{260}105$ and $^{261}105$ by bombarding ^{243}Am with ^{22}Ne. Their evidence was based on time-coincidence measurements of alpha energies. More recently, it was reported that early in 1970 Dubna scientists synthesized Element 105 and that by the end of April 1970 "had investigated all the types of decay of the new element and had determined its chemical properties." In late April 1970, it was announced that Ghiorso, Nurmia, Harris, K. A. Y. Eskola, and P. L. Eskola, working at the University of California at Berkeley, had positively identified Element 105. The discovery was made by bombarding a target of ^{249}Cf with a beam of 84 MeV nitrogen nuclei in the Heavy Ion Linear Accelerator (HILAC). When a ^{15}N nuclear is absorbed by a ^{249}Cf nucleus, four neutrons are emitted and a new atom of $^{260}105$ with a half-life of 1.6 s is formed. While the first atoms of Element 105 are said to have been detected conclusively on March 5, 1970, there is evidence that Element 105 had been formed in Berkeley experiments a year earlier by the method described. Ghiorso and his associates have attempted to confirm Soviet findings by more sophisticated methods without success.

In October 1971, it was announced that two new isotopes of Element 105 were synthesized with the heavy ion linear accelerator by A. Ghiorso and co-workers at Berkeley. Element $^{261}105$ was produced both by bombarding ^{250}Cf with ^{15}N and by bombarding ^{249}Bk with ^{16}O. The isotope emits 8.93-MeV α particles and decays to ^{257}Lr with a half-life of about 1.8 s. Element $^{262}105$ was produced by bombarding ^{249}Bk with ^{18}O. It emits 8.45 MeV α particles and decays to ^{258}Lr with a half-life of about 40 s. Nine isotopes of Dubnium are now recognized. Soon after the discovery the names ***Hahnium*** and ***Joliotium,*** named after Otto Hahn and Jean-Frederic Joliot and Mme. Joliot-Curie, were suggested as names for Element 105. The IUPAC in August 1997 finally resolved the issue, naming Element 105 **Dubnium** with the symbol Db. Dubnium is thought to have properties similar to tantalum.

Dysprosium — (Gr. *dysprositos,* hard to get at), Dy; at. wt. 162.50(3); at. no. 66; m.p. 1412°C; b.p. 2567°C; sp. gr. 8.551 (25°C); valence 3. Dysprosium was discovered in 1886 by Lecoq de Boisbaudran, but not isolated. Neither the oxide nor the metal was available in relatively pure form until the development of ion-exchange separation and metallographic reduction techniques by Spedding and associates about 1950. Dysprosium occurs along with other so-called rare-earth or lanthanide elements in a variety of minerals such as *xenotime, fergusonite, gadolinite, euxenite, polycrase,* and *blomstrandine.* The most important sources, however, are from *monazite* and *bastnasite.* Dysprosium can be prepared by reduction of the trifluoride with calcium. The element has a metallic, bright silver luster. It is relatively stable in air at room temperature, and is readily attacked and dissolved, with the evolution of hydrogen, by dilute and concentrated mineral acids. The metal is soft enough to be cut with a knife and can be machined without sparking if overheating is avoided. Small amounts of impurities can greatly affect its physical properties. While dysprosium has not yet found many applications, its thermal neutron absorption cross-section and high melting point suggest metallurgical uses in nuclear control applications and for alloying with special stainless steels. A dysprosium oxide-nickel cermet has found use in cooling nuclear reactor rods. This cermet absorbs neutrons readily without swelling or contracting under prolonged neutron bombardment. In combination with vanadium and other rare earths, dysprosium has been used in making laser materials. Dysprosium-cadmium chalcogenides, as sources of infrared radiation, have been used for studying chemical reactions. The cost of dysprosium metal has dropped in recent years since the development of ion-exchange and solvent extraction techniques, and the discovery of large ore bodies. Thirty two isotopes and isomers are now known. The metal costs about $5/g (99.9% purity).

Einsteinium — (Albert Einstein [1879–1955]), Es; at. wt. (252); m.p. 860°C (est.); at. no. 99. Einsteinium, the seventh transuranic element of the actinide series to be discovered, was identified by Ghiorso and co-workers at Berkeley in December 1952 in debris from the first large thermonuclear explosion, which took place in the Pacific in November 1952. The isotope produced was the 20-day ^{253}Es isotope. In 1961, a sufficient amount of einsteinium was produced to permit separation of a macroscopic amount of ^{253}Es. This sample weighed about 0.01 µg. A special magnetic-type balance was used in making this determination. ^{253}Es so produced was used to produce mendelevium. About 3 µg of einsteinium has been produced at Oak Ridge National Laboratories by irradiating for several years kilogram quantities of ^{239}Pu in a reactor to produce ^{242}Pu. This was then fabricated into pellets of plutonium oxide and aluminum powder, and loaded into target rods for an initial 1-year irradiation at the Savannah River Plant, followed by irradiation in a HFIR (High Flux Isotopic Reactor). After 4 months in the HFIR the targets were removed for chemical separation of the einsteinium from californium. Nineteen isotopes and isomers of einsteinium are now recognized. ^{254}Es has the longest half-life (276 days). Tracer studies using ^{253}Es show that einsteinium has chemical properties typical of a heavy trivalent, actinide element. Einsteinium is extremely radioactive. Great care must be taken when handling it.

Element 93 — See Neptunium.
Element 94 — See Plutonium.
Element 95 — See Americium.
Element 96 — See Curium.
Element 97 — See Berkelium.
Element 98 — See Californium.
Element 99 — See Einsteinium.
Element 100 — See Fermium (unnilnilium).
Element 101 — See Mendelevium (unnilunium).
Element 102 — See Nobelium (unnilbium).
Element 103 — See Lawrencium (unniltrium).
Element 104 — See Rutherfordium (unnilquadium).
Element 105 — See Dubnium (unnilpentium).
Element 106 — See Seaborgium (unnilhexium).
Element 107 — See Bohrium (unnilseptium).
Element 108 — See Hassium (unniloctium).
Element 109 — See Meitnerium (unnilennium).

Element 110 — In 1987 Oganessian, et al., at Dubna, claimed discovery of this element. Their experiments indicated the spontaneous fissioning nuclide $^{272}110$ with a half-life of 10 ms. More recently a group led by Armbruster at G.S.I. in Darmstadt, Germany, reported evidence of $^{269}110$, which was produced by bombarding lead for many days with more than 10^{18} nickel atoms. A detector searched each collision for Element 110's distinct decay sequence. On November 9, 1994, evidence of 110 was detected. Berkeley scientists, in 1991, performed similar experiments and reported evidence of 110, but this was not confirmed. Workers at Dubna have experiments underway to produce $^{273}110$ by bombarding plutonium with sulfur atoms. Other experiments at G.S.I. and elsewhere are now searching for heavier isotopes. Five isotopes of Element 110 are now recognized. Several years ago the IUPAC suggested the use of the temporary name *ununnilium,* with the symbol Uun, for Element 110 when it was found.

Element 111 — On December 20, 1994, scientists at GSI Darmstadt, Germany announced they had detected three atoms of a new element with 111 protons and 161 neutrons. This element was made by bombarding ^{83}Bi with ^{28}Ni. Signals of Element 111 appeared for less than 0.002 sec, then decayed into lighter elements including Element $^{268}109$ and Element $^{264}107$. These isotopes had not previously been observed. A name for Element 111 has not been suggested although IUPAC has suggested a temporary name of Unununium, with the symbol Uuu. Element 111 is expected to have properties similar to gold.

Element 112—In late February 1996, Siguard Hofmann and his collaborators at GSI Darmstadt announced their discovery of Element 112, having 112 protons and 165 neutrons, with an atomic mass of 277. This element was made by bombarding a lead target with high-energy zinc ions. A single nucleus of Element 112 was detected, which decayed after less than 0.001 sec by emitting an α particle, consisting of two protons and two neutrons. This created Element 110_{273}, which in turn decayed by emitting an α particle to form a new isotope of Element 108 and so on. Evidence indicates that nuclei with 162 neutrons are held together more strongly than nuclei with a smaller or larger number of neutrons. This suggests a narrow "peninsula" of relatively stable isotopes around Element 114. GSI scientists are experimenting to bombard targets with ions heavier than zinc to produce Elements 113 and 114. A name has not yet been suggested for Element 112, although the IUPAC suggested the temporary name of ununbium, with the symbol of Uub, when the element was discovered. Element 112 is expected to have properties similar to mercury.

Element 113 — (ununtrium) As of December 1999 this element remains undiscovered.

Element 114 — (ununquadium) Symbol Uuq. Element 114 is the first new element to be discovered since 1996. This element was found by a Russian–American team, including Livermore researchers, by bombarding a sheet of plutonium with a rare form of calcium hoping to make the atoms stick together in a new element. Radiation showed that the new element broke into smaller pieces. Data of radiation collected at the Russian Joint Institute for Nuclear Research in November and December 1998, were analyzed in January 1999. It was found that some of the heavy atoms created when 114 decayed lived up to 30 seconds, which was longr than ever seen before, for such a heavy element. This isotope decayed into a previously unknown isotope of Element 112, which itself lasted 15 minutes. That isotope, in turn, decayed to a previously undiscovered isotope of Element 108, which survived 17 minutes. Isotopes of these and those with longer life-times have been predicted for some time by theorists. It appears that these isotopes are on the edge of the "island of stability", and that some of the isotopes in this region might last long enough for studies of their nuclear behavior and for a chemical evaluation to be made. No name has yet been suggested for Element 114; however, the temporary name of ununquadium with symbol Uuq may be used.

In June 1999 it was announced that Elements 118 and 116 had been discovered. These disintegrated into Element 114 and other elements. See under Element 118.

Element 115— (ununpentium) As of December 1999, this element remains undiscovered.

Element 116— (ununhexium) Symbol Uuh. This element was discovered along with Element 118 in June 1999. See under Element 118.

Element 117— (ununseptium) As of December 1999, this element remains undiscovered.

Element 118—(ununoctium) Symbol Uuo. Victor Nonov of the Lawrence Berkeley National Laboratory led a team of university and government scientists in experiments that led to the discovery of Elements 118 and 116 in June 1999. A lead target was bombarded for more than 10 days with roughly 1 quintillion krypton ions. The team made three atoms of Element 118, which quickly decayed into Elements 116, 114, and other elements. The isotopes of Element 118 lasted only about 200 microseconds, while the isotope of Element 116 lasted only 1.2 milliseconds. Investigations seem to indicate that the investigations are converging on a cluster of very stable super-heavy isotopes, some of which may survive for very long times. Researchers at GSI, in Russia at Dubna, or at the Berkeley Laboratory may soon fire krypton at bismuth to make Element 119, which would decay into the yet-undiscovered Elements 117, 115, and 113.

Erbium — (*Ytterby*, a town in Sweden), Er; at. wt. 167.26(3); at. no. 68; m.p. 1529°C; b.p. 2868°C; sp. gr. 9.066 (25°C); valence 3, Erbium, one of the so-called rare-earth elements of the lanthanide series, is found in the minerals mentioned under dysprosium above. In 1842 Mosander separated "yttria," found in the mineral *gadolinite,* into three fractions which he called *yttria, erbia,* and *terbia.* The names *erbia* and *terbia* became confused in this early period. After 1860, Mosander's *terbia* was known as *erbia,* and after 1877, the earlier known *erbia* became *terbia.* The *erbia* of this period was later shown to consist of five oxides, now known as *erbia, scandia, holmia, thulia* and *ytterbia.* By 1905 Urbain and James independently succeeded in isolating fairly pure Er_2O_3. Klemm and Bommer first produced reasonably pure erbium metal in 1934 by reducing the anhydrous chloride with potassium vapor. The pure metal is soft and malleable and has a bright, silvery, metallic luster. As with other rare-earth metals, its properties depend to a certain extent on the impurities present. The metal is fairly stable in air and does not oxidize as rapidly as some of the other rare-earth metals. Naturally occurring erbium is a mixture of six isotopes, all of which are stable. Twenty-seven radioactive isotopes of erbium are also recognized. Recent production techniques, using ion-exchange reactions, have resulted in much lower prices of the rare-earth metals and their compounds in recent years. The cost of 99.9% erbium metal is about $10/g. Erbium is finding nuclear and metallurgical uses. Added to vanadium, for example, erbium lowers the hardness and improves workability. Most of the rare-earth oxides have sharp absorption bands in the visible, ultraviolet, and near infrared. This property, associated with the electronic structure, gives beautiful pastel colors to many of the rare-earth salts. Erbium oxide gives a pink color and has been used as a colorant in glasses and porcelain enamel glazes.

Europium — (Europe), Eu; at. wt. 151.964(1); at. no. 63; m.p. 822°C; b.p. 1596°C; sp. gr. 5.244 (25°C); valence 2 or 3. In 1890 Boisbaudran obtained basic fractions from samarium-gadolinium concentrates which had spark spectral lines not accounted for by samarium or gadolinium. These lines subsequently have been shown to belong to europium. The discovery of europium is generally credited to Demarcay, who separated the rare earth in reasonably pure form in 1901. The pure metal was not isolated until recent years. Europium is now prepared by mixing Eu_2O_3 with a 10%-excess of lanthanum metal and heating the mixture in a tantalum crucible under high vacuum. The element is collected as a silvery-white metallic deposit on the walls of the crucible. As with other rare-earth metals, except for lanthanum, europium ignites in air at about 150 to 180°C. Europium is about as hard as lead and is quite ductile. It is the most reactive of the rare-earth metals, quickly oxidizing in air. It resembles calcium in its reaction with water. *Bastnasite* and *monazite* are the principal ores containing europium. Europium has been identified spectroscopically in the sun and certain stars. Europium isotopes are good neutron absorbers and are being studied for use in nuclear control applications. Europium oxide is now widely used as a phosphor activator and europium-activated yttrium vanadate is in commercial use as the red phosphor in color TV tubes. Europium-doped plastic has been used as a laser material. With the development of ion-exchange techniques and special processes, the cost of the metal has been greatly reduced in recent years. Natural europium contains two stable isotopes. Thirty five other radioactive isotopes and isomers are known. Europium is one of the rarest and most costly of the rare-earth metals. It is priced at about $50/g (99.9% pure).

Fermium —(Enrico Fermi [1901–1954], nuclear physicist), Fm; at. wt. [257]; at. no. 100; m.p. 1527°C. Fermium, the eighth transuranium element of the actinide series to be discovered, was identified by Ghiorso and co-workers in 1952 in the debris from a thermonuclear explosion in the Pacific in work involving the University of California Radiation Laboratory, the Argonne National Laboratory, and the Los Alamos Scientific Laboratory. The isotope produced was the 20-hour ^{255}Fm. During 1953 and early 1954, while discovery of elements 99 and 100 was withheld from publication for security reasons, a group from the Nobel Institute of Physics in Stockholm bombarded ^{238}U with ^{16}O ions, and isolated a 30-min α-emitter, which they ascribed to ^{250}100, without claiming discovery of the element. This isotope has since been identified positively, and the 30-min half-life confirmed. The chemical properties of fermium have been studied solely with tracer amounts, and in normal aqueous media only the (III) oxidation state appears to exist. The isotope ^{254}Fm and heavier isotopes can be produced by intense neutron irradiation of lower elements such as plutonium by a process of successive neutron capture interspersed with beta decays until these mass numbers and atomic numbers are reached. Twenty isotopes and isomers of fermium are known to exist. ^{257}Fm, with a half-life of about 100.5 days, is the longest lived. ^{250}Fm, with a half-life of 30 min, has been shown to be a product of decay of Element ^{254}102. It was by chemical identification of ^{250}Fm that production of Element 102 (nobelium) was confirmed. Fermium would probably have chemical properties resembling erbium.

Fluorine — (L. and F. *fluere,* flow, or flux), F; at. wt. 18.9984032(5); at. no. 9; m.p. –219.62°C (1 atm); b.p. –188.12°C (1 atm); t_c -129.02°C; density 1.696 g/L (0°C, 1 atm); liq. den. at b.p. 1.50 g/cm^3; valence 1. In 1529, Georigius Agricola described the use of fluorspar as a flux, and as early as 1670 Schwandhard found that glass was etched when exposed to fluorspar treated with acid. Scheele and many later investigators, including Davy, Gay-Lussac, Lavoisier, and Thenard, experimented with hydrofluoric acid, some experiments ending in tragedy. The element was finally isolated in 1886 by Moisson after nearly 74 years of continuous effort. Fluorine occurs chiefly in *fluorspar* (CaF_2) and *cryolite* (Na_2AlF_6), but is rather widely distributed in other minerals. It is a member of the halogen family of elements, and is obtained by electrolyzing a solution of potassium hydrogen fluoride in anhydrous hydrogen fluoride in a vessel of metal or transparent fluorspar. Modern commercial production methods are essentially variations on the procedures first used by Moisson. Fluorine is the most electronegative and reactive of all elements. It is a pale yellow, corrosive gas, which reacts with practically all organic and inorganic substances. Finely divided metals, glass, ceramics, carbon, and even water burn in fluorine with a bright flame. Until World War II, there was no commercial production of elemental fluorine. The atom bomb project and nuclear energy applications, however, made it necessary to produce large quantities. Safe handling techniques have now been developed and it is possible at present to transport liquid fluorine by the ton. Fluorine and its compounds are used in producing uranium (from the hexafluoride) and more than 100 commercial fluorochemicals, including many well-known high-temperature plastics. Hydrofluoric acid is extensively used for etching the glass of light bulbs, etc. Fluorochloro hydrocarbons have been extensively used in air conditioning and refrigeration. However, in recent years the U.S. and other countries have been phasing out ozone-depleting substances, such as the fluorochloro hydrocarbons that have been used in these applications. It has been suggested that fluorine might be substituted for hydrogen wherever it occurs in organic compounds, which could lead to an astronomical number of new fluorine compounds. The presence of fluorine as a soluble fluoride in drinking water to the extent of 2 ppm may cause mottled enamel in teeth, when used by children acquiring permanent teeth; in smaller amounts, however, fluorides are said to be beneficial and used in water supplies to prevent dental cavities. Elemental fluorine has been studied as a rocket propellant as it has an exceptionally high specific impulse value. Compounds of fluorine with rare gases have now been confirmed. Fluorides of xenon, radon, and krypton are among those known. Elemental fluorine and the fluoride ion are highly toxic. The free element has a characteristic pungent odor, detectable in concentrations as low as 20 ppb, which is below the safe working level. The recommended maximum allowable concentration for a daily 8-hour time-weighted exposure is 1 ppm. Fluorine is known to have fourteen isotopes.

Francium — (France), Fr; at. no. 87; at. wt. [223]; m.p. 27°C; valence 1. Discovered in 1939 by Mlle. Marguerite Perey of the Curie Institute, Paris. Francium, the heaviest known member of the alkali metal series, occurs as a result of an alpha disintegration of actinium. It can also be made artificially by bombarding thorium with protons. While it occurs naturally in uranium minerals, there is probably less than an ounce of francium at any time in the total crust of the earth. It has the highest equivalent weight of any element, and is the most unstable of the first 101 elements of the periodic system. Thirty-six isotopes and isomers of francium are recognized. The longest lived ^{223}Fr(Ac, K), a daughter of ^{227}Ac, has a half-life of 21.8 min. This is the only isotope of francium occurring in nature. Because all known isotopes of francium are highly unstable, knowledge of the chemical properties of this element comes from radiochemical techniques. No weighable quantity of the element has been prepared or isolated. The chemical properties of francium most closely resemble cesium. In 1996, researchers Orozco, Sprouse, and co-workers at the State University of New York, Stony Brook, reported that they had produced francium atoms by bombarding ^{18}O atoms at a gold target heated almost to its melting point. Collisions between gold and oxygen nuclei created atoms of francium-210 which had 87 protons and123 neutrons. This team reported they had generated about 1 million francium-210 ions per second and held 1000 or more atoms at a time for about 20 secs. in a magnetic trap they had devised before the atoms decayed or escaped. Enough francium was trapped so that a video camera could capture the light given off by the atoms as they fluoresced. A cluster of about 10,000 francium atoms appeared as a glowing sphere about 1 mm in diameter. It is thought that the francium atoms could serve as miniature laboratoires for probing interactions between electrons and quarks.

Gadolinium — (*gadolinite,* a mineral named for Gadolin, a Finnish chemist), Gd; at. wt. 157.25(3); at. no. 64; m.p. 1313°C; b.p. 3273°C; sp. gr. 7.901 (25°C); valence 3. Gadolinia, the oxide of gadolinium, was separated by Marignac in 1880 and Lecoq de Boisbaudran independently isolated the element from Mosander's "yttria" in 1886. The element was named for the mineral *gadolinite* from which this rare earth was originally obtained. Gadolinium is found in several other minerals, including *monazite* and *bastnasite,* which are of commercial importance. The element has been isolated only in recent years. With the development of ion-exchange and solvent extraction techniques, the availability and price of gadolinium and the other rare-earth metals have greatly improved. Thirty-one isotopes and isomers of gadolinium are now recognized; seven are stable and occur naturally. The metal can be prepared by the reduction of the anhydrous fluoride with metallic calcium. As with other related rare-earth metals, it is silvery white, has a metallic luster, and is malleable and ductile. At room temperature, gadolinium crystallizes in the hexagonal, close-packed α form. Upon heating to 1235°C, α gadolinium transforms into the β form, which has a body-centered cubic structure. The metal is relatively stable in dry air, but in moist air it tarnishes with the formation of a loosely adhering oxide film which spalls off and exposes more surface to oxidation. The metal reacts slowly with water and is soluble in dilute acid. Gadolinium has the highest thermal neutron capture cross-section of any known element (49,000 barns). Natural gadolinium is a mixture of seven isotopes. Two of these, ^{155}Gd and ^{157}Gd, have excellent capture characteristics, but they are present naturally in low concentrations. As a result, gadolinium has a very fast burnout rate and has limited use as a nuclear control rod material. It has been used in making gadolinium yttrium garnets, which have microwave applications. Compounds of gadolinium are used in making phosphors for color TV tubes. The metal has unusual superconductive properties. As little as 1% gadolinium has been found to improve the workability and resistance of iron, chromium, and related alloys to high temperatures and oxidation. Gadolinium ethyl sulfate has extremely low noise characteristics and may find use in duplicating the performance of amplifiers, such as the maser. The metal is ferromagnetic. Gadolinium is unique for its high magnetic moment and for its special Curie temperature (above which ferromagnetism vanishes) lying just at room temperature. This suggests uses as a magnetic component that senses hot and cold. The price of the metal is about $12/g (99.9% purity).

Gallium — (L. *Gallia,* France; also from Latin, *gallus,* a translation of Lecoq, a cock), Ga; at. wt. 69.723(1); at. no. 31; m.p. 29.76°C; b.p. 2204°C; sp. gr. 5.904 (29.6°C) solid; sp. gr. 6.095 (29.6°C) liquid; valence 2 or 3. Predicted and described by Mendeleev as ekaaluminum, and discovered spectroscopically by Lecoq de Boisbaudran in 1875, who in the same year obtained the free metal by electrolysis of a solution of the hydroxide in KOH. Gallium is often found as a trace element in *diaspore, sphalerite, germanite, bauxite,* and *coal.* Some flue dusts from burning coal have been shown to contain as much as 1.5% gallium. It is the only metal, except for mercury, cesium, and rubidium, which can be liquid near room temperatures; this makes possible its use in high-temperature thermometers. It has one of the longest liquid ranges of any metal and has a low vapor pressure even at high temperatures. There is a strong tendency for gallium to supercool below its freezing point. Therefore, seeding may be necessary to initiate solidification. Ultra-pure gallium has a beautiful, silvery appearance, and the solid metal exhibits a conchoidal fracture similar to glass. The metal expands 3.1% on solidifying; therefore, it should not be stored in glass or metal containers, as they may break as the metal solidifies. Gallium wets glass or porcelain, and forms a brilliant mirror when it is painted on glass. It is widely used in doping semiconductors and producing solid-state devices such as transistors. High-purity gallium is attacked only slowly by mineral acids. Magnesium gallate containing divalent impurities such as Mn^{+2} is finding use in commercial ultraviolet activated powder phosphors. Gallium nitride has been used to produce blue light-emitting diodes. Blue LED's used in compact disc applications can be used to store a 2-h movie, for example, on one 5-in. diameter disc. Extensive use of gallium has found recent application in the **Gallex Detector Experiment** located in the Gran Sasso Underground Laboratory in Italy. This underground facility has been built by the Italian Istituto Nazionale di Fisica Nucleare in the middle of a highway tunnel through the Abruzzese mountains, about 150 km east of Rome. The experiment is shielded by a 3300-m water-equivalent of rock. In this experiment, 30.3 tons of gallium in the form of 110 tons of $GaCl_3$-HCl solution are being used to detect solar neutrinos. The production of ^{71}Ge from gallium is being measured.Gallium arsenide is capable of converting electricity directly into coherent light. Gallium readily alloys with most metals, and has been used as a component in low-melting alloys. Its toxicity appears to be of a low order, but should be handled with care until more data are forthcoming. Natural gallium contains two stable isotopes. Twenty-six other isotopes, one of which is an isomer, are known. The metal can be supplied in ultrapure form (99.99999+%). The cost is about $5/g (99.999%).

Germanium — (L. *Germania,* Germany), Ge; at. wt. 72.61(2); at. no. 32; m.p. 938.25°C; b.p. 2833°C; sp. gr. 5.323 (25°C); valence 2 and 4. Predicted by Mendeleev in 1871 as ekasilicon, and discovered by Winkler in 1886. The metal is found in *argyrodite,* a sulfide of germanium and silver; in *germanite,* which contains 8% of the element; in zinc ores; in coal; and in other minerals. The element is frequently obtained commercially from flue dusts of smelters processing zinc ores, and has been recovered from the by-products of combustion of certain coals. Its presence in coal insures a large reserve of the element in the years to come. Germanium can be separated from other metals by fractional distillation of its volatile tetrachloride. The tetrachloride may then be hydrolyzed to give GeO_2; the dioxide can be reduced with hydrogen to give the metal. Recently developed zone-refining techniques permit the production of germanium of ultra-high purity. The element is a gray-white metalloid, and in its pure state is crystalline and brittle, retaining its luster in air at room temperature. It is a very important semiconductor material. Zone-refining techniques have led to production of crystalline germanium for semiconductor use with an impurity of only one part in 10^{10}. Doped with arsenic, gallium, or other elements, it is used as a transistor element in thousands of electronic applications. Its application in fiber optics and infra-red optical systems now provides the largest use for germanium. Germanium is also finding many other applications including use as an alloying agent, as a phosphor in fluorescent lamps, and as a catalyst. Germanium and germanium oxide are transparent to the infrared and are used in infrared spectroscopes and other optical equipment, including extremely sensitive infrared detectors. Germanium oxide's high index of refraction and dispersion has made it useful as a component of glasses used in wide-angle camera lenses and microscope objectives. The field of organogermanium chemistry is becoming increasingly important. Certain germanium compounds have a low mammalian toxicity, but a marked activity against certain bacteria, which makes them of interest as chemotherapeutic agents. The cost of germanium is about $10/g (99.999% purity). Thirty isotopes and isomers are known, five of which occur naturally.

Gold — (Sanskrit *Jval;* Anglo-Saxon *gold*), Au (L. *aurum,* gold); at. wt. 196.96654(3); at. no. 79; m.p. 1064.18°C; b.p. 2856°C; sp. gr. ~19.3 (20°C); valence 1 or 3. Known and highly valued from earliest times, gold is found in nature as the free metal and in tellurides; it is very widely distributed and is almost always associated with quartz or pyrite. It occurs in veins and alluvial deposits, and is often separated from rocks and other minerals by sluicing and panning operations. About 25% of the world's gold output comes from South Africa, and about two thirds of the total U.S. production now comes from South Dakota and Nevada. The metal is recovered from its ores by cyaniding, amalgamating, and smelting processes. Refining is also frequently done by electrolysis. Gold occurs in sea water to the extent of 0.1 to 2 mg/ton, depending on the location where the sample is taken. As yet, no method has been found for recovering gold from sea water profitably. It is estimated that all the gold in the world, so far refined, could be placed in a single cube 60 ft on a side. Of all the elements, gold in its pure state is undoubtedly the most beautiful. It is metallic, having a yellow color when in a mass, but when finely divided it may be black, ruby, or purple. The Purple of Cassius is a delicate test for auric gold. It is the most malleable and ductile metal; 1 oz. of gold can be beaten out to 300 ft^2. It is a soft metal and is usually alloyed to give it more strength. It is a good conductor of heat and electricity, and is unaffected by air and most reagents. It is used in coinage and is a standard for monetary systems in many countries. It is also extensively used for jewelry, decoration, dental work, and for plating. It is used for coating certain space satellites, as it is a good reflector of infrared and is inert. Gold, like other precious metals, is measured in troy weight; when alloyed with other metals, the term *carat* is used to express the amount of gold present, 24 carats being pure gold. For many years the value of gold was set by the U.S. at $20.67/troy ounce; in 1934 this value was fixed by law at $35.00/troy ounce, 9/10th fine. On March 17, 1968, because of a gold crisis, a two-tiered pricing system was established whereby gold was still used to settle international accounts at the old $35.00/troy ounce price while the price of gold on the private market would be allowed to fluctuate. Since this time, the price of gold on the free market has fluctuated widely. The price of gold on the free market reached a price of $620/ troy oz. in January 1980. More recently, the U.K. and other nations, including the I.M.F. have sold or threatened to sell a sizeable portion of their gold reserves. This has caused wide fluctuations in the price of gold. Because this has damaged the economy of some countires, a moretorium for a few years has been declared. This has tended to stabilize temporarily the price of gold. The most common gold compounds are auric chloride ($AuCl_3$) and chlorauric acid ($HAuCl_4$), the latter being used in photography for toning the silver image. Gold has forty-eight recognized isotopes and isomers; ^{198}Au, with a half-life of 2.7 days, is used for treating cancer and other diseases. Disodium aurothiomalate is administered intramuscularly as a treatment for arthritis. A mixture of one part nitric acid with three of hydrochloric acid is called *aqua regia* (because it dissolved gold, the King of Metals). Gold is available commercially with a purity of 99.999+%. For many years the temperature assigned to the freezing point of gold has been 1063.0°C; this has served as a calibration point for the International Temperature Scales (ITS-27 and ITS-48) and the International Practical Temperature Scale (IPTS-

48). In 1968, a new International Practical Temperature Scale (IPTS-68) was adopted, which demanded that the freezing point of gold be changed to 1064.43°C. In 1990 a new International Temperature Scale (ITS-90) was adopted bringing the t.p. (triple point) of H_2O (t_{90} (°C)) to 0.01°C and the freezing point of gold to 1064.18°C.The specific gravity of gold has been found to vary considerably depending on temperature, how the metal is precipitated, and cold-worked. As of December 1999, gold was priced at about $300/troy oz. ($9.50/g).

Hafnium — (*Hafnia,* Latin name for Copenhagen), Hf; at. wt. 178.49(2); at. no. 72; m.p. 2233°C; b.p. 4603°C; sp. gr. 13.31 (20°C); valence 4. Hafnium was thought to be present in various minerals and concentrations many years prior to its discovery, in 1923, credited to D. Coster and G. von Hevesey. On the basis of the Bohr theory, the new element was expected to be associated with zirconium. It was finally identified in *zircon* from Norway, by means of X-ray spectroscopic analysis. It was named in honor of the city in which the discovery was made. Most zirconium minerals contain 1 to 5% hafnium. It was originally separated from zirconium by repeated recrystallization of the double ammonium or potassium fluorides by von Hevesey and Jantzen. Metallic hafnium was first prepared by van Arkel and deBoer by passing the vapor of the tetraiodide over a heated tungsten filament. Almost all hafnium metal now produced is made by reducing the tetrachloride with magnesium or with sodium (Kroll Process). Hafnium is a ductile metal with a brilliant silver luster. Its properties are considerably influenced by the impurities of zirconium present. Of all the elements, zirconium and hafnium are two of the most difficult to separate. Their chemistry is almost identical, however, the density of zirconium is about half that of hafnium. Very pure hafnium has been produced, with zirconium being the major impurity. Natural hafnium contains six isotopes, one of which is slightly radioactive. Hafnium has a total of 41 recognized isotopes and isomers. Because hafnium has a good absorption cross section for thermal neutrons (almost 600 times that of zirconium), has excellent mechanical properties, and is extremely corrosion resistant, it is used for reactor control rods. Such rods are used in nuclear submarines. Hafnium has been successfully alloyed with iron, titanium, niobium, tantalum, and other metals. Hafnium carbide is the most refractory binary composition known, and the nitride is the most refractory of all known metal nitrides (m.p. 3310°C). Hafnium is used in gas-filled and incandescent lamps, and is an efficient "getter" for scavenging oxygen and nitrogen. Finely divided hafnium is pyrophoric and can ignite spontaneously in air. Care should be taken when machining the metal or when handling hot sponge hafnium. At 700°C hafnium rapidly absorbs hydrogen to form the composition $HfH_{1.86}$. Hafnium is resistant to concentrated alkalis, but at elevated temperatures reacts with oxygen, nitrogen, carbon, boron, sulfur, and silicon. Halogens react directly to form tetrahalides. The price of the metal is in the broad range of $1/g to $5/g, depending on purity and quantity. The yearly demand for hafnium in the U.S. is now in excess of 50,000 kg.

Hahnium — A name previously used for Element 105, now named *dubnium.*

Hassium — (named for the German state, Hesse) Hs, at.wt. [265]; at.no. 108. This element was first synthesized and identified in 1964 by the same G.S.I. Darmstadt Group who first identified *Bohrium* and *Meitnerium.* Presumably this element has chemical properties similar to osmium. Isotope $^{265}108$ was produced using a beam of ^{58}Fe projectiles, produced by the Universal Linear Accelerator (UNILAC) to bombard a ^{208}Pb target. Discovery of *Bohrium* and *Meitnerium* was made using detection of isotopes with odd proton and neutron numbers. Elements having even atomic numbers have been thought to be less stable against spontaneous fusion than odd elements. The production of $^{265}108$ in the same reaction as was used at G.S.I. was confirmed at Dubna with detection of the seventh member of the decay chain ^{253}Es. Isotopes of *Hassium* are believed to decay by spontaneous fission, explaining why 109 was produced before 108. Isotope $^{265}108$ and $^{266}108$ are thought to decay to $^{261}106$, which in turn decay to $^{257}104$ and $^{253}102$. The IUPAC adopted the name *Hassium* after the German state of Hesse in September 1997.

Helium — (Gr. *helios,* the sun), He; at. wt. 4.002602(2); at. no. 2; m.p. below — 272.2°C (26 atm); b.p. — 268.93°C; t_c -267.96°C; density 0.1785 g/l (0°C, 1 atm); liquid density 7.62 lb/ft^3 at. b.p.; valence usually 0. Evidence of the existence of helium was first obtained by Janssen during the solar eclipse of 1868 when he detected a new line in the solar spectrum; Lockyer and Frankland suggested the name *helium* for the new element; in 1895, Ramsay discovered helium in the uranium mineral *clevite,* and it was independently discovered in cleveite by the Swedish chemists Cleve and Langlet about the same time. Rutherford and Royds in 1907 demonstrated that α particles are helium nuclei. Except for hydrogen, helium is the most abundant element found throughout the universe. Helium is extracted from natural gas; all natural gas contains at least trace quantities of helium. It has been detected spectroscopically in great abundance, especially in the hotter stars, and it is an important component in both the proton-proton reaction and the carbon cycle, which account for the energy of the sun and stars. The fusion of hydrogen into helium provides the energy of the hydrogen bomb. The helium content of the atmosphere is about 1 part in 200,000. It is present in various radioactive minerals as a decay product. Much of the world's supply of helium is obtained from wells in Texas, Colorado, and Kansas. The only other known helium extraction plants, outside the United States, in 1999 were inPoland, Russia, China, Algeria, and India. The cost of helium has fallen from $2500/ft^3 in 1915 to about 2.5¢/cu.ft. (.028 cu meters) in 1999. Helium has the lowest melting point of any element and has found wide use in cryogenic research, as its boiling point is close to absolute zero. Its use in the study of superconductivity is vital. Using liquid helium, Kurti and co-workers, and others, have succeeded in obtaining temperatures of a few microkelvins by the adiabatic demagnetization of copper nuclei, starting from about 0.01 K. Liquid helium (He^4) exists in two forms: He^4I and He^4II, with a sharp transition point at 2.174 K (3.83 cm Hg). He^4I (above this temperature) is a normal liquid, but He^4II (below it) is unlike any other known substance. It expands on cooling; its conductivity for heat is enormous; and neither its heat conduction nor viscosity obeys normal rules. It has other peculiar properties. Helium is the only liquid that cannot be solidified by lowering the temperature. It remains liquid down to absolute zero at ordinary pressures, but it can readily be solidified by increasing the pressure. Solid ^3He and ^4He are unusual in that both can readily be changed in volume by more than 30% by application of pressure. The specific heat of helium gas is unusually high. The density of helium vapor at the normal boiling point is also very high, with the vapor expanding greatly when heated to room temperature. Containers filled with helium gas at 5 to 10 K should be treated as though they contained liquid helium due to the large increase in pressure resulting from warming the gas to room temperature. While helium normally has a 0 valence, it seems to have a weak tendency to combine with certain other elements. Means of preparing helium diflouride have been studied, and species such as HeNe and the molecular ions He^+ and He^{++} have been investigated. Helium is widely used as an inert gas shield for arc welding; as a protective gas in growing silicon and germanium crystals, and in titanium and zirconium production; as a cooling medium for nuclear reactors, and as a gas for supersonic wind tunnels. A mixture of helium and oxygen is used as an artificial atmosphere for divers and others working under pressure. Different ratios of He/O_2 are used for different depths at which the diver is operating. Helium is extensively used for filling balloons as it is a much safer gas than hydrogen. One of the recent largest uses for helium has been for pressuring liquid fuel rockets. A Saturn booster such as used on the Apollo lunar missions required about 13 million ft^{3}of helium for a firing, plus more for checkouts. Liquid helium's use in magnetic resonance imaging (MRI) continues to increase as the medical profession accepts and develops new uses for the equipment. This equipment is providing accurate diagnoses of problems where exploratory surgery has previously been required to determine problems. Another medical application that is being developed uses MRI to determine by blood analysis whether a patient has any form of cancer. Lifting gas applications are increasing. Various companies in addition to Goodyear, are now using "blimps" for advertising. The Navy and the Air Force are investigating the use of airships to provide early warning systems to detect low-flying cruise missiles. The Drug Enforcement Agency has used radar-equipped blimps to detect drug smugglers along the southern border of the U.S. In addition, NASA is currently using helium-filled balloons to sample the atmosphere in Antarctica to determine what is depleting the ozone layer that protects Earth from harmful U.V. radiation. Research on and development of materials which become superconductive at temperatures well above the boiling point of helium could have a major impact on the demand for helium. Less costly refrigerants having boiling points considerably higher could replace the present need to cool such superconductive materials to the boiling point of helium. Natural helium contains two stable isotopes ^3He and ^4He. ^3He is present in very small quantities. Six other isotopes of helium are now recognized.

Holmium — (L. *Holmia* , for Stockholm), Ho; at. wt. 164.93032(2); at. no 67; m.p. 1474°C; b.p. 2700°C; sp. gr. 8.795 (25°C); valence + 3. The spectral absorption bands of holmium were noticed in 1878 by the Swiss chemists Delafontaine and Soret, who announced the existence of an "Element X". Cleve, of Sweden, later independently discovered the element while working on erbia earth. The element is named after Cleve's native city. Pure holmia, the yellow oxide, was prepared by Homberg in 1911. Holmium occurs in *gadolinite, monazite,* and in other rare-earth minerals. It is commercially obtained from monazite, occurring in that mineral to the extent of about 0.05%. It has been isolated by the reduction of its anhydrous chloride or fluoride with calcium metal. Pure holmium has a metallic to bright silver luster. It is relatively soft and malleable, and is stable in dry air at room temperature, but rapidly oxidizes in moist air and at elevated temperatures. The metal has unusual magnetic properties. Few uses have yet been found for the element. The element, as with other rare earths, seems to have a low acute toxic rating. Natural holmium consists of one isotope ^{165}Ho, which is not radioactive. Holmium has 49 other isotopes known, all of which are radioactive. The price of 99.9% holmium metal is about $35/g.

Hydrogen — (Gr. *hydro*, water, and *genes,* forming), H; at. wt. 1.00794(7); at. no. 1; m.p. –259.34°C; b.p. –252.87°C; t_c -240.18; density 0.08988 g/l; density (liquid) 70.8 g/l (–253°C); density (solid) 70.6 g/l (–262°C); valence 1. Hydrogen was prepared many years before it was recognized as a distinct substance by Cavendish in 1766. It was named by Lavoisier. Hydrogen is the most abundant of all elements in the universe, and it is thought that the heavier elements were, and still are, being built from hydrogen and helium. It has been estimated that hydrogen makes up more than 90% of all the atoms or three quarters of the mass of the universe. It is found in the sun and most stars, and plays an important part in the proton-proton reaction and carbon-nitrogen cycle, which accounts for the energy of the sun and stars. It is thought that hydrogen is a major component of the planet Jupiter and that at some depth in the planet's interior the pressure is so great that solid molecular hydrogen is converted into solid metallic hydrogen. In 1973, it was reported that a group of Russian experimenters may have produced metallic hydrogen at a pressure of 2.8 Mbar. At the transition the density changed from 1.08 to 1.3 g/cm^3. Earlier, in 1972, a Livermore (California) group also reported on a similar experiment in which they observed a pressure-volume point centered at 2 Mbar. It has been predicted that metallic hydrogen may be metastable; others have predicted it would be a superconductor at room temperature. On earth, hydrogen occurs chiefly in combination with oxygen in water, but it is also present in organic matter such as living plants, petroleum, coal, etc. It is present as the free element in the atmosphere, but only to the extent of less than 1 ppm by volume. It is the lightest of all gases, and combines with other elements, sometimes explosively, to form compounds. Great quantities of hydrogen are required commercially for the fixation of nitrogen from the air in the Haber ammonia process and for the hydrogenation of fats and oils. It is also used in large quantities in methanol production, in hydrodealkylation, hydrocracking, and hydrodesulfurization. It is also used as a rocket fuel, for welding, for production of hydrochloric acid, for the reduction of metallic ores, and for filling balloons. The lifting power of 1 ft^3 of hydrogen gas is about 0.076 lb at 0°C, 760 mm pressure. Production of hydrogen in the U.S. alone now amounts to about 3 billion cubic feet per year. It is prepared by the action of steam on heated carbon, by decomposition of certain hydrocarbons with heat, by the electrolysis of water, or by the displacement from acids by certain metals. It is also produced by the action of sodium or potassium hydroxide on aluminum. Liquid hydrogen is important in cryogenics and in the study of superconductivity, as its melting point is only a 20°C above absolute zero. Hydrogen consists of three isotopes, most of which is ^1H. The ordinary isotope of hydrogen, H, is known as *protium*. In 1932, Urey announced the discovery of a stable isotope, deuterium (^2H or D) with an atomic weight of 2. Deuterium is present in natural hydrogen to the extent of 0.015%. Two years later an unstable isotope, tritium (^3H), with an atomic weight of 3 was discovered. Tritium has a half-life of about 12.32 years. Tritium atoms are also present in natural hydrogen but in much smaller proportion. Tritium is readily produced in nuclear reactors and is used in the production of the hydrogen bomb. It is also used as a radioactive agent in making luminous paints, and as a tracer. Deuterium gas is readily available, without permit, at about $1/l. Heavy water, deuterium oxide (D_2O), which is used as a moderator to slow down neutrons, is available without permit at a cost of 6c to $1/g, depending on quantity and purity. About 1000 tons (4,400,000 kg) of deuterium oxide (heavy water) are now in use at the Sudbury (Ontario) Neutrino Observatory. This observatory is taking data to provide new revolutionary insight into the properties of neutrinos and into the core of the sun. The heavy water is on loan from Atomic Energy of Canada, Ltd. (AECL). The observatory and detectors are located 6800 ft (2072 m) deep in the Creighton mine of the International Nickel Co., near Sudbury. The heavy water is contained in an acrylic vessel, 12 m in diameter. Neutrinos react with the heavy water to produce Cherenkov radiation. This light is then detected with 9600 photomultiplier tubes surrounding the vessel. The detector laboratory is immensely clean to reduce background radiation, which otherwise hide the very weak signals from neutrinos. Quite apart from isotopes, it has been shown that hydrogen gas under ordinary conditions is a mixture of two kinds of molecules, known as *ortho-* and *para*-hydrogen, which differ from one another by the spins of their electrons and nuclei. Normal hydrogen at room temperature contains 25% of the *para* form and 75% of the *ortho* form. The *ortho* form cannot be prepared in the pure state. Since the two forms differ in energy, the physical properties also differ. The melting and boiling points of *para*hydrogen are about 0.1°C lower than those of normal hydrogen. Consideration is being given to an entire economy based on solar- and nuclear-generated hydrogen. Located in remote regions, power plants would electrolyze sea water; the hydrogen produced would travel to distant cities by pipelines. Pollution-free hydrogen could replace natural gas, gasoline, etc., and could serve as a reducing agent in metallurgy, chemical processing, refining, etc. It could also be used to convert trash into methane and ethylene. Public acceptance, high capital investment, and the high present cost of hydrogen with respect to present fuels are but a few of the problems facing establishment of such an economy. Hydrogen is being investigated as a substitute for deep-sea diving applications below 300 m. Hydrogen is readily available from air product suppliers.

Indium — (from the brilliant indigo line in its spectrum), In; at. wt. 114.818(3); at. no. 49; m.p. 156.60°C; b.p. 2072°C; sp. gr. 7.31 (20°C); valence 1, 2, or 3. Discovered by Reich and Richter, who later isolated the metal. Indium is most frequently associated with zinc materials, and it is from these that most commercial indium is now obtained; however, it is also found in iron, lead, and copper ores. Until 1924, a gram or so constituted the world's supply of this element in isolated form. It is probably about as abundant as silver. About 4 million troy ounces of indium are now produced annually in the Free World. Canada is presently producing more than 1,000,000 troy ounces annually. The present cost of indium is about $2 to $10/g, depending on quantity and purity. It is available in ultrapure form. Indium is a very soft, silvery-white metal with a brilliant luster. The pure metal gives a high-pitched "cry" when bent. It wets glass, as does gallium. It has found application in making low-melting alloys; an alloy of 24% indium-76% gallium is liquid at room temperature. Indium is used in making bearing alloys, germanium transistors, rectifiers, thermistors, liquid crystal displays, high definition television, batteries, and photoconductors. It can be plated onto metal and evaporated onto glass, forming a mirror as good as that made with silver but with more resistance to atmospheric corrosion. There is evidence that indium has a low order of toxicity; however, care should be taken until further information is available. Seventy isotopes and isomers are now recognized (more than any other element). Natural indium contains two isotopes. One is stable. The other, ^{115}In, comprising 95.71% of natural indium is slightly radioactive with a very long half-life.

Iodine — (Gr. *iodes,* violet), I; at. wt. 126.90447(3); at. no. 53; m.p. 113.7°C; b.p. 184.4°C; t_c 546°C; density of the gas 11.27 g/l; sp. gr. solid 4.93 (20°C); valence 1, 3, 5, or 7. Discovered by Courtois in 1811. Iodine, a halogen, occurs sparingly in the form of iodides in sea water from which it is assimilated by seaweeds, in Chilean saltpeter and nitrate-bearing earth, known as *caliche* in brines from old sea deposits, and in brackish waters from oil and salt wells. Ultrapure iodine can be obtained from the reaction of potassium iodide with copper sulfate. Several other methods of isolating the element are known. Iodine is a bluish-black, lustrous solid, volatilizing at ordinary temperatures into a blue-violet gas with an irritating odor; it forms compounds with many elements, but is less active than the other halogens, which displace it from iodides. Iodine exhibits some metallic-like properties. It dissolves readily in chloroform, carbon tetrachloride, or carbon disulfide to form beautiful purple solutions. It is only slightly soluble in water. Iodine compounds are important in organic chemistry and very useful in medicine. Forty two isotopes and isomers are recognized. Only one stable isotope, ^{127}I is found in nature. The artificial radioisotope ^{131}I, with a half-life of 8 days, has been used in treating the thyroid gland. The most common compounds are the iodides of sodium and potassium (KI) and the iodates (KIO_3). Lack of iodine is the cause of goiter. Iodides, and thyroxin which contains iodine, are used internally in medicine, and a solution of KI and iodine in alcohol is used for external wounds. Potassium iodide finds use in photography. The deep blue color with starch solution is characteristic of the free element. Care should be taken in handling and using iodine, as contact with the skin can cause lesions; iodine vapor is intensely irritating to the eyes and mucous membranes. Elemental iodine costs about 25 to 75¢/g depending on purity and quantity.

Iridium — (L. *iris,* rainbow), Ir; at. wt. 192.217(3); at. no. 77; m.p. 2446°C; b.p. 4428°C; sp. gr. 22.42 (17°C); valence 3 or 4. Discovered in 1803 by Tennant in the residue left when crude platinum is dissolved by aqua regia. The name iridium is appropriate, for its salts are highly colored. Iridium, a metal of the platinum family, is white, similar to platinum, but with a slight yellowish cast. It is very hard and brittle, making it very hard to machine, form, or work. It is the most corrosion-resistant metal known, and was used in making the standard meter bar of Paris, which is a 90% platinum-10% iridium alloy. This meter bar was replaced in 1960 as a fundamental unit of length (see under Krypton). Iridium is not attacked by any of the acids nor by aqua regia, but is attacked by molten salts, such as NaCl and NaCN. Iridium occurs uncombined in nature with platinum and other metals of this family in alluvial deposits. It is recovered as a by-product from the nickel mining industry. The largest reserves and production of the platinum group of metals, which includes iridium, is in South Africa, followed by Russia and Canada. The U.S. has only one active mine, located at Nye, MT. The presence of iridium has recently been used in examining the Cretaceous-Tertiary (K-T) boundary. Meteorites contain small amounts of iridium. Because iridium is found widely distributed at the K-T boundary, it has been suggested that a large meteorite or asteroid collided with the earth, killing the dinosaurs, and creating a large dust cloud and crater. Searches for such a crater point to one in the Yucatan, known as Chicxulub. Iridium has found use in making crucibles and apparatus for use at high temperatures. It is also used for electrical contacts. Its principal use is as a hardening agent for platinum. With osmium, it forms an alloy which is used for tipping pens and compass bearings. The specific gravity of iridium is only very slightly lower than that of osmium, which has been generally credited as being the heaviest known element. Calculations of the densities of iridium and osmium from the space lattices gives values of 22.65 and 22.61 g/cm^3, respectively. These values may be more reliable than actual physical measurements. At present, therefore, we know that either iridium or osmium is the densest known element, but the data do not yet allow selection between the two. Natural iridium contains two stable isotopes. Forty-five other isotopes, all radioactive, are now recognized. Iridium (99.9%) costs about $100/g.

Iron — (Anglo-Saxon, *iron*), Fe (L. *ferrum*); at. wt. 55.845(2); at. no. 26; m.p. 1538°C; b.p. 2861°C; sp. gr. 7.874 (20°C); valence 2, 3, 4, or 6. The use of iron is prehistoric. Genesis mentions that Tubal-Cain, seven generations from Adam, was "an instructor of every artificer in brass and iron." A remarkable iron pillar, dating to about A.D. 400, remains standing today in Delhi, India. This solid shaft of wrought iron is about 7$^1/_4$ m high by 40 cm in diameter. Corrosion to the pillar has been minimal although it has been exposed to the weather since its erection. Iron is a relatively abundant element in the universe. It is found in the sun and many types of stars in considerable quantity. Its nuclei are very stable. It has been suggested that the iron we have here on earth may have originated in a super-nova. Iron is a very difficult element to produce in ordinary nuclear reactions, such as would take place in the sun. Iron is found native as a principal component of a class of iron-nickel meteorites known as *siderites*, and is a minor constituent of the other two classes of meteorites. The core of the earth, 2150 miles in radius, is thought to be largely composed of iron with about 10% occluded hydrogen. The metal is the fourth most abundant element, by weight, making up the crust of the earth. The most common ore is *hematite* (Fe_2O_3). Magnetite (Fe_3O_4) is frequently seen as *black sands* along beaches and banks of streams. *Lodestone* is another form of magnetite. *Taconite* is becoming increasingly important as a commercial ore. Iron is a vital constituent of plant and animal life, and appears in hemoglobin. The pure metal is not often encountered in commerce, but is usually alloyed with carbon or other metals. The pure metal is very reactive chemically, and rapidly corrodes, especially in moist air or at elevated temperatures. It has four allotropic forms, or ferrites, known as α, β, γ, and δ, with transition points at 700, 928, and 1530°C. The α form is magnetic, but when transformed into the β form, the magnetism disappears although the lattice remains unchanged. The relations of these forms are peculiar. Pig iron is an alloy containing about 3% carbon with varying amounts of S, Si, Mn, and P. It is hard, brittle, fairly fusible, and is used to produce other alloys, including steel. Wrought iron contains only a few tenths of a percent of carbon, is tough, malleable, less fusible, and has usually a "fibrous" structure. Carbon steel is an alloy of iron with carbon, with small amounts of Mn, S, P, and Si. Alloy steels are carbon steels with other additives such as nickel, chromium, vanadium, etc. Iron is the cheapest and most abundant, useful, and important of all metals. Natural iron contains four isotopes and isomers. Twenty-six other isotopes and isomers, all radioactive, are now recognized.

Krypton — (Gr. *kryptos,* hidden), Kr; at. wt. 83.80(1); at. no. 36; m.p. –157.36°C; b.p. –153.22 ± 0.10°C; t_c -63.74°C; density 3.733 g/l (0°C); valence usually 0. Discovered in 1898 by Ramsay and Travers in the residue left after liquid air had nearly boiled away. Krypton is present in the air to the extent of about 1 ppm. The atmosphere of Mars has been found to contain 0.3 ppm of krypton. It is one of the "noble" gases. It is characterized by its brilliant green and orange spectral lines. Naturally occurring krypton contains six stable isotopes. Thirty other unstable isotopes and isomers are now recognized. The spectral lines of krypton are easily produced and some are very sharp. In 1960 it was internationally agreed that the fundamental unit of length, the meter, should be defined in terms of the orange-red spectral line of ^{86}Kr. This replaced the standard meter of Paris, which was defined in terms of a bar made of a platinum-iridium alloy. In October 1983 the meter, which originally was defined as being one ten millionth of a quadrant of the earth's polar circumference, was again redefined by the International Bureau of Weights and Measures as being the length of path traveled by light in a vacuum during a time interval of 1/299,792,458 of a second. Solid krypton is a white crystalline substance with a face-centered cubic structure which is common to all the "rare gases". While krypton is generally thought of as a rare gas that normally does not combine with other elements to form compounds, it now appears that the existence of some krypton compounds is established. Krypton difluoride has been prepared in gram quantities and can be made by several methods. A higher fluoride of krypton and a salt of an oxyacid of krypton also have been reported. Molecule-ions of $ArKr^+$ and KrH^+ have been identified and investigated, and evidence is provided for the formation of KrXe or $KrXe^+$. Krypton clathrates have been prepared with hydroquinone and phenol. ^{85}Kr has found recent application in chemical analysis. By imbedding the isotope in various solids, *kryptonates* are formed. The activity of these kryptonates is sensitive to chemical reactions at the surface. Estimates of the concentration of reactants are therefore made possible. Krypton is used in certain photographic flash lamps for high-speed photography. Uses thus far have been limited because of its high cost. Krypton gas presently costs about $690/100 l.

Kurchatovium — See Rutherfordium.

Lanthanum — (Gr. *lanthanein*, to lie hidden), La; at. wt. 138.9055(2); at. no. 57; m.p. 918°C; b.p. 3464°C; sp. gr. 6.145 (25°C); valence 3. Mosander in 1839 extracted a new earth *lanthana*, from impure cerium nitrate, and recognized the new element. Lanthanum is found in rare-earth minerals such as *cerite, monazite, allanite,* and *bastnasite*. Monazite and bastnasite are principal ores in which lanthanum occurs in percentages up to 25 and 38%, respectively. Misch metal, used in making lighter flints, contains about 25% lanthanum. Lanthanum was isolated in relatively pure form in 1923. Ion-exchange and solvent extraction techniques have led to much easier isolation of the so-called "rare-earth" elements. The availability of lanthanum and other rare earths has improved greatly in recent years. The metal can be produced by reducing the anhydrous fluoride with calcium. Lanthanum is silvery white, malleable, ductile, and soft enough to be cut with a knife. It is one of the most reactive of the rare-earth metals. It oxidizes rapidly when exposed to air. Cold water attacks lanthanum slowly, and hot water attacks it much more rapidly. The metal reacts directly with elemental carbon, nitrogen, boron, selenium, silicon, phosphorus, sulfur, and with halogens. At 310°C, lanthanum changes from a hexagonal to a face-centered cubic structure, and at 865°C it again transforms into a body-centered cubic structure. Natural lanthanum is mixture of two isotopes, one of which is stable and one of which is radioactive with a very long half-life. Thirty other radioactive isotopes are recognized. Rare-earth compounds containing lanthanum are extensively used in carbon lighting applications, especially by the motion picture industry for studio lighting and projection. This application consumes about 25% of the rare-earth compounds produced. La_2O_3 improves the alkali resistance of glass, and is used in making special optical glasses. Small amounts of lanthanum, as an additive, can be used to produce nodular cast iron. There is current interest in hydrogen sponge alloys containing lanthanum. These alloys take up to 400 times their own volume of hydrogen gas, and the process is reversible. Heat energy is released every time they do so; therefore these alloys have possibilities in energy conservation systems. Lanthanum and its compounds have a low to moderate acute toxicity rating; therefore, care should be taken in handling them. The metal costs about $2/g (99.9%).

Lawrencium — (Ernest O. Lawrence [1901–1958], inventor of the cyclotron), Lr; at. no. 103; at. mass no. [262]; valence + 3(?). This member of the 5f transition elements (actinide series) was discovered in March 1961 by A. Ghiorso, T. Sikkeland, A. E. Larsh, and R. M. Latimer.A 3-μg californium target, consisting of a mixture of isotopes of mass number 249, 250, 251, and 252, was bombarded with either ^{10}B or ^{11}B. The electrically charged transmutation nuclei recoiled with an atmosphere of helium and were collected on a thin copper conveyor tape which was then moved to place collected atoms in front of a series of solid-state detectors. The isotope of element 103 produced in this way decayed by emitting an 8.6-MeV alpha particle with a half-life of 8 s. In 1967, Flerov and associates of the Dubna Laboratory reported their inability to detect an alpha emitter with a half-life of 8 s which was assigned by the Berkeley group to $^{257}103$. This assignment has been changed to ^{258}Lr or ^{259}Lr. In 1965, the Dubna workers found a longer-lived lawrencium isotope, ^{256}Lr, with a half-life of 35 s. In 1968, Ghiorso and associates at Berkeley were able to use a few atoms of this isotope to study the oxidation behavior of lawrencium. Using solvent extraction techniques and working very rapidly, they extracted lawrencium ions from a buffered aqueous solution into an organic solvent, completing each extraction in about 30 s. It was found that lawrencium behaves differently from dipositive nobelium and more like the tripositive elements earlier in the actinide series. Ten isotopes of lawrencium are now recognized.

Lead — (Anglo-Saxon *lead*), Pb (L. *plumbum*); at. wt. 207.2(1); at. no. 82; m.p. 327.46°C; b.p. 1749°C; sp. gr. 11.35 (20°C); valence 2 or 4. Long known, mentioned in Exodus. The alchemists believed lead to be the oldest metal and associated it with the planet Saturn. Native lead occurs in nature, but it is rare. Lead is obtained chiefly from *galena* (PbS) by a roasting process. *Anglesite* ($PbSO_4$), *cerussite* ($PbCO_3$), and *minim* (Pb_3O_4) are other common lead minerals. Lead is a bluish-white metal of bright luster, is very soft, highly malleable, ductile, and a poor conductor of electricity. It is very resistant to corrosion; lead pipes bearing the insignia of Roman emperors, used as drains from the baths, are still in service. It is used in containers for corrosive liquids (such as sulfuric acid) and may be toughened by the addition of a small percentage of antimony or other metals. Natural lead is a mixture of four stable isotopes: ^{204}Pb (1.4%), ^{206}Pb (24.1%), ^{207}Pb (22.1%), and ^{208}Pb (52.4%). Lead isotopes are the end products of each of the three series of naturally occurring radioactive elements: ^{206}Pb for the uranium series, ^{207}Pb for the actinium series, and ^{208}Pb for the thorium series. Forty-three other isotopes of lead, all of which are radioactive, are recognized. Its alloys include solder, type metal, and various antifriction metals. Great quantities of lead, both as the metal and as the dioxide, are used in storage batteries. Lead is also used for cable covering, plumbing, and ammunition. The metal is very effective as a sound absorber, is used as a radiation shield around X-ray equipment and nuclear reactors, and is used to absorb vibration. Lead, alloyed with tin, is used in making organ pipes. White lead, the basic carbonate, sublimed white lead ($PbSO_4$), chrome yellow ($PbCrO_4$), red lead (Pb_3O_4), and other lead compounds are used extensively in paints, although in recent years the use of lead in paints has been drastically curtailed to eliminate or reduce health hazards. Lead oxide is used in producing fine "crystal glass" and "flint glass" of a high index of refraction for achromatic lenses. The nitrate and the acetate are soluble salts. Lead salts such as lead arsenate have been used as insecticides, but their use in recent years has been practically eliminated in favor of less harmful organic compounds. Care must be used in handling lead as it is a cumulative poison. Environmental concern with lead poisoning has resulted in a national program to eliminate the lead tetraethyl in gasoline. The U.S. Occupational Safety and Health Administration (OSHA) has recommended that industries limit airborne lead to 50 μgms/cu. meter. Lead is priced at about 90¢/kg (99.9%).

Lithium — (Gr. *lithos*, stone), Li; at. wt. 6.941(2); at. no. 3; m.p. 180.5°C; b.p. 1342°C; sp. gr. 0.534 (20°C); valence 1. Discovered by Arfvedson in 1817. Lithium is the lightest of all metals, with a density only about half that of water. It does not occur free in nature; combined it is found in small amounts in nearly all igneous rocks and in the waters of many mineral springs. *Lepidolite, spodumene, petalite,* and *amblygonite* are the more important minerals containing it. Lithium is presently being recovered from brines of Searles Lake, in California, and from Nevada, Chile, and Argentina. Large deposits of spodumene are found in North Carolina. The metal is produced electrolytically from the fused chloride. Lithium is silvery in appearance, much like Na and K, other members of the alkali metal series. It reacts with water, but not as vigorously as sodium. Lithium imparts a beautiful crimson color to a flame, but when the metal burns strongly the flame is a dazzling white. Since World War II, the production of lithium metal and its compounds has increased greatly. Because the metal has the highest specific heat of any solid element, it has found use in heat transfer applications; however, it is corrosive and requires special handling. The metal has been used as an alloying agent, is of interest in synthesis of organic compounds, and has nuclear applications. It ranks as a leading contender as a battery anode material as it has a high electrochemical potential. Lithium is used in special glasses and ceramics. The glass for the 200-inch telescope at Mt. Palomar contains lithium as a minor ingredient. Lithium chloride is one of the most hygroscopic materials known, and it, as well as lithium bromide, is used in air conditioning and industrial drying systems. Lithium stearate is used as an all-purpose and high-temperature lubricant. Other lithium compounds are used in dry cells and storage batteries. Seven isotopes of lithium are recognized. Natural lithium contains two isotopes. The metal is priced at about $1.50/g (99.9%).

Lutetium — (Lutetia, ancient name for Paris, sometimes called *cassiopeium* by the Germans), Lu; at. wt. 174.967(1); at. no. 71; m.p. 1663°C; b.p. 3402°C; sp. gr. 9.841 (25°C); valence 3. In 1907, Urbain described a process by which Marignac's ytterbium (1879) could be separated into the two elements, ytterbium (neoytterbium)and lutetium. These elements were identical with "aldebaranium" and "cassiopeium," independently discovered by von Welsbach about the same time. Charles James of the University of New Hampshire also independently prepared the very pure oxide, *lutecia*, at this time. The spelling of the element was changed from *lutecium* to *lutetium* in 1949. Lutetium occurs in very small amounts in nearly all minerals containing yttrium, and is present in *monazite* to the extent of about 0.003%, which is a commercial source. The pure metal has been isolated only in recent years and is one of the most difficult to prepare. It can be prepared by the reduction of anhydrous $LuCl_3$ or LuF_3 by an alkali or alkaline earth metal. The metal is silvery white and relatively stable in air. While new techniques, including ion-exchange reactions, have been developed to separate the various rare-earth elements, lutetium is still the most costly of all rare earths. It is priced at about $100/g (99.9%). ^{176}Lu occurs naturally (97.41%) with ^{175}Lu (2.59%), which is radioactive with a very long half-life. It is radioactive with a half-life of about 4×10^{10} years. Lutetium has 50 isotopes and isomers that are now recognized. Stable lutetium nuclides, which emit pure beta radiation after thermal neutron activation, can be used as catalysts in cracking, alkylation, hydrogenation, and polymerization. Virtually no other commercial uses have been found yet for lutetium. While lutetium, like other rare-earth metals, is thought to have a low toxicity rating, it should be handled with care until more information is available.

Magnesium — (*Magnesia,* district in Thessaly) Mg; at. wt. 24.3050(6); at. no. 12; m.p. 650°C; b.p. 1090°C; sp. gr. 1.738 (20°C); valence 2. Compounds of magnesium have long been known. Black recognized magnesium as an element in 1755. It was isolated by Davy in 1808, and prepared in coherent form by Bussy in 1831. Magnesium is the eighth most abundant element in the earth's crust. It does not occur uncombined, but is found in large deposits in the form of *magnesite, dolomite,* and other minerals. The metal is now principally obtained in the U.S. by electrolysis of fused magnesium chloride derived from brines, wells, and sea water. Magnesium is a light, silvery-white, and fairly tough metal. It tarnishes slightly in air, and finely divided magnesium readily ignites upon heating in air and burns with a dazzling white flame. It is used in flashlight photography, flares, and pyrotechnics, including incendiary bombs. It is one third lighter than aluminium, and in alloys is essential for airplane and missile contruction. The metal improves the mechanical, fabrication, and welding characteristics of aluminum when used as an alloying agent. Magnesium is used in producing nodular graphite in cast iron,and is used as an additive to conventional propellants. It is also used as a reducing agent in the production of pure uranium and other metals from their salts. The hydroxide (*milk of magnesia*), chloride, sulfate (*Epsom salts*), and citrate are used in medicine. Dead-burned magnesite is employed for refractory purposes such as brick and liners in furnaces and converters. Calcined magnesia is also used for water treatment and in the manufacture of rubber, paper, etc. Organic magnesium compounds (Grignard's reagents) are important. Magnesium is an important element in both plant and animal life. Chlorophylls are magnesium-centered porphyrins. The adult daily requirement of magnesium is about 300 mg/day, but this is affected by various factors. Great care should be taken in handling magnesium metal, especially in the finely divided state, as serious fires can occur. Water should not be used on burning magnesium or on magnesium fires. Natural magnesium contains three isotopes. Twelve other isotopes are recognized. Magnesium metal costs about $1.70/kg (99.8%).

Manganese — (L. *magnes*, magnet, from magnetic properties of pyrolusite; It. *manganese*, corrupt form of *magnesia*), Mn; at. wt. 54.938049(9); at. no. 25; m.p. 1246°C; b.p. 2061°C; sp. gr. 7.21 to 7.44, depending on allotropic form; valence 1, 2, 3, 4, 6, or 7. Recognized by Scheele, Bergman, and others as an element and isolated by Gahn in 1774 by reduction of the dioxide with carbon. Manganese minerals are widely distributed; oxides, silicates, and carbonates are the most common. The discovery of large quantities of manganese nodules on the floor of the oceans holds promise as a source of manganese. These nodules contain about 24% manganese together with many other elements in lesser abundance. Most manganese today is obtained from ores found in the Ukraine, Brazil, Australia, Republic of So. Africa, Gabon, China, and India. *Pyrolusite* (MnO_2) and *rhodochrosite* ($MnCO_3$) are among the most common manganese minerals. The metal is obtained by reduction of the oxide with sodium, magnesium, aluminum, or by electrolysis. It is gray-white, resembling iron, but is harder and very brittle. The metal is reactive chemically, and decomposes cold water slowly. Manganese is used to form many important alloys. In steel, manganese improves the rolling and forging qualities, strength, toughness, stiffness, wear resistance, hardness, and hardenability. With aluminum and antimony, especially with small amounts of copper, it forms highly ferromagnetic alloys. Manganese metal is ferromagnetic only after special treatment. The pure metal exists in four allotropic forms. The alpha form is stable at ordinary temperature; gamma manganese, which changes to alpha at ordinary temperatures, is said to be flexible, soft, easily cut, and capable of being bent. The dioxide (pyrolusite) is used as a depolarizer in dry cells, and is used to "decolorize" glass that is colored green by impurities of iron. Manganese by itself colors glass an amethyst color, and is responsible for the color of true amethyst. The dioxide is also used in the preparation of oxygen and chlorine, and in drying black paints. The permanganate is a powerful oxidizing agent and is used in quantitative analysis and in medicine. Manganese is widely distributed throughout the animal kingdom. It is an important trace element and may be essential for utilization of vitamin B_1. Twenty-seven isotopes and isomers are known. Manganese metal (99.95%) is priced at about $400/kg. Metal of 99.6% purity is priced at about $60/kg.

Meitnerium — (named for Lise Meitner [1878–1968], Austrian-Swedish physicist and mathematician), Mt; at. wt [266]; at. no. 109. On August 29, 1992, Element 109 was made and identified by physicists at the Heavy Ion Research Laboratory (G.S.I.), Darmstadt, Germany, by bombarding a target of ^{209}Bi with accelerated nuclei of ^{58}Fe. The production of Element 109 has been extremely small. It took a week of target bombardment (10^{11} nuclear encounters) to produce a single atom of 109. Oganessian and his team at Dubna in 1994 repeated the Darmstadt experiment using a tenfold irradiation dose. One fission event from seven alpha decays of 109 was observed, thus indirectly confirming the existence of isotope $^{266}109$. In August 1997, the IUPAC adopted the name *meitnerium* for this element, honoring L. Meitner. Four isotopes of *meitnerium* are now recognized.

Mendelevium — (Dmitri Mendeleev [1834–1907]), Md; at. wt. (258); at. no. 101; m.p. 827°C; valence +2, +3. Mendelevium, the ninth transuranium element of the actinide series to be discovered, was first identified by Ghiorso, Harvey, Choppin, Thompson, and Seaborg early in 1955 as a result of the bombardment of the isotope ^{253}Es with helium ions in the Berkeley 60-inch cyclotron. The isotope produced was ^{256}Md, which has a half-life of 78 min. This first identification was notable in that ^{256}Md was synthesized on a one-atom-at-a-time basis. Nineteen isotopes and isomers are now recognized. ^{258}Md has a half-life of 51.5 days. This isotope has been produced by the bombardment of an isotope of einsteinium with ions of helium. It now appears possible that eventually enough ^{258}Md can be made so that some of its physical properties can be determined. ^{256}Md has been used to elucidate some of the chemical properties of mendelevium in aqueous solution. Experiments seem to show that the element possesses a moderately stable dipositive (II) oxidation state in addition to the tripositive (III) oxidation state,which is characteristic of actinide elements.

Mercury — (Planet *Mercury*), Hg (*hydrargyrum*, liquid silver); at. wt. 200.59(2); at. no. 80; *t*.p. –38.83°C; b.p. 356.73°C; t_c 1447°C; sp. gr. 13.546 (20°C); valence 1 or 2. Known to ancient Chinese and Hindus; found in Egyptian tombs of 1500 B.C. Mercury is the only common metal liquid at ordinary temperatures. It only rarely occurs free in nature. The chief ore is *cinnabar* (HgS). Spain and China produce about 75% of the world's supply of the metal. The commercial unit for handling mercury is the "flask," which weighs 76 lb (34.46 kg). The metal is obtained by heating cinnabar in a current of air and by condensing the vapor. It is a heavy, silvery-white metal; a rather poor conductor of heat, as compared with other metals, and a fair conductor of electricity. It easily forms alloys with many metals, such as gold, silver, and tin, which are called *amalgams*. Its ease in amalgamating with gold is made use of in the recovery of gold from its ores. The metal is widely used in laboratory work for making thermometers, barometers, diffusion pumps, and many other instruments. It is used in making mercury-vapor lamps and advertising signs, etc. and is used in mercury switches and other electrical apparatus. Other uses are in making pesticides, mercury cells for caustic soda and chlorine production, dental preparations, antifouling paint, batteries, and catalysts. The most important salts are mercuric chloride $HgCl_2$ (corrosive sublimate — a violent poison), mercurous chloride Hg_2Cl_2(calomel, occasionally still used in medicine), mercury fulminate ($Hg(ONC)_2$), a detonator widely used in explosives, and mercuric sulfide (HgS, vermillion, a high-grade paint pigment). Organic mercury compounds are important. It has been found that an electrical discharge causes mercury vapor to combine with neon, argon, krypton, and xenon. These products, held together with van der Waals' forces, correspond to HgNe, HgAr, HgKr, and HgXe. Mercury is a virulent poison and is readily absorbed through the respiratory tract, the gastrointestinal tract, or through unbroken skin. It acts as a cumulative poison and dangerous levels are readily attained in air. Air saturated with mercury vapor at 20°C contains a concentration that exceeds the toxic limit many times. The danger increases at higher temperatures. *It is therefore important that mercury be handled with care.* Containers of mercury should be securely covered and spillage should be avoided. If it is necessary to heat mercury or mercury compounds, it should be done in a well-ventilated hood. Methyl mercury is a dangerous pollutant and is now widely found in water and streams. The triple point of mercury, –38.8344°C, is a fixed point on the International Temperature Scale (ITS-90). Mercury (99.98%) is priced at about $110/kg. Native mercury contains seven isotopes. Thirty-six other isotopes and isomers are known.

Molybdenum — (Gr. *molybdos*, lead), Mo; at. wt. 95.94(1); at. no. 42; m.p. 2623°C; b.p. 4639°C; sp. gr. 10.22 (20°C); valence 2, 3, 4?, 5?, or 6. Before Scheele recognized molybdenite as a distinct ore of a new element in 1778, it was confused with graphite and lead ore. The metal was prepared in an impure form in 1782 by Hjelm. Molybdenum does not occur native, but is obtained principally from *molybdenite* (MoS_2). *Wulfenite* ($PbMoO_4$) and *powellite* ($Ca(MoW)O_4$) are also minor commercial ores. Molybdenum is also recovered as a by-product of copper and tungsten mining operations. The U.S., Canada, Chile, and China produce most of the world's molybdenum ores. The metal is prepared from the powder made by the hydrogen reduction of purified molybdic trioxide or ammonium molybdate. The metal is silvery white, very hard, but is softer and more ductile than tungsten. It has a high elastic modulus, and only tungsten and tantalum, of the more readily available metals, have higher melting points. It is a valuable alloying agent, as it contributes to the hardenability and toughness of quenched and tempered steels. It also improves the strength of steel at high temperatures. It is used in certain nickel-based alloys, such as the "Hastelloys®" which are heat-resistant and corrosion-resistant to chemical solutions. Molybdenum oxidizes at elevated temperatures. The metal has found recent application as electrodes for electrically heated glass furnaces and forehearths. The metal is also used in nuclear energy applications and for missile and aircraft parts. Molybdenum is valuable as a catalyst in the refining of petroleum. It has found application as a filament material in electronic and electrical applications. Molybdenum is an essential trace element in plant nutrition. Some lands are barren for lack of this element in the soil. Molybdenum sulfide is useful as a lubricant, especially at high temperatures where oils would decompose. Almost all ultra-high strength steels with minimum yield points up to 300,000 psi(lb/in.^{2}) contain molybdenum in amounts from 0.25 to 8%. Natural molybdenum contains seven isotopes. Thirty other isotopes and isomers are known, all of which are radioactive. Molybdenum metal costs about $1/g (99.999% purity). Commercial molybdenum metal (99.9%) costs about $300/kg.

Neodymium — (Gr. *neos*, new, and *didymos*, twin), Nd; at. wt. 144.24(3); at. no. 60; m.p. 1021°C; b.p. 3074°C; sp. gr. 7.008 (25°C); valence 3. In 1841, Mosander, extracted from *cerite* a new rose-colored oxide, which he believed contained a new element. He named the element *didymium*, as it was *an inseparable twin brother of lanthanum*. In 1885 von Welsbach separated didymium into two new elemental components, *neodymia* and *praseodymia*, by repeated fractionation of ammonium didymium nitrate. While the free metal is in *misch metal*, long known and used as a pyrophoric alloy for light flints, the element was not isolated in relatively pure form until 1925. Neodymium is present in misch metal to the extent of about 18%. It is present in the minerals *monazite* and *bastnasite*, which are principal sources of rare-earth metals. The element may be obtained by separating neodymium salts from other rare earths by ion-exchange or solvent extraction techniques, and by reducing anhydrous halides such as NdF_3 with calcium metal. Other separation techniques are possible. The metal has a bright silvery metallic luster. Neodymium is one of the more reactive rare-earth metals and quickly tarnishes in air, forming an oxide that spalls off and exposes metal to oxidation. The metal, therefore, should be kept under light mineral oil or sealed in a plastic material. Neodymium exists in two allotropic forms, with a transformation from a double hexagonal to a body-centered cubic structure taking place at 863°C. Natural neodymium is a mixture of seven isotopes, one of which has a very long half-life. Twenty seven other radioactive isotopes and isomers are recognized. Didymium, of which neodymium is a component, is used for coloring glass to make welder's goggles. By itself, neodymium colors glass delicate shades ranging from pure violet through wine-red and warm gray. Light transmitted through such glass shows unusually sharp absorption bands. The glass has been used in astronomical work to produce sharp bands by which spectral lines may be calibrated. Glass containing neodymium can be used as a laser material to produce coherent light. Neodymium salts are also used as a colorant for enamels. The element is also being used with iron and boron to produce extremely strong magnets having energy densities as high as 27 to 35 million gauss oersteds. These are the most compact magnets commercially available. The price of the metal is about $4/g. Neodymium has a low-to-moderate acute toxic rating. As with other rare earths, neodymium should be handled with care.

Neon — (Gr. *neos*, new), Ne; at. wt. 20.1797(6); at. no. 10; *t*.p. –248.59°C; b.p. –246.08°C; t_c -228.7°C (1 atm); density of gas 0.89990 g/l (1 atm, 0°C); density of liquid at b.p. 1.207 g/cm^3; valence 0. Discovered by Ramsay and Travers in 1898. Neon is a rare gaseous element present in the atmosphere to the extent of 1 part in 65,000 of air. It is obtained by liquefaction of air and separated from the other gases by fractional distillation. Natural neon is a mixture of three isotopes. Fourteen other unstable isotopes are known. It is very inert element; however, it is said to form a compound with fluorine. It is still questionable if true compounds of neon exist, but evidence is mounting in favor of their existence. The following ions are known from optical and mass spectrometric studies: Ne^+, $(NeAr)^+$, $(NeH)^+$, and $(HeNe^+)$. Neon also forms an unstable hydrate. In a vacuum discharge tube, neon glows reddish orange. Of all the rare gases, the discharge of neon is the most intense at ordinary voltages and currents. Neon is used in making the common neon advertising signs, which accounts for its largest use. It is also used to make high-voltage indicators, lightning arrestors, wave meter tubes, and TV tubes. Neon and helium are used in making gas lasers. Liquid neon is now commercially available and is finding important application as an economical cryogenic refrigerant. It has over 40 times more refrigerating capacity per unit volume than liquid helium and more than three times that of liquid hydrogen. It is compact, inert, and is less expensive than helium when it meets refrigeration requirements. Neon costs about $800/80 cu. ft. (2265 l).

Neptunium — (Planet *Neptune*), Np; at. wt. (237); at. no. 93; m.p. 644°C; sp. gr. 20.25 (20°C); valence 3, 4, 5, and 6. Neptunium was the first synthetic transuranium element of the actinide series discovered; the isotope ^{239}Np was produced by McMillan and Abelson in 1940 at Berkeley, California, as the result of bombarding uranium with cyclotron-produced neutrons. The isotope ^{237}Np (half-life of 2.14×10^6 years) is currently obtained in gram quantities as a by-product from nuclear reactors in the production of plutonium. Twenty-three isotopes and isomers of neptunium are now recognized. Trace quantities of the element are actually found in nature due to transmutation reactions in uranium ores produced by the neutrons which are present. Neptunium is prepared by the reduction of NpF_3 with barium or lithium vapor at about 1200°C. Neptunium metal has a silvery appearance, is chemically reactive, and exists in at least three structural modifications: α-neptunium, orthorhombic, density 20.25 g/cm³, β-neptunium (above 280°C), tetragonal, density (313°C) 19.36 g/cm³; γ-neptunium (above 577°C), cubic, density (600°C) 18.0 g/cm³. Neptunium has four ionic oxidation states in solution: Np^{+3} (pale purple), analogous to the rare earth ion Pm^{+3}, Np^{+4} (yellow green); NpO^+ (green blue); and NpO^{++} (pale pink). These latter oxygenated species are in contrast to the rare earths which exhibit only simple ions of the (II), (III), and (IV) oxidation states in aqueous solution. The element forms tri- and tetrahalides such as NpF_3, NpF_4, $NpCl_4$, $NpBr_3$, NpI_3, and oxides of various compositions such as are found in the uranium-oxygen system, including Np_3O_8 and NpO_2.

Nickel — (Ger. *Nickel,* Satan or Old Nick's and from *kupfernickel,* Old Nick's copper), Ni; at. wt. 58.6934(2); at. no. 28; m.p. 1455°C; b.p. 2913°C; sp. gr. 8.902 (25°C); valence 0, 1, 2, 3. Discovered by Cronstedt in 1751 in kupfernickel (*niccolite*). Nickel is found as a constituent in most meteorites and often serves as one of the criteria for distinguishing a meteorite from other minerals. Iron meteorites, or *siderites,* may contain iron alloyed with from 5 to nearly 20% nickel. Nickel is obtained commercially from *pentlandite* and *pyrrhotite* of the Sudbury region of Ontario, a district that produces much of the world's nickel. It is now thought that the Sudbury deposit is the result of an ancient meteorite impact. Large deposits of nickel, cobalt, and copper have recently been developed at Voisey's Bay, Laborador. Other deposits of nickel are found in Russia, New Caledonia, Australia, Cuba, Indonesia, and elsewhere. Nickel is silvery white and takes on a high polish. It is hard, malleable, ductile, somewhat ferromagnetic, and a fair conductor of heat and electricity. It belongs to the iron-cobalt group of metals and is chiefly valuable for the alloys it forms. It is extensively used for making stainless steel and other corrosion-resistant alloys such as Invar®, Monel®, Inconel®, and the Hastelloys®. Tubing made of a copper-nickel alloy is extensively used in making desalination plants for converting sea water into fresh water. Nickel is also now used extensively in coinage and in making nickel steel for armor plate and burglar-proof vaults, and is a component in Nichrome®, Permalloy®, and constantan. Nickel added to glass gives a green color. Nickel plating is often used to provide a protective coating for other metals, and finely divided nickel is a catalyst for hydrogenating vegetable oils. It is also used in ceramics, in the manufacture of Alnico magnets, and in the Edison® storage battery. The sulfate and the oxides are important compounds. Natural nickel is a mixture of five stable isotopes; twenty-five other unstable isotopes are known. Nickel sulfide fume and dust is recognized as having carcinogenic potential. Nickel metal (99.9%) is priced at about $1.25/g or less in larger quantities.

Nielsbohrium — See Bohrium.

Niobium — (*Niobe,* daughter of Tantalus), Nb; or Columbium (*Columbia,* name for America); at. wt. 92.90638(2); at. no. 41; m.p. 2477°C; b.p. 4744°C, sp. gr. 8.57 (20°C); valence 2, 3, 4?, 5. Discovered in 1801 by Hatchett in an ore sent to England more that a century before by John Winthrop the Younger, first governor of Connecticut. The metal was first prepared in 1864 by Blomstrand, who reduced the chloride by heating it in a hydrogen atmosphere. The name *niobium* was adopted by the International Union of Pure and Applied Chemistry in 1950 after 100 years of controversy. Many leading chemical societies and government organizations refer to it by this name. Most metallurgists, leading metal societies, and all but one of the leading U.S. commercial producers, however, still refer to the metal as "columbium". The element is found in *niobite*(or *columbite), niobite-tantalite, pyrochlore,* and *euxenite*. Large deposits of niobium have been found associated with *carbonatites* (carbon-silicate rocks), as a constituent of *pyrochlore*. Extensive ore reserves are found in Canada, Brazil, Congo-Kinshasa, Rwanda, and Australia. The metal can be isolated from tantalum, and prepared in several ways. It is a shiny, white, soft, and ductile metal, and takes on a bluish cast when exposed to air at room temperatures for a long time. The metal starts to oxidize in air at 200°C, and when processed at even moderate temperatures must be placed in a protective atmosphere. It is used in arc-welding rods for stabilized grades of stainless steel. Thousands of pounds of niobium have been used in advance air frame systems such as were used in the Gemini space program. It has also found use in super-alloys for applications such as jet engine components, rocket subassemblies, and heat-resisting equipment. The element has superconductive properties; superconductive magnets have been made with Nb-Zr wire, which retains its superconductivity in strong magnetic fields. This type of application offers hope of direct large-scale generation of electric power. Natural niobium is composed of only one isotope, ^{93}Nb. Forty-seven other isotopes and isomers of niobium are now recognized. Niobium metal (99.9% pure) is priced at about 50¢/g.

Nitrogen — (L. *nitrum,* Gr. *nitron,* native soda; genes, *forming,* N; at. wt. 14.00674(7); at. no. 7; m.p. –210.00°C; b.p. –198.79°C; t_c -146.94°C; density 1.2506 g/l; sp. gr. liquid 0.808 (–195.8°C), solid 1.026 (–252°C); valence 3 or 5. Discovered by Daniel Rutherford in 1772, but Scheele, Cavendish, Priestley, and others about the same time studied "burnt or dephlogisticated air," as air without oxygen was then called. Nitrogen makes up 78% of the air, by volume. The atmosphere of Mars, by comparison, is 2.6% nitrogen. The estimated amount of this element in our atmosphere is more than 4000 trillion tons. From this inexhaustible source it can be obtained by liquefaction and fractional distillation. Nitrogen molecules give the orange-red, blue-green, blue-violet, and deep violet shades to the aurora. The element is so inert that Lavoisier named it *azote,* meaning without life, yet its compounds are so active as to be most important in foods, poisons, fertilizers, and explosives. Nitrogen can be also easily prepared by heating a water solution of ammonium nitrite. Nitrogen, as a gas, is colorless, odorless, and a generally inert element. As a liquid it is also colorless and odorless, and is similar in appearance to water. Two allotropic forms of solid nitrogen exist, with the transition from the α to the β form taking place at –237°C. When nitrogen is heated, it combines directly with magnesium, lithium, or calcium; when mixed with oxygen and subjected to electric sparks, it forms first nitric oxide (NO) and then the dioxide (NO_2); when heated under pressure with a catalyst with hydrogen, ammonia is formed (Haber process). The ammonia thus formed is of the utmost importance as it is used in fertilizers, and it can be oxidized to nitric acid (Ostwald process). The ammonia industryis the largest consumer of nitrogen. Large amounts of gas are also used by the electronics industry, which uses the gas as a blanketing medium during production of such components as transistors, diodes, etc. Large quantities of nitrogen are used in annealing stainless steel and other steel mill products. The drug industry also uses large quantities. Nitrogen is used as a refrigerant both for the immersion freezing of food products and for transportation of foods. Liquid nitrogen is also used in missile work as a purge for components, insulators for space chambers, etc., and by the oil industry to build up great pressures in wells to force crude oil upward. Sodium and potassium nitrates are formed by the decomposition of organic matter with compounds of the metals present. In certain dry areas of the world these saltpeters are found in quantity. Ammonia, nitric acid, the nitrates, the five oxides (N_2O, NO, N_2O_3, NO_2, and N_2O_5), TNT, the cyanides, etc. are but a few of the important compounds. Nitrogen gas prices vary from 2¢ to $2.75 per 100 ft^3 (2.83 cu. meters), depending on purity, etc. Production of elemental nitrogen in the U.S. is more than 9 million short tons per year. Natural nitrogen contains two isotopes, ^{14}N and ^{15}N. Ten other isotopes are known.

Nobelium — (Alfred Nobel, discoverer of dynamite), No; at. wt. [259]; at. no. 102; valence +2, +3. Nobelium was unambiguously discovered and identified in April 1958 at Berkeley by A. Ghiorso, T. Sikkeland, J. R. Walton, and G. T. Seaborg, who used a new double-recoil technique. A heavy-ion linear accelerator (HILAC) was used to bombard a thin target of curium (95% ^{244}Cm and 4.5% ^{246}Cm) with ^{12}C ions to produce 102^{254} according to the ^{246}Cm (^{12}C, 4*n*) reaction. Earlier in 1957 workers of the U.S., Britain, and Sweden announced the discovery of an isotope of Element 102 with a 10-min half-life at 8.5 MeV, as a result of bombarding ^{244}Cm with ^{13}C nuclei. On the basis of this experiment the name *nobelium* was assigned and accepted by the Commission on Atomic Weights of the International Union of Pure and Applied Chemistry. The acceptance of the name was premature, for both Russian and American efforts now completely rule out the possibility of any isotope of Element 102 having a half-life of 10 min in the vicinity of 8.5 MeV. Early work in 1957 on the search for this element, in Russia at the Kurchatov Institute, was marred by the assignment of 8.9 ± 0.4 MeV alpha radiation with a half-life of 2 to 40 sec, which was too indefinite to support claim to discovery. Confirmatory experiments at Berkeley in 1966 have shown the existence of $^{254}102$ with a 55-s half-life, $^{252}102$ with a 2.3-s half-life, and $^{257}102$ with a 25-s half-life. Twelve isotopes are now recognized, one of which — $^{255}102$ has a half-life of 3.1 min. In view of the discover's traditional right to name an element, the Berkeley group, in 1967, suggested that the hastily given name *nobelium,* along with the symbol No, be retained.

Osmium — (Gr. *osme*, a smell), Os; at. wt. 190.23(3); at. no. 76; m.p. 3033°C; b.p. 5012°C; sp. gr. 22.57; valence 0 to +8, more usually +3, +4, +6, and +8. Discovered in 1803 by Tennant in the residue left when crude platinum is dissolved by *aqua regia.* Osmium occurs in *iridosmine* and in platinum-bearing river sands of the Urals, North America, and South America. It is also found in the nickel-bearing ores of Sudbury, Ontario, region along with other platinum metals. While the quantity of platinum metals in these ores is very small, the large tonnages of nickel ores processed make commercial recovery possible. The metal is lustrous, bluish white, extremely hard, and brittle even at high temperatures. It has the highest melting point and the lowest vapor pressure of the platinum group. The metal is very difficult to fabricate, but the powder can be sintered in a hydrogen atmosphere at a temperature of 2000°C. The solid metal is not affected by air at room temperature, but the powdered or spongy metal slowly gives off osmium tetroxide, which is a powerful oxidizing agent and has a strong smell. The tetroxide is highly toxic, and boils at 130°C (760 mm). Concentrations in air as low as 10^{-7} g/m^3 can cause lung congestion, skin damage, or eye damage. The tetroxide has been used to detect fingerprints and to stain fatty tissue for microscope slides. The metal is almost entirely used to produce very hard alloys, with other metals of the platinum group, for fountain pen tips, instrument pivots, phonograph needles, and electrical contacts. The price of 99.9% pure osmium powder — the form usually supplied commercially — is about $100/g, depending on quantity and supplier. Natural osmium contains seven isotopes, one of which, ^{186}Os, is radioactive with a very long half-life. Thirty four other isotopes and isomers are known, all of which are radioactive.The measured densities of iridium and osmium seem to indicate that osmium is slightly more dense than iridium, so osmium has generally been credited with being the heaviest known element. Calculations of the density from the space lattice, which may be more reliable for these elements than actual measurements, however, give a density of 22.65 for iridium compared to 226.61 for osmium. At present, therefore, we know either iridium or osmium is the heaviest element, but the data do not allow selection between the two.

Oxygen — (Gr. *oxys*, sharp, acid, and *genes*, forming; acid former), O; at. wt. 15.9994(3); at. no. 8; t.p. –218.79°C; t_c -118.56°C; valence 2. For many centuries, workers occasionally realized air was composed of more than one component. The behavior of oxygen and nitrogen as components of air led to the advancement of the phlogiston theory of combustion, which captured the minds of chemists for a century. Oxygen was prepared by several workers, including Bayen and Borch, but they did not know how to collect it, did not study its properties, and did not recognize it as an elementary substance. Priestley is generally credited with its discovery, although Scheele also discovered it independently. Oxygen is the third most abundant element found in the sun, and it plays a part in the carbon-nitrogen cycle, one process thought to give the sun and stars their energy. Oxygen under excited conditions is responsible for the bright red and yellow-green colors of the aurora. Oxygen, as a gaseous element, forms 21% of the atmosphere by volume from which it can be obtained by liquefaction and fractional distillation. The atmosphere of Mars contains about 0.15% oxygen. The element and its compounds make up 49.2%, by weight, of the earth's crust. About two thirds of the human body and nine tenths of water is oxygen. In the laboratory it can be prepared by the electrolysis of water or by heating potassium chlorate with manganese dioxide as a catalyst. The gas is colorless, odorless, and tasteless. The liquid and solid forms are a pale blue color and are strongly paramagnetic. Ozone (O_3), a highly active compound, is formed by the action of an electrical discharge or ultraviolet light on oxygen. Ozone's presence in the atmosphere (amounting to the equivalent of a layer 3 mm thick at ordinary pressures and temperatures) is of vital importance in preventing harmful ultraviolet rays of the sun from reaching the earth's surface. There has been recent concern that pollutants in the atmosphere may have a detrimental effect on this ozone layer. Ozone is toxic and exposure should not exceed 0.2 mg/m^3 (8-hour time-weighted average — 40-hour work week). Undiluted ozone has a bluish color. Liquid ozone is bluish black, and solid ozone is violet-black. Oxygen is very reactive and capable of combining with most elements. It is a component of hundreds of thousands of organic compounds. It is essential for respiration of all plants and animals and for practically all combustion. In hospitals it is frequently used to aid respiration of patients. Its atomic weight was used as a standard of comparison for each of the other elements until 1961 when the International Union of Pure and Applied Chemistry adopted carbon 12 as the new basis. Oxygen has thirteen recognized isotopes. Natural oxygen is a mixture of three isotopes. Oxygen 18 occurs naturally, is stable, and is available commercially. Water (H_2O with 1.5% ^{18}O) is also available. Commercial oxygen consumption in the U.S. is estimated to be 20 million short tons per year and the demand is expected to increase substantially in the next few years. Oxygen enrichment of steel blast furnaces accounts for the greatest use of the gas. Large quantities are also used in making synthesis gas for ammonia and methanol, ethylene oxide, and for oxy-acetylene welding. Air separation plants produce about 99% of the gas, electrolysis plants about 1%. The gas costs 5¢/ft^3 ($1.75/cu. meters) in small quantities.

Palladium — (named after the asteroid *Pallas*, discovered about the same time; Gr. *Pallas*, goddess of wisdom), Pd. at. wt. 106.42(1) at. no. 46; m.p. 1554.9°C; b.p. 2963°C; sp. gr. 12.02 (20°C); valence 2, 3, or 4. Discovered in 1803 by Wollaston. Palladium is found along with platinum and other metals of the platinum group in deposits of Russia, South Africa, Canada (Ontario), and elsewhere. Natural palladium contains six stable isotopes. Twenty-nine other isotopes are recognized, all of which are radioactive. It is frequently found associated with the nickel-copper deposits such as those found in Ontario. Its separation from the platinum metals depends upon the type of ore in which it is found. It is a steel-white metal, does not tarnish in air, and is the least dense and lowest melting of the platinum group of metals. When annealed, it is soft and ductile; cold working greatly increases its strength and hardness. Palladium is attacked by nitric and sulfuric acid. At room temperatures the metal has the unusual property of absorbing up to 900 times its own volume of hydrogen, possibly forming Pd_2H. It is not yet clear if this a true compound. Hydrogen readily diffuses through heated palladium and this provides a means of purifying the gas. Finely divided palladium is a good catalyst and is used for hydrogenation and dehydrogenation reactions. It is alloyed and used in jewelry trades. White gold is an alloy of gold decolorized by the addition of palladium. Like gold, palladium can be beaten into leaf as thin as 1/250,000 in. The metal is used in dentistry, watchmaking, and in making surgical instruments and electrical contacts. Palladium recently has been substituted for higher priced platinum in catalytic converters by some automobile companies. This has caused a large increase in the cost of palladium. The price of the two metals are now, in 1999, about the same. Palladium, however, is less resistant to poisoning by sulfur and lead, than platinum, but it may prove useful in controlling emissions from diesel vehicles. The metal sells for about $425/tr. oz. ($13/g).

Phosphorus — (Gr. *phosphoros*, light bearing; ancient name for the planet Venus when appearing before sunrise), P; at. wt. 30.973762(4); at. no. 15; m.p. (white) 44.15°C; b.p. 280.5°C; sp. gr. (white) 1.82 (red) 2.20, (black) 2.25 to 2.69; valence 3 or 5. Discovered in 1669 by Brand, who prepared it from urine. Phosphorus exists in four or more allotropic forms: white (or yellow), red, and black (or violet). White phosphorus has two modifications: α and β with a transition temperature at –3.8°C. Never found free in nature, it is widely distributed in combination with minerals. Twenty-one isotopes of phosphorus are recognized. *Phosphate* rock, which contains the mineral *apatite*, an impure tri-calcium phosphate, is an important source of the element. Large deposits are found in the Russia,China, Morocco, and in Florida, Tennessee, Utah, Idaho, and elsewhere. Phosphorus in an essential ingredient of all cell protoplasm, nervous tissue, and bones. Ordinary phosphorus is a waxy white solid; when pure it is colorless and transparent. It is insoluble in water, but soluble in carbon disulfide. It takes fire spontaneously in air, burning to the pentoxide. It is very poisonous, 50 mg constituting an approximate fatal dose. Exposure to white phosphorus should not exceed 0.1 mg/m^3 (8-hour time-weighted average — 40-hour work week). White phosphorus should be kept under water, as it is dangerously reactive in air, and it should be handled with forceps, as contact with the skin may cause severe burns. When exposed to sunlight or when heated in its own vapor to 250°C, it is converted to the red variety, which does not phosphoresce in air as does the white variety. This form does not ignite spontaneously and it is not as dangerous as white phosphorus. It should, however, be handled with care as it does convert to the white form at some temperatures and it emits highly toxic fumes of the oxides of phosphorus when heated. The red modification is fairly stable, sublimes with a vapor pressure of 1 atm at 417°C,and is used in the manufacture of safety matches, pyrotechnics, pesticides, incendiary shells, smoke bombs, tracer bullets, etc. White phosphorus may be made by several methods. By one process, tri-calcium phosphate, the essential ingredient of phosphate rock, is heated in the presence of carbon and silica in an electric furnace or fuel-fired furnace. Elementary phosphorus is liberated as vapor and may be collected under water. If desired, the phosphorus vapor and carbon monoxide produced by the reaction can be oxidized at once in the presence of moisture to produce phosphoric acid, an important compound in making super-phosphate fertilizers. In recent years, concentrated phosphoric acids, which may contain as much as 70 to 75% P_2O_5 content, have become of great importance to agriculture and farm production. World-wide demand for fertilizers has caused record phosphate production. Phosphates are used in the production of special glasses, such as those used for sodium lamps. Bone-ash, calcium phosphate, is also used to produce fine chinaware and to produce mono-calcium phosphate used in baking powder. Phosphorus is also important in the production of steels, phosphor bronze, and many other products. Trisodium phosphate is important as a cleaning agent, as a water softener, and for preventing boiler scale and corrosion of pipes and boiler tubes. Organic compounds of phosphorus are important. Amorphous (red) phosphorus costs about $80/kg (99.5%).

Platinum — (It. *platina*, silver), Pt; at. wt. 195.078(2); at. no. 78; m.p. 1768.4°C; b.p. 3825°C; sp. gr. 21.45 (20°C); valence 1?, 2, 3, or 4. Discovered in South America by Ulloa in 1735 and by Wood in 1741. The metal was used by pre-Columbian Indians. Platinum occurs native, accompanied by small quantities of iridium, osmium, palladium, ruthenium, and rhodium, all belonging to the same group of metals. These are found in the alluvial deposits of the Ural mountains and in Columbia. *Sperrylite* ($PtAs_2$), occurring with the nickel-bearing deposits of Sudbury, Ontario, is a source of a considerable amount of metal. The large production of nickel offsets there being only one part of the platinum metals in two million parts of ore. The largest supplier of the platinum group of metals is now South Africa, followed by Russia and Canada. Platinum is a beautiful silvery-white metal, when pure, and is malleable and ductile. It has a coefficient of expansion almost equal to that of soda-lime-silica glass, and is therefore used to make sealed electrodes in glass systems. The metal does not oxidize in air at any temperature, but is corroded by halogens, cyanides, sulfur, and caustic alkalis. It is insoluble in hydrochloric and nitric acid, but dissolves when they are mixed as *aqua regia*, forming chloroplatinic acid (H_2PtCl_6), an important compound. Natural platinum contains six isotopes, one of which, ^{190}Pt, is radioactive with a long half-life. Thirty-seven other radioactive isotopes and isomers are recognized. The metal is extensively used in jewelry, wire, and vessels for laboratory use, and in many valuable instruments including thermocouple elements. It is also used for electrical contacts, corrosion-resistant apparatus, and in dentistry. Platinum-cobalt alloys have magnetic properties. One such alloy made of 76.7% Pt and 23.3% Co, by weight, is an extremely powerful magnet that offers a B-H (max) almost twice that of Alnico V. Platinum resistance wires are used for constructing high-temperature electric furnaces. The metal is used for coating missile nose cones, jet engine fuel nozzles, etc., which must perform reliably for long periods of time at high temperatures. The metal, like palladium, absorbs large volumes, of hydrogen, retaining it at ordinary temperatures but giving it up at red heat. In the finely divided state platinum is an excellent catalyst, having long been used in the contact process for producing sulfuric acid. It is also used as a catalyst in cracking petroleum products. There is also much current interest in the use of platinum as a catalyst in fuel cells and in its use as antipollution devices for automobiles. Platinum anodes are extensively used in cathodic protection systems for large ships and ocean-going vessels, pipelines, steel piers, etc. Pure platinum wire will glow red hot when placed in the vapor of methyl alcohol. It acts here as a catalyst, converting the alcohol to formaldehyde. This phenomenon has been used commercially to produce cigarette lighters and hand warmers. Hydrogen and oxygen explode in the presence of platinum. The price of platinum has varied widely; more than a century ago it was used to adulterate gold. It was nearly eight times as valuable as gold in 1920. The price in December 1999 was about $460/troy oz. ($15/g), somewhat higher than the price of gold.

Plutonium — (Planet *pluto*), Pu; at. wt. (244); at. no. 94; sp. gr. (α modification) 19.84 (25°C); m.p. 640°C; b.p. 3228°C; valence 3, 4, 5, or 6. Plutonium was the second transuranium element of the actinide series to be discovered. The isotope ^{238}Pu was produced in 1940 by Seaborg, McMillan,

Kennedy, and Wahl by deuteron bombardment of uranium in the 60-inch cyclotron at Berkeley, California. Plutonium also exists in trace quantities in naturally occurring uranium ores. It is formed in much the same manner as neptunium, by irradiation of natural uranium with the neutrons which are present. By far of greatest importance is the isotope Pu^{239}, with a half-life of 24,100 years, produced in extensive quantities in nuclear reactors from natural uranium:

$$^{238}U(n,\gamma)\rightarrow{}^{239}U\xrightarrow{\beta}{}^{239}Np\xrightarrow{\beta}{}^{239}Pu$$

Nineteen isotopes of plutonium are now known. Plutonium has assumed the position of dominant importance among the transuranium elements because of its successful use as an explosive ingredient in nuclear weapons and the place which it holds as a key material in the development of industrial use of nuclear power. One kilogram is equivalent to about 22 million kilowatt hours of heat energy. The complete detonation of a kilogram of plutonium produces an explosion equal to about 20,000 tons of chemical explosive. Its importance depends on the nuclear property of being readily fissionable with neutrons and its availability in quantity. The world's nuclear-power reactors are now producing about 20,000 kg of plutonium/yr. By 1982 it was estimated that about 300,000 kg had accumulated. The various nuclear applications of plutonium are well known. ^{238}Pu has been used in the Apollo lunar missions to power seismic and other equipment on the lunar surface. As with neptunium and uranium, plutonium metal can be prepared by reduction of the trifluoride with alkaline-earth metals. The metal has a silvery appearance and takes on a yellow tarnish when slightly oxidized. It is chemically reactive. A relatively large piece of plutonium is warm to the touch because of the energy given off in alpha decay. Larger pieces will produce enough heat to boil water. The metal readily dissolves in concentrated hydrochloric acid, hydroiodic acid, or perchloric acid with formation of the Pu^{+3} ion. The metal exhibits six allotropic modifications having various crystalline structures. The densities of these vary from 16.00 to 19.86 g/cm^3. Plutonium also exhibits four ionic valence states in aqueous solutions: Pu^{+3}(blue lavender), Pu^{+4} (yellow brown), PuO^+ (pink?), and PuO^{+2} (pink orange). The ion PuO^+ is unstable in aqueous solutions, disproportionating into Pu^{+4} and PuO^{+2}. The Pu^{+4} thus formed, however, oxidizes the PuO^+ into PuO^{+2}, itself being reduced to Pu^{+3}, giving finally Pu^{+3} and PuO^{+2}. Plutonium forms binary compounds with oxygen: PuO, PuO_2, and intermediate oxides of variable composition; with the halides: PuF_3, PuF_4, $PuCl_3$, $PuBr_3$, PuI_3; with carbon, nitrogen, and silicon: PuC, PuN, $PuSi_2$. Oxyhalides are also well known: PuOCl, PuOBr, PuOI. Because of the high rate of emission of alpha particles and the element being specifically absorbed by bone marrow, plutonium, as well as all of the other transuranium elements except neptunium, are radiological poisons and must be handled with very special equipment and precautions. Plutonium is a very dangerous radiological hazard. Precautions must also be taken to prevent the unintentional formation of a critical mass. Plutonium in liquid solution is more likely to become critical than solid plutonium. The shape of the mass must also be considered where criticality is concerned. Plutonium-239 is available to authorized users from the O.R.N.L. at a cost of about \$4.80/mg (99.9%) plus packing costs.

Polonium — (Poland, native country of Mme. Curie [1867–1934]), Po; at. wt. (209); at. no. 84; m.p. 254°C; b.p. 962°C; sp. gr. (alpha modification) 9.32; valence –2, 0, +2, +3(?), +4, and +6. Polonium was the first element discovered by Mme. Curie in 1898, while seeking the cause of radioactivity of pitchblende from Joachimsthal, Bohemia. The electroscope showed it separating with bismuth. Polonium is also called Radium F. Polonium is a very rare natural element. Uranium ores contain only about 100 µg of the element per ton. Its abundance is only about 0.2% of that of radium. In 1934, it was found that when natural bismuth (^{209}Bi) was bombarded by neutrons, ^{210}Bi, the parent of polonium, was obtained. Milligram amounts of polonium may now be prepared this way, by using the high neutron fluxes of nuclear reactors. Polonium-210 is a low-melting, fairly volatile metal, 50% of which is vaporized in air in 45 hours at 55°C. It is an alpha emitter with a half-life of 138.39 days. A milligram emits as many alpha particles as 5 g of radium. The energy released by its decay is so large (140 W/g) that a capsule containing about half a gram reaches a temperature above 500°C. The capsule also presents a contact gamma-ray dose rate of 0.012 Gy/h. A few curies (1 curie = 3.7×10^{10} Bq) of polonium exhibit a blue glow, caused by excitation of the surrounding gas. Because almost all alpha radiation is stopped within the solid source and its container, giving up its energy, polonium has attracted attention for uses as a lightweight heat source for thermoelectric power in space satellites. Thirty-eight isotopes and isomers of polonium are known, with atomic masses ranging from 192 to 218. All are radioactive. Polonium-210 is the most readily available. Isotopes of mass 209 (half-life 102 years) and mass 208 (half-life 2.9 years) can be prepared by alpha, proton, or deuteron bombardment of lead or bismuth in a cyclotron, but these are expensive to produce. Metallic polonium has been prepared from polonium hydroxide and some other polonium compounds in the presence of concentrated aqueous or anhydrous liquid ammonia. Two allotropic modifications are known to exist. Polonium is readily dissolved in dilute acids, but is only slightly soluble in alkalis. Polonium salts of organic acids char rapidly; halide amines are reduced to the metal. Polonium can be mixed or alloyed with beryllium to provide a source of neutrons. It has been used in devices for eliminating static charges in textile mills, etc.; however, beta sources are more commonly used and are less dangerous. It is also used on brushes for removing dust from photographic films. The polonium for these is carefully sealed and controlled, minimizing hazards to the user. Polonium-210 is very dangerous to handle in even milligram or microgram amounts, and special equipment and strict control is necessary. Damage arises from the complete absorption of the energy of the alpha particle into tissue. The maximum permissible body burden for ingested polonium is only 0.03 µCi, which represents a particle weighing only 6.8×10^{-12} g. Weight for weight it is about 2.5×10^{11} times as toxic as hydrocyanic acid. The maximum allowable concentration for soluble polonium compounds in air is about 2×10^{11} µCi/cm^3. Polonium-209 is available on special order from the Oak Ridge National Laboratory at a cost of \$3600/µCi plus packing costs..

Potassium — (English, *potash* — pot ashes; L. *kalium*, Arab. *qali*, alkali), K; at. wt. 39.0983(1); at. no. 19; m.p. 63.38°C; b.p. 759°C; sp. gr. 0.862 (20°C); valence 1. Discovered in 1807 by Davy, who obtained it from caustic potash (KOH); this was the first metal isolated by electrolysis. The metal is the seventh most abundant and makes up about 2.4% by weight of the earth's crust. Most potassium minerals are insoluble and the metal is obtained from them only with great difficulty. Certain minerals, however, such as *sylvite, carnallite, langbeinite,* and *polyhalite* are found in ancient lake and sea beds and form rather extensive deposits from which potassium and its salts can readily be obtained. Potash is mined in Germany, New Mexico, California, Utah, and elsewhere. Large deposits of potash, found at a depth of some 1000 m in Saskatchewan, promise to be important in coming years. Potassium is also found in the ocean, but is present only in relatively small amounts, compared to sodium. The greatest demand for potash has been in its use for fertilizers. Potassium is an essential constituent for plant growth and it is found in most soils. Potassium is never found free in nature, but is obtained by electrolysis of the hydroxide, much in the same manner as prepared by Davy. Thermal methods also are commonly used to produce potassium (such as by reduction of potassium compounds with CaC_2, C, Si, or Na). It is one of the most reactive and electropositive of metals. Except for lithium, it is the lightest known metal. It is soft, easily cut with a knife, and is silvery in appearance immediately after a fresh surface is exposed. It rapidly oxidizes in air and should be preserved in a mineral oil. As with other metals of the alkali group, it decomposes in water with the evolution of hydrogen. It catches fire spontaneously on water. Potassium and its salts impart a violet color to flames. Twenty one isotopes, one of which is an isomer, of potassium are known. Ordinary potassium is composed of three isotopes, one of which is ^{40}K (0.0117%), a radioactive isotope with a half-life of 1.26×10^9 years. The radioactivity presents no appreciable hazard. An alloy of sodium and potassium (NaK) is used as a heat-transfer medium. Many potassium salts are of utmost importance, including the hydroxide, nitrate, carbonate, chloride, chlorate, bromide, iodide, cyanide, sulfate, chromate, and dichromate. Metallic potassium is available commercially for about $1200/kg (98% purity) or $55/g (99.95% purity).

Praseodymium — (Gr. *prasios,* green, and *didymos*, twin), Pr; at. wt. 140.90765(2); at. no. 59; m.p. 931°C; b.p. 3520°C; sp. gr. 6.773; valence 3. In 1841 Mosander extracted the rare earth *didymia* from *lanthana*; in 1879, Lecoq de Boisbaudran isolated a new earth, *samaria*, from didymia obtained from the mineral *samarskite*. Six years later, in 1885, von Welsbach separated didymia into two others, *praseodymia* and *neodymia*, which gave salts of different colors. As with other rare earths, compounds of these elements in solution have distinctive sharp spectral absorption bands or lines, some of which are only a few Angstroms wide. The element occurs along with other rare-earth elements in a variety of minerals. *Monazite* and *bastnasite* are the two principal commercial sources of the rare-earth metals. Ion-exchange and solvent extraction techniques have led to much easier isolation of the rare earths and the cost has dropped greatly in the past few years. Thirty-seven isotopes and isomers are now recognized. Praseodymium can be prepared by several methods, such as by calcium reduction of the anhydrous chloride of fluoride. Misch metal, used in making cigarette lighters, contains about 5% praseodymium metal. Praseodymium is soft, silvery, malleable, and ductile. It was prepared in relatively pure form in 1931. It is somewhat more resistant to corrosion in air than europium, lanthanum, cerium, or neodymium, but it does develop a green oxide coating that spalls off when exposed to air. As with other rare-earth metals it should be kept under a light mineral oil or sealed in plastic. The rare-earth oxides, including Pr_2O_3, are among the most refractory substances known. Along with other rare earths, it is widely used as a core material for carbon arcs used by the motion picture industry for studio lighting and projection. Salts of praseodymium are used to color glasses and enamels; when mixed with certain other materials, praseodymium produces an intense and unusually clean yellow color in glass. Didymium glass, of which praseodymium is a component, is a colorant for welder's goggles. The metal (99.9% pure) is priced at about $5/g.

Promethium — (*Prometheus*, who, according to mythology, stole fire from heaven), Pm; at. no. 61; at. wt. (145); m.p. 1042°C; b.p. 3000°C (est.); sp. gr. 7.264 (25°C); valence 3. In 1902 Branner predicted the existence of an element between neodymium and samarium, and this was confirmed by Moseley in 1914. Unsuccessful searches were made for this predicted element over two decades, and various investigators proposed the names "illinium", "florentium", and "cyclonium" for this element. In 1941, workers at Ohio State University irradiated neodymium and praseodymium with neutrons, deuterons, and alpha particles, resp., and produced several new radioactivities, which most likely were those of element 61. Wu and Segre, and Bethe, in 1942, confirmed the formation; however, chemical proof of the production of element 61 was lacking because of the difficulty in separating the rare earths from each other at that time. In 1945, Marinsky, Glendenin, and Coryell made the first chemical identification by use of ion-exchange chromatography. Their work was done by fission of uranium and by neutron bombardment of neodymium. These investigators named the newly discovered element. Searches for the element on earth have been fruitless, and it now appears that promethium is completely missing from the earth's crust. Promethium, however, has been reported to be in the spectrum of the star HR^{465} in Andromeda. This element is being formed recently near the star's surface, for no known isotope of promethium has a half-life longer than 17.7 years. Thirty five isotopes and isomers of promethium, with atomic masses from 130 to 158 are now known. Promethium-145,with a half-life of 17.7 years, is the most useful. Promethium-145 has a specific activity of 940 Ci/g. It is a soft beta emitter; although no gamma rays are emitted, X-radiation can be generated when beta particles impinge on elements of a high atomic number, and great care must be taken in handling it. Promethium salts luminesce in the dark with a pale blue or greenish glow, due to their high radioactivity. Ion-exchange methods led to the preparation of about 10 g of promethium from atomic reactor fuel processing wastes in early 1963. Little is yet generally known about the properties of metallic promethium. Two allotropic modifications exist. The element has applications as a beta source for thickness gages, and it can be absorbed by a phosphor to produce light. Light produced in this manner can be used for signs or signals that require dependable operation; it can be used as a nuclear-powered battery by capturing light in photocells which convert it into electric current. Such a battery, using ^{147}Pm, would have a useful life of about 5 years. It is being used for fluorescent lighting starters and coatings for self-luminous watch dials. Promethium shows promise as a portable X-ray source, and it may become useful as a heat source to provide auxiliary power for space probes and satellites. More than 30 promethium compounds have been prepared. Most are colored. Promethium-147 is available upon special order from the Idaho National Engineering Laboratory, Idaho Falls, ID, or from the Westinghouse Hanford Co., Richland, WA.

Protactinium — (Gr. *protos*, first), Pa; at. wt. 231.03588(2); at. no. 91; m.p. 1572°C; sp. gr. 15.37 (calc.); valence 4 or 5. The first isotope of element 91 to be discovered was ^{234}Pa, also known as UX_2, a short-lived member of the naturally occurring ^{238}U decay series. It was identified by K. Fajans and O. H. Gohring in 1913 and they named the new element *brevium*. When the longer-lived isotope ^{231}Pa was identified by Hahn and Meitner in 1918, the name protoactinium was adopted as being more consistent with the characteristics of the most abundant isotope. Soddy, Cranson, and Fleck were also active in this work. The name *protoactinium* was shortened to *protactinium* in 1949. In 1927, Grosse prepared 2 mg of a white powder, which was shown to be Pa_2O_5. Later, in 1934, from 0.1 g of pure Pa_2O_5 he isolated the element by two methods, one of which was by converting the oxide to an iodide and "cracking" it in a high vacuum by an electrically heated filament by the reaction

$$2PaI_5 \rightarrow 2Pa + 5I_2$$

Protactinium has a bright metallic luster which it retains for some time in air. The element occurs in *pitchblende* to the extent of about 1 part ^{231}Pa to 10 million of ore. Ores from Congo-Kinshasa have about 3 ppm. Protactinium has twenty-eight isotopes and isomers, the most common of which is ^{231}Pr with a half-life of 32,500 years. A number of protactinium compounds are known, some of which are colored. The element is superconductive below 1.4 K. The element is a dangerous toxic material and requires precautions similar to those used when handling plutonium. In 1959 and 1961, it was announced that the Great Britain Atomic Energy Authority extracted by a 12-stage process 125 g of 99.9% protactinium, the world's only stock of the metal for many years to come. The extraction was made from 60 tons of waste material at a cost of about $500,000. Protactinium is one of the rarest and most expensive naturally occurring elements.

Radium — (L. *radius*, ray), Ra; at. wt. (226); at. no. 88; m.p. 700°C; sp. gr. 5; valence 2. Radium was discovered in 1898 by M. and Mme. Curie in the *pitchblende* or *uraninite* of North Bohemia (Czech Republic), where it occurs. There is about 1 g of radium in 7 tons of pitchblende. The element was isolated in 1911 by Mme. Curie and Debierne by the electrolysis of a solution of pure radium chloride, employing a mercury cathode; on distillation in an atmosphere of hydrogen this amalgam yielded the pure metal. Originally, radium was obtained from the rich pitchblende ore found at Joachimsthal, Bohemia. The *carnotite* sands of Colorado furnish some radium, but richer ores are found in the Republic of Congo-Kinshasa and the

Great Bear Lake region of Canada. Radium is present in all uranium minerals, and could be extracted, if desired, from the extensive wastes of uranium processing. Large uranium deposits are located in Ontario, New Mexico, Utah, Australia, and elsewhere. Radium is obtained commercially as the bromide or chloride; it is doubtful if any appreciable stock of the isolated element now exists. The pure metal is brilliant white when freshly prepared, but blackens on exposure to air, probably due to formation of the nitride. It exhibits luminescence, as do its salts; it decomposes in water and is somewhat more volatile than barium. It is a member of the alkaline-earth group of metals. Radium imparts a carmine red color to a flame. Radium emits alpha, beta, and gamma rays and when mixed with beryllium produce neutrons. One gram of ^{226}Ra undergoes 3.7×10^{10} disintegrations per s. The *curie (Ci)* is defined as that amount of radioactivity which has the same disintegration rate as 1 g of ^{226}Ra. Thirty-six isotopes are now known; radium 226, the common isotope, has a half-life of 1599 years. One gram of radium produces about 0.0001 ml (stp) of emanation, or radon gas, per day. This is pumped from the radium and sealed in minute tubes, which are used in the treatment of cancer and other diseases. One gram of radium yields about 4186 kJ per year. Radium is used in producing self-luminous paints, neutron sources, and in medicine for the treatment of disease. Some of the more recently discovered radioisotopes, such as ^{60}Co, are now being used in place of radium. Some of these sources are much more powerful, and others are safer to use. Radium loses about 1% of its activity in 25 years, being transformed into elements of lower atomic weight. Lead is a final product of disintegration. Stored radium should be ventilated to prevent build-up of radon. Inhalation, injection, or body exposure to radium can cause cancer and other body disorders. The maximum permissible burden in the total body for ^{226}Ra is 7400 becquerel.

Radon — (from *radium*; called *niton* at first, L. *nitens*, shining), Rn; at. wt. (222); at. no. 86; m.p. –71°C; b.p. –61.7°C; t_c 104°C; density of gas 9.73 g/l; sp. gr. liquid 4.4 at –62°C, solid 4; valence usually 0. The element was discovered in 1900 by Dorn, who called it *radium emanation*. In 1908 Ramsay and Gray, who named it *niton*, isolated the element and determined its density, finding it to be the heaviest known gas. It is essentially inert and occupies the last place in the zero group of gases in the Periodic Table. Since 1923, it has been called radon. Thirty-seven isotopes and isomers are known. Radon-222, coming from radium, has a half-life of 3.823 days and is an alpha emitter; Radon-220, emanating naturally from thorium and called *thoron*, has a half-life of 55.6 s and is also an alpha emitter. Radon-219 emanates from actinium and is called *actinon*. It has a half-life of 3.96 s and is also on alpha emitter. It is estimated that every square mile of soil to a depth of 6 inch contains about 1 g of radium, which releases radon in tiny amounts to the atmosphere. Radon is present in some spring waters, such as those at Hot Springs, Arkansas. On the average, one part of radon is present to 1×10^{21} part of air. At ordinary temperatures radon is a colorless gas; when cooled below the freezing point, radon exhibits a brilliant phosphorescence which becomes yellow as the temperature is lowered and orange-red at the temperature of liquid air. It has been reported that fluorine reacts with radon, forming radon fluoride. Radon clathrates have also been reported. Radon is still produced for therapeutic use by a few hospitals by pumping it from a radium source and sealing it in minute tubes, called seeds or needles, for application to patients. This practice has now been largely discontinued as hospitals can order the seeds directly from suppliers, who make up the seeds with the desired activity for the day of use. Care must be taken in handling radon, as with other radioactive materials. The main hazard is from inhalation of the element and its solid daughters, which are collected on dust in the air. Good ventilation should be provided where radium, thorium, or actinium is stored to prevent build-up of this element. Radon build-up is a health consideration in uranium mines. Recently radon build-up in homes has been a concern. Many deaths from lung cancer are caused by radon exposure. In the U.S. it is recommended that remedial action be taken if the air from radon in homes exceeds 4 pCi/l.

Rhenium — (L. *Rhenus,* Rhine), Re; at. wt. 186.207(1); at. no. 75; m.p. 3186°C; b.p. 5596°C; sp. gr. 21.02 (20°C); valence –1, +1, 2, 3, 4, 5, 6, 7. Discovery of rhenium is generally attributed to Noddack, Tacke, and Berg, who announced in 1925 they had detected the element in platinum ores and *columbite*. They also found the element in *gadolinite* and *molybdenite*. By working up 660 kg of molybdenite they were able in 1928 to extract 1 g of rhenium. The price in 1928 was $10,000/g. Rhenium does not occur free in nature or as a compound in a distinct mineral species. It is, however, widely spread throughout the earth's crust to the extent of about 0.001 ppm. Commercial rhenium in the U.S. today is obtained from molybdenite roaster-flue dusts obtained from copper-sulfide ores mined in the vicinity of Miami, Arizona, and elsewhere in Arizona and Utah. Some molybdenites contain from 0.002 to 0.2% rhenium. It is estimated that in 1999 about 16,000 kg of rhenium was being produced. The total estimated world reserves of rhenium is 11,000,000 kg. The total estimated Free World reserve of rhenium metal is 3500 tons. Natural rhenium is a mixture of two isotopes, one of which has a very long half-life. Thirty nine other unstable isotopes are recognized. Rhenium metal is prepared by reducing ammonium perrhenate with hydrogen at elevated temperatures. The element is silvery white with a metallic luster; its density is exceeded only by that of platinum, iridium, and osmium, and its melting point is exceeded only by that of tungsten and carbon. It has other useful properties. The usual commercial form of the element is a powder, but it can be consolidated by pressing and resistance-sintering in a vacuum or hydrogen atmosphere. This produces a compact shape in excess of 90% of the density of the metal. Annealed rhenium is very ductile, and can be bent, coiled, or rolled. Rhenium is used as an additive to tungsten and molybdenum-based alloys to impart useful properties. It is widely used for filaments for mass spectrographs and ion gages. Rhenium-molybdenum alloys are superconductive at 10 K. Rhenium is also used as an electrical contact material as it has good wear resistance and withstands arc corrosion. Thermocouples made of Re-W are used for measuring temperatures up to 2200°C, and rhenium wire has been used in photoflash lamps for photography. Rhenium catalysts are exceptionally resistant to poisoning from nitrogen, sulfur, and phosphorus, and are used for hydrogenation of fine chemicals, hydrocracking, reforming, and disproportionation of olefins. Rhenium has recently become especially important as a catalyst for petroleum refining and in making super-alloys for jet engines. Rhenium costs about $16/g (99.99% pure). Little is known of its toxicity; therefore, it should be handled with care until more data are available.

Rhodium — (Gr. *rhodon*, rose), Rh; at. wt. 102.90550(3); at. no. 45; m.p. 1964°C; b.p. 3695°C; sp. gr. 12.41 (20°C); valence 2, 3, 4, 5, and 6. Wollaston discovered rhodium in 1803-4 in crude platinum ore he presumably obtained from South America. Rhodium occurs native with other platinum metals in river sands of the Urals and in North and South America. It is also found with other platinum metals in the copper-nickel sulfide ores of the Sudbury, Ontario region. Although the quantity occurring here is very small, the large tonnages of nickel processed make the recovery commercially feasible. The annual world production of rhodium in 1999 was only about 9000 kg. The metal is silvery white and at red heat slowly changes in air to the sesquioxide. At higher temperatures it converts back to the element. Rhodium has a higher melting point and lower density than platinum. Its major use is as an alloying agent to harden platinum and palladium. Such alloys are used for furnace windings, thermocouple elements, bushings for glass fiber production, electrodes for aircraft spark plugs, and laboratory crucibles. It is useful as an electrical contact material as it has a low electrical resistance, a low and stable contact resistance, and is highly resistant to corrosion. Plated rhodium, produced by electroplating or evaporation, is exceptionally hard and is used for optical instruments. It has a high reflectance and is hard and durable. Rhodium is also used for jewelry, for decoration, and as a catalyst. Fifty-two isotopes and isomers are now known. Soluble salts should not exceed 0.01 mg/m^3. Rhodium metal (powder) costs about $180/g (99.9%).

Rubidium — (L. *rubidus*, deepest red), Rb; at. wt. 85.4678(3); at. no. 37; m.p. 39.31°C; b.p. 688°C; sp. gr. (solid) 1.532 (20°C), (liquid) 1.475 (39°C); valence 1, 2, 3, 4. Discovered in 1861 by Bunsen and Kirchoff in the mineral *lepidolite* by use of the spectroscope. The element is much more abundant than was thought several years ago. It is now considered to be the 16th most abundant element in the earth's crust. Rubidium occurs in *pollucite, carnallite, leucite,* and *zinnwaldite*, which contains traces up to 1%, in the form of the oxide. It is found in lepidolite to the extent of about 1.5%, and is recovered commercially from this source. Potassium minerals, such as those found at Searles Lake, California, and potassium chloride recovered from brines in Michigan also contain the element and are commercial sources. It is also found along with cesium in the extensive deposits of *pollucite* at Bernic Lake, Manitoba. Rubidium can be liquid at room temperature. It is a soft, silvery-white metallic element of the alkali group and is the second most electropositive and alkaline element. It ignites spontaneously in air and reacts violently in water, setting fire to the liberated hydrogen. As with other alkali metals, it forms amalgams with mercury and it alloys with gold, cesium, sodium, and potassium. It colors a flame yellowish violet. Rubidium metal can be prepared by reducing rubidium chloride with calcium, and by a number of other methods. It must be kept under a dry mineral oil or in a vacuum or inert atmosphere. Thirty five isotopes and isomers of rubidium are known. Naturally occurring rubidium is made of two isotopes, ^{85}Rb and ^{87}Rb. Rubidium-87 is present to the extent of 27.83% in natural rubidium and is a beta emitter with a half-life of 4.9×10^{10} years. Ordinary rubidium is sufficiently radioactive to expose a photographic film in about 30 to 60 days. Rubidium forms four oxides: Rb_2O, Rb_2O_2, Rb_2O_3, Rb_2O_4. Because rubidium can be easily ionized, it has been considered for use in "ion engines" for space vehicles; however, cesium is somewhat more efficient for this purpose. It is also proposed for use as a working fluid for vapor turbines and for use in a thermoelectric generator using the magnetohydrodynamic principle where rubidium ions are formed by heat at high temperature and passed through a magnetic field. These conduct electricity and act like an armature of a generator thereby generating an electric current. Rubidium is used as a getter in vacuum tubes and as a photocell component. It has been used in making special glasses. $RbAg_4I_5$ is important, as it has the highest room conductivity of any known ionic crystal. At 20°C its conductivity is about the same as dilute sulfuric acid. This suggests use in thin film batteries and other applications. The present cost in small quantities is about $80/g (99.8% pure).

Ruthenium — (L. *Ruthenia*, Russia), Ru; at. wt. 101.07(2); at. no. 44, m.p. 2334°C; b.p. 4150°C; sp. gr. 12.41 (20°C); valence 0, 1, 2, 3, 4, 5, 6, 7, 8. Berzelius and Osann in 1827 examined the residues left after dissolving crude platinum from the Ural mountains in *aqua regia*. While Berzelius found no unusual metals, Osann thought he found three new metals, one of which he named ruthenium. In 1844 Klaus, generally recognized as the discoverer, showed that Osann's ruthenium oxide was very impure and that it contained a new metal. Klaus obtained 6 g of ruthenium from the portion of crude platinum that is insoluble in *aqua regia*. A member of the platinum group, ruthenium occurs native with other members of the group in ores found in the Ural mountains and in North and South America. It is also found along with other platinum metals in small but commercial quantities in *pentlandite* of the Sudbury, Ontario, nickel-mining region, and in *pyroxinite* deposits of South Africa. Natural ruthenium contains seven isotopes. Twenty-eight other isotopes and isomers are known, all of which are radioactive. The metal is isolated commercially by a complex chemical process, the final stage of which is the hydrogen reduction of ammonium ruthenium chloride, which yields a powder. The powder is consolidated by powder metallurgy techniques or by argon-arc welding. Ruthenium is a hard, white metal and has four crystal modifications. It does not tarnish at room temperatures, but oxidizes in air at about 800°C. The metal is not attacked by hot or cold acids or *aqua regia*, but when potassium chlorate is added to the solution, it oxidizes explosively. It is attacked by halogens, hydroxides, etc. Ruthenium can be plated by electrodeposition or by thermal decomposition methods. The metal is one of the most effective hardeners for platinum and palladium, and is alloyed with these metals to make electrical contacts for severe wear resistance. A ruthenium-molybdenum alloy is said to be superconductive at 10.6 K. The corrosion resistance of titanium is improved a hundredfold by addition of 0.1% ruthenium. It is a versatile catalyst. Hydrogen sulfide can be split catalytically by light using an aqueous suspension of CdS particles loaded with ruthenium dioxide. It is thought this may have application to removal of H_2S from oil refining and other industrial processes. Compounds in at least eight oxidation states have been found, but of these, the +2. +3. and +4 states are the most common. Ruthenium tetroxide, like osmium tetroxide, is highly toxic. In addition, it may explode. Ruthenium compounds show a marked resemblance to those of osmium. The metal is priced at about $30/g (99.95% pure).

Rutherfordium — (named for Ernest Rutherford [1871–1937], New Zealand, Canadian, and British physicist); Rf; at. wt. [261]; at. no. 104. In 1964, workers of the Joint Nuclear Research Institute at Dubna (Russia) bombarded plutonium with accelerated 113 to 115 MeV neon ions. By measuring fission tracks in a special glass with a microscope, they detected an isotope that decays by spontaneous fission. They suggested that this isotope, which has a half-life of 0.3 ± 0.1 s, might be $^{260}104$, produced by the following reaction:

$$^{242}_{94}Pu + ^{22}_{10}Ne \rightarrow ^{260}104 + 4n$$

Element 104, the first *transactinide* element, is expected to have chemical properties similar to those of hafnium. It would, for example, form a relatively volatile compound with chlorine (a tetrachloride). The Soviet scientists have performed experiments aimed at chemical identification, and have attempted to show that the 0.3-s activity is more volatile than that of the relatively nonvolatile actinide trichlorides. This experiment does not fulfill the test of chemically separating the new element from all others, but it provides important evidence for evaluation. New data, reportedly issued by Soviet scientists, have reduced the half-life of the isotope they worked with from 0.3 to 0.15 s. The Dubna scientists suggest the name *kurchatovium* and symbol *Ku* for Element 104, in honor of Igor Vasilevich Kurchatov (1903—1960), late Head of Soviet Nuclear Research. The Dubna Group also has proposed the name *dubnium* for Element 104. In 1969, Ghiorso, Nurmia, Harris, K. A. Y. Eskola, and P. L. Eskola of the University of California at Berkeley reported they had positively identified two, and possibly three, isotopes of Element 104. The group also indicated that after repeated attempts so far they have been unable to produce isotope $^{260}104$ reported by the Dubna groups in 1964. The discoveries at Berkeley were made by bombarding a target of ^{249}Cf with ^{12}C nuclei of 71 MeV, and ^{13}C nuclei of 69 MeV. The combination of ^{12}C with ^{249}Cf followed by instant emission of four neutrons produced Element $^{257}104$. This isotope has a half-life of 4 to 5 s, decaying by emitting an alpha particle into ^{253}No, with a half-life of 105 s. The same reaction, except with the emission of three neutrons, was thought to have produced $^{258}104$ with a half-life of about 1/100 s. Element $^{259}104$ is formed by the merging of a ^{13}C nuclei with ^{249}Cf, followed by emission of three neutrons. This isotope has a half-life of 3 to 4 s, and decays by emitting an alpha particle into ^{255}No, which has a half-life of 185 s. Thousands of atoms of $^{257}104$ and $^{259}104$ have been detected. The Berkeley group believe their identification of $^{258}104$ was correct. Eleven isotopes of Element 104 have now been identified. The Berkeley group proposed for the new element the name *rutherfordium* (symbol Rf), in honor of Ernest Rutherford. This name was formally adapted by IUPAC in August 1997.

Samarium — (*Samarskite* a mineral), Sm; at. wt. 150.36(3); at. no. 62; m.p. 1074°C; b.p. 1794°C; sp. gr (α) 7.520 (25°C); valence 2 or 3. Discovered spectroscopically by its sharp absorption lines in 1879 by Lecoq de Boisbaudran in the mineral *samarskite*, named in honor of a Russian mine official, Col. Samarski. Samarium is found along with other members of the rare-earth elements in many minerals, including *monazite* and *bastnasite*, which are commercial sources. The largest producer of rare earth minerals is now China, followed by the U.S., India, and Russia. It occurs in monazite to the extent of 2.8%. While *misch metal* containing about 1% of samarium metal, has long been used, samarium has not been isolated in relatively pure form until recent years. Ion-exchange and solvent extraction techniques have recently simplified separation of the rare earths from one another; more recently, electrochemical deposition, using an electrolytic solution of lithium citrate and a mercury electrode, is said to be a simple, fast, and highly specific way to separate the rare earths. Samarium metal can be produced by reducing the oxide with barium or lanthanum. Samarium has a bright silver luster and is reasonably stable in air. Three crystal modifications of the metal exist, with transformations at 734 and 922°C. The metal ignites in air at about 150°C. Thirty-three isotopes and isomers of samarium are now recognized. Natural samarium is a mixture of seven isotopes, three of which are unstable but have long half-lives. Samarium, along with other rare earths, is used for carbon-arc lighting for the motion picture industry. The sulfide has excellent high-temperature stability and good thermoelectric efficiencies up to 1100°C. $SmCo_5$ has been used in making a new permanent magnet material with the highest resistance to demagnetization of any known material. It is said to have an intrinsic coercive force as high as 2200 kA/m. Samarium oxide has been used in optical glass to absorb the infrared. Samarium is used to dope calcium fluoride crystals for use in optical masers or lasers. Compounds of the metal act as sensitizers for phosphors excited in the infrared; the oxide exhibits catalytic properties in the dehydration and dehydrogenation of ethyl alcohol. It is used in infrared absorbing glass and as a neutron absorber in nuclear reactors. The metal is priced at about $3.50/g (99.9%). Little is known of the toxicity of samarium; therefore, it should be handled carefully.

Scandium — (L. *Scandia*, Scandinavia), Sc; at. wt. 44.955910(8); at. no. 21; m.p. 1541°C; b.p. 2836°C; sp. gr. 2.989 (25°C); valence 3. On the basis of the Periodic System, Mendeleev predicted the existence of *ekaboron*, which would have an atomic weight between 40 of calcium and 48 of titanium. The element was discovered by Nilson in 1878 in the minerals *euxenite* and *gadolinite*, which had not yet been found anywhere except in Scandinavia. By processing 10 kg of euxenite and other residues of rare-earth minerals, Nilson was able to prepare about 2 g of scandium oxide of high purity. Cleve later pointed out that Nilson's scandium was identical with Mendeleev's ekaboron. Scandium is apparently a much more abundant element in the sun and certain stars than here on earth. It is about the 23rd most abundant element in the sun, compared to the 50th most abundant on earth. It is widely distributed on earth, occurring in very minute quantities in over 800 mineral species. The blue color of beryl (aquamarine variety) is said to be due to scandium. It occurs as a principal component in the rare mineral *thortveitite*, found in Scandinavia and Malagasy. It is also found in the residues remaining after the extraction of tungsten from Zinnwald *wolframite*, and in *wiikite* and *bazzite*. Most scandium is presently being recovered from *thortveitite* or is extracted as a by-product from uranium mill tailings. Metallic scandium was first prepared in 1937 by Fischer, Brunger, and Grieneisen, who electrolyzed a eutectic melt of potassium, lithium, and scandium chlorides at 700 to 800°C. Tungsten wire and a pool of molten zinc served as the electrodes in a graphite crucible. Pure scandium is now produced by reducing scandium fluoride with calcium metal. The production of the first pound of 99% pure scandium metal was announced in 1960. Scandium is a silver-white metal which develops a slightly yellowish or pinkish cast upon exposure to air. It is relatively soft, and resembles yttrium and the rare-earth metals more than it resembles aluminum or titanium. It is a very light metal and has a much higher melting point than aluminum, making it of interest to designers of spacecraft. Scandium is not attacked by a 1:1 mixture of conc. HNO_3 and 48% HF. Scandium reacts rapidly with many acids. Twenty-three isotopes and isomers of scandium are recognized. The metal is expensive, costing about $200/g with a purity of about 99.9%. About 20 kg of scandium (as Sc_2O_3) are now being used yearly in the U.S. to produce high-intensity lights, and the radioactive isotope ^{46}Sc is used as a tracing agent in refinery crackers for crude oil, etc. Scandium iodide added to mercury vapor lamps produces a highly efficient light source resembling sunlight, which is important for indoor or night-time color TV. Little is yet known about the toxicity of scandium; therefore, it should be handled with care.

Seaborgium — (named for Glenn T. Seaborg [1912–1999], American chemist and nuclear physicist). Sg; at. wt. [263]; at no. 106. The discovery of *Seaborgium*, Element 106, took place in 1974 almost simultaneously at the Lawrence-Berkeley Laboratory and at the Joint Institute for Nuclear Research at Dubna, Russia. The Berkeley Group, under direction of Ghiorso, used the Super-Heavy Ion Linear Accelerator (Super HILAC) as a source of heavy ^{18}O ions to bombard a 259-μg target of ^{249}Cf. This resulted in the production and positive identification of $^{263}106$, which decayed with a half-life of 0.9 ± 0.2 s by the emission of alpha particles as follows:

$$^{263}106 \xrightarrow{\alpha} {}^{259}104 \xrightarrow{\alpha} {}^{255}No \xrightarrow{\alpha}.$$

The Dubna Team, directed by Flerov and Organessian, produced heavy ions of ^{54}Cr with their 310-cm heavy-ion cyclotron to bombard ^{207}Pb and ^{208}Pb and found a product that decayed with a half-life of 7 ms. They assigned $^{259}106$ to this isotope. It is now thought seven isotopes of *Seaborgium* have been identified. Two of the isotopes are believed to have half-lives of about 30 s. *Seaborgium* most likely would have properties resembling tungsten. The IUPAC adopted the name *Seaborgium* in August 1997. Normally the naming of an element is not given until after the death of the person for which the element is named; however, in this case, it was named while Dr. Seaborg was still alive.

Selenium — (Gr. *Selene*, moon), Se; at. wt. 78.96(3); at. no. 34; m.p. (gray) 221°C; b.p. (gray) 685°C; sp. gr. (gray) 4.79, (vitreous) 4.28; valence −2, +4, or +6. Discovered by Berzelius in 1817, who found it associated with tellurium, named for the earth. Selenium is found in a few rare minerals, such as *crooksite* and *clausthalite*. In years past it has been obtained from flue dusts remaining from processing copper sulfide ores, but the anode muds from electrolytic copper refineries now provide the source of most of the world's selenium. Selenium is recovered by roasting the muds with soda or sulfuric acid, or by smelting them with soda and niter. Selenium exists in several allotropic forms. Three are generally recognized, but as many as six have been claimed. Selenium can be prepared with either an amorphous or crystalline structure. The color of amorphous selenium is either red, in powder form, or black, in vitreous form. Crystalline monoclinic selenium is a deep red; crystalline hexagonal selenium, the most stable variety, is a metallic gray. Natural selenium contains six stable isotopes. Twenty-nine other isotopes and isomers have been characterized. The element is a member of the sulfur family and resembles sulfur both in its various forms and in its compounds. Selenium exhibits both photovoltaic action, where light is converted directly into electricity, and photoconductive action, where the electrical resistance decreases with increased illumination. These properties make selenium useful in the production of photocells and exposure meters for photographic use, as well as solar cells. Selenium is also able to convert a.c. electricity to d.c., and is extensively used in rectifiers. Below its melting point selenium is a p-type semiconductor and is finding many uses in electronic and solid-state applications. It is used in Xerography for reproducing and copying documents, letters, etc., but recently its use in this application has been decreasing in favor of certain organic compounds. It is used by the glass industry to decolorize glass and to make ruby-colored glasses and enamels. It is also used as a photographic toner, and as an additive to stainless steel. Elemental selenium has been said to be practically nontoxic and is considered to be an essential trace element; however, hydrogen selenide and other selenium compounds are extremely toxic, and resemble arsenic in their physiological reactions. Hydrogen selenide in a concentration of 1.5 ppm is intolerable to man. Selenium occurs in some soils in amounts sufficient to produce serious effects on animals feeding on plants, such as locoweed, grown in such soils. Selenium (99.5%) is priced at about $250/kg. It is also available in high-purity form at a cost of about $750/kg (99.999%).

Silicon — (L. *silex, silicis,* flint), Si; at. wt. 28.0855(3); at. no. 14; m.p. 1414°C; b.p. 3265°C; sp. gr. 2.33 (25°C); valence 4. Davy in 1800 thought silica to be a compound and not an element; later in 1811, Gay Lussac and Thenard probably prepared impure amorphous silicon by heating potassium with silicon tetrafluoride. Berzelius, generally credited with the discovery, in 1824 succeeded in preparing amorphous silicon by the same general method as used earlier, but he purified the product by removing the fluosilicates by repeated washings. Deville in 1854 first prepared crystalline silicon, the second allotropic form of the element. Silicon is present in the sun and stars and is a principal component of a class of meteorites known as "aerolites". It is also a component of *tektites*, a natural glass of uncertain origin. Natural silicon contains three isotopes. Twenty-four other radioactive isotopes are recognized. Silicon makes up 25.7% of the earth's crust, by weight, and is the second most abundant element, being exceeded only by oxygen. Silicon is not found free in nature, but occurs chiefly as the oxide and as silicates. *Sand, quartz, rock crystal, amethyst, agate, flint, jasper,* and *opal* are some of the forms in which the oxide appears. *Granite, hornblende, asbestos, feldspar, clay mica,* etc. are but a few of the numerous silicate minerals. Silicon is prepared commercially by heating silica and carbon in an electric furnace, using carbon electrodes. Several other methods can be used for preparing the element. Amorphous silicon can be prepared as a brown powder, which can be easily melted or vaporized. Crystalline silicon has a metallic luster and grayish color. The Czochralski process is commonly used to produce single crystals of silicon used for solid-state or semiconductor devices. Hyperpure silicon can be prepared by the thermal decomposition of ultra-pure trichlorosilane in a hydrogen atmosphere, and by a vacuum float zone process. This product can be doped with boron, gallium, phosphorus, or arsenic to produce silicon for use in transistors, solar cells, rectifiers, and other solid-state devices which are used extensively in the electronics and space-age industries. Hydrogenated amorphous silicon has shown promise in producing economical cells for converting solar energy into electricity. Silicon is a relatively inert element, but it is attacked by halogens and dilute alkali. Most acids except hydrofluoric, do not affect it. Silicones are important products of silicon. They may be prepared by hydrolyzing a silicon organic chloride, such as dimethyl silicon chloride. Hydrolysis and condensation of various substituted chlorosilanes can be used to produce a very great number of polymeric products, or silicones, ranging from liquids to hard, glasslike solids with many useful properties. Elemental silicon transmits more than 95% of all wavelengths of infrared, from 1.3 to 6.7 μm. Silicon is one of man's most useful elements. In the form of sand and clay it is used to make concrete and brick; it is a useful refractory material for high-temperature work, and in the form of silicates it is used in making enamels, pottery, etc. Silica, as sand, is a principal ingredient of glass, one of the most inexpensive of materials with excellent mechanical, optical, thermal, and electrical properties. Glass can be made in a very great variety of shapes, and is used as containers, window glass, insulators, and thousands of other uses. Silicon tetrachloride can be used to iridize glass. Silicon is important in plant and animal life. Diatoms in both fresh and salt water extract silica from the water to build up their cell walls. Silica is present in ashes of plants and in the human skeleton. Silicon is an important ingredient in steel; silicon carbide is one of the most important abrasives and has been used in lasers to produce coherent light of 4560 Å. A remarkable material, first discovered in 1930, is *Aerogel,* developed and now used by NASA in their *Stardust* mission, which is expected to encounter Comet Wild 2 in 2004, returning cometary and interplanet dust to Earth in 2006. *Aerogel* is a highly insulative material that has the lowest density of any known solid. One form of *Aerogel* is 99.9% air and 0.1% SiO_2, by volume. It is 1000 times less dense than glass. It has been called "blue smoke" or "solid smoke". A block of *Aerogel* as large as a person may weigh less than a pound and yet support the weight of 1000 lbs (455 kg). This material is expected to trap cometary particles traveling at speeds of 32 km/sec. *Aerogel* is said to be non-toxic and non-inflammable. It has high thermal insulating qualities that could be used in home insulation. Its light weight may have aircraft applications. Regular grade silicon (99.5%) costs about $160/kg. Silicon (99.9999%) pure costs about $250/kg; hyperpure silicon is available at a higher cost. Miners, stonecutters, and other engaged in work where siliceous dust is breathed in large quantities often develop a serious lung disease known as *silicosis.*

Silver — (Anglo-Saxon, *Seolfor siolfur*), Ag (L. argentum), at. wt. 107.8682(2); at. no. 47; m.p. 961.78°C; b.p. 2162°C; sp. gr. 10.50 (20°C); valence 1, 2. Silver has been known since ancient times. It is mentioned in Genesis. Slag dumps in Asia Minor and on islands in the Aegean Sea indicate that man learned to separate silver from lead as early as 3000 B.C. Silver occurs native and in ores such as *argentite* (Ag_2S) and *horn silver* (AgCl); lead, lead-zinc, copper, gold, and copper-nickel ores are principal sources. Mexico, Canada, Peru, and the U.S. are the principal silver producers in the western hemisphere. Silver is also recovered during electrolytic refining of copper. Commercial fine silver contains at least 99.9% silver. Purities of 99.999+% are available commercially. Pure silver has a brilliant white metallic luster. It is a little harder than gold and is very ductile and malleable, being exceeded only by gold and perhaps palladium. Pure silver has the highest electrical and thermal conductivity of all metals, and possesses the lowest contact resistance. It is stable in pure air and water, but tarnishes when exposed to ozone, hydrogen sulfide, or air containing sulfur. The alloys of silver are important. Sterling silver is used for jewelry, silverware, etc. where appearance is paramount. This alloy contains 92.5% silver, the remainder being copper or some other metal. Silver is of utmost importance in photography, about 30% of the U.S. industrial consumption going into this application. It is used for dental alloys. Silver is used in making solder and brazing alloys, electrical contacts, and high capacity silver-zinc and silver-cadmium batteries. Silver paints are used for making printed circuits. It is used in mirror production and may be deposited on glass or metals by chemical deposition, electrodeposition, or by evaporation. When freshly deposited, it is the best reflector of visible light known, but is rapidly tarnishes and loses much of its reflectance. It is a poor reflector of ultraviolet. Silver fulminate ($Ag_2C_2N_2O_2$), a powerful explosive, is sometimes formed during the silvering process. Silver iodide is used in seeding clouds to produce rain. Silver chloride has interesting optical properties as it can be made transparent; it also is a cement for glass. Silver nitrate, or *lunar caustic*, the most important silver compound, is used extensively in photography. While silver itself is not considered to be toxic, most of its salts are poisonous. Natural silver contains two stable isotopes. Fifty-six other radioactive isotopes and isomers are known. Silver compounds can be absorbed in the circulatory system and reduced silver deposited in the various tissues of the body. A condition, known as *argyria*, results with a greyish pigmentation of the skin and mucous membranes. Silver has germicidal effects and kills many lower organisms effectively without harm to higher animals. Silver for centuries has been used traditionally for coinage by many countries of the world. In recent times, however, consumption of silver has at times greatly exceeded the output. In 1939, the price of silver was fixed by the U.S. Treasury at 71¢/troy oz., and at 90.5¢/troy oz. in 1946. In November 1961 the U.S. Treasury suspended sales of nonmonetized silver, and the price stabilized for a time at about \$1.29, the melt-down value of silver U.S. coins. The Coinage Act of 1965 authorized a change in the metallic composition of the three U.S. subsidiary denominations to clad or composite type coins. This was the first change in U.S. coinage since the monetary system was established in 1792. Clad dimes and quarters are made of an outer layer of 75% Cu and 25% Ni bonded to a central core of pure Cu. The composition of the one- and five-cent pieces remains unchanged. One-cent coins are 95% Cu and 5% Zn. Five-cent coins are 75% Cu and 25% Ni. Old silver dollars are 90% Ag and 10% Cu. Earlier subsidiary coins of 90% Ag and 10% Cu officially were to circulate alongside the clad coins; however, in practice they have largely disappeared (Gresham's Law), as the value of the silver is now greater than their exchange value. Silver coins of other countries have largely been replaced with coins made of other metals. On June 24, 1968, the U.S. Government ceased to redeem U.S. Silver Certificates with silver. Since that time, the price of silver has fluctuated widely. As of December 1999, the price of silver was about \$5.20/troy oz. (17¢/g); however the price has fluctuated considerably due to market instability. The price of silver in 1999 was only about five times the cost of the metal about 150 years ago. This has largely been caused by Central Banks disposing of some of their silver reserves and the development of more productive mines with better refining methods. Also, silver has been displaced by other metals or processes, such as digital photography.

Sodium — (English, *soda*; Medieval Latin, *sodanum*, headache remedy), Na (L. *natrium*); at. wt. 22.989770(2); at. no. 11; m.p. 97.80°C; b.p. 883°C; sp. gr. 0.971 (20°C); valence 1. Long recognized in compounds, sodium was first isolated by Davy in 1807 by electrolysis of caustic soda. Sodium is present in fair abundance in the sun and stars. The D lines of sodium are among the most prominent in the solar spectrum. Sodium is the sixth most abundant element on earth, comprising about 2.6% of the earth's crust; it is the most abundant of the alkali group of metals of which it is a member. The most common compound is sodium chloride, but it occurs in many other minerals, such as *soda niter, cryolite, amphibole, zeolite, sodalite*,etc. It is a very reactive element and is never found free in nature. It is now obtained commercially by the electrolysis of absolutely dry fused sodium chloride. This method is much cheaper than that of electrolyzing sodium hydroxide, as was used several years ago. Sodium is a soft, bright, silvery metal which floats on water, decomposing it with the evolution of hydrogen and the formation of the hydroxide. It may or may not ignite spontaneously on water, depending on the amount of oxide and metal exposed to the water. It normally does not ignite in air at temperatures below 115°C. Sodium should be handled with respect, as it can be dangerous when improperly handled. Metallic sodium is vital in the manufacture of sodamide and esters, and in the preparation of organic compounds. The metal may be used to improve the structure of certain alloys, to descale metal, to purify molten metals, and as a heat transfer agent. An alloy of sodium with potassium, NaK, is also an important heat transfer agent. Sodium compounds are important to the paper, glass, soap, textile, petroleum, chemical, and metal industries. Soap is generally a sodium salt of certain fatty acids. The importance of common salt to animal nutrition has been recognized since prehistoric times. Among the many compounds that are of the greatest industrial importance are common salt (NaCl), soda ash (Na_2CO_3), baking soda ($NaHCO_3$), caustic soda (NaOH), Chile saltpeter ($NaNO_3$), di- and tri-sodium phosphates, sodium thiosulfate (hypo, $Na_2S_2O_3 \cdot 5H_2O$), and borax ($Na_2B_4O_7 \cdot 10H_2O$). Seventeen isotopes of sodium are recognized. Metallic sodium is priced at about \$130/kg (99.95%). On a volume basis, it is the cheapest of all metals. Sodium metal should be handled with great care. It should be kept in an inert atmosphere and contact with water and other substances with which sodium reacts should be avoided.

Strontium — (*Strontian*, town in Scotland), Sr; at. wt. 87.62(1); at. no. 38; m.p. 777°C; b.p. 1382°C; sp. gr. 2.54; valence 2. Isolated by Davey by electrolysis in 1808; however, Adair Crawford in 1790 recognized a new mineral (strontianite) as differing from other barium minerals (baryta). Strontium is found chiefly as *celestite* ($SrSO_4$) and *strontianite* ($SrCO_3$). *Celestite* is found in Mexico, Turkey, Iran, Spain, Algeria, and in the U.K. The U.S. has no active *celestite* mines. The metal can be prepared by electrolysis of the fused chloride mixed with potassium chloride, or is made by reducing strontium oxide with aluminum in a vacuum at a temperature at which strontium distills off. Three allotropic forms of the metal exist, with transition points at 235 and 540°C. Strontium is softer than calcium and decomposes water more vigorously. It does not absorb nitrogen below 380°C. It should be kept under mineral oil to prevent oxidation. Freshly cut strontium has a silvery appearance, but rapidly turns a yellowish color with the formation of the oxide. The finely divided metal ignites spontaneously in air. Volatile strontium salts impart a beautiful crimson color to flames, and these salts are used in pyrotechnics and in the production of flares. Natural strontium is a mixture of four stable isotopes. Thirty-two other unstable isotopes and isomers are known to exist. Of greatest importance is ^{90}Sr with a half-life of 29 years. It is a product of nuclear fallout and presents a health problem. This isotope is one of the best long-lived high-energy beta emitters known, and is used in SNAP (Systems for Nuclear Auxiliary Power) devices. These devices hold promise for use in space vehicles, remote weather stations, navigational buoys, etc., where a lightweight, long-lived, nuclear-electric power source is needed. The major use for strontium at present is in producing glass for color television picture tubes. All color TV and cathode ray tubes sold in the U.S. are required by law to contain strontium in the face plate glass to block X-ray emission. Strontium also improves the brilliance of the glass and the quality of the picture. It has also found use in producing ferrite magnets and in refining zinc. Strontium titanate is an interesting optical material as it has an extremely high refractive index and an optical dispersion greater than that of diamond. It has been used as a gemstone, but it is very soft. It does not occur naturally. Strontium metal (99% pure) costs about \$220/kg.

Sulfur — (Sanskrit, *sulvere*; L. *sulphurium*), S; at. wt. 32.066(6); at. no. 16; m.p. 115.21°C; b.p. 444.60°C; t_c 1041°C; sp. gr. (rhombic) 2.07, (monoclinic) 1.957 (20°C); valence 2, 4, or 6. Known to the ancients; referred to in Genesis as *brimstone*. Sulfur is found in meteorites. A dark area near the crater Aristarchus on the moon has been studied by R. W. Wood with ultraviolet light. This study suggests strongly that it is a sulfur deposit. Sulfur occurs native in the vicinity of volcanoes and hot springs. It is widely distributed in nature as *iron pyrites, galena, sphalerite, cinnabar, stibnite, gypsum, Epsom salts, celestite, barite*,etc. Sulfur is commercially recovered from wells sunk into the salt domes along the Gulf Coast of the U.S. It is obtained from these wells by the Frasch process, which forces heated water into the wells to melt the sulfur, which is then brought to the surface. Sulfur also occurs in natural gas and petroleum crudes and must be removed from these products. Formerly this was done chemically, which wasted the sulfur. New processes now permit recovery, and these sources promise to be very important. Large amounts of sulfur are being recovered from Alberta gas fields. Sulfur is a pale yellow, odorless, brittle solid, which is insoluble in water but soluble in carbon disulfide. In every state, whether gas, liquid or solid, elemental sulfur occurs in more than one allotropic form or modification; these present a confusing multitude of forms whose relations are not yet fully understood. Amorphous or "plastic" sulfur is obtained by fast cooling of the crystalline form. X-ray studies indicate that amorphous sulfur may have a helical structure with eight atoms per spiral. Crystalline sulfur seems to be made of rings, each containing eight sulfur atoms, which fit together to give a normal X-ray pattern. Twenty-one isotopes of sulfur are now recognized. Four occur in natural sulfur, none of which is radioactive. A finely divided form of sulfur, known as *flowers of sulfur*, is obtained by sublimation. Sulfur readily forms sulfides with many elements. Sulfur is a component of black gunpowder, and is used in the vulcanization of natural rubber and a fungicide. It is also used extensively is making phosphatic fertilizers. A tremendous tonnage is used to produce sulfuric acid, the most important manufactured chemical. It is used in making sulfite paper and other papers, as a fumigant, and in the bleaching of dried fruits. The element is a good electrical insulator. Organic compounds containing sulfur are very important. Calcium sulfate, ammonium sulfate, carbon disulfide, sulfur dioxide, and hydrogen sulfide are but a few of the many other important compounds of sulfur. Sulfur is essential to life. It is a minor constituent of fats, body fluids, and skeletal minerals. Carbon disulfide, hydrogen sulfide, and sulfur dioxide should be handled carefully. Hydrogen sulfide in small concentrations can be metabolized, but in higher concentrations it quickly can cause death by respiratory paralysis. It is insidious in that it quickly deadens the sense of smell. Sulfur dioxide is a dangerous component in atmospheric air pollution. In 1975, University of Pennsylvania scientists reported synthesis of polymeric sulfur nitride, which has the properties of a metal, although it contains no metal atoms. The material has unusual optical and electrical properties.

Tantalum — (Gr. *Tantalos*, mythological character, father of *Niobe*), Ta; at. wt. 180.9479(1); at. no. 73; m.p. 3017°C; b.p. 5458°C; sp. gr. 16.654; valence 2?, 3, 4?, or 5. Discovered in 1802 by Ekeberg, but many chemists thought niobium and tantalum were identical elements until Rose, in 1844, and Marignac, in 1866, showed that niobic and tantalic acids were two different acids. The early investigators only isolated the impure metal. The first relatively pure ductile tantalum was produced by von Bolton in 1903. Tantalum occurs principally in the mineral *columbite-tantalite* $(Fe, Mn)(Nb, Ta)_2O_6$. Tantalum ores are found in Australia, Brazil, Rwanda, Zimbabwe, Congo-Kinshasa, Nigeria, and Canada. Separation of tantalum from niobium requires several complicated steps. Several methods are used to commercially produce the element, including electrolysis of molten potassium fluorotantalate, reduction of potassium fluorotantalate with sodium, or reacting tantalum carbide with tantalum oxide. Thirty four isotopes and isomers of tantalum are known to exist. Natural tantalum contains two isotopes, one of which is radioactive with a very long half-life. Tantalum is a gray, heavy, and very hard metal. When pure, it is ductile and can be drawn into fine wire, which is used as a filament for evaporating metals such as aluminum. Tantalum is almost completely immune to chemical attack at temperatures below 150°C, and is attacked only by hydrofluoric acid, acidic solutions containing the fluoride ion, and free sulfur trioxide. Alkalis attack it only slowly. At high temperatures, tantalum becomes much more reactive. The element has a melting point exceeded only by tungsten and rhenium. Tantalum is used to make a variety of alloys with desirable properties such as high melting point, high strength, good ductility, etc. Scientists at Los Alamos have produced a tantalum carbide graphite composite material, which is said to be one of the hardest materials ever made. The compound has a melting point of 3738°C. Tantalum has good "gettering" ability at high temperatures, and tantalum oxide films are stable and have good rectifying and dielectric properties. Tantalum is used to make electrolytic capacitors and vacuum furnace parts, which account for about 60% of its use. The metal is also widely used to fabricate chemical process equipment, nuclear reactors, and aircraft and missile parts. Tantalum is completely immune to body liquids and is a nonirritating metal. It has, therefore, found wide use in making surgical appliances. Tantalum oxide is used to make special glass with high index of refraction for camera lenses. The metal has many other uses. The price of (99.9%) tantalum is about \$2/g.

Technetium — (Gr. *technetos*, artificial), Tc; at. wt. (98); at. no. 43; m.p. 2157°C; b.p. 4265°C; sp. gr. 11.50 (calc.); valence 0, +2, +4, +5, +6, and +7. Element 43 was predicted on the basis of the periodic table, and was erroneously reported as having been discovered in 1925, at which time it was named *masurium*. The element was actually discovered by Perrier and Segre in Italy in 1937. It was found in a sample of molybdenum, which was bombarded by deuterons in the Berkeley cyclotron, and which E. Lawrence sent to these investigators. Technetium was the first element to be produced artificially. Since its discovery, searches for the element in terrestrial materials have been made without success. If it does exist, the concentration must be very small. Technetium has been found in the spectrum of S-, M-, and N-type stars, and its presence in stellar matter is leading to new theories of the production of heavy elements in the stars. Forty-three isotopes and isomers of technetium, with atomic masses ranging from 86 to 113, are known. ^{97}Tc has a half-life of 2.6×10^6 years. ^{98}Tc has a half-life of 4.2×10^6 years. The isomeric isotope ^{95m}Tc, with a half-life of 61 days, is useful for tracer work, as it produces energetic gamma rays. Technetium metal has been produced in kilogram quantities. The metal was first prepared by passing hydrogen gas at 1100°C over Tc_2S_7. It is now conveniently prepared by the reduction of ammonium pertechnetate with hydrogen. Technetium is a silvery-gray metal that tarnishes slowly in moist air. Until 1960, technetium was available only in small amounts and the price was as high as \$2800/g. ^{99}Tc is now commercially available to holders of O.R.N.L. permits at a price of \$83/g plus packing charges. ^{99}Tc is available at a cost of \$1.56/μCi. The chemistry of technetium is said to be similar to that of rhenium. Technetium dissolves in nitric acid, aqua regia, and conc. sulfuric acid, but is not soluble in hydrochloric acid of any strength. The element is a remarkable corrosion inhibitor for steel. It is reported that mild carbon steels may be effectively protected by as little as 55 ppm of $KTcO_4$ in aerated distilled water at temperatures up to 250°C. This corrosion protection is limited to closed systems, since technetium is radioactive and must be confined. ^{99}Tc has a specific activity of 6.2×10^8 Bq/g. Activity of this level must not be allowed to spread. ^{99}Tc is a contamination hazard and should be handled in a glove box. The metal is an excellent superconductor at 11°K and below.

Tellurium — (L. *tellus*, earth), Te; at. wt. 127.60(3); at. no. 52; m.p. 449.51°C; b.p. 988°C; sp. gr. 6.24 (20°C); valence 2, 4, or 6. Discovered by Muller von Reichenstein in 1782; named by Klaproth, who isolated it in 1798. Tellurium is occasionally found native, but is more often found as the telluride of gold (*calaverite*), and combined with other metals. It is recovered commercially from the anode muds produced during the electrolytic refining of blister copper. The U.S., Canada, Peru, and Japan are the largest Free World producers of the element. Crystalline tellurium has a silvery-white appearance, and when pure exhibits a metallic luster. It is brittle and easily pulverized. Amorphous tellurium is formed by precipitating tellurium from a solution of telluric or tellurous acid. Whether this form is truly amorphous, or made of minute crystals, is open to question. Tellurium is a p-type semiconductor, and shows greater conductivity in certain directions, depending on alignment of the atoms. Its conductivity increases slightly with exposure to light. It can be doped with silver, copper, gold, tin, or other elements. In air, tellurium burns with a greenish-blue flame, forming the dioxide. Molten tellurium corrodes iron, copper, and stainless steel. Tellurium and its compounds are probably toxic and should be handled with care. Workmen exposed to as little as 0.01 mg/m^3 of air, or less, develop "tellurium breath," which has a garlic-like odor. Forty two isotopes and isomers of tellurium are known, with atomic masses ranging from 106 to 138. Natural tellurium consists of eight isotopes, two of which are radioactive with very long half-lives. Tellurium improves the machinability of copper and stainless steel, and its addition to lead decreases the corrosive action of sulfuric acid on lead and improves its strength and hardness. Tellurium catalysts are used in the oxidation of organic compounds and are used in hydrogenation and halogenation reactions. Tellurium is also used in electronic and semi-conductor devices. It is also used as a basic ingredient in blasting caps, and is added to cast iron for chill control. Tellurium is used in ceramics. Bismuth telluride has been used in thermoelectric devices. Tellurium costs about 20¢/g, with a purity of about 99.5%. The metal with a purity of 99.9999% costs about \$6/g.

Terbium — (*Ytterby*, village in Sweden), Tb; at. wt. 158.92534(2); at. no. 65; m.p. 1356°C; b.p. 3230°C; sp. gr. 8.230; valence 3, 4. Discovered by Mosander in 1843. Terbium is a member of the lanthanide or "rare earth" group of elements. It is found in *cerite, gadolinite*, and other minerals along with other rare earths. It is recovered commercially from *monazite* in which it is present to the extent of 0.03%, from *xenotime*, and from *euxenite*, a complex oxide containing 1% of more of terbia. Terbium has been isolated only in recent years with the development of ion-exchange techniques for separating the rare-earth elements. As with other rare earths, it can be produced by reducing the anhydrous chloride or fluoride with calcium metal in a tantalum crucible. Calcium and tantalum impurities can be removed by vacuum remelting. Other methods of isolation are possible. Terbium is reasonably stable in air. It is a silver-gray metal, and is malleable, ductile, and soft enough to be cut with a knife. Two crystal modifications exist, with a transformation temperature of 1289°C. Forty-two isotopes and isomers are recognized. The oxide is a chocolate or dark maroon color. Sodium terbium borate is used as a laser material and emits coherent light at 0.546 μm. Terbium is used to dope calcium fluoride, calcium tungstate, and strontium molybdate, used in solid-state devices. The oxide has potential application as an activator for green phosphors used in color TV tubes. It can be used with ZrO_2 as a crystal stabilizer of fuel cells which operate at elevated temperature. Few other uses have been found. The element is priced at about $40/g (99.9%). Little is known of the toxicity of terbium. It should be handled with care as with other lanthanide elements.

Thallium — (Gr. *thallos*, a green shoot or twig), Tl; at. wt. 204.3833(2); at. no. 81; m.p. 304°C; b.p. 1473°C; sp. gr. 11.85 (20°C); valence 1, or 3. Thallium was discovered spectroscopically in 1861 by Crookes. The element was named after the beautiful green spectral line, which identified the element. The metal was isolated both by Crookes and Lamy in 1862 about the same time. Thallium occurs in *crooksite, lorandite*, and *hutchinsonite*. It is also present in *pyrites* and is recovered from the roasting of this ore in connection with the production of sulfuric acid. It is also obtained from the smelting of lead and zinc ores. Extraction is somewhat complex and depends on the source of the thallium. Manganese nodules, found on the ocean floor, contain thallium. When freshly exposed to air, thallium exhibits a metallic luster, but soon develops a bluish-gray tinge, resembling lead in appearance. A heavy oxide builds up on thallium if left in air, and in the presence of water the hydroxide is formed. The metal is very soft and malleable. It can be cut with a knife. Forty-seven isotopes of thallium, with atomic masses ranging from 179 to 210 are recognized. Natural thallium is a mixture of two isotopes. The element and its compounds are toxic and should be handled carefully. Contact of the metal with skin is dangerous, and when melting the metal adequate ventilation should be provided. Thallium is suspected of carcinogenic potential for man. Thallium sulfate has been widely employed as a rodenticide and ant killer. It is odorless and tasteless, giving no warning of its presence. Its use, however, has been prohibited in the U.S. since 1975 as a household insecticide and rodenticide. The electrical conductivity of thallium sulfide changes with exposure to infrared light, and this compound is used in photocells. Thallium bromide-iodide crystals have been used as infrared optical materials. Thallium has been used, with sulfur or selenium and arsenic, to produce low melting glasses which become fluid between 125 and 150°C. These glasses have properties at room temperatures similar to ordinary glasses and are said to be durable and insoluble in water. Thallium oxide has been used to produce glasses with a high index of refraction. Thallium has been used in treating ringworm and other skin infections; however, its use has been limited because of the narrow margin between toxicity and therapeutic benefits. A mercury-thallium alloy, which forms a eutectic at 8.5% thallium, is reported to freeze at –60°C, some 20° below the freezing point of mercury. Thallium metal (99.999%) costs about $2/g.

Thorium — (*Thor*, Scandinavian god of war), Th; at. wt. 232.0381(1); at. no. 90; m.p. 1750°C; b.p. 4788°C; sp. gr. 11.72; valence +2(?), +3(?), +4. Discovered by Berzelius in 1828. Thorium occurs in *thorite* ($ThSiO_4$) and in *thorianite* ($ThO_2 + UO_2$). Large deposits of thorium minerals have been reported in New England and elsewhere, but these have not yet been exploited. Thorium is now thought to be about three times as abundant as uranium and about as abundant as lead or molybdenum. The metal is a source of nuclear power. There is probably more energy available for use from thorium in the minerals of the earth's crust than from both uranium and fossil fuels. Any sizable demand for thorium as a nuclear fuel is still several years in the future. Work has been done in developing thorium cycle converter-reactor systems. Several prototypes, including the HTGR (high-temperature gas-cooled reactor) and MSRE (molten salt converter reactor experiment), have operated. While the HTGR reactors are efficient, they are not expected to become important commercially for many years because of certain operating difficulties. Thorium is recovered commercially from the mineral *monazite*, which contains from 3 to 9% ThO_2 along with rare-earth minerals. Much of the internal heat the earth produces has been attributed to thorium and uranium. Several methods are available for producing thorium metal: it can be obtained by reducing thorium oxide with calcium, by electrolysis of anhydrous thorium chloride in a fused mixture of sodium and potassium chlorides, by calcium reduction of thorium tetrachloride mixed with anhydrous zinc chloride, and by reduction of thorium tetrachloride with an alkali metal. Thorium was originally assigned a position in Group IV of the periodic table. Because of its atomic weight, valence, etc., it is now considered to be the second member of the *actinide* series of elements. When pure, thorium is a silvery-white metal which is air-stable and retains its luster for several months. When contaminated with the oxide, thorium slowly tarnishes in air, becoming gray and finally black. The physical properties of thorium are greatly influenced by the degree of contamination with the oxide. The purest specimens often contain several tenths of a percent of the oxide. High-purity thorium has been made. Pure thorium is soft, very ductile, and can be cold-rolled, swaged, and drawn. Thorium is dimorphic, changing at 1400°C from a cubic to a body-centered cubic structure. Thorium oxide has a melting point of 3300°C, which is the highest of all oxides. Only a few elements, such as tungsten, and a few compounds, such as tantalum carbide, have higher melting points. Thorium is slowly attacked by water, but does not dissolve readily in most common acids, except hydrochloric. Powdered thorium metal is often pyrophoric and should be carefully handled. When heated in air, thorium turnings ignite and burn brilliantly with a white light. The principal use of thorium has been in the preparation of the Welsbach mantle, used for portable gas lights. These mantles, consisting of thorium oxide with about 1% cerium oxide and other ingredients, glow with a dazzling light when heated in a gas flame. Thorium is an important alloying element in magnesium, imparting high strength and creep resistance at elevated temperatures. Because thorium has a low work-function and high electron emission, it is used to coat tungsten wire used in electronic equipment. The oxide is also used to control the grain size of tungsten used for electric lamps; it is also used for high-temperature laboratory crucibles. Glasses containing thorium oxide have a high refractive index and low dispersion. Consequently, they find application in high quality lenses for cameras and scientific instruments. Thorium oxide has also found use as a catalyst in the conversion of ammonia to nitric acid, in petroleum cracking, and in producing sulfuric acid. Thorium has not found many uses due to its radioactive nature and its handling and disposal problems. Thirty isotopes of thorium are known with atomic masses ranging from 210 to 237. All are unstable. ^{232}Th occurs naturally and has a half-life of 1.4×10^{10} years. It is an alpha emitter. ^{232}Th goes through six alpha and four beta decay steps before becoming the stable isotope ^{208}Pb. ^{232}Th is sufficiently radioactive to expose a photographic plate in a few hours. Thorium disintegrates with the production of "thoron" (^{220}Rn), which is an alpha emitter and presents a radiation hazard. Good ventilation of areas where thorium is stored or handled is therefore essential. Thorium metal (99.8%) costs about $25/g.

Thulium — (*Thule*, the earliest name for Scandinavia), Tm; at. wt. 168.93421(3); at. no. 69; m.p. 1545°C; b.p. 1950°C; sp. gr. 9.321 (25°C); valence 3. Discovered in 1879 by Cleve. Thulium occurs in small quantities along with other rare earths in a number of minerals. It is obtained commercially from *monazite*, which contains about 0.007% of the element. Thulium is the least abundant of the rare earth elements, but with new sources recently discovered, it is now considered to be about as rare as silver, gold, or cadmium. Ion-exchange and solvent extraction techniques have recently permitted much easier separation of the rare earths, with much lower costs. Only a few years ago, thulium metal was not obtainable at any cost; in 1996 the oxide cost \$20/g. Thulium metal powder now costs \$70/g (99.9%). Thulium can be isolated by reduction of the oxide with lanthanum metal or by calcium reduction of the anhydrous fluoride. The pure metal has a bright, silvery luster. It is reasonably stable in air, but the metal should be protected from moisture in a closed container. The element is silver-gray, soft, malleable, and ductile, and can be cut with a knife. Forty-one isotopes and isomers are known, with atomic masses ranging from 146 to 176. Natural thulium, which is 100% ^{169}Tm, is stable. Because of the relatively high price of the metal, thulium has not yet found many practical applications. ^{169}Tm bombarded in a nuclear reactor can be used as a radiation source in portable X-ray equipment. ^{171}Tm is potentially useful as an energy source. Natural thulium also has possible use in *ferrites* (ceramic magnetic materials) used in microwave equipment. As with other lanthanides, thulium has a low-to-moderate acute toxic rating. It should be handled with care.

Tin — (anglo-Saxon, *tin*), Sn (L. *stannum*); at. wt. 118.710(7); at. no. 50; m.p. 231.93°C; b.p. 2602°C; sp. gr. (gray) 5.75, (white) 7.31; valence 2, 4. Known to the ancients. Tin is found chiefly in *cassiterite* (SnO_2). Most of the world's supply comes from China, Indonesia, Peru, Brazil, and Bolivia. The U.S. produces almost none, although occurrences have been found in Alaska and Colorado. Tin is obtained by reducing the ore with coal in a reverberatory furnace. Ordinary tin is composed of ten stable isotopes; thirty-six unstable isotopes and isomers are also known. Ordinary tin is a silver-white metal, is malleable, somewhat ductile, and has a highly crystalline structure. Due to the breaking of these crystals, a "tin cry" is heard when a bar is bent. The element has two allotropic forms at normal pressure. On warming, gray, or α tin, with a cubic structure, changes at 13.2°C into white, or β tin, the ordinary form of the metal. White tin has a tetragonal structure. When tin is cooled below 13.2°C, it changes slowly from white to gray. This change is affected by impurities such as aluminum and zinc, and can be prevented by small additions of antimony or bismuth. This change from the α to β form is called the tin pest. Tin-lead alloys are used to make organ pipes. There are few if any uses for gray tin. Tin takes a high polish and is used to coat other metals to prevent corrosion or other chemical action. Such tin plate over steel is used in the so-called tin can for preserving food. Alloys of tin are very important. Soft solder, type metal, fusible metal, pewter, bronze, bell metal, Babbitt metal, White metal, die casting alloy, and phosphor bronze are some of the important alloys using tin. Tin resists distilled sea and soft tap water, but is attacked by strong acids, alkalis, and acid salts. Oxygen in solution accelerates the attack. When heated in air, tin forms SnO_2, which is feebly acid, forming stannate salts with basic oxides. The most important salt is the chloride ($SnCl_2 \cdot H_2O$), which is used as a reducing agent and as a mordant in calico printing. Tin salts sprayed onto glass are used to produce electrically conductive coatings. These have been used for panel lighting and for frost-free windshields. Most window glass is now made by floating molten glass on molten tin (float glass) to produce a flat surface (Pilkington process). Of recent interest is a crystalline tin-niobium alloy that is superconductive at very low temperatures. This promises to be important in the construction of superconductive magnets that generate enormous field strengths but use practically no power. Such magnets, made of tin-niobium wire, weigh but a few pounds and produce magnetic fields that, when started with a small battery, are comparable to that of a 100 ton electromagnet operated continuously with a large power supply. The small amount of tin found in canned foods is quite harmless. The agreed limit of tin content in U.S. foods is 300 mg/kg. The trialkyl and triaryl tin compounds are used as biocides and must be handled carefully. Over the past 25 years the price of commercial tin has varied from 50¢/lb (\$1.10/kg) to its present price of about \$9/kg as of December 1999.

Titanium — (L. *Titans*, the first sons of the Earth, myth.), Ti; at. wt. 47.867(1); at. no. 22; m.p. 1668°C; b.p. 3287°C; sp. gr. 4.54; valence 2, 3, or 4. Discovered by Gregor in 1791; named by Klaproth in 1795. Impure titanium was prepared by Nilson and Pettersson in 1887; however, the pure metal (99.9%) was not made until 1910 by Hunter by heating $TiCl_4$ with sodium in a steel bomb. Titanium is present in meteorites and in the sun. Rocks obtained during the Apollo 17 lunar mission showed presence of 12.1% TiO_2. Analyses of rocks obtained during earlier Apollo missions show lower percentages. Titanium oxide bands are prominent in the spectra of M-type stars. The element is the ninth most abundant in the crust of the earth. Titanium is almost always present in igneous rocks and in the sediments derived from them. It occurs in the minerals *rutile, ilmenite,* and *sphene*, and is present in titanates and in many iron ores. Deposits of ilmenite and rutile are found in Florida, California, Tennessee, and New York. Australia, Norway, Malaysia, India, and China are also large suppliers of titanium minerals. Titanium is present in the ash of coal, in plants, and in the human body. The metal was a laboratory curiosity until Kroll, in 1946, showed that titanium could be produced commercially by reducing titanium tetrachloride with magnesium. This method is largely used for producing the metal today. The metal can be purified by decomposing the iodide. Titanium, when pure, is a lustrous, white metal. It has a low density, good strength, is easily fabricated, and has excellent corrosion resistance. It is ductile only when it is free of oxygen. The metal burns in air and is the only element that burns in nitrogen. Titanium is resistant to dilute sulfuric and hydrochloric acid, most organic acids, moist chlorine gas, and chloride solutions. Natural titanium consists of five isotopes with atomic masses from 46 to 50. All are stable. Eighteen other unstable isotopes are known. The metal is dimorphic. The hexagonal α form changes to the cubic β form very slowly at about 880°C. The metal combines with oxygen at red heat, and with chlorine at 550°C. Titanium is important as an alloying agent with aluminum, molybdenum, manganese, iron, and other metals. Alloys of titanium are principally used for aircraft and missiles where lightweight strength and ability to withstand extremes of temperature are important. Titanium is as strong as steel, but 45% lighter. It is 60% heavier than aluminum, but twice as strong. Titanium has potential use in desalination plants for converting sea water into fresh water. The metal has excellent resistance to sea water and is used for propeller shafts, rigging, and other parts of ships exposed to salt water. A titanium anode coated with platinum has been used to provide cathodic protection from corrosion by salt water. Titanium metal is considered to be physiologically inert; however, titanium powder may be a carcinogenic hazard. When pure, titanium dioxide is relatively clear and has an extremely high index of refraction with an optical dispersion higher than diamond. It is produced artificially for use as a gemstone, but it is relatively soft. Star sapphires and rubies exhibit their asterism as a result of the presence of TiO_2. Titanium dioxide is extensively used for both house paint and artist's paint, as it is permanent and has good covering power. Titanium oxide pigment accounts for the largest use of the element. Titanium paint is an excellent reflector of infrared, and is extensively used in solar observatories where heat causes poor seeing conditions. Titanium tetrachloride is used to iridize glass. This compound fumes strongly in air and has been used to produce smoke screens. The price of titanium metal (99.8%) is about \$800/kg.

Tungsten — (Swedish, *tung sten*, heavy stone); also known as *wolfram* (from *wolframite*, said to be named from *wolf rahm* or *spumi lupi*, because the ore interfered with the smelting of tin and was supposed to devour the tin), W; at. wt. 183.84(1); at. no. 74; m.p. 3422°C; b.p. 5555°C; sp. gr. 19.3 (20°C); valence 2, 3, 4, 5, or 6. In 1779 Peter Woulfe examined the mineral now known as *wolframite* and concluded it must contain a new substance. Scheele, in 1781, found that a new acid could be made from *tung sten* (a name first applied about 1758 to a mineral now known as *scheelite*). Scheele and Berman suggested the possibility of obtaining a new metal by reducing this acid. The de Elhuyar brothers found an acid in *wolframite* in 1783 that was identical to the acid of *tungsten*(tungstic acid) of Scheele, and in that year they succeeded in obtaining the element by reduction of this acid with charcoal. Tungsten occurs in *wolframite*, (Fe, Mn)WO_4; *scheelite*, $CaWO_4$; *huebnerite*, $MnWO_4$; and *ferberite*, $FeWO_4$. Important deposits of tungsten occur in California, Colorado, Bolivia, Russia, and Portugal. China is reported to have about 75% of the world's tungsten resources. Natural tungsten contains five stable isotopes. Thirty two other unstable isotopes and isomers are recognized. The metal is obtained commercially by reducing tungsten oxide with hydrogen or carbon. Pure tungsten is a steel-gray to tin-white metal. Very pure tungsten can be cut with a hacksaw, and can be forged, spun, drawn, and extruded. The impure metal is brittle and can be worked only with difficulty. Tungsten has the highest melting point of all metals, and at temperatures over 1650°C has the highest tensile strength. The metal oxidizes in air and must be protected at elevated temperatures. It has excellent corrosion resistance and is attacked only slightly by most mineral acids. The thermal expansion is about the same as borosilicate glass, which makes the metal useful for glass-to-metal seals. Tungsten and its alloys are used extensively for filaments for electric lamps, electron and television tubes, and for metal evaporation work; for electrical contact points for automobile distributors; X-ray targets; windings and heating elements for electrical furnaces; and for numerous spacecraft and high-temperature applications. High-speed tool steels, Hastelloy®, Stellite®, and many other alloys contain tungsten. Tungsten carbide is of great importance to the metal-working, mining, and petroleum industries.Calcium and magnesium tungstates are widely used in fluorescent lighting; other salts of tungsten are used in the chemical and tanning industries. Tungsten disulfide is a dry, high-temperature lubricant, stable to 500°C. Tungsten bronzes and other tungsten compounds are used in paints. Zirconium tungstate has found recent applications (see under Zirconium). Tungsten powder (99.999%) costs about $2900/kg.

Uranium — (Planet *Uranus*), U; at. wt. 238.0289(1); at. no. 92; m.p. 1135°C; b.p. 4131°C; sp. gr. ~18.95; valence 2, 3, 4, 5, or 6. Yellow-colored glass, containing more than 1% uranium oxide and dating back to 79 A.D., has been found near Naples, Italy. Klaproth recognized an unknown element in *pitchblende* and attempted to isolate the metal in 1789. The metal apparently was first isolated in 1841 by Peligot, who reduced the anhydrous chloride with potassium. Uranium is not as rare as it was once thought. It is now considered to be more plentiful than mercury, antimony, silver, or cadmium, and is about as abundant as molybdenum or arsenic. It occurs in numerous minerals such as *pitchblende, uraninite, carnotite, autunite, uranophane, davidite,* and *tobernite*. It is also found in *phosphate rock, lignite, monazite sands*, and can be recovered commercially from these sources. Large deposits of uranium ore occur in Utah, Colorado, New Mexico, Canada, and elsewhere. Uranium can be made by reducing uranium halides with alkali or alkaline earth metals or by reducing uranium oxides by calcium, aluminum, or carbon at high temperatures. The metal can also be produced by electrolysis of KUF_5 or UF_4, dissolved in a molten mixture of $CaCl_2$ and NaCl. High-purity uranium can be prepared by the thermal decomposition of uranium halides on a hot filament. Uranium exhibits three crystallographic modifications as follows:

$$\alpha \xrightarrow{688°C} \beta \xrightarrow{776°C} \gamma$$

Uranium is a heavy, silvery-white metal which is pyrophoric when finely divided. It is a little softer than steel, and is attacked by cold water in a finely divided state. It is malleable, ductile, and slightly paramagnetic. In air, the metal becomes coated with a layer of oxide. Acids dissolve the metal, but it is unaffected by alkalis. Uranium has twenty three isotopes, one of which is an isomer and all of which are radioactive. Naturally occurring uranium contains 99.2745% by weight ^{238}U, 0.720% ^{235}U, and 0.0055% ^{234}U. Studies show that the percentage weight of ^{235}U in natural uranium varies by as much as 0.1%, depending on the source. The U.S.D.O.E. has adopted the value of 0.711 as being their "official" percentage of ^{235}U in natural uranium. Natural uranium is sufficiently radioactive to expose a photographic plate in an hour or so. Much of the internal heat of the earth is thought to be attributable to the presence of uranium and thorium. ^{238}U with a half-life of 4.46×10^9 years, has been used to estimate the age of igneous rocks. The origin of uranium, the highest member of the naturally occurring elements — except perhaps for traces of neptunium or plutonium — is not clearly understood, although it has been thought that uranium might be a decay product of elements of higher atomic weight, which may have once been present on earth or elsewhere in the universe. These original elements may have been formed as a result of a primordial "creation," known as "the big bang," in a supernova, or in some other stellar processes. The fact that recent studies show that most trans-uranic elements are extremely rare with very short half-lives indicates that it may be necessary to find some alternative explanation for the very large quantities of radioactive uranium we find on earth. Studies of meteorites from other parts of the solar system show a relatively low radioactive content, compared to terrestrial rocks. Uranium is of great importance as a nuclear fuel. ^{238}U can be converted into fissionable plutonium by the following reactions:

$$^{238}U(n,\gamma) \rightarrow {}^{239}U \xrightarrow{\beta} {}^{239}Np \xrightarrow{\beta} {}^{239}Pu$$

This nuclear conversion can be brought about in "breeder" reactors where it is possible to produce more new fissionable material than the fissionable material used in maintaining the chain reaction. ^{235}U is of even greater importance, for it is the key to the utilization of uranium. ^{235}U, while occurring in natural uranium to the extent of only 0.72%, is so fissionable with slow neutrons that a self-sustaining fission chain reaction can be made to occur in a reactor constructed from natural uranium and a suitable moderator, such as heavy water or graphite, alone. ^{235}U can be concentrated by gaseous diffusion and other physical processes, if desired, and used directly as a nuclear fuel, instead of natural uranium, or used as an explosive. Natural uranium, slightly enriched with ^{235}U by a small percentage, is used to fuel nuclear power reactors for the generation of electricity. Natural thorium can be irradiated with neutrons as follows to produce the important isotope ^{233}U.

$$^{232}Th(n,\gamma) \rightarrow {}^{233}Th \xrightarrow{\beta} {}^{233}Pa \xrightarrow{\beta} {}^{233}U$$

While thorium itself is not fissionable, ^{233}U is, and in this way may be used as a nuclear fuel. One pound of completely fissioned uranium has the fuel value of over 1500 tons of coal. The uses of nuclear fuels to generate electrical power, to make isotopes for peaceful purposes, and to make explosives are well known. The estimated world-wide production of the 437 nuclear power reactors in operation in 1998 amounted to about 352,000 Megawatt hours. In 1998 the U.S. had about 107 commercial reactors with an output of about 100,000 Megawatt-hours. Some nuclear-powered electric generating plants have recently been closed because of safety concerns. There are also serious problems with nuclear waste disposal that have not been completely resolved. Uranium in the U.S.A. is controlled by the U.S. Nuclear Regulatory Commission, under the Department of Energy. Uses are being found for the large quantities of "depleted" uranium, now available, where uranium-235 has been lowered to about 0.2%. Depleted uranium has been used for inertial guidance devices, gyrocompasses, counterweights for aircraft control surfaces, ballast for missile reentry vehicles, and as a shielding material for tanks, etc. Concerns, however, have been raised over its low radioactive properties. Uranium metal is used for X-ray targets for production of high-energy X-rays. The nitrate has been used as photographic toner, and the acetate is used in analytical chemistry. Crystals of uranium nitrate are triboluminescent. Uranium salts have also been used for producing yellow "vaseline" glass and glazes. Uranium and its compounds are highly toxic, both from a chemical and radiological standpoint. Finely divided uranium metal, being pyrophoric, presents a fire hazard. The maximum permissible total body burden of natural uranium (based on radiotoxicity) is 0.2 μCi for soluble compounds. Recently, the natural presence of uranium and thorium in many soils has become of concern to homeowners because of the generation of radon and its daughters (see under Radon). Uranium metal is available commercially at a cost of about $6/g (99.7%) in air-tight glass under argon.

Unilnilium, Unnilbium, etc. — See under the opening paragraphs of this article and also under Elements 110 to 118.

Vanadium — (Scandinavian goddess, *Vanadis*), V; at. wt. 50.9415(1); at. no. 23; m.p. 1910°C; b.p. 3407°C; sp. gr. 6.11 (18.7°C); valence 2, 3, 4, or 5. Vanadium was first discovered by del Rio in 1801. Unfortunately, a French chemist incorrectly declared del Rio's new element was only impure chromium; del Rio thought himself to be mistaken and accepted the French chemist's statement. The element was rediscovered in 1830 by Sefstrom, who named the element in honor of the Scandinavian goddess *Vanadis* because of its beautiful multicolored compounds. It was isolated in nearly pure form by Roscoe, in 1867, who reduced the chloride with hydrogen. Vanadium of 99.3 to 99.8% purity was not produced until 1927. Vanadium is found in about 65 different minerals among which are *carnotite, roscoelite, vanadinite,* and *patronite* important sources of the metal. Vanadium is also found in phosphate rock and certain iron ores, and is present in some crude oils in the form of organic complexes. It is also found in small percentages in meteorites. Commercial production from petroleum ash holds promise as an important source of the element. China, South Africa, and Russia supply much of the world's vanadium ores. High-purity ductile vanadium can be obtained by reduction of vanadium trichloride with magnesium or with magnesium-sodium mixtures. Much of the vanadium metal being produced is now made by calcium reduction of V_2O_5 in a pressure vessel, an adaption of a process developed by McKechnie and Seybolt. Natural vanadium is a mixture of two isotopes, ^{50}V (0.25%) and ^{51}V (99.75%). ^{50}V is slightly radioactive, having a long half-life. Twenty other unstable isotopes are recognized. Pure vanadium is a bright white metal, and is soft and ductile. It has good corrosion resistance to alkalis, sulfuric and hydrochloric acid, and salt water, but the metal oxidizes readily above 660°C. The metal has good structural strength and a low fission neutron cross section, making it useful in nuclear applications. Vanadium is used in producing rust resistant, spring, and highspeed tool steels. It is an important carbide stabilizer in making steels. About 80% of the vanadium now produced is used as ferrovanadium or as a steel additive. Vanadium foil is used as a bonding agent in cladding titanium to steel. Vanadium pentoxide is used in ceramics and as a catalyst. It is also used in producing a superconductive magnet with a field of 175,000 gauss. Vanadium and its compounds are toxic and should be handled with care. Ductile vanadium is commercially available. Vanadium metal (99.7%) costs about $3/g.

Wolfram — see Tungsten.

Xenon — (Gr. *xenon*, stranger), Xe; at. wt. 131.29(2); at. no. 54; m.p. –111.79°C; b.p. –108.12°C; t_c 16.62°C; density (gas) 5.887 ± 0.009 g/l, sp. gr (liquid) 3.52 (–109°C); valence usually 0. Discovered by Ramsay and Travers in 1898 in the residue left after evaporating liquid air components. Xenon is a member of the so-called noble or "inert" gases. It is present in the atmosphere to the extent of about one part in twenty million. Xenon is present in the Martian atmosphere to the extent of 0.08 ppm. The element is found in the gases evolved from certain mineral springs, and is commercially obtained by extraction from liquid air. Natural xenon is composed of nine stable isotopes. In addition to these, thirty five unstable isotopes and isomers have been characterized. Before 1962, it had generally been assumed that xenon and other noble gases were unable to form compounds. Evidence has been mounting in the past few years that xenon, as well as other members of the zero valence elements, do form compounds. Among the "compounds" of xenon now reported are xenon hydrate, sodium perxenate, xenon deuterate, difluoride, tetrafluoride, hexafluoride, and $XePtF_6$ and $XeRhF_6$. Xenon trioxide, which is highly explosive, has been prepared. More than 80 xenon compounds have been made with xenon chemically bonded to fluorine and oxygen. Some xenon compounds are colored. Metallic xenon has been produced, using several hundred kilobars of pressure. Xenon in a vacuum tube produces a beautiful blue glow when excited by an electrical discharge. The gas is used in making electron tubes, stroboscopic lamps, bactericidal lamps, and lamps used to excite ruby lasers for generating coherent light. Xenon is used in the atomic energy field in bubble chambers, probes, and other applications where its high molecular weight is of value. The perxenates are used in analytical chemistry as oxidizing agents. ^{133}Xe and ^{135}Xe are produced by neutron irradiation in air cooled nuclear reactors. ^{133}Xe has useful applications as a radioisotope. The element is available in sealed glass containers for about $20/l of gas at standard pressure. Xenon is not toxic, but its compounds are highly toxic because of their strong oxidizing characteristics.

Ytterbium — (Ytterby, village in Sweden), Yb; at. wt. 173.04(3); at. no. 70; m.p. 819°C; b.p. 1196°C; sp. gr (α) 6.903 (β) 6.966; valence 2, 3. Marignac in 1878 discovered a new component, which he called *ytterbia*, in the earth then known as *erbia*. In 1907, Urbain separated ytterbia into two components, which he called *neoytterbia* and *lutecia*. The elements in these earths are now known as *ytterbium* and *lutetium*, respectively. These elements are identical with *aldebaranium* and *cassiopeium*, discovered independently and at about the same time by von Welsbach. Ytterbium occurs along with other rare earths in a number of rare minerals. It is commercially recovered principally from *monazite sand*, which contains about 0.03%. Ion-exchange and solvent extraction techniques developed in recent years have greatly simplified the separation of the rare earths from one another. The element was first prepared by Klemm and Bonner in 1937 by reducing ytterbium trichloride with potassium. Their metal was mixed, however, with KCl. Daane, Dennison, and Spedding prepared a much purer form in 1953 from which the chemical and physical properties of the element could be determined. Ytterbium has a bright silvery luster, is soft, malleable, and quite ductile. While the element is fairly stable, it should be kept in closed containers to protect it from air and moisture. Ytterbium is readily attacked and dissolved by dilute and concentrated mineral acids and reacts slowly with water. Ytterbium has three allotropic forms with transformation points at –13° and 795°C. The beta form is a room-temperature, face-centered, cubic modification, while the high-temperature gamma form is a body-centered cubic form. Another body-centered cubic phase has recently been found to be stable at high pressures at room temperatures. The beta form ordinarily has metallic-type conductivity, but becomes a semiconductor when the pressure is increased above 16,000 atm. The electrical resistance increases tenfold as the pressure is increased to 39,000 atm and drops to about 80% of its standard temperature-pressure resistivity at a pressure of 40,000 atm. Natural ytterbium is a mixture of seven stable isotopes. Twenty six other unstable isotopes and isomers are known. Ytterbium metal has possible use in improving the grain refinement, strength, and other mechanical properties of stainless steel. One isotope is reported to have been used as a radiation source as a substitute for a portable X-ray machine where electricity is unavailable. Few other uses have been found. Ytterbium metal is available with a purity of about 99.99% for about $20/g. Ytterbium has a low acute toxic rating, but may present a carcinogenic hazard.

Yttrium — (*Ytterby*, village in Sweden near Vauxholm), Y; at. wt. 88.90585(2); at. no. 39; m.p. 1522°C; b.p. 3345°C; sp. gr. 4.469 (25°C); valence 3. *Yttria*, which is an earth containing yttrium, was discovered by Gadolin in 1794. *Ytterby* is the site of a quarry which yielded many unusually minerals containing rare earths and other elements. This small town, near Stockholm, bears the honor of giving names to *erbium, terbium*, and *ytterbium* as well as *yttrium*. In 1843 Mosander showed that yttria could be resolved into the oxides (or earths) of three elements. The name yttria was reserved for the most basic one; the others were named *erbia* and *terbia*. Yttrium occurs in nearly all of the rare-earth minerals. Analysis of lunar rock samples obtained during the Apollo missions show a relatively high yttrium content. It is recovered commercially from *monazite sand*, which contains about 3%, and from *bastnasite*, which contains about 0.2%. Wohler obtained the impure element in 1828 by reduction of the anhydrous chloride with potassium. The metal is now produced commercially by reduction of the fluoride with calcium metal. It can also be prepared by other techniques. Yttrium has a silver-metallic luster and is relatively stable in air. Turnings of the metal, however, ignite in air if their temperature exceeds 400°C, and finely divided yttrium is very unstable in air. Yttrium oxide is one of the most important compounds of yttrium and accounts for the largest use. It is widely used in making YVO_4 europium, and Y_2O_3 europium phosphors to give the red color in color television tubes. Many hundreds of thousands of pounds are now used in this application. Yttrium oxide also is used to produce yttrium-iron-garnets, which are very effective microwave filters. Yttrium iron, aluminum, and gadolinium garnets, with formulas such as $Y_3Fe_5O_{12}$ and $Y_3Al_5O_{12}$, have interesting magnetic properties. Yttrium iron garnet is also exceptionally efficient as both a transmitter and transducer of acoustic energy. Yttrium aluminum garnet, with a hardness of 8.5, is also finding use as a gemstone (simulated diamond). Small amounts of yttrium (0.1 to 0.2%) can be used to reduce the grain size in chromium, molybdenum, zirconium, and titanium, and to increase strength of aluminum and magnesium alloys. Alloys with other useful properties can be obtained by using yttrium as an additive. The metal can be used as a deoxidizer for vanadium and other nonferrous metals. The metal has a low cross section for nuclear capture. ^{90}Y, one of the isotopes of yttrium, exists in equilibrium with its parent ^{90}Sr, a product of atomic explosions. Yttrium has been considered for use as a nodulizer for producing nodular cast iron, in which the graphite forms compact nodules instead of the usual flakes. Such iron has increased ductility. Yttrium is also finding application in laser systems and as a catalyst for ethylene polymerization.It has also potential use in ceramic and glass formulas, as the oxide has a high melting point and imparts shock resistance and low expansion characteristics to glass. Natural yttrium contains but one isotope, ^{89}Y. Forty-three other unstable isotopes and isomers have been characterized. Yttrium metal of 99.9% purity is commercially available at a cost of about $3/g.

Zinc — (Ger. *Zink*, of obscure origin), Zn; at. wt. 65.39(2); at. no. 30; m.p. 419.53°C; b.p. 907°C; sp. gr. 7.133 (25°C); valence 2. Centuries before zinc was recognized as a distinct element, zinc ores were used for making brass. Tubal-Cain, seven generations from Adam, is mentioned as being an "instructor in every artificer in brass and iron." An alloy containing 87% zinc has been found in prehistoric ruins in Transylvania. Metallic zinc was produced in the 13th century A.D. in India by reducing calamine with organic substances such as wool. The metal was rediscovered in Europe by Marggraf in 1746, who showed that it could be obtained by reducing *calamine* with charcoal. The principal ores of zinc are *sphalerite* or *blende* (sulfide), *smithsonite* (carbonate), *calamine* (silicate), and *franklinite* (zinc, manganese, iron oxide). Canada, Japan, Belgium, Germany, and The Netherlands are suppliers of zinc ores. Zinc is also mined in Alaska, Tennessee, Missouri, and elsewhere in the U.S. Zinc can be obtained by roasting its ores to form the oxide and by reduction of the oxide with coal or carbon, with subsequent distillation of the metal. Other methods of extraction are possible. Naturally occurring zinc contains five stable isotopes. Twenty-five other unstable isotopes and isomers are recognized. Zinc is a bluish-white, lustrous metal. It is brittle at ordinary temperatures but malleable at 100 to 150°C. It is a fair conductor of electricity, and burns in air at high red heat with evolution of white clouds of the oxide. The metal is employed to form numerous alloys with other metals. Brass, nickel silver, typewriter metal, commercial bronze, spring brass, German silver, soft solder, and aluminum solder are some of the more important alloys. Large quantities of zinc are used to produce die castings, used extensively by the automotive, electrical, and hardware industries. An alloy called *Prestal®*, consisting of 78% zinc and 22% aluminum is reported to be almost as strong as steel but as easy to mold as plastic. It is said to be so plastic that it can be molded into form by relatively inexpensive die casts made of ceramics and cement. It exhibits superplasticity. Zinc is also extensively used to galvanize other metals such as iron to prevent corrosion. Neither zinc nor zirconium is ferromagnetic; but $ZrZn_2$ exhibits ferromagnetism at temperatures below 35 K. Zinc oxide is a unique and very useful material to modern civilization. It is widely used in the manufacture of paints, rubber products, cosmetics, pharmaceuticals, floor coverings, plastics, printing inks, soap, storage batteries, textiles, electrical equipment, and other products. It has unusual electrical, thermal, optical, and solid-state properties that have not yet been fully investigated. Lithopone, a mixture of zinc sulfide and barium sulfate, is an important pigment. Zinc sulfide is used in making luminous dials, X-ray and TV screens, and fluorescent lights. The chloride and chromate are also important compounds. Zinc is an essential element in the growth of human beings and animals. Tests show that zinc-deficient animals require 50% more food to gain the same weight as an animal supplied with sufficient zinc. Zinc is not considered to be toxic, but when freshly formed ZnO is inhaled a disorder known as the *oxide shakes* or *zinc chills* sometimes occurs. It is recommended that where zinc oxide is encountered good ventilation be provided. The commercial price of zinc was roughly 60¢/lb ($1.30 kg). Zinc metal with a purity of 99.9999% is priced at about $6/g.

Zirconium — (Syriac, *zargun*, color of gold), Zr; at. wt. 91.224(2); at. no. 40; m.p. 1855°C; b.p. 4409°C; sp. gr. 6.506 (20°C); valence +2, +3, and +4. The name *zircon* may have originated from the Syriac word *zargono,* which describes the color of certain gemstones now known as *zircon, jargon, hyacinth, jacinth,* or *ligure.* This mineral, or its variations, is mentioned in biblical writings. These minerals were not known to contain this element until Klaproth, in 1789, analyzed a *jargon* from Sri Lanka and found a new earth, which Werner named zircon (*silex circonius*), and Klaproth called *Zirkonerde (zirconia)*. The impure metal was first isolated by Berzelius in 1824 by heating a mixture of potassium and potassium zirconium fluoride in a small iron tube. Pure zirconium was first prepared in 1914. Very pure zirconium was first produced in 1925 by van Arkel and de Boer by an iodide decomposition process they developed. Zirconium is found in abundance in S-type stars, and has been identified in the sun and meteorites. Analyses of lunar rock samples obtained during the various Apollo missions to the moon show a surprisingly high zirconium oxide content, compared with terrestial rocks. Naturally occurring zirconium contains five isotopes. Thirty-one other radioactive isotopes and isomers are known to exist. *Zircon*, $ZrSiO_4$, the principal ore, is found in deposits in Florida, South Carolina, Australia, South Africa, and elsewhere. *Baddeleyite*, found in Brazil, is an important zirconium mineral. It is principally pure ZrO_2in crystalline form having a hafnium content of about 1%. Zirconium also occurs in some 30 other recognized mineral species. Zirconium is produced commercially by reduction of the chloride with magnesium (the Kroll Process), and by other methods. It is a grayish-white lustrous metal. When finely divided, the metal may ignite spontaneously in air, especially at elevated temperatures. The solid metal is much more difficult to ignite. The inherent toxicity of zirconium compounds is low. Hafnium is invariably found in zirconium ores, and the separation is difficult. Commercial-grade zirconium contains from 1 to 3% hafnium. Zirconium has a low absorption cross section for neutrons, and is therefore used for nuclear energy applications, such as for cladding fuel elements. Commercial nuclear power generation now takes more than 90% of zirconium metal production. Reactors of the size now being made may use as much as a half-million lineal feet of zirconium alloy tubing. Reactor-grade zirconium is essentially free of hafnium. *Zircaloy®* is an important alloy developed specifically for nuclear applications. Zirconium is exceptionally resistant to corrosion by many common acids and alkalis, by sea water, and by other agents. It is used extensively by the chemical industry where corrosive agents are employed. Zirconium is used as a getter in vacuum tubes, as an alloying agent in steel, in surgical appliances, photoflash bulbs, explosive primers, rayon spinnerets, lamp filaments, etc. It is used in poison ivy lotions in the form of the carbonate as it combines with *urushiol*. With niobium, zirconium is superconductive at low temperatures and is used to make superconductive magnets, which offer hope of direct large-scale generation of electric power. Alloyed with zinc, zirconium becomes magnetic at temperatures below 35 K. Zirconium oxide (zircon) has a high index of refraction and is used as a gem material. The impure oxide, zirconia, is used for laboratory crucibles that will withstand heat shock, for linings of metallurgical furnaces, and by the glass and ceramic industries as a refractory material. Its use as a refractory material accounts for a large share of all zirconium consumed. Zirconium tungstate is an unusual material that shrinks, rather than expands, when heated. While this compound has been known for more than 30 years, it is only now that it is being studied to determine the nature of this unusual behavior. A few other compounds are known to possess this property, but they tend to shrink in one direction, while they stretch out in others in order to maintain an overall volume. Zirconium tungstate shrinks in all directions over a wide temperature range of from near absolute zero to +777°C. This material is being considered for use in very small computer chips, which are subject to severe temperature changes. It is also being considered for use in composite materials where thermal expansion may be a problem. Zirconium of about 99.5% purity is available at a cost of about $2100/kg or about $3/g.

PHYSICAL CONSTANTS OF INORGANIC COMPOUNDS

The compounds in this table were selected on the basis of their laboratory and industrial importance, as well as their value in illustrating trends in the variation of physical properties with position in the periodic table. An effort has been made to include the most frequently encountered inorganic substances; a limited number of organometallics are also covered. Many, if not most, of the compounds that are solids at ambient temperature can exist in more than one crystalline modification. The information given here applies to the most stable or common crystalline form. In cases where two or more forms are of practical importance, separate entries will be found in the table.

Compounds are arranged primarily in alphabetical order by the most commonly used name. However, adjustments are made in many instances in order to bring closely related compounds together. For example, hydrides of elements such as boron, silicon, and germanium are grouped together immediately following the entry for the parent element, since they would otherwise be scattered throughout the table. Likewise, the oxoacids of an element are given in one group whenever a strict alphabetical order would separate them (e.g., sulfuric acid and fluorosulfuric acid). The Formula Index following the table provides another means of locating a compound. There is also an index to CAS Registry Numbers.

The following data fields appear in the table:

- **Name:** Systematic name for the substance. The valence state of a metallic element is indicated by a Roman numeral, e.g., copper in the +1 state is written as copper(I) rather than cuprous, iron in the +3 state is iron(III) rather than ferric.
- **Formula:** The simplest descriptive formula is given, but this does not necessarily specify the actual structure of the compound. For example, aluminum chloride is designated as $AlCl_3$, even though a more accurate representation of the structure in the solid phase (and, under some conditions, in the gas phase) is Al_2Cl_6. A few exceptions are made, such as the use of Hg_2^{+2} for the mercury(I) ion.
- **CAS Registry Number:** Chemical Abstracts Service Registry Number. An asterisk (*) following the CAS RN for a hydrate indicates that the number refers to the anhydrous compound. In most cases the generic CAS RN for the compound is given rather than the number for a specific crystalline form or mineral.
- **Mol. Weight:** Molecular weight (relative molar mass) as calculated with the 1997 IUPAC Recommended Atomic Weights. The number of decimal places corresponds to the number of places in the atomic weight of the least accurately known element (e.g., one place for lead compounds, two places for compounds of selenium, germanium, etc.); a maximum of three places is given. For compounds of radioactive elements for which IUPAC makes no recommendation, the mass number of the isotope with longest half-life is used, and the result is rounded to the nearest integer.
- **Physical Form:** The crystal system is given, when available, for compounds that are solid at room temperature, together with color and other descriptive features. Abbreviations are listed below.
- **mp:** Normal melting point in °C. The notation "tp" indicates the temperature at which solid, liquid, and gas are in equilibrium at a pressure greater than one atmosphere (i.e., the normal melting point does not exist). When available, the triple point pressure is listed.
- **bp:** Normal boiling point in °C (referred to 101.325 kPa or 760 mmHg pressure). The notation "sp" following the number indicates the temperature where the pressure of the vapor in equilibrium with the solid reaches as 101.325 kPa. See Reference 8, p. 23, for further discussion of sublimation points and triple points. A notation "sublimes" without a temperature being given indicates that there is a perceptible sublimation pressure above the solid at ambient temperatures.
- **Density:** Density values for solids and liquids are always in units of grams per cubic centimeter and can be assumed to refer to temperatures near room temperature unless otherwise stated. Values for gases are the calculated ideal gas densities in grams per liter at 25°C and 101.325 kPa; the unit is always specified for a gas value.
- **Aqueous Solubility:** Solubility is expressed as the number of grams of the compound (excluding any water of hydration) that will dissolve in 100 g of water. The temperature in °C is given as a superscript. Solubility at other temperatures can be found for many compounds in the table "Aqueous Solubility of Inorganic Compounds at Various Temperatures" in Section 8.
- **Qualitative Solubility:** Qualitative information on the solubility in other solvents (and in water, if quantitative data are unavailable) is given here. The abbreviations are:

i	insoluble
sl	slightly soluble
s	soluble
vs	very soluble

Data were taken from a wide variety of reliable sources, including monographs, treatises, review articles, evaluated compilations and databases, and in some cases the primary literature. Some of the most useful references for the properties covered here are listed below.

List of Abbreviations

Ac	acetyl	brn	brown	dec	decomposes
ace	acetone	bz	benzene	dil	dilute
acid	acid solutions	chl	chloroform	diox	dioxane
alk	alkaline solutions	col	colorless	eth	ethyl ether
amorp	amorphous	conc	concentrated	EtOH	ethanol
anh	anhydrous	cry	crystals, crystalline	exp	explodes, explosive
aq	aqueous	cub	cubic	flam	flammable
blk	black	cyhex	cyclohexane	gl	glass, glassy

List of Abbreviations (continued)

grn	green	peth	petroleum ether	temp	temperature
hc	hydrocarbon solvents	pow	powder	tetr	tetragonal
hex	hexagonal	prec	precipitate	thf	tetrahydrofuran
hp	heptane	pur	purple	tol	toluene
hex	hexane	py	pyridine	tp	triple point
hyd	hydrate	reac	reacts with	trans	transition, transformation
hyg	hygroscopic	refrac	refractory	tricl	triclinic
i	insoluble in	rhom	rhombohedral	trig	trigonal
liq	liquid	s	soluble in	unstab	unstable
MeOH	methanol	silv	silvery	viol	violet
mono	monoclinic	sl	slightly soluble in	visc	viscous
octahed	octahedral	soln	solution	vs	very soluble in
oran	orange	sp	sublimation point	wh	white
orth	orthorhombic	stab	stable	xyl	xylene
os	organic solvents	subl	sublimes	yel	yellow

References

1. Phillips, S. L., and Perry, D.L., *Handbook of Inorganic Compounds*, CRC Press, Boca Raton, FL, 1995.
2. Trotman-Dickenson, A. F., Executive Editor, *Comprehensive Inorganic Chemistry*, Vol. 1-5, Pergamon Press, Oxford, 1973.
3. Greenwood, N. N., and Earnshaw, A., *Chemistry of the Elements, Second Edition*, Butterworth-Heinemann, Oxford, 1997.
4. Budavari, S., Editor, *The Merck Index, Twelfth Edition*, Merck & Co., Rahway, NJ, 1996.
5. *GMELIN Handbook of Inorganic and Organometallic Chemistry*, Springer-Verlag, Heidelberg.
6. Chase, M.W., Davies, C.A., Downey, J.R., Frurip, D. J., McDonald, R.A., and Syverud, A.N., *JANAF Thermochemical Tables, Third Edition, J. Phys. Chem. Ref. Data*, Vol. 14, Suppl. 1, 1985; Chase, M. W., *NIST-JANAF Thermochemical Tables, Fourth Edition, J. Phys. Chem. Ref. Data*, Monograph No. 9, 1998.
7. Donnay, J.D.H., and Ondik, H.M., *Crystal Data Determinative Tables, Third Edition*, Volumes 2 and 4, Inorganic Compounds, Joint Committee on Powder Diffraction Standards, Swarthmore, PA, 1973.
8. Lide, D. R., and Kehiaian, H.V., *CRC Handbook of Thermophysical and Thermochemical Data*, CRC Press, Boca Raton, FL, 1994.
9. *Kirk-Othmer Concise Encyclopedia of Chemical Technology*, Wiley-Interscience, New York, 1985.
10. *Dictionary of Inorganic Compounds*, Chapman & Hall, New York, 1992.
11. Massalski, T. B., Editor, *Binary Alloy Phase Diagrams, 2nd Edition*, ASM International, Metals Park, Ohio, 1990.
12. *Landolt-Börnstein, Numerical Data and Functional Relationships in Science and Technology, Sixth Edition*, II/4, Caloric Quantities of State, Springer-Verlag, Heidelberg, 1961.
13. Deer, W. A., Howie, R.A., and Zussman, J., *An Introduction to the Rock-Forming Minerals*, 2nd Edition, Longman Scientific & Technical, Harlow, Essex, 1992.
14. Carmichael, R. S., *Practical Handbook of Physical Properties of Rocks and Minerals*, CRC Press, Boca Raton, FL, 1989.
15. Dinsdale, A.T., "SGTE Data for Pure Elements", *CALPHAD*, 15, 317-425, 1991.
16. Madelung, O., *Semiconductors: Group IV Elements and III-IV Compounds*, Springer-Verlag, Heidelberg, 1991.
17. Daubert, T.E., Danner, R. P., Sibul, H.M., and Stebbins, C.C., *Physical and Thermodynamic Properties of Pure Compounds: Data Compilation*, extant 1994 (core with 4 supplements), Taylor & Francis, Bristol, PA.
18. Lidin, R. A., Andreeva, L. L., and Molochko, V. A., *Constants of Inorganic Substances*, Begell House, New York, 1995.
19. Gurvich, L. V., Veyts, I. V., and Alcock, C. B., *Thermodynamic Properties of Individual Substances, Fourth Edition*, Hemisphere Publishing Corp., New York, 1989.
20. *The Combined Chemical Dictionary on CDROM*, Chapman & Hall/CRC Press, Boca Raton, FL, 2000.

No.	Name	Formula	CAS Reg No.	Mol. Weight	Physical Form	mp/°C	bp/°C	Density g cm^{-3}	Solubility g/100 g H_2O	Qualitative Solubility
1	Actinium	Ac	7440-34-8	227	silv metal; cub	1051	3198	10		
2	Actinium bromide	$AcBr_3$	33689-81-5	467	wh hex cry		800 subl	5.85		s H_2O
3	Actinium chloride	$AcCl_3$	22986-54-5	333	wh hex cry		960 subl	4.81		
4	Actinium fluoride	AcF_3	33689-80-4	284	wh hex cry			7.88		i H_2O
5	Actinium iodide	AcI_3	33689-82-6	608	wh cry					s H_2O
6	Actinium oxide	Ac_2O_3	12002-61-8	502	wh hex cry	1977		9.19		i H_2O
7	Aluminum	Al	7429-90-5	26.982	silv-wh metal; cub cry	660.32	2519	2.70		i H_2O; s acid, alk
8	Aluminum ammonium sulfate	$AlNH_4(SO_4)_2$	7784-25-0	237.148	wh powder					sl H_2O; i EtOH
9	Aluminum ammonium sulfate dodecahydrate	$AlNH_4(SO_4)_2 \cdot 12H_2O$	7784-26-1	453.331	col cry or powder	94.5	>280 dec	1.65		s H_2O; i EtOH
10	Aluminum antimonide	$AlSb$	25152-52-7	148.742	cub cry	1065		4.26		
11	Aluminum arsenide	$AlAs$	22831-42-1	101.903	oran cub cry; hyg	1740		3.76		
12	Aluminum borate	$2Al_2O_3 \cdot B_2O_3$	11121-16-7	273.543	needles	≈1050				i H_2O
13	Aluminum boride	AlB_2	12041-50-8	48.604	powder	>920 dec		3.19		s dil HCl
14	Aluminum borohydride	$Al(BH_4)_3$	16962-07-5	71.510	flam liq	-64.5	44.5			reac H_2O
15	Aluminum bromate nonahydrate	$Al(BrO_3)_3 \cdot 9H_2O$	11126-81-1*	572.826	wh hyg cry	62	>100 dec			s H_2O
16	Aluminum bromide	$AlBr_3$	7727-15-3	266.694	wh-yel monocl cry; hyg	97.5	255	3.2		reac H_2O; s bz, tol
17	Aluminum bromide hexahydrate	$AlBr_3 \cdot 6H_2O$	7784-11-4	374.785	col-yel hyg cry	93		2.54		s H_2O, EtOH, CS_2
18	Aluminum carbide	Al_4C_3	1299-86-1	143.958	yel hex cry	2100	>2200 dec	2.36		reac H_2O
19	Aluminum chlorate nonahydrate	$Al(ClO_3)_3 \cdot 9H_2O$	15477-33-5	439.472	hyg cry					vs H_2O; s EtOH
20	Aluminum chloride	$AlCl_3$	7446-70-0	133.340	wh hex cry or powder; hyg	192.6	180 sp	2.48	45.1^{25}	s bz, ctc, chl
21	Aluminum chloride hexahydrate	$AlCl_3 \cdot 6H_2O$	7784-13-6	241.431	col hyg cry	100 dec		2.398	45.1^{25}	s EtOH, eth
22	Aluminum diacetate	$Al(OH)(C_2H_3O_2)_2$	142-03-0	162.078	wh amorp powder					i H_2O
23	Aluminum ethanolate	$Al(C_2H_5O)_3$	555-75-9	162.163	liq, condenses to wh solid	140				reac H_2O; sl xyl
24	Aluminum fluoride	AlF_3	7784-18-1	83.977	wh hex cry	≈2250 tp (220 MPa)	1276 sp	3.10	0.50^{25}	
25	Aluminum fluoride monohydrate	$AlF_3 \cdot H_2O$	32287-65-3	101.992	orth cry			2.17	0.50^{25}	
26	Aluminum fluoride trihydrate	$AlF_3 \cdot 3H_2O$	15098-87-0	138.023	wh hyg cry			1.914	0.50^{25}	
27	Aluminum hexafluorosilicate nonahydrate	$Al_2(SiF_6)_3 \cdot 9H_2O$	17099-70-6	642.329	hex prisms	>500 dec				s H_2O
28	Aluminum hydride	AlH_3	7784-21-6	30.006	col hex cry	>150 dec				reac H_2O
29	Aluminum hydroxide	$Al(OH)_3$	21645-51-2	78.004	wh amorp powder			2.42		i H_2O; s alk, acid
30	Aluminum hydroxychloride	$Al_2(OH)_5Cl \cdot 2H_2O$	1327-41-9	210.483	gl solid					s H_2O
31	Aluminum hypophosphite	$Al(H_2PO_2)_3$	7784-22-7	221.948	cry powder	220 dec				i H_2O; s alk, acid
32	Aluminum iodide	AlI_3	7784-23-8	407.695	wh leaflets	188.28	382	3.98		reac H_2O
33	Aluminum iodide hexahydrate	$AlI_3 \cdot 6H_2O$	10090-53-6	515.786	yel hyg cry powder					vs H_2O; s EtOH, eth
34	Aluminum lactate	$Al(C_3H_5O_3)_3$	18917-91-4	294.192	powder					vs H_2O
35	Aluminum nitrate	$Al(NO_3)_3$	13473-90-0	212.997	wh hyg solid	dec			68.9^{25}	vs EtOH; sl ace
36	Aluminum nitrate nonahydrate	$Al(NO_3)_3 \cdot 9H_2O$	7784-27-2	375.134	wh hyg mono cry	73	135 dec	1.72	68.9^{25}	vs EtOH; i pyr
37	Aluminum nitride	AlN	24304-00-5	40.989	blue-wh hex cry	3000		3.255		reac H_2O
38	Aluminum oleate	$Al(C_{18}H_{33}O_2)_3$	688-37-9	871.342	yel solid					i H_2O; s EtOH, bz
39	Aluminum phosphate	$AlPO_4$	7784-30-7	121.953	wh rhomb plates	>1460		2.56		i H_2O; sl acid
40	Aluminum metaphosphate	$Al(PO_3)_3$	32823-06-6	263.898	col powder; tetr	≈1525		2.78		i H_2O
41	Aluminum oxide (corundum)	Al_2O_3	1344-28-1	101.961	wh powder; hex	2053	≈3000	3.97		i H_2O, os; sl alk
42	Aluminum oxyhydroxide	$AlO(OH)$	14457-84-2	59.989	ortho cry			3.44		i H_2O; s acid, alk
43	Aluminum palmitate	$Al(C_{15}H_{31}COO)_3$	555-35-1	793.230	wh-yel powder					i H_2O, EtOH; s peth
44	Aluminum perchlorate nonahydrate	$Al(ClO_4)_3 \cdot 9H_2O$	14452-39-2	487.470	wh hyg cry	82 dec		2.0	182.4^{0}	
45	Aluminum phosphide	AlP	20859-73-8	57.956	grn or yel cub cry	2550		2.40		reac H_2O

No.	Name	Formula	CAS Reg No.	Mol. Weight	Physical Form	mp/°C	bp/°C	Density g cm^{-3}	Solubility g/100 g H_2O	Qualitative Solubility
46	Aluminum selenide	Al_2Se_3	1302-82-5	290.84	yel-brown powder	960		3.437		reac H_2O
47	Aluminum silicate	Al_2SiO_5	12183-80-1	162.046	gray-grn cry			3.145		
48	Aluminum silicate dihydrate	$Al_2O_3 \cdot 2SiO_2 \cdot 2H_2O$	1332-58-7	258.161	wh-yel powder; tricl			2.59		i H_2O, acid, alk
49	Aluminum stearate	$Al(C_{18}H_{35}O_2)_3$	637-12-7	877.390	wh powder	115		1.070		i H_2O, EtOH, eth; s alk
50	Aluminum sulfate	$Al_2(SO_4)_3$	10043-01-3	342.154	wh cry	1040 dec			38.5^{25}	i EtOH
51	Aluminum sulfate octadecahydrate	$Al_2(SO_4)_3 \cdot 18H_2O$	7784-31-8	666.429	col monocl cry	86 dec		1.69	38.5^{25}	
52	Aluminum sulfide	Al_2S_3	1302-81-4	150.161	yel-gray powder	1100		2.02		
53	Aluminum telluride	Al_2Te_3	12043-29-7	436.76	gray-blk hex cry	≈895		4.5		
54	Aluminum thiocyanate	$Al(SCN)_3$	538-17-0	201.232	yel powder					s H_2O; i EtOH, eth
55	Americium	Am	7440-35-9	243	silv metal; hex or cub	1176	2011	12		s acid
56	Americium(III) oxide	Am_2O_3	12254-64-7	534	tan hex cry			11.77		s acid
57	Americium(III) bromide	$AmBr_3$	14933-38-1	483	wh orth cry			6.85		s H_2O
58	Americium(III) chloride	$AmCl_3$	13464-46-5	349	pink hex cry	500		5.87		
59	Americium(III) fluoride	AmF_3	13708-80-0	300	pink hex cry	1393		9.53		
60	Americium(III) iodide	AmI_3	13813-47-3	624	yel ortho cry	≈950		6.9		
61	Americium(IV) fluoride	AmF_4	15947-41-8	319	tan monocl cry			7.23		
62	Americium(IV) oxide	AmO_2	12005-67-3	275	blk cub cry	>1000 dec		11.68		s acid
63	Ammonia	NH_3	7664-41-7	17.031	col gas	-77.73	-33.33	0.696 g/L		vs H_2O; s EtOH, eth
64	Ammonium acetate	$NH_4C_2H_3O_2$	631-61-8	77.083	wh hyg cry	114		1.073	148^{4}	s EtOH; sl ace
65	Ammonium azide	NH_4N_3	12164-94-2	60.059	ortho cry; flam	160	exp	1.346	20.2^{30}	
66	Ammonium benzoate	$NH_4C_7H_5O_2$	1863-63-4	139.152	wh cry or powder	198		1.26		s H_2O; sl EtOH
67	Ammonium hydrogen malate	$NH_4C_4H_4O_5$	5972-71-4	151.118	ortho cry	160		1.15		s H_2O; sl EtOH
68	Ammonium borate tetrahydrate	$(NH_4)_2B_4O_7 \cdot 4H_2O$	12228-87-4	263.377	tetr cry					s H_2O; i EtOH
69	Ammonium bromate	NH_4BrO_3	13843-59-9	145.941	col hex cry	exp				vs H_2O
70	Ammonium bromide	NH_4Br	12124-97-9	97.943	wh hyg tetr cry	542 dec	396 sp	2.429	78.3^{25}	s EtOH, ace; sl eth
71	Ammonium caprylate	$NH_4C_8H_{15}O_2$	5972-76-9	161.243	hyg monocl cry	≈75				reac H_2O; s EtOH; i chl, bz
72	Ammonium carbamate	NH_2COONH_4	1111-78-0	78.071	cry powder					vs H_2O; s EtOH
73	Ammonium carbonate	$(NH_4)_2CO_3$	506-87-6	96.086	col cry powder	58 dec			100^{15}	
74	Ammonium cerium(III) sulfate tetrahydrate	$NH_4Ce(SO_4)_2 \cdot 4H_2O$	21995-38-0*	422.343	monocl cry					s H_2O
75	Ammonium cerium(IV) nitrate	$(NH_4)_2Ce(NO_3)_6$	16774-21-3	548.223	red-oran cry					vs H_2O
76	Ammonium chlorate	NH_4ClO_3	10192-29-7	101.490	wh cry	102 exp		1.80	28.7^{0}	
77	Ammonium chloride	NH_4Cl	12125-02-9	53.492	col cub cry	520 tp (dec)	338 sp	1.519	39.5^{25}	
78	Ammonium chromate	$(NH_4)_2CrO_4$	7788-98-9	152.071	yel cry	185 dec		1.90	37^{25}	sl ace, MeOH; i EtOH
79	Ammonium chromic sulfate dodecahydrate	$NH_4Cr(SO_4)_2 \cdot 12H_2O$	10022-47-6	478.345	blue-viol cry	94 dec		1.72		s H_2O; sl EtOH
80	Ammonium cobalt(II) phosphate	$CoNH_4PO_4$	14590-13-7	171.943	red-viol powder (hyd)					i H_2O; s acid
81	Ammonium cobalt(II) sulfate hexahydrate	$(NH_4)_2Co(SO_4)_2 \cdot 6H_2O$	13586-38-4	395.229	red monocl prisms			1.90		s H_2O; i EtOH
82	Ammonium copper(II) chloride	$CuCl_2 \cdot 2NH_4Cl$	10060-13-6*	241.434	yel hyg orth cry					s H_2O
83	Ammonium copper(II) chloride dihydrate	$CuCl_2 \cdot 2NH_4Cl \cdot 2H_2O$	10060-13-6	277.464	blue-grn tetr cry	110 dec		1.993		s H_2O, EtOH
84	Ammonium cyanide	NH_4CN	12211-52-8	44.056	col tetr cry	dec		1.10		vs H_2O
85	Ammonium dichromate	$(NH_4)_2Cr_2O_7$	7789-09-5	252.065	oran-red monocl cry; hyg	180 dec		2.155	35.6^{20}	
86	Ammonium dihydrogen arsenate	$NH_4H_2AsO_4$	13462-93-6	158.975	tetr cry	300 dec		2.311	52.7^{25}	
87	Ammonium dihydrogen phosphate	$NH_4H_2PO_4$	7722-76-1	115.026	wh tetr cry	190		1.80	40.4^{25}	sl EtOH; i ace
88	Ammonium dithiocarbamate	NH_4NH_2CSS	513-74-6	110.204	yel ortho cry	99 dec		1.45		s H_2O
89	Ammonium ferric chromate	$NH_4Fe(CrO_4)_2$	7789-08-4	305.871	red powder					i H_2O
90	Ammonium ferric oxalate trihydrate	$(NH_4)_3Fe(C_2O_4)_3 \cdot 3H_2O$	13268-42-3	428.063	grn monocl cry; hyg	≈160 dec		1.780		vs H_2O; i EtOH

91	Ammonium ferric sulfate dodecahydrate	$NH_4Fe(SO_4)_2 \cdot 12H_2O$	10138-04-2	482.194	col to viol cry	≈37		1.71		vs H_2O; i EtOH
92	Ammonium ferricyanide trihydrate	$(NH_4)_3Fe(CN)_6 \cdot 3H_2O$	14221-48-8*	320.111	red cry					s H_2O; i EtOH
93	Ammonium ferrocyanide trihydrate	$(NH_4)_4Fe(CN)_6 \cdot 3H_2O$	14481-29-9*	338.149	yel cry	dec				s H_2O; i EtOH
94	Ammonium ferrous sulfate hexahydrate	$(NH_4)_2Fe(SO_4)_2 \cdot 6H_2O$	10045-89-3	392.141	blue-grn monocl cry	≈100 dec		1.86		s H_2O; i EtOH
95	Ammonium fluoride	NH_4F	12125-01-8	37.037	wh hex cry; hyg	dec		1.015	83.5^{25}	sl EtOH
96	Ammonium tetrafluoroborate	NH_4BF_4	13826-83-0	104.844	wh powder; orth	487 dec		1.871	25^{20}	
97	Ammonium fluorosulfonate	NH_4SO_3F	13446-08-7	117.101	col needles	245				s H_2O, EtOH, MeOH
98	Ammonium formate	NH_4CHO_2	540-69-2	63.057	hyg cry	116		1.27	143^{20}	s EtOH
99	Ammonium hexachloroiridate(IV)	$(NH_4)_2IrCl_6$	16940-92-4	441.010	blk cry powder	dec		2.856	1.09^{25}	
100	Ammonium hexachloroosmiate(IV)	$(NH_4)_2OsCl_6$	12125-08-5	439.02	red cry or powder		subl	2.93		s H_2O, EtOH
101	Ammonium hexachloropalladate(IV)	$(NH_4)_2PdCl_6$	19168-23-1	355.21	red-brn hyg cry	dec		2.418		
102	Ammonium hexabromoplatinate(IV)	$(NH_4)_2PtBr_6$	17363-02-9	710.58	powder	145 dec			0.59^{20}	
103	Ammonium hexachloroplatinate(IV)	$(NH_4)_2PtCl_6$	16919-58-7	443.87	red-oran cub cry	380 dec		3.065	0.5^{20}	i EtOH
104	Ammonium hexafluoroaluminate	$(NH_4)_3AlF_6$	7784-19-2	195.087	cub cry			1.78		s H_2O
105	Ammonium hexafluorogallate	$(NH_4)_3GaF_6$	14639-94-2	237.828	col cub cry	>200 dec		2.10		
106	Ammonium hexafluorogermanate	$(NH_4)_2GeF_6$	16962-47-3	222.68	wh cry	380	subl	2.564		s H_2O; i EtOH
107	Ammonium hexafluorophosphate	NH_4PF_6	16941-11-0	163.003	wh cub cry	58 dec		2.180		vs H_2O; s ace, EtOH, MeOH
108	Ammonium hexafluorosilicate	$(NH_4)_2SiF_6$	16919-19-0	178.153	wh cub or trig cry	dec		2.011	22.7^{25}	i EtOH, ace
109	Ammonium hexafluorozirconate(IV)	$(NH_4)_2ZrF_6$	16919-31-6	241.291	wh hex cry			1.154		s H_2O
110	Ammonium hydrogen arsenate	$(NH_4)_2HAsO_4$	7784-44-3	176.004	wh powder			1.99		s H_2O
111	Ammonium hydrogen borate trihydrate	$NH_4HB_4O_7 \cdot 3H_2O$	10135-84-9	228.332	col cry			≈2.5		s H_2O
112	Ammonium hydrogen carbonate	NH_4HCO_3	1066-33-7	79.056	col or wh prisms	107 dec		1.586	24.8^{25}	i EtOH, bz
113	Ammonium hydrogen citrate	$(NH_4)_2HC_6H_5O_7$	3012-65-5	226.184	col cry			1.48		vs H_2O; sl EtOH
114	Ammonium hydrogen fluoride	NH_4HF_2	1341-49-7	57.044	wh orth cry	125	240 dec	1.50	60.2^{20}	
115	Ammonium hydrogen oxalate monohydrate	$NH_4HC_2O_4 \cdot H_2O$	5972-72-5*	125.081	col rhomb cry	dec		1.56		sl H_2O, EtOH
116	Ammonium hydrogen phosphate	$(NH_4)_2HPO_4$	7783-28-0	132.055	wh cry	155 dec		1.619	69.5^{25}	i EtOH, ace
117	Ammonium hydrogen selenate	NH_4HSeO_4	10294-60-7	162.01	rhom cry	dec		2.162		
118	Ammonium hydrogen sulfate	NH_4HSO_4	7803-63-6	115.111	wh hyg cry	147		1.78	100^{20}	i EtOH, ace, py
119	Ammonium hydrogen sulfide	NH_4HS	12124-99-1	51.113	wh tetr or orth cry	dec		1.17	128^{0}	sl ace; i bz, eth
120	Ammonium hydrogen sulfite	NH_4HSO_3	10192-30-0	99.111	col cry	dec		2.03	71.8^{0}	
121	Ammonium hydrogen tartrate	$NH_4C_4H_4O_6$	3095-65-6	167.117	wh cry			1.68		sl H_2O; s alk; i EtOH
122	Ammonium hydroxide	NH_4OH	1336-21-6	35.046	exists only in soln					
123	Ammonium hypophosphite	$NH_4H_2PO_2$	7803-65-8	83.028	wh hyg cry	dec				vs H_2O; sl EtOH; i ace
124	Ammonium iodate	NH_4IO_3	13446-09-8	192.941	wh powder	150		3.3	3.84^{25}	
125	Ammonium iodide	NH_4I	12027-06-4	144.943	wh tetr cry; hyg	551 dec	405 sp	2.514	178^{25}	sl EtOH, MeOH
126	Ammonium lactate	$NH_4C_3H_5O_3$	52003-58-4	107.108	col cry	92				s H_2O, EtOH; sl MeOH; i ace, eth
127	Ammonium metatungstate hexahydrate	$(NH_4)_6W_7O_{24} \cdot 6H_2O$	12028-48-7	1887.19	wh cry					s H_2O; i EtOH
128	Ammonium metavanadate	NH_4VO_3	7803-55-6	116.979	wh-yel cry	200 dec		2.326	4.8^{20}	
129	Ammonium molybdate(VI) tetrahydrate	$(NH_4)_6Mo_7O_{24} \cdot 4H_2O$	12054-85-2	1235.86	col or grn-yel cry	90 dec		2.498	43	i EtOH
130	Ammonium molybdophosphate	$(NH_4)_3PO_4 \cdot 12MoO_3$	54723-94-3*	1876.35	yel cry pow	dec			0.02	s alk; i acid
131	Ammonium nickel chloride hexahydrate	$NH_4NiCl_3 \cdot 6H_2O$	16122-03-5*	291.181	grn hyg cry			1.65		s H_2O
132	Ammonium nickel sulfate hexahydrate	$(NH_4)_2Ni(SO_4)_2 \cdot 6H_2O$	7785-20-8	394.989	blue-grn cry	dec		1.923		sl H_2O; i EtOH
133	Ammonium nitrate	NH_4NO_3	6484-52-2	80.043	wh hyg cry; orth	210 dec		1.72	213^{25}	sl MeOH
134	Ammonium nitrite	NH_4NO_2	13446-48-5	64.044	wh-yel cry	60 exp		1.69	221^{25}	i eth
135	Ammonium nitroferricyanide	$(NH_4)_2Fe(CN)_5NO$	14402-70-1	252.016	red-brn cry					s H_2O, EtOH
136	Ammonium oleate	$NH_4C_{18}H_{33}O_2$	544-60-5	299.493	yel-brn paste	21				s H_2O; sl ace
137	Ammonium oxalate	$(NH_4)_2C_2O_4$	1113-38-8	124.096	col sol			1.5	5.20^{25}	
138	Ammonium oxalate monohydrate	$(NH_4)_2C_2O_4 \cdot H_2O$	6009-70-7	142.110	wh orth cry	dec		1.50	5.20^{25}	sl EtOH

No.	Name	Formula	CAS Reg No.	Mol. Weight	Physical Form	mp/°C	bp/°C	Density g cm^{-3}	Solubility g/100 g H_2O	Qualitative Solubility
139	Ammonium palmitate	$NH_4C_{15}H_{31}CO_2$	593-26-0	273.455	yel-wh powder	22				s H_2O; sl bz, xyl; i ace, EtOH, ctc
140	Ammonium pentaborate tetrahydrate	$NH_4B_5O_8 \cdot 4H_2O$	12007-89-5	272.150	wh cry				7.03^{18}	
141	Ammonium pentachlorozincate	$(NH_4)_3ZnCl_5$	14639-98-6	296.77	hyg orth cry			1.81		vs H_2O
142	Ammonium perchlorate	NH_4ClO_4	7790-98-9	117.490	wh orth cry	dec, exp		1.95	24.5^{25}	s MeOH; sl EtOH, ace; i eth
143	Ammonium permanganate	NH_4MnO_4	13446-10-1	136.975	purp rhomb cry	70 dec		2.22	7.9^{15}	
144	Ammonium peroxydisulfate	$(NH_4)_2S_2O_8$	7727-54-0	228.204	monocl cry or wh powder	dec		1.982	83.5^{25}	
145	Ammonium perrhenate	NH_4ReO_4	13598-65-7	268.244	col powder			3.97	6.23^{20}	
146	Ammonium phosphate trihydrate	$(NH_4)_3PO_4 \cdot 3H_2O$	10361-65-6*	203.133	wh prisms				25.0^{25}	i ace
147	Ammonium phosphite, dibasic, monohydrate	$(NH_4)_2HPO_3 \cdot H_2O$	51503-61-8	134.071	hyg cry					s H_2O
148	Ammonium phosphomolybdate monohydrate	$(NH_4)_3PO_4 \cdot 12MoO_3 \cdot H_2O$	54723-94-3	1894.36	yel cry or powder	dec			0.02	
149	Ammonium phosphotungstate dihydrate	$(NH_4)_3PO_4 \cdot 12WO_3 \cdot 2H_2O$	1311-90-6	2967.18	cry powder					sl H_2O
150	Ammonium picrate	$NH_4C_6H_2N_3O_7$	131-74-8	246.135	yel orth cry	exp		1.72		sl H_2O
151	Ammonium salicylate	$NH_4C_7H_5O_3$	528-94-9	155.151	wh cry powder					vs H_2O; s EtOH
152	Ammonium selenate	$(NH_4)_2SeO_4$	7783-21-3	179.04	wh monocl cry	dec		2.194	117^{25}	i EtOH, ace
153	Ammonium selenite	$(NH_4)_2SeO_3$	7783-19-9	163.04	wh or red hyg cry	dec			121^{25}	
154	Ammonium stearate	$NH_4C_{18}H_{35}O_2$	1002-89-7	301.509	yel-wh powder	22		0.89		sl H_2O, bz; s EtOH, MeOH; i ace
155	Ammonium sulfamate	$NH_4NH_2SO_3$	7773-06-0	114.125	wh hyg cry	131	160 dec			vs H_2O; sl EtOH
156	Ammonium sulfate	$(NH_4)_2SO_4$	7783-20-2	132.141	wh or brn orth cry	280 dec		1.77	76.4^{25}	i EtOH, ace
157	Ammonium sulfide	$(NH_4)_2S$	12135-76-1	68.143	yel-oran cry	≈0 dec				s H_2O, EtOH, alk
158	Ammonium sulfite	$(NH_4)_2SO_3$	17026-44-7	116.141	wh hyg cry				64.2^{25}	
159	Ammonium sulfite monohydrate	$(NH_4)_2SO_3 \cdot H_2O$	7783-11-1	134.156	col cry	dec		1.41	64.2^{25}	i EtOH, ace
160	Ammonium tartrate	$(NH_4)_2C_4H_4O_6$	3164-29-2	184.147	wh cry	dec		1.601		s H_2O
161	Ammonium tellurate	$(NH_4)_2TeO_4$	13453-06-0	227.68	wh powder	dec		3.024		
162	Ammonium tetrachloroaluminate	NH_4AlCl_4	7784-14-7	186.832	wh hyg solid	304				s H_2O, eth
163	Ammonium tetrachloroplatinate(II)	$(NH_4)_2PtCl_4$	13820-41-2	372.97	red cry	dec		2.936		s H_2O; i EtOH
164	Ammonium tetrachlorozincate	$(NH_4)_2ZnCl_4$	14639-97-5	243.28	wh orth plates; hyg	150 dec		1.879		vs H_2O
165	Ammonium tetrathiotungstate	$(NH_4)_2WS_4$	13862-78-7	348.18	oran cry	dec		2.71		s H_2O
166	Ammonium thiocyanate	NH_4SCN	1762-95-4	76.122	col hyg cry	≈149	dec	1.30	181^{25}	vs EtOH; s ace; i chl
167	Ammonium thiosulfate	$(NH_4)_2S_2O_3$	7783-18-8	148.207	wh cry	150 dec		1.678		vs H_2O; i EtOH, eth
168	Ammonium titanium oxalate monohydrate	$(NH_4)_2TiO(C_2O_4)_2 \cdot H_2O$	10580-03-7	293.996	hyg cry					vs H_2O
169	Ammonium tungstate(VI)	$(NH_4)_{10}W_{12}O_{41}$	11120-25-5	3042.44	cry pow			2.3		s H_2O; i EtOH
170	Ammonium uranate(VI)	$(NH_4)_2U_2O_7$	7783-22-4	624.131	red-yel amorp powder					i H_2O, alk; s acid
171	Ammonium uranium fluoride	$UO_2(NH_4)_3F_5$	18433-40-4	419.135	grn-yel monocl cry					s H_2O; i EtOH
172	Ammonium valerate	$NH_4C_4H_9CO_2$	42739-38-8	119.163	hyg cry	108				vs H_2O, EtOH; s eth
173	Ammonium zirconyl carbonate dihydrate	$(NH_4)_3ZrOH(CO_3)_3 \cdot 2H_2O$	12616-24-9*	362.404	prisms; unstable					s H_2O
174	Antimony	Sb	7440-36-0	121.760	silv metal; hex	630.63	1587	6.68		i dil acid
175	Stibine	SbH_3	7803-52-3	124.784	col gas; flam	-88	-17	5.100 g/L		sl H_2O; s EtOH
176	Antimony arsenide	SbAs	12322-34-8	196.682	hex cry	≈680		6.0		
177	Antimony(III) bromide	$SbBr_3$	7789-61-9	361.472	yel orth cry; hyg	96.6	280	4.35		reac H_2O; s ace, bz, chl
178	Antimony(III) chloride	$SbCl_3$	10025-91-9	228.118	col orth cry; hyg	73.4	220.3	3.14	987^{25}	s acid, EtOH, bz, ace
179	Antimony(III) fluoride	SbF_3	7783-56-4	178.755	wh orth cry; hyg	292	≈345	4.38	492^{25}	
180	Antimony(III) iodide	SbI_3	7790-44-5	502.473	red rhomb cry	168	401	4.92		reac H_2O; s EtOH, ace; i ctc
181	Antimony(III) oxide (senarmontite)	Sb_2O_3	1309-64-4	291.518	col cub cry	570 trans	1425	5.58		sl H_2O; i os
182	Antimony(III) oxide (valentinite)	Sb_2O_3	1309-64-4	291.518	wh orth cry	655	1425	5.7		sl H_2O; i os

183	Antimony(III) oxychloride	SbOCl	7791-08-4	173.212	wh momo cry	170 dec				reac H_2O; i EtOH, eth
184	Antimony(III) selenide	Sb_2Se_3	1315-05-5	480.40	grn orth cry	611		5.81		sl H_2O
185	Antimony(III) sulfate	$Sb_2(SO_4)_3$	7446-32-4	531.711	wh cry powder; hyg	dec		3.62		sl H_2O
186	Antimony(III) sulfide	Sb_2S_3	1345-04-6	339.718	gray-blk orth cry	550		4.562		i H_2O; s conc HCl
187	Antimony(III) teluride	Sb_2Te_3	1327-50-0	626.32	gray cry	620		6.5		
188	Antimony(III,V) oxide	Sb_2O_4	1332-81-6	307.518	yel orth cry			6.64		
189	Antimony(V) chloride	$SbCl_5$	7647-18-9	299.024	col or yel liq	4	140 dec	2.34		reac H_2O; s chl, ctc
190	Antimony(V) fluoride	SbF_5	7783-70-2	216.752	hyg visc liq	8.3	141	3.10		reac H_2O
191	Antimony(V) dichlorotrifluoride	$SbCl_2F_3$	7791-16-4	249.660	visc liq					reac H_2O
192	Antimony(V) oxide	Sb_2O_5	1314-60-9	323.517	yel powder; cub	dec		3.78	0.3^{20}	
193	Antimony(V) sulfide	Sb_2S_5	1315-04-4	403.850	oran-yel powder	75 dec		4.120		i H_2O; s acid, alk
194	Argon	Ar	7440-37-1	39.948	col gas	-189.36 tp (69 kPa)	-185.85	1.633 g/L		sl H_2O
195	Arsenic (gray)	As	7440-38-2	74.922	gray metal; rhomb	817 tp (3.70 MPa)	603 sp	5.75		i H_2O
196	Arsine	AsH_3	7784-42-1	77.946	col gas	-116	-62.5	3.186 g/L		sl H_2O
197	Diarsine	As_2H_4	15942-63-9	153.875	unstable liq		≈100			
198	Arsenic acid	H_3AsO_4	7778-39-4	141.944	exists only in soln					
199	Arsenic acid hemihydrate	$H_3AsO_4 \cdot 0.5H_2O$	7778-39-4*	150.951	wh hyg cry	35.5		≈2		vs H_2O, EtOH
200	Arsenious acid	H_3AsO_3	13464-58-9	125.944	exists only in soln					
201	Arsenic diiodide	As_2I_4	13770-56-4	657.461	red cry	137				reac H_2O; s os
202	Arsenic hemiselenide	As_2Se	1303-35-1	228.80	blk cry					i H_2O, os; dec acid, alk
203	Arsenic sulfide	As_4S_4	12279-90-2	427.950	red monocl cry	320	565	3.5		i H_2O; sl bz; s alk
204	Arsenic(III) bromide	$AsBr_3$	7784-33-0	314.634	yel orth cry; hyg	31.1	221	3.40		reac H_2O; s hc, ctc; vs eth, bz
205	Arsenic(III) chloride	$AsCl_3$	7784-34-1	181.280	col liq	-16	130	2.150		reac H_2O; vs chl, ctc, eth
206	Arsenic(III) fluoride	AsF_3	7784-35-2	131.917	col liq	-5.9	57.8	2.7		reac H_2O; s EtOH, eth, bz
207	Arsenic(III) iodide	AsI_3	7784-45-4	455.635	red hex cry	140.9	424	4.73		sl H_2O, EtOH, eth; s bz. tol
208	Arsenic(III) oxide (arsenolite)	As_2O_3	1327-53-3	197.841	wh cub cry	274	460	3.86	2.05^{25}	
209	Arsenic(III) oxide (claudetite)	As_2O_3	1327-53-3	197.841	wh monocl cry	313	460	3.74	2.05^{25}	s dil acid, alk; i EtOH
210	Arsenic(III) selenide	As_2Se_3	1303-36-2	386.72	brn-blk solid	260		4.75		i H_2O; s alk
211	Arsenic(III) sulfide	As_2S_3	1303-33-9	246.041	yel-oran monocl cry	310	707	3.46		i H_2O; s alk
212	Arsenic(III) telluride	As_2Te_3	12044-54-1	532.64	blk monocl cry	621		6.50		
213	Arsenic(V) chloride	$AsCl_5$	22441-45-8	252.186	stable at low temp	≈-50 dec				
214	Arsenic(V) fluoride	AsF_5	7784-36-3	169.914	col gas	-79.8	-52.8	6.945 g/L		reac H_2O; s EtOH, bz, eth
215	Arsenic(V) oxide	As_2O_5	1303-28-2	229.840	wh amorp powder	315		4.32	65.8^{20}	vs EtOH
216	Arsenic(V) selenide	As_2Se_5	1303-37-3	544.64	blk solid	dec				i H_2O, EtOH, eth; s alk
217	Arsenic(V) sulfide	As_2S_5	1303-34-0	310.173	brn-yel amorp solid	dec				i H_2O; s alk
218	Triethyl arsenite	$As(OC_2H_5)_3$	3141-12-6	210.103	liq		166	1.21		
219	Astatine	At	7440-68-8	210	cry	302				s HNO_3, os
220	Barium	Ba	7440-39-3	137.327	silv-yel metal; cub	727	1897	3.62		reac H_2O; sl EtOH
221	Barium acetate	$Ba(C_2H_3O_2)_2$	543-80-6	255.416	wh powder			2.47	79.2^{25}	
222	Barium acetate monohydrate	$Ba(C_2H_3O_2)_2 \cdot H_2O$	5908-64-5	273.431	wh cry	110 dec		2.19	79.2^{25}	sl EtOH
223	Barium aluminate	$BaAl_2O_4$	12004-04-5	255.288	hex cry	1827				
224	Barium azide	$Ba(N_3)_2$	18810-58-7	221.367	monocl cry; exp	≈120 dec		2.936	17.3^{20}	sl EtOH; i eth
225	Barium bromate monohydrate	$Ba(BrO_3)_2 \cdot H_2O$	10326-26-8	411.147	wh monocl cry	260 dec		3.99	0.831^{25}	i EtOH
226	Barium bromide	$BaBr_2$	10553-31-8	297.135	wh orth cry	857	1835	4.781	100^{25}	
227	Barium bromide dihydrate	$BaBr_2 \cdot 2H_2O$	7791-28-8	333.166	wh cry	75 dec		3.7	100^{25}	s MeOH; i EtOH, ace, diox
228	Barium carbide	BaC_2	50813-65-5	161.348	gray tetr cry	dec		3.74		reac H_2O
229	Barium carbonate	$BaCO_3$	513-77-9	197.336	wh orth cry	1555		4.2865	0.0014^{20}	s acid

No.	Name	Formula	CAS Reg No.	Mol. Weight	Physical Form	mp/°C	bp/°C	Density g cm^{-3}	Solubility g/100 g H_2O	Qualitative Solubility
230	Barium chlorate	$Ba(ClO_3)_2$	13477-00-4	304.228	wh cry	414			37.9^{25}	sl EtOH, ace
231	Barium chlorate monohydrate	$Ba(ClO_3)_2 \cdot H_2O$	10294-38-9	322.244	wh monocl cry	120 dec		3.179	37.9^{25}	s acid; sl EtOH, ace
232	Barium chloride	$BaCl_2$	10361-37-2	208.232	wh orth cry; hyg	962	1560	3.9	37.0^{25}	
233	Barium chloride dihydrate	$BaCl_2 \cdot 2H_2O$	10326-27-9	244.263	wh monocl cry	≈120 dec		3.097	37.0^{25}	i EtOH
234	Barium chromate(V)	$Ba_3(CrO_4)_2$	12345-14-1	643.968	grn-blk hex cry			5.25		s H_2O
235	Barium chromate(VI)	$BaCrO_4$	10294-40-3	253.321	yel orth cry	1380		4.50	0.00026^{20}	reac acid
236	Barium citrate monohydrate	$Ba_3(C_6H_5O_7)_2 \cdot H_2O$	512-25-4*	808.195	gray-wh cry					s H_2O, acid
237	Barium copper yttrium oxide	$BaCuY_2O_5$	82642-06-6	458.682	grn cry; not superconductor					
238	Barium copper yttrium oxide	$Ba_2Cu_3YO_7$	109064-29-1	666.194	blk solid; HT superconductor					
239	Barium copper yttrium oxide	$Ba_2Cu_4YO_8$	114104-80-2	745.739	HT superconductor					
240	Barium copper yttrium oxide	$Ba_4Cu_7Y_2O_{15}$	124365-83-9	1411.933	HT superconductor					
241	Barium cyanide	$Ba(CN)_2$	542-62-1	189.361	wh cry powder					vs H_2O; s EtOH
242	Barium dichromate dihydrate	$BaCr_2O_7 \cdot 2H_2O$	10031-16-0	389.346	brn-red needles	dec				reac H_2O
243	Barium dithionate dihydrate	$BaS_2O_6 \cdot 2H_2O$	13845-17-5	333.486	wh cry	140 dec		4.54	22.1^{20}	sl EtOH
244	Barium ferrocyanide hexahydrate	$Ba_2Fe(CN)_6 \cdot 6H_2O$	13821-06-2*	594.694	yel monocl cry	80 dec				i H_2O, EtOH
245	Barium fluoride	BaF_2	7787-32-8	175.324	wh cub cry	1368	2260	4.893	0.161^{25}	
246	Barium formate	$Ba(CHO_2)_2$	541-43-5	227.362	cry			3.21		s H_2O; i EtOH
247	Barium hexaboride	BaB_6	12046-08-1	202.193	blk cub cry	2070		4.36		i H_2O; s acid; i EtOH
248	Barium hexafluorosilicate	$BaSiF_6$	17125-80-3	279.403	wh orth needles	300 dec		4.29		i H_2O, EtOH; sl acid
249	Barium hydride	BaH_2	13477-09-3	139.343	gray orth cry	1200		4.16		reac H_2O
250	Barium hydrogen phosphate	$BaHPO_4$	10048-98-3	233.306	wh cry powder	400 dec		4.16	0.015^{20}	s dil acid
251	Barium hydrosulfide	$Ba(HS)_2$	25417-81-6	203.475	yel hyg cry					s H_2O
252	Barium hydrosulfide tetrahydrate	$Ba(HS)_2 \cdot 4H_2O$	12230-74-9	275.536	yel rhomb cry	50 dec				s H_2O
253	Barium hydroxide	$Ba(OH)_2$	17194-00-2	171.342	wh powder	408			4.91^{25}	
254	Barium hydroxide monohydrate	$Ba(OH)_2 \cdot H_2O$	22326-55-2	189.357	wh powder			3.743	4.91^{25}	s acid
255	Barium hydroxide octahydrate	$Ba(OH)_2 \cdot 8H_2O$	12230-71-6	315.464	wh monocl cry	78 dec		2.18	4.91^{25}	
256	Barium hypophosphite monohydrate	$Ba(H_2PO_2)_2 \cdot H_2O$	14871-79-5*	285.320	monocl plates			2.90		s H_2O; i EtOH
257	Barium iodate	$Ba(IO_3)_2$	10567-69-8	487.132	wh cry powder	476 dec		5.23	0.0396^{25}	
258	Barium iodate monohydrate	$Ba(IO_3)_2 \cdot H_2O$	7787-34-0	505.148	cry	130 dec		5.00	0.0396^{25}	s acid; i EtOH
259	Barium iodide	BaI_2	13718-50-8	391.136	wh orth cry	711		5.15	221^{25}	
260	Barium iodide dihydrate	$BaI_2 \cdot 2H_2O$	7787-33-9	427.167	col cry	740 dec		5.0	221^{25}	s EtOH, ace
261	Barium manganate(VI)	$BaMnO_4$	7787-35-1	256.263	grn-gray hyg cry			4.85	0.00041^{20}	
262	Barium metaborate monohydrate	$Ba(BO_2)_2 \cdot H_2O$	26124-86-7	240.962	wh powder	>900		3.3		sl H_2O
263	Barium molybdate	$BaMoO_4$	7787-37-3	297.27	wh powder	1450		4.975	0.0021^{20}	
264	Barium niobate	$Ba(NbO_3)_2$	12009-14-2	419.136	yel orth cry	1455		5.44		i H_2O
265	Barium nitrate	$Ba(NO_3)_2$	10022-31-8	261.336	wh cub cry	590		3.24	10.3^{25}	sl EtOH, ace
266	Barium nitride	Ba_3N_2	12047-79-9	439.994	yel-brn cry	>500 dec		4.78		reac H_2O
267	Barium nitrite	$Ba(NO_2)_2$	13465-94-6	229.338	col hex cry	267		3.234	79.5^{25}	
268	Barium nitrite monohydrate	$Ba(NO_2)_2 \cdot H_2O$	7787-38-4	247.353	yel-wh hex cry	217 dec		3.18	79.5^{25}	i EtOH
269	Barium oxalate	BaC_2O_4	516-02-9	225.346	wh powder	400 dec		2.658	0.0075	
270	Barium oxalate monohydrate	$BaC_2O_4 \cdot H_2O$	13463-22-4	243.361	wh cry powder			2.66	0.0075^{20}	s acid
271	Barium oxide	BaO	1304-28-5	153.326	wh-yel powder; cub and hex	1972		5.72(cub)	1.5^{20}	s dil acid, EtOH; i ace
272	Barium perchlorate	$Ba(ClO_4)_2$	13465-95-7	336.227	col hex cry	505		3.20	312^{25}	vs EtOH
273	Barium perchlorate trihydrate	$Ba(ClO_4)_2 \cdot 3H_2O$	10294-39-0	390.273	col cry			2.74	312^{25}	s MeOH; sl EtOH, ace; i eth
274	Barium permanganate	$Ba(MnO_4)_2$	7787-36-2	375.198	brn-viol cry	200 dec		3.77	62.5^{20}	reac EtOH

275	Barium peroxide	BaO_2	1304-29-6	169.326	gray-wh tetr cry	450 dec		4.96	0.091^{20}	reac dil acid
276	Barium metaphosphate	$Ba(PO_3)_2$	13466-20-1	295.271	wh powder	1560				i H_2O; sl acid
277	Barium potassium chromate	$BaK_2(CrO_4)_2$	27133-66-0	447.511	yel hex cry			3.63		vs H_2O
278	Barium pyrophosphate	$Ba_2P_2O_7$	13466-21-2	448.597	wh powder	1430		3.9	0.0088^{20}	s acid
279	Barium selenate	$BaSeO_4$	7787-41-9	280.29	wh rhomb cry	dec		4.75	0.015^{20}	
280	Barium selenide	$BaSe$	1304-39-8	216.29	cub cry powder	1780		5.02		reac H_2O
281	Barium selenite	$BaSeO_3$	13718-59-7	264.29	solid					i H_2O
282	Barium disilicate	$BaSi_2O_5$	12650-28-1	273.495	wh orth cry	1420		3.70		
283	Barium metasilicate	$BaSiO_3$	13255-26-0	213.411	col rhomb powder	1605		4.40		i H_2O; s acid
284	Barium silicide	$BaSi_2$	1304-40-1	193.498	gray lumps	1180				reac H_2O
285	Barium sodium niobate	$Ba_2Na(NbO_3)_5$	12323-03-4	1002.167	wh orth cry	1437		5.40		i H_2O
286	Barium stannate	$BaSnO_3$	12009-18-6	304.035	cub cry			7.24		sl H_2O
287	Barium stannate trihydrate	$BaSnO_3 \cdot 3H_2O$	12009-18-6*	358.081	wh cry powder					sl H_2O; s acid
288	Barium stearate	$Ba(C_{18}H_{35}O_2)_2$	6865-35-6	704.266	wh powder	160		1.145		i H_2O, EtOH
289	Barium sulfate	$BaSO_4$	7727-43-7	233.391	wh orth cry	1580		4.49	0.00031^{20}	i EtOH
290	Barium sulfide	BaS	21109-95-5	169.393	col cub cry or gray powder	2229		4.3	8.94^{25}	
291	Barium sulfite	$BaSO_3$	7787-39-5	217.391	wh monocl cry	dec		4.44	0.0011^{25}	i EtOH
292	Barium tartrate	$BaC_4H_4O_6$	5908-81-6	285.398	wh cry			2.98		s H_2O; i EtOH
293	Barium tetracyanoplatinate(II) tetrahydrate	$BaPt(CN)_4 \cdot 4H_2O$	13755-32-3	508.54	yel powder or cry			2.076		sl H_2O; i EtOH
294	Barium tetraiodomercurate(II)	$BaHgI_4$	10048-99-4	845.54	yel-red hyg cry					vs H_2O, EtOH
295	Barium thiocyanate	$Ba(SCN)_2$	2092-17-3	253.493	hyg cry				167^{25}	s ace, MeOH, EtOH
296	Barium thiocyanate dihydrate	$Ba(SCN)_2 \cdot 2H_2O$	2092-17-3*	289.524	hyg wh cry				167^{25}	s EtOH
297	Barium thiocyanate trihydrate	$Ba(SCN)_2 \cdot 3H_2O$	68016-36-4	307.539	wh needles; hyg			2.286	167^{25}	s EtOH
298	Barium thiosulfate	BaS_2O_3	35112-53-9	249.457	wh cry powder	220 dec			0.2^{20}	i EtOH
299	Barium thiosulfate monohydrate	$BaS_2O_3 \cdot H_2O$	7787-40-8	267.473	wh cry powder	dec		3.5	0.2	i EtOH
300	Barium titanate	$BaTiO_3$	12047-27-7	233.192	wh tetr cry	1625		6.02		i H_2O
301	Barium tungstate	$BaWO_4$	7787-42-0	385.17	wh tetr cry	1475	1730	5.04	0.0016^{20}	
302	Barium uranium oxide	BaU_2O_7	10380-31-1	725.381	oran-yel powder					i H_2O; s acid
303	Barium orthovanadate	$Ba_3(VO_4)_2$	39416-30-3	641.859	hex cry	707		5.14		
304	Barium zirconate	$BaZrO_3$	12009-21-1	276.549	gray-wh cub cry	2500		5.52		i H_2O, alk; sl acid
305	Berkelium (α form)	Bk	7440-40-6	247	hex	1050		14.78		
306	Berkelium (β form)	Bk	7440-40-6	247	cub cry	986		13.25		
307	Beryllium	Be	7440-41-7	9.012	hex	1287	2471	1.85		s acid, alk
308	Beryllium acetate	$Be(C_2H_3O_2)_2$	543-81-7	127.101	wh cry	60 dec				i H_2O, EtOH
309	Beryllium 2,4-pentanedioate	$Be(CH_3COCHCOCH_3)_2$	10210-64-7	207.228	monocl cry powder	108	270	1.168		i H_2O; vs EtOH, eth
310	Beryllium aluminate	$BeAl_2O_4$	12004-06-7	126.973	orth cry			3.65		
311	Beryllium aluminum metasilicate	$Be_3Al_2(SiO_3)_6$	1302-52-9	537.502	col or grn-yel cry; hex			2.64		
312	Beryllium basic acetate	$Be_4O(C_2H_3O_2)_6$	1332-52-1	406.312	wh cry	285	330	1.25		i H_2O; s eth, os
313	Beryllium boride	BeB_2	12228-40-9	30.634	refrac solid	>1970				
314	Beryllium borohydride	$Be(BH_4)_2$	17440-85-6	36.682	solid	125 dec	subl			reac H_2O
315	Beryllium bromide	$BeBr_2$	7787-46-4	168.820	orth cry; hyg	508	520	3.465		vs H_2O; s EtOH, pyr
316	Beryllium carbide	Be_2C	506-66-1	30.035	red cub cry	>2100 dec		1.90		reac H_2O
317	Beryllium carbonate tetrahydrate	$BeCO_3 \cdot 4H_2O$	60883-64-9	93.085	wh solid	100 dec			0.36^{0}	
318	Beryllium carbonate, basic	$Be_3(OH)_2(CO_3)_2$	66104-24-3	181.069	wh pow					i H_2O; s acid, alk
319	Beryllium chloride	$BeCl_2$	7787-47-5	79.917	wh-yel orth cry; hyg	415	482	1.90	71.5^{25}	s EtOH, eth, py; i bz, tol
320	Beryllium fluoride	BeF_2	7787-49-7	47.009	tetr cry or gl; hyg	552	1169	2.1		vs H_2O; sl EtOH
321	Beryllium formate	$Be(CHO_2)_2$	1111-71-3	99.047	powder	>250 dec				reac H_2O; i os
322	Beryllium hydride	BeH_2	7787-52-2	11.028	wh amorp solid	250 dec		0.65		reac H_2O; i eth, tol

No.	Name	Formula	CAS Reg No.	Mol. Weight	Physical Form	mp/°C	bp/°C	Density g cm^{-3}	Solubility g/100 g H_2O	Qualitative Solubility
323	Beryllium hydrogen phosphate	$BeHPO_4$	13598-15-7	104.991	cry					i H_2O
324	Beryllium hydroxide	$Be(OH)_2$	13327-32-7	43.027	wh powder or cry	≈200 dec		1.92		sl H_2O, alk; s acid
325	Beryllium iodide	BeI_2	7787-53-3	262.821	hyg needles	470	487	4.32		reac H_2O; s EtOH
326	Beryllium nitrate trihydrate	$Be(NO_3)_2 \cdot 3H_2O$	13597-99-4	187.068	yel-wh hyg cry	≈30	dec		107^{20}	s EtOH
327	Beryllium nitride	Be_3N_2	1304-54-7	55.050	gray refrac cry; cub	2200		2.71		reac acid, alk
328	Beryllium oxide	BeO	1304-56-9	25.011	wh hex cry	2577		3.01		i H_2O; sl acid, alk
329	Beryllium perchlorate tetrahydrate	$Be(ClO_4)_2 \cdot 4H_2O$	7787-48-6	279.974	hyg cry	250 dec			198^{25}	
330	Beryllium selenate tetrahydrate	$BeSeO_4 \cdot 4H_2O$	10039-31-3	224.03	orth cry	100 dec		2.03		vs H_2O
331	Beryllium sulfate	$BeSO_4$	13510-49-1	105.076	col tetr cry; hyg	1127		2.5	41.3^{25}	
332	Beryllium sulfate tetrahydrate	$BeSO_4 \cdot 4H_2O$	7787-56-6	177.137	col tetr cry	≈100 dec		1.71	41.3^{25}	i EtOH
333	Beryllium sulfide	BeS	13598-22-6	41.078	col cub cry	dec		2.36		reac hot H_2O
334	Bismuth	Bi	7440-69-9	208.980	gray-wh soft metal	271.40	1564	9.79		s acid
335	Bismuth arsenate	$BiAsO_4$	13702-38-0	347.900	wh mono cry			7.14		i H_2O; sl conc HNO_3
336	Bismuth basic carbonate	$(BiO)_2CO_3$	5892-10-4	509.969	wh powder			6.86		i H_2O; s acid
337	Bismuth tribromide	$BiBr_3$	7787-58-8	448.692	yel cub cry	218	453	5.72		reac H_2O; s dil acid, ace; i EtOH
338	Bismuth trichloride	$BiCl_3$	7787-60-2	315.338	yel-wh cub cry; hyg	230	447	4.75		reac H_2O; s acid, EtOH, ace
339	Bismuth citrate	$BiC_6H_5O_7$	813-93-4	398.080	wh powder			3.458		i H_2O; sl EtOH
340	Bismuth trifluoride	BiF_3	7787-61-3	265.975	wh-gray cub cry	725	900	8.3		i H_2O
341	Bismuth pentafluoride	BiF_5	7787-62-4	303.972	wh tetr needles; hyg	154	230	5.55		reac H_2O
342	Bismuth hydride	BiH_3	18288-22-7	212.004	col gas; unstable	-67	≈17	8.665 g/L		
343	Bismuth hydroxide	$Bi(OH)_3$	10361-43-0	260.002	wh-yel amorp powder			4.962		i H_2O; s acid
344	Bismuth triiodide	BiI_3	7787-64-6	589.693	blk hex cry	408.6	542	5.778	0.00078^{20}	s EtOH
345	Bismuth hexafluoro-2,4-pentanedioate	$Bi(CF_3COCHCOCF_3)_3$		830.132	pow	96				
346	Bismuth molybdate	$Bi_2(MoO_4)_3$	51898-99-8	897.77	monocl cry			5.95		
347	Bismuth nitrate pentahydrate	$Bi(NO_3)_3 \cdot 5H_2O$	10035-06-0	485.071	col tricl cry; hyg	≈75 dec		2.83		reac H_2O; s ace; i EtOH
348	Bismuth oleate	$Bi(C_{18}H_{33}O_2)_3$	52951-38-9	1053.340	soft yel-brn solid					i H_2O; s eth; sl bz
349	Bismuth oxalate	$Bi_2(C_2O_4)_3$	6591-55-5	682.018	wh powder					i H_2O, EtOH; s dil acid
350	Bismuth oxide	Bi_2O_3	1304-76-3	465.959	yel monocl cry or powder	817	1890	8.9		i H_2O; s acid
351	Bismuth oxybromide	BiOBr	7787-57-7	304.883	col tetr cry			8.08		i H_2O, EtOH; s acid
352	Bismuth oxychloride	BiOCl	7787-59-9	260.432	wh tetr cry			7.72		i H_2O
353	Bismuth oxyiodide	BiOI	7787-63-5	351.883	red tetr cry	>300 dec		7.92		i H_2O, EtOH, chl; s HCl
354	Bismuth oxynitrate	$BiONO_3$	10361-46-3	286.985	wh powder	260 dec		4.93		i H_2O, EtOH; s acid
355	Bismuth phosphate	$BiPO_4$	10049-01-1	303.951	monocl cry			6.32		sl H_2O, dil acid; i EtOH
356	Bismuth potassium iodide	K_4BiI_7	41944-01-8	1253.704	red cry					reac H_2O; s alk iodide soln
357	Bismuth selenide	Bi_2Se_3	12068-69-8	654.84	blk hex cry	710 dec		7.5		i H_2O
358	Bismuth stannate pentahydrate	$Bi_2(SnO_3)_3 \cdot 5H_2O$	12777-45-6	1008.162	wh cry					i H_2O
359	Bismuth subnitrate	$Bi_5O(OH)_9(NO_3)_4$	1304-85-4	1461.987	hyg cry pow	260 dec		4.928		i H_2O, EtOH; s dil acid
360	Bismuth sulfate	$Bi_2(SO_4)_3$	7787-68-0	706.152	wh needles or powder	405 dec		5.08		reac H_2O, EtOH
361	Bismuth sulfide	Bi_2S_3	1345-07-9	514.159	blk-brn orth cry	850		6.78		i H_2O; s acid
362	Bismuth telluride	Bi_2Te_3	1304-82-1	800.76	gray hex plates	580		7.74		i H_2O; s EtOH
363	Bismuth tetroxide	Bi_2O_4	12048-50-9	481.959	red-oran powder	305		5.6		reac H_2O
364	Bismuth titanate	$Bi_4(TiO_4)_3$	12048-51-0	1171.516	wh orth cry			7.85		
365	Bismuth tungstate	$Bi_2(WO_4)_3$	13595-87-4	1161.47	wh pow					
366	Bismuth vanadate	$BiVO_4$	14059-33-7	323.920	orth cry	trans 500		6.25		i H_2O; s acid
367	Boron	B	7440-42-8	10.811	blk rhomb cry	2075	4000	2.34		i H_2O

368	Diborane	B_2H_6	19287-45-7	27.670	col gas; flam	-165.5	-92.4	1.131 g/L		reac H_2O
369	Tetraborane(10)	B_4H_{10}	18283-93-7	53.323	col gas	-121	18	2.180 g/L		reac H_2O
370	Pentaborane(9)	B_5H_9	19624-22-7	63.126	flam liq	-46.6	60	0.60		reac hot H_2O
371	Pentaborane(11)	B_5H_{11}	18433-84-6	65.142	col liq; unstable	-122	65			reac H_2O
372	Hexaborane(10)	B_6H_{10}	23777-80-2	74.945	col liq	-62.3	108 dec	0.67		reac hot H_2O
373	Hexaborane(12)	B_6H_{12}	12008-19-4	76.961	col liq; unstable	-82.3	≈80			reac H_2O
374	Nonaborane(15)	B_9H_{15}	19465-30-6	112.418	col liq	2.6				
375	Decaborane(14)	$B_{10}H_{14}$	17702-41-9	122.221	wh orth cry	99.6	≈213	0.94		sl H_2O; s EtOH, bz, CS_2, ctc
376	Borane carbonyl	BH_3CO	13205-44-2	41.845	col gas	-137	-64	1.710 g/L		reac H_2O
377	Borazine	$B_3N_3H_6$	6569-51-3	80.501	col liq	-58	53	0.824		reac H_2O
378	Boric acid (orthoboric acid)	H_3BO_3	10043-35-3	61.833	col tricl cry	170.9		1.5	5.80^{25}	sl EtOH
379	Metaboric acid (α form)	HBO_2	13460-50-9	43.818	col orth cry; hyg	176		1.784		s H_2O
380	Metaboric acid (β form)	HBO_2	13460-50-9	43.818	col monocl cry; hyg	201		2.045		s H_2O
381	Metaboric acid (γ form)	HBO_2	13460-50-9	43.818	col cub cry	236		2.487		s H_2O
382	Tetrafluoroboric acid	HBF_4	16872-11-0	87.813	col liq		130 dec	≈1.8		vs H_2O, EtOH
383	Boron arsenide	BAs	12005-69-5	85.733	cub cry	920 dec		5.22		
384	Boron tribromide	BBr_3	10294-33-4	250.523	col liq; hyg	-45	91	2.6		reac H_2O, EtOH
385	Boron carbide	B_4C	12069-32-8	55.255	hard blk cry	2350	>3500	2.50		i H_2O, acid
386	Boron trichloride	BCl_3	10294-34-5	117.169	col liq or gas	-107	12.65	4.789 g/L		reac H_2O, EtOH
387	Tetrachlorodiborane	B_2Cl_4	13701-67-2	163.433	col liq; flam	-92.6	65			reac H_2O
388	Boron trifluoride	BF_3	7637-07-2	67.806	col gas	-126.8	-101	2.772 g/L		s H_2O
389	Tetrafluorodiborane	B_2F_4	13965-73-6	97.616	col gas; flam	-56	-34	3.990 g/L		reac H_2O
390	Boron triiodide	BI_3	13517-10-7	391.524	wh needles	49.7	209.5	3.35		i H_2O
391	Boron nitride	BN	10043-11-5	24.818	wh powder; hex or cub cry	2966		2.18		i H_2O, acid
392	Boron oxide	B_2O_3	1303-86-2	69.620	col gl or hex cry; hyg	450		2.55	2.2^{20}	s EtOH
393	Boron phosphide	BP	20205-91-8	41.785	red cub cry or powder	1125 dec				reac H_2O, acid
394	Boron sulfide	B_2S_3	12007-33-9	117.820	yel amorp solid	softens ≈320		≈1.7		
395	Bromine	Br_2	7726-95-6	159.808	red liq	-7.2	58.8	3.1028		sl H_2O
396	Bromic acid	$HBrO_3$	7789-31-3	128.910	stable only in aq soln					s H_2O
397	Bromine oxide	Br_2O	21308-80-5	175.807	brn solid	-17.5 dec				
398	Bromine dioxide	BrO_2	21255-83-4	111.903	unstable yel cry	≈0 dec				
399	Bromine azide	BrN_3	13973-87-0	121.924	red cry; exp	≈45	exp			
400	Bromine chloride	BrCl	13863-41-7	115.357	unstable red-brn gas	≈-66	≈5 dec	4.715 g/L		reac H_2O; s eth, CS_2
401	Bromine fluoride	BrF	13863-59-7	98.902	unstable red-brn gas	≈-33	≈20 dec	4.043 g/L		
402	Bromine trifluoride	BrF_3	7787-71-5	136.899	col hyg liq	8.77	125.8	2.803		reac H_2O
403	Bromine pentafluoride	BrF_5	7789-30-2	174.896	col liq	-60.5	40.76	2.460		reac H_2O (exp)
404	Bromyl fluoride	BrO_2F	22585-64-4	130.901	col liq	-9	50 dec			reac H_2O
405	Cadmium	Cd	7440-43-9	112.411	silv-wh metal	321.07	767	8.69		i H_2O; reac acid
406	Cadmium acetate	$Cd(C_2H_3O_2)_2$	543-90-8	230.500	col cry	255		2.34		s H_2O, EtOH
407	Cadmium acetate dihydrate	$Cd(C_2H_3O_2)_2 \cdot 2H_2O$	5743-04-4	266.529	wh cry	130 dec		2.01		vs H_2O; s EtOH
408	Cadmium antimonide	CdSb	12014-29-8	234.171	orth cry	456		6.92		
409	Cadmium arsenide	Cd_3As_2	12006-15-4	487.076	gray tetr cry	721		6.25		
410	Cadmium azide	$Cd(N_3)_2$	14215-29-3	196.451	yel-wh orth cry; exp	exp		3.24		
411	Cadmium bromide	$CdBr_2$	7789-42-6	272.219	wh hex powder or flakes; hyg	568	844	5.19	115^{25}	sl ace, eth
412	Cadmium bromide tetrahydrate	$CdBr_2 \cdot 4H_2O$	13464-92-1	344.281	wh-yel cry				115^{25}	s ace, EtOH
413	Cadmium carbonate	$CdCO_3$	513-78-0	172.420	wh hex cry	500 dec		4.258		i H_2O; s acid
414	Cadmium chlorate dihydrate	$Cd(ClO_3)_2 \cdot 2H_2O$	22750-54-5*	315.343	col hyg cry	80 dec		2.28	2.64^{0}	
415	Cadmium chloride	$CdCl_2$	10108-64-2	183.316	rhomb cry; hyg	564	960	4.08	120^{25}	s ace; sl EtOH; i eth

No.	Name	Formula	CAS Reg No.	Mol. Weight	Physical Form	mp/°C	bp/°C	Density g cm^{-3}	Solubility g/100 g H_2O	Qualitative Solubility
416	Cadmium chloride hemipentahydrate	$CdCl_2 \cdot 2.5H_2O$	7790-78-5	228.354	wh rhomb leaflets			3.327	120^{25}	s ace
417	Cadmium chloride monohydrate	$CdCl_2 \cdot H_2O$	34330-64-8	201.331	wh cry				120^{25}	
418	Cadmium chromate	$CdCrO_4$	14312-00-6	228.405	yel orth cry			4.5		i H_2O
419	Cadmium cyanide	$Cd(CN)_2$	542-83-6	164.445	wh cub cry			2.23	1.7^{15}	
420	Cadmium 2-ethylhexanoate	$Cd(C_8H_{15}O_2)_2$	2420-98-6	398.818	pow					
421	Cadmium fluoride	CdF_2	7790-79-6	150.408	cub cry	1110	1748	6.33	4.36^{25}	s acid; i EtOH
422	Cadmium hydroxide	$Cd(OH)_2$	21041-95-2	146.426	wh trig or hex cry	130 dec		4.79	0.00015^{20}	s dil acid
423	Cadmium iodate	$Cd(IO_3)_2$	7790-81-0	462.216	wh powder			6.48	0.091^{25}	s HNO_3
424	Cadmium iodide	CdI_2	7790-80-9	366.220	hex flakes	387	742	5.64	86.2^{25}	s EtOH, eth, ace
425	Cadmium metasilicate	$CdSiO_3$	13477-19-5	188.495	grn monocl cry	1252		5.10		
426	Cadmium molybdate	$CdMoO_4$	13972-68-4	272.35	col tetr cry	≈900 dec		5.4		i H_2O; s acid
427	Cadmium niobate	$Cd_2Nb_2O_7$	12187-14-3	522.631	cub cry	≈1410		6.28		i H_2O
428	Cadmium nitrate	$Cd(NO_3)_2$	10325-94-7	236.420	wh cub cry; hyg	350		3.6	156^{25}	s EtOH
429	Cadmium nitrate tetrahydrate	$Cd(NO_3)_2 \cdot 4H_2O$	10022-68-1	308.482	col orth cry; hyg	59.5		2.45	156^{25}	s EtOH, ace
430	Cadmium oxalate	CdC_2O_4	814-88-0	200.430	wh solid			3.32	0.0060^{25}	
431	Cadmium oxalate trihydrate	$CdC_2O_4 \cdot 3H_2O$	20712-42-9	254.476	wh amorp powder	340 dec			0.0060^{25}	i EtOH; s dil acid
432	Cadmium oxide	CdO	1306-19-0	128.410	brn cub cry		1559 sp	8.15		i H_2O; s dil acid
433	Cadmium perchlorate hexahydrate	$Cd(ClO_4)_2 \cdot 6H_2O$	10326-28-0	419.403	wh hex cry			2.37	191.5^{25}	
434	Cadmium phosphate	$Cd_3(PO_4)_2$	13477-17-3	527.176	pow	≈1500				i H_2O
435	Cadmium phosphide	Cd_3P_2	12014-28-7	399.181	gr tetr needles	700		5.96		s dil HCl
436	Cadmium selenate dihydrate	$CdSeO_4 \cdot 2H_2O$	10060-09-0	291.40	orth cry	100 dec		3.62	70.5^{25}	
437	Cadmium selenide	$CdSe$	1306-24-7	191.37	wh cub cry	1240		5.81		i H_2O
438	Cadmium sulfate	$CdSO_4$	10124-36-4	208.475	col orth cry	1000		4.69	76.7^{25}	i EtOH
439	Cadmium sulfate monohydrate	$CdSO_4 \cdot H_2O$	7790-84-3	226.490	monocl cry	105		3.79	76.7^{25}	
440	Cadmium sulfate octahydrate	$CdSO_4 \cdot 8H_2O$	15244-35-6	352.597	col monocl cry	40 dec		3.08	76.7^{25}	
441	Cadmium sulfide	CdS	1306-23-6	144.477	yel-oran cub cry	1750		4.83		i H_2O; s acid
442	Cadmium telluride	$CdTe$	1306-25-8	240.01	brn-blk cub cry	1042		6.2		i H_2O, dil acid
443	Cadmium tetrafluoroborate	$Cd(BF_4)_2$	14486-19-2	286.020	col hyg liq			1.6		vs H_2O, EtOH
444	Cadmium titanate	$CdTiO_3$	12014-14-1	208.276	orth cry			6.5		
445	Cadmium tungstate	$CdWO_4$	7790-85-4	360.25	wh monocl cry			8.0		i H_2O, acid; s NH_4OH
446	Calcium	Ca	7440-70-2	40.078	silv-wh metal	842	1484	1.54		reac H_2O; i bz
447	Calcium acetate	$Ca(C_2H_3O_2)_2$	62-54-4	158.167	wh hyg cry	160 dec		1.50		s H_2O; sl EtOH
448	Calcium acetate monohydrate	$Ca(C_2H_3O_2)_2 \cdot H_2O$	5743-26-0	176.182	wh needles or powder	≈150 dec				s H_2O; sl EtOH
449	Calcium aluminate	$CaAl_2O_4$	12042-68-1	158.039	wh monocl cry	1605		2.98		reac H_2O
450	Calcium aluminate (β form)	$Ca_3Al_2O_6$	12042-78-3	270.193	wh cub cry; refr	1535		3.04		i H_2O
451	Calcium arsenate	$Ca_3(AsO_4)_2$	7778-44-1	398.072	wh powder	dec		3.6	0.0036^{20}	s dil acid
452	Calcium arsenite	$CaAsO_3$	52740-16-6	162.998	wh pow					sl H_2O; s acid
453	Calcium boride	CaB_6	12007-99-7	104.944	refrac solid	2235		2.49		
454	Calcium bromide	$CaBr_2$	7789-41-5	199.886	rhomb cry; hyg	742	1815	3.38	156^{25}	s EtOH, ace
455	Calcium bromide hexahydrate	$CaBr_2 \cdot 6H_2O$	13477-28-6	307.977	wh hyg powder	38 dec		2.29	156^{25}	
456	Calcium carbide	CaC_2	75-20-7	64.099	gray-blk orth cry	2300		2.22		reac H_2O
457	Calcium carbonate (aragonite)	$CaCO_3$	471-34-1	100.087	wh orth cry or powder	825 dec		2.83	0.00066^{20}	s dil acid
458	Calcium carbonate (calcite)	$CaCO_3$	471-34-1	100.087	wh hex cry or powder	1330		2.71	0.00066^{20}	s dil acid
459	Calcium chlorate	$Ca(ClO_3)_2$	10137-74-3	206.979	wh cry	340			197^{25}	
460	Calcium chlorate dihydrate	$Ca(ClO_3)_2 \cdot 2H_2O$	10035-05-9	243.010	wh monocl cry; hyg	100 dec		2.711	197^{25}	s EtOH

461	Calcium chloride	$CaCl_2$	10043-52-4	110.983	wh cub cry or powder; hyg	775	1935.5	2.15	81.3^{25}	vs EtOH
462	Calcium chloride dihydrate	$CaCl_2 \cdot 2H_2O$	10035-04-8	147.014	hyg flakes or powder	175 dec		1.85	81.3^{25}	vs EtOH
463	Calcium chloride hexahydrate	$CaCl_2 \cdot 6H_2O$	7774-34-7	219.074	wh hex cry; hyg	30 dec		1.71	81.3^{25}	
464	Calcium chloride monohydrate	$CaCl_2 \cdot H_2O$	13477-29-7	128.998	wh hyg cry	260 dec		2.24	81.3^{25}	s EtOH
465	Calcium chromate dihydrate	$CaCrO_4 \cdot 2H_2O$	13765-19-0	192.102	yel orth cry			2.50	13.2^{20}	
466	Calcium cyanamide	$CaCN_2$	156-62-7	80.102	col hex cry	≈1340	subl	2.29		reac H_2O
467	Calcium cyanide	$Ca(CN)_2$	592-01-8	92.112	wh rhomb cry; hyg					s H_2O, EtOH
468	Calcium dichromate trihydrate	$CaCr_2O_7 \cdot 3H_2O$	14307-33-6*	310.112	red-oran cry	100 dec		2.37		vs H_2O; reac EtOH; i eth, ctc
469	Calcium 2-ethylhexanoate	$Ca(C_8H_{15}O_2)_2$	136-51-6	326.485	pow					
470	Calcium fluoride	CaF_2	7789-75-5	78.075	wh cub cry or powder	1418	2533.4	3.18	0.0016^{25}	sl acid
471	Calcium formate	$Ca(CHO_2)_2$	544-17-2	130.113	orth cry	300 dec		2.02	16.6^{20}	i EtOH
472	Calcium hexafluoro-2,4-pentanedioate	$Ca(CF_3COCHCOCF_3)_2$	121012-90-6	454.180	pow	135				
473	Calcium hexafluorosilicate dihydrate	$CaSiF_6 \cdot 2H_2O$	16925-39-6	218.185	col tetr cry			2.25	0.52^{20}	i ace; reac hot H_2O
474	Calcium hydride	CaH_2	7789-78-8	42.094	gray orth cry or powder	1000		1.7		reac H_2O, EtOH
475	Calcium hydrogen phosphate	$CaHPO_4$	7757-93-9	136.057	wh tricl cry	dec		2.92	0.02^{25}	i EtOH
476	Calcium hydrogen phosphate dihydrate	$CaHPO_4 \cdot 2H_2O$	7789-77-7	172.088	monocl cry	≈100 dec		2.31	0.02^{25}	i EtOH; s dil acid
477	Calcium hydroxide	$Ca(OH)_2$	1305-62-0	74.093	soft hex cry			≈2.2	0.160^{20}	s acid
478	Calcium hypochlorite	$Ca(OCl)_2$	7778-54-3	142.982	pow	100		2.350		
479	Calcium hypophosphite	$Ca(H_2PO_2)_2$	7789-79-9	170.055	wh monocl cry	300 dec				s H_2O; i EtOH
480	Calcium iodate	$Ca(IO_3)_2$	7789-80-2	389.883	wh monocl cry			4.52	0.306^{25}	s HNO_3; i EtOH
481	Calcium iodide	CaI_2	10102-68-8	293.887	hyg hex cry	783		3.96	215^{25}	s MeOH, EtOH, ace; i eth
482	Calcium iodide hexahydrate	$CaI_2 \cdot 6H_2O$	71626-98-7	401.978	wh hex needles or powder	42 dec		2.55	215^{25}	vs EtOH
483	Calcium metaborate	$Ca(BO_2)_2$	13701-64-9	125.698	pow				0.13^{20}	
484	Calcium molybdate	$CaMoO_4$	7789-82-4	200.02	wh tetr cry	965 dec		4.35	0.0011^{20}	i EtOH; s conc acid
485	Calcium nitrate	$Ca(NO_3)_2$	10124-37-5	164.087	wh cub cry; hyg	561		2.5	144^{25}	s EtOH, MeOH, ace
486	Calcium nitrate tetrahydrate	$Ca(NO_3)_2 \cdot 4H_2O$	13477-34-4	236.149	wh cry	≈40 dec		1.82	144^{25}	s EtOH, ace
487	Calcium nitride	Ca_3N_2	12013-82-0	148.247	red-brn cub cry	1195		2.67		s H_2O, acid; i EtOH
488	Calcium nitrite	$Ca(NO_2)_2$	13780-06-8	132.089	wh-yel hex cry; hyg			2.23	94.6^{25}	sl EtOH
489	Calcium oxalate	CaC_2O_4	563-72-4	128.097	wh cry powder			2.2	0.00061^{20}	
490	Calcium oxalate monohydrate	$CaC_2O_4 \cdot H_2O$	5794-28-5	146.112	cub cry	200 dec		2.2	0.00061^{20}	s dil acid
491	Calcium oxide	CaO	1305-78-8	56.077	gray-wh cub cry	2898		3.34		reac H_2O; s acid
492	Calcium oxide silicate	Ca_3OSiO_4	12168-85-3	228.317	refrac solid	2150				
493	Calcium 2,4-pentanedioate	$Ca(CH_3COCHCOCH_3)_2$	19372-44-2	238.294	cry	175 dec				
494	Calcium perchlorate	$Ca(ClO_4)_2$	13477-36-6	238.978	wh cry	270 dec		2.65	188^{25}	s EtOH
495	Calcium permanganate	$Ca(MnO_4)_2$	10118-76-0	277.949	purp hyg cry			2.4	331^{20}	reac EtOH
496	Calcium peroxide	CaO_2	1305-79-9	72.077	wh-yel tetr cry; hyg	≈200 dec		2.9		sl H_2O; s acid
497	Calcium phosphate	$Ca_3(PO_4)_2$	7758-87-4	310.177	wh amorp powder	1670		3.14	0.00012^{20}	i EtOH; s dil acid
498	Calcium dihydrogen phosphate monohydrate	$Ca(H_2PO_4)_2 \cdot H_2O$	10031-30-8	252.068	col tricl plates	100 dec		2.220		sl H_2O; s dil acid
499	Calcium phosphide	Ca_3P_2	1305-99-3	182.182	red-brn hyg cry	≈1600		2.51		reac H_2O; i EtOH, eth
500	Calcium propanoate	$Ca(C_3H_5O_2)_2$	4075-81-4	186.219	mono cry, pow					s H_2O; sl MeOH, EtOH; i ace, bz
501	Calcium pyrophosphate	$Ca_2P_2O_7$	7790-76-3	254.099	wh powder	1353		3.09		i H_2O; s dil acid
502	Calcium selenate dihydrate	$CaSeO_4 \cdot 2H_2O$	7790-74-1	219.07	wh monocl cry			2.75	8.3^{18}	
503	Calcium selenide	$CaSe$	1305-84-6	119.04	wh-brn cub cry	1400 dec		3.8		reac H_2O
504	Calcium metasilicate	$CaSiO_3$	1344-95-2	116.162	wh monocl cry	1540		2.92		i H_2O
505	Calcium silicide	$CaSi_2$	12013-56-8	96.249	gray hex cry	1040		2.50		i cold H_2O; reac hot H_2O; s acid
506	Calcium silicide	$CaSi$	12013-55-7	68.164	orth cry	1324		2.39		
507	Calcium stearate	$Ca(C_{18}H_{35}O_2)_2$	1592-23-0	607.017	granular pow	180				i H_2O, EtOH

No.	Name	Formula	CAS Reg No.	Mol. Weight	Physical Form	mp/°C	bp/°C	Density g cm^{-3}	Solubility g/100 g H_2O	Qualitative Solubility
508	Calcium sulfate	$CaSO_4$	7778-18-9	136.142	orth cry	1460		2.96	0.205^{25}	
509	Calcium sulfate dihydrate	$CaSO_4 \cdot 2H_2O$	10101-41-4	172.172	monocl cry or powder	150 dec		2.32	0.205^{20}	i os
510	Calcium sulfate hemihydrate	$CaSO_4 \cdot 0.5H_2O$	10034-76-1	145.149	wh powder				0.205^{25}	
511	Calcium sulfide	CaS	20548-54-3	72.144	wh-yel cub cry; hyg	2524		2.59		sl H_2O; i EtOH
512	Calcium sulfite dihydrate	$CaSO_3 \cdot 2H_2O$	10257-55-3	156.173	wh powder				0.0070^{25}	sl EtOH; s acid
513	Calcium telluride	CaTe	12013-57-9	167.68	wh cub cry	1600 dec		4.87		
514	Calcium tetrahydroaluminate	$Ca(AlH_4)_2$	16941-10-9	102.105	gray powder; flam					reac H_2O; s thf; i eth, bz
515	Calcium thiocyanate tetrahydrate	$Ca(SCN)_2 \cdot 4H_2O$	2092-16-2	228.306	hygr cry	160 dec				vs H_2O; s EtOH, ace
516	Calcium thiosulfate hexahydrate	$CaS_2O_3 \cdot 6H_2O$	10124-41-1	260.300	tricl cry	45 dec		1.87		s H_2O; i EtOH
517	Calcium titanate	$CaTiO_3$	12049-50-2	135.943	cub cry	1980		3.98		
518	Calcium tungstate	$CaWO_4$	7790-75-2	287.92	wh tetr cry	1620		6.06	0.2^{18}	s hot acid
519	Calcium zirconate	$CaZrO_3$	12013-47-7	179.300	pow	2550				
520	Californium	Cf	7440-71-3	251	hex or cub metal	900		15.1		
521	Carbon (diamond)	C	7782-40-3	12.011	col cub cry	4440 (12.4 GPa)		3.513		i H_2O
522	Carbon (graphite)	C	7782-42-5	12.011	soft blk hex cry	4489 tp (10.3 MPa)	3825 sp	2.2		i H_2O
523	Carbon (fullerene-C_{60})	C_{60}	99685-96-8	720.642	yel needles or plates	>280				s os
524	Carbon (fullerene-C_{70})	C_{70}	115383-22-7	840.749	red-brn solid	>280				s bz, tol
525	Fullerene fluoride	$C_{60}F_{60}$	134929-59-2	1860.546	col plates	287				vs ace; s thf; i chl
526	Carbon monoxide	CO	630-08-0	28.010	col gas	-205.02	-191.5	1.145 g/L		sl H_2O; s chl, EtOH
527	Carbon dioxide	CO_2	124-38-9	44.010	col gas	-56.56 tp	-78.4 sp	1.799 g/L		s H_2O
528	Carbon diselenide	CSe_2	506-80-9	169.93	yel liq	-43.7	125.5	2.6626		i H_2O; vs ctc, tol
529	Carbon disulfide	CS_2	75-15-0	76.143	col or yel liq	-112.1	46	1.2555		i H_2O; vs EtOH, bz, os
530	Carbon oxyselenide	COSe	1603-84-5	106.97	col gas; unstable	-124.4	-21.7	4.372 g/L		reac H_2O
531	Carbon oxysulfide	COS	463-58-1	60.076	col gas	-138.8	-50	2.456 g/L		s H_2O, EtOH
532	Carbon sulfide selenide	CSSe	5951-19-9	123.04	yel liq	-85	84.5	1.99		i H_2O
533	Carbon sulfide telluride	CSTe	10340-06-4	171.68	red-yel liq; unstable	-54	20 dec			reac H_2O
534	Carbon suboxide	C_3O_2	504-64-3	68.031	col gas	-111.3	6.8	2.781 g/L		reac H_2O
535	Carbon subsulfide	C_3S_2	627-34-9	100.164	red liq	-1	90 dec	1.27		reac H_2O
536	Carbonyl bromide	$COBr_2$	593-95-3	187.818	col liq		64.5	2.5		reac H_2O
537	Carbonyl chloride	$COCl_2$	75-44-5	98.915	col gas	-127.78	8	4.043 g/L		sl H_2O; s bz, tol
538	Carbonyl fluoride	COF_2	353-50-4	66.007	col gas	-111.26	-84.57	2.698 g/L		reac H_2O
539	Cyanogen	C_2N_2	460-19-5	52.034	col gas	-27.83	-21.1	2.127 g/L		sl H_2O, eth; s EtOH
540	Cyanogen bromide	BrCN	506-68-3	105.922	wh hyg needles	52	61.5	2.005		s H_2O, EtOH, eth
541	Cyanogen chloride	ClCN	506-77-4	61.470	col gas	-6.55	13	2.513 g/L		s H_2O, EtOH, eth
542	Cyanogen fluoride	FCN	1495-50-7	45.016	col gas	-82	-46	1.840 g/L		
543	Cyanogen iodide	ICN	506-78-5	152.922	col needles	146.7		1.84		s H_2O, EtOH, eth
544	Cerium	Ce	7440-45-1	140.116	silv metal; cub or hex	798	3443	6.770		s dil acid
545	Cerium boride	CeB_6	12008-02-5	204.982	blue refrac solid; hex	2550		4.87		i H_2O, HCl
546	Cerium carbide	CeC_2	12012-32-7	164.137	red hex cry	2250		5.47		reac H_2O
547	Cerium nitride	CeN	25764-08-3	154.123	refrac cub cry	2557		7.89		
548	Cerium silicide	$CeSi_2$	12014-85-6	196.287	tetr cry	1620		5.31		i H_2O
549	Cerium(II) hydride	CeH_2	13569-50-1	142.132	cub cry			5.45		reac H_2O
550	Cerium(II) iodide	CeI_2	19139-47-0	393.925	bronze cry	808				
551	Cerium(II) sulfide	CeS	12014-82-3	172.182	yel cub cry	2445		5.9		
552	Cerium(III) bromide	$CeBr_3$	14457-87-5	379.828	wh hex cry; hyg	733	1457			s H_2O

553	Cerium(III) bromide heptahydrate	$CeBr_3 \cdot 7H_2O$	7789-56-2	505.935	col hyg needles	732				s H_2O, EtOH
554	Cerium(III) carbide	Ce_2C_3	12115-63-8	316.264	yel-brn cub cry	1505		6.9		
555	Cerium(III) carbonate hydrate	$Ce_2(CO_3)_3 \cdot 5H_2O$	72520-94-6	550.335	wh powder					i H_2O; s dil acid
556	Cerium(III) chloride	$CeCl_3$	7790-86-5	246.474	wh hex cry	817		3.97		s H_2O, EtOH
557	Cerium(III) chloride heptahydrate	$CeCl_3 \cdot 7H_2O$	18618-55-8	372.581	yel orth cry; hyg	90 dec				vs H_2O, EtOH
558	Cerium(III) fluoride	CeF_3	7758-88-5	197.111	wh hex cry; hyg	1430		6.157		i H_2O
559	Cerium(III) iodide	CeI_3	7790-87-6	520.829	yel orth cry; hyg	766				s H_2O
560	Cerium(III) iodide nonahydrate	$CeI_3 \cdot 9H_2O$	7790-87-6*	682.967	wh-red cry					vs H_2O; s EtOH
561	Cerium(III) nitrate hexahydrate	$Ce(NO_3)_3 \cdot 6H_2O$	10108-73-3*	434.222	col-red cry	150 dec			176^{25}	s ace
562	Cerium(III) oxide	Ce_2O_3	1345-13-7	328.230	yel-grn cub cry	2210	3730	6.2		i H_2O; s acid
563	Cerium(III) sulfate octahydrate	$Ce_2(SO_4)_3 \cdot 8H_2O$	13454-94-9	712.545	wh orth cry	≈250 dec		2.87		s H_2O
564	Cerium(III) sulfide	Ce_2S_3	12014-93-6	376.430	red cub cry	2450		5.02		i H_2O
565	Cerium(IV) fluoride	CeF_4	10060-10-3	216.110	wh hyg powder	≈600 dec		4.77		i H_2O
566	Cerium(IV) oxide	CeO_2	1306-38-3	172.115	wh-yel powder; cub	2400		7.65		i H_2O, dil acid
567	Cerium(IV) sulfate tetrahydrate	$Ce(SO_4)_2 \cdot 4H_2O$	10294-42-5	404.305	yel-oran orth cry	180 dec		3.91	9.66^{20}	
568	Cesium	Cs	7440-46-2	132.905	silv-wh metal	28.5	671	1.93		reac H_2O
569	Cesium acetate	$CsC_2H_3O_2$	3396-11-0	191.949	hyg lumps	194			10^{11}	
570	Cesium amide	$CsNH_2$	22205-57-8	148.928	wh tetr cry			3.70		
571	Cesium azide	CsN_3	22750-57-8	174.925	hyg tetr cry; exp	326		≈3.5	224^{0}	
572	Cesium bromate	$CsBrO_3$	13454-75-6	260.807	col hex cry			4.11	3.83^{25}	
573	Cesium bromide	CsBr	7787-69-1	212.809	wh cub cry; hyg	636	≈1300	4.43	123^{25}	s EtOH; i ace
574	Cesium carbonate	Cs_2CO_3	534-17-8	325.820	wh monocl cry; hyg	792		4.24	261^{15}	s EtOH, eth
575	Cesium chlorate	$CsClO_3$	13763-67-2	216.356	col hex cry			3.57	7.78^{25}	
576	Cesium chloride	CsCl	7647-17-8	168.358	wh cub cry; hyg	645	1297	3.988	191^{25}	s EtOH
577	Cesium cyanide	CsCN	21159-32-0	158.923	wh cub cry; hyg	350		3.34		vs H_2O
578	Cesium fluoride	CsF	13400-13-0	151.903	wh cub cry; hyg	703		4.64	573^{25}	s MeOH; i diox, py
579	Cesium formate	$CsCHO_2$	3495-36-1	177.923	wh cry			1.017		vs H_2O
580	Cesium hydride	CsH	58724-12-2	133.913	wh cub cry; flam	≈170 dec		3.42		reac H_2O
581	Cesium hydrogen carbonate	$CsHCO_3$	15519-28-5	193.922	rhom cry	175 dec			209^{15}	s EtOH
582	Cesium hydrogen fluoride	$CsHF_2$	12280-52-3	171.910	tetr cry	170		3.86		
583	Cesium hydrogen sulfate	$CsHSO_4$	7789-16-4	229.977	col rhom prisms	dec		3.352		s H_2O
584	Cesium hydroxide	CsOH	21351-79-1	149.912	wh-yel hyg cry	342.3		3.68	300^{30}	s EtOH
585	Cesium iodate	$CsIO_3$	13454-81-4	307.807	wh mono cry			4.85	2.6^{25}	
586	Cesium iodide	CsI	7789-17-5	259.809	col cub cry; hyg	621	≈1280	4.51	84.8^{25}	s EtOH, MeOH, ace
587	Cesium metaborate	$CsBO_2$	92141-86-1	175.715	cub cry	732		≈3.7		
588	Cesium nitrate	$CsNO_3$	7789-18-6	194.910	wh hex or cub cry	414		3.66	27.9^{25}	s ace; sl EtOH
589	Cesium oxide	Cs_2O	20281-00-9	281.810	yel-oran hex cry	490		4.65		vs H_2O
590	Cesium superoxide	CsO_2	12018-61-0	164.904	yel tetr cry	432		3.77		reac H_2O
591	Cesium perchlorate	$CsClO_4$	13454-84-7	232.356	wh orth cry; hyg	250		3.327	2.00^{25}	
592	Cesium periodate	$CsIO_4$	13478-04-1	323.807	wh rhom prisms			4.26	2.2^{15}	
593	Cesium sulfate	Cs_2SO_4	10294-54-9	361.875	wh orth cry or hex prisms; hyg	1005		4.24	182^{25}	i EtOH, ace, py
594	Cesium sulfide tetrahydrate	$Cs_2S \cdot 4H_2O$	12214-16-3	369.939	wh hyg cry					vs H_2O
595	Chlorine	Cl_2	7782-50-5	70.905	grn-yel gas	-101.5	-34.04	2.898 g/L		sl H_2O
596	Hypochlorous acid	HOCl	7790-92-3	52.460	grn-yel; stable only in aq soln					s H_2O
597	Perchloric acid	$HClO_4$	7601-90-3	100.459	col hyg liq	-112	≈90 dec	1.77		s H_2O
598	Chlorine monoxide	Cl_2O	7791-21-1	86.904	yel-brn gas	-120.6	2.2	3.552 g/L		vs H_2O
599	Chlorine dioxide	ClO_2	10049-04-4	67.452	oran-grn gas	-59	11	2.757 g/L		sl H_2O
600	Chlorine trioxide	Cl_2O_3	17496-59-2	118.903	dark brn solid	exp <25				

No.	Name	Formula	CAS Reg No.	Mol. Weight	Physical Form	mp/°C	bp/°C	Density g cm^{-3}	Solubility g/100 g H_2O	Qualitative Solubility
601	Chlorine hexoxide	Cl_2O_6	12442-63-6	166.901	red liq	3.5	≈200			reac H_2O
602	Chlorine heptoxide	Cl_2O_7	10294-48-1	182.901	col oily liq; exp	-91.5	82	1.9		reac H_2O
603	Chlorine fluoride	ClF	7790-89-8	54.451	col gas	-155.6	-101.1	2.226 g/L		reac H_2O
604	Chlorine trifluoride	ClF_3	7790-91-2	92.448	gas	-76.34	11.75	3.779 g/L		reac H_2O
605	Chlorine trifluoride oxide	$ClOF_3$	30708-80-6	108.447	col liq	-42	29			reac H_2O
606	Chlorine pentafluoride	ClF_5	13637-63-3	130.445	col gas	-103	-13.1	5.332 g/L		
607	Chloryl fluoride	ClO_2F	13637-83-7	86.450	col gas	-15	-6	3.534 g/L		reac H_2O
608	Chloryl trifluoride	ClO_2F_3	38680-84-1	124.447	col gas	-81	-22	5.087 g/L		reac H_2O
609	Perchloryl fluoride	ClO_3F	7616-94-6	102.449	col gas	-147	-46.75	4.187 g/L		
610	Chlorine perchlorate	$ClOClO_3$	27218-16-2	134.903	unstable liq	-117	≈25 dec	1.81^0		
611	Chromium	Cr	7440-47-3	51.996	blue-wh metal; cub	1907	2671	7.15		reac dil acid
612	Chromium antimonide	CrSb	12053-12-2	173.756	hex cry	1110		7.11		
613	Chromium arsenide	Cr_2As	12254-85-2	178.914	tetr cry			7.04		
614	Chromium boride	CrB	12006-79-0	62.807	refrac orth cry	2100		6.1		
615	Chromium boride	CrB_2	12007-16-8	73.618	refrac solid; hex	2200		5.22		
616	Chromium boride	Cr_5B_3	12007-38-4	292.414	tetr cry	1900		6.10		
617	Chromium carbide	Cr_3C_2	12012-35-0	180.009	gray orth cry	1895		6.68		
618	Chromium carbonyl	$Cr(CO)_6$	13007-92-6	220.056	col orth cry	130 dec	subl	1.77		i H_2O, EtOH; s eth, chl
619	Chromium nitride	Cr_2N	12053-27-9	117.999	hex cry	1650		6.8		
620	Chromium nitride	CrN	24094-93-7	66.003	gray cub cry	1080 dec		5.9		
621	Chromium phosphide	CrP	26342-61-0	82.970	orth cry			5.25		
622	Chromium selenide	CrSe	12053-13-3	130.96	hex cry	≈1500		6.1		
623	Chromium silicide	Cr_3Si	12018-36-9	184.074	cub cry	1770		6.4		
624	Chromium silicide	$CrSi_2$	12018-09-6	108.167	gray hex cry	1490		4.91		
625	Chromium(II) acetate monohydrate	$Cr(C_2H_3O_2)_2 \cdot H_2O$	628-52-4*	188.100	red monocl cry			1.79		sl H_2O
626	Chromium(II) bromide	$CrBr_2$	10049-25-9	211.804	wh monocl cry; aq soln blue	842		4.236		s H_2O, EtOH
627	Chromium(II) chloride	$CrCl_2$	10049-05-5	122.901	hyg needles; aq soln blue	814	1300	2.88		s H_2O
628	Chromium(II) chloride tetrahydrate	$Cr(H_2O)_4Cl_2 \cdot 4H_2O$	13931-94-7	267.023	blue hyg cry	51 dec				s H_2O
629	Chromium(II) fluoride	CrF_2	10049-10-2	89.993	blue-grn monocl cry	894		3.79		sl H_2O; i EtOH
630	Chromium(II) iodide	CrI_2	13478-28-9	305.805	red-brn cry; hyg	868		5.1		
631	Chromium(II) oxalate monohydrate	$CrC_2O_4 \cdot H_2O$	814-90-4*	158.030	yel-grn powder			2.468		sl H_2O
632	Chromium(II) sulfate pentahydrate	$CrSO_4 \cdot 5H_2O$	13825-86-0	238.136	blue cry				21^0	s dil acid; sl EtOH; i ace
633	Chromium(II,III) oxide	Cr_3O_4	12018-34-7	219.986	cub cry			6.1		
634	Chromium(III) acetate	$Cr(C_2H_3O_2)_3$	1066-30-4	229.127	bl-grn pwd					sl H_2O
635	Chromium(III) acetate hexahydrate	$Cr(C_2H_3O_2)_3 \cdot 6H_2O$	1066-30-4*	337.220	blue needles					s H_2O
636	Chromium(III) bromide	$CrBr_3$	10031-25-1	291.708	dark grn hex cry	1130		4.68		s hot H_2O
637	Chromium(III) bromide hexahydrate (β)	$Cr(H_2O)_6Br_3$	10031-25-1*	399.799	viol hyg cry					s H_2O; i EtOH, eth
638	Chromium(III) bromide hexahydrate (α)	$CrBr_3(H_2O)_4 \cdot 2H_2O$	18721-05-6	399.799	grn hyg cry					s H_2O, EtOH
639	Chromium(III) chloride	$CrCl_3$	10025-73-7	158.354	purp hex plates	1152	1300 dec	2.87		sl H_2O
640	Chromium(III) chloride hexahydrate	$[CrCl_2(H_2O)_4]Cl \cdot 2H_2O$	10060-12-5	266.445	grn monocl cry; hyg					s H_2O, EtOH; sl ace; i eth
641	Chromium(III) fluoride	CrF_3	7788-97-8	108.991	grn needles	1400		3.8		i H_2O, EtOH
642	Chromium(III) fluoride trihydrate	$CrF_3 \cdot 3H_2O$	16671-27-5	163.037	grn hex cry			2.2		sl H_2O
643	Chromium(III) hydroxide trihydrate	$Cr(OH)_3 \cdot 3H_2O$	1308-14-1	157.063	blue-grn powder					i H_2O; s acid
644	Chromium(III) iodide	CrI_3	13569-75-0	432.709	dark grn hex cry	500 dec		5.32		sl H_2O
645	Chromium(III) nitrate	$Cr(NO_3)_3$	13548-38-4	238.011	grn hyg powder	>60 dec				vs H_2O

646	Chromium(III) nitrate nonahydrate	$Cr(NO_3)_3 \cdot 9H_2O$	7789-02-8	400.148	grn-blk monocl cry	66.3	>100 dec	1.80		vs H_2O
647	Chromium(III) oxide	Cr_2O_3	1308-38-9	151.990	grn hex cry	2329	≈3000	5.22		i H_2O, EtOH; sl acid, alk
648	Chromium(III) 2,4-pentanedioate	$Cr(CH_3COCHCOCH_3)_3$	21679-31-2	349.320	red monocl cry	208	345	1.34		i H_2O; s bz
649	Chromium(III) phosphate	$CrPO_4$	7789-04-0	146.967	blue orth cry	>1800		4.6		i H_2O, acid, aqua regia
650	Chromium(III) phosphate hemiheptahydrate	$CrPO_4 \cdot 3.5H_2O$	84359-31-9	210.021	blue-grn powder			2.15		i H_2O; s acid
651	Chromium(III) phosphate hexahydrate	$CrPO_4 \cdot 6H_2O$	84359-31-9	255.059	viol cry	>500 dec		2.121		i H_2O; s acid, alk
652	Chromium(III) potassium sulfate dodecahydrate	$CrK(SO_4)_2 \cdot 12H_2O$	7788-99-0	499.405	viol-blk cub cry	89 dec		1.83		s H_2O; i EtOH
653	Chromium(III) sulfate	$Cr_2(SO_4)_3$	10101-53-8	392.183	red-brn hex cry			3.1	64^{25}	vs acid
654	Chromium(III) sulfide	Cr_2S_3	12018-22-3	200.190	brn-blk hex cry			3.8		
655	Chromium(III) telluride	Cr_2Te_3	12053-39-3	486.79	hex cry	≈1300		7.0		
656	Chromium(IV) chloride	$CrCl_4$	15597-88-3	193.807	gas, stable at high temp		>600 dec	7.922 g/L		
657	Chromium(IV) fluoride	CrF_4	10049-11-3	127.990	grn cry	277				
658	Chromium(IV) oxide	CrO_2	12018-01-8	83.995	brn-blk tetr powder	≈400 dec		4.89		i H_2O; s acid
659	Chromium(V) fluoride	CrF_5	14884-42-5	146.988	red orth cry	34	117			
660	Chromium(VI) fluoride	CrF_6	13843-28-2	165.986	yel solid; stable at low temp	-100 dec				
661	Chromium(VI) oxide	CrO_3	1333-82-0	99.994	red orth cry	197	≈250 dec	2.7	169^{25}	
662	Chromic acid	H_2CrO_4	7738-94-5	118.010	aq soln only					s H_2O
663	Chromyl chloride	CrO_2Cl_2	14977-61-8	154.900	red liq	-96.5	117	1.91		reac H_2O; s ctc, chl, bz
664	Cobalt	Co	7440-48-4	58.933	gray metal; hex or cub	1495	2927	8.86		s dil acid
665	Cobalt antimonide	CoSb	12052-42-5	180.693	hex cry	1202		8.8		
666	Cobalt arsenic sulfide	CoAsS	12254-82-9	165.921	silv-wh solid			≈6.1		
667	Cobalt arsenide	CoAs	27016-73-5	133.855	orth cry	1180		8.22		
668	Cobalt arsenide	$CoAs_2$	12044-42-7	208.776	monocl cry			7.2		
669	Cobalt arsenide	$CoAs_3$	12256-04-1	283.698	cub cry	942		6.84		
670	Cobalt boride	Co_2B	12045-01-1	128.677	refrac solid	1280		8.1		
671	Cobalt boride	CoB	12006-77-8	69.744	refrac solid	1460		7.25		reac H_2O, HNO_3
672	Cobalt carbonyl	$Co_2(CO)_8$	10210-68-1	341.947	oran cry	51 dec		1.78		i H_2O; s EtOH, eth, CS_2
673	Cobalt phosphide	Co_2P	12134-02-0	148.840	gray needles	1386		6.4		i H_2O; s HNO_3
674	Cobalt silicide	$CoSi_2$	12017-12-8	115.104	gray cub cry	1326		4.9		s hot HCl
675	Cobalt disulfide	CoS_2	12013-10-4	123.065	cub cry			4.3		
676	Cobalt dodecacarbonyl	$Co_4(CO)_{12}$	17786-31-1	571.854	blk cry	60 dec		2.09		
677	Cobalt(II) acetate	$Co(C_2H_3O_2)_2$	71-48-7	177.022	pink cry					vs H_2O; s EtOH
678	Cobalt(II) acetate tetrahydrate	$Co(C_2H_3O_2)_2 \cdot 4H_2O$	6147-53-1	249.082	red monocl cry			1.705		s H_2O, EtOH, dil acid
679	Cobalt(II) aluminate	$CoAl_2O_4$	13820-62-7	176.894	blue cub cry			4.37		i H_2O
680	Cobalt(II) arsenate octahydrate	$Co_3(AsO_4)_2 \cdot 8H_2O$	24719-19-5	598.760	red monocl needles	400 dec	1000 dec	3.0		i H_2O; s dil acid
681	Cobalt(II) bromate hexahydrate	$Co(BrO_3)_2 \cdot 6H_2O$	13476-01-2	422.829	viol cry			≈2.5		vs H_2O
682	Cobalt(II) bromide	$CoBr_2$	7789-43-7	218.741	grn hex cry; hyg	678		4.91	113.2^{20}	s MeOH, EtOH, ace
683	Cobalt(II) bromide hexahydrate	$CoBr_2 \cdot 6H_2O$	13762-12-4	326.832	red hyg cry	47 dec	100 dec	2.46	113.2	
684	Cobalt(II) carbonate	$CoCO_3$	513-79-1	118.942	pink rhomb cry			4.2	0.00014^{20}	i EtOH
685	Cobalt(II) chloride	$CoCl_2$	7646-79-9	129.838	blue hyg leaflets	740	1049	3.36	56.2^{25}	s EtOH, eth, ace, py
686	Cobalt(II) chloride dihydrate	$CoCl_2 \cdot 2H_2O$	16544-92-6	165.869	viol-blue cry			2.477	56.2^{25}	
687	Cobalt(II) chloride hexahydrate	$CoCl_2 \cdot 6H_2O$	7791-13-1	237.929	pink-red monocl cry	87 dec		1.924	56.2^{25}	s EtOH, ace, eth
688	Cobalt(II) chromate	$CoCrO_4$	24613-38-5	174.927	yel-brn orth cry			≈4.0		i H_2O; s acid
689	Cobalt(II) chromite	$CoCr_2O_4$	13455-25-9	226.923	blue-grn cub cry			5.14		i H_2O, conc acid
690	Cobalt(II) cyanide	$Co(CN)_2$	542-84-7	110.967	blue hyg cry			1.872		i H_2O
691	Cobalt(II) cyanide dihydrate	$Co(CN)_2 \cdot 2H_2O$	20427-11-6	146.998	pink-brn needles					i H_2O, acid
692	Cobalt(II) ferricyanide	$Co_3[Fe(CN)_6]_2$	14049-81-1	600.699	red needles					i H_2O, HCl; s NH_4OH

No.	Name	Formula	CAS Reg No.	Mol. Weight	Physical Form	mp/°C	bp/°C	Density g cm^{-3}	Solubility g/100 g H_2O	Qualitative Solubility
693	Cobalt(II) fluoride	CoF_2	10026-17-2	96.930	red tetr cry	1127	≈1400	4.46	1.4^{25}	s acid
694	Cobalt(II) fluoride tetrahydrate	$CoF_2 \cdot 4H_2O$	13817-37-3	168.992	red orth cry	dec		2.22	1.4^{25}	
695	Cobalt(II) formate dihydrate	$Co(CHO_2)_2 \cdot 2H_2O$	6424-20-0	184.998	red cry powder	140 dec		2.13	5.03^{20}	i EtOH
696	Cobalt(II) hexafluoro-2,4-pentanedioate	$Co(CF_3COCHCOCF_3)_2$	19648-83-0	473.035	pow	197				
697	Cobalt(II) hexafluorosilicate hexahydrate	$CoSiF_6 \cdot 6H_2O$	12021-68-0	309.100	pale red cry			2.087	76.8^{22}	
698	Cobalt(II) hydroxide	$Co(OH)_2$	21041-93-0	92.948	blue-grn cry	≈160 dec		3.60		sl H_2O; s acid
699	Cobalt(II) iodate	$Co(IO_3)_2$	13455-28-2	408.738	blk-viol needles	200 dec		5.09	0.46^{20}	
700	Cobalt(II) iodide	CoI_2	15238-00-3	312.742	blk hex cry; hyg	520		5.60	203^{25}	
701	Cobalt(II) iodide hexahydrate	$CoI_2 \cdot 6H_2O$	15238-00-3*	420.833	red hex prisms	130 dec		2.90	203^{25}	s EtOH, eth, ace
702	Cobalt(II) titanate	$CoTiO_3$	12017-01-5	154.798	grn rhomb cry			5.0		
703	Cobalt(II) molybdate	$CoMoO_4$	13762-14-6	218.87	blk monocl cry	1040		4.7		
704	Cobalt(II) nitrate	$Co(NO_3)_2$	10141-05-6	182.942	pale red powder	100 dec		2.49	103^{25}	
705	Cobalt(II) nitrate hexahydrate	$Co(NO_3)_2 \cdot 6H_2O$	10026-22-9	291.034	red monocl cry; hyg	≈55		1.88	103^{25}	s EtOH
706	Cobalt(II) oxalate	CoC_2O_4	814-89-1	146.952	pink powder	250 dec		3.02	0.0037^{20}	s acid, NH_4OH
707	Cobalt(II) oxalate dihydrate	$CoC_2O_4 \cdot 2H_2O$	5965-38-8	182.982	pink needles	dec			0.0037	sl acid; s NH_4OH
708	Cobalt(II) oxide	CoO	1307-96-6	74.932	gray cub cry	1830		6.44		i H_2O; s acid
709	Cobalt(II) perchlorate	$Co(ClO_4)_2$	13455-31-7	257.833	red needles			3.33	113^{25}	i EtOH, ace
710	Cobalt(II) phosphate octahydrate	$Co_3(PO_4)_2 \cdot 8H_2O$	10294-50-5	510.865	pink amorp powder			2.77		i H_2O; s acid
711	Cobalt(II) potassium sulfate hexahydrate	$CoK_2(SO_4)_2 \cdot 6H_2O$	10026-20-7	437.349	red monocl cry	75 dec		2.22		vs H_2O
712	Cobalt(II) selenate pentahydrate	$CoSeO_4 \cdot 5H_2O$	14590-19-3	291.97	red tricl cry	dec		2.51	55^{15}	
713	Cobalt(II) selenide	CoSe	1307-99-9	137.89	yel hex cry	1055		7.65		i H_2O, alk; s aqua regia
714	Cobalt(II) selenite dihydrate	$CoSeO_3 \cdot 2H_2O$	19034-13-0	221.92	blue-red powder					i H_2O
715	Cobalt(II) orthosilicate	Co_2SiO_4	12017-08-2	209.950	red-viol orth cry	1345		4.63		i H_2O; s dil HCl
716	Cobalt(II) stannate	Co_2SnO_4	12139-93-4	300.574	grn-blue cub cry			6.30		i H_2O; s alk
717	Cobalt(II) sulfate	$CoSO_4$	10124-43-3	154.997	red orth cry	>700		3.71	38.3^{25}	
718	Cobalt(II) sulfate heptahydrate	$CoSO_4 \cdot 7H_2O$	10026-24-1	281.103	pink monocl cry	41 dec		2.03	38.3^{25}	sl EtOH, MeOH
719	Cobalt(II) sulfate monohydrate	$CoSO_4 \cdot H_2O$	13455-34-0	173.012	red monocl cry			3.08	38.3^{25}	
720	Cobalt(II) sulfide	CoS	1317-42-6	90.999	blk amorp powder	1182		5.45		i H_2O; s acid
721	Cobalt(II) telluride	CoTe	12017-13-9	186.53	hex cry			≈8.8		
722	Cobalt(II) thiocyanate	$Co(SCN)_2$	3017-60-5	175.099	yel-brn pow				103^{25}	s EtOH, MeOH, ace, eth
723	Cobalt(II) thiocyanate trihydrate	$Co(SCN)_2 \cdot 3H_2O$	97126-35-7	229.145	viol rhomb cry				103^{25}	s EtOH, eth, ace
724	Cobalt(II) tungstate	$CoWO_4$	12640-47-0	306.77	blue monocl cry			≈7.8		i H_2O; s hot conc acid
725	Cobalt(II,III) oxide	Co_3O_4	1308-06-1	240.798	blk cub cry	900 dec		6.11		i H_2O; s acid, alk
726	Cobalt(III) acetate	$Co(C_2H_3O_2)_3$	917-69-1	236.064	grn hyg cry	100 dec				s H_2O, EtOH
727	Cobalt(III) ammonium tetranitrodiammine	$NH_4[Co(NH_3)_2(NO_2)_4]$	13600-89-0	295.054	red-brn orth cry			1.97		s H_2O
728	Cobalt(III) fluoride	CoF_3	10026-18-3	115.928	brn hex cry	927		3.88		reac H_2O
729	Cobalt(III) hexammine chloride	$Co(NH_3)_6Cl_3$	10534-89-1	267.474	red monocl cry			1.71		s H_2O; i EtOH
730	Cobalt(III) hydroxide	$Co(OH)_3$	1307-86-4	109.955	brn powder	dec		≈4		i H_2O; s acid
731	Cobalt(III) nitrate	$Co(NO_3)_3$	15520-84-0	244.948	grn cub cry; hyg			≈3.0		s H_2O; reac os
732	Cobalt(III) oxide	Co_2O_3	1308-04-9	165.864	gray-blk powder	895 dec		5.18		i H_2O; s conc acid
733	Cobalt(III) oxide monohydrate	$Co_2O_3 \cdot H_2O$	12016-80-7	183.880	brn-blk hex cry	150 dec				i H_2O; s acid
734	Cobalt(III) potassium nitrite sesquihydrate	$CoK_3(NO_2)_6 \cdot 1.5H_2O$	13782-01-9*	479.284	yel cub cry			2.6		sl H_2O; reac acid; i EtOH
735	Cobalt(III) sulfide	Co_2S_3	1332-71-4	214.064	blk cub cry			4.8		reac acid
736	Cobalt(III) titanate	Co_2TiO_4	12017-38-8	229.731	grn-blk cub cry			5.1		s conc HCl
737	Copper	Cu	7440-50-8	63.546	red metal; cub	1084.62	2562	8.96		sl dil acid

738	Copper(II) 2,4-pentanedioate	$Cu(CH_3COCHCOCH_3)_2$	13395-16-9	261.762	blue powder	284 dec	subl			sl H_2O; s chl
739	Copper nitride	Cu_3N	1308-80-1	204.645	cub cry	300 dec		5.84		
740	Copper(II) 2-ethylhexanoate	$Cu(C_8H_{15}O_2)_2$	149-11-1	349.953	pow	252 dec				
741	Copper phosphide	CuP_2	12019-11-3	125.494	monocl cry	≈900		4.20		
742	Copper silicide	Cu_5Si	12159-07-8	345.816	solid	825				
743	Copper(I) acetate	$CuC_2H_3O_2$	598-54-9	122.590	col cry	dec	subl			reac H_2O
744	Copper(I) acetylide	Cu_2C_2	1117-94-8	151.113	red amorp powder; exp					
745	Copper(I) azide	CuN_3	14336-80-2	105.566	tetr cry; exp					
746	Copper(I) bromide	$CuBr$	7787-70-4	143.450	wh cub cry; hyg	497	1345	4.98	0.0012^{20}	i ace
747	Copper(I) chloride	$CuCl$	7758-89-6	98.999	wh cub cry	430	≈1400	4.14	0.0047^{20}	i EtOH, ace
748	Copper(I) cyanide	$CuCN$	544-92-3	89.564	wh powder or grn orth cry	474	dec	2.9		i H_2O, EtOH; s KCN soln
749	Copper(I) fluoride	CuF	13478-41-6	82.544	cub cry			7.1		
750	Copper(I) hydride	CuH	13517-00-5	64.554	red-brn solid	60 dec				
751	Copper(I) iodide	CuI	7681-65-4	190.450	wh cub cry	606	≈1290	5.67	0.000020^{20}	i dil acid
752	Copper(I) mercury iodide	Cu_2HgI_4	13876-85-2	835.30	red cry powder	trans ≈60 (brn)				i H_2O, EtOH
753	Copper(I) oxide	Cu_2O	1317-39-1	143.091	red-brn cub cry	1235	1800 dec	6.0		i H_2O
754	Copper(I) selenide	Cu_2Se	20405-64-5	206.05	blue-blk tetr cry	1113		6.84		i H_2O; s acid
755	Copper(I) sulfide	Cu_2S	22205-45-4	159.158	blue-blk orth cry	≈1100		5.6		i H_2O; sl acid
756	Copper(I) sulfite monohydrate	$Cu_2SO_3 \cdot H_2O$		225.172	cry			3.83		sl H_2O; s HCl
757	Copper(I) sulfite hemihydrate	$Cu_2SO_3 \cdot 0.5H_2O$	13982-53-1*	216.164	wh-yel hex cry					sl H_2O; s acid, alk; i EtOH, eth
758	Copper(I) telluride	Cu_2Te	12019-52-2	254.69	blue hex cry	1127		4.6		
759	Copper(I) thiocyanate	$CuSCN$	1111-67-7	121.630	wh-yel amorp powder	1084		2.85		i H_2O, dil acid, EtOH, ace; s eth
760	Copper(I,II) sulfite dihydrate	$Cu_2SO_3 \cdot CuSO_3 \cdot 2H_2O$	13814-81-8	386.797	red prisms or powder					i H_2O, EtOH; s HCl
761	Copper(II) acetate	$Cu(C_2H_3O_2)_2$	142-71-2	181.635	blue-grn hyg powder					
762	Copper(II) acetate metaarsenite	$Cu(C_2H_3O_2)_2 \cdot 3Cu(AsO_2)_2$	12002-03-8	1013.795	grn cry powder					i H_2O; reac acid
763	Copper(II) acetate monohydrate	$Cu(C_2H_3O_2)_2 \cdot H_2O$	6046-93-1	199.650	grn monocl cry	115	240 dec	1.88		s H_2O, EtOH; sl eth
764	Copper(II) acetylide	CuC_2	12540-13-5	87.567	brn-blk solid; exp	exp 100				
765	Copper(II) arsenate	$Cu_3(AsO_4)_2$	10103-61-4	468.476	blue-grn cry					i H_2O, EtOH; s dil acid
766	Copper(II) arsenite	$CuHAsO_3$	10290-12-7	187.474	yel-grn powder					i H_2O, EtOH; s acid
767	Copper(II) azide	$Cu(N_3)_2$	14215-30-6	147.586	brn orth cry; exp			≈2.6		
768	Copper(II) basic acetate	$Cu(C_2H_3O_2)_2 \cdot CuO \cdot 6H_2O$	52503-64-7	369.271	blue-grn cry or powder					sl H_2O, EtOH; s dil acid, NH_4OH
769	Copper(II) borate	$Cu(BO_2)_2$	39290-85-2	149.166	blue-grn powder			3.859		i H_2O; s acid
770	Copper(II) bromide	$CuBr_2$	7789-45-9	223.354	blk monocl cry; hyg	498	900	4.710	126^{25}	s EtOH, ace; i bz, eth
771	Copper(II) butanoate monohydrate	$Cu(C_4H_7O_2)_2 \cdot H_2O$	540-16-9	255.756	grn monocl plates					s H_2O, diox, bz; sl EtOH
772	Copper(II) carbonate hydroxide	$CuCO_3 \cdot Cu(OH)_2$	12069-69-1	221.116	grn monocl cry	200 dec		4.0		i H_2O, EtOH; s dil acid
773	Copper(II) chlorate hexahydrate	$Cu(ClO_3)_2 \cdot 6H_2O$	14721-21-2	338.539	blue-grn hyg cry	65	100 dec		164^{18}	vs EtOH
774	Copper(II) chloride	$CuCl_2$	7447-39-4	134.451	yel-brn monocl cry; hyg	630 dec		3.4	75.7^{25}	s EtOH, ace
775	Copper(II) chloride dihydrate	$CuCl_2 \cdot 2H_2O$	10125-13-0	170.482	grn-blue orth cry; hyg	100 dec		2.51	75.7^{20}	vs EtOH, MeOH; s ace; i eth
776	Copper(II) chloride hydroxide	$Cu_2(OH)_3Cl$	1332-65-6	213.567	pale grn cry					i H_2O; s acid
777	Copper(II) chromate	$CuCrO_4$	13548-42-0	179.540	red-brn cry					i H_2O; s EtOH
778	Copper(II) chromite	$CuCr_2O_4$	12018-10-9	231.536	gray-blk tetr cry			5.4		i H_2O, dil acid
779	Copper(II) citrate hemipentahydrate	$Cu_2C_6H_4O_7 \cdot 2.5H_2O$	10402-15-0	360.221	blue-grn cry	100 dec				sl H_2O; s dil acid
780	Copper(II) cyanide	$Cu(CN)_2$	14763-77-0	115.580	grn powder					i H_2O; s acid, alk
781	Copper(II) cyclohexanebutanoate	$Cu(C_{10}H_{17}O_2)_2$	2218-80-6	402.028	pow	126 dec				
782	Copper(II) dichromate dihydrate	$CuCr_2O_7 \cdot 2H_2O$	13675-47-3	315.565	red-brn tricl cry			2.286		vs H_2O
783	Copper(II) ethanolate	$Cu(C_2H_5O)_2$	2850-65-9	153.667	blue hyg solid	120 dec				i os
784	Copper(II) ethylacetoacetate	$Cu(C_2H_5CO_2CHCOCH_3)_2$	14284-06-1	321.813	pow	192				s EtOH
785	Copper(II) ferrocyanide	$Cu_2Fe(CN)_6$	13601-13-3	339.041	red-br cub cry or powder			2.2		i H_2O, acid, os

No.	Name	Formula	CAS Reg No.	Mol. Weight	Physical Form	mp/°C	bp/°C	Density g cm^{-3}	Solubility g/100 g H_2O	Qualitative Solubility
786	Copper(II) ferrous sulfide	$CuFeS_2$	1308-56-1	183.523	yel tetr cry	950		4.2		i H_2O, HCl; s HNO_3
787	Copper(II) fluoride	CuF_2	7789-19-7	101.543	wh monocl cry	836	1676	4.23	0.075^{25}	
788	Copper(II) fluoride dihydrate	$CuF_2 \cdot 2H_2O$	13454-88-1	137.574	blue monocl cry	130 dec		2.934	0.075^{25}	
789	Copper(II) formate	$Cu(CHO_2)_2$	544-19-4	153.581	blue cry				12.5^{20}	i os
790	Copper(II) formate tetrahydrate	$Cu(CHO_2)_2 \cdot 4H_2O$	5893-61-8	225.641	blue monocl cry				12.5	sl EtOH; i os
791	Copper(II) hexafluoro-2,4-pentanedioate	$Cu(CF_3COCHCOCF_3)_2$	14781-45-4	477.648	cry	98	220 dec			s MeOH, ace, tol
792	Copper(II) hexafluorosilicate tetrahydrate	$CuSiF_6 \cdot 4H_2O$	12062-24-7	277.684	blue monocl cry	dec		2.56	99.7^{17}	sl EtOH
793	Copper(II) hydroxide	$Cu(OH)_2$	20427-59-2	97.561	blue-grn powder			3.37		i H_2O; s acid, conc alk
794	Copper(II) iodate	$Cu(IO_3)_2$	13454-89-2	413.351	grn mono cry	dec		5.241	0.15^{20}	s dil acid
795	Copper(II) iodate monohydrate	$Cu(IO_3)_2 \cdot H_2O$	13454-90-5	431.367	blue tricl cry	248 dec		4.872	0.15^{20}	s dil H_2SO_4
796	Copper(II) molybdate	$CuMoO_4$	13767-34-5	223.48	grn cry	≈500		3.4	0.038	
797	Copper(II) nitrate	$Cu(NO_3)_2$	3251-23-8	187.555	blue-grn orth cry; hyg	255	subl		145^{25}	s diox; reac eth
798	Copper(II) nitrate hexahydrate	$Cu(NO_3)_2 \cdot 6H_2O$	13478-38-1	295.647	blue rhomb cry; hyg			2.07	145^{25}	s EtOH
799	Copper(II) nitrate trihydrate	$Cu(NO_3)_2 \cdot 3H_2O$	10031-43-3	241.602	blue rhomb cry	114	170 dec	2.32	145^{25}	vs EtOH
800	Copper(II) oleate	$Cu(C_{18}H_{33}O_2)_2$	1120-44-1	626.453	blue-grn solid					i H_2O; sl EtOH; s eth
801	Copper(II) oxalate	CuC_2O_4	814-91-5	151.565	blue-wh powder	310 dec			0.0026^{20}	i EtOH, eth; s NH_4OH
802	Copper(II) oxalate hemihydrate	$CuC_2O_4 \cdot 0.5H_2O$	814-91-5*	144.573	blue-wh cry	200 dec			0.0026^{20}	s NH_4OH
803	Copper(II) oxide	CuO	1317-38-0	79.545	blk powder or monocl cry	1446		6.31		i H_2O, EtOH; s dil acid
804	Copper(II) perchlorate	$Cu(ClO_4)_2$	13770-18-8	262.446	grn hyg cry	130 dec			146^{30}	s eth, diox; i bz, ctc
805	Copper(II) perchlorate hexahydrate	$Cu(ClO_4)_2 \cdot 6H_2O$	10294-46-9	370.538	blue monocl cry; hyg	82	120 dec	2.22	146^{30}	vs EtOH, HOAc, ace; sl eth
806	Copper(II) phosphate	$Cu_3(PO_4)_2$	7798-23-4	380.581	blue-grn tricl cry					i H_2O; s acid, NH_4OH
807	Copper(II) phosphate trihydrate	$Cu_3(PO_4)_2 \cdot 3H_2O$	10031-48-8	434.627	blue-grn orth cry					i H_2O; s acid, NH_4OH
808	Copper(II) selenate pentahydrate	$CuSeO_4 \cdot 5H_2O$	10031-45-5	296.58	blue tricl cry	80 dec		2.56	27.4^{25}	s acid, NH_4OH; sl ace; i EtOH
809	Copper(II) selenide	$CuSe$	1317-41-5	142.51	blue-blk needles or plates	550 dec		5.99		reac acid
810	Copper(II) selenite dihydrate	$CuSeO_3 \cdot 2H_2O$	15168-20-4	226.54	blue orth cry			3.31		i H_2O; s acid, NH_4OH
811	Copper(II) stearate	$Cu(C_{18}H_{35}O_2)_2$	660-60-6	630.485	blue-grn amorp powder	≈250				i H_2O, EtOH, eth; s py
812	Copper(II) sulfate	$CuSO_4$	7758-98-7	159.610	wh-grn amorp powder or rhomb cry	560 dec		3.60	22.0^{25}	i EtOH
813	Copper(II) sulfate pentahydrate	$CuSO_4 \cdot 5H_2O$	7758-99-8	249.686	blue tricl cry	110 dec		2.286	22.0^{25}	s MeOH; sl EtOH
814	Copper(II) sulfate, basic	$Cu_3(OH)_4SO_4$	1332-14-5	354.731	grn rhomb cry			3.88		i H_2O
815	Copper(II) sulfide	CuS	1317-40-4	95.612	blk hex cry	trans 507		4.76		i H_2O, EtOH, dil acid, alk
816	Copper(II) tartrate trihydrate	$CuC_4H_4O_6 \cdot 3H_2O$	815-82-7	265.663	blue-grn powder					sl H_2O; s acid, alk
817	Copper(II) telluride	$CuTe$	12019-23-7	191.15	yel orth cry	trans ≈400		7.09		
818	Copper(II) tetrafluoroborate	$Cu(BF_4)_2$	14735-84-3	237.155	solid					s H_2O
819	Copper(II) tungstate	$CuWO_4$	13587-35-4	311.38	yel-brn powder			7.5		
820	Copper(II) tungstate dihydrate	$CuWO_4 \cdot 2H_2O$	13587-35-4*	347.41	grn powder					i H_2O; sl HOAc; reac conc acid
821	Copper(II) vanadate	$Cu(VO_3)_2$	12789-09-2	261.425	pow					
822	Curium	Cm	7440-51-9	247	silv metal; hex or cub	1345	≈3100	13.51		
823	Dysprosium	Dy	7429-91-6	162.50	silv metal; hex	1412	2567	8.55		s dil acid
824	Dysprosium boride	DyB_4	12310-43-9	205.74	tetr cry	2500		6.98		
825	Dysprosium nitride	DyN	12019-88-4	176.51	cub cry			9.93		
826	Dysprosium silicide	$DySi_2$	12133-07-2	218.67	orth cry			5.2		
827	Dysprosium(II) chloride	$DyCl_2$	13767-31-2	233.41	blk cry	721 dec				reac H_2O
828	Dysprosium(II) iodide	DyI_2	36377-94-3	416.31	purp cry	659				reac H_2O
829	Dysprosium(III) bromide	$DyBr_3$	14456-48-5	402.21	wh hyg cry	879				s H_2O
830	Dysprosium(III) chloride	$DyCl_3$	10025-74-8	268.86	yel cry	680		3.67		s H_2O

PHYSICAL CONSTANTS OF INORGANIC COMPOUNDS (CONTINUED)

831	Dysprosium(III) fluoride	DyF_3	13569-80-7	219.50	grn cry	1154				
832	Dysprosium(III) hydride	DyH_3	13537-09-2	165.52	hex cry			7.1		
833	Dysprosium(III) iodide	DyI_3	15474-63-2	543.21	grn cry	978				
834	Dysprosium(III) nitrate pentahydrate	$Dy(NO_3)_3 \cdot 5H_2O$	10143-38-1*	438.59	yel cry	88.6			208.4^{25}	
835	Dysprosium(III) oxide	Dy_2O_3	1308-87-8	373.00	wh cub cry	2228	3900	7.81		s acid
836	Dysprosium(III) sulfide	Dy_2S_3	12133-10-7	421.20	red-brn monocl cry			6.08		
837	Einsteinium	Es	7429-92-7	252	metal; cub	860				
838	Erbium	Er	7440-52-0	167.26	silv metal; hex	1529	2868	9.07		i H_2O; s acid
839	Erbium boride	ErB_4	12310-44-0	210.50	tetr cry	2450		7.0		
840	Erbium bromide	$ErBr_3$	13536-73-7	406.97	viol hyg cry	923				s H_2O
841	Erbium chloride	$ErCl_3$	10138-41-7	273.62	viol monocl cry; hyg	776		4.1		s H_2O
842	Erbium chloride hexahydrate	$ErCl_3 \cdot 6H_2O$	10025-75-9	381.71	pink hyg cry	dec				s H_2O; sl EtOH
843	Erbium fluoride	ErF_3	13760-83-3	224.26	pink orth cry	1147		7.8		i H_2O
844	Erbium hydride	ErH_3	13550-53-3	170.28	hex cry			≈7.6		
845	Erbium iodide	ErI_3	13813-42-8	547.97	viol hex cry; hyg	1014		≈5.5		s H_2O
846	Erbium nitrate pentahydrate	$Er(NO_3)_3 \cdot 5H_2O$	10168-80-6*	443.35	red cry	130 dec			240.8^{25}	s EtOH, ace
847	Erbium nitride	ErN	12020-21-2	181.27	cub cry			10.6		
848	Erbium oxide	Er_2O_3	12061-16-4	382.52	pink powder	2344	3920	8.64		i H_2O; s acid
849	Erbium silicide	$ErSi_2$	12020-28-9	223.43	orth cry			7.26		
850	Erbium sulfate	$Er_2(SO_4)_3$	13478-49-4	622.71	hyg powder	dec		3.68	13^{20}	
851	Erbium sulfate octahydrate	$Er_2(SO_4)_3 \cdot 8H_2O$	10031-52-4	766.83	pink monocl cry	dec		3.20	13^{20}	
852	Erbium sulfide	Er_2S_3	12159-66-9	430.72	red-brn monocl cry	1730		6.07		
853	Erbium telluride	Er_2Te_3	12020-39-2	717.32	orth cry	1213		7.11		
854	Europium	Eu	7440-53-1	151.964	soft silv metal; cub	822	1529	5.24		reac H_2O
855	Europium boride	EuB_6	12008-05-8	216.830	cub cry	≈2600		4.91		
856	Europium nitride	EuN	12020-58-5	165.971	cub cry			8.7		
857	Europium silicide	$EuSi_2$	12434-24-1	208.135	tetr cry	1500		5.46		
858	Europium(II) bromide	$EuBr_2$	13780-48-8	311.772	wh cry	683				s H_2O
859	Europium(II) chloride	$EuCl_2$	13769-20-5	222.869	wh orth cry	731		4.9		s H_2O
860	Europium(II) fluoride	EuF_2	14077-39-5	189.961	grn-yel cub cry	≈1380		6.5		
861	Europium(II) iodide	EuI_2	22015-35-6	405.773	grn cry	580				s H_2O
862	Europium(II) selenide	EuSe	12020-66-5	230.92	brn cub cry			6.45		
863	Europium(II) sulfate	$EuSO_4$	10031-54-6	248.028	col orth cry			4.99		i H_2O
864	Europium(II) sulfide	EuS	12020-65-4	184.030	cub cry			5.7		
865	Europium(II) telluride	EuTe	12020-69-8	279.56	blk cub cry	1526		6.48		
866	Europium(III) bromide	$EuBr_3$	13759-88-1	391.676	gray cry	dec				s H_2O
867	Europium(III) chloride	$EuCl_3$	10025-76-0	258.322	grn-yel needles	623		4.89		
868	Europium(III) chloride hexahydrate	$EuCl_3 \cdot 6H_2O$	13759-92-7	366.413	wh-yel hyg cry	850		4.89		s H_2O
869	Europium(III) fluoride	EuF_3	13765-25-8	208.959	wh hyg cry	1276				i H_2O
870	Europium(III) nitrate hexahydrate	$Eu(NO_3)_3 \cdot 6H_2O$	10031-53-5	446.070	wh-pink hyg cry	85 dec			193^{25}	
871	Europium(III) oxide	Eu_2O_3	1308-96-9	351.926	pink powder	2291	3790	7.42		i H_2O; s acid
872	Europium(III) sulfate	$Eu_2(SO_4)_3$	13537-15-0	592.119	pale pink cry			4.99	2.1^{20}	
873	Europium(III) sulfate octahydrate	$Eu_2(SO_4)_3 \cdot 8H_2O$	10031-52-4	736.241	pink cry	375 dec			2.1^{20}	
874	Fermium	Fm	7440-72-4	257	metal	1527				
875	Fluorine	F_2	7782-41-4	37.997	pale yel gas	-219.67 tp	-188.12	1.553 g/L		reac H_2O
876	Fluorine monoxide	F_2O	7783-41-7	53.996	col gas	-223.8	-144.75	2.207 g/L		sl H_2O
877	Fluorine dioxide	F_2O_2	7783-44-0	69.996	gas, stable only at low temp	-154	-57	2.861 g/L		
878	Fluorine nitrate	FNO_3	7789-26-6	81.003	col gas	-175	-46	3.311 g/L		reac H_2O, EtOH, eth; s ace

No.	Name	Formula	CAS Reg No.	Mol. Weight	Physical Form	mp/°C	bp/°C	Density g cm^{-3}	Solubility g/100 g H_2O	Qualitative Solubility
879	Fluorine perchlorate	$FOClO_3$	10049-03-3	118.449	col gas; exp	-167.3	-16	4.841 g/L		reac H_2O
880	Francium	Fr	7440-73-5	223	short-lived alkali metal	27				
881	Gadolinium	Gd	7440-54-2	157.25	silv metal; hex	1313	3273	7.90		s dil acid
882	Gadolinium boride	GdB_6	12008-06-9	222.12	blk-brn cub cry	2510		5.31		
883	Gadolinium nitride	GdN	25764-15-2	171.26	cub cry			9.10		
884	Gadolinium silicide	$GdSi_2$	12134-75-7	213.42	orth cry			5.9		
885	Gadolinium(II) iodide	GdI_2	13814-72-7	411.06	bronze cry	831				
886	Gadolinium(II) selenide	GdSe	12024-81-6	236.21	cub cry	2170		8.1		
887	Gadolinium(III) bromide	$GdBr_3$	13818-75-2	396.96	wh monocl cry; hyg	770		4.56		
888	Gadolinium(III) chloride	$GdCl_3$	10138-52-0	263.61	wh monocl cry; hyg	609		4.52		s H_2O
889	Gadolinium(III) chloride hexahydrate	$GdCl_3 \cdot 6H_2O$	19423-81-5	371.70	col hyg cry			2.424		s H_2O
890	Gadolinium(III) fluoride	GdF_3	13765-26-9	214.25	wh cry	1231				
891	Gadolinium(III) iodide	GdI_3	13572-98-0	537.96	yel cry	925				
892	Gadolinium(III) nitrate hexahydrate	$Gd(NO_3)_3 \cdot 6H_2O$	19598-90-4	451.36	hyg tricl cry	91 dec		2.33	190^{25}	s EtOH
893	Gadolinium(III) nitrate pentahydrate	$Gd(NO_3)_3 \cdot 5H_2O$	52788-53-1	433.34	wh cry	92 dec		2.41	190^{25}	
894	Gadolinium(III) oxide	Gd_2O_3	12064-62-9	362.50	wh hyg powder	2339	3900	7.07		i H_2O; s acid
895	Gadolinium(III) sulfate octahydrate	$Gd_2(SO_4)_3 \cdot 8H_2O$	13450-87-8	746.81	col monocl cry	400 dec		4.14	2.3^{20}	
896	Gadolinium(III) sulfide	Gd_2S_3	12134-77-9	410.70	yel cub cry			6.1		
897	Gadolinium(III) telluride	Gd_2Te_3	12160-99-5	697.30	orth cry	1255		7.7		
898	Gallium	Ga	7440-55-3	69.723	silv liq or gray orth cry	29.771 tp	2204	5.91		reac alk
899	Gallium antimonide	GaSb	12064-03-8	191.483	cub cry	712		5.6137		
900	Gallium arsenide	GaAs	1303-00-0	144.645	gray cub cry	1238		5.3176		
901	Gallium nitride	GaN	25617-97-4	83.730	gray hex cry	>2500		6.1		
902	Gallium phosphide	GaP	12063-98-8	100.697	yel cub cry	1457		4.138		
903	Gallium suboxide	Ga_2O	12024-20-3	155.445	brn powder	>660	>800 dec	4.77		
904	Gallium(II) chloride	$GaCl_2$	24597-12-4	140.628	wh orth cry	172.4	535	2.74		
905	Gallium(II) selenide	GaSe	12024-11-2	148.68	hex cry	960		5.03		
906	Gallium(II) sulfide	GaS	12024-10-1	101.789	hex cry	965		3.86		
907	Gallium(II) telluride	GaTe	12024-14-5	197.32	monocl cry	824		5.44		
908	Gallium(III) 2,4-pentanedioate	$Ga(CH_3COCHCOCH_3)_3$	14405-43-7	367.047	wh powder	193	subl	1.42		
909	Gallium(III) bromide	$GaBr_3$	13450-88-9	309.435	wh orth cry	121.5	279	3.69		
910	Gallium(III) chloride	$GaCl_3$	13450-90-3	176.081	col needles or gl solid	77.9	201	2.47		
911	Gallium(III) fluoride	GaF_3	7783-51-9	126.718	wh powder or col needles	>1000		4.47		i H_2O
912	Gallium(III) fluoride trihydrate	$GaF_3 \cdot 3H_2O$	22886-66-4	180.764	wh cry	>140 dec				sl H_2O
913	Gallium(III) hydride	GaH_3	13572-93-5	72.747	visc liq	-15	≈0 dec			
914	Gallium(III) hydroxide	$Ga(OH)_3$	12023-99-3	120.745	unstable prec					
915	Gallium(III) iodide	GaI_3	13450-91-4	450.436	monocl cry	212	340	4.5		
916	Gallium(III) nitrate	$Ga(NO_3)_3$	13494-90-1	255.738	wh cry powder					s H_2O, EtOH, eth
917	Gallium(III) oxide	Ga_2O_3	12024-21-4	187.444	wh cry	1806		≈6.0		s hot acid
918	Gallium(III) oxide hydroxide	GaOOH	20665-52-5	102.730	orth cry			5.23		
919	Gallium(III) selenide	Ga_2Se_3	12024-24-7	376.33	cub cry	937		4.92		
920	Gallium(III) sulfate	$Ga_2(SO_4)_3$	13494-91-2	427.637	hex cry					
921	Gallium(III) sulfate octadecahydrate	$Ga_2(SO_4)_3 \cdot 18H_2O$	13780-42-2	751.912	octahed cry					s H_2O, EtOH
922	Gallium(III) sulfide	Ga_2S_3	12024-22-5	235.644	monocl cry	1090		3.7		
923	Gallium(III) telluride	Ga_2Te_3	12024-27-0	522.25	cub cry	790		5.57		

924	Germanium	Ge	7440-56-4	72.61	gray-wh cub cry	938.25	2833	5.3234		i H_2O, dil acid, alk
925	Germane	GeH_4	7782-65-2	76.64	col gas; flam	-165	-88.1	3.133 g/L		i H_2O
926	Digermane	Ge_2H_6	13818-89-8	151.27	col liq; flam	-109	29	1.98^{-109}		
927	Trigermane	Ge_3H_8	14691-44-2	225.89	col liq	-105.6	110.5	2.20^{-105}		i H_2O
928	Tetragermane	Ge_4H_{10}	14691-47-5	300.52	col liq		176.9			i H_2O
929	Pentagermane	Ge_5H_{12}	15587-39-0	375.15	col liq		234			i H_2O
930	Bromogermane	GeH_3Br	13569-43-2	155.54	col liq	-32	52	2.34		reac H_2O
931	Chlorogermane	GeH_3Cl	13637-65-5	111.09	col liq	-52	28	1.75		reac H_2O
932	Chlorotrifluorogermane	GeF_3Cl	14188-40-0	165.06	gas	-66.2	-20.3	6.747 g/L		
933	Dibromogermane	GeH_2Br_2	13769-36-3	234.43	col liq	-15	89	2.80		reac H_2O
934	Dichlorogermane	GeH_2Cl_2	15230-48-5	145.53	col liq	-68	69.5	1.90		reac H_2O
935	Dichlorodifluorogermane	GeF_2Cl_2	24422-21-7	181.51	col gas	-51.8	-2.8	7.419 g/L		
936	Fluorogermane	GeH_3F	13537-30-9	94.63	col gas			3.868 g/L		reac H_2O
937	Iodogermane	GeH_3I	13573-02-9	202.54	liq	-15	≈90			reac H_2O
938	Tribromogermane	$GeHBr_3$	14779-70-5	313.33	col liq	-25	dec			reac H_2O
939	Trichlorogermane	$GeHCl_3$	1184-65-2	179.98	liq	-71	75.3	1.93		reac H_2O
940	Trichlorofluorogermane	$GeCl_3F$	24422-20-6	197.97	liq	-49.8	37.5			
941	Methylgermane	GeH_3CH_3	1449-65-6	90.67	col gas	-158	-23	3.706 g/L		
942	Germanium(II) bromide	$GeBr_2$	24415-00-7	232.42	yel monocl cry	122	150 dec			reac H_2O
943	Germanium(II) chloride	$GeCl_2$	10060-11-4	143.51	wh-yel hyg powder	dec				reac H_2O; s eth, bz
944	Germanium(II) fluoride	GeF_2	13940-63-1	110.61	wh orth cry; hyg	110	130 dec	3.64		reac H_2O
945	Germanium(II) iodide	GeI_2	13573-08-5	326.42	oran-yel hex cry	550 dec		5.4		reac H_2O
946	Germanium(II) oxide	GeO	20619-16-3	88.61	blk solid	700 dec				
947	Germanium(II) selenide	GeSe	12065-10-0	151.57	gray orth cry or brn powder	667		5.6		
948	Germanium(II) sulfide	GeS	12025-32-0	104.68	gray orth cry	615		4.1		
949	Germanium(II) telluride	GeTe	12025-39-7	200.21	cub cry	725		6.16		i H_2O; s conc HNO_3
950	Germanium(IV) bromide	$GeBr_4$	13450-92-5	392.23	wh cry	26.1	186.35	3.132		reac H_2O
951	Germanium(IV) chloride	$GeCl_4$	10038-98-9	214.42	col liq	-51.50	86.55	1.88		reac H_2O; s bz, eth, EtOH, ctc
952	Germanium(IV) fluoride	GeF_4	7783-58-6	148.60	col gas	-15 tp	-36.5 sp	6.074 g/L		reac H_2O
953	Germanium(IV) iodide	GeI_4	13450-95-8	580.23	red-oran cub cry	146	377	4.322		reac H_2O
954	Germanium(IV) nitride	Ge_3N_4	12065-36-0	273.86	orth cry	900 dec				i H_2O, acid, aqua regia
955	Germanium(IV) oxide	GeO_2	1310-53-8	104.61	wh hex cry	1115		4.25		i H_2O
956	Germanium(IV) selenide	$GeSe_2$	12065-11-1	230.53	yel-oran orth ccry	707 dec		4.56		
957	Germanium(IV) sulfide	GeS_2	12025-34-2	136.74	blk orth cry	530		3.01		
958	Gold	Au	7440-57-5	196.967	soft yel metal	1064.18	2856	19.3		s aqua regia
959	Bromoauric acid pentahydrate	$HAuBr_4 \cdot 5H_2O$	17083-68-0	607.667	red-brn hyg cry	27				s H_2O, EtOH
960	Chloroauric acid tetrahydrate	$HAuCl_4 \cdot 4H_2O$	16903-35-8	411.847	yel monocl cry; hyg			≈3.9		vs H_2O, EtOH; s eth
961	Gold(I) bromide	AuBr	10294-27-6	276.871	yel-gray tetr cry	165 dec		8.20		i H_2O
962	Gold(I) chloride	AuCl	10294-29-8	232.420	yel orth cry	289 dec		7.6	0.000031^{20}	
963	Gold(I) cyanide	AuCN	506-65-0	222.985	yel hex cry	dec		7.2		i H_2O, EtOH, eth, dil acid
964	Gold(I) iodide	AuI	10294-31-2	323.871	yel-grn powder; tetr	120 dec		8.25		i H_2O; s CN soln
965	Gold(I) sulfide	Au_2S	1303-60-2	425.999	brn-blk cub cry; unstable	240 dec		≈11		i H_2O, acid; s aqua regia
966	Gold(III) bromide	$AuBr_3$	10294-28-7	436.679	red-br monocl cry	≈160 dec				s H_2O, EtOH
967	Gold(III) chloride	$AuCl_3$	13453-07-1	303.325	red monocl cry	>160 dec		4.7	68^{20}	
968	Gold(III) cyanide trihydrate	$Au(CN)_3 \cdot 3H_2O$	535-37-5*	329.065	wh hyg cry	50 dec				vs H_2O; sl EtOH
969	Gold(III) fluoride	AuF_3	14720-21-9	253.962	oran-yel hex cry	>300	subl	6.75		
970	Gold(III) hydroxide	$Au(OH)_3$	1303-52-2	247.989	brn powder	≈100 dec				i H_2O; s acid
971	Gold(III) iodide	AuI_3	31032-13-0	577.680	unstable grn powder	20 dec				

No.	Name	Formula	CAS Reg No.	Mol. Weight	Physical Form	mp/°C	bp/°C	Density g cm⁻³	Solubility g/100 g H_2O	Qualitative Solubility
972	Gold(III) oxide	Au_2O_3	1303-58-8	441.931	brn powder	≈150 dec				i H_2O; s acid
973	Gold(III) selenate	$Au_2(SeO_4)_3$	10294-32-3	822.81	yel cry					i H_2O; s acid
974	Gold(III) selenide	Au_2Se_3	1303-62-4	630.81	blk amorp solid	dec		4.65		s aqua regia
975	Gold(III) sulfide	Au_2S_3	1303-61-3	490.131	unstable blk powder	200 dec				
976	Hafnium	Hf	7440-58-6	178.49	gray metal; hex	2233	4603	13.3		s HF
977	Hafnium boride	HfB_2	12007-23-7	200.11	gray hex cry	3100		10.5		
978	Hafnium(IV) bromide	$HfBr_4$	13777-22-5	498.11	wh cub cry	424 tp	323 sp	4.90		
979	Hafnium carbide	HfC	12069-85-1	190.50	refrac cub cry	≈3000		12.2		
980	Hafnium(IV) chloride	$HfCl_4$	13499-05-3	320.30	wh monocl cry	432 tp	317 sp			reac H_2O
981	Hafnium fluoride	HfF_4	13709-52-9	254.48	wh monocl cry	>970	970 sp	7.1		
982	Hafnium hydride	HfH_2	12770-26-2	180.51	refrac tetr cry			11.4		
983	Hafnium iodide	HfI_4	13777-23-6	686.11	yel-oran cub cry	449 tp	394 sp	5.6		
984	Hafnium nitride	HfN	25817-87-2	192.50	yel-brn cub cry	3305		13.8		
985	Hafnium oxide	HfO_2	12055-23-1	210.49	wh cub cry	2774		9.68		i H_2O
986	Hafnium oxychloride octahydrate	$HfOCl_2 \cdot 8H_2O$	14456-34-9	409.52	wh tetr cry	dec				s H_2O
987	Hafnium phosphide	HfP	12325-59-6	209.46	hex cry			9.78		
988	Hafnium selenide	$HfSe_2$	12162-21-9	336.41	brn hex cry			7.46		
989	Hafnium orthosilicate	$HfSiO_4$	13870-13-8	270.57	tetr cry			7.0		
990	Hafnium silicide	$HfSi_2$	12401-56-8	234.66	gray orth cry	≈1700		7.6		
991	Hafnium sulfate	$Hf(SO_4)_2$	15823-43-5	370.62	wh cry	>500 dec				
992	Hafnium sulfide	HfS_2	18855-94-2	242.62	purp-brn hex cry			6.03		
993	Helium	He	7440-59-7	4.003	col gas		-268.93	0.164 g/L		sl H_2O; i EtOH
994	Holmium	Ho	7440-60-0	164.930	silv metal; hex	1474	2700	8.80		s dil acid
995	Holmium bromide	$HoBr_3$	13825-76-8	404.642	yel hyg cry	919	1470			
996	Holmium chloride	$HoCl_3$	10138-62-2	271.288	yel monocl cry; hyg	718	1500	3.7		s H_2O
997	Holmium fluoride	HoF_3	13760-78-6	221.925	pink-yel orth cry; hyg	1143	>2200	7.664		s H_2O
998	Holmium iodide	HoI_3	13813-41-7	545.643	yel hex cry	994		5.4		
999	Holmium nitride	HoN	12029-81-1	178.937	cub cry			10.6		
1000	Holmium oxide	Ho_2O_3	12055-62-8	377.859	yel cub cry	2330	3900	8.41		s acid
1001	Holmium silicide	$HoSi_2$	12136-24-2	221.101	hex cry			7.1		
1002	Holmium sulfide	Ho_2S_3	12162-59-3	426.059	yel-oran monocl cry			5.92		
1003	Hydrazine	N_2H_4	302-01-2	32.045	col oily liq	1.4	113.55	1.0036		vs H_2O, EtOH, MeOH
1004	Hydrazine hydrate	$N_2H_4 \cdot H_2O$	7803-57-8	50.060	fuming liq	-51.7	119	1.030		vs H_2O, EtOH; i chl, eth
1005	Hydrazine hydrobromide	$N_2H_4 \cdot HBr$	13775-80-9	112.957	wh monocl cry flakes	84	≈190 dec	2.3		s H_2O, EtOH
1006	Hydrazine hydrochloride	$N_2H_4 \cdot HCl$	2644-70-4	68.506	wh orth cry	89	240 dec	1.5		s H_2O; i os
1007	Hydrazine dihydrochloride	$N_2H_4 \cdot 2HCl$	5341-61-7	104.966	wh orth cry	198 dec		1.42		s H_2O; sl EtOH
1008	Hydrazine hydroiodide	$N_2H_4 \cdot HI$	10039-55-1	159.957	hyg cry	125				s H_2O
1009	Hydrazine nitrate	$N_2H_4 \cdot HNO_3$	13464-97-6	95.058	monocl cry; exp	70				vs H_2O
1010	Hydrazine sulfate	$N_2H_4 \cdot H_2SO_4$	10034-93-2	130.125	col orth cry	254		1.378		sl H_2O; i EtOH
1011	Hydrazoic acid	HN_3	7782-79-8	43.028	col liq; exp	-80	35.7			s H_2O
1012	Hydroxylamine	H_2NOH	7803-49-8	33.030	wh orth flakes or needles	33.1	58	1.21		vs H_2O, MeOH
1013	Hydroxylamine sulfate	$(H_2NOH)_2 \cdot H_2SO_4$	10039-54-0	164.139	cry	170				vs H_2O
1014	Hydrogen	H_2	1333-74-0	2.016	col gas; flam	-259.34	-252.87	0.082 g/L		sl H_2O
1015	Hydrogen bromide	HBr	10035-10-6	80.912	col gas	-86.80	-66.38	3.307 g/L		vs H_2O; s EtOH
1016	Hydrogen chloride	HCl	7647-01-0	36.461	col gas	-114.17	-85	1.490 g/L		vs H_2O

1017	Hydrogen chloride dihydrate	$HCl \cdot 2H_2O$	13465-05-9	72.492	col liq	-17.7		1.46		
1018	Hydrogen cyanide	HCN	74-90-8	27.026	col liq	-13.29	26	0.684		vs H_2O, EtOH; sl eth
1019	Hydrogen fluoride	HF	7664-39-3	20.006	col gas	-83.35	20	0.818 g/L		vs H_2O, EtOH; sl eth
1020	Hydrogen iodide	HI	10034-85-2	127.912	col or yel gas	-50.76	-35.55	5.228 g/L		vs H_2O; s os
1021	Hydrogen peroxide	H_2O_2	7722-84-1	34.015	col liq	-0.43	150.2	1.44		vs H_2O
1022	Hydrogen selenide	H_2Se	7783-07-5	80.98	col gas; flam	-65.73	-41.25	3.310 g/L		s H_2O
1023	Hydrogen sulfide	H_2S	7783-06-4	34.082	col gas; flam	-85.5	-59.55	1.393 g/L		s H_2O
1024	Hydrogen disulfide	H_2S_2	13465-07-1	66.148	col liq		70.7	1.334		
1025	Hydrogen telluride	H_2Te	7783-09-7	129.62	col gas	-49	-2	5.298 g/L		s H_2O, EtOH, alk
1026	Indium	In	7440-74-6	114.818	soft wh metal	156.60	2072	7.31		s acid
1027	Indium antimonide	InSb	1312-41-0	236.578	blk cub cry	525		5.7747		
1028	Indium arsenide	InAs	1303-11-3	189.740	gray cub cry	942		5.67		i acid
1029	Indium nitride	InN	25617-98-5	128.825	hex cry	1100		6.88		
1030	Indium phosphide	InP	22398-80-7	145.792	blk cub cry	1062		4.81		sl acid
1031	Indium(I) bromide	InBr	14280-53-6	194.722	oran-red orth cry	290	656	4.96		reac H_2O
1032	Indium(I) chloride	InCl	13465-10-6	150.271	yel cub cry	211	608	4.19		reac H_2O
1033	Indium(I) iodide	InI	13966-94-4	241.722	orth cry	364.4	712	5.32		
1034	Indium(II) bromide	$InBr_2$	21264-43-7	274.626	orth cry			4.22		reac H_2O
1035	Indium(II) chloride	$InCl_2$	13465-11-7	185.723	col orth cry	235		3.64		reac H_2O
1036	Indium(II) sulfide	InS	12030-14-7	146.884	red-brn orth cry	692		5.2		
1037	Indium(III) bromide	$InBr_3$	13465-09-3	354.530	hyg yel-wh monocl cry	420		4.74	414^{20}	
1038	Indium(III) chloride	$InCl_3$	10025-82-8	221.176	yel monocl cry; hyg	583		4.0	195.1^{22}	s EtOH
1039	Indium(III) fluoride	InF_3	7783-52-0	171.813	wh hex cry; hyg	1170	>1200	4.39		sl H_2O; s dil acid
1040	Indium(III) fluoride trihydrate	$InF_3 \cdot 3H_2O$	14166-78-0	225.859	wh cry	100 dec				s H_2O
1041	Indium(III) hydroxide	$In(OH)_3$	20661-21-6	165.840	cub cry			4.4		
1042	Indium(III) iodide	InI_3	13510-35-5	495.531	yel-red monocl cry; hyg	207		4.69	1308^{22}	
1043	Indium(III) oxide	In_2O_3	1312-43-2	277.634	yel cub cry	1912		7.18		i H_2O; s hot acid
1044	Indium(III) perchlorate octahydrate	$In(ClO_4)_3 \cdot 8H_2O$	13465-15-1	557.291	wh cry	≈80	200 dec			
1045	Indium(III) phosphate	$InPO_4$	14693-82-4	209.789	wh orth cry			4.9		i H_2O
1046	Indium(III) selenide	In_2Se_3	1312-42-1	466.52	blk hex cry	660		5.8		
1047	Indium(III) sulfate	$In_2(SO_4)_3$	13464-82-9	517.827	hyg wh powder			3.44	117^{20}	
1048	Indium(III) sulfide	In_2S_3	12030-24-9	325.834	oran cub cry	1050		4.45		
1049	Indium(III) telluride	In_2Te_3	1312-45-4	612.44	blk cub cry	667		5.75		
1050	Iodine	I_2	7553-56-2	253.809	blue-blk plates	113.7	184.4	4.933	0.03^{20}	s bz, EtOH, eth, ctc, chl
1051	Iodic acid	HIO_3	7782-68-5	175.910	col orth cry	110 dec		4.63	308^{25}	i EtOH, eth
1052	Periodic acid dihydrate	$HIO_4 \cdot 2H_2O$	10450-60-9	227.940	monocl hyg cry	122 dec				s H_2O, EtOH; sl eth
1053	Iodine tetroxide	I_2O_4	12399-08-5	317.807	yel cry	85 dec		4.2		sl H_2O
1054	Iodine pentoxide	I_2O_5	12029-98-0	333.806	hyg wh cry	≈300 dec		4.98	253.4^{20}	i EtOH, eth, CS_2
1055	Iodine nonaoxide	I_4O_9	73560-00-6	651.613	hyg yel powder	75 dec				
1056	Iodine bromide	IBr	7789-33-5	206.808	blk orth cry	40	116 dec	4.3		s H_2O, EtOH, eth
1057	Iodine chloride	ICl	7790-99-0	162.357	red cry or oily liq	27.39	100 dec	3.24		reac H_2O; s EtOH
1058	Iodine trichloride	ICl_3	865-44-1	233.262	yel tricl cry; hyg	101 tp (16 atm)	64 sp dec	3.2		reac H_2O; s EtOH, bz
1059	Iodine fluoride	IF	13873-84-2	145.902	disproportionates at room temp					
1060	Iodine trifluoride	IF_3	22520-96-3	183.899	yel solid, stable at low temp	-28 dec				
1061	Iodine pentafluoride	IF_5	7783-66-6	221.896	yel liq	9.43	100.5	3.19		reac H_2O
1062	Iodine heptafluoride	IF_7	16921-96-3	259.893	col gas	6.5 tp	4.8 sp	10.62 g/L		s H_2O
1063	Iridium	Ir	7439-88-5	192.217	silv-wh metal; cub	2446	4428	22.5		s aqua regia
1064	Iridium(III) sulfide	Ir_2S_3	12136-42-4	480.632	orth cry			10.2		

No.	Name	Formula	CAS Reg No.	Mol. Weight	Physical Form	mp/°C	bp/°C	Density g cm^{-3}	Solubility g/100 g H_2O	Qualitative Solubility
1065	Iridium(III) bromide	$IrBr_3$	10049-24-8	431.929	red-brn monocl cry			6.82		i H_2O, acid, alk
1066	Iridium(III) bromide tetrahydrate	$IrBr_3 \cdot 4H_2O$	10049-24-8*	503.991	grn-brn cry					s H_2O; i EtOH
1067	Iridium(III) chloride	$IrCl_3$	10025-83-9	298.575	brn monocl cry	763 dec		5.30		i H_2O, acid, alk
1068	Iridium(III) fluoride	IrF_3	23370-59-4	249.212	blk hex cry	250 dec		≈8.0		i H_2O, dil acid
1069	Iridium(III) iodide	IrI_3	7790-41-2	572.930	dark brn monocl cry			≈7.4		i H_2O, acid, bz, chl; s alk
1070	Iridium(III) oxide	Ir_2O_3	1312-46-5	432.432	blue-blk cry	1000 dec				i H_2O; sl hot HCl
1071	Iridium(IV) chloride	$IrCl_4$	10025-97-5	334.028	brn hyg solid	≈700 dec				s H_2O, EtOH
1072	Iridium(IV) oxide	IrO_2	12030-49-8	224.216	brn tetr cry	1100 dec		11.7		
1073	Iridium(IV) sulfide	IrS_2	12030-51-2	256.349	orth cry			9.3		
1074	Iridium(VI) fluoride	IrF_6	7783-75-7	306.207	yel cub cry; hyg	44	53.6	4.8		reac H_2O
1075	Iron	Fe	7439-89-6	55.845	silv-wh or gray met	1538	2861	7.87		s dil acid
1076	Ferrocene	$Fe(C_5H_5)_2$	102-54-5	186.031	oran needles	172.5	249			i H_2O; s EtOH, eth, bz, dil HNO_3
1077	Iron pentacarbonyl	$Fe(CO)_5$	13463-40-6	195.896	yel oily liq; flam	-20	103	1.490		i H_2O; s eth, bz, ace
1078	Iron nonacarbonyl	$Fe_2(CO)_9$	15321-51-4	363.781	oran-yel cry	100 dec		2.85		
1079	Iron dodecacarbonyl	$Fe_3(CO)_{12}$	12088-65-2	503.656	blk cry	140		2.00		
1080	Iron hydrocarbonyl	$FeH_2(CO)_4$	17440-90-3	169.902	col liq; unstable	-70	dec			s alk
1081	Iron arsenide	FeAs	12044-16-5	130.767	gray orth cry	1030		7.85		
1082	Iron boride	FeB	12006-84-7	66.656	refr solid; orth	1650		≈7		
1083	Iron boride	Fe_2B	12006-86-9	122.501	refr solid; tetr	1389		7.3		
1084	Iron carbide	Fe_3C	12011-67-5	179.546	gray cub cry	1227		7.694		
1085	Iron phosphide	FeP	26508-33-8	86.819	rhom cry			6.07		
1086	Iron phosphide	Fe_2P	1310-43-6	142.664	gray hex needles	1370		6.8		i H_2O, dil acid, alk
1087	Iron phosphide	Fe_3P	12023-53-9	198.509	gray solid	1100		6.74		i H_2O
1088	Iron disulfide	FeS_2	1317-66-4	119.977	blk cub cry	>600 dec		5.02		i H_2O
1089	Iron silicide	FeSi	12022-95-6	83.931	gray cub cry	1410		6.1		
1090	Iron silicide	$FeSi_2$	12022-99-0	112.016	gray tetr cry	1220		4.74		
1091	Iron(II) aluminate	$Fe(AlO_2)_2$	12068-49-4	173.806	blk cub cry			4.3		
1092	Iron(II) arsenate	$Fe_3(AsO_4)_2$	10102-50-8	445.373	grn pow					i H_2O
1093	Iron(II) arsenate hexahydrate	$Fe_3(AsO_4)_2 \cdot 6H_2O$	10102-50-8*	553.465	grn amorp pow	dec				i H_2O; s acid
1094	Iron(II) bromide	$FeBr_2$	7789-46-0	215.653	yel-brn hex cry; hyg	691	dec	4.636	120^{25}	vs EtOH
1095	Iron(II) bromide hexahydrate	$FeBr_2 \cdot 6H_2O$	13463-12-2	323.744	grn hyg cry	27 dec		4.64	120^{25}	s EtOH
1096	Iron(II) carbonate	$FeCO_3$	563-71-3	115.854	gray-brn hex cry			3.9	0.000062^{20}	
1097	Iron(II) chloride	$FeCl_2$	7758-94-3	126.750	wh hex cry; hyg	677	1023	3.16	65.0^{25}	vs EtOH, ace; sl bz
1098	Iron(II) chloride dihydrate	$FeCl_2 \cdot 2H_2O$	16399-77-2	162.781	wh-grn monocl cry	120 dec		2.39	65.0^{25}	
1099	Iron(II) chloride tetrahydrate	$FeCl_2 \cdot 4H_2O$	13478-10-9	198.812	grn monocl cry	105 dec		1.93	65.0^{25}	s EtOH
1100	Iron(II) chromite	$FeCr_2O_4$	1308-31-2	223.835	blk cub cry			5.0		
1101	Iron(II) fluoride	FeF_2	7789-28-8	93.842	wh tetr cry	1100		4.09		sl H_2O; s dil HF; i EtOH, eth
1102	Iron(II) fluoride tetrahydrate	$FeF_2 \cdot 4H_2O$	13940-89-1	165.904	col hex cry			2.20		
1103	Iron(II) hydroxide	$Fe(OH)_2$	18624-44-7	89.860	wh-grn hex cry			3.4	0.000052^{20}	
1104	Iron(II) iodide	FeI_2	7783-86-0	309.654	red-viol hex cry; hyg	587		5.3		s H_2O, EtOH, eth
1105	Iron(II) iodide tetrahydrate	$FeI_2 \cdot 4H_2O$	7783-86-0*	381.716	blk hyg leaflets	90 dec		2.87		s H_2O, EtOH
1106	Iron(II) molybdate	$FeMoO_4$	13718-70-2	215.78	brn-yel monocl cry	1115		5.6		i H_2O
1107	Iron(II) nitrate	$Fe(NO_3)_2$	14013-86-6	179.854	grn solid				87.5^{25}	
1108	Iron(II) nitrate hexahydrate	$Fe(NO_3)_2 \cdot 6H_2O$	14013-86-6*	287.946	grn solid	60 dec			87.5^{25}	
1109	Iron(II) oxalate dihydrate	$FeC_2O_4 \cdot 2H_2O$	6047-25-2	179.894	yel cry	150 dec		2.28	0.078^{25}	s acid

1110	Iron(II) oxide	FeO	1345-25-1	71.844	blk cub cry	1377		6.0		i H_2O, alk; s acid
1111	Iron(II) perchlorate	$Fe(ClO_4)_2$	13933-23-8	254.745	grn-wh hyg needles	>100 dec			210^{25}	
1112	Iron(II) phosphate octahydrate	$Fe_3(PO_4)_2 \cdot 8H_2O$	14940-41-1	501.600	gray-blue monocl cry; hyg			2.58		i H_2O; s acid
1113	Iron(II) selenide	$FeSe$	1310-32-3	134.81	blk hex cry			6.7		i H_2O
1114	Iron(II) orthosilicate	Fe_2SiO_4	10179-73-4	203.774	brn orth cry			4.30		
1115	Iron(II) sulfate	$FeSO_4$	7720-78-7	151.909	wh orth cry; hyg			3.65	29.5^{25}	
1116	Iron(II) sulfate monohydrate	$FeSO_4 \cdot H_2O$	17375-41-6	169.924	wh-yel monocl cry	300 dec		3.0	29.5^{25}	
1117	Iron(II) sulfate heptahydrate	$FeSO_4 \cdot 7H_2O$	7782-63-0	278.015	blue-grn monocl cry	≈60 dec		1.895	29.5^{25}	i EtOH
1118	Iron(II) sulfide	FeS	1317-37-9	87.911	col hex or tetr cry; hyg	1188	dec	4.7		i H_2O; reac acid
1119	Iron(II) tantalate	$Fe(TaO_3)_2$		513.737	brn tetr cry			7.33		
1120	Iron(II) tartrate	$FeC_4H_4O_6$		203.916	wh cry				0.88	vs acid; s NH_4OH
1121	Iron(II) telluride	$FeTe$	12125-63-2	183.45	tetr cry	914		6.8		
1122	Iron(II) thiocyanate trihydrate	$Fe(SCN)_2 \cdot 3H_2O$	6010-09-9	226.057	grn monocl cry					s H_2O, EtOH, eth
1123	Iron(II) titanate	$FeTiO_3$	12168-52-4	151.710	blk rhomb cry	≈1470		4.72		
1124	Iron(II) tungstate	$FeWO_4$	13870-24-1	303.68	monocl cry			7.51		
1125	Iron(II,III) oxide	Fe_3O_4	1317-61-9	231.533	blk cub cry or amorp powder	1597		5.17		i H_2O; s acid
1126	Iron(III) acetate, basic	$FeOH(C_2H_3O_2)_2$	10450-55-2	190.941	brn-red amorp powder					i H_2O; s EtOH, acid
1127	Iron(III) 2,4-pentanedioate	$Fe(CH_3COCHCOCH_3)_3$	14024-18-1	353.169	red-oran cry	179		5.24		sl H_2O; s os
1128	Iron(III) arsenate dihydrate	$FeAsO_4 \cdot 2H_2O$	10102-49-5	230.795	grn-brn powder	dec		3.18		i H_2O; s dil acid
1129	Iron(III) bromide	$FeBr_3$	10031-26-2	295.557	dark red hex cry; hyg	dec		4.5	455^{25}	s EtOH, eth
1130	Iron(III) chloride	$FeCl_3$	7705-08-0	162.203	grn hex cry; hyg	304	≈316	2.90	91.2^{25}	s EtOH, eth, ace
1131	Iron(III) chloride hexahydrate	$FeCl_3 \cdot 6H_2O$	10025-77-1	270.294	yel-oran monocl cry; hyg	37 dec		1.82	91.2^{25}	s EtOH, eth, ace
1132	Iron(III) chromate	$Fe_2(CrO_4)_3$	10294-52-7	459.671	yel powder					i H_2O, EtOH; s acid
1133	Iron(III) citrate pentahydrate	$FeC_6H_5O_7 \cdot 5H_2O$	3522-50-7	335.021	red-brn cry					s H_2O; i EtOH
1134	Iron(III) dichromate	$Fe_2(Cr_2O_7)_3$	10294-53-8	759.654	red-brn solid					s H_2O, acid
1135	Iron(III) ferrocyanide	$Fe_4[Fe(CN)_6]_3$	14038-43-8	859.229	dark blue powder			1.80		i H_2O, dil acid, os
1136	Iron(III) fluoride	FeF_3	7783-50-8	112.840	grn hex cry	>1000		3.87	5.92^{25}	i EtOH, eth, bz
1137	Iron(III) fluoride trihydrate	$FeF_3 \cdot 3H_2O$	15469-38-2	166.886	yel-brn tetr cry			2.3	5.92^{25}	
1138	Iron(III) formate	$Fe(CHO_2)_3$	555-76-0	190.897	red-yel cry pow					s H_2O; sl EtOH
1139	Iron(III) hydroxide	$Fe(OH)_3$	1309-33-7	106.867	yel monocl cry			3.12		
1140	Iron(III) hydroxide oxide	$FeO(OH)$	20344-49-4	88.852	red-brn orth cry			4.26		i H_2O; s acid
1141	Iron(III) nitrate	$Fe(NO_3)_3$	10421-48-4	241.860	cry				82.5^{20}	
1142	Iron(III) nitrate hexahydrate	$Fe(NO_3)_3 \cdot 6H_2O$	13476-08-9	349.951	viol cub cry	35 dec			82.5^{20}	
1143	Iron(III) nitrate nonahydrate	$Fe(NO_3)_3 \cdot 9H_2O$	7782-61-8	403.997	viol-gray hyg cry	47 dec		1.68	82.5^{20}	vs EtOH, ace
1144	Iron(III) oxalate	$Fe_2(C_2O_4)_3$	19469-07-9	375.747	yel amorp powder	100 dec				s H_2O, acid; i alk
1145	Iron(III) oxide	Fe_2O_3	1309-37-1	159.688	red-brn hex cry	1565		5.25		i H_2O; s acid
1146	Iron(III) phosphate dihydrate	$FePO_4 \cdot 2H_2O$	10045-86-0	186.847	gray-wh orth cry			2.87		i H_2O; s HCl
1147	Iron(III) pyrophosphate nonahydrate	$Fe_4(P_2O_7)_3 \cdot 9H_2O$	10058-44-3	907.348	yel powder					i H_2O; s acid
1148	Iron(III) hypophosphite	$Fe(H_2PO_2)_3$	7783-84-8	250.811	wh-gray powder					i H_2O
1149	Iron(III) sodium pyrophosphate	$FeNaP_2O_7$	10045-87-1	252.778	wh pow			1.5		i H_2O; s HCl
1150	Iron(III) sulfate	$Fe_2(SO_4)_3$	10028-22-5	399.881	gray-wh rhomb cry; hyg			3.10	440^{20}	sl EtOH; i ace
1151	Iron(III) sulfate nonahydrate	$Fe_2(SO_4)_3 \cdot 9H_2O$	13520-56-4	562.018	yel hex cry	400 dec		2.1	440^{20}	
1152	Iron(III) thiocyanate monohydrate	$Fe(SCN)_3 \cdot H_2O$	4119-52-2	248.110	red hyg cry	dec				s H_2O, EtOH, ace; i tol, chl
1153	Iron(III) metavanadate	$Fe(VO_3)_3$	65842-03-7	352.665	gray-brn powder					i H_2O, EtOH; s acid
1154	Krypton	Kr	7439-90-9	83.80	col gas	-157.38 tp (73.2 kPa)	-153.22	3.425 g/L		sl H_2O
1155	Krypton difluoride	KrF_2	13773-81-4	121.80	col tetr cry	≈25 dec		3.24		reac H_2O
1156	Lanthanum	La	7439-91-0	138.906	silv metal; hex	918	3464	6.15		s dil acid
1157	Lanthanum boride	LaB_6	12008-21-8	203.772	blk cub cry; refrac	2715		4.76		

No.	Name	Formula	CAS Reg No.	Mol. Weight	Physical Form	mp/°C	bp/°C	Density g cm⁻³	Solubility g/100 g H_2O	Qualitative Solubility
1158	Lanthanum bromide	$LaBr_3$	13536-79-3	378.618	wh hex cry; hyg	788		5.1		s H_2O
1159	Lanthanum carbide	LaC_2	12071-15-7	162.927	tetr cry	2360		5.29		
1160	Lanthanum carbonate octahydrate	$La_2(CO_3)_3 \cdot 8H_2O$	6487-39-4	601.960	wh cry powder			2.6		i H_2O; s dil acid
1161	Lanthanum chloride	$LaCl_3$	10099-58-8	245.264	wh hex cry; hyg	859		3.84	95.7^{25}	
1162	Lanthanum chloride heptahydrate	$LaCl_3 \cdot 7H_2O$	20211-76-1	371.371	wh tricl cry; hyg	91 dec			95.7^{25}	s EtOH
1163	Lanthanum fluoride	LaF_3	13709-38-1	195.901	wh hex cry; hyg	1493		5.9		i H_2O, acid
1164	Lanthanum hydride	LaH_3	13864-01-2	141.930	blk cub cry			5.36		
1165	Lanthanum hydroxide	$La(OH)_3$	14507-19-8	189.928	wh amorp solid	dec			0.000020^{20}	
1166	Lanthanum iodate	$La(IO_3)_3$	13870-19-4	663.614	col cry				1.7	
1167	Lanthanum iodide	LaI_3	13813-22-4	519.619	wh orth cry; hyg	778		5.6		s H_2O
1168	Lanthanum nitrate hexahydrate	$La(NO_3)_3 \cdot 6H_2O$	10277-43-7	433.012	wh hyg tricl cry	≈40 dec			200^{25}	vs EtOH; s ace
1169	Lanthanum nitride	LaN	25764-10-7	152.913	cub cry			6.73		
1170	Lanthanum oxide	La_2O_3	1312-81-8	325.809	wh amorp powder	2304	3620	6.51		i H_2O; s dil acid
1171	Lanthanum silicide	$LaSi_2$	12056-90-5	195.077	gray tetr cry			5.0		
1172	Lanthanum sulfate nonahydrate	$La_2(SO_4)_3 \cdot 9H_2O$	10294-62-9	728.139	hex cry			2.82	2.7^{20}	i EtOH
1173	Lanthanum sulfide	La_2S_3	12031-49-1	374.009	red cub cry	2110		4.9		
1174	Lanthanum sulfide	LaS	12031-30-0	170.972	yel cub cry	2300		5.61		
1175	Lawrencium	Lr	22537-19-5	262	metal	1627				
1176	Lead	Pb	7439-92-1	207.2	soft silv-gray metal; cub	327.46	1749	11.3		s conc acid
1177	Lead(II) acetate	$Pb(C_2H_3O_2)_2$	301-04-2	325.3	wh cry	280	dec	3.25	44.3^{20}	
1178	Lead(II) acetate trihydrate	$Pb(C_2H_3O_2)_2 \cdot 3H_2O$	6080-56-4	427.3	col cry	75 dec		2.55		vs H_2O; sl EtOH
1179	Lead(II) acetate, basic	$Pb(C_2H_3O_2)2 \cdot 2Pb(OH)_2$	1335-32-6	807.7	wh pow	dec			6.3^{0}	
1180	Lead(II) antimonate	$Pb_3(SbO_4)_2$	13510-89-9	993.1	oran-yel powder			6.58		i H_2O, dil acid
1181	Lead(II) arsenate	$Pb_3(AsO_4)_2$	3687-31-8	899.4	wh cry	1042 dec		5.8		i H_2O; s HNO_3
1182	Lead(II) arsenite	$Pb(AsO_2)_2$	10031-13-7	421.0	wh powder			5.85		i H_2O; s dil HNO_3
1183	Lead(II) azide	$Pb(N_3)_2$	13424-46-9	291.2	col orth needles; exp	exp ≈350		4.7	0.023^{18}	vs HOAc
1184	Lead(II) borate monohydrate	$Pb(BO_2)_2 \cdot H_2O$	10214-39-8	310.8	wh powder	500 dec		5.6		i H_2O; s dil HNO_3
1185	Lead(II) bromate monohydrate	$Pb(BrO_3)_2 \cdot H_2O$	10031-21-7	481.0	col cry	≈180 dec		5.53	1.33^{20}	
1186	Lead(II) bromide	$PbBr_2$	10031-22-8	367.0	wh orth cry	371	892	6.69	0.975^{25}	i EtOH
1187	Lead(II) butanoate	$Pb(C_4H_7O_2)_2$	819-73-8	381.4	col solid	≈90				i H_2O; s dil HNO_3
1188	Lead(II) carbonate	$PbCO_3$	598-63-0	267.2	col orth cry	≈315 dec		6.6		i H_2O
1189	Lead(II) carbonate, basic	$Pb(OH)_2 \cdot 2PbCO_3$	1319-46-6	775.6	wh hex cry	400 dec		≈6.5		i H_2O, EtOH; s acid
1190	Lead(II) chlorate	$Pb(ClO_3)_2$	10294-47-0	374.1	col hyg cry	230 dec		3.9	144^{18}	vs EtOH
1191	Lead(II) chloride	$PbCl_2$	7758-95-4	278.1	wh orth needles or powder	501	951	5.98	1.08^{25}	s alk
1192	Lead(II) chloride fluoride	PbClF	13847-57-9	261.7	tetr cry			7.05	0.035^{20}	
1193	Lead(II) chromate	$PbCrO_4$	7758-97-6	323.2	yel-oran monocl cry	844		6.12	0.000017^{20}	s alk, dil acid
1194	Lead(II) chromate(VI) oxide	$PbCrO_4 \cdot PbO$	18454-12-1	546.4	red powder					i H_2O
1195	Lead(II) citrate trihydrate	$Pb_3(C_6H_5O_7)_2 \cdot 3H_2O$	512-26-5	1053.8	wh cry powder					s H_2O; sl EtOH
1196	Lead(II) cyanide	$Pb(CN)_2$	592-05-2	259.2	wh-yel powder					sl H_2O; reac acid
1197	Lead(II) 2-ethylhexanoate	$Pb(C_7H_{15}CO_2)_2$	301-08-6	493.6	visc liq			1.56		
1198	Lead(II) fluoride	PbF_2	7783-46-2	245.2	wh orth cry	830	1293	8.44	0.0670^{25}	
1199	Lead(II) fluoroborate	$Pb(BF_4)_2$	13814-96-5	380.8	stable only in aq soln					s H_2O
1200	Lead(II) formate	$Pb(CHO_2)_2$	811-54-1	297.2	wh prisms or needles	190 dec		4.63	1.6^{16}	i EtOH
1201	Lead(II) hexafluoro-2,4-pentanedioate	$Pb(CF_3COCHCOCF_3)_2$	19648-88-5	621.3	cry	155	210			
1202	Lead(II) hydrogen arsenate	$PbHAsO_4$	7784-40-9	347.1	wh monocl cry	280 dec		5.943		i H_2O; s HNO_3, alk

1203	Lead(II) hydrogen phosphate	$PbHPO_4$	15845-52-0	303.2	wh monocl cry	dec		5.66		
1204	Lead(II) hydroxide	$Pb(OH)_2$	19783-14-3	241.2	wh powder	145 dec		7.59	0.00012^{20}	s acid
1205	Lead(II) iodate	$Pb(IO_3)_2$	25659-31-8	557.0	wh orth cry			6.50	0.0025^{25}	
1206	Lead(II) iodide	PbI_2	10101-63-0	461.0	yel hex cry or powder	410	872 dec	6.16	0.076^{25}	i EtOH
1207	Lead(II) lactate	$Pb(C_3H_5O_3)_2$	18917-82-3	385.3	wh cry powder					s H_2O, hot EtOH
1208	Lead(II) molybdate	$PbMoO_4$	10190-55-3	367.1	yel tetr cry	≈1060		6.7		i H_2O; s HNO_3, NaOH
1209	Lead(II) niobate	$Pb(NbO_3)_2$	12034-88-7	489.0	rhomb or tetr cry	1343		6.6		i H_2O
1210	Lead(II) nitrate	$Pb(NO_3)_2$	10099-74-8	331.2	col cub cry	470		4.53	59.7^{25}	sl EtOH
1211	Lead(II) oleate	$Pb(C_{18}H_{33}O_2)_2$	1120-46-3	770.1	wax-like solid					i H_2O; s EtOH, bz, eth
1212	Lead(II) oxalate	PbC_2O_4	814-93-7	295.2	wh powder	300 dec		5.28	0.00025^{20}	s dil HNO_3
1213	Lead(II) oxide (litharge)	PbO	1317-36-8	223.2	red tetr cry	trans to massicot 489		9.35		i H_2O, EtOH; s dil HNO_3
1214	Lead(II) oxide (massicot)	PbO	1317-36-8	223.2	yel orth cry	897		9.64		i H_2O, EtOH; s dil HNO_3
1215	Lead(II) oxide hydrate	$3PbO \cdot H_2O$	1311-11-1	687.6	wh powder			7.41		i H_2O; s dil acid
1216	Lead(II) 2,4-pentanedioate	$Pb(CH_3COCHCOCH_3)_2$	15282-88-9	405.4	cry	143				
1217	Lead(II) perchlorate	$Pb(ClO_4)_2$	13453-62-8	406.1	wh cry				441^{25}	
1218	Lead(II) perchlorate trihydrate	$Pb(ClO_4)_2 \cdot 3H_2O$	13637-76-8	460.1	wh cry	100 dec		2.6	441^{25}	s EtOH
1219	Lead(II) phosphate	$Pb_3(PO_4)_2$	7446-27-7	811.5	wh hex cry	1014		7.01		i H_2O, EtOH
1220	Lead(II) hypophosphite	$Pb(H_2PO_2)_2$	10294-58-3	337.2	hyg cry powder	dec				sl H_2O; i EtOH
1221	Lead(II) metasilicate	$PbSiO_3$	10099-76-0	283.3	wh monocl cry powder	764		6.49		i H_2O, os
1222	Lead(II) orthosilicate	Pb_2SiO_4	13566-17-1	506.5	monocl cry	743		7.60		
1223	Lead(II) hexafluorosilicate dihydrate	$PbSiF_6 \cdot 2H_2O$	1310-03-8	385.3	col cry	dec				vs H_2O
1224	Lead(II) selenate	$PbSeO_4$	7446-15-3	350.2	orth cry			6.37	0.013^{25}	s conc acid
1225	Lead(II) selenide	PbSe	12069-00-0	286.2	gray cub cry	1078		8.1		i H_2O; s HNO_3
1226	Lead(II) selenite	$PbSeO_3$	7488-51-9	334.2	wh monocl cry	≈500		7.0		i H_2O
1227	Lead(II) sodium thiosulfate	$Na_4Pb(S_2O_3)_3$	10101-94-7	635.6	wh cry					sl H_2O
1228	Lead(II) stearate	$Pb(C_{18}H_{35}O_2)_2$	1072-35-1	774.1	wh powder	≈100		1.4		i H_2O; s hot EtOH
1229	Lead(II) sulfate	$PbSO_4$	7446-14-2	303.3	orth cry	1087		6.29	0.0044^{25}	i acid; sl alk
1230	Lead(II) sulfide	PbS	1314-87-0	239.3	blk powder or silv cub cry	1113		7.60		i H_2O; s acid
1231	Lead(II) sulfite	$PbSO_3$	7446-10-8	287.3	wh powder	dec				i H_2O; s HNO_3
1232	Lead(II) tantalate	$Pb(TaO_3)_2$	12065-68-8	665.1	orth cry			7.9		i H_2O
1233	Lead(II) telluride	PbTe	1314-91-6	334.8	gray cub cry	924		8.164		i H_2O, acid
1234	Lead(II) thiocyanate	$Pb(SCN)_2$	592-87-0	323.4	wh-yel powder			3.82	0.05^{20}	
1235	Lead(II) thiosulfate	PbS_2O_3	13478-50-7	319.3	wh cry	dec		5.18		i H_2O; s acid
1236	Lead(II) titanate	$PbTiO_3$	12060-00-3	303.1	yel tetr cry			7.9		i H_2O; reac HCl
1237	Lead(II) tungstate (stolzite)	$PbWO_4$	7759-01-5	455.0	yel tetr cry	1130		8.24	0.03^{20}	s alk
1238	Lead(II) tungstate (raspite)	$PbWO_4$	7759-01-5	455.0	monocl cry	trans 400		8.46	0.03^{20}	s alk
1239	Lead(II) metavanadate	$Pb(VO_3)_2$	10099-79-3	405.1	yel powder					i H_2O; reac HNO_3
1240	Lead(II) zirconate	$PbZrO_3$	12060-01-4	346.4	col orth cry			≈8		i H_2O, alk; s acid
1241	Lead(II,IV) oxide	Pb_2O_3	1314-27-8	462.4	blk monocl cry or red amorp powder	530 dec		10.05		i H_2O; s alk; reac conc HCl
1242	Lead(II,II,IV) oxide	Pb_3O_4	1314-41-6	685.6	red tetr cry	830		8.92		i H_2O, EtOH; s hot HCl
1243	Lead(IV) acetate	$Pb(C_2H_3O_2)_4$	546-67-8	443.4	col monocl cry	≈175		2.23		reac H_2O, EtOH; s bz, chl
1244	Lead(IV) bromide	$PbBr_4$	13701-91-2	526.8	unstable liq					
1245	Lead(IV) chloride	$PbCl_4$	13463-30-4	349.0	yel oily liq	-15	≈50 dec			
1246	Lead(IV) fluoride	PbF_4	7783-59-7	283.2	wh tetr cry; hyg	≈600		6.7		
1247	Lead(IV) oxide	PbO_2	1309-60-0	239.2	red tetr cry or brn powder	290 dec		9.64		
1248	Lithium	Li	7439-93-2	6.941	soft silv-wh metal	180.50	1342	0.534		reac H_2O
1249	Lithium acetate	$LiC_2H_3O_2$	546-89-4	65.985	cry	286			45.0^{25}	vs EtOH

No.	Name	Formula	CAS Reg No.	Mol. Weight	Physical Form	mp/°C	bp/°C	Density g cm^{-3}	Solubility g/100 g H_2O	Qualitative Solubility
1250	Lithium acetate dihydrate	$LiC_2H_3O_2 \cdot 2H_2O$	6108-17-4	102.016	wh rhomb cry	58 dec		1.3	45.0^{25}	s EtOH
1251	Lithium aluminum hydride	$LiAlH_4$	16853-85-3	37.955	gray-wh monocl cry	>125 dec		0.917		reac H_2O, EtOH; s eth, thf
1252	Lithium amide	$LiNH_2$	7782-89-0	22.964	tetr cry	380		1.18		reac H_2O
1253	Lithium arsenate	Li_3AsO_4	13478-14-3	159.743	col orth cry			3.07		sl H_2O; s HOAc
1254	Lithium azide	LiN_3	19597-69-4	48.961	hyg monocl cry; exp			1.83		vs H_2O
1255	Lithium metaborate	$LiBO_2$	13453-69-5	49.751	wh monocl cry; hyg	849		2.18		vs H_2O; s EtOH
1256	Lithium borohydride	$LiBH_4$	16949-15-8	21.784	wh-gray orth cry or powder	268	380 dec	0.66		s alk, eth, thf
1257	Lithium bromide	LiBr	7550-35-8	86.845	wh cub cry; hyg	552	≈1300	3.464	181^{25}	s EtOH, eth
1258	Lithium carbonate	Li_2CO_3	554-13-2	73.891	wh monocl cry	723	1300 dec	2.11	1.30^{25}	s acid; i EtOH
1259	Lithium chlorate	$LiClO_3$	13453-71-9	90.392	col hyg rhom needles	127.6	300 dec	1.119	459^{25}	vs EtOH; sl ace
1260	Lithium chloride	LiCl	7447-41-8	42.394	wh cub cry or powder; hyg	610	1383	2.07	84.5^{25}	s EtOH, ace, py
1261	Lithium chromate dihydrate	$Li_2CrO_4 \cdot 2H_2O$	7789-01-7	165.906	yel orth cry; hyg	75 dec		2.15		vs H_2O; s EtOH
1262	Lithium dichromate dihydrate	$Li_2Cr_2O_7 \cdot 2H_2O$	10022-48-7	265.901	yel-red hyg cry	130 dec		2.34		vs H_2O
1263	Lithium dihydrogen phosphate	LiH_2PO_4	13453-80-0	103.928	col hyg cry	>100		2.461	126^{0}	
1264	Lithium ferrosilicon	LiFeSi	64082-35-5	90.872	dark brittle cry					reac H_2O
1265	Lithium fluoride	LiF	7789-24-4	25.939	wh cub cry or powder	848.2	1673	2.640	0.134^{25}	s acid
1266	Lithium formate monohydrate	$Li(CHO_2) \cdot H_2O$	6108-23-2	69.974	col-wh cry			1.46		s H_2O
1267	Lithium hydride	LiH	7580-67-8	7.949	gray cub cry or powder; hyg	688.7		0.78		reac H_2O, EtOH
1268	Lithium hydroxide	LiOH	1310-65-2	23.948	col tetr cry	471.1	1626	1.45	12.5^{25}	sl EtOH
1269	Lithium hydroxide monohydrate	$LiOH \cdot H_2O$	1310-66-3	41.964	wh monocl cry or powder			1.51	12.5^{25}	sl EtOH
1270	Lithium iodate	$LiIO_3$	13765-03-2	181.843	wh hyg hex cry			4.502	77.9^{25}	i EtOH
1271	Lithium iodide	LiI	10377-51-2	133.845	wh cub cry; hyg	469	1171	4.06	165^{25}	
1272	Lithium iodide trihydrate	$LiI \cdot 3H_2O$	7790-22-9	187.891	wh hyg cry	73		3.48	165^{25}	vs EtOH, ace
1273	Lithium niobate	$LiNbO_3$	12031-63-9	147.845	wh hex cry	≈1240		4.30		
1274	Lithium nitrate	$LiNO_3$	7790-69-4	68.946	col hex cry; hyg	253		2.38	102^{25}	s EtOH
1275	Lithium nitride	Li_3N	26134-62-3	34.830	red hex cry	813		1.27		reac H_2O
1276	Lithium nitrite monohydrate	$LiNO_2 \cdot H_2O$	13568-33-7*	70.962	col needles	>100		1.615	139.5^{25}	vs EtOH
1277	Lithium phosphate	Li_3PO_4	10377-52-3	115.794	wh orth cry	1205		2.46	0.027^{25}	
1278	Lithium oxide	Li_2O	12057-24-8	29.881	wh cub cry	1570		2.013		
1279	Lithium perchlorate	$LiClO_4$	7791-03-9	106.392	wh orth cry or powder	236	430 dec	2.428	58.7^{25}	s EtOH, ace, eth
1280	Lithium peroxide	Li_2O_2	12031-80-0	45.881	wh hex cry			2.31		s H_2O; i EtOH
1281	Lithium selenate monohydrate	$Li_2SeO_4 \cdot H_2O$	7790-71-8	174.86	monocl cry			2.56		vs H_2O
1282	Lithium metasilicate	Li_2SiO_3	10102-24-6	89.966	wh orth needles	1201		2.52		i cold H_2O; reac dil acid
1283	Lithium sulfate	Li_2SO_4	10377-48-7	109.946	wh monocl cry; hyg	859		2.21	34.2^{25}	
1284	Lithium sulfate monohydrate	$Li_2SO_4 \cdot H_2O$	10102-25-7	127.961	col cry	130 dec		2.06	34.2^{25}	sl EtOH
1285	Lithium sulfide	Li_2S	12136-58-2	45.948	wh cub cry; hyg	1372		1.64		
1286	Lithium thiocyanate	LiSCN	556-65-0	65.025	wh hyg cry				120^{25}	
1287	Lutetium	Lu	7439-94-3	174.967	silv metal; hex	1663	3402	9.84		s dil acid
1288	Lutetium boride	LuB_4	12688-52-7	218.211	tetr cry	2600		≈7.0		
1289	Lutetium bromide	$LuBr_3$	14456-53-2	414.679	wh hyg cry	1025				vs H_2O
1290	Lutetium chloride	$LuCl_3$	10099-66-8	281.325	wh monocl cry; hyg	925		3.98		s H_2O
1291	Lutetium fluoride	LuF_3	13760-81-1	231.962	orth cry	1182	2200	8.3		i H_2O
1292	Lutetium iodide	LuI_3	13813-45-1	555.680	brn hex cry; hyg	1050		≈5.6		vs H_2O
1293	Lutetium nitride	LuN	12125-25-6	188.974	cub cry			11.6		
1294	Lutetium oxide	Lu_2O_3	12032-20-1	397.932	wh cub cry or powder	2427	3980	9.41		

1295	Lutetium sulfate octahydrate	$Lu_2(SO_4)_3 \cdot 8H_2O$	13473-77-3	782.247	wh cry					vs H_2O
1296	Lutetium sulfide	Lu_2S_3	12163-20-1	446.132	gray rhomb cry	1750 dec		6.26		
1297	Lutetium telluride	Lu_2Te_3	12163-22-3	732.73	orth cry			7.8		
1298	Magnesium	Mg	7439-95-4	24.305	silv-wh metal	650	1090	1.74		s dil acid
1299	Magnesium acetate	$Mg(C_2H_3O_2)_2$	142-72-3	142.394	wh orth/mcl cry	323 dec		1.50	65.6^{25}	
1300	Magnesium acetate tetrahydrate	$Mg(C_2H_3O_2)_2 \cdot 4H_2O$	16674-78-5	214.454	col monocl cry; hyg	80 dec		1.45	65.6^{25}	vs EtOH
1301	Magnesium amide	$Mg(NH_2)_2$	7803-54-5	56.350	wh powder; flam	dec		1.39		reac H_2O
1302	Magnesium antimonide	Mg_3Sb_2	12057-75-9	316.435	hex cry	1245		3.99		
1303	Magnesium boride	MgB_2	12007-25-9	45.927	hex cry	800 dec		2.57		
1304	Magnesium bromate hexahydrate	$Mg(BrO_3)_2 \cdot 6H_2O$	7789-36-8	388.201	col cub cry	200 dec		2.29	98^{25}	
1305	Magnesium bromide	$MgBr_2$	7789-48-2	184.113	wh hex cry; hyg	711		3.72	102^{25}	
1306	Magnesium bromide hexahydrate	$MgBr_2 \cdot 6H_2O$	13446-53-2	292.204	col monocl cry	165 dec		2.0	102^{25}	s EtOH
1307	Magnesium carbonate	$MgCO_3$	546-93-0	84.314	wh hex cry	990		3.05	0.18^{20}	i EtOH; s acid
1308	Magnesium chlorate hexahydrate	$Mg(ClO_3)_2 \cdot 6H_2O$	13446-19-0	299.298	wh hyg cry	≈35 dec		1.80	142^{25}	sl EtOH
1309	Magnesium chloride	$MgCl_2$	7786-30-3	95.210	wh hex leaflets; hyg	714	1412	2.325	56.0^{25}	
1310	Magnesium chloride hexahydrate	$MgCl_2 \cdot 6H_2O$	7791-18-6	203.301	wh hyg cry	≈100 dec		1.56	56.0^{25}	s EtOH
1311	Magnesium chromate heptahydrate	$MgCrO_4 \cdot 7H_2O$	13423-61-5*	266.405	yel rhom cry			1.695	54.8^{25}	
1312	Magnesium fluoride	MgF_2	7783-40-6	62.302	wh tetr cry	1263	2227	3.148	0.013^{25}	
1313	Magnesium formate dihydrate	$Mg(CHO_2)_2 \cdot 2H_2O$	6150-82-9	150.370	wh cry	dec				s H_2O; i EtOH
1314	Magnesium germanide	Mg_2Ge	1310-52-7	121.22	cub cry	1117		3.09		
1315	Magnesium hydride	MgH_2	7693-27-8	26.321	wh tetr cry	327		1.45		reac H_2O
1316	Magnesium hydrogen phosphate trihydrate	$MgHPO_4 \cdot 3H_2O$	7757-86-0	174.331	wh powder	550 dec		2.13		sl H_2O; s dil acid
1317	Magnesium hydroxide	$Mg(OH)_2$	1309-42-8	58.320	wh hex cry	350		2.37	0.00069^{20}	s dil acid
1318	Magnesium iodate tetrahydrate	$Mg(IO_3)_2 \cdot 4H_2O$	7790-32-1*	446.172	col mono cry	210 dec		3.3	11.1^{25}	
1319	Magnesium iodide	MgI_2	10377-58-9	278.114	wh hex cry; hyg	634		4.43	146^{25}	
1320	Magnesium iodide octahydrate	$MgI_2 \cdot 8H_2O$	7790-31-0	422.236	wh orth cry; hyg	41 dec		2.10	146^{25}	s EtOH
1321	Magnesium nitrate	$Mg(NO_3)_2$	10377-60-3	148.314	wh cub cry			≈2.3	71.2^{25}	
1322	Magnesium nitrate dihydrate	$Mg(NO_3)_2 \cdot 2H_2O$	15750-45-5	184.345	wh cry	≈100 dec		1.45	71.2^{25}	s EtOH
1323	Magnesium nitrate hexahydrate	$Mg(NO_3)_2 \cdot 6H_2O$	13446-18-9	256.406	col monocl cry; hyg	≈95 dec		1.46	71.2^{25}	s EtOH
1324	Magnesium nitride	Mg_3N_2	12057-71-5	100.928	yel cub cry	≈1500 dec		2.71		
1325	Magnesium nitrite trihydrate	$Mg(NO_2)_2 \cdot 3H_2O$	15070-34-5	170.362	wh hyg prisms	100 dec			129.9^{25}	s EtOH
1326	Magnesium oxalate	MgC_2O_4	547-66-0	112.324	wh pdw				0.038^{25}	
1327	Magnesium oxalate dihydrate	$MgC_2O_4 \cdot 2H_2O$	6150-88-5	148.354	wh powder				0.038^{25}	i EtOH; s dil acid
1328	Magnesium oxide	MgO	1309-48-4	40.304	wh cub cry	2825	3600	3.6		sl H_2O; i EtOH
1329	Magnesium perchlorate	$Mg(ClO_4)_2$	10034-81-8	223.205	wh hyg powder	250 dec		2.2	100^{25}	
1330	Magnesium perchlorate hexahydrate	$Mg(ClO_4)_2 \cdot 6H_2O$	13446-19-0	331.297	wh hyg cry	190 dec		1.98	100^{25}	s EtOH
1331	Magnesium permanganate hexahydrate	$Mg(MnO_4)_2 \cdot 6H_2O$	10377-62-5	370.268	blue-blk cry	dec		2.18		s H_2O
1332	Magnesium peroxide	MgO_2	1335-26-8	56.304	wh cub cry	100 dec		≈3.0		i H_2O; s dil acid
1333	Magnesium phosphate pentahydrate	$Mg_3(PO_4)_2 \cdot 5H_2O$	7757-87-1*	352.934	wh cry	400 dec			0.00009^{20}	s dil acid
1334	Magnesium phosphate octahydrate	$Mg_3(PO_4)_2 \cdot 8H_2O$	13446-23-6	406.980	wh monocl cry			2.17	0.00009^{20}	s acid
1335	Magnesium pyrophosphate trihydrate	$Mg_2P_2O_7 \cdot 3H_2O$	10102-34-8	276.600	wh powder	100 dec		2.56		i H_2O; s acid
1336	Magnesium phosphide	Mg_3P_2	12057-74-8	134.863	yel cub cry			2.06		reac H_2O
1337	Magnesium selenate hexahydrate	$MgSeO_4 \cdot 6H_2O$	13446-28-1	275.35	wh monocl cry			1.928	55.5^{25}	
1338	Magnesium selenide	MgSe	1313-04-8	103.27	brn cub cry			4.2		reac H_2O
1339	Magnesium selenite hexahydrate	$MgSeO_3 \cdot 6H_2O$	15593-61-0	259.36	col hex cry			2.09		i H_2O; s dil acid
1340	Magnesium metasilicate	$MgSiO_3$	13776-74-4	100.389	wh monocl cry	≈1550 dec		3.19		i H_2O; sl HF
1341	Magnesium orthosilicate	Mg_2SiO_4	26686-77-1	140.694	wh orth cry	1897		3.21		i H_2O
1342	Magnesium trisilicate	$Mg_2Si_3O_8$	14987-04-3	260.862	wh pow					i H_2O, EtOH

No.	Name	Formula	CAS Reg No.	Mol. Weight	Physical Form	mp/°C	bp/°C	Density g cm⁻³	Solubility g/100 g H_2O	Qualitative Solubility
1343	Magnesium hexafluorosilicate hexahydrate	$MgSiF_6 \cdot 6H_2O$	60950-56-3	274.472	wh cry	120 dec		1.79	39.3^{18}	i EtOH
1344	Magnesium silicide	Mg_2Si	22831-39-6	76.696	gray cub cry	1102		1.99		reac H_2O
1345	Magnesium stannide	Mg_2Sn	1313-08-2	167.320	blue cub cry	771		3.60		s H_2O, dil HCl
1346	Magnesium sulfate	$MgSO_4$	7487-88-9	120.369	col orth cry	1127		2.66	35.7^{25}	
1347	Magnesium sulfate monohydrate	$MgSO_4 \cdot H_2O$	14168-73-1	138.384	col monocl cry	150 dec		2.57	35.7^{25}	
1348	Magnesium sulfate heptahydrate	$MgSO_4 \cdot 7H_2O$	10034-99-8	246.475	col orth cry	150 dec		1.67	35.7^{25}	sl EtOH
1349	Magnesium sulfide	MgS	12032-36-9	56.371	red-brn cub cry	2226		2.68		reac H_2O
1350	Magnesium sulfite trihydrate	$MgSO_3 \cdot 3H_2O$	19086-20-5	158.415	col orth cry			2.12	0.79^{25}	
1351	Magnesium sulfite hexahydrate	$MgSO_3 \cdot 6H_2O$	13446-29-2	212.461	wh hex cry	200 dec		1.72	0.79^{25}	i EtOH
1352	Magnesium thiosulfate hexahydrate	$MgS_2O_3 \cdot 6H_2O$	13446-30-5	244.527	col cry	170 dec		1.82	93^{25}	i EtOH
1353	Magnesium titanate	$MgTiO_3$	1312-99-8	120.170	col hex cry	1565		3.85		
1354	Magnesium tungstate	$MgWO_4$	13573-11-0	272.14	wh monocl cry			6.89	0.016^{20}	i EtOH
1355	Manganese	Mn	7439-96-5	54.938	hard gray metal	1246	2061	7.3		s dil acids
1356	Manganese antimonide	$MnSb$	12032-82-5	176.698	hex cry	840		6.9		
1357	Manganese antimonide	Mn_2Sb	12032-97-2	231.636	tetr cry	948		7.0		
1358	Manganese boride	MnB	12045-15-7	65.749	orth cry	1890		6.45		
1359	Manganese boride	MnB_2	12228-50-1	76.560	hex cry	1827		5.3		
1360	Manganese boride	Mn_2B	12045-16-8	120.687	red-brn tetr cry	1580		7.20		
1361	Manganese carbide	Mn_3C	12266-65-8	176.825	refrac solid	1520		6.89		
1362	Manganese carbonyl	$Mn_2(CO)_{10}$	10170-69-1	389.977	yel monocl cry	154		1.75		i H_2O; s os
1363	Manganese phosphide	MnP	12032-78-9	85.912	orth cry	1147		5.49		
1364	Manganese phosphide	Mn_2P	12333-54-9	140.850	hex cry	1327		6.0		
1365	Manganese(II) acetate tetrahydrate	$Mn(C_2H_3O_2)_2 \cdot 4H_2O$	6156-78-1	245.087	red monocl cry	80		1.59		s H_2O, EtOH
1366	Manganese(II) tetraborate octahydrate	$MnB_4O_7 \cdot 8H_2O$	12228-91-0	354.300	red solid					i H_2O, EtOH; s dil acid
1367	Manganese(II) bromide	$MnBr_2$	13446-03-2	214.746	pink hex cry	698		4.385	151^{25}	
1368	Manganese(II) bromide tetrahydrate	$MnBr_2 \cdot 4H_2O$	10031-20-6	286.808	red hyg cry	64 dec			151^{25}	
1369	Manganese(II) carbonate	$MnCO_3$	598-62-9	114.947	pink hex cry	>200 dec		3.70	0.00008^{20}	s dil acid
1370	Manganese(II) chloride	$MnCl_2$	7773-01-5	125.843	pink trig cry; hyg	650	1190	2.977	77.3^{25}	s py, EtOH; i eth
1371	Manganese(II) chloride tetrahydrate	$MnCl_2 \cdot 4H_2O$	13446-34-9	197.905	red monocl cry; hyg	87.5		1.913	77.3^{25}	s EtOH; i eth
1372	Manganese(II) dihydrogen phosphate dihydrate	$Mn(H_2PO_4)_2 \cdot 2H_2O$	18718-07-5	284.944	col hyg cry					s H_2O; i EtOH
1373	Manganese(II) fluoride	MnF_2	7782-64-1	92.935	red tetr cry	930		3.98	1.02^{25}	i EtOH
1374	Manganese(II) hydroxide	$Mn(OH)_2$	18933-05-6	88.953	pink hex cry	dec		3.26	0.00034^{20}	
1375	Manganese(II) iodide	MnI_2	7790-33-2	308.747	wh hex cry; hyg	638		5.04		s H_2O, EtOH
1376	Manganese(II) iodide tetrahydrate	$MnI_2 \cdot 4H_2O$	7790-33-2*	380.809	red cry					vs H_2O; s EtOH
1377	Manganese(II) molybdate	$MnMoO_4$	14013-15-1	214.88	yel monocl cry			4.05		
1378	Manganese(II) nitrate	$Mn(NO_3)_2$	10377-93-2	178.948	col orth cry; hyg			2.2	161^{25}	s diox, thf
1379	Manganese(II) nitrate hexahydrate	$Mn(NO_3)_2 \cdot 6H_2O$	10377-66-9	287.040	rose monocl cry	28 dec		1.8	161^{25}	vs EtOH
1380	Manganese(II) nitrate tetrahydrate	$Mn(NO_3)_2 \cdot 4H_2O$	20694-39-7	251.010	pink hyg cry	37.1 dec		2.13	161^{25}	s EtOH
1381	Manganese(II) oxalate dihydrate	$MnC_2O_4 \cdot 2H_2O$	6556-16-7	178.987	wh cry powder	150 dec		2.45	0.032^{20}	s acid
1382	Manganese(II) oxide	MnO	1344-43-0	70.937	gr cub cry or powder	1839		5.37		i H_2O; s acid
1383	Manganese(II) perchlorate hexahydrate	$Mn(ClO_4)_2 \cdot 6H_2O$	15364-94-0	361.930	pink hex cry			2.10		
1384	Manganese(II) pyrophosphate	$Mn_2P_2O_7$	53731-35-4	283.819	wh monocl cry	1196		3.71		i H_2O
1385	Manganese(II) metasilicate	$MnSiO_3$	7759-00-4	131.022	red orth cry	1291		3.48		i H_2O
1386	Manganese(II) orthosilicate	Mn_2SiO_4	13568-32-6	201.960	orth cry			4.11		i H_2O
1387	Manganese(II) selenide	$MnSe$	1313-22-0	133.90	gray cub cry	1460		5.45		i H_2O

1388	Manganese(II) sulfate	$MnSO_4$	7785-87-7	151.002	wh orth cry	700	850 dec	3.25	63.7^{25}	
1389	Manganese(II) sulfate monohydrate	$MnSO_4 \cdot H_2O$	10034-96-5	169.017	red monocl cry			2.95	63.7^{25}	i EtOH
1390	Manganese(II) sulfate tetrahydrate	$MnSO_4 \cdot 4H_2O$	10101-68-5	223.063	red monocl cry	38 dec		2.26	63.7^{25}	i EtOH
1391	Manganese(II) sulfide (α form)	MnS	18820-29-6	87.004	grn cub cry	1610		4.0		i H_2O; s dil acid
1392	Manganese(II) sulfide (β form)	MnS	18820-29-6	87.004	red cub cry			3.3		i H_2O; s dil acid
1393	Manganese(II) sulfide (γ form)	MnS	18820-29-6	87.004	red hex cry			≈3.3		i H_2O; s dil acid
1394	Manganese(II) telluride	MnTe	12032-88-1	182.54	hex cry	≈1150		6.0		
1395	Manganese(II) titanate	$MnTiO_3$	12032-74-5	150.803	red hex cry	1360		4.55		
1396	Manganese(II) tungstate	$MnWO_4$	13918-22-4	302.78	wh monocl cry			7.2	0.0054^{20}	
1397	Manganese(II,III) oxide	Mn_3O_4	1317-35-7	228.812	brn tetr cry	1567		4.84		i H_2O; s HCl
1398	Manganese(III) fluoride	MnF_3	7783-53-1	111.933	red monocl cry; hyg	>600 dec		3.54		reac H_2O
1399	Manganese(III) hydroxide	MnO(OH)	1332-63-4	87.945	blk monocl cry	250 dec		≈4.3		i H_2O
1400	Manganese(III) oxide	Mn_2O_3	1317-34-6	157.874	blk cub cry	1080 dec		≈5.0		i H_2O
1401	Manganese(IV) oxide	MnO_2	1313-13-9	86.937	blk tetr cry	535 dec		5.08		i H_2O, HNO_3
1402	Manganese(VII) oxide	Mn_2O_7	12057-92-0	221.872	grn oil; exp	5.9	95 exp	2.40		vs H_2O
1403	Mendelevium	Md	7440-11-1	258	Metal	827				
1404	Mercury	Hg	7439-97-6	200.59	heavy silv liq	-38.837 tp	356.73	13.5336		i H_2O
1405	Mercury(I) acetate	$Hg_2(C_2H_3O_2)_2$	631-60-7	519.27	col scales	dec				sl H_2O; i EtOH, eth
1406	Mercury(I) bromate	$Hg_2(BrO_3)_2$	13465-33-3	656.98	col cry	dec				i H_2O; sl acid
1407	Mercury(I) bromide	Hg_2Br_2	15385-58-7	560.99	wh tetr cry or powder	407		7.307		i H_2O, EtOH, eth
1408	Mercury(I) carbonate	Hg_2CO_3	6824-78-8	461.19	yel-brn cry	130 dec			0.0000045	i EtOH
1409	Mercury(I) chlorate	$Hg_2(ClO_3)_2$	10294-44-7	568.08	wh rhom cry	≈250 dec		6.409		sl H_2O; s EtOH
1410	Mercury(I) chloride	Hg_2Cl_2	10112-91-1	472.09	wh tetr cry	525 tp	383 sp	7.16	0.0004^{25}	i EtOH, eth
1411	Mercury(I) fluoride	Hg_2F_2	13967-25-4	439.18	yel cub cry	570 dec	subl	8.73		reac H_2O
1412	Mercury(I) iodide	Hg_2I_2	15385-57-6	654.99	yel amorp powder	290		7.70		i H_2O, EtOH, eth
1413	Mercury(I) nitrate	$Hg_2(NO_3)_2$	10415-75-5	525.19	cry					sl H_2O
1414	Mercury(I) nitrate dihydrate	$Hg_2(NO_3)_2 \cdot 2H_2O$	7782-86-7	561.22	col cry	70 dec		4.8		sl H_2O
1415	Mercury(I) nitrite	$Hg_2(NO_2)_2$	13492-25-6	493.19	yel cry	100 dec		7.3		reac H_2O
1416	Mercury(I) oxalate	$Hg_2C_2O_4$	2949-11-3	489.20	cry					i H_2O; sl HNO_3
1417	Mercury(I) oxide	Hg_2O	15829-53-5	417.18	prob mixture of HgO+Hg	100 dec		9.8		i H_2O; s HNO_3
1418	Mercury(I) perchlorate tetrahydrate	$Hg_2(ClO_4)_2 \cdot 4H_2O$	65202-12-2	672.14	cry	64			442^{25}	
1419	Mercury(I) sulfate	Hg_2SO_4	7783-36-0	497.24	wh-yel cry powder			7.56	0.051^{25}	s dil HNO_3
1420	Mercury(I) thiocyanate	$Hg_2(SCN)_2$		517.35	col pow	dec			0.03^{25}	s HCl, KCNS
1421	Mercury(I) tungstate	Hg_2WO_4	38705-19-0	649.02	yel amorp solid	dec				i H_2O, EtOH
1422	Mercury(II) acetate	$Hg(C_2H_3O_2)_2$	1600-27-7	318.68	wh-yel cry or powder	179 dec		3.28	25^{10}	s EtOH
1423	Mercury(II) amide chloride	$Hg(NH_2)Cl$	10124-48-8	252.07	wh solid		subl	5.38		i H_2O, EtOH; s warm acid
1424	Mercury(II) bromate	$Hg(BrO_3)_2$	26522-91-8	456.39	cry	130 dec			0.15	s acid
1425	Mercury(II) bromide	$HgBr_2$	7789-47-1	360.40	wh rhomb cry or powder	236	322	6.05	0.61^{25}	sl chl; s EtOH, MeOH
1426	Mercury(II) chlorate	$Hg(ClO_3)_2$		367.49	wh needles	dec		4.998	25	
1427	Mercury(II) chloride	$HgCl_2$	7487-94-7	271.50	wh orth cry	276	304	5.6	7.31^{25}	sl bz; s EtOH, MeOH, ace, eth
1428	Mercury(II) chromate	$HgCrO_4$	13444-75-2	316.58	red monocl cry			6.06		sl H_2O
1429	Mercury(II) cyanide	$Hg(CN)_2$	592-04-1	252.62	col tetr cry	320 dec		4.00	11.4^{25}	s EtOH; sl eth
1430	Mercury(II) dichromate	$HgCr_2O_7$	7789-10-8	416.58	red cry powder					i H_2O; s acid
1431	Mercury(II) fluoride	HgF_2	7783-39-3	238.59	wh cub cry; hyg	645 dec		8.95		reac H_2O
1432	Mercury(II) fulminate	$Hg(CNO)_2$	628-86-4	284.62	gray cry	exp		4.42		sl H_2O; s EtOH, NH_4OH
1433	Mercury(II) hydrogen arsenate	$HgHAsO_4$	7784-37-4	340.52	yel powder					i H_2O; s acid
1434	Mercury(II) iodate	$Hg(IO_3)_2$	7783-32-6	550.40	wh powder	175 dec				i H_2O
1435	Mercury(II) iodide	HgI_2	7774-29-0	454.40	red tetr cry or powder	259	354	6.28	0.0055^{25}	sl EtOH, ace, eth

No.	Name	Formula	CAS Reg No.	Mol. Weight	Physical Form	mp/°C	bp/°C	Density g cm^{-3}	Solubility g/100 g H_2O	Qualitative Solubility
1436	Mercury(II) nitrate	$Hg(NO_3)_2$	10045-94-0	324.60	col hyg cry	79		4.3		s H_2O; i EtOH
1437	Mercury(II) nitrate dihydrate	$Hg(NO_3)_2 \cdot 2H_2O$	10045-94-0*	360.63	monocl cry			4.78		s H_2O
1438	Mercury(II) nitrate monohydrate	$Hg(NO_3)_2 \cdot H_2O$	7783-34-8	342.62	wh-yel hyg cry			4.3		s H_2O, dil acid
1439	Mercury(II) oxalate	HgC_2O_4	3444-13-1	288.61	pwd	165 dec				i H_2O
1440	Mercury(II) oxide	HgO	21908-53-2	216.59	red or yel orth cry	500 dec		11.14		i H_2O, EtOH; s dil acid
1441	Mercury(II) oxide sulfate	$(Hg_3O_2)SO_4$	1312-03-4	729.83	yel pow					i H_2O; s acid
1442	Mercury(II) oxycyanide	$Hg(CN)_2 \cdot HgO$	1335-31-5	469.21	wh orth cry	exp		4.44	11.4^{25}	
1443	Mercury(II) perchlorate trihydrate	$Hg(ClO_4)_2 \cdot 3H_2O$	7616-83-3	453.54	cry					
1444	Mercury(II) phosphate	$Hg_3(PO_4)_2$	7782-66-3	791.71	wh-yel powder					i H_2O, EtOH; s acid
1445	Mercury(II) selenide	HgSe	20601-83-6	279.55	gray cub cry		subl	8.21		i H_2O
1446	Mercury(II) sulfate	$HgSO_4$	7783-35-9	296.65	wh monocl cry			6.47		reac H_2O
1447	Mercury(II) sulfide (black)	HgS	1344-48-5	232.66	blk cub cry or powder	850		7.70		i H_2O; s acid, EtOH
1448	Mercury(II) sulfide (red)	HgS	1344-48-5	232.66	red hex cry	trans to blk HgS 344		8.17		i H_2O, acid; s aqua regia
1449	Mercury(II) telluride	HgTe	12068-90-5	328.19	gray cub cry	673		8.63		
1450	Mercury(II) thiocyanate	$Hg(SCN)_2$	592-85-8	316.76	monocl cry	≈165 dec		3.71	0.070^{25}	s dil HCl
1451	Mercury(II) tungstate	$HgWO_4$	37913-38-5	448.43	yel cry	dec				i H_2O, EtOH
1452	Molybdenum	Mo	7439-98-7	95.94	gray-blk metal; cub	2623	4639	10.2		i H_2O, dil acid, alk
1453	Molybdenum boride	Mo_2B	12006-99-4	202.69	refrac tetr cry	2000		9.2		
1454	Molybdenum boride	Mo_2B_5	12007-97-5	245.94	refrac hex cry	1600		≈7.2		
1455	Molybdenum carbide	MoC	12011-97-1	107.95	refrac solid; cub	2577				
1456	Molybdenum carbide	Mo_2C	12069-89-5	203.89	gray orth cry	2687		9.18		
1457	Molybdenum hexacarbonyl	$Mo(CO)_6$	13939-06-5	264.00	wh orth cry	148	155 dec	1.96		i H_2O; s bz; sl eth
1458	Molybdenum nitride	MoN	12033-19-1	109.95	hex cry	1750		9.20		
1459	Molybdenum nitride	Mo_2N	12033-31-7	205.89	gray cub cry	790 dec		9.46		
1460	Molybdenum phosphide	MoP	12163-69-8	126.91	blk hex cry			7.34		
1461	Molybdenum silicide	$MoSi_2$	12136-78-6	152.11	gray tetr cry	≈1900		6.2		i H_2O; s HF
1462	Molybdenum(II) bromide	$MoBr_2$	13446-56-5	255.75	yel-red cry	900 dec				
1463	Molybdenum(II) chloride	$MoCl_2$	13478-17-6	166.85	yel cry	530 dec				
1464	Molybdenum(II) iodide	MoI_2	14055-74-4	349.75	blk hyg cry			5.278		
1465	Molybdenum(III) bromide	$MoBr_3$	13446-57-6	335.65	grn hex cry	977		4.89		i H_2O
1466	Molybdenum(III) chloride	$MoCl_3$	13478-18-7	202.30	dark red monocl cry	1027		3.74		i H_2O
1467	Molybdenum(III) fluoride	MoF_3	20193-58-2	152.94	brn hex cry	>600		4.64		i H_2O
1468	Molybdenum(III) iodide	MoI_3	14055-75-5	476.65	blk solid	927				i H_2O
1469	Molybdenum(III) oxide	Mo_2O_3	1313-29-7	239.88	gray-blk powder					i H_2O; sl acid
1470	Molybdenum(IV) bromide	$MoBr_4$	13520-59-7	415.56	blk cry	dec				reac H_2O
1471	Molybdenum(IV) chloride	$MoCl_4$	13320-71-3	237.75	blk cry	>170 dec				reac H_2O
1472	Molybdenum(IV) fluoride	MoF_4	23412-45-5	171.93	grn cry	dec				reac H_2O
1473	Molybdenum(IV) oxide	MoO_2	18868-43-4	127.94	brn-viol tetr cry	≈1100 dec		6.47		sl H_2O
1474	Molybdenum(IV) selenide	$MoSe_2$	12058-18-3	253.86	gray hex cry	>1200		6.90		
1475	Molybdenum(IV) sulfide	MoS_2	1317-33-5	160.07	blk powder or hex cry	1750		5.06		i H_2O; s conc acid
1476	Molybdenum(IV) telluride	$MoTe_2$	12058-20-7	351.14	gray hex cry			7.7		
1477	Molybdenum(V) chloride	$MoCl_5$	10241-05-1	273.20	gr-blk monocl cry; hyg	194	268	2.93		s EtOH, eth
1478	Molybdenum(V) fluoride	MoF_5	13819-84-6	190.93	yel monocl cry	67	213.6	3.5		
1479	Molybdenum(V) oxytrichloride	$MoOCl_3$	13814-74-9	218.30	blk monocl cry	297	subl			reac H_2O
1480	Molybdenum(VI) acid monohydrate	$H_2MoO_4 \cdot H_2O$	7782-91-4	179.97	wh powder			3.1		sl H_2O; s alk

PHYSICAL CONSTANTS OF INORGANIC COMPOUNDS (CONTINUED)

1481	Molybdenum(VI) fluoride	MoF_6	7783-77-9	209.93	wh cub cry or col liq; hyg	17.5	34.0	2.54		reac H_2O
1482	Molybdenum(VI) oxytetrafluoride	$MoOF_4$	14459-59-7	187.93	volatile solid	98	186.0			
1483	Molybdenum(VI) oxytetrachloride	$MoOCl_4$	13814-75-0	253.75	grn hyg powder	101				
1484	Molybdenum(VI) dioxydichloride	MoO_2Cl_2	13637-68-8	198.84	yel-oran solid	≈175		3.31		reac H_2O
1485	Molybdenum(VI) oxide	MoO_3	1313-27-5	143.94	wh-yel rhomb cry	801	1155	4.70	0.14^{20}	s conc acid
1486	Molybdenum(VI) metaphosphate	$Mo(PO_3)_6$	133863-98-6	569.77	yel powder			3.28		i H_2O, acid
1487	Neodymium	Nd	7440-00-8	144.24	silv metal; hex	1021	3074	7.01		
1488	Neodymium boride	NdB_6	12008-23-0	209.11	blk cub cry	2610		4.93		
1489	Neodymium bromide	$NdBr_3$	13536-80-6	383.95	viol orth cry; hyg	682	1540	5.3		s H_2O
1490	Neodymium chloride	$NdCl_3$	10024-93-8	250.60	viol hex cry	758	1600	4.13	100^{25}	vs EtOH; i eth, chl
1491	Neodymium chloride hexahydrate	$NdCl_3 \cdot 6H_2O$	13477-89-9	358.69	purp cry	124 dec		2.3	100^{25}	s EtOH
1492	Neodymium fluoride	NdF_3	13709-42-7	201.24	viol hex cry; hyg	1377	2300	6.51		i H_2O
1493	Neodymium iodide	NdI_3	13813-24-6	524.95	grn orth cry; hyg	784		5.85		s H_2O
1494	Neodymium nitrate	$Nd(NO_3)_3$	10045-95-1	330.26	viol hyg. cry				152^{25}	s EtOH
1495	Neodymium nitrate hexahydrate	$Nd(NO_3)_3 \cdot 6H_2O$	14517-29-4	438.35	purp hyg cry				152^{25}	s EtOH, ace
1496	Neodymium nitride	NdN	25764-11-8	158.25	blk cub cry			7.69		
1497	Neodymium oxide	Nd_2O_3	1313-97-9	336.48	blue hex cry; hyg	2233	3760	7.24		i H_2O; s dil acid
1498	Neodymium sulfate	$Nd_2(SO_4)_3$	13477-91-3	576.67	pink needles	≈700 dec			7.1^{20}	
1499	Nickel(II) perchlorate hexahydrate	$Ni(ClO_4)_2 \cdot 6H_2O$	13637-71-3*	365.685	grn hex needles	140			158.8^{25}	s EtOH, ace
1500	Neodymium sulfide	Nd_2S_3	12035-32-4	384.68	orth cry	2207		5.46		
1501	Nickel(II) phosphate octahydrate	$Ni_3(PO_4)_2 \cdot 8H_2O$	10381-36-9*	510.145	grn plates					s acid
1502	Nickel(II) selenate hexahydrate	$NiSeO_4 \cdot 6H_2O$	15060-62-5*	309.74	grn tetr cry			2.314	35.5^{20}	
1503	Neodymium telluride	Nd_2Te_3	12035-35-7	671.28	gray orth cry	1377		7.0		
1504	Neon	Ne	7440-01-9	20.180	col gas	-248.61 tp (43 kPa)	-246.08	0.825 g/L		sl H_2O
1505	Neptunium	Np	7439-99-8	237	silv metal	644		20.2		s HCl
1506	Neptunium(IV) oxide	NpO_2	12035-79-9	269	grn cub cry	2547		11.1		
1507	Nickel	Ni	7440-02-0	58.693	wh metal; cub	1455	2913	8.90		i H_2O; sl dil acid
1508	Nickel antimonide	NiSb	12035-52-8	180.453	hex cry	1147		8.74		
1509	Nickel arsenide	NiAs	27016-75-7	133.615	hex cry	967		7.77		
1510	Nickel boride	Ni_3B	12007-02-2	186.891	refrac solid	1156		8.17		
1511	Nickel boride	NiB	12007-00-0	69.504	grn refrac solid	1035		7.13		
1512	Nickel boride	Ni_2B	12007-01-1	128.198	refrac solid	1125		7.90		
1513	Nickel carbonyl	$Ni(CO)_4$	13463-39-3	170.734	col liq	-19.3	43 (exp ≈60)	1.31		i H_2O; s EtOH, bz, ace, ctc
1514	Nickel phosphide	Ni_2P	12035-64-2	148.361	hex cry	1100		7.33		
1515	Nickel silicide	Ni_2Si	12059-14-2	145.473	orth cry	1255		7.40		
1516	Nickel silicide	$NiSi_2$	12201-89-7	114.864	cub cry	993		4.83		
1517	Nickel(II) ammonium sulfate hexahydrate	$Ni(NH_4)_2(SO4)_2 \cdot 6H_2O$	15699-18-0	394.989	blue-grn cry			1.923	6.5^{20}	i EtOH
1518	Nickel(II) arsenate octahydrate	$Ni_3(AsO_4)_2 \cdot 8H_2O$	7784-48-7	598.040	yel-grn powder	dec		4.98		i H_2O; s acid
1519	Nickel(II) bromide	$NiBr_2$	13462-88-9	218.501	yel hex cry; hyg	963	subl	5.10	131^{20}	
1520	Nickel(II) bromide trihydrate	$NiBr_2 \cdot 3H_2O$	13462-88-9*	272.547	yel-grn hyg cry	200 dec				vs H_2O; s EtOH, eth
1521	Nickel(II) carbonate	$NiCO_3$	3333-67-3	118.702	grn rhomb cry			4.39	0.0043^{20}	s dil acid
1522	Nickel(II) chloride	$NiCl_2$	7718-54-9	129.598	yel hex cry; hyg	1009 tp	985 sp	3.51	67.5^{25}	s EtOH
1523	Nickel(II) chloride hexahydrate	$NiCl_2 \cdot 6H_2O$	7791-20-0	237.689	grn monocl cry				67.5^{25}	s EtOH
1524	Nickel(II) cyanide tetrahydrate	$Ni(CN)_2 \cdot 4H_2O$	13477-95-7	182.789	grn plates	200 dec				i H_2O; sl dil acid; s NH_4OH
1525	Nickel(II) fluoride	NiF_2	10028-18-9	96.690	yel tetr cry	1474		4.7	2.56^{25}	i EtOH, eth
1526	Nickel(II) hydroxide	$Ni(OH)_2$	12054-48-7	92.708	grn hex cry	230 dec		4.1	0.00015^{20}	
1527	Nickel(II) hydroxide monohydrate	$Ni(OH)_2 \cdot H_2O$	36897-37-7	110.723	grn powder				0.00015^{20}	s dil acid
1528	Nickel(II) iodate	$Ni(IO_3)_2$	13477-98-0	408.498	yel needles			5.07	1.1^{30}	

No.	Name	Formula	CAS Reg No.	Mol. Weight	Physical Form	mp/°C	bp/°C	Density g cm⁻³	Solubility g/100 g H_2O	Qualitative Solubility
1529	Nickel(II) iodide	NiI_2	13462-90-3	312.502	blk hex cry; hyg	780	subl	5.22	154^{25}	
1530	Nickel(II) iodide hexahydrate	$NiI_2 \cdot 6H_2O$	7790-34-3	420.593	grn monocl cry; hyg				154^{25}	vs EtOH
1531	Nickel(II) nitrate	$Ni(NO_3)_2$	13138-45-9	182.702	grn cry				99.2^{25}	s EtOH
1532	Nickel(II) nitrate hexahydrate	$Ni(NO_3)_2 \cdot 6H_2O$	13478-00-7	290.794	grn monocl cry; hyg	56 dec		2.05	99.2^{25}	s EtOH
1533	Nickel(II) oxide	NiO	1313-99-1	74.692	grn cub cry	1955		6.72		i H_2O; s acid
1534	Nickel(II) selenide	NiSe	1314-05-2	137.65	yel-grn hex cry	980		7.2		
1535	Nickel(II) sulfate	$NiSO_4$	7786-81-4	154.757	grn-yel orth cry	840 dec		4.01	40.4^{25}	
1536	Nickel(II) sulfate heptahydrate	$NiSO_4 \cdot 7H_2O$	10101-98-1	280.863	grn orth cry			1.98	40.4^{25}	s EtOH
1537	Nickel(II) sulfate hexahydrate	$NiSO_4 \cdot 6H_2O$	10101-97-0	262.848	blue-grn tetr cry	≈100 dec		2.07	40.4^{25}	sl EtOH
1538	Nickel(II) sulfide	NiS	16812-54-7	90.759	yel hex cry	976		5.5		i H_2O
1539	Nickel(II) thiocyanate	$Ni(SCN)_2$	13689-92-4	174.859	grn pwd				55.0^{25}	
1540	Nickel(II) titanate	$NiTiO_3$	12035-39-1	154.558	brn hex cry			5.0		
1541	Nickel(II,III) sulfide	Ni_3S_4	12137-12-1	304.344	cub cry	995		4.77		
1542	Nickel(III) oxide	Ni_2O_3	1314-06-3	165.385	gray-blk cub cry	≈600 dec				i H_2O; s hot acid
1543	Nickel(III) sulfide	Ni_3S_2	12035-72-2	240.212	hex cry	787		5.87		
1544	Niobium	Nb	7440-03-1	92.906	gray metal; cub	2477	4744	8.57		i acid
1545	Niobium boride	NbB	12045-19-1	103.717	gray orth cry	2270		7.5		
1546	Niobium boride	NbB_2	12007-29-3	114.528	gray hex cry	3050		6.97		
1547	Niobium carbide	NbC	12069-94-2	104.917	gray cub cry	3608	4300	7.82		i H_2O, acid
1548	Niobium carbide	Nb_2C	12011-99-3	197.824	refrac hex cry	3080		7.8		i H_2O
1549	Niobium nitride	NbN	24621-21-4	106.913	gray cry; cub	2300		8.47		i HCl, acid
1550	Niobium phosphide	NbP	12034-66-1	123.880	tetr cry			6.5		
1551	Niobium silicide	$NbSi_2$	12034-80-9	149.077	gray hex cry	1950		5.7		
1552	Niobium(II) oxide	NbO	12034-57-0	108.905	gray cub cry	1936		7.30		
1553	Niobium(III) bromide	$NbBr_3$	15752-41-7	332.618	dark brn solid		subl			
1554	Niobium(III) chloride	$NbCl_3$	13569-59-0	199.264	blk solid					
1555	Niobium(III) fluoride	NbF_3	15195-53-6	149.901	blue cub cry			4.2		
1556	Niobium(IV) chloride	$NbCl_4$	13569-70-5	234.717	viol-blk monocl cry		275 subl	3.2		
1557	Niobium(IV) fluoride	NbF_4	13842-88-1	168.900	blk tetr cry; hyg	>350 dec		4.01		
1558	Niobium(IV) iodide	NbI_4	13870-21-8	600.524	gray orth cry	503		5.6		
1559	Niobium(IV) oxide	NbO_2	12034-59-2	124.905	wh tetr cry or powder	1901		5.9		
1560	Niobium(IV) selenide	$NbSe_2$	12034-77-4	250.83	gray hex cry	>1300		6.3		
1561	Niobium(IV) sulfide	NbS_2	12136-97-9	157.038	blk rhomb cry			4.4		
1562	Niobium(IV) telluride	$NbTe_2$	12034-83-2	348.11	hex cry			7.6		
1563	Niobium(V) bromide	$NbBr_5$	13478-45-0	492.426	oran orth cry	254	360	4.36		s H_2O, EtOH
1564	Niobium(V) chloride	$NbCl_5$	10026-12-7	270.170	yel monocl cry; hyg	204.7	254.0	2.78		reac H_2O; s HCl, ctc
1565	Niobium(V) fluoride	NbF_5	7783-68-8	187.898	col monocl cry; hyg	80	229	2.70		reac H_2O; sl CS_2, chl
1566	Niobium(V) iodide	NbI_5	13779-92-5	727.428	yel-blk monocl cry	≈200 dec		5.32		
1567	Niobium(V) oxide	Nb_2O_5	1313-96-8	265.810	wh orth cry	1512		4.6		i H_2O; s HF
1568	Niobium(V) oxybromide	$NbOBr_3$	14459-75-7	348.617	yel-brn cry	≈320 dec	subl			
1569	Niobium(V) oxychloride	$NbOCl_3$	13597-20-1	215.263	wh tetr cry		subl	3.72		
1570	Niobium(V) dioxyfluoride	NbO_2F	15195-33-2	143.903	wh cub cry			4.0		
1571	Nitrogen	N_2	7727-37-9	28.013	col gas	-210.00	-195.79	1.145 g/L		sl H_2O; i EtOH
1572	Nitramide	NO_2NH_2	7782-94-7	62.028	unstable wh cry	72 dec				s H_2O, EtOH, ace, eth; i chl
1573	Nitric acid	HNO_3	7697-37-2	63.013	col liq; hyg	-41.6	83	1.55		vs H_2O

1574	Nitrous acid	HNO_2	7782-77-6	47.014	stable only in soln					
1575	Nitrous oxide	N_2O	10024-97-2	44.012	col gas	-90.8	-88.48	1.799 g/L		sl H_2O; s EtOH, eth
1576	Nitric oxide	NO	10102-43-9	30.006	col gas	-163.6	-151.74	1.226 g/L		sl H_2O
1577	Nitrogen dioxide	NO_2	10102-44-0	46.006	brn gas; equil with N_2O_4		see N_2O_4	1.880 g/L		reac H_2O
1578	Nitrogen trioxide	N_2O_3	10544-73-7	76.011	blue solid or liq (low temp)	-101.1	≈3 dec	1.4^{2}		reac H_2O
1579	Nitrogen tetroxide	N_2O_4	10544-72-6	92.011	col liq; equil with NO_2	-9.3	21.15	1.45^{20}		reac H_2O
1580	Nitrogen pentoxide	N_2O_5	10102-03-1	108.010	col hex cry		33 sp	2.0		s chl; sl ctc
1581	Nitrogen tribromide	NBr_3	15162-90-0	253.719	unstable solid	exp -100				
1582	Nitrogen trichloride	NCl_3	10025-85-1	120.365	yel oily liq; exp	-40	71	1.653		i H_2O; s CS_2, bz, ctc
1583	Nitrogen trifluoride	NF_3	7783-54-2	71.002	col gas	-206.79	-128.75	2.902 g/L		i H_2O
1584	Nitrogen triiodide	NI_3	13444-85-4	394.720	unstable blk cry; exp					
1585	Nitrogen chloride difluoride	$NClF_2$	13637-87-1	87.457	col gas	-195	-67	3.575 g/L		
1586	Chloramine	NH_2Cl	10599-90-3	51.476	yel liq	-66				s H_2O, EtOH, eth; sl bz, CCl_4
1587	Fluoramine	NH_2F	15861-05-9	35.021	unstable gas			1.431 g/L		
1588	Difluoramine	NHF_2	10405-27-3	53.012	col gas	-116	-23	2.167 g/L		
1589	cis-Difluorodiazine	N_2F_2	13812-43-6	66.010	col gas	<-195	-105.75	2.698 g/L		
1590	trans-Difluorodiazine	N_2F_2	13776-62-0	66.010	col gas	-172	-111.45	2.698 g/L		
1591	Tetrafluorohydrazine	N_2F_4	10036-47-2	104.007	col gas	-164.5	-74	4.251 g/L		
1592	Nitrosyl bromide	NOBr	13444-87-6	109.910	red gas	-56	≈0	4.492 g/L		reac H_2O
1593	Nitrosyl chloride	NOCl	2696-92-6	65.459	yel gas	-59.6	-5.5	2.676 g/L		reac H_2O
1594	Nitrosyl fluoride	NOF	7789-25-5	49.004	col gas	-132.5	-59.9	2.003 g/L		
1595	Trifluoramine oxide	NOF_3	13847-65-9	87.001	col gas	-161	-87.5	3.556 g/L		
1596	Nitryl chloride	NO_2Cl	13444-90-1	81.459	col gas	-145	-15	3.330 g/L		
1597	Nitryl fluoride	NO_2F	10022-50-1	65.004	col gas	-166	-72.4	2.657 g/L		reac H_2O
1598	Nitrogen selenide	N_4Se_4	12033-88-4	371.87	red monocl cry; hyg	exp		4.2		i H_2O, eth, EtOH; sl bz, CS_2
1599	Nobelium	No	10028-14-5	259	metal	827				
1600	Osmium	Os	7440-04-2	190.23	blue-wh metal; hex	3033	5012	22.59		s aqua regia
1601	Osmium carbonyl	$Os_3(CO)_{12}$	15696-40-9	906.81	yel cry			3.48		
1602	Osmium(III) bromide	$OsBr_3$	59201-51-3	429.94	dark gray cry	340 dec				
1603	Osmium(III) chloride	$OsCl_3$	13444-93-4	296.59	gray cub cry	>450 dec				i H_2O; s HNO_3
1604	Osmium(IV) chloride	$OsCl_4$	10026-01-4	332.04	red-blk orth cry		450 sp	4.38		reac H_2O
1605	Osmium(IV) fluoride	OsF_4	54120-05-7	266.22	yel cry	230				
1606	Osmium(IV) oxide	OsO_2	12036-02-1	222.23	yel-brn tetr cry			11.4		i H_2O, acid
1607	Osmium(V) fluoride	OsF_5	31576-40-6	285.22	blue cry	70	225.9			reac H_2O
1608	Osmium(VI) fluoride	OsF_6	13768-38-2	304.22	yel cub cry	33.2	47.5	4.1		reac H_2O
1609	Osmium(VIII) oxide	OsO_4	20816-12-0	254.23	yel monocl cry	41	135	5.1	6.44^{20}	
1610	Oxygen	O_2	7782-44-7	31.999	col gas	-218.79	-182.95	1.308 g/L		sl H_2O, EtOH, os
1611	Ozone	O_3	10028-15-6	47.998	blue gas	-193	-111.35	1.962 g/L		sl H_2O
1612	Palladium	Pd	7440-05-3	106.42	silv-wh metal; cub	1554.9	2963	12.0		s aqua regia
1613	Palladium(II) sulfide	PdS	12125-22-3	138.49	gray tetr cry			6.7		
1614	Palladium(II) bromide	$PdBr_2$	13444-94-5	266.23	red-blk monocl cry; hyg	250 dec		≈5.2		i H_2O
1615	Palladium(II) chloride	$PdCl_2$	7647-10-1	177.33	red rhomb cry; hyg	679		4.0		s H_2O, EtOH, ace
1616	Palladium(II) fluoride	PdF_2	13444-96-7	144.42	viol tetr cry; hyg	952		5.76		reac H_2O
1617	Palladium(II) iodide	PdI_2	7790-38-7	360.23	blk cry	360 dec		6.0		i H_2O, EtOH, eth
1618	Palladium(II) nitrate	$Pd(NO_3)_2$	10102-05-3	230.43	brn hyg cry	dec				sl H_2O; s dil HNO_3
1619	Palladium(II) oxide	PdO	1314-08-5	122.42	grn-blk tetr cry	750 dec		8.3		i H_2O, acid; sl aqua regia
1620	Phosphorus (white)	P	7723-14-0	30.974	col waxlike cub cry	44.15	280.5	1.823		i H_2O; sl bz, EtOH, chl; s CS_2
1621	Phosphorus (red)	P	7723-14-0	30.974	red-viol amorp powder	590 tp	431 sp	2.16		i H_2O, os

No.	Name	Formula	CAS Reg No.	Mol. Weight	Physical Form	mp/°C	bp/°C	Density g cm^{-3}	Solubility g/100 g H_2O	Qualitative Solubility
1622	Phosphorus (black)	P	7723-14-0	30.974	blk orth cry or amorp solid	610		2.69		i os
1623	Phosphine	PH_3	7803-51-2	33.998	col gas; flam	-133.8	-87.75	1.390 g/L		i H_2O; sl EtOH, eth
1624	Diphosphine	P_2H_4	13445-50-6	65.980	col liq	-99	63.5 dec			reac H_2O
1625	Diphosphorus tetrachloride	P_2Cl_4	13497-91-1	203.759	col oily liq	-28	≈180 dec			
1626	Diphosphorus tetrafluoride	P_2F_4	13824-74-3	137.942	col gas	-86.5	-6.2	5.638 g/L		
1627	Diphosphorus tetraiodide	P_2I_4	13455-00-0	569.566	red tricl needles	125.5	dec	3.89		
1628	Phosphonium chloride	PH_4Cl	24567-53-1	70.459	gas		-27 sp	2.880 g/L		reac H_2O
1629	Phosphonium iodide	PH_4I	12125-09-6	161.910	col tetr cry	18.5	62.5	2.86		reac H_2O, EtOH
1630	Phosphoric acid (orthophosphoric acid)	H_3PO_4	7664-38-2	97.995	col visc liq	42.4	407		548^{20}	s EtOH
1631	Phosphonic acid (phosphorous acid)	H_3PO_3	13598-36-2	81.996	wh hyg cry	74.4	200	1.65	309^{0}	vs EtOH
1632	Phosphinic acid (hypophosphorous acid)	HPH_2O_2	6303-21-5	65.997	hyg cry or col oily liq	26.5	130	1.49		vs H_2O, EtOH, eth
1633	Metaphosphoric acid	HPO_3	37267-86-0	79.980	gl solid; hyg					sl H_2O; s EtOH
1634	Hypophosphoric acid	$H_4P_2O_6$	7803-60-3	161.976	col orth cry	73 dec				vs H_2O
1635	Diphosphoric acid (pyrophosphoric acid)	$H_4P_2O_7$	2466-09-3	177.975	wh cry	71.5			709^{23}	
1636	Difluorophosphoric acid	HPO_2F_2	13779-41-4	101.978	col liq	≈-94	110 dec	1.583		reac H_2O
1637	Hexafluorophosphoric acid	HPF_6	16940-81-1	145.972	col oily liq	25 dec				reac H_2O
1638	Fluorophosphonic acid	H_2PFO_3	13537-32-1	99.986	col visc liq	<-70		1.82		vs H_2O
1639	Phosphorus nitride	P_3N_5	12136-91-3	162.955	yel-brn solid	800 dec				i H_2O; s os
1640	Phosphorus sesquisulfide	P_4S_3	1314-85-8	220.093	yel-grn orth cry	172.5	407	2.03		i H_2O; s bz; vs CS_2
1641	Phosphorus heptasulfide	P_4S_7	12037-82-0	348.357	pale yel monocl cry	312	523	2.19		sl CS_2
1642	Phosphonitrilic chloride trimer	$(PNCl_2)_3$	940-71-6	347.657	wh hyg cry	128.8		1.98		reac H_2O
1643	Phosphorus(III) bromide	PBr_3	7789-60-8	270.686	col liq	-41.5	173.2	2.8		reac H_2O, EtOH; s ace, CS_2
1644	Phosphorus(III) dibromide fluoride	PBr_2F	15597-39-4	209.780	col liq	-115	78.5			
1645	Phosphorus(III) bromide difluoride	$PBrF_2$	15597-40-7	148.875	col gas	-133.8	-16.1	6.085 g/L		
1646	Phosphorus(III) chloride	PCl_3	7719-12-2	137.332	col liq	-93.6	76.1	1.574		reac H_2O, EtOH; s bz, chl, eth
1647	Phosphorus(III) dichloride fluoride	PCl_2F	15597-63-4	120.877	col gas	-144	13.85	4.941 g/L		
1648	Phosphorus(III) chloride difluoride	$PClF_2$	14335-40-1	104.424	col gas	-164.8	-47.3	4.268 g/L		
1649	Phosphorus(III) fluoride	PF_3	7783-55-3	87.969	col gas	-151.5	-101.8	3.596 g/L		reac H_2O
1650	Phosphorus(III) iodide	PI_3	13455-01-1	411.687	red-oran hex cry; hyg	61.2	227 dec	4.18		reac H_2O; s EtOH
1651	Phosphorus(III) oxide	P_2O_3	1314-24-5	109.946	col monocl cry or liq	23.8	173	2.13		reac H_2O
1652	Tetraphosphorus(III) hexoxide	P_4O_6	12440-00-5	219.891	soft wh cry	23.8	175.4			
1653	Phosphorus(III) selenide	P_2Se_3	1314-86-9	298.83	oran-red cry	245	≈380	1.31		reac H_2O; s bz, ctc, CS_2, ace
1654	Phosphorus(III) sulfide	P_2S_3	12165-69-4	158.146	yel solid	290	490			reac H_2O; s EtOH, eth, CS_2
1655	Phosphorus(V) bromide	PBr_5	7789-69-7	430.494	yel orth cry, Donnay	≈100 dec		3.61		reac H_2O, EtOH; s CS_2, ctc
1656	Phosphorus(V) tetrabromide fluoride	PBr_4F		369.588	pale yel cry	87 dec				
1657	Phosphorus(V) dibromide trifluoride	PBr_2F_3	13445-58-4	247.777	yel-red liq	-20	15 dec			
1658	Phosphorus(V) chloride	PCl_5	10026-13-8	208.238	wh-yel tetr cry; hyg	167 tp	160 sp	2.1		reac H_2O; s CS_2, ctc
1659	Phosphorus(V) tetrachloride fluoride	PCl_4F	13498-11-8	191.783	col liq	-59	30 dec			
1660	Phosphorus(V) trichloride difluoride	PCl_3F_2	158704-27-9	175.329	col liq	-63				
1661	Phosphorus(V) dichloride trifluoride	PCl_2F_3	13454-99-4	158.874	col gas	-125	7.1	6.494 g/L		
1662	Phosphorus(V) chloride tetrafluoride	$PClF_4$		142.421	col gas	-132	-43.4	5.821 g/L		
1663	Phosphorus(V) fluoride	PF_5	7647-19-0	125.966	col gas	-93.8	-84.6	5.149 g/L		reac H_2O
1664	Phosphorus(V) oxide	P_2O_5	1314-56-3	141.945	wh orth cry; hyg	562	605	2.30		reac H_2O, EtOH
1665	Phosphorus(V) selenide	P_2Se_5	1314-82-5	456.75	blk-purp amorp solid					reac hot H_2O, ctc; i CS_2
1666	Phosphorus(V) sulfide	P_2S_5	1314-80-3	222.278	grn-yel hyg cry	285	515	2.03		reac H_2O; s CS_2

1667	Phosphonic difluoride	POF_2H	14939-34-5	85.978	volatile liq	>-120	≈60 (gas unstab)			
1668	Phosphoric tribromide (phosphoryl bromide)	$POBr_3$	7789-59-5	286.685	faint oran plates	55	191.7	2.822		reac H_2O; s bz, eth, chl
1669	Phosphoric dibromide chloride	$POBr_2Cl$	13550-31-7	242.234	yel solid	31	165			
1670	Phosphoric dibromide fluoride	$POBr_2F$	14014-19-8	225.779	col liq	-117.2	110.1			
1671	Phosphoric bromide dichloride	$POBrCl_2$	13455-03-3	197.782	col liq	11	136.5	2.104^{14}		
1672	Phosphoric bromide difluoride	$POBrF_2$	14014-18-7	164.874	col liq	-84.8	31.6			
1673	Phosphoric bromide chloride fluoride	POBrClF	14518-81-1	181.328	col liq		79			
1674	Phosphoric trichloride (phosphoryl chloride)	$POCl_3$	10025-87-3	153.331	col liq	1.18	105.5	1.645		reac H_2O, EtOH
1675	Phosphoric dichloride fluoride	$POCl_2F$	13769-76-1	136.876	col liq	-80.1	52.9			
1676	Phosphoric chloride difluoride	$POClF_2$	13769-75-0	120.423	col gas	-96.4	3.1	4.922 g/L		
1677	Phosphoric trifluoride (phosphoryl fluoride)	POF_3	13478-20-1	103.968	col gas	-39.1 tp	-39.7 sp	4.250 g/L		reac H_2O
1678	Phosphoric triodide (phosphoryl iodide)	POI_3	13455-04-4	427.686	viol cry	53				
1679	Phosphorothioc tribromide	$PSBr_3$	3931-89-3	302.752	yel cry	37.8	212 dec	2.85		
1680	Phosphorothioc dibromide fluoride	$PSBr_2F$	13706-10-0	241.846	yel liq	-75.2	125.3			
1681	Phosphorothioc bromide difluoride	$PSBrF_2$	13706-09-7	180.941	yel liq	-136.9	35.5			
1682	Phosphorothioc trichloride	$PSCl_3$	3982-91-0	169.398	fuming liq	-36.2	125	1.635		reac H_2O; s bz, ctc, chl, CS_2
1683	Phosphorothioc dichloride fluoride	$PSCl_2F$	155698-29-6	152.943	col liq	-96.0	64.7			
1684	Phosphorothioc chloride difluoride	$PSClF_2$	2524-02-9	136.490	col gas	-155.2	6.3	5.579 g/L		
1685	Phosphorothioc trifluoride	PSF_3	2404-52-6	120.035	col gas	-148.8	-52.25	4.906 g/L		
1686	Phosphorothioc triiodide	PSI_3	63972-04-3	443.753	yel cry	48	dec			
1687	Platinum	Pt	7440-06-4	195.08	silv-gray metal; cub	1768.4	3825	21.5		i acid; s aqua regia
1688	Platinum(II) bromide	$PtBr_2$	13455-12-4	354.89	red-brn powder	250 dec		6.65		i H_2O
1689	Platinum(II) chloride	$PtCl_2$	10025-65-7	265.98	grn hex cry	581 dec		6.0		i H_2O, EtOH, eth; s HCl
1690	Platinum(II) iodide	PtI_2	7790-39-8	448.89	blk powder	325 dec		6.4		i H_2O
1691	Platinum(II) oxide	PtO	12035-82-4	211.08	blk tetr cry	325 dec		14.1		i H_2O, EtOH; s aqua regia
1692	Platinum(II) sulfide	PtS	12038-20-9	227.14	tetr cry			10.25		
1693	Platinum(III) bromide	$PtBr_3$	25985-07-3	434.79	grn-blk cry	200 dec				
1694	Platinum(III) chloride	$PtCl_3$	25909-39-1	301.44	grn-blk cry	435 dec		5.26		
1695	Platinum(IV) bromide	$PtBr_4$	68938-92-1	514.69	brn-blk cry	180 dec			0.41^{20}	sl EtOH, eth
1696	Platinum(IV) chloride	$PtCl_4$	37773-49-2	336.89	red-brn cub cry	327 dec		4.30	142^{25}	
1697	Platinum(IV) chloride pentahydrate	$PtCl_4 \cdot 5H_2O$	13454-96-1	426.97	red cry			2.43		s H_2O, EtOH
1698	Platinum(IV) fluoride	PtF_4	13455-15-7	271.07	red cry	600				
1699	Platinum(IV) iodide	PtI_4	7790-46-7	702.70	brn-blk powder	130 dec				s H_2O
1700	Platinum(IV) oxide	PtO_2	1314-15-4	227.08	blk hex cry	450		11.8		i H_2O; s conc acid, dil alk
1701	Platinum(IV) sulfide	PtS_2	12038-21-0	259.21	hex cry			7.85		
1702	Platinum(VI) fluoride	PtF_6	13693-05-5	309.07	red cub cry	61.3	69.1	≈4.0		
1703	cis-Diamminedichloroplatinum	$Pt(NH_3)_2Cl_2$	15663-27-1	300.04	yel solid	270 dec			0.253^{25}	
1704	trans-Diamminedichloroplatinum	$Pt(NH_3)_2Cl_2$	14913-33-8	300.04	pale yel solid	270 dec			0.036^{25}	s DMF, DMSO
1705	Hexachloroplatinic acid hexahydrate	$H_2PtCl_6 \cdot 6H_2O$	16941-12-1	517.90	brn-yel hyg cry	60		2.43	140^{18}	vs EtOH
1706	Platinum silicide	PtSi	12137-83-6	223.16	orth cry	1229		12.4		
1707	Plutonium	Pu	7440-07-5	244	silv-wh metal; monocl	640	3228	19.7		
1708	Plutonium nitride	PuN	12033-54-4	258	gray cub cry	2550		14.4		
1709	Plutonium(II) oxide	PuO	12035-83-5	260	cub cry			14.0		
1710	Plutonium(III) bromide	$PuBr_3$	15752-46-2	484	grn orth cry	681		6.75		s H_2O
1711	Plutonium(III) chloride	$PuCl_3$	13569-62-5	350	grn hex cry	760		5.71		s H_2O
1712	Plutonium(III) fluoride	PuF_3	13842-83-6	301	purp hex cry	1396		9.33		i H_2O; sl acid

No.	Name	Formula	CAS Reg No.	Mol. Weight	Physical Form	mp/°C	bp/°C	Density g cm^{-3}	Solubility g/100 g H_2O	Qualitative Solubility
1713	Plutonium(III) iodide	PuI_3	13813-46-2	625	grn orth cry; hyg	777		6.92		s H_2O
1714	Plutonium(III) oxide	Pu_2O_3	12036-34-9	536	blk cub cry			10.5		
1715	Plutonium(IV) fluoride	PuF_4	13709-56-3	320	red-brn monocl cry	1027		7.1		
1716	Plutonium(IV) oxide	PuO_2	12059-95-9	276	yel-brn cub cry	2400		11.5		
1717	Plutonium(VI) fluoride	PuF_6	13693-06-6	358	red-brn orth cry	52		5.08		
1718	Polonium	Po	7440-08-6	209	silv metal; cub	254	962	9.20		
1719	Polonium(IV) chloride	$PoCl_4$	10026-02-5	351	yel hyg cry	≈300	390			s H_2O, EtOH, ace
1720	Polonium(IV) oxide	PoO_2	7446-06-2	241	yel cub cry	500 dec		8.9		
1721	Potassium	K	7440-09-7	39.098	soft silv-wh metal; cub	63.5	759	0.89		reac H_2O
1722	Potassium acetate	$KC_2H_3O_2$	127-08-2	98.142	wh hyg cry	309		1.57	269^{25}	s EtOH; i eth
1723	Potassium aluminate trihydrate	$K_2Al_2O_4 \cdot 3H_2O$	12003-63-3*	250.204	wh orth cry			2.13		vs H_2O; i EtOH
1724	Potassium aluminum silicate	$KAlSi_3O_8$	1327-44-2	278.332	col monocl cry			2.56		i H_2O
1725	Potassium aluminum sulfate	$KAl(SO_4)_2$	10043-67-1	258.207	wh hyg powder				5.9^{20}	
1726	Potassium aluminum sulfate dodecahydrate	$KAl(SO_4)_2 \cdot 12H_2O$	7784-24-9	474.391	col cry	≈100 dec		1.72	5.9^{20}	
1727	Potassium amide	KNH_2	17242-52-3	55.121	wh/yel-grn hyg cry	335				reac H_2O, EtOH
1728	Potassium arsenate	K_3AsO_4	13464-36-3	256.215	col cry			2.8	125^{25}	
1729	Potassium arsenite	$KAsO_2$	13464-35-2	146.019	wh hyg pow					s H_2O; sl EtOH
1730	Potassium azide	KN_3	20762-60-1	81.118	tetr cry; exp			2.04	49.7^{17}	
1731	Potassium borohydride	KBH_4	13762-51-1	53.941	wh cub cry	≈500 dec		1.11		s H_2O
1732	Potassium bromate	$KBrO_3$	7758-01-2	167.000	wh hex cry	434 dec		3.27	8.17^{25}	i EtOH
1733	Potassium bromide	KBr	7758-02-3	119.002	col cub cry; hyg	734	1435	2.74	67.8^{25}	sl EtOH
1734	Potassium carbonate	K_2CO_3	584-08-7	138.206	wh monocl cry; hyg	898	dec	2.29	111^{25}	i EtOH
1735	Potassium carbonate sesquihydrate	$K_2CO_3 \cdot 1.5H_2O$	6381-79-9	165.229	granular cry				111^{20}	
1736	Potassium chlorate	$KClO_3$	3811-04-9	122.549	wh monocl cry	368	dec	2.32	8.61^{25}	
1737	Potassium chloride	KCl	7447-40-7	74.551	wh cub cry	771		1.988	35.5^{25}	i eth, ace
1738	Potassium chromate	K_2CrO_4	7789-00-6	194.191	yel orth cry	975		2.73	65.0^{25}	
1739	Potassium cyanate	KCNO	590-28-3	81.115	wh tetr cry	≈700 dec		2.05	75^{25}	sl EtOH
1740	Potassium cyanide	KCN	151-50-8	65.116	wh cub cry; hyg	634		1.55	69.9^{20}	sl EtOH
1741	Potassium dichromate	$K_2Cr_2O_7$	7778-50-9	294.185	oran-red tricl cry	398	≈500 dec	2.68	15.1^{25}	
1742	Potassium dihydrogen arsenate	KH_2AsO_4	7784-41-0	180.034	col cry	288		2.87	19^{6}	i EtOH
1743	Potassium dihydrogen phosphate	KH_2PO_4	7778-77-0	136.085	wh tetr cry	253		2.34	25.0^{25}	sl EtOH
1744	Potassium ferricyanide	$K_3Fe(CN)_6$	13746-66-2	329.244	red cry	dec		1.89	48.8^{25}	
1745	Potassium ferrocyanide trihydrate	$K_4Fe(CN)_6 \cdot 3H_2O$	14459-95-1	422.388	yel monocl cry	60 dec		1.85	36.0^{25}	i EtOH, eth
1746	Potassium fluoride	KF	7789-23-3	58.096	wh cub cry	858	1502	2.48	102^{25}	
1747	Potassium fluoride dihydrate	$KF \cdot 2H_2O$	13455-21-5	94.127	monocl cry	41 dec		2.5	102^{25}	
1748	Potassium fluoroborate	KBF_4	14075-53-7	125.903	col orth cry	530		2.505	0.55^{25}	sl EtOH
1749	Potassium fluorotantalate	K_2TaF_7	16924-00-8	392.134	col cry	730		5.24	0.5^{0}	
1750	Potassium formate	$KCHO_2$	590-29-4	84.116	col hyg cry	167		1.91	331^{18}	
1751	Potassium hexachloroosmate(IV)	K_2OsCl_6	16871-60-6	481.14	red cub cry					vs H_2O; sl EtOH
1752	Potassium hexachloroplatinate	K_2PtCl_6	16921-30-5	485.99	yel-oran cub cry	250 dec		3.50	0.77^{20}	i EtOH
1753	Potassium hexacyanocobaltate	$K_3Co(CN)_6$	13963-58-1	332.332	yel monocl cry	dec		1.91		vs H_2O; i EtOH
1754	Potassium hexafluoromanganate(IV)	K_2MnF_6	16962-31-5	247.125	yel hex cry					reac H_2O
1755	Potassium hexafluorosilicate	K_2SiF_6	16871-90-2	220.273	wh cry	dec		2.27	0.084^{20}	i EtOH
1756	Potassium hexafluorozirconate(IV)	K_2ZrF_6	16923-95-8	283.411	col mono cry			3.48	0.78^{2}	
1757	Potassium hydride	KH	7693-26-7	40.106	cub cry			1.43		reac H_2O

1758	Potassium hydrogen arsenate	K_2HAsO_4	21093-83-4	218.125	col mono prisms	300 dec			18.7^{6}	i EtOH
1759	Potassium hydrogen arsenite	$KAsO_2 \cdot HAsO_2$	10124-50-2	253.947	wh hyg powder					s H_2O
1760	Potassium hydrogen carbonate	$KHCO_3$	298-14-6	100.115	col monocl cry	≈100 dec		2.17	36.2^{25}	i EtOH
1761	Potassium hydrogen fluoride	KHF_2	7789-29-9	78.103	col tetr cry	238.9		2.37	39.2^{20}	i EtOH
1762	Potassium hydrogen phosphate	K_2HPO_4	7758-11-4	174.176	wh hyg cry	dec			168^{25}	s EtOH
1763	Potassium hydrogen phosphite	K_2HPO_3	13492-26-7	158.177	wh hyg powder	dec			170^{20}	i EtOH
1764	Potassium hydrogen selenite	$KHSeO_3$	7782-70-9	167.06	hyg orth cry	>100 dec				s H_2O; sl EtOH
1765	Potassium hydrogen sulfate	$KHSO_4$	7646-93-7	136.170	wh monocl cry; hyg	≈200		2.32	50.6^{25}	
1766	Potassium hydrogen sulfide	KHS	1310-61-8	72.172	wh hex cry; hyg	≈450		1.69		s H_2O, EtOH
1767	Potassium hydrogen sulfide hemihydrate	$KHS \cdot 0.5H_2O$	1310-61-8*	81.179	wh-yel hyg cry	≈175		1.7		vs H_2O, EtOH
1768	Potassium hydrogen sulfite	$KHSO_3$	7773-03-7	120.170	wh cry powder	190 dec			49^{20}	i EtOH
1769	Potassium hydrogen tartrate	$KHC_4H_4O_6$	868-14-4	188.177	wh cry			1.98	0.57^{20}	s acid, alk; i EtOH
1770	Potassium hydroxide	KOH	1310-58-3	56.105	wh rhomb cry; hyg	406	1327	2.044	121^{25}	s EtOH; s MeOH
1771	Potassium hypophosphite	KH_2PO_2	7782-87-8	104.087	wh hyg cry	dec				vs H_2O; s EtOH
1772	Potassium iodate	KIO_3	7758-05-6	214.001	wh monocl cry	560 dec		3.89	9.22^{25}	
1773	Potassium iodide	KI	7681-11-0	166.003	col cub cry	681	1323	3.12	148^{25}	sl EtOH
1774	Potassium iron(III) oxalate trihydrate	$K_3Fe(C_2O_4)3 \cdot 3H_2O$		491.243	grn mono cry	100	230 dec	2.133	4.7^{0}	i EtOH
1775	Potassium manganate	K_2MnO_4	10294-64-1	197.133	grn cry	190 dec				s H_2O; reac HCl
1776	Potassium metaarsenate	$KAsO_3$	19197-73-0	162.018	wh solid	660				
1777	Potassium metabisulfite	$K_2S_2O_5$	16731-55-8	222.326	wh powder	≈150 dec		2.3	49.5^{25}	reac acid; i EtOH
1778	Potassium metaborate	KBO_2	13709-94-9	81.908	wh hex cry			≈2.3		
1779	Potassium molybdate	K_2MoO_4	13446-49-6	238.14	wh hyg cry	919		2.3	183^{25}	i EtOH
1780	Potassium niobate	$KNbO_3$	12030-85-2	180.002	wh rhomb cry	≈1100		4.64		i H_2O
1781	Potassium nitrate	KNO_3	7757-79-1	101.103	col rhomb cry or powder	337	400 dec	2.11	38.3^{25}	i EtOH
1782	Potassium nitrite	KNO_2	7758-09-0	85.104	wh hyg cry	441	537 exp	1.915	312^{25}	sl EtOH
1783	Potassium oxalate	$K_2C_2O_4$	583-52-8	166.216	wh pwd					sl H_2O
1784	Potassium oxalate monohydrate	$K_2C_2O_4 \cdot H_2O$	6487-48-5	184.231	col cry	160 dec		2.13	36.4^{20}	
1785	Potassium oxide	K_2O	12136-45-7	94.196	gray cub cry	350 dec		2.35		s H_2O, EtOH, eth
1786	Potassium perbromate	$KBrO_4$	22207-96-1	183.000	wh cry	275 dec			4.21^{25}	
1787	Potassium perchlorate	$KClO_4$	7778-74-7	138.549	col orth cry; hyg	525		2.52	2.08^{25}	
1788	Potassium periodate	KIO_4	7790-21-8	230.001	col tetr cry	582	exp	3.618	0.51^{25}	
1789	Potassium permanganate	$KMnO_4$	7722-64-7	158.034	purp orth cry	dec		2.7	7.60^{25}	reac EtOH
1790	Potassium peroxide	K_2O_2	17014-71-0	110.196	yel amorp solid	490				reac H_2O
1791	Potassium persulfate	$K_2S_2O_8$	7727-21-1	270.324	col cry	≈100 dec		2.48	4.7^{20}	
1792	Potassium phosphate	K_3PO_4	7778-53-2	212.266	wh orth cry; hyg	1340		2.564	106^{25}	i EtOH
1793	Potassium pyrophosphate trihydrate	$K_4P_2O_7 \cdot 3H_2O$	7320-34-5*	384.383	col hyg cry	1090		2.33		vs H_2O; i EtOH
1794	Potassium pyrosulfate	$K_2S_2O_7$	7790-62-7	254.325	col needles	≈325		2.28		s H_2O
1795	Potassium selenate	K_2SeO_4	7790-59-2	221.16	wh powder			3.07	114^{25}	
1796	Potassium selenide	K_2Se	1312-74-9	157.16	red cub cry; hyg	800		2.29		s H_2O
1797	Potassium selenite	K_2SeO_3	10431-47-7	205.16	wh hyg cry	875 dec			217^{25}	sl EtOH
1798	Potassium silver cyanide	$KAg(CN)_2$	506-61-6	199.000	wh cry					s H_2O
1799	Potassium stannate trihydrate	$K_2SnO_3 \cdot 3H_2O$	12142-33-5*	298.951	col cry			3.20		vs H_2O; i EtOH
1800	Potassium sulfate	K_2SO_4	7778-80-5	174.261	wh orth cry	1069		2.66	12.0^{25}	i EtOH
1801	Potassium sulfide	K_2S	1312-73-8	110.263	red-yel cub cry; hyg	948		1.74		s H_2O, EtOH; i eth
1802	Potassium sulfide pentahydrate	$K_2S \cdot 5H_2O$	37248-34-3	200.339	col rhomb cry	60				vs H_2O, EtOH; i eth
1803	Potassium sulfite	K_2SO_3	10117-38-1	158.261	col hex cry				106^{25}	sl EtOH
1804	Potassium sulfite dihydrate	$K_2SO_3 \cdot 2H_2O$	7790-56-9	194.292	wh monocl cry	dec			107^{20}	sl EtOH; dec dil acid
1805	Potassium superoxide	KO_2	12030-88-5	71.097	yel tetr cry; hyg	380		2.16		reac H_2O

No.	Name	Formula	CAS Reg No.	Mol. Weight	Physical Form	mp/°C	bp/°C	Density g cm^{-3}	Solubility g/100 g H_2O	Qualitative Solubility
1806	Potassium tellurate(VI) trihydrate	$K_2TeO_4 \cdot 3H_2O$	15571-91-2*	323.84	wh cry powder					s H_2O
1807	Potassium tellurite	K_2TeO_3	7790-58-1	253.80	wh hyg cry	≈460 dec				vs H_2O
1808	Potassium tetraborate pentahydrate	$K_2B_4O_7 \cdot 5H_2O$	1332-77-0	323.513	wh cry powder				16.5^{30}	sl EtOH
1809	Potassium tetrachloroaurate dihydrate	$KAuCl_4 \cdot 2H_2O$	13682-61-6	413.907	yel monocl cry					s H_2O, EtOH, eth
1810	Potassium tetrachloroplatinate	K_2PtCl_4	10025-99-7	415.09	pink-red tetr cry	500 dec		3.38		s H_2O; i EtOH
1811	Potassium tetracyanoplatinate(II) trihydrate	$K_2Pt(CN)_4 \cdot 3H_2O$	562-76-5*	431.39	col rhomb prisms					s H_2O
1812	Potassium tetraiodomercurate(II)	K_2HgI_4	7783-33-7	786.40	yel hyg cry			4.29		vs H_2O; s EtOH, eth, ace
1813	Potassium thiocyanate	KSCN	333-20-0	97.182	col tetr cry; hyg	173	500 dec	1.88	238^{25}	s EtOH
1814	Potassium thiosulfate	$K_2S_2O_3$	10294-66-3	190.327	col hyg cry				165^{25}	i EtOH
1815	Potassium titanate	K_2TiO_3	12030-97-6	174.062	wh orth cry	1515		3.1		reac H_2O
1816	Potassium triiodide monohydrate	$KI_3 \cdot H_2O$	7790-42-3	437.827	brn monocl cry; hyg	225 dec		3.5		s H_2O; reac EtOH, eth
1817	Potassium thiocarbonate	K_2CS_3	26750-66-3	186.406	yel-red hyg cry					vs H_2O
1818	Potassium tungstate	K_2WO_4	7790-60-5	326.04	hyg cry	921		3.12		vs H_2O; i EtOH
1819	Potassium uranate	$K_2U_2O_7$	7790-63-8	666.251	oran cub cry			6.12		i H_2O; s acid
1820	Praseodymium	Pr	7440-10-0	140.908	silv metal; hex	931	3520	6.77		
1821	Praseodymium boride	PrB_6	12008-27-4	205.774	blk cub cry	2610		4.84		
1822	Praseodymium bromide	$PrBr_3$	13536-53-3	380.620	grn hex cry; hyg	693		5.28		s H_2O
1823	Praseodymium chloride	$PrCl_3$	10361-79-2	247.266	grn hex needles; hyg	786		4.0	96.1^{25}	s EtOH
1824	Praseodymium chloride heptahydrate	$PrCl_3 \cdot 7H_2O$	10025-90-8	373.373	grn cry	110 dec			96.1^{25}	s EtOH
1825	Praseodymium fluoride	PrF_3	13709-46-1	197.903	grn hex cry	1395		6.3		
1826	Praseodymium iodide	PrI_3	13813-23-5	521.621	orth hyg cry	737		≈5.8		s H_2O
1827	Praseodymium nitrate	$Pr(NO_3)_3$	10361-80-5	326.923	pale grn hyg cry				165^{25}	s EtOH
1828	Praseodymium nitrate hexahydrate	$Pr(NO_3)_3 \cdot 6H_2O$	15878-77-0	435.014	grn needles				165^{25}	s EtOH, ace
1829	Praseodymium nitride	PrN	25764-09-4	154.915	cub cry			7.46		
1830	Praseodymium oxide	Pr_2O_3	12036-32-7	329.813	wh hex cry	2183	3760	6.9		
1831	Praseodymium silicide	$PrSi_2$	12066-83-0	197.079	tetr cry	1712		5.46		
1832	Praseodymium sulfide	Pr_2S_3	12038-13-0	378.013	cub cry	1765		5.1		
1833	Praseodymium telluride	Pr_2Te_3	12038-12-9	664.62	cub cry	1500		≈7.0		
1834	Promethium	Pm	7440-12-2	145	silv metal; hex	1042	3000	7.26		
1835	Protactinium	Pa	7440-13-3	231.036	shiny metal; tetr or cub	1572		15.4		
1836	Protactinium(V) chloride	$PaCl_5$	13760-41-3	408.300	yel monocl cry	306		3.74		
1837	Radium	Ra	7440-14-4	226	wh metal; cub	700		5		
1838	Radium bromide	$RaBr_2$	10031-23-9	386	wh orth cry	728		5.79	70.6^{20}	s EtOH
1839	Radium chloride	$RaCl_2$	10025-66-8	297	wh orth cry	1000		4.9	24.5^{20}	s EtOH
1840	Radium fluoride	RaF_2	20610-49-5	264	wh cub cry			6.7		
1841	Radium nitrate	$Ra(NO_3)_2$	10213-12-4	350	cry				13.9	
1842	Radium sulfate	$RaSO_4$	7446-16-4	322	wh cry					i H_2O, acid
1843	Radon	Rn	10043-92-2	222	col gas	-71	-61.7	9.074 g/L		sl H_2O
1844	Rhenium	Re	7440-15-5	186.207	silv-gray metal	3186	5596	20.8		i HCl
1845	Perrhenic acid	$HReO_4$	13768-11-1	251.213	exists only in soln					vs H_2O, os
1846	Rhenium carbonyl	$Re_2(CO)_{10}$	14285-68-8	652.515	yel-wh cry	170 dec		2.87		s os
1847	Rhenium(III) bromide	$ReBr_3$	13569-49-8	425.919	red-brn monocl cry		500 subl	6.10		s ace, MeOH, EtOH
1848	Rhenium(III) chloride	$ReCl_3$	13569-63-6	292.565	red-blk hyg cry	500 dec		4.81		s H_2O
1849	Rhenium(III) iodide	ReI_3	15622-42-1	566.920	blk solid	dec				
1850	Rhenium(IV) chloride	$ReCl_4$	13569-71-6	328.018	purp-blk cry; hyg	300 dec		4.9		

1851	Rhenium(IV) fluoride	ReF_4	15192-42-4	262.201	blue tetr cry		>300 subl	7.49		
1852	Rhenium(IV) oxide	ReO_2	12036-09-8	218.206	gray orth cry	900 dec		11.4		
1853	Rhenium(IV) sulfide	ReS_2	12038-63-0	250.339	tricl cry			7.6		
1854	Rhenium(IV) telluride	$ReTe_2$	12067-00-4	441.41	orth cry			8.50		
1855	Rhenium(V) bromide	$ReBr_5$	30937-53-2	585.727	brn solid	110 dec				
1856	Rhenium(V) chloride	$ReCl_5$	39368-69-9	363.471	brn-blk solid	220		4.9		reac H_2O
1857	Rhenium(V) fluoride	ReF_5	30937-52-1	281.199	yel-grn solid	48	221.3			
1858	Rhenium(V) oxide	Re_2O_5	12165-05-8	452.411	blue-blk tetr cry			≈7		
1859	Rhenium(VI) chloride	$ReCl_6$	31234-26-1	398.923	red-grn solid	29				
1860	Rhenium(VI) dioxydifluoride	ReO_2F_2	81155-18-2	256.203	col cry	156				
1861	Rhenium(VI) fluoride	ReF_6	10049-17-9	300.197	yel liq or cub cry	18.5	33.8	4.06(cry)		s HNO_3
1862	Rhenium(VI) oxide	ReO_3	1314-28-9	234.205	redcub cry	400 dec		6.9		i H_2O, acid, alk
1863	Rhenium(VI) oxytetrachloride	$ReOCl_4$	13814-76-1	344.017	brn cry	29.3	223			reac H_2O
1864	Rhenium(VI) oxytetrafluoride	$ReOF_4$	17026-29-8	278.200	blue solid	108	171.7			
1865	Rhenium(VII) fluoride	ReF_7	17029-21-9	319.196	yel cub cry	48.3	73.7	4.32		
1866	Rhenium(VII) oxide	Re_2O_7	1314-68-7	484.410	yel hyg cry	297	360	6.10		s H_2O, EtOH, eth, diox, py
1867	Rhenium(VII) trioxychloride	ReO_3Cl	7791-09-5	269.658	col liq	4.5	128	3.87		reac H_2O
1868	Rhenium(VII) trioxyfluoride	ReO_3F	42246-24-2	253.203	yel solid	147	164			
1869	Rhenium(VII) dioxytrifluoride	ReO_2F_3	57246-89-6	275.201	yel solid	90	185.4			reac H_2O
1870	Rhenium(VII) oxypentafluoride	$ReOF_5$	23377-53-9	297.198	cream solid	43.8	73.0			
1871	Rhenium(VII) sulfide	Re_2S_7	12038-67-4	596.876	brn-blk tetr cry			4.87		i H_2O
1872	Rhodium	Rh	7440-16-6	102.906	silv-wh metal; cub	1964	3695	12.4		i acid, sl aqua regia
1873	Rhodium carbonyl chloride	$[Rh(CO)_2Cl]_2$	14523-22-9	388.757	red-oran cry	124				s os
1874	Rhodium dodecacarbonyl	$Rh_4(CO)_{12}$	19584-30-6	747.743	red hyg cry			2.52		reac H_2O
1875	Rhodium(III) chloride	$RhCl_3$	10049-07-7	209.264	red monocl cry		717	5.38		i H_2O; s alk
1876	Rhodium(III) fluoride	RhF_3	60804-25-3	159.901	red hex cry			5.4		
1877	Rhodium(III) iodide	RhI_3	15492-38-3	483.619	blk monocl cry; hyg			6.4		
1878	Rhodium(III) oxide	Rh_2O_3	12036-35-0	253.809	gray hex cry	1100 dec		8.2		
1879	Rhodium(III) sulfate	$Rh_2(SO_4)_3$	10489-46-0	494.002	red-yel solid	>500 dec				
1880	Rhodium(IV) oxide	RhO_2	12137-27-8	134.905	blk tetr cry			7.2		
1881	Rhodium(VI) fluoride	RhF_6	13693-07-7	216.896	blk cub cry	≈70		3.1		
1882	Rubidium	Rb	7440-17-7	85.468	soft silv metal; cub	39.30	688	1.53		reac H_2O
1883	Rubidium acetate	$RbC_2H_3O_2$	563-67-7	144.512	wh hyg cry	246				vs H_2O
1884	Rubidium aluminum sulfate	$RbAl(SO_4)_2$	13530-57-9	304.577	hex cry			≈3.1	1.60^{20}	i EtOH
1885	Rubidium aluminum sulfate dodecahydrate	$RbAl(SO_4)_2 \cdot 12H_2O$	7784-29-4	520.761	col cub cry	≈100 dec		≈1.9		s H_2O; i EtOH
1886	Rubidium azide	RbN_3	22756-36-1	127.488	tetr cry; exp	317		2.79	107^{16}	
1887	Rubidium bromate	$RbBrO_3$	13446-70-3	213.370	cub cry	430		3.68	2.95^{25}	
1888	Rubidium bromide	RbBr	7789-39-1	165.372	wh cub cry; hyg	682	1340	3.35	116^{25}	
1889	Rubidium carbonate	Rb_2CO_3	584-09-8	230.945	col monocl cry; hyg	837			223^{20}	
1890	Rubidium chlorate	$RbClO_3$	13446-71-4	168.919	col cry			3.19	6.63^{25}	
1891	Rubidium chloride	RbCl	7791-11-9	120.921	wh cub cry; hyg	715	1390	2.76	93.9^{25}	sl EtOH
1892	Rubidium chromate	Rb_2CrO_4	13446-72-5	286.930	yel rhom cry			3.518	76.2^{25}	
1893	Rubidium cyanide	RbCN	19073-56-4	111.486	wh cub cry			2.3		s H_2O; i EtOH, eth
1894	Rubidium fluoride	RbF	13446-74-7	104.466	wh cub cry; hyg	833	1410	3.2	300^{20}	i EtOH
1895	Rubidium hydrogen fluoride	$RbHF_2$	12280-64-7	124.473	tetr cry	188		3.3		
1896	Rubidium formate	$RbCHO_2$	3495-35-0	130.486	wh hyg cry	dec				
1897	Rubidium hydride	RbH	13446-75-8	86.476	wh cub cry; flam	≈170 dec		2.60		reac H_2O
1898	Rubidium hydrogen carbonate	$RbHCO_3$	19088-74-5	146.485	wh rhomb cry	175 dec			116^{20}	

No.	Name	Formula	CAS Reg No.	Mol. Weight	Physical Form	mp/°C	bp/°C	Density g cm⁻³	Solubility g/100 g H_2O	Qualitative Solubility
1899	Rubidium hydrogen sulfate	$RbHSO_4$	15587-72-1	182.540	col monocl cry	208		2.9		s H_2O
1900	Rubidium hydroxide	RbOH	1310-82-3	102.475	gray-wh orth cry; hyg	382		3.2	173^{30}	s EtOH
1901	Rubidium iodate	$RbIO_3$	13446-76-9	260.370	mono or cub cry	dec		4.33	2.44^{25}	vs HCl
1902	Rubidium iodide	RbI	7790-29-6	212.372	wh cub cry	642	1300	3.55	165^{25}	s EtOH
1903	Rubidium nitrate	$RbNO_3$	13126-12-0	147.473	wh hex cry; hyg	305		3.11	65.0^{25}	
1904	Rubidium oxide	Rb_2O	18088-11-4	186.935	yel-brn cub cry; hyg	400 dec		4.0		reac H_2O
1905	Rubidium perchlorate	$RbClO_4$	13510-42-4	184.919	wh hyg cry	281	600 dec	2.8	1.5^{25}	
1906	Rubidium peroxide	Rb_2O_2	23611-30-5	202.935	wh orth cry			3.8		reac H_2O
1907	Rubidium selenide	Rb_2Se	31052-43-4	249.90	wh cub cry	733		3.22		reac H_2O
1908	Rubidium sulfate	Rb_2SO_4	7488-54-2	267.000	wh orth cry	1050		3.6	50.8^{25}	
1909	Rubidium sulfide	Rb_2S	31083-74-6	203.002	wh cub cry	425		2.91		s H_2O
1910	Rubidium superoxide	RbO_2	12137-25-6	117.467	tetr cry	412		≈3.0		
1911	Ruthenium	Ru	7440-18-8	101.07	silv-wh metal; hex	2334	4150	12.1		i acid, aqua regia
1912	Ruthenium dodecacarbonyl	$Ru_3(CO)_{12}$	15243-33-1	639.33	oran cry	150 dec				
1913	Ruthenium(III) 2,4-pentanedioate	$Ru(CH_3COCHCOCH_3)_3$	14284-93-6	398.39	red-brn cry	230				
1914	Ruthenium(III) bromide	$RuBr_3$	14014-88-1	340.78	brn hex cry	>400 dec		5.3		
1915	Ruthenium(III) chloride	$RuCl_3$	10049-08-8	207.43	brn hex cry	>500 dec		3.1		i H_2O; s EtOH
1916	Ruthenium(III) fluoride	RuF_3	51621-05-7	158.07	brn rhomb cry	>600 dec		5.36		
1917	Ruthenium(III) iodide	RuI_3	13896-65-6	481.78	blk hex cry			6.0		
1918	Ruthenium(IV) fluoride	RuF_4	71500-16-8	177.06	yel cry					reac H_2O
1919	Ruthenium(IV) oxide	RuO_2	12036-10-1	133.07	gray-blk tetr cry			7.05		i H_2O, acid
1920	Ruthenium(V) fluoride	RuF_5	14521-18-7	196.06	grn monocl cry	86.5	227	3.90		
1921	Ruthenium(VI) fluoride	RuF_6	13693-08-8	215.06	dark brn orth cry	54		3.54		reac H_2O
1922	Ruthenium(VIII) oxide	RuO_4	20427-56-9	165.07	yel monocl prisms	25.4	40	3.29	171^{0}	vs ctc; reac EtOH
1923	Samarium	Sm	7440-19-9	150.36	silv metal; rhomb	1074	1794	7.52		
1924	Samarium boride	SmB_6	12008-29-6	215.23	refrac solid	2580		5.07		
1925	Samarium silicide	$SmSi_2$	12300-22-0	206.53	orth cry			5.14		
1926	Samarium(II) bromide	$SmBr_2$	50801-97-3	310.17	brn cry	669				reac H_2O
1927	Samarium(II) chloride	$SmCl_2$	13874-75-4	221.27	brn cry	855		3.69		reac H_2O
1928	Samarium(II) fluoride	SmF_2	15192-17-3	188.36	purp cry					reac H_2O
1929	Samarium(II) iodide	SmI_2	32248-43-4	404.17	grn cry	520				reac H_2O
1930	Samarium(III) bromide	$SmBr_3$	13759-87-0	390.07	yel cry	640				reac H_2O
1931	Samarium(III) chloride	$SmCl_3$	10361-82-7	256.72	yel cry	682		4.46	93.8^{25}	
1932	Samarium(III) chloride hexahydrate	$SmCl_3 \cdot 6H_2O$	13465-55-9	364.81	yel cry	dec		2.38	93.8^{25}	
1933	Samarium(III) fluoride	SmF_3	13765-24-7	207.36	wh cry	1306				reac H_2O
1934	Samarium(III) iodide	SmI_3	13813-25-7	531.07	oran cry	850				reac H_2O
1935	Samarium(III) nitrate	$Sm(NO_3)_3$	10361-83-8	336.38	yel-wh hyg solid				144^{25}	s EtOH
1936	Samarium(III) nitrate hexahydrate	$Sm(NO_3)_3 \cdot 6H_2O$	13759-83-6	444.47	pale yel cry	78				s H_2O, MeOH, ace
1937	Samarium(III) oxide	Sm_2O_3	12060-58-1	348.72	yel-wh cub cry	2269	3780	7.6		
1938	Samarium(III) sulfate octahydrate	$Sm_2(SO_4)_3 \cdot 8H_2O$	13465-58-2	733.03	yel cry			2.93	2.67^{20}	
1939	Samarium(III) sulfide	Sm_2S_3	12067-22-0	396.92	gray-brn cub cry	1720		5.87		
1940	Samarium(III) telluride	Sm_2Te_3	12040-00-5	683.52	orth cry			7.31		
1941	Scandium	Sc	7440-20-2	44.956	silv metal; hex	1541	2836	2.99		
1942	Scandium boride	ScB_2	12007-34-0	66.578	refrac solid	2250		3.17		
1943	Scandium bromide	$ScBr_3$	13465-59-3	284.668	wh hyg cry	969		9.33		s H_2O

1944	Scandium chloride	$ScCl_3$	10361-84-9	151.314	wh hyg cry	967		2.4		s H_2O; i EtOH
1945	Scandium fluoride	ScF_3	13709-47-2	101.951	wh powder	1515				sl H_2O
1946	Scandium hydroxide	$Sc(OH)_3$	17674-34-9	95.978	col amorp sol					i H_2O; s dil acid
1947	Scandium nitrate	$Sc(NO_3)_3$	13465-60-6	230.971	wh cry				169^{25}	s EtOH
1948	Scandium oxide	Sc_2O_3	12060-08-1	137.910	wh cub cry	2485		3.864		s conc acid
1949	Scandium sulfide	Sc_2S_3	12166-29-9	186.110	yel orth cry	1775		2.91		
1950	Scandium telluride	Sc_2Te_3	12166-44-8	472.71	blk hex cry			5.29		
1951	Selenium (α form)	Se	7782-49-2	78.96	red monocl cry	221	685	4.39		i H_2O, EtOH; sl eth
1952	Selenium (gray)	Se	7782-49-2	78.96	gray metallic cry; hex	220.5	685	4.81		i H_2O, CS_2
1953	Selenium (vitreous)	Se	7782-49-2	78.96	blk amorp solid	trans to gray Se 180	685	4.28		i H_2O; sl CS_2
1954	Selenic acid	H_2SeO_4	7783-08-6	144.97	wh hyg solid	58	260 dec	2.95		vs H_2O; reac EtOH
1955	Selenous acid	H_2SeO_3	7783-00-8	128.97	wh hyg cry	70 dec		3.0		vs H_2O; s EtOH
1956	Selenium dioxide	SeO_2	7446-08-4	110.96	wh tetr needles or powder	340 tp	315 sp	3.95	264^{22}	s EtOH, MeOH; sl ace
1957	Selenium trioxide	SeO_3	13768-86-0	126.96	wh tetr cry; hyg	118	subl	3.44		s H_2O, os
1958	Selenium bromide	Se_2Br_2	7789-52-8	317.73	red liq		225 dec	3.60		reac H_2O; s CS_2, chl
1959	Selenium chloride	Se_2Cl_2	10025-68-0	228.83	yel-brn oily liq	-85	130 dec	2.774		reac H_2O; s CS_2, bz, ctc, chl
1960	Selenium tetrabromide	$SeBr_4$	7789-65-3	398.58	oran-red cry	123				reac H_2O; s CS_2, chl
1961	Selenium tetrachloride	$SeCl_4$	10026-03-6	220.77	wh-yel cry	305 tp	191.4 sp	2.6		reac H_2O
1962	Selenium tetrafluoride	SeF_4	13465-66-2	154.95	col liq	-10	106	2.75		reac H_2O; vs EtOH, eth
1963	Selenium hexafluoride	SeF_6	7783-79-1	192.95	col gas	-34.6 tp	-46.6 sp	7.887 g/L		i H_2O
1964	Selenium oxybromide	$SeOBr_2$	7789-51-7	254.77	red-yel solid	41.6	220 dec	3.38		reac H_2O; s CS_2, bz, ctc
1965	Selenium oxychloride	$SeOCl_2$	7791-23-3	165.86	col or yel liq	8.5	177	2.44		reac H_2O; s ctc, chl, bz, tol
1966	Selenium oxyfluoride	$SeOF_2$	7783-43-9	132.96	col liq	15	125	2.8		reac H_2O
1967	Selenium dioxydifluoride	SeO_2F_2	14984-81-7	148.96	col gas	-99.5	-8.4	6.089 g/L		reac H_2O
1968	Selenium sulfide	SeS_2	7488-56-4	143.09	red-yel cry	100				i H_2O; s acid
1969	Selenium sulfide	Se_2S_6	75926-26-0	350.32	oran needles	121.5		2.44		s CS_2; sl bz
1970	Selenium sulfide	Se_4S_4	75926-28-2	444.10	red cry	113 dec		3.29		s bz; sl CS_2
1971	Selenium sulfide	Se_6S_2	75926-30-6	537.89	oran cry	121.5				s CS_2
1972	Silicon	Si	7440-21-3	28.086	gray cry or brn amorp solid	1414	3265	2.3290		i H_2O, acid; s alk
1973	Silane	SiH_4	7803-62-5	32.118	col gas; flam	-185	-111.9	1.313 g/L		reac H_2O; i EtOH, bz
1974	Disilane	Si_2H_6	1590-87-0	62.219	col gas; flam	-132.5	-14.3	2.543 g/L		reac H_2O, ctc, chl; s EtOH, bz
1975	Trisilane	Si_3H_8	7783-26-8	92.321	flam liq	-117.4	52.9	0.739		reac H_2O
1976	Tetrasilane	Si_4H_{10}	7783-29-1	122.421	col liq; flam	-89.9	108.1	0.792		reac H_2O
1977	2-Silyltrisilane	Si_4H_{10}	13597-87-0	122.421	col liq	-99.4	101.7	0.792		reac H_2O
1978	Pentasilane	Si_5H_{12}	14868-53-2	152.523	col liq	-72.8	153.2	0.827		reac H_2O
1979	2-Silyltetrasilane	Si_5H_{12}	14868-54-3	152.523	col liq	-109.9	146.2	0.820		reac H_2O
1980	2,2-Disilyltrisilane	Si_5H_{12}	15947-57-6	152.523	col liq	-57.8	134.3	0.815		reac H_2O
1981	Hexasilane	Si_6H_{14}	14693-61-9	182.624	col liq	-44.7	193.6	0.847		reac H_2O
1982	2-Silylpentasilane	Si_6H_{14}	14868-55-4	182.624	col liq	-78.4	185.2	0.840		
1983	3-Silylpentasilane	Si_6H_{14}	14868-55-4	182.624	col liq	-69	179.5	0.843		reac H_2O
1984	Heptasilane	Si_7H_{16}	14693-65-3	212.726	col liq	-30.1	226.8	0.859		reac H_2O
1985	Cyclopentasilane	Si_5H_{10}	289-22-5	150.507	col liq	-10.5	194.3	0.963		reac H_2O
1986	Cyclohexasilane	Si_6H_{12}	291-59-8	180.608	col liq	16.5	226			reac H_2O
1987	Bromosilane	SiH_3Br	13465-73-1	111.014	col gas	-94	1.9	4.538 g/L		
1988	Bromotrichlorosilane	$SiCl_3Br$	13465-74-2	214.348	col liq	-62	80.3	1.826		reac H_2O
1989	Chlorosilane	SiH_3Cl	13465-78-6	66.563	col gas	-118	-30.4	2.721 g/L		
1990	Chlorotrifluorosilane	$SiClF_3$	14049-36-6	120.534	col gas	-138	-70.0	4.927 g/L		reac H_2O
1991	Dibromodichlorosilane	$SiBr_2Cl_2$	13465-75-3	258.799	col liq	-45.5	104	2.172		reac H_2O

No.	Name	Formula	CAS Reg No.	Mol. Weight	Physical Form	mp/°C	bp/°C	Density g cm^{-3}	Solubility g/100 g H_2O	Qualitative Solubility
1992	Dibromosilane	SiH_2Br_2	13768-94-0	189.910	liq	-70.1	66			
1993	Dichlorosilane	SiH_2Cl_2	4109-96-0	101.007	col gas; flam	-122	8.3	4.129 g/L		reac H_2O
1994	Dichlorodifluorosilane	$SiCl_2F_2$	18356-71-3	136.988	col gas	-44	-32	5.599 g/L		reac H_2O
1995	Difluorosilane	SiH_2F_2	13824-36-7	68.099	col gas	-122	-77.8	2.783 g/L		
1996	Diiodosilane	SiH_2I_2	13760-02-6	283.911	col liq	-1	150			
1997	Fluorosilane	SiH_3F	13537-33-2	50.108	col gas		-98.6	2.048 g/L		
1998	Iodosilane	SiH_3I	13598-42-0	158.014	col liq	-57	45.6			
1999	Tetrabromosilane	$SiBr_4$	7789-66-4	347.702	col fuming liq	5.39	154	2.8		reac H_2O
2000	Tetrachlorosilane	$SiCl_4$	10026-04-7	169.897	col fuming liq	-68.74	57.65	1.5		reac H_2O
2001	Tetrafluorosilane	SiF_4	7783-61-1	104.080	col gas	-90.2	-86	4.254 g/L		reac H_2O
2002	Tetraiodosilane	SiI_4	13465-84-4	535.704	wh powder	120.5	287.35	4.1		
2003	Tribromosilane	$SiHBr_3$	7789-57-3	268.806	flam liq	-73	109	2.7		reac H_2O
2004	Tribromochlorosilane	$SiBr_3Cl$	13465-76-4	303.251	col liq	-20.8	127	2.497		reac H_2O
2005	Trichlorosilane	$SiHCl_3$	10025-78-2	135.452	fuming liq	-128.2	33	1.331		reac H_2O
2006	Trichlorofluorosilane	$SiCl_3F$	14965-52-7	153.442	col gas		12.25	6.272 g/L		
2007	Trichloroiodosilane	$SiCl_3I$	13465-85-5	261.348	col liq	-60	113.5			reac H_2O
2008	Trifluorosilane	$SiHF_3$	13465-71-9	86.089	col gas	-131	-95	3.519 g/L		
2009	Triiodosilane	$SiHI_3$	13465-72-0	409.807	liq	8	220 dec			
2010	Disiloxane	$(SiH_3)_2O$	13597-73-4	78.218	gas	-144	-15.2	3.197 g/L		
2011	Metasilicic acid	H_2SiO_3	7699-41-4	78.100	wh amorp powder					i H_2O; s HF
2012	Orthosilicic acid	H_4SiO_4	10193-36-9	96.116	exists only in soln					
2013	Fluorosilicic acid	H_2SiF_6	16961-83-4	144.092	stable only in aq soln					s H_2O
2014	Silicon carbide (hexagonal)	SiC	409-21-2	40.097	hard grn-black hex cry	2830		3.16		i H_2O, EtOH
2015	Silicon nitride	Si_3N_4	12033-89-5	140.284	gray refrac solid; hex	1900		3.17		
2016	Silicon monoxide	SiO	10097-28-6	44.085	blk cub cry, stable >1200			2.18		
2017	Silicon dioxide (α-quartz)	SiO_2	14808-60-7	60.085	col hex cry	trans to beta quartz 573	2950	2.648		i H_2O, acid; s HF
2018	Silicon dioxide (β-quartz)	SiO_2	14808-60-7	60.085	col hex cry	trans to tridymite 867	2950	2.533^{600}		i H_2O, acid; s HF
2019	Silicon dioxide (tridymite)	SiO_2	15468-32-3	60.085	col hex cry	trans cristobalite 1470	2950	2.265		i H_2O, acid; s HF
2020	Silicon dioxide (cristobalite)	SiO_2	14464-46-1	60.085	col hex cry	1722	2950	2.334		i H_2O, acid; s HF
2021	Silicon dioxide (vitreous)	SiO_2	60676-86-0	60.085	col amorp solid	1713	2950	2.196		i H_2O, acid; s HF
2022	Silicon monosulfide	SiS	12504-41-5	60.152	yel-red hyg powder	≈900	940	1.85		reac H_2O
2023	Silicon disulfide	SiS_2	13759-10-9	92.218	wh rhomb cry	1090	subl	2.04		reac H_2O, EtOH; i bz
2024	Silver	Ag	7440-22-4	107.868	silv metal; cub	961.78	2162	10.5		
2025	Silver azide	AgN_3	13863-88-2	149.888	orth cry; exp	exp ≈250		4.9	0.00081^{20}	
2026	Silver subfluoride	Ag_2F	1302-01-8	234.734	yel hex cry	100 dec		8.6		reac H_2O
2027	Silver(I) acetate	$AgC_2H_3O_2$	563-63-3	166.912	wh needles or powder	dec		3.26	1.04^{20}	
2028	Silver(I) acetylide	Ag_2C_2	7659-31-6	239.757	wh powder; exp					
2029	Silver(I) arsenate	Ag_3AsO_4	13510-44-6	462.524	red cub cry	dec		6.657	0.00085	s NH_4OH
2030	Silver(I) acetylide	AgC_2H	13092-75-6	132.897	wh powder; exp					
2031	Silver(I) bromate	$AgBrO_3$	7783-89-3	235.770	wh tetr cry	360 dec		5.21	0.193^{25}	
2032	Silver(I) bromide	AgBr	7785-23-1	187.772	yel cub cry	432	1502	6.47	0.000014^{25}	i acid
2033	Silver(I) carbonate	Ag_2CO_3	534-16-7	275.745	yel monocl cry	218		6.077	0.0036^{20}	s acid
2034	Silver(I) chlorate	$AgClO_3$	7783-92-8	191.319	wh tetr cry	230	270 dec	4.430	17.6^{25}	sl EtOH
2035	Silver(I) chloride	AgCl	7783-90-6	143.321	wh cub cry	455	1547	5.56	0.00019^{25}	
2036	Silver(I) chlorite	$AgClO_2$	7783-91-7	175.320	yel cry	105 exp			0.55^{25}	

2037	Silver(I) chromate	Ag_2CrO_4	7784-01-2	331.730	brn-red monocl cry			5.625	0.000014^{0}	
2038	Silver(I) citrate	$Ag_3C_6H_5O_7$	126-45-4	512.705	wh cry powder					i H_2O; s HNO_3
2039	Silver(I) cyanide	AgCN	506-64-9	133.886	wh-gray hex cry	320 dec		3.95	0.0000011	i EtOH, dil acid
2040	Silver(I) dichromate	$Ag_2Cr_2O_7$	7784-02-3	431.724	red cry			4.770		sl H_2O
2041	Silver(I) diethyldithiocarbamate	$Ag(C_2H_5)_2NCS_2$	1470-61-7	256.140	pow	173				s py
2042	Silver(I) fluoride	AgF	7775-41-9	126.866	yel-brn cub cry; hyg	435	1159	5.852	172^{20}	
2043	Silver(I) hexafluoroantimonate	$AgSbF_6$	26042-64-8	343.618	pow					
2044	Silver(I) hexafluoroarsenate	$AgAsF_6$	12005-82-2	296.780	pow					
2045	Silver(I) hexafluorophosphate	$AgPF_6$	26042-63-7	252.832	pow	102 dec				
2046	Silver(I) hydrogen fluoride	$AgHF_2$	12249-52-4	146.873	hyg cry	dec				
2047	Silver(I) iodate	$AgIO_3$	7783-97-3	282.770	wh orth cry	>200		5.53	0.053^{25}	
2048	Silver(I) iodide	AgI	7783-96-2	234.772	yel powder; hex	558	1506	5.68	0.000003	i acid
2049	Silver(I) lactate monohydrate	$AgC_3H_5O_3 \cdot H_2O$	128-00-7	214.954	gray cry powder					sl H_2O, EtOH
2050	Silver(I) metaphosphate	$AgPO_3$	13465-96-8	186.840	grn glass	490		6.37		i H_2O; s HNO_3, NH_4OH
2051	Silver(I) molybdate	Ag_2MoO_4	13765-74-7	375.67	yel cub cry	483		6.18		sl H_2O
2052	Silver(I) nitrate	$AgNO_3$	7761-88-8	169.873	col rhomb cry	212	440 dec	4.35	234^{25}	sl EtOH, ace
2053	Silver(I) nitrite	$AgNO_2$	7783-99-5	153.874	yel needles	140 dec		4.453	0.415^{25}	i EtOH; reac acid
2054	Silver(I) oxalate	$Ag_2C_2O_4$	533-51-7	303.755	wh cry powder	exp 140		5.03	0.0043^{20}	
2055	Silver(I) oxide	Ag_2O	20667-12-3	231.735	brn-blk cub cry	≈200 dec		7.2	0.0025	i EtOH; s acid, alk
2056	Silver(I) perchlorate	$AgClO_4$	7783-93-9	207.319	col cub cry; hyg	486 dec		2.806	558^{25}	s bz, py, os
2057	Silver(I) perchlorate monohydrate	$AgClO_4 \cdot H_2O$	14242-05-8	225.334	hyg wh cry	43 dec			558^{25}	
2058	Silver(I) permanganate	$AgMnO_4$	7783-98-4	226.804	viol monocl cry	dec		4.49	0.91^{18}	reac EtOH
2059	Silver(I) phosphate	Ag_3PO_4	7784-09-0	418.576	yel powder	849		6.37	0.0064	sl dil acid
2060	Silver(I) picrate monohydrate	$AgC_6H_2N_3O_7 \cdot H_2O$	146-84-9	353.979	yel cry					sl H_2O, EtOH; i chl, eth
2061	Silver(I) selenate	Ag_2SeO_4	7784-07-8	358.69	orth cry			5.72	0.118^{20}	
2062	Silver(I) selenide	Ag_2Se	1302-09-6	294.70	gray hex needles	880		8.216		i H_2O
2063	Silver(I) selenite	Ag_2SeO_3	7784-05-6	342.69	needles	530	>550 dec	5.930		sl H_2O; s acid
2064	Silver(I) sulfate	Ag_2SO_4	10294-26-5	311.800	col cry or powder	652		5.45	0.84^{25}	
2065	Silver(I) sulfide	Ag_2S	21548-73-2	247.802	gray-blk orth powder	825		7.23		i H_2O; s acid
2066	Silver(I) sulfite	Ag_2SO_3	13465-98-0	295.800	wh cry	100 dec			0.00046^{20}	s acid, NH_4OH
2067	Silver(I) telluride	Ag_2Te	12002-99-2	343.34	blk orth cry	955		8.4		
2068	Silver(I) tetraiodomercurate(II)	Ag_2HgI_4	7784-03-4	923.94	yel tetr cry	trans to red cub ≈40		6.1		i H_2O, dil acid
2069	Silver(I) thiocyanate	AgSCN	1701-93-5	165.952	wh powder	dec				i H_2O
2070	Silver(I) thiosulfate	$Ag_2S_2O_3$	23149-52-2	327.866	wh cry	dec				sl H_2O; s NH_4OH
2071	Silver(II) oxide	AgO	1301-96-8	123.867	gray powder; monocl or cub	>100 dec		7.5	0.0027^{25}	s alk; reac acid
2072	Silver(I) tungstate	Ag_2WO_4	13465-93-5	463.57	yel cry	620			0.015	s HNO_3, NH_4OH
2073	Silver(II) fluoride	AgF_2	7783-95-1	145.865	wh or gray hyg cry	690		4.58		reac H_2O
2074	Silver(II) oxide (Ag2O2)	Ag_2O_2	25455-73-6	247.735	gray-blk cub cry	>100		7.44		i H_2O; s acid, NH_4OH
2075	Sodium	Na	7440-23-5	22.990	soft silv met; cub	97.80	883	0.97		reac H_2O
2076	Sodium acetate	$NaC_2H_3O_2$	127-09-3	82.034	col cry	328.2		1.528	50.4^{25}	
2077	Sodium acetate trihydrate	$NaC_2H_3O_2 \cdot 3H_2O$	6131-90-4	136.079	col cry	58 dec		1.45	50.4^{25}	sl EtOH
2078	Sodium aluminate	$NaAlO_2$	1302-42-7	81.971	wh orth cry; hyg	1650		4.63		vs H_2O; i EtOH
2079	Sodium aluminum hydride	$NaAlH_4$	13770-96-2	54.004	wh hyg solid	174 dec		1.24		i eth; s thf
2080	Sodium aluminum sulfate dodecahydrate	$NaAl(SO_4)_2 \cdot 12H_2O$	10102-71-3	458.283	col cry	≈60		1.61	39.7^{20}	i EtOH
2081	Sodium amide	$NaNH_2$	7782-92-5	39.013	wh-grn orth cry	210	500 dec	1.39		reac H_2O
2082	Sodium ammonium phosphate tetrahydrate	$NaNH_4HPO_4 \cdot 4H_2O$	13011-54-6	209.069	monocl cry	≈80 dec		1.54		s H_2O; i EtOH
2083	Sodium arsenite	$NaAsO_2$	7784-46-5	129.911	wh-gray hyg powder			1.87		vs H_2O; i EtOH
2084	Sodium azide	NaN_3	26628-22-8	65.010	col hex cry	300 dec		1.846	40.8^{20}	sl EtOH; i eth

No.	Name	Formula	CAS Reg No.	Mol. Weight	Physical Form	mp/°C	bp/°C	Density g cm^{-3}	Solubility g/100 g H_2O	Qualitative Solubility
2085	Sodium borohydride	$NaBH_4$	16940-66-2	37.833	wh cub cry; hyg	≈400 dec		1.07	55^{20}	reac EtOH
2086	Sodium bromate	$NaBrO_3$	7789-38-0	150.892	col cub cry	381		3.34	39.4^{25}	i EtOH
2087	Sodium bromide	NaBr	7647-15-6	102.894	wh cub cry	747	1390	3.200	94.6^{25}	s EtOH
2088	Sodium bromide dihydrate	$NaBr \cdot 2H_2O$	13466-08-5	138.925	wh cry	36 dec		2.18	94.6^{25}	sl EtOH
2089	Sodium carbonate	Na_2CO_3	497-19-8	105.989	wh hyg powder	858.1		2.54	30.7^{25}	i EtOH
2090	Sodium carbonate decahydrate	$Na_2CO_3 \cdot 10H_2O$	6132-02-1	286.142	col cry	34 dec		1.46	30.7^{25}	i EtOH
2091	Sodium carbonate monohydrate	$Na_2CO_3 \cdot H_2O$	5968-11-6	124.005	col orth cry	100 dec		2.25	30.7^{25}	i EtOH
2092	Sodium chlorate	$NaClO_3$	7775-09-9	106.441	col cub cry	248	>300 dec	2.5	100^{25}	sl EtOH
2093	Sodium chloride	NaCl	7647-14-5	58.443	col cub cry	800.7	1465	2.17	36.0^{25}	sl EtOH
2094	Sodium chlorite	$NaClO_2$	7758-19-2	90.442	wh hyg cry	≈180 dec			64^{17}	
2095	Sodium chromate	Na_2CrO_4	7775-11-3	161.974	yel orth cry	792		2.72	87.6^{25}	sl EtOH
2096	Sodium chromate tetrahydrate	$Na_2CrO_4 \cdot 4H_2O$	10034-82-9	234.035	yel hyg cry	dec			87.6^{25}	sl EtOH
2097	Sodium citrate dihydrate	$Na_3C_6H_5O_7 \cdot 2H_2O$	6132-04-3	294.099	wh cry	150 dec				vs H_2O; i EtOH, eth
2098	Sodium cyanate	NaCNO	917-61-3	65.007	col needles	550		1.89		s H_2O; sl EtOH; i eth
2099	Sodium cyanide	NaCN	143-33-9	49.008	wh cub cry; hyg	563		1.6	58.2^{20}	sl EtOH
2100	Sodium cyanoborohydride	$NaBH_3(CN)$	25895-60-7	62.843	wh hyg powder	240 dec		1.12		vs H_2O; s thf; sl EtOH; i bz, eth
2101	Sodium dichromate	$Na_2Cr_2O_7$	10588-01-9	261.968	red hyg cry	357	400 dec		187^{25}	
2102	Sodium dihydrogen phosphate	NaH_2PO_4	7558-80-7	119.977	col mono cry	200 dec			94.9^{25}	
2103	Sodium dihydrogen phosphate monohydrate	$NaH_2PO_4 \cdot H_2O$	10049-21-5	137.993	wh hyg cry	100 dec			94.9^{25}	i EtOH
2104	Sodium dihydrogen phosphate dihydrate	$NaH_2PO_4 \cdot 2H_2O$	13472-35-0	156.008	col orth cry	60 dec		1.91	94.9^{25}	i EtOH
2105	Sodium dihydrogen hypophosphate hexahydrate	$Na_2H_2P_2O_6 \cdot 6H_2O$	7782-95-8*	314.031	mono plates	110 dec		1.849	2.0^{25}	i EtOH
2106	Sodium dihydrogen pyrophosphate	$Na_2H_2P_2O_7$	7758-16-9	221.939	wh powder	220 dec		≈1.9		s H_2O
2107	Sodium dithionate	$Na_2S_2O_4$	7775-14-6	174.110	gray-wh pow	52 dec			24.1^{20}	sl EtOH
2108	Sodium dithionate dihydrate	$Na_2S_2O_6 \cdot 2H_2O$	7631-94-9*	242.139	col orth cry	110 dec		2.19	15.1^{20}	i EtOH
2109	Sodium ethanolate	NaC_2H_5O	141-52-6	68.050	wh-yel hyg pow					reac H_2O; s EtOH
2110	Sodium ferricyanide monohydrate	$Na_3Fe(CN)_6 \cdot H_2O$	14217-21-1*	298.933	red hyg cry					s H_2O; i EtOH
2111	Sodium ferrocyanide decahydrate	$Na_4Fe(CN)_6 \cdot 10H_2O$	13601-19-9	484.061	yel monocl cry	≈50 dec		1.46	20^{20}	i os
2112	Sodium fluoride	NaF	7681-49-4	41.988	col cub or tetr cry	996	1704	2.78	4.13^{25}	i EtOH
2113	Sodium tetrafluoroborate	$NaBF_4$	13755-29-8	109.795	wh orth prisms	384		2.47	108^{20}	sl EtOH
2114	Sodium fluorophosphate	Na_2PO_3F	10163-15-2	143.950	pow					
2115	Sodium formate	$NaCHO_2$	141-53-7	68.008	wh hyg cry	257.3	dec	1.92	94.9^{25}	sl EtOH
2116	Sodium germanate	Na_2GeO_3	12025-19-3	166.59	wh mono hyg cry	1083		3.31		
2117	Sodium hexabromoplatinate(IV) hexahydrate	$Na_2PtBr_6 \cdot 6H_2O$	39277-13-9	828.57	cry					
2118	Sodium hexachloroiridate(IV) hexahydrate	$Na_2IrCl_6 \cdot 6H_2O$	19567-78-3	559.004	cry	600 dec				
2119	Sodium hexachloroplatinate(IV)	Na_2PtCl_6	16923-58-3	453.77	yel hyg cry				53^{16}	s EtOH
2120	Sodium hexachloroplatinate(IV) hexahydrate	$Na_2PtCl_6 \cdot 6H_2O$	16923-58-3	561.87	yel cry	110 dec		2.50	53^{16}	s EtOH; i eth
2121	Sodium hexafluoroaluminate	Na_3AlF_6	13775-53-6	209.941	col monocl cry; trans cub 560	1009		2.97		i H_2O
2122	Sodium hexafluoroantimonate	$NaSbF_6$	16925-25-0	258.740	wh cub cry			3.375	129^{20}	s EtOH, ace
2123	Sodium hexafluorophosphate monohydrate	$NaPF_6 \cdot H_2O$	20644-15-9	185.969	col orth cry			2.369	103^{0}	s EtOH, MeOH, ace
2124	Sodium hexafluorosilicate	Na_2SiF_6	16893-85-9	188.056	wh hex cry	dec		2.7	0.67^{20}	i EtOH
2125	Sodium hexanitrocobaltate(III)	$Na_3Co(NO_2)_6$	14649-73-1	403.935	yel-brn cry pow					vs H_2O; sl EtOH
2126	Sodium hydride	NaH	7646-69-7	23.998	silv cub cry; flam	425 dec		1.39		reac H_2O, EtOH
2127	Sodium hydrogen arsenate	Na_2HAsO_4	7778-43-0	185.908	wh pow	≈195 dec			51^{20}	sl EtOH

2128	Sodium hydrogen arsenate heptahydrate	$Na_2HAsO_4 \cdot 7H_2O$	10048-95-0	312.014	wh monocl cry	≈50 dec		1.87	51^{20}	sl EtOH
2129	Sodium hydrogen carbonate	$NaHCO_3$	144-55-8	84.007	wh monocl cry	≈50 dec		2.20	10.3^{25}	i EtOH
2130	Sodium hydrogen fluoride	$NaHF_2$	1333-83-1	61.995	wh hex cry	>160 dec		2.08	3.25^{20}	
2131	Sodium hydrogen phosphate	Na_2HPO_4	7558-79-4	141.959	wh hyg powder			1.7	11.8^{25}	
2132	Sodium hydrogen phosphate dodecahydrate	$Na_2HPO_4 \cdot 12H_2O$	10039-32-4	358.143	col cry	≈35 dec		≈1.5	11.8^{25}	i EtOH
2133	Sodium hydrogen phosphate heptahydrate	$Na_2HPO_4 \cdot 7H_2O$	7782-85-6	268.066	col cry			≈1.7	11.8^{25}	i EtOH
2134	Sodium hydrogen sulfate	$NaHSO_4$	7681-38-1	120.062	wh hyg cry	≈315		2.43	28.5^{25}	
2135	Sodium hydrogen sulfate monohydrate	$NaHSO_4 \cdot H_2O$	10034-88-5	138.077	wh monocl cry			2.10	28.5^{25}	reac EtOH
2136	Sodium hydrogen sulfide	NaHS	16721-80-5	56.064	col rhomb cry	350		1.79		s H_2O, EtOh, eth
2137	Sodium hydrogen sulfide dihydrate	$NaHS \cdot 2H_2O$	16721-80-5	92.095	yel hyg needles	55 dec				vs H_2O, EtOH, eth
2138	Sodium hydrogen sulfite	$NaHSO_3$	7631-90-5	104.062	wh cry			1.48		s H_2O; sl EtOH
2139	Sodium hydroxide	NaOH	1310-73-2	39.997	wh orth cry; hyg	323	1388	2.13	100^{25}	s EtOH, MeOH
2140	Sodium hypochlorite	NaClO	7681-52-9	74.442	stable in aq soln	anh form exp			79.9^{25}	
2141	Sodium hypochlorite pentahydrate	$NaOCl \cdot 5H_2O$	10022-70-5	164.518	pale grn orth cry	18		1.6		s H_2O
2142	Sodium iodate	$NaIO_3$	7681-55-2	197.892	wh orth cry	dec		4.28	9.47^{25}	i EtOH
2143	Sodium iodide	NaI	7681-82-5	149.894	wh cub cry; hyg	660	1304	3.67	184^{25}	s EtOH, ace
2144	Sodium bismuthate	$NaBiO_3$	12232-99-4	279.968	yel-brn hyg cry					i cold H_2O, reac acid
2145	Sodium metabisulfite	$Na_2S_2O_5$	7681-57-4	190.109	wh cry				66.7^{25}	sl EtOH
2146	Sodium metaborate	$NaBO_2$	7775-19-1	65.800	wh hex cry	966	1434	2.46		s H_2O
2147	Sodium metasilicate	Na_2SiO_3	6834-92-0	122.064	wh amorp solid; hyg	1089		2.61		s cold H_2O; reac hot H_2O
2148	Sodium molybdate	Na_2MoO_4	7631-95-0	205.92	col cub cry	687		≈3.5	65.0^{25}	
2149	Sodium molybdate dihydrate	$Na_2MoO_4 \cdot 2H_2O$	10102-40-6	241.95	cry powder	100 dec		≈3.5	65.0^{25}	
2150	Sodium niobate	$NaNbO_3$	12034-09-2	163.894	rhomb cry	1422		4.55		i H_2O
2151	Sodium nitrate	$NaNO_3$	7631-99-4	84.995	col hex cry; hyg	307		2.26	91.2^{25}	sl EtOH, MeOH
2152	Sodium nitrite	$NaNO_2$	7632-00-0	68.996	wh orth cry; hyg	271	>320 dec	2.17	84.8^{25}	sl EtOH; reac acid
2153	Sodium nitroprusside dihydrate	$Na_2[Fe(CN)_5NO] \cdot 2H_2O$	13755-38-9	297.949	red cry			1.72	40^{16}	sl EtOH
2154	Sodium orthovanadate	Na_3VO_4	13721-39-6	183.909	col hex prisms	860				s H_2O; i EtOH
2155	Sodium oxalate	$Na_2C_2O_4$	62-76-0	133.999	wh powder	≈250 dec		2.34	3.61^{25}	i EtOH
2156	Sodium oxide	Na_2O	1313-59-3	61.979	wh amorp powder	1132 dec		2.27		reac H_2O
2157	Sodium perborate tetrahydrate	$NaBO_3 \cdot 4H_2O$	7632-04-4	153.861	wh cry	60 dec				reac H_2O
2158	Sodium perchlorate	$NaClO_4$	7601-89-0	122.441	wh orth cry; hyg	480 dec		2.52	205^{25}	
2159	Sodium perchlorate monohydrate	$NaClO_4 \cdot H_2O$	7791-07-3	140.456	wh hyg cry	≈130 dec		2.02	205^{25}	
2160	Sodium periodate	$NaIO_4$	7790-28-5	213.892	wh tetr cry	≈300 dec		3.86	14.4^{25}	s acid
2161	Sodium periodate trihydrate	$NaIO_4 \cdot 3H_2O$	13472-31-6	267.938	wh hex cry	175 dec		3.22	14.4^{25}	
2162	Sodium permanganate trihydrate	$NaMnO_4 \cdot 3H_2O$	10101-50-5*	195.972	red-blk hyg cry	170 dec		2.47	144^{20}	reac EtOH
2163	Sodium peroxide	Na_2O_2	1313-60-6	77.979	yel hyg powder	675		2.805		reac H_2O
2164	Sodium perrhenate	$NaReO_4$	13472-33-8	273.195	cry	300		5.39		
2165	Sodium persulfate	$Na_2S_2O_8$	7775-27-1	238.107	wh hyg cry					vs H_2O; reac EtOH
2166	Sodium phosphate dodecahydrate	$Na_3PO_4 \cdot 12H_2O$	10101-89-0	380.124	col hex cry	≈75		1.62	14.4^{25}	i EtOH
2167	Chlorinated trisodium phosphate	$Na_3PO_4 \cdot NaOCl$	56802-99-4	238.383	wh cry				25^{25}	
2168	Sodium phosphinate	NaH_2PO_2	7681-53-0	87.979	wh cry				100^{25}	
2169	Sodium phosphinate monohydrate	$NaH_2PO_2 \cdot H_2O$	10039-56-2	105.994	col hyg cry	310 dec			100^{25}	s EtOH
2170	Sodium potassium tartrate tetrahydrate	$NaKC_4H_4O_6 \cdot 4H_2O$	304-59-6	282.220	wh cry	≈70 dec	anh at 130	1.79		vs H_2O; i EtOH
2171	Sodium pyrophosphate	$Na_4P_2O_7$	7722-88-5	265.902	col cry	988		2.53	7.09^{25}	
2172	Sodium selenate	Na_2SeO_4	13410-01-0	188.94	col orth cry				58.5^{25}	
2173	Sodium selenate decahydrate	$Na_2SeO_4 \cdot 10H_2O$	10102-23-5	369.09	wh cry			1.61	58.5^{25}	
2174	Sodium selenide	Na_2Se	1313-85-5	124.94	amorp solid	>875		2.62		reac H_2O
2175	Sodium selenite	Na_2SeO_3	10102-18-8	172.94	wh tetr cry				89.8^{25}	i EtOH

No.	Name	Formula	CAS Reg No.	Mol. Weight	Physical Form	mp/°C	bp/°C	Density g cm⁻³	Solubility g/100 g H_2O	Qualitative Solubility
2176	Sodium stearate	$NaC_{18}H_{35}O_2$	822-16-2	306.460	wh pow					sl H_2O, EtOH; vs hot H_2O
2177	Sodium succinate hexahydrate	$Na_2C_4H_4O_4 \cdot 6H_2O$	150-90-3	270.144	cry pow	120 dec			20	i EtOH
2178	Sodium sulfate	Na_2SO_4	7757-82-6	142.044	wh orth cry or powder	884		2.7	28.1^{25}	i EtOH
2179	Sodium sulfate decahydrate	$Na_2SO_4 \cdot 10H_2O$	7727-73-3	322.197	col monocl cry	32 dec		1.46	28.1^{25}	i EtOH
2180	Sodium sulfide	Na_2S	1313-82-2	78.046	wh cub cry; hyg	1172		1.856	20.6^{25}	sl EtOH; i eth
2181	Sodium sulfide nonahydrate	$Na_2S \cdot 9H_2O$	1313-84-4	240.184	wh-yel hyg cry	≈50 dec		1.43	20.6^{25}	sl EtOH; i eth
2182	Sodium sulfide pentahydrate	$Na_2S \cdot 5H_2O$	1313-83-3	168.122	col orth cry	120 dec		1.58	20.6^{25}	s EtOH; i eth
2183	Sodium sulfite	Na_2SO_3	7757-83-7	126.044	wh hex cry	dec		2.63	30.7^{25}	i EtOH
2184	Sodium sulfite heptahydrate	$Na_2SO_3 \cdot 7H_2O$	10102-15-5	252.151	wh monocl cry; unstable			1.56	30.7^{25}	sl EtOH
2185	Sodium superoxide	NaO_2	12034-12-7	54.989	yel cub cry	552		2.2		reac H_2O
2186	Sodium tellurate	Na_2TeO_4	10101-83-4	237.58	wh powder				0.8	
2187	Sodium tellurite	Na_2TeO_3	10102-20-2	221.58	wh rhomb prisms					sl H_2O
2188	Sodium tetraborate	$Na_2B_4O_7$	1330-43-4	201.220	col gl solid; hyg	743	1575	2.4	3.17^{25}	sl MeOH
2189	Sodium tetraborate decahydrate	$Na_2B_4O_7 \cdot 10H_2O$	1303-96-4	381.373	wh monocl cry	75 dec		1.73	3.17^{25}	i EtOH
2190	Sodium tetraborate pentahydrate	$Na_2B_4O_7 \cdot 5H_2O$	12045-88-4	291.296	hex cry	dec		1.88	3.17^{25}	
2191	Sodium tetraborate tetrahydrate	$Na_2B_4O_7 \cdot 4H_2O$	12045-87-3	273.281	wh monocl cry			1.95	3.17^{25}	
2192	Sodium tetrachloroaluminate	$NaAlCl_4$	7784-16-9	191.783	orth cry			2.01		s H_2O
2193	Sodium tetrachloroaurate(III) dihydrate	$NaAuCl_4 \cdot 2H_2O$	13874-02-7	397.799	oran-yel rhom cry	100 dec			150^{10}	s EtOH, eth
2194	Sodium tetrachloropalladate(II) trihydrate	$Na_2PdCl_4 \cdot 3H_2O$	13820-53-6	348.26	brn-red hyg cry					vs H_2O; s EtOH
2195	Sodium tetrachloroplatinate(II) tetrahydrate	$Na_2PtCl_4 \cdot 4H_2O$	10026-00-3	454.93	red prisms	100				s H_2O, EtOH
2196	Sodium tetrafluoroberyllate	Na_2BeF_4	13871-27-7	130.986	orth cry	575		2.47		sl H_2O
2197	Sodium thiocyanate	NaSCN	540-72-7	81.074	col hyg cry	287			151^{25}	
2198	Sodium thiophosphate dodecahydrate	$Na_3PO_3S \cdot 12H_2O$	10101-88-9	396.191	hex hyg leaflets	60				vs hot H_2O
2199	Sodium thiosulfate	$Na_2S_2O_3$	7772-98-7	158.110	col mono cry	100 dec		1.69	76.4^{25}	i EtOH
2200	Sodium thiosulfate pentahydrate	$Na_2S_2O_3 \cdot 5H_2O$	10102-17-7	248.186	col cry	≈50 dec		1.69	76.4^{25}	i EtOH
2201	Sodium trimetaphosphate	$Na_3(PO_3)_3$	7785-84-4	305.885	wh cry			2.49	22	
2202	Sodium trimetaphosphate hexahydrate	$Na_3(PO_3)_3 \cdot 6H_2O$	7785-84-4	413.976	tricl-rhom hyg prisms	53		1.786	22	i EtOH
2203	Sodium tripolyphosphate	$Na_5P_3O_{10}$	7758-29-4	367.864	wh hyg pow	622			20^{25}	
2204	Sodium tungstate	Na_2WO_4	13472-45-2	293.82	wh rhom cry	695		4.18	74.2^{25}	
2205	Sodium tungstate dihydrate	$Na_2WO_4 \cdot 2H_2O$	10213-10-2	329.85	wh orth cry	100 dec		3.25	74.2^{25}	i EtOH
2206	Sodium uranate(VI) monohydrate	$Na_2U_2O_7 \cdot H_2O$	13721-34-1	652.049	yel pow					i H_2O; s acid
2207	Sodium vanadate(V)	$NaVO_3$	13718-26-8	121.930	col mono prisms	630			21^{25}	
2208	Sodium vanadate(V) tetrahydrate	$NaVO_3 \cdot 4H_2O$	13718-26-8	193.992	yel-wh cry pow				21^{25}	
2209	Strontium	Sr	7440-24-6	87.62	silv-wh metal; cub	777	1382	2.64		reac H_2O; s EtOH
2210	Strontium arsenite tetrahydrate	$Sr(AsO_2)_2 \cdot 4H_2O$	10378-48-0	373.52	wh pow					sl H_2O, EtOH; sol dil acid
2211	Strontium bromate monohydrate	$Sr(BrO_3)_2 \cdot H_2O$	14519-18-7	361.44	yel hyg mono cry	120 dec		3.773	39.0^{25}	
2212	Strontium bromide	$SrBr_2$	10476-81-0	247.43	wh tetr cry	657		4.216	107^{25}	
2213	Strontium bromide hexahydrate	$SrBr_2 \cdot 6H_2O$	7789-53-9	355.52	col hyg cry	88 dec			107^{25}	s EtOH; i eth
2214	Strontium carbide	SrC_2	12071-29-3	111.64	blk tetr cry	>1700		3.19		i H_2O
2215	Strontium carbonate	$SrCO_3$	1633-05-2	147.63	wh orth cry; hyg	1494		3.5	0.00034^{20}	s dil acid
2216	Strontium chlorate	$Sr(ClO_3)_2$	7791-10-8	254.52	col cry	120 dec		3.15	176^{25}	sl EtOH
2217	Strontium chloride	$SrCl_2$	10476-85-4	158.53	wh cub cry; hyg	874	1250	3.052	54.7^{25}	
2218	Strontium chloride hexahydrate	$SrCl_2 \cdot 6H_2O$	10025-70-4	266.62	col hyg cry	100 dec		1.96	54.7^{25}	s EtOH
2219	Strontium chromate	$SrCrO_4$	7789-06-2	203.61	yel monocl cry	dec		3.9	0.106^{20}	s dil acid
2220	Strontium cyanide dihydrate	$Sr(CN)_2 \cdot 4H_2O$		211.72	wh hyg cry	dec				vs H_2O

2221	Strontium ferrocyanide pentadecahydrate	$SrFe(CN)_6 \cdot 15H_2O$		569.80	yel mono cry				50	
2222	Strontium fluoride	SrF_2	7783-48-4	125.62	wh cub cry or powder	1477	2460	4.24	0.021^{25}	s dil acid
2223	Strontium formate	$Sr(CHO_2)_2$	592-89-2	177.66	wh cry	71.9		2.693	9.1^{0}	
2224	Strontium formate dihydrate	$Sr(CHO_2)_2 \cdot 2H_2O$	6160-34-5	213.69	col rhom cry	100 dec		2.25	9.1^{37}	i EtOH, eth
2225	Strontium hexaboride	SrB_6	12046-54-7	152.49	blk cub cry	2235		3.39		i H_2O; s HNO_3
2226	Strontium hydride	SrH_2	13598-33-9	89.64	orth cry	1050		3.26		reac H_2O
2227	Strontium hydroxide	$Sr(OH)_2$	18480-07-4	121.64	col orth cry; hyg	535	710 dec	3.625	2.25^{25}	
2228	Strontium iodate	$Sr(IO_3)_2$	13470-01-4	437.43	tricl cry			5.045	0.165^{25}	
2229	Strontium iodide	SrI_2	10476-86-5	341.43	wh hyg cry	538	1773 dec	4.55	177^{25}	
2230	Strontium iodide hexahydrate	$SrI_2 \cdot 6H_2O$	73796-25-5	449.52	wh-yel hex cry; hyg	120 dec		4.4	177^{25}	s EtOH
2231	Strontium niobate	$SrNb_2O_6$	12034-89-8	369.43	monocl cry	1225		5.11		i H_2O
2232	Strontium nitrate	$Sr(NO_3)_2$	10042-76-9	211.63	wh cub cry	570	645	2.99	80.2^{25}	sl EtOH, ace
2233	Strontium nitride	Sr_3N_2	12033-82-8	290.87	refrac solid	1200				reac H_2O; s HCl
2234	Strontium nitrite	$Sr(NO_2)_2$	13470-06-9	179.63	wh-yel hyg needles	240 dec		2.8	72.1^{30}	
2235	Strontium oxide	SrO	1314-11-0	103.62	col cub cry	2531		5.1		reac H_2O
2236	Strontium perchlorate	$Sr(ClO_4)_2$	13450-97-0	286.52	col hyg cry				306^{25}	s EtOH, MeOH
2237	Strontium permanganate trihydrate	$Sr(MnO_4)_2 \cdot 3H_2O$		379.54	pur cub cry	175 dec		2.75	250^{18}	
2238	Strontium peroxide	SrO_2	1314-18-7	119.62	wh tetr cry; unstable	215 dec		4.78		reac H_2O
2239	Strontium phosphate	$Sr_3(PO_4)_2$	7446-28-8	452.80	wh powder				0.0000011^{20}	s acid
2240	Strontium selenate	$SrSeO_4$	7446-21-1	230.58	orth cry			4.25	0.115^{20}	s hot HCl
2241	Strontium selenide	SrSe	1315-07-7	166.58	wh cub cry	1600		4.54		
2242	Strontium orthosilicate	Sr_2SiO_4	13597-55-2	267.32	orth cry			4.5		
2243	Strontium silicide	$SrSi_2$	12138-28-2	143.79	silv-gray cub cry	1100		3.35		
2244	Strontium sulfate	$SrSO_4$	7759-02-6	183.68	wh orth cry	1606		3.96	0.0135^{25}	i EtOH; sl acid
2245	Strontium sulfide	SrS	1314-96-1	119.69	gray cub cry	2226		3.70		sl H_2O; s acid
2246	Strontium sulfite	$SrSO_3$	13451-02-0	167.68	col cry	dec			0.0015^{25}	s H_2SO_4, HCl
2247	Strontium telluride	SrTe	12040-08-3	215.22	wh cub cry			4.83		
2248	Strontium thiosulfate pentahydrate	$SrS_2O_3 \cdot 5H_2O$	15123-90-7	289.83	mono needles	100 dec		2.17	36.3^{25}	i EtOH
2249	Strontium titanate	$SrTiO_3$	12060-59-2	183.49	wh cub cry	2080		5.1		i H_2O
2250	Strontium tungstate	$SrWO_4$	13451-05-3	335.46	col tetr cry	dec		6.187	0.14^{15}	i EtOH
2251	Sulfur (rhombic)	S	7704-34-9	32.066	yel orth cry	95.3 (trans to monocl)	444.60	2.07		i H_2O; sl EtOH, bz, eth; s CS_2
2252	Sulfur (monoclinic)	S	7704-34-9	32.066	yel monocl needles, stable 95.3-120	115.21	444.60	2.07		i H_2O; sl EtOH, bz, eth; s CS_2
2253	Sulfuric acid	H_2SO_4	7664-93-9	98.080	col oily liq	10.31	337	1.8		vs H_2O
2254	Peroxysulfuric acid	H_2SO_5	7722-86-3	114.079	wh cry; unstable	45 dec				vs H_2O
2255	Nitrosylsulfuric acid	$HNOSO_4$	7782-78-7	127.078	prisms	73 dec				reac H_2O; s H_2SO_4
2256	Chlorosulfonic acid	$SO_2(OH)Cl$	7790-94-5	116.525	col-yel liq	-80	152	1.75		reac H_2O; s py
2257	Fluorosulfonic acid	$SO_2(OH)F$	7789-21-1	100.070	col liq	-89	163	1.726		reac H_2O
2258	Sulfurous acid	H_2SO_3	7782-99-2	82.080	exists only in soln					soln of SO_2 in H_2O
2259	Sulfamic acid	H_2NSO_3H	5329-14-6	97.095	orth cry	≈205 dec		2.15	14.7^{0}	sl ace; i eth
2260	Sulfur dioxide	SO_2	7446-09-5	64.065	col gas	-75.5	-10.05	2.619 g/L		s H_2O, EtOh, eth, chl
2261	Sulfur trioxide	SO_3	7446-11-9	80.064	col liq	16.8	45	1.92		reac H_2O
2262	Sulfur bromide	$SSBr_2$	13172-31-1	223.940	red oily liq	-46	>25 dec	2.63		reac H_2O
2263	Sulfur chloride	$SSCl_2$	10025-67-9	135.037	yel-red oily liq	-77	137	1.69		reac H_2O; s EtOH, bz, eth, ctc
2264	Sulfur fluoride	SSF_2	16860-99-4	102.129	col gas	-164.6	-10.6	4.174 g/L		reac H_2O
2265	Sulfur fluoride	FSSF	13709-35-8	102.129	col gas	-133	15	4.174 g/L		reac H_2O
2266	Sulfur dichloride	SCl_2	10545-99-0	102.971	red visc liq	-122	59.6	1.62		reac H_2O
2267	Sulfur tetrafluoride	SF_4	7783-60-0	108.060	col gas	-125	-40.45	4.417 g/L		reac H_2O

No.	Name	Formula	CAS Reg No.	Mol. Weight	Physical Form	mp/°C	bp/°C	Density g cm^{-3}	Solubility g/100 g H_2O	Qualitative Solubility
2268	Sulfur hexafluoride	SF_6	2551-62-4	146.056	col gas	-50.7 tp	-63.8 sp	5.970 g/L		sl H_2O; s EtOH
2269	Sulfur bromide pentafluoride	SF_5Br	15607-89-3	206.962	col gas	-79	3.1	8.459 g/L		
2270	Sulfur chloride pentafluoride	SF_5Cl	13780-57-9	162.511	col gas	-64	-19.05	6.642 g/L		
2271	Sulfur decafluoride	S_2F_{10}	5714-22-7	254.116	liq	-52.7	30; dec 150	2.08		i H_2O
2272	Sulfuryl amide	$(NH_2)_2SO_2$	7803-58-9	96.110	orth plates	93	250 dec			vs H_2O; sl EtOH
2273	Sulfuryl chloride	SO_2Cl_2	7791-25-5	134.970	col liq	-51	69.4	1.680		reac H_2O; s bz, tol, eth
2274	Sulfuryl fluoride	SO_2F_2	2699-79-8	102.062	col gas	-135.8	-55.4	4.172 g/L		sl H_2O, EtOH; s tol, ctc
2275	Pyrosulfuryl chloride	$S_2O_5Cl_2$	7791-27-7	215.034	col fuming liq	-37	151	1.837		reac H_2O
2276	Thionyl bromide	$SOBr_2$	507-16-4	207.873	yel liq	-50	140			reac H_2O
2277	Thionyl chloride	$SOCl_2$	7719-09-7	118.970	yel fuming liq	-101	75.6	1.631		reac H_2O; s bz, ctc, chl
2278	Thionyl fluoride	SOF_2	7783-42-8	86.062	col gas	-129.5	-43.8	3.518 g/L		reac H_2O; s bz, eth
2279	Sulfur fluoride hypofluorite	F_5SOF	15179-32-5	162.055	col gas	-86	-35.1	6.624 g/L		
2280	Tantalum	Ta	7440-25-7	180.948	gray metal; cub	3017	5458	16.4		reac HF
2281	Tantalum aluminide	$TaAl_3$	12004-76-1	261.893	gray refrac powder	≈1400		7.02		i H_2O, acid, alk
2282	Tantalum boride	TaB	12007-07-7	191.759	refrac orth cry	2040		14.2		
2283	Tantalum boride	TaB_2	12007-35-1	202.570	blk hex cry	3140		11.2		i H_2O, acid, alk
2284	Tantalum carbide	TaC	12070-06-3	192.959	gold-brown powder; cub	3880	4780	14.3		s HF-HNO_3 mixture
2285	Tantalum carbide	Ta_2C	12070-07-4	373.907	refrac hex cry	3327		15.1		
2286	Tantalum nitride	TaN	12033-62-4	194.955	blk hex cry	3090		13.7		i H_2O; sl aqua regia; reac alk
2287	Tantalum silicide	$TaSi_2$	12039-79-1	237.119	gray powder	2200		9.14		
2288	Tantalum(IV) oxide	TaO_2	12036-14-5	212.947	tetr cry			10.0		
2289	Tantalum(IV) selenide	$TaSe_2$	12039-55-3	338.87	hex cry			6.7		
2290	Tantalum(IV) sulfide	TaS_2	12143-72-5	245.080	blk hex cry	>3000		6.86		i H_2O
2291	Tantalum(IV) telluride	$TaTe_2$	12067-66-2	436.15	monocl cry			9.4		
2292	Tantalum(V) bromide	$TaBr_5$	13451-11-1	580.468	yel cry powder	265	349	4.99		
2293	Tantalum(V) chloride	$TaCl_5$	7721-01-9	358.212	yel monocl cry; hyg	216	239.35	3.68		reac H_2O; s EtOH
2294	Tantalum(V) fluoride	TaF_5	7783-71-3	275.940	wh monocl cry; hyg	95.1	229.2	5.0		s H_2O, eth; sl CS_2, ctc
2295	Tantalum(V) iodide	TaI_5	14693-81-3	815.470	blk hex cry; hyg	496	543	5.80		
2296	Tantalum(V) oxide	Ta_2O_5	1314-61-0	441.893	wh rhomb cry or powder	1784		8.2		i H_2O, EtOH, acid; s HF
2297	Technetium	Tc	7440-26-8	98	hex cry	2157	4265	11		
2298	Technetium(V) fluoride	TcF_5	31052-14-9	193	yel solid	50	dec			
2299	Technetium(VI) fluoride	TcF_6	13842-93-8	212	yel cub cry	37.4	55.3	3.0		
2300	Tellurium	Te	13494-80-9	127.60	gray-wh rhomb cry	449.51	988	6.24		i H_2O, bz, CS_2
2301	Telluric(VI) acid	H_6TeO_6	7803-68-1	229.64	wh monocl cry	136		3.07	50.1^{30}	
2302	Tellurous acid	H_2TeO_3	10049-23-7	177.61	wh cry	40 dec		3.0		sl H_2O; s dil acid, alk
2303	Tellurium dioxide	TeO_2	7446-07-3	159.60	wh orth cry	733	1245	5.9		i H_2O; s alk, acid
2304	Tellurium trioxide	TeO_3	13451-18-8	175.60	yel-oran cry	430		5.07		i H_2O
2305	Tellurium dibromide	$TeBr_2$	7789-54-0	287.41	grn-brn hyg cry	210	339			reac H_2O; s eth; sl chl
2306	Tellurium dichloride	$TeCl_2$	10025-71-5	198.51	blk amorp solid; hyg	208	328	6.9		reac H_2O; i ctc
2307	Tellurium tetrabromide	$TeBr_4$	10031-27-3	447.22	yel-oran monocl cry	388	≈420 dec	4.3		reac H_2O; s eth
2308	Tellurium tetrachloride	$TeCl_4$	10026-07-0	269.41	wh monocl cry; hyg	224	387	3.0		reac H_2O; s EtOH, tol
2309	Tellurium tetrafluoride	TeF_4	15192-26-4	203.59	col cry	129	195 dec			reac H_2O
2310	Tellurium tetraiodide	TeI_4	7790-48-9	635.22	blk orth cry	280		5.05		reac H_2O; sl ace
2311	Tellurium hexafluoride	TeF_6	7783-80-4	241.59	col gas	-37.6 tp	-38.9 sp	9.875 g/L		reac H_2O
2312	Terbium	Tb	7440-27-9	158.925	silv metal; hex	1356	3230	8.23		

2313	Terbium chloride	$TbCl_3$	10042-88-3	265.283	wh orth cry; hyg	588		4.35		s H_2O
2314	Terbium chloride hexahydrate	$TbCl_3 \cdot 6H_2O$	13798-24-8	373.374	hyg cry			4.35		vs H_2O
2315	Terbium iodide	TbI_3	13813-40-6	539.638	hex cry; hyg	957		≈5.2		s H_2O
2316	Terbium nitrate	$Tb(NO_3)_3$	10043-27-3	344.940	pink hyg solid				157^{25}	s EtOH
2317	Terbium nitrate hexahydrate	$Tb(NO_3)_3 \cdot 6H_2O$	13451-19-9	453.031	col needles	89				s H_2O, EtOH, ace
2318	Terbium nitride	TbN	12033-64-6	172.932	cub cry			9.55		
2319	Terbium oxide	Tb_2O_3	12036-41-8	365.849	wh cub cry	2303		7.91		
2320	Terbium silicide	$TbSi_2$	12039-80-4	215.096	orth cry			6.66		
2321	Terbium sulfide	Tb_2S_3	12138-11-3	414.049	cub cry			6.35		
2322	Thallium	Tl	7440-28-0	204.383	soft blue-wh metal	304	1473	11.8		i H_2O; reac acid
2323	Thallium(I) acetate	$TlC_2H_3O_2$	563-68-8	263.427	hyg wh cry	131		3.68		s H_2O, EtOH
2324	Thallium(I) bromate	$TlBrO_3$	14550-84-6	332.285	col needles	120 dec			0.49^{30}	s EtOH
2325	Thallium(I) bromide	TlBr	7789-40-4	284.287	yel cub cry	460	819	7.5	0.059^{20}	
2326	Thallium(I) carbonate	Tl_2CO_3	6533-73-9	468.776	wh monocl cry	272		7.11	4.69^{20}	i EtOH
2327	Thallium(I) chlorate	$TlClO_3$	13453-30-0	287.834	col hex cry			5.5	3.92^{20}	
2328	Thallium(I) chloride	TlCl	7791-12-0	239.836	wh cub cry	430	720	7.0	0.33^{20}	i EtOH
2329	Thallium(I) chromate	Tl_2CrO_4		524.761	yel cry				0.003^{20}	sl acid, alk
2330	Thallium(I) cyanide	TlCN	13453-34-4	230.401	wh hex plates			6.523		s H_2O, acid, EtOH
2331	Thallium(I) ethanolate	TlC_2H_5O	20398-06-5	249.443	cloudy liq	-3	130 dec	3.49		reac H_2O
2332	Thallium(I) fluoride	TlF	7789-27-7	223.381	wh orth cry	326	826	8.36	245^{25}	
2333	Thallium(I) formate	$TlCHO_2$	992-98-3	249.401	hyg col needles	101		4.97		vs H_2O; s MeOH
2334	Thallium(I) hexafluorophosphate	$TlPF_6$	60969-19-9	349.347	wh cub cry			4.6		
2335	Thallium(I) hydroxide	TlOH	12026-06-1	221.390	yel needles	139 dec		7.44	34.3^{18}	
2336	Thallium(I) iodate	$TlIO_3$	14767-09-0	379.285	wh needles				0.058	sl HNO_3
2337	Thallium(I) iodide	TlI	7790-30-9	331.287	yel cry powder	441.7	824	7.1	0.0085^{20}	i EtOH
2338	Thallium(I) molybdate	Tl_2MoO_4	34128-09-1	568.71	yel-wh cub cry					i H_2O
2339	Thallium(I) nitrate	$TlNO_3$	10102-45-1	266.388	wh cry	206	450 dec	5.55	9.55^{20}	i EtOH
2340	Thallium(I) nitrite	$TlNO_2$	13826-63-6	250.389	cub cry			5.7	32.1^{25}	
2341	Thallium(I) oxalate	$Tl_2C_2O_4$	30737-24-7	496.786	wh powder			6.31	1.83^{20}	
2342	Thallium(I) oxide	Tl_2O	1314-12-1	424.766	blk rhomb cry; hyg	579	≈1080	9.52		s H_2O, EtOH
2343	Thallium(I) perchlorate	$TlClO_4$	13453-40-2	303.834	col orth cry			4.8	19.7^{30}	
2344	Thallium(I) selenate	Tl_2SeO_4	7446-22-2	551.73	orth cry	>400		6.875	2.8^{20}	i EtOH, eth
2345	Thallium(I) selenide	Tl_2Se	15572-25-5	487.73	gray plates	340				i H_2O, acid
2346	Thallium(I) sulfate	Tl_2SO_4	7446-18-6	504.831	wh rhomb prisms	632		6.77	5.47^{25}	
2347	Thallium(I) sulfide	Tl_2S	1314-97-2	440.833	blue-blk cry	448	1367	8.39	0.02^{20}	sl alk; s acid
2348	Thallium(III) bromide tetrahydrate	$TlBr_3 \cdot 4H_2O$	13701-90-1	516.157	yel orth cry			3.65		s H_2O, EtOH
2349	Thallium(III) chloride	$TlCl_3$	13453-32-2	310.741	monocl cry	155		4.7		vs H_2O, EtOH, eth
2350	Thallium(III) chloride tetrahydrate	$TlCl_3 \cdot 4H_2O$	13453-32-2*	382.803	orth cry			3.00		s H_2O
2351	Thallium(III) fluoride	TlF_3	7783-57-5	261.378	wh orth cry; hyg	550 dec		8.65		reac H_2O
2352	Thallium(III) nitrate	$Tl(NO_3)_3$	13746-98-0	390.398	col cry					reac H_2O
2353	Thallium(III) oxide	Tl_2O_3	1314-32-5	456.765	brn cub cry	834		10.2		i H_2O; reac acid
2354	Thallium(III) sulfate	$Tl_2(SO_4)_3$	16222-66-5	696.958	col leaflets					reac H_2O
2355	Thallium selenide	TlSe	12039-52-0	283.34	blk solid	330				i H_2O, acid
2356	Thorium	Th	7440-29-1	232.038	soft gray-wh metal; cub	1750	4788	11.7		s acid
2357	Thorium hydride	ThH_2	16689-88-6	234.054	tetr cry			9.5		
2358	Thorium boride	ThB_6	12229-63-9	296.904	refrac solid	2450		6.99		
2359	Thorium(IV) bromide	$ThBr_4$	13453-49-1	551.654	wh hyg cry	679			65^{20}	
2360	Thorium carbide	ThC	12012-16-7	244.049	cub cry	2500		10.6		reac H_2O

No.	Name	Formula	CAS Reg No.	Mol. Weight	Physical Form	mp/°C	bp/°C	Density g cm^{-3}	Solubility g/100 g H_2O	Qualitative Solubility
2361	Thorium dicarbide	ThC_2	12071-31-7	256.059	yel monocl cry	≈2650		9.0		reac H_2O
2362	Thorium(IV) chloride	$ThCl_4$	10026-08-1	373.849	gray-wh tetr needles; hyg	770	921	4.59		s H_2O, EtOH
2363	Thorium(IV) fluoride	ThF_4	13709-59-6	308.032	wh monocl cry; hyg	1110	1680	6.1		
2364	Thorium(IV) iodide	ThI_4	7790-49-0	739.656	wh-yel monocl cry	570	837			
2365	Thorium(IV) nitrate tetrahydrate	$Th(NO_3)_4 \cdot 4H_2O$	33088-16-3	552.119	wh hyg cry	500 dec			191^{20}	s EtOH
2366	Thorium nitride	ThN	12033-65-7	246.045	refrac cub cry	2820		11.6		reac H_2O
2367	Thorium(IV) oxide	ThO_2	1314-20-1	264.037	wh cub cry	3390	4400	10.0		i H_2O, alk; sl acid
2368	Thorium(IV) selenide	$ThSe_2$	60763-24-8	389.96	orth cry			8.5		
2369	Thorium orthosilicate	$ThSiO_4$	14553-44-7	324.122	brn tetr cry			6.7		
2370	Thorium silicide	$ThSi_2$	12067-54-8	288.209	tetr cry	1850		7.9		
2371	Thorium(IV) sulfate nonahydrate	$Th(SO_4)_2 \cdot 9H_2O$	10381-37-0	586.303	wh monocl cry	dec		2.8	4.2^{20}	
2372	Thorium(IV) sulfide	ThS_2	12138-07-7	296.170	dark brn cry	1905		7.30		i H_2O; s acid
2373	Thulium	Tm	7440-30-4	168.934	silv metal; hex	1545	1950	9.32		s dil acid
2374	Thulium bromide	$TmBr_3$	14456-51-0	408.646	wh hyg cry	954				s H_2O
2375	Thulium chloride	$TmCl_3$	13537-18-3	275.292	yel hyg cry	824				s H_2O
2376	Thulium chloride heptahydrate	$TmCl_3 \cdot 7H_2O$	13778-39-7	401.399	hyg cry					s H_2O, EtOH
2377	Thulium fluoride	TmF_3	13760-79-7	225.929	wh cry	1158				s H_2O
2378	Thulium iodide	TmI_3	13813-43-9	549.647	yel hyg cry	1021				
2379	Thulium nitrate	$Tm(NO_3)_3$	14985-19-4	354.949	grn hyg solid				212^{25}	s EtOH
2380	Thulium nitrate pentahydrate	$Tm(NO_3)_3 \cdot 5H_2O$	36548-87-5	445.025	grn hyg cry					s H_2O, EtOH, ace
2381	Thulium oxide	Tm_2O_3	12036-44-1	385.866	grn-wh cub cry	2341	3945	8.6		sl acid
2382	Tin (gray)	Sn	7440-31-5	118.710	cub cry	trans to wh Sn 13.2	2602	5.769		
2383	Tin (white)	Sn	7440-31-5	118.710	silv tetr cry	231.93	2602	7.265		
2384	Stannane	SnH_4	2406-52-2	122.742	unstable col gas	-146	-51.8	5.017 g/L		
2385	Methylstannane	SnH_3CH_3	1631-78-3	136.769	col gas		0	5.590 g/L		reac H_2O
2386	Tin(II) acetate	$Sn(C_2H_3O_2)_2$	638-39-1	236.799	wh orth cry	183	subl	2.31		s dil HCl
2387	Tin(II) bromide	$SnBr_2$	10031-24-0	278.518	yel powder	215	639	5.12	85^{0}	s EtOH, eth, ace
2388	Tin(II) chloride	$SnCl_2$	7772-99-8	189.615	wh orth cry	247.1	623	3.90	178^{10}	s EtOH, ace, eth; i xyl
2389	Tin(II) chloride dihydrate	$SnCl_2 \cdot 2H_2O$	10025-69-1	225.646	wh monocl cry	37 dec		2.71	178^{10}	s EtOH, NaOH; vs HCl
2390	Tin(II) fluoride	SnF_2	7783-47-3	156.707	wh monocl cry; hyg	213	850	4.57		s H_2O; i EtOH, eth, chl
2391	Tin(II) hexafluorozirconate	$SnZrF_6$	12419-43-1	323.924	cry			4.21		s H_2O
2392	Tin(II) hydroxide	$Sn(OH)_2$	12026-24-3	152.725	wh amorp solid					
2393	Tin(II) iodide	SnI_2	10294-70-9	372.519	red-oran powder	320	714	5.28	0.98^{20}	s bz, chl, CS_2
2394	Tin(II) oxalate	SnC_2O_4	814-94-8	206.729	wh powder	280 dec		3.56		i H_2O; s dil HCl
2395	Tin(II) oxide	SnO	21651-19-4	134.709	blue-blk tetr cry	1080 dec		6.45		i H_2O, EtOH; s acid
2396	Tin(II) pyrophosphate	$Sn_2P_2O_7$	15578-26-4	411.363	wh amorp powder	400 dec		4.009		i H_2O; s conc acid
2397	Tin(II) selenide	SnSe	1315-06-6	197.67	gray orth cry	861		6.18		i H_2O; s aqua regia
2398	Tin(II) sulfate	$SnSO_4$	7488-55-3	214.774	wh orth cry	378 dec		4.15	18.8^{19}	
2399	Tin(II) sulfide	SnS	1314-95-0	150.776	gray orth cry	880	1210	5.08		i H_2O; s conc acid
2400	Tin(II) tartrate	$SnC_4H_4O_6$	815-85-0	266.781	wh cry powder					s H_2O, dil HCl
2401	Tin(II) telluride	SnTe	12040-02-7	246.31	gray cub cry	790		6.5		
2402	Tin(IV) bromide	$SnBr_4$	7789-67-5	438.326	wh cry	29.1	205	3.34		vs H_2O; s EtOH
2403	Tin(IV) chloride	$SnCl_4$	7646-78-8	260.521	col fuming liq	-34.07	114.15	2.234		reac H_2O; s EtOH, ctc, bz, ace
2404	Tin(IV) chloride pentahydrate	$SnCl_4 \cdot 5H_2O$	10026-06-9	350.597	wh-yel cry	56 dec		2.04		vs H_2O; s EtOH
2405	Tin(IV) chromate	$Sn(CrO_4)_2$	38455-77-5	350.697	brn-yel cry powder	dec				s H_2O

2406	Tin(IV) fluoride	SnF_4	7783-62-2	194.704	wh tetr cry		705 subl	4.78	reac H_2O
2407	Tin(IV) iodide	SnI_4	7790-47-8	626.328	yel-brn cub cry	143	364.35	4.46	reac H_2O; s EtOH, bz, chl, eth
2408	Tin(IV) oxide	SnO_2	18282-10-5	150.709	gray tetr cry	1630		6.85	i H_2O, EtOH; s hot conc alk
2409	Tin(IV) selenide	$SnSe_2$	20770-09-6	276.63	red-brn cry	650		≈5.0	i H_2O; s alk, conc acid
2410	Tin(IV) selenite	$Sn(SeO_3)_2$	7446-25-5	372.63	cry powder				i H_2O; s hot HCl
2411	Tin(IV) sulfide	SnS_2	1315-01-1	182.842	gold-yel hex cry	600 dec		4.5	i H_2O; s alk, aqua regia
2412	Titanium	Ti	7440-32-6	47.867	gray metal; hex	1668	3287	4.506	
2413	Titanium hydride	TiH_2	7704-98-5	49.883	gray-blk powder	≈450 dec		3.75	i H_2O
2414	Titanium boride	TiB_2	12045-63-5	69.489	gray refrac solid; hex	3225		4.38	
2415	Titanium carbide	TiC	12070-08-5	59.878	cub cry	3067		4.93	i H_2O; s HNO_3
2416	Titanium nitride	TiN	25583-20-4	61.874	yel-brn cub cry	2950		5.21	i H_2O; s aqua regia
2417	Titanium phosphide	TiP	12037-65-9	78.841	gray hex cry	1990		4.08	
2418	Titanium silicide	$TiSi_2$	12039-83-7	104.038	blk orth cry	1500		4.0	i H_2O, acid, alk; s HF
2419	Titanium(II) bromide	$TiBr_2$	13783-04-5	207.675	blk powder			4.0	reac H_2O
2420	Titanium(II) chloride	$TiCl_2$	10049-06-6	118.772	blk hex cryc	1035	1500	3.13	reac H_2O; s EtOH; i chl, eth
2421	Titanium(II) iodide	TiI_2	13783-07-8	301.676	blk hex cry			5.02	reac H_2O
2422	Titanium(II) oxide	TiO	12137-20-1	63.866	cub cry	1750		4.95	
2423	Titanium(II) sulfide	TiS	12039-07-5	79.933	brn hex cry	1780		3.85	s conc acid
2424	Titanium(III) bromide	$TiBr_3$	13135-31-4	287.579	blue-blk hex cry				s H_2O
2425	Titanium(III) chloride	$TiCl_3$	7705-07-9	154.225	red-viol hex cry; hyg	425 dec	960	2.64	reac H_2O
2426	Titanium(III) fluoride	TiF_3	13470-08-1	104.862	viol hex cry	1200	1400	2.98	i H_2O, dil acid, alk
2427	Titanium(III) oxide	Ti_2O_3	1344-54-3	143.732	viol hex cry	1842		4.486	s hot HF
2428	Titanium(III) sulfate	$Ti_2(SO_4)_3$	10343-61-0	383.925	grn cry				i H_2O, EtOH; s dil HCl
2429	Titanium(III) sulfide	Ti_2S_3	12039-16-6	191.932	blk hex cry			3.56	
2430	Titanium(III,IV) oxide	Ti_3O_5	12065-65-5	223.598	blk monocl cry	1777		4.24	
2431	Titanium(IV) bromide	$TiBr_4$	7789-68-6	367.483	yel-oran cub cry; hyg	39	230	3.37	reac H_2O
2432	Titanium(IV) chloride	$TiCl_4$	7550-45-0	189.678	col or yel liq	-24.12	136.45	1.73	reac H_2O; s EtOH
2433	Titanium(IV) fluoride	TiF_4	7783-63-3	123.861	wh hyg powder	284	subl	2.798	reac H_2O; s EtOH, py
2434	Titanium(IV) iodide	TiI_4	7720-83-4	555.485	red hyg powder	150	377	4.3	reac H_2O
2435	Titanium(IV) oxide	TiO_2	13463-67-7	79.866	wh tetr cry	1843		4.23	i H_2O, dil acid; s conc acid
2436	Titanium(IV) oxysulfate monohydrate	$TiOSO_4 \cdot H_2O$	13825-74-6*	177.945	col orth cry			2.71	reac H_2O
2437	Titanium(IV) sulfate	$Ti(SO_4)_2$	13693-11-3	239.994	wh-yel hyg cry	150 dec			s H_2O
2438	Titanium(IV) sulfide	TiS_2	12039-13-3	111.999	yel-brn hex cry; hyg			3.37	s H_2SO_4
2439	Tungsten	W	7440-33-7	183.84	gray-wh metal; cub	3422	5555	19.3	
2440	Tungstic acid	H_2WO_4	7783-03-1	249.85	yel amorp powder	100 dec		5.5	i H_2O, acid; s alk
2441	Tungsten boride	W_2B	12007-10-2	378.49	refrac blk powder	2670		16.0	i H_2O
2442	Tungsten boride	WB	12007-09-9	194.65	blk refrac powder	2665		15.2	i H_2O
2443	Tungsten boride	W_2B_5	12007-98-6	421.74	refrac solid	2365		11.0	i H_2O
2444	Tungsten carbide	W_2C	12070-13-2	379.69	refrac hex cry	≈2800		14.8	i H_2O
2445	Tungsten carbide	WC	12070-12-1	195.85	gray hex cry	2785		15.6	i H_2O; s HNO_3/HF
2446	Tungsten carbonyl	$W(CO)_6$	14040-11-0	351.90	wh cry	170 dec	subl	2.65	i H_2O; s os
2447	Tungsten nitride	WN_2	60922-26-1	211.85	hex cry	600 dec		7.7	
2448	Tungsten nitride	W_2N	12033-72-6	381.69	gray cub cry	dec		17.8	
2449	Tungsten silicide	WSi_2	12039-88-2	240.01	blue-gray tetr cry	2160		9.3	i H_2O
2450	Tungsten silicide	W_5Si_3	12039-95-1	1003.46	blue-gray refrac solid	2320		14.4	
2451	Tungsten(II) bromide	WBr_2	13470-10-5	343.65	yel powder	400 dec			
2452	Tungsten(II) chloride	WCl_2	13470-12-7	254.75	yel solid	>500 dec			s H_2O
2453	Tungsten(II) iodide	WI_2	13470-17-2	437.65	oran cry			6.79	

No.	Name	Formula	CAS Reg No.	Mol. Weight	Physical Form	mp/°C	bp/°C	Density g cm^{-3}	Solubility g/100 g H_2O	Qualitative Solubility
2454	Tungsten(III) bromide	WBr_3	15163-24-3	423.55	blk hex cry	>80 dec				i H_2O
2455	Tungsten(III) chloride	WCl_3	20193-56-0	290.20	red solid	550 dec	subl			reac H_2O
2456	Tungsten(IV) bromide	WBr_4	14055-81-3	503.46	blk orth cry		240 subl			reac H_2O
2457	Tungsten(IV) chloride	WCl_4	13470-13-8	325.65	blk hyg powder	450 dec		4.62		reac H_2O
2458	Tungsten(IV) fluoride	WF_4	13766-47-7	259.83	red-brn cry	>800 dec				
2459	Tungsten(IV) iodide	WI_4	14055-84-6	691.46	blk powder	dec				reac H_2O; s EtOH; i eth chl
2460	Tungsten(IV) oxide	WO_2	12036-22-5	215.84	blue monocl cry	≈1500 dec		10.8		i H_2O, os
2461	Tungsten(IV) selenide	WSe_2	12067-46-8	341.76	gray hex cry			9.2		
2462	Tungsten(IV) sulfide	WS_2	12138-09-9	247.97	gray hex cry	1250 dec		7.6		i H_2O, HCl, alk
2463	Tungsten(IV) telluride	WTe_2	12067-76-4	439.04	gray orth cry	1020		9.43		
2464	Tungsten(V) bromide	WBr_5	13470-11-6	583.36	brn-blk hyg solid	286	333			
2465	Tungsten(V) chloride	WCl_5	13470-14-9	361.10	blk hyg cry	242	286			reac H_2O
2466	Tungsten(V) ethanolate	$W(C_2H_5O)_5$	62571-53-3	409.14	pow		105(0.05 mmHg)			s EtAc
2467	Tungsten(V) fluoride	WF_5	19357-83-6	278.83	yel solid	>80 dec				
2468	Tungsten(V) oxytribromide	$WOBr_3$	20213-56-3	439.55	dark brn tetr cry			≈5.9		
2469	Tungsten(V) oxytrichloride	$WOCl_3$	14249-98-0	306.20	grn tetr cry			≈4.6		
2470	Tungsten(VI) bromide	WBr_6	13701-86-5	663.26	blue-blk cry	309				
2471	Tungsten(VI) chloride	WCl_6	13283-01-7	396.56	purp hex cry; hyg	275	346.75	3.52		s EtOH, os
2472	Tungsten(VI) dioxydibromide	WO_2Br_2	13520-75-7	375.65	red cry		440 subl			
2473	Tungsten(VI) dioxydichloride	WO_2Cl_2	13520-76-8	286.74	yel orth cry	265		4.67		i H_2O
2474	Tungsten(VI) dioxydiiodide	WO_2I_2	14447-89-3	469.65	grn monocl cry	400 dec		6.39		
2475	Tungsten(VI) fluoride	WF_6	7783-82-6	297.83	col gas	2.3	17.1	12.17 g/L		reac H_2O
2476	Tungsten(VI) oxide	WO_3	1314-35-8	231.84	yel powder	1472		7.2		i H_2O; sl acid; s alk
2477	Tungsten(VI) oxytetrabromide	$WOBr_4$	13520-77-9	519.46	red tetr cry	277	327	≈5.5		reac H_2O
2478	Tungsten(VI) oxytetrachloride	$WOCl_4$	13520-78-0	341.65	red hyg cry	211	227.55	11.92		reac H_2O; s bz, CS_2
2479	Tungsten(VI) oxytetrafluoride	WOF_4	13520-79-1	275.83	wh monocl cry	106	185.9	5.07		reac H_2O
2480	Tungsten(VI) sulfide	WS_3	12125-19-8	280.04	brn powder					sl H_2O; s alk
2481	Uranium	U	7440-61-1	238.029	silv-wh orth cry	1135	4131	19.1		
2482	Uranium boride	UB_2	12007-36-2	259.651	refrac solid	2430		12.7		
2483	Uranium boride	UB_4	12007-84-0	281.273	refrac solid	2530		9.32		i H_2O
2484	Uranium carbide	UC	12070-09-6	250.040	gray cub cry	2790				
2485	Uranium carbide	UC_2	12071-33-9	262.050	gray tetr cry	2350	4370	11.3		reac H_2O; sl EtOH
2486	Uranium carbide	U_2C_3	12076-62-9	512.090	gray cub cry	≈1700 dec		12.7		
2487	Uranium nitride	UN	25658-43-9	252.036	gray cub cry	2805		14.3		i H_2O
2488	Uranium nitride	U_2N_3	12033-83-9	518.078	cub cry	dec		11.3		
2489	Uranium(III) bromide	UBr_3	13470-19-4	477.741	red hyg cry	727				s H_2O
2490	Uranium(III) chloride	UCl_3	10025-93-1	344.387	grn hyg cry	837		5.51		vs H_2O; i bz, ctc
2491	Uranium(III) fluoride	UF_3	13775-06-9	295.024	blk hex cry	dec		8.9		i H_2O; s acid
2492	Uranium(III) hydride	UH_3	13598-56-6	241.053	gray-blk cub cry			11.1		
2493	Uranium(III) iodide	UI_3	13775-18-3	618.742	blk hyg cry	766				s H_2O
2494	Uranium(IV) bromide	UBr_4	13470-20-7	557.645	brn hyg cry	519				s H_2O, EtOH
2495	Uranium(IV) chloride	UCl_4	10026-10-5	379.840	grn octahed cry	590	791	4.72		reac H_2O; s EtOH
2496	Uranium(IV) fluoride	UF_4	10049-14-6	314.023	grn monocl cry	1036	1417	6.7	0.01^{25}	s conc acid, alk
2497	Uranium(IV) iodide	UI_4	13470-22-9	745.647	blk hyg cry	506				s H_2O, EtOH
2498	Uranium(IV) oxide	UO_2	1344-57-6	270.028	brn cub cry	2827		10.97		i H_2O, dil acid; s conc acid

2499	Uranium(IV,V) oxide	U_4O_9	12037-15-9	1096.111	cub cry			11.2		
2500	Uranium(V) bromide	UBr_5	13775-16-1	637.549	brn hyg cry					reac H_2O
2501	Uranium(V) chloride	UCl_5	13470-21-8	415.293	brn hyg cry	287				reac H_2O
2502	Uranium(V) fluoride	UF_5	13775-07-0	333.021	pale blue tetr cry; hyg	348		5.81		s H_2O
2503	Uranium(V,VI) oxide	U_3O_8	1344-59-8	842.082	grn-blk orth cry	1300 dec		8.38		
2504	Uranium(VI) chloride	UCl_6	13763-23-0	450.745	green hex cry	177		3.6		
2505	Uranium(VI) fluoride	UF_6	7783-81-5	352.019	wh monocl solid	64.0 tp	56.5 sp	5.09		reac H_2O; s ctc, chl
2506	Uranium(VI) oxide	UO_3	1344-58-7	286.027	oran-yel cry			≈7.3		i H_2O; s acid
2507	Uranium(VI) oxide monohydrate	$UO_3 \cdot H_2O$	12326-21-5	304.043	yel orth cry	570 dec		7.05		
2508	Uranium peroxide dihydrate	$UO_4 \cdot 2H_2O$	19525-15-6	338.057	yel hyg cry	115 dec				i H_2O
2509	Uranyl chloride	UO_2Cl_2	7791-26-6	340.933	yel orth cry; hyg	577				vs H_2O; s EtOH, ace; i bz
2510	Uranyl fluoride	UO_2F_2	13536-84-0	308.025	yel hyg solid				64.4^{20}	i bz
2511	Uranyl nitrate	$UO_2(NO_3)_2$	10102-06-4	394.037	yel cry				127^{25}	s eth
2512	Uranyl nitrate hexahydrate	$UO_2(NO_3)_2 \cdot 6H_2O$	13520-83-7	502.129	yel orth cry; hyg	60	118 dec	2.81	127^{25}	s EtOH, eth
2513	Uranyl sulfate	UO_2SO_4	1314-64-3	366.091	yel cry					
2514	Uranyl sulfate trihydrate	$UO_2SO_4 \cdot 3H_2O$	20910-28-5	420.138	yel cry			3.28	152^{16}	sl EtOH
2515	Vanadium	V	7440-62-2	50.942	gray-wh metal; cub	1910	3407	6.0		i H_2O; s acid
2516	Vanadium boride	VB	12045-27-1	61.753	refrac solid	2250				i H_2O
2517	Vanadium boride	VB_2	12007-37-3	72.564	refrac solid	2450				
2518	Vanadium carbide	VC	12070-10-9	62.953	refrac blk cry; cub	2810		5.77		i H_2O
2519	Vanadium carbide	V_2C	12012-17-8	113.894	hex cry	2167				
2520	Vanadium carbonyl	$V(CO)_6$	20644-87-5	219.002	blue-grn cry; flam	60 dec	subl			
2521	Vanadium nitride	VN	24646-85-3	64.949	blk powder; cub	2050		6.13		i H_2O; s aqua regia
2522	Vanadium silicide	VSi_2	12039-87-1	107.113	metallic prisms			4.42		s HF
2523	Vanadium silicide	V_3Si	12039-76-8	180.911	cub cry	1935		5.70		
2524	Vanadium(II) bromide	VBr_2	14890-41-6	210.750	oran-brn hex cry		800 subl	4.58		reac H_2O
2525	Vanadium(II) chloride	VCl_2	10580-52-6	121.847	grn hex plates	≈1350	910 subl	3.23		reac H_2O; s EtOH, eth
2526	Vanadium(II) fluoride	VF_2	13842-80-3	88.939	blue hyg cry					reac H_2O
2527	Vanadium(II) iodide	VI_2	15513-84-5	304.751	red-viol hex cry		800 subl	5.44		reac H_2O
2528	Vanadium(II) oxide	VO	12035-98-2	66.941	grn cry	1789		5.758		s acid
2529	Vanadium(II) sulfate heptahydrate	$VSO_4 \cdot 7H_2O$	36907-42-3	273.112	viol cry					
2530	Vanadium(III) 2,4-pentanedioate	$V(CH_3COCHCOCH_3)_3$	13476-99-8	348.266	brn cry	≈185	subl	≈1.0		s MeOH, ace, bz chl
2531	Vanadium(III) bromide	VBr_3	13470-26-3	290.654	gray-brn hyg cry			4.00		reac H_2O
2532	Vanadium(III) chloride	VCl_3	7718-98-1	157.300	red-viol hex cry; hyg	500 dec		3.00		reac H_2O; s EtOH, eth
2533	Vanadium(III) fluoride	VF_3	10049-12-4	107.937	yel-grn hex cry	≈1400	subl	3.363		i H_2O, EtOH
2534	Vanadium(III) fluoride trihydrate	$VF_3 \cdot 3H_2O$	10049-12-4*	161.983	grn rhomb cry	≈100 dec				sl H_2O
2535	Vanadium(III) iodide	VI_3	15513-94-7	431.655	brn-blk rhomb cry; hyg			5.21		reac H_2O
2536	Vanadium(III) oxide	V_2O_3	1314-34-7	149.881	blk powder	2067		4.87		i H_2O
2537	Vanadium(III) sulfate	$V_2(SO_4)_3$	13701-70-7	390.074	yel powder	≈400 dec				sl H_2O
2538	Vanadium(III) sulfide	V_2S_3	1315-03-3	198.081	grn-blk powder	dec		4.7		i H_2O; s hot HCl
2539	Vanadium(IV) bromide	VBr_4	13595-30-7	370.558	unstable magenta cry	-23 dec				
2540	Vanadium(IV) chloride	VCl_4	7632-51-1	192.753	unstable red liq	-25.7	148	1.816		reac H_2O; s EtOH, eth
2541	Vanadium(IV) fluoride	VF_4	10049-16-8	126.936	grn hyg powder	325 dec	subl	3.15		vs H_2O
2542	Vanadium(IV) oxide	VO_2	12036-21-4	82.941	blue-blk powder	1967		4.339		i H_2O; s acid, alk
2543	Vanadium(V) fluoride	VF_5	7783-72-4	145.934	col liq	19.5	48.3	2.50		reac H_2O
2544	Vanadium(V) oxide	V_2O_5	1314-62-1	181.880	yel-brn orth cry	670	1800 dec	3.35	0.07^{25}	s conc acid, alk; i EtOH
2545	Vanadyl bromide	VOBr	13520-88-2	146.845	viol cry	480 dec				
2546	Vanadyl chloride	VOCl	13520-87-1	102.394	brn orth cry		127	1.72		

No.	Name	Formula	CAS Reg No.	Mol. Weight	Physical Form	mp/°C	bp/°C	Density g cm⁻³	Solubility g/100 g H_2O	Qualitative Solubility
2547	Vanadyl dibromide	$VOBr_2$	13520-89-3	226.749	yel-brn cry	180 dec				
2548	Vanadyl dichloride	$VOCl_2$	10213-09-9	137.846	grn hyg cry	380 dec		2.88		reac H_2O; s EtOH
2549	Vanadyl difluoride	VOF_2	13814-83-0	104.938	yel cry					
2550	Vanadyl selenite hydrate	$VOSeO_3 \cdot H_2O$	133578-89-9	211.92	grn tricl plates			3.506		
2551	Vanadyl sulfate dihydrate	$VOSO_4 \cdot 2H_2O$	27774-13-6	199.036	blue cry powder					s H_2O
2552	Vanadyl tribromide	$VOBr_3$	13520-90-6	306.653	deep red liq		180 dec			reac H_2O
2553	Vanadyl trichloride	$VOCl_3$	7727-18-6	173.299	fuming red liq	-79	127	1.829		reac H_2O; s MeOH, eth, ace
2554	Vanadyl trifluoride	VOF_3	13709-31-4	123.936	yel hyg powder	300	480	2.459		reac H_2O
2555	Water	H_2O	7732-18-5	18.015	col liq	0.00	100.0	0.9970		s EtOH, MeOH, ace
2556	Xenon	Xe	7440-63-3	131.29	col gas	-111.79 tp (81.6 kPa)	-108.12	5.366 g/L		sl H_2O
2557	Xenon trioxide	XeO_3	13776-58-4	179.29	col orth cry	exp ≈25		4.55		s H_2O
2558	Xenon tetroxide	XeO_4	12340-14-6	195.29	yel solid; exp	-35.9	≈0 dec			
2559	Xenon difluoride	XeF_2	13709-36-9	169.29	col tetr cry	129.03 tp	114.35 sp	4.32		sl H_2O
2560	Xenon tetrafluoride	XeF_4	13709-61-0	207.28	col monocl cry	117.10 tp	115.75 sp	4.04		reac H_2O
2561	Xenon hexafluoride	XeF_6	13693-09-9	245.28	col monocl cry	49.5	75.6	3.56		reac H_2O
2562	Xenon dioxydifluoride	XeO_2F_2	13875-06-4	201.29	col orth cry	30.8 exp		4.10		
2563	Xenon oxytetrafluoride	$XeOF_4$	13774-85-1	223.28	col liq	-46.2		3.17^{0}		reac H_2O
2564	Xenon fluoride hexafluororuthenate	$XeFRuF_6$	22527-13-5	365.35	yel-grn monocl cry	110		3.78		
2565	Xenon fluoride undecafluoroantimonate	$XeFSb_2F_{11}$	15364-10-0	602.79	yel monocl cry	63		3.69		
2566	Xenon fluoride hexafluoroarsenate	$Xe_2F_3AsF_6$	50432-32-1	508.49	yel-grn monocl cry	99		3.62		reac H_2O
2567	Xenon fluoride hexafluoroantimonate	XeF_3SbF_6	39797-63-2	424.04	yel-grn monocl cry	≈110		3.92		
2568	Xenon trifluoride undecafluoroantimonate	$XeF_3Sb_2F_{11}$	35718-37-7	640.79	yel-grn tricl cry	82		3.98		
2569	Xenon pentafluoride hexafluoroarsenate	XeF_5AsF_6	20328-94-3	415.19	wh monocl cry	130.5		3.51		
2570	Xenon pentafluoride hexafluororuthenate	XeF_5RuF_6	39796-98-0	441.34	grn orth cry	152		3.79		
2571	Ytterbium	Yb	7440-64-4	173.04	silv metal; cub	819	1196	6.90		s dil acid
2572	Ytterbium silicide	$YbSi_2$	12039-89-3	229.21	hex cry			7.54		
2573	Ytterbium(II) bromide	$YbBr_2$	25502-05-0	332.85	yel cry	673				reac H_2O
2574	Ytterbium(II) chloride	$YbCl_2$	13874-77-6	243.95	grn cry	721		5.27		reac H_2O
2575	Ytterbium(II) iodide	YbI_2	19357-86-9	426.85	blk cry	772				reac H_2O
2576	Ytterbium(III) chloride	$YbCl_3$	10361-91-8	279.40	wh hyg powder	875				s H_2O
2577	Ytterbium(III) chloride hexahydrate	$YbCl_3 \cdot 6H_2O$	19423-87-1	387.49	grn hyg cry	150 dec		2.57		vs H_2O
2578	Ytterbium(III) fluoride	YbF_3	13760-80-0	230.04	wh cry	1157		8.2		i H_2O
2579	Ytterbium(III) nitrate	$Yb(NO_3)_3$	13768-67-7	359.06	col hyg solid				239^{25}	s EtOH
2580	Ytterbium(III) oxide	Yb_2O_3	1314-37-0	394.08	col cub cry	2355	4070	9.2		s dil acid
2581	Ytterbium(III) sulfate octahydrate	$Yb_2(SO_4)_3 \cdot 8H_2O$	10034-98-7	778.39	col cry			3.3	38.4^{20}	
2582	Yttrium	Y	7440-65-5	88.906	silv metal; hex	1522	3345	4.47		reac H_2O; s dil acid
2583	Yttrium aluminum oxide	$Y_3Al_5O_{12}$	12005-21-9	593.619	grn cub cry			≈4.5		
2584	Yttrium antimonide	YSb	12186-97-9	210.666	cub cry	2310		5.97		
2585	Yttrium arsenide	YAs	12255-48-0	163.828	cub cry			5.59		
2586	Yttrium boride	YB_6	12008-32-1	153.772	refrac solid	2600		3.72		
2587	Yttrium bromide	YBr_3	13469-98-2	328.618	col hyg cry	904			83.3^{30}	
2588	Yttrium carbide	YC_2	12071-35-1	112.927	refrac solid	≈2400		4.13		
2589	Yttrium carbonate trihydrate	$Y_2(CO_3)_3 \cdot 3H_2O$	5970-44-5	411.885	red-brn powder					i H_2O; s dil acid
2590	Yttrium chloride	YCl_3	10361-92-9	195.264	wh monocl cry; hyg	721		2.61	75.1^{20}	
2591	Yttrium fluoride	YF_3	13709-49-4	145.901	wh hyg powder	≈1150		4.0		i H_2O

2592	Yttrium nitrate	$Y(NO_3)_3$	10361-93-0	274.921	wh hyg solid				149^{25}	s EtOH
2593	Yttrium nitrate tetrahydrate	$Y(NO_3)_3 \cdot 4H_2O$	13773-69-8	346.982	red-wh prisms			2.68	149^{25}	
2594	Yttrium nitrate hexahydrate	$Y(NO_3)_3 \cdot 6H_2O$	13494-98-9	383.012	hyg cry				149^{25}	
2595	Yttrium oxide	Y_2O_3	1314-36-9	225.810	wh cry; cub	2438		5.03		s dil acid
2596	Yttrium phosphide	YP	12294-01-8	119.880	cub cry			≈4.4		
2597	Yttrium sulfate octahydrate	$Y_2(SO_4)_3 \cdot 8H_2O$	7446-33-5	610.125	red monocl cry			2.6	7.47^{16}	
2598	Yttrium sulfide	Y_2S_3	12039-19-9	274.010	yel cub cry	1925		3.87		
2599	Zinc	Zn	7440-66-6	65.39	blue-wh metal; hex	419.53	907	7.14		s acid, alk
2600	Zinc acetate dihydrate	$Zn(C_2H_3O_2)_2 \cdot 2H_2O$	5970-45-6	219.51	wh powder	237 dec		1.735	30.0^{20}	s EtOH
2601	Zinc ammonium sulfate	$Zn(NH_4)_2(SO_4)_2$	7783-24-6	293.59	wh cry				9.2^{20}	
2602	Zinc antimonide	ZnSb	12039-35-9	187.15	silv-wh orth cry	565		6.33		reac H_2O
2603	Zinc arsenate	$Zn_3(AsO_4)_2$	13464-44-3	474.01	wh powder				0.000078^{20}	s acid, alk
2604	Zinc arsenate octahydrate	$Zn_3(AsO_4)_2 \cdot 8H_2O$	13464-45-4	618.13	wh monocl cry			3.33	0.000078^{20}	s acid, alk
2605	Zinc arsenide	Zn_3As_2	12006-40-5	346.01	pow	1015		5.528		
2606	Zinc arsenite	$Zn(AsO_2)_2$	10326-24-6	279.23	col powder					i H_2O; s acid
2607	Zinc borate	$3ZnO \cdot 2B_2O_3$	27043-84-1	383.41	wh amorp powder			3.64		sl H_2O; s dil acid
2608	Zinc borate hemiheptahydrate	$2ZnO \cdot 3B_2O_3 \cdot 3.5H_2O$	12513-27-8	434.69	wh cry	980		4.22		i H_2O
2609	Zinc borate pentahydrate	$2ZnO \cdot 3B_2O_3 \cdot 5H_2O$	12536-65-1	461.72	wh pow			3.64	0.007^{25}	sl HCl
2610	Zinc bromate hexahydrate	$Zn(BrO_3)_2 \cdot 6H_2O$	13517-27-6	429.29	wh hyg solid	100		2.57		vs H_2O
2611	Zinc bromide	$ZnBr_2$	7699-45-8	225.20	wh hex cry; hyg	394	697	4.5	488^{25}	vs EtOH; s eth
2612	Zinc caprylate	$Zn(C_8H_{15}O_2)_2$	557-09-5	351.80	wh hyg cry	136				sl H_2O
2613	Zinc carbonate	$ZnCO_3$	3486-35-9	125.40	wh rhomb cry	140 dec		4.4	0.000091^{20}	s dil acid, alk
2614	Zinc carbonate hydroxide	$3Zn(OH)_2 \cdot 2ZnCO_3$	12070-69-8	549.01	wh pow					
2615	Zinc chlorate	$Zn(ClO_3)_2$	10361-95-2	232.29	yel hyg cry	60 dec		2.15	200^{20}	
2616	Zinc chloride	$ZnCl_2$	7646-85-7	136.29	wh hyg cry	290	732	2.907	408^{25}	s EtOH, ace
2617	Zinc chromate	$ZnCrO_4$	13530-65-9	181.38	yel prisms	316		3.40	3.08	s acid; i ace
2618	Zinc chromite	$ZnCr_2O_4$	12018-19-8	233.38	grn cub cry			5.29		
2619	Zinc citrate dihydrate	$Zn_3(C_6H_5O_7)_2 \cdot 2H_2O$	546-46-3	610.40	col powder					sl H_2O; s dil acid, alk
2620	Zinc cyanide	$Zn(CN)_2$	557-21-1	117.42	wh powder			1.852	0.00047^{20}	reac acid
2621	Zinc dithionate	ZnS_2O_4	7779-86-4	193.52	wh amorp solid	200 dec			40^{20}	
2622	Zinc fluoride	ZnF_2	7783-49-5	103.39	wh tetr needles; hyg	872	1500	4.9	1.55^{25}	
2623	Zinc fluoride tetrahydrate	$ZnF_2 \cdot 4H_2O$	13986-18-0	175.45	wh orth cry			2.30	1.55^{25}	
2624	Zinc fluoroborate hexahydrate	$Zn(BF_4)_2 \cdot 6H_2O$	27860-83-9	347.09	hex cry			2.12		vs H_2O; s EtOH
2625	Zinc formate dihydrate	$Zn(CHO_2)_2 \cdot 2H_2O$	5970-62-7	191.46	wh cry			2.207	5.2^{20}	i EtOH
2626	Zinc hexafluorosilicate hexahydrate	$ZnSiF_6 \cdot 6H_2O$	16871-71-9	315.56	wh cry					s H_2O
2627	Zinc hydroxide	$Zn(OH)_2$	20427-58-1	99.41	col orth cry	125 dec		3.05	0.000042^{20}	
2628	Zinc iodate	$Zn(IO_3)_2$	7790-37-6	415.20	wh cry powder				0.64^{25}	
2629	Zinc iodide	ZnI_2	10139-47-6	319.20	wh hyg cry	446	625	4.74	438^{25}	s EtOH, eth
2630	Zinc laurate	$Zn(C_{12}H_{23}O_2)_2$	2452-01-9	464.01	wh powder	128				sl H_2O
2631	Zinc molybdate	$ZnMoO_4$	13767-32-3	225.33	wh tetr cry	>700		4.3		i H_2O
2632	Zinc nitrate	$Zn(NO_3)_2$	7779-88-6	189.40	wh powder				120^{25}	
2633	Zinc nitrate hexahydrate	$Zn(NO_3)_2 \cdot 6H_2O$	10196-18-6	297.49	col orth cry	36 dec		2.067	120^{25}	vs EtOH
2634	Zinc nitride	Zn_3N_2	1313-49-1	224.18	blue-gray cub cry	700 dec		6.22		i H_2O
2635	Zinc nitrite	$Zn(NO_2)_2$	10102-02-0	157.40	hyg solid					reac H_2O
2636	Zinc oleate	$Zn(C_{18}H_{33}O_2)_2$	557-07-3	628.30	wh powder	70 dec				i H_2O; s EtOH, eth, bz
2637	Zinc oxalate	ZnC_2O_4	547-68-2	153.41	wh pwd				0.0026^{25}	
2638	Zinc oxalate dihydrate	$ZnC_2O_4 \cdot 2H_2O$	4255-07-6	189.44	wh powder	100 dec		2.56	0.0026^{25}	s dil acid
2639	Zinc oxide	ZnO	1314-13-2	81.39	wh powder; hex	1974		5.6		i H_2O; s dil acid

No.	Name	Formula	CAS Reg No.	Mol. Weight	Physical Form	mp/°C	bp/°C	Density /g cm^{-3}	Solubility g/100 g H_2O	Qualitative Solubility
2640	Zinc 2,4-pentanedioate	$Zn(CH_3COCHCOCH_3)_2$	14024-63-6	263.61	cry	137 dec				sl H_2O; s EtOH
2641	Zinc perchlorate hexahydrate	$Zn(ClO_4)_2 \cdot 6H_2O$	10025-64-6	372.38	wh cub cry; hyg	106 dec		2.2	121.3^{25}	s EtOH
2642	Zinc permanganate hexahydrate	$Zn(MnO_4)_2 \cdot 6H_2O$	23414-72-4	411.35	blk orth cry; hyg			2.45		s H_2O; reac EtOH
2643	Zinc peroxide	ZnO_2	1314-22-3	97.39	yel-wh powder	>150 dec	212 exp	1.57		i H_2O; reac acid, EtOH, ace
2644	Zinc phosphate	$Zn_3(PO_4)_2$	7779-90-0	386.11	wh monocl cry	900		4.0		i H_2O
2645	Zinc phosphate tetrahydrate	$Zn_3(PO_4)_2 \cdot 4H_2O$	7543-51-3	458.17	col orth cry			3.04		i H_2O, EtOH; s dil acid, alk
2646	Zinc phosphide	Zn_3P_2	1314-84-7	258.12	gray tetr cry	1160		4.55		i H_2O, EtOH; reac acid; s bz
2647	Zinc pyrophosphate	$Zn_2P_2O_7$	7446-26-6	304.72	wh cry powder			3.75		i H_2O; s dil acid
2648	Zinc selenate pentahydrate	$ZnSeO_4 \cdot 5H_2O$	13597-54-1	298.42	tricl cry	50 dec		2.59	63.4^{25}	
2649	Zinc selenide	ZnSe	1315-09-9	144.35	yel-red cub cry	>1100	subl	5.65		i H_2O; s dil acid
2650	Zinc orthosilicate	Zn_2SiO_4	13597-65-4	222.86	wh hex cry	1509		4.1		i H_2O, dil acid
2651	Zinc selenite	$ZnSeO_3$	13597-46-1	192.35	wh pow					
2652	Zinc stearate	$Zn(C_{18}H_{35}O_2)_2$	557-05-1	632.33	wh powder	130		1.095		i H_2O, EtOH, eth; s bz
2653	Zinc sulfate	$ZnSO_4$	7733-02-0	161.45	col orth cry	680 dec		3.8	57.7^{25}	
2654	Zinc sulfate monohydrate	$ZnSO_4 \cdot H_2O$	7446-19-7	179.47	wh monocl cry	238 dec		3.20	57.7^{25}	i EtOH
2655	Zinc sulfate heptahydrate	$ZnSO_4 \cdot 7H_2O$	7446-20-0	287.56	col orth cry	100 dec		1.97	57.7^{25}	i EtOH
2656	Zinc sulfide (sphalerite)	ZnS	1314-98-3	97.46	gray-wh cub cry	1700		4.04		i H_2O, EtOH; s dil acid
2657	Zinc sulfide (wurtzite)	ZnS	1314-98-3	97.46	wh hex cry	1700		4.09		i H_2O; s dil acid
2658	Zinc sulfite dihydrate	$ZnSO_3 \cdot 2H_2O$	7488-52-0	181.49	wh powder	200 dec			0.224^{25}	i EtOH
2659	Zinc telluride	ZnTe	1315-11-3	192.99	red cub cry	1239		5.9		i H_2O
2660	Zinc thiocyanate	$Zn(SCN)_2$	557-42-6	181.56	wh hyg cry					sl H_2O; s EtOH
2661	Zirconium	Zr	7440-67-7	91.224	gray-wh metal; hex	1855	4409	6.52		s hot conc acid
2662	Zirconium boride	ZrB_2	12045-64-6	112.846	gray refrac solid; hex	3245		6.17		
2663	Zirconium carbide	ZrC	12020-14-3	103.235	gray refrac solid; cub	3532		6.73		s HF
2664	Zirconium(II) chloride	$ZrCl_2$	13762-26-0	162.129	blk cry	772 dec		3.16		reac H_2O
2665	Zirconium(II) hydride	ZrH_2	7704-99-6	93.240	gray tetr cry	800 dec		5.6		i H_2O
2666	Zirconium(IV) bromide	$ZrBr_4$	13777-25-8	410.840	wh cub cry	450 tp	360 sp	3.98		
2667	Zirconium(IV) chloride	$ZrCl_4$	10026-11-6	233.035	wh monocl cry; hyg	437 tp	331 sp	2.80		reac H_2O; s EtOH, eth
2668	Zirconium(IV) fluoride	ZrF_4	7783-64-4	167.218	wh monocl cry	932 tp	912 sp	4.43	1.5^{25}	
2669	Zirconium(IV) hydroxide	$Zr(OH)_4$	14475-63-9	159.254	wh amorp powder	dec		3.25		i H_2O; s acid
2670	Zirconium(IV) iodide	ZrI_4	13986-26-0	598.842	oran cub cry; hyg	499 tp	431 sp	4.85		vs H_2O
2671	Zirconium(IV) nitrate pentahydrate	$Zr(NO_3)_4 \cdot 5H_2O$	13746-89-9	429.320	wh hyg cry	100 dec				vs H_2O; s EtOH
2672	Zirconium(IV) oxide	ZrO_2	1314-23-4	123.223	wh amorp powder	2709		5.68		i H_2O; sl acid
2673	Zirconium(IV) orthosilicate	$ZrSiO_4$	10101-52-7	183.308	wh tetr cry	1540 dec		4.6		i H_2O, acid
2674	Zirconium(IV) sulfate	$Zr(SO_4)_2$	14644-61-2	283.351	wh hyg cry	410 dec		3.22		s H_2O; sl EtOH
2675	Zirconium(IV) sulfate tetrahydrate	$Zr(SO_4)_2 \cdot 4H_2O$	7446-31-3	355.413	wh tetr cry	100 dec		2.80		vs H_2O
2676	Zirconium(IV) sulfide	ZrS_2	12039-15-5	155.356	red-brn hex cry	1480		3.82		i H_2O
2677	Zirconium nitride	ZrN	25658-42-8	105.231	yel cub cry	2960		7.09		s conc HF; sl dil acid
2678	Zirconium phosphide	ZrP_2	12037-80-8	153.172	orth cry			≈5.1		
2679	Zirconium silicide	$ZrSi_2$	12039-90-6	147.395	gray powder	1620		4.88		i H_2O, aqua regia; s HF
2680	Zirconyl chloride	$ZrOCl_2$	7699-43-6	178.128	wh solid	250 dec				s H_2O, EtOH
2681	Zirconyl chloride octahydrate	$ZrOCl_2 \cdot 8H_2O$	13520-92-8	322.251	tetr cry	400 dec		1.91		vs H_2O, EtOH

FORMULA INDEX OF INORGANIC COMPOUNDS

Ac	1
$AcBr_3$	2
$AcCl_3$	3
AcF_3	4
AcI_3	5
Ac_2O_3	6
Ag	2024
$AgAsF_6$	2044
$AgBr$	2032
$AgBrO_3$	2031
$AgCN$	2039
AgC_2H	2030
$AgC_2H_3O_2$	2027
$AgC_3H_5O_3 \cdot H_2O$	2049
$Ag(C_2H_5)_2NCS_2$	2041
$AgC_6H_2N_3O_7 \cdot H_2O$	2060
$AgCl$	2035
$AgClO_2$	2036
$AgClO_3$	2034
$AgClO_4$	2056
$AgClO_4 \cdot H_2O$	2057
AgF	2042
AgF_2	2073
AgF_2H	2046
$AgHF_2$	2046
AgI	2048
$AgIO_3$	2047
$AgMnO_4$	2058
$AgNO_2$	2053
$AgNO_3$	2052
AgN_3	2025
AgO	2071
$AgPF_6$	2045
$AgPO_3$	2050
$AgSCN$	2069
$AgSbF_6$	2043
Ag_2CO_3	2033
Ag_2C_2	2028
$Ag_2C_2O_4$	2054
Ag_2CrO_4	2037
$Ag_2Cr_2O_7$	2040
Ag_2F	2026
Ag_2HgI_4	2068
Ag_2MoO_4	2051
Ag_2O	2055
Ag_2O_2	2074
Ag_2S	2065
Ag_2SO_3	2066
Ag_2SO_4	2064
$Ag_2S_2O_3$	2070
Ag_2Se	2062
Ag_2SeO_3	2063
Ag_2SeO_4	2061
Ag_2Te	2067
Ag_2WO_4	2072
Ag_3AsO_4	2029
$Ag_3C_6H_5O_7$	2038
Ag_3PO_4	2059
Al	7
$AlAs$	11
AlB_2	13
$Al(BH_4)_3$	14
$AlBr_3$	16
$AlBr_3 \cdot 6H_2O$	17
$Al(BrO_3)_3 \cdot 9H_2O$	15
$Al(C_2H_5O)_3$	23
$Al(C_3H_5O_3)_3$	34
$Al(C_{15}H_{31}COO)_3$	43
$Al(C_{18}H_{33}O_2)_3$	38
$Al(C_{18}H_{35}O_2)_3$	49
$AlCl_3$	20
$AlCl_3 \cdot 6H_2O$	21
$Al(ClO_3)_3 \cdot 9H_2O$	19
$Al(ClO_4)_3 \cdot 9H_2O$	44
AlF_3	24
$AlF_3 \cdot H_2O$	25
$AlF_3 \cdot 3H_2O$	26
AlH_3	28
$Al(H_2PO_2)_3$	31
AlI_3	32
$AlI_3 \cdot 6H_2O$	33
AlN	37
$AlNH_4(SO_4)_2$	8
$AlNH_4(SO_4)_2 \cdot 12H_2O$	9
$Al(NO_3)_3$	35
$Al(NO_3)_3 \cdot 9H_2O$	36
$Al(OH)(C_2H_3O_2)_2$	22
$AlO(OH)$	42
$Al(OH)_3$	29
AlP	45
$AlPO_4$	39
$Al(PO_3)_3$	40
$Al(SCN)_3$	54
$AlSb$	10
Al_2O_3	41
$2Al_2O_3 \cdot B_2O_3$	12
$Al_2O_3 \cdot 2SiO_2 \cdot 2H_2O$	48
$Al_2(OH)_5Cl \cdot 2H_2O$	30
Al_2S_3	52
$Al_2(SO_4)_3$	50
$Al_2(SO_4)_3 \cdot 18H_2O$	51
Al_2Se_3	46
Al_2SiO_5	47
$Al_2(SiF_6)_3 \cdot 9H_2O$	27
Al_2Te_3	53
Al_3Ta	2281
Al_4C_3	18
Am	55
$AmBr_3$	57
$AmCl_3$	58
AmF_3	59
AmF_4	61
AmI_3	60
AmO_2	62
Am_2O_3	56
Ar	194
As	195
AsB	383
$AsBr_3$	204
$As(OC_2H_5)_3$	218
$AsCl_3$	205
$AsCl_5$	213
$AsCo$	667
$AsCr_2$	613
AsF_3	206
AsF_5	214
AsF_9Xe_2	2566
$AsF_{11}Xe$	2569
$AsFe$	1081
$AsGa$	900
AsH_3	196
AsH_3O_3	200
AsH_3O_4	198
AsI_3	207
$AsIn$	1028
$AsNi$	1509
$AsSb$	176
AsY	2585
As_2Cd_3	409
As_2Co	668
As_2H_4	197
As_2I_4	201
As_2O_3	209
As_2O_3	208
As_2O_5	215
As_2S_3	211
As_2S_5	217
As_2Se	202
As_2Se_3	210
As_2Se_5	216
As_2Te_3	212
As_2Zn_3	2605
As_3Co	669
As_4S_4	203
At	219
Au	958
$AuBr$	961
$AuBr_3$	966
$AuCN$	963
$Au(CN)_3 \cdot 3H_2O$	968
$AuCl$	962
$AuCl_3$	967
AuF_3	969
AuI	964
AuI_3	971
$Au(OH)_3$	970
Au_2O_3	972
Au_2S	965
Au_2S_3	975
Au_2Se_3	974
$Au_2(SeO_4)_3$	973
B	367
BAs	383
BBr_3	384
BCl_3	386
BCo	671
BCo_2	670
BCr	614
BF_3	388
BF_4H	382
BFe	1082
BFe_2	1083
BHO_2	379
BHO_2	380
BHO_2	381
BH_3CO	376
BH_3O_3	378
BH_4K	1731
BH_4Li	1256
BH_4Na	2085
BI_3	390
BMn	1358
BMn_2	1360
BMo_2	1453
BN	391
BNb	1545
BNi	1511
BNi_2	1512
BNi_3	1510
BP	393
BTa	2282
BV	2516
BW	2442
BW_2	2441
B_2Be	313
B_2BeH_6	314
B_2Cl_4	387
B_2Cr	615
B_2F_4	389
B_2H_6	368
B_2Hf	977
B_2Mg	1303
B_2Mn	1359
B_2Nb	1546
B_2O_3	392
B_2S_3	394
B_2Sc	1942
B_2Ta	2283
B_2Ti	2414
B_2U	2482
B_2V	2517
B_2Zr	2662
B_3Cr_5	616
$B_3H_6N_3$	377
$B_3N_3H_6$	377
B_4C	385
B_4Dy	824
B_4Er	839
B_4H_{10}	369
B_4Lu	1288
B_4U	2483
B_5H_9	370
B_5H_{11}	371
B_5Mo_2	1454
B_5W_2	2443
B_6Ba	247
B_6Ca	453
B_6Ce	545
B_6Eu	855
B_6Gd	882
B_6H_{10}	372
B_6H_{12}	373
B_6La	1157
B_6Nd	1488
B_6Pr	1821
B_6Sm	1924
B_6Sr	2225
B_6Th	2358
B_6Y	2586
B_9H_{15}	374
$B_{10}H_{14}$	375
Ba	220

Formula	No.
$BaAl_2O_4$	223
$Ba(BO_2)_2 \cdot H_2O$	262
BaB_6	247
$BaBr_2$	226
$BaBr_2 \cdot 2H_2O$	227
$Ba(BrO_3)_2 \cdot H_2O$	225
$BaCO_3$	229
BaC_2	228
$Ba(CHO_2)_2$	246
$Ba(CN)_2$	241
BaC_2O_4	269
$BaC_2O_4 \cdot H_2O$	270
$BaC_4H_4O_6$	292
$Ba(C_2H_3O_2)_2$	221
$Ba(C_2H_3O_2)_2 \cdot H_2O$	222
$Ba(C_{18}H_{35}O_2)_2$	288
$BaCl_2$	232
$BaCl_2 \cdot 2H_2O$	233
$Ba(ClO_3)_2$	230
$Ba(ClO_3)_2 \cdot H_2O$	231
$Ba(ClO_4)_2$	272
$Ba(ClO_4)_2 \cdot 3H_2O$	273
$BaCrO_4$	235
$BaCr_2O_7 \cdot 2H_2O$	242
$BaCuY_2O_5$	237
BaF_2	245
$BaHPO_4$	250
BaH_2	249
$Ba(HS)_2$	251
$Ba(HS)_2 \cdot 4H_2O$	252
$Ba(H_2PO_2)_2 \cdot H_2O$	256
$BaHgI_4$	294
BaI_2	259
$BaI_2 \cdot 2H_2O$	260
$Ba(IO_3)_2$	257
$Ba(IO_3)_2 \cdot H_2O$	258
$BaK_2(CrO_4)_2$	277
$BaMnO_4$	261
$Ba(MnO_4)_2$	274
$BaMoO_4$	263
$Ba(NO_2)_2$	267
$Ba(NO_2)_2 \cdot H_2O$	268
$Ba(NO_3)_2$	265
BaN_6	224
$Ba(N_3)_2$	224
$Ba(NbO_3)_2$	264
BaO	271
BaO_2	275
$Ba(OH)_2$	253
$Ba(OH)_2 \cdot H_2O$	254
$Ba(OH)_2 \cdot 8H_2O$	255
$Ba(PO_3)_2$	276
$BaPt(CN)_4 \cdot 4H_2O$	293
BaS	290
$BaSO_3$	291
$BaSO_4$	289
$Ba(SCN)_2$	295
$Ba(SCN)_2 \cdot 2H_2O$	296
$Ba(SCN)_2 \cdot 3H_2O$	297
BaS_2O_3	298
$BaS_2O_3 \cdot H_2O$	299
$BaS_2O_6 \cdot 2H_2O$	243
$BaSe$	280
$BaSeO_3$	281
$BaSeO_4$	279
$BaSiF_6$	248
$BaSiO_3$	283
$BaSi_2$	284
$BaSi_2O_5$	282
$BaSnO_3$	286
$BaSnO_3 \cdot 3H_2O$	287
$BaTiO_3$	300
BaU_2O_7	302
$BaWO_4$	301
$BaZrO_3$	304
$Ba_2Cu_3YO_7$	238
$Ba_2Cu_4YO_8$	239
$Ba_2Fe(CN)_6 \cdot 6H_2O$	244
$Ba_2Na(NbO_3)_5$	285
$Ba_2P_2O_7$	278
$Ba_3(C_6H_5O_7)_2 \cdot H_2O$	236
$Ba_3(CrO_4)_2$	234
Ba_3N_2	266
$Ba_3(VO_4)_2$	303
$Ba_4Cu_7Y_2O_{15}$	240
Be	307
$BeAl_2O_4$	310
BeB_2	313
$Be(BH_4)_2$	314
$BeBr_2$	315
$BeCO_3 \cdot 4H_2O$	317
$Be(CHO_2)_2$	321
$Be(C_2H_3O_2)_2$	308
$Be(CH_3COCHCOCH_3)_2$	309
$BeCl_2$	319
$Be(ClO_4)_2 \cdot 4H_2O$	329
BeF_2	320
$BeHPO_4$	323
BeH_2	322
BeI_2	325
$Be(NO_3)_2 \cdot 3H_2O$	326
BeO	328
$Be(OH)_2$	324
BeS	333
$BeSO_4$	331
$BeSO_4 \cdot 4H_2O$	332
$BeSeO_4 \cdot 4H_2O$	330
Be_2C	316
$Be_3Al_2(SiO_3)_6$	311
Be_3N_2	327
$Be_3(OH)_2(CO_3)_2$	318
$Be_4O(C_2H_3O_2)_6$	312
Bi	334
$BiAsO_4$	335
$BiBr_3$	337
$BiC_6H_5O_7$	339
$Bi(CF_3COCHCOCF_3)_3$	345
$Bi(C_{18}H_{33}O_2)_3$	348
$BiCl_3$	338
BiF_3	340
BiF_5	341
BiH_3	342
BiI_3	344
$Bi(NO_3)_3 \cdot 5H_2O$	347
$BiOBr$	351
$BiOCl$	352
$BiOI$	353
$BiONO_3$	354
$Bi(OH)_3$	343
$BiPO_4$	355
$BiVO_4$	366
$Bi_2(C_2O_4)_3$	349
$Bi_2(MoO_4)_3$	346
$(BiO)_2CO_3$	336
Bi_2O_3	350
Bi_2O_4	363
Bi_2S_3	361
$Bi_2(SO_4)_3$	360
Bi_2Se_3	357
$Bi_2(SnO_3)_3 \cdot 5H_2O$	358
Bi_2Te_3	362
$Bi_2(WO_4)_3$	365
$Bi_4(TiO_4)_3$	364
$Bi_5O(OH)_9(NO_3)_4$	359
Bk	305
Bk	306
$BrCN$	540
$BrCl$	400
$BrClFOP$	1673
$BrCl_2OP$	1671
$BrCl_3Si$	1988
$BrCs$	573
$BrCu$	746
BrF	401
$BrFO_2$	404
BrF_2OP	1672
BrF_2P	1645
BrF_2PS	1681
BrF_3	402
BrF_5	403
BrF_5S	2269
$BrGeH_3$	930
BrH	1015
$BrHO_3$	396
BrH_3Si	1987
BrI	1056
$BrIn$	1031
BrK	1733
$BrLi$	1257
$BrNO$	1592
BrN_3	399
$BrNa$	2087
$BrOV$	2545
BrO_2	398
BrO_2F	404
$BrRb$	1888
$BrTl$	2325
Br_2	395
Br_2Ca	454
Br_2Cd	411
Br_2ClOP	1669
Br_2Cl_2Si	1991
Br_2Co	682
Br_2Cr	626
Br_2Cu	770
Br_2Eu	858
Br_2FOP	1670
Br_2FP	1644
Br_2FPS	1680
Br_2F_3P	1657
Br_2Fe	1094
Br_2Ge	942
Br_2GeH_2	933
Br_2H_2Si	1992
Br_2Hg	1425
Br_2Hg_2	1407
Br_2In	1034
Br_2Mg	1305
Br_2Mn	1367
Br_2Mo	1462
Br_2Ni	1519
Br_2O	397
Br_2OS	2276
Br_2OSe	1964
Br_2OV	2547
Br_2O_2W	2472
Br_2Pb	1186
Br_2Pd	1614
Br_2Pt	1688
Br_2Ra	1838
Br_2S_2	2262
Br_2Se_2	1958
Br_2Sm	1926
Br_2Sn	2387
Br_2Sr	2212
Br_2Te	2305
Br_2Ti	2419
Br_2V	2524
Br_2W	2451
Br_2Yb	2573
Br_2Zn	2611
Br_3Ce	552
Br_3ClSi	2004
Br_3Cr	636
Br_3Dy	829
Br_3Er	840
Br_3Eu	866
Br_3Fe	1129
Br_3Ga	909
Br_3Gd	887
Br_3GeH	938
Br_3HSi	2003
Br_3Ho	995
Br_3In	1037
Br_3Ir	1065
Br_3La	1158
Br_3Lu	1289
Br_3Mo	1465
Br_3N	1581
Br_3Nb	1553
Br_3NbO	1568
Br_3Nd	1489
Br_3OP	1668
Br_3OV	2552
Br_3OW	2468
Br_3Os	1602
Br_3P	1643
Br_3PS	1679
Br_3Pr	1822
Br_3Pt	1693
Br_3Pu	1710
Br_3Re	1847

Br_3Ru	1914
Br_3Sb	177
Br_3Sc	1943
Br_3Sm	1930
Br_3Ti	2424
Br_3Tm	2374
Br_3U	2489
Br_3V	2531
Br_3W	2454
Br_3Y	2587
Br_4FP	1656
Br_4Ge	950
Br_4Hf	978
Br_4Mo	1470
Br_4OW	2477
Br_4Pb	1244
Br_4Pt	1695
Br_4Se	1960
Br_4Si	1999
Br_4Sn	2402
Br_4Te	2307
Br_4Th	2359
Br_4Ti	2431
Br_4U	2494
Br_4V	2539
Br_4W	2456
Br_4Zr	2666
Br_5Nb	1563
Br_5P	1655
Br_5Re	1855
Br_5Ta	2292
Br_5U	2500
Br_5W	2464
Br_6W	2470
C	521
C	522
CB_4	385
CBe_2	316
CBrN	540
CBr_2O	536
$CCaN_2$	466
CClN	541
CFN	542
CF_2O	538
CFe_3	1084
CHN	1018
CH_3SnH_3	2385
CH_6Ge	941
CH_6Sn	2385
CHf	979
CIN	543
CMn_3	1361
CMo	1455
CMo_2	1456
CNb	1547
CNb_2	1548
CO	526
$COBr_2$	536
$COCl_2$	537
COF_2	538
COS	531
COSe	530
CO_2	527
CSSe	532
CSTe	533
CS_2	529
CSe_2	528
CSi	2014
CTa	2284
CTa_2	2285
CTh	2360
CTi	2415
CU	2484
CV	2518
CV_2	2519
CW	2445
CW_2	2444
CZr	2663
C_2Ag_2	2028
C_2Ba	228
C_2Ca	456
C_2Ce	546
C_2Cr_3	617
C_2Cu	764
C_2Cu_2	744
C_2La	1159
C_2N_2	539
C_2Sr	2214
C_2Th	2361
C_2U	2485
C_2Y	2588
C_3Al_4	18
C_3Ce_2	554
C_3O_2	534
C_3S_2	535
C_3U_2	2486
C_4NiO_4	1513
C_5FeO_5	1077
C_6CrO_6	618
C_6O_6V	2520
C_6O_6W	2446
$C_8Co_2O_8$	672
$C_9Fe_2O_9$	1078
C_{60}	523
$C_{60}F_{60}$	525
C_{70}	524
Ca	446
$Ca(AlH_4)_2$	514
$CaAl_2O_4$	449
$CaAsO_3$	452
$Ca(BO_2)_2$	483
CaB_6	453
$CaBr_2$	454
$CaBr_2 \cdot 6H_2O$	455
$CaCN_2$	466
$CaCO_3$	457
$CaCO_3$	458
CaC_2	456
$Ca(CHO_2)_2$	471
$Ca(CN)_2$	467
CaC_2O_4	489
$CaC_2O_4 \cdot H_2O$	490
$Ca(C_2H_3O_2)_2$	447
$Ca(C_2H_3O_2)_2 \cdot H_2O$	448
$Ca(C_3H_5O_2)_2$	500
$Ca(CF_3COCHCOCF_3)_2$	472
$Ca(CH_3COCHCOCH_3)_2$	493
$Ca(C_8H_{15}O_2)_2$	469
$Ca(C_{18}H_{35}O_2)_2$	507
$CaCl_2$	461
$CaCl_2 \cdot H_2O$	464
$CaCl_2 \cdot 2H_2O$	462
$CaCl_2 \cdot 6H_2O$	463
$Ca(ClO_3)_2$	459
$Ca(ClO_3)_2 \cdot 2H_2O$	460
$Ca(ClO_4)_2$	494
$CaCrO_4 \cdot 2H_2O$	465
$CaCr_2O_7 \cdot 3H_2O$	468
CaF_2	470
$CaHPO_4$	475
$CaHPO_4 \cdot 2H_2O$	476
CaH_2	474
$Ca(H_2PO_2)_2$	479
$Ca(H_2PO_4)_2 \cdot H_2O$	498
CaI_2	481
$CaI_2 \cdot 6H_2O$	482
$Ca(IO_3)_2$	480
$Ca(MnO_4)_2$	495
$CaMoO_4$	484
$Ca(NO_2)_2$	488
$Ca(NO_3)_2$	485
$Ca(NO_3)_2 \cdot 4H_2O$	486
CaO	491
CaO_2	496
$Ca(OCl)_2$	478
$Ca(OH)_2$	477
CaS	511
$CaSO_3 \cdot 2H_2O$	512
$CaSO_4$	508
$CaSO_4 \cdot 0.5H_2O$	510
$CaSO_4 \cdot 2H_2O$	509
$Ca(SCN)_2 \cdot 4H_2O$	515
$CaS_2O_3 \cdot 6H_2O$	516
CaSe	503
$CaSeO_4 \cdot 2H_2O$	502
CaSi	506
$CaSiF_6 \cdot 2H_2O$	473
$CaSiO_3$	504
$CaSi_2$	505
CaTe	513
$CaTiO_3$	517
$CaWO_4$	518
$CaZrO_3$	519
$Ca_2P_2O_7$	501
$Ca_3Al_2O_6$	450
$Ca_3(AsO_4)_2$	451
Ca_3N_2	487
Ca_3OSiO_4	492
Ca_3P_2	499
$Ca_3(PO_4)_2$	497
Cd	405
$Cd(BF_4)_2$	443
$CdBr_2$	411
$CdBr_2 \cdot 4H_2O$	412
$CdCO_3$	413
$Cd(CN)_2$	419
CdC_2O_4	430
$CdC_2O_4 \cdot 3H_2O$	431
$Cd(C_2H_3O_2)_2$	406
$Cd(C_2H_3O_2)_2 \cdot 2H_2O$	407
$Cd(C_8H_{15}O_2)_2$	420
$CdCl_2$	415
$CdCl_2 \cdot H_2O$	417
$CdCl_2 \cdot 2.5H_2O$	416
$Cd(ClO_3)_2 \cdot 2H_2O$	414
$Cd(ClO_4)_2 \cdot 6H_2O$	433
$CdCrO_4$	418
CdF_2	421
CdI_2	424
$Cd(IO_3)_2$	423
$CdMoO_4$	426
$Cd(NO_3)_2$	428
$Cd(NO_3)_2 \cdot 4H_2O$	429
CdN_6	410
$Cd(N_3)_2$	410
CdO	432
$Cd(OH)_2$	422
CdS	441
$CdSO_4$	438
$CdSO_4 \cdot H_2O$	439
$CdSO_4 \cdot 8H_2O$	440
CdSb	408
CdSe	437
$CdSeO_4 \cdot 2H_2O$	436
$CdSiO_3$	425
CdTe	442
$CdTiO_3$	444
$CdWO_4$	445
$Cd_2Nb_2O_7$	427
Cd_3As_2	409
Cd_3P_2	435
$Cd_3(PO_4)_2$	434
Ce	544
CeB_6	545
$CeBr_3$	552
$CeBr_3 \cdot 7H_2O$	553
CeC_2	546
$CeCl_3$	556
$CeCl_3 \cdot 7H_2O$	557
CeF_3	558
CeF_4	565
CeH_2	549
CeI_2	550
CeI_3	559
$CeI_3 \cdot 9H_2O$	560
CeN	547
$Ce(NO_3)_3 \cdot 6H_2O$	561
CeO_2	566
CeS	551
$Ce(SO_4)_2 \cdot 4H_2O$	567
$CeSi_2$	548
Ce_2C_3	554
$Ce_2(CO_3)_3 \cdot 5H_2O$	555
Ce_2O_3	562
Ce_2S_3	564
$Ce_2(SO_4)_3 \cdot 8H_2O$	563
Cf	520
ClCN	541
ClCs	576
ClCu	74[illegible]
ClF	[illegible]
$ClFO_2$	[illegible]

$ClFO_3$ 609
$ClFO_4$ 879
ClF_2N 1585
ClF_2OP 1676
ClF_2P 1648
ClF_2PS 1684
ClF_3 604
ClF_3Ge 932
ClF_3O 605
ClF_3O_2 608
ClF_3Si 1990
ClF_4P 1662
ClF_5 606
ClF_5S 2270
$ClGeH_3$ 931
ClH 1016
$ClHO$ 596
$ClHO_3S$ 2256
$ClHO_4$ 597
ClH_2N 1586
ClH_3Si 1989
ClI 1057
$ClIn$ 1032
ClK 1737
$ClLi$ 1260
$ClNO$ 1593
$ClNO_2$ 1596
$ClNa$ 2093
$ClOClO_3$ 610
$ClOF_3$ 605
$ClOSb$ 183
$ClOV$ 2546
ClO_2 599
ClO_2F 607
ClO_2F_3 608
ClO_3F 609
ClO_3Re 1867
$ClRb$ 1891
$ClTl$ 2328
Cl_2 595
Cl_2Co 685
Cl_2Cr 627
Cl_2CrO_2 663
Cl_2Cu 774
Cl_2Dy 827
Cl_2Eu 859
Cl_2FOP 1675
Cl_2FP 1647
Cl_2FPS 1683
Cl_2F_2Ge 935
Cl_2F_2Si 1994
Cl_2F_3P 1661
Cl_2F_3Sb 191
Cl_2Fe 1097
Cl_2Ga 904
Cl_2Ge 943
Cl_2GeH_2 934
Cl_2H_2Si 1993
Cl_2Hg 1427
Cl_2Hg_2 1410
Cl_2In 1035
Cl_2Mg 1309
Cl_2Mn 1370
Cl_2Mo 1463
Cl_2Ni 1522
Cl_2O 598
Cl_2OS 2277
Cl_2OSe 1965
Cl_2OV 2548
Cl_2OZr 2680
Cl_2O_2S 2273
Cl_2O_2U 2509
Cl_2O_2W 2473
Cl_2O_3 600
Cl_2O_4 610
$Cl_2O_5S_2$ 2275
Cl_2O_6 601
Cl_2O_7 602
Cl_2Pb 1191
Cl_2Pd 1615
Cl_2Pt 1689
Cl_2Ra 1839
Cl_2S 2266
Cl_2S_2 2263
Cl_2Se_2 1959
Cl_2Sm 1927
Cl_2Sn 2388
Cl_2Sr 2217
Cl_2Te 2306
Cl_2Ti 2420
Cl_2V 2525
Cl_2W 2452
Cl_2Yb 2574
Cl_2Zn 2616
Cl_2Zr 2664
Cl_3Cr 639
Cl_3Dy 830
Cl_3Er 841
Cl_3Eu 867
Cl_3FGe 940
Cl_3FSi 2006
Cl_3F_2P 1660
Cl_3Fe 1130
Cl_3Ga 910
Cl_3Gd 888
Cl_3GeH 939
Cl_3HSi 2005
Cl_3Ho 996
Cl_3I 1058
Cl_3ISi 2007
Cl_3In 1038
Cl_3Ir 1067
Cl_3La 1161
Cl_3Lu 1290
Cl_3Mo 1466
Cl_3MoO 1479
Cl_3N 1582
Cl_3Nb 1554
Cl_3NbO 1569
Cl_3Nd 1490
Cl_3OP 1674
Cl_3OV 2553
Cl_3OW 2469
Cl_3Os 1603
Cl_3P 1646
Cl_3PS 1682
Cl_3Pr 1823
Cl_3Pt 1694
Cl_3Pu 1711
Cl_3Re 1848
Cl_3Rh 1875
Cl_3Ru 1915
Cl_3Sb 178
Cl_3Sc 1944
Cl_3Sm 1931
Cl_3Tb 2313
Cl_3Ti 2425
Cl_3Tl 2349
Cl_3Tm 2375
Cl_3U 2490
Cl_3V 2532
Cl_3W 2455
Cl_3Y 2590
Cl_3Yb 2576
Cl_4Cr 656
Cl_4FP 1659
Cl_4Ge 951
Cl_4Hf 980
Cl_4Ir 1071
Cl_4Mo 1471
Cl_4MoO 1483
Cl_4Nb 1556
Cl_4ORe 1863
Cl_4OW 2478
Cl_4Os 1604
Cl_4P_2 1625
Cl_4Pb 1245
Cl_4Po 1719
Cl_4Pt 1696
Cl_4Re 1850
Cl_4Se 1961
Cl_4Si 2000
Cl_4Sn 2403
Cl_4Te 2308
Cl_4Th 2362
Cl_4Ti 2432
Cl_4U 2495
Cl_4V 2540
Cl_4W 2457
Cl_4Zr 2667
Cl_5Mo 1477
Cl_5Nb 1564
Cl_5P 1658
Cl_5Pa 1836
Cl_5Re 1856
Cl_5Sb 189
Cl_5Ta 2293
Cl_5U 2501
Cl_5W 2465
$Cl_6N_3P_3$ 1642
Cl_6Re 1859
Cl_6U 2504
Cl_6W 2471
Cm 822
Co 664
$CoAl_2O_4$ 679
$CoAs$ 667
$CoAsS$ 666
$CoAs_2$ 668
$CoAs_3$ 669
CoB 671
$CoBr_2$ 682
$CoBr_2 \cdot 6H_2O$ 683
$Co(BrO_3)_2 \cdot 6H_2O$ 681
$CoCO_3$ 684
$Co(CHO_2)_2 \cdot 2H_2O$ 695
$Co(CN)_2$ 690
$Co(CN)_2 \cdot 2H_2O$ 691
CoC_2O_4 706
$CoC_2O_4 \cdot 2H_2O$ 707
$Co(C_2H_3O_2)_2$ 677
$Co(C_2H_3O_2)_2 \cdot 4H_2O$ 678
$Co(C_2H_3O_2)_3$ 726
$Co(CF_3COCHCOCF_3)_2$ 696
$CoCl_2$ 685
$CoCl_2 \cdot 2H_2O$ 686
$CoCl_2 \cdot 6H_2O$ 687
$Co(ClO_4)_2$ 709
$CoCrO_4$ 688
$CoCr_2O_4$ 689
CoF_2 693
$CoF_2 \cdot 4H_2O$ 694
CoF_3 728
CoI_2 700
$CoI_2 \cdot 6H_2O$ 701
$Co(IO_3)_2$ 699
$CoK_2(SO_4)_2 \cdot 6H_2O$ 711
$CoK_3(NO_2)_6 \cdot 1.5H_2O$ 734
$CoMoO_4$ 703
$CoNH_4PO_4$ 80
$Co(NO_3)_2$ 704
$Co(NO_3)_2 \cdot 6H_2O$ 705
$Co(NO_3)_3$ 731
$Co(NH_3)_6Cl_3$ 729
CoO 708
$Co(OH)_2$ 698
$Co(OH)_3$ 730
CoS 720
$CoSO_4$ 717
$CoSO_4 \cdot H_2O$ 719
$CoSO_4 \cdot 7H_2O$ 718
CoS_2 675
$Co(SCN)_2$ 722
$Co(SCN)_2 \cdot 3H_2O$ 723
$CoSb$ 665
$CoSe$ 713
$CoSeO_3 \cdot 2H_2O$ 714
$CoSeO_4 \cdot 5H_2O$ 712
$CoSiF_6 \cdot 6H_2O$ 697
$CoSi_2$ 674
$CoTe$ 721
$CoTiO_3$ 702
$CoWO_4$ 724
Co_2B 670
$Co_2(CO)_8$ 672
Co_2O_3 732
$Co_2O_3 \cdot H_2O$ 733
Co_2P 673
Co_2S_3 735
Co_2SiO_4 715
Co_2SnO_4 716
Co_2TiO_4 736

$Co_3(AsO_4)_2 \cdot 8H_2O$ 680
$Co_3[Fe(CN)_6]_2$ 692
Co_3O_4 725
$Co_3(PO_4)_2 \cdot 8H_2O$ 710
$Co_4(CO)_{12}$ 676
Cr 611
CrB 614
CrB_2 615
$CrBr_2$ 626
$CrBr_3$ 636
$CrBr_3(H_2O)_4 \cdot 2H_2O$ 638
$CrC_2O_4 \cdot H_2O$ 631
$Cr(C_2H_3O_2)_2 \cdot H_2O$ 625
$Cr(C_2H_3O_2)_3$ 634
$Cr(C_2H_3O_2)_3 \cdot 6H_2O$ 635
$Cr(CO)_6$ 618
$Cr(CH_3COCHCOCH_3)_3$ 648
$CrCl_2$ 627
$[CrCl_2(H_2O)_4]Cl \cdot 2H_2O$ 640
$CrCl_3$ 639
$CrCl_4$ 656
CrF_2 629
CrF_3 641
$CrF_3 \cdot 3H_2O$ 642
CrF_4 657
CrF_5 659
CrF_6 660
$Cr(H_2O)_4Cl_2 \cdot 4H_2O$ 628
$Cr(H_2O)_6Br_3$ 637
CrI_2 630
CrI_3 644
$CrK(SO_4)_2 \cdot 12H_2O$ 652
CrN 620
$Cr(NO_3)_3$ 645
$Cr(NO_3)_3 \cdot 9H_2O$ 646
CrO_2 658
CrO_2Cl_2 663
CrO_3 661
$Cr(OH)_3 \cdot 3H_2O$ 643
CrP 621
$CrPO_4$ 649
$CrPO_4 \cdot 3.5H_2O$ 650
$CrPO_4 \cdot 6H_2O$ 651
$CrSO_4 \cdot 5H_2O$ 632
CrSb 612
CrSe 622
$CrSi_2$ 624
Cr_2As 613
Cr_2N 619
Cr_2O_3 647
Cr_2S_3 654
$Cr_2(SO_4)_3$ 653
Cr_2Te_3 655
Cr_3C_2 617
Cr_3O_4 633
Cr_3Si 623
Cr_5B_3 616
Cs 568
$CsBO_2$ 587
CsBr 573
$CsBrO_3$ 572
$CsCHO_2$ 579
CsCN 577
$CsC_2H_3O_2$ 569
CsCl 576
$CsClO_3$ 575
$CsClO_4$ 591
CsF 578
CsH 580
$CsHCO_3$ 581
$CsHF_2$ 582
$CsHSO_4$ 583
CsI 586
$CsIO_3$ 585
$CsIO_4$ 592
$CsNH_2$ 570
$CsNO_3$ 588
CsN_3 571
CsOH 584
CsO_2 590
Cs_2CO_3 574
Cs_2O 589
$Cs_2S \cdot 4H_2O$ 594
Cs_2SO_4 593
Cu 737
$Cu(BF_4)_2$ 818
$Cu(BO_2)_2$ 769
CuBr 746
$CuBr_2$ 770
CuCN 748
$CuCO_3 \cdot Cu(OH)_2$ 772
CuC_2 764
$Cu(CHO_2)_2$ 789
$Cu(CHO_2)_2 \cdot 4H_2O$ 790
$CuC_2H_3O_2$ 743
$Cu(C_2H_3O_2)_2$ 761
$Cu(C_2H_3O_2)_2 \cdot H_2O$ 763
$Cu(C_2H_3O_2)_2 \cdot CuO \cdot 6H_2O$ 768
$Cu(C_2H_3O_2)_2 \cdot 3Cu(AsO_2)_2$ 762
$Cu(CN)_2$ 780
CuC_2O_4 801
$CuC_2O_4 \cdot 0.5H_2O$ 802
$CuC_4H_4O_6 \cdot 3H_2O$ 816
$Cu(C_2H_5O)_2$ 783
$Cu(C_4H_7O_2)_2 \cdot H_2O$ 771
$Cu(CF_3COCHCOCF_3)_2$ 791
$Cu(CH_3COCHCOCH_3)_2$ 738
$Cu(C_2H_5CO_2CHCOCH_3)_2$ 784
$Cu(C_8H_{15}O_2)_2$ 740
$Cu(C_{10}H_{17}O_2)_2$ 781
$Cu(C_{18}H_{33}O_2)_2$ 800
$Cu(C_{18}H_{35}O_2)_2$ 811
CuCl 747
$CuCl_2$ 774
$CuCl_2 \cdot 2H_2O$ 775
$CuCl_2 \cdot 2NH_4Cl$ 82
$CuCl_2 \cdot 2NH_4Cl \cdot 2H_2O$ 83
$Cu(ClO_3)_2 \cdot 6H_2O$ 773
$Cu(ClO_4)_2$ 804
$Cu(ClO_4)_2 \cdot 6H_2O$ 805
$CuCrO_4$ 777
$CuCr_2O_4$ 778
$CuCr_2O_7 \cdot 2H_2O$ 782
CuF 749
CuF_2 787
$CuF_2 \cdot 2H_2O$ 788
$CuFeS_2$ 786
CuH 750
$CuHAsO_3$ 766
CuI 751
$Cu(IO_3)_2$ 794
$Cu(IO_3)_2 \cdot H_2O$ 795
$CuMoO_4$ 796
$Cu(NO_3)_2$ 797
$Cu(NO_3)_2 \cdot 3H_2O$ 799
$Cu(NO_3)_2 \cdot 6H_2O$ 798
CuN_3 745
CuN_6 767
$Cu(N_3)_2$ 767
CuO 803
$Cu(OH)_2$ 793
CuP_2 741
CuS 815
CuSCN 759
$CuSO_4$ 812
$CuSO_4 \cdot 5H_2O$ 813
CuSe 809
$CuSeO_3 \cdot 2H_2O$ 810
$CuSeO_4 \cdot 5H_2O$ 808
$CuSiF_6 \cdot 4H_2O$ 792
CuTe 817
$Cu(VO_3)_2$ 821
$CuWO_4$ 819
$CuWO_4 \cdot 2H_2O$ 820
Cu_2C_2 744
$Cu_2C_6H_4O_7 \cdot 2.5H_2O$ 779
$Cu_2Fe(CN)_6$ 785
Cu_2HgI_4 752
Cu_2O 753
$Cu_2(OH)_3Cl$ 776
Cu_2S 755
$Cu_2SO_3 \cdot 0.5H_2O$ 757
$Cu_2SO_3 \cdot H_2O$ 756
$Cu_2SO_3 \cdot CuSO_3 \cdot 2H_2O$ 760
Cu_2Se 754
Cu_2Te 758
$Cu_3(AsO_4)_2$ 765
Cu_3N 739
$Cu_3(OH)_4SO_4$ 814
$Cu_3(PO_4)_2$ 806
$Cu_3(PO_4)_2 \cdot 3H_2O$ 807
Cu_5Si 742
Dy 823
DyB_4 824
$DyBr_3$ 829
$DyCl_2$ 827
$DyCl_3$ 830
DyF_3 831
DyH_3 832
DyI_2 828
DyI_3 833
DyN 825
$Dy(NO_3)_3 \cdot 5H_2O$ 834
$DySi_2$ 826
Dy_2O_3 835
Dy_2S_3 836
Er 838
ErB_4 839
$ErBr_3$ 840
$ErCl_3$ 841
$ErCl_3 \cdot 6H_2O$ 842
ErF_3 843
ErH_3 844
ErI_3 845
ErN 847
$Er(NO_3)_3 \cdot 5H_2O$ 846
$ErSi_2$ 849
Er_2O_3 848
Er_2S_3 852
$Er_2(SO_4)_3$ 850
$Er_2(SO_4)_3 \cdot 8H_2O$ 851
Er_2Te_3 853
Es 837
Eu 854
EuB_6 855
$EuBr_2$ 858
$EuBr_3$ 866
$EuCl_2$ 859
$EuCl_3$ 867
$EuCl_3 \cdot 6H_2O$ 868
EuF_2 860
EuF_3 869
EuI_2 861
EuN 856
$Eu(NO_3)_3 \cdot 6H_2O$ 870
EuS 864
$EuSO_4$ 863
EuSe 862
$EuSi_2$ 857
EuTe 865
Eu_2O_3 871
$Eu_2(SO_4)_3$ 872
$Eu_2(SO_4)_3 \cdot 8H_2O$ 873
FCN 542
$FGeH_3$ 936
FH 1019
FHO_3S 2257
FH_2N 1587
FH_3Si 1997
FI 1059
FK 1746
FLi 1265
FNO 1594
FNO_2 1597
FNO_3 878
FNa 2112
$FNbO_2$ 1570
$FOClO_3$ 879
FO_3Re 1868
FRb 1894
FSSF 2265
FTl 2332
F_2 875
F_2Fe 1101
F_2Ge 944
F_2HN 1588
F_2H_2Si 1995
F_2Hg 1431
F_2Hg_2 1411
F_2Kr 1155
F_2Mg 1312
F_2Mn 1373

F_2N_2	1589
F_2N_2	1590
F_2Ni	1525
F_2O	876
F_2OS	2278
F_2OSe	1966
F_2OV	2549
F_2O_2	877
F_2O_2S	2274
F_2O_2Se	1967
F_2O_2U	2510
F_2O_2Xe	2562
F_2Pb	1198
F_2Pd	1616
F_2Ra	1840
F_2S_2	2265
F_2S_2	2264
F_2Sm	1928
F_2Sn	2390
F_2Sr	2222
F_2V	2526
F_2Xe	2559
F_2Zn	2622
F_3Fe	1136
F_3Ga	911
F_3Gd	890
F_3HSi	2008
F_3Ho	997
F_3I	1060
F_3In	1039
F_3Ir	1068
F_3La	1163
F_3Lu	1291
F_3Mn	1398
F_3Mo	1467
F_3N	1583
F_3NO	1595
F_3Nb	1555
F_3Nd	1492
F_3OP	1677
F_3OV	2554
F_3O_2Re	1869
F_3P	1649
F_3PS	1685
F_3Pr	1825
F_3Pu	1712
F_3Rh	1876
F_3Ru	1916
F_3Sb	179
F_3Sc	1945
F_3Sm	1933
F_3Ti	2426
F_3Tl	2351
F_3Tm	2377
F_3U	2491
F_3V	2533
F_3Y	2591
F_3Yb	2578
F_4Ge	952
F_4Hf	981
F_4Mo	1472
F_4MoO	1482
F_4N_2	1591
F_4Nb	1557
F_4ORe	1864
F_4OW	2479
F_4OXe	2563
F_4Os	1605
F_4P_2	1626
F_4Pb	1246
F_4Pt	1698
F_4Pu	1715
F_4Re	1851
F_4Ru	1918
F_4S	2267
F_4Se	1962
F_4Si	2001
F_4Sn	2406
F_4Te	2309
F_4Th	2363
F_4Ti	2433
F_4U	2496
F_4V	2541
F_4W	2458
F_4Xe	2560
F_4Zr	2668
F_5I	1061
F_5Mo	1478
F_5Nb	1565
F_5ORe	1870
F_5Os	1607
F_5P	1663
F_5Re	1857
F_5Ru	1920
F_5SOF	2279
F_5Sb	190
F_5Ta	2294
F_5Tc	2298
F_5U	2502
F_5V	2543
F_5W	2467
F_6HP	1637
F_6H_2Si	2013
F_6Ir	1074
F_6Mo	1481
F_6Os	1608
F_6Pt	1702
F_6Pu	1717
F_6Re	1861
F_6Rh	1881
F_6Ru	1921
F_6S	2268
F_6Se	1963
F_6Tc	2299
F_6Te	2311
F_6U	2505
F_6W	2475
F_6Xe	2561
F_7I	1062
F_7Re	1865
$F_{10}S_2$	2271
Fe	1075
$Fe(AlO_2)_2$	1091
$FeAs$	1081
$FeAsO_4 \cdot 2H_2O$	1128
FeB	1082
$FeBr_2$	1094
$FeBr_2 \cdot 6H_2O$	1095
$FeBr_3$	1129
$FeCO_3$	1096
$FeC_2O_4 \cdot 2H_2O$	1109
$Fe(CHO_2)_3$	1138
$FeC_4H_4O_6$	1120
$Fe(CO)_5$	1077
$FeC_6H_5O_7 \cdot 5H_2O$	1133
$Fe(C_5H_5)_2$	1076
$Fe(CH_3COCHCOCH_3)_3$	1127
$FeCl_2$	1097
$FeCl_2 \cdot 2H_2O$	1098
$FeCl_2 \cdot 4H_2O$	1099
$Fe(ClO_4)_2$	1111
$FeCl_3$	1130
$FeCl_3 \cdot 6H_2O$	1131
$FeCr_2O_4$	1100
FeF_2	1101
$FeF_2 \cdot 4H_2O$	1102
FeF_3	1136
$FeF_3 \cdot 3H_2O$	1137
$FeH_2(CO)_4$	1080
$Fe(H_2PO_2)_3$	1148
FeI_2	1104
$FeI_2 \cdot 4H_2O$	1105
$FeLiSi$	1264
$FeMoO_4$	1106
$Fe(NO_3)_2$	1107
$Fe(NO_3)_2 \cdot 6H_2O$	1108
$Fe(NO_3)_3$	1141
$Fe(NO_3)_3 \cdot 6H_2O$	1142
$Fe(NO_3)_3 \cdot 9H_2O$	1143
$FeNaP_2O_7$	1149
FeO	1110
$FeOH(C_2H_3O_2)_2$	1126
$FeO(OH)$	1140
$Fe(OH)_2$	1103
$Fe(OH)_3$	1139
FeP	1085
$FePO_4 \cdot 2H_2O$	1146
FeS	1118
$FeSO_4$	1115
$FeSO_4 \cdot H_2O$	1116
$FeSO_4 \cdot 7H_2O$	1117
FeS_2	1088
$Fe(SCN)_2 \cdot 3H_2O$	1122
$Fe(SCN)_3 \cdot H_2O$	1152
$FeSe$	1113
$FeSi$	1089
$FeSi_2$	1090
$Fe(TaO_3)_2$	1119
$FeTe$	1121
$FeTiO_3$	1123
$Fe(VO_3)_3$	1153
$FeWO_4$	1124
Fe_2B	1083
$Fe_2(C_2O_4)_3$	1144
$Fe_2(CO)_9$	1078
$Fe_2(CrO_4)_3$	1132
$Fe_2(Cr_2O_7)_3$	1134
Fe_2O_3	1145
Fe_2P	1086
$Fe_2(SO_4)_3$	1150
$Fe_2(SO_4)_3 \cdot 9H_2O$	1151
Fe_2SiO_4	1114
$Fe_3(AsO_4)_2$	1092
$Fe_3(AsO_4)_2 \cdot 6H_2O$	1093
Fe_3C	1084
$Fe_3(CO)_{12}$	1079
Fe_3O_4	1125
Fe_3P	1087
$Fe_3(PO_4)_2 \cdot 8H_2O$	1112
$Fe_4[Fe(CN)_6]_3$	1135
$Fe_4(P_2O_7)_3 \cdot 9H_2O$	1147
Fm	874
Fr	880
Ga	898
$GaAs$	900
$GaBr_3$	909
$Ga(CH_3COCHCOCH_3)_3$	908
$GaCl_2$	904
$GaCl_3$	910
GaF_3	911
$GaF_3 \cdot 3H_2O$	912
GaH_3	913
GaI_3	915
GaN	901
$Ga(NO_3)_3$	916
$GaOOH$	918
$Ga(OH)_3$	914
GaP	902
GaS	906
$GaSb$	899
$GaSe$	905
$GaTe$	907
Ga_2O	903
Ga_2O_3	917
Ga_2S_3	922
$Ga_2(SO_4)_3$	920
$Ga_2(SO_4)_3 \cdot 18H_2O$	921
Ga_2Se_3	919
Ga_2Te_3	923
Gd	881
GdB_6	882
$GdBr_3$	887
$GdCl_3$	888
$GdCl_3 \cdot 6H_2O$	889
GdF_3	890
GdI_2	885
GdI_3	891
GdN	883
$Gd(NO_3)_3 \cdot 5H_2O$	893
$Gd(NO_3)_3 \cdot 6H_2O$	892
$GdSe$	886
$GdSi_2$	884
Gd_2O_3	894
Gd_2S_3	896
$Gd_2(SO_4)_3 \cdot 8H_2O$	895
Gd_2Te_3	897
Ge	924
$GeBr_2$	942
$GeBr_4$	950
$GeCl_2$	943
$GeCl_3F$	940
$GeCl_4$	951

GeF_2	944
GeF_2Cl_2	935
GeF_3Cl	932
GeF_4	952
$GeHBr_3$	938
$GeHCl_3$	939
GeH_2Br_2	933
GeH_2Cl_2	934
GeH_3Br	930
GeH_3CH_3	941
GeH_3Cl	931
GeH_3F	936
GeH_3I	937
GeH_4	925
GeI_2	945
GeI_4	953
$GeMg_2$	1314
GeO	946
GeO_2	955
GeS	948
GeS_2	957
$GeSe$	947
$GeSe_2$	956
$GeTe$	949
Ge_2H_6	926
Ge_3H_8	927
Ge_3N_4	954
Ge_4H_{10}	928
Ge_5H_{12}	929
$HAuBr_4 \cdot 5H_2O$	959
$HAuCl_4 \cdot 4H_2O$	960
HBF_4	382
HBO_2	379
HBO_2	380
HBO_2	381
HBr	1015
$HBrO_3$	396
HCN	1018
HCl	1016
$HCl \cdot 2H_2O$	1017
$HClO_4$	597
HF	1019
HI	1020
HIO_3	1051
$HIO_4 \cdot 2H_2O$	1052
HI_3Si	2009
HK	1757
HLi	1267
$HNOSO_4$	2255
HNO_2	1574
HNO_3	1573
HN_3	1011
HNa	2126
$HOCl$	596
HO_3P	1633
HO_4Re	1845
HPF_6	1637
HPH_2O_2	1632
HPO_2F_2	1636
HPO_3	1633
HRb	1897
$HReO_4$	1845
H_2	1014
H_2CrO_4	662
H_2Hf	982
H_2I_2Si	1996
H_2KN	1727
H_2LiN	1252
H_2Mg	1315
$H_2MoO_4 \cdot H_2O$	1480
H_2NNa	2081
H_2NOH	1012
$(H_2NOH)_2 \cdot H_2SO_4$	1013
H_2NSO_3H	2259
$H_2N_2O_2$	1572
H_2O	2555
H_2O_2	1021
H_2O_3S	2258
H_2O_3Se	1955
H_2O_3Te	2302
H_2O_4S	2253
H_2O_4Se	1954
H_2O_4W	2440
H_2O_5S	2254
H_2PFO_3	1638
$H_2PtCl_6 \cdot 6H_2O$	1705
H_2S	1023
H_2SO_3	2258
H_2SO_4	2253
H_2SO_5	2254
H_2S_2	1024
H_2Se	1022
H_2SeO_3	1955
H_2SeO_4	1954
H_2SiF_6	2013
H_2SiO_3	2011
H_2Sr	2226
H_2Te	1025
H_2TeO_3	2302
H_2Th	2357
H_2Ti	2413
H_2WO_4	2440
H_2Zr	2665
H_3AsO_3	200
H_3AsO_4	198
$H_3AsO_4 \cdot 0.5H_2O$	199
H_3BO_3	378
H_3ISi	1998
H_3La	1164
H_3N	63
H_3NO	1012
H_3O_2P	1632
H_3O_3P	1631
H_3O_4P	1630
H_3P	1623
H_3PO_3	1631
H_3PO_4	1630
H_3Sb	175
H_3U	2492
H_4MgN_2	1301
H_4N_2	1003
H_4N_4	65
$H_4O_6P_2$	1634
$H_4O_7P_2$	1635
H_4P_2	1624
$H_4P_2O_6$	1634
$H_4P_2O_7$	1635
H_4Si	1973
H_4SiO_4	2012
H_4Sn	2384
H_6OSi_2	2010
H_6O_6Te	2301
H_6Si_2	1974
H_6TeO_6	2301
H_8Si_3	1975
$H_{10}Si_4$	1976
$H_{10}Si_4$	1977
$H_{10}Si_5$	1985
$H_{12}Si_5$	1978
$H_{12}Si_5$	1979
$H_{12}Si_5$	1980
$H_{12}Si_6$	1986
$H_{14}Si_6$	1981
$H_{14}Si_6$	1983
$H_{14}Si_6$	1982
$H_{16}Si_7$	1984
He	993
Hf	976
HfB_2	977
$HfBr_4$	978
HfC	979
$HfCl_4$	980
HfF_4	981
HfH_2	982
HfI_4	983
HfN	984
$HfOCl_2 \cdot 8H_2O$	986
HfO_2	985
HfP	987
HfS_2	992
$Hf(SO_4)_2$	991
$HfSe_2$	988
$HfSiO_4$	989
$HfSi_2$	990
Hg	1404
$HgBr_2$	1425
$Hg(BrO_3)_2$	1424
$Hg(CN)_2$	1429
$Hg(CN)_2 \cdot HgO$	1442
$Hg(CNO)_2$	1432
HgC_2O_4	1439
$Hg(C_2H_3O_2)_2$	1422
$HgCl_2$	1427
$Hg(ClO_3)_2$	1426
$Hg(ClO_4)_2 \cdot 3H_2O$	1443
$HgCrO_4$	1428
$HgCr_2O_7$	1430
HgF_2	1431
$HgHAsO_4$	1433
HgI_2	1435
$Hg(IO_3)_2$	1434
$Hg(NH_2)Cl$	1423
$Hg(NO_3)_2$	1436
$Hg(NO_3)_2 \cdot H_2O$	1438
$Hg(NO_3)_2 \cdot 2H_2O$	1437
HgO	1440
HgS	1448
HgS	1447
$HgSO_4$	1446
$Hg(SCN)_2$	1450
$HgSe$	1445
$HgTe$	1449
$HgWO_4$	1451
Hg_2Br_2	1407
$Hg_2(BrO_3)_2$	1406
Hg_2CO_3	1408
$Hg_2C_2O_4$	1416
$Hg_2(C_2H_3O_2)_2$	1405
Hg_2Cl_2	1410
$Hg_2(ClO_3)_2$	1409
$Hg_2(ClO_4)_2 \cdot 4H_2O$	1418
Hg_2F_2	1411
Hg_2I_2	1412
$Hg_2(NO_2)_2$	1415
$Hg_2(NO_3)_2$	1413
$Hg_2(NO_3)_2 \cdot 2H_2O$	1414
Hg_2O	1417
Hg_2SO_4	1419
$Hg_2(SCN)_2$	1420
Hg_2WO_4	1421
$(Hg_3O_2)SO_4$	1441
$Hg_3(PO_4)_2$	1444
Ho	994
$HoBr_3$	995
$HoCl_3$	996
HoF_3	997
HoI_3	998
HoN	999
$HoSi_2$	1001
Ho_2O_3	1000
Ho_2S_3	1002
IBr	1056
ICN	543
ICl	1057
ICl_3	1058
IF	1059
IF_3	1060
IF_5	1061
IF_7	1062
IIn	1033
IK	1773
ILi	1271
INa	2143
IRb	1902
ITl	2337
I_2	1050
I_2Mg	1319
I_2Mn	1375
I_2Mo	1464
I_2Ni	1529
I_2O_2W	2474
I_2O_4	1053
I_2O_5	1054
I_2Pb	1206
I_2Pd	1617
I_2Pt	1690
I_2Sm	1929
I_2Sn	2393
I_2Sr	2229
I_2Ti	2421
I_2V	2527
I_2W	2453

I_2Yb 2575
I_2Zn 2629
I_3In 1042
I_3Ir 1069
I_3La 1167
I_3Lu 1292
I_3Mo 1468
I_3N 1584
I_3Nd 1493
I_3OP 1678
I_3P 1650
I_3PS 1686
I_3Pr 1826
I_3Pu 1713
I_3Re 1849
I_3Rh 1877
I_3Ru 1917
I_3Sb 180
I_3Sm 1934
I_3Tb 2315
I_3Tm 2378
I_3U 2493
I_3V 2535
I_4Nb 1558
I_4O_9 1055
I_4P_2 1627
I_4Pt 1699
I_4Si 2002
I_4Sn 2407
I_4Te 2310
I_4Th 2364
I_4Ti 2434
I_4U 2497
I_4W 2459
I_4Zr 2670
I_5Nb 1566
I_5Ta 2295
In 1026
$InAs$ 1028
$InBr$ 1031
$InBr_2$ 1034
$InBr_3$ 1037
$InCl$ 1032
$InCl_2$ 1035
$InCl_3$ 1038
$In(ClO_4)_3 \cdot 8H_2O$ 1044
InF_3 1039
$InF_3 \cdot 3H_2O$ 1040
InI 1033
InI_3 1042
InN 1029
$In(OH)_3$ 1041
InP 1030
$InPO_4$ 1045
InS 1036
$InSb$ 1027
In_2O_3 1043
In_2S_3 1048
$In_2(SO_4)_3$ 1047
In_2Se_3 1046
In_2Te_3 1049
Ir 1063
$IrBr_3$ 1065
$IrBr_3 \cdot 4H_2O$ 1066
$IrCl_3$ 1067
$IrCl_4$ 1071
IrF_3 1068
IrF_6 1074
IrI_3 1069
IrO_2 1072
IrS_2 1073
Ir_2O_3 1070
Ir_2S_3 1064
K 1721
$KAg(CN)_2$ 1798
$KAl(SO_4)_2$ 1725
$KAl(SO_4)_2 \cdot 12H_2O$ 1726
$KAlSi_3O_8$ 1724
$KAsO_2$ 1729
$KAsO_2 \cdot HAsO_2$ 1759
$KAsO_3$ 1776
$KAuCl_4 \cdot 2H_2O$ 1809
KBF_4 1748
KBH_4 1731
KBO_2 1778
KBr 1733
$KBrO_3$ 1732
$KBrO_4$ 1786
$KCHO_2$ 1750
KCN 1740
$KCNO$ 1739
$KC_2H_3O_2$ 1722
KCl 1737
$KClO_3$ 1736
$KClO_4$ 1787
KF 1746
$KF \cdot 2H_2O$ 1747
KH 1757
$KHCO_3$ 1760
$KHC_4H_4O_6$ 1769
KHF_2 1761
KHS 1766
$KHS \cdot 0.5H_2O$ 1767
$KHSO_3$ 1768
$KHSO_4$ 1765
$KHSeO_3$ 1764
KH_2AsO_4 1742
KH_2PO_2 1771
KH_2PO_4 1743
KI 1773
KIO_3 1772
KIO_4 1788
$KI_3 \cdot H_2O$ 1816
$KMnO_4$ 1789
KNH_2 1727
KNO_2 1782
KNO_3 1781
KN_3 1730
$KNbO_3$ 1780
KOH 1770
KO_2 1805
$KSCN$ 1813
$K_2Al_2O_4 \cdot 3H_2O$ 1723
$K_2B_4O_7 \cdot 5H_2O$ 1808
K_2CO_3 1734
$K_2CO_3 \cdot 1.5H_2O$ 1735
K_2CS_3 1817
$K_2C_2O_4$ 1783
$K_2C_2O_4 \cdot H_2O$ 1784
K_2CrO_4 1738
$K_2Cr_2O_7$ 1741
K_2HAsO_4 1758
K_2HPO_3 1763
K_2HPO_4 1762
K_2HgI_4 1812
K_2MnF_6 1754
K_2MnO_4 1775
K_2MoO_4 1779
K_2O 1785
K_2O_2 1790
K_2OsCl_6 1751
$K_2Pt(CN)_4 \cdot 3H_2O$ 1811
K_2PtCl_4 1810
K_2PtCl_6 1752
K_2S 1801
$K_2S \cdot 5H_2O$ 1802
K_2SO_3 1803
$K_2SO_3 \cdot 2H_2O$ 1804
K_2SO_4 1800
$K_2S_2O_3$ 1814
$K_2S_2O_5$ 1777
$K_2S_2O_7$ 1794
$K_2S_2O_8$ 1791
K_2Se 1796
K_2SeO_3 1797
K_2SeO_4 1795
K_2SiF_6 1755
$K_2SnO_3 \cdot 3H_2O$ 1799
K_2TaF_7 1749
K_2TeO_3 1807
$K_2TeO_4 \cdot 3H_2O$ 1806
K_2TiO_3 1815
$K_2U_2O_7$ 1819
K_2WO_4 1818
K_2ZrF_6 1756
K_3AsO_4 1728
$K_3Co(CN)_6$ 1753
$K_3Fe(CN)_6$ 1744
$K_3Fe(C_2O_4)3 \cdot 3H_2O$ 1774
K_3PO_4 1792
K_4BiI_7 356
$K_4Fe(CN)_6 \cdot 3H_2O$ 1745
$K_4P_2O_7 \cdot 3H_2O$ 1793
Kr 1154
KrF_2 1155
La 1156
LaB_6 1157
$LaBr_3$ 1158
LaC_2 1159
$LaCl_3$ 1161
$LaCl_3 \cdot 7H_2O$ 1162
LaF_3 1163
LaH_3 1164
LaI_3 1167
$La(IO_3)_3$ 1166
LaN 1169
$La(NO_3)_3 \cdot 6H_2O$ 1168
$La(OH)_3$ 1165
LaS 1174
$LaSi_2$ 1171
$La_2(CO_3)_3 \cdot 8H_2O$ 1160
La_2O_3 1170
La_2S_3 1173
$La_2(SO_4)_3 \cdot 9H_2O$ 1172
Li 1248
$LiAlH_4$ 1251
$LiBH_4$ 1256
$LiBO_2$ 1255
$LiBr$ 1257
$Li(CHO_2) \cdot H_2O$ 1266
$LiC_2H_3O_2$ 1249
$LiC_2H_3O_2 \cdot 2H_2O$ 1250
$LiCl$ 1260
$LiClO_3$ 1259
$LiClO_4$ 1279
LiF 1265
$LiFeSi$ 1264
LiH 1267
LiH_2PO_4 1263
LiI 1271
$LiI \cdot 3H_2O$ 1272
$LiIO_3$ 1270
$LiNH_2$ 1252
$LiNO_2 \cdot H_2O$ 1276
$LiNO_3$ 1274
LiN_3 1254
$LiNbO_3$ 1273
$LiOH$ 1268
$LiOH \cdot H_2O$ 1269
$LiSCN$ 1286
Li_2CO_3 1258
$Li_2CrO_4 \cdot 2H_2O$ 1261
$Li_2Cr_2O_7 \cdot 2H_2O$ 1262
Li_2O 1278
Li_2O_2 1280
Li_2S 1285
Li_2SO_4 1283
$Li_2SO_4 \cdot H_2O$ 1284
$Li_2SeO_4 \cdot H_2O$ 1281
Li_2SiO_3 1282
Li_3AsO_4 1253
Li_3N 1275
Li_3PO_4 1277
Lr 1175
Lu 1287
LuB_4 1288
$LuBr_3$ 1289
$LuCl_3$ 1290
LuF_3 1291
LuI_3 1292
LuN 1293
Lu_2O_3 1294
Lu_2S_3 1296
$Lu_2(SO_4)_3 \cdot 8H_2O$ 1295
Lu_2Te_3 1297
Md 1403
Mg 1298
MgB_2 1303
$MgBr_2$ 1305
$MgBr_2 \cdot 6H_2O$ 1306
$Mg(BrO_3)_2 \cdot 6H_2O$ 1304
$MgCO_3$ 1307

$Mg(CHO_2)_2 \cdot 2H_2O$	1313
MgC_2O_4	1326
$MgC_2O_4 \cdot 2H_2O$	1327
$Mg(C_2H_3O_2)_2$	1299
$Mg(C_2H_3O_2)_2 \cdot 4H_2O$	1300
$MgCl_2$	1309
$MgCl_2 \cdot 6H_2O$	1310
$Mg(ClO_3)_2 \cdot 6H_2O$	1308
$Mg(ClO_4)_2$	1329
$Mg(ClO_4)_2 \cdot 6H_2O$	1330
$MgCrO_4 \cdot 7H_2O$	1311
MgF_2	1312
$MgHPO_4 \cdot 3H_2O$	1316
MgH_2	1315
MgI_2	1319
$MgI_2 \cdot 8H_2O$	1320
$Mg(IO_3)_2 \cdot 4H_2O$	1318
$Mg(MnO_4)_2 \cdot 6H_2O$	1331
$Mg(NH_2)_2$	1301
$Mg(NO_2)_2 \cdot 3H_2O$	1325
$Mg(NO_3)_2$	1321
$Mg(NO_3)_2 \cdot 2H_2O$	1322
$Mg(NO_3)_2 \cdot 6H_2O$	1323
MgO	1328
MgO_2	1332
$Mg(OH)_2$	1317
MgS	1349
$MgSO_3 \cdot 3H_2O$	1350
$MgSO_3 \cdot 6H_2O$	1351
$MgSO_4$	1346
$MgSO_4 \cdot H_2O$	1347
$MgSO_4 \cdot 7H_2O$	1348
$MgS_2O_3 \cdot 6H_2O$	1352
$MgSe$	1338
$MgSeO_3 \cdot 6H_2O$	1339
$MgSeO_4 \cdot 6H_2O$	1337
$MgSiF_6 \cdot 6H_2O$	1343
$MgSiO_3$	1340
$MgTiO_3$	1353
$MgWO_4$	1354
Mg_2Ge	1314
$Mg_2P_2O_7 \cdot 3H_2O$	1335
Mg_2Si	1344
Mg_2SiO_4	1341
$Mg_2Si_3O_8$	1342
Mg_2Sn	1345
Mg_3N_2	1324
Mg_3P_2	1336
$Mg_3(PO_4)_2 \cdot 5H_2O$	1333
$Mg_3(PO_4)_2 \cdot 8H_2O$	1334
Mg_3Sb_2	1302
Mn	1355
MnB	1358
MnB_2	1359
$MnB_4O_7 \cdot 8H_2O$	1366
$MnBr_2$	1367
$MnBr_2 \cdot 4H_2O$	1368
$MnCO_3$	1369
$MnC_2O_4 \cdot 2H_2O$	1381
$Mn(C_2H_3O_2)_2 \cdot 4H_2O$	1365
$MnCl_2$	1370
$MnCl_2 \cdot 4H_2O$	1371
$Mn(ClO_4)_2 \cdot 6H_2O$	1383
MnF_2	1373
MnF_3	1398
$Mn(H_2PO_4)_2 \cdot 2H_2O$	1372
MnI_2	1375
$MnI_2 \cdot 4H_2O$	1376
$MnMoO_4$	1377
$Mn(NO_3)_2$	1378
$Mn(NO_3)_2 \cdot 4H_2O$	1380
$Mn(NO_3)_2 \cdot 6H_2O$	1379
MnO	1382
MnO_2	1401
$MnO(OH)$	1399
$Mn(OH)_2$	1374
MnP	1363
MnS	1393
MnS	1391
MnS	1392
$MnSO_4$	1388
$MnSO_4 \cdot H_2O$	1389
$MnSO_4 \cdot 4H_2O$	1390
$MnSb$	1356
$MnSe$	1387
$MnSiO_3$	1385
$MnTe$	1394
$MnTiO_3$	1395
$MnWO_4$	1396
Mn_2B	1360
$Mn_2(CO)_{10}$	1362
Mn_2O_3	1400
Mn_2O_7	1402
Mn_2P	1364
$Mn_2P_2O_7$	1384
Mn_2Sb	1357
Mn_2SiO_4	1386
Mn_3C	1361
Mn_3O_4	1397
Mo	1452
$MoBr_2$	1462
$MoBr_3$	1465
$MoBr_4$	1470
MoC	1455
$Mo(CO)_6$	1457
$MoCl_2$	1463
$MoCl_3$	1466
$MoCl_4$	1471
$MoCl_5$	1477
MoF_3	1467
MoF_4	1472
MoF_5	1478
MoF_6	1481
MoI_2	1464
MoI_3	1468
MoN	1458
$MoOCl_3$	1479
$MoOCl_4$	1483
$MoOF_4$	1482
MoO_2	1473
MoO_2Cl_2	1484
MoO_3	1485
MoP	1460
$Mo(PO_3)_6$	1486
MoS_2	1475
$MoSe_2$	1474
$MoSi_2$	1461
$MoTe_2$	1476
Mo_2B	1453
Mo_2B_5	1454
Mo_2C	1456
Mo_2N	1459
Mo_2O_3	1469
NBr_3	1581
$NClF_2$	1585
NCl_3	1582
NF_3	1583
NHF_2	1588
NH_2COONH_4	72
NH_2Cl	1586
NH_2F	1587
NH_2OH	1012
$NH_2OH \cdot H_2SO_4$	1013
NH_3	63
NH_4AlCl_4	162
NH_4BF_4	96
$NH_4B_5O_8 \cdot 4H_2O$	140
NH_4Br	70
NH_4BrO_3	69
NH_4CHO_2	98
NH_4CN	84
$NH_4C_2H_3O_2$	64
$NH_4C_3H_5O_3$	126
$NH_4C_4H_4O_6$	121
$NH_4C_4H_5O_5$	67
$NH_4C_4H_9COO$	172
$NH_4C_6H_2N_3O_7$	150
$NH_4C_7H_5O_2$	66
$NH_4C_7H_5O_3$	151
$NH_4C_8H_{15}O_2$	71
$NH_4C_{15}H_{31}COO$	139
$NH_4C_{18}H_{33}O_2$	136
$NH_4C_{18}H_{35}O_2$	154
$NH_4Ce(SO_4)_2 \cdot 4H_2O$	74
NH_4Cl	77
NH_4ClO_3	76
NH_4ClO_4	142
$NH_4[Co(NH_3)_2(NO_2)_4]$	727
$NH_4Cr(SO_4)_2 \cdot 12H_2O$	79
NH_4F	95
$NH_4Fe(CrO_4)_2$	89
$NH_4Fe(SO_4)_2 \cdot 12H_2O$	91
$NH_4HB_4O_7 \cdot 3H_2O$	111
NH_4HCO_3	112
$NH_4HC_2O_4 \cdot H_2O$	115
NH_4HF_2	114
NH_4HS	119
NH_4HSO_3	120
NH_4HSO_4	118
NH_4HSeO_4	117
$NH_4H_2AsO_4$	86
$NH_4H_2PO_2$	123
$NH_4H_2PO_4$	87
NH_4I	125
NH_4IO_3	124
NH_4MnO_4	143
NH_4NH_2CSS	88
$NH_4NH_2SO_3$	155
NH_4NO_2	134
NH_4NO_3	133
NH_4N_3	65
$NH_4NiCl_3 \cdot 6H_2O$	131
NH_4OH	122
NH_4PF_6	107
NH_4ReO_4	145
NH_4SCN	166
NH_4SO_3F	97
NH_4VO_3	128
NI_3	1584
NNb	1549
NNd	1496
NO	1576
$NOBr$	1592
$NOCl$	1593
NOF	1594
NOF_3	1595
NO_2	1577
NO_2Cl	1596
NO_2F	1597
NO_2NH_2	1572
NPr	1829
NPu	1708
NTa	2286
NTb	2318
NTh	2366
NTi	2416
NU	2487
NV	2521
NW_2	2448
NZr	2677
N_2	1571
N_2F_2	1589
N_2F_2	1590
N_2F_4	1591
N_2H_4	1003
$N_2H_4 \cdot HBr$	1005
$N_2H_4 \cdot HCl$	1006
$N_2H_4 \cdot 2HCl$	1007
$N_2H_4 \cdot HI$	1008
$N_2H_4 \cdot HNO_3$	1009
$N_2H_4 \cdot H_2O$	1004
$N_2H_4 \cdot H_2SO_4$	1010
$(NH_2)_2SO_2$	2272
$(NH_4)_2B_4O_7 \cdot 4H_2O$	68
$(NH_4)_2CO_3$	73
$(NH_4)_2C_2O_4$	137
$(NH_4)_2C_2O_4 \cdot H_2O$	138
$(NH_4)_2C_4H_4O_6$	160
$(NH_4)_2Ce(NO_3)_6$	75
$(NH_4)_2Co(SO_4)_2 \cdot 6H_2O$	81
$(NH_4)_2CrO_4$	78
$(NH_4)_2Cr_2O_7$	85
$(NH_4)_2Fe(CN)_5NO$	135
$(NH_4)_2Fe(SO_4)_2 \cdot 6H_2O$	94
$(NH_4)_2GeF_6$	106
$(NH_4)_2HAsO_4$	110
$(NH_4)_2HC_6H_5O_7$	113
$(NH_4)_2HPO_3 \cdot H_2O$	147
$(NH_4)_2HPO_4$	116
$(NH_4)_2IrCl_6$	99
$(NH_4)_2Ni(SO_4)_2 \cdot 6H_2O$	132
$(NH_4)_2OsCl_6$	100

Formula	No.
$(NH_4)_2PdCl_6$	101
$(NH_4)_2PtBr_6$	102
$(NH_4)_2PtCl_4$	163
$(NH_4)_2PtCl_6$	103
$(NH_4)_2S$	157
$(NH_4)_2SO_3$	158
$(NH_4)_2SO_3 \cdot H_2O$	159
$(NH_4)_2SO_4$	156
$(NH_4)_2S_2O_3$	167
$(NH_4)_2S_2O_8$	144
$(NH_4)_2SeO_3$	153
$(NH_4)_2SeO_4$	152
$(NH_4)_2SiF_6$	108
$(NH_4)_2TeO_4$	161
$(NH_4)_2TiO(C_2O_4)_2 \cdot H_2O$	168
$(NH_4)_2U_2O_7$	170
$(NH_4)_2WS_4$	165
$(NH_4)_2ZnCl_4$	164
$(NH_4)_2ZrF_6$	109
N_2O	1575
N_2O_3	1578
N_2O_4	1579
N_2O_5	1580
N_2Sr_3	2233
N_2W	2447
N_2Zn_3	2634
$(NH_4)_3AlF_6$	104
$(NH_4)_3Fe(CN)_6 \cdot 3H_2O$	92
$(NH_4)_3Fe(C_2O_4)_3 \cdot 3H_2O$	90
$(NH_4)_3GaF_6$	105
$(NH_4)_3PO_4 \cdot 3H_2O$	146
$(NH_4)_3PO_4 \cdot 12MoO_3$	130
$(NH_4)_3PO_4 \cdot 12MoO_3 \cdot H_2O$	148
$(NH_4)_3PO_4 \cdot 12WO_3 \cdot 2H_2O$	149
$(NH_4)_3ZnCl_5$	141
$(NH_4)_3ZrOH(CO_3)_3 \cdot 2H_2O$	173
N_3Na	2084
N_3Rb	1886
N_3U_2	2488
$(NH_4)_4Fe(CN)_6 \cdot 3H_2O$	93
$(NH_4)_6Mo_7O_{24} \cdot 4H_2O$	129
$(NH_4)_6W_7O_{24} \cdot 6H_2O$	127
N_4Se_4	1598
N_4Si_3	2015
N_5P_3	1639
N_6Pb	1183
$(NH_4)_{10}W_{12}O_{41}$	169
Na	2075
$NaAlCl_4$	2192
$NaAlH_4$	2079
$NaAlO_2$	2078
$NaAl(SO_4)_2 \cdot 12H_2O$	2080
$NaAsO_2$	2083
$NaAuCl_4 \cdot 2H_2O$	2193
$NaBF_4$	2113
$NaBH_3(CN)$	2100
$NaBH_4$	2085
$NaBO_2$	2146
$NaBO_3 \cdot 4H_2O$	2157
$NaBiO_3$	2144
NaBr	2087
$NaBr \cdot 2H_2O$	2088
$NaBrO_3$	2086
$NaCHO_2$	2115
NaCN	2099
NaCNO	2098
$NaC_2H_3O_2$	2076
$NaC_2H_3O_2 \cdot 3H_2O$	2077
NaC_2H_5O	2109
$NaC_{18}H_{35}O_2$	2176
NaCl	2093
NaClO	2140
$NaClO_2$	2094
$NaClO_3$	2092
$NaClO_4$	2158
$NaClO_4 \cdot H_2O$	2159
NaF	2112
NaH	2126
$NaHCO_3$	2129
$NaHF_2$	2130
NaHS	2136
$NaHS \cdot 2H_2O$	2137
$NaHSO_3$	2138
$NaHSO_4$	2134
$NaHSO_4 \cdot H_2O$	2135
NaH_2PO_2	2168
$NaH_2PO_2 \cdot H_2O$	2169
NaH_2PO_4	2102
$NaH_2PO_4 \cdot H_2O$	2103
$NaH_2PO_4 \cdot 2H_2O$	2104
NaI	2143
$NaIO_3$	2142
$NaIO_4$	2160
$NaIO_4 \cdot 3H_2O$	2161
$NaKC_4H_4O_6 \cdot 4H_2O$	2170
$NaMnO_4 \cdot 3H_2O$	2162
$NaNH_2$	2081
$NaNH_4HPO_4 \cdot 4H_2O$	2082
$NaNO_2$	2152
$NaNO_3$	2151
NaN_3	2084
$NaNbO_3$	2150
$NaOCl \cdot 5H_2O$	2141
NaOH	2139
NaO_2	2185
$NaPF_6 \cdot H_2O$	2123
$NaReO_4$	2164
NaSCN	2197
$NaSbF_6$	2122
$NaVO_3$	2207
$NaVO_3 \cdot 4H_2O$	2208
$Na_2B_4O_7$	2188
$Na_2B_4O_7 \cdot 4H_2O$	2191
$Na_2B_4O_7 \cdot 5H_2O$	2190
$Na_2B_4O_7 \cdot 10H_2O$	2189
Na_2BeF_4	2196
Na_2CO_3	2089
$Na_2CO_3 \cdot H_2O$	2091
$Na_2CO_3 \cdot 10H_2O$	2090
$Na_2C_2O_4$	2155
$Na_2C_4H_4O_4 \cdot 6H_2O$	2177
Na_2CrO_4	2095
$Na_2CrO_4 \cdot 4H_2O$	2096
$Na_2Cr_2O_7$	2101
$Na_2[Fe(CN)_5NO] \cdot 2H_2O$	2153
Na_2GeO_3	2116
Na_2HAsO_4	2127
$Na_2HAsO_4 \cdot 7H_2O$	2128
Na_2HPO_4	2131
$Na_2HPO_4 \cdot 7H_2O$	2133
$Na_2HPO_4 \cdot 12H_2O$	2132
$Na_2H_2P_2O_6 \cdot 6H_2O$	2105
$Na_2H_2P_2O_7$	2106
$Na_2IrCl_6 \cdot 6H_2O$	2118
Na_2MoO_4	2148
$Na_2MoO_4 \cdot 2H_2O$	2149
Na_2O	2156
Na_2O_2	2163
Na_2PO_3F	2114
$Na_2PdCl_4 \cdot 3H_2O$	2194
$Na_2PtBr_6 \cdot 6H_2O$	2117
$Na_2PtCl_4 \cdot 4H_2O$	2195
Na_2PtCl_6	2119
$Na_2PtCl_6 \cdot 6H_2O$	2120
Na_2S	2180
$Na_2S \cdot 5H_2O$	2182
$Na_2S \cdot 9H_2O$	2181
Na_2SO_3	2183
$Na_2SO_3 \cdot 7H_2O$	2184
Na_2SO_4	2178
$Na_2SO_4 \cdot 10H_2O$	2179
$Na_2S_2O_3$	2199
$Na_2S_2O_3 \cdot 5H_2O$	2200
$Na_2S_2O_4$	2107
$Na_2S_2O_5$	2145
$Na_2S_2O_6 \cdot 2H_2O$	2108
$Na_2S_2O_8$	2165
Na_2Se	2174
Na_2SeO_3	2175
Na_2SeO_4	2172
$Na_2SeO_4 \cdot 10H_2O$	2173
Na_2SiF_6	2124
Na_2SiO_3	2147
Na_2TeO_3	2187
Na_2TeO_4	2186
$Na_2U_2O_7 \cdot H_2O$	2206
Na_2WO_4	2204
$Na_2WO_4 \cdot 2H_2O$	2205
Na_3AlF_6	2121
$Na_3C_6H_5O_7 \cdot 2H_2O$	2097
$Na_3Co(NO_2)_6$	2125
$Na_3Fe(CN)_6 \cdot H_2O$	2110
$Na_3PO_3S \cdot 12H_2O$	2198
$Na_3PO_4 \cdot 12H_2O$	2166
$Na_3PO_4 \cdot NaOCl$	2167
$Na_3(PO_3)_3$	2201
$Na_3(PO_3)_3 \cdot 6H_2O$	2202
Na_3VO_4	2154
$Na_4Fe(CN)_6 \cdot 10H_2O$	2111
$Na_4P_2O_7$	2171
$Na_4Pb(S_2O_3)_3$	1227
$Na_5P_3O_{10}$	2203
Nb	1544
NbB	1545
NbB_2	1546
$NbBr_3$	1553
$NbBr_5$	1563
NbC	1547
$NbCl_3$	1554
$NbCl_4$	1556
$NbCl_5$	1564
NbF_3	1555
NbF_4	1557
NbF_5	1565
NbI_4	1558
NbI_5	1566
NbN	1549
NbO	1552
$NbOBr_3$	1568
$NbOCl_3$	1569
NbO_2	1559
NbO_2F	1570
NbP	1550
NbS_2	1561
$NbSe_2$	1560
$NbSi_2$	1551
$NbTe_2$	1562
Nb_2C	1548
Nb_2O_5	1567
Nd	1487
NdB_6	1488
$NdBr_3$	1489
$NdCl_3$	1490
$NdCl_3 \cdot 6H_2O$	1491
NdF_3	1492
NdI_3	1493
NdN	1496
$Nd(NO_3)_3$	1494
$Nd(NO_3)_3 \cdot 6H_2O$	1495
Nd_2O_3	1497
Nd_2S_3	1500
$Nd_2(SO_4)_3$	1498
Nd_2Te_3	1503
Ne	1504
Ni	1507
NiAs	1509
NiB	1511
$NiBr_2$	1519
$NiBr_2 \cdot 3H_2O$	1520
$NiCO_3$	1521
$Ni(CN)_2 \cdot 4H_2O$	1524
$Ni(CO)_4$	1513
$NiCl_2$	1522
$NiCl_2 \cdot 6H_2O$	1523
$Ni(ClO_4)_2 \cdot 6H_2O$	1499
NiF_2	1525
NiI_2	1529
$NiI_2 \cdot 6H_2O$	1530
$Ni(IO_3)_2$	1528
$Ni(NH_4)_2(SO4)_2 \cdot 6H_2O$	1517
$Ni(NO_3)_2$	1531
$Ni(NO_3)_2 \cdot 6H_2O$	1532
NiO	1533
$Ni(OH)_2$	1526
$Ni(OH)_2 \cdot H_2O$	1527
NiS	1538
$NiSO_4$	1535
$NiSO_4 \cdot 6H_2O$	1537
$NiSO_4 \cdot 7H_2O$	1536
$Ni(SCN)_2$	1539
NiSb	1508
NiSe	1534

$NiSeO_4 \cdot 6H_2O$	1502
$NiSi_2$	1516
$NiTiO_3$	1540
Ni_2B	1512
Ni_2O_3	1542
Ni_2P	1514
Ni_2Si	1515
$Ni_3(AsO_4)_2 \cdot 8H_2O$	1518
Ni_3B	1510
$Ni_3(PO_4)_2 \cdot 8H_2O$	1501
Ni_3S_2	1543
Ni_3S_4	1541
No	1599
Np	1505
NpO_2	1506
OCS	531
OCSe	530
OPb	1213
OPb	1214
OPd	1619
OPt	1691
OPu	1709
ORb_2	1904
OSi	2016
OSn	2395
OSr	2235
OTi	2422
OTl_2	2342
OV	2528
OZn	2639
O_2	1610
O_2Os	1606
O_2Pb	1247
O_2Po	1720
O_2Pt	1700
O_2Pu	1716
O_2Rb	1910
O_2Rb_2	1906
O_2Re	1852
O_2Rh	1880
O_2Ru	1919
O_2S	2260
O_2Se	1956
O_2Si	2020
O_2Si	2019
O_2Si	2017
O_2Si	2021
O_2Si	2018
O_2Sn	2408
O_2Sr	2238
O_2Ta	2288
O_2Te	2303
O_2Th	2367
O_2Ti	2435
O_2U	2498
O_2V	2542
O_2W	2460
O_2Zn	2643
O_2Zr	2672
O_3	1611
O_3P_2	1651
O_3Pb_2	1241
O_3Pr_2	1830
O_3Pu_2	1714
O_3Re	1862
O_3Rh_2	1878
O_3S	2261
O_3Sb_2	182
O_3Sb_2	181
O_3Sc_2	1948
O_3Se	1957
O_3Sm_2	1937
O_3Tb_2	2319
O_3Te	2304
O_3Ti_2	2427
O_3Tl_2	2353
O_3Tm_2	2381
O_3U	2506
O_3V_2	2536
O_3W	2476
O_3Xe	2557
O_3Y_2	2595
O_3Yb_2	2580
O_4Os	1609
O_4Pb_3	1242
O_4Ru	1922
O_4Sb_2	188
O_4Xe	2558
O_5P_2	1664
O_5Re_2	1858
O_5Sb_2	192
O_5Ta_2	2296
O_5Ti_3	2430
O_5V_2	2544
O_6P_4	1652
O_7Re_2	1866
O_8U_3	2503
O_9U_4	2499
Os	1600
$OsBr_3$	1602
$OsCl_3$	1603
$OsCl_4$	1604
OsF_4	1605
OsF_5	1607
OsF_6	1608
OsO_2	1606
OsO_4	1609
$Os_3(CO)_{12}$	1601
P	1622
P	1620
P	1621
$PBrF_2$	1645
PBr_2F	1644
PBr_2F_3	1657
PBr_3	1643
PBr_4F	1656
PBr_5	1655
$PClF_2$	1648
$PClF_4$	1662
PCl_2F	1647
PCl_2F_3	1661
PCl_3	1646
PCl_3F_2	1660
PCl_4F	1659
PCl_5	1658
PF_3	1649
PF_5	1663
PH_3	1623
PH_4Cl	1628
PH_4I	1629
PI_3	1650
$POBrCl_2$	1671
POBrClF	1673
$POBrF_2$	1672
$POBr_2Cl$	1669
$POBr_2F$	1670
$POBr_3$	1668
$POClF_2$	1676
$POCl_2F$	1675
$POCl_3$	1674
POF_2H	1667
POF_3	1677
POI_3	1678
$PSBrF_2$	1681
$PSBr_2F$	1680
$PSBr_3$	1679
$PSClF_2$	1684
$PSCl_2F$	1683
$PSCl_3$	1682
PSF_3	1685
PSI_3	1686
PTi	2417
PY	2596
P_2Cl_4	1625
P_2F_4	1626
P_2H_4	1624
P_2I_4	1627
P_2O_3	1651
P_2O_5	1664
P_2S_3	1654
P_2S_5	1666
P_2Se_3	1653
P_2Se_5	1665
P_2Zn_3	2646
P_2Zr	2678
$(PNCl_2)_3$	1642
P_3N_5	1639
P_4O_6	1652
P_4S_3	1640
P_4S_7	1641
Pa	1835
$PaCl_5$	1836
Pb	1176
$Pb(AsO_2)_2$	1182
$Pb(BF_4)_2$	1199
$Pb(BO_2)_2 \cdot H_2O$	1184
$PbBr_2$	1186
$Pb(BrO_3)_2 \cdot H_2O$	1185
$PbBr_4$	1244
$PbCO_3$	1188
$Pb(CHO_2)_2$	1200
$Pb(CN)_2$	1196
PbC_2O_4	1212
$Pb(C_2H_3O_2)_2$	1177
$Pb(C_2H_3O_2)_2 \cdot 3H_2O$	1178
$Pb(C_2H_3O_2)2 \cdot 2Pb(OH)_2$	1179
$Pb(C_3H_5O_3)_2$	1207
$Pb(C_4H_7O_2)_2$	1187
$Pb(C_2H_3O_2)_4$	1243
$Pb(CF_3COCHCOCF_3)_2$	1201
$Pb(CH_3COCHCOCH_3)_2$	1216
$Pb(C_7H_{15}CO_2)_2$	1197
$Pb(C_{18}H_{33}O_2)_2$	1211
$Pb(C_{18}H_{35}O_2)_2$	1228
PbClF	1192
$PbCl_2$	1191
$Pb(ClO_3)_2$	1190
$Pb(ClO_4)_2$	1217
$Pb(ClO_4)_2 \cdot 3H_2O$	1218
$PbCl_4$	1245
$PbCrO_4$	1193
$PbCrO_4 \cdot PbO$	1194
PbF_2	1198
PbF_4	1246
$PbHAsO_4$	1202
$PbHPO_4$	1203
$Pb(H_2PO_2)_2$	1220
PbI_2	1206
$Pb(IO_3)_2$	1205
$PbMoO_4$	1208
$Pb(NO_3)_2$	1210
$Pb(N_3)_2$	1183
$Pb(NbO_3)_2$	1209
PbO	1213
PbO	1214
$3PbO \cdot H_2O$	1215
PbO_2	1247
$Pb(OH)_2$	1204
$Pb(OH)_2 \cdot 2PbCO_3$	1189
PbS	1230
$PbSO_3$	1231
$PbSO_4$	1229
$Pb(SCN)_2$	1234
PbS_2O_3	1235
PbSe	1225
$PbSeO_3$	1226
$PbSeO_4$	1224
$PbSiF_6 \cdot 2H_2O$	1223
$PbSiO_3$	1221
$Pb(TaO_3)_2$	1232
PbTe	1233
$PbTiO_3$	1236
$Pb(VO_3)_2$	1239
$PbWO_4$	1237
$PbWO_4$	1238
$PbZrO_3$	1240
Pb_2O_3	1241
Pb_2SiO_4	1222
$Pb_3(AsO_4)_2$	1181
$Pb_3(C_6H_5O_7)_2 \cdot 3H_2O$	1195
Pb_3O_4	1242
$Pb_3(PO_4)_2$	1219
$Pb_3(SbO_4)_2$	1180
Pd	1612
$PdBr_2$	1614
$PdCl_2$	1615
PdF_2	1616
PdI_2	1617
$Pd(NO_3)_2$	1618
PdO	1619
PdS	1613
Pm	1834

Po	1718
$PoCl_4$	1719
PoO_2	1720
Pr	1820
PrB_6	1821
$PrBr_3$	1822
$PrCl_3$	1823
$PrCl_3 \cdot 7H_2O$	1824
PrF_3	1825
PrI_3	1826
PrN	1829
$Pr(NO_3)_3$	1827
$Pr(NO_3)_3 \cdot 6H_2O$	1828
$PrSi_2$	1831
Pr_2O_3	1830
Pr_2S_3	1832
Pr_2Te_3	1833
Pt	1687
$PtBr_2$	1688
$PtBr_3$	1693
$PtBr_4$	1695
$PtCl_2$	1689
$PtCl_3$	1694
$PtCl_4$	1696
$PtCl_4 \cdot 5H_2O$	1697
PtF_4	1698
PtF_6	1702
PtI_2	1690
PtI_4	1699
$Pt(NH_3)_2Cl_2$	1703
$Pt(NH_3)_2Cl_2$	1704
PtO	1691
PtO_2	1700
PtS	1692
PtS_2	1701
PtSi	1706
Pu	1707
$PuBr_3$	1710
$PuCl_3$	1711
PuF_3	1712
PuF_4	1715
PuF_6	1717
PuI_3	1713
PuN	1708
PuO	1709
PuO_2	1716
Pu_2O_3	1714
Ra	1837
$RaBr_2$	1838
$RaCl_2$	1839
RaF_2	1840
$Ra(NO_3)_2$	1841
$RaSO_4$	1842
Rb	1882
$RbAl(SO_4)_2$	1884
$RbAl(SO_4)_2 \cdot 12H_2O$	1885
RbBr	1888
$RbBrO_3$	1887
$RbCHO_2$	1896
RbCN	1893
$RbC_2H_3O_2$	1883
RbCl	1891
$RbClO_3$	1890
$RbClO_4$	1905
RbF	1894
RbH	1897
$RbHCO_3$	1898
$RbHF_2$	1895
$RbHSO_4$	1899
RbI	1902
$RbIO_3$	1901
$RbNO_3$	1903
RbN_3	1886
RbOH	1900
RbO_2	1910
Rb_2CO_3	1889
Rb_2CrO_4	1892
Rb_2O	1904
Rb_2O_2	1906
Rb_2S	1909
Rb_2SO_4	1908
Rb_2Se	1907
Re	1844
$ReBr_3$	1847
$ReBr_5$	1855
$ReCl_3$	1848
$ReCl_4$	1850
$ReCl_5$	1856
$ReCl_6$	1859
ReF_4	1851
ReF_5	1857
ReF_6	1861
ReF_7	1865
ReI_3	1849
$ReOCl_4$	1863
$ReOF_4$	1864
$ReOF_5$	1870
ReO_2	1852
ReO_2F_2	1860
ReO_2F_3	1869
ReO_3	1862
ReO_3Cl	1867
ReO_3F	1868
ReS_2	1853
$ReTe_2$	1854
$Re_2(CO)_{10}$	1846
Re_2O_5	1858
Re_2O_7	1866
Re_2S_7	1871
Rh	1872
$[Rh(CO)_2Cl]_2$	1873
$RhCl_3$	1875
RhF_3	1876
RhF_6	1881
RhI_3	1877
RhO_2	1880
Rh_2O_3	1878
$Rh_2(SO_4)_3$	1879
$Rh_4(CO)_{12}$	1874
Rn	1843
Ru	1911
$RuBr_3$	1914
$Ru(CH_3COCHCOCH_3)_3$	1913
$RuCl_3$	1915
RuF_3	1916
RuF_4	1918
RuF_5	1920
RuF_6	1921
RuI_3	1917
RuO_2	1919
RuO_4	1922
$Ru_3(CO)_{12}$	1912
S	2252
S	2251
SCl_2	2266
SF_4	2267
SF_5Br	2269
SF_5Cl	2270
SF_6	2268
$SOBr_2$	2276
$SOCl_2$	2277
SOF_2	2278
SO_2	2260
SO_2Cl_2	2273
SO_2F_2	2274
$SO_2(OH)Cl$	2256
$SO_2(OH)F$	2257
SO_3	2261
$SSBr_2$	2262
$SSCl_2$	2263
SSF_2	2264
SSi	2022
SSn	2399
SSr	2245
STi	2423
STl_2	2347
SZn	2657
SZn	2656
S_2F_{10}	2271
$S_2O_5Cl_2$	2275
S_2Se	1968
S_2Se_6	1971
S_2Si	2023
S_2Sn	2411
S_2Ta	2290
S_2Th	2372
S_2Ti	2438
S_2W	2462
S_2Zr	2676
S_3Sb_2	186
S_3Sc_2	1949
S_3Sm_2	1939
S_3Tb_2	2321
S_3Ti_2	2429
S_3V_2	2538
S_3W	2480
S_3Y_2	2598
S_4Se_4	1970
S_5Sb_2	193
S_6Se_2	1969
Sb	174
SbAs	176
$SbBr_3$	177
$SbCl_2F_3$	191
$SbCl_3$	178
$SbCl_5$	189
SbF_3	179
SbF_5	190
SbH_3	175
SbI_3	180
SbOCl	183
SbY	2584
SbZn	2602
Sb_2O_3	182
Sb_2O_3	181
Sb_2O_4	188
Sb_2O_5	192
Sb_2S_3	186
$Sb_2(SO_4)_3$	185
Sb_2S_5	193
Sb_2Se_3	184
Sb_2Te_3	187
Sc	1941
ScB_2	1942
$ScBr_3$	1943
$ScCl_3$	1944
ScF_3	1945
$Sc(NO_3)_3$	1947
$Sc(OH)_3$	1946
Sc_2O_3	1948
Sc_2S_3	1949
Sc_2Te_3	1950
Se	1951
Se	1952
Se	1953
$SeBr_4$	1960
$SeCl_4$	1961
SeF_4	1962
SeF_6	1963
$SeOBr_2$	1964
$SeOCl_2$	1965
$SeOF_2$	1966
SeO_2	1956
SeO_2F_2	1967
SeO_3	1957
SeS_2	1968
SeSn	2397
SeSr	2241
SeTl	2355
$SeTl_2$	2345
SeZn	2649
Se_2Br_2	1958
Se_2Cl_2	1959
Se_2S_6	1969
Se_2Sn	2409
Se_2Ta	2289
Se_2Th	2368
Se_2W	2461
Se_4S_4	1970
Se_6S_2	1971
Si	1972
$SiBr_2Cl_2$	1991
$SiBr_3Cl$	2004
$SiBr_4$	1999
SiC	2014
$SiClF_3$	1990
$SiCl_2F_2$	1994
$SiCl_3Br$	1988
$SiCl_3F$	2006
$SiCl_3I$	2007
$SiCl_4$	2000
SiF_4	2001

$SiHBr_3$	2003
$SiHCl_3$	2005
$SiHF_3$	2008
$SiHI_3$	2009
SiH_2Br_2	1992
SiH_2Cl_2	1993
SiH_2F_2	1995
SiH_2I_2	1996
SiH_3Br	1987
SiH_3Cl	1989
SiH_3F	1997
SiH_3I	1998
SiH_4	1973
SiI_4	2002
SiO	2016
SiO_2	2020
SiO_2	2019
SiO_2	2017
SiO_2	2021
SiO_2	2018
SiS	2022
SiS_2	2023
SiV_3	2523
Si_2H_6	1974
$(SiH_3)_2O$	2010
Si_2Sm	1925
Si_2Sr	2243
Si_2Ta	2287
Si_2Tb	2320
Si_2Th	2370
Si_2Ti	2418
Si_2V	2522
Si_2W	2449
Si_2Yb	2572
Si_2Zr	2679
Si_3H_8	1975
Si_3N_4	2015
Si_3W_5	2450
Si_4H_{10}	1976
Si_4H_{10}	1977
Si_5H_{10}	1985
Si_5H_{12}	1978
Si_5H_{12}	1979
Si_5H_{12}	1980
Si_6H_{12}	1986
Si_6H_{14}	1981
Si_6H_{14}	1983
Si_6H_{14}	1982
Si_7H_{16}	1984
Sm	1923
SmB_6	1924
$SmBr_2$	1926
$SmBr_3$	1930
$SmCl_2$	1927
$SmCl_3$	1931
$SmCl_3 \cdot 6H_2O$	1932
SmF_2	1928
SmF_3	1933
SmI_2	1929
SmI_3	1934
$Sm(NO_3)_3$	1935
$Sm(NO_3)_3 \cdot 6H_2O$	1936
$SmSi_2$	1925
Sm_2O_3	1937
Sm_2S_3	1939
$Sm_2(SO_4)_3 \cdot 8H_2O$	1938
Sm_2Te_3	1940
Sn	2382
Sn	2383
$SnBr_2$	2387
$SnBr_4$	2402
SnC_2O_4	2394
$SnC_4H_4O_6$	2400
$Sn(C_2H_3O_2)_2$	2386
$SnCl_2$	2388
$SnCl_2 \cdot 2H_2O$	2389
$SnCl_4$	2403
$SnCl_4 \cdot 5H_2O$	2404
$Sn(CrO_4)_2$	2405
SnF_2	2390
SnF_4	2406
SnH_3CH_3	2385
SnH_4	2384
SnI_2	2393
SnI_4	2407
SnO	2395
SnO_2	2408
$Sn(OH)_2$	2392
SnS	2399
$SnSO_4$	2398
SnS_2	2411
$SnSe$	2397
$SnSe_2$	2409
$Sn(SeO_3)_2$	2410
$SnTe$	2401
$SnZrF_6$	2391
$Sn_2P_2O_7$	2396
Sr	2209
$Sr(AsO_2)_2 \cdot 4H_2O$	2210
SrB_6	2225
$SrBr_2$	2212
$SrBr_2 \cdot 6H_2O$	2213
$Sr(BrO_3)_2 \cdot H_2O$	2211
$SrCO_3$	2215
SrC_2	2214
$Sr(CHO_2)_2$	2223
$Sr(CHO_2)_2 \cdot 2H_2O$	2224
$Sr(CN)_2 \cdot 4H_2O$	2220
$SrCl_2$	2217
$SrCl_2 \cdot 6H_2O$	2218
$Sr(ClO_3)_2$	2216
$Sr(ClO_4)_2$	2236
$SrCrO_4$	2219
SrF_2	2222
$SrFe(CN)_6 \cdot 15H_2O$	2221
SrH_2	2226
SrI_2	2229
$SrI_2 \cdot 6H_2O$	2230
$Sr(IO_3)_2$	2228
$Sr(MnO_4)_2 \cdot 3H_2O$	2237
$Sr(NO_2)_2$	2234
$Sr(NO_3)_2$	2232
$SrNb_2O_6$	2231
SrO	2235
SrO_2	2238
$Sr(OH)_2$	2227
SrS	2245
$SrSO_3$	2246
$SrSO_4$	2244
$SrS_2O_3 \cdot 5H_2O$	2248
$SrSe$	2241
$SrSeO_4$	2240
$SrSi_2$	2243
$SrTe$	2247
$SrTiO_3$	2249
$SrWO_4$	2250
Sr_2SiO_4	2242
Sr_3N_2	2233
$Sr_3(PO_4)_2$	2239
Ta	2280
$TaAl_3$	2281
TaB	2282
TaB_2	2283
$TaBr_5$	2292
TaC	2284
$TaCl_5$	2293
TaF_5	2294
TaI_5	2295
TaN	2286
TaO_2	2288
TaS_2	2290
$TaSe_2$	2289
$TaSi_2$	2287
$TaTe_2$	2291
Ta_2C	2285
Ta_2O_5	2296
Tb	2312
$TbCl_3$	2313
$TbCl_3 \cdot 6H_2O$	2314
TbI_3	2315
TbN	2318
$Tb(NO_3)_3$	2316
$Tb(NO_3)_3 \cdot 6H_2O$	2317
$TbSi_2$	2320
Tb_2O_3	2319
Tb_2S_3	2321
Tc	2297
TcF_5	2298
TcF_6	2299
Te	2300
$TeBr_2$	2305
$TeBr_4$	2307
$TeCl_2$	2306
$TeCl_4$	2308
TeF_4	2309
TeF_6	2311
TeI_4	2310
TeO_2	2303
TeO_3	2304
$TeZn$	2659
Te_2W	2463
Th	2356
ThB_6	2358
$ThBr_4$	2359
ThC	2360
ThC_2	2361
$ThCl_4$	2362
ThF_4	2363
ThH_2	2357
ThI_4	2364
ThN	2366
$Th(NO_3)_4 \cdot 4H_2O$	2365
ThO_2	2367
ThS_2	2372
$Th(SO_4)_2 \cdot 9H_2O$	2371
$ThSe_2$	2368
$ThSiO_4$	2369
$ThSi_2$	2370
Ti	2412
TiB_2	2414
$TiBr_2$	2419
$TiBr_3$	2424
$TiBr_4$	2431
TiC	2415
$TiCl_2$	2420
$TiCl_3$	2425
$TiCl_4$	2432
TiF_3	2426
TiF_4	2433
TiH_2	2413
TiI_2	2421
TiI_4	2434
TiN	2416
TiO	2422
$TiOSO_4 \cdot H_2O$	2436
TiO_2	2435
TiP	2417
TiS	2423
TiS_2	2438
$Ti(SO_4)_2$	2437
$TiSi_2$	2418
Ti_2O_3	2427
Ti_2S_3	2429
$Ti_2(SO_4)_3$	2428
Ti_3O_5	2430
Tl	2322
$TlBr$	2325
$TlBrO_3$	2324
$TlBr_3 \cdot 4H_2O$	2348
$TlCHO_2$	2333
$TlCN$	2330
$TlC_2H_3O_2$	2323
TlC_2H_5O	2331
$TlCl$	2328
$TlClO_3$	2327
$TlClO_4$	2343
$TlCl_3$	2349
$TlCl_3 \cdot 4H_2O$	2350
TlF	2332
TlF_3	2351
TlI	2337
$TlIO_3$	2336
$TlNO_2$	2340
$TlNO_3$	2339
$Tl(NO_3)_3$	2352
$TlOH$	2335
$TlPF_6$	2334
$TlSe$	2355
Tl_2CO_3	2326
$Tl_2C_2O_4$	2341
Tl_2CrO_4	2329
Tl_2MoO_4	2338

Tl_2O	2342
Tl_2O_3	2353
Tl_2S	2347
Tl_2SO_4	2346
$Tl_2(SO_4)_3$	2354
Tl_2Se	2345
Tl_2SeO_4	2344
Tm	2373
$TmBr_3$	2374
$TmCl_3$	2375
$TmCl_3 \cdot 7H_2O$	2376
TmF_3	2377
TmI_3	2378
$Tm(NO_3)_3$	2379
$Tm(NO_3)_3 \cdot 5H_2O$	2380
Tm_2O_3	2381
U	2481
UB_2	2482
UB_4	2483
UBr_3	2489
UBr_4	2494
UBr_5	2500
UC	2484
UC_2	2485
UCl_3	2490
UCl_4	2495
UCl_5	2501
UCl_6	2504
UF_3	2491
UF_4	2496
UF_5	2502
UF_6	2505
UH_3	2492
UI_3	2493
UI_4	2497
UN	2487
UO_2	2498
UO_2Cl_2	2509
UO_2F_2	2510
$UO_2(NO_3)_2$	2511
$UO_2(NO_3)_2 \cdot 6H_2O$	2512
$UO_2(NH_4)_3F_5$	171
UO_2SO_4	2513
$UO_2SO_4 \cdot 3H_2O$	2514
UO_3	2506
$UO_3 \cdot H_2O$	2507
$UO_4 \cdot 2H_2O$	2508
U_2C_3	2486
U_2N_3	2488
U_3O_8	2503
U_4O_9	2499
V	2515
VB	2516
VB_2	2517
VBr_2	2524
VBr_3	2531
VBr_4	2539
VC	2518
$V(CO)_6$	2520
$V(CH_3COCHCOCH_3)_3$	2530
VCl_2	2525
VCl_3	2532
VCl_4	2540
VF_2	2526
VF_3	2533
$VF_3 \cdot 3H_2O$	2534
VF_4	2541
VF_5	2543
VI_2	2527
VI_3	2535
VN	2521
VO	2528
$VOBr$	2545
$VOBr_2$	2547
$VOBr_3$	2552
$VOCl$	2546
$VOCl_2$	2548
$VOCl_3$	2553
VOF_2	2549
VOF_3	2554
$VOSO_4 \cdot 2H_2O$	2551
$VOSeO_3 \cdot H_2O$	2550
VO_2	2542
$VSO_4 \cdot 7H_2O$	2529
VSi_2	2522
V_2C	2519
V_2O_3	2536
V_2O_5	2544
V_2S_3	2538
$V_2(SO_4)_3$	2537
V_3Si	2523
W	2439
WB	2442
WBr_2	2451
WBr_3	2454
WBr_4	2456
WBr_5	2464
WBr_6	2470
WC	2445
$W(CO)_6$	2446
$W(C_2H_5O)_5$	2466
WCl_2	2452
WCl_3	2455
WCl_4	2457
WCl_5	2465
WCl_6	2471
WF_4	2458
WF_5	2467
WF_6	2475
WI_2	2453
WI_4	2459
WN_2	2447
$WOBr_3$	2468
$WOBr_4$	2477
$WOCl_3$	2469
$WOCl_4$	2478
WOF_4	2479
WO_2	2460
WO_2Br_2	2472
WO_2Cl_2	2473
WO_2I_2	2474
WO_3	2476
WS_2	2462
WS_3	2480
WSe_2	2461
WSi_2	2449
WTe_2	2463
W_2B	2441
W_2B_5	2443
W_2C	2444
W_2N	2448
W_5Si_3	2450
Xe	2556
$XeFRuF_6$	2564
$XeFSb_2F_{11}$	2565
XeF_2	2559
XeF_3SbF_6	2567
$XeF_3Sb_2F_{11}$	2568
XeF_4	2560
XeF_5AsF_6	2569
XeF_5RuF_6	2570
XeF_6	2561
$XeOF_4$	2563
XeO_2F_2	2562
XeO_3	2557
XeO_4	2558
$Xe_2F_3AsF_6$	2566
Y	2582
YAs	2585
YB_6	2586
YBr_3	2587
YC_2	2588
YCl_3	2590
YF_3	2591
$Y(NO_3)_3$	2592
$Y(NO_3)_3 \cdot 4H_2O$	2593
$Y(NO_3)_3 \cdot 6H_2O$	2594
YP	2596
YSb	2584
$Y_2(CO_3)_3 \cdot 3H_2O$	2589
Y_2O_3	2595
Y_2S_3	2598
$Y_2(SO_4)_3 \cdot 8H_2O$	2597
$Y_3Al_5O_{12}$	2583
Yb	2571
$YbBr_2$	2573
$YbCl_2$	2574
$YbCl_3$	2576
$YbCl_3 \cdot 6H_2O$	2577
YbF_3	2578
YbI_2	2575
$Yb(NO_3)_3$	2579
$YbSi_2$	2572
Yb_2O_3	2580
$Yb_2(SO_4)_3 \cdot 8H_2O$	2581
Zn	2599
$Zn(AsO_2)_2$	2606
$Zn(BF_4)_2 \cdot 6H_2O$	2624
$ZnBr_2$	2611
$Zn(BrO_3)_2 \cdot 6H_2O$	2610
$ZnCO_3$	2613
$Zn(CHO_2)_2 \cdot 2H_2O$	2625
$Zn(CN)_2$	2620
ZnC_2O_4	2637
$ZnC_2O_4 \cdot 2H_2O$	2638
$Zn(C_2H_3O_2)_2 \cdot 2H_2O$	2600
$Zn(C_8H_{15}O_2)_2$	2612
$Zn(CH_3COCHCOCH_3)_2$	2640
$Zn(C_{12}H_{23}O_2)_2$	2630
$Zn(C_{18}H_{33}O_2)_2$	2636
$Zn(C_{18}H_{35}O_2)_2$	2652
$ZnCl_2$	2616
$Zn(ClO_3)_2$	2615
$Zn(ClO_4)_2 \cdot 6H_2O$	2641
$ZnCrO_4$	2617
$ZnCr_2O_4$	2618
ZnF_2	2622
$ZnF_2 \cdot 4H_2O$	2623
ZnI_2	2629
$Zn(IO_3)_2$	2628
$Zn(MnO_4)_2 \cdot 6H_2O$	2642
$ZnMoO_4$	2631
$Zn(NH_4)_2(SO_4)_2$	2601
$Zn(NO_2)_2$	2635
$Zn(NO_3)_2$	2632
$Zn(NO_3)_2 \cdot 6H_2O$	2633
ZnO	2639
$3ZnO \cdot 2B_2O_3$	2607
$2ZnO \cdot 3B_2O_3 \cdot 3.5H_2O$	2608
$2ZnO \cdot 3B_2O_3 \cdot 5H_2O$	2609
ZnO_2	2643
$Zn(OH)_2$	2627
$3Zn(OH)_2 \cdot 2ZnCO_3$	2614
ZnS	2657
ZnS	2656
$ZnSO_3 \cdot 2H_2O$	2658
$ZnSO_4$	2653
$ZnSO_4 \cdot H_2O$	2654
$ZnSO_4 \cdot 7H_2O$	2655
$Zn(SCN)_2$	2660
ZnS_2O_4	2621
$ZnSb$	2602
$ZnSe$	2649
$ZnSeO_3$	2651
$ZnSeO_4 \cdot 5H_2O$	2648
$ZnSiF_6 \cdot 6H_2O$	2626
$ZnTe$	2659
$Zn_2P_2O_7$	2647
Zn_2SiO_4	2650
Zn_3As_2	2605
$Zn_3(AsO_4)_2$	2603
$Zn_3(AsO_4)_2 \cdot 8H_2O$	2604
$Zn_3(C_6H_5O_7)_2 \cdot 2H_2O$	2619
Zn_3N_2	2634
Zn_3P_2	2646
$Zn_3(PO_4)_2$	2644
$Zn_3(PO_4)_2 \cdot 4H_2O$	2645
Zr	2661
ZrB_2	2662
$ZrBr_4$	2666
ZrC	2663
$ZrCl_2$	2664
$ZrCl_4$	2667
ZrF_4	2668
ZrH_2	2665
ZrI_4	2670
ZrN	2677
$Zr(NO_3)_4 \cdot 5H_2O$	2671
$ZrOCl_2$	2680
$ZrOCl_2 \cdot 8H_2O$	2681
ZrO_2	2672
$Zr(OH)_4$	2669

ZrP_2 2678
ZrS_2 2676
$Zr(SO_4)_2$ 2674
$Zr(SO_4)_2 \cdot 4H_2O$ 2675
$ZrSiO_4$ 2673
$ZrSi_2$ 2679

62-54-4 447
62-76-0 2155
71-48-7 677
74-90-8 1018
75-15-0 529
75-20-7 456
75-44-5 537
102-54-5 1076
124-38-9 527
126-45-4 2038
127-08-2 1722
127-09-3 2076
128-00-7 2049
131-74-8 150
136-51-6 469
141-52-6 2109
141-53-7 2115
142-03-0 22
142-71-2 763
142-71-2 761
142-72-3 1299
143-33-9 2099
144-55-8 2129
146-84-9 2060
149-11-1 740
150-90-3 2177
151-50-8 1740
156-62-7 466
289-22-5 1985
291-59-8 1986
298-14-6 1760
301-04-2 1177
301-08-6 1197
302-01-2 1003
304-59-6 2170
333-20-0 1813
353-50-4 538
409-21-2 2014
460-19-5 539
463-58-1 531
471-34-1 457
471-34-1 458
497-19-8 2089
504-64-3 534
506-61-6 1798
506-64-9 2039
506-65-0 963
506-66-1 316
506-68-3 540
506-77-4 541
506-78-5 543
506-80-9 528
506-87-6 73
507-16-4 2276
512-25-4 236
512-26-5 1195
513-74-6 88
513-77-9 229
513-78-0 413
513-79-1 684
516-02-9 269
528-94-9 151
533-51-7 2054
534-16-7 2033
534-17-8 574
535-37-5 968
538-17-0 54
540-16-9 771
540-69-2 98
540-72-7 2197
541-43-5 246
542-62-1 241
542-83-6 419
542-84-7 690
543-80-6 221
543-81-7 308
543-90-8 406
544-17-2 471
544-18-3 695
544-19-4 789
544-60-5 136
544-92-3 748
546-46-3 2619
546-67-8 1243
546-89-4 1249
546-93-0 1307
547-66-0 1326
547-68-2 2638
547-68-2 2637
554-13-2 1258
555-35-1 43
555-75-9 23
555-76-0 1138
556-63-8 1266
556-65-0 1286
557-05-1 2652
557-07-3 2636
557-09-5 2612
557-19-7 1524
557-21-1 2620
557-39-1 1313
557-42-6 2660
562-76-5 1811
562-81-2 293
563-63-3 2027
563-67-7 1883
563-68-8 2323
563-71-3 1096
563-72-4 489
583-52-8 1783
584-08-7 1734
584-09-8 1889
590-28-3 1739
590-29-4 1750
592-01-8 467
592-04-1 1429
592-05-2 1196
592-85-8 1450
592-87-0 1234
592-89-2 2224
592-89-2 2223
593-26-0 139
593-95-3 536
598-54-9 743
598-62-9 1369
598-63-0 1188
627-34-9 535
628-52-4 625
628-86-4 1432
630-08-0 526
631-60-7 1405
631-61-8 64
637-12-7 49
638-39-1 2386
660-60-6 811
688-37-9 38
811-54-1 1200
813-93-4 339
814-88-0 430
814-89-1 707
814-89-1 706
814-90-4 631
814-91-5 801
814-91-5 802
814-93-7 1212
814-94-8 2394
815-82-7 816
815-85-0 2400
819-73-8 1187
822-16-2 2176
865-44-1 1058
868-14-4 1769
917-61-3 2098
917-69-1 726
940-71-6 1642
992-98-3 2333
1002-89-7 154
1066-30-4 635
1066-30-4 634
1066-33-7 112
1072-35-1 1228
1111-67-7 759
1111-71-3 321
1111-78-0 72
1113-38-8 137
1117-94-8 744
1120-44-1 800
1120-46-3 1211
1184-65-2 939
1299-86-1 18
1301-96-8 2071
1302-01-8 2026
1302-09-6 2062
1302-42-7 2078
1302-52-9 311
1302-74-5 41
1302-81-4 52
1302-82-5 46
1303-00-0 900
1303-11-3 1028
1303-28-2 215
1303-33-9 211
1303-34-0 217
1303-35-1 202
1303-36-2 210
1303-37-3 216
1303-52-2 970
1303-58-8 972
1303-60-2 965
1303-61-3 975
1303-62-4 974
1303-86-2 392
1303-96-4 2189
1304-28-5 271
1304-29-6 275
1304-39-8 280
1304-40-1 284
1304-54-7 327
1304-56-9 328
1304-76-3 350
1304-82-1 362
1304-85-4 359
1305-62-0 477
1305-78-8 491
1305-79-9 496
1305-84-6 503
1305-99-3 499
1306-19-0 432
1306-23-6 441
1306-24-7 437
1306-25-8 442
1306-38-3 566
1307-86-4 730
1307-96-6 708
1307-99-9 713
1308-04-9 732
1308-06-1 725
1308-14-1 643
1308-31-2 1100
1308-38-9 647
1308-56-1 786
1308-80-1 739
1308-87-8 835
1308-96-9 871
1309-33-7 1139
1309-37-1 1145
1309-42-8 1317
1309-48-4 1328
1309-55-3 1397
1309-56-4 1475
1309-60-0 1247
1309-64-4 182
1309-64-4 181
1310-03-8 1223
1310-32-3 1113
1310-43-6 1086
1310-52-7 1314
1310-53-8 955
1310-58-3 1770
1310-61-8 1766
1310-61-8 1767
1310-65-2 1268
1310-66-3 1269
1310-73-2 2139
1310-82-3 1900
1311-11-1 1215
1311-90-6 149
1312-03-4 1441
1312-41-0 1027
1312-42-1 1046
1312-43-2 1043
1312-45-4 1049
1312-46-5 1070
1312-73-8 1802
1312-73-8 1801
1312-74-9 1796
1312-81-8 1170
1312-99-8 1353
1313-04-8 1338
1313-08-2 1345
1313-13-9 1401
1313-22-0 1387
1313-27-5 1485
1313-29-7 1469
1313-49-1 2634
1313-59-3 2156
1313-60-6 2163
1313-82-2 2181
1313-82-2 2180
1313-83-3 2182
1313-84-4 2181
1313-85-5 2174
1313-96-8 1567
1313-97-9 1497
1313-99-1 1533
1314-04-1 1538
1314-05-2 1534
1314-06-3 1542
1314-08-5 1619
1314-11-0 2235
1314-12-1 2342
1314-13-2 2639
1314-15-4 1700
1314-18-7 2238
1314-20-1 2367
1314-22-3 2643
1314-23-4 2672
1314-24-5 1651
1314-27-8 1241
1314-28-9 1862
1314-32-5 2353
1314-34-7 2536
1314-35-8 2476
1314-36-9 2595
1314-37-0 2580
1314-41-6 1242
1314-56-3 1664
1314-60-9 192
1314-61-0 2296
1314-62-1 2544

1314-64-3	2514	1335-26-8	1332	3495-35-0	1896	7429-90-5	7	7440-43-9	405
1314-64-3	2513	1335-31-5	1442	3495-36-1	579	7429-91-6	823	7440-44-0	521
1314-68-7	1866	1335-32-6	1179	3522-50-7	1133	7429-92-7	837	7440-44-0	522
1314-80-3	1666	1336-21-6	122	3687-31-8	1181	7439-88-5	1063	7440-45-1	544
1314-82-5	1665	1341-49-7	114	3811-04-9	1736	7439-89-6	1075	7440-46-2	568
1314-84-7	2646	1343-98-2	2011	3931-89-3	1679	7439-90-9	1154	7440-47-3	611
1314-85-8	1640	1343-98-2	2012	3982-91-0	1682	7439-91-0	1156	7440-48-4	664
1314-86-9	1653	1344-28-1	41	4075-81-4	500	7439-92-1	1176	7440-50-8	737
1314-87-0	1230	1344-43-0	1382	4109-96-0	1993	7439-93-2	1248	7440-51-9	822
1314-91-6	1233	1344-48-5	1448	4119-52-2	1152	7439-94-3	1287	7440-52-0	838
1314-95-0	2399	1344-48-5	1447	4255-07-6	2638	7439-95-4	1298	7440-53-1	854
1314-96-1	2245	1344-54-3	2427	5329-14-6	2259	7439-96-5	1355	7440-54-2	881
1314-97-2	2347	1344-57-6	2498	5341-61-7	1007	7439-97-6	1404	7440-55-3	898
1314-98-3	2657	1344-58-7	2506	5714-22-7	2271	7439-98-7	1452	7440-56-4	924
1314-98-3	2656	1344-59-8	2503	5743-04-4	407	7439-99-8	1505	7440-57-5	958
1315-01-1	2411	1344-95-2	504	5743-26-0	448	7440-00-8	1487	7440-58-6	976
1315-03-3	2538	1345-04-6	186	5794-28-5	490	7440-01-9	1504	7440-59-7	993
1315-04-4	193	1345-07-9	361	5892-10-4	336	7440-02-0	1507	7440-60-0	994
1315-05-5	184	1345-13-7	562	5893-61-8	790	7440-03-1	1544	7440-61-1	2481
1315-06-6	2397	1345-25-1	1110	5908-64-5	222	7440-04-2	1600	7440-62-2	2515
1315-07-7	2241	1449-65-6	941	5908-81-6	292	7440-05-3	1612	7440-63-3	2556
1315-09-9	2649	1470-61-7	2041	5951-19-9	532	7440-06-4	1687	7440-64-4	2571
1315-11-3	2659	1495-50-7	542	5965-38-8	707	7440-07-5	1707	7440-65-5	2582
1317-33-5	1475	1590-87-0	1974	5968-11-6	2091	7440-08-6	1718	7440-66-6	2599
1317-34-6	1400	1592-23-0	507	5970-44-5	2589	7440-09-7	1721	7440-67-7	2661
1317-35-7	1397	1600-27-7	1422	5970-45-6	2600	7440-10-0	1820	7440-68-8	219
1317-36-8	1213	1603-84-5	530	5970-62-7	2625	7440-11-1	1403	7440-69-9	334
1317-36-8	1214	1631-78-3	2385	5972-71-4	67	7440-12-2	1834	7440-70-2	446
1317-37-9	1118	1633-05-2	2215	5972-72-5	115	7440-13-3	1835	7440-71-3	520
1317-38-0	803	1701-93-5	2069	5972-76-9	71	7440-14-4	1837	7440-72-4	874
1317-39-1	753	1762-95-4	166	6009-70-7	138	7440-15-5	1844	7440-73-5	880
1317-40-4	815	1863-63-4	66	6010-09-9	1122	7440-16-6	1872	7440-74-6	1026
1317-41-5	809	2092-16-2	515	6046-93-1	763	7440-17-7	1882	7446-06-2	1720
1317-42-6	720	2092-17-3	297	6047-25-2	1109	7440-18-8	1911	7446-07-3	2303
1317-43-7	1317	2092-17-3	295	6080-56-4	1178	7440-19-9	1923	7446-08-4	1956
1317-58-4	441	2092-17-3	296	6108-17-4	1250	7440-20-2	1941	7446-09-5	2260
1317-61-9	1125	2218-80-6	781	6108-23-2	1266	7440-21-3	1972	7446-10-8	1231
1317-66-4	1088	2404-52-6	1685	6131-90-4	2077	7440-22-4	2024	7446-11-9	2261
1317-98-2	182	2406-52-2	2384	6132-02-1	2090	7440-23-5	2075	7446-14-2	1229
1318-74-7	48	2420-98-6	420	6132-04-3	2097	7440-24-6	2209	7446-15-3	1224
1319-46-6	1189	2452-01-9	2630	6147-53-1	678	7440-25-7	2280	7446-16-4	1842
1319-53-5	772	2466-09-3	1635	6150-82-9	1313	7440-26-8	2297	7446-18-6	2346
1327-41-9	30	2524-02-9	1684	6150-88-5	1327	7440-27-9	2312	7446-19-7	2654
1327-44-2	1724	2551-62-4	2268	6156-78-1	1365	7440-28-0	2322	7446-20-0	2655
1327-50-0	187	2644-70-4	1006	6160-34-5	2224	7440-29-1	2356	7446-21-1	2240
1327-53-3	209	2696-92-6	1593	6303-21-5	1632	7440-30-4	2373	7446-22-2	2344
1327-53-3	208	2699-79-8	2274	6381-79-9	1735	7440-31-5	2382	7446-25-5	2410
1330-43-4	2188	2850-65-9	783	6424-20-0	695	7440-31-5	2383	7446-26-6	2647
1332-14-5	814	2949-11-3	1416	6484-52-2	133	7440-32-6	2412	7446-27-7	1219
1332-52-1	312	3012-65-5	113	6487-39-4	1160	7440-33-7	2439	7446-28-8	2239
1332-58-7	48	3017-60-5	723	6487-48-5	1784	7440-34-8	1	7446-31-3	2675
1332-63-4	1399	3017-60-5	722	6533-73-9	2326	7440-35-9	55	7446-32-4	185
1332-65-6	776	3095-65-6	121	6556-16-7	1381	7440-36-0	174	7446-33-5	2597
1332-71-4	735	3141-12-6	218	6569-51-3	377	7440-37-1	194	7446-70-0	20
1332-77-0	1808	3164-29-2	160	6591-55-5	349	7440-38-2	195	7447-39-4	774
1332-81-6	188	3251-23-8	797	6824-78-8	1408	7440-39-3	220	7447-40-7	1737
1333-24-0	452	3333-67-3	1521	6834-92-0	2147	7440-40-6	305	7447-41-8	1260
1333-74-0	1014	3396-11-0	569	6865-35-6	288	7440-40-6	306	7487-88-9	1346
1333-82-0	661	3444-13-1	1439	7320-34-5	1793	7440-41-7	307	7487-94-7	1427
1333-83-1	2130	3486-35-9	2613	7428-48-0	1228	7440-42-8	367	7488-51-9	1226

CAS REGISTRY NUMBER INDEX OF INORGANIC COMPOUNDS (CONTINUED)

7488-52-0	2658	7699-45-8	2611	7758-99-8	813	7782-94-7	1572	7783-82-6	2475
7488-54-2	1908	7704-34-9	2252	7759-00-4	1385	7782-95-8	2105	7783-84-8	1148
7488-55-3	2398	7704-34-9	2251	7759-01-5	1237	7782-99-2	2258	7783-86-0	1104
7488-56-4	1968	7704-98-5	2413	7759-01-5	1238	7783-00-8	1955	7783-86-0	1105
7543-51-3	2645	7704-99-6	2665	7759-02-6	2244	7783-03-1	2440	7783-89-3	2031
7550-35-8	1257	7705-07-9	2425	7761-88-8	2052	7783-06-4	1023	7783-90-6	2035
7550-45-0	2432	7705-08-0	1130	7772-98-7	2200	7783-07-5	1022	7783-91-7	2036
7553-56-2	1050	7718-54-9	1522	7772-98-7	2199	7783-08-6	1954	7783-92-8	2034
7558-79-4	2132	7718-98-1	2532	7772-99-8	2388	7783-09-7	1025	7783-93-9	2056
7558-79-4	2131	7719-09-7	2277	7773-01-5	1370	7783-11-1	159	7783-95-1	2073
7558-80-7	2103	7719-12-2	1646	7773-03-7	1768	7783-18-8	167	7783-96-2	2048
7558-80-7	2104	7720-78-7	1115	7773-06-0	155	7783-19-9	153	7783-97-3	2047
7558-80-7	2102	7720-83-4	2434	7774-29-0	1435	7783-20-2	156	7783-98-4	2058
7580-67-8	1267	7721-01-9	2293	7774-34-7	463	7783-21-3	152	7783-99-5	2053
7601-54-9	2166	7722-64-7	1789	7775-09-9	2092	7783-22-4	170	7784-01-2	2037
7601-89-0	2158	7722-76-1	87	7775-11-3	2095	7783-24-6	2601	7784-02-3	2040
7601-90-3	597	7722-84-1	1021	7775-14-6	2107	7783-26-8	1975	7784-03-4	2068
7616-83-3	1443	7722-86-3	2254	7775-19-1	2146	7783-28-0	116	7784-05-6	2063
7616-94-6	609	7722-88-5	2112	7775-27-1	2165	7783-29-1	1976	7784-07-8	2061
7631-90-5	2138	7722-88-5	2171	7775-41-9	2042	7783-32-6	1434	7784-09-0	2059
7631-94-9	2108	7723-14-0	1622	7778-18-9	508	7783-33-7	1812	7784-11-4	17
7631-95-0	2148	7723-14-0	1620	7778-39-4	199	7783-34-8	1438	7784-13-6	21
7631-99-4	2151	7723-14-0	1621	7778-39-4	198	7783-35-9	1446	7784-14-7	162
7632-00-0	2152	7726-95-6	395	7778-43-0	2128	7783-36-0	1419	7784-16-9	2192
7632-04-4	2157	7727-15-3	16	7778-43-0	2127	7783-39-3	1431	7784-18-1	24
7632-51-1	2540	7727-18-6	2553	7778-44-1	451	7783-40-6	1312	7784-19-2	104
7637-07-2	388	7727-21-1	1791	7778-50-9	1741	7783-41-7	876	7784-21-6	28
7646-69-7	2126	7727-37-9	1571	7778-53-2	1792	7783-42-8	2278	7784-22-7	31
7646-78-8	2403	7727-43-7	289	7778-54-3	478	7783-43-9	1966	7784-23-8	32
7646-79-9	685	7727-54-0	144	7778-74-7	1787	7783-44-0	877	7784-24-9	1726
7646-85-7	2616	7727-73-3	2179	7778-77-0	1743	7783-46-2	1198	7784-25-0	8
7646-93-7	1765	7732-18-5	2555	7778-80-5	1800	7783-47-3	2390	7784-26-1	9
7647-01-0	1016	7733-02-0	2653	7779-86-4	2621	7783-48-4	2222	7784-27-2	36
7647-10-1	1615	7738-94-5	661	7779-88-6	2632	7783-49-5	2622	7784-29-4	1885
7647-14-5	2093	7738-94-5	662	7779-90-0	2645	7783-50-8	1136	7784-30-7	39
7647-15-6	2087	7745-87-1	1334	7779-90-0	2644	7783-51-9	911	7784-31-8	51
7647-17-8	576	7757-79-1	1781	7782-40-3	521	7783-52-0	1039	7784-33-0	204
7647-18-9	189	7757-82-6	2178	7782-41-4	875	7783-53-1	1398	7784-34-1	205
7647-19-0	1663	7757-83-7	2183	7782-42-5	522	7783-54-2	1583	7784-35-2	206
7659-31-6	2028	7757-86-0	1316	7782-44-7	1610	7783-55-3	1649	7784-36-3	214
7664-38-2	1630	7757-87-1	1333	7782-49-2	1951	7783-56-4	179	7784-37-4	1433
7664-39-3	1019	7757-93-9	476	7782-49-2	1952	7783-57-5	2351	7784-40-9	1202
7664-41-7	63	7757-93-9	475	7782-49-2	1953	7783-58-6	952	7784-41-0	1742
7664-93-9	2253	7758-01-2	1732	7782-50-5	595	7783-59-7	1246	7784-42-1	196
7681-11-0	1773	7758-02-3	1733	7782-61-8	1143	7783-60-0	2267	7784-44-3	110
7681-38-1	2134	7758-05-6	1772	7782-63-0	1117	7783-61-1	2001	7784-45-4	207
7681-49-4	2112	7758-09-0	1782	7782-64-1	1373	7783-62-2	2406	7784-46-5	2083
7681-52-9	2141	7758-11-4	1762	7782-65-2	925	7783-63-3	2433	7784-48-7	1518
7681-52-9	2140	7758-16-9	2106	7782-66-3	1444	7783-64-4	2668	7785-20-8	132
7681-53-0	2169	7758-19-2	2094	7782-68-5	1051	7783-66-6	1061	7785-23-1	2032
7681-53-0	2168	7758-23-8	498	7782-70-9	1764	7783-68-8	1565	7785-24-2	680
7681-55-2	2142	7758-29-4	2203	7782-77-6	1574	7783-70-2	190	7785-84-4	2201
7681-57-4	2145	7758-79-4	2133	7782-78-7	2255	7783-71-3	2294	7785-84-4	2202
7681-65-4	751	7758-87-4	497	7782-79-8	1011	7783-72-4	2543	7785-87-7	1388
7681-82-5	2143	7758-88-5	558	7782-85-6	2133	7783-75-7	1074	7786-30-3	1309
7693-26-7	1757	7758-89-6	747	7782-86-7	1414	7783-77-9	952	7786-81-4	1535
7693-27-8	1315	7758-94-3	1097	7782-87-8	1771	7783-77-9	1481	7787-32-8	245
7697-37-2	1573	7758-95-4	1191	7782-89-0	1252	7783-79-1	1963	7787-33-9	260
7699-41-4	2011	7758-97-6	1193	7782-91-4	1480	7783-80-4	2311	7787-34-0	258
7699-43-6	2680	7758-98-7	812	7782-92-5	2081	7783-81-5	2505	7787-35-1	261

7787-36-2	274
7787-37-3	263
7787-38-4	268
7787-39-5	291
7787-40-8	299
7787-41-9	279
7787-42-0	301
7787-46-4	315
7787-47-5	319
7787-48-6	329
7787-49-7	320
7787-52-2	322
7787-53-3	325
7787-56-6	332
7787-57-7	351
7787-58-8	337
7787-59-9	352
7787-60-2	338
7787-61-3	340
7787-62-4	341
7787-63-5	353
7787-64-6	344
7787-68-0	360
7787-69-1	573
7787-70-4	746
7787-71-5	402
7788-97-8	641
7788-98-9	78
7788-99-0	652
7789-00-6	1738
7789-01-7	1261
7789-02-8	646
7789-04-0	650
7789-04-0	651
7789-04-0	649
7789-06-2	2219
7789-08-4	89
7789-09-5	85
7789-10-8	1430
7789-16-4	583
7789-17-5	586
7789-18-6	588
7789-19-7	787
7789-21-1	2257
7789-23-3	1746
7789-24-4	1265
7789-25-5	1594
7789-26-6	878
7789-27-7	2332
7789-28-8	1101
7789-29-9	1761
7789-30-2	403
7789-31-3	396
7789-33-5	1056
7789-36-8	1304
7789-38-0	2086
7789-39-1	1888
7789-40-4	2325
7789-41-5	454
7789-42-6	411
7789-43-7	683
7789-43-7	682
7789-45-9	770
7789-46-0	1094
7789-47-1	1425
7789-48-2	1305
7789-51-7	1964
7789-52-8	1958
7789-53-9	2213
7789-54-0	2305
7789-56-2	553
7789-57-3	2003
7789-59-5	1668
7789-60-8	1643
7789-61-9	177
7789-65-3	1960
7789-66-4	1999
7789-67-5	2402
7789-68-6	2431
7789-69-7	1655
7789-75-5	470
7789-77-7	476
7789-78-8	474
7789-79-9	479
7789-80-2	480
7789-82-4	484
7790-21-8	1788
7790-22-9	1272
7790-28-5	2160
7790-29-6	1902
7790-30-9	2337
7790-31-0	1320
7790-32-1	1318
7790-33-2	1375
7790-33-2	1376
7790-34-3	1530
7790-37-6	2628
7790-38-7	1617
7790-39-8	1690
7790-41-2	1069
7790-42-3	1816
7790-44-5	180
7790-46-7	1699
7790-47-8	2407
7790-48-9	2310
7790-49-0	2364
7790-56-9	1804
7790-58-1	1807
7790-59-2	1795
7790-60-5	1818
7790-62-7	1794
7790-63-8	1819
7790-69-4	1274
7790-71-8	1281
7790-74-1	502
7790-75-2	518
7790-76-3	501
7790-78-5	416
7790-79-6	421
7790-80-9	424
7790-81-0	423
7790-84-3	439
7790-85-4	445
7790-86-5	556
7790-87-6	559
7790-87-6	560
7790-89-8	603
7790-91-2	604
7790-92-3	596
7790-94-5	2256
7790-98-9	142
7790-99-0	1057
7791-03-9	1279
7791-07-3	2159
7791-08-4	183
7791-09-5	1867
7791-10-8	2216
7791-11-9	1891
7791-12-0	2328
7791-13-1	687
7791-16-4	191
7791-18-6	1310
7791-20-0	1523
7791-21-1	598
7791-23-3	1965
7791-25-5	2273
7791-26-6	2509
7791-27-7	2275
7791-28-8	227
7798-23-4	807
7798-23-4	806
7803-49-8	1012
7803-51-2	1623
7803-52-3	175
7803-54-5	1301
7803-55-6	128
7803-57-8	1004
7803-58-9	2272
7803-60-3	1634
7803-62-5	1973
7803-63-6	118
7803-65-8	123
7803-68-1	2301
8011-62-9	290
10022-31-8	265
10022-47-6	79
10022-48-7	1262
10022-50-1	1597
10022-68-1	429
10022-70-5	2141
10024-93-8	1490
10024-97-2	1575
10025-64-6	2641
10025-65-7	1689
10025-66-8	1839
10025-67-9	2263
10025-68-0	1959
10025-69-1	2389
10025-70-4	2218
10025-71-5	2306
10025-73-7	639
10025-74-8	830
10025-75-9	842
10025-76-0	867
10025-77-1	1131
10025-78-2	2005
10025-82-8	1038
10025-83-9	1067
10025-85-1	1582
10025-87-3	1674
10025-90-8	1824
10025-91-9	178
10025-93-1	2490
10025-97-5	1071
10025-99-7	1810
10026-00-3	2195
10026-01-4	1604
10026-02-5	1719
10026-03-6	1961
10026-04-7	2000
10026-06-9	2404
10026-07-0	2308
10026-08-1	2362
10026-10-5	2495
10026-11-6	2667
10026-12-7	1564
10026-13-8	1658
10026-17-2	693
10026-18-3	728
10026-20-7	711
10026-22-9	705
10026-24-1	718
10028-14-5	1599
10028-15-6	1611
10028-18-9	1525
10028-22-5	1151
10028-22-5	1150
10031-13-7	1182
10031-16-0	242
10031-18-2	1407
10031-20-6	1368
10031-21-7	1185
10031-22-8	1186
10031-23-9	1838
10031-24-0	2387
10031-25-1	636
10031-25-1	637
10031-26-2	1129
10031-27-3	2307
10031-30-8	498
10031-43-3	799
10031-45-5	808
10031-48-8	807
10031-52-4	851
10031-52-4	873
10031-53-5	870
10031-54-6	863
10034-76-1	510
10034-81-8	1329
10034-82-9	2096
10034-85-2	1020
10034-88-5	2135
10034-93-2	1010
10034-96-5	1389
10034-98-7	2581
10034-99-8	1348
10035-04-8	462
10035-05-9	460
10035-06-0	347
10035-10-6	1015
10036-47-2	1591
10038-98-9	951
10039-31-3	330
10039-32-4	2132
10039-54-0	1013
10039-55-1	1008
10039-56-2	2169
10042-76-9	2232
10042-88-3	2313
10043-01-3	50
10043-11-5	391
10043-27-3	2317
10043-27-3	2316
10043-35-3	378
10043-52-4	461
10043-67-1	1725
10043-92-2	1843
10045-86-0	1146
10045-87-1	1149
10045-89-3	94
10045-94-0	1436
10045-94-0	1437
10045-95-1	1494
10048-95-0	2128
10048-98-3	250
10048-99-4	294
10049-01-1	355
10049-03-3	879
10049-04-4	599
10049-05-5	627
10049-06-6	2420
10049-07-7	1875
10049-08-8	1915
10049-10-2	629
10049-11-3	657
10049-12-4	2533
10049-12-4	2534
10049-14-6	2496
10049-16-8	2541
10049-17-9	1861
10049-21-5	2103
10049-23-7	2302
10049-24-8	1066
10049-24-8	1065
10049-25-9	636
10049-25-9	626
10058-44-3	1147
10060-09-0	436
10060-10-3	565
10060-11-4	943

10060-12-5	640
10060-13-6	82
10060-13-6	83
10090-53-6	33
10097-28-6	2016
10099-58-8	1161
10099-59-9	1168
10099-60-2	1172
10099-66-8	1290
10099-74-8	1210
10099-76-0	1221
10099-79-3	1239
10101-41-4	509
10101-50-5	2162
10101-52-7	2673
10101-53-8	653
10101-63-0	1206
10101-68-5	1390
10101-83-4	2186
10101-88-9	2198
10101-89-0	2166
10101-94-7	1227
10101-97-0	1537
10101-98-1	1536
10102-02-0	2635
10102-03-1	1580
10102-05-3	1618
10102-06-4	2511
10102-15-5	2184
10102-17-7	2200
10102-18-8	2175
10102-20-2	2187
10102-23-5	2173
10102-24-6	1282
10102-25-7	1284
10102-34-8	1335
10102-40-6	2149
10102-43-9	1576
10102-44-0	1577
10102-45-1	2339
10102-49-5	1128
10102-50-8	1093
10102-50-8	1092
10102-68-8	481
10102-71-3	2080
10103-61-4	765
10103-62-5	451
10108-64-2	415
10108-73-3	561
10112-91-1	1410
10117-38-1	1803
10118-76-0	495
10124-36-4	440
10124-36-4	438
10124-37-5	485
10124-41-1	516
10124-43-3	717
10124-48-8	1423
10124-50-2	1759
10124-53-5	1352
10125-13-0	775
10135-84-9	111
10137-74-3	460
10137-74-3	459
10138-04-2	91
10138-41-7	841
10138-52-0	888
10138-62-2	996
10139-47-6	2629
10141-00-1	652
10141-05-6	704
10143-38-1	834
10163-15-2	2114
10168-80-6	846
10170-69-1	1362
10179-73-4	1114
10190-55-3	1208
10192-29-7	76
10192-30-0	120
10193-36-9	2012
10196-04-0	159
10196-04-0	158
10196-18-6	2633
10210-64-7	309
10210-68-1	672
10213-09-9	2525
10213-09-9	2548
10213-10-2	2205
10213-12-4	1841
10214-39-8	1184
10214-40-1	810
10241-05-1	1477
10257-55-3	512
10277-43-7	1168
10290-12-7	766
10294-26-5	2064
10294-27-6	961
10294-28-7	966
10294-29-8	962
10294-31-2	964
10294-32-3	973
10294-33-4	384
10294-34-5	386
10294-38-9	231
10294-39-0	273
10294-40-3	235
10294-42-5	567
10294-44-7	1409
10294-46-9	805
10294-47-0	1190
10294-48-1	602
10294-50-5	710
10294-52-7	1132
10294-53-8	1134
10294-54-9	593
10294-58-3	1220
10294-60-7	117
10294-62-9	1172
10294-64-1	1775
10294-66-3	1814
10294-70-9	2393
10325-94-7	428
10326-21-3	1308
10326-24-6	2606
10326-26-8	225
10326-27-9	233
10326-28-0	433
10340-06-4	533
10343-61-0	2428
10361-37-2	232
10361-43-0	343
10361-44-1	347
10361-46-3	354
10361-65-6	146
10361-79-2	1823
10361-80-5	1828
10361-80-5	1827
10361-82-7	1931
10361-83-8	1936
10361-83-8	1935
10361-84-9	1944
10361-91-8	2576
10361-92-9	2590
10361-93-0	2593
10361-93-0	2592
10361-95-2	2615
10377-48-7	1283
10377-51-2	1271
10377-52-3	1277
10377-58-9	1319
10377-60-3	1321
10377-62-5	1331
10377-66-9	1379
10377-93-2	1378
10378-48-0	2210
10380-31-1	302
10381-36-9	1501
10381-37-0	2371
10402-15-0	779
10405-27-3	1588
10415-75-5	1414
10415-75-5	1413
10421-48-4	1142
10421-48-4	1141
10431-47-7	1797
10450-55-2	1126
10450-60-9	1052
10476-81-0	2212
10476-85-4	2217
10476-86-5	2229
10489-46-0	1879
10534-89-1	729
10544-72-6	1579
10544-73-7	1578
10545-99-0	2266
10553-31-8	226
10567-69-8	258
10567-69-8	257
10580-03-7	168
10580-52-6	2525
10588-01-9	2101
10599-90-3	1586
11120-25-5	169
11121-16-7	12
11126-81-1	15
12002-03-8	762
12002-61-8	6
12002-99-2	2067
12003-63-3	1723
12004-04-5	223
12004-06-7	310
12004-76-1	2281
12005-21-9	2583
12005-67-3	62
12005-69-5	383
12005-82-2	2044
12006-15-4	409
12006-40-5	2605
12006-77-8	671
12006-79-0	614
12006-84-7	1082
12006-86-9	1083
12006-99-4	1453
12007-00-0	1511
12007-01-1	1512
12007-02-2	1510
12007-07-7	2282
12007-09-9	2442
12007-10-2	2441
12007-16-8	615
12007-23-7	977
12007-25-9	1303
12007-29-3	1546
12007-33-9	394
12007-34-0	1942
12007-35-1	2283
12007-36-2	2482
12007-37-3	2517
12007-38-4	616
12007-58-8	68
12007-84-0	2483
12007-89-5	140
12007-97-5	1454
12007-98-6	2443
12007-99-7	453
12008-02-5	545
12008-05-8	855
12008-06-9	882
12008-19-4	373
12008-21-8	1157
12008-23-0	1488
12008-27-4	1821
12008-29-6	1924
12008-30-9	1924
12008-32-1	2586
12009-14-2	264
12009-18-6	287
12009-18-6	286
12009-21-1	304
12011-67-5	1084
12011-97-1	1455
12011-99-3	1548
12012-16-7	2360
12012-17-8	2519
12012-32-7	546
12012-35-0	617
12013-10-4	675
12013-47-7	519
12013-55-7	506
12013-56-8	505
12013-57-9	513
12013-82-0	487
12014-14-1	444
12014-28-7	435
12014-29-8	408
12014-82-3	551
12014-85-6	548
12014-93-6	564
12016-80-7	733
12017-01-5	702
12017-08-2	715
12017-12-8	674
12017-13-9	721
12017-38-8	736
12018-01-8	658
12018-09-6	624
12018-10-9	778
12018-19-8	2618
12018-22-3	654
12018-34-7	633
12018-36-9	623
12018-61-0	590
12019-11-3	741
12019-23-7	817
12019-52-2	758
12019-88-4	825
12020-14-3	2663
12020-21-2	847
12020-28-9	849
12020-39-2	853
12020-58-5	856
12020-65-4	864
12020-66-5	862
12020-69-8	865
12021-68-0	697
12022-71-8	1123
12022-95-6	1089
12022-99-0	1090
12023-53-9	1087
12023-99-3	914
12024-10-1	906
12024-11-2	905
12024-14-5	907
12024-20-3	903
12024-21-4	917
12024-22-5	922
12024-24-7	919
12024-27-0	923
12024-81-6	886
12025-19-3	2116

CAS REGISTRY NUMBER INDEX OF INORGANIC COMPOUNDS (CONTINUED)

12025-32-0	948
12025-34-2	957
12025-39-7	949
12026-06-1	2335
12026-24-3	2392
12026-66-3	148
12027-06-4	125
12027-67-7	129
12028-48-7	127
12029-81-1	999
12029-98-0	1054
12030-14-7	1036
12030-24-9	1048
12030-49-8	1072
12030-51-2	1073
12030-85-2	1780
12030-88-5	1805
12030-97-6	1815
12031-30-0	1174
12031-49-1	1173
12031-63-9	1273
12031-80-0	1280
12032-20-1	1294
12032-36-9	1349
12032-74-5	1395
12032-78-9	1363
12032-82-5	1356
12032-88-1	1394
12032-97-2	1357
12033-19-1	1458
12033-31-7	1459
12033-54-4	1708
12033-62-4	2286
12033-64-6	2318
12033-65-7	2366
12033-72-6	2448
12033-82-8	2233
12033-83-9	2488
12033-88-4	1598
12033-89-5	2015
12034-09-2	2150
12034-12-7	2185
12034-57-0	1552
12034-59-2	1559
12034-66-1	1550
12034-77-4	1560
12034-80-9	1551
12034-83-2	1562
12034-88-7	1209
12034-89-8	2231
12035-32-4	1500
12035-35-7	1503
12035-39-1	1540
12035-52-8	1508
12035-64-2	1514
12035-72-2	1543
12035-79-9	1506
12035-82-4	1691
12035-83-5	1709
12035-98-2	2528
12036-02-1	1606
12036-09-8	1852
12036-10-1	1919
12036-14-5	2288
12036-21-4	2542
12036-22-5	2460
12036-32-7	1830
12036-34-9	1714
12036-35-0	1878
12036-41-8	2319
12036-44-1	2381
12037-15-9	2499
12037-65-9	2417
12037-80-8	2678
12037-82-0	1641
12038-12-9	1833
12038-13-0	1832
12038-20-9	1692
12038-21-0	1701
12038-63-0	1853
12038-67-4	1871
12039-07-5	2423
12039-13-3	2438
12039-15-5	2676
12039-16-6	2429
12039-19-9	2598
12039-35-9	2602
12039-52-0	2355
12039-55-3	2289
12039-76-8	2523
12039-79-1	2287
12039-80-4	2320
12039-83-7	2418
12039-87-1	2522
12039-88-2	2449
12039-89-3	2572
12039-90-6	2679
12039-95-1	2450
12040-00-5	1940
12040-02-7	2401
12040-08-3	2247
12041-50-8	13
12042-68-1	449
12042-78-3	450
12043-29-7	53
12044-16-5	1081
12044-30-3	203
12044-42-7	668
12044-54-1	212
12045-01-1	670
12045-15-7	1358
12045-16-8	1360
12045-19-1	1545
12045-27-1	2516
12045-63-5	2414
12045-64-6	2662
12045-87-3	2191
12045-88-4	2190
12046-08-1	247
12046-54-7	2225
12047-27-7	300
12047-79-9	266
12048-50-9	363
12048-51-0	364
12049-50-2	517
12050-27-0	408
12052-42-5	665
12053-12-2	612
12053-13-3	622
12053-27-9	619
12053-39-3	655
12054-48-7	1527
12054-48-7	1526
12054-85-2	129
12055-23-1	985
12055-62-8	1000
12056-90-5	1171
12057-24-8	1278
12057-71-5	1324
12057-74-8	1336
12057-75-9	1302
12057-92-0	1402
12058-18-3	1474
12058-20-7	1476
12059-14-2	1515
12059-95-9	1716
12060-00-3	1236
12060-01-4	1240
12060-08-1	1948
12060-58-1	1937
12060-59-2	2249
12061-16-4	848
12062-24-7	792
12063-98-8	902
12064-03-8	899
12064-62-9	894
12065-10-0	947
12065-11-1	956
12065-36-0	954
12065-65-5	2430
12065-68-8	1232
12066-83-0	1831
12067-00-4	1854
12067-22-0	1939
12067-46-8	2461
12067-54-8	2370
12067-66-2	2291
12067-76-4	2463
12068-49-4	1091
12068-69-8	357
12068-90-5	1449
12069-00-0	1225
12069-32-8	385
12069-69-1	772
12069-85-1	979
12069-89-5	1456
12069-94-2	1547
12070-06-3	2284
12070-07-4	2285
12070-08-5	2415
12070-09-6	2484
12070-10-9	2518
12070-12-1	2445
12070-13-2	2444
12070-69-8	2614
12071-15-7	1159
12071-29-3	2214
12071-31-7	2361
12071-33-9	2485
12071-35-1	2588
12076-62-9	2486
12088-65-2	1079
12115-63-8	554
12124-97-9	70
12124-99-1	119
12125-01-8	95
12125-02-9	77
12125-08-5	100
12125-09-6	1629
12125-19-8	2480
12125-22-3	1613
12125-25-6	1293
12125-63-2	1121
12133-07-2	826
12133-10-7	836
12134-02-0	673
12134-75-7	884
12134-77-9	896
12135-76-1	157
12136-24-2	1001
12136-42-4	1064
12136-44-6	1730
12136-45-7	1785
12136-58-2	1285
12136-78-6	1461
12136-91-3	1639
12136-97-9	1561
12137-12-1	1541
12137-20-1	2422
12137-25-6	1910
12137-27-8	1880
12137-83-6	1706
12138-06-6	2657
12138-07-7	2372
12138-09-9	2462
12138-11-3	2321
12138-28-2	2243
12139-93-4	716
12142-33-5	1799
12143-72-5	2290
12159-07-8	742
12159-66-9	852
12160-99-5	897
12162-21-9	988
12162-59-3	1002
12163-20-1	1296
12163-22-3	1297
12163-69-8	1460
12164-94-2	65
12165-05-8	1858
12165-69-4	1654
12166-29-9	1949
12166-44-8	1950
12168-52-4	1123
12168-85-3	492
12183-80-1	47
12185-10-3	1620
12186-97-9	2584
12187-14-3	427
12201-89-7	1516
12211-52-8	84
12214-16-3	594
12228-40-9	313
12228-50-1	1359
12228-87-4	68
12228-91-0	1366
12229-63-9	2358
12230-71-6	255
12230-74-9	252
12232-99-4	2144
12249-52-4	2046
12254-64-7	56
12254-82-9	666
12254-85-2	613
12255-48-0	2585
12256-04-1	669
12266-65-8	1361
12279-90-2	203
12280-52-3	582
12280-64-7	1895
12294-01-8	2596
12298-70-3	2453
12300-22-0	1925
12310-43-9	824
12310-44-0	839
12322-34-8	176
12323-03-4	285
12325-59-6	987
12326-21-5	2507
12333-54-9	1364
12340-14-6	2558
12345-14-1	234
12393-61-2	211
12399-08-5	1053
12401-56-8	990
12412-52-1	181
12419-43-1	2391
12434-24-1	857
12440-00-5	1652
12442-63-6	601
12504-41-5	2022
12513-27-8	2608
12528-75-5	1106
12536-65-1	2609
12540-13-5	764
12612-73-6	2484
12612-73-6	2486
12616-24-9	173
12640-47-0	724
12650-28-1	282

12688-52-7	1288
12770-26-2	982
12777-45-6	358
12789-09-2	821
13007-92-6	618
13011-54-6	2082
13092-75-6	2030
13126-12-0	1903
13135-31-4	2424
13138-45-9	1532
13138-45-9	1531
13172-31-1	2262
13205-44-2	376
13255-26-0	283
13268-42-3	90
13283-01-7	2471
13320-71-3	1471
13327-32-7	324
13395-16-9	738
13397-24-5	509
13397-26-7	458
13400-13-0	578
13410-01-0	2173
13410-01-0	2172
13423-61-5	1311
13424-46-9	1183
13444-75-2	1428
13444-85-4	1584
13444-87-6	1592
13444-90-1	1596
13444-93-4	1603
13444-94-5	1614
13444-96-7	1616
13445-50-6	1624
13445-58-4	1657
13446-03-2	1367
13446-08-7	97
13446-09-8	124
13446-10-1	143
13446-18-9	1323
13446-19-0	1308
13446-19-0	1330
13446-23-6	1334
13446-24-7	1335
13446-28-1	1337
13446-29-2	1351
13446-30-5	1352
13446-34-9	1371
13446-48-5	134
13446-49-6	1779
13446-53-2	1306
13446-56-5	1462
13446-57-6	1465
13446-70-3	1887
13446-71-4	1890
13446-72-5	1892
13446-74-7	1894
13446-75-8	1897
13446-76-9	1901
13450-87-8	895
13450-88-9	909
13450-90-3	910
13450-91-4	915
13450-92-5	950
13450-95-8	953
13450-97-0	2236
13451-02-0	2246
13451-05-3	2250
13451-11-1	2292
13451-18-8	2304
13451-19-9	2317
13453-06-0	161
13453-07-1	967
13453-30-0	2327
13453-32-2	2349
13453-32-2	2350
13453-34-4	2330
13453-40-2	2343
13453-49-1	2359
13453-62-8	1217
13453-69-5	1255
13453-71-9	1259
13453-80-0	1263
13454-75-6	572
13454-81-4	585
13454-84-7	591
13454-88-1	788
13454-89-2	794
13454-90-5	795
13454-94-9	563
13454-96-1	1697
13454-99-4	1661
13455-00-0	1627
13455-01-1	1650
13455-03-3	1671
13455-04-4	1678
13455-12-4	1688
13455-15-7	1698
13455-21-5	1747
13455-25-9	689
13455-28-2	699
13455-31-7	709
13455-34-0	719
13460-50-9	379
13460-50-9	380
13460-50-9	381
13462-88-9	1519
13462-88-9	1520
13462-90-3	1529
13462-93-6	86
13463-12-2	1095
13463-22-4	270
13463-30-4	1245
13463-39-3	1513
13463-40-6	1077
13463-67-7	2435
13464-35-2	1729
13464-36-3	1728
13464-44-3	2603
13464-45-4	2604
13464-46-5	58
13464-58-9	200
13464-82-9	1047
13464-92-1	412
13464-97-6	1009
13465-05-9	1017
13465-07-1	1024
13465-09-3	1037
13465-10-6	1032
13465-11-7	1035
13465-15-1	1044
13465-33-3	1406
13465-55-9	1932
13465-58-2	1938
13465-59-3	1943
13465-60-6	1947
13465-66-2	1962
13465-71-9	2008
13465-72-0	2009
13465-73-1	1987
13465-74-2	1988
13465-75-3	1991
13465-76-4	2004
13465-78-6	1989
13465-84-4	2002
13465-85-5	2007
13465-93-5	2072
13465-94-6	267
13465-95-7	272
13465-96-8	2050
13465-98-0	2066
13466-08-5	2088
13466-20-1	276
13466-21-2	278
13469-98-2	2587
13470-01-4	2228
13470-06-9	2234
13470-08-1	2426
13470-10-5	2451
13470-11-6	2464
13470-12-7	2452
13470-13-8	2457
13470-14-9	2465
13470-17-2	2453
13470-19-4	2489
13470-20-7	2494
13470-21-8	2501
13470-22-9	2497
13470-26-3	2531
13472-31-6	2161
13472-33-8	2164
13472-35-0	2104
13472-45-2	2204
13473-03-5	209
13473-77-3	1295
13473-90-0	36
13473-90-0	35
13476-01-2	681
13476-08-9	1142
13476-99-8	2530
13477-00-4	231
13477-00-4	230
13477-09-3	249
13477-17-3	434
13477-19-5	425
13477-28-6	455
13477-29-7	464
13477-34-4	486
13477-36-6	494
13477-89-9	1491
13477-91-3	1498
13477-95-7	1524
13477-98-0	1528
13478-00-7	1532
13478-04-1	592
13478-10-9	1099
13478-14-3	1253
13478-17-6	1463
13478-18-7	1466
13478-20-1	1677
13478-28-9	630
13478-38-1	798
13478-41-6	749
13478-45-0	1563
13478-49-4	850
13478-50-7	1235
13492-25-6	1415
13492-26-7	1763
13494-80-9	2300
13494-90-1	916
13494-91-2	920
13494-98-9	2594
13497-91-1	1625
13498-11-8	1659
13499-05-3	980
13510-35-5	1042
13510-42-4	1905
13510-44-6	2029
13510-49-1	331
13510-89-9	1180
13517-00-5	750
13517-10-7	390
13517-27-6	2610
13520-56-4	1151
13520-59-7	1470
13520-75-7	2472
13520-76-8	2473
13520-77-9	2477
13520-78-0	2478
13520-79-1	2479
13520-83-7	2512
13520-87-1	2546
13520-88-2	2545
13520-89-3	2547
13520-90-6	2552
13520-92-8	2681
13530-57-9	1885
13530-57-9	1884
13530-65-9	2617
13536-53-3	1822
13536-73-7	840
13536-79-3	1158
13536-80-6	1489
13536-84-0	2510
13537-09-2	832
13537-15-0	872
13537-18-3	2375
13537-30-9	936
13537-32-1	1638
13537-33-2	1997
13548-38-4	645
13548-42-0	777
13548-43-1	79
13550-31-7	1669
13550-53-3	844
13566-17-1	1222
13568-32-6	1386
13568-33-7	1276
13569-43-2	930
13569-49-8	1847
13569-50-1	549
13569-59-0	1554
13569-62-5	1711
13569-63-6	1848
13569-70-5	1556
13569-71-6	1850
13569-75-0	644
13569-80-7	831
13572-93-5	913
13572-98-0	891
13573-02-9	937
13573-08-5	945
13573-11-0	1354
13586-38-4	81
13587-35-4	819
13587-35-4	820
13590-82-4	567
13595-30-7	2539
13595-87-4	365
13596-22-0	711
13596-46-8	81
13597-20-1	1569
13597-44-9	2658
13597-46-1	2651
13597-54-1	2648
13597-55-2	2242
13597-65-4	2650
13597-73-4	2010
13597-87-0	1977
13597-95-0	329
13597-99-4	326
13598-15-7	323
13598-22-6	333
13598-33-9	2226
13598-36-2	1631
13598-42-0	1998
13598-56-6	2492
13598-65-7	145
13600-89-0	727
13601-13-3	785

13601-19-9	2111
13637-61-1	2641
13637-63-3	606
13637-65-5	931
13637-68-8	1484
13637-71-3	1499
13637-76-8	1218
13637-83-7	607
13637-87-1	1585
13675-47-3	782
13682-61-6	1809
13689-92-4	1539
13693-05-5	1702
13693-06-6	1717
13693-07-7	1881
13693-08-8	1921
13693-09-9	2561
13693-11-3	2437
13701-64-9	483
13701-67-2	387
13701-70-7	2537
13701-86-5	2470
13701-90-1	2348
13701-91-2	1244
13702-38-0	335
13706-09-7	1681
13706-10-0	1680
13708-80-0	59
13709-31-4	2554
13709-35-8	2265
13709-36-9	2559
13709-38-1	1163
13709-42-7	1492
13709-46-1	1825
13709-47-2	1945
13709-49-4	2591
13709-52-9	981
13709-56-3	1715
13709-59-6	2363
13709-61-0	2560
13709-94-9	1778
13718-26-8	2208
13718-26-8	2207
13718-50-8	259
13718-59-7	281
13718-70-2	1106
13721-34-1	2206
13721-39-6	2154
13746-66-2	1744
13746-89-9	2671
13746-98-0	2352
13755-29-8	2113
13755-32-3	293
13755-38-9	2153
13759-10-9	2023
13759-83-6	1936
13759-87-0	1930
13759-88-1	866
13759-92-7	868
13760-02-6	1996
13760-41-3	1836
13760-78-6	997
13760-79-7	2377
13760-80-0	2578
13760-81-1	1291
13760-83-3	843
13762-12-4	683
13762-14-6	703
13762-26-0	2664
13762-51-1	1731
13763-23-0	2504
13763-67-2	575
13765-03-2	1270
13765-19-0	465
13765-24-7	1933
13765-25-8	869
13765-26-9	890
13765-74-7	2051
13766-47-7	2458
13767-31-2	827
13767-32-3	2631
13767-34-5	796
13768-11-1	1845
13768-38-2	1608
13768-67-7	2579
13768-86-0	1957
13768-94-0	1992
13769-20-5	859
13769-36-3	933
13769-75-0	1676
13769-76-1	1675
13770-18-8	804
13770-56-4	201
13770-96-2	2079
13773-69-8	2593
13773-81-4	1155
13774-85-1	2563
13775-06-9	2491
13775-07-0	2502
13775-16-1	2500
13775-18-3	2493
13775-53-6	2121
13775-80-9	1005
13776-58-4	2557
13776-62-0	1590
13776-74-4	1340
13777-22-5	978
13777-23-6	983
13777-25-8	2666
13778-39-7	2376
13778-96-6	1310
13779-41-4	1636
13779-92-5	1566
13780-06-8	488
13780-42-2	921
13780-48-8	858
13780-57-9	2270
13782-01-9	734
13783-04-5	2419
13783-07-8	2421
13798-24-8	2314
13812-43-6	1589
13813-22-4	1167
13813-23-5	1826
13813-24-6	1493
13813-25-7	1934
13813-40-6	2315
13813-41-7	998
13813-42-8	845
13813-43-9	2378
13813-45-1	1292
13813-46-2	1713
13813-47-3	60
13814-62-5	436
13814-72-7	885
13814-74-9	1479
13814-75-0	1483
13814-76-1	1863
13814-81-8	760
13814-83-0	2549
13814-96-5	1199
13817-37-3	694
13818-75-2	887
13818-89-8	926
13819-84-6	1478
13820-41-2	163
13820-53-6	2194
13820-62-7	679
13821-06-2	244
13823-29-5	2365
13824-36-7	1995
13824-74-3	1626
13825-74-6	2436
13825-76-8	995
13825-86-0	632
13826-63-6	2340
13826-83-0	96
13842-80-3	2526
13842-83-6	1712
13842-88-1	1557
13842-93-8	2299
13843-28-2	660
13843-59-9	69
13845-17-5	243
13847-22-8	2644
13847-57-9	1192
13847-65-9	1595
13862-78-7	165
13863-41-7	400
13863-59-7	401
13863-88-2	2025
13864-01-2	1164
13870-13-8	989
13870-19-4	1166
13870-21-8	1558
13870-24-1	1124
13871-27-7	2196
13873-84-2	1059
13874-02-7	2193
13874-75-4	1927
13874-77-6	2574
13875-06-4	2562
13876-85-2	752
13896-65-6	1917
13918-22-4	1396
13931-94-7	628
13933-23-8	1111
13939-06-5	1457
13940-63-1	944
13940-89-1	1102
13943-58-3	1745
13963-58-1	1753
13965-73-6	389
13966-94-4	1033
13967-25-4	1411
13972-68-4	426
13973-87-0	399
13982-53-1	757
13986-18-0	2623
13986-26-0	2670
14013-15-1	1377
14013-86-6	1107
14013-86-6	1108
14014-18-7	1672
14014-19-8	1670
14014-88-1	1914
14019-91-1	502
14024-18-1	1127
14024-63-6	2640
14038-43-8	1135
14040-11-0	2446
14049-36-6	1990
14049-81-1	692
14055-74-4	1464
14055-75-5	1468
14055-81-3	2456
14055-84-6	2459
14059-33-7	366
14075-53-7	1748
14077-39-5	860
14166-78-0	1040
14168-73-1	1347
14177-46-9	1396
14188-40-0	932
14215-29-3	410
14215-30-6	767
14216-75-2	1531
14217-21-1	2110
14221-48-8	92
14242-05-8	2057
14249-98-0	2469
14280-53-6	1031
14284-06-1	784
14284-93-6	1913
14285-68-8	1846
14291-02-2	2244
14307-33-6	468
14312-00-6	418
14335-40-1	1648
14336-80-2	745
14402-70-1	135
14405-43-7	908
14447-89-3	2474
14452-39-2	44
14456-34-9	986
14456-48-5	829
14456-51-0	2374
14456-53-2	1289
14457-84-2	42
14457-87-5	553
14457-87-5	552
14459-59-7	1482
14459-75-7	1568
14459-95-1	1745
14464-46-1	2020
14475-63-9	2669
14476-12-1	1369
14481-29-9	93
14486-19-2	443
14507-19-8	1165
14517-29-4	1495
14518-81-1	1673
14519-07-4	2610
14519-18-7	2211
14521-18-7	1920
14523-22-9	1873
14542-23-5	470
14550-84-6	2324
14553-44-7	2369
14567-54-5	812
14567-59-0	1237
14567-59-0	1238
14590-13-7	80
14590-19-3	712
14639-94-2	105
14639-97-5	164
14639-98-6	141
14644-61-2	2675
14644-61-2	2674
14649-73-1	2125
14676-93-8	631
14691-44-2	927
14691-47-5	928
14693-61-9	1981
14693-65-3	1984
14693-81-3	2295
14693-82-4	1045
14720-21-9	969
14720-53-7	1184
14721-21-2	773
14735-84-3	818
14758-11-3	1202
14763-77-0	780
14767-09-0	2336
14779-70-5	938
14781-45-4	791
14791-73-2	457
14808-60-7	2020
14808-60-7	2019
14808-60-7	2021

14808-60-7	2017	15578-26-4	2396	16940-81-1	1637	19357-86-9	2575	21995-38-0	74
14808-60-7	2018	15587-39-0	929	16940-92-4	99	19372-44-2	493	22015-35-6	861
14859-67-7	1843	15587-72-1	1899	16941-10-9	514	19423-81-5	889	22132-71-4	147
14868-53-2	1978	15593-52-9	1281	16941-11-0	107	19423-87-1	2577	22205-45-4	755
14868-54-3	1979	15593-61-0	1339	16941-12-1	1705	19465-30-6	374	22205-57-8	570
14868-55-4	1983	15597-39-4	1644	16949-15-8	1256	19469-07-9	1144	22207-96-1	1786
14868-55-4	1982	15597-40-7	1645	16961-83-4	2013	19525-15-6	2508	22326-55-2	254
14871-79-5	256	15597-63-4	1647	16962-07-5	14	19567-78-3	2118	22398-80-7	1030
14884-42-5	659	15597-88-3	656	16962-31-5	1754	19584-30-6	1874	22441-45-8	213
14890-41-6	2524	15607-89-3	2269	16962-47-3	106	19597-69-4	1254	22520-96-3	1060
14902-95-5	330	15610-76-1	83	17014-71-0	1790	19598-90-4	892	22527-13-5	2564
14913-33-8	1704	15622-42-1	1849	17026-29-8	1864	19624-22-7	370	22537-19-5	1175
14933-38-1	57	15663-27-1	1703	17026-44-7	158	19648-83-0	696	22585-64-4	404
14939-34-5	1667	15696-40-9	1601	17029-21-9	1865	19648-88-5	1201	22750-54-5	414
14940-41-1	1112	15699-18-0	132	17083-68-0	959	19783-14-3	1204	22750-57-8	571
14965-52-7	2006	15699-18-0	1517	17099-70-6	27	20193-56-0	2455	22756-36-1	1886
14977-61-8	663	15750-45-5	1322	17125-80-3	248	20193-58-2	1467	22831-39-6	1344
14984-81-7	1967	15752-41-7	1553	17194-00-2	253	20205-91-8	393	22831-42-1	11
14985-19-4	2380	15752-46-2	1710	17242-52-3	1727	20211-76-1	1162	22886-66-4	912
14985-19-4	2379	15823-43-5	991	17363-02-9	102	20213-56-3	2468	22986-54-5	3
14986-91-5	1337	15829-53-5	1417	17375-41-6	1116	20281-00-9	589	23032-72-6	1141
14987-04-3	1342	15843-48-8	1212	17440-85-6	314	20328-94-3	2569	23149-52-2	2070
15060-62-5	1502	15845-52-0	1203	17440-90-3	1080	20344-49-4	1140	23370-59-4	1068
15070-34-5	1325	15861-05-9	1587	17496-59-2	600	20398-06-5	2331	23377-53-9	1870
15098-87-0	26	15878-77-0	1828	17674-34-9	1946	20405-64-5	754	23412-45-5	1472
15123-69-0	808	15942-63-9	197	17702-41-9	375	20427-11-6	691	23414-72-4	2642
15123-90-7	2248	15947-41-8	61	17786-31-1	676	20427-56-9	1922	23436-05-7	262
15162-90-0	1581	15947-57-6	1980	18088-11-4	1904	20427-58-1	2627	23611-30-5	1906
15163-24-3	2454	16122-03-5	131	18282-10-5	2408	20427-59-2	793	23777-80-2	372
15168-20-4	810	16222-66-5	2354	18283-93-7	369	20548-54-3	511	24094-93-7	620
15179-32-5	2279	16399-77-2	1098	18288-22-7	342	20601-83-6	1445	24304-00-5	37
15192-17-3	1928	16544-92-6	686	18356-71-3	1994	20610-49-5	1840	24401-69-2	751
15192-26-4	2309	16671-27-5	642	18433-40-4	171	20619-16-3	946	24415-00-7	942
15192-42-4	1851	16674-78-5	1300	18433-84-6	371	20644-15-9	2123	24422-20-6	940
15195-33-2	1570	16689-88-6	2357	18454-12-1	1194	20644-87-5	2520	24422-21-7	935
15195-53-6	1555	16721-80-5	2136	18480-07-4	2227	20661-21-6	1041	24567-53-1	1628
15230-48-5	934	16721-80-5	2137	18618-55-8	557	20665-52-5	918	24597-12-4	904
15238-00-3	700	16731-55-8	1777	18624-44-7	1103	20667-12-3	2055	24613-38-5	688
15238-00-3	701	16774-21-3	75	18718-07-5	1372	20694-39-7	1380	24621-21-4	1549
15243-27-3	1365	16812-54-7	1538	18721-05-6	638	20712-42-9	431	24646-85-3	2521
15243-33-1	1912	16853-85-3	1251	18810-58-7	224	20762-60-1	1730	24719-19-5	680
15244-35-6	440	16860-99-4	2264	18820-29-6	1393	20770-09-6	2409	25152-52-7	10
15282-88-9	1216	16871-60-6	1751	18820-29-6	1391	20816-12-0	1609	25417-81-6	251
15293-86-4	760	16871-71-9	2626	18820-29-6	1392	20859-73-8	45	25455-73-6	2074
15321-51-4	1078	16871-90-2	1755	18855-94-2	992	20910-28-5	2514	25502-05-0	2573
15364-10-0	2565	16872-11-0	382	18868-43-4	1473	21041-93-0	698	25583-20-4	2416
15364-94-0	1383	16893-85-9	2124	18917-82-3	1207	21041-95-2	422	25617-97-4	901
15385-57-6	1412	16903-35-8	960	18917-91-4	34	21093-83-4	1758	25617-98-5	1029
15385-58-7	1407	16919-19-0	108	18933-05-6	1374	21109-95-5	290	25658-42-8	2677
15468-32-3	2019	16919-31-6	109	19034-13-0	714	21159-32-0	577	25658-43-9	2487
15469-38-2	1137	16919-58-7	103	19049-40-2	312	21255-83-4	398	25659-31-8	1205
15474-63-2	833	16921-30-5	1752	19073-56-4	1893	21264-43-7	1034	25764-08-3	547
15477-33-5	19	16921-96-3	1062	19086-20-5	1350	21308-45-2	2566	25764-09-4	1829
15492-38-3	1877	16923-58-3	2119	19088-74-5	1898	21308-80-5	397	25764-10-7	1169
15513-84-5	2527	16923-58-3	2120	19138-68-2	815	21351-79-1	584	25764-11-8	1496
15513-94-7	2535	16923-95-8	1756	19139-47-0	550	21548-73-2	2065	25764-15-2	883
15519-28-5	581	16924-00-8	1749	19168-23-1	101	21645-51-2	29	25808-74-6	1223
15520-84-0	731	16925-25-0	2122	19197-73-0	1776	21651-19-4	2395	25817-87-2	984
15571-91-2	1806	16925-39-6	473	19287-45-7	368	21679-31-2	648	25895-60-7	2100
15572-25-5	2345	16940-66-2	2085	19357-83-6	2467	21908-53-2	1440	25909-39-1	1694

25985-07-3	1693	31052-14-9	2298	37913-38-5	1451	54723-94-3	148	75926-28-2	1970
26042-63-7	2045	31052-43-4	1907	38455-77-5	2405	54723-94-3	130	75926-30-6	1971
26042-64-8	2043	31083-74-6	1909	38680-84-1	608	56802-99-4	2167	81155-18-2	1860
26124-86-7	262	31234-26-1	1859	38705-19-0	1421	57246-89-6	1869	82642-06-6	237
26134-62-3	1275	31576-40-6	1607	39277-13-9	2117	58724-12-2	580	84359-31-9	650
26342-61-0	621	32248-43-4	1929	39290-85-2	769	59201-51-3	1602	84359-31-9	651
26499-65-0	510	32287-65-3	25	39368-69-9	1856	60616-74-2	1315	92141-86-1	587
26500-06-1	2564	32823-06-6	40	39416-30-3	303	60676-86-0	2021	94217-84-2	1246
26508-33-8	1085	33088-16-3	2365	39796-98-0	2570	60763-24-8	2368	97126-35-7	723
26522-91-8	1424	33689-80-4	4	39797-63-2	2567	60804-25-3	1876	99685-96-8	523
26628-22-8	2084	33689-81-5	2	41944-01-8	356	60883-64-9	317	107539-20-8	237
26686-77-1	1341	33689-82-6	5	42246-24-2	1868	60922-26-1	2447	107539-20-8	238
26750-66-3	1817	34018-28-5	1185	42739-38-8	172	60950-56-3	1343	107539-20-8	239
27016-73-5	667	34128-09-1	2338	49756-76-5	2569	60969-19-9	2334	107539-20-8	240
27016-75-7	1509	34330-64-8	417	50432-32-1	2566	62571-53-3	2466	109064-29-1	238
27043-84-1	2607	35112-53-9	298	50801-97-3	1926	63972-04-3	1686	114104-80-2	239
27133-66-0	277	35405-51-7	501	50813-65-5	228	64082-35-5	1264	115383-22-7	524
27218-16-2	610	35718-37-7	2568	50927-81-6	2022	65202-12-2	1418	121012-90-6	472
27774-13-6	2551	36377-94-3	828	51503-61-8	147	65842-03-7	1153	124365-83-9	240
27860-83-9	2624	36478-76-9	2511	51621-05-7	1916	66104-24-3	318	133578-89-9	2550
28380-38-3	208	36539-19-2	2565	51898-99-8	346	68016-36-4	297	133863-98-6	1486
29703-01-3	574	36548-87-5	2380	52003-58-4	126	68938-92-1	1695	134929-59-2	525
29809-42-5	2326	36897-37-7	1527	52503-64-7	768	71500-16-8	1918	155698-29-6	1683
30708-80-6	605	36907-42-3	2529	52740-16-6	452	71626-98-7	482	158704-27-9	1660
30737-24-7	2341	37248-04-7	989	52788-53-1	893	72520-94-6	555		
30937-52-1	1857	37248-34-3	1802	52951-38-9	348	73560-00-6	1055		
30937-53-2	1855	37267-86-0	1633	53731-35-4	1384	73796-25-5	2230		
31032-13-0	971	37773-49-2	1696	54120-05-7	1605	75926-26-0	1969		

PHYSICAL PROPERTIES OF THE RARE EARTH METALS

K.A. Gschneidner, Jr.

Table 1
Data for the Trivalent Ions of the Rare Earth Elements

Rare earth	Symbol	Atomic no.	Atomic wt.[a]	Electronic configuration for R^{3+}: No. 4f electrons	S	L	J	Spectroscopic ground state symbol
Scandium	Sc	21	44.955910	0	—	—	—	—
Yttrium	Y	39	88.90585	0	—	—	—	—
Lanthanum	La	57	138.9055	0	—	—	—	—
Cerium	Ce	58	140.115	1	1/2	3	5/2	$^2F_{5/2}$
Praseodymium	Pr	59	140.90765	2	1	5	4	$^3H_4$
Neodymium	Nd	60	144.24	3	3/2	6	9/2	$^4I_{9/2}$
Promethium	Pm	61	(145)	4	2	6	4	$^5I_4$
Samarium	Sm	62	150.36	5	5/2	5	5/2	$^6H_{5/2}$
Europium	Eu	63	151.965	6	3	3	0	$^7F_0$
Gadolinium	Gd	64	157.25	7	7/2	0	7/2	$^8S_{7/2}$
Terbium	Tb	65	158.92534	8	3	3	6	$^7F_6$
Dysprosium	Dy	66	162.50	9	5/2	5	15/2	$^6H_{15/2}$
Holmium	Ho	67	164.93032	10	2	6	8	$^5I_8$
Erbium	Er	68	167.26	11	3/2	6	15/2	$^4I_{15/2}$
Thulium	Tm	69	168.93421	12	1	5	6	$^3H_6$
Ytterbium	Yb	70	173.04	13	1/2	3	7/2	$^2F_{7/2}$
Lutetium	Lu	71	174.967	14	—	—	—	—

Note: For additional information, see Goldschmidt, Z.B., in *Handbook on the Physics and Chemistry of Rare Earths,* Vol. 1, Gschneidner, K.A., Jr. and Eyring, L., Eds., North-Holland Physics, Amsterdam, 1978; DeLaeter, J.R., and Heumann, K.G., *J. Phys. Chem. Ref. Data,* 20, 1313, 1991; *Pure Appl. Chem.*, 66, 2423, 1994.

[a] 1993 standard atomic weights.

Table 2
Crystallographic Data for the Rare Earth Metals at 24°C (297 K) or Below

Rare earth metal	Crystal structure[a]	Lattice constants(Å): a_o	b_o	c_o	Metallic radius CN = 12 (Å)	Atomic volume (cm^3/mol)	Density (g/cm^3)
αSc	hcp	3.3088	—	5.2680	1.6406	15.039	2.989
αY	hcp	3.6482	—	5.7318	1.8012	19.893	4.469
αLa	dhcp	3.7740	—	12.171	1.8791	22.602	6.146
αCe[b]	fcc	4.85[b]	—	—	1.72[b]	17.2[b]	8.16[b]
βCe	dhcp	3.6810	—	11.857	1.8321	20.947	6.689
γCe[c]	fcc	5.1610	—	—	1.8247	20.696	6.770
αPr	dhcp	3.6721	—	11.8326	1.8279	20.803	6.773
αNd	dhcp	3.6582	—	11.7966	1.8214	20.583	7.008
αPm	dhcp	3.65	—	11.65	1.811	20.24	7.264
αSm	rhomb[d]	3.6290[d]	—	26.207	1.8041	20.000	7.520
Eu	bcc	4.5827	—	—	2.0418	28.979	5.244
αGd	hcp	3.6336	—	5.7810	1.8013	19.903	7.901
α′Tb[e]	ortho	3.605[e]	6.244[e]	5.706[e]	1.784[e]	19.34[e]	8.219[e]
αTb	hcp	3.6055	—	5.6966	1.7833	19.310	8.230
α′Dy[f]	ortho	3.595[f]	6.184[f]	5.678[f]	1.774[f]	19.00[f]	8.551[f]
αDy	hcp	3.5915	—	5.6501	1.7740	19.004	8.551
Ho	hcp	3.5778	—	5.6178	1.7661	18.752	8.795

Table 2
Crystallographic Data for the Rare Earth Metals at 24°C (297 K) or Below (continued)

Rare earth metal	Crystal structure[a]	Lattice constants(Å) a_o	b_o	c_o	Metallic radius CN = 12 (Å)	Atomic volume (cm³/mol)	Density (g/cm³)
Er	hcp	3.5592	—	5.5850	1.7566	18.449	9.066
Tm	hcp	3.5375	—	5.5540	1.7462	18.124	9.321
αYb[g]	hcp	3.8799[g]	—	6.3859[g]	1.9451[g]	25.067[g]	6.903[g]
βYb	fcc	5.4848	—	—	1.9392	24.841	6.966
Lu	hcp	3.5052	—	5.5494	1.7349	17.779	9.841

Note: For additional information, see Gschneidner, K.A., Jr. and Calderwood, F.W., in *Handbook on the Physics and Chemistry of Rare Earths,* Vol. 8, Gschneidner, K.A., Jr. and Eyring, L., Eds., North-Holland Physics, Amsterdam, 1986; Gschneidner, K.A., Jr., Pecharsky, V.K., Cho, Jaephil and Martin, S.W., *Scripta Mater.*, 1996, to be published.

[a] hcp = hexagonal close-packed; $P6_3/mmc$, hP2, A3, Mg-type; dhcp = double-c hexagonal close-packed; $P6_3/mmc$, hP4, A3′, αLa-type; fcc = face-centered cubic; $Fm\bar{3}m$, cF4, A1, Cu-type; rhomb = rhombohedral; $R\bar{3}m$, hR3, αSm-type; bcc = body-centered cubic; $Im\bar{3}m$, cI2, A2, W-type; ortho = orthorhombic; Cmcm, oC4, α′ Dy-type.

[b] At 77 K (–196°C).

[c] Equilibrium room temperature (standard state) phase.

[d] Rhombohedral is the primitive cell. Lattice parameters given are for the nonprimitive hexagonal cell.

[d] At 220 K (–53°C).

[f] At 86 K (–187°C).

[g] At 23°C.

Table 3
Crystallographic Data for Rare Earth Metals at High Temperature

Rare earth metal	Structure	Lattice parameter (Å)	Temp. (°C)	Metallic radius CN = 8 (Å)	CN = 12 (Å)	Atomic volume (cm³/mol)	Density (g/cm³)
βSc	bcc	3.73 (est.)	1337	1.62	1.66	15.6	2.88
βY	bcc	4.10[a]	1478	1.78	1.83	20.8	4.28
βLa	fcc	5.303	325	—	1.875	22.45	6.187
γLa	bcc	4.26	887	1.84	1.90	23.3	5.97
δCe	bcc	4.12	757	1.78	1.84	21.1	6.65
βPr	bcc	4.13	821	1.79	1.84	21.2	6.64
βNd	bcc	4.13	883	1.79	1.84	21.2	6.80
βPm	bcc	4.10 (est.)	890	1.78	1.83	20.8	6.99
βSm	hcp	a = 3.6630 c = 5.8448	450[b]	—	1.8176	20.450	7.353
γSm	bcc	4.10 (est.)	922	1.77	1.82	20.8	7.25
βGd	bcc	4.06	1265	1.76	1.81	20.2	7.80
βTb	bcc	4.07[a]	1289	1.76	1.81	20.3	7.82
βDy	bcc	4.03[a]	1381	1.75	1.80	19.7	8.23
γYb	bcc	4.44	763[c]	1.92	1.98	26.4	6.57

Note: The rare earths Eu, Ho, Er, Tm, and Lu are monomorphic. For additional information, see Gschneidner, K.A., Jr. and Calderwood, F.W., in *Handbook on the Physics and Chemistry of Rare Earths,* Vol. 8, Gschneidner, K.A., Jr. and Eyring, L., Eds., North-Holland Physics, Amsterdam, 1986, 1.

[a] Determined by extrapolation to 0% solute of a vs. composition data for R-Mg alloys at 24°C and corrected for thermal expansion to temperature given.

[b] The hcp phase was stabilized by impurities and the temperature of measurement was below the equilibrium transition temperature (see Table 4).

[c] The bcc phase was stabilized by impurities and the temperature of measurement was below the equilibrium transition temperature (see Table 4).

Table 4
High Temperature Transition Temperatures and Melting Point of Rare Earth Metals

Rare earth metal	Transition I (α – β)[a] Temp. (°C)	Transition I (α – β)[a] Phases	Transition II (β – γ)[a] Temp. (C°)	Transition II (β – γ)[a] Phases	Melting point (C°)
Sc	1337	hcp ⇌ bcc	—	—	1541
Y	1478	hcp ⇌ bcc	—	—	1522
La[b]	310	dhcp → fcc	865	fcc ⇌ bcc	918
Ce[c,d]	139	dhcp → fcc (β - γ)	726	fcc ⇌ bcc (γ - δ)	798
Pr	795	dhcp ⇌ bcc	—	—	931
Nd	863	dhcp ⇌ bcc	—	—	1021
Pm	890	dhcp ⇌ bcc	—	—	1042
Sm[e]	734	rhom → hcp	922	hcp ⇌ bcc	1074
Eu	—	—	—	—	822
Gd	1235	hcp ⇌ bcc	—	—	1313
Tb	1289	hcp ⇌ bcc	—	—	1356
Dy	1381	hcp ⇌ bcc	—	—	1412
Ho	—	—	—	—	1474
Er	—	—	—	—	1529
Tm	—	—	—	—	1545
Yb	795	fcc ⇌ bcc (β - γ)	—	—	819
Lu	—	—	—	—	1663

Note: For additional information, see Gschneidner, K.A., Jr. and Calderwood, F.W., in *Handbook on the Physics and Chemistry of Rare Earths,* Vol. 8, Gschneidner, K.A., Jr. and Eyring, L., Eds., North-Holland Physics, Amsterdam, 1986; Gschneidner, K.A., Jr., Pecharsky, V.K., Cho, Jaephil and Martin, S.W., *Scripta Mater.*, 34, 1717, 1996.

[a] For all the transformations listed, unless otherwise noted.
[b] On cooling, fcc → dhcp (β → α), 260°C.
[c] The β ⇌ γ equilibrium transition temperature is 10 ± 5°C.
[d] On cooling, fcc → dhcp (γ → β), –16°C.
[e] On cooling, hcp → rhomb (β → α), 727°C.

Table 5
Low Temperature Transition Temperatures of the Rare Earth Metals

Rare earth metal	Cooling: Transformation	°C	K	Rare earth metal	Heating: Transformation	°C	K
Ce	γ → β[a]	–16	257	Ce	α → β	–148	125
	γ → α	–172	101		α → β + γ	–104	169
	β → α	–228	45		β → γ[a]	139	412
Tb	α → α′	–53	220	Yb	α → β	7	280
Dy	α → α′	–187	86				
Yb	β → α	–13	260				

Note: For additional information, see Beaudry, B.J. and Gschneidner, K.A., Jr., in *Handbook on the Physics and Chemistry of Rare Earths,* Vol. 1, Gschneidner, K.A., Jr. and Eyring, L., Eds., North-Holland Physics, Amsterdam, 1978, 173. Koskenmaki, D.C. and Gschneidner, K.A., Jr., 1978, in *Handbook on the Physics and Chemistry of Rare Earths,* Vol. 1, Gschneidner, K.A., Jr. and Eyring, L., Eds., North-Holland Physics, Amsterdam, 1978, 337. Gschneidner, K.A., Jr., Pecharsky, V.K., Cho, Jaephil and Martin, S.W., *Scripta Mater.*, 34, 1717, 1996.

[a] The β ⇌ γ equilibrium transition temperature is 10 ± 5°C (283 ± 5K).

Table 6
Heat Capacity, Standard Entropy, Heats of Transformation, and Fusion of the Rare Earth Metals

Rare earth metal	Heat capacity at 298 K (J/mol K)	Standard entropy S°_{298} (J/mol K)	Heat of transformation (kJ/mol) trans. 1	ΔH_{tr}^1	trans. 2	ΔH_{tr}^2	Heat of fusion (kJ/mol)
Sc	25.5	34.6	$\alpha \rightleftharpoons \beta$	4.00	—	—	14.1
Y	26.5	44.4	$\alpha \rightleftharpoons \beta$	4.99	—	—	11.4
La	27.1	56.9	$\alpha \rightleftharpoons \beta$	0.36	$\beta \rightleftharpoons \gamma$	3.12	6.20
Ce	26.9	72.0	$\beta \rightleftharpoons \gamma$	0.05	$\gamma \rightleftharpoons \delta$	2.99	5.46
Pr	27.2	73.2	$\alpha \rightleftharpoons \beta$	3.17	—	—	6.89
Nd	27.5	71.5	$\alpha \rightleftharpoons \beta$	3.03	—	—	7.14
Pm	27.3[a]	71.6[a]	$\alpha \rightleftharpoons \beta$	3.0[a]	—	—	7.7[a]
Sm	29.5	69.6	$\alpha \rightleftharpoons \beta$	0.2[a]	$\beta \rightleftharpoons \gamma$	3.11	8.62
Eu	27.7	77.8	—	—	—	—	9.21
Gd	37.0	68.1	$\alpha \rightleftharpoons \beta$	3.91	—	—	10.0
Tb	28.9	73.2	$\alpha \rightleftharpoons \beta$	5.02	—	—	10.79
Dy	27.7	75.6	$\alpha \rightleftharpoons \beta$	4.16	—	—	11.06
Ho	27.2	75.3	—	—	—	—	17.0[a]
Er	28.1	73.2	—	—	—	—	19.9
Tm	27.0	74.0	—	—	—	—	16.8
Yb	26.7	59.9	$\beta \rightleftharpoons \gamma$	1.75	—	—	7.66
Lu	26.9	51.0	—	—	—	—	22[a]

Note: For additional information, see Hultgren, R., Desai, P.D., Hawkins, D.T., Gleiser, M., Kelley, K.K., and Wagman, D.D., *Selected Values of the Thermodynamic Properties of the Elements,* ASM International, Metals Park, Ohio, 1973; Wagman, D.D., Evans, W.H., Parker, V.B., Schumm, R.H., Halow, I., Bailey, S.M., Churney, K.L., and Nuttall, R.L., *The NBS Tables of Chemical Thermodynamic Properties, J. Phys. Chem. Ref. Data,* Vol. 11, Suppl 2, 1982; Amitin, E.B., Bessergenev, W.G., Kovalevskaya, Yu. A., and Paukov, I.E., *J. Chem. Thermodyn.*, 15, 181, 1983; Amitin, E.B., Bessergenev, W.G., Kovalevskaya, Yu. A., and Paukov, I.E., *J. Chem. Thermodyn.*, 15, 181, 1983.

[a] Estimated.

Table 7
Vapor Pressures, Boiling Points, and Heats of Sublimation of Rare Earth Metals

Rare earth metal	Temperature in °C[a] for a vapor pressure of 10^{-8} atm (0.001 Pa)	10^{-6} atm (0.101 Pa)	10^{-4} atm (10.1Pa)	10^{-2} atm (1013 Pa)	Boiling point[a] (°C)	Heat of sublimation at 25°C (kJ/mol)
Sc	1036	1243	1533	1999	2836	377.8
Y	1222	1460	1812	2360	3345	424.7
La	1301	1566	1938	2506	3464	431.0
Ce	1290	1554	1926	2487	3443	422.6
Pr	1083	1333	1701	2305	3520	355.6
Nd	955	1175	1500	2029	3074	327.6
Pm	—	—	—	—	3000[b]	348[b]
Sm	508	642	835	1150	1794	206.7
Eu	399	515	685	964	1529	175.3
Gd	1167	1408	1760	2306	3273	397.5
Tb	1124	1354	1698	2237	3230	388.7
Dy	804	988	1252	1685	2567	290.4
Ho	845	1036	1313	1771	2700	300.8
Er	908	1113	1405	1896	2868	317.1
Tm	599	748	964	1300	1950	232.2
Yb	301	400	541	776	1196	152.1
Lu	1241	1483	1832	2387	3402	427.6

Note: For additional information, see Hultgren, R., Desai, P.D., Hawkins, D.T., Gleiser, M., Kelley, K.K., and Wagman, D.D., *Selected Values of the Thermodynamic Properties of the Elements,* ASM International, Metals Park, Ohio, 1973 and Beaudry, B.J. and Gschneidner, K.A., Jr., in *Handbook on the Physics and Chemistry of Rare Earths,* Vol. 1, Gschneidner, K.A., Jr. and Eyring, L., Eds., North-Holland Physics, Amsterdam, 1978, 173.

[a] International Temperature Scale of 1990 (ITS-90) values.
[b] Estimated.

Table 8
Magnetic Properties of the Rare Earth Metals

Rare earth metal	$\chi_A \times 10^6$ at 298 K (emu/mol)	Effective magnetic moment: Paramagnetic at ~298 K, Theory[a]	Paramagnetic at ~298 K, Obs.	Ferromagnetic at ~0 K, Theory[b]	Ferromagnetic at ~0 K, Obs.	Easy axis	Néel temp. T_N (K): Hex sites	Néel temp. T_N (K): Cubic sites	Curie temp. T_C (K)	θ_p (K): ∥c	θ_p (K): ⊥c	θ_p (K): Polycryst. or avg.
αSc	295.2	—	—	—	—	—	—	—	—	—	—	—
αY	187.7	—	—	—	—	—	—	—	—	—	—	—
αLa	95.9	—	—	—	—	—	—	—	—	—	—	—
βLa	105	—	—	—	—	—	—	—	—	—	—	—
γCe	2,270	2.54	2.52	2.14	—	—	—	14.4	—	—	—	−50
βCe	2,500	2.54	2.61	2.14	—	—	13.7	12.5	—	—	—	−41
αPr	5,530	3.58	3.56	3.20	2.7[c]	**a**	0.03	—	—	—	—	0
αNd	5,930	3.62	3.45	3.27	2.2[c]	**b**	19.9	7.5	—	0	5	3.3
αPm	—	2.68	—	2.40	—	—	—	—	—	—	—	—
αSm	1,278[d]	0.85	1.74	0.71	0.5[c]	**a**	109	14.0	—	—	—	—
Eu	30,900	7.94	8.48	7.0	5.9	<110>	—	90.4	—	—	—	100
αGd	185,000[e]	7.94	7.98	7.0	7.63	30° to **c**	—	—	293.4	317	317	317
αTb	170,000	9.72	9.77	—	—	—	230.0	—	—	195	239	224
α′Tb	—	—	—	9.0	9.34	**b**	—	—	219.5	—	—	—
αDy	98,000	10.64	10.83	—	—	—	180.2	—	—	121	169	153
α′Dy	—	—	—	10.0	10.33	**a**	—	—	90.5[g]	—	—	—
Ho	72,900	10.60	11.2	10.0	10.34	**b**	132	—	19.5	73.0	88.0	83.0
Er	48,000	9.58	9.9	9.0	9.1	30° to **c**	85	—	18.7	61.7	32.5	42.2
Tm	24,700	7.56	7.61	7.0	7.14	**c**	58	—	32.0	41.0	−17.0	2.3
βYb	67[d]	—	—	—	—	—	—	—	—	—	—	—
Lu	182.9	—	—	—	—	—	—	—	—	—	—	—

Note: For additional information, see McEwen, K.A., in *Handbook on the Physics and Chemistry of Rare Earths,* Vol. 1, Gschneidner, K.A., Jr. and Eyring, L., Eds., North-Holland Physics, Amsterdam, 1978, 411 and Legvold, S., in *Ferromagnetic Materials,* Vol. 1, Wohlfarth, E.P., Ed., North-Holland Physics, Amsterdam, 1980, 183; Pecharsky, V.K., Gschneidner, K.A., Jr. and Fort, D., *Phys. Rev. B,* 47, 5063, 1993; Pecharsky, V.K., Gschneidner, K.A., Jr. and Fort, D., 1996, to be published; Steward, A.M. and Collocott, S.J., *J. Phys.: Condens. Matter,* 1, 677, 1988.

[a] $g[J(J+1)]^{1/2}$.
[b] gJ.
[c] At 38 T and 4.2 K.
[d] At 290 K.
[e] At 350 K.
[g] On cooling $T_C = 89.6$ K and on warming $T_C = 91.5$ K.

Table 9
Room Temperature Coefficient of Thermal Expansion, Thermal Conductivity, Electrical Resistance, and Hall Coefficient

Rare earth metal	Expansion $\alpha_i \times 10^6$ (°C⁻¹)			Thermal conductivity (W/cm-K)	Electrical resistance (μΩ·cm)			Hall coefficient ($R_i \times 10^{12}$) (V·cm/A·Oe)		
	α_a	α_c	α_{poly}		ρ_a	ρ_c	ρ_{poly}	R_a	R_c	R_{poly}
αSc	7.6	15.3	10.2	0.158	70.9	26.9	56.2[a]	—	—	–0.13
αY	6.0	19.7	10.6	0.172	72.5	35.5	59.6	–0.27	–1.6	—
aLa	4.5	27.2	12.1	0.134	—	—	61.5	—	—	–0.35
bCe	—	—	—	—	—	—	82.8	—	—	—
γCe	6.3	—	6.3	0.113	—	—	74.4	—	—	+1.81
αPr	4.5	11.2	6.7	0.125	—	—	70.0	—	—	+0.709
αNd	7.6	13.5	9.6	0.165	—	—	64.3	—	—	+0.971
αPm	9[b]	16[b]	11[b]	0.15[b]	—	—	75[b]	—	—	—
αSm	9.6	19.0	12.7	0.133	—	—	94.0	—	—	–0.21
Eu	35.0	—	35.0	0.139[b]	—	—	90.0	—	—	+24.4
αGd	9.1[c]	10.0[c]	9.4[c]	0.105	135.1	121.7	131.0	–10	–54	–4.48[d]
αTb	9.3	12.4	10.3	0.111	123.5	101.5	115.0	–1.0	–3.7	—
αDy	7.1	15.6	9.9	0.107	111.0	76.6	92.6	–0.3	–3.7	—
Ho	7.0	19.5	11.2	0.162	101.5	60.5	81.4	+0.2	–3.2	—
Er	7.9	20.9	12.2	0.145	94.5	60.3	86.0	+0.3	–3.6	—
Tm	8.8	22.2	13.3	0.169	88.0	47.2	67.6	—	—	–1.8
βYb	26.3	—	26.3	0.385	—	—	25.0	—	—	+3.77
Lu	4.8	20.0	9.9	0.164	76.6	34.7	58.2	+0.45	–2.6	–0.535

Note: For additional information, see Beaudry, B. J. and Gschneidner, K.A., Jr., in *Handbook on the Physics and Chemistry of Rare Earths,* Vol. 1, Gschneidner, K.A., Jr. and Eyring, L., Eds., North-Holland Physics, Amsterdam, 1978, 173 and McEwen, K.A., in *Handbook on the Physics and Chemistry of Rare Earths,* Vol. 1, Gschneidner, K.A., Jr. and Eyring, L., Eds., North-Holland Physics, Amsterdam, 1978, 411.

[a] Calculated from single crystal values.
[b] Estimated.
[c] At 100°C.
[d] At 77°C.

Table 10
Electronic Specific Heat Constant (γ), Electron-Electron (Coulomb) Coupling Constant (μ^*), Electron-Phonon Coupling Constant (λ), Debye Temperature at 0 K(θ_D), and Superconducting Transition Temperature

Rare earth metal	γ (mJ/mol·K^2)	μ^*	λ	θ_D (K) from		Superconducting temperature (K)
				Heat capacity	Elastic constants	
αSc	10.334	0.16	0.30	345.3	—	0.050[a]
αY	7.878	0.15	0.30	244.4	258	1.3[b]
αLa	9.45	0.08	0.76	150	154	5.10
βLa	11.5	—	—	140	—	6.00
αCe	12.8	—	—	179	—	0.022[c]
αPr	20	—	1.07[d]	155[e]	153	—
αNd	f	—	0.86[d]	157[e]	163	—
αPm	—	—	—	159[e]	—	—
αSm	8.1 ± 1.5[g]	—	0.81[d]	162[e,f]	169	—
Eu	f	—	—	f	118	—
αGd	4.48	—	0.30	169	182	—
α′Tb	3.71	—	0.34[d]	169.6	177	—
α′Dy	4.9	—	0.32[d]	192	183	—
Ho	2.1	—	0.30[d]	175[e]	190	—
Er	8.7	—	0.33[d]	176.9	188	—
Tm	f	—	0.36[d]	179[e]	200	—
αYb	3.30	—	—	117.6	118	—
βYb	8.36	—	—	109	—	—
Lu	8.194	0.14	0.31	183.2	185	0.022[h]

Note: For additional information, see Sundström, L.J., in *Handbook on the Physics and Chemistry of Rare Earths,* Vol. 1, Gschneidner, K.A., Jr., and Eyring, L., Eds., North-Holland Physics, Amsterdam, 1978, 379, Scott, T., in *Handbook on the Physics and Chemistry of Rare Earths,* Vol. 1, Gschneidner, K.A., Jr. and Eyring, L., Eds., North-Holland Physics, Amsterdam, 1978, 591, Probst, C. and Wittig, J., in *Handbook on the Physics and Chemistry of Rare Earths,* Vol. 1, Gschneidner, K.A., Jr. and Eyring, L., Eds., North-Holland Physics, Amsterdam, 1978, 749, and Tsang, T.-W.E., Gschneidner, K.A., Jr., Schmidt, F.A., and Thome, D.K., *Phys. Rev.,* B, 31, 235, 1985. Collocott, S.J., Hill, R.W. and Stewart, A.M., *J. Phys. F,* 18, L223, 1988. Hill, R.W. and Gschneidner, K.A., Jr., *J. Phys. F,* 18, 2545, 1988. Skriver, H.L. and Mertig, I., *Phys. Rev. B,* 41, 6553, 1990. Collocott, S.J. and Stewart, A.M., *J. Phys.: Condens. Matter,* 4, 6743, 1992. Pecharsky, V.K., Gschneidner, K.A., Jr. and Fort, D., *Phys. Rev. B,* 47, 5063, 1993.

[a] At 18.6 GPa.
[b] At 11 GPa.
[c] At 2.2 GPa.
[d] Calculated value.
[e] Estimated.
[f] Heat capacity results have been reported, but the resultant γ and θ_D values are unreliable because of the presence of impurities and/or there was no reliable procedure or model to correct for the magnetic contribution to the heat capacity.
[g] Based on the values reported for the purer Sm sample (IV).
[h] At 4.5 GPa.

Table 11
Room Temperature Elastic Moduli and Mechanical Properties

Rare earth metal	Elastic moduli (GPa)				Mechanical properties (MPa)				
	Young's (elastic) modulus	Shear modulus	Bulk modulus	Poisson's ratio	Yield strength 0.2% offset	Ultimate tensile strength	Uniform elongation (%)	Reduction in area (%)	Recryst. temp. (°C)
Sc	74.4	29.1	56.6	0.279	173[a]	255[a]	5.0[a]	8.0[a]	550
Y	63.5	25.6	41.2	0.243	42	129	34.0	—	550
αLa	36.6	14.3	27.9	0.280	126[a]	130	7.9[a]	—	300
βCe	—	—	—	—	86	138	—	24.0	—
γCe	33.6	13.5	21.5	0.24	28	117	22.0	30.0	325
αPr	37.3	14.8	28.8	0.281	73	147	15.4	67.0	400
αNd	41.4	16.3	31.8	0.281	71	164	25.0	72.0	400
αPm	46[b]	18[b]	33[b]	0.28[b]	—	—	—	—	400[b]
αSm	49.7	19.5	37.8	0.274	68	156	17.0	29.5	440
Eu	18.2	7.9	8.3	0.152	—	—	—	—	300
αGd	54.8	21.8	37.9	0.259	15	118	37.0	56.0	500
αTb	55.7	22.1	38.7	0.261	—	—	—	—	500
αDy	61.4	24.7	40.5	0.247	43	139	30.0	30.0	550
Ho	64.8	26.3	40.2	0.231	—	—	—	—	520
Er	69.9	28.3	44.4	0.237	60	136	11.5	11.9	520
Tm	74.0	30.5	44.5	0.213	—	—	—	—	600
βYb	23.9	9.9	30.5	0.207	7	58	43.0	92.0	300
Lu	68.6	27.2	47.6	0.261	—	—	—	—	600

Note: For additional information, see Scott, T., in *Handbook on the Physics and Chemistry of Rare Earths,* Vol. 1, Gschneidner, K.A., Jr. and Eyring, L., Eds., North-Holland Physics, Amsterdam, 1978, 591.

[a] Value is questionable.
[b] Estimated.

Table 12
Liquid Metal Properties Near the Melting Point

Rare earth metal	Density (g/cm³)	Surface tension (N/m)	Viscosity (centipoise)	Heat capacity (J/mol K)	Thermal conductivity (W/cm K)	Magnetic susceptibility $\chi \times 10^4$ (emu/mol)	Electrical resistivity (μΩ·cm)	$\Delta V_{L \to s}$[a] (%)	Spectral emittance at λ = 645 nm: ε (%)	Spectral emittance at λ = 645 nm: Temp. range (°C)
Sc	2.80	0.954	—	44.2[b]	—	—	—	—	—	—
Y	4.24	0.871	—	43.1	—	—	—	—	36.8	1522–1647
La	5.96	0.718	2.65	34.3	0.238	1.20	133	–0.6	25.4	920–1287
Ce	6.68	0.706	3.20	37.7	0.210	9.37	130	+1.1	32.2	877–1547
Pr	6.59	0.707	2.85	43.0	0.251	17.3	139	–0.02	28.4	931–1537
Nd	6.72	0.687	—	48.8	0.195	18.7	151	–0.9	39.4	1021–1567
Pm	6.9[b]	0.680[b]	—	50[b]	—	—	160[b]	—	—	—
Sm	7.16	0.431	—	50.2[b]	—	18.3	182	–3.6	43.7	1075
Eu	4.87	0.264	—	38.1	—	97	242	–4.8	—	—
Gd	7.4	0.664	—	37.2	0.149	67	195	–2.0	34.2	1313–1600
Tb	7.65	0.669	—	46.5	—	82	193	–3.1	—	—
Dy	8.2	0.648	—	49.9	0.187	95	210	–4.5	29.7	1412–1437
Ho	8.34	0.650	—	43.9	—	88	221	–7.4	—	—
Er	8.6	0.637	—	38.7	—	69	226	–9.0	37.2	1529–1587
Tm	9.0[b]	—	—	41.4	—	41	235[b]	–6.9	—	—
Yb	6.21	0.320	2.67	36.8	—	—	113	–5.1	—	—
Lu	9.3	0.940	—	47.9[b]	—	—	224	–3.6	—	—

Note: For additional information, see Van Zytveld, J., in *Handbook on the Physics and Chemistry of Rare Earths,* Vol. 12, Gschneidner, K.A., Jr. and Eyring, L., Eds., North-Holland Physics, Amsterdam, 1989, 357. Stretz, L.A. and Bautista, R.G., in *Temperature, Its Measurement and Control in Science and Industry,* Vol. 4, part I, H.H. Plumb, Ed., Instrument Society of America, Pittsburgh, 1972, 489. King, T.S., Baria, D.N., and Bautista, R.G., *Met. Trans. B,* 7, 411, 1976. Baria, D.N., King, T.S., and Bautista, R.G., *Met. Trans. B,* 7, 577, 1976.

[a] Volume change on freezing.
[b] Estimated.

Table 13
Ionization Potentials (Electronvolts)

Rare earth	I Neutral atom	II Singly ionized	III Doubly ionized	IV Triply ionized	V Quadruply ionized
Sc	6.56144	12.79967	24.75666	73.4894	91.65
Y	6.217	12.24	20.52	60.597	77.0
La	5.5770	11.060	19.1773	49.95	61.6
Ce	5.5387	10.85	20.198	36.758	65.55
Pr	5.464	10.55	21.624	38.98	57.53
Nd	5.5250	10.73	22.1	40.41	—
Pm	5.554	10.90	22.3	41.1	—
Sm	5.6437	11.07	23.4	41.4	—
Eu	5.6704	11.241	24.92	42.7	—
Gd	6.1500	12.09	20.63	44.0	—
Tb	5.8639	11.52	21.91	39.79	—
Dy	5.9389	11.67	22.8	41.47	—
Ho	6.0216	11.80	22.84	42.5	—
Er	6.1078	11.93	22.74	42.7	—
Tm	6.18431	12.05	23.68	42.7	—
Yb	6.25416	12.1761	25.05	43.56	—
Lu	5.42585	13.9	20.9594	45.25	66.8

Note: For references, see the table "Ionization Potentials of Atoms and Atomic Ions" in Section 10.

Table 14
Effective Ionic Radii (Å)[a]

Rare earth ion	R^{2+} CN = 6	R^{2+} CN = 8	R^{3+} CN = 6	R^{3+} CN = 8	R^{3+} CN = 12	R^{4+} CN = 6	R^{4+} CN = 8
Sc	—	—	0.745	0.87	1.116	—	—
Y	—	—	0.900	1.015	1.220	—	—
La	—	—	1.045	1.18	1.320	—	—
Ce	—	—	1.010	1.14	1.290	0.80	0.97
Pr	—	—	0.997	1.14	1.286	0.78	0.96
Nd	—	—	0.983	1.12	1.276	—	—
Pm	—	—	0.97	1.10	1.267	—	—
Sm	1.19	1.27	0.958	1.09	1.260	—	—
Eu	1.17	1.25	0.947	1.07	1.252	—	—
Gd	—	—	0.938	1.06	1.246	—	—
Tb	—	—	0.923	1.04	1.236	0.76	0.88
Dy	—	—	0.912	1.03	1.228	—	—
Ho	—	—	0.901	1.02	1.221	—	—
Er	—	—	0.890	1.00	1.214	—	—
Tm	—	—	0.880	0.99	1.207	—	—
Yb	1.00	1.07	0.868	0.98	1.199	—	—
Lu	—	—	0.861	0.97	1.194	—	—

Note: For additional information, see Shannon, R.D. and Prewitt, C.T., *Acta Cryst.*, 25, 925, 1969 and Shannon, R.D. and Prewitt, C.T., *Acta Cryst.*, 26, 1046, 1970.

[a] Radius of O^{2-} is 1.40 Å for a coordination number (CN) of 6.

MELTING, BOILING, AND CRITICAL TEMPERATURES OF THE ELEMENTS

This table summarizes the melting point t_m, normal boiling point t_b, and critical temperature t_c (on the ITS-90 scale) for the elements for which data are available. A "tp" after a value indicates a solid-liquid-gas triple point, and "sp" indicates a sublimation point, where the vapor pressure of the solid phase reaches 101.325 kPa (1 atm). Transition temperatures between allotropic forms are included for several elements. References may be found in the tables *Physical Constants of Inorganic Compounds* and *Critical Constants*.

Name	t_m/°C	t_b/°C	t_c/°C
Actinium	1051	3198	
Aluminum	660.32	2519	
Americium	1176	2011	
Antimony	630.63	1587	
Argon	-189.35	-185.85	-122.28
Arsenic (gray)	817 tp (3.70 MPa)	603 sp	1400
Astatine	302		
Barium	727	1897	
Berkelium (α form)	1050		
Berkelium (β form)	986		
Beryllium	1287	2471	
Bismuth	271.40	1564	
Boron	2075	4000	
Bromine	-7.2	58.8	315
Cadmium	321.07	767	
Calcium	842	1484	
Californium	900		
Carbon (graphite)	4489 tp (10.3 MPa)	3825 sp	
Carbon (diamond)	4440 (12.4 GPa)		
Cerium	798	3443	
Cesium	28.5	671	1665
Chlorine	-101.5	-34.04	143.8
Chromium	1907	2671	
Cobalt	1495	2927	
Copper	1084.62	2562	
Curium	1345	≈3100	
Dysprosium	1412	2567	
Einsteinium	860		
Erbium	1529	2868	
Europium	822	1529	
Fermium	1527		
Fluorine	-219.67 tp	-188.12	-129.02
Francium	27		
Gadolinium	1313	3273	
Gallium	29.771 tp	2204	
Germanium	938.25	2833	
Gold	1064.18	2856	
Hafnium	2233	4603	
Helium		-268.93	-267.96
Holmium	1474	2700	
Hydrogen	-259.34	-252.87	-240.18
Indium	156.60	2072	
Iodine	113.7	184.4	546
Iridium	2446	4428	
Iron	1538	2861	
Krypton	-157.38 tp (73.2 kPa)	-153.22	-63.74
Lanthanum	918	3464	
Lawrencium	1627		
Lead	327.46	1749	
Lithium	180.50	1342	2950
Lutetium	1663	3402	
Magnesium	650	1090	
Manganese	1246	2061	
Mendelevium	827		
Mercury	-38.837 tp	356.73	1477
Molybdenum	2623	4639	
Neodymium	1021	3074	
Neon	-248.59	-246.08	-228.7
Neptunium	644		
Nickel	1455	2913	
Niobium	2477	4744	
Nitrogen	-210.00	-195.79	-146.94
Nobelium	827		
Osmium	3033	5012	
Oxygen	-218.79	-182.95	-118.56
Palladium	1554.9	2963	
Phosphorus (white)	44.15	280.5	721
Phosphorus (red)	590 tp	431 sp	721
Phosphorus (black)	610		
Platinum	1768.4	3825	
Plutonium	640	3228	
Polonium	254	962	
Potassium	63.5	759	1950
Praseodymium	931	3520	
Promethium	1042	3000	
Protactinium	1572		
Radium	700		
Radon	-71	-61.7	104
Rhenium	3186	5596	
Rhodium	1964	3695	
Rubidium	39.30	688	1820
Ruthenium	2334	4150	
Samarium	1074	1794	
Scandium	1541	2836	
Selenium(vitreous)	180 (trans to gray)	685	
Selenium (gray)	220.5	685	1493
Silicon	1414	3265	
Silver	961.78	2162	
Sodium	97.80	883	2300
Strontium	777	1382	
Sulfur (rhombic)	95.3 (trans to mono)	444.60	1041
Sulfur (monoclinic)	119.6	444.60	1041
Tantalum	3017	5458	
Technetium	2157	4265	
Tellurium	449.51	988	
Terbium	1356	3230	
Thallium	304	1473	
Thorium	1750	4788	
Thulium	1545	1950	
Tin (gray)	13.2 (trans to white)	2602	
Tin (white)	231.93	2602	
Titanium	1668	3287	
Tungsten	3422	5555	
Uranium	1135	4131	
Vanadium	1910	3407	
Xenon	-111.79 tp (81.6 kPa)	-108.12	16.62
Ytterbium	819	1196	
Yttrium	1522	3345	
Zinc	419.53	907	
Zirconium	1855	4409	

HEAT CAPACITY OF THE ELEMENTS AT 25°C

This table gives the specific heat capacity (c_p) in J/g K and the molar heat capacity (C_p) in J/mol K at a temperature of 25°C and a pressure of 100 kPa (1 bar or 0.987 standard atmospheres) for all the elements for which reliable data are available.

Name	c_p J/g K	C_p J/mol K
Actinium	0.120	27.2
Aluminum	0.897	24.200
Antimony	0.207	25.23
Argon	0.520	20.786
Arsenic	0.329	24.64
Barium	0.204	28.07
Beryllium	1.825	16.443
Bismuth	0.122	25.52
Boron	1.026	11.087
Bromine (Br_2)	0.226	36.057
Cadmium	0.232	26.020
Calcium	0.647	25.929
Carbon (graphite)	0.709	8.517
Cerium	0.192	26.94
Cesium	0.242	32.210
Chlorine (Cl_2)	0.479	33.949
Chromium	0.449	23.35
Cobalt	0.421	24.81
Copper	0.385	24.440
Dysprosium	0.170	27.7
Erbium	0.168	28.12
Europium	0.182	27.66
Fluorine (F_2)	0.824	31.304
Gadolinium	0.236	37.03
Gallium	0.371	25.86
Germanium	0.320	23.222
Gold	0.129	25.418
Hafnium	0.144	25.73
Helium	5.193	20.786
Holmium	0.165	27.15
Hydrogen (H_2)	14.304	28.836
Indium	0.233	26.74
Iodine (I_2)	0.145	36.888
Iridium	0.131	25.10
Iron	0.449	25.10
Krypton	0.248	20.786
Lanthanum	0.195	27.11
Lead	0.129	26.650
Lithium	3.582	24.860
Lutetium	0.154	26.86
Magnesium	1.023	24.869
Manganese	0.479	26.32
Mercury	0.140	27.983
Molybdenum	0.251	24.06
Neodymium	0.190	27.45
Neon	1.030	20.786
Nickel	0.444	26.07
Niobium	0.265	24.60
Nitrogen (N_2)	1.040	29.124
Osmium	0.130	24.7
Oxygen (O_2)	0.918	29.378
Palladium	0.246	25.98
Phosphorous (white)	0.769	23.824
Platinum	0.133	25.86
Potassium	0.757	29.600
Praseodymium	0.193	27.20
Radon	0.094	20.786
Rhenium	0.137	25.48
Rhodium	0.243	24.98
Rubidium	0.363	31.060
Ruthenium	0.238	24.06
Samarium	0.197	29.54
Scandium	0.568	25.52
Selenium	0.321	25.363
Silicon	0.705	19.789
Silver	0.235	25.350
Sodium	1.228	28.230
Strontium	0.301	26.4
Sulfur (rhombic)	0.710	22.75
Tantalum	0.140	25.36
Tellurium	0.202	25.73
Terbium	0.182	28.91
Thallium	0.129	26.32
Thorium	0.113	26.230
Thulium	0.160	27.03
Tin (white)	0.228	27.112
Titanium	0.523	25.060
Tungsten	0.132	24.27
Uranium	0.116	27.665
Vanadium	0.489	24.89
Xenon	0.158	20.786
Ytterbium	0.155	26.74
Yttrium	0.298	26.53
Zinc	0.388	25.390
Zirconium	0.278	25.36

VAPOR PRESSURE OF THE METALLIC ELEMENTS

C. B. Alcock

This table gives coefficients in an equation for the vapor pressure of 65 metallic elements in both the solid and liquid state. Vapor pressures in the range 10^{-10} to 10^{2} Pa (10^{-15} to 10^{-3} atm) are covered. The equation is:

for p in pascals: $\log(p/\text{Pa}) = 5.006 + A + BT^{-1} + C\log T + DT^{-3}$

for p in atmospheres: $\log(p/\text{atm}) = A + BT^{-1} + C\log T + DT^{-3}$, where T is the temperature in K

This equation reproduces the observed vapor pressures to an accuracy of ±5% or better. Reprinted with permission of the publisher, Pergamon Press.

REFERENCE

Alcock, C. B., Itkin, V. P., and Horrigan, M. K., *Canadian Metallurgical Quarterly*, 23, 309, 1984.

Element, state	*A*	*B*	*C*	*D*	Temperature range
Li sol	5.667	-8310			298-m.p.
Li liq	5.055	-8023			m.p.-1000
Na sol	5.298	-5603			298-m.p.
Na liq	4.704	-5377			m.p.-700
K sol	4.961	-4646			298-m.p.
K liq	4.402	-4453			m.p.-600
Rb sol	4.857	-4215			298-m.p.
Rb liq	4.312	-4040			m.p.-550
Cs sol	4.711	-3999			298-m.p.
Cs liq	4.165	-3830			m.p.-550
Be sol	8.042	-17020	-0.4440		298-m.p.
Be liq	5.786	-15731			m.p.-1800
Mg sol	8.489	-7813	-0.8253		298-m.p.
Ca sol	10.127	-9517	-1.4030		298-m.p.
Sr sol	9.226	-8572	-1.1926		298-m.p.
Ba sol	12.405	-9690	-2.2890		298-m.p.
Ba liq	4.007	-8163			m.p.-1200
Al sol	9.459	-17342	-0.7927		298-m.p.
Al liq	5.911	-16211			m.p.-1800
Ga sol	6.657	-14208			298-m.p.
Ga liq	6.754	-13984	-0.3413		m.p.-1600
In sol	5.991	-12548			298-m.p.
In liq	5.374	-12276			m.p.-1500
Tl sol	5.971	-9447			298-m.p.
Tl liq	5.259	-9037			m.p.-1100
Sn sol	6.036	-15710			298-m.p.
Sn liq	5.262	-15332			m.p.-1850
Pb sol	5.643	-10143			298-m.p.
Pb liq	4.911	-9701			m.p.-1200
Sc sol	6.650	-19721	0.2885	-0.3663	298-m.p.
Sc liq	5.795	-17681			m.p.-2000
Y sol	9.735	-22306	-0.8705		298-m.p.
Y liq	5.795	-20341			m.p.-2300
La sol	7.463	-22551	-0.3142		298-m.p.
La liq	5.911	-21855			m.p.-2450
Ti sol	11.925	-24991	-1.3376		298-m.p.
Ti liq	6.358	-22747			m.p.-2400
Zr sol	10.008	-31512	-0.7890		298-m.p
Zr liq	6.806	-30295			m.p.-2500
Hf sol	9.445	-32482	-0.6735		298-m.p.
V sol	9.744	-27132	-0.5501		298-m.p.

Element, state	*A*	*B*	*C*	*D*	Temperature range
V liq	6.929	-25011			m.p.-2500
Nb sol	8.822	-37818	-0.2575		298-2500
Ta sol	16.807	-41346	-3.2152	0.7437	248-2500
Cr sol	6.800	-20733	0.4391	-0.4094	298-2000
Mo sol	11.529	-34626	-1.1331		298-2500
W sol	2.945	-44094	1.3677		298-2350
W sol	-54.527	-57687	-12.2231		2200-2500
Mn sol	12.805	-15097	-1.7896		298-m.p.
Re sol	11.543	-40726	-1.1629		298-2500
Fe sol	7.100	-21723	0.4536	-0.5846	298-m.p.
Fe liq	6.347	-19574			m.p.-2100
Ru sol	9.755	-34154	-0.4723		298-m.p.
Os sol	9.419	-41198	-0.3896		298-2500
Co sol	10.976	-22576	-1.0280		298-m.p.
Co liq	6.488	-20578			m.p.-2150
Rh sol	10.168	-29010	-0.7068		298-m.p.
Rh liq	6.802	-26792			m.p.-2500
Ir sol	10.506	-35099	-0.7500		298-2500
Ni sol	10.557	-22606	-0.8717		298-m.p.
Ni liq	6.666	-20765			m.p.-2150
Pd sol	9.502	-19813	-0.9258		298-m.p.
Pd liq	5.426	-17899			m.p.-2100
Pt sol	4.882	-29387	1.1039	-0.4527	298-m.p.
Pt liq	6.386	-26856			m.p.-2500
Cu sol	9.123	-17748	-0.7317		298-m.p.
Cu liq	5.849	-16415			m.p.-1850
Ag sol	9.127	-14999	-0.7845		298-m.p.
Ag liq	5.752	-13827			m.p.-1600
Au sol	9.152	-19343	-0.7479		298-m.p.
Au liq	5.832	-18024			m.p.-2050
Zn sol	6.102	-6776			298-m.p.
Zn liq	5.378	-6286			m.p.-750
Cd sol	5.939	-5799			298-m.p.
Cd liq	5.242	-5392			m.p.-650
Hg liq	5.116	-3190			298-400
Ce sol	6.139	-21752			298-m.p.
Ce liq	5.611	-21200			m.p.-2450
Pr sol	8.859	-18720	-0.9512		298-m.p.
Pr liq	4.772	-17315			m.p.-2200
Nd sol	8.996	-17264	-0.9519		298-m.p.
Nd liq	4.912	-15824			m.p.-2000
Sm sol	9.988	-11034	-1.3287		298-m.p.
Eu sol	9.240	-9459	-1.1661		298-m.p.
Gd sol	8.344	-20861	-0.5775		298-m.p.
Gd liq	5.557	-19389			m.p.-2250
Tb sol	9.510	-20457	-0.9247		298-m.p.
Tb liq	5.411	-18639			m.p.-2200
Dy sol	9.579	-15336	-1.1114		298-m.p.
Ho sol	9.785	-15899	-1.1753		298-m.p.
Er sol	9.916	-16642	-1.2154		298-m.p.
Er liq	4.668	-14380			m.p.-1900
Tm sol	8.882	-12270	-0.9564		298-1400
Yb sol	9.111	-8111	-1.0849		298-900
Lu sol	8.793	-22423	-0.6200		298-m.p.
Lu liq	5.648	-20302			m.p.-2350
Th sol	8.668	-31483	-0.5288		298-m.p.
Th liq	-18.453	-24569	6.6473		m.p.-2500
Pa sol	10.552	-34869	-1.0075		298-m.p.

Element, state	*A*	*B*	*C*	*D*	Temperature range
Pa liq	6.177	-32874			m.p.-2500
U sol	0.770	-27729	2.6982	-1.5471	298-m.p.
U liq	20.735	-28776	-4.0962		m.p.-2500
Np sol	19.643	-24886	-3.9991		298-m.p.
Np liq	10.076	-23378	-1.3250		m.p.-2500
Pu sol	26.160	-19162	-6.6675		298-600
Pu sol	18.858	-18460	-4.4720		500-m.p.
Pu liq	3.666	-16658			m.p.-2450
Am sol	11.311	-15059	-1.3449		298-m.p.
Cm sol	8.369	-20364	-0.5770		298-m.p.
Cm liq	5.223	-18292			m.p.-2200

DENSITY OF MOLTEN ELEMENTS AND REPRESENTATIVE SALTS

This table lists the liquid density at the melting point, ρ_m, for elements that are solid at room temperature, as well as for some representative salts of these elements. Densities at higher temperatures (up to the t_{max} given in the last column) may be estimated from the equation

$$\rho(t) = \rho_m - k(t - t_m)$$

where t_m is the melting point and k is given in the fifth column of the table. If a value of t_{max} is not given, the equation should not be used to extrapolate more than about 20°C beyond the melting point.

Data for the elements were selected from the primary literature; the assistance of Gernot Lang in compiling these data is gratefully acknowledged. The molten salt data were derived from Reference 1.

REFERENCE

1. Janz, G. J., Thermodynamic and Transport Properties of Molten Salts: Correlation Equations for Critically Evaluated Density, Surface Tension, Electrical Conductance, and Viscosity Data, *J. Phys. Chem. Ref. Data*, 17, Suppl. 2, 1988.
2. Nasch, P. M., and Steinemann, S. G., *Phys. Chem. Liq.*, 29, 43, 1995.

Formula	Name	t_m/°C	ρ_m/g cm^{-3}	k/g cm^{-3} °C^{-1}	t_{max}
Ag	Silver	961.78	9.320	0.0009	1500
$AgBr$	Silver(I) bromide	432	5.577	0.001035	667
$AgCl$	Silver(I) chloride	455	4.83	0.00094	627
AgI	Silver(I) iodide	558	5.58	0.00101	802
$AgNO_3$	Silver(I) nitrate	212	3.970	0.001098	360
Ag_2SO_4	Silver(I) sulfate	652	4.84	0.001089	770
Al	Aluminum	660.32	2.375	0.000233	1340
$AlBr_3$	Aluminum bromide	97.5	2.647	0.002435	267
$AlCl_3$	Aluminum chloride	192.6	1.302	0.002711	296
AlI_3	Aluminum iodide	188.32	3.223	0.0025	240
As	Arsenic	817	5.22	0.000544	
Au	Gold	1064.18	17.31	0.001343	1200
B	Boron	2075	2.08		
Ba	Barium	727	3.338	0.000299	1550
$BaBr_2$	Barium bromide	857	3.991	0.000924	900
$BaCl_2$	Barium chloride	962	3.174	0.000681	1081
BaF_2	Barium fluoride	1368	4.14	0.000999	1727
BaI_2	Barium iodide	711	4.26	0.000977	975
Be	Beryllium	1287	1.690	0.00011	
$BeCl_2$	Beryllium chloride	415	1.54	0.0011	473
BeF_2	Beryllium fluoride	552	1.96	0.000015	850
Bi	Bismuth	271.40	10.05	0.00135	800
$BiBr_3$	Bismuth bromide	218	4.76	0.002637	927
$BiCl_3$	Bismuth chloride	230	3.916	0.0023	350
Ca	Calcium	842	1.378	0.000230	1484
$CaBr_2$	Calcium bromide	742	3.111	0.0005	791
$CaCl_2$	Calcium chloride	775	2.085	0.000422	950
CaF_2	Calcium fluoride	1418	2.52	0.000391	2027
CaI_2	Calcium iodide	783	3.443	0.000751	1028
Cd	Cadmium	321.07	7.996	0.001218	500
$CdBr_2$	Cadmium bromide	568	4.075	0.00108	720
$CdCl_2$	Cadmium chloride	564	3.392	0.00082	807
CdI_2	Cadmium iodide	387	4.396	0.001117	700
Ce	Cerium	799	6.55	0.000710	1460
$CeCl_3$	Cerium(III) chloride	817	3.25	0.00092	950
CeF_3	Cerium(III) fluoride	1430	4.659	0.000936	1927
Co	Cobalt	1495	7.75	0.00165	1580
Cr	Chromium	1907	6.3	0.0011	2100
Cs	Cesium	28.44	1.843	0.000556	510
$CsBr$	Cesium bromide	636	3.133	0.001223	860
$CsCl$	Cesium chloride	645	2.79	0.001065	906
CsF	Cesium fluoride	703	3.649	0.001282	912
CsI	Cesium iodide	621	3.197	0.001183	907

Formula	Name	t_m/°C	ρ_m/g cm^{-3}	k/g cm^{-3} °C^{-1}	t_{max}
$CsNO_3$	Cesium nitrate	414	2.820	0.001166	491
Cs_2SO_4	Cesium sulfate	1005	3.1	0.00095	1530
Cu	Copper	1084.62	8.02	0.000609	1630
CuCl	Copper(I) chloride	430	3.692	0.00076	585
Dy	Dysprosium	1411	8.37	0.00143	1540
$DyCl_3$	Dysprosium(III) chloride	680	3.62	0.00068	987
Er	Erbium	1529	8.86	0.00157	1700
Eu	Europium	822	5.13	0.0028	980
Fe	Iron	1538	6.98	0.000572	1680
$FeCl_2$	Iron(II) chloride	677	2.348	0.000555	877
Ga	Gallium	29.76	6.08	0.00062	400
$GaBr_3$	Gallium(III) bromide	121.5	3.116	0.00246	135
$GaCl_3$	Gallium(III) chloride	77.9	2.053	0.002083	141
GaI_3	Gallium(III) iodide	212	3.630	0.002377	252
Gd	Gadolinium	1314	7.4		
$GdCl_3$	Gadolinium(III) chloride	609	3.56	0.000671	1007
GdI_3	Gadolinium(III) iodide	925	4.12	0.000908	1032
Ge	Germanium	938.25	5.60	0.00055	1600
Hf	Hafnium	2233	12		
$HgBr_2$	Mercury(II) bromide	236	5.126	0.003233	319
$HgCl_2$	Mercury(II) chloride	276	4.368	0.002862	304
HgI_2	Mercury(II) iodide	259	5.222	0.003235	354
Ho	Holmium	1472	8.34		
In	Indium	156.60	7.02	0.000836	500
$InBr_3$	Indium(III) bromide	420	3.121	0.0015	528
$InCl_3$	Indium(III) chloride	583	2.140	0.0021	666
InI_3	Indium(III) iodide	207	3.820	0.0015	360
Ir	Iridium	2446	19		
K	Potassium	63.38	0.828	0.000232	500
KBr	Potassium bromide	734	2.127	0.000825	930
KCl	Potassium chloride	771	1.527	0.000583	939
KF	Potassium fluoride	858	1.910	0.000651	1037
KI	Potassium iodide	681	2.448	0.000956	904
KNO_3	Potassium nitrate	337	1.865	0.000723	457
La	Lanthanum	920	5.94	0.00061	1600
$LaBr_3$	Lanthanum bromide	788	4.933	0.000096	912
$LaCl_3$	Lanthanum chloride	859	3.209	0.000777	973
LaF_3	Lanthanum fluoride	1493	4.589	0.000682	2177
LaI_3	Lanthanum iodide	778	4.29	0.001110	907
Li	Lithium	180.5	0.512	0.00052	285
LiBr	Lithium bromide	552	2.528	0.000652	739
LiCl	Lithium chloride	610	1.502	0.000432	781
LiF	Lithium fluoride	848.2	1.81	0.000490	1047
LiI	Lithium iodide	469	3.109	0.000917	667
$LiNO_3$	Lithium nitrate	253	1.781	0.000546	441
Li_2SO_4	Lithium sulfate	859	2.003	0.000407	1214
Lu	Lutetium	1663	9.3		
Mg	Magnesium	650	1.584	0.000234	900
$MgBr_2$	Magnesium bromide	711	2.62	0.000478	935
$MgCl_2$	Magnesium chloride	714	1.68	0.000271	826
MgI_2	Magnesium iodide	634	3.05	0.000651	888
Mn	Manganese	1246	5.95	0.00105	1590
$MnCl_2$	Manganese(II) chloride	650	2.353	0.000437	850
Mo	Molybdenum	2623	9.33		
Na	Sodium	97.80	0.927	0.00023	600
NaBr	Sodium bromide	747	2.342	0.000816	945
Na_2CO_3	Sodium carbonate	858.1	1.972	0.000448	1004
NaCl	Sodium chloride	800.7	1.556	0.000543	1027
NaF	Sodium fluoride	996	1.948	0.000636	1097
NaI	Sodium iodide	660	2.742	0.000949	912

DENSITY OF MOLTEN ELEMENTS AND REPRESENTATIVE SALTS (continued)

Formula	Name	t_m/°C	ρ_m/g cm^{-3}	k/g cm^{-3} °C^{-1}	t_{max}
$NaNO_3$	Sodium nitrate	307	1.90	0.000715	370
Na_2SO_4	Sodium sulfate	884	2.069	0.000483	1077
Nd	Neodymium	1016	6.89	0.00076	1350
Ni	Nickel	1455	7.81	0.000726	1700
$NiCl_2$	Nickel(II) chloride	1009	2.653	0.00066	1057
Os	Osmium	3033	20		
Pb	Lead	327.46	10.66	0.00122	700
$PbBr_2$	Lead(II) bromide	371	5.73	0.00165	600
$PbCl_2$	Lead(II) chloride	501	4.951	0.0015	710
PbI_2	Lead(II) iodide	410	5.691	0.001594	697
Pd	Palladium	1554.9	10.38	0.001169	1700
Pr	Praseodymium	931	6.50	0.00093	1460
$PrCl_3$	Praseodymium chloride	786	3.23	0.00074	977
Pt	Platinum	1768.4	19.77	0.0024	2200
Pu	Plutonium	640	16.63	0.001419	950
Rb	Rubidium	39.31	1.46	0.000451	800
RbBr	Rubidium bromide	682	2.715	0.001072	907
Rb_2CO_3	Rubidium carbonate	837	2.84	0.000640	1007
RbCl	Rubidium chloride	715	2.248	0.000883	923
RbF	Rubidium fluoride	833	2.87	0.00102	1067
RbI	Rubidium iodide	642	2.904	0.001143	902
$RbNO_3$	Rubidium nitrate	305	2.519	0.001068	417
Rb_2SO_4	Rubidium sulfate	1050	2.56	0.000665	1545
Re	Rhenium	3186	18.9		
Rh	Rhodium	1964	10.7	0.000895	2200
Ru	Ruthenium	2334	10.65		
S	Sulfur	115.21	1.819	0.00080	160
Sb	Antimony	630.63	6.53	0.00067	745
$SbCl_3$	Antimony(III) chloride	73.4	2.681	0.002293	77
$SbCl_5$	Antimony(V) chloride	4	2.37	0.001869	77
SbI_3	Antimony(III) iodide	168	4.171	0.002483	322
Sc	Scandium	1541	2.80		
Se	Selenium	221	3.99		
Si	Silicon	1414	2.57	0.000936	1500
Sm	Samarium	1072	7.16		
Sn	Tin	231.93	6.99	0.000601	1200
$SnCl_2$	Tin(II) chloride	247	3.36	0.001253	480
$SnCl_4$	Tin(IV) chloride	-33	2.37	0.002687	138
Sr	Strontium	777	6.980		
$SrBr_2$	Strontium bromide	657	3.70	0.000745	1004
$SrCl_2$	Strontium chloride	874	2.727	0.000578	1037
SrF_2	Strontium fluoride	1477	3.470	0.000751	1927
SrI_2	Strontium iodide	538	4.085	0.000885	1026
Ta	Tantalum	3017	15		
$TaCl_5$	Tantalum(V) chloride	216	2.700	0.004316	457
Tb	Terbium	1359	7.65		
Te	Tellurium	449.51	5.70	0.00035	600
$ThCl_4$	Thorium chloride	770	3.363	0.0014	847
ThF_4	Thorium fluoride	1110	6.058	0.000759	1378
Ti	Titanium	1668	4.11		
$TiCl_4$	Titanium(IV) chloride	-25	1.807	0.001735	137
Tl	Thallium	304	11.22	0.00144	600
TlBr	Thallium(I) bromide	460	5.98	0.001755	647
TlCl	Thallium(I) chloride	430	5.628	0.0018	642
TlI	Thallium(I) iodide	441.8	6.15	0.001761	737
$TlNO_3$	Thallium(I) nitrate	206	4.91	0.001873	279
Tl_2SO_4	Thallium(I) sulfate	632	5.62	0.00130	927
Tm	Thulium	1545	8.56	0.00050	1675
U	Uranium	1135	17.3		
UCl_3	Uranium(III) chloride	837	4.84	0.007943	1057

Formula	Name	t_m/°C	ρ_m/g cm^{-3}	k/g cm^{-3} °C^{-1}	t_{max}
UCl_4	Uranium(IV) chloride	590	3.572	0.001945	667
UF_4	Uranium(IV) fluoride	1036	6.485	0.000992	1341
V	Vanadium	1910	5.5		
W	Tungsten	3422	17.6		
Y	Yttrium	1526	4.24		
YCl_3	Yttrium chloride	721	2.510	0.0005	845
Yb	Ytterbium	824	6.21		
Zn	Zinc	419.53	6.57	0.0011	700
$ZnBr_2$	Zinc bromide	394	3.47	0.000959	602
$ZnCl_2$	Zinc chloride	290	2.54	0.00053	557
ZnI_2	Zinc iodide	446	3.878	0.00136	588
$ZnSO_4$	Zinc sulfate	680	3.14	0.00047	987
Zr	Zirconium	1855	5.8		
$ZrCl_4$	Zirconium chloride	437	1.643	0.007464	492

MAGNETIC SUSCEPTIBILITY OF THE ELEMENTS AND INORGANIC COMPOUNDS

When a material is placed in a magnetic field H, a magnetization (magnetic moment per unit volume) M is induced in the material which is related to H by $M = \kappa H$, where κ is called the volume susceptibility. Since H and M have the same dimensions, κ is dimensionless. A more useful parameter is the molar susceptibility χ_m, defined by

$$\chi_m = \kappa V_m = \kappa M/\rho$$

where V_m is the molar volume of the substance, M the molar mass, and ρ the mass density. When the cgs system is used, the customary units for χ_m are $cm^3\ mol^{-1}$; the corresponding SI units are $m^3\ mol^{-1}$.

Substances that have no unpaired electron orbital or spin angular momentum generally have negative values of χ_m and are called diamagnetic. Their molar susceptibility varies only slightly with temperature. Substances with unpaired electrons, which are termed paramagnetic, have positive χ_m and show a much stronger temperature dependence, varying roughly as $1/T$. The net susceptibility of a paramagnetic substance is the sum of the paramagnetic and diamagnetic contributions, but the former almost always dominates.

This table gives values of χ_m for the elements and selected inorganic compounds. All values refer to nominal room temperature (285 to 300 K) unless otherwise indicated. When the physical state (s = solid, l = liquid, g = gas, aq = aqueous solution) is not given, the most common crystalline form is understood. An entry of "Ferro." indicates a ferromagnetic substance.

Substances are arranged in alphabetical order by the most common name, except that compounds such as hydrides, oxides, and acids are grouped with the parent element (the same ordering used in the table "Physical Constants of Inorganic Compounds").

In keeping with customary practice, the molar susceptibility is given here in units appropriate to the cgs system. These values should be multiplied by 4π to obtain values for use in SI equations (where the magnetic field strength H has units of $A\ m^{-1}$).

REFERENCES

1. *Landolt-Börnstein, Numerical Data and Functional Relationships in Science and Technology, New Series*, II/16, *Diamagnetic Susceptibility*, Springer-Verlag, Heidelberg, 1986.
2. *Landolt-Börnstein, Numerical Data and Functional Relationships in Science and Technology, New Series*, III/19, Subvolumes a to i2, *Magnetic Properties of Metals*, Springer-Verlag, Heidelberg, 1986-1992.
3. *Landolt-Börnstein, Numerical Data and Functional Relationships in Science and Technology, New Series*, II/2, II/8, II/10, II/11, and II/12a, *Coordination and Organometallic Transition Metal Compounds*, Springer-Verlag, Heidelberg, 1966-1984.
4. *Tables de Constantes et Données Numérique, Volume 7, Relaxation Paramagnetique*, Masson, Paris, 1957.

Name	Formula	$\chi_m/10^{-6}\ cm^3\ mol^{-1}$	Name	Formula	$\chi_m/10^{-6}\ cm^3\ mol^{-1}$
Aluminum	Al	+16.5	Arsenic (yellow)	As	-23.2
Aluminum trifluoride	AlF_3	-13.9	Arsine (g)	AsH_3	-35.2
Aluminum oxide	Al_2O_3	-37	Arsenic(III) bromide	$AsBr_3$	-106
Aluminum sulfate	$Al_2(SO_4)_3$	-93	Arsenic(III) chloride	$AsCl_3$	-72.5
Ammonia (g)	NH_3	-16.3	Arsenic(III) iodide	AsI_3	-142.2
Ammonia (aq)	NH_3	-18.3	Arsenic(III) oxide	As_2O_3	-30.34
Ammonium acetate	$NH_4C_2H_3O_2$	-41.1	Arsenic(III) sulfide	As_2S_3	-70
Ammonium bromide	NH_4Br	-47	Barium	Ba	+20.6
Ammonium carbonate	$(NH_4)_2CO_3$	-42.5	Barium bromide	$BaBr_2$	-92
Ammonium chlorate	NH_4ClO_3	-42.1	Barium bromide dihydrate	$BaBr_2 \cdot 2H_2O$	-119.3
Ammonium chloride	NH_4Cl	-36.7	Barium carbonate	$BaCO_3$	-58.9
Ammonium fluoride	NH_4F	-23	Barium chloride	$BaCl_2$	-72.6
Ammonium iodate	NH_4IO_3	-62.3	Barium chloride dihydrate	$BaCl_2 \cdot 2H_2O$	-100
Ammonium iodide	NH_4I	-66	Barium fluoride	BaF_2	-51
Ammonium nitrate	NH_4NO_3	-33	Barium hydroxide	$Ba(OH)_2$	-53.2
Ammonium sulfate	$(NH_4)_2SO_4$	-67	Barium iodate	$Ba(IO_3)_2$	-122.5
Ammonium thiocyanate	NH_4SCN	-48.1	Barium iodide	BaI_2	-124.4
Antimony	Sb	-99	Barium iodide dihydrate	$BaI_2 \cdot 2H_2O$	-163
Stibine (g)	SbH_3	-34.6	Barium nitrate	$Ba(NO_3)_2$	-66.5
Antimony(III) bromide	$SbBr_3$	-111.4	Barium oxide	BaO	-29.1
Antimony(III) chloride	$SbCl_3$	-86.7	Barium peroxide	BaO_2	-40.6
Antimony(III) fluoride	SbF_3	-46	Barium sulfate	$BaSO_4$	-65.8
Antimony(III) iodide	SbI_3	-147.2	Beryllium	Be	-9.0
Antimony(III) oxide	Sb_2O_3	-69.4	Beryllium chloride	$BeCl_2$	-26.5
Antimony(III) sulfide	Sb_2S_3	-86	Beryllium hydroxide	$Be(OH)_2$	-23.1
Antimony(V) chloride	$SbCl_5$	-120.5	Beryllium oxide	BeO	-11.9
Argon (g)	Ar	-19.32	Beryllium sulfate	$BeSO_4$	-37
Arsenic (gray)	As	-5.6	Bismuth	Bi	-280.1

Name	Formula	$\chi_m/10^{-6}\ cm^3\ mol^{-1}$
Bismuth tribromide	$BiBr_3$	-147
Bismuth trichloride	$BiCl_3$	-26.5
Bismuth fluoride	BiF_3	-61.2
Bismuth hydroxide	$Bi(OH)_3$	-65.8
Bismuth triiodide	BiI_3	-200.5
Bismuth nitrate pentahydrate	$Bi(NO_3)_3 \cdot 5H_2O$	-159
Bismuth oxide	Bi_2O_3	-83
Bismuth phosphate	$BiPO_4$	-77
Bismuth sulfate	$Bi_2(SO_4)_3$	-199
Bismuth sulfide	Bi_2S_3	-123
Boron	B	-6.7
Diborane (g)	B_2H_6	-21.0
Boric acid (orthoboric acid)	H_3BO_3	-34.1
Boron trichloride	BCl_3	-59.9
Boron oxide	B_2O_3	-38.7
Bromine (l)	Br_2	-56.4
Bromine (g)	Br_2	-73.5
Bromine trifluoride	BrF_3	-33.9
Bromine pentafluoride	BrF_5	-45.1
Cadmium	Cd	-19.7
Cadmium bromide	$CdBr_2$	-87.3
Cadmium bromide tetrahydrate	$CdBr_2 \cdot 4H_2O$	-131.5
Cadmium carbonate	$CdCO_3$	-46.7
Cadmium chloride	$CdCl_2$	-68.7
Cadmium chromate	$CdCrO_4$	-16.8
Cadmium cyanide	$Cd(CN)_2$	-54
Cadmium fluoride	CdF_2	-40.6
Cadmium hydroxide	$Cd(OH)_2$	-41
Cadmium iodate	$Cd(IO_3)_2$	-108.4
Cadmium iodide	CdI_2	-117.2
Cadmium nitrate	$Cd(NO_3)_2$	-55.1
Cadmium nitrate tetrahydrate	$Cd(NO_3)_2 \cdot 4H_2O$	-140
Cadmium oxide	CdO	-30
Cadmium sulfate	$CdSO_4$	-59.2
Cadmium sulfide	CdS	-50
Calcium	Ca	+40
Calcium bromide	$CaBr_2$	-73.8
Calcium carbonate	$CaCO_3$	-38.2
Calcium chloride	$CaCl_2$	-54.7
Calcium fluoride	CaF_2	-28
Calcium hydroxide	$Ca(OH)_2$	-22
Calcium iodate	$Ca(IO_3)_2$	-101.4
Calcium iodide	CaI_2	-109
Calcium oxide	CaO	-15.0
Calcium sulfate	$CaSO_4$	-49.7
Calcium sulfate dihydrate	$CaSO_4 \cdot 2H_2O$	-74
Carbon (diamond)	C	-5.9
Carbon (graphite)	C	-6.0
Carbon monoxide (g)	CO	-11.8
Carbon dioxide (g)	CO_2	-21.0
Cerium (β)	Ce	+2500
Cerium(II) sulfide	CeS	+2110
Cerium(III) chloride	$CeCl_3$	+2490
Cerium(III) fluoride	CeF_3	+2190
Cerium(III) sulfide	Ce_2S_3	+5080
Cerium(IV) oxide	CeO_2	+26
Cerium(IV) sulfate tetrahydrate	$Ce(SO_4)_2 \cdot 4H_2O$	-97
Cesium	Cs	+29
Cesium bromate	$CsBrO_3$	-75.1
Cesium bromide	CsBr	-67.2
Cesium carbonate	Cs_2CO_3	-103.6
Cesium chlorate	$CsClO_3$	-65
Cesium chloride	CsCl	-56.7
Cesium fluoride	CsF	-44.5
Cesium iodide	CsI	-82.6
Cesium superoxide	CsO_2	+1534
Cesium sulfate	Cs_2SO_4	-116
Chlorine (l)	Cl_2	-40.4
Chlorine trifluoride (g)	ClF_3	-26.5
Chromium	Cr	+167
Chromium(II) chloride	$CrCl_2$	+7230
Chromium(III) chloride	$CrCl_3$	+6350
Chromium(III) fluoride	CrF_3	+4370
Chromium(III) oxide	Cr_2O_3	+1960
Chromium(III) sulfate	$Cr_2(SO_4)_3$	+11800
Chromium(VI) oxide	CrO_3	+40
Cobalt	Co	Ferro.
Cobalt(II) bromide	$CoBr_2$	+13000
Cobalt(II) chloride	$CoCl_2$	+12660
Cobalt(II) chloride hexahydrate	$CoCl_2 \cdot 6H_2O$	+9710
Cobalt(II) cyanide	$Co(CN)_2$	+3825
Cobalt(II) fluoride	CoF_2	+9490
Cobalt(II) iodide	CoI_2	+10760
Cobalt(II) sulfate	$CoSO_4$	+10000
Cobalt(II) sulfide	CoS	+225
Cobalt(II,III) oxide	Co_3O_4	+7380
Cobalt(III) fluoride	CoF_3	+1900
Cobalt(III) oxide	Co_2O_3	+4560
Copper	Cu	-5.46
Copper(I) bromide	CuBr	-49
Copper(I) chloride	CuCl	-40
Copper(I) cyanide	CuCN	-24
Copper(I) iodide	CuI	-63
Copper(I) oxide	Cu_2O	-20
Copper(II) bromide	$CuBr_2$	+685
Copper(II) chloride	$CuCl_2$	+1080
Copper(II) chloride dihydrate	$CuCl_2 \cdot 2H_2O$	+1420
Copper(II) fluoride	CuF_2	+1050
Copper(II) fluoride dihydrate	$CuF_2 \cdot 2H_2O$	+1600
Copper(II) hydroxide	$Cu(OH)_2$	+1170
Copper(II) nitrate trihydrate	$Cu(NO_3)_2 \cdot 3H_2O$	+1570
Copper(II) nitrate hexahydrate	$Cu(NO_3)_2 \cdot 6H_2O$	+1625
Copper(II) oxide	CuO	+238
Copper(II) sulfate	$CuSO_4$	+1330
Copper(II) sulfate pentahydrate	$CuSO_4 \cdot 5H_2O$	+1460
Copper(II) sulfide	CuS	-2.0
Dysprosium (α)	Dy	+98000
Dysprosium(III) oxide	Dy_2O_3	+89600
Dysprosium(III) sulfide	Dy_2S_3	+95200
Erbium	Er	+48000
Erbium oxide	Er_2O_3	+73920
Erbium sulfate octahydrate	$Er_2(SO_4)_3 \cdot 8H_2O$	+74600
Erbium sulfide	Er_2S_3	+77200
Europium	Eu	+30900
Europium(II) bromide	$EuBr_2$	+26800

Name	Formula	$\chi_m/10^{-6}$ cm^3 mol^{-1}
Europium(II) chloride	$EuCl_2$	+26500
Europium(II) fluoride	EuF_2	+23750
Europium(II) iodide	EuI_2	+26000
Europium(II) sulfide	EuS	+23800
Europium(III) oxide	Eu_2O_3	+10100
Europium(III) sulfate	$Eu_2(SO_4)_3$	+10400
Fluorine	F_2	-9.63
Gadolinium (350 K)	Gd	+185000
Gadolinium(III) chloride	$GdCl_3$	+27930
Gadolinium(III) oxide	Gd_2O_3	+53200
Gadolinium(III) sulfate octahydrate	$Gd_2(SO_4)_3 \cdot 8H_2O$	+53280
Gadolinium(III) sulfide	Gd_2S_3	+55500
Gallium	Ga	-21.6
Gallium suboxide	Ga_2O	-34
Gallium(II) sulfide	GaS	-23
Gallium(III) chloride	$GaCl_3$	-63
Gallium(III) sulfide	Ga_2S_3	-80
Germanium	Ge	-11.6
Germane (g)	GeH_4	-29.7
Germanium(II) oxide	GeO	-28.8
Germanium(II) sulfide	GeS	-40.9
Germanium(IV) chloride	$GeCl_4$	-72
Germanium(IV) fluoride	GeF_4	-50
Germanium(IV) iodide	GeI_4	-171
Germanium(IV) oxide	GeO_2	-34.3
Germanium(IV) sulfide	GeS_2	-53.9
Gold	Au	-28
Gold(I) bromide	AuBr	-61
Gold(I) chloride	AuCl	-67
Gold(I) iodide	AuI	-91
Gold(III) chloride	$AuCl_3$	-112
Hafnium	Hf	+71
Hafnium oxide	HfO_2	-23
Helium (g)	He	-2.02
Holmium	Ho	+72900
Holmium oxide	Ho_2O_3	+88100
Hydrazine (l)	N_2H_4	-201
Hydrogen (l, 20.3 K)	H_2	-5.44
Hydrogen (g)	H_2	-3.99
Hydrogen chloride (l)	HCl	-22.6
Hydrogen chloride (aq)	HCl	-22
Hydrogen fluoride (l)	HF	-8.6
Hydrogen fluoride (aq)	HF	-9.3
Hydrogen iodide (s, 195 K)	HI	-47.3
Hydrogen iodide (l, 233 K)	HI	-48.3
Hydrogen iodide (aq)	HI	-50.2
Hydrogen peroxide (l)	H_2O_2	-17.3
Hydrogen sulfide (g)	H_2S	-25.5
Indium	In	-10.2
Indium(I) chloride	InCl	-30
Indium(II) chloride	$InCl_2$	-56
Indium(II) sulfide	InS	-28
Indium(III) bromide	$InBr_3$	-107
Indium(III) chloride	$InCl_3$	-86
Indium(III) oxide	In_2O_3	-56
Indium(III) sulfide	In_2S_3	-98
Iodine	I_2	-90
Iodic acid	HIO_3	-48
Iodine pentoxide	I_2O_5	-79.4
Iodine chloride	ICl	-54.6
Iodine trichloride	ICl_3	-90.2
Iodine pentafluoride	IF_5	-58.1
Iridium	Ir	+25
Iridium(III) chloride	$IrCl_3$	-14.4
Iridium(IV) oxide	IrO_2	+224
Iron	Fe	Ferro.
Iron(II) bromide	$FeBr_2$	+13600
Iron(II) carbonate	$FeCO_3$	+11300
Iron(II) chloride	$FeCl_2$	+14750
Iron(II) chloride tetrahydrate	$FeCl_2 \cdot 4H_2O$	+12900
Iron(II) fluoride	FeF_2	+9500
Iron(II) iodide	FeI_2	+13600
Iron(II) oxide	FeO	+7200
Iron(II) sulfate	$FeSO_4$	+12400
Iron(II) sulfate monohydrate	$FeSO_4 \cdot H_2O$	+10500
Iron(II) sulfate heptahydrate	$FeSO_4 \cdot 7H_2O$	+11200
Iron(II) sulfide	FeS	+1074
Iron(III) chloride	$FeCl_3$	+13450
Iron(III) chloride hexahydrate	$FeCl_3 \cdot 6H_2O$	+15250
Iron(III) fluoride	FeF_3	+13760
Iron(III) fluoride trihydrate	$FeF_3 \cdot 3H_2O$	+7870
Iron(III) nitrate nonahydrate	$Fe(NO_3)_3 \cdot 9H_2O$	+15200
Krypton (g)	Kr	-29.0
Lanthanum (α)	La	+95.9
Lanthanum oxide	La_2O_3	-78
Lanthanum sulfate nonahydrate	$La_2(SO_4)_3 \cdot 9H_2O$	-262
Lanthanum sulfide	La_2S_3	-37
Lead	Pb	-23
Lead(II) acetate	$Pb(C_2H_3O_2)_2$	-89.1
Lead(II) bromide	$PbBr_2$	-90.6
Lead(II) carbonate	$PbCO_3$	-61.2
Lead(II) chloride	$PbCl_2$	-73.8
Lead(II) chromate	$PbCrO_4$	-18
Lead(II) fluoride	PbF_2	-58.1
Lead(II) iodate	$Pb(IO_3)_2$	-131
Lead(II) iodide	PbI_2	-126.5
Lead(II) nitrate	$Pb(NO_3)_2$	-74
Lead(II) oxide	PbO	-42
Lead(II) phosphate	$Pb_3(PO_4)_2$	-182
Lead(II) sulfate	$PbSO_4$	-69.7
Lead(II) sulfide	PbS	-83.6
Lithium	Li	+14.2
Lithium bromide	LiBr	-34.3
Lithium carbonate	Li_2CO_3	-27
Lithium chloride	LiCl	-24.3
Lithium fluoride	LiF	-10.1
Lithium hydride	LiH	-4.6
Lithium hydroxide (aq)	LiOH	-12.3
Lithium iodide	LiI	-50
Lithium sulfate	Li_2SO_4	-41.6
Lutetium	Lu	+182.9
Magnesium	Mg	+13.1
Magnesium bromide	$MgBr_2$	-72
Magnesium carbonate	$MgCO_3$	-32.4
Magnesium chloride	$MgCl_2$	-47.4
Magnesium fluoride	MgF_2	-22.7

Name	Formula	$\chi_m/10^{-6}\ cm^3\ mol^{-1}$
Magnesium hydroxide	$Mg(OH)_2$	-22.1
Magnesium iodide	MgI_2	-111
Magnesium oxide	MgO	-10.2
Magnesium sulfate	$MgSO_4$	-42
Magnesium sulfate monohydrate	$MgSO_4 \cdot H_2O$	-61
Magnesium sulfate heptahydrate	$MgSO_4 \cdot 7H_2O$	-135.7
Manganese	Mn	+511
Manganese(II) bromide	$MnBr_2$	+13900
Manganese(II) carbonate	$MnCO_3$	+11400
Manganese(II) chloride	$MnCl_2$	+14350
Manganese(II) chloride tetrahydrate	$MnCl_2 \cdot 4H_2O$	+14600
Manganese(II) fluoride	MnF_2	+10700
Manganese(II) hydroxide	$Mn(OH)_2$	+13500
Manganese(II) iodide	MnI_2	+14400
Manganese(II) oxide	MnO	+4850
Manganese(II) sulfate	$MnSO_4$	+13660
Manganese(II) sulfate monohydrate	$MnSO_4 \cdot H_2O$	+14200
Manganese(II) sulfate tetrahydrate	$MnSO_4 \cdot 4H_2O$	+14600
Manganese(II) sulfide (α form)	MnS	+5630
Manganese(II) sulfide (β form)	MnS	+3850
Manganese(II,III) oxide	Mn_3O_4	+12400
Manganese(III) fluoride	MnF_3	+10500
Manganese(III) oxide	Mn_2O_3	+14100
Manganese(IV) oxide	MnO_2	+2280
Mercury (s, 234 K)	Hg	-24.1
Mercury (l)	Hg	-33.5
Mercury(I) bromide	Hg_2Br_2	-105
Mercury(I) chloride	Hg_2Cl_2	-120
Mercury(I) fluoride	Hg_2F_2	-106
Mercury(I) iodide	Hg_2I_2	-166
Mercury(I) nitrate	$Hg_2(NO_3)_2$	-121
Mercury(I) oxide	Hg_2O	-76.3
Mercury(I) sulfate	Hg_2SO_4	-123
Mercury(II) bromide	$HgBr_2$	-94.2
Mercury(II) chloride	$HgCl_2$	-82
Mercury(II) cyanide	$Hg(CN)_2$	-67
Mercury(II) fluoride	HgF_2	-57.3
Mercury(II) iodide	HgI_2	-165
Mercury(II) nitrate	$Hg(NO_3)_2$	-74
Mercury(II) oxide	HgO	-46
Mercury(II) sulfate	$HgSO_4$	-78.1
Mercury(II) sulfide	HgS	-55.4
Mercury(II) thiocyanate	$Hg(SCN)_2$	-96.5
Molybdenum	Mo	+72
Molybdenum(III) bromide	$MoBr_3$	+525
Molybdenum(III) chloride	$MoCl_3$	+43
Molybdenum(III) oxide	Mo_2O_3	-42.0
Molybdenum(IV) bromide	$MoBr_4$	+520
Molybdenum(IV) chloride	$MoCl_4$	+1750
Molybdenum(IV) oxide	MoO_2	+41
Molybdenum(V) chloride	$MoCl_5$	+990
Molybdenum(VI) fluoride	MoF_6	-26.0

Name	Formula	$\chi_m/10^{-6}\ cm^3\ mol^{-1}$
Molybdenum(VI) oxide	MoO_3	+3
Neodymium (α)	Nd	+5930
Neodymium fluoride	NdF_3	+4980
Neodymium oxide	Nd_2O_3	+10200
Neodymium sulfate	$Nd_2(SO_4)_3$	+9990
Neodymium sulfide	Nd_2S_3	+5550
Neon (g)	Ne	-6.96
Neptunium	Np	+575
Nickel	Ni	Ferro.
Nickel(II) bromide	$NiBr_2$	+5600
Nickel(II) chloride	$NiCl_2$	+6145
Nickel(II) chloride hexahydrate	$NiCl_2 \cdot 6H_2O$	+4240
Nickel(II) fluoride	NiF_2	+2410
Nickel(II) hydroxide	$Ni(OH)_2$	+4500
Nickel(II) iodide	NiI_2	+3875
Nickel(II) nitrate hexahydrate	$Ni(NO_3)_2 \cdot 6H_2O$	+4300
Nickel(II) oxide	NiO	+660
Nickel(II) sulfate	$NiSO_4$	+4005
Nickel(II) sulfide	NiS	+190
Nickel(III) sulfide	Ni_3S_2	+1030
Niobium	Nb	+208
Niobium(V) oxide	Nb_2O_5	-10
Nitrogen (g)	N_2	-12.0
Nitric acid (l)	HNO_3	-19.9
Nitrous oxide (g)	N_2O	-18.9
Nitric oxide (s, 90 K)	NO	+19.8
Nitric oxide (l, 118 K)	NO	+114.2
Nitric oxide (g)	NO	+1461
Nitrogen dioxide (g, 408 K)	NO_2	+150
Nitrogen trioxide (g)	N_2O_3	-16
Nitrogen tetroxide (g)	N_2O_4	-23.0
Osmium	Os	+11
Oxygen (s, 54 K)	O_2	+10200
Oxygen (l, 90 K)	O_2	+7699
Oxygen (g)	O_2	+3449
Ozone (l)	O_3	+6.7
Palladium	Pd	+540
Palladium(II) chloride	$PdCl_2$	-38
Phosphorus (white)	P	-26.66
Phosphorus (red)	P	-20.77
Phosphine (g)	PH_3	-26.2
Phosphoric acid (aq)	H_3PO_4	-43.8
Phosphorous acid (aq)	H_3PO_3	-42.5
Phosphorus(III) chloride (l)	PCl_3	-63.4
Platinum	Pt	+193
Platinum(II) chloride	$PtCl_2$	-54
Platinum(III) chloride	$PtCl_3$	-66.7
Platinum(IV) chloride	$PtCl_4$	-93
Platinum(IV) fluoride	PtF_4	+445
Plutonium	Pu	+525
Plutonium(IV) fluoride	PuF_4	+1760
Plutonium(IV) oxide	PuO_2	+730
Plutonium(VI) fluoride	PuF_6	+173
Potassium	K	+20.8
Potassium bromate	$KBrO_3$	-52.6
Potassium bromide	KBr	-49.1
Potassium carbonate	K_2CO_3	-59
Potassium chlorate	$KClO_3$	-42.8

Name	Formula	$\chi_m/10^{-6}$ cm³ mol⁻¹
Potassium chloride	KCl	-38.8
Potassium chromate	K_2CrO_4	-3.9
Potassium cyanide	KCN	-37
Potassium ferricyanide	$K_3Fe(CN)_6$	+2290
Potassium ferrocyanide trihydrate	$K_4Fe(CN)_6 \cdot 3H_2O$	-172.3
Potassium fluoride	KF	-23.6
Potassium hydrogen sulfate	$KHSO_4$	-49.8
Potassium hydroxide (aq)	KOH	-22
Potassium iodate	KIO_3	-63.1
Potassium iodide	KI	-63.8
Potassium nitrate	KNO_3	-33.7
Potassium nitrite	KNO_2	-23.3
Potassium permanganate	$KMnO_4$	+20
Potassium sulfate	K_2SO_4	-67
Potassium sulfide	K_2S	-60
Potassium superoxide	KO_2	+3230
Potassium thiocyanate	$KSCN$	-48
Praseodymium (α)	Pr	+5530
Praseodymium chloride	$PrCl_3$	+44.5
Praseodymium oxide	Pr_2O_3	+8994
Praseodymium sulfide	Pr_2S_3	+10770
Protactinium	Pa	+277
Rhenium	Re	+67
Rhenium(IV) oxide	ReO_2	+44
Rhenium(IV) sulfide	ReS_2	+38
Rhenium(V) chloride	$ReCl_5$	+1225
Rhenium(VI) oxide	ReO_3	+16
Rhenium(VII) oxide	Re_2O_7	-16
Rhodium	Rh	+102
Rhodium(III) chloride	$RhCl_3$	-7.5
Rhodium(III) oxide	Rh_2O_3	+104
Rubidium	Rb	+17
Rubidium bromide	$RbBr$	-56.4
Rubidium carbonate	Rb_2CO_3	-75.4
Rubidium chloride	$RbCl$	-46
Rubidium fluoride	RbF	-31.9
Rubidium iodide	RbI	-72.2
Rubidium nitrate	$RbNO_3$	-41
Rubidium sulfate	Rb_2SO_4	-88.4
Rubidium superoxide	RbO_2	+1527
Ruthenium	Ru	+39
Ruthenium(III) chloride	$RuCl_3$	+1998
Ruthenium(IV) oxide	RuO_2	+162
Samarium (α)	Sm	+1278
Samarium(II) bromide	$SmBr_2$	+5337
Samarium(III) bromide	$SmBr_3$	+972
Samarium(III) oxide	Sm_2O_3	+1988
Samarium(III) sulfate octahydrate	$Sm_2(SO_4)_3 \cdot 8H_2O$	+1710
Samarium(III) sulfide	Sm_2S_3	+3300
Scandium (α)	Sc	+295.2
Selenium	Se	-25
Selenium dioxide	SeO_2	-27.2
Selenium bromide	Se_2Br_2	-113
Selenium chloride (l)	Se_2Cl_2	-94.8
Selenium hexafluoride (g)	SeF_6	-51
Silicon	Si	-3.12
Silane (g)	SiH_4	-20.4
Disilane (g)	Si_2H_6	-37.3
Tetramethylsilane (l)	$(CH_3)_4Si$	-74.80
Tetraethylsilane (l)	$(C_2H_5)_4Si$	-120.2
Tetrabromosilane (l)	$SiBr_4$	-126
Tetrachlorosilane (l)	$SiCl_4$	-87.5
Silicon carbide	SiC	-12.8
Silicon dioxide	SiO_2	-29.6
Silver	Ag	-19.5
Silver(I) bromide	$AgBr$	-61
Silver(I) carbonate	Ag_2CO_3	-80.90
Silver(I) chloride	$AgCl$	-49
Silver(I) chromate	Ag_2CrO_4	-40
Silver(I) cyanide	$AgCN$	-43.2
Silver(I) fluoride	AgF	-36.5
Silver(I) iodide	AgI	-80
Silver(I) nitrate	$AgNO_3$	-45.7
Silver(I) nitrite	$AgNO_2$	-42
Silver(I) oxide	Ag_2O	-134
Silver(I) phosphate	Ag_3PO_4	-120
Silver(I) sulfate	Ag_2SO_4	-92.90
Silver(I) thiocyanate	$AgSCN$	-61.8
Silver(II) oxide	AgO	-19.6
Sodium	Na	+16
Sodium acetate	$NaC_2H_3O_2$	-37.6
Sodium bromate	$NaBrO_3$	-44.2
Sodium bromide	$NaBr$	-41
Sodium carbonate	Na_2CO_3	-41
Sodium chlorate	$NaClO_3$	-34.7
Sodium chloride	$NaCl$	-30.2
Sodium dichromate	$Na_2Cr_2O_7$	+55
Sodium fluoride	NaF	-15.6
Sodium hydrogen phosphate	Na_2HPO_4	-56.6
Sodium hydroxide (aq)	$NaOH$	-15.8
Sodium iodate	$NaIO_3$	-53
Sodium iodide	NaI	-57
Sodium nitrate	$NaNO_3$	-25.6
Sodium nitrite	$NaNO_2$	-14.5
Sodium oxide	Na_2O	-19.8
Sodium peroxide	Na_2O_2	-28.10
Sodium sulfate	Na_2SO_4	-52
Sodium sulfate decahydrate	$Na_2SO_4 \cdot 10H_2O$	-184
Sodium sulfide	Na_2S	-39
Sodium tetraborate	$Na_2B_4O_7$	-85
Strontium	Sr	+92
Strontium bromide	$SrBr_2$	-86.6
Strontium bromide hexahydrate	$SrBr_2 \cdot 6H_2O$	-160
Strontium carbonate	$SrCO_3$	-47
Strontium chlorate	$Sr(ClO_3)_2$	-73
Strontium chloride	$SrCl_2$	-61.5
Strontium chloride hexahydrate	$SrCl_2 \cdot 6H_2O$	-145
Strontium chromate	$SrCrO_4$	-5.1
Strontium fluoride	SrF_2	-37.2
Strontium hydroxide	$Sr(OH)_2$	-40
Strontium iodate	$Sr(IO_3)_2$	-108
Strontium iodide	SrI_2	-112
Strontium nitrate	$Sr(NO_3)_2$	-57.2
Strontium oxide	SrO	-35

Name	Formula	$\chi_m/10^{-6}$ cm³ mol⁻¹
Strontium peroxide	SrO_2	-32.3
Strontium sulfate	$SrSO_4$	-57.9
Sulfur (rhombic)	S	-15.5
Sulfur (monoclinic)	S	-14.9
Sulfuric acid (l)	H_2SO_4	-39
Sulfur dioxide (g)	SO_2	-18.2
Sulfur trioxide (l)	SO_3	-28.54
Sulfur chloride (l)	$SSCl_2$	-62.2
Sulfur dichloride (l)	SCl_2	-49.4
Sulfur hexafluoride (g)	SF_6	-44
Thionyl chloride (l)	$SOCl_2$	-44.3
Tantalum	Ta	+154
Tantalum(V) chloride	$TaCl_5$	+140
Tantalum(V) oxide	Ta_2O_5	-32
Technetium	Tc	+115
Tellurium	Te	-38
Tellurium dibromide	$TeBr_2$	-106
Tellurium dichloride	$TeCl_2$	-94
Tellurium hexafluoride (g)	TeF_6	-66
Terbium (α)	Tb	+170000
Terbium oxide	Tb_2O_3	+78340
Thallium	Tl	-50
Thallium(I) bromate	$TlBrO_3$	-75.9
Thallium(I) bromide	TlBr	-63.9
Thallium(I) carbonate	Tl_2CO_3	-101.6
Thallium(I) chlorate	$TlClO_3$	-65.5
Thallium(I) chloride	TlCl	-57.8
Thallium(I) chromate	Tl_2CrO_4	-39.3
Thallium(I) cyanide	TlCN	-49
Thallium(I) fluoride	TlF	-44.4
Thallium(I) iodate	$TlIO_3$	-86.8
Thallium(I) iodide	TlI	-82.2
Thallium(I) nitrate	$TlNO_3$	-56.5
Thallium(I) nitrite	$TlNO_2$	-50.8
Thallium(I) sulfate	Tl_2SO_4	-112.6
Thallium(I) sulfide	Tl_2S	-88.8
Thorium	Th	+97
Thorium(IV) oxide	ThO_2	-16
Thulium	Tm	+24700
Thulium oxide	Tm_2O_3	+51444
Tin (gray)	Sn	-37.4
Tin(II) chloride	$SnCl_2$	-69
Tin(II) chloride dihydrate	$SnCl_2·2H_2O$	-91.4
Tin(II) oxide	SnO	-19
Tin(IV) bromide	$SnBr_4$	-149
Tin(IV) chloride (l)	$SnCl_4$	-115
Tin(IV) oxide	SnO_2	-41
Titanium	Ti	+151
Titanium(II) bromide	$TiBr_2$	+720
Titanium(II) chloride	$TiCl_2$	+484
Titanium(II) iodide	TiI_2	+1790
Titanium(II) sulfide	TiS	+432
Titanium(III) bromide	$TiBr_3$	+660
Titanium(III) chloride	$TiCl_3$	+1110
Titanium(III) fluoride	TiF_3	+1300
Titanium(III) oxide	Ti_2O_3	+132
Titanium(IV) chloride	$TiCl_4$	-54
Titanium(IV) oxide	TiO_2	+5.9
Tungsten	W	+53
Tungsten carbide	WC	+10
Tungsten(II) chloride	WCl_2	-25
Tungsten(IV) oxide	WO_2	+57
Tungsten(IV) sulfide	WS_2	+5850
Tungsten(V) bromide	WBr_5	+270
Tungsten(V) chloride	WCl_5	+387
Tungsten(VI) chloride	WCl_6	-71
Tungsten(VI) fluoride (g)	WF_6	-53
Tungsten(VI) oxide	WO_3	-15.8
Uranium	U	+409
Uranium(III) bromide	UBr_3	+4740
Uranium(III) chloride	UCl_3	+3460
Uranium(III) hydride	UH_3	+6244
Uranium(III) iodide	UI_3	+4460
Uranium(IV) bromide	UBr_4	+3530
Uranium(IV) chloride	UCl_4	+3680
Uranium(IV) fluoride	UF_4	+3530
Uranium(IV) oxide	UO_2	+2360
Uranium(VI) fluoride	UF_6	+43
Uranium(VI) oxide	UO_3	+128
Vanadium	V	+285
Vanadium(II) bromide	VBr_2	+3230
Vanadium(II) chloride	VCl_2	+2410
Vanadium(III) bromide	VBr_3	+2910
Vanadium(III) chloride	VCl_3	+3030
Vanadium(III) fluoride	VF_3	+2757
Vanadium(III) oxide	V_2O_3	+1976
Vanadium(III) sulfide	V_2S_3	+1560
Vanadium(IV) chloride	VCl_4	+1215
Vanadium(IV) oxide	VO_2	+99
Vanadium(V) oxide	V_2O_5	+128
Water (s, 273 K)	H_2O	-12.63
Water (l, 293 K)	H_2O	-12.96
Water (l, 373 K)	H_2O	-13.09
Water (g, 373 K))	H_2O	-13.1
Xenon (g)	Xe	-45.5
Ytterbium (β)	Yb	+67
Yttrium (α)	Y	+187.7
Yttrium oxide	Y_2O_3	+44.4
Yttrium sulfide	Y_2S_3	+100
Zinc	Zn	-9.15
Zinc carbonate	$ZnCO_3$	-34
Zinc chloride	$ZnCl_2$	-55.33
Zinc cyanide	$Zn(CN)_2$	-46
Zinc fluoride	ZnF_2	-34.3
Zinc hydroxide	$Zn(OH)_2$	-67
Zinc iodide	ZnI_2	-108
Zinc oxide	ZnO	-27.2
Zinc phosphate	$Zn_3(PO_4)_2$	-141
Zinc sulfate	$ZnSO_4$	-47.8
Zinc sulfate monohydrate	$ZnSO_4·H_2O$	-63
Zinc sulfate heptahydrate	$ZnSO_4·7H_2O$	-138
Zinc sulfide	ZnS	-25
Zirconium	Zr	+120
Zirconium carbide	ZrC	-26
Zirconium nitrate pentahydrate	$Zr(NO_3)_4·5H_2O$	-77
Zirconium(IV) oxide	ZrO_2	-13.8

INDEX OF REFRACTION OF INORGANIC LIQUIDS

This table gives the index of refraction *n* of several inorganic substances in the liquid state at specified temperatures. The measurements refer to ambient atmospheric pressure except for substances whose normal boiling points are greater than the indicated temperature; in this case the pressure is the saturated vapor pressure of the substance. All values refer to a wavelength of 589 nm unless otherwise indicated. Entries are arranged in alphabetical order by chemical formula as normally written.

Data on the index of refraction at other temperatures and wavelengths may be found in Reference 1.

REFERENCES

1. Wohlfarth, C., and Wohlfarth, B., *Landolt-Börnstein, Numerical Data and Functional Relationships in Science and Technology, New Series*, III/38A, Martienssen, W., Editor, Springer-Verlag, Heidelberg, 1996.
2. Francis, A.W., *J. Chem. Eng. Data*, 5, 534, 1960.

Formula	Name	t/°C	n
Ar	Argon	-188	1.2312
$AsCl_3$	Arsenic(III) chloride	16	1.604
BBr_3	Boron tribromide	16	1.312
BrF_3	Bromine trifluoride	25	1.4536
BrF_5	Bromine pentafluoride	25	1.3529
Br_2	Bromine	15	1.659
COS	Carbon oxysulfide	25	1.3506
CO_2	Carbon dioxide	24	1.6630
CS_2	Carbon disulfide	20	1.62774
C_3O_2	Carbon suboxide	0	1.453
Cl_2	Chlorine	20	1.3834
CrO_2Cl_2	Chromyl chloride	23	1.524
$Fe(CO)_5$	Iron pentacarbonyl	14	1.523
$GeBr_4$	Germanium(IV) bromide	26	1.6269
$GeCl_4$	Germanium(IV) chloride	25	1.4614
HBr	Hydrogen bromide	10	1.325
HCN	Hydrogen cyanide	20	1.26136
HCl	Hydrogen chloride	18	1.3287 [a]
$HClO_4$	Perchloric acid	50	1.3819
HF	Hydrogen fluoride	25	1.1574
HI	Hydrogen iodide	16	1.466
HNO_3	Nitric acid	25	1.393
H_2	Hydrogen	-253	1.1096
H_2O	Water	20	1.33336
H_2O_2	Hydrogen peroxide	28	1.4061
H_2S	Hydrogen sulfide	-80	1.460
		20	1.3682
H_2SO_4	Sulfuric acid	20	1.4183
H_2S_2	Hydrogen disulfide	20	1.630
He	Helium	-269	1.02451 [c]
Kr	Krypton	-157	1.3032 [c]
NH_3	Ammonia	-77	1.3944 [b]
		20	1.3327
NO	Nitric oxide	-90	1.330
N_2	Nitrogen	-196	1.19876 [b]
N_2H_4	Hydrazine	22	1.470
N_2O	Nitrous oxide	25	1.238
O_2	Oxygen	-183	1.2243 [c]
PBr_3	Phosphorus(III) bromide	25	1.687
PCl_3	Phosphorus(III) chloride	21	1.5122
PH_3	Phosphine	17	1.317
P_2O_3	Phosphorus(III) oxide	27	1.540
S	Sulfur	125	1.9170
SCl_2	Sulfur dichloride	14	1.557
SF_6	Sulfur hexafluoride	25	1.167
$SOCl_2$	Thionyl chloride	10	1.527
SO_2	Sulfur dioxide	25	1.3396
SO_2Cl_2	Sulfuryl chloride	12	1.444

INDEX OF REFRACTION OF INORGANIC LIQUIDS (continued)

Formula	Name	t/°C	n
SO_3	Sulfur trioxide	20	1.40965
$SSCl_2$	Sulfur chloride	20	1.671
$SbCl_5$	Antimony(V) chloride	22	1.5925
$SiBr_4$	Tetrabromosilane	31	1.5685
$SiCl_4$	Tetrachlorosilane	25	1.41156
$SnBr_4$	Tin(IV) bromide	31	1.6628
$SnCl_4$	Tin(IV) chloride	25	1.5086
$TiCl_4$	Titanium(IV) chloride	18	1.6076
Xe	Xenon	-112	1.3918 [c]

[a] At 581 nm
[b] At 578 nm
[c] At 546 nm

PHYSICAL AND OPTICAL PROPERTIES OF MINERALS

The chemical formula, crystal system, density, hardness, and index of refraction of some common minerals are given in this table. Entries are arranged alphabetically by mineral name. The columns are:

- **Formula:** Chemical formula for a typical sample of the mineral. Composition often varies considerably with the origin of the sample.
- **Crystal system:** tricl = triclinic; monocl = monoclinic; orth = orthorhombic; tetr = tetragonal; hex = hexagonal; rhomb = rhombohedral; cub = cubic.
- **Density:** Typical density in g/cm^3. Individual samples may vary by a few percent.
- **Hardness:** On the Mohs' scale (range of 1 to 10, with talc = 1 and diamond = 10).
- **Index of refraction:** Values are given for the three coordinate axes in the order of least, intermediate, and greatest index. For cubic crystals there is only a single value. See Reference 1 for details on the axis systems. Variations of several percent, depending on the origin and exact composition of the sample, are common.

REFERENCES

1. Deer, W.A., Howie, R.A., and Zussman, J., *An Introduction to the Rock-Forming Minerals*, 2nd Edition, Longman Scientific & Technical, Harlow, Essex, 1992.
2. Carmichael, R.S., *Practical Handbook of Physical Properties of Rocks and Minerals*, CRC Press, Boca Raton, FL, 1989.
3. Donnay, J.D.H., and Ondik, H.M., *Crystal Data Determinative Tables, Third Edition, Volume 2, Inorganic Compounds*, Joint Committee on Powder Diffraction Standards, Swarthmore, PA, 1973.

Name	Formula	Crystal system	Density g/cm^3	Hard-ness	Index of refraction n_α	n_β	n_γ
Acanthite	Ag_2S	orth	7.2	2.3			
Actinolite	$Ca_2(Mg,Fe)_5Si_8O_{22}(OH,F)_2$	monocl	3.23	5.5	1.624	1.655	1.664
Aegirine	$NaFe(SiO_3)_2$	monocl	3.58	6	1.763	1.800	1.815
Akermanite	$Ca_2MgSi_2O_7$	tetr	2.94	5.5	1.632	1.640	
Alabandite	MnS	cub	4.0	3.8			
Albite	$NaAlSi_3O_8$	tricl	2.63	6.3	1.527	1.531	1.538
Allanite	$(Ca,Mn,Ce,La,Y,Th)_2(Fe,Ti)(Al,Fe)$ $O{\cdot}OH(Si_2O_7)(SiO_4)$	monocl	3.8	5.8	1.75	1.78	1.80
Allemontite	SbAs	hex	6.0	3.5			
Almandine	$Fe_3Al_2Si_3O_{12}$	cub	4.32	6.8	1.830		
Altaite	PbTe	cub	8.16	3			
Aluminite	$Al_2(SO_4)(OH)_4{\cdot}7H_2O$	monocl	1.74	1.5	1.459	1.464	1.470
Alunite	$(K,Na)Al_3(SO_4)_2(OH)_6$	rhomb	2.8	3.8	1.572	1.592	
Alunogen	$Al_2(SO_4)_3{\cdot}18H_2O$	monocl	1.69	1.8	1.467	1.47	1.478
Amblygonite	$(Li,Na)Al(PO_4)(F,OH)$	tricl	3.1	5.8	1.591	1.604	1.613
Analcite	$NaAlSi_2O_6{\cdot}H_20$	cub	2.27	5.5	1.486		
Anatase	TiO_2	tetr	4.23	5.8	2.488	2.561	
Andalusite	Al_2OSiO_4	orth	3.15	7.5	1.635	1.639	1.644
Andesine	$NaAlSi_3O_8{\cdot}CaAl_2Si_2O_8$	tricl	2.67	6.3	1.550	1.553	1.557
Andorite	$PbAgSb_3S_6$	rhomb	5.35	3.3			
Andradite	$Ca_3(Fe,Ti)_2Si_3O_{12}$	cub	3.86	6.8	1.887		
Anglesite	$PbSO_4$	orth	6.29	2.8	1.877	1.883	1.894
Anhydrite	$CaSO_4$	orth	2.96	3.5	1.570	1.575	1.614
Ankerite	$Ca(Fe,Mg,Mn)(CO_3)_2$	rhomb	3.0	3.8	1.529	1.720	
Anorthite	$CaAl_2Si_2O_8$	tricl	2.76	6.3	1.577	1.585	1.590
Anorthoclase	$(Na,K)AlSi_3O_8$	tricl	2.58	6	1.523	1.528	1.529
Anthophyllite	$(Mg,Fe)_7Si_8O_{22}(OH,F)_2$	rhomb	3.21	5.8	1.645	1.658	1.668
Apatite	$Ca_5(PO_4)_3(OH,F,Cl)$	hex	3.2	5	1.645	1.648	
Apophyllite	$KFCa_4Si_8O_{20}{\cdot}8H_2O$	tetr	2.35	4.8	1.535	1.536	
Aragonite	$CaCO_3$	orth	2.83	3.5	1.531	1.680	1.686
Arcanite	K_2SO_4	orth	2.66		1.494	1.494	1.497
Argentite	Ag_2S	orth	7.2	2.3			
Arsenolite	As_2O_3	cub	3.86	1.5	1.755		
Arsenopyrite	FeAsS	monocl	6.1	5.8			
Atacamite	$Cu_2(OH)_3Cl$	rhomb	3.76	3.3	1.831	1.861	1.880
Augelite	$Al_2(PO_4)(OH)_3$	monocl	2.70	4.8	1.574	1.576	1.588
Augite	$(Ca,Mg,Fe,Ti,Al)_2(Si,Al)_2O_6$	monocl	3.38	6	1.703	1.707	1.738
Autunite	$Ca(UO_{22})(PO_4)_2{\cdot}10H20$	tetr	3.2	2.3	1.553	1.577	
Axinite	$(Ca,Mn,Fe)_3Al_2BO_3Si_4O_{12}(OH)$	tricl	3.31	6.8	1.684	1.691	1.694

PHYSICAL AND OPTICAL PROPERTIES OF MINERALS (continued)

Name	Formula	Crystal system	Density g/cm³	Hard-ness	Index of refraction n_α	n_β	n_γ
Azurite	$Cu_3(OH)_2(CO_3)_2$	monocl	3.77	3.8	1.730	1.758	1.838
Baddeleyite	ZrO_2	monocl	5.7	6.5	2.13	2.19	2.20
Barite	$BaSO_4$	orth	4.49	3.3	1.636	1.637	1.648
Benitoite	$BaTi(SiO_3)_3$	rhomb	3.65	6.3	1.757	1.804	
Bertrandite	$Be_4Si_2O_7(OH)_2$	rhomb	2.6	6	1.589	1.602	1.613
Beryl	$Be_3Al_2(SiO_3)_6$	hex	2.64	7.8	1.582	1.589	
Beryllonite	$NaBe(PO)_4$	monocl	2.81	5.8	1.552	1.558	1.561
Biotite	$K(Mg,Fe)_3AlSi_3O_{10}(OH,F)_2$	monocl	3.0	2.8	1.595	1.651	1.651
Bismuthinite	Bi_2S_3	orth	6.78	2			
Bixbyite	$(Mn,Fe)_2O_3$	cub	4.95	6.3			
Bloedite	$Na_2Mg(SO_4)_2 \cdot 4H_2O$	monocl	2.25	2.8	1.483	1.486	1.487
Boehmite	$AlO(OH)$	orth	3.44	3.8	1.64	1.65	1.66
Boracite	$Mg_3B_7O_{13}Cl$	rhomb	2.94	7.3	1.66	1.66	1.67
Borax	$Na_2B_4O_7 \cdot 10H_2O$	monocl	1.73	2.3	1.447	1.469	1.472
Bornite	Cu_5FeS_4	cub	5.07	3			
Boulangerite	$Pb_5Sb_4S_{11}$	monocl	6.1	2.8			
Bournonite	$PbCuSbS_3$	rhomb	5.83	2.8			
Braggite	PtS	tetr	10.2				
Braunite	$(Mn,Si)_2O_3$	tetr	4.78	6.3			
Bravoite	$(Ni,Fe)S_2$	cub	4.62	5.8			
Breithauptite	$NiSb$	hex	≈8.7	5.5			
Brochantite	$Cu_4(SO_4)(OH)_6$	monocl	3.79	3.8	1.728	1.771	1.800
Bromyrite	$AgBr$	cub	6.47	2.5	2.253		
Brookite	TiO_2	orth	4.23	5.8	2.583	2.584	2.700
Brucite	$Mg(OH)_2$	hex	2.37	2.5	1.575	1.59	
Bunsenite	NiO	cub	6.72	5.5			
Cacoxenite	$Fe_4(PO_4)_3(OH)_3 \cdot 12H_20$	hex	2.3	3.5	1.580	1.646	
Calcite	$CaCO_3$	hex	2.71	3	1.486	1.658	
Caledonite	$Cu_2Pb_5(SO_4)_3(CO_3)(OH)_6$	rhomb	5.76	2.8	1.818	1.866	1.909
Calomel	Hg_2Cl_2	tetr	7.16	1.5	1.973	2.656	
Cancrinite	$(Na,Ca,K)_7[Al_6Si_6O_{24}]$ $(CO_3,SO_4,Cl,OH)_2 \cdot H_20$	hex	2.42	5.5	1.495	1.509	
Carnalite	$KMgCl_3 \cdot 6H_20$	rhomb	1.60	2.5	1.466	1.475	1.494
Carnotite	$K_2(UO_2)_2(VO_4)_2 \cdot 3H_2O$	rhomb		1.5	1.75	1.92	1.95
Cassiterite	SnO_2	tetr	6.85	6.5	2.006	2.097	
Celestite	$SrSO_4$	orth	3.96	3.3	1.622	1.624	1.631
Celsian	$BaAl_2Si_2O_8$	monocl	3.25	6.3	1.583	1.588	1.594
Cerargyrite	$AgCl$	cub	5.56	2.5	2.071		
Cerussite	$PbCO_3$	orth	6.6	3.3	1.804	2.076	2.079
Cervantite	Sb_2O_4	orth	6.64	4.5			
Chabazite	$Ca[Al_2Si_4O_{12}] \cdot 6H_2O$	trig	2.08	4.5	1.482		
Chalcanthite	$CuSO_4 \cdot 5H_2O$	tricl	2.29	2.5	1.514	1.537	1.543
Chalcocite	Cu_2S	orth	5.6	2.8			
Chalcopyrite	$CuFeS_2$	tetr	4.2	3.8			
Chiolite	$Na_5Al_3F_{14}$	tetr	3.00	3.8	1.342	1.349	
Chlorite	$(Mg,Al,Fe)_{12}(Si,Al)_8O_{20}(OH)_{16}$	monocl	3.0	2.5	1.61	1.62	1.62
Chloritoid	$FeAl_4O_2(SiO_4)_2(OH)_4$	monocl	3.66	6.5	1.717	1.721	1.726
Chondrodite	$Mg(OH,F)_2 \cdot 2Mg_2SiO_4$	monocl	3.21	6.5	1.604	1.615	1.634
Chromite	$FeCr_2O_4$	cub	5.0	5.5	2.16		
Chrysoberyl	$BeAl_2O_4$	orth	3.65	8.5	1.746	1.748	1.756
Chrysocolla	$CuSiO_3 \cdot 2H_2O$	rhomb	2.4	2	1.575	1.597	1.598
Cinnabar	HgS	hex	8.17	2.3	2.814	3.143	
Claudetite	As_2O_3	monocl	3.74	2.5	1.87	1.92	2.01
Clinohumite	$Mg(OH,F)_2 \cdot 4Mg_2SiO_4$	monocl	3.21	6	1.633	1.647	1.668
Clinozoisite	$Ca_2Al_3Si_3O_{12}(OH)$	monocl	3.30	6.5	1.693	1.700	1.712
Cobaltite	$CoAsS$	cub	≈6.1	5.5			
Colemanite	$Ca_2B_6O_{11} \cdot 5H_2O$	monocl	2.42	4.5	1.586	1.592	1.614
Columbite	$(Fe,Mn)(Nb,Ta)_2O_6$	rhomb	5.20	6			
Connellite	$Cu_{19}(SO_4)Cl_4(OH)_{32} \cdot 3H_2O$	hex	3.36	3	1.731	1.752	

PHYSICAL AND OPTICAL PROPERTIES OF MINERALS (continued)

Name	Formula	Crystal system	Density g/cm³	Hard-ness	Index of refraction n_α	n_β	n_γ
Copiapite	$(Fe,Mg)Fe_4(SO_4)_6(OH)_2 \cdot 20H_2O$	tricl	2.13	2.8	1.52	1.54	1.59
Coquimbite	$Fe_2(SO_4)_3 \cdot 9H_2O$	hex	2.1	2.5	1.54	1.56	
Cordierite	$Al_3(Mg,Fe)_2Si_5AlO_{18}$	rhomb	2.66	7	1.540	1.549	1.553
Corundum	Al_2O_3	hex	3.97	9	1.761	1.769	
Cotunnite	$PbCl_2$	orth	5.98	2.5	2.199	2.217	2.260
Covellite	CuS	hex	4.8	1.8			
Cristobalite	SiO_2	hex	2.33	6.5	1.484	1.487	
Crocoite	$PbCrO_4$	monocl	6.12	2.8	2.29	2.36	2.66
Cryolite	Na_3AlF_6	monocl	2.97	2.5	1.338	1.338	1.339
Cryolithionite	$Na_3Li_3Al_2F_{12}$	cub	2.77	2.8	1.340		
Cubanite	$CuFe_2S_3$	rhomb	4.11	3.5			
Cummingtonite	$(Mg,Fe)_7Si_8O_{22}(OH)_2$	monocl	3.4	5.5	1.650	1.660	1.676
Cuprite	Cu_2O	cub	6.0	3.8			
Danburite	$CaSi_2B_2O_8$	rhomb	3.0	7	1.63	1.63	1.63
Datolite	$CaBSiO_4(OH)$	monocl	2.98	5.3	1.624	1.652	1.668
Daubreelite	Cr_2FeS_4	cub	3.81				
Derbylite	$Fe_6Ti_6Sb_2O_{23}$	rhomb	4.53	5	2.45	2.45	2.51
Diamond	C	cub	3.51	10	2.418		
Diaspore	$AlO(OH)$	orth	3.4	6.8	1.694	1.715	1.741
Digenite	$Cu_{2-x}S$	cub	5.55	2.8			
Diopside	$CaMgSi_2O_6$	monocl	3.30	6	1.680	1.687	1.708
Dioptase	$CuSiO_2(OH)_2$	rhomb	3.5	5	1.65	1.70	
Dolomite	$CaMg(CO_3)_2$	rhomb	2.86	3.5	1.500	1.679	
Douglasite	$K_2FeCl_4 \cdot 2H_2O$	orth	2.16		1.488	1.500	
Dyscrasite	Ag_3Sb	rhomb	9.74	3.8			
Eddingtonite	$BaAl_2Si_3O_{10} \cdot 4H_2O$	rhomb	2.8		1.541	1.553	1.557
Eglestonite	Hg_4OCl_2	cub	8.4	2.5	2.49		
Emplectite	$CuBiS_2$	rhomb	6.38	2			
Enargite	Cu_3AsS_4	rhomb	4.5	3			
Enstatite	$MgSiO_3$	monocl	3.19	5.5	1.656	1.662	1.669
Epidote	$Ca_2Al_2(Al,Fe)OH(SiO_4)_3$	monocl	3.44	6	1.733	1.755	1.765
Epsomite	$MgSO_4 \cdot 7H_2O$	orth	1.67	2.3	1.433	1.455	1.461
Erythrite	$(Co,Ni)_3(AsO_4)_2 \cdot 8H_2O$	monocl	3.06	2	1.626	1.661	1.699
Eucairite	$CuAgSe$	orth	7.7	2.5			
Euclasite	$BeAlSiO_4(OH)$	monocl	3.1	7.5	1.651	1.655	1.671
Eudialite	$(Na,Ca,Ce)_5(Fe,Mn)(Zr,Ti)(Si_3O_9)_2(OH,Cl)$	hex	3.0	5.5	1.623	1.600	1.615
Eulytite	$Bi_4Si_3O_{12}$	cub	6.6	4.5	2.05		
Euxenite	$(Y,Ca,Ce,U,Th)(Nb,Ta,Ti)_2O_6$	rhomb	5.5	6	2.2		
Fayalite	Fe_2SiO_4	orth	4.30	6.5	1.827	1.869	1.879
Ferberite	$FeWO_4$	monocl	7.51	4.3			
Fergussonite	$(Y,Er,Ce,Fe)(Nb,Ta,Ti)O_4$	tetr	5.7	6	2.1		
Fluorite	CaF_2	cub	3.18	4	1.434		
Forsterite	Mg_2SiO_4	orth	3.21	7	1.635	1.651	1.670
Franklinite	$ZnFe_2O_4$	cub	5.21	6	2.36		
Gahnite	$ZnAl_2O_4$	cub	4.62	7.8	1.805		
Galaxite	$MnAl_2O_4$	cub	4.04	7.8	1.92		
Galena	PbS	cub	7.60	2.5	3.91		
Galenabismuthite	$PbBi_2S_4$	rhomb	7.04	3			
Ganomalite	$(Ca,Pb)_{10}(OH,Cl)_2(Si_2O_7)_3$	hex	5.6	3.5	1.910	1.945	
Gaylussite	$Na_2Ca(CO_3)_2 \cdot 5H_2O$	monocl	1.99	2.8	1.444	1.516	1.523
Gehlenite	$Ca_2Al_2SiO_7$	tetr	3.04	5.5	1.658	1.669	
Geikielite	$MgTiO_3$	hex	3.85	5.5	1.95	2.31	
Gibbsite	$Al(OH)_3$	monocl	2.42	3	1.57	1.57	1.59
Glauberite	$Na_2Ca(SO_4)_2$	monocl	2.80	2.8	1.515	1.535	1.536
Glauconite	$(K,Na,Ca)_{1.6}(Fe,Al,Mg)_{4.0}Si_{7.3}Al_{0.7}O_{20}(OH)_4$	monocl	2.7	2	1.60	1.63	1.63
Glaucophane	$Na_2Mg_3Al_2Si_8O_{22}(OH)_2$	monocl	3.19	6	1.634	1.645	1.648
Gmelinite	$(Ca,Na_2)[Al_2Si_4O_{12}] \cdot 6H_2O$	hex	2.10	4.5	1.477	1.485	

Name	Formula	Crystal system	Density g/cm^3	Hard- ness	Index of refraction n_α	n_β	n_γ
Goethite	$FeO(OH)$	orth	4.3	5.3	2.268	2.401	2.457
Goslarite	$ZnSO_4{\cdot}7H_2O$	orth	1.97	2.3	1.457	1.480	1.484
Greenockite	CdS	cub	4.8	3.3	2.506	2.529	
Grossularite	$Ca_3Al_2Si_3O_{12}$	cub	3.59	6.8	1.734		
Gummite	$UO_3{\cdot}H_2O$	orth	7.05	3.8			
Gypsum	$CaSO_4{\cdot}2H_2O$	monocl	2.32	2	1.520	1.525	1.530
Halite	$NaCl$	cub	2.17	2	1.544		
Hambergite	$Be_2(OH)(BO_3)$	rhomb	2.36	7.5	1.56	1.59	1.63
Hanksite	$Na_{22}K(SO_4)_9(CO_3)_2Cl$	hex	2.56	3.3	1.461	1.481	
Harmotome	$Ba[Al_2Si_6O_{16}]{\cdot}6H_2O$	monocl	2.44	4.5	1.506	1.507	1.511
Hausmannite	Mn_3O_4	tetr	4.84	5.5	2.15	2.46	
Haüyne	$(Na,Ca)_{4\text{-}8}Al_6Si_6O_{24}(SO_4,S)_{1\text{-}2}$	cub	2.47	5.8	1.502		
Hedenbergite	$CaFeSi_2O_6$	monocl	3.53	6	1.721	1.727	1.746
Helvite	$Mn_4Be_3Si_3O_{12}S$	cub	3.32	6	1.739		
Hematite	Fe_2O_3	hex	5.25	6	2.91	3.19	
Hemimorphite	$Zn_4Si_2O_7(OH)_2{\cdot}H_2O$	rhomb	3.45	5	1.614	1.617	1.636
Hercynite	$Fe(AlO_2)_2$	cub	4.3	7.8	1.835		
Herderite	$CaBe(PO_4)(Fe,OH)$	monocl	2.98	5.3	1.592	1.612	1.621
Hessite	Ag_2Te	orth	8.4	2.5			
Heulandite	$(Ca,Na_2,K_2)[Al_2Si_7O_{18}]{\cdot}6H_2O$	monocl	2.2	3.8	1.498	1.498	1.506
Hopeite	$Zn_3(PO_4)_2{\cdot}4H_2O$	orth	3.0	3.2	1.58	1.59	1.59
Hornblende	$Ca_2(Mg,Fe)_4Al(Si_7AlO_{22})(OH)_2$	monocl	3.24	5.5	1.67	1.67	1.69
Huebnerite	$MnWO_4$	monocl	7.2	4.3	2.17	2.22	2.32
Humite	$Mg(OH,F)_2{\cdot}3Mg_2SiO_4$	orth	3.3	6	1.625	1.636	1.657
Huntite	$Mg_3Ca(CO_3)_4$	trig	2.70				
Hydrogrossularite	$Ca_3Al_2Si_2O_8(SiO_4)_{1\text{-}m}(OH)_{4m}$	cub	3.4	6.8	1.70		
Hydromagnesite	$3MgCO_3{\cdot}Mg(OH)_2{\cdot}3H_2O$	monocl	2.24	3.5	1.523	1.527	1.545
Illite	$KAl_4[Si_7AlO_{20}](OH)_4$	monocl	2.8	1.5	1.56	1.59	1.59
Ilmenite	$FeTiO_3$	rhomb	4.72	5.5			
Iodyrite	AgI	hex	5.68	1.5	2.21	2.22	
Jacobsite	$MnFe_2O_4$	cub	4.87	7.8	2.3		
Jadeite	$NaAlSi_2O_6$	monocl	3.34	6	1.649	1.654	1.663
Jamesonite	$Pb_4FeSb_6S_{14}$	monocl	5.63	2.5			
Jarosite	$KFe_3(SO_4)_2(OH)_6$	rhomb	3.09	3	1.715	1.820	
Kainite	$KMg(SO_4)Cl{\cdot}3H_2O$	monocl	2.15	2.8	1.494	1.505	1.516
Kaliophylite	$KAlSiO_4$	hex	2.61	6	1.532	1.537	
Kaolinite	$Al_4Si_4O_{10}(OH)_8$	tricl	2.65	2.3	1.549	1.564	1.565
Kernite	$Na_2B_4O_7{\cdot}4H_2O$	monocl	1.95	2.5	1.454	1.472	1.488
Kieserite	$MgSO_4{\cdot}H_2O$	monocl	2.57	3.5	1.520	1.533	1.584
Kyanite	Al_2OSiO_4	tricl	3.59	6.3	1.715	1.722	1.731
Lanarkite	$Pb_2(SO_4)O$	monocl	6.92	2.3	1.928	2.007	2.036
Lanthanite	$(La,Ce)_2(CO_3)_3{\cdot}8H_2O$	rhomb	2.72	2.8	1.52	1.587	1.613
Laumontite	$Ca_4[Al_8Si_{16}O_{48}]{\cdot}16H_2O$	monocl	2.3	3.3	1.508	1.517	1.519
Laurionite	$Pb(OH)Cl$	rhomb	6.24	3.3	2.08	2.12	2.16
Lawsonite	$CaAl_2(OH)_2Si_2O_7{\cdot}H_2O$	rhomb	3.08	6	1.655	1.675	1.685
Lazulite	$(Mg,Fe)Al_2(PO_4)_2(OH)_2$	monocl	3.23	5.8	1.615	1.64	1.650
Lazurite	$Na_4SSi_3Al_3O_{12}$	cub	2.42	5.3	1.500		
Leadhillite	$Pb_4(SO_4)(CO_3)_2(OH)_2$	monocl	6.55	2.8	1.87	2.00	2.01
Lepidocrocite	$FeO(OH)$	orth	4.26	5	1.94	2.20	2.51
Lepidolite	$K_2(Li,Al)_{5\text{-}6}[Si_{6\text{-}7}Al_{2\text{-}1}O_{20}](OH,F)_4$	monocl	2.85	3.3	1.536	1.565	1.566
Leucite	$KAlSi_2O_6$	tetr	2.49	5.8	1.510		
Levyne	$(Ca,Na_2)Al_2Si_4O_{12}{\cdot}6H_2O$	rhomb	2.10	4.5	1.496	1.501	
Litharge	PbO	tetr	9.35	2	2.535	2.665	
Loellingite	$FeAs_2$	rhomb	7.40	5.3			
Maghemite	Fe_2O_3	cub	4.88	7.8	2.63		
Magnesite	$MgCO_3$	hex	3.05	4	1.536	1.741	
Magnetite	Fe_3O_4	cub	5.17	6	2.42		
Malachite	$Cu_2(OH)_2(CO_3)$	monocl	4.05	3.8	1.655	1.875	1.909
Manganite	$MnO(OH)$	monocl	≈4.3	4	2.25	2.25	2.53

Name	Formula	Crystal system	Density g/cm³	Hard-ness	Index of refraction n_α	n_β	n_γ
Manganosite	MnO	cub	5.37	5.5			
Marcasite	FeS_2	cub	5.02	6.3			
Marialite	$Na_4Al_3Si_9O_{24}Cl$	tetr	2.56	5.5	1.541	1.548	
Marshite	CuI	cub	5.67	2.5	2.346		
Mascagnite	$(NH_4)_2SO_4$	orth	1.77	2.3	1.520	1.523	1.533
Matlockite	$PbClF$	tetr	7.05	2.8	2.006	2.145	
Meionite	$Ca_4Al_6Si_6O_{24}CO_3$	tetr	2.78	5.5	1.559	1.595	
Melanterite	$FeSO_4 \cdot 7H_2O$	monocl	1.89	2	1.47	1.48	1.49
Melilite	$(Ca,Na)_2(Mg,Fe,Al,Si)_3O_7$	tetr	3.00	5.5	1.639	1.645	
Mellite	$Al_2C_{12}O_{12} \cdot 18H_2O$	tetr	1.64	2.3	1.511	1.539	
Mendipite	$Pb_3O_2Cl_2$	rhomb	7.24	2.5	2.24	2.27	2.31
Mesolite	$Na_2Ca_2(Al_2Si_3O_{10})_3 \cdot 8H_2O$	orth	2.26	5	1.506		
Metacinnabar	HgS	cub	7.70	3			
Microcline	$KAlSi_3O_8$	monocl	2.56	6.3	1.522	1.526	1.530
Miersite	AgI	hex	5.68	2.5	2.20		
Millerite	NiS	hex	5.5	3.3			
Mimetite	$Pb_5(AsO_4,PO_4)_3Cl$	hex	7.24	3.8	2.128	2.147	
Minium	Pb_3O_4	tetr	8.9	2.5			
Mirabilite	$Na_2SO_4 \cdot 10H_2O$	monocl	1.46	1.8	1.394	1.396	1.398
Moissanite	SiC	hex	3.16	9.5	2.648	2.691	
Molybdenite	MoS_2	hex	5.06	1.3			
Monazite	$(Ce,La,Th)PO_4$	monocl	5.2	5	1.787	1.789	1.840
Monetite	$CaHPO_4$	tricl	2.92	3.5	1.587	1.61	1.640
Monticellite	$Ca(Mg,Fe)SiO_4$	orth	3.18	5.5	1.647	1.655	1.664
Montmorillonite	$(0.5Ca,Na)_{0.7}(Al,Mg,Fe)_4[(Si,Al)_8O_{20}](OH)_4 \cdot nH_2O$	monocl	2.5	1.5	1.55	1.57	1.57
Montroydite	HgO	orth	11.14	2.5	2.37	2.50	2.65
Mordenite	$(Na,K,Ca)[Al_2Si_{10}O_{24}] \cdot 7H_2O$	orth	2.13	3.5	1.478	1.480	1.482
Muscovite	$KAl_2Si_3AlO_{10}(OH,F)_2$	monocl	2.83	2.8	1.563	1.596	1.602
Nantokite	$CuCl$	cub	4.14	2.5	1.930		
Natrolite	$Na_2Al_2Si_3O_{10} \cdot 2H_2O$	orth	2.23	5	1.478	1.481	1.491
Nepheline	$Na_3KAl_4Si_4O_{16}$	hex	2.61	5.8	1.534	1.538	
Newberyite	$MgHPO_4 \cdot 3H_2O$	orth	2.13	3.3	1.514	1.517	1.533
Niccolite	$NiAs$	hex	7.77	5.3			
Norbergite	$Mg(OH,F)_2 \cdot Mg_2SiO_4$	orth	3.21	6.5	1.565	1.573	1.592
Nosean	$Na_8Al_6Si_6O_{24}SO_4$	cub	2.35	5.5	1.495		
Oldhamite	CaS	cub	2.59	4	2.137		
Oligoclase	$([NaSi]_{0.9\text{-}0.7}[CaAl]_{0.1\text{-}0.3})AlSi_2O_8$	tricl	2.64	6.3	1.539	1.543	1.547
Olivenite	$Cu_2(AsO_4)(OH)$	rhomb	4.2	3	1.77	1.80	1.85
Olivine	$(Mg,Fe)SiO_4$	rhomb	3.81	6.8	1.73	1.76	1.78
Opal	$SiO_2 \cdot nH_2O$	amorp	1.9	5	1.44		
Orpiment	As_2S_3	monocl	3.46	1.8	2.40	2.81	3.02
Orthoclase	$KAlSi_3O_8$	monocl	2.56	6	1.523	1.527	1.531
Orthopyroxene	$(Mg,Fe)SiO_3$	rhomb	3.6	5.5	1.709	1.712	1.723
Paragonite	$NaAl_2Si_3AlO_{10}(OH)_2$	monocl	2.85	2.5	1.572	1.602	1.605
Parisite	$(Ce,La,Na)FCO_3 \cdot CaCO_3$	hex	4.42	4.5	1.672	1.771	
Pectolite	$Ca_2NaH(SiO_3)_3$	tricl	2.88	4.8	1.603	1.610	1.639
Penfieldite	$Pb_4Cl_6(OH)_2$	hex	6.6		2.13	2.21	
Pentlandite	$(Fe,Ni)_9S_8$	cub	4.8	3.8			
Percylite	$PbCuCl_2(OH)_2$	cub		2.5	2.05		
Periclase	MgO	cub	3.6	5.5	1.735		
Perovskite	$CaTiO_3$	cub	3.98	5.5	2.34		
Petalite	$LiAlSi_4O_{10}$	monocl	2.42	6.5	1.506	1.511	1.519
Pharmacosiderite	$Fe_3(AsO_4)_2(OH)_3 \cdot 5H_2O$	cub	2.80	2.5	1.690		
Phenakite	Be_2SiO_4	rhomb	2.98	7.5	1.654	1.670	
Phillipsite	$K(Ca_{0.5},Na)_2[Al_3Si_5O_{16}] \cdot 6H_2O$	monocl	2.2	4.3	1.494	1.497	1.505
Phlogopite	$KMg_3AlSi_3O_{10}(OH,F)_2$	monocl	2.83	2.3	1.560	1.597	1.598
Phosgenite	$Pb_2(CO_3)Cl_2$	tetr	6.13	2.5	2.118	2.145	
Piemontite	$Ca_2(Mn,Fe,Al)_3O(Si_2O_7)(SiO_4)(OH)$	monocl	3.49	6	1.762	1.773	1.796

Name	Formula	Crystal system	Density g/cm³	Hard-ness	Index of refraction n_α	n_β	n_γ
Pigeonite	$(Mg,Fe,Ca)(Mg,Fe)Si_2O_6$	monocl	3.38	6	1.702	1.703	1.728
Pollucite	$CsAlSi_2O_6$	tetr	2.9	6.5	1.517		
Polybasite	$(Ag,Cu)_{16}Sb_2S_{11}$	monocl	6.1	2.5			
Powellite	$Ca(Mo,W)O_4$	tetr	4.35	3.8	1.971	1.980	
Prehnite	$Ca_2Al_2Si_3O_{10}(OH)_2$	rhomb	2.93	6.3	1.622	1.628	1.648
Proustite	Ag_3AsS_3	rhomb	5.57	2.3	2.792	3.088	
Pseudobrookite	Fe_2TiO_5	rhomb	4.36	6	2.38	2.39	2.42
Psilomelane	$BaMn_9O_{16}(OH)_4$	rhomb	4.71	5.5			
Pumpellyite	$Ca_2Al_2(Al,Fe,Mg)[Si_2(O,OH)_7](SiO_4)(OH,O)_3$	monocl	3.21	5.5	1.688	1.695	1.705
Pyrargyrite	Ag_3SbS_3	rhomb	5.85	2.5	2.88	3.08	
Pyrite	FeS_2	cub	5.02	6.3			
Pyrochlore	$NaCaNb_2O_6F$	cub	5.3	5.3			
Pyrochroite	$Mn(OH)_2$	hex	3.26	2.5	1.68	1.72	
Pyrolusite	MnO_2	tetr	5.08	6.3			
Pyromorphite	$Pb_5(PO_4,AsO_4)_3Cl$	hex	7.04	3.8	2.048	2.058	
Pyrope	$Mg_3Al_2Si_3O_{12}$	cub	3.58	6.8	1.714		
Pyrophyllite	$Al_2Si_4O_{10}(OH)_2$	monocl	2.78	1.5	1.545	1.579	1.599
Pyrrhotite	Fe_7S_8	hex	4.62	4			
Quartz	SiO_2	hex	2.65	7	1.544	1.553	
Rammelsbergite	$NiAs_2$	orth	7.1	5.8			
Raspite	$PbWO_4$	monocl	8.46	2.8	1.27	1.27	1.30
Realgar	As_4S_4	monocl	3.5	1.8	2.538	2.684	2.704
Rhodochrosite	$MnCO_3$	hex	3.70	3.8	1.597	1.816	
Rhodonite	$(Mn,Fe,Ca)SiO_3$	orth	3.48	6	1.725	1.729	1.737
Riebeckite	$Na_2Fe_5(Si_8O_{22})(OH)_2$	monocl	3.3	5	1.675	1.683	1.694
Rutile	TiO_2	tetr	4.23	6.2	2.609	2.900	
Safflorite	$(Co,Fe)As_2$	rhomb	7.3	4.8			
Samarskite	$(Y,Er,Ce,U,Ca,Fe,Pb,Th)(Nb,Ta,Ti,Sn)_2O_6$	rhomb	5.69	5.5	2.200		
Sapphirine	$(Mg,Fe)_2Al_4O_6SiO_4$	monocl	3.49	7.5	1.709	1.712	1.715
Scapolite	$(Na,Ca)_4Al_3(Al,Si)_3Si_6O_{24}(Cl,F,OH,CO_3,SO_4)$	tetr	2.64	5.5	1.551	1.573	
Scheelite	$CaWO_4$	tetr	6.06	4.8	1.920	1.936	
Scolecite	$CaAl_2Si_3O_{10}\cdot 3H_2O$	monocl	2.27	5	1.510	1.518	1.519
Scorodite	$Fe(AsO_4)\cdot 2H_2O$	rhomb	3.28	3.8	1.784	1.795	1.814
Sellaite	MgF_2	tetr	3.15	5	1.378	1.390	
Senarmontite	Sb_2O_3	cub	5.58	2.3	2.087		
Serpentine	$Mg_3Si_2O_5(OH)_4$	monocl	2.55	3	1.55	1.56	1.56
Siderite	$FeCO_3$	hex	3.9	4.3	1.635	1.875	
Sillimanite	Al_2OSiO_4	rhomb	3.25	7	1.658	1.660	1.660
Skutterudite	$(Co,Ni)As_3$	cub	6.8	5.8			
Smithsonite	$ZnCO_3$	rhomb	4.4	4.3	1.621	1.848	
Sodalite	$Na_8Al_6Si_6O_{24}Cl_2$	cub	2.30	5.8	1.485		
Sperrylite	$PtAs_2$	cub	10.58	6.5			
Spessartite	$Mn_3Al_2Si_3O_{12}$	cub	4.19	6.8	1.800		
Sphalerite	ZnS	cub	4.0	3.8	2.369		
Sphene	$CaTiSiO_4(O,OH,F)$	monocl	3.50	5	1.90	1.95	2.03
Spinel	$MgAl_2O_4$	cub	3.55	7.8	1.719		
Spodumene	$LiAlSi_2O_6$	monocl	3.13	6.8	1.656	1.662	1.671
Stannite	Cu_2FeSn_4	tetr	4.4	4			
Staurolite	$(Fe,Mg,Zn)_2(Al,Fe,Ti)_9O_6[(Si,Al)O_4]_4(O,OH)_2$	monocl	3.79	7.5	1.743	1.747	1.755
Stercorite	$Na(NH_4)H(PO_4)\cdot 4H_2O$	tricl	1.62	2	1.439	1.442	1.469
Stibiotantalite	$Sb(Ta,Nb)O_4$	rhomb	6.6	5.5	2.38	2.41	2.46
Stibnite	Sb_2S_3	orth	4.56	2			
Stilbite	$NaCa_2[Al_5Si_{13}O_{36}]\cdot 14H_2O$	monocl	2.2	3.8	1.492	1.499	1.503

Name	Formula	Crystal system	Density g/cm³	Hard-ness	Index of refraction		
					n_α	n_β	n_γ
Stilpnomelane	$(K,Na,Ca)_{0.6}(Fe,Mg)_6Si_8Al(O,OH)_{27}\cdot 2H_2O$	monocl	2.8	3.5	1.585	1.665	1.665
Stolzite	$PbWO_4$	tetr	8.2	2.8	2.19	2.27	
Strengite	$FePO_4\cdot 2H_2O$	orth	2.87	4	1.707	1.719	1.741
Strontianite	$SrCO_3$	orth	3.5	3.5	1.518	1.666	1.668
Struvite	$Mg(NH_4)(PO_4)\cdot 6H_2O$	rhomb	1.71	2	1.495	1.496	1.504
Sulfur	S	orth	2.07	2	1.958	2.038	2.245
Sylvanite	$(Ag,Au)Te_2$	monocl	8.16	1.8			
Sylvite	KCl	cub	1.99	2	1.490		
Talc	$Mg_3Si_4O_{10}(OH)_2$	monocl	2.71	1	1.545	1.592	1.595
Tantalite	$(Fe,Mn)(Ta,Nb)_2O_6$	rhomb	7.95	6.5	2.26	2.32	2.43
Tapiolite	$FeTa_2O_6$	tetr	7.9	6.3	2.27	2.42	
Tellurobismuthite	Bi_2Te_3	hex	7.74	1.8			
Terlinguaite	Hg_2OCl	monocl	8.73	2.5	2.35	2.64	2.66
Tetrahedrite	$(Cu,Fe)_{12}Sb_4S_{13}$	cub	4.9	3.8			
Thenardite	Na_2SO_4	orth	2.7	2.8	1.468	1.475	1.483
Thermonatrite	$Na_2CO_3\cdot H_2O$	orth	2.25	1.3	1.420	1.506	1.524
Thomsenolite	$NaCaAlF_6\cdot H_2O$	monocl	2.98	2	1.407	1.414	1.415
Thorianite	ThO_2	cub	10.0	6.5	2.200		
Thorite	$ThSiO_4$	tetr	6.7	4.8	1.8		
Topaz	$Al_2SiO_3(OH,F)_2$	rhomb	3.53	8	1.618	1.620	1.627
Torbernite	$Cu(UO_2)_2(PO_4)_2\cdot 8H_2O$	tetr	3.22	2.3	1.582	1.592	
Tourmaline	$Na(Mg,Fe,Mn,Li,Al)_3Al_6Si_6O_{18}(BO_3)_3$	rhomb	3.14	7	1.62	1.65	
Tremolite	$Ca_2Mg_5Si_8O_{22}(OH,F)_2$	monocl	3.0	5.5	1.599	1.612	1.622
Trevorite	$NiFe_2O_4$	cub	5.33	7.8	2.3		
Tridymite	SiO_2	hex	2.27	7	1.475	1.476	1.479
Triphyllite-Lithiophyllite	$Li(Fe,Mn)PO_4$	rhomb	3.46	4.5	1.68	1.68	1.69
Troegerite	$(UO_2)_3(AsO_4)_2\cdot 12H_2O$	tetr		2.5	1.59	1.630	
Troilite	FeS	hex	4.7	4			
Trona	$Na_3H(CO_3)_2\cdot 2H_2O$	monocl	2.14	2.8	1.412	1.492	1.540
Turquois	$Cu(Al,Fe)_6(PO_4)_4(OH)_8\cdot 4H_2O$	tricl	2.9	5.3	1.70	1.73	1.75
Ullmannite	$NiSbS$	cub	6.65	5.3			
Uraninite	UO_2	cub	11.0	5.5			
Uvarovite	$Ca_3Cr_2Si_3O_{12}$	cub	3.83	6.8	1.865		
Valentinite	Sb_2O_3	orth	5.7	2.8	2.18	2.35	2.35
Vanadinite	$Pb_5(VO_4)_3Cl$	hex	6.8	2.9	2.350	2.416	
Variseite-Strengite	$(Al,Fe)(PO_4)\cdot 2H_2O$	rhomb	2.72	4	1.635	1.654	1.668
Vaterite	$CaCO_3$	hex	2.71		1.550	1.645	
Vermiculite	$(Mg,Ca)_{0.7}(Mg,Fe,Al)_6[(Al,Si)_8O_{20}](OH)_4\cdot 8H_2O$	monocl	2.3	1.5	1.542	1.556	1.556
Vesuvianite	$Ca_{10}(Mg,Fe)_2Al_4(Si_2O_7)_2(SiO_4)_5(OH,F)_4$	tetr	3.33	6.5	1.72	1.73	
Villiaumite	NaF	cub	2.78	2.3	1.327		
Vivianite	$Fe_3(PO_4)_2\cdot 8H_2O$	monocl	2.58	1.8	1.598	1.629	1.652
Wagnerite	$Mg_2(PO_4)F$	monocl	3.15	5.3	1.568	1.572	1.582
Wavellite	$Al_3(OH)_3(PO_4)_2\cdot 5H_2O$	rhomb	2.36	3.6	1.527	1.535	1.553
Whewellite	$CaC_2O_4\cdot H_2O$	cub	2.2	2.8	1.491	1.554	1.650
Willemite	Zn_2SiO_4	hex	4.1	5.5	1.691	1.719	
Witherite	$BaCO_3$	orth	4.29	3.5	1.529	1.676	1.677
Wolframite	$(Fe,Mn)WO_4$	monocl	7.3	4.3	2.26	2.32	2.42
Wollastonite	$CaSiO_3$	monocl	2.92	4.8	1.628	1.639	1.642
Wulfenite	$PbMoO_4$	tetr	6.7	2.9	2.283	2.403	
Wurtzite	ZnS	hex	4.09	3.8	2.356	2.378	
Xenotime	YPO_4	tetr	4.8	4.5	1.721	1.816	
Zeunerite	$Cu(UO_2)_2(AsO_4)_2\cdot 10H_2O$	tetr			1.606		
Zincite	ZnO	hex	5.6	4	2.013	2.029	
Zircon	$ZrSiO_4$	tetr	4.6	7.5	1.94	1.99	
Zoisite	$Ca_2Al_3Si_3O_{12}(OH)$	rhomb	3.26	6	1.695	1.699	1.711

CRYSTALLOGRAPHIC DATA ON MINERALS

This table contains x-ray crystallographic data on about 400 common minerals, as well as selected crystalline elements. Entries are arranged alphabetically by mineral name. The columns are:

Name: Common name of the mineral.
Formula: Chemical formula for a typical sample of the mineral. Composition often varies considerably with the origin of the sample.
Crystal system: tricl = triclinic; monocl = monoclinic; orth = orthorhombic; tetr = tetragonal; hex = hexagonal; rhomb = rhombohedral; cubic = cubic.
Structure type: Prototype for the structural arrangement of the crystallographic cell.
Z: Number of formula units per the unit cell.
a, *b*, *c*: Lengths of the cell edges in Å (1Å = 10^{-8} cm).
α, β, γ: Angles between cell axes.

REFERENCES

1. Robie, R.A., Bethke, P.M., and Beardsley, K.M., *U. S. Geological Survey Bulletin 1248*, U. S. Government Printing Office, Washington, D.C.
2. Donnay, J.D.H., and Ondik, H.M., *Crystal Data Determinative Tables, Third Edition, Volume 2, Inorganic Compounds*, Joint Committee on Powder Diffraction Standards, Swarthmore, PA, 1973.
3. Deer, W.A., Howie, R.A., and Zussman, J., *An Introduction to the Rock-Forming Minerals, 2nd Edition*, Longman Scientific & Technical, Harlow, Essex, 1992.

Name	Formula	Crystal system	Structure type	Z	a/Å	b/Å	c/Å	α	β	γ
Acanthite	Ag_2S	monocl		4	4.228	6.928	7.862		99.58°	
Acmite (Aegirine)	$NaFe(SiO_3)_2$	monocl	diopside	4	9.658	8.795	5.294		107.42°	
Akermanite	$Ca_2MgSi_2O_7$	tetr	melilite	2	7.8435		5.010			
Alabandite	MnS	cubic	rock salt	4	5.223					
Almandine (Almandite)	$Fe_3Al_2Si_3O_{12}$	cubic	garnet	8	11.526					
Altaite	PbTe	cubic	rock salt	4	6.4606					
Aluminum	Al	cubic	copper	4	4.049					
Alunite	$KAl_3(SO_4)_2(OH)_6$	rhomb		3	6.982		17.32			
Analcite	$NaAlSi_2O_6 \cdot H_20$	cubic		16	13.733					
Anatase	TiO_2	tetr		4	3.785		9.514			
Andalusite	Al_2OSiO_4	orth		4	7.7959	7.8983	5.5583			
Andradite	$Ca_3Fe_2Si_3O_{12}$	cubic	garnet	8	12.048					
Anglesite	$PbSO_4$	orth	barite	4	8.480	5.398	6.958			
Anhydrite	$CaSO_4$	orth	anhydrite	4	6.991	6.996	6.238			
Annite	$KFe_3[AlSi_3O_{10}](OH)_2$	monocl	1M mica	2	10.29	9.33	5.39		105.1°	
Anorthite	$CaAl_2Si_2O_8$	tricl	primitive cell	8	8.177	12.877	14.169	93.17°	115.85°	91.22°
Anthophyllite	$Mg_7Si_8O_{22}(OH)_2$	orth		4	18.61	18.01	5.24			
Antimony	Sb	rhomb	arsenic	6	4.2996		11.2516			
Aragonite	$CaCO_3$	orth	aragonite	4	5.741	7.968	4.959			
Arcanite	K_2SO_4	orth	arcanite	4	5.772	10.072	7.483			
Argentite	Ag_2S	cubic		2	4.870					
Argentopyrite	$AgFe_2S_3$	orth		4	6.64	11.47	6.45			
Arsenic	As	rhomb	arsenic	6	3.760		10.555			
Arsenolite	As_2O_3	cubic	diamond	16	11.074					
Arsenopyrite	FeAsS	tricl		4	5.760	5.690	5.785	90.00°	112.23°	90.00°
Azurite	$Cu_3(OH)_2(CO_3)_2$	monocl		2	5.008	5.844	10.336		92.45°	
Baddeleyite	ZrO_2	monocl	baddeleyite	4	5.1454	5.2075	5.3107		99.23°	
Banalsite	$BaNa_2Al_4Si_4O_{16}$	orth		4	8.50	9.97	16.72			
Barite	$BaSO_4$	orth	barite	4	8.878	5.450	7.152			
Berlinite	$AlPO_4$	hex	α-quartz	3	4.942		10.97			
Beryl	$Be_3Al_2(SiO_3)_6$	hex	beryl	2	9.215		9.192			
Berzelianite	Cu_2Se	cubic		4	5.85					
Bismite	Bi_2O_3	monocl	pseudo-orth	4	7.48	8.14	5.83		112.9°	
Bismuth	Bi	rhomb	arsenic	6	4.5367		11.8383			
Bismuthinite	Bi_2S_3	orth	stibnite	4	11.150	11.300	3.981			
Bixbyite	Mn_2O_3	cubic	thallium trioxide	16	9.411					
Boehmite	AlO(OH)	orth	lepidocrocite	4	2.868	12.227	3.700			
Borax	$Na_2B_4O_7 \cdot 10H_2O$	monocl		4	11.858	10.674	12.197		106.68°	
Bornite (metastable)	Cu_5FeS_4	cubic		8	10.94					
Breithauptite	NiSb	hex	niccolite	2	3.942		5.155			
Brochantite	$Cu_4SO_4(OH)_6$	monocl		4	13.066	9.85	6.022		103.27°	
Bromargyrite	AgBr	cubic	rock salt	4	5.7745					
Bromellite	BeO	hex	zincite	2	2.6979		4.3772			
Brookite	TiO_2	orth		8	5.456	9.182	5.143			
Brucite	$Mg(OH)_2$	hex	cadmium iodide	1	3.147		4.769			

Name	Formula	Crystal system	Structure type	Z	a/Å	b/Å	c/Å	α	β	γ
Bunsenite	NiO	cubic	rock salt	4	4.177					
Bustamite	$CaMn(SiO_3)_2$	tricl		6	7.736	7.157	13.824	90.52°	94.58°	103.87°
Cadmium telluride	$CdTe$	cubic	sphalerite	4	6.4805					
Cadmoselite	$CdSe$	hex	zincite	2	4.2977		7.0021			
Calcite	$CaCO_3$	rhomb	calcite	6	4.9899		17.064			
Calomel	Hg_2Cl_2	tetr		4	4.478		10.910			
Carbonate-apatite	$Ca_{10}(PO_4)_6CO_3{\cdot}H_2O$	hex	apatite	1	9.436		6.883			
Cassiterite	SnO_2	tetr	rutile	2	4.738		3.188			
Cattierite	CoS_2	cubic	pyrite	4	5.5345					
Celestite	$SrSO_4$	orth	barite	4	8.359	5.352	6.866			
Celsian	$BaAl_2Si_2O_8$	monocl		8	8.627	13.045	14.408		115.20°	
Cerianite	CeO_2	cubic	fluorite	4	5.4110					
Cerussite	$PbCO_3$	orth	aragonite	4	6.152	8.436	5.195			
Cervantite	Sb_2O_4	orth		4	5.424	11.76	4.804			
Chalcanthite	$CuSO_4{\cdot}5H_2O$	tricl		2	6.1045	10.72	5.949	97.57°	107.28°	77.43°
Chalcocite	Cu_2S	orth		96	11.881	27.323	13.491			
Chalcopyrite	$CuFeS_2$	tetr		4	5.2988		10.434			
Chlorapatite	$Ca_5(PO_4)_3Cl$	hex	apatite	2	9.629		6.777			
Chlorargyrite	$AgCl$	cubic	rock salt	4	5.5491					
Chloritoid	$FeAl_4O_2(SiO_4)_2(OH)_4$	monocl		8	9.48	5.48	18.18		101.77°	
Chloromagnesite	$MgCl_2$	rhomb		3	3.632		17.795			
Chondrodite	$2Mg_2SiO_4{\cdot}MgF_2$	monocl		2	7.89	4.743	10.29		109.03°	
Chrysoberyl	$BeAl_2O_4$	orth	olivine	4	5.4756	9.4041	4.4267			
Cinnabar	HgS	hex	cinnabar	3	4.149		9.495			
Claudetite	As_2O_3	monocl		4	5.339	12.984	4.5405		94.27°	
Clausthalite	$PbSe$	cubic	rock salt	4	6.1255					
Clinoenstatite	$MgSiO_3$	monocl		8	9.620	8.825	5.188		108.33°	
Clinoferrosilite	$FeSiO_3$	monocl		8	9.7085	9.0872	5.2284		108.43°	
Clinohumite	$4Mg_2SiO_4{\cdot}MgF_2$	monocl		2	13.68	4.75	10.27		100.83°	
Clinozoisite	$Ca_2Al_3(SiO_4)_3OH$	monocl		2	8.887	5.581	10.14		115.93°	
Cobalt olivine	Co_2SiO_4	orth	olivine	4	4.782	10.301	6.003			
Cobalt oxide	CoO	cubic	rock salt	4	4.260					
Cobalt sulfide	CoS	cubic	sphalerite	4	5.339					
Cobalt titanate	$CoTiO_3$	rhomb	ilmenite	6	5.066		13.918			
Cobalticalcite	$CoCO_3$	rhomb	calcite	6	4.6581		14.958			
Cobaltite	$CoAsS$	cubic	NiSbS	4	5.60					
Coesite	SiO_2	monocl		16	7.152	12.379	7.152		120.00°	
Coffinite	$USiO_4$	tetr	zircon	4	6.995		6.263			
Colemanite	$Ca_2B_6O_{11}{\cdot}5H_2O$	monocl		4	8.743	11.264	6.102		110.12°	
Coloradoite	$HgTe$	cubic	sphalerite	4	6.4600					
Cooperite	PtS	tetr		2	3.4699		6.1098			
Copper	Cu	cubic	face-centered cubic	4	3.6150					
Corundum	Al_2O_3	rhomb	corundum	6	4.7591		12.9894			
Cotunnite	$PbCl_2$	orth		4	4.535	7.62	9.05			
Covellite	CuS	hex		6	3.792		16.34			

Name	Formula	Crystal system	Structure type	Z	a/Å	b/Å	c/Å	α	β	γ
Cristobalite (α)	SiO_2	tetr		4	4.971		6.918			
Cristobalite (β)	SiO_2	cubic		8	7.1382					
Cryolite	Na_3AlF_6	monocl		2	5.40	5.60	7.776		90.18°	
Cubanite	$CuFe_2S_3$	orth		4	6.46	11.12	6.23			
Cummingtonite	$(Mg,Fe,Mn)_7(Si_4O_{11})_2(OH)_2$	monocl	tremolite	2	9.522	18.223	5.332		101.92°	
Cuprite	Cu_2O	cubic		2	4.2696					
Danburite	$CaB_2Si_2O_8$	orth		4	8.04	8.77	7.74			
Datolite	$CaBSiO_4(OH)$	monocl		4	9.62	7.60	4.84		90.15°	
Daubreeite	$FeCr_2S_4$	cubic	spinel	8	9.966					
Diamond	C	cubic	diamond	8	3.5670					
Diaspore	$AlO(OH)$	orth		4	4.401	9.421	2.845			
Dickite	$Al_2Si_2O_5(OH)_4$	monocl		4	5.150	8.940	14.736		103.58°	
Digenite	$Cu_{1.79}S$	cubic	deformed fluorite	4	5.5695					
Diopside	$CaMg(SiO_3)_2$	monocl	diopside	4	9.743	8.923	5.251		105.93°	
Dioptase	$CuSiO_2(OH)_2$	rhomb	phenacite	18	14.61		7.80			
Dolerophanite	$Cu_2O(SO_4)$	monocl		4	8.334	6.312	7.628		108.4°	
Dolomite	$CaMg(CO_3)_2$	rhomb	calcite	3	4.8079		16.010			
Dravite	$NaMg_3Al_6B_3Si_6O_{27}(OH)_4$	rhomb	tourmaline	3	15.942		7.224			
Elbaite	$NaLiAl_{7.67}B_3Si_6O_{27}(OH)_4$	rhomb	tourmaline	3	15.842		7.009			
Enargite	Cu_3AsS_4	orth		2	6.426	7.422	6.144			
Enstatite	$MgSiO_3$	orth		16	8.829	18.22	5.192			
Epidote	$Ca_2Al_2(Al,Fe)OH(SiO_4)_3$	monocl		2	8.89	5.63	10.19		115.40°	
Epsomite	$MgSO_4{\cdot}7H_2O$	orth		4	11.86	11.99	6.858			
Eskolaite	Cr_2O_3	rhomb	corundum	6	4.9607		13.599			
Eucairite	AgCuSe	orth		10	4.105	20.35	6.31			
Euclase	$AlBeSiO_4(OH)$	monocl		4	4.763	14.29	4.618		100.25°	
Famatimite	Cu_3SbS_4	tetr		2	5.384		10.770			
Fayalite	Fe_2SiO_4	orth	olivine	4	4.817	10.477	6.105			
Fe-Cordierite	$Fe_2Al_3(AlSi_5O_{18})$	orth	cordierite	4	9.726	17.065	9.287			
Fe-Gehlenite	$Ca_2Fe_2SiO_7$	tetr	melilite	2	7.54		4.855			
Fe-Indialite	$Fe_2Al_3(AlSi_5O_{18})$	hex	beryl	2	9.860		9.285			
Fe-Leucite	$KFeSi_2O_6$	tetr		16	13.205		13.970			
Fe-Microcline	$KFeSi_3O_8$	tricl		4	8.68	13.10	7.340	90.75°	116.05°	86.23°
Fe-Sanidine	$KFeSi_3O_8$	monocl		4	8.689	13.12	7.319		116.10°	
Fe-Skutterudite	$FeAs_{2.95}$	cubic		8	8.1814					
Ferberite	$FeWO_4$	monocl	wolframite	2	4.732	5.708	4.965		90.00°	
Ferriannite	$KFe_3[FeSi_3O_{10}](OH)_2$	monocl		2	5.430	9.404	10.341		100.07°	
Ferroselite	$FeSe_2$	orth	marcasite	2	4.801	5.778	3.587			
Ferrotremolite	$Ca_2Fe_5[Si_8O_{22}](OH)_2$	monocl	tremolite	2	9.97	18.34	5.30		104.50°	
Fluor-edenite	$NaCa_2Mg_5[AlSi_7O_{22}]F_2$	monocl	tremolite	2	9.847	18.00	5.282		104.83°	
Fluor-humite	$3Mg_2SiO_4{\cdot}MgF_2$	orth		4	10.243	20.72	4.735			
Fluor-norbergite	$Mg_2SiO_4{\cdot}MgF_2$	orth		4	8.727	10.271	4.709			
Fluor-phlogopite	$KMg_3[AlSi_3O_{10}]F_2$	monocl	1M mica	2	5.299	9.188	10.135		99.92°	
Fluor-richterite	$Na_2CaMg_5[Si_8O_{22}]F_2$	monocl	tremolite	2	9.823	17.96	5.268		104.33°	
Fluor-tremolite	$Ca_2Mg_5[Si_8O_{22}]F_2$	monocl	tremolite	2	9.781	18.01	5.267		104.52°	

Name	Formula	Crystal system	Structure type	Z	a/Å	b/Å	c/Å	α	β	γ
Fluorapatite	$Ca_5(PO_4)_3F$	hex	apatite	2	9.3684		6.8841			
Fluorite	CaF_2	cubic	fluorite	4	5.4638					
Forsterite	Mg_2SiO_4	orth	olivine	4	4.758	10.214	5.984			
Frohbergite	$FeTe_2$	orth	marcasite	2	5.265	6.265	3.869			
Gahnite	$ZnAl_2O_4$	cubic	spinel	8	8.0848					
Galaxite	$MnAl_2O_4$	cubic	spinel	8	8.258					
Galena	PbS	cubic	rock salt	4	5.9360					
Gallium oxide	Ga_2O_3	rhomb	corundum	6	4.9793		13.429			
Gehlenite	$Ca_2Al_2SiO_7$	tetr	melilite	2	7.690		5.0675			
Geikielite	$MgTiO_3$	rhomb	ilmenite	6	5.054		13.898			
Gerhardite	$Cu_2(NO_3)(OH)_3$	orth		4	6.075	13.812	5.592			
Gersdorfite	NiAsS	cubic		4	5.693					
Gibbsite	$Al(OH)_3$	monocl		8	9.719	5.0705	8.6412		94.57°	
Glauchroite	$CaMnSiO_4$	orth	olivine	4	4.944	11.19	6.529			
Glaucodot	(Co,Fe)AsS	orth		24	6.64	28.39	5.64			
Glaucophane I	$Na_2Mg_3Al_2[Si_8O_{22}](OH)_2$	monocl	tremolite	2	9.748	17.915	5.273		102.78°	
Glaucophane II	$Na_2Mg_3Al_2[Si_8O_{22}](OH)_2$	monocl	tremolite	2	9.663	17.696	5.277		103.67°	
Goethite	FeO(OH)	orth		4	4.596	9.957	3.021			
Gold	Au	cubic	face-centered cubic	4	4.0786					
Goldmanite	$Ca_3V_2Si_3O_{12}$	cubic	garnet	8	12.070					
Goslarite	$ZnSO_4 \cdot 7H_2O$	orth	epsomite	4	11.779	12.050	6.822			
Graphite	C	hex	graphite	4	2.4612		6.7079			
Greenockite	CdS	hex	zincite	2	4.1354		6.7120			
Greigite	Fe_3S_4	cubic	spinel	8	9.876					
Grossularite	$Ca_3Al_2Si_3O_{12}$	cubic	garnet	8	11.851					
Grunerite	$Fe_7[Si_8O_{22}](OH)_2$	monocl	tremolite	2	9.572	18.44	5.342		101.77°	
Gudmundite	FeSbS	monocl		8	10.00	5.93	6.73		90.00°	
Gypsum	$CaSO_4 \cdot 2H_2O$	monocl		4	5.68	15.18	6.29		113.83°	
Hafnia	HfO_2	monocl	baddeleyite	4	5.1156	5.1722	5.2948		99.18°	
Halite	NaCl	cubic	rock salt	4	5.6402					
Hambergite	$Be_2(OH,F)BO_3$	orth		8	9.755	12.201	4.426			
Hardystonite	$Ca_2ZnSi_2O_7$	tetr	melilite	2	7.87		5.01			
Hauerite	MnS_2	cubic	pyrite	4	6.1014					
Hausmannite	Mn_3O_4	tetr		8	8.136		9.422			
Hawleyite	CdS	cubic	sphalerite	4	5.833					
Heazelwoodite	Ni_3S_2	rhomb		3	5.746		7.134			
Hedenbergite	$CaFe(SiO_3)_2$	monocl	diopside	4	9.854	9.024	5.263		104.23°	
Hematite	Fe_2O_3	rhomb	corundum	6	5.0329				13.749	
Hemimorphite	$Zn_4(OH)_2Si_2O_7 \cdot H_2O$	orth		2	8.370	10.719	5.120			
Hercynite	$Fe(AlO_2)_2$	cubic	spinel	8	8.150					
Herzenbergite	SnS	orth	germanium sulfide	4	4.328	11.190	3.978			
Hessite	Ag_2Te	monocl		4	8.13	4.48	8.09		111.9°	
Hexahydrite	$MgSO_4 \cdot 6H_2O$	monocl		8	10.110	7.212	24.41		98.30°	
High albite (Analbite)	$NaAlSi_3O_8$	tricl		4	8.160	12.870	7.106	93.54°	116.36°	90.19°
High argentite	Ag_2S	cubic		4	6.269					

Name	Formula	Crystal system	Structure type	Z	a/Å	b/Å	c/Å	α	β	γ
High bornite	Cu_5FeS_4	cubic		1	5.50					
High carnegeite	$NaAlSiO_4$	cubic		4	7.325					
High chalcocite	Cu_2S	hex		2	3.961		6.722			
High clinoenstatite	$MgSiO_3$	tricl		8	10.000	8.934	5.170	88.27°	70.03°	91.01°
High digenite	Cu_2S	cubic		4	5.725					
High germania	GeO_2	hex	α–quartz	3	4.987		5.652			
High leucite	$KAlSi_2O_6$	cubic		16	13.43					
High naumanite	Ag_2Se	cubic		2	4.993					
High sanidine	$KAlSi_3O_8$	monocl		4	8.615	13.031	7.177		115.98°	
Huebnerite	$MnWO_4$	monocl	wolframite	2	4.834	5.758	4.999		91.18°	
Huntite	$Mg_3Ca(CO_3)_4$	rhomb	calcite	3	9.498		7.816			
Hydroxylapatite	$Ca_5(PO_4)_3OH$	hex	apatite	2	9.418		6.883			
Ice	H_2O	hex		4	4.5212		7.3666			
Ilmenite	$FeTiO_3$	rhomb	ilmenite	6	5.093		14.055			
Indialite (Cordierite)	$Mg_2Al_3(AlSi_5O_{18})$	hex	beryl	2	9.7698		9.3517			
Iodargyrite	AgI	hex	zincite	2	4.5955		7.5005			
Iron (α)	Fe	cubic	body-centered cubic	2	2.8664					
Jacobsite	$MnFe_2O_4$	cubic	spinel	8	8.499					
Jadeite	$NaAl(SiO_3)_2$	monocl	diopside	4	9.409	8.564	5.220		107.50°	
Jalpaite	$Ag_{1.55}Cu_{0.45}S$	tetr		16	8.673		11.756			
Johannsenite	$CaMn(SiO_3)_2$	monocl	diopside	4	9.83	9.04	5.27		105.00°	
Kaliophilite	$KAlSiO_4$	hex		54	26.930		8.522			
Kalsilite	$KAlSiO_4$	hex		2	5.1597		8.7032			
Kaolinite	$Al_2Si_2O_5(OH)_4$	tricl		2	5.155	8.959	7.407	91.68°	104.87°	89.93°
Karelianite	V_2O_3	rhomb	corundum	6	4.952		14.002			
Keatite	SiO_2	tetr		12	7.456		8.604			
Kernite	$Na_2B_4O_7 \cdot 4H_2O$	monocl		4	7.022	9.151	15.676		108.83°	
Kerschsteinite	$CaFeSiO_4$	orth	olivine	4	4.886	11.146	6.434			
Klockmannite	CuSe	hex	deformed covellite	78	14.206		17.25			
Knebelite	$MnFeSiO_4$	orth	olivine	4	4.854	10.602	6.162			
Kyanite	Al_2OSiO_4	tricl		4	7.123	7.848	5.564	89.92°	101.25°	105.97°
Larnite	Ca_2SiO_4	monocl		4	5.48	6.76	9.28		94.55°	
Laurite	RuS_2	cubic	pyrite	4	5.60					
Lawrencite	$FeCl_2$	rhomb		3	3.593		17.58			
Lawsonite	$CaAl_2Si_2O_7(OH)_2 \cdot H_2O$	orth		4	8.787	5.836	13.123			
Lead	Pb	cubic	face-centered cubic	4	4.9505					
Leonhardtite	$MgSO_4 \cdot 4H_2O$	monocl		4	5.922	13.604	7.905		90.85°	
Lepidocrocite	FeO(OH)	orth		4	3.868	12.525	3.066			
Lepidolite	$K_2Al_3Li_2AlSi_7O_{20}(OH)_4$	monocl	2M2 mica	2	9.2	5.3	20.0		98.00°	
Leucite	$KAlSi_2O_6$	tetr		16	13.074		13.738			
Lime	CaO	cubic	rock salt	4	4.8108					
Lime olivine	Ca_2SiO_4	orth	olivine	4	5.091	11.371	6.782			
Linnaeite	Co_3S_4	cubic	spinel	8	9.401					
Litharge	PbO	tetr		2	3.9759		5.023			
Loellingite	$FeAs_2$	orth	marcasite	2	5.300	5.981	2.882			

Name	Formula	Crystal system	Structure type	Z	a/Å	b/Å	c/Å	α	β	γ
Low albite	$NaAlSi_3O_8$	tricl		4	8.139	12.788	7.160	94.27°	116.57°	87.68°
Low bornite	Cu_5FeS_4	tetr		16	10.94		21.88			
Low cordierite	$Mg_2Al_3(AlSi_5O_{18})$	orth		4	9.721	17.062	9.339			
Low germania	GeO_2	tetr	rutile	2	4.3963		2.8626			
Low nepheline	$NaAlSiO_4$	hex		8	9.986		8.330			
Luzonite	Cu_3AsS_4	tetr		2	5.289		10.440			
Mackinawite	FeS	tetr		2	3.675		5.030			
Magnesioriebeckite	$Na_2Mg_3Fe_2[Si_8O_{22}](OH)_2$	monocl	tremolite	2	9.733	17.946	5.299		103.30°	
Magnesite	$MgCO_3$	rhomb	calcite	6	4.6330		15.016			
Magnetite	Fe_3O_4	cubic	spinel	8	8.3940					
Malachite	$Cu_2(OH)_2CO_3$	monocl		4	9.502	11.974	3.240		98.75°	
Maldonite	Au_2Bi	cubic		8	7.958					
Manganese sulfide (γ)	MnS	hex	zincite	2	3.976		6.432			
Manganese sulfide (β)	MnS	cubic	sphalerite	4	5.611					
Manganosite	MnO	cubic	rock salt	4	4.4448					
Marcasite	FeS_2	orth	marcasite	2	4.443	5.423	3.3876			
Margarite	$CaAl_2[AlSi_2O_{10}](OH)_2$	monocl	2M mica	4	5.13	8.92	19.50		95.00°	
Marialite	$Na_4Al_3Si_9O_{24}Cl$	tetr		2	12.064		7.514			
Marshite	CuI	cubic	sphalerite	4	6.0507					
Mascagnite	$(NH_4)_2SO_4$	orth	arcanite	4	7.782	5.993	10.636			
Massicot	PbO	orth		4	5.489	4.755	5.891			
Matlockite	PbClF	tetr		2	4.106		7.23			
Maucherite	$Ni_{11}As_8$	tetr		4	6.870		21.81			
Meionite	$Ca_4Al_6Si_6O_{24}CO_3$	tetr		2	12.174		7.652			
Melanophlogite	SiO_2	cubic	clathrate type	46	13.402					
Melanterite	$FeSO_4{\cdot}7H_2O$	monocl		4	14.072	6.503	11.041		105.57°	
Melonite	$NiTe_2$	hex	cadmium iodide	1	3.869		5.308			
Metacinnabar	HgS	cubic	sphalerite	4	5.8517					
Miargyrite	$AgSbS_2$	monocl		8	12.862	4.111	13.220		98.63°	
Microcline	$KAlSi_3O_8$	tricl		4	8.582	12.964	7.222	90.62°	115.92°	87.68°
Miersite	AgI	cubic	sphalerite	4	6.4963					
Millerite	NiS	rhomb		9	9.616		3.152			
Minium	Pb_3O_4	tetr		4	8.815		6.565			
Minnesotaite	$Fe_3Si_4O_{10}(OH)_2$	monocl		4	5.4	9.42	19.4		100.00°	
Mirabilite	$Na_2SO_4{\cdot}10H_2O$	monocl		4	11.51	10.38	12.83		107.75°	
Mn-Indialite	$Mn_2Al_3(AlSi_5O_{18})$	hex	beryl	2	9.925		9.297			
Molybdenite	MoS_2	hex	molybdenite	2	3.1604		12.295			
Molybdenum	Mo	cubic		2	3.1653					
Molybdite	MoO_3	orth		4	3.962	13.858	3.697			
Monteponite	CdO	cubic	rock salt	4	4.6953					
Monticellite	$CaMgSiO_4$	orth	olivine	4	4.827	11.084	6.376			
Montroydite	HgO	orth		4	6.608	5.518	3.519			
Mullite (2:1)	$2Al_2O_3{\cdot}SiO_2$	orth		6	7.5788	7.6909	2.8883			
Mullite (3:2)	$3Al_2O_3{\cdot}2SiO_2$	orth		3	7.557	7.6876	2.8842			
Muscovite	$KAl_2AlSi_3O_{10}(OH)_2$	monocl	2M2 mica	4	5.203	8.995	20.030		94.47°	

Name	Formula	Crystal system	Structure type	Z	a/Å	b/Å	c/Å	α	β	γ
Nacrite	$Al_2Si_2O_5(OH)_4$	monocl		4	8.909	5.146	15.697		113.70°	
Nantokite	CuCl	cubic	sphalerite	4	5.416					
Natroalunite	$NaAl_3(SO_4)_2(OH)_6$	rhomb		3	6.974		16.69			
Natrolite	$Na_2Al_2Si_3O_{10}\cdot 2H_2O$	orth		8	18.30	18.63	6.60			
Neighborite	$NaMgF_3$	orth	perovskite	4	5.363	7.676	5.503			
Ni-Skutterudite	$NiAs_{2.95}$	cubic		8	8.3300					
Niccolite	NiAs	hex	niccolite	2	3.618		5.034			
Nickel	Ni	cubic	face-centered cubic	4	3.5238					
Nickel carbonate	$NiCO_3$	rhomb	calcite	6	4.5975		14.723			
Nickel olivine	Ni_2SiO_4	orth	olivine	4	4.727	10.121	5.915			
Nickel selenide	$NiSe_2$	cubic	pyrite	4	5.9604					
Niter	KNO_3	orth	aragonite	4	6.431	9.164	5.414			
Norsethite	$BaMg(CO_3)_2$	rhomb	calcite	3	5.020		16.75			
Oldhamite	CaS	cubic	rock salt	4	5.689					
Orpiment	As_2S_3	monocl		4	11.49	9.59	4.25		90.45°	
Orthoclase	$KAlSi_3O_8$	monocl		4	8.562	12.996	7.193		116.02°	
Orthoferrosilite	$FeSiO_3$	orth	enstatite	16	9.080	18.431	5.238			
Otavite	$CdCO_3$	rhomb	calcite	6	4.9204		16.298			
Paracelsian	$BaAl_2Si_2O_8$	monocl		4	8.58	9.583	9.08		90.00°	
Paragonite	$NaAl_2AlSi_3O_{10}(OH)_2$	monocl	2M1 mica	4	5.13	8.89	19.32		95.17°	
Pararammelsbergite	$NiAs_2$	orth		8	5.75	5.82	11.428			
Paratellurite	TeO_2	tetr		4	4.810		7.613			
Parawollastonite	$CaSiO_3$	monocl		12	15.417	7.321	7.066		95.40°	
Pectolite	$Ca_2NaH(SiO_3)_3$	tricl		2	7.99	7.04	7.02	90.05°	95.27°	102.47°
Pentlandite	$Fe_{5.25}Ni_{3.75}S_8$	cubic		4	10.196					
Pentlandite	$Fe_{4.75}Ni_{5.25}S_8$	cubic		4	10.095					
Periclase	MgO	cubic	rock salt	4	4.2117					
Perovskite	$CaTiO_3$	orth	perovskite	4	5.3670	7.6438	5.4439			
Petalite	$LiAlSi_4O_{10}$	monocl		2	11.32	5.14	7.62		105.90°	
Petzite	Ag_3AuTe_2	cubic		8	10.38					
Phenacite	Be_2SiO_4	rhomb	phenacite	18	12.472		8.252			
Phlogopite	$KMg_3AlSi_3O_{10}(OH)_2$	monocl	1M mica	2	5.326	9.210	10.311		100.17°	
Picrochromite	$MgCr_2O_4$	cubic	spinel	8	8.333					
Piemontite	$Ca_2Al_{1.5}Mn_{1.5}(SiO_4)_3OH$	monocl		2	8.95	5.70	9.41		115.70°	
Platinum	Pt	cubic	face-centered cubic	4	3.9231					
Polymidite	Ni_3S_4	cubic	spinel	8	9.480					
Portlandite	$Ca(OH)_2$	hex	cadmium iodide	1	3.5933		4.9086			
Powellite	$CaMoO_4$	tetr	scheelite	4	5.226		11.43			
Protoenstatite	$MgSiO_3$	orth		8	9.25	8.74	5.32			
Proustite	Ag_3AsS_3	rhomb		6	10.816		8.6948			
Pseudowollastonite	$CaSiO_3$	tricl		24	6.90	11.78	19.65	90.00°	90.80°	90.00°
Pyrargyrite	Ag_3SbS_3	rhomb		6	11.052		8.7177			
Pyrite	FeS_2	cubic	pyrite	4	5.4175					
Pyrolusite	MnO_2	tetr	rutile	2	4.388		2.865			
Pyrope	$Mg_3Al_2Si_3O_{12}$	cubic	garnet	8	11.459					

Name	Formula	Crystal system	Structure type	Z	a/Å	b/Å	c/Å	α	β	γ
Pyrophanite	$MnTiO_3$	rhomb	ilmenite	6	5.155		14.18			
Pyrophyllite	$Al_2Si_4O_{10}(OH)_2$	monocl	2M1 mica	4	5.14	8.90	18.55		99.92°	
Pyroxmangite	$MnFe(SiO_3)_2$	tricl		7	7.56	17.45	6.67	84.00°	94.30°	113.70°
Pyrrhotite	$Fe_{0.980}S$	hex	defect niccolite	2	3.446		5.848			
Pyrrhotite	$Fe_{0.885}S$	hex	defect niccolite	2	3.440		5.709			
Quartz (α)	SiO_2	hex		3	4.9136		5.4051			
Quartz (β)	SiO_2	hex		3	4.999		5.4592			
Rammelsbergite	$NiAs_2$	orth	marcasite	2	4.757	5.797	3.542			
Realgar	AsS	monocl		16	9.29	13.53	6.57		106.55°	
Retgersite	$NiSO_4 \cdot 4H_2O$	tetr		4	6.782		18.28			
Rhodochrosite	$MnCO_3$	rhomb	calcite	6	4.7771		15.664			
Rhodonite	$MnSiO_3$	tricl		10	7.682	11.818	6.707	92.36°	93.95°	105.66°
Riebeckite	$Na_2Fe_5FSi_8O_{22}(OH)_2$	monocl	tremolite	2	9.729	18.065	5.334		103.31°	
Rutile	TiO_2	tetr		2	4.5937		2.9618			
Safflorite	$Co_{0.5}Fe_{0.5}As_2$	orth	marcasite	2	5.231	5.953	2.962			
Sanmartinite	$ZnWO_4$	monocl	wolframite	2	4.691	5.720	4.925		89.36°	
Sapphirine	$Mg_2Al_4O_6SiO_4$	monocl		8	9.96	28.60	9.85		110.5°	
Scacchite	$MnCl_2$	rhomb		3	3.711		17.59			
Scheelite	$CaWO_4$	tetr	scheelite	4	5.242		11.372			
Schorl	$NaFe_3Al_6B_3Si_6O_{27}(OH)_4$	rhomb	tourmaline	3	16.032		7.149			
Selenium	Se	hex		3	4.3642		4.9588			
Selenolite	SeO_2	tetr		8	8.35		5.05			
Sellaite	MgF_2	tetr	rutile	2	4.621		3.050			
Senarmontite	Sb_2O_3	cubic	arsenic trioxide	16	11.152					
Shandite	$Ni_3Pb_2S_2$	rhomb		3	5.576		13.658			
Shortite	$Na_2Ca_2(CO_3)_3$	orth		2	4.961	11.03	7.12			
Siderite	$FeCO_3$	rhomb	calcite	6	4.6887		15.373			
Silicon	Si	cubic	diamond	8	5.4305					
Sillimanite	Al_2OSiO_4	orth		4	7.4843	7.6730	5.7711			
Silver	Ag	cubic	face-centered cubic	4	4.0862					
Silver telluride I	Ag_2Te	cubic		2	5.29					
Silver telluride II	Ag_2Te	cubic		4	6.585					
Smithsonite	$ZnCO_3$	rhomb	calcite	6	4.6528		15.025			
Soda niter	$NaNO_3$	rhomb	calcite	6	5.0696		16.829			
Sodium melilite	$NaCaAlSi_2O_7$	tetr	melilite	2	8.511		4.809			
Sperrylite	$PtAs_2$	cubic	pyrite	4	5.968					
Spessartite	$Mn_3Al_2Si_3O_{12}$	cubic	garnet	8	11.621					
Sphalerite	ZnS	cubic	sphalerite	4	5.4093					
Sphene	$CaTiSiO_5$	monocl		4	7.07	8.72	6.56		113.95°	
Spinel	$MgAl_2O_4$	cubic	spinel	8	8.080					
Spodumene	$LiAl(SiO_3)_2$	monocl	diopside	4	9.451	8.387	5.208		110.07°	
Spodumene (β)	$LiAl(SiO_3)_2$	tetr		4	7.5332		9.1540			
Staurolite	$Fe_2Al_9Si_4O_{22}(OH)_2$	monocl		2	7.90	16.65	5.63		90.00°	
Sternbergite	$AgFe_2S_3$	orth		8	11.60	12.675	6.63			
Stibnite	Sb_2S_3	orth	stibnite	4	11.229	11.310	3.8389			

Name	Formula	Crystal system	Structure type	Z	a/Å	b/Å	c/Å	α	β	γ
Stilleite	ZnSe	cubic	sphalerite	4	5.6685					
Stishovite	SiO_2	tetr	rutile	2	4.1790		2.6649			
Stolzite	$PbWO_4$	tetr	scheelite	4	5.4616		12.046			
Stromeyerite	$Ag_{0.93}Cu_{1.07}S$	orth		4	4.066	6.628	7.972			
Strontianite	$SrCO_3$	orth	aragonite	4	6.029	8.414	5.107			
Sulfur (monoclinic)	S	monocl	S8 ring molecules	48	11.04	10.98	10.92		96.73°	
Sulfur (orthorhombic)	S	orth	S8 ring molecules	128	10.4646	12.8660	24.4860			
Sulfur (rhombohedral)	S	rhomb	S6 ring molecules	18	10.818		4.280			
Sylvite	KCl	cubic	rock salt	4	6.2931					
Syngenite	$K_2Ca(SO_4)_2{\cdot}H_2O$	monocl		2	9.775	7.156	6.251		104.00°	
Synthetic anorthite	$CaAl_2Si_2O_8$	hex		2	5.10		14.72			
Synthetic anorthite	$CaAl_2Si_2O_8$	orth		2	8.22	8.60	4.83			
Talc	$Mg_3Si_4O_{10}(OH)_2$	monocl	2M1 mica	4	5.287	9.158	18.95		99.50°	
Tantalum	Ta	cubic	tungsten	2	3.3058					
Teallite	$PbSnS_2$	orth	germanium sulfide	2	4.266	11.419	4.090			
Tellurite	TeO_2	orth	tellurite	8	5.607	12.034	5.463			
Tellurium	Te	hex	selenium	3	4.4570		5.9290			
Tellurobismuthite	Bi_2Te_3	rhomb		3	4.3835		30.487			
Tennantite	$Cu_{12}As_4S_{13}$	cubic	tetrahedrite	2	10.190					
Tenorite	CuO	monocl		4	4.684	3.425	5.129		99.47°	
Tephroite	Mn_2SiO_4	orth	olivine	4	4.871	10.636	6.232			
Tetrahedrite	$Cu_{12}Sb_4S_{13}$	cubic	tetrahedrite	2	10.327					
Thenardite	Na_2SO_4	orth	thenardite	8	5.863	12.304	9.821			
Thorianite	ThO_2	cubic	fluorite	4	5.5952					
Thorite	$ThSiO_4$	tetr	zircon	4	7.143		6.327			
Tiemannite	HgSe	cubic	sphalerite	4	6.0853					
Tin	Sn	tetr		4	5.8315		3.1813			
Titanium	Ti	hex		2	2.953		4.729			
Titanium(III) oxide	Ti_2O_3	rhomb	corundum	6	5.149		13.642			
Topaz	$Al_2(SiO_4)(OH)$	orth		4	8.394	8.792	4.649			
Tremolite	$Ca_2Mg_5Si_8O_{22}(OH)_2$	monocl	tremolite	2	9.840	18.052	5.275		104.70°	
Trevorite	$NiFe_2O_4$	cubic	spinel	8	8.339					
Tridymite (β)	SiO_2	hex		4	5.0463		8.2563			
Trogtalite	$CoSe_2$	cubic	pyrite	4	5.8588					
Troilite	FeS	hex	niccolite	2	3.446		5.877			
Tschermakite	$CaAl_2SiO_6$	monocl	diopside	4	9.615	8.661	5.272		106.12°	
Tungsten	W	cubic		2	3.1653					
Tungstenite	WS_2	hex	molybdenite	2	3.154		12.362			
Turquois	$CuAl_6(PO_4)_4(OH)_8{\cdot}4H_2O$	tricl		1	7.424	7.629	9.910	68.61°	69.71°	65.08°
Umangite	Cu_3Se_2	tetr		2	6.402		4.276			
Uraninite	UO_2	cubic	fluorite	4	5.4682					
Ureyite	$NaCr(SiO_3)_2$	monocl	diopside	4	9.550	8.712	5.273		107.44°	
Uvarovite	$Ca_3Cr_2Si_3O_{12}$	cubic	garnet	8	11.999					
Uvite	$CaMg_4Al_5B_3Si_6O_{27}(OH)_4$	rhomb	tourmaline	3	15.86		7.19			
Vaesite	NiS_2	cubic	pyrite	4	5.6873					

Name	Formula	Crystal system	Structure type	Z	a/Å	b/Å	c/Å	α	β	γ
Valentinite	Sb_2O_3	orth	antimony trioxide	4	4.914	12.468	5.421			
Vanthoffite	$MgSO_4 \cdot 3Na_2SO_4$	monocl		2	9.797	9.217	8.199		113.50°	
Vaterite	$CaCO_3$	hex		6	7.135		8.524			
Villiaumite	NaF	cubic	rock salt	4	4.6342					
Violarite	$FeNi_2S_4$	cubic	spinel	8	9.464					
Willemite	Zn_2SiO_4	rhomb	phenacite	18	13.94		9.309			
Witherite	$BaCO_3$	orth	aragonite	4	6.430	8.904	5.314			
Wolframite	$Fe_{0.5}Mn_{0.5}WO_4$	monocl	wolframite	2	4.782	5.731	4.982		90.57°	
Wollastonite	$CaSiO_3$	tricl		6	7.94	7.32	7.07	90.03°	95.37°	103.43°
Wulfenite	$PbMoO_4$	tetr	scheelite	4	5.435		12.110			
Wurtzite	ZnS	hex	zincite	2	3.8230		6.2565			
Wustite	$Fe_{0.953}O$	cubic	defect rock salt	4	4.3088					
Xenotime	YPO_4	tetr	zircon	4	6.885		5.982			
Zinc	Zn	hex	hexagonal close pack	2	2.665		4.947			
Zinc telluride	ZnTe	cubic	sphalerite	4	6.1020					
Zincite	ZnO	hex	zincite	2	3.2495		5.2069			
Zinkosite	$ZnSO_4$	orth	barite	4	8.588	6.740	4.770			
Zircon	$ZrSiO_4$	tetr	zircon	4	6.604		5.979			
Zoisite	$Ca_2Al_3(SiO_4)_3OH$	orth		4	16.15	5.581	10.06			

Section 5
Thermochemistry, Electrochemistry, and Kinetics

CODATA Key Values for Thermodynamics 5-1
Standard Thermodynamic Properties of Chemical Substances 5-4
Thermodynamic Properties as a Function of Temperature 5-61
Thermodynamic Properties of Aqueous Systems 5-85
Heat of Combustion 5-89
Electrical Conductivity of Water 5-90
Standard KCl Solutions for Calibrating Conductivity Cells 5-91
Molar Conductivity of Aqueous HF, HCl, HBr, and HI 5-92
Equivalent Conductivity of Electrolytes in Aqueous Solution 5-94
Ionic Conductivity and Diffusion at Infinite Dilution 5-95
Activity Coefficients of Acids, Bases, and Salts 5-97
Mean Activity Coefficients of Electrolytes as a Function of Concentration 5-99
Enthalpy of Dilution of Acids 5-104
Enthalpy of Solution of Electrolytes 5-105
Chemical Kinetic Data for Stratospheric Modeling 5-106

CODATA KEY VALUES FOR THERMODYNAMICS

The Committee on Data for Science and Technology (CODATA) has conducted a project to establish internationally agreed values for the thermodynamic properties of key chemical substances. This table presents the final results of the project. Use of these recommended, internally consistent values is encouraged in the analysis of thermodynamic measurements, data reduction, and preparation of other thermodynamic tables.

The table includes the standard enthalpy of formation at 298.15 K, the entropy at 298.15 K, and the quantity $H°$ (298.15 K)–$H°$ (0). A value of 0 in the $\Delta_f H°$ column for an element indicates the reference state for that element. The standard state pressure is 100000 Pa (1 bar). See the reference for information on the dependence of gas-phase entropy on the choice of standard state pressure.

Substances are listed in alphabetical order of their chemical formulas when written in the most common form.

The table is reprinted with permission of CODATA.

REFERENCE

Cox, J. D., Wagman, D. D., and Medvedev, V. A., *CODATA Key Values for Thermodynamics,* Hemisphere Publishing Corp., New York, 1989.

Substance	State	$\Delta_f H°$ (298.15 K) kJ·mol^{-1}	$S°$ (298.15 K) J·K^{-1}·mol^{-1}	$H°$ (298.15 K)–$H°$ (0) kJ·mol^{-1}
Ag	cr	0	42.55 ± 0.20	5.745 ± 0.020
Ag	g	284.9 ± 0.8	172.997 ± 0.004	6.197 ± 0.001
Ag^+	aq	105.79 ± 0.08	73.45 ± 0.40	
AgCl	cr	-127.01 ± 0.05	96.25 ± 0.20	12.033 ± 0.020
Al	cr	0	28.30 ± 0.10	4.540 ± 0.020
Al	g	330.0 ± 4.0	164.554 ± 0.004	6.919 ± 0.001
Al^{+3}	aq	-538.4 ± 1.5	-325 ± 10	
AlF_3	cr	-1510.4 ± 1.3	66.5 ± 0.5	11.62 ± 0.04
Al_2O_3	cr, corundum	-1675.7 ± 1.3	50.92 ± 0.10	10.016 ± 0.020
Ar	g	0	154.846 ± 0.003	6.197 ± 0.001
B	cr, rhombic	0	5.90 ± 0.08	1.222 ± 0.008
B	g	565 ± 5	153.436 ± 0.015	6.316 ± 0.002
BF_3	g	-1136.0 ± 0.8	254.42 ± 0.20	11.650 ± 0.020
B_2O_3	cr	-1273.5 ± 1.4	53.97 ± 0.30	9.301 ± 0.040
Be	cr	0	9.50 ± 0.08	1.950 ± 0.020
Be	g	324 ± 5	136.275 ± 0.003	6.197 ± 0.001
BeO	cr	-609.4 ± 2.5	13.77 ± 0.04	2.837 ± 0.008
Br	g	111.87 ± 0.12	175.018 ± 0.004	6.197 ± 0.001
Br^-	aq	-121.41 ± 0.15	82.55 ± 0.20	
Br_2	l	0	152.21 ± 0.30	24.52 ± 0.01
Br_2	g	30.91 ± 0.11	245.468 ± 0.005	9.725 ± 0.001
C	cr, graphite	0	5.74 ± 0.10	1.050 ± 0.020
C	g	716.68 ± 0.45	158.100 ± 0.003	6.536 ± 0.001
CO	g	-110.53 ± 0.17	197.660 ± 0.004	8.671 ± 0.001
CO_2	g	-393.51 ± 0.13	213.785 ± 0.010	9.365 ± 0.003
CO_2	aq, undissoc.	-413.26 ± 0.20	119.36 ± 0.60	
CO_3^{-2}	aq	-675.23 ± 0.25	-50.0 ± 1.0	
Ca	cr	0	41.59 ± 0.40	5.736 ± 0.040
Ca	g	177.8 ± 0.8	154.887 ± 0.004	6.197 ± 0.001
Ca^{+2}	aq	-543.0 ± 1.0	-56.2 ± 1.0	
CaO	cr	-634.92 ± 0.90	38.1 ± 0.4	6.75 ± 0.06
Cd	cr	0	51.80 ± 0.15	6.247 ± 0.015
Cd	g	111.80 ± 0.20	167.749 ± 0.004	6.197 ± 0.001
Cd^{+2}	aq	-75.92 ± 0.60	-72.8 ± 1.5	
CdO	cr	-258.35 ± 0.40	54.8 ± 1.5	8.41 ± 0.08
$CdSO_4 \cdot 8/3H_2O$	cr	-1729.30 ± 0.80	229.65 ± 0.40	35.56 ± 0.04
Cl	g	121.301 ± 0.008	165.190 ± 0.004	6.272 ± 0.001
Cl^-	aq	-167.080 ± 0.10	56.60 ± 0.20	
ClO_4^-	aq	-128.10 ± 0.40	184.0 ± 1.5	
Cl_2	g	0	223.081 ± 0.010	9.181 ± 0.001
Cs	cr	0	85.23 ± 0.40	7.711 ± 0.020
Cs	g	76.5 ± 1.0	175.601 ± 0.003	6.197 ± 0.001
Cs^+	aq	-258.00 ± 0.50	132.1 ± 0.5	

Substance	State	$\Delta_f H°$ (298.15 K) kJ·mol^{-1}	$S°$ (298.15 K) J·K^{-1}·mol^{-1}	$H°$ (298.15 K)–$H°$ (0) kJ·mol^{-1}
Cu	cr	0	33.15 ± 0.08	5.004 ± 0.008
Cu	g	337.4 ± 1.2	166.398 ± 0.004	6.197 ± 0.001
Cu^{+2}	aq	64.9 ± 1.0	-98 ± 4	
$CuSO_4$	cr	-771.4 ± 1.2	109.2 ± 0.4	16.86 ± 0.08
F	g	79.38 ± 0.30	158.751 ± 0.004	6.518 ± 0.001
F^-	aq	-335.35 ± 0.65	-13.8 ± 0.8	
F_2	g	0	202.791 ± 0.005	8.825 ± 0.001
Ge	cr	0	31.09 ± 0.15	4.636 ± 0.020
Ge	g	372 ± 3	167.904 ± 0.005	7.398 ± 0.001
GeF_4	g	-1190.20 ± 0.50	301.9 ± 1.0	17.29 ± 0.10
GeO_2	cr, tetragonal	-580.0 ± 1.0	39.71 ± 0.15	7.230 ± 0.020
H	g	217.998 ± 0.006	114.717 ± 0.002	6.197 ± 0.001
H^+	aq	0	0	
HBr	g	-36.29 ± 0.16	198.700 ± 0.004	8.648 ± 0.001
HCO_3^-	aq	-689.93 ± 0.20	98.4 ± 0.5	
HCl	g	-92.31 ± 0.10	186.902 ± 0.005	8.640 ± 0.001
HF	g	-273.30 ± 0.70	173.779 ± 0.003	8.599 ± 0.001
HI	g	26.50 ± 0.10	206.590 ± 0.004	8.657 ± 0.001
HPO_4^{-2}	aq	-1299.0 ± 1.5	-33.5 ± 1.5	
HS^-	aq	-16.3 ± 1.5	67 ± 5	
HSO_4^-	aq	-886.9 ± 1.0	131.7 ± 3.0	
H_2	g	0	130.680 ± 0.003	8.468 ± 0.001
H_2O	l	-285.830 ± 0.040	69.95 ± 0.03	13.273 ± 0.020
H_2O	g	-241.826 ± 0.040	188.835 ± 0.010	9.905 ± 0.005
$H_2PO_4^-$	aq	-1302.6 ± 1.5	92.5 ± 1.5	
H_2S	g	-20.6 ± 0.5	205.81 ± 0.05	9.957 ± 0.010
H_2S	aq, undissoc.	-38.6 ± 1.5	126 ± 5	
H_3BO_3	cr	-1094.8 ± 0.8	89.95 ± 0.60	13.52 ± 0.04
H_3BO_3	aq, undissoc.	-1072.8 ± 0.8	162.4 ± 0.6	
He	g	0	126.153 ± 0.002	6.197 ± 0.001
Hg	l	0	75.90 ± 0.12	9.342 ± 0.008
Hg	g	61.38 ± 0.04	174.971 ± 0.005	6.197 ± 0.001
Hg^{+2}	aq	170.21 ± 0.20	-36.19 ± 0.80	
HgO	cr, red	-90.79 ± 0.12	70.25 ± 0.30	9.117 ± 0.025
Hg_2^{+2}	aq	166.87 ± 0.50	65.74 ± 0.80	
Hg_2Cl_2	cr	-265.37 ± 0.40	191.6 ± 0.8	23.35 ± 0.20
Hg_2SO_4	cr	-743.09 ± 0.40	200.70 ± 0.20	26.070 ± 0.030
I	g	106.76 ± 0.04	180.787 ± 0.004	6.197 ± 0.001
I^-	aq	-56.78 ± 0.05	106.45 ± 0.30	
I_2	cr	0	116.14 ± 0.30	13.196 ± 0.040
I_2	g	62.42 ± 0.08	260.687 ± 0.005	10.116 ± 0.001
K	cr	0	64.68 ± 0.20	7.088 ± 0.020
K	g	89.0 ± 0.8	160.341 ± 0.003	6.197 ± 0.001
K^+	aq	-252.14 ± 0.08	101.20 ± 0.20	
Kr	g	0	164.085 ± 0.003	6.197 ± 0.001
Li	cr	0	29.12 ± 0.20	4.632 ± 0.040
Li	g	159.3 ± 1.0	138.782 ± 0.010	6.197 ± 0.001
Li^+	aq	-278.47 ± 0.08	12.24 ± 0.15	
Mg	cr	0	32.67 ± 0.10	4.998 ± 0.030
Mg	g	147.1 ± 0.8	148.648 ± 0.003	6.197 ± 0.001
Mg^{+2}	aq	-467.0 ± 0.6	-137 ± 4	
MgF_2	cr	-1124.2 ± 1.2	57.2 ± 0.5	9.91 ± 0.06
MgO	cr	-601.60 ± 0.30	26.95 ± 0.15	5.160 ± 0.020
N	g	472.68 ± 0.40	153.301 ± 0.003	6.197 ± 0.001
NH_3	g	-45.94 ± 0.35	192.77 ± 0.05	10.043 ± 0.010
NH_4^+	aq	-133.26 ± 0.25	111.17 ± 0.40	
NO_3^-	aq	-206.85 ± 0.40	146.70 ± 0.40	

CODATA KEY VALUES FOR THERMODYNAMICS (continued)

Substance	State	$\Delta_f H°$ (298.15 K) kJ·mol^{-1}	$S°$ (298.15 K) J·K^{-1}·mol^{-1}	$H°$ (298.15 K)–$H°$ (0) kJ·mol^{-1}
N_2	g	0	191.609 ± 0.004	8.670 ± 0.001
Na	cr	0	51.30 ± 0.20	6.460 ± 0.020
Na	g	107.5 ± 0.7	153.718 ± 0.003	6.197 ± 0.001
Na^+	aq	-240.34 ± 0.06	58.45 ± 0.15	
Ne	g	0	146.328 ± 0.003	6.197 ± 0.001
O	g	249.18 ± 0.10	161.059 ± 0.003	6.725 ± 0.001
OH^-	aq	-230.015 ± 0.040	-10.90 ± 0.20	
O_2	g	0	205.152 ± 0.005	8.680 ± 0.002
P	cr, white	0	41.09 ± 0.25	5.360 ± 0.015
P	g	316.5 ± 1.0	163.199 ± 0.003	6.197 ± 0.001
P_2	g	144.0 ± 2.0	218.123 ± 0.004	8.904 ± 0.001
P_4	g	58.9 ± 0.3	280.01 ± 0.50	14.10 ± 0.20
Pb	cr	0	64.80 ± 0.30	6.870 ± 0.030
Pb	g	195.2 ± 0.8	175.375 ± 0.005	6.197 ± 0.001
Pb^{+2}	aq	0.92 ± 0.25	18.5 ± 1.0	
$PbSO_4$	cr	-919.97 ± 0.40	148.50 ± 0.60	20.050 ± 0.040
Rb	cr	0	76.78 ± 0.30	7.489 ± 0.020
Rb	g	80.9 ± 0.8	170.094 ± 0.003	6.197 ± 0.001
Rb^+	aq	-251.12 ± 0.10	121.75 ± 0.25	
S	cr, rhombic	0	32.054 ± 0.050	4.412 ± 0.006
S	g	277.17 ± 0.15	167.829 ± 0.006	6.657 ± 0.001
SO_2	g	-296.81 ± 0.20	248.223 ± 0.050	10.549 ± 0.010
SO_4^{-2}	aq	-909.34 ± 0.40	18.50 ± 0.40	
S_2	g	128.60 ± 0.30	228.167 ± 0.010	9.132 ± 0.002
Si	cr	0	18.81 ± 0.08	3.217 ± 0.008
Si	g	450 ± 8	167.981 ± 0.004	7.550 ± 0.001
SiF_4	g	-1615.0 ± 0.8	282.76 ± 0.50	15.36 ± 0.05
SiO_2	cr, alpha quartz	-910.7 ± 1.0	41.46 ± 0.20	6.916 ± 0.020
Sn	cr, white	0	51.18 ± 0.08	6.323 ± 0.008
Sn	g	301.2 ± 1.5	168.492 ± 0.004	6.215 ± 0.001
Sn^{+2}	aq	-8.9 ± 1.0	-16.7 ± 4.0	
SnO	cr, tetragonal	-280.71 ± 0.20	57.17 ± 0.30	8.736 ± 0.020
SnO_2	cr, tetragonal	-577.63 ± 0.20	49.04 ± 0.10	8.384 ± 0.020
Th	cr	0	51.8 ± 0.5	6.35 ± 0.05
Th	g	602 ± 6	190.17 ± 0.05	6.197 ± 0.003
ThO_2	cr	-1226.4 ± 3.5	65.23 ± 0.20	10.560 ± 0.020
Ti	cr	0	30.72 ± 0.10	4.824 ± 0.015
Ti	g	473 ± 3	180.298 ± 0.010	7.539 ± 0.002
$TiCl_4$	g	-763.2 ± 3.0	353.2 ± 4.0	21.5 ± 0.5
TiO_2	cr, rutile	-944.0 ± 0.8	50.62 ± 0.30	8.68 ± 0.05
U	cr	0	50.20 ± 0.20	6.364 ± 0.020
U	g	533 ± 8	199.79 ± 0.10	6.499 ± 0.020
UO_2	cr	-1085.0 ± 1.0	77.03 ± 0.20	11.280 ± 0.020
UO_2^{+2}	aq	-1019.0 ± 1.5	-98.2 ± 3.0	
UO_3	cr, gamma	-1223.8 ± 1.2	96.11 ± 0.40	14.585 ± 0.050
U_3O_8	cr	-3574.8 ± 2.5	282.55 ± 0.50	42.74 ± 0.10
Xe	g	0	169.685 ± 0.003	6.197 ± 0.001
Zn	cr	0	41.63 ± 0.15	5.657 ± 0.020
Zn	g	130.40 ± 0.40	160.990 ± 0.004	6.197 ± 0.001
Zn^{+2}	aq	-153.39 ± 0.20	-109.8 ± 0.5	
ZnO	cr	-350.46 ± 0.27	43.65 ± 0.40	6.933 ± 0.040

STANDARD THERMODYNAMIC PROPERTIES OF CHEMICAL SUBSTANCES

This table gives the standard state chemical thermodynamic properties of about 2400 individual substances in the crystalline, liquid, and gaseous states. Substances are listed by molecular formula in a modified Hill order; all compounds not containing carbon appear first, followed by those that contain carbon. The properties tabulated are:

$\Delta_f H°$	Standard molar enthalpy (heat) of formation at 298.15 K in kJ/mol
$\Delta_f G°$	Standard molar Gibbs energy of formation at 298.15 K in kJ/mol
$S°$	Standard molar entropy at 298.15 K in J/mol K
C_p	Molar heat capacity at constant pressure at 298.15 K in J/mol K

The standard state pressure is 100 kPa (1 bar). The standard states are defined for different phases by:

- The standard state of a pure gaseous substance is that of the substance as a (hypothetical) ideal gas at the standard state pressure.
- The standard state of a pure liquid substance is that of the liquid under the standard state pressure.
- The standard state of a pure crystalline substance is that of the crystalline substance under the standard state pressure.

An entry of 0.0 for $\Delta_f H°$ for an element indicates the reference state of that element. See References 1 and 2 for further information on reference states. A blank means no value is available.

We are indebted to M.V. Korobov for providing data on fullerene compounds.

REFERENCES

1. Cox, J.D., Wagman, D.D., and Medvedev, V.A., *CODATA Key Values for Thermodynamics*, Hemisphere Publishing Corp., New York, 1989.
2. Wagman, D.D., Evans, W.H., Parker, V.B., Schumm, R.H., Halow, I., Bailey, S.M., Churney, K.L., and Nuttall, R.L., *The NBS Tables of Chemical of Chemical Thermodynamic Properties, J. Phys. Chem. Ref. Data*, Vol. 11, Suppl. 2, 1982.
3. Chase M.W., Davies, C.A., Downey, J.R., Frurip, D.J., McDonald, R.A., and Syverud, A.N., *JANAF Thermochemical Tables, Third Edition, J. Phys. Chem. Ref. Data*, Vol. 14, Suppl. 1, 1985.
4. Daubert, T.E., Danner, R.P., Sibul, H.M., and Stebbins, C.C., *Physical and Thermodynamic Properties of Pure Compounds: Data Compilation*, extant 1994 (core with 4 supplements), Taylor & Francis, Bristol, PA.
5. Pedley, J.B., Naylor, R.D., and Kirby, S.P., *Thermochemical Data of Organic Compounds, Second Edition*, Chapman & Hall, London, 1986.
6. Pedley, J.B., *Thermochemical Data and Structures of Organic Compounds*, Thermodynamic Research Center, Texas A & M University, College Station, TX, 1994.
7. Dolmalski, E.S., Evans, W.H., Hearing E.D., *Heat Capacities and Entropies of Organic Compounds in the Condensed Phase, J. Phys. Chem. Ref. Data*, Vol. 13, Suppl. 1, 1984; Vol. 19, No. 4, 881—1047, 1990.
8. Gurvich, L.V., Veyts, I.V., and Alcock, C.B., *Thermodynamic Properties of Individual Substances, Fourth Edition*, Vol. 1, Hermisphere Publishing Corp., New York, 1989.
9. Gurvich, L.V., Veyts, I.V., and Alcock, C.B., *Thermodynamic Properties of Individual Substances, Fourth Edition*, Vol. 3, CRC Press, Boca Raton, FL, 1994.

Molecular formula	Name	$\Delta_f H^\circ$/kJ mol^{-1} Crys.	$\Delta_f H^\circ$/kJ mol^{-1} Liq.	$\Delta_f H^\circ$/kJ mol^{-1} Gas	$\Delta_f G^\circ$/kJ mol^{-1} Crys.	$\Delta_f G^\circ$/kJ mol^{-1} Liq.	$\Delta_f G^\circ$/kJ mol^{-1} Gas	S°/J mol^{-1} K^{-1} Crys.	S°/J mol^{-1} K^{-1} Liq.	S°/J mol^{-1} K^{-1} Gas	C_p/J mol^{-1} K^{-1} Crys.	C_p/J mol^{-1} K^{-1} Liq.	C_p/J mol^{-1} K^{-1} Gas
Ac	Actinium	0.0		406.0			366.0	56.5		188.1	27.2		20.8
Ag	Silver	0.0		284.9			246.0	42.6		173.0	25.4		20.8
AgBr	Silver(I) bromide	-100.4			-96.9			107.1			52.4		
$AgBrO_3$	Silver(I) bromate	-10.5			71.3			151.9					
AgCl	Silver(I) chloride	-127.0			-109.8			96.3			50.8		
$AgClO_3$	Silver(I) chlorate	-30.3			64.5			142.0					
$AgClO_4$	Silver(I) perchlorate	-31.1											
AgF	Silver(I) fluoride	-204.6											
AgF_2	Silver(II) fluoride	-360.0											
AgI	Silver(I) iodide	-61.8			-66.2			115.5			56.8		
$AgIO_3$	Silver(I) iodate	-171.1			-93.7			149.4			102.9		
$AgNO_3$	Silver(I) nitrate	-124.4			-33.4			140.9			93.1		
Ag_2	Disilver			410.0			358.8			257.1			37.0
Ag_2CrO_4	Silver(I) chromate	-731.7			-641.8			217.6			142.3		
Ag_2O	Silver(I) oxide	-31.1			-11.2			121.3			65.9		
Ag_2O_2	Silver(II) oxide	-24.3			27.6			117.0			88.0		
Ag_2O_3	Silver(III) oxide	33.9			121.4			100.0					
Ag_2O_4S	Silver(I) sulfate	-715.9			-618.4			200.4			131.4		
Ag_2S	Silver(I) sulfide (argentite)	-32.6			-40.7			144.0			76.5		
Al	Aluminum	0.0		330.0			289.4	28.3		164.6	24.4		21.4
AlB_3H_{12}	Aluminum borohydride		-16.3	13.0		145.0	147.0		289.1	379.2		194.6	
AlBr	Aluminum monobromide			-4.0			-42.0			239.5			35.6
$AlBr_3$	Aluminum tribromide	-527.2		-425.1				180.2			100.6		
AlCl	Aluminum monochloride			-47.7			-74.1			228.1			35.0
$AlCl_2$	Aluminum dichloride			-331.0									
$AlCl_3$	Aluminum trichloride	-704.2		-583.2	-628.8			109.3			91.1		
AlF	Aluminum monofluoride			-258.2			-283.7			215.0			31.9
AlF_3	Aluminum trifluoride	-1510.4		-1204.6	-1431.1		-1188.2	66.5		277.1	75.1		62.6
AlF_4Na	Sodium tetrafluoroaluminate			-1869.0			-1827.5			345.7			105.9
AlH	Aluminum hydride			259.2			231.2			187.9			29.4
AlH_3	Aluminum hydride	-46.0						30.0			40.2		
AlH_4K	Potassium aluminum hydride	-183.7											
AlH_4Li	Lithium aluminum hydride	-116.3			-44.7			78.7			83.2		
AlH_4Na	Sodium aluminum hydride	-15.5											
AlI	Aluminum monoiodide			65.5									36.0
AlI_3	Aluminum triiodide	-313.8		-207.5	-300.8			159.0			98.7		
AlN	Aluminum nitride	-318.0			-287.0			20.2			30.1		
AlO	Aluminum monoxide			91.2			65.3			218.4			30.9
AlO_4P	Aluminum phosphate	-1733.8			-1617.9			90.8			93.2		
AlP	Aluminum phosphide	-166.5											
AlS	Aluminum monosulfide			200.9			150.1			230.6			33.4
Al_2	Dialuminum			485.9			433.3			233.2			36.4
Al_2Br_6	Aluminum hexabromide			-970.7									
Al_2Cl_6	Aluminum hexachloride			-1290.8			-1220.4			490.0			

Molecular formula	Name	$\Delta_f H°$/kJ mol^{-1} Crys.	Liq.	Gas	$\Delta_f G°$/kJ mol^{-1} Crys.	Liq.	Gas	$S°$/J mol^{-1} K^{-1} Crys.	Liq.	Gas	C_p/J mol^{-1} K^{-1} Crys.	Liq.	Gas
Al_2F_6	Aluminum hexafluoride			-2628.0									
Al_2I_6	Aluminum hexaiodide			-516.7									
Al_2O	Aluminum oxide (Al_2O)			-130.0			-159.0			259.4			45.7
Al_2O_3	Aluminum oxide (corundum)	-1675.7			-1582.3			50.9			79.0		
Al_2S_3	Aluminum sulfide	-724.0						116.9			105.1		
Am	Americium	0.0											
Ar	Argon			0.0						154.8			20.8
As	Arsenic (gray)	0.0		302.5			261.0	35.1		174.2	24.6		20.8
As	Arsenic (yellow)	14.6											
$AsBr_3$	Arsenic(III) bromide	-197.5		-130.0			-159.0			363.9			79.2
$AsCl_3$	Arsenic(III) chloride		-305.0	-261.5		-259.4	-248.9		216.3	327.2			75.7
AsF_3	Arsenic(III) fluoride		-821.3	-785.8		-774.2	-770.8		181.2	289.1		126.6	65.6
AsGa	Gallium arsenide	-71.0			-67.8			64.2			46.2		
AsH_3	Arsine			66.4			68.9			222.8			38.1
AsH_3O_4	Arsenic acid	-906.3											
AsI_3	Arsenic(III) iodide	-58.2			-59.4			213.1		388.3	105.8		80.6
AsIn	Indium arsenide	-58.6			-53.6			75.7			47.8		
AsO	Arsenic monoxide			70.0									
As_2	Diarsenic			222.2			171.9			239.4			35.0
As_2O_5	Arsenic(V) oxide	-924.9			-782.3			105.4			116.5		
As_2S_3	Arsenic(III) sulfide	-169.0			-168.6			163.6			116.3		
At	Astatine	0.0											
Au	Gold	0.0		366.1			326.3	47.4		180.5	25.4		20.8
AuBr	Gold(I) bromide	-14.0											
$AuBr_3$	Gold(III) bromide	-53.3											
AuCl	Gold(I) chloride	-34.7											
$AuCl_3$	Gold(III) chloride	-117.6											
AuF_3	Gold(III) fluoride	-363.6											
AuH	Gold hydride			295.0			265.7			211.2			29.2
AuI	Gold(I) iodide	0.0											
Au_2	Digold			515.1									36.9
B	Boron (rhombic)	0.0		565.0			521.0	5.9		153.4	11.1		20.8
BBr	Bromoborane			238.1			195.4			225.0			32.9
BBr_3	Boron tribromide		-239.7	-205.6		-238.5	-232.5		229.7	324.2			67.8
BCl	Chloroborane			149.5			120.9			213.2			31.7
BClO	Chloroxyborane			-314.0									
BCl_3	Boron trichloride		-427.2	-403.8		-387.4	-388.7		206.3	290.1		106.7	62.7
$BCsO_2$	Cesium metaborate	-972.0			-915.0			104.4			80.6		
BF	Fluoroborane			-122.2			-149.8			200.5			29.6
BFO	Fluorooxyborane			-607.0									
BF_3	Boron trifluoride			-1136.0			-1119.4			254.4			
BF_3H_3N	Aminetrifluoroboron	-1353.9											
BF_3H_3P	Trihydro(phosphorus trifluoride)boron			-854.0									
BF_4Na	Sodium tetrafluoroborate	-1844.7			-1750.1			145.3			120.3		

Molecular formula	Name	$\Delta_f H°$/kJ mol^{-1} Crys.	Liq.	Gas	$\Delta_f G°$/kJ mol^{-1} Crys.	Liq.	Gas	$S°$/J mol^{-1} K^{-1} Crys.	Liq.	Gas	C_p/J mol^{-1} K^{-1} Crys.	Liq.	Gas
BH	Boron monohydride			442.7			419.6			171.9			29.2
BHO_2	Metaboric acid (monoclinic)	-794.3		-561.9	-723.4		-551.0	38.0		240.1			42.2
BH_3	Borane (3)			89.2						188.2			36.0
BH_3O_3	Boric acid (orthoboric acid)	-1094.3		-994.1	-968.9			90.0			86.1		
BH_4K	Potassium borohydride	-227.4			-160.3			106.3			96.1		
BH_4Li	Lithium borohydride	-190.8			-125.0			75.9			82.6		
BH_4Na	Sodium borohydride	-188.6			-123.9			101.3			86.8		
BI_3	Boron triiodide			71.1			20.7			349.2			70.8
BKO_2	Potassium metaborate	-981.6			-923.4			80.0			66.7		
$BLiO_2$	Lithium metaborate	-1032.2			-976.1			51.5			59.8		
BN	Boron nitride	-254.4		647.5	-228.4		614.5	14.8		212.3	19.7		29.5
$BNaO_2$	Sodium metaborate	-977.0			-920.7			73.5			65.9		
BO	Boron monoxide			25.0			-4.0			203.5			29.2
BO_2	Boron dioxide			-300.4			-305.9			229.6			43.0
BO_2Rb	Rubidium metaborate	-971.0			-913.0			94.3			74.1		
BS	Boron monosulfide			342.0			288.8			216.2			30.0
B_2	Diboron			830.5			774.0			201.9			30.5
B_2Cl_4	Tetrachlorodiborane		-523.0	-490.4		-464.8	-460.6		262.3	357.4		137.7	95.4
B_2F_4	Tetrafluorodiborane			-1440.1			-1410.4			317.3			79.1
B_2H_6	Diborane			36.4			86.7			232.1			56.7
B_2O_2	Diboron dioxide			-454.8			-462.3			242.5			57.3
B_2O_3	Boron oxide	-1273.5		-843.8	-1194.3		-832.0	54.0		279.8	62.8		66.9
B_2S_3	Boron sulfide	-240.6		67.0				100.0			111.7		
$B_3H_6N_3$	Borazine		-541.0			-392.7			199.6				
B_4H_{10}	Tetraborane (10)			66.1						280.3			93.2
$B_4Na_2O_7$	Sodium tetraborate	-3291.1			-3096.0			189.5			186.8		
B_5H_9	Pentaborane(9)		42.7	73.2		171.8	175.0		184.2	280.6		151.1	99.6
B_5H_{11}	Pentaborane (11)		73.2	103.3						321.0			130.3
B_6H_{10}	Hexaborane (10)		56.3	94.6						296.8			125.7
Ba	Barium	0.0		180.0			146.0	62.5		170.2	28.1		20.8
$BaBr_2$	Barium bromide	-757.3			-736.8			146.0					
$BaCl_2$	Barium chloride	-855.0			-806.7			123.7			75.1		
$BaCl_2H_4O_2$	Barium chloride dihydrate	-1456.9			-1293.2			203.0					
BaF_2	Barium fluoride	-1207.1			-1156.8			96.4			71.2		
BaH_2	Barium hydride	-177.0			-138.2			63.0			46.0		
BaH_2O_2	Barium hydroxide	-944.7											
BaI_2	Barium iodide	-602.1											
BaN_2O_4	Barium nitrite	-768.2											
BaN_2O_6	Barium nitrate	-988.0			-792.6			214.0			151.4		
BaO	Barium oxide	-548.0			-520.3			72.1			47.3		
BaO_4S	Barium sulfate	-1473.2			-1362.2			132.2			101.8		
BaS	Barium sulfide	-460.0			-456.0			78.2			49.4		
Be	Beryllium	0.0		324.0			286.6	9.5		136.3	16.4		20.8
$BeBr_2$	Beryllium bromide	-353.5						108.0			69.4		
$BeCl_2$	Beryllium chloride	-490.4			-445.6			75.8			62.4		

Molecular formula	Name	$\Delta_f H°$/kJ mol^{-1} Crys.	Liq.	Gas	$\Delta_f G°$/kJ mol^{-1} Crys.	Liq.	Gas	$S°$/J mol^{-1} K^{-1} Crys.	Liq.	Gas	C_p/J mol^{-1} K^{-1} Crys.	Liq.	Gas
BeF_2	Beryllium fluoride	-1026.8			-979.4			53.4			51.8		
BeH_2O_2	Beryllium hydroxide	-902.5			-815.0			45.5			62.1		
BeI_2	Beryllium iodide	-192.5						121.0			71.1		
BeO	Beryllium oxide	-609.4			-580.1			13.8			25.6		
BeO_4S	Beryllium sulfate	-1205.2			-1093.8			77.9			85.7		
BeS	Beryllium sulfide	-234.3						34.0			34.0		
Bi	Bismuth	0.0		207.1			168.2	56.7		187.0	25.5		20.8
BiClO	Bismuth oxychloride	-366.9			-322.1			120.5					
$BiCl_3$	Bismuth trichloride	-379.1		-265.7	-315.0		-256.0	177.0		358.9	105.0		79.7
BiH_3O_3	Bismuth hydroxide	-711.3											
BiI_3	Bismuth triiodide				-175.3								
Bi_2	Dibismuth			219.7									36.9
Bi_2O_3	Bismuth oxide	-573.9			-493.7			151.5			113.5		
$Bi_2O_{12}S_3$	Bismuth sulfate	-2544.3											
Bi_2S_3	Bismuth sulfide	-143.1			-140.6			200.4			122.2		
Bk	Berkelium	0.0											
Br	Bromine (atomic)			111.9			82.4			175.0			20.8
BrCl	Bromine chloride			14.6			-1.0			240.1			35.0
$BrCl_3Si$	Bromotrichlorosilane									350.1			90.9
BrCs	Cesium bromide	-405.8			-391.4			113.1			52.9		
BrCu	Copper(I) bromide	-104.6			-100.8			96.1			54.7		
BrF	Bromine fluoride			-93.8			-109.2			229.0			33.0
BrF_3	Bromine trifluoride		-300.8	-255.6		-240.5	-229.4		178.2	292.5		124.6	66.6
BrF_5	Bromine pentafluoride		-458.6	-428.9		-351.8	-350.6		225.1	320.2			99.6
BrGe	Germanium monobromide			235.6									37.1
$BrGeH_3$	Bromogermane									274.8			56.4
BrH	Hydrogen bromide			-36.3			-53.4			198.7			29.1
BrHSi	Bromosilylene			-464.4									
BrH_3Si	Bromosilane									262.4			52.8
BrH_4N	Ammonium bromide	-270.8			-175.2			113.0			96.0		
BrI	Iodine bromide			40.8			3.7			258.8			36.4
BrIn	Indium(I) bromide	-175.3		-56.9	-169.0		-94.3	113.0		259.5			36.7
BrK	Potassium bromide	-393.8			-380.7			95.9			52.3		
$BrKO_3$	Potassium bromate	-360.2			-271.2			149.2			105.2		
$BrKO_4$	Potassium perbromate	-287.9			-174.4			170.1			120.2		
BrLi	Lithium bromide	-351.2			-342.0			74.3					
BrNO	Nitrosyl bromide			82.2			82.4			273.7			45.5
BrNa	Sodium bromide	-361.1		-143.1	-349.0		-177.1	86.8		241.2	51.4		36.3
$BrNaO_3$	Sodium bromate	-334.1			-242.6			128.9					
BrO	Bromine monoxide			125.8			108.2			237.6			32.1
BrO_2	Bromine dioxide	48.5											
BrRb	Rubidium bromide	-394.6			-381.8			110.0			52.8		
BrSi	Bromosilyldyne			209.0									38.6
BrTl	Thallium(I) bromide	-173.2		-37.7	-167.4			120.5					
Br_2	Bromine		0.0	30.9			3.1		152.2	245.5		75.7	36.0

Molecular formula	Name	$\Delta_f H°$/kJ mol^{-1}			$\Delta_f G°$/kJ mol^{-1}			$S°$/J mol^{-1} K^{-1}			C_p/J mol^{-1} K^{-1}		
		Crys.	Liq.	Gas	Crys.	Liq.	Gas	Crys.	Liq.	Gas	Crys.	Liq.	Gas
Br_2Ca	Calcium bromide	-682.8			-663.6			130.0					
Br_2Cd	Cadmium bromide	-316.2			-296.3			137.2			76.7		
Br_2Co	Cobalt(II) bromide	-220.9									79.5		
Br_2Cr	Chromium(II) bromide	-302.1											
Br_2Cu	Copper(II) bromide	-141.8											
Br_2Fe	Iron(II) bromide	-249.8			-238.1			140.6					
Br_2H_2Si	Dibromosilane									309.7			65.5
Br_2Hg	Mercury(II) bromide	-170.7			-153.1			172.0					
Br_2Hg_2	Mercury(I) bromide	-206.9			-181.1			218.0					
Br_2Mg	Magnesium bromide	-524.3			-503.8			117.2					
Br_2Mn	Manganese(II) bromide	-384.9											
Br_2Ni	Nickel(II) bromide	-212.1											
Br_2Pb	Lead(II) bromide	-278.7			-261.9			161.5			80.1		
Br_2Pt	Platinum(II) bromide	-82.0											
Br_2S_2	Sulfur bromide		-13.0										
Br_2Se	Selenium bromide			-21.0									
Br_2Sn	Tin(II) bromide	-243.5											
Br_2Sr	Strontium bromide	-717.6			-697.1			135.1			75.3		
Br_2Ti	Titanium(II) bromide	-402.0											
Br_2Zn	Zinc bromide	-328.7			-312.1			138.5					
Br_3ClSi	Tribromochlorosilane									377.1			95.3
Br_3Fe	Iron(III) bromide	-268.2											
Br_3Ga	Gallium(III) bromide	-386.6			-359.8			180.0					
Br_3HSi	Tribromosilane		-355.6	-317.6		-336.4	-328.5		248.1	348.6			80.8
Br_3In	Indium(III) bromide	-428.9		-282.0									
Br_3OP	Phosphorus(V) oxybromide	-458.6								359.8			89.9
Br_3P	Phosphorus(III) bromide		-184.5	-139.3		-175.7	-162.8		240.2	348.1			76.0
Br_3Pt	Platinum(III) bromide	-120.9											
Br_3Re	Rhenium(III) bromide	-167.0											
Br_3Ru	Ruthenium(III) bromide	-138.0											
Br_3Sb	Antimony(III) bromide	-259.4		-194.6	-239.3		-223.9	207.1		372.9			80.2
Br_3Sc	Scandium bromide	-743.1											
Br_3Ti	Titanium(III) bromide	-548.5			-523.8			176.6			101.7		
Br_4Ge	Germanium(IV) bromide		-347.7	-300.0		-331.4	-318.0		280.7	396.2			101.8
Br_4Pa	Protactinium(IV) bromide	-824.0			-787.8			234.0					
Br_4Pt	Platinum(IV) bromide	-156.5											
Br_4Si	Tetrabromosilane		-457.3	-415.5		-443.9	-431.8		277.8	377.9			97.1
Br_4Sn	Tin(IV) bromide	-377.4		-314.6	-350.2		-331.4	264.4		411.9			103.4
Br_4Te	Tellurium tetrabromide	-190.4											
Br_4Ti	Titanium(IV) bromide	-616.7		-549.4	-589.5		-568.2	243.5		398.4	131.5		100.8
Br_4V	Vanadium(IV) bromide			-336.8									
Br_4Zr	Zirconium(IV) bromide	-760.7											
Br_5P	Phosphorus(V) bromide	-269.9											
Br_5Ta	Tantalum(V) bromide	-598.3											
Br_6W	Tungsten(VI) bromide	-348.5											

Molecular formula	Name	$\Delta_f H°$/kJ mol^{-1}			$\Delta_f G°$/kJ mol^{-1}			$S°$/J mol^{-1} K^{-1}			C_p/J mol^{-1} K^{-1}		
		Crys.	Liq.	Gas	Crys.	Liq.	Gas	Crys.	Liq.	Gas	Crys.	Liq.	Gas
Ca	Calcium	0.0		177.8			144.0	41.6		154.9	25.9		20.8
$CaCl_2$	Calcium chloride	-795.4			-748.8			108.4			72.9		
CaF_2	Calcium fluoride	-1228.0			-1175.6			68.5			67.0		
CaH_2	Calcium hydride	-181.5			-142.5			41.4			41.0		
CaH_2O_2	Calcium hydroxide	-985.2			-897.5			83.4			87.5		
CaI_2	Calcium iodide	-533.5			-528.9			142.0					
CaN_2O_6	Calcium nitrate	-938.2			-742.8			193.2			149.4		
CaO	Calcium oxide	-634.9			-603.3			38.1			42.0		
CaO_4S	Calcium sulfate	-1434.5			-1322.0			106.5			99.7		
CaS	Calcium sulfide	-482.4			-477.4			56.5			47.4		
$Ca_3O_8P_2$	Calcium phosphate	-4120.8			-3884.7			236.0			227.8		
Cd	Cadmium	0.0		111.8				51.8		167.7	26.0		20.8
$CdCl_2$	Cadmium chloride	-391.5			-343.9			115.3			74.7		
CdF_2	Cadmium fluoride	-700.4			-647.7			77.4					
CdH_2O_2	Cadmium hydroxide	-560.7			-473.6			96.0					
CdI_2	Cadmium iodide	-203.3			-201.4			161.1			80.0		
CdO	Cadmium oxide	-258.4			-228.7			54.8			43.4		
CdO_4S	Cadmium sulfate	-933.3			-822.7			123.0			99.6		
CdS	Cadmium sulfide	-161.9			-156.5			64.9					
CdTe	Cadmium telluride	-92.5			-92.0			100.0					
Ce	Cerium (γ, fcc)	0.0		423.0			385.0	72.0		191.8	26.9		23.1
$CeCl_3$	Cerium(III) chloride	-1060.5			-984.8			151.0			87.4		
CeI_3	Cerium(III) iodide	-669.3											
CeO_2	Cerium(IV) oxide	-1088.7			-1024.6			62.3			61.6		
CeS	Cerium(II) sulfide	-459.4			-451.5			78.2			50.0		
Ce_2O_3	Cerium(III) oxide	-1796.2			-1706.2			150.6			114.6		
Cf	Californium	0.0											
Cl	Chlorine (atomic)			121.3			105.3			165.2			21.8
ClCs	Cesium chloride	-443.0			-414.5			101.2			52.5		
$ClCsO_4$	Cesium perchlorate	-443.1			-314.3			175.1			108.3		
ClCu	Copper(I) chloride	-137.2			-119.9			86.2			48.5		
ClF	Chlorine fluoride			-50.3			-51.8			217.9			32.1
$ClFO_3$	Perchloryl fluoride			-23.8			48.2			279.0			64.9
ClF_3	Chlorine trifluoride		-189.5	-163.2			-123.0			281.6			63.9
ClF_5S	Sulfur chloride pentafluoride		-1065.7										
ClGe	Germanium monochloride			155.2			124.2			247.0			36.9
$ClGeH_3$	Chlorogermane									263.7			54.7
ClH	Hydrogen chloride			-92.3			-95.3			186.9			29.1
ClHO	Hypochlorous acid			-78.7			-66.1			236.7			37.2
$ClHO_4$	Perchloric acid		-40.6										
ClH_3Si	Chlorosilane									250.7			51.0
ClH_4N	Ammonium chloride	-314.4			-202.9			94.6			84.1		
ClH_4NO_4	Ammonium perchlorate	-295.3			-88.8			186.2					
ClH_4P	Phosphonium chloride	-145.2											
ClI	Iodine chloride		-23.9	17.8		-13.6	-5.5		135.1	247.6			35.6

Molecular formula	Name	$\Delta_f H°$/kJ mol^{-1} Crys.	Liq.	Gas	$\Delta_f G°$/kJ mol^{-1} Crys.	Liq.	Gas	$S°$/J mol^{-1} K^{-1} Crys.	Liq.	Gas	C_p/J mol^{-1} K^{-1} Crys.	Liq.	Gas
ClIn	Indium(I) chloride	-186.2		-75.0									
ClK	Potassium chloride	-436.5		-214.6	-408.5		-233.3	82.6		239.1	51.3		36.5
$ClKO_3$	Potassium chlorate	-397.7			-296.3			143.1			100.3		
$ClKO_4$	Potassium perchlorate	-432.8			-303.1			151.0			112.4		
ClLi	Lithium chloride	-408.6			-384.4			59.3			48.0		
$ClLiO_4$	Lithium perchlorate	-381.0											
ClNO	Nitrosyl chloride			51.7			66.1			261.7			44.7
$ClNO_2$	Nitryl chloride			12.6			54.4			272.2			53.2
ClNa	Sodium chloride	-411.2			-384.1			72.1			50.5		
$ClNaO_2$	Sodium chlorite	-307.0											
$ClNaO_3$	Sodium chlorate	-365.8			-262.3			123.4					
$ClNaO_4$	Sodium perchlorate	-383.3			-254.9			142.3					
ClO	Chlorine monoxide			101.8			98.1			226.6			31.5
ClOV	Vanadyl chloride	-607.0			-556.0			75.0					
ClO_2	Chlorine dioxide			102.5			120.5			256.8			42.0
ClO_2	Chlorine superoxide (ClOO)			89.1			105.0			263.7			46.0
ClO_4Rb	Rubidium perchlorate	-437.2			-306.9			161.1					
ClRb	Rubidium chloride	-435.4			-407.8			95.9			52.4		
ClSi	Chlorosilylidyne			189.9									36.9
ClTl	Thallium(I) chloride	-204.1		-67.8	-184.9			111.3			50.9		
Cl_2	Chlorine			0.0						223.1			33.9
Cl_2Co	Cobalt(II) chloride	-312.5			-269.8			109.2			78.5		
Cl_2Cr	Chromium(II) chloride	-395.4			-356.0			115.3			71.2		
Cl_2CrO_2	Chromyl chloride		-579.5	-538.1		-510.8	-501.6		221.8	329.8			84.5
Cl_2Cu	Copper(II) chloride	-220.1			-175.7			108.1			71.9		
Cl_2Fe	Iron(II) chloride	-341.8			-302.3			118.0			76.7		
Cl_2H_2Si	Dichlorosilane									285.7			60.5
Cl_2Hg	Mercury(II) chloride	-224.3			-178.6			146.0					
Cl_2Hg_2	Mercury(I) chloride	-265.4			-210.7			191.6					
Cl_2Mg	Magnesium chloride	-641.3			-591.8			89.6			71.4		
Cl_2Mn	Manganese(II) chloride	-481.3			-440.5			118.2			72.9		
Cl_2Ni	Nickel(II) chloride	-305.3			-259.0			97.7			71.7		
Cl_2O	Chlorine oxide			80.3			97.9			266.2			45.4
Cl_2OS	Thionyl chloride		-245.6	-212.5			-198.3			309.8		121.0	66.5
Cl_2O_2S	Sulfuryl chloride		-394.1	-364.0			-320.0			311.9		134.0	77.0
Cl_2O_2U	Uranyl chloride	-1243.9			-1146.4			150.5			107.9		
Cl_2Pb	Lead(II) chloride	-359.4			-314.1			136.0					
Cl_2Pt	Platinum(II) chloride	-123.4											
Cl_2S	Sulfur dichloride		-50.0										
Cl_2S_2	Sulfur chloride		-59.4										
Cl_2Sn	Tin(II) chloride	-325.1											
Cl_2Sr	Strontium chloride	-828.9			-781.1			114.9			75.6		
Cl_2Ti	Titanium(II) chloride	-513.8			-464.4			87.4			69.8		
Cl_2Zn	Zinc chloride	-415.1		-266.1	-369.4			111.5			71.3		
Cl_2Zr	Zirconium(II) chloride	-502.0											

Molecular formula	Name	$\Delta_f H°$/kJ mol^{-1}			$\Delta_f G°$/kJ mol^{-1}			$S°$/J mol^{-1} K^{-1}			C_p/J mol^{-1} K^{-1}		
		Crys.	Liq.	Gas	Crys.	Liq.	Gas	Crys.	Liq.	Gas	Crys.	Liq.	Gas
Cl_3Cr	Chromium(III) chloride	-556.5			-486.1			123.0			91.8		
Cl_3Dy	Dysprosium(III) chloride	-1000.0											
Cl_3Er	Erbium chloride	-998.7									100.0		
Cl_3Eu	Europium(III) chloride	-936.0											
Cl_3Fe	Iron(III) chloride	-399.5			-334.0			142.3			96.7		
Cl_3Ga	Gallium(III) chloride	-524.7			-454.8			142.0					
Cl_3Gd	Gadolinium(III) chloride	-1008.0									88.0		
Cl_3HSi	Trichlorosilane		-539.3	-513.0		-482.5	-482.0		227.6	313.9			75.8
Cl_3Ho	Holmium chloride	-1005.4									88.0		
Cl_3In	Indium(III) chloride	-537.2		-374.0									
Cl_3Ir	Iridium(III) chloride	-245.6											
Cl_3La	Lanthanum chloride	-1072.2									108.8		
Cl_3Lu	Lutetium chloride	-945.6		-649.0									
Cl_3N	Nitrogen trichloride		230.0										
Cl_3Nd	Neodymium chloride	-1041.0									113.0		
Cl_3OP	Phosphorus(V) oxychloride		-597.1	-558.5		-520.8	-512.9		222.5	325.5		138.8	84.9
Cl_3OV	Vanadyl trichloride		-734.7	-695.6		-668.5	-659.3		244.3	344.3			89.9
Cl_3Os	Osmium(III) chloride	-190.4											
Cl_3P	Phosphorus(III) chloride		-319.7	-287.0		-272.3	-267.8		217.1	311.8			71.8
Cl_3Pr	Praseodymium chloride	-1056.9									100.0		
Cl_3Pt	Platinum(III) chloride	-182.0											
Cl_3Re	Rhenium(III) chloride	-264.0			-188.0			123.8			92.4		
Cl_3Rh	Rhodium(III) chloride	-299.2											
Cl_3Ru	Ruthenium(III) chloride	-205.0											
Cl_3Sb	Antimony(III) chloride	-382.2			-323.7			184.1			107.9		
Cl_3Sc	Scandium chloride	-925.1											
Cl_3Sm	Samarium(III) chloride	-1025.9											
Cl_3Tb	Terbium chloride	-997.0											
Cl_3Ti	Titanium(III) chloride	-720.9			-653.5			139.7			97.2		
Cl_3Tl	Thallium(III) chloride	-315.1											
Cl_3Tm	Thulium chloride	-986.6											
Cl_3U	Uranium(III) chloride	-866.5			-799.1			159.0			102.5		
Cl_3V	Vanadium(III) chloride	-580.7			-511.2			131.0			93.2		
Cl_3Y	Yttrium chloride	-1000.0		-750.2									75.0
Cl_3Yb	Ytterbium(III) chloride	-959.8											
Cl_4Ge	Germanium(IV) chloride		-531.8	-495.8		-462.7	-457.3		245.6	347.7			96.1
Cl_4Hf	Hafnium(IV) chloride	-990.4		-884.5	-901.3			190.8			120.5		
Cl_4Pa	Protactinium(IV) chloride	-1043.0			-953.0			192.0					
Cl_4Pb	Lead(IV) chloride		-329.3										
Cl_4Pt	Platinum(IV) chloride	-231.8											
Cl_4Si	Tetrachlorosilane		-687.0	-657.0		-619.8	-617.0		239.7	330.7		145.3	90.3
Cl_4Sn	Tin(IV) chloride		-511.3	-471.5		-440.1	-432.2		258.6	365.8		165.3	98.3
Cl_4Te	Tellurium tetrachloride	-326.4									138.5		
Cl_4Th	Thorium(IV) chloride	-1186.2		-964.4	-1094.1		-932.0	190.4		390.7	120.3		107.5
Cl_4Ti	Titanium(IV) chloride		-804.2	-763.2		-737.2	-726.3		252.3	353.2		145.2	95.4

Molecular formula	Name	$\Delta_f H°$/kJ mol^{-1} Crys.	Liq.	Gas	$\Delta_f G°$/kJ mol^{-1} Crys.	Liq.	Gas	$S°$/J mol^{-1} K^{-1} Crys.	Liq.	Gas	C_p/J mol^{-1} K^{-1} Crys.	Liq.	Gas
Cl_4U	Uranium(IV) chloride	-1019.2		-809.6	-930.0		-786.6	197.1		419.0	122.0		
Cl_4V	Vanadium(IV) chloride		-569.4	-525.5		-503.7	-492.0		255.0	362.4			96.2
Cl_4Zr	Zirconium(IV) chloride	-980.5			-889.9			181.6			119.8		
Cl_5Nb	Niobium(V) chloride	-797.5		-703.7	-683.2		-646.0	210.5		400.6	148.1		120.8
Cl_5P	Phosphorus(V) chloride	-443.5		-374.9			-305.0			364.6			112.8
Cl_5Pa	Protactinium(V) chloride	-1145.0			-1034.0			238.0					
Cl_5Ta	Tantalum(V) chloride	-859.0											
Cl_6U	Uranium(VI) chloride	-1092.0		-1013.0	-962.0		-928.0	285.8		431.0	175.7		
Cl_6W	Tungsten(VI) chloride	-602.5		-513.8									
Cm	Curium	0.0											
Co	Cobalt	0.0		424.7			380.3	30.0		179.5	24.8		23.0
CoF_2	Cobalt(II) fluoride	-692.0			-647.2			82.0			68.8		
CoH_2O_2	Cobalt(II) hydroxide	-539.7			-454.3			79.0					
CoI_2	Cobalt(II) iodide	-88.7											
CoN_2O_6	Cobalt(II) nitrate	-420.5											
CoO	Cobalt(II) oxide	-237.9			-214.2			53.0			55.2		
CoO_4S	Cobalt(II) sulfate	-888.3			-782.3			118.0					
CoS	Cobalt(II) sulfide	-82.8											
Co_2S_3	Cobalt(III) sulfide	-147.3											
Co_3O_4	Cobalt(II,III) oxide	-891.0			-774.0			102.5			123.4		
Cr	Chromium	0.0		396.6			351.8	23.8		174.5	23.4		20.8
CrF_2	Chromium(II) fluoride	-778.0											
CrF_3	Chromium(III) fluoride	-1159.0			-1088.0			93.9			78.7		
CrI_2	Chromium(II) iodide	-156.9											
CrI_3	Chromium(III) iodide	-205.0											
CrO_2	Chromium(IV) oxide	-598.0											
CrO_4Pb	Lead(II) chromate	-930.9											
Cr_2FeO_4	Chromium iron oxide	-1444.7			-1343.8			146.0			133.6		
Cr_2O_3	Chromium(III) oxide	-1139.7			-1058.1			81.2			118.7		
Cr_3O_4	Chromium(II,III) oxide	-1531.0											
Cs	Cesium	0.0		76.5			49.6	85.2		175.6	32.2		20.8
CsF	Cesium fluoride	-553.5			-525.5			92.8			51.1		
CsF_2H	Cesium hydrogen fluoride	-923.8			-858.9			135.2			87.3		
CsH	Cesium hydride	-54.2											
CsHO	Cesium hydroxide	-417.2											
$CsHO_4S$	Cesium hydrogen sulfate	-1158.1											
CsH_2N	Cesium amide	-118.4											
CsI	Cesium iodide	-346.6			-340.6			123.1			52.8		
$CsNO_3$	Cesium nitrate	-506.0			-406.5			155.2					
CsO_2	Cesium superoxide	-286.2											
Cs_2O	Cesium oxide	-345.8			-308.1			146.9			76.0		
Cs_2O_3S	Cesium sulfite	-1134.7											
Cs_2O_4S	Cesium sulfate	-1443.0			-1323.6			211.9			134.9		
Cs_2S	Cesium sulfide	-359.8											
Cu	Copper	0.0		337.4			297.7	33.2		166.4	24.4		20.8

Molecular formula	Name	$\Delta_f H°$/kJ mol^{-1} Crys.	$\Delta_f H°$/kJ mol^{-1} Liq.	$\Delta_f H°$/kJ mol^{-1} Gas	$\Delta_f G°$/kJ mol^{-1} Crys.	$\Delta_f G°$/kJ mol^{-1} Liq.	$\Delta_f G°$/kJ mol^{-1} Gas	$S°$/J mol^{-1} K^{-1} Crys.	$S°$/J mol^{-1} K^{-1} Liq.	$S°$/J mol^{-1} K^{-1} Gas	C_p/J mol^{-1} K^{-1} Crys.	C_p/J mol^{-1} K^{-1} Liq.	C_p/J mol^{-1} K^{-1} Gas
CuF_2	Copper(II) fluoride	-542.7											
CuH_2O_2	Copper(II) hydroxide	-449.8											
CuI	Copper(I) iodide	-67.8			-69.5			96.7			54.1		
CuN_2O_6	Copper(II) nitrate	-302.9											
CuO	Copper(II) oxide	-157.3			-129.7			42.6			42.3		
CuO_4S	Copper(II) sulfate	-771.4			-662.2			109.2					
CuO_4W	Copper(II) tungstate	-1105.0											
CuS	Copper(II) sulfide	-53.1			-53.6			66.5			47.8		
$CuSe$	Copper(II) selenide	-39.5											
Cu_2	Dicopper			484.2			431.9			241.6			36.6
Cu_2O	Copper(I) oxide	-168.6			-146.0			93.1			63.6		
Cu_2S	Copper(I) sulfide	-79.5			-86.2			120.9			76.3		
Dy	Dysprosium	0.0		290.4			254.4	75.6		196.6	27.7		20.8
Dy_2O_3	Dysprosium(III) oxide	-1863.1			-1771.5			149.8			116.3		
Er	Erbium	0.0		317.1			280.7	73.2		195.6	28.1		20.8
ErF_3	Erbium fluoride	-1711.0											
Er_2O_3	Erbium oxide	-1897.9			-1808.7			155.6			108.5		
Es	Einsteinium	0.0											
Eu	Europium	0.0		175.3			142.2	77.8		188.8	27.7		20.8
Eu_2O_3	Europium(III) oxide	-1651.4			-1556.8			146.0			122.2		
Eu_3O_4	Europium(II,III) oxide	-2272.0			-2142.0			205.0					
F	Fluorine (atomic)			79.4			62.3			158.8			22.7
FGa	Gallium monofluoride			-251.9									33.3
FGe	Germanium monofluoride			-33.4									34.7
$FGeH_3$	Fluorogermane									252.8			51.6
FH	Hydrogen fluoride		-299.8	-273.3			-275.4			173.8			
FH_3Si	Fluorosilane									238.4			47.4
FH_4N	Ammonium fluoride	-464.0			-348.7			72.0			65.3		
FI	Iodine fluoride			-95.7			-118.5			236.2			33.4
FIn	Indium monofluoride			-203.4									
FK	Potassium fluoride	-567.3			-537.8			66.6			49.0		
FLi	Lithium fluoride	-616.0			-587.7			35.7			41.6		
FNO	Nitrosyl fluoride			-66.5			-51.0			248.1			41.3
FNO_2	Nitryl fluoride									260.4			49.8
FNS	Thionitrosyl fluoride (NSF)									259.8			44.1
FNa	Sodium fluoride	-576.6			-546.3			51.1			46.9		
FO	Fluorine monoxide			109.0			105.0			216.8			30.5
FRb	Rubidium fluoride	-557.7											
FSi	Fluorosilylidyne			7.1			-24.3			225.8			32.6
FTl	Thallium(I) fluoride	-324.7		-182.4									
F_2	Fluorine			0.0						202.8			31.3
F_2Fe	Iron(II) fluoride	-711.3			-668.6			87.0			68.1		
F_2HK	Potassium hydrogen fluoride	-927.7			-859.7			104.3			76.9		
F_2HN	Difluoramine									252.8			43.4
F_2HNa	Sodium hydrogen fluoride	-920.3			-852.2			90.9			75.0		

Molecular formula	Name	$\Delta_f H°$/kJ mol^{-1} Crys.	Liq.	Gas	$\Delta_f G°$/kJ mol^{-1} Crys.	Liq.	Gas	$S°$/J mol^{-1} K^{-1} Crys.	Liq.	Gas	C_p/J mol^{-1} K^{-1} Crys.	Liq.	Gas
F_2HRb	Rubidium hydrogen fluoride	-922.6			-855.6			120.1			79.4		
F_2Mg	Magnesium fluoride	-1124.2			-1071.1			57.2			61.6		
F_2N	Difluoroamidogen			43.1			57.8			249.9			41.0
F_2N_2	*cis*-Difluorodiazine			69.5									
F_2N_2	*trans*-Difluorodiazine			82.0									
F_2Ni	Nickel(II) fluoride	-651.4			-604.1			73.6			64.1		
F_2O	Fluorine monoxide			24.7			41.9			247.4			43.3
F_2OS	Thionyl fluoride									278.7			56.8
F_2O_2	Fluorine dioxide			18.0									
F_2O_2S	Sulfuryl fluoride									284.0			66.0
F_2O_2U	Uranyl fluoride	-1653.5			-1557.4			135.6			103.2		
F_2Pb	Lead(II) fluoride	-664.0			-617.1			110.5					
F_2Si	Difluorosilylene			-619.0			-628.0			252.7			43.9
F_2Sr	Strontium fluoride	-1216.3			-1164.8			82.1			70.0		
F_2Zn	Zinc fluoride	-764.4			-713.3			73.7			65.7		
F_3Ga	Gallium(III) fluoride	-1163.0			-1085.3			84.0					
F_3Gd	Gadolinium(III) fluoride			-1297.0									
F_3HSi	Trifluorosilane									271.9			60.5
F_3Ho	Holmium fluoride	-1707.0											
F_3N	Nitrogen trifluoride			-132.1			-90.6			260.8			53.4
F_3Nd	Neodymium fluoride	-1657.0											
F_3OP	Phosphorus(V) oxyfluoride			-1254.3			-1205.8			285.4			68.8
F_3P	Phosphorus(III) fluoride			-958.4			-936.9			273.1			58.7
F_3Sb	Antimony(III) fluoride	-915.5											
F_3Sc	Scandium fluoride	-1629.2		-1247.0	-1555.6		-1234.0	92.0		300.5			67.8
F_3Sm	Samarium(III) fluoride	-1778.0											
F_3Th	Thorium(III) fluoride			-1166.1			-1160.6			339.2			73.3
F_3U	Uranium(III) fluoride	-1502.1		-1058.5	-1433.4		-1051.9	123.4		331.9	95.1		74.3
F_3Y	Yttrium fluoride	-1718.8		-1288.7	-1644.7		-1277.8	100.0		311.8			70.3
F_4Ge	Germanium(IV) fluoride			-1190.2			-1150.0			301.9			
F_4Hf	Hafnium fluoride	-1930.5		-1669.8	-1830.4			113.0					
F_4N_2	Tetrafluorohydrazine			-8.4			79.9			301.2			79.2
F_4Pb	Lead(IV) fluoride	-941.8											
F_4S	Sulfur tetrafluoride			-763.2			-722.0			299.6			77.6
F_4Si	Tetrafluorosilane			-1615.0			-1572.8			282.8			73.6
F_4Th	Thorium(IV) fluoride	-2097.8		-1759.0	-2003.4		-1724.0	142.0		341.7	110.7		93.0
F_4U	Uranium(IV) fluoride	-1914.2		-1598.7	-1823.3		-1572.7	151.7		368.0	116.0		91.2
F_4V	Vanadium(IV) fluoride	-1403.3											
F_4Xe	Xenon tetrafluoride	-261.5											
F_4Zr	Zirconium(IV) fluoride	-1911.3			-1809.9			104.6			103.7		
F_5I	Iodine pentafluoride		-864.8	-822.5			-751.7			327.7			99.2
F_5Nb	Niobium(V) fluoride	-1813.8		-1739.7	-1699.0		-1673.6	160.2		321.9	134.7		97.1
F_5P	Phosphorus(V) fluoride			-1594.4			-1520.7			300.8			84.8
F_5Ta	Tantalum(V) fluoride	-1903.6											
F_5V	Vanadium(V) fluoride		-1480.3	-1433.9		-1373.1	-1369.8		175.7	320.9			98.6

Molecular formula	Name	$\Delta_f H°$/kJ mol^{-1} Crys.	$\Delta_f H°$/kJ mol^{-1} Liq.	$\Delta_f H°$/kJ mol^{-1} Gas	$\Delta_f G°$/kJ mol^{-1} Crys.	$\Delta_f G°$/kJ mol^{-1} Liq.	$\Delta_f G°$/kJ mol^{-1} Gas	$S°$/J mol^{-1} K^{-1} Crys.	$S°$/J mol^{-1} K^{-1} Liq.	$S°$/J mol^{-1} K^{-1} Gas	C_p/J mol^{-1} K^{-1} Crys.	C_p/J mol^{-1} K^{-1} Liq.	C_p/J mol^{-1} K^{-1} Gas
$F_6H_8N_2Si$	Ammonium hexafluorosilicate	-2681.7			-2365.3			280.2			228.1		
F_6Ir	Iridium(VI) fluoride	-579.7		-544.0	-461.6		-460.0	247.7		357.8			121.1
F_6K_2Si	Potassium hexafluorosilicate	-2956.0			-2798.6			226.0					
F_6Mo	Molybdenum(VI) fluoride		-1585.5	-1557.7		-1473.0	-1472.2		259.7	350.5		169.8	120.6
F_6Na_2Si	Sodium hexafluorosilicate	-2909.6			-2754.2			207.1			187.1		
F_6Os	Osmium(VI) fluoride							246.0		358.1			120.8
F_6Pt	Platinum(VI) fluoride							235.6		348.3			122.8
F_6S	Sulfur hexafluoride			-1220.5			-1116.5			291.5			97.0
F_6Se	Selenium hexafluoride			-1117.0			-1017.0			313.9			110.5
F_6Te	Tellurium hexafluoride			-1318.0									
F_6U	Uranium(VI) fluoride	-2197.0		-2147.4	-2068.5		-2063.7	227.6		377.9	166.8		129.6
F_6W	Tungsten(VI) fluoride		-1747.7	-1721.7		-1631.4	-1632.1		251.5	341.1			119.0
Fe	Iron	0.0		416.3			370.7	27.3		180.5	25.1		25.7
FeI_2	Iron(II) iodide	-113.0											
FeI_3	Iron(III) iodide			71.0									
$FeMoO_4$	Iron(II) molybdate	-1075.0			-975.0			129.3			118.5		
FeO	Iron(II) oxide	-272.0											
FeO_4S	Iron(II) sulfate	-928.4			-820.8			107.5			100.6		
FeO_4W	Iron(II) tungstate	-1155.0			-1054.0			131.8			114.6		
FeS	Iron(II) sulfide	-100.0			-100.4			60.3			50.5		
FeS_2	Iron disulfide	-178.2			-166.9			52.9			62.2		
Fe_2O_3	Iron(III) oxide	-824.2			-742.2			87.4			103.9		
Fe_2O_4Si	Iron(II) orthosilicate	-1479.9			-1379.0			145.2			132.9		
Fe_3O_4	Iron(II,III) oxide	-1118.4			-1015.4			146.4			143.4		
Fm	Fermium	0.0											
Fr	Francium	0.0						95.4					
Ga	Gallium	0.0	5.6	272.0	0.0		233.7	40.8		169.0	26.1		25.3
GaH_3O_3	Gallium(III) hydroxide	-964.4			-831.3			100.0					
GaI_3	Gallium(III) iodide	-238.9						205.0			100.0		
GaN	Gallium nitride	-110.5											
GaO	Gallium monoxide			279.5			253.5			231.1			32.1
GaP	Gallium phosphide	-88.0											
GaSb	Gallium antimonide	-41.8			-38.9			76.1			48.5		
Ga_2	Digallium			438.5									
Ga_2O	Gallium suboxide	-356.0											
Ga_2O_3	Gallium(III) oxide	-1089.1			-998.3			85.0			92.1		
Gd	Gadolinium	0.0		397.5			359.8	68.1		194.3	37.0		27.5
Gd_2O_3	Gadolinium(III) oxide	-1819.6									106.7		
Ge	Germanium	0.0		372.0			331.2	31.1		167.9	23.3		30.7
GeH_3I	Iodogermane									283.2			57.5
GeH_4	Germane			90.8			113.4			217.1			45.0
GeI_4	Germanium(IV) iodide	-141.8		-56.9	-144.3		-106.3	271.1		428.9			104.1
GeO	Germanium(II) oxide	-261.9		-46.2	-237.2		-73.2	50.0		224.3			30.9
GeO_2	Germanium(IV) oxide	-580.0			-521.4			39.7			52.1		

Molecular formula	Name	$\Delta_f H°$/kJ mol^{-1} Crys.	Liq.	Gas	$\Delta_f G°$/kJ mol^{-1} Crys.	Liq.	Gas	$S°$/J mol^{-1} K^{-1} Crys.	Liq.	Gas	C_p/J mol^{-1} K^{-1} Crys.	Liq.	Gas
GeP	Germanium phosphide	-21.0			-17.0			63.0					
GeS	Germanium(II) sulfide	-69.0		92.0	-71.5		42.0	71.0		234.0			33.7
GeTe	Germanium(II) telluride	20.0											
Ge_2	Digermanium			473.1			416.3			252.8			35.6
Ge_2H_6	Digermane		137.3	162.3									
Ge_3H_8	Trigermane		193.7	226.8									
H	Hydrogen (atomic)			218.0			203.3			114.7			20.8
HI	Hydrogen iodide			26.5			1.7			206.6			29.2
HIO_3	Iodic acid	-230.1											
HK	Potassium hydride	-57.7											
HKO	Potassium hydroxide	-424.6		-228.0	-378.7		-229.7	78.9		238.3	64.9		49.2
HKO_4S	Potassium hydrogen sulfate	-1160.6			-1031.3			138.1					
HLi	Lithium hydride	-90.5			-68.3			20.0			27.9		
HLiO	Lithium hydroxide	-484.9			-439.0			42.8			49.7		
HN	Imidogen			351.5			345.6			181.2			29.2
HNO_2	Nitrous acid			-79.5			-46.0			254.1			45.6
HNO_3	Nitric acid		-174.1	-133.9		-80.7	-73.5		155.6	266.9		109.9	54.1
HN_3	Hydrazoic acid		264.0	294.1		327.3	328.1		140.6	239.0			43.7
HNa	Sodium hydride	-56.3			-33.5			40.0			36.4		
HNaO	Sodium hydroxide	-425.6			-379.5			64.5			59.5		
$HNaO_4S$	Sodium hydrogen sulfate	-1125.5			-992.8			113.0					
HNa_2O_4P	Sodium hydrogen phosphate	-1748.1			-1608.2			150.5			135.3		
HO	Hydroxyl			39.0			34.2			183.7			29.9
HORb	Rubidium hydroxide	-418.2											
HOTl	Thallium(I) hydroxide	-238.9			-195.8			88.0					
HO_2	Hydroperoxy			10.5			22.6			229.0			34.9
HO_3P	Metaphosphoric acid	-948.5											
HO_4RbS	Rubidium hydrogen sulfate	-1159.0											
HO_4Re	Perrhenic acid	-762.3			-656.4			158.2					
HRb	Rubidium hydride	-52.3											
HS	Mercapto			142.7			113.3			195.7			32.3
HSi	Silylidyne			361.0									
HTa_2	Tantalum hydride	-32.6			-69.0			79.1			90.8		
H_2	Hydrogen			0.0						130.7			28.8
H_2KN	Potassium amide	-128.9											
H_2KO_4P	Potassium dihydrogen phosphate	-1568.3			-1415.9			134.9			116.6		
H_2LiN	Lithium amide	-179.5											
H_2Mg	Magnesium hydride	-75.3			-35.9			31.1			35.4		
H_2MgO_2	Magnesium hydroxide	-924.5			-833.5			63.2			77.0		
H_2N	Amidogen			184.9			194.6			195.0			33.9
H_2NNa	Sodium amide	-123.8			-64.0			76.9			66.2		
H_2NRb	Rubidium amide	-113.0											
$H_2N_2O_2$	Nitramide	-89.5											
H_2NiO_2	Nickel(II) hydroxide	-529.7			-447.2			88.0					

Molecular formula	Name	$\Delta_f H°$/kJ mol^{-1} Crys.	$\Delta_f H°$/kJ mol^{-1} Liq.	$\Delta_f H°$/kJ mol^{-1} Gas	$\Delta_f G°$/kJ mol^{-1} Crys.	$\Delta_f G°$/kJ mol^{-1} Liq.	$\Delta_f G°$/kJ mol^{-1} Gas	$S°$/J mol^{-1} K^{-1} Crys.	$S°$/J mol^{-1} K^{-1} Liq.	$S°$/J mol^{-1} K^{-1} Gas	C_p/J mol^{-1} K^{-1} Crys.	C_p/J mol^{-1} K^{-1} Liq.	C_p/J mol^{-1} K^{-1} Gas
H_2O	Water		-285.8	-241.8		-237.1	-228.6		70.0	188.8		75.3	33.6
H_2O_2	Hydrogen peroxide		-187.8	-136.3		-120.4	-105.6		109.6	232.7		89.1	43.1
H_2O_2Sn	Tin(II) hydroxide	-561.1			-491.6			155.0					
H_2O_2Sr	Strontium hydroxide	-959.0											
H_2O_2Zn	Zinc hydroxide	-641.9			-553.5			81.2					
H_2O_3Si	Metasilicic acid	-1188.7			-1092.4			134.0					
H_2O_4S	Sulfuric acid		-814.0			-690.0			156.9			138.9	
H_2O_4Se	Selenic acid	-530.1											
H_2S	Hydrogen sulfide			-20.6			-33.4			205.8			34.2
H_2S_2	Hydrogen disulfide		-18.1	15.5								84.1	51.5
H_2Se	Hydrogen selenide			29.7			15.9			219.0			34.7
H_2Sr	Strontium hydride	-180.3											
H_2Te	Hydrogen telluride			99.6									
H_2Th	Thorium hydride	-139.7			-100.0			50.7			36.7		
H_2Zr	Zirconium(II) hydride	-169.0			-128.8			35.0			31.0		
H_3ISi	Iodosilane									270.9			54.4
H_3N	Ammonia			-45.9			-16.4			192.8			35.1
H_3NO	Hydroxylamine	-114.2											
H_3O_2P	Hypophosphorous acid	-604.6	-595.4										
H_3O_3P	Phosphorous acid	-964.4											
H_3O_4P	Phosphoric acid (orthophosphoric acid)	-1284.4	-1271.7		-1124.3	-1123.6		110.5	150.8		106.1	145.0	
H_3P	Phosphine			5.4			13.4			210.2			37.1
H_3Sb	Stibine			145.1			147.8			232.8			41.1
H_3U	Uranium(III) hydride	-127.2			-72.8			63.7			49.3		
H_4IN	Ammonium iodide	-201.4			-112.5			117.0					
H_4N_2	Hydrazine		50.6	95.4		149.3	159.4		121.2	238.5		98.9	48.4
$H_4N_2O_2$	Ammonium nitrite	-256.5											
$H_4N_2O_3$	Ammonium nitrate	-365.6			-183.9			151.1			139.3		
H_4N_4	Ammonium azide	115.5			274.2			112.5					
H_4O_4Si	Orthosilicic acid	-1481.1			-1332.9			192.0					
$H_4O_7P_2$	Pyrophosphoric acid	-2241.0	-2231.7										
H_4P_2	Diphosphine		-5.0	20.9									
H_4Si	Silane			34.3			56.9			204.6			42.8
H_4Sn	Stannane			162.8			188.3			227.7			49.0
H_5NO	Ammonium hydroxide		-361.2			-254.0			165.6			154.9	
H_5NO_3S	Ammonium hydrogen sulfite	-768.6											
H_5NO_4S	Ammonium hydrogen sulfate	-1027.0											
H_6Si_2	Disilane			80.3			127.3			272.7			80.8
$H_8N_2O_4S$	Ammonium sulfate	-1180.9			-901.7			220.1			187.5		
H_8Si_3	Trisilane		92.5	120.9									
$H_9N_2O_4P$	Diammonium hydrogen phosphate	-1566.9									188.0		
$H_{12}N_3O_4P$	Ammonium phosphate	-1671.9											
He	Helium			0.0						126.2			20.8

STANDARD THERMODYNAMIC PROPERTIES OF CHEMICAL SUBSTANCES (continued)

Molecular formula	Name	$\Delta_f H°$/kJ mol^{-1} Crys.	$\Delta_f H°$/kJ mol^{-1} Liq.	$\Delta_f H°$/kJ mol^{-1} Gas	$\Delta_f G°$/kJ mol^{-1} Crys.	$\Delta_f G°$/kJ mol^{-1} Liq.	$\Delta_f G°$/kJ mol^{-1} Gas	$S°$/J mol^{-1} K^{-1} Crys.	$S°$/J mol^{-1} K^{-1} Liq.	$S°$/J mol^{-1} K^{-1} Gas	C_p/J mol^{-1} K^{-1} Crys.	C_p/J mol^{-1} K^{-1} Liq.	C_p/J mol^{-1} K^{-1} Gas
Hf	Hafnium	0.0		619.2			576.5	43.6		186.9	25.7		20.8
HfO_2	Hafnium oxide	-1144.7			-1088.2			59.3			60.3		
Hg	Mercury		0.0	61.4			31.8		75.9	175.0		28.0	20.8
HgI_2	Mercury(II) iodide	-105.4			-101.7			180.0					
HgO	Mercury(II) oxide	-90.8			-58.5			70.3			44.1		
HgO_4S	Mercury(II) sulfate	-707.5											
HgS	Mercury(II) sulfide	-58.2			-50.6			82.4			48.4		
HgTe	Mercury(II) telluride	-42.0											
Hg_2	Dimercury			108.8			68.2			288.1			37.4
Hg_2I_2	Mercury(I) iodide	-121.3			-111.0			233.5					
Hg_2O_4S	Mercury(I) sulfate	-743.1			-625.8			200.7			132.0		
Ho	Holmium	0.0		300.8			264.8	75.3		195.6	27.2		20.8
Ho_2O_3	Holmium oxide	-1880.7			-1791.1			158.2			115.0		
I	Iodine (atomic)			106.8			70.2			180.8			20.8
IIn	Indium(I) iodide	-116.3		7.5	-120.5		-37.7	130.0		267.3			36.8
IK	Potassium iodide	-327.9			-324.9			106.3			52.9		
IKO_3	Potassium iodate	-501.4			-418.4			151.5			106.5		
IKO_4	Potassium periodate	-467.2			-361.4			175.7					
ILi	Lithium iodide	-270.4			-270.3			86.8			51.0		
INa	Sodium iodide	-287.8			-286.1			98.5			52.1		
$INaO_3$	Sodium iodate	-481.8									92.0		
$INaO_4$	Sodium periodate	-429.3			-323.0			163.0					
IO	Iodine monoxide			175.1			149.8			245.5			32.9
IRb	Rubidium iodide	-333.8			-328.9			118.4			53.2		
ITl	Thallium(I) iodide	-123.8		7.1	-125.4			127.6					
I_2	Iodine (rhombic)	0.0		62.4			19.3	116.1		260.7	54.4		36.9
I_2Mg	Magnesium iodide	-364.0			-358.2			129.7					
I_2Ni	Nickel(II) iodide	-78.2											
I_2Pb	Lead(II) iodide	-175.5			-173.6			174.9			77.4		
I_2Sn	Tin(II) iodide	-143.5											
I_2Sr	Strontium iodide	-558.1									81.6		
I_2Zn	Zinc iodide	-208.0			-209.0			161.1					
I_3In	Indium(III) iodide	-238.0		-120.5									
I_3La	Lanthanum iodide	-668.9											
I_3Lu	Lutetium iodide	-548.0											
I_3P	Phosphorus(III) iodide	-45.6								374.4			78.4
I_3Ru	Ruthenium(III) iodide	-65.7											
I_3Sb	Antimony(III) iodide	-100.4											
I_4Pt	Platinum(IV) iodide	-72.8											
I_4Si	Tetraiodosilane	-189.5											
I_4Sn	Tin(IV) iodide									446.1	84.9		105.4
I_4Ti	Titanium(IV) iodide	-375.7		-277.8	-371.5			249.4			125.7		
I_4V	Vanadium(IV) iodide			-122.6									
I_4Zr	Zirconium(IV) iodide	-481.6											
In	Indium	0.0		243.3			208.7	57.8		173.8	26.7		20.8

Molecular formula	Name	$\Delta_f H°$/kJ mol^{-1} Crys.	$\Delta_f H°$/kJ mol^{-1} Liq.	$\Delta_f H°$/kJ mol^{-1} Gas	$\Delta_f G°$/kJ mol^{-1} Crys.	$\Delta_f G°$/kJ mol^{-1} Liq.	$\Delta_f G°$/kJ mol^{-1} Gas	$S°$/J mol^{-1} K^{-1} Crys.	$S°$/J mol^{-1} K^{-1} Liq.	$S°$/J mol^{-1} K^{-1} Gas	C_p/J mol^{-1} K^{-1} Crys.	C_p/J mol^{-1} K^{-1} Liq.	C_p/J mol^{-1} K^{-1} Gas
InO	Indium monoxide			387.0			364.4			236.5			32.6
InP	Indium phosphide	-88.7			-77.0			59.8			45.4		
InS	Indium(II) sulfide	-138.1		238.0	-131.8			67.0					
$InSb$	Indium antimonide	-30.5		344.3	-25.5			86.2			49.5		
In_2	Diindium			380.9									
In_2O_3	Indium(III) oxide	-925.8			-830.7			104.2			92.0		
In_2S_3	Indium(III) sulfide	-427.0			-412.5			163.6			118.0		
Ir	Iridium	0.0		665.3			617.9	35.5		193.6	25.1		20.8
IrO_2	Iridium(IV) oxide	-274.1									57.3		
IrS_2	Iridium(IV) sulfide	-138.0											
Ir_2S_3	Iridium(III) sulfide	-234.0											
K	Potassium	0.0		89.0			60.5	64.7		160.3	29.6		20.8
$KMnO_4$	Potassium permanganate	-837.2			-737.6			171.7			117.6		
KNO_2	Potassium nitrite	-369.8			-306.6			152.1			107.4		
KNO_3	Potassium nitrate	-494.6			-394.9			133.1			96.4		
KNa	Potassium sodium		6.3										
KO_2	Potassium superoxide	-284.9			-239.4			116.7			77.5		
K_2	Dipotassium			123.7			87.5			249.7			37.9
K_2O	Potassium oxide	-361.5											
K_2O_2	Potassium peroxide	-494.1			-425.1			102.1					
K_2O_4S	Potassium sulfate	-1437.8			-1321.4			175.6			131.5		
K_2S	Potassium sulfide	-380.7			-364.0			105.0					
K_3O_4P	Potassium phosphate	-1950.2											
Kr	Krypton			0.0						164.1			20.8
La	Lanthanum	0.0		431.0			393.6	56.9		182.4	27.1		22.8
LaS	Lanthanum sulfide	-456.0			-451.5			73.2			59.0		
La_2O_3	Lanthanum oxide	-1793.7			-1705.8			127.3			108.8		
Li	Lithium	0.0		159.3			126.6	29.1		138.8	24.8		20.8
$LiNO_2$	Lithium nitrite	-372.4			-302.0			96.0					
$LiNO_3$	Lithium nitrate	-483.1			-381.1			90.0					
Li_2	Dilithium			215.9			174.4			197.0			36.1
Li_2O	Lithium oxide	-597.9			-561.2			37.6			54.1		
Li_2O_2	Lithium peroxide	-634.3											
Li_2O_3Si	Lithium metasilicate	-1648.1			-1557.2			79.8			99.1		
Li_2O_4S	Lithium sulfate	-1436.5			-1321.7			115.1			117.6		
Li_2S	Lithium sulfide	-441.4											
Li_3O_4P	Lithium phosphate	-2095.8											
Lr	Lawrencium	0.0											
Lu	Lutetium	0.0		427.6			387.8	51.0		184.8	26.9		20.9
Lu_2O_3	Lutetium oxide	-1878.2			-1789.0			110.0			101.8		
Md	Mendelevium	0.0											
Mg	Magnesium	0.0		147.1			112.5	32.7		148.6	24.9		20.8
MgN_2O_6	Magnesium nitrate	-790.7			-589.4			164.0			141.9		
MgO	Magnesium oxide	-601.6			-569.3			27.0			37.2		
MgO_4S	Magnesium sulfate	-1284.9			-1170.6			91.6			96.5		

STANDARD THERMODYNAMIC PROPERTIES OF CHEMICAL SUBSTANCES (continued)

Molecular formula	Name	$\Delta_f H°$/kJ mol^{-1} Crys.	$\Delta_f H°$/kJ mol^{-1} Liq.	$\Delta_f H°$/kJ mol^{-1} Gas	$\Delta_f G°$/kJ mol^{-1} Crys.	$\Delta_f G°$/kJ mol^{-1} Liq.	$\Delta_f G°$/kJ mol^{-1} Gas	$S°$/J mol^{-1} K^{-1} Crys.	$S°$/J mol^{-1} K^{-1} Liq.	$S°$/J mol^{-1} K^{-1} Gas	C_p/J mol^{-1} K^{-1} Crys.	C_p/J mol^{-1} K^{-1} Liq.	C_p/J mol^{-1} K^{-1} Gas
MgO_4Se	Magnesium selenate	-968.5											
MgS	Magnesium sulfide	-346.0			-341.8			50.3			45.6		
Mg_2	Dimagnesium			287.7									
Mg_2O_4Si	Magnesium orthosilicate	-2174.0			-2055.1			95.1			118.5		
Mn	Manganese	0.0		280.7			238.5	32.0		173.7	26.3		20.8
MnN_2O_6	Manganese(II) nitrate	-576.3											
$MnNa_2O_4$	Sodium permanganate	-1156.0											
MnO	Manganese(II) oxide	-385.2			-362.9			59.7			45.4		
MnO_2	Manganese(IV) oxide	-520.0			-465.1			53.1			54.1		
MnO_3Si	Manganese(II) metasilicate	-1320.9			-1240.5			89.1			86.4		
MnS	Manganese(II) sulfide	-214.2			-218.4			78.2			50.0		
$MnSe$	Manganese(II) selenide	-106.7			-111.7			90.8			51.0		
Mn_2O_3	Manganese(III) oxide	-959.0			-881.1			110.5			107.7		
Mn_2O_4Si	Manganese(II) orthosilicate	-1730.5			-1632.1			163.2			129.9		
Mn_3O_4	Manganese(II,III) oxide	-1387.8			-1283.2			155.6			139.7		
Mo	Molybdenum	0.0		658.1			612.5	28.7		182.0	24.1		20.8
$MoNa_2O_4$	Sodium molybdate	-1468.1			-1354.3			159.7			141.7		
MoO_2	Molybdenum(IV) oxide	-588.9			-533.0			46.3			56.0		
MoO_3	Molybdenum(VI) oxide	-745.1			-668.0			77.7			75.0		
MoO_4Pb	Lead(II) molybdate	-1051.9			-951.4			166.1			119.7		
MoS_2	Molybdenum(IV) sulfide	-235.1			-225.9			62.6			63.6		
N	Nitrogen (atomic)			472.7			455.5			153.3			20.8
$NNaO_2$	Sodium nitrite	-358.7			-284.6			103.8					
$NNaO_3$	Sodium nitrate	-467.9			-367.0			116.5			92.9		
NO	Nitric oxide			91.3			87.6			210.8			29.9
NO_2	Nitrogen dioxide			33.2			51.3			240.1			37.2
NO_2Rb	Rubidium nitrite	-367.4			-306.2			172.0					
NO_3Rb	Rubidium nitrate	-495.1			-395.8			147.3			102.1		
NO_3Tl	Thallium(I) nitrate	-243.9			-152.4			160.7			99.5		
NP	Phosphorus nitride	-63.0											
N_2	Nitrogen			0.0						191.6			29.1
N_2O	Nitrous oxide			81.6			103.7			220.0			38.6
N_2O_3	Nitrogen trioxide		50.3	86.6			142.4			314.7			72.7
N_2O_4	Nitrogen tetroxide		-19.5	11.1		97.5	99.8		209.2	304.4		142.7	79.2
N_2O_4Sr	Strontium nitrite	-762.3											
N_2O_5	Nitrogen pentoxide	-43.1		13.3	113.9		117.1	178.2		355.7	143.1		95.3
N_2O_6Pb	Lead(II) nitrate	-451.9											
N_2O_6Ra	Radium nitrate	-992.0			-796.1			222.0					
N_2O_6Sr	Strontium nitrate	-978.2			-780.0			194.6			149.9		
N_2O_6Zn	Zinc nitrate	-483.7											
N_3Na	Sodium azide	21.7			93.8			96.9			76.6		
N_4Si_3	Silicon nitride	-743.5			-642.6			101.3					
Na	Sodium	0.0		107.5			77.0	51.3		153.7	28.2		20.8
NaO_2	Sodium superoxide	-260.2			-218.4			115.9			72.1		
Na_2	Disodium			142.1			103.9			230.2			37.6

Molecular formula	Name	$\Delta_f H°$/kJ mol^{-1} Crys.	$\Delta_f H°$/kJ mol^{-1} Liq.	$\Delta_f H°$/kJ mol^{-1} Gas	$\Delta_f G°$/kJ mol^{-1} Crys.	$\Delta_f G°$/kJ mol^{-1} Liq.	$\Delta_f G°$/kJ mol^{-1} Gas	$S°$/J mol^{-1} K^{-1} Crys.	$S°$/J mol^{-1} K^{-1} Liq.	$S°$/J mol^{-1} K^{-1} Gas	C_p/J mol^{-1} K^{-1} Crys.	C_p/J mol^{-1} K^{-1} Liq.	C_p/J mol^{-1} K^{-1} Gas
Na_2O	Sodium oxide	-414.2			-375.5			75.1			69.1		
Na_2O_2	Sodium peroxide	-510.9			-447.7			95.0			89.2		
Na_2O_3S	Sodium sulfite	-1100.8			-1012.5			145.9			120.3		
Na_2O_3Si	Sodium metasilicate	-1554.9			-1462.8			113.9					
Na_2O_4S	Sodium sulfate	-1387.1			-1270.2			149.6			128.2		
Na_2S	Sodium sulfide	-364.8			-349.8			83.7					
Nb	Niobium	0.0		725.9			681.1	36.4		186.3	24.6		30.2
NbO	Niobium(II) oxide	-405.8			-378.6			48.1			41.3		
NbO_2	Niobium(IV) oxide	-796.2			-740.5			54.5			57.5		
Nb_2O_5	Niobium(V) oxide	-1899.5			-1766.0			137.2			132.1		
Nd	Neodymium	0.0		327.6			292.4	71.5		189.4	27.5		22.1
Nd_2O_3	Neodymium oxide	-1807.9			-1720.8			158.6			111.3		
Ne	Neon			0.0						146.3			20.8
Ni	Nickel	0.0		429.7			384.5	29.9		182.2	26.1		23.4
NiO_4S	Nickel(II) sulfate	-872.9			-759.7			92.0			138.0		
NiS	Nickel(II) sulfide	-82.0			-79.5			53.0			47.1		
Ni_2O_3	Nickel(III) oxide	-489.5											
No	Nobelium	0.0											
O	Oxygen (atomic)			249.2			231.7			161.1			21.9
OP	Phosphorus monoxide			-28.5			-51.9			222.8			31.8
OPb	Lead(II) oxide (massicot)	-217.3			-187.9			68.7			45.8		
OPb	Lead(II) oxide (litharge)	-219.0			-188.9			66.5			45.8		
OPd	Palladium(II) oxide	-85.4		348.9			325.9			218.0	31.4		
ORa	Radium oxide	-523.0											
ORb_2	Rubidium oxide	-339.0											
ORh	Rhodium monoxide			385.0									
OS	Sulfur monoxide			6.3			-19.9			222.0			30.2
OSe	Selenium monoxide			53.4			26.8			234.0			31.3
OSi	Silicon monoxide			-99.6			-126.4			211.6			29.9
OSn	Tin(II) oxide	-280.7		15.1	-251.9		-8.4	57.2		232.1	44.3		31.6
OSr	Strontium oxide	-592.0			-561.9			54.4			45.0		
OTi	Titanium(II) oxide	-519.7			-495.0			50.0			40.0		
OTl_2	Thallium(I) oxide	-178.7			-147.3			126.0					
OU	Uranium(II) oxide			21.0									
OV	Vanadium(II) oxide	-431.8			-404.2			38.9			45.4		
OZn	Zinc oxide	-350.5			-320.5			43.7			40.3		
O_2	Oxygen			0.0						205.2			29.4
O_2P	Phosphorus dioxide			-279.9			-281.6			252.1			39.5
O_2Pb	Lead(IV) oxide	-277.4			-217.3			68.6			64.6		
O_2Rb	Rubidium superoxide	-278.7											
O_2Rb_2	Rubidium peroxide	-472.0											
O_2Ru	Ruthenium(IV) oxide	-305.0											
O_2S	Sulfur dioxide		-320.5	-296.8			-300.1			248.2			39.9
O_2Se	Selenium dioxide	-225.4											
O_2Si	Silicon dioxide (α-quartz)	-910.7		-322.0	-856.3			41.5			44.4		

Molecular formula	Name	$\Delta_f H°$/kJ mol^{-1} Crys.	$\Delta_f H°$/kJ mol^{-1} Liq.	$\Delta_f H°$/kJ mol^{-1} Gas	$\Delta_f G°$/kJ mol^{-1} Crys.	$\Delta_f G°$/kJ mol^{-1} Liq.	$\Delta_f G°$/kJ mol^{-1} Gas	$S°$/J mol^{-1} K^{-1} Crys.	$S°$/J mol^{-1} K^{-1} Liq.	$S°$/J mol^{-1} K^{-1} Gas	C_p/J mol^{-1} K^{-1} Crys.	C_p/J mol^{-1} K^{-1} Liq.	C_p/J mol^{-1} K^{-1} Gas
O_2Sn	Tin(IV) oxide	-577.6			-515.8			49.0			52.6		
O_2Te	Tellurium dioxide	-322.6			-270.3			79.5					
O_2Th	Thorium(IV) oxide	-1226.4			-1169.2			65.2			61.8		
O_2Ti	Titanium(IV) oxide	-944.0			-888.8			50.6			55.0		
O_2U	Uranium(IV) oxide	-1085.0		-465.7	-1031.8		-471.5	77.0		274.6	63.6		51.4
O_2W	Tungsten(IV) oxide	-589.7			-533.9			50.5			56.1		
O_2Zr	Zirconium(IV) oxide	-1100.6			-1042.8			50.4			56.2		
O_3	Ozone			142.7			163.2			238.9			39.2
O_3PbS	Lead(II) sulfite	-669.9											
O_3PbSi	Lead(II) metasilicate	-1145.7			-1062.1			109.6			90.0		
O_3Pr_2	Praseodymium oxide	-1809.6									117.4		
O_3Rh_2	Rhodium(III) oxide	-343.0									103.8		
O_3S	Sulfur trioxide	-454.5	-441.0	-395.7	-374.2	-373.8	-371.1	70.7	113.8	256.8			50.7
O_3Sc_2	Scandium oxide	-1908.8			-1819.4			77.0			94.2		
O_3SiSr	Strontium metasilicate	-1633.9			-1549.7			96.7			88.5		
O_3Sm_2	Samarium(III) oxide	-1823.0			-1734.6			151.0			114.5		
O_3Tb_2	Terbium oxide	-1865.2									115.9		
O_3Ti_2	Titanium(III) oxide	-1520.9			-1434.2			78.8			97.4		
O_3Tm_2	Thulium oxide	-1888.7			-1794.5			139.7			116.7		
O_3U	Uranium(VI) oxide	-1223.8			-1145.7			96.1			81.7		
O_3V_2	Vanadium(III) oxide	-1218.8			-1139.3			98.3			103.2		
O_3W	Tungsten(VI) oxide	-842.9			-764.0			75.9			73.8		
O_3Y_2	Yttrium oxide	-1905.3			-1816.6			99.1			102.5		
O_3Yb_2	Ytterbium(III) oxide	-1814.6			-1726.7			133.1			115.4		
O_4Os	Osmium(VIII) oxide	-394.1		-337.2	-304.9		-292.8	143.9		293.8			74.1
O_4PbS	Lead(II) sulfate	-920.0			-813.0			148.5			103.2		
O_4PbSe	Lead(II) selenate	-609.2			-504.9			167.8					
O_4Pb_2Si	Lead(II) orthosilicate	-1363.1			-1252.6			186.6			137.2		
O_4Pb_3	Lead(II,II,IV) oxide	-718.4			-601.2			211.3			146.9		
O_4RaS	Radium sulfate	-1471.1			-1365.6			138.0					
O_4Rb_2S	Rubidium sulfate	-1435.6			-1316.9			197.4			134.1		
O_4Ru	Ruthenium(VIII) oxide	-239.3			-152.2			146.4					
O_4SSr	Strontium sulfate	-1453.1			-1340.9			117.0					
O_4STl_2	Thallium(I) sulfate	-931.8			-830.4			230.5					
O_4SZn	Zinc sulfate	-982.8			-871.5			110.5			99.2		
O_4SiSr_2	Strontium orthosilicate	-2304.5			-2191.1			153.1			134.3		
O_4SiZn_2	Zinc orthosilicate	-1636.7			-1523.2			131.4			123.3		
O_4SiZr	Zirconium(IV) orthosilicate	-2033.4			-1919.1			84.1			98.7		
O_5Sb_2	Antimony(V) oxide	-971.9			-829.2			125.1					
O_5Ta_2	Tantalum(V) oxide	-2046.0			-1911.2			143.1			135.1		
O_5Ti_3	Titanium(III,IV) oxide	-2459.4			-2317.4			129.3			154.8		
O_5V_2	Vanadium(V) oxide	-1550.6			-1419.5			131.0			127.7		
O_5V_3	Vanadium(III,IV) oxide	-1933.0			-1803.0			163.0					
O_7Re_2	Rhenium(VII) oxide	-1240.1		-1100.0	-1066.0		-994.0	207.1		452.0	166.1		
O_7U_3	Uranium(IV,VI) oxide	-3427.1			-3242.9			250.5			215.5		

Molecular formula	Name	$\Delta_f H°$/kJ mol^{-1} Crys.	$\Delta_f H°$/kJ mol^{-1} Liq.	$\Delta_f H°$/kJ mol^{-1} Gas	$\Delta_f G°$/kJ mol^{-1} Crys.	$\Delta_f G°$/kJ mol^{-1} Liq.	$\Delta_f G°$/kJ mol^{-1} Gas	$S°$/J mol^{-1} K^{-1} Crys.	$S°$/J mol^{-1} K^{-1} Liq.	$S°$/J mol^{-1} K^{-1} Gas	C_p/J mol^{-1} K^{-1} Crys.	C_p/J mol^{-1} K^{-1} Liq.	C_p/J mol^{-1} K^{-1} Gas
O_8S_2Zr	Zirconium(IV) sulfate	-2217.1									172.0		
O_8U_3	Uranium(V,VI) oxide	-3574.8			-3369.5			282.6			238.4		
O_9U_4	Uranium(IV,V) oxide	-4510.4			-4275.1			334.1			293.3		
Os	Osmium	0.0		791.0			745.0	32.6		192.6	24.7		20.8
P	Phosphorus (white)	0.0		316.5			280.1	41.1		163.2	23.8		20.8
P	Phosphorus (red)	-17.6						22.8			21.2		
P	Phosphorus (black)	-39.3											
P_2	Diphosphorus			144.0			103.5			218.1			32.1
P_4	Tetraphosphorus			58.9			24.4			280.0			67.2
Pa	Protactinium	0.0		607.0			563.0	51.9		198.1			22.9
Pb	Lead	0.0		195.2			162.2	64.8		175.4	26.4		20.8
PbS	Lead(II) sulfide	-100.4			-98.7			91.2			49.5		
PbSe	Lead(II) selenide	-102.9			-101.7			102.5			50.2		
PbTe	Lead(II) telluride	-70.7			-69.5			110.0			50.5		
Pd	Palladium	0.0		378.2			339.7	37.6		167.1	26.0		20.8
PdS	Palladium(II) sulfide	-75.0			-67.0			46.0					
Pm	Promethium	0.0								187.1			24.3
Po	Polonium	0.0											
Pr	Praseodymium	0.0		355.6			320.9	73.2		189.8	27.2		21.4
Pt	Platinum	0.0		565.3			520.5	41.6		192.4	25.9		25.5
PtS	Platinum(II) sulfide	-81.6			-76.1			55.1			43.4		
PtS_2	Platinum(IV) sulfide	-108.8			-99.6			74.7			65.9		
Pu	Plutonium	0.0											
Ra	Radium	0.0		159.0			130.0	71.0		176.5			20.8
Rb	Rubidium	0.0		80.9			53.1	76.8		170.1	31.1		20.8
Re	Rhenium	0.0		769.9			724.6	36.9		188.9	25.5		20.8
Rh	Rhodium	0.0		556.9			510.8	31.5		185.8	25.0		21.0
Rn	Radon			0.0						176.2			20.8
Ru	Ruthenium	0.0		642.7			595.8	28.5		186.5	24.1		21.5
S	Sulfur (rhombic)	0.0		277.2			236.7	32.1		167.8	22.6		23.7
S	Sulfur (monoclinic)	0.3											
SSi	Silicon monosulfide			112.5			60.9			223.7			32.3
SSn	Tin(II) sulfide	-100.0			-98.3			77.0			49.3		
SSr	Strontium sulfide	-472.4			-467.8			68.2			48.7		
STl_2	Thallium(I) sulfide	-97.1			-93.7			151.0					
SZn	Zinc sulfide (wurtzite)	-192.6											
SZn	Zinc sulfide (sphalerite)	-206.0			-201.3			57.7			46.0		
S_2	Disulfur			128.6			79.7			228.2			32.5
Sb	Antimony	0.0		262.3			222.1	45.7		180.3	25.2		20.8
Sb_2	Diantimony			235.6			187.0			254.9			36.4
Sc	Scandium	0.0		377.8			336.0	34.6		174.8	25.5		22.1
Se	Selenium	0.0		227.1			187.0	42.4		176.7	25.4		20.8
SeSr	Strontium selenide	-385.8											
$SeTl_2$	Thallium(I) selenide	-59.0			-59.0			172.0					
SeZn	Zinc selenide	-163.0			-163.0			84.0					

Molecular formula	Name	$\Delta_f H°$/kJ mol^{-1}			$\Delta_f G°$/kJ mol^{-1}			$S°$/J mol^{-1} K^{-1}			C_p/J mol^{-1} K^{-1}		
		Crys.	Liq.	Gas	Crys.	Liq.	Gas	Crys.	Liq.	Gas	Crys.	Liq.	Gas
Se_2	Diselenium			146.0			96.2			252.0			35.4
Si	Silicon	0.0		450.0			405.5	18.8		168.0	20.0		22.3
Si_2	Disilicon			594.0			536.0			229.9			34.4
Sm	Samarium	0.0		206.7			172.8	69.6		183.0	29.5		30.4
Sn	Tin (white)	0.0		301.2			266.2	51.2		168.5	27.0		21.3
Sn	Tin (gray)	-2.1			0.1			44.1			25.8		
Sr	Strontium	0.0		164.4			130.9	55.0		164.6	26.8		20.8
Ta	Tantalum	0.0		782.0			739.3	41.5		185.2	25.4		20.9
Tb	Terbium	0.0		388.7			349.7	73.2		203.6	28.9		24.6
Tc	Technetium	0.0		678.0						181.1			20.8
Te	Tellurium	0.0		196.7			157.1	49.7		182.7	25.7		20.8
Te_2	Ditellurium			168.2			118.0			268.1			36.7
Th	Thorium	0.0		602.0			560.7	51.8		190.2	27.3		20.8
Ti	Titanium	0.0		473.0			428.4	30.7		180.3	25.0		24.4
Tl	Thallium	0.0		182.2			147.4	64.2		181.0	26.3		20.8
Tm	Thulium	0.0		232.2			197.5	74.0		190.1	27.0		20.8
U	Uranium	0.0		533.0			488.4	50.2		199.8	27.7		23.7
V	Vanadium	0.0		514.2			754.4	28.9		182.3	24.9		26.0
W	Tungsten	0.0		849.4			807.1	32.6		174.0	24.3		21.3
Xe	Xenon			0.0						169.7			20.8
Y	Yttrium	0.0		421.3			381.1	44.4		179.5	26.5		25.9
Yb	Ytterbium	0.0		152.3			118.4	59.9		173.1	26.7		20.8
Zn	Zinc	0.0		130.4			94.8	41.6		161.0	25.4		20.8
Zr	Zirconium	0.0		608.8			566.5	39.0		181.4	25.4		26.7
C	Carbon (graphite)	0.0		716.7			671.3	5.7		158.1	8.5		20.8
C	Carbon (diamond)	1.9			2.9			2.4			6.1		
CAgN	Silver(I) cyanide	146.0			156.9			107.2			66.7		
CAg_2O_3	Silver(I) carbonate	-505.8			-436.8			167.4			112.3		
$CBaO_3$	Barium carbonate	-1213.0			-1134.4			112.1			86.0		
$CBeO_3$	Beryllium carbonate	-1025.0						52.0			65.0		
$CBrClF_2$	Bromochlorodifluoromethane									318.5			74.6
$CBrCl_2F$	Bromodichlorofluoromethane									330.6			80.0
$CBrCl_3$	Bromotrichloromethane			-41.1									85.3
$CBrF_3$	Bromotrifluoromethane			-648.3									69.3
CBrN	Cyanogen bromide	140.5		186.2			165.3			248.3			46.9
$CBrN_3O_6$	Bromotrinitromethane		32.5	80.3									
CBr_2ClF	Dibromochlorofluoromethane									342.8			82.4
CBr_2Cl_2	Dibromodichloromethane									347.8			87.1
CBr_2F_2	Dibromodifluoromethane									325.3			77.0
CBr_2O	Carbonyl bromide		-127.2	-96.2			-110.9			309.1			61.8
CBr_3Cl	Tribromochloromethane									357.8			89.4
CBr_3F	Tribromofluoromethane									345.9			84.4

Molecular formula	Name	$\Delta_f H°$/kJ mol^{-1}			$\Delta_f G°$/kJ mol^{-1}			$S°$/J mol^{-1} K^{-1}			C_p/J mol^{-1} K^{-1}		
		Crys.	Liq.	Gas	Crys.	Liq.	Gas	Crys.	Liq.	Gas	Crys.	Liq.	Gas
CBr_4	Tetrabromomethane	29.4		83.9	47.7		67.0	212.5		358.1	144.3		91.2
$CCaO_3$	Calcium carbonate (calcite)	-1207.6			-1129.1			91.7			83.5		
$CCaO_3$	Calcium carbonate (aragonite)	-1207.8			-1128.2			88.0			82.3		
$CCdO_3$	Cadmium carbonate	-750.6			-669.4			92.5					
$CClFO$	Carbonyl chloride fluoride									276.7			52.4
$CClF_3$	Chlorotrifluoromethane			-706.3									66.9
$CClN$	Cyanogen chloride		112.1	138.0			131.0			236.2			45.0
$CClN_3O_6$	Chlorotrinitromethane		-27.1	18.4									
CCl_2F_2	Dichlorodifluoromethane			-477.4			-439.4			300.8			72.3
CCl_2O	Carbonyl chloride			-219.1			-204.9			283.5			57.7
CCl_3	Trichloromethyl			59.0									
CCl_3F	Trichlorofluoromethane		-301.3	-268.3		-236.8			225.4			121.6	78.1
CCl_4	Tetrachloromethane		-128.2	-95.7								130.7	83.3
$CCoO_3$	Cobalt(II) carbonate	-713.0											
CCs_2O_3	Cesium carbonate	-1139.7			-1054.3			204.5			123.9		
$CCuN$	Copper(I) cyanide	96.2			111.3			84.5					
CFN	Cyanogen fluoride									224.7			41.8
CF_2O	Carbonyl fluoride			-639.8									46.8
CF_3	Trifluoromethyl			-477.0			-464.0			264.5			49.6
CF_3I	Trifluoroiodomethane			-587.8						307.4			70.9
CF_4	Tetrafluoromethane			-933.6						261.6			61.1
$CFeO_3$	Iron(II) carbonate	-740.6			-666.7			92.9			82.1		
CFe_3	Iron carbide	25.1			20.1			104.6			105.9		
CH	Methylidyne			595.8									
$CHBrClF$	Bromochlorofluoromethane									304.3			63.2
$CHBrCl_2$	Bromodichloromethane									316.4			67.4
$CHBrF_2$	Bromodifluoromethane			-424.9						295.1			58.7
$CHBr_2Cl$	Chlorodibromomethane									327.7			69.2
$CHBr_2F$	Dibromofluoromethane									316.8			65.1
$CHBr_3$	Tribromomethane		-22.3	23.8		-5.0	8.0		220.9	330.9		130.7	71.2
$CHClF_2$	Chlorodifluoromethane			-482.6						280.9			55.9
$CHCl_2F$	Dichlorofluoromethane									293.1			60.9
$CHCl_3$	Trichloromethane		-134.1	-102.7		-73.7	6.0		201.7	295.7		114.2	65.7
$CHCsO_3$	Cesium hydrogen carbonate	-966.1											
$CHFO$	Formyl fluoride									246.6			39.9
CHF_3	Trifluoromethane			-695.4						259.7			51.0
CHI_3	Triiodomethane	-181.1		251.0						356.2			75.0
$CHKO_2$	Potassium formate	-679.7											
$CHKO_3$	Potassium hydrogen carbonate	-963.2			-863.5			115.5					
CHN	Hydrogen cyanide		108.9	135.1		125.0	124.7		112.8	201.8		70.6	35.9
$CHNO$	Isocyanic acid (HNCO)									238.0			44.9
$CHNS$	Isothiocyanic acid			127.6			113.0			247.8			46.9
CHN_3O_6	Trinitromethane		-32.8	-0.2									
$CHNaO_2$	Sodium formate	-666.5			-599.9			103.8			82.7		

STANDARD THERMODYNAMIC PROPERTIES OF CHEMICAL SUBSTANCES (continued)

Molecular formula	Name	$\Delta_f H°$/kJ mol^{-1} Crys.	$\Delta_f H°$/kJ mol^{-1} Liq.	$\Delta_f H°$/kJ mol^{-1} Gas	$\Delta_f G°$/kJ mol^{-1} Crys.	$\Delta_f G°$/kJ mol^{-1} Liq.	$\Delta_f G°$/kJ mol^{-1} Gas	$S°$/J mol^{-1} K^{-1} Crys.	$S°$/J mol^{-1} K^{-1} Liq.	$S°$/J mol^{-1} K^{-1} Gas	C_p/J mol^{-1} K^{-1} Crys.	C_p/J mol^{-1} K^{-1} Liq.	C_p/J mol^{-1} K^{-1} Gas
$CHNaO_3$	Sodium hydrogen carbonate	-950.8			-851.0			101.7			87.6		
CHO	Oxomethyl (HCO)			43.1			28.0			224.7			34.6
CH_2	Methylene			390.4			372.9			194.9			33.8
CH_2BrCl	Bromochloromethane									287.6			52.7
CH_2BrF	Bromofluoromethane									276.3			49.2
CH_2Br_2	Dibromomethane									293.2			54.7
CH_2ClF	Chlorofluoromethane									264.4			47.0
CH_2Cl_2	Dichloromethane		-124.2	-95.4					177.8	270.2		101.2	51.0
CH_2F_2	Difluoromethane			-452.3						246.7			42.9
CH_2I_2	Diiodomethane		68.5	119.5		90.4	95.8		174.1	309.7		134.0	57.7
CH_2N_2	Diazomethane									242.9			52.5
CH_2N_2	Cyanamide	58.8											
$CH_2N_2O_4$	Dinitromethane		-104.9	-58.9									
CH_2O	Formaldehyde			-108.6			-102.5			218.8			35.4
$(CH_2O)_x$	Paraformaldehyde	-177.6											
CH_2O_2	Formic acid		-425.0	-378.7		-361.4			129.0			99.0	
CH_2S_3	Trithiocarbonic acid		24.0										
CH_3	Methyl			145.7			147.9			194.2			38.7
CH_3BO	Borane carbonyl			-111.2			-92.9			249.4			59.5
CH_3Br	Bromomethane		-59.8	-35.4			-26.3			246.4			42.4
CH_3Cl	Chloromethane			-81.9						234.6			40.8
CH_3F	Fluoromethane									222.9			37.5
CH_3I	Iodomethane		-13.6	14.4					163.2	254.1		126.0	44.1
CH_3NO	Formamide		-254.0	-193.9									
CH_3NO_2	Nitromethane		-112.6	-74.3		-14.4	-6.8		171.8	275.0		106.6	57.3
CH_3NO_2	Methyl nitrite			-66.1									
CH_3NO_3	Methyl nitrate		-156.3	-122.4		-43.4	-39.2		217.1	318.5		157.3	
CH_4	Methane			-74.6			-50.5			186.3			35.7
CH_4N_2	Ammonium cyanide	0.4									134.0		
CH_4N_2O	Urea	-333.1		-245.8									
CH_4N_2S	Thiourea	-89.1		22.9									
$CH_4N_4O_2$	Nitroguanidine	-92.4											
CH_4O	Methanol		-239.2	-201.0		-166.6	-162.3		126.8	239.9		81.1	44.1
CH_4S	Methanethiol		-46.7	-22.9		-7.7	-9.3		169.2	255.2		90.5	50.3
CH_5N	Methylamine		-47.3	-22.5		35.7	32.7		150.2	242.9		102.1	50.1
CH_5NO_3	Ammonium hydrogen carbonate	-849.4			-665.9			120.9					
CH_5N_3	Guanidine	-56.0											
CH_5N_3S	Hydrazinecarbothioamide	24.7											
$CH_5N_5O_2$	3-Amino-1-nitroguanidine	22.1											
CH_6ClN	Methylamine hydrochloride	-298.1											
CH_6N_2	Methylhydrazine		54.2	94.7		180.0	187.0		165.9	278.8		134.9	71.1
CH_6Si	Methylsilane									256.5			65.9
CHg_2O_3	Mercury(I) carbonate	-553.5			-468.1			180.0					
CIN	Cyanogen iodide	166.2		225.5	185.0		196.6	96.2		256.8			48.3

Molecular formula	Name	$\Delta_f H°$/kJ mol^{-1}			$\Delta_f G°$/kJ mol^{-1}			$S°$/J mol^{-1} K^{-1}			C_p/J mol^{-1} K^{-1}		
		Crys.	Liq.	Gas	Crys.	Liq.	Gas	Crys.	Liq.	Gas	Crys.	Liq.	Gas
CI_4	Tetraiodomethane	-392.9		474.0						391.9			95.9
CKN	Potassium cyanide	-113.0			-101.9			128.5			66.3		
CKNS	Potassium thiocyanate	-200.2			-178.3			124.3			88.5		
CK_2O_3	Potassium carbonate	-1151.0			-1063.5			155.5			114.4		
CLi_2O_3	Lithium carbonate	-1215.9			-1132.1			90.4			99.1		
$CMgO_3$	Magnesium carbonate	-1095.8			-1012.1			65.7			75.5		
$CMnO_3$	Manganese(II) carbonate	-894.1			-816.7			85.8			81.5		
CN	Cyanide			437.6			407.5			202.6			29.2
CNNa	Sodium cyanide	-87.5			-76.4			115.6			70.4		
CNNaO	Sodium cyanate	-405.4			-358.1			96.7			86.6		
CN_4O_8	Tetranitromethane		38.4	88.3									
CNa_2O_3	Sodium carbonate	-1130.7			-1044.4			135.0			112.3		
CO	Carbon monoxide			-110.5			-137.2			197.7			29.1
COS	Carbon oxysulfide			-142.0			-169.2			231.6			41.5
CO_2	Carbon dioxide			-393.5			-394.4			213.8			37.1
CO_3Pb	Lead(II) carbonate	-699.1			-625.5			131.0			87.4		
CO_3Rb_2	Rubidium carbonate	-1136.0			-1051.0			181.3			117.6		
CO_3Sr	Strontium carbonate	-1220.1			-1140.1			97.1			81.4		
CO_3Tl_2	Thallium(I) carbonate	-700.0			-614.6			155.2					
CO_3Zn	Zinc carbonate	-812.8			-731.5			82.4			79.7		
CS	Carbon sulfide			234.0			184.0			210.6			29.8
CS_2	Carbon disulfide		89.0	116.7		64.6	67.1		151.3	237.8		76.4	45.4
CSe_2	Carbon diselenide		164.8										
CSi	Silicon carbide (cubic)	-65.3			-62.8			16.6			26.9		
CSi	Silicon carbide (hexagonal)	-62.8			-60.2			16.5			26.7		
C_2	Dicarbon			831.9			775.9			199.4			43.2
C_2BrF_5	Bromopentafluoroethane			-1064.4									
$C_2Br_2ClF_3$	1,2-Dibromo-1-chloro-1,2,2-trifluoroethane		-691.7	-656.6									
$C_2Br_2F_4$	1,2-Dibromotetrafluoroethane		-817.7	-789.1									
C_2Br_4	Tetrabromoethylene									387.1			102.7
C_2Br_6	Hexabromoethane									441.9			139.3
C_2Ca	Calcium carbide	-59.8			-64.9			70.0			62.7		
C_2CaN_2	Calcium cyanide	-184.5											
C_2CaO_4	Calcium oxalate	-1360.6											
C_2ClF_3	Chlorotrifluoroethylene		-522.7	-505.5			-523.8			322.1			83.9
C_2ClF_5	Chloropentafluoroethane			-1118.8									184.2
$C_2Cl_2F_4$	1,2-Dichlorotetrafluoroethane		-960.2	-937.0								111.7	
$C_2Cl_2O_2$	Ethanedioyl dichloride		-367.6	-335.8									
$C_2Cl_3F_3$	1,1,2-Trichlorotrifluoroethane		-745.0	-716.8								170.1	
C_2Cl_3N	Trichloroacetonitrile									336.6			96.1
C_2Cl_4	Tetrachloroethylene		-50.6	-10.9		3.0			266.9			143.4	

STANDARD THERMODYNAMIC PROPERTIES OF CHEMICAL SUBSTANCES (continued)

Molecular formula	Name	$\Delta_f H°$/kJ mol^{-1} Crys.	Liq.	Gas	$\Delta_f G°$/kJ mol^{-1} Crys.	Liq.	Gas	$S°$/J mol^{-1} K^{-1} Crys.	Liq.	Gas	C_p/J mol^{-1} K^{-1} Crys.	Liq.	Gas
$C_2Cl_4F_2$	1,1,1,2-Tetrachloro-2,2-difluoroethane			-489.9			-407.0			382.9			123.4
$C_2Cl_4F_2$	1,1,2,2-Tetrachloro-1,2-difluoroethane											173.6	
C_2Cl_4O	Trichloroacetyl chloride		-280.8	-239.8									
C_2Cl_6	Hexachloroethane	-202.8		-143.6				237.3			198.2		
C_2F_3N	Trifluoroacetonitrile			-497.9						298.1			77.9
C_2F_4	Tetrafluoroethylene	-820.5		-658.9						300.1			80.5
C_2F_6	Hexafluoroethane			-1344.2						332.3			106.7
C_2HBr	Bromoacetylene									253.7			55.7
$C_2HBrClF_3$	1-Bromo-2-chloro-1,1,2-trifluoroethane		-675.3	-644.8									
$C_2HBrClF_3$	2-Bromo-2-chloro-1,1,1-trifluoroethane		-720.0	-690.4									
C_2HCl	Chloroacetylene									242.0			54.3
C_2HClF_2	1-Chloro-2,2-difluoroethylene			-315.5			-289.1			303.0			72.1
C_2HCl_2F	1,1-Dichloro-2-fluoroethylene									313.9			76.5
$C_2HCl_2F_3$	2,2-Dichloro-1,1,1-trifluoroethane									352.8			102.5
C_2HCl_3	Trichloroethylene		-43.6	-9.0					228.4	324.8		124.4	80.3
C_2HCl_3O	Trichloroacetaldehyde		-234.5	-196.6								151.0	
C_2HCl_3O	Dichloroacetyl chloride		-280.4	-241.0									
$C_2HCl_3O_2$	Trichloroacetic acid	-503.3											
C_2HCl_5	Pentachloroethane		-187.6	-142.0								173.8	
C_2HF	Fluoroacetylene									231.7			52.4
C_2HF_3	Trifluoroethylene			-490.5									
$C_2HF_3O_2$	Trifluoroacetic acid		-1069.9	-1031.4									
C_2HF_5	Pentafluoroethane			-1100.4									
C_2H_2	Acetylene			227.4			209.9			200.9			44.0
$C_2H_2BrF_3$	2-Bromo-1,1,1-trifluoroethane			-694.5									
$C_2H_2Br_2$	*cis*-1,2-Dibromoethylene									311.3			68.8
$C_2H_2Br_2$	*trans*-1,2-Dibromoethylene									313.5			70.3
$C_2H_2Br_2Cl_2$	1,2-Dibromo-1,2-dichloroethane			-36.9									
$C_2H_2Br_4$	1,1,2,2-Tetrabromoethane											165.7	
$C_2H_2ClF_3$	2-Chloro-1,1,1-trifluoroethane									326.5			89.1
$C_2H_2Cl_2$	1,1-Dichloroethylene		-23.9	2.8		24.1	25.4		201.5	289.0		111.3	67.1
$C_2H_2Cl_2$	*cis*-1,2-Dichloroethylene		-26.4	4.6					198.4	289.6		116.4	65.1
$C_2H_2Cl_2$	*trans*-1,2-Dichloroethylene		-24.3	5.0		27.3	28.6		195.9	290.0		116.8	66.7
$C_2H_2Cl_2O$	Chloroacetyl chloride		-283.7	-244.8									
$C_2H_2Cl_2O_2$	Dichloroacetic acid		-496.3										

Molecular formula	Name	$\Delta_f H°$/kJ mol^{-1}			$\Delta_f G°$/kJ mol^{-1}			$S°$/J mol^{-1} K^{-1}			C_p/J mol^{-1} K^{-1}		
		Crys.	Liq.	Gas	Crys.	Liq.	Gas	Crys.	Liq.	Gas	Crys.	Liq.	Gas
$C_2H_2Cl_3NO$	2,2,2-Trichloroacetamide	-358.0											
$C_2H_2Cl_4$	1,1,1,2-Tetrachloroethane									356.0			102.7
$C_2H_2Cl_4$	1,1,2,2-Tetrachloroethane		-195.0	-149.2					246.9	362.8		162.3	100.8
$C_2H_2F_2$	1,1-Difluoroethylene			-335.0						266.2			60.1
$C_2H_2F_2$	*cis*-1,2-Difluoroethylene									268.3			58.2
$C_2H_2F_3I$	1,1,1-Trifluoro-2-iodoethane			-644.5									
$C_2H_2I_2$	*cis*-1,2-Diiodoethylene			-207.4									
C_2H_2O	Ketene		-67.9	-47.5			-48.3			247.6			51.8
$C_2H_2O_2$	Glyoxal			-212.0									
$C_2H_2O_4$	Oxalic acid	-821.7		-723.7				109.8			91.0		
$C_2H_2O_4Sr$	Strontium formate	-1393.3											
$C_2H_2S_2$	Thiirene			300.0			275.8			255.3			54.7
C_2H_3Br	Bromoethylene			79.2			81.8			275.8			55.5
C_2H_3BrO	Acetyl bromide		-223.5	-190.4									
C_2H_3Cl	Chloroethylene	-94.1	14.6	37.2			53.6			264.0	59.4		53.7
$C_2H_3ClF_2$	1-Chloro-1,1-difluoroethane									307.2			82.5
C_2H_3ClO	Acetyl chloride		-272.9	-242.8		-208.0	-205.8		200.8	295.1		117.0	67.8
$C_2H_3ClO_2$	Chloroacetic acid	-510.5		-435.2									
$C_2H_3Cl_2F$	1,1-Dichloro-1-fluoroethane									320.2			88.7
$C_2H_3Cl_3$	1,1,1-Trichloroethane		-177.4	-144.4					227.4	323.1		144.3	93.3
$C_2H_3Cl_3$	1,1,2-Trichloroethane		-190.8	-151.3					232.6	337.2		150.9	89.0
C_2H_3F	Fluoroethylene			-138.8									
C_2H_3FO	Acetyl fluoride		-467.2	-442.1									
$C_2H_3F_3$	1,1,1-Trifluoroethane			-744.6						279.9			78.2
$C_2H_3F_3$	1,1,2-Trifluoroethane			-730.7									
$C_2H_3F_3O$	2,2,2-Trifluoroethanol		-932.4	-888.4									
C_2H_3I	Iodoethylene									285.0			57.9
C_2H_3IO	Acetyl iodide		-163.5	-126.4									
$C_2H_3KO_2$	Potassium acetate	-723.0											
C_2H_3N	Acetonitrile		40.6	74.0		86.5	91.9		149.6	243.4		91.5	52.2
C_2H_3N	Isocyanomethane		130.8	163.5		159.5	165.7		159.0	246.9			52.9
C_2H_3NO	Methylisocyanate		-92.0										
$C_2H_3NO_3$	Oxamic acid	-661.2		-552.3									
C_2H_3NS	Methyl isothiocyanate	79.4											
$C_2H_3NaO_2$	Sodium acetate	-708.8			-607.2			123.0			79.9		
C_2H_4	Ethylene			52.4			68.4			219.3			42.9
C_2H_4BrCl	1-Bromo-2-chloroethane											130.1	
$C_2H_4Br_2$	1,1-Dibromoethane		-66.2							327.7			80.8
$C_2H_4Br_2$	1,2-Dibromoethane		-79.2	-37.5					223.3			136.0	
C_2H_4ClF	1-Chloro-1-fluoroethane			-313.4									
$C_2H_4Cl_2$	1,1-Dichloroethane		-158.4	-127.7		-73.8	-70.8		211.8	305.1		126.3	76.2
$C_2H_4Cl_2$	1,2-Dichloroethane		-166.8	-126.4						308.4		128.4	78.7

Molecular formula	Name	$\Delta_fH°$/kJ mol^{-1}			$\Delta_fG°$/kJ mol^{-1}			$S°$/J mol^{-1} K^{-1}			C_p/J mol^{-1} K^{-1}		
		Crys.	Liq.	Gas	Crys.	Liq.	Gas	Crys.	Liq.	Gas	Crys.	Liq.	Gas
$C_2H_4F_2$	1,1-Difluoroethane			-497.0						282.5			67.8
$C_2H_4I_2$	1,2-Diiodoethane	9.3		75.0									
$C_2H_4N_2O_2$	Oxamide	-504.4		-387.1									
$C_2H_4N_2O_2$	Ethanedial, dioxime	-90.5											
$C_2H_4N_2O_4$	1,1-Dinitroethane		-148.2										
$C_2H_4N_2O_4$	1,2-Dinitroethane		-165.2										
$C_2H_4N_2S_2$	Ethanedithioamide	-20.8		83.0									
$C_2H_4N_4$	1*H*-1,2,4-Triazol-3-amine	76.8											
C_2H_4O	Acetaldehyde		-192.2	-166.2		-127.6	-133.0		160.2	263.8		89.0	55.3
C_2H_4O	Ethylene oxide		-78.0	-52.6		-11.8	-13.0		153.9	242.5		88.0	47.9
C_2H_4OS	Thioacetic acid		-216.9	-175.1									
$C_2H_4O_2$	Acetic acid		-484.3	-432.2		-389.9	-374.2		159.8	283.5		123.3	63.4
$C_2H_4O_2$	Methyl formate		-386.1	-357.4						285.3		119.1	64.4
C_2H_4S	Thiirane		51.6	82.0			96.8			255.2			53.3
C_2H_4Si	Ethynylsilane									269.4			72.6
C_2H_5Br	Bromoethane		-90.5	-61.9		-25.8	-23.9		198.7	286.7		100.8	64.5
C_2H_5Cl	Chloroethane		-136.8	-112.1		-59.3	-60.4		190.8	276.0		104.3	62.8
C_2H_5ClO	Ethylene chlorohydrin		-295.4										
C_2H_5F	Fluoroethane									264.5			58.6
C_2H_5I	Iodoethane		-40.0	-8.1		14.7	19.2		211.7	306.0		115.1	66.9
C_2H_5N	Ethyleneimine		91.9	126.5									
C_2H_5NO	Acetamide	-317.0		-238.3				115.0			91.3		
C_2H_5NO	*N*-Methylformamide											123.8	
$C_2H_5NO_2$	Nitroethane		-143.9	-102.3								134.4	
$C_2H_5NO_2$	Glycine	-528.5		-392.1									
$C_2H_5NO_3$	2-Nitroethanol		-350.7										
$C_2H_5NO_3$	Ethyl nitrate		-190.4	-154.1									
C_2H_5NS	Thioacetamide	-71.7		11.4									
C_2H_6	Ethane			-84.0			-32.0			229.2			52.5
C_2H_6Cd	Dimethyl cadmium		63.6	101.6		139.0	146.9		201.9	303.0		132.0	
C_2H_6Hg	Dimethyl mercury		59.8	94.4		140.3	146.1		209.0	306.0			83.3
$C_2H_6N_2O$	*N*-Methylurea	-332.8											
$C_2H_6N_4O_2$	1,2-Hydrazinedicarboxamide	-498.7											
$C_2H_6N_4O_2$	Ethanedioic acid, dihydrazide	-295.2											
C_2H_6O	Ethanol		-277.6	-234.8		-174.8	-167.9		160.7	281.6		112.3	65.6
C_2H_6O	Dimethyl ether		-203.3	-184.1			-112.6			266.4			64.4
C_2H_6OS	Dimethyl sulfoxide		-204.2	-151.3		-99.9			188.3			153.0	
$C_2H_6O_2$	Ethylene glycol		-460.0	-392.2					163.2	303.8		148.6	82.7
$C_2H_6O_2S$	Dimethyl sulfone	-450.1		-373.1	-302.4		-272.7	142.0		310.6			100.0
$C_2H_6O_3S$	Dimethyl sulfite		-523.6	-483.4									
$C_2H_6O_4S$	Dimethyl sulfate		-735.5	-687.0									
C_2H_6S	Ethanethiol		-73.6	-46.1		-5.5	-4.8		207.0	296.2		117.9	72.7
C_2H_6S	Dimethyl sulfide		-65.3	-37.4					196.4	286.0		118.1	74.1
$C_2H_6S_2$	1,2-Ethanedithiol		-54.3	-9.7									
$C_2H_6S_2$	Dimethyl disulfide		-62.6	-24.7					235.4			146.1	

Molecular formula	Name	$\Delta_f H°$/kJ mol^{-1}			$\Delta_f G°$/kJ mol^{-1}			$S°$/J mol^{-1} K^{-1}			C_p/J mol^{-1} K^{-1}		
		Crys.	Liq.	Gas	Crys.	Liq.	Gas	Crys.	Liq.	Gas	Crys.	Liq.	Gas
C_2H_6Zn	Dimethyl zinc		23.4	53.0					201.6			129.2	
C_2H_7N	Ethylamine		-74.1	-47.5			36.3			283.8		130.0	71.5
C_2H_7N	Dimethylamine		-43.9	-18.8		70.0	68.5		182.3	273.1		137.7	70.7
C_2H_7NO	Ethanolamine											195.5	
C_2H_8ClN	Dimethylamine hydrochloride	-289.3											
$C_2H_8N_2$	1,2-Ethanediamine		-63.0	-18.0								172.6	
$C_2H_8N_2$	1,1-Dimethylhydrazine		48.9	84.1		206.4			198.0			164.1	
$C_2H_8N_2$	1,2-Dimethylhydrazine		52.7	92.2									
$C_2H_8N_2O_4$	Ammonium oxalate	-1123.0									226.0		
C_2HgO_4	Mercury(II) oxalate	-678.2											
C_2I_2	Diiodoacetylene									313.1			70.3
C_2I_4	Tetraiodoethylene	305.0											
$C_2K_2O_4$	Potassium oxalate	-1346.0											
C_2MgO_4	Magnesium oxalate	-1269.0											
C_2N_2	Cyanogen		285.9	306.7						241.9			56.8
$C_2N_4O_6$	Trinitroacetonitrile		183.7										
$C_2Na_2O_4$	Sodium oxalate			-1318.0									
C_2O_4Pb	Lead(II) oxalate	-851.4			-750.1			146.0			105.4		
C_3F_8	Perfluoropropane			-1783.2									
$C_3H_2N_2$	Malononitrile	186.4		265.5									
$C_3H_2O_2$	2-Propynoic acid		-193.2										
$C_3H_2O_3$	1,3-Dioxol-2-one		-459.9	-418.6									
$C_3H_3Cl_3$	1,2,3-Trichloropropene		-101.8										
$C_3H_3F_3$	3,3,3-Trifluoropropene			-614.2									
C_3H_3N	2-Propenenitrile		147.1	180.6									
C_3H_3NO	Oxazole		-48.0	-15.5									
C_3H_3NO	Isoxazole		42.1	78.6									
C_3H_4	Allene			190.5									
C_3H_4	Propyne			184.9									
C_3H_4	Cyclopropene			277.1									
$C_3H_4Cl_2$	2,3-Dichloropropene		-73.3										
$C_3H_4Cl_4$	1,1,1,3-Tetrachloropropane		-208.7										
$C_3H_4Cl_4$	1,2,2,3-Tetrachloropropane		-251.8										
$C_3H_4F_4O$	2,2,3,3-Tetrafluoro-1-propanol		-1114.9	-1061.3									
$C_3H_4N_2$	1*H*-Pyrazole		105.4	179.4									
$C_3H_4N_2$	Imidazole	49.8		132.9									
$C_3H_4O_2$	1,2-Propanedione		-309.1	-271.0									
$C_3H_4O_2$	Propenoic acid		-383.8									145.7	
$C_3H_4O_2$	2-Oxetanone		-329.9	-282.9					175.3			122.1	
$C_3H_4O_3$	Ethylene carbonate		-571.5	-508.4								133.9	
C_3H_5Br	*cis*-1-Bromopropene		7.9	40.8									
C_3H_5Br	3-Bromopropene		12.2	45.2									
C_3H_5BrO	Bromoacetone			-181.0									

Molecular formula	Name	$\Delta_f H°$/kJ mol^{-1} Crys.	$\Delta_f H°$/kJ mol^{-1} Liq.	$\Delta_f H°$/kJ mol^{-1} Gas	$\Delta_f G°$/kJ mol^{-1} Crys.	$\Delta_f G°$/kJ mol^{-1} Liq.	$\Delta_f G°$/kJ mol^{-1} Gas	$S°$/J mol^{-1} K^{-1} Crys.	$S°$/J mol^{-1} K^{-1} Liq.	$S°$/J mol^{-1} K^{-1} Gas	C_p/J mol^{-1} K^{-1} Crys.	C_p/J mol^{-1} K^{-1} Liq.	C_p/J mol^{-1} K^{-1} Gas
C_3H_5Cl	2-Chloropropene			-21.0									
C_3H_5Cl	3-Chloropropene											125.1	
C_3H_5ClO	Epichlorohydrin		-148.4	-107.8								131.6	
$C_3H_5ClO_2$	2-Chloropropanoic acid		-522.5	-475.8									
$C_3H_5ClO_2$	3-Chloropropanoic acid	-549.3											
$C_3H_5ClO_2$	Ethyl chloroformate		-505.3	-462.9									
$C_3H_5Cl_3$	1,2,3-Trichloropropane		-230.6	-182.9								183.6	
C_3H_5I	3-Iodopropene		53.7	91.5									
C_3H_5IO	Iodoacetone			-130.5									
$C_3H_5IO_2$	3-Iodopropanoic acid	-460.0											
C_3H_5N	Propanenitrile		15.5	51.7								119.3	
C_3H_5N	2-Propyn-1-amine		205.7										
C_3H_5N	Ethyl isocyanide		108.6	141.7									
$C_3H_5NO_3$	Nitroacetone		-278.6										
$C_3H_5NO_4$	Methyl nitroacetate		-464.0										
$C_3H_5N_3O_9$	Trinitroglycerol		-370.9	-270.9									
C_3H_6	Propene		4.0	20.0									
C_3H_6	Cyclopropane		35.2	53.3			104.5			237.5			55.6
$C_3H_6Br_2$	1,2-Dibromopropane		-113.6	-71.6									
$C_3H_6Cl_2$	1,2-Dichloropropane		-198.8	-162.8								149.1	
$C_3H_6Cl_2$	1,3-Dichloropropane		-199.9	-159.2									
$C_3H_6Cl_2$	2,2-Dichloropropane		-205.8	-173.2									
$C_3H_6Cl_2O$	2,3-Dichloro-1-propanol		-381.5	-316.3									
$C_3H_6Cl_2O$	1,3-Dichloro-2-propanol		-385.3	-318.4									
$C_3H_6I_2$	1,2-Diiodopropane			35.6									
$C_3H_6I_2$	1,3-Diiodopropane		-9.0										
$C_3H_6N_2O_2$	Propanediamide	-546.1											
$C_3H_6N_2O_2$	*N*-(Aminocarbonyl) acetamide	-544.2		-441.2									
$C_3H_6N_2O_4$	1,1-Dinitropropane		-163.2	-100.7									
$C_3H_6N_2O_4$	1,3-Dinitropropane		-207.1										
$C_3H_6N_2O_4$	2,2-Dinitropropane		-181.2										
C_3H_6O	Allyl alcohol		-171.8	-124.5								138.9	
C_3H_6O	Propanal		-215.6	-185.6						304.5			80.7
C_3H_6O	Acetone		-248.4	-217.1			-152.7		199.8	295.3		126.3	74.5
C_3H_6O	Methyloxirane		-123.0	-94.7					196.5	286.9		120.4	72.6
C_3H_6O	Oxetane		-110.8	-80.5									
$C_3H_6O_2$	Propanoic acid		-510.7	-455.7					191.0			152.8	
$C_3H_6O_2$	Ethyl formate											149.3	
$C_3H_6O_2$	Methyl acetate		-445.9	-413.3						324.4		141.9	86.0
$C_3H_6O_2$	1,3-Dioxolane		-333.5	-298.0								118.0	
$C_3H_6O_2S$	Thiolactic acid		-468.4										
$C_3H_6O_3$	Trioxane	-522.5		-465.9				133.0			111.4		
C_3H_6S	Thiacyclobutane		24.7	60.6			107.1		184.9	285.0			68.3
C_3H_6S	Methylthiirane		11.3	45.8									

Molecular formula	Name	$\Delta_f H°$/kJ mol^{-1} Crys.	$\Delta_f H°$/kJ mol^{-1} Liq.	$\Delta_f H°$/kJ mol^{-1} Gas	$\Delta_f G°$/kJ mol^{-1} Crys.	$\Delta_f G°$/kJ mol^{-1} Liq.	$\Delta_f G°$/kJ mol^{-1} Gas	$S°$/J mol^{-1} K^{-1} Crys.	$S°$/J mol^{-1} K^{-1} Liq.	$S°$/J mol^{-1} K^{-1} Gas	C_p/J mol^{-1} K^{-1} Crys.	C_p/J mol^{-1} K^{-1} Liq.	C_p/J mol^{-1} K^{-1} Gas
$C_3H_6S_2$	1,2-Dithiolane			0.0			47.7			313.5			86.5
$C_3H_6S_2$	1,3-Dithiolane			10.0			54.7			323.3			84.7
C_3H_7Br	1-Bromopropane		-121.9	-87.0									
C_3H_7Br	2-Bromopropane		-130.5	-99.4									
C_3H_7Cl	1-Chloropropane		-160.5	-131.9									
C_3H_7Cl	2-Chloropropane		-172.3	-144.9									
$C_3H_7ClO_2$	3-Chloro-1,2-propanediol		-525.3										
$C_3H_7ClO_2$	2-Chloro-1,3-propanediol		-517.5										
C_3H_7F	1-Fluoropropane			-285.9									
C_3H_7F	2-Fluoropropane			-293.5									
C_3H_7I	1-Iodopropane		-66.0	-30.0									
C_3H_7I	2-Iodopropane		-74.8	-40.3									
C_3H_7N	Allylamine		-10.0										
C_3H_7N	Cyclopropylamine		45.8	77.0					187.7			147.1	
C_3H_7NO	*N,N*-Dimethylformamide		-239.3	-192.4								150.6	
C_3H_7NO	Propanamide	-338.2		-259.0									
$C_3H_7NO_2$	1-Nitropropane		-167.2	-123.8									
$C_3H_7NO_2$	2-Nitropropane		-180.3	-138.9								170.3	
$C_3H_7NO_2$	*DL*-Alanine	-563.6											
$C_3H_7NO_2$	*D*-Alanine	-561.2											
$C_3H_7NO_2$	*L*-Alanine	-604.0		-465.9									
$C_3H_7NO_2$	β-Alanine	-558.0		-424.0									
$C_3H_7NO_2$	*N*-Methylglycine	-513.3		-367.3									
$C_3H_7NO_2S$	*L*-Cysteine	-534.1											
$C_3H_7NO_3$	Propyl nitrate		-214.5	-173.9									
$C_3H_7NO_3$	Isopropyl nitrate		-229.7	-191.0									
$C_3H_7NO_3$	*DL*-Serine	-739.0											
$C_3H_7NO_3$	*L*-Serine	-732.7											
C_3H_8	Propane		-120.9	-103.8			-23.4			270.3			73.6
$C_3H_8N_2O$	*N*-Ethylurea	-357.8											
$C_3H_8N_2O$	*N,N*-Dimethylurea	-319.1											
$C_3H_8N_2O$	*N,N'*-Dimethylurea	-312.1											
$C_3H_8N_2O_3$	Oxymethurea	-717.0											
C_3H_8O	1-Propanol		-302.6	-255.1					193.6	322.6		143.9	85.6
C_3H_8O	2-Propanol		-318.1	-272.6					181.1	309.2		156.5	89.3
C_3H_8O	Ethyl methyl ether			-216.4						309.2			93.3
$C_3H_8O_2$	1,2-Propylene glycol		-501.0	-429.8								190.8	
$C_3H_8O_2$	1,3-Propylene glycol		-480.8	-408.0									
$C_3H_8O_2$	Ethylene glycol monomethyl ether											171.1	
$C_3H_8O_2$	Dimethoxymethane		-377.8	-348.5					244.0			162.0	
$C_3H_8O_3$	Glycerol		-669.6	-577.9					206.3			218.9	
C_3H_8S	1-Propanethiol		-99.9	-67.8					242.5			144.6	
C_3H_8S	2-Propanethiol		-105.9	-76.2					233.5			145.3	
C_3H_8S	Ethyl methyl sulfide		-91.6	-59.6					239.1			144.6	

Molecular formula	Name	$\Delta_f H°$/kJ mol^{-1} Crys.	Liq.	Gas	$\Delta_f G°$/kJ mol^{-1} Crys.	Liq.	Gas	$S°$/J mol^{-1} K^{-1} Crys.	Liq.	Gas	C_p/J mol^{-1} K^{-1} Crys.	Liq.	Gas
$C_3H_8S_2$	1,3-Propanedithiol		-79.4	-29.8									
C_3H_9Al	Trimethyl aluminum		-136.4	-74.1		-9.9			209.4			155.6	
C_3H_9B	Trimethylborane		-143.1	-124.3		-32.1	-35.9		238.9	314.7			88.5
C_3H_9ClSi	Trimethylchlorosilane		-382.8	-352.8		-246.4	-243.5		278.2	369.1			
C_3H_9N	Propylamine		-101.5	-70.1			39.9			325.4		164.1	91.2
C_3H_9N	Isopropylamine		-112.3	-83.7			32.2		218.3	312.2		163.8	97.5
C_3H_9N	Trimethylamine		-45.7	-23.6					208.5	287.1		137.9	91.8
$C_3H_{10}ClN$	Propylamine hydrochloride	-354.7											
$C_3H_{10}ClN$	Trimethylamine hydrochloride	-282.9											
$C_3H_{10}N_2$	1,2-Propanediamine, (±)-		-97.8	-53.6									
$C_3H_{10}Si$	Trimethylsilane									331.0			117.9
$C_3H_{12}BN$	Trimethylamine borane	-142.5			70.7			187.0					
$C_3H_{12}BN$	Aminetrimethylboron	-284.1			-79.3			218.0					
C_4F_8	Perfluorocyclobutane			-1542.6									
C_4F_{10}	Perfluorobutane											127.2	
$C_4H_2N_2$	*trans*-2-Butenedinitrile	268.2		340.2									
$C_4H_2O_3$	Maleic anhydride	-469.8		-398.3									
$C_4H_2O_4$	2-Butynedioic acid	-577.3											
$C_4H_3NO_3$	2-Nitrofuran	-104.1		-28.8									
$C_4H_4BrNO_2$	*N*-Bromosuccinimide	-335.9											
$C_4H_4ClNO_2$	*N*-Chlorosuccinimide	-357.9											
$C_4H_4N_2$	Succinonitrile		139.7	209.7					191.6			145.6	
$C_4H_4N_2$	Pyrazine	139.8		196.1									
$C_4H_4N_2$	Pyrimidine		145.9	195.7									
$C_4H_4N_2$	Pyridazine		224.9	278.3									
$C_4H_4N_2O_2$	Uracil	-429.4		-302.9							120.5		
$C_4H_4N_2O_3$	2,4,6-Trihydroxypyrimidine	-634.7											
C_4H_4O	Furan		-62.3	-34.8					177.0	267.2		114.8	65.4
$C_4H_4O_2$	Diketene		-233.1	-190.3									
$C_4H_4O_3$	Succinic anhydride	-608.6		-527.9									
$C_4H_4O_4$	Maleic acid	-789.4		-679.4				160.8			137.0		
$C_4H_4O_4$	Fumaric acid	-811.7		-675.8				168.0			142.0		
C_4H_4S	Thiophene		80.2	114.9			126.1		181.2	278.8		123.8	72.8
C_4H_5N	*trans*-2-Butenenitrile		95.1	134.3									
C_4H_5N	3-Butenenitrile		117.8	159.7									
C_4H_5N	Pyrrole		63.1	108.2					156.4			127.7	
C_4H_5N	Cyclopropanecarbonitrile		140.8	182.8									
$C_4H_5NO_2$	Succinimide	-459.0		-375.4									
C_4H_5NS	4-Methylthiazole		67.9	111.8									
$C_4H_5N_3O$	Cytosine	-221.3									132.6		
C_4H_6	1,2-Butadiene		138.6	162.3									
C_4H_6	1,3-Butadiene		88.5	110.0					199.0			123.6	
C_4H_6	1-Butyne		141.4	165.2									
C_4H_6	2-Butyne		119.1	145.7									

Molecular formula	Name	$\Delta_f H°$/kJ mol^{-1} Crys.	Liq.	Gas	$\Delta_f G°$/kJ mol^{-1} Crys.	Liq.	Gas	$S°$/J mol^{-1} K^{-1} Crys.	Liq.	Gas	C_p/J mol^{-1} K^{-1} Crys.	Liq.	Gas
C_4H_6	Cyclobutene			156.7									
$C_4H_6N_2O_2$	2,5-Piperazinedione	-446.5											
C_4H_6O	Divinyl ether		-39.8	-13.6									
C_4H_6O	*trans*-2-Butenal		-138.7	-100.6									
$C_4H_6O_2$	*trans*-Crotonic acid												
$C_4H_6O_2$	Methacrylic acid											161.1	
$C_4H_6O_2$	Vinyl acetate		-349.2	-314.4									
$C_4H_6O_2$	Methyl acrylate		-362.2	-333.0					239.5			158.8	
$C_4H_6O_2$	γ-Butyrolactone		-420.9	-366.5								141.4	
$C_4H_6O_3$	Acetic anhydride		-624.4	-572.5									
$C_4H_6O_3$	Propylene carbonate		-613.2	-582.5								218.6	
$C_4H_6O_4$	Succinic acid	-940.5		-823.0				167.3			153.1		
$C_4H_6O_4$	Dimethyl oxalate	-756.3		-708.9									
C_4H_6S	2,3-Dihydrothiophene		52.9	90.7			133.5			303.5			79.8
C_4H_6S	2,5-Dihydrothiophene		47.0	86.9			131.6			297.1			83.3
C_4H_7ClO	2-Chloroethyl vinyl ether		-208.1	-170.1									
$C_4H_7ClO_2$	2-Chlorobutanoic acid		-575.5										
$C_4H_7ClO_2$	3-Chlorobutanoic acid		-556.3										
$C_4H_7ClO_2$	4-Chlorobutanoic acid		-566.3										
$C_4H_7ClO_2$	Propyl chlorocarbonate		-533.4	-492.7									
C_4H_7N	Butanenitrile		-5.8	33.6									
C_4H_7N	2-Methylpropanenitrile		-13.8	23.4									
C_4H_7NO	2-Pyrrolidone		-286.2										
C_4H_7NO	2-Methyl-2-oxazoline		-169.5	-130.5									
$C_4H_7NO_4$	Iminodiacetic acid	-932.6											
$C_4H_7NO_4$	Ethyl nitroacetate		-487.1										
$C_4H_7NO_4$	*L*-Aspartic acid	-973.3											
$C_4H_7N_3O$	4*H*-Imidazol-4-one, 2-amino-1,5-dihydro-1-methyl-	-238.5											
C_4H_8	1-Butene		-20.8	0.1					227.0			118.0	
C_4H_8	*cis*-2-Butene		-29.8	-7.1					219.9			127.0	
C_4H_8	*trans*-2-Butene		-33.3	-11.4									
C_4H_8	Isobutene		-37.5	-16.9									
C_4H_8	Cyclobutane		3.7	27.7									
C_4H_8	Methylcyclopropane		1.7										
$C_4H_8Br_2$	1,2-Dibromobutane		-142.1	-91.6									
$C_4H_8Br_2$	1,3-Dibromobutane		-148.0										
$C_4H_8Br_2$	1,4-Dibromobutane		-140.3	-87.8									
$C_4H_8Br_2$	2,3-Dibromobutane		-139.6	-102.0									
$C_4H_8Br_2$	1,2-Dibromo-2-methylpropane		-156.6	-113.3									
$C_4H_8Cl_2$	1,3-Dichlorobutane		-237.3	-195.0									
$C_4H_8Cl_2$	1,4-Dichlorobutane		-229.8	-183.4									
$C_4H_8Cl_2O$	Bis(2-chloroethyl) ether											220.9	

Molecular formula	Name	$\Delta_f H°$/kJ mol^{-1} Crys.	Liq.	Gas	$\Delta_f G°$/kJ mol^{-1} Crys.	Liq.	Gas	$S°$/J mol^{-1} K^{-1} Crys.	Liq.	Gas	C_p/J mol^{-1} K^{-1} Crys.	Liq.	Gas
$C_4H_8I_2$	1,4-Diiodobenzene		-30.0										
$C_4H_8N_2O_2$	Succinamide	-581.2											
$C_4H_8N_2O_2$	Dimethylglyoxime	-199.7											
$C_4H_8N_2O_3$	*L*-Asparagine	-789.4											
$C_4H_8N_2O_3$	*N*-Glycylglycine	-747.7											
$C_4H_8N_2O_4$	1,4-Dinitrobutane		-237.5										
C_4H_8O	Ethyl vinyl ether		-167.4	-140.8									
C_4H_8O	1,2-Epoxybutane		-168.9						230.9			147.0	
C_4H_8O	Butanal		-239.2	-204.8					246.6	343.7		163.7	103.4
C_4H_8O	Isobutanal		-247.3	-215.7									
C_4H_8O	2-Butanone		-273.3	-238.5					239.1	339.9		158.7	101.7
C_4H_8O	Tetrahydrofuran		-216.2	-184.1					204.3	302.4		124.0	76.3
C_4H_8OS	*S*-Ethyl thioacetate		-268.2	-228.1									
$C_4H_8O_2$	Butanoic acid		-533.8	-475.9					222.2			178.6	
$C_4H_8O_2$	2-Methylpropanoic acid											173.0	
$C_4H_8O_2$	Propyl formate		-500.3	-462.7									
$C_4H_8O_2$	Ethyl acetate		-479.3	-443.6					257.7			170.7	
$C_4H_8O_2$	1,3-Dioxane		-379.7	-340.6								143.9	
$C_4H_8O_2$	1,4-Dioxane		-353.9	-315.3					270.2			152.1	
$C_4H_8O_2$	2-Methyl-1,3-dioxolane		-386.9	-352.0									
$C_4H_8O_2S$	Sulfolane											180.0	
C_4H_8S	Tetrahydrothiophene		-72.9	-34.1			45.8			309.6			92.5
$C_4H_8S_2$	1,3-Dithiane			-10.0			72.4			333.5			110.4
C_4H_9Br	1-Bromobutane		-143.8	-107.1									
C_4H_9Br	2-Bromobutane		-154.9	-120.3									
C_4H_9Br	2-Bromo-2-methylpropane		-164.4	-132.4									
C_4H_9Cl	1-Chlorobutane		-188.1	-154.4									
C_4H_9Cl	2-Chlorobutane		-192.8	-161.1									
C_4H_9Cl	1-Chloro-2-methylpropane		-191.1	-159.3									
C_4H_9Cl	2-Chloro-2-methylpropane		-211.3	-182.2									
C_4H_9ClO	2-Chloroethyl ethyl ether		-335.6	-301.3									
C_4H_9I	1-Iodo-2-methylpropane											162.3	
C_4H_9I	2-Iodo-2-methylpropane		-107.5	-72.1									
C_4H_9N	Cyclobutanamine		5.6	41.2									
C_4H_9N	Pyrrolidine		-41.1	-3.6					204.1			156.6	
C_4H_9NO	Butanamide		-346.9	-282.0									
C_4H_9NO	*N*-Methylpropanamide											179.0	
C_4H_9NO	2-Methylpropanamide	-368.6		-282.6									
C_4H_9NO	*N,N*-Dimethylacetamide		-278.3	-228.0								175.6	
C_4H_9NO	Morpholine											164.8	
$C_4H_9NO_2$	1-Nitrobutane		-192.5	-143.9									
$C_4H_9NO_2$	2-Nitroisobutane		-217.2	-177.1									
$C_4H_9NO_2$	Propyl carbamate	-552.6		-471.4									
$C_4H_9NO_2$	4-Aminobutanoic acid	-581.0		-441.0									
$C_4H_9NO_3$	3-Nitro-2-butanol		-390.0										

Molecular formula	Name	$\Delta_f H°$/kJ mol^{-1}			$\Delta_f G°$/kJ mol^{-1}			$S°$/J mol^{-1} K^{-1}			C_p/J mol^{-1} K^{-1}		
		Crys.	Liq.	Gas	Crys.	Liq.	Gas	Crys.	Liq.	Gas	Crys.	Liq.	Gas
$C_4H_9NO_3$	2-Methyl-2-nitro-1-propanol	-410.1											
$C_4H_9NO_3$	*DL*-Threonine	-758.8											
$C_4H_9NO_3$	*L*-Threonine	-807.2											
$C_4H_9N_3O_2$	Creatine	-537.2											
C_4H_{10}	Butane		-147.3	-125.7								140.9	
C_4H_{10}	Isobutane		-154.2	-134.2									
$C_4H_{10}Hg$	Diethyl mercury		30.1	75.3								182.8	
$C_4H_{10}N_2$	Piperazine	-45.6											
$C_4H_{10}N_2O$	Trimethylurea	-330.5											
$C_4H_{10}N_2O_2$	*N*-Nitrodiethylamine		-106.2	-53.0									
$C_4H_{10}N_2O_4$	*L*-Asparagine, monohydrate	-1086.6											
$C_4H_{10}O$	1-Butanol		-327.3	-274.9					225.8			177.2	
$C_4H_{10}O$	2-Butanol		-342.6	-292.8					214.9	359.5		196.9	112.7
$C_4H_{10}O$	2-Methyl-1-propanol		-334.7	-283.8					214.7			181.5	
$C_4H_{10}O$	2-Methyl-2-propanol		-359.2	-312.5					193.3	326.7		218.6	113.6
$C_4H_{10}O$	Diethyl ether		-279.5	-252.1					172.4	342.7		175.6	119.5
$C_4H_{10}O$	Methyl propyl ether		-266.0	-238.1					262.9			165.4	
$C_4H_{10}O$	Isopropyl methyl ether		-278.8	-252.0					253.8			161.9	
$C_4H_{10}OS$	Diethyl sulfoxide		-268.0	-205.6									
$C_4H_{10}O_2$	1,2-Butanediol, (±)-		-523.6										
$C_4H_{10}O_2$	1,3-Butanediol		-501.0	-433.2									
$C_4H_{10}O_2$	1,4-Butanediol		-505.3	-428.7					223.4			200.1	
$C_4H_{10}O_2$	2,3-Butanediol		-541.5	-482.3								213.0	
$C_4H_{10}O_2$	2-Methyl-1,2-propanediol		-539.7										
$C_4H_{10}O_2$	Ethylene glycol monoethyl ether											210.8	
$C_4H_{10}O_2$	Ethylene glycol dimethyl ether		-376.6									193.3	
$C_4H_{10}O_2$	Dimethylacetal		-420.6	-389.7									
$C_4H_{10}O_2$	*tert*-Butyl hydroperoxide		-293.6	-245.9									
$C_4H_{10}O_3$	Diethylene glycol		-628.5	-571.2								244.8	
$C_4H_{10}O_3S$	Diethyl sulfite		-600.7	-552.2									
$C_4H_{10}O_4S$	Sulfuric acid, diethyl ester		-813.2	-756.3									
$C_4H_{10}S$	1-Butanethiol		-124.7	-88.0								171.2	
$C_4H_{10}S$	2-Butanethiol		-131.0	-96.9									
$C_4H_{10}S$	2-Methyl-1-propanethiol		-132.0	-97.3									
$C_4H_{10}S$	2-Methyl-2-propanethiol		-140.5	-109.6									
$C_4H_{10}S$	Diethyl sulfide		-119.4	-83.5					269.3	368.1		171.4	117.0
$C_4H_{10}S$	Methyl propyl sulfide		-118.5	-82.2					272.5			171.6	
$C_4H_{10}S$	Isopropyl methyl sulfide		-124.7	-90.5					263.1			172.4	
$C_4H_{10}S_2$	1,4-Butanedithiol		-105.7	-50.6									
$C_4H_{10}S_2$	Diethyl disulfide		-120.1	-79.4					269.3			171.4	
$C_4H_{11}N$	Butylamine		-127.6	-91.9								179.2	
$C_4H_{11}N$	*sec*-Butylamine		-137.5	-104.6									
$C_4H_{11}N$	*tert*-Butylamine		-150.6	-121.0								192.1	
$C_4H_{11}N$	Isobutylamine		-132.6	-98.7								183.2	
$C_4H_{11}N$	Diethylamine		-103.7	-72.2								169.2	

Molecular formula	Name	$\Delta_f H°$/kJ mol^{-1}			$\Delta_f G°$/kJ mol^{-1}			$S°$/J mol^{-1} K^{-1}			C_p/J mol^{-1} K^{-1}		
		Crys.	Liq.	Gas	Crys.	Liq.	Gas	Crys.	Liq.	Gas	Crys.	Liq.	Gas
$C_4H_{11}NO$	*N,N*-Dimethylethanolamine		-253.7	-203.6									
$C_4H_{11}NO_2$	Diethanolamine	-493.8		-397.1							233.5		
$C_4H_{11}NO_3$	2-Amino-2-(hydroxymethyl)-1,3-propanediol	-717.8											
$C_4H_{12}BrN$	Tetramethylammonium bromide	-251.0											
$C_4H_{12}ClN$	Diethylamine hydrochloride	-358.6											
$C_4H_{12}ClN$	Tetramethylammonium chloride	-276.4											
$C_4H_{12}IN$	Tetramethylammonium iodide	-203.9											
$C_4H_{12}N_2$	2-Methyl-1,2-propanediamine		-133.9	-90.3									
$C_4H_{12}Pb$	Tetramethyl lead		97.9	135.9									
$C_4H_{12}Si$	Tetramethylsilane		-264.0	-239.1		-100.0	-99.9		277.3	359.0		204.1	143.9
$C_4H_{12}Sn$	Tetramethylstannane		-52.3	-18.8									
$C_4H_{13}N_3$	Diethylenetriamine											254.0	
C_4N_2	2-Butynedinitrile		500.4	529.2									
C_4NiO_4	Nickel carbonyl		-633.0	-602.9		-588.2	-587.2		313.4	410.6		204.6	145.2
C_5FeO_5	Iron pentacarbonyl		-774.0			-705.3			338.1			240.6	
$C_5H_2F_6O_2$	Hexafluoroacetylacetone	-2286.7											
$C_5H_3NO_5$	5-Nitro-2-furancarboxylic acid	-516.8											
$C_5H_4N_4$	1*H*-Purine	169.4											
$C_5H_4N_4O$	Hypoxanthine	-110.8						145.6			134.5		
$C_5H_4N_4O_2$	Xanthine	-379.6						161.1			151.3		
$C_5H_4N_4O_3$	Uric acid	-618.8						173.2			166.1		
$C_5H_4O_2$	Furfural		-201.6	-151.0								163.2	
$C_5H_4O_3$	2-Furancarboxylic acid	-498.4		-390.0									
$C_5H_4O_3$	3-Methyl-2,5-furandione		-504.5	-447.2									
$C_5H_5F_3O_2$	1,1,1-Trifluoro-2,4-pentanedione		-1040.2	-993.3									
C_5H_5N	Pyridine		100.2	140.4								132.7	
C_5H_5NO	1*H*-Pyrrole-2-carboxaldehyde	-106.4											
$C_5H_5N_5$	Adenine	96.9		205.7							147.0		
$C_5H_5N_5O$	Guanine	-183.9											
C_5H_6	*cis*-3-Penten-1-yne		226.5										
C_5H_6	*trans*-3-Penten-1-yne		228.2										
C_5H_6	1,3-Cyclopentadiene		105.9	134.3									
$C_5H_6N_2O_2$	Thymine	-462.8		-328.7							150.8		
$C_5H_6O_2$	Furfuryl alcohol		-276.2	-211.8								204.0	
$C_5H_6O_4$	*trans*-1-Propene-1,2-dicarboxylic acid	-824.4											
C_5H_6S	2-Methylthiophene		44.6	83.5					218.5			149.8	
C_5H_6S	3-Methylthiophene		43.1	82.5									
C_5H_7N	*trans*-3-Pentenenitrile		80.9	125.7									
C_5H_7N	Cyclobutanecarbonitrile		103.0	147.4									

Molecular formula	Name	$\Delta_f H°$/kJ mol^{-1} Crys.	Liq.	Gas	$\Delta_f G°$/kJ mol^{-1} Crys.	Liq.	Gas	$S°$/J mol^{-1} K^{-1} Crys.	Liq.	Gas	C_p/J mol^{-1} K^{-1} Crys.	Liq.	Gas
C_5H_7N	1-Methylpyrrole		62.4	103.1									
C_5H_7N	2-Methylpyrrole		23.3	74.0									
C_5H_7N	3-Methylpyrrole		20.5	70.2									
$C_5H_7NO_2$	Ethyl cyanoacetate											220.2	
C_5H_8	1,2-Pentadiene			140.7									
C_5H_8	*cis*-1,3-Pentadiene			81.4									
C_5H_8	*trans*-1,3-Pentadiene			76.1									
C_5H_8	1,4-Pentadiene			105.7									
C_5H_8	2,3-Pentadiene			133.1									
C_5H_8	3-Methyl-1,2-butadiene		101.2										
C_5H_8	2-Methyl-1,3-butadiene		48.2	75.5					229.3			152.6	
C_5H_8	Cyclopentene		4.3	34.0					201.2			122.4	
C_5H_8	Methylenecyclobutane		93.8	121.6									
C_5H_8	Spiropentane		157.5	185.2					193.7			134.5	
$C_5H_8N_4O_{12}$	Pentaerythritol tetranitrate	-538.6											
C_5H_8O	Cyclopentanone		-235.9	-192.1									
$C_5H_8O_2$	4-Pentenoic acid	-430.6											
$C_5H_8O_2$	Methyl *trans*-2-butenoate		-382.9	-341.9									
$C_5H_8O_2$	Methyl methacrylate											191.2	
$C_5H_8O_2$	Allyl acetate											184.1	
$C_5H_8O_2$	2,4-Pentanedione		-423.8	-382.0									
$C_5H_8O_2$	4-Methyl-gamma-butyrolactone		-461.3	-406.5									
$C_5H_8O_2$	Tetrahydro-2*H*-pyran-2-one		-436.7	-379.6									
$C_5H_8O_4$	Glutaric acid	-960.0											
$C_5H_9ClO_2$	Propyl chloroacetate		-515.5	-467.0									
C_5H_9N	Pentanenitrile		-33.1	10.5									
C_5H_9N	2,2-Dimethylpropanenitrile		-39.8	-2.3					232.0			179.4	
C_5H_9N	1,2,5,6-Tetrahydropyridine		33.5										
C_5H_9NO	2-Piperidinone	-306.6											
C_5H_9NO	*N*-Methyl-2-pyrrolidone		-262.2									307.8	
$C_5H_9NO_2$	*L*-Proline	-515.2		-366.2									
$C_5H_9NO_4$	*D*-Glutamic acid	-1005.3											
$C_5H_9NO_4$	*L*-Glutamic acid	-1009.7											
C_5H_{10}	1-Pentene		-46.9	-21.1					262.6			154.0	
C_5H_{10}	*cis*-2-Pentene		-53.7	-27.6					258.6			151.7	
C_5H_{10}	*trans*-2-Pentene		-58.2	-31.9					256.5			157.0	
C_5H_{10}	2-Methyl-1-butene		-61.1	-35.2					254.0			157.2	
C_5H_{10}	3-Methyl-1-butene		-51.5	-27.5					253.3			156.1	
C_5H_{10}	2-Methyl-2-butene		-68.6	-41.7					251.0			152.8	
C_5H_{10}	Cyclopentane		-105.1	-76.4					204.5			128.8	
C_5H_{10}	Methylcyclobutane		-44.5										
C_5H_{10}	Ethylcyclopropane		-24.8										
C_5H_{10}	1,1-Dimethylcyclopropane		-33.3	-8.2									
C_5H_{10}	*cis*-1,2-Dimethylcyclopropane		-26.3										

Molecular formula	Name	$\Delta_f H°$/kJ mol^{-1} Crys.	Liq.	Gas	$\Delta_f G°$/kJ mol^{-1} Crys.	Liq.	Gas	$S°$/J mol^{-1} K^{-1} Crys.	Liq.	Gas	C_p/J mol^{-1} K^{-1} Crys.	Liq.	Gas
C_5H_{10}	*trans*-1,2-Dimethyl-cyclopropane		-30.7										
$C_5H_{10}Br_2$	1,3-Dibromo-2-methylpropane			-137.6									
$C_5H_{10}N_2O$	*N*-Nitrosopiperidine		-31.1	16.6									
$C_5H_{10}N_2O_2$	*N*-Nitropiperidine		-93.0	-44.5									
$C_5H_{10}N_2O_3$	*L*-Glutamine	-826.4											
$C_5H_{10}O$	Cyclopentanol		-300.1	-242.5					206.3			184.1	
$C_5H_{10}O$	Pentanal		-267.2	-228.4									
$C_5H_{10}O$	2-Pentanone		-297.3	-258.8								184.1	
$C_5H_{10}O$	3-Pentanone		-296.5	-257.9					266.0			190.9	
$C_5H_{10}O$	3-Methyl-2-butanone		-299.5	-262.6					268.5			179.9	
$C_5H_{10}O$	3,3-Dimethyloxetane		-182.2	-148.2									
$C_5H_{10}O$	Tetrahydropyran		-258.3	-223.4									
$C_5H_{10}OS$	S-Propyl thioacetate		-294.5	-250.4									
$C_5H_{10}O_2$	Pentanoic acid		-559.4	-491.9					259.8			210.3	
$C_5H_{10}O_2$	2-Methylbutanoic acid		-554.5										
$C_5H_{10}O_2$	3-Methylbutanoic acid		-561.6	-510.0									
$C_5H_{10}O_2$	2,2-Dimethylpropanoic acid	-564.5		-491.3									
$C_5H_{10}O_2$	Butyl formate											200.2	
$C_5H_{10}O_2$	Propyl acetate											196.2	
$C_5H_{10}O_2$	Isopropyl acetate		-518.9	-481.6								199.4	
$C_5H_{10}O_2$	Ethyl propanoate		-502.7	-463.4									
$C_5H_{10}O_2$	*cis*-1,2-Cyclopentanediol	-485.0											
$C_5H_{10}O_2$	*trans*-1,2-Cyclopentanediol	-490.1											
$C_5H_{10}O_2$	4-Methyl-1,3-dioxane	-416.1		-376.9									
$C_5H_{10}O_2$	(Ethoxymethyl)oxirane		-296.5										
$C_5H_{10}O_2$	Tetrahydrofurfuryl alcohol		-435.7	-369.1									
$C_5H_{10}O_3$	Diethyl carbonate		-681.5	-637.9									
$C_5H_{10}O_3$	Ethylene glycol momomethyl ether acetate											310.0	
$C_5H_{10}O_3$	Ethyl lactate											254.0	
$C_5H_{10}O_4$	1,2,3-Propanetriol, 1-acetate, (±)-		-909.2										
$C_5H_{10}O_5$	*D*-Ribose	-1047.2											
$C_5H_{10}O_5$	*D*-Xylose	-1057.8											
$C_5H_{10}O_5$	α-*D*-Arabinopyranose	-1057.9											
$C_5H_{10}S$	Thiacyclohexane		-106.3	-63.5			53.1		218.2	323.0		163.3	109.7
$C_5H_{10}S$	Cyclopentanethiol		-89.5	-48.1					256.9			165.2	
$C_5H_{11}Br$	1-Bromopentane		-170.2	-128.9									
$C_5H_{11}Cl$	1-Chloropentane		-213.2	-174.9									
$C_5H_{11}Cl$	2-Chloro-2-methylbutane		-235.7	-202.2									
$C_5H_{11}Cl$	1-Chloro-3-methylbutane		-216.0	-179.7									
$C_5H_{11}Cl$	2-Chloro-3-methylbutane		-226.6	-185.1									
$C_5H_{11}N$	Cyclopentylamine		-95.1	-54.9					241.0			181.2	

Molecular formula	Name	$\Delta_f H°$/kJ mol^{-1}			$\Delta_f G°$/kJ mol^{-1}			$S°$/J mol^{-1} K^{-1}			C_p/J mol^{-1} K^{-1}		
		Crys.	Liq.	Gas	Crys.	Liq.	Gas	Crys.	Liq.	Gas	Crys.	Liq.	Gas
$C_5H_{11}N$	Piperidine		-86.4	-47.1					210.0			179.9	
$C_5H_{11}NO$	Pentanamide	-379.5		-290.2									
$C_5H_{11}NO$	2,2-Dimethylpropanamide	-399.7		-313.1									
$C_5H_{11}NO_2$	1-Nitropentane		-215.4										
$C_5H_{11}NO_2$	*DL*-Valine	-628.9											
$C_5H_{11}NO_2$	*L*-Valine	-617.9		-455.1									
$C_5H_{11}NO_2$	5-Aminopentanoic acid	-604.1		-460.0									
$C_5H_{11}NO_2S$	*L*-Methionine	-577.5		-413.5									
$C_5H_{11}NO_4$	2-Ethyl-2-nitro-1,3-propanediol	-606.4											
C_5H_{12}	Pentane		-173.5	-146.9								167.2	
C_5H_{12}	Isopentane		-178.4	-153.6					260.4			164.8	
C_5H_{12}	Neopentane		-190.2	-168.0									
$C_5H_{12}N_2O$	Butylurea	-419.5											
$C_5H_{12}N_2O$	*tert*-Butylurea	-417.4											
$C_5H_{12}N_2O$	*N,N*-Diethylurea	-372.2											
$C_5H_{12}N_2O$	Tetramethylurea	-262.2											
$C_5H_{12}N_2S$	Tetramethylthiourea	-38.1		44.9									
$C_5H_{12}O$	1-Pentanol		-351.6	-294.6								208.1	
$C_5H_{12}O$	2-Pentanol		-365.2	-311.0									
$C_5H_{12}O$	3-Pentanol		-368.9	-314.9								239.7	
$C_5H_{12}O$	2-Methyl-1-butanol		-356.6	-301.4									
$C_5H_{12}O$	3-Methyl-1-butanol		-356.4	-300.7									
$C_5H_{12}O$	2-Methyl-2-butanol		-379.5	-329.3								247.1	
$C_5H_{12}O$	3-Methyl-2-butanol		-366.6	-313.5									
$C_5H_{12}O$	2,2-Dimethyl-1-propanol		-399.4										
$C_5H_{12}O$	Butyl methyl ether		-290.6	-258.1					295.3			192.7	
$C_5H_{12}O$	*tert*-Butyl methyl ether		-313.6	-283.7					265.3			187.5	
$C_5H_{12}O$	Ethyl propyl ether		-303.6	-272.0					295.0			197.2	
$C_5H_{12}O_2$	1,5-Pentanediol		-528.8	-450.8									
$C_5H_{12}O_2$	2,2-Dimethyl-1,3-propanediol	-551.2											
$C_5H_{12}O_2$	Diethoxymethane		-450.5	-414.7									
$C_5H_{12}O_2$	1,1-Dimethoxypropane		-443.6										
$C_5H_{12}O_2$	2,2-Dimethoxypropane		-459.4	-429.9									
$C_5H_{12}O_3$	Diethylene glycol monomethyl ether											271.1	
$C_5H_{12}O_3$	2-(Hydroxymethyl)-2-methyl-1,3-propanediol	-744.6											
$C_5H_{12}O_4$	Pentaerythritol	-920.6		-776.7									
$C_5H_{12}O_5$	Xylitol	-1118.5											
$C_5H_{12}S$	1-Pentanethiol		-151.3	-110.0									
$C_5H_{12}S$	2-Methyl-1-butanethiol		-154.4	-114.9									
$C_5H_{12}S$	3-Methyl-1-butanethiol		-154.4	-114.9									
$C_5H_{12}S$	2-Methyl-2-butanethiol		-162.8	-127.1					290.1			198.1	
$C_5H_{12}S$	3-Methyl-2-butanethiol		-158.8	-121.3									

Molecular formula	Name	$\Delta_f H°$/kJ mol^{-1}			$\Delta_f G°$/kJ mol^{-1}			$S°$/J mol^{-1} K^{-1}			C_p/J mol^{-1} K^{-1}		
		Crys.	Liq.	Gas	Crys.	Liq.	Gas	Crys.	Liq.	Gas	Crys.	Liq.	Gas
$C_5H_{12}S$	2,2-Dimethyl-1-propanethiol		-165.4	-129.0									
$C_5H_{12}S$	Butyl methyl sulfide		-142.9	-102.4					307.5			200.9	
$C_5H_{12}S$	*tert*-Butyl methyl sulfide		-157.1	-121.3					276.1			199.9	
$C_5H_{12}S$	Ethyl propyl sulfide		-144.8	-104.8					309.5			198.4	
$C_5H_{12}S$	Ethyl isopropyl sulfide		-156.1	-118.3									
$C_5H_{13}N$	Pentylamine											218.0	
$C_5H_{14}N_2$	*N,N,N′,N′*-Tetramethyl-methanediamine		-51.1	-18.2									
C_6ClF_5	Chloropentafluorobenzene	-858.4		-809.3									
C_6Cl_6	Hexachlorobenzene	-127.6		-35.5				260.2			201.2		
C_6F_6	Hexafluorobenzene		-991.3	-955.4					280.8			221.6	
C_6F_{10}	Perfluorocyclohexene		-1963.5	-1932.7									
C_6F_{12}	Perfluorocyclohexane		-2406.3	-2370.4									
C_6HF_5	Pentafluorobenzene	-852.7	-841.8	-806.5									
C_6HF_5O	Pentafluorophenol	-1024.1	-1007.7										
$C_6H_2F_4$	1,2,4,5-Tetrafluorobenzene		-683.8										
$C_6H_3Cl_3$	1,2,3-Trichlorobenzene	-70.8		3.8									
$C_6H_3Cl_3$	1,2,4-Trichlorobenzene		-63.1	-8.1									
$C_6H_3Cl_3$	1,3,5-Trichlorobenzene	-78.4		-13.4									
$C_6H_3N_3O_8$	2,4,6-Trinitro-1,3-benzenediol	-467.5											
$C_6H_4Cl_2$	*o*-Dichlorobenzene		-17.5	30.2								162.4	
$C_6H_4Cl_2$	*m*-Dichlorobenzene		-20.7	25.7									
$C_6H_4Cl_2$	*p*-Dichlorobenzene	-42.3		22.5				175.4			147.8		
$C_6H_4F_2$	*o*-Difluorobenzene		-330.0	-293.8					222.6			159.0	
$C_6H_4F_2$	*m*-Difluorobenzene		-343.9	-309.2					223.8			159.1	
$C_6H_4F_2$	*p*-Difluorobenzene		-342.3	-306.7								157.5	
$C_6H_4N_2O_5$	2,4-Dinitrophenol	-232.7		-128.1									
$C_6H_4O_2$	*p*-Benzoquinone	-185.7		-122.9							129.0		
C_6H_5Br	Bromobenzene		60.9						219.2			154.3	
C_6H_5Cl	Chlorobenzene		11.1	52.0								150.1	
C_6H_5ClO	*m*-Chlorophenol	-206.4	-189.3										
C_6H_5ClO	*p*-Chlorophenol	-197.7	-181.3										
C_6H_5F	Fluorobenzene		-150.6	-115.9					205.9			146.4	
C_6H_5I	Iodobenzene		117.2	164.9					205.4			158.7	
$C_6H_5NO_2$	Nitrobenzene		12.5	67.5								185.8	
$C_6H_5NO_2$	3-Pyridinecarboxylic acid	-344.9		-221.5									
$C_6H_5NO_3$	2-Nitrophenol	-202.4											
$C_6H_5N_3$	1*H*-Benzotriazole	236.5		335.5									
$C_6H_5N_3O_4$	2,3-Dinitroaniline	-11.7											
$C_6H_5N_3O_4$	2,4-Dinitroaniline	-67.8											
$C_6H_5N_3O_4$	2,5-Dinitroaniline	-44.3											
$C_6H_5N_3O_4$	2,6-Dinitroaniline	-50.6											
$C_6H_5N_3O_4$	3,5-Dinitroaniline	-38.9											
C_6H_6	1,5-Hexadiyne		384.2										
C_6H_6	Benzene		49.1	82.9		124.5	129.7		173.4	269.2		136.0	82.4

Molecular formula	Name	$\Delta_f H°$/kJ mol^{-1}			$\Delta_f G°$/kJ mol^{-1}			$S°$/J mol^{-1} K^{-1}			C_p/J mol^{-1} K^{-1}		
		Crys.	Liq.	Gas	Crys.	Liq.	Gas	Crys.	Liq.	Gas	Crys.	Liq.	Gas
$C_6H_6N_2O_2$	*o*-Nitroaniline	-26.1	-9.4	63.8							166.0		
$C_6H_6N_2O_2$	*m*-Nitroaniline	-38.3	-14.4	58.4							158.8		
$C_6H_6N_2O_2$	*p*-Nitroaniline	-42.0	-20.7	58.8							167.0		
C_6H_6O	Phenol	-165.1		-96.4				144.0			127.4		
C_6H_6O	2-Vinylfuran		-10.3	27.8									
$C_6H_6O_2$	*p*-Hydroquinone	-364.5		-265.3							136.0		
$C_6H_6O_2$	Pyrocatechol	-354.1		-267.5									
$C_6H_6O_2$	Resorcinol	-368.0		-274.7									
$C_6H_6O_3$	1,2,3-Benzenetriol	-551.1		-434.2									
$C_6H_6O_3$	1,2,4-Benzenetriol	-563.8		-444.0									
$C_6H_6O_3$	1,3,5-Benzenetriol	-584.6		-452.9									
$C_6H_6O_3$	3,4-Dimethyl-2,5-furandione	-581.4											
$C_6H_6O_6$	*cis*-1-Propene-1,2,3-tricarboxylic acid	-1224.4											
$C_6H_6O_6$	*trans*-1-Propene-1,2,3-tricarboxylic acid	-1232.7											
C_6H_6S	Benzenethiol		63.7	111.3					222.8			173.2	
C_6H_7N	Aniline		31.6	87.5			-7.0			317.9		191.9	107.9
C_6H_7N	2-Methylpyridine		56.7	99.2								158.6	
C_6H_7N	3-Methylpyridine		61.9	106.4					216.3			158.7	
C_6H_7N	4-Methylpyridine		59.2	103.8					209.1			159.0	
C_6H_7N	1-Cyclopentenecarbonitrile		111.5	156.5									
$C_6H_8N_2$	Adiponitrile		85.1	149.5								128.7	
$C_6H_8N_2$	*o*-Phenylenediamine	-0.3											
$C_6H_8N_2$	*m*-Phenylenediamine	-7.8						154.5			159.6		
$C_6H_8N_2$	*p*-Phenylenediamine	3.0											
$C_6H_8N_2$	Phenylhydrazine		141.0	202.9								217.0	
$C_6H_8N_2S$	Bis(2-cyanoethyl) sulfide		96.3										
$C_6H_8O_4$	Dimethyl maleate											263.2	
$C_6H_8O_6$	*L*-Ascorbic acid	-1164.6											
$C_6H_8O_7$	2-Hydroxy-1,2,3-propanetricarboxylic acid	-1543.8											
$C_6H_9Cl_3O_2$	Butyl trichloroacetate		-545.8	-492.3									
$C_6H_9Cl_3O_2$	Isobutyl trichloroacetate		-553.4	-500.2									
C_6H_9N	Cyclopentanecarbonitrile		0.7	44.1									
C_6H_9N	2,4-Dimethylpyrrole	-422.3											
C_6H_9N	2,5-Dimethylpyrrole		-16.7	39.8									
$C_6H_9NO_3$	Triacetamide		-610.5	-550.1									
$C_6H_9NO_6$	Nitrilotriacetic acid	-1311.9											
$C_6H_9N_3O_2$	*L*-Histidine	-466.7											
C_6H_{10}	1,5-Hexadiene		54.1	84.2									
C_6H_{10}	3,3-Dimethyl-1-butyne		78.4										
C_6H_{10}	Cyclohexene		-38.5	-5.0					214.6			148.3	
C_6H_{10}	1-Methylcyclopentene		-36.4	-3.8									
C_6H_{10}	3-Methylcyclopentene		-23.7	7.4									

Molecular formula	Name	$\Delta_f H°$/kJ mol^{-1}			$\Delta_f G°$/kJ mol^{-1}			$S°$/J mol^{-1} K^{-1}			C_p/J mol^{-1} K^{-1}		
		Crys.	Liq.	Gas	Crys.	Liq.	Gas	Crys.	Liq.	Gas	Crys.	Liq.	Gas
C_6H_{10}	4-Methylcyclopentene		-17.6	14.6									
$C_6H_{10}Cl_2O_2$	Butyl dichloroacetate		-550.1	-497.8									
$C_6H_{10}O$	Cyclohexanone		-271.2	-226.1								182.2	
$C_6H_{10}O$	2-Methylcyclopentanone		-265.2										
$C_6H_{10}O$	Mesityl oxide											212.5	
$C_6H_{10}O_2$	Ethyl *trans*-2-butenoate		-420.0	-375.6									
$C_6H_{10}O_2$	Methyl cyclobutane-carboxylate		-395.0	-350.2									
$C_6H_{10}O_3$	Ethyl acetoacetate											248.0	
$C_6H_{10}O_3$	Propanoic anhydride		-679.1	-626.5									
$C_6H_{10}O_4$	Adipic acid	-994.3											
$C_6H_{10}O_4$	Diethyl oxalate		-805.5	-742.0									
$C_6H_{10}O_4$	Ethylene glycol diacetate											310.0	
$C_6H_{11}Cl$	Chlorocyclohexane		-207.2	-163.7									
$C_6H_{11}ClO_2$	Ethyl 4-chlorobutanoate		-566.5	-513.8									
$C_6H_{11}ClO_2$	Propyl 3-chloropropanoate		-537.6	-485.7									
$C_6H_{11}ClO_2$	Butyl chloroacetate		-538.4	-487.4									
$C_6H_{11}NO$	Caprolactam	-329.4		-239.6							156.8		
$C_6H_{11}NO$	1-Methyl-2-piperidinone		-293.0										
C_6H_{12}	1-Hexene		-74.2	-43.5					295.2			183.3	
C_6H_{12}	*cis*-2-Hexene		-83.9	-52.3									
C_6H_{12}	*trans*-2-Hexene		-85.5	-53.9									
C_6H_{12}	*cis*-3-Hexene		-78.9	-47.6									
C_6H_{12}	*trans*-3-Hexene		-86.1	-54.4									
C_6H_{12}	2-Methyl-1-pentene		-90.0	-59.4									
C_6H_{12}	3-Methyl-1-pentene		-78.2	-49.5									
C_6H_{12}	4-Methyl-1-pentene		-80.0	-51.3									
C_6H_{12}	3-Methyl-*cis*-2-pentene		-94.5	-62.3									
C_6H_{12}	4-Methyl-*cis*-2-pentene		-87.0	-57.5									
C_6H_{12}	3-Methyl-*trans*-2-pentene		-94.6	-63.1									
C_6H_{12}	4-Methyl-*trans*-2-pentene		-91.6	-61.5									
C_6H_{12}	2-Ethyl-1-butene		-87.1	-56.0									
C_6H_{12}	2,3-Dimethyl-1-butene		-93.2	-62.4									
C_6H_{12}	3,3-Dimethyl-1-butene		-87.5	-60.3									
C_6H_{12}	2,3-Dimethyl-2-butene		-101.4	-68.1					270.2			174.7	
C_6H_{12}	Cyclohexane		-156.4	-123.4								154.9	
C_6H_{12}	Methylcyclopentane		-137.9	-106.2									
C_6H_{12}	Ethylcyclobutane		-59.0	-27.5									
C_6H_{12}	1,1,2-Trimethylcyclo-propane		-96.2										
C_6H_{12}	2-Methyl-2-pentene		-98.5	-66.9									
$C_6H_{12}N_2O_4S_2$	*L*-Cystine	-1032.7											
$C_6H_{12}O$	Cyclohexanol		-348.2	-286.2								208.2	
$C_6H_{12}O$	*cis*-2-Methylcyclopentanol		-345.5										
$C_6H_{12}O$	Butyl vinyl ether		-218.8	-182.6								232.0	

Molecular formula	Name	$\Delta_f H°$/kJ mol^{-1} Crys.	$\Delta_f H°$/kJ mol^{-1} Liq.	$\Delta_f H°$/kJ mol^{-1} Gas	$\Delta_f G°$/kJ mol^{-1} Crys.	$\Delta_f G°$/kJ mol^{-1} Liq.	$\Delta_f G°$/kJ mol^{-1} Gas	$S°$/J mol^{-1} K^{-1} Crys.	$S°$/J mol^{-1} K^{-1} Liq.	$S°$/J mol^{-1} K^{-1} Gas	C_p/J mol^{-1} K^{-1} Crys.	C_p/J mol^{-1} K^{-1} Liq.	C_p/J mol^{-1} K^{-1} Gas
$C_6H_{12}O$	2-Hexanone		-322.0	-278.9								213.3	
$C_6H_{12}O$	3-Hexanone		-320.2	-277.6					305.3			216.9	
$C_6H_{12}O$	Methyl isobutyl ketone											213.3	
$C_6H_{12}O$	2-Methyl-3-pentanone		-325.9	-286.0									
$C_6H_{12}O$	3,3-Dimethyl-2-butanone		-328.6	-290.6									
$C_6H_{12}O_2$	Hexanoic acid		-583.8	-511.9									
$C_6H_{12}O_2$	Butyl acetate		-529.2	-485.3								227.8	
$C_6H_{12}O_2$	Isobutyl acetate											233.8	
$C_6H_{12}O_2$	Ethyl butanoate											228.0	
$C_6H_{12}O_2$	Methyl pentanoate		-514.2	-471.1								229.3	
$C_6H_{12}O_2$	Methyl 2,2-dimethylpropanoate		-530.0	-491.2								257.9	
$C_6H_{12}O_2$	Diacetone alcohol											221.3	
$C_6H_{12}O_3$	Ethylene glycol ethyl ether acetate											376.0	
$C_6H_{12}O_3$	Paraldehyde		-673.1	-631.7									
$C_6H_{12}O_6$	β-*D*-Fructose	-1265.6											
$C_6H_{12}O_6$	*D*-Galactose	-1286.3											
$C_6H_{12}O_6$	α-*D*-Glucose	-1273.3											
$C_6H_{12}O_6$	*D*-Mannose	-1263.0											
$C_6H_{12}O_6$	*L*-Sorbose	-1271.5											
$C_6H_{12}S$	Cyclohexanethiol		-140.7	-96.2					255.6			192.6	
$C_6H_{12}S$	Cyclopentyl methyl sulfide		-109.8	-64.7									
$C_6H_{13}Br$	1-Bromohexane		-194.2	-148.3					453.0			203.5	
$C_6H_{13}Cl$	2-Chlorohexane		-246.1	-204.3									
$C_6H_{13}N$	Cyclohexylamine		-147.6	-104.0									
$C_6H_{13}N$	(±)-2-Methylpiperidine		-124.9	-84.4									
$C_6H_{13}NO$	Hexanamide		-397.9	-324.2									
$C_6H_{13}NO$	*N*-Butylacetamide		-380.9	-305.9									
$C_6H_{13}NO_2$	*DL*-Leucine	-640.6											
$C_6H_{13}NO_2$	*D*-Leucine	-637.3											
$C_6H_{13}NO_2$	*L*-Leucine	-637.4		-486.8							200.1		
$C_6H_{13}NO_2$	*DL*-Isoleucine	-635.3											
$C_6H_{13}NO_2$	*L*-Isoleucine	-637.8											
$C_6H_{13}NO_2$	Norleucine	-639.1											
$C_6H_{13}NO_2$	6-Aminohexanoic acid	-637.3											
C_6H_{14}	Hexane		-198.7	-166.9								195.6	
C_6H_{14}	2-Methylpentane		-204.6	-174.6					290.6			193.7	
C_6H_{14}	3-Methylpentane		-202.4	-171.9					292.5			190.7	
C_6H_{14}	2,2-Dimethylbutane		-213.8	-185.9					272.5			191.9	
C_6H_{14}	2,3-Dimethylbutane		-207.4	-178.1					287.8			189.7	
$C_6H_{14}N_2$	Azopropane		11.5	51.3									
$C_6H_{14}N_2O_2$	Lysine	-678.7											
$C_6H_{14}N_4O_2$	*D*-Arginine	-623.5						250.6			232.0		
$C_6H_{14}O$	1-Hexanol		-377.5	-315.9					287.4			240.4	

Molecular formula	Name	$\Delta_f H°$/kJ mol^{-1}			$\Delta_f G°$/kJ mol^{-1}			$S°$/J mol^{-1} K^{-1}			C_p/J mol^{-1} K^{-1}		
		Crys.	Liq.	Gas	Crys.	Liq.	Gas	Crys.	Liq.	Gas	Crys.	Liq.	Gas
$C_6H_{14}O$	2-Hexanol		-392.0	-333.5									
$C_6H_{14}O$	3-Hexanol		-392.4									286.2	
$C_6H_{14}O$	2-Methyl-1-pentanol											248.0	
$C_6H_{14}O$	3-Methyl-2-pentanol											275.9	
$C_6H_{14}O$	4-Methyl-2-pentanol		-394.7									273.0	
$C_6H_{14}O$	2-Methyl-3-pentanol		-396.4										
$C_6H_{14}O$	3-Methyl-3-pentanol											293.4	
$C_6H_{14}O$	Dipropyl ether		-328.8	-293.0					323.9			221.6	
$C_6H_{14}O$	Diisopropyl ether		-351.5	-319.2								216.8	
$C_6H_{14}O$	Butyl ethyl ether											159.0	
$C_6H_{14}O$	Ethyl *tert*-butyl ether			-313.9									
$C_6H_{14}OS$	Dipropyl sulfoxide		-329.4	-254.9									
$C_6H_{14}O_2$	1,2-Hexanediol		-577.1	-490.1									
$C_6H_{14}O_2$	1,6-Hexanediol	-569.9	-548.6	-461.2									
$C_6H_{14}O_2$	Hexylene glycol											336.0	
$C_6H_{14}O_2$	Ethylene glycol monobutyl ether											281.0	
$C_6H_{14}O_2$	1,1-Diethoxyethane		-491.4	-453.5									
$C_6H_{14}O_2$	1,2-Diethoxyethane		-451.4	-408.1								259.4	
$C_6H_{14}O_3$	Diethylene glycol monoethyl ether											301.0	
$C_6H_{14}O_3$	Diethylene glycol dimethyl ether											274.1	
$C_6H_{14}O_3$	Trimethylolpropane	-750.9											
$C_6H_{14}O_4$	Triethylene glycol		-804.3	-725.0									
$C_6H_{14}O_4S$	Dipropyl sulfate		-859.0	-792.0									
$C_6H_{14}O_6$	Galactitol		-1317.0										
$C_6H_{14}O_6$	*D*-Mannitol		-1314.5										
$C_6H_{14}S$	1-Hexanethiol		-175.7	-129.9									
$C_6H_{14}S$	2-Methyl-2-pentanethiol		-188.3	-148.3									
$C_6H_{14}S$	2,3-Dimethyl-2-butanethiol		-187.1	-147.9									
$C_6H_{14}S$	Diisopropyl sulfide		-181.6	-142.0					313.0			232.0	
$C_6H_{14}S$	Butyl ethyl sulfide		-172.3	-127.8									
$C_6H_{14}S$	Methyl pentyl sulfide		-167.1	-121.8									
$C_6H_{14}S_2$	Dipropyl disulfide		-171.5	-118.3									
$C_6H_{15}B$	Triethylborane		-194.6	-157.7		9.4	16.1		336.7	437.8		241.2	
$C_6H_{15}N$	Dipropylamine		-156.1	-116.0									
$C_6H_{15}N$	Diisopropylamine		-178.5	-143.8									
$C_6H_{15}N$	Triethylamine		-127.7	-92.7								219.9	
$C_6H_{15}NO$	2-Diethylaminoethanol		-305.9										
$C_6H_{15}NO_3$	Triethanolamine	-664.2		-558.3							389.0		
$C_6H_{18}N_3OP$	Hexamethylphosphoric triamide											321.0	
$C_6H_{18}OSi_2$	Hexamethyldisiloxane		-815.0	-777.7		-541.5	-534.5		433.8	535.0		311.4	238.5
C_6MoO_6	Molybdenum hexacarbonyl	-982.8		-912.1	-877.7		-856.0	325.9		490.0	242.3		205.0

Molecular formula	Name	$\Delta_f H°$/kJ mol^{-1}			$\Delta_f G°$/kJ mol^{-1}			$S°$/J mol^{-1} K^{-1}			C_p/J mol^{-1} K^{-1}		
		Crys.	Liq.	Gas	Crys.	Liq.	Gas	Crys.	Liq.	Gas	Crys.	Liq.	Gas
C_6N_4	Tetracyanoethylene	623.8		705.0									
C_7F_8	Perfluorotoluene		-1311.1						355.5			262.3	
C_7F_{14}	Perfluoromethylcyclohexane		-2931.1	-2897.2								353.1	
C_7F_{16}	Perfluoroheptane		-3420.0	-3383.6					561.8			419.0	
$C_7H_3F_5$	2,3,4,5,6-Pentafluorotoluene		-883.8	-842.7					306.4			225.8	
$C_7H_4Cl_2O$	*m*-Chlorobenzoyl chloride		-189.7										
$C_7H_4N_2O_6$	3,5-Dinitrobenzoic acid	-409.8											
C_7H_5ClO	Benzoyl chloride		-158.0	-103.2									
$C_7H_5ClO_2$	*o*-Chlorobenzoic acid	-404.5		-325.0									
$C_7H_5ClO_2$	*o*-Chlorobenzoic acid	-424.3		-342.3									
$C_7H_5ClO_2$	*p*-Chlorobenzoic acid	-428.9		-341.0							163.2		
$C_7H_5F_3$	(Trifluoromethyl)benzene											188.4	
C_7H_5N	Benzonitrile		163.2	215.7					209.1			165.2	
C_7H_5NO	Benzoxazole	-24.2		44.8									
$C_7H_5NO_4$	2-Nitrobenzoic acid	-378.8											
$C_7H_5NO_4$	3-Nitrobenzoic acid	-394.7											
$C_7H_5NO_4$	4-Nitrobenzoic acid	-392.2											
$C_7H_6N_2$	1*H*-Benzimidazole	79.5		181.7									
$C_7H_6N_2$	1*H*-Indazole	151.9		243.0									
C_7H_6O	Benzaldehyde		-87.0	-36.7					221.2			172.0	
$C_7H_6O_2$	Benzoic acid	-385.2		-294.0				167.6			146.8		
$C_7H_6O_2$	3-(2-Furanyl)-2-propenal	-182.0		-105.9									
$C_7H_6O_2$	Salicylaldehyde											222.0	
$C_7H_6O_3$	Salicylic acid	-589.9		-494.8									
C_7H_7Cl	*o*-Chlorotoluene											166.8	
C_7H_7Cl	(Chloromethyl)benzene		-32.5	18.9									
C_7H_7F	*p*-Fluorotoluene		-186.9	-147.4								171.2	
C_7H_7NO	Benzamide	-202.6		-100.9									
$C_7H_7NO_2$	2-Aminobenzoic acid		-380.4	-296.0									
$C_7H_7NO_2$	Aniline-3-carboxylic acid		-389.8	-283.6									
$C_7H_7NO_2$	Aniline-4-carboxylic acid		-391.9	-296.7									
$C_7H_7NO_2$	*o*-Nitrotoluene		-9.7										
$C_7H_7NO_2$	*m*-Nitrotoluene		-31.5										
$C_7H_7NO_2$	*p*-Nitrotoluene	-48.1		31.0							172.3		
$C_7H_7NO_2$	(Nitromethyl)benzene		-22.8	30.7									
$C_7H_7NO_2$	Salicylaldoxime	-183.7											
C_7H_8	Toluene		12.4	50.5								157.3	
$C_7H_8N_2O$	Phenylurea	-218.6											
C_7H_8O	*o*-Cresol	-204.6		-128.6				165.4			154.6		
C_7H_8O	*m*-Cresol		-194.0	-132.3					212.6			224.9	
C_7H_8O	*p*-Cresol	-199.3		-125.4				167.3			150.2		
C_7H_8O	Benzyl alcohol		-160.7	-100.4					216.7			217.9	
C_7H_8O	Anisole		-114.8	-67.9									
C_7H_9N	Benzylamine		34.2	94.4									
C_7H_9N	*o*-Methylaniline		-6.3	56.4			167.6			351.0			130.2

Molecular formula	Name	$\Delta_f H°$/kJ mol^{-1} Crys.	$\Delta_f H°$/kJ mol^{-1} Liq.	$\Delta_f H°$/kJ mol^{-1} Gas	$\Delta_f G°$/kJ mol^{-1} Crys.	$\Delta_f G°$/kJ mol^{-1} Liq.	$\Delta_f G°$/kJ mol^{-1} Gas	$S°$/J mol^{-1} K^{-1} Crys.	$S°$/J mol^{-1} K^{-1} Liq.	$S°$/J mol^{-1} K^{-1} Gas	C_p/J mol^{-1} K^{-1} Crys.	C_p/J mol^{-1} K^{-1} Liq.	C_p/J mol^{-1} K^{-1} Gas
C_7H_9N	*m*-Methylaniline		-8.1	54.6			165.4			352.5			125.5
C_7H_9N	*p*-Methylaniline	-23.5		55.3			167.7			347.0			126.2
C_7H_9N	*N*-Methylaniline											207.1	
C_7H_9N	1-Cyclohexenecarbonitrile		48.1	101.6									
C_7H_9N	2,3-Dimethylpyridine		19.4	67.1					243.7			189.5	
C_7H_9N	2,4-Dimethylpyridine		16.1	63.6					248.5			184.8	
C_7H_9N	2,5-Dimethylpyridine		18.7	66.5					248.8			184.7	
C_7H_9N	2,6-Dimethylpyridine		12.7	58.1					244.2			185.2	
C_7H_9N	3,4-Dimethylpyridine		18.3	68.8					240.7			191.8	
C_7H_9N	3,5-Dimethylpyridine		22.5	72.0					241.7			184.5	
$C_7H_{10}O_2$	Ethyl 2-pentynoate		-301.8	-250.3									
$C_7H_{10}O_2$	Methyl 2-hexynoate		-242.7										
$C_7H_{11}Cl_3O_2$	3-Methylbutyl trichloroacetate		-580.9	-523.1									
$C_7H_{11}N$	Cyclohexanecarbonitrile		-47.2	4.8									
C_7H_{12}	Methylenecyclohexane		-61.3	-25.2									
C_7H_{12}	Vinylcyclopentane		-34.8										
C_7H_{12}	1-Ethylcyclopentene		-53.3	-19.8									
C_7H_{12}	Bicyclo[2.2.1]heptane	-95.1		-54.8									
C_7H_{12}	1-Methylbicyclo(3,1,0) hexane		-33.2	1.7									
$C_7H_{12}O$	2-Methylenecyclohexanol	-277.6											
$C_7H_{12}O_4$	Diethyl malonate											285.0	
$C_7H_{13}ClO_2$	Butyl 2-chloropropanoate		-571.7	-517.3									
$C_7H_{13}ClO_2$	Isobutyl 2-chloropropanoate		-603.1	-549.6									
$C_7H_{13}ClO_2$	Butyl 3-chloropropanoate		-557.9	-502.3									
$C_7H_{13}ClO_2$	Isobutyl 3-chloropropanoate		-572.6	-517.3									
$C_7H_{13}ClO_2$	Propyl 2-chlorobutanoate		-630.7	-578.4									
$C_7H_{13}N$	Heptanenitrile		-82.8	-31.0									
C_7H_{14}	1-Heptene		-97.9	-62.3					327.6			211.8	
C_7H_{14}	*cis*-2-Heptene		-105.1										
C_7H_{14}	*trans*-2-Heptene		-109.5										
C_7H_{14}	*cis*-3-Heptene		-104.3										
C_7H_{14}	*trans*-3-Heptene		-109.3										
C_7H_{14}	5-Methyl-1-hexene		-100.0	-65.7									
C_7H_{14}	3-Methyl-*cis*-3-hexene		-115.9	-79.4									
C_7H_{14}	3-Methyl-*trans*-3-hexene		-112.7	-76.8									
C_7H_{14}	2,4-Dimethyl-1-pentene		-117.0	-83.8									
C_7H_{14}	4,4-Dimethyl-1-pentene		-110.6	-81.6									
C_7H_{14}	2,4-Dimethyl-2-pentene		-123.1	-88.7									
C_7H_{14}	*cis*-4,4-Dimethyl-2-pentene		-105.3	-72.6									
C_7H_{14}	*trans*-4,4-Dimethyl-2-pentene		-121.7	-88.8									
C_7H_{14}	2-Ethyl-3-methyl-1-butene		-114.1	-79.5									
C_7H_{14}	2,3,3-Trimethyl-1-butene		-117.7	-85.5									
C_7H_{14}	Cycloheptane		-156.6	-118.1									
C_7H_{14}	Methylcyclohexane		-190.1	-154.7								184.8	

Molecular formula	Name	$\Delta_f H°$/kJ mol^{-1}			$\Delta_f G°$/kJ mol^{-1}			$S°$/J mol^{-1} K^{-1}			C_p/J mol^{-1} K^{-1}		
		Crys.	Liq.	Gas	Crys.	Liq.	Gas	Crys.	Liq.	Gas	Crys.	Liq.	Gas
C_7H_{14}	Ethylcyclopentane		-163.4	-126.9					279.9				
C_7H_{14}	1,1-Dimethylcyclopentane		-172.1	-138.2									
C_7H_{14}	*cis*-1,2-Dimethylcyclopentane		-165.3	-129.5					269.2				
C_7H_{14}	*trans*-1,2-Dimethylcyclopentane		-171.2	-136.6									
C_7H_{14}	*cis*-1,3-Dimethylcyclopentane		-170.1	-135.8									
C_7H_{14}	*trans*-1,3-Dimethylcyclopentane		-168.1	-133.6									
C_7H_{14}	1,1,2,2-Tetramethylcyclopropane		-119.8										
$C_7H_{14}Br_2$	1,2-Dibromoheptane		-212.3	-157.9									
$C_7H_{14}O$	1-Heptanal		-311.5	-263.8					335.4			230.1	
$C_7H_{14}O$	2-Heptanone											232.6	
$C_7H_{14}O$	2,2-Dimethyl-3-pentanone		-356.1	-313.6									
$C_7H_{14}O$	2,4-Dimethyl-3-pentanone		-352.9	-311.3					318.0			233.7	
$C_7H_{14}O$	*cis*-2-Methylcyclohexanol		-390.2	-327.0									
$C_7H_{14}O$	*trans*-2-Methylcyclohexanol		-415.7	-352.5									
$C_7H_{14}O$	*cis*-3-Methylcyclohexanol		-416.1	-350.9									
$C_7H_{14}O$	*trans*-3-Methylcyclohexanol		-394.4	-329.1									
$C_7H_{14}O$	*cis*-4-Methylcyclohexanol		-413.2	-347.5									
$C_7H_{14}O$	*trans*-4-Methylcyclohexanol		-433.3	-367.2									
$C_7H_{14}O_2$	Heptanoic acid		-610.2	-536.2								265.4	
$C_7H_{14}O_2$	Pentyl acetate											261.0	
$C_7H_{14}O_2$	Isopentyl acetate											248.5	
$C_7H_{14}O_2$	Ethyl pentanoate		-553.0	-505.9									
$C_7H_{14}O_2$	Ethyl 3-methylbutanoate		-571.0	-527.0									
$C_7H_{14}O_2$	Ethyl 2,2-dimethylpropanoate		-577.2	-536.0									
$C_7H_{14}O_2$	Methyl hexanoate		-540.2	-492.2									
$C_7H_{14}O_6$	α-Methylglucoside	-1233.3											
$C_7H_{15}Br$	1-Bromoheptane		-218.4	-167.8									
C_7H_{16}	Heptane		-224.2	-187.6								224.7	
C_7H_{16}	2-Methylhexane		-229.5	-194.5					323.3			222.9	
C_7H_{16}	3-Methylhexane		-226.4	-191.3									
C_7H_{16}	3-Ethylpentane		-224.9	-189.5					314.5			219.6	
C_7H_{16}	2,2-Dimethylpentane		-238.3	-205.7					300.3			221.1	
C_7H_{16}	2,3-Dimethylpentane		-233.1	-198.7									
C_7H_{16}	2,4-Dimethylpentane		-234.6	-201.6					303.2			224.2	
C_7H_{16}	3,3-Dimethylpentane		-234.2	-201.0									
C_7H_{16}	2,2,3-Trimethylbutane		-236.5	-204.4					292.2			213.5	
$C_7H_{16}O$	1-Heptanol		-403.3	-336.5								272.1	
$C_7H_{16}O$	*tert*-Butyl isopropyl ether		-392.8	-358.1									
$C_7H_{16}O_2$	1,7-Heptanediol		-574.2										

Molecular formula	Name	$\Delta_f H°$/kJ mol^{-1}			$\Delta_f G°$/kJ mol^{-1}			$S°$/J mol^{-1} K^{-1}			C_p/J mol^{-1} K^{-1}		
		Crys.	Liq.	Gas	Crys.	Liq.	Gas	Crys.	Liq.	Gas	Crys.	Liq.	Gas
$C_7H_{16}O_2$	2,2-Diethoxypropane		-538.9	-506.9									
$C_7H_{16}S$	1-Heptanethiol		-200.5	-149.9									
$C_8H_4O_3$	Phthalic anhydride	-460.1		-371.4				180.0			160.0		
$C_8H_5NO_2$	1*H*-Indole-2,3-dione	-268.2											
$C_8H_6O_4$	Phthalic acid	-782.0						207.9			188.1		
$C_8H_6O_4$	Isophthalic acid	-803.0		-696.3									
$C_8H_6O_4$	Terephthalic acid	-816.1		-717.9									
C_8H_6S	Benzo[b]thiophene	100.6		166.3									
C_8H_7N	1*H*-Indole	86.6		156.5									
C_8H_8	Styrene		103.8	147.9								182.0	
C_8H_8O	Phenyl vinyl ether		-26.2	22.7									
C_8H_8O	Acetophenone		-142.5	-86.7									
$C_8H_8O_2$	*o*-Toluic acid	-416.5									174.9		
$C_8H_8O_2$	*m*-Toluic acid	-426.1									163.6		
$C_8H_8O_2$	*p*-Toluic acid	-429.2									169.0		
$C_8H_8O_2$	Methyl benzoate		-343.5	-287.9								221.3	
$C_8H_8O_3$	Methyl salicylate											249.0	
C_8H_{10}	1,7-Octadiyne		334.4										
C_8H_{10}	Ethylbenzene		-12.3	29.9								183.2	
C_8H_{10}	*o*-Xylene		-24.4	19.1								186.1	
C_8H_{10}	*m*-Xylene		-25.4	17.3								183.0	
C_8H_{10}	*p*-Xylene		-24.4	18.0								181.5	
$C_8H_{10}O$	*o*-Ethylphenol		-208.8	-145.2									
$C_8H_{10}O$	*m*-Ethylphenol		-214.3	-146.1									
$C_8H_{10}O$	*p*-Ethylphenol	-224.4		-144.1							206.9		
$C_8H_{10}O$	2,3-Xylenol	-241.1		-157.2									
$C_8H_{10}O$	2,4-Xylenol		-228.7	-163.8									
$C_8H_{10}O$	2,5-Xylenol	-246.6		-161.6									
$C_8H_{10}O$	2,6-Xylenol	-237.4		-162.1									
$C_8H_{10}O$	3,4-Xylenol	-242.3		-157.3									
$C_8H_{10}O$	3,5-Xylenol	-244.4		-162.4									
$C_8H_{10}O$	Phenetole		-152.6	-101.6								228.5	
$C_8H_{10}O_2$	Veratrole		-290.3	-223.3									
$C_8H_{11}N$	*N*-Ethylaniline		4.0	56.3									
$C_8H_{11}N$	*N,N*-Dimethylaniline		47.7	100.5									
$C_8H_{11}N$	2,6-Xylidine											238.9	
C_8H_{12}	1-Octen-3-yne		140.7										
C_8H_{12}	*cis*-1,2-Divinylcyclobutane		124.3	166.5									
C_8H_{12}	*trans*-1,2-Divinylcyclobutane		101.3	143.5									
$C_8H_{12}N_4$	2,2′-Azobis[isobutyronitrile]	228.9											
$C_8H_{12}O_2$	2,2,4,4-Tetramethyl-1,3-cyclobutanedione	-379.9		-307.6									
C_8H_{14}	Ethylidenecyclohexane		-103.5	-59.5									

Molecular formula	Name	$\Delta_f H°$/kJ mol^{-1}			$\Delta_f G°$/kJ mol^{-1}			$S°$/J mol^{-1} K^{-1}			C_p/J mol^{-1} K^{-1}		
		Crys.	Liq.	Gas	Crys.	Liq.	Gas	Crys.	Liq.	Gas	Crys.	Liq.	Gas
C_8H_{14}	Allylcyclopentane		-64.5	-24.1									
$C_8H_{14}O_3$	Butanoic anhydride											283.7	
$C_8H_{15}ClO_2$	3-Methylbutyl 2-chloropropanoate		-627.3	-575.0									
$C_8H_{15}ClO_2$	3-Methylbutyl 3-chloropropanoate		-593.4	-539.4									
$C_8H_{15}N$	Octanenitrile		-107.3	-50.5									
C_8H_{16}	1-Octene		-121.8	-81.3								241.0	
C_8H_{16}	*cis*-2-Octene		-135.7									239.0	
C_8H_{16}	*trans*-2-Octene		-135.7									239.0	
C_8H_{16}	2,2-Dimethyl-*cis*-3-hexene		-126.4	-89.3									
C_8H_{16}	2,2-Dimethyl-*trans*-3-hexene		-144.9	-107.7									
C_8H_{16}	3-Ethyl-2-methyl-1-pentene		-137.9	-100.3									
C_8H_{16}	2,4,4-Trimethyl-1-pentene		-145.9	-110.5									
C_8H_{16}	2,4,4-Trimethyl-2-pentene		-142.4	-104.9									
C_8H_{16}	Cyclooctane		-167.7	-124.4									
C_8H_{16}	Ethylcyclohexane		-212.1	-171.5					280.9			211.8	
C_8H_{16}	1,1-Dimethylcyclohexane		-218.7	-180.9					267.2			209.2	
C_8H_{16}	*cis*-1,2-Dimethylcyclohexane		-211.8	-172.1					274.1			210.2	
C_8H_{16}	*trans*-1,2-Dimethylcyclohexane		-218.2	-179.9					273.2			209.4	
C_8H_{16}	*cis*-1,3-Dimethylcyclohexane		-222.9	-184.6					272.6			209.4	
C_8H_{16}	*trans*-1,3-Dimethylcyclohexane		-215.7	-176.5					276.3			212.8	
C_8H_{16}	*cis*-1,4-Dimethylcyclohexane		-215.6	-176.6					271.1			212.1	
C_8H_{16}	*trans*-1,4-Dimethylcyclohexane		-222.4	-184.5					268.0			210.2	
C_8H_{16}	Propylcyclopentane		-188.8	-147.7					310.8			216.3	
C_8H_{16}	1-Ethyl-1-methylcyclopentane		-193.8										
C_8H_{16}	*cis*-1-Ethyl-2-methylcyclopentane		-190.8										
C_8H_{16}	*trans*-1-Ethyl-2-methylcyclopentane		-195.1	-156.2									
C_8H_{16}	*cis*-1-Ethyl-3-methylcyclopentane		-194.4										
C_8H_{16}	*trans*-1-Ethyl-3-methylcyclopentane		-196.0										
$C_8H_{16}O$	2-Ethylhexanal		-348.5	-299.6									
$C_8H_{16}O$	2-Octanone											273.3	
$C_8H_{16}O$	2,2,4-Trimethyl-3-pentanone		-381.6	-338.3									
$C_8H_{16}O_2$	Octanoic acid		-636.0	-554.3								297.9	
$C_8H_{16}O_2$	2-Ethylhexanoic acid		-635.1	-559.5									
$C_8H_{16}O_2$	Hexyl acetate											282.8	
$C_8H_{16}O_2$	Propyl pentanoate		-583.0	-533.6									
$C_8H_{16}O_2$	Isopropyl pentanoate		-592.2	-544.9									

Molecular formula	Name	$\Delta_f H°$/kJ mol^{-1} Crys.	$\Delta_f H°$/kJ mol^{-1} Liq.	$\Delta_f H°$/kJ mol^{-1} Gas	$\Delta_f G°$/kJ mol^{-1} Crys.	$\Delta_f G°$/kJ mol^{-1} Liq.	$\Delta_f G°$/kJ mol^{-1} Gas	$S°$/J mol^{-1} K^{-1} Crys.	$S°$/J mol^{-1} K^{-1} Liq.	$S°$/J mol^{-1} K^{-1} Gas	C_p/J mol^{-1} K^{-1} Crys.	C_p/J mol^{-1} K^{-1} Liq.	C_p/J mol^{-1} K^{-1} Gas
$C_8H_{16}O_2$	Methyl heptanoate		-567.1	-515.5								285.1	
$C_8H_{17}Br$	1-Bromooctane		-245.1	-189.3									
$C_8H_{17}Cl$	1-Chlorooctane		-291.3	-238.9									
$C_8H_{17}NO$	Octanamide	-473.2		-362.7									
C_8H_{18}	Octane		-250.1	-208.5								254.6	
C_8H_{18}	2-Methylheptane		-255.0	-215.3					356.4			252.0	
C_8H_{18}	3-Methylheptane		-252.3	-212.5					362.6			250.2	
C_8H_{18}	4-Methylheptane		-251.6	-211.9								251.1	
C_8H_{18}	3-Ethylhexane		-250.4	-210.7									
C_8H_{18}	2,2-Dimethylhexane		-261.9	-224.5									
C_8H_{18}	2,3-Dimethylhexane		-252.6	-213.8									
C_8H_{18}	2,4-Dimethylhexane		-257.0	-219.2									
C_8H_{18}	2,5-Dimethylhexane		-260.4	-222.5								249.2	
C_8H_{18}	3,3-Dimethylhexane		-257.5	-219.9								246.6	
C_8H_{18}	3,4-Dimethylhexane		-251.8	-212.8									
C_8H_{18}	3-Ethyl-2-methylpentane		-249.6	-211.0									
C_8H_{18}	3-Ethyl-3-methylpentane		-252.8	-214.8									
C_8H_{18}	2,2,3-Trimethylpentane		-256.9	-220.0									
C_8H_{18}	2,2,4-Trimethylpentane		-259.2	-224.0								239.1	
C_8H_{18}	2,3,3-Trimethylpentane		-253.5	-216.3								245.6	
C_8H_{18}	2,3,4-Trimethylpentane		-255.0	-217.3					329.3			247.3	
C_8H_{18}	2,2,3,3-Tetramethylbutane	-269.0		-226.0				273.7			239.2		
$C_8H_{18}N_2$	Azobutane		-40.1	9.2									
$C_8H_{18}O$	1-Octanol		-426.5	-355.6								305.2	
$C_8H_{18}O$	2-Octanol											330.1	
$C_8H_{18}O$	2-Ethyl-1-hexanol		-432.8	-365.3					347.0			317.5	
$C_8H_{18}O$	Dibutyl ether		-377.9	-332.8								278.2	
$C_8H_{18}O$	Di-*sec*-butyl ether		-401.5	-360.6									
$C_8H_{18}O$	Di-*tert*-butyl ether		-399.6	-362.0								276.1	
$C_8H_{18}O$	*tert*-Butyl isobutyl ether		-409.1	-369.0									
$C_8H_{18}O_2$	1,8-Octanediol	-626.6											
$C_8H_{18}O_2$	2,5-Dimethyl-2,5-hexanediol	-681.7											
$C_8H_{18}O_3$	Diethylene glycol diethyl ether											341.4	
$C_8H_{18}O_3S$	Dibutyl sulfite		-693.1	-625.3									
$C_8H_{18}O_5$	Tetraethylene glycol		-981.7	-883.0								428.8	
$C_8H_{18}S$	Dibutyl sulfide		-220.7	-167.7					405.1			284.3	
$C_8H_{18}S$	Di-*sec*-butyl sulfide		-220.7	-167.7									
$C_8H_{18}S$	Di-*tert*-butyl sulfide		-232.6	-188.8									
$C_8H_{18}S$	Diisobutyl sulfide		-229.2	-180.5									
$C_8H_{18}S_2$	Dibutyl disulfide		-222.9	-160.6									
$C_8H_{18}S_2$	Di-*tert*-butyl disulfide		-255.2	-201.0									
$C_8H_{19}N$	Dibutylamine		-206.0	-156.6								292.9	
$C_8H_{19}N$	Diisobutylamine		-218.5	-179.2									

Molecular formula	Name	$\Delta_f H°$/kJ mol^{-1} Crys.	$\Delta_f H°$/kJ mol^{-1} Liq.	$\Delta_f H°$/kJ mol^{-1} Gas	$\Delta_f G°$/kJ mol^{-1} Crys.	$\Delta_f G°$/kJ mol^{-1} Liq.	$\Delta_f G°$/kJ mol^{-1} Gas	$S°$/J mol^{-1} K^{-1} Crys.	$S°$/J mol^{-1} K^{-1} Liq.	$S°$/J mol^{-1} K^{-1} Gas	C_p/J mol^{-1} K^{-1} Crys.	C_p/J mol^{-1} K^{-1} Liq.	C_p/J mol^{-1} K^{-1} Gas
$C_8H_{20}BrN$	Tetraethylammonium bromide	-342.7											
$C_8H_{20}Pb$	Tetraethyl lead		52.7	109.6					464.6			307.4	
$C_8H_{20}Si$	Tetraethylsilane											298.1	
C_9H_7N	Quinoline		141.2	200.5									
C_9H_7N	Isoquinoline		144.3	204.6					216.0			196.2	
C_9H_7NO	2-Hydroxyquinoline	-144.9		-25.5									
C_9H_8	Indene		110.6	163.4					215.3			186.9	
$C_9H_8O_4$	2-(Acetyloxy)benzoic acid	-815.6											
C_9H_{10}	Cyclopropylbenzene		100.3	150.5									
C_9H_{10}	Indan		11.5	60.3					56.0			190.2	
$C_9H_{10}N_2$	2,2′-Dipyrrolylmethane	126.2											
$C_9H_{10}O_2$	Benzyl acetate											148.5	
$C_9H_{10}O_2$	Ethyl benzoate											246.0	
$C_9H_{11}NO_2$	*L*-Phenylalanine	-466.9		-312.9				213.6			203.0		
$C_9H_{11}NO_3$	*L*-Tyrosine	-685.1						214.0			216.4		
C_9H_{12}	Propylbenzene		-38.3	7.9					287.8			214.7	
C_9H_{12}	Cumene		-41.1	4.0								210.7	
C_9H_{12}	*o*-Ethyltoluene		-46.4	1.3									
C_9H_{12}	*m*-Ethyltoluene		-48.7	-1.8									
C_9H_{12}	*p*-Ethyltoluene		-49.8	-3.2									
C_9H_{12}	1,2,3-Trimethylbenzene		-58.5	-9.5					267.9			216.4	
C_9H_{12}	1,2,4-Trimethylbenzene		-61.8	-13.8								215.0	
C_9H_{12}	Mesitylene		-63.4	-15.9								209.3	
$C_9H_{12}O$	2-Isopropylphenol		-233.7	-182.2									
$C_9H_{12}O$	3-Isopropylphenol		-252.5	-196.0									
$C_9H_{12}O$	4-Isopropylphenol		-265.9	-209.4									
$C_9H_{12}O_2$	Hydroperoxide, 1-methyl-1-phenylethyl		-148.3	-78.4									
$C_9H_{13}NO_2$	Ethyl 3,5-dimethylpyrrole-2-carboxylate	-474.5											
$C_9H_{13}NO_2$	Ethyl 2,4-dimethylpyrrole-3-carboxylate	-463.2											
$C_9H_{13}NO_2$	Ethyl 2,5-dimethylpyrrole-3-carboxylate	-478.7											
$C_9H_{13}NO_2$	Ethyl 4,5-dimethylpyrrole-3-carboxylate	-470.3											
$C_9H_{14}O$	Isophorone											253.5	
$C_9H_{14}O_6$	Triacetin		-1330.8	-1245.0					458.3			384.7	
$C_9H_{15}N$	3-Ethyl-2,4,5-trimethyl-pyrrole	-89.2											
$C_9H_{16}O_4$	Nonanedioic acid	-1054.3											
$C_9H_{17}NO$	2,2,6,6-Tetramethyl-4-piperidinone	-334.2		-273.4									
C_9H_{18}	Propylcyclohexane		-237.4	-192.3					311.9			242.0	

Molecular formula	Name	$\Delta_f H°$/kJ mol^{-1}			$\Delta_f G°$/kJ mol^{-1}			$S°$/J mol^{-1} K^{-1}			C_p/J mol^{-1} K^{-1}		
		Crys.	Liq.	Gas	Crys.	Liq.	Gas	Crys.	Liq.	Gas	Crys.	Liq.	Gas
C_9H_{18}	1,3,5-Trimethylcyclo-hexane (1′,3′,5′)			-212.1									
$C_9H_{18}O$	2-Nonanone		-397.2	-340.7									
$C_9H_{18}O$	5-Nonanone		-398.2	-344.9					401.4			303.6	
$C_9H_{18}O$	2,6-Dimethyl-4-heptanone		-408.5	-357.6								297.3	
$C_9H_{18}O_2$	Nonanoic acid		-659.7	-577.3								362.4	
$C_9H_{18}O_2$	Butyl pentanoate	-613.3		-560.2									
$C_9H_{18}O_2$	*sec*-Butyl pentanoate		-624.2	-573.2									
$C_9H_{18}O_2$	Isobutyl pentanoate		-620.0	-568.6									
$C_9H_{18}O_2$	Methyl octanoate		-590.3	-533.9									
$C_9H_{19}N$	*N*-Butylpiperidine		-171.8										
$C_9H_{19}N$	2,2,6,6-Tetramethyl-piperidine		-206.9	-159.9									
C_9H_{20}	Nonane		-274.7	-228.2								284.4	
C_9H_{20}	2,2-Dimethylheptane		-288.1										
C_9H_{20}	2,2,3-Trimethylhexane		-282.7										
C_9H_{20}	2,2,4-Trimethylhexane		-282.8										
C_9H_{20}	2,2,5-Trimethylhexane		-293.3										
C_9H_{20}	2,3,3-Trimethylhexane		-281.1										
C_9H_{20}	2,3,5-Trimethylhexane		-284.0	-242.6									
C_9H_{20}	2,4,4-Trimethylhexane		-280.2										
C_9H_{20}	3,3,4-Trimethylhexane		-277.5										
C_9H_{20}	3,3-Diethylpentane		-275.4	-233.3								278.2	
C_9H_{20}	3-Ethyl-2,2-dimethylpentane		-272.7										
C_9H_{20}	3-Ethyl-2,4-dimethylpentane		-269.7										
C_9H_{20}	2,2,3,3-Tetramethylpentane		-278.3	-237.1								271.5	
C_9H_{20}	2,2,3,4-Tetramethylpentane		-277.7	-236.9									
C_9H_{20}	2,2,4,4-Tetramethylpentane		-280.0	-241.6								266.3	
C_9H_{20}	2,3,3,4-Tetramethylpentane		-277.9	-236.1									
$C_9H_{20}N_2O$	Tetraethylurea	-403.0											
$C_9H_{20}O$	1-Nonanol		-453.4	-376.5									
$C_9H_{20}O_2$	1,9-Nonanediol	-657.6											
$C_9H_{21}N$	Tripropylamine		-207.1	-161.0									
$C_{10}H_6N_2O_4$	1,5-Dinitronaphthalene	29.8											
$C_{10}H_6N_2O_4$	1,8-Dinitronaphthalene	39.7											
$C_{10}H_7Cl$	1-Chloronaphthalene		54.6	119.8								212.6	
$C_{10}H_7Cl$	2-Chloronaphthalene	55.4		137.4									
$C_{10}H_7I$	1-Iodonaphthalene		161.5	233.8									
$C_{10}H_7I$	2-Iodonaphthalene	144.3		235.1									
$C_{10}H_7NO_2$	1-Nitronaphthalene	42.6		111.2									
$C_{10}H_8$	Naphthalene	78.5		150.6	201.6		224.1	167.4		333.1	165.7		131.9
$C_{10}H_8$	Azulene	212.3		289.1									
$C_{10}H_8O$	1-Naphthol	-121.5		-30.4							166.9		
$C_{10}H_8O$	2-Naphthol		-124.1	-29.9									
$C_{10}H_9N$	1-Naphthalenamine	67.8		132.8									

Molecular formula	Name	$\Delta_f H°$/kJ mol^{-1} Crys.	$\Delta_f H°$/kJ mol^{-1} Liq.	$\Delta_f H°$/kJ mol^{-1} Gas	$\Delta_f G°$/kJ mol^{-1} Crys.	$\Delta_f G°$/kJ mol^{-1} Liq.	$\Delta_f G°$/kJ mol^{-1} Gas	$S°$/J mol^{-1} K^{-1} Crys.	$S°$/J mol^{-1} K^{-1} Liq.	$S°$/J mol^{-1} K^{-1} Gas	C_p/J mol^{-1} K^{-1} Crys.	C_p/J mol^{-1} K^{-1} Liq.	C_p/J mol^{-1} K^{-1} Gas
$C_{10}H_9N$	2-Naphthalenamine	60.2		134.3									
$C_{10}H_{10}$	1,2-Dihydronaphthalene		71.6										
$C_{10}H_{10}$	1,4-Dihydronaphthalene		84.2										
$C_{10}H_{10}O$	1-Tetralone	-209.6											
$C_{10}H_{10}O_4$	Dimethyl phthalate											303.1	
$C_{10}H_{10}O_4$	Dimethyl isophthalate	-730.9											
$C_{10}H_{10}O_4$	Dimethyl terephthalate	-732.6									261.1		
$C_{10}H_{12}$	1,2,3,4-Tetrahydrona- phthalene		-29.2	26.0								217.5	
$C_{10}H_{14}$	Butylbenzene		-63.2	-11.8					321.2			243.4	
$C_{10}H_{14}$	*sec*-Butylbenzene		-66.4	-18.4									
$C_{10}H_{14}$	*tert*-Butylbenzene		-70.7	-23.0									
$C_{10}H_{14}$	Isobutylbenzene		-69.8	-21.9									
$C_{10}H_{14}$	*o*-Cymene		-73.3										
$C_{10}H_{14}$	*m*-Cymene		-78.6										
$C_{10}H_{14}$	*p*-Cymene		-78.0									236.4	
$C_{10}H_{14}$	*o*-Diethylbenzene		-68.5										
$C_{10}H_{14}$	*m*-Diethylbenzene		-73.5										
$C_{10}H_{14}$	*p*-Diethylbenzene		-72.8										
$C_{10}H_{14}$	3-Ethyl-*o*-xylene		-80.5										
$C_{10}H_{14}$	4-Ethyl-1,2-dimethylbenzene		-86.0										
$C_{10}H_{14}$	2-Ethyl-1,3-dimethylbenzene		-80.1										
$C_{10}H_{14}$	2-Ethyl-1,4-dimethylbenzene		-84.8										
$C_{10}H_{14}$	1-Ethyl-2,4-dimethylbenzene		-84.1										
$C_{10}H_{14}$	1-Ethyl-3,5-dimethylbenzene		-87.8										
$C_{10}H_{14}$	1,2,4,5-Tetramethylbenzene	-119.9						245.6			215.1		
$C_{10}H_{14}O$	Thymol	-309.7		-218.5									
$C_{10}H_{16}$	*cis, cis*-2,6-Dimethyl- 2,4,6-octatriene		-24.0										
$C_{10}H_{16}$	Dipentene		-50.8	-2.6								249.4	
$C_{10}H_{16}$	*D*-Limonene		-54.5									249.0	
$C_{10}H_{16}$	β-Myrcene		14.5										
$C_{10}H_{16}$	α-Pinene		-16.4	28.3									
$C_{10}H_{16}$	β-Pinene		-7.7	38.7									
$C_{10}H_{16}$	α-Terpinene			-20.6									
$C_{10}H_{16}N_2O_8$	Glycine, *N,N'*-1,2-ethane- diylbis[*N*-(carboxymethyl)-	-1759.5											
$C_{10}H_{16}O$	Camphor	-319.4		-267.5							271.2		
$C_{10}H_{18}$	1,1'-Bicyclopentyl		-178.9										
$C_{10}H_{18}$	*cis*-Decahydronaphthalene		-219.4	-169.2					265.0			232.0	
$C_{10}H_{18}$	*trans*-Decahydronaphthalene		-230.6	-182.1					264.9			228.5	
$C_{10}H_{18}O_4$	Sebacic acid	-1082.6		-921.9									
$C_{10}H_{19}N$	Decanenitrile		-158.4	-91.5									
$C_{10}H_{20}$	1-Decene		-173.8	-123.3					425.0			300.8	
$C_{10}H_{20}$	*cis*-1,2-Di-*tert*-butylethylene		-163.6										

Molecular formula	Name	$\Delta_f H°$/kJ mol^{-1}			$\Delta_f G°$/kJ mol^{-1}			$S°$/J mol^{-1} K^{-1}			C_p/J mol^{-1} K^{-1}		
		Crys.	Liq.	Gas	Crys.	Liq.	Gas	Crys.	Liq.	Gas	Crys.	Liq.	Gas
$C_{10}H_{20}$	Butylcyclohexane		-263.1	-213.7					345.0			271.0	
$C_{10}H_{20}O_2$	Decanoic acid	-713.7	-684.3	-594.9									
$C_{10}H_{20}O_2$	Methyl nonanoate		-616.2	-554.2									
$C_{10}H_{21}NO_2$	1-Nitrodecane		-351.5										
$C_{10}H_{22}$	Decane		-300.9	-249.5								314.4	
$C_{10}H_{22}$	2-Methylnonane		-309.8	-260.2					420.1			313.3	
$C_{10}H_{22}$	5-Methylnonane		-307.9	-258.6					423.8			314.4	
$C_{10}H_{22}O$	1-Decanol		-478.1	-396.6								370.6	
$C_{10}H_{22}O$	Dipentyl ether											250.0	
$C_{10}H_{22}O$	Diisopentyl ether											379.0	
$C_{10}H_{22}O_2$	1,10-Decanediol	-678.9											
$C_{10}H_{22}O_2$	Ethylene glycol dibutyl ether											350.0	
$C_{10}H_{22}S$	1-Decanethiol	-309.9	-276.5	-211.5					476.1			350.4	
$C_{10}H_{22}S$	Dipentyl sulfide		-266.4	-204.9									
$C_{10}H_{22}S$	Diisopentylsulfide		-281.8	-221.5									
$C_{11}H_8O_2$	1-Naphthalenecarboxylic acid	-333.5		-223.1									
$C_{11}H_8O_2$	2-Naphthoic acid	-346.1		-232.5									
$C_{11}H_{10}$	1-Methylnaphthalene		56.3						254.8			224.4	
$C_{11}H_{10}$	2-Methylnaphthalene	44.9		106.7				220.0			196.0		
$C_{11}H_{12}N_2O_2$	*L*-Tryptophan	-415.3						251.0			238.1		
$C_{11}H_{14}$	1,1-Dimethylindan		-53.6	-1.6									
$C_{11}H_{16}$	Pentamethylbenzene	-144.6		-67.2									
$C_{11}H_{20}$	Spiro[5.5]undecane		-244.5	-188.3									
$C_{11}H_{22}$	1-Undecene											344.9	
$C_{11}H_{22}O_2$	Methyl decanoate		-640.5	-573.8									
$C_{11}H_{24}$	Undecane		-327.2	-270.8								344.9	
$C_{11}H_{24}O$	1-Undecanol		-504.8										
$C_{12}F_{27}N$	Trinonafluorobutylamine											418.4	
$C_{12}H_8$	Acenaphthylene	186.7		259.7							166.4		
$C_{12}H_8N_2$	Phenazine	237.0		328.8									
$C_{12}H_8O$	Dibenzofuran	-5.3		83.4									
$C_{12}H_8S$	Dibenzothiophene	120.0		205.1									
$C_{12}H_8S_2$	Thianthrene	182.0		286.0									
$C_{12}H_9N$	Carbazole	101.7		200.7									
$C_{12}H_{10}$	Acenaphthene	70.3		156.0				188.9			190.4		
$C_{12}H_{10}$	Biphenyl	99.4		181.4				209.4			198.4		
$C_{12}H_{10}N_2O$	*trans*-Azoxybenzene	243.4		342.0									
$C_{12}H_{10}N_2O$	*N*-Nitrosodiphenylamine	227.2											
$C_{12}H_{10}O$	Diphenyl ether	-32.1	-14.9	52.0				233.9			216.6		
$C_{12}H_{10}O_2$	1-Naphthaleneacetic acid	-359.2											
$C_{12}H_{10}O_2$	2-Naphthaleneacetic acid	-371.9											
$C_{12}H_{11}N$	2-Aminobiphenyl	93.8		184.4									
$C_{12}H_{11}N$	*p*-Biphenylamine	81.0											
$C_{12}H_{11}N$	Diphenylamine	130.2		219.3									
$C_{12}H_{12}N_2$	*p*-Benzidine	70.7											

Molecular formula	Name	$\Delta_f H°$/kJ mol^{-1}			$\Delta_f G°$/kJ mol^{-1}			$S°$/J mol^{-1} K^{-1}			C_p/J mol^{-1} K^{-1}		
		Crys.	Liq.	Gas	Crys.	Liq.	Gas	Crys.	Liq.	Gas	Crys.	Liq.	Gas
$C_{12}H_{14}O_4$	Diethyl phthalate		-776.6	-688.4					425.1			366.1	
$C_{12}H_{16}$	Cyclohexylbenzene		-76.6	-16.7									
$C_{12}H_{17}NO_4$	Diethyl 3,5-dimethyl-pyrrole-2,4-dicarboxylate	-916.7											
$C_{12}H_{18}$	3,9-Dodecadiyne		197.8										
$C_{12}H_{18}$	5,7-Dodecadiyne		181.5										
$C_{12}H_{18}$	Hexamethylbenzene	-162.4		-77.4				306.3			245.6		
$C_{12}H_{22}$	Cyclohexylcyclohexane		-273.7	-215.7									
$C_{12}H_{22}O_4$	Dodecanedioic acid	-1130.0		-976.9									
$C_{12}H_{22}O_{11}$	Sucrose	-2226.1											
$C_{12}H_{22}O_{11}$	β-Lactose	-2236.7											
$C_{12}H_{24}$	1-Dodecene		-226.2	-165.4					484.8			360.7	
$C_{12}H_{24}O_2$	Dodecanoic acid	-774.6	-737.9	-642.0							404.3		
$C_{12}H_{24}O_2$	Methyl undecanoate		-665.2	-593.8									
$C_{12}H_{24}O_{12}$	α-Lactose monohydrate	-2484.1											
$C_{12}H_{25}Br$	1-Bromododecane		-344.7	-269.9									
$C_{12}H_{25}Cl$	1-Chlorododecane		-392.3	-321.1									
$C_{12}H_{26}$	Dodecane		-350.9	-289.4								375.8	
$C_{12}H_{26}O$	1-Dodecanol		-528.5	-436.6								438.1	
$C_{12}H_{26}O_3$	Diethylene glycol dibutyl ether											452.0	
$C_{12}H_{27}N$	Tributylamine		-281.6										
$C_{13}H_8O_2$	Xanthone	-191.5											
$C_{13}H_9N$	Acridine	179.4		273.9									
$C_{13}H_9N$	Phenanthridine	141.9		240.5									
$C_{13}H_9N$	Benzo[f]quinoline	150.6		233.7									
$C_{13}H_{10}N_2$	9-Acridinamine	159.2											
$C_{13}H_{10}O$	Benzophenone	-34.5		54.9							224.8		
$C_{13}H_{11}N$	9-Methylcarbazole	105.5		201.0									
$C_{13}H_{12}$	Diphenylmethane	71.5	89.7	139.0				239.3					
$C_{13}H_{24}O_4$	Tridecanedioic acid	-1148.3											
$C_{13}H_{26}$	1-Tridecene											391.8	
$C_{13}H_{26}O_2$	Methyl dodecanoate		-693.0	-614.9									
$C_{13}H_{28}$	Tridecane											406.7	
$C_{13}H_{28}O$	1-Tridecanol	-599.4											
$C_{14}H_8O_2$	9,10-Anthracenedione	-188.5		-75.7									
$C_{14}H_8O_2$	9,10-Phenanthrenedione	-154.7		-46.6									
$C_{14}H_8O_4$	1,4-Dihydroxy-9,10-anthracenedione	-595.8		-471.7									
$C_{14}H_{10}$	Anthracene	129.2		230.9				207.5			210.5		
$C_{14}H_{10}$	Phenanthrene	116.2		207.5				215.1			220.6		
$C_{14}H_{10}$	Diphenylacetylene	312.4									225.9		
$C_{14}H_{10}O_2$	Benzil	-153.9		-55.5									
$C_{14}H_{12}$	*cis*-Stilbene		183.3	252.3									
$C_{14}H_{12}$	*trans*-Stilbene	136.9		236.1									

Molecular formula	Name	$\Delta_f H°$/kJ mol^{-1} Crys.	Liq.	Gas	$\Delta_f G°$/kJ mol^{-1} Crys.	Liq.	Gas	$S°$/J mol^{-1} K^{-1} Crys.	Liq.	Gas	C_p/J mol^{-1} K^{-1} Crys.	Liq.	Gas
$C_{14}H_{14}$	1,1-Diphenylethane		48.7										
$C_{14}H_{14}$	1,2-Diphenylethane	51.5		142.9									
$C_{14}H_{23}N_3O_{10}$	Glycine, *N,N*-bis[2-[bis(carboxymethyl) amino]ethyl]	-2225.2											
$C_{14}H_{27}N$	Tetradecanenitrile		-260.2	-174.9									
$C_{14}H_{28}O_2$	Tetradecanoic acid	-833.5	-788.8	-693.7							432.0		
$C_{14}H_{28}O_2$	Methyl tridecanoate		-717.9	-635.3									
$C_{14}H_{30}O$	1-Tetradecanol	-629.6	-580.6								388.0		
$C_{15}H_{16}O_2$	2,2-Bis(4-hydroxyphenyl) propane	-368.6											
$C_{15}H_{30}$	Decylcyclopentane		-367.3										
$C_{15}H_{30}O_2$	Pentadecanoic acid	-861.7	-811.7	-699.0							443.3		
$C_{15}H_{30}O_2$	Methyl tetradecanoate		-743.9	-656.9									
$C_{15}H_{32}O$	1-Pentadecanol	-658.2											
$C_{16}H_{10}$	Fluoranthene	189.9		289.0				230.6			230.2		
$C_{16}H_{10}$	Pyrene	125.5		225.7				224.9			229.7		
$C_{16}H_{22}O_4$	Dibutyl phthalate		-842.6	-750.9									
$C_{16}H_{22}O_{11}$	α-*D*-Glucose pentaacetate	-2249.4											
$C_{16}H_{22}O_{11}$	β-*D*-Glucose pentaacetate	-2232.6											
$C_{16}H_{26}$	Decylbenzene		-218.3	-138.6									
$C_{16}H_{32}$	1-Hexadecene		-328.7	-248.4					587.9			488.9	
$C_{16}H_{32}O_2$	Hexadecanoic acid	-891.5	-838.1	-737.1				452.4			460.7		
$C_{16}H_{32}O_2$	Methyl pentadecanoate		-771.0	-680.0									
$C_{16}H_{33}Br$	1-Bromohexadecane		-444.5	-350.2									
$C_{16}H_{34}$	Hexadecane		-456.1	-374.8								501.6	
$C_{16}H_{34}O$	1-Hexadecanol	-686.5		-517.0							422.0		
$C_{16}H_{36}IN$	Tetrabutyl ammonium iodide	-498.6											
$C_{17}H_{34}O_2$	Margaric acid	-924.4	-865.6								475.7		
$C_{18}H_{12}$	Benz[a]anthracene	170.8		293.0									
$C_{18}H_{12}$	Chrysene	145.3		269.8									
$C_{18}H_{15}N$	Triphenylamine	234.7		326.8									
$C_{18}H_{34}O_2$	Oleic acid											577.0	
$C_{18}H_{34}O_4$	Dibutyl sebacate											619.0	
$C_{18}H_{36}O_2$	Stearic acid	-947.7	-884.7	-781.2							501.5		
$C_{18}H_{37}Cl$	1-Chlorooctadecane		-544.1	-446.0									
$C_{18}H_{38}$	Octadecane	-567.4		-414.6				480.2			485.6		
$C_{18}H_{39}N$	Trihexylamine		-433.0										
$C_{19}H_{16}O$	Triphenylmethanol	-2.5											
$C_{19}H_{36}O_2$	Methyl oleate		-734.5	-649.9									
$C_{19}H_{36}O_2$	Methyl *trans*-9-octadecenoate		-737.0										
$C_{20}H_{12}$	Perylene	182.8						264.6			274.9		
$C_{20}H_{14}O_4$	1,2-Benzenedicarboxylic acid, diphenyl ester	-489.2											
$C_{20}H_{38}O_2$	Ethyl *cis*-9-octadecenoate		-775.8										

Molecular formula	Name	$\Delta_f H°$/kJ mol^{-1}			$\Delta_f G°$/kJ mol^{-1}			$S°$/J mol^{-1} K^{-1}			C_p/J mol^{-1} K^{-1}		
		Crys.	Liq.	Gas	Crys.	Liq.	Gas	Crys.	Liq.	Gas	Crys.	Liq.	Gas
$C_{20}H_{38}O_2$	Ethyl *trans*-9-octa-decenoate		-773.3										
$C_{20}H_{40}O_2$	Arachidic acid	-1011.9	-940.0	-812.4							545.1		
$C_{22}H_{42}O_2$	Brassidic acid	-960.7											
$C_{22}H_{42}O_2$	Butyl oleate		-816.9										
$C_{22}H_{44}O_2$	Butyl stearate												
$C_{24}H_{38}O_4$	Bis(2-ethylhexyl) phthalate											704.7	
$C_{24}H_{51}N$	Trioctylamine		-585.0										
$C_{26}H_{18}$	9,10-Diphenylanthracene	308.7		465.6									
$C_{26}H_{54}$	5-Butyldocosane		-713.5	-587.6									
$C_{26}H_{54}$	11-Butyldocosane		-716.0	-593.4									
$C_{28}H_{18}$	9,9′-Bianthracene	326.2		454.3									
$C_{31}H_{64}$	11-Decylheneicosane		-848.0	-705.8									
$C_{32}H_{66}$	Dotriacontane		-968.3	-697.2									
C_{60}	Fullerene-C_{60}	2327.0		2502.0	2302.0		2442.0	426.0		544.0	520.0		512.0
C_{70}	Fullerene-C_{70}	2555.0		2755.0	2537.0		2692.0	464.0		614.0	650.0		585.0

THERMODYNAMIC PROPERTIES AS A FUNCTION OF TEMPERATURE

L. V. Gurvich, V. S. Iorish, V. S. Yungman, and O. V. Dorofeeva

The thermodynamic properties $C_p^\circ(T)$, $S^\circ(T)$, $H^\circ(T)-H^\circ(T_r)$, $-[G^\circ(T)-H^\circ(T_r)]/T$ and formation properties $\Delta_f H^\circ(T)$, $\Delta_f G^\circ(T)$, $\log K_f^\circ(T)$ are tabulated as functions of temperature in the range 298.15 to 1500 K for 80 substances in the standard state.The reference temperature, T_r, is equal to 298.15 K. The standard state pressure is taken as 1 bar (100,000 Pa). The tables are presented in the JANAF Thermochemical Tables format (Reference 2). The numerical data are extracted from IVTANTHERMO databases except for C_2H_4O, C_3H_6O, C_6H_6, C_6H_6O, $C_{10}H_8$, and CH_5N, which are based upon TRC Tables. See the references for information on standard states and other details.

REFERENCES

1. Gurvich, L. V., Veyts, I. V., and Alcock, C. B., Eds., *Thermodynamic Properties of Individual Substances, 4th ed.*, Hemisphere Publishing Corp., New York, 1989.
2. Chase, M. W., et al., *JANAF Thermochemical Tables, 3rd ed., J. Phys. Chem. Ref. Data*, 14, Suppl. 1, 1985.

Order of Listing of Tables

No.	Formula	Name	State
1	Ar	Argon	g
2	Br	Bromine	g
3	Br_2	Dibromine	g
4	BrH	Hydrogen bromide	g
5	C	Carbon (graphite)	cr
6	C	Carbon (diamond)	cr
7	C_2	Dicarbon	g
8	C_3	Tricarbon	g
9	CO	Carbon oxide	g
10	CO_2	Carbon dioxide	g
11	CH_4	Methane	g
12	C_2H_2	Acetylene	g
13	C_2H_4	Ethylene	g
14	C_2H_6	Ethane	g
15	C_3H_6	Cyclopropane	g
16	C_3H_8	Propane	g
17	C_6H_6	Benzene	l
18	C_6H_6	Benzene	g
19	$C_{10}H_8$	Naphthalene	cr, l
20	$C_{10}H_8$	Naphthalene	g
21	CH_2O	Formaldehyde	g
22	CH_4O	Methanol	g
23	C_2H_4O	Acetaldehyde	g
24	C_2H_6O	Ethanol	g
25	$C_2H_4O_2$	Acetic acid	g
26	C_3H_6O	Acetone	g
27	C_6H_6O	Phenol	g
28	CF_4	Carbon tetrafluoride	g
29	CHF_3	Trifluoromethane	g
30	$CClF_3$	Chlorotrifluoromethane	g
31	CCl_2F_2	Dichlorodifluoromethane	g
32	$CHClF_2$	Chlorodifluoromethane	g
33	CH_5N	Methylamine	g
34	Cl	Chlorine	g
35	Cl_2	Dichlorine	g
36	ClH	Hydrogen chloride	g
37	Cu	Copper	cr, l
38	Cu	Copper	g
39	CuO	Copper oxide	cr
40	Cu_2O	Dicopper oxide	cr
41	$CuCl_2$	Copper dichloride	cr, l
42	$CuCl_2$	Copper dichloride	g
43	F	Fluorine	g
44	F_2	Difluorine	g
45	FH	Hydrogen fluoride	g
46	Ge	Germanium	cr, l
47	Ge	Germanium	g
48	GeO_2	Germanium dioxide	cr, l
49	$GeCl_4$	Germanium tetrachloride	g
50	H	Hydrogen	g
51	H_2	Dihydrogen	g
52	HO	Hydroxyl	g
53	H_2O	Water	l
54	H_2O	Water	g
55	I	Iodine	g
56	I_2	Diiodine	cr, l
57	I_2	Diiodine	g
58	IH	Hydrogen iodide	g
59	K	Potassium	cr, l
60	K	Potassium	g
61	K_2O	Dipotassium oxide	cr, l
62	KOH	Potassium hydroxide	cr, l
63	KOH	Potassium hydroxide	g
64	KCl	Potassium chloride	cr, l
65	KCl	Potassium chloride	g
66	N_2	Dinitrogen	g
67	NO	Nitric oxide	g
68	NO_2	Nitrogen dioxide	g
69	NH_3	Ammonia	g
70	O	Oxygen	g
71	O_2	Dioxygen	g
72	S	Sulfur	cr, l
73	S	Sulfur	g
74	S_2	Disulfur	g
75	S_8	Octasulfur	g
76	SO_2	Sulfur dioxide	g
77	Si	Silicon	cr
78	Si	Silicon	g
79	SiO_2	Silicon dioxide	cr
80	$SiCl_4$	Silicon tetrachloride	g

	J/K·mol			kJ/mol			
T/K	C_p°	S°	$-(G^\circ - H^\circ(T_r))/T$	$H^\circ - H^\circ(T_r)$	$\Delta_f H^\circ$	$\Delta_f G^\circ$	Log K_f
1. ARGON Ar (g)							
298.15	20.786	154.845	154.845	0.000	0.000	0.000	0.000
300	20.786	154.973	154.845	0.038	0.000	0.000	0.000
400	20.786	160.953	155.660	2.117	0.000	0.000	0.000
500	20.786	165.591	157.200	4.196	0.000	0.000	0.000
600	20.786	169.381	158.924	6.274	0.000	0.000	0.000
700	20.786	172.585	160.653	8.353	0.000	0.000	0.000
800	20.786	175.361	162.322	10.431	0.000	0.000	0.000
900	20.786	177.809	163.909	12.510	0.000	0.000	0.000
1000	20.786	179.999	165.410	14.589	0.000	0.000	0.000
1100	20.786	181.980	166.828	16.667	0.000	0.000	0.000
1200	20.786	183.789	168.167	18.746	0.000	0.000	0.000
1300	20.786	185.453	169.434	20.824	0.000	0.000	0.000
1400	20.786	186.993	170.634	22.903	0.000	0.000	0.000
1500	20.786	188.427	171.773	24.982	0.000	0.000	0.000
2. BROMINE Br (g)							
298.15	20.786	175.017	175.017	0.000	111.870	82.379	–14.432
300	20.786	175.146	175.018	0.038	111.838	82.196	–14.311
400	20.787	181.126	175.833	2.117	96.677	75.460	–9.854
500	20.798	185.765	177.373	4.196	96.910	70.129	–7.326
600	20.833	189.559	179.097	6.277	97.131	64.752	–5.637
700	20.908	192.776	180.827	8.364	97.348	59.338	–4.428
800	21.027	195.575	182.499	10.461	97.568	53.893	–3.519
900	21.184	198.061	184.093	12.571	97.796	48.420	–2.810
1000	21.365	200.302	185.604	14.698	98.036	42.921	–2.242
1100	21.559	202.347	187.034	16.844	98.291	37.397	–1.776
1200	21.752	204.231	188.390	19.010	98.560	31.850	–1.386
1300	21.937	205.980	189.676	21.195	98.844	26.279	–1.056
1400	22.107	207.612	190.900	23.397	99.141	20.686	–0.772
1500	22.258	209.142	192.065	25.615	99.449	15.072	–0.525
3. DIBROMINE Br_2 (g)							
298.15	36.057	245.467	245.467	0.000	30.910	3.105	–0.544
300	36.074	245.690	245.468	0.067	30.836	2.933	–0.511
332.25	36.340	249.387	245.671	1.235		pressure = 1 bar	
400	36.729	256.169	246.892	3.711	0.000	0.000	0.000
500	37.082	264.406	249.600	7.403	0.000	0.000	0.000
600	37.305	271.188	252.650	11.123	0.000	0.000	0.000
700	37.464	276.951	255.720	14.862	0.000	0.000	0.000
800	37.590	281.962	258.694	18.615	0.000	0.000	0.000
900	37.697	286.396	261.530	22.379	0.000	0.000	0.000
1000	37.793	290.373	264.219	26.154	0.000	0.000	0.000
1100	37.883	293.979	266.763	29.938	0.000	0.000	0.000
1200	37.970	297.279	269.170	33.730	0.000	0.000	0.000
1300	38.060	300.322	271.451	37.532	0.000	0.000	0.000
1400	38.158	303.146	273.615	41.343	0.000	0.000	0.000
1500	38.264	305.782	275.673	45.164	0.000	0.000	0.000
4. HYDROGEN BROMIDE HBr (g)							
298.15	29.141	198.697	198.697	0.000	–36.290	–53.360	9.348
300	29.141	198.878	198.698	0.054	–36.333	–53.466	9.309
400	29.220	207.269	199.842	2.971	–52.109	–55.940	7.305
500	29.454	213.811	202.005	5.903	–52.484	–56.854	5.939
600	29.872	219.216	204.436	8.868	–52.844	–57.694	5.023
700	30.431	223.861	206.886	11.882	–53.168	–58.476	4.363
800	31.063	227.965	209.269	14.957	–53.446	–59.214	3.866

	J/K·mol			kJ/mol			
T/K	C_p°	S°	$-(G^\circ-H^\circ(T_r))/T$	$H^\circ-H^\circ(T_r)$	$\Delta_f H^\circ$	$\Delta_f G^\circ$	Log K_f
4. HYDROGEN BROMIDE HBr (g) (continued)							
900	31.709	231.661	211.555	18.095	−53.677	−59.921	3.478
1000	32.335	235.035	213.737	21.298	−53.864	−60.604	3.166
1100	32.919	238.145	215.816	24.561	−54.012	−61.271	2.909
1200	33.454	241.032	217.799	27.880	−54.129	−61.925	2.696
1300	33.938	243.729	219.691	31.250	−54.220	−62.571	2.514
1400	34.374	246.261	221.499	34.666	−54.291	−63.211	2.358
1500	34.766	248.646	223.230	38.123	−54.348	−63.846	2.223
5. CARBON (GRAPHITE) C (cr; graphite)							
298.15	8.536	5.740	5.740	0.000	0.000	0.000	0.000
300	8.610	5.793	5.740	0.016	0.000	0.000	0.000
400	11.974	8.757	6.122	1.054	0.000	0.000	0.000
500	14.537	11.715	6.946	2.385	0.000	0.000	0.000
600	16.607	14.555	7.979	3.945	0.000	0.000	0.000
700	18.306	17.247	9.113	5.694	0.000	0.000	0.000
800	19.699	19.785	10.290	7.596	0.000	0.000	0.000
900	20.832	22.173	11.479	9.625	0.000	0.000	0.000
1000	21.739	24.417	12.662	11.755	0.000	0.000	0.000
1100	22.452	26.524	13.827	13.966	0.000	0.000	0.000
1200	23.000	28.502	14.968	16.240	0.000	0.000	0.000
1300	23.409	30.360	16.082	18.562	0.000	0.000	0.000
1400	23.707	32.106	17.164	20.918	0.000	0.000	0.000
1500	23.919	33.749	18.216	23.300	0.000	0.000	0.000
6. CARBON (DIAMOND) C (cr; diamond)							
298.15	6.109	2.362	2.362	0.000	1.850	2.857	−0.501
300	6.201	2.400	2.362	0.011	1.846	2.863	−0.499
400	10.321	4.783	2.659	0.850	1.645	3.235	−0.422
500	13.404	7.431	3.347	2.042	1.507	3.649	−0.381
600	15.885	10.102	4.251	3.511	1.415	4.087	−0.356
700	17.930	12.709	5.274	5.205	1.361	4.537	−0.339
800	19.619	15.217	6.361	7.085	1.338	4.993	−0.326
900	21.006	17.611	7.479	9.118	1.343	5.450	−0.316
1000	22.129	19.884	8.607	11.277	1.372	5.905	−0.308
1100	23.020	22.037	9.731	13.536	1.420	6.356	−0.302
1200	23.709	24.071	10.842	15.874	1.484	6.802	−0.296
1300	24.222	25.990	11.934	18.272	1.561	7.242	−0.291
1400	24.585	27.799	13.003	20.714	1.646	7.675	−0.286
1500	24.824	29.504	14.047	23.185	1.735	8.103	−0.282
7. DICARBON C_2 (g)							
298.15	43.548	197.095	197.095	0.000	830.457	775.116	−135.795
300	43.575	197.365	197.096	0.081	830.506	774.772	−134.898
400	42.169	209.809	198.802	4.403	832.751	755.833	−98.700
500	39.529	218.924	201.959	8.483	834.170	736.423	−76.933
600	37.837	225.966	205.395	12.342	834.909	716.795	−62.402
700	36.984	231.726	208.758	16.078	835.148	697.085	−52.016
800	36.621	236.637	211.943	19.755	835.020	677.366	−44.227
900	36.524	240.943	214.931	23.411	834.618	657.681	−38.170
1000	36.569	244.793	217.728	27.065	834.012	638.052	−33.328
1100	36.696	248.284	220.349	30.728	833.252	618.492	−29.369
1200	36.874	251.484	222.812	34.406	832.383	599.006	−26.074
1300	37.089	254.444	225.133	38.104	831.437	579.596	−23.288
1400	37.329	257.201	227.326	41.824	830.445	560.261	−20.903
1500	37.589	259.785	229.405	45.570	829.427	540.997	−18.839

	J/K·mol			kJ/mol			
T/K	C_p°	S°	$-(G^\circ-H^\circ(T_r))/T$	$H^\circ-H^\circ(T_r)$	$\Delta_f H^\circ$	$\Delta_f G^\circ$	Log K_f
8. TRICARBON C_3 (g)							
298.15	42.202	237.611	237.611	0.000	839.958	774.249	–135.643
300	42.218	237.872	237.611	0.078	839.989	773.841	–134.736
400	43.383	250.164	239.280	4.354	841.149	751.592	–98.147
500	44.883	260.003	242.471	8.766	841.570	729.141	–76.172
600	46.406	268.322	246.104	13.331	841.453	706.659	–61.519
700	47.796	275.582	249.807	18.042	840.919	684.230	–51.057
800	48.997	282.045	253.440	22.884	840.053	661.901	–43.217
900	50.006	287.876	256.948	27.835	838.919	639.698	–37.127
1000	50.844	293.189	260.310	32.879	837.572	617.633	–32.261
1100	51.535	298.069	263.524	37.999	836.059	595.711	–28.288
1200	52.106	302.578	266.593	43.182	834.420	573.933	–24.982
1300	52.579	306.768	269.524	48.417	832.690	552.295	–22.191
1400	52.974	310.679	272.326	53.695	830.899	530.793	–19.804
1500	53.307	314.346	275.006	59.010	829.068	509.421	–17.739
9. CARBON OXIDE CO (g)							
298.15	29.141	197.658	197.658	0.000	–110.530	–137.168	24.031
300	29.142	197.838	197.659	0.054	–110.519	–137.333	23.912
400	29.340	206.243	198.803	2.976	–110.121	–146.341	19.110
500	29.792	212.834	200.973	5.930	–110.027	–155.412	16.236
600	30.440	218.321	203.419	8.941	–110.157	–164.480	14.319
700	31.170	223.067	205.895	12.021	–110.453	–173.513	12.948
800	31.898	227.277	208.309	15.175	–110.870	–182.494	11.915
900	32.573	231.074	210.631	18.399	–111.378	–191.417	11.109
1000	33.178	234.538	212.851	21.687	–111.952	–200.281	10.461
1100	33.709	237.726	214.969	25.032	–112.573	–209.084	9.928
1200	34.169	240.679	216.990	28.426	–113.228	–217.829	9.482
1300	34.568	243.430	218.920	31.864	–113.904	–226.518	9.101
1400	34.914	246.005	220.763	35.338	–114.594	–235.155	8.774
1500	35.213	248.424	222.527	38.845	–115.291	–243.742	8.488
10. CARBON DIOXIDE CO_2 (g)							
298.15	37.135	213.783	213.783	0.000	–393.510	–394.373	69.092
300	37.220	214.013	213.784	0.069	–393.511	–394.379	68.667
400	41.328	225.305	215.296	4.004	–393.586	–394.656	51.536
500	44.627	234.895	218.280	8.307	–393.672	–394.914	41.256
600	47.327	243.278	221.762	12.909	–393.791	–395.152	34.401
700	49.569	250.747	225.379	17.758	–393.946	–395.367	29.502
800	51.442	257.492	228.978	22.811	–394.133	–395.558	25.827
900	53.008	263.644	232.493	28.036	–394.343	–395.724	22.967
1000	54.320	269.299	235.895	33.404	–394.568	–395.865	20.678
1100	55.423	274.529	239.172	38.893	–394.801	–395.984	18.803
1200	56.354	279.393	242.324	44.483	–395.035	–396.081	17.241
1300	57.144	283.936	245.352	50.159	–395.265	–396.159	15.918
1400	57.818	288.196	248.261	55.908	–395.488	–396.219	14.783
1500	58.397	292.205	251.059	61.719	–395.702	–396.264	13.799
11. METHANE CH_4 (g)							
298.15	35.695	186.369	186.369	0.000	–74.600	–50.530	8.853
300	35.765	186.590	186.370	0.066	–74.656	–50.381	8.772
400	40.631	197.501	187.825	3.871	–77.703	–41.827	5.462
500	46.627	207.202	190.744	8.229	–80.520	–32.525	3.398
600	52.742	216.246	194.248	13.199	–82.969	–22.690	1.975
700	58.603	224.821	198.008	18.769	–85.023	–12.476	0.931
800	64.084	233.008	201.875	24.907	–86.693	–1.993	0.130

	J/K·mol			kJ/mol			
T/K	C_p°	S°	$-(G^\circ - H^\circ(T_r))/T$	$H^\circ - H^\circ(T_r)$	$\Delta_f H^\circ$	$\Delta_f G^\circ$	Log K_f
11. METHANE CH_4 (g) (continued)							
900	69.137	240.852	205.773	31.571	–88.006	8.677	–0.504
1000	73.746	248.379	209.660	38.719	–88.996	19.475	–1.017
1100	77.919	255.607	213.511	46.306	–89.698	30.358	–1.442
1200	81.682	262.551	217.310	54.289	–90.145	41.294	–1.797
1300	85.067	269.225	221.048	62.630	–90.367	52.258	–2.100
1400	88.112	275.643	224.720	71.291	–90.390	63.231	–2.359
1500	90.856	281.817	228.322	80.242	–90.237	74.200	–2.584
12. ACETYLENE C_2H_2 (g)							
298.15	44.036	200.927	200.927	0.000	227.400	209.879	–36.769
300	44.174	201.199	200.927	0.082	227.397	209.770	–36.524
400	50.388	214.814	202.741	4.829	227.161	203.928	–26.630
500	54.751	226.552	206.357	10.097	226.846	198.154	–20.701
600	58.121	236.842	210.598	15.747	226.445	192.452	–16.754
700	60.970	246.021	215.014	21.704	225.968	186.823	–13.941
800	63.511	254.331	219.418	27.931	225.436	181.267	–11.835
900	65.831	261.947	223.726	34.399	224.873	175.779	–10.202
1000	67.960	268.995	227.905	41.090	224.300	170.355	–8.898
1100	69.909	275.565	231.942	47.985	223.734	164.988	–7.835
1200	71.686	281.725	235.837	55.067	223.189	159.672	–6.950
1300	73.299	287.528	239.592	62.317	222.676	154.400	–6.204
1400	74.758	293.014	243.214	69.721	222.203	149.166	–5.565
1500	76.077	298.218	246.709	77.264	221.774	143.964	–5.013
13. ETHYLENE C_2H_4 (g)							
298.15	42.883	219.316	219.316	0.000	52.400	68.358	–11.976
300	43.059	219.582	219.317	0.079	52.341	68.457	–11.919
400	53.045	233.327	221.124	4.881	49.254	74.302	–9.703
500	62.479	246.198	224.864	10.667	46.533	80.887	–8.450
600	70.673	258.332	229.441	17.335	44.221	87.982	–7.659
700	77.733	269.770	234.393	24.764	42.278	95.434	–7.121
800	83.868	280.559	239.496	32.851	40.655	103.142	–6.734
900	89.234	290.754	244.630	41.512	39.310	111.036	–6.444
1000	93.939	300.405	249.730	50.675	38.205	119.067	–6.219
1100	98.061	309.556	254.756	60.280	37.310	127.198	–6.040
1200	101.670	318.247	259.688	70.271	36.596	135.402	–5.894
1300	104.829	326.512	264.513	80.599	36.041	143.660	–5.772
1400	107.594	334.384	269.225	91.223	35.623	151.955	–5.669
1500	110.018	341.892	273.821	102.107	35.327	160.275	–5.581
14. ETHANE C_2H_6 (g)							
298.15	52.487	229.161	229.161	0.000	–84.000	–32.015	5.609
300	52.711	229.487	229.162	0.097	–84.094	–31.692	5.518
400	65.459	246.378	231.379	5.999	–88.988	–13.473	1.759
500	77.941	262.344	235.989	13.177	–93.238	5.912	–0.618
600	89.188	277.568	241.660	21.545	–96.779	26.086	–2.271
700	99.136	292.080	247.835	30.972	–99.663	46.800	–3.492
800	107.936	305.904	254.236	41.334	–101.963	67.887	–4.433
900	115.709	319.075	260.715	52.525	–103.754	89.231	–5.179
1000	122.552	331.628	267.183	64.445	–105.105	110.750	–5.785
1100	128.553	343.597	273.590	77.007	–106.082	132.385	–6.286
1200	133.804	355.012	279.904	90.131	–106.741	154.096	–6.708
1300	138.391	365.908	286.103	103.746	–107.131	175.850	–7.066
1400	142.399	376.314	292.178	117.790	–107.292	197.625	–7.373
1500	145.905	386.260	298.121	132.209	–107.260	219.404	–7.640

	J/K·mol			kJ/mol			
T/K	C_p°	S°	$-(G^\circ-H^\circ(T_r))/T$	$H^\circ-H^\circ(T_r)$	$\Delta_f H^\circ$	$\Delta_f G^\circ$	Log K_f
15. CYCLOPROPANE C_3H_6 (g)							
298.15	55.571	237.488	237.488	0.000	53.300	104.514	–18.310
300	55.941	237.832	237.489	0.103	53.195	104.832	–18.253
400	76.052	256.695	239.924	6.708	47.967	122.857	–16.043
500	93.859	275.637	245.177	15.230	43.730	142.091	–14.844
600	108.542	294.092	251.801	25.374	40.405	162.089	–14.111
700	120.682	311.763	259.115	36.854	37.825	182.583	–13.624
800	130.910	328.564	266.755	49.447	35.854	203.404	–13.281
900	139.658	344.501	274.516	62.987	34.384	224.441	–13.026
1000	147.207	359.616	282.277	77.339	33.334	245.618	–12.830
1100	153.749	373.961	289.965	92.395	32.640	266.883	–12.673
1200	159.432	387.588	297.538	108.060	32.249	288.197	–12.545
1300	164.378	400.549	304.967	124.257	32.119	309.533	–12.437
1400	168.689	412.892	312.239	140.915	32.215	330.870	–12.345
1500	172.453	424.662	319.344	157.976	32.507	352.193	–12.264
16. PROPANE C_3H_8 (g)							
298.15	73.597	270.313	270.313	0.000	–103.847	–23.458	4.110
300	73.931	270.769	270.314	0.136	–103.972	–22.959	3.997
400	94.014	294.739	273.447	8.517	–110.33	15.029	–0.657
500	112.591	317.768	280.025	18.872	–115.658	34.507	–3.605
600	128.700	339.753	288.162	30.955	–119.973	64.961	–5.655
700	142.674	360.668	297.039	44.540	–123.384	96.065	–7.168
800	154.766	380.528	306.245	59.427	–126.016	127.603	–8.331
900	165.352	399.381	315.555	75.444	–127.982	159.430	–9.253
1000	174.598	417.293	324.841	92.452	–129.380	191.444	–10.000
1100	182.673	434.321	334.026	110.325	–130.296	223.574	–10.617
1200	189.745	450.526	343.064	128.954	–130.802	255.770	–11.133
1300	195.853	465.961	351.929	148.241	–130.961	287.993	–11.572
1400	201.209	480.675	360.604	168.100	–130.829	320.217	–11.947
1500	205.895	494.721	369.080	188.460	–130.445	352.422	–12.272
17. BENZENE C_6H_6 (l)							
298.15	135.950	173.450	173.450	0.000	49.080	124.521	–21.815
300	136.312	174.292	173.453	.252	49.077	124.989	–21.762
400	161.793	216.837	179.082	15.102	48.978	150.320	–19.630
500	207.599	257.048	190.639	33.204	50.330	175.559	–18.340
18. BENZENE C_6H_6 (g)							
298.15	82.430	269.190	269.190	0.000	82.880	129.750	–22.731
300	83.020	269.700	269.190	0.153	82.780	130.040	–22.641
400	113.510	297.840	272.823	10.007	77.780	146.570	–19.140
500	139.340	326.050	280.658	22.696	73.740	164.260	–17.160
600	160.090	353.360	290.517	37.706	70.490	182.680	–15.903
700	176.790	379.330	301.360	54.579	67.910	201.590	–15.042
800	190.460	403.860	312.658	72.962	65.910	220.820	–14.418
900	201.840	426.970	324.084	92.597	64.410	240.280	–13.945
1000	211.430	448.740	335.473	113.267	63.340	259.890	–13.575
1100	219.580	469.280	346.710	134.827	62.620	277.640	–13.184
1200	226.540	488.690	357.743	157.137	62.200	299.320	–13.029
1300	232.520	507.070	368.534	180.097	62.000	319.090	–12.821
1400	237.680	524.490	379.056	203.607	61.990	338.870	–12.643
1500	242.140	541.040	389.302	227.607	62.110	358.640	–12.489

	J/K·mol			kJ/mol			
T/K	C_p°	S°	$-(G^\circ-H^\circ(T_r))/T$	$H^\circ-H^\circ(T_r)$	$\Delta_f H^\circ$	$\Delta_f G^\circ$	Log K_f
19. NAPHTHALENE $C_{10}H_8$ (cr, l)							
298.15	165.720	167.390	167.390	0.000	78.530	201.585	–35.316
300	167.001	168.419	167.393	0.308	78.466	202.349	–35.232
353.43	208.722	198.948	169.833	10.290	96.099	224.543	–33.186
PHASE TRANSITION: $\Delta_{trs}H$ = 18.980 kJ/mol, $\Delta_{trs}S$ = 53.702 J/K·mol, cr–l							
353.43	217.200	252.650	169.833	29.270	96.099	224.543	–33.186
400	241.577	280.916	181.124	39.917	96.067	241.475	–31.533
470	276.409	322.712	199.114	58.091	97.012	266.859	–29.658
20. NAPHTHALENE $C_{10}H_8$ (g)							
298.15	131.920	333.150	333.150	0.000	150.580	224.100	–39.260
300	132.840	333.970	333.157	0.244	150.450	224.560	–39.098
400	180.070	378.800	338.950	15.940	144.190	250.270	–32.681
500	219.740	423.400	351.400	36.000	139.220	277.340	–28.973
600	251.530	466.380	367.007	59.624	135.350	305.330	–26.581
700	277.010	507.140	384.146	86.096	132.330	333.950	–24.919
800	297.730	545.520	401.935	114.868	130.050	362.920	–23.696
900	314.850	581.610	419.918	145.523	128.430	392.150	–22.759
1000	329.170	615.550	437.806	177.744	127.510	421.700	–22.027
1100	341.240	647.500	455.426	211.281	127.100	450.630	–21.398
1200	351.500	677.650	472.707	245.932	126.960	480.450	–20.913
1300	360.260	706.130	489.568	281.531	127.060	509.770	–20.482
1400	367.780	733.110	506.009	317.941	127.390	539.740	–20.137
1500	374.270	758.720	522.019	355.051	127.920	568.940	–19.812
21. FORMALDEHYDE H_2CO (g)							
298.15	35.387	218.760	218.760	0.000	–108.700	–102.667	17.987
300	35.443	218.979	218.761	0.066	–108.731	–102.630	17.869
400	39.240	229.665	220.192	3.789	–110.438	–100.340	13.103
500	43.736	238.900	223.028	7.936	–112.073	–97.623	10.198
600	48.181	247.270	226.381	12.534	–113.545	–94.592	8.235
700	52.280	255.011	229.924	17.560	–114.833	–91.328	6.815
800	55.941	262.236	233.517	22.975	–115.942	–87.893	5.739
900	59.156	269.014	237.088	28.734	–116.889	–84.328	4.894
1000	61.951	275.395	240.603	34.792	–117.696	–80.666	4.213
1100	64.368	281.416	244.042	41.111	–118.382	–76.929	3.653
1200	66.453	287.108	247.396	47.655	–118.966	–73.134	3.183
1300	68.251	292.500	250.660	54.392	–119.463	–69.294	2.784
1400	69.803	297.616	253.833	61.297	–119.887	–65.418	2.441
1500	71.146	302.479	256.915	68.346	–120.249	–61.514	2.142
22. METHANOL CH_3OH (g)							
298.15	44.101	239.865	239.865	0.000	–201.000	–162.298	28.434
300	44.219	240.139	239.866	0.082	–201.068	–162.057	28.216
400	51.713	253.845	241.685	4.864	–204.622	–148.509	19.393
500	59.800	266.257	245.374	10.442	–207.750	–134.109	14.010
600	67.294	277.835	249.830	16.803	–210.387	–119.125	10.371
700	73.958	288.719	254.616	23.873	–212.570	–103.737	7.741
800	79.838	298.987	259.526	31.569	–214.350	–88.063	5.750
900	85.025	308.696	264.455	39.817	–215.782	–72.188	4.190
1000	89.597	317.896	269.343	48.553	–216.916	–56.170	2.934
1100	93.624	326.629	274.158	57.718	–217.794	–40.050	1.902
1200	97.165	334.930	278.879	67.262	–218.457	–23.861	1.039
1300	100.277	342.833	283.497	77.137	–218.936	–7.624	0.306
1400	103.014	350.367	288.007	87.304	–219.261	8.644	–0.322
1500	105.422	357.558	292.405	97.729	–219.456	24.930	–0.868

	J/K·mol			kJ/mol			
T/K	C_p°	S°	$-(G^\circ - H^\circ(T_r))/T$	$H^\circ - H^\circ(T_r)$	$\Delta_f H^\circ$	$\Delta_f G^\circ$	Log K_f
23. ACETALDEHYDE C_2H_4O (g)							
298.15	55.318	263.840	263.840	0.000	–166.190	–133.010	23.302
300	55.510	264.180	263.837	0.103	–166.250	–132.800	23.122
400	66.282	281.620	266.147	6.189	–169.530	–121.130	15.818
500	76.675	297.540	270.850	13.345	–172.420	–108.700	11.356
600	85.942	312.360	276.550	21.486	–174.870	–95.720	8.334
700	94.035	326.230	282.667	30.494	–176.910	–82.350	6.145
800	101.070	339.260	288.938	40.258	–178.570	–68.730	4.487
900	107.190	351.520	295.189	50.698	–179.880	–54.920	3.187
1000	112.490	363.100	301.431	61.669	–180.850	–40.930	2.138
1100	117.080	374.040	307.537	73.153	–181.560	–27.010	1.283
1200	121.060	384.400	313.512	85.065	–182.070	–12.860	0.560
1300	124.500	394.230	319.350	97.344	–182.420	1.240	–0.050
1400	127.490	403.570	325.031	109.954	–182.640	15.470	–0.577
1500	130.090	412.460	330.571	122.834	–182.750	29.580	–1.030
24. ETHANOL C_2H_5OH (g)							
298.15	65.652	281.622	281.622	0.000	–234.800	–167.874	29.410
300	65.926	282.029	281.623	0.122	–234.897	–167.458	29.157
400	81.169	303.076	284.390	7.474	–239.826	–144.216	18.832
500	95.400	322.750	290.115	16.318	–243.940	–119.820	12.517
600	107.656	341.257	297.112	26.487	–247.260	–94.672	8.242
700	118.129	358.659	304.674	37.790	–249.895	–69.023	5.151
800	127.171	375.038	312.456	50.065	–251.951	–43.038	2.810
900	135.049	390.482	320.276	63.185	–253.515	–16.825	0.976
1000	141.934	405.075	328.033	77.042	–254.662	9.539	–0.498
1100	147.958	418.892	335.670	91.543	–255.454	36.000	–1.709
1200	153.232	431.997	343.156	106.609	–255.947	62.520	–2.721
1300	157.849	444.448	350.473	122.168	–256.184	89.070	–3.579
1400	161.896	456.298	357.612	138.160	–256.206	115.630	–4.314
1500	165.447	467.591	364.571	154.531	–256.044	142.185	–4.951
25. ACETIC ACID $C_2H_4O_2$ (g)							
298.15	63.438	283.470	283.470	0.000	–432.249	–374.254	65.567
300	63.739	283.863	283.471	0.118	–432.324	–373.893	65.100
400	79.665	304.404	286.164	7.296	–436.006	–353.840	46.206
500	93.926	323.751	291.765	15.993	–438.875	–332.950	34.783
600	106.181	341.988	298.631	26.014	–440.993	–311.554	27.123
700	116.627	359.162	306.064	37.169	–442.466	–289.856	21.629
800	125.501	375.331	313.722	49.287	–443.395	–267.985	17.497
900	132.989	390.558	321.422	62.223	–443.873	–246.026	14.279
1000	139.257	404.904	329.060	75.844	–443.982	–224.034	11.702
1100	144.462	418.429	336.576	90.039	–443.798	–202.046	9.594
1200	148.760	431.189	343.933	104.707	–443.385	–180.086	7.839
1300	152.302	443.240	351.113	119.765	–442.795	–158.167	6.355
1400	155.220	454.637	358.105	135.146	–442.071	–136.299	5.085
1500	157.631	465.432	364.903	150.793	–441.247	–114.486	3.987
26. ACETONE C_3H_6O (g)							
298.15	74.517	295.349	295.349	0.000	–217.150	–152.716	26.757
300	74.810	295.809	295.349	0.138	–217.233	–152.339	26.521
400	91.755	319.658	298.498	8.464	–222.212	–129.913	16.962
500	107.864	341.916	304.988	18.464	–226.522	–106.315	11.107
600	122.047	362.836	312.873	29.978	–230.120	–81.923	7.133
700	134.306	382.627	321.470	42.810	–233.049	–56.986	4.252
800	144.934	401.246	330.265	56.785	–235.350	–31.673	2.068

	J/K·mol			kJ/mol			
T/K	C_p°	S°	$-(G^\circ-H^\circ(T_r))/T$	$H^\circ-H^\circ(T_r)$	$\Delta_f H^\circ$	$\Delta_f G^\circ$	Log K_f
26. ACETONE C_3H_6O (g) (continued)							
900	154.097	418.860	339.141	71.747	–237.149	–6.109	0.353
1000	162.046	435.513	347.950	87.563	–238.404	19.707	–1.030
1100	168.908	451.286	356.617	104.136	–239.283	45.396	–2.157
1200	174.891	466.265	365.155	121.332	–239.827	71.463	–3.110
1300	180.079	480.491	373.513	139.072	–240.120	97.362	–3.912
1400	184.556	493.963	381.596	157.314	–240.203	123.470	–4.607
1500	188.447	506.850	389.533	175.975	–240.120	149.369	–5.202
27. PHENOL C_6H_6O (g)							
298.15	103.220	314.810	314.810	0.000	–96.400	–32.630	5.720
300	103.860	315.450	314.810	0.192	–96.490	–32.230	5.610
400	135.790	349.820	319.278	12.217	–100.870	–10.180	1.330
500	161.910	383.040	328.736	27.152	–104.240	12.970	–1.360
600	182.480	414.450	340.430	44.412	–106.810	36.650	–3.190
700	198.840	443.860	353.134	63.508	–108.800	60.750	–4.530
800	212.140	471.310	366.211	84.079	–110.300	85.020	–5.550
900	223.190	496.950	379.327	105.861	–111.370	109.590	–6.360
1000	232.490	520.960	392.302	128.658	–111.990	134.280	–7.010
1100	240.410	543.500	405.033	152.314	–112.280	158.620	–7.530
1200	247.200	564.720	417.468	176.703	–112.390	183.350	–7.980
1300	253.060	584.740	429.568	201.723	–112.330	208.070	–8.360
1400	258.120	603.680	441.331	227.288	–112.120	233.050	–8.700
1500	262.520	621.650	452.767	253.325	–111.780	257.540	–8.970
28. CARBON TETRAFLUORIDE CF_4 (g)							
298.15	61.050	261.455	261.455	0.000	–933.200	–888.518	155.663
300	61.284	261.833	261.456	0.113	–933.219	–888.240	154.654
400	72.399	281.057	264.001	6.822	–933.986	–873.120	114.016
500	80.713	298.153	269.155	14.499	–934.372	–857.852	89.618
600	86.783	313.434	275.284	22.890	–934.490	–842.533	73.348
700	91.212	327.162	281.732	31.801	–934.431	–827.210	61.726
800	94.479	339.566	288.199	41.094	–934.261	–811.903	53.011
900	96.929	350.842	294.542	50.670	–934.024	–796.622	46.234
1000	98.798	361.156	300.695	60.460	–933.745	–781.369	40.814
1100	100.250	370.643	306.629	70.416	–933.442	–766.146	36.381
1200	101.396	379.417	312.334	80.500	–933.125	–750.952	32.688
1300	102.314	387.571	317.811	90.687	–932.800	–735.784	29.564
1400	103.059	395.181	323.069	100.957	–932.470	–720.641	26.887
1500	103.671	402.313	328.116	111.295	–932.137	–705.522	24.568
29. TRIFLUOROMETHANE CHF_3 (g)							
298.15	51.069	259.675	259.675	0.000	–696.700	–662.237	116.020
300	51.258	259.991	259.676	0.095	–696.735	–662.023	115.267
400	61.148	276.113	261.807	5.722	–698.427	–650.186	84.905
500	69.631	290.700	266.149	12.275	–699.715	–637.969	66.647
600	76.453	304.022	271.368	19.593	–700.634	–625.528	54.456
700	81.868	316.230	276.917	27.519	–701.253	–612.957	45.739
800	86.201	327.455	282.542	35.930	–701.636	–600.315	39.196
900	89.719	337.818	288.116	44.732	–701.832	–587.636	34.105
1000	92.617	347.426	293.572	53.854	–701.879	–574.944	30.032
1100	95.038	356.370	298.879	63.240	–701.805	–562.253	26.699
1200	97.084	364.730	304.022	72.849	–701.629	–549.574	23.922
1300	98.833	372.571	308.997	82.647	–701.368	–536.913	21.573
1400	100.344	379.952	313.804	92.607	–701.033	–524.274	19.561
1500	101.660	386.921	318.449	102.709	–700.635	–511.662	17.817

	J/K·mol			kJ/mol			
T/K	C_p°	S°	$-(G^\circ-H^\circ(T_r))/T$	$H^\circ-H^\circ(T_r)$	$\Delta_f H^\circ$	$\Delta_f G^\circ$	Log K_f
30. CHLOROTRIFLUOROMETHANE $CClF_3$ (g)							
298.15	66.886	285.419	285.419	0.000	–707.800	–667.238	116.896
300	67.111	285.834	285.421	0.124	–707.810	–666.986	116.131
400	77.528	306.646	288.187	7.383	–708.153	–653.316	85.313
500	85.013	324.797	293.734	15.532	–708.170	–639.599	66.818
600	90.329	340.794	300.271	24.314	–707.975	–625.901	54.489
700	94.132	355.020	307.096	33.547	–707.654	–612.246	45.686
800	96.899	367.780	313.897	43.106	–707.264	–598.642	39.087
900	98.951	379.317	320.536	52.903	–706.837	–585.090	33.957
1000	100.507	389.827	326.947	62.880	–706.396	–571.586	29.856
1100	101.708	399.465	333.108	72.993	–705.950	–558.126	26.503
1200	102.651	408.357	339.013	83.213	–705.505	–544.707	23.710
1300	103.404	416.604	344.668	93.517	–705.064	–531.326	21.349
1400	104.012	424.290	350.084	103.889	–704.628	–517.977	19.326
1500	104.512	431.484	355.273	114.316	–704.196	–504.660	17.574
31. DICHLORODIFLUOROMETHANE CCl_2F_2 (g)							
298.15	72.476	300.903	300.903	0.000	–486.000	–447.030	78.317
300	72.691	301.352	300.905	0.134	–486.002	–446.788	77.792
400	82.408	323.682	303.883	7.919	–485.945	–433.716	56.637
500	89.063	342.833	309.804	16.514	–485.618	–420.692	43.949
600	93.635	359.500	316.729	25.663	–485.136	–407.751	35.497
700	96.832	374.189	323.909	35.196	–484.576	–394.897	29.467
800	99.121	387.276	331.027	44.999	–483.984	–382.126	24.950
900	100.801	399.053	337.942	55.000	–483.388	–369.429	21.441
1000	102.062	409.742	344.596	65.146	–482.800	–356.799	18.637
1100	103.030	419.517	350.969	75.402	–482.226	–344.227	16.346
1200	103.786	428.515	357.061	85.745	–481.667	–331.706	14.439
1300	104.388	436.847	362.882	96.154	–481.121	–319.232	12.827
1400	104.874	444.602	368.445	106.618	–480.588	–306.799	11.447
1500	105.270	451.851	373.767	117.126	–480.065	–294.404	10.252
32. CHLORODIFLUOROMETHANE $CHClF_2$ (g)							
298.15	55.853	280.915	280.915	0.000	–475.000	–443.845	77.759
300	56.039	281.261	280.916	0.104	–475.028	–443.652	77.246
400	65.395	298.701	283.231	6.188	–476.390	–432.978	56.540
500	73.008	314.145	287.898	13.123	–477.398	–422.001	44.086
600	78.940	328.003	293.448	20.733	–478.103	–410.851	35.767
700	83.551	340.533	299.294	28.867	–478.574	–399.603	29.818
800	87.185	351.936	305.172	37.411	–478.870	–388.299	25.353
900	90.100	362.379	310.956	46.280	–479.031	–376.967	21.878
1000	92.475	371.999	316.586	55.413	–479.090	–365.622	19.098
1100	94.433	380.908	322.033	64.761	–479.068	–354.276	16.823
1200	96.066	389.196	327.289	74.289	–478.982	–342.935	14.927
1300	97.438	396.941	332.352	83.966	–478.843	–331.603	13.324
1400	98.601	404.206	337.228	93.769	–478.661	–320.283	11.950
1500	99.593	411.044	341.923	103.681	–478.443	–308.978	10.759
33. METHYLAMINE CH_5N (g)							
298.15	50.053	242.881	242.881	0.000	–22.529	32.734	–5.735
300	50.227	243.196	242.893	0.091	–22.614	33.077	–5.759
400	60.171	258.986	244.975	5.604	–26.846	52.294	–6.829
500	70.057	273.486	249.244	12.121	–30.431	72.510	–7.575
600	78.929	287.063	254.431	19.579	–33.364	93.382	–8.129
700	86.711	299.826	260.008	27.873	–35.712	114.702	–8.559
800	93.545	311.865	265.749	36.893	–37.548	136.316	–8.900

T/K	C_p° (J/K·mol)	S° (J/K·mol)	$-(G^\circ-H^\circ(T_r))/T$ (J/K·mol)	$H^\circ-H^\circ(T_r)$ (kJ/mol)	Δ_fH° (kJ/mol)	Δ_fG° (kJ/mol)	Log K_f
33. METHYLAMINE CH_5N (g) (continued)							
900	99.573	323.239	271.511	46.555	–38.949	158.138	–9.178
1000	104.886	334.006	277.220	56.786	–39.967	180.098	–9.407
1100	109.576	344.233	282.861	67.509	–40.681	201.822	–9.584
1200	113.708	353.944	288.374	78.685	–41.136	224.240	–9.761
1300	117.341	363.190	293.775	90.239	–41.376	246.364	–9.899
1400	120.542	372.012	299.061	102.131	–41.451	268.504	–10.018
1500	123.353	380.426	304.209	114.326	–41.381	290.639	–10.121
34. CHLORINE Cl (g)							
298.15	21.838	165.190	165.190	0.000	121.302	105.306	–18.449
300	21.852	165.325	165.190	0.040	121.311	105.207	–18.318
400	22.467	171.703	166.055	2.259	121.795	99.766	–13.028
500	22.744	176.752	167.708	4.522	122.272	94.203	–9.841
600	22.781	180.905	169.571	6.800	122.734	88.546	–7.709
700	22.692	184.411	171.448	9.074	123.172	82.813	–6.179
800	22.549	187.432	173.261	11.337	123.585	77.019	–5.029
900	22.389	190.079	174.986	13.584	123.971	71.175	–4.131
1000	22.233	192.430	176.615	15.815	124.334	65.289	–3.410
1100	22.089	194.542	178.150	18.031	124.675	59.368	–2.819
1200	21.959	196.458	179.597	20.233	124.996	53.416	–2.325
1300	21.843	198.211	180.963	22.423	125.299	47.439	–1.906
1400	21.742	199.826	182.253	24.602	125.587	41.439	–1.546
1500	21.652	201.323	183.475	26.772	125.861	35.418	–1.233
35. DICHLORINE Cl_2 (g)							
298.15	33.949	223.079	223.079	0.000	0.000	0.000	0.000
300	33.981	223.290	223.080	0.063	0.000	0.000	0.000
400	35.296	233.263	224.431	3.533	0.000	0.000	0.000
500	36.064	241.229	227.021	7.104	0.000	0.000	0.000
600	36.547	247.850	229.956	10.736	0.000	0.000	0.000
700	36.874	253.510	232.926	14.408	0.000	0.000	0.000
800	37.111	258.450	235.815	18.108	0.000	0.000	0.000
900	37.294	262.832	238.578	21.829	0.000	0.000	0.000
1000	37.442	266.769	241.203	25.566	0.000	0.000	0.000
1100	37.567	270.343	243.692	29.316	0.000	0.000	0.000
1200	37.678	273.617	246.052	33.079	0.000	0.000	0.000
1300	37.778	276.637	248.290	36.851	0.000	0.000	0.000
1400	37.872	279.440	250.416	40.634	0.000	0.000	0.000
1500	37.961	282.056	252.439	44.426	0.000	0.000	0.000
36. HYDROGEN CHLORIDE HCl (g)							
298.15	29.136	186.902	186.902	0.000	–92.310	–95.298	16.696
300	29.137	187.082	186.902	0.054	–92.314	–95.317	16.596
400	29.175	195.468	188.045	2.969	–92.587	–96.278	12.573
500	29.304	201.990	190.206	5.892	–92.911	–97.164	10.151
600	29.576	207.354	192.630	8.835	–93.249	–97.983	8.530
700	29.988	211.943	195.069	11.812	–93.577	–98.746	7.368
800	30.500	215.980	197.435	14.836	–93.879	–99.464	6.494
900	31.063	219.604	199.700	17.913	–94.149	–100.145	5.812
1000	31.639	222.907	201.858	21.049	–94.384	–100.798	5.265
1100	32.201	225.949	203.912	24.241	–94.587	–101.430	4.816
1200	32.734	228.774	205.867	27.488	–94.760	–102.044	4.442
1300	33.229	231.414	207.732	30.786	–94.908	–102.645	4.124
1400	33.684	233.893	209.513	34.132	–95.035	–103.235	3.852
1500	34.100	236.232	211.217	37.522	–95.146	–103.817	3.615

	J/K·mol			kJ/mol			
T/K	C_p°	S°	$-(G^\circ-H^\circ(T_r))/T$	$H^\circ-H^\circ(T_r)$	$\Delta_f H^\circ$	$\Delta_f G^\circ$	Log K_f
37. COPPER Cu (cr, l)							
298.15	24.440	33.150	33.150	0.000	0.000	0.000	0.000
300	24.460	33.301	33.150	0.045	0.000	0.000	0.000
400	25.339	40.467	34.122	2.538	0.000	0.000	0.000
500	25.966	46.192	35.982	5.105	0.000	0.000	0.000
600	26.479	50.973	38.093	7.728	0.000	0.000	0.000
700	26.953	55.090	40.234	10.399	0.000	0.000	0.000
800	27.448	58.721	42.322	13.119	0.000	0.000	0.000
900	28.014	61.986	44.328	15.891	0.000	0.000	0.000
1000	28.700	64.971	46.245	18.726	0.000	0.000	0.000
1100	29.553	67.745	48.075	21.637	0.000	0.000	0.000
1200	30.617	70.361	49.824	24.644	0.000	0.000	0.000
1300	31.940	72.862	51.501	27.769	0.000	0.000	0.000
1358	32.844	74.275	52.443	29.647	0.000	0.000	0.000
PHASE TRANSITION: $\Delta_{trs} H$ = 13.141 kJ/mol, $\Delta_{trs} S$ = 9.676 J/K·mol, cr–l							
1358	32.800	83.951	52.443	42.788	0.000	0.000	0.000
1400	32.800	84.950	53.403	44.166	0.000	0.000	0.000
1500	32.800	87.213	55.583	47.446	0.000	0.000	0.000
38. COPPER Cu (g)							
298.15	20.786	166.397	166.397	0.000	337.600	297.873	–52.185
300	20.786	166.525	166.397	0.038	337.594	297.626	–51.821
400	20.786	172.505	167.213	2.117	337.179	284.364	–37.134
500	20.786	177.143	168.752	4.196	336.691	271.215	–28.333
600	20.786	180.933	170.476	6.274	336.147	258.170	–22.475
700	20.786	184.137	172.205	8.353	335.554	245.221	–18.298
800	20.786	186.913	173.874	10.431	334.913	232.359	–15.171
900	20.786	189.361	175.461	12.510	334.219	219.581	–12.744
1000	20.786	191.551	176.963	14.589	333.463	206.883	–10.806
1100	20.788	193.532	178.380	16.667	332.631	194.265	–9.225
1200	20.793	195.341	179.719	18.746	331.703	181.726	–7.910
1300	20.803	197.006	180.986	20.826	330.657	169.270	–6.801
1400	20.823	198.548	182.186	22.907	316.342	157.305	–5.869
1500	20.856	199.986	183.325	24.991	315.146	145.987	–5.084
39. COPPER OXIDE CuO (cr)							
298.15	42.300	42.740	42.740	0.000	–162.000	–134.277	23.524
300	42.417	43.002	42.741	0.078	–161.994	–134.105	23.349
400	46.783	55.878	44.467	4.564	–161.487	–124.876	16.307
500	49.190	66.596	47.852	9.372	–160.775	–115.803	12.098
600	50.827	75.717	51.755	14.377	–159.973	–106.883	9.305
700	52.099	83.651	55.757	19.526	–159.124	–98.102	7.320
800	53.178	90.680	59.691	24.791	–158.247	–89.444	5.840
900	54.144	97.000	63.491	30.158	–157.356	–80.897	4.695
1000	55.040	102.751	67.134	35.617	–156.462	–72.450	3.784
1100	55.890	108.037	70.615	41.164	–155.582	–64.091	3.043
1200	56.709	112.936	73.941	46.794	–154.733	–55.812	2.429
1300	57.507	117.507	77.118	52.505	–153.940	–47.601	1.913
1400	58.288	121.797	80.158	58.295	–166.354	–39.043	1.457
1500	59.057	125.845	83.070	64.163	–165.589	–29.975	1.044
40. DICOPPER OXIDE Cu_2O (cr)							
298.15	62.600	92.550	92.550	0.000	–173.100	–150.344	26.339
300	62.721	92.938	92.551	0.116	–173.102	–150.203	26.152
400	67.587	111.712	95.078	6.654	–173.036	–142.572	18.618
500	70.784	127.155	99.995	13.580	–172.772	–134.984	14.101

	J/K·mol			kJ/mol			
T/K	C_p°	S°	$-(G^\circ-H^\circ(T_r))/T$	$H^\circ-H^\circ(T_r)$	$\Delta_f H^\circ$	$\Delta_f G^\circ$	Log K_f
40. DICOPPER OXIDE Cu_2O (cr) (continued)							
600	73.323	140.291	105.643	20.789	–172.389	–127.460	11.096
700	75.552	151.764	111.429	28.235	–171.914	–120.009	8.955
800	77.616	161.989	117.121	35.894	–171.363	–112.631	7.354
900	79.584	171.245	122.629	43.755	–170.750	–105.325	6.113
1000	81.492	179.729	127.920	51.809	–170.097	–98.091	5.124
1100	83.360	187.584	132.992	60.052	–169.431	–90.922	4.317
1200	85.202	194.917	137.850	68.480	–168.791	–83.814	3.648
1300	87.026	201.808	142.507	77.092	–168.223	–76.756	3.084
1400	88.836	208.324	146.978	85.885	–194.030	–68.926	2.572
1500	90.636	214.515	151.276	94.858	–193.438	–60.010	2.090
41. COPPER DICHLORIDE $CuCl_2$ (cr, l)							
298.15	71.880	108.070	108.070	0.000	–218.000	–173.826	30.453
300	71.998	108.515	108.071	0.133	–217.975	–173.552	30.218
400	76.338	129.899	110.957	7.577	–216.494	–158.962	20.758
500	78.654	147.204	116.532	15.336	–214.873	–144.765	15.123
600	80.175	161.687	122.884	23.282	–213.182	–130.901	11.396
675	81.056	171.183	127.732	29.329	–211.185	–120.693	9.340
PHASE TRANSITION: $\Delta_{trs}H$ = 0.700 kJ/mol, $\Delta_{trs}S$ = 1.037 J/K·mol, crII–crI							
675	82.400	172.220	127.732	30.029	–211.185	–120.693	9.340
700	82.400	175.216	129.375	32.089	–210.719	–117.350	8.757
800	82.400	186.219	135.808	40.329	–208.898	–104.137	6.799
871	82.400	193.226	140.207	46.179	–192.649	–94.893	5.691
PHASE TRANSITION: $\Delta_{trs}H$ = 15.001 kJ/mol, $\Delta_{trs}S$ = 17.221 J/K·mol, crI–l							
871	100.000	210.447	140.207	61.180	–192.649	–94.893	5.691
900	100.000	213.723	142.523	64.080	–191.640	–91.655	5.319
1000	100.000	224.259	150.179	74.080	–188.212	–80.730	4.217
1100	100.000	233.790	157.353	84.080	–184.873	–70.144	3.331
1130.75	100.000	236.547	159.470	87.155	–183.867	–66.951	3.093
42. COPPER DICHLORIDE $CuCl_2$ (g)							
298.15	56.814	278.418	278.418	0.000	–43.268	–49.883	8.739
300	56.869	278.769	278.419	0.105	–43.271	–49.924	8.692
400	58.992	295.456	280.679	5.911	–43.428	–52.119	6.806
500	60.111	308.752	285.010	11.871	–43.606	–54.271	5.670
600	60.761	319.774	289.911	17.918	–43.814	–56.385	4.909
700	61.168	329.173	294.865	24.015	–44.060	–58.462	4.362
800	61.439	337.360	299.677	30.147	–44.349	–60.500	3.950
900	61.630	344.608	304.274	36.301	–44.688	–62.499	3.627
1000	61.776	351.109	308.638	42.471	–45.088	–64.457	3.367
1100	61.900	357.003	312.771	48.655	–45.566	–66.372	3.152
1200	62.022	362.394	316.685	54.851	–46.139	–68.239	2.970
1300	62.159	367.364	320.395	61.060	–46.829	–70.053	2.815
1400	62.325	371.976	323.916	67.284	–60.784	–71.404	2.664
1500	62.531	376.283	327.265	73.526	–61.613	–72.133	2.512
43. FLUORINE F (g)							
298.15	22.746	158.750	158.750	0.000	79.380	62.280	–10.911
300	22.742	158.891	158.750	0.042	79.393	62.173	–10.825
400	22.432	165.394	159.639	2.302	80.043	56.332	–7.356
500	22.100	170.363	161.307	4.528	80.587	50.340	–5.259
600	21.832	174.368	163.161	6.724	81.046	44.246	–3.852
700	21.629	177.717	165.008	8.897	81.442	38.081	–2.842
800	21.475	180.595	166.780	11.052	81.792	31.862	–2.080
900	21.357	183.117	168.458	13.193	82.106	25.601	–1.486

	J/K·mol			kJ/mol			
T/K	C_p°	S°	$-(G^\circ-H^\circ(T_r))/T$	$H^\circ-H^\circ(T_r)$	$\Delta_f H^\circ$	$\Delta_f G^\circ$	Log K_f
43. FLUORINE F (g) (continued)							
1000	21.266	185.362	170.039	15.324	82.391	19.308	–1.009
1100	21.194	187.386	171.525	17.447	82.654	12.986	–0.617
1200	21.137	189.227	172.925	19.563	82.897	6.642	–0.289
1300	21.091	190.917	174.245	21.675	83.123	0.278	–0.011
1400	21.054	192.479	175.492	23.782	83.335	–6.103	0.228
1500	21.022	193.930	176.673	25.886	83.533	–12.498	0.435
44. DIFLUORINE F_2 (g)							
298.15	31.304	202.790	202.790	0.000	0.000	0.000	0.000
300	31.337	202.984	202.790	0.058	0.000	0.000	0.000
400	32.995	212.233	204.040	3.277	0.000	0.000	0.000
500	34.258	219.739	206.453	6.643	0.000	0.000	0.000
600	35.171	226.070	209.208	10.117	0.000	0.000	0.000
700	35.839	231.545	212.017	13.669	0.000	0.000	0.000
800	36.343	236.365	214.765	17.279	0.000	0.000	0.000
900	36.740	240.669	217.409	20.934	0.000	0.000	0.000
1000	37.065	244.557	219.932	24.625	0.000	0.000	0.000
1100	37.342	248.103	222.334	28.346	0.000	0.000	0.000
1200	37.588	251.363	224.619	32.093	0.000	0.000	0.000
1300	37.811	254.381	226.794	35.863	0.000	0.000	0.000
1400	38.019	257.191	228.866	39.654	0.000	0.000	0.000
1500	38.214	259.820	230.843	43.466	0.000	0.000	0.000
45. HYDROGEN FLUORIDE HF (g)							
298.15	29.137	173.776	173.776	0.000	–273.300	–275.399	48.248
300	29.137	173.956	173.776	0.054	–273.302	–275.412	47.953
400	29.149	182.340	174.919	2.968	–273.450	–276.096	36.054
500	29.172	188.846	177.078	5.884	–273.679	–276.733	28.910
600	29.230	194.169	179.496	8.804	–273.961	–277.318	24.142
700	29.350	198.683	181.923	11.732	–274.277	–277.852	20.733
800	29.549	202.614	184.269	14.676	–274.614	–278.340	18.174
900	29.827	206.110	186.505	17.645	–274.961	–278.785	16.180
1000	30.169	209.270	188.626	20.644	–275.309	–279.191	14.583
1100	30.558	212.163	190.636	23.680	–275.652	–279.563	13.275
1200	30.974	214.840	192.543	26.756	–275.988	–279.904	12.184
1300	31.403	217.336	194.355	29.875	–276.315	–280.217	11.259
1400	31.831	219.679	196.081	33.037	–276.631	–280.505	10.466
1500	32.250	221.889	197.729	36.241	–276.937	–280.771	9.777
46. GERMANIUM Ge (cr, l)							
298.15	23.222	31.090	31.090	0.000	0.000	0.000	0.000
300	23.249	31.234	31.090	0.043	0.000	0.000	0.000
400	24.310	38.083	32.017	2.426	0.000	0.000	0.000
500	24.962	43.582	33.798	4.892	0.000	0.000	0.000
600	25.452	48.178	35.822	7.414	0.000	0.000	0.000
700	25.867	52.133	37.876	9.980	0.000	0.000	0.000
800	26.240	55.612	39.880	12.586	0.000	0.000	0.000
900	26.591	58.723	41.804	15.227	0.000	0.000	0.000
1000	26.926	61.542	43.639	17.903	0.000	0.000	0.000
1100	27.252	64.124	45.386	20.612	0.000	0.000	0.000
1200	27.571	66.509	47.048	23.353	0.000	0.000	0.000
1211.4	27.608	66.770	47.232	23.668	0.000	0.000	0.000
PHASE TRANSITION: $\Delta_{trs} H$ = 37.030 kJ/mol, $\Delta_{trs} S$ = 30.568 J/K·mol, cr–l							
1211.4	27.600	97.338	47.232	60.698	0.000	0.000	0.000
1300	27.600	99.286	50.714	63.143	0.000	0.000	0.000

	J/K·mol			kJ/mol			
T/K	C_p°	S°	$-(G^\circ-H^\circ(T_r))/T$	$H^\circ-H^\circ(T_r)$	$\Delta_f H^\circ$	$\Delta_f G^\circ$	Log K_f
46. GERMANIUM Ge (cr, l) (continued)							
1400	27.600	101.331	54.258	65.903	0.000	0.000	0.000
1500	27.600	103.236	57.460	68.663	0.000	0.000	0.000
47. GERMANIUM Ge (g)							
298.15	30.733	167.903	167.903	0.000	367.800	327.009	–57.290
300	30.757	168.094	167.904	0.057	367.814	326.756	–56.893
400	31.071	177.025	169.119	3.162	368.536	312.959	–40.868
500	30.360	183.893	171.415	6.239	369.147	298.991	–31.235
600	29.265	189.334	173.965	9.222	369.608	284.914	–24.804
700	28.102	193.758	176.487	12.090	369.910	270.773	–20.205
800	27.029	197.439	178.882	14.845	370.060	256.598	–16.754
900	26.108	200.567	181.122	17.501	370.073	242.414	–14.069
1000	25.349	203.277	183.205	20.072	369.969	228.234	–11.922
1100	24.741	205.664	185.141	22.575	369.763	214.069	–10.165
1200	24.264	207.795	186.941	25.025	369.471	199.928	–8.703
1300	23.898	209.722	188.621	27.432	332.088	188.521	–7.575
1400	23.624	211.483	190.192	29.807	331.704	177.492	–6.622
1500	23.426	213.105	191.666	32.159	331.296	166.491	–5.798
48. GERMANIUM DIOXIDE GeO_2 (cr, l)							
298.15	50.166	39.710	39.710	0.000	–580.200	–521.605	91.382
300	50.475	40.021	39.711	0.093	–580.204	–521.242	90.755
400	61.281	56.248	41.850	5.759	–579.893	–501.610	65.503
500	66.273	70.519	46.191	12.164	–579.013	–482.134	50.368
600	69.089	82.872	51.299	18.943	–577.915	–462.859	40.295
700	70.974	93.671	56.597	25.952	–576.729	–443.776	33.115
800	72.449	103.247	61.841	33.125	–575.498	–424.866	27.741
900	73.764	111.857	66.928	40.436	–574.235	–406.113	23.570
1000	75.049	119.696	71.819	47.877	572.934	–387.502	20.241
1100	76.378	126.910	76.504	55.447	–571.582	–369.024	17.523
1200	77.796	133.616	80.987	63.155	–570.166	–350.671	15.264
1300	79.332	139.903	85.279	71.010	–605.685	–329.732	13.249
1308	79.460	140.390	85.615	71.646	–584.059	–328.034	13.100
PHASE TRANSITION: $\Delta_{trs}H$ = 21.500 kJ/mol, $\Delta_{trs}S$ = 16.437 J/K·mol, crII–crI							
1308	80.075	156.827	85.615	93.146	–584.059	–328.034	13.100
1388	81.297	161.617	89.858	99.601	–565.504	–312.415	11.757
PHASE TRANSITION: $\Delta_{trs}H$ = 17.200 kJ/mol, $\Delta_{trs}S$ = 12.392 J/K·mol, crI–l							
1388	78.500	174.009	89.858	116.801	–565.504	–312.415	11.757
1400	78.500	174.685	90.582	117.743	–565.328	–310.228	11.575
1500	78.500	180.100	96.372	125.593	–563.882	–292.057	10.170
49. GERMANIUM TETRACHLORIDE $GeCl_4$ (g)							
298.15	95.918	348.393	348.393	0.000	–500.000	–461.582	80.866
300	96.041	348.987	348.395	0.178	–499.991	–461.343	80.326
400	100.750	377.342	352.229	10.045	–499.447	–448.540	58.573
500	103.206	400.114	359.604	20.255	–498.845	–435.882	45.536
600	104.624	419.067	367.980	30.652	–498.234	–423.347	36.855
700	105.509	435.266	376.463	41.162	–497.634	–410.914	30.662
800	106.096	449.396	384.715	51.744	–497.057	–398.565	26.023
900	106.504	461.917	392.611	62.375	–496.509	–386.287	22.419
1000	106.799	473.155	400.113	73.041	–495.993	–374.068	19.539
1100	107.020	483.344	407.224	83.733	–495.512	–361.899	17.185
1200	107.189	492.664	413.961	94.444	–495.067	–349.772	15.225
1300	107.320	501.249	420.349	105.169	–531.677	–334.973	13.459
1400	107.425	509.206	426.416	115.907	–531.265	–319.857	11.934
1500	107.509	516.621	432.185	126.654	–530.861	–304.771	10.613

	J/K·mol			kJ/mol			
T/K	C_p°	S°	$-(G^\circ-H^\circ(T_r))/T$	$H^\circ-H^\circ(T_r)$	$\Delta_f H^\circ$	$\Delta_f G^\circ$	Log K_f
50. HYDROGEN H (g)							
298.15	20.786	114.716	114.716	0.000	217.998	203.276	–35.613
300	20.786	114.845	114.716	0.038	218.010	203.185	–35.377
400	20.786	120.824	115.532	2.117	218.635	198.149	–25.875
500	20.786	125.463	117.071	4.196	219.253	192.956	–20.158
600	20.786	129.252	118.795	6.274	219.867	187.639	–16.335
700	20.786	132.457	120.524	8.353	220.476	182.219	–13.597
800	20.786	135.232	122.193	10.431	221.079	176.712	–11.538
900	20.786	137.680	123.780	12.510	221.670	171.131	–9.932
1000	20.786	139.870	125.282	14.589	222.247	165.485	–8.644
1100	20.786	141.852	126.700	16.667	222.806	159.781	–7.587
1200	20.786	143.660	128.039	18.746	223.345	154.028	–6.705
1300	20.786	145.324	129.305	20.824	223.864	148.230	–5.956
1400	20.786	146.864	130.505	22.903	224.360	142.393	–5.313
1500	20.786	148.298	131.644	24.982	224.835	136.522	–4.754
51. DIHYDROGEN H_2 (g)							
298.15	28.836	130.680	130.680	0.000	0.000	0.000	0.000
300	28.849	130.858	130.680	0.053	0.000	0.000	0.000
400	29.181	139.217	131.818	2.960	0.000	0.000	0.000
500	29.260	145.738	133.974	5.882	0.000	0.000	0.000
600	29.327	151.078	136.393	8.811	0.000	0.000	0.000
700	29.440	155.607	138.822	11.749	0.000	0.000	0.000
800	29.623	159.549	141.172	14.702	0.000	0.000	0.000
900	29.880	163.052	143.412	17.676	0.000	0.000	0.000
1000	30.204	166.217	145.537	20.680	0.000	0.000	0.000
1100	30.580	169.113	147.550	23.719	0.000	0.000	0.000
1200	30.991	171.791	149.460	26.797	0.000	0.000	0.000
1300	31.422	174.288	151.275	29.918	0.000	0.000	0.000
1400	31.860	176.633	153.003	33.082	0.000	0.000	0.000
1500	32.296	178.846	154.653	36.290	0.000	0.000	0.000
52. HYDROXYL OH (g)							
298.15	29.886	183.737	183.737	0.000	39.349	34.631	–6.067
300	29.879	183.922	183.738	0.055	39.350	34.602	–6.025
400	29.604	192.476	184.906	3.028	39.384	33.012	–4.311
500	29.495	199.067	187.104	5.982	39.347	31.422	–3.283
600	29.513	204.445	189.560	8.931	39.252	29.845	–2.598
700	29.655	209.003	192.020	11.888	39.113	28.287	–2.111
800	29.914	212.979	194.396	14.866	38.945	26.752	–1.747
900	30.265	216.522	196.661	17.874	38.763	25.239	–1.465
1000	30.682	219.731	198.810	20.921	38.577	23.746	–1.240
1100	31.135	222.677	200.848	24.012	38.393	22.272	–1.058
1200	31.603	225.406	202.782	27.149	38.215	20.814	–0.906
1300	32.069	227.954	204.621	30.332	38.046	19.371	–0.778
1400	32.522	230.347	206.374	33.562	37.886	17.941	–0.669
1500	32.956	232.606	208.048	36.836	37.735	16.521	–0.575
53. WATER H_2O (l)							
298.15	75.300	69.950	69.950	0.000	–285.830	–237.141	41.546
300	75.281	70.416	69.951	0.139	–285.771	–236.839	41.237
373.21	76.079	86.896	71.715	5.666	–283.454	–225.160	31.513
54. WATER H_2O (g)							
298.15	33.598	188.832	188.832	0.000	–241.826	–228.582	40.046

54. WATER H_2O (g) (continued)

	J/K·mol			kJ/mol			
T/K	C_p°	S°	$-(G^\circ - H^\circ(T_r))/T$	$H^\circ - H^\circ(T_r)$	$\Delta_f H^\circ$	$\Delta_f G^\circ$	Log K_f
300	33.606	189.040	188.833	0.062	–241.844	–228.500	39.785
400	34.283	198.791	190.158	3.453	–242.845	–223.900	29.238
500	35.259	206.542	192.685	6.929	–243.822	–219.050	22.884
600	36.371	213.067	195.552	10.509	–244.751	–214.008	18.631
700	37.557	218.762	198.469	14.205	–245.620	–208.814	15.582
800	38.800	223.858	201.329	18.023	–246.424	–203.501	13.287
900	40.084	228.501	204.094	21.966	–247.158	–198.091	11.497
1000	41.385	232.792	206.752	26.040	–247.820	–192.603	10.060
1100	42.675	236.797	209.303	30.243	–248.410	–187.052	8.882
1200	43.932	240.565	211.753	34.574	–248.933	–181.450	7.898
1300	45.138	244.129	214.108	39.028	–249.392	–175.807	7.064
1400	46.281	247.516	216.374	43.599	–249.792	–170.132	6.348
1500	47.356	250.746	218.559	48.282	–250.139	–164.429	5.726

55. IODINE I (g)

T/K	C_p°	S°	$-(G^\circ - H^\circ(T_r))/T$	$H^\circ - H^\circ(T_r)$	$\Delta_f H^\circ$	$\Delta_f G^\circ$	Log K_f
298.15	20.786	180.787	180.787	0.000	106.760	70.172	–12.294
300	20.786	180.915	180.787	0.038	106.748	69.945	–12.178
400	20.786	186.895	181.602	2.117	97.974	58.060	–7.582
500	20.786	191.533	183.142	4.196	75.988	50.202	–5.244
600	20.786	195.323	184.866	6.274	76.190	45.025	–3.920
700	20.786	198.527	186.594	8.353	76.385	39.816	–2.971
800	20.787	201.303	188.263	10.432	76.574	34.579	–2.258
900	20.789	203.751	189.851	12.510	76.757	29.319	–1.702
1000	20.795	205.942	191.352	14.589	76.936	24.038	–1.256
1100	20.806	207.924	192.770	16.669	77.109	18.740	–0.890
1200	20.824	209.735	194.110	18.751	77.277	13.426	–0.584
1300	20.851	211.403	195.377	20.835	77.440	8.098	–0.325
1400	20.889	212.950	196.577	22.921	77.596	2.758	–0.103
1500	20.936	214.392	197.717	25.013	77.745	–2.592	0.090

56. DIIODINE I_2 (cr, l)

T/K	C_p°	S°	$-(G^\circ - H^\circ(T_r))/T$	$H^\circ - H^\circ(T_r)$	$\Delta_f H^\circ$	$\Delta_f G^\circ$	Log K_f
298.15	54.440	116.139	116.139	0.000	0.000	0.000	0.000
300	54.518	116.476	116.140	0.101	0.000	0.000	0.000
386.75	61.531	131.039	117.884	5.088	0.000	0.000	0.000
PHASE TRANSITION: $\Delta_{trs} H$ = 15.665 kJ/mol, $\Delta_{trs} S$ = 40.504 J/K·mol, cr–l							
386.75	79.555	171.543	117.884	20.753	0.000	0.000	0.000
400	79.555	174.223	119.706	21.807	0.000	0.000	0.000
457.67	79.555	184.938	127.266	26.395	0.000	0.000	0.000

57. DIIODINE I_2 (g)

T/K	C_p°	S°	$-(G^\circ - H^\circ(T_r))/T$	$H^\circ - H^\circ(T_r)$	$\Delta_f H^\circ$	$\Delta_f G^\circ$	Log K_f
298.15	36.887	260.685	260.685	0.000	62.420	19.324	–3.385
300	36.897	260.913	260.685	0.068	62.387	19.056	–3.318
400	37.256	271.584	262.138	3.778	44.391	5.447	–0.711
457.67	37.385	276.610	263.652	5.931		pressure = 1 bar	
500	37.464	279.921	264.891	7.515	0.000	0.000	0.000
600	37.613	286.765	267.983	11.269	0.000	0.000	0.000
700	37.735	292.573	271.092	15.037	0.000	0.000	0.000
800	37.847	297.619	274.099	18.816	0.000	0.000	0.000
900	37.956	302.083	276.965	22.606	0.000	0.000	0.000
1000	38.070	306.088	279.681	26.407	0.000	0.000	0.000
1100	38.196	309.722	282.249	30.220	0.000	0.000	0.000
1200	38.341	313.052	284.679	34.047	0.000	0.000	0.000
1300	38.514	316.127	286.981	37.890	0.000	0.000	0.000
1400	38.719	318.989	289.166	41.751	0.000	0.000	0.000
1500	38.959	321.668	291.245	45.635	0.000	0.000	0.000

58. HYDROGEN IODIDE HI (g)

T/K	C_p° (J/K·mol)	S° (J/K·mol)	$-(G^\circ-H^\circ(T_r))/T$ (J/K·mol)	$H^\circ-H^\circ(T_r)$ (kJ/mol)	$\Delta_f H^\circ$ (kJ/mol)	$\Delta_f G^\circ$ (kJ/mol)	Log K_f
298.15	29.157	206.589	206.589	0.000	26.500	1.700	–0.298
300	29.158	206.769	206.589	0.054	26.477	1.546	–0.269
400	29.329	215.176	207.734	2.977	17.093	–6.289	0.821
500	29.738	221.760	209.904	5.928	–5.481	–9.946	1.039
600	30.351	227.233	212.348	8.931	–5.819	–10.806	0.941
700	31.070	231.965	214.820	12.002	–6.101	–11.614	0.867
800	31.807	236.162	217.230	15.145	–6.323	–12.386	0.809
900	32.511	239.950	219.548	18.362	–6.489	–13.133	0.762
1000	33.156	243.409	221.763	21.646	–6.608	–13.865	0.724
1100	33.735	246.597	223.878	24.991	–6.689	–14.586	0.693
1200	34.249	249.555	225.896	28.391	–6.741	–15.302	0.666
1300	34.703	252.314	227.823	31.839	–6.775	–16.014	0.643
1400	35.106	254.901	229.666	35.330	–6.797	–16.723	0.624
1500	35.463	257.336	231.430	38.858	–6.814	–17.432	0.607

59. POTASSIUM K (cr, l)

T/K	C_p° (J/K·mol)	S° (J/K·mol)	$-(G^\circ-H^\circ(T_r))/T$ (J/K·mol)	$H^\circ-H^\circ(T_r)$ (kJ/mol)	$\Delta_f H^\circ$ (kJ/mol)	$\Delta_f G^\circ$ (kJ/mol)	Log K_f
298.15	29.600	64.680	64.680	0.000	0.000	0.000	0.000
300	29.671	64.863	64.681	0.055	0.000	0.000	0.000
336.86	32.130	68.422	64.896	1.188	0.000	0.000	0.000
PHASE TRANSITION: $\Delta_{trs} H$ = 2.321 kJ/mol, $\Delta_{trs} S$ = 6.891 J/K·mol, cr–l							
336.86	32.129	75.313	64.896	3.509	0.000	0.000	0.000
400	31.552	80.784	66.986	5.519	0.000	0.000	0.000
500	30.741	87.734	70.469	8.632	0.000	0.000	0.000
600	30.158	93.283	73.824	11.675	0.000	0.000	0.000
700	29.851	97.905	76.943	14.673	0.000	0.000	0.000
800	29.838	101.887	79.818	17.655	0.000	0.000	0.000
900	30.130	105.415	82.470	20.651	0.000	0.000	0.000
1000	30.730	108.618	84.927	23.691	0.000	0.000	0.000
1039.4	31.053	109.812	85.847	24.908	0.000	0.000	0.000

60. POTASSIUM K (g)

T/K	C_p° (J/K·mol)	S° (J/K·mol)	$-(G^\circ-H^\circ(T_r))/T$ (J/K·mol)	$H^\circ-H^\circ(T_r)$ (kJ/mol)	$\Delta_f H^\circ$ (kJ/mol)	$\Delta_f G^\circ$ (kJ/mol)	Log K_f
298.15	20.786	160.340	160.340	0.000	89.000	60.479	–10.596
300	20.786	160.468	160.340	0.038	88.984	60.302	–10.499
400	20.786	166.448	161.155	2.117	85.598	51.332	–6.703
500	20.786	171.086	162.695	4.196	84.563	42.887	–4.480
600	20.786	174.876	164.419	6.274	83.599	34.643	–3.016
700	20.786	178.080	166.148	8.353	82.680	26.557	–1.982
800	20.786	180.856	167.817	10.431	81.776	18.601	–1.215
900	20.786	183.304	169.404	12.510	80.859	10.759	–0.624
1000	20.786	185.494	170.905	14.589	79.897	3.021	–0.158
1039.4	20.786	186.297	171.474	15.408		pressure = 1 bar	
1100	20.786	187.475	172.323	16.667	0.000	0.000	0.000
1200	20.786	189.284	173.662	18.746	0.000	0.000	0.000
1300	20.789	190.948	174.929	20.825	0.000	0.000	0.000
1400	20.793	192.489	176.129	22.904	0.000	0.000	0.000
1500	20.801	193.923	177.268	24.983	0.000	0.000	0.000

61. DIPOTASSIUM OXIDE K_2O (cr, l)

T/K	C_p° (J/K·mol)	S° (J/K·mol)	$-(G^\circ-H^\circ(T_r))/T$ (J/K·mol)	$H^\circ-H^\circ(T_r)$ (kJ/mol)	$\Delta_f H^\circ$ (kJ/mol)	$\Delta_f G^\circ$ (kJ/mol)	Log K_f
298.15	72.000	96.000	96.000	0.000	–361.700	–321.171	56.267
300	72.130	96.446	96.001	0.133	–361.704	–320.920	55.876
400	79.154	118.158	98.914	7.698	–366.554	–306.416	40.013
500	86.178	136.575	104.647	15.964	–366.043	–291.423	30.444
590	92.500	151.348	110.662	24.005	–364.204	–278.079	24.619
PHASE TRANSITION: $\Delta_{trs} H$ = 0.700 kJ/mol, $\Delta_{trs} S$ = 1.186 J/K·mol, crIII–crII							

61. DIPOTASSIUM OXIDE K_2O (cr, l) (continued)

	J/K·mol			kJ/mol			
T/K	C_p°	S°	$-(G^\circ-H^\circ(T_r))/T$	$H^\circ-H^\circ(T_r)$	$\Delta_f H^\circ$	$\Delta_f G^\circ$	Log K_f
590	100.000	152.534	110.662	24.705	–364.204	–278.079	24.619
600	100.000	154.215	111.374	25.705	–363.968	–276.621	24.082
645	100.000	161.447	114.618	30.205	–358.901	–270.109	21.874
PHASE TRANSITION: $\Delta_{trs}H$ = 4.000 kJ/mol, $\Delta_{trs}S$ = 6.202 J/K·mol, crII–crI							
645	100.000	167.649	114.618	34.205	–358.901	–270.109	21.874
700	100.000	175.832	119.111	39.705	–357.592	–262.592	19.595
800	100.000	189.185	127.054	49.705	–355.224	–249.183	16.270
900	100.000	200.963	134.625	59.705	–352.919	–236.067	13.701
1000	100.000	211.499	141.794	69.705	–350.732	–223.202	11.659
1013	100.000	212.791	142.697	71.005	–323.459	–221.546	11.424
PHASE TRANSITION: $\Delta_{trs}H$ = 27.000 kJ/mol, $\Delta_{trs}S$ =26.654 J/K·mol, crI–l							
1013	100.000	239.444	142.697	98.005	–323.459	–221.546	11.424
1100	100.000	247.684	150.679	106.705	–479.439	–203.633	9.670
1200	100.000	256.385	159.131	116.705	–475.371	–178.740	7.780
1300	100.000	264.389	166.924	126.705	–471.321	–154.185	6.195
1400	100.000	271.800	174.154	136.705	–467.287	–129.941	4.848
1500	100.000	278.699	180.896	146.705	–463.268	–105.986	3.691

62. POTASSIUM HYDROXIDE KOH (cr, l)

T/K	C_p°	S°	$-(G^\circ-H^\circ(T_r))/T$	$H^\circ-H^\circ(T_r)$	$\Delta_f H^\circ$	$\Delta_f G^\circ$	Log K_f
298.15	64.900	78.870	78.870	0.000	–424.580	–378.747	66.354
300	65.038	79.272	78.871	0.120	–424.569	–378.463	65.895
400	72.519	99.007	81.512	6.998	–426.094	–362.765	47.372
500	80.000	115.993	86.745	14.624	–424.572	–347.093	36.260
520	81.496	119.159	87.931	16.239	–417.725	–344.002	34.555
PHASE TRANSITION: $\Delta_{trs}H$ = 6.450 kJ/mol, $\Delta_{trs}S$ = 12.404 J/K·mol, crII–crI							
520	79.000	131.563	87.931	22.689	–417.725	–344.002	34.555
600	79.000	142.868	94.520	29.009	–416.274	–332.766	28.969
678	79.000	152.523	100.649	35.171	–405.464	–321.998	24.807
PHASE TRANSITION: $\Delta_{trs}H$ = 9.400 kJ/mol, $\Delta_{trs}S$ = 13.865 J/K·mol, crI–l							
678	83.000	166.388	100.649	44.571	–405.464	–321.998	24.807
700	83.000	169.038	102.757	46.397	–404.981	–319.297	23.826
800	83.000	180.121	111.750	54.697	–402.808	–307.206	20.058
900	83.000	189.897	119.901	62.997	–400.694	–295.383	17.143
1000	83.000	198.642	127.345	71.297	–398.668	–283.791	14.824
1100	83.000	206.553	134.192	79.597	–475.618	–267.780	12.716
1200	83.000	213.775	140.527	87.897	–472.711	–249.014	10.839
1300	83.000	220.418	146.421	96.197	–469.843	–230.490	9.261
1400	83.000	226.569	151.929	104.497	–467.011	–212.184	7.917
1500	83.000	232.296	157.098	112.797	–464.217	–194.080	6.758

63. POTASSIUM HYDROXIDE KOH (g)

T/K	C_p°	S°	$-(G^\circ-H^\circ(T_r))/T$	$H^\circ-H^\circ(T_r)$	$\Delta_f H^\circ$	$\Delta_f G^\circ$	Log K_f
298.15	49.184	238.283	238.283	0.000	–227.989	–229.685	40.239
300	49.236	238.588	238.284	0.091	–228.007	–229.696	39.993
400	51.178	253.053	240.243	5.124	–231.377	–229.667	29.991
500	52.178	264.591	243.998	10.296	–232.309	–229.129	23.937
600	52.804	274.163	248.251	15.547	–233.145	–228.413	19.885
700	53.296	282.340	252.551	20.853	–233.934	–227.562	16.981
800	53.758	289.487	256.730	26.206	–234.708	–226.599	14.795
900	54.229	295.846	260.730	31.605	–235.495	–225.538	13.090
1000	54.713	301.585	264.533	37.052	–236.322	–224.388	11.721
1100	55.203	306.823	268.143	42.548	–316.077	–218.535	10.377
1200	55.686	311.647	271.570	48.092	–315.925	–209.674	9.127
1300	56.153	316.122	274.827	53.684	–315.764	–200.826	8.069
1400	56.598	320.300	277.927	59.322	–315.595	–191.991	7.163
1500	57.016	324.220	280.884	65.003	–315.420	–183.169	6.378

	J/K·mol			kJ/mol			
T/K	C_p°	S°	$-(G^\circ-H^\circ(T_r))/T$	$H^\circ-H^\circ(T_r)$	$\Delta_f H^\circ$	$\Delta_f G^\circ$	Log K_f
64. POTASSIUM CHLORIDE KCl (cr, l)							
298.15	51.300	82.570	82.570	0.000	–436.490	–408.568	71.579
300	51.333	82.887	82.571	0.095	–436.481	–408.395	71.107
400	52.977	97.886	84.605	5.312	–438.463	–398.651	52.058
500	54.448	109.867	88.498	10.685	–437.990	–388.749	40.612
600	55.885	119.921	92.919	16.201	–437.332	–378.960	32.991
700	57.425	128.649	97.413	21.865	–436.502	–369.295	27.557
800	59.205	136.430	101.812	27.694	–435.505	–359.760	23.490
900	61.361	143.523	106.058	33.719	–434.337	–350.360	20.334
1000	64.032	150.121	110.138	39.983	–432.981	–341.100	17.817
1044	65.405	152.908	111.882	42.830	–485.450	–336.720	16.847
PHASE TRANSITION: $\Delta_{trs}H$ = 26.320 kJ/mol, $\Delta_{trs}S$ = 25.210 J/K·mol, cr–l							
1044	72.000	178.118	111.882	69.150	–485.450	–336.720	16.847
1100	72.000	181.880	115.351	73.182	–483.633	–328.790	15.613
1200	72.000	188.145	121.160	80.382	–480.393	–314.856	13.705
1300	72.000	193.908	126.537	87.582	–477.158	–301.192	12.102
1400	72.000	199.244	131.542	94.782	–473.928	–287.778	10.737
1500	72.000	204.211	136.223	101.982	–470.704	–274.594	9.562
65. POTASSIUM CHLORIDE KCl (g)							
298.15	36.505	239.091	239.091	0.000	–214.575	–233.320	40.876
300	36.518	239.317	239.092	0.068	–214.594	–233.436	40.644
400	37.066	249.904	240.532	3.749	–218.112	–239.107	31.224
500	37.384	258.212	243.267	7.473	–219.287	–244.219	25.513
600	37.597	265.048	246.344	11.222	–220.396	–249.100	21.686
700	37.769	270.857	249.441	14.991	–221.461	–253.799	18.938
800	37.907	275.910	252.441	18.775	–222.509	–258.347	16.868
900	38.041	280.382	255.302	22.572	–223.568	–262.764	15.250
1000	38.162	284.397	258.014	26.383	–224.667	–267.061	13.950
1100	38.279	288.039	260.581	30.205	–304.696	–266.627	12.661
1200	38.401	291.375	263.010	34.039	–304.821	–263.161	11.455
1300	38.518	294.454	265.312	37.885	–304.941	–259.684	10.434
1400	38.639	297.313	267.496	41.743	–305.053	–256.199	9.559
1500	38.761	299.983	269.574	45.613	–305.159	–252.706	8.800
66. DINITROGEN N_2 (g)							
298.15	29.124	191.608	191.608	0.000	0.000	0.000	0.000
300	29.125	191.788	191.608	0.054	0.000	0.000	0.000
400	29.249	200.180	192.752	2.971	0.000	0.000	0.000
500	29.580	206.738	194.916	5.911	0.000	0.000	0.000
600	30.109	212.175	197.352	8.894	0.000	0.000	0.000
700	30.754	216.864	199.812	11.936	0.000	0.000	0.000
800	31.433	221.015	202.208	15.046	0.000	0.000	0.000
900	32.090	224.756	204.509	18.222	0.000	0.000	0.000
1000	32.696	228.169	206.706	21.462	0.000	0.000	0.000
1100	33.241	231.311	208.802	24.759	0.000	0.000	0.000
1200	33.723	234.224	210.801	28.108	0.000	0.000	0.000
1300	34.147	236.941	212.708	31.502	0.000	0.000	0.000
1400	34.517	239.485	214.531	34.936	0.000	0.000	0.000
1500	34.842	241.878	216.275	38.404	0.000	0.000	0.000
67. NITRIC OXIDE NO (g)							
298.15	29.862	210.745	210.745	0.000	91.277	87.590	–15.345
300	29.858	210.930	210.746	0.055	91.278	87.567	–15.247
400	29.954	219.519	211.916	3.041	91.320	86.323	–11.272
500	30.493	226.255	214.133	6.061	91.340	85.071	–8.887

T/K	C_p° (J/K·mol)	S° (J/K·mol)	$-(G^\circ-H^\circ(T_r))/T$ (J/K·mol)	$H^\circ-H^\circ(T_r)$ (kJ/mol)	Δ_fH° (kJ/mol)	Δ_fG° (kJ/mol)	Log K_f
67. NITRIC OXIDE NO (g) (continued)							
600	31.243	231.879	216.635	9.147	91.354	83.816	–7.297
700	32.031	236.754	219.168	12.310	91.369	82.558	–6.160
800	32.770	241.081	221.642	15.551	91.386	81.298	–5.308
900	33.425	244.979	224.022	18.862	91.405	80.036	–4.645
1000	33.990	248.531	226.298	22.233	91.426	78.772	–4.115
1100	34.473	251.794	228.469	25.657	91.445	77.505	–3.680
1200	34.883	254.811	230.540	29.125	91.464	76.237	–3.318
1300	35.234	257.618	232.516	32.632	91.481	74.967	–3.012
1400	35.533	260.240	234.404	36.170	91.495	73.697	–2.750
1500	35.792	262.700	236.209	39.737	91.506	72.425	–2.522
68. NITROGEN DIOXIDE NO_2 (g)							
298.15	37.178	240.166	240.166	0.000	34.193	52.316	–9.165
300	37.236	240.397	240.167	0.069	34.181	52.429	–9.129
400	40.513	251.554	241.666	3.955	33.637	58.600	–7.652
500	43.664	260.939	244.605	8.167	33.319	64.882	–6.778
600	46.383	269.147	248.026	12.673	33.174	71.211	–6.199
700	48.612	276.471	251.575	17.427	33.151	77.553	–5.787
800	50.405	283.083	255.107	22.381	33.213	83.893	–5.478
900	51.844	289.106	258.555	27.496	33.334	90.221	–5.236
1000	53.007	294.631	261.891	32.741	33.495	96.534	–5.042
1100	53.956	299.729	265.102	38.090	33.686	102.828	–4.883
1200	54.741	304.459	268.187	43.526	33.898	109.105	–4.749
1300	55.399	308.867	271.148	49.034	34.124	115.363	–4.635
1400	55.960	312.994	273.992	54.603	34.360	121.603	–4.537
1500	56.446	316.871	276.722	60.224	34.604	127.827	–4.451
69. AMMONIA NH_3 (g)							
298.15	35.630	192.768	192.768	0.000	–45.940	–16.407	2.874
300	35.678	192.989	192.769	0.066	–45.981	–16.223	2.825
400	38.674	203.647	194.202	3.778	–48.087	–5.980	0.781
500	41.994	212.633	197.011	7.811	–49.908	4.764	–0.498
600	45.229	220.578	200.289	12.174	–51.430	15.846	–1.379
700	48.269	227.781	203.709	16.850	–52.682	27.161	–2.027
800	51.112	234.414	207.138	21.821	–53.695	38.639	–2.523
900	53.769	240.589	210.516	27.066	–54.499	50.231	–2.915
1000	56.244	246.384	213.816	32.569	–55.122	61.903	–3.233
1100	58.535	251.854	217.027	38.309	–55.589	73.629	–3.496
1200	60.644	257.039	220.147	44.270	–55.920	85.392	–3.717
1300	62.576	261.970	223.176	50.432	–56.136	97.177	–3.905
1400	64.339	266.673	226.117	56.779	–56.251	108.975	–4.066
1500	65.945	271.168	228.971	63.295	–56.282	120.779	–4.206
70. OXYGEN O (g)							
298.15	21.911	161.058	161.058	0.000	249.180	231.743	–40.600
300	21.901	161.194	161.059	0.041	249.193	231.635	–40.331
400	21.482	167.430	161.912	2.207	249.874	225.677	–29.470
500	21.257	172.197	163.511	4.343	250.481	219.556	–22.937
600	21.124	176.060	165.290	6.462	251.019	213.319	–18.571
700	21.040	179.310	167.067	8.570	251.500	206.997	–15.446
800	20.984	182.115	168.777	10.671	251.932	200.610	–13.098
900	20.944	184.584	170.399	12.767	252.325	194.171	–11.269
1000	20.915	186.789	171.930	14.860	252.686	187.689	–9.804
1100	20.893	188.782	173.372	16.950	253.022	181.173	–8.603
1200	20.877	190.599	174.733	19.039	253.335	174.628	–7.601
1300	20.864	192.270	176.019	21.126	253.630	168.057	–6.753

	J/K·mol			kJ/mol			
T/K	C_p°	S°	$-(G^\circ-H^\circ(T_r))/T$	$H^\circ-H^\circ(T_r)$	$\Delta_f H^\circ$	$\Delta_f G^\circ$	Log K_f
70. OXYGEN O (g) (continued)							
1400	20.853	193.815	177.236	23.212	253.908	161.463	–6.024
1500	20.845	195.254	178.389	25.296	254.171	154.851	–5.392
71. DIOXYGEN O_2 (g)							
298.15	29.378	205.148	205.148	0.000	0.000	0.000	0.000
300	29.387	205.330	205.148	0.054	0.000	0.000	0.000
400	30.109	213.873	206.308	3.026	0.000	0.000	0.000
500	31.094	220.695	208.525	6.085	0.000	0.000	0.000
600	32.095	226.454	211.045	9.245	0.000	0.000	0.000
700	32.987	231.470	213.612	12.500	0.000	0.000	0.000
800	33.741	235.925	216.128	15.838	0.000	0.000	0.000
900	34.365	239.937	218.554	19.244	0.000	0.000	0.000
1000	34.881	243.585	220.878	22.707	0.000	0.000	0.000
1100	35.314	246.930	223.096	26.217	0.000	0.000	0.000
1200	35.683	250.019	225.213	29.768	0.000	0.000	0.000
1300	36.006	252.888	227.233	33.352	0.000	0.000	0.000
1400	36.297	255.568	229.162	36.968	0.000	0.000	0.000
1500	36.567	258.081	231.007	40.611	0.000	0.000	0.000
72. SULFUR S (cr, l)							
298.15	22.690	32.070	32.070	0.000	0.000	0.000	0.000
300	22.737	32.210	32.070	0.042	0.000	0.000	0.000
368.3	24.237	37.030	32.554	1.649	0.000	0.000	0.000
PHASE TRANSITION: $\Delta_{trs} H$ = 0.401 kJ/mol, $\Delta_{trs} S$ = 1.089 J/K·mol, crII–crI							
368.3	24.773	38.119	32.553	2.050	0.000	0.000	0.000
388.36	25.180	39.444	32.875	2.551	0.000	0.000	0.000
PHASE TRANSITION: $\Delta_{trs} H$ = 1.722 kJ/mol, $\Delta_{trs} S$ = 4.431 J/K·mol, crI–l							
388.36	31.710	43.875	32.872	4.273	0.000	0.000	0.000
400	32.369	44.824	33.206	4.647	0.000	0.000	0.000
500	38.026	53.578	36.411	8.584	0.000	0.000	0.000
600	34.371	60.116	39.842	12.164	0.000	0.000	0.000
700	32.451	65.278	43.120	15.511	0.000	0.000	0.000
800	32.000	69.557	46.163	18.715	0.000	0.000	0.000
882.38	32.000	72.693	48.496	21.351	0.000	0.000	0.000
73. SULFUR S (g)							
298.15	23.673	167.828	167.828	0.000	277.180	236.704	–41.469
300	23.669	167.974	167.828	0.044	277.182	236.453	–41.170
400	23.233	174.730	168.752	2.391	274.924	222.962	–29.115
500	22.741	179.860	170.482	4.689	273.286	210.145	–21.953
600	22.338	183.969	172.398	6.942	271.958	197.646	–17.206
700	22.031	187.388	174.302	9.160	270.829	185.352	–13.831
800	21.800	190.314	176.125	11.351	269.816	173.210	–11.309
900	21.624	192.871	177.847	13.522	215.723	162.258	–9.417
1000	21.489	195.142	179.465	15.677	216.018	156.301	–8.164
1100	21.386	197.185	180.985	17.821	216.284	150.317	–7.138
1200	21.307	199.043	182.413	19.955	216.525	144.309	–6.282
1300	21.249	200.746	183.759	22.083	216.743	138.282	–5.556
1400	21.209	202.319	185.029	24.206	216.940	132.239	–4.934
1500	21.186	203.781	186.231	26.325	217.119	126.182	–4.394
74. DISULFUR S_2 (g)							
298.15	32.505	228.165	228.165	0.000	128.600	79.696	–13.962
300	32.540	228.366	228.165	0.060	128.576	79.393	–13.823
400	34.108	237.956	229.462	3.398	122.703	63.380	–8.276
500	35.133	245.686	231.959	6.863	118.296	49.031	–5.122

T/K	C_p° (J/K·mol)	S° (J/K·mol)	$-(G^\circ-H^\circ(T_r))/T$ (J/K·mol)	$H^\circ-H^\circ(T_r)$ (kJ/mol)	Δ_fH° (kJ/mol)	Δ_fG° (kJ/mol)	Log K_f
74. DISULFUR S_2 (g) (continued)							
600	35.815	252.156	234.800	10.413	114.685	35.530	–3.093
700	36.305	257.715	237.686	14.020	111.599	22.588	–1.685
800	36.697	262.589	240.501	17.671	108.841	10.060	–0.657
882.38	36.985	266.200	242.734	20.706		pressure = 1 bar	
900	37.045	266.932	243.201	21.358	0.000	0.000	0.000
1000	37.377	270.852	245.773	25.079	0.000	0.000	0.000
1100	37.704	274.430	248.218	28.833	0.000	0.000	0.000
1200	38.030	277.725	250.541	32.620	0.000	0.000	0.000
1300	38.353	280.781	252.751	36.439	0.000	0.000	0.000
1400	38.669	283.635	254.856	40.290	0.000	0.000	0.000
1500	38.976	286.314	256.865	44.173	0.000	0.000	0.000
75. OCTASULFUR S_8 (g)							
298.15	156.500	432.536	432.536	0.000	101.277	48.810	–8.551
300	156.768	433.505	432.539	0.290	101.231	48.484	–8.442
400	167.125	480.190	438.834	16.542	80.642	32.003	–4.179
500	173.181	518.176	451.022	33.577	66.185	21.409	–2.237
600	177.936	550.180	464.951	51.137	55.101	13.549	–1.180
700	182.441	577.948	479.152	69.157	46.349	7.343	–0.548
800	186.764	602.596	493.071	87.620	39.177	2.263	–0.148
900	190.595	624.821	506.495	106.494	–392.062	6.554	–0.380
1000	193.618	645.067	519.355	125.712	–387.728	50.614	–2.644
1100	195.684	663.625	531.639	145.185	–383.272	94.233	–4.475
1200	196.825	680.707	543.359	164.817	–378.786	137.444	–5.983
1300	197.195	696.480	554.539	184.524	–374.356	180.283	–7.244
1400	196.988	711.089	565.206	204.237	–370.048	222.785	–8.312
1500	196.396	724.662	575.389	223.909	–365.905	264.984	–9.227
76. SULFUR DIOXIDE SO_2 (g)							
298.15	39.842	248.219	248.219	0.000	–296.810	–300.090	52.574
300	39.909	248.466	248.220	0.074	–296.833	–300.110	52.253
400	43.427	260.435	249.828	4.243	–300.240	–300.935	39.298
500	46.490	270.465	252.978	8.744	–302.735	–300.831	31.427
600	48.938	279.167	256.634	13.520	–304.699	–300.258	26.139
700	50.829	286.859	260.413	18.513	–306.308	–299.386	22.340
800	52.282	293.746	264.157	23.671	–307.691	–298.302	19.477
900	53.407	299.971	267.796	28.958	–362.075	–295.987	17.178
1000	54.290	305.646	271.301	34.345	–362.012	–288.647	15.077
1100	54.993	310.855	274.664	39.810	–361.934	–281.314	13.358
1200	55.564	315.665	277.882	45.339	–361.849	–273.989	11.926
1300	56.033	320.131	280.963	50.920	–361.763	–266.671	10.715
1400	56.426	324.299	283.911	56.543	–361.680	–259.359	9.677
1500	56.759	328.203	286.735	62.203	–361.605	–252.053	8.777
77. SILICON Si (cr)							
298.15	19.789	18.810	18.810	0.000	0.000	0.000	0.000
300	19.855	18.933	18.810	0.037	0.000	0.000	0.000
400	22.301	25.023	19.624	2.160	0.000	0.000	0.000
500	23.610	30.152	21.231	4.461	0.000	0.000	0.000
600	24.472	34.537	23.092	6.867	0.000	0.000	0.000
700	25.124	38.361	25.006	9.348	0.000	0.000	0.000
800	25.662	41.752	26.891	11.888	0.000	0.000	0.000
900	26.135	44.802	28.715	14.478	0.000	0.000	0.000
1000	26.568	47.578	30.464	17.114	0.000	0.000	0.000
1100	26.974	50.130	32.138	19.791	0.000	0.000	0.000
1200	27.362	52.493	33.737	22.508	0.000	0.000	0.000

	J/K·mol			kJ/mol			
T/K	C_p°	S°	$-(G^\circ-H^\circ(T_r))/T$	$H^\circ-H^\circ(T_r)$	$\Delta_f H^\circ$	$\Delta_f G^\circ$	Log K_f
77. SILICON Si (cr) (continued)							
1300	27.737	54.698	35.265	25.263	0.000	0.000	0.000
1400	28.103	56.767	36.728	28.055	0.000	0.000	0.000
1500	28.462	58.719	38.130	30.883	0.000	0.000	0.000
78. SILICON Si (g)							
298.15	22.251	167.980	167.980	0.000	450.000	405.525	–71.045
300	22.234	168.117	167.980	0.041	450.004	405.249	–70.559
400	21.613	174.416	168.843	2.229	450.070	390.312	–50.969
500	21.316	179.204	170.456	4.374	449.913	375.388	–39.216
600	21.153	183.074	172.246	6.497	449.630	360.508	–31.385
700	21.057	186.327	174.032	8.607	449.259	345.682	–25.795
800	21.000	189.135	175.748	10.709	448.821	330.915	–21.606
900	20.971	191.606	177.375	12.808	448.329	316.205	–18.352
1000	20.968	193.815	178.911	14.904	447.791	301.553	–15.751
1100	20.989	195.815	180.358	17.002	447.211	286.957	–13.626
1200	21.033	197.643	181.723	19.103	446.595	272.416	–11.858
1300	21.099	199.329	183.014	21.209	445.946	257.927	–10.364
1400	21.183	200.895	184.236	23.323	445.268	243.489	–9.085
1500	21.282	202.360	185.396	25.446	444.563	229.101	–7.978
79. SILICON DIOXIDE SiO_2 (cr)							
298.15	44.602	41.460	41.460	0.000	–910.700	–856.288	150.016
300	44.712	41.736	41.461	0.083	–910.708	–855.951	149.032
400	53.477	55.744	43.311	4.973	–910.912	–837.651	109.385
500	60.533	68.505	47.094	10.705	–910.540	–819.369	85.598
600	64.452	79.919	51.633	16.971	–909.841	–801.197	69.749
700	68.234	90.114	56.414	23.590	–908.958	–783.157	58.439
800	76.224	99.674	61.226	30.758	–907.668	–765.265	49.966
848	82.967	104.298	63.533	34.569	–906.310	–756.747	46.613
PHASE TRANSITION: $\Delta_{trs} H$ = 0.411 kJ/mol, $\Delta_{trs} S$ = 0.484 J/K·mol, crII–crII′							
848	67.446	104.782	63.532	34.980	–906.310	–756.747	46.613
900	67.953	108.811	66.033	38.500	–905.922	–747.587	43.388
1000	68.941	116.021	70.676	45.345	–905.176	–730.034	38.133
1100	69.940	122.639	75.104	52.289	–904.420	–712.557	33.836
1200	70.947	128.768	79.323	59.333	–901.382	–695.148	30.259
PHASE TRANSITION: $\Delta_{trs} H$ = 2.261 kJ/mol, $\Delta_{trs} S$ = 1.883 J/K·mol, crII′–crI							
1200	71.199	130.651	79.323	61.594	–901.382	–695.148	30.259
1300	71.743	136.372	83.494	68.742	–900.574	–677.994	27.242
1400	72.249	141.707	87.463	75.941	–899.782	–660.903	24.658
1500	72.739	146.709	91.248	83.191	–899.004	–643.867	22.421
80. SILICON TETRACHLORIDE $SiCl_4$ (g)							
298.15	90.404	331.446	331.446	0.000	–662.200	–622.390	109.039
300	90.562	332.006	331.448	0.167	–662.195	–622.143	108.323
400	96.893	359.019	335.088	9.572	–661.853	–608.841	79.505
500	100.449	381.058	342.147	19.456	–661.413	–595.637	62.225
600	102.587	399.576	350.216	29.616	–660.924	–582.527	50.713
700	103.954	415.500	358.432	39.948	–660.417	–569.501	42.496
800	104.875	429.445	366.455	50.392	–659.912	–556.548	36.338
900	105.523	441.837	374.155	60.914	–659.422	–543.657	31.553
1000	105.995	452.981	381.490	71.491	–658.954	–530.819	27.727
1100	106.349	463.101	388.456	82.109	–658.515	–518.027	24.599
1200	106.620	472.366	395.068	92.758	–658.107	–505.274	21.994
1300	106.834	480.909	401.347	103.431	–657.735	–492.553	19.791
1400	107.003	488.833	407.316	114.123	–657.400	–479.860	17.904
1500	107.141	496.220	413.000	124.830	–657.104	–467.189	16.269

THERMODYNAMIC PROPERTIES OF AQUEOUS SYSTEMS

This table contains standard state thermodynamic properties of ions and neutral species in aqueous solution. It includes enthalpy and Gibbs energy of formation, entropy, and heat capacity, and thus serves as a companion to the preceding table, "Standard Thermodynamic Properties of Chemical Substances". The standard state is the hypothetical ideal solution with molality $m = 1$ mol/kg (mean ionic molality $m_{\pm}$ in the case of a species which is assumed to dissociate at infinite dilution). Further details on conventions may be found in Reference 1.

Cations are listed by formula in the first part of the table, followed by anions and finally neutral species. All values refer to standard conditions of 25°C and 100 kPa pressure.

REFERENCES

1. Wagman, D. D., Evans, W. H., Parker, V. B., Schumm, R. H., Halow, I., Bailey, S. M., Churney, K. L., and Nuttall, R. L., *The NBS Tables of Chemical Thermodynamic Properties, J. Phys. Chem. Ref. Data*, Vol. 11, Suppl. 2, 1982.
2. Zemaitis, J. F., Clark, D. M., Rafal, M., and Scrivner, N. C., *Handbook of Aqueous Electrolyte Thermodynamics*, American Institute of Chemical Engineers, New York, 1986.

Species	$\Delta_f H°$/ kJ mol^{-1}	$\Delta_f G°$/ kJ mol^{-1}	$S°$/ J mol^{-1}K^{-1}	C_p/ J mol^{-1}K^{-1}
Cations				
Ag^{+}	105.6	77.1	72.7	21.8
Al^{+3}	-531.0	-485.0	-321.7	
$AlOH^{+2}$		-694.1		
Ba^{+2}	-537.6	-560.8	9.6	
$BaOH^{+}$		-730.5		
Be^{+2}	-382.8	-379.7	-129.7	
Bi^{+3}		82.8		
$BiOH^{+2}$		-146.4		
Ca^{+2}	-542.8	-553.6	-53.1	
$CaOH^{+}$		-718.4		
Cd^{+2}	-75.9	-77.6	-73.2	
$CdOH^{+}$		-261.1		
Ce^{+3}	-696.2	-672.0	-205.0	
Ce^{+4}	-537.2	-503.8	-301.0	
Co^{+2}	-58.2	-54.4	-113.0	
Co^{+3}	92.0	134.0	-305.0	
Cr^{+2}	-143.5			
Cs^{+}	-258.3	-292.0	133.1	-10.5
Cu^{+}	71.7	50.0	40.6	
Cu^{+2}	64.8	65.5	-99.6	
Dy^{+3}	-699.0	-665.0	-231.0	21.0
Er^{+3}	-705.4	-669.1	-244.3	21.0
Eu^{+2}	-527.0	-540.2	-8.0	
Eu^{+3}	-605.0	-574.1	-222.0	8.0
Fe^{+2}	-89.1	-78.9	-137.7	
Fe^{+3}	-48.5	-4.7	-315.9	
$FeOH^{+}$	-324.7	-277.4	-29.0	
$FeOH^{+2}$	-290.8	-229.4	-142.0	
$Fe(OH)_2^{+}$		-438.0		
Ga^{+2}		-88.0		
Ga^{+3}	-211.7	-159.0	-331.0	
$GaOH^{+2}$		-380.3		
$Ga(OH)_2^{+}$		-597.4		
Gd^{+3}	-686.0	-661.0	-205.9	
H^{+}	0	0	0	0
Hg^{+2}	171.1	164.4	-32.2	
Hg_2^{+2}	172.4	153.5	84.5	
$HgOH^{+}$	-84.5	-52.3	71.0	
Ho^{+3}	-705.0	-673.7	-226.8	17.0
In^{+}		-12.1		
In^{+2}		-50.7		
In^{+3}	-105.0	-98.0	-151.0	
$InOH^{+2}$	-370.3	-313.0	-88.0	
$In(OH)_2^{+}$	-619.0	-525.0	25.0	
K^{+}	-252.4	-283.3	102.5	21.8
La^{+3}	-707.1	-683.7	-217.6	-13.0
Li^{+}	-278.5	-293.3	13.4	68.6
Lu^{+3}	-665.0	-628.0	-264.0	25.0
LuF^{+2}		-931.4		
Mg^{+2}	-466.9	-454.8	-138.1	
$MgOH^{+}$		-626.7		
Mn^{+2}	-220.8	-228.1	-73.6	50.0
$MnOH^{+}$	-450.6	-405.0	-17.0	
NH_4^{+}	-132.5	-79.3	113.4	79.9
$N_2H_5^{+}$	-7.5	82.5	151.0	70.3
Na^{+}	-240.1	-261.9	59.0	46.4
Nd^{+3}	-696.2	-671.6	-206.7	-21.0
Ni^{+2}	-54.0	-45.6	-128.9	
$NiOH^{+}$	-287.9	-227.6	-71.0	
PH_4^{+}		92.1		
Pa^{+4}	-619.0			
Pb^{+2}	-1.7	-24.4	10.5	
$PbOH^{+}$		-226.3		
Pd^{+2}	149.0	176.5	-184.0	
Po^{+2}		71.0		
Po^{+4}		293.0		
Pr^{+3}	-704.6	-679.1	-209.0	-29.0
Pt^{+2}		254.8		
Ra^{+2}	-527.6	-561.5	54.0	
Rb^{+}	-251.2	-284.0	121.5	
Re^{+}		-33.0		
Sc^{+3}	-614.2	-586.6	-255.0	
$ScOH^{+2}$	-861.5	-801.2	-134.0	
Sm^{+2}		-497.5		
Sm^{+3}	-691.6	-666.6	-211.7	-21.0
Sn^{+2}	-8.8	-27.2	-17.0	
$SnOH^{+}$	-286.2	-254.8	50.0	
Sr^{+2}	-545.8	-559.5	-32.6	
$SrOH^{+}$		-721.3		
Tb^{+3}	-682.8	-651.9	-226.0	17.0
$Te(OH)_3^{+}$	-608.4	-496.1	111.7	
Th^{+4}	-769.0	-705.1	-422.6	
$Th(OH)^{+3}$	-1030.1	-920.5	-343.0	
$Th(OH)_2^{+2}$	-1282.4	-1140.9	-218.0	
Tl^{+}	5.4	-32.4	125.5	
Tl^{+3}	196.6	214.6	-192.0	

Species	$\Delta_f H°$/ kJ mol^{-1}	$\Delta_f G°$/ kJ mol^{-1}	$S°$/ J mol^{-1}K^{-1}	C_p/ J mol^{-1}K^{-1}
$TlOH^{+2}$		-15.9		
$Tl(OH)_2^+$		-244.7		
Tm^{+3}	-697.9	-662.0	-243.0	25.0
U^{+3}	-489.1	-476.2	-188.0	
U^{+4}	-591.2	-531.9	-410.0	
Y^{+3}	-723.4	-693.8	-251.0	
$Y_2(OH)_2^{+4}$		-1780.3		
Yb^{+2}		-527.0		
Yb^{+3}	-674.5	-644.0	-238.0	25.0
$Y(OH)^{+2}$		-879.1		
Zn^{+2}	-153.9	-147.1	-112.1	46.0
$ZnOH^+$		-330.1		
Anions				
AlO_2^-	-930.9	-830.9	-36.8	
$Al(OH)_4^-$	-1502.5	-1305.3	102.9	
AsO_2^-	-429.0	-350.0	40.6	
AsO_4^{-3}	-888.1	-648.4	-162.8	
BF_4^-	-1574.9	-1486.9	180.0	
BH_4^-	48.2	114.4	110.5	
BO_2^-	-772.4	-678.9	-37.2	
$B_4O_7^{-2}$		-2604.8		
BeO_2^{-2}	-790.8	-640.1	-159.0	
Br^-	-121.6	-104.0	82.4	-141.8
BrO^-	-94.1	-33.4	42.0	
BrO_3^-	-67.1	18.6	161.7	
BrO_4^-	13.0	118.1	199.6	
$CHOO^-$	-425.6	-351.0	92.0	-87.9
CH_3COO^-	-486.0	-369.3	86.6	-6.3
$C_2O_4^{-2}$	-825.1	-673.9	45.6	
$C_2O_4H^-$	-818.4	-698.3	149.4	
Cl^-	-167.2	-131.2	56.5	-136.4
ClO^-	-107.1	-36.8	42.0	
ClO_2^-	-66.5	17.2	101.3	
ClO_3^-	-104.0	-8.0	162.3	
ClO_4^-	-129.3	-8.5	182.0	
CN^-	150.6	172.4	94.1	
CO_3^{-2}	-677.1	-527.8	-56.9	
CrO_4^{-2}	-881.2	-727.8	50.2	
$Cr_2O_7^{-2}$	-1490.3	-1301.1	261.9	
F^-	-332.6	-278.8	-13.8	-106.7
$Fe(CN)_6^{-3}$	561.9	729.4	270.3	
$Fe(CN)_6^{-4}$	455.6	695.1	95.0	
$HB_4O_7^-$		-2685.1		
HCO_3^-	-692.0	-586.8	91.2	
HF_2^-	-649.9	-578.1	92.5	
HPO_3F^-		-1198.2		
HPO_4^{-2}	-1292.1	-1089.2	-33.5	
$HP_2O_7^{-3}$	-2274.8	-1972.2	46.0	
HS^-	-17.6	12.1	62.8	
HSO_3^-	-626.2	-527.7	139.7	
HSO_4^-	-887.3	-755.9	131.8	-84.0
$HS_2O_4^-$		-614.5		
HSe^-	15.9	44.0	79.0	
$HSeO_3^-$	-514.6	-411.5	135.1	
$HSeO_4^-$	-581.6	-452.2	149.4	
$H_2AsO_3^-$	-714.8	-587.1	110.5	
$H_2AsO_4^-$	-909.6	-753.2	117.0	

Species	$\Delta_f H°$/ kJ mol^{-1}	$\Delta_f G°$/ kJ mol^{-1}	$S°$/ J mol^{-1}K^{-1}	C_p/ J mol^{-1}K^{-1}
$H_2PO_4^-$	-1296.3	-1130.2	90.4	
$H_2P_2O_7^{-2}$	-2278.6	-2010.2	163.0	
I^-	-55.2	-51.6	111.3	-142.3
IO^-	-107.5	-38.5	-5.4	
IO_3^-	-221.3	-128.0	118.4	
IO_4^-	-151.5	-58.5	222.0	
MnO_4^-	-541.4	-447.2	191.2	-82.0
MnO_4^{-2}	-653.0	-500.7	59.0	
MoO_4^{-2}	-997.9	-836.3	27.2	
NO_2^-	-104.6	-32.2	123.0	-97.5
NO_3^-	-207.4	-111.3	146.4	-86.6
N_3^-	275.1	348.2	107.9	
OCN^-	-146.0	-97.4	106.7	
OH^-	-230.0	-157.2	-10.8	-148.5
PO_4^{-3}	-1277.4	-1018.7	-220.5	
$P_2O_7^{-4}$	-2271.1	-1919.0	-117.0	
Re^-	46.0	10.1	230.0	
S^{-2}	33.1	85.8	-14.6	
SCN^-	76.4	92.7	144.3	-40.2
SO_3^{-2}	-635.5	-486.5	-29.0	
SO_4^{-2}	-909.3	-744.5	20.1	-293.0
S_2^{-2}	30.1	79.5	28.5	
$S_2O_3^{-2}$	-652.3	-522.5	67.0	
$S_2O_4^{-2}$	-753.5	-600.3	92.0	
$S_2O_8^{-2}$	-1344.7	-1114.9	244.3	
Se^{-2}		129.3		
SeO_3^{-2}	-509.2	-369.8	13.0	
SeO_4^{-2}	-599.1	-441.3	54.0	
VO_3^-	-888.3	-783.6	50.0	
VO_4^{-3}		-899.0		
WO_4^{-2}	-1075.7			
Neutral species				
AgBr	-16.0	-26.9	155.2	-120.1
AgCl	-61.6	-54.1	129.3	-114.6
AgF	-227.1	-201.7	59.0	-84.9
AgI	50.4	25.5	184.1	-120.5
$AgNO_3$	-101.8	-34.2	219.2	-64.9
Ag_2SO_4	-698.1	-590.3	165.7	-251.0
$AlBr_3$	-895.0	-799.0	-74.5	
$AlCl_3$	-1033.0	-879.0	-152.3	
AlF_3	-1531.0	-1322.0	-363.2	
AlI_3	-699.0	-640.0	12.1	
$Al_2(SO_4)_3$	-3791.0	-3205.0	-583.2	
$BaBr_2$	-780.7	-768.7	174.5	
$BaCO_3$	-1214.8	-1088.6	-47.3	
$BaCl_2$	-872.0	-823.2	122.6	
BaF_2	-1202.9	-1118.4	-18.0	
$Ba(HCO_3)_2$	-1921.6	-1734.3	192.0	
BaI_2	-648.0	-663.9	232.2	
$Ba(NO_3)_2$	-952.4	-783.3	302.5	
$BaSO_4$	-1446.9	-1305.3	29.7	
$BeSO_4$	-1292.0	-1124.3	-109.6	
CCl_3COOH	-516.3			
$CHCl_2COOH$	-512.1			
CHOOCs	-683.8	-643.0	226.0	
CHOOH	-425.6	-351.0	92.0	-87.9
CHOOK	-677.9	-634.2	192.0	-66.1

Species	$\Delta_f H°$/ kJ mol^{-1}	$\Delta_f G°$/ kJ mol^{-1}	$S°$/ J mol^{-1}K^{-1}	C_p/ J mol^{-1}K^{-1}
$CHOONH_4$	-558.1	-430.4	205.0	-7.9
$CHOONa$	-665.7	-612.9	151.0	-41.4
$CHOORb$	-676.7	-635.1	213.0	
$CH_2ClCOOH$	-501.3			
CH_3COOCs	-744.3	-661.3	219.7	
CH_3COOH	-486.0	-369.3	86.6	-6.3
CH_3COOK	-738.4	-652.6	189.1	15.5
CH_3COONH_4	-618.5	-448.6	200.0	73.6
CH_3COONa	-726.1	-631.2	145.6	40.2
CH_3COORb	-737.2	-653.3	207.9	
$(COOH)_2$	-825.1	-673.9	45.6	
$(CH_3)_3N$	-76.0	93.1	133.5	
$CaBr_2$	-785.9	-761.5	111.7	
$CaCO_3$	-1220.0	-1081.4	-110.0	
$CaCl_2$	-877.1	-816.0	59.8	
CaF_2	-1208.1	-1111.2	-80.8	
CaI_2	-653.2	-656.7	169.5	
$Ca(NO_3)_2$	-957.6	-776.1	239.7	
$CaSO_4$	-1452.1	-1298.1	-33.1	
$CdBr_2$	-319.0	-285.5	91.6	
$CdCl_2$	-410.2	-340.1	39.7	
CdF_2	-741.2	-635.2	-100.8	
CdI_2	-186.3	-180.8	149.4	
$Cd(NO_3)_2$	-490.6	-300.1	219.7	
$CdSO_4$	-985.2	-822.1	-53.1	
$CeCl_3$	-1197.5	-1065.6	-38.0	
$CoBr_2$	-301.2	-262.3	50.0	
$CoCl_2$	-392.5	-316.7		
CoI_2	-168.6	-157.7	109.0	
$Co(NO_3)_2$	-472.8	-276.9	180.0	
$CoSO_4$	-967.3	-799.1	-92.0	
$CsBr$	-379.8	-396.0	215.5	
$CsCl$	-425.4	-423.2	189.5	-146.9
CsF	-590.9	-570.8	119.2	
$CsHCO_3$	-950.3	-878.8	224.3	
$CsHSO_4$	-1145.6	-1047.9	264.8	
CsI	-313.5	-343.6	244.3	-152.7
$CsNO_3$	-465.6	-403.3	279.5	-99.0
Cs_2CO_3	-1193.7	-1111.9	209.2	
Cs_2S	-483.7	-498.3	251.0	
Cs_2SO_4	-1425.8	-1328.6	286.2	
Cs_2Se		-454.8		
$Cu(NO_3)_2$	-350.0	-157.0	193.3	
$CuSO_4$	-844.5	-679.0	-79.5	
$DyCl_3$	-1197.0	-1059.0	-61.9	-389.0
$ErCl_3$	-1207.1	-1062.7	-75.3	-389.0
$EuCl_2$	-862.0			
$EuCl_3$	-1106.2	-967.7	-54.0	-402.0
$FeBr_2$	-332.2	-286.8	27.2	
$FeBr_3$	-413.4	-316.7	-68.6	
$FeCl_2$	-423.4	-341.3	-24.7	
$FeCl_3$	-550.2	-398.3	-146.4	
FeF_2	-754.4	-636.5	-165.3	
FeF_3	-1046.4	-840.9	-357.3	
FeI_2	-199.6	-182.1	84.9	
FeI_3	-214.2	-159.4	18.0	
$Fe(NO_3)_3$	-670.7	-338.3	123.4	
$FeSO_4$	-998.3	-823.4	-117.6	
$Fe_2(SO_4)_3$	-2825.0	-2242.8	-571.5	
$GdCl_3$	-1188.0	-1059.0	-36.8	-410.0
HBr	-121.6	-104.0	82.4	-141.8
HCN	150.6	172.4	94.1	
HCl	-167.2	-131.2	56.5	-136.4
HF	-332.6	-278.8	-13.8	-106.7
HI	-55.2	-51.6	111.3	-142.3
HNO_3	-207.4	-111.3	146.4	-86.6
$HSCN$	76.4	92.7	144.3	-40.2
H_2SO_4	-909.3	-744.5	20.1	-293.0
$HoCl_3$	-1206.7	-1067.3	-57.7	-393.0
KBr	-373.9	-387.2	184.9	-120.1
KCl	-419.5	-414.5	159.0	-114.6
KF	-585.0	-562.1	88.7	-84.9
$KHCO_3$	-944.4	-870.0	193.7	
$KHSO_4$	-1139.7	-1039.2	234.3	-63.0
KI	-307.6	-334.9	213.8	-120.5
KNO_3	-459.7	-394.5	248.9	-64.9
K_2CO_3	-1181.9	-1094.4	148.1	
K_2S	-471.5	-480.7	190.4	
K_2SO_4	-1414.0	-1311.1	225.1	-251.0
K_2Se		-437.2		
$LaCl_3$	-1208.8	-1077.3	-50.0	-423.0
$LiBr$	-400.0	-397.3	95.8	-73.2
$LiCl$	-445.6	-424.6	69.9	-67.8
LiF	-611.1	-571.9	-0.4	-38.1
LiI	-333.7	-344.8	124.7	-73.6
$LiNO_3$	-485.9	-404.5	160.2	-18.0
Li_2CO_3	-1234.1	-1114.6	-29.7	
Li_2SO_4	-1466.2	-1331.2	47.3	-155.6
$LuCl_3$	-1167.0	-1021.0	-96.0	-385.0
$MgBr_2$	-709.9	-662.7	26.8	
$MgCl_2$	-801.2	-717.1	-25.1	
MgI_2	-577.2	-558.1	84.5	
$Mg(NO_3)_2$	-881.6	-677.3	154.8	
$MgSO_4$	-1376.1	-1199.5	-118.0	
$MnBr_2$	-464.0			
$MnCl_2$	-555.1	-490.8	38.9	-222.0
MnI_2	-331.0			
$Mn(NO_3)_2$	-635.5	-450.9	218.0	-121.0
$MnSO_4$	-1130.1	-972.7	-53.6	-243.0
NH_4Br	-254.1	-183.3	195.8	-61.9
NH_4BrO_3	-199.6	-60.7	275.1	
NH_4CN	18.0	93.0	207.5	
NH_4Cl	-299.7	-210.5	169.9	-56.5
NH_4ClO_3	-236.5	-87.3	275.7	
NH_4ClO_4	-261.8	-87.8	295.4	
NH_4F	-465.1	-358.1	99.6	-26.8
NH_4HCO_3	-824.5	-666.1	204.6	
NH_4HS	-150.2	-67.2	176.1	
NH_4HSO_3	-758.7	-607.0	253.1	
NH_4HSO_4	-1019.9	-835.2	245.2	-3.8
NH_4HSeO_4	-714.2	-531.6	262.8	
$NH_4H_2AsO_3$	-847.3	-666.4	223.8	
$NH_4H_2AsO_4$	-1042.1	-832.5	230.5	
$NH_4H_2PO_4$	-1428.8	-1209.6	203.8	
$NH_4H_3P_2O_7$	-2409.1	-2102.6	326.0	
NH_4I	-187.7	-130.9	224.7	-62.3
NH_4IO_3	-354.0	-207.4	231.8	
NH_4NO_2	-237.2	-111.6	236.4	-17.6

Species	$\Delta_f H°$/ kJ mol^{-1}	$\Delta_f G°$/ kJ mol^{-1}	$S°$/ J mol^{-1}K^{-1}	C_p/ J mol^{-1}K^{-1}
NH_4NO_3	-339.9	-190.6	259.8	-6.7
NH_4OH	-362.5	-236.5	102.5	-68.6
NH_4SCN	-56.1	13.4	257.7	39.7
$(NH_4)_2CO_3$	-942.2	-686.4	169.9	
$(NH_4)_2CrO_4$	-1146.2	-886.4	277.0	
$(NH_4)_2Cr_2O_7$	-1755.2	-1459.5	488.7	
$(NH_4)_2HAsO_4$	-1171.4	-873.2	225.1	
$(NH_4)_2HPO_4$	-1557.2	-1247.8	193.3	
$(NH_4)_2S$	-231.8	-72.6	212.1	
$(NH_4)_2SO_3$	-900.4	-645.0	197.5	
$(NH_4)_2SO_4$	-1174.3	-903.1	246.9	-133.1
$(NH_4)_2SeO_4$	-864.0	-599.8	280.7	
$(NH_4)_3PO_4$	-1674.9	-1256.6	117.0	
NaBr	-361.7	-365.8	141.4	-95.4
NaCl	-407.3	-393.1	115.5	-90.0
NaF	-572.8	-540.7	45.2	-60.2
$NaHCO_3$	-932.1	-848.7	150.2	
$NaHSO_4$	-1127.5	-1017.8	190.8	-38.0
NaI	-295.3	-313.5	170.3	-95.8
$NaNO_3$	-447.5	-373.2	205.4	-40.2
Na_2CO_3	-1157.4	-1051.6	61.1	
Na_2S	-447.3	-438.1	103.3	
Na_2SO_4	-1389.5	-1268.4	138.1	-201.0
Na_2Se		-394.6		
$NdCl_3$	-1197.9	-1065.6	-37.7	-431.0
$NiBr_2$	-297.1	-253.6	36.0	
$NiCl_2$	-388.3	-307.9	-15.1	
NiF_2	-719.2	-603.3	-156.5	
NiI_2	-164.4	-149.0	93.7	
$Ni(NO_3)_2$	-468.6	-268.5	164.0	
$NiSO_4$	-963.2	-790.3	-108.8	
$PbBr_2$	-244.8	-232.3	175.3	
$PbCl_2$	-336.0	-286.9	123.4	
PbF_2	-666.9	-582.0	-17.2	
PbI_2	-112.1	-127.6	233.0	
$Pb(NO_3)_2$	-416.3	-246.9	303.3	
$PrCl_3$	-1206.2	-1072.7	-42.0	-439.0
$RaCl_2$	-861.9	-823.8	167.0	
$Ra(NO_3)_2$	-942.2	-784.0	347.0	
$RaSO_4$	-1436.8	-1306.2	75.0	
RbBr	-372.7	-387.9	203.9	
RbCl	-418.3	-415.2	178.0	
RbF	-583.8	-562.8	107.5	
$RbHCO_3$	-943.2	-870.8	212.7	
$RbHSO_4$	-1138.5	-1039.9	253.1	
RbI	-306.4	-335.6	232.6	
$RbNO_3$	-458.5	-395.2	267.8	
Rb_2CO_3	-1179.5	-1095.8	186.2	
Rb_2S	-469.4	-482.0	228.4	
Rb_2SO_4	-1411.6	-1312.5	263.2	
$SmCl_3$	-1193.3	-1060.2	-42.7	-431.0
$SrBr_2$	-788.9	-767.4	132.2	
$SrCO_3$	-1222.9	-1087.3	-89.5	
$SrCl_2$	-880.1	-821.9	80.3	
SrI_2	-656.2	-662.6	190.0	
$Sr(NO_3)_2$	-960.5	-782.0	260.2	
$SrSO_4$	-1455.1	-1304.0	-12.6	
$TbCl_3$	-1184.1	-1045.5	-59.0	-393.0
TlBr	-116.2	-136.4	207.9	
$TlBr_3$	-168.2	-97.1	54.0	
TlCl	-161.8	-163.6	182.0	
$TlCl_3$	-305.0	-179.0	-23.0	
TlF	-327.3	-311.2	111.7	
TlI	-49.8	-84.0	236.8	
$TlNO_3$	-202.0	-143.7	272.0	
Tl_2SO_4	-898.6	-809.3	271.1	
$TmCl_3$	-1199.1	-1055.6	-75.0	-385.0
UCl_4	-1259.8	-1056.8	-184.0	
UO_2CO_3	-1696.6	-1481.5	-154.4	
$UO_2(NO_3)_2$	-1434.3	-1176.0	195.4	
UO_2SO_4	-1928.8	-1698.2	-77.4	
$YbCl_3$	-1176.1	-1037.6	-71.0	-385.0
$ZnBr_2$	-397.0	-355.0	52.7	-238.0
$ZnCl_2$	-488.2	-409.5	0.8	-226.0
ZnF_2	-819.1	-704.6	-139.7	-167.0
ZnI_2	-264.3	-250.2	110.5	-238.0
$Zn(NO_3)_2$	-568.6	-369.6	180.7	-126.0
$ZnSO_4$	-1063.2	-891.6	-92.0	-247.0

HEAT OF COMBUSTION

The heat of combustion of a substance at 25°C can be calculated from the enthalpy of formation ($\Delta_f H°$) data in the table "Standard Thermodynamic Properties of Chemical Substances" in this Section. We can write the general combustion reaction as

$$X + O_2 \rightarrow CO_2(g) + H_2O(l) + \text{other products}$$

For a compound containing only carbon, hydrogen, and oxygen, the reaction is simply

$$C_aH_bO_c + \left(a + \frac{1}{4}b - \frac{1}{2}c\right)O_2 \rightarrow a\ CO_2(g) + \frac{1}{2}b\ H_2O(l)$$

and the standard heat of combustion $\Delta_c H°$, which is defined as the negative of the enthalpy change for the reaction (i.e., the heat released in the combustion process), is given by

$$\Delta_c H° = -a\Delta_f H°(CO_2,g) - \frac{1}{2}b\,\Delta_f H°(H_2O,l) + \Delta_f H°(C_aH_b,O_c)$$

$$= 393.51\,a + 142.915\,b + \Delta_f H°(C_aH_b,O_c)$$

This equation applies if the reactants start in their standard states (25°C and one atmosphere pressure) and the products return to the same conditions. The same equation applies to a compound containing another element if that element ends in its standard reference state (e.g., nitrogen, if the product is N_2); in general, however, the exact products containing the other elements must be known in order to calculate the heat of combustion.

The following table gives the standard heat of combustion calculated in this manner for a few representative substances.

Molecular formula	Name	$\Delta_c H°$/kJ mol^{-1}
	Inorganic substances	
C	Carbon (graphite)	393.5
CO	Carbon monoxide (g)	283.0
H_2	Hydrogen (g)	285.8
H_3N	Ammonia (g)	382.8
H_4N_2	Hydrazine (g)	667.1
N_2O	Nitrous oxide (g)	82.1
	Hydrocarbons	
CH_4	Methane (g)	890.8
C_2H_2	Acetylene (g)	1301.1
C_2H_4	Ethylene (g)	1411.2
C_2H_6	Ethane (g)	1560.7
C_3H_6	Propylene (g)	2058.0
C_3H_6	Cyclopropane (g)	2091.3
C_3H_8	Propane (g)	2219.2
C_4H_6	1,3-Butadiene (g)	2541.5
C_4H_{10}	Butane (g)	2877.6
C_5H_{12}	Pentane (l)	3509.0
C_6H_6	Benzene (l)	3267.6
C_6H_{12}	Cyclohexane (l)	3919.6
C_6H_{14}	Hexane (l)	4163.2
C_7H_8	Toluene (l)	3910.3
C_7H_{16}	Heptane (l)	4817.0
$C_{10}H_8$	Naphthalene (s)	5156.3
	Alcohols and ethers	
CH_4O	Methanol (l)	726.1
C_2H_6O	Ethanol (l)	1366.8
C_2H_6O	Dimethyl ether (g)	1460.4
$C_2H_6O_2$	Ethylene glycol (l)	1189.2
C_3H_8O	1-Propanol (l)	2021.3
$C_3H_8O_3$	Glycerol (l)	1655.4
$C_4H_{10}O$	Diethyl ether (l)	2723.9
$C_5H_{12}O$	1-Pentanol (l)	3330.9
C_6H_6O	Phenol (s)	3053.5
	Carbonyl compounds	
CH_2O	Formaldehyde (g)	570.7
C_2H_2O	Ketene (g)	1025.4
C_2H_4O	Acetaldehyde (l)	1166.9
C_3H_6O	Acetone (l)	1789.9
C_3H_6O	Propanal (l)	1822.7
C_4H_8O	2-Butanone (l)	2444.1
	Acids and esters	
CH_2O_2	Formic acid (l)	254.6
$C_2H_4O_2$	Acetic acid (l)	874.2
$C_2H_4O_2$	Methyl formate (l)	972.6
$C_3H_6O_2$	Methyl acetate (l)	1592.2
$C_4H_8O_2$	Ethyl acetate (l)	2238.1
$C_7H_6O_2$	Benzoic acid (s)	3226.9
	Nitrogen compounds	
CHN	Hydrogen cyanide (g)	671.5
CH_3NO_2	Nitromethane (l)	709.2
CH_5N	Methylamine (g)	1085.6
C_2H_3N	Acetonitrile (l)	1247.2
C_2H_5NO	Acetamide (s)	1184.6
C_3H_9N	Trimethylamine (g)	2443.1
C_5H_5N	Pyridine (l)	2782.3
C_6H_7N	Aniline (l)	3392.8

ELECTRICAL CONDUCTIVITY OF WATER

This table gives the electrical conductivity of highly purified water over a range of temperature and pressure. The first column of conductivity data refers to water at its own vapor pressure. Equations for calculating the conductivity at any temperature and pressure may be found in the reference.

REFERENCE

Marshall, W. L., *J. Chem. Eng. Data* 32, 221, 1987.

Conductivity in μS/cm at the indicated pressure

t/°C	Sat. vapor	50 MPa	100 MPa	200 MPa	400 MPa	600 MPa
0	0.0115	0.0150	0.0189	0.0275	0.0458	0.0667
25	0.0550	0.0686	0.0836	0.117	0.194	0.291
100	0.765	0.942	1.13	1.53	2.45	3.51
200	2.99	4.08	5.22	7.65	13.1	19.5
300	2.41	4.87	7.80	14.1	28.9	46.5
400		1.17	4.91	14.3	39.2	71.3
600			0.134	4.65	33.8	85.7

STANDARD KCl SOLUTIONS FOR CALIBRATING CONDUCTIVITY CELLS

This table presents recommended electrolytic conductivity (κ) values for aqueous potassium chloride solutions with molalities of 0.01 mol/kg, 0.1 mol/kg and 1.0 mol/kg at temperatures from 0°C to 50°C. The values, which are based on measurements at the National Institute of Standards and Technology, provide primary standards for the calibration of conductivity cells. The measurements at 0.01 and 0.1 molal are described in Reference 1, while those at 1.0 molal are in Reference 2. Temperatures are given on the ITS-90 scale. The uncertainty in the conductivity is about 0.03% for the 0.01 molal values and about 0.04% for the 0.1 and 1.0 molal values. The conductivity of water saturated with atmospheric CO_2 is given in the last column. These values were subtracted from the original measurements to give the values in the second, third, and fourth columns. All κ values are given in units of 10^{-4} S/m (numerically equal to μS/cm).

The assistance of Kenneth W. Pratt is appreciated.

REFERENCES

1. Wu, Y.C., Koch, W.F., and Pratt, K.W., *J. Res. Natl. Inst. Stand. Technol.* 96, 191, 1991.
2. Wu, Y.C., Koch, W.F., Feng, D., Holland, L.A., Juhasz, E., Arvay, E., and Tomek, A., *J. Res. Natl. Inst. Stand. Technol.* 99, 241, 1994.

	10^4 κ/S m^{-1}			
t/°C	0.01 m KCl	0.1 m KCl	1.0 m KCl	H_2O (CO_2 sat.)
0	772.92	7 116.85	63 488	0.58
5	890.96	8 183.70	72 030	0.68
10	1 013.95	9 291.72	80 844	0.79
15	1 141.45	10 437.1	89 900	0.89
18	1 219.93	11 140.6	—	0.95
20	1 273.03	11 615.9	99 170	0.99
25	1 408.23	12 824.6	108 620	1.10
30	1 546.63	14 059.2	118 240	1.20
35	1 687.79	15 316.0	127 970	1.30
40	1 831.27	16 591.0	137 810	1.40
45	1 976.62	17 880.6	147 720	1.51
50	2 123.43	19 180.9	157 670	1.61

MOLAR CONDUCTIVITY OF AQUEOUS HF, HCl, HBr, AND HI

The molar conductivity Λ of an electrolyte solution is defined as the conductivity divided by amount-of-substance concentration. The customary unit is S $cm^2 mol^{-1}$ (i.e., Ω^{-1} $cm^2 mol^{-1}$). The first part of this table gives the molar conductivity of the hydrohalogen acids at 25°C as a function of the concentration in mol/L. The second part gives the temperature dependence of Λ for HCl and HBr. More extensive tables and mathematical representations may be found in the reference.

REFERENCE

Hamer, W.J., and DeWane, H.J., *Electrolytic Conductance and the Conductances of the Hydrohalogen Acids in Water*, Natl. Stand. Ref. Data Sys.-Natl. Bur. Standards (U.S.), No. 33, 1970.

c/mol L^{-1}	HF	HCl	HBr	HI
Inf. dil.	405.1	426.1	427.7	426.4
0.0001		424.5	425.9	424.6
0.0005		422.6	424.3	423.0
0.001		421.2	422.9	421.7
0.005	128.1	415.7	417.6	416.4
0.01	96.1	411.9	413.7	412.8
0.05	50.1	398.9	400.4	400.8
0.10	39.1	391.1	391.9	394.0
0.5	26.3	360.7	361.9	369.8
1.0	24.3	332.2	334.5	343.9
1.5		305.8	307.6	316.4
2.0		281.4	281.7	288.9
2.5		258.9	257.8	262.5
3.0		237.6	236.8	237.9
3.5		218.3	217.5	215.4
4.0		200.0	199.4	195.1
4.5		183.1	182.4	176.8
5.0		167.4	166.5	160.4
5.5		152.9	151.8	145.5
6.0		139.7	138.2	131.7
6.5		127.7	125.7	118.6
7.0		116.9	114.2	105.7
7.5		107.0	103.8	
8.0		98.2	94.4	
8.5		90.3	85.8	
9.0		83.1		
9.5		76.6		
10.0		70.7		

c/mol L^{-1}	−20°C	−10°C	0°C	10°C	20°C	30°C	40°C	50°C
HCl								
0.5			228.7	283.0	336.4	386.8	436.9	482.4
1.0			211.7	261.6	312.2	359.0	402.9	445.3
1.5			196.2	241.5	287.5	331.1	371.6	410.8
2.0			182.0	222.7	262.9	303.3	342.4	378.2
2.5		131.7	168.5	205.1	239.8	277.0	315.2	347.6
3.0		120.8	154.6	188.5	219.3	253.3	289.3	319.0
3.5	85.5	111.3	139.6	172.2	201.6	232.9	263.9	292.1
4.0	79.3	102.7	129.2	158.1	185.6	214.2	242.2	268.2
4.5	73.7	94.9	119.5	145.4	170.6	196.6	222.5	246.7
5.0	68.5	87.8	110.3	133.5	156.6	180.2	204.1	226.5
5.5	63.6	81.1	101.7	122.5	143.6	165.0	187.1	207.7
6.0	58.9	74.9	93.7	112.3	131.5	151.0	171.3	190.3
6.5	54.4	69.1	86.2	103.0	120.4	138.2	156.9	174.3
7.0	50.2	63.7	79.3	94.4	110.2	126.4	143.3	159.7
7.5	46.3	58.6	73.0	86.5	100.9	115.7	131.6	146.2
8.0	42.7	54.0	67.1	79.4	92.4	106.1	120.6	134.0
8.5	39.4	49.8	61.7	72.9	84.7	97.3	110.7	123.0
9.0	36.4	45.9	56.8	67.1	77.8	89.4	101.7	112.9
9.5	33.6	42.3	52.3	61.8	71.5	82.3	93.6	103.9
10.0	31.2	39.1	48.2	57.0	65.8	75.9	86.3	95.7
10.5	28.9	36.1	44.5	52.7	60.7	70.1	79.6	88.4
11.0	26.8	33.4	41.1	48.8	56.1	64.9	73.6	81.7
11.5	24.9	31.0	38.0	45.3	51.9	60.1	68.0	75.6
12.0	23.1	28.7	35.3	42.0	48.0	55.6	62.8	70.0
12.5	21.4	26.7	32.7	39.0	44.4	51.4	57.9	64.8

MOLAR CONDUCTIVITY OF AQUEOUS HF, HCl, HBr, AND HI (continued)

c/mol L^{-1}	–20°C	–10°C	0°C	10°C	20°C	30°C	40°C	50°C
				HBr				
0.5			240.9	295.9	347.0	398.9	453.6	496.8
1.0			229.6	276.0	329.0	380.4	418.6	465.2
1.5			209.5	254.9	298.9	340.6	381.8	421.4
2.0		150.8	188.6	231.3	271.8	314.1	350.5	387.4
2.5		136.8	171.7	208.3	244.8	281.7	316.0	349.1
3.0		125.7	157.2	189.5	222.2	255.0	287.8	318.6
3.5		116.1	144.1	174.6	203.2	234.4	263.7	291.9
4.0	84.0	107.5	132.3	160.2	186.8	214.2	239.7	266.9
4.5	78.0	99.0	123.0	146.4	171.2	195.1	218.8	242.6
5.0	72.3	91.4	112.6	134.0	155.7	178.2	199.6	221.3
5.5	67.0	84.2	103.1	122.7	142.1	162.8	181.4	201.8
6.0	61.8	77.2	94.3	112.0	129.6	148.0	165.4	183.4
6.5	56.8	70.7	86.0	102.0	118.0	134.1	150.5	166.3
7.0	51.9	64.6	78.4	92.6	107.1	121.4	136.3	150.8

EQUIVALENT CONDUCTIVITY OF ELECTROLYTES IN AQUEOUS SOLUTION

Petr Vanýsek

This table gives the equivalent (molar) conductivity Λ at 25°C for some common electrolytes in aqueous solution at concentrations up to 0.1 mol/L. The units of Λ are 10^{-4} m^2 S mol^{-1}.

For very dilute solutions, the equivalent conductivity for any electrolyte of concentration c can be approximately calculated using the Debye-Hückel-Onsager equation, which can be written for a symmetrical (equal charge on cation and anion) electrolyte as

$$\Lambda = \Lambda^\circ - (A + B\Lambda^\circ)c^{1/2}$$

For a solution at 25°C and both cation and anion with charge $|1|$, the constants are $A = 60.20$ and $B = 0.229$. Λ° can be found from the next table, "Ionic Conductivity and Diffusion at Infinite Dilution". The equation is reliable for $c < 0.001$ mol/L; with higher concentration the error increases.

		Concentration (mol/L)						
Compound	**Infinite dilution**	**0.0005**	**0.001**	**0.005**	**0.01**	**0.02**	**0.05**	**0.1**
	Λ°				**Λ**			
$AgNO_3$	133.29	131.29	130.45	127.14	124.70	121.35	115.18	109.09
$1/2BaCl_2$	139.91	135.89	134.27	127.96	123.88	119.03	111.42	105.14
$1/2CaCl_2$	135.77	131.86	130.30	124.19	120.30	115.59	108.42	102.41
$1/2Ca(OH)_2$	258	—	—	233	226	214	—	—
$1/2CuSO_4$	133.6	121.6	115.20	94.02	83.08	72.16	59.02	50.55
HCl	425.95	422.53	421.15	415.59	411.80	407.04	398.89	391.13
KBr	151.9	149.8	148.9	146.02	143.36	140.41	135.61	131.32
KCl	149.79	147.74	146.88	143.48	141.20	138.27	133.30	128.90
$KClO_4$	139.97	138.69	137.80	134.09	131.39	127.86	121.56	115.14
$1/3K_3Fe(CN)_6$	174.5	166.4	163.1	150.7	—	—	—	—
$1/4K_4Fe(CN)_6$	184	—	167.16	146.02	134.76	122.76	107.65	97.82
$KHCO_3$	117.94	116.04	115.28	112.18	110.03	107.17	—	—
KI	150.31	148.2	143.32	144.30	142.11	139.38	134.90	131.05
KIO_4	127.86	125.74	124.88	121.18	118.45	114.08	106.67	98.2
KNO_3	144.89	142.70	141.77	138.41	132.75	132.34	126.25	120.34
$KMnO_4$	134.8	132.7	131.9	—	126.5	—	—	113
KOH	271.5	—	234	230	228	—	219	213
$KReO_4$	128.20	126.03	125.12	121.31	118.49	114.49	106.40	97.40
$1/3LaCl_3$	145.9	139.6	137.0	127.5	121.8	115.3	106.2	99.1
LiCl	114.97	113.09	112.34	109.35	107.27	104.60	100.06	95.81
$LiClO_4$	105.93	104.13	103.39	100.52	98.56	96.13	92.15	88.52
$1/2MgCl_2$	129.34	125.55	124.15	118.25	114.49	109.99	103.03	97.05
NH_4Cl	149.6	147.5	146.7	134.4	141.21	138.25	133.22	128.69
NaCl	126.39	124.44	123.68	120.59	118.45	115.70	111.01	106.69
$NaClO_4$	117.42	115.58	114.82	111.70	109.54	106.91	102.35	98.38
NaI	126.88	125.30	124.19	121.19	119.18	116.64	112.73	108.73
$NaOOCCH_3$	91.0	89.2	88.5	85.68	83.72	81.20	76.88	72.76
NaOH	247.7	245.5	244.6	240.7	237.9	—	—	—
Na picrate	80.45	78.7	78.6	75.7	73.7	—	66.3	61.8
$1/2Na_2SO_4$	129.8	125.68	124.09	117.09	112.38	106.73	97.70	89.94
$1/2SrCl_2$	135.73	131.84	130.27	124.18	120.23	115.48	108.20	102.14
$1/2ZnSO_4$	132.7	121.3	114.47	95.44	84.87	74.20	61.17	52.61

IONIC CONDUCTIVITY AND DIFFUSION AT INFINITE DILUTION

Petr Vanýsek

This table gives the molar (equivalent) conductivity λ for common ions at infinite dilution. All values refer to aqueous solutions at 25°C. It also lists the diffusion coefficient D of the ion in dilute aqueous solution, which is related to λ through the equation

$$D = \left(RT/F^2\right)\left(\lambda/|z|\right)$$

where R is the molar gas constant, T the temperature, F the Faraday constant, and z the charge on the ion. The variation with temperature is fairly sharp; for typical ions, λ and D increase by 2 to 3% per degree as the temperature increases from 25°C.

The diffusion coefficient for a salt, D_{salt}, may be calculated from the D_+ and D_- values of the constituent ions by the relation

$$D_{salt} = \frac{\left(z_+ + |z_-|\right)D_+D_-}{z_+D_+ + |z_-|D_-}$$

For solutions of simple, pure electrolytes (one positive and one negative ionic species), such as NaCl, equivalent ionic conductivity $\Lambda°$, which is the conductivity per unit concentration of charge, is defined as

$$\Lambda° = \lambda_+ + \lambda_-$$

where λ_+ and λ_- are equivalent ionic conductivities of the cation and anion. The more general formula is

$$\Lambda° = \nu_+\lambda_+ + \nu_-\lambda_-$$

where ν_+ and ν_- refer to the number of moles of cations and anions to which one mole of the electrolyte gives a rise in the solution.

REFERENCES

1. Gray, D. E., Ed., *American Institute of Physics Handbook,* McGraw-Hill, New York, 1972, 2—226.
2. Robinson, R. A., and Stokes, R. H., *Electrolyte Solutions,* Butterworths, London, 1959.
3. Lobo, V. M. M., and Quaresma, J. L., *Handbook of Electrolyte Solutions*, Physical Science Data Series 41, Elsevier, Amsterdam, 1989.
4. Conway, B. E., *Electrochemical Data,* Elsevier, Amsterdam, 1952.
5. Milazzo, G., *Electrochemistry: Theoretical Principles and Practical Applications,* Elsevier, Amsterdam, 1963.

Ion	λ 10^{-4} m² S mol⁻¹	D 10^{-5} cm² s⁻¹
Inorganic Cations		
Ag^+	61.9	1.648
$1/3Al^{3+}$	61	0.541
$1/2Ba^{2+}$	63.6	0.847
$1/2Be^{2+}$	45	0.599
$1/2Ca^{2+}$	59.47	0.792
$1/2Cd^{2+}$	54	0.719
$1/3Ce^{3+}$	69.8	0.620
$1/2Co^{2+}$	55	0.732
$1/3[Co(NH_3)_6]^{3+}$	101.9	0.904
$1/3[Co(en)_3]^{3+}$	74.7	0.663
$1/6[Co_2(trien)_3]^{6+}$	69	0.306
$1/3Cr^{3+}$	67	0.595
Cs^+	77.2	2.056
$1/2Cu^{2+}$	53.6	0.714
D^+	249.9	6.655
$1/3Dy^{3+}$	65.6	0.582
$1/3Er^{3+}$	65.9	0.585
$1/3Eu^{3+}$	67.8	0.602
$1/2Fe^{2+}$	54	0.719
$1/3Fe^{3+}$	68	0.604
$1/3Gd^{3+}$	67.3	0.597
H^+	349.65	9.311
$1/2Hg^{2+}$	68.6	0.913
$1/2Hg^{2+}$	63.6	0.847
$1/3Ho^{3+}$	66.3	0.589
K^+	73.48	1.957
$1/3La^{3+}$	69.7	0.619
Li^+	38.66	1.029
$1/2Mg^{2+}$	53.0	0.706
$1/2Mn^{2+}$	53.5	0.712
NH_4^+	73.5	1.957
$N_2H_5^+$	59	1.571
Na^+	50.08	1.334
$1/3Nd^{3+}$	69.4	0.616
$1/2Ni^{2+}$	49.6	0.661
$1/4[Ni_2(trien)_3]^{4+}$	52	0.346
$1/2Pb^{2+}$	71	0.945
$1/3Pr^{3+}$	69.5	0.617
$1/2Ra^{2+}$	66.8	0.889
Rb^+	77.8	2.072
$1/3Sc^{3+}$	64.7	0.574
$1/3Sm^{3+}$	68.5	0.608
$1/2Sr^{2+}$	59.4	0.791
Tl^+	74.7	1.989
$1/3Tm^{3+}$	65.4	0.581
$1/2UO_2^{2+}$	32	0.426
$1/3Y^{3+}$	62	0.550
$1/3Yb^{3+}$	65.6	0.582
$1/2Zn^{2+}$	52.8	0.703

Ion	λ 10^{-4} m^2 S mol^{-1}	*D* 10^{-5} cm^2 s^{-1}
Inorganic Anions		
$Au(CN)_2^-$	50	1.331
$Au(CN)_4^-$	36	0.959
$B(C_6H_5)_4^-$	21	0.559
Br^-	78.1	2.080
Br_3^-	43	1.145
BrO_3^-	55.7	1.483
CN^-	78	2.077
CNO^-	64.6	1.720
$1/2CO_3^{2-}$	69.3	0.923
Cl^-	76.31	2.032
ClO_2^-	52	1.385
ClO_3^-	64.6	1.720
ClO_4^-	67.3	1.792
$1/3[Co(CN)_6]^{3-}$	98.9	0.878
$1/2CrO_4^{2-}$	85	1.132
F^-	55.4	1.475
$1/4[Fe(CN)_6]^{4-}$	110.4	0.735
$1/3[Fe(CN)_6]^{3-}$	100.9	0.896
$H_2AsO_4^-$	34	0.905
HCO_3^-	44.5	1.185
HF_2^-	75	1.997
$1/2HPO_4^{2-}$	57	0.759
$H_2PO_4^-$	36	0.959
$H_2PO_2^-$	46	1.225
HS^-	65	1.731
HSO_3^-	58	1.545
HSO_4^-	52	1.385
$H_2SbO_4^-$	31	0.825
I^-	76.8	2.045
IO_3^-	40.5	1.078
IO_4^-	54.5	1.451
MnO_4^-	61.3	1.632
$1/2MoO_4^{2-}$	74.5	1.984
$N(CN)_2^-$	54.5	1.451
NO_2^-	71.8	1.912
NO_3^-	71.42	1.902
$NH_2SO_3^-$	48.3	1.286
N_3^-	69	1.837
OCN^-	64.6	1.720
OD^-	119	3.169
OH^-	198	5.273
PF_6^-	56.9	1.515
$1/2PO_3F^{2-}$	63.3	0.843
$1/3PO_4^{3-}$	92.8	0.824
$1/4P_2O_7^{4-}$	96	0.639
$1/3P_3O_9^{3-}$	83.6	0.742
$1/5P_3O_{10}^{5-}$	109	0.581
ReO_4^-	54.9	1.462
SCN^-	66	1.758
$1/2SO_3^{2-}$	72	0.959
$1/2SO_4^{2-}$	80.0	1.065
$1/2S_2O_3^{2-}$	85.0	1.132
$1/2S_2O_4^{2-}$	66.5	0.885
$1/2S_2O_6^{2-}$	93	1.238
$1/2S_2O_8^{2-}$	86	1.145
$Sb(OH)_6^-$	31.9	0.849
$SeCN^-$	64.7	1.723
$1/2SeO_4^{2-}$	75.7	1.008
$1/2WO_4^{2-}$	69	0.919
Organic Cations		
Benzyltrimethylammonium$^+$	34.6	0.921
Isobutylammonium$^+$	38	1.012
Butyltrimethylammonium$^+$	33.6	0.895
Decylpyridinium$^+$	29.5	0.786
Decyltrimethylammonium$^+$	24.4	0.650
Diethylammonium$^+$	42.0	1.118
Dimethylammonium$^+$	51.8	1.379
Dipropylammonium$^+$	30.1	0.802
Dodecylammonium$^+$	23.8	0.634
Dodecyltrimethylammonium$^+$	22.6	0.602
Ethanolammonium$^+$	42.2	1.124
Ethylammonium$^+$	47.2	1.257
Ethyltrimethylammonium$^+$	40.5	1.078
Hexadecyltrimethylammonium$^+$	20.9	0.557
Hexyltrimethylammonium$^+$	29.6	0.788
Histidyl$^+$	23.0	0.612
Hydroxyethyltrimethylarsonium$^+$	39.4	1.049
Methylammonium$^+$	58.7	1.563
Octadecylpyridinium$^+$	20	0.533
Octadecyltributylammonium$^+$	16.6	0.442
Octadecyltriethylammonium$^+$	17.9	0.477
Octadecyltrimethylammonium$^+$	19.9	0.530
Octadecyltripropylammonium$^+$	17.2	0.458
Octyltrimethylammonium$^+$	26.5	0.706
Pentylammonium$^+$	37	0.985
Piperidinium$^+$	37.2	0.991
Propylammonium$^+$	40.8	1.086
Pyrilammonium$^+$	24.3	0.647
Tetrabutylammonium$^+$	19.5	0.519
Tetradecyltrimethylammonium$^+$	21.5	0.573
Tetraethylammonium$^+$	32.6	0.868
Tetramethylammonium$^+$	44.9	1.196
Tetraisopentylammonium$^+$	17.9	0.477
Tetrapentylammmonium$^+$	17.5	0.466
Tetrapropylammonium$^+$	23.4	0.623
Triethylammonium$^+$	34.3	0.913
Triethylsulfonium$^+$	36.1	0.961
Trimethylammonium$^+$	47.23	1.258
Trimethylhexylammonium$^+$	34.6	0.921
Trimethylsulfonium$^+$	51.4	1.369
Tripropylammonium$^+$	26.1	0.695
Organic Anions		
Acetate$^-$	40.9	1.089
p-Anisate$^-$	29.0	0.772
1/2Azelate^{2-}	40.6	0.541
Benzoate$^-$	32.4	0.863
Bromoacetate$^-$	39.2	1.044
Bromobenzoate$^-$	30	0.799
Butyrate$^-$	32.6	0.868
Chloroacetate$^-$	39.8	1.060
m-Chlorobenzoate$^-$	31	0.825
o-Chlorobenzoate$^-$	30.2	0.804

IONIC CONDUCTIVITY AND DIFFUSION AT INFINITE DILUTION (continued)

Ion	λ 10^{-4} m^2 S mol^{-1}	D 10^{-5} cm^2 s^{-1}
1/3Citrate^{3-}	70.2	0.623
Crotonate^{-}	33.2	0.884
Cyanoacetate^{-}	43.4	1.156
Cyclohexane carboxylate^{-}	28.7	0.764
1/2 1,1-Cyclopropanedicarboxylate^{2-}	53.4	0.711
Decylsulfate^{-}	26	0.692
Dichloroacetate^{-}	38.3	1.020
1/2Diethylbarbiturate^{2-}	26.3	0.350
Dihydrogencitrate^{-}	30	0.799
1/2Dimethylmalonate^{2-}	49.4	0.658
3,5-Dinitrobenzoate^{-}	28.3	0.754
Dodecylsulfate^{-}	24	0.639
Ethylmalonate^{-}	49.3	1.313
Ethylsulfate^{-}	39.6	1.055
Fluoroacetate^{-}	44.4	1.182
Fluorobenzoate^{-}	33	0.879
Formate^{-}	54.6	1.454
1/2Fumarate^{2-}	61.8	0.823
1/2Glutarate^{2-}	52.6	0.700
Hydrogenoxalate^{-}	40.2	1.070
Isovalerate^{-}	32.7	0.871
Iodoacetate^{-}	40.6	1.081
Lactate^{-}	38.8	1.033
1/2Malate^{2-}	58.8	0.783
1/2Maleate^{2-}	61.9	0.824
1/2Malonate^{2-}	63.5	0.845
Methylsulfate^{-}	48.8	1.299
Naphthylacetate^{-}	28.4	0.756
1/2Oxalate^{2-}	74.11	0.987
Octylsulfate^{-}	29	0.772
Phenylacetate^{-}	30.6	0.815
1/2*o*-Phthalate^{2-}	52.3	0.696
1/2*m*-Phthalate^{2-}	54.7	0.728
Picrate^{-}	30.37	0.809
Pivalate^{-}	31.9	0.849
Propionate^{-}	35.8	0.953
Propylsulfate^{-}	37.1	0.988
Salicylate^{-}	36	0.959
1/2Suberate^{2-}	36	0.479
1/2Succinate^{2-}	58.8	0.783
p-Sulfonate	29.3	0.780
1/2Tartarate^{2-}	59.6	0.794
Trichloroacetate^{-}	35	0.932

ACTIVITY COEFFICIENTS OF ACIDS, BASES, AND SALTS

Petr Vanýsek

This table gives mean activity coefficients at 25°C for molalities in the range 0.1 to 1.0. See the following table for definitions, references, and data over a wider concentration range.

	0.1	0.2	0.3	0.4	0.5	0.6	0.7	0.8	0.9	1.0
$AgNO_3$	0.734	0.657	0.606	0.567	0.536	0.509	0.485	0.464	0.446	0.429
$AlCl_3$	0.337	0.305	0.302	0.313	0.331	0.356	0.388	0.429	0.479	0.539
$Al_2(SO_4)_3$	0.035	0.0225	0.0176	0.0153	0.0143	0.014	0.0142	0.0149	0.0159	0.0175
$BaCl_2$	0.500	0.444	0.419	0.405	0.397	0.391	0.391	0.391	0.392	0.395
$BeSO_4$	0.150	0.109	0.0885	0.0769	0.0692	0.0639	0.0600	0.0570	0.0546	0.0530
$CaCl_2$	0.518	0.472	0.455	0.448	0.448	0.453	0.460	0.470	0.484	0.500
$CdCl_2$	0.2280	0.1638	0.1329	0.1139	0.1006	0.0905	0.0827	0.0765	0.0713	0.0669
$Cd(NO_3)_2$	0.513	0.464	0.442	0.430	0.425	0.423	0.423	0.425	0.428	0.433
$CdSO_4$	0.150	0.103	0.0822	0.0699	0.0615	0.0553	0.0505	0.0468	0.0438	0.0415
$CoCl_2$	0.522	0.479	0.463	0.459	0.462	0.470	0.479	0.492	0.511	0.531
$CrCl_3$	0.331	0.298	0.294	0.300	0.314	0.335	0.362	0.397	0.436	0.481
$Cr(NO_3)_3$	0.319	0.285	0.279	0.281	0.291	0.304	0.322	0.344	0.371	0.401
$Cr_2(SO_4)_3$	0.0458	0.0300	0.0238	0.0207	0.0190	0.0182	0.0181	0.0185	0.0194	0.0208
CsBr	0.754	0.694	0.654	0.626	0.603	0.586	0.571	0.558	0.547	0.538
CsCl	0.756	0.694	0.656	0.628	0.606	0.589	0.575	0.563	0.553	0.544
CsI	0.754	0.692	0.651	0.621	0.599	0.581	0.567	0.554	0.543	0.533
$CsNO_3$	0.733	0.655	0.602	0.561	0.528	0.501	0.478	0.458	0.439	0.422
CsOH	0.795	0.761	0.744	0.739	0.739	0.742	0.748	0.754	0.762	0.771
CsOAc	0.799	0.771	0.761	0.759	0.762	0.768	0.776	0.783	0.792	0.802
Cs_2SO_4	0.456	0.382	0.338	0.311	0.291	0.274	0.262	0.251	0.242	0.235
$CuCl_2$	0.508	0.455	0.429	0.417	0.411	0.409	0.409	0.410	0.413	0.417
$Cu(NO_3)_2$	0.511	0.460	0.439	0.429	0.426	0.427	0.431	0.437	0.445	0.455
$CuSO_4$	0.150	0.104	0.0829	0.0704	0.0620	0.0559	0.0512	0.0475	0.0446	0.0423
$FeCl_2$	0.5185	0.473	0.454	0.448	0.450	0.454	0.463	0.473	0.488	0.506
HBr	0.805	0.782	0.777	0.781	0.789	0.801	0.815	0.832	0.850	0.871
HCl	0.796	0.767	0.756	0.755	0.757	0.763	0.772	0.783	0.795	0.809
$HClO_4$	0.803	0.778	0.768	0.766	0.769	0.776	0.785	0.795	0.808	0.823
HI	0.818	0.807	0.811	0.823	0.839	0.860	0.883	0.908	0.935	0.963

	0.1	0.2	0.3	0.4	0.5	0.6	0.7	0.8	0.9	1.0
HNO_3	0.791	0.754	0.735	0.725	0.720	0.717	0.717	0.718	0.721	0.724
H_2SO_4	0.2655	0.2090	0.1826	—	0.1557	—	0.1417	—	—	0.1316
KBr	0.772	0.722	0.693	0.673	0.657	0.646	0.636	0.629	0.622	0.617
KCl	0.770	0.718	0.688	0.666	0.649	0.637	0.626	0.618	0.610	0.604
$KClO_3$	0.749	0.681	0.635	0.599	0.568	0.541	0.518	—	—	—
K_2CrO_4	0.456	0.382	0.340	0.313	0.292	0.276	0.263	0.253	0.243	0.235
KF	0.775	0.727	0.700	0.682	0.670	0.661	0.654	0.650	0.646	0.645
$K_3Fe(CN)_6$	0.268	0.212	0.184	0.167	0.155	0.146	0.140	0.135	0.131	0.128
$K_4Fe(CN)_6$	0.139	0.0993	0.0808	0.0693	0.0614	0.0556	0.0512	0.0479	0.0454	—
KH_2PO_4	0.731	0.653	0.602	0.561	0.529	0.501	0.477	0.456	0.438	0.421
KI	0.778	0.733	0.707	0.689	0.676	0.667	0.660	0.654	0.649	0.645
KNO_3	0.739	0.663	0.614	0.576	0.545	0.519	0.496	0.476	0.459	0.443
KOAc	0.796	0.766	0.754	0.750	0.751	0.754	0.759	0.766	0.774	0.783
KOH	0.798	0.760	0.742	0.734	0.732	0.733	0.736	0.742	0.749	0.756
KSCN	0.769	0.716	0.685	0.663	0.646	0.633	0.623	0.614	0.606	0.599
K_2SO_4	0.441	0.360	0.316	0.286	0.264	0.246	0.232	—	—	—
LiBr	0.796	0.766	0.756	0.752	0.753	0.758	0.767	0.777	0.789	0.803
LiCl	0.790	0.757	0.744	0.740	0.739	0.743	0.748	0.755	0.764	0.774
$LiClO_4$	0.812	0.794	0.792	0.798	0.808	0.820	0.834	0.852	0.869	0.887
LiI	0.815	0.802	0.804	0.813	0.824	0.838	0.852	0.870	0.888	0.910
$LiNO_3$	0.788	0.752	0.736	0.728	0.726	0.727	0.729	0.733	0.737	0.743
LiOH	0.760	0.702	0.665	0.638	0.617	0.599	0.585	0.573	0.563	0.554
LiOAc	0.784	0.742	0.721	0.709	0.700	0.691	0.689	0.688	0.688	0.689
Li_2SO_4	0.468	0.398	0.361	0.337	0.319	0.307	0.297	0.289	0.282	0.277
$MgCl_2$	0.529	0.489	0.477	0.475	0.481	0.491	0.506	0.522	0.544	0.570
$MgSO_4$	0.150	0.107	0.0874	0.0756	0.0675	0.0616	0.0571	0.0536	0.0508	0.0485
$MnCl_2$	0.516	0.469	0.450	0.442	0.440	0.443	0.448	0.455	0.466	0.479
$MnSO_4$	0.150	0.105	0.0848	0.0725	0.0640	0.0578	0.0530	0.0493	0.0463	0.0439
NH_4Cl	0.770	0.718	0.687	0.665	0.649	0.636	0.625	0.617	0.609	0.603
NH_4NO_3	0.740	0.677	0.636	0.606	0.582	0.562	0.545	0.530	0.516	0.504
$(NH_4)_2SO_4$	0.439	0.356	0.311	0.280	0.257	0.240	0.226	0.214	0.205	0.196
NaBr	0.782	0.741	0.719	0.704	0.697	0.692	0.689	0.687	0.687	0.687
NaCl	0.778	0.735	0.710	0.693	0.681	0.673	0.667	0.662	0.659	0.657
$NaClO_3$	0.772	0.720	0.688	0.664	0.645	0.630	0.617	0.606	0.597	0.589
$NaClO_4$	0.775	0.729	0.701	0.683	0.668	0.656	0.648	0.641	0.635	0.629
Na_2CrO_4	0.464	0.394	0.353	0.327	0.307	0.292	0.280	0.269	0.261	0.253
NaF	0.765	0.710	0.676	0.651	0.632	0.616	0.603	0.592	0.582	0.573
NaH_2PO_4	0.744	0.675	0.629	0.593	0.563	0.539	0.517	0.499	0.483	0.468
NaI	0.787	0.751	0.735	0.727	0.723	0.723	0.724	0.727	0.731	0.736
$NaNO_3$	0.762	0.703	0.666	0.638	0.617	0.599	0.583	0.570	0.558	0.548
NaOAc	0.791	0.757	0.744	0.737	0.735	0.736	0.740	0.745	0.752	0.757
NaOH	0.766	0.727	0.708	0.697	0.690	0.685	0.681	0.679	0.678	0.678
NaSCN	0.787	0.750	—	0.720	0.715	0.712	0.710	0.710	0.711	0.712
Na_2SO_4	0.445	0.365	0.320	0.289	0.266	0.248	0.233	0.221	0.210	0.201
$NiCl_2$	0.522	0.479	0.463	0.460	0.464	0.471	0.482	0.496	0.515	0.563
$NiSO_4$	0.150	0.105	0.0841	0.0713	0.0627	0.0562	0.0515	0.0478	0.0448	0.0425
$Pb(NO_3)_2$	0.395	0.308	0.260	0.228	0.205	0.187	0.172	0.160	0.150	0.141
RbBr	0.763	0.706	0.673	0.650	0.632	0.617	0.605	0.595	0.586	0.578
RbCl	0.764	0.709	0.675	0.652	0.634	0.620	0.608	0.599	0.590	0.583
RbI	0.762	0.705	0.671	0.647	0.629	0.614	0.602	0.591	0.583	0.575
$RbNO_3$	0.734	0.658	0.606	0.565	0.534	0.508	0.485	0.465	0.446	0.430
RbOAc	0.796	0.767	0.756	0.753	0.755	0.759	0.766	0.773	0.782	0.792
Rb_2SO_4	0.451	0.374	0.331	0.301	0.279	0.263	0.249	0.238	0.228	0.219
$SrCl_2$	0.511	0.462	0.442	0.433	0.430	0.431	0.434	0.441	0.449	0.461
$TlClO_4$	0.730	0.652	0.599	0.559	0.527	—	—	—	—	—
$TlNO_3$	0.702	0.606	0.545	0.500	—	—	—	—	—	—
UO_2Cl_2	0.544	0.510	0.520	0.505	0.517	0.532	0.549	0.571	0.595	0.620
UO_2SO_4	0.150	0.102	0.0807	0.0689	0.0611	0.0566	0.0515	0.0483	0.0458	0.0439
$ZnCl_2$	0.515	0.462	0.432	0.411	0.394	0.380	0.369	0.357	0.348	0.339
$Zn(NO_3)_2$	0.531	0.489	0.474	0.469	0.473	0.480	0.489	0.501	0.518	0.535
$ZnSO_4$	0.150	0.140	0.0835	0.0714	0.0630	0.0569	0.0523	0.0487	0.0458	0.0435

MEAN ACTIVITY COEFFICIENTS OF ELECTROLYTES AS A FUNCTION OF CONCENTRATION

The mean activity coefficient γ of an electrolyte X_aY_b is defined as

$$\gamma = \left(\gamma_+^a \gamma_-^b\right)^{1/(a+b)}$$

where γ_+ and γ_- are activity coefficients of the individual ions (which cannot be directly measured). This table gives the mean activity coefficients of about 100 electrolytes in aqueous solution as a function of concentration, expressed in molality terms. All values refer to a temperature of 25°C. Substances are arranged in alphabetical order by formula.

REFERENCES

1. Hamer, W. J., and Wu, Y. C., *J. Phys. Chem. Ref. Data*, 1, 1047, 1972.
2. Staples, B. R., *J. Phys. Chem. Ref. Data*, 6, 385, 1977; 10, 767, 1981; 10, 779, 1981.
3. Goldberg, R. N. et al., *J. Phys. Chem. Ref. Data*, 7, 263, 1978; 8, 923, 1979; 8, 1005, 1979; 10, 1, 1981; 10, 671, 1981.

Mean Activity Coefficient at 25°C

m/mol kg^{-1}	$AgNO_3$	$BaBr_2$	$BaCl_2$	BaI_2	$CaBr_2$	$CaCl_2$	CaI_2
0.001	0.964	0.881	0.887	0.890	0.890	0.888	0.890
0.002	0.950	0.850	0.849	0.853	0.853	0.851	0.853
0.005	0.924	0.785	0.782	0.792	0.791	0.787	0.791
0.010	0.896	0.727	0.721	0.737	0.735	0.727	0.736
0.020	0.859	0.661	0.653	0.678	0.674	0.664	0.677
0.050	0.794	0.573	0.559	0.600	0.594	0.577	0.600
0.100	0.732	0.517	0.492	0.551	0.540	0.517	0.552
0.200	0.656	0.463	0.436	0.520	0.502	0.469	0.524
0.500	0.536	0.435	0.391	0.536	0.500	0.444	0.554
1.000	0.430	0.470	0.393	0.664	0.604	0.495	0.729
2.000	0.316	0.654		1.242	1.125	0.784	
5.000	0.181				18.7	5.907	
10.000	0.108					43.1	
15.000	0.085						

m/mol kg^{-1}	$Cd(NO_2)_2$	$Cd(NO_3)_2$	$CoBr_2$	$CoCl_2$	CoI_2	$Co(NO_3)_2$	$CsBr$
0.001	0.881	0.888	0.890	0.889	0.887	0.888	0.965
0.002	0.837	0.851	0.854	0.852	0.849	0.850	0.951
0.005	0.759	0.787	0.794	0.789	0.783	0.786	0.925
0.010	0.681	0.728	0.740	0.732	0.724	0.728	0.898
0.020	0.589	0.664	0.681	0.670	0.661	0.663	0.864
0.050	0.451	0.576	0.605	0.586	0.582	0.576	0.806
0.100	0.344	0.515	0.556	0.528	0.540	0.516	0.752
0.200	0.247	0.465	0.523	0.483	0.527	0.469	0.691
0.500	0.148	0.428	0.538	0.465	0.596	0.446	0.605
1.000	0.098	0.437	0.685	0.532	0.845	0.492	0.540
2.000	0.069	0.517	1.421	0.864	2.287	0.722	0.485
5.000	0.054		13.9		55.3	3.338	0.454
10.000					196		

m/mol kg^{-1}	$CsCl$	CsF	CsI	$CsNO_3$	$CsOH$	Cs_2SO_4	$CuBr_2$
0.001	0.965	0.965	0.965	0.964	0.966	0.885	0.889
0.002	0.951	0.952	0.951	0.951	0.953	0.845	0.853
0.005	0.925	0.929	0.925	0.924	0.930	0.775	0.791
0.010	0.898	0.905	0.898	0.897	0.906	0.709	0.735
0.020	0.864	0.876	0.863	0.860	0.878	0.634	0.674
0.050	0.805	0.830	0.804	0.796	0.836	0.526	0.594
0.100	0.751	0.792	0.749	0.733	0.802	0.444	0.541

MEAN ACTIVITY COEFFICIENTS OF ELECTROLYTES AS A FUNCTION OF CONCENTRATION (continued)

m/mol kg^{-1}	CsCl	CsF	CsI	$CsNO_3$	CsOH	Cs_2SO_4	$CuBr_2$
0.200	0.691	0.755	0.688	0.655	0.772	0.369	0.504
0.500	0.607	0.721	0.601	0.529	0.755	0.285	0.503
1.000	0.546	0.726	0.534	0.421	0.782	0.233	0.591
2.000	0.496	0.803	0.470				0.859
5.000	0.474						
10.000	0.508						

m/mol kg^{-1}	$CuCl_2$	$Cu(ClO_4)_2$	$Cu(NO_3)_2$	$FeCl_2$	HBr	HCl	$HClO_4$
0.001	0.887	0.890	0.888	0.888	0.966	0.965	0.966
0.002	0.849	0.854	0.851	0.850	0.953	0.952	0.953
0.005	0.783	0.795	0.787	0.785	0.930	0.929	0.929
0.010	0.722	0.741	0.729	0.725	0.907	0.905	0.906
0.020	0.654	0.685	0.664	0.659	0.879	0.876	0.878
0.050	0.561	0.613	0.577	0.570	0.837	0.832	0.836
0.100	0.495	0.572	0.516	0.509	0.806	0.797	0.803
0.200	0.441	0.553	0.466	0.462	0.783	0.768	0.776
0.500	0.401	0.617	0.431	0.443	0.790	0.759	0.769
1.000	0.405	0.892	0.456	0.500	0.872	0.811	0.826
2.000	0.453	2.445	0.615	0.782	1.167	1.009	1.055
5.000	0.601		2.083		3.800	2.380	3.100
10.000					33.4	10.4	30.8
15.000							323

m/mol kg^{-1}	HF	HI	HNO_3	H_2SO_4	KBr	KCNS	KCl
0.001	0.551	0.966	0.965	0.804	0.965	0.965	0.965
0.002	0.429	0.953	0.952	0.740	0.952	0.951	0.951
0.005	0.302	0.931	0.929	0.634	0.927	0.927	0.927
0.010	0.225	0.909	0.905	0.542	0.902	0.901	0.901
0.020	0.163	0.884	0.875	0.445	0.870	0.869	0.869
0.050	0.106	0.847	0.829	0.325	0.817	0.815	0.816
0.100	0.0766	0.823	0.792	0.251	0.771	0.768	0.768
0.200	0.0550	0.811	0.756	0.195	0.772	0.716	0.717
0.500	0.0352	0.845	0.725	0.146	0.658	0.647	0.649
1.000	0.0249	0.969	0.730	0.125	0.617	0.598	0.604
2.000	0.0175	1.363	0.788	0.119	0.593	0.556	0.573
5.000	0.0110	4.760	1.063	0.197	0.626	0.525	0.593
10.000	0.0085	49.100	1.644	0.527			
15.000	0.0077		2.212	1.077			
20.000	0.0075		2.607	1.701			

m/mol kg^{-1}	$KClO_3$	K_2CrO_4	KF	KH_2PO_4*	K_2HPO_4**	KI	KNO_3
0.001	0.965	0.886	0.965	0.964	0.886	0.965	0.964
0.002	0.951	0.847	0.952	0.950	0.847	0.952	0.950
0.005	0.926	0.779	0.927	0.924	0.779	0.927	0.924
0.010	0.899	0.715	0.902	0.896	0.715	0.902	0.896
0.020	0.865	0.643	0.870	0.859	0.643	0.871	0.860
0.050	0.805	0.539	0.818	0.793	0.538	0.820	0.797
0.100	0.749	0.460	0.773	0.730	0.457	0.776	0.735
0.200	0.681	0.385	0.726	0.652	0.379	0.731	0.662
0.500	0.569	0.296	0.670	0.529	0.283	0.676	0.546
1.000		0.239	0.645	0.422		0.646	0.444

m/mol kg^{-1}	$KClO_3$	K_2CrO_4	KF	KH_2PO_4*	K_2HPO_4**	KI	KNO_3
2.000		0.199	0.658			0.638	0.332
5.000			0.871				
10.000			1.715				
15.000			3.120				

m/mol kg^{-1}	KOH	K_2SO_4	LiBr	LiCl	$LiClO_4$	LiI	$LiNO_3$
0.001	0.965	0.885	0.965	0.965	0.966	0.966	0.965
0.002	0.952	0.844	0.952	0.952	0.953	0.953	0.952
0.005	0.927	0.772	0.929	0.928	0.931	0.930	0.928
0.010	0.902	0.704	0.905	0.904	0.908	0.908	0.904
0.020	0.871	0.625	0.877	0.874	0.882	0.882	0.874
0.050	0.821	0.511	0.832	0.827	0.843	0.843	0.827
0.100	0.779	0.424	0.797	0.789	0.815	0.817	0.788
0.200	0.740	0.343	0.767	0.756	0.795	0.802	0.753
0.500	0.710	0.251	0.754	0.739	0.806	0.824	0.726
1.000	0.733		0.803	0.775	0.887	0.912	0.743
2.000	0.860		1.012	0.924	1.161	1.197	0.837
5.000	1.697		2.696	2.000			1.298
10.000	6.110		20.0	9.600			2.500
15.000	19.9		147	30.9			3.960
20.000	46.4		486				4.970

m/mol kg^{-1}	LiOH	Li_2SO_4	$MgBr_2$	$MgCl_2$	MgI_2	$MnBr_2$	$MnCl_2$
0.001	0.964	0.887	0.889	0.889	0.889	0.889	0.888
0.002	0.950	0.847	0.852	0.852	0.853	0.853	0.850
0.005	0.923	0.780	0.790	0.790	0.791	0.791	0.786
0.010	0.895	0.716	0.733	0.734	0.736	0.735	0.727
0.020	0.858	0.645	0.672	0.672	0.677	0.674	0.662
0.050	0.794	0.544	0.593	0.590	0.602	0.595	0.574
0.100	0.735	0.469	0.543	0.535	0.556	0.543	0.513
0.200	0.668	0.400	0.512	0.493	0.535	0.508	0.464
0.500	0.579	0.325	0.540	0.485	0.594	0.519	0.437
1.000	0.522	0.284	0.715	0.577	0.858	0.650	0.477
2.000	0.484	0.270	1.590	1.065	2.326	1.224	0.661
5.000	0.493		36.1	14.40	109.8	6.697	1.539

m/mol kg^{-1}	$Mn(ClO_4)_2$	NH_4Cl	NII_4ClO_4	$(NH_4)_2HPO_4$**	NH_4NO_3	NaBr	$NaBrO_3$
0.001	0.892	0.965	0.964	0.882	0.964	0.965	0.965
0.002	0.858	0.952	0.950	0.839	0.951	0.952	0.951
0.005	0.801	0.927	0.924	0.763	0.925	0.928	0.926
0.010	0.752	0.901	0.895	0.688	0.897	0.903	0.900
0.020	0.700	0.869	0.859	0.600	0.862	0.873	0.867
0.050	0.637	0.816	0.794	0.469	0.801	0.824	0.811
0.100	0.604	0.769	0.734	0.367	0.744	0.783	0.759
0.200	0.596	0.718	0.663	0.273	0.678	0.742	0.698
0.500	0.686	0.649	0.560	0.171	0.582	0.697	0.605
1.000	1.030	0.603	0.479	0.114	0.502	0.687	0.528
2.000	3.072	0.569	0.399	0.074	0.419	0.730	0.449
5.000		0.563			0.303	1.083	
10.000					0.220		
15.000					0.179		
20.000					0.154		

MEAN ACTIVITY COEFFICIENTS OF ELECTROLYTES AS A FUNCTION OF CONCENTRATION (continued)

m/mol kg^{-1}	Na_2CO_3	NaCl	$NaClO_3$	$NaClO_4$	Na_2CrO_4	NaF	Na_2HPO_4*
0.001	0.887	0.965	0.965	0.965	0.887	0.965	0.887
0.002	0.847	0.952	0.952	0.952	0.849	0.951	0.848
0.005	0.780	0.928	0.927	0.928	0.783	0.926	0.780
0.010	0.716	0.903	0.902	0.903	0.722	0.901	0.717
0.020	0.644	0.872	0.870	0.872	0.653	0.868	0.644
0.050	0.541	0.822	0.818	0.821	0.554	0.813	0.539
0.100	0.462	0.779	0.771	0.777	0.479	0.764	0.456
0.200	0.385	0.734	0.719	0.729	0.406	0.710	0.373
0.500	0.292	0.681	0.646	0.668	0.318	0.633	0.266
1.000	0.229	0.657	0.590	0.630	0.261	0.573	0.191
2.000	0.182	0.668	0.537	0.608	0.231		0.133
5.000		0.874		0.648			

m/mol kg^{-1}	NaI	$NaNO_3$	NaOH	Na_2SO_3	Na_2SO_4	Na_2WO_4	$NiBr_2$
0.001	0.965	0.965	0.965	0.887	0.886	0.886	0.889
0.002	0.952	0.951	0.952	0.847	0.846	0.846	0.853
0.005	0.928	0.926	0.927	0.779	0.777	0.777	0.791
0.010	0.904	0.900	0.902	0.716	0.712	0.712	0.735
0.020	0.874	0.866	0.870	0.644	0.637	0.638	0.675
0.050	0.827	0.810	0.819	0.540	0.529	0.534	0.596
0.100	0.789	0.759	0.775	0.462	0.446	0.457	0.546
0.200	0.753	0.701	0.731	0.386	0.366	0.388	0.514
0.500	0.722	0.617	0.685	0.296	0.268	0.320	0.535
1.000	0.734	0.550	0.674	0.237	0.204	0.291	0.692
2.000	0.823	0.480	0.714	0.196	0.155	0.291	1.476
5.000	1.402	0.388	1.076				
10.000	4.011	0.329	3.258				
15.000			9.796				
20.000			19.410				

m/mol kg^{-1}	$NiCl_2$	$Ni(ClO_4)_2$	$Ni(NO_3)_2$	$Pb(ClO_4)_2$	$Pb(NO_3)_2$	RbBr	RbCl
0.001	0.889	0.891	0.889	0.889	0.882	0.965	0.965
0.002	0.852	0.855	0.851	0.851	0.840	0.951	0.951
0.005	0.789	0.797	0.787	0.787	0.764	0.926	0.926
0.010	0.732	0.745	0.730	0.729	0.690	0.900	0.900
0.020	0.669	0.690	0.666	0.666	0.604	0.866	0.867
0.050	0.584	0.621	0.581	0.580	0.476	0.811	0.811
0.100	0.527	0.582	0.524	0.522	0.379	0.760	0.761
0.200	0.482	0.567	0.481	0.476	0.291	0.705	0.707
0.500	0.465	0.639	0.467	0.458	0.195	0.630	0.633
1.000	0.538	0.946	0.528	0.516	0.136	0.578	0.583
2.000	0.915	2.812	0.797	0.799		0.535	0.546
5.000	4.785			4.043		0.514	0.544
10.000				33.8			

m/mol kg^{-1}	RbF	RbI	$RbNO_3$	Rb_2SO_4	$SrBr_2$	$SrCl_2$	SrI_2
0.001	0.965	0.965	0.964	0.886	0.889	0.888	0.890
0.002	0.952	0.951	0.950	0.845	0.852	0.850	0.854
0.005	0.927	0.926	0.924	0.776	0.790	0.785	0.793
0.010	0.902	0.900	0.896	0.710	0.734	0.725	0.740
0.020	0.871	0.866	0.859	0.635	0.673	0.659	0.681

m/mol kg^{-1}	RbF	RbI	$RbNO_3$	Rb_2SO_4	$SrBr_2$	$SrCl_2$	SrI_2
0.050	0.821	0.810	0.795	0.526	0.591	0.569	0.606
0.100	0.780	0.759	0.733	0.443	0.535	0.506	0.557
0.200	0.739	0.703	0.657	0.365	0.492	0.455	0.526
0.500	0.701	0.627	0.536	0.274	0.476	0.421	0.542
1.000	0.697	0.574	0.430	0.217	0.545	0.451	0.686
2.000	0.724	0.532	0.320		0.921	0.650	
5.000		0.517					

m/mol kg^{-1}	UO_2Cl_2	$UO_2(NO_3)_2$	$ZnBr_2$	$ZnCl_2$	ZnI_2
0.001	0.888	0.888	0.890	0.887	0.893
0.002	0.851	0.849	0.854	0.847	0.859
0.005	0.787	0.784	0.794	0.781	0.804
0.010	0.729	0.726	0.741	0.719	0.757
0.020	0.666	0.663	0.683	0.652	0.708
0.050	0.583	0.583	0.606	0.561	0.644
0.100	0.529	0.535	0.553	0.499	0.601
0.200	0.493	0.509	0.515	0.447	0.574
0.500	0.501	0.532	0.516	0.384	0.635
1.000	0.601	0.673	0.558	0.330	0.836
2.000	0.948	1.223	0.578	0.283	1.062
5.000		3.020	0.788	0.342	1.546
10.000			2.317	0.876	4.698
15.000			5.381	1.914	
20.000			7.965	2.968	

* The anion is $H_2PO_4^-$.

** The anion is HPO_4^{-2}.

ENTHALPY OF DILUTION OF ACIDS

The quantity given in this table is $-\Delta_{dil}H$, the negative of the enthalpy (heat) of dilution to infinite dilution for aqueous solutions of several common acids; i.e., the negative of the enthalpy change when a solution of molality *m* at a temperature of 25°C is diluted with an infinite amount of water. The tabulated numbers thus represent the heat produced (or, if the value is negative, the heat absorbed) when the acid is diluted. The initial molality *m* is given in the first column. The second column gives the dilution ratio, which is the number of moles of water that must be added to one mole of the acid to produce a solution of the molality in the first column.

REFERENCE

Parker, V. B., *Thermal Properties of Aqueous Uni-Univalent Electrolytes*, Natl. Stand. Ref. Data Ser. - Natl. Bur. Stand. (U.S.) 2, U.S. Government Printing Office, 1965.

$-\Delta_{dil}H$ in kJ/mol at 25°C

m	Dil. ratio	HF	HCl	$HClO_4$	HBr	HI	HNO_3	CH_2O_2	$C_2H_4O_2$
55.506	1.0		45.61		48.83		19.73	0.046	2.167
20	2.775	14.88	19.87	13.81	19.92	21.71	9.498	0.038	2.075
15	3.700	14.34	15.40	7.920	14.29	14.02	6.883	0.109	1.962
10	5.551	13.87	10.24	2.013	8.694	7.615	3.933	0.205	1.824
9	6.167	13.81	9.213	1.280	7.719	6.569	3.368	0.230	1.782
8	6.938	13.77	8.201	0.611	6.786	5.607	2.791	0.255	1.724
7	7.929	13.73	7.217	0.046	5.925	4.728	2.251	0.272	1.648
6	9.251	13.69	6.268	-0.351	5.004	3.975	1.749	0.280	1.540
5.5506	10	13.66	5.841	-0.490	4.590	3.577	1.540	0.285	1.477
5	11.10	13.62	5.318	-0.628	4.113	3.197	1.310	0.289	1.393
4.5	12.33	13.58	4.899	-0.732	3.711	2.828	1.109	0.289	1.310
4	13.88	13.53	4.402	-0.787	3.330	2.460	0.958	0.289	1.218
3.5	15.86	13.47	3.958	-0.820	2.966	2.105	0.791	0.289	1.121
3	18.50	13.45	3.506	-0.782	2.611	1.787	0.665	0.289	1.025
2.5	22.20	13.43	3.063	-0.724	2.301	1.527	0.582	0.285	0.912
2	27.75	13.40	2.623	-0.623	1.996	1.318	0.527	0.276	0.803
1.5	37.00	13.36	2.167	-0.431	1.665	1.125	0.506	0.259	0.678
1	55.51	13.30	1.695	-0.201	1.314	0.933	0.506	0.226	0.544
0.5551	100	13.22	1.234	0.050	0.983	0.736	0.502	0.184	0.423
0.5	111.0	13.20	1.172	0.075	0.941	0.711	0.498	0.176	0.406
0.2	277.5	13.09	0.761	0.247	0.649	0.536	0.439	0.146	0.331
0.1	555.1	12.80	0.556	0.272	0.498	0.439	0.372	0.134	0.289
0.0925	600	12.79	0.540	0.272	0.481	0.427	0.368	0.134	0.285
0.0793	700	12.70	0.502	0.272	0.452	0.402	0.351	0.134	0.285
0.0694	800	12.61	0.473	0.268	0.427	0.385	0.339	0.130	0.280
0.0617	900	12.50	0.448	0.264	0.406	0.368	0.326	0.126	0.276
0.05551	1000	12.42	0.427	0.259	0.385	0.351	0.318	0.121	0.272
0.05	1110	12.24	0.406	0.259	0.372	0.339	0.305	0.121	0.272
0.02775	2000	11.29	0.310	0.226	0.285	0.264	0.247	0.117	0.264
0.01850	3000	10.66	0.251	0.197	0.234	0.218	0.213	0.117	0.259
0.01388	4000	10.25	0.226	0.180	0.205	0.192	0.192	0.113	0.259
0.01110	5000	9.874	0.197	0.167	0.184	0.172	0.176	0.109	0.255
0.00555	10000	8.912	0.142	0.126	0.130	0.121	0.130	0.105	0.243
0.00278	20000	7.531	0.105	0.092	0.092	0.084	0.096	0.096	0.230
0.00111	50000	5.439	0.067	0.059	0.054	0.050	0.063	0.084	0.222
0.000555	100000	3.766	0.042	0.042	0.038	0.038	0.046	0.054	0.209
0.000111	500000	1.255	0.021	0.021	0.021	0.021	0.021	0.038	0.167
0	∞	0	0	0	0	0	0	0	0

ENTHALPY OF SOLUTION OF ELECTROLYTES

This table gives the molar enthalpy (heat) of solution at infinite dilution for some common uni-univalent electrolytes. This is the enthalpy change when 1 mol of solute in its standard state is dissolved in an infinite amount of water. Values are given in kilojoules per mole at 25°C.

REFERENCE

Parker, V. B., *Thermal Properties of Uni-Univalent Electrolytes*, Natl. Stand. Ref. Data Series — Natl. Bur. Stand.(U.S.), No.2, 1965.

Solute	State	$\Delta_{sol}H°$ kJ/mol
HF	g	–61.50
HCl	g	–74.84
$HClO_4$	l	–88.76
$HClO_4 \cdot H_2O$	c	–32.95
HBr	g	–85.14
HI	g	–81.67
HIO_3	c	8.79
HNO_3	l	–33.28
HCOOH	l	–0.86
CH_3COOH	l	–1.51
NH_3	g	–30.50
NH_4Cl	c	14.78
NH_4ClO_4	c	33.47
NH_4Br	c	16.78
NH_4I	c	13.72
NH_4IO_3	c	31.80
NH_4NO_2	c	19.25
NH_4NO_3	c	25.69
$NH_4C_2H_3O_2$	c	–2.38
NH_4CN	c	17.57
NH_4CNS	c	22.59
CH_3NH_3Cl	c	5.77
$(CH_3)_3NHCl$	c	1.46
$N(CH_3)_4Cl$	c	4.08
$N(CH_3)_4Br$	c	24.27
$N(CH_3)_4I$	c	42.07
$AgClO_4$	c	7.36
$AgNO_2$	c	36.94
$AgNO_3$	c	22.59
LiOH	c	–23.56
$LiOH \cdot H_2O$	c	–6.69
LiF	c	4.73
LiCl	c	–37.03
$LiCl \cdot H_2O$	c	–19.08
$LiClO_4$	c	–26.55
$LiClO_4 \cdot 3H_2O$	c	32.61
LiBr	c	–48.83
$LiBr \cdot H_2O$	c	–23.26
$LiBr \cdot 2H_2O$	c	–9.41
$LiBrO_3$	c	1.42
LiI	c	–63.30
$LiI \cdot H_2O$	c	–29.66
$LiI \cdot 2H_2O$	c	–14.77
$LiI \cdot 3H_2O$	c	0.59
$LiNO_2$	c	–11.00
$LiNO_2 \cdot H_2O$	c	7.03
$LiNO_3$	c	–2.51
NaOH	c	–44.51
$NaOH \cdot H_2O$	c	–21.41
NaF	c	0.91
NaCl	c	3.88
$NaClO_2$	c	0.33
$NaClO_2 \cdot 3H_2O$	c	28.58
$NaClO_3$	c	21.72
$NaClO_4$	c	13.88
$NaClO_4 \cdot H_2O$	c	22.51
NaBr	c	–0.60
$NaBr \cdot 2H_2O$	c	18.64
$NaBrO_3$	c	26.90
NaI	c	–7.53
$NaI \cdot 2H_2O$	c	16.13
$NaIO_3$	c	20.29
$NaNO_2$	c	13.89
$NaNO_3$	c	20.50
$NaC_2H_3O_2$	c	–17.32
$NaC_2H_3O_2 \cdot 3H_2O$	c	19.66
NaCN	c	1.21
$NaCN \cdot 0.5H_2O$	c	3.31
$NaCN \cdot 2H_2O$	c	18.58
NaCNO	c	19.20
NaCNS	c	6.83
KOH	c	–57.61
$KOH \cdot H_2O$	c	–14.64
$KOH \cdot 1.5H_2O$	c	–10.46
KF	c	–17.73
$KF \cdot 2H_2O$	c	6.97
KCl	c	17.22
$KClO_3$	c	41.38
$KClO_4$	c	51.04
KBr	c	19.87
$KBrO_3$	c	41.13
KI	c	20.33
KIO_3	c	27.74
KNO_2	c	13.35
KNO_3	c	34.89
$KC_2H_3O_2$	c	–15.33
KCN	c	11.72
KCNO	c	20.25
KCNS	c	24.23
$KMnO_4$	c	43.56
RbOH	c	–62.34
$RbOH \cdot H_2O$	c	–17.99
$RbOH \cdot 2H_2O$	c	0.88
RbF	c	–26.11
$RbF \cdot H_2O$	c	–0.42
$RbF \cdot 1.5H_2O$	c	1.34
RbCl	c	17.28
$RbClO_3$	c	47.74
$RbClO_4$	c	56.74
RbBr	c	21.88
$RbBrO_3$	c	48.95
RbI	c	25.10
$RbNO_3$	c	36.48
CsOH	c	–71.55
$CsOH \cdot H_2O$	c	–20.50
CsF	c	–36.86
$CsF \cdot H_2O$	c	–10.46
$CsF \cdot 1.5H_2O$	c	–5.44
CsCl	c	17.78
$CsClO_4$	c	55.44
CsBr	c	25.98
$CsBrO_3$	c	50.46
CsI	c	33.35
$CsNO_3$	c	40.00

CHEMICAL KINETIC DATA FOR STRATOSPHERIC MODELING

The present compilation of kinetic data represents the 12th evaluation prepared by the NASA Panel for Data Evaluation. The Panel was established in 1977 by the NASA Upper Atmosphere Research Program Office for the purpose of providing a critical tabulation of the latest kinetic and photochemical data for use by modelers in computer simulations of stratospheric chemistry. The recommended rate data and cross sections are based on laboratory measurements. The major use of theoretical extrapolation of data is in connection with three-body reactions, in which the required pressure or temperature dependence is sometimes unavailable from laboratory measurements, and can be estimated by use of appropriate theoretical treatment. In the case of important rate constants for which no experimental data are available, the panel may provide estimates of rate constant parameters based on analogy to similar reactions for which data are available.

Rate constants are expressed in the form $k(T) = A \exp(-E/RT)$, where A is the pre-exponential factor, E the activation energy, R the gas constant, and T the absolute temperature. Uncertainties are expressed by the factor f, e.g., a value of 4.2×10^{-10} with $f = 2$ indicates that the true value is believed to lie between 2.1×10^{-10} and 8.4×10^{-10}. The value of f at other temperatures may be calculated from $f(298)$, given in the last column, by:

$$f(T) = f(298) \exp[(\Delta E/R)(1/T - 1/298)],$$

where $\Delta E/R$ is the uncertainty in E/R.

Table 1 covers rate constant data on second order reactions, grouped by class, while Table 2 covers association reactions. Relevant equilibrium constant data are given in Table 3. All concentrations are measured in molecules cm^{-3}. Notes on each reaction, as well as related photochemical data, may be found in the reference.

The assistance of Robert Hampson is gratefully acknowledged.

REFERENCE

DeMore, W. B., Sander, S. P., Golden, D. M., Hampson, R. F., Kurylo, M. J., Howard, C. J., Ravishankara, A. R., Kolb, C. E., and Molina, M. J., *Chemical Kinetics and Photochemical Data for use in Atmospheric Modeling. Evaluation Number 12*, Jet Propulsion Laboratory Publication 97-4, Pasadena CA, 1997.

The report is also available at the World Wide Web site < http://remus.jpl.nasa.gov/pub/jpl97>.

Table 1. Rate Constants for Second Order Reactions

Reaction	*A* cm^3 molecule^{-1} s^{-1}	*E/R* K	*k* (298 K) cm^3 molecule^{-1} s^{-1}	*f*(298)
O_x Reactions				
$O + O_3 \rightarrow O_2 + O_2$	8.0×10^{-12}	2060±250	8.0×10^{-15}	1.15
$O(^1D)$ Reactions				
$O(^1D) + O_2 \rightarrow O + O_2$	3.2×10^{-11}	-(70±100)	4.0×10^{-11}	1.2
$O(^1D) + O_3 \rightarrow O_2 + O_2$	1.2×10^{-10}	0±100	1.2×10^{-10}	1.3
$\rightarrow O_2 + O + O$	1.2×10^{-10}	0±100	1.2×10^{-10}	1.3
$O(^1D) + H_2 \rightarrow OH + H$	1.1×10^{-10}	0±100	1.1×10^{-10}	1.1
$O(^1D) + H_2O \rightarrow OH + OH$	2.2×10^{-10}	0±100	2.2×10^{-10}	1.2
$O(^1D) + N_2 \rightarrow O + N_2$	1.8×10^{-11}	-(110±100)	2.6×10^{-11}	1.2
$O(^1D) + N_2O \rightarrow N_2 + O_2$	4.9×10^{-11}	0±100	4.9×10^{-11}	1.3
$\rightarrow NO + NO$	6.7×10^{-11}	0±100	6.7×10^{-11}	1.3
$O(^1D) + NH_3 \rightarrow OH + NH_2$	2.5×10^{-10}	0±100	2.5×10^{-10}	1.3
$O(^1D) + CO_2 \rightarrow O + CO_2$	7.4×10^{-11}	-(120±100)	1.1×10^{-10}	1.2
$O(^1D) + CH_4 \rightarrow$ products	1.5×10^{-10}	0±100	1.5×10^{-10}	1.2
$O(^1D) + HCl \rightarrow$ products	1.5×10^{-10}	0±100	1.5×10^{-10}	1.2
$O(^1D) + HF \rightarrow OH + F$	1.4×10^{-10}	0±100	1.4×10^{-10}	2.0
$O(^1D) + HBr \rightarrow$ products	1.5×10^{-10}	0±100	1.5×10^{-10}	2.0
$O(^1D) + Cl_2 \rightarrow$ products	2.8×10^{-10}	0±100	2.8×10^{-10}	2.0
$O(^1D) + CCl_2O \rightarrow$ products	3.6×10^{-10}	0±100	3.6×10^{-10}	2.0
$O(^1D) + CClFO \rightarrow$ products	1.9×10^{-10}	0±100	1.9×10^{-10}	2.0
$O(^1D) + CF_2O \rightarrow$ products	7.4×10^{-11}	0±100	7.4×10^{-11}	2.0
$O(^1D) + CCl_4 \rightarrow$ products (CFC-10)	3.3×10^{-10}	0±100	3.3×10^{-10}	1.2
$O(^1D) + CH_3Br \rightarrow$ products	1.8×10^{-10}	0±100	1.8×10^{-10}	1.3
$O(^1D) + CH_2Br_2 \rightarrow$ products	2.7×10^{-10}	0±100	2.7×10^{-10}	1.3
$O(^1D) + CHBr_3 \rightarrow$ products	6.6×10^{-10}	0±100	6.6×10^{-10}	1.5
$O(^1D) + CH_3F \rightarrow$ products (HFC-41)	1.5×10^{-10}	0±100	1.5×10^{-10}	1.2

Table 1. Rate Constants for Second Order Reactions (continued)

Reaction	*A* cm^3 $molecule^{-1}$ s^{-1}	*E/R* K	*k* (298 K) cm^3 $molecule^{-1}$ s^{-1}	*f*(298)
$O(^1D) + CH_2F_2 \rightarrow$ products (HFC-32)	5.1×10^{-11}	0±100	5.1×10^{-11}	1.3
$O(^1D) + CHF_3 \rightarrow$ products (HFC-23)	9.1×10^{-12}	0±100	9.1×10^{-12}	1.2
$O(^1D) + CHCl_2F \rightarrow$ products (HCFC-21)	1.9×10^{-10}	0±100	1.9×10^{-10}	1.3
$O(^1D) + CHClF_2 \rightarrow$ products (HCFC-22)	1.0×10^{-10}	0±100	1.0×10^{-10}	1.2
$O(^1D) + CCl_3F \rightarrow$ products (CFC-11)	2.3×10^{-10}	0±100	2.3×10^{-10}	1.2
$O(^1D) + CCl_2F_2 \rightarrow$ products (CFC-12)	1.4×10^{-10}	0±100	1.4×10^{-10}	1.3
$O(^1D) + CClF_3 \rightarrow$ products (CFC-13)	8.7×10^{-11}	0±100	8.7×10^{-11}	1.3
$O(^1D) + CClBrF_2 \rightarrow$ products (Halon-1211)	1.5×10^{-10}	0±100	1.5×10^{-10}	1.3
$O(^1D) + CBr_2F_2 \rightarrow$ products (Halon-1202)	2.2×10^{-10}	0±100	2.2×10^{-10}	1.3
$O(^1D) + CBrF_3 \rightarrow$ products (Halon-1301)	1.0×10^{-10}	0±100	1.0×10^{-10}	1.3
$O(^1D) + CF_4 \rightarrow CF_4 + O$ (CFC-14)	-	-	2.0×10^{-14}	1.5
$O(^1D) + CH_3CH_2F \rightarrow$ products (HFC-161)	2.6×10^{-10}	0±100	2.6×10^{-10}	1.3
$O(^1D) + CH_3CHF_2 \rightarrow$ products (HFC-152a)	2.0×10^{-10}	0±100	2.0×10^{-10}	1.3
$O(^1D) + CH_3CCl_2F \rightarrow$ products (HCFC-141b)	2.6×10^{-10}	0±100	2.6×10^{-10}	1.3
$O(^1D) + CH_3CClF_2 \rightarrow$ products (HCFC-142b)	2.2×10^{-10}	0±100	2.2×10^{-10}	1.3
$O(^1D) + CH_3CF_3 \rightarrow$ products (HFC-143a)	1.0×10^{-10}	0±100	1.0×10^{-10}	3.0
$O(^1D) + CH_2ClCClF_2 \rightarrow$ products (HCFC-132b)	1.6×10^{-10}	0±100	1.6×10^{-10}	2.0
$O(^1D) + CH_2ClCF_3 \rightarrow$ products (HCFC-133a)	1.2×10^{-10}	0±100	1.2×10^{-10}	1.3
$O(^1D) + CH_2FCF_3 \rightarrow$ products (HFC-134a)	4.9×10^{-11}	0±100	4.9×10^{-11}	1.3
$O(^1D) + CHCl_2CF_3 \rightarrow$ products (HCFC-123)	2.0×10^{-10}	0±100	2.0×10^{-10}	1.3
$O(^1D) + CHClFCF_3 \rightarrow$ products (HCFC-124)	8.6×10^{-11}	0±100	8.6×10^{-11}	1.3
$O(^1D) + CHF_2CF_3 \rightarrow$ products (HFC-125)	1.2×10^{-10}	0±100	1.2×10^{-10}	2.0
$O(^1D) + CCl_3CF_3 \rightarrow$ products (CFC-113a)	2×10^{-10}	0±100	2×10^{-10}	2.0
$O(^1D) + CCl_2FCClF_2 \rightarrow$ products (CFC-113)	2×10^{-10}	0±100	2×10^{-10}	2.0
$O(^1D) + CCl_2FCF_3 \rightarrow$ products (CFC-114a)	1×10^{-10}	0±100	1×10^{-10}	2.0
$O(^1D) + CClF_2CClF_2 \rightarrow$ products (CFC-114)	1.3×10^{-10}	0±100	1.3×10^{-10}	1.3
$O(^1D) + CClF_2CF_3 \rightarrow$ products (CFC-115)	5×10^{-11}	0±100	5×10^{-11}	1.3
$O(^1D) + CBrF_2CBrF_2 \rightarrow$ products (Halon-2402)	1.6×10^{-10}	0±100	1.6×10^{-10}	1.3

Table 1. Rate Constants for Second Order Reactions (continued)

Reaction	*A* cm^3 molecule^{-1} s^{-1}	*E/R* K	*k* (298 K) cm^3 molecule^{-1} s^{-1}	*f*(298)
$O(^1D) + CF_3CF_3 \rightarrow O + CF_3CF_3$ (CFC-116)	-	-	$1.5x10^{-13}$	1.5
$O(^1D) + CHF_2CF_2CF_2CHF_2 \rightarrow$ products (HFC-338pcc)	$1.8x10^{-11}$	0±100	$1.8x10^{-11}$	1.5
$O(^1D) + c\text{-}C_4F_8 \rightarrow$ products	-	-	$8x10^{-13}$	1.3
$O(^1D) + CF_3CHFCHFCF_2CF_3 \rightarrow$ products (HFC-43-10mee)	$2.1x10^{-10}$	0±100	$2.1x10^{-10}$	4
$O(^1D) + C_5F_{12} \rightarrow$ products (CFC-41-12)	-	-	$3.9x10^{-13}$	2
$O(^1D) + C_6F_{14} \rightarrow$ products (CFC-51-14)	-	-	$1x10^{-12}$	2
$O(^1D) + 1,2\text{-}(CF_3)_2c\text{-}C_4F_6 \rightarrow$ products	-	-	$2.8x10^{-13}$	2
$O(^1D) + SF_6 \rightarrow$ products	-	-	$1.8x10^{-14}$	1.5
Singlet O_2 Reactions				
$O_2(^1\Delta) + O \rightarrow$ products	-	-	$<2x10^{-16}$	-
$O_2(^1\Delta) + O_2 \rightarrow$ products	$3.6x10^{-18}$	220±100	$1.7x10^{-18}$	1.2
$O_2(^1\Delta) + O_3 \rightarrow O + 2O_2$	$5.2x10^{-11}$	2840±500	$3.8x10^{-15}$	1.2
$O_2(^1\Delta) + H_2O \rightarrow$ products	-	-	$4.8x10^{-18}$	1.5
$O_2(^1\Delta) + N \rightarrow NO + O$	-	-	$<9x10^{-17}$	-
$O_2(^1\Delta) + N_2 \rightarrow$ products	-	-	$<10^{-20}$	-
$O_2(^1\Delta) + CO_2 \rightarrow$ products	-	-	$<2x10^{-20}$	-
$O_2(^1\Sigma) + O \rightarrow$ products	-	-	$8x10^{-14}$	5.0
$O_2(^1\Sigma) + O_2 \rightarrow$ products	-	-	$3.9x10^{-17}$	1.5
$O_2(^1\Sigma) + O_3 \rightarrow$ products	$2.2x10^{-11}$	0±200	$2.2x10^{-11}$	1.2
$O_2(^1\Sigma) + H_2O \rightarrow$ products	-	-	$5.4x10^{-12}$	1.3
$O_2(^1\Sigma) + N \rightarrow$ products	-	-	$<10^{-13}$	-
$O_2(^1\Sigma) + N_2 \rightarrow$ products	$2.1x10^{-15}$	0±200	$2.1x10^{-15}$	1.2
$O_2(^1\Sigma) + CO_2 \rightarrow$ products	$4.2x10^{-13}$	0±200	$4.2x10^{-13}$	1.2
HO_x Reactions				
$O + OH \rightarrow O_2 + H$	$2.2x10^{-11}$	-(120±100)	$3.3x10^{-11}$	1.2
$O + HO_2 \rightarrow OH + O_2$	$3.0x10^{-11}$	-(200±100)	$5.9x10^{-11}$	1.2
$O + H_2O_2 \rightarrow OH + HO_2$	$1.4x10^{-12}$	2000±1000	$1.7x10^{-15}$	2.0
$H + O_3 \rightarrow OH + O_2$	$1.4x10^{-10}$	470±200	$2.9x10^{-11}$	1.25
$H + HO_2 \rightarrow$ products	$8.1x10^{-11}$	0±100	$8.1x10^{-11}$	1.3
$OH + O_3 \rightarrow HO_2 + O_2$	$1.6x10^{-12}$	940±300	$6.8x10^{-14}$	1.3
$OH + H_2 \rightarrow H_2O + H$	$5.5x10^{-12}$	2000±100	$6.7x10^{-15}$	1.1
$OH + HD \rightarrow$ products	$5.0x10^{-12}$	2130±200	$4.0x10^{-15}$	1.2
$OH + OH \rightarrow H_2O + O$	$4.2x10^{-12}$	240±240	$1.9x10^{-12}$	1.4
$OH + HO_2 \rightarrow H_2O + O_2$	$4.8x10^{-11}$	-(250±200)	$1.1x10^{-10}$	1.3
$OH + H_2O_2 \rightarrow H_2O + HO_2$	$2.9x10^{-12}$	160±100	$1.7x10^{-12}$	1.2
$HO_2 + O_3 \rightarrow OH + 2O_2$	$1.1x10^{-14}$	500±	$2.0x10^{-15}$	1.3
$HO_2 + HO_2 \rightarrow H_2O_2 + O_2$	$2.3x10^{-13}$	-(600±200)	$1.7x10^{-12}$	1.3
$H_2O_2 + O_2$	$1.7x10^{-33}$[M]	-(1000±400)	$4.9x10^{-32}$[M]	1.3
NO_x Reactions				
$O + NO_2 \rightarrow NO + O_2$	$6.5x10^{-12}$	-(120±120)	$9.7x10^{-12}$	1.1
$O + NO_3 \rightarrow O_2 + NO_2$	$1.0x10^{-11}$	0±150	$1.0x10^{-11}$	1.5
$O + N_2O_5 \rightarrow$ products			$<3.0x10^{-16}$	
$O + HNO_3 \rightarrow OH + NO_3$			$<3.0x10^{-17}$	
$O + HO_2NO_2 \rightarrow$ products	$7.8x10^{-11}$	3400±750	$8.6x10^{-16}$	3.0

Table 1. Rate Constants for Second Order Reactions (continued)

Reaction	*A* cm^3 molecule^{-1} s^{-1}	*E/R* K	*k* (298 K) cm^3 molecule^{-1} s^{-1}	*f*(298)
$H + NO_2 \rightarrow OH + NO$	4.0x10^{-10}	340±300	1.3x10^{-10}	1.3
$OH + NO_3 \rightarrow$ products			2.2x10^{-11}	1.5
$OH + HONO \rightarrow H_2O + NO_2$	1.8x10^{-11}	390±	4.5x10^{-12}	1.5
$OH + HNO_3 \rightarrow H_2O + NO_3$	See reference	1.3		
$OH + HO_2NO_2 \rightarrow$ products	1.3x10^{-12}	-(380±)	4.6x10^{-12}	1.5
$OH + NH_3 \rightarrow H_2O + NH_2$	1.7x10^{-12}	710±200	1.6x10^{-13}	1.2
$HO_2 + NO \rightarrow NO_2 + OH$	3.5x10^{-12}	-(250±50)	8.1x10^{-12}	1.15
$HO_2 + NO_2 \rightarrow HONO + O_2$	See reference			
$HO_2 + NO_3 \rightarrow$ products			3.5x10^{-12}	1.5
$HO_2 + NH_2 \rightarrow$ products			3.4x10^{-11}	2.0
$N + O_2 \rightarrow NO + O$	1.5x10^{-11}	3600±400	8.5x10^{-17}	1.25
$N + O_3 \rightarrow NO + O_2$			<2.0x10^{-16}	
$N + NO \rightarrow N_2 + O$	2.1x10^{-11}	-(100±100)	3.0x10^{-11}	1.3
$N + NO_2 \rightarrow N_2O + O$	5.8x10^{-12}	-(220±100)	1.2x10^{-11}	1.5
$NO + O_3 \rightarrow NO_2 + O_2$	2.0x10^{-12}	1400±200	1.8x10^{-14}	1.1
$NO + NO_3 \rightarrow 2NO_2$	1.5x10^{-11}	-(170±100)	2.6x10^{-11}	1.3
$NO_2 + O_3 \rightarrow NO_3 + O_2$	1.2x10^{-13}	2450±150	3.2x10^{-17}	1.15
$NO_2 + NO_3 \rightarrow NO + NO_2 + O_2$	See reference			
$NO_3 + NO_3 \rightarrow 2NO_2 + O_2$	8.5x10^{-13}	2450±500	2.3x10^{-16}	1.5
$NH_2 + O_2 \rightarrow$ products			<6.0x10^{-21}	
$NH_2 + O_3 \rightarrow$ products	4.3x10^{-12}	930±500	1.9x10^{-13}	3.0
$NH_2 + NO \rightarrow$ products	4.0x10^{-12}	-(450±150)	1.8x10^{-11}	1.3
$NH_2 + NO_2 \rightarrow$ products	2.1x10^{-12}	-(650±250)	1.9x10^{-11}	3.0
$NH + NO \rightarrow$ products	4.9x10^{-11}	0±300	4.9x10^{-11}	1.5
$NH + NO_2 \rightarrow$ products	3.5x10^{-13}	-(1140±500)	1.6x10^{-11}	2.0
$O_3 + HNO_2 \rightarrow O_2 + HNO_3$			<5.0x10^{-19}	
$N_2O_5 + H_2O \rightarrow 2HNO_3$			<2.0x10^{-21}	
$N_2(A,v) + O_2 \rightarrow$ products			2.5x10^{-12}, v=0	1.5
$N_2(A,v) + O_3 \rightarrow$ products			4.1x10^{-11}, v=0	2.0
Reactions of Organic Compounds				
$O + CH_3 \rightarrow$ products	1.1x10^{-10}	0±250	1.1x10^{-10}	1.3
$O + HCN \rightarrow$ products	1.0x10^{-11}	4000±1000	1.5x10^{-17}	10
$O + C_2H_2 \rightarrow$ products	3.0x10^{-11}	1600±250	1.4x10^{-13}	1.3
$O + H_2CO \rightarrow$ products	3.4x10^{-11}	1600±250	1.6x10^{-13}	1.25
$O + CH_3CHO \rightarrow CH_3CO + OH$	1.8x10^{-11}	1100±200	4.5x10^{-13}	1.25
$O_3 + C_2H_2 \rightarrow$ products	1.0x10^{-14}	4100±500	1.0x10^{-20}	3
$O_3 + C_2H_4 \rightarrow$ products	1.2x10^{-14}	2630±100	1.7x10^{-18}	1.25
$O_3 + C_3H_6 \rightarrow$ products	6.5x10^{-15}	1900±200	1.1x10^{-17}	1.2
$OH + CO \rightarrow$ Products	1.5x10^{-13} x (1+0.6P_{atm})	0±300	1.5x10^{-13} x (1+0.6P_{atm})	1.3
$OH + CH_4 \rightarrow CH_3 + H_2O$	2.45x10^{-12}	1775±100	6.3x10^{-15}	1.1
$OH + {}^{13}CH_4 \rightarrow {}^{13}CH_3 + H_2O$	See reference			
$OH + CH_3D \rightarrow$ products	3.5x10^{-12}	1950 ± 200	5.0x10^{-15}	1.15
$OH + H_2CO \rightarrow H_2O + HCO$	1.0x10^{-11}	0±200	1.0x10^{-11}	1.25
$OH + CH_3OH \rightarrow$ products	6.7x10^{-12}	600±300	8.9x10^{-13}	1.2
$OH + CH_3OOH \rightarrow$ Products	3.8x10^{-12}	-(200±200)	7.4x10^{-12}	1.5
$OH + HC(O)OH \rightarrow$ products	4.0x10^{-13}	0±200	4.0x10^{-13}	1.3
$OH + HCN \rightarrow$ products	1.2x10^{-13}	400±150	3.1x10^{-14}	3
$OH + C_2H_6 \rightarrow H_2O + C_2H_5$	8.7 x 10^{-12}	1070±100	2.4x10^{-13}	1.1
$OH + C_3H_8 \rightarrow H_2O + C_3H_7$	1.0 x 10^{-11}	660±100	1.1x10^{-12}	1.2
$OH + CH_3CHO \rightarrow CH_3CO + H_2O$	5.6x10^{-12}	-(270±200)	1.4x10^{-11}	1.2
$OH + C_2H_5OH \rightarrow$ products	7.0x10^{-12}	235±100	3.2x10^{-12}	1.3
$OH + CH_3C(O)OH \rightarrow$ products	4.0x10^{-13}	-(200±400)	8.0x10^{-13}	1.3

Table 1. Rate Constants for Second Order Reactions (continued)

Reaction	*A* cm^3 molecule^{-1} s^{-1}	*E/R* K	*k* (298 K) cm^3 molecule^{-1} s^{-1}	*f*(298)
$OH + CH_3C(O)CH_3 \rightarrow CH_3C(O)CH_2 + H_2O$	2.2 x 10^{-12}	685±100	2.2x10^{-13}	1.15
$OH + CH_3CN \rightarrow$ products	7.8x10^{-13}	1050±200	2.3x10^{-14}	1.5
$OH + CH_3ONO_2 \rightarrow$ products	5.0x10^{-13}	890±500	2.4x10^{-14}	3
$OH + CH_3C(O)O_2NO_2$ (PAN) $\rightarrow$ products			<4 x 10^{-14}	
$OH + C_2H_5ONO_2 \rightarrow$ products	8.2x10^{-13}	450±300	1.8x10^{-13}	3
$HO_2 + CH_2O \rightarrow$ adduct	6.7x10^{-15}	-(600±600)	5.0x10^{-14}	5
$HO_2 + CH_3O_2 \rightarrow CH_3OOH + O_2$	3.8x10^{-13}	-(800±400)	5.6x10^{-12}	2
$HO_2 + C_2H_5O_2 \rightarrow C_2H_5OOH + O_2$	7.5x10^{-13}	-(700±250)	8.0x10^{-12}	1.5
$HO_2 + CH_3C(O)O_2 \rightarrow$ products	4.5x10^{-13}	-(1000±600)	1.3x10^{-11}	2
$NO_3 + CO \rightarrow$ products			<4.0x10^{-19}	
$NO_3 + CH_2O \rightarrow$ products			5.8x10^{-16}	1.3
$NO_3 + CH_3CHO \rightarrow$ products	1.4x10^{-12}	1900±300	2.4x10^{-15}	1.3
$CH_3 + O_2 \rightarrow$ products			<3.0x10^{-16}	
$CH_3 + O_3 \rightarrow$ products	5.4x10^{-12}	220±150	2.6x10^{-12}	2
$HCO + O_2 \rightarrow CO + HO_2$	3.5x10^{-12}	-(140±140)	5.5x10^{-12}	1.3
$CH_2OH + O_2 \rightarrow CH_2O + HO_2$	9.1x10^{-12}	0±200	9.1x10^{-12}	1.3
$CH_3O + O_2 \rightarrow CH_2O + HO_2$	3.9x10^{-14}	900±300	1.9x10^{-15}	1.5
$CH_3O + NO \rightarrow CH_2O + HNO$	See reference			
$CH_3O + NO_2 \rightarrow CH_2O + HONO$	1.1 x 10^{-11}	1200±600	2.0 x 10^{-13}	5
$CH_3O_2 + O_3 \rightarrow$ products			<3.0x10^{-17}	
$CH_3O_2 + CH_3O_2 \rightarrow$ products	2.5x10^{-13}	-(190±190)	4.7x10^{-13}	1.5
$CH_3O_2 + NO \rightarrow CH_3O + NO_2$	3.0x10^{-12}	-(280±60)	7.7x10^{-12}	1.15
$CH_3O_2 + CH_3C(O)O_2 \rightarrow$ products	1.3x10^{-12}	-(640±200)	1.1x10^{-11}	1.5
$C_2H_5 + O_2 \rightarrow C_2H_4 + HO_2$			<2.0x10^{-14}	
$C_2H_5O + O_2 \rightarrow CH_3CHO + HO_2$	6.3 x 10^{-14}	550±200	1.0x10^{-14}	1.5
$C_2H_5O_2 + C_2H_5O_2 \rightarrow$ products	6.8x10^{-14}	0±300	6.8x10^{-14}	2
$C_2H_5O_2 + NO \rightarrow$ products	2.6x10^{-12}	-(365±150)	8.7x10^{-12}	1.2
$CH_3C(O)O_2 + CH_3C(O)O_2 \rightarrow$ products	2.9x10^{-12}	-(500±150)	1.5x10^{-11}	1.5
$CH_3C(O)O_2 + NO \rightarrow$ products	5.3x10^{-12}	-(360±150)	1.8x10^{-11}	1.4
FO_x Reactions				
$O + FO \rightarrow F + O_2$	2.7x10^{-11}	0±250	2.7x10^{-11}	3.0
$O + FO_2 \rightarrow FO + O_2$	5.0x10^{-11}	0±250	5.0x10^{-11}	5.0
$OH + CH_3F \rightarrow CH_2F + H_2O$ (HFC-41)	3.0x10^{-12}	1500±300	2.0x10^{-14}	1.1
$OH + CH_2F_2 \rightarrow CHF_2 + H_2O$ (HFC-32)	1.9x10^{-12}	1550±200	1.0x10^{-14}	1.2
$OH + CHF_3 \rightarrow CF_3 + H_2O$ (HFC-23)	1.0x10^{-12}	2440±200	2.8x10^{-16}	1.3
$OH + CF_3OH \rightarrow CF_3O + H_2O$			<2x10^{-17}	
$OH + CH_3CH_2F \rightarrow$ products (HFC-161)	7.0x10^{-12}	1100±300	1.7x10^{-13}	1.4
$OH + CH_3CHF_2 \rightarrow$ products (HFC-152a)	2.4x10^{-12}	1260±200	3.5x10^{-14}	1.2
$OH + CH_2FCH_2F \rightarrow CHFCH_2F + H_2O$ (HFC-152)	1.7x10^{-11}	1500±500	1.1x10^{-13}	2.0
$OH + CH_3CF_3 \rightarrow CH_2CF_3 + H_2O$ (HFC-143a)	1.8x10^{-12}	2170±150	1.2x10^{-15}	1.1
$OH + CH_2FCHF_2 \rightarrow$ products (HFC-143)	4.0x10^{-12}	1650±300	1.6x10^{-14}	1.5
$OH + CH_2FCF_3 \rightarrow CHFCF_3 + H_2O$ (HFC-134a)	1.5x10^{-12}	1750±200	4.2x10^{-15}	1.1

Table 1. Rate Constants for Second Order Reactions (continued)

Reaction	*A* cm^3 molecule^{-1} s^{-1}	*E/R* K	*k* (298 K) cm^3 molecule^{-1} s^{-1}	*f*(298)
$OH + CHF_2CHF_2 \rightarrow CF_2CHF_2 + H_2O$ (HFC-134)	$1.6x10^{-12}$	1680±300	$5.7x10^{-15}$	2.0
$OH + CHF_2CF_3 \rightarrow CF_2CF_3 + H_2O$ (HFC-125)	$5.6x10^{-13}$	1700±300	$1.9x10^{-15}$	1.3
$OH + CH_3OCHF_2 \rightarrow$ products (HFOC-152a)	$6.0x10^{-12}$	1530±150	$3.5x10^{-14}$	1.2
$OH + CF_3OCH_3 \rightarrow CF_3OCH_2 + H_2O$ (HFOC-143a)	$1.5x10^{-12}$	1450±150	$1.2x10^{-14}$	1.1
$OH + CF_2HOCF_2H \rightarrow CF_2OCF_2H + H_2O$ (HFOC-134)	$1.9x10^{-12}$	2000±150	$2.3x10^{-15}$	1.2
$OH + CF_3OCHF_2 \rightarrow CF_3OCF_2 + H_2O$ (HFOC-125)	$4.7x10^{-13}$	2100±300	$4.1x10^{-16}$	1.2
$OH + CF_3CH_2CH_3 \rightarrow$ products (HFC-263fb)	-	-	$4.2x10^{-14}$	1.5
$OH + CH_2FCF_2CHF_2 \rightarrow$ products (HFC-245ca)	$2.4x10^{-12}$	1660±150	$9.1x10^{-15}$	1.3
$OH + CHF_2CHFCHF_2 \rightarrow$ products (HFC-245ea)	-	-	$1.6x10^{-14}$	2.0
$OH + CF_3CHFCH_2F \rightarrow$ products (HFC-245eb)	-	-	$1.5x10^{-14}$	2.0
$OH + CHF_2CH_2CF_3 \rightarrow$ products (HFC-245fa)	$6.1x10^{-13}$	1330±150	$7.0x10^{-15}$	1.2
$OH + CF_3CF_2CH_2F \rightarrow CF_3CF_2CHF + H_2O$ (HFC-236cb)	$1.5x10^{-12}$	1750±500	$4.2x10^{-15}$	2.0
$OH + CF_3CHFCHF_2 \rightarrow$ products (HFC-236ea)	$1.1x10^{-12}$	1590±150	$5.3x10^{-15}$	1.1
$OH + CF_3CH_2CF_3 \rightarrow CF_3CHCF_3 + H_2O$ (HFC-236fa)	$1.3x10^{-12}$	2480±150	$3.2x10^{-16}$	1.1
$OH + CF_3CHFCF_3 \rightarrow CF_3CFCF_3 + H_2O$ (HFC-227ea)	$5.0x10^{-13}$	1700±300	$1.7x10^{-15}$	1.1
$OH + CHF_2OCH_2CF_3 \rightarrow$ products (HFOC-245fa)	$2.6x10^{-12}$	1610±150	$1.2x10^{-14}$	2.0
$OH + CF_3CH_2CF_2CH_3 \rightarrow$ products (HFC-365mfc)	$2.0x10^{-12}$	1750±200	$5.7x10^{-15}$	1.3
$OH + CF_3CH_2CH_2CF_3 \rightarrow$ products (HFC-356mff)	$3.0x10^{-12}$	1800±300	$7.1x10^{-15}$	1.3
$OH + CF_3CF_2CH_2CH_2F \rightarrow$ products (HFC-356mcf)	$1.7x10^{-12}$	1110±200	$4.2x10^{-14}$	2.0
$OH + CHF_2CF_2CF_2CF_2H \rightarrow$ products (HFC-338pcc)	$7.8x10^{-13}$	1530±200	$4.6x10^{-15}$	1.5
$OH + CF_3CH_2CF_2CH_2CF_3 \rightarrow$ products (HFC-458mfcf)	$1.2x10^{-12}$	1830±200	$2.6x10^{-15}$	2.0
$OH + CF_3CHFCHFCF_2CF_3 \rightarrow$ products (HFC-43-10mee)	$5.2x10^{-13}$	1500±300	$3.4x10^{-15}$	1.3
$OH + CF_3CF_2CH_2CH_2CF_2CF_3 \rightarrow$ products (HFC-55-10-mcff)	-	-	$8.3x10^{-15}$	1.5
$F + O_3 \rightarrow FO + O_2$	$2.2x10^{-11}$	230±200	$1.0x10^{-11}$	1.5
$F + H_2 \rightarrow HF + H$	$1.4x10^{-10}$	500±200	$2.6x10^{-11}$	1.2
$F + H_2O \rightarrow HF + OH$	$1.4x10^{-11}$	0±200	$1.4x10^{-11}$	1.3
$F + HNO_3 \rightarrow HF + NO_3$	$6.0x10^{-12}$	-(400±200)	$2.3x10^{-11}$	1.3
$F + CH_4 \rightarrow HF + CH_3$	$1.6x10^{-10}$	260±200	$6.7x10^{-11}$	1.4
$FO + O_3 \rightarrow$ products			$<1 x 10^{-14}$	
$FO + NO \rightarrow NO_2 + F$	$8.2x10^{-12}$	-(300±200)	$2.2x10^{-11}$	1.5
$FO + FO \rightarrow 2 F + O_2$	$1.0x10^{-11}$	0±250	$1.0x10^{-11}$	1.5
$FO_2 + O_3 \rightarrow$ products			$<3.4x10^{-16}$	
$FO_2 + NO \rightarrow FNO + O_2$	$7.5x10^{-12}$	690±400	$7.5x10^{-13}$	2.0

Table 1. Rate Constants for Second Order Reactions (continued)

Reaction	*A* cm^3 molecule^{-1} s^{-1}	*E/R* K	*k* (298 K) cm^3 molecule^{-1} s^{-1}	*f*(298)
$FO_2 + NO_2 \rightarrow$ products	3.8x10^{-11}	2040±500	4.0x10^{-14}	2.0
$FO_2 + CO \rightarrow$ products			<5.1x10^{-16}	
$FO_2 + CH_4 \rightarrow$ products			<2x10^{-16}	
$CF_3O + O_2 \rightarrow FO_2 + CF_2O$	<3 x 10^{-11}	5000	<1.5 x 10^{-18}	
$CF_3O + O_3 \rightarrow CF_3O_2 + O_2$	2 x 10^{-12}	1400±600	1.8 x 10^{-14}	1.3
$CF_3O + H_2O \rightarrow OH + CF_3OH$	3 x 10^{-12}	>3600	<2 x 10^{-17}	
$CF_3O + NO \rightarrow CF_2O + FNO$	3.7 x 10^{-11}	-(110±70)	5.4 x 10^{-11}	1.2
$CF_3O + NO_2 \rightarrow$ products	See reference			
$CF_3O + CO \rightarrow$ products			<2 x 10^{-15}	
$CF_3O + CH_4 \rightarrow CH_3 + CF_3OH$	2.6 x 10^{-12}	1420±200	2.2 x 10^{-14}	1.1
$CF_3O + C_2H_6 \rightarrow C_2H_5 + CF_3OH$	4.9 x 10^{-12}	400±100	1.3 x 10^{-12}	1.2
$CF_3O_2 + O_3 \rightarrow CF_3O + 2O_2$			<3 x 10^{-15}	
$CF_3O_2 + CO \rightarrow CF_3O + CO_2$			<5 x 10^{-16}	
$CF_3O_2 + NO \rightarrow CF_3O + NO_2$	5.4 x 10^{-12}	-(320±150)	1.6 x 10^{-11}	1.1
ClO_x Reactions				
$O + ClO \rightarrow Cl + O_2$	3.0x10^{-11}	-(70±70)	3.8x10^{-11}	1.2
$O + OClO \rightarrow ClO + O_2$	2.4x10^{-12}	960±300	1.0x10^{-13}	2.0
$O + Cl_2O \rightarrow ClO + ClO$	2.7x10^{-11}	530±150	4.5x10^{-12}	1.3
$O + HCl \rightarrow OH + Cl$	1.0x10^{-11}	3300±350	1.5x10^{-16}	2.0
$O + HOCl \rightarrow OH + ClO$	1.7x10^{-13}	0±300	1.7x10^{-13}	3.0
$O + ClONO_2 \rightarrow$ products	2.9x10^{-12}	800±200	2.0x10^{-13}	1.5
$O_3 + OClO \rightarrow$ products	2.1x10^{-12}	4700±1000	3.0x10^{-19}	2.5
$O_3 + Cl_2O_2 \rightarrow$ products	-	-	<1.0x10^{-19}	-
$OH + Cl_2 \rightarrow HOCl + Cl$	1.4x10^{-12}	900±400	6.7x10^{-14}	1.2
$OH + ClO \rightarrow$ products	1.1x10^{-11}	-(120±150)	1.7x10^{-11}	1.5
$OH + OClO \rightarrow HOCl + O_2$	4.5x10^{-13}	-(800±200)	6.8x10^{-12}	2.0
$OH + HCl \rightarrow H_2O + Cl$	2.6x10^{-12}	350±100	8.0x10^{-13}	1.2
$OH + HOCl \rightarrow H_2O + ClO$	3.0x10^{-12}	500±500	5.0x10^{-13}	3.0
$OH + ClNO_2 \rightarrow HOCl + NO_2$	2.4x10^{-12}	1250±300	3.6x10^{-14}	2.0
$OH + ClONO_2 \rightarrow$ products	1.2x10^{-12}	330±200	3.9x10^{-13}	1.5
$OH + CH_3Cl \rightarrow CH_2Cl + H_2O$	4.0x10^{-12}	1400±250	3.6x10^{-14}	1.2
$OH + CH_2Cl_2 \rightarrow CHCl_2 + H_2O$	3.8x10^{-12}	1050±150	1.1x10^{-13}	1.4
$OH + CHCl_3 \rightarrow CCl_3 + H_2O$	2.0x10^{-12}	900±150	1.0x10^{-13}	1.2
$OH + CCl_4 \rightarrow$ products	~1.0x10^{-12}	>2300	<5.0x10^{-16}	-
$OH + CFCl_3 \rightarrow$ products (CFC-11)	~1.0x10^{-12}	>3700	<5.0x10^{-18}	-
$OH + CF_2Cl_2 \rightarrow$ products (CFC-12)	~1.0x10^{-12}	>3600	<6.0x10^{-18}	-
$OH + CH_2ClF \rightarrow CHClF + H_2O$ (HCFC-31)	2.8x10^{-12}	1270±200	3.9x10^{-14}	1.2
$OH + CHFCl_2 \rightarrow CFCl_2 + H_2O$ (HCFC-21)	1.7x10^{-12}	1250±150	2.6x10^{-14}	1.2
$OH + CHF_2Cl \rightarrow CF_2Cl + H_2O$ (HCFC-22)	1.0x10^{-12}	1600±150	4.7x10^{-15}	1.1
$OH + CH_3OCl \rightarrow$ products	2.4x10^{-12}	360±200	7.2x10^{-13}	3.0
$OH + CH_3CCl_3 \rightarrow CH_2CCl_3 + H_2O$ (HCC-140)	1.8x10^{-12}	1550±150	1.0x10^{-14}	1.1
$OH + C_2HCl_3 \rightarrow$ products	4.9x10^{-13}	-(450±200)	2.2x10^{-12}	1.25
$OH + C_2Cl_4 \rightarrow$ products	9.4x10^{-12}	1200±200	1.7x10^{-13}	1.25
$OH + CCl_3CHO \rightarrow H_2O + CCl_3CO$	8.2x10^{-12}	600±300	1.1x10^{-12}	1.5
$OH + CH_3CFCl_2 \rightarrow CH_2CFCl_2 + H_2O$ (HCFC-141b)	1.7x10^{-12}	1700±150	5.7x10^{-15}	1.2
$OH + CH_3CF_2Cl \rightarrow CH_2CF_2Cl + H_2O$ (HCFC-142b)	1.3x10^{-12}	1800±150	3.1x10^{-15}	1.2

Table 1. Rate Constants for Second Order Reactions (continued)

Reaction	*A* cm³ molecule⁻¹ s⁻¹	*E/R* K	*k* (298 K) cm³ molecule⁻¹ s⁻¹	*f*(298)
$OH + CH_2ClCF_2Cl \rightarrow CHClCF_2Cl + H_2O$ (HCFC-132b)	$3.6x10^{-12}$	1600±400	$1.7x10^{-14}$	2.0
$OH + CHCl_2CF_2Cl \rightarrow CCl_2CF_2Cl + H_2O$ (HCFC-122)	$1.0x10^{-12}$	900±150	$4.9x10^{-14}$	1.2
$OH + CHFClCFCl_2 \rightarrow CFClCFCl_2 + H_2O$ (HCFC-122a)	$1.0x10^{-12}$	1250±150	$1.5x10^{-14}$	1.1
$OH + CH_2ClCF_3 \rightarrow CHClCF_3 + H_2O$ (HCFC-133a)	$5.2x10^{-13}$	1100±300	$1.3x10^{-14}$	1.3
$OH + CHCl_2CF_3 \rightarrow CCl_2CF_3 + H_2O$ (HCFC-123)	$7.0x10^{-13}$	900±150	$3.4x10^{-14}$	1.2
$OH + CHFClCF_2Cl \rightarrow CFClCF_2Cl$ + H2O (HCFC-123a)	$9.2x10^{-13}$	1280±150	$1.3x10^{-14}$	1.2
$OH + CHFClCF_3 \rightarrow CFClCF_3 + H_2O$ (HCFC-124)	$8.0x10^{-13}$	1350±150	$8.6x10^{-15}$	1.2
$OH + CH_3CF_2CFCl_2 \rightarrow$ products (HCFC-243cc)	$7.7x10^{-13}$	1700±300	$2.6x10^{-15}$	2.0
$OH + CF_3CF_2CHCl_2 \rightarrow$ products (HCFC-225ca)	$1.0x10^{-12}$	1100±200	$2.5x10^{-14}$	1.3
$OH + CF_2ClCF_2CHFCl \rightarrow$ products (HCFC-225cb)	$5.5x10^{-13}$	1250±200	$8.3x10^{-15}$	1.3
$HO_2 + Cl \rightarrow HCl + O_2$	$1.8x10^{-11}$	-(170±200)	$3.2x10^{-11}$	1.5
$\rightarrow OH + ClO$	$4.1x10^{-11}$	450±200	$9.1x10^{-12}$	2.0
$HO_2 + ClO \rightarrow HOCl + O_2$	$4.8x10^{-13}$	-(700±)	$5.0x10^{-12}$	1.4
$H_2O + ClONO_2 \rightarrow$ products	-	-	$<2.0x10^{-21}$	-
$NO + OClO \rightarrow NO_2 + ClO$	$2.5x10^{-12}$	600±300	$3.4x10^{-13}$	2.0
$NO + Cl_2O_2 \rightarrow$ products	-	-	$<2.0x10^{-14}$	-
$NO_3 + HCl \rightarrow HNO_3 + Cl$	-	-	$<5.0x10^{-17}$	-
$HO_2NO_2 + HCl \rightarrow$ products	-	-	$<1.0x10^{-21}$	-
$Cl + O_3 \rightarrow ClO + O_2$	$2.9x10^{-11}$	260±100	$1.2x10^{-11}$	1.15
$Cl + H_2 \rightarrow HCl + H$	$3.7x10^{-11}$	2300±200	$1.6x10^{-14}$	1.25
$Cl + H_2O_2 \rightarrow HCl + HO_2$	$1.1x10^{-11}$	980±500	$4.1x10^{-13}$	1.5
$Cl + NO_3 \rightarrow ClO + NO_2$	$2.4x10^{-11}$	0±400	$2.4x10^{-11}$	1.5
$Cl + N_2O \rightarrow ClO + N_2$	See reference			
$Cl + HNO_3 \rightarrow$ products	-	-	$<2.0x10^{-16}$	-
$Cl + CH_4 \rightarrow HCl + CH_3$	$1.1x10^{-11}$	1400±150	$1.0x10^{-13}$	1.1
$Cl + CH_3D \rightarrow$ products	-	-	$7.4x10^{-14}$	2.0
$Cl + H_2CO \rightarrow HCl + HCO$	$8.1x10^{-11}$	30±100	$7.3x10^{-11}$	1.15
$Cl + CH_3O_2 \rightarrow$ products	-	-	$1.6x10^{-10}$	1.5
$Cl + CH_3OH \rightarrow CH_2OH + HCl$	$5.4x10^{-11}$	0±250	$5.4x10^{-11}$	1.5
$Cl + C_2H_6 \rightarrow HCl + C_2H_5$	$7.7x10^{-11}$	90±90	$5.7x10^{-11}$	1.1
$Cl + C_2H_5O_2 \rightarrow ClO + C_2H_5O$	-	-	$7.4x10^{-11}$	2.0
$\rightarrow HCl + C_2H_4O_2$	-	-	$7.7x10^{-11}$	2.0
$Cl + CH_3CN \rightarrow$ products	$1.6x10^{-11}$	2140±300	$1.2x10^{-14}$	2.0
$Cl + CH_3CO_3NO_2 \rightarrow$ products	-	-	$<1x10^{-14}$	
$Cl + C_3H_8 \rightarrow HCl + C_3H_7$	$1.2x10^{-10}$	-(40±250)	$1.4x10^{-10}$	1.3
$Cl + OClO \rightarrow ClO + ClO$	$3.4x10^{-11}$	-(160±200)	$5.8x10^{-11}$	1.25
$Cl + ClOO \rightarrow Cl_2 + O_2$	$2.3x10^{-10}$	0±250	$2.3x10^{-10}$	3.0
$\rightarrow ClO + ClO$	$1.2x10^{-11}$	0±250	$1.2x10^{-11}$	3.0
$Cl + Cl_2O \rightarrow Cl_2 + ClO$	$6.2x10^{-11}$	-(130±130)	$9.6x10^{-11}$	1.2
$Cl + Cl_2O_2 \rightarrow$ products	-	-	$1.0x10^{-10}$	2.0
$Cl + HOCl \rightarrow$ products	$2.5x10^{-12}$	130±250	$1.6x10^{-12}$	1.5
$Cl + ClNO \rightarrow NO + Cl_2$	$5.8x10^{-11}$	-(100±200)	$8.1x10^{-11}$	1.5
$Cl + ClONO_2 \rightarrow$ products	$6.5x10^{-12}$	-(135±50)	$1.0x10^{-11}$	1.2
$Cl + CH_3Cl \rightarrow CH_2Cl + HCl$	$3.2x10^{-11}$	1250±200	$4.8x10^{-13}$	1.2
$Cl + CH_2Cl_2 \rightarrow HCl + CHCl_2$	$3.1x10^{-11}$	1350±500	$3.3x10^{-13}$	1.5
$Cl + CHCl_3 \rightarrow HCl + CCl_3$	$8.2x10^{-12}$	1325±300	$9.6x10^{-14}$	1.3

Table 1. Rate Constants for Second Order Reactions (continued)

Reaction	*A* cm³ molecule⁻¹ s⁻¹	*E/R* K	*k* (298 K) cm³ molecule⁻¹ s⁻¹	*f*(298)
$Cl + CH_3F \rightarrow HCl + CH_2F$ (HFC-41)	2.0×10^{-11}	1200±500	3.5×10^{-13}	1.3
$Cl + CH_2F_2 \rightarrow HCl + CHF_2$ (HFC-32)	1.2×10^{-11}	1630±500	5.0×10^{-14}	1.5
$Cl + CF_3H \rightarrow HCl + CF_3$ (HFC-23)	-	-	3.0×10^{-18}	5.0
$Cl + CH_2FCl \rightarrow HCl + CHFCl$ (HCFC-31)	1.2×10^{-11}	1390±500	1.1×10^{-13}	2.0
$Cl + CHFCl_2 \rightarrow HCl + CFCl_2$ (HCFC-21)	5.5×10^{-12}	1675±200	2.0×10^{-14}	1.3
$Cl + CHF_2Cl \rightarrow HCl + CF_2Cl$ (HCFC-22)	5.9×10^{-12}	2430±200	1.7×10^{-15}	1.3
$Cl + CH_3CCl_3 \rightarrow CH_2CCl_3 + HCl$	2.8×10^{-12}	1790±400	7.0×10^{-15}	2.0
$Cl + CH_3CH_2F \rightarrow HCl + CH_3CHF$ (HFC-161)	1.8×10^{-11}	290±500	6.8×10^{-12}	3.0
$\rightarrow HCl + CH_2CH_2F$	1.4×10^{-11}	880±500	7.3×10^{-13}	3.0
$Cl + CH_3CHF_2 \rightarrow HCl + CH_3CF_2$ (HFC-152a)	6.4×10^{-12}	950±500	2.6×10^{-13}	1.3
$\rightarrow HCl + CH_2CHF_2$	7.2×10^{-12}	2390±500	2.4×10^{-15}	3.0
$Cl + CH_2FCH_2F \rightarrow HCl + CHFCH_2F$ (HFC-152)	2.6×10^{-11}	1060±500	7.5×10^{-13}	3.0
$Cl + CH_3CFCl_2 \rightarrow HCl + CH_2CFCl_2$ (HCFC-141b)	1.8×10^{-12}	2000±300	2.2×10^{-15}	1.2
$Cl + CH_3CF_2Cl \rightarrow HCl + CH_2CF_2Cl$ (HCFC-142b)	1.4×10^{-12}	2420±500	4.2×10^{-16}	1.2
$Cl + CH_3CF_3 \rightarrow HCl + CH_2CF_3$ (HFC-143a)	1.2×10^{-11}	3880±500	2.6×10^{-17}	5.0
$Cl + CH_2FCHF_2 \rightarrow HCl + CH_2FCF_2$ (HFC-143)	5.5×10^{-12}	1610±500	2.5×10^{-14}	3.0
$\rightarrow HCl + CHFCHF_2$	7.7×10^{-12}	1720±500	2.4×10^{-14}	3.0
$Cl + CH_2ClCF_3 \rightarrow HCl + CHClCF_3$ (HCFC-133a)	1.8×10^{-12}	1710±500	5.9×10^{-15}	3.0
$Cl + CH_2FCF_3 \rightarrow HCl + CHFCF_3$ (HFC-134a)	-	-	1.5×10^{-15}	1.2
$Cl + CHF_2CHF_2 \rightarrow HCl + CF_2CHF_2$ (HCF-134)	7.5×10^{-12}	2430±500	2.2×10^{-15}	1.5
$Cl + CHCl_2CF_3 \rightarrow HCl + CCl_2CF_3$ (HCFC-123)	4.4×10^{-12}	1750±500	1.2×10^{-14}	1.3
$Cl + CHFClCF_3 \rightarrow HCl + CFClCF_3$ (HCFC-124)	1.1×10^{-12}	1800±500	2.7×10^{-15}	1.3
$Cl + CHF_2CF_3 \rightarrow HCl + CF_2CF_3$ (HFC-125)	-	-	2.4×10^{-16}	1.3
$ClO + O_3 \rightarrow ClOO + O_2$	-	-	$<1.4\times10^{-17}$	-
$\rightarrow OClO + O_2$	1.0×10^{-12}	>4000	$<1.0\times10^{-18}$	-
$ClO + H_2 \rightarrow$ products	$\sim1.0\times10^{-12}$	>4800	$<1.0\times10^{-19}$	-
$ClO + NO \rightarrow NO_2 + Cl$	6.4×10^{-12}	-(290±100)	1.7×10^{-11}	1.15
$ClO + NO_3 \rightarrow ClOO + NO_2$	4.7×10^{-13}	0±400	4.7×10^{-13}	1.5
$ClO + N_2O \rightarrow$ products	$\sim1.0\times10^{-12}$	>4300	$<6.0\times10^{-19}$	-
$ClO + CO \rightarrow$ products	$\sim1.0\times10^{-12}$	>3700	$<4.0\times10^{-18}$	-
$ClO + CH_4 \rightarrow$ products	$\sim1.0\times10^{-12}$	>3700	$<4.0\times10^{-18}$	-
$ClO + H_2CO \rightarrow$ products	$\sim1.0\times10^{-12}$	>2100	$<1.0\times10^{-15}$	-
$ClO + CH_3O_2 \rightarrow$ products	3.3×10^{-12}	115±115	2.2×10^{-12}	1.5
$ClO + ClO \rightarrow Cl_2 + O_2$	1.0×10^{-12}	1590±300	4.8×10^{-15}	1.5
$\rightarrow ClOO + Cl$	3.0×10^{-11}	2450±500	8.0×10^{-15}	1.5
$\rightarrow OClO + Cl$	3.5×10^{-13}	1370±300	3.5×10^{-15}	1.5
$HCl + ClONO_2 \rightarrow$ products	-	-	$<1.0\times10^{-20}$	-

Table 1. Rate Constants for Second Order Reactions (continued)

Reaction	*A* cm^3 molecule^{-1} s^{-1}	*E/R* K	*k* (298 K) cm^3 molecule^{-1} s^{-1}	*f*(298)
$CH_2ClO + O_2 \rightarrow CHClO + HO_2$	-	-	6 x 10^{-14}	5
$CH_2ClO_2 + HO_2 \rightarrow CH_2ClO_2H + O_2$	3.3 x 10^{-13}	-(820±200)	5.2 x 10^{-12}	1.5
$CH_2ClO_2 + NO \rightarrow CH_2ClO + NO_2$	7 x 10^{-12}	-(300±200)	1.9 x 10^{-11}	1.5
$CCl_3O_2 + NO \rightarrow CCl_2O + NO_2 + Cl$	7.3 x 10^{-12}	-(270±200)	1.8 x 10^{-11}	1.3
$CCl_2FO_2 + NO \rightarrow CClFO + NO_2 + Cl$	4.5 x 10^{-12}	-(350±200)	1.5 x 10^{-11}	1.3
$CClF_2O_2 + NO \rightarrow CF_2O + NO_2 + Cl$	3.8 x 10^{-12}	-(400±200)	1.5 x 10^{-11}	1.2
BrO_x Reactions				
$O + BrO \rightarrow Br + O_2$	1.9x10^{-11}	-(230±150)	4.1x10^{-11}	1.5
$O + HBr \rightarrow OH + Br$	5.8x10^{-12}	1500±200	3.8x10^{-14}	1.3
$O + HOBr \rightarrow OH + BrO$	1.2x10^{-10}	430±300	2.8x10^{-11}	3.0
$OH + Br_2 \rightarrow HOBr + Br$	4.2x10^{-11}	0±600	4.2x10^{-11}	1.3
$OH + BrO \rightarrow$ products	-	-	7.5x10^{-11}	3.0
$OH + HBr \rightarrow H_2O + Br$	1.1x10^{-11}	0±250	1.1x10^{-11}	1.2
$OH + CH_3Br \rightarrow CH_2Br + H_2O$	4.0x10^{-12}	1470±150	2.9x10^{-14}	1.1
$OH + CH_2Br_2 \rightarrow CHBr_2 + H_2O$	2.4x10^{-12}	900±300	1.2x10^{-13}	1.1
$OH + CHBr_3 \rightarrow CBr_3 + H_2O$	1.6x10^{-12}	710±200	1.5x10^{-13}	2.0
$OH + CHF_2Br \rightarrow CF_2Br + H_2O$	1.1x10^{-12}	1400±200	1.0x10^{-14}	1.1
$OH + CH_2ClBr \rightarrow CHClBr + H_2O$	2.3x10^{-12}	930±150	1.0x10^{-13}	1.2
$OH + CF_2ClBr \rightarrow$ products	-	-	<1.5x10^{-16}	-
$OH + CF_2Br_2 \rightarrow$ products	-	-	<5.0x10^{-16}	-
$OH + CF_3Br \rightarrow$ products	-	-	<1.2x10^{-16}	-
$OH + CH_2BrCF_3 \rightarrow CHBrCF_3 + H_2O$	1.4x10^{-12}	1340±200	1.6x10^{-14}	1.3
$OH + CHFBrCF_3 \rightarrow CFBrCF_3$	7.2x10^{-13}	1110±150	1.8x10^{-14}	1.5
$OH + CHClBrCF_3 \rightarrow CClBrCF_3 + H_2O$	1.3x10^{-12}	995±150	4.5x10^{-14}	1.5
$OH + CF_2BrCHFCl \rightarrow CF_2BrCFCl + H_2O$	9.3x10^{-13}	1250±150	1.4x10^{-14}	1.5
$OH + CF_2BrCF_2Br \rightarrow$ products	-	-	<1.5x10^{-16}	-
$HO_2 + Br \rightarrow HBr + O_2$	1.5x10^{-11}	600±600	2.0x10^{-12}	2.0
$HO_2 + BrO \rightarrow$ products	3.4x10^{-12}	-(540±200)	2.1x10^{-11}	1.5
$NO_3 + HBr \rightarrow HNO_3 + Br$	-	-	<1.0x10^{-16}	-
$Cl + CH_2ClBr \rightarrow HCl + CHClBr$	4.3x10^{-11}	1370±500	4.3x10^{-13}	3.0
$Cl + CH_3Br \rightarrow HCl + CH_2Br$	1.5x10^{-11}	1060±100	4.3x10^{-13}	1.2
$Cl + CH_2Br_2 \rightarrow HCl + CHBr_2$	6.4x10^{-12}	810±100	4.2x10^{-13}	1.2
$Br + O_3 \rightarrow BrO + O_2$	1.7x10^{-11}	800±200	1.2x10^{-12}	1.2
$Br + H_2O_2 \rightarrow HBr + HO_2$	1.0x10^{-11}	>3000	<5.0x10^{-16}	-
$Br + NO_3 \rightarrow BrO + NO_2$	-	-	1.6x10^{-11}	2.0
$Br + H_2CO \rightarrow HBr + HCO$	1.7x10^{-11}	800±200	1.1x10^{-12}	1.3
$Br + OClO \rightarrow BrO + ClO$	2.6x10^{-11}	1300±300	3.4x10^{-13}	2.0
$Br + Cl_2O \rightarrow BrCl + ClO$	2.1x10^{-11}	470±150	4.3x10^{-12}	1.3
$Br + Cl_2O_2 \rightarrow$ products	-	-	3.0x10^{-12}	2.0
$BrO + O_3 \rightarrow$ products	~1.0x10^{-12}	>3200	<2.0x10^{-17}	-
$BrO + NO \rightarrow NO_2 + Br$	8.8x10^{-12}	-(260±130)	2.1x10^{-11}	1.15
$BrO + NO_3 \rightarrow$ products	-	-	1.0x10^{-12}	3.0
$BrO + ClO \rightarrow Br + OClO$	1.6x10^{-12}	-(430±200)	6.8x10^{-12}	1.25
$\rightarrow Br + ClOO$	2.9x10^{-12}	-(220±200)	6.1x10^{-12}	1.25
$\rightarrow BrCl + O_2$	5.8x10^{-13}	-(170±200)	1.0x10^{-12}	1.25
$BrO + BrO \rightarrow$ products	1.5x10^{-12}	-(230±150)	3.2x10^{-12}	1.15
$CH_2BrO_2 + NO \rightarrow CH_2O + NO_2 + Br$	4x10^{-12}	-(300±200)	1.1 x 10^{-11}	1.5

Table 1. Rate Constants for Second Order Reactions (continued)

Reaction	*A* cm^3 molecule^{-1} s^{-1}	*E/R* K	*k* (298 K) cm^3 molecule^{-1} s^{-1}	*f*(298)
IO_x Reactions				
$O + I_2 \rightarrow IO + I$	1.4x10^{-10}	0±250	1.4x10^{-10}	1.4
$O + IO \rightarrow O_2 + I$			1.2x10^{-10}	2.0
$OH + I_2 \rightarrow HOI + I$			1.8x10^{-10}	2.0
$OH + HI \rightarrow H_2O + I$			3.0x10^{-11}	2.0
$OH + CH_3I \rightarrow H_2O + CH_2I$	3.1x10^{-12}	1120±500	7.2x10^{-14}	3.0
$OH + CF_3I \rightarrow HOI + CF_3$			3.1x10^{-14}	5.0
$HO_2 + I \rightarrow HI + O_2$	1.5x10^{-11}	1090±500	3.8x10^{-13}	2.0
$HO_2 + IO \rightarrow HOI + O_2$			8.4x10^{-11}	1.5
$NO_3 + HI \rightarrow HNO_3 + I$	See reference			
$I + O_3 \rightarrow IO + O_2$	2.3x10^{-11}	870±200	1.2x10^{-12}	1.2
$I + BrO \rightarrow IO + Br$	-	-	1.2x10^{-11}	2.0
$IO + NO \rightarrow I + NO_2$	9.1x10^{-12}	-(240±150)	2.0x10^{-11}	1.2
$IO + ClO \rightarrow$ products	5.1x10^{-12}	-(280±200)	1.3x10^{-11}	2.0
$IO + BrO \rightarrow$ products	-	-	6.9x10^{-11}	1.5
$IO + IO \rightarrow$ products	1.5x10^{-11}	-(500±500)	8.0x10^{-11}	1.5
$INO + INO \rightarrow I_2 + 2NO$	8.4x10^{-11}	2620±600	1.3x10^{-14}	2.5
$INO_2 + INO_2 \rightarrow I_2 + 2NO_2$	2.9x10^{-11}	2600±1000	4.7x10^{-15}	3.0
SO_x Reactions				
$O + SH \rightarrow SO + H$	-	-	1.6x10^{-10}	5.0
$O + CS \rightarrow CO + S$	2.7x10^{-10}	760±250	2.1x10^{-11}	1.1
$O + H_2S \rightarrow OH + SH$	9.2x10^{-12}	1800±550	2.2x10^{-14}	1.7
$O + OCS \rightarrow CO + SO$	2.1x10^{-11}	2200±150	1.3x10^{-14}	1.2
$O + CS_2 \rightarrow CS + SO$	3.2x10^{-11}	650±150	3.6x10^{-12}	1.2
$O + CH_3SCH_3 \rightarrow CH_3SO + CH_3$	1.3x10^{-11}	-(410±100)	5.0x10^{-11}	1.1
$O + CH_3SSCH_3 \rightarrow CH_3SO + CH_3S$	5.5x10^{-11}	-(250±100)	1.3x10^{-10}	1.3
$O_3 + H_2S \rightarrow$ products	-	-	<2.0x10^{-20}	-
$O_3 + CH_3SCH_3 \rightarrow$ products	-	-	<1.0x10^{-18}	-
$O_3 + SO_2 \rightarrow SO_3 + O_2$	3.0x10^{-12}	>7000	<2.0x10^{-22}	-
$OH + H_2S \rightarrow SH + H_2O$	6.0x10^{-12}	75±75	4.7x10^{-12}	1.2
$OH + OCS \rightarrow$ products	1.1x10^{-13}	1200±500	1.9x10^{-15}	2.0
$OH + CS_2 \rightarrow$ products	See reference	-	-	-
$OH + CH_3SH \rightarrow CH_3S + H_2O$	9.9x10^{-12}	-(360±100)	3.3x10^{-11}	1.2
$OH + CH_3SCH_3 \rightarrow H_2O + CH_2SCH_3$	1.2x10^{-11}	260±100	5.0x10^{-12}	1.15
$OH + CH_3SSCH_3 \rightarrow$ products	6.0x10^{-11}	-(400±200)	2.3x10^{-10}	1.2
$OH + S \rightarrow H + SO$	-	-	6.6x10^{-11}	3.0
$OH + SO \rightarrow H + SO_2$	-	-	8.6x10^{-11}	2.0
$HO_2 + H_2S \rightarrow$ products	-	-	<3.0x10^{-15}	-
$HO_2 + CH_3SH \rightarrow$ products	-	-	<4.0x10^{-15}	-
$HO_2 + CH_3SCH_3 \rightarrow$ products	-	-	<5.0x10^{-15}	-
$HO_2 + SO_2 \rightarrow$ products	-	-	<1.0x10^{-18}	-
$NO_2 + SO_2 \rightarrow$ products	-	-	<2.0x10^{-26}	-
$NO_3 + H_2S \rightarrow$ products	-	-	<8.0x10^{-16}	-
$NO_3 + OCS \rightarrow$ products	-	-	<1.0x10^{-16}	-
$NO_3 + CS_2 \rightarrow$ products	-	-	<4.0x10^{-16}	-
$NO_3 + CH_3SH \rightarrow$ products	4.4x10^{-13}	-(210±210)	8.9x10^{-13}	1.25
$NO_3 + CH_3SCH_3 \rightarrow CH_3SCH_2 + HNO_3$	1.9x10^{-13}	-(500±200)	1.0x10^{-12}	1.2
$NO_3 + CH_3SSCH_3 \rightarrow$ products	1.3x10^{-12}	270±270	5.3x10^{-13}	1.4
$NO_3 + SO_2 \rightarrow$ products	-	-	<7.0x10^{-21}	-
$N_2O_5 + CH_3SCH_3 \rightarrow$ products	-	-	<1.0x10^{-17}	-
$CH_3O_2 + SO_2 \rightarrow$ products	-	-	<5.0x10^{-17}	-
$F + CH_3SCH_3 \rightarrow$ products	-	-	2.4.x10^{-10}	2.0

Table 1. Rate Constants for Second Order Reactions (continued)

Reaction	*A* cm³ molecule⁻¹ s⁻¹	*E/R* K	*k* (298 K) cm³ molecule⁻¹ s⁻¹	*f*(298)
$Cl + H_2S \rightarrow HCl + SH$	3.7×10^{-11}	-(210±100)	7.4×10^{-11}	1.25
$Cl + OCS \rightarrow$ products	-	-	$<1.0 \times 10^{-16}$	-
$Cl + CS_2 \rightarrow$ products	-	-	$<4.0 \times 10^{-15}$	-
$Cl + CH_3SH \rightarrow CH_3S + HCl$	1.2×10^{-10}	-(150±50)	2.0×10^{-10}	1.25
$Cl + CH_3SCH_3 \rightarrow$ products	See reference	-	-	-
$ClO + OCS \rightarrow$ products	-	-	$<2.0 \times 10^{-16}$	-
$ClO + CH_3SCH_3 \rightarrow$ products	-	-	9.5×10^{-15}	2.0
$ClO + SO \rightarrow Cl + SO_2$	2.8×10^{-11}	0±50	2.8×10^{-11}	1.3
$ClO + SO_2 \rightarrow Cl + SO_3$	-	-	$<4.0 \times 10^{-18}$	-
$Br + H_2S \rightarrow HBr + SH$	1.4×10^{-11}	2750±300	1.4×10^{-15}	2.0
$Br + CH_3SH \rightarrow CH_3S + HBr$	9.2×10^{-12}	390±100	2.5×10^{-12}	2.0
$Br + CH_3SCH_3 \rightarrow$ products	See reference			
$BrO + CH_3SCH_3 \rightarrow$ products	1.5×10^{-14}	-(850±200)	2.6×10^{-13}	1.3
$BrO + SO \rightarrow Br + SO_2$			5.7×10^{-11}	1.4
$IO + CH_3SH \rightarrow$ products			6.6×10^{-16}	2.0
$IO + CH_3SCH_3 \rightarrow$ products			1.2×10^{-14}	1.5
$S + O_2 \rightarrow SO + O$	2.3×10^{-12}	0±200	2.3×10^{-12}	1.2
$S + O_3 \rightarrow SO + O_2$			1.2×10^{-11}	2.0
$SO + O_2 \rightarrow SO_2 + O$	2.6×10^{-13}	2400±500	8.4×10^{-17}	2.0
$SO + O_3 \rightarrow SO_2 + O_2$	3.6×10^{-12}	1100±200	9.0×10^{-14}	1.2
$SO + NO_2 \rightarrow SO_2 + NO$	1.4×10^{-11}	0±50	1.4×10^{-11}	1.2
$SO + OClO \rightarrow SO_2 + ClO$			1.9×10^{-12}	3.0
$SO_3 + H_2O \rightarrow$ products	See reference		-	-
$SO_3 + NO_2 \rightarrow$ products			1.0×10^{-19}	10.0
$SH + O_2 \rightarrow OH + SO$			$<4.0 \times 10^{-19}$	-
$SH + O_3 \rightarrow HSO + O_2$	9.0×10^{-12}	280±200	3.5×10^{-12}	1.3
$SH + H_2O_2 \rightarrow$ products			$<5.0 \times 10^{-15}$	-
$SH + NO_2 \rightarrow HSO + NO$	2.9×10^{-11}	-(240±50)	6.5×10^{-11}	1.2
$SH + Cl_2 \rightarrow ClSH + Cl$	1.7×10^{-11}	690±200	1.7×10^{-12}	2.0
$SH + BrCl \rightarrow$ products	2.3×10^{-11}	-(350±200)	7.4×10^{-11}	2.0
$SH + Br_2 \rightarrow BrSH + Br$	6.0×10^{-11}	-(160±160)	1.0×10^{-10}	2.0
$SH + F_2 \rightarrow FSH + F$	4.3×10^{-11}	1390±200	4.0×10^{-13}	2.0
$HSO + O_2 \rightarrow$ products			$<2.0 \times 10^{-17}$	-
$HSO + O_3 \rightarrow$ products			1.0×10^{-13}	1.3
$HSO + NO \rightarrow$ products			$<1.0 \times 10^{-15}$	-
$HSO + NO_2 \rightarrow HSO_2 + NO$			9.6×10^{-12}	2.0
$HSO_2 + O_2 \rightarrow HO_2 + SO_2$			3.0×10^{-13}	3.0
$HOSO_2 + O_2 \rightarrow HO_2 + SO_3$	1.3×10^{-12}	330±200	4.4×10^{-13}	1.2
$CS + O_2 \rightarrow OCS + O$			2.9×10^{-19}	2.0
$CS + O_3 \rightarrow OCS + O_2$			3.0×10^{-16}	3.0
$CS + NO_2 \rightarrow OCS + NO$			7.6×10^{-17}	3.0
$CH_3S + O_2 \rightarrow$ products			$<3.0 \times 10^{-18}$	-
$CH_3S + O_3 \rightarrow$ products	2.0×10^{-12}	-(290±100)	5.3×10^{-12}	1.15
$CH_3S + NO \rightarrow$ products			$<1.0 \times 10^{-13}$	-
$CH_3S + NO_2 \rightarrow CH_3SO + NO$	2.1×10^{-11}	-(320±100)	6.1×10^{-11}	1.15
$CH_2SH + O_2 \rightarrow$ products			6.5×10^{-12}	2.0
$CH_2SH + O_3 \rightarrow$ products			3.5×10^{-11}	2.0
$CH_2SH + NO \rightarrow$ products			1.9×10^{-11}	2.0
$CH_2SH + NO_2 \rightarrow$ products			5.2×10^{-11}	2.0
$CH_3SO + O_3 \rightarrow$ products			6.0×10^{-13}	1.5
$CH_3SO + NO_2 \rightarrow CH_3SO_2 + NO$			1.2×10^{-11}	1.4
$CH_3SOO + O_3 \rightarrow$ products			$<8.0 \times 10^{-13}$	-
$CH_3SOO + NO \rightarrow$ products	1.1×10^{-11}	0±100	1.1×10^{-11}	2.0
$CH_3SO_2 + NO_2 \rightarrow$ products	2.2×10^{-11}	0±100	2.2×10^{-11}	2.0
$CH_3SCH_2 + NO_3 \rightarrow$ products			3.0×10^{-10}	2.0
$CH_3SCH_2O_2 + NO \rightarrow CH_3SCH_2O + NO_2$			1.9×10^{-11}	2.0

Table 1. Rate Constants for Second Order Reactions (continued)

Reaction	*A* cm³ molecule⁻¹ s⁻¹	*E/R* K	*k* (298 K) cm³ molecule⁻¹ s⁻¹	*f*(298)
$CH_3SS + O_3 \rightarrow$ products			$4.6x10^{-13}$	2.0
$CH_3SS + NO_2 \rightarrow$ products			$1.8x10^{-11}$	2.0
$CH_3SSO + NO_2 \rightarrow$ products			$4.5x10^{-12}$	2.0
Metal Reactions				
$Na + O_3 \rightarrow NaO + O_2$	$1.0x10^{-9}$	95±50	$7.3x10^{-10}$	1.2
$\rightarrow NaO_2 + O$	-	-	$<4.0x10^{-11}$	-
$Na + N_2O \rightarrow NaO + N_2$	$2.8x10^{-10}$	1600±400	$1.3x10^{-12}$	1.2
$Na + Cl_2 \rightarrow NaCl + Cl$	$7.3x10^{-10}$	0±200	$7.3x10^{-10}$	1.3
$NaO + O \rightarrow Na + O_2$	$3.7x10^{-10}$	0±400	$3.7x10^{-10}$	3.0
$NaO + O_3 \rightarrow NaO_2 + O_2$	$1.1x10^{-9}$	570±300	$1.6x10^{-10}$	1.5
$\rightarrow Na + 2O_2$	$6.0x10^{-11}$	0±800	$6.0x10^{-11}$	3.0
$NaO + H_2 \rightarrow NaOH + H$	$2.6x10^{-11}$	0±600	$2.6x10^{-11}$	2.0
$NaO + H_2O \rightarrow NaOH + OH$	$2.2x10^{-10}$	0±400	$2.2x10^{-10}$	2.0
$NaO + NO \rightarrow Na + NO_2$	$1.5x10^{-10}$	0±400	$1.5x10^{-10}$	4.0
$NaO + HCl \rightarrow$ products	$2.8x10^{-10}$	0±400	$2.8x10^{-10}$	3.0
$NaO_2 + O \rightarrow NaO + O_2$	$2.2x10^{-11}$	0±600	$2.2x10^{-11}$	5.0
$NaO_2 + NO \rightarrow NaO + NO_2$	-	-	$<10^{-14}$	-
$NaO_2 + HCl \rightarrow$ products	$2.3x10^{-10}$	0±400	$2.3x10^{-10}$	3.0
$NaOH + HCl \rightarrow NaCl + H_2O$	$2.8x10^{-10}$	0±400	$2.8x10^{-10}$	3.0

Table 2. Rate Constants for Association Reactions

The values quoted are suitable for air as the third body, M. The integer in parentheses is the power of ten.

Reaction	Low pressure limit $k_0(T) = k_0(300)\,(T/300)^{-n}$ cm⁶ molecule⁻² s⁻¹ $k_0(300)$	*n*	High pressure limit $k_\infty(T) = k_\infty(300)\,(T/300)^{-m}$ cm³ molecule⁻¹ s⁻¹ $k_\infty(300)$	*m*
O_x Reactions				
$O + O_2 \rightarrow O_3$	(6.0±0.5) (-34)	2.3±0.5	-	-
$O(^1D)$ Reactions				
$O(^1D) + N_2 \rightarrow N_2O$	(3.5±3.0) (-37)	0.6	.	
HO_x Reactions				
$H + O_2 \rightarrow HO_2$	(5.7±0.5) (-32)	1.6±0.5	(7.5±4.0) (-11)	0±1.0
$OH + OH \rightarrow H_2O_2$	(6.2±1.2) (-31)	1.0	(2.6±1.0) (-11)	0±0.5
NO_x Reactions				
$O + NO \rightarrow NO_2$	(9.0±2.0) (-32)	1.5±0.3	(3.0±1.0) (-11)	0±1.0
$O + NO_2 \rightarrow NO_3$	(9.0±1.0) (-32)	2.0±1.0	(2.2±0.3) (-11)	0±1.0
$OH + NO \rightarrow HONO$	(7.0±1.0) (-31)	2.6±0.3	(3.6±1.0) (-11)	0.1±0.5
$OH + NO_2 \rightarrow HNO_3$	(2.5±0.1) (-30)	4.4±0.3	(1.6±0.2) (-11)	1.7±0.2
$HO_2 + NO_2 \rightarrow HO_2NO_2$	(1.8±0.3) (-31)	3.2±0.4	(4.7±1.0) (-12)	1.4±1.4
$NO_2 + NO_3 \rightarrow N_2O_5$	(2.2±0.5) (-30)	3.9±1.0	(1.5±0.8) (-12)	0.7±0.4
$NO_3 \rightarrow NO + O_2$	See reference			
Hydrocarbon Reactions				
$CH_3 + O_2 \rightarrow CH_3O_2$	(4.5±1.5) (-31)	3.0±1.0	(1.8±0.2) (-12)	1.7±1.7
$C_2H_5 + O_2 \rightarrow C_2H_5O_2$	(1.5±1.0) (-28)	3.0±1.0	(8.0±1.0) (-12)	0±1.0
$OH + C_2H_2 \rightarrow HOCHCH$	(5.5±2.0) (-30)	0.0±0.2	(8.3±1.0) (-13)	-2
$OH + C_2H_4 \rightarrow HOCH_2CH_2$	(1.0±0.6) (-28)	0.8±2.0	(8.8±0.9) (-12)	0

Table 2. Rate Constants for Association Reactions (continued)

The values quoted are suitable for air as the third body, M. The integer in parentheses is the power of ten.

Reaction	Low pressure limit $k_0(T) = k_0(300)\,(T/300)^{-n}$ cm^6 molecule^{-2} s^{-1} $k_0(300)$	n	High pressure limit $k_\infty(T) = k_\infty(300)\,(T/300)^{-m}$ cm^3 molecule^{-1} s^{-1} $k_\infty(300)$	m
$CH_3O + NO \rightarrow CH_3ONO$	(1.4±0.5) (-29)	3.8±1.0	(3.6±1.6) (-11)	0.6±1.0
$CH_3O + NO_2 \rightarrow CH_3ONO_2$	(1.1±0.4) (-28)	4.0±2.0	(1.6±0.5) (-11)	1.0±1.0
$C_2H_5O + NO \rightarrow C_2H_5ONO$	(2.8±1.0) (-27)	4.0±2.0	(5.0±1.0) (-11)	1.0±1.0
$C_2H_5O + NO_2 \rightarrow C_2H_5ONO_2$	(2.0±1.0) (-27)	4.0±2.0	(2.8±0.4) (-11)	1.0±1.0
$CH_3O_2 + NO_2 \rightarrow CH_3O_2NO_2$	(1.5±0.8) (-30)	4.0±2.0	(6.5±3.2) (-12)	2.0±2.0
$CH_3C(O)O_2 + NO_2 \rightarrow$ $CH_3C(O)O_2NO_2$	(9.7±3.8) (-29)	5.6±2.8	(9.3±0.4)(-12)	1.5±0.3
FO_x Reactions				
$F + O_2 \rightarrow FO_2$	(4.4±0.4) (-33)	1.2±0.5	-	-
$F + NO \rightarrow FNO$	(1.8±0.3) (-31)	1.0±10	(2.8±1.4) (-10)	0.0±1.0
$F + NO_2 \rightarrow FNO_2$	(6.3±3.0) (-32)	2.0±2.0	(2.6±1.3) (-10)	0.0±1.0
$FO + NO_2 \rightarrow FONO_2$	(2.6±2.0) (-31)	1.3±1.3	(2.0±1.0) (-11)	1.5±1.5
$CF_3 + O_2 \rightarrow CF_3O_2$	(3.0±0.3) (-29)	4.0±2.0	(4.0±1.0) (-12)	1.0±1.0
$CF_3O + NO_2 \rightarrow CF_3ONO_2$	See reference			
$CF_3O_2 + NO_2CF_3O_2NO_2$	(2.2±0.5) (-29)	5.0±1.0	(6.0±1.0) (-12)	2.5±1.0
$CF_3O + CO \rightarrow CF_3OCO$	(2.5±0.2) (-31)	-	(6.8±0.4) (-14)	-1.2
$CF_3O \rightarrow CF_2O + F$	See reference			
ClO_x Reactions				
$Cl + O_2 \rightarrow ClOO$	(2.7±1.0) (-33)	1.5±0.5	-	-
$Cl + NO \rightarrow ClNO$	(9.0±2.0) (-32)	1.6±0.5	-	-
$Cl + NO_2$ $ClONO \rightarrow$	(1.3±0.2) (-30)	2.0±1.0	(1.0±0.5) (-10)	1.0±1.0
$ClNO_2$	(1.8±0.3) (-31)	2.0±1.0	(1.0±0.5) (-10)	1.0±1.0
$Cl + CO \rightarrow ClCO$	(1.3±0.5) (-33)	3.8±0.5	-	-
$Cl + C_2H_2 \rightarrow ClC_2H_2$	((5.9±1.0) (-30)	2.1±1.0	(2.1±0.4) (-10)	1.0±0.5
$Cl + C_2H_4 \rightarrow ClC_2H_4$	(1.6±1) (-29)	3.3±1.0	(3.1±2) (-10)	1.0±0.5
$Cl + C_2Cl_4 \rightarrow C_2Cl_5$	(1.4±0.6) (-28)	8.5±1.0	(4.0±1.0) (-11)	1.2±0.5
$ClO + NO_2 \rightarrow ClONO_2$	(1.8±0.3) (-31)	3.4±1.0	(1.5±0.7) (-11)	1.9±1.9
$OClO + NO_3 \rightarrow O_2ClONO_2$	See reference			
$ClO + ClO \rightarrow Cl_2O_2$	(2.2±0.4) (-32)	3.1±0.5	(3.5±2) (-12)	1.0±1.0
$ClO + OClO \rightarrow Cl_2O_3$	(6.2±1.0) (-32)	4.7±0.6	(2.4±1.2) (-11)	0±1.0
$OClO + O \rightarrow ClO_3$	(1.9±0.5) (-31)	1.1±1.0	(3.1±0.8) (-11)	0±1.0
$CH_2Cl + O_2 \rightarrow CH_2ClO_2$	(1.9±0.1) (-30)	3.2±0.2	(2.9±0.2) (-12)	1.2±0.6
$CHCl_2 + O_2 \rightarrow CHCl_2O_2$	(1.3±0.1) (-30)	4.0±0.2	(2.8±0.2) (-12)	1.4±0.6
$CCl_3 + O_2 \rightarrow CCl_3O_2$	(6.9±0.2) (-31)	6.4±0.3	(2.4±0.2) (-12)	2.1±0.6
$CFCl_2 + O_2 \rightarrow CFCl_2O_2$	(5.0±0.8) (-30)	4.0±2.0	(6.0±1.0) (-12)	1.0±1.0
$CF_2Cl + O_2 \rightarrow CF_2ClO_2$	(3.0±1.5) (-30)	4.0±2.0	(3±2) (-12)	1.0±1.0
$CCl_3O_2 + NO_2 \rightarrow CCl_3O_2NO_2$	(5.0±1.0) (-29)	5.0±1.0	(6.0±1.0) (-12)	2.5±1.0
$CFCl_2O_2 + NO_2 \rightarrow CFCl_2O_2NO_2$	(3.5±0.5) (-29)	5.0±1.0	(6.0±1.0) (-12)	2.5±1.0
$CF_2ClO_2 + NO_2 \rightarrow CF_2ClO_2NO_2$	(3.3±0.7) (-29)	6.7±1.3	(4.1±1.9) (-12)	2.8±0.7
BrO_x Reactions				
$Br + NO_2 \rightarrow BrNO_2$	(4.2±0.8) (-31)	2.4±0.5	(2.7±0.5) (-11)	0±1.0
$BrO + NO_2 \rightarrow BrONO_2$	(5.2±0.6) (-31)	3.2±0.8	(6.9±1.0) (-12)	2.9±1.0
IO_x Reactions				
$I + NO \rightarrow INO$	(1.8±0.5) (-32)	1.0±0.5	(1.7±1.0) (-11)	0±1.0
$I + NO_2 \rightarrow INO_2$	(3.0±1.5) (-31)	1.0±1.0	(6.6±5.0) (-11)	0±1.0
$IO + NO_2 \rightarrow IONO_2$	(5.9±2.0) (-31)	3.5±1.0	(9.0±1.0) (-12)	1.5±1.0

Table 2. Rate Constants for Association Reactions (continued)

The values quoted are suitable for air as the third body, M. The integer in parentheses is the power of ten.

Reaction	Low pressure limit $k_0(T) = k_0(300)\ (T/300)^{-n}$ cm^6 molecule^{-2} s^{-1} $k_0(300)$	n	High pressure limit $k_\infty(T) = k_\infty(300)\ (T/300)^{-m}$ cm^3 molecule^{-1} s^{-1} $k_\infty(300)$	m
SO_x Reactions				
$HS + NO \rightarrow HSNO$	(2.4±0.4) (-31)	3.0±1.0	(2.7±0.5) (-11)	0
$CH_3S + NO \rightarrow CH_3SNO$	(3.2±0.4) (-29)	4.0±1.0	(3.9±0.6) (-11)	2.7±1.0
$O + SO_2 \rightarrow SO_3$	(1.3±)(-33)	-3.6±0.7		
$OH + SO_2 \rightarrow HOSO_2$	(3.0±1.0) (-31)	3.3±1.5	(1.5±0.5) (-12)	0
$CH_3SCH_2 + O_2 \rightarrow CH_3SCH_2O_2$	See reference			
$SO_3 + NH_3 \rightarrow H_3NSO_3$	(3.9±0.8) (-30)	3.0±3.0	(4.7±1.3) (-11)	0±1.0
Metal Reactions				
$Na + O_2 \rightarrow NaO_2$	(3.2±0.3) (-30)	1.4±0.3	(6.0±2.0) (-10)	0±1.0
$NaO + O_2 \rightarrow NaO_3$	(3.5±0.7) (-30)	2.0±2.0	(5.7±3.0) (-10)	0±1.0
$NaO + CO_2 \rightarrow NaCO_3$	(8.7±2.6) (-28)	2.0±2.0	(6.5±3.0) (-10)	0±1.0
$NaOH + CO_2 \rightarrow NaHCO_3$	(1.3±0.3) (-28)	2.0±2.0	(6.8±4.0) (-10)	0±1.0

Table 3. Equilibrium Constants

$K(T)$/cm^3 molecule^{-1} = $A \exp(B/T)$ [200 < T/K < 300]

Reaction	A/cm^3 molecule^{-1}	B/K	K (298 K)	f (298 K)
$HO_2 + NO_2 \rightarrow HO_2NO_2$	2.1×10^{-27}	10900±1000	1.6×10^{-11}	5
$NO + NO_2 \rightarrow N_2O_3$	3.3×10^{-27}	4667±100	2.1×10^{-20}	2
$NO_2 + NO_2 \rightarrow N_2O_4$	5.2×10^{-29}	6643±250	2.5×10^{-19}	2
$NO_2 + NO_3 \rightarrow N_2O_5$	2.7×10^{-27}	11000±500	2.9×10^{-11}	1.3
$CH_3O_2 + NO_2 \rightarrow CH_3O_2NO_2$	1.3×10^{-28}	11200±1000	2.7×10^{-12}	2
$CH_3C(O)O_2 + NO_2 \rightarrow CH_3C(O)O_2NO_2$	9.0×10^{-29}	14000±200	2.3×10^{-8}	2
$F + O_2 \rightarrow FOO$	3.2×10^{-25}	6100±1200	2.5×10^{-16}	1.0
$Cl + O_2 \rightarrow ClOO$	5.7×10^{-25}	2500±750	2.5×10^{-21}	2
$Cl + CO \rightarrow ClCO$	1.6×10^{-25}	4000±500	1.1×10^{-19}	5
$ClO + O_2 \rightarrow ClO{\cdot}O_2$	2.9×10^{-26}	<3700	$<7.2 \times 10^{-21}$	-
$ClO + ClO \rightarrow Cl_2O_2$	1.3×10^{-27}	8744±850	7.2×10^{-15}	1.5
$ClO + OClO \rightarrow Cl_2O_3$	1.1×10^{-24}	5455±300	9.8×10^{-17}	3
$OClO + NO_3 \rightarrow O_2ClONO_2$	1×10^{-28}	9300±1000	3.6×10^{-15}	5
$OH + CS_2 \rightarrow CS_2OH$	4.5×10^{-25}	5140±500	1.4×10^{-17}	1.4
$CH_3S + O_2 \rightarrow CH_3SO_2$	1.8×10^{-27}	5545±300	2.2×10^{-19}	1.4

Section 6
Fluid Properties

Thermodynamic Properties of Air 6-1
Properties of Water in the Range 0-100°C 6-3
Enthalpy of Vaporization of Water 6-3
Fixed Point Properties of H_2O and D_2O 6-4
Thermal Conductivity of Saturated H_2O and D_2O 6-4
Standard Density of Water 6-5
Properties of Ice and Supercooled Water 6-6
Volumetric Properties of Aqueous Sodium Chloride Solutions 6-8
Density of D_2O 6-9
Vapor Pressure of Ice 6-9
Vapor Pressure of Water from 0 to 370°C 6-10
Boiling Point of Water at Various Pressures 6-12
Melting Point of Ice as a Function of Pressure 6-12
Properties of Water and Steam as a Function of Temperature and Pressure 6-13
Permittivity (Dielectric Constant) of Water at Various Frequencies 6-15
Thermophysical Properties of Fluids 6-16
Virial Coefficients of Selected Gases 6-25
Van der Waals Constants for Gases 6-45
Mean Free Path and Related Properties of Gases 6-46
Influence of Pressure on Freezing Points 6-47
Critical Constants 6-48
Sublimation Pressure of Solids 6-62
Vapor Pressure 6-65
Vapor Pressure of Fluids at Temperatures below 300 K 6-93
Vapor Pressure of Saturated Salt Solutions 6-104
IUPAC Recommended Data for Vapor Pressure Calibration 6-105
Enthalpy of Vaporization 6-106
Enthalpy of Fusion 6-122
Pressure and Temperature Dependence of Liquid Density 6-137
Properties of Cryogenic Fluids 6-139
Properties of Liquid Helium 6-140
Halocarbon Refrigerants 6-141
Density and Specific Volume of Mercury 6-144
Thermal Properties of Mercury 6-145
Surface Tension of Common Liquids 6-146
Surface Tension of Aqueous Mixtures 6-150
Permittivity (Dielectric Constant) of Liquids 6-151
Permittivity (Dielectric Constant) of Gases 6-174
Azeotropic Data for Binary Mixtures 6-176
Viscosity of Gases 6-180
Viscosity of Liquids 6-182
Viscosity of Carbon Dioxide along the Saturation Line 6-187
Thermal Conductivity of Gases 6-188
Thermal Conductivity of Liquids 6-190
Diffusion in Gases 6-192
Diffusion of Gases in Water 6-194
Diffusion Coefficients in Liquids at Infinite Dilution 6-195

THERMODYNAMIC PROPERTIES OF AIR

These tables summarize the thermodynamic properties of air in the liquid and gaseous states, as well as along the saturation line. In the table for the saturation state, *P*(boil) is the bubble point temperature (i.e., the pressure at which boiling begins as the temperature of the liquid is raised), and *P*(con) is the dew point temperature (pressure at which condensation begins as the temperature of the gas is lowered). The other properties tabulated are density (ρ), enthalpy (*H*), entropy (*S*), and isobaric heat capacity (C_p). More detailed tables may be found in the references.

REFERENCES

1. Vasserman, A.A., and Rabinovich, V.A., *Thermophysical Properties of Liquid Air and its Components*, Izdatel'stvo Komiteta, Standartov, Moscow, 1968.
2. Vasserman, A.A., et al., *Thermophysical Properties of Air and Air Components,* Izdatel'stvo Nauka, Moscow, 1966.

Properties in the saturation state:

T K	*P*(boil) bar	*P*(con) bar	ρ (liq) g/cm³	ρ (gas) g/L
65	0.1468	0.0861	0.939	0.464
70	0.3234	0.2052	0.917	1.033
75	0.6366	0.4321	0.894	2.048
80	1.146	0.8245	0.871	3.709
85	1.921	1.453	0.845	6.258
90	3.036	2.397	0.819	9.980
95	4.574	3.748	0.792	15.21
100	6.621	5.599	0.763	22.39
110	12.59	11.22	0.699	45.15
120	21.61	20.14	0.622	87.34
130	34.16	33.32	0.487	184.33
132.55	37.69	37.69	0.313	312.89

Properties of liquid air:

P bar	*T* K	ρ g/cm³	*H* J/g	*S* J/g K	C_p J/g K
1	75	0.8935	−131.7	2.918	1.843
5	75	0.8942	−131.4	2.916	1.840
5	80	0.8718	−122.3	3.031	1.868
5	85	0.8482	−112.9	3.143	1.901
5	90	0.8230	−103.3	3.250	1.941
5	95	0.7962	−93.5	3.356	1.991
10	75	0.8952	−131.1	2.913	1.836
10	80	0.8729	−122.0	3.028	1.863
10	90	0.8245	−103.1	3.246	1.932
10	100	0.7695	−83.2	3.452	2.041
50	75	0.9025	−128.2	2.892	1.806
50	100	0.7859	−81.8	3.415	1.939
50	125	0.6222	−28.3	3.889	2.614
50	150	0.1879	91.9	4.764	2.721
100	75	0.9111	−124.5	2.867	1.774
100	100	0.8033	−79.4	3.376	1.852
100	125	0.6746	−31.4	3.805	2.062
100	150	0.4871	32.8	4.271	2.832

Properties of air in the gaseous state:

P bar	*T* K	ρ g/L	*H* J/g	*S* J/g K	C_p J/g K
1	100	3.556	98.3	5.759	1.032
1	200	1.746	199.7	6.463	1.007
1	300	1.161	300.3	6.871	1.007

THERMODYNAMIC PROPERTIES OF AIR (continued)

P bar	*T* K	ρ g/L	*H* J/g	*S* J/g K	C_p J/g K
1	500	0.696	503.4	7.389	1.030
1	1000	0.348	1046.6	8.138	1.141
10	200	17.835	195.2	5.766	1.049
10	300	11.643	298.3	6.204	1.021
10	500	6.944	502.9	6.727	1.034
10	1000	3.471	1047.2	7.477	1.142
100	200	213.950	148.8	4.949	1.650
100	300	116.945	279.9	5.486	1.158
100	500	66.934	499.0	6.048	1.073
100	1000	33.613	1052.4	6.812	1.151

PROPERTIES OF WATER IN THE RANGE 0 — 100 °C

This table summarizes the best available values of the density, specific heat capacity at constant pressure (C_p), vapor pressure, viscosity, thermal conductivity, dielectric constant, and surface tension for liquid water in the range 0 — 100 °C. All values (except vapor pressure) refer to a pressure of 100 kPa (1 bar). The temperature scale is IPTS-68.

t °C	Density g/cm³	C_p J/g K	Vap. pres. kPa	Visc. µPa s	Ther. cond. mW/K m	Diel. const.	Surf. ten. mN/m
0	0.99984	4.2176	0.6113	1793	561.0	87.90	75.64
10	0.99970	4.1921	1.2281	1307	580.0	83.96	74.23
20	0.99821	4.1818	2.3388	1002	598.4	80.20	72.75
30	0.99565	4.1784	4.2455	797.7	615.4	76.60	71.20
40	0.99222	4.1785	7.3814	653.2	630.5	73.17	69.60
50	0.98803	4.1806	12.344	547.0	643.5	69.88	67.94
60	0.98320	4.1843	19.932	466.5	654.3	66.73	66.24
70	0.97778	4.1895	31.176	404.0	663.1	63.73	64.47
80	0.97182	4.1963	47.373	354.4	670.0	60.86	62.67
90	0.96535	4.2050	70.117	314.5	675.3	58.12	60.82
100	0.95840	4.2159	101.325	281.8	679.1	55.51	58.91
Ref.	1—3	2	1, 3	3	3	4	5

REFERENCES

1. L. Harr, J. S. Gallagher, and G. S. Kell, *NBS/NRC Steam Tables*, Hemisphere Publishing Corp., 1984.
2. K. N. Marsh, Ed., *Recommended Reference Materials for the Realization of Physicochemical Properties*, Blackwell Scientific Publications, Oxford, 1987.
3. J. V. Sengers and J. T. R. Watson, Improved international formulations for the viscosity and thermal conductivity of water substance, *J. Phys. Chem. Ref. Data*, 15, 1291, 1986.
4. D. G. Archer and P. Wang, The dielectric constant of water and Debye-Hückel limiting law slopes, *J. Phys. Chem. Ref. Data*, 19, 371, 1990.
5. N. B. Vargaftik, et al., International tables of the surface tension of water, *J. Phys. Chem. Ref. Data*, 12, 817, 1983.

ENTHALPY OF VAPORIZATION OF WATER

The enthalpy (heat) of vaporization of water is tabulated as a function of temperature on the IPTS-68 scale.

REFERENCE

Marsh, K. N., Ed., *Recommended Reference Materials for the Realization of Physicochemical Properties*, Blackwell, Oxford, 1987.

t °C	$\Delta_{vap}H$ kJ/mol	*t* °C	$\Delta_{vap}H$ kJ/mol
0	45.054	200	34.962
25	43.990	220	33.468
40	43.350	240	31.809
60	42.482	260	29.930
80	41.585	280	27.795
100	40.657	300	25.300
120	39.684	320	22.297
140	38.643	340	18.502
160	37.518	360	12.966
180	36.304	374	2.066

FIXED POINT PROPERTIES OF H_2O AND D_2O

Temperatures are given on the IPTS-68 scale.

REFERENCES

1. Haar, L., Gallagher, J.S., and Kell, G.S., *NBS/NRC Steam Tables*, Hemisphere Publishing Corp., 1984.
2. Levelt Sengers, J.M.H., Straub, J., Watanabe, K., and Hill, P.G., Assessment of critical parameter values for H_2O and D_2O, *J. Phys. Chem. Ref. Data,* 14, 193, 1985.
3. Kestin, J. et. al., Thermophysical properties of fluid D_2O, *J. Phys. Chem. Ref. Data,* 13, 601, 1984.
4. Kestin, J. et. al., Thermophysical properties of fluid H_2O, *J. Phys. Chem. Ref. Data,* 13, 175, 1984.
5. Hill, P.G., MacMillan, R.D.C., and Lee, V., A fundamental equation of state for heavy water, *J. Phys. Chem. Ref. Data,* 11, 1, 1982.

	Unit	H_2O	D_2O
Molar mass	g/mol	18.01528	20.02748
Melting point(101.325 kPa)	°C	0.00	3.82
Boiling point(101.325 kPa)	°C	100.00	101.42
Triple point temperature	°C	0.01	3.82
Triple point pressure	Pa	611.73	661
Triple point density(l)	g/cm³	0.99978	1.1055
Triple point density(g)	mg/L	4.885	5.75
Critical temperature	°C	373.99	370.74
Critical pressure	MPa	22.064	21.671
Critical density	g/cm³	0.322	0.356
Critical specific volume	cm³/g	3.11	2.81
Maximum density(saturated liquid)	g/cm³	0.99995	1.1053
Temperature of maximum density	°C	4.0	11.2

THERMAL CONDUCTIVITY OF SATURATED H_2O AND D_2O

This table gives the thermal conductivity λ for water (H_2O or D_2O) in equilibrium with its vapor. Values for the liquid (λ_l) and vapor (λ_v) are listed, as well as the vapor pressure.

REFERENCES

1. Sengers, J.V. and Watson, J.T.R., Improved international formulations for the viscosity and thermal conductivity of water substance, *J. Phys. Chem. Ref. Data,* 15, 1291, 1986.
2. Matsunaga, N. and Nagashima, A., Transport properties of liquid and gaseous D_2O over a wide range of temperature and pressure, *J. Phys. Chem. Ref. Data,* 12, 933, 1983.

	H_2O			D_2O		
t/°C	*P*/kPa	λ_l/(mW/K m)	λ_v/(mW/K m)	*P*/kPa	λ_l/(mW/K m)	λ_v/(mW/K m)
0	0.6	561.0	16.49			
10	1.2	580.0	17.21	1.0	575	17.0
20	2.3	598.4	17.95	2.0	589	17.8
30	4.2	615.4	18.70	3.7	600	18.5
40	7.4	630.5	19.48	6.5	610	19.3
50	12.3	643.5	20.28	11.1	618	20.2
60	19.9	654.3	21.10	18.2	625	21.0
70	31.2	663.1	21.96	28.8	629	21.9
80	47.4	670.0	22.86	44.2	633	22.8
90	70.1	675.3	23.80	66.1	635	23.8
100	101.3	679.1	24.79	96.2	636	24.8
150	476	682.1	30.77	465	625	30.8
200	1555	663.4	39.10	1546	592	39.0
250	3978	621.4	51.18	3995	541	52.0
300	8593	547.7	71.78	8688	473	75.2
350	16530	447.6	134.59	16820	391	143.0

STANDARD DENSITY OF WATER

This table gives the density ρ of standard mean ocean water (SMOW), free from dissolved salts and gases, at a pressure of 101325 Pa. SMOW is a standard water sample of high purity and known isotopic composition. Methods of correcting for different isotopic compositions are discussed in the reference. The table below is reprinted with the permission of IUPAC. Note that the temperature scale is IPTS-68.

REFERENCE

Marsh, K. N., Ed., *Recommended Reference Materials for the Realization of Physicochemical Properties,* Blackwell Scientific Publications, Oxford, 1987.

t_{68}/°C	ρ/kg m^{-3} 0.0	0.1	0.2	0.3	0.4	0.5	0.6	0.7	0.8	0.9
0	999.8426	8493	8558	8622	8683	8743	8801	8857	8912	8964
1	999.9015	9065	9112	9158	9202	9244	9284	9323	9360	9395
2	999.9429	9461	9491	9519	9546	9571	9595	9616	9636	9655
3	999.9672	9687	9700	9712	9722	9731	9738	9743	9747	9749
4	999.9750	9748	9746	9742	9736	9728	9719	9709	9696	9683
5	999.9668	9651	9632	9612	9591	9568	9544	9518	9490	9461
6	999.9430	9398	9365	9330	9293	9255	9216	9175	9132	9088
7	999.9043	8996	8948	8898	8847	8794	8740	8684	8627	8569
8	999.8509	8448	8385	8321	8256	8189	8121	8051	7980	7908
9	999.7834	7759	7682	7604	7525	7444	7362	7279	7194	7108
10	999.7021	6932	6842	6751	6658	6564	6468	6372	6274	6174
11	999.6074	5972	5869	5764	5658	5551	5443	5333	5222	5110
12	999.4996	4882	4766	4648	4530	4410	4289	4167	4043	3918
13	999.3792	3665	3536	3407	3276	3143	3010	2875	2740	2602
14	999.2464	2325	2184	2042	1899	1755	1609	1463	1315	1166
15	999.1016	0864	0712	0558	0403	0247	0090	9932*	9772*	9612*
16	998.9450	9287	9123	8957	8791	8623	8455	8285	8114	7942
17	998.7769	7595	7419	7243	7065	6886	6706	6525	6343	6160
18	998.5976	5790	5604	5416	5228	5038	4847	4655	4462	4268
19	998.4073	3877	3680	3481	3282	3081	2880	2677	2474	2269
20	998.2063	1856	1649	1440	1230	1019	0807	0594	0380	0164
21	997.9948	9731	9513	9294	9073	8852	8630	8406	8182	7957
22	997.7730	7503	7275	7045	6815	6584	6351	6118	5883	5648
23	997.5412	5174	4936	4697	4456	4215	3973	3730	3485	3240
24	997.2994	2747	2499	2250	2000	1749	1497	1244	0990	0735
25	997.0480	0223	9965*	9707*	9447*	9186*	8925*	8663*	8399*	8135*
26	996.7870	7604	7337	7069	6800	6530	6259	5987	5714	5441
27	996.5166	4891	4615	4337	4059	3780	3500	3219	2938	2655
28	996.2371	2087	1801	1515	1228	0940	0651	0361	0070	9778*
29	995.9486	9192	8898	8603	8306	8009	7712	7413	7113	6813
30	995.6511	6209	5906	5602	5297	4991	4685	4377	4069	3760
31	995.3450	3139	2827	2514	2201	1887	1572	1255	0939	0621
32	995.0302	9983*	9663*	9342*	9020*	8697*	8373*	8049*	7724*	7397*
33	994.7071	6743	6414	6085	5755	5423	5092	4759	4425	4091
34	994.3756	3420	3083	2745	2407	2068	1728	1387	1045	0703
35	994.0359	0015	9671*	9325*	8978*	8631*	8283*	7934*	7585*	7234*
36	993.6883	6531	6178	5825	5470	5115	4759	4403	4045	3687
37	993.3328	2968	2607	2246	1884	1521	1157	0793	0428	0062
38	992.9695	9328	8960	8591	8221	7850	7479	7107	6735	6361
39	992.5987	5612	5236	4860	4483	4105	3726	3347	2966	2586
40	992.2204									

* The leading figure decreases by 1.

PROPERTIES OF ICE AND SUPERCOOLED WATER

The common form of ice at ambient temperature and pressure is hexagonal ice, designated as ice I_h (see phase diagram on p. 12-174). The data given here refer to that form. Data have been taken from the references indicated; values have been interpolated and smoothed in some cases. All properties are sensitive to the method of preparation of the sample, since air or other gases are sometimes occluded. For this reason there is often disagreement among values found in the literature.

Density values (except at 0°C) and the thermal expansion coefficient were calculated from the temperature variation in the crystal lattice constants of ice (see Ref. 1). The thermal expansion coefficient appears to become negative around -200°C, but there is considerable scatter in the data.

Density of ice I_h and supercooled water in g cm^{-3}

t/°C	ρ (ice)	ρ (supercooled water)
0	0.9167	0.9998
-10	0.9187	0.9982
-20	0.9203	0.9935
-30	0.9216	0.9839
-40	0.9228	
-50	0.9240	
-60	0.9252	
-80	0.9274	
-100	0.9292	
-120	0.9305	
-140	0.9314	
-160	0.9331	
-180	0.9340	
Ref.	1	8

Phase transition properties:

$\Delta_{fus}H(0°C) = 333.6$ J/g (Ref. 2)

$\Delta_{sub}H(0°C) = 2838$ J/g (Ref. 2)

Other properties of ice I_h :

α_V: cubic thermal expansion coefficient, $\alpha_V = -(1/V)(\partial V/\partial t)_p$

κ: adiabatic compressibility, $\kappa = -(1/V)(\partial V/\partial p)_S$

ε: relative permittivity (dielectric constant)

k: thermal conductivity

c_p: specific heat capacity at constant pressure

t/°C	$\alpha_V/10^{-6}$ °C^{-1}	$\kappa/10^{-5}$ MPa^{-1}	ε	k/W cm^{-1} °C^{-1}	c_p/J g^{-1} °C^{-1}
0	159	13.0	91.6	0.0214	2.11
–10	155	12.8	94.4	0.023	2.03
–20	149	12.7	97.5	0.024	1.96
–30	143	12.5	99.7	0.025	1.88
–40	137	12.4	101.9	0.026	1.80
–50	130	12.2	106.9	0.028	1.72
–60	122	12.1	119.5	0.030	1.65
–80	105	11.9		0.033	1.50
–100	85	11.6		0.037	1.36
–120	77	11.4		0.042	1.23
–140	60	11.3		0.049	1.10
–160	45	11.2		0.057	0.97
–180	30	11.1		0.070	0.83
–200		11.0		0.087	0.67
–220		10.9		0.118	0.50
–240		10.9		0.20	0.29
–250		10.9		0.32	0.17
Ref.	1,2,3,5	1,5	6	7	1

REFERENCES

1. Eisenberg, D., and Kauzmann, W., *The Structure and Properties of Water*, Oxford University Press, Oxford, 1969.
2. *Landolt-Börnstein, Numerical Data and Functional Relationships in Science and Technology, New Series*, V/1b, Springer-Verlag, Heidelberg, 1982.
3. LaPlaca, S., and Post, B., *Acta Cryst.*, 13, 503, 1960. [Thermal expansion of lattice]
4. Brill, R., and Tippe, A., *Acta Cryst.*, 23, 343, 1967. [Thermal expansion of lattice]
5. Leadbetter, A. J., *Proc. Roy. Soc. A* 287, 403, 1965. [Compressibility and thermal expansion]
6. Auty, R. P., and Cole, R. H., *J. Chem. Phys.*, 20, 1309, 1952. [Dielectric constant]
7. Slack, G. A., *Phys. Rev. B*, 22, 3065, 1980. [Thermal conductivity]
8. Hare, D. E., and Sorensen, C. M., *J. Chem. Phys.*, 87, 4840, 1987. [Supercooled water]
9. Hobbs, P. V., *Ice Physics*, Clarendon Press, Oxford, 1974.

VOLUMETRIC PROPERTIES OF AQUEOUS SODIUM CHLORIDE SOLUTIONS

This table gives the following properties of aqueous solutions of NaCl as a function of temperature and concentration:

Specific volume v (reciprocal of density) in cm^3/g
Isothermal compressibility $\kappa_T = -(1/v)(\partial v/\partial P)_T$ in GPa^{-1}
Cubic expansion coefficient $\alpha_v = (1/v)(\partial v/\partial T)_P$ in kK^{-1}

All data refer to a pressure of 100 kPa (1 bar). The reference gives properties over a wider range of temperature and pressure.

REFERENCE

Rogers, P. S. Z., and Pitzer, K. S., *J. Phys. Chem. Ref. Data,* 11, 15, 1982.

T/°C	Molality in mol/kg 0.100	0.250	0.500	0.750	1.000	2.000	3.000	4.000	5.000
	Specific volume v in cm^3/g								
0	0.995732	0.989259	0.978889	0.968991	0.959525	0.925426	0.896292	0.870996	0.848646
10	0.995998	0.989781	0.979804	0.970256	0.961101	0.927905	0.899262	0.874201	0.851958
20	0.997620	0.991564	0.981833	0.972505	0.963544	0.930909	0.902565	0.877643	0.855469
25	0.998834	0.992832	0.983185	0.973932	0.965038	0.932590	0.904339	0.879457	0.857301
30	1.000279	0.994319	0.984735	0.975539	0.966694	0.934382	0.906194	0.881334	0.859185
40	1.003796	0.997883	0.988374	0.979243	0.970455	0.938287	0.910145	0.885276	0.863108
50	1.008064	1.002161	0.992668	0.983551	0.974772	0.942603	0.914411	0.889473	0.867241
60	1.0130	1.0071	0.9976	0.9885	0.9797	0.9474	0.9191	0.8940	0.8716
70	1.0186	1.0127	1.0031	0.9939	0.9851	0.9526	0.9240	0.8987	0.8762
80	1.0249	1.0188	1.0092	0.9999	0.9909	0.9581	0.9293	0.9037	0.8809
90	1.0317	1.0256	1.0157	1.0063	0.9972	0.9640	0.9348	0.9089	0.8858
100	1.0391	1.0329	1.0228	1.0133	1.0040	0.9703	0.9406	0.9144	0.8910
	Compressibility κ_T in GPa^{-1}								
0	0.503	0.492	0.475	0.459	0.443	0.389	0.346	0.315	0.294
10	0.472	0.463	0.449	0.436	0.423	0.377	0.341	0.313	0.294
20	0.453	0.446	0.433	0.422	0.411	0.371	0.338	0.313	0.294
25	0.447	0.440	0.428	0.417	0.407	0.369	0.337	0.313	0.294
30	0.443	0.436	0.425	0.414	0.404	0.367	0.337	0.313	0.294
40	0.438	0.432	0.421	0.411	0.401	0.367	0.338	0.315	0.296
50	0.438	0.431	0.421	0.411	0.402	0.369	0.340	0.317	0.299
60	0.44	0.44	0.43	0.42	0.41	0.38	0.35	0.32	0.30
70	0.45	0.44	0.43	0.42	0.42	0.38	0.36	0.33	0.31
80	0.46	0.45	0.44	0.43	0.43	0.39	0.37	0.34	0.32
90	0.47	0.47	0.46	0.45	0.44	0.41	0.38	0.35	0.33
100	0.49	0.48	0.47	0.46	0.45	0.42	0.39	0.37	0.34
	Cubic expansion coefficient α_V in kK^{-1}								
0	-0.058	-0.026	0.024	0.069	0.110	0.237	0.313	0.355	
10	0.102	0.123	0.156	0.186	0.213	0.297	0.349	0.380	
20	0.218	0.232	0.254	0.274	0.292	0.349	0.384	0.406	
25	0.267	0.278	0.296	0.312	0.327	0.373	0.401	0.420	
30	0.311	0.320	0.334	0.347	0.359	0.395	0.418	0.433	
40	0.389	0.394	0.402	0.410	0.417	0.438	0.451	0.460	
50	0.458	0.460	0.464	0.467	0.470	0.479	0.484	0.486	
60	0.52	0.52	0.52	0.52	0.52	0.52	0.52	0.52	
70	0.58	0.58	0.58	0.57	0.57	0.56	0.55	0.54	
80	0.64	0.63	0.63	0.62	0.61	0.60	0.58	0.56	
90	0.69	0.68	0.67	0.67	0.66	0.63	0.61	0.59	
100	0.74	0.73	0.72	0.71	0.70	0.66	0.64	0.61	

DENSITY OF D_2O

Density of liquid D_2O in g/cm³ at a pressure of 100 kPa (1 bar).

REFERENCE

Kirillin, V.A., Ed., *Heavy Water: Thermophysical Properties*, Gosudarstvennoe Energeticheskoe Izdatel'stvo, Moscow, 1963.

t/°C	3.8	5	10	15	20	25	30
Density	1.1053	1.1055	1.1057	1.1056	1.105	1.1044	1.1034
t/°C	35	40	45	50	55	60	65
Density	1.1019	1.1001	1.0979	1.0957	1.0931	1.0905	1.0875
t/°C	70	75	80	85	90	95	100
Density	1.0847	1.0815	1.0783	1.0748	1.0712	1.0673	1.0635

VAPOR PRESSURE OF ICE

The values of the vapor (sublimation) pressure of ice in this table were calculated from the equation recommended by the International Association for the Properties of Steam (IAPS) in 1993. Temperature values correspond to the ITS-90 temperature scale. The uncertainty in the pressure is estimated to be 0.1% for $t > -25$°C and 0.5% for $t < -25$°C. The first entry in the table is the triple point of water.

REFERENCE

Wagner, W., Saul, A., and Pruss, A., *J. Phys. Chem. Ref. Data,* 23, 515, 1994.

t/°C	*p*/Pa	*t*/°C	*p*/Pa	*t*/°C	*p*/Pa
0.01	611.657	–16	150.68	–33	27.71
0	611.15	–17	137.25	–34	24.90
–1	562.67	–18	124.92	–35	22.35
–2	517.72	–19	113.62	–36	20.04
–3	476.06	–20	103.26	–37	17.96
–4	437.47	–21	93.77	–38	16.07
–5	401.76	–22	85.10	–39	14.37
–6	368.73	–23	77.16	–40	12.84
–7	338.19	–24	69.91	–45	7.202
–8	309.98	–25	63.29	–50	3.936
–9	283.94	–26	57.25	–55	2.093
–10	259.90	–27	51.74	–60	1.080
–11	237.74	–28	46.73	–65	0.540
–12	217.32	–29	42.16	–70	0.261
–13	198.52	–30	38.01	–75	0.122
–14	181.22	–31	34.24	–80	0.055
–15	165.30	–32	30.82		

VAPOR PRESSURE OF WATER FROM 0 TO 370° C

This table gives the vapor pressure of water at intervals of 1° C from the melting point to the critical point.

REFERENCE

Haar, L., Gallagher, J.S., and Kell, G.S., *NBS/NRC Steam Tables*, Hemisphere Publishing Corp., New York, 1984.

t/°C	*P*/kPa	*t*/°C	*P*/kPa	*t*/°C	*P*/kPa	*t*/°C	*P*/kPa
0	0.61129	53	14.303	106	125.03	159	602.11
1	0.65716	54	15.012	107	129.39	160	617.66
2	0.70605	55	15.752	108	133.88	161	633.53
3	0.75813	56	16.522	109	138.50	162	649.73
4	0.81359	57	17.324	110	143.24	163	666.25
5	0.87260	58	18.159	111	148.12	164	683.10
6	0.93537	59	19.028	112	153.13	165	700.29
7	1.0021	60	19.932	113	158.29	166	717.83
8	1.0730	61	20.873	114	163.58	167	735.70
9	1.1482	62	21.851	115	169.02	168	753.94
10	1.2281	63	22.868	116	174.61	169	772.52
11	1.3129	64	23.925	117	180.34	170	791.47
12	1.4027	65	25.022	118	186.23	171	810.78
13	1.4979	66	26.163	119	192.28	172	830.47
14	1.5988	67	27.347	120	198.48	173	850.53
15	1.7056	68	28.576	121	204.85	174	870.98
16	1.8185	69	29.852	122	211.38	175	891.80
17	1.9380	70	31.176	123	218.09	176	913.03
18	2.0644	71	32.549	124	224.96	177	934.64
19	2.1978	72	33.972	125	232.01	178	956.66
20	2.3388	73	35.448	126	239.24	179	979.09
21	2.4877	74	36.978	127	246.66	180	1001.9
22	2.6447	75	38.563	128	254.25	181	1025.2
23	2.8104	76	40.205	129	262.04	182	1048.9
24	2.9850	77	41.905	130	270.02	183	1073.0
25	3.1690	78	43.665	131	278.20	184	1097.5
26	3.3629	79	45.487	132	286.57	185	1122.5
27	3.5670	80	47.373	133	295.15	186	1147.9
28	3.7818	81	49.324	134	303.93	187	1173.8
29	4.0078	82	51.342	135	312.93	188	1200.1
30	4.2455	83	53.428	136	322.14	189	1226.9
31	4.4953	84	55.585	137	331.57	190	1254.2
32	4.7578	85	57.815	138	341.22	191	1281.9
33	5.0335	86	60.119	139	351.09	192	1310.1
34	5.3229	87	62.499	140	361.19	193	1338.8
35	5.6267	88	64.958	141	371.53	194	1368.0
36	5.9453	89	67.496	142	382.11	195	1397.6
37	6.2795	90	70.117	143	392.92	196	1427.8
38	6.6298	91	72.823	144	403.98	197	1458.5
39	6.9969	92	75.614	145	415.29	198	1489.7
40	7.3814	93	78.494	146	426.85	199	1521.4
41	7.7840	94	81.465	147	438.67	200	1553.6
42	8.2054	95	84.529	148	450.75	201	1586.4
43	8.6463	96	87.688	149	463.10	202	1619.7
44	9.1075	97	90.945	150	475.72	203	1653.6
45	9.5898	98	94.301	151	488.61	204	1688.0
46	10.094	99	97.759	152	501.78	205	1722.9
47	10.620	100	101.32	153	515.23	206	1758.4
48	11.171	101	104.99	154	528.96	207	1794.5
49	11.745	102	108.77	155	542.99	208	1831.1
50	12.344	103	112.66	156	557.32	209	1868.4
51	12.970	104	116.67	157	571.94	210	1906.2
52	13.623	105	120.79	158	586.87	211	1944.6

VAPOR PRESSURE OF WATER FROM 0 TO 370° C (continued)

t/°C	P/kPa	t/°C	P/kPa	t/°C	P/kPa	t/°C	P/kPa
212	1983.6	253	4178.9	294	7881.3	335	13701
213	2023.2	254	4249.1	295	7995.2	336	13876
214	2063.4	255	4320.2	296	8110.3	337	14053
215	2104.2	256	4392.2	297	8226.8	338	14232
216	2145.7	257	4465.1	298	8344.5	339	14412
217	2187.8	258	4539.0	299	8463.5	340	14594
218	2230.5	259	4613.7	300	8583.8	341	14778
219	2273.8	260	4689.4	301	8705.4	342	14964
220	2317.8	261	4766.1	302	8828.3	343	15152
221	2362.5	262	4843.7	303	8952.6	344	15342
222	2407.8	263	4922.3	304	9078.2	345	15533
223	2453.8	264	5001.8	305	9205.1	346	15727
224	2500.5	265	5082.3	306	9333.4	347	15922
225	2547.9	266	5163.8	307	9463.1	348	16120
226	2595.9	267	5246.3	308	9594.2	349	16320
227	2644.6	268	5329.8	309	9726.7	350	16521
228	2694.1	269	5414.3	310	9860.5	351	16725
229	2744.2	270	5499.9	311	9995.8	352	16931
230	2795.1	271	5586.4	312	10133	353	17138
231	2846.7	272	5674.0	313	10271	354	17348
232	2899.0	273	5762.7	314	10410	355	17561
233	2952.1	274	5852.4	315	10551	356	17775
234	3005.9	275	5943.1	316	10694	357	17992
235	3060.4	276	6035.0	317	10838	358	18211
236	3115.7	277	6127.9	318	10984	359	18432
237	3171.8	278	6221.9	319	11131	360	18655
238	3228.6	279	6317.0	320	11279	361	18881
239	3286.3	280	6413.2	321	11429	362	19110
240	3344.7	281	6510.5	322	11581	363	19340
241	3403.9	282	6608.9	323	11734	364	19574
242	3463.9	283	6708.5	324	11889	365	19809
243	3524.7	284	6809.2	325	12046	366	20048
244	3586.3	285	6911.1	326	12204	367	20289
245	3648.8	286	7014.1	327	12364	368	20533
246	3712.1	287	7118.3	328	12525	369	20780
247	3776.2	288	7223.7	329	12688	370	21030
248	3841.2	289	7330.2	330	12852	371	21283
249	3907.0	290	7438.0	331	13019	372	21539
250	3973.6	291	7547.0	332	13187	373	21799
251	4041.2	292	7657.2	333	13357	373.98	22055
252	4109.6	293	7768.6	334	13528		

BOILING POINT OF WATER AT VARIOUS PRESSURES

Data are based on the equation of state recommended by the International Association for the Properties of Steam in 1984, as presented in Haar, Gallagher, and Kell, *NBS-NRC Steam Tables* (Hemisphere Publishing Corp., New York, 1984). The temperature scale is IPTS-68.
Note that: 1 mbar = 100 Pa = 0.000986923 atmos = 0.750062 mmHg.

P/mbar	*T*/°C	*P*/mbar	*T*/°C	*P*/mbar	*T*/°C	*P*/mbar	*T*/°C
50	32.88	915	97.17	1013.25	100.00	1200	104.81
100	45.82	920	97.32	1015	100.05	1250	105.99
150	53.98	925	97.47	1020	100.19	1300	107.14
200	60.07	930	97.62	1025	100.32	1350	108.25
250	64.98	935	97.76	1030	100.46	1400	109.32
300	69.11	940	97.91	1035	100.60	1450	110.36
350	72.70	945	98.06	1040	100.73	1500	111.38
400	75.88	950	98.21	1045	100.87	1550	112.37
450	78.74	955	98.35	1050	101.00	1600	113.33
500	81.34	960	98.50	1055	101.14	1650	114.26
550	83.73	965	98.64	1060	101.27	1700	115.18
600	85.95	970	98.78	1065	101.40	1750	116.07
650	88.02	975	98.93	1070	101.54	1800	116.94
700	89.96	980	99.07	1075	101.67	1850	117.79
750	91.78	985	99.21	1080	101.80	1900	118.63
800	93.51	990	99.35	1085	101.93	1950	119.44
850	95.15	995	99.49	1090	102.06	2000	120.24
900	96.71	1000	99.63	1095	102.19	2050	121.02
905	96.87	1005	99.77	1100	102.32	2100	121.79
910	97.02	1010	99.91	1150	103.59	2150	122.54

MELTING POINT OF ICE AS A FUNCTION OF PRESSURE

This table gives values of the melting temperature of ice at various pressures, as calculated from the equation for the ice I - liquid water phase boundary recommended by the International Association for the Properties of Steam (IAPS). Temperatures are on the ITS-90 scale. See the Reference for information on forms of ice that exist at higher pressures. The transition points for transformations of the various forms of ice (in each case in equilibrium with liquid water) are:

ice I - ice III	209.9 MPa	–21.985°C
ice III - ice V	350.1	–16.986
ice V - ice VI	632.4	0.16
ice VI - ice VII	2216	82

REFERENCE

Wagner, W., Saul, A., and Pruss, A., *J. Phys. Chem. Ref. Data,* 23, 515, 1994.

p/MPa	*t*/°C	*p*/MPa	*t*/°C	*p*/MPa	*t*/°C
0.1	0.00	40	-3.15	130	-12.07
1	-0.06	50	-4.02	140	-13.22
2	-0.14	60	-4.91	150	-14.40
3	-0.21	70	-5.83	160	-15.62
4	-0.29	80	-6.79	170	-16.85
5	-0.36	90	-7.78	180	-18.11
10	-0.74	100	-8.80	190	-19.39
20	-1.52	110	-9.86	200	-20.69
30	-2.32	120	-10.95	210	-22.00

PROPERTIES OF WATER AND STEAM AS A FUNCTION OF TEMPERATURE AND PRESSURE

This table gives properties of compressed water and superheated steam at selected pressures and temperatures. The properties included are density ρ, enthalpy H, entropy S, heat capacity at constant pressure C_p, and static dielectric constant (relative permittivity). The table was generated from the formulation approved by the International Association for the Properties of Water and Steam for general and scientific use. The reference state for this table is the liquid at the triple point, at which the internal energy and entropy are taken as zero. A duplicate entry in the temperature column indicates a phase transition (liquid-vapor) at that temperature; property values are then given for both phases. In the 100 MPa section of the table, an entry is given at the critical temperature, 647.10 K. Temperatures refer to the ITS-90 scale, on which the normal boiling point of water is 373.12 K (99.97°C).

REFERENCES

1. Release on the IAPWS Formulation 1995 for the Thermodynamic Properties of Ordinary Water Substance for General and Scientific Use, September 1996; available from Executive Secretary of IAPWS, Electric Power Research Institute, 3412 Hillview Ave., Palo Alto, CA 94304-1395.
2. NIST Chemistry WebBook, NIST Standard Reference Database Number 69, Mallard, W. G., and Linstrom, P. J., Eds., March 1998, National Institute of Standards and Technology, Gaithersburg, MD, 20899 (http://webbook.nist.gov).
3. Pruss, A. and Wagner, W., to be published.
4. Fernandez, D. P., Goodwin, A. R. H., Lemmon, E. W., Levelt Sengers, J. M. H., and Williams, R. C., *J. Phys. Chem. Ref. Data,* 26, 1125, 1997. [Dielectric constant]

p/MPa	T/K	ρ/kg m^{-3}	H/J g^{-1}	S/J g^{-1}K^{-1}	C_p/J g^{-1}K^{-1}	Diel. const.
0.1	273.16	999.84	0.10	0.0000	4.2194	87.90
0.1	300	996.56	112.65	0.3931	4.1806	77.75
0.1	325	987.19	217.15	0.7276	4.1819	69.32
0.1	350	973.73	321.84	1.0380	4.1945	61.79
0.1	372.76	958.63	417.50	1.3028	4.2152	55.61
0.1	372.76	0.59034	2674.9	7.3588	2.0784	1.006
0.1	375	0.58653	2679.6	7.3713	2.0686	1.006
0.1	400	0.54761	2730.4	7.5025	2.0078	1.005
0.1	450	0.48458	2829.7	7.7365	1.9752	1.004
0.1	500	0.43514	2928.6	7.9447	1.9813	1.003
0.1	550	0.39507	3028.1	8.1344	2.0010	1.003
0.1	600	0.36185	3128.8	8.3096	2.0268	1.002
0.1	650	0.33384	3230.8	8.4730	2.0557	1.002
0.1	700	0.30988	3334.4	8.6264	2.0867	1.002
0.1	750	0.28915	3439.5	8.7715	2.1191	1.002
0.1	800	0.27102	3546.3	8.9093	2.1525	1.001
0.1	850	0.25504	3654.8	9.0408	2.1868	1.001
0.1	900	0.24085	3765.0	9.1668	2.2216	1.001
0.1	950	0.22815	3876.9	9.2879	2.2568	1.001
0.1	1000	0.21673	3990.7	9.4045	2.2921	1.001
0.1	1050	0.20640	4106.1	9.5172	2.3273	1.001
0.1	1100	0.19701	4223.4	9.6263	2.3621	1.001
0.1	1150	0.18844	4342.3	9.7321	2.3965	1.001
0.1	1200	0.18058	4463.0	9.8348	2.4302	1.001
1	273.16	1000.3	1.02	0.0000	4.2150	87.93
1	300	996.96	113.48	0.3928	4.1781	77.78
1	325	987.58	217.93	0.7272	4.1798	69.36
1	350	974.13	322.56	1.0374	4.1925	61.82
1	375	957.43	427.64	1.3274	4.2158	55.09
1	400	937.87	533.47	1.6005	4.2535	49.06
1	450	890.39	749.20	2.1086	4.3924	38.81
1	453.03	887.13	762.51	2.1381	4.4045	38.23
1	453.03	5.1450	2777.1	6.5850	2.7114	1.042
1	500	4.5323	2891.2	6.8250	2.2795	1.034

PROPERTIES OF WATER AND STEAM AS A FUNCTION OF TEMPERATURE AND PRESSURE (continued)

p/MPa	*T*/K	ρ/kg m^{-3}	*H*/J g^{-1}	*S*/J g^{-1}K^{-1}	C_p/J g^{-1}K^{-1}	Diel. const.
1	550	4.0581	3001.8	7.0359	2.1647	1.028
1	600	3.6871	3109.0	7.2224	2.1292	1.024
1	650	3.3843	3215.2	7.3925	2.1254	1.020
1	700	3.1305	3321.7	7.5504	2.1368	1.017
1	750	2.9140	3429.0	7.6984	2.1566	1.015
1	800	2.7265	3537.5	7.8384	2.1816	1.013
1	850	2.5624	3647.3	7.9715	2.2098	1.012
1	900	2.4174	3758.5	8.0986	2.2402	1.011
1	950	2.2882	3871.3	8.2206	2.2721	1.010
1	1000	2.1723	3985.7	8.3380	2.3048	1.009
1	1050	2.0678	4101.8	8.4512	2.3380	1.008
1	1100	1.9729	4219.5	8.5608	2.3713	1.007
1	1150	1.8865	4338.9	8.6669	2.4044	1.007
1	1200	1.8074	4460.0	8.7699	2.4371	1.006
10	273.16	1004.8	10.1	0.000	4.173	88.30
10	300	1001.0	121.7	0.390	4.153	78.11
10	325	991.46	225.6	0.723	4.160	69.67
10	350	978.09	329.7	1.031	4.173	62.13
10	375	961.62	434.4	1.320	4.195	55.40
10	400	942.42	539.6	1.592	4.230	49.39
10	450	896.16	753.9	2.096	4.355	39.17
10	500	838.02	977.1	2.566	4.602	30.79
10	550	761.82	1218	3.027	5.140	23.53
10	584.15	688.42	1408	3.360	6.123	18.70
10	584.15	55.463	2725	5.616	7.140	1.404
10	600	49.773	2820	5.775	5.136	1.365
10	650	40.479	3022	6.100	3.396	1.267
10	700	35.355	3177	6.330	2.874	1.214
10	750	31.810	3314	6.520	2.645	1.179
10	800	29.107	3443	6.686	2.531	1.154
10	850	26.933	3568	6.838	2.473	1.134
10	900	25.123	3691	6.978	2.445	1.118
10	950	23.580	3813	7.110	2.436	1.105
10	1000	22.241	3935	7.235	2.439	1.095
10	1050	21.063	4057	7.354	2.450	1.086
10	1100	20.017	4180	7.469	2.466	1.078
10	1150	19.078	4304	7.579	2.485	1.072
10	1200	18.230	4429	7.685	2.507	1.066
100	273.16	1045.3	95.4	-0.008	3.905	91.83
100	300	1037.2	201.4	0.362	3.979	81.22
100	325	1026.6	301.3	0.682	4.008	72.58
100	350	1013.6	401.7	0.979	4.025	64.95
100	375	998.59	502.6	1.258	4.040	58.19
100	400	981.82	603.7	1.518	4.056	52.20
100	450	943.51	807.8	1.999	4.110	42.15
100	500	899.21	1015	2.436	4.196	34.15
100	550	848.78	1228	2.842	4.323	27.67
100	600	791.49	1448	3.225	4.501	22.29
100	647.10	730.24	1665	3.573	4.733	17.97
100	650	726.21	1679	3.595	4.750	17.72
100	700	651.77	1925	3.958	5.083	13.75
100	750	568.52	2188	4.322	5.449	10.34
100	800	482.23	2466	4.681	5.610	7.562
100	850	404.66	2742	5.016	5.380	5.571
100	900	343.61	3000	5.310	4.887	4.284
100	950	298.61	3231	5.560	4.382	3.477
100	1000	265.45	3440	5.774	3.978	2.956
100	1050	240.32	3631	5.961	3.683	2.601
100	1100	220.62	3809	6.127	3.471	2.347
100	1150	204.71	3979	6.278	3.319	2.158
100	1200	191.53	4142	6.417	3.209	2.011

PERMITTIVITY (DIELECTRIC CONSTANT) OF WATER AT VARIOUS FREQUENCIES

The permittivity of liquid water in the radiofrequency and microwave regions can be represented by the Debye equation (References 1 and 2):

$$\varepsilon' = \varepsilon_\infty + \frac{\varepsilon_s - \varepsilon_\infty}{1+\omega^2\tau^2}$$

$$\varepsilon'' = \frac{(\varepsilon_s - \varepsilon_\infty)\omega\tau}{1+\omega^2\tau^2}$$

where $\varepsilon = \varepsilon' + i\,\varepsilon''$ is the (complex) relative permittivity (i.e., the absolute permittivity divided by the permittivity of free space $\varepsilon_0 = 8.854 \cdot 10^{-12}$ F m^{-1}). Here ε_s is the static permittivity (see Reference 3 and the table "Properties of Water in the Range 0—100°C" in this Section); ε_∞ is a parameter describing the permittivity in the high frequency limit; τ is the relaxation time for molecular orientation; and $\omega = 2\pi f$ is the angular frequency. The values in this table have been calculated from parameters given in Reference 2:

	0°C	25°C	50°C
ε_∞	5.7	5.2	4.0
τ/ps	17.67	8.27	4.75

Other useful quantities that can be calculated from the values in the table are the loss tangent:

$$\tan\delta = \varepsilon''/\varepsilon'$$

and the absorption coefficient α which describes the power attentuation per unit length ($P = P_0\, e^{-\alpha l}$):

$$\alpha = \frac{\pi f \varepsilon''}{c\sqrt{\varepsilon'}}$$

and c is the speed of light. The last equation is valid when $\varepsilon''/\varepsilon' << 1$.

REFERENCES

1. Fernendez, D.P., Mulev, Y., Goodwin, A.R.H., and Levelt Sengers, J.M.H., *J. Phys. Chem. Ref. Data*, 24, 33, 1995.
2. Kaatze, U., *J. Chem. Eng. Data*, 34, 371, 1989.
3. Archer, D.G., and Wang, P., *J. Phys. Chem. Ref. Data*, 12, 817, 1983.

	0°C		25°C		50°C	
Frequency	ε′	ε″	ε′	ε″	ε′	ε″
0	87.90	0.00	78.36	0.00	69.88	0.00
1 kHz	87.90	0.00	78.36	0.00	69.88	0.00
1 MHz	87.90	0.01	78.36	0.00	69.88	0.00
10 MHz	87.90	0.09	78.36	0.04	69.88	0.02
100 MHz	87.89	0.91	78.36	0.38	69.88	0.20
200 MHz	87.86	1.82	78.35	0.76	69.88	0.39
500 MHz	87.65	4.55	78.31	1.90	69.87	0.98
1 GHz	86.90	9.01	78.16	3.79	69.82	1.96
2 GHz	84.04	17.39	77.58	7.52	69.65	3.92
3 GHz	79.69	24.64	76.62	11.13	69.36	5.85
4 GHz	74.36	30.49	75.33	14.58	68.95	7.75
5 GHz	68.54	34.88	73.73	17.81	68.45	9.62
10 GHz	42.52	40.88	62.81	29.93	64.49	18.05
20 GHz	19.56	30.78	40.37	36.55	52.57	28.99
30 GHz	12.50	22.64	26.53	33.25	40.57	32.74
40 GHz	9.67	17.62	18.95	28.58	31.17	32.43
50 GHz	8.28	14.34	14.64	24.53	24.42	30.47

THERMOPHYSICAL PROPERTIES OF FLUIDS

These tables give thermodynamic and transport properties of some important fluids, as generated from the equations of state presented in the references below. The properties tabulated are density (ρ), energy (E), enthalpy (H), entropy (S), isochoric heat capacity (C_v), isobaric heat capacity (C_p), speed of sound (v_s), viscosity (η), thermal conductivity (λ), and dielectric constant (D). All extensive properties are given on a molar basis. Not all properties are included for every substance. The references should be consulted for information on the uncertainties and the reference states for E, H, and S.

Values are given as a function of temperature for several isobars. The phase can be determined by noting the sharp decrease in density between two successive temperature entries; all lines above this point refer to the liquid phase, and all lines below refer to the gas phase. If there is no sharp discontinuity in density, all data in the table refer to the supercritical region (i.e., the isobar is above the critical pressure).

REFERENCES

1. Younglove, B.A., *Thermophysical Properties of Fluids. Part I, J. Phys. Chem. Ref. Data,* 11, Suppl. 1, 1982.
2. Younglove, B.A., and Ely, J.F., *Thermophysical Properties of Fluids. Part II, J. Phys. Chem. Ref. Data,* 16, 577, 1987.
3. McCarty, R.D., *Thermodynamic Properties of Helium, J. Phys. Chem. Ref. Data,* 2, 923, 1973.

Nitrogen (N_2)

P = 0.1 MPa (1 bar)

T K	ρ mol/L	E J/mol	H J/mol	S J/mol K	C_v J/mol K	C_p J/mol K	η μPa s	λ mW/m K	D
70	30.017	−3828	−3824	73.8	28.5	57.2	203.9	143.5	1.45269
77.25	28.881	−3411	−3407	79.5	27.8	57.8	152.2	133.8	1.43386
77.25	0.163	1546	2161	151.6	21.6	31.4	5.3	7.6	1.00215
100	0.123	2041	2856	159.5	21.1	30.0	6.8	9.6	1.00162
200	0.060	4140	5800	179.9	20.8	29.2	12.9	18.4	1.00079
300	0.040	6223	8717	191.8	20.8	29.2	18.0	25.8	1.00053
400	0.030	8308	11635	200.2	20.9	29.2	22.2	32.3	1.00040
500	0.024	10414	14573	206.7	21.2	29.6	26.1	38.5	1.00032
600	0.020	12563	17554	212.2	21.8	30.1	29.5	44.5	1.00026
700	0.017	14770	20593	216.8	22.4	30.7	32.8	50.5	1.00023
800	0.015	17044	23698	221.0	23.1	31.4	35.8	56.3	1.00020
900	0.013	19383	26869	224.7	23.7	32.0	38.7	62.0	1.00017
1000	0.012	21786	30103	228.1	24.3	32.6	41.5	67.7	1.00016
1500	0.008	34530	47004	241.8	26.4	34.7	54.0	93.3	1.00010

P = 1 MPa

T K	ρ mol/L	E J/mol	H J/mol	S J/mol K	C_v J/mol K	C_p J/mol K	η μPa s	λ mW/m K	D
70	30.070	−3838	−3805	73.6	28.9	56.9	205.9	144.1	1.45355
80	28.504	−3267	−3232	81.3	27.8	57.7	139.5	130.7	1.42760
90	26.721	−2685	−2648	88.2	26.7	59.4	100.1	115.3	1.39824
100	24.634	−2073	−2032	94.6	26.2	64.4	73.1	98.5	1.36417
103.75	23.727	−1828	−1786	97.1	26.2	67.8	64.8	91.8	1.34947
103.75	1.472	1788	2467	138.1	24.1	45.0	7.6	12.5	1.01954
200	0.614	4048	5675	160.3	21.0	30.4	13.2	19.3	1.00812
300	0.402	6171	8661	172.5	20.9	29.6	18.1	26.3	1.00529
400	0.300	8273	11609	180.9	20.9	29.5	22.4	32.7	1.00395
500	0.240	10389	14563	187.5	21.3	29.7	26.1	38.8	1.00315
600	0.200	12544	17554	193.0	21.8	30.2	29.6	44.8	1.00262
700	0.171	14756	20600	197.7	22.4	30.8	32.8	50.7	1.00224
800	0.150	17032	23709	201.8	23.1	31.4	35.9	56.5	1.00196
900	0.133	19374	26884	205.6	23.7	32.1	38.8	62.2	1.00174
1000	0.120	21778	30121	209.0	24.3	32.7	41.5	67.8	1.00157
1500	0.080	34527	47029	222.7	26.4	34.8	54.0	93.4	1.00104

P = 10 MPa

T K	ρ mol/L	E J/mol	H J/mol	S J/mol K	C_v J/mol K	C_p J/mol K	η μPa s	λ mW/m K	D
65.32	31.120	−4176	−3855	68.6	31.8	53.8	275.7	153.8	1.47067
100	26.201	−2328	−1946	92.0	27.4	56.3	90.2	112.3	1.38942
200	7.117	3037	4442	136.4	22.7	45.5	17.6	30.4	1.09698
300	3.989	5667	8174	151.7	21.4	33.4	20.1	31.9	1.05347
400	2.898	7941	11392	161.0	21.3	31.3	23.7	36.7	1.03860
500	2.302	10148	14492	167.9	21.5	30.8	27.1	42.0	1.03055

T K	ρ mol/L	*E* J/mol	*H* J/mol	*S* J/mol K	C_v J/mol K	C_p J/mol K	η μPa s	λ mW/m K	*D*
600	1.918	12361	17575	173.5	21.9	30.9	30.4	47.4	1.02538
700	1.647	14613	20683	178.3	22.5	31.3	33.5	53.0	1.02175
800	1.445	16919	23837	182.5	23.2	31.8	36.4	58.6	1.01904
900	1.288	19283	27046	186.3	23.8	32.4	39.3	64.1	1.01694
1000	1.162	21705	30308	189.8	24.4	32.9	42.0	69.6	1.01526
1500	0.783	34504	47283	203.5	26.5	34.8	54.3	94.7	1.01020

Oxygen (O_2)

T K	ρ mol/L	*E* J/mol	*H* J/mol	*S* J/mol K	C_v J/mol K	C_p J/mol K	η μPa s	λ mW/m K	*D*
P = 0.1 MPa (1 bar)									
60	40.049	–5883	–5880	72.4	34.9	53.4	425.2	188.2	1.55619
80	37.204	–4814	–4812	87.7	31.0	53.6	251.7	166.1	1.51114
100	0.123	2029	2840	172.9	21.4	30.5	7.5	9.3	1.00146
120	0.102	2458	3442	178.4	21.0	29.8	9.0	11.2	1.00121
140	0.087	2881	4035	182.9	20.9	29.5	10.5	13.1	1.00103
160	0.076	3301	4624	186.9	20.9	29.4	11.9	15.0	1.00090
180	0.067	3720	5210	190.3	20.8	29.3	13.3	16.7	1.00080
200	0.060	4138	5796	193.4	20.8	29.3	14.6	18.4	1.00072
220	0.055	4556	6381	196.2	20.8	29.3	15.9	20.1	1.00065
240	0.050	4974	6966	198.8	20.9	29.3	17.2	21.7	1.00060
260	0.046	5393	7552	201.1	20.9	29.3	18.4	23.2	1.00055
280	0.043	5812	8138	203.3	21.0	29.4	19.5	24.8	1.00051
300	0.040	6234	8726	205.3	21.1	29.4	20.6	26.3	1.00048
320	0.038	6657	9316	207.2	21.2	29.5	21.7	27.8	1.00045
340	0.035	7082	9908	209.0	21.3	29.7	22.8	29.3	1.00042
360	0.033	7510	10503	210.7	21.5	29.8	23.8	30.8	1.00040
380	0.032	7941	11100	212.3	21.6	30.0	24.8	32.2	1.00038
P = 1 MPa									
60	40.084	–5887	–5863	72.3	34.9	53.3	428.5	188.4	1.55674
80	37.254	–4822	–4795	87.6	31.0	53.5	253.8	166.4	1.51192
100	34.153	–3741	–3712	99.7	28.5	55.2	155.6	137.9	1.46381
120	1.198	2163	2997	156.7	24.0	40.6	9.4	13.9	1.01429
140	0.950	2683	3735	162.4	22.2	34.4	10.8	14.9	1.01133
160	0.802	3151	4398	166.8	21.5	32.2	12.2	16.3	1.00955
180	0.698	3598	5030	170.5	21.2	31.2	13.5	17.7	1.00831
200	0.620	4035	5647	173.8	21.1	30.6	14.8	19.3	1.00738
220	0.559	4466	6255	176.7	21.0	30.3	16.1	20.8	1.00665
240	0.509	4894	6858	179.3	21.0	30.1	17.3	22.3	1.00606
260	0.468	5321	7458	181.7	21.0	29.9	18.5	23.8	1.00556
280	0.433	5748	8056	183.9	21.1	29.9	19.6	25.2	1.00515
300	0.403	6174	8654	186.0	21.1	29.9	20.7	26.7	1.00479
320	0.377	6602	9252	187.9	21.2	29.9	21.8	28.2	1.00448
340	0.355	7032	9851	189.7	21.4	30.0	22.8	29.6	1.00421
360	0.335	7463	10452	191.4	21.5	30.1	23.9	31.1	1.00397
380	0.317	7898	11056	193.1	21.7	30.2	24.9	32.6	1.00376
P = 10 MPa									
60	40.419	–5931	–5684	71.5	35.1	53.0	461.8	189.9	1.56210
80	37.727	–4893	–4628	86.7	31.6	52.7	274.4	168.6	1.51936
100	34.881	–3856	–3570	98.5	29.1	53.4	171.0	141.2	1.47500
120	31.721	–2796	–2481	108.4	27.3	55.9	113.0	115.1	1.42677
140	27.890	–1662	–1304	117.5	26.2	62.9	76.3	91.8	1.36972
160	22.379	-322	125	127.0	26.1	84.8	48.6	71.2	1.29037
180	13.232	1489	2245	139.5	26.6	105.9	26.2	46.8	1.16560
200	8.666	2681	3835	147.9	24.0	60.6	21.2	34.0	1.10650

T K	ρ mol/L	*E* J/mol	*H* J/mol	*S* J/mol K	C_v J/mol K	C_p J/mol K	η μPa s	λ mW/m K	*D*
220	6.868	3424	4880	152.9	22.6	46.4	20.5	30.8	1.08380
240	5.836	4029	5742	156.6	22.0	40.6	20.8	30.1	1.07090
260	5.134	4573	6521	159.7	21.8	37.6	21.4	30.2	1.06219
280	4.613	5086	7254	162.5	21.6	35.8	22.1	30.8	1.05575
300	4.205	5581	7959	164.9	21.6	34.7	22.9	31.6	1.05073
320	3.874	6063	8645	167.1	21.7	33.9	23.7	32.6	1.04667
340	3.598	6538	9318	169.1	21.8	33.4	24.6	33.7	1.04329
360	3.363	7009	9982	171.0	21.9	33.0	25.4	34.9	1.04043
380	3.161	7477	10641	172.8	22.0	32.8	26.3	36.1	1.03796

Hydrogen (H_2)

T K	ρ mol/L	*E* J/mol	*H* J/mol	*S* J/mol K	C_v J/mol K	C_p J/mol K	v_s m/s	*D*
P = 0.1 MPa (1 bar)								
15	37.738	–605	–603	11.2	9.7	14.4	1319	1.24827
20	35.278	–524	–521	15.8	11.3	19.1	1111	1.23093
40	0.305	491	818	75.6	12.5	21.3	521	1.00186
60	0.201	748	1244	84.3	13.1	21.6	636	1.00122
80	0.151	1030	1694	90.7	15.3	23.7	714	1.00091
100	0.120	1370	2202	96.4	18.7	27.1	773	1.00073
120	0.100	1777	2776	101.6	21.8	30.2	827	1.00061
140	0.086	2237	3401	106.4	23.8	32.2	883	1.00052
160	0.075	2723	4054	110.8	24.6	33.0	940	1.00046
180	0.067	3216	4714	114.7	24.6	32.9	998	1.00041
200	0.060	3703	5367	118.1	24.1	32.4	1054	1.00037
220	0.055	4179	6009	121.2	23.4	31.8	1110	1.00033
240	0.050	4641	6638	123.9	22.8	31.2	1163	1.00030
260	0.046	5093	7256	126.4	22.3	30.6	1214	1.00028
280	0.043	5535	7865	128.6	21.9	30.2	1263	1.00026
300	0.040	5970	8466	130.7	21.6	29.9	1310	1.00024
400	0.030	8093	11421	139.2	21.0	29.3	1518	1.00018
P = 1 MPa								
15	38.109	–609	–583	10.9	10.1	14.1	1315	1.25089
20	35.852	–532	–504	15.5	11.4	18.4	1155	1.23496
40	3.608	399	676	54.1	12.9	28.4	498	1.02209
60	2.098	697	1173	64.3	13.2	23.5	635	1.01280
80	1.523	994	1651	71.1	15.4	24.7	719	1.00928
100	1.204	1343	2174	77.0	18.8	27.7	779	1.00733
120	0.999	1756	2758	82.3	21.9	30.6	835	1.00608
140	0.854	2219	3390	87.1	23.9	32.5	891	1.00520
160	0.747	2709	4048	91.5	24.7	33.2	949	1.00454
180	0.663	3204	4712	95.4	24.6	33.1	1006	1.00404
200	0.597	3693	5368	98.9	24.1	32.5	1063	1.00363
220	0.543	4170	6012	102.0	23.5	31.9	1118	1.00330
240	0.498	4634	6643	104.7	22.9	31.2	1171	1.00303
260	0.460	5087	7263	107.2	22.3	30.7	1222	1.00279
280	0.427	5530	7873	109.5	21.9	30.3	1271	1.00259
300	0.399	5966	8475	111.5	21.6	30.0	1317	1.00242
400	0.299	8091	11433	120.1	21.0	29.4	1525	1.00182
P = 10 MPa								
20	39.669	–568	–316	13.0	10.9	15.0	1458	1.26198
40	31.344	–209	110	27.3	13.2	27.0	1171	1.20354
60	21.273	255	725	39.7	13.8	32.5	931	1.13527
80	14.830	686	1360	48.8	15.9	31.1	886	1.09303
100	11.417	1110	1986	55.8	19.3	31.9	904	1.07109

T K	ρ mol/L	*E* J/mol	*H* J/mol	*S* J/mol K	C_v J/mol K	C_p J/mol K	v_s m/s	*D*
120	9.357	1571	2640	61.8	22.4	33.5	941	1.05801
140	7.969	2068	3323	67.0	24.3	34.6	989	1.04925
160	6.963	2583	4020	71.7	25.0	34.9	1042	1.04294
180	6.195	3099	4713	75.7	24.9	34.4	1096	1.03814
200	5.588	3604	5393	79.3	24.4	33.6	1150	1.03436
220	5.094	4094	6057	82.5	23.7	32.8	1203	1.03129
240	4.683	4569	6704	85.3	23.1	32.0	1254	1.02874
260	4.336	5030	7336	87.8	22.6	31.3	1302	1.02659
280	4.038	5481	7958	90.1	22.1	30.8	1349	1.02475
300	3.780	5924	8570	92.3	21.8	30.4	1394	1.02315
400	2.869	8073	11559	100.9	21.2	29.6	1592	1.01753

Helium (He-4)

T K	ρ mol/L	*E* J/mol	*H* J/mol	*S* J/mol K	C_v J/mol K	C_p J/mol K	v_s m/s	η μPa s	*D*
P = 0.1 MPa (1 bar)									
3	35.794	–39	–36	9.8	7.6	9.4	222	3.85	1.05646
4	32.477	–27	–24	13.3	9.1	16.3	185	3.33	1.05114
5	2.935	52	86	39.1	12.7	27.1	120	1.39	1.00456
10	1.238	120	201	55.2	12.5	21.7	185	2.26	1.00192
20	0.602	247	413	69.9	12.5	21.0	264	3.58	1.00093
50	0.240	623	1039	89.0	12.5	20.8	417	6.36	1.00037
100	0.120	1247	2079	103.4	12.5	20.8	589	9.78	1.00019
200	0.060	2494	4158	117.8	12.5	20.8	833	15.14	1.00009
300	0.040	3741	6237	126.3	12.5	20.8	1020	19.93	1.00006
400	0.030	4988	8315	132.3	12.5	20.8	1177	24.29	1.00005
500	0.024	6236	10394	136.9	12.5	20.8	1316	28.36	1.00004
600	0.020	7483	12472	140.7	12.5	20.8	1441	32.22	1.00003
700	0.017	8730	14551	143.9	12.5	20.8	1557	35.89	1.00003
800	0.015	9977	16630	146.7	12.5	20.8	1664	39.43	1.00002
900	0.013	11224	18708	149.1	12.5	20.8	1765	42.85	1.00002
1000	0.012	12471	20787	151.3	12.5	20.8	1861	46.16	1.00002
1500	0.008	18707	31179	159.7	12.5	20.8	2279	61.55	1.00001
P = 1 MPa									
3	39.703	–42	–16	8.6	7.1	7.8	300	5.63	1.06274
4	38.210	–34	–7	11.2	8.3	10.9	290	5.01	1.06034
5	35.818	–22	6	14.0	9.7	15.1	269	4.38	1.05650
10	15.378	78	143	32.2	12.3	30.5	198	3.07	1.02402
20	6.067	228	393	49.8	12.6	22.9	274	3.94	1.00943
50	2.353	617	1042	69.8	12.5	21.1	428	6.53	1.00365
100	1.186	1245	2089	84.3	12.5	20.9	597	9.89	1.00184
200	0.597	2495	4170	98.7	12.5	20.8	838	15.21	1.00093
300	0.399	3742	6249	107.1	12.5	20.8	1024	19.96	1.00062
400	0.300	4990	8327	113.1	12.5	20.8	1180	24.32	1.00046
500	0.240	6237	10406	117.8	12.5	20.8	1319	28.38	1.00037
600	0.200	7485	12484	121.5	12.5	20.8	1444	32.23	1.00031
700	0.172	8732	14562	124.7	12.5	20.8	1559	35.91	1.00027
800	0.150	9979	16641	127.5	12.5	20.8	1666	39.44	1.00023
900	0.133	11227	18719	130.0	12.5	20.8	1767	42.86	1.00021
1000	0.120	12474	20798	132.2	12.5	20.8	1862	46.17	1.00019
1500	0.080	18710	31190	140.6	12.5	20.8	2280	61.55	1.00012
P = 10 MPa									
4	51.978	–24	169	6.7	6.0	7.3	586	24.27	1.08262
5	51.118	–18	177	8.5	7.9	9.3	576	18.16	1.08122
10	46.872	23	236	16.6	11.0	14.5	546	9.31	1.07432

T K	ρ mol/L	*E* J/mol	*H* J/mol	*S* J/mol K	C_v J/mol K	C_p J/mol K	v_s m/s	η μPa s	*D*
20	37.092	154	423	29.5	12.6	20.7	498	6.99	1.05854
50	19.192	572	1093	49.9	12.9	22.4	541	8.07	1.03003
100	10.525	1231	2181	65.0	12.8	21.3	674	10.93	1.01640
200	5.605	2500	4284	79.6	12.6	20.9	889	15.82	1.00871
300	3.829	3755	6367	88.0	12.6	20.8	1063	20.25	1.00595
400	2.908	5006	8445	94.0	12.6	20.8	1212	24.54	1.00452
500	2.344	6256	10522	98.6	12.5	20.8	1346	28.56	1.00364
600	1.963	7505	12599	102.4	12.5	20.8	1467	32.38	1.00305
700	1.689	8754	14676	105.6	12.5	20.8	1580	36.04	1.00262
800	1.481	10003	16753	108.4	12.5	20.8	1685	39.56	1.00230
900	1.320	11252	18830	110.9	12.5	20.8	1784	42.96	1.00205
1000	1.189	12500	20907	113.0	12.5	20.8	1877	46.26	1.00185
1500	0.797	18742	31294	121.5	12.5	20.8	2289	61.62	1.00124

Argon (Ar)

T K	ρ mol/L	*E* J/mol	*H* J/mol	*S* J/mol K	C_v J/mol K	C_p J/mol K	v_s m/s	η μPa s	λ mW/m K
P = 0.1 MPa (1 bar)									
85	35.243	–4811	–4808	53.6	23.1	44.7	820	278.8	132.4
90	0.138	1077	1802	129.4	13.1	22.5	174	7.5	6.0
100	0.123	1211	2024	131.8	12.9	21.9	184	8.2	6.6
120	0.102	1471	2456	135.7	12.6	21.4	203	9.8	7.8
140	0.087	1727	2881	139.0	12.6	21.1	220	11.4	9.0
160	0.076	1980	3302	141.8	12.5	21.0	235	13.0	10.2
180	0.067	2232	3722	144.3	12.5	21.0	250	14.5	11.4
200	0.060	2483	4141	146.5	12.5	20.9	263	16.0	12.5
220	0.055	2734	4559	148.5	12.5	20.9	276	17.5	13.7
240	0.050	2984	4976	150.3	12.5	20.9	289	18.9	14.8
260	0.046	3234	5394	152.0	12.5	20.9	300	20.3	15.8
280	0.043	3484	5811	153.5	12.5	20.8	312	21.6	16.9
300	0.040	3734	6227	155.0	12.5	20.8	323	22.9	17.9
320	0.038	3984	6644	156.3	12.5	20.8	333	24.2	18.9
340	0.035	4234	7060	157.6	12.5	20.8	344	25.4	19.9
360	0.033	4484	7477	158.7	12.5	20.8	354	26.6	20.8
380	0.032	4734	7893	159.9	12.5	20.8	363	27.8	21.7
P = 1 MPa									
85	35.307	–4820	–4792	53.5	23.1	44.6	823	281.3	133.0
90	34.542	–4598	–4569	56.1	21.6	44.7	808	242.7	124.2
100	32.909	–4145	–4115	60.9	19.9	46.2	753	185.0	109.2
120	1.181	1210	2057	114.3	14.7	30.1	189	10.3	9.3
140	0.945	1544	2603	118.5	13.5	25.4	212	11.8	10.1
160	0.799	1838	3089	121.8	13.0	23.6	231	13.3	11.1
180	0.697	2116	3551	124.5	12.8	22.7	247	14.8	12.1
200	0.619	2384	3999	126.9	12.7	22.2	262	16.3	13.2
220	0.559	2648	4438	128.9	12.6	21.8	275	17.7	14.2
240	0.509	2908	4873	130.8	12.6	21.6	288	19.1	15.3
260	0.468	3167	5304	132.6	12.6	21.5	301	20.4	16.3
280	0.433	3423	5732	134.2	12.6	21.4	312	21.8	17.3
300	0.403	3679	6159	135.6	12.5	21.3	324	23.1	18.3
320	0.377	3934	6583	137.0	12.5	21.2	334	24.3	19.2
340	0.355	4188	7007	138.3	12.5	21.2	345	25.5	20.2
360	0.335	4441	7429	139.5	12.5	21.1	355	26.7	21.1
380	0.317	4694	7851	140.6	12.5	21.1	365	27.9	22.0
P = 10 MPa									
90	35.208	–4694	–4410	55.0	21.9	43.2	846	265.2	129.5

T K	ρ mol/L	*E* J/mol	*H* J/mol	*S* J/mol K	C_v J/mol K	C_p J/mol K	v_s m/s	η μPa s	λ mW/m K
100	33.744	–4271	–3974	59.6	20.4	44.0	800	205.0	115.1
120	30.525	–3396	–3069	67.8	18.8	46.9	672	131.2	92.1
140	26.609	–2447	–2072	75.5	17.6	54.1	526	85.9	71.7
160	20.816	–1279	–799	83.9	17.4	78.6	357	51.3	52.8
180	12.296	228	1042	94.8	17.3	83.6	257	27.8	32.0
200	8.442	1118	2302	101.4	15.3	48.6	268	23.3	23.6
220	6.776	1661	3137	105.4	14.2	36.8	284	22.8	21.6
240	5.787	2087	3815	108.4	13.7	31.6	300	23.2	21.3
260	5.105	2458	4416	110.8	13.4	28.8	314	23.9	21.4
280	4.596	2798	4974	112.9	13.2	27.1	327	24.8	21.8
300	4.195	3119	5503	114.7	13.1	25.9	339	25.7	22.3
320	3.869	3427	6012	116.3	13.0	25.0	350	26.7	22.9
340	3.596	3726	6506	117.8	13.0	24.4	361	27.7	23.5
360	3.364	4017	6989	119.2	12.9	23.9	372	28.7	24.2
380	3.164	4303	7464	120.5	12.9	23.5	381	29.7	24.9

Methane (CH_4)

T K	ρ mol/L	*E* J/mol	*H* J/mol	*S* J/mol K	C_v J/mol K	C_p J/mol K	η μPa s	λ mW/m K	*D*
P = 0.1 MPa (1 bar)									
100	27.370	–5258	–5254	73.0	33.4	54.1	156.3	208.1	1.65504
125	0.099	3026	4039	156.5	25.4	34.6	5.0	13.4	1.00193
150	0.081	3667	4896	162.7	25.2	34.0	5.9	16.2	1.00159
175	0.069	4301	5743	168.0	25.2	33.8	6.9	19.1	1.00136
200	0.061	4935	6587	172.5	25.3	33.8	7.8	21.9	1.00119
225	0.054	5571	7434	176.5	25.5	34.0	8.7	24.8	1.00105
250	0.048	6216	8288	180.1	26.0	34.4	9.6	27.8	1.00095
275	0.044	6875	9156	183.4	26.6	35.0	10.4	30.9	1.00086
300	0.040	7552	10042	186.4	27.5	35.9	11.2	34.1	1.00079
325	0.037	8252	10951	189.4	28.5	36.9	12.0	37.6	1.00073
350	0.034	8979	11887	192.1	29.7	38.0	12.8	41.2	1.00068
375	0.032	9737	12853	194.8	30.9	39.3	13.5	45.1	1.00063
400	0.030	10528	13852	197.4	32.3	40.7	14.3	49.1	1.00059
425	0.028	11354	14886	199.9	33.7	42.1	15.0	53.3	1.00056
450	0.027	12215	15956	202.3	35.2	43.5	15.7	57.6	1.00053
500	0.024	14047	18204	207.1	38.0	46.4	17.0	66.5	1.00047
600	0.020	18111	23101	216.0	42.9	51.3	19.4	84.1	1.00039
P = 1 MPa									
100	27.413	–5268	–5231	72.9	33.4	54.0	158.1	208.9	1.65617
125	25.137	–3882	–3842	85.3	32.4	57.4	89.2	168.2	1.59261
150	0.969	3282	4315	140.9	27.9	45.2	6.2	18.4	1.01911
175	0.765	4041	5348	147.3	26.4	38.9	7.1	20.6	1.01507
200	0.644	4736	6289	152.3	25.9	36.8	8.0	23.1	1.01268
225	0.560	5410	7197	156.6	25.9	36.0	8.9	25.8	1.01102
250	0.497	6081	8093	160.4	26.2	35.8	9.7	28.7	1.00979
275	0.448	6758	8991	163.8	26.8	36.1	10.6	31.7	1.00882
300	0.408	7449	9901	167.0	27.6	36.7	11.4	34.9	1.00803
325	0.375	8160	10829	169.9	28.6	37.6	12.1	38.3	1.00738
350	0.347	8897	11781	172.8	29.7	38.6	12.9	41.9	1.00683
375	0.323	9662	12760	175.5	31.0	39.8	13.6	45.7	1.00636
400	0.302	10460	13770	178.1	32.4	41.1	14.4	49.6	1.00595
425	0.284	11291	14814	180.6	33.8	42.4	15.1	53.8	1.00559
450	0.268	12157	15892	183.1	35.2	43.8	15.7	58.1	1.00527
500	0.241	13997	18153	187.8	38.1	46.6	17.0	66.9	1.00474
600	0.200	18073	23070	196.8	43.0	51.4	19.5	84.5	1.00394

T K	ρ mol/L	E J/mol	H J/mol	S J/mol K	C_v J/mol K	C_p J/mol K	η μPa s	λ mW/m K	D
P = 10 MPa									
100	27.815	–5362	–5003	72.0	33.8	53.2	175.4	217	1.66668
125	25.754	–4036	–3648	84.1	32.7	55.3	100.4	178.8	1.60895
150	23.441	–2655	–2229	94.4	31.4	58.6	65.7	144.6	1.54553
175	20.613	–1175	–689	103.9	30.3	65.5	44.9	113.4	1.47021
200	16.602	542	1144	113.6	30.1	84.7	29.4	85.8	1.36789
225	10.547	2680	3628	125.3	30.8	102.2	17.6	61.0	1.22352
250	7.013	4289	5714	134.1	29.3	67.4	14.3	47.6	1.14481
275	5.530	5387	7195	139.8	28.7	53.4	13.8	44.1	1.11297
300	4.685	6320	8454	144.2	28.9	48.0	13.9	44.6	1.09513
325	4.115	7192	9622	147.9	29.6	45.8	14.3	46.6	1.08322
350	3.695	8047	10753	151.3	30.5	44.9	14.7	49.2	1.07450
375	3.366	8903	11874	154.4	31.7	44.8	15.2	52.3	1.06773
400	3.101	9774	12999	157.3	32.9	45.2	15.8	55.7	1.06227
425	2.880	10666	14138	160.0	34.3	46.0	16.3	59.4	1.05775
450	2.692	11584	15298	162.7	35.7	46.9	16.9	63.3	1.05392
500	2.389	13507	17692	167.7	38.5	48.9	18.0	71.6	1.04775
600	1.963	17700	22795	177.0	43.3	52.9	20.2	88.3	1.03911

Ethane (C_2H_6)

T K	ρ mol/L	E J/mol	H J/mol	S J/mol K	C_v J/mol K	C_p J/mol K	v_s m/s	D
P = 0.1 MPa (1 bar)								
95	21.50	–14555	–14550	80.2	47.2	68.7	1970	1.93480
100	21.32	–14210	–14205	83.8	47.1	69.3	1943	1.92500
125	20.41	–12468	–12463	99.3	45.0	69.8	1775	1.87634
150	19.47	–10717	–10712	112.1	43.4	70.4	1587	1.82726
175	18.49	-8938	-8933	123.1	42.7	72.1	1396	1.77671
200	0.062	5503	7123	210.1	34.5	43.8	258	1.00208
225	0.054	6401	8238	215.4	36.5	45.5	273	1.00183
250	0.049	7349	9401	220.3	38.9	47.7	287	1.00164
275	0.044	8360	10624	224.9	41.6	50.2	300	1.00148
300	0.040	9439	11914	229.4	44.5	53.1	312	1.00136
325	0.037	10592	13278	233.8	47.6	56.1	324	1.00125
350	0.035	11823	14719	238.1	50.7	59.2	335	1.00116
375	0.032	13133	16240	242.3	54.0	62.4	345	1.00108
400	0.030	14525	17841	246.4	57.2	65.6	355	1.00101
450	0.027	17548	21282	254.5	63.6	72.0	375	1.00090
500	0.024	20883	25035	262.4	69.7	78.1	393	1.00081
600	0.020	28429	33415	277.6	80.9	89.3	428	1.00067
P = 1 MPa								
95	21.514	–14562	–14515	80.2	47.3	68.7	1972	1.93537
100	21.334	–14217	–14170	83.7	47.2	69.3	1946	1.92560
125	20.427	–12478	–12429	99.2	45.0	69.8	1778	1.87709
150	19.494	–10731	–10679	112.0	43.4	70.3	1592	1.82823
175	18.515	–8957	–8903	123.0	42.7	72.0	1402	1.77800
200	17.464	–7127	–7070	132.7	42.9	74.9	1209	1.72513
225	16.288	–5199	–5137	141.8	43.8	80.2	1008	1.66733
250	0.564	6762	8534	198.7	41.6	57.5	260	1.01909
275	0.489	7902	9949	204.1	43.2	56.2	280	1.01650
300	0.435	9063	11363	209.0	45.5	57.2	297	1.01467
325	0.393	10273	12815	213.7	48.3	59.1	311	1.01327
350	0.360	11546	14321	218.1	51.3	61.5	325	1.01214
375	0.333	12889	15893	222.5	54.4	64.2	337	1.01121
400	0.310	14306	17534	226.7	57.5	67.1	349	1.01043

T K	ρ mol/L	*E* J/mol	*H* J/mol	*S* J/mol K	C_v J/mol K	C_p J/mol K	v_s m/s	*D*
450	0.272	17367	21038	234.9	63.8	73.0	370	1.00917
500	0.244	20730	24836	242.9	69.9	78.9	390	1.00819
600	0.201	28313	33278	258.3	81.0	89.8	427	1.00677
P = 10 MPa								
95	21.624	–14626	–14163	79.5	47.4	68.5	2000	1.94104
100	21.448	–14286	–13819	83.0	47.4	69.1	1974	1.93146
125	20.570	–12572	–12086	98.5	45.5	69.3	1814	1.88436
150	19.678	–10858	–10350	111.1	43.9	69.6	1637	1.83753
175	18.758	–9130	–8596	121.9	43.3	70.8	1459	1.79010
200	17.793	–7363	–6801	131.5	43.5	73.0	1284	1.74134
225	16.760	–5535	–4938	140.3	44.3	76.4	1110	1.69017
250	15.620	–3609	–2969	148.6	45.8	81.5	935	1.63488
275	14.301	–1539	–839	156.7	47.9	89.4	758	1.57249
300	12.666	757	1547	165.0	50.8	102.7	577	1.49740
325	10.398	3443	4404	174.1	54.7	129.1	399	1.39745
350	7.292	6643	8015	184.8	58.8	150.1	290	1.26832
375	5.182	9419	11349	194.1	60.0	115.7	289	1.18570
400	4.182	11577	13968	200.8	61.4	96.9	310	1.14797
450	3.204	15379	18500	211.5	65.8	87.5	347	1.11193
500	2.677	19135	22870	220.7	71.2	88.0	378	1.09288
600	2.076	27160	31978	237.3	81.8	94.7	427	1.07142

Propane (C_3H_8)

T K	ρ mol/L	*E* J/mol	*H* J/mol	*S* J/mol K	C_v J/mol K	C_p J/mol K	v_s m/s	*D*
P = 0.1 MPa (1 bar)								
90	16.526	–21486	–21426	87.3	59.2	84.5	2126	2.07988
100	16.295	–20639	–20577	96.2	59.6	85.2	2041	2.05806
125	15.726	–18495	–18432	115.4	59.2	86.5	1856	2.00674
150	15.156	–16319	–16253	131.3	58.9	88.0	1685	1.95796
175	14.577	–14096	–14028	145.0	59.5	90.3	1521	1.91036
200	13.982	–11806	–11735	157.3	61.0	93.5	1359	1.86300
225	13.339	–9395	–9387	168.5	63.4	97.9	1197	1.81487
250	0.050	9194	11213	257.6	57.2	66.8	228	1.00238
275	0.045	10691	12930	264.1	61.6	70.7	239	1.00215
300	0.041	12297	14752	270.5	66.2	75.1	249	1.00195
325	0.037	14019	16689	276.7	71.1	79.8	259	1.00179
350	0.035	15862	18744	282.8	76.0	84.6	269	1.00166
375	0.032	17827	20921	288.8	80.9	89.5	278	1.00154
400	0.030	19912	23217	294.7	85.7	94.3	286	1.00144
450	0.027	24441	28166	306.4	95.2	103.6	303	1.00128
500	0.024	29428	33573	317.7	104.1	112.6	318	1.00115
600	0.020	40677	45658	339.7	120.4	128.8	347	1.00095
P = 1 MPa								
90	16.526	–21486	–21426	87.2	59.3	84.5	2128	2.08034
100	16.295	–20639	–20577	96.2	59.7	85.2	2043	2.05856
125	15.726	–18495	–18432	115.3	59.2	86.4	1859	2.00736
150	15.156	–16319	–16253	131.2	59.0	88.0	1690	1.95873
175	14.577	–14096	–14028	144.9	59.6	90.2	1526	1.91132
200	13.982	–11806	–11735	157.2	61.1	93.4	1365	1.86421
225	13.361	–9424	–9349	168.4	63.4	97.7	1205	1.81642
250	12.696	–6919	–6840	179.0	66.4	103.3	1045	1.76672
275	11.962	–4252	–4169	189.1	70.0	110.8	881	1.71316
300	11.102	–1360	–1270	199.2	74.1	121.9	708	1.65216
325	0.428	13278	15614	255.2	74.1	89.6	233	1.02067

T K	ρ mol/L	E J/mol	H J/mol	S J/mol K	C_v J/mol K	C_p J/mol K	v_s m/s	D
350	0.383	15259	17869	261.9	78.0	91.2	248	1.01846
375	0.349	17318	20183	268.3	82.2	94.2	261	1.01678
400	0.322	19472	22582	274.4	86.7	97.8	272	1.01544
450	0.279	24092	27672	286.4	95.7	105.9	293	1.01337
500	0.248	29137	33172	298.0	104.4	114.1	312	1.01184
600	0.203	40455	45374	320.2	120.5	129.7	344	1.00968
P = 10 MPa								
90	16.590	–21553	–20951	86.5	59.9	84.4	2146	2.08489
100	16.364	–20714	–20103	95.4	60.1	85.1	2068	2.06350
125	15.810	–18595	–17962	114.5	59.6	86.1	1895	2.01342
150	15.259	–16448	–15793	130.3	59.3	87.5	1733	1.96617
175	14.705	–14261	–13581	144.0	59.9	89.5	1577	1.92048
200	14.141	–12016	–11309	156.1	61.4	92.4	1425	1.87557
225	13.562	–9692	–8955	167.2	63.7	96.1	1277	1.83076
250	12.960	–7268	–6496	177.5	66.7	100.7	1133	1.78529
275	12.322	–4721	–3909	187.4	70.2	106.4	991	1.73826
300	11.631	–2027	–1167	196.9	74.1	113.2	851	1.68849
325	10.860	843	1764	206.3	78.4	121.5	715	1.63437
350	9.973	3924	4927	215.7	82.9	132.0	582	1.57361
375	8.905	7270	8393	225.2	87.7	146.1	455	1.50271
400	7.561	10957	12279	235.3	93.0	165.7	339	1.41671
450	4.614	18845	21013	255.8	101.8	167.8	249	1.24060
500	3.241	25567	28652	272.0	107.8	142.7	276	1.16439
600	2.242	38131	42591	297.4	121.7	140.5	332	1.11122

VIRIAL COEFFICIENTS OF SELECTED GASES

Henry V. Kehiaian

This table gives second virial coefficients of about 110 inorganic and organic gases as a function of temperature. Selected data from the literature have been fitted by least squares to the equation

$$B/\text{cm}^3\,\text{mol}^{-1} = \sum_{i=1}^{n} a(i)\left[(T_o/T)-1\right]^{i-1}$$

where $T_o = 298.15$ K. The table gives the coefficients $a(i)$ and values of B at fixed temperature increments, as calculated from this smoothing equation.

The equation may be used with the tabulated coefficients for interpolation within the indicated temperature range. It should not be used for extrapolation beyond this range.

Compounds are listed in the modified Hill order (see Introduction), with carbon-containing compounds following those compounds not containing carbon.

A useful compilation of virial coefficient data from the literature may be found in:

J. H. Dymond and E. B. Smith, *The Virial Coefficients of Pure Gases and Mixtures, A Critical Compilation*, Oxford University Press, Oxford, 1980.

Compounds Not Containing Carbon

Mol. form.	Name	T/K	B/cm³ mol⁻¹
Ar	Argon	100	–184
		120	–131
		140	–98
	a(1) = –16	160	–76
	a(2) = –60	80	–60
	a(3) = –10	200	–48
		300	–16
		400	–1
		500	7
		600	12
		700	15
		800	18
		900	20
		1000	22
BF_3	Boron trifluoride	200	–338
		240	–202
		280	–129
	a(1) = –106	320	–85
	a(2) = –330	360	–56
	a(3) = –251	400	–37
	a(4) = –80	440	–23
ClH	Hydrogen chloride	190	–451
		230	–269
		270	–181
	a(1) = –144	310	–132
	a(2) = –325	350	–102
	a(3) = –277	390	–81
	a(4) = –170	430	–66
		470	–54
Cl_2	Chlorine	210	–508
		220	–483

VIRIAL COEFFICIENTS OF SELECTED GASES (continued)

Mol. form.	Name	*T*/K	*B*/cm^3 mol^{-1}
		230	–457
	a(1) = –303	240	–432
	a(2) = –555	250	–407
	a(3) = 9	260	–383
	a(4) = 329	270	–360
	a(5) = 68	280	–339
		290	–318
		300	–299
		350	–221
		400	–166
		450	–126
		500	–97
		600	–59
		700	–36
		800	–22
		900	–12
F_2	Fluorine	80	–386
		110	–171
		140	–113
	a(1) = –25	170	–73
	a(2) = 21	200	–47
	a(3) = –185	230	–32
	a(4) = 113	260	–25
F_4Si	Silicon tetrafluoride	210	–268
		240	–213
		270	–170
	a(1) = –138	300	–136
	a(2) = –312	330	–108
		360	–84
		390	–64
		420	–47
		450	–32
F_5I	Iodine pentafluoride	320	–2540
		330	–2344
		340	–2172
	a(1) = –3077	350	–2021
	a(2) = –8474	360	–1890
	a(3) = –9116	370	–1775
		380	–1674
		390	–1587
		400	–1510
		410	–1443
F_5P	Phosphorus pentafluoride	320	–162
		340	–143
		360	–127
	a(1) = –186	380	–112
	a(2) = –345	400	–98
		420	–86
		440	–75
		460	–64
F_6Mo	Molybdenum hexafluoride	300	–896
		310	–810
		320	–737

Mol. form.	Name	*T*/K	*B*/cm³ mol⁻¹
	a(1) = –914	330	–677
	a(2) = –2922	340	–627
	a(3) = –4778	350	–586
		360	–553
		370	–527
		380	–506
		390	–491
F_6S	Sulfur hexafluoride	200	–685
		250	–416
		300	–275
	a(1) = –279	350	–190
	a(2) = –647	400	–135
	a(3) = –335	450	–96
	a(4) = –72	500	–68
F_6U	Uranium hexafluoride	320	–1030
		340	–905
		360	–805
	a(1) = –1204	380	–724
	a(2) = –2690	400	–658
	a(3) = –2144	420	–604
		440	–560
F_6W	Tungsten hexafluoride	320	–641
		340	–578
		360	–523
	a(1) = –719	380	–473
	a(2) = –1143	400	–428
		420	–387
		440	–350
		460	–317
H_2	Hydrogen	15	–230
		20	–151
		25	–108
	a(1) = 15.4	30	–82
	a(2) = –9.0	35	–64
	a(3) = –0.2	40	–52
		45	–42
		50	–35
		60	–24
		70	–16
		80	–11
		90	–7
		100	–3
		200	11
		300	15
		400	18
H_2O	Water	300	–1126
		320	–850
		340	–660
	a(1) = –1158	360	–526
	a(2) = –5157	380	–428
	a(3) = –10301	400	–356
	a(4) =–10597	420	–301
	a(5) = –4415	440	–258

Mol. form.	Name	*T*/K	*B*/cm³ mol⁻¹
		460	–224
		480	–197
		500	–175
		600	–104
		700	–67
		800	–44
		900	–30
		1000	–20
		1100	–14
		1200	–11
H_3N	Ammonia	290	–302
		300	–265
		310	–236
	a(1) = –271	320	–213
	a(2) = –1022	330	–194
	a(3) = –2715	340	–179
	a(4) = –4189	350	–166
		360	–154
		370	–144
		380	–135
		400	–118
		420	–101
H_3P	Phosphine	190	–457
		200	–404
		210	–364
	a(1) = –146	220	–332
	a(2) = –733	230	–305
	a(3) = 1022	240	–281
	a(4) = –1220	250	–258
		260	–235
		270	–213
		280	–190
		290	–166
He	Helium	2	–172
		6	–48
		10	–24
	a(1) = 12	14	–13
	a(2) = –1	18	–7
		22	–3
		26	–1
		30	1
		50	6
		70	8
		90	10
		110	10
		150	11
		250	12
		650	13
		700	13
Kr	Krypton	110	–363
		120	–307
		130	–263
	a(1) = –51	140	–229
	a(2) = –118	150	–201

Mol. form.	Name	*T*/K	*B*/cm³ mol⁻¹
	a(3) = –29	160	–178
	a(4) = –5	170	–159
		180	–143
		190	–129
		200	–117
		250	–75
		300	–51
		400	–23
		500	–8
		600	2
		700	8
NO	Nitric oxide	120	–232
		130	–176
		140	–138
	a(1) = –12	150	–113
	a(2) = –119	160	–96
	a(3) = 89	170	–83
	a(4) = –73	180	–73
		190	–65
		200	–58
		210	–52
		230	–42
		250	–32
		270	–24
N_2	Nitrogen	75	–274
		100	–161
		125	–104
	a(1) = –4	150	–71
	a(2) = –56	175	–49
	a(3) = –12	200	–34
		225	–24
		250	–15
		300	–4
		400	9
		500	16
		600	21
		700	24
N_2O	Nitrous oxide	240	–219
		260	–181
		280	–151
	a(1) = –130	300	–128
	a(2) = –307	320	–110
	a(3) = –248	340	–96
		360	–85
		380	–76
		400	–68
Ne	Neon	60	–25
		80	–13
		100	–6
	a(1) = 10.8	120	–1
	a(2) = –7.5	140	2
	a(3) = 0.4	160	4
		180	6
		200	7

Mol. form.	Name	*T*/K	*B*/cm^3 mol^{-1}
		300	11
		400	13
		500	14
		600	15
O_2	Oxygen	90	–241
		110	–161
		130	–117
	a(1) = –16	150	–88
	a(2) = –62	170	–69
	a(3) = –8	190	–55
	a(4) = –3	210	–44
		230	–36
		250	–29
		270	–23
		290	–18
		310	–14
		330	–10
		350	–7
		400	–1
O_2S	Sulfur dioxide	290	–465
		320	–354
		350	–276
	a(1) = –430	380	–221
	a(2) = –1193	410	–181
	a(3) = –1029	440	–153
		470	–132
Xe	Xenon	160	–421
		170	–377
		180	–340
	a(1) = –130	190	–307
	a(2) = –262	200	–280
	a(3) = –87	210	–255
		220	–234
		230	–215
		240	–199
		250	–184
		300	–129
		350	–93
		400	–69
		500	–39
		600	–21
		650	–14

Compounds Containing Carbon

Mol. form.	Name	*T*/K	*B*/cm^3 mol^{-1}
$CClF_3$	Chlorotrifluoromethane	240	–369
		290	–237
		340	–165
	a(1) = –223	390	–119
	a(2) = –504	440	–86
	a(3) = –340	490	–60
	a(4) = –291	540	–39

Mol. form.	Name	*T*/K	*B*/cm³ mol⁻¹
CCl_2F_2	Dichlorodifluoromethane	250	–769
		280	–570
		310	–441
	a(1) = –486	340	–353
	a(2) = –1217	370	–289
	a(3) = –1188	400	–241
	a(4) = –698	430	–204
		460	–174
CCl_3F	Trichlorofluoromethane	240	–1140
		280	–879
		320	–689
	a(1) = –786	360	–545
	a(2) = –1428	400	–431
	a(3) = –142	440	–340
		480	–265
CCl_4	Tetrachloromethane	320	–1345
		340	–1171
		360	–1040
	a(1) = –1600	380	–942
	a(2) = –4059	400	–868
	a(3) = –4653	420	–814
CF_4	Tetrafluoromethane	250	–137
		300	–87
		350	–55
	a(1) = –88	400	–32
	a(2) = –238	450	–16
	a(3) = –70	500	–4
		600	14
		700	25
		800	33
$CHClF_2$	Chlorodifluoromethane	300	–343
		325	–298
		350	–257
	a(1) = –347	375	–221
	a(2) = –575	400	–188
	a(3) = 187	425	–158
$CHCl_2F$	Dichlorofluoromethane	250	–728
		275	–634
		300	–557
	a(1) = –562	325	–491
	a(2) = –862	350	–434
		375	–385
		400	–343
		425	–305
		450	–271
$CHCl_3$	Trichloromethane	320	–1001
		330	–926
		340	–858
	a(1) = –1193	350	–797
	a(2) = –2936	360	–740
	a(3) = –1751	370	–689
		380	–642

Mol. form.	Name	*T*/K	*B*/cm³ mol⁻¹
		390	–599
		400	–559
CHF_3	Trifluoromethane	200	–433
		220	–350
		240	–288
	a(1) = –177	260	–241
	a(2) = –399	280	–204
	a(3) = –250	300	–174
		320	–151
		340	–132
		360	–116
		380	–103
		400	–91
CH_2Cl_2	Dichloromethane	320	–706
		330	–634
		340	–574
	a(1) = –913	350	–524
	a(2) = –3371	360	–482
	a(3) = –5013	370	–447
		380	–420
		400	–380
		420	–357
CH_2F_2	Difluoromethane	280	–375
		290	–343
		300	–316
	a(1) = –321	310	–294
	a(2) = –754	320	–275
	a(3) = –1300	330	–260
		340	–248
		350	–238
CH_3Br	Bromomethane	280	–645
		290	–596
		300	–551
	a(1) = –559	310	–509
	a(2) = –1324	320	–469
		340	–396
		360	–332
		380	–274
CH_3Cl	Chloromethane	280	–466
		300	–402
		320	–348
	a(1) = –407	340	–304
	a(2) = –887	360	–266
	a(3) = –385	380	–234
		400	–206
		420	–182
		440	–161
		460	–142
		480	–126
		500	–112
		600	–58
CH_3F	Fluoromethane	280	–244

Mol. form.	Name	*T*/K	*B*/cm³ mol⁻¹
		300	−205
		320	−174
	a(1) = −209	340	−150
	a(2) = −525	360	−129
	a(3) = −365	380	−112
		400	−99
		420	−87
CH_3I	Iodomethane	310	−725
		320	−646
		330	−582
	a(1) = −844	340	−531
	a(2) = −3353	350	−492
	a(3) = −6590	360	−462
		370	−441
		380	−427
CH_4	Methane	110	−328
		120	−276
		130	−237
	a(1) = −43	140	−206
	a(2) = −114	150	−181
	a(3) = −19	160	−160
	a(4) = −7	170	−143
		180	−128
		190	−116
		200	−105
		250	−66
		300	−43
		350	−27
		400	−16
		500	0
		600	10
CH_4O	Methanol	320	−1431
		330	−1299
		340	−1174
	a(1) = −1752	350	−1056
	a(2) = −4694	360	−945
		370	−840
		380	−741
		390	−646
		400	−557
CH_5N	Methylamine	300	−451
		325	−367
		350	−304
	a(1) = −459	375	−257
	a(2) = −1191	400	−220
	a(3) = −995	425	−192
		450	−170
		500	−140
		550	−122
CO	Carbon monoxide	210	−36
		240	−24
		270	−15
	a(1) = −9	300	−8

Mol. form.	Name	*T*/K	*B*/cm³ mol⁻¹
	a(2) = −58	330	−3
	a(3) = −18	360	1
		420	7
		480	11
CO_2	Carbon dioxide	220	−244
		240	−204
		260	−172
	a(1) = −127	280	−146
	a(2) = −288	300	−126
	a(3) = −118	320	−108
		340	−94
		360	−81
		380	−71
		400	−62
		500	−30
		600	−13
		700	−1
		800	7
		900	12
		1000	16
		1100	19
CS_2	Carbon disulfide	280	−932
		310	−740
		340	−603
	a(1) = −807	370	−504
	a(2) = −1829	400	−431
	a(3) = −1371	430	−375
$C_2Cl_2F_4$	1,2-Dichloro-1,1,2,2-tetrafluoroethane	300	−801
		320	−695
		340	−608
	a(1) = −812	360	−536
	a(2) = −1773	380	−475
	a(3) = −963	400	−423
		420	−379
		440	−341
		460	−307
		480	−279
		500	−253
$C_2Cl_3F_3$	1,1,2-Trichloro-1,2,2-trifluoroethane	290	−1041
		310	−943
		330	−856
	a(1) = −999	350	−780
	a(2) = −1479	370	−712
		390	−651
		410	−596
		430	−546
		450	−500
C_2H_2	Ethyne	200	−573
		210	−500
		220	−440
	a(1) = −216	230	−390
	a(2) = −375	240	−349
	a(3) = −716	250	−315

Mol. form.	Name	*T*/K	*B*/cm³ mol⁻¹
		260	–287
		270	–263
C_2H_3N	Ethanenitrile	330	–3468
		340	–2971
		350	–2563
	a(1) = –5840	360	–2233
	a(2) = –29175	370	–1970
	a(3) = –47611	380	–1765
		390	–1610
		400	–1499
		410	–1425
C_2H_4	Ethene	240	–218
		270	–172
		300	–139
	a(1) = –140	330	–113
	a(2) = –296	360	–92
	a(3) = –101	390	–76
		420	–63
		450	–52
$C_2H_4Cl_2$	1,2-Dichloroethane	370	–812
		390	–716
		410	–635
	a(1) = –1362	430	–566
	a(2) = –3240	450	–508
	a(3) = –2100	470	–458
		490	–416
		510	–379
		530	–347
		550	–319
		570	–295
C_2H_4O	Ethanal	290	–1352
		320	–927
		350	–654
	a(1) = –1217	380	–482
	a(2) = –4647	410	–375
	a(3) = –5725	440	–314
		470	–283
$C_2H_4O_2$	Methyl methanoate	320	–821
		330	–744
		340	–677
	a(1) = –1035	350	–620
	a(2) = –3425	360	–571
	a(3) = –4203	370	–528
		380	–492
		390	–461
		400	–435
C_2H_5Cl	Chloroethane	320	–634
		360	–450
		400	–330
	a(1) = –777	440	–249
	a(2) = –2205	480	–195
	a(3) = –1764	520	–157

Mol. form.	Name	T/K	B/cm³ mol⁻¹
		560	–131
		600	–114
C_2H_6	Ethane	200	–409
		220	–337
		240	–284
	a(1) = –184	260	–242
	a(2) = –376	280	–209
	a(3) = –143	300	–181
	a(4) = –54	320	–159
		340	–140
		360	–123
		380	–109
		400	–96
		500	–52
		600	–24
C_2H_6O	Ethanol	320	–2710
		330	–2135
		340	–1676
	a(1) = –4475	350	–1317
	a(2) = –29719	360	–1043
	a(3) = –56716	370	–843
		380	–705
		390	–622
C_2H_6O	Dimethyl ether	275	–536
		280	–517
		285	–499
	a(1) = –455	290	–482
	a(2) = –965	295	–465
		300	–449
		305	–433
		310	–418
C_2H_7N	Dimethylamine	310	–606
		320	–563
		330	–523
	a(1) = –662	340	–487
	a(2) = –1504	350	–454
	a(3) = –667	360	–423
		370	–395
		380	–369
		390	–345
		400	–322
C_2H_7N	Ethylamine	300	–773
		310	–710
		320	–654
	a(1) = –785	330	–604
	a(2) = –2012	340	–558
	a(3) = –1397	350	–517
		360	–480
		370	–447
		380	–416
		390	–389
		400	–363
C_3H_6	Cyclopropane	300	–383

Mol. form.	Name	*T*/K	*B*/cm³ mol⁻¹
		310	–356
		320	–332
	a(1) = –388	330	–310
	a(2) = –861	340	–290
	a(3) = –538	350	–272
		360	–256
		370	–241
		380	–227
		390	–215
		400	–204
C_3H_6	Propene	280	–395
		300	–342
		320	–299
	a(1) = –347	340	–262
	a(2) = –727	360	–232
	a(3) = –325	380	–205
		400	–183
		420	–163
		440	–146
		460	–131
		480	–118
		500	–106
C_3H_6O	2-Propanone	300	–1996
		320	–1522
		340	–1198
	a(1) = –2051	360	–971
	a(2) = –8903	380	–806
	a(3) = –18056	400	–683
	a(4) = –16448	420	–586
		440	–506
		460	–437
		480	–375
C_3H_6O	Ethyl methanoate	330	–1003
		340	–916
		350	–839
	a(1) = –1371	360	–771
	a(2) = –4231	370	–712
	a(3) = –4312	380	–660
		390	–614
C_3H_6O	Methyl ethanoate	320	–1320
		330	–1186
		340	–1074
	a(1) = –1709	350	–980
	a(2) = –6348	360	–903
	a(3) = –9650	370	–840
		380	–789
		390	–749
C_3H_7Cl	1-Chloropropane	310	–1001
		340	–772
		370	–614
	a(1) = –1121	400	–501
	a(2) = –3271	430	–417
	a(3) = –3786	460	–352
	a(4) = –1974	490	–302

Mol. form.	Name	*T*/K	*B*/cm³ mol⁻¹
		520	–261
		550	–227
		580	–198
C_3H_8	Propane	240	–641
		260	–527
		280	–444
	a(1) = –386	300	–381
	a(2) = –844	320	–331
	a(3) = –720	340	–292
	a(4) = –574	360	–259
		380	–232
		400	–208
		440	–169
		480	–138
		520	–112
		560	–90
C_3H_8O	1-Propanol	380	–873
		385	–826
		390	–783
	a(1) = –2690	395	–744
	a(2) = –12040	400	–709
	a(3) = –16738	405	–679
		410	–651
		415	–627
		420	–606
C_3H_8O	2-Propanol	380	–821
		385	–766
		390	–717
	a(1) = –3165	395	–674
	a(2) = –16092	400	–636
	a(3) = –24197	405	–604
		410	–576
		415	–552
		420	–533
C_3H_9N	Trimethylamine	310	–675
		320	–628
		330	–585
	a(1) = –737	340	–547
	a(2) = –1669	350	–512
	a(3) = –986	360	–480
		370	–450
C_4H_8	1-Butene	300	–624
		320	–539
		340	–470
	a(1) = –633	360	–413
	a(2) = –1442	380	–366
	a(3) = –932	400	–327
		420	–294
C_4H_8O	2-Butanone	310	–2056
		320	–1878
		330	–1712
	a(1) = –2282	340	–1555

Mol. form.	Name	*T*/K	*B*/cm³ mol⁻¹
	a(2) = –5907	350	–1407
		360	–1267
		370	–1135
$C_4H_8O_2$	Propyl methanoate	330	–1496
		340	–1354
		350	–1231
	a(1) = –2118	360	–1126
	a(2) = –7299	370	–1035
	a(3) = –8851	380	–957
		390	–890
		400	–834
$C_4H_8O_2$	Ethyl ethanoate	330	–1543
		340	–1385
		350	–1254
	a(1) = –2272	360	–1144
	a(2) = –8818	370	–1055
	a(3) = –13130	380	–982
		390	–923
		400	–878
$C_4H_8O_2$	Methyl propanoate	330	–1588
		340	–1444
		350	–1319
	a(1) = –2216	360	–1211
	a(2) = –7339	370	–1117
	a(3) = –8658	380	–1037
		390	–968
		400	–908
C_4H_9Cl	1-Chlorobutane	330	–1224
		370	–898
		410	–691
	a(1) = –1643	450	–551
	a(2) = –4897	490	–449
	a(3) = –6178	530	–371
	a(4) = –3718	570	–309
C_4H_{10}	Butane	250	–1170
		280	–863
		310	–668
	a(1) = –735	340	–536
	a(2) = –1835	370	–442
	a(3) = –1922	400	–371
	a(4) = –1330	430	–315
		460	–270
		490	–232
		520	–199
		550	–171
C_4H_{10}	2-Methylpropane	270	–900
		300	–697
		330	–553
	a(1) = –707	360	–450
	a(2) = –1719	390	–374
	a(3) = –1282	420	–317
		450	–273

Mol. form.	Name	*T*/K	*B*/cm³ mol⁻¹
		480	–240
		510	–215
$C_4H_{10}O$	1-Butanol	350	–1693
		360	–1544
		370	–1402
	a(1) = –2629	380	–1268
	a(2) = –6315	390	–1141
		400	–1021
		420	–796
		440	–593
$C_4H_{10}O$	2-Methyl-1-propanol	390	–1076
		400	–979
		410	–887
	a(1) = –2269	420	–800
	a(2) = –5065	430	–716
		440	–636
$C_4H_{10}O$	2-Butanol	380	–1110
		390	–1005
		400	–906
	a(1) = –2232	410	–811
	a(2) = –5209	420	–721
$C_4H_{10}O$	2-Methyl-2-propanol	380	–924
		390	–827
		400	–736
	a(1) = –1952	410	–649
	a(2) = –4775	420	–567
$C_4H_{10}O$	Diethyl ether	280	–1550
		300	–1199
		320	–954
	a(1) = –1226	340	–776
	a(2) = –4458	360	–638
	a(3) = –7746	380	–525
	a(4) = –10005	400	–428
		420	–340
$C_4H_{11}N$	Diethylamine	320	–1228
		330	–1134
		340	–1056
	a(1) = –1522	350	–988
	a(2) = –5204	360	–926
	a(3) = –15047	370	–868
	a(4) = –28835	380	–812
		390	–755
		400	–697
C_5H_5N	Pyridine	350	–1257
		360	–1176
		370	–1099
	a(1) = –1765	380	–1026
	a(2) = –3431	390	–957
		400	–892
		420	–770
		440	–659

Mol. form.	Name	*T*/K	*B*/cm³ mol⁻¹
C_5H_{10}	Cyclopentane	300	–1049
		305	–1015
		310	–981
	a(1) = –1062	315	–949
	a(2) = –2116	320	–918
C_5H_{10}	1-Pentene	310	–966
		320	–898
		330	–836
	a(1) = –1055	340	–780
	a(2) = –2377	350	–729
	a(3) = –1189	360	–681
		370	–638
		380	–598
		390	–561
		400	–527
		410	–495
$C_5H_{10}O$	2-Pentanone	330	–2850
		340	–2420
		350	–2076
	a(1) = –4962	360	–1804
	a(2) = –26372	370	–1595
	a(3) = –46537	380	–1440
		390	–1332
C_5H_{12}	Pentane	300	–1234
		310	–1130
		320	–1038
	a(1) = –1254	330	–957
	a(2) = –3345	340	–884
	a(3) = –2726	350	–818
		400	–579
		450	–436
		500	–348
		550	–294
C_5H_{12}	2-Methylbutane	280	–1263
		290	–1166
		300	–1079
	a(1) = –1095	310	–1001
	a(2) = –2503	320	–931
	a(3) = –1534	330	–867
		340	–810
		350	–757
		400	–557
		450	–424
C_5H_{12}	2,2-Dimethylpropane	300	–916
		310	–843
		320	–780
	a(1) = –931	330	–724
	a(2) = –2387	340	–674
	a(3) = –2641	350	–629
	a(4) = –1810	360	–590
		370	–554
		380	–521
		390	–492

Mol. form.	Name	*T*/K	*B*/cm³ mol⁻¹
		400	–464
		450	–357
		500	–279
		550	–218
C_6H_6	Benzene	290	–1588
		300	–1454
		310	–1335
	a(1) = –1477	320	–1231
	a(2) = –3851	330	–1139
	a(3) = –3683	340	–1056
	a(4) = –1423	350	–983
		400	–712
		450	–542
		500	–429
		550	–349
		600	–291
C_6H_7N	2-Methylpyridine	360	–1656
		370	–1523
		380	–1404
	a(1) = –2940	390	–1297
	a(2) = –8813	400	–1202
	a(3) = –7809	410	–1117
		420	–1040
		430	–972
C_6H_7N	3-Methylpyridine	380	–1819
		390	–1612
		400	–1448
	a(1) = –6304	410	–1322
	a(2) = –30415	420	–1230
	a(3) = –44549	430	–1166
C_6H_7N	4-Methylpyridine	380	–1787
		390	–1578
		400	–1417
	a(1) = –6553	410	–1297
	a(2) = –32873	420	–1214
	a(3) = –49874	430	–1163
C_6H_{12}	Cyclohexane	300	–1698
		320	–1391
		340	–1170
	a(1) = –1733	360	–1007
	a(2) = –5618	380	–883
	a(3) = –9486	400	–786
	a(4) = –7936	420	–707
		440	–641
		460	–584
		480	–534
		500	–488
		520	–446
		540	–406
		560	–368
C_6H_{12}	Methylcyclopentane	305	–1447
		315	–1357

Mol. form.	Name	*T*/K	*B*/cm³ mol⁻¹
		325	–1272
	a(1) = –1512	335	–1192
	a(2) = –2910	345	–1117
C_6H_{14}	Hexane	300	–1920
		310	–1724
		320	–1561
	a(1) = –1961	330	–1424
	a(2) = –6691	340	–1309
	a(3) = –13167	350	–1209
	a(4) = –15273	360	–1123
		370	–1046
		380	–978
		390	–916
		400	–859
		410	–806
		430	–707
		450	–616
$C_6H_{15}N$	Triethylamine	330	–1562
		340	–1444
		350	–1340
	a(1) = –2061	360	–1249
	a(2) = –5735	370	–1169
	a(3) = –5899	380	–1099
		390	–1037
		400	–983
C_7H_8	Toluene	350	–1641
		360	–1511
		370	–1394
	a(1) = –2620	380	–1289
	a(2) = –7548	390	–1195
	a(3) = –6349	400	–1110
		410	–1034
		420	–965
		430	–903
C_7H_{14}	1-Heptene	340	–1781
		350	–1651
		360	–1532
	a(1) = –2491	370	–1424
	a(2) = –6230	380	–1324
	a(3) = –3780	390	–1233
		400	–1150
		410	–1073
C_7H_{16}	Heptane	300	–2782
		320	–2297
		340	–1928
	a(1) = –2834	360	–1641
	a(2) = –8523	380	–1415
	a(3) = –10068	400	–1233
	a(4) = –5051	420	–1085
		440	–963
		460	–862
		480	–775
		500	–702

Mol. form.	Name	*T*/K	*B*/cm³ mol⁻¹
		540	–583
		580	–490
		620	–416
		660	–355
		700	–304
C_8H_{10}	1,2-Dimethylbenzene	380	–2046
		390	–1848
		400	–1681
	a(1) = –5632	410	–1543
	a(2) = –22873	420	–1428
	a(3) = –28900	430	–1335
		440	–1261
C_8H_{10}	1,3-Dimethylbenzene	380	–2082
		390	–1865
		400	–1679
	a(1) = –5808	410	–1521
	a(2) = –23244	420	–1388
	a(3) = –27607	430	–1276
		440	–1184
C_8H_{10}	1,4-Dimethylbenzene	380	–2043
		390	–1851
		400	–1680
	a(1) = –4921	410	–1529
	a(2) = –16843	420	–1395
	a(3) = –16159	430	–1276
		440	–1171
C_8H_{16}	1-Octene	360	–2147
		370	–2000
		380	–1861
	a(1) = –3273	390	–1729
	a(2) = –6557	400	–1604
		410	–1485
C_8H_{18}	Octane	300	–4042
		350	–2511
		400	–1704
	a(1) = –4123	450	–1234
	a(2) = –13120	500	–936
	a(3) = –16408	550	–732
	a(4) = –8580	600	–583
		650	–468
		700	–375

VAN DER WAALS CONSTANTS FOR GASES

The van der Waals equation of state for a real gas is

$$(P + n^2a/V^2)(V - nb) = nRT$$

where P is the pressure, V the volume, T the temperature, n the amount of substance (in moles), and R the gas constant. The van der Waals constants a and b are characteristic of the substance and are independent of temperature. They are related to the critical temperature and pressure, T_c and P_c, by

$$a = 27R^2T_c^2/64P_c \qquad b = RT_c/8P_c$$

This table gives values of a and b for some common gases. Most of the values have been calculated from the critical temperature and pressure values given in the table "Critical Constants" in this section. Van der Waals constants for other gases may easily be calculated from the data in that table.

To convert the van der Waals constants to SI units, note that 1 bar L^2/mol^2 = 0.1 Pa m^6/mol^2 and 1 L/mol = 0.001 m^3/mol.

REFERENCE

Reid, R.C, Prausnitz, J. M., and Poling, B.E., *The Properties of Gases and Liquids, Fourth Edition*, McGraw-Hill, New York, 1987.

Substance	*a* bar L^2/mol^2	*b* L/mol
Acetic acid	17.71	0.1065
Acetone	16.02	0.1124
Acetylene	4.516	0.0522
Ammonia	4.225	0.0371
Aniline	29.14	0.1486
Argon	1.355	0.0320
Benzene	18.82	0.1193
Bromine	9.75	0.0591
Butane	13.89	0.1164
1-Butanol	20.94	0.1326
2-Butanone	19.97	0.1326
Carbon dioxide	3.658	0.0429
Carbon disulfide	11.25	0.0726
Carbon monoxide	1.472	0.0395
Chlorine	6.343	0.0542
Chlorobenzene	25.80	0.1454
Chloroethane	11.66	0.0903
Chloromethane	7.566	0.0648
Cyclohexane	21.92	0.1411
Cyclopropane	8.34	0.0747
Decane	52.74	0.3043
1-Decanol	59.51	0.3086
Diethyl ether	17.46	0.1333
Dimethyl ether	8.690	0.0774
Dodecane	69.38	0.3758
1-Dodecanol	75.70	0.3750
Ethane	5.580	0.0651
Ethanol	12.56	0.0871
Ethylene	4.612	0.0582
Fluorine	1.171	0.0290
Furan	12.74	0.0926
Helium	0.0346	0.0238
Heptane	31.06	0.2049
1-Heptanol	38.17	0.2150
Hexane	24.84	0.1744
1-Hexanol	31.79	0.1856
Hydrazine	8.46	0.0462
Hydrogen	0.2452	0.0265
Hydrogen bromide	4.500	0.0442
Hydrogen chloride	3.700	0.0406
Hydrogen cyanide	11.29	0.0881
Hydrogen fluoride	9.565	0.0739
Hydrogen iodide	6.309	0.0530
Hydrogen sulfide	4.544	0.0434
Isobutane	13.32	0.1164
Krypton	5.193	0.0106
Methane	2.303	0.0431
Methanol	9.476	0.0659
Methylamine	7.106	0.0588
Neon	0.208	0.0167
Neopentane	17.17	0.1411
Nitric oxide	1.46	0.0289
Nitrogen	1.370	0.0387
Nitrogen dioxide	5.36	0.0443
Nitrogen trifluoride	3.58	0.0545
Nitrous oxide	3.852	0.0444
Octane	37.88	0.2374
1-Octanol	44.71	0.2442
Oxygen	1.382	0.0319
Ozone	3.570	0.0487
Pentane	19.09	0.1449
1-Pentanol	25.88	0.1568
Phenol	22.93	0.1177
Propane	9.39	0.0905
1-Propanol	16.26	0.1079
2-Propanol	15.82	0.1109
Propene	8.442	0.0824
Pyridine	19.77	0.1137
Pyrrole	18.82	0.1049
Silane	4.38	0.0579
Sulfur dioxide	6.865	0.0568
Sulfur hexafluoride	7.857	0.0879
Tetrachloromethane	20.01	0.1281
Tetrachlorosilane	20.96	0.1470
Tetrafluoroethylene	6.954	0.0809
Tetrafluoromethane	4.040	0.0633
Tetrafluorosilane	5.259	0.0724
Tetrahydrofuran	16.39	0.1082
Thiophene	17.21	0.1058
Toluene	24.86	0.1497
1,1,1-Trichloroethane	20.15	0.1317
Trichloromethane	15.34	0.1019
Trifluoromethane	5.378	0.0640
Trimethylamine	13.37	0.1101
Water	5.537	0.0305
Xenon	4.192	0.0516

MEAN FREE PATH AND RELATED PROPERTIES OF GASES

In the simplest version of the kinetic theory of gases, molecules are treated as hard spheres of diameter d which make binary collisions only. In this approximation the mean distance traveled by a molecule between successive collisions, the mean free path l, is related to the collision diameter by:

$$l = \frac{kT}{\pi\sqrt{2}Pd^2}$$

where P is the pressure, T the absolute temperature, and k the Boltzmann constant. At standard conditions ($P = 100\,000$ Pa and $T = 298.15$ K) this relation becomes:

$$l = \frac{9.27 \cdot 10^{-27}}{d^2}$$

where l and d are in meters.

Using the same model and the same standard pressure, the collision diameter can be calculated from the viscosity η by the kinetic theory relation:

$$\eta = \frac{2.67 \cdot 10^{-20}(MT)^{1/2}}{d^2}$$

where η is in units of μPa s and M is the molar mass in g/mol. Kinetic theory also gives a relation for the mean velocity $\bar{v}$ of molecules of mass m:

$$\bar{v} = \left(\frac{8kT}{\pi m}\right)^{1/2} = 145.5(T/M)^{1/2} \text{ m/s}$$

Finally, the mean time τ between collisions can be calculated from the relation $\tau\bar{v} = l$.

The table below gives values of l, $\bar{v}$, and τ for some common gases at 25°C and atmospheric pressure, as well as the value of d, all calculated from measured gas viscosities (see References 2 and 3 and the table "Viscosity of Gases" in this section). It is seen from the above equations that the mean free path varies directly with T and inversely with P, while the mean velocity varies as the square root of T and, in this approximation, is independent of P.

A more accurate model, in which molecular interactions are described by a Lennard-Jones potential, gives mean free path values about 5% lower than this table (see Reference 4).

REFERENCES

1. Reid, R.C., Prausnitz, J.M., and Poling, B.E., *The Properties of Gases and Liquids, Fourth Edition* , McGraw-Hill, New York, 1987.
2. Lide, D.R., and Kehiaian, H.V., *CRC Handbook of Thermophysical and Thermochemical Data*, CRC Press, Boca Raton, FL, 1994.
3. Vargaftik, N.B., *Tables of Thermophysical Properties of Liquids and Gases, Second Edition*, John Wiley, New York, 1975.
4. Kaye, G.W.C., and Laby, T.H., *Tables of Physical and Chemical Constants, 15th Edition*, Longman, London, 1986.

Gas	d	l	$\bar{v}$	τ
Air	$3.66 \cdot 10^{-10}$ m	$6.91 \cdot 10^{-8}$ m	467 m/s	148 ps
Ar	3.58	7.22	397	182
CO_2	4.53	4.51	379	119
H_2	2.71	12.6	1769	71
He	2.15	20.0	1256	159
Kr	4.08	5.58	274	203
N_2	3.70	6.76	475	142
NH_3	4.32	4.97	609	82
Ne	2.54	14.3	559	256
O_2	3.55	7.36	444	166
Xe	4.78	4.05	219	185

INFLUENCE OF PRESSURE ON FREEZING POINTS

This table illustrates the variation of the freezing point of representative types of liquids with pressure. Substances are listed in alphabetical order. Note that 1 MPa = 0.01 kbar = 9.87 atm.

REFERENCES

1. Isaacs, N.S., *Liquid Phase High Pressure Chemistry*, John Wiley, New York, 1981.
2. Merrill, L., *J. Phys. Chem. Ref. Data*, 6, 1205, 1977; 11, 1005, 1982.

Substance	Molecular formula	Freezing point in °C at: 0.1 MPa	100 MPa	1000 MPa
Acetic acid	$C_2H_4O_2$	16.6	37	
Acetophenone	C_8H_8O	20.0	41.2	
Aniline	C_6H_7N	−6.0	13.5	140
Benzene	C_6H_6	5.5	33.4	
Benzonitrile	C_7H_5N	−12.8	7.6	
Benzyl alcohol	C_7H_8O	−15.2	0.2	
Bromobenzene	C_6H_5Br	−30.6	−12	108
Bromoethane	C_2H_5Br	−118.6	−108	
1-Bromonaphthalene	$C_{10}H_7Br$	−1.8	6.1	
1-Bromopropane	C_3H_7Br	−110	−98	
p-Bromotoluene	C_7H_7Br	28.0	56.7	
Butanoic acid	$C_4H_8O_2$	−5.7	13.8	
1-Butanol	$C_4H_{10}O$	−89.8	−77.2	
Carbon disulfide	CS_2	−111.5	−98	
Chlorobenzene	C_6H_5Cl	−45.2	−28	84
p-Chlorotoluene	C_7H_7Cl	6.9	33.1	
o-Cresol	C_7H_8O	29.8	47.7	
m-Cresol	C_7H_8O	11.8	25.6	
p-Cresol	C_7H_8O	35.8	56.2	
Cyclohexane	C_6H_{12}	6.6	32.5	
Cyclohexanol	$C_6H_{12}O$	25.5	62.3	
1,2-Dibromoethane	$C_2H_4Br_2$	9.9	34.0	
p-Dichlorobenzene	$C_6H_4Cl_2$	52.7	79.1	
Dichloromethane	CH_2Cl_2	−95.1	−83	
N,N-Dimethylaniline	$C_8H_{11}N$	2.5	26.3	
1,4-Dioxane	$C_4H_8O_2$	11	23	
Ethanol	C_2H_6O	−114.1	−108	
Formamide	CH_3NO	−15.5	10.8	
Formic acid	CH_2O_2	8.3	20.6	
Furan	C_4H_4O	−85.6	−73	
Hexamethyldisiloxane	$C_6H_{18}OSi_2$	−66	−37	
Menthol	$C_{10}H_{20}O$	42	60	
Methyl benzoate	$C_8H_8O_2$	−15	31.8	
2-Methyl-2-butanol	$C_5H_{12}O$	−8.8	13.4	
2-Methyl-2-propanol	$C_4H_{10}O$	25.4	58.1	
Naphthalene	$C_{10}H_8$	78.2	115.7	
Nitrobenzene	$C_6H_5NO_2$	5.7	13.5	
m-Nitrotoluene	$C_7H_7NO_2$	15.5	40.6	
Pentachloroethane	C_2HCl_5	−29.0	−6.3	
Potassium	K	63.7	78	170
Potassium chloride	ClK	771		945
Propanoic acid	$C_3H_6O_2$	−20.7	−1.2	
Silver chloride	AgCl	455		545
Sodium	Na	97.8	106	167
Sodium chloride	ClNa	800.7		997
Sodium fluoride	FNa	996		1115
Tetrachloromethane	CCl_4	−23.0	14.2	
Tribromomethane	$CHBr_3$	8.1	31.5	
Trichloromethane	$CHCl_3$	−63.6	−45.2	
Water	H_2O	0.0	−9.0	
o-Xylene	C_8H_{10}	−25.2	−3.5	
m-Xylene	C_8H_{10}	−47.8	−25.2	
p-Xylene	C_8H_{10}	13.2	46.0	

CRITICAL CONSTANTS

This table gives the liquid–gas critical constants and normal boiling points of about 750 inorganic and organic substances. The properties tabulated are:

T_b: Normal boiling point in K at a pressure of 101.325 kPa (1 atmosphere); an "s" following the value indicates a sublimation point (temperature at which the solid is in equilibrium with the gas at a pressure of 101.325 kPa)

T_c: Critical temperature in K

P_c: Critical pressure in MPa

V_c: Critical molar volume in cm^3/mol

The number of digits given for T_b, T_c, and P_c indicates the estimated accuracy of these quantities; however, values of T_c greater than 750 K may be in error by 10 K or more. Although most V_c values are given to three figures, they cannot be assumed accurate to better than a few percent.

Many of the critical constants in this table are taken from reviews produced by the IUPAC Commission on Thermodynamics (References 2-6) and from the publications of the Design Institute for Physical Property Data (References 8-9). Compounds are listed by molecular formula in modified Hill order, with compounds not containing carbon preceding those that do contain carbon.

* An asterisk following an entry indicates the value was obtained by extrapolation.

REFERENCES

1. Ambrose, D., "Vapor-Liquid Constants of Fluids", in Stevenson, R. M., and Malanowski, S., *Handbook of the Thermodynamics of Organic Compounds*, Elsevier, New York, 1987.
2. Ambrose, D., and Tsonopoulos, C., *J. Chem. Eng. Data*, 40, 531, 1995.
3. Tsonopoulos, C., and Ambrose, D., *J. Chem. Eng. Data*, 40, 547, 1995.
4. Gude, M., and Teja, A. S., *J. Chem. Eng. Data*, 40, 1025, 1995.
5. Daubert, T. E., *J. Chem. Eng. Data*, 41, 365, 1966.
6. Tsonopoulos, C., and Ambrose, D., *J. Chem. Eng. Data*, 41, 645, 1996.
7. Das, A., Frenkel, M., Gadalla, N. A. M., Kudchadker, S., Marsh, K. N., Rodgers, A. S., and Wilhoit, R. C., *J. Phys. Chem. Ref. Data*, 22, 659, 1993.
8. Wilson, L. C., Wilson, H. L., Wilding, W. V., and Wilson, G. M., *J. Chem. Eng. Data*, 41, 1252, 1997.
9. Daubert, T. E., Danner, R. P., Sibul, H. M., and Stebbins, C. C., *Physical and Thermodynamic Properties of Pure Compounds: Data Compilation*, extant 1996 (core with 6 supplements), Taylor & Francis, Bristol, PA.
10. Reid, R. C., Prausnitz, J. M., and Poling, B. E., *The Properties of Gases and Liquids, Fourth Edition,* McGraw-Hill, New York, 1987.
11. Vargaftik, N.B., Vinogradov, Y. K., and Yargin, V. S., *Handbook of Physical Properties of Liquids and Gases, Third Edition*, Begell House, New York, 1996.
12. Morton, D. W., Lui, M. P. W., Tran, C. A., and Young, C. L., *J. Chem. Eng. Data*, 45, 437, 2000.
13. VonNiederhausern, D. M., Wilson, L. C., Giles, N. F., and Wilson, G. M., *J. Chem. Eng. Data*, 45, 154, 2000.
14. VonNiederhausern, D. M., Wilson, G. M., and Giles, N. F., *J. Chem. Eng. Data*, 45, 157, 2000.

Molecular Formula	Name	T_b/K	T_c/K	P_c/MPa	V_c/cm^3 mol^{-1}
$AlBr_3$	Aluminum bromide	528	763	2.89	310
$AlCl_3$	Aluminum chloride	453 s	620	2.63	257
AlI_3	Aluminum iodide	655	983		408
Ar	Argon	87.30	150.87	4.898	75
As	Arsenic	876 s	1673		35
$AsCl_3$	Arsenic(III) chloride	403	654		252
AsH_3	Arsine	210.7	373.1		
BBr_3	Boron tribromide	364	581		272
BCl_3	Boron trichloride	285.80	455	3.87	239
BF_3	Boron trifluoride	172	260.8	4.98	115
BI_3	Boron triiodide	482.7	773		356
B_2H_6	Diborane	180.8	289.8	4.05	
$BiBr_3$	Bismuth tribromide	726	1220		301
$BiCl_3$	Bismuth trichloride	720	1179	12.0	261
BrH	Hydrogen bromide	206.77	363.2	8.55	
BrI	Iodine bromide	389	719		139
Br_2	Bromine	332.0	588	10.34	127

Molecular Formula	Name	T_b/K	T_c/K	P_c/MPa	V_c/cm^3 mol^{-1}
Br_2Hg	Mercury(II) bromide	595	1012		
Br_3Ga	Gallium(III) bromide	552	806.7		303
Br_3HSi	Tribromosilane	382	610.0		305
Br_3P	Phosphorus(III) bromide	446.4	711		300
Br_3Sb	Antimony(III) bromide	553	904		300
Br_4Ge	Germanium(IV) bromide	459.50	718		392
Br_4Hf	Hafnium(IV) bromide	596 s	746		415
Br_4Si	Tetrabromosilane	427	663		382
Br_4Sn	Tin(IV) bromide	478	744		417
Br_4Ti	Titanium(IV) bromide	503	795.7		391
Br_4Zr	Zirconium(IV) bromide	633 s	805		424
Br_5Ta	Tantalum(V) bromide	622	974		461
$ClFO_3$	Perchloryl fluoride	226.40	368.4	5.37	161
ClF_2N	Nitrogen chloride difluoride	206	337.5	5.15	
ClF_2P	Phosphorus(III) chloride difluoride	225.9	362.4	4.52	
ClF_2PS	Phosphorothioc chloride difluoride	279.5	439.2	4.14	
ClF_3Si	Chlorotrifluorosilane	203.2	307.7	3.46	
ClF_5	Chlorine pentafluoride	260.1	416	5.27	233
ClF_5S	Sulfur chloride pentafluoride	254.10	390.9		
ClH	Hydrogen chloride	188	324.7	8.31	81
ClH_4N	Ammonium chloride	611 s	1155	163.5	
ClH_4P	Phosphonium chloride	246	322.3	7.37	
$ClNO$	Nitrosyl chloride	267.7	440		
$ClOV$	Vanadyl chloride	400	636		171
Cl_2	Chlorine	239.11	416.9	7.991	123
Cl_2FP	Phosphorus(III) dichloride fluoride	287.00	463.0	4.96	
Cl_2F_2Si	Dichlorodifluorosilane	241	369.0	3.5	
Cl_2Hg	Mercury(II) chloride	577	973		174
Cl_2OSe	Selenium oxychloride	450	730	7.09	235
Cl_3FSi	Trichlorofluorosilane	285.40	438.6	3.58	
Cl_3Ga	Gallium(III) chloride	474	694		263
Cl_3HSi	Trichlorosilane	306	479		268
Cl_3P	Phosphorus(III) chloride	349.3	563		264
Cl_3Sb	Antimony(III) chloride	493.5	794		272
Cl_4Ge	Germanium(IV) chloride	359.70	553.2	3.861	330
Cl_4Hf	Hafnium(IV) chloride	590 s	725.7	5.42	314
Cl_4ORe	Rhenium(VI) oxytetrachloride	496	781		362
Cl_4OW	Tungsten(VI) oxytetrachloride	500.70	782		338
Cl_4Si	Tetrachlorosilane	330.80	508.1	3.593	326
Cl_4Sn	Tin(IV) chloride	387.30	591.9	3.75	351
Cl_4Te	Tellurium tetrachloride	660	1002	8.56	310
Cl_4Ti	Titanium(IV) chloride	409.60	638	4.66	339
Cl_4Zr	Zirconium(IV) chloride	604 s	778	5.77	319
Cl_5Mo	Molybdenum(V) chloride	541	850		369
Cl_5Nb	Niobium(V) chloride	527.2	803.5	4.88	397
Cl_5P	Phosphorus(V) chloride	433 s	646		
Cl_5Ta	Tantalum(V) chloride	512.50	767		402
Cl_6W	Tungsten(VI) chloride	619.90	923		422
Cs	Cesium	944	1938	9.4	341
FH	Hydrogen fluoride	293	461	6.48	69
FNO_2	Nitryl fluoride	200.8	349.5		
F_2	Fluorine	85.03	144.13	5.172	66
F_2HN	Difluoramine	250	403		
F_2N_2	*cis*-Difluorodiazine	167.40	272	7.09	
F_2N_2	*trans*-Difluorodiazine	161.70	260	5.57	
F_2O	Fluorine monoxide	128.40	215		
F_2Xe	Xenon difluoride	387.50 s	631	9.32	148

CRITICAL CONSTANTS (continued)

Molecular Formula	Name	T_b/K	T_c/K	P_c/MPa	V_c/cm³ mol⁻¹
F_3N	Nitrogen trifluoride	144.40	234.0	4.46	126
F_3NO	Trifluoramine oxide	185.7	303	6.43	147
F_3P	Phosphorus(III) fluoride	171.4	271.2	4.33	
F_3PS	Phosphorothioc trifluoride	220.90	346.0	3.82	
F_4N_2	Tetrafluorohydrazine	199	309	3.75	
F_4S	Sulfur tetrafluoride	232.70	364		
F_4Si	Tetrafluorosilane	187	259.0	3.72	
F_4Xe	Xenon tetrafluoride	388.90 s	612	7.04	188
F_5Nb	Niobium(V) fluoride	502	737	6.28	155
F_6Mo	Molybdenum(VI) fluoride	307.2	473	4.75	226
F_6S	Sulfur hexafluoride	209.35 s	318.69	3.77	199
F_6Se	Selenium hexafluoride	226.55 s	345.5		
F_6Te	Tellurium hexafluoride	234.25 s	356		
F_6U	Uranium(VI) fluoride	329.65 s	505.8	4.66	250
F_6W	Tungsten(VI) fluoride	290.3	444	4.34	233
GaI_3	Gallium(III) iodide	613	951		395
GeH_4	Germane	185.1	312.2	4.95	147
GeI_4	Germanium(IV) iodide	650	973		500
HI	Hydrogen iodide	237.60	424.0	8.31	
H_2	Hydrogen	20.28	32.97	1.293	65
H_2O	Water	373.2	647.14	22.06	56
H_2O_2	Hydrogen peroxide	423.4	728*	22*	
H_2S	Hydrogen sulfide	213.60	373.2	8.94	99
H_2Se	Hydrogen selenide	231.90	411	8.92	
H_3N	Ammonia	239.82	405.5	11.35	72
H_3P	Phosphine	185.40	324.5	6.54	
H_4N_2	Hydrazine	386.70	653	14.7	
He	Helium	4.22	5.19	0.227	57
HfI_4	Hafnium iodide	667 s	916		528
Hg	Mercury	629.88	1750	172.00	43
HgI_2	Mercury(II) iodide	627	1072		
I_2	Iodine	457.6	819		155
I_3Sb	Antimony(III) iodide	674	1102		
I_4Si	Tetraiodosilane	560.50	944		558
I_4Sn	Tin(IV) iodide	637.50	968		531
I_4Ti	Titanium(IV) iodide	650	1040		505
I_4Zr	Zirconium(IV) iodide	704 s	960		530
K	Potassium	1032	2223*	16*	209*
Kr	Krypton	119.93	209.41	5.50	91
Li	Lithium	1615	3223*	67*	66*
NO	Nitric oxide	121.41	180	6.48	58
N_2	Nitrogen	77.36	126.21	3.39	90
N_2O	Nitrous oxide	184.67	309.57	7.255	97
N_2O_4	Nitrogen tetroxide	294.30	431	10.1	167
Na	Sodium	1156	2573*	35*	116*
Ne	Neon	27.07	44.4	2.76	42
O_2	Oxygen	90.20	154.59	5.043	73
O_2S	Sulfur dioxide	263.10	430.8	7.884	122
O_3	Ozone	161.80	261.1	5.57	89
O_3S	Sulfur trioxide	318	491.0	8.2	127
O_4Os	Osmium(VIII) oxide	408	678		
O_7Re_2	Rhenium(VII) oxide	633	942		334
P	Phosphorus	553.7	994		
Rb	Rubidium	961	2093*	16*	247*
Rn	Radon	211.5	377	6.28	
S	Sulfur	717.75	1314	20.7	
Se	Selenium	958	1766	27.2	
Xe	Xenon	165.03	289.77	5.841	118

Molecular Formula	Name	T_b/K	T_c/K	P_c/MPa	V_c/cm³ mol⁻¹
$CBrClF_2$	Bromochlorodifluoromethane	269.5	426.88	4.254	246
$CBrF_3$	Bromotrifluoromethane	215.4	340.2	3.97	196
CBr_2F_2	Dibromodifluoromethane	298	471.3		
$CClF_3$	Chlorotrifluoromethane	191.8	302	3.870	180
CCl_2F_2	Dichlorodifluoromethane	243.4	384.95	4.136	217
CCl_2O	Carbonyl chloride	281	455	5.67	190
CCl_3F	Trichlorofluoromethane	296.9	471.1	4.47	247
CCl_4	Tetrachloromethane	350.0	556.6	4.516	276
CF_3I	Trifluoroiodomethane	250.7	396.44	3.953	
CF_4	Tetrafluoromethane	145.2	227.6	3.74	140
$CHClF_2$	Chlorodifluoromethane	232.5	369.5	5.035	164
$CHCl_2F$	Dichlorofluoromethane	282.1	451.58	5.18	196
$CHCl_3$	Trichloromethane	334.32	536.4	5.47	239
CHF_3	Trifluoromethane	191.1	299.3	4.858	133
CHN	Hydrogen cyanide	299	456.7	5.39	139
CH_2Cl_2	Dichloromethane	313	510	6.10	
CH_2F_2	Difluoromethane	221.6	351.6	5.830	121
CH_2O_2	Formic acid	374	588		
CH_3Cl	Chloromethane	249.06	416.25	6.679	139
CH_3Cl_3Si	Methyltrichlorosilane	338.8	517	3.28	348
CH_3F	Fluoromethane	194.8	317.8	5.88	113
CH_3I	Iodomethane	315.58	528		
CH_3NO_2	Nitromethane	374.34	588	5.87	173
CH_4	Methane	111.67	190.56	4.599	98.60
CH_4O	Methanol	337.8	512.5	8.084	117
CH_4S	Methanethiol	279.1	470.0	7.23	145
CH_5ClSi	Chloromethylsilane	280	517.8		
CH_5N	Methylamine	266.83	430.7	7.614	
CH_6N_2	Methylhydrazine	360.7	567	8.24	271
CH_6Si	Methylsilane	215.7	352.5		
CO	Carbon monoxide	81.7	132.91	3.499	93
COS	Carbon oxysulfide	223	375	5.88	137
CO_2	Carbon dioxide	194.65 s	304.13	7.375	94
CS_2	Carbon disulfide	319	552	7.90	173
$C_2Br_2ClF_3$	1,2-Dibromo-1-chloro-1,2,2-trifluoroethane	366	560.7	3.61	368
$C_2Br_2F_4$	1,2-Dibromotetrafluoroethane	320.50	487.8	3.393	341
C_2ClF_3	Chlorotrifluoroethene	245.4	379	4.05	212
C_2ClF_5	Chloropentafluoroethane	235.20	353.2	3.229	252
$C_2Cl_2F_4$	1,1-Dichloro-1,2,2,2-tetrafluoroethane	277	418.6	3.30	294
$C_2Cl_2F_4$	1,2-Dichloro-1,1,2,2-tetrafluoroethane	277.0	418.78	3.252	297
$C_2Cl_3F_3$	1,1,2-Trichloro-1,2,2-trifluoroethane	320.9	487.3	3.42	325
C_2Cl_4	Tetrachloroethene	394.5	620.2		
$C_2Cl_4F_2$	1,1,2,2-Tetrachloro-1,2-difluoroethane	366.0	551		
C_2F_3N	Trifluoroacetonitrile	204.4	311.11	3.618	202
C_2F_4	Tetrafluoroethene	197.3	306.5	3.94	172
C_2F_6	Hexafluoroethane	195.1	293		222
C_2HClF_2	1-Chloro-2,2-difluoroethene	254.7	400.6	4.46	197
C_2HClF_4	1-Chloro-1,1,2,2-tetrafluoroethane	263	399.9	3.72	244
C_2HClF_4	1-Chloro-1,2,2,2-tetrafluoroethane	261	395.4	3.66	
$C_2HCl_2F_3$	1,2-Dichloro-1,1,2-trifluoroethane	301	461.6		278
$C_2HCl_2F_3$	2,2-Dichloro-1,1,1-trifluoroethane	300.97	456.83	3.661	278
C_2HCl_3	Trichloroethene	360.36	544.2	5.02	
$C_2HF_3O_2$	Trifluoroacetic acid	346	491.3	3.258	204
C_2HF_5	Pentafluoroethane	224.7	339.41	3.639	209.7
C_2HF_5O	Trifluoromethyl difluoromethyl ether	237	354.49	3.33	
C_2H_2	Acetylene	188.45 s	308.3	6.138	112.2
$C_2H_2Cl_2$	*cis*-1,2-Dichloroethene	333.3	544.2		
$C_2H_2Cl_2$	*trans*-1,2-Dichloroethene	321.9	516.5	5.51	

Molecular Formula	Name	T_b/K	T_c/K	P_c/MPa	V_c/cm^3 mol^{-1}
$C_2H_2Cl_4$	1,1,2,2-Tetrachloroethane	419.7	661.15		
$C_2H_2F_2$	1,1-Difluoroethene	187.5	302.9	4.46	154
$C_2H_2F_4$	1,1,1,2-Tetrafluoroethane	247.07	374.18	4.065	198
$C_2H_2F_4$	1,1,2,2-Tetrafluoroethane	253.3	391.8		191
$C_2H_3ClF_2$	1-Chloro-1,1-difluoroethane	263.5	410.34	4.048	225
$C_2H_3Cl_2F$	1,1-Dichloro-1-fluoroethane	305.2	477.5	4.194	255
$C_2H_3Cl_3$	1,1,1-Trichloroethane	347.24	545	4.30	
C_2H_3F	Fluoroethene	201	327.9	5.24	144
$C_2H_3F_3$	1,1,1-Trifluoroethane	225.90	345.86	3.764	194
$C_2H_3F_3O$	Methyl trifluoromethyl ether	249	378.0	3.680	228
C_2H_3N	Acetonitrile	354.80	545.5	4.85	173
C_2H_3N	Acetonitrile	354.80	545.6	4.884	173
C_2H_4	Ethylene	169.38	282.34	5.041	131.1
$C_2H_4Br_2$	1,2-Dibromoethane	404.8	583.0	7.2	
$C_2H_4Cl_2$	1,1-Dichloroethane	330.5	523	5.07	236
$C_2H_4Cl_2$	1,2-Dichloroethane	356.7	561	5.4	225
$C_2H_4F_2$	1,1-Difluoroethane	248.20	386.7	4.50	181
C_2H_4O	Acetaldehyde	293.3	466		154
C_2H_4O	Ethylene oxide	283.8	469	7.19	140
$C_2H_4O_2$	Acetic acid	391.1	592.71	5.786	171
$C_2H_4O_2$	Methyl formate	304.9	487.2	5.998	172
C_2H_5Br	Bromoethane	311.7	503.9	6.23	215
C_2H_5Cl	Chloroethane	285.5	460.4	5.3	
C_2H_5F	Fluoroethane	235.5	375.31	5.028	
C_2H_6	Ethane	184.6	305.32	4.872	145.5
$C_2H_6Cl_2Si$	Dichlorodimethylsilane	343.5	520.4	3.49	350
C_2H_6O	Ethanol	351.44	514.0	6.137	168
C_2H_6O	Dimethyl ether	248.4	400.0	5.37	190
$C_2H_6O_2$	Ethylene glycol	470.5	719	8.2	
C_2H_6S	Ethanethiol	308.3	499	5.49	207
C_2H_6S	Dimethyl sulfide	310.48	503.0	5.53	201
C_2H_7N	Ethylamine	289.7	456	5.62	182
C_2H_7N	Dimethylamine	280.03	437.22	5.340	
$C_2H_8N_2$	1,2-Ethanediamine	390	613.1	6.707	
C_2N_2	Cyanogen	252.1	400	5.98	
C_3ClF_5O	Chloropentafluoroacetone	281	410.6	2.878	
C_3F_6O	Perfluoroacetone	245.8	357.14	2.84	329
C_3F_6O	Perfluorooxetane	244	361.8	3.094	272
C_3F_8	Perfluoropropane	236.6	345.1	2.680	299
$C_3F_8O_2$	Perfluorodimethoxymethane	263	372.3	2.333	363
C_3HF_7O	Trifluoromethyl 1,1,2,2-tetrafluoroethyl ether	270	387.7	2.640	337
$C_3H_2F_6$	1,1,1,2,3,3-Hexafluoropropane	279.3	412.38	3.412	269
$C_3H_2F_6O$	1,2,2,2-Tetrafluoroethyl difluoromethyl ether	296.50	428.95	3.050	315
$C_3H_3F_3$	3,3,3-Trifluoropropene	256	376.2	3.80	211
$C_3H_3F_5$	1,1,1,2,2-Pentafluoropropane	255.8	380.11	3.137	273
$C_3H_3F_5O$	Perfluoroethyl methyl ether	278.74	406.80	2.887	301
C_3H_3N	Acrylonitrile	350.5	540	4.660	
C_3H_3NO	Isoxazole	368	552.0		
C_3H_4	Allene	238.8	394	5.25	
C_3H_4	Propyne	250.0	402.4	5.63	163.5
C_3H_5Cl	3-Chloropropene	318.3	514		
$C_3H_5F_3O$	2,2,2-Trifluoroethyl methyl ether	304.77	448.98	3.513	277
C_3H_5N	Propanenitrile	370.29	561.3	4.26	229
C_3H_6	Propene	225.46	364.9	4.60	184.6
C_3H_6	Cyclopropane	240.34	398.0	5.54	162
$C_3H_6Cl_2$	1,2-Dichloropropane	369.6	578.5		
$C_3H_6Cl_2$	1,3-Dichloropropane	394.1	614.6		
C_3H_6O	Allyl alcohol	370.2	545.1		

CRITICAL CONSTANTS (continued)

Molecular Formula	Name	T_b/K	T_c/K	P_c/MPa	V_c/cm³ mol⁻¹
C_3H_6O	Propanal	321	504.4	5.27	204
C_3H_6O	Acetone	329.20	508.1	4.700	209
C_3H_6O	Methyloxirane	308	482.2	4.92	186
$C_3H_6O_2$	Propanoic acid	414.30	604	4.53	222
$C_3H_6O_2$	Ethyl formate	327.6	508.5	4.74	229
$C_3H_6O_2$	Methyl acetate	330.02	506.55	4.75	228
C_3H_7Cl	1-Chloropropane	319.7	503	4.58	
C_3H_7Cl	2-Chloropropane	308.9	484		
C_3H_7NO	*N,N*-Dimethylformamide	426	649.6		262
C_3H_8	Propane	231.1	369.83	4.248	200
C_3H_8O	1-Propanol	370.4	536.8	5.169	218
C_3H_8O	2-Propanol	355.5	508.3	4.764	222
C_3H_8O	Ethyl methyl ether	280.6	437.8	4.40	221
$C_3H_8O_2$	1,2-Propylene glycol	460.8	676.4	5.941	
$C_3H_8O_2$	1,3-Propylene glycol	487.6	722	6.3	
$C_3H_8O_2$	Ethylene glycol monomethyl ether	397.3	597.6	5.285	263
C_3H_8S	1-Propanethiol	341.0	536.6		286
C_3H_8S	Ethyl methyl sulfide	339.9	533	4.26	
$C_3H_9BO_3$	Trimethyl borate	340.7	501.7	3.59	
C_3H_9ClSi	Trimethylchlorosilane	333	497.8	3.20	366
C_3H_9N	Propylamine	320.37	497.0	4.72	
C_3H_9N	Isopropylamine	304.91	471.8	4.54	221
C_3H_9N	Trimethylamine	276.02	432.79	4.087	254
$C_4Br_2F_8$	1,4-Dibromooctafluorobutane	370	532.5	2.39	
C_4F_8	Perfluorocyclobutane	267.3	388.46	2.784	324
C_4F_{10}	Perfluorobutane	271.3	386.4	2.323	378
C_4F_{10}	Perfluoroisobutane	273	395.4		
$C_4H_2F_8O$	Perfluoroethyl 2,2,2-trifluoroethyl ether	301.04	421.68	2.330	409
$C_4H_3F_7O$	Perfluoropropyl methyl ether	307.38	437.70	2.481	377
$C_4H_3F_7O$	Perfluoroisopropyl methyl ether	302.56	433.30	2.553	369
C_4H_4O	Furan	304.7	490.2	5.50	218
C_4H_4S	Thiophene	357.2	579.4	5.69	219
$C_4H_5F_5O$	Perfluoroethyl ethyl ether	301.26	431.23	2.533	366
C_4H_5N	Pyrrole	402.94	639.7	6.34	200
C_4H_6	1,3-Butadiene	268.74	425	4.32	221
C_4H_6	1-Butyne	281.23	440	4.60	208
C_4H_6	2-Butyne	300.1	488.7		
$C_4H_6O_2$	γ-Butyrolactone	477	731.0	5.131	
$C_4H_6O_3$	Acetic anhydride	412.7	606	4.0	
C_4H_7N	Butanenitrile	390.8	585.4	3.88	
C_4H_8	1-Butene	266.89	419.5	4.02	240.8
C_4H_8	*cis*-2-Butene	276.86	435.5	4.21	233.8
C_4H_8	*trans*-2-Butene	274.03	428.6	4.10	237.7
C_4H_8	Isobutene	266.3	417.9	4.000	238.8
C_4H_8	Cyclobutane	285.8	460.0	4.98	210
C_4H_8O	Ethyl vinyl ether	308.7	475	4.07	
C_4H_8O	Butanal	348.0	537.2	4.32	258
C_4H_8O	2-Butanone	352.74	536.78	4.207	267
C_4H_8O	Tetrahydrofuran	338	540.1	5.19	224
C_4H_8OS	Ethyl thioacetate	389.6	590.55	4.075	319
$C_4H_8O_2$	Butanoic acid	436.90	624	4.03	290
$C_4H_8O_2$	2-Methylpropanoic acid	427.60	605	3.7	292
$C_4H_8O_2$	Propyl formate	354.1	538.0	4.06	285
$C_4H_8O_2$	Ethyl acetate	350.26	523.3	3.882	286
$C_4H_8O_2$	Methyl propanoate	353.0	530.6	4.004	282
$C_4H_8O_2$	1,4-Dioxane	374.7	587	5.21	238
C_4H_8S	Tetrahydrothiophene	394.2	632.0		
C_4H_9Cl	1-Chlorobutane	351.6	539.2		

Molecular Formula	Name	T_b/K	T_c/K	P_c/MPa	V_c/cm³ mol⁻¹
C_4H_9Cl	2-Chlorobutane	341.4	518.6		
C_4H_9Cl	2-Chloro-2-methylpropane	324.1	500		
C_4H_9N	Pyrrolidine	359.71	568	6.00	238
C_4H_{10}	Butane	272.7	425.12	3.796	255
C_4H_{10}	Isobutane	261.42	407.8	3.640	259
$C_4H_{10}O$	1-Butanol	390.88	563.0	4.414	274
$C_4H_{10}O$	2-Butanol	372.66	536.2	4.202	269
$C_4H_{10}O$	2-Methyl-1-propanol	381.04	547.8	4.295	274
$C_4H_{10}O$	2-Methyl-2-propanol	355.6	506.2	3.972	275
$C_4H_{10}O$	Diethyl ether	307.7	466.74	3.638	280
$C_4H_{10}O$	Methyl propyl ether	312.3	476.25	3.801	
$C_4H_{10}O$	Isopropyl methyl ether	303.92	464.48	3.762	
$C_4H_{10}O_2$	1,2-Butanediol	463.7	680	5.21	303
$C_4H_{10}O_2$	1,3-Butanediol	480.7	676	4.02	305
$C_4H_{10}O_2$	1,4-Butanediol	508	727	6.22	
$C_4H_{10}O_2$	Ethylene glycol dimethyl ether	358	536	3.87	271
$C_4H_{10}O_2$	Propylene glycol monomethyl ether	392	579.8	4.113	
$C_4H_{10}S$	1-Butanethiol	371.7	570.1		324
$C_4H_{10}S$	Diethyl sulfide	365.3	557.8	3.897	318
$C_4H_{11}N$	Butylamine	350.15	531.9	4.25	277
$C_4H_{11}N$	*sec*-Butylamine	335.88	514.3	4.20	278
$C_4H_{11}N$	*tert*-Butylamine	317.19	483.9	3.84	292
$C_4H_{11}N$	Isobutylamine	340.90	519	4.07	278
$C_4H_{11}N$	Diethylamine	328.7	499.99	3.758	
$C_4H_{12}Si$	Tetramethylsilane	299.8	448.64	2.821	362
$C_4H_{12}Sn$	Tetramethylstannane	351	521.8	2.981	
$C_4H_{13}N_3$	Bis(2-amimoethyl)amine	480	709	4.3	
C_5F_{12}	Perfluoropentane	302.4	420.59	2.045	473
$C_5H_2F_6O_2$	Hexafluoroacetylacetone	327.30	485.1	2.767	
$C_5H_4O_2$	Furfural	434.9	670	5.89	
C_5H_5N	Pyridine	388.38	620.0	5.67	243
$C_5H_6N_2$	2-Methylpyrazine	410	634.3	5.01	283
C_5H_6O	2-Methylfuran	337.9	527	4.72	247
C_5H_7N	1-Methylpyrrole	385.96	596.0	4.86	271
C_5H_7N	2-Methylpyrrole	420.8	654	5.08	266
C_5H_7N	3-Methylpyrrole	416.1	647	5.08	266
C_5H_8	1-Pentyne	313.3	493.5		
C_5H_8	Cyclopentene	317.4	506.5	4.80	245
C_5H_8O	Cyclopentanone	403.72	624.5	4.60	
C_5H_8O	3,4-Dihydro-2*H*-pyran	359	561.7	4.56	268
C_5H_9N	Pentanenitrile	414.5	610.3	3.58	
C_5H_9NO	*N*-Methyl-2-pyrrolidone	475	721.8		311
C_5H_{10}	1-Pentene	303.11	464.8	3.56	298.4
C_5H_{10}	*cis*-2-Pentene	310.08	475	3.69	
C_5H_{10}	*trans*-2-Pentene	309.49	471	3.52	
C_5H_{10}	2-Methyl-1-butene	304.4	470	3.8	
C_5H_{10}	3-Methyl-1-butene	293.3	452.7	3.53	304.9
C_5H_{10}	2-Methyl-2-butene	311.71	470	3.42	
C_5H_{10}	Cyclopentane	322.5	511.7	4.51	259
$C_5H_{10}O$	Cyclopentanol	413.57	619.5	4.9	
$C_5H_{10}O$	Allyl ethyl ether	340.8	518		
$C_5H_{10}O$	Pentanal	376	566.1	3.97	313
$C_5H_{10}O$	2-Pentanone	375.41	561.08	3.694	301
$C_5H_{10}O$	3-Pentanone	374.9	561.46	3.729	336
$C_5H_{10}O$	3-Methyl-2-butanone	367.48	553.4	3.85	310
$C_5H_{10}O$	Tetrahydropyran	361	572.2	4.77	263
$C_5H_{10}O$	2-Methyltetrahydrofuran	351	537	3.76	267
$C_5H_{10}O_2$	Pentanoic acid	459.3	643	3.58	340

Molecular Formula	Name	T_b/K	T_c/K	P_c/MPa	V_c/cm^3 mol^{-1}
$C_5H_{10}O_2$	3-Methylbutanoic acid	449.7	629	3.40	
$C_5H_{10}O_2$	Isobutyl formate	371.4	551	3.88	352
$C_5H_{10}O_2$	Propyl acetate	374.69	549.7	3.36	345
$C_5H_{10}O_2$	Isopropyl acetate	361.8	531		
$C_5H_{10}O_2$	Ethyl propanoate	372.3	546.0	3.362	345
$C_5H_{10}O_2$	Methyl butanoate	376.0	554.4	3.47	340
$C_5H_{10}O_2$	Methyl isobutanoate	365.7	540.8	3.43	339
$C_5H_{11}Cl$	1-Chloropentane	381.6	571.2		
$C_5H_{11}Cl$	2-Chloro-2-methylbutane	358.8	509.1		
$C_5H_{11}N$	Piperidine	379.37	594	4.94	288
C_5H_{12}	Pentane	309.21	469.7	3.370	311
C_5H_{12}	Isopentane	301.03	460.4	3.38	306
C_5H_{12}	Neopentane	282.63	433.8	3.196	307
$C_5H_{12}O$	1-Pentanol	411.13	588.1	3.897	326
$C_5H_{12}O$	2-Pentanol	392.5	560.3	3.675	329
$C_5H_{12}O$	3-Pentanol	389.40	559.6		325
$C_5H_{12}O$	2-Methyl-1-butanol	400.7	575.4	3.94	
$C_5H_{12}O$	3-Methyl-1-butanol	404.3	577.2	3.93	
$C_5H_{12}O$	2-Methyl-2-butanol	375.6	543.7	3.71	
$C_5H_{12}O$	3-Methyl-2-butanol	386.1	556.1	3.87	
$C_5H_{12}O$	Butyl methyl ether	343.31	512.78	3.371	329
$C_5H_{12}O$	Methyl *tert*-butyl ether	328.4	497.1	3.430	
$C_5H_{12}O$	Ethyl propyl ether	336.36	500.23	3.370	339
$C_5H_{12}O_2$	Ethylene glycol monopropyl ether	423.0	614.6		364
$C_5H_{12}O_3$	Diethylene glycol monomethyl ether	466	672	3.67	
$C_5H_{12}S$	3-Methyl-1-butanethiol	393	604		
C_6BrF_5	Bromopentafluorobenzene	410	601	3.0	
C_6ClF_5	Chloropentafluorobenzene	391.11	570.81	3.238	376
$C_6Cl_2F_4$	1,2-Dichloro-3,4,5,6-tetrafluorobenzene	430.9	626	5.32	
$C_6Cl_3F_3$	1,3,5-Trichloro-2,4,6-trifluorobenzene	471.6	684.8	3.27	448
C_6F_6	Hexafluorobenzene	353.41	516.73	3.273	335
C_6F_{10}	Perfluorocyclohexene	325.2	461.8		
C_6F_{12}	Perfluoro-1-hexene	330.2	454.4		
C_6F_{12}	Perfluorocyclohexane	325.95 s	457.2	2.43	
C_6F_{14}	Perfluorohexane	329.8	448.77	1.868	606
C_6F_{14}	Perfluoro-2-methylpentane	330.8	455.3	1.923	532
C_6F_{14}	Perfluoro-3-methylpentane	331.6	450	1.69	
C_6F_{14}	Perfluoro-2,3-dimethylbutane	333.0	463	1.87	525
C_6HF_5	Pentafluorobenzene	358.89	530.97	3.531	324
C_6HF_5O	Pentafluorophenol	418.8	609	4.0	348
C_6HF_{11}	Undecafluorocyclohexane	335.2	477.7		
$C_6H_2F_4$	1,2,3,4-Tetrafluorobenzene	367.5	550.83	3.791	313
$C_6H_2F_4$	1,2,3,5-Tetrafluorobenzene	357.6	535.25	3.747	
$C_6H_2F_4$	1,2,4,5-Tetrafluorobenzene	363.4	543.35	3.801	
$C_6H_3ClF_2$	1-Chloro-2,4-difluorobenzene	400	609.6		
$C_6H_3ClF_2$	1-Chloro-2,5-difluorobenzene	401	612.5		
$C_6H_3ClF_2$	1-Chloro-3,4-difluorobenzene	400	609.2		
$C_6H_3ClF_2$	1-Chloro-3,5-difluorobenzene	391.7	592.0		
$C_6H_3F_3$	1,2,3-Trifluorobenzene		560.3		
$C_6H_3F_3$	1,2,4-Trifluorobenzene	363	551.1		
$C_6H_3F_3$	1,3,5-Trifluorobenzene	348.7	530.9		
C_6H_4BrF	1-Bromo-2-fluorobenzene	427	669.6		
C_6H_4BrF	1-Bromo-3-fluorobenzene	423	652.0		
C_6H_4BrF	1-Bromo-4-fluorobenzene	424.7	654.8		
C_6H_4ClF	1-Chloro-2-fluorobenzene	410.8	633.8		
C_6H_4ClF	1-Chloro-3-fluorobenzene	400.8	615.9		
C_6H_4ClF	1-Chloro-4-fluorobenzene	403	620.1		
$C_6H_4Cl_2$	*m*-Dichlorobenzene	446	685.7		

Molecular Formula	Name	T_b/K	T_c/K	P_c/MPa	V_c/cm³ mol⁻¹
$C_6H_4F_2$	*o*-Difluorobenzene	367	566.0		
$C_6H_4F_2$	*m*-Difluorobenzene	355.8	548.4		
$C_6H_4F_2$	*p*-Difluorobenzene	362	556.9	4.40	
C_6H_5Br	Bromobenzene	429.21	670	4.52	324
C_6H_5Cl	Chlorobenzene	404.87	633.4	4.52	308
C_6H_5F	Fluorobenzene	357.88	560.09	4.551	269
C_6H_5I	Iodobenzene	461.6	721	4.52	351
C_6H_6	Benzene	353.24	562.05	4.895	256
C_6H_6O	Phenol	455.02	694.2	6.13	
C_6H_7N	Aniline	457.32	699	4.89	287
C_6H_7N	2-Methylpyridine	402.53	621.0	4.60	292
C_6H_7N	3-Methylpyridine	417.29	645.0	4.65	288
C_6H_7N	4-Methylpyridine	418.51	645.7	4.70	292
C_6H_{10}	1,5-Hexadiene	332.6	508		
C_6H_{10}	Cyclohexene	356.13	560.4		
$C_6H_{10}O$	Cyclohexanone	428.58	653.0	4.0	
$C_6H_{10}O_2$	Ethyl *trans*-2-butenoate	411	599		
$C_6H_{10}S$	Diallyl sulfide	411.8	653		
$C_6H_{11}Cl$	Chlorocyclohexane	415	586		
$C_6H_{11}N$	Hexanenitrile	436.80	633.8	3.30	
C_6H_{12}	1-Hexene	336.63	504.0	3.21	355.1
C_6H_{12}	Cyclohexane	353.88	553.8	4.08	308
C_6H_{12}	Methylcyclopentane	345.0	532.7	3.79	318
$C_6H_{12}O$	Hexanal	404	591	3.46	
$C_6H_{12}O$	2-Hexanone	400.8	587.0	3.32	
$C_6H_{12}O$	3-Hexanone	396.7	582.82	3.320	
$C_6H_{12}O$	4-Methyl-2-pentanone	389.7	571	3.27	
$C_6H_{12}O$	Cyclohexanol	433.99	647.1	4.401	
$C_6H_{12}O_2$	Hexanoic acid	478.4	662	3.20	
$C_6H_{12}O_2$	Pentyl formate	403.6	576	3.46	
$C_6H_{12}O_2$	Isopentyl formate	396.7	578		
$C_6H_{12}O_2$	Butyl acetate	399.3	579		
$C_6H_{12}O_2$	Isobutyl acetate	389.7	561	3.16	
$C_6H_{12}O_2$	Propyl propanoate	395.7	578		
$C_6H_{12}O_2$	Ethyl butanoate	394.7	566	3.06	421
$C_6H_{12}O_2$	Ethyl 2-methylpropanoate	383.3	553	3.07	421
$C_6H_{12}O_2$	Methyl pentanoate	400.6	567	3.19	
$C_6H_{12}O_3$	Ethylene glycol monoethyl ether acetate	429.6	611	3.18	441
$C_6H_{12}O_3$	Paraldehyde	397.5	563		
$C_6H_{13}Cl$	1-Chlorohexane	408.3	599		
$C_6H_{13}Cl$	3-Chloro-3-methylpentane	389	528		
C_6H_{14}	Hexane	341.88	507.6	3.025	368
C_6H_{14}	2-Methylpentane	333.41	497.7	3.04	368
C_6H_{14}	3-Methylpentane	336.42	504.6	3.12	368
C_6H_{14}	2,2-Dimethylbutane	322.88	489.0	3.10	358
C_6H_{14}	2,3-Dimethylbutane	331.08	500.0	3.15	361
$C_6H_{14}O$	1-Hexanol	430.8	610.3	3.417	387
$C_6H_{14}O$	2-Hexanol	413	585.9	3.31	384
$C_6H_{14}O$	3-Hexanol	408	582.4	3.36	383
$C_6H_{14}O$	2-Methyl-1-pentanol	422	604.4	3.45	
$C_6H_{14}O$	4-Methyl-1-pentanol	425.1	603.5		
$C_6H_{14}O$	2-Methyl-2-pentanol	394.3	559.5		
$C_6H_{14}O$	4-Methyl-2-pentanol	404.8	574.4		
$C_6H_{14}O$	2-Methyl-3-pentanol	399.7	576.0	3.46	
$C_6H_{14}O$	3-Methyl-3-pentanol	395.6	575.6	3.52	
$C_6H_{14}O$	Dipropyl ether	363.23	530.6	3.028	
$C_6H_{14}O$	Diisopropyl ether	341.66	500.32	2.832	386
$C_6H_{14}O$	Methyl pentyl ether	372	546.53	3.042	391

Molecular Formula	Name	T_b/K	T_c/K	P_c/MPa	V_c/cm^3 mol^{-1}
$C_6H_{14}O_2$	1-Propoxy-2-propanol	423	605.1	3.051	
$C_6H_{14}O_2$	Ethylene glycol monobutyl ether	441.6	633.9		424
$C_6H_{14}O_2$	1,1-Diethoxyethane	375.40	527		
$C_6H_{14}O_3$	Diethylene glycol monoethyl ether	469	670	3.167	
$C_6H_{15}N$	Dipropylamine	382.5	555.8	3.63	
$C_6H_{15}N$	Diisopropylamine	357.1	523.1	3.02	
$C_6H_{15}N$	Triethylamine	362	535.6	3.032	389
C_7F_8	Perfluorotoluene	377.7	534.47	2.705	428
C_7F_{14}	Perfluoro-1-heptene	354.2	478.2		
C_7F_{14}	Perfluoromethylcyclohexane	349.5	485.91	2.019	570
C_7F_{16}	Perfluoroheptane	355.7	474.8	1.62	664
C_7HF_{15}	1*H*-Pentadecafluoroheptane	369.2	495.8		
$C_7H_3F_5$	2,3,4,5,6-Pentafluorotoluene	390.7	566.52	3.126	384
$C_7H_4BrF_3$	1-Bromo-2-(trifluoromethyl)benzene	440.7	656.5		
$C_7H_4BrF_3$	1-Bromo-3-(trifluoromethyl)benzene	424.7	627.1		
$C_7H_4BrF_3$	1-Bromo-4-(trifluoromethyl)benzene	433	629.8		
C_7H_5N	Benzonitrile	464.3	699.4	4.21	
$C_7H_6F_2$	2,4-Difluorotoluene	390	581.4		
$C_7H_6F_2$	2,5-Difluorotoluene	391	587.8		
$C_7H_6F_2$	2,6-Difluorotoluene	385	581.8		
$C_7H_6F_2$	3,4-Difluorotoluene	385	598.5		
C_7H_6O	Benzaldehyde	452.0	695	4.65	
C_7H_7F	2-Fluorotoluene	388	591.2		
C_7H_7F	3-Fluorotoluene	388	591.8		
C_7H_7F	4-Fluorotoluene	389.8	592.1		
C_7H_8	Toluene	383.78	591.80	4.110	316
C_7H_8O	*o*-Cresol	464.19	697.6	5.01	
C_7H_8O	*m*-Cresol	475.42	705.8	4.56	309
C_7H_8O	*p*-Cresol	475.13	704.6	5.15	
C_7H_8O	Benzyl alcohol	478.46	715	4.3	
C_7H_8O	Anisole	426.9	646.1	4.222	341
C_7H_9N	2-Methylaniline	473.5	707	4.37	
C_7H_9N	3-Methylaniline	476.5	707	4.28	
C_7H_9N	4-Methylaniline	473.6	706	4.58	
C_7H_9N	*N*-Methylaniline	469.4	701	5.20	
C_7H_9N	2,3-Dimethylpyridine	434.27	655.4	4.10	356
C_7H_9N	2,4-Dimethylpyridine	431.53	647	3.95	361
C_7H_9N	2,5-Dimethylpyridine	430.13	645	3.85	361
C_7H_9N	2,6-Dimethylpyridine	417.16	624	3.85	361
C_7H_9N	3,4-Dimethylpyridine	452.25	684	4.20	355
C_7H_9N	3,5-Dimethylpyridine	444.99	668	4.05	361
C_7H_{14}	1-Heptene	366.79	537.3	2.92	409
C_7H_{14}	Cycloheptane	391.6	604.2	3.82	353
C_7H_{14}	Methylcyclohexane	374.08	572.1	3.48	369
C_7H_{14}	Ethylcyclopentane	376.7	569.5	3.40	375
C_7H_{14}	1,1-Dimethylcyclopentane	360.7	547	3.45	
C_7H_{14}	*cis*-1,2-Dimethylcyclopentane	372.7	565	3.45	
C_7H_{14}	*trans*-1,2-Dimethylcyclopentane	365.1	553	3.45	
C_7H_{14}	*cis*-1,3-Dimethylcyclopentane	364.0	551	3.45	
C_7H_{14}	*trans*-1,3-Dimethylcyclopentane	364.9	553	3.45	
$C_7H_{14}O$	2-Heptanone	424.20	611.5	3.436	
$C_7H_{14}O_2$	Heptanoic acid	495.4	679	2.90	
$C_7H_{14}O_2$	Pentyl acetate	422.4	600	2.68	470
$C_7H_{14}O_2$	Isopentyl acetate	415.7	599		
$C_7H_{14}O_2$	Isobutyl propanoate	410	592		
$C_7H_{14}O_2$	Propyl butanoate	416.2	600		
$C_7H_{14}O_2$	Propyl isobutanoate	408.6	589		
$C_7H_{14}O_2$	Ethyl pentanoate	419.3	570		

Molecular Formula	Name	T_b/K	T_c/K	P_c/MPa	V_c/cm^3 mol^{-1}
$C_7H_{14}O_2$	Ethyl 3-methylbutanoate	408.2	588		
$C_7H_{15}Cl$	1-Chloroheptane	433.6	614		
C_7H_{16}	Heptane	371.6	540.2	2.74	428
C_7H_{16}	2-Methylhexane	363.19	530.4	2.74	421
C_7H_{16}	3-Methylhexane	365	535.2	2.81	404
C_7H_{16}	3-Ethylpentane	366.7	540.6	2.89	416
C_7H_{16}	2,2-Dimethylpentane	352.4	520.5	2.77	416
C_7H_{16}	2,3-Dimethylpentane	362.93	537.3	2.91	393
C_7H_{16}	2,4-Dimethylpentane	353.64	519.8	2.74	418
C_7H_{16}	3,3-Dimethylpentane	359.21	536.4	2.95	414
C_7H_{16}	2,2,3-Trimethylbutane	354.01	531.1	2.95	398
$C_7H_{16}O$	1-Heptanol	449.60	632.6	3.058	435
$C_7H_{16}O$	2-Heptanol	432	608.3	3.021	442
$C_7H_{16}O$	3-Heptanol	430	605.4		434
$C_7H_{16}O$	4-Heptanol	429	602.6		432
$C_7H_{16}O_2$	1-Butoxy-2-propanol	444.7	624.9	2.739	
$C_8F_{16}O$	Perfluoro-2-butyltetrahydrofuran	375.8	500.2	1.607	588
C_8F_{18}	Perfluorooctane	379.1	502	1.66	
C_8H_7N	4-Methylbenzonitrile	490.2	723		
C_8H_7N	Indole	526.8	794	4.8	356
C_8H_8	Styrene	418	635.2	3.87	
C_8H_8O	Acetophenone	475	713	4.40	380
$C_8H_8O_3$	Methyl salicylate	496.1	709		
C_8H_{10}	Ethylbenzene	409.34	617.15	3.609	374
C_8H_{10}	*o*-Xylene	417.7	630.3	3.732	370
C_8H_{10}	*m*-Xylene	412.27	617.0	3.541	375
C_8H_{10}	*p*-Xylene	411.52	616.2	3.511	378
$C_8H_{10}O$	2-Ethylphenol	477.7	703.0		
$C_8H_{10}O$	3-Ethylphenol	491.6	718.8		
$C_8H_{10}O$	4-Ethylphenol	491.1	716.4		
$C_8H_{10}O$	2,3-Xylenol	490.1	722.8		
$C_8H_{10}O$	2,4-Xylenol	484.13	707.6		
$C_8H_{10}O$	2,5-Xylenol	484.3	706.9		
$C_8H_{10}O$	2,6-Xylenol	474.22	701.0		
$C_8H_{10}O$	3,4-Xylenol	500	729.8		
$C_8H_{10}O$	3,5-Xylenol	494.89	715.6		
$C_8H_{10}O$	1-Phenylethanol	478	699	3.77	
$C_8H_{10}O$	Phenetole	442.96	647	3.42	
$C_8H_{11}N$	*N*-Ethylaniline	476.2	698		
$C_8H_{11}N$	*N,N*-Dimethylaniline	467.30	687	3.63	
$C_8H_{14}O_4$	Diethyl succinate	490.9	663		
$C_8H_{15}N$	Octanenitrile	478.40	674.4	2.85	
C_8H_{16}	1-Octene	394.44	567.0	2.68	468
C_8H_{16}	Cyclooctane	422	647.2	3.56	410
C_8H_{16}	Ethylcyclohexane	405.1	609	3.04	
C_8H_{16}	*cis*-1,2-Dimethylcyclohexane	403.0	606	2.95	
C_8H_{16}	*trans*-1,2-Dimethylcyclohexane	396.7	596	2.94	
C_8H_{16}	*cis*-1,3-Dimethylcyclohexane	393.3	591	2.94	
C_8H_{16}	*trans*-1,3-Dimethylcyclohexane	397.7	598	2.94	
C_8H_{16}	*trans*-1,4-Dimethylcyclohexane	392.6	587.7		
$C_8H_{16}O_2$	Octanoic acid	512	695	2.64	
$C_8H_{16}O_2$	2-Ethylhexanoic acid	501	674.6	2.778	528
$C_8H_{16}O_2$	Isopentyl propanoate	433.4	611		
$C_8H_{16}O_2$	Isobutyl butanoate	430.1	611		
$C_8H_{16}O_2$	Isobutyl isobutanoate	421.8	602		
$C_8H_{16}O_2$	Propyl 3-methylbutanoate	429.1	609		
$C_8H_{16}O_4$	Diethylene glycol monoethyl ether acetate	491.7	663	2.73	
$C_8H_{17}Cl$	1-Chlorooctane	456.7	643		

Molecular Formula	Name	T_b/K	T_c/K	P_c/MPa	V_c/cm³ mol⁻¹
C_8H_{18}	Octane	398.82	568.7	2.49	492
C_8H_{18}	2-Methylheptane	390.81	559.7	2.50	488
C_8H_{18}	3-Methylheptane	392.1	563.6	2.55	464
C_8H_{18}	4-Methylheptane	390.87	561.7	2.54	476
C_8H_{18}	3-Ethylhexane	391.8	565.5	2.61	455
C_8H_{18}	2,2-Dimethylhexane	380.01	549.8	2.53	478
C_8H_{18}	2,3-Dimethylhexane	388.77	563.5	2.63	468
C_8H_{18}	2,4-Dimethylhexane	382.7	553.5	2.56	472
C_8H_{18}	2,5-Dimethylhexane	382.27	550.0	2.49	482
C_8H_{18}	3,3-Dimethylhexane	385.12	562.0	2.65	443
C_8H_{18}	3,4-Dimethylhexane	390.88	568.8	2.69	466
C_8H_{18}	3-Ethyl-2-methylpentane	388.81	567.1	2.70	442
C_8H_{18}	3-Ethyl-3-methylpentane	391.42	576.5	2.81	455
C_8H_{18}	2,2,3-Trimethylpentane	383	563.5	2.73	436
C_8H_{18}	2,2,4-Trimethylpentane	372.37	543.8	2.57	468
C_8H_{18}	2,3,3-Trimethylpentane	388.0	573.5	2.82	455
C_8H_{18}	2,3,4-Trimethylpentane	386.7	566.4	2.73	460
C_8H_{18}	2,2,3,3-Tetramethylbutane	379.60	567.8	2.87	461
$C_8H_{18}O$	1-Octanol	468.31	652.5	2.777	497
$C_8H_{18}O$	2-Octanol	452.5	629.6	2.754	519
$C_8H_{18}O$	3-Octanol	444	628.5		515
$C_8H_{18}O$	4-Octanol	449.5	625.1		515
$C_8H_{18}O$	4-Methyl-3-heptanol	443	623.5		
$C_8H_{18}O$	5-Methyl-3-heptanol	445	621.2		
$C_8H_{18}O$	2-Ethyl-1-hexanol	457.8	640.6	2.8	
$C_8H_{18}O$	Dibutyl ether	413.43	584.1	3.01	
$C_8H_{18}O$	Di-*tert*-butyl ether	380.38	550		
$C_8H_{19}N$	Dibutylamine	432.8	607.5	3.11	
$C_8H_{19}N$	Diisobutylamine	412.8	584.4	3.20	
$C_8H_{20}Si$	Tetraethylsilane	427.9	603.7	2.602	
C_9F_{20}	Perfluorononane	398.5	524	1.56	
C_9H_7N	Quinoline	510.31	782	4.86	371
C_9H_7N	Isoquinoline	516.37	803	5.10	374
C_9H_{10}	Indan	451.12	684.9	3.95	
C_9H_{12}	Propylbenzene	432.39	638.35	3.200	440
C_9H_{12}	Isopropylbenzene	425.56	631.0	3.209	
C_9H_{12}	2-Ethyltoluene	438.4	651	3.38	
C_9H_{12}	3-Ethyltoluene	434.5	637	3.25	
C_9H_{12}	4-Ethyltoluene	435	640.2	3.23	
C_9H_{12}	1,2,3-Trimethylbenzene	449.27	664.5	3.454	
C_9H_{12}	1,2,4-Trimethylbenzene	442.53	649.1	3.232	
C_9H_{12}	1,3,5-Trimethylbenzene	437.89	637.3	3.127	
$C_9H_{13}N$	2-Methyl-*N,N*-dimethylaniline	467.3	668	3.12	
C_9H_{18}	1-Nonene	420.1	594.0		526
C_9H_{18}	Cyclononane	451.6	682	3.34	
C_9H_{18}	1α,3α,5β-1,3,5-Trimethylcyclohexane	413.7	602.2		
$C_9H_{18}O$	2-Nonanone	468.5	651.9	2.482	
$C_9H_{18}O$	5-Nonanone	461.60	640		
$C_9H_{18}O_2$	Nonanoic acid	527.7	711	2.40	
$C_9H_{18}O_2$	Isopentyl butanoate	452	619		
$C_9H_{18}O_2$	Isobutyl 3-methylbutanoate	441.7	621		
C_9H_{20}	Nonane	423.97	594.6	2.29	555
C_9H_{20}	2-Methyloctane	416.4	582.8	2.31	
C_9H_{20}	2,2-Dimethylheptane	405.9	576.7	2.35	
C_9H_{20}	2,2,5-Trimethylhexane	397.24	569.8		
C_9H_{20}	2,2,3,3-Tetramethylpentane	413.4	607.5	2.74	
C_9H_{20}	2,2,3,4-Tetramethylpentane	406.2	592.6	2.60	
C_9H_{20}	2,2,4,4-Tetramethylpentane	395.44	574.6	2.49	

Molecular Formula	Name	T_b/K	T_c/K	P_c/MPa	V_c/cm³ mol⁻¹
C_9H_{20}	2,3,3,4-Tetramethylpentane	414.7	607.5	2.72	
$C_9H_{20}O$	1-Nonanol	486.52	670.7	2.528	572
$C_9H_{20}O$	2-Nonanol	466.7	649.6	2.53	575
$C_9H_{20}O$	3-Nonanol	468	648.0		577
$C_9H_{20}O$	4-Nonanol	465.7	645.1		575
$C_{10}F_8$	Perfluoronaphthalene	482	673.1		
$C_{10}F_{18}$	Perfluorodecalin	415	566	1.52	
$C_{10}F_{22}$	Perfluorodecane	417.4	542	1.45	
$C_{10}H_8$	Naphthalene	491.1	748.4	4.05	407
$C_{10}H_9N$	1-Naphthylamine	573.9	850	5.0	438
$C_{10}H_9N$	2-Naphthylamine	579.4	850	4.9	438
$C_{10}H_{12}$	1,2,3,4-Tetrahydronaphthalene	480.8	720	3.65	408
$C_{10}H_{14}$	Butylbenzene	456.46	660.5	2.89	497
$C_{10}H_{14}$	Isobutylbenzene	445.94	650	3.05	
$C_{10}H_{14}$	1-Isopropyl-4-methylbenzene	450.3	652	2.8	
$C_{10}H_{14}$	*p*-Diethylbenzene	456.9	657.9	2.803	
$C_{10}H_{14}$	1,2,4,5-Tetramethylbenzene	470.0	676	2.9	
$C_{10}H_{14}O$	Thymol	505.7	698		
$C_{10}H_{16}$	*d*-Limonene	451	653		470
$C_{10}H_{16}$	α-Pinene	429.4	644		454
$C_{10}H_{16}$	3-Carene	444	658		487
$C_{10}H_{18}$	1,3-Decadiene	442	615		
$C_{10}H_{18}$	*cis*-Decahydronaphthalene	469.0	702.3	3.20	
$C_{10}H_{18}$	*trans*-Decahydronaphthalene	460.5	687.1		
$C_{10}H_{20}$	1-Decene	443.7	617	2.22	584
$C_{10}H_{20}O$	Decanal	481.7	674.2		
$C_{10}H_{20}O$	Menthol	489	694		
$C_{10}H_{20}O_2$	Decanoic acid	541.9	726	2.23	
$C_{10}H_{20}O_2$	Ethyl octanoate	481.7	659		
$C_{10}H_{20}O_4$	Diethylene glycol monobutyl ether acetate	518	681	3.15	
$C_{10}H_{22}$	Decane	447.30	617.7	2.11	624
$C_{10}H_{22}$	3,3,5-Trimethylheptane	428.9	609.5	2.32	
$C_{10}H_{22}$	2,2,3,3-Tetramethylhexane	433.5	623.0	2.51	
$C_{10}H_{22}$	2,2,5,5-Tetramethylhexane	410.6	581.4	2.19	
$C_{10}H_{22}O$	1-Decanol	504.3	687.3	2.315	649
$C_{10}H_{22}O$	2-Decanol	484	668.6		646
$C_{10}H_{22}O$	3-Decanol	486	666.1		643
$C_{10}H_{22}O$	4-Decanol	483.7	663.7		643
$C_{10}H_{22}O$	5-Decanol	474	663.2		646
$C_{10}H_{22}S$	Diisopentyl sulfide	484	664		
$C_{11}H_{10}$	1-Methylnaphthalene	517.9	772	3.60	
$C_{11}H_{10}$	2-Methylnaphthalene	514.3	761		
$C_{11}H_{22}O_2$	Ethyl nonanoate	500.2	674		
$C_{11}H_{24}$	Undecane	469.1	639	1.98	689
$C_{11}H_{24}O$	1-Undecanol	518	703.6	2.147	718
$C_{12}H_8$	Acenaphthylene	553	792	3.20	
$C_{12}H_9N$	Carbazole	627.84	901.8	3.13	454
$C_{12}H_{10}$	Biphenyl	529.3	773	3.38	497
$C_{12}H_{10}O$	Diphenyl ether	531.2	766.8		
$C_{12}H_{12}$	2,7-Dimethylnaphthalene	538	775	3.23	601
$C_{12}H_{18}$	Hexamethylbenzene	536.6	758		
$C_{12}H_{24}$	1-Dodecene	487.0	658	1.93	
$C_{12}H_{26}$	Dodecane	489.47	658	1.82	754
$C_{12}H_{26}O$	1-Dodecanol	533	719.4	1.994	
$C_{13}H_9N$	Acridine	618.01	891.1	3.21	548
$C_{13}H_9N$	Phenanthridine	622.1	895	3.6	548
$C_{13}H_{11}N$	9-Methylcarbazole	616.79	890	3.38	572
$C_{13}H_{12}$	Diphenylmethane	538.2	760	2.71	563

Molecular Formula	Name	T_b/K	T_c/K	P_c/MPa	V_c/cm³ mol⁻¹
$C_{13}H_{26}O_2$	Methyl dodecanoate	540	712		
$C_{13}H_{28}$	Tridecane	508.62	675	1.68	823
$C_{13}H_{28}O$	1-Tridecanol	547	734	1.935	
$C_{14}H_{10}$	Anthracene	613.1	869.3		554
$C_{14}H_{10}$	Phenanthrene	613	869		
$C_{14}H_{30}$	Tetradecane	526.73	693	1.57	894
$C_{14}H_{30}O$	1-Tetradecanol	560	747	1.81	
$C_{15}H_{32}$	Pentadecane	543.8	708	1.48	966
$C_{16}H_{34}$	Hexadecane	560.01	723	1.40	1034
$C_{16}H_{34}$	2,2,4,4,6,8,8-Heptamethylnonane	519.5	692		
$C_{16}H_{34}O$	1-Hexadecanol	585	770	1.61	
$C_{17}H_{36}$	Heptadecane	575.2	736	1.34	1103
$C_{17}H_{36}O$	1-Heptadecanol	597	780	1.50	
$C_{18}H_{14}$	*o*-Terphenyl	605	857	2.99	731
$C_{18}H_{14}$	*m*-Terphenyl	636	883	2.48	724
$C_{18}H_{14}$	*p*-Terphenyl	649	908	2.99	729
$C_{18}H_{38}$	Octadecane	589.5	747	1.29	1189
$C_{18}H_{38}O$	1-Octadecanol	608	790	1.44	
$C_{19}H_{40}$	Nonadecane	603.1	755	1.16	
$C_{20}H_{42}$	Eicosane	616	768	1.07	
$C_{20}H_{42}O$	1-Eicosanol	629	809	1.30	
$C_{21}H_{44}$	Heneicosane	629.7	778	1.03	
$C_{22}H_{46}$	Docosane	641.8	786	0.98	
$C_{23}H_{48}$	Tricosane	653	790	0.92	
$C_{24}H_{50}$	Tetracosane	664.5	800	0.87	
$C_{30}H_{50}$	Squalene	694.5	795.9	0.59	

SUBLIMATION PRESSURE OF SOLIDS

This table gives the sublimation (vapor) pressure of some representative solids as a function of temperature. Entries include simple inorganic and organic substances in their solid phase below room temperature, as well as polycyclic organic compounds which show measurable sublimation pressure only at elevated temperatures. Substances are listed by molecular formula in the Hill order. Values marked by * represent the solid-liquid-gas triple point. Note that some pressure values are in pascals (Pa) and others are in kilopascals (kPa). For conversion, 1 kPa = 7.506 mmHg = 0.0098692 atm.

REFERENCES

1. Lide, D.R. and Kehiaian, H.V., *CRC Handbook of Thermophysical and Thermochemical Data,* CRC Press, Boca Raton, FL, 1994.
2. *TRC Thermodynamic Tables,* Thermodynamic Research Center, Texas A&M University, College Station, TX.
3. Oja, V. and Suuberg, E.M., *J. Chem. Eng. Data,*43, 486, 1998.

Ar	*T*/K	55	60	65	70	75	80	83.81*	
Argon	*p*/kPa	0.2	0.8	2.8	7.7	18.7	40.7	68.8*	
BrH	*T*/K	135	140	150	160	170	180	185.1*	
Hydrogen bromide	*p*/kPa	0.1	0.3	1.1	3.3	8.7	20.1	27.4*	
Br_2	*T*/K	170	180	190	200	210	220	230	240*
Bromine	*p*/Pa	0.069	0.416	2.04	8.45	30.3	96.0	273	710*
ClH	*T*/K	120	130	140	150	155	159.0*		
Hydrogen chloride	*p*/kPa	0.1	0.5	1.9	5.8	9.5	13.5*		
Cl_2	*T*/K	120	130	140	150	160	170*		
Chlorine	*p*/Pa	0.144	1.52	11.2	63.1	283	1054*		
F_4Si	*T*/K	130	140	150	160	170	175	180	186.3*
Tetrafluorosilane	*p*/kPa	0.2	0.9	3.9	14.0	43.8	74.2	122.4	220.8*
F_6S	*T*/K	150	165	180	190	200	210	220	223.1*
Sulfur hexafluoride	*p*/kPa	0.4	2.6	11.3	25.9	54.5	106.1	195.1	232.7*
HI	*T*/K	160	170	180	190	200	210	220	222.4*
Hydrogen iodide	*p*/kPa	0.2	0.8	2.2	5.3	11.7	23.6	44.1	49.3*
H_2O	*T*/K	190	210	225	240	250	260	270	273.16*
Water	*p*/Pa	0.032	0.702	4.942	27.28	76.04	195.8	470.1	611.66*
H_2S	*T*/K	140	150	160	165	170	175	180	187.6*
Hydrogen sulfide	*p*/kPa	0.2	0.6	1.9	3.2	5.2	8.3	12.7	22.7*
H_3N	*T*/K	160	170	180	190	195	195.4*		
Ammonia	*p*/kPa	0.1	0.4	1.2	3.5	5.8	6.12*		
I_2	*T*/K	240	250	260	270	280	290	300	310*
Iodine	*p*/Pa	0.081	0.297	0.971	2.89	7.92	20.1	47.9	107*
Kr	*T*/K	80	90	95	100	105	110	115.8*	
Krypton	*p*/kPa	0.4	2.7	6.0	12.1	22.8	40.4	73.1*	
NO	*T*/K	85	90	95	100	105	109.5*		
Nitric oxide	*p*/kPa	0.1	0.4	1.3	3.8	10.0	21.9*		
Xe	*T*/K	110	120	130	140	150	155	160	161.4*
Xenon	*p*/kPa	0.3	1.5	4.9	14.0	34.2	51.1	74.2	81.7*
CHN	*T*/K	200	210	220	230	240	250	255	259.83*
Hydrogen cyanide	*p*/kPa	0.2	0.4	1.0	2.2	4.8	9.7	13.6	18.62*
CH_4	*T*/K	65	70	75	80	85	90.69*		
Methane	*p*/kPa	0.1	0.3	0.8	2.1	4.9	11.70*		

CO	T/K	50	55	60	65	68.13*			
Carbon monoxide	p/kPa	0.1	0.6	2.6	8.2	15.4*			
CO_2	T/K	130	140	155	170	185	194.7	205	216.58*
Carbon dioxide	p/kPa	0.032	0.187	1.674	9.987	44.02	101.3	227.1	518.0*
C_2Cl_6	T/K	275	300	325	350	375	400	425	459.9*
Hexachloroethane	p/Pa	0.004	0.056	0.383	1.62	5.30	14.8	36.4	107.4*
C_2H_2	T/K	130	140	150	160	170	180	190	192.4*
Acetylene	p/kPa	0.2	0.7	2.6	7.8	20.6	49.0	106.3	126.0*
$C_2H_4O_2$	T/K	250	260	270	280	289.7*			
Acetic acid	p/kPa	0.092	0.199	0.406	0.79	1.29*			
C_5H_{12}	T/K	200	210	220	230	240	250	255	256.58*
Neopentane	p/kPa	0.7	1.6	3.6	7.3	13.9	24.8	32.4	35.8*
$C_6H_6Cl_6$	T/K	300	320	330	340	350	360	370	380
1,2,3,4,5,6-Hexa-chlorocyclohexane (Lindane)	p/Pa	0.01	0.13	0.39	1.04	2.66	6.42	14.8	32.7
$C_6H_6O_2$	T/K	330	340	350	360	370	380		
Resorcinol	p/Pa	1.03	2.78	7.09	17.2	39.6	87.6		
$C_6H_6O_2$	T/K	350	360	370	380	390	400		
p-Hydroquinone	p/Pa	1.20	3.18	7.96	19.0	43.4	95.1		
$C_{10}H_8$	T/K	250	270	280	290	300	310	330	353.43*
Naphthalene	p/Pa	0.036	0.514	1.662	4.918	13.43	34.15	182.9	999.6*
$C_{12}H_8N_2$	T/K	290	300	310	320				
Phenazine	p/Pa	0.0013	0.0046	0.0150	0.0448				
$C_{12}H_8O$	T/K	300	310	320	330	340	350		
Dibenzofuran	p/Pa	0.408	1.21	3.35	8.71	21.4	50.0		
$C_{12}H_9N$	T/K	350	355	360					
Carbazole	p/Pa	0.086	0.140	0.245					
$C_{13}H_7NO_2$	T/K	330	340	350	360	370	380		
Benz[g]isoquinoline-5,10-dione	p/Pa	0.006	0.018	0.053	0.148	0.394	0.994		
$C_{13}H_8O$	T/K	330	340	350					
1H-Phenalen-1-one	p/Pa	0.040	0.113	0.302					
$C_{13}H_8O_2$	T/K	400	410	420	430				
3-Hydroxy-1H-phenalen-1-one	p/Pa	0.006	0.018	0.053	0.144				
$C_{13}H_9N$	T/K	290	300	310	320				
Acridine	p/Pa	0.0024	0.0085	0.0278	0.0845				
$C_{13}H_9N$	T/K	310	320	330	340				
Phenanthridine	p/Pa	0.020	0.066	0.206	0.603				
$C_{14}H_{10}$	T/K	320	330	340	350	360	370	380	390
Anthracene	p/Pa	0.014	0.043	0.125	0.342	1.01	2.38	5.35	11.5

$C_{14}H_{10}$	*T*/K	300	310	320	330	340	350	360	
Phenanthrene	*p*/Pa	0.025	0.085	0.270	0.796	2.02	4.89	11.2	
$C_{16}H_{10}$	*T*/K	320	330	340	350	360	370	380	390
Pyrene	*p*/Pa	0.008	0.024	0.073	0.208	0.556	1.32	2.86	6.30
$C_{16}H_{10}O$	*T*/K	360	370	380	390	400			
1-Pyrenol	*p*/Pa	0.005	0.016	0.047	0.135	0.364			
$C_{16}H_{12}S$	*T*/K	330	340	350	360	370	380	390	
Benzo[b]naphtho-(2,1-d)thiophene	*p*/Pa	0.001	0.004	0.012	0.036	0.098	0.255	0.631	
$C_{17}H_{12}$	*T*/K	340	350	360	370	380	390	400	
11*H*-Benzo[b]fluorene	*p*/Pa	0.003	0.009	0.029	0.085	0.235	0.619	1.55	
$C_{18}H_{10}O_4$	*T*/K	420	430	440	450				
6,11-Dihydroxy-5,12-naphthacenedione	*p*/Pa	0.008	0.022	0.055	0.131				
$C_{18}H_{12}$	*T*/K	390	400	410	420				
Chrysene	*p*/Pa	0.087	0.221	0.539	1.26				
$C_{18}H_{12}$	*T*/K	390	400	410	420	430	440	450	460
Naphthacene	*p*/Pa	0.005	0.014	0.035	0.084	0.194	0.432	0.928	1.929
$C_{20}H_{12}$	*T*/K	390	400	410	420	430			
Perylene	*p*/Pa	0.006	0.015	0.040	0.102	0.246			
$C_{22}H_{14}$	*T*/K	450	460	470	480	490			
Pentacene	*p*/Pa	0.002	0.006	0.013	0.031	0.069			
$C_{24}H_{12}$	*T*/K	430	440	450	460	470	480	490	500
Coronene	*p*/Pa	0.004	0.010	0.021	0.046	0.097	0.197	0.389	0.747

VAPOR PRESSURE

This table gives vapor pressure data for about 1800 inorganic and organic substances. In order to accommodate elements and compounds ranging from refractory to highly volatile in a single table, the temperature at which the vapor pressure reaches specified pressure values is listed. The pressure values run in decade steps from 1 Pa (about 7.5 μm Hg) to 100 kPa (about 750 mm Hg). All temperatures are given in °C.

The data used in preparing the table came from a large number of sources; the main references used for each substance are indicated in the last column. Since the data were refit in most cases, values appearing in this table may not be identical with values in the source cited. The temperature entry in the 100 kPa column is close to, but not identical with, the normal boiling point (which is defined as the temperature at which the vapor pressure reaches 101.325 kPa). Although some temperatures are quoted to 0.1°C, uncertainties of several degrees should generally be assumed. Values followed by an "e" were obtained by extrapolating (usually with an Antoine equation) beyond the region for which experimental measurements were available and are thus subject to even greater uncertainty.

Compounds are listed by molecular formula following the Hill convention. Substances not containing carbon are listed first, followed by those that contain carbon. To locate an organic compound by name or CAS Registry Number when the molecular formula is not known, use the table *Physical Constants of Organic Compounds* in Section 3 and its indexes to determine the molecular formula. The indexes to *Physical Constants of Inorganic Compounds* in Section 4 can be used in a similar way.

More extensive and detailed vapor pressure data on selected important substances appear in other tables in this section of the *Handbook*. These substances are flagged by a symbol following the name as follows:

*	See *Vapor Pressure of Fluids below 300 K*
**	See *IUPAC Recommended Data for Vapor Pressure Calibration*
***	See *Vapor Pressure of Ice* and *Vapor Pressure of Water from 0 to 370°C*

The following notations appear after individual temperature entries:

s — Indicates the substance is a solid at this temperature.
e — Indicates an extrapolation beyond the region where experimental measurements exist.
i — Indicates the value was calculated from ideal gas thermodynamic functions, such as those in the *JANAF Thermochemical Tables* (see Reference 8).

REFERENCES

1. Lide, D.R., and Kehiaian, H.V., *CRC Handbook of Thermophysical and Thermochemical Data*, CRC Press, Boca Raton, FL, 1994.
2. Stull, D., in *American Institute of Physics Handbook, Third Edition*, Gray, D.E., Ed., McGraw Hill, New York, 1972.
3. Hultgren, R., Desai, P.D., Hawkins, D.T., Gleiser, M., Kelley, K.K., and Wagman, D.D., *Selected Values of Thermodynamic Properties of the Elements*, American Society for Metals, Metals Park, OH, 1973.
4. Stull, D., *Ind. Eng. Chem.*, 39, 517, 1947.
5. *TRCVP, Vapor Pressure Database, Version 2.2P*, Thermodynamic Research Center, Texas A&M University, College Station, TX.
6. *TRC Thermodynamic Tables*, Thermodynamic Research Center, Texas A&M University, College Station, TX.
7. Ohe, S., *Computer Aided Data Book of Vapor Pressure*, Data Book Publishing Co., Tokyo, 1976.
8. Chase, M.W., Davies, C.A., Downey, J.R., Frurip, D.J., McDonald, R.A., and Syverud, A.N., *JANAF Thermochemical Tables, Third Edition*, *J. Phys. Chem. Ref. Data*, Vol. 14, Suppl. 1, 1985.
9. Barin, I., *Thermochemical Data of Pure Substances*, VCH Publishers, New York, 1993.
10. Jacobsen, R.T., et. al, *International Thermodynamic Tables of the Fluid State, No. 10. Ethylene*, Blackwell Scientific Publications, Oxford, 1988.
11. Wakeham, W.A., *International Thermodynamic Tables of the Fluid State, No. 12. Methanol*, Blackwell Scientific Publications, Oxford, 1993.
12. Janz, G.J., *Molten Salts Handbook*, Academic Press, New York, 1967.
13. Ohse, R.W. *Handbook of Thermodynamic and Transport Properties of Alkali Metals*, Blackwell Scientific Publications, Oxford, 1994.
14. Gschneidner, K.A., in *CRC Handbook of Chemistry and Physics, 77th Edition*, p. 4-112, CRC Press, Boca Raton, FL, 1996.
15. Leider, H.R., Krikorian, O.H., and Young, D.A., *Carbon*, 11, 555, 1973.
16. Ruzicka, K., and Majer, V., *J. Phys. Chem. Ref. Data*, 23, 1, 1994.
17. Tillner-Roth, R., and Baehr, H.D., *J. Phys. Chem. Ref. Data*, 23, 657, 1994.
18. Younglove, B.A., and McLinden, M.O., *J. Phys. Chem. Ref. Data*, 23, 731, 1994.
19. Outcalt, S.L., and McLinden, M.O., *J. Phys. Chem. Ref. Data*, 25, 605, 1996.
20. Weber, L.A., and Defibaugh, D.R., *J. Chem. Eng. Data*, 41, 382, 1996.
21. Rodrigues, M.F., and Bernardo-Gil, M.G., *J. Chem. Eng. Data*, 41, 581, 1996.
22. Piacente, V., Gigli, G., Scardala, P., and Giustini, A., *J. Phys. Chem.*, 100, 9815, 1996.
23. Barton, J.L., and Bloom, H., *J. Phys. Chem.*, 60, 1413, 1956.
24. Sense, K.A., Alexander, C.A., Bowman, R.E., and Filbert, R.B., *J. Phys. Chem.*, 61, 337, 1957.
25. Ewing, C.T., and Stern, K.H., *J. Phys. Chem.* 78, 1998, 1974.
26. Cady, G.H., and Hargreaves, G.B., *J. Chem. Soc.*, 1961, 1563; 1961, 1568.
27. Skudlarski, K., Dudek, J., and Kapala, J., *J. Chem. Thermodynamics*, 19, 857, 1987.
28. Wagner, W., and de Reuck, K.M., *International Thermodynamic Tables of the Fluid State, No. 9. Oxygen*, Blackwell Scientific Publications, Oxford, 1987.

29. Marsh, K.N., Editor, *Recommended Reference Materials for the Realization of Physicochemical Properties*, Blackwell Scientific Publications, Oxford, 1987.
30. Alcock, C.B., Itkin, V.P., and Horrigan, M.K., *Canadian Metallurgical Quarterly*, 23, 309, 1984.
31. Stewart, R.B., and Jacobsen, R.T., *J. Phys. Chem. Ref. Data*, 18, 639, 1989.
32. Sifner, O., and Klomfar, J., *J. Phys. Chem. Ref. Data*, 23, 63, 1994.
33. Bah, A., and Dupont-Pavlovsky, N., *J. Chem. Eng. Data*, 40, 869, 1995.
34. Behrens, R.G., and Rosenblatt, G., *J. Chem. Thermodynamics*, 4, 175, 1972.
35. Behrens, R.G., and Rosenblatt, G., *J. Chem. Thermodynamics*, 5, 173, 1973.
36. Haar, L., Gallagher, J.S., and Kell, G.S., *NBS/NRC Steam Tables*, Hemisphere Publishing Corp., New York, 1984.
37. Wagner, W., Saul, A., and Pruss, A., *J. Phys. Chem. Ref. Data*, 23, 515. 1994.
38. Behrens, R.G., Lemons, R.S., and Rosenblatt, G., *J. Chem. Thermodynamics*, 6, 457, 1974.
39. Boublik, T., Fried, V., and Hala, E., *The Vapor Pressure of Pure Substances, Second Edition*, Elsevier, Amsterdam, 1984.
40. Goodwin, R.D., *J. Phys. Chem. Ref. Data*, 14, 849, 1985.
41. Younglove, B.A., and Ely, J.F., *J. Phys. Chem. Ref. Data*, 16, 577, 1987.

		Temperature in °C for the indicated pressure						
Mol. Form.	**Name**	**1 Pa**	**10 Pa**	**100 Pa**	**1 kPa**	**10 kPa**	**100 kPa**	**Ref.**
Substances not containing carbon:								
Ag	Silver	1010	1140	1302	1509	1782	2160	2
AgBr	Silver(I) bromide	569 i	656 i	765 i	905 i	1093 i	1359 i	9
AgCl	Silver(I) chloride	670	769	873	1052	1264	1561	4
AgI	Silver(I) iodide	594	686	803	959	1177	1503	4
Al	Aluminum	1209	1359	1544	1781	2091	2517	2
AlB_3H_{12}	Aluminum borohydride				-46.8	-9.4	45.5	4
$AlCl_3$	Aluminum trichloride	58.4 s	76.5 s	97.1 s	120.7 s	148.2 s	180.5 s	4
AlF_3	Aluminum trifluoride	744 s	819 s	906 s	1008 s	1130 s	1276 s	8
AlI_3	Aluminum triiodide				218	285	385	4
Al_2O_3	Aluminum oxide			2122	2351	2629	2975	4
Ar	Argon*		-226.4 s	-220.3 s	-212.4 s	-201.7 s	-186.0	1,5,31
As	Arsenic	280 s	323 s	373 s	433 s	508 s	601 s	3
$AsCl_3$	Arsenic(III) chloride			-8 e	21.3	63.1	129.4	1
AsF_3	Arsenic(III) fluoride					8.1	56.0	4
AsI_3	Arsenic(III) iodide				187	261	367 e	7
As_2O_3	Arsenic(III) oxide (arsenolite)	133.7 s	163.0 s	196.8 s	236.2 s	283.0		34
At	Astatine	88 s	119 s	156 s	202 s	258 s	334	2
Au	Gold	1373	1541	1748	2008	2347	2805	2
B	Boron	2075	2289	2549	2868	3272	3799	2
BBr_3	Boron tribromide			-45 e	-15 e	27.5	90.4	1
BCl_3	Boron trichloride*			-94.0	-70.5	-37.4	12.3	4
BF_3	Boron trifluoride*	-173.9 s	-166.0 s	-156.0 s	-143.0 s	-125.9	-101.1	4
B_2F_4	Tetrafluorodiborane						-34	1
B_2H_6	Diborane			-162 e	-147.0	-125.8	-92.6	1
B_5H_9	Pentaborane(9)				-34.8	3.8	57.6	4
Ba	Barium	638 s	765	912	1115	1413	1897	9
Be	Beryllium	1189 s	1335	1518	1750	2054	2469	2
$BeBr_2$	Beryllium bromide	203 s	240 s	283 s	335 s	397 s	473 s	4
$BeCl_2$	Beryllium chloride	196 s	237 s	284 s	339 s	402 s	487	4
BeF_2	Beryllium fluoride		686 e	767 e	869	999	1172 e	7
BeI_2	Beryllium iodide	188 s	229 s	276 s	333 s	402 s	487	4
Bi	Bismuth	668	768	892	1052	1265	1562	2
$BiBr_3$	Bismuth tribromide			217 s	273 i	348 i	455 i	4,9
$BiCl_3$	Bismuth trichloride				248.9	328.6	438.7	1,4
BrCs	Cesium bromide	531 s	601 s	701 i	834 i	1019 i	1293 e	9
BrH	Hydrogen bromide*		-153.3 s	-140.4 s	-123.8 s	-101.5 s	-67.0	5
BrH_3Si	Bromosilane				-81.0	-47.3	2.2	4
BrH_4N	Ammonium bromide	121 s	154 s	195 s	246 s	310.4 s	395.1 s	5
BrK	Potassium bromide	597 s	674 s	773				25
BrLi	Lithium bromide		630	733	868	1049	1308	4
BrNa	Sodium bromide			791	931	1120	1389	4
BrRb	Rubidium bromide			766	903	1087	1350	4
BrTl	Thallium(I) bromide				509	635	817	4
Br_2	Bromine*	-87.7 s	-71.8 s	-52.7 s	-29.3 s	2.5	58.4	1
Br_2Cd	Cadmium bromide	373 s	435 s	509 s				27
Br_2Hg	Mercury(II) bromide	71 s	98 s	132 s	174 s	227 s	318	4
Br_2OS	Thionyl bromide	-49 e	-29 e	-5 e	27.8	72.9	139.6	5

VAPOR PRESSURE (continued)

Mol. Form.	Name	1 Pa	10 Pa	100 Pa	1 kPa	10 kPa	100 kPa	Ref.
		Temperature in °C for the indicated pressure						
Br_2Pb	Lead(II) bromide	374	431	502	597	726	914	4
Br_2S_2	Sulfur bromide	-7 e	15 e	42 e	78.4	128.1	200.9	5
Br_3In	Indium(III) bromide			304.6 s	328.7 s	364.8 s		1
Br_3OP	Phosphorus(V) oxybromide				64 e	115.5	191.4	5
Br_3P	Phosphorus(III) bromide		-23 e	5 e	42.3	94.6	172.6	5
Br_3Sb	Antimony(III) bromide				136.5	196.9	286.5	1
Br_4Ge	Germanium(IV) bromide				51	105	188	4
Br_4Sn	Tin(IV) bromide				67	122	204	4
Br_4Zr	Zirconium(IV) bromide	136 s	167 s	203 s	245 s	295 s	356 s	4
Br_5P	Phosphorus(V) bromide		-19 s	4 s	31 s	65.5 s	110.1	5
Ca	Calcium	591 s	683 s	798 s	954	1170	1482	2
Cd	Cadmium	257 s	310 s	381	472	594	767	2
$CdCl_2$	Cadmium chloride	412 s	471 s	541 s	634	768	959	23, 27
CdF_2	Cadmium fluoride				1257	1461	1742	4
CdI_2	Cadmium iodide	296 s	344 s	406	498	622	795	4,27
CdO	Cadmium oxide	770 s	866 s	983 s	1128 s	1314 s	1558 s	4
Ce	Cerium	1719	1921	2169	2481	2886	3432	14
ClCs	Cesium chloride			730	864	1043	1297	4
ClCu	Copper(I) chloride		459	543	675	914	1477	4
ClF	Chlorine fluoride*				-144.4	-122.6	-90.2	5
ClF_2P	Phosphorus(III) chloride difluoride				-119.5	-91.1	-47.6	5
ClF_3	Chlorine trifluoride				-63.7	-33.0	11.4	5
ClF_5	Chlorine pentafluoride				-88 e	-59	-14	7
ClH	Hydrogen chloride*				-138.2 s	-118.0	-85.2	1,5
$ClHO_3S$	Chlorosulfonic acid	-40 e	-20 e	5 e	38.7	85.0	153.6	5
ClH_4N	Ammonium chloride	91 s	121 s	159 s	204.7 s	263.1 s	339.5 s	5
ClK	Potassium chloride	625 s	704 s	804	945	1137	1411	23,25
ClLi	Lithium chloride		649 i	761 i	905 i	1101 i	1381 i	8
ClNO	Nitrosyl chloride		-116 s	-100 s	-78.7 s	-50.2	-5.7	5
$ClNO_2$	Nitryl chloride	-121 e	-113 e	-102 e	-86.1	-60.9	-15.7	5
ClNa	Sodium chloride	653 s	733 s	835	987	1182	1461	23,25
ClO_2	Chlorine dioxide*					-34.3	10.5	5
ClRb	Rubidium chloride			777	916	1105	1379	4
ClTl	Thallium(I) chloride				504	626	806	4
Cl_2	Chlorine*	-145 s	-133.7 s	-120.2 s	-103.6 s	-76.1	-34.2	1
Cl_2Co	Cobalt(II) chloride					818	1048	4
Cl_2FP	Phosphorus(III) dichloride fluoride				-71.1	-37.4	13.5	5
Cl_2F_3P	Phosphorus(V) dichloride trifluoride		-120 e	-101 e	-77.1	-44.3	3 e	7
Cl_2Fe	Iron(II) chloride				685	821	1025	4
Cl_2Hg	Mercury(II) chloride	64.4 s	94.7 s	130.8 s	174.5 s	228.5 s	304.0	4
Cl_2Mg	Magnesium chloride			762	908	1111	1414	4
Cl_2Mn	Manganese(II) chloride				760	933	1189	4
Cl_2Ni	Nickel(II) chloride	534 s	592 s	662 s	747 s	852 s	985 s	4
Cl_2OS	Thionyl chloride	-99 e	-81 e	-58 e	-27.1	14.6	75.2	5
Cl_2O_2S	Sulfuryl chloride				-27 e	11.8	69.0	5
Cl_2Pb	Lead(II) chloride			541 e	637	765	949	23
Cl_2S	Sulfur dichloride	-76 e	-61 e	-41 e	-16.7	15.3	58.7	5
Cl_2S_2	Sulfur chloride	-55 e	-36 e	-12 e	21.0	67.2	137.1	5
Cl_2Sn	Tin(II) chloride		253	308	381	479	622	4
Cl_2Zn	Zinc chloride	305 i	356 i	419 i	497 i	596 i	726 i	4,9,12
Cl_3Fe	Iron(III) chloride	118 s	153 s	190 s	229 s	268 s	319	4
Cl_3HSi	Trichlorosilane			-81 e	-56 e	-21 e	31.6	7
Cl_3N	Nitrogen trichloride				-25 e	13.2	70.6	5
Cl_3OP	Phosphorus(V) oxychloride					39.9	105.0	5
Cl_3P	Phosphorus(III) chloride	-93 e	-77 e	-55 e	-26.0	14.5	75.7	5
Cl_4Po	Polonium(IV) chloride					300.6	389.4	5
Cl_4Se	Selenium tetrachloride	23 s	45 s	71 s	102 s	141.4 s	191.1 s	5
Cl_4Si	Tetrachlorosilane*				-39 e	0 e	57.3	1
Cl_4Te	Tellurium tetrachloride				237 e	299.4	387.8	5
Cl_4Zr	Zirconium(IV) chloride	117 s	146 s	181 s	222 s	272 s	336 s	9
Cl_5P	Phosphorus(V) chloride	-2 s	19 s	44 s	74 s	111.4 s	158.9 s	5
Co	Cobalt	1517	1687	1892	2150	2482	2925	2
Cr	Chromium	1383 s	1534 s	1718 s	1950	2257	2669	2
Cs	Cesium	144.5	195.6	260.9	350.0	477.1	667.0	13,30

Mol. Form.	Name	Temperature in °C for the indicated pressure						Ref.
		1 Pa	10 Pa	100 Pa	1 kPa	10 kPa	100 kPa	
CsF	Cesium fluoride				825	999	1249	4
CsI	Cesium iodide	523 s	595 s	692	854	1029	1278	4,25
Cu	Copper	1236	1388	1577	1816	2131	2563	2
CuI	Copper(I) iodide				636	864	1331	4
Dy	Dysprosium	1105 s	1250 s	1431 i	1681 i	2031 i	2558 i	3
Er	Erbium	1231 s	1390 s	1612 i	1890 i	2279 i	2859 i	3
Eu	Europium	590 s	684 s	799 s	961	1179	1523	14
FH	Hydrogen fluoride*				-71.1	-33.7	19.2	1,5
FHO_3S	Fluorosulfonic acid	-14 e	4 e	28 e	59.1	101.3	162.2	5
FK	Potassium fluoride			869	1017	1216	1499	4
FLi	Lithium fluoride	801 s	896	1024	1188	1395	1672	4,12,25
FNO	Nitrosyl fluoride			-131 e	-116.1	-94.3	-60.1	5
FNO_2	Nitryl fluoride		-156 e	-144 e	-128.1	-106.0	-72.6	5
FNO_3	Fluorine nitrate	-160 e	-149 e	-135 e	-115.1	-87.4	-45.0	5
FNa	Sodium fluoride		920 s	1058	1218	1426	1702	4,12,24
FRb	Rubidium fluoride			910	1001	1145	1409	4,12
F_2	Fluorine*	-235 s	-229.5 s	-222.9 s	-214.8	-204.3	-188.3	1,5
F_2O	Fluorine monoxide*	-211.7	-204.7	-195.9	-184.2	-168.2	-144.9	5
F_2OS	Thionyl fluoride			-124 e	-106.5	-81.5	-44.1	5
F_2O_2Re	Rhenium(VI) dioxydifluoride				89.2	131.9	185 e	26
F_2Pb	Lead(II) fluoride				865	1054	1292	4
F_2Xe	Xenon difluoride			2.9 s	31.8 s	67.9 s	114 s	1,5
F_2Zn	Zinc fluoride	731 s	813 s	911 i	1048 i	1237 i	1503 i	9
F_3N	Nitrogen trifluoride*	-201 e	-194 e	-185 e	-172.8	-155.5	-129.2	5
F_3OP	Phosphorus(V) oxyfluoride	-124 s	-113 s	-100 s	-83.7 s	-64.1 s	-39.7 s	5
F_3P	Phosphorus(III) fluoride*				-152 e	-132.6	-101.4	5
F_4MoO	Molybdenum(VI) oxytetrafluoride	-21 s	3 s	33 s	69.3 s	117.3	184.1	26
F_4ORe	Rhenium(VI) oxytetrafluoride	5 s	26 s	50.7 s	80.1 s	117.1	171.2	26
F_4OW	Tungsten(VI) oxytetrafluoride	2 s	25 s	52.1 s	84.3 s	126.7	185.4	26
F_4S	Sulfur tetrafluoride				-110.0	-82.1	-40.3	5
F_4Se	Selenium tetrafluoride				13.6	51.6	104.7	5
F_4Si	Tetrafluorosilane*	-166 s	-157 s	-145.6 s	-132.3 s	-115.7 s	-94.9 s	4,7
F_5Mo	Molybdenum(V) fluoride				86.6	140.3	213 e	26
F_5Nb	Niobium(V) fluoride				80	140	224	4
F_5ORe	Rhenium(VII) oxypentafluoride	-103 s	-84 s	-59 s	-28 s	13.7 s	72.8	26
F_5Os	Osmium(V) fluoride			74.1	113.2	162.3	226 e	26
F_5P	Phosphorus(V) fluoride	-157 s	-148 s	-137 s	-124.5 s	-108.6 s	-84.8	5
F_5Re	Rhenium(V) fluoride			58.8	99.5	152 e	221 e	26
F_5Ta	Tantalum(V) fluoride					119	229	4
F_6Ir	Iridium(VI) fluoride	-88 s	-71 s	-51 s	-27 s	3.8 s	53.1	26
F_6Mo	Molybdenum(VI) fluoride	-98 s	-82 s	-64 s	-41.2 s	-13.4 s	33.5	26
F_6Os	Osmium(VI) fluoride	-89 s	-73 s	-54 s	-30.6 s	-1.7 s	47.4	26
F_6Re	Rhenium(VI) fluoride	-97 s	-82 s	-63 s	-40.2 s	-11.9 s	33.4	26
F_6S	Sulfur hexafluoride*	-158 s	-147 s	-133.6 s	-116.6 s	-94.4 s	-64.1 s	5
F_6Se	Selenium hexafluoride	-143 s	-132 s	-118 s	-100.7 s	-77.8 s	-46.5 s	5
F_6Te	Tellurium hexafluoride	-142 s	-130 s	-115 s	-96 s	-71.8 s	-39.1 s	5
F_6W	Tungsten(VI) fluoride	-107 s	-92 s	-74 s	-52.1 s	-24.8 s	16.9	26
$F_{10}S_2$	Sulfur decafluoride					-22.0	28.5	5
Fe	Iron	1455 s	1617	1818	2073	2406	2859	2
Fr	Francium	131 e	181 e	246 e	335 e	465 e	673 e	2
Ga	Gallium	1037	1175	1347	1565	1852	2245	2
Gd	Gadolinium	1563 i	1755 i	1994 i	2300 i	2703 i	3262 i	3
Ge	Germanium	1371	1541	1750	2014	2360	2831	2
HI	Hydrogen iodide*	-146 s	-135.2 s	-120.8 s	-101.9 s	-75.9 s	-35.9	5
HKO	Potassium hydroxide	520 e	601 e	704	842	1035	1325	4
HNO_3	Nitric acid			-37 e	-9 e	28.4	82.2	5
HN_3	Hydrazoic acid			-79 e	-54 e	-18.0	35.7	5
HNaO	Sodium hydroxide	513	605	722	874	1080	1377	4
H_2	Hydrogen*					-258.6	-252.8	1
H_2I_2Si	Diiodosilane				11.8	70.5	149.4	4
H_2O	Water***	-60.7 s	-42.2 s	-20.3 s	7.0	45.8	99.6	36,37
H_2O_2	Hydrogen peroxide			13 e	45 e	89.0	149.8	5
H_2O_4S	Sulfuric acid	72	103	140	187	248	330	4
H_2S	Hydrogen sulfide*		-149 s	-136 s	-118.9 s	-95.9 s	-60.5	1,5
H_2S_2	Hydrogen disulfide				-27 e	12.2	70.7	5
H_2Se	Hydrogen selenide	-145 s	-134 s	-120 s	-102.8 s	-78.9 s	-41.5	5

VAPOR PRESSURE (continued)

Mol. Form.	Name	Temperature in °C for the indicated pressure 1 Pa	10 Pa	100 Pa	1 kPa	10 kPa	100 kPa	Ref.
H_2Te	Hydrogen telluride					-46.6	-2.3	5
H_3ISi	Iodosilane				-47.7	-10.1	45.2	4
H_3N	Ammonia*	-139 s	-127 s	-112 s	-94.5 s	-71.3	-33.6	1,5,6
H_3NO	Hydroxylamine				43.7	73.3	109.8	4
H_3P	Phosphine*	-182 s	-173 s	-161 s	-145 s	-122.7	-88.0	5
H_4IN	Ammonium iodide	125 s	159 s	201 s	253 s	318.4 s	405.2 s	5
H_4N_2	Hydrazine				14.7	55.6	113 e	5
H_4Si	Silane*			-181	-165.4	-143.7	-111.8	4
He	Helium*					-270.6	-268.9	2
Hf	Hafnium	2416	2681	3004	3406	3921	4603	9
Hg	Mercury**	42.0	76.6	120.0	175.6	250.3	355.9	29,30
HgI_2	Mercury(II) iodide	85.1 s	115.6 s	152.4 s	197.8 s	255.1 s	353.6	4
Ho	Holmium	1159 s	1311 s	1502 i	1767 i	2137 i	2691 i	3
IK	Potassium iodide			731	866	1052	1322	4
ILi	Lithium iodide	545	619	710	824	972	1170	4
INa	Sodium iodide			753	883	1058	1301	4
IRb	Rubidium iodide			733	866	1045	1302	4
ITl	Thallium(I) iodide				520	644	821	4
I_2	Iodine (rhombic)	-12.8 s	9.3 s	35.9 s	68.7 s	108 s	184.0	1,2
I_2Pb	Lead(II) iodide			470	558	682	869	4
I_2Zn	Zinc iodide	301 s	351 s	409 s	488 i	598 i	750 i	9
I_3Sb	Antimony(III) iodide				214.9	292.0	401.2	4
I_4Sn	Tin(IV) iodide				167.1	242.7	347.7	4
I_4Zr	Zirconium(IV) iodide	187 s	220 s	259 s	305 s	361 s	430 s	4
In	Indium	923	1052	1212	1417	1689	2067	2
Ir	Iridium	2440 s	2684	2979	3341	3796	4386	2
K	Potassium	200.2	256.5	328	424	559	756.2	13,30
Kr	Krypton*	-214.0 s	-208.0 s	-199.4 s	-188.9 s	-174.6 s	-153.6	5
La	Lanthanum	1732 i	1935 i	2185 i	2499 i	2905 i	3453 i	3
Li	Lithium	524.3	612.3	722.1	871.2	1064.3	1337.1	13,30
Lu	Lutetium	1633 s	1829.8	2072.8	2380 i	2799 i	3390 i	3
Mg	Magnesium	428 s	500 s	588 s	698	859	1088	2
Mn	Manganese	955 s	1074 s	1220 s	1418	1682	2060	2
Mo	Molybdenum	2469 s	2721	3039	3434	3939	4606	2
MoO_3	Molybdenum(VI) oxide				801	935	1151	4
NO	Nitric oxide*	-201 s	-195 s	-188 s	-179.3 s	-168.1 s	-151.9	5
N_2	Nitrogen*	-236 s	-232 s	-226.8 s	-220.2 s	-211.1 s	-195.9	1,5
N_2O	Nitrous oxide*	-167 s	-157 s	-145.4 s	-131.1 s	-112.9 s	-88.7	5
N_2O_4	Nitrogen tetroxide	-92 s	-78 s	-61 s	-41.1 s	-16.6 s	28.7	5
N_2O_5	Nitrogen pentoxide	-71 s	-56 s	-40 s	-19.9 s	3.9 s	33.2	5
Na	Sodium	280.6	344.2	424.3	529	673	880.2	13,30
Nb	Niobium	2669	2934	3251	3637	4120	4740	2
Nd	Neodymium	1322.3	1501.2	1725.3	2023 i	2442 i	3063 i	3
Ne	Neon*	-261 s	-260 s	-258 s	-255 s	-252 s	-246.1	2
Ni	Nickel	1510	1677	1881	2137	2468	2911	2
OPb	Lead(II) oxide	724	816	928	1065	1241	1471	4
OSr	Strontium oxide	1789 s	1903 s	2047 s	2235 s	2488 s		4
O_2	Oxygen*				-211.9	-200.5	-183.1	1,28
O_2S	Sulfur dioxide*			-98 s	-80 s	-52.2	-10.3	1,5
O_2Se	Selenium dioxide	124.5 s	153.9 s	188 s	228 s	275 s	315 s	38
O_2Si	Silicon dioxide	1966 i	2149 i	2368 i				8
O_3	Ozone*	-189 e	-182 e	-172 e	-158 e	-139.7	-111.5	5
O_3P_2	Phosphorus(III) oxide				47.3	100.3	172.8	4
O_3S	Sulfur trioxide				-20 s	6.6 s	44.5	5
O_3Sb_2	Antimony(III) oxide (valentinite)	426.1 s	478 s	539 s	610 s	907	1420	4,35
O_5P_2	Phosphorus(V) oxide	285 s	328 s	377.5 s	434.4 s	500.5 s	591	4
O_7Re_2	Rhenium(VII) oxide	147 s	176 s	208 s	244 s	284 s	362	4
Os	Osmium	2887 s	3150	3478	3875	4365	4983	2
P	Phosphorus (white)	6 s	34 s	69	115	180	276	3,9
P	Phosphorus (red)	182 s	216 s	256 s	303 s	362 s	431 s	2,3
Pb	Lead	705	815	956	1139	1387	1754	2
PbS	Lead(II) sulfide	656 s	741 s	838 s	953 s	1088 s	1280	4
Pd	Palladium	1448 s	1624	1844	2122	2480	2961	2
Po	Polonium				573 e	730.2	963.3	5
Pr	Praseodymium	1497.7	1699.4	1954 i	2298 i	2781 i	3506 i	3
Pt	Platinum	2057	2277 e	2542	2870	3283	3821	2

Mol. Form.	Name	Temperature in °C for the indicated pressure 1 Pa	10 Pa	100 Pa	1 kPa	10 kPa	100 kPa	Ref.
Pu	Plutonium	1483	1680	1925	2238	2653	3226	2
Ra	Radium	546 s	633 s	764	936	1173	1526	2
Rb	Rubidium	160.4	212.5	278.9	368	496.1	685.3	13,30
Re	Rhenium	3030 s	3341	3736	4227	4854	5681	2
Rh	Rhodium	2015	2223	2476	2790	3132	3724	2
Rn	Radon*	-163 s	-152 s	-139 s	-121.4 s	-97.6 s	-62.3	5
Ru	Ruthenium	2315 s	2538	2814	3151	3572	4115	2
S	Sulfur	102 s	135	176	235	318	444	3
Sb	Antimony	534 s	603 s	738	946	1218	1585	2,3
Sc	Scandium	1372 s	1531 s	1733 i	1993 i	2340 i	2828 i	3
Se	Selenium	227	279	344	431	540	685	3
Si	Silicon	1635	1829	2066	2363	2748	3264	2
Sm	Samarium	728 s	833 s	967 s	1148 i	1402 i	1788 i	3
Sn	Tin	1224	1384	1582	1834	2165	2620	2
Sr	Strontium	523 s	609 s	717 s	866	1072	1373	2
Ta	Tantalum	3024	3324	3684	4122	4666	5361	2
Tb	Terbium	1516.1	1706.1	1928 i	2232 i	2640 i	3218 i	3
Tc	Technetium	2454 e	2725 e	3051 e	3453 e	3961 e	4621 e	2
Te	Tellurium			502 e	615 e	768.8	992.4	5
Th	Thorium	2360	2634	2975	3410	3986	4782	2
Ti	Titanium	1709	1898	2130 e	2419	2791	3285	2
Tl	Thallium	609	704	824	979	1188	1485	2
Tm	Thulium	844 s	962 s	1108 s	1297 s	1548 i	1944 i	3
U	Uranium	2052	2291	2586	2961	3454	4129	2
V	Vanadium	1828 s	2016	2250	2541	2914	3406	2
W	Tungsten	3204 s	3500	3864	4306	4854	5550	2
Xe	Xenon*	-190 s	-181 s	-170 s	-155.8 s	-136.6 s	-108.4	5,32
Y	Yttrium	1610.1	1802.3	2047 i	2354 i	2763 i	3334 i	3
Yb	Ytterbium	463 s	540 s	637 s	774 s	993 i	1192 i	3
Zn	Zinc	337 s	397 s	477	579	717	912 e	2
Zr	Zirconium	2366	2618	2924	3302	3780	4405	2
Substances containing carbon:								
C	Carbon (graphite)		2566 s	2775 s	3016 s	3299 s	3635 s	15
$CBrClF_2$	Bromochloro-difluoromethane	-136 e	-123 e	-106 e	-83.4	-51.8	-4.3	1
$CBrCl_3$	Bromotrichloromethane				-6 e	38.9	104.4	5
$CBrF_3$	Bromotrifluoromethane*	-168 e	-156 e	-142 e	-122.8	-96.6	-58.1	5
CBrN	Cyanogen bromide				-13 s	17.7 s	61.0	1
CBr_2F_2	Dibromodifluoromethane		-110 e	-91 e	-66 e	-30 e	22.5	1
CBr_4	Tetrabromomethane			25.6 s	65.8 s	111.6	188.9	5
$CClF_3$	Chlorotrifluoromethane	-176 e	-167 e	-155 e	-139 e	-116 e	-81.7	5
CClN	Cyanogen chloride		-94.6 s	-78.1 s	-57 s	-29 s	13.0	5
CCl_2F_2	Dichlorodifluoromethane*	-150 e	-138 e	-122 e	-101.8	-73.1	-30.0	5
CCl_2O	Carbonyl chloride	-127 e	-113 e	-96 e	-73 e	-40.6	7.2	5
CCl_3F	Trichlorofluoromethane*		-107 e	-89 e	-63 e	-28.5	23.3	1,5
CCl_3NO_2	Trichloronitromethane		-59 e	-30 e	4.4	47.8	112.0	5
CCl_4	Tetrachloromethane*	-79.4 s	-70.8 s	-53.5 s	-24.4 s	15.8	76.2	1,5
CFN	Cyanogen fluoride		-135 s	-121.2 s	-104.1 s	-82.8 s	-46.2	1,5
CF_4	Tetrafluoromethane*	-199.9 s	-193 s	-183.9 s	-171.6	-153.9	-128.3	1,5
$CHBrF_2$	Bromodifluoromethane		-128 s	-111.4 s	-89.7 s	-59.7 s	-16 s	5
$CHBr_3$	Tribromomethane				30.5	78.3	148.8	1
$CHClF_2$	Chlorodifluoromethane*	-152 e	-141 e	-126 e	-107.1	-80.5	-41.1	5
$CHCl_2F$	Dichlorofluoromethane	-76 e	-70 e	-61 e	-49 e	-28.7	8.6	1
$CHCl_3$	Trichloromethane*			-61 e	-34 e	4.3	60.8	1
CHF_3	Trifluoromethane*			-152 e	-136 e	-114.4	-82.3	1
CHI_3	Triiodomethane	51.1 s	82.7 s	121 e			218.0	5
CHN	Hydrogen cyanide*			-77 s	-52.6 s	-22.7 s	25.4	1,5
CHNO	Cyanic acid			-81.1	-56.8	-23.9	23 e	5
CH_2BrCl	Bromochloromethane	-83 e	-69 e	-50 e	-25 e	11.4	67.7	1
CH_2Br_2	Dibromomethane			-37 e	-7 e	35.2	96.5	5
CH_2ClF	Chlorofluoromethane		-124 e	-108 e	-86.2	-55.7	-9.4	5
CH_2Cl_2	Dichloromethane*		-92 e	-73 e	-48 e	-12.5	39.3	1
CH_2F_2	Difluoromethane*	-156.7	-145.8	-131.9	-113.6	-88.6	-51.9	1
CH_2I_2	Diiodomethane			17 e	55 e	106.1	181.6	5
CH_2O	Formaldehyde*				-91 e	-61.7	-19.3	1

VAPOR PRESSURE (continued)

Mol. Form.	Name	Temperature in °C for the indicated pressure 1 Pa	10 Pa	100 Pa	1 kPa	10 kPa	100 kPa	Ref.
CH_2O_2	Formic acid	-56 s	-40.4 s	-22.3 s	-0.8 s	37.0	100.2	1,5
CH_3AsF_2	Methyldifluoroarsine				-15 e	22.1	76.1	5
CH_3BO	Borane carbonyl				-124	-99	-64	4
CH_3Br	Bromomethane				-77 e	-44.3	3.3	1
CH_3Cl	Chloromethane*	-140.2 s	-128.6 s	-114.7 s	-96 e	-67.1	-24.4	1,33
CH_3Cl_3Si	Methyltrichlorosilane		-83 e	-61 e	-33 e	7 e	65.7	1
CH_3F	Fluoromethane*				-130 e	-111 e	-78.6	1
CH_3I	Iodomethane				-49 e	-12.4	42.1	1
CH_3NO	Formamide		22 e	53 e	93 e	145.0	218 e	5
CH_3NO_2	Nitromethane				-2 e	40 e	100.8	1
CH_3NO_3	Methyl nitrate		-75 e	-55 e	-27 e	9.8	63 e	5
CH_4	Methane*	-220 s	-214.2 s	-206.8 s	-197 s	-183.6 s	-161.7	5,41
CH_4Cl_2Si	Dichloromethylsilane			-77 e	-51 e	-14 e	40.5	1
CH_4O	Methanol*	-87 e	-69 e	-47.5	-20.4	15.2	64.2	11
CH_4S	Methanethiol		-115 e	-97 e	-74 e	-41.7	5.7	1
CH_5ClSi	Chloromethylsilane	-129 e	-115 e	-97.9	-74.4	-41.5	8.3	5
CH_5N	Methylamine				-76.7	-48.1	-6.6	1
CH_6N_2	Methylhydrazine			-31 e	-4.7	32.9	91 e	1
CH_6OSi	Methyl silyl ether				-90.2	-61.8	-18 e	1
CH_6Si	Methylsilane			-144 e	-124.6	-97.5	-57.5	5
CIN	Cyanogen iodide						153.8	5
$CNNa$	Sodium cyanide		672 e	798	961	1182	1497	4
CN_4O_8	Tetranitromethane				18.0	61.8	124 e	5
CO	Carbon monoxide*			-223 s	-216.5 s	-207.2 s	-191.7	40
COS	Carbon oxysulfide*			-136 e	-117 e	-90.0	-50.4	1
$COSe$	Carbon oxyselenide			-120	-98	-67	-22	4
CO_2	Carbon dioxide*	-159.1 s	-148.9 s	-136.7 s	-121.6 s	-103.1 s	-78.6 s	5
CS_2	Carbon disulfide		-96 e	-76 e	-49 e	-10.9	45.9	1
CSe_2	Carbon diselenide			-24 e	9.4	56.2	127 e	1
$C_2Br_2ClF_3$	1,2-Dibromo-1-chloro-1,2,2-trifluoroethane						92.3	5
$C_2Br_2F_4$	1,2-Dibromotetrafluoroethane		-97 e	-75 e	-46 e	-7.2	47.1	5
C_2Br_4	Tetrabromoethylene		-54.5 s	-31.7 s	-3.5 s	32.2 s	226.0	5
C_2ClF_3	Chlorotrifluoroethylene	-146 e	-134 e	-119 e	-99 e	-71 e	-28.4	1
C_2ClF_5	Chloropentafluoroethane					-80.3	-39.4	1
$C_2Cl_2F_4$	1,1-Dichlorotetrafluoroethane					-45.4	2.7	5
$C_2Cl_2F_4$	1,2-Dichlorotetrafluoroethane				-76.8	-44.9	3.2	5
$C_2Cl_3F_3$	1,1,1-Trichlorotrifluoroethane						45.6	1,5
$C_2Cl_3F_3$	1,1,2-Trichlorotrifluoroethane					-8.2	47.3	1,5
C_2Cl_3N	Trichloroacetonitrile				-16 e	25.3	85.1	1
C_2Cl_4	Tetrachloroethylene			-22 e	10 e	54.4	120.7	1
$C_2Cl_4F_2$	1,1,1,2-Tetrachloro-2,2-difluoroethane				-7 e	31.0	91.1	5
$C_2Cl_4F_2$	1,1,2,2-Tetrachloro-1,2-difluoroethane					32.3	92.5	1
C_2Cl_4O	Trichloroacetyl chloride			-25 e	7 e	51.7	117.8	1,5
C_2Cl_6	Hexachloroethane	-7.6 s	9.9 s	33.6 s	67.7 s	116.9 s	184.2 s	5
C_2F_3N	Trifluoroacctonitrile				-126.1	-102.5	-67.8	1
C_2F_4	Tetrafluoroethylene				-132.3	-109.7	-75.8	1
$C_2F_4N_2O_4$	1,1,2,2-Tetrafluoro-1,2-dinitroethane				-30 e	6.4	59.5	5
C_2F_6	Hexafluoroethane**			-155.2 s	-137.5 s	-113.4 s	-78.4 s	1,5
$C_2HBrClF_3$	2-Bromo-2-chloro-1,1,1-trifluoroethane				-41.4	-4.8	49.8	1
C_2HBr_3O	Tribromoacetaldehyde			15.0	52.7	103.0	173.5	5
C_2HClF_4	1-Chloro-1,1,2,2-tetrafluoroethanc			-110 e	-87.6	-57.0	-12.1	5
$C_2HCl_2F_3$	2,2-Dichloro-1,1,1-trifluoroethane		-101.0	-82.2	-57.4	-23.3	26.7	18
C_2HCl_3	Trichloroethylene	-74 e	-59 e	-39 e	-12 e	26.7	86.8	1
C_2HCl_3O	Trichloroacetaldehyde			-41.6	-9.8	33.8	97.4	5
$C_2HCl_3O_2$	Trichloroacetic acid				83.8	130.0	197.2	1,5
C_2HCl_5	Pentachloroethane		-23 e	3 e	37.4	86.0	159.4	1
$C_2HF_3O_2$	Trifluoroacetic acid					16.8	71.4	1,5
C_2HF_5O	Trifluoromethyl difluoromethyl ether	-147 e	-136 e	-121 e	-102 e	-75.0	-35.4	20
C_2H_2	Acetylene*			-146.6 s	-130.7 s	-110.6 s	-84.8 s	5

VAPOR PRESSURE (continued)

Mol. Form.	Name	Temperature in °C for the indicated pressure						Ref.
		1 Pa	10 Pa	100 Pa	1 kPa	10 kPa	100 kPa	
$C_2H_2Br_2$	*cis*-1,2-Dibromoethylene		-45 e	-21 e	10 e	52.2	114.8	1
$C_2H_2Br_2$	*trans*-1,2-Dibromoethylene				-4 e	42.2	107.4	5
$C_2H_2Br_2Cl_2$	1,2-Dibromo-1,1-dichloroethane					103.6	177.8	5
$C_2H_2Br_2Cl_2$	1,2-Dibromo-1,2-dichloroethane		-11 e	22 e	64.1	119 e	193 e	5
$C_2H_2Br_4$	1,1,2,2-Tetrabromoethane	14 e	38 e	69 e	109 e	163.7	242.9	5
$C_2H_2Cl_2$	1,1-Dichloroethylene	-116 e	-101 e	-82 e	-57 e	-21.4	31.2	1
$C_2H_2Cl_2$	*cis*-1,2-Dichloroethylene			-62 e	-34 e	3.8	60.3	1
$C_2H_2Cl_2$	*trans*-1,2-Dichloroethylene				-44 e	-7.5	47.3	1
$C_2H_2Cl_2F_2$	1,2-Dichloro-1,1-difluoroethane	-101 e	-87 e	-68 e	-42.2	-6.8	46.3	5
$C_2H_2Cl_2O$	Chloroacetyl chloride			-23.7	5.6	46.1	105.6	5
$C_2H_2Cl_4$	1,1,1,2-Tetrachloroethane	-58 e	-40 e	-15 e	17 e	62.2	129.7	1
$C_2H_2Cl_4$	1,1,2,2-Tetrachloroethane		-22 e	1 e	32.4	76.9	144.7	1
$C_2H_2F_4$	1,1,1,2-Tetrafluoroethane				-94.3	-66.8	-26.4	17
$C_2H_2F_4$	1,1,2,2-Tetrafluoroethane				-96.0	-66.9	-23.3	5
C_2H_2O	Ketene		-151 e	-135 e	-115 e	-88.2	-50.0	1
C_2H_3Br	Bromoethylene	-124 e	-110 e	-92 e	-68 e	-34.5	15.4	5
C_2H_3BrO	Acetyl bromide	-78 e	-65 e	-49 e	-25 e	13.9	84 e	5
$C_2H_3Br_3$	1,1,2-Tribromoethane	-18 e	4 e	32 e	68 e	117.1	188.4	5
C_2H_3Cl	Chloroethylene	-139 e	-127 e	-110 e	-89 e	-59.0	-14.1	1
$C_2H_3ClF_2$	1-Chloro-1,1-difluoroethane		-123 e	-107 e	-85.3	-55.4	-10.5	5
C_2H_3ClO	Acetyl chloride	-100 e	-85 e	-66 e	-40 e	-3.6	50.4	1
$C_2H_3ClO_2$	Chloroacetic acid				78.4	123.9	188.9	1
$C_2H_3Cl_2F$	1,1-Dichloro-1-fluoroethane		-101 e	-83 e	-57.9	-22.7	31.4	5
$C_2H_3Cl_2F$	1,2-Dichloro-1-fluoroethane			-50 e	-23.8	14.1	73.4	5
$C_2H_3Cl_3$	1,1,1-Trichloroethane				-25.3	14.2	73.7	5
$C_2H_3Cl_3$	1,1,2-Trichloroethane			-23 e	7 e	49.9	113.4	1
C_2H_3F	Fluoroethylene			-153.3	-135.2	-109.9	-72.2	5
C_2H_3FO	Acetyl fluoride					-64.1	17.0	5
$C_2H_3F_3$	1,1,1-Trifluoroethane				-113 e	-86.6	-47.8	1
$C_2H_3F_3O$	2,2,2-Trifluoroethanol			-33 e	-8 e	26.0	74 e	5
C_2H_3I	Iodoethylene				-41 e	-3 e	55.6	5
C_2H_3IO	Acetyl iodide				-0.6	47 e	107.0	5
C_2H_3N	Acetonitrile				-20 e	21.4	81.2	1
C_2H_3NO	Methylisocyanate				-43.5	-10.2	38.8	1
C_2H_3NS	Methyl thiocyanate			-18.4	16.2	63.5	132.5	5
C_2H_4	Ethylene*				-155.6	-135.1	-104.0	1,10
C_2H_4BrCl	1-Bromo-2-chloroethane				-0.4	41.7	105.7	6
$C_2H_4Br_2$	1,1-Dibromoethane		-49 e	-26 e	5 e	46.4	107.6	5
$C_2H_4Br_2$	1,2-Dibromoethane				18 e	62.2	130.9	1
C_2H_4ClF	1-Chloro-1-fluoroethane				-69.9	-36.1	15.8	5
$C_2H_4Cl_2$	1,1-Dichloroethane		-84 e	-64 e	-36.7	1.0	56.9	1
$C_2H_4Cl_2$	1,2-Dichloroethane				-16.4	23.7	83.1	1
$C_2H_4F_2$	1,1-Difluoroethane			-115.2	-94.6	-66.1	-24.3	19
$C_2H_4N_2O_6$	Ethylene glycol dinitrate	4 e	25.6	51.0	81 e	117 e	162 e	5
C_2H_4O	Acetaldehyde		-105 e	-87 e	-62.8	-29.4	20.0	5
C_2H_4O	Ethylene oxide		-111 e	-93 e	-70 e	-37.0	10.2	1
$C_2H_4O_2$	Acetic acid	-42.8 s	-26.7 s	-8 s	14.2 s	55.9	117.5	1,5
$C_2H_4O_2$	Methyl formate		-95 e	-76 e	-51.8	-18.1	31.4	5
$C_2H_4O_3$	Peroxyacetic acid				14.4	55.3	109.7	5
$C_2H_4O_3$	Glycolic acid						99.9	5
$C_2H_5AsF_2$	Ethyldifluoroarsine			-36 e	-6.0	35.0	93.1	5
C_2H_5Br	Bromoethane	-111 e	-96 e	-77 e	-51.3	-15.5	38.0	5
C_2H_5Cl	Chloroethane	-126 e	-112 e	-94 e	-70 e	-37.0	12.0	1
C_2H_5ClO	2-Chloroethanol	-61 e	-39 e	-12 e	23 e	67.1	127.3	5
C_2H_5ClO	Chloromethyl methyl ether	-96 e	-80 e	-59 e	-32 e	6 e	61 e	5
$C_2H_5Cl_3OSi$	Trichloroethoxysilane	-78 e	-60 e	-36.0	-4.6	38.7	102.0	5
$C_2H_5Cl_3Si$	Trichloroethylsilane	-79 e	-61 e	-38 e	-8 e	34.9	98.7	5
C_2H_5F	Fluoroethane		-142 e	-127 e	-106.3	-78.7	-37.9	1
C_2H_5FO	2-Fluoroethanol			-22 e	8.3	47.5	99 e	5
C_2H_5I	Iodoethane	-94 e	-78 e	-56 e	-27.9	11.9	71.9	5
C_2H_5N	Ethyleneimine		-74 e	-55 e	-30 e	4.1	55 e	5
C_2H_5NO	Acetamide	16.7 s	39.1 s	65.2 s	102.8	150.8	218.2	5
C_2H_5NO	*N*-Methylformamide		13 e	41 e	78 e	127.9	199.1	1
$C_2H_5NO_2$	Nitroethane	-61 e	-44 e	-21 e	8.3	50.1	113.5	5
$C_2H_5NO_3$	Ethyl nitrate	-81 e	-63 e	-41 e	-12 e	28.2	87 e	1
C_2H_6	Ethane*	-183.3 s	-173.2	-161.3	-145.3	-122.8	-88.8	41
$C_2H_6Cl_2Si$	Dichlorodimethylsilane					11.1	70.1	5

Mol. Form.	Name	Temperature in °C for the indicated pressure 1 Pa	10 Pa	100 Pa	1 kPa	10 kPa	100 kPa	Ref.
C_2H_6Hg	Dimethyl mercury				-13.5	29.0	92.1	5
$C_2H_6N_2O$	*N*-Nitrosodimethylamine				30.7	80.5	149.8	5
C_2H_6O	Ethanol	-73 e	-56 e	-34 e	-7 e	29.2	78.0	1,5
C_2H_6O	Dimethyl ether*		-135 e	-118 e	-96.8	-67.6	-25.1	1,5
C_2H_6OS	Dimethyl sulfoxide			27.4	65.0	115.9	188.6	1
$C_2H_6O_2$	Ethylene glycol	2 e	24 e	51.1	86.1	132.5	196.9	1
$C_2H_6O_2$	Ethyl hydroperoxide	-70 e	-49 e	-25 e	6.8	47.0	101 e	5
$C_2H_6O_2S$	Dimethyl sulfone				109 e	166.8	248.9	5
C_2H_6S	Ethanethiol	-112 e	-97 e	-78 e	-53 e	-18 e	34.7	1
C_2H_6S	Dimethyl sulfide		-96 e	-77 e	-51.2	-16.0	37.0	1,5
$C_2H_6S_2$	Dimethyl disulfide	-71 e	-53 e	-29 e	1.7	45.0	109.3	5
$C_2H_7BO_2$	Dimethoxyborane	-116 e	-101.9	-83.5	-59.2	-25.4	25 e	5
C_2H_7N	Ethylamine			-71 e	-53 e	-27 e	16.4	1
C_2H_7N	Dimethylamine			-88 e	-66.9	-37.2	6.6	1
C_2H_7NO	Ethanolamine		11 e	35 e	66.2	109.0	170.6	1
$C_2H_8N_2$	1,2-Ethanediamine				17.0	57.5	116.6	1,5
$C_2H_8N_2$	1,1-Dimethylhydrazine			-52 e	-25.6	10.5	63 e	5
$C_2H_8N_2$	1,2-Dimethylhydrazine		-49 e	-33 e	-9 e	26.4	88 e	1
C_2N_2	Cyanogen	-127 s	-114.1 s	-98.5 s	-79.2 s	-54.9 s	-21.4	5
C_3ClF_5O	Chloropentafluoroacetone	-122 e	-109 e	-93 e	-71 e	-39.4	7.4	5
C_3Cl_6	Hexachloropropene	-12 e	11 e	40 e	79 e	132.8	213.6	5
C_3F_6	Perfluoropropene	-150 e	-138 e	-122 e	-101 e	-72 e	-30.6	5
C_3F_6O	Perfluoroacetone			-113 e	-94 e	-67.8	-27.6	5
C_3F_8	Perfluoropropane		-139 e	-124 e	-105 e	-77.5	-37.0	1
C_3HN	Cyanoacetylene			-58.7 s	-35.6 s	-7 s	42.0	5
$C_3H_2F_6O$	1,1,1,3,3,3-Hexafluoro-2-propanol					12.7	57.1	5
$C_3H_3F_5$	1,1,1,2,2-Pentafluoropropane					-60 e	-17.9	5
C_3H_3N	2-Propenenitrile		-72 e	-50 e	-22 e	17.7	77.0	1
C_3H_3NS	Thiazole					54.4	117.8	5
C_3H_4	Allene*		-129 e	-118 e	-101.4	-76.7	-34.7	5
C_3H_4	Propyne				-94 e	-65.3	-23.2	1
$C_3H_4ClF_3$	3-Chloro-1,1,1-trifluoropropane	-102 e	-87 e	-68 e	-43 e	-8 e	45.3	5
$C_3H_4Cl_2O$	1,1-Dichloroacetone				1 e	47.8	118.0	5
$C_3H_4Cl_2O_2$	Methyl dichloroacetate	-44 e	-25 e	0 e	33 e	77.7	142.3	5
$C_3H_4Cl_4$	1,1,1,2-Tetrachloropropane	-48 e	-28 e	-2 e	32 e	79.1	149.5	5
$C_3H_4F_4O$	2,2,3,3-Tetrafluoro-1-propanol			-10 e	17 e	53.9	107.2	5
C_3H_4O	Acrolein		-87 e	-67 e	-40 e	-3.0	52.8	1
$C_3H_4O_2$	Propenoic acid				35 e	78.0	140.7	1
$C_3H_4O_2$	Vinyl formate			-58 e	-34 e	-1.6	46.2	1
$C_3H_4O_2$	2-Oxetanone		-21 e	8 e	45.5	93.8	159.3	5
$C_3H_4O_3$	Ethylene carbonate	12.7 s	37 e				247	5
C_3H_5Br	*cis*-1-Bromopropene	-100 e	-84 e	-64 e	-37 e	1.0	57.4	5
C_3H_5Br	2-Bromopropene	-112 e	-95 e	-75 e	-47 e	-9 e	48.0	5
C_3H_5Br	3-Bromopropene	-98 e	-80 e	-58 e	-28 e	12 e	69.6	5
C_3H_5Cl	*cis*-1-Chloropropene	-114 e	-100 e	-81 e	-55 e	-20.1	32.4	5
C_3H_5Cl	*trans*-1-Chloropropene		-97 e	-77 e	-52 e	-16.2	37.0	5
C_3H_5Cl	2-Chloropropene	-120 e	-106 e	-87 e	-63 e	-28.7	22.3	5
C_3H_5Cl	3-Chloropropene	-107 e	-92 e	-72.4	-46.3	-9.8	44.6	5
C_3H_5ClO	Epichlorohydrin			-21 e	11 e	53.8	115.5	5
$C_3H_5ClO_2$	Methyl chloroacetate		-28 e	-5 e	25 e	66.9	129.1	5
$C_3H_5Cl_3$	1,1,3-Trichloropropane	-51 e	-31 e	-5 e	28 e	75.3	145.1	5
$C_3H_5Cl_3$	1,2,3-Trichloropropane			2 e	37 e	84.9	156.3	5
$C_3H_5Cl_3Si$	Trichloro-2-propenylsilane					53.0	116.5	5
C_3H_5I	3-Iodopropene	80 c	-62 e	-39 e	-8 e	36 e	101.5	5
C_3H_5N	Propanenitrile	-69.4	-55.3	-36.0	-7.9	35.2	97.4	1,5
C_3H_5NO	Acrylamide			109.6	161 e			5
C_3H_5NO	3-Hydroxypropanenitrile	-11 e	18 e	53 e	96.1	150.3	220.8	5
C_3H_5NS	Ethyl thiocyanate	-39 e	-20 e	4 e	35 e	79.1	143.4	5
C_3H_5NS	Ethyl isothiocyanate				17.4	66 e	136 e	5
$C_3H_5N_3O_9$	Trinitroglycerol	48.6	75.7	118 e	191 e	353 e	1007 e	5
C_3H_6	Propene*	-160.6	-149.0	-134.3	-114.9	-88.2	-47.9	1,5
C_3H_6	Cyclopropane			-124 e	-104 e	-75.7	-33.1	1
C_3H_6BrCl	1-Bromo-3-chloropropane	-51 e	-31 e	-6 e	28 e	74.1	142.9	5
$C_3H_6Br_2$	1,2-Dibromopropane	-46 e	-26 e	-2 e	31 e	75.3	139.5	5

VAPOR PRESSURE (continued)

Mol. Form.	Name	Temperature in °C for the indicated pressure 1 Pa	10 Pa	100 Pa	1 kPa	10 kPa	100 kPa	Ref.
$C_3H_6Br_2$	1,3-Dibromopropane	-30 e	-9 e	17 e	52 e	98.7	166.8	5
$C_3H_6Cl_2$	1,1-Dichloropropane				-14 e	27.0	87.7	5
$C_3H_6Cl_2$	1,2-Dichloropropane	-78 e	-61 e	-38.1	-8.1	33.7	95.9	5
$C_3H_6Cl_2$	1,3-Dichloropropane	-65 e	-46 e	-22 e	10 e	54.0	119.9	5
$C_3H_6Cl_2$	2,2-Dichloropropane				-28 e	10.8	68.9	5
$C_3H_6Cl_2O$	1,3-Dichloro-2-propanol			21.8	59.0	107.6	173.9	5
$C_3H_6N_2O_4$	1,1-Dinitropropane	-9 e	12 e	39 e	73.2	120 e	187 e	5
C_3H_6O	Allyl alcohol	-63 e	-48 e	-21.9	6.8	44.5	96.2	5
C_3H_6O	Methyl vinyl ether			-114 e	-89 e	-52.7	4.6	1
C_3H_6O	Propanal			-69 e	-42 e	-6 e	47.7	1
C_3H_6O	Acetone	-95	-81.8	-62.8	-35.6	1.3	55.7	1,5
C_3H_6O	Methyloxirane	-109 e	-95 e	-76 e	-51.5	-17.2	33.9	5
$C_3H_6O_2$	Propanoic acid			0 e	35.1	79.9	140.8	1,5
$C_3H_6O_2$	Ethyl formate		-80 e	-61 e	-35 e	1 e	54.0	1
$C_3H_6O_2$	Methyl acetate	-95 e	-79 e	-59 e	-33 e	3.3	56.6	1
$C_3H_6O_2$	1,3-Dioxolane		-72 e	-50 e	-22 e	17.0	75.3	1
$C_3H_6O_3$	1,3,5-Trioxane					53 e	113.7	1
C_3H_6S	Thiacyclobutane		-62 e	-40 e	-9 e	32.5	94.5	5
C_3H_7Br	1-Bromopropane	-95 e	-78 e	-57 e	-28 e	11.6	70.6	1
C_3H_7Br	2-Bromopropane		-84 e	-65 e	-39.6	-1.7	59.1	1,5
C_3H_7Cl	1-Chloropropane	-106 e	-90 e	-71 e	-44.5	-8.1	46.2	1
C_3H_7Cl	2-Chloropropane		-91 e	-74 e	-51.1	-17.8	35.4	1,5
C_3H_7ClO	2-Chloro-1-propanol				23 e	63.8	125.7	5
C_3H_7F	1-Fluoropropane	-133 e	-120 e	-103 e	-80.7	-49.4	-2.8	5
C_3H_7I	1-Iodopropane	-78 e	-60 e	-37 e	-6 e	36.9	102.0	5
C_3H_7I	2-Iodopropane	-89 e	-71 e	-47 e	-16.3	26.5	89.2	5
C_3H_7N	Allylamine		-88 e	-65 e	-37 e	0.4	52 e	5
C_3H_7NO	*N,N*-Dimethylformamide	-39 e	-20 e	5 e	38.0	83.9	152.6	1
C_3H_7NO	*N*-Methylacetamide	-13.3 s	13 s	43 e	83.8	136.1	206.3	5
$C_3H_7NO_2$	1-Nitropropane	-56 e	-37 e	-13 e	20 e	64.8	130.8	1
$C_3H_7NO_2$	2-Nitropropane		-48 e	-22 e	10.7	55.6	119.8	1
$C_3H_7NO_3$	Propyl nitrate			-23.9	6.1	48.1	111 e	5
C_3H_8	Propane*	-156.9	-145.6	-130.9	-111.4	-83.8	-42.3	1,41
C_3H_8O	1-Propanol	-54 e	-38 e	-16 e	10 e	47 e	96.9	1,5
C_3H_8O	2-Propanol	-65 e	-49 e	-28 e	-1.3	33.6	82.0	1,5
C_3H_8O	Ethyl methyl ether	-98 e	-89 e	-77 e	-60 e	-34.8	7.0	5
$C_3H_8O_2$	1,2-Propylene glycol	-11 e	13 e	42 e	78 e	125.0	187.2	5
$C_3H_8O_2$	1,3-Propylene glycol	4 e	30 e	62 e	101 e	149.9	214.0	5
$C_3H_8O_2$	Ethylene glycol monomethyl ether	-57 e	-37 e	-12 e	21 e	63.8	124.3	1
$C_3H_8O_2$	Dimethoxymethane	-93 e	-81 e	-64 e	-42 e	-9.3	41.7	5
$C_3H_8O_3$	Glycerol	96 e	113 e	136 e	168 e	213.4	287 e	1
C_3H_8S	1-Propanethiol	-94 e	-78 e	-57 e	-29.1	9.6	67.4	1,5
C_3H_8S	2-Propanethiol	-102 e	-87 e	-67 e	-41 e	-3 e	52.2	1
C_3H_8S	Ethyl methyl sulfide	-94 e	-78 e	-57 e	-29.7	8.8	66.3	1
$C_3H_8S_2$	1,3-Propanedithiol	-53 e	-28 e	3 e	43 e	97 e	172.4	5
C_3H_9As	Trimethylarsine			-74 e	-45 e	-5.4	52.0	5
$C_3H_9BO_3$	Trimethyl borate				-14 e	15.6	67.9	5
C_3H_9BS	Methyl dimethylthioborane			-62 e	-30.4	11.4	70.7	5
C_3H_9ClSi	Trimethylchlorosilane				-37.8	0.4	57.3	5
C_3H_9N	Propylamine		-81 e	-63 e	-38.3	-4.1	46.9	1,5
C_3H_9N	Isopropylamine		-91 e	-74 e	-50.4	-17.6	31.5	1,5
C_3H_9N	Trimethylamine		-114 e	-97 e	-75.0	-43.8	2.6	1,5
C_3H_9NO	1-Amino-2-propanol			18 e	53.2	98.2	157.9	5
$C_3H_9O_4P$	Trimethyl phosphate	-31 e	-7 e	23.6	62.8	116.0	192.0	5
C_3H_9P	Trimethylphosphine			-81 e	-53 e	-15.0	37.1	5
C_3H_9Sb	Trimethylstibine			-56 e	-23.8	19 e	80 e	5
$C_3H_{10}N_2$	1,2-Propanediamine		-35.4	-12.0	18.8	61 e	119 e	5
C_3N_2O	Carbonyl dicyanide				-21.7	15.3	65.2	5
C_4Cl_6	Hexachloro-1,3-butadiene	-1 e	22 e	50 e	86.7	137.0	209.7	5
$C_4F_6O_3$	Trifluoroacetic acid anhydride			-63 e	-39 e	-7.1	38.8	5
C_4F_8	Perfluorocyclobutane						-6.2	1
C_4F_{10}	Perfluorobutane		-122 e	-105 e	-82 e	-49.8	-2.5	1,5
$C_4H_2Cl_2O_2$	*trans*-2-Butenedioyl dichloride			8.0	45.6	94.3	159.8	5
$C_4H_2Cl_2S$	2,5-Dichlorothiophene			-20 e	22 e	81.4	171 e	5

Mol. Form.	Name	Temperature in °C for the indicated pressure						Ref.
		1 Pa	10 Pa	100 Pa	1 kPa	10 kPa	100 kPa	
$C_4H_2O_3$	Maleic anhydride				73.7	127.9	201.7	5
C_4H_3ClS	2-Chlorothiophene		-62 e	-35 e	2 e	51.8	123 e	5
C_4H_3IS	2-Iodothiophene			-25 e	23 e	94.9	181.0	5
C_4H_4	1-Buten-3-yne			-96.1	-73.4	-41.8	4.9	5
$C_4H_4N_2$	Succinonitrile	24.8 s					266.0	5
C_4H_4O	Furan			-78 e	-54 e	-20 e	31.0	1
$C_4H_4O_2$	Diketene				19.3	63.3	126 e	5
$C_4H_4O_3$	Succinic anhydride				121 e	180.8	260.8	5
$C_4H_4O_4$	Fumaric acid	123.9 s	150 s	180 s				5
C_4H_4S	Thiophene				-17 e	23.7	83.7	5
C_4H_5Cl	2-Chloro-1,3-butadiene	-113 e	-95 e	-71 e	-41 e	0.3	59.0	5
C_4H_5ClO	2-Methyl-2-propenoyl chloride		-57 e	-35 e	-5 e	36.4	98.2	5
$C_4H_5Cl_3O_2$	Ethyl trichloroacetate			15.3	51.9	100.1	166.6	5
C_4H_5N	3-Butenenitrile	-67 e	-48 e	-23.1	9.3	53.7	118.4	5
C_4H_5N	Methylacrylonitrile				-12 e	29.0	89.8	5
C_4H_5N	Pyrrole			-8 e	24 e	66.7	129.4	1
$C_4H_5NO_2$	Methyl cyanoacetate	-3 e	19 e	48 e	84 e	134.0	204.6	5
C_4H_5NS	Allyl isothiocyanate	-45 e	-27 e	-3 e	32.1	89 e	198 e	5
C_4H_5NS	4-Methylthiazole						67.0	5
C_4H_6	1,2-Butadiene	-132 e	-117 e	-98 e	-72.8	-38.9	10.5	5
C_4H_6	1,3-Butadiene*			-106 e	-83 e	-51.9	-4.7	1
C_4H_6	1-Butyne	-125 e	-111 e	-94 e	-71.2	-39.4	7.8	1
C_4H_6	2-Butyne		-89.2 s	-73.8 s	-53.5 s	-23.9	26.6	5
$C_4H_6Cl_2O_2$	Ethyl dichloroacetate			2.6	40.1	89.1	156.3	5
C_4H_6O	Divinyl ether		-99 e	-80 e	-56 e	-22.1	28.0	5
C_4H_6O	*trans*-2-Butenal	-74 e	-56 e	-33 e	-3 e	39.7	102.4	5
C_4H_6O	3-Buten-2-one					21 e	81.0	5
C_4H_6O	Cyclobutanone			-34 e	-4 e	37.1	97 e	5
$C_4H_6O_2$	*cis*-Crotonic acid			30 e	63 e	106.7	168.9	5
$C_4H_6O_2$	*trans*-Crotonic acid				74 e	120.8	184.9	5
$C_4H_6O_2$	3-Butenoic acid	-19 e	2 e	27 e	61 e	105.6	168.6	5
$C_4H_6O_2$	Methacrylic acid			22 e	56 e	99.9	161.5	5
$C_4H_6O_2$	Vinyl acetate	-88 e	-71 e	-50 e	-22 e	16.2	72.2	1
$C_4H_6O_2$	Methyl acrylate		-71 e	-48 e	-18 e	22 e	79.9	5
$C_4H_6O_2$	2,3-Butanedione					30.7	84.8	5
$C_4H_6O_2$	gamma-Butyrolactone		-17 e	24 e	72 e	130.2	203 e	5
$C_4H_6O_3$	Acetic anhydride	-44 e	-25 e	-1 e	31 e	75.1	139.7	1
$C_4H_6O_3$	Propylene carbonate	-40 e	-5 e	43 e	112 e	220 e	410 e	5
$C_4H_6O_4$	Dimethyl oxalate				50.5	98.1	163.0	5
C_4H_7Br	*trans*-1-Bromo-1-butene	-87 e	-68 e	-43.3	-11.4	31.9	94.4	5
C_4H_7Br	2-Bromo-1-butene	-87 e	-70 e	-48 e	-20 e	20.7	80.6	5
C_4H_7Br	*cis*-2-Bromo-2-butene	-90 e	-72 e	-49.0	-18.5	23.5	85.2	5
C_4H_7Br	*trans*-2-Bromo-2-butene	-86 e	-67 e	-43.4	-12.0	31.0	93.5	5
$C_4H_7Br_3$	1,2,3-Tribromobutane	0 e	23 e	53 e	91 e	143.7	219.5	5
$C_4H_7Br_3$	1,2,4-Tribromobutane	-3 e	20 e	49 e	87 e	139.4	214.5	5
C_4H_7Cl	3-Chloro-1-butene			-64 e	-36 e	4 e	63.6	5
C_4H_7Cl	*cis*-2-Chloro-2-butene	-100 e	-83 e	-62 e	-34 e	6 e	66.4	5
C_4H_7Cl	*trans*-2-Chloro-2-butene	-102 e	-86 e	-65 e	-37 e	3 e	62.2	5
C_4H_7Cl	3-Chloro-2-methylpropene		-75 e	-54 e	-25 e	13.8	71.5	5
$C_4H_7ClO_2$	Ethyl chloroacetate			-2.6	32.6	79.1	143.8	5
C_4H_7N	Butanenitrile	-67 e	-48 e	-24 e	8 e	52.3	117.2	1
C_4H_8	1-Butene	-139.0	-125.2	-107.8	-85.3	-53.7	-6.6	1,5
C_4H_8	*cis*-2-Butene	-131.2	-117.4	-99.8	-76.7	-44.8	3.4	1,5
C_4H_8	*trans*-2-Butene			-102 e	-80 e	-47.6	0.6	1
C_4H_8	Isobutene	-139.1	-125.5	-108.2	-85.5	-54.5	-7.3	1,5
C_4H_8	Cyclobutane				-71.8	-38.1	12.1	5
C_4H_8	Methylcyclopropane	-130 e	-116 e	-99.3	-76.3	-44.2	4.2	5
$C_4H_8Br_2$	1,2-Dibromobutane	-54 e	-30 e	0.4	39.6	92.1	166.1	5
$C_4H_8Br_2$	1,4-Dibromobutane	-13 e	9 e	37 e	74 e	124.0	196.5	5
$C_4H_8Cl_2$	1,1-Dichlorobutane			-25 e	6 e	49.3	113.4	5
$C_4H_8Cl_2$	1,2-Dichlorobutane			-28.4	5.8	53.1	123.1	5
$C_4H_8Cl_2$	1,4-Dichlorobutane		-26 e	0 e	35 e	82.4	153.4	5
$C_4H_8Cl_2$	2,2-Dichlorobutane		-58 e	-35 e	-5 e	37.8	102.1	5
$C_4H_8Cl_2O$	Bis(2-chloroethyl) ether	-32 e	-9 e	19.8	56.9	106.9	177.9	5
C_4H_8O	Ethyl vinyl ether		-102 e	-81 e	-53.1	-16.5	34.7	5
C_4H_8O	1,2-Epoxybutane	-135 e	-114 e	-87 e	-53 e	-5.5	62.1	5

Mol. Form.	Name	Temperature in °C for the indicated pressure 1 Pa	10 Pa	100 Pa	1 kPa	10 kPa	100 kPa	Ref.
C_4H_8O	Butanal	-88 e	-72 e	-50 e	-22 e	16.6	74.5	1,5
C_4H_8O	Isobutanal			-56 e	-29 e	8 e	63.8	1
C_4H_8O	2-Butanone	-85 e	-68 e	-46 e	-18.1	21.2	79.2	1
C_4H_8O	Tetrahydrofuran	-94 e	-78 e	-57.3	-29.8	9 e	65.6	1
$C_4H_8O_2$	Butanoic acid			12.9	52.2	101.4	163.3	1,5
$C_4H_8O_2$	2-Methylpropanoic acid	-30.1	-8.2	18.1	50.5	92.9	154.0	5
$C_4H_8O_2$	Propyl formate	-78 e	-62 e	-42 e	-15.1	23.0	80.4	1,5
$C_4H_8O_2$	Isopropyl formate	-80 e	-65 e	-47 e	-22.2	13.2	67.7	5
$C_4H_8O_2$	Ethyl acetate	-83 e	-66 e	-45 e	-18 e	20.4	76.8	1
$C_4H_8O_2$	Methyl propanoate	-80 e	-64 e	-43 e	-15.8	22.2	79.0	1
$C_4H_8O_2$	*cis*-2-Butene-1,4-diol	17 e	44 e	77 e	117.4	168.5	234.9	5
$C_4H_8O_2$	1,3-Dioxane			-37 e	-3 e	43.4	106.0	5
$C_4H_8O_2$	1,4-Dioxane					39.6	101.0	1
$C_4H_8O_2S$	Sulfolane		49 e	87 e	135 e	198.0	283.5	5
C_4H_8S	Tetrahydrothiophene	-66 e	-47 e	-23 e	9.4	54.1	120.5	1
C_4H_9Br	1-Bromobutane	-68.4	-53.9	-34.1	-5.4	37.6	101.1	1,5
C_4H_9Br	2-Bromobutane	-86 e	-68 e	-46 e	-16 e	26.6	90.7	5
C_4H_9Br	1-Bromo-2-methylpropane	-85 e	-68 e	-46 e	-16 e	26.8	91.1	5
C_4H_9Br	2-Bromo-2-methylpropane					11.7	72.4	1,5
C_4H_9Cl	1-Chlorobutane	-87 e	-71 e	-49 e	-21 e	18.4	78.1	1
C_4H_9Cl	2-Chlorobutane	-96 e	-80 e	-59 e	-31.0	8.5	67.9	1
C_4H_9Cl	1-Chloro-2-methylpropane	-94 e	-78 e	-56.6	-28.7	10.2	68.5	5
C_4H_9Cl	2-Chloro-2-methylpropane					-4.2	50.3	5
$C_4H_9Cl_3Si$	Butyltrichlorosilane					77.2	148.4	5
C_4H_9F	1-Fluorobutane	-114 e	-99 e	-80 e	-55 e	-20.0	32.1	5
C_4H_9F	2-Fluorobutane	-117 e	-103 e	-85 e	-60.7	-26.7	24.7	5
C_4H_9I	1-Iodobutane	-62 e	-43 e	-19 e	14 e	60.5	130.0	5
C_4H_9I	2-Iodobutane	-70 e	-51 e	-27 e	5 e	50 e	119.5	5
C_4H_9I	1-Iodo-2-methylpropane		-47 e	-21.4	12.0	56.8	120.0	5
C_4H_9I	2-Iodo-2-methylpropane	-75.1 s	-58.8 s	-39.5 s	-5.2	41 e	100.0	5
C_4H_9N	Pyrrolidine		-59 e	-38 e	-10 e	28.5	86.2	1
C_4H_9NO	*N*-Methylpropanamide				81.1	105 e		5
C_4H_9NO	*N,N*-Dimethylacetamide	-8 e	8 e	28.0	56.4	98.2	165.7	1
C_4H_9NO	2-Butanone oxime		-18 e	7 e	38.9	81.9	142.9	5
C_4H_9NO	Morpholine				21 e	64.5	128.5	1
$C_4H_9NO_3$	Isobutyl nitrate			-18 e	15.1	59.2	123.0	5
C_4H_{10}	Butane*	-134.3	-121.0	-103.9	-81.1	-49.1	-0.8	1,41
C_4H_{10}	Isobutane*		-129.0	-113.0	-90.9	-59.4	-12.0	1,41
$C_4H_{10}O$	1-Butanol	-37 e	-20 e	0 e	28 e	64 e	117.4	1
$C_4H_{10}O$	2-Butanol	-50 e	-34 e	-14 e	12.6	48.2	99.2	1,5
$C_4H_{10}O$	2-Methyl-1-propanol	-39 e	-24 e	-5 e	20.9	56.0	107.6	1,5
$C_4H_{10}O$	2-Methyl-2-propanol					34.4	82.1	1,5
$C_4H_{10}O$	Diethyl ether	-111 e	-96 e	-77 e	-52.6	-17.8	34.1	1
$C_4H_{10}O$	Methyl propyl ether				-40 e	-11.3	38.7	5
$C_4H_{10}O$	Isopropyl methyl ether				-56 e	-21.2	30.4	5
$C_4H_{10}O_2$	1,3-Butanediol	-4 e	23 e	55 e	94 e	142.9	206.1	5
$C_4H_{10}O_2$	1,4-Butanediol		45 e	77 e	116 e	164.7	227.6	5
$C_4H_{10}O_2$	2,3-Butanediol		15 e	43 e	77 e	121.2	180.3	5
$C_4H_{10}O_2$	Ethylene glycol monoethyl ether	-49 e	-29 e	-3 e	30 e	73.6	135.3	1
$C_4H_{10}O_2$	Ethylene glycol dimethyl ether			-44 e	-15 e	25.2	85.2	1
$C_4H_{10}O_2$	Dimethylacetal	-89 e	-74 e	-55 e	-29 e	7.7	64.1	5
$C_4H_{10}O_2$	Diethylperoxide				-39 e	3.6	65.0	5
$C_4H_{10}O_2S$	Bis(2-hydroxyethyl) sulfide			31 e	114.2		282.0	5
$C_4H_{10}O_3$	Diethylene glycol	35 e	58 e	86 e	123 e	173.6	245.2	1
$C_4H_{10}O_4S$	Diethyl sulfate		3 e	36 e	79 e	134 e	208.3	5
$C_4H_{10}S$	1-Butanethiol	-77 e	-59 e	-37 e	-6 e	35.4	98.0	5
$C_4H_{10}S$	2-Butanethiol	-86 e	-69 e	-47 e	-17 e	23.4	84.5	5
$C_4H_{10}S$	2-Methyl-1-propanethiol		-66 e	-44 e	-15 e	26.5	88.1	5
$C_4H_{10}S$	2-Methyl-2-propanethiol					5.8	63.8	5
$C_4H_{10}S$	Diethyl sulfide	-80 e	-62 e	-40 e	-10.8	30.3	91.7	1
$C_4H_{10}S$	Methyl propyl sulfide	-78 e	-61 e	-38 e	-8 e	33.1	95.1	5
$C_4H_{10}S$	Isopropyl methyl sulfide	-85 e	-68 e	-46 e	-17 e	23.4	84.3	5
$C_4H_{10}S_2$	1,4-Butanedithiol	-17 e	5 e	32 e	69.1	119.9	195.1	5
$C_4H_{10}S_2$	Diethyl disulfide	-46 e	-26 e	0 e	35 e	82.4	153.5	5

Mol. Form.	Name	Temperature in °C for the indicated pressure 1 Pa	10 Pa	100 Pa	1 kPa	10 kPa	100 kPa	Ref.
$C_4H_{11}N$	Butylamine			-46 e	-18.1	20.0	75.9	5
$C_4H_{11}N$	*sec*-Butylamine			-55 e	-29.1	7.5	62.3	5
$C_4H_{11}N$	*tert*-Butylamine			-67 e	-42.4	-8.1	43.7	5
$C_4H_{11}N$	Isobutylamine	-85 e	-70 e	-50 e	-24.5	12.0	67.3	5
$C_4H_{11}N$	Diethylamine			-46 e	-26 e	5 e	55.2	1
$C_4H_{11}NO$	*N,N*-Dimethylethanolamine	-52 e	-31 e	-6 e	27 e	70.9	133 e	5
$C_4H_{11}NO_2$	Diethanolamine	53 e	77 e	107 e	146 e	197.3	268 e	5
$C_4H_{12}BN$	(Dimethylamino)dimethylborane		-81 e	-60.1	-31.9	7.0	64.2	5
$C_4H_{12}Cl_2OSi_2$	1,3-Dichloro-1,1,3,3-tetramethyldisiloxane		-33 e	-9 e	23.8	69.1	136.5	5
$C_4H_{12}O_4Si$	Tetramethyl silicate				14.4	59.3	119.7	5
$C_4H_{12}Si$	Tetramethylsilane			-83 e	-59 e	-25 e	26.7	5
$C_4H_{12}Sn$	Tetramethylstannane			-55.0	-25.6	16.6	77.7	5
$C_4H_{13}N_3$	Diethylenetriamine	-10 e	13 e	43 e	80 e	129.6	198 e	5
C_4NiO_4	Nickel carbonyl					-12	42	4
C_5F_{12}	Perfluoropentane				-54.7	-20.9	28.6	5
C_5FeO_5	Iron pentacarbonyl				0	44	105	4
C_5H_4ClN	2-Chloropyridine			7.4	45.8	97.3	169.9	5
$C_5H_4O_2$	Furfural	-26 e	-8 e	16 e	47 e	92.4	161.4	1
C_5H_5N	Pyridine			-23 e	8 e	51.0	114.9	1
C_5H_6	1,3-Cyclopentadiene			-77 e	-51 e	-14 e	39.8	5
$C_5H_6N_2$	Pentanedinitrile	24.1	52 e	85 e	126 e	178 e	245 e	5
C_5H_6O	2-Methylfuran			-66 e	-35 e	6 e	64.5	1
$C_5H_6O_2$	Furfuryl alcohol	-30 e	-5 e	25 e	62.6	109.3	169.7	5
C_5H_6S	2-Methylthiophene		-58 e	-32 e	2 e	47.9	112.2	1
C_5H_6S	3-Methylthiophene		-53 e	-28 e	6 e	50.6	115.1	1
C_5H_7N	1-Methylpyrrole				8 e	49.9	112.3	5
$C_5H_7NO_2$	Ethyl cyanoacetate	16 e	39 e	67.0	102.1	146.7	205.6	5
C_5H_8	1,2-Pentadiene	-109 e	-93 e	-73 e	-46.1	-9.7	44.5	5
C_5H_8	*cis*-1,3-Pentadiene	-109 e	-93 e	-73 e	-47.0	-10.5	43.7	1,5
C_5H_8	*trans*-1,3-Pentadiene			-75 e	-49.0	-13 e	42 e	1
C_5H_8	1,4-Pentadiene	-120 e	-105 e	-86 e	-60.9	-26.2	25.6	5
C_5H_8	2,3-Pentadiene	-106 e	-90 e	-70 e	-42.9	-6.3	47.9	5
C_5H_8	3-Methyl-1,2-butadiene	-111 e	-95 e	-75 e	-49.2	-13.1	40.4	5
C_5H_8	2-Methyl-1,3-butadiene	-115 e	-100 e	-81 e	-55.4	-19.7	33.7	1,5
C_5H_8	1-Pentyne			-75 e	-49.1	-13.5	39.9	5
C_5H_8	2-Pentyne	-100 e	-85 e	-65 e	-37.9	-0.5	55.7	5
C_5H_8	3-Methyl-1-butyne			-82 e	-57.5	-23.1	28.6	5
C_5H_8	Cyclopentene	-109 e	-94 e	-74 e	-48 e	-11.1	43.8	5
C_5H_8	Spiropentane	-110 e	-95 e	-76 e	-51 e	-15 e	38.6	5
C_5H_8O	3-Methyl-3-buten-2-one			-35 e	-5 e	36.0	97.3	5
C_5H_8O	Cyclopropyl methyl ketone		-57 e	-31 e	3 e	49 e	112 e	5
C_5H_8O	Cyclopentanone		-39 e	-14 e	19 e	64 e	130.3	1
C_5H_8O	3,4-Dihydro-2H-pyran				-22 e	22.0	84.9	5
$C_5H_8O_2$	4-Pentenoic acid	0 e	19 e	44 e	77 e	122.0	187.5	5
$C_5H_8O_2$	Vinyl propanoate					31.2	94 e	5
$C_5H_8O_2$	Ethyl acrylate		-55 e	-32.7	-2.8	38.5	99.2	5
$C_5H_8O_2$	Methyl methacrylate			-31 e	-1 e	39.7	100.0	1
$C_5H_8O_2$	2,4-Pentanedione			-5 e	24.7	67.8	137.4	1
$C_5H_8O_2$	Tetrahydro-2*H*-pyran-2-one		5 e	35.1	74.4	128.3	207.0	5
$C_5H_8O_3$	Methyl acetoacetate				50.1	101.1	171.3	5
$C_5H_8O_4$	Glutaric acid		121 e	153.2	191.9	240.3	302.5	5
$C_5H_8O_4$	Dimethyl malonate	-22 e	1 e	30.0	66.7	114.7	180.2	5
$C_5H_9ClO_2$	Ethyl 2-chloropropanoate			1.4	36.4	82.5	146.0	5
$C_5H_9ClO_2$	Isopropyl chloroacetate			-2 e	35.0	83.3	148.1	5
C_5H_9N	Pentanenitrile	-54 e	-34 e	-8 e	26 e	72.2	140.9	1
C_5H_9N	2,2-Dimethylpropanenitrile					41.1	104.8	5
C_5H_9NO	*N*-Methyl-2-pyrrolidone	1 e	24 e	53.1	92.3	147.2	229 e	5
C_5H_{10}	1-Pentene	-118.9	-103.4	-84.0	-58.8	-23.3	29.6	1,5
C_5H_{10}	*cis*-2-Pentene	-113.8	-98.1	-78.4	-52.7	-16.8	36.6	1,5
C_5H_{10}	*trans*-2-Pentene	-114.5	-98.9	-79.1	-53.3	-17.5	36.0	1,5
C_5H_{10}	2-Methyl-1-butene	-117.7	-102.2	-82.7	-57.2	-21.9	30.8	1,5
C_5H_{10}	3-Methyl-1-butene	-125.0	-110.1	-91.2	-66.7	-32.1	19.7	1,5
C_5H_{10}	2-Methyl-2-butene	-113.4	-97.6	-77.7	-51.6	-15.8	38.2	1,5

Mol. Form.	Name	1 Pa	10 Pa	100 Pa	1 kPa	10 kPa	100 kPa	Ref.
		Temperature in °C for the indicated pressure						
C_5H_{10}	Cyclopentane			-77.0	-45.4	-7.1	48.8	5
C_5H_{10}	Ethylcyclopropane	-118 e	-102 e	-83 e	-57 e	-20 e	35.5	5
C_5H_{10}	*cis*-1,2-Dimethylcyclopropane	-118 e	-103 e	-83 e	-57 e	-20 e	36.6	5
C_5H_{10}	*trans*-1,2-Dimethylcyclopropane	-122 e	-108 e	-89 e	-63 e	-27 e	27.8	5
$C_5H_{10}Br_2$	1,5-Dibromopentane	1 e	25 e	54 e	93 e	145.6	221.8	5
$C_5H_{10}Cl_2$	1,2-Dichloropentane				30 e	77.4	147.8	5
$C_5H_{10}Cl_2$	1,5-Dichloropentane	-31 e	-10 e	17 e	54 e	104.1	178.9	5
$C_5H_{10}N_2$	3-(Dimethylamino)-propanenitrile				51.1	101.8	171.4	5
$C_5H_{10}O$	Cyclopentanol		-13 e	11.5	42.2	82.5	140.0	5
$C_5H_{10}O$	Allyl ethyl ether			-56 e	-28.7	9.8	67.2	5
$C_5H_{10}O$	Pentanal	-71 e	-53 e	-31 e	-1 e	40.8	102.6	5
$C_5H_{10}O$	2-Pentanone				-1 e	40.3	101.9	1,5
$C_5H_{10}O$	3-Pentanone			-31 e	-1 e	40 e	101.6	1
$C_5H_{10}O$	3-Methyl-2-butanone	-69 e	-54 e	-34 e	-6.9	32.2	94.0	1,5
$C_5H_{10}O$	Tetrahydropyran				-15 e	26.0	88 e	5
$C_5H_{10}O$	2-Methyltetrahydrofuran				-20 e	19.7	79.8	5
$C_5H_{10}O_2$	Pentanoic acid	-7.4	15.3	42.7	76.3	122.1	185.7	5
$C_5H_{10}O_2$	2-Methylbutanoic acid	-10 e	10 e	36 e	69 e	112.8	175.2	5
$C_5H_{10}O_2$	3-Methylbutanoic acid	-15.8	4 e	30.0	64.7	110.6	176.1	5
$C_5H_{10}O_2$	Butyl formate			-29 e	2 e	44.4	105.7	5
$C_5H_{10}O_2$	Isobutyl formate	-69 e	-53 e	-31 e	-3 e	37.4	97.6	5
$C_5H_{10}O_2$	Propyl acetate	-69 e	-51 e	-29 e	0 e	40.9	101.2	1
$C_5H_{10}O_2$	Isopropyl acetate		-61 e	-40 e	-11 e	29.8	88.2	5
$C_5H_{10}O_2$	Ethyl propanoate	-69 e	-52 e	-30 e	-1 e	38.9	98.7	1
$C_5H_{10}O_2$	Methyl butanoate	-68 e	-50 e	-28 e	0.9	41.7	102.3	5
$C_5H_{10}O_2$	Methyl isobutanoate	-83 e	-65 e	-41 e	-11 e	31 e	92.1	5
$C_5H_{10}O_2$	Tetrahydrofurfuryl alcohol	-40 e	-16 e	15 e	55 e	106 e	176.8	5
$C_5H_{10}O_3$	Diethyl carbonate		-42 e	-17 e	17 e	61.6	125.9	5
$C_5H_{10}O_3$	Ethylene glycol monomethyl ether acetate	-47 e	-26 e	0 e	34 e	79.4	144.1	5
$C_5H_{10}S$	Thiacyclohexane				24 e	71.1	141.2	5
$C_5H_{10}S$	Cyclopentanethiol				18 e	64 e	131.7	5
$C_5H_{11}Br$	1-Bromopentane	-60 e	-41 e	-16 e	16 e	61.5	129.1	5
$C_5H_{11}Br$	2-Bromopentane	-69 e	-51 e	-27 e	5 e	49.7	116.9	5
$C_5H_{11}Br$	3-Bromopentane	-68 e	-50 e	-26 e	6 e	50.8	118.1	5
$C_5H_{11}Br$	1-Bromo-3-methylbutane	-67 e	-49 e	-25 e	8 e	52.4	119.9	5
$C_5H_{11}Cl$	1-Chloropentane	-73 e	-55 e	-32 e	-1 e	42.5	107.9	5
$C_5H_{11}Cl$	2-Chloropentane	-80 e	-62 e	-39 e	-9 e	33.2	96.1	5
$C_5H_{11}Cl$	3-Chloropentane	-77 e	-60 e	-37 e	-7 e	34.9	97.3	5
$C_5H_{11}Cl$	2-Chloro-2-methylbutane			-52 e	-21 e	21.8	85.2	5
$C_5H_{11}Cl$	1-Chloro-2,2-dimethyl-propane				-17 e	23.5	83.9	5
$C_5H_{11}F$	1-Fluoropentane	-97 e	-80 e	-60 e	-32 e	5.7	62.4	5
$C_5H_{11}I$	1-Iodopentane	-47 e	-27 e	-1 e	34 e	83.0	156.5	5
$C_5H_{11}I$	1-Iodo-3-methylbutane		-34 e	-6.6	28.8	77.3	147.8	5
$C_5H_{11}N$	Cyclopentylamine	-66 e	-48 e	-26 e	4 e	45.8	108 e	5
$C_5H_{11}N$	Piperidine				2 e	43.3	105.8	5
$C_5H_{11}N$	*N*-Methylpyrrolidine				-23 e	18.5	78 e	5
$C_5H_{11}NO_3$	3-Methylbutyl nitrate		-26 e	1.0	35.5	81.7	147.0	5
C_5H_{12}	Pentane**	-115.5	-99.8	-80.0	-54.0	-18.1	35.7	16
C_5H_{12}	Isopentane	-119 e	-105 e	-86 e	-61 e	-26 e	27.5	1
C_5H_{12}	Neopentane*		-107.5 s	-90.8 s	-68.8 s	-38.5 s	9.2	1,5
$C_5H_{12}N_2O$	Tetramethylurea			20.7	58.0	106.7	179.5	5
$C_5H_{12}O$	1-Pentanol	-27 e	-10 e	12 e	41 e	79.8	137.4	5
$C_5H_{12}O$	2-Pentanol	-35 e	-19 e	1 e	28.0	64.9	118.7	1
$C_5H_{12}O$	3-Pentanol	-41 e	-25 e	-4 e	24 e	61.1	114.9	5
$C_5H_{12}O$	2-Methyl-1-butanol	-27 e	-11 e	9 e	36.2	73.4	128.3	1
$C_5H_{12}O$	3-Methyl-1-butanol	-22 e	-7 e	13 e	39.1	75.7	130.1	5
$C_5H_{12}O$	2-Methyl-2-butanol			-5 e	17.7	50.6	101.7	1,5
$C_5H_{12}O$	3-Methyl-2-butanol			-3 e	22.7	58.2	111.1	5
$C_5H_{12}O$	2,2-Dimethyl-1-propanol					59.2	112.7	5
$C_5H_{12}O$	Butyl methyl ether			-54 e	-27 e	12 e	69.8	1
$C_5H_{12}O$	Methyl *tert*-butyl ether			-66 e	-39 e	-2 e	54.8	1
$C_5H_{12}O$	Ethyl propyl ether	-92 e	-77 e	-57 e	-30.5	6.7	63.4	1,5

Mol. Form.	Name	Temperature in °C for the indicated pressure						Ref.
		1 Pa	10 Pa	100 Pa	1 kPa	10 kPa	100 kPa	
$C_5H_{12}O_2$	1,5-Pentanediol	25 e	52 e	85 e	125 e	175.1	238.9	5
$C_5H_{12}O_2$	Ethylene glycol monopropyl ether				40 e	85.6	149.3	5
$C_5H_{12}O_2$	Diethoxymethane		-65 e	-43 e	-14 e	27.3	87.7	5
$C_5H_{12}O_3$	Diethylene glycol monomethyl ether		12 e	40 e	76 e	124.2	193.7	1
$C_5H_{12}S$	1-Pentanethiol	-60 e	-41 e	-17 e	15 e	60 e	126.2	1
$C_5H_{12}S$	2-Pentanethiol	-70 e	-52 e	-28 e	3 e	46.6	111.9	5
$C_5H_{12}S$	3-Pentanethiol	-70 e	-51 e	-28 e	4 e	47.7	113.4	5
$C_5H_{12}S$	2-Methyl-1-butanethiol				8.0	52.3	118.5	5
$C_5H_{12}S$	3-Methyl-1-butanethiol				7.8	51.9	117.9	5
$C_5H_{12}S$	2-Methyl-2-butanethiol				-8.0	34.6	98.7	5
$C_5H_{12}S$	Butyl methyl sulfide		-43 e	-19 e	13 e	57 e	123.0	1
$C_5H_{12}S$	*tert*-Butyl methyl sulfide				-7.8	34.7	98.4	5
$C_5H_{12}S$	Ethyl propyl sulfide	-64 e	-46 e	-23 e	9 e	52.7	118.0	5
$C_5H_{12}S$	Ethyl isopropyl sulfide	-72 e	-54 e	-31 e	0 e	42.7	106.9	5
$C_5H_{13}N$	Pentylamine		-52 e	-29 e	1 e	42.8	104.0	5
C_6BrF_5	Bromopentafluorobenzene			-10 e	23 e	68 e	136.0	5
C_6ClF_5	Chloropentafluorobenzene		-44 e	-21 e	11 e	53.8	117.6	1
$C_6Cl_3F_3$	1,3,5-Trichloro-2,4,6-trifluorobenzene	-19 e	4 e	32 e	70 e	121.7	197.9	1
C_6F_6	Hexafluorobenzene		-56.9 s	-36 s	-11.5 s	22.6	79.9	1,5
C_6F_{12}	Perfluorocyclohexane				-46.2 s	-7.6 s	48.9 s	5
C_6F_{14}	Perfluorohexane		-75 e	-57 e	-32 e	2.8	56.8	5
C_6F_{14}	Perfluoro-2-methylpentane				-33 e	2.9	57.1	5
C_6F_{14}	Perfluoro-3-methylpentane	-95 e	-80 e	-60 e	-34 e	2.8	57.9	5
C_6F_{14}	Perfluoro-2,3-dimethylbutane					4.3	59.3	5
C_6HF_5	Pentafluorobenzene			-41 e	-13 e	27 e	85.3	5
C_6HF_5O	Pentafluorophenol				39 e	82 e	145.2	5
$C_6H_2F_4$	1,2,3,4-Tetrafluorobenzene			-36 e	-7 e	33.8	94.0	1
$C_6H_2F_4$	1,2,3,5-Tetrafluorobenzene			-43 e	-14 e	25.5	84.1	1
$C_6H_2F_4$	1,2,4,5-Tetrafluorobenzene					30.7	89.9	1
$C_6H_3Cl_3O$	2,4,6-Trichlorophenol			71.8	114.0	169.5	245.7	5
$C_6H_3F_3$	1,3,5-Trifluorobenzene					18.2	75.0	5
$C_6H_4Br_2$	*m*-Dibromobenzene	-7 e	16 e	44 e	83 e	137.0	218.2	5
$C_6H_4ClNO_2$	1-Chloro-4-nitrobenzene	15.4 s	35.8 s		97 e	156.0	238 e	5
$C_6H_4Cl_2$	*o*-Dichlorobenzene		-13 e	16.3	53.9	104.6	180.0	1,5
$C_6H_4Cl_2$	*m*-Dichlorobenzene		-22 e	8.0	46.7	97.8	172.5	1,5
$C_6H_4Cl_2$	*p*-Dichlorobenzene	-45.5 s	-21.8 s	8 s	46.7 s	99.0	173.6	1,5
$C_6H_4O_2$	*p*-Benzoquinone	-4.1 s	17.8 s	43.5 s	74.3 s	111.6 s		5
$C_6H_5AsCl_2$	Dichlorophenylarsine	6.9	35.2	70 e	113 e	170 e	245 e	5
C_6H_5Br	Bromobenzene		-25 e	1 e	34.9	83.1	155.4	1
C_6H_5Cl	Chlorobenzene		-43 e	-17 e	16.8	62.9	131.3	1,5
C_6H_5ClO	*o*-Chlorophenol				45.8	97.9	173.9	5
C_6H_5ClO	*m*-Chlorophenol			39.7	80.2	135.1	213.4	5
C_6H_5ClO	*p*-Chlorophenol			45.0	86.5	142.0	219.9	5
$C_6H_5Cl_3Si$	Trichlorophenylsilane			33 e	70.2	122.6	201 e	5
C_6H_5F	Fluorobenzene				-16.9	24.2	84.4	1
C_6H_5I	Iodobenzene	-30 e	-7 e	20.9	58.5	110.6	187.8	1
$C_6H_5NO_2$	Nitrobenzene		10 e	40 e	78 e	132 e	210.3	1
$C_6H_5NO_3$	*p*-Nitrophenol	72.6 s	97.4 s					5
C_6H_6	1,5-Hexadien-3-yne	-82 e	-66 e	-44.3	-16.0	23.7	83.6	5
C_6H_6	Benzene**			-40 s	-15.1 s	20.0	79.7	1,5
C_6H_6ClN	*o*-Chloroaniline		10 e	39.0	75.2	131.4	208.3	5
C_6H_6ClN	*m*-Chloroaniline	-5 e	19.7	49.4	94.2	162 e	1069 e	5
$C_6H_6N_2O_2$	*p*-Nitroaniline	87.8 s			192.0	252.6	331.2	5
C_6H_6O	Phenol	-9.7 s	9.6 s	34.1 s	68.9	113.7	181.4	1,5
$C_6H_6O_3$	1,2,3-Benzenetriol				162.0	222.8	308.3	5
C_6H_6S	Benzenethiol		-15 e	12 e	47 e	96.0	168.6	5
C_6H_7N	Aniline		-2.5	26.7	63.5	112.5	183.5	1,5
C_6H_7N	2-Methylpyridine	-56.5	-37.8	-13.9	18.3	62.9	129.0	1,5
C_6H_7N	3-Methylpyridine			-5 e	28.8	75.2	143.7	1
C_6H_7N	4-Methylpyridine	-58.2 s	-43.1 s	-3.9 s	29.6	76.1	144.9	1,5
C_6H_8	*cis*-1,3,5-Hexatriene					21 e	78 e	5
C_6H_8	1,3-Cyclohexadiene	-88 e	-71 e	-50 e	-21 e	19 e	79.9	5
C_6H_8	1,4-Cyclohexadiene				-15 e	27.3	85.0	5
$C_6H_8N_2$	Adiponitrile	30 e	61 e	100 e	148.6	211.8	297 e	5

Mol. Form.	Name	Temperature in °C for the indicated pressure						Ref.
		1 Pa	10 Pa	100 Pa	1 kPa	10 kPa	100 kPa	
$C_6H_8N_2$	*m*-Phenylenediamine			94.5	140.2	200.8	285.0	5
$C_6H_8N_2$	Phenylhydrazine		38 e	69 e	109 e	163.9	242.5	5
$C_6H_8O_4$	Dimethyl maleate		5 e	36 e	76 e	127.3	197 e	5
C_6H_8S	2,5-Dimethylthiophene		-43 e	-16 e	20 e	67.5	134.8	5
C_6H_{10}	*trans*-1,3-Hexadiene	-86 e	-70 e	-51 e	-24 e	14 e	72 e	5
C_6H_{10}	*trans*-1,4-Hexadiene	-98 e	-81 e	-60 e	-33 e	7 e	65 e	5
C_6H_{10}	1,5-Hexadiene	-99 e	-84 e	-64 e	-37 e	0.9	59.2	5
C_6H_{10}	*cis,cis*-2,4-Hexadiene					18 e	79.6	5
C_6H_{10}	*trans,cis*-2,4-Hexadiene	-89 e	-73 e	-52 e	-23 e	18 e	79.6	5
C_6H_{10}	*trans,trans*-2,4-Hexadiene				-23 e	18 e	79.6	5
C_6H_{10}	*trans*-2-Methyl-1,3-pentadiene	-92 e	-75 e	-54 e	-26 e	14 e	75.6	5
C_6H_{10}	2,3-Dimethyl-1,3-butadiene			-59 e	-30 e	9.7	68.1	5
C_6H_{10}	1-Hexyne	-91 e	-75 e	-54 e	-26 e	12.8	71.0	5
C_6H_{10}	2-Hexyne	-84 e	-67 e	-46 e	-17 e	23.6	84.1	5
C_6H_{10}	3-Hexyne	-86 e	-69 e	-48 e	-19.1	21.0	81.0	1,5
C_6H_{10}	4-Methyl-1-pentyne	-97 e	-81 e	-61 e	-34 e	4.1	60.7	5
C_6H_{10}	4-Methyl-2-pentyne	-91 e	-74 e	-54 e	-26 e	13.8	72.7	5
C_6H_{10}	Cyclohexene	-87 e	-70 e	-49 e	-19 e	21 e	82.6	1
$C_6H_{10}Cl_2$	1,1-Dichlorocyclohexane	-39 e	-19 e	8 e	43 e	93.5	170.5	5
$C_6H_{10}Cl_2$	*cis*-1,2-Dichlorocyclohexane			27 e	69 e	125.7	206.2	5
$C_6H_{10}O$	4-Methyl-4-penten-2-one	-59 e	-41 e	-17 e	14 e	57.0	121.0	5
$C_6H_{10}O$	Cyclohexanone		-25 e	1 e	36 e	84 e	155.2	1
$C_6H_{10}O$	Mesityl oxide	-56 e	-37 e	-13 e	19 e	63.5	129.3	5
$C_6H_{10}O_2$	Vinyl butanoate					53 e	114.5	5
$C_6H_{10}O_2$	Ethyl methacrylate				8 e	53.2	116.8	5
$C_6H_{10}O_2$	Allyl glycidyl ether				40.1	85.7	152.8	5
$C_6H_{10}O_3$	Ethyl acetoacetate	-25 e	-3 e	25.7	62.3	111.3	180.2	5
$C_6H_{10}O_3$	Propanoic anhydride	-32 e	-15 e	6 e	36 e	77.6	142.9	5
$C_6H_{10}O_4$	Diethyl oxalate	-5 e	18 e	44.9	79.4	124.3	185.2	5
$C_6H_{10}O_4$	Dimethyl succinate			30 e	70.4	123.3	195.4	5
$C_6H_{10}O_4$	Ethylene glycol diacetate	-17 e	6 e	35.0	71.9	121.1	190.0	5
$C_6H_{10}S$	Diallylsulfide	-58 e	-38 e	-12.4	21.7	68.8	138.1	5
$C_6H_{11}Cl$	Chlorocyclohexane		-35 e	-9 e	25 e	71.6	142.1	5
$C_6H_{11}N$	Hexanenitrile	-40 e	-19 e	8 e	43 e	91.5	163.2	1,5
$C_6H_{11}N$	4-Methylpentanenitrile		-50 e	-20 e	20 e	75.2	155.2	5
$C_6H_{11}NO$	Caprolactam	36.8 s	58.9 s	86.6 s			270	5
C_6H_{12}	1-Hexene	-99.8	-82.8	-61.4	-33.7	5.2	63.1	1,5
C_6H_{12}	*cis*-2-Hexene	-97 e	-80 e	-58 e	-30 e	9.9	68.5	5
C_6H_{12}	*trans*-2-Hexene	-94 e	-78 e	-57 e	-30 e	9.3	67.5	5
C_6H_{12}	*cis*-3-Hexene	-96 e	-79 e	-59 e	-30.8	7.9	66.0	5
C_6H_{12}	*trans*-3-Hexene	-95 e	-79 e	-58 e	-30.0	8.8	66.7	5
C_6H_{12}	2-Methyl-1-pentene	-98 e	-82 e	-62 e	-34.2	4.1	61.7	5
C_6H_{12}	3-Methyl-1-pentene	-104 e	-88 e	-68 e	-41.5	-3.6	53.8	5
C_6H_{12}	4-Methyl-1-pentene	-105 e	-89 e	-69 e	-41.6	-3.6	53.5	5
C_6H_{12}	2-Methyl-2-pentene	-95 e	-78 e	-58 e	-30 e	9.0	66.9	5
C_6H_{12}	3-Methyl-*cis*-2-pentene	-95 e	-79 e	-58 e	-30 e	8.9	67.3	5
C_6H_{12}	3-Methyl-*trans*-2-pentene	-93 e	-77 e	-55 e	-27.4	11.7	70.0	5
C_6H_{12}	4-Methyl-*cis*-2-pentene	-102 e	-86 e	-66 e	-38.7	-0.9	56.0	5
C_6H_{12}	4-Methyl-*trans*-2-pentene	-100 e	-84 e	-64 e	-36.8	1.2	58.2	5
C_6H_{12}	2-Ethyl-1-butene	-98 e	-81 e	-60 e	-32 e	6.6	64.3	5
C_6H_{12}	2,3-Dimethyl-1-butene	-103 e	-87 e	-67 e	-39.9	-1.9	55.2	5
C_6H_{12}	3,3-Dimethyl-1-butene	-110 e	-95 e	-76 e	-50.8	-14.5	40.8	5
C_6H_{12}	2,3-Dimethyl-2-butene		-75 e	-54 e	-25 e	14 e	72.9	1
C_6H_{12}	Cyclohexane	-85.6 s	-68.9 s	-47.6 s	-19.8 s	19.3	80.4	1,5
C_6H_{12}	Methylcyclopentane	-97 e	-80 e	-58 e	-28.8	11.6	71.4	1,5
C_6H_{12}	Ethylcyclobutane	-99 e	-82 e	-61 e	-32 e	9 e	70.2	5
C_6H_{12}	Isopropylcyclopropane	-104 e	-88 e	-68 e	-40 e	-1 e	57.9	5
C_6H_{12}	1-Ethyl-1-methylcyclopropane	-105 e	-89 e	-69 e	-41 e	-3 e	56.3	5
C_6H_{12}	1,1,2-Trimethylcyclopropane	-109 e	-94 e	-73 e	-46 e	-7 e	52.0	5
$C_6H_{12}Cl_2$	1,2-Dichlorohexane				49 e	98.1	171.7	5
$C_6H_{12}Cl_2O$	2,2′-Dichlorodiisopropyl ether		-1 e	27.3	63.4	112.3	182.1	5
$C_6H_{12}O$	Butyl vinyl ether	-87 e	-67 e	-42 e	-9.3	33.6	93.2	5
$C_6H_{12}O$	Isobutyl vinyl ether	-87 e	-68 e	-44 e	-13 e	26.5	80.7	5
$C_6H_{12}O$	Hexanal	-56 e	-37 e	-13 e	19 e	62.6	127.8	5
$C_6H_{12}O$	2-Hexanone	-43 e	-21 e	4.2	34.5	61.9	127.2	1,5
$C_6H_{12}O$	3-Hexanone		-40 e	-16 e	15 e	58.5	123.1	1
$C_6H_{12}O$	3-Methyl-2-pentanone				8.5	52.7	117.0	5

VAPOR PRESSURE (continued)

Mol. Form.	Name	Temperature in °C for the indicated pressure 1 Pa	10 Pa	100 Pa	1 kPa	10 kPa	100 kPa	Ref.
$C_6H_{12}O$	4-Methyl-2-pentanone	-61 e	-43 e	-21 e	9 e	51.5	116.1	5
$C_6H_{12}O$	2-Methyl-3-pentanone					50.2	113.0	5
$C_6H_{12}O$	3,3-Dimethyl-2-butanone			-30 e	0 e	42.5	105.7	1
$C_6H_{12}O$	Cyclohexanol			34 e	61 e	99.2	160.7	1
$C_6H_{12}O_2$	Hexanoic acid		33 e	59 e	93 e	139.3	204.5	1
$C_6H_{12}O_2$	4-Methylpentanoic acid	36 e	49 e	67.1	92.9	133.6	206.8	5
$C_6H_{12}O_2$	Diethylacetic acid	-9 e	16 e	46 e	83 e	130.7	192.5	5
$C_6H_{12}O_2$	Isopentyl formate	-60 e	-41 e	-17 e	15 e	59.1	124 e	5
$C_6H_{12}O_2$	Butyl acetate	-63 e	-43 e	-19 e	14 e	61.0	125.6	1,5
$C_6H_{12}O_2$	Isobutyl acetate	-63 e	-45 e	-21 e	10 e	53.4	116 e	5
$C_6H_{12}O_2$	Propyl propanoate	-62 e	-42 e	-18 e	14 e	58.3	122.0	5
$C_6H_{12}O_2$	Ethyl butanoate	-49 e	-34 e	-14 e	14.3	55.2	121.1	5
$C_6H_{12}O_2$	Ethyl 2-methylpropanoate	-65 e	-47 e	-24.6	5.4	47.3	109.8	5
$C_6H_{12}O_2$	Methyl pentanoate				19.2	63.7	127.4	5
$C_6H_{12}O_2$	Methyl isopentanoate					53.3	116.3	5
$C_6H_{12}O_2$	Diacetone alcohol	-41 e	-17 e	13 e	50.1	98.5	164 e	5
$C_6H_{12}O_3$	Ethylene glycol monoethyl ether acetate	-25 e	-8 e	14 e	44.6	88.0	155.6	5
$C_6H_{12}O_3$	Paraldehyde				17 e	62.2	124 e	5
$C_6H_{12}S$	Cyclohexanethiol					84.8	158.3	5
$C_6H_{12}S$	*cis*-Tetrahydro-2,5-dimethylthiophene	-53 e	-34 e	-8 e	25 e	72.0	142.1	5
$C_6H_{12}S$	Tetrahydro-3-methyl-2H-thiopyran	-48 e	-27 e	0 e	35 e	84.1	157.5	5
$C_6H_{13}Br$	1-Bromohexane	-45 e	-25 e	2 e	36 e	83.7	154.8	5
$C_6H_{13}Cl$	1-Chlorohexane	-55 e	-36 e	-11 e	21 e	66.7	134.6	5
$C_6H_{13}F$	1-Fluorohexane	-80 e	-62 e	-40 e	-11 e	30.4	91.1	5
$C_6H_{13}I$	1-Iodohexane	-33 e	-11 e	16 e	53 e	104.0	180.8	5
$C_6H_{13}N$	Cyclohexylamine			-9 e	22 e	66.6	133.5	1
C_6H_{14}	Hexane	-96.4 s	-79.2	-57.6	-29.3	9.8	68.3	16
C_6H_{14}	2-Methylpentane	-100 e	-84 e	-64 e	-36 e	2 e	59.9	1
C_6H_{14}	3-Methylpentane	-99 e	-83 e	-62 e	-34.3	4.6	62.9	1
C_6H_{14}	2,2-Dimethylbutane		-90 e	-71.5	-45.5	-7.7	49.4	1
C_6H_{14}	2,3-Dimethylbutane	-103 e	-87 e	-66 e	-39.0	-0.4	57.6	1
$C_6H_{14}O$	1-Hexanol		5 e	28 e	56.8	97.3	157.1	1
$C_6H_{14}O$	2-Hexanol	-28 e	-10 e	12 e	41.4	81.5	139.6	1
$C_6H_{14}O$	3-Hexanol	-43 e	-23 e	1 e	33 e	75.4	135.1	1
$C_6H_{14}O$	2-Methyl-1-pentanol			14 e	45.9	88.3	147.6	5
$C_6H_{14}O$	4-Methyl-1-pentanol			24 e	53 e	92.4	151.4	5
$C_6H_{14}O$	2-Methyl-2-pentanol	-29 e	-15 e	3 e	27.1	63.0	120.9	5
$C_6H_{14}O$	3-Methyl-2-pentanol				36.5	76.1	133.8	5
$C_6H_{14}O$	4-Methyl-2-pentanol	-43 e	-24 e	0 e	30 e	71.9	131.3	5
$C_6H_{14}O$	2-Methyl-3-pentanol				29.8	68.8	126.0	5
$C_6H_{14}O$	3-Methyl-3-pentanol		-23 e	-4 e	22.9	61.1	121.1	5
$C_6H_{14}O$	2-Ethyl-1-butanol		-5 e	17 e	46 e	85.7	146.1	5
$C_6H_{14}O$	3,3-Dimethyl-1-butanol	-37 e	-16 e	9 e	42 e	84.3	142.5	5
$C_6H_{14}O$	2,3-Dimethyl-2-butanol			-5 e	23 e	61.3	118.2	5
$C_6H_{14}O$	Dipropyl ether	-80 e	-63 e	-41 e	-12 e	28.8	89.7	1
$C_6H_{14}O$	Diisopropyl ether		-76 e	-55 e	-28 e	11 e	68.1	1
$C_6H_{14}O$	Butyl ethyl ether	-78 e	-61 e	-39 e	-10 e	31.0	91.9	1
$C_6H_{14}O$	*tert*-Butyl ethyl ether	-90 e	-74 e	-53 e	-24.6	14.4	72.6	5
$C_6H_{14}O_2$	2-Methyl-2,4-pentanediol	-8 e	17 e	48 e	86 e	134.4	197.5	5
$C_6H_{14}O_2$	Ethylene glycol monobutyl ether	-31 e	-8 e	20 e	55 e	103.2	170.2	5
$C_6H_{14}O_2$	1,1-Diethoxyethane	-68 e	-49 e	-26 e	3.7	44.2	101.9	5
$C_6H_{14}O_2$	Ethylene glycol diethyl ether		-59 e	-35.3	-2.8	44.4	118.8	5
$C_6H_{14}O_3$	1,2,6-Hexanetriol	92 e	114.8	146.0	191 e			5
$C_6H_{14}O_3$	Dipropylene glycol				110 e	162.6	231.4	5
$C_6H_{14}O_3$	Diethylene glycol monoethyl ether			40 e	80.3	132.4	201.4	5
$C_6H_{14}O_3$	Diethylene glycol dimethyl ether	-42 e	-20 e	8.3	44.3	92.3	159.4	5
$C_6H_{14}O_3$	Trimethylolpropane	73 e	98 e	128 e	167.8	220.5	295 e	5
$C_6H_{14}O_4$	Triethylene glycol	44 e	74 e	109.0	152.6	207.2	277.9	5
$C_6H_{14}S$	1-Hexanethiol	-45 e	-25 e	1 e	35 e	81.7	152.2	5
$C_6H_{14}S$	2-Hexanethiol	-50 e	-32 e	-8 e	25 e	69.9	138.4	5
$C_6H_{14}S$	Dipropyl sulfide	-50 e	-30 e	-6 e	28 e	73.6	142.4	5

VAPOR PRESSURE (continued)

Mol. Form.	Name	Temperature in °C for the indicated pressure 1 Pa	10 Pa	100 Pa	1 kPa	10 kPa	100 kPa	Ref.
$C_6H_{14}S$	Diisopropyl sulfide	-65 e	-47 e	-23 e	9 e	53.1	119.6	5
$C_6H_{14}S$	Isopropyl propyl sulfide				18.5	63.8	131.6	5
$C_6H_{14}S$	Butyl ethyl sulfide	-49 e	-30 e	-5 e	29 e	74.8	143.8	5
$C_6H_{15}N$	Hexylamine			-10 e	22 e	66.0	130.6	5
$C_6H_{15}N$	Butylethylamine				6.1	47.7	107.0	5
$C_6H_{15}N$	Dipropylamine		-48 e	-25 e	6 e	47.5	108.8	5
$C_6H_{15}N$	Diisopropylamine			-47 e	-17.5	23.5	84.0	5
$C_6H_{15}N$	Triethylamine	-58 e	-45 e	-29 e	-5 e	29.9	88.5	1
$C_6H_{15}NO$	2-Diethylaminoethanol					97 e	160.6	5
$C_6H_{15}NO_3$	Triethanolamine	75 e	108 e	148 e	196 e	256.7	334 e	5
$C_6H_{15}O_4P$	Triethyl phosphate			34	76	132	211	4
$C_6H_{16}N_2$	Hexamethylenediamine				76.0	128.2	199.0	5
$C_6H_{16}O_2Si$	Diethoxydimethylsilane	-62 e	-44 e	-21.2	9.1	51.0	113.0	5
$C_6H_{18}Cl_2O_2Si_3$	1,5-Dichloro-1,1,3,3,5,5-hexamethyltrisiloxane	-29 e	-7 e	22.2	59.7	110.5	183.4	5
$C_6H_{18}OSi_2$	Hexamethyldisiloxane		-56 e	-34 e	-5 e	37.1	100.1	5
C_6MoO_6	Molybdenum hexacarbonyl		17.4 s	42.8 s	73.1 s	109.9 s	155.4 s	5
C_7F_{14}	Perfluoromethylcyclohexane				-21 e	18 e	75.9	1
C_7F_{16}	Perfluoroheptane		-62 e	-41 e	-14 e	24.7	82.1	1
C_7HF_{15}	1H-Pentadecafluoroheptane				-7 e	35.9	96.0	5
$C_7H_3ClF_3NO_2$	1-Chloro-2-nitro-4-(trifluoromethyl)benzene	3 e	26 e	55 e	92.8	145.2	222.0	5
$C_7H_3F_5$	2,3,4,5,6-Pentafluorotoluene			-20 e	11 e	53.6	117.0	5
$C_7H_4ClF_3$	1-Chloro-2-(trifluoromethyl)benzene			1 e	34.5	81.8	151.8	5
$C_7H_4ClF_3$	1-Chloro-3-(trifluoromethyl)benzene	-53 e	-34 e	-9 e	24.2	69.8	137.2	5
$C_7H_4ClF_3$	1-Chloro-4-(trifluoromethyl)benzene			-9 e	24.2	70.4	138.1	5
$C_7H_4Cl_2O$	*o*-Chlorobenzoyl chloride				93 e	149 e	237.0	5
$C_7H_4Cl_2O$	*m*-Chlorobenzoyl chloride				87.8	147 e	225.0	5
$C_7H_4F_3NO_2$	1-Nitro-3-(trifluoromethyl)benzene		11 e	39 e	76.2	127.3	202.2	5
$C_7H_4F_4$	1-Fluoro-4-(trifluoromethyl)benzene			-38 e	-6 e	38.6	102.3	5
C_7H_5BrO	Benzoyl bromide	-15 e	11 e	42.6	83.9	139.5	218.0	5
C_7H_5ClO	Benzoyl chloride			27.5	67.0	120.4	196.7	5
$C_7H_5Cl_3$	(Trichloromethyl)benzene		9 e	40.6	81.5	136.2	213.0	5
$C_7H_5F_3$	(Trifluoromethyl)benzene				-3 e	39 e	101.6	5
C_7H_5N	Benzonitrile		-6 e	23.9	63.1	115.7	190.0	5
C_7H_5NS	Phenyl isothiocyanate				79.4	105 e	117 e	5
$C_7H_6Cl_2$	2,4-Dichlorotoluene		6 e	33 e	68.3	119.5	199.1	5
$C_7H_6Cl_2$	3,4-Dichlorotoluene	-13 e	9 e	38 e	76 e	129.3	208.4	5
$C_7H_6Cl_2$	(Dichloromethyl)benzene			31	72	130	213	4
C_7H_6O	Benzaldehyde		-9 e	19 e	54.6	104.6	178.3	1
$C_7H_6O_2$	Salicylaldehyde		-1 e	29 e	68 e	120.7	196.2	5
C_7H_7Br	*o*-Bromotoluene		-10 e	17 e	54 e	104.8	181.1	5
C_7H_7Br	*m*-Bromotoluene	-34 e	-11 e	19.4	58.1	109.9	183.1	5
C_7H_7Br	*p*-Bromotoluene				57 e	107.8	183.8	5
C_7H_7Br	(Bromomethyl)benzene			25.4	66.8	121.7	198.3	5
C_7H_7Cl	*o*-Chlorotoluene		-24 e	3 e	38 e	86.3	158.7	1,5
C_7H_7Cl	*m*-Chlorotoluene	-41 e	-21 e	6 e	41 e	89 e	161.8	5
C_7H_7Cl	*p*-Chlorotoluene				40 e	88.9	161.5	1,5
C_7H_7Cl	(Chloromethyl)benzene	-34 e	-11 e	17.7	55.4	106.3	178.9	5
C_7H_7ClO	1-Chloro-2-methoxybenzene	-22 e	2 e	33 e	72 e	125.2	201 e	5
C_7H_7F	*o*-Fluorotoluene		-50 e	-26 e	5 e	49.0	113.9	5
C_7H_7F	*m*-Fluorotoluene	-67 e	-48 e	-25 e	7 e	51.0	116.1	5
C_7H_7F	*p*-Fluorotoluene		-48 e	-24 e	7 e	51 e	116.2	5
$C_7H_7NO_2$	*o*-Nitrotoluene	23 e	40 e	62 e	94 e	141.9	221.9	5
$C_7H_7NO_2$	*m*-Nitrotoluene			45 e	89.7	148.7	231.3	5
$C_7H_7NO_3$	2-Nitroanisole	15 e	45 e	82 e	129 e	189.4	271.8	5
C_7H_8	Toluene	-78.1	-57.1	-31.3	1.5	45.2	110.1	5
C_7H_8	Bicyclo[2.2.1]hepta-2,5-diene				-15 e	27.4	91 e	5
$C_7H_8Cl_2Si$	Dichloromethylphenylsilane			32.4	71.8	126.0	205.0	5
C_7H_8O	*o*-Cresol	-6.4 s	12.8 s	40.2	72.3	120.3	190.5	1,5
C_7H_8O	*m*-Cresol	20.8	33.6	52.4	82.6	130.6	201.8	1,5

Mol. Form.	Name	Temperature in °C for the indicated pressure						Ref.
		1 Pa	10 Pa	100 Pa	1 kPa	10 kPa	100 kPa	
C_7H_8O	*p*-Cresol	-0.2 s	20.7 s	52.7	83.1	130.7	201.5	1,5
C_7H_8O	Benzyl alcohol	8 e	28 e	54 e	88 e	134.7	204.9	1
C_7H_8O	Anisole		-21 e	4 e	38 e	84 e	153.2	1,5
C_7H_8S	3-Methylbenzenethiol		0 e	29 e	66 e	117.9	194.6	5
C_7H_9N	Benzylamine			25.6	62.6	112.7	183.9	5
C_7H_9N	*o*-Methylaniline	1.0	18.8	42.6	76.1	125.6	199.9	1,5
C_7H_9N	*m*-Methylaniline	3.8	22.0	46.2	80.1	128.8	202.9	1,5
C_7H_9N	*p*-Methylaniline				77.1	126.2	199.9	5
C_7H_9N	*N*-Methylaniline	-16 e	6 e	34 e	70.3	121.1	195.8	1
C_7H_9N	2-Ethylpyridine	-46 e	-26 e	-1 e	33 e	79.3	149.0	5
C_7H_9N	3-Ethylpyridine	-38 e	-17 e	9 e	44 e	92.7	166.5	5
C_7H_9N	4-Ethylpyridine	-35 e	-15 e	11 e	46 e	94.4	168.6	5
C_7H_9N	2,3-Dimethylpyridine				42 e	89.9	160.6	5
C_7H_9N	2,4-Dimethylpyridine		-25 e	3.7	40.0	87.5	157.9	1,5
C_7H_9N	2,5-Dimethylpyridine			4 e	39 e	86.2	156.6	1
C_7H_9N	2,6-Dimethylpyridine			-3 e	29.9	75.8	143.6	1
C_7H_9N	3,4-Dimethylpyridine		-9 e	19 e	55 e	104.8	178.6	5
C_7H_9N	3,5-Dimethylpyridine			11 e	48 e	98 e	171.5	1
$C_7H_{10}N_2$	Toluene-2,4-diamine			100.4	145.3	202.9	279.5	5
C_7H_{12}	1-Heptyne	-75 e	-57 e	-35 e	-5 e	37.1	99.5	5
C_7H_{12}	2-Heptyne		-51 e	-27 e	4 e	46.9	111.5	5
C_7H_{12}	3-Heptyne	-71 e	-53 e	-31 e	0 e	42.7	106.4	5
C_7H_{12}	5-Methyl-1-hexyne	-80 e	-62 e	-40 e	-11 e	30.1	91.4	5
C_7H_{12}	5-Methyl-2-hexyne	-75 e	-57 e	-34 e	-4 e	38.6	102.0	5
C_7H_{12}	2-Methyl-3-hexyne	-78 e	-61 e	-39 e	-9 e	32.6	94.8	5
C_7H_{12}	4,4-Dimethyl-1-pentyne		-73 e	-52 e	-24 e	15.9	75.6	5
C_7H_{12}	4,4-Dimethyl-2-pentyne		-70 e	-48 e	-19 e	21.4	82.6	5
C_7H_{12}	Bicyclo[4.1.0]heptane					49.9	116.3	5
C_7H_{12}	Cycloheptene			-30.0	3.4	47.5	108 e	5
C_7H_{12}	1-Methylbicyclo(3,1,0)hexane					29.8	92.6	5
C_7H_{12}	Methylenecyclohexane	-76 e	-58 e	-35 e	-5 e	38 e	103.0	5
C_7H_{12}	1-Methylcyclohexene	-72 e	-53 e	-30 e	1 e	45 e	109.8	5
C_7H_{12}	4-Methylcyclohexene	-76 e	-59 e	-36 e	-5 e	37.9	102.3	5
C_7H_{12}	1-Ethylcyclopentene	-75 e	-57 e	-34 e	-3 e	40.7	105.8	5
C_7H_{12}	1,2-Dimethylcyclopentene	-75 e	-57 e	-34 e	-3 e	40.2	105.3	5
C_7H_{12}	1,5-Dimethylcyclopentene	-77 e	-59 e	-36 e	-5.5	37.3	101.5	5
$C_7H_{12}O$	Cycloheptanone			18 e	53.7	104.0	178.7	5
$C_7H_{12}O_2$	Butyl acrylate	-52 e	-31 e	-4.5	30.4	78.0	146.9	5
$C_7H_{12}O_2$	Propyl methacrylate				26 e	73.8	139.7	5
$C_7H_{12}O_3$	Ethyl levulinate		17 e	45.3	82.6	133.2	205.7	5
$C_7H_{12}O_4$	Diethyl malonate	-23 e	4 e	36.0	76.4	128.5	198.3	5
$C_7H_{12}O_4$	Dimethyl glutarate	-11 e	15 e	47 e	87.7	139.8	209.5	5
$C_7H_{13}ClO$	Heptanoyl chloride	-17 e	4 e	29.4	59.7	96.9	144.0	5
C_7H_{14}	1-Heptene	-82.1	-63.8	-40.6	-10.7	31.1	93.2	1,5
C_7H_{14}	*cis*-2-Heptene	-79 e	-61 e	-38 e	-8 e	34.3	98.0	5
C_7H_{14}	*trans*-2-Heptene	-79 e	-61 e	-39 e	-8 e	34.0	97.5	5
C_7H_{14}	*cis*-3-Heptene	-80 e	-62 e	-40 e	-10 e	32.3	95.3	5
C_7H_{14}	*trans*-3-Heptene	-80 e	-62 e	-40 e	-10 e	32.2	95.2	5
C_7H_{14}	2-Methyl-1-hexene	-81 e	-64 e	-42 e	-12 e	29.3	91.6	5
C_7H_{14}	4-Methyl-1-hexene	-84 e	-67 e	-45 e	-16 e	25.3	86.3	5
C_7H_{14}	2-Methyl-2-hexene	-80 e	-63 e	-40 e	-10 e	32.0	95.0	5
C_7H_{14}	*cis*-3-Methyl-2-hexene	-79 e	-62 e	-39 e	-9 e	33.4	96.8	5
C_7H_{14}	*trans*-4-Methyl-2-hexene	-83 e	-66 e	-44 e	-15 e	25.9	87.1	5
C_7H_{14}	*trans*-5-Methyl-2-hexene	-83 e	-66 e	-44 e	-15 e	26.3	87.7	5
C_7H_{14}	*trans*-2-Methyl-3-hexene	-84 e	-67 e	-45 e	-16 e	24.6	85.5	5
C_7H_{14}	3-Ethyl-1-pentene	-85 e	-68 e	-46 e	-17 e	23.2	83.7	5
C_7H_{14}	2,3-Dimethyl-1-pentene	-85 e	-68 e	-46 e	-17 e	23.4	83.8	5
C_7H_{14}	2,4-Dimethyl-1-pentene	-88 e	-71 e	-50 e	-21 e	20.0	81.2	5
C_7H_{14}	3,3-Dimethyl-1-pentene	-87 e	-71 e	-50 e	-21 e	18.1	77.1	5
C_7H_{14}	4,4-Dimethyl-1-pentene	-94 e	-78 e	-57 e	-28 e	11.5	72.1	5
C_7H_{14}	2,3-Dimethyl-2-pentene	-79 e	-62 e	-39 e	-9 e	33.5	96.9	5
C_7H_{14}	2,4-Dimethyl-2-pentene	-84 e	-68 e	-46 e	-18 e	22.6	82.9	5
C_7H_{14}	*cis*-3,4-Dimethyl-2-pentene	-83 e	-65 e	-43 e	-14 e	27.2	88.8	5
C_7H_{14}	*trans*-3,4-Dimethyl-2-pentene	-82 e	-64 e	-42 e	-13 e	29.0	91.1	5
C_7H_{14}	*cis*-4,4-Dimethyl-2-pentene	-90 e	-73 e	-51 c	-22 e	18.6	80.0	5
C_7H_{14}	*trans*-4,4-Dimethyl-2-pentene	-90 e	-73 e	-52 e	-23 e	16.6	76.3	5
C_7H_{14}	2,3,3-Trimethyl-1-butene	-91 e	-75 e	-53 e	-24.2	16.3	77.5	5
C_7H_{14}	Cycloheptane				6 e	51.1	118.4	1

Mol. Form.	Name	1 Pa	10 Pa	100 Pa	1 kPa	10 kPa	100 kPa	Ref.
		Temperature in °C for the indicated pressure						
C_7H_{14}	Methylcyclohexane	-79 e	-62 e	-39 e	-7.9	35.5	100.5	1
C_7H_{14}	Ethylcyclopentane	-76 e	-59 e	-35 e	-5 e	38.4	103.0	5
C_7H_{14}	1,1-Dimethylcyclopentane		-69 e	-47 e	-17 e	24.8	87.4	5
C_7H_{14}	*cis*-1,2-Dimethylcyclopentane			-38 e	-8 e	34.9	99.0	5
C_7H_{14}	*trans*-1,2-Dimethylcyclopentane	-83 e	-66 e	-43 e	-13 e	28.4	91.4	5
C_7H_{14}	*cis*-1,3-Dimethylcyclopentane	-84 e	-66 e	-44 e	-14 e	28.2	91.1	5
C_7H_{14}	*trans*-1,3-Dimethylcyclopentane	-84 e	-67 e	-44 e	-14 e	27.4	90.3	5
$C_7H_{14}O$	1-Heptanal	-41 e	-21 e	4 e	37 e	83.7	152.3	5
$C_7H_{14}O$	2-Heptanone		-22 e	3 e	36 e	82.2	150.6	1
$C_7H_{14}O$	3-Heptanone		-28 e	0 e	36 e	83.2	147.0	5
$C_7H_{14}O$	4-Heptanone	-27 e	-6 e	18.8	50.2	90.3	143.4	5
$C_7H_{14}O$	5-Methyl-2-hexanone		-27 e	-2 e	31.0	76.6	144.4	5
$C_7H_{14}O$	2,4-Dimethyl-3-pentanone	-61 e	-42 e	-18 e	14 e	58.5	124.8	1
$C_7H_{14}O_2$	Heptanoic acid	24 e	46 e	72 e	107 e	154.6	222.6	5
$C_7H_{14}O_2$	Pentyl acetate	-58 e	-39 e	-14 e	20 e	70.1	149 e	5
$C_7H_{14}O_2$	Isopentyl acetate	-51 e	-30 e	-4 e	30.3	76.2	141.4	5
$C_7H_{14}O_2$	Isobutyl propanoate	-35 e	-19 e	2 e	31 e	72.0	136.1	5
$C_7H_{14}O_2$	Propyl butanoate	-35 e	-19 e	3 e	32.0	74.9	142.8	5
$C_7H_{14}O_2$	Propyl isobutanoate		-28 e	-5.7	24.5	67.5	133.3	5
$C_7H_{14}O_2$	Isopropyl isobutanoate		-44 e	-19.7	12.2	56.0	120.1	5
$C_7H_{14}O_2$	Ethyl 3-methylbutanoate	-57 e	-36 e	-10 e	23.9	69.5	134.4	5
$C_7H_{14}O_2$	Methyl hexanoate	-47 e	-26 e	2 e	36.6	83.3	149 e	5
$C_7H_{14}O_2$	4-Methoxy-4-methyl-2-pentanone				43 e	89.8	160 e	5
$C_7H_{15}Br$	1-Bromoheptane	-30 e	-9 e	18 e	54 e	104.4	178.4	5
$C_7H_{15}Cl$	1-Chloroheptane	-39 e	-19 e	7 e	41 e	88.6	159.9	5
$C_7H_{15}F$	1-Fluoroheptane	-64 e	-45 e	-22 e	10 e	53.3	117.4	5
$C_7H_{15}I$	1-Iodoheptane	-19 e	3 e	32 e	71 e	123.8	203.4	5
C_7H_{16}	Heptane	-78.6	-60.2	-37.0	-6.6	35.4	98.0	16
C_7H_{16}	2-Methylhexane	-82 e	-65 e	-43 e	-13 e	27.8	89.7	1
C_7H_{16}	3-Methylhexane	-81 e	-64 e	-42 e	-12 e	29.2	91.5	1
C_7H_{16}	3-Ethylpentane	-81 e	-63 e	-41 e	-11 e	30.5	93.1	1
C_7H_{16}	2,2-Dimethylpentane	-90 e	-73 e	-52 e	-22.9	17.6	78.8	1
C_7H_{16}	2,3-Dimethylpentane	-87 e	-68.4	-45.3	-14.9	26.8	89.3	5
C_7H_{16}	2,4-Dimethylpentane	-89 e	-72 e	-50 e	-21.3	19.2	80.1	1
C_7H_{16}	3,3-Dimethylpentane	-88 e	-71 e	-49 e	-18.8	22.9	85.6	1
C_7H_{16}	2,2,3-Trimethylbutane				-23.2	18.1	80.4	5
$C_7H_{16}O$	1-Heptanol		17 e	40 e	70.1	112.5	176 e	1
$C_7H_{16}O$	2-Heptanol	-9 e	7 e	27 e	55.0	95.2	158.7	5
$C_7H_{16}O$	3-Heptanol	-8 e	7 e	27 e	54.5	93.9	156.3	5
$C_7H_{16}O$	4-Heptanol	-16 e	1 e	22 e	51 e	91.9	154.6	5
$C_7H_{16}O$	2,2-Dimethyl-3-pentanol			9 e	35 e	73.1	135.5	5
$C_7H_{16}S$	1-Heptanethiol	-30 e	-9 e	18 e	53 e	102.7	176.4	5
$C_7H_{17}N$	Heptylamine			5 e	39 e	86.7	156.4	5
$C_7H_{18}N_2$	*N,N*-Diethyl-1,3-propanediamine				50.1	99.9	167.7	5
C_8F_{18}	Perfluorooctane				5 e	45.0	105.6	5
$C_8H_4O_3$	Phthalic anhydride	48.2 s	72.4 s			192.7	284.2	5
C_8H_6O	Benzofuran		-16 e	12 e	47.9	97.7	170.7	5
C_8H_7Cl	*o*-Chlorostyrene	-33 e	-10 e	20 e	58 e	110.8	188 e	5
C_8H_7N	2-Methylbenzonitrile		1 e	32.1	72.2	126.6	204.7	5
C_8H_7N	4-Methylbenzonitrile			40.1	78.7	134.3	221.3	5
C_8H_7N	Benzeneacetonitrile	-3 e	23 e	55.3	97.4	153.7	233.1	5
C_8H_7N	Indole	20.6 s	44.5 s				254.0	5
$C_8H_7NO_4$	Methyl 2-nitrobenzoate	17 e	49 e	89 e	140 e	208 e	302 e	5
C_8H_8	Styrene		-31 e	-5 e	28.6	75.4	144.7	1
C_8H_8	1,3,5,7-Cyclooctatetraene				24.3	71.0	140.1	5
C_8H_8O	Acetophenone			36 e	73 e	125.3	201.5	5
$C_8H_8O_2$	Phenyl acetate		3 e	33.1	72.2	123.9	195.5	5
$C_8H_8O_2$	Methyl benzoate		-1 e	29 e	68 e	121.2	198.9	5
$C_8H_8O_2$	4-Methoxybenzaldehyde	9 e	35 e	68.1	110.8	167.9	248.5	5
$C_8H_8O_3$	Methyl salicylate	-1 e	22 e	51 e	88.8	141.8	219.9	5
C_8H_9Cl	1-Chloro-2-ethylbenzene	-30 e	-9 e	18 e	54 e	103.7	177.9	5
C_8H_9Cl	1-Chloro-4-ethylbenzene	-27 e	-6 e	22 e	58 e	108.7	183.9	5
$C_8H_9NO_2$	1-Ethyl-4-nitrobenzene	10 e	36 e	69 e	111.6	168 e	245 e	5
C_8H_{10}	Ethylbenzene	-56.2	-36.8	-12.0	21.1	67.1	135.7	1
C_8H_{10}	*o*-Xylene			-7 e	27 e	74.2	143.9	1
C_8H_{10}	*m*-Xylene		-35 e	-10 e	23.4	69.8	138.7	1
C_8H_{10}	*p*-Xylene				22.4	68.9	137.9	1

Mol. Form.	Name	Temperature in °C for the indicated pressure						Ref.
		1 Pa	10 Pa	100 Pa	1 kPa	10 kPa	100 kPa	
$C_8H_{10}O$	*o*-Ethylphenol		16.9	44.5	81.1	130.9	204.0	5
$C_8H_{10}O$	*m*-Ethylphenol	5.6	29.2	57.5	91.9	144.8	217.9	5
$C_8H_{10}O$	*p*-Ethylphenol			60 e	95.5	144.6	217.5	5
$C_8H_{10}O$	2,3-Xylenol	14.3 s	34.3 s	57.2 s	91.4	141.7	216.4	1,5
$C_8H_{10}O$	2,4-Xylenol			50.2	85.5	137.2	210.5	1,5
$C_8H_{10}O$	2,5-Xylenol	13.4 s	33.2 s	55.9 s	87.4	137.0	210.6	5
$C_8H_{10}O$	2,6-Xylenol	-3.1 s	16.7 s	39.6 s	75.3	125.9	200.6	1,5
$C_8H_{10}O$	3,4-Xylenol	19.7 s	40.2 s	63.7 s	102.1	152.3	226.4	1,5
$C_8H_{10}O$	3,5-Xylenol	16.5 s	37.2 s	61.1 s	98.0	147.9	221.3	1,5
$C_8H_{10}O$	Benzeneethanol	2 e	25 e	54 e	92 e	143.6	217.7	5
$C_8H_{10}O$	Phenetole		-9 e	17 e	51 e	99 e	169.3	5
$C_8H_{10}O_2$	2-Phenoxyethanol	21 e	46 e	75.9	115.4	168.7	244.8	5
$C_8H_{10}O_2$	1,3-Dimethoxybenzene	18 e	34 e	56 e	86.7	135.5	223 e	5
$C_8H_{11}N$	*p*-Ethylaniline	-2 e	21 e	49 e	87 e	139.4	216.7	5
$C_8H_{11}N$	*N*-Ethylaniline	-15 e	8 e	38 e	76.4	128.8	204.2	5
$C_8H_{11}N$	*N,N*-Dimethylaniline			28 e	66 e	118.1	193.6	1
$C_8H_{11}N$	2,4-Xylidine	-2 e	21 e	51 e	88 e	139.1	210.9	5
$C_8H_{11}N$	2,6-Xylidine			37 e	80 e	137.7	217.7	5
$C_8H_{11}N$	5-Ethyl-2-picoline	-33 e	-9.3	20 e			178.0	5
$C_8H_{11}NO$	*o*-Phenetidine	0 e	27 e	60 e	102.2	156.0	228.1	5
C_8H_{12}	1,5-Cyclooctadiene		-37 e	-8 e	30 e	80.2	150 e	5
C_8H_{12}	4-Vinylcyclohexene	-62 e	-43 e	-19 e	14.1	59.9	129 e	5
$C_8H_{12}O_4$	Diethyl maleate	-6 e	20 e	52.2	93.5	148.4	224.8	5
C_8H_{14}	2,5-Dimethyl-1,5-hexadiene	-38 e	-26 e	-10 e	14 e	50.8	115.1	5
C_8H_{14}	1-Octyne	-59 e	-40 e	-16 e	16 e	60.3	125.8	1
C_8H_{14}	2-Octyne	-52 e	-33 e	-8 e	25 e	70.6	137.8	1
C_8H_{14}	3-Octyne	-55 e	-35 e	-11 e	22 e	66.8	132.8	1
C_8H_{14}	4-Octyne	-56 e	-36 e	-12 e	21 e	65.6	131.4	1
C_8H_{14}	1-Ethylcyclohexene	-55 e	-35 e	-11 e	22 e	68 e	136.5	5
$C_8H_{14}O_2$	Cyclohexyl acetate					103.1	172.9	5
$C_8H_{14}O_2$	Butyl methacrylate				47 e	93.3	159.0	5
$C_8H_{14}O_3$	Butanoic anhydride	-28 e	-2 e	30 e	71 e	123.8	196.5	5
$C_8H_{14}O_4$	Ethyl succinate	-6 e	20 e	51.0	91.1	143.7	216.1	5
$C_8H_{14}O_4$	Dipropyl oxalate	-4 e	20 e	49.9	88.6	140.4	213.0	5
$C_8H_{14}O_4$	Dimethyl adipate		28 e	61 e	103 e	156.1	227.3	5
$C_8H_{15}Br$	(2-Bromoethyl)cyclohexane	-14 e	8 e	36.9	75.3	129.7	212.5	5
$C_8H_{15}ClO$	Octanoyl chloride	1 e	22 e	46 e	74.7	109 e	150 e	5
$C_8H_{15}N$	Octanenitrile	-15 e	8 e	37 e	75 e	127.7	204.4	5
C_8H_{16}	1-Octene	-65.7	-46.1	-21.4	10.5	54.9	120.9	1,5
C_8H_{16}	*cis*-2-Octene	-59 e	-41 e	-17 e	15 e	59 e	125.2	5
C_8H_{16}	*trans*-2-Octene	-59 e	-41 e	-17 e	14 e	59 e	124.5	5
C_8H_{16}	*cis*-3-Octene	-65 e	-46 e	-22 e	10 e	55.1	122.4	5
C_8H_{16}	*trans*-3-Octene	-61 e	-43 e	-19 e	13 e	57 e	122.8	5
C_8H_{16}	*cis*-4-Octene	-63 e	-44 e	-20 e	11 e	56 e	122.1	5
C_8H_{16}	*trans*-4-Octene	-65 e	-46 e	-22 e	10 e	54.6	121.8	5
C_8H_{16}	2-Methyl-1-heptene	-66 e	-48 e	-24 e	8 e	52.3	118.7	5
C_8H_{16}	2,2-Dimethyl-*cis*-3-hexene	-74 e	-56 e	-33 e	-3 e	40.1	105.0	5
C_8H_{16}	2,3-Dimethyl-2-hexene	-65 e	-47 e	-23 e	10 e	54.3	121.3	5
C_8H_{16}	2,3,3-Trimethyl-1-pentene		-53 e	-30 e	1 e	43.8	107.9	5
C_8H_{16}	2,4,4-Trimethyl-1-pentene	-79 e	-61 e	-38 e	-7 e	36.2	101.0	5
C_8H_{16}	2,3,4-Trimethyl-2-pentene	-68 e	-49 e	-26 e	6 e	50.0	115.8	5
C_8H_{16}	2,4,4-Trimethyl-2-pentene	-73 e	-56 e	-33 e	-2 e	40.4	104.5	5
C_8H_{16}	Cyclooctane				30 e	78 e	150.7	1
C_8H_{16}	Ethylcyclohexane	-61 e	-42 e	-17 e	15.8	61.9	131.3	5
C_8H_{16}	1,1-Dimethylcyclohexane			-27 e	5 e	50.6	119.1	5
C_8H_{16}	*cis*-1,2-Dimethylcyclohexane		-44 e	-20 e	14 e	59.7	129.2	5
C_8H_{16}	*trans*-1,2-Dimethylcyclohexane	-68 e	-49 e	-25 e	8 e	53.9	122.9	5
C_8H_{16}	*cis*-1,3-Dimethylcyclohexane	-68 e	-48 e	-23 e	10 e	55.6	123.1	5
C_8H_{16}	*trans*-1,3-Dimethylcyclohexane	-62 e	-45 e	-23 e	8 e	51.5	120.9	5
C_8H_{16}	*cis*-1,4-Dimethylcyclohexane	-66 e	-47 e	-23 e	10 e	55.3	123.8	5
C_8H_{16}	*trans*-1,4-Dimethylcyclohexane			-27 e	5 e	50.6	118.9	5
C_8H_{16}	Propylcyclopentane	-60 e	-41 e	-16 e	16.5	62.1	130.5	5
C_8H_{16}	Isopropylcyclopentane	-65 e	-46 e	-21 e	12 e	57.3	125.9	5
C_8H_{16}	1-Ethyl-1-methylcyclopentane	-67 e	-49 e	-24 e	8 e	53.2	121.0	5
C_8H_{16}	*cis*-1-Ethyl-2-methylcyclopentane	-63 e	-44 e	-19 e	13.3	59.1	127.6	5
C_8H_{16}	1,1,2-Trimethylcyclopentane				2 e	46.2	113.2	5
C_8H_{16}	1,1,3-Trimethylcyclopentane	-77 e	-59 e	-36 e	-5 e	38.7	104.4	5

Mol. Form.	Name	Temperature in °C for the indicated pressure						
		1 Pa	10 Pa	100 Pa	1 kPa	10 kPa	100 kPa	Ref.
C_8H_{16}	1′,2′,4a-1,2,4-Trimethylcyclopentane	-70 e	-52 e	-28 e	4 e	48.9	116.2	5
C_8H_{16}	1′,2a,4′-1,2,4-Trimethylcyclopentane	-74 e	-56 e	-33 e	-1 e	42.8	108.8	5
$C_8H_{16}O$	1-Propylcyclopentanol	9 e	24 e	43 e	69.0	108.4	173.5	5
$C_8H_{16}O$	Octanal			6 e	45.7	97.8	170.2	5
$C_8H_{16}O$	2-Octanone		-3 e	23 e	57 e	103.8	172.1	5
$C_8H_{16}O$	3-Octanone			8 e	47.7	97 e	161 e	5
$C_8H_{16}O$	2,2,4-Trimethyl-3-pentanone			11.3	42.1	81.7	134.6	5
$C_8H_{16}O_2$	Octanoic acid	37 e	58 e	85 e	120 e	165.5	238.4	1,5
$C_8H_{16}O_2$	2-Ethylhexanoic acid				108 e	159.6	226.6	5
$C_8H_{16}O_2$	Hexyl acetate	-37 e	-13 e	16 e	52.8	100.4	164 e	5
$C_8H_{16}O_2$	Isopentyl propanoate			3.1	40.7	90.6	159.8	5
$C_8H_{16}O_2$	Isobutyl isobutanoate	-47 e	-26 e	0.4	34.8	81.1	147.0	5
$C_8H_{16}O_2$	Propyl 3-methylbutanoate			1.8	38.9	87.9	155.6	5
$C_8H_{16}O_2$	Ethyl hexanoate	-31 e	-9 e	18.7	53.9	100.7	166.2	5
$C_8H_{16}O_2$	Methyl heptanoate	-30 e	-9 e	19 e	54.2	102.4	172 e	5
$C_8H_{16}O_4$	Diethylene glycol monoethyl ether acetate	-16 e	10.6	43.9	86.2	141.3	216.6	5
$C_8H_{17}Br$	1-Bromooctane	-17 e	6 e	34 e	72 e	123.8	200.3	5
$C_8H_{17}Cl$	1-Chlorooctane	-25 e	-4 e	23 e	59 e	108.8	182.9	5
$C_8H_{17}Cl$	3-(Chloromethyl)heptane					100.3	172.4	5
$C_8H_{17}F$	1-Fluorooctane				29 e	74.6	141.8	5
$C_8H_{17}I$	1-Iodooctane	-6 e	18 e	48 e	87 e	142.5	224.5	5
C_8H_{18}	Octane		-42.6	-17.9	14.4	58.9	125.3	16
C_8H_{18}	2-Methylheptane	-69 e	-49.1	-24.5	7.6	51.6	117.2	1,5
C_8H_{18}	3-Methylheptane	-67 e	-48.1	-23.6	8.5	52.7	118.5	1,5
C_8H_{18}	4-Methylheptane	-65 e	-47 e	-24 e	7.8	51.6	117.2	5
C_8H_{18}	3-Ethylhexane				8 e	52.1	118.1	5
C_8H_{18}	2,2-Dimethylhexane	-73 e	-55 e	-32 e	-1.5	41.6	106.4	5
C_8H_{18}	2,3-Dimethylhexane				5 e	49.2	115.1	5
C_8H_{18}	2,4-Dimethylhexane				0.6	43.9	109.0	5
C_8H_{18}	2,5-Dimethylhexane	-71 e	-53 e	-30 e	0.7	43.8	108.6	5
C_8H_{18}	3,3-Dimethylhexane	-72 e	-54 e	-30 e	1.4	45.4	111.5	5
C_8H_{18}	3,4-Dimethylhexane				7 e	50.9	117.3	5
C_8H_{18}	3-Ethyl-2-methylpentane	-69 e	-50 e	-27 e	5 e	48.9	115.2	5
C_8H_{18}	3-Ethyl-3-methylpentane	-70 e	-51 e	-27 e	5 e	50.2	117.8	5
C_8H_{18}	2,2,3-Trimethylpentane	-74 e	-56 e	-32 e	-0.8	43.1	109.4	5
C_8H_{18}	2,2,4-Trimethylpentane	-81.9	-63.4	-39.8	-8.9	34.0	98.8	5
C_8H_{18}	2,3,3-Trimethylpentane	-72 e	-54 e	-30 e	2.1	46.9	114.3	5
C_8H_{18}	2,3,4-Trimethylpentane	-74 e	-54.5	-30.0	2.2	46.7	113.1	1,5
C_8H_{18}	2,2,3,3-Tetramethylbutane	-62.5 s	-44 s	-20.9 s	8.9 s	48.8 s	105.8	5
$C_8H_{18}O$	1-Octanol	12 e	30 e	53 e	84 e	128.2	194.8	1,39
$C_8H_{18}O$	2-Octanol			40 e	69.9	112.5	179.4	1,39
$C_8H_{18}O$	3-Octanol	12 e	24 e	40 e	64 e	102.8	174.1	1
$C_8H_{18}O$	4-Octanol			40 e	66.9	107.3	176.0	1,39
$C_8H_{18}O$	4-Methyl-3-heptanol	-52 e	-28 e	1 e	39 e	87.6	155.0	5
$C_8H_{18}O$	5-Methyl-3-heptanol	-35 e	-16 e	8 e	40 e	84.8	153.0	5
$C_8H_{18}O$	4-Methyl-4-heptanol	-17 e	1 e	24 e	55 e	97.2	160.7	5
$C_8H_{18}O$	2-Ethyl-1-hexanol			45 e	75 e	118.3	184.2	1
$C_8H_{18}O$	2-Ethyl-2-hexanol	-13 e	4 e	26 e	55 e	96.3	160.3	5
$C_8H_{18}O$	2,4,4-Trimethyl-2-pentanol		-7 e	13 e	40 e	79.8	146.1	5
$C_8H_{18}O$	2,2,4-Trimethyl-3-pentanol	-2 e	9 e	24 e	47 e	82.6	150.4	5
$C_8H_{18}O$	Dibutyl ether	-55 e	-35 e	-8 e	26 e	73.0	141.2	5
$C_8H_{18}O$	Di-*sec*-butyl ether			-19 e	12.1	55.4	120.6	5
$C_8H_{18}O$	Di-*tert*-butyl ether			-33 e	-2 e	41.7	106.8	1
$C_8H_{18}O_2$	Ethylene glycol monohexyl ether	-13 e	14 e	46 e	86 e	137.7	206.9	5
$C_8H_{18}O_2$	1,2-Dipropoxyethane			-44.2	-2.0	63.6	179.2	5
$C_8H_{18}O_2$	Di-*tert*-butyl peroxide			-26 e	4.3	46.6	110.5	5
$C_8H_{18}O_3$	Diethylene glycol monobutyl ether	14 e	37 e	66.8	104.9	153 e	230.4	5
$C_8H_{18}O_3$	Diethylene glycol diethyl ether	-32 e	-7 e	25 e	64.9	117.1	189 e	5
$C_8H_{18}O_5$	Tetraethylene glycol	89 e	117 e	151.1	192.2	242.9	307.3	5
$C_8H_{18}S$	1-Octanethiol	-15 e	6 e	34 e	71 e	122.1	198.5	5
$C_8H_{18}S$	Dibutyl sulfide	-22 e	0 e	27 e	63 e	113.5	188.4	5

Mol. Form.	Name	1 Pa	10 Pa	100 Pa	1 kPa	10 kPa	100 kPa	Ref.
		Temperature in °C for the indicated pressure						
$C_8H_{19}N$	Dibutylamine	-37 e	-16 e	10 e	44 e	90.8	159.1	5
$C_8H_{19}N$	Diisobutylamine	-57 e	-36 e	-9.0	25.5	72.2	139.0	5
$C_8H_{20}O_4Si$	Ethyl silicate	-77 e	-52 e	-21 e	21.6	80.5	164.1	5
$C_8H_{20}Si$	Tetraethylsilane			-6.5	30.5	80.6	152.6	5
C_9F_{20}	Perfluorononane					40 e	114.7	5
$C_9H_6N_2O_2$	Toluene-2,4-diisocyanate		39 e	72 e	113.9	169.7	247 e	5
C_9H_7N	Quinoline	-1.3	23.7	55.4	96.8	153.4	236.5	1,5
C_9H_7N	Isoquinoline		30.2	60.7	101.3	157.9	242.7	1,5
C_9H_8	Indene			12 e	53.0	106.8	181.0	5
C_9H_{10}	*cis*-1-Propenylbenzene	-38 e	-15.4	13.3	51.4	103.7	178.4	5
C_9H_{10}	*trans*-1-Propenylbenzene		-16 e	13.3	51.6	103.7	178.4	5
C_9H_{10}	Isopropenylbenzene			3.2	41.5	92.8	164.9	5
C_9H_{10}	Indan	-33 e	-12 e	16 e	52 e	102.3	177.5	1
$C_9H_{10}O$	2,4-Dimethylbenzaldehyde	-3 e	23 e	54 e	93.2	144.6	214.5	5
$C_9H_{10}O_2$	Ethyl benzoate	-18 e	8 e	39 e	80.1	135.1	212.8	5
$C_9H_{10}O_2$	Benzyl acetate	-11 e	15 e	46.6	86.9	139.5	211 e	5
$C_9H_{11}Br$	1-Bromo-4-isopropylbenzene	-8 e	15 e	45 e	84 e	138.1	218.5	5
$C_9H_{11}Cl$	1-Chloro-2-isopropylbenzene	-23 e	-1 e	27 e	64 e	114.6	190.5	5
$C_9H_{11}Cl$	1-Chloro-4-isopropylbenzene		3 e	31 e	69 e	120.5	197.8	5
C_9H_{12}	Propylbenzene	-43 e	-23 e	4 e	38 e	86.7	158.8	1
C_9H_{12}	Isopropylbenzene	-46 e	-26 e	-1 e	33 e	80.9	152.0	1
C_9H_{12}	*o*-Ethyltoluene	-40 e	-19 e	8 e	43 e	92.1	164.7	5
C_9H_{12}	*m*-Ethyltoluene	-42 e	-21 e	5 e	40.4	88.9	160.8	5
C_9H_{12}	*p*-Ethyltoluene	-41 e	-21 e	6 e	41 e	89.2	161.5	5
C_9H_{12}	1,2,3-Trimethylbenzene		-12 e	15 e	52 e	101.5	175.6	1
C_9H_{12}	1,2,4-Trimethylbenzene	-37 e	-16 e	11 e	47 e	95.9	168.9	1
C_9H_{12}	1,3,5-Trimethylbenzene	-39 e	-18 e	9 e	43.7	92.4	164.3	1
$C_9H_{12}O$	Benzyl ethyl ether		-10 e	20.4	59.3	111.3	184.5	5
$C_9H_{12}O$	Phenyl propyl ether		-10 e	21 e	61 e	113.9	189.3	5
$C_9H_{12}O$	Phenyl isopropyl ether	-20 e	-1 e	23 e	56 e	103.7	176.9	5
$C_9H_{13}N$	2,4,6-Trimethylaniline	12 e	36 e	66 e	104.1	154.9	226 e	5
$C_9H_{13}N$	*N,N*-Dimethyl-*o*-toluidine	-25 e	-3 e	24.4	60.6	110.7	184.5	5
$C_9H_{13}N$	Amphetamine			33 e	70.1	118 e	202.0	5
$C_9H_{14}O$	Isophorone		1 e	33.1	75.1	132.4	215.1	5
$C_9H_{14}O_6$	Triacetin	37.6	62 e	90 e	124 e	165 e	214 e	5
$C_9H_{16}O_4$	Diethyl glutarate	-1 e	26 e	60.2	103.3	159.6	236.5	5
$C_9H_{17}N$	Nonanenitrile	-3 e	21 e	50.9	90.7	145.4	225.1	5
C_9H_{18}	1-Nonene	-50.1	-29.4	-3.3	30.4	77.1	146.4	1,5
C_9H_{18}	2-Methyl-1-octene	-53 e	-34 e	-9 e	25 e	72 e	144.1	5
C_9H_{18}	Butylcyclopentane	-45 e	-24 e	1 e	36 e	84 e	156.1	5
C_9H_{18}	Propylcyclohexane	-46 e	-26 e	0 e	35.1	83.6	156.2	5
C_9H_{18}	Isopropylcyclohexane	-48 e	-28 e	-2 e	33 e	81.3	154.0	5
C_9H_{18}	*trans*-1-Ethyl-4-methylcyclohexane	-53 e	-33 e	-8 e	25 e	71.8	141.5	5
C_9H_{18}	1,1,2-Trimethylcyclohexane			-12 e	23 e	71.5	145.5	5
C_9H_{18}	1,1,3-Trimethylcyclohexane	-60 e	-41 e	-16 e	18 e	65.2	136.1	5
C_9H_{18}	1′,2a,4a-1,2,4-Trimethylcyclohexane	-71 e	-50 e	-22 e	15 e	65.7	140.7	5
C_9H_{18}	1′,3′,5′-1,3,5-Trimethylcyclohexane	-72 e	-50 e	-22 e	14 e	65.1	140.0	5
C_9H_{18}	Isobutylcyclopentane	-105 e	-88 e	-64 e	-28 e	31 e	147.0	5
C_9H_{18}	*cis*-1-Methyl-2-propylcyclopentane	-52 e	-33 e	-7 e	28 e	77 e	152.0	5
C_9H_{18}	*trans*-1-Methyl-2-propylcyclopentane	-56 e	-36 e	-11 e	23 e	72 e	145.8	5
C_9H_{18}	1,1,3,3-Tetramethylcyclopentane	-72 e	-54 e	-30 e	2 e	47 e	117.4	5
$C_9H_{18}O$	Nonanal		-3 e	27.4	65.5	115.6	184.6	5
$C_9H_{18}O$	2-Nonanone		8 e	35 e	71 e	121.0	194.0	5
$C_9H_{18}O$	5-Nonanone			-1 e	39.1	94 e	188 e	5
$C_9H_{18}O$	2,6-Dimethyl-4-heptanone	-32 e	-12 e	14 e	48 e	96.2	167.7	5
$C_9H_{18}O_2$	Nonanoic acid	48 e	69 e	97 e	133 e	182.7	255.1	5
$C_9H_{18}O_2$	Heptyl acetate	-16 e	6 e	34 e	70 e	119.9	191.9	5
$C_9H_{18}O_2$	Isopentyl butanoate				55 e	105.6	178.4	5
$C_9H_{18}O_2$	Isobutyl 3-methylbutanoate			11.3	48.3	97.9	168.3	5
$C_9H_{18}O_2$	Propyl hexanoate	-26 e	-2 e	28 e	65.1	113.4	178 e	5
$C_9H_{18}O_2$	Methyl octanoate	-26 e	-9 e	13 e	40 e	76 e	127.9	5

Mol. Form.	Name	Temperature in °C for the indicated pressure						Ref.
		1 Pa	10 Pa	100 Pa	1 kPa	10 kPa	100 kPa	
$C_9H_{19}Cl$	1-Chlorononane	-11 e	11 e	39 e	76 e	127.8	204.7	5
C_9H_{20}	Nonane	-46.8	-26.0	0.0	34.0	80.8	150.3	16
C_9H_{20}	2-Methyloctane	-49 e	-30 e	-5 e	28 e	73.9	142.8	5
C_9H_{20}	3-Methyloctane	-49 e	-29 e	-5 e	29 e	74.7	143.7	5
C_9H_{20}	4-Methyloctane	-50 e	-30 e	-6 e	27 e	73.2	141.9	5
C_9H_{20}	2,2-Dimethylheptane	-58 e	-39 e	-15 e	18 e	63.6	132.3	5
C_9H_{20}	2,3-Dimethylheptane	-53 e	-33 e	-9 e	25 e	70.8	140.0	5
C_9H_{20}	2,6-Dimethylheptane	-55 e	-36 e	-12 e	21 e	66.4	134.7	5
C_9H_{20}	3-Ethyl-4-methylhexane			-9 e	24 e	70.6	139.9	5
C_9H_{20}	2,2,4-Trimethylhexane	-66.1	-46.4	-21.3	11.8	57.7	126.0	5
C_9H_{20}	2,2,5-Trimethylhexane	-65.1	-45.8	-21.2	11.2	56.2	123.7	1,5
C_9H_{20}	2,3,3-Trimethylhexane	-58 e	-38 e	-13 e	20 e	66.7	137.2	5
C_9H_{20}	2,3,5-Trimethylhexane	-60 e	-41 e	-16 e	17 e	62.3	130.9	5
C_9H_{20}	2,4,4-Trimethylhexane	-62 e	-43 e	-18 e	15 e	61.0	130.2	5
C_9H_{20}	3,3,4-Trimethylhexane	-53 e	-33 e	-7 e	28 e	76.3	148.9	5
C_9H_{20}	3,3-Diethylpentane			-9 e	26 e	73.7	145.7	1
C_9H_{20}	3-Ethyl-2,4-dimethylpentane	-58 e	-38 e	-13 e	20 e	66.7	136.2	5
C_9H_{20}	2,2,3,3-Tetramethylpentane				21 e	68.5	139.8	1
C_9H_{20}	2,2,3,4-Tetramethylpentane	-61 e	-42 e	-17 e	16 e	62.5	132.6	1
C_9H_{20}	2,2,4,4-Tetramethylpentane		-49 e	-25 e	8 e	53.2	121.8	1
C_9H_{20}	2,3,3,4-Tetramethylpentane	-57 e	-37 e	-12 e	22 e	69.7	141.1	1
$C_9H_{20}O$	1-Nonanol		40 e	64 e	96.9	141.0	213.0	5,39
$C_9H_{20}O$	3-Nonanol		24 e	47 e	78 e	123.0	194.2	5
$C_9H_{20}O$	4-Nonanol			45 e	76.4	121.3	192.0	5
$C_9H_{20}O$	5-Nonanol	13 e	31 e	54 e	84.5	128.1	194.7	5
$C_9H_{20}O$	2,2,4,4-Tetramethyl-3-pentanol				58	100	167	5
$C_9H_{20}S$	1-Nonanethiol	-2 e	21 e	49 e	87 e	140.4	219.2	5
$C_9H_{21}BO_3$	Triisopropyl borate					73.1	139.0	5
$C_9H_{21}N$	Nonylamine		9 e	37 e	75 e	126.2	202.1	5
$C_9H_{21}N$	Tripropylamine	-39 e	-18 e	8 e	42 e	88.2	156.0	5
$C_{10}F_8$	Perfluoronaphthalene	5.2 s	25.1 s	48.1 s				5
$C_{10}F_{22}$	Perfluorodecane				52 e		132.9	5
$C_{10}H_7Br$	1-Bromonaphthalene	17 e	45 e	80.3	126.7	189.8	280.5	5
$C_{10}H_7Cl$	1-Chloronaphthalene	14 e	39 e	70.5	112.8	171.6	258.6	5
$C_{10}H_8$	Naphthalene**	3.2 s	24.1 s	49.3 s	80.7	135.6	217.5	1,5
$C_{10}H_8$	Azulene	24.1 s	46 s	71.5 s	103.3	162.6	244.0	5
$C_{10}H_8O$	1-Naphthol				137.2	196.7	281.8	5
$C_{10}H_8O$	2-Naphthol				140.7	200.5	286.8	5
$C_{10}H_9N$	1-Naphthalenamine		62 e	99.0	146.9	210.7	300.1	5
$C_{10}H_9N$	2-Naphthalenamine	36.3 s	65.9 s	103 s	150.9	215.1	305.5	5
$C_{10}H_9N$	2-Methylquinoline	5.3	31.9	63.8	102.9	165.8	247.2	5
$C_{10}H_9N$	4-Methylquinoline	29 e	54 e	85 e	127 e	183.0	265.1	5
$C_{10}H_9N$	6-Methylquinoline	27 e	51 e	81 e	122 e	179.2	264.5	5
$C_{10}H_9N$	8-Methylquinoline	15 e	40 e	70 e	111 e	166.1	247.3	5
$C_{10}H_{10}$	*m*-Divinylbenzene	-29 e	-4 e	27.1	67.6	122.1	199 e	5
$C_{10}H_{10}O_4$	Dimethyl phthalate	27 e	56 e	92.7	137.8	195.8	272.7	5
$C_{10}H_{10}O_4$	Dimethyl isophthalate			85 e	129.5	189.2	273 e	5
$C_{10}H_{10}O_4$	Dimethyl terephthalate	56.6 s	79.4 s	106.1 s	137.9 s	197.9	282 e	5
$C_{10}H_{12}$	1,2,3,4-Tetrahydronaphthalene	-21 e	3 e	33.2	74.1	127.4	207.8	5
$C_{10}H_{12}$	2-Ethylstyrene	-31 e	-8 e	21 e	60 e	111.7	187 e	5
$C_{10}H_{12}$	3-Ethylstyrene	-28 e	-5.3	24.1	62.6	116 e	193 e	5
$C_{10}H_{12}$	4-Ethylstyrene	-31 e	-8.2	21.3	60.5	115 e	196 e	5
$C_{10}H_{12}O$	Estragole			48.5	88.0	140.7	214.6	5
$C_{10}H_{12}O$	4-Isopropylbenzaldehyde			54.1	96.0	152.2	231.5	5
$C_{10}H_{12}O_2$	4-Allyl-2-methoxyphenol	9 e	37 e	72 e	115.9	173.8	252.9	5
$C_{10}H_{12}O_2$	2-Phenylethyl acetate	-4 e	22 e	54 e	96 e	152.3	232.0	5
$C_{10}H_{12}O_2$	Propyl benzoate	-8 e	18 e	50.2	92.3	149.2	230.5	5
$C_{10}H_{12}O_2$	Ethyl phenylacetate	-9 e	19 e	52 e	95 e	150.2	225 e	5
$C_{10}H_{12}O_2$	Isoeugenol				125 e	185.3	267.1	5
$C_{10}H_{14}$	Butylbenzene	-28 e	-7 e	21 e	56.9	107.6	182.8	1,5
$C_{10}H_{14}$	*sec*-Butylbenzene	-35 e	-14 e	13 e	48 e	98.3	172.8	5
$C_{10}H_{14}$	*tert*-Butylbenzene	-37 e	-16 e	10 e	46 e	94.9	168.6	5
$C_{10}H_{14}$	Isobutylbenzene	-36 e	-15 e	12 e	47.9	97.8	172.3	5
$C_{10}H_{14}$	*o*-Cymene	-39 e	-16 e	13 e	51 e	103.1	177.8	5
$C_{10}H_{14}$	*m*-Cymene	-34 e	-13 e	14 e	50 e	99.9	174.6	5
$C_{10}H_{14}$	*p*-Cymene	-33 e	-12 e	16 e	52 e	102.2	176.6	5
$C_{10}H_{14}$	*o*-Diethylbenzene	-28 e	-6 e	21 e	58 e	107.9	182.9	5

VAPOR PRESSURE (continued)

Mol. Form.	Name	Temperature in °C for the indicated pressure						Ref.
		1 Pa	10 Pa	100 Pa	1 kPa	10 kPa	100 kPa	
$C_{10}H_{14}$	*m*-Diethylbenzene	-28 e	-7 e	20 e	56 e	106.2	180.6	5
$C_{10}H_{14}$	*p*-Diethylbenzene	-28 e	-6 e	21 e	57 e	108.1	183.3	5
$C_{10}H_{14}$	3-Ethyl-1,2-dimethylbenzene	-22 e	0 e	28 e	66 e	117.2	193.4	5
$C_{10}H_{14}$	4-Ethyl-1,2-dimethylbenzene	-24 e	-2 e	26 e	63 e	113.6	189.2	5
$C_{10}H_{14}$	2-Ethyl-1,3-dimethylbenzene		-2 e	26 e	63 e	113.7	189.5	5
$C_{10}H_{14}$	2-Ethyl-1,4-dimethylbenzene	-27 e	-5 e	23 e	60 e	110.6	186.4	5
$C_{10}H_{14}$	1-Ethyl-2,4-dimethylbenzene	-25 e	-4 e	24 e	61 e	112.2	187.9	5
$C_{10}H_{14}$	1-Ethyl-3,5-dimethylbenzene	-28 e	-6 e	21 e	58 e	108.3	183.2	5
$C_{10}H_{14}$	1-Methyl-2-propylbenzene	-27 e	-6 e	22 e	58.2	108.9	184.3	5
$C_{10}H_{14}$	1-Methyl-3-propylbenzene	-29 e	-8 e	20 e	56.1	106.5	181.3	5
$C_{10}H_{14}$	1-Methyl-4-propylbenzene	-29 e	-7 e	20 e	56.6	107.4	182.8	5
$C_{10}H_{14}$	1,2,3,4-Tetramethylbenzene		7 e	36 e	74 e	126.6	204.5	5
$C_{10}H_{14}$	1,2,3,5-Tetramethylbenzene	-19 e	3 e	32 e	69 e	120.9	197.5	5
$C_{10}H_{14}$	1,2,4,5-Tetramethylbenzene					119.9	196.3	5
$C_{10}H_{14}O$	2-Butylphenol	7 e	31 e	61 e	101 e	155.2	234.4	5
$C_{10}H_{14}O$	Butyl phenyl ether	-16 e	8 e	38 e	77 e	131.3	209.7	5
$C_{10}H_{14}O$	Thymol	18.9 s	37.9 s	59.5	101.2	155.0	230.4	5
$C_{10}H_{15}N$	2-Methyl-5-isopropylaniline	19 e	43 e	72 e	107.4	150 e	204 e	5
$C_{10}H_{15}N$	*N*-Butylaniline	11 e	35 e	66 e	106 e	160.9	241.0	5
$C_{10}H_{15}N$	*N,N*-Diethylaniline	-11 e	14 e	44.3	84.2	138.4	216.3	5
$C_{10}H_{16}$	Dipentene	-42 e	-19 e	10.6	48.7	100.2	173.9	5
$C_{10}H_{16}$	*d*-Limonene	-45 e	-21 e	9.1	48.0	100.4	174.5	5
$C_{10}H_{16}$	*l*-Limonene	-33 e	-12 e	16 e	52.0	102.3	177.0	21
$C_{10}H_{16}$	β-Myrcene			9.4	47.3	98.3	171.0	5
$C_{10}H_{16}$	α-Pinene	-48 e	-27 e	-1 e	33.6	82.2	155.1	21
$C_{10}H_{16}$	β-Pinene	-43 e	-22 e	5.0	40.6	90.5	165.5	21
$C_{10}H_{16}$	Camphene					90.7	160.1	4
$C_{10}H_{16}$	Terpinolene			26.5	64.9	115.4	184.6	5
$C_{10}H_{16}$	β-Phellandrene			16 e	53.2	104 e	171.0	5
$C_{10}H_{16}O$	(+)-Camphor	-15.8 s	10 s	41.5 s	80.8 s	131.4 s	207.6	5
$C_{10}H_{16}O$	Pulegone	37 e	49.1	66.4	92.2	135.1	220.2	5
$C_{10}H_{18}$	1-Decyne	-34 e	-13 e	14 e	51 e	100.3	173.5	5
$C_{10}H_{18}$	*cis*-Decahydronaphthalene	-26 e	-4 e	24 e	62.4	115.5	195.3	1
$C_{10}H_{18}$	*trans*-Decahydronaphthalene		-10 e	18 e	55.3	107.9	186.8	1
$C_{10}H_{18}O$	α-Terpineol			48	89	142	217	4
$C_{10}H_{18}O$	Eucalyptol			10.6	48.5	100.3	175.4	5
$C_{10}H_{18}O$	*trans*-Geraniol	4 e	31 e	63.2	104.3	157.7	229.6	5
$C_{10}H_{18}O_4$	Sebacic acid	125.9 s						5
$C_{10}H_{18}O_4$	Dipropyl succinate	11 e	38 e	72.1	115.4	172.3	250.4	5
$C_{10}H_{18}O_4$	Diethyl adipate	4 e	35 e	72 e	116.6	171.2	239.5	5
$C_{10}H_{19}N$	Decanenitrile	13 e	36 e	66 e	105.8	160.6	241.6	5
$C_{10}H_{20}$	1-Decene	-35.5	-13.7	13.7	49.0	97.9	170.1	1,5
$C_{10}H_{20}$	Cyclodecane			29 e	68 e	121.3	201.8	1
$C_{10}H_{20}$	Butylcyclohexane	-31 e	-9 e	18 e	54 e	104.7	180.4	5
$C_{10}H_{20}$	Isobutylcyclohexane	-37 e	-16 e	10 e	46 e	95.9	170.8	5
$C_{10}H_{20}$	*tert*-Butylcyclohexane	-39 e	-18 e	9 e	45 e	95.3	171.1	5
$C_{10}H_{20}O$	Decanal		16 e	47.2	86.3	137.7	208.0	5
$C_{10}H_{20}O_2$	Decanoic acid	58 e	80 e	108 e	145 e	195.2	269.5	5
$C_{10}H_{20}O_2$	Octyl acetate	-26 e	-3 e	27 e	66.3	120.0	198.2	5
$C_{10}H_{20}O_2$	2-Ethylhexyl acetate	-11 e	5 e	26 e	57.6	107.1	197.2	5
$C_{10}H_{20}O_2$	Isopentyl isopentanoate			22 e	62.8	116.9	193.6	5
$C_{10}H_{20}O_2$	Ethyl octanoate	-17 e	9 e	41 e	81.4	133.2	203 e	5
$C_{10}H_{20}O_4$	Diethylene glycol monobutyl ether acetate	6 e	34 e	69 e	112.6	169.2	245.4	5
$C_{10}H_{21}Br$	1-Bromodecane	9 e	33 e	63 e	104 e	159.2	240.0	5
$C_{10}H_{21}Cl$	1-Chlorodecane	2 e	25 e	54 e	92 e	145.7	225.3	5
$C_{10}H_{21}F$	1-Fluorodecane	-22 e	0 e	27 e	64 e	113.3	185.7	5
$C_{10}H_{22}$	Decane		-10.6	16.7	52.3	101.1	173.7	16
$C_{10}H_{22}$	2-Methylnonane	-34 e	-14 e	12 e	47 e	94.8	166.5	5
$C_{10}H_{22}$	3-Methylnonane	-34 e	-14 e	12 e	47 e	95.1	167.3	5
$C_{10}H_{22}$	4-Methylnonane	-36 e	-16 e	10 e	45 e	93.1	165.2	5
$C_{10}H_{22}$	5-Methylnonane	-36 e	-16 e	10 e	45 e	92.6	164.6	5
$C_{10}H_{22}$	2,4-Dimethyloctane				38 e	84.9	155.4	5
$C_{10}H_{22}$	2,7-Dimethyloctane	-39 e	-19 e	7 e	41 e	88.4	159.4	5
$C_{10}H_{22}$	2,2,6-Trimethylheptane	-46 e	-27 e	-2 e	32 e	78.5	148.4	5
$C_{10}H_{22}$	3,3,5-Trimethylheptane			0 e	35 e	82.7	155.2	5
$C_{10}H_{22}$	2,2,3,3-Tetramethylhexane	-46 e	-25 e	1 e	36 e	85.6	159.8	5

Mol. Form.	Name	1 Pa	10 Pa	100 Pa	1 kPa	10 kPa	100 kPa	Ref.
		Temperature in °C for the indicated pressure						
$C_{10}H_{22}$	2,2,5,5-Tetramethylhexane			-10 e	22 e	68.3	137.0	5
$C_{10}H_{22}$	2,4-Dimethyl-3-isopropylpentane	-46 e	-26 e	0 e	35 e	83.2	156.5	5
$C_{10}H_{22}$	2,2,3,3,4-Pentamethylpentane		-24 e	3 e	39 e	89.1	165.5	5
$C_{10}H_{22}$	2,2,3,4,4-Pentamethylpentane		-29 e	-3 e	33 e	82.8	158.7	5
$C_{10}H_{22}O$	1-Decanol	30 e	50 e	75 e	109 e	157.3	230.6	1,39
$C_{10}H_{22}O$	4-Decanol	18 e	37 e	61 e	93 e	139 e	210 e	5
$C_{10}H_{22}O$	Dipentyl ether	-31 e	-8 e	22 e	60 e	111.6	186.2	5
$C_{10}H_{22}O$	Diisopentyl ether			14.0	51.5	101.8	172.8	5
$C_{10}H_{22}O_2$	Ethylene glycol dibutyl ether	0 e	20 e	44 e	78.4	127.1	202.9	5
$C_{10}H_{22}O_5$	Tetraethylene glycol dimethyl ether				138 e	200.9	275.3	5
$C_{10}H_{22}S$	1-Decanethiol	11 e	34 e	64 e	103 e	157.5	238.6	5
$C_{10}H_{22}S$	Diisopentylsulfide			7 e	82 e	118 e	139 e	5
$C_{10}H_{23}N$	Dipentylamine				77 e	127.7	202.0	5
$C_{10}H_{30}O_3Si_4$	Decamethyltetrasiloxane	-31 e	-6 e	26 e	66.8	118.8	193.9	5
$C_{10}H_{30}O_5Si_5$	Decamethylcyclopentasiloxane	-2 e	19 e	46 e	82 e	132.9	210.4	5
$C_{11}H_8O_2$	1-Naphthalenecarboxylic acid				191.9	239.3	299.6	5
$C_{11}H_8O_2$	2-Naphthalenecarboxylic acid				197.9	246.0	308.1	5
$C_{11}H_{10}$	1-Methylnaphthalene	5 e	29 e	60 e	102 e	159.1	244.1	1
$C_{11}H_{10}$	2-Methylnaphthalene			57 e	99 e	156.0	240.5	1
$C_{11}H_{12}O_2$	Ethyl *trans*-cinnamate			79	125	187	271	4
$C_{11}H_{12}O_3$	Myristicin	23 e	53 e	88.9	135.2	196.0	279.4	5
$C_{11}H_{14}$	4-Isopropylstyrene	-25 e	-1 e	30.2	70.3	124.5	202.1	5
$C_{11}H_{14}$	1,2,3,4-Tetrahydro-5-methylnaphthalene	9 e	31 e	60 e	99 e	153.1	233.8	5
$C_{11}H_{14}$	1,2,3,4-Tetrahydro-6-methylnaphthalene	17 e	36 e	62 e	97 e	147.8	228.5	5
$C_{11}H_{14}O_2$	Butyl benzoate	6 e	34 e	67.9	110.3	165 e	237 e	5
$C_{11}H_{16}$	Pentylbenzene	-14 e	8 e	37 e	74 e	126.7	204.9	5
$C_{11}H_{16}$	*p-tert*-Butyltoluene	-24 e	-2 e	27 e	64.1	115.5	190.8	5
$C_{11}H_{16}$	1,3-Diethyl-5-methylbenzene	-26 e	-1 e	29.5	69.5	123.5	200.2	5
$C_{11}H_{16}$	2-Ethyl-1,3,5-trimethylbenzene		6 e	36 e	75.7	129.6	207.6	5
$C_{11}H_{16}$	1-Ethyl-2,4,5-trimethylbenzene	-13 e	11 e	40 e	79.4	132.1	207.7	5
$C_{11}H_{20}$	1-Undecyne	-22 e	0 e	29 e	67 e	118.5	194.5	5
$C_{11}H_{20}$	2-Undecyne	-17 e	6 e	35 e	74 e	127.4	205.4	5
$C_{11}H_{20}O_2$	10-Undecenoic acid	35 e	67 e	105 e	150.0	205.4	274.5	5
$C_{11}H_{20}O_4$	Ethyl diethylmalonate			74 e	105 e	149.4	219 e	5
$C_{11}H_{21}N$	Undecanenitrile			78.6	120.3	177.3	259.9	5
$C_{11}H_{22}$	1-Undecene	-21.6	1.2	29.7	66.4	117.1	192.2	5
$C_{11}H_{22}$	*cis*-2-Undecene	-14 e	7 e	34 e	70.2	120.6	196 e	5
$C_{11}H_{22}$	*trans*-2-Undecene	-14 e	7 e	33 e	69.3	119.6	195 e	5
$C_{11}H_{22}$	*cis*-4-Undecene	-19 e	3 e	30 e	66.6	117.1	192 e	5
$C_{11}H_{22}$	*trans*-4-Undecene	-17 e	4 e	31 e	67.1	117.4	193 e	5
$C_{11}H_{22}$	*cis*-5-Undecene	-19 e	2 e	30 e	66.2	116.7	191 e	5
$C_{11}H_{22}$	*trans*-5-Undecene	-18 e	3 e	31 e	67.0	117.4	192 e	5
$C_{11}H_{22}$	Pentylcyclohexane	-17 e	6 e	34 e	72 e	124.2	202.7	5
$C_{11}H_{22}$	Hexylcyclopentane	-15 e	7 e	36 e	73 e	125.0	202.5	5
$C_{11}H_{22}O$	2-Undecanone	17 e	37 e	64.3	103.0	153.6	232.6	1,5
$C_{11}H_{22}O$	6-Undecanone		28 e	57 e	95 e	148.4	226.9	1
$C_{11}H_{22}O_2$	Undecanoic acid	68 e	90 e	118 e	156 e	207.2	283.6	5
$C_{11}H_{22}O_2$	Heptyl butanoate	2 e	29 e	62 e	102.6	155.1	224.7	5
$C_{11}H_{22}O_2$	Propyl octanoate	-2 e	23 e	55 e	94.0	145.2	215 e	5
$C_{11}H_{22}O_2$	Methyl decanoate	10 e	33 e	62 e	100.9	154.0	232 e	5
$C_{11}H_{24}$	Undecane	-18.4	4.3	32.6	69.5	120.2	195.4	16
$C_{11}H_{24}$	2-Methyldecane	-20 e	1 e	28 e	64 e	114.0	188.7	5
$C_{11}H_{24}$	3-Methyldecane	-35 e	-10 e	22 e	61.9	115.6	190.4	5
$C_{11}H_{24}$	4-Methyldecane	-38 e	-12 e	20 e	60.8	113.9	186.4	5
$C_{11}H_{24}$	2,4,7-Trimethyloctane				43 e	94 e	170.4	5
$C_{11}H_{24}O$	1-Undecanol	52.2	80.0	82 e	118 e	167.6	244.1	5
$C_{11}H_{24}S$	1-Undecanethiol	23 e	47 e	77 e	118 e	173.6	256.8	5
$C_{12}F_{27}N$	Trinonafluorobutylamine		3 e	29.0	63.3	109.9	176.8	5
$C_{12}H_8$	Acenaphthylene	24 s	49.8 s	80.6 s				5
$C_{12}H_9N$	Carbazole					254.7	354.0	5
$C_{12}H_{10}$	Acenaphthene				126.2	187 e	276 e	1
$C_{12}H_{10}$	Biphenyl			69.0	111.1	169.5	254.7	1
$C_{12}H_{10}N_2$	Azobenzene			98.1	144.8	206.7	292.7	4
$C_{12}H_{10}O$	Diphenyl ether		44 e	75 e	116 e	173 e	257.4	5

Mol. Form.	Name	Temperature in °C for the indicated pressure: 1 Pa	10 Pa	100 Pa	1 kPa	10 kPa	100 kPa	Ref.
$C_{12}H_{10}O$	1-Acetonaphthone	37 e	69 e	107.0	154.6	215.2	294.9	5
$C_{12}H_{10}O$	2-Acetonaphthone	48.3 s		118.7	163.0	221.1	300.3	5
$C_{12}H_{10}S$	Diphenyl sulfide	20 e	51 e	88.7	137.5	202.2	291.8	5
$C_{12}H_{11}N$	Diphenylamine	48 s		102.8	150.5	213.7	301.4	5
$C_{12}H_{12}$	1-Ethylnaphthalene	16 e	41 e	72 e	114 e	171.8	257.7	5
$C_{12}H_{12}$	2-Ethylnaphthalene	14 e	39 e	71 e	113 e	171.2	257.3	5
$C_{12}H_{12}$	1,2-Dimethylnaphthalene	26 e	51 e	82 e	123 e	180.5	265.7	5
$C_{12}H_{12}$	2,7-Dimethylnaphthalene	31.5 s	53.1 s	78.8 s	115.9	175 e	260 e	5
$C_{12}H_{14}O_4$	Diethyl phthalate	12 e	51 e	96 e	150.5	215.9	296.2	5
$C_{12}H_{16}$	*p*-Isopropenylisopropylbenzene	-11 e	15 e	46 e	87 e	142.4	221 e	5
$C_{12}H_{16}$	Cyclohexylbenzene		28 e	58 e	98 e	154.7	239.5	5
$C_{12}H_{16}O_2$	3-Methylbutyl benzoate			66 e	115.0	177.7	261.4	5
$C_{12}H_{18}$	Hexylbenzene	-2 e	22 e	51 e	90 e	144.5	225.5	5
$C_{12}H_{18}$	1,2-Diisopropylbenzene	-14 e	9 e	37 e	74 e	125.9	203.2	5
$C_{12}H_{18}$	1,3-Diisopropylbenzene	-14 e	8 e	36 e	74 e	125.5	202.6	5
$C_{12}H_{18}$	1,4-Diisopropylbenzene	-6 e	18 e	49 e	90 e	148.8	238 e	5
$C_{12}H_{18}$	Hexamethylbenzene	46.3 s	72.5 s	81.7 s	121.8 s	178.3	263.7	5
$C_{12}H_{18}$	1,5,9-Cyclododecatriene	-14 e	11 e	44 e	87 e	145.0	229.8	5
$C_{12}H_{20}O_2$	Geranyl acetate			67.7	110.8	166.9	242.9	5
$C_{12}H_{20}O_4$	Dibutyl maleate	12.3	50.4	94.0	144.2	203 e	272 e	5
$C_{12}H_{22}$	1-Dodecyne	-11 e	13 e	43 e	82 e	135.8	214.4	5
$C_{12}H_{22}$	Cyclohexylcyclohexane		20 e	53.1	96.0	154.1	237.2	5
$C_{12}H_{22}O_2$	Methyl 10-undecenoate	10 e	38 e	73 e	116 e	172.2	247.1	5
$C_{12}H_{22}O_4$	Dimethyl sebacate		53 e	97	150	214	293	4
$C_{12}H_{23}N$	Dodecanenitrile	36 e	60 e	92 e	133 e	190.5	275.5	5
$C_{12}H_{24}$	1-Dodecene	-8.3	15.2	44.8	82.9	135.4	212.8	5
$C_{12}H_{24}$	Hexylcyclohexane	-3 e	20 e	50 e	89 e	143.1	224.2	5
$C_{12}H_{24}$	Heptylcyclopentane	-1 e	22 e	51 e	90 e	143.5	223.5	5
$C_{12}H_{24}O$	Dodecanal			70 e	116.2	175.9	256.6	5
$C_{12}H_{24}O_2$	Dodecanoic acid	78 e	100 e	128 e	166 e	219.1	298.1	5
$C_{12}H_{24}O_2$	Decyl acetate	12 e	40 e	74 e	115.1	168.1	238 e	5
$C_{12}H_{24}O_2$	Ethyl decanoate	8 e	35 e	69 e	111.8	166.1	238 e	5
$C_{12}H_{25}Br$	1-Bromododecane	31 e	57 e	90 e	132 e	190.8	275.3	5
$C_{12}H_{25}Cl$	1-Chlorododecane	27 e	51 e	81 e	122 e	178.7	262.6	5
$C_{12}H_{26}$	Dodecane	-5.4	18.2	47.6	85.8	138.2	215.8	16
$C_{12}H_{26}O$	1-Dodecanol				133 e	185.0	264.1	1
$C_{12}H_{26}O_3$	Diethylene glycol dibutyl ether	5 e	34.4	70.2	115.3	174.1	253.8	5
$C_{12}H_{27}N$	Tributylamine	-26 e	1 e	35 e	77.7	134.5	213.4	5
$C_{12}H_{27}N$	Triisobutylamine		1 e	28.9	64.9	112.5	178.5	5
$C_{12}H_{27}O_4P$	Tributyl phosphate					205 e	288.3	5
$C_{12}H_{36}O_6Si_6$	Dodecamethylcyclohexasiloxane	18 e	41 e	69 e	108 e	162.2	244.7	5
$C_{13}H_9N$	Acridine			124.4	176.2	246.0	345.4	5
$C_{13}H_9N$	Phenanthridine	79 s						5
$C_{13}H_{10}$	Fluorene	48.4 s			137.4	205.4	295 e	5
$C_{13}H_{10}O_2$	Phenyl benzoate			102.3	151.4	217.9	313.3	5
$C_{13}H_{10}O_3$	Phenyl salicylate				166.0	224.8	312.4	5
$C_{13}H_{12}$	Diphenylmethane		45 e	77 e	119.3	177.7	263.6	1,5
$C_{13}H_{13}N$	Methyldiphenylamine	35 e	63 e	98.4	143.1	201.6	281.6	5
$C_{13}H_{14}$	1-Isopropylnaphthalene	27 e	51 e	82 e	123.2	180.8	267.3	5
$C_{13}H_{20}$	Heptylbenzene	12 e	36 e	66 e	107 e	162.7	246.2	5
$C_{13}H_{24}O_2$	Ethyl 10-undecenoate	32 e	55 e	86 e	125.2	179.5	258.4	5
$C_{13}H_{26}$	1-Tridecene	4.1	28.5	59.0	98.3	152.5	232.3	5
$C_{13}H_{26}$	Heptylcyclohexane	11 e	34 e	65 e	105 e	160.9	244.3	5
$C_{13}H_{26}$	Octylcyclopentane	13 e	36 e	66 e	106 e	160.9	243.1	5
$C_{13}H_{26}O_2$	Tridecanoic acid	87 e	109 e	138 e	176 e	230.3	311.5	5
$C_{13}H_{26}O_2$	Methyl dodecanoate	38 e	61 e	90 e	130 e	184.9	269 e	5
$C_{13}H_{28}$	Tridecane	7.2	31.5	61.8	101.1	155.1	234.9	16
$C_{13}H_{28}O$	1-Tridecanol	71.6	101.0	103 e	140 e	192.3	273.1	5
$C_{14}H_{10}$	Anthracene	89.2 s	125.9 s	151.5 s	165 s	238.8	340.2	1,5
$C_{14}H_{10}$	Phenanthrene	53 s	83 s	120.8	170.4	238.4	337.7	5
$C_{14}H_{10}O_2$	Benzil			123	175	246	346	4
$C_{14}H_{12}$	*cis*-Stilbene	26 e	54 e	88 e	130.4	183 e	253 e	5
$C_{14}H_{12}$	*trans*-Stilbene				155.6	218.1	305.8	5
$C_{14}H_{12}O_2$	Benzoin				181	248	342	4
$C_{14}H_{14}$	1,1-Diphenylethane	19 e	47 e	82.0	125.3	181 e	254 e	5
$C_{14}H_{15}N$	Dibenzylamine	48 e	77 e	113.1	158.9	218.5	299.4	5
$C_{14}H_{16}$	1-Butylnaphthalene	67 e	82 e	103 e	135 e	186.7	288.6	5

VAPOR PRESSURE (continued)

Mol. Form.	Name	Temperature in °C for the indicated pressure 1 Pa	10 Pa	100 Pa	1 kPa	10 kPa	100 kPa	Ref.
$C_{14}H_{16}$	2-Butylnaphthalene	44 e	67 e	98 e	139 e	197.5	287.4	5
$C_{14}H_{22}$	Octylbenzene	20.1	46.2	79.1	121.9	178.1	263.8	5
$C_{14}H_{26}O_4$	Diethyl sebacate		83 e	120	166	225	305	4
$C_{14}H_{27}N$	Tetradecanenitrile	52 e	79 e	114.0	159.0	219.7	306.3	5
$C_{14}H_{28}$	1-Tetradecene	16.1	41.3	72.7	113.2	168.7	250.6	5
$C_{14}H_{28}$	Octylcyclohexane	16.9	44.3	77.8	120.0	177.6	263.2	5
$C_{14}H_{28}$	Nonylcyclopentane	25 e	49 e	80 e	120 e	177.2	261.5	5
$C_{14}H_{28}O_2$	Tetradecanoic acid	96 e	118 e	147 e	186 e	241.3	325.6	5
$C_{14}H_{30}$	Tetradecane	19.1	44.1	75.3	115.7	171.1	253.0	16
$C_{14}H_{30}O$	1-Tetradecanol	80.0	110.5	149.6	152 e	205.3	286.7	5
$C_{14}H_{31}N$	Tetradecylamine			104 e	147 e	206.1	290.9	5
$C_{14}H_{42}O_5Si_6$	Tetradecamethylhexasiloxane	6 e	36 e	72 e	117 e	176.0	259.1	5
$C_{15}H_{18}$	1-Pentylnaphthalene	34 e	62 e	96 e	141.3	202.2	289 e	5
$C_{15}H_{24}$	Nonylbenzene	33.0	58.9	92.0	135.4	193.7	281.4	5
$C_{15}H_{30}$	Nonylcyclohexane	35 e	60 e	92 e	134 e	193.4	280.9	5
$C_{15}H_{30}$	Decylcyclopentane	37 e	61 e	93 e	134 e	192.5	278.8	5
$C_{15}H_{30}O_2$	Methyl tetradecanoate		75 e	110	155	214	295	4
$C_{15}H_{32}$	Pentadecane	30.5	56.1	88.1	129.6	186.3	270.1	16
$C_{16}H_{22}O_4$	Dibutyl phthalate		104.0	142.7	191.5	254.5	339.4	4
$C_{16}H_{32}$	1-Hexadecene	38.4	65.0	98.1	140.5	198.8	284.3	5
$C_{16}H_{32}O_2$	Hexadecanoic acid		136 e	165 e	205 e	261.9	350.2	5
$C_{16}H_{34}$	Hexadecane	41.1	67.4	100.3	142.7	200.7	286.3	16
$C_{16}H_{34}O$	1-Hexadecanol	99.5	130.6	171.9	175 e	229.0	311.7	5
$C_{16}H_{35}N$	Hexadecylamine	63 e	91 e	126 e	171 e	232.6	320.5	5
$C_{17}H_{10}O$	Benzanthrone		184 e	229.3	290.3	377.2	511 e	5
$C_{17}H_{34}O_2$	Methyl hexadecanoate	65 e	93	129	177			4
$C_{17}H_{36}$	Heptadecane	51.5	78.5	112.0	155.3	214.5	302 e	16
$C_{17}H_{36}O$	1-Heptadecanol	94 e	117 e	146 e	185 e	240.1	323.3	5
$C_{18}H_{14}$	*o*-Terphenyl	66 e	94 e	129 e	176 e	241.3	336.3	5
$C_{18}H_{14}$	*m*-Terphenyl	87 e	118 e	156 e	206.6	275.3	374.6	5
$C_{18}H_{14}$	*p*-Terphenyl	127.1 s	154.7 s		217.2	284.0	383.0	5
$C_{18}H_{30}$	Hexaethylbenzene				144.1	206.8	297.5	5
$C_{18}H_{34}O_2$	Oleic acid	94 e	126 e	165.5	214.5	277.0	359.7	5
$C_{18}H_{34}O_2$	Elaidic acid		124 e	166	216	280	361	4
$C_{18}H_{36}O$	Stearaldehyde			142 e	186 e	246.9	336.7	5
$C_{18}H_{36}O_2$	Stearic acid		153 e	183 e	223 e	281.6	374.5	5
$C_{18}H_{38}$	Octadecane	61.5	89.0	123.1	167.3	227.6	316 e	16
$C_{18}H_{38}O$	1-Octadecanol	106 e	130 e	160 e	200.5	257.3	343.0	5
$C_{19}H_{16}$	Triphenylmethane	81 s		112 e	175 e	254.6	360.0	5
$C_{19}H_{36}O_2$	Methyl oleate	85 e	114 e	149.7	195.6	256 e	340 e	5
$C_{19}H_{40}$	Nonadecane	71.1	99.1	133.8	178.8	240.1	330 e	16
$C_{20}H_{42}$	Eicosane	80.4	108.9	144.2	189.8	252.1	344 e	16
$C_{20}H_{42}O$	1-Eicosanol	119 e	143 e	173 e	213 e	270.0	355.1	5
$C_{20}H_{60}O_8Si_9$	Eicosamethylnonasiloxane			141 e	183.1	236.7	307.1	5
$C_{21}H_{21}O_4P$	Tri-*o*-cresyl phosphate	119.0	156.1	201.0	256.3	326.3	418 e	5
$C_{21}H_{21}O_4P$	Tri-*m*-cresyl phosphate	147.8	177.3	211.4	251.3	298 e	355 e	5
$C_{21}H_{21}O_4P$	Tri-*p*-cresyl phosphate	140.6	174 e	214 e	262 e	320 e	392 e	5
$C_{21}H_{44}$	Heneicosane	82.3	113.5	152.2	201.6	263.8	355.9	5
$C_{22}H_{42}O_2$	Brassidic acid	134 e	166 e	203.6	249.8	307.6	382.0	5
$C_{22}H_{42}O_2$	Erucic acid	126 e	160 e	199.4	247.4	306.5	381.1	5
$C_{22}H_{42}O_2$	Butyl oleate	95.5	124.2	158 e	198 e	245 e	304 e	5
$C_{22}H_{44}O_2$	Behenic acid	145.4	176.5	213.7	259.3	316.2	390 e	5
$C_{22}H_{44}O_2$	Butyl stearate	99.6	128 e	162 e	201 e	249 e	307 e	5
$C_{22}H_{46}$	Docosane	83.5	115.0	154.0	203.6	274.8	368.0	5
$C_{23}H_{48}$	Tricosane	102.9	135.1	174.8	221 e	285.3	379.5	5
$C_{24}H_{38}O_4$	Dioctyl phthalate	130 e	163.7	203.8	252 e	311 e	385 e	5
$C_{24}H_{38}O_4$	Bis(2-ethylhexyl) phthalate	122.0	153.2	189.2	231.3	281.1	341.1	5
$C_{24}H_{50}$	Tetracosane	115.0	148.1	188.5	239.1	295.4	390.6	5
$C_{25}H_{52}$	Pentacosane	119.7	152.7	193.2	244.4	305.0	401.1	5
$C_{26}H_{54}$	Hexacosane	125.1	158.8	200.1	252.1	314.3	411.3	5
$C_{27}H_{56}$	Heptacosane	136.7	168.8	206.5	255.8	323.3	421.2	5
$C_{28}H_{58}$	Octacosane	136.5	169.8	210.9	263.1	332.0	430.6	5
$C_{29}H_{60}$	Nonacosane	148.2	182.8	221.2	271.5	340.2	439.7	5
$C_{30}H_{62}$	Squalane	66 e	84 e	105.8	131.9	163.7	203.2	5
C_{70}	Carbon (fullerene-C_{70})	598 s	662 s					22

VAPOR PRESSURE OF FLUIDS AT TEMPERATURES BELOW 300 K

This table gives vapor pressures of 67 important fluids in the temperature range 2 to 300 K. Helium (^4He), hydrogen (H_2), and neon (Ne) are covered on this page. The remaining fluids are listed on subsequent pages by molecular formula in the Hill order (see Introduction). The data have been taken from evaluated sources; references are listed at the end of the table.

Pressures are given in kilopascals (kPa). Note that:

1 kPa = 7.50062 Torr
100 kPa = 1 bar
101.325 kPa = 1 atmos

s following an entry indicates that the compound is solid at that temperature.

Helium		Hydrogen		Neon	
T/K	*P*/kPa	*T*/K	*P*/kPa	*T*/K	*P*/kPa
2.2	5.3	14.0	7.90	25.0	51.3
2.3	6.7	14.5	10.38	26.0	71.8
2.4	8.3	15.0	13.43	27.0	98.5
2.5	10.2	15.5	17.12	28.0	132.1
2.6	12.4	16.0	21.53	29.0	173.5
2.7	14.8	16.5	26.74	30.0	223.8
2.8	17.5	17.0	32.84	31.0	284.0
2.9	20.6	17.5	39.92	32.0	355.2
3.0	24.0	18.0	48.08	33.0	438.6
3.1	27.8	18.5	57.39	34.0	535.2
3.2	32.0	19.0	67.96	35.0	646.2
3.3	36.5	19.5	79.89	36.0	772.8
3.4	41.5	20.0	93.26	37.0	916.4
3.5	47.0	20.5	108.2	38.0	1078
3.6	52.9	21.0	124.7	39.0	1260
3.7	59.3	21.5	143.1	40.0	1462
3.8	66.1	22.0	163.2	41.0	1688
3.9	73.5	22.5	185.3	42.0	1939
4.0	81.5	23.0	209.4	43.0	2216
4.1	90.0	23.5	235.7	44.0	2522
4.2	99.0	24.0	264.2		
4.3	108.7	24.5	295.1		
4.4	119.0	25.0	328.5		
4.5	129.9	25.5	364.3		
4.6	141.6	26.0	402.9		
4.7	153.9	26.5	444.3		
4.8	167.0	27.0	488.5		
4.9	180.8	27.5	535.7		
5.0	195.4	28.0	586.1		
5.1	210.9	28.5	639.7		
		29.0	696.7		
		29.5	757.3		
		30.0	821.4		
		30.5	889.5		
		31.0	961.5		
		31.5	1038.0		
		32.0	1119.0		
		32.5	1204.0		
Ref.	17,18		1		13

T/K	Ar Argon	BCl_3 Boron trichloride	BF_3 Boron trifluoride	BrH Hydrogen bromide	Br_2 Bromine	ClF Chlorine fluoride	ClH Hydrogen chloride
50	0.1 s						
55	0.2 s						
60	0.8 s						
65	2.8 s						
70	7.7 s						
75	18.7 s						
80	40.7 s						
85	79.0						
90	134						
95	213						
100	324						
105	473						
110	666						
115	910					0.1	
120	1214					0.3	0.1 s
125	1584					0.6	0.3 s
130	2027					1.2	0.5 s
135	2553			0.1 s		2.1	1.0 s
140	3170			0.3 s		3.6	1.9 s
145	3892		7.7	0.6 s		6.0	3.4 s
150	4736		13.4	1.1 s		9.5	5.8 s
155			22.3	1.9 s		14.6	9.5 s
160			35.2	3.3 s		21.8	14.7
165			53.7	5.4 s		31.7	22.0
170			79.1	8.7 s		44.8	31.9
175			113	13.4 s		62.0	45.1
180		0.1	157	20.1 s		84.2	62.5
185		0.2	214	29.5 s		112	84.7
190		0.3	285	37.9		147	113
195		0.5	372	51.8		190	148
200		0.8	479	69.5		242	190
205		1.2	608	91.8		304	242
210		1.8	762	119		378	304
215		2.6	944	153		464	377
220		3.8	1160	194	0.1 s	564	463
225		5.2	1413	242	0.2 s	680	563
230		7.2	1709	299	0.3 s	812	678
235		9.7	2056	366	0.4 s	961	811
240		12.9	2460	443	0.7 s	1130	961
245		17.0	2913	532	1.1 s	1319	1132
250		22.0	3481	633	1.7 s	1529	1325
255		28.1	4123	748	2.6 s	1762	1542
260		35.6	4874	878	3.8 s	2019	1784
265		44.5		1023	5.5 s	2301	2054
270		55.1		1185	7.3	2608	2354
275		67.6		1364	9.5	2941	2686
280		82.2		1562	12.3	3303	3053
285		99.1		1780	15.6	3693	3457
290		119		2018	19.7	4111	3901
295		141		2278	24.6	4560	4388
300		166		2561	30.5	5039	4921
Ref.	8,15	12	12	12	12	12	12

T/K	ClO_2 Chlorine dioxide	Cl_2 Chlorine	Cl_4Si Silicon tetrachloride	FH Hydrogen fluoride	F_2 Fluorine	F_2O Difluorine oxide	F_3N Nitrogen trifluoride
50							
55					0.4		
60					1.5		
65					4.8		
70					12.3		
75					27.6	0.1	
80					55.3	0.2	
85					101	0.5	0.1
90					172	1.2	0.2
95					276	2.6	0.4
100					420	5.3	0.9
105					615	10.1	2.0
110					870	18.0	4.0
115					1196	30.5	7.3
120					1605	49.3	12.8
125					2108	76.7	21.1
130					2721	115	33.5
135					3458	168	51.1
140					4339	237	75.4
145						328	108
150						444	150
155						588	205
160						766	273
165						981	357
170						1238	459
175		1.8				1541	581
180		2.8				1895	726
185		4.2				2303	896
190		6.1		0.3		2771	1092
195	0.1	8.7		0.5		3302	1319
200	0.3	12.3		0.8		3899	1578
205	0.5	16.9		1.2		4567	1871
210	0.9	22.9	0.1	1.7		5308	2203
215	1.4	30.5	0.2	2.3			2577
220	2.3	40.1	0.3	3.2			2995
225	3.5	51.9	0.5	4.4			3464
230	5.3	66.4	0.7	5.9			3991
235	7.6	84.0	1.0	7.9			
240	10.8	105	1.5	10.3			
245	14.9	130	2.0	13.4			
250	20.1	160	2.8	17.2			
255	26.6	194	3.8	21.8			
260	34.6	234	5.0	27.4			
265	44.4	280	6.6	34.2			
270	56.1	332	8.6	42.2			
275	69.9	392	11.1	51.8			
280	86.2	459	14.2	63.1			
285	105	535	17.9	76.3			
290	127	619	22.3	91.7			
295	151	714	27.7	110			
300	179	818	34.0	130			
Ref.	12	5	12	12	12	12	1

T/K	F_3P Phosphorous trifluoride	F_4Si Silicon tetrafluoride	F_6S Sulfur hexafluoride	HI Hydrogen iodide	H_2S Hydrogen sulfide	H_3N Ammonia	H_3P Phosphine
50							
55							
60							
65							
70							
75							
80							
85							
90							
95							
100							
105	0.1						
110	0.2						0.1
115	0.5						0.2
120	1.0						0.4
125	1.9	0.1 s					0.7
130	3.5	0.2 s					1.3
135	5.9	0.4 s			0.1 s		2.3
140	9.5	0.9 s	0.1 s		0.2 s		3.9
145	14.9	1.9 s	0.2 s		0.3 s		6.2
150	22.5	3.8 s	0.4 s		0.6 s		9.6
155	33.1	7.5 s	0.8 s	0.1 s	1.1 s		14.5
160	47.3	14.0 s	1.5 s	0.2 s	1.9 s	0.1 s	21.1
165	66.0	25.2 s	2.6 s	0.4 s	3.2 s	0.2 s	30.0
170	90.1	43.8 s	4.4 s	0.8 s	5.2 s	0.3 s	41.6
175	121	74.2 s	7.1 s	1.3 s	8.3 s	0.6 s	56.6
180	159	122 s	11.3 s	2.2 s	12.7 s	1.2 s	75.6
185	206	197 s	17.3 s	3.4 s	18.9 s	2.1 s	99.2
190	262	280	25.9 s	5.3 s	26.6	3.5 s	128
195	330	376	38.0 s	8.0 s	36.7	5.8 s	163
200	410	488	54.4 s	11.7 s	49.8	8.7	205
205	503	618	76.6 s	16.8 s	66.4	12.6	254
210	611	766	106 s	23.6 s	87.1	17.9	312
215	736	932	145 s	32.5 s	113	24.9	379
220	877	1117	195 s	44.0 s	144	34.1	456
225	1037	1324	249	56.2	182	45.9	544
230	1217	1555	305	71.4	227	60.8	644
235	1418	1816	371	89.7	281	79.6	756
240	1640	2111	448	112	344	103	881
245	1885	2449	536	137	416	131	1019
250	2154	2841	636	168	500	165	1172
255	2448	3301	750	203	597	207	1341
260	2767		878	244	706	256	1525
265	3112		1021	290	830	313	1725
270			1181	343	969	381	1942
275			1358	404	1124	460	2176
280			1554	472	1297	552	2428
285			1768	548	1488	655	2699
290			2003	633	1698	774	2987
295			2258	727	1929	909	3295
300			2534	831	2181	1062	3621
Ref.	12	12	12,15	12	12,15	11	12

T/K	H_4Si Silane	Kr Krypton	NO Nitric oxide	N_2 Nitrogen	N_2O Nitrous oxide	O_2 Oxygen	O_2S Sulfur dioxide
50				0.4 s			
55				1.8 s		0.2	
60				6.3 s		0.7	
65				17.4		2.3	
70				38.6		6.3	
75		0.1 s		76.1		14.5	
80		0.4 s		137		30.1	
85		1.1 s	0.1 s	229		56.8	
90		2.7 s	0.4 s	361		99.3	
95	0.1	6.0 s	1.3 s	541		163	
100	0.2	12.1 s	3.8 s	779		254	
105	0.4	22.8 s	10.0 s	1084		379	
110	1.0	40.4 s	23.5	1467		543	
115	1.9	68.0 s	46.8	1939	0.1	756	
120	3.5	103	86.5	2513	0.1	1022	
125	6.1	150	151	3209	0.3	1351	
130	10.0	211	248		0.7	1749	
135	15.8	290	391		1.3	2225	
140	24.1	388	592		2.5	2788	
145	35.3	509	867		4.3	3448	
150	50.3	655	1231		7.1	4219	
155	69.8	830	1703		11.4		
160	94.6	1037	2302		17.6		
165	126	1278	3050		26.4		
170	164	1557	3971		38.5		0.1
175	210	1877	5089		54.7		0.2
180	265	2241	6433		75.9		0.3
185	331	2655			103		0.5
190	408	3120			138		0.8
195	498	3641			181		1.3
200	602	4223			234		2.0
205	722	4870			298		3.0
210	859				374		4.4
215	1017				465		6.3
220	1196				571		9.0
225	1398				694		12.6
230	1628				835		17.3
235	1888				996		23.3
240	2180				1179		31.1
245	2509				1385		40.9
250	2880				1615		53.2
255	3296				1870		68.3
260	3763				2152		86.7
265	4288				2462		109
270					2802		136
275					3172		168
280					3573		205
285					4006		249
290					4473		300
295					4973		359
300					5508		426
Ref.	12	13, 15	12, 15	1	12	3	12

VAPOR PRESSURE OF FLUIDS AT TEMPERATURES BELOW 300 K (continued)

T/K	O_3 Ozone	Rn Radon	Xe Xenon	$CBrF_3$ Bromotri-fluoromethane	$CClF_3$ Chlorotri-fluoromethane	CCl_2F_2 Dichlorodi-fluoromethane	CCl_3F Trichloro-fluoromethane
50							
55							
60							
65							
70							
75							
80							
85							
90							
95							
100	0.1		0.1 s				
105	0.2		0.1 s				
110	0.4		0.3 s				
115	1.0		0.7 s		0.1		
120	2.0		1.5 s		0.2		
125	3.8		2.7 s		0.3		
130	6.8	0.1	4.9 s		0.6		
135	11.5	0.3	8.5 s	0.1	1.1		
140	18.7	0.5	14.0 s	0.3	2.0		
145	29.1	0.9	22.2 s	0.5	3.3		
150	43.7	1.5	34.2 s	0.9	5.3		
155	63.6	2.4	51.1 s	1.5	8.3	0.1	
160	89.9	3.8	74.2 s	2.5	12.6	0.3	
165	124	5.8	101	3.9	18.6	0.5	
170	168	8.6	134	5.9	26.8	0.8	
175	222	12.5	173	8.8	37.6	1.3	
180	289	17.7	222	12.8	51.7	2.1	
185	367	24.5	280	18.1	69.7	3.2	
190	468	33.2	348	25.1	92.3	4.8	0.2
195	584	44.4	428	34.1	120	6.9	0.3
200	721	58.2	521	45.6	155	9.9	0.4
205	881	75.3	628	60.0	196	13.7	0.6
210	1068	96	750	77.8	246	18.8	1.0
215	1285	121	889	99.5	304	25.2	1.4
220	1536	151	1045	126	372	33.3	2.0
225	1824	185	1220	157	451	43.3	2.9
230	2155		1416	194	542	55.5	4.1
235	2534		1633	237	646	70.4	5.6
240	2968		1872	287	763	88.1	7.6
245	3464		2136	344	896	109	10.1
250	4031		2425	410	1044	134	13.3
255	4678		2742	485	1210	163	17.2
260	5417		3087	570	1394	196	22.1
265			3462	665	1598	234	28.0
270			3869	771	1823	278	35.1
275			4310	889	2071	327	43.7
280			4786	1021	2343	383	53.8
285			5299	1166	2641	445	65.7
290				1325	2968	515	79.6
295				1501	3325	593	95.6
300				1692	3716	679	114.1
Ref.	12	15	12,13	12	12	12	12

T/K	CCl_4 Tetrachloro-methane	CF_4 Tetrafluoro-methane	CO Carbon monoxide	COS Carbon oxysulfide	CO_2 Carbon dioxide	$CHClF_2$ Chlorodifluo-methane	$CHCl_3$ Trichloro-methane
50			0.1 s				
55			0.6 s				
60			2.6 s				
65			8.2 s				
70			21.0				
75			44.4				
80			83.7				
85			147				
90		0.1	239				
95		0.3	371				
100		0.8	545				
105		1.7	771				
110		3.4	1067				
115		6.5	1428				
120		11.5	1877				
125		19.3	2400				
130		30.8	3064				
135		47.4			0.1 s		
140		70.2		0.1	0.2 s		
145		101		0.2	0.4 s		
150		141		0.4	0.8 s	0.1	
155		191		0.8	1.7 s	0.3	
160		254		1.3	3.1 s	0.5	
165		332		2.2	5.7 s	0.8	
170		425		3.4	9.9 s	1.4	
175		537		5.2	16.8 s	2.3	
180		669		7.8	27.6 s	3.6	
185		824		11.3	44.0 s	5.5	
190		1005		15.9	68.4 s	8.1	
195		1216		22.1	104 s	11.8	
200		1460		30.0	155 s	16.7	
205		1743		40.1	227 s	23.1	
210		2073		52.7	327 s	31.5	
215		2457		68.2	465 s	42.1	0.1
220		2907		87.2	600	55.3	0.2
225		3438		110	735	71.7	0.3
230				137	894	91.6	0.4
235				169	1075	116	0.7
240				207	1283	144	1.0
245				250	1519	178	1.4
250				301	1786	218	2.0
255	1.5			358	2085	264	2.7
260	2.1			423	2419	317	3.7
265	2.8			497	2790	377	5.0
270	3.7			580	3203	446	6.6
275	4.9			673	3658	525	8.7
280	6.4			777	4160	613	11.3
285	8.2			892	4712	711	14.4
290	10.5			1019	5315	821	18.3
295	13.2			1159	5984	944	22.9
300	16.5			1313	6710	1080	28.5
Ref.	12	12	9	12	6	12	12

T/K	CHF_3 Trifluoro-methane	CHN Hydrogen cyanide		CH_2Cl_2 Dichloro-methane	CH_2F_2 Difluoro methane	CH_2O Formaldehyde	CH_3Cl Chloromethane	CH_3F Fluoromethane
50								
55								
60								
65								
70								
75								
80								
85								
90								
95								
100								
105								
110								
115								
120	0.1							
125	0.2							
130	0.4							
135	0.7							0.6
140	1.4				0.1			1.2
145	2.5				0.2			2.1
150	4.3				0.3			3.6
155	7.1				0.6			5.9
160	11.1				1.0			9.3
165	17.0				1.7			14.1
170	25.3				2.8			20.9
175	36.5				4.4			29.9
180	51.4				6.8			42.0
185	70.9				10.2	1.3	2.1	57.6
190	95.8				14.8	2.0	3.1	77.4
195	127				21.2	3.0	4.6	102
200	166	0.1	s	0.1	29.5	4.4	6.7	133
205	214	0.2	s	0.2	40.5	6.4	9.5	171
210	271	0.4	s	0.3	54.5	9.1	13.1	216
215	340	0.6	s	0.4	72.1	12.7	17.9	270
220	421	1	s	0.6	94.1	17.4	24.0	333
225	516	1.5	s	0.9	121	23.4	31.8	408
230	626	2.2	s	1.4	154	31.0	41.4	495
235	754	3.3	s	2.0	193	40.6	53.3	595
240	900	4.7	s	2.8	240	52.5	67.7	711
245	1067	6.8	s	3.8	295	67.0	85.1	843
250	1257	9.7	s	5.3	360	84.6	106	993
255	1472	13.6	s	7.1	434	106	131	1163
260	1713	18.8		9.5	521	131	159	1355
265	1984	24.1		12.4	620	161	193	1571
270	2287	30.5		16.1	732	196	232	1813
275	2624	38.3		20.7	860	236	277	2084
280	3000	47.7		26.3	1004	283	327	2387
285	3418	58.8		33.0	1165	337	385	2724
290	3881	72.1		41.1	1346	399	450	3099
295	4393	87.6		50.8	1547	470	524	3516
300		105.9		62.1	1770	549	606	3978
Ref.	12	12,16		12	12	12	12	12

VAPOR PRESSURE OF FLUIDS AT TEMPERATURES BELOW 300 K (continued)

T/K	CH_4 Methane	CH_4O Methanol	C_2H_2 Acetylene	C_2H_4 Ethylene	C_2H_6 Ethane	C_2H_6O Dimethyl ether	C_3H_4 Propadiene
50							
55							
60							
65	0.1						
70	0.3						
75	0.8						
80	2.1						
85	4.9						
90	10.6						
95	20.0						
100	34.5						
105	57.0						
110	88.4			0.3			
115	133			0.8	0.1		
120	192			1.4	0.4		
125	269			2.7	0.7		
130	368		0.1 s	4.5	1.3		
135	491		0.3 s	7.7	2.2		
140	642		0.7 s	11.9	3.8		
145	824		1.3 s	18.3	6.0		
150	1041		2.6 s	27.5	9.7		0.1
155	1297		4.6 s	39.9	15.0	0.1	0.2
160	1594		7.8 s	56.4	21.5	0.2	0.3
165	1937		12.8 s	77.9	31.0	0.3	0.6
170	2331		20.6 s	105	42.9	0.5	1.0
175	2779		32.2 s	140	59.0	0.9	1.7
180	3288		49.0 s	182	78.7	1.4	2.7
185	3865		72.9 s	234	104	2.1	4.1
190	4520		106 s	296	135	3.2	6.1
195			146	369	172	4.7	8.9
200			190	456	217	6.8	12.5
205			244	557	271	9.6	17.4
210			309	673	334	13.3	23.7
215			385	806	407	18.1	31.6
220			475	958	492	24.3	41.4
225			579	1128	590	32.1	53.5
230		0.1	699	1321	700	41.9	68.2
235		0.2	837	1535	826	53.9	85.8
240		0.4	993	1774	967	68.6	107
245		0.5	1170	2039	1125	86.3	131
250		0.8	1370	2331	1301	108	160
255		1.2	1593	2652	1496	133	193
260		1.7	1843	3005	1712	162	230
265		2.4	2121	3391	1949	197	273
270		3.3	2429	3813	2210	237	322
275		4.5	2771	4275	2495	283	376
280		6.2	3150		2806	335	438
285		8.3	3567		3146	395	506
290		11	4028		3515	463	582
295		14.4	4535		3917	538	666
300		18.7	5093		4355	623	759
Ref.	2,16	12	12,16	4	2	12	12

VAPOR PRESSURE OF FLUIDS AT TEMPERATURES BELOW 300 K (continued)

T/K	C_3H_6 Propylene	C_3H_8 Propane	C_4H_6 Buta-1,3-diene	C_4H_{10} *n*-Butane	C_4H_{10} Isobutane	C_5H_{12} *n*-Pentane	C_5H_{12} Neopentane
50							
55							
60							
65							
70							
75							
80							
85							
90							
95							
100							
105							
110							
115							
120							
125							
130							
135							
140	0.1						
145	0.2						
150	0.4						
155	0.7						
160	1.2	0.8			0.1		
165	2.0	1.4			0.1		
170	3.1	2.2	0.1	0.1	0.3		
175	4.7	3.3	0.2	0.2	0.4		
180	7.0	5.0	0.4	0.3	0.7		
185	10.1	7.3	0.6	0.5	1.1		0.1 s
190	14.2	10.5	1.0	0.8	1.7		0.2 s
195	19.7	15.0	1.5	1.3	2.5		0.4 s
200	26.9	20.1	2.3	1.9	3.7		0.7 s
205	35.9	27.0	3.4	2.8	5.3		1.1 s
210	47.3	36.0	4.8	4.0	7.4		1.6 s
215	61.3	47.0	6.7	5.7	10.2		2.4 s
220	78.5	60.0	9.2	7.8	13.8	1.0	3.6 s
225	99.2	77.0	12.5	10.6	18.3	1.5	5.2 s
230	124	97.0	16.7	14.1	24.0	2.1	7.3 s
235	153	120	21.9	18.5	31.1	3.0	10.2 s
240	188	148	28.4	24.1	39.8	4.2	13.9 s
245	228	180	36.3	30.9	50.3	5.7	18.7 s
250	274	218	46.0	39.1	62.9	7.6	24.8 s
255	327	261	57.6	49.1	77.8	10.0	32.4 s
260	387	311	71.3	61.0	95.4	13.0	41.6
265	456	367	87.6	75.0	116	16.6	51.4
270	533	431	107	91.5	140	21.1	63.0
275	619	502	129	111	167	26.6	76.6
280	715	582	154	133	198	33.1	92.3
285	822	671	184	159	234	40.8	111
290	940	769	217	188	274	50.0	131
295	1069	878	255	221	319	60.7	155
300	1212	998	297	258	370	73.2	182
Ref.	7	2	12	2	2	14	12,16

REFERENCES

1. B. A. Younglove, Thermophysical properties of fluids. I. Ethylene, parahydrogen, nitrogen trifluoride, and oxygen, *J. Phys. Chem. Ref. Data*, 11, Supp. 1, 1982.
2. B. A. Younglove and J. F. Ely, Thermophysical properties of fluids. II. Methane, ethane, propane, isobutane, and normal butane, *J. Phys. Chem. Ref. Data*, 16, 577, 1987.
3. W. Wagner, et al., *International Tables for the Fluid State: Oxygen*, Blackwell Scientific Publications, Oxford, 1987.
4. R. T. Jacobsen, et al., *International Tables for the Fluid State: Ethylene*, Blackwell Scientific Publications, Oxford, 1988.
5. S. Angus, et al., *International Tables for the Fluid State: Chlorine*, Pergamon Press, Oxford, 1985.
6. S. Angus, et al., *International Tables for the Fluid State: Carbon Dioxide*, Pergamon Press, Oxford, 1976.
7. S. Angus, et al., *International Tables for the Fluid State: Propylene*, Pergamon Press, Oxford, 1980.
8. R. B. Stewart and R. T. Jacobsen, Thermophysical properties of argon, *J. Phys. Chem. Ref. Data*, 18, 639, 1989.
9. R. D. Goodwin, Carbon monoxide thermophysical properties, *J. Phys. Chem. Ref. Data*, 14, 849, 1985.
10. R. D. Goodwin, Methanol thermophysical properties, *J. Phys. Chem. Ref. Data*, 16, 799, 1987.
11. L. Haar, Thermodynamic properties of ammonia, *J. Phys. Chem. Ref. Data*, 7, 635, 1978.
12. DIPPR Data Compilation of Pure Compound Properties, Design Institute for Physical Properties Data, American Institute of Chemical Engineers, 1987.
13. V. A. Rabinovich, et al., *Thermophysical Properties of Neon, Argon, Krypton, and Xenon*, Hemisphere Publishing Corp., New York, 1987.
14. K. N. Marsh, *Recommended Reference Methods for the Realization of Physicochemical Properties*, Blackwell Scientific Publications, Oxford, 1987.
15. TRC Thermodynamic Tables: Non-Hydrocarbons, Thermodynamic Research Center, Texas A & M University, College Station, Texas, 1985.
16. R. M. Stevenson and S. Malanowski, *Handbook of the Thermodynamics of Organic Compounds*, Elsevier, New York, 1987.
17. S. Angus and K. M. de Reuck, *International Tables of the Fluid State: Helium-4*, Pergamon Press, Oxford, 1977.
18. R. D. McCarty, *J. Phys. Chem. Ref. Data*, 2, 923, 1973.

VAPOR PRESSURE OF SATURATED SALT SOLUTIONS

This table gives the vapor pressure of water above saturated solutions of some common salts at ambient temperatures. Data on pure water are given on the last line for comparison.

The references provide additional information on water activity, osmotic coefficient, and enthalpy of vaporization.

REFERENCES

1. Apelblat, A., *J. Chem. Thermodynamics*, 24, 619, 1992.
2. Apelblat, A., *J. Chem. Thermodynamics*, 25, 63, 1993.
3. Apelblat, A., *J. Chem. Thermodynamics*, 25, 1513, 1993.
4. Apelblat, A. and Korin, E., *J. Chem. Thermodynamics*, 30, 59, 1998.

Vapor Pressure in kPa

Salt	10°C	15°C	20°C	25°C	30°C	35°C	40°C	Ref.
$BaCl_2$	0.971	1.443	2.073	2.887	3.903	5.133	6.576	1
$Ca(NO_3)_2$	0.701	1.015	1.381	1.772	2.154	2.487		1
$CuSO_4$	1.113	1.574	2.189	2.996	4.037	5.363		3
$FeSO_4$	0.978	1.516	2.208	3.035	3.950	4.884		3
KBr	0.953	1.338	1.853	2.533	3.419	4.563		3
KIO_3	1.100	1.564	2.177	2.970	3.979	5.236	6.778	4
K_2CO_3	0.541	0.802	1.134	1.536	1.997	2.499	3.016	1
LiCl	0.128	0.193	0.279	0.384				2
$Mg(NO_3)_2$	0.726	0.999	1.339	1.749	2.231	2.782	3.397	1
$MnCl_2$	0.697	1.064	1.515	2.020	2.535	3.002		3
NH_4Cl	0.971	1.328	1.836	2.481				2
NH_4NO_3	0.853	1.152	1.524	1.972				2
$(NH_4)_2SO_4$	0.901	1.319	1.871	2.573	3.439	4.474		3
NaBr	0.722	1.004	1.376	1.858	2.475	3.255	4.229	4
NaCl	0.921	1.285	1.768	2.401	3.218	4.262	5.581	4
$NaNO_2$	0.703	0.994	1.381	1.888	2.540	3.368	4.403	4
$NaNO_3$	0.884	1.244	1.719	2.335	3.121	4.109	5.333	4
RbCl	0.862	1.215	1.684	2.298	3.088	4.089	5.343	4
$ZnSO_4$	0.945	1.401	1.986	2.698	3.523	4.431	5.382	1
Water	1.228	1.706	2.339	3.169	4.246	5.627	7.381	

IUPAC RECOMMENDED DATA FOR VAPOR PRESSURE CALIBRATION

Vapor pressures are given in kPa (1 kPa = 0.0098692 atmos = 7.5006 Torr). Reprinted with permission if IUPAC.

REFERENCE

Marsh, K.N., Ed., *Recommended Reference Materials for the Realization of Physicochemical Properties*, Blackwell Scientific Publications,Oxford (1987).

T/K	CO_2(s)	H_2O(s)	$C_{10}H_8$(s)	*n*-C_5H_{12}	C_6H_6	C_6F_6	H_2O	Hg
180	27.62							
190	68.44							
200	155.11	0.0002						
210	327.17	0.0007						
220		0.0026						
230		0.0089						
240		0.0273						
250		0.0760		7.60				
260		0.1958	0.0001	12.98				
270		0.4701	0.0005	21.15			0.485	
280			0.0017	33.11	5.148	4.322	0.991	
290			0.0049	50.01	8.606	7.463	1.919	
300			0.0134	73.17	13.816	12.328	3.535	
310			0.0341	104.07	21.389	19.576	6.228	
320			0.0814	144.3	32.054	30.009	10.540	
330			0.1829	195.7	46.656	44.578	17.202	
340			0.3899	260.1	66.152	64.380	27.167	
350			0.7920	339.4	91.609	90.664	41.647	
360				435.9	124.192	124.816	62.139	
370				551.5	165.2	168.4	90.453	
380				688.8	215.9	223.0	128.74	
390				850.2	277.7	290.4	179.48	
400				1038	353.2	372.6	245.54	0.138
410				1256	441.0	471.5	330.15	0.215
420				1507	545.5	589.3	436.89	0.329
430				1793	667.6	728.3	569.73	0.493
440				2120	808.8	890.9	732.99	0.724
450				2490	971.1	1080	931.36	1.045
460				2910	1156	1297	1169.9	1.485
470					1366	1547	1453.9	2.078
480					1602	1833	1789.0	2.866
490					1868	2159	2181.4	3.899
500					2164	2530	2637.3	5.239
510					2494	2954	3163.3	6.955
520					2861		3766.4	9.131
530					3267		4453.9	11.861
540					3717		5233.5	15.256
550					4216		6113.4	19.438
560					4770		7102.0	25.547
570							8208.6	30.74
580							9443.0	38.19
590							10816	47.09
600							12339	57.64

ENTHALPY OF VAPORIZATION

The molar enthalpy (heat) of vaporization $\Delta_{vap}H$, which is defined as the enthalpy change in the conversion of one mole of liquid to gas at constant temperature, is tabulated here for approximately 850 inorganic and organic compounds. Values are given, when available, both at the normal boiling point t_b, referred to a pressure of 101.325 kPa (760 mmHg), and at 25°C. Substances are listed by molecular formula in the modified Hill order (see Preface).

The values in this table were measured either by calorimetric techniques or by application of the Claperyon equation to the variation of vapor pressure with temperature. See Reference 1 for a discussion of the accuracy of different experimental techniques and for methods of estimating enthalpy of vaporization at other temperatures.

REFERENCES

1. Majer, V. and Svoboda, V., *Enthalpies of Vaporization of Organic Compounds,* Blackwell Scientific Publications, Oxford, 1985.
2. Chase, M. W., Davies, C. A., Downey, J. R., Frurip, D. J., McDonald, R. A., and Syverud, A. N., *JANAF Thermochemical Tables, Third Edition, J. Phys. Chem. Ref. Data,* Vol. 14, Suppl. 1, 1985.
3. *Landolt-Börnstein, Numerical Data and Functional Relationships in Science and Technology, Sixth Edition,* II/4, *Caloric Quantities of State,* Springer-Verlag, Heidelberg, 1961.
4. Daubert, T. E., Danner, R. P., Sibul, H. M., and Stebbins, C. C., *Physical and Thermodynamic Properties of Pure Compounds: Data Compilation,* extant 1994 (core with 4 supplements), Taylor & Francis, Bristol, PA.
5. Ruzicka, K. and Majer, V., "Simultaneous Treatment of Vapor Pressures and Related Thermal Data Between the Triple and Normal Boiling Temperatures for *n*-Alkanes C_5 - C_{20}", *J. Phys. Chem. Ref. Data,* 23, 1, 1994.
6. Verevkin, S. P., "Thermochemistry of Amines: Experimental Standard Molar Enthalpies of Formation of Some Aliphatic and Aromatic Amines", *J. Chem. Thermodynamics,* 29, 891, 1997.
7. Cady, G. H. and Hargreaves, G. B., "The Vapor Pressure of Some Heavy Transition Metal Hexafluorides", *J. Chem. Soc.,* 1961, 1563; 1961, 1578.
8. Steele, W. V., Chirico, R. D., Knipmeyer, S. E., and Nguyen, A., *J. Chem. Eng. Data,* 41, 1255, 1996.

Mol. Form.	Name	t_b/°C	$\Delta_{vap}H(t_b)$ kJ/mol	$\Delta_{vap}H$(25°C) kJ/mol
AgBr	Silver(I) bromide	1502	198	
AgCl	Silver(I) chloride	1547	199	
AgI	Silver(I) iodide	1506	143.9	
Al	Aluminum	2519	294	
AlB_3H_{12}	Aluminum borohydride	44.5	30	
$AlBr_3$	Aluminum tribromide	255	23.5	
AlI_3	Aluminum triiodide	382	32.2	
Ar	Argon	-185.85	6.43	
$AsBr_3$	Arsenic(III) bromide	221	41.8	
$AsCl_3$	Arsenic(III) chloride	130	35.01	
AsF_3	Arsenic(III) fluoride	57.8	29.7	
AsF_5	Arsenic(V) fluoride	-52.8	20.8	
AsH_3	Arsine	-62.5	16.69	
AsI_3	Arsenic(III) iodide	424	59.3	
Au	Gold	2856	324	
B	Boron	4000	480	
BBr_3	Boron tribromide	91	30.5	
BCl_3	Boron trichloride	12.65	23.77	23.1
BF_3	Boron trifluoride	-101	19.33	
BI_3	Boron triiodide	210	40.5	
B_2F_4	Tetrafluorodiborane	-34	28	
B_2H_6	Diborane	-92.4	14.28	
B_4H_{10}	Tetraborane	18	27.1	
B_5H_{11}	Pentaborane(11)	63	31.8	
Ba	Barium	1897	140	
$BeCl_2$	Beryllium chloride	482	105	
BeI_2	Beryllium iodide	487	70.5	
Bi	Bismuth	1564	151	
$BiBr_3$	Bismuth tribromide	453	75.4	
$BiCl_3$	Bismuth trichloride	447	72.61	
BrF	Bromine fluoride	20	25.1	
BrF_3	Bromine trifluoride	125.8	47.57	

Mol. Form.	Name	t_b/°C	$\Delta_{vap}H(t_b)$ kJ/mol	$\Delta_{vap}H(25°C)$ kJ/mol
BrF_5	Bromine pentafluoride	40.76	30.6	
BrH	Hydrogen bromide	-66.38		12.69
BrH_3Si	Bromosilane	1.9	24.4	
$BrIn$	Indium(I) bromide	656	92	
$BrTl$	Thallium(I) bromide	819	99.56	
Br_2	Bromine	58.8	29.96	30.91
Br_2Cd	Cadmium bromide	844	115	
Br_2H_2Si	Dibromosilane	66	31	
Br_2Hg	Mercury(II) bromide	322	58.89	
Br_2Pb	Lead(II) bromide	892	133	
Br_2Sn	Tin(II) bromide	639	102	
Br_2Zn	Zinc bromide	697	118	
Br_3Ga	Gallium(III) bromide	279	38.9	
Br_3HSi	Tribromosilane	109	34.8	
Br_3OP	Phosphorus(V) oxybromide	191.7	38	
Br_3P	Phosphorus(III) bromide	172.95	38.8	
Br_3Sb	Antimony(III) bromide	280	59	
Br_4Ge	Germanium(IV) bromide	186.35	41.4	
Br_4Si	Tetrabromosilane	154	37.9	
Br_4Sn	Tin(IV) bromide	205	43.5	
Br_4Ti	Titanium(IV) bromide	230	44.37	
Br_5Ta	Tantalum(V) bromide	349	62.3	
Cd	Cadmium	767	99.87	
$CdCl_2$	Cadmium chloride	960	124.3	
CdF_2	Cadmium fluoride	1748	214	
CdI_2	Cadmium iodide	742	115	
ClF	Chlorine fluoride	-101.1	24	
$ClFO_3$	Perchloryl fluoride	-46.75	19.33	
ClF_2P	Phosphorus(III) chloride difluoride	-47.25	17.6	
ClF_3	Chlorine trifluoride	11.75	27.53	
ClF_3Si	Chlorotrifluorosilane	-70.0	18.7	
ClH	Hydrogen chloride	-85	16.15	9.08
ClH_3Si	Chlorosilane	-30.4	21	
$ClNO$	Nitrosyl chloride	-5.5	25.78	
$ClNO_2$	Nitryl chloride	-15	25.7	
ClO_2	Chlorine dioxide	11	30	
$ClTl$	Thallium(I) chloride	720	102.2	
Cl_2	Chlorine	-34.04	20.41	17.65
Cl_2Cr	Chromium(II) chloride	1300	197	
Cl_2CrO_2	Chromyl chloride	117	35.1	
Cl_2FP	Phosphorus(III) dichloride fluoride	14	24.9	
Cl_2F_2Si	Dichlorodifluorosilane	-32	21.2	
Cl_2H_2Si	Dichlorosilane	8.3	25	24.2
Cl_2Hg	Mercury(II) chloride	304	58.9	
Cl_2O	Chlorine monoxide	2.2	25.9	
Cl_2OS	Thionyl chloride	75.6	31.7	31
Cl_2O_2S	Sulfuryl chloride	69.4	31.4	30.1
Cl_2Pb	Lead(II) chloride	951	127	
Cl_2Sn	Tin(II) chloride	623	86.8	
Cl_2Ti	Titanium(II) chloride	1500	232	
Cl_2Zn	Zinc chloride	732	126	
Cl_3Ga	Gallium(III) chloride	201	23.9	
Cl_3HSi	Trichlorosilane	33		25.7
Cl_3OP	Phosphorus(V) oxychloride	105.5	34.35	38.6
Cl_3OV	Vanadyl trichloride	127	36.78	
Cl_3P	Phosphorus(III) chloride	75.95	30.5	32.1
Cl_3Sb	Antimony(III) chloride	220.3	45.19	
Cl_3Ti	Titanium(III) chloride	960	124	

Mol. Form.	Name	t_b/°C	$\Delta_{vap}H(t_b)$ kJ/mol	$\Delta_{vap}H(25°C)$ kJ/mol
Cl_4Ge	Germanium(IV) chloride	86.55	27.9	
Cl_4OW	Tungsten(VI) oxytetrachloride	227.55	67.8	
Cl_4Si	Tetrachlorosilane	57.65	28.7	29.7
Cl_4Sn	Tin(IV) chloride	114.15	34.9	
Cl_4Te	Tellurium tetrachloride	387	77	
Cl_4Th	Thorium(IV) chloride	921	146.4	
Cl_4Ti	Titanium(IV) chloride	136.45	36.2	
Cl_4V	Vanadium(IV) chloride	148	41.4	42.5
Cl_5Mo	Molybdenum(V) chloride	268	62.8	
Cl_5Nb	Niobium(V) chloride	254.0	52.7	
Cl_5Ta	Tantalum(V) chloride	239.35	54.8	
Cl_6W	Tungsten(VI) chloride	346.75	52.7	
FH_3Si	Fluorosilane	-98.6	18.8	
FLi	Lithium fluoride	1673	147	
FNO	Nitrosyl fluoride	-59.9	19.28	
FNO_2	Nitryl fluoride	-72.4	18.05	
FNS	Thionitrosyl fluoride (NSF)	4.8	22.2	
F_2	Fluorine	-188.12	6.62	
F_2H_2Si	Difluorosilane	-77.8	16.3	
F_2O	Fluorine monoxide	-144.75	11.09	
F_2OS	Thionyl fluoride	-43.8	21.8	
F_2O_2	Fluorine dioxide	-57	19.1	
F_2Pb	Lead(II) fluoride	1293	160.4	
F_2Zn	Zinc fluoride	1500	190.1	
F_3HSi	Trifluorosilane	-95	16.2	
F_3N	Nitrogen trifluoride	-128.75	11.56	
F_3O_2Re	Rhenium(VII) dioxytrifluoride	185.4	65.7	
F_3P	Phosphorus(III) fluoride	-101.5	16.5	
F_3PS	Phosphorus(V) sulfide trifluoride	-52.25	19.6	
F_4MoO	Molybdenum(VI) oxytetrafluoride	186.0	50.6	
F_4N_2	Tetrafluorohydrazine	-74	13.27	
F_4ORe	Rhenium(VI) oxytetrafluoride	171.7	61.0	
F_4OW	Tungsten(VI) oxytetrafluoride	185.9	59.5	
F_4S	Sulfur tetrafluoride	-40.45	26.44	
F_4Se	Selenium tetrafluoride	106	47.2	
F_4Th	Thorium(IV) fluoride	1680	258	
F_5I	Iodine pentafluoride	100.5	41.3	
F_5Mo	Molybdenum(V) fluoride	213.6	51.8	
F_5Nb	Niobium(V) fluoride	229	52.3	
F_5Os	Osmium(V) fluoride	225.9	65.6	
F_5P	Phosphorus(V) fluoride	-84.6	17.2	
F_5Re	Rhenium(V) fluoride	221.3	58.1	
F_5Ta	Tantalum(V) fluoride	229.2	56.9	
F_5V	Vanadium(V) fluoride	48.3	44.52	
F_6Ir	Iridium(VI) fluoride	53.6	30.9	
F_6Mo	Molybdenum(VI) fluoride	34.0	29.0	
F_6Os	Osmium(VI) fluoride	47.5	28.1	
F_6Re	Rhenium(VI) fluoride	33.8	28.7	
F_6S	Sulfur hexafluoride			8.99
F_6W	Tungsten(VI) fluoride	17.1	26.5	
Ga	Gallium	2204	254	
GaI_3	Gallium(III) iodide	340	56.5	
Ge	Germanium	2833	334	
GeH_4	Germane	-88.1	14.06	
Ge_2H_6	Digermane	30.8	25.1	
Ge_3H_8	Trigermane	110.5	32.2	
HI	Hydrogen iodide	-35.55	19.76	17.36
HLiO	Lithium hydroxide	1626	188	

Mol. Form.	Name	t_b/°C	$\Delta_{vap}H(t_b)$ kJ/mol	$\Delta_{vap}H$(25°C) kJ/mol
HNO_3	Nitric acid	83		39.1
HN_3	Hydrazoic acid	35.7	30.5	
HNaO	Sodium hydroxide	1388	175	
H_2	Hydrogen	-252.87	0.90	
H_2O	Water	100.0	40.65	43.98
H_2O_2	Hydrogen peroxide	150.2		51.6
H_2S	Hydrogen sulfide	-59.55	18.67	14.08
H_2S_2	Hydrogen disulfide	70.7		33.78
H_2Se	Hydrogen selenide	-41.25	19.7	
H_2Te	Hydrogen telluride	-2	19.2	
H_3N	Ammonia	-33.33	23.33	19.86
H_3P	Phosphine	-87.75	14.6	
H_3Sb	Stibine	-17	21.3	
H_4N_2	Hydrazine	113.55	41.8	44.7
H_4P_2	Diphosphine	63.5	28.8	
H_4Si	Silane	-111.9	12.1	
H_4Sn	Stannane	-51.8	19.05	
H_6Si_2	Disilane	-14.3	21.2	
H_8Si_3	Trisilane	52.9	28.5	
He	Helium	-268.93	0.08	
Hg	Mercury	356.73	59.11	
HgI_2	Mercury(II) iodide	354	59.2	
IIn	Indium(I) iodide	712	90.8	
ITl	Thallium(I) iodide	824	104.7	
I_2	Iodine	184.4	41.57	
I_2Pb	Lead(II) iodide	872	104	
I_2Sn	Tin(II) iodide	714	105	
I_3P	Phosphorus(III) iodide	227	43.9	
I_3Sb	Antimony(III) iodide	401	68.6	
I_4Si	Tetraiodosilane	287.35	50.2	
I_4Sn	Tin(IV) iodide	364.35	56.9	
I_4Ti	Titanium(IV) iodide	377	58.4	
Kr	Krypton	-153.22	9.08	
MoO_3	Molybdenum(VI) oxide	1155	138	
NO	Nitric oxide	-151.74	13.83	
N_2	Nitrogen	-195.79	5.57	
N_2O	Nitrous oxide	-88.48	16.53	
N_2O_4	Nitrogen tetroxide	21.15	38.12	
Ne	Neon	-246.08	1.71	
O_2	Oxygen	-182.95	6.82	
O_2S	Sulfur dioxide	-10.05	24.94	22.92
O_3S	Sulfur trioxide	45	40.69	43.14
P	Phosphorus	280.5	12.4	14.2
Pb	Lead	1749	179.5	
S	Sulfur	444.60	45	
STl_2	Thallium(I) sulfide	1367	154	
Se	Selenium	685	95.48	
Te	Tellurium	988	114.1	
Xe	Xenon	-108.11	12.57	
$CClF_3$	Chlorotrifluoromethane	-81.4	15.8	
CCl_2F_2	Dichlorodifluoromethane	-29.8	20.1	
CCl_3F	Trichlorofluoromethane	23.7	25.1	
CCl_4	Tetrachloromethane	76.8	29.82	32.43
$CHBr_3$	Tribromomethane	149.1	39.66	46.05
$CHClF_2$	Chlorodifluoromethane	-40.7	20.2	
$CHCl_2F$	Dichlorofluoromethane	8.9	25.2	
$CHCl_3$	Trichloromethane	61.17	29.24	31.28
CH_2BrCl	Bromochloromethane	68.0	30.0	

Mol. Form.	Name	t_b/°C	$\Delta_{vap}H(t_b)$ kJ/mol	$\Delta_{vap}H$(25°C) kJ/mol
CH_2Br_2	Dibromomethane	97	32.92	36.97
CH_2Cl_2	Dichloromethane	40	28.06	28.82
CH_2I_2	Diiodomethane	182	42.5	
CH_2O_2	Formic acid	101	22.69	20.10
CH_3Br	Bromomethane	3.5	23.91	22.81
CH_3Cl	Chloromethane	-24.09	21.40	18.92
CH_3I	Iodomethane	42.55	27.34	27.97
CH_3NO	Formamide	220		60.15
CH_3NO_2	Nitromethane	101.19	33.99	38.27
CH_4	Methane	-161.48	8.19	
CH_4O	Methanol	64.6	35.21	37.43
CH_5N	Methylamine	-6.32	25.60	23.37
CH_6N_2	Methylhydrazine	87.5	36.12	40.37
CN_4O_8	Tetranitromethane	126.1	40.74	49.93
CO	Carbon monoxide	-191.5	6.04	
CS_2	Carbon disulfide	46	26.74	27.51
$C_2Br_2ClF_3$	1,2-Dibromo-1-chloro-1,2,2-trifluoroethane	93	31.17	35.04
$C_2Br_2F_4$	1,2-Dibromotetrafluoroethane	47.35	27.03	28.39
C_2ClF_5	Chloropentafluoroethane	-37.95	19.41	
$C_2Cl_2F_4$	1,2-Dichlorotetrafluoroethane	3.8	23.3	
$C_2Cl_3F_3$	1,1,1-Trichlorotrifluoroethane	46.1	26.85	28.08
$C_2Cl_3F_3$	1,1,2-Trichloro-1,2,2-trifluoroethane	47.7	27.04	28.40
C_2Cl_4	Tetrachloroethylene	121.3	34.68	39.68
C_2F_6	Hexafluoroethane	-78.1	16.15	
$C_2HBrClF_3$	2-Bromo-2-chloro-1,1,1-trifluoroethane	50.2	28.08	29.61
C_2HCl_3	Trichloroethylene	87.21	31.40	34.54
C_2HCl_5	Pentachloroethane	159.8	36.9	
$C_2HF_3O_2$	Trifluoroacetic acid	73	33.3	
$C_2H_2Br_4$	1,1,2,2-Tetrabromoethane	243.5	48.7	
$C_2H_2Cl_2$	1,1-Dichloroethylene	31.6	26.14	26.48
$C_2H_2Cl_2$	*cis*-1,2-Dichloroethylene	60.1	30.2	
$C_2H_2Cl_2$	*trans*-1,2-Dichloroethylene	48.7	28.9	
$C_2H_2Cl_4$	1,1,2,2-Tetrachloroethane	146.5	37.64	45.71
C_2H_3Br	Bromoethylene	15.8	23.4	
C_2H_3Cl	Chloroethylene	-13.3	20.8	
$C_2H_3Cl_2F$	1,1-Dichloro-1-fluoroethane	32.0	26.06	26.48
$C_2H_3Cl_3$	1,1,1-Trichloroethane	74.09	29.86	32.50
$C_2H_3Cl_3$	1,1,2-Trichloroethane	113.8	34.82	40.24
$C_2H_3F_3$	1,1,1-Trifluoroethane	-47.25	18.99	
C_2H_3N	Acetonitrile	81.65	29.75	32.94
C_2H_4	Ethylene	-103.77	13.53	
$C_2H_4Br_2$	1,2-Dibromoethane	131.6	34.77	41.73
$C_2H_4Cl_2$	1,1-Dichloroethane	57.4	28.85	30.62
$C_2H_4Cl_2$	1,2-Dichloroethane	83.5	31.98	35.16
$C_2H_4F_2$	1,1-Difluoroethane	-24.95	21.56	19.08
C_2H_4O	Acetaldehyde	20.1	25.76	25.47
C_2H_4O	Ethylene oxide	10.6	25.54	24.75
$C_2H_4O_2$	Acetic acid	117.9	23.70	23.36
$C_2H_4O_2$	Methyl formate	31.7	27.92	28.35
C_2H_5Br	Bromoethane	38.5	27.04	28.03
C_2H_5Cl	Chloroethane	12.3	24.65	
C_2H_5ClO	2-Chloroethanol	128.6	41.4	
C_2H_5I	Iodoethane	72.5	29.44	31.93
C_2H_5NO	*N*-Methylformamide	199.51		56.19
$C_2H_5NO_2$	Nitroethane	114.0	38.0	
C_2H_6	Ethane	-88.6	14.69	5.16
C_2H_6O	Ethanol	78.29	38.56	42.32
C_2H_6O	Dimethyl ether	-24.8	21.51	18.51

Mol. Form.	Name	t_b/°C	$\Delta_{vap}H(t_b)$ kJ/mol	$\Delta_{vap}H$(25°C) kJ/mol
C_2H_6OS	Dimethyl sulfoxide	189	43.1	
$C_2H_6O_2$	Ethylene glycol	197.3	50.5	
C_2H_6S	Ethanethiol	35.1	26.79	27.30
C_2H_6S	Dimethyl sulfide	37.33	27.0	27.65
$C_2H_6S_2$	1,2-Ethanedithiol	146.1	37.93	44.68
$C_2H_6S_2$	Dimethyl disulfide	109.8	33.78	37.86
C_2H_7N	Dimethylamine	6.88	26.40	25.05
C_2H_7NO	Ethanolamine	171	49.83	
$C_2H_8N_2$	1,2-Ethanediamine	117	37.98	44.98
$C_2H_8N_2$	1,1-Dimethylhydrazine	63.9	32.55	35.0
C_2N_2	Cyanogen	-21.1	23.33	19.75
$C_3Cl_2F_6$	1,2-Dichlorohexafluoropropane	34.1	26.28	26.93
$C_3H_3Cl_3O_2$	Methyl trichloroacetate	153.8		48.33
C_3H_3N	Acrylonitrile	77.3	32.6	
$C_3H_4Cl_2O_2$	Methyl dichloroacetate	142.9	39.28	47.72
C_3H_4O	Acrolein	52.6	28.3	
$C_3H_4O_2$	2-Oxetanone	162		47.03
C_3H_5Br	3-Bromopropene	70.1	30.24	32.73
C_3H_5Cl	3-Chloropropene	45.1	29.0	
$C_3H_5ClO_2$	Methyl chloroacetate	129.5	39.23	46.73
$C_3H_5Cl_3$	1,2,3-Trichloropropane	157	37.1	
C_3H_5N	Propanenitrile	97.14	31.81	36.03
C_3H_6	Propene	-47.69	18.42	14.24
C_3H_6	Cyclopropane	-32.81	20.05	16.93
$C_3H_6Br_2$	1,2-Dibromopropane	141.9	35.61	41.67
$C_3H_6Br_2$	1,3-Dibromopropane	167.3		47.45
$C_3H_6Cl_2$	1,3-Dichloropropane	120.9	35.18	40.75
C_3H_6O	Allyl alcohol	97.0	40.0	
C_3H_6O	Propanal	48	28.31	29.62
C_3H_6O	Acetone	56.05	29.10	30.99
C_3H_6O	Methyloxirane	35	27.35	27.89
C_3H_6O	Oxetane	47.6	28.67	29.85
$C_3H_6O_2$	Propanoic acid	141.15		32.14
$C_3H_6O_2$	Ethyl formate	54.4	29.91	31.96
$C_3H_6O_2$	Methyl acetate	56.87	30.32	32.29
C_3H_6S	Thiacyclobutane	95	32.32	35.97
C_3H_7Br	1-Bromopropane	71.1	29.84	32.01
C_3H_7Br	2-Bromopropane	59.5	28.33	30.17
C_3H_7Cl	1-Chloropropane	46.5	27.18	28.35
C_3H_7Cl	2-Chloropropane	35.7	26.30	26.90
C_3H_7I	1-Iodopropane	102.6	32.08	36.25
C_3H_7I	2-Iodopropane	89.5	30.68	34.06
C_3H_7NO	*N*-Ethylformamide	198		58.44
C_3H_7NO	*N*,*N*-Dimethylformamide	153		46.89
$C_3H_7NO_2$	1-Nitropropane	131.1	38.5	
$C_3H_7NO_2$	2-Nitropropane	120.2	36.8	
C_3H_8	Propane	-42.1	19.04	14.79
C_3H_8O	1-Propanol	97.2	41.44	47.45
C_3H_8O	2-Propanol	82.3	39.85	45.39
$C_3H_8O_2$	1,2-Propylene glycol	187.6	52.4	
$C_3H_8O_2$	1,3-Propylene glycol	214.4	57.9	
$C_3H_8O_2$	Ethylene glycol monomethyl ether	124.1	37.54	45.17
$C_3H_8O_3$	Glycerol	290	61.0	
C_3H_8S	1-Propanethiol	67.8	29.54	31.89
C_3H_8S	2-Propanethiol	52.6	27.91	29.45
C_3H_8S	Ethyl methyl sulfide	66.7	29.53	31.85
$C_3H_8S_2$	1,3-Propanedithiol	172.9		49.66
C_3H_9N	Propylamine	47.22	29.55	31.27

Mol. Form.	Name	t_b/°C	$\Delta_{vap}H(t_b)$ kJ/mol	$\Delta_{vap}H$(25°C) kJ/mol
C_3H_9N	Isopropylamine	31.76	27.83	28.36
C_3H_9N	Trimethylamine	2.87	22.94	21.66
$C_3H_{10}N_2$	1,3-Propanediamine	139.8	40.85	50.16
C_4F_8	Perfluorocyclobutane	-5.9	23.2	
C_4F_{10}	Perfluorobutane	-1.9	22.9	
$C_4H_4N_2$	Succinonitrile	266	48.5	
$C_4H_4N_2$	Pyrimidine	123.8	43.09	49.79
$C_4H_4N_2$	Pyridazine	208		53.47
C_4H_4O	Furan	31.5	27.10	27.45
$C_4H_4O_2$	Diketene	126.1	36.80	42.89
C_4H_4S	Thiophene	84.0	31.48	34.70
$C_4H_5Cl_3O_2$	Ethyl trichloroacetate	167.5		50.97
C_4H_5N	2-Methylacrylonitrile	90.3	31.8	
C_4H_5N	Pyrrole	129.79	38.75	45.09
C_4H_5N	Cyclopropanecarbonitrile	135.1	35.55	41.94
$C_4H_5NO_2$	Methyl cyanoacetate	200.5	48.2	
C_4H_5NS	4-Methylthiazole	133.3	37.58	43.85
C_4H_6	1,2-Butadiene	10.9	24.02	23.21
C_4H_6	1,3-Butadiene	-4.41	22.47	20.86
C_4H_6	1-Butyne	8.08	24.52	23.35
$C_4H_6Cl_2O_2$	Ethyl dichloroacetate	155		50.60
$C_4H_6O_2$	Vinyl acetate	72.5	34.6	
$C_4H_6O_2$	Methyl acrylate	80.7	33.1	
$C_4H_6O_2$	γ-Butyrolactone	204	52.2	
$C_4H_6O_3$	Acetic anhydride	139.5	38.2	
C_4H_6S	2,3-Dihydrothiophene	112.1	33.24	37.74
C_4H_6S	2,5-Dihydrothiophene	122.4	34.83	39.95
$C_4H_7ClO_2$	Ethyl chloroacetate	144.3	40.43	49.47
C_4H_7N	Butanenitrile	117.6	33.68	39.33
C_4H_7N	2-Methylpropanenitrile	103.9	32.39	37.13
C_4H_8	1-Butene	-6.26	22.07	20.22
C_4H_8	*cis*-2-Butene	3.71	23.34	22.16
C_4H_8	*trans*-2-Butene	0.88	22.72	21.40
C_4H_8	Cyclobutane	12.6	24.19	23.51
$C_4H_8Br_2$	1,4-Dibromobutane	197		53.09
$C_4H_8Cl_2$	1,2-Dichlorobutane	124.1	33.90	39.58
$C_4H_8Cl_2$	1,4-Dichlorobutane	161		46.36
$C_4H_8Cl_2O$	Bis(2-chloroethyl) ether	178.5	45.2	
C_4H_8O	Ethyl vinyl ether	35.5	26.2	
C_4H_8O	1,2-Epoxybutane	63.4	30.3	
C_4H_8O	Butanal	74.8	31.5	
C_4H_8O	2-Butanone	79.59	31.30	34.79
C_4H_8O	Tetrahydrofuran	65	29.81	31.99
$C_4H_8O_2$	Butanoic acid	163.75		40.45
$C_4H_8O_2$	2-Methylpropanoic acid	154.45		35.30
$C_4H_8O_2$	Propyl formate	80.9	33.61	37.53
$C_4H_8O_2$	Ethyl acetate	77.11	31.94	35.60
$C_4H_8O_2$	Methyl propanoate	79.8	32.24	35.85
$C_4H_8O_2$	1,3-Dioxane	106.1	34.37	39.09
$C_4H_8O_2$	1,4-Dioxane	101.5	34.16	38.60
C_4H_8S	Tetrahydrothiophene	121.0	34.66	39.43
C_4H_9Br	1-Bromobutane	101.6	32.51	36.64
C_4H_9Br	2-Bromobutane	91.3	30.77	34.41
C_4H_9Br	1-Bromo-2-methylpropane	91.1	31.33	34.82
C_4H_9Br	2-Bromo-2-methylpropane	73.3	29.23	31.81
C_4H_9Cl	1-Chlorobutane	78.6	30.39	33.51
C_4H_9Cl	2-Chlorobutane	68.2	29.17	31.53
C_4H_9Cl	1-Chloro-2-methylpropane	68.5	29.22	31.67

Mol. Form.	Name	t_b/°C	$\Delta_{vap}H(t_b)$ kJ/mol	$\Delta_{vap}H(25°C)$ kJ/mol
C_4H_9Cl	2-Chloro-2-methylpropane	50.9	27.55	28.98
C_4H_9I	1-Iodobutane	130.6	34.66	40.63
C_4H_9I	2-Iodobutane	120.1	33.27	38.46
C_4H_9I	1-Iodo-2-methylpropane	121.1	33.54	38.83
C_4H_9I	2-Iodo-2-methylpropane	100.1	31.43	35.41
C_4H_9N	Pyrrolidine	86.56	33.01	37.52
C_4H_9NO	*N*-Ethylacetamide	205		64.89
C_4H_9NO	*N,N*-Dimethylacetamide	165		50.24
C_4H_9NO	Morpholine	128	37.1	
C_4H_{10}	Butane	-0.5	22.44	21.02
C_4H_{10}	Isobutane	-11.73	21.30	19.23
$C_4H_{10}O$	1-Butanol	117.73	43.29	52.35
$C_4H_{10}O$	2-Butanol	99.51	40.75	49.72
$C_4H_{10}O$	2-Methyl-1-propanol	107.89	41.82	50.82
$C_4H_{10}O$	2-Methyl-2-propanol	82.4	39.07	46.69
$C_4H_{10}O$	Diethyl ether	34.5	26.52	27.10
$C_4H_{10}O$	Methyl propyl ether	39.1	26.75	27.60
$C_4H_{10}O$	Isopropyl methyl ether	30.77	26.05	26.41
$C_4H_{10}O_2$	1,2-Butanediol	190.5	52.84	71.55
$C_4H_{10}O_2$	1,3-Butanediol	207.5	54.31	74.46
$C_4H_{10}O_2$	Ethylene glycol monoethyl ether	135	39.22	48.21
$C_4H_{10}O_2$	Ethylene glycol dimethyl ether	85	32.42	36.39
$C_4H_{10}O_3$	Diethylene glycol	245.8	52.3	
$C_4H_{10}S$	1-Butanethiol	98.5	32.23	36.63
$C_4H_{10}S$	2-Butanethiol	85	30.59	33.99
$C_4H_{10}S$	2-Methyl-1-propanethiol	88.5	31.01	34.63
$C_4H_{10}S$	2-Methyl-2-propanethiol	64.3	28.45	30.78
$C_4H_{10}S$	Diethyl sulfide	92.1	31.77	35.80
$C_4H_{10}S$	Methyl propyl sulfide	95.6	32.08	36.24
$C_4H_{10}S$	Isopropyl methyl sulfide	84.8	30.71	34.15
$C_4H_{10}S_2$	1,4-Butanedithiol	195.5		55.10
$C_4H_{10}S_2$	Diethyl disulfide	154.1	37.58	45.18
$C_4H_{11}N$	Butylamine	77.00	31.81	35.72
$C_4H_{11}N$	*sec*-Butylamine	62.73	29.92	32.85
$C_4H_{11}N$	*tert*-Butylamine	44.04	28.27	29.64
$C_4H_{11}N$	Isobutylamine	67.75	30.61	33.85
$C_4H_{11}N$	Diethylamine	55.5	29.06	31.31
$C_4H_{11}N$	Isopropylmethylamine	50.4	28.71	30.69
$C_4H_{11}NO$	2-Amino-2-methyl-1-propanol	165.5	50.6	
$C_4H_{11}NO_2$	Diethanolamine	268.8	65.2	
$C_5H_2F_6O_2$	Hexafluoroacetylacetone	54.15	27.05	30.58
$C_5H_4O_2$	Furfural	161.7	43.2	
C_5H_5N	Pyridine	115.23	35.09	40.21
$C_5H_6O_2$	Furfuryl alcohol	171	53.6	
C_5H_6S	2-Methylthiophene	112.6	33.90	38.87
C_5H_6S	3-Methylthiophene	115.5	34.24	39.43
C_5H_7N	*trans*-3-Pentenenitrile	142.6	37.09	44.77
C_5H_7N	Cyclobutanecarbonitrile	149.6	36.88	44.34
C_5H_8	Spiropentane	39	26.76	27.49
C_5H_8O	Cyclopropyl methyl ketone	111.3	34.07	39.41
C_5H_8O	Cyclopentanone	130.57	36.35	42.72
$C_5H_8O_2$	Methyl cyclopropanecarboxylate	114.9	35.25	41.27
$C_5H_8O_2$	Allyl acetate	103.5	36.3	
$C_5H_8O_2$	Ethyl acrylate	99.4	34.7	
$C_5H_8O_2$	Methyl methacrylate	100.5	36.0	
$C_5H_8O_2$	2,4-Pentanedione	138	34.30	41.77
C_5H_9N	Pentanenitrile	141.3	36.09	43.60
C_5H_9N	3-Methylbutanenitrile	127.5	35.10	41.64

Mol. Form.	Name	t_b/°C	$\Delta_{vap}H(t_b)$ kJ/mol	$\Delta_{vap}H$(25°C) kJ/mol
C_5H_9N	2,2-Dimethylpropanenitrile	106.1	32.40	37.35
C_5H_{10}	1-Pentene	29.96	25.20	25.47
C_5H_{10}	*cis*-2-Pentene	36.93		26.86
C_5H_{10}	*trans*-2-Pentene	36.34		26.76
C_5H_{10}	2-Methyl-1-butene	31.2	25.50	25.92
C_5H_{10}	3-Methyl-1-butene	20.1		23.77
C_5H_{10}	2-Methyl-2-butene	38.56	26.31	27.06
C_5H_{10}	Cyclopentane	49.3	27.30	28.52
$C_5H_{10}Cl_2$	1,2-Dichloropentane	148.3	36.45	43.89
$C_5H_{10}Cl_2$	1,5-Dichloropentane	179		50.71
$C_5H_{10}O$	Cyclopentanol	140.42		57.05
$C_5H_{10}O$	2-Pentanone	102.26	33.44	38.40
$C_5H_{10}O$	3-Pentanone	101.96	33.45	38.52
$C_5H_{10}O$	3-Methyl-2-butanone	94.33	32.35	36.78
$C_5H_{10}O$	3,3-Dimethyloxetane	80.6	30.85	33.94
$C_5H_{10}O$	Tetrahydropyran	88	31.17	34.58
$C_5H_{10}O_2$	Pentanoic acid	186.1	44.1	
$C_5H_{10}O_2$	2-Methylbutanoic acid	177		46.91
$C_5H_{10}O_2$	Butyl formate	106.1	36.58	41.11
$C_5H_{10}O_2$	Isobutyl formate	98.2	33.6	
$C_5H_{10}O_2$	Propyl acetate	101.54	33.92	39.72
$C_5H_{10}O_2$	Isopropyl acetate	88.6	32.93	37.20
$C_5H_{10}O_2$	Ethyl propanoate	99.1	33.88	39.21
$C_5H_{10}O_2$	Methyl butanoate	102.8	33.79	39.28
$C_5H_{10}O_2$	Methyl isobutanoate	92.5	32.61	37.32
$C_5H_{10}O_2$	Tetrahydrofurfuryl alcohol	178	45.2	
$C_5H_{10}O_3$	Diethyl carbonate	126		43.60
$C_5H_{10}O_3$	Ethylene glycol monomethyl ether acetate	143	43.9	
$C_5H_{10}S$	Thiacyclohexane	141.8	35.96	42.58
$C_5H_{10}S$	Cyclopentanethiol	132.1	35.32	41.42
$C_5H_{11}Br$	1-Bromopentane	129.8	35.01	41.28
$C_5H_{11}Cl$	1-Chloropentane	107.8	33.15	38.24
$C_5H_{11}Cl$	2-Chloropentane	97.0	31.79	36.03
$C_5H_{11}Cl$	1-Chloro-3-methylbutane	98.9	32.02	36.24
$C_5H_{11}I$	1-Iodopentane	155		45.27
$C_5H_{11}N$	Piperidine	106.22		39.29
C_5H_{12}	Pentane	36.06	25.79	26.43
C_5H_{12}	Isopentane	27.88	24.69	24.85
C_5H_{12}	Neopentane	9.48	22.74	21.84
$C_5H_{12}O$	1-Pentanol	137.98	44.36	57.02
$C_5H_{12}O$	2-Pentanol	119.3	41.40	54.21
$C_5H_{12}O$	3-Pentanol	116.25		54.0
$C_5H_{12}O$	2-Methyl-1-butanol	128		55.16
$C_5H_{12}O$	3-Methyl-1-butanol	131.1	44.07	55.61
$C_5H_{12}O$	2-Methyl-2-butanol	102.4	39.04	50.10
$C_5H_{12}O$	3-Methyl-2-butanol	112.9		53.0
$C_5H_{12}O$	Butyl methyl ether	70.16	29.55	32.37
$C_5H_{12}O$	*sec*-Butyl methyl ether	59.1	28.09	30.23
$C_5H_{12}O$	Methyl *tert*-butyl ether	55.2	27.94	29.82
$C_5H_{12}O$	Isobutyl methyl ether	58.6	28.02	30.13
$C_5H_{12}O$	Ethyl propyl ether	63.21	28.94	31.43
$C_5H_{12}O$	Ethyl isopropyl ether	54.1	28.21	30.08
$C_5H_{12}O_2$	1-Ethoxy-2-methoxyethane	102.1	34.33	39.83
$C_5H_{12}O_2$	1,5-Pentanediol	239	60.7	
$C_5H_{12}O_2$	Ethylene glycol monopropyl ether	149.8	41.40	52.12
$C_5H_{12}O_2$	Diethoxymethane	88	31.33	35.65
$C_5H_{12}O_3$	Diethylene glycol monomethyl ether	193	46.6	
$C_5H_{12}S$	1-Pentanethiol	126.6	34.88	41.24

Mol. Form.	Name	t_b/°C	$\Delta_{vap}H(t_b)$ kJ/mol	$\Delta_{vap}H$(25°C) kJ/mol
$C_5H_{12}S$	2-Methyl-1-butanethiol	119.1	33.79	39.45
$C_5H_{12}S$	2-Methyl-2-butanethiol	99.1	31.37	35.67
$C_5H_{12}S$	Butyl methyl sulfide	123.5	34.47	40.46
$C_5H_{12}S$	*tert*-Butyl methyl sulfide	99	31.47	35.84
$C_5H_{12}S$	Ethyl propyl sulfide	118.6	34.24	39.97
$C_5H_{12}S$	Ethyl isopropyl sulfide	107.5	32.74	37.78
$C_5H_{13}N$	Pentylamine	104.3	34.01	40.08
$C_5H_{13}N$	Ethylisopropylamine	69.6	29.94	33.13
C_6ClF_5	Chloropentafluorobenzene	117.96	34.76	41.07
C_6F_6	Hexafluorobenzene	80.26	31.66	35.71
C_6HF_5	Pentafluorobenzene	85.74	32.15	36.27
$C_6H_4Cl_2$	*o*-Dichlorobenzene	180	39.66	50.21
$C_6H_4Cl_2$	*m*-Dichlorobenzene	173	38.62	48.58
$C_6H_4Cl_2$	*p*-Dichlorobenzene	174	38.79	49.0
$C_6H_4F_2$	*o*-Difluorobenzene	94	32.21	36.18
$C_6H_4F_2$	*m*-Difluorobenzene	82.6	31.10	34.59
$C_6H_4F_2$	*p*-Difluorobenzene	89	31.77	35.54
C_6H_5Br	Bromobenzene	156.06		44.54
C_6H_5Cl	Chlorobenzene	131.72	35.19	40.97
C_6H_5F	Fluorobenzene	84.73	31.19	34.58
C_6H_5I	Iodobenzene	188.4	39.5	
$C_6H_5NO_2$	Nitrobenzene	210.8		55.01
C_6H_6	Benzene	80.09	30.72	33.83
C_6H_6ClN	*o*-Chloroaniline	208.8	44.4	
C_6H_6O	Phenol	181.87	45.69	57.82
C_6H_6S	Benzenethiol	169.1	39.93	47.56
C_6H_7N	Aniline	184.17	42.44	55.83
C_6H_7N	2-Methylpyridine	129.38	36.17	42.48
C_6H_7N	3-Methylpyridine	144.14	37.35	44.44
C_6H_7N	4-Methylpyridine	145.36	37.51	44.56
C_6H_7N	1-Cyclopentenecarbonitrile			44.98
C_6H_9N	Cyclopentanecarbonitrile			43.43
$C_6H_9NO_3$	Triacetamide			60.41
C_6H_{10}	Cyclohexene	82.98	30.46	33.47
$C_6H_{10}O$	Cyclohexanone	155.43		45.06
$C_6H_{10}O$	Mesityl oxide	130	36.1	
$C_6H_{10}O_2$	Methyl cyclobutanecarboxylate	135.5	37.13	44.72
$C_6H_{10}O_3$	Propanoic anhydride	170	41.7	
$C_6H_{10}O_4$	Diethyl oxalate	185.7	42.0	
$C_6H_{10}O_4$	Ethylene glycol diacetate	190		61.44
$C_6H_{11}N$	Hexanenitrilc	163.65		47.91
C_6H_{12}	1-Hexene	63.48		30.61
C_6H_{12}	*cis*-2-Hexene	68.8		32.19
C_6H_{12}	*trans*-2-Hexene	67.9		31.60
C_6H_{12}	*cis*-3-Hexene	66.4		31.23
C_6H_{12}	*trans*-3-Hexene	67.1		31.55
C_6H_{12}	2-Methyl-1-pentene	62.1		30.48
C_6H_{12}	3-Methyl-1-pentene	54.2		28.62
C_6H_{12}	4-Methyl-1-pentene	53.9		28.71
C_6H_{12}	2-Methyl-2-pentene	67.3		31.60
C_6H_{12}	3-Methyl-*cis*-2-pentene	67.7		32.09
C_6H_{12}	3-Methyl-*trans*-2-pentene	70.4		31.35
C_6H_{12}	4-Methyl-*cis*-2-pentene	56.3		29.48
C_6H_{12}	4-Methyl-*trans*-2-pentene	58.6		29.97
C_6H_{12}	2-Ethyl-1-butene	64.7		31.13
C_6H_{12}	2,3-Dimethyl-1-butene	55.6		29.18
C_6H_{12}	3,3-Dimethyl-1-butene	41.2		26.61
C_6H_{12}	2,3-Dimethyl-2-butene	73.3	29.64	32.51

Mol. Form.	Name	t_b/°C	$\Delta_{vap}H(t_b)$ kJ/mol	$\Delta_{vap}H$(25°C) kJ/mol
C_6H_{12}	Cyclohexane	80.73	29.97	33.01
C_6H_{12}	Methylcyclopentane	71.8	29.08	31.64
C_6H_{12}	Ethylcyclobutane	70.8	28.67	31.24
$C_6H_{12}Cl_2$	1,2-Dichlorohexane	173		48.16
$C_6H_{12}O$	Butyl vinyl ether	94	31.58	36.17
$C_6H_{12}O$	2-Hexanone	127.6	36.35	43.14
$C_6H_{12}O$	3-Hexanone	123.5	35.36	42.47
$C_6H_{12}O$	3-Methyl-2-pentanone	117.5	34.16	40.53
$C_6H_{12}O$	4-Methyl-2-pentanone	116.5	34.49	40.61
$C_6H_{12}O$	2-Methyl-3-pentanone	113.5	33.84	39.79
$C_6H_{12}O$	3,3-Dimethyl-2-butanone	106.1	33.39	37.91
$C_6H_{12}O$	Cyclohexanol	160.84		62.01
$C_6H_{12}O_2$	Butyl acetate	126.1	36.28	43.86
$C_6H_{12}O_2$	*tert*-Butyl acetate	95.1	33.07	38.03
$C_6H_{12}O_2$	Isobutyl acetate	116.5	35.9	
$C_6H_{12}O_2$	Propyl propanoate	122.5	35.54	43.45
$C_6H_{12}O_2$	Ethyl butanoate	121.5	35.47	42.68
$C_6H_{12}O_2$	Ethyl 2-methylpropanoate	110.1	33.67	39.83
$C_6H_{12}O_2$	Methyl pentanoate	127.4	35.36	43.10
$C_6H_{12}O_2$	Methyl 2,2-dimethylpropanoate	101.1	33.42	38.76
$C_6H_{12}O_3$	Ethylene glycol monoethyl ether acetate	156.4	40.76	52.61
$C_6H_{12}S$	Cyclohexanethiol	158.9	37.06	44.57
$C_6H_{13}Br$	1-Bromohexane	155.3		45.89
$C_6H_{13}Cl$	1-Chlorohexane	135	35.67	42.83
$C_6H_{13}I$	1-Iodohexane	181		49.75
$C_6H_{13}N$	Cyclohexylamine	134	36.14	43.67
C_6H_{14}	Hexane	68.73	28.85	31.56
C_6H_{14}	2-Methylpentane	60.26	27.79	29.89
C_6H_{14}	3-Methylpentane	63.27	28.06	30.28
C_6H_{14}	2,2-Dimethylbutane	49.73	26.31	27.68
C_6H_{14}	2,3-Dimethylbutane	57.93	27.38	29.12
$C_6H_{14}N_2$	Azopropane	114		39.88
$C_6H_{14}O$	1-Hexanol	157.6	44.50	61.61
$C_6H_{14}O$	2-Hexanol	140	41.01	58.46
$C_6H_{14}O$	2-Methyl-1-pentanol	149	50.2	
$C_6H_{14}O$	4-Methyl-1-pentanol	151.9	44.46	60.47
$C_6H_{14}O$	2-Methyl-2-pentanol	121.1	39.59	54.77
$C_6H_{14}O$	4-Methyl-2-pentanol	131.6	44.2	
$C_6H_{14}O$	2-Ethyl-1-butanol	147	43.2	
$C_6H_{14}O$	Dipropyl ether	90.08	31.31	35.69
$C_6H_{14}O$	Diisopropyl ether	68.51	29.10	32.12
$C_6H_{14}O$	Butyl ethyl ether	92.3	31.63	36.32
$C_6H_{14}O$	Methyl pentyl ether	99	32.02	36.85
$C_6H_{14}O_2$	2-Methyl-2,4-pentanediol	197.1	57.3	
$C_6H_{14}O_2$	Ethylene glycol monobutyl ether	168.4		56.59
$C_6H_{14}O_2$	1,1-Diethoxyethane	102.25	36.28	43.20
$C_6H_{14}O_2$	Ethylene glycol diethyl ether	119.4	36.28	43.20
$C_6H_{14}O_3$	Bis(ethoxymethyl) ether	140.6	36.17	44.69
$C_6H_{14}O_3$	Diethylene glycol monoethyl ether	196	47.5	
$C_6H_{14}O_3$	Diethylene glycol dimethyl ether	162	36.17	44.69
$C_6H_{14}O_4$	Triethylene glycol	285	71.4	
$C_6H_{14}S$	Dipropyl sulfide	142.9	36.60	44.21
$C_6H_{14}S$	Diisopropyl sulfide	120.1	33.80	39.60
$C_6H_{14}S$	Isopropyl propyl sulfide	132.1	35.11	41.78
$C_6H_{14}S$	Butyl ethyl sulfide	144.3	37.01	44.51
$C_6H_{14}S$	Methyl pentyl sulfide	145.1	37.41	45.24
$C_6H_{15}N$	Hexylamine	132.8	36.54	45.10
$C_6H_{15}N$	Butylethylamine	107.5	33.97	40.15

Mol. Form.	Name	t_b/°C	$\Delta_{vap}H(t_b)$ kJ/mol	$\Delta_{vap}H$(25°C) kJ/mol
$C_6H_{15}N$	Dipropylamine	109.3	33.47	40.04
$C_6H_{15}N$	Diisopropylamine	83.9	30.40	34.61
$C_6H_{15}N$	Isopropylpropylamine	96.9	32.14	37.23
$C_6H_{15}N$	Triethylamine	89	31.01	34.84
C_6MoO_6	Molybdenum hexacarbonyl	701	72.51	
$C_7H_3F_5$	2,3,4,5,6-Pentafluorotoluene	117.5	34.75	41.12
$C_7H_5F_3$	(Trifluoromethyl)benzene	102.1	32.63	37.60
C_7H_5N	Benzonitrile	191.1	45.9	
C_7H_6O	Benzaldehyde	179.0	42.5	
$C_7H_6O_2$	Salicylaldehyde	197	38.2	
C_7H_7Cl	*o*-Chlorotoluene	159.0	37.5	
C_7H_7Cl	*p*-Chlorotoluene	162.4	38.7	
C_7H_7F	*o*-Fluorotoluene	115	35.4	
C_7H_7F	*p*-Fluorotoluene	116.6	34.08	39.42
C_7H_8	Toluene	110.63	33.18	38.01
C_7H_8O	*o*-Cresol	191.04	45.19	
C_7H_8O	*m*-Cresol	202.27	47.40	61.71
C_7H_8O	*p*-Cresol	201.98	47.45	
C_7H_8O	Benzyl alcohol	205.31	50.48	
C_7H_8O	Anisole	153.7	38.97	46.90
C_7H_9N	Benzylamine	185		60.16
C_7H_9N	*o*-Methylaniline	200.3	44.6	
C_7H_9N	*m*-Methylaniline	203.3	44.9	
C_7H_9N	*p*-Methylaniline	200.4	44.3	
C_7H_9N	1-Cyclohexenecarbonitrile			53.55
C_7H_9N	2,3-Dimethylpyridine	161.12	39.08	47.82
C_7H_9N	2,4-Dimethylpyridine	158.38	38.53	47.49
C_7H_9N	2,5-Dimethylpyridine	156.98	38.68	47.04
C_7H_9N	2,6-Dimethylpyridine	144.01	37.46	45.34
C_7H_9N	3,4-Dimethylpyridine	179.10	39.99	50.50
C_7H_9N	3,5-Dimethylpyridine	171.84	39.46	49.33
$C_7H_{10}O$	Dicyclopropyl ketone	161		53.70
$C_7H_{11}N$	Cyclohexanecarbonitrile			51.92
C_7H_{12}	1-Methylbicyclo(3,1,0)hexane	93.1	31.07	34.77
$C_7H_{12}O_4$	Diethyl malonate	200	54.8	
C_7H_{14}	1-Heptene	93.64		35.49
C_7H_{14}	*cis*-2-Heptene	98.4		36.26
C_7H_{14}	*trans*-2-Heptene	98		36.27
C_7H_{14}	*cis*-3-Heptene	95.8		35.81
C_7H_{14}	*trans*-3-Heptene	95.7		35.84
C_7H_{14}	*cis*-3-Methyl-3-hexene	95.4		36.31
C_7H_{14}	*trans*-3-Methyl-3-hexene	93.5		35.70
C_7H_{14}	2,4-Dimethyl-1-pentene	81.6		33.03
C_7H_{14}	4,4-Dimethyl-1-pentene	72.5		31.13
C_7H_{14}	2,4-Dimethyl-2-pentene	83.4		34.19
C_7H_{14}	*cis*-4,4-Dimethyl-2-pentene	80.4		32.56
C_7H_{14}	*trans*-4,4-Dimethyl-2-pentene	76.7		32.81
C_7H_{14}	2-Ethyl-3-methyl-1-butene	89		34.35
C_7H_{14}	2,3,3-Trimethyl-1-butene	77.9		32.09
C_7H_{14}	Methylcyclohexane	100.93	31.27	35.36
C_7H_{14}	Ethylcyclopentane	103.5	31.96	36.40
C_7H_{14}	*cis*-1,3-Dimethylcyclopentane	90.8	30.40	34.20
$C_7H_{14}O$	2-Heptanone	151.05		47.24
$C_7H_{14}O$	2,2-Dimethyl-3-pentanone	125.6	36.09	42.34
$C_7H_{14}O$	2,4-Dimethyl-3-pentanone	125.4	34.64	41.51
$C_7H_{14}O$	1-Methylcyclohexanol	155	79.0	
$C_7H_{14}O$	*cis*-2-Methylcyclohexanol	165	48.5	
$C_7H_{14}O$	*trans*-2-Methylcyclohexanol	167.5	53.0	

Mol. Form.	Name	t_b/°C	$\Delta_{vap}H(t_b)$ kJ/mol	$\Delta_{vap}H$(25°C) kJ/mol
$C_7H_{14}O_2$	Pentyl acetate	149.2	38.42	48.56
$C_7H_{14}O_2$	Isopentyl acetate	142.5	37.5	
$C_7H_{14}O_2$	Ethyl pentanoate	146.1	36.96	47.01
$C_7H_{14}O_2$	Ethyl 3-methylbutanoate	135.0	37.0	
$C_7H_{14}O_2$	Ethyl 2,2-dimethylpropanoate	118.4	34.51	41.25
$C_7H_{14}O_2$	Methyl hexanoate	149.5	38.55	48.04
$C_7H_{15}Br$	1-Bromoheptane	179		50.60
$C_7H_{15}Cl$	1-Chloroheptane	159		47.66
C_7H_{16}	Heptane	98.5	31.77	36.57
C_7H_{16}	2-Methylhexane	90.04	30.62	34.87
C_7H_{16}	3-Methylhexane	92	30.9	
C_7H_{16}	3-Ethylpentane	93.5	31.12	35.22
C_7H_{16}	2,2-Dimethylpentane	79.2	29.23	32.42
C_7H_{16}	2,3-Dimethylpentane	89.78	30.46	34.26
C_7H_{16}	2,4-Dimethylpentane	80.49	29.55	32.88
C_7H_{16}	3,3-Dimethylpentane	86.06	29.62	33.03
C_7H_{16}	2,2,3-Trimethylbutane	80.86	28.90	32.05
$C_7H_{16}O$	Hexyl methyl ether	126.1	34.93	42.07
$C_7H_{16}O$	1-Heptanol	176.45		66.81
$C_7H_{16}O$	3-Heptanol	157	42.5	
$C_7H_{16}O$	Butyl propyl ether	118.1	33.72	40.22
$C_7H_{16}O$	Ethyl pentyl ether	117.6	34.41	41.01
$C_7H_{17}N$	Heptylamine	156		49.96
C_8F_{18}	Perfluorooctane	105.9	33.38	41.13
C_8H_8	Styrene	145	38.7	
C_8H_8O	Acetophenone	202	43.98	55.40
$C_8H_8O_2$	Methyl benzoate	199		55.57
$C_8H_8O_3$	Methyl salicylate	222.9	46.7	
C_8H_{10}	Ethylbenzene	136.19	35.57	42.24
C_8H_{10}	*o*-Xylene	144.5	36.24	43.43
C_8H_{10}	*m*-Xylene	139.12	35.66	42.65
C_8H_{10}	*p*-Xylene	138.37	35.67	42.40
$C_8H_{10}O$	2,4-Xylenol	210.98		64.96
$C_8H_{10}O$	2,5-Xylenol	211.1	46.9	
$C_8H_{10}O$	2,6-Xylenol	201.07		75.31
$C_8H_{10}O$	3,4-Xylenol	227		85.03
$C_8H_{10}O$	3,5-Xylenol	221.74		82.01
$C_8H_{10}O$	Phenetole	169.81		51.04
$C_8H_{11}N$	*N*-Ethylaniline	203.0		58.3
$C_8H_{11}N$	*N,N*-Dimethylaniline	194.15		52.83
$C_8H_{11}N$	2,4-Dimethylaniline	214		61.3
$C_8H_{11}N$	2,5-Dimethylaniline	214		61.7
$C_8H_{11}N$	2,3,6-Trimethylpyridine	171.6	39.95	50.61
$C_8H_{11}N$	2,4,6-Trimethylpyridine	170.6	39.87	50.33
C_8H_{14}	1-Octyne	126.3	35.83	42.30
C_8H_{14}	2-Octyne	137.6	37.26	44.49
C_8H_{14}	3-Octyne	133.1	36.94	43.92
C_8H_{14}	4-Octyne	131.6	36.0	42.73
$C_8H_{14}O_3$	Butanoic anhydride	200	50.0	
$C_8H_{15}N$	Octanenitrile	205.25		56.80
C_8H_{16}	1-Octene	121.29	34.07	40.34
C_8H_{16}	*cis*-2,2-Dimethyl-3-hexene	105.5		36.86
C_8H_{16}	*trans*-2,2-Dimethyl-3-hexene	100.8		37.03
C_8H_{16}	3-Ethyl-2-methyl-1-pentene	109.5		37.27
C_8H_{16}	2,4,4-Trimethyl-1-pentene	101.4		35.59
C_8H_{16}	2,4,4-Trimethyl-2-pentene	104.9		37.23
C_8H_{16}	Ethylcyclohexane	131.9	34.04	40.56
C_8H_{16}	1,1-Dimethylcyclohexane	119.6	32.51	37.92

Mol. Form.	Name	t_b/°C	$\Delta_{vap}H(t_b)$ kJ/mol	$\Delta_{vap}H$(25°C) kJ/mol
C_8H_{16}	*cis*-1,2-Dimethylcyclohexane	129.8	33.47	39.70
C_8H_{16}	*trans*-1,2-Dimethylcyclohexane	123.5	32.96	38.36
C_8H_{16}	*cis*-1,3-Dimethylcyclohexane	120.1	32.91	38.26
C_8H_{16}	*trans*-1,3-Dimethylcyclohexane	124.5	33.39	39.16
C_8H_{16}	*cis*-1,4-Dimethylcyclohexane	124.4	33.28	39.02
C_8H_{16}	*trans*-1,4-Dimethylcyclohexane	119.4	32.56	37.90
C_8H_{16}	Propylcyclopentane	131	34.70	41.08
C_8H_{16}	Isopropylcyclopentane	126.5	33.56	39.44
C_8H_{16}	1-Ethyl-1-methylcyclopentane	121.6	33.20	38.85
$C_8H_{16}O$	2,2,4-Trimethyl-3-pentanone	135.1	35.64	43.30
$C_8H_{16}O_2$	Octanoic acid	239	58.5	
$C_8H_{16}O_2$	2-Ethylhexanoic acid	228		75.60
$C_8H_{16}O_2$	Isobutyl isobutanoate	148.6	38.2	
$C_8H_{16}O_2$	Ethyl hexanoate	167		51.72
$C_8H_{16}O_2$	Methyl heptanoate	174		51.62
$C_8H_{17}Br$	1-Bromooctane	200		55.77
$C_8H_{17}Cl$	1-Chlorooctane	181.5		52.42
$C_8H_{17}F$	1-Fluorooctane	142.4	40.43	49.65
C_8H_{18}	Octane	125.67	34.41	41.49
C_8H_{18}	2-Methylheptane	117.66	33.26	39.67
C_8H_{18}	3-Methylheptane	118.9	33.66	39.83
C_8H_{18}	4-Methylheptane	117.72	33.35	39.69
C_8H_{18}	3-Ethylhexane	118.6	33.59	39.64
C_8H_{18}	2,2-Dimethylhexane	106.86	32.07	37.28
C_8H_{18}	2,3-Dimethylhexane	115.62	33.17	38.78
C_8H_{18}	2,4-Dimethylhexane	109.5	32.51	37.76
C_8H_{18}	2,5-Dimethylhexane	109.12	32.54	37.85
C_8H_{18}	3,3-Dimethylhexane	111.97	32.31	37.53
C_8H_{18}	3,4-Dimethylhexane	117.73	33.24	38.97
C_8H_{18}	3-Ethyl-2-methylpentane	115.66	32.93	38.52
C_8H_{18}	3-Ethyl-3-methylpentane	118.27	32.78	37.99
C_8H_{18}	2,2,3-Trimethylpentane	110	31.94	36.91
C_8H_{18}	2,2,4-Trimethylpentane	99.22	30.79	35.14
C_8H_{18}	2,3,3-Trimethylpentane	114.8	32.12	37.27
C_8H_{18}	2,3,4-Trimethylpentane	113.5	32.36	37.75
C_8H_{18}	2,2,3,3-Tetramethylbutane	106.45		42.90
$C_8H_{18}N_2$	Azobutane			49.31
$C_8H_{18}O$	1-Octanol	195.16		70.98
$C_8H_{18}O$	2-Octanol	180	44.4	
$C_8H_{18}O$	2-Ethyl-1-hexanol	184.6	54.2	
$C_8H_{18}O$	Dibutyl ether	140.28	36.49	44.97
$C_8H_{18}O$	Di-*sec*-butyl ether	121.1	34.06	40.84
$C_8H_{18}O$	Di-*tert*-butyl ether	107.23	32.15	37.61
$C_8H_{18}O_2$	1,2-Dipropoxyethane			50.62
$C_8H_{18}O_3$	Diethylene glycol diethyl ether	188		58.40
$C_8H_{18}S$	Dibutyl sulfide	185		52.96
$C_8H_{18}S$	Di-*tert*-butyl sulfide	149.1	33.26	43.76
$C_8H_{18}S$	Diisobutyl sulfide	171		48.71
$C_8H_{19}N$	Dibutylamine	159.6	38.44	49.45
$C_8H_{19}N$	2-Ethylhexylamine	169.2	40.0	
C_9H_7N	Quinoline	237.16	49.7	59.30
C_9H_7N	Isoquinoline	243.22	49.0	60.26
C_9H_{10}	Cyclopropylbenzene	173.6		50.22
C_9H_{10}	Indan	177.97	39.63	48.79
$C_9H_{10}O_2$	Benzyl acetate	213	49.4	
C_9H_{12}	Propylbenzene	159.24		46.22
C_9H_{12}	Isopropylbenzene	152.41		45.13
C_9H_{12}	1,2,3-Trimethylbenzene	176.12		49.05

Mol. Form.	Name	t_b/°C	$\Delta_{vap}H(t_b)$ kJ/mol	$\Delta_{vap}H$(25°C) kJ/mol
C_9H_{12}	1,2,4-Trimethylbenzene	169.38		47.93
C_9H_{12}	1,3,5-Trimethylbenzene	164.74		47.50
$C_9H_{14}O_6$	Triacetin	259		85.74
C_9H_{18}	Butylcyclopentane	156.6	36.16	45.89
C_9H_{18}	Propylcyclohexane	156.7		45.08
C_9H_{18}	Isopropylcyclohexane	154.8		44.02
$C_9H_{18}O$	2-Nonanone	195.3		56.44
$C_9H_{18}O$	5-Nonanone	188.45		53.30
$C_9H_{18}O$	2,6-Dimethyl-4-heptanone	169.4		50.92
$C_9H_{18}O_2$	Methyl octanoate	192.9		56.41
C_9H_{20}	Nonane	150.82	37.18	46.55
C_9H_{20}	2,2,5-Trimethylhexane	124.09	33.65	40.16
C_9H_{20}	2,3,5-Trimethylhexane	131.4	34.43	41.41
C_9H_{20}	3,3-Diethylpentane	146.3	34.61	42.0
C_9H_{20}	2,2,4,4-Tetramethylpentane	122.29	32.51	38.49
$C_9H_{20}O$	1-Nonanol	213.37		76.86
$C_{10}H_7Br$	1-Bromonaphthalene	281	39.3	
$C_{10}H_7Cl$	1-Chloronaphthalene	259	52.1	
$C_{10}H_8$	Naphthalene	217.9	43.2	
$C_{10}H_9N$	2-Methylquinoline	246.5		66.1
$C_{10}H_9N$	4-Methylquinoline	262		67.6
$C_{10}H_9N$	6-Methylquinoline	258.6		67.7
$C_{10}H_9N$	8-Methylquinoline	247.5		65.7
$C_{10}H_{12}$	1,2,3,4-Tetrahydronaphthalene	207.6	43.9	
$C_{10}H_{14}$	Butylbenzene	183.31	38.87	51.36
$C_{10}H_{14}$	*sec*-Butylbenzene	173.3		47.98
$C_{10}H_{14}$	*tert*-Butylbenzene	169.1		47.71
$C_{10}H_{14}$	Isobutylbenzene	172.79		47.86
$C_{10}H_{14}$	1-Isopropyl-4-methylbenzene	177.1	38.2	
$C_{10}H_{16}O$	(+)-Camphor	207.4	59.5	
$C_{10}H_{18}$	*cis*-Decahydronaphthalene	195.8	41.0	
$C_{10}H_{18}$	*trans*-Decahydronaphthalene	187.3	40.2	
$C_{10}H_{19}N$	Decanenitrile	243		66.84
$C_{10}H_{20}$	1-Decene	170.5		50.43
$C_{10}H_{20}$	Butylcyclohexane	180.9		49.36
$C_{10}H_{20}O_2$	2-Ethylhexyl acetate	199	43.5	
$C_{10}H_{20}O_2$	Isopentyl isopentanoate	190.4	45.9	
$C_{10}H_{22}$	Decane	174.15	39.58	51.42
$C_{10}H_{22}$	2-Methylnonane	167.1	38.23	49.63
$C_{10}H_{22}$	3-Methylnonane	167.9	38.26	49.71
$C_{10}H_{22}$	5-Methylnonane	165.1	38.14	49.34
$C_{10}H_{22}$	2,4-Dimethyloctane	156	36.47	47.13
$C_{10}H_{22}O$	1-Decanol	231.1		81.50
$C_{10}H_{22}O$	Diisopentyl ether	172.5	35.1	
$C_{10}H_{22}S$	1-Decanethiol	240.6		65.48
$C_{11}H_{10}$	1-Methylnaphthalene	244.7	45.5	
$C_{11}H_{21}N$	Undecanenitrile	253		71.14
$C_{11}H_{22}$	Pentylcyclohexane	203.7		53.88
$C_{11}H_{24}$	Undecane	195.9	41.91	56.58
$C_{11}H_{24}$	2-Methyldecane	189.3	40.25	54.28
$C_{11}H_{24}$	4-Methyldecane	187	40.70	53.76
$C_{11}H_{24}$	2,4,7-Trimethyloctane	168.1	38.22	49.91
$C_{12}F_{27}N$	Tris(perfluorobutyl)amine	178	46.4	
$C_{12}H_{10}O$	Diphenyl ether	258.0	48.2	
$C_{12}H_{16}$	Cyclohexylbenzene	240.1		59.94
$C_{12}H_{22}$	Cyclohexylcyclohexane	238		57.98
$C_{12}H_{23}N$	Dodecanenitrile	277		76.12
$C_{12}H_{24}$	1-Dodecene	213.8		60.78

Mol. Form.	Name	t_b/°C	$\Delta_{vap}H(t_b)$ kJ/mol	$\Delta_{vap}H(25°C)$ kJ/mol
$C_{12}H_{26}$	2,2,4,6,6-Pentamethylheptane	177.8		48.97
$C_{12}H_{26}$	Dodecane	216.32	44.09	61.52
$C_{12}H_{26}O$	1-Dodecanol	259		91.96
$C_{12}H_{27}BO_3$	Tributyl borate	234	56.1	
$C_{12}H_{27}N$	Tributylamine	216.5	46.9	
$C_{13}H_{13}N$	*N*-Benzylaniline	306.5		79.6
$C_{13}H_{26}O_2$	Methyl dodecanoate	267		77.17
$C_{13}H_{28}$	Tridecane	235.47	46.20	66.68
$C_{14}H_{10}$	Phenanthrene	340		75.50
$C_{14}H_{12}O_2$	Benzyl benzoate	323.5	53.6	
$C_{14}H_{27}N$	Tetradecanenitrile			85.29
$C_{14}H_{30}$	Tetradecane	253.58	48.16	71.73
$C_{14}H_{30}O$	1-Tetradecanol	289		102.20
$C_{15}H_{32}$	Pentadecane	270.6	50.08	76.77
$C_{16}H_{22}O_4$	Dibutyl phthalate	340	79.2	
$C_{16}H_{32}$	1-Hexadecene	284.9		80.25
$C_{16}H_{34}$	Hexadecane	286.86	51.84	81.35
$C_{17}H_{36}$	Heptadecane	302.0	53.58	86.47
$C_{18}H_{34}O_2$	Oleic acid	360	67.4	
$C_{18}H_{38}$	Octadecane	316.3	55.23	91.44
$C_{19}H_{40}$	Nonadecane	329.9	56.93	96.4
$C_{20}H_{42}$	Eicosane	343	58.49	101.81

ENTHALPY OF FUSION

This table lists the molar enthalpy (heat) of fusion, $\Delta_{fus}H$, of over 800 inorganic and organic compounds. All values refer to the enthalpy change at equilibrium between the liquid phase and the most stable solid phase at the transition temperature. Most values of $\Delta_{fus}H$ are given at the normal melting point t_m. However, a "t" following the entry in the melting point column indicate a triple-point temperature, where the solid, liquid, and gas phases are in equilibrium. Substances are listed by molecular formula in the Hill order, with substances containing carbon (except graphite) following those that do not contain carbon.

All temperatures are given on the ITS-90 scale.

A * following an entry indicates that the value includes the enthalpy of transition between crystalline phases whose transformation occurs within 1°C of the melting point.

REFERENCES

1. Chase, M. W., Davies, C. A., Downey, J. R., Frurip, D. J., McDonald, R. A., and Syverud, A. N., *JANAF Thermochemical Tables, Third Edition, J. Phys. Chem. Ref. Data*, Vol. 14, Suppl. 1, 1985.
2. Gurvich, L. V., Veyts, I. V., and Alcock, C. B., *Thermodynamic Properties of Individual Substances, Fourth Edition*; Vol. 2, Hemisphere Publishing Corp., New York, 1991; Vol. 3, CRC Press, Boca Raton, FL, 1994.
3. Dinsdale, A. T., *CALPHAD*, 15, 317, 1991
4. *Landolt-Börnstein, Numerical Data and Functional Relationships in Science and Technology, New Series,* IV/8A, *Enthalpies of Fusion and Transition of Organic Compounds*, Springer-Verlag, Heidelberg, 1995.
5. *Landolt-Börnstein, Numerical Values and Functions for Physics, Chemistry, Astronomy, Geophysics, and Technology, Sixth Edition,* Vol. 2, Part 4, Springer-Verlag, Heidelberg, 1961.
6. Janz, G. J., et al., *Physical Properties Data Compilations Relevant to Energy Storage. II. Molten Salts,* Nat. Stand. Ref. Data Sys.- Nat. Bur. Standards (U.S.), No. 61, Part 2, 1979.
7. *TRC Thermodynamic Tables,* Thermodynamic Research Center, Texas A&M University, College Station, TX.

Molecular formula	Name	t_m/°C	$\Delta_{fus}H$/kJ mol^{-1}
Ag	Silver	961.78	11.28
AgBr	Silver(I) bromide	432	9.12
AgCl	Silver(I) chloride	455	13.2
AgI	Silver(I) iodide	558	9.41
$AgNO_3$	Silver(I) nitrate	212	11.5
Ag_2S	Silver(I) sulfide	825	14.1
Al	Aluminum	660.32	10.789
$AlBr_3$	Aluminum bromide	97.5	11.25
$AlCl_3$	Aluminum chloride	192.6	35.4
AlF_3	Aluminum fluoride	2250 t	98
AlI_3	Aluminum iodide	188.28	15.9
Al_2O_3	Aluminum oxide	2053	111.4
Al_2S_3	Aluminum sulfide	1100	55
Am	Americium	1176	14.39
Ar	Argon	-189.36 t	1.18
As	Arsenic (gray)	817 t	24.44
$AsBr_3$	Arsenic(III) bromide	31.1	11.7
$AsCl_3$	Arsenic(III) chloride	-16	10.1
AsF_3	Arsenic(III) fluoride	-5.9	10.4
Au	Gold	1064.18	12.72
B	Boron	2075	50.2
BCl_3	Boron trichloride	-107	2.10
BF_3	Boron trifluoride	-126.8	4.20
BHO_2	Metaboric acid (γ form)	236	14.3
BH_3O_3	Boric acid (orthoboric acid)	170.9	22.3
BN	Boron nitride	2966	81

Molecular formula	Name	t_m/°C	$\Delta_{fus}H$/kJ mol^{-1}
$BNaO_2$	Sodium metaborate	966	36.2
B_2O_3	Boron oxide	450	24.56
Ba	Barium	727	7.12
$BaBr_2$	Barium bromide	857	32.2
$BaCl_2$	Barium chloride	962	15.85
BaF_2	Barium fluoride	1368	17.8
BaH_2	Barium hydride	1200	25
BaH_2O_2	Barium hydroxide	408	16
BaI_2	Barium iodide	711	26.5
BaO	Barium oxide	1972	46
BaO_4S	Barium sulfate	1580	40
BaS	Barium sulfide	2229	63
Be	Beryllium	1287	7.895
$BeBr_2$	Beryllium bromide	508	18
$BeCl_2$	Beryllium chloride	415	8.66
BeF_2	Beryllium fluoride	552	4.77
BeI_2	Beryllium iodide	470	18
BeO	Beryllium oxide	2577	86
BeO_4S	Beryllium sulfate	1127	6
Bi	Bismuth	271.40	11.145
$BiCl_3$	Bismuth trichloride	230	10.9
BrF_5	Bromine pentafluoride	-60.5	5.67
BrH	Hydrogen bromide	-86.80	2.41
BrIn	Indium(I) bromide	290	15
BrK	Potassium bromide	734	25.5
BrLi	Lithium bromide	552	17.6
BrNa	Sodium bromide	747	26.11
$BrNaO_3$	Sodium bromate	381	28.11
BrRb	Rubidium bromide	682	15.5
BrTl	Thallium(I) bromide	460	16.4
Br_2	Bromine	-7.2	10.57
Br_2Ca	Calcium bromide	742	29.1
Br_2Cd	Cadmium bromide	568	20.9
Br_2Fe	Iron(II) bromide	691	50.2
Br_2Hg	Mercury(II) bromide	236	17.9
Br_2Mg	Magnesium bromide	711	39.3
Br_2Pb	Lead(II) bromide	371	16.44
Br_2Sr	Strontium bromide	657	10.1
Br_2Zn	Zinc bromide	394	16.7
Br_3Ga	Gallium(III) bromide	121.5	12.1
Br_3In	Indium(III) bromidc	420	26
Br_3Pu	Plutonium(III) bromide	681	55.2
Br_3U	Uranium(III) bromide	727	43.9
Br_4Sn	Tin(IV) bromide	29.1	12.2
Br_4Th	Thorium(IV) bromide	679	66.9
Br_4Ti	Titanium(IV) bromide	39	12.9
Br_4U	Uranium(IV) bromide	519	55.2
Br_5Ta	Tantalum(V) bromide	265	45.6
C	Carbon (graphite)	4489 t	117
Ca	Calcium	842	8.54
$CaCl_2$	Calcium chloride	775	28.05
CaF_2	Calcium fluoride	1418	30
CaH_2	Calcium hydride	1000	6.7
CaI_2	Calcium iodide	783	41.8
CaO	Calcium oxide	2898	80
CaO_4S	Calcium sulfate	1460	28
CaS	Calcium sulfide	2524	70
Cd	Cadmium	321.07	6.21

Molecular formula	Name	t_m/°C	$\Delta_{fus}H$/kJ mol^{-1}
$CdCl_2$	Cadmium chloride	564	48.58
CdF_2	Cadmium fluoride	1110	22.6
CdI_2	Cadmium iodide	387	15.3
Ce	Cerium	798	5.46
$CeCl_3$	Cerium(III) chloride	817	54.4
ClCs	Cesium chloride	645	15.9
ClCu	Copper(I) chloride	430	10.2
ClH	Hydrogen chloride	-114.17	2.00
ClI	Iodine chloride	27.39	11.6
ClIn	Indium(I) chloride	211	21.3
ClK	Potassium chloride	771	26.53
ClLi	Lithium chloride	610	19.9
$ClLiO_4$	Lithium perchlorate	236	29
ClNa	Sodium chloride	800.7	28.16
$ClNaO_3$	Sodium chlorate	248	22.1
ClRb	Rubidium chloride	715	18.4
ClTl	Thallium(I) chloride	430	15.56
Cl_2	Chlorine	-101.5	6.40
Cl_2Co	Cobalt(II) chloride	740	45
Cl_2Cr	Chromium(II) chloride	814	32.2
Cl_2Cu	Copper(II) chloride	630	20.4
Cl_2Fe	Iron(II) chloride	677	43.01
Cl_2Hg	Mercury(II) chloride	276	19.41
Cl_2Mg	Magnesium chloride	714	43.1
Cl_2Mn	Manganese(II) chloride	650	30.7
Cl_2Ni	Nickel(II) chloride	1009	71.2
Cl_2Pb	Lead(II) chloride	501	21.75
Cl_2Sn	Tin(II) chloride	247.1	14.52
Cl_2Sr	Strontium chloride	874	17.5
Cl_3Fe	Iron(III) chloride	304	43.1
Cl_3Ga	Gallium(III) chloride	77.9	11.13
Cl_3In	Indium(III) chloride	583	27
Cl_3La	Lanthanum chloride	859	43.1
Cl_3OP	Phosphorus(V) oxychloride	1.18	13.1
Cl_3P	Phosphorus(III) chloride	-112	7.10
Cl_3Sb	Antimony(III) chloride	73.4	12.7
Cl_4OW	Tungsten(VI) oxytetrachloride	211	45
Cl_4Si	Tetrachlorosilane	-68.74	7.60
Cl_4Sn	Tin(IV) chloride	-34.07	9.20
Cl_4Th	Thorium(IV) chloride	770	40.2
Cl_4Ti	Titanium(IV) chloride	-24.12	9.97
Cl_4U	Uranium(IV) chloride	590	45
Cl_4V	Vanadium(IV) chloride	-25.7	2.30
Cl_4Zr	Zirconium(IV) chloride	437 t	50
Cl_5Mo	Molybdenum(V) chloride	194	19
Cl_5Nb	Niobium(V) chloride	204.7	38.3
Cl_5Ta	Tantalum(V) chloride	216	41.6
Cl_6W	Tungsten(VI) chloride	275	6.60
Co	Cobalt	1495	16.06
CoF_2	Cobalt(II) fluoride	1127	59
Cr	Chromium	1907	21.0
Cr_2O_3	Chromium(III) oxide	2329	130
Cs	Cesium	28.5	2.09
CsF	Cesium fluoride	703	21.7
CsHO	Cesium hydroxide	342.3	7.78
Cs_2O_4S	Cesium sulfate	1005	35.7
Cu	Copper	1084.62	12.93
CuF_2	Copper(II) fluoride	836	55

Molecular formula	Name	t_m/°C	$\Delta_{fus}H$/kJ mol^{-1}
CuO	Copper(II) oxide	1446	11.8
Dy	Dysprosium	1412	11.06
Er	Erbium	1529	19.9
Eu	Europium	822	9.21
FH	Hydrogen fluoride	-83.35	4.58
FK	Potassium fluoride	858	27.2
FLi	Lithium fluoride	848.2	27.09
FNa	Sodium fluoride	996	33.35
FRb	Rubidium fluoride	833	17.3
FTl	Thallium(I) fluoride	326	13.87
F_2	Fluorine	-219.66	0.51
F_2Fe	Iron(II) fluoride	1100	52
F_2HK	Potassium hydrogen fluoride	238.9	6.62
F_2Mg	Magnesium fluoride	1263	58.5
F_2Pb	Lead(II) fluoride	830	14.7
F_2Sr	Strontium fluoride	1477	28.5
F_3In	Indium(III) fluoride	1170	64
F_3Pu	Plutonium(III) fluoride	1396	59.8
F_4Pu	Plutonium(IV) fluoride	1027	65.3
F_4Th	Thorium(IV) fluoride	1110	44.0
F_4U	Uranium(IV) fluoride	1036	42.7
F_4Zr	Zirconium(IV) fluoride	932 t	64.2
F_5Nb	Niobium(V) fluoride	80	12.2
F_5V	Vanadium(V) fluoride	19.5	49.96
F_6Ir	Iridium(VI) fluoride	44	8.40
F_6Mo	Molybdenum(VI) fluoride	17.5	4.33
F_6Pu	Plutonium(VI) fluoride	52	17.6
F_6S	Sulfur hexafluoride	-50.7 t	5.02
F_6U	Uranium(VI) fluoride	64.0 t	19.1
F_6W	Tungsten(VI) fluoride	2.3	4.10
Fe	Iron	1538	13.81
FeI_2	Iron(II) iodide	587	45
FeO	Iron(II) oxide	1377	24
FeS	Iron(II) sulfide	1188	31.5
Fe_3O_4	Iron(II,III) oxide	1597	138
Ga	Gallium	29.76	5.576
GaI_3	Gallium(III) iodide	212	12.9
GaSb	Gallium antimonide	712	25.1
Ga_2O_3	Gallium(III) oxide	1806	100
Gd	Gadolinium	1313	10.0
Ge	Germanium	938.25	36.94
HI	Hydrogen iodide	-50.76	2.87
HKO	Potassium hydroxide	406	7.9
HLi	Lithium hydride	688.7	22.59
HLiO	Lithium hydroxide	471.1	20.88
HNO_3	Nitric acid	-41.6	10.5
HNaO	Sodium hydroxide	323	6.60
HORb	Rubidium hydroxide	382	8.0
H_2	Hydrogen	-259.34	0.12
H_2Mg	Magnesium hydride	327	14
H_2O	Water	0.00	6.01
H_2O_2	Hydrogen peroxide	-0.43	12.50
H_2O_2Sr	Strontium hydroxide	535	23
H_2O_4S	Sulfuric acid	10.31	10.71
H_2S	Hydrogen sulfide	-85.5	2.38
H_2Sr	Strontium hydride	1050	23
H_3N	Ammonia	-77.73	5.66
H_3O_2P	Hypophosphorous acid	26.5	9.7

Molecular formula	Name	t_m/°C	$\Delta_{fus}H$/kJ mol^{-1}
H_3O_3P	Phosphorous acid	74.4	12.8
H_3O_4P	Phosphoric acid	42.4	13.4
H_4IN	Ammonium iodide	551	21
H_4N_2	Hydrazine	1.4	12.6
$H_4N_2O_3$	Ammonium nitrate	210	6.40
Hf	Hafnium	2233	27.2
Hg	Mercury	-38.83	2.29
HgI_2	Mercury(II) iodide	259	18.9
Hg_2I_2	Mercury(I) iodide	290	27
Ho	Holmium	1474	17.0
IIn	Indium(I) iodide	364.4	17.26
IK	Potassium iodide	681	24
ILi	Lithium iodide	469	14.6
INa	Sodium iodide	660	23.6
IRb	Rubidium iodide	642	12.5
ITl	Thallium(I) iodide	441.7	14.73
I_2	Iodine	113.7	15.52
I_2Mg	Magnesium iodide	634	26
I_2Pb	Lead(II) iodide	410	23.4
I_2Sr	Strontium iodide	538	19.7
I_3In	Indium(III) iodide	207	18.48
I_4Si	Tetraiodosilane	120.5	19.7
I_4Th	Thorium(IV) iodide	570	61.4
I_4Ti	Titanium(IV) iodide	150	19.8
I_4U	Uranium(IV) iodide	506	70.7
In	Indium	156.60	3.281
InSb	Indium antimonide	525	25.5
In_2O_3	Indium(III) oxide	1912	105
Ir	Iridium	2446	41.12
K	Potassium	63.5	2.33
KNO_3	Potassium nitrate	337	10.1
K_2O_4S	Potassium sulfate	1069	36.4
K_2S	Potassium sulfide	948	16.15
Kr	Krypton	-157.38 t	1.64
La	Lanthanum	918	6.20
Li	Lithium	180.50	3.00
$LiNO_3$	Lithium nitrate	253	24.9
Li_2O_3Si	Lithium metasilicate	1201	28
Li_2O_4S	Lithium sulfate	859	7.50
Lu	Lutetium	1663	22
Mg	Magnesium	650	8.48
MgO	Magnesium oxide	2825	77
MgO_4S	Magnesium sulfate	1127	14.6
MgS	Magnesium sulfide	2226	63
Mg_2O_4Si	Magnesium orthosilicate	1897	71
Mn	Manganese	1246	12.91
MnO	Manganese(II) oxide	1839	54.4
Mo	Molybdenum	2623	37.48
MoO_3	Molybdenum(VI) oxide	801	48
$NNaO_3$	Sodium nitrate	307	15
NO	Nitric oxide	-163.6	2.30
NO_3Rb	Rubidium nitrate	305	5.60
NO_3Tl	Thallium(I) nitrate	206	9.6
N_2	Nitrogen	-210.0	0.71
N_2O	Nitrous oxide	-90.8	6.54
N_2O_4	Nitrogen tetroxide	-9.3	14.65
Na	Sodium	97.80	2.60
Na_2O	Sodium oxide	1132	48

Molecular formula	Name	t_m/°C	$\Delta_{fus}H$/kJ mol^{-1}
Na_2O_3Si	Sodium metasilicate	1089	52
Na_2O_4S	Sodium sulfate	884	23.6
Na_2S	Sodium sulfide	1172	19
Nb	Niobium	2477	30
NbO	Niobium(II) oxide	1936	85
NbO_2	Niobium(IV) oxide	1901	92
Nb_2O_5	Niobium(V) oxide	1512	104.3
Nd	Neodymium	1021	7.14
Ne	Neon	-248.61 t	0.328
Ni	Nickel	1455	17.04
NiS	Nickel(II) sulfide	976	30.1
Np	Neptunium	644	3.20
OSr	Strontium oxide	2531	81
OTl_2	Thallium(I) oxide	579	30.3
OV	Vanadium(II) oxide	1789	63
OZn	Zinc oxide	1974	52.3
O_2	Oxygen	-218.79	0.44
O_2Si	Silicon dioxide (cristobalite)	1722	9.6
O_2Zr	Zirconium(IV) oxide	2709	87
O_3S	Sulfur trioxide	16.8	8.60
O_3Tl_2	Thallium(III) oxide	834	53
O_3W	Tungsten(VI) oxide	1472	73
O_3Y_2	Yttrium oxide	2438	105
O_4Os	Osmium(VIII) oxide	41	9.8
O_4SSr	Strontium sulfate	1606	36
O_4STl_2	Thallium(I) sulfate	632	23
O_5P_2	Phosphorus(V) oxide	562	27.2
O_5Ta_2	Tantalum(V) oxide	1784	120
O_5V_2	Vanadium(V) oxide	670	64.5
O_7Re_2	Rhenium(VII) oxide	297	64.2
Os	Osmium	3033	57.85
P	Phosphorus (white)	44.15	0.66
Pa	Protactinium	1572	12.34
Pb	Lead	327.46	4.782
PbS	Lead(II) sulfide	1113	49.4
Pd	Palladium	1554.9	16.74
Pr	Praseodymium	931	6.89
Pt	Platinum	1768.4	22.17
Pu	Plutonium	640	2.82
Rb	Rubidium	39.3	2.19
Re	Rhenium	3186	60.43
Rh	Rhodium	1964	26.59
Ru	Ruthenium	2334	38.59
S	Sulfur (monoclinic)	115.21	1.72
SSr	Strontium sulfide	2226	63
STl_2	Thallium(I) sulfide	448	12
Sb	Antimony	630.63	19.79
Sc	Scandium	1541	14.1
Se	Selenium (gray)	220.5	6.69
Si	Silicon	1414	50.21
Sm	Samarium	1074	8.62
Sn	Tin (white)	231.93	7.173
Sr	Strontium	777	7.43
Ta	Tantalum	3017	36.57
Tb	Terbium	1356	10.15
Tc	Technetium	2157	33.29
Te	Tellurium	449.51	17.49
Th	Thorium	1750	13.81

Molecular formula	Name	t_m/°C	$\Delta_{fus}H$/kJ mol^{-1}
Ti	Titanium	1668	14.15
Tl	Thallium	304	4.14
Tm	Thulium	1545	16.84
U	Uranium	1135	9.14
V	Vanadium	1910	21.5
W	Tungsten	3422	52.31
Xe	Xenon	-111.79 t	2.27
Y	Yttrium	1522	11.4
Yb	Ytterbium	819	7.66
Zn	Zinc	419.53	7.068
Zr	Zirconium	1855	21.00
$CBaO_3$	Barium carbonate	1555	40
$CBrCl_3$	Bromotrichloromethane	-5.65	2.53
CBr_4	Tetrabromomethane	92.3	3.76
$CCaO_3$	Calcium carbonate (calcite)	1330	36
CCl_2O	Carbonyl chloride	-127.78	5.74
CCl_3F	Trichlorofluoromethane	-110.44	6.89
CCl_4	Tetrachloromethane	-22.62	2.56
CF_4	Tetrafluoromethane	-183.60	0.704
$CHBr_3$	Tribromomethane	8.69	11.05
$CHClF_2$	Chlorodifluoromethane	-157.42	4.12
$CHCl_3$	Trichloromethane	-63.41	9.5
CHF_3	Trifluoromethane	-155.2	4.06
CHI_3	Triiodomethane	121.2	16.44
CHN	Hydrogen cyanide	-13.29	8.41
$CHNaO_2$	Sodium formate	257.3	17.7
CHO_2Tl	Thallium(I) formate	101	10.9
CH_2Cl_2	Dichloromethane	-97.2	4.60
CH_2N_2	Cyanamide	45.56	7.27
CH_2N_4	Tetrazole	157.3	18.2
CH_2O_2	Formic acid	8.3	12.68
CH_3Br	Bromomethane	-93.68	5.98
CH_3Cl	Chloromethane	-97.7	6.43
CH_3NO	Formamide	2.49	8.44
CH_3NO_2	Nitromethane	-28.38	9.70
CH_3NO_3	Methyl nitrate	-83.0	8.24
CH_4	Methane	-182.47	0.94
CH_4N_2O	Urea	133.3	13.9
CH_4N_2S	Thiourea	178	14.0
CH_4O	Methanol	-97.53	3.215
CH_4S	Methanethiol	-123	5.91
CH_5N	Methylamine	-93.5	6.13
CH_6N_2	Methylhydrazine	-52.36	10.42
CK_2O_3	Potassium carbonate	898	27.6
CLi_2O_3	Lithium carbonate	723	41
$CMgO_3$	Magnesium carbonate	990	59
CNa_2O_3	Sodium carbonate	858.1	29.7
CO	Carbon monoxide	-205.02	0.833
COS	Carbon oxysulfide	-138.8	4.73
CO_2	Carbon dioxide	-56.56 t	9.02
CO_3Sr	Strontium carbonate	1494	40
CO_3Tl_2	Thallium(I) carbonate	272	18.4
CS_2	Carbon disulfide	-112.1	4.39
CSe_2	Carbon diselenide	-43.7	6.36
$C_2Br_2F_4$	1,2-Dibromotetrafluoroethane	-110.32	7.04
C_2ClF_3	Chlorotrifluoroethene	-158.2	5.55
C_2ClF_5	Chloropentafluoroethane	-99.4	1.86
$C_2Cl_2F_4$	1,2-Dichloro-1,1,2,2-tetrafluoroethane	-92.53	1.51

Molecular formula	Name	t_m/°C	$\Delta_{fus}H$/kJ mol^{-1}
$C_2Cl_3F_3$	1,1,2-Trichloro-1,2,2-trifluoroethane	-36.22	2.47
C_2Cl_4	Tetrachloroethene	-22.3	10.88
$C_2Cl_4F_2$	1,1,2,2-Tetrachloro-1,2-difluoroethane	24.8	3.67
C_2Cl_6	Hexachloroethane	186.8 t	9.75
C_2F_4	Tetrafluoroethene	-131.15	7.72
C_2F_6	Hexafluoroethane	-100.05	2.69
C_2HCl_3	Trichloroethylene	-84.7	8.45
$C_2HCl_3O_2$	Trichloroacetic acid	59.2	5.90
C_2HCl_5	Pentachloroethane	-28.78	11.3
$C_2H_2Cl_2$	1,1-Dichloroethene	-122.56	6.51
$C_2H_2Cl_2$	*cis*-1,2-Dichloroethene	-80.0	7.2
$C_2H_2Cl_4$	1,1,2,2-Tetrachloroethane	-42.4	9.17
C_2H_3Br	Bromoethene	-139.54	5.12
C_2H_3Cl	Chloroethene	-153.84	4.92
$C_2H_3ClO_2$	Chloroacetic acid	63	12.28
$C_2H_3Cl_3$	1,1,1-Trichloroethane	-30.01	2.35
$C_2H_3Cl_3$	1,1,2-Trichloroethane	-36.3	11.46
$C_2H_3F_3$	1,1,1-Trifluoroethane	-111.3	6.19
$C_2H_3KO_2$	Potassium acetate	309	7.65
C_2H_3N	Acetonitrile	-43.82	8.16
$C_2H_3NaO_2$	Sodium acetate	328.2	17.9
C_2H_4	Ethylene	-169.15	3.35
$C_2H_4Br_2$	1,2-Dibromoethane	9.84	10.89
$C_2H_4Cl_2$	1,1-Dichloroethane	-96.9	7.87
$C_2H_4Cl_2$	1,2-Dichloroethane	-35.7	8.84
C_2H_4O	Acetaldehyde	-123.37	2.31
C_2H_4O	Ethylene oxide	-112.5	5.17
$C_2H_4O_2$	Acetic acid	16.64	11.73
C_2H_5Br	Bromoethane	-118.6	7.47
C_2H_5Cl	Chloroethane	-138.4	4.45
C_2H_5NO	Acetamide	80.16	15.59
$C_2H_5NO_2$	Nitroethane	-89.5	9.85
C_2H_6	Ethane	-182.79	2.72*
$C_2H_6N_2O$	*N*-Methylurea	104.9	14.0
C_2H_6O	Ethanol	-114.14	4.931
C_2H_6O	Dimethyl ether	-141.5	4.94
C_2H_6OS	Dimethyl sulfoxide	17.89	14.37
$C_2H_6O_2$	Ethylene glycol	-12.69	9.96
$C_2H_6O_2S$	Dimethyl sulfone	108.9	18.30
C_2H_6S	Ethanethiol	-147.88	4.98
C_2H_6S	Dimethyl sulfide	-98.24	7.99
$C_2H_6S_2$	Dimethyl disulfide	-84.67	9.19
C_2H_6Zn	Dimethyl zinc	-43.0	6.83
C_2H_7N	Dimethylamine	-92.18	5.94
$C_2H_8N_2$	1,2-Ethanediamine	11.14	22.58
$C_2H_8N_2$	1,1-Dimethylhydrazine	-57.20	10.07
$C_2H_8N_2$	1,2-Dimethylhydrazine	-8.9	13.64
C_2N_2	Cyanogen	-27.83	8.11
C_3F_6O	Perfluoroacetone	-125.45	8.38
C_3F_8	Perfluoropropane	-147.70	0.477
C_3H_3N	Acrylonitrile	-83.48	6.23
C_3H_3NS	Thiazole	-33.62	9.57
$C_3H_3N_3$	1,3,5-Triazine	80.3	14.56
C_3H_4	Allene	-136.6	4.40
$C_3H_4N_2$	1*H*-Pyrazole	70.7	14.0
$C_3H_4N_2$	Imidazole	89.5	12.82
$C_3H_4O_2$	Acrylic acid	12.5	9.51
C_3H_5N	Propanenitrile	-92.78	5.03

Molecular formula	Name	t_m/°C	$\Delta_{fus}H$/kJ mol^{-1}
$C_3H_5N_3O_9$	Trinitroglycerol	13.5	21.87
C_3H_6	Propene	-185.24	3.003
C_3H_6	Cyclopropane	-127.58	5.44
$C_3H_6Br_2$	1,2-Dibromopropane	-55.49	8.94
$C_3H_6Br_2$	1,3-Dibromopropane	-34.5	14.6
$C_3H_6Cl_2$	1,2-Dichloropropane, (±)	-100.53	6.40
$C_3H_6Cl_2$	2,2-Dichloropropane	-33.9	2.30
C_3H_6O	Acetone	-94.7	5.77
C_3H_6O	Methyloxirane	-111.9	6.53
C_3H_6O	Oxetane	-97	6.5
$C_3H_6O_2$	Propanoic acid	-20.5	10.66
$C_3H_6O_2$	Methyl acetate	-98.25	7.49
$C_3H_6O_2$	1,3-Dioxolane	-97.22	6.57
$C_3H_6O_3$	1,3,5-Trioxane	60.29	15.11
C_3H_6S	Thiacyclobutane	-73.24	8.25
C_3H_7Br	1-Bromopropane	-110.3	6.44
C_3H_7Br	2-Bromopropane	-89.0	6.53
C_3H_7Cl	1-Chloropropane	-122.9	5.54
C_3H_7Cl	2-Chloropropane	-117.18	7.39
C_3H_7N	Cyclopropylamine	-35.39	13.18
C_3H_7NO	*N,N*-Dimethylformamide	-60.48	7.90
C_3H_8	Propane	-187.63	3.50
$C_3H_8N_2O$	*N,N*-Dimethylurea	182.1	23.0
$C_3H_8N_2O$	N,N'-Dimethylurea	106.6	13.0
C_3H_8O	1-Propanol	-124.39	5.37
C_3H_8O	2-Propanol	-87.9	5.41
$C_3H_8O_2$	1,3-Propylene glycol	-27.7	7.1
$C_3H_8O_2$	Dimethoxymethane	-105.1	8.33
$C_3H_8O_3$	Glycerol	18.1	18.3
C_3H_8S	1-Propanethiol	-113.13	5.48
C_3H_8S	2-Propanethiol	-130.5	5.74
C_3H_8S	Ethyl methyl sulfide	-105.93	9.76
C_3H_9N	Propylamine	-84.75	10.97
C_3H_9N	Isopropylamine	-95.13	7.33
C_3H_9N	Trimethylamine	-117.1	7
C_3H_9NO	3-Amino-1-propanol	12.4	19.7
C_4F_8	Perfluorocyclobutane	-40.19	2.77
C_4F_{10}	Perfluorobutane	-129.1	7.66
$C_4H_2O_3$	Maleic anhydride	52.56	13.60
$C_4H_4N_2$	Succinonitrile	57.98	3.70
$C_4H_4N_2$	Pyrazine	51.0	12.9
C_4H_4O	Furan	-85.61	3.80
$C_4H_4O_3$	Succinic anhydride	119	20.4
C_4H_4S	Thiophene	-38.21	5.07
C_4H_5N	Pyrrole	-23.39	7.91
C_4H_6	1,2-Butadiene	-136.2	6.96
C_4H_6	1,3-Butadiene	-108.91	7.98
C_4H_6	1-Butyne	-125.7	6.03
C_4H_6	2-Butyne	-32.2	9.23
C_4H_6O	Divinyl ether	-100.6	7.9
$C_4H_6O_2$	*cis*-Crotonic acid	15	12.6
$C_4H_6O_2$	*trans*-Crotonic acid	71.5	13.0
$C_4H_6O_2$	γ-Butyrolactone	-43.61	9.57
$C_4H_6O_3$	Acetic anhydride	-74.1	10.5
$C_4H_6O_4$	Succinic acid	187.9	32.4
$C_4H_6O_4$	Dimethyl oxalate	54.8	21.1
C_4H_8	1-Butene	-185.34	3.96
C_4H_8	*cis*-2-Butene	-138.88	7.31

Molecular formula	Name	t_m/°C	$\Delta_{fus}H$/kJ mol^{-1}
C_4H_8	*trans*-2-Butene	-105.52	9.76
C_4H_8	Isobutene	-140.7	5.92
C_4H_8	Cyclobutane	-90.7	1.09
C_4H_8	Methylcyclopropane	-177.6	2.8
C_4H_8O	Butanal	-96.86	10.77
C_4H_8O	2-Butanone	-86.64	8.39
C_4H_8O	Tetrahydrofuran	-108.44	8.54
$C_4H_8O_2$	Butanoic acid	-5.1	11.59
$C_4H_8O_2$	Ethyl acetate	-83.8	10.48
$C_4H_8O_2$	1,4-Dioxane	11.85	12.84
C_4H_8S	Tetrahydrothiophene	-96.2	7.35
C_4H_9Br	1-Bromobutane	-112.6	9.23
C_4H_9Br	2-Bromobutane, (±)	-112.65	6.89
C_4H_9Cl	2-Chloro-2-methylpropane	-25.60	2.07
C_4H_9N	Pyrrolidine	-57.79	8.58
C_4H_9NO	Morpholine	-4.8	14.5
C_4H_{10}	Butane	-138.3	4.66
C_4H_{10}	Isobutane	-159.4	4.54
$C_4H_{10}O$	1-Butanol	-88.6	9.37
$C_4H_{10}O$	2-Butanol	-88.5	5.97
$C_4H_{10}O$	2-Methyl-1-propanol	-101.9	6.32
$C_4H_{10}O$	2-Methyl-2-propanol	25.69	6.70
$C_4H_{10}O$	Diethyl ether	-116.2	7.19
$C_4H_{10}O_2$	1,4-Butanediol	20.4	18.70
$C_4H_{10}O_2$	Ethylene glycol dimethyl ether	-69.20	12.6
$C_4H_{10}S$	1-Butanethiol	-115.7	10.46
$C_4H_{10}S$	Diethyl sulfide	-103.91	10.90
$C_4H_{11}N$	*tert*-Butylamine	-66.94	0.882
$C_4H_{12}Pb$	Tetramethyl lead	-30.2	10.80
$C_4H_{12}Si$	Tetramethylsilane	-99.06	6.87
$C_4H_{12}Sn$	Tetramethylstannane	-55.1	9.30
$C_5H_4O_2$	Furfural	-38.1	14.37
C_5H_5N	Pyridine	-41.70	8.28
C_5H_6O	2-Methylfuran	-91.3	8.55
$C_5H_6O_2$	Furfuryl alcohol	-14.6	13.13
C_5H_8	*cis*-1,3-Pentadiene	-140.8	5.64
C_5H_8	*trans*-1,3-Pentadiene	-87.4	7.14
C_5H_8	1,4-Pentadiene	-148.2	6.12
C_5H_8	2-Methyl-1,3-butadiene	-145.9	4.93
C_5H_8	Cyclopentene	-135.0	3.36
C_5H_8	Spiropentane	-107.0	6.43
$C_5H_8O_2$	Methyl methacrylate	-47.55	14.4
$C_5H_8O_3$	4-Oxopentanoic acid	33	9.22
$C_5H_8O_4$	Glutaric acid	97.8	20.3
C_5H_9N	Pentanenitrile	-96.2	9
C_5H_{10}	1-Pentene	-165.12	5.94
C_5H_{10}	*cis*-2-Pentene	-151.36	7.11
C_5H_{10}	*trans*-2-Pentene	-140.21	8.35
C_5H_{10}	2-Methyl-1-butene	-137.53	7.91
C_5H_{10}	3-Methyl-1-butene	-168.43	5.36
C_5H_{10}	2-Methyl-2-butene	-133.72	7.60
C_5H_{10}	Cyclopentane	-93.4	0.61
$C_5H_{10}O$	Cyclopentanol	-17.5	1.535
$C_5H_{10}O$	2-Pentanone	-76.8	10.63
$C_5H_{10}O$	3-Pentanone	-39	11.59
$C_5H_{10}O$	3-Methyl-2-butanone	-93.1	9.34
$C_5H_{10}O$	Tetrahydropyran	-49.1	1.8
$C_5H_{10}O_2$	Pentanoic acid	-33.6	14.16

Molecular formula	Name	t_m/°C	$\Delta_{fus}H$/kJ mol^{-1}
$C_5H_{11}Br$	1-Bromopentane	-88.0	14.37
$C_5H_{11}N$	Cyclopentylamine	-82.7	8.31
$C_5H_{11}N$	Piperidine	-11.02	14.85
C_5H_{12}	Pentane	-129.67	8.40
C_5H_{12}	Isopentane	-159.77	5.15
C_5H_{12}	Neopentane	-16.4	3.10
$C_5H_{12}O$	1-Pentanol	-77.6	10.50
$C_5H_{12}O$	2-Methyl-2-butanol	-9.1	4.46
$C_5H_{12}O$	Butyl methyl ether	-115.7	10.85
$C_5H_{12}O$	Methyl *tert*-butyl ether	-108.6	7.60
$C_5H_{12}O_4$	Pentaerythritol	258	4.8
$C_5H_{12}S$	1-Pentanethiol	-75.65	17.53
C_6Cl_6	Hexachlorobenzene	228.83	25.2
C_6F_6	Hexafluorobenzene	5.03	11.59
C_6F_{14}	Perfluorohexane	-88.2	6.84
C_6HF_5	Pentafluorobenzene	-47.4	10.87
C_6HF_5O	Pentafluorophenol	37.5	16.41
$C_6H_2F_4$	1,2,3,5-Tetrafluorobenzene	-46.25	6.36
$C_6H_2F_4$	1,2,4,5-Tetrafluorobenzene	3.88	15.05
$C_6H_3Cl_3$	1,2,3-Trichlorobenzene	51.3	17.9
$C_6H_3Cl_3$	1,2,4-Trichlorobenzene	16.92	16.4
$C_6H_3Cl_3$	1,3,5-Trichlorobenzene	62.8	18.1
$C_6H_3N_3O_6$	1,3,5-Trinitrobenzene	122.9	15.4
$C_6H_4ClNO_2$	1-Chloro-2-nitrobenzene	32.1	17.9
$C_6H_4ClNO_2$	1-Chloro-3-nitrobenzene	44.4	19.4
$C_6H_4ClNO_2$	1-Chloro-4-nitrobenzene	82	14.1
$C_6H_4Cl_2$	*o*-Dichlorobenzene	-17.0	12.4
$C_6H_4Cl_2$	*m*-Dichlorobenzene	-24.8	12.6
$C_6H_4Cl_2$	*p*-Dichlorobenzene	53.09	18.19
$C_6H_4F_2$	*o*-Difluorobenzene	-47.1	11.05
$C_6H_4F_2$	*m*-Difluorobenzene	-69.12	8.58
$C_6H_4O_2$	*p*-Benzoquinone	115	18.5
C_6H_5Br	Bromobenzene	-30.72	10.70
C_6H_5Cl	Chlorobenzene	-45.31	9.6
C_6H_5ClO	*o*-Chlorophenol	9.4	13.0
C_6H_5ClO	*m*-Chlorophenol	32.6	14.9
C_6H_5ClO	*p*-Chlorophenol	42.8	14.1
C_6H_5F	Fluorobenzene	-42.18	11.31
C_6H_5I	Iodobenzene	-31.3	9.75
C_6H_5NO	Nitrosobenzene	67	31.0
$C_6H_5NO_2$	Nitrobenzene	5.7	12.12
$C_6H_5NO_3$	*o*-Nitrophenol	44.8	17.7
$C_6H_5NO_3$	*m*-Nitrophenol	96.8	20.6
$C_6H_5NO_3$	*p*-Nitrophenol	113.6	18.8
C_6H_6	Benzene	5.49	9.87
C_6H_6ClN	*o*-Chloroaniline	-1.9	11.9
C_6H_6ClN	*m*-Chloroaniline	-10.28	10.15
C_6H_6ClN	*p*-Chloroaniline	70.5	20.0
$C_6H_6N_2O_2$	*o*-Nitroaniline	71.0	16.1
$C_6H_6N_2O_2$	*m*-Nitroaniline	113.4	23.6
$C_6H_6N_2O_2$	*p*-Nitroaniline	147.5	21.2
C_6H_6O	Phenol	40.89	11.51
$C_6H_6O_2$	*p*-Hydroquinone	172.4	26.8
$C_6H_6O_2$	Pyrocatechol	104.6	22.8
$C_6H_6O_2$	Resorcinol	109.4	20.4
C_6H_6S	Benzenethiol	-14.93	11.48
C_6H_7N	Aniline	-6.02	10.54
C_6H_7N	2-Methylpyridine	-66.68	9.72

Molecular formula	Name	t_m/°C	$\Delta_{fus}H$/kJ mol^{-1}
C_6H_7N	3-Methylpyridine	-18.14	14.18
C_6H_7N	4-Methylpyridine	3.67	12.58
$C_6H_8N_2$	*o*-Phenylenediamine	102.1	23.1
$C_6H_8N_2$	*m*-Phenylenediamine	66.0	15.57
$C_6H_8N_2$	*p*-Phenylenediamine	141.1	23.8
$C_6H_8N_2$	Phenylhydrazine	20.6	14.05
C_6H_{10}	Cyclohexene	-103.5	3.29
$C_6H_{10}O$	Cyclohexanone	-27.9	1.328
$C_6H_{10}O_2$	2-Oxepanone	-1.0	13.83
$C_6H_{10}O_4$	Adipic acid	152.5	36.3
$C_6H_{11}Cl$	Chlorocyclohexane	-43.81	2.043
C_6H_{12}	1-Hexene	-139.76	9.35
C_6H_{12}	*cis*-2-Hexene	-141.11	8.88
C_6H_{12}	2,3-Dimethyl-2-butene	-74.19	6.45
C_6H_{12}	Cyclohexane	6.59	2.68
C_6H_{12}	Methylcyclopentane	-142.42	6.93
$C_6H_{12}O$	Hexanal	-56	13.3
$C_6H_{12}O$	2-Hexanone	-55.5	14.9
$C_6H_{12}O$	3-Hexanone	-55.4	13.49
$C_6H_{12}O$	Cyclohexanol	25.93	1.78
$C_6H_{12}O_3$	Paraldehyde	12.6	13.5
$C_6H_{13}Br$	1-Bromohexane	-83.7	18.1
$C_6H_{13}N$	Cyclohexylamine	-17.8	17.5
C_6H_{14}	Hexane	-95.35	13.08
C_6H_{14}	2-Methylpentane	-153.6	6.27
C_6H_{14}	3-Methylpentane	-162.90	5.30
C_6H_{14}	2,2-Dimethylbutane	-98.8	0.58
C_6H_{14}	2,3-Dimethylbutane	-128.10	0.79
$C_6H_{14}O$	1-Hexanol	-47.4	15.38
$C_6H_{14}O$	Dipropyl ether	-114.8	10.8
$C_6H_{14}O$	Diisopropyl ether	-85.4	12.04
$C_6H_{14}O_2$	1,6-Hexanediol	41.5	22.2
C_7F_8	Perfluorotoluene	-65.49	11.54
C_7F_{16}	Perfluoroheptane	-51.2	6.95
$C_7H_3F_5$	2,3,4,5,6-Pentafluorotoluene	-29.78	13.1
C_7H_5ClO	Benzoyl chloride	-0.4	19.2
$C_7H_5ClO_2$	*o*-Chlorobenzoic acid	140.2	25.6
C_7H_5N	Benzonitrile	-13.99	9.1
$C_7H_5N_3O_6$	2,4,6-Trinitrotoluene	80.5	22.9
$C_7H_6O_2$	Benzoic acid	122.35	18.02
$C_7H_6O_3$	*o*-Hydroxybenzoic acid	159.0	14.2
C_7H_7Cl	*o*-Chlorotoluene	-35.8	9.6
C_7H_7NO	Benzamide	127.3	19.5
$C_7H_7NO_2$	*p*-Nitrotoluene	51.63	16.81
C_7H_8	Toluene	-94.95	6.64
C_7H_8O	*o*-Cresol	31.03	15.82
C_7H_8O	*m*-Cresol	12.24	10.71
C_7H_8O	*p*-Cresol	34.77	12.71
C_7H_8O	Benzyl alcohol	-15.4	8.97
C_7H_8O	Anisole	-37.13	12.9
C_7H_9N	*o*-Methylaniline	-14.41	11.66
C_7H_9N	*m*-Methylaniline	-31.3	7.9
C_7H_9N	*p*-Methylaniline	43.6	18.9
C_7H_{14}	1-Heptene	-118.9	12.41
C_7H_{14}	Cycloheptane	-8.46	1.88
C_7H_{14}	Methylcyclohexane	-126.6	6.75
$C_7H_{14}O$	1-Heptanal	-43.4	23.2
$C_7H_{14}O$	Cycloheptanol	7.2	1.60

Molecular formula	Name	t_m/°C	$\Delta_{fus}H$/kJ mol^{-1}
$C_7H_{14}O_2$	Heptanoic acid	-7.17	15.13
$C_7H_{15}Br$	1-Bromoheptane	-56.1	21.8
C_7H_{16}	Heptane	-90.55	14.03
C_7H_{16}	2-Methylhexane	-118.2	9.19
C_7H_{16}	3-Ethylpentane	-118.55	9.55
C_7H_{16}	2,2-Dimethylpentane	-123.7	5.82
C_7H_{16}	2,4-Dimethylpentane	-119.2	6.85
C_7H_{16}	3,3-Dimethylpentane	-134.4	6.85
C_7H_{16}	2,2,3-Trimethylbutane	-24.6	2.26
$C_7H_{16}O$	1-Heptanol	-33.2	18.17
C_8H_8	Styrene	-30.65	10.9
$C_8H_8O_2$	*o*-Toluic acid	103.5	19.5
$C_8H_8O_2$	*m*-Toluic acid	109.9	15.7
$C_8H_8O_2$	*p*-Toluic acid	179.6	22.7
$C_8H_8O_2$	Benzeneacetic acid	76.5	16.3
$C_8H_8O_2$	Methyl benzoate	-12.4	9.74
C_8H_{10}	Ethylbenzene	-94.96	9.18
C_8H_{10}	*o*-Xylene	-25.2	13.6
C_8H_{10}	*m*-Xylene	-47.8	11.6
C_8H_{10}	*p*-Xylene	13.25	17.12
$C_8H_{10}O$	2,3-Xylenol	72.5	21.0
$C_8H_{10}O$	2,5-Xylenol	74.8	23.4
$C_8H_{10}O$	2,6-Xylenol	45.8	18.9
$C_8H_{10}O$	3,4-Xylenol	65.1	18.1
$C_8H_{10}O$	3,5-Xylenol	63.4	17.4
C_8H_{16}	1-Octene	-101.7	15.31
C_8H_{16}	Cyclooctane	14.59	2.41
C_8H_{16}	Ethylcyclohexane	-111.3	8.33
C_8H_{16}	1,1-Dimethylcyclohexane	-33.3	2.07
C_8H_{16}	*cis*-1,2-Dimethylcyclohexane	-49.8	1.64
C_8H_{16}	*trans*-1,2-Dimethylcyclohexane	-88.15	10.49
C_8H_{16}	*cis*-1,3-Dimethylcyclohexane	-75.53	10.82
C_8H_{16}	*trans*-1,3-Dimethylcyclohexane	-90.07	9.87
C_8H_{16}	*cis*-1,4-Dimethylcyclohexane	-87.39	9.31
C_8H_{16}	*trans*-1,4-Dimethylcyclohexane	-36.93	12.33
$C_8H_{16}O_2$	Octanoic acid	16.5	21.35
$C_8H_{17}Br$	1-Bromooctane	-55.0	24.7
C_8H_{18}	Octane	-56.82	20.73
C_8H_{18}	2-Methylheptane	-109.02	11.92
C_8H_{18}	3-Methylheptane	-120.48	11.69
C_8H_{18}	4-Methylheptane	-121.0	10.8
C_8H_{18}	2,2,4-Trimethylpentane	-107.3	9.20
$C_8H_{18}O$	1-Octanol	-14.8	23.7
C_9H_7N	Quinoline	-14.78	10.66
C_9H_7N	Isoquinoline	26.47	13.54
C_9H_8	Indene	-1.5	10.20
C_9H_{10}	Indan	-51.38	8.60
C_9H_{12}	Propylbenzene	-99.6	9.27
C_9H_{12}	Isopropylbenzene	-96.02	7.33
C_9H_{12}	*o*-Ethyltoluene	-79.83	9.96
C_9H_{12}	*m*-Ethyltoluene	-95.6	7.6
C_9H_{12}	*p*-Ethyltoluene	-62.35	12.7
C_9H_{12}	1,2,3-Trimethylbenzene	-25.4	8.18
C_9H_{12}	1,2,4-Trimethylbenzene	-43.77	13.19
C_9H_{12}	1,3,5-Trimethylbenzene	-44.72	9.51
C_9H_{18}	Propylcyclohexane	-94.9	10.37
$C_9H_{18}O$	Nonanal	-19.3	30.5
$C_9H_{18}O$	5-Nonanone	-3.8	24.93

Molecular formula	Name	t_m/°C	$\Delta_{fus}H$/kJ mol^{-1}
$C_9H_{18}O_2$	Nonanoic acid	12.4	19.82
C_9H_{20}	Nonane	-53.46	15.47
C_9H_{20}	3,3-Diethylpentane	-33.1	10.09
C_9H_{20}	2,2,3,3-Tetramethylpentane	-9.75	2.33
C_9H_{20}	2,2,4,4-Tetramethylpentane	-66.54	9.74
$C_{10}H_7Br$	1-Bromonaphthalene	6.1	15.2
$C_{10}H_7Br$	2-Bromonaphthalene	55.9	14.4
$C_{10}H_7Cl$	1-Chloronaphthalene	-2.5	12.9
$C_{10}H_7Cl$	2-Chloronaphthalene	58.0	14.0
$C_{10}H_8$	Naphthalene	80.26	19.01
$C_{10}H_8O$	1-Naphthol	95.0	23.1
$C_{10}H_8O$	2-Naphthol	121.5	18.1
$C_{10}H_{14}$	Butylbenzene	-87.85	11.22
$C_{10}H_{14}$	1-Isopropyl-4-methylbenzene	-67.94	9.66
$C_{10}H_{14}$	1,2,4,5-Tetramethylbenzene	79.3	21
$C_{10}H_{14}O$	Thymol	49.5	21.3
$C_{10}H_{18}$	*cis*-Decahydronaphthalene	-42.9	9.49
$C_{10}H_{18}$	*trans*-Decahydronaphthalene	-30.4	14.41
$C_{10}H_{18}O_4$	Sebacic acid	130.9	40.8
$C_{10}H_{20}$	1-Decene	-66.3	13.81
$C_{10}H_{20}$	Butylcyclohexane	-74.73	14.16
$C_{10}H_{20}O$	Decanal	-4.0	34.5
$C_{10}H_{20}O_2$	Decanoic acid	31.4	27.8
$C_{10}H_{22}$	Decane	-29.6	28.72
$C_{10}H_{22}O$	1-Decanol	6.9	43
$C_{11}H_{10}$	1-Methylnaphthalene	-30.43	6.95
$C_{11}H_{10}$	2-Methylnaphthalene	34.6	12.13
$C_{11}H_{24}$	Undecane	-25.5	22.2
$C_{12}H_8$	Acenaphthylene	91.8	6.9
$C_{12}H_9N$	Carbazole	246.3	24.1
$C_{12}H_{10}$	Acenaphthene	93.4	21.49
$C_{12}H_{10}$	Biphenyl	68.93	18.57
$C_{12}H_{10}N_2$	Azobenzene	67.88	22.52
$C_{12}H_{10}N_2O$	*trans*-Azoxybenzene	34.6	17.9
$C_{12}H_{10}O$	Diphenyl ether	26.87	17.22
$C_{12}H_{11}N$	Diphenylamine	53.2	18.5
$C_{12}H_{16}$	Cyclohexylbenzene	7.07	15.6
$C_{12}H_{18}$	Hexamethylbenzene	165.5	20.6
$C_{12}H_{24}$	1-Dodecene	-35.2	19.9
$C_{12}H_{24}O_2$	Dodecanoic acid	43.8	36.3
$C_{12}H_{26}$	Dodecane	-9.57	36.8
$C_{12}H_{26}O$	1-Dodecanol	23.9	40.2
$C_{13}H_{10}$	9*H*-Fluorene	114.77	19.58
$C_{13}H_{10}O$	Benzophenone	47.9	18.19
$C_{13}H_{12}$	Diphenylmethane	25.4	18.6
$C_{13}H_{28}$	Tridecane	-5.4	28.50
$C_{13}H_{28}O$	1-Tridecanol	31.7	41.4
$C_{14}H_{10}$	Anthracene	215.76	29.4
$C_{14}H_{10}$	Phenanthrene	99.24	16.46
$C_{14}H_{10}O_2$	Benzil	94.87	23.5
$C_{14}H_{12}$	*trans*-Stilbene	124.2	27.7
$C_{14}H_{12}O_2$	α-Phenylbenzeneacetic acid	147.29	31.3
$C_{14}H_{28}O_2$	Tetradecanoic acid	54.2	45.1
$C_{14}H_{30}$	Tetradecane	5.82	45.07
$C_{14}H_{30}O$	1-Tetradecanol	38.2	25.1*
$C_{15}H_{32}$	Pentadecane	9.95	34.6
$C_{16}H_{10}$	Fluoranthene	110.19	18.69
$C_{16}H_{10}$	Pyrene	150.62	17.36

Molecular formula	Name	t_m/°C	$\Delta_{fus}H$/kJ mol^{-1}
$C_{16}H_{32}O_2$	Hexadecanoic acid	62.5	53.7
$C_{16}H_{34}$	Hexadecane	18.12	53.36
$C_{16}H_{34}O$	1-Hexadecanol	49.2	33.6
$C_{17}H_{36}$	Heptadecane	22.0	40.16
$C_{18}H_{12}$	Benz[a]anthracene	160.5	21.4
$C_{18}H_{12}$	Benzo[c]phenanthrene	68	16.3
$C_{18}H_{12}$	Chrysene	255.5	26.2
$C_{18}H_{12}$	Triphenylene	197.8	24.74
$C_{18}H_{14}$	*o*-Terphenyl	56.20	17.19
$C_{18}H_{14}$	*p*-Terphenyl	213.9	35.3
$C_{18}H_{15}N$	Triphenylamine	126.5	24.9
$C_{18}H_{36}O_2$	Stearic acid	69.3	61.2
$C_{18}H_{38}$	Octadecane	28.2	61.7
$C_{18}H_{38}O$	1-Octadecanol	57.9	45
$C_{19}H_{40}$	Nonadecane	32.0	45.8
$C_{20}H_{12}$	Perylene	277.76	31.9
$C_{20}H_{12}$	Benzo[a]pyrene	181.1	17.3
$C_{20}H_{12}$	Benzo[e]pyrene	181.4	16.6
$C_{20}H_{14}$	2,2'-Binaphthalene	187.9	38.9
$C_{20}H_{42}$	Eicosane	36.6	69.9
$C_{20}H_{42}O$	1-Eicosanol	65.4	42
$C_{24}H_{12}$	Coronene	437.4	19.2

PRESSURE AND TEMPERATURE DEPENDENCE OF LIQUID DENSITY

This table gives data on the variation of the density of some common liquids with pressure and temperature. The pressure dependence is described to first order by the isothermal compressibility coefficient κ defined as

$$\kappa = -(1/V)\ (\partial V/\partial P)_T$$

where V is the volume, and the temperature dependence by the cubic expansion coefficient α,

$$\alpha = (1/V)\ (\partial V/\partial T)_P$$

Substances are listed by molecular formula in the Hill order. More precise data on the variation of density with temperature over a wide temperature range can be found in Reference 1.

REFERENCES

1. Lide, D. R., and Kehiaian, H. V., *CRC Handbook of Thermophysical and Thermochemical Data*, CRC Press, Boca Raton, FL, 1994.
2. Le Neindre, B., *Effets des Hautes et Très Hautes Pressions*, in *Techniques de l'Ingénieur*, Paris, 1991.
3. *Landolt-Börnstein, Numerical Data and Functional Relationships in Science and Technology, New Series*, IV/4, *High-pressure Properties of Matter*, Springer-Verlag, Heidelberg, 1980.
4. Riddick, J.A., Bunger, W.B., and Sakano, T.K., *Organic Solvents, Fourth Edition*, John Wiley & Sons, New York, 1986.
5. Isaacs, N. S., *Liquid Phase High Pressure Chemistry*, John Wiley, New York, 1981.

Molecular formula	Name	Isothermal Compressibility		Cubic Thermal Expansion	
		t/°C	$\kappa \times 10^4$/MPa^{-1}	t/°C	$\alpha \times 10^3$/°C^{-1}
Cl_3P	Phosphorus trichloride	20	9.45	20	1.9
H_2O	Water	20	4.591	20	0.206
		25	4.524	25	0.256
		30	4.475	30	0.302
Hg	Mercury	20	0.401	20	1.811
CCl_4	Tetrachloromethane	20	10.50	20	1.14
		40	12.20	40	1.21
		70	15.6	70	1.33
$CHBr_3$	Tribromomethane	50	8.76	25	0.91
$CHCl_3$	Trichloromethane	20	9.96	20	1.21
		50	12.9	50	1.33
CH_2Br_2	Dibromomethane	27	6.85		
CH_2Cl_2	Dichloromethane	25	10.3	25	1.39
CH_3I	Iodomethane	27	10.3	25	1.26
CH_4O	Methanol	20	12.14	20	1.49
		40	13.83	40	1.59
CS_2	Carbon disulfide	20	9.38	20	1.12
		40	10.6	35	1.16
C_2Cl_4	Tetrachloroethylene	25	7.56	25	1.02
C_2HCl_3	Trichloroethylene	25	8.57	25	1.17
$C_2H_2Cl_2$	trans-1,2-Dichloroethylene	25	11.2	25	1.36
$C_2H_4Cl_2$	1,1-Dichloroethane	20	7.97	25	0.93
$C_2H_4Cl_2$	1,2-Dichloroethane	30	8.46	20	1.14
$C_2H_4O_2$	Acetic acid	20	9.08	20	1.08
		80	13.7	80	1.38
C_2H_5Br	Bromoethane	20	11.53	20	1.31
C_2H_5I	Iodoethane	20	9.82	25	1.17
C_2H_6O	Ethanol	20	11.19	20	1.40
		70	15.93	70	1.67
$C_2H_6O_2$	Ethylene glycol	20	3.64	20	0.626
C_3H_6O	Acetone	20	12.62	20	1.46
		40	15.6	40	1.57
C_3H_7Br	1-Bromopropane	0	10.22	25	1.2
C_3H_7Cl	1-Chloropropane	0	12.09	20	1.4
C_3H_7I	1-Iodopropane	0	10.22	25	1.09
C_3H_8O	1-Propanol	0	8.43	0	1.22
C_3H_8O	2-Propanol	40	13.32	40	1.55
$C_3H_8O_2$	1,2-Propanediol	0	4.45	20	0.695

Molecular formula	Name	Isothermal Compressibility		Cubic Thermal Expansion	
		t/°C	$\kappa \times 10^4$/MPa^{-1}	t/°C	$\alpha \times 10^3$/°C^{-1}
$C_3H_8O_2$	1,3-Propanediol	0	4.09	20	0.61
$C_3H_8O_3$	Glycerol	0	2.54	20	0.520
$C_4H_8O_2$	Ethyl acetate	20	11.32	20	1.35
		60	16.2	60	1.54
C_4H_9Br	1-Bromobutane	25	10.26	20	1.13
C_4H_9I	1-Iodobutane	0	7.73	25	1.02
$C_4H_{10}O$	1-Butanol	0	8.10	0	1.12
$C_4H_{10}O$	Diethyl ether	20	18.65	20	1.65
		30	20.85	30	1.72
$C_4H_{10}O_3$	Diethylene glycol	0	3.34	20	0.635
C_5H_{10}	Cyclopentane	20	13.31	20	1.35
$C_5H_{11}Br$	1-Bromopentane	0	8.42	25	1.04
$C_5H_{11}I$	1-Iodopentane	0	7.56		
C_5H_{12}	Pentane	25	21.80	25	1.64
$C_5H_{12}O$	1-Pentanol	0	7.71	0	1.02
C_6H_5Br	Bromobenzene	20	6.46	20	0.86
C_6H_5Cl	Chlorobenzene	20	7.45	20	0.94
$C_6H_5NO_2$	Nitrobenzene	20	4.93	25	0.833
C_6H_6	Benzene	25	9.66	25	1.14
		45	11.28	45	1.21
C_6H_6O	Phenol	60	6.05	60	0.82
C_6H_7N	Aniline	20	4.53	20	0.81
		80	6.32	80	0.91
C_6H_{12}	Cyclohexane	20	11.30	20	1.15
		60	15.2	60	1.29
C_6H_{14}	Hexane	25	16.69	25	1.41
		45	20.27	45	1.52
C_6H_{14}	2-Methylpentane	0	13.97	25	1.43
C_6H_{14}	3-Methylpentane	0	14.57	25	1.40
C_6H_{14}	2,3-Dimethylbutane	20	17.97	25	1.39
$C_6H_{14}O$	1-Hexanol	25	8.24	25	1.03
$C_6H_{15}NO_3$	Triethanolamine	0	3.61	55	0.53
C_7H_8	Toluene	20	8.96	20	1.05
		50	11.0	50	1.13
C_7H_8O	Anisole	20	6.60	20	0.951
C_7H_{14}	Cycloheptane	20	9.22		
C_7H_{16}	Heptane	25	14.38	25	1.26
C_8H_{10}	*o*-Xylene	25	8.10	25	0.96
C_8H_{10}	*m*-Xylene	20	8.46	20	0.99
C_8H_{10}	*p*-Xylene	25	8.59	25	1.00
C_8H_{16}	Cyclooctane	20	8.03		
C_8H_{18}	Octane	25	12.82	25	1.16
		45	15.06	45	1.23
$C_8H_{18}O$	1-Octanol	25	7.64	25	0.827
C_9H_{12}	Mesitylene	25	8.14	25	0.94
$C_9H_{14}O_6$	Triacetin	0	4.49	25	0.94
C_9H_{20}	Nonane	25	11.75	25	1.08
$C_{10}H_{22}$	Decane	25	10.94	25	1.02
$C_{11}H_{24}$	Undecane	25	10.31	25	0.97
$C_{12}H_{26}$	Dodecane	25	9.88	25	0.93
$C_{13}H_{28}$	Tridecane	25	9.48	25	0.90
$C_{14}H_{30}$	Tetradecane	25	9.10	25	0.87
$C_{15}H_{32}$	Pentadecane	25	8.82		
$C_{16}H_{22}O_4$	Butyl phthalate	0	5.0	25	0.86
$C_{16}H_{34}$	Hexadecane	25	8.57		
		45	9.78		
$C_{19}H_{36}O_2$	Methyl oleate	0	6.18	60	0.85

PROPERTIES OF CRYOGENIC FLUIDS

This table gives physical and thermodynamic properties of eight cryogenic fluids. The properties are:

M	Molar mass in grams per mole
T_t	Triple point temperature in kelvins
P_t	Triple point pressure in kilopascals
ρ_t (1)	Liquid density at the triple point in grams per milliliter
$\Delta_{fus}H$ @ T_t	Enthalpy of fusion at the triple point in joules per gram
T_b	Normal boiling point in kelvins at a pressure of 101325 pascals (760 mmHg)
$\Delta_{vap}H$ @ T_b	Enthalpy of vaporization at the normal boiling point in joules per gram
ρ (l) @ T_b	Liquid density at the normal boiling point in grams per milliliter
ρ (g) @ T_b	Vapor density at the normal boiling point in grams per liter
C_p (l) @ T_b	Liquid heat capacity at constant pressure at the normal boiling point in joules per gram kelvin
C_p (g) @ T_b	Vapor heat capacity at constant pressure at the normal boiling point in joules per gram kelvin
T_c	Critical temperature in kelvins
P_c	Critical pressure in megapascals
ρ_c	Critical density in grams per milliliter

In the case of air, the value given for the triple point temperature is the incipient solidification temperature, and the normal boiling point value is the incipient boiling (bubble) point. See Reference 3 for more details.

REFERENCES

1. Younglove, B. A., *J. Phys. Chem. Ref. Data*, 11, Suppl. 1, 1982.
2. Daubert, T. E., Danner, R. P., Sibul, H. M., and Stebbins, C. C., *Physical and Thermodynamic Properties of Pure Compounds: Data Compilation*, extant 1994 (core with 4 supplements), Taylor & Francis, Bristol, PA (also available as database).
3. Sytchev, V. V., et al., *Thermodynamic Properties of Air*, Hemisphere Publishing, New York, 1987.
4. Jacobsen, R. T., Stewart, R. B., and Jahangiri, M., *J. Phys. Chem. Ref. Data*, 15, 735, 1986. [Nitrogen]
5. Stewart, R. B., Jacobsen, R. T., and Wagner, W., *J. Phys. Chem. Ref. Data*, 20, 917, 1991. [Oxygen]
6. McCarty, R. D., *J. Phys. Chem. Ref. Data*, 2, 923, 1973. [Helium] Also, Donnelly, R. J., private communication.
7. Stewart, R. B. and Jacobsen, R. T., *J. Phys. Chem. Ref. Data*, 18, 639, 1989. [Argon]
8. Setzmann, U. and Wagner, W., *J. Phys. Chem. Ref. Data*, 20, 1061, 1991. [Methane]
9. Vargaftik, N. B., *Thermophysical Properties of Liquids and Gases*, 2nd ed., John Wiley, New York, 1975.

Property	Units	Air	N_2	O_2	H_2	He	Ne	Ar	Kr	Xe	CH_4
M	g/mol	28.96	28.014	31.999	2.0159	4.0026	20.180	39.948	83.800	131.290	16.043
T_t	K	59.75	63.15	54.3584	13.8		24.5561	83.8058	115.8	161.4	90.694
P_t	kPa		12.463	0.14633	7.042		50	68.95	72.92	81.59	11.696
ρ_t (l)	g/mL	0.959	0.870	1.306	0.0770		1.251	1.417	2.449	2.978	0.4515
$\Delta_{fus}H$ @ T_t	J/g		25.3	13.7	59.5		16.8	28.0	16.3	13.8	58.41
T_b	K	78.67	77.35	90.188	20.28	4.2221	27.07	87.293	119.92	165.10	111.668
$\Delta_{vap}H$ @ T_b	J/g	198.7	198.8	213.1	445	20.7	84.8	161.0	108.4	96.1	510.83
ρ (l) @ T_b	g/mL	0.8754	0.807	1.141	0.0708	0.124901	1.204	1.396	2.418	2.953	0.4224
ρ (g) @ T_b	g/L	3.199	4.622	4.467	1.3390	16.89	9.51	5.79	8.94		1.816
C_p (l) @ T_b	J/g K	1.865	2.042	1.699	9.668	4.545	1.877	1.078	0.533	0.340	3.481
C_p (g) @ T_b	J/g K		1.341	0.980	12.24	9.78		0.570	0.248	0.158	2.218
T_c	K	132.5	126.20	154.581	32.98	5.1953	44.40	150.663	209.40	289.73	190.56
P_c	MPa	3.766	3.390	5.043	1.293	0.227460	2.760	4.860	5.500	5.840	4.592
ρ_c	g/mL	0.316	0.313	0.436	0.031	0.06964	0.484	0.531	0.919	1.110	0.1627

PROPERTIES OF LIQUID HELIUM

The following data were obtained by a critical evaluation of all existing experimental measurements on liquid helium, using a fitting procedure described in the reference. All values refer to liquid helium at saturated vapor pressure; temperatures are on the ITS-90 scale. Several properties show a singularity at the lambda point (2.1768 K).

p : vapor pressure
ρ : density
C_s : molar heat capacity
$\Delta_{vap}H$: molar enthalpy of vaporization
ε : relative permittivity (dielectric constant)
σ : surface tension
α : coefficient of linear expansion
η : viscosity
λ : thermal conductivity

REFERENCE

Donnelly, R. J., and Barenghi, C. F., *J. Phys. Chem. Reference Data* 27, 1217, 1998.

T/K	p/kPa	ρ/g cm^{-3}	C_s/J mol^{-1}K^{-1}	$\Delta_{vap}H$/J mol^{-1}	ε	σ/mN m^{-1}	$10^3\alpha$/K^{-1}	η/μPa s	λ/W cm^{-1}K^{-1}
0.0		0.1451397	0	59.83	1.057255		0.000		
0.5		0.1451377	0.010	70.24	1.057254	0.3530	0.107		
1.0	0.01558	0.1451183	0.415	80.33	1.057246	0.3471	0.309	3.873	
1.5	0.4715	0.1451646	4.468	89.35	1.057265	0.3322	-2.36	1.346	
2.0	3.130	0.1456217	21.28	93.07	1.057449	0.3021	-12.2	1.468	
2.5	10.23	0.1448402	9.083	92.50	1.057135	0.2623	39.4	3.259	0.1497
3.0	24.05	0.1412269	9.944	94.11	1.055683	0.2161	61.5	3.517	0.1717
3.5	47.05	0.1360736	12.37	92.84	1.053615	0.1626	88.7	3.509	0.1868
4.0	81.62	0.1289745	15.96	87.00	1.050770	0.1095	129	3.319	0.1965
4.5	130.3	0.1188552	21.8	75.86	1.046725	0.0609	211		
5.0	196.0		44.7	47.67		0.0157			

HALOCARBON REFRIGERANTS

Halogen derivatives of the lower aliphatic hydrocarbons, especially fluorocarbons and chlorocarbons, are widely used as refrigerants, solvents, and cleaning agents. This table lists the most important halocarbons in current use or being considered as possible substitutes for those compounds that are potential environmental hazards. The code number for each compound appears in the first column; this number is frequently used in expressions like R 11 or CFC 11 or HFC 134.

In addition to name, molecular formula, CAS Registry Number, and molecular weight, the table lists the following properties:

t_m: normal melting point
t_b: normal boiling point (at 101.324 kPa)
t_c: critical temperature
TLV: Threshold Limit Value, which is the maximum safe concentration in air in the workplace, expressed as the time-weighted average (TWA) in parts per million by volume over an 8-hr workday and 40-hr workweek (see Reference 3). An * following the TLV indicates that the substance in a confirmed or suspected human carcinogen.

REFERENCES

1. *1989 ASHRAE Handbook: Fundamentals*, American Society of Heating, Refrigerating, and Air Conditioning Engineers, Atlanta, GA, 1989.
2. Platzer, B., Polt, A., and Mauer,G.,*Thermophysical Properties of Refrigerants*, Springer, Berlin, 1990.
3. 1991-1992 Threshold Limit Values and Biological Exposure Indexes, American Conference of Government Industrial Hygienists, Cincinnati, OH, 1991.

Code no.	Name	Molecular formula	CAS reg. no.	Molecular weight	t_m/°C	t_b/°C	t_c/°C	TLV ppm
10	Tetrachloromethane	CCl_4	56-23-5	153.822	-23	76.8	283.5	5*
11	Trichlorofluoromethane	CCl_3F	75-69-4	137.368	-111.11	23.75	198.1	1000
12	Dichlorodifluoromethane	CCl_2F_2	75-71-8	120.913	-158	-29.8	111.80	1000
12B1	Bromochlorocifluoromethane	$CBrClF_2$	353-59-3	165.365	-159.5	-3.72	153.73	
13	Chlorotrifluoromethane	$CClF_3$	75-72-9	104.459	-181	-81.4	29	
13B1	Bromotrifluoromethane	$CBrF_3$	75-63-8	148.910	-172	-57.89	67.1	
14	Tetrafluoromethane	CF_4	75-73-0	88.005	-183.59	-128.02	-45.5	
20	Trichloromethane	$CHCl_3$	67-66-3	119.377	-63.6	61.17	263.3	10*
21	Dichlorofluoromethane	$CHCl_2F$	75-43-4	102.923	-135	8.95	178.43	10
22	Chlorodifluoromethane	$CHClF_2$	75-45-6	86.468	-157.42	-40.75	96.2	1000
23	Trifluoromethane	CHF_3	75-46-7	70.014	-155.18	-82.1	26.2	
30	Dichloromethane	CH_2Cl_2	75-09-2	84.932	-95.14	40	237	50*
31	Chlorofluoromethane	CH_2ClF	593-70-4	68.478	-133	-9.1		
32	Difluoromethane	CH_2F_2	75-10-5	52.024	-136	-51.69	78.5	
40	Chloromethane	CH_3Cl	74-87-3	50.488	-97.7	-24.09	143.10	50*
41	Fluoromethane	CH_3F	593-53-3	34.033	-141.8	-78.41	44.7	
110	Hexachloroethane	C_2Cl_6	67-72-1	236.738	187	187		1*
111	Pentachlorofluoroethane	C_2Cl_5F	354-56-3	220.284				
112	Tetrachloro-1,2-difluoroethane	$C_2Cl_4F_2$	76-12-0	203.830	26	93	278	500
112a	1,1,1,2-Tetrachloro-2,2-difluoroethane	$C_2Cl_4F_2$	76-11-9	203.830	40.6	91.5		
113	1,1,2-Trichlorotrifluoroethane	$C_2Cl_3F_3$	76-13-1	187.375	-35	47.7	214.2	1000
113a	1,1,1-Trichlorotrifluoroethane	$C_2Cl_3F_3$	354-58-5	187.375	14.2	46.1		
114	1,2-Dichlorotetrafluoroethane	$C_2Cl_2F_4$	76-14-2	170.921	-94	3.8	145.63	1000
114a	1,1-Dichlorotetrafluoroethane	$C_2Cl_2F_4$	374-07-2	170.921	-56.6	4	145.5	

HALOCARBON REFRIGERANTS (continued)

Code no.	Name	Molecular formula	CAS reg. no.	Molecular weight	t_m/°C	t_b/°C	t_c/°C	TLV ppm
114B2	1,2-Dibromotetrafluoroethane	$C_2Br_2F_4$	124-73-2	259.824	-110.4	47.35	214.7	
115	Chloropentafluoroethane	C_2ClF_5	76-15-3	154.467	-99.44	-37.95	80.1	1000
116	Hexafluoroethane	C_2F_6	76-16-4	138.012	-100.7	-78.1	20	
120	Pentachloroethane	C_2HCl_5	76-01-7	202.293	-29	159.88		
121	1,1,2,2-Tetrachloro-1-fluoroethane	C_2HCl_4F	354-14-3	185.839	-82.6	116.6		
121a	1,1,1,2-Tetrachloro-2-fluoroethane	C_2HCl_4F	354-11-0	185.840	-95.35	116.5		
122	1,2,2-Trichloro-1,1-difluoroethane	$C_2HCl_3F_2$	354-21-2	169.385	-140	71.9		
122a	1,1,2-Trichloro-1,2-difluoroethane	$C_2HCl_3F_2$	354-15-4	169.385		72.5		
122b	1,1,1-Trichloro-2,2-difluoroethane	$C_2HCl_3F_2$	354-12-1	169.385		73		
123	2,2-Dichloro-1,1,1-trifluoroethane	$C_2HCl_2F_3$	306-83-2	152.931	-107	27.1		
123a	1,2-Dichloro-1,2,2-trifluoroethane	$C_2HCl_2F_3$	354-23-4	152.931		28.2		
124	2-Chloro-1,1,1,2-tetrafluoroethane	C_2HClF_4	2837-89-0	136.476		-12		
124a	1-Chloro-1,1,2,2-tetrafluoroethane	C_2HClF_4	354-25-6	136.476	-117	-10.2	126.8	
125	Pentafluoroethane	C_2HF_5	354-33-6	120.022	-103	-48.5		
130	1,1,2,2-Tetrachloroethane	$C_2H_2Cl_4$	79-34-5	167.849	-43.8	146.5	388.00	1*
131	1,1,2-Trichloro-2-fluoroethane	$C_2H_2Cl_3F$	359-28-4	151.394		102.5		
132	1,2-Dichloro-1,2-difluoroethane	$C_2H_2Cl_2F_2$	431-06-1	134.940	-155	59		
132b	1,2-Dichloro-1,1-difluoroethane	$C_2H_2Cl_2F_2$	1649-08-7	134.940				
133	1-Chloro-1,2,2-trifluoroethane	$C_2H_2ClF_3$	431-07-2	118.486		17		
133a	2-Chloro-1,1,1-trifluoroethane	$C_2H_2ClF_3$	75-88-7	118.486	-105.5	6.1		
133b	1-Chloro-1,1,2-trifluoroethane	$C_2H_2ClF_3$	421-04-5	118.486		12		
134	1,1,2,2-Tetrafluoroethane	$C_2H_2F_4$	359-35-3	102.032	-89	-19.9	119	
134a	1,1,1,2-Tetrafluoroethane	$C_2H_2F_4$	811-97-2	102.032	-101	-26.5	101.2	
140	1,1,2-Trichloroethane	$C_2H_3Cl_3$	79-00-5	133.404	-36.6	113.8		10
140a	1,1,1-Trichloroethane	$C_2H_3Cl_3$	71-55-6	133.404	-30.4	74.09	272	350
141	1,2-Dichloro-1-fluoroethane	$C_2H_3Cl_2F$	430-57-9	116.950		75.7		
141b	1,1-Dichloro-1-fluoroethane	$C_2H_3Cl_2F$	1717-00-6	116.950	-103.5	32.11		
142	2-Chloro-1,1-difluoroethane	$C_2H_3ClF_2$	338-65-8	100.495		35.1		
142b	1-Chloro-1,1-difluoroethane	$C_2H_3ClF_2$	75-68-3	100.495	-130.8	-9.75	137.14	
143	1,1,2-Trifluoroethane	$C_2H_3F_3$	430-66-0	84.041	-84	5		
143a	1,1,1-Trifluoroethane	$C_2H_3F_3$	420-46-2	84.041	-111.3	-47.55	73.2	
150	1,2-Dichloroethane	$C_2H_4Cl_2$	107-06-2	98.959	-35.5	83.5	288	10*
150a	1,1-Dichloroethane	$C_2H_4Cl_2$	75-34-3	98.959	-96.96	57.4	250	100
151	1-Chloro-2-fluoroethane	C_2H_4ClF	762-50-5	82.505		53.2		
151a	1-Chloro-1-fluoroethane	C_2H_4ClF	1615-75-4	82.505		16.1		
152	1,2-Difluoroethane	$C_2H_4F_2$	624-72-6	66.051		30.7		
152a	1,1-Difluoroethane	$C_2H_4F_2$	75-37-6	66.051	-117	-24.95	113.6	
160	Chloroethane	C_2H_5Cl	75-00-3	64.514	-138.7	13.1	187.3	1000
161	Fluoroethane	C_2H_5F	353-36-6	48.060	-143.2	-37.65	102.16	
218	Perfluoropropane	C_3F_8	76-19-7	188.020	-147.69	-36.65	72.0	
1112a	1,1-Dichloro-2,2-difluoroethylene	$C_2Cl_2F_2$	79-35-6	132.924				
1113	Chlorotrifluoroethylene	C_2ClF_3	79-38-9	116.470	-158	-27.85	106	

HALOCARBON REFRIGERANTS (continued)

Code no.	Name	Molecular formula	CAS reg. no.	Molecular weight	t_m/°C	t_b/°C	t_c/°C	TLV ppm
1114	Tetrafluoroethylene	C_2F_4	116-14-3	100.016	-142.5	-75.95	33.4	
1120	Trichloroethylene	C_2HCl_3	79-01-6	131.388	-84.75	87.21	271.1	50*
1130	*cis*-1,2-Dichloroethylene	$C_2H_2Cl_2$	156-59-2	96.943	-80	60.19	271.1	200
1130	*trans*-1,2-Dichloroethylene	$C_2H_2Cl_2$	156-60-5	96.943	-49.8	48.73	243.4	200
1132a	1,1-Difluoroethylene	$C_2H_2F_2$	75-38-7	64.035	-144	-85.7	29.8	
1140	Chloroethylene	C_2H_3Cl	75-01-4	62.499	-153.79	-13.37		5*
1141	Fluoroethylene	C_2H_3F	75-02-5	46.044	-160.5	-72	54.8	
C316	1,2-Dichlorohexafluorocyclobutane	$C_4Cl_2F_6$	356-18-3	232.940				
C317	Chloroheptafluorocyclobutane	C_4ClF_7	377-41-3	216.486				
C318	Perfluorocyclobutane	C_4F_8	115-25-3	200.031	-40.19	-5.99	115.31	

* Confirmed or suspected human carcinogen.

DENSITY AND SPECIFIC VOLUME OF MERCURY

The data in this table have been adjusted to the ITS-90 temperature scale. The uncertainty in density values is 0.0003 g/mL between –20 and –10°C; 0.0001 or less between –10 and 200°C; and 0.0002 between 200 and 300°C.

REFERENCE

Ambrose, D., *Metrologia*, 27, 245, 1990.

t/°C	ρ/(g/mL)	*v*/(mL/kg)	*t*/°C	ρ/(g/mL)	*v*/(mL/kg)	*t*/°C	ρ/(g/mL)	*v*/(mL/kg)
–20	13.64461	73.2890	27	13.52869	73.9170	74	13.41423	74.5477
–19	13.64212	73.3024	28	13.52624	73.9304	75	13.41181	74.5612
–18	13.63964	73.3157	29	13.52379	73.9438	76	13.40939	74.5746
–17	13.63716	73.3291	30	13.52134	73.9572	77	13.40697	74.5881
–16	13.63468	73.3424	31	13.51889	73.9705	78	13.40455	74.6016
–15	13.63220	73.3558	32	13.51645	73.9839	79	13.40213	74.6150
–14	13.62972	73.3691	33	13.51400	73.9973	80	13.39971	74.6285
–13	13.62724	73.3824	34	13.51156	74.0107	81	13.39729	74.6420
–12	13.62476	73.3958	35	13.50911	74.0241	82	13.39487	74.6554
–11	13.62228	73.4091	36	13.50667	74.0375	83	13.39245	74.6689
–10	13.61981	73.4225	37	13.50422	74.0509	84	13.39003	74.6824
–9	13.61733	73.4358	38	13.50178	74.0643	85	13.38762	74.6959
–8	13.61485	73.4492	39	13.49934	74.0777	86	13.38520	74.7094
–7	13.61238	73.4625	40	13.49690	74.0911	87	13.38278	74.7229
–6	13.60991	73.4759	41	13.49446	74.1045	88	13.38037	74.7364
–5	13.60743	73.4892	42	13.49202	74.1179	89	13.37795	74.7498
–4	13.60496	73.5026	43	13.48958	74.1313	90	13.37554	74.7633
–3	13.60249	73.5160	44	13.48714	74.1447	91	13.37313	74.7768
–2	13.60002	73.5293	45	13.48470	74.1581	92	13.37071	74.7903
–1	13.59755	73.5427	46	13.48226	74.1715	93	13.36830	74.8038
0	13.59508	73.5560	47	13.47982	74.1850	94	13.36589	74.8173
1	13.59261	73.5694	48	13.47739	74.1984	95	13.36347	74.8308
2	13.59014	73.5827	49	13.47495	74.2118	96	13.36106	74.8443
3	13.58768	73.5961	50	13.47251	74.2252	97	13.35865	74.8579
4	13.58521	73.6095	51	13.47008	74.2386	98	13.35624	74.8714
5	13.58275	73.6228	52	13.46765	74.2520	99	13.35383	74.8849
6	13.58028	73.6362	53	13.46521	74.2655	100	13.35142	74.8984
7	13.57782	73.6495	54	13.46278	74.2789	110	13.3273	75.0337
8	13.57535	73.6629	55	13.46035	74.2923	120	13.3033	75.1693
9	13.57289	73.6763	56	13.45791	74.3057	130	13.2793	75.3052
10	13.57043	73.6896	57	13.45548	74.3192	140	13.2553	75.4413
11	13.56797	73.7030	58	13.45305	74.3326	150	13.2314	75.5778
12	13.56551	73.7164	59	13.45062	74.3460	160	13.2075	75.7147
13	13.56305	73.7297	60	13.44819	74.3594	170	13.1836	75.8519
14	13.56059	73.7431	61	13.44576	74.3729	180	13.1597	75.9895
15	13.55813	73.7565	62	13.44333	74.3863	190	13.1359	76.1274
16	13.55567	73.7698	63	13.44090	74.3998	200	13.1120	76.2659
17	13.55322	73.7832	64	13.43848	74.4132	210	13.0882	76.4047
18	13.55076	73.7966	65	13.43605	74.4266	220	13.0644	76.5440
19	13.54831	73.8100	66	13.43362	74.4401	230	13.0406	76.6838
20	13.54585	73.8233	67	13.43120	74.4535	240	13.0167	76.8241
21	13.54340	73.8367	68	13.42877	74.4670	250	12.9929	76.9650
22	13.54094	73.8501	69	13.42635	74.4804	260	12.9691	77.1064
23	13.53849	73.8635	70	13.42392	74.4939	270	12.9453	77.2484
24	13.53604	73.8769	71	13.42150	74.5073	280	12.9214	77.3909
25	13.53359	73.8902	72	13.41908	74.5208	290	12.8975	77.5341
26	13.53114	73.9036	73	13.41665	74.5342	300	12.8736	77.6779

THERMAL PROPERTIES OF MERCURY

Lev R. Fokin

The first of these tables gives the molar heat capacity at constant pressure of liquid and gaseous mercury as a function of temperature. To convert to specific heat in units of J/g K, divide these values by 200.59, the atomic weight of mercury.

REFERENCE

Douglas, T. B., Ball, A. T., and Ginnings, D. C., *J. Res. Natl. Bur. Stands.*, 46, 334, 1951.

t/°C	C_p/(J/mol K) Liquid	C_p/(J/mol K) Gas	t/°C	C_p/(J/mol K) Liquid	C_p/(J/mol K) Gas	t/°C	C_p/(J/mol K) Liquid	C_p/(J/mol K) Gas
−38.84	28.2746	20.786	140	27.3675	20.786	340	27.1500	20.836
−20	28.1466	20.786	160	27.3090	20.786	356.73	27.1677	20.849
0	28.0190	20.786	180	27.2588	20.790	360	27.1709	20.853
20	27.9002	20.786	200	27.2169	20.790	380	27.1981	20.870
25	27.8717	20.786	220	27.1834	20.794	400	27.2324	20.891
40	27.7897	20.786	240	27.1583	20.794	420	27.2738	20.916
60	27.6880	20.786	260	27.1412	20.799	440	27.3207	20.941
80	27.5952	20.786	280	27.1320	20.807	460	27.3742	20.974
100	27.5106	20.786	300	27.1303	20.815	480	27.4332	21.008
120	27.4349	20.786	320	27.1366	20.824	500	27.4985	21.046

The second table gives the molar heat capacity of solid mercury in its rhombohedral (α-mercury) form.

REFERENCES

1. Busey and Giaque, *J. Am. Chem. Soc.*, 75, 806, 1953.
2. Amitin, Lebedeva, and Paukov, *Rus. J. Phys. Chem.*, 2666, 1979.

t/°C	C_p/(J/mol K)	t/°C	C_p/(J/mol K)	t/°C	C_p/(J/mol K)	t/°C	C_p/(J/mol K)
−268.99	0.99*	−248.15	12.74	−193.15	23.16	−113.15	26.15
−268.99	0.97**	−243.15	14.78	−183.15	23.76	−93.15	26.69
−268.15	1.6	−233.15	17.90	−173.15	24.24	−73.15	27.28
−263.15	4.6	−223.15	19.94	−153.15	25.00	−53.15	27.96
−258.15	7.6	−213.15	21.40	−133.15	25.61	−38.87	28.5
−253.15	10.33	−203.15	22.42				

* Superconducting state
** Normal state

The final table gives the cubic thermal expansion coefficient α, the isothermal compressibility coefficient κ_T, and the speed of sound U for liquid mercury as a function of temperature. These properties are defined as follows:

$$\alpha = \frac{1}{v}\left(\frac{\partial v}{\partial T}\right)_p \qquad \kappa_T = -\frac{1}{v}\left(\frac{\partial v}{\partial P}\right)_T \qquad U^2 = \left(\frac{\partial P}{\partial \rho}\right)_S \qquad \rho = v^{-1}$$

where v is the specific volume (given in the table on the preceding page).

REFERENCE

Vukalovich, M. P., et al., *Thermophysical Properties of Mercury,* Moscow Standard Press, 1971.

t/°C	$\alpha \times 10^4$/K^{-1}	$\kappa_T \times 10^6$/bar^{-1} At 1 bar	$\kappa_T \times 10^6$/bar^{-1} At 1000 bar	U/m s^{-1}	t/°C	$\alpha \times 10^4$/K^{-1}	$\kappa_T \times 10^6$/bar^{-1} At 1 bar	$\kappa_T \times 10^6$/bar^{-1} At 1000 bar	U/m s^{-1}
−20	1.818	3.83		1470	120	1.8058	4.513	4.33	1404.7
0	1.8144	3.918	3.78	1460.8	140	1.8074	4.622		1395.4
20	1.8110	4.013	3.87	1451.4	160	1.8100	4.731	4.53	1386.1
40	1.8083	4.109	3.96	1442.0	180	1.8136	4.844		1376.7
60	1.8064	4.207		1432.7	200	1.818	4.96		1367
80	1.8053	4.308	4.14	1423.4	250	1.834	5.26		1344
100	1.8051	4.410		1414.1	300	1.856	5.59		1321

SURFACE TENSION OF COMMON LIQUIDS

The surface tension γ of about 200 liquids is tabulated here as a function of temperature. Values of γ are given in units of millinewtons per meter (mN/m), which is equivalent to dyn/cm in cgs units. The values refer to a nominal pressure of one atmosphere (about 100 kPa) except in cases where the indicated temperature is above the normal boiling point of the substance; in those cases, the applicable pressure is the saturation vapor pressure at the temperature in question.

The uncertainty of the values is 0.1 to 0.2 mN/m or less in most cases. Values at temperatures between the points tabulated can be obtained by linear interpolation to a good approximation.

Substances are listed by molecular formula in the modified Hill order, with substances not containing carbon appearing before those that do contain carbon. A more extensive compilation of surface tension may be found in the Reference.

REFERENCE

Jasper, J. J., *J. Phys. Chem. Ref. Data*, 1, 841, 1972.

Mol. form.	Name	γ in mN/m 10°C	25°C	50°C	75°C	100°C
Br_2	Bromine	43.68	40.95	36.40		
Cl_2O_2S	Sulfuryl chloride		28.78			
Cl_3OP	Phosphoryl chloride		32.03	28.85	25.66	
Cl_3P	Phosphorus trichloride		27.98	24.81		
Cl_4Si	Silicon tetrachloride	19.78	18.29	15.80		
H_2O	Water	74.23	71.99	67.94	63.57	58.91
H_4N_2	Hydrazine		66.39			
Hg	Mercury	488.55	485.48	480.36	475.23	470.11
CCl_4	Tetrachloromethane		26.43	23.37	20.31	17.25
CS_2	Carbon disulfide	33.81	31.58	27.87		
$CHBr_3$	Tribromomethane		44.87	41.60	38.33	
$CHCl_3$	Trichloromethane		26.67	23.44	20.20	
CH_2Br_2	Dibromomethane		39.05	35.33	31.61	
CH_2Cl_2	Dichloromethane		27.20			
CH_2O_2	Formic acid		37.13	34.38	31.64	
CH_3I	Iodomethane	32.19	30.34			
CH_3NO	Formamide		57.03	54.92	52.82	50.71
CH_3NO_2	Nitromethane	39.04	36.53	32.33		
CH_4O	Methanol	23.23	22.07	20.14		
CH_5N	Methylamine		19.15			
C_2HCl_5	Pentachloroethane		34.15	31.20	28.26	
$C_2HF_3O_2$	Trifluoroacetic acid		13.53	11.42		
$C_2H_2Cl_4$	1,1,2,2-Tetrachloroethane		35.58	32.41	29.24	26.07
$C_2H_3Cl_3$	1,1,1-Trichloroethane		25.18	22.07		
$C_2H_3Cl_3$	1,1,2-Trichloroethane		34.02	30.65	27.27	23.89
C_2H_3N	Acetonitrile		28.66	25.51		
$C_2H_4Br_2$	1,2-Dibromoethane		39.55	36.25	32.95	
$C_2H_4Cl_2$	1,1-Dichloroethane		24.07			
$C_2H_4Cl_2$	1,2-Dichloroethane		31.86	28.29	24.72	
C_2H_4O	Acetaldehyde	22.54	20.50	17.10		
$C_2H_4O_2$	Acetic acid		27.10	24.61	22.13	
$C_2H_4O_2$	Methyl formate	26.72	24.36	20.43	16.50	12.57
C_2H_5Br	Bromoethane	25.36	23.62			
C_2H_5I	Iodoethane	30.38	28.46	25.24		
$C_2H_5NO_2$	Nitroethane	34.02	32.13	29.00		
C_2H_6O	Ethanol	23.22	21.97	19.89		
C_2H_6OS	Dimethyl sulfoxide		42.92	40.06		
$C_2H_6O_2$	Ethylene glycol		47.99	45.76	43.54	41.31
C_2H_6S	Dimethyl sulfide	25.27	24.06			
C_2H_6S	Ethanethiol		23.08			
$C_2H_6S_2$	Dimethyl disulfide		33.39	30.04		
C_2H_7N	Dimethylamine		26.34			
C_2H_7N	Ethylamine		19.20			

SURFACE TENSION OF COMMON LIQUIDS (continued)

Mol. form.	Name	γ in mN/m 10°C	25°C	50°C	75°C	100°C
C_2H_7NO	Ethanolamine		48.32	45.53	42.73	
C_3H_5Br	3-Bromopropene		26.31	23.17		
C_3H_5Cl	3-Chloropropene		23.14			
C_3H_5ClO	Epichlorohydrin	38.40	36.36	32.96	29.56	26.16
C_3H_5N	Propanenitrile		26.75	23.87		
$C_3H_6Cl_2$	1,2-Dichloropropane		28.32	25.22	22.12	
C_3H_6O	Acetone		23.46	20.66		
C_3H_6O	Allyl alcohol	26.63	25.28	23.02	20.77	
$C_3H_6O_2$	Ethyl formate	25.16	23.18			
$C_3H_6O_2$	Methyl acetate	26.66	24.73	21.51		
$C_3H_6O_2$	Propanoic acid		26.20	23.72	21.23	
C_3H_7Br	1-Bromopropane	27.08	25.26	22.21		
C_3H_7Br	2-Bromopropane	25.03	23.25	20.30		
C_3H_7Cl	1-Chloropropane	23.16	21.30			
C_3H_7Cl	2-Chloropropane	20.49	19.16			
$C_3H_7NO_2$	2-Nitropropane	31.02	29.29	26.39		
C_3H_8O	1-Propanol	24.48	23.32	21.38	19.43	
C_3H_8O	2-Propanol	22.11	20.93	18.96	16.98	
$C_3H_8O_2$	2-Methoxyethanol	32.32	30.84	28.38	25.92	23.46
C_3H_8S	1-Propanethiol		24.20	21.02		
C_3H_8S	2-Propanethiol		21.33	18.39		
C_3H_9N	Propylamine		21.75			
C_3H_9N	Trimethylamine		13.41			
$C_4H_4N_2$	Pyridazine	49.51	47.96	45.37	42.78	40.19
$C_4H_4N_2$	Pyrimidine		30.33	27.80	25.28	22.75
C_4H_4S	Thiophene		30.68	27.36		
C_4H_5N	Pyrrole	38.71	37.06	34.31		
$C_4H_6O_3$	Acetic anhydride	34.08	31.93	28.34	24.75	21.16
C_4H_7N	Butanenitrile		26.92	24.33	21.73	
C_4H_8O	2-Butanone		23.97	21.16		
$C_4H_8O_2$	1,4-Dioxane		32.75	29.28	25.80	22.32
$C_4H_8O_2$	Ethyl acetate	25.13	23.39	20.49	17.58	14.68
$C_4H_8O_2$	Methyl propanoate	26.32	24.44	21.29		
$C_4H_8O_2$	Butanoic acid		26.05	23.75	21.45	
C_4H_9Br	1-Bromobutane	27.58	25.90	23.08	20.27	17.45
C_4H_9Cl	1-Chlorobutane	24.85	23.18	20.39		
C_4H_9I	1-Iodobutane	29.79	28.24	25.67	23.09	20.51
C_4H_9N	Pyrrolidine	30.58	29.23	26.98		
$C_4H_{10}O$	1-Butanol	26.28	24.93	22.69	20.44	18.20
$C_4H_{10}O$	2-Butanol	23.74	22.54	20.56	18.57	16.58
$C_4H_{10}O$	2-Methyl-2-propanol		19.96	17.71		
$C_4H_{10}O$	Diethyl ether		16.65			
$C_4H_{10}O_2$	2-Ethoxyethanol		28.35	26.11	23.86	21.62
$C_4H_{10}O_3$	Diethylene glycol		44.77	42.57	40.37	38.17
$C_4H_{10}S$	Diethyl sulfide	26.22	24.57	21.80		
$C_4H_{11}N$	Butylamine		23.44	20.63		
$C_4H_{11}N$	Isobutylamine		21.75	19.02		
$C_4H_{11}N$	*tert*-Butylamine		16.87			
$C_4H_{11}N$	Diethylamine		19.85			
$C_5H_4O_2$	Furfural	45.08	43.09	39.78	36.46	33.14
C_5H_5N	Pyridine		36.56	33.29	30.03	
C_5H_8	Cyclopentene	24.45	22.20			
C_5H_8O	Cyclopentanone	34.45	32.80	30.05	27.30	24.55
C_5H_{10}	1-Pentene	17.10	15.45			
C_5H_{10}	2-Methyl-2-butene	18.61	17.15			
C_5H_{10}	Cyclopentane	24.07	21.88	18.22		
$C_5H_{10}O$	2-Pentanone		23.25	21.62		

Mol. form.	Name	γ in mN/m 10°C	25°C	50°C	75°C	100°C
$C_5H_{10}O$	3-Pentanone		24.74	22.13		
$C_5H_{10}O$	Pentanal	26.95	25.44	22.91		
$C_5H_{10}O_2$	Butyl formate	26.05	24.52	21.95	19.39	16.82
$C_5H_{10}O_2$	Propyl acetate	25.48	23.80	21.00	18.20	15.40
$C_5H_{10}O_2$	Isopropyl acetate	23.37	21.76	19.08	16.40	
$C_5H_{10}O_2$	Ethyl propanoate	25.55	23.80	20.88	17.96	
$C_5H_{10}O_2$	Methyl butanoate	26.34	24.62	21.76	18.89	16.03
$C_5H_{11}Cl$	1-Chloropentane	26.01	24.40	21.71	19.02	16.33
$C_5H_{11}N$	Piperidine	30.64	28.91	26.03	23.14	20.26
C_5H_{12}	Pentane	17.15	15.49			
$C_5H_{12}O$	1-Pentanol	26.67	25.36	23.17	20.99	18.80
$C_5H_{12}O$	2-Pentanol	24.96	23.45	20.94	18.43	15.92
$C_5H_{12}O$	3-Methyl-1-butanol	24.94	23.71	21.66	19.61	17.56
$C_5H_{13}N$	Pentylamine		24.69	22.14	19.58	
$C_6H_4Cl_2$	*m*-Dichlorobenzene	37.15	35.43	32.57	29.70	26.83
C_6H_5Br	Bromobenzene	36.98	35.24	32.34	29.44	26.54
C_6H_5Cl	Chlorobenzene	34.78	32.99	30.02	27.04	24.06
C_6H_5ClO	*o*-Chlorophenol		39.70	36.89	34.09	31.28
C_6H_5ClO	*m*-Chlorophenol		41.18	38.66	36.13	33.61
C_6H_5F	Fluorobenzene	28.47	26.66	23.65	20.64	
C_6H_5I	Iodobenzene	40.40	38.71	35.91	33.10	30.29
$C_6H_5NO_2$	Nitrobenzene			40.56	37.66	34.77
C_6H_6	Benzene		28.22	25.00	21.77	
C_6H_6O	Phenol			38.20	35.53	32.86
C_6H_7N	Aniline		42.12	39.41	36.69	
C_6H_7N	2-Methylpyridine		33.00	29.90	26.79	
$C_6H_8N_2$	Adiponitrile		45.45	43.02	40.58	
C_6H_{10}	Cyclohexene	28.01	26.17	23.12		
$C_6H_{10}O$	Cyclohexanone	36.43	34.57	31.46	28.36	25.25
$C_6H_{11}N$	Hexanenitrile		27.37	25.11	22.84	
C_6H_{12}	Cyclohexane	26.43	24.65	21.68		
C_6H_{12}	Methylcyclopentane	23.47	21.72	18.82		
C_6H_{12}	1-Hexene	19.44	17.90	15.33		
$C_6H_{12}O$	Cyclohexanol		32.92	30.50	28.09	25.67
$C_6H_{12}O$	2-Hexanone		25.45	22.72		
$C_6H_{12}O_2$	Butyl acetate	26.48	24.88	22.21	19.54	16.87
$C_6H_{12}O_2$	Isobutyl acetate	24.58	23.06	20.53	17.99	15.46
$C_6H_{12}O_2$	Ethyl butanoate	25.51	23.94	21.33	18.71	16.10
$C_6H_{12}O_3$	Paraldehyde	27.22	25.63	22.97	20.32	17.66
$C_6H_{13}Cl$	1-Chlorohexane	27.28	25.73	23.13	20.54	17.94
$C_6H_{13}N$	Cyclohexylamine		31.22	28.25	25.28	
C_6H_{14}	Hexane	19.42	17.89	15.33		
C_6H_{14}	2-Methylpentane	18.37	16.88	14.39		
C_6H_{14}	3-Methylpentane	19.20	17.61	14.96		
$C_6H_{14}O$	Diisopropyl ether		17.27	14.65		
$C_6H_{14}O$	1-Hexanol		25.81	23.81	21.80	19.80
$C_6H_{14}O_2$	1,1-Diethoxyethane		20.89	18.31	15.74	
$C_6H_{14}O_2$	2-Butoxyethanol	27.36	26.14	24.10	22.06	20.02
$C_6H_{15}N$	Triethylamine		20.22	17.74		
$C_6H_{15}N$	Dipropylamine		22.31	19.75	17.20	
$C_6H_{15}N$	Diisopropylamine		19.14	16.45		
C_7H_5N	Benzonitrile		38.79	35.90	33.00	
C_7H_6O	Benzaldehyde	39.63	38.00	35.27	32.55	29.82
C_7H_8	Toluene	29.71	27.93	24.96	21.98	19.01
C_7H_8O	*o*-Cresol		36.90	34.38	31.85	29.32
C_7H_8O	*m*-Cresol		35.69	33.38	31.07	28.76
C_7H_8O	Benzyl alcohol				27.89	24.44

SURFACE TENSION OF COMMON LIQUIDS (continued)

Mol. form.	Name	γ in mN/m 10°C	25°C	50°C	75°C	100°C
C_7H_8O	Anisole		35.10	32.09	29.08	
C_7H_9N	*N*-Methylaniline		36.90	34.47	32.05	
C_7H_9N	2,3-Dimethylpyridine		32.71	30.04	27.36	
C_7H_9N	Benzylamine		39.30	36.27	33.23	
C_7H_{14}	Methylcyclohexane	24.98	23.29	20.46		
C_7H_{14}	1-Heptene	21.29	19.80	17.33	14.85	
$C_7H_{14}O$	2-Heptanone		26.12	23.48		
$C_7H_{14}O_2$	Pentyl acetate	26.67	25.17	22.69	20.20	17.72
$C_7H_{14}O_2$	Heptanoic acid		27.76	25.64		
C_7H_{16}	Heptane	21.12	19.65	17.20	14.75	
C_7H_{16}	3-Methylhexane	20.76	19.31	16.88	14.46	
C_8H_8O	Acetophenone		39.04	36.15	33.27	
$C_8H_8O_2$	Methyl benzoate		37.17	34.25	31.32	
$C_8H_8O_3$	Methyl salicylate	40.98	39.22	36.28	33.35	30.41
C_8H_{10}	Ethylbenzene	30.39	28.75	26.01	23.28	20.54
C_8H_{10}	*o*-Xylene	31.41	29.76	27.01	24.25	21.50
C_8H_{10}	*m*-Xylene	30.13	28.47	25.71	22.95	20.19
C_8H_{10}	*p*-Xylene		28.01	25.32	22.64	19.95
$C_8H_{10}O$	Phenetole		32.41	29.65	26.89	
$C_8H_{11}N$	*N,N*-Dimethylaniline		35.52	32.90	30.27	
$C_8H_{11}N$	*N*-Ethylaniline		36.33	33.65	30.98	
C_8H_{16}	Ethylcyclohexane	26.73	25.15	22.51		
C_8H_{18}	Octane	22.57	21.14	18.77	16.39	14.01
C_8H_{18}	2,5-Dimethylhexane	20.77	19.40	17.12	14.84	12.56
$C_8H_{18}O$	1-Octanol	28.30	27.10	25.12		
$C_8H_{19}N$	Dibutylamine		24.12	21.74	19.36	
$C_8H_{19}N$	Diisobutylamine		21.72	19.44	17.16	
C_9H_7N	Quinoline	44.19	42.59	39.94	37.28	34.62
C_9H_{12}	Cumene	29.27	27.69	25.05	22.42	19.78
C_9H_{12}	1,2,4-Trimethylbenzene	30.74	29.20	26.64	24.07	21.51
C_9H_{12}	Mesitylene	28.89	27.55	25.31	23.07	20.82
$C_9H_{18}O$	5-Nonanone		26.28	23.85		
C_9H_{20}	Nonane	23.79	22.38	20.05	17.71	15.37
$C_9H_{20}O$	1-Nonanol	29.03	27.89	26.00	24.10	22.20
$C_{10}H_{12}$	1,2,3,4-Tetrahydronaphthalene		33.17	30.78	28.40	
$C_{10}H_{22}$	Decane	24.75	23.37	21.07	18.77	16.47
$C_{10}H_{22}O$	1-Decanol	29.61	28.51	26.68	24.85	23.02
$C_{11}H_{24}$	Undecane	25.56	24.21	21.96	19.70	17.45
$C_{12}H_{10}O$	Diphenyl ether		26.75	24.80		
$C_{12}H_{27}N$	Tributylamine		24.39	22.32	20.24	
$C_{13}H_{28}$	Tridecane	26.86	25.55	23.37	21.19	19.01
$C_{14}H_{12}O_2$	Benzyl benzoate	44.47	42.82	40.06	37.31	34.55
$C_{14}H_{30}$	Tetradecane	27.43	26.13	23.96	21.78	19.61
$C_{16}H_{34}$	Hexadecane		27.05	24.91	22.78	20.64
$C_{18}H_{38}$	Octadecane		27.87	25.77	23.66	21.55

SURFACE TENSION OF AQUEOUS MIXTURES

The composition dependence of the surface tension of binary mixtures of several compounds with water is given in this table. The data are tabulated as a function of the mass percent of the non-aqueous component. Data for methanol, ethanol, 1-propanol, and 2-propanol are taken from Reference 1, which also gives values at other temperatures.

REFERENCES

1. Vazquez, G., Alvarez, E., and Navaza, J. M., *J. Chem. Eng. Data*, 40, 611, 1995.
2. *Landolt-Börnstein, Numerical Data and Functional Relationships in Science and Technology, New Series*, IV/16, *Surface Tension*, Springer-Verlag, Heidelberg, 1997.

Surface Tension in mN/m² for the Specified Mass %

Compound	*t*/°C	0%	10%	20%	30%	40%	50%	60%	70%	80%	90%	100%
Acetic acid	30	71.2	51.4	43.3	41.2	38.2	37.4	36.1	33.5	31.5	30.2	26.3
Acetone	25	72.0	44.9	40.5	36.7	33.0	30.1	29.4	29.4	27.6	24.5	23.1
Acetonitrile	20	72.8	48.5	40.2	34.1	31.6	30.6	30.0	29.6	29.1	28.7	28.4
1,2-Butanediol	25	72.0	66.1	60.4	55.1	50.1	45.6	43.3	41.9	40.8	39.2	35.8
1,3-Butanediol	30	71.2	58.1	51.6	48.7	45.8	43.9	42.4	41.2	40.0	39.0	37.0
1,4-Butanediol	30	71.2	61.2	56.9	54.2	52.0	50.7	49.5	47.9	46.6	45.2	43.8
Butanoic acid	30	71.2	42.4	37.5	35.5	34.8	32.2	30.8	29.2	27.4	26.3	25.5
2-Butanone	20	72.8	41.6	32.2				25.2				24.6
γ-Butyrolactone	30	71.2	64	58	53	50	48	46	45	44	42.8	42.7
Chloroacetic acid	25	72.0	59.8	53.6	51.3	49.7	48.3	47.5	46.1			
Diethanolamine	25	72.0	66.8	63.2	60.7	58.8	57.2	55.7	54.3	52.7	50.6	47.2
N,N-Dimethylacetamide	25	72.0	72.0	72.0	72.4	73.5	74.9	75.4	73.0	65.7	54.7	36.4
N,N-Dimethylformamide	25	72.0	65.4	59.2	53.8	49.6	47.3	46.9	44.9	42.3	38.4	35.2
1,4-Dioxane	25	72.0					41.2	39.6	37.9	36.2	34.5	33.7
Ethanol	25	72.01	47.53	37.97	32.98	30.16	27.96	26.23	25.01	23.82	22.72	21.82
Ethylene glycol	20	72.8	68.5	64.9	61.9		57.0					48.2
Formic acid	20	72.8	66	60	55.7	52.2	50.3	48.8	47.1	44.7	40.9	38.0
Glycerol	25	72.0	70.5	69.5	68.5	67.9	67.4	66.9	66.5	65.7	64.5	62.5
Methanol	25	72.01	56.18	47.21	41.09	36.51	32.86	29.83	27.48	25.54	23.93	22.51
Morpholine	20	72.8	65.1	60.7	58.9	56.7	53.0	49.6	47.0	43.7	41.8	38.7
Nitric acid	20	72.8	71.9	70.7	68.9	66.6	63.8	60.6	56.8	52.6	47.9	42.6
Propanoic acid	30	71.2	46.6	42.2	37.7	35.6	33.1	31.7	30.2	28.2	27.4	25.8
1-Propanol	25	72.01	34.32	27.84	25.98	25.26	24.80	24.49	24.08	23.86	23.59	23.28
2-Propanol	25	72.01	40.42	30.57	26.82	25.27	24.26	23.51	22.68	22.14	21.69	21.22
1,2-Propylene glycol	30	71.2	60.5	54.9	50.7	47.2	44.5	41.5	38.6	37.6	36.3	35.5
1,3-Propylene glycol	30	71.2	62.6	58.8	55.7	53.8	52.8	51.7	50.8	49.6	48.2	47.0
Pyridine	25	72.0	52.8	51.2	48.0	46.8	46.6	45.8	45.0	43.6	40.9	37.0
Sulfolane	20	72.8					62.5	61.6	59.6	57.1	54.9	50.9
Sulfuric acid	50	67.9	73.5	75.1	73.6	71.2	68.0	64.1	60.0	56.4	53.6	51.7
Trichloroacetaldehyde	25	72.0	56.7	51.0	46.7	44.1	43.0	42.5	41.5	38.9	34.7	29.4
Trichloroacetic acid	25	72.0	55.8	46.5	42.8	41.6	40.6	39.4	38.3	37.4	36.5	

PERMITTIVITY (DIELECTRIC CONSTANT) OF LIQUIDS

Christian Wohlfarth

The permittivity of a substance (often called the dielectric constant) is the ratio of the electric displacement D to the electric field strength E when an external field is applied to the substance. The quantity tabulated here is the relative permittivity, which is the ratio of the actual permittivity to the permittivity of a vacuum; it is a dimensionless number.

The table gives the static relative permittivity ε_r, i.e., the relative permittivity measured in static fields or at low frequencies where no relaxation effects occur. The fourth column of the table lists the value of ε_r at the temperature specified in the third column, usually 293.15 or 298.15 K. Otherwise, the temperature closest to 293.15 K was chosen, or (as it is the case for many of the substances included here) ε_r is given at the only temperature for which data are available.

The static permittivity refers to nominal atmospheric pressure as long as the corresponding temperature is below the normal boiling point. Otherwise, at temperatures above the normal boiling point, the pressure is understood to be the saturated vapor pressure of the substance considered.

For substances where information on the temperature dependence of the permittivity is available, the table gives the coefficients of a simple polynomial fitting of permittivity to temperature with an equation of the form

$$\varepsilon_r(T) = a + bT + cT^2 + dT^3$$

where T is the absolute temperature in K. Since the parameter d was used in only a few cases where the quadratic fit was not satisfactory, only a, b, and c are listed as columns in the table, while the d values are given at the end of this introduction. For all other substances, $d = 0$. The temperature range of the fit is given in the last column. The coefficients of the fitting equation can be used to calculate dielectric constants within the fitted temperature range but should not be used for extrapolation outside this range. The user who needs dielectric constant data with more accuracy than can be provided by this equation is referred to Reference 1, which gives the original data together with their literature source.

Substances are listed by molecular formula in modified Hill order, with substances not containing carbon preceding those that do contain carbon.

* Indicates that the isomer was not specified in the original reference.
** Indicates a compound for which the cubic term is needed:

Ethanol	d = -0.15512E-05
N-Methylacetamide	d = -0.12998E-04
1,2-Propylene glycol	d = -0.32544E-05
1-Butanol	d = -0.48841E-06
2-Butanol	d = -0.89512E-06
2-Methyl-1-propanol	d = -0.45229E-06
2-Methyl-2-propanol	d = -0.25968E-05
N-Butylacetamide	d = -0.48716E-05

REFERENCES

1. Wohlfarth, Ch., "Static Dielectric Constants of Pure Liquids and Binary Liquid Mixtures", *Landolt-Börnstein, Numerical Data and Functional Relationships in Science and Technology*, New Series, Editor in Chief, O. Madelung, Group IV, Macroscopic and Technical Properties of Matter, Volume 6, Springer-Verlag, Berlin, Heidelberg, New York, 1991.
2. Marsh, K. N., Ed., *Recommended Reference Materials for the Realization of Physicochemical Properties*, Blackwell Scientific Publications, Oxford, 1987.

Mol. Form.	Name	T/K	ε	*a*	*b*	*c*	Range/K
$AlBr_3$	Aluminum tribromide	373.2	3.38				
Ar	Argon	140.00	1.3247	0.12408E+01	0.68755E-02	-0.45344E-04	87-149
AsH_3	Arsine	200.9	2.40	0.37674E+01	-0.97454E-02	0.14537E-04	157-201
BBr_3	Boron tribromide	273.2	2.58				
B_2H_6	Diborane	180.66	1.8725	0.23848E+01	-0.29501E-02	0.64189E-06	108-181
B_5H_9	Pentaborane(9)	298.2	21.1	0.40952E+03	-0.24414E+01	0.38225E-02	226-298
BrF_3	Bromine trifluoride	298.2	106.8				
BrF_5	Bromine pentafluoride	297.7	7.91	0.11428E+02	-0.11822E-01		262-298
BrH	Hydrogen bromide	186.8	8.23				
BrNO	Nitrosyl bromide	288.4	13.4				
Br_2	Bromine	297.9	3.1484	0.32701E+01	-0.12535E-03		273-327
Br_2OS	Thionyl bromide	293.2	9.06				
Br_3OV	Vanadyl tribromide	298.2	3.6	0.61112E+01	-0.84211E-02		203-298
Br_4Ge	Germanium(IV) bromide	299.9	2.955	0.34450E+01	-0.16083E-02		300-316
Br_4Sn	Tin(IV) bromide	303.45	3.169	0.50001E+01	-0.60383E-02		304-316
$ClFO_3$	Perchloryl fluoride	150.2	2.194	0.23808E+01	-0.38629E-03	-0.57143E-05	125-150
ClF_3	Chlorine trifluoride	293.2	4.394	0.96716E+01	-0.18000E-01		273-313

Mol. Form.	Name	*T*/K	ε	*a*	*b*	*c*	Range/K
ClF_5	Chlorine pentafluoride	193.2	4.28	0.78192E+01	-0.20860E-01	0.13132E-04	193-256
ClH	Hydrogen chloride	158.9	14.3	0.47316E+02	-0.28455E+00	0.48650E-03	159-258
$ClNO$	Nitrosyl chloride	285.2	18.2				
Cl_2	Chlorine	208.0	2.147	0.29440E+01	-0.44649E-02	0.30388E-05	208-240
Cl_2F_3P	Phosphorus(V) dichloride trifluoride	228.63	2.8129	0.46501E+01	-0.80358E-02		172-229
Cl_2OS	Thionyl chloride	298.2	8.675				
Cl_2OSe	Selenium oxychloride	293.2	46.2				
Cl_2O_2S	Sulfuryl chloride	293.2	9.1				
Cl_2S	Sulfur dichloride	298.2	2.915				
Cl_2S_2	Sulfur chloride	288.2	4.79				
Cl_3F_2P	Phosphorus(V) trichloride difluoride	268.0	2.3752	0.28905E+01	-0.19228E-02		215-268
Cl_3OP	Phosphorus(V) oxychloride	293.2	14.1				
Cl_3OV	Vanadyl trichloride	298.2	3.4				
Cl_3P	Phosphorus(III) chloride	290.2	3.498	0.59098E+01	-0.83322E-02		290-333
Cl_3PS	Phosphorus(V) sulfide trichloride	298.2	4.94				
Cl_4FP	Phosphorus(V) tetrachloride fluoride	272.64	2.6499	0.33503E+01	-0.29651E-02		244-273
Cl_4Ge	Germanium(IV) chloride	273.2	2.463	-0.55078E+01	0.64881E-01	-0.13091E-03	246-273
Cl_4Pb	Lead(IV) chloride	293.2	2.78				
Cl_4Si	Tetrachlorosilane	273.2	2.248	0.58041E+01	-0.27129E-01	0.51678E-04	207-273
Cl_4Sn	Tin(IV) chloride	273.2	3.014	0.43951E+01	-0.48805E-02		234-273
Cl_4Ti	Titanium(IV) chloride	257.4	2.843	0.33668E+01	-0.19675E-02		237-257
Cl_4V	Vanadium(IV) chloride	298.2	3.05				
Cl_5P	Phosphorus(V) chloride	433.2	2.85				
Cl_5Sb	Antimony(V) chloride	293.0	3.222	0.45413E+01	-0.45078E-02		276-320
FH	Hydrogen fluoride	273.2	83.6	0.50352E+03	-0.19297E+01	0.14372E-02	200-273
F_2	Fluorine	53.48	1.4913	0.14144E+01	0.26387E-02	-0.28356E-04	54-144
F_5I	Iodine pentafluoride	293.2	37.13	0.95184E+02	-0.19800E+00		273-313
F_6S	Sulfur hexafluoride	223.2	1.81				
F_6Xe	Xenon hexafluoride	398.2	4.10				
F_7I	Iodine heptafluoride	298.2	1.75				
$F_{10}S_2$	Sulfur decafluoride	293.2	2.0202				
HI	Hydrogen iodide	220.2	3.87	0.51557E+03	-0.44552E+01	0.96795E-02	220-236
H_2	Hydrogen	13.52	1.2792	0.13327E+01	-0.51946E-02		14-19
H_2O	Water	293.2	80.100	0.24921E+03	-0.79069E+00	0.72997E-03	273-372
H_2O_2	Hydrogen peroxide	290.2	74.6	0.48511E+03	-0.23145E+01	0.31020E-02	233-303
H_2S	Hydrogen sulfide	283.2	5.93	0.14736E+02	-0.33675E-01	0.96740E-05	212-363
H_3N	Ammonia	293.2	16.61	0.66756E+02	-0.24696E+00	0.25913E-03	238-323
H_4N_2	Hydrazine	298.2	51.7	0.22061E+03	-0.89633E+00	0.11066E-02	278-323
He	Helium	2.055	1.0555	0.10640E+01	-0.35584E-02		2-4
I_2	Iodine	391.25	11.08	0.64730E+02	-0.29266E+00	0.39759E-03	391-441
Kr	Krypton	119.80	1.664				
Mn_2O_7	Manganese(VII) oxide	293.2	3.28	0.37655E+01	-0.16463E-02		283-312
NO	Nitric oxide	1.997					
N_2	Nitrogen	63.15	1.4680	0.12550E+01	0.67949E-02	-0.56704E-04	63-126
N_2O_3	Nitrogen trioxide	203.2	31.13	0.92287E+02	-0.43306E+00	0.65000E-03	203-243
N_2O_4	Nitrogen tetroxide	293.2	2.44	0.28212E+01	-0.13000E-02		253-293
Ne	Neon	26.11	1.1907	0.12667E+01	-0.29064E-02		26-29
O_2	Oxygen	54.478	1.5684	0.15434E+01	0.14615E-02	-0.21964E-04	55-154
O_2S	Sulfur dioxide	298.2	16.3	0.52045E+02	-0.16125E+00	0.11042E-03	213-449
O_3	Ozone	90.2	4.75	0.86344E+01	-0.54807E-01	0.12596E-03	90-185
O_3S	Sulfur trioxide	291.2	3.11				
P	Phosphorus	307.2	4.096	0.79018E+00	0.23911E-01	-0.42826E-04	307-358
S	Sulfur	407.2	3.4991	0.51651E+01	-0.77381E-02	0.89120E-05	407-479
Se	Selenium	510.65	5.44	0.67569E+01	-0.25829E-02		511-575
Xe	Xenon	161.35	1.880				
$CBrClF_2$	Bromochlorodifluoromethane	123.2	3.920	0.52442E+01	-0.11000E-01		123-223
$CBrCl_3$	Bromotrichloromethane	293.2	2.405	0.29249E+01	-0.17650E-02		273-333
$CBrF_3$	Bromotrifluoromethane	123.2	3.730	0.54154E+01	-0.13680E-01		123-173
CBr_2Cl_2	Dibromodichloromethane	298.2	2.542	0.32330E+01	-0.23162E-02		298-333

Mol. Form.	Name	T/K	ε	a	b	c	Range/K
CBr_2F_2	Dibromodifluoromethane	273.2	2.939	0.67296E+01	-0.22133E-01	0.30213E-04	139-273
CBr_3Cl	Tribromochloromethane	333.2	2.601				
CBr_3F	Tribromofluoromethane	293.2	3.00	0.53203E+01	-0.11061E-01	0.10688E-04	206-323
CBr_3NO_2	Tribromonitromethane	298.2	9.034	0.16079E+02	-0.23630E-01		298-328
$CClF_3$	Chlorotrifluoromethane	123.2	3.010	0.43677E+01	-0.11020E-01		123-173
CCl_2F_2	Dichlorodifluoromethane	123.2	3.500	0.46984E+01	-0.97600E-02		123-223
CCl_2O	Carbonyl chloride	295.2	4.30				
CCl_3D	Trichloromethane-*d*	298.2	4.67				
CCl_3F	Trichlorofluoromethane	293.2	3.00	0.53203E+01	-0.11061E-01	0.10688E-04	206-323
CCl_3NO_2	Trichloronitromethane	293.2	7.319	0.14403E+02	-0.24178E-01		276-333
CCl_4	Tetrachloromethane	293.2	2.2379	0.28280E+01	-0.20339E-02	0.71795E-07	283-333
CF_4	Tetrafluoromethane	126.3	1.685	0.20350E+01	-0.27616E-02		126-142
$CHBr_3$	Tribromomethane	283.2	4.404	0.71707E+01	-0.98000E-02		283-343
$CHCl_3$	Trichloromethane	293.2	4.8069	0.15115E+02	-0.51830E-01	0.56803E-04	218-323
CHF_3	Trifluoromethane	294.0	5.2	0.11442E+03	-0.75600E+00	0.13562E-02	130-263
CHN	Hydrogen cyanide	293.2	114.9	0.37331E+04	-0.23180E+02	0.36963E-01	258-299
CH_2Br_2	Dibromomethane	283.2	7.77	0.18060E+02	-0.36333E-01		283-313
CH_2Cl_2	Dichloromethane	298.0	8.93	0.40452E+02	-0.17748E+00	0.23942E-03	184-306
CH_2F_2	Difluoromethane	152.2	53.74	0.19428E+03	-0.12939E+01	0.24280E-02	152-224
CH_2I_2	Diiodomethane	298.2	5.32				
CH_2O_2	Formic acid	298.2	51.1	0.14040E+03	-0.24673E+00	-0.17151E-03	287-358
CH_3Br	Bromomethane	275.7	9.71	0.40580E+02	-0.18418E+00	0.26219E-03	195-276
CH_3Cl	Chloromethane	295.2	10.0	0.42775E+02	-0.16175E+00	0.17108E-03	190-392
CH_3ClO_2S	Methanesulfonyl chloride	293.2	34.0	0.10384E+03	-0.33838E+00	0.34156E-03	293-373
CH_3DO	Methan-d_1-ol	297.5	31.68	0.20839E+03	-0.10318E+01	0.14740E-02	176-298
CH_3F	Fluoromethane	131.0	51.0	0.11338E+03	-0.63979E+00	0.96983E-03	150-299
CH_3I	Iodomethane	293.2	6.97	0.24264E+02	-0.93914E-01	0.11926E-03	223-303
CH_3NO	Formamide	293.2	111.0	0.26076E+03	-0.61145E+00	0.34296E-03	278-333
CH_3NO_2	Nitromethane	293.2	37.27	0.11227E+03	-0.35591E+00	0.34206E-03	288-343
CH_3NO_2	Methyl nitrite	200.0	20.77	0.11071E+03	-0.73428E+00	0.14054E-02	110-260
CH_3NO_3	Methyl nitrate	293.2	23.9				
CH_4	Methane	91.0	1.6761	0.15996E+01	0.27434E-02	-0.22086E-04	91-184
CH_4O	Methanol	293.2	33.0	0.19341E+03	-0.92211E+00	0.12839E-02	177-293
CH_5N	Methylamine	215.2	16.7	0.34398E+02	-0.73630E-01	-0.41279E-04	198-258
CN_4O_8	Tetranitromethane	293.2	2.317				
COS	Carbon oxysulfide	185.0	4.47	0.84702E+01	-0.21488E-01		143-185
$COSe$	Carbon oxyselenide	283.2	3.47	0.48740E+01	-0.49425E-02		219-283
CO_2	Carbon dioxide	295.0	1.4492	0.79062E+00	0.10639E-01	-0.28510E-04	220-300
CS_2	Carbon disulfide	293.2	2.6320	0.45024E+01	-0.12054E-01	0.19147E-04	154-319
$C_2Br_2F_4$	1,2-Dibromotetrafluoroethane	298.2	2.34				
$C_2Cl_2F_4$	1,2-Dichlorotetrafluoroethane	273.2	2.4842	0.36663E+01	-0.42271E-02	-0.36255E-06	193-273
$C_2Cl_2O_2$	Oxalyl chloride	294.35	3.470				
C_2Cl_3N	Trichloroacetonitrile	292.2	7.85				
C_2Cl_4	Tetrachloroethylene	303.2	2.268				
$C_2Cl_4F_2$	1,1,2,2-Tetrachloro-1,2-difluoroethane	308.2	2.52				
C_2HBr_3O	Tribromoacetaldehyde	293.2	7.6				
C_2HCl_3	Trichloroethylene	301.5	3.390	0.58319E+01	-0.80828E-02		302-338
$C_2HCl_3F_2$	1,2,2-Trichloro-1,1-difluoroethane	303.2	4.01	0.75423E+01	-0.11667E-01		303-333
C_2HCl_3O	Trichloroacetaldehyde	298.2	6.8				
$C_2HCl_3O_2$	Trichloroacetic acid	333.2	4.34	0.13412E+01	0.90000E-02	-0.24130E-14	333-393
C_2HCl_5	Pentachloroethane	298.2	3.716	0.65972E+01	-0.96800E-02		298-338
$C_2HF_3O_2$	Trifluoroacetic acid	293.2	8.42	0.21652E+02	-0.68146E-01	0.78571E-04	263-323
C_2H_2	Acetylene	195.0	2.4841				
$C_2H_2Br_2$	*cis*-1,2-Dibromoethylene	298.2	7.08				
$C_2H_2Br_2$	*trans*-1,2-Dibromoethylene	298.2	2.88				
$C_2H_2Br_4$	1,1,2,2-Tetrabromoethane	303.2	6.72	0.16246E+02	-0.31500E-01		303-333
$C_2H_2Cl_2$	1,1-Dichloroethylene	293.2	4.60				
$C_2H_2Cl_2$	*cis*-1,2-Dichloroethylene	298.2	9.20				
$C_2H_2Cl_2$	*trans*-1,2-Dichloroethylene	293.2	2.14				

Mol. Form.	Name	*T*/K	ε	*a*	*b*	*c*	Range/K
$C_2H_2Cl_2O_2$	Dichloroacetic acid	293.2	8.33	0.11014E+02	-0.10859E-01	0.49242E-05	284-363
$C_2H_2Cl_4$	1,1,1,2-Tetrachloroethane	207.2	9.22	0.19606E+02	-0.49847E-01		207-233
$C_2H_2Cl_4$	1,1,2,2-Tetrachloroethane	293.2	8.50				
$C_2H_2I_2$	*cis*-1,2-Diiodoethylene	345.65	4.46				
C_2H_3ClO	Acetyl chloride	295.2	15.8				
$C_2H_3ClO_2$	Chloroacetic acid	338.2	12.35	0.17310E+02	-0.14674E-01		338-393
$C_2H_3Cl_2NO_2$	1,1-Dichloro-1-nitroethane	303.2	16.3	0.37576E+02	-0.70400E-01		303-333
$C_2H_3Cl_3$	1,1,1-Trichloroethane	293.2	7.243	0.27705E+02	-0.10621E+00	0.12424E-03	258-318
$C_2H_3Cl_3$	1,1,2-Trichloroethane	298.2	7.1937	0.17147E+02	-0.33371E-01		288-318
$C_2H_3F_3O$	2,2,2-Trifluoroethanol	293.2	27.68	0.90593E+02	-0.21421E+00		293-318
C_2H_3N	Acetonitrile	293.2	36.64	0.29724E+03	-0.15508E+01	0.22591E-02	288-333
C_2H_3NO	Methyl isocyanate	288.7	21.75				
C_2H_4	Ethylene	270.0	1.4833	0.13546E+01	0.62614E-02	-0.21374E-04	200-270
C_2H_4BrCl	1-Bromo-2-chloroethane	283.2	7.41	0.19493E+02	-0.59054E-01	0.58036E-04	263-363
$C_2H_4Br_2$	1,2-Dibromoethane	293.2	4.9612	0.67142E+01	-0.59800E-02		293-313
$C_2H_4Cl_2$	1,1-Dichloroethane	298.2	10.10	0.24429E+02	-0.48000E-01		288-318
$C_2H_4Cl_2$	1,2-Dichloroethane	293.2	10.42	0.24404E+02	-0.47892E-01		293-343
$C_2H_4Cl_2O$	Bis(chloromethyl) ether	293.2	3.51				
$C_2H_4N_2O_6$	Ethylene glycol dinitrate	293.2	28.26				
C_2H_4O	Acetaldehyde	291.2	21.0				
C_2H_4O	Ethylene oxide	293.2	12.42	0.52661E+02	-0.21337E+00	0.25947E-03	293-243
C_2H_4OS	Thioacetic acid	298.2	14.30				
$C_2H_4O_2$	Acetic acid	293.2	6.20	-0.15731E+02	0.12662E+00	-0.17738E-03	293-363
$C_2H_4O_2$	Methyl formate	288.2	9.20	0.19699E+02	-0.36429E-01		288-302
$C_2H_4O_3S$	Ethylene glycol sulfite	298.2	39.6	0.85483E+02	-0.15400E+00		298-328
C_2H_5Br	Bromoethane	298.2	9.01	0.28473E+02	-0.85495E-01	0.67971E-04	243-308
C_2H_5Cl	Chloroethane	293.2	9.45	0.60693E+02	-0.31290E+00	0.47154E-03	237-293
C_2H_5ClO	2-Chloroethanol	293.2	25.80	0.11155E+03	-0.30149E+00		140-175
C_2H_5I	Iodoethane	293.2	7.82	0.25598E+02	-0.94367E-01	0.11424E-03	183-343
C_2H_5N	Ethyleneimine	298.2	18.3	0.61405E+02	-0.14474E+00		273-298
C_2H_5NO	Acetamide	363.7	67.6	-0.20055E+03	0.15515E+01	-0.22392E-02	364-448
C_2H_5NO	*N*-Methylformamide	293.2	189.0	0.10383E+04	-0.43165E+01	0.48398E-02	276-353
C_2H_5NO	Acetaldoxime	298.2	4.70				
$C_2H_5NO_2$	Nitroethane	288.2	29.11	0.57406E+02	-0.97657E-01		276-333
$C_2H_5NO_2$	Methyl carbamate	328.2	18.48	0.36773E+02	-0.55700E-01		328-368
$C_2H_5NO_3$	Ethyl nitrate	293.2	19.7				
C_2H_6	Ethane	95.0	1.9356	0.20185E+01	-0.51493E-03	-0.48148E-05	95-295
C_2H_6O	Ethanol	293.2	25.3	0.15145E+03	-0.87020E+00	0.19570E-02	163-523
C_2H_6O	Dimethyl ether	258.0	6.18	0.22389E+02	-0.86524E-01	0.91291E-04	155-258
C_2H_6OS	Dimethyl sulfoxide	293.2	47.24	0.38478E+02	0.16939E+00	-0.47423E-03	288-343
$C_2H_6O_2$	Ethylene glycol	293.2	41.4	0.14355E+03	-0.48573E+00	0.46703E-03	293-423
$C_2H_6O_2S$	Dimethyl sulfone	383.2	47.39	0.10830E+03	-0.15900E+00		383-398
$C_2H_6O_4S$	Dimethyl sulfate	298.2	55.0				
C_2H_6S	Ethanethiol	298.2	6.667				
C_2H_6S	Dimethyl sulfide	294.2	6.70				
$C_2H_6S_2$	1,2-Ethanedithiol	293.2	7.26	0.11228E+02	-0.13500E-01		293-333
$C_2H_6S_2$	Dimethyl disulfide	298.2	9.6	0.19109E+02	-0.32000E-01		298-323
C_2H_7N	Ethylamine	273.2	8.7	0.30163E+02	-0.79000E-01		233-273
C_2H_7NO	Ethanolamine	293.2	31.94	0.14890E+03	-0.62491E+00	0.77143E-03	253-293
$C_2H_8N_2$	1,2-Ethanediamine	293.2	13.82	0.48922E+02	-0.17021E+00	0.17262E-03	273-333
C_3Cl_6O	Hexachloroacetone	291.9	3.925	0.76423E+01	-0.15838E-01	0.10618E-04	269-303
C_3F_6O	Perfluoroacetone	202.2	2.104	0.34809E+01	-0.92883E-02	0.12282E-04	151-238
C_3HN	Cyanoacetylene	291.9	72.3	0.91803E+03	-0.49149E+01	0.69104E-02	281-314
$C_3H_2F_6O$	1,1,1,3,3,3-Hexafluoro-2-propanol	293.2	16.70				
$C_3H_3ClO_3$	4-Chloro-1,3-dioxolan-2-one	313.2	62.0				
C_3H_3N	Acrylonitrile	293.2	33.0	0.11109E+03	-0.36806E+00	0.34879E-03	233-413
$C_3H_3NO_2$	Cyanoacetic acid	277.2	33.4				
C_3H_4	Allene	269.0	2.025	0.26049E+01	-0.44147E-03	-0.63420E-05	156-269
C_3H_4	Propyne	246.0	3.218	0.60871E+01	-0.11730E-01		185-246

Mol. Form.	Name	*T*/K	ε	*a*	*b*	*c*	Range/K
$C_3H_4ClF_3$	3-Chloro-1,1,1-trifluoropropane	295.2	7.32	0.22361E+02	-0.68840E-01	0.60594E-04	275-313
C_3H_4ClNO	2-Chloroethyl isocyanate	288.2	29.1	0.64311E+02	-0.12217E+00		288-403
$C_3H_4Cl_2O$	1,1-Dichloroacetone	293.2	14.6				
$C_3H_4F_4O$	2,2,3,3-Tetrafluoro-1-propanol	298.2	21.03				
C_3H_4O	Propargyl alcohol	293.2	20.8	0.99895E+02	-0.38911E+00	0.40776E-03	213-293
$C_3H_4O_3$	Ethylene carbonate	313.2	89.78	0.20746E+03	-0.37610E+00		313-343
C_3H_5Br	3-Bromopropene	293.2	7.0				
$C_3H_5BrO_2$	2-Bromopropanoic acid	294.2	11.0				
$C_3H_5Br_3$	1,2,3-Tribromopropane	303.2	6.00	0.11024E+02	-0.16596E-01		303-358
C_3H_5Cl	2-Chloropropene	299.25	8.92				
C_3H_5Cl	3-Chloropropene	293.2	8.2				
$C_3H_5ClN_2O_6$	3-Chloro-1,2-propanediol dinitrate	293.2	17.50				
C_3H_5ClO	Epichlorohydrin	293.2	22.6				
$C_3H_5ClO_2$	Ethyl chloroformate	308.7	9.736	0.15356E+02	-0.18250E-01		309-349
$C_3H_5ClO_2$	Methyl chloroacetate	293.2	12.0				
$C_3H_5Cl_3$	1,2,3-Trichloropropane	293.2	7.5				
C_3H_5I	3-Iodopropene	292.2	6.1				
C_3H_5N	Propanenitrile	293.2	29.7	0.82222E+02	-0.22937E+00	0.17424E-03	213-473
C_3H_5NO	Ethyl isocyanate	293.2	19.7				
C_3H_5NS	Ethyl isothiocyanate	293.2	19.6				
$C_3H_5N_3O_9$	Trinitroglycerol	293.2	19.25				
C_3H_6	Propene	220.0	2.1365	0.29623E+01	-0.37564E-02		220-250
$C_3H_6Br_2$	1,2-Dibromopropane	283.2	4.60	0.54973E+01	-0.31695E-02		283-333
$C_3H_6Br_2$	1,3-Dibromopropane	293.2	9.482	0.29193E+02	-0.94450E-01	0.92800E-04	293-368
$C_3H_6ClNO_2$	2-Chloro-2-nitropropane	250.4	31.90				
$C_3H_6Cl_2$	1,2-Dichloropropane	293.2	8.37	0.18915E+02	-0.35907E-01		281-323
$C_3H_6Cl_2$	1,3-Dichloropropane	303.2	10.27	0.21609E+02	-0.37333E-01		303-333
$C_3H_6Cl_2$	2,2-Dichloropropane	293.2	11.37	0.32421E+02	-0.72188E-01		245-293
$C_3H_6N_2O_4$	2,2-Dinitropropane	325.1	42.4				
C_3H_6O	Allyl alcohol	293.2	19.7	0.62714E+02	-0.14771E+00	0.37879E-05	213-303
C_3H_6O	Propanal	290.2	18.5				
C_3H_6O	Acetone	293.2	21.01	0.88157E+02	-0.34300E+00	0.38925E-03	273-323
$C_3H_6O_2$	Propanoic acid	298.2	3.44	0.18793E+01	0.46841E-02	0.19983E-05	289-408
$C_3H_6O_2$	Ethyl formate	288.2	8.57	0.15884E+02	-0.25333E-01		288-318
$C_3H_6O_2$	Methyl acetate	288.2	7.07	0.13190E+02	-0.21226E-01		276-318
$C_3H_6O_3$	3-Hydroxypropanoic acid	296.2	30.0				
$C_3H_6O_3$	Dimethyl carbonate	298.2	3.087				
$C_3H_6O_3$	1,3,5-Trioxane	338.2	15.55				
C_3H_7Br	1-Bromopropane	293.2	8.09	0.17769E+02	-0.32599E-01		274-328
C_3H_7Br	2-Bromopropane	293.2	9.46	0.26195E+02	-0.72995E-01	0.55454E-04	186-328
C_3H_7Cl	1-Chloropropane	293.2	8.588	0.21214E+02	-0.43130E-01		273-313
C_3H_7ClO	3-Chloro-1-propanol	215.2	36.0	0.12436E+03	-0.60841E+00	0.92060E-03	145-215
C_3H_7ClO	1-Chloro-2-propanol	153.2	59.0	-0.19169E+02	0.13605E+01	-0.55567E-02	153-177
$C_3H_7ClO_2$	3-Chloro-1,2-propanediol	293.2	31.0				
C_3H_7I	1-Iodopropane	293.2	7.07	0.13744E+02	-0.22745E-01		293-323
C_3H_7I	2-Iodopropane	298.2	8.19				
C_3H_7NO	*N*-Ethylformamide	298.2	102.7	0.64764E+03	-0.28499E+01	0.34286E-02	298-338
C_3H_7NO	*N,N*-Dimethylformamide	293.2	38.25	0.15364E+03	-0.60367E+00	0.71505E-03	213-353
C_3H_7NO	*N*-Methylacetamide	303.2	179.0	0.15975E+04	-0.90451E+01	0.18345E-01	303-473
$C_3H_7NO_2$	1-Nitropropane	288.2	24.70	0.94999E+02	-0.38358E+00	0.48480E-03	276-333
$C_3H_7NO_2$	2-Nitropropane	288.2	26.74	0.60138E+02	-0.11566E+00		276-303
$C_3H_7NO_2$	Propyl nitrite	250.0	12.35	0.70552E+02	-0.40362E+00	0.66687E-03	110-310
$C_3H_7NO_2$	Isopropyl nitrite	260.0	13.92	0.74578E+02	-0.38283E+00	0.57071E-03	150-300
$C_3H_7NO_2$	Ethyl carbamate	328.2	14.14	0.32431E+02	-0.65097E-01	0.28571E-04	328-368
C_3H_8	Propane	293.19	1.6678	0.22883E+01	-0.23276E-02	0.84710E-06	90-300
C_3H_8O	1-Propanol	293.2	20.8	0.98045E+02	-0.36860E+00	0.36422E-03	193-493
C_3H_8O	2-Propanol	293.2	20.18	0.10416E+03	-0.41011E+00	0.42049E-03	193-493
$C_3H_8O_2$	1,2-Propylene glycol	303.2	27.5	0.24546E+03	-0.15738E+01	0.38068E-02	193-403
$C_3H_8O_2$	1,3-Propylene glycol	293.2	35.1	0.11365E+03	-0.36680E+00	0.33766E-03	288-328

Mol. Form.	Name	*T*/K	ε	*a*	*b*	*c*	Range/K
$C_3H_8O_2$	Ethylene glycol monomethyl ether	298.2	17.2	0.11803E+03	-0.58000E+00	0.81001E-03	254-318
$C_3H_8O_2$	Dimethoxymethane	293.2	2.644	0.25877E+01	-0.93019E-03	0.38472E-05	171-293
$C_3H_8O_3$	Glycerol	293.2	46.53	0.77503E+02	-0.37984E-01	-0.23107E-03	288-343
C_3H_8S	1-Propanethiol	288.2	5.937	0.11602E+02	-0.19580E-01		273-318
C_3H_8S	2-Propanethiol	298.2	5.952				
$C_3H_8S_2$	1,2-Propanedithiol	293.2	7.24	0.14667E+02	-0.32660E-01	0.25000E-04	293-333
$C_3H_8S_2$	1,3-Propanedithiol	303.2	8.11	0.66607E+01	0.31310E-01	-0.87500E-04	303-343
$C_3H_9BO_3$	Trimethyl borate	293.2	2.2762				
C_3H_9ClSi	Trimethylchlorosilane	273.2	10.21	-0.19492E+02	0.29806E+00	-0.69284E-03	223-273
C_3H_9N	Propylamine	296.2	5.08	0.17719E+02	-0.59022E-01	0.54780E-04	204-296
C_3H_9N	Isopropylamine	293.2	5.6268	0.40429E+02	-0.21441E+00	0.32634E-03	213-298
C_3H_9N	Trimethylamine	298.2	2.440	0.39745E+01	-0.51331E-02		273-298
$C_3H_9O_4P$	Trimethyl phosphate	293.2	20.6				
C_4Cl_6	Hexachloro-1,3-butadiene	293.2	2.55				
$C_4Cl_6O_3$	Trichloroacetic anhydride	298.2	5.0				
$C_4F_6O_3$	Trifluoroacetic acid anhydride	298.2	2.7				
$C_4H_2Cl_4O_3$	Dichloroacetic anhydride	298.2	15.8				
$C_4H_2O_3$	Maleic anhydride	326.2	52.75				
$C_4H_3F_7O$	2,2,3,3,4,4,4-Heptafluoro-1-butanol	298.2	14.4				
$C_4H_4N_2$	Succinonitrile	298.2	62.6	0.17724E+03	-0.54654E+00	0.54046E-03	236-351
$C_4H_4N_2$	Pyrazine	323.2	2.80				
C_4H_4O	Furan	277.1	2.88	0.13636E+01	0.12864E-01	-0.22701E-04	188-277
C_4H_4S	Thiophene	293.2	2.739	0.32941E+01	-0.19019E-02		253-293
C_4H_5Cl	2-Chloro-1,3-butadiene	293.2	4.914				
$C_4H_5Cl_3O_2$	Ethyl trichloroacetate	293.2	8.428				
C_4H_5N	Pyrrole	293.0	8.00	0.12672E+02	-0.14075E-01	-0.62671E-05	293-357
C_4H_5NO	Allyl isocynate	288.2	15.15	0.34299E+02	-0.66444E-01		288-333
C_4H_6	1,3-Butadiene	265.0	2.050	0.27674E+01	-0.26738E-02		185-265
C_4H_6O	Divinyl ether	288.2	3.94				
C_4H_6O	Ethoxyacetylene	298.2	8.05				
C_4H_6O	Cyclobutanone	298.2	14.27	0.43974E+02	-0.15712E+00	0.19264E-03	220-317
$C_4H_6O_2$	Methyl acrylate	303.2	7.03	0.11968E+02	-0.16500E-01		303-333
$C_4H_6O_2$	2,3-Butanedione	298.2	4.04	0.46907E+01	-0.22302E-02		278-348
$C_4H_6O_2$	γ-Butyrolactone	293.2	39.0				
$C_4H_6O_3$	Acetic anhydride	293.2	22.45				
$C_4H_6O_3$	Propylene carbonate	293.0	66.14	0.15940E+03	-0.39530E+00	0.26284E-03	273-333
C_4H_7Br	*cis*-2-Bromo-2-butene	293.2	5.38				
C_4H_7Br	*trans*-2-Bromo-2-butene	293.2	6.76				
$C_4H_7BrO_2$	2-Bromobutanoic acid	293.2	7.2				
$C_4H_7BrO_2$	Ethyl bromoacetate	303.2	9.75	0.15627E+02	-0.19600E-01		303-333
$C_4H_7BrO_2$	Methyl 3-bromopropanoate	303.2	5.81	0.36001E+01	0.72500E-02		303-343
$C_4H_7ClO_2$	Propyl chlorocarbonate	293.2	11.2				
$C_4H_7ClO_2$	Methyl 2-chloropropanoate	303.2	11.45	0.22449E+02	-0.36250E-01		303-343
C_4H_7N	Butanenitrile	293.2	24.83	0.53884E+02	-0.99257E-01		293-333
C_4H_7N	2-Methylpropanenitrile	293.2	24.42	0.52554E+02	-0.96000E-01		293-313
C_4H_7NO	2-Pyrrolidone	298.2	28.18	0.11054E+03	-0.47945E+00	0.68182E-03	298-338
C_4H_8	1-Butene	220.0	2.2195	0.29354E+01	-0.32580E-02		220-250
C_4H_8	*cis*-2-Butene	296.0	1.960	0.28802E+01	-0.31064E-02		197-296
C_4H_8	Isobutene	288.7	2.1225	0.33701E+01	-0.43295E-02		220-289
$C_4H_8Br_2$	1,2-Dibromobutane	293.2	4.74	0.11199E+03	-0.63334E+00	0.91250E-03	293-333
$C_4H_8Br_2$	1,3-Dibromobutane	293.2	9.14	0.34031E+02	-0.13254E+00	0.16250E-03	293-333
$C_4H_8Br_2$	1,4-Dibromobutane	303.2	8.68	0.20944E+02	-0.55620E-01	0.50000E-04	303-333
$C_4H_8Br_2$	2,3-Dibromobutane	298.2	6.245	0.23849E+02	-0.96300E-01	0.12500E-03	293-333
$C_4H_8Br_2$	1,2-Dibromo-2-methylpropane	293.2	4.1				
$C_4H_8Cl_2$	1,2-Dichlorobutane	293.2	7.74	0.31925E+02	-0.13232E+00	0.17007E-03	293-356
$C_4H_8Cl_2$	1,4-Dichlorobutane	308.2	9.30	0.59766E+01	0.49300E-01	-0.12500E-03	308-338
$C_4H_8Cl_2$	1,2-Dichloro-2-methylpropane	296.0	7.15	0.39429E+02	-0.20028E+00	0.30917E-03	165-296
$C_4H_8Cl_2O$	Bis(2-chloroethyl) ether	293.2	21.20				
C_4H_8O	Butanal	298.2	13.45				

Mol. Form.	Name	*T*/K	ε	*a*	*b*	*c*	Range/K
C_4H_8O	2-Butanone	293.2	18.56	0.15457E+02	0.90152E-01	-0.27100E-03	293-333
C_4H_8O	Tetrahydrofuran	295.2	7.52	0.30739E+02	-0.12946E+00	0.17195E-03	224-295
$C_4H_8O_2$	Butanoic acid	287.2	2.98	0.15010E+01	0.50046E-02		287-403
$C_4H_8O_2$	2-Methylpropanoic acid	293.2	2.58				
$C_4H_8O_2$	Propyl formate	303.2	6.92				
$C_4H_8O_2$	Ethyl acetate	293.2	6.0814	0.15646E+02	-0.44066E-01	0.39137E-04	293-433
$C_4H_8O_2$	Methyl propanoate	293.2	6.200	0.12798E+02	-0.22540E-01		293-333
$C_4H_8O_2$	1,4-Dioxane	293.2	2.2189	0.27299E+01	-0.17440E-02		293-313
$C_4H_8O_3$	2-Hydroxybutanoic acid	296.2	37.7				
$C_4H_8O_3$	3-Hydroxybutanoic acid	296.2	31.5				
$C_4H_8O_3$	Ethyl methyl carbonate	293.2	2.985				
$C_4H_8O_3$	Ethylene glycol monoacetate	303.2	12.95				
C_4H_9Br	1-Bromobutane	283.2	7.315	0.22542E+02	-0.79306E-01	0.89867E-04	183-363
C_4H_9Br	2-Bromobutane	298.2	8.64	0.18461E+02	-0.32933E-01		274-328
C_4H_9Br	1-Bromo-2-methylpropane	273.2	7.70	0.37558E+02	-0.20571E+00	0.35496E-03	112-273
C_4H_9Br	2-Bromo-2-methylpropane	293.0	10.98	0.35085E+02	-0.14075E+00	0.19960E-03	258-293
C_4H_9Cl	1-Chlorobutane	293.2	7.276	0.13565E+02	-0.10161E-01	-0.38750E-04	273-323
C_4H_9Cl	2-Chlorobutane	293.2	8.564	0.30376E+02	-0.11377E+00	0.13429E-03	273-323
C_4H_9Cl	1-Chloro-2-methylpropane	293.2	7.027	0.14945E+02	-0.33747E-01	0.23036E-04	273-323
C_4H_9Cl	2-Chloro-2-methylpropane	293.2	9.663	0.35077E+02	-0.12867E+00	0.14304E-03	273-323
C_4H_9I	1-Iodobutane	293.2	6.27	0.16493E+02	-0.50262E-01	0.52485E-04	293-323
C_4H_9I	2-Iodobutane	293.2	7.873	0.10883E+02	-0.14680E-02	-0.30000E-04	293-323
C_4H_9I	2-Iodo-2-methylpropane	283.2	6.65	0.76780E+01	0.69900E-02	-0.37500E-04	283-323
C_4H_9N	Pyrrolidine	293.0	8.30	0.38191E+02	-0.15462E+00	0.17941E-03	274-333
C_4H_9NO	*N*-Methylpropanamide	293.2	170.0				
C_4H_9NO	*N*-Ethylacetamide	293.2	135.0	0.74494E+03	-0.31400E+01	0.36131E-02	213-353
C_4H_9NO	*N,N*-Dimethylacetamide	294.2	38.85	0.15420E+03	-0.57506E+00	0.61911E-03	294-433
C_4H_9NO	2-Butanone oxime	293.2	3.4				
C_4H_9NO	Morpholine	298.2	7.42				
$C_4H_9NO_2$	*tert*-Butyl nitrite	298.2	11.47				
$C_4H_9NO_2$	Propyl carbamate	338.2	12.06	0.24356E+02	-0.36400E-01		338-378
$C_4H_9NO_2$	Ethyl-*N*-methyl carbamate	298.2	21.10	0.11477E+03	-0.47568E+00	0.54127E-03	298-373
$C_4H_9NO_2$	*N*-Acetylethanolamine	298.2	96.6	0.37016E+03	-0.13113E+01	0.13214E-02	298-348
$C_4H_9NO_3$	Butyl nitrate	293.2	13.10				
C_4H_{10}	Butane	295.0	1.7697	0.22379E+01	-0.13884E-02	-0.66711E-06	135-303
C_4H_{10}	Isobutane	295.0	1.7518	0.23295E+01	-0.19953E-02	0.14197E-06	115-303
$C_4H_{10}O$	1-Butanol	293.2	17.84	0.10578E+03	-0.50587E+00	0.84733E-03	193-553
$C_4H_{10}O$	2-Butanol	293.2	17.26	0.13850E+03	-0.75146E+00	0.14086E-02	172-533
$C_4H_{10}O$	2-Methyl-1-propanol	293.2	17.93	0.10762E+03	-0.51398E+00	0.83702E-03	173-533
$C_4H_{10}O$	2-Methyl-2-propanol	298.2	12.47	0.22541E+03	-0.14990E+01	0.34050E-02	298-503
$C_4H_{10}O$	Diethyl ether	293.2	4.2666	0.79725E+01	-0.12519E-01		283-301
$C_4H_{10}O_2$	1,2-Butanediol	298.2	22.4	0.63702E+02	-0.13807E+00		278-323
$C_4H_{10}O_2$	1,3-Butanediol	298.2	28.8	0.72883E+02	-0.14770E+00		278-323
$C_4H_{10}O_2$	1,4-Butanediol	298.2	31.9	0.13079E+03	-0.46985E+00	0.46320E-03	288-328
$C_4H_{10}O_2$	Ethylene glycol monoethyl ether	298.2	13.38				
$C_4H_{10}O_2$	Ethylene glycol dimethyl ether	296.7	7.30	0.48832E+02	-0.24218E+00	0.34413E-03	256-318
$C_4H_{10}O_2S$	Bis(2-hydroxyethyl) sulfide	293.2	28.61	0.13128E+03	-0.52719E+00	0.60465E-03	253-333
$C_4H_{10}O_3$	Diethylene glycol	293.2	31.82	0.13973E+03	-0.54725E+00	0.61149E-03	288-343
$C_4H_{10}O_3S$	Diethyl sulfite	293.2	15.6				
$C_4H_{10}O_4$	1,2,3,4-Butanetetrol	393.2	28.2				
$C_4H_{10}O_4S$	Diethyl sulfate	293.2	29.2				
$C_4H_{10}S$	1-Butanethiol	288.2	5.204	0.11201E+02	-0.20767E-01		273-318
$C_4H_{10}S$	2-Butanethiol	288.2	5.645	0.10866E+02	-0.17993E-01		273-318
$C_4H_{10}S$	2-Methyl-1-propanethiol	298.2	4.961				
$C_4H_{10}S$	2-Methyl-2-propanethiol	293.2	5.475	0.10597E+02	-0.17500E-01		283-313
$C_4H_{10}S$	Diethyl sulfide	298.2	5.723				
$C_4H_{11}N$	Butylamine	293.2	4.71	0.13322E+02	-0.44176E-01	0.50250E-04	223-333
$C_4H_{11}N$	Diethylamine	293.2	3.680	0.26462E+02	-0.13750E+00	0.20373E-03	243-323
$C_4H_{11}NO_2$	Diethanolamine	293.2	25.75	0.73435E+02	-0.21377E+00	0.17500E-03	273-323

PERMITTIVITY (DIELECTRIC CONSTANT) OF LIQUIDS (continued)

Mol. Form.	Name	*T*/K	ε	*a*	*b*	*c*	Range/K
$C_4H_{12}O_2Si$	Dimethoxydimethylsilane	298.2	3.663				
$C_4H_{12}O_3Si$	Trimethoxymethylsilane	298.2	4.9				
$C_4H_{12}O_4Si$	Tetramethyl silicate	293.2	6.0				
$C_4H_{12}Si$	Diethylsilane	293.2	2.544				
$C_4H_{12}Si$	Tetramethylsilane	293.2	1.921				
$C_4H_{13}N_3$	Diethylenetriamine	293.2	12.62	0.57840E+02	-0.23873E+00	0.28841E-03	213-333
C_5FeO_5	Iron pentacarbonyl	293.2	2.602				
C_5H_4BrN	2-Bromopyridine	298.2	23.18	0.73391E+02	-0.23678E+00	0.22930E-03	298-398
C_5H_4ClN	2-Chloropyridine	298.2	27.32	0.98702E+02	-0.34237E+00	0.34502E-03	298-398
$C_5H_4F_8O$	2,2,3,3,4,4,5,5-Octafluoro-1-pentanol	298.2	15.30				
$C_5H_4O_2$	Furfural	293.2	42.1				
C_5H_5N	Pyridine	293.2	13.260	0.43991E+02	-0.15150E+00	0.15925E-03	293-323
C_5H_5NO	Pyridine-1-oxide	343.0	35.94	0.20878E+02	0.16450E+00	-0.35269E-03	343-398
C_5H_6O	2-Methylfuran	293.2	2.76				
$C_5H_6O_2$	Furfuryl alcohol	298.2	16.85				
$C_5H_7Cl_3O_2$	Propyl trichloroacetate	298.2	8.32				
$C_5H_7NO_2$	Ethyl cyanoacetate	263.2	31.62				
C_5H_8	1,3-Pentadiene*	298.2	2.319				
C_5H_8	1,4-Pentadiene	294.0	2.054	0.29994E+01	-0.34578E-02	0.85300E-06	178-294
C_5H_8	2-Methyl-1,3-butadiene	293.2	2.098	0.28170E+01	-0.23147E-02	-0.43975E-06	198-293
C_5H_8	Cyclopentene	295.0	2.083	0.28177E+01	-0.27597E-02	0.89346E-06	171-319
C_5H_8O	Cyclopentanone	298.2	13.58	0.24083E+02	-0.30286E-01	-0.16802E-04	219-298
$C_5H_8O_2$	Ethyl acrylate	303.2	6.05	0.47827E+02	-0.24394E+00	0.35000E-03	303-343
$C_5H_8O_2$	Methyl *trans*-2-butenoate	293.2	6.6645				
$C_5H_8O_2$	Methyl methacrylate	303.2	6.32	0.32098E+02	-0.14568E+00	0.20000E-03	303-343
$C_5H_8O_2$	2,4-Pentanedione	303.2	26.524				
$C_5H_8O_4$	Dimethyl malonate	293.2	9.82	0.26470E+02	-0.76656E-01	0.67888E-04	293-433
$C_5H_9BrO_2$	Ethyl 2-bromopropanoate	293.2	9.4				
$C_5H_9ClO_2$	Isobutyl chlorocarbonate	293.2	9.1				
$C_5H_9ClO_2$	Ethyl 2-chloropropanoate	303.2	11.95	0.25965E+02	-0.46250E-01		303-343
$C_5H_9ClO_2$	Ethyl 3-chloropropanoate	303.2	10.19	0.21951E+02	-0.38750E-01		303-343
$C_5H_9ClO_2$	Methyl 4-chlorobutanoate	303.2	9.51	0.17127E+02	-0.25000E-01		303-343
C_5H_9N	Pentanenitrile	293.2	20.04	0.55793E+02	-0.15750E+00	0.12432E-03	183-333
C_5H_9N	2,2-Dimethylpropanenitrile	293.2	21.1	0.58418E+02	-0.16884E+00	0.14131E-03	293-453
C_5H_9NO	Isobutyl isocyanate	293.2	11.638	0.38026E+02	-0.12714E+00	0.12679E-03	293-353
C_5H_9NO	*N*-Methyl-2-pyrrolidone	293.2	32.55				
C_5H_{10}	1-Pentene	293.2	2.011	-0.11438E+01	0.25420E-01	-0.50000E-04	273-293
C_5H_{10}	2-Methyl-1-butene	293.2	2.180				
C_5H_{10}	2-Methyl-2-butene	296.0	1.979	0.26064E+01	-0.19578E-02	-0.53908E-06	225-296
C_5H_{10}	Cyclopentane	293.2	1.9687	0.24287E+01	-0.15304E-02	-0.13095E-06	278-313
C_5H_{10}	Ethylcyclopropane	293.2	1.933				
$C_5H_{10}Br_2$	1,2-Dibromopentane	298.2	4.39				
$C_5H_{10}Br_2$	1,4-Dibromopentane	293.2	9.05	0.26443E+02	-0.88640E-01	0.10000E-03	293-333
$C_5H_{10}Br_2$	1,5-Dibromopentane	303.2	9.14	0.38192E+02	-0.15648E+00	0.20000E-03	303-333
$C_5H_{10}Cl_2$	1,2-Dichloropentane	293.2	6.89	0.19016E+02	-0.57954E-01	0.56801E-04	293-356
$C_5H_{10}Cl_2$	1,5-Dichloropentane	298.2	9.92				
$C_5H_{10}O$	Cyclopentanol	288.2	18.5	0.10565E+03	-0.44244E+00	0.48657E-03	258-323
$C_5H_{10}O$	Pentanal	293.2	10.00				
$C_5H_{10}O$	2,2-Dimethylpropanal	293.2	9.051	0.18645E+02	-0.32395E-01	-0.16157E-05	280-333
$C_5H_{10}O$	2-Pentanone	293.2	15.45	0.40893E+02	-0.10423E+00	0.60557E-04	204-353
$C_5H_{10}O$	3-Pentanone	293.2	17.00	0.12690E+02	0.95177E-01	-0.27321E-03	233-353
$C_5H_{10}O$	3-Methyl-2-butanone	293.2	10.37	0.30695E+02	-0.10962E+00	0.13810E-03	293-328
$C_5H_{10}O$	Tetrahydropyran	293.2	5.66	0.19793E+02	-0.76071E-01	0.94852E-04	234-333
$C_5H_{10}O$	2-Methyltetrahydrofuran	298.2	6.97				
$C_5H_{10}O_2$	Pentanoic acid	294.4	2.661	0.33491E+01	-0.75156E-02	0.17820E-04	250-344
$C_5H_{10}O_2$	Butyl formate	303.2	6.10	0.21532E+02	-0.84106E-01	0.10952E-03	288-323
$C_5H_{10}O_2$	Isobutyl formate	293.2	6.41				
$C_5H_{10}O_2$	Propyl acetate	293.2	5.62	0.17677E+02	-0.61404E-01	0.69196E-04	253-353
$C_5H_{10}O_2$	Ethyl propanoate	293.2	5.76				

Mol. Form.	Name	T/K	ε	a	b	c	Range/K
$C_5H_{10}O_2$	Methyl butanoate	301.2	5.48	0.38604E+02	-0.19171E+00	0.27128E-03	301-343
$C_5H_{10}O_2$	Tetrahydrofurfuryl alcohol	303.2	13.48				
$C_5H_{10}O_2S$	3-Methyl sulfolane	298.2	29.4	0.53158E+02	-0.93730E-01	0.47275E-04	298-398
$C_5H_{10}O_3$	Diethyl carbonate	297.2	2.820				
$C_5H_{10}O_3$	Ethyl lactate	303.2	15.4	0.31225E+02	-0.43531E-01	-0.28571E-04	273-373
$C_5H_{10}O_4$	1,2,3-Propanetriol-1-acetate	242.2	38.57	0.10653E+03	-0.26439E+00	-0.62371E-04	215-242
$C_5H_{11}Br$	2-Bromo-2-methylbutane	298.2	9.21				
$C_5H_{11}Br$	1-Bromopentane	299.2	6.31	0.20954E+02	-0.78743E-01	0.98908E-04	183-328
$C_5H_{11}Br$	3-Bromopentane	298.2	8.37				
$C_5H_{11}Br$	1-Bromo-3-methylbutane	291.5	6.33	0.27743E+02	-0.13927E+00	0.22627E-03	123-292
$C_5H_{11}Cl$	1-Chloropentane	293.2	6.654	0.18626E+02	-0.54719E-01	0.47143E-04	273-323
$C_5H_{11}Cl$	1-Chloro-3-methylbutane	292.0	6.10	0.22228E+02	-0.93189E-01	0.12991E-03	171-297
$C_5H_{11}Cl$	2-Chloro-2-methylbutane	222.75	12.31	0.55104E+02	-0.29866E+00	0.47840E-03	201-223
$C_5H_{11}F$	1-Fluoropentane	293.2	3.931				
$C_5H_{11}I$	1-Iodopentane	293.2	5.78	0.15753E+02	-0.50543E-01	0.56401E-04	293-323
$C_5H_{11}I$	3-Iodopentane	293.2	7.432				
$C_5H_{11}I$	1-Iodo-3-methylbutane	292.2	5.6				
$C_5H_{11}I$	2-Iodo-2-methylbutane	293.2	8.192				
$C_5H_{11}N$	Piperidine	293.0	4.33	0.82317E+01	-0.11229E-01	-0.71429E-05	293-333
$C_5H_{11}N$	*N*-Methylpyrrolidine	298.2	32.2				
$C_5H_{11}NO$	2,2-Dimethylpropanamide	298.2	20.13	0.10400E+03	-0.46017E+00	0.60000E-03	298-328
$C_5H_{11}NO$	*N,N*-Diethylformamide	293.2	29.6				
$C_5H_{11}NO$	2-Pentanone oxime	293.2	3.3				
$C_5H_{11}NO_2$	Pentyl nitrite	298.2	7.21				
C_5H_{12}	Pentane	293.2	1.8371				
C_5H_{12}	Isopentane	293.2	1.845	0.22384E+01	-0.12985E-02	-0.16182E-06	143-293
C_5H_{12}	Neopentane	296.0	1.769	0.10949E+02	-0.63057E-01	0.10835E-03	251-296
$C_5H_{12}N_2O$	Tetramethylurea	293.2	23.10				
$C_5H_{12}O$	1-Pentanol	298.2	15.13	0.73397E+02	-0.28165E+00	0.28427E-03	213-513
$C_5H_{12}O$	2-Pentanol	298.2	13.71	0.16437E+03	-0.86506E+00	0.11955E-02	273-323
$C_5H_{12}O$	3-Pentanol	298.2	13.35	0.12838E+03	-0.60980E+00	0.75000E-03	288-318
$C_5H_{12}O$	2-Methyl-1-butanol	298.2	15.63	0.14020E+02	0.13948E+00	-0.45000E-03	288-318
$C_5H_{12}O$	3-Methyl-1-butanol	293.2	15.63	0.79733E+02	-0.31272E+00	0.32014E-03	173-513
$C_5H_{12}O$	2-Methyl-2-butanol	298.2	5.78	0.11662E+03	-0.69756E+00	0.10920E-02	268-318
$C_5H_{12}O$	3-Methyl-2-butanol	298.2	12.1				
$C_5H_{12}O$	2,2-Dimethyl-1-propanol	333.2	8.35	0.92350E+02	-0.41870E+00	0.50000E-03	333-373
$C_5H_{12}O_2$	1,2-Pentanediol	296.8	17.31	0.18436E+03	-0.10682E+01	0.17037E-02	197-297
$C_5H_{12}O_2$	1,4-Pentanediol	295.7	26.74	0.13568E+03	-0.59198E+00	0.75398E-03	193-318
$C_5H_{12}O_2$	1,5-Pentanediol	293.2	26.2	0.11858E+03	-0.45920E+00	0.49341E-03	243-343
$C_5H_{12}O_2$	2,3-Pentanediol	296.9	17.37	0.95876E+02	-0.46463E+00	0.67434E-03	238-297
$C_5H_{12}O_2$	2,4-Pentanediol	294.2	24.69	0.11914E+03	-0.52569E+00	0.69607E-03	224-294
$C_5H_{12}O_2$	Diethoxymethane	293.2	2.527	0.25294E+01	0.73988E-04	-0.28331E-06	227-293
$C_5H_{12}O_4$	Tetramethoxymethane	293.2	2.40				
$C_5H_{12}O_5$	Xylitol	293.2	40.0				
$C_5H_{12}S$	1-Pentanethiol	293.2	4.847	0.71131E+01	-0.30228E-02	-0.16414E-04	273-333
$C_5H_{12}S$	2-Methyl-2-butanethiol	293.2	5.087	0.15116E+02	-0.50700E-01	0.56250E-04	273-333
$C_5H_{12}S_4$	Tetrakis(methylthio)methane	343.2	2.818				
$C_5H_{13}N$	Pentylamine	293.2	4.27	0.11274E+02	-0.34965E-01	0.37706E-04	223-353
$C_5H_{13}N_3$	1,1,3,3-Tetramethylguanidine	298.2	11.5				
$C_5H_{14}OSi$	Ethoxytrimethylsilane	298.2	3.013				
C_6F_6	Hexafluorobenzene	298.2	2.029	0.24041E+01	-0.83086E-03	-0.14286E-05	298-338
C_6F_{14}	Perfluorohexane	298.2	1.76				
$C_6H_3N_3O_7$	2,4,6-Trinitrophenol	294.2	4.0				
C_6H_4BrF	1-Bromo-2-fluorobenzene	298.2	4.72				
C_6H_4BrF	1-Bromo-3-fluorobenzene	298.2	4.85				
C_6H_4BrF	1-Bromo-4-fluorobenzene	298.2	2.60				
$C_6H_4BrNO_2$	1-Bromo-3-nitrobenzene	328.2	20.2	0.81413E+02	-0.27645E+00	0.27367E-03	328-413
$C_6H_4Br_2$	*o*-Dibromobenzene	293.2	7.86	-0.81849E-02	0.62671E-01	-0.12222E-03	293-353
$C_6H_4Br_2$	*m*-Dibromobenzene	293.2	4.81	0.93214E+01	-0.20273E-01	0.16667E-04	293-353

Mol. Form.	Name	*T*/K	ε	*a*	*b*	*c*	Range/K
$C_6H_4Br_2$	*p*-Dibromobenzene	368.2	2.57				
C_6H_4ClF	1-Chloro-2-fluorobenzene	298.2	6.10				
C_6H_4ClF	1-Chloro-3-fluorobenzene	298.2	4.96				
C_6H_4ClF	1-Chloro-4-fluorobenzene	298.2	3.34				
$C_6H_4ClNO_2$	1-Chloro-2-nitrobenzene	323.2	37.7	0.16800E+03	-0.59708E+00	0.59957E-03	323-436
$C_6H_4ClNO_2$	1-Chloro-3-nitrobenzene	323.2	20.9	0.77193E+02	-0.25118E+00	0.23798E-03	323-433
$C_6H_4ClNO_2$	1-Chloro-4-nitrobenzene	393.2	8.09				
$C_6H_4Cl_2$	*o*-Dichlorobenzene	293.2	10.12	0.13629E+02	0.10622E-02	-0.44444E-04	293-353
$C_6H_4Cl_2$	*m*-Dichlorobenzene	293.2	5.02	0.77565E+01	-0.93333E-02	-0.26880E-14	293-353
$C_6H_4Cl_2$	*p*-Dichlorobenzene	328.2	2.3943	0.26999E+01	-0.35325E-03	-0.17619E-05	328-363
C_6H_4FI	1-Fluoro-2-iodobenzene	298.2	8.22				
C_6H_4FI	1-Fluoro-4-iodobenzene	298.2	3.12				
$C_6H_4F_2$	*o*-Difluorobenzene	301.2	13.38	0.59107E+02	-0.23611E+00	0.27987E-03	273-323
$C_6H_4F_2$	*m*-Difluorobenzene	301.2	5.01	0.14448E+02	-0.46982E-01	0.51948E-04	273-323
$C_6H_4I_2$	*o*-Diiodobenzene	323.2	5.41	0.31150E+02	-0.14428E+00	0.20000E-03	323-353
$C_6H_4I_2$	*m*-Diiodobenzene	323.2	4.11				
$C_6H_4I_2$	*p*-Diiodobenzene	393.2	2.88				
$C_6H_4N_2$	2-Pyridinecarbonitrile	303.2	93.77	0.45596E+03	-0.17746E+01	0.19105E-02	303-398
$C_6H_4N_2$	3-Pyridinecarbonitrile	323.2	20.54	0.60484E+02	-0.17280E+00	0.15218E-03	323-398
$C_6H_4N_2$	4-Pyridinecarbonitrile	353.2	5.23	0.12533E+02	-0.30115E-01	0.26674E-04	353-398
$C_6H_4N_2O_4$	1,3-Dinitrobenzene	365.2	22.9	0.10406E+03	-0.34133E+00	0.32609E-03	365-413
C_6H_5Br	Bromobenzene	293.2	5.45	0.94100E+01	-0.12537E-01	-0.31127E-05	234-333
C_6H_5Cl	Chlorobenzene	293.2	5.6895	0.19471E+02	-0.70786E-01	0.82466E-04	293-430
C_6H_5ClO	*o*-Chlorophenol	296.2	7.40	0.29755E+02	-0.11256E+00	0.12390E-03	296-448
C_6H_5ClO	*m*-Chlorophenol	293.2	6.255				
C_6H_5ClO	*p*-Chlorophenol	314.2	11.18	0.31997E+02	-0.94241E-01	0.88392E-04	314-453
$C_6H_5ClO_2S$	Benzenesulfonyl chloride	323.2	28.90	0.83886E+02	-0.23405E+00	0.19713E-03	323-473
C_6H_5ClS	4-Chlorobenzenethiol	338.2	3.59				
C_6H_5F	Fluorobenzene	293.2	5.465				
C_6H_5I	Iodobenzene	293.2	4.59	0.89442E+01	-0.20008E-01	0.17641E-04	243-323
C_6H_5NOS	*N*-Sulfinylaniline	298.2	6.97				
$C_6H_5NO_2$	Nitrobenzene	293.0	35.6	0.11212E+03	-0.35211E+00	0.31128E-03	279-533
$C_6H_5NO_3$	*o*-Nitrophenol	323.2	16.50	0.33827E+02	-0.62123E-01	0.26774E-04	323-453
$C_6H_5NO_3$	*m*-Nitrophenol	373.2	35.45	0.18967E+03	-0.66144E+00	0.66532E-03	373-458
$C_6H_5NO_3$	*p*-Nitrophenol	393.2	42.20	0.22901E+03	-0.74264E+00	0.68006E-03	393-463
C_6H_6	Benzene	293.2	2.2825	0.26706E+01	-0.91648E-03	-0.14257E-05	293-513
C_6H_6BrN	*m*-Bromoaniline	293.2	13.0				
C_6H_6ClN	*o*-Chloroaniline	293.2	13.40				
C_6H_6ClN	*m*-Chloroaniline	293.2	13.3				
$C_6H_6N_2O_2$	*o*-Nitroaniline	353.0	47.3	0.18900E+03	-0.56977E+00	0.47484E-03	353-468
$C_6H_6N_2O_2$	*m*-Nitroaniline	398.0	35.6	0.20352E+03	-0.66582E+00	0.61310E-03	398-468
$C_6H_6N_2O_2$	*p*-Nitroaniline	428.0	78.5	0.48673E+03	-0.15040E+01	0.12857E-02	428-468
C_6H_6O	Phenol	303.2	12.40	0.63391E+02	-0.24988E+00	0.26930E-03	303-433
$C_6H_6O_2$	Pyrocatechol	388.2	17.57	0.74930E+02	-0.22142E+00	0.18919E-03	388-463
$C_6H_6O_2$	Resorcinol	393.2	13.55	0.30252E+02	-0.56443E-01	0.35578E-04	393-463
C_6H_6S	Benzenethiol	303.2	4.26	0.57155E+01	-0.70336E-02	0.73617E-05	303-358
C_6H_7N	Aniline	293.2	7.06	0.89534E+01	0.38990E-02	-0.36310E-04	293-413
C_6H_7N	2-Methylpyridine	293.2	10.18	0.34560E+02	-0.11980E+00	0.12500E-03	293-333
C_6H_7N	3-Methylpyridine	303.0	11.10	0.19643E+03	-0.11167E+01	0.16667E-02	303-333
C_6H_7N	4-Methylpyridine	293.0	12.2	0.33765E+02	-0.10113E+00	0.93860E-04	274-333
C_6H_7NO	2-Methylpyridine-1-oxide	323.2	36.4	0.11705E+03	-0.35301E+00	0.32000E-03	323-398
C_6H_7NO	3-Methylpyridine-1-oxide	318.2	28.26	0.59851E+02	-0.12682E+00	0.86622E-04	318-398
C_6H_8	1,3-Cyclohexadiene	184.2	2.68				
C_6H_8	1,4-Cyclohexadiene	296.0	2.211	0.27459E+01	-0.16975E-02	-0.36461E-06	232-356
$C_6H_8N_2$	Phenylhydrazine	293.2	7.15				
$C_6H_8N_2$	2,5-Dimethylpyrazine	293.2	2.436				
$C_6H_8N_2$	2,6-Dimethylpyrazine	308.2	2.653				
$C_6H_8O_2$	1,4-Cyclohexanedione	351.2	4.40				
$C_6H_9Cl_3O_2$	Butyl trichloroacetate	293.2	7.480				

Mol. Form.	Name	T/K	ε	a	b	c	Range/K
$C_6H_9Cl_3O_2$	Isobutyl trichloroacetate	293.2	7.667				
C_6H_9N	Cyclopentanecarbonitrile	293.2	22.68	0.69830E+02	-0.25303E+00	0.31491E-03	201-293
C_6H_{10}	1,5-Hexadiene	294.0	2.125	0.30014E+01	-0.28668E-02	-0.31026E-06	151-294
C_6H_{10}	*cis,cis*-2,4-Hexadiene	297.0	2.163	0.27284E+01	-0.17178E-02	-0.62926E-06	234-351
C_6H_{10}	*trans,trans*-2,4-Hexadiene	297.0	2.123	0.26774E+01	-0.16977E-02	-0.55637E-06	232-353
C_6H_{10}	2-Methyl-1,3-pentadiene*	298.2	2.422				
C_6H_{10}	3-Methyl-1,3-pentadiene	298.2	2.426				
C_6H_{10}	4-Methyl-1,3-pentadiene	293.2	2.599	0.51328E+01	-0.12774E-01	0.14215E-04	198-323
C_6H_{10}	2,3-Dimethyl-1,3-butadiene	293.2	2.102	0.26258E+01	-0.17990E-02	0.12035E-06	223-323
C_6H_{10}	1-Hexyne	296.0	2.621	0.58591E+01	-0.17099E-01	0.20856E-04	184-296
C_6H_{10}	Cyclohexene	293.2	2.2176	0.30598E+01	-0.39841E-02	0.37554E-05	141-313
$C_6H_{10}O$	Butoxyacetylene	298.2	6.62				
$C_6H_{10}O$	Cyclohexanone	293.0	16.1	0.41577E+02	-0.11463E+00	0.92454E-04	253-423
$C_6H_{10}O$	Mesityl oxide	273.2	15.6				
$C_6H_{10}O_2$	Ethyl 2-butenoate	293.2	5.4				
$C_6H_{10}O_2$	Ethyl methacrylate	303.2	5.68	0.40962E+02	-0.20520E+00	0.29286E-03	303-343
$C_6H_{10}O_3$	Ethyl acetoacetate	293.2	14.0				
$C_6H_{10}O_3$	Propanoic anhydride	293.2	18.30				
$C_6H_{10}O_4$	Monomethyl glutarate	293.2	8.37	0.16779E+02	-0.39839E-01	0.38095E-04	293-363
$C_6H_{10}O_4$	Diethyl oxalate	293.2	8.266	0.21938E+02	-0.66226E-01	0.66800E-04	293-368
$C_6H_{10}O_4$	Dimethyl succinate	293.2	7.19	0.13551E+02	-0.23109E-01	0.55440E-05	293-433
$C_6H_{10}O_4$	Ethylene glycol diacetate	290.2	7.7	0.25093E+02	-0.95171E-01	0.12224E-03	223-290
$C_6H_{11}Br$	Bromocyclohexane	303.2	8.0026				
$C_6H_{11}BrO_2$	Ethyl 2-bromobutanoate	303.2	8.57	0.49005E+02	-0.23193E+00	0.32500E-03	303-333
$C_6H_{11}BrO_2$	Ethyl 2-bromo-2-methylpropanoate	303.2	8.55	0.77044E+02	-0.40784E+00	0.60000E-03	303-333
$C_6H_{11}Cl$	Chlorocyclohexane	303.2	7.9505				
$C_6H_{11}N$	Hexanenitrile	298.2	17.26				
$C_6H_{11}N$	4-Methylpentanenitrile	295.2	17.5				
$C_6H_{11}NO$	Cyclohexanone oxime	362.2	3.04				
C_6H_{12}	1-Hexene	294.0	2.077	0.31476E+01	-0.50003E-02	0.46673E-05	149-294
C_6H_{12}	*trans*-2-Hexene	295.0	1.978	0.24338E+01	-0.11323E-02	-0.13720E-05	157-295
C_6H_{12}	*cis*-3-Hexene	296.0	2.069	0.30691E+01	-0.45458E-02	0.39898E-05	155-296
C_6H_{12}	*trans*-3-Hexene	293.2	1.954				
C_6H_{12}	Cyclohexane	293.2	2.0243	0.24293E+01	-0.12095E-02	-0.58741E-06	283-333
C_6H_{12}	Methylcyclopentane	293.2	1.9853	0.21587E+01	-0.22450E-03	-0.12500E-05	293-323
C_6H_{12}	Ethylcyclobutane	293.2	1.965				
$C_6H_{12}Br_2$	1,6-Dibromohexane	298.2	8.52	-0.55185E+01	0.11746E+00	-0.23658E-03	274-328
$C_6H_{12}Br_2$	3,4-Dibromohexane	298.2	6.732				
$C_6H_{12}Cl_2$	1,6-Dichlorohexane	308.2	8.60	0.11277E+02	0.67200E-02	-0.50000E-04	308-338
$C_6H_{12}O$	1-Methylcyclopentanol	310.1	7.11	0.75444E+02	-0.36617E+00	0.47021E-03	310-333
$C_6H_{12}O$	Isobutyl vinyl ether	293.2	3.34	0.48060E+01	-0.50000E-02	-0.41495E-14	293-323
$C_6H_{12}O$	2-Hexanone	293.2	14.56	0.70378E+02	-0.29385E+00	0.35289E-03	243-293
$C_6H_{12}O$	4-Methyl-2-pentanone	293.2	13.11	0.36341E+02	-0.97119E-01	0.61896E-04	204-373
$C_6H_{12}O$	3,3-Dimethyl-2-butanone	293.2	12.73	0.66857E+02	-0.28552E+00	0.34422E-03	243-293
$C_6H_{12}O$	Cyclohexanol	293.2	16.40	0.10173E+03	-0.43072E+00	0.47926E-03	293-423
$C_6H_{12}O_2$	Hexanoic acid	298.2	2.600	0.21730E+01	0.14840E-02	-0.16526E-06	298-433
$C_6H_{12}O_2$	2-Ethylbutanoic acid	296.2	2.72				
$C_6H_{12}O_2$	*tert*-Butylacetic acid	296.2	2.85				
$C_6H_{12}O_2$	Pentyl formate	292.2	5.7				
$C_6H_{12}O_2$	Isopentyl formate	288.2	5.44	0.29257E+02	-0.14028E+00	0.20000E-03	288-323
$C_6H_{12}O_2$	Butyl acetate	293.2	5.07	0.13825E+02	-0.43994E-01	0.48214E-04	253-353
$C_6H_{12}O_2$	*sec*-Butyl acetate	293.2	5.135	0.12427E+02	-0.32035E-01	0.24286E-04	273-323
$C_6H_{12}O_2$	*tert*-Butyl acetate	293.2	5.672	0.55435E+02	-0.30494E+00	0.46107E-03	273-323
$C_6H_{12}O_2$	Isobutyl acetate	293.2	5.068	0.14323E+02	-0.46048E-01	0.49286E-04	273-323
$C_6H_{12}O_2$	Propyl propanoate	293.2	5.249				
$C_6H_{12}O_2$	Ethyl butanoate	301.2	5.18	0.48698E+02	-0.25660E+00	0.37237E-03	301-343
$C_6H_{12}O_2$	Methyl pentanoate	293.2	4.992				
$C_6H_{12}O_2$	Diacetone alcohol	298.2	18.2				
$C_6H_{12}O_3$	Ethylene glycol monoethyl ether acetate	303.2	7.567	0.23290E+02	-0.71566E-01	0.65000E-04	303-323

Mol. Form.	Name	*T*/K	ε	*a*	*b*	*c*	Range/K
$C_6H_{12}S$	Cyclohexanethiol	298.2	5.420				
$C_6H_{13}Br$	1-Bromohexane	298.2	5.82	0.15233E+02	-0.44385E-01	0.43039E-04	274-328
$C_6H_{13}Cl$	1-Chlorohexane	293.2	6.104	0.15994E+02	-0.43647E-01	0.33393E-04	273-323
$C_6H_{13}ClO$	6-Chloro-1-hexanol	242.2	21.6	-0.73364E+01	0.46377E+00	-0.14202E-02	195-242
$C_6H_{13}I$	1-Iodohexane	293.3	5.35	0.16685E+02	-0.61309E-01	0.77262E-04	293-323
$C_6H_{13}N$	Cyclohexylamine	293.2	4.547				
$C_6H_{13}NO$	*N*-Propylpropanamide	298.2	118.1	0.58846E+03	-0.22012E+01	0.20870E-02	298-328
$C_6H_{13}NO$	*N*-Butylacetamide	293.2	104.0	0.70739E+03	-0.37369E+01	0.71585E-02	253-493
$C_6H_{13}NO$	*N,N*-Diethylacetamide	293.2	32.1				
C_6H_{14}	Hexane	293.2	1.8865	0.19768E+01	0.70933E-03	-0.34470E-05	293-473
C_6H_{14}	2-Methylpentane	293.2	1.886	0.20745E+01	0.50871E-03	-0.39286E-05	273-323
C_6H_{14}	3-Methylpentane	293.2	1.886	0.24739E+01	-0.23190E-02	0.10714E-05	273-323
C_6H_{14}	2,2-Dimethylbutane	293.2	1.869	0.22740E+01	-0.96229E-03	-0.14286E-05	273-313
C_6H_{14}	2,3-Dimethylbutane	293.2	1.889	0.24305E+01	-0.20081E-02	0.53571E-06	273-323
$C_6H_{14}O$	1-Hexanol	293.2	13.03	0.62744E+02	-0.24214E+00	0.24704E-03	233-513
$C_6H_{14}O$	2-Hexanol	298.2	11.06				
$C_6H_{14}O$	3-Hexanol	298.2	9.66				
$C_6H_{14}O$	3-Methyl-1-pentanol	298.2	15.2				
$C_6H_{14}O$	3-Methyl-3-pentanol	293.2	4.322				
$C_6H_{14}O$	2-Ethyl-1-butanol	362.2	6.19				
$C_6H_{14}O$	2,2-Dimethyl-1-butanol	293.2	10.5	0.14054E+03	-0.72925E+00	0.97821E-03	243-393
$C_6H_{14}O$	Dipropyl ether	297.0	3.38	0.14600E+02	-0.72670E-01	0.11742E-03	161-297
$C_6H_{14}O$	Diisopropyl ether	303.2	3.805				
$C_6H_{14}OS$	Dipropyl sulfoxide	303.2	30.37	0.84868E+02	-0.23486E+00	0.18198E-03	303-373
$C_6H_{14}O_2$	2-Methyl-2,4-pentanediol	293.2	25.86	0.14531E+03	-0.65285E+00	0.83503E-03	203-333
$C_6H_{14}O_2$	Ethylene glycol diethyl ether	293.2	3.90	0.99099E+01	-0.33403E-01	0.44048E-04	223-303
$C_6H_{14}O_2S$	Dipropyl sulfone	303.2	32.62	0.70195E+02	-0.15008E+00	0.86506E-04	303-398
$C_6H_{14}O_3$	1,2,6-Hexanetriol	285.3	31.5	0.26127E+03	-0.14552E+01	0.22765E-02	261-285
$C_6H_{14}O_3$	Diethylene glycol dimethyl ether	298.2	7.23	0.28291E+02	-0.11236E+00	0.14000E-03	298-333
$C_6H_{14}O_4$	Triethylene glycol	293.2	23.69	0.91845E+02	-0.33827E+00	0.36062E-03	253-333
$C_6H_{14}O_6$	*D*-Glucitol	353.2	35.5				
$C_6H_{14}O_6$	*D*-Mannitol	443.2	24.6				
$C_6H_{14}S$	1-Hexanethiol	293.2	4.436	0.11774E+02	-0.37298E-01	0.41875E-04	273-333
$C_6H_{15}B$	Triethylborane	293.2	1.974				
$C_6H_{15}N$	Hexylamine	293.2	4.08	0.80244E+01	-0.16627E-01	0.10874E-04	253-373
$C_6H_{15}N$	Dipropylamine	293.2	2.923	0.11376E+02	-0.49796E-01	0.71792E-04	243-323
$C_6H_{15}N$	Triethylamine	293.2	2.418	0.29205E+01	-0.14007E-02	-0.13469E-05	233-323
$C_6H_{15}OP$	Triethylphosphine oxide	323.2	35.5				
$C_6H_{15}O_4P$	Triethyl phosphate	298.2	13.20	0.61230E+02	-0.26047E+00	0.33333E-03	298-333
$C_6H_{15}PS$	Triethylphosphine sulfide	371.2	39.0				
$C_6H_{16}O_2Si$	Diethoxydimethylsilane	298.2	3.216				
$C_6H_{16}Si$	Triethylsilane	293.2	2.323				
$C_6H_{18}N_3OP$	Hexamethylphosphoric triamide	293.2	31.3	0.95666E+02	-0.29769E+00	0.26407E-03	283-363
$C_6H_{18}N_4$	*N,N'*-Bis(2-aminoethyl)-1,2-ethanediamine	293.2	10.76	0.50699E+02	-0.21730E+00	0.27582E-03	213-333
$C_6H_{18}OSi_2$	Hexamethyldisiloxane	293.2	2.179	0.34537E+01	-0.61530E-02	0.61544E-05	213-313
$C_6H_{18}O_3Si_3$	Hexamethylcyclotrisiloxane	343.2	2.139				
$C_6H_{19}NSi_2$	Hexamethyldisilazane	294.2	2.273	0.23358E+01	0.16127E-02	-0.62078E-05	294-333
C_7F_{14}	Perfluoromethylcyclohexane	298.2	1.82				
C_7F_{16}	Perfluoroheptane	289.2	1.847				
$C_7H_3Cl_5$	2,3,4,5,6-Pentachlorotoluene	293.2	4.8				
C_7H_4ClNO	4-Chlorophenyl isocyanate	288.2	3.177	0.40896E+01	-0.31667E-02		288-348
C_7H_5BrO	Benzoyl bromide	293.2	21.33	0.84231E+02	-0.31089E+00	0.32857E-03	283-313
C_7H_5ClO	Benzoyl chloride	293.2	23.0				
C_7H_5FO	Benzoyl fluoride	293.2	22.7				
$C_7H_5F_3$	(Trifluoromethyl)benzene	298.2	9.22				
C_7H_5N	Benzonitrile	293.2	25.9	0.57605E+02	-0.13354E+00	0.87767E-04	273-453
C_7H_5NO	Phenyl isocyanate	293.2	8.940	0.17541E+02	-0.29790E-01	0.15476E-05	293-353
$C_7H_6ClNO_2$	4-Chloro-3-nitrotoluene	301.2	28.07				

PERMITTIVITY (DIELECTRIC CONSTANT) OF LIQUIDS (continued)

Mol. Form.	Name	T/K	ε	a	b	c	Range/K
$C_7H_6Cl_2$	2,4-Dichlorotoluene	301.2	5.68				
$C_7H_6Cl_2$	2,6-Dichlorotoluene	301.2	3.36				
$C_7H_6Cl_2$	3,4-Dichlorotoluene	301.2	9.39				
$C_7H_6Cl_2$	(Dichloromethyl)benzene	293.2	6.9				
C_7H_6O	Benzaldehyde	293.2	17.85	0.35046E+02	-0.61271E-01	0.16222E-04	301-346
$C_7H_6O_2$	Salicylaldehyde	293.2	18.35	0.51315E+02	-0.15379E+00	0.14111E-03	289-453
C_7H_7Br	*o*-Bromotoluene	293.2	4.641	0.10229E+02	-0.25050E-01	0.20357E-04	273-323
C_7H_7Br	*m*-Bromotoluene	293.2	5.566	0.11522E+02	-0.24946E-01	0.15714E-04	273-323
C_7H_7Br	*p*-Bromotoluene	293.2	5.503	0.10014E+02	-0.13918E-01	-0.50000E-05	273-293
C_7H_7Br	(Bromomethyl)benzene	293.2	6.658	0.18482E+02	-0.57207E-01	0.57321E-04	273-323
C_7H_7BrO	*o*-Bromoanisole	303.2	8.96	0.12023E+02	-0.59116E-02	-0.13787E-04	303-358
C_7H_7BrO	*p*-Bromoanisole	303.2	7.40	0.74367E+01	0.12648E-01	-0.42128E-04	303-358
C_7H_7Cl	*o*-Chlorotoluene	293.2	4.721	0.11507E+02	-0.31148E-01	0.27143E-04	273-323
C_7H_7Cl	*m*-Chlorotoluene	293.2	5.763	0.13921E+02	-0.37186E-01	0.31786E-04	273-323
C_7H_7Cl	*p*-Chlorotoluene	293.2	6.25	0.20265E+01	0.40060E-01	-0.87500E-04	293-333
C_7H_7Cl	(Chloromethyl)benzene	293.2	6.854	0.17108E+02	-0.45285E-01	0.35000E-04	273-323
C_7H_7ClO	*p*-Chloroanisole	293.2	7.84	0.64019E+01	0.30560E-01	-0.87500E-04	293-333
$C_7H_7ClO_2S$	*p*-Toluenesulfonyl chloride	343.2	22.6				
$C_7H_7ClO_3S$	4-Methoxybenzenesulfonyl chloride	314.2	27.2				
C_7H_7F	*o*-Fluorotoluene	298.2	4.23				
C_7H_7F	*m*-Fluorotoluene	298.2	5.41				
C_7H_7F	*p*-Fluorotoluene	298.2	5.88				
C_7H_7I	*p*-Iodotoluene	308.2	4.4				
C_7H_7N	2-Vinylpyridine	293.2	9.126				
C_7H_7N	4-Vinylpyridine	293.2	10.50				
$C_7H_7NO_2$	Benzyl nitrite	298.2	7.78				
$C_7H_7NO_2$	*o*-Nitrotoluene	293.0	26.26	0.10420E+03	-0.41726E+00	0.51607E-03	273-323
$C_7H_7NO_2$	*m*-Nitrotoluene	303.2	24.95	0.62492E+02	-0.16235E+00	0.12844E-03	303-403
$C_7H_7NO_2$	*p*-Nitrotoluene	331.2	22.2				
$C_7H_7NO_2S$	4-Nitrothioanisole	346.0	21.7				
$C_7H_7NO_3$	2-Nitroanisole	293.2	45.75	0.16684E+03	-0.58196E+00	0.57382E-03	293-423
$C_7H_7NO_3$	3-Nitroanisole	318.2	25.7	0.65402E+02	-0.16460E+00	0.12560E-03	318-443
$C_7H_7NO_3$	4-Nitroanisole	338.2	26.95	0.59811E+02	-0.10955E+00	0.36042E-04	338-443
C_7H_8	Toluene	296.35	2.379	0.32584E+01	-0.34410E-02	0.15937E-05	207-316
C_7H_8O	*o*-Cresol	298.2	6.76	0.21633E+02	-0.71069E-01	0.70590E-04	298-453
C_7H_8O	*m*-Cresol	298.2	12.44	0.81716E+02	-0.35039E+00	0.39878E-03	274-463
C_7H_8O	*p*-Cresol	298.2	13.05	0.70253E+02	-0.28870E+00	0.31979E-03	298-453
C_7H_8O	Benzyl alcohol	303.2	11.916	0.13661E+03	-0.72127E+00	0.10225E-02	303-333
C_7H_8O	Anisole	294.2	4.30	0.10887E+02	-0.32372E-01	0.33629E-04	294-413
$C_7H_8O_2$	2-Methoxyphenol	298.2	11.95	0.31751E+02	-0.88173E-01	0.72953E-04	291-448
$C_7H_8O_2$	3-Methoxyphenol	298.2	11.59	0.37279E+02	-0.12113E+00	0.11698E-03	298-433
$C_7H_8O_2$	4-Methoxyphenol	333.7	11.05	0.39483E+02	-0.12142E+00	0.10841E-03	334-453
$C_7H_8O_2S$	Ethyl thiophene-2-carboxylate	293.2	6.18				
$C_7H_8O_2S$	Methyl phenyl sulfone	373.2	37.9				
C_7H_8S	Benzenemethanethiol	298.2	4.705	0.16628E+02	-0.68276E-01	0.94636E-04	298-358
C_7H_8S	4-Methylbenzenethiol	323.2	4.74	0.87052E+01	-0.15347E-01	0.95238E-05	323-358
C_7H_8S	(Methylthio)benzene	303.2	4.88	0.21841E+02	-0.97630E-01	0.13750E-03	303-343
C_7H_9N	Benzylamine	293.2	5.18				
C_7H_9N	*o*-Methylaniline	298.2	6.138	0.10988E+02	-0.18976E-01	0.91958E-05	298-398
C_7H_9N	*m*-Mcthylaniline	298.2	5.816	0.13477E+02	-0.35551E-01	0.33135E-04	298-398
C_7H_9N	*p*-Methylaniline	333.2	5.058	0.78897E+01	-0.10196E-01	0.51190E-05	333-403
C_7H_9N	*N*-Methylaniline	293.2	5.96				
C_7H_9N	2-Ethylpyridine	293.2	8.33	0.36397E+02	-0.15070E+00	0.18750E-03	293-333
C_7H_9N	4-Ethylpyridine	293.2	10.98	-0.73831E+01	0.14326E+00	-0.27500E-03	293-333
C_7H_9N	2,4-Dimethylpyridinc	293.2	9.60	0.25895E+02	-0.73900E-01	0.62500E-04	293-333
C_7H_9N	2,6-Dimethylpyridine	293.2	7.33	0.17714E+02	-0.39080E-01	0.12500E-04	293-333
C_7H_9NO	2,6-Dimethylpyridine-1-oxide	298.2	46.11	0.22765E+03	-0.90760E+00	0.10011E-02	298-398
C_7H_9NO	*o*-Methoxyaniline	303.2	5.230	0.79911E+01	-0.92183E-02	0.37879E-06	303-393
C_7H_9NO	*m*-Methoxyaniline	298.2	8.76	0.28179E+02	-0.97840E-01	0.11027E-03	289-393

Mol. Form.	Name	*T*/K	ε	*a*	*b*	*c*	Range/K
C_7H_9NO	*p*-Methoxyaniline	333.2	7.85	0.30149E+02	-0.10523E+00	0.11467E-03	333-453
$C_7H_{10}N_2$	1-Methyl-1-phenylhydrazine	292.2	7.3				
$C_7H_{11}Cl_3O_2$	Isopentyl trichloroacetate	293.2	7.287				
C_7H_{12}	1,6-Heptadiene	293.0	2.161	0.30815E+01	-0.36095E-02	0.16354E-05	184-293
C_7H_{12}	Cycloheptene	295.0	2.265	0.32309E+01	-0.42373E-02	0.32572E-05	227-363
$C_7H_{12}O$	Cycloheptanone	298.2	13.16	0.17511E+03	-0.11221E+01	0.19417E-02	258-298
$C_7H_{12}O$	2-Methylcyclohexanone	293.2	14.0				
$C_7H_{12}O$	3-Methylcyclohexanone	293.2	12.4				
$C_7H_{12}O$	4-Methylcyclohexanone	293.2	12.35				
$C_7H_{12}O_2$	Cyclohexanecarboxylic acid	304.2	2.67				
$C_7H_{12}O_2$	Cyclohexyl formate	293.2	6.47				
$C_7H_{12}O_2$	Butyl acrylate	301.2	5.25	0.38296E+02	-0.19109E+00	0.27006E-03	301-343
$C_7H_{12}O_4$	Monomethyl adipate	293.2	6.69	0.11962E+02	-0.23973E-01	0.20608E-04	293-433
$C_7H_{12}O_4$	Diethyl malonate	304.2	7.550	0.14809E+02	-0.31207E-01	0.24066E-04	304-393
$C_7H_{12}O_4$	Dimethyl glutarate	293.2	7.87	0.20697E+02	-0.57794E-01	0.48405E-04	293-433
$C_7H_{12}O_5$	1,2,3-Propanetriol-1,3-diacetate	288.2	9.80	0.28321E+02	-0.89073E-01	0.86891E-04	258-374
C_7H_{14}	1-Heptene	293.2	2.092	0.21755E+01	0.13896E-02	-0.57049E-05	273-323
C_7H_{14}	2-Methyl-2-hexene	293.2	2.962				
C_7H_{14}	3-Ethyl-2-pentene	293.2	2.051				
C_7H_{14}	Cycloheptane	293.2	2.0784	0.25136E+01	-0.15089E-02	0.84915E-07	278-333
C_7H_{14}	Methylcyclohexane	293.2	2.024				
$C_7H_{14}Br_2$	1,2-Dibromoheptane	298.2	3.77				
$C_7H_{14}Br_2$	2,3-Dibromoheptane	298.2	5.08				
$C_7H_{14}Br_2$	3,4-Dibromoheptane	298.2	4.70				
$C_7H_{14}Cl_2$	1,7-Dichloroheptane	298.2	8.34				
$C_7H_{14}O$	1-Heptanal	295.2	9.07				
$C_7H_{14}O$	2-Heptanone	293.2	11.95	0.38348E+02	-0.12531E+00	0.12005E-03	253-413
$C_7H_{14}O$	3-Heptanone	293.2	12.7				
$C_7H_{14}O$	4-Heptanone	293.2	12.60	0.41520E+02	-0.13839E+00	0.13497E-03	253-393
$C_7H_{14}O$	5-Methyl-2-hexanone	293.2	13.53	0.52353E+02	-0.17695E+00	0.15195E-03	293-333
$C_7H_{14}O$	Cyclohexanemethanol	333.2	9.70	0.10164E+03	-0.45839E+00	0.54762E-03	333-368
$C_7H_{14}O$	2-Methylcyclohexanol*	293.2	9.375	0.17315E+03	-0.98794E+00	0.14634E-02	273-323
$C_7H_{14}O$	3-Methylcyclohexanol*	293.2	13.79	0.65896E+02	-0.21954E+00	0.14107E-03	273-323
$C_7H_{14}O$	4-Methylcyclohexanol*	293.2	13.45	0.65021E+02	-0.22896E+00	0.17946E-03	273-323
$C_7H_{14}O_2$	Heptanoic acid	288.2	3.04	0.36423E+01	-0.31996E-02	0.39362E-05	288-423
$C_7H_{14}O_2$	Pentyl acetate	293.2	4.79	0.12091E+02	-0.36536E-01	0.39732E-04	253-353
$C_7H_{14}O_2$	Isopentyl acetate	293.2	4.72				
$C_7H_{14}O_2$	Butyl propanoate	293.2	4.838				
$C_7H_{14}O_2$	Propyl butanoate	293.2	4.3				
$C_7H_{14}O_2$	Ethyl pentanoate	291.2	4.71				
$C_7H_{14}O_2$	Ethyl 3-methylbutanoate	293.2	4.71				
$C_7H_{14}O_2$	Methyl hexanoate	293.2	4.615				
$C_7H_{15}Br$	1-Bromoheptane	303.2	5.255	0.15289E+02	-0.50621E-01	0.57753E-04	203-343
$C_7H_{15}Br$	2-Bromoheptane	295.2	6.46				
$C_7H_{15}Br$	4-Bromoheptane	295.2	6.81				
$C_7H_{15}Cl$	1-Chloroheptane	293.2	5.521	0.14279E+02	-0.39431E-01	0.32321E-04	273-323
$C_7H_{15}Cl$	2-Chloroheptane	295.2	6.52				
$C_7H_{15}Cl$	3-Chloroheptane	295.2	6.70				
$C_7H_{15}Cl$	4-Chloroheptane	295.2	6.54				
$C_7H_{15}I$	1-Iodoheptane	298.2	4.92	0.11856E+02	-0.33493E-01	0.34368E-04	294-323
$C_7H_{15}I$	3-Iodoheptane	295.2	6.39				
C_7H_{16}	Heptane	293.2	1.9209	0.24740E+01	-0.22577E-02	0.12428E-05	273-373
C_7H_{16}	2-Methylhexane	293.2	1.9221	0.24759E+01	-0.22535E-02	0.12500E-05	293-323
C_7H_{16}	3-Methylhexane	293.2	1.920	0.27089E+01	-0.37908E-02	0.37500E-05	273-323
C_7H_{16}	3-Ethylpentane	293.2	1.942	0.23771E+01	-0.15140E-02	0.10093E-06	163-363
C_7H_{16}	2,2-Dimethylpentane	293.2	1.915	0.23414E+01	-0.14362E-02	-0.51322E-07	153-353
C_7H_{16}	2,3-Dimethylpentane	293.2	1.929	0.25637E+01	-0.26328E-02	0.16071E-05	273-323
C_7H_{16}	2,4-Dimethylpentane	293.2	1.902	0.23979E+01	-0.17436E-02	0.17857E-06	273-323
C_7H_{16}	3,3-Dimethylpentane	291.3	1.9419	0.24007E+01	-0.16802E-02	0.36069E-06	291-322

Mol. Form.	Name	*T*/K	ε	*a*	*b*	*c*	Range/K
C_7H_{16}	2,2,3-Trimethylbutane	293.2	1.930				
$C_7H_{16}O$	1-Heptanol	293.2	11.75	0.60662E+02	-0.24049E+00	0.25155E-03	239-513
$C_7H_{16}O$	2-Heptanol	293.7	9.72	0.10050E+03	-0.49793E+00	0.64504E-03	207-365
$C_7H_{16}O$	3-Heptanol	296.1	7.07	0.19586E+03	-0.11465E+01	0.17175E-02	248-349
$C_7H_{16}O$	4-Heptanol	296.2	6.18	0.28995E+03	-0.18499E+01	0.30109E-02	270-301
$C_7H_{16}O$	2-Methyl-2-hexanol	297.0	3.257				
$C_7H_{16}O$	3-Methyl-2-hexanol	297.2	4.990	0.59724E+02	-0.32417E+00	0.47058E-03	244-372
$C_7H_{16}O$	3-Methyl-3-hexanol	298.2	3.248				
$C_7H_{16}O$	3-Ethyl-3-pentanol	293.2	3.158				
$C_7H_{16}O$	2,2-Dimethyl-1-pentanol	293.2	6.020	0.37318E+02	-0.17095E+00	0.22022E-03	283-393
$C_7H_{16}O$	Ethyl pentyl ether	296.2	3.6				
$C_7H_{16}O$	Ethyl isopentyl ether	293.2	3.955	0.66541E+01	-0.55450E-02	-0.12500E-04	293-323
$C_7H_{16}O_3$	Triethoxymethane	293.2	4.779				
$C_7H_{16}S$	1-Heptanethiol	293.2	4.194	0.71333E+01	-0.97320E-02	-0.12500E-05	273-333
$C_7H_{17}N$	Heptylamine	293.2	3.81	0.87794E+01	-0.24363E-01	0.25325E-04	253-373
$C_7H_{18}O_3Si$	Triethoxymethylsilane	298.2	3.845				
$C_8H_4F_6$	1,3-Bis(trifluoromethyl)benzene	303.2	5.98				
C_8H_6	Phenylacetylene	298.2	2.98				
$C_8H_6Cl_2$	2,5-Dichlorostyrene	298.2	2.58				
$C_8H_6Cl_4$	1,2,3,4-Tetrachloro-5,6-dimethylbenzene	293.2	8.0				
$C_8H_6Cl_4$	1,2,3,5-Tetrachloro-4,6-dimethylbenzene	293.2	5.4				
C_8H_6O	Phenoxyacetylene	298.2	4.76				
C_8H_7N	Benzeneacetonitrile	299.2	17.87	0.82175E+02	-0.37416E+00	0.53220E-03	299-343
$C_8H_7NO_2$	4-Methoxyphenyl isocyanate	333.2	10.26	0.20780E+02	-0.31571E-01		333-403
$C_8H_7NO_4$	Methyl 2-nitrobenzoate	300.1	27.76				
C_8H_8	Styrene	293.2	2.4737	0.44473E+01	-0.11422E-01	0.16000E-04	293-313
C_8H_8O	Acetophenone	298.2	17.44	0.26099E+02	0.64048E-02	-0.11905E-03	298-333
$C_8H_8O_2$	Benzeneacetic acid	353.2	3.47	0.24104E+01	0.30000E-02		353-393
$C_8H_8O_2$	Benzyl formate	303.2	6.34	0.26162E+02	-0.11026E+00	0.14787E-03	303-358
$C_8H_8O_2$	Phenyl acetate	298.2	5.403	0.11327E+02	-0.26707E-01	0.22938E-04	298-404
$C_8H_8O_2$	Methyl benzoate	302.7	6.642	0.17486E+02	-0.51027E-01	0.50222E-04	303-393
$C_8H_8O_2$	(Hydroxyacetyl)benzene	298.2	21.33	0.42286E+02	-0.69215E-01	-0.35714E-05	298-368
$C_8H_8O_2$	4-Methoxybenzaldehyde	303.2	22.0				
$C_8H_8O_3$	Methyl salicylate	314.4	8.80	0.20501E+02	-0.39045E-01	0.68298E-05	223-398
C_8H_9Br	1-Bromo-2-ethylbenzene	298.2	5.55				
C_8H_9Br	1-Bromo-3-ethylbenzene	298.2	5.56				
C_8H_9Br	1-Bromo-4-ethylbenzene	298.2	5.42				
C_8H_9BrO	1-Bromo-2-ethoxybenzene	313.2	7.04	0.23146E+02	-0.75753E-01	0.77778E-04	313-358
C_8H_9Cl	1-Chloro-2-ethylbenzene	298.2	4.36				
C_8H_9Cl	1-Chloro-3-ethylbenzene	298.2	5.18				
C_8H_9Cl	1-Chloro-4-ethylbenzene	298.2	5.16				
$C_8H_9NO_2$	1-Ethyl-2-nitrobenzene	273.4	21.9				
$C_8H_9NO_2$	Methyl 2-aminobenzoate	298.2	21.9				
$C_8H_9NO_2$	Ethyl 4-pyridinecarboxylate	293.2	8.95				
C_8H_{10}	Ethylbenzene	293.2	2.4463	0.35969E+01	-0.53169E-02	0.47500E-05	293-323
C_8H_{10}	*o*-Xylene	293.2	2.562	0.36163E+01	-0.40177E-02	0.14286E-05	273-323
C_8H_{10}	*m*-Xylene	293.2	2.359	0.28421E+01	-0.10191E-02	-0.21429E-05	273-323
C_8H_{10}	*p*-Xylene	293.2	2.2735	0.23140E+01	0.97221E-03	-0.37500E-05	293-363
$C_8H_{10}O$	2,3-Xylenol	343.2	4.81	0.14399E+02	-0.41438E-01	0.39244E-04	343-433
$C_8H_{10}O$	2,4-Xylenol	303.2	5.060	0.22125E+02	-0.85543E-01	0.96548E-04	303-363
$C_8H_{10}O$	2,5-Xylenol	338.2	5.36	0.18049E+02	-0.54991E-01	0.51656E-04	338-455
$C_8H_{10}O$	2,6-Xylenol	313.2	4.90	0.12284E+02	-0.32996E-01	0.29867E-04	313-453
$C_8H_{10}O$	3,4-Xylenol	333.2	9.02	0.54423E+02	-0.21153E+00	0.22508E-03	333-453
$C_8H_{10}O$	3,5-Xylenol	323.2	9.06	0.54251E+02	-0.21647E+00	0.23542E-03	323-453
$C_8H_{10}O$	Benzeneethanol	293.2	12.31	0.12170E+03	-0.63124E+00	0.87776E-03	278-333
$C_8H_{10}O$	1-Phenylethanol	293.2	8.77	0.32971E+02	-0.12042E+00	0.12809E-03	293-423
$C_8H_{10}O$	Phenetole	293.2	4.216	-0.15043E+02	0.13752E+00	-0.24500E-03	293-313

Mol. Form.	Name	*T*/K	ε	*a*	*b*	*c*	Range/K
$C_8H_{10}O$	2-Methylanisole	293.2	3.502	0.50825E+01	-0.62297E-02	0.28571E-05	293-333
$C_8H_{10}O$	3-Methylanisole	293.2	3.967	0.12830E+02	-0.49701E-01	0.66429E-04	293-333
$C_8H_{10}O$	4-Methylanisole	293.2	3.914	0.86608E+01	-0.23510E-01	0.25000E-04	293-333
$C_8H_{10}O_2$	1,2-Dimethoxybenzene	293.2	4.45	0.74604E+01	-0.13445E-01	0.10737E-04	293-443
$C_8H_{10}O_2$	1,3-Dimethoxybenzene	298.2	5.363	0.11911E+02	-0.30804E-01	0.29643E-04	298-358
$C_8H_{10}O_2$	1,4-Dimethoxybenzene	333.7	5.60	0.11289E+02	-0.20765E-01	0.11987E-04	334-463
$C_8H_{10}O_2S$	Ethyl phenyl sulfone	348.2	39.0				
$C_8H_{10}S$	(Ethylthio)benzene	298.2	4.95				
$C_8H_{11}N$	*p*-Ethylaniline	298.2	4.84				
$C_8H_{11}N$	*N*-Ethylaniline	293.2	5.87				
$C_8H_{11}N$	*N,N*-Dimethylaniline	298.2	4.90	0.84052E+01	-0.13549E-01	0.62835E-05	289-453
$C_8H_{11}N$	2,4,6-Trimethylpyridine	298.2	7.807	0.20990E+02	-0.57419E-01	0.44286E-04	298-358
$C_8H_{11}NO$	4-Ethoxyaniline	298.2	7.43				
$C_8H_{12}N_2O_2$	Hexamethylene diisocyanate	288.2	14.41	0.26715E+02	-0.42696E-01		288-403
$C_8H_{12}O_4$	Diethyl maleate	298.2	7.560	0.13953E+02	-0.21969E-01	0.17817E-05	298-343
$C_8H_{12}O_4$	Diethyl fumarate	296.2	6.56				
C_8H_{14}	1,7-Octadiene	293.0	2.186	0.28376E+01	-0.17442E-02	-0.16141E-05	214-293
C_8H_{14}	*cis*-Cyclooctene	296.0	2.306	0.31115E+01	-0.32058E-02	0.16713E-05	269-406
C_8H_{14}	1,2-Dimethylcyclohexene	296.0	2.144	0.26443E+01	-0.17973E-02	0.35815E-06	211-374
C_8H_{14}	1,3-Dimethylcyclohexene	296.0	2.182	0.29951E+01	-0.34615E-02	0.24026E-05	213-373
$C_8H_{14}O_2$	Methyl cyclohexanecarboxylate	293.2	4.87				
$C_8H_{14}O_2$	Cyclohexyl acetate	293.2	5.08				
$C_8H_{14}O_3$	Butanoic anhydride	293.2	12.8				
$C_8H_{14}O_3$	2-Methylpropanoic anhydride	292.2	13.6				
$C_8H_{14}O_4$	Diisopropyl oxalate	293.2	6.403	0.10709E+02	-0.16328E-01	0.56000E-05	293-368
$C_8H_{14}O_4$	Diethyl succinate	293.2	6.098	0.80213E+01	0.11810E-02	-0.26400E-04	293-343
$C_8H_{14}O_4$	Dimethyl adipate	293.2	6.84	0.11739E+02	-0.17281E-01	0.11447E-05	293-433
$C_8H_{15}N$	Octanenitrile	293.2	13.90				
C_8H_{16}	1-Octene	293.2	2.113	0.24348E+01	0.34200E-03	-0.50000E-05	273-323
C_8H_{16}	*cis*-3-Octene	298.2	2.062				
C_8H_{16}	*trans*-3-Octene	298.2	2.002				
C_8H_{16}	*cis*-4-Octene	298.2	2.053				
C_8H_{16}	*trans*-4-Octene	298.2	2.004				
C_8H_{16}	3-Methyl-2-heptene*	293.2	2.436				
C_8H_{16}	2,5-Dimethyl-2-hexene	293.2	2.431				
C_8H_{16}	2,4,4-Trimethyl-1-pentene	298.2	2.0908				
C_8H_{16}	Cyclooctane	295.0	2.116	0.25036E+01	-0.12460E-02	-0.23175E-06	295-411
$C_8H_{16}Br_2$	1,8-Dibromooctane	298.2	7.43	0.94117E+00	0.61520E-01	-0.13333E-03	298-328
$C_8H_{16}Cl_2$	1,8-Dichlorooctane	298.2	7.64				
$C_8H_{16}O$	2-Octanone	293.2	9.51	-0.16219E+02	0.18799E+00	-0.34156E-03	293-333
$C_8H_{16}O$	3-Octanone	303.2	10.50				
$C_8H_{16}O_2$	Octanoic acid	288.2	2.85	0.29391E+01	-0.38721E-03		288-423
$C_8H_{16}O_2$	2-Ethylhexanoic acid	296.2	2.64				
$C_8H_{16}O_2$	Hexyl acetate	293.2	4.42				
$C_8H_{16}O_2$	Pentyl propanoate	293.2	4.552				
$C_8H_{16}O_2$	Isopentyl propanoate	273.2	5.21	0.17665E+02	-0.71718E-01	0.95635E-04	273-373
$C_8H_{16}O_2$	Butyl butanoate	298.2	4.39	0.79684E+01	-0.12000E-01	0.15266E-13	298-318
$C_8H_{16}O_2$	Propyl pentanoate	292.2	4.0				
$C_8H_{16}O_2$	Ethyl hexanoate	293.2	4.45	0.11007E+02	-0.32800E-01	0.35714E-04	253-353
$C_8H_{16}O_2$	Methyl heptanoate	293.2	4.355				
$C_8H_{16}O_3$	Isopentyl lactate	273.2	11.2	0.48649E+02	-0.21253E+00	0.27619E-03	273-373
$C_8H_{17}Br$	1-Bromooctane	293.2	5.0957	0.12404E+02	-0.35050E-01	0.34542E-04	283-353
$C_8H_{17}Br$	2-Bromooctane	293.2	5.44				
$C_8H_{17}Cl$	1-Chlorooctane	298.2	5.05	0.11346E+02	-0.25120E-01	0.13450E-04	274-328
$C_8H_{17}Cl$	2-Chlorooctane	293.2	5.42				
$C_8H_{17}F$	1-Fluorooctane	293.2	3.89				
$C_8H_{17}I$	1-Iodooctane	293.2	4.67	0.12452E+02	-0.41229E-01	0.50108E-04	233-313
$C_8H_{17}NO_2$	1-Nitrooctane	293.2	11.46				
C_8H_{18}	Octane	293.2	1.948	0.22590E+01	-0.84212E-03	-0.75758E-06	233-393

Mol. Form.	Name	*T*/K	ε	*a*	*b*	*c*	Range/K
C_8H_{18}	2-Methylheptane	293.2	1.9519				
C_8H_{18}	3-Ethylhexane	293.2	1.9617				
C_8H_{18}	2,2-Dimethylhexane	293.2	1.9498				
C_8H_{18}	2,5-Dimethylhexane	293.95	1.9619	0.25821E+01	-0.26804E-02	0.19404E-05	294-324
C_8H_{18}	3,3-Dimethylhexane	293.2	1.9645				
C_8H_{18}	3,4-Dimethylhexane	292.1	1.9814	0.26849E+01	-0.33712E-02	0.32949E-05	292-324
C_8H_{18}	3-Ethyl-3-methylpentane	291.49	1.9869	0.25983E+01	-0.28027E-02	0.24195E-05	292-324
C_8H_{18}	2,2,3-Trimethylpentane	293.2	1.960				
C_8H_{18}	2,2,4-Trimethylpentane	293.2	1.943	0.23677E+01	-0.14768E-02	0.94261E-07	173-373
C_8H_{18}	2,3,3-Trimethylpentane	293.2	1.9780				
C_8H_{18}	2,3,4-Trimethylpentane	293.2	1.9738				
$C_8H_{18}O$	1-Octanol	293.2	10.30	0.51647E+02	-0.20371E+00	0.21320E-03	258-513
$C_8H_{18}O$	2-Octanol	293.2	8.13	0.63760E+02	-0.27643E+00	0.31075E-03	213-513
$C_8H_{18}O$	3-Octanol	293.2	5.55	0.12505E+03	-0.70646E+00	0.10245E-02	223-383
$C_8H_{18}O$	4-Octanol	293.2	4.48	0.51049E+02	-0.26664E+00	0.37280E-03	243-403
$C_8H_{18}O$	2-Methyl-1-heptanol	293.1	5.16	0.61698E+02	-0.33647E+00	0.49066E-03	236-328
$C_8H_{18}O$	3-Methyl-1-heptanol	290.3	2.884	0.84687E+01	-0.33712E-01	0.49793E-04	241-316
$C_8H_{18}O$	4-Methyl-1-heptanol	290.6	4.63	0.48612E+02	-0.26773E+00	0.39972E-03	237-332
$C_8H_{18}O$	5-Methyl-1-heptanol	290.4	7.68	0.54581E+02	-0.24772E+00	0.29734E-03	235-328
$C_8H_{18}O$	6-Methyl-1-heptanol	290.3	10.54	0.57997E+02	-0.23517E+00	0.24663E-03	265-328
$C_8H_{18}O$	2-Methyl-2-heptanol	292.2	3.43				
$C_8H_{18}O$	3-Methyl-2-heptanol	289.6	7.47	0.39178E+02	-0.17976E+00	0.24218E-03	229-329
$C_8H_{18}O$	4-Methyl-2-heptanol	290.0	3.59	0.39715E+02	-0.23115E+00	0.36771E-03	240-333
$C_8H_{18}O$	5-Methyl-2-heptanol	278.5	7.5	0.68568E+02	-0.40706E+00	0.67433E-03	230-279
$C_8H_{18}O$	6-Methyl-2-heptanol	290.1	6.41	0.77520E+02	-0.41724E+00	0.59448E-03	239-329
$C_8H_{18}O$	2-Methyl-3-heptanol	293.2	3.260	-0.59739E+01	0.56700E-01	-0.83125E-04	343-403
$C_8H_{18}O$	3-Methyl-3-heptanol	293.2	3.013	-0.38440E+01	0.42327E-01	-0.61250E-04	343-403
$C_8H_{18}O$	4-Methyl-3-heptanol	293.2	3.312	-0.48003E+01	0.50740E-01	-0.75000E-04	343-403
$C_8H_{18}O$	5-Methyl-3-heptanol	293.2	3.832	0.61967E+01	-0.63750E-02		343-383
$C_8H_{18}O$	6-Methyl-3-heptanol	293.2	4.992	0.23037E+02	-0.98029E-01	0.12479E-03	283-383
$C_8H_{18}O$	2-Methyl-4-heptanol	296.3	3.338	0.42102E+00	0.10427E-01	-0.20438E-05	230-333
$C_8H_{18}O$	3-Methyl-4-heptanol	290.0	7.46	0.33354E+02	-0.14077E+00	0.17750E-03	230-330
$C_8H_{18}O$	4-Methyl-4-heptanol	296.2	2.902				
$C_8H_{18}O$	2-Ethyl-1-hexanol	298.2	7.58	0.86074E+02	-0.42636E+00	0.55078E-03	208-318
$C_8H_{18}O$	2,2-Dimethyl-1-hexanol	293.2	4.50	0.91244E+01	-0.21785E-01	0.21018E-04	283-393
$C_8H_{18}O$	Dibutyl ether	293.2	3.0830	0.65383E+01	-0.16172E-01	0.14969E-04	293-314
$C_8H_{18}OS$	Dibutyl sulfoxide	313.2	24.73	0.67156E+02	-0.16448E+00	0.92275E-04	313-393
$C_8H_{18}O_2$	2-Ethyl-1,3-hexanediol	293.2	18.73	0.57919E+02	-0.17128E+00	0.12949E-03	233-333
$C_8H_{18}O_2S$	Dibutyl sulfone	323.2	25.72	0.66248E+02	-0.16417E+00	0.12001E-03	323-398
$C_8H_{18}O_4$	Triethylene glycol dimethyl ether	298.2	7.62				
$C_8H_{18}O_5$	Tetraethylene glycol	293.2	20.44	0.83547E+02	-0.31691E+00	0.34689E-03	253-333
$C_8H_{18}S$	1-Octanethiol	293.2	3.949	0.63667E+01	-0.87920E-02	0.18750E-05	273-333
$C_8H_{18}S$	Dibutyl sulfide	298.2	4.29				
$C_8H_{19}N$	Octylamine	293.2	3.58	0.77931E+01	-0.20015E-01	0.19347E-04	273-373
$C_8H_{19}N$	Dibutylamine	293.2	2.765	0.52504E+01	-0.10538E-01	0.71485E-05	243-323
$C_8H_{20}O_4Si$	Ethyl silicate	293.2	2.50				
$C_8H_{20}Si$	Tetraethylsilane	293.2	2.090				
$C_8H_{20}Sn$	Tetraethylstannane	293.2	2.241				
$C_8H_{23}N_5$	Tetraethylenepentamine	293.2	9.40	0.40553E+02	-0.16681E+00	0.20659E-03	213-333
$C_8H_{24}O_4Si_4$	Octamethylcyclotetrasiloxane	296.2	2.390	0.36286E+01	-0.56885E-02	0.50874E-05	296-333
$C_9H_6N_2O_2$	Toluene-2,4-diisocyanate	293.2	8.433	0.22174E+02	-0.66982E-01	0.68571E-04	293-353
$C_9H_6O_2$	2*H*-1-Benzopyran-2-one	343.2	34.04	0.11311E+03	-0.33804E+00	0.31324E-03	343-423
C_9H_7N	Quinoline	293.2	9.16	0.33432E+02	-0.13497E+00	0.17788E-03	258-323
C_9H_7N	Isoquinoline	298.2	11.0	0.14412E+03	-0.79935E+00	0.11839E-02	298-323
C_9H_8O	Cinnamaldehyde	305.8	17.72	0.41837E+02	-0.11060E+00	0.10401E-03	306-354
$C_9H_8O_4$	2-(Acetyloxy)benzoic acid	333.2	6.55	0.69994E+01	-0.14553E-02		333-416
C_9H_{10}	1-Propenylbenzene	293.2	2.73				
C_9H_{10}	Allylbenzene	293.2	2.63				
C_9H_{10}	Isopropenylbenzene	293.2	2.28				

Mol. Form.	Name	T/K	ε	a	b	c	Range/K
$C_9H_{10}OS$	4-Acetylthioanisole	355.2	11.34				
$C_9H_{10}O_2$	Ethyl benzoate	293.2	6.20	0.18216E+02	-0.62361E-01	0.72884E-04	288-343
$C_9H_{10}O_2$	Methyl 4-methylbenzoate	306.2	4.3				
$C_9H_{10}O_2$	Benzyl acetate	303.2	5.34	0.11727E+02	-0.30869E-01	0.32340E-04	303-358
$C_9H_{10}O_2$	Phenyl propanoate	293.2	4.77				
$C_9H_{10}O_2$	4-Acetylanisole	313.2	17.3				
$C_9H_{10}O_3$	Ethyl salicylate	308.2	8.48	0.18910E+02	-0.35623E-01	0.46529E-05	225-321
$C_9H_{10}O_3$	Methyl 2-methoxybenzoate	294.2	7.7				
$C_9H_{11}Br$	(3-Bromopropyl)benzene	302.2	5.41	0.11360E+02	-0.27471E-01	0.25775E-04	302-358
$C_9H_{11}NO$	*N*-Ethylbenzamide	352.7	42.6	-0.20109E+03	0.17866E+01	-0.31065E-02	353-389
$C_9H_{11}NO$	*N,N*-Dimethylbenzamide	318.2	20.77	0.76725E+02	-0.26908E+00	0.29409E-03	318-443
$C_9H_{11}NO_2$	Ethyl 2-aminobenzoate	298.2	4.14				
C_9H_{12}	Propylbenzene	293.2	2.370	0.26933E+01	0.21679E-03	-0.44643E-05	273-323
C_9H_{12}	Isopropylbenzene	293.2	2.381	0.31149E+01	-0.30801E-02	0.19643E-05	273-323
C_9H_{12}	*o*-Ethyltoluene	293.2	2.595				
C_9H_{12}	*m*-Ethyltoluene	293.2	2.365				
C_9H_{12}	*p*-Ethyltoluene	293.2	2.265				
C_9H_{12}	1,2,3-Trimethylbenzene	293.2	2.656	0.76006E+01	-0.29118E-01	0.41786E-04	273-323
C_9H_{12}	1,2,4-Trimethylbenzene	293.2	2.377	0.31517E+01	-0.30634E-02	0.14286E-05	273-323
C_9H_{12}	1,3,5-Trimethylbenzene	293.2	2.279	0.38998E+01	-0.88072E-02	0.11149E-04	288-358
$C_9H_{12}O$	Benzenepropanol	293.2	11.97	0.94482E+02	-0.45540E+00	0.59307E-03	213-303
$C_9H_{12}O$	α-Ethylbenzenemethanol	293.2	6.68	0.44520E+02	-0.21505E+00	0.29443E-03	233-373
$C_9H_{12}O$	α,α-Dimethylbenzenemethanol	303.2	5.61	0.57072E+01	0.86568E-02	-0.29580E-04	303-373
$C_9H_{12}O$	1-Phenyl-2-propanol	293.2	9.35	0.10762E+03	-0.56026E+00	0.76915E-03	233-373
$C_9H_{12}O$	Benzyl ethyl ether	298.2	3.90				
$C_9H_{12}O$	2,6-Dimethylanisole	293.2	3.780	0.76700E+01	-0.18298E-01	0.17143E-04	293-333
$C_9H_{12}O$	3,5-Dimethylanisole	293.2	3.711	0.54981E+01	-0.56651E-02	-0.14286E-05	293-333
$C_9H_{12}O_2S$	Butyl thiophene-2-carboxylate	293.2	6.40				
$C_9H_{12}S$	Benzenepropanethiol	303.2	4.36	0.82411E+01	-0.15034E-01	0.73617E-05	303-358
$C_9H_{13}N$	Benzylethylamine	293.2	4.3				
$C_9H_{13}N$	*N*-Propylaniline	293.2	5.48				
$C_9H_{13}N$	2-Methyl-*N,N*-dimethylaniline	293.2	3.4				
$C_9H_{13}N$	4-Methyl-*N,N*-dimethylaniline	293.2	3.9				
$C_9H_{14}OSi$	Trimethylphenoxysilane	298.2	3.3953				
$C_9H_{14}O_6$	Triacetin	293.6	7.11	0.17819E+02	-0.53656E-01	0.57759E-04	219-304
$C_9H_{14}Si$	Trimethylphenylsilane	298.2	2.3533	0.21463E+01	0.32711E-02	-0.86264E-05	288-323
$C_9H_{16}O_2$	2-Nonenoic acid	296.2	2.5				
$C_9H_{16}O_2$	Cyclohexyl propanoate	293.2	4.82				
$C_9H_{16}O_2$	Ethyl cyclohexanecarboxylate	293.2	4.64				
$C_9H_{16}O_4$	Diethyl glutarate	303.2	6.659				
$C_9H_{17}N$	Nonanenitrile	293.2	12.08				
C_9H_{18}	1-Nonene	293.2	2.180	0.22710E+01	0.15797E-02	-0.64286E-05	273-323
$C_9H_{18}Br_2$	1,9-Dibromononane	293.2	7.153	0.18931E+02	-0.57764E-01	0.60000E-04	293-343
$C_9H_{18}O$	2-Nonanone	295.2	9.14				
$C_9H_{18}O$	5-Nonanone	293.2	10.6				
$C_9H_{18}O$	Di-*tert*-butyl ketone	287.65	10.0				
$C_9H_{18}O$	2,6-Dimethyl-4-heptanone	293.2	9.91	0.33178E+02	-0.11290E+00	0.11454E-03	273-393
$C_9H_{18}O_2$	Nonanoic acid	294.9	2.475	0.25039E+01	0.67274E-03	-0.24180E-05	295-365
$C_9H_{18}O_2$	2-Methyloctanoic acid	293.2	2.39				
$C_9H_{18}O_2$	2-Ethylheptanoic acid	293.2	1.98				
$C_9H_{18}O_2$	Heptyl acetate	293.2	4.2				
$C_9H_{18}O_2$	Pentyl butanoate	301.2	4.08	0.59029E+01	-0.49905E-02	-0.34292E-05	301-343
$C_9H_{18}O_2$	Isopentyl butanoate	293.2	4.0				
$C_9H_{18}O_2$	Isobutyl pentanoate	292.2	3.8				
$C_9H_{18}O_2$	Methyl octanoate	293.2	4.101				
$C_9H_{19}Br$	1-Bromononane	298.2	4.74	0.79870E+01	-0.10488E-01	-0.13450E-05	274-328
$C_9H_{19}Cl$	1-Chlorononane	293.2	4.803	0.95528E+01	-0.16200E-01	-0.16365E-13	293-323
$C_9H_{19}NO$	*N,N*-Dibutylformamide	293.2	18.4				
C_9H_{20}	Nonane	293.2	1.9722	0.23894E+01	-0.14830E-02	0.14881E-06	253-393

Mol. Form.	Name	*T*/K	ε	*a*	*b*	*c*	Range/K
C_9H_{20}	2-Methyloctane	293.2	1.967				
C_9H_{20}	4-Methyloctane	293.2	1.967				
C_9H_{20}	2,4-Dimethylheptane	293.2	1.89				
C_9H_{20}	2,5-Dimethylheptane	293.2	1.89				
C_9H_{20}	2,6-Dimethylheptane	293.2	1.987				
$C_9H_{20}N_2O$	Tetraethylurea	296.8	14.29	0.52820E+02	-0.18790E+00	0.19580E-03	205-411
$C_9H_{20}O$	1-Nonanol	293.2	8.83	0.97467E+02	-0.51103E+00	0.71429E-03	288-343
$C_9H_{20}O$	2-Nonanol	298.2	6.66	0.10136E+03	-0.55612E+00	0.80000E-03	288-308
$C_9H_{20}O$	3-Nonanol	298.2	4.49	0.55214E+02	-0.31920E+00	0.50000E-03	288-308
$C_9H_{20}O$	4-Nonanol	298.2	3.69	0.27954E+01	0.30000E-02	-0.52375E-13	288-308
$C_9H_{20}O$	5-Nonanol	298.2	3.54	-0.25463E+01	0.35320E-01	-0.50000E-04	288-308
$C_9H_{21}B$	Tripropylborane	293.2	2.026				
$C_9H_{21}N$	Nonylamine	293.2	3.42	0.53575E+01	-0.71982E-02	0.19481E-05	293-373
$C_9H_{21}N$	Tripropylamine	293.2	2.380	0.33380E+01	-0.86332E-02	0.18322E-04	243-293
$C_9H_{21}O_4P$	Tripropyl phosphate	293.2	10.93	0.33166E+02	-0.10514E+00	0.10000E-03	293-373
$C_{10}H_7Br$	1-Bromonaphthalene	298.2	4.768	0.10561E+02	-0.27671E-01	0.27655E-04	293-323
$C_{10}H_7Cl$	1-Chloronaphthalene	298.2	5.04	0.84861E+01	-0.12357E-01	0.26899E-05	274-328
$C_{10}H_7NO_2$	1-Nitronaphthalene	333.2	19.68	0.36267E+02	-0.41283E-01	-0.25595E-04	333-403
$C_{10}H_8$	Naphthalene	363.2	2.54				
$C_{10}H_8O$	1-Naphthol	373.0	5.03	0.16489E+02	-0.46700E-01	0.42857E-04	373-453
$C_{10}H_8O$	2-Naphthol	413.0	4.95	0.92865E+01	-0.10500E-01	0.42501E-15	413-453
$C_{10}H_9N$	1-Naphthylamine	333.2	5.20	0.10577E+02	-0.22114E-01	0.17857E-04	333-453
$C_{10}H_9N$	2-Naphthylamine	393.0	5.26	0.19722E+02	-0.60679E-01	0.60714E-04	393-473
$C_{10}H_9N$	2-Methylquinoline	293.2	7.24	0.11688E+02	-0.78400E-02	-0.25000E-04	293-333
$C_{10}H_9N$	4-Methylquinoline	293.2	9.31	0.17788E+02	-0.32580E-01	0.12500E-04	293-333
$C_{10}H_9N$	6-Methylquinoline	293.2	8.48	0.21696E+02	-0.63400E-01	0.62500E-04	293-333
$C_{10}H_9N$	8-Methylquinoline	293.2	6.58	0.19356E+02	-0.61900E-01	0.62500E-04	293-333
$C_{10}H_{10}O_4$	Methyl 2-(acetyloxy)benzoate	328.9	5.31	0.19579E+02	-0.69970E-01	0.80889E-04	329-371
$C_{10}H_{10}O_4$	Dimethyl phthalate	293.2	8.66				
$C_{10}H_{12}$	1,2,3,4-Tetrahydronaphthalene	298.2	2.771	0.29172E+01	0.12832E-02	-0.59453E-05	298-343
$C_{10}H_{12}$	4-Ethylstyrene	298.2	3.350				
$C_{10}H_{12}$	Dicyclopentadiene	313.2	2.43	0.30564E+01	-0.20000E-02	0.82443E-15	313-373
$C_{10}H_{12}O$	Tetrahydro-2-naphthol*	293.2	11.70	0.98978E+02	-0.48267E+00	0.63008E-03	293-363
$C_{10}H_{12}O$	4-Isopropylbenzaldehyde	288.2	10.68				
$C_{10}H_{12}O_2$	4-Allyl-2-methoxyphenol	293.2	9.55	0.52377E+02	-0.24380E+00	0.33333E-03	273-323
$C_{10}H_{12}O_2$	2-Phenylethyl acetate	297.2	4.93				
$C_{10}H_{12}O_2$	Benzyl propanoate	303.0	5.11	0.42301E+01	0.13962E-01	-0.36426E-04	303-358
$C_{10}H_{12}O_2$	Phenyl butanoate	293.2	4.48				
$C_{10}H_{12}O_2$	Propyl benzoate	303.2	5.78	0.10927E+02	-0.20535E-01	0.11745E-04	303-358
$C_{10}H_{12}O_2$	Ethyl phenylacetate	293.2	5.320				
$C_{10}H_{14}$	Butylbenzene	293.2	2.359				
$C_{10}H_{14}$	*sec*-Butylbenzene	293.2	2.357	0.28348E+01	-0.68586E-03	-0.32143E-05	273-323
$C_{10}H_{14}$	*tert*-Butylbenzene	293.2	2.359	0.27924E+01	-0.38350E-03	-0.37500E-05	273-323
$C_{10}H_{14}$	Isobutylbenzene	293.2	2.318	0.28055E+01	-0.92614E-03	-0.25000E-05	273-323
$C_{10}H_{14}$	1-Isopropyl-4-methylbenzene	298.2	2.2322	0.25266E+01	-0.25121E-03	-0.24867E-05	277-333
$C_{10}H_{14}$	*o*-Diethylbenzene	293.2	2.594				
$C_{10}H_{14}$	*m*-Diethylbenzene	293.2	2.369				
$C_{10}H_{14}$	*p*-Diethylbenzene	293.2	2.259				
$C_{10}H_{14}$	1-Ethyl-3,5-dimethylbenzene	293.2	2.275				
$C_{10}H_{14}$	1,2,3,4-Tetramethylbenzene	296.0	2.538	0.33822E+01	-0.33630E-02	0.17475E-05	273-412
$C_{10}H_{14}$	1,2,4,5-Tetramethylbenzene	356.0	2.223	0.26834E+01	-0.10327E-02	-0.73533E-06	356-430
$C_{10}H_{14}N_2$	*L*-Nicotine	293.2	8.937	0.21347E+02	-0.57177E-01	0.50655E-04	293-363
$C_{10}H_{14}O$	1-Phenyl-2-methyl-2-propanol	298.2	5.71	0.21922E+02	-0.84231E-01	0.99475E-04	298-423
$C_{10}H_{14}O$	Butyl phenyl ether	293.2	3.734				
$C_{10}H_{14}O$	Thymol	333.2	4.259				
$C_{10}H_{15}N$	*N,N*-Diethylaniline	303.2	5.15	0.50773E+01	0.15399E-01	-0.50000E-04	303-328
$C_{10}H_{16}$	γ-Terpinene	298.2	2.2738				
$C_{10}H_{16}$	*d*-Limonene	298.2	2.3746				
$C_{10}H_{16}$	*l*-Limonene	298.2	2.3738				

Mol. Form.	Name	*T*/K	ε	*a*	*b*	*c*	Range/K
$C_{10}H_{16}$	Terpinolene	298.2	2.2918				
$C_{10}H_{16}$	α-Pinene	298.2	2.1787				
$C_{10}H_{16}$	β-Pinene	298.2	2.4970				
$C_{10}H_{16}$	α-Terpinene	298.2	2.4526				
$C_{10}H_{16}$	β-Myrcene	298.2	2.3				
$C_{10}H_{16}O$	Carvenone	293.2	18.8				
$C_{10}H_{16}O$	*d*-Fenchone	294.2	12.8				
$C_{10}H_{17}Cl$	2-Chlorobornane	368.2	5.21				
$C_{10}H_{18}$	Pinane	298.2	2.1456				
$C_{10}H_{18}$	*cis*-Decahydronaphthalene	293.2	2.219	0.25410E+01	-0.11420E-02	0.15092E-06	293-373
$C_{10}H_{18}$	*trans*-Decahydronaphthalene	293.2	2.184	0.26615E+01	-0.21241E-02	0.16864E-05	293-373
$C_{10}H_{18}O$	Eucalyptol	298.2	4.57				
$C_{10}H_{18}O_2$	Cyclohexyl butanoate	293.2	4.58				
$C_{10}H_{18}O_4$	Diethyl adipate	293.2	6.109	0.14824E+02	-0.40749E-01	0.37600E-04	293-343
$C_{10}H_{20}$	1-Decene	293.2	2.136	0.19091E+01	0.33442E-02	-0.87500E-05	273-323
$C_{10}H_{20}$	*cis*-5-Decene	298.2	2.071				
$C_{10}H_{20}$	*trans*-5-Decene	298.2	2.030				
$C_{10}H_{20}$	5-Methyl-4-nonene	293.2	2.175				
$C_{10}H_{20}$	2,4,6-Trimethyl-3-heptene	293.2	2.293				
$C_{10}H_{20}Br_2$	1,10-Dibromodecane	303.2	6.56	0.17350E+02	-0.50328E-01	0.48633E-04	303-368
$C_{10}H_{20}Cl_2$	1,10-Dichlorodecane	308.2	6.68	-0.57423E+01	0.94220E-01	-0.17500E-03	308-338
$C_{10}H_{20}O$	2-Decanone	287.2	8.3				
$C_{10}H_{20}O$	Menthol	309.3	3.90	0.68202E+01	-0.15894E-01	0.20837E-04	309-358
$C_{10}H_{20}O_2$	2,2-Dimethyloctanoic acid	296.2	2.8				
$C_{10}H_{20}O_2$	Octyl acetate	288.2	4.18	-0.34691E+01	0.58106E-01	-0.10952E-03	288-323
$C_{10}H_{20}O_2$	2-Methylheptyl acetate	288.2	4.27	0.23285E+02	-0.11538E+00	0.17143E-03	288-323
$C_{10}H_{20}O_2$	Pentyl pentanoate	305.6	4.076	0.77641E+01	-0.14335E-01	0.73740E-05	306-393
$C_{10}H_{20}O_2$	Isopentyl pentanoate	292.2	3.6				
$C_{10}H_{20}O_2$	Isopentyl isopentanoate	288.2	4.39	0.14698E+02	-0.57726E-01	0.76190E-04	288-323
$C_{10}H_{20}O_2$	Methyl nonanoate	293.2	3.943				
$C_{10}H_{21}Br$	1-Bromodecane	298.2	4.44	0.11202E+02	-0.33491E-01	0.36314E-04	274-328
$C_{10}H_{21}Cl$	1-Chlorodecane	293.2	4.581	0.68741E+01	-0.12210E-02	-0.22500E-04	293-323
$C_{10}H_{21}NO$	*N,N*-Dibutylacetamide	293.2	19.1				
$C_{10}H_{22}$	Decane	293.2	1.9853	0.24054E+01	-0.15445E-02	0.44643E-06	253-393
$C_{10}H_{22}$	2,7-Dimethyloctane	293.2	1.98				
$C_{10}H_{22}$	4-Propylheptane	293.2	1.9955				
$C_{10}H_{22}O$	1-Decanol	293.2	7.93	0.47195E+02	-0.20740E+00	0.24942E-03	293-343
$C_{10}H_{22}O$	2-Decanol	298.2	5.82	0.13621E+03	-0.81000E+00	0.12500E-02	288-308
$C_{10}H_{22}O$	3-Decanol	298.2	4.05	0.52090E+02	-0.31020E+00	0.50000E-03	288-308
$C_{10}H_{22}O$	4-Decanol	298.2	3.42	-0.11260E+02	0.93960E-01	-0.15000E-03	288-308
$C_{10}H_{22}O$	5-Decanol	298.2	3.24	-0.25832E+01	0.31456E-01	-0.40000E-04	288-308
$C_{10}H_{22}O$	2,2-Dimethyl-1-octanol	293.2	7.86	0.69536E+02	-0.34596E+00	0.46250E-03	293-333
$C_{10}H_{22}O$	Dipentyl ether	298.2	2.798				
$C_{10}H_{22}O$	Diisopentyl ether	293.2	2.817	0.44690E+01	-0.63710E-02	0.25000E-05	293-323
$C_{10}H_{22}OS$	Dipentyl sulfoxide	348.2	18.8				
$C_{10}H_{22}O_5$	Tetraethylene glycol dimethyl ether	298.2	7.68				
$C_{10}H_{22}S$	Dipentyl sulfide	298.2	3.826				
$C_{10}H_{23}N$	Decylamine	293.2	3.31	0.61497E+01	-0.12801E-01	0.10606E-04	293-373
$C_{10}H_{30}O_3Si_4$	Decamethyltetrasiloxane	293.2	2.370				
$C_{10}H_{30}O_5Si_5$	Decamethylcyclopentasiloxane	293.2	2.50				
$C_{11}H_{10}$	1-Methylnaphthalene	293.2	2.915	0.45126E+01	-0.76480E-02	0.75000E-05	293-333
$C_{11}H_{10}$	2-Methylnaphthalene	313.2	2.747				
$C_{11}H_{10}O$	1-Methoxynaphthalene	293.2	4.020	0.71885E+01	-0.14838E-01	0.13750E-04	293-333
$C_{11}H_{10}O$	2-Methoxynaphthalene	353.2	3.563	0.56702E+01	-0.69754E-02	0.28571E-05	353-373
$C_{11}H_{12}O_2$	Ethyl *trans*-cinnamate	293.2	5.63				
$C_{11}H_{12}O_3$	Ethyl benzoylacetate	303.2	13.50	0.93644E+01	0.74280E-01	-0.20000E-03	303-323
$C_{11}H_{14}O_2$	Benzyl butanoate	301.2	4.55				
$C_{11}H_{14}O_2$	Phenyl pentanoate	293.2	4.30				
$C_{11}H_{14}O_2$	Butyl benzoate	303.2	5.52	0.77854E+01	-0.34972E-02	-0.13149E-04	303-358

Mol. Form.	Name	*T*/K	ε	*a*	*b*	*c*	Range/K
$C_{11}H_{14}O_2$	Isobutyl benzoate	291.2	5.39				
$C_{11}H_{16}$	1,3-Diethyl-5-methylbenzene	293.2	2.264				
$C_{11}H_{16}$	Pentamethylbenzene	334.0	2.358	0.30196E+01	-0.22619E-02	0.83831E-06	334-413
$C_{11}H_{22}$	1-Undecene	293.2	2.137	0.22132E+01	0.13121E-02	-0.53571E-05	273-323
$C_{11}H_{22}O$	2-Undecanone	285.3	8.3				
$C_{11}H_{22}O_2$	Nonyl acetate	293.2	3.87				
$C_{11}H_{22}O_2$	Pentyl hexanoate	288.2	4.22	0.83503E+01	-0.18449E-01	0.14286E-04	288-323
$C_{11}H_{23}Br$	1-Bromoundecane	272.6	4.61				
$C_{11}H_{24}$	Undecane	293.2	1.9972	0.23637E+01	-0.12500E-02	-0.85869E-16	283-363
$C_{11}H_{24}O$	1-Undecanol	313.2	5.98				
$C_{11}H_{25}N$	Undecylamine	293.2	3.25	0.54945E+01	-0.96161E-02	0.66017E-05	293-373
$C_{12}F_{27}N$	Tris(perfluorobutyl)amine	293.2	2.15				
$C_{12}H_8O$	Dibenzofuran	373.2	3.00				
$C_{12}H_{10}$	Biphenyl	348.2	2.53	0.26869E+01	0.63072E-03	-0.30995E-05	348-428
$C_{12}H_{10}N_2O$	*trans*-Azoxybenzene	311.2	5.2				
$C_{12}H_{10}O$	Diphenyl ether	283.2	3.726				
$C_{12}H_{10}O$	2-Acetonaphthone	333.2	13.03	0.14538E+03	-0.73040E+00	0.10000E-02	333-363
$C_{12}H_{10}OS$	Diphenyl sulfoxide	344.7	16.6				
$C_{12}H_{10}O_2S$	Diphenyl sulfone	406.2	21.1				
$C_{12}H_{10}S$	Diphenyl sulfide	298.2	5.43				
$C_{12}H_{11}N$	Diphenylamine	323.2	3.73				
$C_{12}H_{11}NO$	*N*-1-Naphthylenylacetamide	433.2	24.3	0.84739E+02	-0.12391E+00	-0.35714E-04	433-533
$C_{12}H_{12}$	1,6-Dimethylnaphthalene	293.2	2.7250				
$C_{12}H_{12}O$	1-Ethoxynaphthalene	292.2	3.3				
$C_{12}H_{14}O_2$	Propyl cinnamate	293.2	5.45				
$C_{12}H_{14}O_4$	Diethyl phthalate	293.2	7.86				
$C_{12}H_{16}O$	2-Cyclohexylphenol	328.2	3.97				
$C_{12}H_{16}O$	4-Cyclohexylphenol	404.2	4.42				
$C_{12}H_{16}O_2$	Pentyl benzoate	293.2	5.07				
$C_{12}H_{16}O_3$	Pentyl salicylate	301.2	6.25				
$C_{12}H_{16}O_3$	Isopentyl salicylate	293.12	7.26	0.13129E+02	-0.19190E-01	-0.36060E-05	225-397
$C_{12}H_{17}NO$	*N*-Butyl-*N*-phenylacetamide	298.2	11.66				
$C_{12}H_{18}$	Hexylbenzene	293.2	2.3				
$C_{12}H_{18}$	1,3,5-Triethylbenzene	293.2	2.256				
$C_{12}H_{18}$	Hexamethylbenzene	449.0	2.172	0.35710E+01	-0.46912E-02	0.35088E-05	449-489
$C_{12}H_{20}O_2$	*l*-Bornyl acetate	303.2	4.46	0.60791E+01	0.98200E-02	-0.50000E-04	303-323
$C_{12}H_{22}O$	Dicyclohexyl ether	293.2	3.45	0.95324E+01	-0.31740E-01	0.37500E-04	293-333
$C_{12}H_{22}O$	Cyclododecanone	303.2	11.4	0.39327E+02	-0.13248E+00	0.13298E-03	303-423
$C_{12}H_{22}O_6$	Dibutyl tartrate	314.2	9.4				
$C_{12}H_{24}$	1-Dodecene	293.2	2.152	0.22581E+01	0.11106E-02	-0.50000E-05	273-323
$C_{12}H_{24}O_2$	Decyl acetate	293.2	3.75				
$C_{12}H_{24}O_2$	Ethyl decanoate	293.2	3.75	0.70969E+01	-0.15080E-01	0.12500E-04	293-353
$C_{12}H_{24}O_2$	Methyl undecanoate	293.2	3.671				
$C_{12}H_{25}Br$	1-Bromododecane	298.2	4.07	0.86103E+01	-0.20891E-01	0.18994E-04	274-328
$C_{12}H_{25}Cl$	1-Chlorododecane	298.2	4.17	0.10002E+02	-0.27798E-01	0.27559E-04	274-328
$C_{12}H_{25}I$	1-Iodododecane	298.2	3.91	0.34641E+01	0.97404E-02	-0.27602E-04	293-323
$C_{12}H_{26}$	Dodecane	293.2	2.0120	0.23697E+01	-0.12200E-02	-0.36375E-16	283-363
$C_{12}H_{26}O$	1-Dodecanol	303.2	5.82	0.18518E+02	-0.44859E-01	0.99900E-05	303-358
$C_{12}H_{26}O$	2-Butyl-1-octanol	363.2	3.28				
$C_{12}H_{27}BO_3$	Tributyl borate	293.2	2.23				
$C_{12}H_{27}N$	Dodecylamine	303.2	3.07	0.27999E+01	0.44810E-02	-0.11905E-04	303-373
$C_{12}H_{27}N$	Tributylamine	293.2	2.340	0.19846E+01	0.28108E-02	-0.54545E-05	233-293
$C_{12}H_{27}O_4P$	Tributyl phosphate	293.2	8.34	0.26304E+02	-0.88480E-01	0.92857E-04	293-373
$C_{12}H_{28}O_4Si$	Tetrapropoxysilane	298.2	3.21				
$C_{12}H_{28}Sn$	Tetrapropylstannane	293.2	2.267				
$C_{12}H_{30}OSi_2$	Hexaethyldisiloxane	298.2	2.259	0.36559E+01	-0.72406E-02	0.85714E-05	298-333
$C_{13}H_{10}O$	Benzophenone	300.2	12.62	0.34130E+02	-0.10249E+00	0.10268E-03	300-420
$C_{13}H_{10}O_3$	Phenyl salicylate	290.2	6.92	0.26545E+02	-0.11180E+00	0.15220E-03	290-358
$C_{13}H_{12}$	Diphenylmethane	303.2	2.540	0.30638E+01	-0.17286E-02		303-333

Mol. Form.	Name	*T*/K	ε	*a*	*b*	*c*	Range/K
$C_{13}H_{12}O$	Benzyl phenyl ether	313.2	3.748				
$C_{13}H_{18}O_2$	Hexyl benzoate	293.2	4.80				
$C_{13}H_{20}$	Heptylbenzene	293.2	2.26				
$C_{13}H_{20}O$	α-Ionone*	292.4	10.78				
$C_{13}H_{20}O$	β-Ionone*	297.65	11.66				
$C_{13}H_{24}O_4$	Diethyl nonanedioate	303.2	5.133				
$C_{13}H_{26}$	1-Tridecene	293.2	2.139	0.14154E+01	0.66514E-02	-0.14286E-04	273-323
$C_{13}H_{26}O$	7-Tridecanone	303.2	7.6				
$C_{13}H_{26}O_2$	Ethyl undecanoate	293.2	3.55				
$C_{13}H_{26}O_2$	Methyl dodecanoate	293.2	3.539				
$C_{13}H_{27}Br$	1-Bromotridecane	281.15	4.19				
$C_{13}H_{28}$	Tridecane	293.2	2.0213	0.23731E+01	-0.12000E-02	-0.21841E-15	283-363
$C_{13}H_{28}$	5-Butylnonane	293.2	2.0319				
$C_{13}H_{28}O$	1-Tridecanol	333.2	4.02				
$C_{14}H_{10}$	Anthracene	502.0	2.649	0.20571E+02	-0.69169E-01	0.66667E-04	502-516
$C_{14}H_{10}$	Phenanthrene	383.2	2.72				
$C_{14}H_{10}O_2$	Benzil	368.2	13.04	-0.23599E+02	0.22715E+00	-0.34667E-03	368-393
$C_{14}H_{12}O_2$	Benzyl benzoate	303.2	5.26	0.76856E+01	-0.80000E-02	-0.80361E-15	303-358
$C_{14}H_{12}O_3$	Benzyl salicylate	301.2	4.12				
$C_{14}H_{14}$	1,2-Diphenylethane	331.2	2.47	0.31178E+01	-0.21572E-02	0.59800E-06	331-451
$C_{14}H_{14}O$	Dibenzyl ether	293.2	3.821	0.80154E+01	-0.20536E-01	0.21250E-04	293-333
$C_{14}H_{15}N$	Dibenzylamine	293.2	3.446				
$C_{14}H_{16}O_2Si$	Dimethyldiphenoxysilane	298.2	3.500	0.51669E+01	-0.77001E-02	0.70156E-05	283-353
$C_{14}H_{18}O_2$	Pentyl cinnamate	293.2	4.89				
$C_{14}H_{22}$	Octylbenzene	293.2	2.26				
$C_{14}H_{26}O_4$	Diisobutyl adipate	293.2	5.19				
$C_{14}H_{26}O_4$	Diethyl sebacate	303.2	4.995	0.39143E+02	-0.20965E+00	0.32000E-03	303-313
$C_{14}H_{28}O_2$	Dodecyl acetate	293.2	3.6				
$C_{14}H_{28}O_2$	Ethyl laurate	273.2	3.94				
$C_{14}H_{28}O_2$	Methyl tridecanoate	293.2	3.442				
$C_{14}H_{29}Br$	1-Bromotetradecane	293.2	3.84	0.10058E+02	-0.33905E-01	0.43528E-04	274-328
$C_{14}H_{30}$	Tetradecane	293.2	2.0343	0.23832E+01	-0.11900E-02	-0.51229E-16	283-363
$C_{14}H_{30}O$	1-Tetradecanol	318.2	4.42	0.12272E+02	-0.24667E-01	-0.13168E-13	318-358
$C_{14}H_{31}N$	Tetradecylamine	312.55	2.90				
$C_{15}H_{12}O_4$	Phenyl 2-(acetyloxy)benzoate	384.2	4.33				
$C_{15}H_{26}O_6$	Tributyrin	282.8	5.72	0.13152E+02	-0.36684E-01	0.36795E-04	199-283
$C_{15}H_{30}O_2$	Methyl tetradecanoate	293.2	3.352				
$C_{15}H_{31}Br$	1-Bromopentadecane	293.35	3.88				
$C_{15}H_{32}$	Pentadecane	293.2	2.0391	0.23792E+01	-0.11600E-02	-0.71069E-16	283-363
$C_{15}H_{32}O$	1-Pentadecanol	333.2	3.70				
$C_{15}H_{33}N$	Pentadecylamine	313.25	2.85				
$C_{15}H_{33}N$	Triisopentylamine	294.2	2.29				
$C_{16}H_{22}O_4$	Dibutyl phthalate	293.2	6.58	0.12444E+02	-0.20000E-01		293-333
$C_{16}H_{32}O_2$	Hexadecanoic acid	338.2	2.417				
$C_{16}H_{32}O_2$	Ethyl myristate	293.2	3.50	0.52642E+01	-0.60000E-02	-0.47358E-15	293-353
$C_{16}H_{32}O_2$	Methyl pentadecanoate	293.2	3.296				
$C_{16}H_{33}Br$	1-Bromohexadecane	298.2	3.68	0.58668E+01	-0.73333E-02	-0.52666E-14	298-328
$C_{16}H_{33}I$	1-Iodohexadecane	293.2	3.57	0.79531E+01	-0.22859E-01	0.26955E-04	293-323
$C_{16}H_{34}$	Hexadecane	293.2	2.0460	0.23861E+01	-0.11600E-02	0.25555E-15	293-363
$C_{16}H_{34}O$	1-Hexadecanol	333.2	3.69	0.85935E+01	-0.14714E-01	-0.45533E-13	333-363
$C_{16}H_{35}N$	Hexadecylamine	328.35	2.71				
$C_{16}H_{36}Sn$	Tetrabutylstannane	293.2	9.74	0.56115E+02	-0.24812E+00	0.30682E-03	293-313
$C_{17}H_{12}O_3$	2-Naphthyl salicylate	293.0	6.30	0.11229E+02	-0.18857E-01	0.70332E-05	293-353
$C_{17}H_{34}O$	9-Heptadecanone	328.2	5.43	0.44176E+02	-0.21183E+00	0.28571E-03	328-363
$C_{17}H_{34}O_2$	Methyl palmitate	313.2	3.124				
$C_{17}H_{36}$	Heptadecane	293.2	2.0578	0.23627E+01	-0.10400E-02	-0.10397E-12	293-308
$C_{17}H_{36}O$	1-Heptadecanol	333.2	3.41				
$C_{18}H_{26}O_4$	Dipentyl phthalate	293.2	6.00				
$C_{18}H_{28}O_2$	Phenyl laurate	293.2	3.28				

Mol. Form.	Name	*T*/K	ε	*a*	*b*	*c*	Range/K
$C_{18}H_{30}O_2$	Linolenic acid	293.2	2.825	0.33867E+01	-0.19181E-02		274-368
$C_{18}H_{30}O_4$	Dicyclohexyl adipate	308.2	4.84				
$C_{18}H_{32}O_2$	Linoleic acid	293.2	2.754	0.32073E+01	-0.15477E-02		275-368
$C_{18}H_{34}O_2$	Oleic acid	293.2	2.336	0.25385E+01	-0.69448E-03		275-368
$C_{18}H_{34}O_4$	Dibutyl sebacate	293.2	4.54				
$C_{18}H_{36}O_2$	Stearic acid	293.2	2.314	0.27159E+01	-0.13300E-02		293-373
$C_{18}H_{36}O_2$	Hexadecyl acetate	308.2	3.19	0.47310E+01	-0.50000E-02	0.41338E-14	308-348
$C_{18}H_{36}O_2$	Ethyl palmitate	303.2	3.07	0.57938E+01	-0.12294E-01	0.10919E-04	303-455
$C_{18}H_{36}O_2$	Methyl heptadecanoate	313.2	3.07				
$C_{18}H_{37}Br$	1-Bromooctadecane	303.35	3.53	0.46790E+01	-0.30355E-02	-0.24798E-05	303-332
$C_{18}H_{38}O$	1-Octadecanol	333.2	3.38	0.73784E+01	-0.12000E-01	-0.22871E-13	333-363
$C_{18}H_{39}BO_3$	Trihexyl borate	293.2	2.22				
$C_{18}H_{39}N$	Octadecylamine	326.35	2.67				
$C_{19}H_{16}$	Triphenylmethane	367.2	2.46	0.40201E+01	-0.66507E-02	0.65329E-05	367-448
$C_{19}H_{18}O_3Si$	Methyltriphenoxysilane	298.2	3.628				
$C_{19}H_{32}O_2$	Methyl linolenate	293.2	3.355				
$C_{19}H_{34}O_2$	Methyl linoleate	293.2	3.466				
$C_{19}H_{36}O_2$	Methyl oleate	293.2	3.211				
$C_{19}H_{38}O$	10-Nonadecanone	353.2	5.37				
$C_{19}H_{38}O_2$	Methyl stearate	313.2	3.021				
$C_{19}H_{40}$	Nonadecane	293.2	2.0706				
$C_{20}H_{30}O_4$	Dihexyl phthalate	293.2	5.62				
$C_{20}H_{38}O_2$	Ethyl oleate	301.2	3.17	0.57033E+01	-0.11223E-01	0.93447E-05	301-423
$C_{20}H_{40}O_2$	Octadecyl acetate	308.2	3.07	0.44569E+01	-0.45000E-02	0.33923E-14	308-348
$C_{20}H_{40}O_2$	Ethyl stearate	313.2	2.958	0.70930E+01	-0.19081E-01	0.19555E-04	331-440
$C_{20}H_{40}O_2$	Methyl nonadecanoate	313.2	2.982				
$C_{20}H_{42}O$	1-Eicosanol	338.2	3.13	0.21700E+01	0.12497E-01	-0.28571E-04	338-363
$C_{20}H_{42}O$	Didecyl ether	293.2	2.644	0.41465E+01	-0.62240E-02	0.37500E-05	293-333
$C_{20}H_{60}O_8Si_9$	Eicosamethylnonasiloxane	293.2	2.645	0.57840E+01	-0.16568E-01	0.20000E-04	293-323
$C_{21}H_{21}O_4P$	Tricresyl phosphate*	298.2	6.7				
$C_{21}H_{38}O_6$	1,2,3-Propanetriyl hexanoate	293.2	4.476				
$C_{22}H_{42}O_2$	Butyl oleate	298.2	4.00				
$C_{22}H_{44}O_2$	Butyl stearate	298.2	3.120	0.73894E+02	-0.46261E+00	0.75500E-03	298-343
$C_{22}H_{46}$	Docosane	293.2	2.0840				
$C_{22}H_{46}O$	1-Docosanol	348.2	2.94	0.82062E+01	-0.25069E-01	0.28571E-04	348-373
$C_{24}H_{20}O_4Si$	Tetraphenoxysilane	333.2	3.4915				
$C_{24}H_{38}O_4$	Dioctyl phthalate	293.2	5.22				
$C_{26}H_{50}O_4$	Dioctyl sebacate	299.2	4.01				
$C_{27}H_{50}O_6$	1,2,3-Propanetriyl octanoate	293.2	3.931				
$C_{30}H_{58}O_4$	Ethylene glycol ditetradecanoate	343.2	2.98				
$C_{30}H_{62}$	Triacontane	373.2	1.9112				
$C_{30}H_{62}$	2,6,10,15,19,23-Hexamethyltetracosane	373.2	1.9106				
$C_{34}H_{66}O_4$	Ethylene glycol dipalmitate	348.2	2.89				
$C_{34}H_{68}O_2$	Hexadecyl stearate	333.2	2.61				
$C_{38}H_{74}O_4$	Ethylene glycol distearate	353.2	2.79				
$C_{39}H_{74}O_6$	Glycerol trilaurate	313.2	3.287				
$C_{51}H_{98}O_6$	Glycerol tripalmitate	328.2	2.901	-0.29131E+01	0.32206E-01	-0.44154E-04	328-393
$C_{57}H_{104}O_6$	Glycerol trioleate	293.2	3.109				
$C_{57}H_{104}O_6$	Glycerol trielaidate	313.2	2.980				
$C_{57}H_{110}O_6$	Glycerol tristearate	353.2	2.740				

PERMITTIVITY (DIELECTRIC CONSTANT) OF GASES

This table gives the relative permittivity ε (often called the dielectric constant) of some common gases at a temperature of 20°C and pressure of one atmosphere (101.325 kPa). Values of the permanent dipole moment μ in Debye Units (1 D = 3.33564 × 10^{-30} C m) are also included.

The density dependence of the permittivity is given by the equation

$$\frac{\varepsilon-1}{\varepsilon+2}=\rho_m\left(\frac{4\pi N\alpha}{3}+\frac{4\pi N\mu^2}{9kT}\right)$$

where ρ_m is the molar density, N is Avogadro's number, k is the Boltzmann constant, T is the temperature, and α is the molecular polarizability. Therefore, in regions where the gas can be considered ideal, ε – 1 is approximately proportional to the pressure at constant temperature. For nonpolar gases (μ = 0), ε –1 is inversely proportional to temperature at constant pressure.

The number of significant figures indicates the accuracy of the values given. The values of ε for air, Ar, H_2, He, N_2, O_2, and CO_2 are recommended as reference values; these are accurate to 1 ppm or better.

The second part of the table gives the permittivity of water vapor in equilibrium with liquid water as a function of temperature (derived from Reference 4).

REFERENCE

1. A. A. Maryott and F. Buckley, *Table of Dielectric Constants and Electric Dipole Moments of Substances in the Gaseous State*, National Bureau of Standards Circular 537, 1953.
2. B. A. Younglove, *J. Phys. Chem. Ref. Data*, 11, Suppl. 1, 1982; 16, 577, 1987 (for data on N_2, H_2, O_2, and hydrocarbons over a range of pressure and temperature).
3. Landolt-Börnstein, *Numerical Data and Functional Relationships in Science and Technology*, New Series, Group IV, Vol. 4, Springer-Verlag, Heidelberg, 1980 (for data at high pressures).
4. G. Birnbaum and S. K. Chatterjee, *J. Appl. Phys.*, 23, 220, 1952 (for data on water vapor).

Mol. form.	Name	ε	μ/D
	Compounds not containing carbon		
	Air (dry,CO_2 free)	1.0005364	
Ar	Argon	1.0005172	0
BF_3	Boron trifluoride	1.0011	0
BrH	Hydrogen bromide	1.00279	0.827
ClH	Hydrogen chloride	1.00390	1.109
F_3N	Nitrogen trifluoride	1.0013	0.235
F_6S	Sulfur hexafluoride	1.00200	0
HI	Hydrogen iodide	1.00214	0.448
H_2	Hydrogen	1.0002538	0
H_2S	Hydrogen sulfide	1.00344	0.97
H_3N	Ammonia	1.00622	1.471
He	Helium	1.0000650	0
Kr	Krypton	1.00078	0
NO	Nitric oxide	1.00060	0.159
N_2	Nitrogen	1.0005480	0
N_2O	Nitrous oxide	1.00104	0.161
Ne	Neon	1.00013	0
O_2	Oxygen	1.0004947	0
O_2S	Sulfur dioxide	1.00825	1.633
O_3	Ozone	1.0017	0.534
Xe	Xenon	1.00126	0
	Compounds containing carbon		
CF_4	Tetrafluoromethane	1.00121	0
CO	Carbon monoxide	1.00065	0.110
CO_2	Carbon dioxide	1.000922	0

PERMITTIVITY (DIELECTRIC CONSTANT) OF GASES (continued)

Mol. form.	Name	ε	μ/D
CH_3Br	Bromomethane	1.01028	1.822
CH_3Cl	Chloromethane	1.01080	1.892
CH_3F	Fluoromethane	1.00973	1.858
CH_3I	Iodomethane	1.00914	1.62
CH_4	Methane	1.00081	0
C_2H_2	Acetylene	1.00124	0
C_2H_3Cl	Chloroethylene	1.0075	1.45
C_2H_4	Ethylene	1.00134	0
C_2H_5Cl	Chloroethane	1.01325	2.05
C_2H_6	Ethane	1.00140	0
C_2H_6O	Dimethyl ether	1.0062	1.30
C_3H_6	Propene	1.00228	0.366
C_3H_6	Cyclopropane	1.00178	0
C_3H_8	Propane	1.00200	0.084
C_4H_{10}	Butane	1.00258	0
C_4H_{10}	Isobutane	1.00260	0.132

PERMITTIVITY OF SATURATED WATER VAPOR

t/°C	ε	*t*/°C	ε
0	1.00007	60	1.00144
10	1.00012	70	1.00213
20	1.00022	80	1.00305
30	1.00037	90	1.00428
40	1.00060	100	1.00587
50	1.00095		

AZEOTROPIC DATA FOR BINARY MIXTURES

Liquid mixtures having an extremum (maximum or minimum) vapor pressure at constant temperature, as a function of composition, are called azeotropic mixtures, or simply azeotropes. Mixtures that do not show a maximum or minimum are called zeotropic. Azeotropes in which the pressure is a maximum are often called positive azeotropes, while pressure-minimum azeotropes are called negative azeotropes. The coordinates of an azeotropic point are the azeotropic temperature t_{az}, pressure P_{az}, and liquid-phase composition, usually expressed as mole fractions. At the azeotropic point, the vapor-phase composition is the same as the liquid-phase composition.

This table gives azeotropic data for a number of binary mixtures at normal atmospheric pressure (P_{az} =101.3 kPa). Component 1 of each mixture is given in bold face. The temperature t_{az} and mole fraction x_1 of component 1 are listed for each choice of component 2.

The components are arranged in a modified Hill order, with substances that do not contain carbon preceding those that do contain carbon.

REFERENCES

1. Lide, D.R., and Kehiaian, H.V., *CRC Handbook of Thermophysical and Thermochemical Data*, CRC Press, Boca Raton, FL, 1994.
2. Horsley, L.H., *Azeotropic Data, III*, American Chemical Society, Washington, D.C., 1973.

Molecular formula	Name	t_{az}/°C	x_1
	Water H_2O		
$CHCl_3$	Trichloromethane	56.1	0.160
CH_2O_2	Formic acid	107.2	0.427
CH_3NO_2	Nitromethane	83.6	0.511
CS_2	Carbon disulfide	42.6	0.109
C_2H_3N	Acetonitrile	76.5	0.307
$C_2H_5NO_2$	Nitroethane	87.2	0.624
C_2H_6O	Ethanol	78.2	0.096
$C_4H_8O_2$	Ethyl acetate	70.4	0.312
$C_4H_{10}O$	1-Butanol	92.7	0.753
$C_4H_{10}O$	2-Butanol	87	0.601
C_5H_5N	Pyridine	93.6	0.755
$C_5H_{11}N$	Piperidine	92.8	0.718
C_5H_{12}	Pentane	34.6	0.054
C_6H_5Cl	Chlorobenzene	90.2	0.712
C_6H_6	Benzene	69.3	0.295
C_6H_6O	Phenol	99.5	0.981
C_6H_{10}	Cyclohexene	70.8	0.308
C_6H_{12}	Cyclohexane	69.5	0.300
C_6H_{14}	Hexane	61.6	0.221
C_7H_8	Toluene	84.1	0.444
C_7H_{16}	Heptane	79.2	0.452
C_8H_{10}	1,3-Dimethylbenzene	92	0.767
C_8H_{10}	Ethylbenzene	92	0.744
C_8H_{18}	Octane	89.6	0.673
$C_8H_{18}O$	Dibutyl ether	92.9	0.781
C_9H_{20}	Nonane	94.8	0.970
$C_{12}H_{27}N$	Tributylamine	99.7	0.976
	Tetrachloromethane CCl_4		
CH_2O_2	Formic acid	66.7	0.569
CH_3NO_2	Nitromethane	71.3	0.660
CH_4O	Methanol	55.7	0.445
C_2H_3N	Acetonitrile	65.1	0.566
C_2H_6O	Ethanol	65.0	0.615
C_3H_6O	Acetone	56.1	0.047
C_3H_8O	1-Propanol	73.4	0.820
$C_4H_{10}O$	1-Butanol	76.6	0.951
	Formic acid CH_2O_2		
CS_2	Carbon disulfide	42.6	0.253
	Nitromethane CH_3NO_2		
CS_2	Carbon disulfide	41.2	0.845
	Methanol CH_4O		
C_3H_6O	Acetone	55.5	0.198

Molecular formula	Name	t_{az}/°C	x_1
$C_3H_6O_2$	Methyl acetate	53.5	0.352
C_5H_{10}	Cyclopentane	38.8	0.263
C_5H_{12}	Pentane	30.9	0.145
$C_5H_{12}O$	*tert*-Butyl methyl ether	51.3	0.315
C_6H_6	Benzene	57.5	0.610
C_6H_{12}	Cyclohexane	53.9	0.601
C_7H_8	Toluene	63.5	0.883
C_7H_{16}	Heptane	59.1	0.769
C_8H_{18}	Octane	62.8	0.881
C_9H_{20}	Nonane	64.1	0.953
	Carbon disulfide CS_2		
C_2H_6O	Ethanol	42.6	0.860
C_3H_6O	Acetone	39.3	0.608
C_3H_8O	1-Propanol	45.7	0.931
$C_4H_8O_2$	Ethyl acetate	46.1	0.974
	Acetonitrile C_2H_3N		
C_2H_6O	Ethanol	72.5	0.469
C_7H_8	Toluene	81.4	0.900
	Acetic acid $C_2H_4O_2$		
$C_4H_8O_2$	1,4-Dioxane	119.5	0.831
C_5H_5N	Pyridine	138.1	0.579
C_6H_6	Benzene	80.1	0.026
C_6H_{12}	Cyclohexane	78.8	0.130
C_6H_{14}	Hexane	68.3	0.084
$C_6H_{15}N$	Triethylamine	163	0.774
C_7H_8	Toluene	100.7	0.375
C_7H_{16}	Heptane	91.7	0.451
C_8H_{10}	Ethylbenzene	114.7	0.774
C_8H_{18}	Octane	105.7	0.688
C_9H_{20}	Nonane	112.9	0.826
	Iodoethane C_2H_5I		
C_6H_{14}	Hexane	64.7	0.420
	Ethanol C_2H_6O		
C_5H_{10}	Cyclopentane	44.7	0.110
C_5H_{12}	Pentane	34.3	0.076
C_6H_6	Benzene	67.9	0.440
C_6H_{12}	Cyclohexane	64.8	0.430
C_6H_{14}	Hexane	58.7	0.332
C_7H_8	Toluene	76.7	0.810
C_8H_{18}	Octane	77	0.898
	Ethylene glycol $C_2H_6O_2$		
C_7H_8	Toluene	110.1	0.034
C_7H_{16}	Heptane	97.9	0.048
$C_8H_{18}O$	Dibutyl ether	139.5	0.125
$C_{10}H_{22}$	Decane	161	0.406
	Dimethyl sulfide C_2H_6S		
C_5H_{12}	Pentane	31.8	0.503
	1,2-Ethanediamine $C_2H_8N_2$		
C_7H_8	Toluene	104	0.406
	Propanenitrile C_3H_5N		
C_6H_{14}	Hexane	63.5	0.134
	Acetone C_3H_6O		
$C_3H_6O_2$	Methyl acetate	55.8	0.544
C_5H_{10}	Cyclopentane	41	0.404
C_6H_{12}	Cyclohexane	53	0.751
	Ethyl formate $C_3H_6O_2$		
C_5H_{12}	Pentane	32.5	0.294
	Methyl acetate $C_3H_6O_2$		
C_6H_{12}	Cyclohexane	55.5	0.801

Molecular formula	Name	t_{az}/°C	x_1
C_6H_{14}	Hexane	51.8	0.642
	Propanoic acid $C_3H_6O_2$		
C_5H_5N	Pyridine	148.6	0.686
C_7H_{16}	Heptane	97.8	0.027
C_9H_{12}	Propylbenzene	139.5	0.830
	1-Nitropropane $C_3H_7NO_2$		
C_3H_8O	1-Propanol	97.0	0.061
C_7H_{16}	Heptane	96.6	0.149
	1-Propanol C_3H_8O		
$C_4H_8O_2$	1,4-Dioxane	95.3	0.642
C_6H_6	Benzene	77.1	0.209
C_6H_{12}	Cyclohexane	74.7	0.241
C_7H_{16}	Heptane	84.6	0.470
	2-Propanol C_3H_8O		
$C_4H_{11}N$	Butylamine	74.7	0.646
C_5H_{12}	Pentane	35.5	0.071
C_6H_{12}	Cyclohexane	69.4	0.397
C_7H_8	Toluene	80.6	0.773
	Ethyl methyl sulfide C_3H_8S		
C_6H_{12}	Methylcyclopentane	65.6	0.664
C_7H_{16}	2,2-Dimethylpentane	66.4	0.908
	1-Propanethiol C_3H_8S		
C_6H_{12}	Cyclohexane	67.8	0.978
C_6H_{14}	Hexane	64.4	0.557
$C_6H_{14}O$	Diisopropyl ether	65.9	0.714
	Thiophene C_4H_4S		
C_6H_{12}	Cyclohexane	77.9	0.412
C_6H_{14}	Hexane	68.5	0.114
	Butanal C_4H_8O		
C_6H_{14}	Hexane	60	0.296
	2-Butanone C_4H_8O		
C_4H_9Cl	1-Chlorobutane	77	0.440
$C_4H_{11}N$	Butylamine	74	0.353
C_6H_6	Benzene	78.3	0.460
C_6H_{12}	Cyclohexane	71.8	0.438
C_7H_{16}	Heptane	77	0.764
	Butanoic acid $C_4H_8O_2$		
C_5H_5N	Pyridine	163.2	0.912
C_6H_5Cl	Chlorobenzene	131.8	0.035
C_8H_{10}	1,2-Dimethylbenzene	143	0.118
	1,4-Dioxane $C_4H_8O_2$		
C_4H_9Br	1-Bromobutane	98	0.580
	Ethyl acetate $C_4H_8O_2$		
C_6H_{14}	Hexane	65.2	0.394
	Methyl propanoate $C_4H_8O_2$		
C_4H_9Cl	1-Chlorobutane	76.8	0.392
	Propyl formate $C_4H_8O_2$		
C_4H_9Cl	1-Chlorobutane	76.1	0.392
C_6H_6	Benzene	78.5	0.440
C_6H_{12}	Cyclohexane	75	0.469
	1-Butanol $C_4H_{10}O$		
C_5H_5N	Pyridine	118.6	0.704
C_6H_5Cl	Chlorobenzene	115.3	0.659
C_6H_{10}	Cyclohexene	82	0.055
C_7H_8	Toluene	105.5	0.324
C_7H_{16}	Heptane	93.9	0.229
C_8H_{10}	1,2-Dimethylbenzene	116.8	0.811
$C_8H_{18}O$	Dibutyl ether	117.7	0.892

Molecular formula	Name	t_{az}/°C	x_1
	2-Butanol $C_4H_{10}O$		
C_6H_6	Benzene	78.5	0.161
C_7H_{16}	Heptane	88.1	0.439
	Diethyl ether $C_4H_{10}O$		
C_5H_{12}	Pentane	33.7	0.553
	***tert*-Butyl alcohol $C_4H_{10}O$**		
C_6H_6	Benzene	74.0	0.378
C_7H_{16}	Heptane	78	0.688
	Methyl propyl ether $C_4H_{10}O$		
C_5H_{12}	Pentane	35.6	0.215
	2-Ethoxyethanol $C_4H_{10}O_2$		
C_7H_{16}	Heptane	96.5	0.153
C_9H_{12}	Propylbenzene	134.6	0.842
	2-Furaldehyde $C_5H_4O_2$		
C_7H_{16}	Heptane	98.3	0.055
C_9H_{12}	Propylbenzene	151.4	0.475
	Pyridine C_5H_5N		
C_7H_8	Toluene	110.1	0.249
	Benzene C_6H_6		
C_6H_{10}	Cyclohexene	78.9	0.635
C_6H_{12}	Cyclohexane	77.6	0.538
	Phenol C_6H_6O		
C_6H_7N	2-Methylpyridine	185.5	0.752
C_7H_9N	2,4-Dimethylpyridine	193.4	0.601
C_9H_{12}	1,3,5-Trimethylbenzene	163.5	0.253
$C_{10}H_{22}$	Decane	168	0.449
	Aniline C_6H_7N		
C_9H_{12}	1,3,5-Trimethylbenzene	164.4	0.150
$C_{10}H_{22}O$	Dipentyl ether	177.5	0.675
$C_{12}H_{26}$	Dodecane	180.4	0.821
	2-Methylpyridine C_6H_7N		
C_8H_{18}	Octane	121.1	0.470
	Cyclohexanol $C_6H_{12}O$		
C_8H_{10}	1,2-Dimethylbenzene	143	0.147

VISCOSITY OF GASES

The following table gives the viscosity of some common gases as a function of temperature. Unless otherwise noted, the viscosity values refer to a pressure of 100 kPa (1 bar). The notation P=0 indicates the low pressure limiting value is given. The difference between the viscosity at 100 kPa and the limiting value is generally less than 1%. Viscosity is given in units of μPa s; note that 1 μPa s = 10^{-5} poise. Substances are listed in the modified Hill order (see Introduction).

		Viscosity in micropascal seconds (μPa s)						
		100 K	200 K	300 K	400 K	500 K	600 K	Ref.
	Air	7.1	13.3	18.6	23.1	27.1	30.8	1
Ar	Argon	8.0	15.9	22.9	28.8	34.2	39.0	2,8
BF_3	Boron trifluoride		12.3	17.1	21.7	26.1	30.2	13
ClH	Hydrogen chloride			14.6	19.7	24.3		13
F_6S	Sulfur hexafluoride (P=0)			15.3	19.8	23.9	27.7	10
H_2	Hydrogen (P=0)	4.2	6.8	9.0	10.9	12.7	14.4	4
D_2	Deuterium (P=0)	5.9	9.6	12.6	15.4	17.9	20.3	11
H_2O	Water			10.0	13.3	17.3	21.4	6
D_2O	Deuterium oxide			11.1	13.7	17.7	22.0	7
He	Helium (P=0)	9.7	15.3	20.0	24.4	28.4	32.3	8
Kr	Krypton (P=0)	8.8	17.1	25.6	33.1	39.8	45.9	8
NO	Nitric oxide		13.8	19.2	23.8	28.0	31.9	13
N_2	Nitrogen (P=0)		12.9	17.9	22.2	26.1	29.6	12
N_2O	Nitrous oxide		10.0	15.0	19.4	23.6	27.4	13
Ne	Neon (P=0)	14.4	24.3	32.1	38.9	45.0	50.8	8
O_2	Oxygen (P=0)	7.5	14.6	20.8	26.1	30.8	35.1	12
O_2S	Sulfur dioxide		8.6	12.9	17.5	21.7		13
Xe	Xenon (P=0)	8.3	15.4	23.2	30.7	37.6	44.0	8
CO	Carbon monoxide	6.7	12.9	17.8	22.1	25.8	29.1	13
CO_2	Carbon dioxide		10.0	15.0	19.7	24.0	28.0	9,10
$CHCl_3$	Chloroform			10.2	13.7	16.9	20.1	13
CH_4	Methane		7.7	11.2	14.3	17.0	19.4	10
CH_4O	Methanol				13.2	16.5	19.6	13
C_2H_2	Acetylene			10.4	13.5	16.5		13
C_2H_4	Ethylene		7.0	10.4	13.6	16.5	19.1	3
C_2H_6	Ethane		6.4	9.5	12.3	14.9	17.3	5
C_2H_6O	Ethanol				11.6	14.5	17.0	13
C_3H_8	Propane			8.3	10.9	13.4	15.8	5
C_4H_{10}	Butane			7.5	10.0	12.3	14.6	5
C_4H_{10}	Isobutane			7.6	10.0	12.3	14.6	5
$C_4H_{10}O$	Diethyl ether			7.6	10.1	12.4		13
C_5H_{12}	Pentane			6.7	9.2	11.4	13.4	13
C_6H_{14}	Hexane				8.6	10.8	12.8	13

REFERENCES

1. K. Kadoya, N. Matsunaga, and A. Nagashima, Viscosity and thermal conductivity of dry air in the gaseous phase, *J. Phys. Chem. Ref. Data*, 14, 947, 1985.
2. B. A. Younglove and H. J. M. Hanley, The viscosity and thermal conductivity coefficients of gaseous and liquid argon, *J. Phys. Chem. Ref. Data*, 15, 1323, 1986.
3. P. M. Holland, B. E. Eaton, and H. J. M. Hanley, A Correlation of the viscosity and thermal conductivity data of gaseous and liquid ethylene, *J. Phys. Chem. Ref. Data*, 12, 917, 1983.
4. M. J. Assael, S. Mixafendi, and W. A. Wakeham, The viscosity and thermal conductivity of normal hydrogen in the limit zero density, *J. Phys. Chem. Ref. Data*, 15, 1315, 1986.
5. B. A. Younglove and J. F. Ely, Thermophysical properties of fluids. II. Methane, ethane, propane, isobutane, and normal butane, *J. Phys. Chem. Ref. Data*, 16, 577, 1987.

6. J. V. Sengers and J. T. R. Watson, Improved international formulations for the viscosity and thermal conductivity of water substance, *J. Phys. Chem. Ref. Data*, 15, 1291, 1986.
7. N. Matsunaga and A. Nagashima, Transport properties of liquid and gaseous D_2O over a wide range of temperature and pressure, *J. Phys. Chem. Ref. Data*, 12, 933, 1983.
8. J. Kestin, et al., Equilibrium and transport properties of the noble gases and their mixtures at low density, *J. Phys. Chem. Ref. Data*, 13, 299, 1984.
9. V. Vescovic, et al., The transport properties of carbon dioxide, *J. Phys. Chem. Ref. Data*, 19, 1990.
10. R. D. Trengove and W. A. Wakeham, The viscosity of carbon dioxide, methane, and sulfur hexafluoride in the limit of zero density, *J. Phys. Chem. Ref. Data*, 16, 175, 1987.
11. M. J. Assael, S. Mixafendi, and W. A. Wakeham, The viscosity of normal deuterium in the limit of zero density, *J. Phys. Chem. Ref. Data*, 16, 189, 1987.
12. W. A. Cole and W. A. Wakeham, The viscosity of nitrogen, oxygen, and their binary mixtures in the limit of zero density, *J. Phys. Chem. Ref. Data*, 14, 209, 1985.
13. C. Y. Ho, Ed., *Properties of Inorganic and Organic Fluids, CINDAS Data Series on Materials Properties*, Vol. V-1, Hemisphere Publishing Corp., New York, 1988.

VISCOSITY OF LIQUIDS

The absolute viscosity of some common liquids at temperatures between –25 and 100°C is given in this table. Values were derived by fitting experimental data to suitable expressions for the temperature dependence. The substances are arranged by molecular formula in the modified Hill order (see Preface). All values are given in units of millipascal seconds (mPa s); this unit is identical to centipoise (cp).

Viscosity values correspond to a nominal pressure of 1 atmosphere. If a value is given at a temperature above the normal boiling point, the applicable pressure is understood to be the vapor pressure of the liquid at that temperature. A few values are given at a temperature slightly below the normal freezing point; these refer to the supercooled liquid.

The accuracy ranges from 1% in the best cases to 5 to 10% in the worst cases. Additional significant figures are included in the table to facilitate interpolation.

REFERENCES

1. Viswanath, D. S. and Natarajan, G., *Data Book on the Viscosity of Liquids*, Hemisphere Publishing Corp., New York, 1989.
2. Daubert, T. E., Danner, R. P., Sibul, H. M., and Stebbins, C. C., *Physical and Thermodynamic Properties of Pure Compounds: Data Compilation*, extant 1994 (core with 4 supplements), Taylor & Francis, Bristol, PA (also available as database).
3. Ho, C. Y., Ed., *CINDAS Data Series on Material Properties*, Vol. V-1, *Properties of Inorganic and Organic Fluids*, Hemisphere Publishing Corp., New York, 1988.
4. Stephan, K. and Lucas, K., *Viscosity of Dense Fluids*, Plenum Press, New York, 1979.
5. Vargaftik, N. B., *Tables of Thermophysical Properties of Liquids and Gases*, 2nd ed., John Wiley, New York, 1975.

Molecular formula	Name	Viscosity in mPa s					
		–25°C	0°C	25°C	50°C	75°C	100°C
	Compounds not containing carbon						
Br_2	Bromine		1.252	0.944	0.746		
Cl_3HSi	Trichlorosilane		0.415	0.326			
Cl_3P	Phosphorous trichloride	0.870	0.662	0.529	0.439		
Cl_4Si	Tetrachlorosilane			99.4	96.2		
H_2O	Water		1.793	0.890	0.547	0.378	0.282
H_4N_2	Hydrazine			0.876	0.628	0.480	0.384
Hg	Mercury			1.526	1.402	1.312	1.245
NO_2	Nitrogen dioxide		0.532	0.402			
	Compounds containing carbon						
CCl_3F	Trichlorofluoromethane	0.740	0.539	0.421			
CCl_4	Tetrachloromethane		1.321	0.908	0.656	0.494	
CS_2	Carbon disulfide		0.429	0.352			
$CHBr_3$	Tribromomethane			1.857	1.367	1.029	
$CHCl_3$	Trichloromethane	0.988	0.706	0.537	0.427		
CHN	Hydrogen cyanide		0.235	0.183			
CH_2Br_2	Dibromomethane	1.948	1.320	0.980	0.779	0.652	
CH_2Cl_2	Dichloromethane	0.727	0.533	0.413			
CH_2O_2	Formic acid			1.607	1.030	0.724	0.545
CH_3I	Iodomethane		0.594	0.469			
CH_3NO	Formamide		7.114	3.343	1.833		
CH_3NO_2	Nitromethane	1.311	0.875	0.630	0.481	0.383	0.317
CH_4O	Methanol	1.258	0.793	0.544			
CH_5N	Methylamine	0.319	0.231				
$C_2Cl_3F_3$	1,1,2-Trichlorotrifluoro-ethane	1.465	0.945	0.656	0.481		
C_2Cl_4	Tetrachloroethylene		1.114	0.844	0.663	0.535	0.442
C_2HCl_3	Trichloroethylene		0.703	0.545	0.444	0.376	
C_2HCl_5	Pentachloroethane		3.761	2.254	1.491	1.061	
$C_2HF_3O_2$	Trifluoroacetic acid			0.808	0.571		
$C_2H_2Cl_2$	*cis*-1,2-Dichloroethylene	0.786	0.575	0.445			
$C_2H_2Cl_2$	*trans*-1,2-Dichloroethylene	0.522	0.398	0.317	0.261		
$C_2H_2Cl_4$	1,1,1,2-Tetrachloroethane	3.660	2.200	1.437	1.006	0.741	0.570
$C_2H_3ClF_2$	1-Chloro-1,1-difluoro-ethane	0.477	0.376				

VISCOSITY OF LIQUIDS (continued)

Molecular formula	Name	Viscosity in mPa s –25°C	0°C	25°C	50°C	75°C	100°C
C_2H_3ClO	Acetyl chloride			0.368	0.294		
$C_2H_3Cl_3$	1,1,1-Trichloroethane	1.847	1.161	0.793	0.578	0.428	
C_2H_3N	Acetonitrile		0.400	0.369	0.284	0.234	
$C_2H_4Br_2$	1,2-Dibromoethane			1.595	1.116	0.837	0.661
$C_2H_4Cl_2$	1,1-Dichloroethane			0.464	0.362		
$C_2H_4Cl_2$	1,2-Dichloroethane		1.125	0.779	0.576	0.447	
$C_2H_4O_2$	Acetic acid			1.056	0.786	0.599	0.464
$C_2H_4O_2$	Methyl formate		0.424	0.325			
C_2H_5Br	Bromoethane	0.635	0.477	0.374			
C_2H_5Cl	Chloroethane	0.416	0.319				
C_2H_5I	Iodoethane		0.723	0.556	0.444	0.365	
C_2H_5NO	*N*-Methylformamide		2.549	1.678	1.155	0.824	0.606
$C_2H_5NO_2$	Nitroethane	1.354	0.940	0.688	0.526	0.415	0.337
C_2H_6O	Ethanol	3.262	1.786	1.074	0.694	0.476	
C_2H_6OS	Dimethyl sulfoxide			1.987	1.290		
$C_2H_6O_2$	Ethylene glycol			16.1	6.554	3.340	1.975
C_2H_6S	Dimethyl sulfide		0.356	0.284			
C_2H_6S	Ethanethiol		0.364	0.287			
C_2H_7N	Dimethylamine	0.300	0.232				
C_2H_7NO	Ethanolamine			21.1	8.560	3.935	1.998
C_3H_5Br	3-Bromopropene		0.620	0.471	0.373		
C_3H_5Cl	3-Chloropropene		0.408	0.314			
C_3H_5ClO	Epichlorohydrin	2.492	1.570	1.073	0.781	0.597	0.474
C_3H_5N	Propanenitrile			0.294	0.240	0.202	
C_3H_6O	Acetone	0.540	0.395	0.306	0.247		
C_3H_6O	Allyl alcohol			1.218	0.759	0.505	
C_3H_6O	Propanal			0.321	0.249		
$C_3H_6O_2$	Ethyl formate		0.506	0.380	0.300		
$C_3H_6O_2$	Methyl acetate		0.477	0.364	0.284		
$C_3H_6O_2$	Propanoic acid		1.499	1.030	0.749	0.569	0.449
C_3H_7Br	1-Bromopropane		0.645	0.489	0.387		
C_3H_7Br	2-Bromopropane		0.612	0.458	0.359		
C_3H_7Cl	1-Chloropropane		0.436	0.334			
C_3H_7Cl	2-Chloropropane		0.401	0.303			
C_3H_7I	1-Iodopropane		0.970	0.703	0.541	0.436	0.363
C_3H_7I	2-Iodopropane		0.883	0.653	0.506	0.407	
C_3H_7NO	*N,N*-Dimethylformamide		1.176	0.794	0.624		
$C_3H_7NO_2$	1-Nitropropane	1.851	1.160	0.798	0.589	0.460	0.374
C_3H_8O	1-Propanol	8.645	3.815	1.945	1.107	0.685	
C_3H_8O	2-Propanol		4.619	2.038	1.028	0.576	
$C_3H_8O_2$	1,2-Propylene glycol		248	40.4	11.3	4.770	2.750
$C_3H_8O_3$	Glycerol			934	152	39.8	14.8
C_3H_8S	1-Propanethiol		0.503	0.385			
C_3H_8S	2-Propanethiol		0.477	0.357	0.280		
C_3H_9N	Propylamine			0.376			
C_3H_9N	Isopropylamine		0.454	0.325			
C_4H_4O	Furan	0.661	0.475	0.361			
C_4H_5N	Pyrrole		2.085	1.225	0.828	0.612	
$C_4H_6O_3$	Acetic anhydride		1.241	0.843	0.614	0.472	0.377
C_4H_7N	Butanenitrile			0.553	0.418	0.330	0.268
C_4H_8O	2-Butanone	0.720	0.533	0.405	0.315	0.249	
C_4H_8O	Tetrahydrofuran	0.849	0.605	0.456	0.359		
$C_4H_8O_2$	1,4-Dioxane			1.177	0.787	0.569	
$C_4H_8O_2$	Ethyl acetate		0.578	0.423	0.325	0.259	
$C_4H_8O_2$	Methyl propionate		0.581	0.431	0.333	0.266	
$C_4H_8O_2$	Propyl formate		0.669	0.485	0.370	0.293	
$C_4H_8O_2$	Butanoic acid		2.215	1.426	0.982	0.714	0.542

Molecular formula	Name	Viscosity in mPa s					
		–25°C	0°C	25°C	50°C	75°C	100°C
$C_4H_8O_2$	2-Methylpropanoic acid		1.857	1.226	0.863	0.639	0.492
$C_4H_8O_2S$	Sulfolane				6.280	3.818	2.559
C_4H_8S	Tetrahydrothiophene			0.973	0.912		
C_4H_9Br	1-Bromobutane		0.815	0.606	0.471	0.379	
C_4H_9Cl	1-Chlorobutane		0.556	0.422	0.329	0.261	
C_4H_9N	Pyrrolidine	1.914	1.071	0.704	0.512		
C_4H_9NO	*N,N*-Dimethylacetamide			1.956	1.279	0.896	0.661
C_4H_9NO	Morpholine			2.021	1.247	0.850	0.627
$C_4H_{10}O$	1-Butanol	12.19	5.185	2.544	1.394	0.833	0.533
$C_4H_{10}O$	2-Butanol			3.096	1.332	0.698	0.419
$C_4H_{10}O$	2-Methyl-2-propanol			4.312	1.421	0.678	
$C_4H_{10}O$	Diethyl ether		0.283	0.224			
$C_4H_{10}O_3$	Diethylene glycol			30.200	11.130	4.917	2.505
$C_4H_{10}S$	Diethyl sulfide		0.558	0.422	0.331	0.267	
$C_4H_{11}N$	Butylamine		0.830	0.574	0.409	0.298	
$C_4H_{11}N$	Isobutylamine		0.770	0.571	0.367		
$C_4H_{11}N$	Diethylamine			0.319	0.239		
$C_4H_{11}NO_2$	Diethanolamine				109.5	28.7	9.100
$C_5H_4O_2$	Furfural		2.501	1.587	1.143	0.906	0.772
C_5H_5N	Pyridine		1.361	0.879	0.637	0.497	0.409
C_5H_{10}	1-Pentene	0.313	0.241	0.195			
C_5H_{10}	2-Methyl-2-butene		0.255	0.203			
C_5H_{10}	Cyclopentane		0.555	0.413	0.321		
$C_5H_{10}O$	Mesityl oxide	1.291	0.838	0.602	0.465	0.381	0.326
$C_5H_{10}O$	2-Pentanone		0.641	0.470	0.362	0.289	0.238
$C_5H_{10}O$	3-Pentanone		0.592	0.444	0.345	0.276	0.227
$C_5H_{10}O_2$	Butyl formate		0.937	0.644	0.472	0.362	0.289
$C_5H_{10}O_2$	Propyl acetate		0.768	0.544	0.406	0.316	0.255
$C_5H_{10}O_2$	Ethyl propanoate		0.691	0.501	0.380	0.299	0.242
$C_5H_{10}O_2$	Methyl butanoate		0.759	0.541	0.406	0.318	0.257
$C_5H_{10}O_2$	Methyl isobutanoate		0.672	0.488	0.373	0.296	
$C_5H_{11}N$	Piperidine			1.573	0.958	0.649	0.474
C_5H_{12}	Pentane	0.351	0.274	0.224			
C_5H_{12}	Isopentane	0.376	0.277	0.214			
$C_5H_{12}O$	1-Pentanol	25.4	8.512	3.619	1.820	1.035	0.646
$C_5H_{12}O$	2-Pentanol			3.470	1.447	0.761	0.465
$C_5H_{12}O$	3-Pentanol			4.149	1.473	0.727	0.436
$C_5H_{12}O$	2-Methyl-1-butanol			4.453	1.963	1.031	0.612
$C_5H_{12}O$	3-Methyl-1-butanol		8.627	3.692	1.842	1.031	0.631
$C_5H_{13}N$	Pentylamine		1.030	0.702	0.493	0.356	
C_6F_6	Hexafluorobenzene			2.789	1.730	1.151	
$C_6H_4Cl_2$	*o*-Dichlorobenzene		1.958	1.324	0.962	0.739	0.593
$C_6H_4Cl_2$	*m*-Dichlorobenzene		1.492	1.044	0.787	0.628	0.525
C_6H_5Br	Bromobenzene		1.560	1.074	0.798	0.627	0.512
C_6H_5Cl	Chlorobenzene	1.703	1.058	0.753	0.575	0.456	0.369
C_6H_5ClO	*o*-Chlorophenol			3.589	1.835	1.131	0.786
C_6H_5ClO	*m*-Chlorophenol				4.041		
C_6H_5F	Fluorobenzene		0.749	0.550	0.423	0.338	
C_6H_5I	Iodobenzene		2.354	1.554	1.117	0.854	0.683
$C_6H_5NO_2$	Nitrobenzene		3.036	1.863	1.262	0.918	0.704
C_6H_6	Benzene			0.604	0.436	0.335	
C_6H_6ClN	*o*-Chloroaniline			3.316	1.913	1.248	0.887
C_6H_6O	Phenol				3.437	1.784	1.099
C_6H_7N	Aniline			3.847	2.029	1.247	0.850
$C_6H_8N_2$	Phenylhydrazine			13.0	4.553	1.850	0.848
C_6H_{10}	Cyclohexene		0.882	0.625	0.467	0.364	
$C_6H_{10}O$	Cyclohexanone			2.017	1.321	0.919	0.671

Molecular formula	Name	Viscosity in mPa s –25°C	0°C	25°C	50°C	75°C	100°C
$C_6H_{11}N$	Hexanenitrile			0.912	0.650	0.488	0.382
C_6H_{12}	Cyclohexane			0.894	0.615	0.447	
C_6H_{12}	Methylcyclopentane	0.927	0.653	0.479	0.364		
C_6H_{12}	1-Hexene	0.441	0.326	0.252	0.202		
$C_6H_{12}O$	Cyclohexanol			57.5	12.3	4.274	1.982
$C_6H_{12}O$	2-Hexanone	1.300	0.840	0.583	0.429	0.329	0.262
$C_6H_{12}O$	4-Methyl-2-pentanone			0.545	0.406		
$C_6H_{12}O_2$	Butyl acetate		1.002	0.685	0.500	0.383	0.305
$C_6H_{12}O_2$	Isobutyl acetate			0.676	0.493	0.370	0.286
$C_6H_{12}O_2$	Ethyl butanoate			0.639	0.453		
$C_6H_{12}O_2$	Diacetone alcohol	28.7	6.621	2.798	1.829	1.648	
$C_6H_{12}O_3$	Paraldehyde			1.079	0.692	0.485	0.362
$C_6H_{13}N$	Cyclohexylamine			1.944	1.169	0.782	0.565
C_6H_{14}	Hexane		0.405	0.300	0.240		
C_6H_{14}	2-Methylpentane		0.372	0.286	0.226		
C_6H_{14}	3-Methylpentane		0.395	0.306			
$C_6H_{14}O$	Dipropyl ether		0.542	0.396	0.304	0.242	
$C_6H_{14}O$	1-Hexanol			4.578	2.271	1.270	0.781
$C_6H_{15}N$	Triethylamine		0.455	0.347	0.273	0.221	
$C_6H_{15}N$	Dipropylamine		0.751	0.517	0.377	0.288	0.228
$C_6H_{15}N$	Diisopropylamine			0.393	0.300	0.237	
$C_6H_{15}NO_3$	Triethanolamine			609	114	31.5	11.7
C_7H_5N	Benzonitrile			1.267	0.883	0.662	0.524
C_7H_7Cl	*o*-Chlorotoluene		1.390	0.964	0.710	0.547	0.437
C_7H_7Cl	*m*-Chlorotoluene		1.165	0.823	0.616	0.482	0.391
C_7H_7Cl	*p*-Chlorotoluene			0.837	0.621	0.483	0.390
C_7H_8	Toluene	1.165	0.778	0.560	0.424	0.333	0.270
C_7H_8O	*o*-Cresol				3.035	1.562	0.961
C_7H_8O	*m*-Cresol			12.9	4.417	2.093	1.207
C_7H_8O	Benzyl alcohol			5.474	2.760	1.618	1.055
C_7H_8O	Anisole			1.056	0.747	0.554	0.427
C_7H_9N	*N*-Methylaniline		4.120	2.042	1.222	0.825	0.606
C_7H_9N	*o*-Methyl aniline		10.3	3.823	1.936	1.198	0.839
C_7H_9N	*m*-Methyl aniline		8.180	3.306	1.679	1.014	0.699
C_7H_9N	Benzylamine			1.624	1.080	0.769	0.577
C_7H_{14}	Methylcyclohexane		0.991	0.679	0.501	0.390	0.316
C_7H_{14}	1-Heptene		0.441	0.340	0.273	0.226	
$C_7H_{14}O$	2-Heptanone			0.714	0.407	0.297	
$C_7H_{14}O_2$	Heptanoic acid			3.840	2.282	1.488	1.041
C_7H_{16}	Heptane	0.757	0.523	0.387	0.301	0.243	
C_7H_{16}	3-Methylhexane			0.350			
$C_7H_{16}O$	1-Heptanol			5.810	2.603	1.389	0.849
$C_7H_{16}O$	2-Heptanol			3.955	1.799	0.987	0.615
$C_7H_{16}O$	3-Heptanol				1.957	0.976	0.584
$C_7H_{16}O$	4-Heptanol			4.207	1.695	0.882	0.539
$C_7H_{17}N$	Heptylamine			1.314	0.865	0.600	0.434
C_8H_8	Styrene		1.050	0.695	0.507	0.390	0.310
C_8H_8O	Acetophenone			1.681			0.634
$C_8H_8O_2$	Methyl benzoate			1.857			
$C_8H_8O_3$	Methyl salicylate					1.102	0.815
C_8H_{10}	Ethylbenzene		0.872	0.631	0.482	0.380	0.304
C_8H_{10}	*o*-Xylene		1.084	0.760	0.561	0.432	0.345
C_8H_{10}	*m*-Xylene		0.795	0.581	0.445	0.353	0.289
C_8H_{10}	*p*-Xylene			0.603	0.457	0.359	0.290
$C_8H_{10}O$	Phenetole			1.197	0.817	0.594	0.453
$C_8H_{11}N$	*N,N*-Dimethylaniline		1.996	1.300	0.911	0.675	0.523
$C_8H_{11}N$	*N*-Ethylaniline		3.981	2.047	1.231	0.825	0.596

Molecular formula	Name	Viscosity in mPa s –25°C	0°C	25°C	50°C	75°C	100°C
C_8H_{16}	Ethylcyclohexane		1.139	0.784	0.579		
$C_8H_{16}O_2$	Octanoic acid			5.020	2.656	1.654	1.147
C_8H_{18}	Octane		0.700	0.508	0.385	0.302	0.243
$C_8H_{18}O$	1-Octanol			7.288	3.232	1.681	0.991
$C_8H_{18}O$	4-Methyl-3-heptanol		1.904	1.085	0.702	0.497	0.375
$C_8H_{18}O$	5-Methyl-3-heptanol		2.052	1.178	0.762	0.536	0.401
$C_8H_{18}O$	2-Ethyl-1-hexanol		20.7	6.271	2.631	1.360	0.810
$C_8H_{18}O$	Dibutyl ether	1.417	0.918	0.637	0.466	0.356	0.281
$C_8H_{19}N$	Dibutylamine		1.509	0.918	0.619	0.449	0.345
$C_8H_{19}N$	Diisobutylamine		1.115	0.723	0.511	0.384	0.303
C_9H_7N	Quinoline			3.337	1.892	1.201	0.833
C_9H_{10}	Indane		2.230	1.357	0.931	0.692	0.545
C_9H_{12}	Cumene		1.075	0.737	0.547		
$C_9H_{14}O$	Isophorone		4.201	2.329	1.415	0.923	0.638
$C_9H_{18}O$	5-Nonanone			1.199	0.834	0.619	0.484
$C_9H_{18}O_2$	Nonanoic acid			7.011	3.712	2.234	1.475
C_9H_{20}	Nonane		0.964	0.665	0.488	0.375	0.300
$C_9H_{20}O$	1-Nonanol			9.123	4.032		
$C_{10}H_{10}O_4$	Dimethyl phthalate		63.2	14.4	5.309	2.824	1.980
$C_{10}H_{14}$	Butylbenzene			0.950	0.683	0.515	
$C_{10}H_{18}$	*cis*-Decahydronaphthalene	12.8	5.645	3.042	1.875	1.271	0.924
$C_{10}H_{18}$	*trans*-Decahydronaphthalene	6.192	3.243	1.948	1.289	0.917	0.689
$C_{10}H_{20}O_2$	Decanoic acid				4.327	2.651	
$C_{10}H_{22}$	Decane	2.188	1.277	0.838	0.598	0.453	0.359
$C_{10}H_{22}O$	1-Decanol			10.9	4.590		
$C_{11}H_{24}$	Undecane		1.707	1.098	0.763	0.562	0.433
$C_{12}H_{10}O$	Diphenyl ether				2.130	1.407	1.023
$C_{12}H_{26}$	Dodecane		2.277	1.383	0.930	0.673	0.514
$C_{13}H_{12}$	Diphenylmethane					1.265	0.929
$C_{13}H_{28}$	Tridecane		2.909	1.724	1.129	0.796	0.594
$C_{14}H_{30}$	Tetradecane			2.128	1.376	0.953	0.697
$C_{16}H_{22}O_4$	Dibutyl phthalate	483	66.4	16.6	6.470	3.495	2.425
$C_{16}H_{34}$	Hexadecane			3.032	1.879	1.260	0.899
$C_{18}H_{38}$	Octadecane				2.487	1.609	1.132

VISCOSITY OF CARBON DIOXIDE ALONG THE SATURATION LINE

The table below gives the viscosity of gas and liquid CO_2 along the liquid-vapor saturation line.

REFERENCES

1. Fenghour, A., Wakeham, W. A., and Vesovic, V., *J. Phys. Chem. Ref. Data*, 27, 31, 1998.
2. Angus, S., et al., *International Tables for the Fluid State: Carbon Dioxide*, Pergamon Press, Oxford, 1976.

T/K	*P*/kPa	Gas η/μPa s	Liquid η/μPa s
205	227	10.33	
210	327	10.60	
215	465	10.87	
220	600	11.13	241.68
225	735	11.41	221.72
230	894	11.69	203.75
235	1075	11.98	187.48
240	1283	12.27	172.67
245	1519	12.58	159.13
250	1786	12.90	146.69
255	2085	13.24	135.20
260	2419	13.61	124.30
265	2790	14.02	114.63
270	3203	14.47	105.21
275	3658	14.99	96.44
280	4160	15.61	87.89
285	4712	16.37	79.64
290	5315	17.36	71.47
295	5984	18.79	63.01
300	6710	21.29	53.33
302	6997	23.52	48.30

THERMAL CONDUCTIVITY OF GASES

This table gives the thermal conductivity of several gases as a function of temperature. Unless otherwise noted, the values refer to a pressure of 100 kPa (1 bar) or to the saturation vapor pressure if that is less than 100 kPa. The notation $P = 0$ indicates the low pressure limiting value is given. In general, the $P = 0$ and $P = 100$ kPa values differ by less than 1%. Units are milliwatts per meter kelvin. Substances are listed in the modified Hill order.

		Thermal conductivity in mW/m K						
MF	**Name**	**100 K**	**200 K**	**300 K**	**400 K**	**500 K**	**600 K**	**Ref.**
	Air	9.4	18.4	26.2	33.3	39.7	45.7	1
Ar	Argon	6.2	12.4	17.9	22.6	26.8	30.6	2,8
BF_3	Boron trifluoride			19.0	24.6			11
ClH	Hydrogen chloride		9.2	14.5	19.5	24.0	28.1	11
F_6S	Sulfur hexafluoride ($P = 0$)			13.0	20.6	27.5	33.8	16
H_2	Hydrogen ($P = 0$)	68.6	131.7	186.9	230.4			4
H_2O	Water			18.7	27.1	35.7	47.1	6
	Deuterium oxide				27.0	36.5	47.6	7
H_2S	Hydrogen sulfide			14.6	20.5	26.4	32.4	11
H_3N	Ammonia			24.4	37.4	51.6	66.8	11
He	Helium ($P = 0$)	75.5	119.3	156.7	190.6	222.3	252.4	8
Kr	Krypton ($P = 0$)	3.3	6.4	9.5	12.3	14.8	17.1	8
NO	Nitric oxide		17.8	25.9	33.1	39.6	46.2	11
N_2	Nitrogen	9.8	18.7	26.0	32.3	38.3	44.0	12
N_2O	Nitrous oxide		9.8	17.4	26.0	34.1	41.8	11
Ne	Neon ($P = 0$)	22.3	37.6	49.8	60.3	69.9	78.7	8
O_2	Oxygen	9.3	18.4	26.3	33.7	41.0	48.1	10
O_2S	Sulfur dioxide			9.6	14.3	20.0	25.6	11
Xe	Xenon ($P = 0$)	2.0	3.6	5.5	7.3	8.9	10.4	8
CCl_2F_2	Dichlorodifluoromethane			9.9	15.0	20.1	25.2	13
CF_4	Tetrafluoromethane ($P = 0$)			16.0	24.1	32.2	39.9	16
CO	Carbon monoxide ($P = 0$)			25.0	32.3	39.2	45.7	14
CO_2	Carbon dioxide		9.6	16.8	25.1	33.5	41.6	9
$CHCl_3$	Trichloromethane			7.5	11.1	15.1		11
CH_4	Methane		22.5	34.1	49.1	66.5	84.1	5,15
CH_4O	Methanol				26.2	38.6	53.0	11
$C_2Cl_2F_4$	1,2-Dichlorotetrafluoroethane			10.25	15.7	21.1		13
$C_2Cl_3F_3$	1,1,2-Trichlorotrifluoroethane			9.0	13.6	18.3		13
C_2H_2	Acetylene			21.4	33.3	45.4	56.8	11
C_2H_4	Ethylene		11.1	20.5	34.6	49.9	68.6	3
C_2H_6	Ethane		11.0	21.3	35.4	52.2	70.5	5
C_2H_6O	Ethanol			14.4	25.8	38.4	53.2	11
C_3H_6O	Acetone			11.5	20.2	30.6	42.7	11
C_3H_8	Propane			18.0	30.6	45.5	61.9	5
C_4F_8	Perfluorocyclobutane			12.5	19.5			13
C_4H_{10}	Butane			16.4	28.4	43.0	59.1	5
C_4H_{10}	Isobutane			16.1	27.9	42.1	57.6	5
$C_4H_{10}O$	Diethyl ether			15.1	25.0	37.1		11
C_5H_{12}	Pentane			14.4	24.9	37.8	52.7	11
C_6H_{14}	Hexane				23.4	35.4	48.7	11

REFERENCES

1. Kadoya, K. Matsunaga, N., and Nagashima, A., Viscosity and thermal conductivity of dry air in the gaseous phase, *J. Phys. Chem. Ref. Data*, 14, 947, 1985.
2. Younglove, B. A. and Hanley, H. J. M., The viscosity and thermal conductivity coefficients of gaseous and liquid argon, *J. Phys. Chem. Ref. Data*, 15, 1323, 1986.
3. Holland, P. M., Eaton, B. E., and Hanley, H. J. M., A correlation of the viscosity and thermal conductivity data of gaseous and liquid ethylene, *J. Phys. Chem. Ref. Data*, 12, 917, 1983.

4. Assael, M. J., Mixafendi, S., and Wakeham, W. A., The viscosity and thermal conductivity of normal hydrogen in the limit of zero density, *J. Phys. Chem. Ref. Data*, 15, 1315, 1986.
5. Younglove, B. A. and Ely, J. F., Thermophysical properties of fluids. II. Methane, ethane, propane, isobutane, and normal butane, *J. Phys. Chem. Ref. Data*, 16, 577, 1987.
6. Sengers, J. V. and Watson, J. T. R., Improved international formulations for the viscosity and thermal conductivity of water substance, *J. Phys. Chem. Ref. Data*, 15, 1291, 1986.
7. Matsunaga, N. and Nagashima, A., Transport properties of liquid and gaseous D_2O over a wide range of temperature and pressure, *J. Phys. Chem. Ref. Data*, 12, 933, 1983.
8. Kestin, J. et al., Equilibrium and transport properties of the noble gases and their mixtures at low density, *J. Phys. Chem. Ref. Data*, 13, 229, 1984.
9. Vescovic, V. et al., The transport properties of carbon dioxide, *J. Phys. Chem. Ref. Data*, 19, 1990.
10. Younglove, B. A., Thermophysical properties of fluids. I. Argon, ethylene, parahydrogen, nitrogen, nitrogen trifluoride, and oxygen, *J. Phys. Chem. Ref. Data*, 11, Suppl. 1, 1982.
11. Ho, C. Y., Ed., *Properties of Inorganic and Organic Fluids, CINDAS Data Series on Materials Properties*, Volume V-1, Hemisphere Publishing Corp., New York, 1988.
12. Stephen, K., Krauss, R., and Laesecke, A., Viscosity and thermal conductivity of nitrogen for a wide range of fluid states, *J. Phys. Chem. Ref. Data*, 16, 993, 1987.
13. Krauss, R. and Stephan, K., Thermal conductivity of refrigerants in a wide range of temperature and pressure, *J. Phys. Chem. Ref. Data*, 18, 43, 1989.
14. Millat, J. and Wakeham, W. A., The thermal conductivity of nitrogen and carbon monoxide in the limit of zero density, *J. Phys. Chem. Ref. Data*, 18, 565, 1989.
15. Friend, D. G., Ely, J. F., and Ingham, H., Thermophysical properties of methane, *J. Phys. Chem. Ref. Data*, 18, 583, 1989.
16. Uribe, F. J., Mason, E. A., and Kestin, J., Thermal conductivity of nine polyatomic gases at low density, *J. Phys. Chem. Ref. Data*, 19, 1123, 1990.

THERMAL CONDUCTIVITY OF LIQUIDS

This table gives the thermal conductivity of some common liquids at temperatures between -25 and 100°C. All values are given in units of watts per meter kelvin (W/m K). Values refer to nominal atmospheric pressure (about 100 kPa); when an entry is given at a temperature above the normal boiling point of the substance, the pressure is understood to be the saturation vapor pressure at that temperature.

Substances are arranged by molecular formula in the modified Hill order, with compounds not containing carbon preceding those that do contain carbon.

The values for water, benzene, toluene, heptane, and dimethyl phthalate are particularly well determined and can be used for calibration purposes.

REFERENCES

1. Daubert, T. E., Danner, R. P., Sibul, H. M., and Stebbins, C. C., *Physical and Thermodynamic Properties of Pure Compounds: Data Compilation*, extant 1994 (core with 4 supplements), Taylor & Francis, Bristol, PA (also available as database).
2. Marsh, K. N., Ed., *Recommended Reference Materials for the Realization of Physicochemical Properties,* Blackwell Scientific Publications, Oxford, 1987.

Molecular formula	Name	Thermal conductivity in W/m K					
		–25°C	0°C	25°C	50°C	75°C	100°C
Cl_4Si	Silicon tetrachloride			0.099	0.096		
H_2O	Water		0.5610	0.6071	0.6435	0.6668	0.6791
Hg	Mercury	7.25	7.77	8.25	8.68	9.07	9.43
CCl_4	Tetrachloromethane		0.104	0.099	0.093	0.088	
CS_2	Carbon disulfide		0.154	0.149			
$CHCl_3$	Trichloromethane	0.127	0.122	0.117	0.112	0.107	0.102
CH_2Br_2	Dibromomethane	0.120	0.114	0.108	0.103	0.097	
CH_4O	Methanol	0.214	0.207	0.200	0.193		
C_2Cl_4	Tetrachloroethylene		0.117	0.110	0.104	0.097	0.091
C_2HCl_3	Trichloroethylene	0.133	0.124	0.116	0.108	0.100	
$C_2H_3Cl_3$	1,1,1-Trichloroethane		0.106	0.101	0.096		
C_2H_3N	Acetonitrile	0.208	0.198	0.188	0.178	0.168	
$C_2H_4O_2$	Acetic acid			0.158	0.153	0.149	0.144
C_2H_5Cl	Chloroethane	0.145	0.132	0.119	0.106	0.093	
C_2H_5NO	*N*-Methylformamide			0.203	0.201	0.199	0.196
C_2H_6O	Ethanol		0.176	0.169	0.162		
$C_2H_6O_2$	Ethylene glycol		0.256	0.256	0.256	0.256	0.256
C_2H_7NO	Ethanolamine			0.299	0.286	0.274	0.261
C_3H_5ClO	Epichlorohydrin	0.142	0.137	0.131	0.125	0.119	0.114
C_3H_6O	Acetone		0.169	0.161			
$C_3H_6O_2$	Methyl acetate	0.174	0.164	0.153	0.143	0.133	0.122
C_3H_7NO	*N,N*-Dimethylformamide			0.184	0.178	0.171	0.165
C_3H_8O	1-Propanol	0.162	0.158	0.154	0.149	0.145	0.141
C_3H_8O	2-Propanol	0.146	0.141	0.135	0.129	0.124	0.118
$C_3H_8O_2$	1,2-Propanediol		0.202	0.200	0.199	0.198	0.197
$C_3H_8O_3$	Glycerol			0.292	0.295	0.297	0.300
C_3H_9N	Trimethylamine	0.143	0.133				
C_4H_4O	Furan	0.142	0.134	0.126			
C_4H_4S	Thiophene			0.199	0.195	0.191	0.186
C_4H_6	2-Butyne	0.137	0.129	0.121			
C_4H_8O	2-Butanone	0.158	0.151	0.145	0.139	0.133	
C_4H_8O	Tetrahydrofuran	0.132	0.126	0.120	0.114		
$C_4H_8O_2$	1,4-Dioxane			0.159	0.147	0.135	0.123
$C_4H_8O_2$	Ethyl acetate	0.162	0.153	0.144	0.135	0.126	
$C_4H_{10}O$	1-Butanol		0.158	0.154	0.149		
$C_4H_{10}O$	Diethyl ether	0.150	0.140	0.130	0.120	0.110	0.100
C_5H_5N	Pyridine		0.169	0.165	0.161	0.158	
C_5H_8	Cyclopentene	0.143	0.136	0.129			
C_5H_{10}	1-Pentene	0.131	0.124	0.116			
C_5H_{10}	Cyclopentane	0.140	0.133	0.126			
C_5H_{12}	Pentane	0.132	0.122	0.113	0.103	0.095	0.087
$C_5H_{12}O$	1-Pentanol		0.157	0.153	0.149	0.145	
C_6H_5Cl	Chlorobenzene	0.136	0.131	0.127	0.122	0.117	0.112

Molecular formula	Name	Thermal conductivity in W/m K					
		–25°C	0°C	25°C	50°C	75°C	100°C
C_6H_6	Benzene			0.1411	0.1329	0.1247	
C_6H_6O	Phenol				0.156	0.153	0.151
C_6H_{10}	Cyclohexene	0.142	0.136	0.130	0.124	0.118	
$C_6H_{10}O$	Mesityl oxide	0.170	0.163	0.156	0.149	0.142	0.134
C_6H_{12}	Cyclohexane			0.123	0.117	0.111	
C_6H_{12}	1-Hexene	0.137	0.129	0.121	0.113		
$C_6H_{12}O$	Cyclohexanol			0.134	0.131		
$C_6H_{12}O$	2-Hexanone	0.151	0.145	0.139	0.133	0.127	0.121
C_6H_{14}	Hexane	0.137	0.128	0.120	0.111	0.102	0.093
$C_6H_{14}O$	1-Hexanol	0.159	0.154	0.150	0.145	0.141	0.137
C_7H_6O	Benzaldehyde			0.151	0.141	0.131	0.121
C_7H_8	Toluene	0.1461	0.1386	0.1311	0.1236	0.1161	
C_7H_8O	Anisole	0.170	0.163	0.156	0.150	0.143	0.136
C_7H_{16}	Heptane	0.1378	0.1303	0.1228	0.1152	0.1077	
$C_7H_{16}O$	1-Heptanol		0.166	0.159	0.153	0.147	0.141
C_8H_8	Styrene	0.148	0.142	0.137	0.131	0.126	0.120
C_8H_{10}	Ethylbenzene			0.130	0.124	0.118	0.112
C_8H_{10}	*o*-Xylene			0.131	0.126	0.120	0.114
C_8H_{10}	*m*-Xylene			0.130	0.124	0.118	0.113
C_8H_{10}	*p*-Xylene			0.130	0.124	0.118	0.112
C_8H_{18}	Octane	0.143	0.135	0.128	0.120	0.113	0.106
$C_8H_{18}O$	1-Octanol		0.168	0.161	0.154	0.147	0.141
C_9H_{12}	Cumene			0.128	0.120	0.112	0.107
C_9H_{12}	Mesitylene	0.147	0.141	0.136	0.130	0.124	0.118
C_9H_{20}	Nonane	0.144	0.138	0.131	0.124	0.118	0.111
$C_9H_{20}O$	1-Nonanol		0.166	0.161	0.155	0.149	0.143
$C_{10}H_{10}O_4$	Dimethyl phthalate		0.1501	0.1473	0.1443	0.1409	0.1373
$C_{10}H_{14}$	*p*-Cymene	0.132	0.127	0.122	0.117	0.112	0.107
$C_{10}H_{22}$	Decane	0.144	0.138	0.132	0.126	0.119	0.113
$C_{10}H_{22}O$	1-Decanol			0.162	0.156	0.150	0.145
$C_{11}H_{24}$	Undecane			0.140	0.135	0.129	0.123
$C_{12}H_{10}O$	Diphenyl ether				0.139	0.135	0.131
$C_{12}H_{26}$	Dodecane		0.157	0.152	0.146	0.140	0.135
$C_{12}H_{26}O$	1-Dodecanol			0.146	0.142	0.139	0.135
$C_{13}H_{28}$	Tridecane			0.137	0.132	0.127	0.122
$C_{14}H_{30}$	Tetradecane			0.136	0.131	0.126	0.121
$C_{14}H_{30}O$	1-Tetradecanol				0.167	0.162	0.157
$C_{16}H_{22}O_4$	Dibutyl phthalate	0.144	0.140	0.136	0.133	0.129	0.125
$C_{16}H_{34}$	Hexadecane			0.140	0.135	0.130	0.125
$C_{18}H_{38}$	Octadecane				0.146	0.142	0.137

DIFFUSION IN GASES

This table gives binary diffusion coefficients D_{12} for a number of common gases as a function of temperature. Values refer to atmospheric pressure. The diffusion coefficient is inversely proportional to pressure as long as the gas is in a regime where binary collisions dominate. See Reference 1 for a discussion of the dependence of D_{12} on temperature and composition.

The first part of the table gives data for several gases in the presence of a large excess of air. The remainder applies to equimolar mixtures of gases. Each gas pair is ordered alphabetically according to the most common way of writing the formula. The listing of pairs then follows alphabetical order by the first constituent.

REFERENCES

1. Marrero, T. R., and Mason, E. A., *J. Phys. Chem. Ref. Data*, 1, 1, 1972.
2. Kestin, J., et al., *J. Phys. Chem. Ref. Data*, 13, 229, 1984.

D_{12}/cm^2 s^{-1} for p = 101.325 kPa and the Specified T/K

System	200	273.15	293.15	373.15	473.15	573.15	673.15
Large Excess of Air							
Ar-air		0.167	0.148	0.289	0.437	0.612	0.810
CH_4-air			0.106	0.321	0.485	0.678	0.899
CO-air			0.208	0.315	0.475	0.662	0.875
CO_2-air			0.160	0.252	0.390	0.549	0.728
H_2-air		0.668	0.627	1.153	1.747	2.444	3.238
H_2O-air			0.242	0.399	0.638	0.873	1.135
He-air		0.617	0.580	1.057	1.594	2.221	2.933
SF_6-air				0.150	0.233	0.329	0.438
Equimolar Mixture							
Ar-CH_4				0.306	0.467	0.657	0.876
Ar-CO		0.168	0.187	0.290	0.439	0.615	0.815
Ar-CO_2		0.129	0.078	0.235	0.365	0.517	0.689
Ar-H_2		0.698	0.794	1.228	1.876	2.634	3.496
Ar-He	0.381	0.645	0.726	1.088	1.617	2.226	2.911
Ar-Kr	0.064	0.117	0.134	0.210	0.323	0.456	0.605
Ar-N_2		0.168	0.190	0.290	0.439	0.615	0.815
Ar-Ne	0.160	0.277	0.313	0.475	0.710	0.979	1.283
Ar-O_2		0.166	0.189	0.285	0.430	0.600	0.793
Ar-SF_6				0.128	0.202	0.290	0.389
Ar-Xe	0.052	0.095	0.108	0.171	0.264	0.374	0.498
CH_4-H_2			0.782	1.084	1.648	2.311	3.070
CH_4-He			0.723	0.992	1.502	2.101	2.784
CH_4-N_2			0.220	0.317	0.480	0.671	0.890
CH_4-O_2			0.210	0.341	0.523	0.736	0.978
CH_4-SF_6				0.167	0.257	0.363	0.482
CO-CO_2			0.162	0.250	0.384		
CO-H_2	0.408	0.686	0.772	1.162	1.743	2.423	3.196
CO-He	0.365	0.619	0.698	1.052	1.577	2.188	2.882
CO-Kr		0.131	0.581	0.227	0.346	0.485	0.645
CO-N_2	0.133	0.208	0.231	0.336	0.491	0.673	0.878
CO-O_2			0.202	0.307	0.462	0.643	0.849
CO-SF_6				0.144	0.226	0.323	0.432
CO_2-C_3H_8			0.084	0.133	0.209		
CO_2-H_2	0.315	0.552	0.412	0.964	1.470	2.066	2.745
CO_2-H_2O			0.162	0.292	0.496	0.741	1.021
CO_2-He	0.300	0.513	0.400	0.878	1.321		
CO_2-N_2			0.160	0.253	0.392	0.553	0.733
CO_2-N_2O	0.055	0.099	0.113	0.177	0.276		
CO_2-Ne	0.131	0.227	0.199	0.395	0.603	0.847	

System	200	273.15	293.15	373.15	473.15	573.15	673.15
CO_2-O_2			0.159	0.248	0.380	0.535	0.710
CO_2-SF_6				0.099	0.155		
D_2-H_2	0.631	1.079	1.219	1.846	2.778	3.866	5.103
H_2-He	0.775	1.320	1.490	2.255	3.394	4.726	6.242
H_2-Kr	0.340	0.601	0.682	1.053	1.607	2.258	2.999
H_2-N_2	0.408	0.686	0.772	1.162	1.743	2.423	3.196
H_2-Ne	0.572	0.982	0.317	1.684	2.541	3.541	4.677
H_2-O_2		0.692	0.756	1.188	1.792	2.497	3.299
H_2-SF_6			0.208	0.649	0.998	1.400	1.851
H_2-Xe		0.513	0.122	0.890	1.349	1.885	2.493
H_2O-N_2			0.242	0.399			
H_2O-O_2			0.244	0.403	0.645	0.882	1.147
He-Kr	0.330	0.559	0.629	0.942	1.404	1.942	2.550
He-N_2	0.365	0.619	0.698	1.052	1.577	2.188	2.882
He-Ne	0.563	0.948	1.066	1.592	2.362	3.254	4.262
He-O_2		0.641	0.697	1.092	1.640	2.276	2.996
He-SF_6			1.109	0.592	0.871	1.190	1.545
He-Xe	0.282	0.478	0.538	0.807	1.201	1.655	2.168
Kr-N_2		0.131	0.149	0.227	0.346	0.485	0.645
Kr-Ne	0.131	0.228	0.258	0.392	0.587	0.812	1.063
Kr-Xe	0.035	0.064	0.073	0.116	0.181	0.257	0.344
N_2-Ne			0.258	0.483	0.731	1.021	1.351
N_2-O_2			0.202	0.307	0.462	0.643	0.849
N_2-SF_6				0.148	0.231	0.328	0.436
N_2-Xe		0.107	0.123	0.188	0.287	0.404	0.539
Ne-Xe	0.111	0.193	0.219	0.332	0.498	0.688	0.901
O_2-SF_6			0.097	0.154	0.238	0.334	0.441

DIFFUSION OF GASES IN WATER

This table gives values of the diffusion coefficient, *D*, for diffusion of several common gases in water at various temperatures. For simple one-dimensional transport, the diffusion coefficient describes the time-rate of change of concentration, d*c*/d*t*, through the equation

$$dc/dt = D\, d^2c/dx^2$$

where *x* is, for example, the perpendicular distance from a gas-liquid interface. The values below have been selected from the references indicated; in some cases data have been refitted to permit interpolation in temperature.

Gas-liquid diffusion coefficients are difficult to measure, and large differences are found between values obtained by different authors and through different experimental methods. See References 1 and 2 for a discussion of measurement techniques.

REFERENCES

1. Jähne, B., Heinz, G., and Dietrich, W., *J. Geophys. Res.*, 92, 10767, 1987.
2. Himmelblau, D. M., *Chem. Rev.* 64, 527, 1964.
3. Boerboom, A. J. H., and Kleyn, G., *J. Chem. Phys.*, 50, 1086, 1969.
4. O'Brien, R. N., and Hyslop, W. F., *Can. J. Chem.*, 55, 1415, 1977.
5. Maharajh, D. M., and Walkley, J., *Can. J. Chem.*, 51, 944, 1973.
6. *Landolt-Börnstein, Numerical Data and Functional Relationships in Science and Technology, Sixth Edition*, II/5a, *Transport Phenomena I (Viscosity and Diffusion)*, Springer-Verlag, Heidelberg, 1969.

	$D/10^{-5}$ cm^2 s^{-1}						
	10°C	**15°C**	**20°C**	**25°C**	**30°C**	**35°C**	**Ref.**
Ar				2.5			3,4
$CHCl_2F$				1.80			5
CH_3Br				1.35			5
CH_3Cl				1.40			5
CH_4	1.24	1.43	1.62	1.84	2.08	2.35	1
CO_2	1.26	1.45	1.67	1.91	2.17	2.47	1
C_2H_2	1.43	1.59	1.78	1.99	2.23		2
Cl_2		1.13	1.5	1.89			2,6
HBr				3.15			6
HCl				3.07			6
H_2	3.62	4.08	4.58	5.11	5.69	6.31	1
H_2S				1.36			2,6
He	5.67	6.18	6.71	7.28	7.87	8.48	1,3
Kr	1.20	1.39	1.60	1.84	2.11	2.40	1,3
NH_3		1.3	1.5				2
NO_2			1.23	1.4	1.59		2,6
N_2				2.0			2
N_2O		1.62	2.11	2.57			2,6
Ne	2.93	3.27	3.64	4.03	4.45	4.89	1,3
O_2		1.67	2.01	2.42			2,6
Rn	0.81	0.96	1.13	1.33	1.55	1.80	1
SO_2			1.62	1.83	2.07	2.32	2
Xe	0.93	1.08	1.27	1.47	1.70	1.95	1,3

DIFFUSION COEFFICIENTS IN LIQUIDS AT INFINITE DILUTION

This table lists diffusion coefficients D_{AB} at infinite dilution for some binary liquid mixtures. Although values are given to two decimal places, measurements in the literature are often in poor agreement. Therefore most values in the table cannot be relied upon to better than 10%.

Solvents are listed in alphabetical order, as are the solutes within each solvent group.

REFERENCE

Landolt-Börnstein, *Numerical Data and Functional Relationships in Science and Technology*, Sixth Edition, Vol. II/5a, 1969.

Solute	Solvent	t/°C	D_{AB} 10^{-5} cm^2 s^{-1}
Acetic acid	Acetone	25	3.31
Benzoic acid	Acetone	25	2.62
Formic acid	Acetone	25	3.77
Nitrobenzene	Acetone	20	2.94
Tetrachloromethane	Acetone	25	3.29
Trichloromethane	Acetone	25	3.64
Water	Acetone	25	4.56
Acetic acid	Benzene	25	2.09
Aniline	Benzene	25	1.96
Benzoic acid	Benzene	25	1.38
Bromobenzene	Benzene	8	1.45
2-Butanone	Benzene	30	2.09
Chloroethylene	Benzene	8	1.77
Cyclohexane	Benzene	25	2.25
Ethanol	Benzene	25	3.02
Formic acid	Benzene	25	2.28
Heptane	Benzene	25	1.78
Methanol	Benzene	25	3.80
Toluene	Benzene	25	1.85
1,2,4-Trichlorobenzene	Benzene	8	1.34
Trichloromethane	Benzene	25	2.26
Adipic acid	1-Butanol	30	0.40
Benzene	1-Butanol	25	1.00
Biphenyl	1-Butanol	25	0.63
Butyric acid	1-Butanol	30	0.51
p-Dichlorobenzene	1-Butanol	25	0.82
Methanol	1-Butanol	30	0.59
Oleic acid	1-Butanol	30	0.25
Propane	1-Butanol	25	1.57
Water	1-Butanol	25	0.56
Benzene	Cyclohexane	25	1.41
Tetrachloromethane	Cyclohexane	25	1.49
Toluene	Cyclohexane	25	1.57
Allyl alcohol	Ethanol	20	0.98
Benzene	Ethanol	25	1.81
Iodine	Ethanol	25	1.32
Iodobenzene	Ethanol	20	1.00
3-Methyl-1-butanol	Ethanol	20	0.81
Pyridine	Ethanol	20	1.10
Tetrachloromethane	Ethanol	25	1.50
Water	Ethanol	25	1.24
Acetic acid	Ethyl acetate	20	2.18
Acetone	Ethyl acetate	20	3.18
2-Butanone	Ethyl acetate	30	2.93
Ethyl benzoate	Ethyl acetate	20	1.85
Nitrobenzene	Ethyl acetate	20	2.25
Water	Ethyl acetate	25	3.20
Benzene	Heptane	25	3.91
Toluene	Heptane	25	3.72
Bromobenzene	Hexane	8	2.60
2-Butanone	Hexane	30	3.74
Dodecane	Hexane	25	2.73
Iodine	Hexane	25	4.45
Methane	Hexane	25	0.09
Propane	Hexane	25	4.87
Tetrachloromethane	Hexane	25	3.70
Toluene	Hexane	25	4.21
Acetone	Tetrachloromethane	25	1.75
Benzene	Tetrachloromethane	25	1.42
Cyclohexane	Tetrachloromethane	25	1.30
Ethanol	Tetrachloromethane	25	1.90
Iodine	Tetrachloromethane	30	1.63
Trichloromethane	Tetrachloromethane	25	1.66
Acetic acid	Toluene	25	2.26
Benzene	Toluene	25	2.54
Benzoic acid	Toluene	25	1.49
Cyclohexane	Toluene	25	2.42
Formic acid	Toluene	25	2.65
Water	Toluene	25	6.19
Acetone	Trichloromethane	25	2.55
Benzene	Trichloromethane	25	2.89
2-Butanone	Trichloromethane	25	2.13
Butyl acetate	Trichloromethane	25	1.71
Diethyl ether	Trichloromethane	25	2.15
Ethanol	Trichloromethane	15	2.20
Ethyl acetate	Trichloromethane	25	2.02
Acetic acid	Water	25	1.29
Acetone	Water	25	1.28
Acetonitrile	Water	15	1.26
Alanine	Water	25	0.91
Allyl alcohol	Water	15	0.90
Aniline	Water	20	0.92
Arabinose	Water	20	0.69
Benzene	Water	20	1.02
1-Butanol	Water	25	0.56
Caprolactam	Water	25	0.87
Chloroethylene	Water	25	1.34
Cyclohexane	Water	20	0.84
Diethylamine	Water	20	0.97
Ethanol	Water	25	1.24
Ethanolamine	Water	25	1.08
Ethyl acetate	Water	20	1.00
Ethylbenzene	Water	20	0.81
Ethylene glycol	Water	25	1.16
Glucose	Water	25	0.67
Glycerol	Water	25	1.06
Glycine	Water	25	1.05
Lactose	Water	15	0.38
Maltose	Water	15	0.38
Mannitol	Water	15	0.50
Methane	Water	25	1.49
Methanol	Water	15	1.28
3-Methyl 1 butanol	Water	10	0.69
Methylcyclopentane	Water	20	0.85
Phenol	Water	20	0.89
1-Propanol	Water	15	0.87
Propene	Water	25	1.44
Pyridine	Water	25	0.58
Raffinose	Water	15	0.33
Sucrose	Water	25	0.52
Toluene	Water	20	0.85
Urea	Water	25	1.38
Urethane	Water	15	0.80

Section 7
Biochemistry

Properties of Amino Acids 7-1
Structures of Common Amino Acids 7-3
Properties of Purine and Pyrimidine Bases 7-4
The Genetic Code 7-5
Properties of Fatty Acids 7-6
Carbohydrate Names and Symbols 7-7
Standard Transformed Gibbs Energy of Formation for Important Biochemical Species 7-9
Biological Buffers 7-12
Typical pH Values of Biological Materials and Foods 7-13
Chemical Composition of the Human Body 7-14

PROPERTIES OF AMINO ACIDS

This table gives selected properties of some important amino acids and closely related compounds. The first part of the table lists the 20 "standard" amino acids that are the basic constituents of proteins (structures of these amino acids may be found in the following table). The second part includes other amino acids and related compounds of biochemical importance. Within each part of the table the compounds are listed by name in alphabetical order.

Symbol — Three-letter symbol for the standard amino acids
M_r — Molecular weight
t_m — Melting point
pK_a, pK_b, pK_c, pK_d — Negative of the logarithm of the acid dissociation constants for the COOH and NH_2 groups (and, in some cases, other groups) in the molecule (at 25°C)
pI — pH at the isoelectric point
S — Solubility in water at 25°C in units of grams of compound per kilogram of water; when quantitative data are not available, the notations sl.s. (for slightly soluble) and v.s. (for very soluble) are used.

Data on the enthalpy of formation of many of these compounds are included in the table "Standard Thermodynamic Properties of Chemical Substances" in Section 5 of this Handbook. Absorption spectra and optical rotation data can be found in Reference 3. Partial molar volume and other thermodynamic properties, including solubility as a function of temperature, are given in References 3 and 5. Most of the pK values come from Reference 7.

REFERENCES

1. Dawson, R. M. C., Elliott, D. C., Elliott, W. H., and Jones, K. M., *Data for Biochemical Research*, 3rd ed., Clarendon Press, Oxford, 1986.
2. Budavari, S., Ed., *The Merck Index, Twelfth Edition*, Merck & Co., Rahway, NJ, 1996.
3. Sober, H. A., Ed., *CRC Handbook of Biochemistry. Selected Data for Molecular Biology*, CRC Press, Boca Raton, FL, 1968.
4. Voet, D. and Voet, J. G., *Biochemistry, Second Edition*, John Wiley & Sons, New York, 1995.
5. Hinz, H. J., Ed., *Thermodynamic Data for Biochemistry and Biotechnology,* Springer-Verlag, Heidelberg, 1986.
6. Fasman, G. D., Ed., *Practical Handbook of Biochemistry and Molecular Biology,* CRC Press, Boca Raton, FL, 1989.
7. Smith, R. M., and Martell, A. E., *NIST Standard Reference Database 46: Critically Selected Stability Constants of Metal Complexes Database, Version 3.0*, National Institute of Standards and Technology, Gaithersburg, MD, 1997.
8. Jin, Z. and Chao, K. C., *J. Chem. Eng. Data*, 37, 199, 1992.

The standard amino acids:

Symbol	Name	Mol. form.	M_r	t_m/°C	pK_a	pK_b	pK_c	pI	S/g kg^{-1}
Ala	Alanine	$C_3H_7NO_2$	89.09	297	2.33	9.71		6.00	165.0
Arg	Arginine	$C_6H_{14}N_4O_2$	174.20	244	2.03	9.00	12.10	10.76	182.6
Asn	Asparagine	$C_4H_8N_2O_3$	132.12	235	2.16	8.73		5.41	25.1
Asp	Aspartic acid	$C_4H_7NO_4$	133.10	270	1.95	9.66	3.71	2.77	4.95
Cys	Cysteine	$C_3H_7NO_2S$	121.16	240	1.91	10.28	8.14	5.07	v.s.
Glu	Glutamic acid	$C_5H_9NO_4$	147.13	160	2.16	9.58	4.15	3.22	8.61
Gln	Glutamine	$C_5H_{10}N_2O_3$	146.15	185	2.18	9.00		5.65	42
Gly	Glycine	$C_2H_5NO_2$	75.07	290	2.34	9.58		5.97	250.9
His	Histidine	$C_6H_9N_3O_2$	155.16	287	1.70	9.09	6.04	7.59	43.5
Ile	Isoleucine	$C_6H_{13}NO_2$	131.17	284	2.26	9.60		6.02	34.2
Leu	Leucine	$C_6H_{13}NO_2$	131.17	293	2.32	9.58		5.98	22.0
Lys	Lysine	$C_6H_{14}N_2O_2$	146.19	224	2.15	9.16	10.67	9.74	5.8
Met	Methionine	$C_5H_{11}NO_2S$	149.21	281	2.16	9.08		5.74	56
Phe	Phenylalanine	$C_9H_{11}NO_2$	165.19	283	2.18	9.09		5.48	27.9
Pro	Proline	$C_5H_9NO_2$	115.13	221	1.95	10.47		6.30	1623
Ser	Serine	$C_3H_7NO_3$	105.09	228	2.13	9.05		5.68	50.2
Thr	Threonine	$C_4H_9NO_3$	119.12	256	2.20	8.96		5.60	98.1
Trp	Tryptophan	$C_{11}H_{12}N_2O_2$	204.23	289	2.38	9.34		5.89	13.2
Tyr	Tyrosine	$C_9H_{11}NO_3$	181.19	343	2.24	9.04	10.10	5.66	0.46
Val	Valine	$C_5H_{11}NO_2$	117.15	315	2.27	9.52		5.96	88.5

Other amino acids and related compounds:

Name	Mol. form.	M_r	t_m/°C	pK_a	pK_b	pK_c	pK_d	S/g kg^{-1}
N-Acetylglutamic acid	$C_7H_{11}NO_5$	189.17	199					
*N*6-Acetyl-*L*-lysine	$C_8H_{16}N_2O_3$	188.23	265	2.12	9.51			
β-Alanine	$C_3H_7NO_2$	89.09	200	3.51	10.08			891
DL-2-Aminobutanoic acid	$C_4H_9NO_2$	103.12	304	2.30	9.63			210
DL-3-Aminobutanoic acid	$C_4H_9NO_2$	103.12	194	3.43	10.05			1250
4-Aminobutanoic acid	$C_4H_9NO_2$	103.12	203	4.02	10.35			v.s.
5-Aminolevulinic acid	$C_5H_9NO_3$	131.13	118	4.05	8.90			
L-3-Amino-2-methylpropanoic acid	$C_4H_9NO_2$	103.12	183					
Azaserine	$C_5H_7N_3O_4$	173.13	150		8.55			v.s.
L-γ-Carboxyglutamic acid	$C_6H_9NO_6$	191.14	167	1.70	9.90	4.75	3.20	
Carnosine	$C_9H_{14}N_4O_3$	226.24	260	2.51	9.35	6.76		322
Citrulline	$C_6H_{13}N_3O_3$	175.19	222	2.32	9.30			
Creatine	$C_4H_9N_3O_2$	131.13	303	2.63	14.30			16
L-Cysteic acid	$C_3H_7NO_5S$	169.16	260	1.89	8.70	1.30		v.s.
L-Cystine	$C_6H_{12}N_2O_4S_2$	240.30	260	1.50	8.80	2.05	8.03	0.11
L-3,5-Diiodotyrosine	$C_9H_9I_2NO_3$	432.98	213	2.12	9.10	6.16		0.62
Dopamine	$C_8H_{11}NO_2$	153.18			10.36	8.88		
L-Ethionine	$C_6H_{13}NO_2S$	163.24	273	2.18	9.05	13.10		
Glycocyamine	$C_3H_7N_3O_2$	117.11	282	2.82				5
N-Glycylglycine	$C_4H_8N_2O_3$	132.12	263	3.13	8.10			
Histamine	$C_5H_9N_3$	111.15	83		9.83	6.11		v.s.
Homocysteine	$C_4H_9NO_2S$	135.19	232	2.15	8.57	10.38		
Homocystine	$C_8H_{16}N_2O_4S_2$	268.36	264	1.59	9.44	2.54	8.52	0.2
L-Homoserine	$C_4H_9NO_3$	119.12	203	2.27	9.28			1100
5-Hydroxylysine	$C_6H_{14}N_2O_3$	162.19		2.13	8.85	9.83		
trans-4-Hydroxyproline	$C_5H_9NO_3$	131.13	274	1.82	9.47			361
L-3-Iodotyrosine	$C_9H_{10}INO_3$	307.09	205	2.20	9.10	8.70		sl.s.
L-Kynurenine	$C_{10}H_{12}N_2O_3$	208.22	185					sl.s.
L-Lanthionine	$C_6H_{12}N_2O_4S$	208.24	295					1.5
Levodopa	$C_9H_{11}NO_4$	197.19	277	2.20	8.75	9.81	13.40	1650
2-Methylalanine	$C_4H_9NO_2$	103.12	335	2.36	10.21			137
L-1-Methylhistidine	$C_7H_{11}N_3O_2$	169.18	250	1.69	8.85	6.48		
L-Norleucine	$C_6H_{13}NO_2$	131.17	301	2.31	9.68			15
L-Norvaline	$C_5H_{11}NO_2$	117.15	305	2.31	9.65			107
L-Ornithine	$C_5H_{12}N_2O_2$	132.16	140	1.94	8.78	10.52		v.s.
O-Phosphoserine	$C_3H_8NO_6P$	185.07	166	2.14	9.80	5.70		
L-Pyroglutamic acid	$C_5H_7NO_3$	129.12	162	3.32				
Sarcosine	$C_3H_7NO_2$	89.09	212	2.18	9.97			428
L-Thyroxine	$C_{15}H_{11}I_4NO_4$	776.87	235	2.20	10.01	6.45		sl.s.

STRUCTURES OF COMMON AMINO ACIDS

Alanine
(Ala, A)

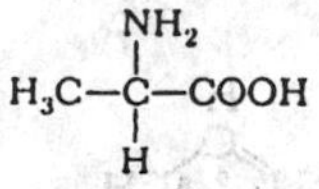

Arginine
(Arg, R)

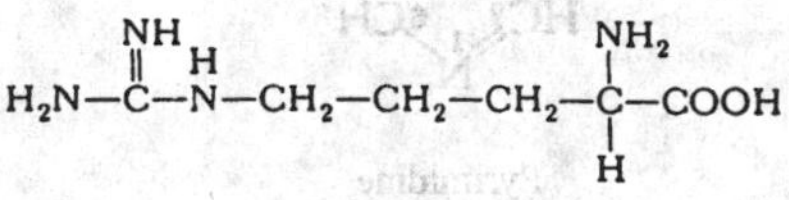

Asparagine
(Asn, N)

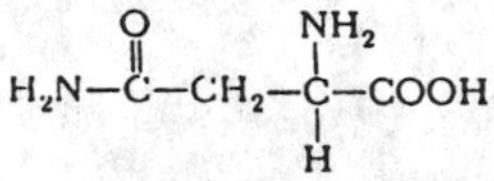

Aspartic acid
(Asp, D)

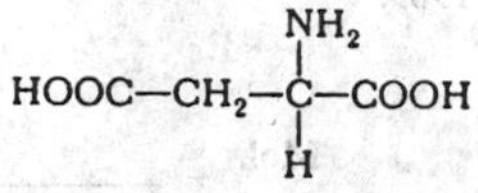

Cysteine
(Cys, C)

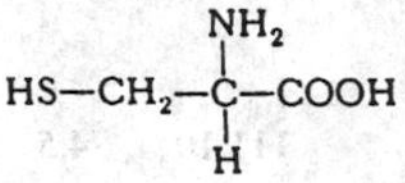

Glutamine
(Gln, Q)

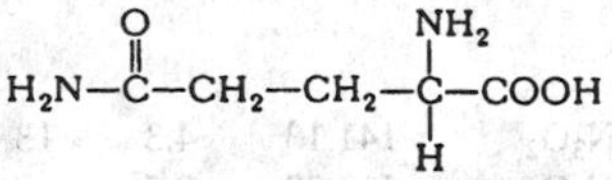

Glutamic acid
(Glu, E)

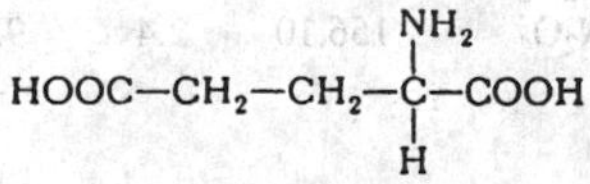

Glycine
(Gly, G)

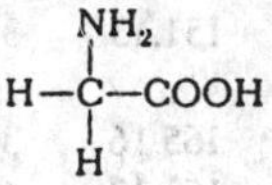

Histidine
(His, H)

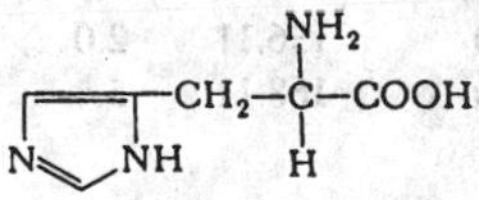

Isoleucine
(Ile, I)

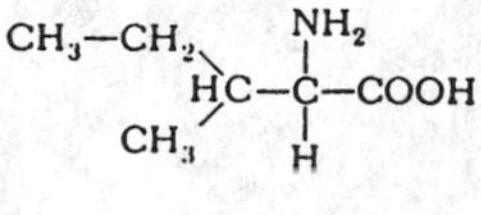

Leucine
(Leu, L)

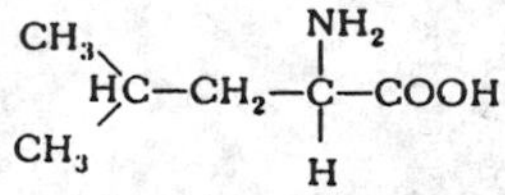

Lysine
(Lys, K)

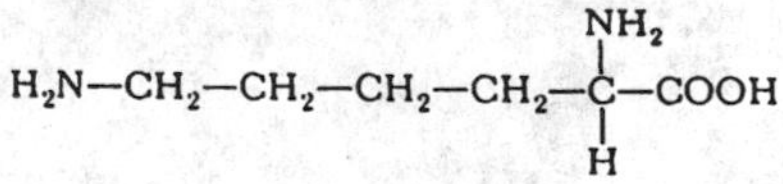

Methionine
(Met, M)

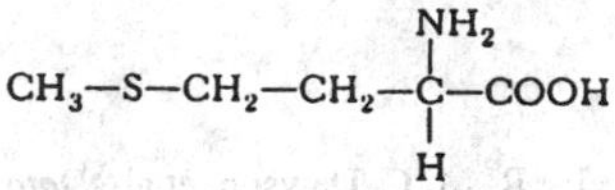

Phenylalanine
(Phe, F)

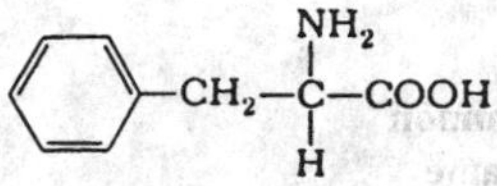

Proline
(Pro, P)

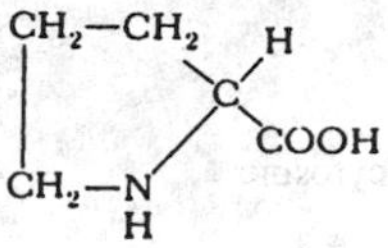

Serine
(Ser, S)

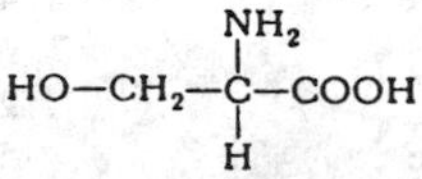

Threonine
(Thr, T)

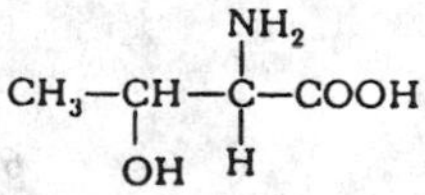

Tryptophan
(Trp, W)

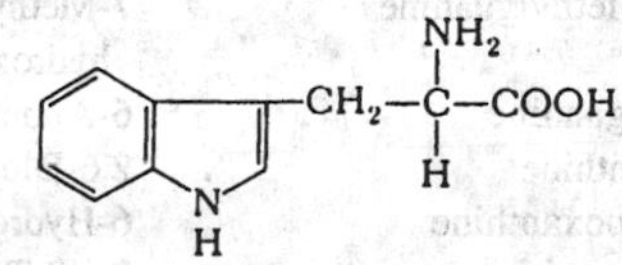

Tyrosine
(Tyr, Y)

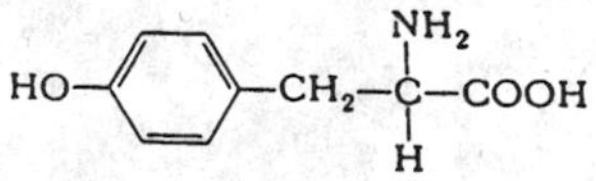

Valine
(Val, V)

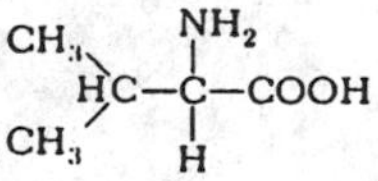

PROPERTIES OF PURINE AND PYRIMIDINE BASES

This table lists some of the important purine and pyrimidine bases that occur in nucleic acids. The pK_a values (negative logarithm of the acid dissociation constant) are given for each ionization stage. The last column gives the aqueous solubility S at the indicated temperature in units of grams per 100 grams of solution.

The numbering system in the rings is:

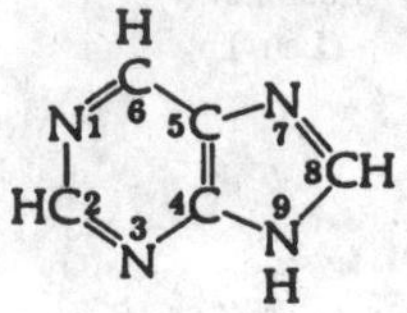

Purine

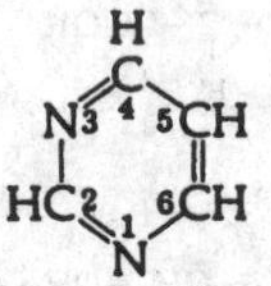

Pyrimidine

REFERENCES

1. R. M. C. Dawson, et al., *Data for Biochemical Research*, 3rd Ed., Clarendon Press, Oxford, 1986.
2. S. Budavari, Ed., *The Merck Index*, 11th Ed., Merk and Co., Rahway, NJ., 1989.

Common name	Systematic name	Mol form.	Mol. wt.	pK_a values			S/mass % (temp.)
Pyrimidines							
Cytosine	4-Amino-2-hydroxypyrimidine	$C_4H_5N_3O$	111.10	4.5	12.2		0.76 (25°C)
5-Methylcytosine	4-Amino-2-hydroxy-5-methylpyrimidine	$C_5H_7N_3O$	125.13	4.6	12.4		0.45 (25°C)
5-Hydroxymethyl-cytosine	4-Amino-2-hydroxy-5-hydroxy-methylpyrimidine	$C_5H_7N_3O_2$	141.13	4.3	13		
Uracil	2,4-Dihydroxypyrimidine	$C_4H_4N_2O_2$	112.09	0.5	9.5	>13	0.36 (25°C)
Thymine	5-Methyluracil	$C_5H_6N_2O_2$	126.11	9.9	>13		0.4 (25°C)
Orotic acid	Uracil-6-carboxylic acid	$C_5H_4N_2O_4$	156.10	2.4	9.5	>13	0.18 (18°C)
Purines							
Adenine	6-Aminopurine	$C_5H_5N_5$	135.14	<1	4.1	9.8	0.09 (25°C)
Guanine	2-Amino-6-hydroxypurine	$C_5H_5N_5O$	151.13	3.3	9.2	12.3	0.004 (40°C)
7-Methylguanine	7-Methyl-2-amino-6-hydroxypurine	$C_6H_7N_5O$	165.16	3.5	9.9		
Isoguanine	6-Amino-2-hydroxypurine	$C_5H_5N_5O$	151.13	4.5	9.0		0.006 (25°C)
Xanthine	2,6-Dioxopurine	$C_5H_4N_4O_2$	152.11	0.8	7.4	11.1	0.05 (20°C)
Hypoxanthine	6-Hydroxypurine	$C_5H_4N_4O$	136.11	2.0	8.9	12.1	0.07 (19°C)
Uric acid	2,6,8-Trihydroxypurine	$C_5H_4N_4O_3$	168.11	5.4	11.3		0.002 (20°C)

THE GENETIC CODE

This table gives the correspondence between a messenger RNA codon and the amino acid which it specifies. The symbols for bases in the codon are:

U: uracil
C: cytosine
A: adenine
G: guanine

The amino acid symbols are given in the table entitled "Structures of Common Amino Acids". A chain-initiating codon is indicated by **init** and a chain-terminating codon by **term**.

Example: UCA codes for **Ser**, UAC codes for **Tyr**, etc.

First position	Second position U	Second position C	Second position A	Second position G	Third position
U	Phe	Ser	Tyr	Cys	U
	Phe	Ser	Tyr	Cys	C
	Leu	Ser	**term**	**term**	A
	Leu	Ser	**term**	Trp	G
C	Leu	Pro	His	Arg	U
	Leu	Pro	His	Arg	C
	Leu	Pro	Gln	Arg	A
	Leu	Pro	Gln	Arg	G
A	Ile	Thr	Asn	Ser	U
	Ile	Thr	Asn	Ser	C
	Ile	Thr	Lys	Arg	A
	Met (init)	Thr	Lys	Arg	G
G	Val	Ala	Asp	Gly	U
	Val	Ala	Asp	Gly	C
	Val	Ala	Glu	Gly	A
	Val (init)	Ala	Glu	Gly	G

PROPERTIES OF FATTY ACIDS

This table gives the systematic names and selected properties of some of the more important fatty acids of five or more carbon atoms. Compounds are listed first by degree of saturation and, secondly, by number of carbon atoms. The following data are included:

M_r: Molecular weight
t_m: Melting point in °C
S: Aqueous solubility at 20°C in units of grams of solute per 100 grams of water

REFERENCES

1. Dawson, R. M. C., Elliott, D. C., Elliott, W. H., and Jones, K. M., *Data for Biochemical Research*, Third Edition, Clarendon Press, Oxford, 1986.
2. Fasman, G. D., Ed., *Practical Handbook of Biochemistry and Molecular Biology*, CRC Press, Boca Raton, FL, 1989.

Common name	Systematic name	Mol. form.	M_r	t_m/°C	S
		Saturated			
Valeric acid	Pentanoic acid	$C_5H_{10}O_2$	102.13	-34	2.5
Isovaleric acid	3-Methylbutanoic acid	$C_5H_{10}O_2$	102.13	-29.3	4.3
Caproic acid	Hexanoic acid	$C_6H_{12}O_2$	116.16	-3	0.967
Enanthic acid	Heptanoic acid	$C_7H_{14}O_2$	130.19	-7.5	0.24
Caprylic acid	Octanoic acid	$C_8H_{16}O_2$	144.21	16.3	0.080
Pelargonic acid	Nonanoic acid	$C_9H_{18}O_2$	158.24	12.3	0.0284
Capric acid	Decanoic acid	$C_{10}H_{20}O_2$	172.27	31.9	0.015
Lauric acid	Dodecanoic acid	$C_{12}H_{24}O_2$	200.32	43.2	0.0055
Tridecylic acid	Tridecanoic acid	$C_{13}H_{26}O_2$	214.35	41.5	0.0033
Myristic acid	Tetradecanoic acid	$C_{14}H_{28}O_2$	228.38	53.9	0.0020
Pentadecylic acid	Pentadecanoic acid	$C_{15}H_{30}O_2$	242.40	52.3	0.0012
Palmitic acid	Hexadecanoic acid	$C_{16}H_{32}O_2$	256.43	63.1	0.00072
Margaric acid	Heptadecanoic acid	$C_{17}H_{34}O_2$	270.46	61.3	0.00042
Stearic acid	Octadecanoic acid	$C_{18}H_{36}O_2$	284.48	69.6	0.00029
Arachidic acid	Eicosanoic acid	$C_{20}H_{40}O_2$	312.54	76.5	
Phytanic acid	3,7,11,15-Tetramethylhexadecanoic acid	$C_{20}H_{40}O_2$	312.54	-65	
Behenic acid	Docosanoic acid	$C_{22}H_{44}O_2$	340.59	81.5	
Lignoceric acid	Tetracosanoic acid	$C_{24}H_{48}O_2$	368.64	87.5	
Cerotic acid	Hexacosanoic acid	$C_{26}H_{52}O_2$	396.70	88.5	
Montanic acid	Octacosanoic acid	$C_{28}H_{56}O_2$	424.75	90.9	
		Monounsaturated			
Caproleic acid	9-Decenoic acid	$C_{10}H_{18}O_2$	170.25	26.5	
Palmitoleic acid	*cis*-9-Hexadecenoic acid	$C_{16}H_{30}O_2$	254.41	-0.1	
Oleic acid	*cis*-9-Octadecenoic acid	$C_{18}H_{34}O_2$	282.47	13.4	
Elaidic acid	*trans*-9-Octadecenoic acid	$C_{18}H_{34}O_2$	282.47	45	
Vaccenic acid	*trans*-11-Octadecenoic acid	$C_{18}H_{34}O_2$	282.47	44	
Erucic acid	*cis*-13-Docosenoic acid	$C_{22}H_{42}O_2$	338.57	34.7	
Brassidic acid	*trans*-13-Docosenoic acid	$C_{22}H_{42}O_2$	338.57	61.9	
Nervonic acid	*cis*-15-Tetracosenoic acid	$C_{24}H_{46}O_2$	366.63	43	
		Diunsaturated			
Linoleic acid	*cis,cis*-9,12-Octadecadienoic acid	$C_{18}H_{32}O_2$	280.45	-12	
		Triunsaturated			
cis-Eleostearic acid	*trans,cis,trans*-9,11,13-Octadecatrienoic acid	$C_{18}H_{30}O_2$	278.44	49	
trans-Eleostearic acid	*trans,trans,trans*-9,11,13-Octadecatrienoic acid	$C_{18}H_{30}O_2$	278.44	71.5	
Linolenic acid	*cis,cis,cis*-9,12,15-Octadecatrienoic acid	$C_{18}H_{30}O_2$	278.44	-11	
		Tetraunsaturated			
Arachidonic acid	5,8,11,14-Eicosatetraenoic acid, (all-*trans*)	$C_{20}H_{32}O_2$	304.47	-49.5	

CARBOHYDRATE NAMES AND SYMBOLS

The following table lists the systematic names and symbols for selected carbohydrates and some of their derivatives. The symbols for monosaccharide residues and derivatives are recommended by IUPAC for use in describing the structures of oligosaccharide chains. A more complete list can be found in the reference.

REFERENCE

McNaught, A. D., *Pure Appl. Chem.*, 68, 1919-2008, 1996.

Common Name	Symbol	Systematic Name
Abequose	Abe	3,6-Dideoxy-D-*xylo*-hexose
N-Acetyl-2-deoxyneur-2-enaminic acid	Neu2en5Ac	
N-Acetylgalactosamine	GalNAc	
N-Acetylglucosamine	GlcNAc	
N-Acetylneuraminic acid	Neu5Ac	
Allose	All	*allo*-Hexose
Altrose	Alt	*altro*-Hexose
Apiose	Api	3-*C*-(Hydroxymethyl)-*glycero*-tetrose
Arabinitol	Ara-ol	Arabinitol
Arabinose	Ara	*arabino*-Pentose
Arcanose		2,6-Dideoxy-3-*C*-methyl-3-*O*-methyl-*xylo*-hexose
Ascarylose		3,6-Dideoxy-L-*arabino*-hexose
Boivinose		2,6-Dideoxy-D-gulose
Chalcose		4,6-Dideoxy-3-*O*-methyl-D-*xylo*-hexose
Cladinose		2,6-Dideoxy-3-*C*-methyl-3-*O*-methyl-L-*ribo*-hexose
Colitose		3,6-Dideoxy-L-*xylo*-hexose
Cymarose		6-Deoxy-3-*O*-methyl-*ribo*-hexose
3-Deoxy-D-*manno*-oct-2-ulosonic acid	Kdo	
2-Deoxyribose	dRib	2-Deoxy-*erythro*-pentose
2,3-Diamino-2,3-dideoxy-D-glucose	GlcN3N	
Diginose		2,6-Dideoxy-3-*O*-methyl-*lyxo*-hexose
Digitalose		6-Deoxy-3-*O*-methyl-D-galactose
Digitoxose		2,6-Dideoxy-D-*ribo*-hexose
3,4-Di-*O*-methylrhamnose	Rha3,4Me_2	
Ethyl glucopyranuronate	Glc*p*A6Et	
Evalose		6-Deoxy-3-*C*-methyl-D-mannose
Fructose	Fru	*arabino*-Hex-2-ulose
Fucitol	Fuc-ol	6-Deoxy-D-galactitol
Fucose	Fuc	6-Deoxygalactose
β-D-Galactopyranose 4-sulfate	β-D-Gal*p*4*S*	
Galactosamine	GalN	2-Amino-2-deoxygalactose
Galactose	Gal	*galacto*-Hexose
Glucitol	Glc-ol	
Glucosamine	GlcN	2-Amino-2-deoxyglucose
Glucose	Glc	*gluco*-Hexose
Glucuronic acid	GlcA	
N-Glycoloylneuraminic acid	Neu5Gc	
Gulose	Gul	*gulo*-Hexose
Hamamelose		2-*C*-(Hydroxymethyl)-D-ribose
Idose	Ido	*ido*-Hexose
Iduronic acid	IdoA	
Lactose	Lac	β-D-Galactopyranosyl-(1→4)-D-glucose
Lyxose	Lyx	*lyxo*-Pentose
Maltose		α-D-Glucopyranosyl-(1→4)-D-glucose
Mannose	Man	*manno*-Hexose
2-*C*-Methylxylose	Xyl2*C*Me	
Muramic acid	Mur	2-Amino-3-*O*-[(R)-1-carboxyethyl]-2-deoxy-D-glucose
Mycarose		2,6-Dideoxy-3-*C*-methyl-L-*ribo*-hexose
Mycinose		6-Deoxy-2,3-di-*O*-methyl-D-allose
Neuraminic acid	Neu	5-Amino-3,5-dideoxy-D-*glycero*-D-*galacto*-non-2-ulosonic acid

CARBOHYDRATE NAMES AND SYMBOLS (continued)

Common Name	Symbol	Systematic Name
Panose		α-D-Glucopyranosyl-(1→6)-α-D-glucopyranosyl-(1→ 4)-D-glucose
Paratose		3,6-Dideoxy-D-*ribo*-hexose
Primeverose		β-D-Xylopyranosyl-(1→6)-D-glucose
Psicose	Psi	*ribo*-Hex-2-ulose
Quinovose	Qui	6-Deoxyglucose
Raffinose		β-D-Fructofuranosyl-α-D-galactopyranosyl-(1→6)-α-D-glucopyranoside
Rhamnose	Rha	6-Deoxymannose
Rhodinose		2,3,6-Trideoxy-L-*threo*-hexose
Ribose	Rib	*ribo*-Pentose
Ribose 5-phosphate	Rib5*P*	
Ribulose	Ribulo (Rul)	*erythro*-Pent-2-ulose
Rutinose		α-L-Rhamnopyranosyl-(1→6)-D-glucose
Sarmentose		2,6-Dideoxy-3-*O*-methyl-D-*xylo*-hexose
Sedoheptulose		D-*altro*-Hept-2-ulose
Sorbose	Sor	*xylo*-Hex-2-ulose
Streptose		5-Deoxy-3-*C*-formyl-L-lyxose
Sucrose		β-D-Fructofuranosyl-α-D-glucopyranoside
Tagatose	Tag	*lyxo*-Hex-2-ulose
Talose	Tal	*talo*-Hexose
Turanose		α-D-Glucopyranosyl-(1→3)-D-fructose
Tyvelose	Tyv	3,6-Dideoxy-D-*arabino*-hexose
Xylose	Xyl	*xylo*-Pentose
Xylulose	Xylulo (Xul)	*threo*-Pent-2-ulose

STANDARD TRANSFORMED GIBBS ENERGY OF FORMATION FOR IMPORTANT BIOCHEMICAL SPECIES

Petr Vanýsek

This table lists transformed values of the standard Gibbs energy of formation for several molecules and ions of biochemical importance. Values of $\Delta_f G'^o$ are given at pH 7, 298.15 K, and 100 kPa for infinite dilution and for two finite ionic strengths, $I = 0.1$ mol/L and $I = 0.25$ mol/L. The charge of the species (z_i) is also given.

The table can be used for calculating practical (pH 7) reduction potentials for important biological processes. Such listing is more compact than offering reduction potentials, which would require tabulating a large number of reactant-product combinations.

To calculate the standard apparent reduction potential E'^o for reduction of acetaldehyde to ethanol at infinite dilution, for example, write first the reaction:

$$CH_3CHO + 2H^+ + 2e^- \rightleftharpoons CH_3CH_2OH$$

The change in hydrogen count can be accomplished by adding H^+, which in turn has to be compensated by adding the appropriate number of electrons (reduction). The correct count of electrons is needed in the subsequent equation for the reduction potential:

$$E'^o = [-1/nF] \cdot [\Delta_f G'^o(\text{product}) - \Delta_f G'^o(\text{reactant})]$$

where n is the number of electrons to be added and F is the Faraday constant.

Specifically, for the above reaction:

$$E'^o = [-1/(2 \cdot 9.6485 \cdot 10^4\ \text{C mol}^{-1})] \cdot [58.1 \cdot 10^3\ \text{J mol}^{-1} - 20.83 \cdot 10^3\ \text{J mol}^{-1}] = -0.193\ \text{V}.$$

REFERENCE

Alberty, R. A., *Arch. Biochem. Biophys.*, **353**, 116-130, 1998; **358**, 25-39, 1998.

	$\Delta_f G'^o$ /kJ mol^{-1}			
Compound	**$I = 0$**	**$I = 0.1$ mol/L**	**$I = 0.25$ mol/L**	z_i
Acetaldehyde	20.83	23.27	24.06	0
Acetate	-249.44	-248.22	-247.82	-1
Acetone	80.03	83.71	84.89	0
cis-Aconitate	-797.26	-800.94	-802.12	-3
Adenine	512.07	515.13	516.12	0
Adenosine	519.43	527.39	529.96	0
Adenosine diphosphate (ADP)	-1234.36	-1230.97	-1230.12	-3
Adenosine monophosphate (AMP)	-367.5	-361.99	-360.29	-2
Adenosine triphosphate (ATP)	-2098	-2097.55	-2097.89	-4
Alanine	-91.31	-87.02	-85.64	0
Ammonium	80.52	82.35	82.94	1
Arabinose	-342.67	-336.55	-334.57	0
L-Asparagine	-206.28	-201.38	-199.8	0
Aspartate	-456.15	-453.09	-452.1	-1
1-Butanol	227.72	233.84	235.82	0
Butyrate	-72.94	-69.26	-68.08	-1
Carbonate	-547.33	-547.15	-547.1	-2
iso-Citrate	-956.82	-958.84	-959.58	-3
Citrate	-963.46	-965.49	-966.23	-3
CO(aq)	-119.9	-119.9	-119.9	0
CO(g)	-137.17	-137.17	-137.17	0
CO_2(g)	-394.36	-394.36	-394.36	0
Creatine	100.41	105.92	107.69	0
Creatinine	256.55	260.84	262.22	0
L-Cysteine	-59.23	-55.01	-53.65	0
L-Cystine	-187.03	-179.69	-177.32	0
Cytochrome c [oxidized]	0	-5.51	-7.29	3
Cytochrome c [reduced]	-24.54	-26.96	-27.75	2
Ethanol	58.1	61.77	62.96	0

STANDARD TRANSFORMED GIBBS ENERGY OF FORMATION FOR IMPORTANT BIOCHEMICAL SPECIES (continued)

	$\Delta_f G'^o$ /kJ mol^{-1}			
Compound	$I = 0$	$I = 0.1$ mol/L	$I = 0.25$ mol/L	z_i
Ethyl acetate	-18	-13.1	-11.52	0
Ferredoxin [oxidized]	0	-0.61	-0.81	1
Ferredoxin [reduced]	38.07	38.07	38.07	0
Flavin adenine dinucleotide [oxidized]	1238.65	1255.17	1260.51	-2
Flavin adenine dinucleotide [reduced]	1279.68	1297.43	1303.16	-2
Flavin mononucleotide [oxidized]	759.17	768.35	771.32	-2
Flavin mononucleotide [reduced]	800.2	810.61	813.97	-2
Formate	-311.04	-311.04	-311.04	-1
Fructose	-436.03	-428.69	-426.32	0
Fructose-6-phosphate	-1321.71	-1317.16	-1315.74	-1
Fumarate	-521.96	-523.18	-523.58	-2
Galactose	-429.45	-422.11	-419.74	0
Galactose-1-phosphate	-1317.5	-1313.01	-1311.6	-2
Glucose	-436.42	-429.08	-426.71	0
Glucose-1-phosphate	-1318.03	-1313.34	-1311.89	-2
Glucose-6-phosphate	-1325	-1320.37	-1318.92	-2
Glutamate	-377.82	-373.54	-372.16	-1
Glutamine	-128.46	-122.34	-120.36	0
Glutathione [oxidized]	1198.69	1214.6	1219.74	-2
Glutathione [reduced]	625.56	633.52	636.09	-1
Glycerol	-177.83	-172.93	-171.35	0
Glycerol-3-phosphate	-1080.22	-1077.83	-1077.14	-1
Glycine	-180.13	-177.07	-176.08	0
Glycolate	-411.08	-409.86	-409.46	-1
Glycylglycine	-200.55	-195.65	-194.07	0
Glyoxylate	-428.64	-428.64	-428.64	-1
$H_2(aq)$	97.51	98.74	99.13	0
$H_2(g)$	79.91	81.17	81.53	0
H_2O	-157.28	-156.05	-155.66	0
β-Hydroxypropionate	-318.62	-316.17	-315.38	-1
Hydrogen peroxide	-54.12	-52.89	-52.5	0
Hypoxanthine	249.33	251.77	252.56	0
Indole	503.49	507.78	509.16	0
L-Isoleucine	175.53	183.49	186.06	0
Lactate	-316.94	-314.49	-313.7	-1
Lactose	-688.29	-674.82	-670.48	0
L-Leucine	-167.18	-175.14	-177.71	0
Lyxose	-349.58	-343.46	-341.48	0
Malate	-682.83	-682.83	-682.83	-2
Maltose	-695.65	-682.19	-677.84	0
D-Mannitol	-383.22	-374.65	-371.89	0
Mannose	-431.51	-424.17	-421.8	0
Methane	109.11	111.55	112.34	0
Methanol	-15.45	-13.04	-12.25	0
L-Methionine	-63.4	-56.67	-54.49	0
Methylamine	199.88	202.94	203.93	1
$N_2(aq)$	18.07	18.07	18.07	0
$N_2(g)$	0	0	0	0
Nicotinamide adenine dinucleotide (NAD) [reduced]	1101.47	1115.55	1120.09	-2
Nicotinamide adenine dinucleotide (NAD) [oxidized]	1038.86	1054.17	1059.11	-1
Nicotinamide adenine dinucleotide phosphate (NADP) [reduced]	1064.85	1070.97	1072.95	-4
Nicotinamide adenine dinucleotide phosphate (NADP) [oxidized]	998.91	1008.7	1011.86	-3

STANDARD TRANSFORMED GIBBS ENERGY OF FORMATION FOR IMPORTANT BIOCHEMICAL SPECIES (continued)

Compound	$\Delta_f G'^{o}$ /kJ mol^{-1}			
	$I = 0$	$I = 0.1$ mol/L	$I = 0.25$ mol/L	z_i
$O_2(aq)$	16.4	16.4	16.4	0
$O_2(g)$	0	0	0	0
Oxalate	-673.9	-676.35	-677.14	-2
Oxaloacetate	-713.37	-714.6	-714.99	-2
Oxalosuccinate	-979.06	-979.06	-979.06	-2
2-Oxoglutarate	-633.59	-633.59	-633.59	-2
Palmitate	979.25	997.6	1003.54	-1
L-Phenylalanine	232.42	239.15	241.33	0
Phosphate	-1058.56	-1059.17	-1059.49	-2
1-Propanol	143.84	148.74	150.32	0
2-Propanol	134.43	139.32	140.9	0
Pyrophosphate	-1937.66	-1941.82	-1943.35	-1
Pyruvate	-352.4	-351.18	-350.78	-1
Retinal	1118.78	1135.91	1141.45	0
Retinol	1170.77	1189.13	1195.06	0
Ribose	-339.23	-333.11	-331.13	0
Ribose-1-phosphate	-1215.87	-1212.24	-1211.14	-2
Ribose-5-phosphate	-1223.95	-1220.32	-1219.22	-2
Ribulose	-336.38	-330.26	-328.28	0
L-Serine	-231.18	-226.89	225.81	0
D-Sorbose	-432.47	-425.13	-422.76	0
Succinate	-530.62	-530.62	-530.62	-2
Sucrose	-685.66	-672.2	-667.85	0
Thioredoxin [oxidized]	0	0	0	0
Thioredoxin [reduced]	54.03	55.26	55.65	0
L-Tryptophane	366.88	374.22	376.59	0
L-Tyrosine	68.82	75.55	77.73	0
Ubiquinone [oxidized]	3596.07	3651.15	3668.94	0
Ubiquinone [reduced]	3586.16	3642.47	3660.65	0
Urate	-206.1	-204.85	204.45	-1
Urea	-42.92	-40.53	-39.73	0
L-Valine	-80.87	-87.6	-89.78	0
Xylose	-350.93	-344.81	-342.83	0
D-Xylulose	-346.59	-340.47	-338.49	0

BIOLOGICAL BUFFERS

This table of frequently used buffers gives the pK_a value at 25°C and the useful pH range of each buffer. The buffers are listed in order of increasing pH.

The table is reprinted with permission of Sigma Chemical Company, St. Louis, Mo.

Acronym	Name	Mol. wt.	pK_a	Useful pH range
MES	2-(*N*-Morpholino)ethanesulfonic acid	195.2	6.1	5.5—6.7
BIS TRIS	*Bis*(2-hydroxyethyl)iminotris(hydroxymethyl)methane	209.2	6.5	5.8—7.2
ADA	*N*-(2-Acetamido)-2-iminodiacetic acid	190.2	6.6	6.0—7.2
ACES	2-[(2-Amino-2-oxoethyl)amino]ethanesulfonic acid	182.2	6.8	6.1—7.5
PIPES	Piperazine-*N,N'-bis*(2-ethanesulfonic acid)	302.4	6.8	6.1—7.5
MOPSO	3-(*N*-Morpholino)-2-hydroxypropanesulfonic acid	225.3	6.9	6.2—7.6
BIS TRIS PROPANE	1,3-*Bis*[*tris*(hydroxymethyl)methylamino]propane	282.3	6.8[a]	6.3—9.5
BES	*N,N-Bis*(2-hydroxyethyl)-2-aminoethanesulfonic acid	213.2	7.1	6.4—7.8
MOPS	3-(*N*-Morpholino)propanesulfonic acid	209.3	7.2	6.5—7.9
HEPES	*N*-(2-Hydroxyethyl)piperazine-*N'*-(2-ethanesulfonic acid)	238.3	7.5	6.8—8.2
TES	*N-Tris*(hydroxymethyl)methyl-2-aminoethanesulfonic acid	229.2	7.5	6.8—8.2
DIPSO	3-[*N,N-Bis*(2-hydroxyethyl)amino]-2-hydroxypropanesulfonic acid	243.3	7.6	7.0—8.2
TAPSO	3-[*N-Tris*(hydroxymethyl)methylamino)-2-hydroxypropanesulfonic acid	259.3	7.6	7.0—8.2
TRIZMA	Tris(hydroxymethyl)aminomethane	121.1	8.1	7.0—9.1
HEPPSO	*N*-(2-hydroxyethyl)piperazine-*N'*-(2-hydroxypropanesulfonic acid)	268.3	7.8	7.1—8.5
POPSO	Piperazine-*N,N'-bis*(2-hydroxypropanesulfonic acid)	362.4	7.8	7.2—8.5
EPPS	*N*-(2-Hydroxyethyl)piperazine-*N'*-(3-propanesulfonic acid)	252.3	8.0	7.3—8.7
TEA	Triethanolamine	149.2	7.8	7.3—8.3
TRICINE	*N-Tris*(hydroxymethyl)methylglycine	179.2	8.1	7.4—8.8
BICINE	*N,N-Bis*(2-hydroxyethyl)glycine	163.2	8.3	7.6—9.0
TAPS	*N-Tris*(hydroxymethyl)methyl-3-aminopropanesulfonic acid	243.3	8.4	7.7—9.1
AMPSO	3-[(1,1-Dimethyl-2-hydroxyethyl)amino]-2-hydroxypropanesulfonic acid	227.3	9.0	8.3—9.7
CHES	2-(*N*-Cyclohexylamino)ethanesulfonic acid	207.3	9.3	8.6—10.0
CAPSO	3-(Cyclohexylamino)-2-hydroxy-1-propanesulfonic acid	237.3	9.6	8.9—10.3
AMP	2-Amino-2-methyl-1-propanol	89.1	9.7	9.0—10.5
CAPS	3-(Cyclohexylamino)-1-propanesulfonic acid	221.3	10.4	9.7—11.1

[a] pK_a = 9.0 for the second dissociation stage.

TYPICAL pH VALUES OF BIOLOGICAL MATERIALS AND FOODS

This table gives typical pH ranges for various biological fluids and common foods. All values refer to 25°C.

Biological Materials

Blood, human	7.35-7.45
Blood, dog	6.9-7.2
Spinal fluid, human	7.3-7.5
Saliva, human	6.5-7.5
Gastric contents, human	1.0-3.0
Duodenal contents, human	4.8-8.2
Feces, human	4.6-8.4
Urine, human	4.8-8.4
Milk, human	6.6-7.6
Bile, human	6.8-7.0

Foods

Apples	2.9-3.3
Apricots	3.6-4.0
Asparagus	5.4-5.8
Bananas	4.5-4.7
Beans	5.0-6.0
Beers	4.0-5.0
Beets	4.9-5.5
Blackberries	3.2-3.6
Bread, white	5.0-6.0
Butter	6.1-6.4
Cabbage	5.2-5.4
Carrots	4.9-5.3
Cheese	4.8-6.4
Cherries	3.2-4.0
Cider	2.9-3.3
Corn	6.0-6.5
Crackers	6.5-8.5
Dates	6.2-6.4
Eggs, fresh white	7.6-8.0
Flour, wheat	5.5-6.5
Gooseberries	2.8-3.0
Grapefruit	3.0-3.3
Grapes	3.5-4.5
Hominy (lye)	6.8-8.0
Jams, fruit	3.5-4.0
Jellies, fruit	2.8-3.4
Lemons	2.2-2.4
Limes	1.8-2.0
Maple syrup	6.5-7.0
Milk, cows	6.3-6.6
Olives	3.6-3.8
Oranges	3.0-4.0
Oysters	6.1-6.6
Peaches	3.4-3.6
Pears	3.6-4.0
Peas	5.8-6.4
Pickles, dill	3.2-3.6
Pickles, sour	3.0-3.4
Pimento	4.6-5.2
Plums	2.8-3.0
Potatoes	5.6-6.0
Pumpkin	4.8-5.2
Raspberries	3.2-3.6
Rhubarb	3.1-3.2
Salmon	6.1-6.3
Sauerkraut	3.4-3.6
Shrimp	6.8-7.0
Soft drinks	2.0-4.0
Spinach	5.1-5.7
Squash	5.0-5.4
Strawberries	3.0-3.5
Sweet potatoes	5.3-5.6
Tomatoes	4.0-4.4
Tuna	5.9-6.1
Turnips	5.2-5.6
Vinegar	2.4-3.4
Water, drinking	6.5-8.0
Wines	2.8-3.8

CHEMICAL COMPOSITION OF THE HUMAN BODY

The elemental composition of the "standard man" of mass 70 kg is given below.

REFERENCES

1. Padikal, T.N., and Fivozinsky, S.P., *Medical Physics Data Book, National Bureau of Standards Handbook 138*, U. S. Government Printing Office, Washington, DC, 1981.
2. Snyde, W.S., et al., *Reference Man: Anatomical, Physiological, and Metabolic Characteristics*, Pergamon, New York, 1975.

Element	Amount (g)	Percent of total body mass
Oxygen	43,000	61
Carbon	16,000	23
Hydrogen	7000	10
Nitrogen	1800	2.6
Calcium	1000	1.4
Phosphorus	780	1.1
Sulfur	140	0.20
Potassium	140	0.20
Sodium	100	0.14
Chlorine	95	0.12
Magnesium	19	0.027
Silicon	18	0.026
Iron	4.2	0.006
Fluorine	2.6	0.0037
Zinc	2.3	0.0033
Rubidium	0.32	0.00046
Strontium	0.32	0.00046
Bromine	0.20	0.00029
Lead	0.12	0.00017
Copper	0.072	0.00010
Aluminum	0.061	0.00009
Cadmium	0.050	0.00007
Boron	<0.048	0.00007
Barium	0.022	0.00003
Tin	<0.017	0.00002
Manganese	0.012	0.00002
Iodine	0.013	0.00002
Nickel	0.010	0.00001
Gold	<0.010	0.00001
Molybdenum	<0.0093	0.00001
Chromium	<0.0018	0.000003
Cesium	0.0015	0.000002
Cobalt	0.0015	0.000002
Uranium	0.00009	0.0000001
Beryllium	0.000036	
Radium	$3.1 \cdot 10^{-11}$	

Section 8
Analytical Chemistry

Preparation of Special Analytical Reagents 8-1
Standard Solutions of Acids, Bases, and Salts 8-5
Standard Solutions of Oxidation and Reduction Reagents 8-7
Organic Analytical Reagents for the Determination of Inorganic Substances 8-8
Acid-Base Indicators 8-16
Fluorescent Indicators 8-19
Conversion Formulas for Concentration of Solutions 8-20
Electrochemical Series 8-21
Reduction and Oxidation Potentials for Certain Ion Radicals 8-32
pH Scale for Aqueous Solutions 8-35
Practical pH Measurements on Natural Waters 8-41
Buffer Solutions Giving Round Values of pH at 25°C 8-43
Dissociation Constants of Inorganic Acids and Bases 8-44
Dissociation Constants of Organic Acids and Bases 8-46
Concentrative Properties of Aqueous Solutions: Density, ... 8-57
Refractive Index, Freezing Point Depression, and Viscosity Ion Product of
Water Substance 8-84
Ionization Constant of Normal and Heavy Water 8-85
Solubility of Selected Gases in Water 8-86
Solubility of Carbon Dioxide in Water at Various Temperatures and Pressures 8-90
Aqueous Solubility and Henry's Law Constants of Organic Compounds 8-91
Aqueous Solubility of Inorganic Compounds at Various Temperatures 8-108
Solubility Product Constants 8-117
Solubility of Common Salts at Ambient Temperatures 8-121
Solubility Chart 8-122
Reduction of Weighings in Air to Vacuo 8-125
Volume of One Gram of Water 8-125
Properties of Carrier Gases for Gas Chromatography 8-126
Solvents for Ultraviolet Spectrophotometry 8-127
^{13}C Chemical Shifts of Useful NMR Solvents 8-128
Mass Spectral Peaks of Common Organic Solvents 8-129

PREPARATION OF SPECIAL ANALYTICAL REAGENTS

Aluminon (qualitative test for aluminum). Aluminon is a trade name for the ammonium salt of aurintricarboxylic acid. Dissolve 1 g of the salt in 1 L of distilled water. Shake the solution well to insure thorough mixing.

Bang's reagent (for glucose estimation). Dissolve 100 g of K_2CO_3, 66 g of KCl and 160 g of $KHCO_3$ in the order given in about 700 mL of water at 30°C. Add 4.4 g of $CuSO_4$ and dilute to 1 L after the CO_2 is evolved. This solution should be shaken only in such a manner as not to allow entry of air. After 24 hours 300 mL are diluted to 1 L with saturated KCl solution, shaken gently and used after 24 hours; 50 mL is equivalent to 10 mg glucose.

Barfoed's reagent (test for glucose). See Cupric acetate.

Baudisch's reagent. See Cupferron.

Benedict's solution (qualitative reagent for glucose). With the aid of heat, dissolve 173 g of sodium citrate and 100 g of Na_2CO_3 in 800 mL of water. Filter, if necessary, and dilute to 850 mL. Dissolve 17.3 g of $CuSO_4{\cdot}5H_2O$ in 100 mL of water. Pour the latter solution, with constant stirring, into the carbonate-citrate solution, and dilute to 1 L.

Benzidine hydrochloride solution (for sulfite determination). Make a paste of 8 g of benzidine hydrochloride ($C_{12}H_8(NH_3)_2{\cdot}2HCl$) and 20 mL of water, add 20 mL of HCl (sp. gr. 1.12) and dilute to 1 L with water. Each mL of this solution is equivalent to 0.00357 g of H_2SO_4.

Bertrand's reagent (glucose estimation). Consists of the following solutions:

1. Dissolve 200 g of Rochelle salt and 150 g of NaOH in sufficient water to make 1 L of solution.
2. Dissolve 40 g of $CuSO_4$ in enough water to make 1 L of solution.
3. Dissolve 50 g of $Fe_2(SO_4)_3$ and 200 g of H_2SO_4 (sp. gr. 1.84) in sufficient water to make 1 L of solution.
4. Dissolve 5 g of $KMnO_4$ in sufficient water to make 1 L of solution.

Bial's reagent (for pentose). Dissolve 1 g of orcinol (5-methyl-1,3-benzenediol) in 500 mL of 30% HCl to which 30 drops of a 10% solution of $FeCl_3$ has been added.

Boutron — Boudet soap solution:

1. Dissolve 100 g of pure castile soap in about 2.5 L of 56% ethanol.
2. Dissolve 0.59 g of $Ba(NO_3)_2$ in 1 L of water.

Adjust the castile soap solution so that 2.4 mL of it will give a permanent lather with 40 mL of solution (b). When adjusted, 2.4 mL of soap solution is equivalent to 220 parts per million of hardness (as $CaCO_3$) for a 40 mL sample. See also Soap solution.

Brucke's reagent (protein precipitation). See Potassium iodide-mercuric iodide.

Clarke's soap solution (estimation of hardness in water).

1. Dissolve 100 g of pure powdered castile soap in 1 L of 80% ethanol and allow to stand over night.
2. Prepare a solution of $CaCl_2$ by dissolving 0.5 g of $CaCO_3$ in HCl (sp. gr. 1.19), neutralize with NH_4OH and make slightly alkaline to litmus, and dilute to 500 mL. One mL is equivalent to 1 mg of $CaCO_3$.

Titrate (1) against (2) and dilute (1) with 80% ethanol until 1 mL of the resulting solution is equivalent to 1 mL of (2) after making allowance for the lather factor (the amount of standard soap solution required to produce a permanent lather in 50 mL of distilled water). One mL of the adjusted solution after subtracting the lather factor is equivalent to 1 mg of $CaCO_3$. See also Soap solution.

Cobalticyanide paper (Rinnmann's test for Zn). Dissolve 4 g of $K_3Co(CN)_6$ and 1 g of $KClO_3$ in 100 mL of water. Soak filter paper in solution and dry at 100°C. Apply drop of zinc solution and burn in an evaporating dish. A green disk is obtained if zinc is present.

Cochineal. Extract 1 g of cochineal for 4 days with 20 mL of alcohol and 60 mL of distilled water. Filter.

Congo red. Dissolve 0.5 g of congo red in 90 mL of distilled water and 10 mL of alcohol.

Cupferron (Baudisch's reagent for iron analysis). Dissolve 6 g of the ammonium salt of *N*-hydroxy-*N*-nitrosoaniline (cupferron) in 100 mL of H_2O. Reagent good for 1 week only and must be kept in the dark.

Cupric acetate (Barfoed's reagent for reducing monosaccharides). Dissolve 66 g of cupric acetate and 10 mL of glacial acetic acid in water and dilute to 1 L.

Cupric oxide, ammoniacal; Schweitzer's reagent (dissolves cotton, linen, and silk, but not wool).

1. Dissolve 5 g of cupric sulfate in 100 mL of boiling water, and add sodium hydroxide until precipitation is complete. Wash the precipitate well, and dissolve it in a minimum quantity of ammonium hydroxide.
2. Bubble a slow stream of air through 300 mL of strong ammonium hydroxide containing 50 g of fine copper turnings. Continue for 1 hour.

Cupric sulfate in glycerin-potassium hydroxide (reagent for silk). Dissolve 10 g of cupric sulfate, $CuSO_4{\cdot}5H_2O$, in 100 mL of water and add 5 g of glycerol. Add KOH solution slowly until a deep blue solution is obtained.

Cupron (precipitates copper). Dissolve 5 g of benzoinoxime in 100 mL of 95% ethanol.

Cuprous chloride, acidic (reagent for CO in gas analysis).

1. Cover the bottom of a 2-L flask with a layer of cupric oxide about 0.5 inch deep, suspend a coil of copper wire so as to reach from the bottom to the top of the solution, and fill the flask with hydrochloric acid (sp. gr. 1.10). Shake occasionally. When the solution becomes nearly colorless, transfer to reagent bottles, which should also contain copper wire. The stock bottle may be refilled with dilute hydrochloric acid until either the cupric oxide or the copper wire is used up. Copper sulfate may be substituted for copper oxide in the above procedure.
2. Dissolve 340 g of $CuCl_2{\cdot}2H_2O$ in 600 mL of conc. HCl and reduce the cupric chloride by adding 190 mL of a saturated solution of stannous chloride or until the solution is colorless. The stannous chloride is prepared by treating 300 g of metallic tin in a 500 mL flask with conc. HCl until no more tin goes into solution.
3. (Winkler method). Add a mixture of 86 g of CuO and 17 g of finely divided metallic Cu, made by the reduction of CuO with hydrogen, to a solution of HCl, made by diluting 650 mL of conc. HCl with 325 mL of water. After the mixture has been added slowly and with frequent stirring, a spiral of copper wire is suspended in the bottle, reaching all the way to the bottom. Shake occasionally, and when the solution becomes colorless, it is ready for use.

Cuprous chloride, ammoniacal (reagent for CO in gas analysis).

1. The acid solution of cuprous chloride as prepared above is neutralized with ammonium hydroxide until an ammonia odor persists. An excess of metallic copper must be kept in the solution.

2. Pour 800 mL of acidic cuprous chloride, prepared by the Winkler method, into about 4 L of water. Transfer the precipitate to a 250 mL graduate. After several hours, siphon off the liquid above the 50 mL mark and refill with 7.5% NH_4OH solution which may be prepared by diluting 50 mL of conc. NH_4OH with 150 mL of water. The solution is well shaken and allowed to stand for several hours. It should have a faint odor of ammonia.

Dichlorofluorescein indicator. Dissolve 1 g in 1 L of 70% alcohol or 1 g of the sodium salt in 1 L of water.

Dimethyglyoxime, 0.01 N. Dissolve 0.6 g of dimethylglyoxime (2,3-butanedione oxime) in 500 mL of 95% ethanol. This is an especially sensitive test for nickel, a very definite crimson color being produced.

Diphenylamine (reagent for rayon). Dissolve 0.2 g in 100 mL of concentrated sulfuric acid.

Diphenylamine sulfonate (for titration of iron with $K_2Cr_2O_7$). Dissolve 0.32 g of the barium salt of diphenylamine sulfonic acid in 100 mL of water, add 0.5 g of sodium sulfate and filter off the precipitate of $BaSO_4$.

Diphenylcarbazide. Dissolve 0.2 g of diphenylcarbazide in 10 mL of glacial acetic acid and dilute to 100 mL with 95% ethanol.

Esbach's reagent (estimation of protein). To a water solution of 10 g of picric acid and 20 g of citric acid, add sufficient water to make 1 L of solution.

Eschka's compound. Two parts of calcined ("light") magnesia are thoroughly mixed with 1 part of anhydrous sodium carbonate.

Fehling's solution (reagent for reducing sugars.)

1. Copper sulfate solution. Dissolve 34.66 g of $CuSO_4 \cdot 5H_2O$ in water and dilute to 500 mL.

2. Alkaline tartrate solution. Dissolve 173 g of potassium sodium tartrate (Rochelle salt, $KNaC_4H_4O_6 \cdot 4H_2O$) and 50 g of NaOH in water and dilute when cold to 500 mL.

Mix equal volumes of the two solutions at the time of using.

Ferric-alum indicator. Dissolve 140 g of ferric ammonium sulfate crystals in 400 mL of hot water. When cool, filter, and make up to a volume of 500 mL with dilute nitric acid.

Folin's mixture (for uric acid). To 650 mL of water add 500 g of $(NH_4)_2SO_4$, 5 g of uranium acetate, and 6 g of glacial acetic acid. Dilute to 1 L.

Formaldehyde — sulfuric acid (Marquis' reagent for alkaloids). Add 10 mL of formaldehyde solution to 50 mL of sulfuric acid.

Froehde's reagent. See Sulfomolybdic acid.

Fuchsin (reagent for linen). Dissolve 1 g of fuchsin in 100 mL of alcohol.

Fuchsin — sulfurous acid (Schiff's reagent for aldehydes). Dissolve 0.5 g of fuchsin and 9 g of sodium bisulfite in 500 mL of water, and add 10 mL of HCl. Keep in well-stoppered bottles and protect from light.

Gunzberg's reagent (detection of HCl in gastric juice). Prepare as needed a solution containing 4 g of phloroglucinol (1,3,5-benzenetriol) and 2 g of vanillin in 100 mL of absolute ethanol.

Hager's reagent. See Picric acid.

Hanus solution (for iodine number). Dissolve 13.2 g of resublimed iodine in 1 L of glacial acetic acid which will pass the dichromate test for reducible matter. Add sufficient bromine to double the halogen content, determined by titration (3 mL is about the proper amount). The iodine may be dissolved by the aid of heat, but the solution should be cold when the bromine is added.

Iodine, tincture of. To 50 mL of water add 70 g of I_2 and 50 g of KI. Dilute to 1 L with alcohol.

Iodo-potassium iodide (Wagner's reagent for alkaloids). Dissolve 2 g of iodine and 6 g of KI in 100 mL of water.

Litmus (indicator). Extract litmus powder three times with boiling alcohol, each treatment consuming an hour. Reject the alcoholic extract. Treat residue with an equal weight of cold water and filter; then exhaust with five times its weight of boiling water, cool and filter. Combine the aqueous extracts.

Magnesia mixture (reagent for phosphates and arsenates). Dissolve 55 g of magnesium chloride and 105 g of ammonium chloride in water, barely acidify with hydrochloric acid, and dilute to 1 L. The ammonium hydroxide may be omitted until just previous to use. The reagent, if completely mixed and stored for any period of time, becomes turbid

Magnesium uranyl acetate. Dissolve 100 g of $UO_2(C_2H_3O_2)_2 \cdot 2H_2O$ in 60 mL of glacial acetic acid and dilute to 500 mL. Dissolve 330 g of $Mg(C_2H_3O_2)_2 \cdot 4H_2O$ in 60 mL of glacial acetic acid and dilute to 200 mL. Heat solutions to the boiling point until clear, pour the magnesium solution into the uranyl solution, cool and dilute to 1 L. Let stand over night and filter if necessary.

Marme's reagent. See Potassium-cadmium iodide.

Marquis' reagent. See Formaldehyde-sulfuric acid.

Mayer's reagent (white precipitate with most alkaloids in slightly acid solutions). Dissolve 1.358 g of $HgCl_2$ in 60 mL of water and pour into a solution of 5 g of KI in 10 mL of H_2O. Add sufficient water to make 100 mL.

Methyl orange indicator. Dissolve 1 g of methyl orange in 1 L of water. Filter, if necessary.

Methyl orange, modified. Dissolve 2 g of methyl orange and 2.8 g of xylene cyanole FF in 1 L of 50% alcohol.

Methyl red indicator. Dissolve 1 g of methyl red in 600 mL of alcohol and dilute with 400 mL of water.

Methyl red, modified. Dissolve 0.50 g of methyl red and 1.25 g of xylene cyanole FF in 1 L of 90% alcohol. Or, dissolve 1.25 g of methyl red and 0.825 g of methylene blue in 1 L of 90% alcohol.

Millon's reagent (for albumins and phenols). Dissolve 1 part of mercury in 1 part of cold fuming nitric acid. Dilute with twice the volume of water and decant the clear solution after several hours.

Molisch's reagent. See 1-Naphthol.

1-Naphthol (Molisch's reagent for wool). Dissolve 15 g of 1-naphthol in 100 mL of alcohol or chloroform.

Nessler's reagent (for ammonia). Dissolve 50 g of KI in the smallest possible quantity of cold water (50 mL). Add a saturated solution of mercuric chloride (about 22 g in 350 mL of water will be needed) until an excess is indicated by the formation of a precipitate. Then add 200 mL of 5 N NaOH and dilute to 1 L. Let settle, and draw off the clear liquid.

Nickel oxide, ammoniacal (reagent for silk). Dissolve 5 g of nickel sulfate in 100 mL of water, and add sodium hydroxide solution until nickel hydroxide is completely precipitated. Wash the precipitate well and dissolve in 25 mL of concentrated ammonium hydroxide and 25 mL of water.

Nitron (detection of nitrate radical). Dissolve 10 g of nitron (1,4-diphenyl-3-(phenylamino)-1,2,4-triazolium hydroxide) in 5 mL of glacial acetic acid and 95 mL of water. The solution may be filtered with slight suction through an alumdum crucible and kept in a dark bottle.

1-Nitroso-2-naphthol. Make a saturated solution in 50% acetic acid (1 part of glacial acetic acid with 1 part of water). Does not keep well.

Nylander's solution (carbohydrates). Dissolve 20 g of bismuth subnitrate and 40 g of Rochelle salt in 1 L of 8% NaOH solution. Cool and filter.

Obermayer's reagent (for indoxyl in urine). Dissolve 4 g of $FeCl_3$ in 1 L of HCl (sp. gr. 1.19).

Oxine. Dissolve 14 g of 8-hydroxyquinoline in 30 mL of glacial acetic acid. Warm slightly, if necessary. Dilute to 1 L.

Oxygen absorbent. Dissolve 300 g of ammonium chloride in 1 L of water and add 1 L of concentrated ammonium hydroxide solution. Shake the solution thoroughly. For use as an oxygen absorbent, a bottle half full of copper turnings is filled nearly full with the NH_4Cl-NH_4OH solution and the gas passed through.

Pasteur's salt solution. To 1 L of distilled water add 2.5 g of potassium phosphate, 0.25 g of calcium phosphate, 0.25 g of magnesium sulfate, and 12.00 g of ammonium tartrate.

Pavy's solution (glucose reagent). To 120 mL of Fehling's solution, add 300 mL of NH_4OH (sp. gr. 0.88) and dilute to 1 L with water.

Phenanthroline ferrous ion indicator. Dissolve 1.485 g of 1,10-phenanthroline monohydrate in 100 mL of 0.025 M ferrous sulfate solution.

Phenolphthalein. Dissolve 1 g of phenolphthalein in 50 mL of alcohol and add 50 mL of water.

Phenolsulfonic acid (determination of nitrogen as nitrate). Dissolve 25 g of phenol in 150 mL of conc. H_2SO_4, add 75 mL of fuming H_2SO_4 (15% SO_3), stir well and heat for 2 hours at 100°C.

Phloroglucinol solution (pentosans). Make a 3% phloroglucinol (1,3,5-benzenetriol) solution in alcohol. Keep in a dark bottle.

Phosphomolybdic acid (Sonnenschein's reagent for alkaloids).

1. Prepare ammonium phosphomolybdate and after washing with water, boil with nitric acid and expel NH_3; evaporate to dryness and dissolve in 2 M nitric acid.
2. Dissolve ammonium molybdate in HNO_3 and treat with phosphoric acid. Filter, wash the precipitate, and boil with aqua regia until the ammonium salt is decomposed. Evaporate to dryness. The residue dissolved in 10 % HNO_3 constitutes Sonnenschein's reagent.

Phosphoric acid — sulfuric acid mixture. Dilute 150 mL of conc. H_2SO_4 and 100 mL of conc. H_3PO_4 (85%) with water to a volume of 1 L.

Phosphotungstic acid (Schcibicr's reagent for alkaloids).

1. Dissolve 20 g of sodium tungstate and 15 g of sodium phosphate in 100 mL of water containing a little nitric acid.
2. The reagent is a 10% solution of phosphotungstic acid in water. Thc.phosphotungstic acid is prepared by evaporating a mixture of 10 g of sodium tungstate dissolved in 5 g of phosphoric acid (sp. gr. 1.13) and enough boiling water to effect solution. Crystals of phosphotungstic acid separate.

Picric acid (Hager's reagent for alkaloids, wool and silk). Dissolve 1 g of picric acid in 100 mL of water.

Potassium antimonate (reagent for sodium). Boil 22 g of potassium antimonate with 1 L of water until nearly all of the salt has dissolved, cool quickly, and add 35 mL of 10% potassium hydroxide. Filter after standing overnight.

Potassium-cadmium iodide (Marme's reagent for alkaloids). Add 2 g of CdI_2 to a boiling solution of 4 g of KI in 12 mL of water, and then mix with 12 mL of saturated KI solution.

Potassium hydroxide (for CO_2 absorption). Dissolve 360 g of KOH in water and dilute to 1 L.

Potassium iodide — mercuric iodide (Brucke's reagent for proteins). Dissolve 50 g of KI in 500 mL of water, and saturate with mercuric iodide (about 120 g). Dilute to 1 L.

Potassium pyrogallate (for oxygen absorption). For mixtures of gases containing less than 28% oxygen, add 100 mL of KOH solution (50 g of KOH to 100 mL of water) to 5 g of pyrogallol. For mixtures containing more than 28% oxygen the KOH solution should contain 120 g of KOH to 100 mL of water.

Pyrogallol, alkaline.

1. Dissolve 75 g of pyrogallic acid in 75 mL of water.
2. Dissolve 500 g of KOH in 250 mL of water. When cool, adjust until sp. gr. is 1.55.

For use, add 270 mL of solution (2) to 30 mL of solution (1).

Rosolic acid (indicator). Dissolve 1 g of rosolic acid in 10 mL of alcohol and add 100 mL of water.

Scheibler's reagent. See Phosphotungstic acid.

Schiff's reagent. See Fuchsin-sulfurous acid.

Schweitzer's reagent. See Cupric oxide, ammoniacal.

Soap solution (reagent for hardness in water). Dissolve 100 g of dry castile soap in 1 L of 80% alcohol (5 parts alcohol to 1 part water). Allow to stand several days and dilute with 70% to 80% alcohol until 6.4 mL produces a permanent lather with 20 mL of standard calcium solution. The latter solution is made by dissolving 0.2 g of $CaCO_3$ in a small amount of dilute HCl, evaporating to dryness and making up to 1 L.

Sodium bismuthate (oxidation of manganese). Heat 20 parts of NaOH nearly to redness in an iron or nickel crucible and add slowly 10 parts of basic bismuth nitrate which has been previously dried. Add 2 parts of sodium peroxide, and pour the brownish-yellow fused mass onto an iron plate to cool. When cold, break up in a mortar, extract with water, and collect on an asbestos filter.

Sodium hydroxide (for CO_2 absorption). Dissolve 330 g of NaOH in water and dilute to 1 L.

Sodium nitroprusside (reagent for hydrogen sulfide and wool). Use a freshly prepared solution of 1 g of sodium nitroferricyanide in 10 mL of water.

Sodium oxalate (primary standard). Dissolve 30 g of the commercial salt in 1 L of water, make slightly alkaline with sodium hydroxide, and let stand until perfectly clear. Filter and evaporate the filtrate to 100 mL. Cool and filter. Pulverize the residue and wash it several times with small volumes of water. The procedure is repeated until the mother liquor is free from sulfate and is neutral to phenolphthalein.

Sodium plumbite (reagent for wool). Dissolve 5 g of sodium hydroxide in 100 mL of water. Add 5 g of litharge (PbO) and boil until dissolved.

Sodium polysulfide. Dissolve 480 g of $Na_2S{\cdot}9H_2O$ in 500 mL of water, add 40 g of NaOH and 18 g of sulfur. Stir thoroughly and dilute to 1 L with water.

Sonnenschein's reagent. See Phosphomolybdic acid.

Starch solution.

1. Make a paste with 2 g of soluble starch and 0.01 g of HgI_2 with a small amount of water. Add the mixture slowly to 1 L of boiling water and boil for a few minutes. Keep in a glass stoppered bottle. If other than soluble starch is used, the solution will not clear on boiling; it should be allowed to stand and the clear liquid decanted.
2. A solution of starch which keeps indefinitely is made as follows: Mix 500 mL of saturated NaCl solution (filtered), 80 mL of glacial acetic acid, 20 mL of water and 3 g of starch. Bring slowly to a boil and boil for 2 minutes.
3. Make a paste with 1 g of soluble starch and 5 mg of HgI_2, using as little cold water as possible. Then pour about 200 mL of boiling water on the paste and stir immediately. This will give a clear solution if the paste is prepared correctly and the water actually boiling. Cool and add 4 g of KI. Starch solution decomposes on standing due to bacterial action, but this solution will keep well if stored under a layer of toluene.

Stoke's reagent. Dissolve 30 g of $FeSO_4$ and 20 g of tartaric acid in water and dilute to 1 L. Just before using, add concentrated NH_4OH until the precipitate first formed is redissolved.

Sulfanilic acid (reagent for nitrites). Dissolve 0.5 g of sulfanilic acid in a mixture of 15 mL of glacial acetic acid and 135 mL of recently boiled water.

Sulfomolybdic acid (Froehde's reagent for alkaloids and glucosides). Dissolve 10 g of molybdic acid or sodium molybdate in 100 mL of conc. H_2SO_4.

Tannic acid (reagent for albumin, alkaloids, and gelatin). Dissolve 10 g of tannic acid in 10 mL of alcohol and dilute with water to 100 mL.

Titration mixture. (residual chlorine in water analyasis). Prepare 1 L of dilute HCl (100 mL of HCl (sp. gr. 1.19) in sufficient water to make 1 L). Dissolve 1 g of *o*-tolidine in 100 mL of the dilute HCl and dilute to 1 L with dilute HCl solution.

Trinitrophenol solution. See Picric acid.

Turmeric tincture (reagent for borates). Digest ground turmeric root with several quantities of water which are discarded. Dry the residue and digest it several days with six times its weight of alcohol. Filter.

Uffelmann's reagent (turns yellow in presence of lactic acid). To a 2% solution of pure phenol in water, add a water solution of $FeCl_3$ until the phenol solution becomes violet in color.

Wagner's reagent. See Iodo-potassium iodide.

Wagner's solution (used in phosphate rock analysis to prevent precipitation of iron and aluminum). Dissolve 25 g of citric acid and 1 g of salicylic acid in water and dilute to 1 L. Use 50 mL of the reagent.

Wij's iodine monochloride solution (for iodine number). Dissolve 13 g of resublimed iodine in 1 L of glacial acetic acid which will pass the dichromate test for reducible matter. Set aside 25 mL of this solution. Pass into the remainder of the solution dry chlorine gas (dried and washed by passing through H_2SO_4 (sp. gr. 1.84)) until the characteristic color of free iodine has been discharged. Now add the iodine solution which was reserved, until all free chlorine has been destroyed. A slight excess of iodine does little or no harm, but an excess of chlorine must be avoided. Preserve in well stoppered, amber colored bottles. Avoid use of solutions which have been prepared for more than 30 days.

Wij's special solution (for iodine number). To 200 mL of glacial acetic acid that will pass the dichromate test for reducible matter, add 12 g of dichloramine T (*N,N*-dichloro-4-methyl-benzenesulfonamide), and 16.6 g of dry KI (in small quantities with continual shaking until all the KI has dissolved). Make up to 1 L with the same quality of acetic acid used above and preserve in a dark colored bottle.

Zimmermann-Reinhardt reagent (determination of iron). Dissolve 70 g of $MnSO_4{\cdot}4H_2O$ in 500 mL of water, add 125 mL of conc. H_2SO_4 and 125 mL of 85% H_3P0_4, and dilute to 1 L.

Zinc chloride solution, basic (reagent for silk). Dissolve 1000 g of zinc chloride in 850 mL of water, and add 40 g of zinc oxide. Heat until solution is complete.

Zinc uranyl acetate (reagent for sodium). Dissolve 10 g of $UO_2(C_2H_3O_2)_2{\cdot}2H_2O$ in 6 g of 30% acetic acid with heat, if necessary, and dilute to 50 mL. Dissolve 30 g of $Zn(C_2H_3O_2)_2{\cdot}H_2O$ in 3 g of 30% acetic acid and dilute to 50 mL. Mix the two solutions, add 50 mg of NaCl, allow to stand overnight and filter.

STANDARD SOLUTIONS OF ACIDS, BASES, AND SALTS

For each compound listed, the last column of this table gives the mass in grams which is contained in 1 liter of a solution whose amount-of-substance concentration divided by the equivalence factor of the compound equals 0.1 mol/L. In the older literature such a solution is often referred to as a "decinormal solution" (0.1 N).

REFERENCE

Compendium of Analytical Nomenclature (IUPAC), Pergamon Press, Oxford, 1978.

Name	Formula	Atomic or molecular weight	Equivalence factor	Mass in grams
Acetic acid	$HC_2H_3O_2$	60.0530	1	6.0053
Ammonia	NH_3	17.0306	1	1.7031
Ammonium ion	NH_4^+	18.0386	1	1.8039
Ammonium chloride	NH_4Cl	53.4916	1	5.3492
Ammonium sulfate	$(NH_4)_2SO_4$	132.1388	1/2	6.6069
Ammonium thiocyanate	NH_4CNS	76.1204	1	7.6120
Barium	Ba	137.34	1/2	6.867
Barium carbonate	$BaCO_3$	197.3494	1/2	9.8675
Barium chloride hydrate	$BaCl_2 \cdot 2H_2O$	244.2767	1/2	12.2138
Barium hydroxide	$Ba(OH)_2$	171.3547	1/2	8.5677
Barium oxide	BaO	153.3394	1/2	7.6670
Bromine	Br	79.909	1	7.9909
Calcium	Ca	40.08	1/2	2.004
Calcium carbonate	$CaCO_3$	100.0894	1/2	5.0045
Calcium chloride	$CaCl_2$	110.9860	1/2	5.5493
Calcium chloride hydrate	$CaCl_2 \cdot 6H_2O$	219.0150	1/2	10.9508
Calcium hydroxide	$Ca(OH)_2$	74.0947	1/2	3.7047
Calcium oxide	CaO	56.0794	1/2	2.8040
Chlorine	Cl	35.453	1	3.5453
Citric acid	$C_6H_8O_7 \cdot H_2O$	210.1418	1/3	7.0047
Cobalt	Co	58.9332	1/2	2.9466
Copper	Cu	63.54	1/2	3.177
Copper oxide (cupric)	CuO	79.5394	1/2	3.9770
Copper sulfate hydrate	$CuSO_4 \cdot 5H_2O$	249.6783	1/2	12.4839
Hydrochloric acid	HCl	36.4610	1	3.6461
Hydrocyanic acid	HCN	27.0258	1	2.7026
Iodine	I	126.9044	1	12.6904
Lactic acid	$C_3H_6O_3$	90.0795	1	9.0080
Malic acid	$C_4H_6O_5$	134.0894	1/2	6.7045
Magnesium	Mg	24.312	1/2	1.2156
Magnesium carbonate	$MgCO_3$	84.3214	1/2	4.2161
Magnesium chloride	$MgCl_2$	95.2180	1/2	4.7609
Magnesium chloride hydrate	$MgCl_2 \cdot 6H_2O$	203.2370	1/2	10.1623
Magnesium oxide	MgO	40.3114	1/2	2.0156
Manganese	Mn	54.938	1/2	2.7469
Manganese sulfate	$MnSO_4$	150.9996	1/2	7.5500
Mercuric chloride	$HgCl_2$	271.4960	1/2	13.5748
Nickel	Ni	58.71	1/2	2.9356
Nitric acid	HNO_3	63.0129	1	6.3013
Oxalic acid	$H_2C_2O_4$	90.0358	1/2	4.5018
Oxalic acid hydrate	$H_2C_2O_4 \cdot 2H_2O$	126.0665	1/2	6.3033
Oxalic acid anhydride	C_2O_3	72.0205	1/2	3.6010
Phosphoric acid	H_3PO_4	97.9953	1/3	3.2665
Potassium	K	39.102	1	3.9102
Potassium bicarbonate	$KHCO_3$	100.1193	1	10.0119
Potassium carbonate	K_2CO_3	138.2134	1/2	6.9106
Potassium chloride	KCl	74.5550	1	7.4555

Name	Formula	Atomic or molecular weight	Equivalence factor	Mass in grams
Potassium cyanide	KCN	65.1199	1	6.5120
Potassium hydroxide	KOH	56.1094	1	5.6109
Potassium oxide	K_2O	94.2034	1/2	4.7102
Potassium tartrate	$K_2H_4C_4O_6$	226.2769	1/2	11.3139
Silver	Ag	107.87	1	10.787
Silver nitrate	$AgNO_3$	169.8749	1	16.9875
Sodium	Na	22.9898	1	2.2990
Sodium bicarbonate	$NaHCO_3$	84.0071	1	8.4007
Sodium carbonate	Na_2CO_3	105.9890	1/2	5.2995
Sodium chloride	NaCl	58.4428	1	5.8443
Sodium hydroxide	NaOH	39.9972	1	3.9997
Sodium oxide	Na_2O	61.9790	1/2	3.0990
Sodium sulfide	Na_2S	78.0436	1/2	3.9022
Succinic acid	$H_2C_4H_4O_4$	118.0900	1/2	5.9045
Sulfuric acid	H_2SO_4	98.0775	1/2	4.9039
Tartaric acid	$C_4H_6O_6$	150.0888	1/2	7.5044
Zinc	Zn	65.37	1/2	3.269
Zinc sulfate hydrate	$ZnSO_4 \cdot 7H_2O$	287.5390	1/2	14.3769

STANDARD SOLUTIONS OF OXIDATION AND REDUCTION REAGENTS

For each reagent listed, the last column of this table gives the mass in grams which is contained in a solution whose amount-of-substance concentration divided by the equivalence factor of the compound equals 0.1 mol/L. The equivalence factor given refers to the most common reactions of the reagent. In the older literature such a solution is often called a "decinormal solution" (0.1 N).

REFERENCE

Compendium of Analytical Nomenclature (IUPAC), Pergamon Press, Oxford, 1978.

Name	Formula	Atomic or molecular weight	Equivalence factor	Mass in grams
Antimony	Sb	121.75	1/2	6.0875
Arsenic	As	74.9216	1/2	3.7461
Arsenic trisulfide	As_2S_3	246.0352	1/4	6.1509
Arsenous oxide	As_2O_3	197.8414	1/4	4.9460
Barium peroxide	BaO_2	169.3388	1/2	8.4669
Barium peroxide hydrate	$BaO_2 \cdot 8H_2O$	313.4615	1/2	15.6730
Calcium	Ca	40.08	1/2	2.004
Calcium carbonate	$CaCO_3$	100.0894	1/2	5.0045
Calcium hypochlorite	$Ca(OCl)_2$	142.9848	1/4	3.5746
Calcium oxide	CaO	56.0794	1/2	2.8040
Chlorine	Cl	35.453	1	3.5453
Chromium trioxide	CrO_3	99.9942	1/3	3.3331
Ferrous ammonium sulfate	$FeSO_4(NH_4)SO_4 \cdot 6H_2O$	392.0764	1	39.2076
Hydroferrocyanic acid	$H_4Fe(CN)_6$	215.9860	1	21.5986
Hydrogen peroxide	H_2O_2	34.0147	1/2	1.7007
Hydrogen sulfide	H_2S	34.0799	1/2	1.7040
Iodine	I	126.9044	1	12.6904
Iron	Fe	55.847	1	5.5847
Iron oxide (ferrous)	FeO	71.8464	1	7.1846
Iron oxide (ferric)	Fe_2O_3	159.6922	1/2	7.9846
Lead peroxide	PbO_2	239.1888	1/2	11.9594
Manganese dioxide	MnO_2	86.9368	1/2	4.3468
Nitric acid	HNO_3	63.0129	1/3	2.1004
Nitrogen trioxide	N_2O_3	76.0116	1/4	1.9002
Nitrogen pentoxide	N_2O_5	108.0104	1/6	1.8001
Oxalic acid	$C_2H_2O_4$	90.0358	1/2	4.5018
Oxalic acid hydrate	$C_2H_2O_4 \cdot 2H_2O$	126.0665	1/2	6.3033
Oxygen	O	15.9994	1/2	0.8000
Potassium dichromate	$K_2Cr_2O_7$	294.1918	1/6	4.9032
Potassium chlorate	$KClO_3$	122.5532	1/6	2.0425
Potassium chromate	K_2CrO_4	194.1076	1/3	6.4733
Potassium ferrocyanide	$K_4Fe(CN)_6$	368.3621	1	36.8362
Potassium ferrocyanide hydrate	$K_4Fe(CN)_6 \cdot 3H_2O$	422.4081	1	42.2408
Potassium iodide	KI	166.0064	1	16.6006
Potassium nitrate	KNO_3	101.1069	1/3	3.3702
Potassium perchlorate	$KClO_4$	138.5526	1/8	1.7319
Potassium permanganate	$KMnO_4$	158.0376	1/5	3.1608
Sodium chlorate	$NaClO_3$	106.4410	1/6	1.7740
Sodium nitrate	$NaNO_3$	84.9947	1/3	2.8332
Sodium thiosulfate hydrate	$Na_2S_2O_3 \cdot 5H_2O$	248.1825	1	24.8183
Stannous chloride	$SnCl_2$	189.5960	1/2	9.4798
Stannous oxide	SnO	134.6894	1/2	6.7345
Sulfur dioxide	SO_2	64.0628	1/2	3.2031
Tin	Sn	118.69	1/2	5.935

ORGANIC ANALYTICAL REAGENTS FOR THE DETERMINATION OF INORGANIC SUBSTANCES

G. Ackermann, L. Sommer, and D. Thorburn Burns

Determination	Reagents	Ref.
Aluminium	Alizarin Red S	Onishi, Part II a, p 28. (5), Snell, *Metals I,* p 587. (7)
	Aluminon	Fries/Getrost, p 16. (2), Onishi, IIa, p 21. (5), Snell, *Metals I,* p 590. (7)
	Aluminon + Cetyltrimethylammonium bromide	Huaxue Shiji, *8*, 85, (1986)
	Chrome Azurol S	Onishi, Part IIa, p 26. (5), Snell, *Metals I,* p 605. (7)
	Chrome Azurol S + Cetyltrimethylammonium bromide	Marczenko, p 133 (3), Snell, *Metals I,* p 606. (7)
	Chromazol KS + Cetylpyridinium bromide	*Analyst, 107*, 428, (1982).
	Eriochrome Cyanine R	Fries/Getrost, p 19 (2), Onishi, Part IIa p 25. (5), Snell, *Metals I,* p 611. (7)
	Eriochrome Cyanine R + Cetyltrimethylammonium bromide	Snell, *Metals I,* p 613. (7), *Analyst, 107*, 1431, (1982).
	8-Hydroxyquinoline	Fries/Getrost, p 22 (2), Marczenko, p 131 (3), Onishi, Part IIa, p 31. (5), Snell, *Metals I , p* 622 (7)
Ammonia	Phenol + Sodium hypochlorite	Boltz, p 210 (1), Marczenko, p 413 (3), Snell, *Nonmetals*, p 604 (9)
Antimony	Brilliant Green	Onishi, Part IIa, p 102. (5), Snell, *Metals I,* p 384. (7)
	Bromopyrogallol Red	*Talanta, 13*, 507, (1966).
	Rhodamine B	Fries/Getrost, p 32, (2), Marczenko, p 141. (3), Onishi, Part IIa, p 93. (5), Snell, *Metals I,* p 404. (7)
	Silver diethyldithiocarbamate	Fries/Getrost, p 36. (2)
Arsenic	Silver diethyldithiocarbamate	Fries/Getrost, p 41. (2), Marczenko, p 153. (3), Onishi, Part IIa, p 153. (5), Snell, *Metals I,* p 370. (7)
Barium	Sulfonazo III	Fries/Getrost, p 46. (2), Snell, *Metals II,* p 1782. (8), Onishi, Part IIa, p 202. (5)
Beryllium	Beryllon II	Snell, *Metals I,* p 667. (7)
	Chrome Azurol S	Marczenko, p 163. (3), Snell, *Metals I,* p 672. (7)
	Chrome Azurol S + Cetyltrimethylammonium bromide	Marczenko, p 164. (3), Snell, *Metals I,* p 673. (7)
	Eriochrome Cyanine R	Snell, *Metals I,* p 675. (7), *Talanta, 31*, 249, (1984).
	Eriochrome Cyanine R + Cetyltrimethylammonium bromide	Zh, *Anal. Khim.*, *33*, 1298, (1978).
Bismuth	Dithizone	Onishi, Part IIa, p 262. (5), Snell, *Metals I,* p 303. (7)
	Pyrocatechol Violet	Fres. Z. *Anal. Chem.*, *186*, 418, (1962).
	Pyrocatechol Violet + Cetyltrimethylammonium bromide	Zh. *Anal. Khim.*, *38*, 216, (1983).
	Thiourea	Onishi, Part IIa, p 260. (5), Snell, *Metals I,* p 317. (7)
	Xylenol Orange	Friez/Getrost, p 57. (2), Marczenko, p 172. (3), Snell, *Metals I,* p 320. (7)
Boron	Azomethine H	Snell, Nonmetals, p 165. (9)
	Carminic acid	Boltz, p 14. (1), Fries/Getrost, p 65. (2), Snell, Nonmetals, p 170. (9), Williams, p 35. (11)
	Curcumin	Boltz, p 8. (1), Fries/Getrost, p 68. (2), Marczenko, p 180. (3), Snell, Nonmetals, p 180. (9), Fres. Z. *Anal. Chem.*, *323*, 266, (1986).

Determination	Reagents	Ref.
	Methylene Blue	Boltz, p 21. (1), Marczenko, p 183. (3), Snell, Nonmetals, p 205. (9), *Talanta, 31*, 547, (1984).
Bromide	Fluorescein	Boltz, p 48. (1), Snell, Nonmetals, p 276., Fres. Z. *Anal. Chem., 301*, 28 (1980).
	Phenol Red	Boltz, p 44. (1), Marczenko, p 190. (3), Snell, Nonmetals, p 28. (9)
Cadmium	2-(5-Bromo-2-pyridylazo)-5-diethylaminophenol	Marczenko, p 197. (3)
	Cadion	Onishi, Part IIa, p 323. (5)
	Dithizone	Fries/Getrost, p 78. (2), Onishi, Part IIa, p 315. (5), Snell, *Metals I*, p 279. (7), West, p 25. (10).
	4–(2-Pyridylazo)resorcinol	Fres. Z. *Anal. Chem., 310*, 51, (1982).
Calcium	Chlorophosphonazo III	Marczenko, p 207. (3), Snell, *Metals II*, p 1744. (8)
	Glyoxal-bis(2-hydroxyanil)	Fries/Getrost, p 86. (2), Onishi, Part IIa, p 352. (5), Snell, *Metals I*, p 1762. (8)
	Murexide	Onishi, Part IIa, p 357. (5), Snell, *Metals II*, p 1769. (8)
	Phthalein Purple	*Anal. Chim. Acta, 34*, 71 (1966).
Cerium	*N*-benzoyl-*N*-phenylhydroxylamine	*Anal. Chim. Acta, 48*, 155, (1969).
	8-Hydroxyquinoline	Fries/Getrost, p 93. (2), Marczenko, p 220. (3), Onishi, Part IIa, p 383. (7)
Chlorine	*N,N*-Diethyl-1,4-phenylenediamine	Boltz, p 92. (1), Fries/Getrost, p 101. (2), Snell, Nonmetals, p 225. (9), *Analyst, 90*, 187, (1965).
Chromium	1,5-Diphenylcarbazide	Fries/Getrost, p 105. (2), Onishi, Part IIa, p 412. (5), Snell, *Metals I*, p 714. (7), West, p 12. (10)
	4-(2-Pyridylazo)resorcinol	Snell, *Metals I*, p 736. (7), West, p 17. (10)
	4-(2-Pyridylazo)resorcinol + Tetradecyldimethyl-benzylammonium chloride	West, p 17. (10), *Anal. Chim. Acta, 67*, 297, (1973).
	4-(2-Pyridylazo)resorcinol + Hydrogen peroxide	Fres. Z. *Anal. Chem., 304*, 382, (1980).
Cobalt	Nitroso-R salt	Fries/Getrost, p 118. (2), Onishi, Part IIa, p 454. (5), Snell, *Metals I*, p 953. (7)
	1-Nitroso-2-naphthol	Fries/Getrost, p 111. (2), Marczenko, p 246. (3), Snell, *Metals I*, p 947. (5)
	2-Nitroso-1-naphthol	Fries/Getrost, p 113. (2), Onishi, Part IIa, p 459. (5), Snell, *Metals I*, p 949. (7), West, p 45. (10)
	4-(2-Pyridylazo)resorcinol	Snell, *Metals I*, p 969. (7), West, p 44. (10)
	4-(2-Pyridylazo)resorcinol + Diphenylguanidine	Zh. *Anal. Khim., 35*, 1306, (1980).
Copper	Bathocuproine	Fries/Getrost, p 135. (2), Snell, *Metals I*, p 148. (7)
	Bathocuproine disulfonic acid	Fries/Getrost, p 137. (2), West, p 52. (10)
	Dithizone	Marczenko, p 258. (3), Onishi, Part IIa, p 529. (5), Snell, *Metals I*, p 199. (7)
	Neocuproine	Snell, *Metals I*, p 217. (5), West, p 51. (10)
	Cuprizone	Onishi, Part IIa, p 534. (5), Snell, *Metals I*, p 157. (7), West, p 53. (10)
	4-(2-pyridylazo)resorcinol + Tetradecyldimethyl-benzylammonium chloride	*Anal. Chim.* Acta, *138*, 321, (1982).
Cyanide	Barbituric Acid + Pyridine	Fries/Getrost, p 153. (2), Snell, Nonmetals, p 653. (9)
	Barbituric Acid + Pyridine-4-carboxylic acid	*Anal. Chim. Acta, 99*, 197, (1978).

ORGANIC ANALYTICAL REAGENTS FOR THE DETERMINATION OF INORGANIC SUBSTANCES (continued)

Determination	Reagents	Ref.
Fluoride	Alizarin Fluorine blue + Lanthanum(III) ion	Boltz, p 129. (1), Fries/Getrost, p 158. (2), Snell, *Nonmetals,* p 333. (9), Williams, p 354. (11)
	Eriochrome Cyanine R + Zirconium(IV) ion	Boltz, p 119. (1), Snell, *Nonmetals*, p 359. (2), Williams, p 357. (10)
Gallium	Pyrocatechol violet + Diphenylguanidine	Snell, *Metals I,* p 500. (7)
	8-Hydroxyquinoline	Onishi Pt IIa, p 582. (5), Snell, *Metals I,* p 505. (7)
	1-(2-Pyridylazo)-2-naphthol	Snell, *Metals I,* p 512. (7)
	4-(2-Pyridylazo)resorcinol	Snell, *Metals I,* p 513. (7)
	Rhodamine B	Marczenko, p 284. (3), Onishi, Part IIa, p 578. (5), Snell, *Metals I,* p 515. (7)
	Xylenol Orange	Fries/Getrost, p 166. (2), Snell, *Metals I,* p 523. (7)
	Xylenol Orange + 8-Hydroxyquinoline	*Zh. Anal. Khim., 26,* 75, (1971).
Germanium	Brilliant Green + Molybdate	Snell, *Metals I,* p 562. (7)
	Phenylfluorone	Fries/Getrost, p 168. (2), Marczenko, p 292. (3), Onishi, Part IIa, p 607. (5), Snell, *Metals I,* p 570. (7)
Gold	5-(4-Diethylaminobenzylidene) rhodanine	Fries/Getrost, p 173. (2), Onishi, Part IIa, p 631. (5), Snell, *Metals II,* p 1516. (8)
	Rhodamine B	Fries/Getrost, p 175. (2), Marczenko, p 301. (3), Onishi, Part IIa, p 637. (5), Snell, *Metals II,* p 513. (8)
Hafnium	Arsenazo III	Snell, *Metals II,* p 1184. (8), *Talanta, 19,* 807, (1972).
Indium	Bromopyrogallol Red	Snell, *Metals I,* p 469. (7)
	Chrome Azurol S	Snell, *Metals I,* p 474. (7)
	Chrome Azurol S + Cetyltrimethylammonium bromide	*Anal. Chim. Acta, 67,* 107, (1973).
	Dithizone	Fries/Getrost, p 179. (2), Onishi, Part IIa, p 672. (5), Snell, *Metals I,* p 474. (7)
	8-Hydroxyquinoline	Onishi, Part IIa, p 670. (5), Snell, *Metals I,* p 475. (7)
	1-(2-Pyridylazo)-2-naphthol	Snell, *Metals I,* p 480. (7)
	4-(2-Pyridylazo)resorcinol	Marczenko, p 309. (3), Snell, Metals I, p 480. (7)
Iodide	Neocuproine + Copper(II)	*Anal. Chim. Acta, 69,* 321, (1974).
Iodine	Starch	Boltz, p 162. (1), Marczenko, p 316. (3), Snell, *Nonmetals,* p 307. (9)
Iridium	Rhodamine 6G + Tin(II)	Marczenko, p 323. (3)
	N,N-Dimethyl-4-nitrosoaniline	*Anal. Chem., 27,* 1776, (1955).
Iron	Bathophenanthroline	Fries/Getrost, p 189. (2), Onishi, Part IIa, p 729. (5), Snell, *Metals I,* p 763. (7)
	Bathophenanthroline disulfonic acid	Fries/Getrost, p 191. (2), Snell, *Metals I,* p 772. (7)
	2,2′-Bipyridyl	Snell, *Metals I,* p 750. (7)
	Chrome Azurol S + Cetyltrimethylammonium bromide	Snell, *Metals I,* p 757. (7), *Coll. Czech. Chem. Comm., 45,* 2656, (1980).
	1,10-Phenanthroline	Fries/Getrost, p 199. (2), Marczenko, p 331. (3), Onishi, Part IIa, p 725. (5), Snell, *Metals I,* p 795. (7)
	1,10-Phenanthroline + Bromothymol Blue	*Zh. Anal. Khim., 25,* 1348, (1970).
	Ferrozine	Onishi, Part IIa, p 730. (5), Snell, *Metals I,* p 783. (7)
Lanthanum	Arsenazo III	Marczenko, p 468. (3), Snell, *Metals II,* p 1910. (8)

ORGANIC ANALYTICAL REAGENTS FOR THE DETERMINATION OF INORGANIC SUBSTANCES (continued)

Determination	Reagents	Ref.
Lead	Dithizone	Fries/Getrost, p 207. (2), Onishi, Part IIa, p 824. (5), Snell, *Metals I,* p 2. (7), West, p 34. (10)
	Sodium diethyldithiocarbamate	Fries/Getrost, p 214. (2), Snell, *Metals I,* p 27. (7)
	4-(2-Pyridylazo)resorcinol	Fries/Getrost, p 220. (2), Marczenko, p 347. (3), Snell, *Metals I,* p 34. (7)
Lithium	Thoron	Onishi, Part IIa, p 863. (5), Snell, *Metals II,* p 1726. (8), T*alanta, 30,* 587, (1983).
Magnesium	Eriochrome Black T	Fries/Getrost, p 226. (2), Marczenko, p 355. (3), Onishi, Part IIb, p 13. (6), Snell, *Metals II,* p 1932. (8)
	8-Hydroxyquinoline	Onishi, Part IIb, p 11. (6), Snell, *Metals II,* p 1938. (8)
	8-Hydroxyquinoline + Butylamine	Fries/Getrost, p 228. (2), Snell, *Metals II,* p 1938. (8)
	Titan Yellow	Fries/Getrost, p 234. (2), Marczenko, p 352. (3), Snell, *Metals II,* p 1945. (8)
	Xylidyl Blue	Fries/Getrost, p 231. (2), Onishi, Part IIb, p 14. (6), Snell, *Metals II,* p 1950. (8)
Manganese	Formaldoxime	Fries/Getrost, p 236. (2), Marczenko, p 364. (3), Onishi Part IIb, p 38. (6), Snell, *Metals II,* 1010. (8)
Mercury	Dithizone	Fries/Getrost, p 243. (2), Marczenko, p 373. (3), Onishi, Part IIb, p 66. (6), Snell, *Metals I,* p 107. (7), West, p 29. (10)
	Michler's thioketone	Marczenko, p 375. (3), Snell, *Metals I,* p 126. (7)
	Xylenol Orange	*Talanta, 16,* 1023, (1969)
Molybdenum	Bromopyrogallol Red + Cetylpyridium chloride	West, p 58. (10)
	Phenylfluorone	Snell, *Metals II,* p 1311., *Microchem. J., 31,* 56, (1985).
	Toluene-3,4-dithiol	Fries/Getrost, p 251. (2), Marczenko, p 384. (3), Onishi, Part IIb, p 96. (6), Snell, *Metals II,* p 1301. (8)
Nickel	2-(5-Bromo-2-pyridylazo)-5-diethylaminophenol	Marczenko, p 397. (3), *Talanta 28,* 189, (1981).
	Dimethylglyoxime	Fries/Getrost, p 263. (2), Marczenko, p 393. (3), Onishi, Part IIb, p 125. (6), Snell, *Metals I,* p 887. (7)
	Dimethylglyoxime + Oxidant	Fries/Getrost, p 263. (2), Onishi, Part IIb, p 125. (6), Snell, *Metals I,* p 887. (7)
	2,2′-Furildioxime	Marczenko, p 396. (3), Snell, *Metals I,* p 904. (7)
	2-(2-Pyridylazo)-2-naphthol	Snell, *Metals I,* p 910. (7)
	4-(2-Pyridylazo)resorcinol	Snell, *Metals I,* p 911. (7), West, p 39. (10), *Anal. Chim. Acta, 82,* 431, (1976).
Niobium	*N*-Benzoyl-*N*-phenylhydroxylamine	Snell, *Metals II,* p 1425. (8)
	Pyrocatechol + EDTA or 2,2′Bipyridyl or 1-(2-thenoyl)-3,3,3,-trifluoroacetone	Snell, *Metals II,* p 1427. (8)
	Bromopyrogallol red	Marczenko, p 407. (3), Snell, *Metals II,* p 1426. (8)
	Bromopyrogallol red + Cetylpyridinium chloride	*Talanta, 32,* 189, (1985).
	4-(2-Pyridylazo)resorcinol	Fries/Getrost, p 274. (2), Marczenko, p 406. (3), Onishi, Part IIb, p 160. (7), Snell, *Metals II,* p 1447. (8)

ORGANIC ANALYTICAL REAGENTS FOR THE DETERMINATION OF INORGANIC SUBSTANCES (continued)

Determination	Reagents	Ref.
	Sulfochlorophenol S	Onishi, Part IIb, p 161. (7), Snell, *Metals II,* p 1430. (8)
	Xylenol Orange	Onishi, Part IIb, p 164. (7)
Nitrate	Brucine	Boltz, p 227. (1), Fries/Getrost, p 280. (2), Snell, *Nonmetals,* p 546. (9)
	Chromotropic acid	Boltz, p 229. (1), Fries/Getrost, p 281. (2), Snell, *Nonmetals,* p 548. (9), Williams, p 132. (11), *Fres. Z. Anal. Chem., 320,* 490, (1985).
	Sulfanilamide + *N*-(1-Naphthyl)ethylenediamine dihydrochloride	Fries/Getrost, p 279. (2), Snell, *Nonmetals,* p 559. (9)
Nitrite	Sulfanilamide + *N*-(1-Naphthyl)ethylenediamine dihydrochlorine	Boltz, p 241. (1), Snell, *Nonmetals, p* 585. (8), *Analyst, 109,* 1281, (1984).
	Sulfanilic acid + 1-Naphthylamine	Boltz, p 237. (1), Fries/Getrost, p 285. (2), Marczenko, p 419. (3), Snell, *Nonmetals,* p 586. (9)
Osmium	1,5-Diphenylcarbazide	Marczenko, p 428. (3)
Palladium	2-(5 Bromo-2-pyridylazo)-5-diethylaminophenol	*Talanta, 33,* 939, (1986).
	Dithizone	Marczenko, p 440. (3), Onishi, Part IIb, p 227. (6), Snell, *Metals II,* p 1577. (8)
	2-Nitroso-1-naphthol	Fries/Getrost, p 294. (2), Onishi, Part IIb, p 226. (6), Snell, *Metals II,* p 1581. (8)
	4-(2-Pyridylazo)resorcinol	Snell, *Metals II,* p 1583. (8) *Analyst, 107,* 708, (1982).
Phosphate	Rhodamine B + Molybdate	Snell, *Nonmetals,* p 103. (9)
	Malachite Green + Molybdate	Snell, *Nonmetals,* p 12. (9), Analyst, *108,* 361, (1983).
Platinum	Sulfochlorophenolazorhodamine	Onishi, Part IIb, p 253. (6), *Talanta, 34,* 87, (1987).
	Dithizone	Fries/Getrost, p 300. (2), Onishi, Part IIb, p 253. (6), Snell, *Metals II,* p 1534. (8)
	2-Mercaptobenzothiazole	Fries/Getrost, p 302. (2), *Zh. Anal. Khim., 24,* 1172, (1969).
Rare Earths	Arsenazo I	Marczenko, p 470. (3), Onishi, Part IIa, p 785. (5), Snell, *Metals II,* p 1857. (8)
	Arsenazo III	Fries/Getrost, p 309. (2), Marczenko, p 468. (3), Onishi, Part IIa, p 786. (5), Snell, *Metals II,* p 1862. (8)
	Xylenol Orange	Onishi, Part IIa, p 787. (5), Snell, *Metals II,* p 1874. (8)
Rhenium	2,2′-Furildioxime	Fries/Getrost, p 310. (2), Marczenko, p 481. (3), Onishi, Part IIb, p 288. (6), Snell, *Metals II,* p 1659. (8)
Rhodium	1-(2-Pyridylazo)-2-naphthol	Fries/Getrost, p 311. (2), Snell, *Metals II,* p 1553. (8)
Ruthenium	1,10-Phenanthroline	Onishi, Part IIb, p 331. (6), Snell, *Metals II,* p 1623. (8)
	Thiourea	Fries/Getrost, p 318. (2), Onishi, Part IIb, 329. (6), Snell, *Metals II,* p 1626. (8)
	1,4-Diphenylthiosemicarbazide	Marczenko, p 493. (3), Onishi, Part IIb, p 330. (8)
Scandium	Alizarin red S	Fries/Getrost, p 319. (2), Onishi, Part IIb, p 360. (6), Snell, *Metals I,* p 536. (7)
	Arsenazo III	Onishi, Part IIb, p 359. (6), Snell, *Metals I,* p 539. (7)

ORGANIC ANALYTICAL REAGENTS FOR THE DETERMINATION OF INORGANIC SUBSTANCES (continued)

Determination	Reagents	Ref.
	Chrome Azurol S	Snell, Metals I, p 551. (7), *Anal. Chim. Acta, 159*, 309, (1984).
	Xylenol Orange	Marczenko, p 501. (3), Onishi, Part IIb, p 357. (6), Snell, *Metals I,* p 547. (7)
Selenium	3,3′-Diaminobenzidine	Boltz, p 391. (1), Fries/Getrost, p 323. (2), Marczenko, p 508. (3), Snell, *Nonmetals,* p 490. (9), West, p 4. (10).
	2,3-Diaminonaphthaline	Snell, *Nonmetals,* p 501. (9)
Silver	Dithizone	Fries/Getrost, p 328. (2), Marczenko, p 524. (3), Onishi, Part IIb, p 379. (6), Snell, *Metals I,* p 82. (7)
	Eosin + 1,10-Phenanthroline	Snell, *Metals I,* p 93. (7)
Sulfate	Methylthymol blue + Barium (II)	Snell, *Nonmetals,* p 457. (9)
Sulfide	*N,N,*-Dimethyl-1,4-phenylenediamine	Boltz, p 483. (1), Fries/Getrost, p 344. (2), Snell, *Nonmetals,* p 400. (9), Williams, p 578. (11)
Sulfite	Pararosaniline + Formaldehyde	Boltz, p 478. (1), Marczenko, p 540. (3), Snell, *Nonmetals,* p 430. (9), Williams, p 591. (11)
Tantalum	Methyl Violet	Marczenko, p 551. (3), Snell, *Metals II,* p 1485. (8)
	4-(2-Pyridylazo)resorcinol	Snell, *Metals II,* p 1488. (8)
	Phenylfluoronc	Onishi, Part IIb, p 166. (6), Snell, *Metals II,* p 1486. (8)
Tellurium	Diethyldithiocarbamate	Boltz, p 402. (1), Fries/Getrost, p 348. (2), Snell, *Nonmetals,* p 533. (9), Williams, p 220. (10)
	Bismuthiol II	Boltz, p 401. (1), Marczenko, p 557. (3), Snell, *Nonmetals,* p 524. (9)
Thallium	Brilliant green	Fries/Getrost, p 352. (2), Marczenko, p 567. (3), Onishi, Part IIb, p 426. (6), Snell, *Metals I,* p 45. (7)
	Dithizone	Fries/Getrost, p 355. (2), Onishi, Part IIb, p 426. (6), Snell, *Metals I,* p 54. (7)
	Rhodamine B	Fries/Getrost, p 354. (2), Marczenko, p 566. (3), Onishi, Part IIb, p 424. (6), Snell, *Metals I,* p 63. (7)
Thorium	Arsenazo III	Fries/Getrost, p 360. (2), Marczenko, p 575. (3), Onishi, Part IIb, p 460. (6), Snell, *Metals II,* p 1820. (8)
	Thoron	Marczenko, p 574. (3), Onishi, Part IIb, p 463. (6), Snell, *Metals I,* p 1835. (7)
	Xylenol Orange	Snell, *Metals I,* p 1852. (7)
	Xylenol Orange + Cetyltrimethylammonium bromide	*Talanta, 26*, 499, (1979).
Tin	Pyrocatechol violet (and + Cetyltrimethylammonium bromide)	Marczenko, p 585. (3), Onishi, Part IIb, p 501. (6), Snell, *Metals I, p* 422. (7)
	Gallein	Onishi, Part IIb, p 507, 510. (6), Snell, *Metals I,* p 432. (7)
	Phenylfluorone	Fries/Getrost, p 368. (2), Marczenko, p 582. (3), Onishi, Part IIb, p 497. (6), Snell, *Metals I,* p 444. (7)
	Toluene-3,4-dithiol + Dispersant	Fries/Getrost, p 366. (2), Onishi, Part IIb, p 502. (6), Snell, *Metals I,* p 427. (7)
Titanium	Chromotropic acid	Marczenko, p 593. (3), Onishi, Part IIb, p 551. (6), Snell, *Metals II,* p 1080. (8)

ORGANIC ANALYTICAL REAGENTS FOR THE DETERMINATION OF INORGANIC SUBSTANCES (continued)

Determination	Reagents	Ref.
	Diantipyrinylmethane	Onishi, Part IIb, p 545. (6), Snell, *Metals II,* 1085. (8)
	Tiron	Fries/Getrost, p 376. (2), Onishi, Part IIb, p 549. (6), Snell, *Metals II,* p 1114. (8)
Tungsten	Pyrocatechol Violet	Snell, Metals II, p 1265. (8)
	Tetraphenylarsonium chloride + Thiocyanate	Onishi, Part IIb, p 596. (6), Snell, *Metals II,* p 1278. (8)
	Toluene-3,5-dithiol	Marczenko, p 605. (3), Onishi, Part IIb, p 590. (6), Snell, *Metals II,* p 1267. (8)
Uranium	Arsenazo III	Marczenko, p 611. (3), Onishi, Part IIb, p 627. (6), Snell, *Metals II,* p 1356. (8)
	2-(5-Bromo-2-pyridylazo)diethylaminophenol	Fries/Getrost, p 388. (2), Onishi, Part IIb, p 625. (6)
	Chlorophosphonazo III	Snell, Metals II, p 1367. (8), *Fres. Z. Anal. Chem., 306,* 110, (1981).
	1-(2-Pyridylazo)-2-naphthol	Fries/Getrost, p 386. (2), Onishi, Part IIb, p 625. (6), Snell, *Metals II,* p 1387. (8)
Vanadium	*N*-Benzoyl-*N*-phenylhydroxylamine	Fries/Getrost, p 395. (2), Marczenko, p 625. (3), Snell, *Metals II,* p 1196. (8)
	8-Hydroxyquinoline	Marczenko, p 623. (3), Snell, *Metals II,* p 1209. (8)
	4-(2-pyridylazo)resorcinol	Fries/Getrost, p 404. (23), Marczenko, p 628. (3), Onishi, Part IIb, p 625. (6), Snell, *Metals II,* p 1226. (8)
Yttrium	Alizarin Red S	Fries/Getrost, p 406. (2), Onishi, Part IIa, p 784. (5), Snell, *Metals II,* p 1919. (8)
	Arsenazo III	Marczenko, p 468. (3), Onishi, Part IIa, p 786. (5), Snell, *Metals II,* p 1921. (8)
	Xylenol Orange	Fries/Getrost, p 406. (2), Onishi, Part IIa, p 787. (5), Snell, *Metals II,* p 1923. (8)
Zinc	Dithizone	Fries/Getrost, p 408. (2), Marczenko, p 637. (3), Onishi, Part IIb, p 708. (6), Snell, *Metals II,* p 1042. (8)
	1-(2-Pyridylazo)-2-naphthol	Marczenko, p 639. (3), Onishi, Part IIb, p 719. (6), Snell, *Metals II,* p 1056. (8)
	Xylenol Orange	Fries/Getrost, p 417. (2), Snell, *Metals II,* p 1062. (8), Talanta, *26,* 693, (1979).
	Zircon	Fries/Getrost, p 412. (2), Onishi, Part IIb, p 719. (6), Snell, *Metals II,* p 1063. (8), West, p 23. (10)
Zirconium	Alizarin Red S	Fries/Getrost, p 421. (2), Marczenko, p 647. (3), Onishi, Part IIb, p 763. (6), Snell, *Metals II,* p 1136. (8)
	Arsenazo III	Fries/Getrost, p 421. (2), Onishi, Part IIb, p 770. (6), Snell, *Metals II,* p 1143. (8)
	Pyrocatechol Violet	Onishi, Part IIb, p 771. (6), Snell, *Metals I I,* p 1149. (8)
	Morin	Fries/Getrost, p 424. (2), Onishi, Part IIb, p 765. (6), Snell, *Metals II,* p 1158. (8)
	Xylenol Orange	Fries/Getrost, p 419. (2), Marczenko, p 648. (3), Onishi, Part IIb, p 767. (6), Snell, *Metals II,* p 1167. (8)

ORGANIC ANALYTICAL REAGENTS FOR THE DETERMINATION OF INORGANIC SUBSTANCES (continued)

REVIEWS

Sommer, L, Ackermann, G., Thorburn Burns, D., and Savvin, S. B., *Pure and Applied Chem.*, 62, 2147, 1990.

Sommer, L., Ackermann, G., and Thorburn Burns, D., *Pure and Applied Chem.*, 62, 2323, 1990)

Sommer, L., Komarek, J., and Thorburn Burns, D., *Pure and Applied Chem., 64,* 213, 1992.

Savvin, S. B., *Crit. Rev. Anal. Chem.,* 8, 55, 1979.

MONOGRAPHS

1. Boltz, D. F., and Howell, J. A., *Colorimetric Determination of Nonmetals,* 2nd ed, Wiley, New York, 1978.
2. Fries, J. and Getrost, H., *Organic Reagents for Trace Analysis,* E Merck, Darmstadt, 1977.
3. Marczenko, Z., *Separation and Spectrophotometric Determination of Elements,* Ellis Horwood, Chichester, 1986.
4. Sandell, E. B. and Onishi, H., *Photometric Determination of Traces of Metals. General Aspects, Part I,* 4th ed, J. Wiley, New York, 1978.
5. Onishi, H., *Photometric Determination of Traces of Metals. Part IIa: Individual Metals, Aluminium to Lithium,* 4th ed, J. Wiley, New York, 1986.
6. Onishi, H., *Photometric Determination of Traces of Metals. Part IIb: Individual Metals, Magnesium to Zinc,* 4th ed, J. Wiley, New York, 1989.
7. Snell, F. D., *Photometric and Fluorimetric Methods of Analysis, Metals Part 1,* J. Wiley, New York, 1978.
8. Snell, F. D., *Photometric and Fluorimetric Methods of Analysis, Metals Part 2,* J. Wiley, New York, 1978.
9. Snell, F. D., *Photometric and Fluorimetric Methods of Analysis, Nonmetals,* J. Wiley, New York, 1981.
10. West, T. S. and Nürnberg, H. W., Eds., *The Determination of Trace Metals in Natural Waters,* Blackwell, Oxford, 1988.
11. Williams, W. J.,, *Handbook of Anion Determination,* Butterworth, London, 1979.
12. Townshend, A., Burns, D. T., Guilbault, G. G., Lobinski, R., Mavenzenko, Z., Newman, E., and Onishi, H., *Dictionary of Analytical Reagents,* Chapman & Hall, London, 1993.

ACID-BASE INDICATORS

A. K. Covington

The first part of this table lists some common acid-base indicators in alphabetical order along with the approximate pH range(s) at which a color change occurs. Following this is a table of the same indicators ordered by pH range, which includes the nature of the color change, instructions on preparation of the indicator solution, and the acid dissociation constant p*K*, when available. The color code is:

C = colorless A = amber B/G = blue-green Pk = pink Y = yellow V = violet R = red B = blue
P = purple O = orange

REFERENCE

Bishop, E., Ed., *Indicators*, Pergamon, Oxford, 1972.

Indicator	pH Range
Alizarin	5.6-7.2; 11.0-12.4
Alizarin Red S	4.6-6.0
Alizarin Yellow R	10.1-12.0
Benzopurpurine 4B	2.2-4.2
4,4'-Bis(2-amino-1-naphthylazo)-2,2'-stilbenedisulfonic acid	3.0-4.0
4,4'-Bis(4-amino-1-naphthylazo)-2,2'-stilbenedisulfonic acid	8.0-9.0
Brilliant Yellow	6.6-7.8
Bromocresol Green	3.8-5.4
Bromocresol Purple	5.2-6.8
Bromophenol Blue	3.0-4.6
Bromothymol Blue	6.0-7.6
Chlorophenol Red	5.2-6.8
Clayton Yellow	12.2-13.2
Congo Red	3.0-5.0
o-Cresolphthalein	8.2-9.8
Cresol Red	0.0-1.0; 7.0-8.8
Crystal Violet	0.0-1.8
Curcumin (Turmaric)	7.4-8.6
p-(2,4-Dihydroxyphenylazo) benzenesulfonic acid, sodium salt	11.4-12.6
p-Dimethylaminoazobenzene	2.8-4.4
4-(4-Dimethylamino-1-naphylazo)-3-methoxybenzenesulfonic acid	3.5-4.8
2-(*p*-Dimethylamino-phenylazo)pyridine	0.2-1.8; 4.4-5.6
N,N-Dimethyl-*p*-(*m*-tolylazo)aniline	2.6-4.8
2,4-Dinitrophenol	2.0-4.7
2-(2,4 Dinitrophenylazo)-1-naphthol-3,6-disulfonic acid, disodium salt	6.0-7.0
6,8-Dinitro-2,4-(1*H*)quinazolinedione	6.4-8.0
Erythrosin, disodium salt	2.2-3.6
4-(*p*-Ethoxyphenylazo)-*m*-phenylene-diamine monohydrochloride	4.4-5.8
Ethyl bis(2,4-dimethylphenyl) ethanoate	8.4-9.6
Ethyl Orange	3.4-4.8
Ethyl Red	4.0-5.8
Ethyl Violet	0.0-2.4
5,5'-Indigodisulfonic acid, disodium salt	11.4-13.0
Malachite Green	0.2-1.8
Metacresol Purple	1.2-2.8; 7.4-9.0
Metanil Yellow	1.2-2.4
Methyl Green	0.2-1.8
Methyl Orange	3.2-4.4
Methyl Red	4.8-6.0
Methyl Violet	0.0-1.6
p-Naphtholbenzein	8.2-10.0
Neutral Red	6.8-8.0
p-Nitrophenol	5.4-6.6
m-Nitrophenol	6.8-8.6
Orange IV	1.4-2.8
Paramethyl Red	1.0-3.0
Phenolphthalein	8.2-10.0
Phenol Red	6.6-8.0
4-Phenylazodiphenylamine	1.2-2.6
4-Phenylazo-1-naphthylamine	4.0-5.6
Propyl Red	4.8-6.6
Quinaldine Red	1.4-3.2
Resazurin	3.8-6.4
Resorcin Blue	4.4-6.2
Tetrabromophenolphthalein ethyl ester, potassium salt	3.0-4.2
Thymol Blue	1.2-2.8; 8.0-9.6
Thymolphthalein	9.4-10.6
4-*o*-Tolylazo-*o*-toluidine	1.4-2.8
1,3,5-Trinitrobenzene	12.0-14.0
2,4,6-Trinitrotoluene	11.5-13.0
Turmaric	7.4-8.6

pH range	Color change	Indicator	pK	Preparation
0.0-1.0	R-Y	Cresol Red		0.1 g in 26.2 mL 0.01 M NaOH + 223.8 mL water
0.0-1.6	Y-B	Methyl Violet		0.01-0.05% in water
0.0-1.8	Y-B	Crystal Violet		0.02% in water
0.0-2.4	Y-B	Ethyl Violet		0.1 g in 50 mL 50% v/v methanol-water
0.2-1.8	Y-B/G	Malachite Green	1.3	water
0.2-1.8	Y-B	Methyl Green		0.1% in water
0.2-1.8	Y-R	2-(*p*-Dimethylaminophenylazo)pyridine		0.1% in ethanol
1.0-3.0	R-Y	Paramethyl Red		ethanol
1.2-2.4	R-Y	Metanil Yellow		0.01% in water
1.2-2.6	R-Y	4-Phenylazodiphenylamine		0.01 g in 1 mL 1 M HCl + 50 mL ethanol + 49 mL water
1.2-2.8	R-Y	Thymol Blue	1.65	0.1 g in 21.5 mL 0.01 M NaOH + 228.5 mL water
1.2-2.8	R-Y	Metacresol Purple	1.51	0.1 g in 26.2 mL 0.01 M NaOH + 223.8 mL water
1.4-2.8	R-Y	Orange IV		0.01% in water
1.4-2.8	O-Y	4-*o*-Tolylazo-*o*-toluidine		water
1.4-3.2	C-R	Quinaldine Red	2.63	1% in ethanol
2.0-4.7	C-Y	2,4-Dinitrophenol	3.96	sat. solution in water
2.2-3.6	O-R	Erythrosin, disodium salt		0.1% in water
2.2-4.2	V-R	Benzopurpurine 4B		0.1% in water
2.6-4.8	R-Y	*N,N*-Dimethyl-*p*-(*m*-tolylazo)aniline		0.1% in water
2.8-4.4	R-Y	*p*-Dimethylaminoazobenzene		0.1 g in 100 mL 90% v/v ethanol-water
3.0-4.0	P-R	4,4'-Bis(2-amino-1-naphthylazo)-2,2'-stilbenedisulfonic acid		0.1 g in 5.9 mL 0.05 M NaOH + 94.1 mL water
3.0-4.2	Y-B	Tetrabromophenolphthalein ethyl ester, potassium salt		0.1% in ethanol
3.0-4.6	Y-B	Bromophenol Blue	4.10	0.1 g in 14.9 mL 0.01 M NaOH + 235.1 mL water
3.0-5.0	B-R	Congo Red		0.1% in water
3.2-4.4	R-Y	Methyl Orange	3.46	0.1% in water
3.4-4.8	R-Y	Ethyl Orange	4.34	0.05-0.2% in water or aqueous ethanol
3.5-4.8	V-Y	4-(4-Dimethylamino-1-naphylazo)-3-methoxybenzenesulfonic acid		0.1% in 60% ethanol-water
3.8-5.4	Y-B	Bromocresol Green	4.90	0.1 g in 14.3 mL 0.01 M NaOH + 235.7 mL water
3.8-6.4	O-V	Resazurin		water
4.0-5.6	R-Y	4-Phenylazo-1-naphthylamine		0.1% in ethanol
4.0-5.8	C-R	Ethyl Red	5.42	0.1 g in 100 mL 50% v/v methanol-water
4.4-5.6	R-Y	2-(*p*-Dimethylaminophenylazo)pyridine		0.1% in ethanol
4.4-5.8	O-Y	4-(*p*-Ethoxyphenylazo)-*m*-phenylene-diamine monohydrochloride		0.1% in water
4.4-6.2	R-B	Resorcin Blue		0.2% in ethanol
4.6-6.0	Y-R	Alizarin Red S		water
4.8-6.0	R-Y	Methyl Red	5.00	0.02 g in 100 mL 60% v/v ethanol-water
4.8-6.6	R-Y	Propyl Red	5.48	ethanol
5.2-6.8	Y-P	Bromocresol Purple	6.40	0.1 g in 18.5 mL 0.01 M NaOH + 231.5 mL water
5.2-6.8	Y-R	Chlorophenol Red	6.25	0.1 g in 23.6 mL 0.01 M NaOH + 226.4 mL water
5.4-6.6	C-Y	*p*-Nitrophenol	7.15	0.1% in water
5.6-7.2	Y-R	Alizarin		0.1% in methanol
6.0-7.0	Y-B	2-(2,4-Dinitrophenylazo)-1-naphthol-3,6-disulfonic acid, disodium salt		0.1% in water
6.0-7.6	Y-B	Bromothymol Blue	7.30	0.1 g in 16 mL 0.01 M NaOH + 234 mL water
6.4-8.0	C-Y	6,8-Dinitro-2,4-(1*H*)quinazolinedione		25 g in 115 mL 1 M NaOH + 50 mL water at 100°C
6.6-7.8	Y-R	Brilliant Yellow		1% in water
6.6-8.0	Y-R	Phenol Red	8.00	0.1 g in 28.2 mL 0.01 M NaOH + 221.8 mL water
6.8-8.0	R-A	Neutral Red		0.01 g in 100 mL 50% v/v ethanol-water
6.8-8.6	C-Y	*m*-Nitrophenol	8.28	0.3% in water
7.0-8.8	Y-R	Cresol Red	8.46	0.1 g in 26.2 mL 0.01 M NaOH + 223.8 mL water
7.4-8.6	Y-R	Turmaric (Curcumin)		ethanol
7.4-9.0	Y-P	Metacresol Purple	8.3	0.1 g in 26.2 mL 0.01 M NaOH + 223.8 mL water

pH range	Color change	Indicator	p*K*	Preparation
8.0-9.0	B-R	4,4'-Bis(4-amino-1-naphthylazo)-2,2'-stilbenedisulfonic acid		0.1 g in 5.9 mL 0.05 M NaOH + 94.1 mL water
8.0-9.6	Y-B	Thymol Blue	9.20	0.1 g in 21.5 mL 0.01 M NaOH + 228.5 mL water
8.2-10.0	O-B	*p*-Naphtholbenzein		1% in dil. alkali
8.2-10.0	C-Pk	Phenolphthalein	9.5	0.5 g in 100 mL 50% v/v ethanol-water
8.2-9.8	C-R	*o*-Cresolphthalein		0.04% in ethanol
8.4-9.6	C-B	Ethyl bis(2,4-dimethylphenyl)ethanoate		sat. solution in 50% acetone-ethanol
9.4-10.6	C-B	Thymolphthalein		0.04 g in 100 mL 50% v/v ethanol-water
10.1-12.0	Y-R	Alizarin Yellow R		0.01% in water
11.0-12.4	R-P	Alizarin		0.1% in methanol
11.4-12.6	Y-O	*p*-(2,4-Dihydroxyphenylazo) benzenesulfonic acid, sodium salt		0.1% in water
11.4-13.0	B-Y	5,5'-Indigodisulfonic acid, disodium salt		water
11.5-13.0	C-O	2,4,6-Trinitrotoluene		0.1-0.5% in ethanol
12.0-14.0	C-O	1,3,5-Trinitrobenzene		0.1-0.5% in ethanol
12.2-13.2	Y-A	Clayton Yellow		0.1% in water

FLUORESCENT INDICATORS

Jack DeMent

Fluorescent indicators are substances which show definite changes in fluorescence with change in pH. Some fluorescent materials are not suitable for indicators since their change in fluorescence is too gradual. Fluorescent indicators find greatest utility in the titration of opaque, highly turbid or deeply colored solutions. A long wavelength ultraviolet ("black light") lamp in a dimly lighted room provides the best environment for titrations involving fluorescent indicators, although bright daylight is sometimes sufficient to evoke a response in the bright green, yellow and orange fluorescent indicators. Titrations are carried out in non-fluorescent glassware. One should check the glassware prior to use to make certain that it does not fluoresce due to the wavelengths of light involved in the titration. The meniscus of the liquid in the burette can be followed when a few particles of an insoluble fluorescent solid are dropped onto its surface.

In this table the indicators are arranged by approximate pH range covered. In the case of some of the dyestuffs the end point may vary slightly with the source or manufacturer.

Indicator	C.I.	From pH	To pH
pH 0 to 2			
Benzoflavine	—	0.3, yellow fl.	1.7, green fl.
3,6-Dioxyphthalimide	—	0, blue fl.	2.4, green fl.
Eosine YS	768	0, yellow colored	3.0, yellow fl.
Erythrosine	772	0, yellow colored	3.6, yellow fl.
Esculin	—	1.5, colorless	2, blue fl.
4-Ethoxyacridone	—	1.2, green fl.	3.2, blue fl.
3,6-Tetramethyldiaminooxanthone	—	1.2, green fl.	3.4, blue fl.
pH 2 to 4			
Chromotropic acid	—	3.5, colorless	4.5, blue fl.
Fluorescein	766	4, colorless	4.5, green fl.
Magdala Red	—	3.0, purple colored	4.0, fl.
α-Naphthylamine	—	3.4, colorless	4.8, blue fl.
β-Naphthylamine	—	2.8, colorless	4.4, violet fl.
Phloxine	774	3.4, colorless	5.0, bright yellow fl.
Salicylic acid	—	2.5, colorless	3.5, blue fl.
pH 4 to 6			
Acridine	788	4.9, green fl.	5.1, violet colored
Dichlorofluorescein	—	4.0, colorless	5.0, green fl.
3,6-Dioxyxanthone	—	5.4, colorless	7.6, blue-violet fl.
Erythrosine	772	4.0, colorless	4.5, yellow-green fl.
β-Methylesculetin	—	4.0, colorless	6.2, blue fl.
Neville-Winther acid	—	6.0, colorless	6.5, blue fl.
Resorufin	—	4.4, yellow fl.	6.4, weak orange fl.
Quininic acid	—	4.0, yellow colored	5.0, blue fl.
Quinine [first end point]	—	5.0, blue fl.	6.1, violet fl.
pH 6 to 8			
Acid R Phosphine	—	(claimed for range pH 6.0–7.0)	
Brilliant Diazol Yellow	—	6.5, colorless	7.5, violet fl.
Cleves acid	—	6.5, colorless	7.5, green fl.
Coumaric acid	—	7.2, colorless	9.0, green fl.
3,6-Dioxyphthalic dinitrile	—	5.8, blue fl.	8.2, green fl.
Magnesium 8-hydroxyquinolinate	—	6.5, colorless	7.5, golden fl.
β-Methylumbelliferone	—	7.0, colorless	7.5, blue fl.
1-Naphthol-4-sulfonic acid	—	6.0, colorless	6.5, blue fl.
Orcinaurine	—	6.5, colorless	8.0, green fl.
Patent Phosphine	789	(for the range pH 6.0–7.0, green-yellow fl.)	
Thioflavine	816	(for the region pH 6.5–7.0, yellow fl.)	
Umbelliferone	—	6.5, colorless	7.6, blue fl.

pH 8 to 10

Indicator	C.I.	From pH	To pH
Acridine Orange	788	8.4, orange colored	10.4, green fl.
Ethoxyphenylnaphthostilbazonium chloride	—	9, green fl.	11, non-fl.
G Salt	—	9.0, dull blue fl.	9.5, bright blue fl.
Naphthazol derivatives	—	8.2, colorless	10.0, yellow or green fl.
α-Naphthionic acid	—	9, blue fl.	11, green fl.
2-Naphthol-3,6-disulfonic acid	—	9.5, dark blue fl.	Light blue fl. at higher pH
β-Naphthol	—	8.6, colorless	Blue fl. at higher pH
α-Naphtholsulfonic acid	—	8.0, dark blue fl.	9.0, bright violet fl.
1,4-Naphtholsulfonic acid	—	8.2, dark blue fl.	Light blue fl. at higher pH
Orcinsulfonphthalein	—	8.6, yellow colored	10.0 fl.
Quinine [second end point]	—	9.5, violet fl.	10.0, colorless
R-Salt	—	9.0, dull blue fl.	9.5, bright blue fl.
Sodium 1-naphthol-2-sulfonate	—	9.0, dark blue fl.	10.0, bright violet fl.

pH 10 to 12

Indicator	C.I.	From pH	To pH
Coumarin	—	9.8, deep green fl.	12, light green fl.
Eosine BN	771	10.5, colorless	14.0, yellow fl.
Papaverine (permanganate oxidized)	—	9.5, yellow fl.	11.0, blue fl.
Schaffers Salt	—	5.0, violet fl.	11.0, green-blue fl.
SS-Acid (sodium salt)	—	10.0, violet fl.	12.0, yellow colored

pH 12 to 14

Indicator	C.I.	From pH	To pH
Cotarnine	—	12.0, yellow fl.	13.0, white fl.
α-Naphthionic acid	—	12, blue fl.	13, green fl.
β-Naphthionic acid	—	12, blue fl.	13, violet fl.

CONVERSION FORMULAS FOR CONCENTRATION OF SOLUTIONS

A = Weight per cent of solute
B = Molecular weight of solvent
E = Molecular weight of solute
F = Grams of solute per liter of solution
G = Molality
M = Molarity
N = Mole fraction
R = Density of solution in grams per milliliter

Concentration of solute—SOUGHT	Concentration of solute—GIVEN: A	N	G	M	F
A	—	$\frac{100N \times E}{N \times E + (1 - N)B}$	$\frac{100G \times E}{1000 + G \times E}$	$\frac{M \times E}{10R}$	$\frac{F}{10R}$
N	$\frac{\frac{A}{E}}{\frac{A}{E} + \frac{100 - A}{B}}$	—	$\frac{B \times G}{B \times G + 1000}$	$\frac{B \times M}{M(B - E) + 1000R}$	$\frac{B \times F}{F(B - E) + 1000R \times E}$
G	$\frac{1000A}{E(100 - A)}$	$\frac{1000N}{B - N \times B}$	—	$\frac{1000M}{1000R - (M \times E)}$	$\frac{1000F}{E(1000R - F)}$
M	$\frac{10R \times A}{E}$	$\frac{1000R \times N}{N \times E + (1 - N)B}$	$\frac{1000R \times G}{1000 + E \times G}$	—	$\frac{F}{E}$
F	$10AR$	$\frac{1000R \times N \times E}{N \times E + (1 - N)B}$	$\frac{1000R \times G \times E}{1000 + G \times E}$	$M \times E$	—

ELECTROCHEMICAL SERIES

Petr Vanýsek

There are three tables for this electrochemical series. Each table lists standard reduction potentials, $E°$ values, at 298.15 K (25°C), and at a pressure of 101.325 kPa (1 atm). Table 1 is an alphabetical listing of the elements, according to the symbol of the elements. Thus, data for silver (Ag) precedes those for aluminum (Al). Table 2 lists only those reduction reactions which have $E°$ values positive in respect to the standard hydrogen electrode. In Table 2, the reactions are listed in the order of increasing positive potential, and they range from 0.0000 V to + 3.4 V. Table 3 lists only those reduction potentials which have $E°$ negative with respect to the standard hydrogen electrode. In Table 3, the reactions are listed in the order of decreasing potential and range from 0.0000 V to –4.10 V. The reliability of the potentials is not the same for all the data. Typically, the values with fewer significant figures have lower reliability. The values of reduction potentials, in particular those of less common reactions, are not definite; they are subject to occasional revisions.

Abbreviations: ac = acetate; bipy = 2,2′-dipyridine, or bipyridine; en = ethylenediamine; phen = 1,10-phenanthroline.

REFERENCES

1. G. Milazzo, S. Caroli, and V. K. Sharma, *Tables of Standard Electrode Potentials*, Wiley, Chichester, 1978.
2. A. J. Bard, R. Parsons, and J. Jordan, *Standard Potentials in Aqueous Solutions*, Marcel Dekker, New York, 1985.
3. S. G. Bratsch, *J. Phys. Chem. Ref. Data*, 18, 1—21, 1989.

TABLE 1
Alphabetical Listing

Reaction	$E°$/V
$Ac^{3+} + 3e \rightleftharpoons Ac$	–2.20
$Ag^{+} + e \rightleftharpoons Ag$	0.7996
$Ag^{2+} + e \rightleftharpoons Ag^{+}$	1.980
$Ag(ac) + e \rightleftharpoons Ag + (ac)^{-}$	0.643
$AgBr + e \rightleftharpoons Ag + Br^{-}$	0.07133
$AgBrO_3 + e \rightleftharpoons Ag + BrO_3^{-}$	0.546
$Ag_2C_2O_4 + 2e \rightleftharpoons 2Ag + C_2O_4^{2-}$	0.4647
$AgCl + e \rightleftharpoons Ag + Cl^{-}$	0.22233
$AgCN + e \rightleftharpoons Ag + CN^{-}$	–0.017
$Ag_2CO_3 + 2e \rightleftharpoons 2Ag + CO_3^{2-}$	0.47
$Ag_2CrO_4 + 2e \rightleftharpoons 2Ag + CrO_4^{2-}$	0.4470
$AgF + e \rightleftharpoons Ag + F^{-}$	0.779
$Ag_4[Fe(CN)_6] + 4e \rightleftharpoons 4Ag + [Fe(CN)_6]^{4-}$	0.1478
$AgI + e \rightleftharpoons Ag + I^{-}$	–0.15224
$AgIO_3 + e \rightleftharpoons Ag + IO_3^{-}$	0.354
$Ag_2MoO_4 + 2e \rightleftharpoons 2Ag + MoO_4^{2-}$	0.4573
$AgNO_2 + e \rightleftharpoons Ag + 2NO_2^{-}$	0.564
$Ag_2O + H_2O + 2e \rightleftharpoons 2Ag + 2OH^{-}$	0.342
$Ag_2O_3 + H_2O + 2e \rightleftharpoons 2AgO + 2OH^{-}$	0.739
$Ag^{3+} + 2e \rightleftharpoons Ag^{+}$	1.9
$Ag^{3+} + e \rightleftharpoons Ag^{2+}$	1.8
$Ag_2O_2 + 4H^{+} + e \rightleftharpoons 2Ag + 2H_2O$	1.802
$2AgO + H_2O + 2e \rightleftharpoons Ag_2O + 2OH^{-}$	0.607
$AgOCN + e \rightleftharpoons Ag + OCN^{-}$	0.41
$Ag_2S + 2e \rightleftharpoons 2Ag + S^{2-}$	–0.691
$Ag_2S + 2H^{+} + 2e \rightleftharpoons 2Ag + H_2S$	–0.0366
$AgSCN + e \rightleftharpoons Ag + SCN^{-}$	0.08951
$Ag_2SeO_3 + 2e \rightleftharpoons 2Ag + SeO_4^{2-}$	0.3629
$Ag_2SO_4 + 2e \rightleftharpoons 2Ag + SO_4^{2-}$	0.654
$Ag_2WO_4 + 2e \rightleftharpoons 2Ag + WO_4^{2-}$	0.4660
$Al^{3+} + 3e \rightleftharpoons Al$	–1.662
$Al(OH)_3 + 3e \rightleftharpoons Al + 3OH^{-}$	–2.31
$Al(OH)_4^{-} + 3e \rightleftharpoons Al + 4OH^{-}$	–2.328
$H_2AlO_3^{-} + H_2O + 3e \rightleftharpoons Al + 4OH^{-}$	–2.33
$AlF_6^{3-} + 3e \rightleftharpoons Al + 6F^{-}$	–2.069
$Am^{4+} + e \rightleftharpoons Am^{3+}$	2.60
$Am^{2+} + 2e \rightleftharpoons Am$	–1.9
$Am^{3+} + 3e \rightleftharpoons Am$	–2.048
$Am^{3+} + e \rightleftharpoons Am^{2+}$	–2.3
$As + 3H^{+} + 3e \rightleftharpoons AsH_3$	–0.608
$As_2O_3 + 6H^{+} + 6e \rightleftharpoons 2As + 3H_2O$	0.234
$HAsO_2 + 3H^{+} + 3e \rightleftharpoons As + 2H_2O$	0.248
$AsO_2^{-} + 2H_2O + 3e \rightleftharpoons As + 4OH^{-}$	–0.68
$H_3AsO_4 + 2H^{+} + 2e^{-} \rightleftharpoons HAsO_2 + 2H_2O$	0.560
$AsO_4^{3-} + 2H_2O + 2e \rightleftharpoons AsO_2^{-} + 4OH^{-}$	–0.71
$At_2 + 2e \rightleftharpoons 2At^{-}$	0.3
$Au^{+} + e \rightleftharpoons Au$	1.692
$Au^{3+} + 2e \rightleftharpoons Au^{+}$	1.401
$Au^{3+} + 3e \rightleftharpoons Au$	1.498
$Au^{2+} + e^{-} \rightleftharpoons Au^{+}$	1.8
$AuOH^{2+} + H^{+} + 2e \rightleftharpoons Au^{+} + H_2O$	1.32
$AuBr_2^{-} + e \rightleftharpoons Au + 2Br^{-}$	0.959
$AuBr_4^{-} + 3e \rightleftharpoons Au + 4Br^{-}$	0.854
$AuCl_4^{-} + 3e \rightleftharpoons Au + 4Cl^{-}$	1.002
$Au(OH)_3 + 3H^{+} + 3e \rightleftharpoons Au + 3H_2O$	1.45
$H_2BO_3^{-} + 5H_2O + 8e \rightleftharpoons BH_4^{-} + 8OH^{-}$	–1.24
$H_2BO_3^{-} + H_2O + 3e \rightleftharpoons B + 4OH^{-}$	–1.79
$H_3BO_3 + 3H^{+} + 3e \rightleftharpoons B + 3H_2O$	–0.8698
$B(OH)_3 + 7H^{+} + 8e \rightleftharpoons BH_4^{-} + 3H_2O$	–0.481
$Ba^{2+} + 2e \rightleftharpoons Ba$	–2.912
$Ba^{2+} + 2e \rightleftharpoons Ba(Hg)$	–1.570
$Ba(OH)_2 + 2e \rightleftharpoons Ba + 2OH^{-}$	–2.99
$Be^{2+} + 2e \rightleftharpoons Be$	–1.847
$Be_2O_3^{2-} + 3H_2O + 4e \rightleftharpoons 2Be + 6OH^{-}$	–2.63

TABLE 1
Alphabetical Listing (continued)

Reaction	$E°/V$
p–benzoquinone + $2\,H^{+} + 2\,e \rightleftharpoons$ hydroquinone	0.6992
$Bi^{+} + e \rightleftharpoons Bi$	0.5
$Bi^{3+} + 3\,e \rightleftharpoons Bi$	0.308
$Bi^{3+} + 2\,e \rightleftharpoons Bi^{+}$	0.2
$Bi + 3\,H^{+} + 3\,e \rightleftharpoons BiH_3$	–0.8
$BiCl_4^{-} + 3\,e \rightleftharpoons Bi + 4\,Cl^{-}$	0.16
$Bi_2O_3 + 3\,H_2O + 6\,e \rightleftharpoons 2\,Bi + 6\,OH^{-}$	–0.46
$Bi_2O_4 + 4\,H^{+} + 2\,e \rightleftharpoons 2\,BiO^{+} + 2\,H_2O$	1.593
$BiO^{+} + 2\,H^{+} + 3\,e \rightleftharpoons Bi + H_2O$	0.320
$BiOCl + 2\,H^{+} + 3\,e \rightleftharpoons Bi + Cl^{-} + H_2O$	0.1583
$Bk^{4+} + e \rightleftharpoons Bk^{3+}$	1.67
$Bk^{2+} + 2\,e \rightleftharpoons Bk$	–1.6
$Bk^{3+} + e \rightleftharpoons Bk^{2+}$	–2.8
$Br_2(aq) + 2\,e \rightleftharpoons 2\,Br^{-}$	1.0873
$Br_2(l) + 2\,e \rightleftharpoons 2\,Br^{-}$	1.066
$HBrO + H^{+} + 2\,e \rightleftharpoons Br^{-} + H_2O$	1.331
$HBrO + H^{+} + e \rightleftharpoons 1/2\,Br_2(aq) + H_2O$	1.574
$HBrO + H^{+} + e \rightleftharpoons 1/2\,Br_2(l) + H_2O$	1.596
$BrO^{-} + H_2O + 2\,e \rightleftharpoons Br^{-} + 2\,OH^{-}$	0.761
$BrO_3^{-} + 6\,H^{+} + 5\,e \rightleftharpoons 1/2\,Br_2 + 3\,H_2O$	1.482
$BrO_3^{-} + 6\,H^{+} + 6\,e \rightleftharpoons Br^{-} + 3\,H_2O$	1.423
$BrO_3^{-} + 3\,H_2O + 6\,e \rightleftharpoons Br^{-} + 6\,OH^{-}$	0.61
$(CN)_2 + 2\,H^{+} + 2\,e \rightleftharpoons 2\,HCN$	0.373
$2\,HCNO + 2\,H^{+} + 2\,e \rightleftharpoons (CN)_2 + 2\,H_2O$	0.330
$(CNS)_2 + 2\,e \rightleftharpoons 2\,CNS^{-}$	0.77
$CO_2 + 2\,H^{+} + 2\,e \rightleftharpoons HCOOH$	–0.199
$Ca^{+} + e \rightleftharpoons Ca$	–3.80
$Ca^{2+} + 2\,e \rightleftharpoons Ca$	–2.868
$Ca(OH)_2 + 2\,e \rightleftharpoons Ca + 2\,OH^{-}$	–3.02
Calomel electrode, 1 molal KCl	0.2800
Calomel electrode, 1 molar KCl (NCE)	0.2801
Calomel electrode, 0.1 molar KCl	0.3337
Calomel electrode, saturated KCl (SCE)	0.2412
Calomel electrode, saturated NaCl (SSCE)	0.2360
$Cd^{2+} + 2\,e \rightleftharpoons Cd$	–0.4030
$Cd^{2+} + 2\,e \rightleftharpoons Cd(Hg)$	–0.3521
$Cd(OH)_2 + 2\,e \rightleftharpoons Cd(Hg) + 2\,OH^{-}$	–0.809
$CdSO_4 + 2\,e \rightleftharpoons Cd + SO_4^{2-}$	–0.246
$Cd(OH)_4^{2-} + 2\,e \rightleftharpoons Cd + 4\,OH^{-}$	–0.658
$CdO + H_2O + 2\,e \rightleftharpoons Cd + 2\,OH^{-}$	–0.783
$Ce^{3+} + 3\,e \rightleftharpoons Ce$	–2.336
$Ce^{3+} + 3\,e \rightleftharpoons Ce(Hg)$	–1.4373
$Ce^{4+} + e \rightleftharpoons Ce^{3+}$	1.72
$CeOH^{3+} + H^{+} + e \rightleftharpoons Ce^{3+} + H_2O$	1.715
$Cf^{4+} + e \rightleftharpoons Cf^{3+}$	3.3
$Cf^{3+} + e \rightleftharpoons Cf^{2+}$	–1.6
$Cf^{3+} + 3\,e \rightleftharpoons Cf$	–1.94
$Cf^{2+} + 2\,e \rightleftharpoons Cf$	–2.12
$Cl_2(g) + 2\,e \rightleftharpoons 2\,Cl^{-}$	1.35827
$HClO + H^{+} + e \rightleftharpoons 1/2\,Cl_2 + H_2O$	1.611
$HClO + H^{+} + 2\,e \rightleftharpoons Cl^{-} + H_2O$	1.482
$ClO^{-} + H_2O + 2\,e \rightleftharpoons Cl^{-} + 2\,OH^{-}$	0.81
$ClO_2 + H^{+} + e \rightleftharpoons HClO_2$	1.277
$HClO_2 + 2\,H^{+} + 2\,e \rightleftharpoons HClO + H_2O$	1.645
$HClO_2 + 3\,H^{+} + 3\,e \rightleftharpoons 1/2\,Cl_2 + 2\,H_2O$	1.628
$HClO_2 + 3\,H^{+} + 4\,e \rightleftharpoons Cl^{-} + 2\,H_2O$	1.570
$ClO_2^{-} + H_2O + 2\,e \rightleftharpoons ClO^{-} + 2\,OH^{-}$	0.66
$ClO_2^{-} + 2\,H_2O + 4\,e \rightleftharpoons Cl^{-} + 4\,OH^{-}$	0.76
$ClO_2(aq) + e \rightleftharpoons ClO_2^{-}$	0.954
$ClO_3^{-} + 2\,H^{+} + e \rightleftharpoons ClO_2 + H_2O$	1.152
$ClO_3^{-} + 3\,H^{+} + 2\,e \rightleftharpoons HClO_2 + H_2O$	1.214
$ClO_3^{-} + 6\,H^{+} + 5\,e \rightleftharpoons 1/2\,Cl_2 + 3\,H_2O$	1.47
$ClO_3^{-} + 6\,H^{+} + 6\,e \rightleftharpoons Cl^{-} + 3\,H_2O$	1.451
$ClO_3^{-} + H_2O + 2\,e \rightleftharpoons ClO_2^{-} + 2\,OH^{-}$	0.33
$ClO_3^{-} + 3\,H_2O + 6\,e \rightleftharpoons Cl^{-} + 6\,OH^{-}$	0.62
$ClO_4^{-} + 2\,H^{+} + 2\,e \rightleftharpoons ClO_3^{-}\,H_2O$	1.189
$ClO_4^{-} + 8\,H^{+} + 7\,e \rightleftharpoons 1/2\,Cl_2 + 4\,H_2O$	1.39
$ClO_4^{-} + 8\,H^{+} + 8\,e \rightleftharpoons Cl^{-} + 4\,H_2O$	1.389
$ClO_4^{-} + H_2O + 2\,e \rightleftharpoons ClO_3^{-} + 2\,OH^{-}$	0.36
$Cm^{4+} + e \rightleftharpoons Cm^{3+}$	3.0
$Cm^{3+} + 3\,e \rightleftharpoons Cm$	–2.04
$Co^{2+} + 2\,e \rightleftharpoons Co$	–0.28
$Co^{3+} + e \rightleftharpoons Co^{2+}$	1.92
$[Co(NH_3)_6]^{3+} + e \rightleftharpoons [Co(NH_3)_6]^{2+}$	0.108
$Co(OH)_2 + 2\,e \rightleftharpoons Co + 2\,OH^{-}$	–0.73
$Co(OH)_3 + e \rightleftharpoons Co(OH)_2 + OH^{-}$	0.17
$Cr^{2+} + 2\,e \rightleftharpoons Cr$	–0.913
$Cr^{3+} + e \rightleftharpoons Cr^{2+}$	–0.407
$Cr^{3+} + 3\,e \rightleftharpoons Cr$	–0.744
$Cr_2O_7^{2-} + 14\,H^{+} + 6\,e \rightleftharpoons 2\,Cr^{3+} + 7\,H_2O$	1.232
$CrO_2^{-} + 2\,H_2O + 3\,e \rightleftharpoons Cr + 4\,OH^{-}$	–1.2
$HCrO_4^{-} + 7\,H^{+} + 3\,e \rightleftharpoons Cr^{3+} + 4\,H_2O$	1.350
$CrO_2 + 4\,H^{+} + e \rightleftharpoons Cr^{3+} + 2H_2O$	1.48
$Cr(V) + e \rightleftharpoons Cr(IV)$	1.34
$CrO_4^{2-} + 4\,H_2O + 3\,e \rightleftharpoons Cr(OH)_3 + 5\,OH^{-}$	–0.13
$Cr(OH)_3 + 3\,e \rightleftharpoons Cr + 3\,OH^{-}$	–1.48
$Cs^{+} + e \rightleftharpoons Cs$	–3.026
$Cu^{+} + e \rightleftharpoons Cu$	0.521
$Cu^{2+} + e \rightleftharpoons Cu^{+}$	0.153
$Cu^{2+} + 2\,e \rightleftharpoons Cu$	0.3419
$Cu^{2+} + 2\,e \rightleftharpoons Cu(Hg)$	0.345
$Cu^{3+} + e \rightleftharpoons Cu^{2+}$	2.4
$Cu_2O_3 + 6\,H^{+} + 2e \rightleftharpoons 2Cu^{2+} + 3\,H_2O$	2.0
$Cu^{2+} + 2\,CN^{-} + e \rightleftharpoons [Cu(CN)_2]^{-}$	1.103
$CuI_2^{-} + e \rightleftharpoons Cu + 2\,I^{-}$	0.00
$Cu_2O + H_2O + 2\,e \rightleftharpoons 2\,Cu + 2\,OH^{-}$	–0.360
$Cu(OH)_2 + 2\,e \rightleftharpoons Cu + 2\,OH^{-}$	–0.222
$2\,Cu(OH)_2 + 2\,e \rightleftharpoons Cu_2O + 2\,OH^{-} + H_2O$	–0.080
$2\,D^{+} + 2\,e \rightleftharpoons D_2$	–0.013
$Dy^{2+} + 2\,e \rightleftharpoons Dy$	–2.2
$Dy^{3+} + 3\,e \rightleftharpoons Dy$	–2.295
$Dy^{3+} + e \rightleftharpoons Dy^{2+}$	–2.6
$Er^{2+} + 2\,e \rightleftharpoons Er$	–2.0
$Er^{3+} + 3\,e \rightleftharpoons Er$	–2.331
$Er^{3+} + e \rightleftharpoons Er^{2+}$	–3.0
$Es^{3+} + e \rightleftharpoons Es^{2+}$	–1.3
$Es^{3+} + 3\,e \rightleftharpoons Es$	–1.91
$Es^{2+} + 2\,e \rightleftharpoons Es$	–2.23
$Eu^{2+} + 2\,e \rightleftharpoons Eu$	–2.812
$Eu^{3+} + 3\,e \rightleftharpoons Eu$	–1.991

TABLE 1
Alphabetical Listing (continued)

Reaction	$E°/V$
$Eu^{3+} + e \rightleftharpoons Eu^{2+}$	–0.36
$F_2 + 2\,H^+ + 2\,e \rightleftharpoons 2\,HF$	3.053
$F_2 + 2\,e \rightleftharpoons 2\,F^-$	2.866
$F_2O + 2\,H^+ + 4\,e \rightleftharpoons H_2O + 2\,F^-$	2.153
$Fe^{2+} + 2\,e \rightleftharpoons Fe$	–0.447
$Fe^{3+} + 3\,e \rightleftharpoons Fe$	–0.037
$Fe^{3+} + e \rightleftharpoons Fe^{2+}$	0.771
$2\,HFeO_4^- + 8\,H^+ + 6\,e \rightleftharpoons Fe_2O_3 + 5\,H_2O$	2.09
$HFeO_4^- + 4\,H^+ + 3\,e \rightleftharpoons FeOOH + 2\,H_2O$	2.08
$HFeO_4^- + 7\,H^+ + 3\,e \rightleftharpoons Fe^{3+} + 4\,H_2O$	2.07
$Fe_2O_3 + 4\,H^+ + 2\,e \rightleftharpoons 2\,FeOH^+ + H_2O$	0.16
$[Fe(CN)_6]^{3-} + e \rightleftharpoons [Fe(CN)_6]^{4-}$	0.358
$FeO_4^{2-} + 8\,H^+ + 3\,e \rightleftharpoons Fe^{3+} + 4\,H_2O$	2.20
$[Fe(bipy)_2]^{3+} + e \rightleftharpoons Fe(bipy)_2]^{2+}$	0.78
$[Fe(bipy)_3]^{3+} + e \rightleftharpoons Fe(bipy)_3]^{2+}$	1.03
$Fe(OH)_3 + e \rightleftharpoons Fe(OH)_2 + OH^-$	–0.56
$[Fe(phen)_3]^{3+} + e \rightleftharpoons [Fe(phen)_3]^{2+}$	1.147
$[Fe(phen)_3]^{3+} + e \rightleftharpoons [Fe(phen)_3]^{2+}$ (1 molar H_2SO_4)	1.06
$[Ferricinium]^+ + e \rightleftharpoons$ ferrocene	0.400
$Fm^{3+} + e \rightleftharpoons Fm^{2+}$	–1.1
$Fm^{3+} + 3\,e \rightleftharpoons Fm$	–1.89
$Fm^{2+} + 2\,e \rightleftharpoons Fm$	–2.30
$Fr^+ + e \rightleftharpoons Fr$	–2.9
$Ga^{3+} + 3\,e \rightleftharpoons Ga$	–0.549
$Ga^+ + e \rightleftharpoons Ga$	–0.2
$GaOH^{2+} + H^+ + 3\,e \rightleftharpoons Ga + H_2O$	–0.498
$H_2GaO^-_3 + H_2O + 3\,e \rightleftharpoons Ga + 4\,OH^-$	–1.219
$Gd^{3+} + 3\,e \rightleftharpoons Gd$	–2.279
$Ge^{2+} + 2\,e \rightleftharpoons Ge$	0.24
$Ge^{4+} + 4\,e \rightleftharpoons Ge$	0.124
$Ge^{4+} + 2\,e \rightleftharpoons Ge^{2+}$	0.00
$GeO_2 + 2\,H^+ + 2\,e \rightleftharpoons GeO + H_2O$	–0.118
$H_2GeO_3 + 4\,H^+ + 4\,e \rightleftharpoons Ge + 3\,H_2O$	–0.182
$2\,H^+ + 2\,e \rightleftharpoons H_2$	0.00000
$H_2 + 2\,e \rightleftharpoons 2\,H^-$	–2.23
$HO_2 + H^+ + e \rightleftharpoons H_2O_2$	1.495
$2\,H_2O + 2\,e \rightleftharpoons H_2 + 2\,OH^-$	–0.8277
$H_2O_2 + 2\,H^+ + 2\,e \rightleftharpoons 2\,H_2O$	1.776
$Hf^{4+} + 4\,e \rightleftharpoons Hf$	–1.55
$HfO^{2+} + 2\,H^+ + 4\,e \rightleftharpoons Hf + H_2O$	–1.724
$HfO_2 + 4\,H^+ + 4\,e \rightleftharpoons Hf + 2\,H_2O$	–1.505
$HfO(OH)_2 + H_2O + 4\,e \rightleftharpoons Hf + 4\,OH^-$	–2.50
$Hg^{2+} + 2\,e \rightleftharpoons Hg$	0.851
$2\,Hg^{2+} + 2\,e \rightleftharpoons Hg_2^{2+}$	0.920
$Hg_2^{2+} + 2\,e \rightleftharpoons 2\,Hg$	0.7973
$Hg_2(ac)_2 + 2\,e \rightleftharpoons 2\,Hg + 2(ac)^-$	0.51163
$Hg_2Br_2 + 2\,e \rightleftharpoons 2\,Hg + 2\,Br^-$	0.13923
$Hg_2Cl_2 + 2\,e \rightleftharpoons 2\,Hg + 2\,Cl^-$	0.26808
$Hg_2HPO_4 + 2\,e \rightleftharpoons 2\,Hg + HPO_4^{2-}$	0.6359
$Hg_2I_2 + 2\,e \rightleftharpoons 2\,Hg + 2\,I^-$	–0.0405
$Hg_2O + H_2O + 2\,e \rightleftharpoons 2\,Hg + 2\,OH^-$	0.123
$HgO + H_2O + 2\,e \rightleftharpoons Hg + 2\,OH^-$	0.0977
$Hg(OH)_2 + 2\,H^+ + 2\,e \rightleftharpoons Hg + 2\,H_2O$	1.034
$Hg_2SO_4 + 2\,e \rightleftharpoons 2\,Hg + SO_4^{2-}$	0.6125
$Ho^{2+} + 2\,e \rightleftharpoons Ho$	–2.1

Reaction	$E°/V$
$Ho^{3+} + 3\,e \rightleftharpoons Ho$	–2.33
$Ho^{3+} + e \rightleftharpoons Ho^{2+}$	–2.8
$I_2 + 2\,e \rightleftharpoons 2\,I^-$	0.5355
$I_3^- + 2\,e \rightleftharpoons 3\,I^-$	0.536
$H_3IO_6^{2-} + 2\,e \rightleftharpoons IO^-_3 + 3\,OH^-$	0.7
$H_5IO_6 + H^+ + 2\,e \rightleftharpoons IO_3^- + 3\,H_2O$	1.601
$2\,HIO + 2\,H^+ + 2\,e \rightleftharpoons I_2 + 2\,H_2O$	1.439
$HIO + H^+ + 2\,e \rightleftharpoons I^- + H_2O$	0.987
$IO^- + H_2O + 2\,e \rightleftharpoons I^- + 2\,OH^-$	0.485
$2\,IO_3^- + 12\,H^+ + 10\,e \rightleftharpoons I_2 + 6\,H_2O$	1.195
$IO_3^- + 6\,H^+ + 6\,e \rightleftharpoons I^- + 3\,H_2O$	1.085
$IO_3^- + 2\,H_2O + 4\,e \rightleftharpoons IO^- + 4\,OH^-$	0.15
$IO_3^- + 3\,H_2O + 6\,e \rightleftharpoons IO^- + 6\,OH^-$	0.26
$In^+ + e \rightleftharpoons In$	–0.14
$In^{2+} + e \rightleftharpoons In^+$	–0.40
$In^{3+} + e \rightleftharpoons In^{2+}$	–0.49
$In^{3+} + 2\,e \rightleftharpoons In^+$	–0.443
$In^{3+} + 3\,e \rightleftharpoons In$	–0.3382
$In(OH)_3 + 3\,e \rightleftharpoons In + 3\,OH^-$	–0.99
$In(OH)_4^- + 3\,e \rightleftharpoons In + 4\,OH^-$	–1.007
$In_2O_3 + 3\,H_2O + 6\,e \rightleftharpoons 2\,In + 6\,OH^-$	–1.034
$Ir^{3+} + 3\,e \rightleftharpoons Ir$	1.156
$[IrCl_6]^{2-} + e \rightleftharpoons [IrCl_6]^{3-}$	0.8665
$[IrCl_6]^{3-} + 3\,e \rightleftharpoons Ir + 6\,Cl^-$	0.77
$Ir_2O_3 + 3\,H_2O + 6\,e \rightleftharpoons 2\,Ir + 6\,OH^-$	0.098
$K^+ + e \rightleftharpoons K$	–2.931
$La^{3+} + 3\,e \rightleftharpoons La$	–2.379
$La(OH)_3 + 3\,e \rightleftharpoons La + 3\,OH^-$	–2.90
$Li^+ + e \rightleftharpoons Li$	–3.0401
$Lr^{3+} + 3\,e \rightleftharpoons Lr$	–1.96
$Lu^{3+} + 3\,e \rightleftharpoons Lu$	–2.28
$Md^{3+} + e \rightleftharpoons Md^{2+}$	–0.1
$Md^{3+} + 3\,e \rightleftharpoons Md$	–1.65
$Md^{2+} + 2\,e \rightleftharpoons Md$	–2.40
$Mg^+ + e \rightleftharpoons Mg$	–2.70
$Mg^{2+} + 2\,e \rightleftharpoons Mg$	–2.372
$Mg(OH)_2 + 2\,e \rightleftharpoons Mg + 2\,OH^-$	–2.690
$Mn^{2+} + 2\,e \rightleftharpoons Mn$	–1.185
$Mn^{3+} + 3e \rightleftharpoons Mn^{2+}$	1.5415
$MnO_2 + 4\,H^+ + 2\,e \rightleftharpoons Mn^{2+} + 2\,H_2O$	1.224
$MnO_4^- + e \rightleftharpoons MnO_4^{2-}$	0.558
$MnO_4^- + 4\,H^+ + 3\,e \rightleftharpoons MnO_2 + 2\,H_2O$	1.679
$MnO_4^- + 8\,H^+ + 5\,e \rightleftharpoons Mn^{2+} + 4\,H_2O$	1.507
$MnO_4^- + 2\,H_2O + 3\,e \rightleftharpoons MnO_2 + 4\,OH^-$	0.595
$MnO_4^{2-} + 2\,H_2O + 2\,e \rightleftharpoons MnO_2 + 4\,OH^-$	0.60
$Mn(OH)_2 + 2\,e \rightleftharpoons Mn + 2\,OH^-$	–1.56
$Mn(OH)_3 + e \rightleftharpoons Mn(OH)_2 + OH^-$	0.15
$Mn_2O_3 + 6\,H^+ + e \rightleftharpoons 2\,Mn^{2+} + 3\,H_2O$	1.485
$Mo^{3+} + 3\,e \rightleftharpoons Mo$	–0.200
$MoO_2 + 4\,H^+ + 4\,e \rightleftharpoons Mo + 4\,H_2O$	–0.152
$H_3Mo_7O_{24}^{3-} + 45\,H^+ + 42\,e \rightleftharpoons 7\,Mo + 24\,H_2O$	0.082
$MoO_3 + 6\,H^+ + 6\,e \rightleftharpoons Mo + 3\,H_2O$	0.075
$N_2 + 2\,H_2O + 6\,H^+ + 6\,e \rightleftharpoons 2\,NH_4OH$	0.092
$3\,N_2 + 2\,H^+ + 2\,e \rightleftharpoons 2\,HN_3$	–3.09
$N_5^+ + 3\,H^+ + 2\,e \rightleftharpoons 2\,NH_4^+$	1.275

TABLE 1
Alphabetical Listing (continued)

Reaction	$E°/V$
$N_2O + 2 H^+ + 2 e \rightleftharpoons N_2 + H_2O$	1.766
$H_2N_2O_2 + 2 H^+ + 2 e \rightleftharpoons N_2 + 2 H_2O$	2.65
$N_2O_4 + 2 e \rightleftharpoons 2 NO_2^-$	0.867
$N_2O_4 + 2 H^+ + 2 e \rightleftharpoons 2 NHO_2$	1.065
$N_2O_4 + 4 H^+ + 4 e \rightleftharpoons 2 NO + 2 H_2O$	1.035
$2 NH_3OH^+ + H^+ + 2 e \rightleftharpoons N_2H_5^+ + 2 H_2O$	1.42
$2 NO + 2 H^+ + 2 e \rightleftharpoons N_2O + H_2O$	1.591
$2 NO + H_2O + 2 e \rightleftharpoons N_2O + 2 OH^-$	0.76
$HNO_2 + H^+ + e \rightleftharpoons NO + H_2O$	0.983
$2 HNO_2 + 4 H^+ + 4 e \rightleftharpoons H_2N_2O_2 + 2 H_2O$	0.86
$2 HNO_2 + 4 H^+ + 4 e \rightleftharpoons N_2O + 3 H_2O$	1.297
$NO_2^- + H_2O + e \rightleftharpoons NO + 2 OH^-$	−0.46
$2 NO_2^- + 2 H_2O + 4 e \rightleftharpoons N_2O_2^{2-} + 4 OH^-$	−0.18
$2 NO_2^- + 3 H_2O + 4 e \rightleftharpoons N_2O + 6 OH^-$	0.15
$NO_3^- + 3 H^+ + 2 e \rightleftharpoons HNO_2 + H_2O$	0.934
$NO_3^- + 4 H^+ + 3 e \rightleftharpoons NO + 2 H_2O$	0.957
$2 NO_3^- + 4 H^+ + 2 e \rightleftharpoons N_2O_4 + 2 H_2O$	0.803
$NO_3^- + H_2O + 2 e \rightleftharpoons NO_2^- + 2 OH^-$	0.01
$2 NO_3^- + 2 H_2O + 2 e \rightleftharpoons N_2O_4 + 4 OH^-$	−0.85
$Na^+ + e \rightleftharpoons Na$	−2.71
$Nb^{3+} + 3 e \rightleftharpoons Nb$	−1.099
$NbO_2 + 2 H^+ + 2 e \rightleftharpoons NbO + H_2O$	−0.646
$NbO_2 + 4 H^+ + 4 e \rightleftharpoons Nb + 2 H_2O$	−0.690
$NbO + 2 H^+ + 2 e \rightleftharpoons Nb + H_2O$	−0.733
$Nb_2O_5 + 10 H^+ + 10 e \rightleftharpoons 2 Nb + 5 H_2O$	−0.644
$Nd^{3+} + 3 e \rightleftharpoons Nd$	−2.323
$Nd^{2+} + 2 e \rightleftharpoons Nd$	−2.1
$Nd^{3+} + e \rightleftharpoons Nd^{2+}$	−2.7
$Ni^{2+} + 2 e \rightleftharpoons Ni$	−0.257
$Ni(OH)_2 + 2 e \rightleftharpoons Ni + 2 OH^-$	−0.72
$NiO_2 + 4 H^+ + 2 e \rightleftharpoons Ni^{2+} + 2 H_2O$	1.678
$NiO_2 + 2 H_2O + 2 e \rightleftharpoons Ni(OH)_2 + 2 OH^-$	−0.490
$No^{3+} + e \rightleftharpoons No^{2+}$	1.4
$No^{3+} + 3 e \rightleftharpoons No$	−1.20
$No^{2+} + 2 e \rightleftharpoons No$	−2.50
$Np^{3+} + 3 e \rightleftharpoons Np$	−1.856
$Np^{4+} + e \rightleftharpoons Np^{3+}$	0.147
$NpO_2 + H_2O + H^+ + e \rightleftharpoons Np(OH)_3$	−0.962
$O_2 + 2 H^+ + 2 e \rightleftharpoons H_2O_2$	0.695
$O_2 + 4 H^+ + 4 e \rightleftharpoons 2 H_2O$	1.229
$O_2 + H_2O + 2 e \rightleftharpoons HO_2^- + OH^-$	−0.076
$O_2 + 2 H_2O + 2 e \rightleftharpoons H_2O_2 + 2 OH^-$	−0.146
$O_2 + 2 H_2O + 4 e \rightleftharpoons 4 OH^-$	0.401
$O_3 + 2 H^+ + 2 e \rightleftharpoons O_2 + H_2O$	2.076
$O_3 + H_2O + 2 e \rightleftharpoons O_2 + 2 OH^-$	1.24
$O(g) + 2 H^+ + 2 e \rightleftharpoons H_2O$	2.421
$OH + e \rightleftharpoons OH^-$	2.02
$HO_2^- + H_2O + 2 e \rightleftharpoons 3 OH^-$	0.878
$OsO_4 + 8 H^+ + 8 e \rightleftharpoons Os + 4 H_2O$	0.838
$OsO_4 + 4 H^+ + 4 e \rightleftharpoons OsO_2 + 2 H_2O$	1.02
$[Os(bipy)_2]^{3+} + e \rightleftharpoons [Os(bipy)_2]^{2+}$	0.81
$[Os(bipy)_3]^{3+} + e \rightleftharpoons [Os(bipy)_3]^{2+}$	0.80
$P(red) + 3 H^+ + 3 e \rightleftharpoons PH_3(g)$	−0.111
$P(white) + 3 H^+ + 3 e \rightleftharpoons PH_3(g)$	−0.063
$P + 3 H_2O + 3 e \rightleftharpoons PH_3(g) + 3 OH^-$	−0.87
$H_2P_2^- + e \rightleftharpoons P + 2 OH^-$	−1.82
$H_3PO_2 + H^+ + e \rightleftharpoons P + 2 H_2O$	−0.508
$H_3PO_3 + 2 H^+ + 2 e \rightleftharpoons H_3PO_2 + H_2O$	−0.499
$H_3PO_3 + 3 H^+ + 3 e \rightleftharpoons P + 3 H_2O$	−0.454
$HPO_3^{2-} + 2 H_2O + 2 e \rightleftharpoons H_2PO_2^- + 3 OH^-$	−1.65
$HPO_3^{2-} + 2 H_2O + 3 e \rightleftharpoons P + 5 OH^-$	−1.71
$H_3PO_4 + 2 H^+ + 2 e \rightleftharpoons H_3PO_3 + H_2O$	−0.276
$PO_4^{3-} + 2 H_2O + 2 e \rightleftharpoons HPO_3^{2-} + 3 OH^-$	−1.05
$Pa^{3+} + 3 e \rightleftharpoons Pa$	−1.34
$Pa^{4+} + 4 e \rightleftharpoons Pa$	−1.49
$Pa^{4+} + e \rightleftharpoons Pa^{3+}$	−1.9
$Pb^{2+} + 2 e \rightleftharpoons Pb$	−0.1262
$Pb^{2+} + 2 e \rightleftharpoons Pb(Hg)$	−0.1205
$PbBr_2 + 2 e \rightleftharpoons Pb + 2 Br^-$	−0.284
$PbCl_2 + 2 e \rightleftharpoons Pb + 2 Cl^-$	−0.2675
$PbF_2 + 2 e \rightleftharpoons Pb + 2 F^-$	−0.3444
$PbHPO_4 + 2 e \rightleftharpoons Pb + HPO_4^{2-}$	−0.465
$PbI_2 + 2 e \rightleftharpoons Pb + 2 I^-$	−0.365
$PbO + H_2O + 2 e \rightleftharpoons Pb + 2 OH^-$	−0.580
$PbO_2 + 4 H^+ + 2 e \rightleftharpoons Pb^{2+} + 2 H_2O$	1.455
$HPbO_2^- + H_2O + 2 e \rightleftharpoons Pb + 3 OH^-$	−0.537
$PbO_2 + H_2O + 2 e \rightleftharpoons PbO + 2 OH^-$	0.247
$PbO_2 + SO_4^{2-} + 4 H^+ + 2 e \rightleftharpoons PbSO_4 + 2 H_2O$	1.6913
$PbSO_4 + 2 e \rightleftharpoons Pb + SO_4^{2-}$	−0.3588
$PbSO_4 + 2 e \rightleftharpoons Pb(Hg) + SO_4^{2-}$	−0.3505
$Pd^{2+} + 2 e \rightleftharpoons Pd$	0.951
$[PdCl_4]^{2-} + 2 e \rightleftharpoons Pd + 4 Cl^-$	0.591
$[PdCl_6]^{2-} + 2 e \rightleftharpoons [PdCl_4]^{2-} + 2 Cl^-$	1.288
$Pd(OH)_2 + 2 e \rightleftharpoons Pd + 2 OH^-$	0.07
$Pm^{2+} + 2 e \rightleftharpoons Pm$	−2.2
$Pm^{3+} + 3 e \rightleftharpoons Pm$	−2.30
$Pm^{3+} + e \rightleftharpoons Pm^{2+}$	−2.6
$Po^{4+} + 2 e \rightleftharpoons Po^{2+}$	0.9
$Po^{4+} + 4 e \rightleftharpoons Po$	0.76
$Pr^{4+} + e \rightleftharpoons Pr^{3+}$	3.2
$Pr^{2+} + 2 e \rightleftharpoons Pr$	−2.0
$Pr^{3+} + 3 e \rightleftharpoons Pr$	−2.353
$Pr^{3+} + e \rightleftharpoons Pr^{2+}$	−3.1
$Pt^{2+} + 2 e \rightleftharpoons Pt$	1.18
$[PtCl_4]^{2-} + 2 e \rightleftharpoons Pt + 4 Cl^-$	0.755
$[PtCl_6]^{2-} + 2 e \rightleftharpoons [PtCl_4]^{2-} + 2 Cl^-$	0.68
$Pt(OH)_2 + 2 e \rightleftharpoons Pt + 2 OH^-$	0.14
$PtO_3 + 2 H^+ + 2 e \rightleftharpoons PtO_2 + H_2O$	1.7
$PtO_3 + 4 H^+ + 2 e \rightleftharpoons Pt(OH)_2^{2+} + H_2O$	1.5
$PtOH^+ + H^+ + 2 e \rightleftharpoons Pt + H_2O$	1.2
$PtO_2 + 2 H^+ + 2 e \rightleftharpoons PtO + H_2O$	1.01
$PtO_2 + 4 H^+ + 4 e \rightleftharpoons Pt + 2 H_2O$	1.00
$Pu^{3+} + 3 e \rightleftharpoons Pu$	−2.031
$Pu^{4+} + e \rightleftharpoons Pu^{3+}$	1.006
$Pu^{5+} + e \rightleftharpoons Pu^{4+}$	1.099
$PuO_2(OH)_2 + 2 H^+ + 2 e \rightleftharpoons Pu(OH)_4$	1.325
$PuO_2(OH)_2 + H^+ + e \rightleftharpoons PuO_2OH + H_2O$	1.062
$Ra^{2+} + 2 e \rightleftharpoons Ra$	−2.8
$Rb^+ + e \rightleftharpoons Rb$	−2.98
$Re^{3+} + 3 e \rightleftharpoons Re$	0.300

TABLE 1
Alphabetical Listing (continued)

Reaction	$E°/V$
$ReO_4^- + 4 H^+ + 3 e \rightleftharpoons ReO_2 + 2 H_2O$	0.510
$ReO_2 + 4 H^+ + 4 e \rightleftharpoons Re + 2 H_2O$	0.2513
$ReO_4^- + 2 H^+ + e \rightleftharpoons ReO_3 + H_2O$	0.768
$ReO_4^- + 4 H_2O + 7 e \rightleftharpoons Re + 8 OH^-$	–0.584
$ReO_4^- + 8 H^+ + 7 e \rightleftharpoons Re + 4 H_2O$	0.368
$Rh^+ + e \rightleftharpoons Rh$	0.600
$Rh^+ + 2e \rightleftharpoons Rh$	0.600
$Rh^{3+} + 3 e \rightleftharpoons Rh$	0.758
$[RhCl_6]^{3-} + 3 e \rightleftharpoons Rh + 6 Cl^-$	0.431
$RhOH^{2+} + H^+ + 3 e \rightleftharpoons Rh + H_2O$	0.83
$Ru^{2+} + 2 e \rightleftharpoons Ru$	0.455
$Ru^{3+} + e \rightleftharpoons Ru^{2+}$	0.2487
$RuO_2 + 4 H^+ + 2 e \rightleftharpoons Ru^{2+} + 2 H_2O$	1.120
$RuO_4^- + e \rightleftharpoons RuO_4^{2-}$	0.59
$RuO_4 + e \rightleftharpoons RuO_4^-$	1.00
$RuO_4 + 6 H^+ + 4 e \rightleftharpoons Ru(OH)_2^{2+} + 2 H_2O$	1.40
$RuO_4 + 8 H^+ + 8 e \rightleftharpoons Ru + 4 H_2O$	1.038
$[Ru(bipy)_3]^{3+} + e^- \rightleftharpoons [Ru(bipy)_3]^{2+}$	1.24
$[Ru(H_2O)_6]^{3+} + e^- \rightleftharpoons [Ru(H_2O)_6]^{2+}$	0.23
$[Ru(NH_3)_6]^{3+} + e^- \rightleftharpoons [Ru(NH_3)_6]^{2+}$	0.10
$[Ru(en)_3]^{3+} + e^- \rightleftharpoons [Ru(en)_3]^{2+}$	0.210
$[Ru(CN)_6]^{3-} + e^- \rightleftharpoons [Ru(CN)_6]^{4-}$	0.86
$S + 2 e \rightleftharpoons S^{2-}$	–0.47627
$S + 2H^+ + 2 e \rightleftharpoons H_2S(aq)$	0.142
$S + H_2O + 2 e \rightleftharpoons SH^- + OH^-$	–0.478
$2 S + 2 e \rightleftharpoons S_2^{2-}$	–0.42836
$S_2O_6^{2-} + 4 H^+ + 2 e \rightleftharpoons 2 H_2SO_3$	0.564
$S_2O_8^{2-} + 2 e \rightleftharpoons 2 SO_4^{2-}$	2.010
$S_2O_8^{2-} + 2 H^+ + 2 e \rightleftharpoons 2 HSO_4^-$	2.123
$S_4O_6^{2-} + 2 e \rightleftharpoons 2 S_2O_3^{2-}$	0.08
$2 H_2SO_3 + H^+ + 2 e \rightleftharpoons HS_2O_4^- + 2 H_2O$	–0.056
$H_2SO_3 + 4 H^+ + 4 e \rightleftharpoons S + 3 H_2O$	0.449
$2 SO_3^{2-} + 2 H_2O + 2 e \rightleftharpoons S_2O_4^{2-} + 4 OH^-$	–1.12
$2 SO_3^{2-} + 3 H_2O + 4 e \rightleftharpoons S_2O_3^{2-} + 6 OH^-$	–0.571
$SO_4^{2-} + 4 H^+ + 2 e \rightleftharpoons H_2SO_3 + H_2O$	0.172
$2 SO_4^{2-} + 4 H^+ + 2 e \rightleftharpoons S_2O_6^{2-} + H_2O$	–0.22
$SO_4^{2-} + H_2O + 2 e \rightleftharpoons SO_3^{2-} + 2 OH^-$	–0.93
$Sb + 3 H^+ + 3 e \rightleftharpoons SbH_3$	–0.510
$Sb_2O_3 + 6 H^+ + 6 e \rightleftharpoons 2 Sb + 3 H_2O$	0.152
Sb_2O_5 (senarmontite) $+ 4 H^+ + 4 e \rightleftharpoons Sb_2O_3 + 2 H_2O$	0.671
Sb_2O_5 (valentinite) $+ 4 H^+ + 4 e \rightleftharpoons Sb_2O_3 + 2 H_2O$	0.649
$Sb_2O_5 + 6 H^+ + 4 e \rightleftharpoons 2 SbO^+ + 3 H_2O$	0.581
$SbO^+ + 2 H^+ + 3 e \rightleftharpoons Sb + 2 H_2O$	0.212
$SbO_2^- + 2 H_2O + 3 e \rightleftharpoons Sb + 4 OH^-$	–0.66
$SbO_3^- + H_2O + 2 e \rightleftharpoons SbO_2^- + 2 OH^-$	–0.59
$Sc^{3+} + 3 e \rightleftharpoons Sc$	–2.077
$Se + 2 e \rightleftharpoons Se^{2-}$	–0.924
$Se + 2 H^+ + 2 e \rightleftharpoons H_2Se(aq)$	–0.399
$H_2SeO_3 + 4 H^+ + 4 e \rightleftharpoons Se + 3 H_2O$	0.74
$Se + 2 H^+ + 2 e \rightleftharpoons H_2Se$	–0.082
$SeO_3^{2-} + 3 H_2O + 4 e \rightleftharpoons Se + 6 OH^-$	–0.366
$SeO_4^{2-} + 4 H^+ + 2 e \rightleftharpoons H_2SeO_3 + H_2O$	1.151
$SeO_4^{2-} + H_2O + 2 e \rightleftharpoons SeO_3^{2-} + 2 OH^-$	0.05
$SiF_6^{2-} + 4 e \rightleftharpoons Si + 6 F^-$	–1.24
$SiO + 2 H^+ + 2 e \rightleftharpoons Si + H_2O$	–0.8

Reaction	$E°/V$
SiO_2 (quartz) $+ 4 H^+ + 4 e \rightleftharpoons Si + 2 H_2O$	0.857
$SiO_3^{2-} + 3 H_2O + 4 e \rightleftharpoons Si + 6 OH^-$	–1.697
$Sm^{3+} + e \rightleftharpoons Sm^{2+}$	–1.55
$Sm^{3+} + 3 e \rightleftharpoons Sm$	–2.304
$Sm^{2+} + 2 e \rightleftharpoons Sm$	–2.68
$Sn^{2+} + 2 e \rightleftharpoons Sn$	–0.1375
$Sn^{4+} + 2 e \rightleftharpoons Sn^{2+}$	0.151
$Sn(OH)_3^+ + 3 H^+ + 2 e \rightleftharpoons Sn^{2+} + 3 H_2O$	0.142
$SnO_2 + 4 H^+ + 2 e^- \rightleftharpoons Sn^{2+} + 2 H_2O$	–0.094
$SnO_2 + 4 H^+ + 4 e \rightleftharpoons Sn + 2 H_2O$	–0.117
$SnO_2 + 3 H^+ + 2 e \rightleftharpoons SnOH^+ + H_2O$	–0.194
$SnO_2 + 2 H_2O + 4 e \rightleftharpoons Sn + 4 OH^-$	–0.945
$HSnO_2^- + H_2O + 2 e \rightleftharpoons Sn + 3 OH^-$	–0.909
$Sn(OH)_6^{2-} + 2 e \rightleftharpoons HSnO_2^- + 3 OH^- + H_2O$	–0.93
$Sr^+ + e \rightleftharpoons Sr$	–4.10
$Sr^{2+} + 2 e \rightleftharpoons Sr$	–2.899
$Sr^{2+} + 2 e \rightleftharpoons Sr(Hg)$	–1.793
$Sr(OH)_2 + 2 e \rightleftharpoons Sr + 2 OH^-$	–2.88
$Ta_2O_5 + 10 H^+ + 10 e \rightleftharpoons 2 Ta + 5 H_2O$	–0.750
$Ta^{3+} + 3 e \rightleftharpoons Ta$	–0.6
$Tc^{2+} + 2 e \rightleftharpoons Tc$	0.400
$TcO_4^- + 4 H^+ + 3 e \rightleftharpoons TcO_2 + 2 H_2O$	0.782
$Tc^{3+} + e \rightleftharpoons Tc^{2+}$	0.3
$TcO_4^- + 8 H^+ + 7 e \rightleftharpoons Tc + 4 H_2O$	0.472
$Tb^{4+} + e \rightleftharpoons Tb^{3+}$	3.1
$Tb^{3+} + 3 e \rightleftharpoons Tb$	–2.28
$Te + 2 e \rightleftharpoons Te^{2-}$	–1.143
$Te + 2 H^+ + 2 e \rightleftharpoons H_2Te$	–0.793
$Te^{4+} + 4 e \rightleftharpoons Te$	0.568
$TeO_2 + 4 H^+ + 4 e \rightleftharpoons Te + 2 H_2O$	0.593
$TeO_3^{2-} + 3 H_2O + 4 e \rightleftharpoons Te + 6 OH^-$	–0.57
$TeO_4^- + 8 H^+ + 7 e \rightleftharpoons Te + 4 H_2O$	0.472
$H_6TeO_6 + 2 H^+ + 2 e \rightleftharpoons TeO_2 + 4 H_2O$	1.02
$Th^{4+} + 4 e \rightleftharpoons Th$	–1.899
$ThO_2 + 4 H^+ + 4 e \rightleftharpoons Th + 2 H_2O$	–1.789
$Th(OH)_4 + 4 e \rightleftharpoons Th + 4 OH^-$	–2.48
$Ti^{2+} + 2 e \rightleftharpoons Ti$	–1.630
$Ti^{3+} + e \rightleftharpoons Ti^{2+}$	–0.9
$TiO_2 + 4 H^+ + 2 e \rightleftharpoons Ti^{2+} + 2 H_2O$	–0.502
$Ti^{3+} + 3 e \rightleftharpoons Ti$	–1.37
$TiOH^{3+} + H^+ + e \rightleftharpoons Ti^{3+} + H_2O$	–0.055
$Tl^+ + e \rightleftharpoons Tl$	–0.336
$Tl^+ + e \rightleftharpoons Tl(Hg)$	–0.3338
$Tl^{3+} + 2 e \rightleftharpoons Tl^+$	1.252
$Tl^{3+} + 3 e \rightleftharpoons Tl$	0.741
$TlBr + e \rightleftharpoons Tl + Br^-$	–0.658
$TlCl + e \rightleftharpoons Tl + Cl^-$	–0.5568
$TlI + e \rightleftharpoons Tl + I^-$	–0.752
$Tl_2O_3 + 3 H_2O + 4 e \rightleftharpoons 2 Tl^+ + 6 OH^-$	0.02
$TlOH + e \rightleftharpoons Tl + OH^-$	–0.34
$Tl(OH)_3 + 2 e \rightleftharpoons TlOH + 2 OH^-$	–0.05
$Tl_2SO_4 + 2 e \rightleftharpoons Tl + SO_4^{2-}$	–0.4360
$Tm^{3+} + e \rightleftharpoons Tm^{2+}$	–2.2
$Tm^{3+} + 3 e \rightleftharpoons Tm$	–2.319
$Tm^{2+} + 2 e \rightleftharpoons Tm$	–2.4

TABLE 1
Alphabetical Listing (continued)

Reaction	$E°$/V
$U^{3+} + 3\,e \rightleftharpoons U$	–1.798
$U^{4+} + e \rightleftharpoons U^{3+}$	–0.607
$UO_2^{+} + 4\,H^{+} + e \rightleftharpoons U^{4+} + 2\,H_2O$	0.612
$UO_2^{2+} + e \rightleftharpoons UO^{+}_2$	0.062
$UO_2^{2+} + 4\,H^{+} + 2\,e \rightleftharpoons U^{4+} + 2\,H_2O$	0.327
$UO_2^{2+} + 4\,H^{+} + 6\,e \rightleftharpoons U + 2\,H_2O$	–1.444
$V^{2+} + 2\,e \rightleftharpoons V$	–1.175
$V^{3+} + e \rightleftharpoons V^{2+}$	–0.255
$VO^{2+} + 2\,H^{+} + e \rightleftharpoons V^{3+} + H_2O$	0.337
$VO_2^{+} + 2\,H^{+} + e \rightleftharpoons VO^{2+} + H_2O$	0.991
$V_2O_5 + 6\,H^{+} + 2\,e \rightleftharpoons 2\,VO^{2+} + 3\,H_2O$	0.957
$V_2O_5 + 10\,H^{+} + 10\,e \rightleftharpoons 2\,V + 5\,H_2O$	–0.242
$V(OH)_4^{+} + 2\,H^{+} + e \rightleftharpoons VO^{2+} + 3\,H_2O$	1.00
$V(OH)_4^{+} + 4\,H^{+} + 5\,e \rightleftharpoons V + 4\,H_2O$	–0.254
$[V(phen)_3]^{3+} + e \rightleftharpoons [V(phen)_3]^{2+}$	0.14
$W^{3+} + 3\,e \rightleftharpoons W$	0.1
$W_2O_5 + 2\,H^{+} + 2\,e \rightleftharpoons 2\,WO_2 + H_2O$	–0.031
$WO_2 + 4\,H^{+} + 4\,e \rightleftharpoons W + 2\,H_2O$	–0.119
$WO_3 + 6\,H^{+} + 6\,e \rightleftharpoons W + 3\,H_2O$	–0.090
$WO_3 + 2\,H^{+} + 2\,e \rightleftharpoons WO_2 + H_2O$	0.036

Reaction	$E°$/V
$2\,WO_3 + 2\,H^{+} + 2\,e \rightleftharpoons W_2O_5 + H_2O$	–0.029
$H_4XeO_6 + 2\,H^{+} + 2\,e \rightleftharpoons XeO_3 + 3\,H_2O$	2.42
$XeO_3 + 6\,H^{+} + 6\,e \rightleftharpoons Xe + 3\,H_2O$	2.10
$XeF + e \rightleftharpoons Xe + F^{-}$	3.4
$Y^{3+} + 3\,e \rightleftharpoons Y$	–2.372
$Yb^{3+} + e \rightleftharpoons Yb^{2+}$	–1.05
$Yb^{3+} + 3\,e \rightleftharpoons Yb$	–2.19
$Yb^{2+} + 2\,e \rightleftharpoons Yb$	–2.76
$Zn^{2+} + 2\,e \rightleftharpoons Zn$	–0.7618
$Zn^{2+} + 2\,e \rightleftharpoons Zn(Hg)$	–0.7628
$ZnO_2^{2-} + 2\,H_2O + 2\,e \rightleftharpoons Zn + 4\,OH^{-}$	–1.215
$ZnSO_4 \cdot 7\,H_2O + 2\,e = Zn(Hg) + SO_4^{2-} + 7\,H_2O$ (Saturated $ZnSO_4$)	–0.7993
$ZnOH^{+} + H^{+} + 2\,e \rightleftharpoons Zn + H_2O$	–0.497
$Zn(OH)_4^{2-} + 2\,e \rightleftharpoons Zn + 4\,OH^{-}$	–1.199
$Zn(OH)_2 + 2\,e \rightleftharpoons Zn + 2\,OH^{-}$	–1.249
$ZnO + H_2O + 2\,e \rightleftharpoons Zn + 2\,OH^{-}$	–1.260
$ZrO_2 + 4\,H^{+} + 4\,e \rightleftharpoons Zr + 2\,H_2O$	–1.553
$ZrO(OH)_2 + H_2O + 4\,e \rightleftharpoons Zr + 4\,OH^{-}$	–2.36
$Zr^{4+} + 4\,e \rightleftharpoons Zr$	–1.45

TABLE 2
Reduction Reactions Having $E°$ Values More Positive than that of the Standard Hydrogen Electrode

Reaction	$E°$/V
$2\,H^{+} + 2\,e \rightleftharpoons H_2$	0.00000
$CuI_2^{-} + e \rightleftharpoons Cu + 2\,I^{-}$	0.00
$Ge^{4+} + 2\,e \rightleftharpoons Ge^{2+}$	0.00
$NO_3^{-} + H_2O + 2\,e \rightleftharpoons NO_2^{-} + 2\,OH^{-}$	0.01
$Tl_2O_3 + 3\,H_2O + 4\,e \rightleftharpoons 2\,Tl^{+} + 6\,OH^{-}$	0.02
$SeO_4^{2-} + H_2O + 2\,e \rightleftharpoons SeO_3^{2-} + 2\,OH^{-}$	0.05
$WO_3 + 2\,H^{+} + 2\,e \rightleftharpoons WO_2 + H_2O$	0.036
$UO_2^{2+} + e = UO_2^{+}$	0.062
$Pd(OH)_2 + 2\,e \rightleftharpoons Pd + 2\,OH^{-}$	0.07
$AgBr + e \rightleftharpoons Ag + Br^{-}$	0.07133
$MoO_3 + 6\,H^{+} + 6\,e \rightleftharpoons Mo + 3\,H_2O$	0.075
$S_4O_6^{2-} + 2\,e \rightleftharpoons 2\,S_2O_3^{2-}$	0.08
$H_3Mo_7O_{24}^{3-} + 45\,H^{+} + 42\,e \rightleftharpoons 7\,Mo + 24\,H_2O$	0.082
$AgSCN + e \rightleftharpoons Ag + SCN^{-}$	0.8951
$N_2 + 2\,H_2O + 6\,H^{+} + 6\,e \rightleftharpoons 2\,NH_4OH$	0.092
$HgO + H_2O + 2\,e \rightleftharpoons Hg + 2\,OH^{-}$	0.0977
$Ir_2O_3 + 3\,H_2O + 6\,e \rightleftharpoons 2\,Ir + 6\,OH^{-}$	0.098
$2\,NO + 2\,e \rightleftharpoons N_2O_2^{2-}$	0.10
$[Ru(NH_3)_6]^{3+} + e \rightleftharpoons [Ru(NH_3)_6]^{2+}$	0.10
$W^{3+} + 3\,e \rightleftharpoons W$	0.1
$[Co(NH_3)_6]^{3+} + e \rightleftharpoons [Co(NH_3)_6]^{2+}$	0.108
$Hg_2O + H_2O + 2\,e \rightleftharpoons 2\,Hg + 2\,OH^{-}$	0.123
$Ge^{4+} + 4\,e \rightleftharpoons Ge$	0.124
$Hg_2Br_2 + 2\,e \rightleftharpoons 2\,Hg + 2\,Br^{-}$	0.13923
$Pt(OH)_2 + 2\,e \rightleftharpoons Pt + 2\,OH^{-}$	0.14
$[V(phen)_3]^{3+} + e \rightleftharpoons [V(phen)_3]^{2+}$	0.14
$S + 2H^{+} + 2\,e \rightleftharpoons H_2S(aq)$	0.142

Reaction	$E°$/V
$Sn(OH)_3^{+} + 3\,H^{+} + 2\,e \rightleftharpoons Sn^{2+} + 3\,H_2O$	0.142
$Np^{4+} + e \rightleftharpoons Np^{3+}$	0.147
$Ag_4[Fe(CN)_6] + 4\,e \rightleftharpoons 4\,Ag + [Fe(CN)_6]^{4-}$	0.1478
$IO_3^{-} + 2\,H_2O + 4\,e \rightleftharpoons IO^{-} + 4\,OH^{-}$	0.15
$Mn(OH)_3 + e \rightleftharpoons Mn(OH)_2 + OH^{-}$	0.15
$2\,NO_2^{-} + 3\,H_2O + 4\,e \rightleftharpoons N_2O + 6\,OH^{-}$	0.15
$Sn^{4+} + 2\,e \rightleftharpoons Sn^{2+}$	0.151
$Sb_2O_3 + 6\,H^{+} + 6\,e \rightleftharpoons 2\,Sb + 3\,H_2O$	0.152
$Cu^{2+} + e \rightleftharpoons Cu^{+}$	0.153
$BiOCl + 2\,H^{+} + 3\,e \rightleftharpoons Bi + Cl^{-} + H_2O$	0.1583
$BiCl_4^{-} + 3\,e \rightleftharpoons Bi + 4\,Cl^{-}$	0.16
$Fe_2O_3 + 4\,H^{+} + 2\,e \rightleftharpoons 2\,FeOH^{+} + H_2O$	0.16
$Co(OH)_3 + e \rightleftharpoons Co(OH)_2 + OH^{-}$	0.17
$SO_4^{2-} + 4\,H^{+} + 2\,e \rightleftharpoons H_2SO_3 + H_2O$	0.172
$Bi^{3+} + 2\,e \rightleftharpoons Bi^{+}$	0.2
$[Ru(en)_3]^{3+} + e \rightleftharpoons [Ru(en)_3]^{2+}$	0.210
$SbO^{+} + 2\,H^{+} + 3\,e \rightleftharpoons Sb + 2\,H_2O$	0.212
$AgCl + e \rightleftharpoons Ag + Cl^{-}$	0.22233
$[Ru(H_2O)_6]^{3+} + e \rightleftharpoons [Ru(H_2O)_6]^{2+}$	0.23
$As_2O_3 + 6\,H^{+} + 6\,e \rightleftharpoons 2\,As + 3\,H_2O$	0.234
Calomel electrode, saturated NaCl (SSCE)	0.2360
$Ge^{2+} + 2\,e \rightleftharpoons Ge$	0.24
$Ru^{3+} + e \rightleftharpoons Ru^{2+}$	0.24
Calomel electrode, saturated KCl	0.2412
$PbO_2 + H_2O + 2\,e \rightleftharpoons PbO + 2\,OH^{-}$	0.247
$HAsO_2 + 3\,H^{+} + 3\,e \rightleftharpoons As + 2\,H_2O$	0.248
$Ru^{3+} + e \rightleftharpoons Ru^{2+}$	0.2487

TABLE 2
Reduction Reactions Having $E°$ Values More Positive than that of the Standard Hydrogen Electrode (continued)

Reaction	$E°$/V
$ReO_2 + 4 H^+ + 4 e \rightleftharpoons Re + 2 H_2O$	0.2513
$IO_3^- + 3 H_2O + 6 e \rightleftharpoons I^- + OH^-$	0.26
$Hg_2Cl_2 + 2 e \rightleftharpoons 2 Hg + 2 Cl^-$	0.26808
Calomel electrode, 1 molal KCl	0.2800
Calomel electrode, 1 molar KCl (NCE)	0.2801
$At_2 + 2 e \rightleftharpoons 2 At^-$	0.3
$Re^{3+} + 3 e \rightleftharpoons Re$	0.300
$Tc^{3+} + e \rightleftharpoons Tc^{2+}$	0.3
$Bi^{3+} + 3 e \rightleftharpoons Bi$	0.308
$BiO^+ + 2 H^+ + 3 e \rightleftharpoons Bi + H_2O$	0.320
$UO_2^{2+} + 4 H^+ + 2 e \rightleftharpoons U^{4+} + 2 H_2O$	0.327
$ClO_3^- + H_2O + 2 e \rightleftharpoons ClO_2^- + 2 OH^-$	0.33
$2 HCNO + 2 H^+ + 2 e \rightleftharpoons (CN)_2 + 2 H_2O$	0.330
Calomel electrode, 0.1 molar KCl	0.3337
$VO^{2+} + 2 H^+ + e \rightleftharpoons V^{3+} + H_2O$	0.337
$Cu^{2+} + 2 e \rightleftharpoons Cu$	0.3419
$Ag_2O + H_2O + 2 e \rightleftharpoons 2 Ag + 2 OH^-$	0.342
$Cu^{2+} + 2 e \rightleftharpoons Cu(Hg)$	0.345
$AgIO_3 + e \rightleftharpoons Ag + IO_3^-$	0.354
$[Fe(CN)_6]^{3-} + e \rightleftharpoons [Fe(CN)_6]^{4-}$	0.358
$ClO_4^- + H_2O + 2 e \rightleftharpoons ClO^-_3 + 2 OH^-$	0.36
$Ag_2SeO_3 + 2 e \rightleftharpoons 2 Ag + SeO_3^{2-}$	0.3629
$ReO_4^- + 8 H^+ + 7 e \rightleftharpoons Re + 4 H_2O$	0.368
$(CN)_2 + 2 H^+ + 2 e \rightleftharpoons 2 HCN$	0.373
$[Ferricinium]^+ + e \rightleftharpoons$ ferrocene	0.400
$Tc^{2+} + 2 e \rightleftharpoons Tc$	0.400
$O_2 + 2 H_2O + 4 e \rightleftharpoons 4 OH^-$	0.401
$AgOCN + e \rightleftharpoons Ag + OCN^-$	0.41
$[RhCl_6]^{3-} + 3 e \rightleftharpoons Rh + 6 Cl^-$	0.431
$Ag_2CrO_4 + 2 e \rightleftharpoons 2 Ag + CrO_4^{2-}$	0.4470
$H_2SO_3 + 4 H^+ + 4 e \rightleftharpoons S + 3 H_2O$	0.449
$Ru^{2+} + 2 e \rightleftharpoons Ru$	0.455
$Ag_2MoO_4 + 2 e \rightleftharpoons 2 Ag + MoO_4^{2-}$	0.4573
$Ag_2C_2O_4 + 2 e \rightleftharpoons 2 Ag + C_2O_4^{2-}$	0.4647
$Ag_2WO_4 + 2 e \rightleftharpoons 2 Ag + WO_4^{2-}$	0.4660
$Ag_2CO_3 + 2 e \rightleftharpoons 2 Ag + CO_3^{2-}$	0.47
$TcO_4^- + 8 H^+ + 7 e \rightleftharpoons Tc + 4 H_2O$	0.472
$TeO_4^- + 8 H^+ + 7 e \rightleftharpoons Te + 4 H_2O$	0.472
$IO^- + H_2O + 2 e \rightleftharpoons I^- + 2 OH^-$	0.485
$NiO_2 + 2 H_2O + 2 e \rightleftharpoons Ni(OH)_2 + 2 OH^-$	0.490
$Bi^+ + e \rightleftharpoons Bi$	0.5
$ReO_4^- + 4 H^+ + 3 e \rightleftharpoons ReO_2 + 2 H_2O$	0.510
$Hg_2(ac)_2 + 2 e \rightleftharpoons 2 Hg + 2(ac)^-$	0.51163
$Cu^+ + e \rightleftharpoons Cu$	0.521
$I_2 + 2 e \rightleftharpoons 2 I^-$	0.5355
$I_3^- + 2 e \rightleftharpoons 3 I^-$	0.536
$AgBrO_3 + e \rightleftharpoons Ag + BrO_3^-$	0.546
$MnO_4^- + e \rightleftharpoons MnO_2^-$	0.558
$H_3AsO_4 + 2 H^+ + 2 e \rightleftharpoons HAsO_2 + 2 H_2O$	0.560
$S_2O_6^{2-} + 4 H^+ + 2 e \rightleftharpoons 2 H_2SO_3$	0.564
$AgNO_2 + e \rightleftharpoons Ag + NO_2^-$	0.564
$Te^{4+} + 4 e \rightleftharpoons Te$	0.568
$Sb_2O_5 + 6 H^+ + 4 e \rightleftharpoons 2 SbO^+ + 3 H_2O$	0.581
$RuO_4^- + e \rightleftharpoons RuO_4^{2-}$	0.59

Reaction	$E°$/V
$[PdCl_4]^{2-} + 2 e \rightleftharpoons Pd + 4 Cl^-$	0.591
$TeO_2 + 4 H^+ + 4 e \rightleftharpoons Te + 2 H_2O$	0.593
$MnO_4^- + 2 H_2O + 3 e \rightleftharpoons MnO_2 + 4 OH^-$	0.595
$Rh^{2+} + 2 e \rightleftharpoons Rh$	0.600
$Rh^+ + e \rightleftharpoons Rh$	0.600
$MnO_4^{2-} + 2 H_2O + 2 e \rightleftharpoons MnO_2 + 4 OH^-$	0.60
$2 AgO + H_2O + 2 e \rightleftharpoons Ag_2O + 2 OH^-$	0.607
$BrO_3^- + 3 H_2O + 6 e \rightleftharpoons Br^- + 6 OH^-$	0.61
$UO_2^+ + 4 H^+ + e \rightleftharpoons U^{4+} + 2 H_2O$	0.612
$Hg_2SO_4 + 2 e \rightleftharpoons 2 Hg + SO_4^{2-}$	0.6125
$ClO_3^- + 3 H_2O + 6 e \rightleftharpoons Cl^- + 6 OH^-$	0.62
$Hg_2HPO_4 + 2 e \rightleftharpoons 2 Hg + HPO_4^{2-}$	0.6359
$Ag(ac) + e \rightleftharpoons Ag + (ac)^-$	0.643
Sb_2O_5(valentinite) $+ 4 H^+ + 4 e \rightleftharpoons Sb_2O_3 + 2 H_2O$	0.649
$Ag_2SO_4 + 2 e \rightleftharpoons 2 Ag + SO_4^{2-}$	0.654
$ClO_2^- + H_2O + 2 e \rightleftharpoons ClO^- + 2 OH^-$	0.66
Sb_2O_5(senarmontite) $+ 4 H^+ + 4 e \rightleftharpoons Sb_2O_5 + 2 H_2O$	0.671
$[PtCl_6]^{2-} + 2 e \rightleftharpoons [PtCl_4]^{2-} + 2 Cl^-$	0.68
$O_2 + 2 H^+ + 2 e \rightleftharpoons H_2O_2$	0.695
p-benzoquinone $+ 2 H^+ + 2 e \rightleftharpoons$ hydroquinone	0.6992
$H_3IO_6^{2-} + 2 e \rightleftharpoons IO_3^- + 3 OH^-$	0.7
$Ag_2O_3 + H_2O + 2 e \rightleftharpoons 2 AgO + 2 OH^-$	0.739
$Tl^{3+} + 3 e \rightleftharpoons Tl$	0.741
$[PtCl_4]^{2-} + 2 e \rightleftharpoons Pt + 4 Cl^-$	0.755
$Rh^{3+} + 3 e \rightleftharpoons Rh$	0.758
$ClO_2^- + 2 H_2O + 4 e \rightleftharpoons Cl^- + 4 OH^-$	0.76
$2 NO + H_2O + 2 e \rightleftharpoons N_2O + 2 OH^-$	0.76
$Po^{4+} + 4 e \rightleftharpoons Po$	0.76
$BrO^- + H_2O + 2 e \rightleftharpoons Br^- + 2 OH^-$	0.761
$ReO_4^- + 2 H^+ + e \rightleftharpoons ReO_3 + H_2O$	0.768
$(CNS)_2 + 2 e \rightleftharpoons 2 CNS^-$	0.77
$[IrCl_6]^{3-} + 3 e \rightleftharpoons Ir + 6 Cl^-$	0.77
$Fe^{3+} + e \rightleftharpoons Fe^{2+}$	0.771
$AgF + e \rightleftharpoons Ag + F^-$	0.779
$[Fe(bipy)_2]^{3+} + e \rightleftharpoons [Fe(bipy)_2]^{2+}$	0.78
$TcO_4^- + 4 H^+ + 3 e \rightleftharpoons TcO_2 + 2 H_2O$	0.782
$Hg_2^{2+} + 2 e \rightleftharpoons 2 Hg$	0.7973
$Ag^+ + e \rightleftharpoons Ag$	0.7996
$[Os(bipy)_3]^{3+} + e \rightleftharpoons [Os(bipy)_3]^{2+}$	0.80
$2 NO_3^- + 4 H^+ + 2 e \rightleftharpoons N_2O_4 + 2 H_2O$	0.803
$[Os(bipy)_2]^{3+} + e \rightleftharpoons [Os(bipy)_2]^{2+}$	0.81
$RhOH^{2+} + H + 3 e \rightleftharpoons Rh + H_2O$	0.83
$OsO_4 + 8 H^+ + 8 e \rightleftharpoons Os + 4 H_2O$	0.838
$ClO^- + H_2O + 2 e \rightleftharpoons Cl^- + 2 OH^-$	0.841
$Hg^{2+} + 2 e \rightleftharpoons Hg$	0.851
$AuBr_4^- + 3 e \rightleftharpoons Au + 4 Br^-$	0.854
SiO_2(quartz) $+ 4 H^+ + 4 e \rightleftharpoons Si + 2 H_2O$	0.857
$2 HNO_2 + 4 H^+ + 4 e \rightleftharpoons H_2N_2O_2 + H_2O$	0.86
$[Ru(CN)_6]^{3-} + e^- \rightleftharpoons [Ru(CN)_6]^{4-}$	0.86
$[IrCl_6]^{2-} + e \rightleftharpoons [IrCl_6]^{3-}$	0.8665
$N_2O_4 + 2 e \rightleftharpoons 2 NO_2^-$	0.867
$HO_2^- + H_2O + 2 e \rightleftharpoons 3 OH^-$	0.878
$Po^{4+} + 2 e \rightleftharpoons Po^{2+}$	0.9
$2 Hg^+ + 2 e \rightleftharpoons Hg_2^{2+}$	0.920

TABLE 2
Reduction Reactions Having $E°$ Values More Positive than that of the Standard Hydrogen Electrode (continued)

Reaction	$E°$/V
$NO_3^- + 3H^+ + 2e \rightleftharpoons HNO_2 + H_2O$	0.934
$Pd^{2+} + 2e \rightleftharpoons Pd$	0.951
$ClO_2(aq) + e \rightleftharpoons ClO_2^-$	0.954
$NO_3^- + 4H^+ + 3e \rightleftharpoons NO + 2H_2O$	0.957
$V_2O_5 + 6H^+ + 2e \rightleftharpoons 2VO^{2+} + 3H_2O$	0.957
$AuBr_2^- + e \rightleftharpoons Au + 2Br^-$	0.959
$HNO_2 + H^+ + e \rightleftharpoons NO + H_2O$	0.983
$HIO + H^+ + 2e \rightleftharpoons I^- + H_2O$	0.987
$VO_2^+ + 2H^+ + e \rightleftharpoons VO^{2+} + H_2O$	0.991
$PtO_2 + 4H^+ + 4e \rightleftharpoons Pt + 2H_2O$	1.00
$RuO_4 + e \rightleftharpoons RuO_4^-$	1.00
$V(OH)_4^+ + 2H^+ + e \rightleftharpoons VO^{2+} + 3H_2O$	1.00
$AuCl_4^- + 3e \rightleftharpoons Au + 4Cl^-$	1.002
$Pu^{4+} + e \rightleftharpoons Pu^{3+}$	1.006
$PtO_2 + 2H^+ + 2e \rightleftharpoons PtO + H_2O$	1.01
$OsO_4 + 4H + 4e \rightleftharpoons OsO_2 + 2H_2O$	1.02
$H_6TeO_6 + 2H^+ + 2e \rightleftharpoons TeO_2 + 4H_2O$	1.02
$[Fe(bipy)_3]^{3+} + e \rightleftharpoons [Fe(bipy)_3]^{2+}$	1.03
$Hg(OH)_2 + 2H^+ + 2e \rightleftharpoons Hg + 2H_2O$	1.034
$N_2O_4 + 4H^+ + 4e \rightleftharpoons 2NO + 2H_2O$	1.035
$RuO_4 + 8H^+ + 8e \rightleftharpoons Ru + 4H_2O$	1.038
$[Fe(phen)_3]^{3+} + e \rightleftharpoons [Fe(phen)_3]^{2+}$ (1 molar H_2SO_4)	1.06
$PuO_2(OH)_2 + H^+ + e \rightleftharpoons PuO_2OH + H_2O$	1.062
$N_2O_4 + 2H^+ + 2e \rightleftharpoons 2HNO_2$	1.065
$Br_2(l) + 2e \rightleftharpoons 2Br^-$	1.066
$IO_3^- + 6H^+ + 6e \rightleftharpoons I^- + 3H_2O$	1.085
$Br_2(aq) + 2e \rightleftharpoons 2Br^-$	1.0873
$Pu^{5+} + e \rightleftharpoons Pu^{4+}$	1.099
$Cu^{2+} + 2CN^- + e \rightleftharpoons [Cu(CN)_2]^-$	1.103
$RuO_2 + 4H^+ + 2e \rightleftharpoons Ru^{2+} + 2H_2O$	1.120
$[Fe(phen)_3]^{3+} + e \rightleftharpoons [Fe(phen)_3]^{2+}$	1.147
$SeO_4^{2-} + 4H^+ + 2e \rightleftharpoons H_2SeO_3 + H_2O$	1.151
$ClO_3^- + 2H^+ + e \rightleftharpoons ClO_2 + H_2O$	1.152
$Ir^{3+} + 3e \rightleftharpoons Ir$	1.156
$Pt^{2+} + 2e \rightleftharpoons Pt$	1.18
$ClO_4^- + 2H^+ + 2e \rightleftharpoons ClO_3^- + H_2O$	1.189
$2IO_3^- + 12H^+ + 10e \rightleftharpoons I_2 + 6H_2O$	1.195
$PtOH^+ + H^+ + 2e \rightleftharpoons Pt + H_2O$	1.2
$ClO_3^- + 3H^+ + 2e \rightleftharpoons HClO_2 + H_2O$	1.214
$MnO_2 + 4H^+ + 2e \rightleftharpoons Mn^{2+} + 2H_2O$	1.224
$O_2 + 4H^+ + 4e \rightleftharpoons 2H_2O$	1.229
$Cr_2O_7^{2-} + 14H^+ + 6e \rightleftharpoons 2Cr^{3+} + 7H_2O$	1.232
$O_3 + H_2O + 2e \rightleftharpoons O_2 + 2OH^-$	1.24
$[Ru(bipy)_3]^{3+} + e \rightleftharpoons [Ru(bipy)_3]^{2+}$	1.24
$Tl^{3+} + 2e \rightleftharpoons Tl^+$	1.252
$N_2H_5^+ + 3H^+ + 2e \rightleftharpoons 2NH_4^+$	1.275
$ClO_2 + H^+ + e \rightleftharpoons HClO_2$	1.277
$[PdCl_6]^{2-} + 2e \rightleftharpoons [PdCl_4]^{2-} + 2Cl^-$	1.288
$2HNO_2 + 4H^+ + 4e \rightleftharpoons N_2O + 3H_2O$	1.297
$AuOH^{2+} + H^+ + 2e \rightleftharpoons Au^+ + H_2O$	1.32
$PuO_2(OH)_2 + 2H^- + 2e \rightleftharpoons Pu(OH)_4$	1.325
$HBrO + H^+ + 2e \rightleftharpoons Br^- + H_2O$	1.331
$Cr(V) + e \rightleftharpoons Cr(IV)$	1.34
$HCrO_4^- + 7H^+ + 3e \rightleftharpoons Cr^{3+} + 4H_2O$	1.350
$Cl_2(g) + 2e \rightleftharpoons 2Cl^-$	1.35827
$ClO_4^- + 8H^+ + 8e \rightleftharpoons Cl^- + 4H_2O$	1.389
$ClO_4^- + 8H^+ + 7e \rightleftharpoons 1/2\,Cl_2 + 4H_2O$	1.39
$No^{3+} + e \rightleftharpoons No^{2+}$	1.4
$RuO_4 + 6H^+ + 4e \rightleftharpoons Ru(OH)_2^{2+} + 2H_2O$	1.40
$Au^{3+} + 2e \rightleftharpoons Au^+$	1.401
$2NH_3OH^+ + H^+ + 2e \rightleftharpoons N_2H_5^+ + 2H_2O$	1.42
$BrO_3^- + 6H^+ + 6e \rightleftharpoons Br^- + 3H_2O$	1.423
$2HIO + 2H^+ + 2e \rightleftharpoons I_2 + 2H_2O$	1.439
$Au(OH)_3 + 3H^+ + 3e \rightleftharpoons Au^- + 3H_2O$	1.45
$3IO_3^- + 6H^+ + 6e \rightleftharpoons Cl^- + 3H_2O$	1.451
$PbO_2 + 4H^+ + 2e \rightleftharpoons Pb^{2+} + 2H_2O$	1.455
$ClO_3^- + 6H^+ + 5e \rightleftharpoons 1/2\,Cl_2 + 3H_2O$	1.47
$CrO_2 + 4H^+ + e \rightleftharpoons Cr^{3+} + 2H_2O$	1.48
$BrO_3^- + 6H^+ + 5e \rightleftharpoons 1/2\,Br_2 + 3H_2O$	1.482
$HClO + H^+ + 2e \rightleftharpoons Cl^- + H_2O$	1.482
$Mn_2O_3 + 6H^+ + e \rightleftharpoons 2Mn^{2+} + 3H_2O$	1.485
$HO_2 + H^+ + e \rightleftharpoons H_2O_2$	1.495
$Au^{3+} + 3e \rightleftharpoons Au$	1.498
$PtO_3 + 4H^+ + 2e \rightleftharpoons Pt(OH)_2^{2+} + H_2O$	1.5
$MnO_4^- + 8H^+ + 5e \rightleftharpoons Mn^{2+} + 4H_2O$	1.507
$Mn^{3+} + e \rightleftharpoons Mn^{2-}$	1.5415
$HClO_2 + 3H^+ + 4e \rightleftharpoons Cl^- + 2H_2O$	1.570
$HBrO + H^+ + e \rightleftharpoons 1/2\,Br_2(aq) + H_2O$	1.574
$2NO + 2H^+ + 2e \rightleftharpoons N_2O + H_2O$	1.591
$Bi_2O_4 + 4H^+ + 2e \rightleftharpoons 2BiO^+ + 2H_2O$	1.593
$HBrO + H^+ + e \rightleftharpoons 1/2\,Br_2(l) + H_2O$	1.596
$H_5IO_6 + H^+ + 2e \rightleftharpoons IO_3^- + 3H_2O$	1.601
$HClO + H^+ + e \rightleftharpoons 1/2\,Cl_2 + H_2O$	1.611
$HClO_2 + 3H^+ + 3e \rightleftharpoons 1/2\,Cl_2 + 2H_2O$	1.628
$HClO_2 + 2H^+ + 2e \rightleftharpoons HClO + H_2O$	1.645
$Bk^{4+} + e \rightleftharpoons Bk^{3+}$	1.67
$NiO_2 + 4H^+ + 2e \rightleftharpoons Ni^{2+} + 2H_2O$	1.678
$MnO_4^- + 4H^+ + 3e \rightleftharpoons MnO_2 + 2H_2O$	1.679
$PbO_2 + SO_4^{2-} + 4H^+ + 2e \rightleftharpoons PbSO_4 + 2H_2O$	1.6913
$Au^+ + e \rightleftharpoons Au$	1.692
$PtO_3 + 2H^+ + 2e \rightleftharpoons PtO_2 + H_2O$	1.7
$CeOH^{3+} + H^+ + e \rightleftharpoons Ce^{3+} + H_2O$	1.715
$Ce^{4+} + e \rightleftharpoons Ce^{3+}$	1.72
$N_2O + 2H^+ + 2e \rightleftharpoons N_2 + H_2O$	1.766
$H_2O_2 + 2H^+ + 2e \rightleftharpoons 2H_2O$	1.776
$Ag^{3+} + e \rightleftharpoons Ag^{2+}$	1.8
$Au^{2+} + e^- \rightleftharpoons Au^+$	1.8
$Ag_2O_2 + 4H^+ + e \rightleftharpoons 2Ag + 2H_2O$	1.802
$Co^{3+} + e \rightleftharpoons Co^{2-}$(2 molar H_2SO_4)	1.83
$Ag^{3+} + 2e \rightleftharpoons Ag^+$	1.9
$Co^{3+} + e \rightleftharpoons Co^{2+}$	1.92
$Ag^{2+} + e \rightleftharpoons Ag^+$	1.980
$Cu_2O_3 + 6H^+ + 2e \rightleftharpoons 2Cu^{2+} + 3H_2O$	2.0
$S_2O_8^{2-} + 2e \rightleftharpoons 2SO_4^{2-}$	2.010
$OH + e \rightleftharpoons OH^-$	2.02
$HFeO_4^- + 7H^+ + 3e \rightleftharpoons Fe^{3+} + 4H_2O$	2.07
$O_3 + 2H^+ + 2e \rightleftharpoons O_2 + H_2O$	2.076
$HFeO_4^- + 4H^+ + 3e \rightleftharpoons FeOOH + 2H_2O$	2.08

TABLE 2
Reduction Reactions Having $E°$ Values More Positive than that of the Standard Hydrogen Electrode (continued)

Reaction	$E°$/V
$2\,HFeO_4^- + 8\,H^+ + 6\,e \rightleftharpoons Fe_2O_3 + 5\,H_2O$	2.09
$XeO_3 + 6\,H^+ + 6\,e \rightleftharpoons Xe + 3\,H_2O$	2.10
$S_2O_8^{2-} + 2\,H^+ + 2\,e \rightleftharpoons 2\,HSO_4^-$	2.123
$F_2O + 2\,H^+ + 4\,e \rightleftharpoons H_2O + 2\,F^-$	2.153
$FeO_4^{2-} + 8\,H^+ + 3\,e \rightleftharpoons Fe^{3+} + 4\,H_2O$	2.20
$Cu^{3+} + e \rightleftharpoons Cu^{2+}$	2.4
$H_4XeO_6 + 2\,H^+ + 2\,e \rightleftharpoons XeO_3 + 3\,H_2O$	2.42
$O(g) + 2\,H^+ + 2\,e \rightleftharpoons H_2O$	2.421
$Am^{4+} + e \rightleftharpoons Am^{3+}$	2.60
$H_2N_2O_2 + 2\,H^+ + 2\,e \rightleftharpoons N_2 + 2\,H_2O$	2.65
$F_2 + 2\,e \rightleftharpoons 2\,F^-$	2.866
$Cm^{4+} + e \rightleftharpoons Cm^{3+}$	3.0
$F_2 + 2\,H^+ + 2\,e \rightleftharpoons 2\,HF$	3.053
$Tb^{4+} + e \rightleftharpoons Tb^{3+}$	3.1
$Pr^{4+} + e \rightleftharpoons Pr^{3+}$	3.2
$Cf^{4+} + e \rightleftharpoons Cf^{3+}$	3.3
$XeF + e \rightleftharpoons Xe + F^-$	3.4

TABLE 3
Reduction Reactions Having $E°$ Values More Negative than that of the Standard Hydrogen Electrode

Reaction	$E°$/V
$2\,H^+ + 2\,e \rightleftharpoons H_2$	0.00000
$2\,D^+ + 2\,e \rightleftharpoons D_2$	–0.013
$AgCN + e \rightleftharpoons Ag + CN^-$	–0.017
$2\,WO_3 + 2\,H^+ + 2\,e \rightleftharpoons W_2O_5 + H_2O$	–0.029
$W_2O_5 + 2\,H^+ + 2\,e \rightleftharpoons 2\,WO_2 + H_2O$	–0.031
$Ag_2S + 2\,H^+ + 2\,e \rightleftharpoons 2\,Ag + H_2S$	–0.0366
$Fe^{3+} + 3\,e \rightleftharpoons Fe$	–0.037
$Hg_2I_2 + 2\,e \rightleftharpoons 2\,Hg + 2\,I^-$	–0.0405
$Tl(OH)_3 + 2\,e \rightleftharpoons TlOH + 2\,OH^-$	–0.05
$TiOH^{3+} + H^+ + e \rightleftharpoons Ti^{3+} + H_2O$	–0.055
$2\,H_2SO_3 + H^+ + 2\,e \rightleftharpoons HS_2O_4^- + 2\,H_2O$	–0.056
$P(white) + 3\,H^+ + 3\,e \rightleftharpoons PH_3(g)$	–0.063
$O_2 + H_2O + 2\,e \rightleftharpoons HO_2^- + OH^-$	–0.076
$2\,Cu(OH)_2 + 2\,e \rightleftharpoons Cu_2O + 2\,OH^- + H_2O$	–0.080
$Se + 2\,H^+ + 2\,e \rightleftharpoons H_2Se$	–0.082
$WO_3 + 6\,H^+ + 6\,e \rightleftharpoons W + 3\,H_2O$	–0.090
$SnO_2 + 4\,H^+ + 2\,e \rightleftharpoons Sn^{2+} + 2\,H_2O$	–0.094
$Md^{3+} + e \rightleftharpoons Md^{2+}$	–0.1
$P(red) + 3\,H^+ + 3\,e \rightleftharpoons PH_3(g)$	–0.111
$SnO_2 + 4\,H^+ + 4\,e \rightleftharpoons Sn + 2\,H_2O$	–0.117
$GeO_2 + 2\,H^+ + 2\,e \rightleftharpoons GeO + H_2O$	–0.118
$WO_2 + 4\,H^+ + 4\,e \rightleftharpoons W + 2\,H_2O$	–0.119
$Pb^{2+} + 2\,e \rightleftharpoons Pb(Hg)$	–0.1205
$Pb^{2+} + 2\,e \rightleftharpoons Pb$	–0.1262
$CrO_4^{2-} + 4\,H_2O + 3\,e \rightleftharpoons Cr(OH)_3 + 5\,OH^-$	–0.13
$Sn^{2-} + 2\,e \rightleftharpoons Sn$	–0.1375
$In^+ + e \rightleftharpoons In$	–0.14
$O_2 + 2\,H_2O + 2\,e \rightleftharpoons H_2O_2 + 2\,OH^-$	–0.146
$MoO_2 + 4\,H^+ + 4\,e \rightleftharpoons Mo + 4\,H_2O$	–0.152
$AgI + e \rightleftharpoons Ag + I^-$	–0.15224
$2\,NO_2^- + 2\,H_2O + 4\,e \rightleftharpoons N_2O_2^{2-} + 4\,OH^-$	–0.18
$H_2GeO_3 + 4\,H^+ + 4\,e \rightleftharpoons Ge + 3\,H_2O$	–0.182
$SnO_2 + 3\,H^+ + 2\,e \rightleftharpoons SnOH^+ + H_2O$	–0.194
$CO_2 + 2\,H^+ + 2\,e \rightleftharpoons HCOOH$	–0.199
$Mo^{3+} + 3\,e \rightleftharpoons Mo$	–0.200
$Ga^+ + e \rightleftharpoons Ga$	–0.2
$2\,SO_2^{2-} + 4\,H^+ + 2\,e \rightleftharpoons S_2O_6^{2-} + H_2O$	–0.22
$Cu(OH)_2 + 2\,e \rightleftharpoons Cu + 2\,OH^-$	–0.222
$V_2O_5 + 10\,H^+ + 10\,e \rightleftharpoons 2\,V + 5\,H_2O$	–0.242
$CdSO_4 + 2\,e \rightleftharpoons Cd + SO_4^{2-}$	–0.246
$V(OH)_4^+ + 4\,H^+ + 5\,e \rightleftharpoons V + 4\,H_2O$	–0.254
$V^{3+} + e \rightleftharpoons V^{2+}$	–0.255
$Ni^{2+} + 2\,e \rightleftharpoons Ni$	–0.257
$PbCl_2 + 2\,e \rightleftharpoons Pb + 2\,Cl^-$	–0.2675
$H_3PO_4 + 2\,H^+ + 2\,e \rightleftharpoons H_3PO_3 + H_2O$	–0.276
$Co^{2+} + 2\,e \rightleftharpoons Co$	–0.28
$PbBr_2 + 2\,e \rightleftharpoons Pb + 2\,Br^-$	–0.284
$Tl^+ + e \rightleftharpoons Tl(Hg)$	–0.3338
$Tl^+ + e \rightleftharpoons Tl$	–0.336
$In^{3+} + 3\,e \rightleftharpoons In$	–0.3382
$TlOH + e \rightleftharpoons Tl + OH^-$	–0.34
$PbF_2 + 2\,e \rightleftharpoons Pb + 2\,F^-$	–0.3444
$PbSO_4 + 2\,e \rightleftharpoons Pb(Hg) + SO_4^{2-}$	–0.3505
$Cd^{2+} + 2\,e \rightleftharpoons Cd(Hg)$	–0.3521
$PbSO_4 + 2\,e \rightleftharpoons Pb + SO_4^{2-}$	–0.3588
$Cu_2O + H_2O + 2\,e \rightleftharpoons 2\,Cu + 2\,OH^-$	–0.360
$Eu^{3+} + e \rightleftharpoons Eu^{2+}$	–0.36
$PbI_2 + 2\,e \rightleftharpoons Pb + 2\,I^-$	–0.365
$SeO_3^{2-} + 3\,H_2O + 4\,e \rightleftharpoons Se + 6\,OH^-$	–0.366
$Se + 2\,H^+ + 2\,e \rightleftharpoons H_2Se(aq)$	–0.399
$In^{2+} + e \rightleftharpoons In^+$	–0.40
$Cd^{2+} + 2\,e \rightleftharpoons Cd$	–0.4030
$Cr^{3+} + e \rightleftharpoons Cr^{2+}$	–0.407
$2\,S + 2\,e \rightleftharpoons S_2^{2-}$	–0.42836
$Tl_2SO_4 + 2\,e \rightleftharpoons Tl + SO_4^{2-}$	–0.4360
$In^{3+} + 2\,e \rightleftharpoons In^+$	–0.443
$Fe^{2+} + 2\,e \rightleftharpoons Fe$	–0.447
$H_3PO_3 + 3\,H^+ + 3\,e \rightleftharpoons P + 3\,H_2O$	–0.454
$Bi_2O_3 + 3\,H_2O + 6\,e \rightleftharpoons 2\,Bi + 6\,OH^-$	–0.46
$NO_2^- + H_2O + e \rightleftharpoons NO + 2\,OH$	–0.46
$PbHPO_4 + 2\,e \rightleftharpoons Pb + HPO_4^{2-}$	–0.465
$S + 2\,e \rightleftharpoons S^{2-}$	–0.47627
$S + H_2O + 2\,e \rightleftharpoons HS^- + OH^-$	–0.478
$B(OH)_3 + 7\,H^+ + 8\,e \rightleftharpoons BH_4^- + 3\,H_2O$	–0.481

TABLE 3
Reduction Reactions Having $E°$ Values More Negative than that of the Standard Hydrogen Electrode (continued)

Reaction	$E°/V$
$In^{3+} + e \rightleftharpoons In^{2+}$	–0.49
$ZnOH^+ + H^+ + 2e \rightleftharpoons Zn + H_2O$	–0.497
$GaOH^{2+} + H^+ + 3e \rightleftharpoons Ga + H_2O$	–0.498
$H_3PO_3 + 2H^+ + 2e \rightleftharpoons H_3PO_2 + H_2O$	–0.499
$TiO_2 + 4H^+ + 2e \rightleftharpoons Ti^{2+} + 2H_2O$	–0.502
$H_3PO_2 + H^+ + e \rightleftharpoons P + 2H_2O$	–0.508
$Sb + 3H^+ + 3e \rightleftharpoons SbH_3$	–0.510
$HPbO_2^- + H_2O + 2e \rightleftharpoons Pb + 3OH^-$	–0.537
$Ga^{3+} + 3e \rightleftharpoons Ga$	–0.549
$TlCl + e \rightleftharpoons Tl + Cl^-$	–0.5568
$Fe(OH)_3 + e \rightleftharpoons Fe(OH)_2 + OH^-$	–0.56
$TeO_3^{2-} + 3H_2O + 4e \rightleftharpoons Te + 6OH^-$	–0.57
$2SO_3^{2-} + 3H_2O + 4e \rightleftharpoons S_2O_3^{2-} + 6OH^-$	–0.571
$PbO + H_2O + 2e \rightleftharpoons Pb + 2OH^-$	–0.580
$ReO_2^- + 4H_2O + 7e \rightleftharpoons Re + 8OH^-$	–0.584
$SbO_3^- + H_2O + 2e \rightleftharpoons SbO_2^- + 2OH^-$	–0.59
$Ta^{3+} + 3e \rightleftharpoons Ta$	–0.6
$U^{4+} + e \rightleftharpoons U^{3+}$	–0.607
$As + 3H^+ + 3e \rightleftharpoons AsH_3$	–0.608
$Nb_2O_5 + 10H^+ + 10e \rightleftharpoons 2Nb + 5H_2O$	–0.644
$NbO_2 + 2H^+ + 2e \rightleftharpoons NbO + H_2O$	–0.646
$Cd(OH)_4^{2-} + 2e \rightleftharpoons Cd + 4OH^-$	–0.658
$TlBr + e \rightleftharpoons Tl + Br^-$	–0.658
$SbO_2^- + 2H_2O + 3e \rightleftharpoons Sb + 4OH^-$	–0.66
$AsO_2^- + 2H_2O + 3e \rightleftharpoons As + 4OH^-$	–0.68
$NbO_2 + 4H^+ + 4e \rightleftharpoons Nb + 2H_2O$	–0.690
$Ag_2S + 2e \rightleftharpoons 2Ag + S^{2-}$	–0.691
$AsO_4^{3-} + 2H_2O + 2e \rightleftharpoons AsO_2^- + 4OH^-$	–0.71
$Ni(OH)_2 + 2e \rightleftharpoons Ni + 2OH^-$	–0.72
$Co(OH)_2 + 2e \rightleftharpoons Co + 2OH^-$	–0.73
$NbO + 2H^+ + 2e \rightleftharpoons Nb + H_2O$	–0.733
$H_2SeO_3 + 4H^+ + 4e \rightleftharpoons Se + 3H_2O$	–0.74
$Cr^{3+} + 3e \rightleftharpoons Cr$	–0.744
$Ta_2O_5 + 10H^+ + 10e \rightleftharpoons 2Ta + 5H_2O$	–0.750
$TlI + e \rightleftharpoons Tl + I^-$	–0.752
$Zn^{2+} + 2e \rightleftharpoons Zn$	–0.7618
$Zn^{2+} + 2e \rightleftharpoons Zn(Hg)$	–0.7628
$CdO + H_2O + 2e \rightleftharpoons Cd + 2OH^-$	–0.783
$Te + 2H^+ + 2e \rightleftharpoons H_2Te$	–0.793
$ZnSO_4 \cdot 7H_2O + 2e \rightleftharpoons Zn(Hg) + SO_4^{2-} + 7H_2O$ (Saturated $ZnSO_4$)	–0.7993
$Bi + 3H^+ + 3e \rightleftharpoons BiH_3$	–0.8
$SiO + 2H^+ + 2e \rightleftharpoons Si + H_2O$	–0.8
$Cd(OH)_2 + 2e \rightleftharpoons Cd(Hg) + 2OH^-$	–0.809
$2H_2O + 2e \rightleftharpoons H_2 + 2OH^-$	–0.8277
$2NO^-_3 + 2H_2O + 2e \rightleftharpoons N_2O_4 + 4OH^-$	–0.85
$H_3BO_3 + 3H^+ + 3e \rightleftharpoons B + 3H_2O$	–0.8698
$P + 3H_2O + 3e \rightleftharpoons PH_3(g) + 3OH^-$	–0.87
$Ti^{3+} + e \rightleftharpoons Ti^{2+}$	–0.9
$HSnO_2^- + H_2O + 2e \rightleftharpoons Sn + 3OH^-$	–0.909
$Cr^{2+} + 2e \rightleftharpoons Cr$	–0.913
$Se + 2e \rightleftharpoons Se^{2-}$	–0.924
$SO_4^{2-} + H_2O + 2e \rightleftharpoons SO_3^{2-} + 2OH^-$	–0.93
$Sn(OH)_6^{2-} + 2e \rightleftharpoons HSnO_2^- + 3OH^- + H_2O$	–0.93
$SnO_2 + 2H_2O + 4e \rightleftharpoons Sn + 4OH^-$	–0.945
$In(OH)_3 + 3e \rightleftharpoons In + 3OH^-$	–0.99
$NpO_2 + H_2O + H^+ + e \rightleftharpoons Np(OH)_3$	–0.962
$In(OH)_4^- + 3e \rightleftharpoons In + 4OH^-$	–1.007
$In_2O_3 + 3H_2O + 6e \rightleftharpoons 2In + 6OH^-$	–1.034
$PO_4^{3-} + 2H_2O + 2e \rightleftharpoons HPO_3^{2-} + 3OH^-$	–1.05
$Yb^{3+} + e \rightleftharpoons Yb^{2+}$	–1.05
$Nb^{3+} + 3e \rightleftharpoons Nb$	–1.099
$Fm^{3+} + e \rightleftharpoons Fm^{2+}$	–1.1
$2SO_3^{2-} + 2H_2O + 2e \rightleftharpoons S_2O_4^{2-} + 4OH^-$	–1.12
$Te + 2e \rightleftharpoons Te^{2-}$	–1.143
$V^{2+} + 2e \rightleftharpoons V$	–1.175
$Mn^{2+} + 2e \rightleftharpoons Mn$	–1.185
$Zn(OH)_4^{2-} + 2e \rightleftharpoons Zn + 4OH^-$	–1.199
$CrO_2 + 2H_2O + 3e \rightleftharpoons Cr + 4OH^-$	–1.2
$No^{3+} + 3e \rightleftharpoons No$	–1.20
$ZnO_2^- + 2H_2O + 2e \rightleftharpoons Zn + 4OH^-$	–1.215
$H_2GaO_3^- + H_2O + 3e \rightleftharpoons Ga + 4OH^-$	–1.219
$H_2BO_3^- + 5H_2O + 8e \rightleftharpoons BH_4^- + 8OH^-$	–1.24
$SiF_6^{2-} + 4e \rightleftharpoons Si + 6F^-$	–1.24
$Zn(OH)_2 + 2e \rightleftharpoons Zn + 2OH^-$	–1.249
$ZnO + H_2O + 2e \rightleftharpoons Zn + 2OH^-$	–1.260
$Es^{3+} + e \rightleftharpoons Es^{2+}$	–1.3
$Pa^{3+} + 3e \rightleftharpoons Pa$	–1.34
$Ti^{3+} + 3e \rightleftharpoons Ti$	–1.37
$Ce^{3+} + 3e \rightleftharpoons Ce(Hg)$	–1.4373
$UO_2^{2+} + 4H^+ + 6e \rightleftharpoons U + 2H_2O$	–1.444
$Zr^{4+} + 4e \rightleftharpoons Zr$	–1.45
$Cr(OH)_3 + 3e \rightleftharpoons Cr + 3OH^-$	–1.48
$Pa^{4+} + 4e \rightleftharpoons Pa$	–1.49
$HfO_2 + 4H^+ + 4e \rightleftharpoons Hf + 2H_2O$	–1.505
$Hf^{4+} + 4e \rightleftharpoons Hf$	–1.55
$Sm^{3+} + e \rightleftharpoons Sm^{2+}$	–1.55
$ZrO_2 + 4H^+ + 4e \rightleftharpoons Zr + 2H_2O$	–1.553
$Mn(OH)_2 + 2e \rightleftharpoons Mn + 2OH^-$	–1.56
$Ba^{2+} + 2e \rightleftharpoons Ba(Hg)$	–1.570
$Bk^{2+} + 2e \rightleftharpoons Bk$	–1.6
$Cf^{3+} + e \rightleftharpoons Cf^{2+}$	–1.6
$Ti^{2+} + 2e \rightleftharpoons Ti$	–1.630
$Md^{3+} + 3e \rightleftharpoons Md$	–1.65
$HPO_3^{2-} + 2H_2O + 2e \rightleftharpoons H_2PO_2^- + 3OH^-$	–1.65
$Al^{3+} + 3e \rightleftharpoons Al$	–1.662
$SiO_3^{2-} + H_2O + 4e \rightleftharpoons Si + 6OH^-$	–1.697
$HPO_3^{2-} + 2H_2O + 3e \rightleftharpoons P + 5OH^-$	–1.71
$HfO^{2+} + 2H^+ + 4e \rightleftharpoons Hf + H_2O$	–1.724
$ThO_2 + 4H^+ + 4e \rightleftharpoons Th + 2H_2O$	–1.789
$H_2BO_3^- + H_2O + 3e \rightleftharpoons B + 4OH^-$	–1.79
$Sr^{2+} + 2e \rightleftharpoons Sr(Hg)$	–1.793
$U^{3+} + 3e \rightleftharpoons U$	–1.798
$H_2PO^-_2 + e \rightleftharpoons P + 2OH^-$	–1.82
$Be^{2+} + 2e \rightleftharpoons Be$	–1.847
$Np^{3+} + 3e \rightleftharpoons Np$	–1.856
$Fm^{3+} + 3e \rightleftharpoons Fm$	–1.89
$Th^{4+} + 4e \rightleftharpoons Th$	–1.899

TABLE 3
Reduction Reactions Having $E°$ Values More Negative than that of the Standard Hydrogen Electrode (continued)

Reaction	$E°$/V
$Am^{2+} + 2e \rightleftharpoons Am$	–1.9
$Pa^{4+} + e \rightleftharpoons Pa^{3+}$	–1.9
$Es^{3+} + 3e \rightleftharpoons Es$	–1.91
$Cf^{3+} + 3e \rightleftharpoons Cf$	–1.94
$Lr^{3+} + 3e \rightleftharpoons Lr$	–1.96
$Eu^{3+} + 3e \rightleftharpoons Eu$	–1.991
$Er^{2+} + 2e \rightleftharpoons Er$	–2.0
$Pr^{2+} + 2e \rightleftharpoons Pr$	–2.0
$Pu^{3+} + 3e \rightleftharpoons Pu$	–2.031
$Cm^{3+} + 3e \rightleftharpoons Cm$	–2.04
$Am^{3+} + 3e \rightleftharpoons Am$	–2.048
$AlF_6^{3-} + 3e \rightleftharpoons Al + 6F^-$	–2.069
$Sc^{3+} + 3e \rightleftharpoons Sc$	–2.077
$Ho^{2+} + 2e \rightleftharpoons Ho$	–2.1
$Nd^{2+} + 2e \rightleftharpoons Nd$	–2.1
$Cf^{2+} + 2e \rightleftharpoons Cf$	–2.12
$Yb^{3+} + 3e \rightleftharpoons Yb$	–2.19
$Ac^{3+} + 3e \rightleftharpoons Ac$	–2.20
$Dy^{2+} + 2e \rightleftharpoons Dy$	–2.2
$Tm^{3+} + e \rightleftharpoons Tm^{2+}$	–2.2
$Pm^{2+} + 2e \rightleftharpoons Pm$	–2.2
$Es^{2+} + 2e \rightleftharpoons Es$	–2.23
$H_2 + 2e \rightleftharpoons 2H^-$	–2.23
$Gd^{3+} + 3e \rightleftharpoons Gd$	–2.279
$Tb^{3+} + 3e \rightleftharpoons Tb$	–2.28
$Lu^{3+} + 3e \rightleftharpoons Lu$	–2.28
$Dy^{3+} + 3e \rightleftharpoons Dy$	–2.295
$Am^{3+} + e \rightleftharpoons Am^{2+}$	–2.3
$Fm^{2+} + 2e \rightleftharpoons Fm$	–2.30
$Pm^{3+} + 3e \rightleftharpoons Pm$	–2.30
$Sm^{3+} + 3e \rightleftharpoons Sm$	–2.304
$Al(OH)_3 + 3e \rightleftharpoons Al + 3OH^-$	–2.31
$Tm^{3+} + 3e \rightleftharpoons Tm$	–2.319
$Nd^{3+} + 3e \rightleftharpoons Nd$	–2.323
$Al(OH)^- + 3e \rightleftharpoons Al + 4OH^-$	–2.328
$H_2AlO_3^- + H_2O + 3e \rightleftharpoons Al + 4OH^-$	–2.33
$Ho^{3+} + 3e \rightleftharpoons Ho$	–2.33
$Er^{3+} + 3e \rightleftharpoons Er$	–2.331
$Ce^{3+} + 3e \rightleftharpoons Ce$	–2.336
$Pr^{3+} + 3e \rightleftharpoons Pr$	–2.353
$ZrO(OH)_2 + H_2O + 4e \rightleftharpoons Zr + 4OH^-$	–2.36
$Mg^{2+} + 2e \rightleftharpoons Mg$	–2.372
$Y^{3+} + 3e \rightleftharpoons Y$	–2.372
$La^{3+} + 3e \rightleftharpoons La$	–2.379
$Tm^{2+} + 2e \rightleftharpoons Tm$	–2.4
$Md^{2+} + 2e \rightleftharpoons Md$	–2.40
$Th(OH)_4 + 4e \rightleftharpoons Th + 4OH^-$	–2.48
$HfO(OH)_2 + H_2O + 4e \rightleftharpoons Hf + 4OH^-$	–2.50
$No^{2+} + 2e \rightleftharpoons No$	–2.50
$Dy^{3+} + e \rightleftharpoons Dy^{2+}$	–2.6
$Pm^{3+} + e \rightleftharpoons Pm^{2+}$	–2.6
$Be_2O_3^{2-} + 3H_2O + 4e \rightleftharpoons 2Be + 6OH^-$	–2.63
$Sm^{2+} + 2e \rightleftharpoons Sm$	–2.68
$Mg(OH)_2 + 2e \rightleftharpoons Mg + 2OH^-$	–2.690
$Nd^{3+} + e \rightleftharpoons Nd^{2+}$	–2.7
$Mg^+ + e \rightleftharpoons Mg$	–2.70
$Na^+ + e \rightleftharpoons Na$	–2.71
$Yb^{2+} + 2e \rightleftharpoons Yb$	–2.76
$Bk^{3+} + e \rightleftharpoons Bk^{2+}$	–2.8
$Ho^{3+} + e \rightleftharpoons Ho^{2+}$	–2.8
$Ra^{2+} + 2e \rightleftharpoons Ra$	–2.8
$Eu^{2+} + 2e \rightleftharpoons Eu$	–2.812
$Ca^{2+} + 2e \rightleftharpoons Ca$	–2.868
$Sr(OH)_2 + 2e \rightleftharpoons Sr + 2OH^-$	–2.88
$Sr^{2+} + 2e \rightleftharpoons Sr$	–2.89
$Fr^+ + e \rightleftharpoons Fr$	–2.9
$La(OH)_3 + 3e \rightleftharpoons La + 3OH^-$	–2.90
$Ba^{2+} + 2e \rightleftharpoons Ba$	–2.912
$K^+ + e \rightleftharpoons K$	–2.931
$Rb^+ + e \rightleftharpoons Rb$	–2.98
$Ba(OH)_2 + 2e \rightleftharpoons Ba + 2OH^-$	–2.99
$Er^{3+} + e \rightleftharpoons Er^{2+}$	–3.0
$Ca(OH)_2 + 2e \rightleftharpoons Ca + 2OH^-$	–3.02
$Cs^+ + e \rightleftharpoons Cs$	–3.026
$Li^+ + e \rightleftharpoons Li$	–3.0401
$3N_2 + 2H^+ + 2e \rightleftharpoons 2HN_3$	–3.09
$Pr^{3+} + e \rightleftharpoons Pr^{2+}$	–3.1
$Ca^+ + e \rightleftharpoons Ca$	–3.80
$Sr^+ + e \rightleftharpoons Sr$	–4.10

REDUCTION AND OXIDATION POTENTIALS FOR CERTAIN ION RADICALS

Petr Vanýsek

There are two tables for ion radicals. The first table lists reduction potentials for organic compounds which produce anion radicals during reduction, a process described as $A + e^- \rightleftharpoons A^{-\cdot}$. The second table lists oxidation potentials for organic compounds which produce cation radicals during oxidation, a process described as $A \rightleftharpoons A^{+\cdot} + e^-$. To obtain reduction potential for a reverse reaction, the sign for the potential is changed.

Unlike the table of the Electrochemical Series, which lists *standard* potentials, values for radicals are experimental values with experimental conditions given in the second column. Since the measurements leading to potentials for ion radicals are very dependent on conditions, an attempt to report standard potentials for radicals would serve no useful purpose. For the same reason, the potentials are also reported as experimental values, usually a half-wave potential ($E_{1/2}$ in polarography) or a peak potential (E_p in cyclic voltammetry). Unless otherwise stated, the values are reported vs. SCE (saturated calomel electrode). To obtain a value vs. normal hydrogen electrode, 0.241 V has to be added to the SCE values. All the ion radicals chosen for inclusion in the tables result from electrochemically reversible reactions. More detailed data on ion radicals can be found in the *Encyclopedia of Electrochemistry of Elements,* (A. J. Bard, Ed.), Vol. XI and XII in particular, Marcel Dekker, New York, 1978.

Abbreviations are: CV — cyclic voltammetry; DMF — *N,N*-Dimethylformamide; *E* swp — potential sweep; E° — standard potential; E_p — peak potential; $E_{p/2}$ — half-peak potential; $E_{1/2}$ — half wave potential; *M* — mol/L; MeCN — acetonitrile; pol — polarography; rot Pt dsk — rotated Pt disk; SCE — saturated calomel electrode; $TBABF_4$ — tetrabutylammonium tetrafluoroborate; TBAI — tetrabutylammonium iodide; TBAP — tetrabutylammonium perchlorate; TEABr — tetraethylammonium bromide; TEAP — tetraethylammonium perchlorate; THF — tetrahydrofuran; $TPACF_3SO_3$ — tetrapropylammonium trifluoromethanesulfite; TPAP — tetrapropylammonium perchlorate; and wr — wire.

Reduction Potentials (Products are Anion Radicals)

Substance	Conditions/electrode/technique	Potential V (vs. SCE)
Acetone	DMF, 0.1 *M* TEABr/Hg/pol	$E_{1/2} = -2.84$
1-Naphthyphenylacetylene	DMF, 0.03 *M* TBAI/Hg/pol	$E_{1/2} = -1.91$
1-Naphthalenecarboxyaldehyde	-/Hg/pol	$E_{1/2} = -0.91$
2-Naphthalenecarboxyaldehyde	-/Hg/pol	$E_{1/2} = -0.96$
2-Phenanthrenecarboxaldehyde	-/Hg/pol	$E_{1/2} = -1.00$
3-Phenanthrenecarboxaldehyde	-/Hg/pol	$E_{1/2} = -0.94$
9-Phenanthrenecarboxaldehyde	-/Hg/pol	$E_{1/2} = -0.83$
1-Anthracenecarboxaldehyde	-/Hg/pol	$E_{1/2} = -0.75$
1-Pyrenecarboxaldehyde	-/Hg/pol	$E_{1/2} = -0.76$
2-Pyrenecarboxaldehyde	-/Hg/pol	$E_{1/2} = -1.00$
Anthracene	DMF, 0.1 *M* TBAP/Pt dsk/CV	$E_p = -2.00$
	DMF, 0.5 *M* $TBABF_4$/Hg/CV	$E_{1/2} = -1.93$
	MeCN, 0.1 *M* TEAP/Hg/CV	$E_{1/2} = -2.07$
	DMF, 0.1 *M* TBAI/Hg/pol	$E_{1/2} = -1.92$
9,10-Dimethylanthracene	DMF, 0.1 *M* TBAP/Pt/CV	$E_p = -2.08$
	MeCN, 0.1 *M* TBAP/Pt/CV	$E_p = -2.10$
1-Phenylanthracene	DMF, 0.5 *M* $TBABF_2$/Hg/CV	$E_{1/2} = -1.91$
	DMF, 0.1 *M* TBAI/Hg/pol	$E_{1/2} = -1.878$
2-Phenylanthracene	DMF, 0.1 *M* TBAI/Hg/pol	$E_{1/2} = -1.875$
8-Phenylanthracene	DMF, 0.5 *M* $TBABF_4$/Hg/CV	$E_{1/2} = -1.91$
9-Phenylanthracene	DMF, 0.5 *M* $TBABF_4$/Hg/CV	$E_{1/2} = -1.93$
	DMF, 0.1 *M* TBAI/Hg/pol	$E_{1/2} = -1.863$
1,8-Diphenylanthracene	DMF, 0.5 *M* $TBABF_4$/Hg/CV	$E_{1/2} = -1.88$
1,9-Diphenylanthracene	DMF, 0.1 *M* TBAI/Hg/pol	$E_{1/2} = -1.846$
1,10-Diphenylanthracene	DMF, 0.1 *M* TBAI/Hg/pol	$E_{1/2} = -1.786$
8,9-Diphenylanthracene	DMF, 0.5 *M* $TBABF_4$/Hg/CV	$E_{1/2} = -1.90$
9,10-Diphenylanthracene	MeCN, 0.1 *M* TBAP/rot Pt/E swp	$E_{1/2} = -1.83$
	DMF, 0.1 *M* TBAI/Hg/pol	$E_{1/2} = -1.835$
1,8,9-Triphenylanthracene	DMF, 0.5 *M* $TBABF_4$/Hg/CV	$E_{1/2} = -1.85$
1,8,10-Triphenylanthracene	DMF, 0.5 *M* $TBABF_4$/Hg/CV	$E_{1/2} = -1.81$
9,10-Dibiphenylanthracene	MeCN, 0.1 *M* TBAP/rot Pt/E swp	$E_{1/2} = -1.94$
Benz(a)anthracene	MeCN, 0.1 *M* TEAP/Hg/CV	$E_{1/2} = -2.11$
	MeCN, 0.1 *M* TEAP/Hg/pol	$E_{1/2} = -2.40$[a]
Azulene	DMF, 0.1 *M* TBAI/Hg/pol	$E_{1/2} = -1.10$[c]
Annulene	DMF, 0.5 *M* TBAP 0°C/Hg/pol	$E_{1/2} = -1.23$
Benzaldehyde	DMF, 0.1 *M* TBAP/Hg/pol	$E_{1/2} = -1.67$
Benzil	DMSO, 0.1 *M* TBAP/Hg/pol	$E_{1/2} = -1.04$

Reduction Potentials (Products are Anion Radicals) (continued)

Substance	Conditions/electrode/technique	Potential V (vs. SCE)
Benzophenone	-/Hg/pol	$E_{1/2} = -1.80$
	DMF/Pt dsk/CV	$E° = -1.72$
Chrysene	MeCN, 0.1 *M* TEAP/Hg/pol	$E_{1/2} = -2.73$[a]
Fluoranthrene	DMF, 0.1 *M* TBAP/Pt dsk/CV	$E_p = -1.76$
Cyclohexanone	DMF, 0.1 *M* TEABr/Hg/pol	$E_{1/2} = -2.79$
5,5-Dimethyl-3-phenyl-2-cyclohexen-1-one	DMF, 0.5 *M*/Hg/pol	$E_{1/2} = -1.71$
1,2,3-Indanetrione hydrate (ninhydrin)	DMF, 0.2 *M* $NaNO_3$/Hg/pol	$E_{1/2} = -0.039$
Naphthacene	DMF, 0.1 *M* TBAI/Hg/pol	$E_{1/2} = -1.53$
Naphthalene	DMF, 0.1 *M* TBAP/Pt dsk/CV	$E_p = -2.55$
	DMF, 0.5 *M* $TBABF_4$/Hg/CV	$E_{1/2} = -2.56$
	DMF, MeCN, 0.1 *M* TEAP/Hg/CV	$E_{1/2} = -2.63$
	DMF, 0.1 *M* TBAI/Hg/pol	$E_{1/2} = -2.50$
1-Phenylnaphthalene	DMF, 0.5 *M* $TBABF_4$/Hg/CV	$E_{1/2} = -2.36$
1,2-Diphenylnaphthalene	DMF, 0.5 *M* $TBABF_4$/Hg/CV	$E_{1/2} = -2.25$
Cyclopentanone	DMF, 0.1 *M* TEABr/Hg/pol	$E_{1/2} = -2.82$
Phenanthrene	MeCN, 0.1 *M* TBAP/Pt wr/CV	$E_{1/2} = -2.47$
	MeCN, 0.1 *M* TEAP/Hg/pol	$E_{1/2} = -2.88$[a]
Pentacene	THF, 0.1 *M* TBAP/rot Pt dsk/E swp	$E_{1/2} = -1.40$
Perylene	MeCN, 0.1 *M* TEAP/Hg/CV	$E_{1/2} = -1.73$
1,3-Diphenyl-1,3-propanedione	DMSO, 0.2 *M* TBAP/Hg/CV	$E_{1/2} = -1.42$
2,2-Dimethyl-1,3-diphenyl-1,3 propanedione	DMSO, TBAP/Hg/CV	$E_{1/2} = -1.80$
Pyrene	DMF, 0.1 *M* TBAP/Pt/CV	$E_p = -2.14$
	MeCN, 0.1 *M* TEAP/Hg/pol	$E_{1/2} = -2.49$[a]
Diphenylsulfone	DMF, TEABr	$E_{1/2} = -2.16$
Triphenylene	MeCN, 0.1 *M* TEAP/Hg/pol	$E_{1/2} = -2.87$[a]
9,10-Anthraquinone	DMF, 0.5 *M* TBAP, 20°/Pt dsk/CV	$E_{1/2} = -1.01$
1,4-Benzoquinone	MeCN, 0.1 *M* TEAP/Pt/CV	$E_p = -0.54$
1,4-Naphthohydroquinone, dipotassium salt	DMF, 0.5 *M* TBAP, 20°/Pt dsk/CV	$E_{1/2} = -1.55$
Rubrene	DMF, 0.1 *M* TBAP/Pt dsk/CV	$E_p = -1.48$
	DMF, 0.1 *M* TBAI/Hg/pol	$E_{1/2} = -1.410$
Benzocyclooctatetraene	THF, 0.1 *M* TBAP/Hg/pol	$E_{1/2} = -2.13$
sym-Dibenzocyclooctatetraene	THF, 0.1 *M* TBAP/Hg/pol	$E_{1/2} = -2.29$
Ubiquinone-6	MeCN, 0.1 *M* TEAP/Pt/CV	$E_p = -1.05$[e]
(9-Phenyl-fluorenyl)$^+$	10.2 *M* H_2SO_4/Hg/CV	$E_p = -0.01$[b]
(Triphenylcyclopropenyl)$^+$	MeCN, 0.1 *M* TEAP/Hg/CV	$E_p = -1.87$
(Triphenylmethyl)$^+$	MeCN, 0.1 *M* TBAP/Hg/pol	$E_{1/2} = 0.27$
	H_2SO_4, 10.2 *M*/Hg/CV	$E_p = -0.58$[b]
(Tribiphenylmethyl)$^+$	MeCN, 0.1 *M* TBAP/Hg/pol	$E_{1/2} = 0.19$
(Tri-4-*t*-butyl-5-phenylmethyl)$^+$	MeCN, 0.1 *M* TBAP/Hg/pol	$E_{1/2} = 0.13$
(Tri-4-isopropylphenylmethyl)$^+$	MeCN, 0.1 *M* TBAP/Hg/pol	$E_{1/2} = 0.07$
(Tri-4-methylphenylmethyl)$^+$	MeCN, 0.1 *M* TBAP/Hg/pol	$E_{1/2} = 0.05$
(Tri-4-cyclopropylphenylmethyl)$^+$	MeCN, 0.1 *M* TBAP/Hg/pol	$E_{1/2} = 0.01$
(Tropylium)$^+$	MeCN, 0.1 *M* TBAP/Hg/pol	$E_{1/2} = -0.17$
	DMF, 0.15 *M* TBAI/Hg/pol	$E_{1/2} = -1.55$
	DMF, 0.15 *M* TBAI/Hg/pol	$E_{1/2} = -1.55$
	DMF, 0.15 *M* TBAI/Hg/pol	$E_{1/2} = -1.57$
	DMF, 0.15 *M* TBAI/Hg/pol	$E_{1/2} = -1.60$
	DMF, 0.15 *M* TBAI/Hg/pol	$E_{1/2} = -1.87$
	DMF, 0.15 *M* TBAI/Hg/pol	$E_{1/2} = -1.96$
	DMF, 0.15 *M* TBAI/Hg/pol	$E_{1/2} = -2.05$

Oxidation Potentials (Products are Cation Radicals)

Substance	Conditions/electrode/technique	Potential V (vs. SCE)
Anthracene	CH_2Cl_2, 0.2 *M* $TBABF_4$, –70°C/Pt dsk/CV	$E_p = +0.73$[d]
9,10-Dimethylanthracene	MeCN, 0.1 *M* $LiClO_4$/Pt wr/CV	$E_p = +1.0$
9,10-Dipropylanthracene	MeCN, 0.1 *M* TEAP/Pt/CV	$E_p = +1.08$
1,8-Diphenylanthracene	CH_2Cl_2, 0.2 *M* $TPrACF_3SO_3$/rot Pt wr/E swp	$E_{1/2} = +1.34$
8,9-Diphenylanthracene	CH_2Cl_2, 0.2 *M* $TPrACF_3SO_3$/rot Pt wr/E swp	$E_{1/2} = +1.30$
9,10-Diphenylanthracene	MeCN/Pt/CV	$E_p = +1.22$
Perylene	MeCN, 0.1 *M* TBAP/Pt/CV	$E_p = +1.34$
Pyrene	DMF, 0.1 *M* TBAP/Pt dsk/CV	$E_p = +1.25$
Rubrene	DMF, 0.1 *M* TBAP/Pt dsk/CV	$E_p = +1.10$
Tetracene	CH_2Cl_2, 0.2 *M* $TBABF_4$, –70°C/Pt wr/CV	$E_p = +0.35$[d]
1,4-Dithiabenzene	MeCN, 0.1 *M* TEAP/Pt dsk/rot	$E_{1/2} = +0.69$
1,4-Dithianaphthalene	MeCN, 0.1 *M* TEAP/Pt dsk/rot	$E_{1/2} = +0.80$
Thianthrene	0.1 *M* TPAP/Pt/CV	$E_{1/2} = +1.28$

[a] vs 0.01 *M* $Ag/AgClO_4$
[b] vs. Hg/Hg_2SO_4, 17 *M* H_2SO_4
[c] vs Hg pool
[d] vs Ag/saturated $AgNO_3$
[e] vs Ag/0.01 *M* Ag^+

pH SCALE FOR AQUEOUS SOLUTIONS

A. K. Covington

The pH value is the negative decadic logarithm of the (relative) ion activity of the hydrogen ion in the solution.

$$pH = -\log a_H \tag{1}$$

This is only a notional definition since Equation 1 involves a single ion activity, which is immeasurable, and has to be attained through a nonthermodynamic assumption such as that described in Equation 5 below. In terms of substance concentration, molarity, Equation 1 may be rewritten

$$pH = -\log (c_H y_H / c^o) \tag{2}$$

where c^o is an arbitrary constant representing the standard state condition and equal to 1 mol dm^{-3}, c_H is the concentration of hydrogen ion and y_H is the single ion activity of the hydrogen ion. In terms of molality, Equation 1 may be rewritten

$$pH = -\log (m_H \gamma_H / m^o) \tag{3}$$

where m^o is an arbitrary constant representing the standard state condition and equal to 1 mol kg^{-1}, m_H is the concentration of hydrogen ion and γ_H is the single ion activity of the hydrogen ion. For most purposes the difference between these two scales can be ignored for dilute aqueous solutions; the difference is 0.001 at 25° C and 0.02 at 100° C. Arising from the nonexperimental determinability of single ion activities, the definition and determination of pH have an operational basis, and depend on the assignment of pH values to a standard solution (or solutions) together with the determination of pH difference by a cell with liquid junction called the operational cell.

The Operational Definition of pH Difference[1,2]

The electromotive force, EMF, E(X) of the cell with liquid junction:

Reference electrode | KCl (aq., concentrated) || Solution X | H_2 | Pt (I)

is measured, and likewise that, E(S), of the cell:

Reference electrode | KCl (aq., concentrated) || Solution S | H_2 | Pt (II)

The temperature of both cells (I and II) must be equal and uniform throughout, and the hydrogen gas pressures identical. The two bridge solutions may be any molality of KCl not less than 3.5 mol kg^{-1} provided they are the same.

The pH of the solution X, pH(X), is then related to the assigned pH of the solution S, pH(S) by the definition:

$$pH(X) = pH(S) + [E(S) - E(X)]/[(RT/F) \ln 10] \tag{4}$$

where R is the gas constant, T the thermodynamic temperature, F the Faraday constant. The quantity $k = (RT/F) \ln 10$ is called the slope factor whose values are given as a function of temperature in Table 1. As a consequence of this definition any difference in liquid junction potential between cells I and II is subsumed into the value of pH(X).

The pH Scale

The pH scale at a particular temperature is defined by Equation 4 as a straight line, on the plot of pH against E(X), having a slope of k drawn through the pH value assigned to the Reference Value Standard (RVS) solution (as given in Table 2) and the value of E(S) for cell II when it contains the Reference Value Standard solution. The solution chosen for the RVS is 0.05 mol kg^{-1} aqueous potassium hydrogen phthalate. The procedure by which pH(RVS) values have been assigned to the Reference Value Standard (RVS) is the cell III without transference:[1,3]

Pt(Pd) | H_2 (g, p=1 atm = 101 325 Pa) | RVS, Cl^- | AgCl | Ag (III)

The palladised-platinum hydrogen electrode is used to reduce the catalytised chemical reduction of the phthalate by hydrogen gas. The calculation involves a non-thermodynamic assumption, the Bates-Guggenheim Convention, for the single ion activity of the chloride ion[1,2] as

$$\log (\gamma_{Cl})^o = -A(I/m^o)^{1/2}/[1 + 1.5\,(I/m^o)^{1/2}] \tag{5}$$

where I is the ionic strength = $(1/2)\Sigma m_i z_i^2$ = 0.0534 mol kg^{-1} for the RVS solution and A is a known function of temperature (Table 1).

To prepare the RVS solution, dry the sample at 110° C for 2 h before use. The water should have a conductivity of less than 0.1 mS m^{-1}. The required solution contains 10.211 g kg^{-1} water. It can be prepared on a volume basis by dissolving 10.138 g potassium hydrogen phthalate in water and making up to 1 L at 20° C. This solution is 0.04964 mol/L with a density of 1.00300 g/L at 20° C.

Primary Standards[2]

pH values may be assigned by the cell without transference method (cell III) to six other buffer solutions which meet certain criteria of reproducibility of preparation and properties. These solutions are called primary pH standards (PS), and details and the pH(PS) values assigned to them are given in Table 3. When these PS solutions are used in the operational cell I, the experimental value of the slope will not be in accord with the slope factor values of Table 1, and, moreover, the experimental value could change if additional primary solutions were to be defined. Hence the pH value determined for an unknown solution can be slightly dependent (±0.02) on the choice of primary standard.[2,4,5] Some useful data for standard buffers are given in Table 5.

Operational Standards[2,6]

Operational standards (OS) are also defined which are traceable to the Reference Value Standard (RVS). Values are assigned by means of the operational cells I and II where the liquid junctions are the free diffusion type reproducibly formed in 1 mm vertical capillary tubes. These operational standards are not restricted in number provided certain preparation criteria are met, and pH(OS) values for 16 solutions are given in Table 4.[2] These OS represent an alternative procedure and are in no way to be regarded as inferior to the primary standards. As a consequence of their definition, all pH(OS) values fall on the line with slope given by the slope factor value for the appropriate temperature in Table 1. Any difference in liquid junction potential between the solutions of cells I and II and KCl is subsumed into the assigned value of pH(OS).

Measurement of pH. Choice of Standard Reference Solution

1a. If pH is not required to better than ±0.05 any standard reference solution may be selected.

1b. If pH is required to ±0.002 and interpretation in terms of hydrogen ion concentration or activity is desired, choose a standard reference solution, pH(PS) or pH(OS), to match X as closely as possible in terms of pH, composition and ionic strength.

2. Alternatively, a bracketting procedure may be adopted whereby two standard reference solutions are chosen whose pH values, pH(S1), pH(S2) are on either side of pH(X). Then if the corresponding potential difference measurements are $E(S1)$, $E(S2)$, $E(X)$, then pH(X) is obtained from

$$\mathrm{pH(X)} = \mathrm{pH(S1)} + [E(\mathrm{X}) - E(\mathrm{S1})]/\ \%k$$

where $\%k = 100[E(\mathrm{S2}) - E(\mathrm{S1})]/[\mathrm{pH(S2)} - \mathrm{pH(S1)}]$ is the apparent percentage slope. This procedure is very easily done on some pH meters simply by adjusting downwards the slope factor control with the electrodes in S2. The purpose of the bracketting procedure is to compensate for deficiencies in the electrodes and measuring system.

Information to be Given about the Measurement of pH(X)

The standard solutions selected for calibration of the pH meter system should be reported with the measurement as follows,

1. System calibrated with pH(RVS) = ... at ...K.
2. System calibrated with two primary standards pH(PS1) = ... and pH(PS2) = ... at ... K.
3. System calibrated with two operational standards pH(OS1) =..... and pH(OS2) = ... at .... K.

Interpretation of pH(X) in Terms of Hydrogen Ion Concentration

The operationally defined pH has no simple interpretation in terms of hydrogen ion concentration but the mean ionic activity coefficient of a typical 1:1 electrolyte can be substituted into equation 2 or 3 to obtain hydrogen ion concentration subject to an uncertainty of 3.9% in concentration corresponding to 0.02 in pH.

REFERENCES

1. R. G. Bates, *Measurement of pH. Theory and Practice*, 2nd ed., John Wiley & Sons,, New York, 1973.
2. A. K. Covington, R. G. Bates, and R. A. Durst, *Pure Appl. Chem.*, 57, 531, 1985.
3. H. P. Butikofer and A. K. Covington, *Anal. Chim. Acta*, 108, 179, 1979.
4. R. G. Bates, *Crit. Rev. Anal. Chem.*, 10, 247, 1981.
5. A. K. Covington, *Anal. Chim. Acta*, 127, 1, 1981.
6. A. K. Covington and M. J. Rebelo, *Anal. Chim. Acta.*, 200, 245, 1987.
7. V. E. Bower and R. G. Bates, *J. Res. Natl. Bur. Stand.*, 39, 263, 1954.

TABLE 1
Standard EMF, Slope Factor and Debye-Huckel Constant *A* (Unit Weight of Solvent) as Functions of Temperature

Temperature/°C	$E°$/mV[7]	Slope Factor k/mV	A[1]
0	236.55	54.199	0.4918
5	234.13	55.191	0.4952
10	231.42	56.183	0.4988
15	228.57	57.175	0.5026
20	225.57	58.167	0.5066
25	222.34	59.159	0.5108
30	219.04	60.152	0.5150
35	215.65	61.144	0.5196
40	212.08	62.136	0.5242
45	208.35	63.128	0.5291
50	204.49	64.120	0.5341
55	200.56	65.112	0.5393
60	196.49	66.104	0.5448
70	187.82	68.088	0.5562
80	178.73	70.073	0.5685
90	169.52	72.057	0.5817
95	165.11	73.049	0.5886

TABLE 2
Values of pH(RVS) for the Reference Value Standard of 0.05 mol kg^{-1} Potassium Hydrogen Phthalate at Various Temperatures

t/°C	pH(RVS)	t/°C	pH(RVS)	t/°C	pH(RVS)
0	4.000	35	4.018	65	4.097
5	3.998	37	4.022	70	4.116
10	3.997	40	4.027	75	4.137
15	3.998	45	4.038	80	4.159
20	4.001	50	4.050	85	4.183
25	4.005	55	4.064	90	4.21
30	4.011	60	4.080	95	4.24

TABLE 3
Values of pH(PS) for Primary Standard Reference Solutions

Primary ref. standard	t/°C 0	5	10	15	20	25	30	35	37	40	50	60	70	80	90	95
Saturated (at 25° C) Potassium hydrogen tartrate	—	—	—	—	—	3.557	3.552	3.549	3.548	3.547	3.549	3.560	3.580	3.610	3.650	3.674
0.1 mol/kg Potassium dihydrogen citrate	3.863	3.840	3.820	3.802	3.788	3.776	3.766	3.759	3.756	3.754	3.749	—	—	—	—	—
0.025 mol/kg Disodium hydrogen phosphate +0.025 mol/kg Potassium dihydrogen phosphate	6.984	6.951	6.923	6.900	6.881	6.865	6.853	6.844	6.841	6.838	6.833	6.836	6.845	6.859	6.876	6.886
0.03043 mol/kg Disodium hydrogen phosphate +0.008695 mol/kg Potassium dihydrogen phosphate	7.534	7.500	7.472	7.448	7.429	7.413	7.400	7.389	7.386	7.380	7.367	—	—	—	—	—
0.01 mol/kg Disodium tetraborate	9.464	9.395	9.332	9.276	9.225	9.180	9.139	9.102	9.088	9.068	9.011	8.962	8.921	8.884	8.850	8.833
0.025 mol/kg Sodium hydrogen carbonate +0.025 mol/kg sodium carbonate	10.317	10.245	10.179	10.118	10.062	10.012	9.966	9.926	9.910	9.889	9.828	—	—	—	—	—

Note: Based on an uncertainty of ±0.2 mV in determined (E-E^{θ}), the uncertainty is ±0.003 in pH in the range 0—50° C.

TABLE 4
pH (OS) Values for Operational Reference Solutions

Operational standard ref. solution	t/°C														
	0	5	10	15	20	25	30	37	40	50	60	70	80	90	95
0.1 mol/kg Potassium tetroxalate[a]	—	—	—	—	1.475	1.479	1.483	1.490	1.493	1.503	1.513	1.52	1.53	1.53	1.53
0.05 mol/kg potassium tetroxalate[a]	—	—	1.638	1.642	1.644	1.646	1.648	1.649	1.650	1.653	1.660	1.671	1.689	1.72	1.73
0.05 mol/kg sodium hydrogen diglycolate[b]	—	3.466	3.470	3.476	3.484	3.492	3.502	3.519	3.527	3.558	3.595	—	—	—	—
Saturated (at 25° C) potassium hydrogen tartrate	—	—	—	—	—	3.556	3.549	3.544	3.542	3.544	3.553	3.570	3.596	3.627	3.649
0.05 mol/kg Potassium hydrogen phthalate (RVS)	4.000	3.998	3.997	3.998	4.000	4.005	4.011	4.022	4.027	4.050	4.080	4.115	4.159	4.21	4.24
0.1 mol/dm³ Acetic acid + 0.1 mol/dm³ sodium acetate	4.664	4.657	4.652	4.647	4.645	4.644	4.643	4.647	4.650	4.663	4.684	4.713	4.75	4.80	4.83
0.01 mol/dm³ Acetic acid + 0.1 mol/dm³ sodium acetate	4.729	4.722	4.717	4.714	4.712	4.713	4.715	4.722	4.726	4.743	4.768	4.800	4.839	4.88	4.91
0.02 mol/kg Piperazine phosphate[c]	—	6.477	6.419	6.364	6.310	6.259	6.209	6.143	6.116	6.030	5.952	—	—	—	—
0.025 mol/kg Disodium hydrogen phosphate + 0.025 mol/kg potassium dihydrogen phosphate	6.961	6.935	6.912	6.891	6.873	6.857	6.843	6.828	6.823	6.814	6.817	6.830	6.85	6.90	6.92
0.03043 mol/kg Disodium hydrogen phosphate + 0.008695 mol/kg potassium disodium phosphate	7.506	7.482	7.460	7.441	7.423	7.406	7.390	7.369	—	—	—	—	—	—	—
0.04 mol/kg Disodium hydrogen phosphate + 0.01 mol/kg potassium dihydrogen phosphate	—	7.512	7.488	7.466	7.445	7.428	7.414	7.404	—	—	—	—	—	—	—
0.05 mol/kg Tris hydrochloride + 0.01667 mol/kg Tris[d]	8.399	8.238	8.083	7.933	7.788	7.648	7.513	7.332	7.257	7.018	6.794	—	—	—	—
0.05 mol/kg Disodium tetraborate ($Na_2B_4O_7$)	9.475	9.409	9.347	9.288	9.233	9.182	9.134	9.074	9.051	8.983	8.932	8.898	8.88	8.84	8.89
0.01 mol/kg Disodium tetraborate ($Na_2B_4O_7$)	9.451	9.388	9.329	9.275	9.225	9.179	9.138	9.086	9.066	9.009	8.965	8.932	8.91	8.90	8.89

TABLE 4
pH(OS) Values for Operational Standard Reference Solutions (continued)

Operational standard ref. solution	t/°C 0	5	10	15	20	25	30	37	40	50	60	70	80	90	95
0.025 mol/kg Sodium hydrogen carbonate + 0.025 mol/kg sodium carbonate	10.273	10.212	10.154	10.098	10.045	9.995	9.948	9.889	9.866	9.800	9.753	9.728	9.725	9.75	9.77
Saturated (at 20° C) calcium hydroxide	13.360	13.159	12.965	12.780	12.602	12.431	12.267	12.049	11.959	11.678	11.423	11.192	10.984	10.80	10.71

Note: Uncertainty is ±0.003 in pH between 0 and 60° C rising to ±0.01 above 70° C.

[a] Potassium trihydrogen dioxalate ($KH_3C_4O_8$).
[b] Sodium hydrogen 2,2′-oxydiethanoate.
[c] $C_4H_{10}N_2 \cdot H_3PO_4$.
[d] 2-Amino-2-(hydroxymethyl)-1,3 propanediol or tris(hydroxymethyl)aminomethane.

TABLE 5
Useful Data on Some Standard Buffer Solutions

	Molecular formula	Molality (mol/kg)	Relative molar mass	Density at 20° C (g/cm³)	Molarity at at 20° C (mol/L)	Mass of 1 L at 20° C (g)	Mass tolerance for ±0.001 pH[a] (g)	Mass tolerance expressed as a percentage (%)
Potassium tetraoxalate	$KH_3C_4O_8 \cdot 2H_2O$	0.1	254.1913	1.0091	0.09875	25.1017	0.07	0.27
Potassium tetraoxalate	$KH_3C_4O_8 \cdot 2H_2O$	0.05	254.1913	1.0038	0.04965	12.6202	0.034	0.26
Disodium hydrogen orthophosphate	Na_2HPO_4	0.025	141.9588 }	1.0038	0.02492	3.5379	0.02	0.56
Potassium dihydrogen orthophosphate	KH_2PO_4	0.025	136.0852 }			3.3912	0.02	0.58
Disodium tetraborate	$Na_2B_4O_7 \cdot 10H_2O$	0.05	381.367	1.0075	0.04985	19.0117	0.9	4.73
Disodium tetraborate	$Na_2B_4O_7 \cdot 10H_2O$	0.01	381.367	1.0001	0.009981	3.8064	0.19	0.49
Sodium carbonate	Na_2CO_3	0.025	105.9887 }	1.0021	0.02494	2.6428	0.017	0.064
Sodium hydrogen carbonate	$NaHCO_3$	0.025	84.0069 }			2.0947	0.013	0.62

a Calculated from known dilution value of solution.

PRACTICAL pH MEASUREMENTS ON NATURAL WATERS

A. K. Covington and W. Davison

(1) Dilute solutions and freshwater including 'acid-rain' samples ($I < 0.02$ mol kg^{-1})

Major problems could be encountered due to errors associated with the liquid junction. It is recommended that either a free diffusion junction is used or it is verified that the junction is working correctly using dilute solutions as follows. For commercial electrodes calibrated with IUPAC aqueous RVS or PS standards, the pH(X) of dilute solutions should be within ±0.02 of those given in Table 1. The difference in determined pH(X) between a stirred and unstirred dilute solution should be < 0.02. The characteristics of glass electrodes are such that below pH 5 the readings should be stable within 2 min, but for pH 5 to 8, 8 or so minutes may be necessary to attain stability. Interpretation of pH(X) measured in this way in terms of activity of hydrogen ion, a_{H+} is subject[1] to an uncertainty of ±0.02 in pH.

(2) Seawater

Measurements made by calibration of electrodes with IUPAC aqueous RVS or PS standards to obtain pH(X) are perfectly valid. However, the interpretation of pH(X) in terms of the activity of hydrogen ion is complicated by the non zero residual liquid junction potential as well as by systematic differences between electrode pairs, principally attributable to the reference electrode. For 35‰ salinity seawater ($S = 0.035$) a_{H+} calculated from pH(X) is typically 12% too low. Special seawater pH scales have been devised to overcome this problem:
(i) The total hydrogen ion scale, pH_T, is defined in terms of the sum of free and complexed (total) hydrogen ion concentrations, where

$$^{T}C_H = [H^+] + [HSO_4^-] + [HF].$$

$$\text{So, } pH_T = -\log {}^{T}C_H$$

Calibration of the electrodes with a buffer having a composition similar to that of seawater, to which pH_T has been assigned, results in values of pHT(X) (Tables 2, 3) which are accurately interpretable in terms of $^{T}C_H$.
(ii) The free hydrogen ion scale, pH_F, is defined, and fully interpretable, in terms of the concentration of free hydrogen ions.

$$pH_F = -\log [H^+]$$

Values of pH_F as a function of temperature have been assigned to the same set of pH_T seawater buffers,and so alternatively can be used for calibration (Tables 2, 3) [2,3]

(3) Estuarine water

Prescriptions for seawater scale buffers are available for a range of salinities. Reliable estuarine pH measurements can be made by calibrating with a buffer of the same salinity as the sample. However, these buffers are difficult to prepare and their use presumes prior knowledge of salinity of the sample. Interpretable measurements of estuarine pH can be made by calibration with IUPAC aqueous RVS or PS standards if the electrode pair is additionally calibrated using a 20‰ salinity seawater buffer.[4] The difference between the assigned pH_{SWS} of the seawater buffer and its measured pH(X) value using RVS or PS standards is

$$\Delta pH = pH_{SWS} - pH(X)$$

Values of ΔpH should be in the range of 0.08 to 0.18. It empirically corrects for differences between the two pH scales and for measurement errors associated with the electrode pair. The pH(X) of samples measured using IUPAC aqueous buffers, can be converted to pH_T or pH_F using the appropriate measured ΔpH:

$$pH_T = pH(X) - \Delta pH$$
$$\text{or } pH_F = pH(X) - \Delta pH$$

This simple procedure is appropriate to pH measurement at salinities from 2‰ to 35‰. For salinities lower than 2‰ the procedures for freshwaters should be adopted.

REFERENCES

1. Davison, W. and Harbinson,T. R., *Analyst*, 113, 709, 1988.
2. Culberson, C. H., in *Marine Electrochemistry*, Whitfield, M. and Jagner,D., Eds., Wiley, 1981.
3. Millero, F. J., *Limnol. Oceanogr.*, 31, 839, 1986.
4. Covington, A. K., Whalley, P. D., Davison, W., and Whitfield, M., in *The Determination of Trace Metals in Natural Waters*, West, T. S. and Nurnberg, H. W., Eds., Blackwell, Oxford, 1988.
5. Koch, W. F., Marinenko, G., and Paule, R. C., *J. Res. NBS*, 91, 33, 1986.

Table 1
pH of Dilute Solutions at 25°C, Degassed and Equilibrated with Air, Suitable as Quality Control Standards

	Ionic strength mmol kg^{-1}	Concentration(x) mmol kg^{-1}	pH p_{CO_2} = 0	pH p_{CO_2} = air
Potassium hydrogen phthalate	10.7	10	4.12	4.12
	1.1	1	4.33	4.33
$xKH_2PO_4 + xNa_2HPO_4$	9.9	2.5	7.07	7.05
$xKH_2PO_4 + 3.5xNa_2HPO_4$	10	0.87	7.61	7.58
$Na_2B_4O_7 \cdot 10H_2O$	10	5	9.20	—
HCl	0.1	0.1	4.03	4.03
SRM2694-I[a]	—	—	4.30	—
SRM2694-II[a]	—	—	3.59	—

Note: The pH of solutions near to pH 4 is virtually independent of temperature over the range of 5 to 30°C.

[a] Simulated rainwater samples are available (Reference 5) from NIST containing sulfate, nitrate, chloride, fluoride, sodium, potassium, calcium and magnesium

Table 2
Composition of Seawater Buffer of Salinity *S* = 35‰ at 25°C (Reference 3)

Solute	mol dm^{-3}	mol kg^{-1}	g kg^{-1}	g dm^{-3}
NaCl	0.3666	0.3493	20.416	20.946
Na_2SO_4	0.02926	0.02788	3.96	4.063
KCl	0.01058	0.01008	0.752	0.772
$CaCl_2$	0.01077	0.01026	1.139	1.169
$MgCl_2$	0.05518	0.05258	5.006	5.139
Tris	0.06	0.05717	6.926	7.106
Tris · HCl	0.06	0.05717	9.010	9.244

Tris = tris(hydroxymethyl)aminomethane $(HOCH_2)_3CNH_2$.
A 20‰ buffer is made by diluting the 35‰ in the ratio 20:35.

Table 3
Assigned Values of 20‰ and 35‰ Buffers on Free and Total Hydrogen Ion Scales. Calculated from Equations Provided by Millero (Reference 3)

Temp (°C)	pH_T *S* = 20‰	pH_T *S* = 35‰	pH_F *S* = 20‰	pH_F *S* = 35‰
5	8.683	8.718	8.759	8.81
10	8.513	8.542	8.597	8.647
15	8.351	8.374	8.442	8.491
20	8.195	8.212	8.292	8.341
25	8.045	8.057	8.149	8.197
30	7.901	7.908	8.011	8.059
35	7.762	7.764	7.879	7.926

BUFFER SOLUTIONS GIVING ROUND VALUES OF pH AT 25°C

A

pH	*x*
1.00	67.0
1.10	52.8
1.20	42.5
1.30	33.6
1.40	26.6
1.50	20.7
1.60	16.2
1.70	13.0
1.80	10.2
1.90	8.1
2.00	6.5
2.10	5.10
2.20	3.9

B

pH	*x*
2.20	49.5
2.30	45.8
2.40	42.2
2.50	38.8
2.60	35.4
2.70	32.1
2.80	28.9
2.90	25.7
3.00	22.3
3.10	18.8
3.20	15.7
3.30	12.9
3.40	10.4
3.50	8.2
3.60	6.3
3.70	4.5
3.80	2.9
3.90	1.4
4.00	0.1

C

pH	*x*
4.10	1.3
4.20	3.0
4.30	4.7
4.40	6.6
4.50	8.7
4.60	11.1
4.70	13.6
4.80	16.5
4.90	19.4
5.00	22.6
5.10	25.5
5.20	28.8
5.30	31.6
5.40	34.1
5.50	36.6
5.60	38.8
5.70	40.6
5.80	42.3
5.90	43.7

D

pH	*x*
5.80	3.6
5.90	4.6
6.00	5.6
6.10	6.8
6.20	8.1
6.30	9.7
6.40	11.6
6.50	13.9
6.60	16.4
6.70	19.3
6.80	22.4
6.90	25.9
7.00	29.1
7.10	32.1
7.20	34.7
7.30	37.0
7.40	39.1
7.50	40.9
7.60	42.4
7.70	43.5
7.80	44.5
7.90	45.3
8.00	46.1

E

pH	*x*
7.00	46.6
7.10	45.7
7.20	44.7
7.30	43.4
7.40	42.0
7.50	40.3
7.60	38.5
7.70	36.6
7.80	34.5
7.90	32.0
8.00	29.2
8.10	26.2
8.20	22.9
8.30	19.9
8.40	17.2
8.50	14.7
8.60	12.2
8.70	10.3
8.80	8.5
8.90	7.0
9.00	5.7

F

pH	*x*
8.00	20.5
8.10	19.7
8.20	18.8
8.30	17.7
8.40	16.6
8.50	15.2
8.60	13.5
8.70	11.6
8.80	9.6
8.90	7.1
9.00	4.6
9.10	2.0

G

pH	*x*
9.20	0.9
9.30	3.6
9.40	6.2
9.50	8.8
9.60	11.1
9.70	13.1
9.80	15.0
9.90	16.7
10.00	18.3
10.10	19.5
10.20	20.5
10.30	21.3
10.40	22.1
10.50	22.7
10.60	23.3
10.70	23.8
10.80	24.25

H

pH	*x*
9.60	5.0
9.70	6.2
9.80	7.6
9.90	9.1
10.00	10.7
10.10	12.2
10.20	13.8
10.30	15.2
10.40	16.5
10.50	17.8
10.60	19.1
10.70	20.2
10.80	21.2
10.90	22.0
11.00	22.7

I

pH	*x*
10.90	3.3
11.00	4.1
11.10	5.1
11.20	6.3
11.30	7.6
11.40	9.1
11.50	11.1
11.60	13.5
11.70	16.2
11.80	19.4
11.90	23.0
12.00	26.9

J

pH	*x*
12.00	6.0
12.10	8.0
12.20	10.2
12.30	12.8
12.40	16.2
12.50	20.4
12.60	25.6
12.70	32.2
12.80	41.2
12.90	53.0
13.00	66.0

A. 25 ml of 0.2 molar KCl + *x* ml of 0.2 molar HCl.
B. 50 ml of 0.1 molar potassium hydrogen phthalate + *x* ml of 0.1 molar HCl.
C. 50 ml of 0.1 molar potassium hydrogen phthalate + *x* ml of 0.1 molar NaOH.
D. 50 ml of 0.1 molar potassium dihydrogen phosphate + *x* ml of 0.1 molar NaOH.
E. 50 ml of 0.1 molar tris(hydroxymethyl)aminomethane + *x* ml of 0.1 M HCl.
F. 50 ml of 0.025 molar borax + *x* ml of 0.1 molar HCl.
G. 50 ml of 0.025 molar borax + *x* ml of 0.1 molar NaOH.
H. 50 ml of 0.05 molar sodium bicarbonate + *x* ml of 0.1 molar NaOH.
I. 50 ml of 0.05 molar disodium hydrogen phosphate + *x* ml of 0.1 molar NaOH.
J. 25 ml of 0.2 molar KCl + *x* ml of 0.2 molar NaOH.

Final volume of mixtures = 100 ml.

REFERENCES

1. Bower, V.E., and Bates, R.G., *J. Res. Natl. Bur. Stand.*, 55, 197, 1955 (A–D).
2. Bates, R.G., and Bower, V.E., *Anal. Chem.*, 28, 1322, 1956 (E–J).

DISSOCIATION CONSTANTS OF INORGANIC ACIDS AND BASES

The data in this table are presented as values of pK_a, defined as the negative logarithm of the acid dissociation constant K_a for the reaction

$$BH \rightleftharpoons B^- + H^+$$

Thus pK_a = $-\log K_a$, and the hydrogen ion concentration [H^+] can be calculated from

$$K_a = \frac{[H^+][B^-]}{[BH]}$$

In the case of bases, the entry in the table is for the conjugate acid; e.g., ammonium ion for ammonia. The OH^- concentration in the system

$$NH_3 + H_2O \rightleftharpoons NH_4^+ + OH^-$$

can be calculated from the equation

$$K_b = K_{water} / K_a = \frac{[OH^-][NH_4^+]}{[NH_3]}$$

where $K_{water} = 1.01 \times 10^{-14}$ at 25 °C. Note that pK_a + pK_b = pK_{water}.

All values refer to dilute aqueous solutions at zero ionic strength at the temperature indicated. The table is arranged alphabetically by compound name.

REFERENCE

1. Perrin, D. D., *Ionization Constants of Inorganic Acids and Bases in Aqueous Solution, Second Edition,* Pergamon, Oxford, 1982.

Name	Formula	Step	t/°C	pK_a
Aluminum(III) ion	Al^{+3}		25	5.0
Ammonia	NH_3		25	9.25
Arsenic acid	H_3AsO_4	1	25	2.26
		2	25	6.76
		3	25	11.29
Arsenious acid	H_2AsO_3		25	9.29
Barium(II) ion	Ba^{+2}		25	13.4
Boric acid	H_3BO_3	1	20	9.27
		2	20	>14
Calcium(II) ion	Ca^{+2}		25	12.6
Carbonic acid	H_2CO_3	1	25	6.35
		2	25	10.33
Chlorous acid	$HClO_2$		25	1.94
Chromic acid	H_2CrO_4	1	25	0.74
		2	25	6.49
Cyanic acid	HCNO		25	3.46
Germanic acid	H_2GeO_3	1	25	9.01
		2	25	12.3
Hydrazine	N_2H_4		25	8.1
Hydrazoic acid	HN_3		25	4.6
Hydrocyanic acid	HCN		25	9.21
Hydrofluoric acid	HF		25	3.20
Hydrogen peroxide	H_2O_2		25	11.62
Hydrogen selenide	H_2Se	1	25	3.89
		2	25	11.0
Hydrogen sulfide	H_2S	1	25	7.05
		2	25	19
Hydrogen telluride	H_2Te	1	18	2.6
		2	25	11
Hydroxylamine	NH_2OH		25	5.94
Hypobromous acid	HBrO		25	8.55

Name	Formula	Step	t/°C	pK_a
Hypochlorous acid	$HClO$		25	7.40
Hypoiodous acid	HIO		25	10.5
Iodic acid	HIO_3		25	0.78
Lithium ion	Li^+		25	13.8
Magnesium(II) ion	Mg^{+2}		25	11.4
Nitrous acid	HNO_2		25	3.25
Perchloric acid	$HClO_4$		20	-1.6
Periodic acid	HIO_4		25	1.64
Phosphoric acid	H_3PO_4	1	25	2.16
		2	25	7.21
		3	25	12.32
Phosphorous acid	H_3PO_3	1	20	1.3
		2	20	6.70
Pyrophosphoric acid	$H_4P_2O_7$	1	25	0.91
		2	25	2.10
		3	25	6.70
		4	25	9.32
Selenic acid	H_2SeO_4	2	25	1.7
Selenious acid	H_2SeO_3	1	25	2.62
		2	25	8.32
Silicic acid	H_4SiO_4	1	30	9.9
		2	30	11.8
		3	30	12
		4	30	12
Sodium ion	Na^+		25	14.8
Strontium(II) ion	Sr^{+2}		25	13.2
Sulfamic acid	NH_2SO_3H		25	1.05
Sulfuric acid	H_2SO_4	2	25	1.99
Sulfurous acid	H_2SO_3	1	25	1.85
		2	25	7.2
Telluric acid	H_2TeO_4	1	18	7.68
		2	18	11.0
Tellurous acid	H_2TeO_3	1	25	6.27
		2	25	8.43
Tetrafluoroboric acid	HBF_4		25	0.5
Thiocyanic acid	$HSCN$		25	-1.8
Water	H_2O		25	13.995

DISSOCIATION CONSTANTS OF ORGANIC ACIDS AND BASES

This table lists the dissociation (ionization) constants of over 1070 organic acids, bases, and amphoteric compounds. All data apply to dilute aqueous solutions and are presented as values of pK_a, which is defined as the negative of the logarithm of the equilibrium constant K_a for the reaction

$$HA \rightleftharpoons H^+ + A^-$$

i.e.,

$$K_a = [H^+][A^-]/[HA]$$

where $[H^+]$, etc. represent the concentrations of the respective species in mol/L. It follows that $pK_a = pH + \log[HA] - \log[A^-]$, so that a solution with 50% dissociation has pH equal to the pK_a of the acid.

Data for bases are presented as pK_a values for the conjugate acid, i.e., for the reaction

$$BH^+ \rightleftharpoons H^+ + B$$

In older literature, an ionization constant K_b was used for the reaction $B + H_2O \rightleftharpoons BH^+ + OH^-$. This is related to K_a by

$$pK_a + pK_b = pK_{water} = 14.00 \quad \text{(at 25°C)}$$

Compounds are listed by molecular formula in Hill order.

REFERENCES

1. Perrin, D.D., *Dissociation Constants of Organic Bases in Aqueous Solution*, Butterworths, London, 1965; Supplement, 1972.
2. Serjeant, E.P., and Dempsey, B., *Ionization Constants of Organic Acids in Aqueous Solution*, Pergamon, Oxford, 1979.
3. Albert, A., "Ionization Constants of Heterocyclic Substances", in Katritzky, A.R., Ed., *Physical Methods in Heterocyclic Chemistry*, Academic Press, New York, 1963.
4. Sober, H.A., Ed., *CRC Handbook of Biochemistry*, CRC Press, Boca Raton, FL, 1968.
5. Perrin, D.D., Dempsey, B., and Serjeant, E.P., *pK_a Prediction for Organic Acids and Bases*, Chapman and Hall, London, 1981.
6. Albert, A., and Serjeant, E. P., *The Determination of Ionization Constants, Third Edition*, Chapman and Hall, London, 1984.
7. Budavari, S., Editor, *The Merck Index, Twelth Edition*, Merck & Co., Whitehouse Station, NJ, 1996.

Mol. Form.	Name	Step	t/°C	pK_a
$CHNO$	Cyanic acid		25	3.7
CH_2N_2	Cyanamide		29	1.1
CH_2O	Formaldehyde		25	13.27
CH_2O_2	Formic acid		25	3.75
CH_3NO_2	Nitromethane		25	10.21
CH_3NS_2	Carbamodithioic acid		25	2.95
CH_4N_2O	Urea		25	0.10
CH_4N_2S	Thiourea		25	-1
CH_4O	Methanol		25	15.5
CH_4S	Methanethiol		25	10.33
CH_5N	Methylamine		25	10.66
CH_5NO	*O*-Methylhydroxylamine			12.5
CH_5N_3	Guanidine		25	13.6
C_2HCl_3O	Trichloroacetaldehyde		25	10.04
$C_2HCl_3O_2$	Trichloroacetic acid		20	0.66
$C_2HF_3O_2$	Trifluoroacetic acid		25	0.52
$C_2H_2Cl_2O_2$	Dichloroacetic acid		25	1.35
$C_2H_2O_3$	Glyoxylic acid		25	3.18
$C_2H_2O_4$	Oxalic acid	1	25	1.25
		2	25	3.81
$C_2H_3BrO_2$	Bromoacetic acid		25	2.90
$C_2H_3ClO_2$	Chloroacetic acid		25	2.87
$C_2H_3Cl_3O$	2,2,2-Trichloroethanol		25	12.24
$C_2H_3FO_2$	Fluoroacetic acid		25	2.59
$C_2H_3F_3O$	2,2,2-Trifluoroethanol		25	12.37
$C_2H_3IO_2$	Iodoacetic acid		25	3.18
$C_2H_3NO_4$	Nitroacetic acid		24	1.48
$C_2H_3N_3$	1H-1,2,3-Triazole		20	1.17
$C_2H_3N_3$	1H-1,2,4-Triazole		20	2.27
$C_2H_4N_2$	Aminoacetonitrile		25	5.34
C_2H_4O	Acetaldehyde		25	13.57
C_2H_4OS	Thioacetic acid		25	3.33
$C_2H_4O_2$	Acetic acid		25	4.756
$C_2H_4O_2S$	Thioglycolic acid		25	3.68
$C_2H_4O_3$	Glycolic acid		25	3.83
C_2H_5N	Ethyleneimine		25	8.04
C_2H_5NO	Acetamide		25	15.1
$C_2H_5NO_2$	Acetohydroxamic acid			8.70
$C_2H_5NO_2$	Nitroethane		25	8.46
$C_2H_5NO_2$	Glycine	1	25	2.35
		2	25	9.78
$C_2H_6N_2$	Ethanimidamide		25	12.1
C_2H_6O	Ethanol		25	15.5
C_2H_6OS	2-Mercaptoethanol		25	9.72
$C_2H_6O_2$	Ethyleneglycol		25	15.1
$C_2H_7AsO_2$	Dimethylarsinic acid	1	25	1.57
		2	25	6.27
C_2H_7N	Ethylamine		25	10.65
C_2H_7N	Dimethylamine		25	10.73
C_2H_7NO	Ethanolamine		25	9.50
$C_2H_7NO_3S$	2-Aminoethanesulfonic acid	1	25	1.5
		2	25	9.06
C_2H_7NS	Cysteamine	1	25	8.27
		2	25	10.53
$C_2H_7N_5$	Biguanide	1		11.52
		2		2.93
$C_2H_8N_2$	1,2-Ethanediamine	1	25	9.92
		2	25	6.86

Mol. Form.	Name	Step	t/°C	pK_a
$C_2H_8O_7P_2$	1-Hydroxy-1,1-diphosphonoethane	1		1.35
		2		2.87
		3		7.03
		4		11.3
$C_3H_2O_2$	2-Propynoic acid		25	1.84
C_3H_3NO	Oxazole		33	0.8
C_3H_3NO	Isoxazole		25	-2.0
$C_3H_3NO_2$	Cyanoacetic acid		25	2.47
C_3H_3NS	Thiazole		25	2.52
$C_3H_3N_3O_3$	Cyanuric acid	1		6.88
		2		11.40
		3		13.5
$C_3H_4N_2$	1*H*-Pyrazole		25	2.49
$C_3H_4N_2$	Imidazole		25	6.99
$C_3H_4N_2S$	2-Thiazolamine		20	5.36
C_3H_4O	Propargyl alcohol		25	13.6
$C_3H_4O_2$	Acrylic acid		25	4.25
$C_3H_4O_3$	Pyruvic acid		25	2.39
$C_3H_4O_4$	Malonic acid	1	25	2.85
		2	25	5.70
$C_3H_4O_5$	Hydroxypropanedioic acid	1		2.42
		2		4.54
$C_3H_5BrO_2$	3-Bromopropanoic acid		25	4.00
$C_3H_5ClO_2$	2-Chloropropanoic acid		25	2.83
$C_3H_5ClO_2$	3-Chloropropanoic acid		25	3.98
$C_3H_6N_2$	3-Aminopropanenitrile		20	7.80
$C_3H_6N_6$	1,3,5-Triazine-2,4,6-triamine		25	5.00
C_3H_6O	Allyl alcohol		25	15.5
$C_3H_6O_2$	Propanoic acid		25	4.87
$C_3H_6O_2S$	(Methylthio)acetic acid		25	3.66
$C_3H_6O_3$	Lactic acid		25	3.86
$C_3H_6O_3$	3-Hydroxypropanoic acid		25	4.51
$C_3H_6O_4$	Glyceric acid		25	3.52
C_3H_7N	Allylamine		25	9.49
C_3H_7N	Azetidine		25	11.29
C_3H_7NO	2-Propanone oxime		25	12.42
$C_3H_7NO_2$	*L*-Alanine	1	25	2.34
		2	25	9.87
$C_3H_7NO_2$	β-Alanine	1	25	3.55
		2	25	10.24
$C_3H_7NO_2$	Sarcosine	1	25	2.21
		2	25	10.1
$C_3H_7NO_2S$	*L*-Cysteine	1	25	1.5
		2	25	8.7
		3	25	10.2
$C_3H_7NO_3$	*L*-Serine	1	25	2.19
		2	25	9.21
$C_3H_7NO_5S$	*DL*-Cysteic acid	1	25	1.3
		2	25	1.9
		3	25	8.70
$C_3H_7N_3O_2$	Glycocyamine		25	2.82
$C_3H_8O_2$	Ethylene glycol monomethyl ether		25	14.8
$C_3H_8O_3$	Glycerol		25	14.15
C_3H_9N	Propylamine		25	10.54
C_3H_9N	Isopropylamine		25	10.63
C_3H_9N	Trimethylamine		25	9.80
C_3H_9NO	2-Methoxyethylamine		25	9.40
C_3H_9NO	Trimethylamine oxide		20	4.65
$C_3H_{10}N_2$	1,2-Propanediamine, (±)	1	25	9.82
		2	25	6.61
$C_3H_{10}N_2$	1,3-Propanediamine	1	25	10.55
		2	25	8.88
$C_3H_{10}N_2O$	1,3-Diamino-2-propanol	1	20	9.69
		2	20	7.93
$C_3H_{11}N_3$	1,2,3-Triaminopropane	1	20	9.59
		2	20	7.95
$C_4H_4FN_3O$	Flucytosine			3.26
$C_4H_4N_2$	Pyrazine		20	0.65
$C_4H_4N_2$	Pyrimidine		20	1.23
$C_4H_4N_2$	Pyridazine		20	2.24
$C_4H_4N_2O_2$	Uracil		25	9.45
$C_4H_4N_2O_3$	Barbituric acid		25	4.01
$C_4H_4N_2O_5$	Alloxanic acid		25	6.64
$C_4H_4N_4O_2$	5-Nitropyrimidinamine		20	0.35
$C_4H_4O_2$	2-Butynoic acid		25	2.62
$C_4H_4O_4$	Maleic acid	1	25	1.92
		2	25	6.23
$C_4H_4O_4$	Fumaric acid	1	25	3.02
		2	25	4.38
$C_4H_4O_5$	Oxaloacetic acid	1	25	2.55
		2	25	4.37
		3	25	13.03
C_4H_5N	Pyrrole		25	-3.8
$C_4H_5NO_2$	Succinimide		25	9.62
$C_4H_5N_3$	2-Pyrimidinamine		20	3.45
$C_4H_5N_3$	4-Pyrimidinamine		20	5.71
$C_4H_5N_3O$	Cytosine	1		4.60
		2		12.16
$C_4H_5N_3O_2$	6-Methyl-1,2,4-triazine-3,5(2H,4H)-dione			7.6
$C_4H_6N_2$	1-Methylimidazol		25	6.95
$C_4H_6N_4O_3$	Allantoin		25	8.96
$C_4H_6N_4O_3S_2$	Acetazolamide			7.2
$C_4H_6O_2$	*trans*-Crotonic acid		25	4.69
$C_4H_6O_2$	3-Butenoic acid		25	4.34
$C_4H_6O_2$	Cyclopropanecarboxylic acid		25	4.83
$C_4H_6O_3$	2-Oxobutanoic acid		25	2.50
$C_4H_6O_3$	Acetoacetic acid		25	3.6
$C_4H_6O_4$	Succinic acid	1	25	4.21
		2	25	5.64
$C_4H_6O_4$	Methylmalonic acid	1	25	3.07
		2	25	5.76
$C_4H_6O_5$	Malic acid	1	25	3.40
		2	25	5.11
$C_4H_6O_6$	*DL*-Tartaric acid	1	25	3.03
		2	25	4.37
$C_4H_6O_6$	*meso*-Tartaric acid	1	25	3.17
		2	25	4.91
$C_4H_6O_6$	*L*-Tartaric acid	1	25	2.98
		2	25	4.34
$C_4H_6O_8$	Dihydroxytartaric acid		25	1.92
$C_4H_7ClO_2$	2-Chlorobutanoic acid			2.86
$C_4H_7ClO_2$	3-Chlorobutanoic acid			4.05
$C_4H_7ClO_2$	4-Chlorobutanoic acid			4.52

Mol. Form.	Name	Step	t/°C	pK_a
$C_4H_7NO_2$	4-Cyanobutanoic acid		25	2.42
$C_4H_7NO_3$	*N*-Acetylglycine		25	3.67
$C_4H_7NO_4$	Iminodiacetic acid	1		2.98
		2		9.89
$C_4H_7NO_4$	*L*-Aspartic acid	1	25	1.99
		2	25	3.90
		3	25	9.90
$C_4H_7N_3O$	Creatinine	1	25	4.8
		2		9.2
$C_4H_7N_5$	2,4,6-Pyrimidinetriamine		20	6.84
$C_4H_8N_2O_3$	*L*-Asparagine	1	20	2.1
		2	20	8.80
$C_4H_8N_2O_3$	*N*-Glycylglycine	1	25	3.14
		2		8.17
$C_4H_8O_2$	Butanoic acid		25	4.83
$C_4H_8O_2$	2-Methylpropanoic acid		20	4.84
$C_4H_8O_3$	3-Hydroxybutanoic acid, (±)		25	4.70
$C_4H_8O_3$	4-Hydroxybutanoic acid		25	4.72
$C_4H_8O_3$	Ethoxyacetic acid		18	3.65
C_4H_9N	Pyrrolidine		25	11.31
C_4H_9NO	Morpholine		25	8.50
$C_4H_9NO_2$	2-Methylalanine	1	25	2.36
		2	25	10.21
$C_4H_9NO_2$	*N,N*-Dimethylglycine		25	9.89
$C_4H_9NO_2$	*DL*-2-Aminobutanoic acid	1	25	2.29
		2	25	9.83
$C_4H_9NO_2$	4-Aminobutanoic acid	1	25	4.031
		2	25	10.556
$C_4H_9NO_2S$	*DL*-Homocysteine	1	25	2.22
		2	25	8.87
		3	25	10.86
$C_4H_9NO_3$	*L*-Threonine	1	25	2.09
		2	25	9.10
$C_4H_9NO_3$	*L*-Homoserine	1	25	2.71
		2	25	9.62
$C_4H_9N_3O_2$	Creatine	1	25	2.63
		2	25	14.3
$C_4H_{10}N_2$	Piperazine	1	25	9.73
		2	25	5.33
$C_4H_{10}N_2O_2$	2,4-Diaminobutanoic acid	1	25	1.85
		2	25	8.24
		3	25	10.44
$C_4H_{10}O_4$	1,2,3,4-Butanetetrol			13.9
$C_4H_{11}N$	Butylamine		25	10.60
$C_4H_{11}N$	*sec*-Butylamine		25	10.56
$C_4H_{11}N$	*tert*-Butylamine		25	10.68
$C_4H_{11}N$	Diethylamine		25	10.84
$C_4H_{11}NO_3$	Tris(hydroxymethyl) methylamine		20	8.3
$C_4H_{12}N_2$	1,4-Butanediamine	1	25	10.80
		2	25	9.63
C_5H_4BrN	3-Bromopyridine		25	2.84
C_5H_4ClN	2-Chloropyridine		25	0.49
C_5H_4ClN	3-Chloropyridine		25	2.81
C_5H_4ClN	4-Chloropyridine		25	3.83
C_5H_4FN	2-Fluoropyridine		25	-0.44
$C_5H_4N_2O_2$	4-Nitropyridine		25	1.61
$C_5H_4N_4$	1*H*-Purine	1	20	2.30
		2	20	8.96

Mol. Form.	Name	Step	t/°C	pK_a
$C_5H_4N_4O$	Hypoxanthine		25	8.7
$C_5H_4N_4O$	Allopurinol			10.2
$C_5H_4N_4O_3$	Uric acid		12	3.89
$C_5H_4N_4S$	1,7-Dihydro-6H-	1		7.77
	purine-6-thione	2		11.17
$C_5H_4O_2S$	2-Thiophenecarboxylic acid		25	3.49
$C_5H_4O_2S$	3-Thiophenecarboxylic acid		25	4.1
$C_5H_4O_3$	2-Furancarboxylic acid		25	3.16
$C_5H_4O_3$	3-Furancarboxylic acid		25	3.9
C_5H_5N	Pyridine		25	5.23
C_5H_5NO	2-Pyridinol	1	20	0.75
		2	20	11.65
C_5H_5NO	3-Pyridinol	1	20	4.79
		2	20	8.75
C_5H_5NO	4-Pyridinol	1	20	3.20
		2	20	11.12
C_5H_5NO	2(1H)-Pyridinone	1	20	0.75
		2	20	11.65
C_5H_5NO	Pyridine-1-oxide		24	0.79
$C_5H_5NO_2$	1*H*-Pyrrole-2-carboxylic acid		20	4.45
$C_5H_5NO_2$	1*H*-Pyrrole-3-carboxylic acid		20	5.00
$C_5H_5N_3O$	Pyrazinecarboxamide			0.5
$C_5H_5N_5$	Adenine	1		4.3
		2		9.83
$C_5H_5N_5O$	Guanine		40	9.92
$C_5H_6N_2$	2-Pyridinamine		20	6.82
$C_5H_6N_2$	3-Pyridinamine		25	6.04
$C_5H_6N_2$	4-Pyridinamine		25	9.11
$C_5H_6N_2$	2-Methylpyrazine		27	1.45
$C_5H_6N_2O_2$	Thymine		25	9.94
$C_5H_6O_4$	1,1-Cyclopropanedi-	1	25	1.82
	carboxylic acid	2	25	7.43
$C_5H_6O_4$	*trans*-1-Propene-1,2-	1	25	3.09
	dicarboxylic acid	2	25	4.75
$C_5H_6O_4$	1-Propene-2,3-	1	25	3.85
	dicarboxylic acid	2	25	5.45
$C_5H_6O_5$	2-Oxoglutaric acid	1	25	2.47
		2	25	4.68
$C_5H_7NO_3$	5,5-Dimethyl-2,4-oxazolidinedione		37	6.13
$C_5H_7NO_3$	*L*-Pyroglutamic acid		25	3.32
$C_5H_7N_3$	2,5-Pyridinediamine		20	6.48
$C_5H_7N_3$	Methylaminopyrazine		25	3.39
$C_5H_7N_3O_4$	Azaserine			8.55
$C_5H_8N_2$	2,4-Dimethylimidazole		25	8.36
$C_5H_8N_4O_3S_2$	Methazolamide			7.30
$C_5H_8O_2$	*trans*-3-Pentenoic acid		25	4.51
$C_5H_8O_4$	Dimethylmalonic acid		25	3.15
$C_5H_8O_4$	Glutaric acid	1	18	4.32
		2	25	5.42
$C_5H_8O_4$	Methylsuccinic acid	1	25	4.13
		2	25	5.64
$C_5H_9NO_2$	*L*-Proline	1	25	1.95
		2	25	10.64
$C_5H_9NO_3$	5-Amino-4-oxopentanoic	1	25	4.05
	acid	2	25	8.90
$C_5H_9NO_3$	*trans*-4-Hydroxyproline	1	25	1.82

Mol. Form.	Name	Step	t/°C	pK_a
		2	25	9.66
$C_5H_9NO_4$	*L*-Glutamic acid	1	25	2.13
		2	25	4.31
		3		9.67
$C_5H_9N_3$	Histamine	1	25	6.04
		2	25	9.75
$C_5H_{10}N_2O_3$	Glycylalanine		25	3.15
$C_5H_{10}N_2O_3$	*L*-Glutamine	1	25	2.17
		2	25	9.13
$C_5H_{10}N_2O_4$	Glycylserine	1	25	2.98
		2	25	8.38
$C_5H_{10}O_2$	Pentanoic acid		20	4.83
$C_5H_{10}O_2$	2-Methylbutanoic acid		25	4.80
$C_5H_{10}O_2$	3-Methylbutanoic acid		25	4.77
$C_5H_{10}O_2$	2,2-Dimethylpropanoic acid		20	5.03
$C_5H_{10}O_4$	*D*-2-Deoxyribose		25	12.61
$C_5H_{10}O_5$	*L*-Ribose		25	12.22
$C_5H_{10}O_5$	*D*-Xylose		18	12.14
$C_5H_{11}N$	Piperidine		25	11.123
$C_5H_{11}N$	*N*-Methylpyrrolidine		25	10.46
$C_5H_{11}NO$	4-Methylmorpholine		25	7.38
$C_5H_{11}NO_2$	*L*-Valine	1	25	2.29
		2	25	9.74
$C_5H_{11}NO_2$	*DL*-Norvaline	1		2.36
		2		9.72
$C_5H_{11}NO_2$	*L*-Norvaline	1	25	2.32
		2	25	9.81
$C_5H_{11}NO_2$	*N*-Propylglycine	1	25	2.35
		2	25	10.19
$C_5H_{11}NO_2$	5-Aminopentanoic acid	1	25	4.27
		2	25	10.77
$C_5H_{11}NO_2$	Betaine		0	1.83
$C_5H_{11}NO_2S$	*L*-Methionine	1	25	2.13
		2	25	9.27
$C_5H_{12}N_2O$	Tetramethylurea			2
$C_5H_{12}N_2O_2$	*L*-Ornithine	1	25	1.71
		2	25	8.69
		3	25	10.76
$C_5H_{13}N$	Pentylamine		25	10.63
$C_5H_{13}N$	3-Pentanamine		17	10.59
$C_5H_{13}N$	3-Methyl-1-butanamine		25	10.60
$C_5H_{13}N$	2-Methyl-2-butanamine		19	10.85
$C_5H_{13}N$	2,2-Dimethylpropylamine		25	10.15
$C_5H_{13}N$	Diethylmethylamine		25	10.35
$C_5H_{14}NO$	Choline		25	13.9
$C_5H_{14}N_2$	1,5-Pentanediamine	1	25	10.05
		2	25	10.93
$C_6H_3Cl_3N_2O_2$	4-Amino-3,5,6-trichloro-2-pyridinecarboxlic acid			3.6
$C_6H_3N_3O_7$	2,4,6-Trinitrophenol		24	0.42
$C_6H_4Cl_2O$	2,3-Dichlorophenol		25	7.44
$C_6H_4N_2O_5$	2,4-Dinitrophenol		25	4.07
$C_6H_4N_2O_5$	2,5-Dinitrophenol		15	5.15
$C_6H_4N_4$	Pteridine		20	4.05
C_6H_5BrO	2-Bromophenol		25	8.45
C_6H_5BrO	3-Bromophenol		25	9.03
C_6H_5BrO	4-Bromophenol		25	9.37
$C_6H_5Br_2N$	3,5-Dibromoaniline		25	2.34
C_6H_5ClO	2-Chlorophenol		25	8.56
C_6H_5ClO	3-Chlorophenol		25	9.12
C_6H_5ClO	4-Chlorophenol		25	9.41
$C_6H_5Cl_2N$	2,4-Dichloroaniline		22	2.05
C_6H_5FO	2-Fluorophenol		25	8.73
C_6H_5FO	3-Fluorophenol		25	9.29
C_6H_5FO	4-Fluorophenol		25	9.89
C_6H_5IO	2-Iodophenol		25	8.51
C_6H_5IO	3-Iodophenol		25	9.03
C_6H_5IO	4-Iodophenol		25	9.33
C_6H_5NO	2-Pyridinecarboxaldehyde		25	12.68
C_6H_5NO	4-Pyridinecarboxaldehyde		30	12.05
$C_6H_5NO_2$	Nitrobenzene		0	3.98
$C_6H_5NO_2$	2-Pyridinecarboxylic acid	1	20	0.99
		2	20	5.39
$C_6H_5NO_2$	3-Pyridinecarboxylic acid	1	25	2.00
		2	25	4.82
$C_6H_5NO_2$	4-Pyridinecarboxylic acid	1	25	1.77
		2	25	4.84
$C_6H_5NO_3$	2-Nitrophenol		25	7.23
$C_6H_5NO_3$	3-Nitrophenol		25	8.36
$C_6H_5NO_3$	4-Nitrophenol		25	7.15
$C_6H_5N_3$	1*H*-Benzotriazole		20	1.6
$C_6H_5N_5O$	2-Amino-4-hydroxypteridine	1	20	2.27
		2	20	7.96
$C_6H_5N_5O_2$	Xanthopterin	2	20	6.59
		3	20	9.31
C_6H_6BrN	2-Bromoaniline		25	2.53
C_6H_6BrN	3-Bromoaniline		25	3.53
C_6H_6BrN	4-Bromoaniline		25	3.89
C_6H_6ClN	2-Chloroaniline		25	2.66
C_6H_6ClN	3-Chloroaniline		25	3.52
C_6H_6ClN	4-Chloroaniline		25	3.98
C_6H_6FN	2-Fluoroaniline		25	3.20
C_6H_6FN	3-Fluoroaniline		25	3.59
C_6H_6FN	4-Fluoroaniline		25	4.65
C_6H_6IN	2-Iodoaniline		25	2.54
C_6H_6IN	3-Iodoaniline		25	3.58
C_6H_6IN	4-Iodoaniline		25	3.81
$C_6H_6N_2O$	3-Pyridinecarboxamide		20	3.3
$C_6H_6N_2O$	2-Pyridinecarboxaldehyde oxime	1	20	3.59
		2	20	10.18
$C_6H_6N_2O_2$	2-Nitroaniline		25	-0.25
$C_6H_6N_2O_2$	3-Nitroaniline		25	2.46
$C_6H_6N_2O_2$	4-Nitroaniline		25	1.02
C_6H_6O	Phenol		25	9.99
$C_6H_6O_2$	*p*-Hydroquinone	1	25	9.85
		2	25	11.4
$C_6H_6O_2$	Pyrocatechol	1	25	9.34
		2	25	12.6
$C_6H_6O_2$	Resorcinol	1	25	9.32
		2	25	11.1
$C_6H_6O_2S$	Benzenesulfinic acid		20	1.3
$C_6H_6O_3S$	Benzenesulfonic acid		25	0.70
$C_6H_6O_4$	5-Hydroxy-2-(hydroxymethyl)-4H-pyran-4-one			7.9
$C_6H_6O_4S$	3-Hydroxybenzenesulfonic acid		25	9.07

Mol. Form.	Name	Step	t/°C	pK_a
$C_6H_6O_4S$	4-Hydroxybenzene-sulfonic acid		25	9.11
$C_6H_6O_6$	*cis*-1-Propene-1,2,3-tricarboxylic acid		25	1.95
$C_6H_6O_6$	*trans*-1-Propene-1,2,3-tricarboxylic acid	1	25	2.80
		2	25	4.46
C_6H_6S	Benzenethiol		25	6.62
$C_6H_7BO_2$	Benzeneboronic acid			8.83
C_6H_7N	Aniline		25	4.87
C_6H_7N	2-Methylpyridine		25	6.00
C_6H_7N	3-Methylpyridine		25	5.70
C_6H_7N	4-Methylpyridine		25	5.99
C_6H_7NO	2-Aminophenol	1	20	4.78
		2	20	9.97
C_6H_7NO	3-Aminophenol	1	20	4.37
		2	20	9.82
C_6H_7NO	4-Aminophenol	1	25	5.48
		2	25	10.30
C_6H_7NO	2-Methoxypyridine		20	3.28
C_6H_7NO	3-Methoxypyridine		25	4.78
C_6H_7NO	4-Methoxypyridine		25	6.58
$C_6H_7NO_3S$	2-Aminobenzenesulfonic acid		25	2.46
$C_6H_7NO_3S$	3-Aminobenzenesulfonic acid		25	3.74
$C_6H_7NO_3S$	4-Aminobenzenesulfonic acid		25	3.23
$C_6H_8N_2$	*N*-Methylpyridinamine		20	9.65
$C_6H_8N_2$	*o*-Phenylenediamine	1	20	4.57
		2	20	0.80
$C_6H_8N_2$	*m*-Phenylenediamine	1	20	5.11
		2	20	2.50
$C_6H_8N_2$	*p*-Phenylenediamine	1	20	6.31
		2	20	2.97
$C_6H_8N_2$	Phenylhydrazine		15	8.79
$C_6H_8O_2$	2,4-Hexadienoic acid		25	4.76
$C_6H_8O_2$	1,3-Cyclohexanedione		25	5.26
$C_6H_8O_4$	2,2-Dimethyl-1,3-dioxane-4,6-dione			5.1
$C_6H_8O_6$	*L*-Ascorbic acid	1	25	4.04
		2	16	11.7
$C_6H_8O_7$	Citric acid	1	25	3.13
		2	25	4.76
		3	25	6.40
$C_6H_8O_7$	Isocitric acid	1	25	3.29
		2	25	4.71
		3	25	6.40
$C_6H_9NO_6$	Nitrilotriacetic acid	1	20	3.03
		2	20	3.07
		3	20	10.70
$C_6H_9NO_6$	*L*-γ-Carboxyglutamic acid	1	25	1.7
		2	25	3.2
		3	25	4.75
		4	25	9.9
$C_6H_9N_3$	4,6-Dimethylpyrimidinamine		20	4.82
$C_6H_9N_3O_2$	*L*-Histidine	1	25	1.80
		2	25	6.04
		3	25	9.33

Mol. Form.	Name	Step	t/°C	pK_a
$C_6H_{10}O_2$	Cyclopentanecarboxylic acid		25	4.99
$C_6H_{10}O_3$	Ethyl acetoacetate		25	10.68
$C_6H_{10}O_4$	3-Methylglutaric acid		25	4.24
$C_6H_{10}O_4$	Adipic acid	1	18	4.41
		2	18	5.41
$C_6H_{11}NO_2$	2-Piperidinecarboxylic acid	1	25	2.28
		2	25	10.72
$C_6H_{11}NO_3$	Adipamic acid		25	4.63
$C_6H_{11}NO_4$	2-Aminoadipic acid	1	25	2.14
		2	25	4.21
		3	25	9.77
$C_6H_{11}N_3O_4$	*N*-(*N*-Glycylglycyl)glycine	1	25	3.225
		2	25	8.09
$C_6H_{11}N_3O_4$	Glycylasparagine	1	25	2.942
		2	18	8.44
$C_6H_{12}N_2$	Triethylenediamine	1		3.0
		2		8.7
$C_6H_{12}N_2O_4S_2$	*L*-Cystine	1		1
		2		2.1
		3		8.02
		4		8.71
$C_6H_{12}O_2$	Hexanoic acid		25	4.85
$C_6H_{12}O_2$	4-Methylpentanoic acid		18	4.84
$C_6H_{12}O_6$	β-*D*-Fructose		25	12.27
$C_6H_{12}O_6$	α-*D*-Glucose		25	12.46
$C_6H_{12}O_6$	*D*-Mannose		25	12.08
$C_6H_{13}N$	Cyclohexylamine		25	10.64
$C_6H_{13}N$	1-Methylpiperidine		25	10.38
$C_6H_{13}N$	1,2-Dimethylpyrrolidine		26	10.20
$C_6H_{13}NO$	*N*-Ethylmorpholine		25	7.67
$C_6H_{13}NO_2$	*L*-Leucine	1	25	2.33
		2	25	9.74
$C_6H_{13}NO_2$	*L*-Isoleucine	1	25	2.32
		2	25	9.76
$C_6H_{13}NO_2$	*L*-Norleucine	1	25	2.34
		2	25	9.83
$C_6H_{13}NO_2$	6-Aminohexanoic acid	1	25	4.37
		2	25	10.80
$C_6H_{13}NO_4$	*N,N*-Bis(2-hydroxyethyl)glycine	2	20	8.35
$C_6H_{13}N_3O_3$	Citrulline	1	25	2.43
		2	25	9.69
$C_6H_{14}N_2$	*cis*-1,2-Cyclohexanediamine	1	20	9.93
		2	20	6.13
$C_6H_{14}N_2$	*trans*-1,2-Cyclohexanediamine	1	20	9.94
		2	20	6.47
$C_6H_{14}N_2$	*cis*-2,5-Dimethylpiperazine	1	25	9.66
		2	25	5.20
$C_6H_{14}N_2O_2$	*L*-Lysine	1	25	2.16
		2	25	9.06
		3	25	10.54
$C_6H_{14}N_4O_2$	*L*-Arginine	1	25	1.82
		2	25	8.99
		3	25	12.5
$C_6H_{14}O_6$	*D*-Mannitol		18	13.5
$C_6H_{15}N$	Hexylamine		25	10.56
$C_6H_{15}N$	Diisopropylamine		25	11.05
$C_6H_{15}N$	Triethylamine		25	10.75

Mol. Form.	Name	Step	t/°C	pK_a
$C_6H_{15}NO_3$	Triethanolamine		25	7.76
$C_6H_{16}N_2$	1,6-Hexanediamine	1	0	11.86
		2	0	10.76
$C_6H_{16}N_2$	*N,N,N′,N′*-Tetramethyl-	1	25	10.40
	1,2-ethanediamine	2	25	8.26
$C_6H_{19}NSi_2$	Hexamethyldisilazane			7.55
$C_7HF_5O_2$	Pentafluorobenzoic acid		25	1.75
$C_7H_3Br_2NO$	3,5-Dibromo-4-hydroxybenzonitrile			4.06
$C_7H_3N_3O_8$	2,4,6-Trinitrobenzoic acid		25	0.65
$C_7H_4Cl_3NO_3$	Triclopyr			2.68
$C_7H_4N_2O_6$	2,4-Dinitrobenzoic acid		25	1.43
$C_7H_5BrO_2$	2-Bromobenzoic acid		25	2.85
$C_7H_5BrO_2$	3-Bromobenzoic acid		25	3.81
$C_7H_5BrO_2$	4-Bromobenzoic acid		25	3.96
$C_7H_5ClO_2$	2-Chlorobenzoic acid		25	2.90
$C_7H_5ClO_2$	3-Chlorobenzoic acid		25	3.84
$C_7H_5ClO_2$	4-Chlorobenzoic acid		25	4.00
$C_7H_5FO_2$	2-Fluorobenzoic acid		25	3.27
$C_7H_5FO_2$	3-Fluorobenzoic acid		25	3.86
$C_7H_5FO_2$	4-Fluorobenzoic acid		25	4.15
$C_7H_5F_3O$	2-(Trifluoromethyl)phenol		25	8.95
$C_7H_5F_3O$	3-(Trifluoromethyl)phenol		25	8.68
$C_7H_5IO_2$	2-Iodobenzoic acid		25	2.86
$C_7H_5IO_2$	3-Iodobenzoic acid		25	3.87
$C_7H_5IO_2$	4-Iodobenzoic acid		25	4.00
C_7H_5NO	2-Hydroxybenzonitrile		25	6.86
C_7H_5NO	3-Hydroxybenzonitrile		25	8.61
C_7H_5NO	4-Hydroxybenzonitrile		25	7.97
$C_7H_5NO_3S$	Saccharin		18	11.68
$C_7H_5NO_4$	2-Nitrobenzoic acid		25	2.17
$C_7H_5NO_4$	3-Nitrobenzoic acid		25	3.46
$C_7H_5NO_4$	4-Nitrobenzoic acid		25	3.43
$C_7H_5NO_4$	2,3-Pyridinedicarboxylic	1	25	2.43
	acid	2	25	4.78
$C_7H_5NO_4$	2,4-Pyridinedicarboxylic acid	1	25	2.15
$C_7H_5NO_4$	2,6-Pyridinedicarboxylic	1	25	2.16
	acid	2	25	4.76
$C_7H_5NO_4$	3,5-Pyridinedicarboxylic acid	1	25	2.80
$C_7H_6ClN_3O_4S_2$	Chlorothiazide	1		6.85
		2		9.45
$C_7H_6F_3N$	3-(Trifluoromethyl)aniline		25	3.49
$C_7H_6F_3N$	4-(Trifluoromethyl)aniline		25	2.45
$C_7H_6N_2$	1*H*-Benzimidazole		25	5.53
$C_7H_6N_2$	2-Aminobenzonitrile		25	0.77
$C_7H_6N_2$	3-Aminobenzonitrile		25	2.75
$C_7H_6N_2$	4-Aminobenzonitrile		25	1.74
C_7H_6O	Benzaldehyde		25	14.90
$C_7H_6O_2$	Benzoic acid		25	4.204
$C_7H_6O_2$	Salicylaldehyde		25	8.37
$C_7H_6O_2$	3-Hydroxybenzaldehyde		25	8.98
$C_7H_6O_2$	4-Hydroxybenzaldehyde		25	7.61
$C_7H_6O_3$	2-Hydroxybenzoic acid	1	20	2.98
		2	20	13.6
$C_7H_6O_3$	3-Hydroxybenzoic acid	1	25	4.08
		2	19	9.92
$C_7H_6O_3$	4-Hydroxybenzoic acid	1	25	4.57
		2	25	9.46
$C_7H_6O_4$	2,4-Dihydroxybenzoic acid	1	25	3.11
		2	25	8.55
		3	25	14.0
$C_7H_6O_4$	2,5-Dihydroxybenzoic acid	1	25	2.97
$C_7H_6O_4$	3,4-Dihydroxybenzoic acid	1	25	4.48
		2	25	8.83
		3	25	12.6
$C_7H_6O_4$	3,5-Dihydroxybenzoic acid	1	25	4.04
$C_7H_6O_5$	2,4,6-Trihydroxybenzoic acid		25	1.68
$C_7H_6O_5$	3,4,5-Trihydroxybenzoic acid		25	4.41
C_7H_7NO	Benzamide		25	~13
$C_7H_7NO_2$	Aniline-2-carboxylic acid	1	25	2.17
		2	25	4.85
$C_7H_7NO_2$	Aniline-3-carboxylic acid	1	25	3.07
		2	25	4.79
$C_7H_7NO_2$	Aniline-4-carboxylic acid	1	25	2.50
		2	25	4.87
$C_7H_7NO_3$	4-Amino-2-hydroxybenzoic acid			3.25
$C_7H_8ClN_3O_4S_2$	Hydrochlorothiazide	1		7.9
		2		9.2
$C_7H_8N_4O_2$	Theobromine		18	7.89
$C_7H_8N_4O_2$	Theophylline	1	25	8.77
C_7H_8O	*o*-Cresol		25	10.29
C_7H_8O	*m*-Cresol		25	10.09
C_7H_8O	*p*-Cresol		25	10.26
C_7H_8OS	4-(Methylthio)phenol		25	9.53
$C_7H_8O_2$	2-Methoxyphenol		25	9.98
$C_7H_8O_2$	3-Methoxyphenol		25	9.65
$C_7H_8O_2$	4-Methoxyphenol		25	10.21
C_7H_8S	Benzenemethanethiol		25	9.43
C_7H_9N	Benzylamine		25	9.34
C_7H_9N	2-Methylaniline		25	4.45
C_7H_9N	3-Methylaniline		25	4.71
C_7H_9N	4-Methylaniline		25	5.08
C_7H_9N	*N*-Methylaniline		25	4.85
C_7H_9N	2-Ethylpyridine		25	5.89
C_7H_9N	2,3-Dimethylpyridine		25	6.57
C_7H_9N	2,4-Dimethylpyridine		25	6.99
C_7H_9N	2,5-Dimethylpyridine		25	6.40
C_7H_9N	2,6-Dimethylpyridine		25	6.65
C_7H_9N	3,4-Dimethylpyridine		25	6.46
C_7H_9N	3,5-Dimethylpyridine		25	6.15
C_7H_9NO	2-Methoxyaniline		25	4.53
C_7H_9NO	3-Methoxyaniline		25	4.20
C_7H_9NO	4-Methoxyaniline		25	5.36
C_7H_9NS	2-(Methylthio)aniline		25	3.45
C_7H_9NS	4-(Methylthio)aniline		25	4.35
$C_7H_9N_5$	2-Dimethylaminopurine	1	20	4.00
		2	20	10.24
$C_7H_{11}N_3O_2$	*L*-1-Methylhistidine	1	25	1.69
		2	25	6.48
		3	25	8.85
$C_7H_{11}N_3O_2$	*L*-3-Methylhistidine	1	25	1.92
		2	25	6.56
		3	25	8.73

DISSOCIATION CONSTANTS OF ORGANIC ACIDS AND BASES (continued)

Mol. Form.	Name	Step	t/°C	pK_a
$C_7H_{12}O_2$	Cyclohexanecarboxylic acid		25	4.91
$C_7H_{12}O_4$	Heptanedioic acid	1	25	4.71
		2	25	5.58
$C_7H_{12}O_4$	Butylpropanedioic acid	1	5	2.96
$C_7H_{13}NO_4$	α-Ethylglutamic acid	1	25	3.846
		2	25	7.838
$C_7H_{14}O_2$	Heptanoic acid		25	4.89
$C_7H_{14}O_6$	α-Methylglucoside		25	13.71
$C_7H_{15}N$	1-Ethylpiperidine		23	10.45
$C_7H_{15}N$	1,2-Dimethylpiperidine,(±)		25	10.22
$C_7H_{15}NO_3$	Carnitine		25	3.80
$C_7H_{17}N$	Heptylamine		25	10.67
$C_7H_{17}N$	2-Heptanamine		19	10.7
$C_8H_5NO_2$	3-Cyanobenzoic acid		25	3.60
$C_8H_5NO_2$	4-Cyanobenzoic acid		25	3.55
$C_8H_6N_2$	Cinnoline		20	2.37
$C_8H_6N_2$	Quinazoline		29	3.43
$C_8H_6N_2$	Quinoxaline		20	0.56
$C_8H_6N_2$	Phthalazine		20	3.47
$C_8H_6N_4O_5$	Nitrofurantoin			7.2
$C_8H_6O_3$	3-Formylbenzoic acid		25	3.84
$C_8H_6O_3$	4-Formylbenzoic acid		25	3.77
$C_8H_6O_4$	Phthalic acid	1	25	2.943
		2	25	5.432
$C_8H_6O_4$	Isophthalic acid	1	25	3.70
		2	25	4.60
$C_8H_6O_4$	Terephthalic acid	1	25	3.54
		2	25	4.34
$C_8H_7ClO_2$	2-Chlorobenzeneacetic acid		25	4.07
$C_8H_7ClO_2$	3-Chlorobenzeneacetic acid		25	4.14
$C_8H_7ClO_2$	4-Chlorobenzeneacetic acid		25	4.19
$C_8H_7ClO_3$	2-Chlorophenoxyacetic acid		25	3.05
$C_8H_7ClO_3$	3-Chlorophenoxyacetic acid		25	3.10
$C_8H_7NO_4$	2-Nitrobenzeneacetic acid		25	4.00
$C_8H_7NO_4$	3-Nitrobenzeneacetic acid		25	3.97
$C_8H_7NO_4$	4-Nitrobenzeneacetic acid		25	3.85
$C_8H_8F_3N_3O_4S_2$	Hydroflumethiazide	1		8.9
		2		9.7
$C_8H_8N_2$	2-Methyl-1*H*-benzimidazole		25	6.19
$C_8H_8O_2$	*o*-Toluic acid		25	3.91
$C_8H_8O_2$	*m*-Toluic acid		25	4.25
$C_8H_8O_2$	*p*-Toluic acid		25	4.37
$C_8H_8O_2$	Benzeneacetic acid		25	4.31
$C_8H_8O_2$	1-(2-Hydroxyphenyl)ethanone		25	10.06
$C_8H_8O_2$	1-(3-Hydroxyphenyl)ethanone		25	9.19
$C_8H_8O_2$	1-(4-Hydroxyphenyl)ethanone		25	8.05
$C_8H_8O_3$	2-Methoxybenzoic acid		25	4.08
$C_8H_8O_3$	3-Methoxybenzoic acid		25	4.10
$C_8H_8O_3$	4-Methoxybenzoic acid		25	4.50
$C_8H_8O_3$	Phenoxyacetic acid		25	3.17
$C_8H_8O_3$	Mandelic acid		25	3.37
$C_8H_8O_4$	2,5-Hydroxybenzeneacetic acid		25	4.40
C_8H_9NO	Acetanilide		25	0.5
$C_8H_9NO_2$	2-(Methylamino)benzoic acid		25	5.34
$C_8H_9NO_2$	3-(Methylamino)benzoic acid		25	5.10
$C_8H_9NO_2$	4-(Methylamino)benzoic acid		25	5.04
$C_8H_9NO_2$	*N*-Phenylglycine	1	25	1.83
		2		4.39
$C_8H_{10}BrN$	4-Bromo-*N,N*-dimethylaniline		25	4.23
$C_8H_{10}ClN$	3-Chloro-*N,N*-dimethylaniline		20	3.83
$C_8H_{10}ClN$	4-Chloro-*N,N*-dimethylaniline		20	4.39
$C_8H_{10}N_2O_2$	*N,N*-Dimethyl-3-nitroaniline		25	2.62
$C_8H_{11}N$	*N*-Ethylaniline		25	5.12
$C_8H_{11}N$	*N,N*-Dimethylaniline		25	5.07
$C_8H_{11}N$	2,6-Dimethylaniline		25	3.89
$C_8H_{11}N$	Benzeneethanamine		25	9.83
$C_8H_{11}N$	2,4,6-Trimethylpyridine		25	7.43
$C_8H_{11}NO$	2-Ethoxyaniline		28	4.43
$C_8H_{11}NO$	3-Ethoxyaniline		25	4.18
$C_8H_{11}NO$	4-Ethoxyaniline		28	5.20
$C_8H_{11}NO$	4-(2-Aminoethyl)phenol	1	25	9.74
		2	25	10.52
$C_8H_{11}NO$	2-(2-Methoxyethyl)pyridine			5.5
$C_8H_{11}NO_2$	Dopamine	1	25	8.9
		2	25	10.6
$C_8H_{11}NO_3$	Norepinephrine	1	25	8.64
		2	25	9.70
$C_8H_{11}N_3O_6$	6-Azauridine			6.70
$C_8H_{11}N_5$	Phenylbiguanide	1		10.76
		2		2.13
$C_8H_{12}N_2O_3$	Barbital		25	7.43
$C_8H_{12}O_2$	5,5-Dimethyl-1,3-cyclohexanedione		25	5.15
$C_8H_{13}NO_2$	Arecoline			6.84
$C_8H_{14}O_2S_2$	Thioctic acid			5.4
$C_8H_{14}O_4$	Octanedioic acid	1	25	4.52
$C_8H_{15}NO$	Tropine		15	3.80
$C_8H_{15}NO$	Pseudotropine		15	3.80
$C_8H_{16}N_2O_3$	*N*-Glycylleucine		25	3.18
$C_8H_{16}N_2O_3$	*N*-Leucylglycine	1	25	3.25
		2	25	8.2
$C_8H_{16}N_2O_4S_2$	Homocystine	1	25	1.59
		2	25	2.54
		3	25	8.52
		4	25	9.44
$C_8H_{16}O_2$	Octanoic acid		25	4.89
$C_8H_{16}O_2$	2-Propylpentanoic acid			4.6
$C_8H_{17}N$	2-Propylpiperidine,(*S*)			10.9
$C_8H_{17}N$	2,2,4-Trimethylpiperidine		30	11.04
$C_8H_{17}NO$	*trans*-6-Propyl-3-piperidinol,(3*S*)			10.3
$C_8H_{19}N$	Octylamine		25	10.65
$C_8H_{19}N$	*N*-Methyl-2-heptanamine		17	10.99
$C_8H_{19}N$	Dibutylamine		21	11.25
$C_8H_{20}N_2$	1,8-Octanediamine	1	20	11.00
		2	20	10.1
C_9H_6BrN	3-Bromoquinoline		25	2.69
$C_9H_7ClO_2$	*trans*-*o*-Chlorocinnamic acid		25	4.23

Mol. Form.	Name	Step	t/°C	pK_a
$C_9H_7ClO_2$	*trans-m*-Chlorocinnamic acid		25	4.29
$C_9H_7ClO_2$	*trans-p*-Chlorocinnamic acid		25	4.41
C_9H_7N	Quinoline		20	4.90
C_9H_7N	Isoquinoline		20	5.40
C_9H_7NO	2-Quinolinol	1	20	-0.31
		2	20	11.76
C_9H_7NO	3-Quinolinol	1	20	4.28
		2	20	8.08
C_9H_7NO	4-Quinolinol	1	20	2.23
		2	20	11.28
C_9H_7NO	6-Quinolinol	1	20	5.15
		2	20	8.90
C_9H_7NO	8-Quinolinol	1	25	4.91
		2	25	9.81
C_9H_7NO	7-Isoquinolinol	1	20	5.68
		2	20	8.90
$C_9H_7NO_3$	2-Cyanophenoxyacetic acid		25	2.98
$C_9H_7NO_3$	3-Cyanophenoxyacetic acid		25	3.03
$C_9H_7NO_3$	4-Cyanophenoxyacetic acid		25	2.93
$C_9H_7N_7O_2S$	Azathioprine			8.2
$C_9H_8N_2$	2-Quinolinamine		20	7.34
$C_9H_8N_2$	3-Quinolinamine		20	4.91
$C_9H_8N_2$	4-Quinolinamine		20	9.17
$C_9H_8N_2$	1-Isoquinolinamine		20	7.62
$C_9H_8N_2$	3-Isoquinolinamine		20	5.05
$C_9H_8O_2$	*cis*-Cinnamic acid		25	3.88
$C_9H_8O_2$	*trans*-Cinnamic acid		25	4.44
$C_9H_8O_2$	α-Methylenebenezene-acetic acid			4.35
$C_9H_8O_4$	2-(Acetyloxy)benzoic acid		25	3.48
$C_9H_9Br_2NO_3$	3,5-Dibromo-*L*-tyrosine	1		2.17
		2		6.45
		3		7.60
$C_9H_9ClO_2$	3-(2-Chlorophenyl)-propanoic acid		25	4.58
$C_9H_9ClO_2$	3-(3-Chlorophenyl)-propanoic acid		25	4.59
$C_9H_9ClO_2$	3-(4-Chlorophenyl)-propanoic acid		25	4.61
$C_9H_9I_2NO_3$	*L*-3,5-Diiodotyrosine	1	25	2.12
		2	25	5.32
		3	25	9.48
$C_9H_9NO_3$	*N*-Benzoylglycine		25	3.62
$C_9H_9NO_4$	3-(2-Nitrophenyl)-propanoic acid		25	4.50
$C_9H_9NO_4$	3-(4-Nitrophenyl)-propanoic acid		25	4.47
$C_9H_9N_3O_2$	Carbendazim			4.48
$C_9H_9N_3O_2S_2$	Sulfathiazole			7.2
$C_9H_{10}INO_3$	*L*-3-Iodotyrosine	1	25	2.2
		2	25	8.7
		3	25	9.1
$C_9H_{10}N_2$	2-Ethylbenzimidazole		25	6.18
$C_9H_{10}O_2$	3,5-Dimethylbenzoic acid		25	4.32
$C_9H_{10}O_2$	Benzenepropanoic acid		25	4.66
$C_9H_{10}O_2$	α-Methylbenzeneacetic acid		25	4.64
$C_9H_{10}O_3$	α-Hydroxy-α-methyl-benezeneacetic acid		25	3.47
$C_9H_{11}Cl_2N_3O_4S_2$	Methylclothiazide			9.4
$C_9H_{11}N$	*N*-Allylaniline		25	4.17
$C_9H_{11}N$	1-Indanamine		22	9.21
$C_9H_{11}NO_2$	4-(Dimethylamino)-benzoic acid	1		6.03
		2		11.49
$C_9H_{11}NO_2$	Ethyl 4-aminobenzoate			2.5
$C_9H_{11}NO_2$	*L*-Phenylalanine	1	25	2.20
		2	25	9.31
$C_9H_{11}NO_3$	*L*-Tyrosine	1	25	2.20
		2	25	9.11
		3	25	10.1
$C_9H_{11}NO_4$	Levodopa	1	25	2.32
		2	25	8.72
		3	25	9.96
		4	25	11.79
$C_9H_{12}N_2O_2$	Tyrosineamide		25	7.33
$C_9H_{13}N$	*N*-Isopropylaniline		25	5.77
$C_9H_{13}NO_3$	Epinephrine	1	25	8.66
		2	25	9.95
$C_9H_{13}N_2O_9P$	5'-Uridylic acid	1		6.4
		2		9.5
$C_9H_{13}N_3O_5$	Cytidine	1		4.22
		2		12.5
$C_9H_{14}ClNO$	Phenylpropanolamine hydrochloride			9.44
$C_9H_{14}N_2O_3$	Metharbital			8.45
$C_9H_{14}N_3O_8P$	3'-Cytidylic acid	1		0.8
		2		4.28
		3		6.0
$C_9H_{14}N_4O_3$	Carnosine	1	20	2.73
		2	20	6.87
		3	20	9.73
$C_9H_{15}NO_3S$	Captopril	1		3.7
		2		9.8
$C_9H_{15}N_5O$	Minoxidil			4.61
$C_9H_{16}O_4$	Nonanedioic acid	1	25	4.53
		2	25	5.33
$C_9H_{18}O_2$	Nonanoic acid		25	4.96
$C_9H_{19}N$	*N*-Butylpiperidine		23	10.47
$C_9H_{19}N$	2,2,6,6-Tetramethyl-piperidine		25	11.07
$C_9H_{21}N$	Nonylamine		25	10.64
$C_{10}H_7NO_2$	8-Quinolinecarboxylic acid		25	1.82
$C_{10}H_8O$	1-Naphthol		25	9.39
$C_{10}H_8O$	2-Naphthol		25	9.63
$C_{10}H_9N$	1-Naphthylamine		25	3.92
$C_{10}H_9N$	2-Naphthylamine		25	4.16
$C_{10}H_9N$	2-Methylquinoline		20	5.83
$C_{10}H_9N$	4-Methylquinoline		20	5.67
$C_{10}H_9N$	5-Methylquinoline		20	5.20
$C_{10}H_9NO$	5-Amino-1-naphthol		25	3.97
$C_{10}H_9NO$	6-Methoxyquinoline		20	5.03
$C_{10}H_9NO_2$	1H-Indole-3-acetic acid			4.75
$C_{10}H_{10}O_2$	*o*-Methylcinnamic acid		25	4.50
$C_{10}H_{10}O_2$	*m*-Methylcinnamic acid		25	4.44
$C_{10}H_{10}O_2$	*p*-Methylcinnamic acid		25	4.56
$C_{10}H_{12}N_2$	Tryptamine		25	10.2
$C_{10}H_{12}N_2O$	5-Hydroxytryptamine	1	25	9.8
		2	25	11.1

Mol. Form.	Name	Step	t/°C	pK_a
$C_{10}H_{12}N_2O_5$	Dinoseb			4.62
$C_{10}H_{12}N_4O_3$	Dideoxyinosine			9.12
$C_{10}H_{12}O$	5,6,7,8-Tetrahydro-2-naphthalenol		25	10.48
$C_{10}H_{12}O_2$	Benzenebutanoic acid		25	4.76
$C_{10}H_{12}O_5$	Propyl 3,4,5-trihydroxybenzoate			8.11
$C_{10}H_{13}N_5O_4$	Adenosine	1	25	3.6
		2	25	12.4
$C_{10}H_{14}N_2$	*L*-Nicotine	1		8.02
		2		3.12
$C_{10}H_{14}N_5O_7P$	5'-Adenylic acid	1		3.8
		2		6.2
$C_{10}H_{14}O$	2-*tert*-Butylphenol		25	10.62
$C_{10}H_{14}O$	3-*tert*-Butylphenol		25	10.12
$C_{10}H_{14}O$	4-*tert*-Butylphenol		25	10.23
$C_{10}H_{15}N$	*N*-*tert*-Butylaniline		25	7.00
$C_{10}H_{15}N$	*N,N*-Diethylaniline		25	6.57
$C_{10}H_{15}NO$	*d*-Ephedrine		10	10.139
$C_{10}H_{15}NO$	*l*-Ephedrine		10	9.958
$C_{10}H_{17}N_3O_6S$	*l*-Glutathione	1	25	2.12
		2	25	3.59
		3	25	8.75
		4	25	9.65
$C_{10}H_{18}N_4O_5$	*L*-Argininosuccinic acid	1	25	1.62
		2	25	2.70
		3	25	4.26
		4	25	9.58
$C_{10}H_{18}O_4$	Sebacic acid	1		4.59
		2		5.59
$C_{10}H_{19}N$	Bornylamine		25	10.17
$C_{10}H_{19}N$	Neobornylamine		25	10.01
$C_{10}H_{21}N$	Butylcyclohexylamine		25	11.23
$C_{10}H_{21}N$	1,2,2,6,6-Pentamethylpiperidine		30	11.25
$C_{10}H_{23}N$	Decylamine		25	10.64
$C_{11}H_8N_2$	1*H*-Perimidine		20	6.35
$C_{11}H_8O_2$	1-Naphthalenecarboxylic acid		25	3.69
$C_{11}H_8O_2$	2-Naphthalenecarboxylic acid		25	4.16
$C_{11}H_{11}N$	Methyl-1-naphthylamine		27	3.67
$C_{11}H_{12}I_3NO_2$	Iopanoic acid			4.8
$C_{11}H_{12}N_2O_2$	*L*-Tryptophan	1	25	2.46
		2	25	9.41
$C_{11}H_{12}N_4O_3S$	Sulfamethoxypyridazine			6.7
$C_{11}H_{13}F_3N_2O_3S$	Mefluidide			4.6
$C_{11}H_{13}NO_3$	Hydrastinine			11.38
$C_{11}H_{13}N_3O_3S$	Sulfisoxazole			5
$C_{11}H_{14}N_2O$	Cytisine	1		6.11
		2		13.08
$C_{11}H_{14}O_2$	2-*tert*-Butylbenzoic acid		25	3.54
$C_{11}H_{14}O_2$	3-*tert*-Butylbenzoic acid		25	4.20
$C_{11}H_{14}O_2$	4-*tert*-Butylbenzoic acid		25	4.38
$C_{11}H_{16}N_2O_2$	Pilocarpine	1	25	1.6
		2	25	6.9
$C_{11}H_{16}N_4O_4$	Pentostatin			5.2
$C_{11}H_{17}N$	*N,N*-Diethyl-2-methylaniline		25	7.24
$C_{11}H_{17}NO_3$	Isoproterenol			8.64
$C_{11}H_{17}N_3O_8$	Tetrodotoxin			8.76
$C_{11}H_{18}ClNO_3$	Methoxamine hydrochloride		25	9.2
$C_{11}H_{18}N_2O_3$	Amobarbital		25	8.0
$C_{11}H_{25}N$	Undecylamine		25	10.63
$C_{11}H_{26}NO_2PS$	Methylphosphonothioic acid *S*[2-[bis(1-isopropyl)amino]-ethyl],*O*-ethylester			7.9
$C_{12}H_6Cl_4O_2S$	Bithionol	1		4.82
		2		10.50
$C_{12}H_8N_2$	1,10-Phenanthroline		25	4.84
$C_{12}H_8N_2$	Phenazine		20	1.20
$C_{12}H_{10}O$	2-Hydroxybiphenyl		25	10.01
$C_{12}H_{10}O$	3-Hydroxybiphenyl		25	9.64
$C_{12}H_{10}O$	4-Hydroxybiphenyl		25	9.55
$C_{12}H_{11}N$	Diphenylamine		25	0.79
$C_{12}H_{11}N$	2-Aminobiphenyl		25	3.83
$C_{12}H_{11}N$	3-Aminobiphenyl		18	4.25
$C_{12}H_{11}N$	4-Aminobiphenyl		18	4.35
$C_{12}H_{11}N$	2-Benzylpyridine		25	5.13
$C_{12}H_{11}N_3$	4-Aminoazobenzene		25	2.82
$C_{12}H_{12}N_2$	*p*-Benzidine	1	20	4.65
		2	20	3.43
$C_{12}H_{12}N_2O_3$	Phenobarbital	1		7.3
		2		11.8
$C_{12}H_{13}I_3N_2O_3$	Iocetamic acid			4
$C_{12}H_{13}N$	*N,N*-Dimethyl-1-naphthylamine		25	4.83
$C_{12}H_{13}N$	*N,N*-Dimethyl-2-naphthylamine		25	4.566
$C_{12}H_{14}N_4O_2S$	Sulfamethazine	1		7.4
		2		2.65
$C_{12}H_{14}N_4O_3S$	Sulfacytine			6.9
$C_{12}H_{17}N_3O_4$	Agaritine	1		3.4
		2		8.86
$C_{12}H_{20}N_2O_2$	Aspergillic acid			5.5
$C_{12}H_{21}N_5O_2S_2$	Nizatidine	1		2.1
		2		6.8
$C_{12}H_{22}O_{11}$	Sucrose		25	12.7
$C_{12}H_{22}O_{11}$	α-Maltose		21	12.05
$C_{12}H_{23}N$	Dicyclohexylamine			10.4
$C_{12}H_{27}N$	Dodecylamine		25	10.63
$C_{13}H_9N$	Acridine		20	5.58
$C_{13}H_9N$	Phenanthridine		20	5.58
$C_{13}H_{10}N_2$	9-Acridinamine		20	9.99
$C_{13}H_{10}N_2$	2-Phenylbenzimidazole	1	25	5.23
		2	25	11.91
$C_{13}H_{10}O_2$	2-Phenylbenzoic acid		25	3.46
$C_{13}H_{10}O_3$	2-Phenoxybenzoic acid		25	3.53
$C_{13}H_{10}O_3$	3-Phenoxybenzoic acid		25	3.95
$C_{13}H_{10}O_3$	4-Phenoxybenzoic acid		25	4.57
$C_{13}H_{11}N_3$	3,6-Acridinediamine		20	9.65
$C_{13}H_{12}Cl_2O_4$	Ethacrynic acid			3.50
$C_{13}H_{12}N_2O$	Harmine			7.70
$C_{13}H_{12}N_2O_3S$	Sulfabenzamide		25	4.57
$C_{13}H_{13}N$	4-Benzylaniline		25	2.17
$C_{13}H_{14}N_2O_{13}$	Harmaline			4.2
$C_{13}H_{15}N_3O_3$	Imazapyr	1		1.9
		2		3.6

Mol. Form.	Name	Step	t/°C	pK_a
$C_{13}H_{16}ClNO$	Ketamine			7.5
$C_{13}H_{19}NO_4S$	4-[(Dipropylamino)-sulfonyl]benzoic acid			5.8
$C_{13}H_{21}N$	2,6-Di-*tert*-butylpyridine			3.58
$C_{13}H_{29}N$	(Tridecyl)amine		25	10.63
$C_{14}H_{12}F_3NO_4S_2$	Perfluidone			2.5
$C_{14}H_{12}O_2$	α-Phenylbenzeneacetic acid		25	3.94
$C_{14}H_{12}O_3$	α-Hydroxy-α-phenyl-benezeneacetic acid		25	3.04
$C_{14}H_{18}N_4O_3$	Trimethoprim			6.6
$C_{14}H_{19}NO_2$	Methylphenidate			8.9
$C_{14}H_{21}N_3O_3S$	Tolazamide		25	3.6
$C_{14}H_{22}N_2O_3$	Atenolol			9.6
$C_{14}H_{31}N$	Tetradecylamine		25	10.62
$C_{15}H_{10}ClN_3O_3$	Clonazepam	1		1.5
		2		10.5
$C_{15}H_{11}I_4NO_4$	*L*-Thyroxine	1	25	2.2
		2	25	6.45
		3	25	10.1
$C_{15}H_{14}O_3$	Fenoprofen			7.3
$C_{15}H_{15}NO_2$	Mefenamic acid			4.2
$C_{15}H_{15}N_3O_2$	Methyl Red	1		2.5
		2		9.5
$C_{15}H_{17}ClN_4$	NeutralRed			6.7
$C_{15}H_{19}NO_2$	Tropacocaine		15	4.32
$C_{15}H_{19}N_3O_3$	Imazethapyr	1		2.1
		2		3.9
$C_{15}H_{21}N_3O_2$	Physostigmine	1		6.12
		2		12.24
$C_{15}H_{26}N_2$	Sparteine	1	20	2.24
		2	20	9.46
$C_{15}H_{33}N$	Pentadecylamine		25	10.61
$C_{16}H_{13}ClN_2O$	Valium			3.4
$C_{16}H_{14}ClN_3O$	Chlorodiazepoxide			4.8
$C_{16}H_{16}N_2O_2$	Lysergic acid	1		3.44
		2		7.68
$C_{16}H_{17}N_3O_4S$	Cephalexin	1		5.2
		2		7.3
$C_{16}H_{19}N_3O_4S$	Cephradine	1		2.63
		2		7.27
$C_{16}H_{22}N_2$	Lycodine	1		3.97
		2		8.08
$C_{16}H_{35}N$	Hexadecylamine		25	10.61
$C_{17}H_{17}NO_2$	Apomorphine	1		7.0
		2		8.92
$C_{17}H_{19}NO_3$	Piperine		18	12.22
$C_{17}H_{19}NO_3$	Morphine	1	25	8.21
		2	20	9.85
$C_{17}H_{20}N_4O_6$	Riboflavin	1		1.7
		2	25	9.69
$C_{17}H_{20}O_6$	Mycophenolic acid			4.5
$C_{17}H_{23}NO_3$	Hyoscyamine		21	9.7
$C_{17}H_{27}NO_4$	Nadolol			9.67
$C_{18}H_{19}ClN_4$	Clozapine	1		3.70
		2		7.60
$C_{18}H_{21}NO_3$	Codeine			8.21
$C_{18}H_{21}N_3O$	Dibenzepin			8.25
$C_{18}H_{32}O_2$	Linoleic acid			7.6
$C_{18}H_{33}ClN_2O_5S$	Clindamycin			7.6
$C_{18}H_{39}N$	Octadecylamine		25	10.60
$C_{19}H_{10}Br_4O_5S$	Bromophenol Blue			4.0
$C_{19}H_{14}O_5S$	Phenol Red			7.9
$C_{19}H_{16}ClNO_4$	Indomethacin			4.5
$C_{19}H_{17}N_3O_4S_2$	Cephaloridine			3.2
$C_{19}H_{20}N_2O_2$	Phenylbutazone			4.5
$C_{19}H_{21}N$	Protriptyline			8.2
$C_{19}H_{21}NO_3$	Thebaine		15	6.05
$C_{19}H_{22}N_2O$	Cinchonine	1		5.85
		2		9.92
$C_{19}H_{22}N_2O$	Cinchonidine	1		5.80
		2		10.03
$C_{19}H_{22}N_2O_2$	Cupreine			6.57
$C_{19}H_{22}O_6$	Gibberellic acid			4.0
$C_{19}H_{23}N_3O_2$	Ergometrinine			7.3
$C_{19}H_{23}N_3O_2$	Ergonovine			6.8
$C_{20}H_{14}O_4$	Phenolphthalein		25	9.7
$C_{20}H_{21}NO_4$	Papaverine			6.4
$C_{20}H_{23}N$	Amitriptyline			9.4
$C_{20}H_{23}N_7O_7$	Folinic acid	1		3.1
		2		4.8
		3		10.4
$C_{20}H_{24}N_2O_2$	Quinine	1	25	8.52
		2	25	4.13
$C_{20}H_{24}N_2O_2$	Quinidine	1	20	5.4
		2	20	10.0
$C_{20}H_{26}N_2O_2$	Hydroquinine			5.33
$C_{21}H_{14}Br_4O_5S$	Bromocresol Green			4.7
$C_{21}H_{16}Br_2O_5S$	Bromocresol Purple			6.3
$C_{21}H_{18}O_5S$	CresolRed			8.3
$C_{21}H_{21}NO_6$	Hydrastine			7.8
$C_{21}H_{22}N_2O_2$	Strychnine		25	8.26
$C_{21}H_{23}ClFNO_2$	Haloperidol			8.3
$C_{21}H_{31}NO_4$	Furethidine			7.48
$C_{21}H_{35}N_3O_7$	Lisinopril	1		2.5
		2		4.0
		3		6.7
		4		10.1
$C_{22}H_{18}O_4$	*o*-Cresolphthalein			9.4
$C_{22}H_{22}FN_3O_2$	Droperidol			7.64
$C_{22}H_{23}NO_7$	Noscapine			7.8
$C_{22}H_{25}NO_6$	Colchicine		20	12.36
$C_{22}H_{25}N_3O$	Benzpiperylon	1		6.73
		2		9.13
$C_{22}H_{33}NO_2$	Atisine			12.2
$C_{23}H_{26}N_2O_4$	Brucine	1		6.04
		2		11.07
$C_{24}H_{40}O_4$	Deoxycholic acid			6.58
$C_{24}H_{40}O_5$	Cholic acid			6.4
$C_{25}H_{29}I_2NO_3$	Amiodarone		25	6.56
$C_{25}H_{41}NO_9$	Aconine			9.52
$C_{26}H_{43}NO_6$	Glycocholic acid			4.4
$C_{26}H_{45}NO_7S$	Taurocholic acid			1.4
$C_{27}H_{28}Br_2O_5S$	Bromothymol Blue			7.0
$C_{27}H_{38}N_2O_4$	Verapamil			8.6
$C_{29}H_{32}O_{13}$	Etoposide			9.8
$C_{29}H_{40}N_2O_4$	Emetine	1		5.77
		2		6.64

Mol. Form.	Name	Step	t/°C	pK_a
$C_{30}H_{23}BrO_4$	Bromadiolone		21	4.04
$C_{30}H_{48}O_3$	Oleanolic acid			2.52
$C_{31}H_{36}N_2O_{11}$	Novobiocin	1		4.3
		2		9.1
$C_{32}H_{32}O_{13}S$	Teniposide			10.13
$C_{33}H_{40}N_2O_9$	Reserpine			6.6
$C_{34}H_{47}NO_{11}$	Aconitine			5.88
$C_{36}H_{51}NO_{11}$	Veratridine			9.54
$C_{37}H_{67}NO_{13}$	Erythromycin			8.8
$C_{43}H_{58}N_4O_{12}$	Rifampin	1		1.7
		2		7.9
$C_{45}H_{73}NO_{15}$	Solanine		15	6.66
$C_{46}H_{56}N_4O_{10}$	Vincristine			5.4
$C_{46}H_{58}N_4O_9$	Vinblastine	1		5.4
		2		7.4

CONCENTRATIVE PROPERTIES OF AQUEOUS SOLUTIONS: DENSITY, REFRACTIVE INDEX, FREEZING POINT DEPRESSION, AND VISCOSITY

This table gives properties of aqueous solutions of 66 substances as a function of concentration. All data refer to a temperature of 20°C. The properties are:

Mass %:	Mass of solute divided by total mass of solution, expressed as percent.
m	Molality (moles of solute per kg of water).
c	Molarity (moles of solute per liter of solution).
ρ	Density of solution in g/cm^3.
n	Index of refraction, relative to air, at a wavelength of 589 nm (sodium D line); the index of pure water at 20°C is 1.3330.
Δ	Freezing point depression in °C relative to pure water.
η	Absolute (dynamic) viscosity in mPa s (equal to centipoise, cP); the viscosity of pure water at 20°C is 1.002 mPa s.

Density data for aqueous solutions over a wider range of temperatures and pressures (and for other compounds) may be found in Reference 2. Solutes are listed in the following order:

Acetic acid
Acetone
Ammonia
Ammonium chloride
Ammonium sulfate
Barium chloride
Calcium chloride
Cesium chloride
Citric acid
Copper sulfate
Disodium ethylenediamine tetraacetate (EDTA sodium)
Ethanol
Ethylene glycol
Ferric chloride
Formic acid
D-Fructose
D-Glucose
Glycerol
Hydrochloric acid
Lactic acid
Lactose
Lithium chloride
Magnesium chloride
Magnesium sulfate
Maltose
Manganese(II) sulfate
D-Mannitol
Methanol
Nitric acid
Oxalic acid
Phosphoric acid
Potassium bicarbonate
Potassium bromide
Potassium carbonate
Potassium chloride
Potassium hydroxide
Potassium iodide
Potassium nitrate
Potassium permanganate
Potassium hydrogen phosphate
Potassium dihydrogen phosphate
Potassium sulfate
1-Propanol
2-Propanol
Silver nitrate
Sodium acetate
Sodium bicarbonate
Sodium bromide
Sodium carbonate
Sodium chloride
Sodium citrate
Sodium hydroxide
Sodium nitrate
Sodium phosphate
Sodium hydrogen phosphate
Sodium dihydrogen phosphate
Sodium sulfate
Sodium thiosulfate
Strontium chloride
Sucrose
Sulfuric acid
Trichloroacetic acid
Tris(hydroxymethyl)methylamine
Urea
Zinc sulfate

REFERENCES

1. Wolf, A. V., Aqueous Solutions and Body Fluids, Hoeber, 1966.
2. Söhnel, O., and Novotny, P., *Densities of Aqueous Solutions of Inorganic Substances*, Elsevier, Amsterdam, 1985.

Solute	Mass %	m/mol kg^{-1}	c/mol L^{-1}	ρ/g cm^{-3}	n	Δ/°C	η/mPa s
Acetic acid	0.5	0.084	0.083	0.9989	1.3334	0.16	1.012
CH_3COOH	1.0	0.168	0.166	0.9996	1.3337	0.32	1.022
	2.0	0.340	0.333	1.0011	1.3345	0.63	1.042
	3.0	0.515	0.501	1.0025	1.3352	0.94	1.063
	4.0	0.694	0.669	1.0038	1.3359	1.26	1.084
	5.0	0.876	0.837	1.0052	1.3366	1.58	1.105
	6.0	1.063	1.006	1.0066	1.3373	1.90	1.125
	7.0	1.253	1.175	1.0080	1.3381	2.23	1.143
	8.0	1.448	1.345	1.0093	1.3388	2.56	1.162
	9.0	1.647	1.515	1.0107	1.3395	2.89	1.186
	10.0	1.850	1.685	1.0121	1.3402	3.23	1.210
	12.0	2.271	2.028	1.0147	1.3416	3.91	1.253
	14.0	2.711	2.372	1.0174	1.3430	4.61	1.298
	16.0	3.172	2.718	1.0200	1.3444	5.33	1.341
	18.0	3.655	3.065	1.0225	1.3458	6.06	1.380
	20.0	4.163	3.414	1.0250	1.3472	6.81	1.431

Solute	Mass %	m/mol kg^{-1}	c/mol L^{-1}	ρ/g cm^{-3}	n	Δ/°C	η/mPa s
	22.0	4.697	3.764	1.0275	1.3485	7.57	1.478
	24.0	5.259	4.116	1.0299	1.3498	8.36	1.525
	26.0	5.851	4.470	1.0323	1.3512	9.17	1.572
	28.0	6.476	4.824	1.0346	1.3525	10.00	1.613
	30.0	7.137	5.180	1.0369	1.3537	10.84	1.669
	32.0	7.837	5.537	1.0391	1.3550	11.70	1.715
	34.0	8.579	5.896	1.0413	1.3562	12.55	1.762
	36.0	9.367	6.255	1.0434	1.3574	13.38	1.812
	38.0	10.207	6.615	1.0454	1.3586		1.852
	40.0	11.102	6.977	1.0474	1.3598		1.912
	50.0	16.653	8.794	1.0562	1.3653		2.158
	60.0	24.979	10.620	1.0629	1.3700		2.409
	70.0	38.857	12.441	1.0673	1.3738		2.629
	80.0	66.611	14.228	1.0680	1.3767		2.720
	90.0	149.875	15.953	1.0644	1.3771		2.386
	92.0	191.507	16.284	1.0629	1.3766		2.240
	94.0	260.894	16.602	1.0606	1.3759		2.036
	96.0	399.667	16.911	1.0578	1.3748		1.813
	98.0	815.987	17.198	1.0538	1.3734		1.535
	100.0		17.447	1.0477	1.3716		1.223
Acetone	0.5	0.087	0.086	0.9975	1.3334	0.16	1.013
$(CH_3)_2CO$	1.0	0.174	0.172	0.9968	1.3337	0.32	1.024
	2.0	0.351	0.343	0.9954	1.3344	0.65	1.047
	3.0	0.533	0.513	0.9940	1.3352	0.97	1.072
	4.0	0.717	0.684	0.9926	1.3359	1.30	1.099
	5.0	0.906	0.853	0.9912	1.3366	1.63	1.125
	6.0	1.099	1.023	0.9899	1.3373	1.96	1.150
	7.0	1.296	1.191	0.9886	1.3381	2.29	1.174
	8.0	1.497	1.360	0.9874	1.3388	2.62	1.198
	9.0	1.703	1.528	0.9861	1.3395	2.95	1.221
	10.0	1.913	1.696	0.9849	1.3402	3.29	1.244
Ammonia	0.5	0.295	0.292	0.9960	1.3332	0.55	1.009
NH_3	1.0	0.593	0.584	0.9938	1.3335	1.14	1.015
	2.0	1.198	1.162	0.9895	1.3339	2.32	1.029
	3.0	1.816	1.736	0.9853	1.3344	3.53	1.043
	4.0	2.447	2.304	0.9811	1.3349	4.78	1.057
	5.0	3.090	2.868	0.9770	1.3354	6.08	1.071
	6.0	3.748	3.428	0.9730	1.3359	7.43	1.085
	7.0	4.420	3.983	0.9690	1.3365	8.95	1.099
	8.0	5.106	4.533	0.9651	1.3370	10.34	1.113
	9.0	5.807	5.080	0.9613	1.3376	11.90	1.127
	10.0	6.524	5.622	0.9575	1.3381	13.55	1.141
	12.0	8.007	6.695	0.9502	1.3393	17.13	1.169
	14.0	9.558	7.753	0.9431	1.3404	21.13	1.195
	16.0	11.184	8.794	0.9361	1.3416	25.63	1.218
	18.0	12.889	9.823	0.9294	1.3428	30.70	1.237
	20.0	14.679	10.837	0.9228	1.3440	36.42	1.254
	22.0	16.561	11.838	0.9164	1.3453	43.36	1.268
	24.0	18.542	12.826	0.9102	1.3465	51.38	1.280
	26.0	20.630	13.801	0.9040	1.3477	60.77	1.288
	28.0	22.834	14.764	0.8980	1.3490	71.66	
	30.0	25.164	15.713	0.8920	1.3502	84.06	
Ammonium	0.5	0.094	0.093	0.9998	1.3340	0.32	0.999
chloride	1.0	0.189	0.187	1.0014	1.3349	0.64	0.996
NH_4Cl	2.0	0.382	0.376	1.0045	1.3369	1.27	0.992

Solute	Mass %	m/mol kg^{-1}	c/mol L^{-1}	ρ/g cm^{-3}	n	Δ/°C	η/mPa s
	3.0	0.578	0.565	1.0076	1.3388	1.91	0.988
	4.0	0.779	0.756	1.0107	1.3407	2.57	0.985
	5.0	0.984	0.948	1.0138	1.3426	3.25	0.982
	6.0	1.193	1.141	1.0168	1.3445	3.94	0.979
	7.0	1.407	1.335	1.0198	1.3464	4.66	0.976
	8.0	1.626	1.529	1.0227	1.3483	5.40	0.974
	9.0	1.849	1.726	1.0257	1.3502	6.16	0.972
	10.0	2.077	1.923	1.0286	1.3521	6.95	0.970
	12.0	2.549	2.320	1.0344	1.3559	8.60	0.969
	14.0	3.043	2.722	1.0401	1.3596		0.969
	16.0	3.561	3.128	1.0457	1.3634		0.971
	18.0	4.104	3.537	1.0512	1.3671		0.973
	20.0	4.674	3.951	1.0567	1.3708		0.978
	22.0	5.273	4.368	1.0621	1.3745		0.986
	24.0	5.903	4.789	1.0674	1.3782		0.996
Ammonium	0.5	0.038	0.038	1.0012	1.3338	0.17	1.008
sulfate	1.0	0.076	0.076	1.0042	1.3346	0.33	1.014
$(NH_4)_2SO_4$	2.0	0.154	0.153	1.0101	1.3363	0.63	1.027
	3.0	0.234	0.231	1.0160	1.3379	0.92	1.041
	4.0	0.315	0.309	1.0220	1.3395	1.21	1.057
	5.0	0.398	0.389	1.0279	1.3411	1.49	1.073
	6.0	0.483	0.469	1.0338	1.3428	1.77	1.090
	7.0	0.570	0.551	1.0397	1.3444	2.05	1.108
	8.0	0.658	0.633	1.0456	1.3460	2.33	1.127
	9.0	0.748	0.716	1.0515	1.3476	2.61	1.147
	10.0	0.841	0.800	1.0574	1.3492	2.89	1.168
	12.0	1.032	0.971	1.0691	1.3523	3.47	1.210
	14.0	1.232	1.145	1.0808	1.3555	4.07	1.256
	16.0	1.441	1.323	1.0924	1.3586	4.69	1.305
	18.0	1.661	1.504	1.1039	1.3616		1.359
	20.0	1.892	1.688	1.1154	1.3647		1.421
	22.0	2.134	1.876	1.1269	1.3677		1.490
	24.0	2.390	2.067	1.1383	1.3707		1.566
	26.0	2.659	2.262	1.1496	1.3737		1.650
	28.0	2.943	2.460	1.1609	1.3766		1.743
	30.0	3.243	2.661	1.1721	1.3795		1.847
	32.0	3.561	2.866	1.1833	1.3824		1.961
	34.0	3.898	3.073	1.1945	1.3853		2.086
	36.0	4.257	3.284	1.2056	1.3881		2.222
	38.0	4.638	3.499	1.2166	1.3909		2.371
	40.0	5.045	3.716	1.2277	1.3938		2.530
Barium	0.5	0.024	0.024	1.0026	1.3337	0.12	1.009
chloride	1.0	0.049	0.048	1.0070	1.3345	0.23	1.016
$BaCl_2$	2.0	0.098	0.098	1.0159	1.3360	0.46	1.026
	3.0	0.149	0.148	1.0249	1.3375	0.69	1.037
	4.0	0.200	0.199	1.0341	1.3391	0.93	1.049
	5.0	0.253	0.251	1.0434	1.3406	1.18	1.062
	6.0	0.307	0.303	1.0528	1.3422	1.44	1.075
	7.0	0.361	0.357	1.0624	1.3438	1.70	1.087
	8.0	0.418	0.412	1.0721	1.3454	1.98	1.101
	9.0	0.475	0.468	1.0820	1.3470	2.27	1.114
	10.0	0.534	0.524	1.0921	1.3487	2.58	1.129
	12.0	0.655	0.641	1.1128	1.3520	3.22	1.161
	14.0	0.782	0.763	1.1342	1.3555	3.92	1.195
	16.0	0.915	0.889	1.1564	1.3591	4.69	1.234

Solute	Mass %	m/mol kg^{-1}	c/mol L^{-1}	ρ/g cm^{-3}	n	Δ/°C	η/mPa s
	18.0	1.054	1.019	1.1793	1.3627		1.277
	20.0	1.201	1.156	1.2031	1.3664		1.325
	22.0	1.355	1.297	1.2277	1.3703		1.378
	24.0	1.517	1.444	1.2531	1.3741		1.437
	26.0	1.687	1.597	1.2793	1.3781		1.503
Calcium chloride $CaCl_2$	0.5	0.045	0.045	1.0024	1.3342	0.22	1.015
	1.0	0.091	0.091	1.0065	1.3354	0.44	1.028
	2.0	0.184	0.183	1.0148	1.3378	0.88	1.050
	3.0	0.279	0.277	1.0232	1.3402	1.33	1.078
	4.0	0.375	0.372	1.0316	1.3426	1.82	1.110
	5.0	0.474	0.469	1.0401	1.3451	2.35	1.143
	6.0	0.575	0.567	1.0486	1.3475	2.93	1.175
	7.0	0.678	0.667	1.0572	1.3500	3.57	1.208
	8.0	0.784	0.768	1.0659	1.3525	4.28	1.242
	9.0	0.891	0.872	1.0747	1.3549	5.04	1.279
	10.0	1.001	0.976	1.0835	1.3575	5.86	1.319
	12.0	1.229	1.191	1.1014	1.3625	7.70	1.408
	14.0	1.467	1.413	1.1198	1.3677	9.83	1.508
	16.0	1.716	1.641	1.1386	1.3730	12.28	1.625
	18.0	1.978	1.878	1.1579	1.3784	15.11	1.764
	20.0	2.253	2.122	1.1775	1.3839	18.30	1.930
	22.0	2.541	2.374	1.1976	1.3895	21.70	2.127
	24.0	2.845	2.634	1.2180	1.3951	25.30	2.356
	26.0	3.166	2.902	1.2388	1.4008	29.70	2.645
	28.0	3.504	3.179	1.2600	1.4066	34.70	3.000
	30.0	3.862	3.464	1.2816	1.4124	41.00	3.467
	32.0	4.240	3.759	1.3036	1.4183	49.70	4.035
	34.0	4.642	4.062	1.3260	1.4242		4.820
	36.0	5.068	4.375	1.3488	1.4301		5.807
	38.0	5.522	4.698	1.3720	1.4361		7.321
	40.0	6.007	5.030	1.3957	1.4420		8.997
Cesium chloride CsCl	0.5	0.030	0.030	1.0020	1.3334	0.10	1.000
	1.0	0.060	0.060	1.0058	1.3337	0.20	0.997
	2.0	0.121	0.120	1.0135	1.3345	0.40	0.992
	3.0	0.184	0.182	1.0214	1.3353	0.61	0.988
	4.0	0.247	0.245	1.0293	1.3361	0.81	0.984
	5.0	0.313	0.308	1.0374	1.3369	1.02	0.980
	6.0	0.379	0.373	1.0456	1.3377	1.22	0.977
	7.0	0.447	0.438	1.0540	1.3386	1.43	0.974
	8.0	0.516	0.505	1.0625	1.3394	1.64	0.971
	9.0	0.587	0.573	1.0711	1.3403	1.85	0.969
	10.0	0.660	0.641	1.0798	1.3412	2.06	0.966
	12.0	0.810	0.782	1.0978	1.3430	2.51	0.961
	14.0	0.967	0.928	1.1163	1.3448	2.97	0.955
	16.0	1.131	1.079	1.1355	1.3468	3.46	0.950
	18.0	1.304	1.235	1.1552	1.3487	3.96	0.945
	20.0	1.485	1.397	1.1756	1.3507	4.49	0.939
	22.0	1.675	1.564	1.1967	1.3528		0.934
	24.0	1.876	1.737	1.2185	1.3550		0.930
	26.0	2.087	1.917	1.2411	1.3572		0.926
	28.0	2.310	2.103	1.2644	1.3594		0.924
	30.0	2.546	2.296	1.2885	1.3617		0.922
	32.0	2.795	2.497	1.3135	1.3641		0.922
	34.0	3.060	2.705	1.3393	1.3666		0.924
	36.0	3.341	2.921	1.3661	1.3691		0.926

Solute	Mass %	*m*/mol kg⁻¹	*c*/mol L⁻¹	ρ/g cm⁻³	*n*	Δ/°C	η/mPa s
	38.0	3.640	3.146	1.3938	1.3717		0.930
	40.0	3.960	3.380	1.4226	1.3744		0.934
	42.0	4.301	3.624	1.4525	1.3771		0.940
	44.0	4.667	3.877	1.4835	1.3800		0.947
	46.0	5.060	4.142	1.5158	1.3829		0.956
	48.0	5.483	4.418	1.5495	1.3860		0.967
	50.0	5.940	4.706	1.5846	1.3892		0.981
	60.0	8.910	6.368	1.7868	1.4076		1.120
	64.0	10.560	7.163	1.8842	1.4167		1.238
Citric acid	0.5	0.026	0.026	1.0002	1.3336	0.05	1.013
$(HO)C(COOH)_3$	1.0	0.053	0.052	1.0022	1.3343	0.11	1.024
	2.0	0.106	0.105	1.0063	1.3356	0.21	1.048
	3.0	0.161	0.158	1.0105	1.3368	0.32	1.073
	4.0	0.217	0.211	1.0147	1.3381	0.43	1.098
	5.0	0.274	0.265	1.0189	1.3394	0.54	1.125
	6.0	0.332	0.320	1.0232	1.3407	0.65	1.153
	7.0	0.392	0.374	1.0274	1.3420	0.76	1.183
	8.0	0.453	0.430	1.0316	1.3433	0.88	1.214
	9.0	0.515	0.485	1.0359	1.3446	1.00	1.247
	10.0	0.578	0.541	1.0402	1.3459	1.12	1.283
	12.0	0.710	0.655	1.0490	1.3486	1.38	1.357
	14.0	0.847	0.771	1.0580	1.3514	1.66	1.436
	16.0	0.991	0.889	1.0672	1.3541	1.95	1.525
	18.0	1.143	1.008	1.0764	1.3569	2.26	1.625
	20.0	1.301	1.130	1.0858	1.3598	2.57	1.740
	22.0	1.468	1.254	1.0953	1.3626	2.88	1.872
	24.0	1.644	1.380	1.1049	1.3655	3.21	2.017
	26.0	1.829	1.508	1.1147	1.3684	3.55	2.178
	28.0	2.024	1.639	1.1246	1.3714	3.89	2.356
	30.0	2.231	1.772	1.1346	1.3744	4.25	2.549
Copper	0.5	0.031	0.031	1.0033	1.3339	0.08	1.017
sulfate	1.0	0.063	0.063	1.0085	1.3348	0.14	1.036
$CuSO_4$	2.0	0.128	0.128	1.0190	1.3367	0.26	1.084
	3.0	0.194	0.194	1.0296	1.3386	0.37	1.129
	4.0	0.261	0.261	1.0403	1.3405	0.48	1.173
	5.0	0.330	0.329	1.0511	1.3424	0.59	1.221
	6.0	0.400	0.399	1.0620	1.3443	0.70	1.276
	7.0	0.472	0.471	1.0730	1.3462	0.82	1.336
	8.0	0.545	0.543	1.0842	1.3481	0.93	1.400
	9.0	0.620	0.618	1.0955	1.3501	1.05	1.469
	10.0	0.696	0.694	1.1070	1.3520	1.18	1.543
	12.0	0.854	0.850	1.1304	1.3560	1.45	1.701
	14.0	1.020	1.013	1.1545	1.3601	1.75	1.889
	16.0	1.193	1.182	1.1796	1.3644		2.136
	18.0	1.375	1.360	1.2059	1.3689		2.449
Disodium	0.5	0.015	0.015	1.0009	1.3339	0.07	1.017
ethylenediamine	1.0	0.030	0.030	1.0036	1.3348	0.14	1.032
tetraacetate	1.5	0.045	0.045	1.0062	1.3356	0.21	1.046
(EDTA sodium)	2.0	0.061	0.060	1.0089	1.3365	0.27	1.062
$Na_2C_{10}H_{14}N_2O_8$	2.5	0.076	0.075	1.0115	1.3374	0.33	1.077
	3.0	0.092	0.090	1.0142	1.3383	0.40	1.093
	3.5	0.108	0.106	1.0169	1.3392	0.46	1.109
	4.0	0.124	0.121	1.0196	1.3400	0.52	1.125
	4.5	0.140	0.137	1.0223	1.3409	0.58	1.142

Solute	Mass %	m/mol kg^{-1}	c/mol L^{-1}	ρ/g cm^{-3}	n	Δ/°C	η/mPa s
	5.0	0.157	0.152	1.0250	1.3418	0.65	1.160
	5.5	0.173	0.168	1.0277	1.3427	0.71	1.178
	6.0	0.190	0.184	1.0305	1.3436	0.77	1.197
Ethanol	0.5	0.109	0.108	0.9973	1.3333	0.20	1.023
CH_3CH_2OH	1.0	0.219	0.216	0.9963	1.3336	0.40	1.046
	2.0	0.443	0.432	0.9945	1.3342	0.81	1.095
	3.0	0.671	0.646	0.9927	1.3348	1.23	1.140
	4.0	0.904	0.860	0.9910	1.3354	1.65	1.183
	5.0	1.142	1.074	0.9893	1.3360	2.09	1.228
	6.0	1.385	1.286	0.9878	1.3367	2.54	1.279
	7.0	1.634	1.498	0.9862	1.3374	2.99	1.331
	8.0	1.887	1.710	0.9847	1.3381	3.47	1.385
	9.0	2.147	1.921	0.9833	1.3388	3.96	1.442
	10.0	2.412	2.131	0.9819	1.3395	4.47	1.501
	12.0	2.960	2.551	0.9792	1.3410	5.56	1.627
	14.0	3.534	2.967	0.9765	1.3425	6.73	1.761
	16.0	4.134	3.382	0.9739	1.3440	8.01	1.890
	18.0	4.765	3.795	0.9713	1.3455	9.40	2.019
	20.0	5.427	4.205	0.9687	1.3469	10.92	2.142
	22.0	6.122	4.613	0.9660	1.3484	12.60	2.259
	24.0	6.855	5.018	0.9632	1.3498	14.47	2.370
	26.0	7.626	5.419	0.9602	1.3511	16.41	2.476
	28.0	8.441	5.817	0.9571	1.3524	18.43	2.581
	30.0	9.303	6.212	0.9539	1.3535	20.47	2.667
	32.0	10.215	6.601	0.9504	1.3546	22.44	2.726
	34.0	11.182	6.987	0.9468	1.3557	24.27	2.768
	36.0	12.210	7.370	0.9431	1.3566	25.98	2.803
	38.0	13.304	7.747	0.9392	1.3575	27.62	2.829
	40.0	14.471	8.120	0.9352	1.3583	29.26	2.846
	42.0	15.718	8.488	0.9311	1.3590	30.98	2.852
	44.0	17.055	8.853	0.9269	1.3598	32.68	2.850
	46.0	18.490	9.213	0.9227	1.3604	34.36	2.843
	48.0	20.036	9.568	0.9183	1.3610	36.04	2.832
	50.0	21.706	9.919	0.9139	1.3616	37.67	2.813
	60.0	32.559	11.605	0.8911	1.3638	44.93	2.547
	70.0	50.648	13.183	0.8676	1.3652		2.214
	80.0	86.824	14.649	0.8436	1.3658		1.881
	90.0	195.355	15.980	0.8180	1.3650		1.542
	92.0	249.620	16.225	0.8125	1.3646		1.475
	94.0	340.062	16.466	0.8070	1.3642		1.407
	96.0	520.946	16.697	0.8013	1.3636		1.342
	98.0		16.920	0.7954	1.3630		1.273
	100.0		17.133	0.7893	1.3614		1.203
Ethylene	0.5	0.081	0.080	0.9988	1.3335	0.15	1.010
glycol	1.0	0.163	0.161	0.9995	1.3339	0.30	1.020
$(CH_2OH)_2$	2.0	0.329	0.322	1.0007	1.3348	0.61	1.048
	3.0	0.498	0.484	1.0019	1.3358	0.92	1.074
	4.0	0.671	0.646	1.0032	1.3367	1.24	1.099
	5.0	0.848	0.809	1.0044	1.3377	1.58	1.125
	6.0	1.028	0.972	1.0057	1.3386	1.91	1.153
	7.0	1.213	1.136	1.0070	1.3396	2.26	1.182
	8.0	1.401	1.299	1.0082	1.3405	2.62	1.212
	9.0	1.593	1.464	1.0095	1.3415	2.99	1.243
	10.0	1.790	1.628	1.0108	1.3425	3.37	1.277
	12.0	2.197	1.959	1.0134	1.3444	4.16	1.348

Solute	Mass %	m/mol kg^{-1}	c/mol L^{-1}	ρ/g cm^{-3}	n	Δ/°C	η/mPa s
	14.0	2.623	2.292	1.0161	1.3464	5.01	1.424
	16.0	3.069	2.626	1.0188	1.3484	5.91	1.500
	18.0	3.537	2.962	1.0214	1.3503	6.89	1.578
	20.0	4.028	3.300	1.0241	1.3523	7.93	1.661
	24.0	5.088	3.981	1.0296	1.3564	10.28	1.843
	28.0	6.265	4.669	1.0350	1.3605	13.03	2.047
	32.0	7.582	5.364	1.0405	1.3646	16.23	2.280
	36.0	9.062	6.067	1.0460	1.3687	19.82	2.537
	40.0	10.741	6.776	1.0514	1.3728	23.84	2.832
	44.0	12.659	7.491	1.0567	1.3769	28.32	3.166
	48.0	14.872	8.212	1.0619	1.3811	33.30	3.544
	52.0	17.453	8.939	1.0670	1.3851	38.81	3.981
	56.0	20.505	9.671	1.0719	1.3892	44.83	4.475
	60.0	24.166	10.406	1.0765	1.3931	51.23	5.026
Ferric	0.5	0.031	0.031	1.0025	1.3344	0.21	1.024
chloride	1.0	0.062	0.062	1.0068	1.3358	0.39	1.047
$FeCl_3$	2.0	0.126	0.125	1.0153	1.3386	0.75	1.093
	3.0	0.191	0.189	1.0238	1.3413	1.15	1.139
	4.0	0.257	0.255	1.0323	1.3441	1.56	1.187
	5.0	0.324	0.321	1.0408	1.3468	2.00	1.238
	6.0	0.394	0.388	1.0493	1.3496	2.48	1.292
	7.0	0.464	0.457	1.0580	1.3524	2.99	1.350
	8.0	0.536	0.526	1.0668	1.3552	3.57	1.412
	9.0	0.610	0.597	1.0760	1.3581	4.19	1.480
	10.0	0.685	0.669	1.0853	1.3611	4.85	1.553
	12.0	0.841	0.817	1.1040	1.3670	6.38	1.707
	14.0	1.004	0.969	1.1228	1.3730	8.22	1.879
	16.0	1.174	1.126	1.1420		10.45	2.080
	18.0	1.353	1.289	1.1615		13.08	2.311
	20.0	1.541	1.457	1.1816		16.14	2.570
	24.0	1.947	1.810	1.2234		23.79	3.178
	28.0	2.398	2.189	1.2679		33.61	4.038
	32.0	2.901	2.595	1.3153		49.16	5.274
	36.0	3.468	3.030	1.3654			7.130
	40.0	4.110	3.496	1.4176			9.674
Formic acid	0.5	0.109	0.109	0.9994	1.3333	0.21	1.006
HCOOH	1.0	0.219	0.217	1.0006	1.3336	0.42	1.011
	2.0	0.443	0.436	1.0029	1.3342	0.82	1.017
	3.0	0.672	0.655	1.0053	1.3348	1.24	1.195
	4.0	0.905	0.876	1.0077	1.3354	1.67	1.032
	5.0	1.143	1.097	1.0102	1.3359	2.10	1.039
	6.0	1.387	1.320	1.0126	1.3365	2.53	1.046
	7.0	1.635	1.544	1.0150	1.3371	2.97	1.052
	8.0	1.889	1.768	1.0175	1.3376	3.40	1.058
	9.0	2.149	1.994	1.0199	1.3382	3.84	1.064
	10.0	2.414	2.221	1.0224	1.3387	4.27	1.070
	12.0	2.962	2.678	1.0273	1.3397	5.19	1.082
	14.0	3.537	3.139	1.0322	1.3408	6.11	1.094
	16.0	4.138	3.605	1.0371	1.3418	7.06	1.106
	18.0	4.769	4.074	1.0419	1.3428	8.08	1.119
	20.0	5.431	4.548	1.0467	1.3437	9.11	1.132
	28.0	8.449	6.481	1.0654	1.3475	13.10	1.179
	36.0	12.220	8.477	1.0839	1.3511	17.65	1.227
	44.0	17.070	10.529	1.1015	1.3547	22.93	1.281
	52.0	23.535	12.633	1.1183	1.3581	29.69	1.340

Solute	Mass %	*m*/mol kg⁻¹	*c*/mol L⁻¹	ρ/g cm⁻³	*n*	Δ/°C	η/mPa s
	60.0	32.587	14.813	1.1364	1.3612	38.26	1.410
	68.0	46.166	17.054	1.1544	1.3641		1.490
D-Fructose	0.5	0.028	0.028	1.0002	1.3337	0.05	1.015
$C_6H_{12}O_6$	1.0	0.056	0.056	1.0021	1.3344	0.10	1.028
	2.0	0.113	0.112	1.0061	1.3358	0.21	1.054
	3.0	0.172	0.168	1.0101	1.3373	0.32	1.080
	4.0	0.231	0.225	1.0140	1.3387	0.43	1.106
	5.0	0.292	0.283	1.0181	1.3402	0.54	1.134
	6.0	0.354	0.340	1.0221	1.3417	0.66	1.165
	7.0	0.418	0.399	1.0262	1.3431	0.78	1.198
	8.0	0.483	0.458	1.0303	1.3446	0.90	1.232
	9.0	0.549	0.517	1.0344	1.3461	1.03	1.270
	10.0	0.617	0.576	1.0385	1.3476	1.16	1.309
	12.0	0.757	0.697	1.0469	1.3507	1.43	1.391
	14.0	0.904	0.820	1.0554	1.3538	1.71	1.483
	16.0	1.057	0.945	1.0640	1.3569	2.01	1.587
	18.0	1.218	1.072	1.0728	1.3601	2.32	1.703
	20.0	1.388	1.201	1.0816	1.3634	2.64	1.837
	22.0	1.566	1.332	1.0906	1.3667	3.05	1.986
	24.0	1.753	1.465	1.0996	1.3700	3.43	2.154
	26.0	1.950	1.600	1.1089	1.3734	3.82	2.348
	28.0	2.159	1.738	1.1182	1.3768	4.20	2.562
	30.0	2.379	1.878	1.1276	1.3803		2.817
	32.0	2.612	2.020	1.1372	1.3839		3.112
	34.0	2.859	2.164	1.1469	1.3874		3.462
	36.0	3.122	2.312	1.1568	1.3911		3.899
	38.0	3.402	2.461	1.1668	1.3948		4.418
	40.0	3.700	2.613	1.1769	1.3985		5.046
	42.0	4.019	2.767	1.1871	1.4023		5.773
	44.0	4.361	2.925	1.1975	1.4062		6.644
	46.0	4.728	3.084	1.2080	1.4101		7.753
	48.0	5.124	3.247	1.2187	1.4141		9.060
D-Glucose	0.5	0.028	0.028	1.0001	1.3337	0.05	1.010
$C_6H_{12}O_6$	1.0	0.056	0.056	1.0020	1.3344	0.11	1.021
	2.0	0.113	0.112	1.0058	1.3358	0.21	1.052
	3.0	0.172	0.168	1.0097	1.3373	0.32	1.083
	4.0	0.231	0.225	1.0136	1.3387	0.43	1.113
	5.0	0.292	0.282	1.0175	1.3402	0.55	1.145
	6.0	0.354	0.340	1.0214	1.3417	0.67	1.179
	7.0	0.418	0.398	1.0254	1.3432	0.79	1.214
	8.0	0.483	0.457	1.0294	1.3447	0.91	1.250
	9.0	0.549	0.516	1.0334	1.3462	1.04	1.289
	10.0	0.617	0.576	1.0375	1.3477	1.17	1.330
	12.0	0.757	0.697	1.0457	1.3508	1.44	1.416
	14.0	0.904	0.819	1.0540	1.3539	1.73	1.512
	16.0	1.057	0.944	1.0624	1.3571	2.03	1.625
	18.0	1.218	1.070	1.0710	1.3603	2.35	1.757
	20.0	1.388	1.199	1.0797	1.3635	2.70	1.904
	22.0	1.566	1.329	1.0884	1.3668	3.07	2.063
	24.0	1.753	1.462	1.0973	1.3702	3.48	2.242
	26.0	1.950	1.597	1.1063	1.3736	3.90	2.458
	28.0	2.159	1.734	1.1154	1.3770	4.34	2.707
	30.0	2.379	1.873	1.1246	1.3805	4.79	2.998
	32.0	2.612	2.014	1.1340	1.3840		3.324
	34.0	2.859	2.158	1.1434	1.3876		3.704

Solute	Mass %	m/mol kg^{-1}	c/mol L^{-1}	ρ/g cm^{-3}	n	Δ/°C	η/mPa s
	36.0	3.122	2.304	1.1529	1.3912		4.193
	38.0	3.402	2.452	1.1626	1.3949		4.786
	40.0	3.700	2.603	1.1724	1.3986		5.493
	42.0	4.019	2.756	1.1823	1.4024		6.288
	44.0	4.361	2.912	1.1924	1.4062		7.235
	46.0	4.728	3.071	1.2026	1.4101		8.454
	48.0	5.124	3.232	1.2130	1.4141		9.883
	50.0	5.551	3.396	1.2235	1.4181		11.884
	52.0	6.013	3.562	1.2342	1.4222		14.489
	54.0	6.516	3.732	1.2451	1.4263		17.916
	56.0	7.064	3.905	1.2562	1.4306		22.886
	58.0	7.665	4.081	1.2676	1.4349		29.389
	60.0	8.326	4.261	1.2793	1.4394		37.445
Glycerol	0.5	0.055	0.054	0.9994	1.3336	0.07	1.011
$CH_2OHCHOHCH_2OH$	1.0	0.110	0.109	1.0005	1.3342	0.18	1.022
	2.0	0.222	0.218	1.0028	1.3353	0.41	1.048
	3.0	0.336	0.327	1.0051	1.3365	0.63	1.074
	4.0	0.452	0.438	1.0074	1.3376	0.85	1.100
	5.0	0.572	0.548	1.0097	1.3388	1.08	1.127
	6.0	0.693	0.659	1.0120	1.3400	1.32	1.157
	7.0	0.817	0.771	1.0144	1.3412	1.56	1.188
	8.0	0.944	0.883	1.0167	1.3424	1.81	1.220
	9.0	1.074	0.996	1.0191	1.3436	2.06	1.256
	10.0	1.207	1.109	1.0215	1.3448	2.32	1.291
	12.0	1.481	1.337	1.0262	1.3472	2.88	1.365
	14.0	1.768	1.568	1.0311	1.3496	3.47	1.445
	16.0	2.068	1.800	1.0360	1.3521	4.09	1.533
	18.0	2.384	2.035	1.0409	1.3547	4.76	1.630
	20.0	2.715	2.271	1.0459	1.3572	5.46	1.737
	24.0	3.429	2.752	1.0561	1.3624	7.01	1.988
	28.0	4.223	3.242	1.0664	1.3676	8.77	2.279
	32.0	5.110	3.742	1.0770	1.3730	10.74	2.637
	36.0	6.108	4.252	1.0876	1.3785	12.96	3.088
	40.0	7.239	4.771	1.0984	1.3841	15.50	3.653
	44.0	8.532	5.300	1.1092	1.3897		4.443
	48.0	10.024	5.838	1.1200	1.3954		5.413
	52.0	11.764	6.385	1.1308	1.4011		6.666
	56.0	13.820	6.944	1.1419	1.4069		8.349
	60.0	16.288	7.512	1.1530	1.4129		10.681
	64.0	19.305	8.092	1.1643	1.4189		13.657
	68.0	23.075	8.680	1.1755	1.4249		18.457
	72.0	27.923	9.277	1.1866	1.4310		27.625
	76.0	34.387	9.884	1.1976	1.4370		40.571
	80.0	43.436	10.498	1.2085	1.4431		59.900
	84.0	57.009	11.121	1.2192	1.4492		84.338
	88.0	79.632	11.753	1.2299	1.4553		147.494
	92.0	124.878	12.392	1.2404	1.4613		384.467
	96.0	260.615	13.039	1.2508	1.4674		780.458
	100.0		13.694	1.2611	1.4735		
Hydrochloric	0.5	0.138	0.137	1.0007	1.3341	0.49	1.008
acid	1.0	0.277	0.275	1.0031	1.3353	0.99	1.015
HCl	2.0	0.560	0.553	1.0081	1.3376	2.08	1.029
	3.0	0.848	0.833	1.0130	1.3399	3.28	1.044
	4.0	1.143	1.117	1.0179	1.3422	4.58	1.059
	5.0	1.444	1.403	1.0228	1.3445	5.98	1.075

Solute	Mass %	*m*/mol kg^{-1}	*c*/mol L^{-1}	ρ/g cm^{-3}	*n*	Δ/°C	η/mPa s
	6.0	1.751	1.691	1.0278	1.3468	7.52	1.091
	7.0	2.064	1.983	1.0327	1.3491	9.22	1.108
	8.0	2.385	2.277	1.0377	1.3515	11.10	1.125
	9.0	2.713	2.574	1.0426	1.3538	13.15	1.143
	10.0	3.047	2.873	1.0476	1.3561	15.40	1.161
	12.0	3.740	3.481	1.0576	1.3607	20.51	1.199
	14.0	4.465	4.099	1.0676	1.3653		1.239
	16.0	5.224	4.729	1.0777	1.3700		1.282
	18.0	6.020	5.370	1.0878	1.3746		1.326
	20.0	6.857	6.023	1.0980	1.3792		1.374
	22.0	7.736	6.687	1.1083	1.3838		1.426
	24.0	8.661	7.362	1.1185	1.3884		1.483
	26.0	9.636	8.049	1.1288	1.3930		1.547
	28.0	10.666	8.748	1.1391	1.3976		1.620
	30.0	11.754	9.456	1.1492	1.4020		1.705
	32.0	12.907	10.175	1.1594	1.4066		1.799
	34.0	14.129	10.904	1.1693	1.4112		1.900
	36.0	15.427	11.642	1.1791	1.4158		2.002
	38.0	16.810	12.388	1.1886	1.4204		2.105
	40.0	18.284	13.140	1.1977	1.4250		
Lactic acid	0.5	0.056	0.055	0.9992	1.3335	0.10	1.014
$CH_3CHOHCOOH$	1.0	0.112	0.111	1.0002	1.3340	0.19	1.027
	2.0	0.227	0.223	1.0023	1.3350	0.38	1.056
	3.0	0.343	0.334	1.0043	1.3360	0.57	1.084
	4.0	0.463	0.447	1.0065	1.3370	0.76	1.110
	5.0	0.584	0.560	1.0086	1.3380	0.95	1.138
	6.0	0.709	0.673	1.0108	1.3390	1.16	1.167
	7.0	0.836	0.787	1.0131	1.3400	1.36	1.198
	8.0	0.965	0.902	1.0153	1.3410	1.57	1.229
	9.0	1.098	1.017	1.0176	1.3420	1.79	1.262
	10.0	1.233	1.132	1.0199	1.3430	2.02	1.296
	12.0	1.514	1.365	1.0246	1.3450	2.49	1.366
	14.0	1.807	1.600	1.0294	1.3470	2.99	1.441
	16.0	2.115	1.837	1.0342	1.3491	3.48	1.522
	18.0	2.437	2.076	1.0390	1.3511	3.96	1.607
	20.0	2.775	2.318	1.0439	1.3532	4.44	1.699
	24.0	3.506	2.807	1.0536	1.3573		1.902
	28.0	4.317	3.305	1.0632	1.3615		2.136
	32.0	5.224	3.811	1.0728	1.3657		2.414
	36.0	6.244	4.325	1.0822	1.3700		2.730
	40.0	7.401	4.847	1.0915	1.3743		3.114
	44.0	8.722	5.377	1.1008	1.3786		3.566
	48.0	10.247	5.917	1.1105	1.3828		4.106
	52.0	12.026	6.466	1.1201	1.3871		4.789
	56.0	14.129	7.023	1.1297	1.3914		5.579
	60.0	16.652	7.588	1.1392	1.3958		6.679
	64.0	19.736	8.161	1.1486	1.4001		8.024
	68.0	23.590	8.741	1.1579	1.4045		9.863
	72.0	28.546	9.328	1.1670	1.4088		12.866
	76.0	35.154	9.922	1.1760	1.4131		16.974
	80.0	44.405	10.522	1.1848	1.4173		22.164
Lactose	0.5	0.015	0.015	1.0002	1.3337	0.03	1.013
$C_{12}H_{22}O_{11}$	1.0	0.030	0.029	1.0021	1.3345	0.06	1.026
	2.0	0.060	0.059	1.0061	1.3359	0.11	1.058
	3.0	0.090	0.089	1.0102	1.3375	0.17	1.089
	4.0	0.122	0.119	1.0143	1.3390	0.23	1.120

Solute	Mass %	m/mol kg^{-1}	c/mol L^{-1}	ρ/g cm^{-3}	n	Δ/°C	η/mPa s
	5.0	0.154	0.149	1.0184	1.3406	0.29	1.154
	6.0	0.186	0.179	1.0225	1.3421	0.35	1.191
	7.0	0.220	0.210	1.0267	1.3437	0.42	1.232
	8.0	0.254	0.241	1.0308	1.3453	0.50	1.276
	9.0	0.289	0.272	1.0349	1.3468		1.321
	10.0	0.325	0.304	1.0390	1.3484		1.370
	12.0	0.398	0.367	1.0473	1.3515		1.476
	14.0	0.476	0.432	1.0558	1.3548		1.593
	16.0	0.556	0.498	1.0648	1.3582		1.724
	18.0	0.641	0.565	1.0746	1.3619		1.869
Lithium	0.5	0.119	0.118	1.0012	1.3341	0.42	1.019
chloride	1.0	0.238	0.237	1.0041	1.3351	0.84	1.037
LiCl	2.0	0.481	0.476	1.0099	1.3373	1.72	1.072
	3.0	0.730	0.719	1.0157	1.3394	2.68	1.108
	4.0	0.983	0.964	1.0215	1.3415	3.73	1.146
	5.0	1.241	1.211	1.0272	1.3436	4.86	1.185
	6.0	1.506	1.462	1.0330	1.3457	6.14	1.226
	7.0	1.775	1.715	1.0387	1.3478	7.56	1.269
	8.0	2.051	1.971	1.0444	1.3499	9.11	1.313
	9.0	2.333	2.230	1.0502	1.3520	10.79	1.360
	10.0	2.621	2.491	1.0560	1.3541	12.61	1.411
	12.0	3.217	3.022	1.0675	1.3583	16.59	1.522
	14.0	3.840	3.564	1.0792	1.3625	21.04	1.647
	16.0	4.493	4.118	1.0910	1.3668		1.787
	18.0	5.178	4.683	1.1029	1.3711		1.942
	20.0	5.897	5.260	1.1150	1.3755		2.128
	22.0	6.653	5.851	1.1274	1.3799		2.341
	24.0	7.449	6.453	1.1399	1.3844		2.600
	26.0	8.288	7.069	1.1527	1.3890		2.925
	28.0	9.173	7.700	1.1658	1.3936		3.318
	30.0	10.109	8.344	1.1791	1.3983		3.785
Magnesium	0.5	0.053	0.053	1.0022	1.3343	0.26	1.024
chloride	1.0	0.106	0.106	1.0062	1.3356	0.52	1.046
$MgCl_2$	2.0	0.214	0.213	1.0144	1.3381	1.06	1.091
	3.0	0.325	0.322	1.0226	1.3406	1.65	1.139
	4.0	0.438	0.433	1.0309	1.3432	2.30	1.188
	5.0	0.553	0.546	1.0394	1.3457	3.01	1.241
	6.0	0.670	0.660	1.0479	1.3483		1.298
	7.0	0.791	0.777	1.0564	1.3508		1.358
	8.0	0.913	0.895	1.0651	1.3534		1.423
	9.0	1.039	1.015	1.0738	1.3560		1.493
	10.0	1.167	1.137	1.0826	1.3587		1.570
	12.0	1.432	1.387	1.1005	1.3641		1.745
	14.0	1.710	1.645	1.1189	1.3695		1.956
	16.0	2.001	1.911	1.1372	1.3749		2.207
	18.0	2.306	2.184	1.1553	1.3804		2.507
	20.0	2.626	2.467	1.1742	1.3859		2.867
	22.0	2.962	2.758	1.1938	1.3915		3.323
	24.0	3.317	3.060	1.2140	1.3972		3.917
	26.0	3.690	3.371	1.2346	1.4030		4.694
	28.0	4.085	3.692	1.2555	1.4089		5.709
	30.0	4.501	4.022	1.2763	1.4148		7.017
Magnesium	0.5	0.042	0.042	1.0033	1.3340	0.10	1.027
sulfate	1.0	0.084	0.084	1.0084	1.3350	0.19	1.054
$MgSO_4$	2.0	0.170	0.169	1.0186	1.3371	0.36	1.112

Solute	Mass %	m/mol kg-1	c/mol L-1	ρ/g cm-3	n	Δ/°C	η/mPa s
	3.0	0.257	0.256	1.0289	1.3391	0.52	1.177
	4.0	0.346	0.345	1.0392	1.3411	0.69	1.249
	5.0	0.437	0.436	1.0497	1.3431	0.87	1.328
	6.0	0.530	0.528	1.0602	1.3451	1.05	1.411
	7.0	0.625	0.623	1.0708	1.3471	1.24	1.498
	8.0	0.722	0.719	1.0816	1.3492	1.43	1.593
	9.0	0.822	0.817	1.0924	1.3512	1.64	1.702
	10.0	0.923	0.917	1.1034	1.3532	1.85	1.829
	12.0	1.133	1.122	1.1257	1.3572	2.31	2.104
	14.0	1.352	1.336	1.1484	1.3613	2.86	2.412
	16.0	1.582	1.557	1.1717	1.3654	3.67	2.809
	18.0	1.824	1.788	1.1955	1.3694		3.360
	20.0	2.077	2.027	1.2198	1.3735		4.147
	22.0	2.343	2.275	1.2447	1.3776		5.199
	24.0	2.624	2.532	1.2701	1.3817		6.498
	26.0	2.919	2.800	1.2961	1.3858		8.066
Maltose	0.5	0.015	0.015	1.0003	1.3337	0.03	1.016
$C_{12}H_{22}O_{11}$	1.0	0.030	0.029	1.0023	1.3345	0.06	1.030
	2.0	0.060	0.059	1.0063	1.3359	0.11	1.060
	3.0	0.090	0.089	1.0104	1.3374	0.17	1.092
	4.0	0.122	0.119	1.0144	1.3389	0.23	1.126
	5.0	0.154	0.149	1.0184	1.3404	0.29	1.162
	6.0	0.186	0.179	1.0224	1.3420	0.35	1.200
	7.0	0.220	0.210	1.0265	1.3435	0.42	1.239
	8.0	0.254	0.241	1.0305	1.3450	0.48	1.281
	9.0	0.289	0.272	1.0345	1.3466	0.55	1.325
	10.0	0.325	0.303	1.0385	1.3482	0.62	1.372
	12.0	0.398	0.367	1.0465	1.3513	0.77	1.474
	14.0	0.476	0.431	1.0545	1.3546	0.92	1.588
	16.0	0.556	0.497	1.0629	1.3578	1.08	1.715
	18.0	0.641	0.564	1.0716	1.3612	1.25	1.859
	20.0	0.730	0.631	1.0801	1.3644	1.43	2.021
	22.0	0.824	0.700	1.0894	1.3678	1.64	2.216
	24.0	0.923	0.770	1.0984	1.3714	1.85	2.463
	26.0	1.026	0.842	1.1080	1.3749	2.08	2.753
	28.0	1.136	0.914	1.1171	1.3785	2.34	3.066
	30.0	1.252	0.988	1.1269	1.3821	2.62	3.427
	40.0	1.948	1.375	1.1769	1.4013	4.41	6.926
	50.0	2.921	1.797	1.2304	1.4217		17.786
	52.0	3.165	1.886	1.2416	1.4260		22.034
	54.0	3.429	1.976	1.2528	1.4308		28.757
	56.0	3.718	2.068	1.2638	1.4350		38.226
	58.0	4.034	2.159	1.2740	1.4394		49.298
	60.0	4.382	2.253	1.2855	1.4440		
Manganese(II)	1.0	0.067	0.067	1.0080	1.3348	0.16	1.046
sulfate	2.0	0.135	0.135	1.0178	1.3366	0.31	1.090
$MnSO_4$	3.0	0.205	0.204	1.0277	1.3384	0.44	1.137
	4.0	0.276	0.275	1.0378	1.3402	0.57	1.187
	5.0	0.349	0.347	1.0480	1.3420	0.70	1.242
	6.0	0.423	0.421	1.0583	1.3438	0.84	1.301
	7.0	0.498	0.495	1.0688	1.3457	0.98	1.363
	8.0	0.576	0.572	1.0794	1.3475	1.12	1.431
	9.0	0.655	0.650	1.0902	1.3494	1.28	1.505
	10.0	0.736	0.729	1.1012	1.3513	1.44	1.587
	12.0	0.903	0.893	1.1236	1.3551	1.80	1.779

Solute	Mass %	m/mol kg^{-1}	c/mol L^{-1}	ρ/g cm^{-3}	n	Δ/°C	η/mPa s
	14.0	1.078	1.063	1.1467	1.3589	2.21	2.005
	16.0	1.261	1.240	1.1705	1.3629	2.67	2.272
	18.0	1.454	1.424	1.1950	1.3668	3.19	2.580
	20.0	1.656	1.616	1.2203	1.3708	3.80	2.938
D-Mannitol	0.5	0.028	0.027	1.0000	1.3337	0.05	1.019
$CH_2(CHOH)_4CH_2OH$	1.0	0.055	0.055	1.0017	1.3345	0.10	1.032
	2.0	0.112	0.110	1.0053	1.3359	0.21	1.057
	3.0	0.170	0.166	1.0088	1.3374	0.32	1.081
	4.0	0.229	0.222	1.0124	1.3389	0.43	1.107
	5.0	0.289	0.279	1.0159	1.3403	0.54	1.135
	6.0	0.350	0.336	1.0195	1.3418	0.66	1.166
	7.0	0.413	0.393	1.0230	1.3433	0.77	1.200
	8.0	0.477	0.451	1.0266	1.3447	0.90	1.236
	9.0	0.543	0.509	1.0302	1.3462	1.02	1.275
	10.0	0.610	0.567	1.0338	1.3477	1.15	1.314
	11.0	0.678	0.626	1.0375	1.3491	1.28	1.355
	12.0	0.749	0.686	1.0412	1.3506	1.41	1.398
	13.0	0.820	0.746	1.0450	1.3521	1.55	1.443
	14.0	0.894	0.806	1.0489	1.3536	1.69	1.489
	15.0	0.969	0.867	1.0529	1.3552	1.84	1.537
Methanol	0.5	0.157	0.156	0.9973	1.3331	0.28	1.022
CH_3OH	1.0	0.315	0.311	0.9964	1.3332	0.56	1.040
	2.0	0.637	0.621	0.9947	1.3334	1.14	1.070
	3.0	0.965	0.930	0.9930	1.3336	1.75	1.100
	4.0	1.300	1.238	0.9913	1.3339	2.37	1.131
	5.0	1.643	1.544	0.9896	1.3341	3.02	1.163
	6.0	1.992	1.850	0.9880	1.3343	3.71	1.196
	7.0	2.349	2.155	0.9864	1.3346	4.41	1.229
	8.0	2.714	2.459	0.9848	1.3348	5.13	1.264
	9.0	3.087	2.762	0.9832	1.3351	5.85	1.297
	10.0	3.468	3.064	0.9816	1.3354	6.60	1.329
	12.0	4.256	3.665	0.9785	1.3359	8.14	1.389
	14.0	5.081	4.262	0.9755	1.3365	9.72	1.446
	16.0	5.945	4.856	0.9725	1.3370	11.36	1.501
	18.0	6.851	5.447	0.9695	1.3376	13.13	1.554
	20.0	7.803	6.034	0.9666	1.3381	15.02	1.604
	22.0	8.803	6.616	0.9636	1.3387	16.98	1.652
	24.0	9.856	7.196	0.9606	1.3392	19.04	1.697
	26.0	10.966	7.771	0.9576	1.3397	21.23	1.735
	28.0	12.138	8.341	0.9545	1.3402	23.59	1.769
	30.0	13.376	8.908	0.9514	1.3407	25.91	1.795
	32.0	14.688	9.470	0.9482	1.3411	28.15	1.814
	34.0	16.078	10.028	0.9450	1.3415	30.48	1.827
	36.0	17.556	10.580	0.9416	1.3419	32.97	1.835
	38.0	19.129	11.127	0.9382	1.3422	35.60	1.839
	40.0	20.807	11.669	0.9347	1.3425	38.60	1.837
	50.0	31.211	14.288	0.9156	1.3431	54.50	1.761
	60.0	46.816	16.749	0.8944	1.3426	74.50	1.600
	70.0	72.826	19.040	0.8715	1.3411		1.368
	80.0	124.844	21.144	0.8468	1.3385		1.128
	90.0	280.899	23.045	0.8204	1.3348		0.861
	100.0		24.710	0.7917	1.3290		0.586
Nitric acid	0.5	0.080	0.079	1.0009	1.3336	0.28	1.004
HNO_3	1.0	0.160	0.159	1.0037	1.3343	0.56	1.005
	2.0	0.324	0.320	1.0091	1.3356	1.12	1.007

Solute	Mass %	m/mol kg^{-1}	c/mol L^{-1}	ρ/g cm^{-3}	n	Δ/°C	η/mPa s
	3.0	0.491	0.483	1.0146	1.3368	1.70	1.010
	4.0	0.661	0.648	1.0202	1.3381	2.32	1.014
	5.0	0.835	0.814	1.0257	1.3394	2.96	1.018
	6.0	1.013	0.982	1.0314	1.3407	3.63	1.022
	7.0	1.194	1.152	1.0370	1.3421	4.33	1.027
	8.0	1.380	1.324	1.0427	1.3434	5.05	1.032
	9.0	1.570	1.498	1.0485	1.3447	5.81	1.038
	10.0	1.763	1.673	1.0543	1.3460	6.60	1.044
	12.0	2.164	2.030	1.0660	1.3487	8.27	1.058
	14.0	2.583	2.395	1.0780	1.3514	10.08	1.075
	16.0	3.023	2.768	1.0901	1.3541	12.04	1.094
	18.0	3.484	3.149	1.1025	1.3569	14.16	1.116
	20.0	3.967	3.539	1.1150	1.3596		1.141
	22.0	4.476	3.937	1.1277	1.3624		1.169
	24.0	5.011	4.344	1.1406	1.3652		1.199
	26.0	5.576	4.760	1.1536	1.3680		1.233
	28.0	6.172	5.185	1.1668	1.3708		1.271
	30.0	6.801	5.618	1.1801	1.3736		1.311
	32.0	7.468	6.060	1.1934	1.3763		1.354
	34.0	8.175	6.512	1.2068	1.3790		1.400
	36.0	8.927	6.971	1.2202	1.3817		1.450
	38.0	9.727	7.439	1.2335	1.3842		1.504
	40.0	10.580	7.913	1.2466	1.3867		1.561
Oxalic acid $(COOH)_2$	0.5	0.056	0.056	1.0006	1.3336	0.16	1.013
	1.0	0.112	0.111	1.0030	1.3342	0.30	1.023
	1.5	0.169	0.167	1.0054	1.3347	0.44	1.033
	2.0	0.227	0.224	1.0079	1.3353	0.57	1.044
	2.5	0.285	0.281	1.0103	1.3359	0.71	1.055
	3.0	0.343	0.337	1.0126	1.3364	0.84	1.065
	3.5	0.403	0.395	1.0150	1.3370	0.97	1.076
	4.0	0.463	0.452	1.0174	1.3375	1.09	1.086
	4.5	0.523	0.510	1.0197	1.3381		1.097
	5.0	0.585	0.568	1.0220	1.3386		1.108
	6.0	0.709	0.684	1.0265	1.3397		1.129
	7.0	0.836	0.802	1.0310	1.3407		1.150
	8.0	0.966	0.920	1.0355	1.3418		1.172
Phosphoric acid H_3PO_4	0.5	0.051	0.051	1.0010	1.3335	0.12	1.010
	1.0	0.103	0.102	1.0038	1.3340	0.24	1.020
	2.0	0.208	0.206	1.0092	1.3349	0.46	1.050
	3.0	0.316	0.311	1.0146	1.3358	0.69	1.079
	4.0	0.425	0.416	1.0200	1.3367	0.93	1.108
	5.0	0.537	0.523	1.0254	1.3376	1.16	1.138
	6.0	0.651	0.631	1.0309	1.3385	1.38	1.169
	7.0	0.768	0.740	1.0363	1.3394	1.62	1.200
	8.0	0.887	0.850	1.0418	1.3403	1.88	1.232
	9.0	1.009	0.962	1.0474	1.3413	2.16	1.267
	10.0	1.134	1.075	1.0531	1.3422	2.45	1.303
	12.0	1.392	1.304	1.0647	1.3441	3.01	1.382
	14.0	1.661	1.538	1.0765	1.3460	3.76	1.469
	16.0	1.944	1.777	1.0885	1.3480	4.45	1.565
	18.0	2.240	2.022	1.1009	1.3500	5.25	1.671
	20.0	2.551	2.273	1.1135	1.3520	6.23	1.788
	22.0	2.878	2.529	1.1263	1.3540	7.38	1.914
	24.0	3.223	2.791	1.1395	1.3561	8.69	2.049
	26.0	3.585	3.059	1.1528	1.3582	10.12	2.198

Solute	Mass %	*m*/mol kg^{-1}	*c*/mol L^{-1}	ρ/g cm^{-3}	*n*	Δ/°C	η/mPa s
	28.0	3.968	3.333	1.1665	1.3604	11.64	2.365
	30.0	4.373	3.614	1.1804	1.3625	13.23	2.553
	32.0	4.802	3.901	1.1945	1.3647	14.94	2.766
	34.0	5.257	4.194	1.2089	1.3669	16.81	3.001
	36.0	5.740	4.495	1.2236	1.3691	18.85	3.260
	38.0	6.254	4.803	1.2385	1.3713	21.09	3.544
	40.0	6.803	5.117	1.2536	1.3735	23.58	3.856
Potassium bicarbonate $KHCO_3$	0.5	0.050	0.050	1.0014	1.3335	0.18	1.009
	1.0	0.101	0.100	1.0046	1.3341	0.34	1.015
	2.0	0.204	0.202	1.0114	1.3353	0.67	1.027
	3.0	0.309	0.305	1.0181	1.3365	0.98	1.040
	4.0	0.416	0.409	1.0247	1.3376	1.29	1.053
	5.0	0.526	0.515	1.0310	1.3386	1.60	1.067
	6.0	0.638	0.622	1.0379	1.3397	1.91	1.081
	7.0	0.752	0.730	1.0446	1.3409	2.22	1.096
	8.0	0.869	0.840	1.0514	1.3419	2.53	1.112
	9.0	0.988	0.951	1.0581	1.3430	2.84	1.128
	10.0	1.110	1.064	1.0650	1.3441	3.16	1.145
	12.0	1.362	1.293	1.0788	1.3462	3.79	1.183
	14.0	1.626	1.528	1.0929	1.3484	4.41	1.224
	16.0	1.903	1.770	1.1073	1.3506		1.270
	18.0	2.193	2.017	1.1221	1.3528		1.319
	20.0	2.497	2.272	1.1372	1.3550		1.373
	22.0	2.817	2.533	1.1527	1.3572		1.432
	24.0	3.154	2.801	1.1685	1.3595		1.497
Potassium bromide KBr	0.5	0.042	0.042	1.0018	1.3336	0.15	1.000
	1.0	0.085	0.084	1.0054	1.3342	0.29	0.998
	2.0	0.171	0.170	1.0127	1.3354	0.59	0.994
	3.0	0.260	0.257	1.0200	1.3366	0.88	0.990
	4.0	0.350	0.345	1.0275	1.3379	1.18	0.985
	5.0	0.442	0.435	1.0350	1.3391	1.48	0.981
	6.0	0.536	0.526	1.0426	1.3403	1.78	0.977
	7.0	0.633	0.618	1.0503	1.3416	2.10	0.974
	8.0	0.731	0.711	1.0581	1.3429	2.42	0.970
	9.0	0.831	0.806	1.0660	1.3441	2.74	0.967
	10.0	0.934	0.903	1.0740	1.3454	3.07	0.964
	12.0	1.146	1.099	1.0903	1.3481	3.76	0.958
	14.0	1.368	1.302	1.1070	1.3507	4.49	0.953
	16.0	1.601	1.512	1.1242	1.3535	5.25	0.949
	18.0	1.845	1.727	1.1419	1.3562	6.04	0.946
	20.0	2.101	1.950	1.1601	1.3591	6.88	0.944
	22.0	2.370	2.179	1.1788	1.3620	7.76	0.943
	24.0	2.654	2.416	1.1980	1.3650	8.70	0.943
	26.0	2.952	2.661	1.2179	1.3680	9.68	0.944
	28.0	3.268	2.914	1.2383	1.3711	10.72	0.947
	30.0	3.601	3.175	1.2593	1.3743	11.82	0.952
	32.0	3.954	3.445	1.2810	1.3776	12.98	0.959
	34.0	4.329	3.724	1.3033	1.3809		0.968
	36.0	4.727	4.012	1.3263	1.3843		0.979
	38.0	5.150	4.311	1.3501	1.3878		0.993
	40.0	5.602	4.620	1.3746	1.3914		1.010
Potassium carbonate K_2CO_3	0.5	0.036	0.036	1.0027	1.3339	0.18	1.013
	1.0	0.073	0.073	1.0072	1.3347	0.34	1.025
	2.0	0.148	0.147	1.0163	1.3365	0.66	1.048

Solute	Mass %	*m*/mol kg⁻¹	*c*/mol L⁻¹	ρ/g cm⁻³	*n*	Δ/°C	η/mPa s
	3.0	0.224	0.223	1.0254	1.3382	0.99	1.071
	4.0	0.301	0.299	1.0345	1.3399	1.32	1.094
	5.0	0.381	0.378	1.0437	1.3416	1.67	1.119
	6.0	0.462	0.457	1.0529	1.3433	2.03	1.146
	7.0	0.545	0.538	1.0622	1.3450	2.40	1.174
	8.0	0.629	0.620	1.0715	1.3467	2.77	1.204
	9.0	0.716	0.704	1.0809	1.3484	3.17	1.235
	10.0	0.804	0.789	1.0904	1.3501	3.57	1.269
	12.0	0.987	0.963	1.1095	1.3535	4.45	1.339
	14.0	1.178	1.144	1.1291	1.3569	5.39	1.414
	16.0	1.378	1.330	1.1490	1.3603	6.42	1.497
	18.0	1.588	1.523	1.1692	1.3637	7.55	1.594
	20.0	1.809	1.722	1.1898	1.3671	8.82	1.707
	24.0	2.285	2.139	1.2320	1.3739	11.96	1.978
	28.0	2.814	2.584	1.2755	1.3807	16.01	2.331
	32.0	3.405	3.057	1.3204	1.3874	21.46	2.834
	36.0	4.070	3.559	1.3665	1.3940	28.58	3.503
	40.0	4.824	4.093	1.4142	1.4006	37.55	4.360
	50.0	7.236	5.573	1.5404	1.4168		9.369
Potassium chloride KCl	0.5	0.067	0.067	1.0014	1.3337	0.23	1.000
	1.0	0.135	0.135	1.0046	1.3343	0.46	0.999
	2.0	0.274	0.271	1.0110	1.3357	0.92	0.999
	3.0	0.415	0.409	1.0174	1.3371	1.38	0.998
	4.0	0.559	0.549	1.0239	1.3384	1.85	0.997
	5.0	0.706	0.691	1.0304	1.3398	2.32	0.996
	6.0	0.856	0.835	1.0369	1.3411	2.80	0.994
	7.0	1.010	0.980	1.0434	1.3425	3.29	0.992
	8.0	1.166	1.127	1.0500	1.3438	3.80	0.990
	9.0	1.327	1.276	1.0566	1.3452	4.30	0.989
	10.0	1.490	1.426	1.0633	1.3466	4.81	0.988
	12.0	1.829	1.733	1.0768	1.3493	5.88	0.990
	14.0	2.184	2.048	1.0905	1.3521		0.994
	16.0	2.555	2.370	1.1043	1.3549		0.999
	18.0	2.944	2.701	1.1185	1.3577		1.004
	20.0	3.353	3.039	1.1328	1.3606		1.012
	22.0	3.783	3.386	1.1474	1.3635		1.024
	24.0	4.236	3.742	1.1623	1.3665		1.040
Potassium hydroxide KOH	0.5	0.090	0.089	1.0025	1.3340	0.30	1.010
	1.0	0.180	0.179	1.0068	1.3350	0.61	1.019
	2.0	0.364	0.362	1.0155	1.3369	1.24	1.038
	3.0	0.551	0.548	1.0242	1.3388	1.89	1.058
	4.0	0.743	0.736	1.0330	1.3408	2.57	1.079
	5.0	0.938	0.929	1.0419	1.3427	3.36	1.102
	6.0	1.138	1.124	1.0509	1.3445	4.14	1.126
	7.0	1.342	1.322	1.0599	1.3464	4.92	1.151
	8.0	1.550	1.524	1.0690	1.3483		1.177
	9.0	1.763	1.729	1.0781	1.3502		1.205
	10.0	1.980	1.938	1.0873	1.3520		1.233
	12.0	2.431	2.365	1.1059	1.3558		1.294
	14.0	2.902	2.806	1.1246	1.3595		1.361
	16.0	3.395	3.261	1.1435	1.3632		1.436
	18.0	3.913	3.730	1.1626	1.3670		1.521
	20.0	4.456	4.213	1.1818	1.3707		1.619
	22.0	5.027	4.711	1.2014	1.3744		1.732
	24.0	5.629	5.223	1.2210	1.3781		1.861

Solute	Mass %	m/mol kg^{-1}	c/mol L^{-1}	ρ/g cm^{-3}	n	Δ/°C	η/mPa s
	26.0	6.262	5.750	1.2408	1.3818		2.006
	28.0	6.931	6.293	1.2609	1.3854		2.170
	30.0	7.639	6.851	1.2813	1.3889		2.357
	40.0	11.882	9.896	1.3881	1.4068		3.879
	50.0	17.824	13.389	1.5024	1.4247		7.892
Potassium	0.5	0.030	0.030	1.0019	1.3337	0.11	0.999
iodide	1.0	0.061	0.061	1.0056	1.3343	0.22	0.997
KI	2.0	0.123	0.122	1.0131	1.3357	0.43	0.991
	3.0	0.186	0.184	1.0206	1.3370	0.64	0.986
	4.0	0.251	0.248	1.0282	1.3384	0.86	0.981
	5.0	0.317	0.312	1.0360	1.3397	1.08	0.976
	6.0	0.385	0.377	1.0438	1.3411	1.30	0.969
	7.0	0.453	0.443	1.0517	1.3425	1.53	0.963
	8.0	0.524	0.511	1.0598	1.3440	1.77	0.957
	9.0	0.596	0.579	1.0679	1.3454	2.01	0.951
	10.0	0.669	0.648	1.0762	1.3469	2.26	0.946
	12.0	0.821	0.790	1.0931	1.3498	2.77	0.937
	14.0	0.981	0.937	1.1105	1.3529	3.30	0.929
	16.0	1.147	1.088	1.1284	1.3560	3.87	0.921
	18.0	1.322	1.244	1.1469	1.3593	4.46	0.915
	20.0	1.506	1.405	1.1659	1.3626	5.09	0.910
	22.0	1.699	1.571	1.1856	1.3661	5.76	0.905
	24.0	1.902	1.744	1.2060	1.3696	6.46	0.901
	26.0	2.117	1.922	1.2270	1.3733	7.21	0.898
	28.0	2.343	2.106	1.2487	1.3771	8.01	0.895
	30.0	2.582	2.297	1.2712	1.3810	8.86	0.892
	32.0	2.835	2.495	1.2944	1.3851	9.76	0.891
	34.0	3.103	2.700	1.3185	1.3893	10.72	0.890
	36.0	3.388	2.913	1.3434	1.3936	11.73	0.890
	38.0	3.692	3.134	1.3692	1.3981	12.81	0.893
	40.0	4.016	3.364	1.3959	1.4027	13.97	0.897
Potassium	0.5	0.050	0.050	1.0014	1.3335	0.17	0.999
nitrate	1.0	0.100	0.099	1.0045	1.3339	0.33	0.996
KNO_3	2.0	0.202	0.200	1.0108	1.3349	0.64	0.990
	3.0	0.306	0.302	1.0171	1.3358	0.94	0.986
	4.0	0.412	0.405	1.0234	1.3368	1.22	0.983
	5.0	0.521	0.509	1.0298	1.3377	1.50	0.980
	6.0	0.631	0.615	1.0363	1.3386	1.76	0.977
	7.0	0.744	0.722	1.0428	1.3396	2.02	0.975
	8.0	0.860	0.830	1.0494	1.3405	2.27	0.973
	9.0	0.978	0.940	1.0560	1.3415	2.52	0.971
	10.0	1.099	1.051	1.0627	1.3425	2.75	0.970
	12.0	1.349	1.277	1.0762	1.3444		0.970
	14.0	1.610	1.509	1.0899	1.3463		0.972
	16.0	1.884	1.747	1.1039	1.3482		0.976
	18.0	2.171	1.991	1.1181	1.3502		0.982
	20.0	2.473	2.240	1.1326	1.3521		0.990
	22.0	2.790	2.497	1.1473	1.3541		0.999
	24.0	3.123	2.759	1.1623	1.3561		1.010
Potassium	0.5	0.032	0.032	1.0017		0.11	1.001
permanganate	1.0	0.064	0.064	1.0051		0.22	1.000
$KMnO_4$	1.5	0.096	0.096	1.0085		0.32	0.999
	2.0	0.129	0.128	1.0118		0.43	0.998
	3.0	0.196	0.193	1.0186			0.995

Solute	Mass %	m/mol kg⁻¹	c/mol L⁻¹	ρ/g cm⁻³	n	Δ/°C	η/mPa s
	4.0	0.264	0.260	1.0254			0.992
	5.0	0.333	0.327	1.0322			0.989
	6.0	0.404	0.394	1.0390			0.985
Potassium hydrogen phosphate K_2HPO_4	0.5	0.029	0.029	1.0025	1.3338	0.13	1.013
	1.0	0.058	0.058	1.0068	1.3345	0.25	1.023
	1.5	0.087	0.087	1.0110	1.3353	0.37	1.034
	2.0	0.117	0.117	1.0153	1.3361	0.49	1.046
	2.5	0.147	0.146	1.0195	1.3368	0.61	1.057
	3.0	0.178	0.176	1.0238	1.3376	0.73	1.069
	3.5	0.208	0.207	1.0281	1.3384	0.86	1.081
	4.0	0.239	0.237	1.0324	1.3392	0.97	1.094
	4.5	0.271	0.268	1.0368	1.3399	1.10	1.107
	5.0	0.302	0.299	1.0412	1.3407	1.22	1.120
	6.0	0.366	0.362	1.0500	1.3422	1.46	1.147
	7.0	0.432	0.426	1.0590	1.3438	1.70	1.177
	8.0	0.499	0.491	1.0680	1.3453	1.95	1.209
Potassium dihydrogen phosphate KH_2PO_4	0.5	0.037	0.037	1.0018	1.3336	0.13	1.010
	1.0	0.074	0.074	1.0053	1.3342	0.25	1.019
	1.5	0.112	0.111	1.0089	1.3348	0.37	1.028
	2.0	0.150	0.149	1.0125	1.3354	0.49	1.038
	3.0	0.227	0.225	1.0197	1.3365	0.72	1.060
	4.0	0.306	0.302	1.0269	1.3377	0.96	1.083
	5.0	0.387	0.380	1.0342	1.3388	1.19	1.108
	6.0	0.469	0.459	1.0414	1.3400	1.41	1.133
	7.0	0.553	0.539	1.0486	1.3411	1.63	1.160
	8.0	0.639	0.621	1.0558	1.3422	1.84	1.187
	9.0	0.727	0.703	1.0630	1.3434	2.04	1.215
	10.0	0.816	0.786	1.0703	1.3445	2.23	1.245
Potassium sulfate K_2SO_4	0.5	0.029	0.029	1.0022	1.3336	0.14	1.006
	1.0	0.058	0.058	1.0062	1.3343	0.26	1.011
	2.0	0.117	0.116	1.0143	1.3355	0.50	1.021
	3.0	0.177	0.176	1.0224	1.3368	0.73	1.033
	4.0	0.239	0.237	1.0306	1.3380	0.95	1.045
	5.0	0.302	0.298	1.0388	1.3393	1.17	1.058
	6.0	0.366	0.360	1.0470	1.3405		1.072
	7.0	0.432	0.424	1.0553	1.3417		1.087
	8.0	0.499	0.488	1.0637	1.3428		1.102
	9.0	0.568	0.554	1.0721	1.3440		1.117
	10.0	0.638	0.620	1.0806	1.3452		1.132
1-Propanol $CH_3CH_2CH_2OH$	1.0	0.168	0.166	0.9963	1.3339	0.31	1.051
	2.0	0.340	0.331	0.9946	1.3348	0.61	1.100
	3.0	0.515	0.496	0.9928	1.3357	0.93	1.152
	4.0	0.693	0.660	0.9911	1.3366	1.24	1.208
	5.0	0.876	0.823	0.9896	1.3376	1.57	1.267
	6.0	1.062	0.987	0.9882	1.3385	1.91	1.325
	7.0	1.252	1.149	0.9868	1.3394	2.26	1.387
	8.0	1.447	1.312	0.9855	1.3404	2.61	1.449
	9.0	1.646	1.474	0.9842	1.3414	2.99	1.514
	10.0	1.849	1.635	0.9829	1.3423	3.36	1.577
	12.0	2.269	1.958	0.9804	1.3442	4.09	1.710
	14.0	2.709	2.278	0.9779	1.3460	4.91	1.849
	16.0	3.169	2.595	0.9749	1.3477	5.78	1.986
	18.0	3.652	2.911	0.9719	1.3494	6.67	2.106

Solute	Mass %	*m*/mol kg^{-1}	*c*/mol L^{-1}	ρ/g cm^{-3}	*n*	Δ/°C	η/mPa s
	20.0	4.160	3.223	0.9686	1.3510	7.76	2.218
	24.0	5.254	3.838	0.9612	1.3539	9.12	2.432
	28.0	6.471	4.441	0.9533	1.3566	10.17	2.612
	32.0	7.830	5.033	0.9452	1.3592	10.66	2.765
	36.0	9.359	5.613	0.9370	1.3614		2.900
	40.0	11.093	6.182	0.9288	1.3635		3.010
	60.0	24.958	8.860	0.8875	1.3734		3.186
	80.0	66.556	11.275	0.8470	1.3812		2.822
	100.0		13.368	0.8034	1.3852		2.227
2-Propanol	1.0	0.168	0.166	0.9960	1.3338	0.30	1.056
$CH_3CHOHCH_3$	2.0	0.340	0.331	0.9939	1.3346	0.60	1.112
	3.0	0.515	0.495	0.9920	1.3355	0.93	1.166
	4.0	0.693	0.659	0.9902	1.3364	1.26	1.225
	5.0	0.876	0.822	0.9884	1.3373	1.61	1.287
	6.0	1.062	0.985	0.9871	1.3382	1.96	1.352
	7.0	1.252	1.148	0.9855	1.3392	2.32	1.417
	8.0	1.447	1.310	0.9843	1.3400	2.68	1.485
	9.0	1.646	1.472	0.9831	1.3410	3.06	1.553
	10.0	1.849	1.633	0.9816	1.3420	3.48	1.629
	12.0	2.269	1.955	0.9793	1.3439	4.43	1.794
	14.0	2.709	2.276	0.9772	1.3459	5.29	1.970
	16.0	3.169	2.596	0.9751	1.3478	6.36	2.160
	18.0	3.652	2.913	0.9725	1.3496	7.40	2.352
	20.0	4.160	3.227	0.9696	1.3514	8.52	2.550
	40.0	11.093	6.191	0.9302	1.3642		
	60.0	24.958	8.809	0.8824	1.3717		
	80.0	66.556	11.103	0.8341	1.3765		
	100.0		13.058	0.7848	1.3742		
Silver	0.5	0.030	0.030	1.0027	1.3336	0.10	1.003
nitrate	1.0	0.059	0.059	1.0070	1.3342	0.20	1.005
$AgNO_3$	2.0	0.120	0.120	1.0154	1.3352	0.40	1.009
	3.0	0.182	0.181	1.0239	1.3363	0.59	1.013
	4.0	0.245	0.243	1.0327	1.3374	0.78	1.016
	5.0	0.310	0.307	1.0417	1.3385	0.96	1.020
	6.0	0.376	0.371	1.0506	1.3396	1.15	1.024
	7.0	0.443	0.437	1.0597	1.3407	1.33	1.027
	8.0	0.512	0.503	1.0690	1.3419	1.51	1.031
	9.0	0.582	0.571	1.0785	1.3431	1.69	1.035
	10.0	0.654	0.641	1.0882	1.3443	1.87	1.039
	12.0	0.803	0.783	1.1079	1.3467	2.21	1.049
	14.0	0.958	0.930	1.1284	1.3493	2.55	1.060
	16.0	1.121	1.083	1.1496	1.3519	2.86	1.072
	18.0	1.292	1.241	1.1715	1.3546		1.086
	20.0	1.472	1.406	1.1942	1.3574		1.101
	22.0	1.660	1.577	1.2177	1.3602		1.117
	24.0	1.859	1.755	1.2420	1.3632		1.135
	26.0	2.068	1.940	1.2672	1.3662		1.154
	28.0	2.289	2.132	1.2933	1.3694		1.176
	30.0	2.523	2.332	1.3204	1.3726		1.200
	32.0	2.770	2.541	1.3487	1.3760		1.227
	34.0	3.033	2.758	1.3780	1.3795		1.257
	36.0	3.311	2.985	1.4087	1.3832		1.290
	38.0	3.608	3.223	1.4407	1.3871		1.326
	40.0	3.925	3.472	1.4743	1.3911		1.366

Solute	Mass %	m/mol kg^{-1}	c/mol L^{-1}	ρ/g cm^{-3}	n	Δ/°C	η/mPa s
Sodium	0.5	0.061	0.061	1.0008	1.3337	0.22	1.021
acetate	1.0	0.123	0.122	1.0034	1.3344	0.43	1.040
CH_3COONa	2.0	0.249	0.246	1.0085	1.3358	0.88	1.080
	3.0	0.377	0.371	1.0135	1.3372	1.34	1.124
	4.0	0.508	0.497	1.0184	1.3386	1.82	1.171
	5.0	0.642	0.624	1.0234	1.3400	2.32	1.222
	6.0	0.778	0.752	1.0283	1.3414	2.85	1.278
	7.0	0.918	0.882	1.0334	1.3428	3.40	1.337
	8.0	1.060	1.013	1.0386	1.3442	3.98	1.401
	9.0	1.206	1.145	1.0440	1.3456	4.57	1.468
	10.0	1.354	1.279	1.0495	1.3470		1.539
	12.0	1.662	1.552	1.0607	1.3498		1.688
	14.0	1.984	1.829	1.0718	1.3526		1.855
	16.0	2.322	2.112	1.0830	1.3554		2.054
	18.0	2.676	2.400	1.0940	1.3583		2.284
	20.0	3.047	2.694	1.1050	1.3611		2.567
	22.0	3.438	2.993	1.1159	1.3639		2.948
	24.0	3.849	3.297	1.1268	1.3666		3.400
	26.0	4.283	3.606	1.1377	1.3693		3.877
	28.0	4.741	3.921	1.1488	1.3720		4.388
	30.0	5.224	4.243	1.1602	1.3748		4.940
Sodium	0.5	0.060	0.060	1.0018	1.3337	0.20	1.015
bicarbonate	1.0	0.120	0.120	1.0054	1.3344	0.40	1.028
$NaHCO_3$	1.5	0.181	0.180	1.0089	1.3351	0.59	1.042
	2.0	0.243	0.241	1.0125	1.3357	0.78	1.057
	2.5	0.305	0.302	1.0160	1.3364	0.98	1.071
	3.0	0.368	0.364	1.0196	1.3370	1.16	1.086
	3.5	0.432	0.426	1.0231	1.3377	1.35	1.102
	4.0	0.496	0.489	1.0266	1.3383	1.54	1.118
	4.5	0.561	0.552	1.0301	1.3390	1.72	1.134
	5.0	0.627	0.615	1.0337	1.3396	1.90	1.151
	5.5	0.693	0.679	1.0372	1.3403	2.08	1.168
	6.0	0.760	0.743	1.0408	1.3409	2.26	1.185
Sodium	0.5	0.049	0.049	1.0021	1.3337	0.17	1.004
bromide	1.0	0.098	0.098	1.0060	1.3344	0.34	1.007
NaBr	2.0	0.198	0.197	1.0139	1.3358	0.69	1.012
	3.0	0.301	0.298	1.0218	1.3372	1.04	1.017
	4.0	0.405	0.400	1.0298	1.3386	1.39	1.022
	5.0	0.512	0.504	1.0380	1.3401	1.76	1.028
	6.0	0.620	0.610	1.0462	1.3415	2.14	1.034
	7.0	0.732	0.717	1.0546	1.3430	2.53	1.040
	8.0	0.845	0.826	1.0630	1.3445	2.93	1.046
	9.0	0.961	0.937	1.0716	1.3460	3.34	1.053
	10.0	1.080	1.050	1.0803	1.3475	3.77	1.060
	12.0	1.325	1.281	1.0981	1.3506	4.67	1.077
	14.0	1.582	1.519	1.1164	1.3538	5.65	1.096
	16.0	1.851	1.765	1.1352	1.3570	6.74	1.119
	18.0	2.133	2.020	1.1546	1.3604		1.144
	20.0	2.430	2.283	1.1745	1.3638		1.174
	22.0	2.741	2.555	1.1951	1.3673		1.207
	24.0	3.069	2.837	1.2163	1.3708		1.244
	26.0	3.415	3.129	1.2382	1.3745		1.287
	28.0	3.780	3.431	1.2608	1.3783		1.336
	30.0	4.165	3.744	1.2842	1.3822		1.395
	32.0	4.574	4.069	1.3083	1.3862		1.465

CONCENTRATIVE PROPERTIES OF AQUEOUS SOLUTIONS: DENSITY, REFRACTIVE INDEX, FREEZING POINT DEPRESSION, AND VISCOSITY (continued)

Solute	Mass %	m/mol kg^{-1}	c/mol L^{-1}	ρ/g cm^{-3}	n	Δ/°C	η/mPa s
	34.0	5.007	4.406	1.3333	1.3903		1.546
	36.0	5.467	4.755	1.3592	1.3946		1.639
	38.0	5.957	5.119	1.3860	1.3990		1.745
	40.0	6.479	5.496	1.4138	1.4035		1.866
Sodium carbonate Na_2CO_3	0.5	0.047	0.047	1.0034	1.3341	0.22	1.025
	1.0	0.095	0.095	1.0086	1.3352	0.43	1.049
	2.0	0.193	0.192	1.0190	1.3375	0.75	1.102
	3.0	0.292	0.291	1.0294	1.3397	1.08	1.159
	4.0	0.393	0.392	1.0398	1.3419	1.42	1.222
	5.0	0.497	0.495	1.0502	1.3440	1.77	1.292
	6.0	0.602	0.600	1.0606	1.3462	2.13	1.367
	7.0	0.710	0.707	1.0711	1.3483		1.448
	8.0	0.820	0.816	1.0816	1.3504		1.538
	9.0	0.933	0.927	1.0922	1.3525		1.638
	10.0	1.048	1.041	1.1029	1.3547		1.754
	11.0	1.166	1.156	1.1136	1.3568		1.884
	12.0	1.287	1.273	1.1244	1.3589		2.028
	13.0	1.410	1.392	1.1353	1.3610		2.186
	14.0	1.536	1.514	1.1463	1.3631		2.361
	15.0	1.665	1.638	1.1574	1.3652		2.551
Sodium chloride NaCl	0.5	0.086	0.086	1.0018	1.3339	0.30	1.011
	1.0	0.173	0.172	1.0053	1.3347	0.59	1.020
	2.0	0.349	0.346	1.0125	1.3365	1.19	1.036
	3.0	0.529	0.523	1.0196	1.3383	1.79	1.052
	4.0	0.713	0.703	1.0268	1.3400	2.41	1.068
	5.0	0.901	0.885	1.0340	1.3418	3.05	1.085
	6.0	1.092	1.069	1.0413	1.3435	3.70	1.104
	7.0	1.288	1.256	1.0486	1.3453	4.38	1.124
	8.0	1.488	1.445	1.0559	1.3470	5.08	1.145
	9.0	1.692	1.637	1.0633	1.3488	5.81	1.168
	10.0	1.901	1.832	1.0707	1.3505	6.56	1.193
	12.0	2.333	2.229	1.0857	1.3541	8.18	1.250
	14.0	2.785	2.637	1.1008	1.3576	9.94	1.317
	16.0	3.259	3.056	1.1162	1.3612	11.89	1.388
	18.0	3.756	3.486	1.1319	1.3648	14.04	1.463
	20.0	4.278	3.928	1.1478	1.3684	16.46	1.557
	22.0	4.826	4.382	1.1640	1.3721	19.18	1.676
	24.0	5.403	4.847	1.1804	1.3757		1.821
	26.0	6.012	5.326	1.1972	1.3795		1.990
Sodium citrate $(HO)C(COONa)_3$	1.0	0.039	0.039	1.0049	1.3348	0.20	1.043
	2.0	0.079	0.078	1.0120	1.3366	0.39	1.081
	3.0	0.120	0.118	1.0186	1.3383	0.59	1.122
	4.0	0.161	0.159	1.0260	1.3401	0.79	1.166
	5.0	0.204	0.200	1.0331	1.3419	0.97	1.210
	6.0	0.247	0.242	1.0405	1.3437	1.17	1.263
	7.0	0.292	0.284	1.0482	1.3455	1.36	1.314
	8.0	0.337	0.327	1.0557	1.3473	1.57	1.371
	9.0	0.383	0.371	1.0632	1.3491	1.77	1.427
	10.0	0.431	0.415	1.0708	1.3509	1.96	1.499
	12.0	0.528	0.505	1.0861	1.3546	2.38	1.649
	14.0	0.631	0.598	1.1019	1.3583	2.82	1.832
	16.0	0.738	0.693	1.1173	1.3618	3.27	2.045
	18.0	0.851	0.790	1.1327	1.3656	3.82	2.290
	20.0	0.969	0.891	1.1492	1.3693	4.39	2.596

Solute	Mass %	m/mol kg^{-1}	c/mol L^{-1}	ρ/g cm^{-3}	n	Δ/°C	η/mPa s
	24.0	1.224	1.099	1.1813	1.3767		3.409
	28.0	1.507	1.318	1.2151	1.3845		4.586
	32.0	1.823	1.548	1.2487	1.3923		6.541
	36.0	2.180	1.792	1.2843	1.4001		9.788
Sodium	0.5	0.126	0.125	1.0039	1.3344	0.43	1.027
hydroxide	1.0	0.253	0.252	1.0095	1.3358	0.86	1.054
NaOH	2.0	0.510	0.510	1.0207	1.3386	1.74	1.112
	3.0	0.773	0.774	1.0318	1.3414	2.64	1.176
	4.0	1.042	1.043	1.0428	1.3441	3.59	1.248
	5.0	1.316	1.317	1.0538	1.3467	4.57	1.329
	6.0	1.596	1.597	1.0648	1.3494	5.60	1.416
	7.0	1.882	1.883	1.0758	1.3520	6.69	1.510
	8.0	2.174	2.174	1.0869	1.3546	7.87	1.616
	9.0	2.473	2.470	1.0979	1.3572	9.12	1.737
	10.0	2.778	2.772	1.1089	1.3597	10.47	1.882
	12.0	3.409	3.393	1.1309	1.3648	13.42	2.201
	14.0	4.070	4.036	1.1530	1.3697	16.76	2.568
	15.0	4.412	4.365	1.1640	1.3722		2.789
	16.0	4.762	4.701	1.1751	1.3746		3.043
	18.0	5.488	5.387	1.1971	1.3793		3.698
	20.0	6.250	6.096	1.2192	1.3840		4.619
	22.0	7.052	6.827	1.2412	1.3885		5.765
	24.0	7.895	7.579	1.2631	1.3929		7.100
	26.0	8.784	8.352	1.2848	1.3971		8.744
	28.0	9.723	9.145	1.3064	1.4012		10.832
	30.0	10.715	9.958	1.3277	1.4051		13.517
	32.0	11.766	10.791	1.3488	1.4088		16.844
	34.0	12.880	11.643	1.3697	1.4123		20.751
	36.0	14.064	12.512	1.3901	1.4156		25.290
	38.0	15.324	13.398	1.4102	1.4186		30.461
	40.0	16.668	14.300	1.4299	1.4215		36.312
Sodium	0.5	0.059	0.059	1.0016	1.3336	0.20	1.004
nitrate	1.0	0.119	0.118	1.0050	1.3341	0.40	1.007
$NaNO_3$	2.0	0.240	0.238	1.0117	1.3353	0.79	1.012
	3.0	0.364	0.359	1.0185	1.3364	1.18	1.018
	4.0	0.490	0.483	1.0254	1.3375	1.56	1.025
	5.0	0.619	0.607	1.0322	1.3387	1.94	1.032
	6.0	0.751	0.734	1.0392	1.3398	2.32	1.040
	7.0	0.886	0.862	1.0462	1.3409	2.70	1.049
	8.0	1.023	0.991	1.0532	1.3421	3.08	1.059
	9.0	1.164	1.123	1.0603	1.3432	3.46	1.069
	10.0	1.307	1.256	1.0674	1.3443	3.84	1.081
	12.0	1.604	1.527	1.0819	1.3466	4.60	1.107
	14.0	1.915	1.806	1.0967	1.3489	5.37	1.138
	18.0	2.583	2.387	1.1272	1.3536	6.98	1.215
	20.0	2.941	2.689	1.1429	1.3559	7.81	1.263
	30.0	5.042	4.326	1.2256	1.3678		1.609
	40.0	7.844	6.200	1.3175	1.3802		2.226
Sodium	0.5	0.031	0.031	1.0042	1.3343	0.19	1.033
phosphate	1.0	0.062	0.062	1.0100	1.3356	0.37	1.064
Na_3PO_4	1.5	0.093	0.093	1.0158	1.3369	0.53	1.094
	2.0	0.124	0.125	1.0216	1.3381	0.67	1.126
	2.5	0.156	0.157	1.0275	1.3394	0.79	1.161
	3.0	0.189	0.189	1.0335	1.3406		1.198

Solute	Mass %	m/mol kg⁻¹	c/mol L⁻¹	ρ/g cm⁻³	n	Δ/°C	η/mPa s
	3.5	0.221	0.222	1.0395	1.3419		1.238
	4.0	0.254	0.255	1.0456	1.3432		1.281
	4.5	0.287	0.289	1.0517	1.3444		1.327
	5.0	0.321	0.323	1.0579	1.3457		1.375
	5.5	0.355	0.357	1.0642	1.3470		1.426
	6.0	0.389	0.392	1.0705	1.3482		1.480
	6.5	0.424	0.427	1.0768	1.3495		1.538
	7.0	0.459	0.462	1.0832	1.3507		1.598
	7.5	0.495	0.498	1.0896	1.3519		1.662
	8.0	0.530	0.535	1.0961	1.3532		1.729
Sodium hydrogen phosphate Na_2HPO_4	0.5	0.035	0.035	1.0032	1.3340	0.17	1.021
	1.0	0.071	0.071	1.0082	1.3349	0.32	1.042
	1.5	0.107	0.107	1.0131	1.3358	0.46	1.064
	2.0	0.144	0.143	1.0180	1.3368		1.088
	2.5	0.181	0.180	1.0229	1.3377		1.113
	3.0	0.218	0.217	1.0279	1.3386		1.138
	3.5	0.255	0.255	1.0328	1.3396		1.165
	4.0	0.293	0.292	1.0378	1.3405		1.193
	4.5	0.332	0.331	1.0428	1.3414		1.223
	5.0	0.371	0.369	1.0478	1.3424		1.254
	5.5	0.410	0.408	1.0528	1.3433		1.286
Sodium dihydrogen phosphate NaH_2PO_4	0.5	0.042	0.042	1.0019	1.3336	0.14	1.018
	1.0	0.084	0.084	1.0056	1.3343	0.28	1.035
	1.5	0.127	0.126	1.0094	1.3349	0.42	1.051
	2.0	0.170	0.169	1.0131	1.3356	0.56	1.068
	2.5	0.214	0.212	1.0168	1.3362	0.70	1.085
	3.0	0.258	0.255	1.0206	1.3369	0.84	1.103
	3.5	0.302	0.299	1.0244	1.3375	0.98	1.121
	4.0	0.347	0.343	1.0281	1.3382	1.12	1.140
	4.5	0.393	0.387	1.0319	1.3388	1.25	1.160
	5.0	0.439	0.432	1.0358	1.3395	1.39	1.180
	6.0	0.532	0.522	1.0434	1.3408	1.65	1.223
	7.0	0.627	0.613	1.0511	1.3421	1.89	1.270
	8.0	0.725	0.706	1.0589	1.3434	2.12	1.319
	9.0	0.824	0.800	1.0668	1.3447	2.35	1.371
	10.0	0.926	0.896	1.0747	1.3460	2.58	1.428
	12.0	1.137	1.091	1.0907	1.3486	3.06	1.552
	14.0	1.357	1.292	1.1070	1.3512	3.53	1.694
	16.0	1.588	1.499	1.1236	1.3538	4.03	1.861
	18.0	1.830	1.711	1.1404	1.3565	4.55	2.050
	20.0	2.084	1.930	1.1576	1.3592	5.10	2.283
	22.0	2.351	2.155	1.1752	1.3618		2.550
	24.0	2.632	2.387	1.1931	1.3646		2.850
	26.0	2.929	2.625	1.2113	1.3673		3.214
	28.0	3.242	2.870	1.2299	1.3700		3.682
	30.0	3.572	3.123	1.2488	1.3728		4.300
	32.0	3.923	3.383	1.2682	1.3756		5.079
	34.0	4.294	3.650	1.2879	1.3784		6.008
	36.0	4.689	3.925	1.3080	1.3812		7.098
	38.0	5.109	4.208	1.3285	1.3840		8.363
	40.0	5.557	4.499	1.3493	1.3869		9.814
Sodium sulfate Na_2SO_4	0.5	0.035	0.035	1.0027	1.3338	0.17	1.013
	1.0	0.071	0.071	1.0071	1.3345	0.32	1.026
	2.0	0.144	0.143	1.0161	1.3360	0.61	1.058

Solute	Mass %	m/mol kg^{-1}	c/mol L^{-1}	ρ/g cm^{-3}	n	Δ/°C	η/mPa s
	3.0	0.218	0.217	1.0252	1.3376	0.87	1.091
	4.0	0.293	0.291	1.0343	1.3391	1.13	1.126
	5.0	0.371	0.367	1.0436	1.3406	1.36	1.163
	6.0	0.449	0.445	1.0526	1.3420	1.56	1.202
	7.0	0.530	0.523	1.0619	1.3435		1.244
	8.0	0.612	0.603	1.0713	1.3449		1.289
	9.0	0.696	0.685	1.0808	1.3464		1.337
	10.0	0.782	0.768	1.0905	1.3479		1.390
	12.0	0.960	0.938	1.1101	1.3509		1.508
	14.0	1.146	1.114	1.1301	1.3539		1.646
	16.0	1.341	1.296	1.1503	1.3567		1.812
	18.0	1.545	1.483	1.1705	1.3595		2.005
	20.0	1.760	1.677	1.1907	1.3620		2.227
	22.0	1.986	1.875	1.2106	1.3643		2.481
Sodium	0.5	0.032	0.032	1.0024	1.3340	0.14	1.012
thiosulfate	1.0	0.064	0.064	1.0065	1.3351	0.28	1.023
$Na_2S_2O_3$	2.0	0.129	0.128	1.0148	1.3371	0.57	1.044
	3.0	0.196	0.194	1.0231	1.3392	0.84	1.066
	4.0	0.264	0.261	1.0315	1.3413	1.09	1.090
	5.0	0.333	0.329	1.0399	1.3434	1.34	1.115
	6.0	0.404	0.398	1.0483	1.3454	1.59	1.141
	7.0	0.476	0.468	1.0568	1.3475	1.83	1.169
	8.0	0.550	0.539	1.0654	1.3496	2.06	1.199
	9.0	0.626	0.611	1.0740	1.3517	2.30	1.231
	10.0	0.703	0.685	1.0827	1.3538	2.55	1.267
	12.0	0.862	0.835	1.1003	1.3581	3.06	1.345
	14.0	1.030	0.990	1.1182	1.3624	3.60	1.435
	16.0	1.205	1.150	1.1365	1.3667	4.17	1.537
	18.0	1.388	1.315	1.1551	1.3711	4.76	1.657
	20.0	1.581	1.485	1.1740	1.3756	5.37	1.798
	30.0	2.711	2.417	1.2739	1.3987		2.903
	40.0	4.216	3.498	1.3827	1.4229		5.758
Strontium	0.5	0.032	0.032	1.0027	1.3339	0.16	1.012
chloride	1.0	0.064	0.064	1.0071	1.3348	0.31	1.021
$SrCl_2$	2.0	0.129	0.128	1.0161	1.3366	0.62	1.039
	3.0	0.195	0.194	1.0252	1.3384	0.93	1.057
	4.0	0.263	0.261	1.0344	1.3402	1.26	1.076
	5.0	0.332	0.329	1.0437	1.3421	1.61	1.096
	6.0	0.403	0.399	1.0532	1.3440	1.98	1.116
	7.0	0.475	0.469	1.0628	1.3459	2.38	1.136
	8.0	0.549	0.541	1.0726	1.3478	2.80	1.157
	9.0	0.624	0.615	1.0825	1.3498	3.25	1.180
	10.0	0.701	0.689	1.0925	1.3518	3.74	1.204
	12.0	0.860	0.843	1.1131	1.3558	4.81	1.258
	14.0	1.027	1.002	1.1342	1.3599	6.03	1.317
	16.0	1.202	1.167	1.1558	1.3641	7.41	1.383
	18.0	1.385	1.338	1.1780	1.3684	8.98	1.460
	20.0	1.577	1.515	1.2008	1.3728	10.74	1.549
	22.0	1.779	1.699	1.2241	1.3772	12.74	1.650
	24.0	1.992	1.890	1.2481	1.3817	14.99	1.765
	26.0	2.216	2.087	1.2728	1.3864		1.897
	28.0	2.453	2.293	1.2983	1.3911		2.056
	30.0	2.703	2.507	1.3248	1.3961		2.245
	32.0	2.968	2.730	1.3523	1.4013		2.527

Solute	Mass %	*m*/mol kg⁻¹	*c*/mol L⁻¹	ρ/g cm⁻³	*n*	Δ/°C	η/mPa s
	34.0	3.250	2.962	1.3811	1.4067		2.846
	36.0	3.548	3.205	1.4114	1.4124		3.206
Sucrose	0.5	0.015	0.015	1.0002	1.3337	0.03	1.015
$C_{12}H_{22}O_{11}$	1.0	0.030	0.029	1.0021	1.3344	0.06	1.028
	2.0	0.060	0.059	1.0060	1.3359	0.11	1.055
	3.0	0.090	0.089	1.0099	1.3373	0.17	1.084
	4.0	0.122	0.118	1.0139	1.3388	0.23	1.114
	5.0	0.154	0.149	1.0178	1.3403	0.29	1.146
	6.0	0.186	0.179	1.0218	1.3418	0.35	1.179
	7.0	0.220	0.210	1.0259	1.3433	0.42	1.215
	8.0	0.254	0.241	1.0299	1.3448	0.49	1.254
	9.0	0.289	0.272	1.0340	1.3463	0.55	1.294
	10.0	0.325	0.303	1.0381	1.3478	0.63	1.336
	12.0	0.398	0.367	1.0465	1.3509	0.77	1.429
	14.0	0.476	0.431	1.0549	1.3541	0.93	1.534
	16.0	0.556	0.497	1.0635	1.3573	1.10	1.653
	18.0	0.641	0.564	1.0722	1.3606	1.27	1.790
	20.0	0.730	0.632	1.0810	1.3639	1.47	1.945
	22.0	0.824	0.700	1.0899	1.3672	1.67	2.124
	24.0	0.923	0.771	1.0990	1.3706	1.89	2.331
	26.0	1.026	0.842	1.1082	1.3741	2.12	2.573
	28.0	1.136	0.914	1.1175	1.3776	2.37	2.855
	30.0	1.252	0.988	1.1270	1.3812	2.64	3.187
	32.0	1.375	1.063	1.1366	1.3848	2.94	3.762
	34.0	1.505	1.139	1.1464	1.3885	3.27	4.052
	36.0	1.643	1.216	1.1562	1.3922	3.63	4.621
	38.0	1.791	1.295	1.1663	1.3960	4.02	5.315
	40.0	1.948	1.375	1.1765	1.3999	4.45	6.162
	42.0	2.116	1.456	1.1868	1.4038	4.93	7.234
	44.0	2.295	1.539	1.1972	1.4078		8.596
	46.0	2.489	1.623	1.2079	1.4118		10.301
	48.0	2.697	1.709	1.2186	1.4159		12.515
	50.0	2.921	1.796	1.2295	1.4201		15.431
	60.0	4.382	2.255	1.2864	1.4419		58.487
	70.0	6.817	2.755	1.3472	1.4654		481.561
	80.0	11.686	3.299	1.4117	1.4906		
Sulfuric acid	0.5	0.051	0.051	1.0016	1.3336	0.21	1.010
H_2SO_4	1.0	0.103	0.102	1.0049	1.3342	0.42	1.019
	2.0	0.208	0.206	1.0116	1.3355	0.80	1.036
	3.0	0.315	0.311	1.0183	1.3367	1.17	1.059
	4.0	0.425	0.418	1.0250	1.3379	1.60	1.085
	5.0	0.537	0.526	1.0318	1.3391	2.05	1.112
	6.0	0.651	0.635	1.0385	1.3403	2.50	1.136
	7.0	0.767	0.746	1.0453	1.3415	2.95	1.159
	8.0	0.887	0.858	1.0522	1.3427	3.49	1.182
	9.0	1.008	0.972	1.0591	1.3439	4.08	1.206
	10.0	1.133	1.087	1.0661	1.3451	4.64	1.230
	12.0	1.390	1.322	1.0802	1.3475	5.93	1.282
	14.0	1.660	1.563	1.0947	1.3500	7.49	1.337
	16.0	1.942	1.810	1.1094	1.3525	9.26	1.399
	18.0	2.238	2.064	1.1245	1.3551	11.29	1.470
	20.0	2.549	2.324	1.1398	1.3576	13.64	1.546
	22.0	2.876	2.592	1.1554	1.3602	16.48	1.624
	24.0	3.220	2.866	1.1714	1.3628	19.85	1.706
	26.0	3.582	3.147	1.1872	1.3653	24.29	1.797
	28.0	3.965	3.435	1.2031	1.3677	29.65	1.894

CONCENTRATIVE PROPERTIES OF AQUEOUS SOLUTIONS: DENSITY, REFRACTIVE INDEX, FREEZING POINT DEPRESSION, AND VISCOSITY (continued)

Solute	Mass %	m/mol kg^{-1}	c/mol L^{-1}	ρ/g cm^{-3}	n	Δ/°C	η/mPa s
	30.0	4.370	3.729	1.2191	1.3701	36.21	2.001
	32.0	4.798	4.030	1.2353	1.3725	44.76	2.122
	34.0	5.252	4.339	1.2518	1.3749	55.28	2.255
	36.0	5.735	4.656	1.2685	1.3773		2.392
	38.0	6.249	4.981	1.2855	1.3797		2.533
	40.0	6.797	5.313	1.3028	1.3821		2.690
	42.0	7.383	5.655	1.3205	1.3846		2.872
	44.0	8.011	6.005	1.3386	1.3870		3.073
	46.0	8.685	6.364	1.3570	1.3895		3.299
	48.0	9.411	6.734	1.3759	1.3920		3.546
	50.0	10.196	7.113	1.3952	1.3945		3.826
	52.0	11.045	7.502	1.4149	1.3971		4.142
	54.0	11.969	7.901	1.4351	1.3997		4.499
	56.0	12.976	8.312	1.4558	1.4024		4.906
	58.0	14.080	8.734	1.4770	1.4050		5.354
	60.0	15.294	9.168	1.4987	1.4077		5.917
	70.0	23.790	11.494	1.6105			
	80.0	40.783	14.088	1.7272			
	90.0	91.762	16.649	1.8144			
	92.0	117.251	17.109	1.8240			
	94.0	159.734	17.550	1.8312			
	96.0	244.698	17.966	1.8355			
	98.0	499.592	18.346	1.8361			
	100.0		18.663	1.8305			
Trichloroacetic	0.5	0.031	0.031	1.0008	1.3337	0.11	1.011
acid	1.0	0.062	0.061	1.0034	1.3343	0.21	1.021
CCl_3COOH	2.0	0.125	0.123	1.0083	1.3356	0.42	1.044
	3.0	0.189	0.186	1.0133	1.3369	0.64	1.069
	4.0	0.255	0.249	1.0182	1.3381	0.86	1.096
	5.0	0.322	0.313	1.0230	1.3394	1.08	1.123
	6.0	0.391	0.377	1.0279	1.3406	1.30	1.150
	7.0	0.461	0.442	1.0328	1.3418	1.53	1.177
	8.0	0.532	0.508	1.0378	1.3431	1.76	1.204
	9.0	0.605	0.574	1.0428	1.3444	1.99	1.233
	10.0	0.680	0.641	1.0479	1.3456	2.23	1.263
	12.0	0.835	0.777	1.0583	1.3483	2.73	1.326
	14.0	0.996	0.916	1.0692	1.3510	3.26	1.393
	16.0	1.166	1.058	1.0806	1.3539	3.82	1.462
	18.0	1.343	1.203	1.0921	1.3568		1.533
	20.0	1.530	1.351	1.1035	1.3597		1.608
	24.0	1.933	1.654	1.1260	1.3652		1.768
	28.0	2.380	1.968	1.1485	1.3705		1.935
	32.0	2.880	2.294	1.1713	1.3759		2.118
	36.0	3.443	2.632	1.1947	1.3813		2.320
	40.0	4.080	2.984	1.2188	1.3868		1.543
	44.0	4.809	3.349	1.2435	1.3923		2.797
	48.0	5.650	3.726	1.2682	1.3977		3.076
Tris	0.5	0.041	0.041	0.9994	1.3337	0.08	1.014
(hydroxymethyl)-	1.0	0.083	0.083	1.0006	1.3344	0.16	1.027
methylamine	2.0	0.168	0.166	1.0030	1.3359	0.31	1.054
$H_2NC(CH_2OH)_3$	3.0	0.255	0.249	1.0054	1.3374	0.47	1.083
	4.0	0.344	0.333	1.0078	1.3388	0.64	1.115
	5.0	0.434	0.417	1.0103	1.3403	0.80	1.148
	6.0	0.527	0.502	1.0128	1.3418	0.97	1.182

Solute	Mass %	m/mol kg^{-1}	c/mol L^{-1}	ρ/g cm^{-3}	n	Δ/°C	η/mPa s
	7.0	0.621	0.587	1.0153	1.3433	1.15	1.218
	8.0	0.718	0.672	1.0179	1.3448	1.33	1.256
	9.0	0.816	0.758	1.0204	1.3463	1.51	1.295
	10.0	0.917	0.844	1.0230	1.3478	1.70	1.337
	12.0	1.126	1.019	1.0282	1.3508	2.08	1.427
	14.0	1.344	1.194	1.0335	1.3539	2.47	1.527
	16.0	1.572	1.372	1.0389	1.3570	2.90	1.642
	18.0	1.812	1.552	1.0443	1.3601	3.36	1.772
	20.0	2.064	1.733	1.0498	1.3633	3.85	1.920
	30.0	3.538	2.670	1.0781	1.3797		2.998
	40.0	5.503	3.657	1.1076	1.3970		5.208
Urea	0.5	0.084	0.083	0.9995	1.3337	0.16	1.007
$(NH_2)_2CO$	1.0	0.168	0.167	1.0007	1.3344	0.31	1.010
	2.0	0.340	0.334	1.0033	1.3358	0.62	1.012
	3.0	0.515	0.502	1.0058	1.3372	0.93	1.017
	4.0	0.694	0.672	1.0085	1.3387	1.24	1.025
	5.0	0.876	0.842	1.0111	1.3401	1.55	1.033
	6.0	1.063	1.013	1.0138	1.3416	1.88	1.041
	7.0	1.253	1.185	1.0165	1.3431	2.22	1.049
	8.0	1.448	1.358	1.0192	1.3446	2.56	1.057
	9.0	1.647	1.531	1.0220	1.3461	2.91	1.065
	10.0	1.850	1.706	1.0248	1.3476	3.26	1.074
	12.0	2.270	2.059	1.0304	1.3506	3.95	1.091
	14.0	2.710	2.415	1.0360	1.3537	4.66	1.109
	16.0	3.171	2.775	1.0417	1.3568	5.40	1.130
	18.0	3.655	3.139	1.0473	1.3599	6.19	1.153
	20.0	4.163	3.506	1.0530	1.3629	7.00	1.178
	22.0	4.696	3.878	1.0586	1.3661	7.81	1.205
	24.0	5.258	4.253	1.0643	1.3692	8.64	1.235
	26.0	5.850	4.632	1.0699	1.3723	9.52	1.266
	28.0	6.475	5.014	1.0756	1.3754	10.45	1.298
	30.0	7.136	5.401	1.0812	1.3785	11.40	1.332
	32.0	7.835	5.791	1.0869	1.3817	12.34	1.371
	34.0	8.577	6.185	1.0926	1.3848	13.27	1.413
	36.0	9.366	6.584	1.0984	1.3881	14.20	1.459
	38.0	10.205	6.988	1.1044	1.3913	15.11	1.509
	40.0	11.100	7.397	1.1106	1.3947	15.99	1.565
	42.0	12.057	7.812	1.1171	1.3982	16.83	1.629
	44.0	13.082	8.234	1.1239	1.4018	17.62	1.700
	46.0	14.183	8.665	1.1313	1.4056		1.780
Zinc sulfate	0.5	0.031	0.031	1.0034	1.3339	0.08	1.021
$ZnSO_4$	1.0	0.063	0.062	1.0085	1.3348	0.15	1.040
	2.0	0.126	0.126	1.0190	1.3366	0.28	1.081
	3.0	0.192	0.191	1.0296	1.3384	0.41	1.126
	4.0	0.258	0.258	1.0403	1.3403	0.53	1.175
	5.0	0.326	0.326	1.0511	1.3421	0.65	1.227
	6.0	0.395	0.395	1.0620	1.3439	0.77	1.283
	7.0	0.466	0.465	1.0730	1.3457	0.89	1.341
	8.0	0.539	0.537	1.0842	1.3475	1.01	1.403
	9.0	0.613	0.611	1.0956	1.3494	1.14	1.470
	10.0	0.688	0.686	1.1071	1.3513	1.27	1.545
	12.0	0.845	0.840	1.1308	1.3551	1.55	1.716
	14.0	1.008	1.002	1.1553	1.3590	1.89	1.918
	16.0	1.180	1.170	1.1806	1.3630	2.31	2.152

ION PRODUCT OF WATER SUBSTANCE

William L. Marshall and E. U. Franck

Pressure (bars)	Temperature (°C) 0	25	50	75	100	150	200	250	300
Saturated vapor	14.938	13.995	13.275	12.712	12.265	11.638	11.289	11.191	11.406
250	14.83	13.90	13.19	12.63	12.18	11.54	11.16	11.01	11.14
500	14.72	13.82	13.11	12.55	12.10	11.45	11.05	10.85	10.86
750	14.62	13.73	13.04	12.48	12.03	11.36	10.95	10.72	10.66
1,000	14.53	13.66	12.96	12.41	11.96	11.29	10.86	10.60	10.50
1,500	14.34	13.53	12.85	12.29	11.84	11.16	10.71	10.43	10.26
2,000	14.21	13.40	12.73	12.18	11.72	11.04	10.57	10.27	10.08
2,500	14.08	13.28	12.62	12.07	11.61	10.92	10.45	10.12	9.91
3,000	13.97	13.18	12.53	11.98	11.53	10.83	10.34	9.99	9.76
3,500	13.87	13.09	12.44	11.90	11.44	10.74	10.24	9.88	9.63
4,000	13.77	13.00	12.35	11.82	11.37	10.66	10.16	9.79	9.52
5,000	13.60	12.83	12.19	11.66	11.22	10.52	10.00	9.62	9.34
6,000	13.44	12.68	12.05	11.53	11.09	10.39	9.87	9.48	9.18
7,000	13.31	12.55	11.93	11.41	10.97	10.27	9.75	9.35	9.04
8,000	13.18	12.43	11.82	11.30	10.86	10.17	9.64	9.24	8.93
9,000	13.04	12.31	11.71	11.20	10.77	10.07	9.54	9.13	8.82
10,000	12.91	12.21	11.62	11.11	10.68	9.98	9.45	9.04	8.71

Pressure (bars)	Temperature (°C) 350	400	450	500	600	700	800	900	1000
Saturated vapor	12.30	—	—	—	—	—	—	—	—
250	11.77	19.43	21.59	22.40	23.27	23.81	24.23	24.59	24.93
500	11.14	11.88	13.74	16.13	18.30	19.29	19.92	20.39	20.80
750	10.79	11.17	11.89	13.01	15.25	16.55	17.35	17.93	18.39
1,000	10.54	10.77	11.19	11.81	13.40	14.70	15.58	16.22	16.72
1,500	10.22	10.29	10.48	10.77	11.59	12.50	13.30	13.97	14.50
2,000	9.98	9.98	10.07	10.23	10.73	11.36	11.98	12.54	12.97
2,500	9.79	9.74	9.77	9.86	10.18	10.63	11.11	11.59	12.02
3,000	9.61	9.54	9.53	9.57	9.78	10.11	10.49	10.89	11.24
3,500	9.47	9.37	9.33	9.34	9.48	9.71	10.02	10.35	10.62
4,000	9.34	9.22	9.16	9.15	9.23	9.41	9.65	9.93	10.13
5,000	9.13	8.99	8.90	8.85	8.85	8.95	9.11	9.30	9.42
6,000	8.96	8.80	8.69	8.62	8.57	8.61	8.72	8.86	8.97
7,000	8.81	8.64	8.51	8.42	8.34	8.34	8.40	8.51	8.64
8,000	8.68	8.50	8.36	8.25	8.13	8.10	8.13	8.21	8.38
9,000	8.57	8.37	8.22	8.10	7.95	7.89	7.89	7.95	8.12
10,000	8.46	8.25	8.09	7.96	7.78	7.70	7.68	7.70	7.85

Data in this table were calculated from the equation, $\log_{10} K_w^* = A + B/T + C/T^2 + D/T^3 + (E + F/T + G/T^2) \log_{10} \rho_w^*$, where $K_w^* = K_w/(\text{mol kg}^{-1})$, and $\rho_w^* = \rho_w/(\text{g cm}^{-3})$. The parameters are:

$A = -4.098$
$B = -3245.2$ K
$C = +2.2362 \times 10^5$ K^2
$D = -3.984 \times 10^7$ K^3
$E = +13.957$
$F = 1262.3$ K
$G = +8.5641 \times 10^5$ K^2

Reprinted with permission from W. L. Marshall and E. U. Franck, *J. Phys. Chem. Ref. Data*, 10, 295, 1981.

IONIZATION CONSTANT OF NORMAL AND HEAVY WATER

This table gives the ionization constant in molality terms for H_2O and D_2O at temperatures from 0 to 100°C at the saturated vapor pressure. The quantity tabulated is $-\log K_W$, where K_W is defined by

$$K_W = m_+ \times m_-$$

and m_+ and m_- are the molalities, in mol/kg of water, for H^+ and OH^-, respectively.

REFERENCES

1. W.L. Marshall and E.U. Franck, *J. Phys. Chem. Ref. Data*, 10, 295, 1981.
2. R.E. Mesmer and D.L. Herting, *J. Solution Chem.*, 7, 901, 1978.

	$-\log K_W$	
t/°C	H_2O	D_2O
0	14.938	15.972
5	14.727	15.743
10	14.528	15.527
15	14.340	15.324
20	14.163	15.132
25	13.995	14.951
30	13.836	14.779
35	13.685	14.616
40	13.542	14.462
45	13.405	14.316
50	13.275	14.176
55	13.152	14.044
60	13.034	13.918
65	12.921	13.798
70	12.814	13.683
75	12.712	13.574
80	12.613	13.470
85	12.520	13.371
90	12.428	13.276
95	12.345	13.186
100	12.265	13.099

SOLUBILITY OF SELECTED GASES IN WATER

L. H. Gevantman

The values in this table are taken almost exclusively from the International Union of Pure and Applied Chemistry "Solubility Data Series". Unless noted, they comprise evaluated data fitted to a smoothing equation. The data at each temperature are then derived from the smoothing equation which expresses the mole fraction solubility X_1 of the gas in solution as:

$$\ln X_1 = A + B/T^* + C \ln T^*$$

where

$$T^* = T/100\ \text{K}$$

All values refer to a partial pressure of the gas of 101.325 kPa (one atmosphere).

The equation constants, the standard deviation for $\ln X_1$ (except where noted), and the temperature range over which the equation applies are given in the column headed Equation constants. There are two exceptions. The equation for methane has an added term, DT^*. The equation for H_2Se and H_2S takes the form,

$$\ln X_1 = A + B/T + C \ln T + DT$$

where T is the temperature in kelvin.

Solubilities given for those gases which react with water, namely ozone, nitrogen oxides, chlorine and its oxides, carbon dioxide, hydrogen sulfide, hydrogen selenide and sulfur dioxide, are recorded as bulk solubilities; i.e., all chemical species of the gas and its reaction products with water are included.

Gas	T/K	Solubility (X_1)	Equation constants	Ref.
Hydrogen (H_2)	288.15	1.510×10^{-5}	$A = -48.1611$	1
$M_r = 2.01588$	293.15	1.455×10^{-5}	$B = 55.2845$	
	298.15	1.411×10^{-5}	$C = 16.8893$	
	303.15	1.377×10^{-5}	Std. dev. = ± 0.54%	
	308.15	1.350×10^{-5}	Temp.range = 273.15—353.15	
Deuterium (D_2)	283.15	$1.675 \times 10^{-5} \pm 0.57\%$	Averaged experimental	1
$M_r = 4.0282$	288.15	$1.595 \times 10^{-5} \pm 0.57\%$	values	
	293.15	$1.512 \times 10^{-5} \pm 0.78\%$	Temp. range = 278.15—303.15	
	298.15	$1.460 \times 10^{-5} \pm 0.52\%$		
	303.15	$1.395 \times 10^{-5} \pm 0.37\%$		
Helium (He)	288.15	7.123×10^{-6}	$A = -41.4611$	2
$A_r = 4.0026$	293.15	7.044×10^{-6}	$B = 42.5962$	
	298.15	6.997×10^{-6}	$C = 14.0094$	
	303.15	6.978×10^{-6}	Std. dev. = ±0.54%	
	308.15	6.987×10^{-6}	Temp.range = 273.15—348.15	
Neon (Ne)	288.15	8.702×10^{-6}	$A = -52.8573$	2
$A_r = 20.1797$	293.15	8.395×10^{-6}	$B = 61.0494$	
	298.15	8.152×10^{-6}	$C = 18.9157$	
	303.15	7.966×10^{-6}	Std. dev. = ±0.47%	
	308.15	7.829×10^{-6}	Temp.range = 273.15—348.15	
Argon (Ar)	288.15	3.025×10^{-5}	$A = -57.6661$	3
$A_r = 39.948$	293.15	2.748×10^{-5}	$B = 74.7627$	
	298.15	2.519×10^{-5}	$C = 20.1398$	
	303.15	2.328×10^{-5}	Std. dev. = ±0.26%	
	308.15	2.169×10^{-5}	Temp.range = 273.15—348.15	
Krypton (Kr)	288.15	5.696×10^{-5}	$A = -66.9928$	4
$A_r = 83.80$	293.15	5.041×10^{-5}	$B = 91.0166$	
	298.15	4.512×10^{-5}	$C = 24.2207$	

Gas	T/K	Solubility (X_1)	Equation constants	Ref.
	303.15	4.079×10^{-5}	Std. dev. = ±0.32%	
	308.15	3.725×10^{-5}	Temp.range = 273.15—353.15	
Xenon (Xe)	288.15	10.519×10^{-5}	$A = -74.7398$	4
$A_r = 131.29$	293.15	9.051×10^{-5}	$B = 105.210$	
	298.15	7.890×10^{-5}	$C = 27.4664$	
	303.15	6.961×10^{-5}	Std. dev. = ±0.35%	
	308.15	6.212×10^{-5}	Temp.range = 273.15—348.15	
Radon-222(^{222}Rn)	288.15	2.299×10^{-4}	$A = -90.5481$	
$A_r = 222$	293.15	1.945×10^{-4}	$B = 130.026$	
	298.15	1.671×10^{-4}	$C = 35.0047$	
	303.15	1.457×10^{-4}	Std. dev. = ±1.02%	
	308.15	1.288×10^{-4}	Temp.range = 273.15—373.15	
Oxygen (O_2)	288.15	2.756×10^{-5}	$A = -66.7354$	5
$M_r = 31.9988$	293.15	2.501×10^{-5}	$B = 87.4755$	
	298.15	2.293×10^{-5}	$C = 24.4526$	
	303.15	2.122×10^{-5}	Std. dev. = ±0.36%	
	308.15	1.982×10^{-5}	Temp.range = 273.15—348.15	
Ozone (O_3)	293.15	$1.885 \times 10^{-6} \pm 10\%$	Experimental value derived	5
$M_r = 47.9982$		pH = 7.0	from Henry's Law Constant	
Nitrogen (N_2)	288.15	1.386×10^{-5}	$A = -67.3877$	6
$M_r = 28.0134$	293.15	1.274×10^{-5}	$B = 86.3213$	
	298.15	1.183×10^{-5}	$C = 24.7981$	
	303.15	1.108×10^{-5}	Std. dev. = ±0.72%	
	308.15	1.047×10^{-5}	Temp.range = 273.15—348.15	
Nitrous oxide (N_2O)	288.15	5.948×10^{-4}	$A = -60.7467$	7
$M_r = 44.0129$	293.15	5.068×10^{-4}	$B = 88.8280$	
	298.15	4.367×10^{-4}	$C = 21.2531$	
	303.15	3.805×10^{-4}	Std. dev. = ±1.2%	
	308.15	3.348×10^{-4}	Temp.range = 273.15—313.15	
Nitric oxide (NO)	288.15	4.163×10^{-5}	$A = -62.8086$	7
$M_r = 30.0061$	293.15	3.786×10^{-5}	$B = 82.3420$	
	298.15	3.477×10^{-5}	$C = 22.8155$	
	303.15	3.222×10^{-5}	Std. dev. = ±0.76%	
	308.15	3.012×10^{-5}	Temp.range = 273.15—358.15	
Carbon monoxide (CO)	288.15	2.095×10^{-5}	Derived from Henry's	8
$M_r = 28.0104$	293.15	1.918×10^{-5}	Law Constant Equation	
	298.15	1.774×10^{-5}	Std. dev. = ±0.043%	
	303.15	1.657×10^{-5}	Temp.range = 273.15—328.15	
	308.15	1.562×10^{-5}		
Carbon dioxide (CO_2)	288.15	8.21×10^{-4}	Derived from Henry's	9
$M_r = 44.0098$	293.15	7.07×10^{-4}	Law Constant Equation	
	298.15	6.15×10^{-4}	Std. dev. = ±1.1%	
	303.15	5.41×10^{-4}	Temp.range = 273.15—353.15	
	308.15	4.80×10^{-4}		
Hydrogen selenide (H_2Se)	288.15	1.80×10^{-3}	$A = 9.15$	10
$M_r = 80.976$	298.15	1.49×10^{-3}	$B = 974$	
	308.15	1.24×10^{-3}	$C = -3.542$	
			$D = 0.0042$	

Gas	T/K	Solubility (X_1)	Equation constants	Ref.
			Std. dev. = ±2.3 × 10^{-5}	
			Temp. range = 288.15—343.15	
Hydrogen sulfide (H_2S)	288.15	2.335 × 10^{-3}	A = –24.912	10
M_r = 34.082	293.15	2.075 × 10^{-3}	B = 3477	
	298.15	1.85 × 10^{-3}	C = 0.3993	
	303.15	1.66 × 10^{-3}	D = 0.0157	
	308.15	1.51 × 10^{-3}	Std. dev. = ±6.5 × 10^{-5}	
			Temp. range = 283.15—603.15	
Sulfur dioxide (SO_2)	288.15	3.45 × 10^{-2}	A = –25.2629	11
M_r = 64.0648	293.15	2.90 × 10^{-2}	B = 45.7552	
	298.15	2.46 × 10^{-2}	C = 5.6855	
	303.15	2.10 × 10^{-2}	Std. dev. = ±1.8%	
	308.15	1.80 × 10^{-2}	Temp.range = 278.15—328.15	
Chlorine (Cl_2)	283.15	2.48 × 10^{-3} ± 2%	Experimental data	11
M_r = 70.9054	293.15	1.88 × 10^{-3} ± 2%	Temp.range = 283.15—333.15	
	303.15	1.50 × 10^{-3} ± 2%		
	313.15	1.23 × 10^{-3} ± 2%		
Chlorine monoxide (Cl_2O)	273.15	5.25 × 10^{-1} ± 1%	Experimental data	11
M_r = 86.9048	276.61	4.54 × 10^{-1} ± 1%	Temp. range = 273.15—293.15	
	283.15	4.273 × 10^{-1} ± 1%		
	293.15	3.353 × 10^{-1} ± 1%		
Chlorine dioxide (ClO_2)	288.15	2.67 × 10^{-2}	A = 7.9163	11
M_r = 67.4515	293.15	2.20 × 10^{-2}	B = 0.4791	
	298.15	1.823 × 10^{-2}	C = 11.0593	
	303.15	1.513 × 10^{-2}	Std. dev. = ±4.6%	
	308.15	1.259 × 10^{-2}	Temp.range = 283.15—333.15	
Methane (CH_4)	288.15	3.122 × 10^{-5}	A = –115.6477	12
M_r = 16.0428	293.15	2.806 × 10^{-5}	B = 155.5756	
	298.15	2.552 × 10^{-5}	C = 65.2553	
	303.15	2.346 × 10^{-5}	D = –6.6170	
	308.15	2.180 × 10^{-5}	Std. dev. = ±0.056%	
			Temp.range = 273.15—328.15	
Ethane (C_2H_6)	288.15	4.556 × 10^{-5}	A = –90.8225	13
M_r = 30.0696	293.15	3.907 × 10^{-5}	B = 126.9559	
	298.15	3.401 × 10^{-5}	C = 34.7413	
	303.15	3.002 × 10^{-5}	Std. dev. = ±0.13%	
	308.15	2.686 × 10^{-5}	Temp.range = 273.15—323.15	
Propane (C_3H_8)	288.15	3.813 × 10^{-5}	A = –102.044	14
M_r = 44.097	293.15	3.200 × 10^{-5}	B = 144.345	
	298.15	2.732 × 10^{-5}	C = 39.4740	
	303.15	2.370 × 10^{-5}	Std. dev. = ±0.012%	
	308.15	2.088 × 10^{-5}	Temp.range = 273.15—347.15	
Butane (C_4H_{10})	288.15	3.274 × 10^{-5}	A = –102.029	14
M_r = 58.123	293.15	2.687 × 10^{-5}	B = 146.040	
	298.15	2.244 × 10^{-5}	C = 38.7599	
	303.15	1.906 × 10^{-5}	Std. dev. = ±0.026%	
	308.15	1.645 × 10^{-5}	Temp.range = 273.15—349.15	
2-Methyl propane (Isobutane)	288.15	2.333 × 10^{-5}	A = –129.714	14

Gas	T/K	Solubility (X_1)	Equation constants	Ref.
(C_4H_{10})	293.15	1.947×10^{-5}	B = 183.044	
M_r = 58.123	298.15	1.659×10^{-5}	C = 53.4651	
	303.15	1.443×10^{-5}	Std. dev. = ±0.034%	
	308.15	1.278×10^{-5}	Temp.range = 278.15—318.15	

REFERENCES

1. C. L. Young, Ed., *IUPAC Solubility Data Series*, Vol. 5/6, Hydrogen and Deuterium, Pergamon Press, Oxford, England, 1981.
2. H. L. Clever, Ed., *IUPAC Solubility Data Series*, Vol. 1, Helium and Neon, Pergamon Press, Oxford, England, 1979.
3. H. L. Clever, Ed., *IUPAC Solubility Data Series*, Vol. 4, Argon, Pergamon Press, Oxford, England, 1980.
4. H. L. Clever, Ed., *IUPAC Solubility Data Series*, Vol. 2, Krypton, Xenon and Radon, Pergamon Press, Oxford, England, 1979.
5. R. Battino, Ed., *IUPAC Solubility Data Series*, Vol. 7, Oxygen and Ozone, Pergamon Press, Oxford, England, 1981.
6. R. Battino, Ed., *IUPAC Solubility Data Series*, Vol. 10, Nitrogen and Air, Pergamon Press, Oxford, England, 1982.
7. C. L. Young, Ed., *IUPAC Solubility Data Series*, Vol. 8, Oxides of Nitrogen, Pergamon Press, Oxford, England, 1981.
8. R. W. Cargill, Ed., *IUPAC Solubility Data Series*, Vol. 43, Carbon Monoxide, Pergamon Press, Oxford, England, 1990.
9. R. Crovetto, Evaluation of Solubility Data for the System CO_2-H_2O, *J. Phys. Chem. Ref. Data*, 20, 575, 1991.
10. P. G. T. Fogg and C. L. Young, Eds., *IUPAC Solubility Data Series*, Vol. 32, Hydrogen Sulfide, Deuterium Sulfide, and Hydrogen Selenide, Pergamon Press, Oxford, England, 1988.
11. C. L. Young, Ed., *IUPAC Solubility Data Series*, Vol. 12, Sulfur Dioxide, Chlorine, Fluorine and Chlorine Oxides, Pergamon Press, Oxford, England, 1983.
12. H. L. Clever and C. L. Young, Eds., *IUPAC Solubility Data Series*, Vol. 27/28, Methane, Pergamon Press, Oxford, England, 1987.
13. W. Hayduk, Ed., *IUPAC Solubility Data Series*, Vol. 9, Ethane, Pergamon Press, Oxford, England, 1982.
14. W. Hayduk, Ed., *IUPAC Solubility Data Series*, Vol. 24, Propane, Butane and 2-Methylpropane, Pergamon Press, Oxford, England, 1986.

SOLUBILITY OF CARBON DIOXIDE IN WATER AT VARIOUS TEMPERATURES AND PRESSURES

The solubility of CO_2 in water, expressed as mole fraction of CO_2 in the liquid phase, is given for pressures up to atmospheric and temperatures of 0 to 100°C. Note that 1 standard atmosphere equals 101.325 kPa. The references give data over a wider range of temperature and pressure. The estimated accuracy is about 2%.

REFERENCES

1. Carroll, J. J., Slupsky, J. D., and Mather, A. E., *J. Phys. Chem. Ref. Data*, 20, 1201, 1991.
2. Fernandez-Prini, R. and Crovetto, R., *J. Phys. Chem. Ref. Data*, 18, 1231, 1989.
3. Crovetto, R., *J. Phys. Chem. Ref. Data*, 20, 575,1991

1000 × mole fraction of CO_2 in liquid phase

	Partial pressure of CO_2 in kPa						
t/°C	**5**	**10**	**20**	**30**	**40**	**50**	**100**
0	0.067	0.135	0.269	0.404	0.538	0.671	1.337
5	0.056	0.113	0.226	0.338	0.451	0.564	1.123
10	0.048	0.096	0.191	0.287	0.382	0.477	0.950
15	0.041	0.082	0.164	0.245	0.327	0.409	0.814
20	0.035	0.071	0.141	0.212	0.283	0.353	0.704
25	0.031	0.062	0.123	0.185	0.247	0.308	0.614
30	0.027	0.054	0.109	0.163	0.218	0.271	0.541
35	0.024	0.048	0.097	0.145	0.193	0.242	0.481
40	0.022	0.043	0.087	0.130	0.173	0.216	0.431
45	0.020	0.039	0.078	0.117	0.156	0.196	0.389
50	0.018	0.036	0.071	0.107	0.142	0.178	0.354
55	0.016	0.033	0.065	0.098	0.131	0.163	0.325
60	0.015	0.030	0.060	0.090	0.121	0.150	0.300
65	0.014	0.028	0.056	0.084	0.112	0.140	0.279
70	0.013	0.026	0.052	0.079	0.105	0.131	0.261
75	0.012	0.025	0.049	0.074	0.099	0.123	0.245
80	0.012	0.023	0.047	0.070	0.093	0.116	0.232
85	0.011	0.022	0.044	0.067	0.089	0.111	0.221
90	0.011	0.021	0.042	0.064	0.085	0.106	0.211
95	0.010	0.020	0.041	0.061	0.082	0.102	0.203
100	0.010	0.020	0.039	0.059	0.079	0.098	0.196

AQUEOUS SOLUBILITY AND HENRY'S LAW CONSTANTS OF ORGANIC COMPOUNDS

The solubility in water of about 800 organic compounds, including many compounds of environmental interest, is tabulated here. Values are given at 25°C or at the nearest temperature to this where data are available. In some cases solubility values are given at other temperatures as well.

Solubility of a solid is defined as the concentration of the compound in a solution that is in equilibrium with the solid phase at the specified temperature and one atmosphere pressure. For liquids whose water mixtures separate into two phases, the solubility given here is the concentration of the compound in the water-rich phase at equilibrium. In the case of gases (i.e., compounds whose vapor pressure at the specified temperature exceeds one atmosphere) the solubility is defined here as the concentration in the water phase when the partial pressure of the compound above the solution is 101.325 kPa (1 atm). Values for gases are marked with an asterisk.

All solubility values are expressed as mass percent of solute, $S = 100\ w_2$, where the mass fraction w_2 is given by

$$w_2 = m_2/(m_1 + m_2)\ .$$

In these equations m_2 is the mass of solute and m_1 the mass of water. This quantity is related to other common measures of solubility as follows:

Molality: $m_2 = 1000w_2/M_2(1-w_2)$

Mole fraction: $x_2 = (w_2/M_2)/\{(w_2/M_2) + (1-w_2)/M_1\}$

Mass of solute per 100 g of H_2O: $r_2 = 100w_2/(1-w_2)$

Here M_2 is the molar mass of the solute and $M_1 = 18.015$ g/mol is the molar mass of water. For small values of S the amount of substance concentration c_2 in moles per liter is approximately $10S/M_2$.

Data have been selected from evaluated sources wherever possible, in particular the *IUPAC Solubility Data Series* (References 1, 2, 3, 4, 25). The primary source for each value is listed in the column following the solubility values. The user is cautioned that wide variations of data are found in the literature for the lower solubility compounds.

The table also contains values of the Henry's Law constant k_H, which provides a measure of the partition of a substance between the atmosphere and the aqueous phase. Here k_H is defined as the limit of p_2/c_2 as the concentration approaches zero, where p_2 is the partial pressure of the solute above the solution and c_2 is the solute concentration (other formulations of Henry's Law are often used; see Reference 5). The values of k_H listed here are based on direct experimental measurement whenever available, but many of them are simply calculated as the ratio of the pure compound vapor pressure to the solubility. This approximation is reliable only for compounds of very low solubility. In fact, values of k_H found in the literature frequently differ by a factor of two or three, and variations over an order of magnitude are not unusual (Reference 5). Therefore the data given here should be taken only as a rough indication of the true Henry's Law constant, which is difficult to measure precisely.

All values of k_H refer to 25°C. If the vapor pressure of the compound at 25°C is greater than one atmosphere, it can be assumed that the k_H value has been calculated as $101.325/c_2$ kPa m^3/mol. The source of the Henry's Law data is given in the last column. The air-water partition coefficient (i.e., ratio of air concentration to water concentration when both are expressed in the same units) is equal to k_H/RT or $k_H/2.48$ in the units used here.

Compounds are listed by molecular formula following the Hill convention. To locate a compound by name or CAS Registry Number when the molecular formula is not known, use the "Physical Constants of Organic Compounds" table in Section 3 and its indexes to determine the molecular formula.

* Indicates a value of S for a gas at a partial pressure of 101.325 kPa (1 atm) in equilibrium with the solution.

REFERENCES

1. *Solubility Data Series, International Union of Pure and Applied Chemistry, Vol. 15*, Pergamon Press, Oxford, 1982.
2. *Solubility Data Series, International Union of Pure and Applied Chemistry, Vol. 20*, Pergamon Press, Oxford, 1985.
3. *Solubility Data Series, International Union of Pure and Applied Chemistry, Vol. 37*, Pergamon Press, Oxford, 1988.
4. *Solubility Data Series, International Union of Pure and Applied Chemistry, Vol. 38*, Pergamon Press, Oxford, 1988.
5. Mackay, D., and Shiu, W. Y., *J. Phys. Chem. Ref. Data*, 10, 1175, 1981.
6. Pearlman, R. S., and Yalkowsky, S. H., *J. Phys. Chem. Ref. Data*, 13, 975, 1984.
7. Shiu, W. Y., and Mackay, D., *J. Phys. Chem. Ref. Data*, 15, 911, 1986.
8. Varhanickova, D., Lee, S. C., Shiu, W. Y., and Mackay, D., *J. Chem. Eng. Data*, 40, 620, 1995.
9. Miller, M. M., Ghodbane, S., Wasik, S. P., Tewari, Y. B., and Martire, D. E., *J. Chem. Eng. Data*, 29, 184, 1984.
10. Riddick, J. A., Bunger, W. B., and Sakano, T. K., *Organic Solvents, Fourth Edition*, John Wiley & Sons, New York, 1986.
11. Mackay, D., Shiu, W. Y., and Ma, K. C., *Illustrated Handbook of Physical-Chemical Properties and Environmental Fate for Organic Chemicals, Vol. I*, Lewis Publishers/CRC Press, Boca Raton, FL, 1992.
12. Mackay, D., Shiu, W. Y., and Ma, K. C., *Illustrated Handbook of Physical-Chemical Properties and Environmental Fate for Organic Chemicals, Vol. II*, Lewis Publishers/CRC Press, Boca Raton, FL, 1992.
13. Mackay, D., Shiu, W. Y., and Ma, K. C., *Illustrated Handbook of Physical-Chemical Properties and Environmental Fate for Organic Chemicals, Vol. III*, Lewis Publishers/CRC Press, Boca Raton, FL, 1993.
14. Horvath, A. L., *Halogenated Hydrocarbons*, Marcel Dekker, New York, 1982.
15. Howard, P. H., *Handbook of Environmental Fate and Exposure Data for Organic Chemicals, Vol. I*, Lewis Publishers/CRC Press, Boca Raton, FL, 1989.
16. Howard, P. H., *Handbook of Environmental Fate and Exposure Data for Organic Chemicals, Vol. II*, Lewis Publishers/CRC Press, Boca Raton, FL, 1990.

17. Banergee, S., Yalkowsky, S. H., and Valvani, S. C., *Environ. Sci. Technol.*, 14, 1227, 1980.
18. Gevantman, L., in *CRC Handbook of Chemistry and Physics, 82nd Edition*, Section 8, CRC Press, Boca Raton, FL, 2001.
19. Wilhelm, E., Battino, R., and Wilcock, R. J., *Chem. Rev.* 77, 219, 1977.
20. Stephenson, R. M., *J. Chem. Eng. Data*, 37, 80, 1992.
21. Stephenson, R. M., Stuart, J., and Tabak, M., *J. Chem. Eng. Data*, 29, 287, 1984.
22. Shiu, W.-Y., and Ma, K.-C, *J. Phys. Chem. Ref. Data*, 29, 41, 2000.
23. Lun, R., Varhanickova, D., Shiu, W.-Y., and Mackay, D., *J. Chem. Eng. Data*, 42, 951, 1997.
24. Huang, G.-L., Xiao, H., Chi, J., Shiu, W.-Y., and Mackay, D., *J. Chem. Eng. Data*, 45, 411, 2000.
25. Horvath, A. L., Getzen, F. W., and Maczynska, Z., *J. Phys. Chem. Ref. Data*, 28, 395, 1999.
26. Dawson, R. M. C., Elliott, D. C., Elliott, W. H., and Jones, K. M., *Data for Biochemical Research*, Third Edition, Clarendon Press, Oxford, 1986.
27. Stephen, H., and Stephen, T., *Solubilities of Organic and Inorganic Compounds*, MacMillan, New York, 1963.
28. Shiu, W.-Y., and Mackay, D., *J. Chem. Eng. Data*, 42, 27, 1997.
29. Hinz, H.-J., ed., *Thermodynamic Data for Biochemistry and Biotechnology*, Springer-Verlag, Berlin, 1986.
30. Budavari, S., ed., *The Merck Index, Twelfth Edition*, Merck & Co., Rahway, NJ, 1996.
31. Bamford, H. A., Poster, D. L., and Baker, J. E., *J. Chem. Eng. Data*, 45, 1069, 2000.

Mol. Form	Name	t/°C	S/mass %	Ref.	k_H/kPa m³mol⁻¹	Ref.
$CBrF_3$	Bromotrifluoromethane	25	0.032*	14		
CBr_3F	Tribromofluoromethane	25	0.040	14		
CBr_4	Tetrabromomethane	30	0.024	14		
$CClF_3$	Chlorotrifluoromethane	25	0.009*	10	6.9	13
CCl_2F_2	Dichlorodifluoromethane	20	0.028*	5	41	13
CCl_3F	Trichlorofluoromethane	20	0.11	5	10.2	13
CCl_4	Tetrachloromethane	25	0.065	20	2.99	13
CF_4	Tetrafluoromethane	25	0.00187*	19		
$CHBr_3$	Tribromomethane	25	0.30	5	0.047	13
$CHClF_2$	Chlorodifluoromethane	25	0.30*	10	3.0	13
$CHCl_2F$	Dichlorofluoromethane	25	0.95*	10		
$CHCl_3$	Trichloromethane	25	0.80	20	0.43	13
CHF_3	Trifluoromethane	25	0.09*	14		
CHI_3	Triiodomethane	25	0.012	14		
CH_2BrCl	Bromochloromethane	25	1.7	10	0.18	13
CH_2Br_2	Dibromomethane	25	1.14	14	0.086	13
CH_2ClF	Chlorofluoromethane	25	1.05*	14		
CH_2Cl_2	Dichloromethane	25	1.73	20	0.30	13
CH_2I_2	Diiodomethane	30	0.124	10	0.032	13
CH_3Br	Bromomethane	20	1.80*	5	0.63	13
CH_3Cl	Chloromethane	25	0.535*	5	0.98	13
CH_3F	Fluoromethane	30	0.177*	5		
CH_3I	Iodomethane	20	1.4	10	0.54	13
CH_3NO_2	Nitromethane	25	11.1	10		
CH_4	Methane	25	0.00227*	18	67.4	5
CO	Carbon monoxide	25	0.00276*	18		
CO_2	Carbon dioxide	25	0.1501	18		
CS_2	Carbon disulfide	20	0.210	10		
$C_2Br_2F_4$	1,2-Dibromotetrafluoroethane	25	0.00030	25		
C_2ClF_5	Chloropentafluoroethane	25	0.006*	10	260	13
$C_2Cl_2F_4$	1,2-Dichlorotetrafluoroethane	25	0.013*	10	127	13
$C_2Cl_3F_3$	1,1,2-Trichlorotrifluoroethane	25	0.017	25	32	13
C_2Cl_4	Tetrachloroethylene	0	0.024	25		
		25	0.021	25	1.73	13
		50	0.020	25		
$C_2Cl_4F_2$	1,1,2,2-Tetrachloro-1,2-difluoroethane	27	0.016	25		
C_2Cl_6	Hexachloroethane	25	0.005	25	0.85	13
C_2F_4	Tetrafluoroethylene	25	0.0158*	19		
$C_2HBrClF_3$	2-Bromo-2-chloro-1,1,1-trifluoroethane	10	0.52	25		
		25	0.41	25		

Mol. Form	Name	t/°C	S/mass %	Ref.	k_H/kPa m^{3}mol^{-1}	Ref.
		40	0.40	25		
$C_2HCl_2F_3$	2,2-Dichloro-1,1,1-trifluoroethane	25	0.46	25		
C_2HCl_3	Trichloroethylene	0	0.145	25		
		25	0.128	25	1.03	13
		60	0.133	25		
$C_2HCl_3O_2$	Trichloroacetic acid	25	92.3	27		
C_2HCl_5	Pentachloroethane	25	0.049	25	0.25	13
C_2H_2	Acetylene	25	0.1081*	19		
$C_2H_2Br_2Cl_2$	1,2-Dibromo-1,2-dichloroethane	20	0.070	25		
$C_2H_2Br_4$	1,1,2,2-Tetrabromoethane	0	0.052	25		
		25	0.068	25		
		50	0.106	25		
		100	0.307	25		
$C_2H_2Cl_2$	1,1-Dichloroethylene	5	0.310	25		
		25	0.242	25	2.62	13
		50	0.225	25		
		90	0.355	25		
$C_2H_2Cl_2$	*cis*-1,2-Dichloroethylene	10	0.76	25		
		25	0.64	25	0.46	13
		40	0.66	25		
$C_2H_2Cl_2$	*trans*-1,2-Dichloroethylene	10	0.53	25		
		25	0.45	25	0.96	13
		40	0.41	25		
$C_2H_2Cl_2F_2$	1,2-Dichloro-1,1-difluoroethane	24	0.49	25		
$C_2H_2Cl_4$	1,1,1,2-Tetrachloroethane	0	0.120	25		
		25	0.107	25	0.24	13
		50	0.123	25		
$C_2H_2Cl_4$	1,1,2,2-Tetrachloroethane	5	0.302	25		
		25	0.283	25	0.026	13
		50	0.318	25		
$C_2H_2I_2$	*cis*-1,2-Diiodoethene	25	0.046	25		
$C_2H_2I_2$	*trans*-1,2-Diiodoethene	25	0.015	25		
$C_2H_2O_4$	Oxalic acid	20	8.69	27		
		80	45.8	27		
$C_2H_3Br_2Cl$	1,2-Dibromo-1-chloroethane	20	0.060	25		
$C_2H_3Br_3$	1,1,2-Tribromoethane	20	0.050	25		
C_2H_3Cl	Chloroethylene	25	0.27*	5	2.68	13
$C_2H_3Cl_2F$	1,1-Dichloro-1-fluoroethane	25	0.042	25		
$C_2H_3Cl_3$	1,1,1-Trichloroethane	0	0.134	25		
		25	0.129	25	1.76	13
		50	0.138	25		
$C_2H_3Cl_3$	1,1,2-Trichloroethane	0	0.425	25		
		25	0.459	25	0.092	13
		50	0.536	25		
$C_2H_3KO_2$	Potassium acetate	25	72.9			
$C_2H_3NaO_2$	Sodium acetate	25	33.5			
C_2H_4	Ethylene	25	0.01336*	19	21.7	5
C_2H_4BrCl	1-Bromo-2-chloroethane	30	0.683	25		
$C_2H_4Br_2$	1,2-Dibromoethane	0	0.31	25		
		25	0.39	25	0.066	13
		50	0.54	25		
		75	0.76	25		
$C_2H_4Cl_2$	1,1-Dichloroethane	0	0.62	25		
		25	0.50	25	0.63	13
		50	0.50	25		
$C_2H_4Cl_2$	1,2-Dichloroethane	0	0.92	25		
		25	0.86	25	0.14	13
		50	1.05	25		
		100	2.17	25		

Mol. Form	Name	t/°C	S/mass %	Ref.	k_H/kPa m^{3}mol^{-1}	Ref.
$C_2H_4O_2$	Methyl formate	25	23	10		
C_2H_5Br	Bromoethane	0	1.05	25		
		25	0.90	25	1.23	13
C_2H_5Cl	Chloroethane	0	0.45	25		
		25	0.67	25	1.02	13
C_2H_5F	Fluoroethane	25	0.216*	14		
C_2H_5I	Iodoethane	0	0.44	25		
		25	0.40	25	0.52	13
C_2H_5NO	Acetamide	20	40.8	10		
$C_2H_5NO_2$	Nitroethane	25	4.68	10		
$C_2H_5NO_2$	Methyl carbamate	15	69	27		
$C_2H_5NO_2$	Glycine	25	20.06	26		
C_2H_6	Ethane	25	0.00568*	18	50.6	5
C_2H_6O	Dimethyl ether	24	35.3*	10	0.077	13
C_2H_6OS	Dimethyl sulfoxide	25	25.3	10		
$C_2H_6O_4S$	Dimethyl sulfate	18	2.7	27		
C_2H_6S	Dimethyl sulfide	25	2	10		
C_2N_2	Cyanogen	25	0.8	30		
$C_3Br_2F_6$	1,2-Dibromo-1,1,2,3,3,3-hexafluoropropane	21	0.0068	25		
$C_3Cl_2F_6$	1,2-Dichlorohexafluoropropane	21	0.0096	25		
$C_3Cl_3F_5$	1,1,1-Trichloro-2,2,3,3,3-pentafluoropropane	21	0.0058	25		
$C_3Cl_4F_4$	1,1,1,3-Tetrachloro-2,2,3,3-tetrafluoropropane	21	0.0052	25		
C_3Cl_6	Hexachloropropene	20	0.00118	25		
C_3F_6	Perfluoropropene	25	0.0194*	14		
C_3F_8	Perfluoropropane	15	0.0015*	14		
C_3H_3N	2-Propenenitrile	20	7.35	10		
C_3H_4	Propyne	25	0.364*	5	1.11	5
$C_3H_4ClF_3$	3-Chloro-1,1,1-trifluoropropane	20	0.133	25		
$C_3H_4Cl_2$	*cis*-1,3-Dichloropropene	20	0.27	5	0.24	5
$C_3H_4Cl_2$	*trans*-1,3-Dichloropropene	20	0.28	5	0.18	5
$C_3H_4Cl_2$	2,3-Dichloropropene	25	0.215	5	0.36	5
$C_3H_4N_2O_2$	2,4-Imidazolidinedione	25	3.93	29		
C_3H_4O	Acrolein	20	20.8	10		
C_3H_5Br	3-Bromopropene	25	0.38	25		
$C_3H_5Br_2Cl$	1,2-Dibromo-3-chloropropane	20	0.123	25		
C_3H_5Cl	3-Chloropropene	25	0.40	25	1.10	5
		50	0.13	25		
C_3H_5ClO	Epichlorohydrin	20	6.58	10	0.003	13
$C_3H_5Cl_3$	1,2,3-Trichloropropane	10	0.14	25		
		25	0.20	25	0.038	13
C_3H_5N	Propanenitrile	25	10.3	10		
C_3H_6	Propene	25	0.0200*	5	21.3	5
C_3H_6	Cyclopropane	25	0.0484*	19		
C_3H_6BrCl	1-Bromo-3-chloropropane	25	0.223	25		
$C_3H_6Br_2$	1,2-Dibromopropane	25	0.143	10		
$C_3H_6Br_2$	1,3-Dibromopropane	25	0.169	25		
$C_3H_6Cl_2$	1,2-Dichloropropane	5	0.270	25		
		25	0.274	25	0.29	13
		40	0.297	25		
$C_3H_6Cl_2$	1,3-Dichloropropane	5	0.218	25		
		25	0.280	25		
$C_3H_6N_6O_6$	Hexahydro-1,3,5-trinitro-1,3,5-triazine	25	0.0060	17		
C_3H_6O	Propanal	25	30.6	10		
C_3H_6O	Methyloxirane	20	40.5	10	0.0087	13
$C_3H_6O_2$	Ethyl formate	25	11.8	10		
$C_3H_6O_2$	Methyl acetate	20	24.5	10		

Mol. Form	Name	t/°C	S/mass %	Ref.	k_H/kPa m^{3}mol^{-1}	Ref.
$C_3H_6O_3$	1,3,5-Trioxane	25	17.4	30		
C_3H_7Br	1-Bromopropane	0	0.298	25		
		25	0.234	25	3.8	13
C_3H_7Br	2-Bromopropane	20	0.32	25	1.27	13
C_3H_7Cl	1-Chloropropane	25	0.250	25	1.41	13
C_3H_7Cl	2-Chloropropane	0	0.44	25		
		20	0.30	25		
C_3H_7F	1-Fluoropropane	14	0.386*	14		
C_3H_7F	2-Fluoropropane	15	0.366	14		
C_3H_7I	1-Iodopropane	0	0.114	25		
		20	0.100	25	0.93	13
C_3H_7I	2-Iodopropane	0	0.167	25		
		20	0.140	25		
$C_3H_7NO_2$	1-Nitropropane	25	1.50	10		
$C_3H_7NO_2$	2-Nitropropane	25	1.71	10		
$C_3H_7NO_2$	Ethyl carbamate	15	48	27		
$C_3H_7NO_2$	Alanine	25	14.30	26		
$C_3H_7NO_2$	β-Alanine	25	47.1	26		
$C_3H_7NO_2$	Sarcosine [N-Methylglycine]	25	30.0	26		
$C_3H_7NO_3$	Serine	25	4.76	26		
$C_3H_7N_3O_2$	Glycocyamine	25	0.5	26		
C_3H_8	Propane	25	0.00669*	18	71.6	5
$C_3H_8O_2$	Dimethoxymethane	16	24.4	10		
C_4Cl_6	Hexachloro-1,3-butadiene	25	0.41	25		
C_4F_8	Perfluorocyclobutane	21	0.014*	14		
$C_4H_4N_2$	Succinonitrile	25	11.5	10		
$C_4H_4N_2O_2$	Uracil	25	0.27	29		
C_4H_4O	Furan	25	1	10	0.54	13
C_4H_5N	Methylacrylonitrile	20	2.57	10		
C_4H_5N	Pyrrole	25	4.5	10		
$C_4H_5N_3O$	Cytosine	25	0.73	29		
C_4H_6	1,3-Butadiene	25	0.0735*	5	20.7	13
C_4H_6	1-Butyne	25	0.287*	5	1.91	5
$C_4H_6BaO_4$	Barium acetate	25	44.2			
$C_4H_6N_2O_2$	2,5-Piperazinedione	25	1.64	29		
C_4H_6O	*trans*-2-Butenal	20	15.6	10		
$C_4H_6O_2$	*trans*-Crotonic acid	25	9	10		
$C_4H_6O_2$	Methacrylic acid	20	8.9	10		
$C_4H_6O_2$	Vinyl acetate	20	2.0	10		
$C_4H_6O_2$	Methyl acrylate	25	4.94	10		
$C_4H_6O_4$	Succinic acid	25	7.71	27		
		75	37.6	27		
$C_4H_6O_4$	Dimethyl oxalate	20	5.82	27		
$C_4H_6O_5$	Malic acid	25	58	27		
C_4H_7Br	4-Bromo-1-butene	25	0.076	25		
C_4H_7Cl	1-Chloro-2-methylpropene	25	0.916	5	0.12	5
C_4H_7N	Butanenitrile	20	3.3	10		
$C_4H_7NO_4$	Aspartic acid	25	0.501	26		
C_4H_8	1-Butene	25	0.0222*	5	25.6	13
C_4H_8	Isobutene	25	0.0263*	5	21.6	13
$C_4H_8Br_2$	1,4-Dibromobutane		0.035	25		
$C_4H_8Cl_2$	1,1-Dichlorobutane	25	0.050	25		
$C_4H_8Cl_2$	1,4-Dichlorobutane	25	0.16	25		
$C_4H_8Cl_2$	2,3-Dichlorobutane	20	0.056	25		
$C_4H_8Cl_2O$	Bis(2-chloroethyl) ether	25	1.03	10	0.003	13
$C_4H_8N_2O_2$	Succinamide	50	18.4	27		
$C_4H_8N_2O_3$	Asparagine	25	2.45	26		
$C_4H_8N_2O_3$	*N*-Glycylglycine	25	18.4	29		
C_4H_8O	*cis*-Crotonyl alcohol	20	16.6	10		

Mol. Form	Name	t/°C	S/mass %	Ref.	k_H/kPa m^{3}mol^{-1}	Ref.
C_4H_8O	Ethyl vinyl ether	20	0.9	10		
C_4H_8O	Butanal	25	7.1	10		
C_4H_8O	Isobutanal	20	9.1	10		
C_4H_8O	2-Butanone	25	25.9	20		
$C_4H_8O_2$	2-Methylpropanoic acid	20	22.8	10		
$C_4H_8O_2$	Propyl formate	22	2.05	10		
$C_4H_8O_2$	Ethyl acetate	25	8.08	10		
$C_4H_8O_2$	Methyl propanoate		6	30		
C_4H_9Br	1-Bromobutane	25	0.087	25	1.2	13
C_4H_9Br	1-Bromo-2-methylpropane	18	0.051	25		
C_4H_9Cl	1-Chlorobutane	1	0.062	25		
		25	0.087	25	1.54	13
C_4H_9Cl	2-Chlorobutane	0	0.107	25		
		25	0.092	25		
C_4H_9Cl	1-Chloro-2-methylpropane	25	0.92	25		
C_4H_9Cl	2-Chloro-2-methylpropane	15	0.29	25		
C_4H_9I	1-Iodobutane	17	0.021	10	1.87	13
$C_4H_9NO_2$	Ethyl-*N*-methyl carbamate	15	69	27		
$C_4H_9NO_2$	2-Methylalanine	25	12.0	26		
$C_4H_9NO_2$	*DL*-2-Aminobutanoic acid	25	17.4	26		
$C_4H_9NO_2$	*DL*-3-Aminobutanoic acid	25	55.6	26		
$C_4H_9NO_3$	Threonine	25	8.93	26		
$C_4H_9NO_3$	L-Homoserine	25	52.4	26		
$C_4H_9N_3O_2$	Creatine	25	1.6	26		
C_4H_{10}	Butane	25	0.00724*	18	95.9	5
C_4H_{10}	Isobutane	25	0.00535*	18	120	5
$C_4H_{10}O$	1-Butanol	0	10.4	1		
		25	7.4	1		
		50	6.4	1		
$C_4H_{10}O$	2-Butanol	10	23.9	1		
		25	18.1	1		
		50	14.0	1		
$C_4H_{10}O$	2-Methyl-1-propanol	0	11.5	1		
		25	8.1	1	0.00273	28
		50	6.5	1		
$C_4H_{10}O$	Diethyl ether	25	6.04	10	0.088	13
$C_4H_{10}O$	Methyl propyl ether	25	3.5	30		
$C_4H_{10}O_4$	1,2,3,4-Butanetetrol	20	38.0	27		
$C_4H_{10}S$	1-Butanethiol	20	0.0597	10		
$C_4H_{11}NO_2$	Diethanolamine	20	95.4	10		
$C_4H_{12}Si$	Tetramethylsilane	25	0.00196	10		
C_5Cl_8	Octachloro-1,3-pentadiene	20	0.000020	25		
C_5F_{12}	Perfluoropentane	25	0.00012	25		
$C_5H_4N_2O_4$	Orotic acid	18	0.18	26		
$C_5H_4N_4O$	Hypoxanthine	25	0.070	29		
$C_5H_4N_4O_2$	Xanthine	20	0.05	26		
$C_5H_4N_4O_3$	Uric acid	20	0.002	26		
$C_5H_4O_2$	Furfural	20	8.2	10		
$C_5H_5N_5$	Adenine	25	0.104	29		
$C_5H_5N_5O$	Guanine	25	0.0068	29		
$C_5H_5N_5O$	Isoguanine	25	0.006	26		
C_5H_6	1,3-Cyclopentadiene	25	0.068	3		
$C_5H_6N_2O_2$	Thymine	25	0.35	29		
$C_5H_7NO_2$	Ethyl cyanoacetate	20	25.9	10		
$C_5H_7N_3O$	5-Methylcytosine	25	0.45	26		
C_5H_8	1,4-Pentadiene	25	0.056	3	12	5
C_5H_8	2-Methyl-1,3-butadiene	25	0.061	3	7.78	5
		50	0.076*	3		
C_5H_8	1-Pentyne	25	0.157	3	2.5	5

Mol. Form	Name	t/°C	S/mass %	Ref.	k_H/kPa m^{3}mol^{-1}	Ref.
C_5H_8	Cyclopentene	25	0.054	3	6.56	13
$C_5H_8O_2$	Ethyl acrylate	25	1.50	10		
$C_5H_8O_2$	Methyl methacrylate	20	1.56	10		
$C_5H_8O_2$	2,4-Pentanedione	20	16.6	10		
$C_5H_9NO_2$	Proline	25	61.9	26		
$C_5H_9NO_3$	trans-4-Hydroxyproline	25	26.5	26		
$C_5H_9NO_4$	*DL*-Glutamic acid	25	2.30	29		
$C_5H_9NO_4$	Glutamic acid	25	0.85	26		
C_5H_{10}	1-Pentene	25	0.0148	3	40.3	5
C_5H_{10}	*cis*-2-Pentene	25	0.0203	3	22.8	5
C_5H_{10}	3-Methyl-1-butene	25	0.013*	3	54.7	5
C_5H_{10}	2-Methyl-2-butene	25	0.041	3		
C_5H_{10}	Cyclopentane	25	0.0157	3	19.1	13
$C_5H_{10}Cl_2$	2,3-Dichloro-2-methylbutane	25	0.029	25		
$C_5H_{10}Cl_2$	2,3-Dichloropentane	25	0.029	25		
$C_5H_{10}Cl_2$	1,2-Dichloropentane	25	0.029	25		
$C_5H_{10}Cl_2$	1,5-Dichloropentane	19	0.02	25		
$C_5H_{10}N_2O_3$	Glutamine	25	4.0	26		
$C_5H_{10}O$	2-Pentanone	25	5.5	20	0.00847	28
$C_5H_{10}O$	3-Pentanone	25	4.72	20		
$C_5H_{10}O$	Tetrahydropyran	25	8.02	10		
$C_5H_{10}O$	2-Methyltetrahydrofuran	25	13.9	10	0.67	13
$C_5H_{10}O_2$	Pentanoic acid	20	2.5	26		
$C_5H_{10}O_2$	3-Methylbutanoic acid	21	4.3	26		
$C_5H_{10}O_2$	Isobutyl formate	22	1.0	10		
$C_5H_{10}O_2$	Propyl acetate	20	2.3	10		
$C_5H_{10}O_2$	Isopropyl acetate	20	2.9	10		
$C_5H_{10}O_2$	Ethyl propanoate	20	1.92	10		
$C_5H_{10}O_2$	Methyl butanoate		1.6	30		
$C_5H_{11}Br$	1-Bromopentane	25	0.0127	25		
$C_5H_{11}Br$	1-Bromo-3-methylbutane	16	0.020	25		
$C_5H_{11}Cl$	1-Chloropentane	5	0.020	25		
		25	0.021	25	2.37	13
$C_5H_{11}Cl$	3-Chloropentane	25	0.025	25		
$C_5H_{11}NO_2$	Valine	25	8.13	26		
$C_5H_{11}NO_2$	L-Norvaline	25	9.7	26		
$C_5H_{11}NO_2S$	Methionine	25	5.3	26		
C_5H_{12}	Pentane	25	0.0041	3	128	13
C_5H_{12}	Isopentane	25	0.00485	3	479	13
C_5H_{12}	Neopentane	25	0.00332*	3	220	13
$C_5H_{12}O$	1-Pentanol	0	3.1	1		
		25	2.20	1		
		50	1.8	1		
$C_5H_{12}O$	2-Pentanol	25	4.3	21		
$C_5H_{12}O$	3-Pentanol	25	5.6	21		
$C_5H_{12}O$	2-Methyl-1-butanol	25	3.0	3		
$C_5H_{12}O$	3-Methyl-1-butanol	25	2.7	1		
$C_5H_{12}O$	2-Methyl-2-butanol	25	11.0	1		
$C_5H_{12}O$	3-Methyl-2-butanol	25	5.6	1		
$C_5H_{12}O$	2,2-Dimethyl-1-propanol	25	3.5	1		
$C_5H_{12}O$	Methyl *tert*-butyl ether	25	3.62		0.070	13
$C_5H_{12}O_4$	Pentaerythritol	15	5.3	30		
C_6Cl_6	Hexachlorobenzene	25	0.0000005	2	0.131	11
C_6F_{14}	Perfluorohexane	25	0.0000098	25		
C_6F_{14}	Perfluoro-2-methylpentane	25	0.000017	25		
C_6HCl_5	Pentachlorobenzene	25	0.000055	2	0.085	11
C_6HCl_5O	Pentachlorophenol	25	0.0013	24		
$C_6H_2Br_4$	1,2,4,5-Tetrabromobenzene	25	0.00000434	2		
$C_6H_2Cl_4$	1,2,3,4-Tetrachlorobenzene	25	0.0000433	2	0.144	11

Mol. Form	Name	t/°C	S/mass %	Ref.	k_H/kPa m^{3}mol^{-1}	Ref.
$C_6H_2Cl_4$	1,2,3,5-Tetrachlorobenzene	25	0.000346	2	0.59	11
$C_6H_2Cl_4$	1,2,4,5-Tetrachlorobenzene	25	0.0000606	2	0.122	11
$C_6H_2Cl_4O$	2,3,4,6-Tetrachlorophenol	25	0.017	24		
$C_6H_2Cl_4O_2$	3,4,5,6-Tetrachloro-1,2-benzenediol	25	0.071	8		
$C_6H_3Br_3$	1,2,4-Tribromobenzene	25	0.0010	2		
$C_6H_3Br_3$	1,3,5-Tribromobenzene	25	0.0000789	2		
$C_6H_3Br_3O$	2,4,6-Tribromophenol	15	0.0007	2		
$C_6H_3Cl_3$	1,2,3-Trichlorobenzene	25	0.00309	2	0.242	11
$C_6H_3Cl_3$	1,2,4-Trichlorobenzene	25	0.00379	2	0.277	11
$C_6H_3Cl_3$	1,3,5-Trichlorobenzene	25	0.000655	2	1.1	11
$C_6H_3Cl_3O$	2,4,5-Trichlorophenol	25	0.1	2		
$C_6H_3Cl_3O$	2,4,6-Trichlorophenol	25	0.050	24		
$C_6H_3Cl_3O_2$	3,4,5-Trichloro-1,2-benzenediol	25	0.051	8		
$C_6H_3N_3O_7$	2,4,6-Trinitrophenol	20	1.43	27		
C_6H_4BrCl	1-Bromo-2-chlorobenzene	25	0.0124	2		
C_6H_4BrCl	1-Bromo-3-chlorobenzene	25	0.0118	2		
C_6H_4BrCl	1-Bromo-4-chlorobenzene	25	0.00442	2		
C_6H_4BrI	1-Bromo-4-iodobenzene	25	0.000794	2		
$C_6H_4Br_2$	*o*-Dibromobenzene	25	0.00748	2		
$C_6H_4Br_2$	*m*-Dibromobenzene	25	0.0064	2		
$C_6H_4Br_2$	*p*-Dibromobenzene	25	0.0020	2		
$C_6H_4Br_2O$	2,4-Dibromophenol	25	0.2	2		
C_6H_4ClI	1-Chloro-2-iodobenzene	25	0.00689	2		
C_6H_4ClI	1-Chloro-3-iodobenzene	25	0.00674	2		
C_6H_4ClI	1-Chloro-4-iodobenzene	25	0.00311	2		
$C_6H_4Cl_2$	*o*-Dichlorobenzene	0	0.0142	2		
		25	0.0147	2	0.195	28
		50	0.0212	2		
$C_6H_4Cl_2$	*m*-Dichlorobenzene	10	0.0103	2		
		25	0.0106	2	0.376	11
		50	0.0165	2		
$C_6H_4Cl_2$	*p*-Dichlorobenzene	10	0.00512	2		
		25	0.00829	2	0.244	28
		50	0.0167	2		
$C_6H_4Cl_2O$	2,4-Dichlorophenol	20	0.49	24		
$C_6H_4Cl_2O_2$	3,5-Dichloro-1,2-benzenediol	25	0.78	8		
$C_6H_4Cl_2O_2$	4,5-Dichloro-1,2-benzenediol	25	1.19	8		
$C_6H_4F_2$	*o*-Difluorobenzene	25	0.114	2		
$C_6H_4F_2$	*m*-Difluorobenzene	25	0.114	2		
$C_6H_4F_2$	*p*-Difluorobenzene	25	0.122	2		
$C_6H_4I_2$	*o*-Diiodobenzene	25	0.00192	2		
$C_6H_4I_2$	*m*-Diiodobenzene	25	0.000185	2		
$C_6H_4I_2$	*p*-Diiodobenzene	25	0.000893	2		
$C_6H_4N_2O_4$	1,2-Dinitrobenzene	20	0.21	27		
$C_6H_4N_2O_4$	1,3-Dinitrobenzene	20	2.09	27		
$C_6H_4N_2O_4$	1,4-Dinitrobenzene	20	1.30	27		
$C_6H_4O_2$	*p*-Benzoquinone	25	1.36	27		
C_6H_5Br	Bromobenzene	10	0.0387	2		
		25	0.0445	2	0.250	28
		40	0.0516	2		
C_6H_5BrO	*p*-Bromophenol	25	1.86	2		
C_6H_5Cl	Chlorobenzene	10	0.0387	2		
		25	0.0495	2	0.32	28
		50	0.0882	2		
C_6H_5ClO	*o*-Chlorophenol	25	2.0	2		
C_6H_5ClO	*m*-Chlorophenol	25	2.2	2		
C_6H_5ClO	*p*-Chlorophenol	25	2.7	2		
C_6H_5F	Fluorobenzene	27	0.154	2	0.70	11
C_6H_5I	Iodobenzene	10	0.0193	2		

Mol. Form	Name	t/°C	S/mass %	Ref.	k_H/kPa m³mol⁻¹	Ref.
		25	0.0226	2	0.078	11
		45	0.0279	2		
$C_6H_5NO_2$	Nitrobenzene	25	0.21	17		
$C_6H_5NO_3$	2-Nitrophenol	20	0.21	27		
$C_6H_5NO_3$	3-Nitrophenol	20	2.14	27		
$C_6H_5NO_3$	4-Nitrophenol	20	1.32	27		
C_6H_6	Benzene	10	0.178	3		
		25	0.178	22	0.557	22
		50	0.208	3		
C_6H_6ClN	*o*-Chloroaniline	25	0.876	10		
$C_6H_6N_2O_2$	2-Nitroaniline	30	1.47	27		
$C_6H_6N_2O_2$	3-Nitroaniline	30	0.121	27		
$C_6H_6N_2O_2$	4-Nitroaniline	30	0.073	27		
C_6H_6O	Phenol	25	8.66	10		
$C_6H_6O_2$	*p*-Hydroquinone	25	7.42	27		
$C_6H_6O_2$	Pyrocatechol	20	31.1	27		
$C_6H_6O_2$	Resorcinol	20	63.7	27		
$C_6H_6O_3$	1,2,3-Benzenetriol	25	38.5	27		
$C_6H_6O_3$	1,3,5-Benzenetriol	20	1.12	27		
$C_6H_6O_6$	Aconitic acid	15	58.5	27		
C_6H_7N	Aniline	25	3.38	10	14	15
$C_6H_7NO_3S$	4-Aminobenzenesulfonic acid	7	0.59	27		
C_6H_8	1,4-Cyclohexadiene	25	0.08	3	1.03	13
C_6H_8ClN	Aniline hydrochloride	15	15.1	27		
$C_6H_8N_2$	Adiponitrile	20	0.80	16		
$C_6H_8N_2$	*m*-Phenylenediamine	20	19.2	27		
$C_6H_8N_2$	*p*-Phenylenediamine		1	30		
$C_6H_8O_4$	Dimethyl maleate	25	8.0	10		
$C_6H_9N_3O_2$	Histidine	25	4.17	26		
C_6H_{10}	1,5-Hexadiene	25	0.017	3		
C_6H_{10}	1-Hexyne	25	0.036	3	4.14	13
C_6H_{10}	Cyclohexene	25	0.016	3	4.57	13
$C_6H_{10}O$	Cyclohexanone	25	8.8	20		
$C_6H_{10}O$	Mesityl oxide	20	2.89	10		
$C_6H_{10}O_3$	Ethyl acetoacetate	25	12	10		
$C_6H_{10}O_4$	Adipic acid	15	1.42	27		
$C_6H_{11}NO$	Caprolactam	25	84.0	10		
C_6H_{12}	1-Hexene	25	0.0053	3	41.8	5
C_6H_{12}	*trans*-2-Hexene	25	0.0067	3		
C_6H_{12}	2-Methyl-1-pentene	25	0.0078	3	28.1	5
C_6H_{12}	4-Methyl-1-pentene	25	0.0048	3	63.2	5
C_6H_{12}	2,3-Dimethyl-1-butene	30	0.046	3		
C_6H_{12}	Cyclohexane	25	0.0058	3	19.4	13
C_6H_{12}	Methylcyclopentane	25	0.0043	3	36.7	5
$C_6H_{12}N_2O_4S$	*L*-Lanthionine	25	0.15	26		
$C_6H_{12}N_2O_4S_2$	*L*-Cystine	25	0.011	26		
$C_6H_{12}N_4$	Methenamine	12	44.8	27		
$C_6H_{12}O$	1-Hexen-3-ol	25	2.52	1		
$C_6H_{12}O$	4-Hexen-2-ol	25	3.81	1		
$C_6H_{12}O$	Butyl vinyl ether	20	0.3	10		
$C_6H_{12}O$	2-Hexanone	20	1.75	10		
$C_6H_{12}O$	4-Methyl-2-pentanone	25	1.7	10		
$C_6H_{12}O$	Cyclohexanol	10	4.62	1		
		25	3.8	1		
		40	3.30	1		
$C_6H_{12}O_2$	Hexanoic acid	20	0.97	26		
$C_6H_{12}O_2$	Isopentyl formate	22	0.3	27		
$C_6H_{12}O_2$	Butyl acetate	20	0.68	10		
$C_6H_{12}O_2$	*sec*-Butyl acetate	20	0.62	10		

Mol. Form	Name	t/°C	S/mass %	Ref.	k_H/kPa m^{3}mol^{-1}	Ref.
$C_6H_{12}O_2$	Isobutyl acetate	20	0.63	10		
$C_6H_{12}O_2$	Propyl propanoate	25	0.6	27		
$C_6H_{12}O_2$	Ethyl butanoate	20	0.49	10		
$C_6H_{12}O_3$	Ethylene glycol monoethyl ether acetate		14	30		
$C_6H_{12}O_3$	Paraldehyde	25	11	30		
$C_6H_{12}O_6$	*D*-Galactose	20	40.6	27		
$C_6H_{12}O_6$	α-*D*-Glucose	15	45.0	27		
		30	54.6	27		
		80	81.5	27		
$C_6H_{13}Br$	1-Bromohexane	25	0.00258	25		
$C_6H_{13}Cl$	1-Chlorohexane	5	0.0047	25		
		25	0.0064	25		
$C_6H_{13}NO_2$	Leucine	25	2.15	26		
$C_6H_{13}NO_2$	Isoleucine	25	3.31	26		
$C_6H_{13}NO_2$	L-Norleucine	25	1.5	26		
$C_6H_{13}NO_2$	Ethyl *N*-propylcarbamate	15	7.70	27		
C_6H_{14}	Hexane	25	0.0011	3	183	13
		60	0.00136	3		
C_6H_{14}	2-Methylpentane	25	0.00137	3	176	13
C_6H_{14}	3-Methylpentane	25	0.00129	3	170	13
C_6H_{14}	2,2-Dimethylbutane	25	0.0021	3	199	13
C_6H_{14}	2,3-Dimethylbutane	25	0.0021	3	144	13
$C_6H_{14}N_2O_2$	Lysine	25	0.58	26		
$C_6H_{14}N_4O_2$	Arginine	25	15.44	26		
$C_6H_{14}O$	2-Methoxy-2-methylbutane	20	1.25	27		
$C_6H_{14}O$	1-Hexanol	0	0.79	1		
		25	0.60	1		
		50	0.51	1		
$C_6H_{14}O$	2-Hexanol	25	1.4	1		
$C_6H_{14}O$	3-Hexanol	25	1.6	1		
$C_6H_{14}O$	2-Methyl-1-pentanol	25	0.81	1		
$C_6H_{14}O$	4-Methyl-1-pentanol	25	0.76	1		
$C_6H_{14}O$	2-Methyl-2-pentanol	25	3.2	1		
$C_6H_{14}O$	3-Methyl-2-pentanol	25	1.9	1		
$C_6H_{14}O$	4-Methyl-2-pentanol	27	1.5	1		
$C_6H_{14}O$	2-Methyl-3-pentanol	25	2.0	1		
$C_6H_{14}O$	3-Methyl-3-pentanol	25	4.3	1		
$C_6H_{14}O$	2-Ethyl-1-butanol	25	1.0	1		
$C_6H_{14}O$	2,2-Dimethyl-1-butanol	25	0.8	1		
$C_6H_{14}O$	2,3-Dimethyl-2-butanol	25	4.2	1		
$C_6H_{14}O$	3,3-Dimethyl-2-butanol	25	2.4	1		
$C_6H_{14}O$	Dipropyl ether	25	0.49	10	0.26	13
$C_6H_{14}O$	Diisopropyl ether	20	1.2	10	0.26	13
$C_6H_{14}O_2$	1,1-Diethoxyethane	25	5	10		
$C_6H_{14}O_6$	*D*-Mannitol	25	17.7	27		
$C_6H_{15}N$	Dipropylamine	20	2.5	10		
$C_6H_{15}N$	Triethylamine	20	5.5	10		
$C_6H_{16}ClN$	Triethylamine hydrochloride	25	57.8	27		
C_7F_{16}	Perfluoroheptane	25	0.0000013	25		
$C_7H_4ClNO_4$	3-Chloro-2-nitrobenzoic acid	25	0.047	27		
$C_7H_4ClNO_4$	5-Chloro-2-nitrobenzoic acid	25	0.96	27		
$C_7H_4Cl_4O$	2,3,4,6-Tetrachloro-5-methylphenol	25	0.00061	2		
$C_7H_4N_2O_6$	3,5-Dinitrobenzoic acid	25	0.134	27		
$C_7H_4O_6$	4-Oxo-4H-pyran-2,6-dicarboxylic acid	25	1.45	27		
$C_7H_4O_7$	3-Hydroxy-4-oxo-4H-pyran-2,6-dicarboxylic acid	25	0.84	27		
$C_7H_5BrO_2$	2-Bromobenzoic acid	25	0.185	27		
$C_7H_5BrO_2$	3-Bromobenzoic acid	25	0.040	27		

Mol. Form	Name	t/°C	S/mass %	Ref.	k_H/kPa m^{3}mol^{-1}	Ref.
$C_7H_5BrO_2$	4-Bromobenzoic acid	25	0.0056	27		
$C_7H_5ClO_2$	2-Chlorobenzoic acid	25	0.209	27		
$C_7H_5ClO_2$	3-Chlorobenzoic acid	25	0.040	27		
$C_7H_5ClO_2$	4-Chlorobenzoic acid	25	0.072	27		
$C_7H_5Cl_3$	(Trichloromethyl)benzene	5	0.0053	10		
$C_7H_5Cl_3O$	2,4,6-Trichloro-3-methylphenol	25	0.0112	2		
$C_7H_5FO_2$	2-Fluorobenzoic acid	25	0.72	27		
$C_7H_5FO_2$	3-Fluorobenzoic acid	25	0.15	27		
$C_7H_5FO_2$	4-Fluorobenzoic acid	25	0.12	27		
$C_7H_5IO_2$	2-Iodobenzoic acid	25	0.095	27		
$C_7H_5IO_2$	3-Iodobenzoic acid	25	0.016	27		
$C_7H_5IO_2$	4-Iodobenzoic acid	25	0.0027	27		
C_7H_5N	Benzonitrile	25	0.2	10		
C_7H_5NO	Benzoxazole	20	0.834	6		
$C_7H_5NO_3$	3-Nitrobenzaldehyde	25	0.16	27		
$C_7H_5NO_3$	4-Nitrobenzaldehyde	25	0.23	27		
$C_7H_5NO_3S$	Saccharin	25	0.40	27		
		100	4.0	27		
$C_7H_5N_3O_6$	2,4,6-Trinitrotoluene	25	0.015	27		
		100	0.015	27		
$C_7H_6Cl_2$	(Dichloromethyl)benzene	30	0.025	10		
$C_7H_6Cl_2O$	2,4-Dichloro-6-methylphenol	25	0.0283	2		
$C_7H_6Cl_2O$	2,6-Dichloro-4-methylphenol	25	0.0673	2		
$C_7H_6N_2$	1*H*-Benzimidazole	20	0.201	6		
$C_7H_6N_2$	1*H*-Indazole	20	0.0827	6		
C_7H_6O	Benzaldehyde	20	0.3	10		
$C_7H_6O_2$	Benzoic acid	25	0.34	27		
$C_7H_6O_2$	Salicylaldehyde	86	1.68	10		
$C_7H_6O_3$	4-Hydroxybenzoic acid	25	0.64	27		
		80	12.0	27		
$C_7H_6O_5$	3,4,5-Trihydroxybenzoic acid	15	0.94	27		
		100	25.0	27		
C_7H_7Br	*p*-Bromotoluene	25	0.011	2		
C_7H_7Cl	(Chloromethyl)benzene	20	0.0493	10		
C_7H_7ClO	2-Chloro-6-methylphenol	25	0.36	2		
C_7H_7ClO	4-Chloro-2-methylphenol	25	0.68	2		
C_7H_7ClO	4-Chloro-3-methylphenol	25	0.40	2		
C_7H_7NO	Benzamide	12	0.577	27		
$C_7H_7NO_2$	2-Nitrotoluene	30	0.065	27		
$C_7H_7NO_2$	3-Nitrotoluene	30	0.050	27		
$C_7H_7NO_2$	4-Nitrotoluene	30	0.044	27		
$C_7H_7NO_3$	2-Nitroanisole	30	0.169	10		
$C_7H_7NO_3$	4-Nitroanisole	30	0.059	27		
C_7H_8	Toluene	5	0.063	3		
		25	0.0531	22	0.660	22
C_7H_8	1,3,5-Cycloheptatriene	25	0.064	3	0.47	13
C_7H_8	1,6-Heptadiyne	25	0.125	3		
$C_7H_8N_2S$	Phenylthiourea	25	2.55	27		
$C_7H_8N_4O_2$	Theophylline	20	0.52	29		
C_7H_8O	*o*-Cresol	40	3.08	10		
C_7H_8O	*m*-Cresol	40	2.51	10		
C_7H_8O	*p*-Cresol	40	2.26	10		
C_7H_8O	Benzyl alcohol	20	0.08	10		
C_7H_8O	Anisole	25	0.19	20	0.025	13
C_7H_9N	*o*-Methylaniline	20	1.66	10		
C_7H_9N	*p*-Methylaniline	21	7.35	10		
$C_7H_9NO_2S$	2-Methylbenzenesulfonamide	25	0.162	27		
$C_7H_9NO_2S$	3-Methylbenzenesulfonamide	25	0.78	27		
$C_7H_9NO_2S$	4-Methylbenzenesulfonamide	25	0.316	27		

Mol. Form	Name	t/°C	S/mass %	Ref.	k_H/kPa m^{3}mol^{-1}	Ref.
C_7H_{12}	1-Heptyne	25	0.0094	3	4.47	13
C_7H_{12}	Cycloheptene	25	0.0066	3	4.9	13
C_7H_{12}	1-Methylyclohexene	25	0.0052	3		
$C_7H_{12}O_2$	Cyclohexanecarboxylic acid	15	0.201	27		
C_7H_{14}	1-Heptene	25	0.032	3	40.3	13
C_7H_{14}	*trans*-2-Heptene	25	0.015	3	42.2	13
C_7H_{14}	Cycloheptane	25	0.0030	3	9.59	13
C_7H_{14}	Methylcyclohexane	25	0.00151	3	43.3	13
		50	0.0019	3		
C_7H_{14}	Ethylcyclopentane	20	0.012	3		
$C_7H_{14}O$	1-Heptanal	11	0.124	27		
$C_7H_{14}O$	2-Heptanone	25	0.43	10	0.0171	28
$C_7H_{14}O$	3-Heptanone	20	1.43	10		
$C_7H_{14}O$	2,4-Dimethyl-3-pentanone	20	0.59	10		
$C_7H_{14}O_2$	Heptanoic acid	15	0.24	27		
$C_7H_{14}O_2$	Pentyl acetate	20	0.17	10		
$C_7H_{14}O_2$	Isopentyl acetate	20	0.2	10		
$C_7H_{14}O_2$	*sec*-Pentyl acetate	25	0.2	27		
$C_7H_{14}O_2$	Butyl propanoate	22	0.572	27		
$C_7H_{14}O_2$	Propyl butanoate	17	0.162	27		
$C_7H_{14}O_2$	Ethyl pentanoate	25	0.3	27		
$C_7H_{14}O_2$	Ethyl 3-methylbutanoate	20	0.2	10		
$C_7H_{15}Br$	1-Bromoheptane	25	0.00067	25		
$C_7H_{15}Cl$	1-Chloroheptane	25	0.00136	25		
$C_7H_{15}I$	1-Iodoheptane	25	0.00035	25		
C_7H_{16}	Heptane	0	0.0003	3		
		25	0.00024	3	209	13
		40	0.00025	3		
C_7H_{16}	2-Methylhexane	25	0.00025	3	346	5
C_7H_{16}	3-Methylhexane	25	0.00026	3	249	13
C_7H_{16}	2,2-Dimethylpentane	25	0.00044	3	318	5
C_7H_{16}	2,3-Dimethylpentane	25	0.00052	3	175	5
C_7H_{16}	2,4-Dimethylpentane	25	0.00042	3	323	13
C_7H_{16}	3,3-Dimethylpentane	25	0.00059	3	186	5
$C_7H_{16}O$	1-Heptanol	10	0.25	1		
		25	0.174	1	0.00562	28
		50	0.12	1		
$C_7H_{16}O$	2-Heptanol	30	0.33	1		
$C_7H_{16}O$	3-Heptanol	25	0.43	1		
$C_7H_{16}O$	4-Heptanol	25	0.47	1		
$C_7H_{16}O$	2-Methyl-2-hexanol	25	1.0	1		
$C_7H_{16}O$	5-Methyl-2-hexanol	25	0.49	1		
$C_7H_{16}O$	3-Methyl-3-hexanol	25	1.2	1		
$C_7H_{16}O$	3-Ethyl-3-pentanol	25	1.7	1		
$C_7H_{16}O$	2,3-Dimethyl-2-pentanol	25	1.5	1		
$C_7H_{16}O$	2,4-Dimethyl-2-pentanol	25	1.3	1		
$C_7H_{16}O$	2,2-Dimethyl-3-pentanol	25	0.82	1		
$C_7H_{16}O$	2,3-Dimethyl-3-pentanol	25	1.6	1		
$C_7H_{16}O$	2,4-Dimethyl-3-pentanol	25	0.70	1		
$C_7H_{16}O$	2,3,3-Trimethyl-2-butanol	40	2.2	1		
C_8F_{18}	Perfluorooctane	25	0.00000017	25		
$C_8H_4F_6$	1,3-Bis(trifluoromethyl)benzene	25	0.0041	2		
$C_8H_6N_2$	Quinoxaline	50	54	6		
$C_8H_6O_4$	Phthalic acid	14	0.54	27		
$C_8H_6O_4$	Isophthalic acid	25	0.013	27		
C_8H_6S	Benzo[b]thiophene	20	0.0130	6		
$C_8H_7ClO_3$	3-Chloro-4-hydroxy-5-methoxybenzaldehyde	25	0.093	8		
$C_8H_7ClO_3$	2-Chloro-4-hydroxy-5-methoxybenzaldehyde	25	0.013	8		

Mol. Form	Name	t/°C	S/mass %	Ref.	k_H/kPa m^{3}mol^{-1}	Ref.
$C_8H_7Cl_3O$	2,4,6-Trichloro-3,5-dimethylphenol	25	0.00050	2		
C_8H_7N	Indole	20	0.187	6		
C_8H_8	Styrene	25	0.0321	22	0.286	22
		50	0.046	4	0.30	13
$C_8H_8HgO_2$	Mercury(II) phenyl acetate		0.2	30		
$C_8H_8N_2$	2-Methyl-1H-benzimidazole	20	0.145	6		
C_8H_8O	Acetophenone	25	0.55	28	0.00108	28
$C_8H_8O_2$	*o*-Toluic acid	25	0.118	27		
$C_8H_8O_2$	*m*-Toluic acid	25	0.098	27		
$C_8H_8O_2$	*p*-Toluic acid	25	0.345	27		
$C_8H_8O_2$	Benzeneacetic acid	25	1.71	27		
$C_8H_8O_2$	Methyl benzoate	20	0.21	10		
$C_8H_8O_3$	4-Methoxybenzoic acid	25	0.023	27		
$C_8H_8O_3$	Mandelic acid	25	11.3	27		
$C_8H_8O_3$	Methyl salicylate	30	0.74	10		
$C_8H_8O_3$	4-Hydroxy-3-methoxybenzaldehyde	25	0.247	8		
C_8H_9ClO	4-Chloro-2,5-dimethylphenol	25	0.89	2		
C_8H_9ClO	4-Chloro-2,6-dimethylphenol	25	0.52	2		
C_8H_9ClO	4-Chloro-3,5-dimethylphenol	25	0.34	2		
C_8H_9NO	Acetanilide	20	0.52	27		
		70	2.7	27		
C_8H_{10}	Ethylbenzene	0	0.020	4		
		25	0.0161	22	0.843	22
		40	0.0200	4		
C_8H_{10}	*o*-Xylene	25	0.0171	22	0.551	22
		45	0.021	4		
C_8H_{10}	*m*-Xylene	0	0.0203	4		
		25	0.0161	22	0.730	22
		40	0.022	4		
C_8H_{10}	*p*-Xylene	0	0.0160	4		
		25	0.0181	22	0.690	22
		40	0.022	4		
$C_8H_{10}N_4O_2$	Caffeine	25	2.12	29		
$C_8H_{10}O$	2,4-Xylenol	25	0.787	10		
$C_8H_{10}O$	3,5-Xylenol	29	0.62	10		
$C_8H_{10}O$	Phenetole	25	0.12	10		
$C_8H_{11}N$	2,5-Dimethylaniline	20	0.66	27		
C_8H_{12}	4-Vinylcyclohexene	25	0.005	4		
C_8H_{14}	1-Octyne	25	0.0024	4	7.87	13
C_8H_{16}	1-Octene	25	0.00027	4	96.3	13
C_8H_{16}	Cyclooctane	25	0.00079	4	10.7	13
C_8H_{16}	Ethylcyclohexane	40	0.00066	4		
C_8H_{16}	*cis*-1,2-Dimethylcyclohexane	25	0.00060	4	36	5
C_8H_{16}	*trans*-1,4-Dimethylcyclohexane	25	0.000384	4	88.2	5
C_8H_{16}	Propylcyclopentane	25	0.00020	4	90.2	5
C_8H_{16}	1,1,3-Trimethylcyclopentane	25	0.00037	4	159	5
$C_8H_{16}N_2O_4S_2$	Homocystine	25	0.02	26		
$C_8H_{16}O$	2-Octanone	25	0.113	10		
$C_8H_{16}O_2$	Octanoic acid	25	0.080	26		
$C_8H_{16}O_2$	Hexyl acetate	20	0.02	10		
$C_8H_{16}O_2$	*sec*-Hexyl acetate	20	0.13	10		
$C_8H_{16}O_2$	Pentyl propanoate	20	0.1	27		
$C_8H_{16}O_2$	Isobutyl isobutanoate	20	0.5	10		
$C_8H_{16}O_2$	Ethyl hexanoate	20	0.063	27		
$C_8H_{17}Br$	1-Bromooctane	25	0.000167	25		
$C_8H_{17}Cl$	1-Chlorooctane	25	0.0345	25		
$C_8H_{17}Cl$	3-(Chloromethyl)heptane	20	0.01	10		
C_8H_{18}	Octane	25	0.000071	4	311	13
		50	0.00010	4		

Mol. Form	Name	t/°C	S/mass %	Ref.	k_H/kPa m^{3}mol^{-1}	Ref.
C_8H_{18}	3-Methylheptane	25	0.000079	4	376	5
C_8H_{18}	2,2,4-Trimethylpentane	25	0.00022	4	307	13
C_8H_{18}	2,3,4-Trimethylpentane	25	0.00018	4	206	13
$C_8H_{18}O$	1-Octanol	25	0.054	1		
$C_8H_{18}O$	2-Octanol	25	0.4	1		
$C_8H_{18}O$	2-Methyl-2-heptanol	30	0.25	1		
$C_8H_{18}O$	2-Ethyl-1-hexanol	25	0.01	1		
$C_8H_{18}O$	Dibutyl ether	20	0.03	10	0.48	13
$C_8H_{19}N$	Dibutylamine	20	0.47	10		
$C_8H_{19}N$	2-Ethylhexylamine	20	0.25	10		
$C_8H_{20}Si$	Tetraethylsilane	25	0.0000325	10		
$C_9H_7BrO_4$	2-(Acetyloxy)-5-bromobenzoic acid		0.07	30		
C_9H_7N	Quinoline	20	0.633	6		
C_9H_7N	Isoquinoline	20	0.452	6		
$C_9H_8O_2$	*trans*-Cinnamic acid	25	0.056	27		
$C_9H_8O_4$	2-(Acetyloxy)benzoic acid		0.25	27		
$C_9H_9I_2NO_3$	L-3,5-Diiodotyrosine	25	0.062	26		
C_9H_9N	3-Methyl-1H-indole	20	0.050	6		
$C_9H_9NO_3$	*N*-Benzoylglycine	25	0.37	29		
C_9H_{10}	Indan	25	0.010	4		
$C_9H_{10}O_2$	Ethyl benzoate	25	0.083	20		
$C_9H_{11}NO_2$	*DL*-Phenylalanine	25	1.40	29		
$C_9H_{11}NO_2$	Phenylalanine	25	2.71	26		
$C_9H_{11}NO_3$	*L*-Tyrosine	25	0.046	26		
$C_9H_{11}NO_3$	*DL*-Tyrosine	25	0.35	30		
$C_9H_{11}NO_4$	Levodopa [3-Hydroxy-*L*-tyrosine]	25	62.3	26		
C_9H_{12}	1,8-Nonadiyne	25	0.0125	4		
C_9H_{12}	Propylbenzene	25	0.0052	22	1.041	22
C_9H_{12}	Isopropylbenzene	25	0.0050	22	1.466	22
C_9H_{12}	*o*-Ethyltoluene	25	0.0093	5	0.529	13
C_9H_{12}	*p*-Ethyltoluene	25	0.0094	5	0.500	13
C_9H_{12}	1,2,3-Trimethylbenzene	25	0.0070	22	0.343	22
C_9H_{12}	1,2,4-Trimethylbenzene	25	0.0057	22	0.569	22
C_9H_{12}	1,3,5-Trimethylbenzene	25	0.0050	22	0.781	22
$C_9H_{14}N_4O_3$	Carnosine	25	24.4	26		
$C_9H_{14}O_6$	Triacetin	25	5.8	10		
C_9H_{16}	1-Nonyne	25	0.00072	4		
C_9H_{18}	1,1,3-Trimethylcyclohexane	25	0.000177	4	105	13
$C_9H_{18}O$	Diisobutyl ketone	25	0.043	10		
$C_9H_{18}O_2$	Nonanoic acid	20	0.0284	26		
$C_9H_{18}O_2$	Ethyl heptanoate	20	0.029	27		
C_9H_{20}	Nonane	25	0.000017	4	333	13
		50	0.000022	4		
C_9H_{20}	4-Methyloctane	25	0.0000115	4	1000	5
C_9H_{20}	2,2,5-Trimethylhexane	25	0.00008	4	246	13
$C_9H_{20}O$	3,5-Dimethyl-4-heptanol	15	0.072	1		
$C_9H_{20}O$	1-Nonanol	25	0.014	1		
$C_9H_{20}O$	2-Nonanol	15	0.026	1		
$C_9H_{20}O$	3-Nonanol	15	0.032	1		
$C_9H_{20}O$	4-Nonanol	15	0.0026	1		
$C_9H_{20}O$	5-Nonanol	15	0.0032	1		
$C_{10}F_{22}$	Perfluorodecane	20	0.000031	25		
$C_{10}H_7Cl$	1-Chloronaphthalene	25	0.00224	5	0.0363	28
$C_{10}H_7Cl$	2-Chloronaphthalene	25	0.00117	5	0.0335	28
$C_{10}H_8$	Naphthalene	10	0.0019	4		
		25	0.00316	22	0.043	22
		50	0.0082	4		
$C_{10}H_8O$	2-Naphthol		0.1	30		
$C_{10}H_9N$	3-Methylisoquinoline	20	0.092	6		

Mol. Form	Name	t/°C	S/mass %	Ref.	k_H/kPa m^{3}mol^{-1}	Ref.
$C_{10}H_{10}O_4$	Dimethyl phthalate	25	0.40	15		
$C_{10}H_{12}N_4O_5$	Inosine	20	1.6	29		
$C_{10}H_{13}N_5O_3$	2'-Deoxyadenosine	25	0.67	29		
$C_{10}H_{13}N_5O_4$	Adenosine	25	0.51	29		
$C_{10}H_{13}N_5O_5$	Guanosine	25	0.0500	29		
$C_{10}H_{14}$	Butylbenzene	25	0.00138	22	1.33	22
$C_{10}H_{14}$	*sec*-Butylbenzene	25	0.0014	4	1.89	11
$C_{10}H_{14}$	*tert*-Butylbenzene	25	0.0032	4	1.28	11
$C_{10}H_{14}$	Isobutylbenzene	25	0.0010	4	3.32	11
$C_{10}H_{14}$	*p*-Cymene	25	0.0051	23	0.80	5
$C_{10}H_{14}$	1,2,4,5-Tetramethylbenzene	25	0.000348	4	2.55	11
$C_{10}H_{14}N_2O_5$	Thymidine	25	5.1	29		
$C_{10}H_{14}O$	Carvone	15	0.13	27		
$C_{10}H_{14}O$	Thymol		0.1	30		
$C_{10}H_{16}$	*d*-Limonene	0	0.00097	4		
		25	0.00138	4		
$C_{10}H_{16}O$	Camphor	20	0.01	10		
$C_{10}H_{16}O$	Carvenone	15	0.22	27		
$C_{10}H_{16}O_4$	*trans*-Camphoric acid	25	0.8	27		
$C_{10}H_{18}$	*trans*-Decahydronaphthalene	25	0.000089	4	3	13
$C_{10}H_{18}O$	Borneol	25	0.074	27		
$C_{10}H_{18}O$	α-Terpineol	15	0.20	27		
$C_{10}H_{20}$	1-Decene	25	0.00057	4		
$C_{10}H_{20}$	Pentylcyclopentanc	25	0.0000115	4	185	5
$C_{10}H_{20}O_2$	Decanoic acid	20	0.015	26		
$C_{10}H_{20}O_2$	Ethyl octanoate	20	0.007	27		
$C_{10}H_{22}$	Decane	0	0.0000015	4	479	13
$C_{10}H_{22}O$	1-Decanol	25	0.0037	1		
$C_{10}H_{22}O$	Diisopentyl ether	20	0.02	10		
$C_{11}H_8O_2$	1-Naphthalenecarboxylic acid	25	0.0058	27		
$C_{11}H_{10}$	1-Methylnaphthalene	25	0.00281	22	0.045	22
$C_{11}H_{10}$	2-Methylnaphthalene	25	0.0025	4	0.051	12
$C_{11}H_{12}N_2O_2$	Tryptophan	25	1.30	26		
$C_{11}H_{16}$	Pentylbenzene	25	0.00105	5	1.69	11
$C_{11}H_{22}O_2$	Ethyl nonanoate	20	0.003	27		
$C_{12}Cl_{10}$	Decachlorobiphenyl	25	0.00000000012	7	0.0208	7
$C_{12}F_{26}$	Hexacosafluorododecane	20	0.00000096	25		
$C_{12}HCl_9$	2,2',3,3',4,5,5',6,6'-Nonachlorobiphenyl	25	0.0000000018	7		
$C_{12}H_2Cl_8$	2,2',3,3',5,5',6,6'-Octachlorobiphenyl	25	0.0000003	7	0.0381	7
$C_{12}H_3Cl_7$	2,2',3,3',4,4',6-Heptachlorobiphenyl	25	0.0000002	7	0.0054	7
$C_{12}H_4Cl_6$	2,2',3,3',4,4'-Hexachlorobiphenyl	25	0.00000006	7	0.0354	31
$C_{12}H_4Cl_6$	2,2',4,4',6,6'-Hexachlorobiphenyl	25	0.00000007	7	0.818	7
$C_{12}H_4Cl_6$	2,2',3,3',6,6'-Hexachlorobiphenyl	25	0.00000008	7		
$C_{12}H_5Cl_5$	2,3,4,5,6-Pentachlorobiphenyl	25	0.0000008	7		
$C_{12}H_5Cl_5$	2,2',4,5,5'-Pentachlorobiphenyl	25	0.000001	7	0.0421	31
$C_{12}H_6Cl_4$	2,3,4,5-Tetrachlorobiphenyl	25	0.000002	7		
$C_{12}H_6Cl_4$	2,2',4',5-Tetrachlorobiphenyl	25	0.0000016	9		
$C_{12}H_7Cl_3$	2,4,5-Trichlorobiphenyl	25	0.000014	7	0.0379	31
$C_{12}H_7Cl_3$	2,4,6-Trichlorobiphenyl	25	0.00002	7	0.0495	7
$C_{12}H_8$	Acenaphthylene	20	0.0016	28	0.012	28
$C_{12}H_8Cl_2$	2,5-Dichlorobiphenyl	25	0.0002	7	0.0201	7
$C_{12}H_8Cl_2$	2,6-Dichlorobiphenyl	25	0.00014	7		
$C_{12}H_8O$	Dibenzofuran	25	0.000656	6	0.011	12
$C_{12}H_8S$	Dibenzothiophene	25	0.000103	6		
$C_{12}H_9Cl$	2-Chlorobiphenyl	25	0.00055	7	0.0701	7
$C_{12}H_9N$	Carbazole	22	0.000120	6		
$C_{12}H_{10}$	Acenaphthene	0	0.00015	4		
		25	0.000380	22	0.01217	22
		50	0.00092	4		

Mol. Form	Name	t/°C	S/mass %	Ref.	k_H/kPa m^{3}mol^{-1}	Ref.
$C_{12}H_{10}$	Biphenyl	0	0.000272	4		
		25	0.00072	22	0.0280	22
		50	0.0022	4		
$C_{12}H_{10}N_2$	Azobenzene	20	0.03	27		
$C_{12}H_{10}N_2O$	*N*-Nitrosodiphenylamine	25	0.0035	17		
$C_{12}H_{10}O$	Diphenyl ether	25	0.00180	6	0.027	13
$C_{12}H_{12}$	1-Ethylnaphthalene	25	0.00101	4	0.039	12
$C_{12}H_{12}$	2-Ethylnaphthalene	25	0.00080	4	0.078	12
$C_{12}H_{12}$	1,3-Dimethylnaphthalene	25	0.0008	4		
$C_{12}H_{12}$	1,4-Dimethylnaphthalene	25	0.00114	4		
$C_{12}H_{12}$	1,5-Dimethylnaphthalene	25	0.00031	4	0.036	28
$C_{12}H_{12}$	2,3-Dimethylnaphthalene	25	0.00025	4		
$C_{12}H_{12}$	2,6-Dimethylnaphthalene	25	0.00017	4		
$C_{12}H_{18}$	Hexylbenzene	25	0.00021	4		
$C_{12}H_{22}O_{11}$	Sucrose	20	67.1	27		
		50	72.3	27		
		100	83.0	27		
$C_{12}H_{22}O_{11}$	α-Maltose	20	51.9	27		
$C_{12}H_{24}O_2$	Dodecanoic acid	20	0.0055	26		
$C_{12}H_{24}O_2$	Ethyl decanoate	20	0.0015	27		
$C_{12}H_{26}$	Dodecane	25	0.00000037	4	750	5
$C_{12}H_{26}O$	1-Dodecanol	25	0.0004	1		
$C_{12}H_{27}O_4P$	Tributyl phosphate	25	0.039	10		
$C_{13}H_9N$	Acridine	25	0.00466	6		
$C_{13}H_9N$	Benzo[f]quinoline	25	0.0079	6		
$C_{13}H_{10}$	9H-Fluorene	0	0.00007	4		
		25	0.00019	22	0.00787	22
		50	0.00063	4		
$C_{13}H_{12}$	Diphenylmethane	25	0.000141	4	0.001	12
$C_{13}H_{14}$	1,4,5-Trimethylnaphthalene	25	0.00021	4		
$C_{13}H_{26}O_2$	Tridecanoic acid	20	0.0033	26		
$C_{14}H_{10}$	Anthracene	0	0.0000022	4		
		25	0.0000045	22	0.00396	22
$C_{14}H_{10}$	Phenanthrene	10	0.000050	4		
		25	0.00011	22	0.00324	22
		50	0.00041	4		
$C_{14}H_{12}$	*trans*-Stilbene	25	0.000029	4	0.040	12
$C_{14}H_{14}$	1,2-Diphenylethane	25	0.00044	6	0.017	12
$C_{14}H_{14}O$	Dibenzyl ether	35	0.0040	10		
$C_{14}H_{28}O_2$	Tetradecanoic acid	20	0.0020	26		
$C_{14}H_{29}Cl$	1-Chlorotetradecane	25	0.0232	25		
$C_{14}H_{30}$	Tetradecane	25	0.000012	5		
$C_{14}H_{30}O$	1-Tetradecanol	25	0.000031	1		
$C_{15}H_{12}$	1-Methylphenanthrene	25	0.0000269	4		
$C_{15}H_{12}$	2-Methylanthracene	25	0.00003	22		
$C_{15}H_{12}$	9-Methylanthracene	25	0.000026	4		
$C_{15}H_{30}O_2$	Pentadecanoic acid	20	0.0012	26		
$C_{15}H_{32}O$	1-Pentadecanol	25	0.000010	1		
$C_{16}H_{10}$	Fluoranthene	25	0.000026	22	0.00096	22
$C_{16}H_{10}$	Pyrene	25	0.000013	22	0.00092	22
		50	0.00009	4		
$C_{16}H_{14}$	9,10-Dimethylanthracene	25	0.0000056	4		
$C_{16}H_{15}NO_3$	*N*-Benzoyl-*L*-phenylalanine	25	0.085	29		
$C_{16}H_{22}O_4$	Dibutyl phthalate	25	0.00112	15		
$C_{16}H_{32}O_2$	Hexadecanoic acid	20	0.00072	26		
$C_{16}H_{34}O$	1-Hexadecanol	25	0.000003	1		
$C_{17}H_{12}$	11H-Benzo[a]fluorene	25	0.0000045	4		
$C_{17}H_{12}$	11H-Benzo[b]fluorene	25	0.0000002	4		
$C_{17}H_{19}NO_3$	Morphine	20	0.015	27		

Mol. Form	Name	t/°C	S/mass %	Ref.	k_H/kPa m³mol⁻¹	Ref.
$C_{17}H_{21}NO_4$	Cocaine	25	0.17	27		
$C_{17}H_{34}O_2$	Heptadecanoic acid	20	0.00042	26		
$C_{18}H_{12}$	Benzo[a]anthracene	25	0.0000011	22	0.00058	22
$C_{18}H_{12}$	Chrysene	25	0.0000002	22	0.000065	22
$C_{18}H_{12}$	Naphthacene	25	0.00000006	4	0.000004	12
$C_{18}H_{12}$	Triphenylene	25	0.0000041	4	0.00001	12
$C_{18}H_{12}N_2$	2,2'-Biquinoline	24	0.000102	6		
$C_{18}H_{21}NO_3$	Codeine	25	0.79	27		
$C_{18}H_{32}O_{16}$	Raffinose	20	12.5	27		
$C_{18}H_{34}O_4$	Dibutyl sebacate	20	0.004	10		
$C_{18}H_{36}O_2$	Octadecanoic acid	20	0.00029	26		
$C_{18}H_{38}O$	1-Octadecanol	34	0.000011	1		
$C_{19}H_{14}$	9-Methylbenz[a]anthracene	27	0.0000066	4		
$C_{19}H_{14}$	10-Methylbenz[a]anthracene	25	0.0000055	4		
$C_{19}H_{14}$	5-Methylchrysene	27	0.0000062	4		
$C_{20}H_{12}$	Perylene	25	0.00000004	4	0.000003	12
$C_{20}H_{12}$	Benzo[a]pyrene	25	0.0000003	22	0.0000465	22
$C_{20}H_{12}$	Benzo[e]pyrene	20	0.0000005	22	0.0000467	22
$C_{20}H_{12}O_5$	Fluorescein	20	0.005	27		
$C_{20}H_{13}N$	13H-Dibenzo[a,i]carbazole	24	0.00000104	6		
$C_{20}H_{14}$	1,2-Dihydrobenz[j]aceanthrylene	25	0.00000036	6		
$C_{20}H_{14}O_4$	Phenolphthalein	20	0.018	27		
$C_{20}H_{24}N_2O_2$	Quinine	25	0.057	27		
$C_{20}H_{24}N_2O_2$	Quinidine	20	0.020	27		
$C_{20}H_{42}$	Eicosane	25	0.00000019	4		
$C_{21}H_{13}N$	Dibenz[a,j]acridine	25	0.000016	6		
$C_{21}H_{16}$	1,2-Dihydro-3-methylbenz[j]aceanthrylene	25	0.00000022	6		
$C_{21}H_{22}N_2O_2$	Strychnine	20	0.013	27		
$C_{21}H_{28}O_5$	17,21-Dihydroxypregn-4-ene-3,11,20-trione	25	0.028	30		
$C_{22}H_{12}$	Benzo[ghi]perylene	25	0.000000026	4	0.000075	12
$C_{22}H_{14}$	Picene	25	0.00000025	4		
$C_{22}H_{14}$	Benzo[b]triphenylene	25	0.0000027	4		
$C_{22}H_{14}$	Dibenz[a,h]anthracene	25	0.00000006	4		
$C_{22}H_{14}$	Dibenz[a,j]anthracene	25	0.0000012	4		
$C_{22}H_{44}O_2$	Butyl stearate	25	0.2	10		
$C_{23}H_{26}N_2O_4$	Brucine	20	0.012	27		
$C_{23}H_{27}NO_8$	Narceine	13	0.078	27		
$C_{24}H_{12}$	Coronene	25	0.000000014	4		

AQUEOUS SOLUBILITY OF INORGANIC COMPOUNDS AT VARIOUS TEMPERATURES

The solubility of over 300 common inorganic compounds in water is tabulated here as a function of temperature. Solubility is defined as the concentration of the compound in a solution that is in equilibrium with a solid phase at the specified temperature. In this table the solid phase is generally the most stable crystalline phase at the temperature in question. An asterisk * on solubility values in adjacent columns indicates that the solid phase changes between those two temperatures (usually from one hydrated phase to another or from a hydrate to the anhydrous solid). In such cases the slope of the solubility vs. temperature curve may show a discontinuity.

All solubility values are expressed as mass percent of solute, $100 \cdot w_2$, where

$$w_2 = m_2/(m_1 + m_2)$$

and m_2 is the mass of solute and m_1 the mass of water. This quantity is related to other common measures of solubility as follows:

Molality: $m_2 = 1000w_2/M_2(1-w_2)$
Mole fraction: $x_2 = (w_2/M_2)/\{(w_2/M_2) + (1-w_2)/M_1\}$
Mass of solute per 100 g of H_2O: $r_2 = 100w_2/(1-w_2)$

Here M_2 is the molar mass of the solute and M_1 = 18.015 g/mol is the molar mass of water.

The data in the table have been derived from the references indicated; in many cases the data have been refitted or interpolated in order to present solubility at rounded values of temperature. Where available, values were taken from the IUPAC *Solubility Data Series* (Reference 1) or the related papers in the *Journal of Physical and Chemical Reference Data* (References 2 to 5), which present carefully evaluated data.

The solubility of sparingly soluble compounds that do not appear in this table may be calculated from the data in the table "Solubility Product Constants". Solubility of inorganic gases may be found in the table "Solubility of Selected Gases in Water".

Compounds are listed alphabetically by chemical formula in the most commonly used form (e.g., NaCl, NH_4NO_3, etc.).

REFERENCES

1. *Solubility Data Series*, International Union of Pure and Applied Chemistry. Volumes 1 to 53 were published by Pergamon Press, Oxford, from 1979 to 1994; subsequent volumes were published by Oxford University Press, Oxford. The number following the colon is the volume number in the series.
2. Clever, H.L., and Johnston, F.J., *J. Phys. Chem. Ref. Data*, 9, 751, 1980.
3. Marcus, Y., *J. Phys. Chem. Ref. Data*, 9, 1307, 1980.
4. Clever, H.L., Johnson, S.A., and Derrick, M.E., *J. Phys. Chem. Ref. Data*, 14, 631, 1985.
5. Clever, H.L., Johnson, S.A., and Derrick, M.E., *J. Phys. Chem. Ref. Data*, 21, 941, 1992.
6. Söhnel, O., and Novotny, P., *Densities of Aqueous Solutions of Inorganic Substances*, Elsevier, Amsterdam, 1985.
7. Krumgalz, B.S., *Mineral Solubility in Water at Various Temperatures*, Israel Oceanographic and Limnological Research Ltd., Haifa, 1994.
8. Potter, R.W., and Clynne, M.A., *J. Research U.S. Geological Survey*, 6, 701, 1978; Clynne, M.A., and Potter, R.W., *J. Chem. Eng. Data*, 24, 338, 1979.
9. Marshal, W.L., and Slusher, R., *J. Phys. Chem.*, 70, 4015, 1966; Knacke, O., and Gans, W., *Zeit. Phys. Chem.*, NF, 104, 41, 1977.
10. Stephen, H., and Stephen, T., *Solubilities of Inorganic and Organic Compounds, Vol. 1*, Macmillan, New York, 1963.

Compound	0°C	10°C	20°C	25°C	30°C	40°C	50°C	60°C	70°C	80°C	90°C	100°C	Ref.
$AgBrO_3$				0.193							1.32		7
$AgClO_2$	0.17	0.31	0.47	0.55	0.64	0.82	1.02	1.22	1.44	1.66	1.88	2.11	7
$AgClO_3$				15									7
$AgClO_4$	81.6	83.0	84.2	84.8	85.3	86.3	86.9	87.5	87.9	88.3	88.6	88.8	6
$AgNO_2$	0.155			0.413									7
$AgNO_3$	55.9	62.3	67.8	70.1	72.3	76.1	79.2	81.7	83.8	85.4	86.7	87.8	6
Ag_2SO_4	0.56	0.67	0.78	0.83	0.88	0.97	1.05	1.13	1.20	1.26	1.32	1.39	7
$AlCl_3$	30.84	30.91	31.03	31.10	31.18	31.37	31.60	31.87	32.17	32.51	32.90	33.32	7
$Al(ClO_4)_3$	54.9										64.4		7
AlF_3	0.25	0.34	0.44	0.50	0.56	0.68	0.81	0.96	1.11	1.28	1.45	1.64	7
$Al(NO_3)_3$	37.0	38.2	39.9	40.8	42.0	44.5	47.3	50.4	53.8*			61.5*	6
$Al_2(SO_4)_3$	27.5			27.8	28.2	29.2	30.7	32.6	34.9	37.6	40.7	44.2	7
As_2O_3	1.19	1.48	1.80	2.01	2.27	2.86	3.43	4.11	4.89	5.77	6.72	7.71	10
$BaBr_2$	47.6	48.5	49.5	50.0	50.4	51.4	52.5	53.5	54.5	55.5	56.6	57.6	6
$Ba(BrO_3)_2$	0.285	0.442	0.656	0.788	0.935	1.30	1.74	2.27	2.90	3.61	4.40	5.25	1:14
$Ba(C_2H_3O_2)_2$	37.0			44.2									7
$BaCl_2$	23.30	24.88	26.33	27.03	27.70	29.00	30.27	31.53	32.81	34.14	35.54	37.05	8
$Ba(ClO_2)_2$	30.5			31.3								44.7	7
$Ba(ClO_3)_2$	16.90	21.23	23.66	27.50	29.43	33.16	36.69	40.05	43.04	45.90	48.70	51.17	1:14
$Ba(ClO_4)_2$	67.30	70.96	74.30	75.75	77.05	79.23	80.92	82.21	83.16	83.88	84.43	84.90	7
BaF_2		0.158		0.161									7
BaI_2	62.5	64.7	67.3	68.8	69.1	69.5	70.1	70.7	71.3	72.0	72.7	73.4	6
$Ba(IO_3)_2$	0.0182	0.0262	0.0342	0.0396	0.045*	0.058*	0.073	0.090	0.109	0.131	0.156	0.182	1:14
$Ba(NO_2)_2$	31.1	36.6	41.8	44.3	46.8	51.6	56.2	60.5	64.6	68.5	72.1	75.6	10
$Ba(NO_3)_2$	4.7	6.3	8.2	9.3	10.2	12.4	14.7	17.0	19.3	21.5	23.5	25.5	6
$Ba(OH)_2$	1.67			4.68	8.4	19	33	52	74	101			7
BaS	2.79	4.78	6.97	8.21	9.58	12.67	16.18	20.05	24.19	28.55	33.04	37.61	7
$Ba(SCN)_2$				62.6									7
$BaSO_3$				0.0011									1:26
$BeCl_2$	40.5			41.7									7
$Be(ClO_4)_2$				59.5									7
$BeSO_4$	26.69	27.58	28.61	29.22	29.90	31.51	33.39	35.50	37.78	40.21	42.72	45.28	7
$CaBr_2$	55	56	59	61	63	68	71	73					10
$CaCl_2$	36.70	39.19	42.13	44.83*	49.12*	52.85*	56.05*	56.73	57.44	58.21	59.04	59.94	8
$Ca(ClO_3)_2$	63.2	64.2	65.5	66.3	67.2	69.0	71.0	73.2	75.5*	77.4*	77.7	78.0	1:14
$Ca(ClO_4)_2$				65.3									7
CaF_2	0.0013			0.0016									10
CaI_2	64.6	66.0	67.6	68.3	69.0	70.8	72.4	74.0	76.0	78.0	79.6	81.0	7
$Ca(IO_3)_2$	0.082	0.155	0.243	0.305	0.384*	0.517*	0.590	0.652	0.811*	0.665*	0.668		1:14
$Ca(NO_2)_2$	38.6	39.5	44.5	48.6									7
$Ca(NO_3)_2$	50.1	53.1	56.7	59.0	60.9	65.4	77.8	78.1	78.2	78.3	78.4	78.5	6
$CaSO_3$			0.0059	0.0054	0.0049	0.0041	0.0035	0.0030	0.0026	0.0023	0.0020	0.0019	1:26
$CaSO_4$	0.174	0.191	0.202	0.205	0.208	0.210	0.207	0.201	0.193	0.184	0.173	0.163	9

AQUEOUS SOLUBILITY OF INORGANIC COMPOUNDS AT VARIOUS TEMPERATURES (continued)

Compound	0°C	10°C	20°C	25°C	30°C	40°C	50°C	60°C	70°C	80°C	90°C	100°C	Ref.
$CdBr_2$	36.0	43.0	49.9	53.4	56.4	60.3*	60.3*	60.5	60.7	60.9	61.3	61.6	6
CdC_2O_4				0.0060									5
$CdCl_2$	47.2	50.1	53.2	54.6	56.3*	57.3*	57.5	57.8	58.1	58.51	58.98	59.5	6
$Cd(ClO_4)_2$				58.7								66.9	7
CdF_2		5.82	4.65	4.18	3.76								5
CdI_2	44.1	44.9	45.8	46.3	46.8	47.9	49.0	50.2	51.5	52.7	54.1	55.4	6
$Cd(IO_3)_2$				0.091									5
$Cd(NO_3)_2$	55.4	57.1	59.6	61.0	62.8	66.5	70.6	86.1	86.5	86.8	87.1	87.4	6
$CdSO_4$	43.1	43.1	43.2	43.4	43.6	44.1	43.5	42.5	41.4	40.2	38.5	36.7	6
$CdSeO_4$	42.04	40.59	39.02	38.18	37.29	35.35	33.15	30.65	27.84	24.69	21.24	17.49	5
$Ce(NO_3)_3$	57.99	59.80	61.89	63.05	64.31*	67.0*	68.6	71.1*	74.9*	79.2	80.9	83.1	1:13
$CoCl_2$	30.30	32.60	34.87	35.99	37.10	39.27	41.38	43.46	45.50	47.51	49.51	51.50	7
$Co(ClO_4)_2$	50.0			53.0									7
CoF_2				1.4									7
CoI_2	58.00	61.78	65.35	66.99	68.51	71.17	73.41	75.29	76.89	78.28	79.52	80.70	7
$Co(NO_2)_2$	0.076			0.49									7
$Co(NO_3)_2$	45.5	47.0	49.4	50.8	52.4	56.0	60.1	62.6	64.9	67.7			6
$CoSO_4$	19.9	23.0	26.1	27.7	29.2	32.3	34.4	35.9	35.5	33.2	30.6	27.8	6
$Co(SCN)_2$				50.7									7
CrO_3	62.2	62.3	62.6	62.8	63.0	63.5	64.1	64.7	65.5	66.2	67.1	67.9	6
CsBr				55.2									7
$CsBrO_3$	1.16	1.93	3.01	3.69	4.46	6.32	8.60	11.32	14.45	17.96	21.83	25.98	1:30
CsCl	61.83	63.48	64.96	65.64	66.29	67.50	68.60	69.61	70.54	71.40	72.21	72.96	1:47
$CsClO_3$	2.40	3.87	5.94	7.22	8.69	12.15	16.33	21.14	26.45	32.10	37.89	43.42	1:30
$CsClO_4$	0.79	1.01	1.51	1.96	2.57	4.28	6.55	9.29	12.41	15.80	19.39	23.07	7
CsI	30.9	37.2	43.2	45.9	48.6	53.3	57.3	60.7	63.6	65.9	67.7	69.2	6
$CsIO_3$	1.08	1.58	2.21	2.59	3.02	3.96	5.06	6.29	7.70	9.20	10.79	12.45	1:30
$CsNO_3$	8.46	13.0	18.6	21.8	25.1	32.0	39.0	45.7	51.9	57.3	62.1	66.2	6
CsOH					75								7
Cs_2SO_4	62.6	63.4	64.1	64.5	64.8	65.5	66.1	66.7	67.3	67.8	68.3	68.8	6
$CuBr_2$				55.8									7
$CuCl_2$	40.8	41.7	42.6	43.1	43.7	44.8	46.0	47.2	48.5	49.9	51.3	52.7	6
$Cu(ClO_4)_2$	54.3				59.3								7
CuF_2				0.075									7
$Cu(NO_3)_2$	45.2	49.8	56.3	59.2	61.1	62.0	63.1	64.5	65.9	67.5	69.2	71.0	6
$CuSO_4$	12.4	14.4	16.7	18.0	19.3	22.2	25.4	28.8	32.4	36.3	40.3	43.5	6
$CuSeO_4$	10.6			16.0									7
$Dy(NO_3)_3$	58.79	59.99	61.49	62.35	63.29	65.43	68.04	71.58					1:13
$Er(NO_3)_3$	61.58	63.15	64.84	65.75	66.69	68.70	70.96	73.64	77.75				1:13
$Eu(NO_3)_3$	55.2	56.7	58.5	59.4	60.4	62.5	64.6						1:13
$FeBr_2$				54.6								64.8*	7
$FeCl_2$	33.2*			39.4*								48.7*	7
$FeCl_3$	42.7	44.9	47.9	47.7	51.6	74.8	76.7	84.6	84.3	84.3	84.4	84.7	6

AQUEOUS SOLUBILITY OF INORGANIC COMPOUNDS AT VARIOUS TEMPERATURES (continued)

Compound	0°C	10°C	20°C	25°C	30°C	40°C	50°C	60°C	70°C	80°C	90°C	100°C	Ref.
$Fe(ClO_4)_2$	63.39			67.76									7
FeF_3				5.59									7
$Fe(NO_3)_3$	40.15			46.57									7
$Fe(NO_3)_2$	41.44			46.67									7
$FeSO_4$	13.5	17.0	20.8	22.8	24.8	28.8	32.8	35.5	33.6	30.4	27.1	24.0	6
$Gd(NO_3)_3$	56.3	57.7	59.2	60.1	61.0	62.9	65.2	67.9	71.5				1:13
HIO_3	73.45	74.10	74.98	75.48	76.03	77.20	78.46	79.78	81.13	82.48	83.82	85.14	1:30
H_3BO_3	2.61	3.57	4.77	5.48	6.27	8.10	10.3	12.9	15.9	19.3	23.1	27.3	6
$HgBr_2$	0.26	0.37	0.52	0.61	0.72	0.96	1.26	1.63	2.08	2.61	3.23	3.95	4
$Hg(CN)_2$	6.57	7.83	9.33	10.2	11.1	13.1	15.5	18.2	21.2	24.6	28.3	32.3	6
$HgCl_2$	4.24	5.05	6.17	6.81	7.62	9.53	12.02	15.18	19.16	24.06	29.90	36.62	4
HgI_2			0.0041	0.0055	0.0072	0.0122	0.0199						4
$Hg(SCN)_2$				0.070									4
Hg_2Cl_2				0.0004									3
$Hg_2(ClO_4)_2$	73.8			79.8*								85.3*	7
Hg_2SO_4	0.038	0.043	0.048	0.051	0.054	0.059	0.065	0.070	0.076	0.082	0.088	0.093	4
$Ho(NO_3)_3$				63.8									1:13
KBF_4	0.28	0.34	0.45	0.55	0.75	1.38	2.09	2.82	3.58	4.34	5.12	5.90	10
KBr	35.0	37.3	39.4	40.4	41.4	43.2	44.8	46.2	47.6	48.8	49.8	50.8	6
$KBrO_3$	2.97	4.48	6.42	7.55	8.79	11.57	14.71	18.14	21.79	25.57	29.42	33.28	1:30
$KC_2H_3O_2$	68.40	70.29	72.09	72.92	73.70	75.08	76.27	77.31	78.22	79.04	79.80	80.55	7
KCl	21.74	23.61	25.39	26.22	27.04	28.59	30.04	31.40	32.66	33.86	34.99	36.05	1:47
$KClO_3$	3.03	4.67	6.74	7.93	9.21	12.06	15.26	18.78	22.65	26.88	31.53	36.65	1:30
$KClO_4$	0.70	1.10	1.67	2.04	2.47	3.54	4.94	6.74	8.99	11.71	14.94	18.67	6
KF	30.90	39.8	47.3	50.41	53.2					60.0			7
$KHCO_3$	18.62	21.73	24.92	26.6	28.13	31.32	34.46	37.51	40.45				6
$KHSO_4$	27.1	29.7	32.3	33.6	35.0	37.8	40.5	43.4	46.2	49.02	51.82	54.6	6
KH_2PO_4	11.74	14.91	18.25	19.97	21.77	25.28	28.95	32.76	36.75	40.96	45.41	50.12	1:31
KI	56.0	57.6	59.0	59.7	60.4	61.6	62.8	63.8	64.8	65.7	66.6	67.4	6
KIO_3	4.53	5.96	7.57	8.44	9.34	11.09	13.22	15.29	17.41	19.58	21.78	24.03	1:30
KIO_4	0.16	0.22	0.37	0.51	0.70	1.24	1.96	2.83	3.82	4.89	6.02	7.17	7
$KMnO_4$	2.74	4.12	5.96	7.06	8.28	11.11	14.42	18.16					6
KNO_2	73.7	74.6	75.3	75.7	76.0	76.7	77.4	78.0	78.5	79.1	79.6	80.1	6
KNO_3	12.0	17.6	24.2	27.7	31.3	38.6	45.7	52.2	58.0	63.0	67.3	70.8	6
KOH	48.7	50.8	53.2	54.7	56.1	57.9	58.6	59.5	60.6	61.8	63.1	64.6	6
$KSCN$	63.8	66.4	69.1	70.4	71.6	74.1	76.5	78.9	81.1	83.3	85.3	87.3	6
K_2CO_3	51.3	51.7	52.3	52.7	53.1	54.0	54.9	56.0	57.2	58.4	59.6	61.0	6
K_2CrO_4	37.1	38.1	38.9	39.4	39.8	40.5	41.3	41.9	42.6	43.2	43.8	44.3	6
$K_2Cr_2O_7$	4.30	7.12	10.9	13.1	15.5	20.8	26.3	31.7	36.9	41.5	45.5	48.9	6
K_2HAsO_4	48.5*			63.6*								79.8*	7
K_2HPO_4	57.0	59.1	61.5	62.7	64.1	67.7*		72.7*					1:31
K_2MoO_4				64.7							66.5		7
K_2SO_3	51.30	51.39	51.49	51.55	51.62	51.76	51.93	52.11	52.32	52.54	52.79	53.06	1:26

Compound	0°C	10°C	20°C	25°C	30°C	40°C	50°C	60°C	70°C	80°C	90°C	100°C	Ref.
K_2SO_4	7.11	8.46	9.95	10.7	11.4	12.9	14.2	15.5	16.7	17.7	18.6	19.3	6
$K_2S_2O_3$	49.0*			62.3*							75.7*		7
$K_2S_2O_5$	22.1	26.7	31.1	33.1	35.2	39.0	42.6	46.0	49.1	52.0	54.6		1:26
K_2SeO_3	68.4*			68.5*								68.5*	7
K_2SeO_4	52.70	52.93	53.17	53.30	53.43	53.70	53.99	54.30	54.61	54.94	55.26	55.60	7
K_3AsO_4	51.5*			55.6*								73*	7
$K_3Fe(CN)_6$	23.9	27.6	31.1	32.8	34.3	37.2	39.6	41.7	43.5	45.0	46.1	47.0	6
K_3PO_4	44.3			51.4									7
$K_4Fe(CN)_6$	12.5	17.3	22.0	23.9	25.6	29.2	32.5	35.5	38.2	40.6	41.4	43.1	6
$LaCl_3$	49.0	48.5	48.6	48.9	49.3	50.5	52.1	54.0	56.3	58.9	61.7		6
$La(NO_3)_3$	55.0	56.9	58.9	60.0	61.1	63.6	66.3	69.9*	74.1*				1:13
LiBr	58.4	60.1	62.7	64.4	65.9	67.8	68.3	69.0	69.8	70.7	71.7	72.8	6
$LiBrO_3$	61.03	62.62	64.44	65.44	66.51	68.90	71.68*	73.24*	74.43	75.66	76.93	78.32	1:30
$LiC_2H_3O_2$	23.76	26.49	29.42	31.02	32.72	36.48	40.65	45.15	49.93	54.91	60.04	65.26	7
LiCl	40.45	42.46*	45.29*	45.81	46.25	47.30	48.47	49.78	51.27	52.98	54.98*	56.34*	1:47
$LiClO_3$	73.2	75.6*	80.8*	82.1	83.4	85.9*	87.1*	88.2	89.6	91.3	93.4	95.7	1:30
$LiClO_4$	30.1	32.6	35.5	37.0	38.6	41.9	45.5	49.2	53.2	57.2	61.3	71.4	6
LiF	0.120	0.126	0.131	0.134									7
LiH_2PO_4	55.8												7
LiI	59.4	60.5	61.7	62.3	63.0	64.3	65.8	67.3	68.8	81.3	81.7	82.6	6
$LiIO_3$				43.8									1:30
$LiNO_2$	41	45	49	51	53	56	60	63	66	68			10
$LiNO_3$	34.8	37.6	42.7	50.5	57.9	60.1	62.2	64.0	65.7	67.2	68.5	69.7	6
LiOH	10.8	10.8	11.0	11.1	11.3	11.7	12.2	12.7	13.4	14.2	15.1	16.1	6
LiSCN				54.5									7
Li_2CO_3	1.54	1.43	1.33	1.28	1.24	1.15	1.07	0.99	0.92	0.85	0.78	0.72	7
$Li_2C_2O_4$				5.87									7
Li_2HPO_3	9.07	8.40	7.77	7.47	7.18	6.64	6.16	5.71	5.30	4.91	4.53	4.16	7
Li_2SO_4	26.3	25.9	25.6	25.5	25.3	25.0	24.8	24.5	24.3	24.0	23.8	23.6	6
Li_3PO_4				0.027									1:31
$Lu(NO_3)_3$				71.1									1:13
$MgBr_2$	49.3	49.8	50.3	50.6	50.9	51.5	52.1	52.8	53.5	54.2	55.0	55.7	6
$Mg(BrO_3)_2$	43.0	45.2	48.0	49.4	51.0	54.3	57.9	61.6	65.3	69.0*	70.9*	71.7	1:14
$Mg(C_2H_3O_2)_2$	36.18	37.55	38.92	39.61									7
MgC_2O_4				0.038									7
$MgCl_2$	33.96	34.85	35.58	35.90	36.20	36.77	37.34	37.97	38.71	39.62	40.75	42.15	8
$Mg(ClO_3)_2$	53.35	54.40	56.81	58.66	60.91*	65.46*	67.33	69.27	71.01	72.44	73.48		1:14
$Mg(ClO_4)_2$	47.8	48.7	49.6	50.1	50.5	51.3	52.1						6
$MgCrO_4$	32.06*			35.39*									7
$MgCr_2O_7$				58.9						67.0			7
MgF_2				0.013									7
MgI_2	54.7	56.1	58.2	59.4	60.8	63.9	65.0	65.0	65.0	65.0	65.1	65.2	6
$Mg(IO_3)_2$	3.19*	6.70*	7.92	8.52	9.11	10.45	11.99	13.7	15.6	17.6	19.6		1:14

Compound	0°C	10°C	20°C	25°C	30°C	40°C	50°C	60°C	70°C	80°C	90°C	100°C	Ref.
$Mg(NO_2)_2$				47									7
$Mg(NO_3)_2$	38.4	39.5	40.8	41.6	42.4	44.1	45.9	47.9	50.0	52.2	70.6	72.0	6
$MgSO_3$	0.32	0.37	0.46	0.52	0.61	0.87*	0.85*	0.76	0.69	0.64	0.62	0.60	1:26
$MgSO_4$	18.2	21.7	25.1	26.3	28.2	30.9	33.4	35.6	36.9	35.9	34.7	33.3	6
MgS_2O_3	30.7			34.1									7
$MgSeO_4$	31.4*			35.7*								47*	7
$MnBr_2$	56.00	57.72	59.39	60.19	60.96	62.41	63.75	65.01	66.19	67.32	68.42	69.50	7
$MnCl_2$	38.7	40.6	42.5	43.6	44.7	47.0	49.4	54.1	54.7	55.2	55.7	56.1	6
MnF_2	0.80*			1.01*								0.48	7
$Mn(IO_3)_2$				0.27							0.34		7
$Mn(NO_3)_2$	50.5			61.7									7
$MnSO_4$	34.6	37.3	38.6	38.9	38.9	37.7	36.3	34.6	32.8	30.8	28.8	26.7	6
NH_4Br	37.5	40.2	42.7	43.9	45.1	47.3	49.4	51.3	53.0	54.6	56.1	57.4	7
NH_4Cl	22.92	25.12	27.27	28.34	29.39	31.46	33.50	35.49	37.46	39.40	41.33	43.24	1:47
NH_4ClO_4	10.8	14.1	17.8	19.7	21.7	25.8	29.8	33.6	37.3	40.7	43.8	46.6	6
NH_4F	41.7	43.2	44.7	45.5	46.3	47.8	49.3	50.9	52.5	54.1			7
NH_4HCO_3	10.6	13.7	17.6	19.9	22.4	27.9	34.2	41.4	49.3	58.1	67.6	78.0	7
$NH_4H_2AsO_4$	25.2	29.0	32.7	34.5	36.3	39.7	43.1	46.2	49.3	52.2	55.0		7
$NH_4H_2PO_4$	17.8	22.0	26.4	28.8	31.2	36.2	41.6	47.2	53.0	59.2	65.7	72.4	7
NH_4I	60.7	62.1	63.4	64.0	64.6	65.8	66.8	67.8	68.7	69.6	70.4	71.1	6
NH_4IO_3				3.70	4.20	5.64	7.63						1:30
NH_4NO_2	55.7	59.0	64.9	68.8									7
NH_4NO_3	54.0	60.1	65.5	68.0	70.3	74.3	77.7	80.8	83.4	85.8	88.2	90.3	6
NH_4SCN				64.4					81.1				7
$(NH_4)_2C_2O_4$	2.31	3.11	4.25	4.94	5.73	7.56	9.73	12.2	15.1	18.3	21.8	25.7	7
$(NH_4)_2HPO_4$	36.4	38.2	40.0	41.0	42.0	44.1	46.2	48.5	50.9	53.3	55.9	58.6	7
$(NH_4)_2S_2O_5$	65.5	67.9	69.8	70.5	71.3	72.3	72.9	73.1					1:26
$(NH_4)_2S_2O_8$	37.00	40.45	43.84	45.49	47.11	50.25	53.28	56.23	59.13	62.00			7
$(NH_4)_2SO_3$	32.2	34.9	37.7	39.1	40.6	43.7	47.0	50.6	54.5	58.9			1:26
$(NH_4)_2SO_4$	41.3	42.1	42.9	43.3	43.8	44.7	45.6	46.6	47.5	48.5	49.5	50.5	6
$(NH_4)_2SeO_3$	49.0	51.1	53.4	54.7	56.0	58.9	62.0	65.4	69.1				7
$(NH_4)_2SeO_4$				54.02									7
$(NH_4)_3PO_4$				15.5									7
$NaBr$	44.4	45.9	47.7	48.6	49.6	51.6	53.7	54.1	54.3	54.5	54.7	54.9	6
$NaBrO_3$	20.0	23.22	26.65	28.28	29.86	32.83	35.55	38.05	40.37	42.52			1:30
$NaCHO_2$	30.8	37.9	45.7	48.7	50.6	52.0	53.5	55.0					6
$NaC_2H_3O_2$	26.5	28.8	31.8	33.5	35.5	39.9	45.1	58.3	59.3	60.5	61.7	62.9	6
$NaCl$	26.28	26.32	26.41	26.45	26.52	26.67	26.84	27.03	27.25	27.50	27.78	28.05	1:47
$NaClO$	22.7			44.4									7
$NaClO_2$				97.0*				95.3*					7
$NaClO_3$	44.27	46.67	49.3	50.1	51.2	53.6	55.5	57.0	58.5	60.5	63.3	67.1	1:30
$NaClO_4$	61.9	64.1	66.2	67.2	68.3	70.4	72.5	74.1	74.7	75.4	76.1	76.7	6
NaF	3.52	3.72	3.89	3.97	4.05	4.20	4.34	4.46	4.57	4.66	4.75	4.82	6

AQUEOUS SOLUBILITY OF INORGANIC COMPOUNDS AT VARIOUS TEMPERATURES (continued)

Compound	0°C	10°C	20°C	25°C	30°C	40°C	50°C	60°C	70°C	80°C	90°C	100°C	Ref.
$NaHCO_3$	6.48	7.59	8.73	9.32	9.91	11.13	12.40	13.70	15.02	16.37	17.73	19.10	7
$NaHSO_4$				22.2								33.3	10
NaH_2PO_4	36.54	41.07	46.00	48.68	51.54	57.89*	61.7*	62.3*	65.9	68.7			1:31
NaI	61.2	62.4	63.9	64.8	65.7	67.7	69.8	72.0	74.7	74.8	74.9	75.1	6
$NaIO_3$	2.43	4.40	7.78*	8.65*	9.60	11.67	13.99	16.52	19.25*	21.1*	22.9	24.7	1:30
$NaIO_4$				12.62									7
$NaNO_2$	41.9	43.4	45.1	45.9	46.8	48.7	50.7	52.8	55.0	57.2	59.5	61.8	6
$NaNO_3$	42.2	44.4	46.6	47.7	48.8	51.0	53.2	55.3	57.5	59.6	61.7	63.8	6
NaOH	30	39	46	50	53	58	63	67	71	74	76	79	10
NaSCN		52.9	57.1	60.2	62.7	63.5	64.2	65.0	65.9	66.9	67.9	69.0	6
$Na_2B_4O_7$	1.23	1.71	2.50	3.07	3.82	6.02	9.7	14.9	17.1	19.9	23.5	28.0	6
Na_2CO_3	6.44	10.8	17.9	23.5	28.7	32.8	32.2	31.7	31.3	31.1	30.9	30.9	6
$Na_2C_2O_4$	2.62	2.95	3.30	3.48	3.65	4.00	4.36	4.71	5.06	5.41	5.75	6.08	6
Na_2CrO_4	22.6	32.3	44.6	46.7	46.9	48.9	51.0	53.4	55.3	55.5	55.8	56.1	6
$Na_2Cr_2O_7$	62.1	63.1	64.4	65.2	66.1	68.0	70.1	72.3	74.6	77.0	79.6	80.7	6
Na_2HAsO_4	5.6*			29.3*								67*	7
Na_2HPO_4	1.66	4.19	7.51	10.55	16.34*	35.17*	44.64*	45.20	46.81	48.78	50.52	51.53	1:31
Na_2MoO_4	30.6	38.8	39.4	39.4	39.8	40.3	41.0	41.7	42.6	43.5	44.5	45.5	6
Na_2S	11.1	13.2	15.7	17.1	18.6	22.1	26.7	28.1	30.2	33.0	36.4	41.0	6
Na_2SO_3	12.0	16.1	20.9	23.5	26.3*	27.3*	25.9	24.8	23.7	22.8	22.1	21.5	1:26
Na_2SO_4			16.13	21.94	29.22*	32.35*	31.55	30.90	30.39	30.02	29.79	29.67	8
$Na_2S_2O_3$	33.1	36.3	40.6	43.3	45.9	52.0	62.3	65.7	68.8	69.4	70.1	71.0	6
$Na_2S_2O_5$		38.4	39.5	40.0	40.6	41.8	43.0	44.2	45.5	46.8	48.1	49.5	1:26
Na_2SeO_3				47.3*								45*	7
Na_2SeO_4	11.7			36.9*								42.1*	7
Na_2WO_4	41.6	41.9	42.3	42.6	42.9	43.6	44.4	45.3	46.2	47.3	48.4	49.5	6
Na_3PO_4	4.28	7.30	10.8	12.6	14.1	16.6	22.9	28.4	32.4	37.6	40.4	43.5	6
$Na_4P_2O_7$	2.23	3.28	4.81	6.62	7.00	10.10	14.38	20.07	27.31	36.03	32.37	30.67	6
$NdCl_3$	49.0	49.3	49.7	50.0	50.4	51.2	52.2	53.3	54.5	55.8	57.1	58.5	6
$Nd(NO_3)_3$	55.76	57.49	59.37	60.38	61.43	63.69	66.27	69.47					1:13
$NiCl_2$	34.7	36.1	38.5	40.3	41.7	42.1	43.2	45.0	46.1	46.2	46.4	46.6	6
$Ni(ClO_4)_2$	51.1			52.8									7
NiF_2				2.50							2.52		7
NiI_2	55.40	57.68	59.78	60.69	61.50	62.80	63.73	64.38	64.80	65.09	65.30		7
$Ni(NO_3)_2$	44.1	46.0	48.4	49.8	51.3	54.6	58.3	61.0	63.1	65.6	67.9	69.0	6
$NiSO_4$	21.4	24.4	27.4	28.8	30.3*	32.0*	34.1	35.8	37.7	39.9	42.3	44.8	6
$Ni(SCN)_2$				35.48									7
$NiSeO_4$	21.6		26.2*									45.6*	7
$PbBr_2$	0.449	0.620	0.841	0.966	1.118	1.46	1.89						2
$PbCl_2$	0.66	0.81	0.98	1.07	1.17	1.39	1.64	1.93	2.24	2.60	2.99	3.42	2
$Pb(ClO_4)_2$				81.5									7
PbF_2		0.0603	0.0649	0.0670	0.0693								2
PbI_2	0.041	0.052	0.067	0.076	0.086	0.112	0.144	0.187	0.243	0.315			2

Compound	0°C	10°C	20°C	25°C	30°C	40°C	50°C	60°C	70°C	80°C	90°C	100°C	Ref.
$Pb(IO_3)_2$				0.0025									7
$Pb(NO_3)_2$	28.46	32.13	35.67	37.38	39.05	42.22	45.17	47.90	50.42	52.72	54.82	56.75	2
$PbSO_4$	0.0033	0.0038	0.0042	0.0044	0.0047	0.0052	0.0058						2
$PrCl_3$	48.0	48.1	48.6	49.0	49.5	50.8	52.3	54.1	56.1	58.3			6
$Pr(NO_3)_3$	57.50	59.20	61.16	62.24	63.40*	65.7*	67.8	70.2	73.4				1:13
RbBr	47.4	50.1	52.6	53.8	54.9	57.0	58.8	60.6	62.1	63.5	64.8	65.9	6
$RbBrO_3$	0.97	1.55	2.36	2.87	3.45	4.87	6.64	3.78	11.29	14.15	17.32	20.76	1:30
RbCl	43.58	45.65	47.53	48.42	49.27	50.86	52.34	53.67	54.92	56.08	57.16	58.15	1:47
$RbClO_3$	2.10	3.38	5.14	6.22	7.45	10.35	13.85	17.93	22.53	27.57	32.96	38.60	1:30
$RbClO_4$	1			1.5								17	7
RbF			75										7
$RbHCO_3$			53.7										7
RbI	55.8	58.6	61.1	62.3	63.4	65.4	67.2	68.8	70.3	71.6	72.7	73.8	6
$RbIO_3$	1.09	1.53	2.07	2.38	2.74	3.52	4.41	5.42	6.52	7.74	9.00	10.36	1:30
$RbNO_3$	16.4	25.0	34.6	39.4	44.2	53.1	60.8	67.2	72.2	76.1	79.0	81.2	6
RbOH					63.4								7
Rb_2CrO_4	38.27			43.26									7
Rb_2SO_4	27.3	30.0	32.5	33.7	34.8	36.9	38.7	40.3	41.8	43.0	44.1	44.9	6
$SbCl_3$	85.7			90.8									7
SbF_3	79.4			83.1									7
$Sc(NO_3)_3$	57.0	59.3	61.6	62.8	63.9	66.2	68.5						1:13
$Sm(NO_3)_3$	54.83	56.33	58.08	59.05	60.08	62.38	65.05*	68.1*	70.8	74.2			1:13
$SmCl_3$		48.0	48.2	48.4	48.6	49.2	50.0						6
$SnCl_2$	46	64											7
SnI_2			0.97									3.87	7
$SrBr_2$	46.0	48.3	50.6	51.7	52.9	55.2	57.6	59.9	62.3	64.6	66.8	69.0	6
$Sr(BrO_3)_2$	18.53	22.00	25.39	27.02	28.59	31.55	34.21	36.57	38.64*	40.2*	40.8	41.0	1:14
$SrCl_2$	31.94	32.93	34.43	35.37	36.43	38.93	41.94	45.44*	46.81*	47.69	48.70	49.87	8
$Sr(ClO_2)_2$	13.0	13.6	14.1	14.3	14.5	14.9	15.3	15.6	15.9				7
$Sr(ClO_3)_2$	63.29	63.42	63.64	63.77	63.93	64.29	64.70	65.16	65.65	66.18	66.74	67.31	1:14
$Sr(ClO_4)_2$	70.04*			75.35*		78.44*							7
SrF_2	0.011			0.021									7
SrI_2	62.5	62.8	63.5	63.9	64.5	65.8	67.3	69.0	70.8	72.7	74.7	79.2	6
$Sr(IO_3)_2$	0.102	0.126	0.152	0.165	0.179	0.206	0.233	0.259	0.284	0.307	0.328	0.346	1:14
$Sr(MnO_4)_2$	2.5												7
$Sr(NO_2)_2$					41.9	44.3						58.6	7
$Sr(NO_3)_2$	28.2	34.6	41.0	44.5	47.0	47.4	47.9	48.4	48.9	49.5	50.1	50.7	6
$Sr(OH)_2$	0.9			2.2									7
$SrSO_3$				0.0015									1:26
$SrSO_4$				0.0135									7
SrS_2O_3	8.8	13.2	17.7	20.0	22.2	26.8							7
$Tb(NO_3)_3$			60.6	61.02									1:13
Tl_2SO_4	2.65	3.56	4.61	5.19	5.80	7.09	8.46	9.89	11.33	12.77	14.18	15.53	6

Compound	0°C	10°C	20°C	25°C	30°C	40°C	50°C	60°C	70°C	80°C	90°C	100°C	Ref.
$Tm(NO_3)_3$				67.9									1:13
$UO_2(NO_3)_2$	49.52	51.82	54.42	55.85	57.55	61.59	67.07						1:55
$Y(NO_3)_3$	55.57	56.93	58.75	59.86	61.11*	63.3*	64.9	67.9	72.5				1:13
$Yb(NO_3)_3$				70.5									1:13
$ZnBr_2$	79.3	80.1	81.8	83.0	84.1	85.6	85.8	86.1	86.3	86.6	86.8	87.1	6
ZnC_2O_4		0.0010	0.0019	0.0026									5
$ZnCl_2$		76.6	79.0	80.3	81.4	81.8	82.4	83.0	83.7	84.4	85.2	86.0	6
$Zn(ClO_4)_2$	44.29*			46.27*			48.70						7
ZnF_2				1.53									5
ZnI_2	81.1	81.2	81.3	81.4	81.5	81.7	82.0	82.3	82.6	83.0	83.3	83.7	6
$Zn(IO_3)_2$			0.58	0.64	0.69	0.77	0.82						5
$Zn(NO_3)_2$	47.8	50.8	54.4	54.6	58.5	79.1	80.1	87.5	89.9				6
$ZnSO_3$			0.1786	0.1790	0.1794	0.1803	0.1812						5
$ZnSO_4$	29.1	32.0	35.0	36.6	38.2	41.3	43.0	42.1	41.0	39.9	38.8	37.6	6
$ZnSeO_4$	33.06	34.98	37.38	38.79	40.34								5

SOLUBILITY PRODUCT CONSTANTS

The solubility product constant K_{sp} is a useful parameter for calculating the aqueous solubility of sparingly soluble compounds under various conditions. It may be determined by direct measurement or calculated from the standard Gibbs energies of formation $\Delta_f G°$ of the species involved at their standard states. Thus if $K_{sp} = [M^+]^m [A^-]^n$ is the equilibrium constant for the reaction

$$M_mA_n(s) \rightleftharpoons mM^+(aq) + nA^-(aq),$$

where M_mA_n is the slightly soluble substance and M^+ and A^- are the ions produced in solution by the dissociation of M_mA_n, then the Gibbs energy change is

$$\Delta G° = m\,\Delta_f G°\,(M^+,aq) + n\;\Delta_f G°\,(A^-,aq) - \Delta_f G°\,(M_mA_n,\,s)$$

The solubility product constant is calculated from the equation

$$\ln K_{sp} = -\Delta\,G°/RT$$

The first table below gives selected values of K_{sp} at 25°C. Many of these have been calculated from standard state thermodynamic data in References 1 and 2; other values are taken from publications of the IUPAC Solubility Data Project (References 3 to 7).

The above formulation is not convenient for treating sulfides because the S^{-2} ion is usually not present in significant concentrations (see Reference 8). This is due to the hydrolysis reaction

$$S^{-2} + H_2O \rightleftharpoons HS^- + OH^-$$

which is strongly shifted to the right except in very basic solutions. Furthermore, the equilibrium constant for this reaction, which depends on the second ionization constant of H_2S, is poorly known. Therefore it is more useful in the case of sulfides to define a different solubility product K_{spa} based on the reaction

$$M_mS_n(s) + 2H^+ \rightleftharpoons mM^+ + nH_2S\,(aq)$$

Values of K_{spa}, taken from Reference 8, are given for several sulfides in the auxiliary table following the main table. Additional discussion of sulfide equilibria may be found in References 7 and 9.

REFERENCES

1. Wagman, D.D., Evans, W.H., Parker, V.B., Schumm, R.H., Halow, I., Bailey, S.M., Churney, K.L., and Nuttall, R L., *The NBS Tables of Chemical Thermodynamic Properties, J. Phys. Chem. Ref. Data*, Vol. 11, Suppl. 2, 1982.
2. Garvin, D., Parker, V.B., and White, H.J., *CODATA Thermodynamic Tables*, Hemisphere, New York, 1987.
3. *Solubility Data Series* (53 Volumes), International Union of Pure and Applied Chemistry, Pergamon Press, Oxford, 1979—1992.
4. Clever, H.L., and Johnston, F.J., *J. Phys. Chem. Ref. Data*, 9, 751, 1980.
5. Marcus, Y., *J. Phys. Chem. Ref. Data*, 9, 1307, 1980.
6. Clever, H.L., Johnson, S.A., and Derrick, M.E., *J. Phys. Chem. Ref. Data*, 14, 631, 1985.
7. Clever, H.L., Johnson, S.A., and Derrick, M.E., *J. Phys. Chem. Ref. Data*, 21, 941, 1992.
8. Myers, R.J., *J. Chem. Educ*., 63, 687, 1986.
9. Licht, S., *J. Electrochem. Soc*.,135, 2971, 1988.

Compound	Formula	K_{sp}
Aluminum phosphate	$AlPO_4$	$9.84 \cdot 10^{-21}$
Barium bromate	$Ba(BrO_3)_2$	$2.43 \cdot 10^{-4}$
Barium carbonate	$BaCO_3$	$2.58 \cdot 10^{-9}$
Barium chromate	$BaCrO_4$	$1.17 \cdot 10^{-10}$
Barium fluoride	BaF_2	$1.84 \cdot 10^{-7}$
Barium hydroxide octahydrate	$Ba(OH)_2 \cdot 8H_2O$	$2.55 \cdot 10^{-4}$
Barium iodate	$Ba(IO_3)_2$	$4.01 \cdot 10^{-9}$
Barium iodate monohydrate	$Ba(IO_3)_2 \cdot H_2O$	$1.67 \cdot 10^{-9}$
Barium molybdate	$BaMoO_4$	$3.54 \cdot 10^{-8}$
Barium nitrate	$Ba(NO_3)_2$	$4.64 \cdot 10^{-3}$
Barium selenate	$BaSeO_4$	$3.40 \cdot 10^{-8}$
Barium sulfate	$BaSO_4$	$1.08 \cdot 10^{-10}$
Barium sulfite	$BaSO_3$	$5.0 \cdot 10^{-10}$
Beryllium hydroxide	$Be(OH)_2$	$6.92 \cdot 10^{-22}$
Bismuth arsenate	$BiAsO_4$	$4.43 \cdot 10^{-10}$

SOLUBILITY PRODUCT CONSTANTS (continued)

Compound	Formula	K_{sp}
Bismuth iodide	BiI_3	$7.71 \cdot 10^{-19}$
Cadmium arsenate	$Cd_3(AsO_4)_2$	$2.2 \cdot 10^{-33}$
Cadmium carbonate	$CdCO_3$	$1.0 \cdot 10^{-12}$
Cadmium fluoride	CdF_2	$6.44 \cdot 10^{-3}$
Cadmium hydroxide	$Cd(OH)_2$	$7.2 \cdot 10^{-15}$
Cadmium iodate	$Cd(IO_3)_2$	$2.5 \cdot 10^{-8}$
Cadmium oxalate trihydrate	$CdC_2O_4 \cdot 3H_2O$	$1.42 \cdot 10^{-8}$
Cadmium phosphate	$Cd_3(PO_4)_2$	$2.53 \cdot 10^{-33}$
Calcium carbonate (calcite)	$CaCO_3$	$3.36 \cdot 10^{-9}$
Calcium fluoride	CaF_2	$3.45 \cdot 10^{-11}$
Calcium hydroxide	$Ca(OH)_2$	$5.02 \cdot 10^{-6}$
Calcium iodate	$Ca(IO_3)_2$	$6.47 \cdot 10^{-6}$
Calcium iodate hexahydrate	$Ca(IO_3)_2 \cdot 6H_2O$	$7.10 \cdot 10^{-7}$
Calcium molybdate	$CaMoO_4$	$1.46 \cdot 10^{-8}$
Calcium oxalate monohydrate	$CaC_2O_4 \cdot H_2O$	$2.32 \cdot 10^{-9}$
Calcium phosphate	$Ca_3(PO_4)_2$	$2.07 \cdot 10^{-33}$
Calcium sulfate	$CaSO_4$	$4.93 \cdot 10^{-5}$
Calcium sulfate dihydrate	$CaSO_4 \cdot 2H_2O$	$3.14 \cdot 10^{-5}$
Calcium sulfite hemihydrate	$CaSO_3 \cdot 0.5H_2O$	$3.1 \cdot 10^{-7}$
Cesium perchlorate	$CsClO_4$	$3.95 \cdot 10^{-3}$
Cesium periodate	$CsIO_4$	$5.16 \cdot 10^{-6}$
Cobalt(II) arsenate	$Co_3(AsO_4)_2$	$6.80 \cdot 10^{-29}$
Cobalt(II) hydroxide (blue)	$Co(OH)_2$	$5.92 \cdot 10^{-15}$
Cobalt(II) iodate dihydrate	$Co(IO_3)_2 \cdot 2H_2O$	$1.21 \cdot 10^{-2}$
Cobalt(II) phosphate	$Co_3(PO_4)_2$	$2.05 \cdot 10^{-35}$
Copper(I) bromide	$CuBr$	$6.27 \cdot 10^{-9}$
Copper(I) chloride	$CuCl$	$1.72 \cdot 10^{-7}$
Copper(I) cyanide	$CuCN$	$3.47 \cdot 10^{-20}$
Copper(I) iodide	CuI	$1.27 \cdot 10^{-12}$
Copper(I) thiocyanate	$CuSCN$	$1.77 \cdot 10^{-13}$
Copper(II) arsenate	$Cu_3(AsO_4)_2$	$7.95 \cdot 10^{-36}$
Copper(II) iodate monohydrate	$Cu(IO_3)_2 \cdot H_2O$	$6.94 \cdot 10^{-8}$
Copper(II) oxalate	CuC_2O_4	$4.43 \cdot 10^{-10}$
Copper(II) phosphate	$Cu_3(PO_4)_2$	$1.40 \cdot 10^{-37}$
Europium(III) hydroxide	$Eu(OH)_3$	$9.38 \cdot 10^{-27}$
Gallium(III) hydroxide	$Ga(OH)_3$	$7.28 \cdot 10^{-36}$
Iron(II) carbonate	$FeCO_3$	$3.13 \cdot 10^{-11}$
Iron(II) fluoride	FeF_2	$2.36 \cdot 10^{-6}$
Iron(II) hydroxide	$Fe(OH)_2$	$4.87 \cdot 10^{-17}$
Iron(III) hydroxide	$Fe(OH)_3$	$2.79 \cdot 10^{-39}$
Iron(III) phosphate dihydrate	$FePO_4 \cdot 2H_2O$	$9.91 \cdot 10^{-16}$
Lanthanum iodate	$La(IO_3)_3$	$7.50 \cdot 10^{-12}$
Lead(II) bromide	$PbBr_2$	$6.60 \cdot 10^{-6}$
Lead(II) carbonate	$PbCO_3$	$7.40 \cdot 10^{-14}$
Lead(II) chloride	$PbCl_2$	$1.70 \cdot 10^{-5}$
Lead(II) fluoride	PbF_2	$3.3 \cdot 10^{-8}$
Lead(II) hydroxide	$Pb(OH)_2$	$1.43 \cdot 10^{-20}$
Lead(II) iodate	$Pb(IO_3)_2$	$3.69 \cdot 10^{-13}$
Lead(II) iodide	PbI_2	$9.8 \cdot 10^{-9}$
Lead(II) selenate	$PbSeO_4$	$1.37 \cdot 10^{-7}$
Lead(II) sulfate	$PbSO_4$	$2.53 \cdot 10^{-8}$
Lithium carbonate	Li_2CO_3	$8.15 \cdot 10^{-4}$
Lithium fluoride	LiF	$1.84 \cdot 10^{-3}$
Lithium phosphate	Li_3PO_4	$2.37 \cdot 10^{-11}$
Magnesium carbonate	$MgCO_3$	$6.82 \cdot 10^{-6}$
Magnesium carbonate trihydrate	$MgCO_3 \cdot 3H_2O$	$2.38 \cdot 10^{-6}$
Magnesium carbonate pentahydrate	$MgCO_3 \cdot 5H_2O$	$3.79 \cdot 10^{-6}$
Magnesium fluoride	MgF_2	$5.16 \cdot 10^{-11}$
Magnesium hydroxide	$Mg(OH)_2$	$5.61 \cdot 10^{-12}$
Magnesium oxalate dihydrate	$MgC_2O_4 \cdot 2H_2O$	$4.83 \cdot 10^{-6}$

Compound	Formula	K_{sp}
Magnesium phosphate	$Mg_3(PO_4)_2$	$1.04 \cdot 10^{-24}$
Manganese(II) carbonate	$MnCO_3$	$2.24 \cdot 10^{-11}$
Manganese(II) iodate	$Mn(IO_3)_2$	$4.37 \cdot 10^{-7}$
Manganese(II) oxalate dihydrate	$MnC_2O_4 \cdot 2H_2O$	$1.70 \cdot 10^{-7}$
Mercury(I) bromide	Hg_2Br_2	$6.40 \cdot 10^{-23}$
Mercury(I) carbonate	Hg_2CO_3	$3.6 \cdot 10^{-17}$
Mercury(I) chloride	Hg_2Cl_2	$1.43 \cdot 10^{-18}$
Mercury(I) fluoride	Hg_2F_2	$3.10 \cdot 10^{-6}$
Mercury(I) iodide	Hg_2I_2	$5.2 \cdot 10^{-29}$
Mercury(I) oxalate	$Hg_2C_2O_4$	$1.75 \cdot 10^{-13}$
Mercury(I) sulfate	Hg_2SO_4	$6.5 \cdot 10^{-7}$
Mercury(I) thiocyanate	$Hg_2(SCN)_2$	$3.2 \cdot 10^{-20}$
Mercury(II) bromide	$HgBr_2$	$6.2 \cdot 10^{-20}$
Mercury(II) iodide	HgI_2	$2.9 \cdot 10^{-29}$
Neodymium carbonate	$Nd_2(CO_3)_3$	$1.08 \cdot 10^{-33}$
Nickel(II) carbonate	$NiCO_3$	$1.42 \cdot 10^{-7}$
Nickel(II) hydroxide	$Ni(OH)_2$	$5.48 \cdot 10^{-16}$
Nickel(II) iodate	$Ni(IO_3)_2$	$4.71 \cdot 10^{-5}$
Nickel(II) phosphate	$Ni_3(PO_4)_2$	$4.74 \cdot 10^{-32}$
Palladium(II) thiocyanate	$Pd(SCN)_2$	$4.39 \cdot 10^{-23}$
Potassium hexachloroplatinate	K_2PtCl_6	$7.48 \cdot 10^{-6}$
Potassium perchlorate	$KClO_4$	$1.05 \cdot 10^{-2}$
Potassium periodate	KIO_4	$3.71 \cdot 10^{-4}$
Praseodymium hydroxide	$Pr(OH)_3$	$3.39 \cdot 10^{-24}$
Radium iodate	$Ra(IO_3)_2$	$1.16 \cdot 10^{-9}$
Radium sulfate	$RaSO_4$	$3.66 \cdot 10^{-11}$
Rubidium perchlorate	$RbClO_4$	$3.00 \cdot 10^{-3}$
Scandium fluoride	ScF_3	$5.81 \cdot 10^{-24}$
Scandium hydroxide	$Sc(OH)_3$	$2.22 \cdot 10^{-31}$
Silver(I) acetate	$AgCH_3COO$	$1.94 \cdot 10^{-3}$
Silver(I) arsenate	Ag_3AsO_4	$1.03 \cdot 10^{-22}$
Silver(I) bromate	$AgBrO_3$	$5.38 \cdot 10^{-5}$
Silver(I) bromide	$AgBr$	$5.35 \cdot 10^{-13}$
Silver(I) carbonate	Ag_2CO_3	$8.46 \cdot 10^{-12}$
Silver(I) chloride	$AgCl$	$1.77 \cdot 10^{-10}$
Silver(I) chromate	Ag_2CrO_4	$1.12 \cdot 10^{-12}$
Silver(I) cyanide	$AgCN$	$5.97 \cdot 10^{-17}$
Silver(I) iodate	$AgIO_3$	$3.17 \cdot 10^{-8}$
Silver(I) iodide	AgI	$8.52 \cdot 10^{-17}$
Silver(I) oxalate	$Ag_2C_2O_4$	$5.40 \cdot 10^{-12}$
Silver(I) phosphate	Ag_3PO_4	$8.89 \cdot 10^{-17}$
Silver(I) sulfate	Ag_2SO_4	$1.20 \cdot 10^{-5}$
Silver(I) sulfite	Ag_2SO_3	$1.50 \cdot 10^{-14}$
Silver(I) thiocyanate	$AgSCN$	$1.03 \cdot 10^{-12}$
Strontium arsenate	$Sr_3(AsO_4)_2$	$4.29 \cdot 10^{-19}$
Strontium carbonate	$SrCO_3$	$5.60 \cdot 10^{-10}$
Strontium fluoride	SrF_2	$4.33 \cdot 10^{-9}$
Strontium iodate	$Sr(IO_3)_2$	$1.14 \cdot 10^{-7}$
Strontium iodate monohydrate	$Sr(IO_3)_2 \cdot H_2O$	$3.77 \cdot 10^{-7}$
Strontium iodate hexahydrate	$Sr(IO_3)_2 \cdot 6H_2O$	$4.55 \cdot 10^{-7}$
Strontium sulfate	$SrSO_4$	$3.44 \cdot 10^{-7}$
Thallium(I) bromate	$TlBrO_3$	$1.10 \cdot 10^{-4}$
Thallium(I) bromide	$TlBr$	$3.71 \cdot 10^{-6}$
Thallium(I) chloride	$TlCl$	$1.86 \cdot 10^{-4}$
Thallium(I) chromate	Tl_2CrO_4	$8.67 \cdot 10^{-13}$
Thallium(I) iodate	$TlIO_3$	$3.12 \cdot 10^{-6}$
Thallium(I) iodide	TlI	$5.54 \cdot 10^{-8}$
Thallium(I) thiocyanate	$TlSCN$	$1.57 \cdot 10^{-4}$
Thallium(III) hydroxide	$Tl(OH)_3$	$1.68 \cdot 10^{-44}$
Tin(II) hydroxide	$Sn(OH)_2$	$5.45 \cdot 10^{-27}$

SOLUBILITY PRODUCT CONSTANTS (continued)

Compound	Formula	K_{sp}
Yttrium carbonate	$Y_2(CO_3)_3$	$1.03 \cdot 10^{-31}$
Yttrium fluoride	YF_3	$8.62 \cdot 10^{-21}$
Yttrium hydroxide	$Y(OH)_3$	$1.00 \cdot 10^{-22}$
Yttrium iodate	$Y(IO_3)_3$	$1.12 \cdot 10^{-10}$
Zinc arsenate	$Zn_3(AsO_4)_2$	$2.8 \cdot 10^{-28}$
Zinc carbonate	$ZnCO_3$	$1.46 \cdot 10^{-10}$
Zinc carbonate monohydrate	$ZnCO_3 \cdot H_2O$	$5.42 \cdot 10^{-11}$
Zinc fluoride	ZnF_2	$3.04 \cdot 10^{-2}$
Zinc hydroxide	$Zn(OH)_2$	$3 \cdot 10^{-17}$
Zinc iodate dihydrate	$Zn(IO_3)_2 \cdot 2H_2O$	$4.1 \cdot 10^{-6}$
Zinc oxalate dihydrate	$ZnC_2O_4 \cdot 2H_2O$	$1.38 \cdot 10^{-9}$
Zinc selenide	ZnSe	$3.6 \cdot 10^{-26}$
Zinc selenite monohydrate	$ZnSeO_3 \cdot H_2O$	$1.59 \cdot 10^{-7}$

Sulfides

Compound	Formula	K_{spa}
Cadmium sulfide	CdS	$8 \cdot 10^{-7}$
Copper(II) sulfide	CuS	$6 \cdot 10^{-16}$
Iron(II) sulfide	FeS	$6 \cdot 10^{2}$
Lead(II) sulfide	PbS	$3 \cdot 10^{-7}$
Manganese(II) sulfide (green)	MnS	$3 \cdot 10^{7}$
Mercury(II) sulfide (red)	HgS	$4 \cdot 10^{-33}$
Mercury(II) sulfide (black)	HgS	$2 \cdot 10^{-32}$
Silver(I) sulfide	Ag_2S	$6 \cdot 10^{-30}$
Tin(II) sulfide	SnS	$1 \cdot 10^{-5}$
Zinc sulfide (sphalerite)	ZnS	$2 \cdot 10^{-4}$
Zinc sulfide (wurtzite)	ZnS	$3 \cdot 10^{-2}$

SOLUBILITY OF COMMON SALTS AT AMBIENT TEMPERATURES

This table gives the aqueous solubility of selected salts at temperatures from 10°C to 40°C. Values are given in molality terms.

REFERENCES

1. Apelblat, A., *J. Chem. Thermodynamics*, 24, 619, 1992.
2. Apelblat, A., *J. Chem. Thermodynamics*, 25, 63, 1993.
3. Apelblat, A., *J. Chem. Thermodynamics*, 25, 1513, 1993.
4. Apelblat, A. and Korin, E., *J. Chem. Thermodynamics*, 30, 59, 1998.

Salt	10°C	15°C	20°C	25°C	30°C	35°C	40°C	Ref.
$BaCl_2$	1.603	1.659	1.716	1.774	1.834	1.895	1.958	1
$Ca(NO_3)_2$	6.896	7.398	7.986	8.675	9.480	10.421		1
$CuSO_4$	1.055	1.153	1.260	1.376	1.502	1.639		3
$FeSO_4$	1.352	1.533	1.729	1.940	2.165	2.405		3
KBr	5.002	5.237	5.471	5.703	5.932	6.157		3
KIO_3	0.291	0.333	0.378	0.426	0.478	0.534	0.593	4
K_2CO_3	7.756	7.846	7.948	8.063	8.191	8.331	8.483	1
LiCl	19.296	19.456	19.670	19.935				2
$Mg(NO_3)_2$	4.403	4.523	4.656	4.800	4.958	5.130	5.314	1
$MnCl_2$	5.421	5.644	5.884	6.143	6.422	6.721		3
NH_4Cl	6.199	6.566	6.943	7.331				2
NH_4NO_3	18.809	21.163	23.721	26.496				2
$(NH_4)_2SO_4$	5.494	5.589	5.688	5.790	5.896	6.005		3
NaBr	8.258	8.546	8.856	9.191	9.550	9.937	10.351	4
NaCl	6.110	6.121	6.136	6.153	6.174	6.197	6.222	4
$NaNO_2$	11.111	11.484	11.883	12.310	12.766	13.253	13.772	4
$NaNO_3$	9.395	9.819	10.261	10.723	11.204	11.706	12.230	4
RbCl	6.911	7.180	7.449	7.717	7.986	8.253	8.520	4
$ZnSO_4$	2.911	3.116	3.336	3.573	3.827	4.099	4.194	1

SOLUBILITY CHART

Abbreviations: **W**, soluble in water; **A**, insoluble in water but soluble in acids; **w**, sparingly soluble in water but soluble in acids; a, insoluble in water and only sparingly soluble in acids; **I**, insoluble in water and acids; **d**, decomposes in water. * Indicates two modifications of the salt

No.		Al	NH4	Sb	Ba	Bi	Cd	Ca	Cr	Co	Cu	Au (I)	Au (II)	H	Fe (II)	Fe (III)
1	Acetate —($C_2H_3O_2$)	W $Al(—)_3$	W $NH_4(—)$		W $Ba(—)_2$	W $Bi(—)_3$	W $Cd(—)_2$	W $Ca(—)_2$	W $Cr(—)_3$	W $Co(—)_2$	W $Cu(—)_2$	W	W	W $C_2H_4O_2$	$Fe(—)_2$	$Fe_2(—)_6$
2	Arsenate —(AsO_4)	a $Al(—)$	W $(NH_4)_3(—)$	A $Sb(—)$	w $Ba_3(—)_2$	A $Bi(—)$	A $Cd_3(—)_2$	w $Ca_3(—)_2$		A $Co_3(—)_2$	A $Cu_3(—)_2$			W H_3AsO_4	A $Fe_3(—)_2$	A $Fe(—)$
3	Arsenite —(AsO_3)		W NH_4AsO_2	A $Sb(—)$				w $Ca_3(—)_2$		A $Co_3H_6(—)_4$	A $CuH(—)$					
4	Benzoate —($C_7H_5O_2$)		W $NH_4(—)$		W $Ba(—)_2$	A $Bi(—)_3$	W $Cd(—)_2$	W $Ca(—)_2$		W $Co(—)_2$	w $Cu(—)_2$			W $C_7H_6O_2$	W $Fe(—)_2$	A $Fe_2(—)_6$
5	Bromide	W $AlBr_2$	W NH_4Br	d $SbBr_3$	W $BaBr_2$	d $BiBr_3$	W $CdBr_2$	W $CaBr_2$	W(I)* $CrBr_3$	W $CoBr_2$	W $CuBr_2$	w AuBr	W $AuBr_3$	W HBr	W $FeBr_2$	W $FeBr_3$
6	Carbonate		W $(NH_4)_2CO_3$		w $BaCO_3$		A $CdCO_3$	w $CaCO_3$	W $CrCO_3$	A $CoCO_3$					w $FeCO_3$	
7	Chlorate —(ClO_3)	W $Al(—)_3$	W $NH_4(—)$		W $Ba(—)_2$	W $Bi(—)_3$	W $Cd(—)_2$	W $Ca(—)_2$		W $Co(—)_2$	W $Cu(—)_2$			W $HClO_3$	W $Fe(—)_2$	W $Fe(—)_3$
8	Chloride	W $AlCl_3$	W NH_4Cl	W $SbCl_3$	W $BaCl_2$	d $BiCl_3$	W $CdCl_2$	W $CaCl_2$	I $CrCl_3$	W $CoCl_2$	W $CuCl_2$	w AuCl	W $AuCl_3$	W HCl	W $FeCl_2$	W $FeCl_3$
9	Chromate —(CrO_4)		W $(NH_4)_2(—)$		A $Ba(—)$		A $Cd(—)$	W $Ca(—)$		A $Co(—)$						A $Fe_2(—)_3$
10	Citrate —($C_6H_5O_7$)	W $Al(—)$	W $(NH_4)_3(—)$		w $Ba_3(—)_2$	A $Bi(—)$	A $Cd_3(—)_2$	w $Ca_3(—)_2$		w $Co_3(—)_2$				W $C_6H_8O_7$		W $Fe(—)$
11	Cyanide		W NH_4CN		W $Ba(CN)_2$	w $Bi(CN)_3$	W $Cd(CN)_2$	W $Ca(CN)_2$	A $Cr(CN)_3$	A $Co(CN)_2$	A $Cu(CN)_2$	w AuCN	W $Au(CN)_3$	W HCN	a $Fe(CN)_2$	
12	Ferricy'de —($Fe(CN)_6$)		W $(NH_4)_3(—)$		w $Ba_3(—)_2$		A $Cd_3(—)_2$	W $Ca_3(—)_2$		I $Co_3(—)_2$	I $Cu_3(—)_2$			W $H_3(—)$	I $Fe_3(—)_2$	
13	Ferrocy'de —($Fe(CN)_6$)	w $Al_4(—)_3$	W $(NH_4)_4(—)$		W $Ba_2(—)$		A $Cd_2(—)$	W $Ca_2(—)$		I $Co_2(—)$	I $Cu_2(—)$			W $H_4(—)$	I $Fe_2(—)$	a $Fe_4(—)_3$
14	Fluoride	W AlF_3	W NH_4F	W SbF_3	w BaF_2	W BiF_3	W CdF_2	w CaF_2	W(a)* CrF_3	W CoF_2	w CuF_2			W HF	w FeF_2	w FeF_3
15	Formate —(CHO_2)	W $Al(—)_3$	W $NH_4(—)$		W $Ba(—)_2$	W $Bi(—)_3$	W $Cd(—)_2$	W $Ca(—)_2$		W $Co(—)_2$	W $Cu(—)_2$			W CH_2O_2	W $Fe(—)_2$	W $Fe(—)_3$
16	Hydroxide	A $Al(OH)_3$	W NH_4OH		W $Ba(OH)_2$	A $Bi(OH)_3$	A $Cd(OH)_2$	W $Ca(OH)_2$	A $Cr(OH)_3$	A $Co(OH)_2$	A $Cu(OH)_2$	W AuOH	A $Au(OH)_3$		A $Fe(OH)_2$	A $Fe(OH)_3$
17	Iodide	W AlI_3	W NH_4I	d SbI_3	W BaI_2	A BiI_3	W CdI_2	W CaI_2	W CrI_3	W CoI_2	a CuI	a AuI	a AuI_3	W HI	W FeI_2	W FeI_3
18	Nitrate	W $Al(NO_3)_3$	W NH_4NO_3		W $Ba(NO_3)_2$	d $Bi(NO_3)_3$	W $Cd(NO_3)_2$	W $Ca(NO_3)_2$	W $Cr(NO_3)_3$	W $Co(NO_3)_2$	W $Cu(NO_3)_2$			W HNO_3	W $Fe(NO_3)_2$	W $Fe(NO)_3$
19	Oxalate —(C_2O_4)	A $Al_2(—)_3$	W $(NH_4)_2(—)$		w $Ba(—)$	A $Bi_2(—)_3$	w $Cd(—)$	A $Ca(—)$	W $Cr(—)$	A $Co(—)$	A $Cu(—)$			W $C_2H_2O_4$	A $Fe(—)$	W $Fe_2(—)_3$
20	Oxide	a Al_2O_3		w Sb_2O_3	W BaO	A Bi_2O_3	A CdO	w CaO	a Cr_2O_3	A CoO	A CuO	Au_2O	A Au_2O_3	W H_2O_2	A FeO	A Fe_2O_3
21	Phosphate	A $AlPO_4$	W $NH_4H_2PO_4$		A $Ba_3(PO_4)_2$	A $BiPO_4$	A $Cd_3(PO_4)_2$	w $Ca_3(PO_4)_2$	w $Cr_2(PO_4)_2$	A $Co_3(PO_4)_2$	A $Cu_3(PO_4)_2$		H_3PO_4	W $Fe_3(PO_4)_2$	A $FePO_4$	w
22	Silicate, —(SiO_3)	I $Al_2(—)_3$			W $Ba(—)$		A $Cd(—)$	w $Ca(—)$		A Co_2SiO_4	A $Cu(—)$			I H_2SiO_3		
23	Sulfate	W $Al_2(SO_4)_3$	W $(NH_4)_2SO_4$	A $Sb_2(SO_4)_3$	a $BaSO_4$	d $Bi_2(SO_4)_3$	W $CdSO_4$	w $CaSO_4$	W(I)* $Cr_2(SO_4)_3$	W $CoSO_4$	W $CuSO_4$			W H_2SO_4	W $FeSO_4$	w $Fe(SO_4)_3$
24	Sulfide	d Al_2S_3	W $(NH_4)_2S$	A Sb_2S_3	d BaS	A Bi_2S_3	A CdS	w CaS	d Cr_2S_3	A CoS	A CuS	I Au_2S	I Au_2S_3	W H_2S	A FeS	d Fe_2S_3
25	Tartrate —($C_4H_4O_6$)	w $Al_2(—)_3$	W $(NH_4)_2(—)$	W $Sb_2(—)_3$	w $Ba(—)$	A $Bi_2(—)_3$	A $Cd(—)$	w $Ca(—)$	d	A $Co(—)$	A $Cu(—)$	I	I	W $C_4H_6O_6$	A $Fe(—)$	d $Fe_2(—)_3$

SOLUBILITY CHART (continued)

No.		Al	NH_4	Sb	Ba	Bi	Cd	Ca	Cr	Co	Cu	Au (I)	Au (II)	H	Fe (II)	Fe (III)
26	Thiocy'te		W NH_4CNS		W $Ba(CNS)_2$			W Ca(CNS)		W $Co(CNS)_2$	d CuCNS			W CNSH	W $Fe(CNS)_2$	W $Fe(CNS)_3$

No.		Pb	Mg	Mn	Hg (I)	Hg (II)	Ni	K	Pt	Ag	Na	Sn (IV)	Sn (II)	Sr	Zn
1	Acetate —($C_2H_3O_2$)	W $Pb(—)_2$	W $Mg(—)_2$	W $Mn(—)_2$	w Hg(—)	W $Hg(—)_2$	W $Ni(—)_2$	W K(—)		w Ag(—)	W Na(—)	W $Sn(—)_4$	d $Sn(—)_2$	W $Sr(—)_2$	W $Zn(—)_2$
2	Arsenate —(AsO_4)	A PbH(—)	A $Mg_3(—)$	w MnH(—)	A $Hg_3(—)$	w $Hg_3(—)_2$	A $Ni_3(—)_2$	W $K_3(—)$		A $Ag_3(—)$	W $Na_3(—)$			w SrH(—)	A $Zn_3(—)_2$
3	Arsenite —(AsO_3)		W $Mg_3(—)_2$	A $Mn_3H_6(—)_4$	A $Hg_3(—)$	A $Hg_3(—)$	A $Ni_3H_6(—)_4$	W K_3AsO_3		A $Ag_3(—)$	W $Na_2H(—)$		A $Sn_3(—)_2$	w $Sr_3(—)_2$	
4	Benzoate —($C_7H_5O_2$)	w $Pb(—)_2$	W $Mg(—)_2$	W $Mn(—)_2$	A $Hg_2(—)_2$	w $Hg(—)_2$	w $Ni(—)_2$	W K(—)		w Ag(—)	W Na(—)				W $Zn(—)_2$
5	Bromide	W $PbBr_2$	W $MgBr_2$	W $MnBr_2$	A HgBr	W $HgBr_2$	W $NiBr_2$	W KBr	w $PtBr_4$	a AgBr	W NaBr	W $SnBr_4$	W $SnBr_2$	W $SrBr_2$	W $ZnBr_2$
6	Carbonate	A $PbCO_3$	w $MgCO_3$	w $MnCO_3$	A Hg_2CO_3		w $NiCO_3$	W K_2CO_3		A Ag_2CO_3	W Na_2CO_3			w $SrCO_3$	w $ZnCO_3$
7	Chlorate —(ClO_3)	W $Pb(—)_2$	W $Mg(—)_2$	W $Mn(—)_2$	W Hg(—)	W $Hg(—)_2$	W $Ni(—)_2$	W K(—)		W Ag(—)	W Na(—)		W $Sn(—)_2$	W $Sr(—)_2$	W $Zn(—)_2$
8	Chloride	W $PbCl_2$	W $MgCl_2$	W $MnCl_2$	a HgCl	W $HgCl_2$	W $NiCl_2$	W KCl	W $PtCl_4$	a AgCl	W NaCl	W $SnCl_4$	W $SnCl_2$	W $SrCl_2$	W $ZnCl_2$
9	Chromate —(CrO_4)	A Pb(—)	W Mg(—)		w $Hg_2(—)$	w Hg(—)	A Ni(—)	W $K_2(—)$		w $Ag_2(—)$	W $Ma_2(—)$	W $Sn(—)_2$	A Sn(—)	w Sr(—)	w Zn(—)
10	Citrate —($C_6H_5O_7$)	W $Pb_3(—)_2$	W $Mg_3(—)_2$	w MnH(—)	w $Hg_3(—)$		W $Ni_3(—)_2$	W $K_3(—)$		w $Ag_3(—)$	W $Na_3(—)$			A SrH(—)	w $Zn_3(—)_2$
11	Cyanide	w $Pb(CN)_2$	W $Mg(CN)_2$		A HgCN	W $Hg(CN)_2$	a $Ni(CN)_2$	W KCN	I $Pt(CN)_2$	a AgCN	W NaCN			W $Sr(CN)_2$	A $Zn(CN)_2$
12	Ferricy'de —$Fe(CN)_6$	w $Pb_3(—)_2$	W $Mg_3(—)_2$			A $Hg_3(—)_2$	I $Ni_3(—)_2$	W $K_3(—)$		I $Ag_3(—)$	W $Na_3(—)$		A $Sn_3(—)_2$	W $Sr_3(—)_2$	A $Zn_3(—)_2$
13	Ferrocy'de —$Fe(CN)_6$	a $Pb_2(—)$	W $Mg_2(—)$	A $Mn_2(—)$		I $Hg_2(—)$	I $Ni_2(—)$	W $K_4(—)$		I $Ag_4(—)$	W $Na_4(—)$		a $Sn_2(—)$	W $Sr_2(—)$	I $Zn_2(—)$
14	Fluoride	w PbF_2	w MgF_2	A MnF_2	d HgF	d HgF_2	w NiF_2	W KF	W PtF_4	W AgF	W NaF	W SnF_4	W SnF_2	w SrF_2	w ZnF_2
15	Formate —(CHO_2)	W $Pb(—)_2$	W $Mg(—)_2$	W $Mn(—)_2$	w Hg(—)	W $Hg(—)_2$	W $Ni(—)_2$	W K(—)		W Ag(—)	W Na(—)			W $Sr(—)_2$	W $Zn(—)_2$
16	Hydroxide	w $Pb(OH)_2$	A $Mg(OH)_2$	A $Mn(OH)_2$		A $Hg(OH)_2$	w $Ni(OH)_2$	W KOH	A $Pt(OH)_4$		W NaOH	w $Sn(OH)_4$	A $Sn(OH)_2$	W $Sr(OH)_2$	A $Zn(OH)_2$
17	Iodide	w PbI_2	W MgI_2	W MnI_2	A HgI	w HgI_2	W NiI_2	W KI	I PtI_2	I AgI	W NaI	d SnI_4	W SnI_2	W SrI_2	W ZnI_2
18	Nitrate	W $Pb(NO_3)_2$	W $Mg(NO_3)_2$	W $Mn(NO_3)_2$	W $HgNO_3$	W $Hg(NO_3)_2$	W $Ni(NO_3)_2$	W KNO_3	W $Pt(NO_3)_4$	W $AgNO_3$	W $NaNO_3$		d $Sn(NO_3)_2$	W $Sr(NO_3)_2$	W $Zn(NO_3)_2$
19	Oxalate —(C_2O_4)	A Pb(—)	w Mg(—)	w Mn(—)	a $Hg_2(—)$	A Hg(—)	A Ni(—)	W $K_2(—)$		a $Ag_2(—)$	W $Na_2(—)$		A Sn(—)	w Sr(—)	A Zn(—)
20	Oxide	w PbO	A MgO	A MnO	A Hg_2O	w HgO	A NiO	W K_2O	A PtO	w Ag_2O	d Na_2O	A SnO_2	A SnO	W SrO	w ZnO
21	Phosphate	A $Pb_3(PO_4)_2$	w $Mg_3(PO_4)_2$	w $Mn_3(PO_4)_2$	A Hg_3PO_4	A $Hg_3(PO_4)_2$	A $Ni_3(PO_4)_2$	W K_3PO_4		A Ag_3PO_4	W Na_3PO_4		A $Sn_3(PO_4)$	A $Sr_3(PO_4)_2$	A $Zn_3(PO_4)_2$
22	Silicate —(SiO_3)	A Pb(—)	A Mg(—)	I Mn(—)				W $K_2(—)$			W $Na_2(—)$			A Sr(—)	A Zn(—)
23	Sulfate	w $PbSO_4$	W $MgSO_4$	W $MnSO_4$	w Hg_2SO_4	d $HgSO_4$	W $NiSO_4$	W K_2SO_4	W $Pt(SO_4)_2$	w Ag_2SO_4	W Na_2SO_4	W $Sn(SO_4)_2$	W $SnSO_4$	w $SrSO_4$	W $ZnSO_4$

No.		Pb	Mg	Mn	Hg (I)	Hg (II)	Ni	K	Pt	Ag	Na	Sn (IV)	Sn (II)	Sr	Zn
24	Sulfide	A PbS	d MgS	A MnS	I Hg_2S	I HgS	A NiS	W K_2S	I PtS	A Ag_2S	W Na_2S	A SnS_2	A SnS	W SrS	A ZnS
25	Tartrate —$(C_4H_4O_6)$	A Pb(—)	w Mg(—)	w Mn(—)	I Hg_2(—)	I	A Ni(—)	W K_2(—)	I	A Ag_2(—)	W Na_2(—)	A	A Sn(—)	W Sr(—)	A Zn(—)
26	Thiocy'te	w $Pb(CNS)_2$	W $Mg(CNS)_2$	W $Mn(CNS)_2$	A HgCNS	w $Hg(CNS)_2$		W KCNS		I AgCNS	W NaCNS			W $Sr(CNS)_2$	W $Zn(CNS)_2$

REDUCTION OF WEIGHINGS IN AIR TO VACUO

When the mass M of a body is determined in air, a correction is necessary for the buoyancy of the air. The corrected mass is given by $M + kM/1000$, where k is a function of the material used for the weights, given by

$$k = 1000\rho_{air}(1/\rho_{body} - 1/\rho_{weight})$$

and ρ is density. The table below is computed for an air density of 0.0012 g/cm³ and for densities of three common weights: platinum-iridium (21.6 g/cm³), brass (8.5 g/cm³), and aluminum or quartz (2.65 g/cm³).

REFERENCES

1. Kaye, G. W. C., and Laby, T. H., *Tables of Physical and Chemical Constants, 16th Edition,* pp. 25-28, Longman, London, 1995.
2. Giacomo, P., *Metrologia* 18, 33, 1982.
3. Davis, R. S., *Metrologia* 29, 67, 1992.

Density of body (g/cm³)	Value of *k* for weights of: Pt-Ir	Brass	Quartz or Al
0.5	2.34	2.26	1.95
0.6	1.94	1.86	1.55
0.7	1.66	1.57	1.26
0.8	1.44	1.36	1.05
0.9	1.28	1.19	0.88
1.0	1.14	1.06	0.75
1.1	1.04	0.95	0.64
1.2	0.94	0.86	0.55
1.3	0.87	0.78	0.47
1.4	0.80	0.72	0.40
1.5	0.74	0.66	0.35
1.6	0.69	0.61	0.30
1.7	0.65	0.56	0.25
1.8	0.61	0.53	0.21
1.9	0.58	0.49	0.18
2.0	0.54	0.46	0.15
2.5	0.42	0.34	0.03
3.0	0.34	0.26	-0.05
4.0	0.24	0.16	-0.15
6.0	0.14	0.06	-0.25
8.0	0.09	0.01	-0.30
10.0	0.06	-0.02	-0.33
15.0	0.02	-0.06	-0.37
20.0	0.00	-0.08	-0.39
22.0	0.00	-0.09	-0.40

For a more accurate calculation, use the following values of the density of air (assuming 50% relative humidity and 0.04% CO_2):

	Air temperature		
P/kPa	10°C	20°C	30°C
85	0.001043	0.001005	0.000968
90	0.001105	0.001065	0.001025
95	0.001166	0.001124	0.001083
100	0.001228	0.001184	0.001140
105	0.001290	0.001243	0.001198

Formulas for calculating the density of air over more extended ranges of temperature, pressure, and humidity may be found in the references.

VOLUME OF ONE GRAM OF WATER

The following table, which is designed for gravimetric calibration of volumetric apparatus, gives the specific volume of water at standard atmospheric pressure as a function of temperature.

REFERENCE

Marsh, K. N., Editor, *Recommended Reference Materials for the Realization of Physicochemical Properties*, pp. 25-27, Blackwell Scientific Publications, Oxford, 1987.

t/°C	Volume of 1 g H_2O in cm³
10	1.0002980
11	1.0003928
12	1.0005007
13	1.0006212
14	1.0007542
15	1.0008992
16	1.0010561
17	1.0012246
18	1.0014044
19	1.0015952
20	1.0017969
21	1.0020092
22	1.0022320
23	1.0024649
24	1.0027079
25	1.0029607
26	1.0032234
27	1.0034956
28	1.0037771
29	1.0040679
30	1.0043679

PROPERTIES OF CARRIER GASES FOR GAS CHROMATOGRAPHY

The following is a list of carrier gases sometimes used in gas chromatography, with properties relevant to the design of chromatographic systems. All data refer to normal atmospheric pressure (101.325 kPa).

M_r : Molecular weight (relative molar mass)
ρ_{25} : Density at 25°C in g/L
λ : Thermal conductivity in mW/m °C
η : Viscosity in μPa s (equal to 10^{-3} cp)
c_p : Specific heat at 25°C in J/g °C

REFERENCES

1. Lide, D. R., and Kehiaian, H. V., *CRC Handbook of Thermophysical and Thermochemical Data*, CRC Press, Boca Raton, FL, 1994.
2. Bruno, T. J., and Svoronos, P. D. N., *CRC Handbook of Basic Tables for Chemical Analysis*, CRC Press, Boca Raton, FL, 1989

			At 25°C		At 250°C		
Gas	M_r	ρ_{25} g L⁻¹	λ mW/m °C	η μPa s	λ mW/m °C	η μPa s	c_p(25 °C) J/g °C
Hydrogen	2.016	0.0824	185.9	8.9	280	13.1	14.3
Helium	4.003	0.1636	154.6	19.9	230	29.5	5.20
Argon	39.95	1.6329	17.8	22.7	27.7	35.3	0.521
Nitrogen	28.01	1.1449	25.9	17.9	39.6	26.8	1.039
Oxygen	32.00	1.3080	26.2	20.7	42.6	31.8	0.919
Carbon monoxide	28.01	1.1449	24.8	17.8	40.7	26.5	1.039
Carbon dioxide	44.01	1.7989	16.7	14.9	35.5	24.9	0.843
Sulfur hexafluoride	146.05	5.9696	13.1	28.1	15.3	24.8	0.664
Methane	16.04	0.6556	34.5	11.1	75.0	17.6	2.23
Ethane	30.07	1.2291	20.9	9.4	57.7	15.5	1.75
Ethylene	28.05	1.1465	20.5	10.3	53.8	17.2	1.53
Propane	44.10	1.8025	17.9	8.3	49.2	14.0	1.67

SOLVENTS FOR ULTRAVIOLET SPECTROPHOTOMETRY

This table lists some solvents commonly used for sample preparation for ultraviolet spectrophotometry. The properties given are:

λ_c: cutoff wavelength, below which the solvent absorption becomes excessive.
ε: dielectric constant (relative permittivity); the temperature in °C is given as a superscript.
t_b: normal boiling point.

REFERENCES

1. Bruno, T. J., and Svoronos, P. D. N., *CRC Handbook of Basic Tables for Chemical Analysis*, CRC Press, Boca Raton, FL, 1989.
2. *Landolt-Börnstein, Numerical Data and Functional Relationships in Science and Technology, New Series,* IV/6, *Static Dielectric Constants of Pure Liquids and Binary Liquid Mixtures,* Springer-Verlag, Heidelberg, 1991.

Name	λ_c/nm	ε	t_b/°C
Acetic acid	260	6.20^{20}	117.9
Acetone	330	21.01^{20}	56.0
Acetonitrile	190	36.64^{20}	81.6
Benzene	280	2.28^{20}	80.0
2-Butanol	260	17.26^{20}	99.5
Butyl acetate	254	5.07^{20}	126.1
Carbon disulfide	380	2.63^{20}	46
Carbon tetrachloride	265	2.24^{20}	76.8
1-Chlorobutane	220	7.28^{20}	78.6
Chloroform	245	4.81^{20}	61.1
Cyclohexane	210	2.02^{20}	80.7
1,2-Dichloroethane	226	10.42^{20}	83.5
Dichloromethane	235	8.93^{25}	40
Diethyl ether	218	4.27^{20}	34.5
N,N-Dimethylacetamide	268	38.85^{21}	165
N,N-Dimethylformamide	270	38.25^{20}	153
Dimethyl sulfoxide	265	47.24^{20}	189
1,4-Dioxane	215	2.22^{20}	101.5
Ethanol	210	25.3^{20}	78.2
Ethyl acetate	255	6.08^{20}	77.1
Ethylene glycol dimethyl ether	240	7.30^{24}	85
Ethylene glycol monoethyl ether	210	13.38^{25}	135
Ethylene glycol monomethyl ether	210	17.2^{25}	124.1
Glycerol	207	46.53^{20}	290
Heptane	197	1.92^{20}	98.5
Hexadecane	200	2.05^{20}	286.8
Hexane	210	1.89^{20}	68.7
Methanol	210	33.0^{20}	64.6
Methylcyclohexane	210	2.02^{20}	100.9
Methyl ethyl ketone	330	18.56^{20}	79.5
Methyl isobutyl ketone	335	13.11^{20}	116.5
2-Methyl-1-propanol	230	17.93^{20}	107.8
N-Methyl-2-pyrrolidone	285	32.55^{20}	202
Nitromethane	380	37.27^{20}	101.1
Pentane	210	1.84^{20}	36.0
Pentyl acetate	212	4.79^{20}	149.2
1-Propanol	210	20.8^{20}	97.2
2-Propanol	210	20.18^{20}	82.3
Pyridine	330	13.26^{20}	115.2
Tetrachloroethylene	290	2.27^{30}	121.3
Tetrahydrofuran	220	7.52^{22}	65
Toluene	286	2.38^{23}	110.6
1,1,2-Trichloro-1,2,2-trifluoroethane	231	2.41^{25}	47.7
2,2,4-Trimethylpentane	215	1.94^{20}	99.2
Water	191	80.10^{20}	100.0
o-Xylene	290	2.56^{20}	144.5
m-Xylene	290	2.36^{20}	139.1
p-Xylene	290	2.27^{20}	138.3

^{13}C CHEMICAL SHIFTS OF USEFUL NMR SOLVENTS

The following table gives the expected carbon-13 chemical shifts, relative to tetramethylsilane, for various useful NMR solvents. In some solvents, slight changes can occur with change of concentration.[2,3]

REFERENCES

1. Bruno, T. J. and Svoronos, P. D. N., *CRC Handbook of Basic Tables for Chemical Analysis*, CRC Press, Boca Raton, FL, 1989.
2. Silverstein, R. M., Bassler, G. C., and Morrill, T. C., *Spectrometric Identification of Organic Compounds,* John Wiley & Sons, Now York, 1981.
3. Rahman, A. U., *Nuclear Magnetic Resonance. Basic Principles,* Springer-Verlag, New York, 1986.
4. Pretsch, E., Clerc, T., Seibl, J., and Simon, W., *Spectral Data for Structure Determination of Organic Compounds, Second Edition*, Springer-Verlag, Heidelberg, 1989.

Solvent	Formula	Chemical shift (ppm)
Acetic acid-d_4	CD_3COOD	20.0 (CD_3) 205.8 (C=O)
Acetone	$(CH_3)_2C{=}O$	30.7 (CH_3) 206.7 (C=O)
Acetone-d_6	$(CD_3)_2C{=}O$	29.2 (CD_3) 204.1 (C=O)
Acetonitrile-d_3	$CD_3C{\equiv}N$	1.3 (CD_3) 117.1 (C≡N)
Benzene	C_6H_6	128.5
Benzene-d_6	C_6D_6	128.4
Carbon disulfide	CS_2	192.3
Carbon tetrachloride	CCl_4	96.0
Chloroform	$CHCl_3$	77.2
Chloroform-d_3	$CDC1_3$	77.05
Cyclohexane-d_{12}	C_6D_{12}	27.5
Dichloromethane-d_2	CD_2C1_2	53.6
Dimethylformamide-d_7	$(CD_3)_2NCDO$	31 (CD_3) 36 (CD_3) 162.4 (C=O)
Dimethylsulfoxide-d_6	$(CD_3)_2S{=}O$	39.6
Dioxane-d_8	$C_4D_3O_2$	67.4
Formic acid-d_2	DCOOD	165.5
Methanol-d_4	CD_3OD	49.3
Nitromethane-d_3	CD_3NO_2	57.3
Pyridine	C_5H_5N	123.6 (C_3) 135.7 (C_4) 149.8 (C_2)
Pyridine-d_5	C_5D_5N	123.9 (C_3) 135.9 (C_4) 150.2 (C_2)
1,1,2,2-Tetrachloroethane-d_2	$CDC1_2CDC1_2$	75.5
Tetrahydrofuran-d_8	C_4D_8O	25.8 (C_2) 67.9 (C_1)
Trichlorofluoromethane	$CFC1_3$	117.6

MASS SPECTRAL PEAKS OF COMMON ORGANIC SOLVENTS

The strongest peaks in the mass spectra of 200 important solvents are listed in this table. The *e/m* value for each peak is followed by the relative intensity in parentheses, with the strongest peak assigned an intensity of 100. The peaks for each compound are listed in order of decreasing intensity. Solvents are listed in alphabetical order by common name.

Data on the physical properties of the same compounds may be found in Section 15 in the table "Properties of Common Laboratory Solvents".

REFERENCES

1. NIST/EPA/NIH Mass Spectral Database, National Institute of Standards and Technology, Gaithersburg, MD, 20899.
2. Lide, D.R., and Milne, G.W.A., Editors, *Handbook of Data on Organic Compounds, Third Edition*, CRC Press, Boca Raton, FL, 1994. (Also available as a CD ROM database.)

Compound	e/m (intensity)									
Acetal (1,1-Diethoxyethane)	44(100)	43(92)	29(77)	31(76)	45(74)	27(52)	72(48)	73(23)	28(17)	46(15)
Acetic acid	43(100)	45(87)	60(57)	15(42)	42(14)	29(13)	14(13)	28(7)	18(6)	16(6)
Acetone	43(100)	15(34)	58(23)	27(9)	14(9)	42(8)	26(7)	29(5)	28(5)	39(4)
Acetonitrile	41(100)	40(46)	39(13)	14(9)	38(6)	28(4)	26(4)	25(3)	42(2)	27(2)
Acetylacetone	43(100)	85(31)	100(20)	27(12)	42(10)	29(10)	41(7)	39(7)	31(5)	26(5)
Acrylonitrile	53(100)	26(85)	52(79)	51(34)	27(13)	50(8)	25(7)	38(5)	54(3)	37(3)
Adiponitrile	41(100)	68(50)	54(42)	40(21)	55(20)	27(17)	39(16)	28(13)	52(7)	42(6)
Allyl alcohol	57(100)	31(34)	29(32)	28(31)	58(25)	39(22)	27(20)	30(16)	32(14)	26(11)
Allylamine	30(100)	56(80)	28(76)	57(33)	39(21)	29(20)	27(18)	26(13)	41(8)	18(8)
2-Aminoisobutanol	58(100)	41(18)	18(17)	42(13)	28(11)	56(10)	30(10)	29(8)	43(6)	59(5)
Benzal chloride	125(100)	127(32)	160(14)	89(13)	162(9)	63(9)	126(8)	62(7)	105(5)	39(5)
Benzaldehyde	51(100)	77(81)	50(55)	106(44)	105(43)	52(26)	78(16)	39(13)	27(10)	74(8)
Benzene	78(100)	77(20)	52(19)	51(17)	50(15)	39(12)	79(6)	76(5)	74(4)	38(4)
Benzonitrile	103(100)	76(34)	50(13)	104(9)	75(7)	51(7)	77(5)	52(4)	39(4)	74(3)
Benzyl chloride	91(100)	126(20)	65(14)	92(9)	39(9)	63(8)	128(6)	45(6)	89(5)	125(3)
Bromochloromethane	49(100)	130(67)	128(52)	51(31)	93(23)	81(20)	79(20)	95(17)	132(16)	47(8)
Bromoform (Tribromomethane)	173(100)	171(50)	175(49)	93(22)	91(22)	79(18)	81(17)	94(13)	92(13)	254(11)
Butyl acetate	43(100)	56(34)	41(17)	27(16)	29(15)	73(11)	61(10)	28(7)	55(6)	39(6)
Butyl alcohol	31(100)	56(81)	41(62)	43(60)	27(50)	42(31)	29(31)	28(17)	39(16)	55(12)
sec-Butyl alcohol	45(100)	31(22)	27(22)	59(20)	29(18)	43(13)	41(12)	44(8)	18(8)	28(5)
tert-Butyl alcohol	59(100)	31(33)	41(22)	43(18)	29(13)	27(11)	57(10)	42(4)	60(3)	28(3)
Butylamine	30(100)	73(10)	28(5)	41(3)	27(3)	18(3)	44(2)	42(2)	31(2)	29(2)
tert-Butylamine	58(100)	41(21)	42(15)	18(9)	30(8)	15(8)	39(7)	57(6)	28(6)	59(4)
Butyl methyl ketone	43(100)	58(60)	57(17)	100(16)	29(15)	41(13)	85(8)	27(8)	71(7)	59(5)
p-tert-Butyltoluene	133(100)	105(38)	41(23)	148(18)	93(16)	91(14)	115(13)	134(11)	39(11)	116(10)
γ-Butyrolactone	28(100)	42(74)	29(48)	27(33)	41(27)	56(25)	86(24)	26(18)	85(10)	39(10)
Caprolactam	55(100)	113(87)	30(81)	56(66)	84(60)	85(57)	42(51)	41(33)	28(26)	43(17)
Carbon disulfide	76(100)	32(22)	44(17)	78(9)	38(6)	28(5)	77(3)	64(1)	46(1)	39(1)
Carbon tetrachloride	117(100)	119(98)	121(31)	82(24)	47(23)	84(16)	35(14)	49(8)	28(8)	36(6)
1-Chloro-1,1-difluoroethane	65(100)	45(31)	85(14)	31(10)	64(8)	44(7)	35(6)	26(6)	87(5)	81(4)
Chlorobenzene	112(100)	77(63)	114(33)	51(29)	50(14)	75(8)	113(7)	78(5)	76(5)	28(4)
Chloroform	83(100)	85(64)	47(35)	35(19)	48(16)	49(12)	87(10)	37(6)	50(5)	84(4)
Chloropentafluoroethane	85(100)	69(61)	31(38)	87(32)	50(17)	35(8)	119(6)	66(4)	100(3)	47(3)
Cumene (Isopropylbenzene)	105(100)	120(25)	77(13)	51(12)	79(10)	106(9)	39(9)	27(8)	103(6)	91(5)
Cyclohexane	56(100)	84(71)	41(70)	27(37)	55(36)	39(35)	42(30)	69(23)	28(18)	43(14)
Cyclohexanol	57(100)	44(68)	41(68)	39(51)	32(40)	43(38)	31(32)	42(22)	67(18)	82(16)
Cyclohexanone	55(100)	42(85)	41(34)	27(33)	98(31)	39(27)	69(26)	70(20)	43(14)	28(14)
Cyclohexylamine	56(100)	43(23)	28(17)	99(10)	70(8)	57(6)	30(6)	93(5)	54(4)	41(4)
Cyclopentane	42(100)	70(30)	55(29)	41(29)	39(22)	27(15)	40(7)	29(5)	28(4)	43(3)
Cyclopentanone	55(100)	28(50)	84(42)	41(38)	56(29)	27(24)	39(19)	42(15)	26(9)	29(7)
p-Cymene (1-Methyl-4-isopropylbenzene)	119(100)	91(42)	134(33)	39(27)	41(20)	117(18)	65(18)	77(17)	27(16)	120(15)
cis-Decalin	67(100)	81(87)	41(81)	138(67)	96(62)	82(62)	39(50)	55(45)	27(44)	95(42)
trans-Decalin	41(100)	68(91)	67(88)	82(67)	27(65)	96(61)	95(55)	138(51)	81(51)	29(51)

MASS SPECTRAL PEAKS OF COMMON ORGANIC SOLVENTS (continued)

Compound	e/m (intensity)									
Diacetone alcohol	43(100)	59(41)	58(17)	101(10)	41(9)	31(9)	83(6)	56(6)	55(6)	29(6)
1,2-Dibromoethane	27(100)	107(77)	109(72)	26(24)	28(10)	81(5)	79(5)	25(5)	95(4)	93(4)
Dibromofluoromethane	111(100)	113(98)	192(29)	43(16)	41(16)	190(15)	194(14)	81(9)	79(9)	122(7)
Dibromomethane	174(100)	93(96)	95(84)	172(53)	176(50)	91(11)	81(9)	79(9)	94(5)	65(5)
1,2-Dibromotetrafluoroethane	179(100)	181(97)	129(34)	131(33)	100(17)	31(13)	260(12)	50(8)	69(7)	262(6)
Dibutylamine	86(100)	72(52)	30(48)	44(40)	29(31)	57(24)	41(21)	73(15)	28(15)	43(13)
o-Dichlorobenzene	146(100)	148(64)	111(38)	75(23)	113(12)	74(12)	50(11)	150(10)	73(9)	147(7)
1,1-Dichloroethane (Ethylidene dichloride)	63(100)	27(71)	65(31)	26(19)	83(11)	85(7)	61(7)	35(6)	98(5)	62(5)
1,2-Dichloroethane (Ethylene dichloride)	62(100)	27(91)	49(40)	64(32)	26(31)	63(19)	98(14)	51(13)	61(12)	100(9)
1,1-Dichloroethylene	61(100)	96(61)	98(38)	63(32)	26(16)	60(15)	62(7)	25(7)	100(6)	35(6)
cis-1,2-Dichloroethylene	61(100)	96(73)	98(47)	63(32)	26(30)	60(21)	25(13)	35(12)	62(9)	100(8)
trans-1,2-Dichloroethylene	61(100)	96(67)	98(43)	26(34)	63(32)	60(24)	25(15)	62(10)	100(7)	47(7)
Dichloroethyl ether	93(100)	63(74)	27(38)	95(32)	65(24)	31(9)	49(4)	28(4)	94(3)	62(3)
Dichloromethane (Methylene chloride)	49(100)	84(64)	86(39)	51(31)	47(14)	48(8)	88(6)	50(3)	85(2)	83(2)
1,2-Dichloropropane	63(100)	62(71)	27(57)	41(49)	39(32)	65(31)	76(27)	64(25)	49(13)	77(12)
1,2-Dichlorotetrafluoroethane	85(100)	135(52)	87(33)	137(17)	101(9)	31(9)	103(6)	100(6)	50(5)	69(4)
Diethanolamine	30(100)	74(82)	28(77)	56(69)	18(50)	42(46)	29(36)	27(34)	45(30)	43(19)
Diethylamine	30(100)	58(81)	44(28)	73(18)	29(18)	28(17)	72(12)	42(11)	27(11)	59(4)
Diethyl carbonate	29(100)	45(70)	31(53)	27(39)	91(24)	28(15)	63(11)	26(10)	30(6)	43(5)
Diethylene glycol	45(100)	75(23)	31(20)	44(16)	27(14)	76(12)	29(12)	43(11)	42(9)	41(4)
Diethylene glycol dimethyl ether (Diglyme)	59(100)	58(43)	31(34)	29(32)	45(28)	28(19)	89(15)	43(9)	27(5)	60(4)
Diethylene glycol monoethyl ether (Carbitol)	45(100)	59(56)	72(37)	73(22)	60(14)	31(13)	75(11)	44(9)	104(8)	103(7)
Diethylene glycol monoethyl ether acetate	43(100)	29(51)	31(42)	45(40)	59(24)	72(18)	44(10)	73(9)	42(9)	30(6)
Diethylene glycol monomethyl ether	45(100)	31(42)	59(41)	29(38)	28(32)	58(21)	43(14)	27(13)	44(11)	32(10)
Diethylenetriamine	44(100)	73(59)	30(35)	19(18)	56(16)	28(16)	27(16)	42(11)	99(8)	43(8)
Diethyl ether	31(100)	29(63)	59(40)	27(35)	45(33)	74(23)	15(17)	43(9)	28(9)	26(9)
Diisobutyl ketone (Isovalerone)	57(100)	85(82)	41(46)	43(39)	58(33)	28(30)	26(30)	39(22)	42(12)	142(11)
Diisopropyl ether	45(100)	43(39)	87(15)	41(12)	59(10)	27(8)	39(4)	69(3)	42(3)	31(3)
N,N-Dimethylacetamide	44(100)	87(69)	43(46)	45(23)	42(19)	72(15)	15(11)	30(8)	28(5)	88(4)
Dimethylamine	44(100)	45(81)	18(32)	28(30)	43(19)	42(15)	15(9)	46(5)	41(5)	27(5)
Dimethyl disulfide	94(100)	45(63)	79(59)	46(38)	47(26)	15(18)	48(14)	61(12)	64(11)	96(9)
N,N-Dimethylformamide	73(100)	44(86)	42(36)	30(22)	28(20)	29(8)	43(7)	72(6)	58(5)	74(4)
Dimethyl sulfoxide	63(100)	78(70)	15(40)	45(35)	29(16)	61(13)	46(12)	31(11)	48(10)	47(10)
1,4-Dioxane	28(100)	29(37)	88(31)	58(24)	31(17)	15(17)	27(15)	30(13)	43(11)	26(9)
1,3-Dioxolane	73(100)	29(56)	44(53)	45(28)	28(21)	43(20)	27(13)	31(7)	74(5)	42(3)
Dipentene	68(100)	93(50)	67(44)	94(22)	39(22)	107(18)	92(18)	53(18)	136(16)	79(16)
Epichlorohydrin	57(100)	27(39)	29(32)	49(25)	31(22)	62(18)	28(16)	92(1)		

Compound	e/m (intensity)									
Ethanolamine (Glycinol)	30(100)	18(30)	28(15)	42(7)	31(6)	17(6)	61(5)	15(5)	43(3)	29(3)
Ethyl acetate	43(100)	29(46)	27(33)	45(32)	61(28)	28(25)	42(18)	73(11)	88(10)	70(10)
Ethyl acetoacetate	43(100)	29(24)	88(18)	28(16)	85(14)	27(12)	42(11)	60(9)	130(6)	45(6)
Ethyl alcohol	31(100)	45(44)	46(18)	27(18)	29(15)	43(14)	30(6)	42(3)	19(3)	14(3)
Ethylamine	30(100)	28(32)	44(20)	45(19)	27(13)	15(10)	42(9)	29(8)	41(5)	40(5)
Ethylbenzene	91(100)	106(31)	51(14)	39(10)	77(8)	65(8)	105(7)	92(7)	78(7)	27(6)
Ethyl bromide (Bromoethane)	108(100)	110(97)	29(62)	27(51)	28(35)	26(14)	93(6)	32(6)	95(5)	81(5)
Ethyl chloride (Cloroethane)	64(100)	28(91)	29(84)	27(75)	66(32)	26(28)	49(25)	51(8)	63(6)	65(4)
Ethylene carbonate	29(100)	44(62)	43(54)	88(40)	30(16)	28(11)	45(7)	58(6)	42(6)	73(4)
Ethylenediamine (1,2-Ethane-diamine)	30(100)	18(13)	42(6)	43(5)	27(5)	44(4)	29(4)	17(4)	15(4)	41(3)
Ethylene glycol	31(100)	33(35)	29(13)	32(11)	43(6)	27(5)	28(4)	62(3)	30(3)	44(2)
Ethylene glycol diethyl ether	31(100)	59(71)	29(58)	45(43)	27(33)	74(27)	43(15)	15(14)	28(12)	44(10)
Ethylene glycol dimethyl ether	45(100)	60(13)	29(13)	90(7)	58(6)	31(5)	28(5)	43(4)	59(3)	46(2)
Ethylene glycol monobutyl ether	57(100)	45(38)	29(35)	41(31)	87(16)	27(12)	56(11)	31(9)	75(7)	28(7)
Ethylene glycol monoethyl ether (Cellosolve)	31(100)	29(52)	59(50)	27(27)	45(26)	72(14)	43(14)	15(14)	28(8)	26(6)
Ethylene glycol monoethyl ether acetate	43(100)	31(34)	59(31)	72(28)	44(25)	29(24)	45(12)	27(11)	15(11)	87(7)
Ethylene glycol monomethyl ether	45(100)	31(15)	29(14)	28(11)	47(9)	76(6)	43(6)	58(4)	46(4)	27(4)
Ethylene glycol monomethyl ether acetate	43(100)	45(48)	58(42)	29(10)	42(4)	31(4)	73(3)	27(3)	59(2)	26(2)
Ethyl formate	31(100)	28(73)	27(51)	29(38)	45(34)	26(17)	74(11)	43(9)	47(8)	56(4)
Furan	68(100)	39(64)	40(9)	38(9)	42(6)	29(6)	37(5)	69(4)	34(2)	67(1)
Furfural	39(100)	96(55)	95(52)	38(38)	29(35)	37(29)	40(11)	97(9)	50(7)	42(7)
Furfuryl alcohol	98(100)	41(65)	39(59)	81(55)	53(53)	97(51)	42(49)	69(39)	70(36)	29(28)
Glycerol	61(100)	43(90)	31(57)	44(54)	29(38)	18(32)	27(12)	42(11)	60(10)	45(10)
Heptane	43(100)	41(56)	29(49)	57(47)	27(46)	71(45)	56(27)	42(26)	39(23)	70(18)
1-Heptanol	41(100)	70(87)	56(86)	31(78)	43(72)	29(70)	55(67)	27(65)	42(54)	69(41)
Hexane	57(100)	43(78)	41(77)	29(61)	27(57)	56(45)	42(39)	39(27)	28(16)	86(14)
1-Hexanol (Caproyl alcohol)	56(100)	43(78)	31(74)	41(71)	27(64)	29(59)	55(58)	42(53)	39(37)	69(27)
Hexylene glycol	59(100)	43(61)	56(25)	45(17)	41(16)	57(13)	42(13)	85(11)	61(10)	31(10)
Hexyl methyl ketone	43(100)	58(79)	41(56)	59(52)	71(49)	27(46)	29(36)	39(27)	57(18)	55(17)
Isobutyl acetate	43(100)	56(26)	73(15)	41(10)	29(5)	71(3)	57(3)	39(3)	27(3)	86(2)
Isobutyl alcohol	43(100)	33(73)	31(72)	41(66)	42(60)	27(43)	29(18)	39(17)	28(8)	74(6)
Isobutylamine	30(100)	28(9)	41(6)	73(5)	27(5)	39(4)	29(3)	15(3)	58(2)	56(2)
Isopentyl acetate	43(100)	70(49)	55(38)	61(15)	42(15)	41(14)	27(12)	87(11)	29(10)	73(9)
Isophorone	82(100)	39(20)	138(17)	54(13)	27(12)	41(10)	53(8)	83(7)	29(7)	55(6)
Isopropyl acetate	43(100)	61(17)	41(14)	87(9)	59(8)	27(8)	42(7)	39(4)	45(3)	44(2)
Isopropyl alcohol	45(100)	43(19)	27(17)	29(12)	41(7)	31(6)	19(6)	42(5)	44(4)	59(3)
Isoquinoline	129(100)	102(26)	51(20)	128(18)	50(11)	130(10)	75(10)	76(9)	103(8)	74(7)
d-Limonene (Citrene)	68(100)	93(50)	67(49)	41(22)	94(21)	79(21)	39(21)	136(20)	53(19)	121(16)
2,6-Lutidine (2,6-Dimethyl-pyridine)	107(100)	39(39)	106(29)	66(22)	92(18)	65(18)	38(12)	27(11)	79(9)	63(9)

Compound	*e/m* (intensity)									
Mesitylene (1,3,5-Trimethylbenzene)	105(100)	120(64)	119(15)	77(13)	39(11)	106(9)	91(9)	51(8)	27(7)	121(6)
Mesityl oxide	55(100)	83(89)	43(73)	29(42)	98(36)	39(32)	27(28)	53(11)	41(10)	56(5)
Methyl acetate	43(100)	74(52)	28(38)	42(19)	59(17)	44(8)	32(8)	29(6)	31(4)	75(2)
Methylal (Dimethoxymethane)	45(100)	75(61)	29(59)	31(13)	30(6)	15(6)	47(5)	76(2)	46(2)	44(2)
Methyl alcohol	31(100)	29(72)	32(67)	15(42)	28(12)	14(10)	30(9)	13(6)	12(3)	16(2)
Methylamine	30(100)	31(87)	28(56)	29(19)	32(15)	15(12)	27(9)			
Methyl benzoate	105(100)	77(81)	51(45)	136(24)	50(18)	106(8)	78(6)	28(6)	39(5)	27(5)
Methylcyclohexane	83(100)	55(82)	41(60)	98(44)	42(35)	56(30)	27(29)	39(27)	69(23)	70(22)
Methyl ethyl ketone	43(100)	72(24)	29(19)	27(12)	57(7)	42(5)	26(4)	28(3)	44(2)	39(2)
N-Methylformamide	59(100)	30(54)	28(34)	29(13)	58(8)	15(7)	60(3)	41(3)	27(3)	31(2)
Methyl formate	31(100)	29(63)	32(34)	60(28)	30(7)	28(7)	44(2)	18(2)	61(1)	59(1)
Methyl iodide (Iodomethane)	142(100)	127(38)	141(14)	15(13)	139(5)	140(4)	128(3)	14(1)	13(1)	71(0)
Methyl isobutyl ketone	43(100)	58(84)	29(65)	41(56)	57(44)	27(42)	39(31)	85(19)	100(14)	42(14)
Methyl isopentyl ketone	43(100)	58(34)	27(14)	41(13)	15(13)	57(11)	39(9)	71(8)	59(8)	29(8)
2-Methylpentane	43(100)	42(53)	41(35)	27(31)	71(29)	39(20)	29(18)	57(11)	15(10)	70(7)
4-Methyl-2-pentanol	45(100)	43(47)	69(30)	41(27)	27(19)	39(13)	29(12)	87(11)	84(10)	57(10)
Methyl pentyl ketone	43(100)	58(60)	71(14)	41(11)	27(11)	59(9)	39(8)	29(8)	42(5)	114(4)
Methyl propyl ketone	43(100)	41(17)	86(12)	42(12)	27(11)	39(8)	71(7)	58(7)	45(7)	44(3)
N-Methyl-2-pyrrolidone	99(100)	44(89)	98(80)	42(60)	41(38)	43(17)	28(17)	71(13)	39(11)	70(10)
Morpholine	57(100)	29(100)	87(69)	28(69)	30(38)	56(33)	86(28)	31(28)	27(12)	15(7)
Nitrobenzene	77(100)	51(59)	123(42)	50(25)	30(15)	65(14)	39(10)	93(9)	74(7)	78(6)
Nitroethane	29(100)	30(12)	28(11)	26(9)	27(8)	43(5)	41(5)	14(5)	15(3)	46(2)
Nitromethane	30(100)	61(64)	46(39)	28(30)	45(8)	27(8)	44(7)	29(7)	60(5)	43(4)
1-Nitropropane	43(100)	27(93)	41(90)	39(34)	30(25)	44(20)	42(20)	26(20)	28(13)	54(12)
2-Nitropropane	43(100)	41(73)	27(71)	39(30)	30(18)	15(11)	42(9)	28(8)	26(8)	38(6)
Octane	43(100)	57(30)	85(25)	41(25)	71(19)	29(17)	56(14)	70(10)	42(10)	27(10)
1-Octanol	41(100)	56(85)	43(82)	55(81)	31(69)	27(69)	29(68)	42(62)	70(53)	69(48)
Pentachloroethane	167(100)	165(91)	117(90)	119(89)	83(58)	169(54)	130(43)	132(42)	60(40)	85(37)
Pentamethylene glycol (1,5-Pentanediol)	31(100)	56(85)	41(67)	57(59)	55(51)	44(45)	29(37)	43(31)	68(29)	27(26)
Pentane	43(100)	42(55)	41(45)	27(42)	29(26)	39(19)	57(13)	28(9)	15(9)	72(8)
1-Pentanol (Amyl alcohol)	42(100)	70(72)	55(65)	41(56)	31(47)	29(41)	27(26)	57(22)	28(22)	43(21)
Pentyl acetate (Amyl acetate)	43(100)	70(90)	42(52)	28(51)	61(50)	55(41)	73(21)	41(20)	29(14)	69(11)
2-Picoline (2-Methylpyridine)	93(100)	66(41)	39(31)	92(20)	78(19)	51(19)	65(16)	38(13)	50(12)	52(11)
α-Pinene	93(100)	92(30)	39(24)	41(23)	77(22)	91(21)	27(21)	79(18)	121(13)	53(10)
β-Pinene	93(100)	41(64)	69(47)	39(33)	27(31)	79(20)	77(18)	53(14)	94(13)	91(13)
Piperidine (Hexahydropyridine)	84(100)	85(53)	56(46)	57(43)	28(41)	29(37)	44(34)	42(30)	30(30)	43(25)
Propanenitrile	28(100)	54(63)	26(20)	27(17)	52(11)	55(10)	51(9)	15(9)	53(7)	25(7)
Propyl acetate	43(100)	61(19)	31(18)	27(15)	42(11)	73(9)	41(9)	29(9)	59(5)	39(5)
Propyl alcohol	31(100)	27(19)	29(18)	59(11)	42(9)	60(7)	41(7)	28(7)	43(3)	32(3)
Propylamine	30(100)	28(13)	59(8)	27(7)	41(5)	42(3)	39(3)	29(3)	26(3)	18(3)
Propylbenzene	91(100)	120(21)	92(10)	38(10)	65(9)	78(6)	51(6)	27(5)	63(4)	105(3)
1,2-Propylene glycol	45(100)	18(46)	29(21)	43(19)	31(18)	27(17)	28(11)	19(8)	44(6)	61(5)

Compound	*e/m* (intensity)									
Pseudocumene (1,2,4-Trimethylbenzene)	105(100)	120(56)	119(17)	77(15)	39(15)	51(11)	91(10)	27(10)	106(9)	79(7)
Pyridine	79(100)	52(62)	51(31)	50(19)	78(11)	53(7)	39(7)	80(6)	27(3)	77(2)
Pyrrole	67(100)	41(58)	39(58)	40(51)	28(42)	38(20)	37(12)	66(7)	68(5)	27(3)
Pyrrolidine	43(100)	28(52)	70(33)	71(26)	42(22)	41(20)	27(16)	39(15)	29(10)	30(9)
2-Pyrrolidone	85(100)	42(43)	41(36)	28(33)	30(29)	56(16)	84(14)	40(12)	27(12)	29(9)
Quinoline	129(100)	51(28)	76(25)	128(24)	44(24)	50(20)	32(19)	75(18)	74(12)	103(11)
Styrene	104(100)	103(41)	78(32)	51(28)	77(23)	105(12)	50(12)	52(11)	39(11)	102(10)
Sulfolane	41(100)	28(94)	56(82)	55(72)	120(37)	27(32)	39(19)	29(17)	26(11)	48(5)
α-Terpinene	121(100)	93(85)	136(43)	91(40)	77(34)	39(33)	27(33)	79(27)	41(26)	43(18)
1,1,1,2-Tetrachloro-2,2-difluoroethane	167(100)	169(96)	117(85)	119(82)	171(31)	85(29)	121(26)	82(14)	47(14)	101(13)
Tetrachloro-1,2-difluoroethane	101(100)	103(64)	167(54)	169(52)	117(19)	119(18)	171(17)	105(11)	31(11)	132(9)
1,1,1,2-Tetrachloroethane	131(100)	133(96)	117(76)	119(73)	95(34)	135(31)	121(23)	97(23)	61(19)	60(18)
1,1,2,2-Tetrachloroethane	83(100)	85(63)	95(11)	87(10)	168(8)	133(8)	131(8)	96(8)	61(8)	60(8)
Tetrachloroethylene	166(100)	164(82)	131(71)	129(71)	168(45)	94(38)	47(31)	96(24)	133(20)	59(17)
Tetraethylene glycol	45(100)	89(10)	44(8)	43(6)	31(6)	29(6)	27(6)	101(5)	75(5)	28(5)
Tetrahydrofuran	42(100)	41(52)	27(33)	72(29)	71(27)	39(24)	43(22)	29(22)	40(13)	15(10)
1,2,3,4-Tetrahydronaphthalene	104(100)	132(53)	91(43)	51(17)	39(17)	131(15)	117(15)	115(14)	78(13)	77(13)
Tetrahydropyran	41(100)	28(64)	56(57)	45(57)	29(51)	27(49)	85(47)	86(42)	39(28)	55(23)
Tetramethylsilane	73(100)	43(14)	45(12)	74(8)	29(7)	15(5)	75(4)	44(4)	42(4)	31(4)
Toluene	91(100)	92(73)	39(20)	65(14)	63(11)	51(11)	50(7)	27(6)	93(5)	90(5)
o-Toluidine	106(100)	107(83)	77(17)	79(13)	39(12)	53(10)	52(10)	54(9)	51(9)	28(9)
Triacetin	43(100)	103(44)	145(34)	116(17)	115(13)	44(10)	86(9)	28(8)	73(7)	42(7)
Tributylamine	142(100)	100(19)	143(11)	29(8)	185(7)	57(6)	44(6)	41(6)	30(5)	86(4)
1,1,1-Trichloroethane	97(100)	99(64)	61(58)	26(31)	27(24)	117(19)	63(19)	119(18)	35(17)	62(11)
1,1,2-Trichloroethane	97(100)	83(95)	99(62)	85(60)	61(58)	26(23)	96(21)	63(19)	27(17)	98(15)
Trichloroethylene	95(100)	130(90)	132(85)	60(65)	97(64)	35(40)	134(27)	47(26)	62(21)	59(13)
Trichlorofluoromethane	101(100)	103(66)	66(13)	105(11)	35(11)	47(9)	31(8)	82(4)	68(4)	37(4)
1,1,2-Trichlorotrifluoroethane	101(100)	151(68)	103(64)	85(45)	31(45)	153(44)	35(20)	66(19)	47(18)	87(14)
Triethanolamine	118(100)	56(69)	45(60)	42(56)	44(27)	43(25)	41(14)	116(8)	57(8)	86(7)
Triethylamine	86(100)	30(68)	58(37)	28(24)	29(23)	27(19)	44(18)	101(17)	42(16)	56(8)
Triethylene glycol	45(100)	58(11)	89(9)	31(8)	29(8)	75(7)	44(7)	43(7)	27(7)	28(5)
Triethyl phosphate	99(100)	81(71)	155(56)	82(45)	45(45)	109(44)	127(41)	43(24)	125(16)	111(14)
Trimethylamine	58(100)	59(47)	30(29)	42(26)	44(17)	15(14)	28(10)	18(10)	43(8)	57(7)
Trimethylene glycol (1,3-Propanediol)	28(100)	58(93)	31(76)	57(70)	29(40)	27(26)	45(24)	43(23)	19(18)	30(17)
Trimethyl phosphate	110(100)	109(35)	79(34)	95(25)	80(23)	15(20)	140(18)	47(10)	31(7)	139(5)
Veratrole	138(100)	95(65)	77(48)	123(44)	52(42)	41(33)	65(30)	51(29)	39(19)	63(17)
o-Xylene	91(100)	106(40)	39(21)	105(17)	51(17)	77(15)	27(12)	65(10)	92(8)	79(8)
m-Xylene	91(100)	106(65)	105(29)	39(18)	51(15)	77(14)	27(10)	92(8)	79(8)	78(8)
p-Xylene	91(100)	106(62)	105(30)	51(16)	39(16)	77(13)	27(11)	92(7)	78(7)	65(7)

Section 9
Molecular Structure and Spectroscopy

Bond Lengths in Crystalline Organic Compounds 9-1
Bond Lengths in Organometallic Compounds 9-15
Bond Lengths and Angles in Gas-Phase Molecules 9-17
Dipole Moments 9-44
Strengths of Chemical Bonds 9-51
Electronegativity 9-75
Force Constants for Bond Stretching 9-76
Fundamental Vibrational Frequencies of Small Molecules 9-77
Spectroscopic Constants of Diatomic Molecules 9-81
Infrared Correlation Charts 9-87
Nuclear Spins, Moments, and Other Data Related to NMR Spectroscopy 9-92
Proton NMR Chemical Shifts for Characteristic Organic Structures 9-95
^{13}C NMR Absorptions of Major Functional Groups 9-96

BOND LENGTHS IN CRYSTALLINE ORGANIC COMPOUNDS

The following table gives average interatomic distances for bonds between the elements H, B, C, N, O, F, Si, P, S, Cl, As, Se, Br, Te, and I as determined from X-ray and neutron diffraction measurements on organic crystals. The table has been derived from an analysis of high-precision structure data on about 10,000 crystals contained in the 1985 version of the Cambridge Structural Database, which is maintained by the Cambridge Crystallographic Data Center. The explanation of the columns is:

Column 1: Specification of elements in the bond, with coordination number given in parentheses, and bond type (single, double, etc.). For carbon, the hybridization state is given.
Column 2: Substructure in which the bond is found. The target bond is set in boldface. Where X is not specified, it denotes any element type. C# indicates any sp^3 carbon atom, and C* denotes an sp^3 carbon whose bonds, in addition to those specified in the linear formulation, are to C and H atoms only.
Column 3: *d* is the unweighted mean in Å units of all the values for that bond length found in the sample.
Column 4: *m* is the median in Å units of all values.
Column 5: σ is the standard deviation in the sample.
Column 6: q_l is the lower quartile for the sample (i.e., 25% of values are less than q_l and 75% exceed it).
Column 7: q_u is the upper quartile for the sample.
Column 8: *n* is number of observations in the sample.
Column 9: Notes refer to the footnotes in Appendix 1.

References to special cases are given in a shorthand form and listed in Appendix 2. Further information on the method of analysis of the data may be found in the reference cited below.

The table is reprinted with permission of the authors, the Royal Society of Chemistry, and the International Union of Crystallography.

REFERENCE

Frank H. Allen, Olga Kennard, David G. Watson, Lee Brammer, A. Guy Orpen, and Robin Taylor, *J. Chem. Soc. Perkin Trans. II*, S1—S19, 1987.

Bond	Substructure	*d*	*m*	σ	q_l	q_u	*n*	Note
As(3)–As(3)	X_2–**As–As**–X_2	2.459	2.457	0.011	2.456	2.466	8	
As–B	see CUDLOC (2.065), CUDLUI (2.041)							
As–Br	see CODDEE, CODDII (2.346—3.203)							
As(4)–C	X_3–**As–C**H_3	1.903	1.907	0.016	1.893	1.916	12	
	$(X)_2$(C,O,S=)**As–C***sp*^{3}	1.927	1.929	0.017	1.921	1.937	16	
	As–C*ar* in Ph_4As^+	1.905	1.909	0.012	1.897	1.912	108	
	$(X)_2$(C,O,S=)**As–C***ar*	1.922	1.927	0.016	1.908	1.934	36	
As(3)–C	X_2–**As–C***sp*^{3}	1.963	1.965	0.017	1.948	1.978	6	
	X_2–**As–C***ar*	1.956	1.956	0.015	1.944	1.964	41	
As(3)–Cl	X_2–**As–Cl**	2.268	2.256	0.039	2.247	2.281	10	
As(6)–F	in AsF_6^-	1.678	1.676	0.020	1.659	1.695	36	
As(3)–I	see OPIMAS (2.579, 2.590)							
As(3)–N(3)	X_2–**As–N**–X_2	1.858	1.858	0.029	1.839	1.873	19	
As(4)=N(2)	see TPASSN (1.837)							
As(4)–O	$(X)_2$(O=)**As–O**H	1.710	1.712	0.017	1.695	1.726	6	
As(3)–O	see ASAZOC, PHASOC01 (1.787—1.845)							
As(4)=O	X_3–**As=O**	1.661	1.661	0.016	1.652	1.667	9	
As(3)=P(3)	see BELNIP (2.350, 2.362)							†
As(3)–P(3)	see BUTHAZ10 (2.124)							†
As(3)–S	X_2–**As–S**	2.275	2.266	0.032	2.247	2.298	14	
As(4)=S	X_3–**As=S**	2.083	2.082	0.004	2.080	2.086	9	
As(3)–Se(2)	see COSDIX, ESEARS (2.355—2.401)							†
As(3)–Si(4)	see BICGEZ, MESIAD (2.351—2.365)							†
As(3)–Te(2)	see ETEARS (2.571, 2.576)							†
B(n)–B(n)	n = 5—7 in boron cages	1.775	1.773	0.031	1.763	1.786	688	
B(4)–B(4)	see CETTAW (2.041)							
B(4)–B(3)	see COFVOI (1.698)							
B(3)–B(3)	X_2–**B–B**–X_2	1.701	1.700	0.014	1.691	1.712	8	
B(6)–Br		1.967	1.971	0.014	1.954	1.979	7	†
B(4)–Br		2.017	2.008	0.031	1.990	2.044	15	†
B(*n*)–C	*n* = 5—7: **B–C** in cages	1.716	1.717	0.020	1.707	1.728	96	
	n = 3—4: **B–C***sp*^{3} not cages	1.597	1.599	0.022	1.585	1.611	29	1
	n = 4: **B–C***ar*	1.606	1.607	0.012	1.596	1.615	41	
	n = 4: **B–C***ar* in Ph_4B^-	1.643	1.643	0.006	1.641	1.645	16	
B(*n*)–C	*n* = 3: **B–C***ar*	1.556	1.552	0.015	1.546	1.566	24	
B(*n*)–Cl	B(5)–Cl and B(3)–Cl	1.751	1.751	0.011	1.743	1.761	14	
	B(4)–Cl	1.833	1.833	0.013	1.821	1.843	22	
B(4)–F	**B–F** (B neutral)	1.366	1.368	0.017	1.356	1.375	25	
	B^-–F in BF_4^-	1.365	1.372	0.029	1.352	1.390	84	
B(4)–I	see TMPBTI (2.220, 2.253)							
B(4)–N(3)	X_3–**B–N**(=C)(X)	1.611	1.617	0.013	1.601	1.625	8	
	in pyrazaboles	1.549	1.552	0.015	1.536	1.560	10	

Bond	Substructure	*d*	*m*	σ	q_1	q_u	*n*	Note
B(3)–N(3)	X_2–**B**–N–C_2: all coplanar	1.404	1.404	0.014	1.389	1.408	40	2
	for τ(BN) > 30° see BOGSUL, BUSHAY, CILRUK (1.434—1.530)							
	S_2–**B**–N–X_2	1.447	1.443	0.013	1.435	1.470	14	
B(4)–O	B^-–**O** in BO_4^-	1.468	1.468	0.022	1.453	1.479	24	
	for neutral B–O see Note 3							3
B(3)–O(2)	X_2–**B–O**–X	1.367	1.367	0.024	1.349	1.382	35	
B(*n*)–P	*n* = 4: **B–P**	1.922	1.927	0.027	1.900	1.954	10	
	n = 3: see BUPSIB10 (1.892, 1.893)							
B(4)–S	**B**(4)–**S**(3)	1.930	1.927	0.009	1.925	1.934	10	
	B(4)–**S**(2)	1.896	1.896	0.004	1.893	1.899	6	
B(3)–S	N–**B–S**$_2$	1.806	1.806	0.010	1.799	1.816	28	
	(=X–)(N–)**B–S**	1.851	1.854	0.013	1.842	1.859	10	
Br–Br	see BEPZEB, TPASTB	2.542	2.548	0.015	2.526	2.551	4	
Br–C	**Br–C***	1.966	1.967	0.029	1.951	1.983	100	4
	Br–C*sp*^{3} (cyclopropane)	1.910	1.910	0.010	1.900	1.914	8	
	Br–C*sp*^{2}	1.883	1.881	0.015	1.874	1.894	31	4
	Br–C*ar* (mono-Br + *m,p*-Br_2)	1.899	1.899	0.012	1.892	1.906	119	4
	Br–C*ar* (*o*-Br_2)	1.875	1.872	0.011	1.864	1.884	8	4
$^-$Br(2)–Cl	see TEACBR (2.362—2.402)							†
Br–I	see DTHIBR10 (2.646), TPHOSI (2.695)							
Br–N	see NBBZAM (1.843)							
Br–O	see CIYFOF	1.581	1.581	0.007	1.574	1.587	4	
Br–P	see CISTED (2.366)							
Br–S(2)	see BEMLIO (2.206)							†
Br–S(3)	see CIWYIQ (2.435, 2.453)							†
Br–S(3)$^+$	see THINBR (2.321)							†
Br–Se	see CIFZUM (2.508, 2.619)							
Br–Si	see BIZJAV (2.284)							
Br–Te	In Br_6Te^{2-} see CUGBAH (2.692—2.716)							
	Br–Te(4) see BETUTE10 (3.079, 3.015)							
	Br–Te(3) see BTUPTE (2.835)							
C*sp*^{3}–C*sp*^{3}	C#–CH_2–CH_3	1.513	1.514	0.014	1.507	1.523	192	
	(C#)$_2$–CH–CH_3	1.524	1.526	0.015	1.518	1.534	226	
	(C#)$_3$–C–CH_3	1.534	1.534	0.011	1.527	1.541	825	
	C#–CH_2–CH_2–C#	1.524	1.524	0.014	1.516	1.532	2 459	
	(C#)$_2$–CH–CH_2–C#	1.531	1.531	0.012	1.524	1.538	1 217	
	(C#)$_3$–C–CH_2–C#	1.538	1.539	0.010	1.533	1.544	330	
	(C#)$_2$–CH–CH–(C#)$_2$	1.542	1.542	0.011	1.536	1.549	321	
	(C#)$_3$–C–CH–(C#)$_2$	1.556	1.556	0.011	1.549	1.562	215	
	(C#)$_3$–C–C–(C#)$_3$	1.588	1.580	0.025	1.566	1.610	21	
	C*–C* (overall)	1.530	1.530	0.015	1.521	1.539	5 777	5,6
	in cyclopropane (any subst.)	1.510	1.509	0.026	1.497	1.523	888	7
	in cyclobutane (any subst.)	1.554	1.553	0.021	1.540	1.567	679	8
	in cyclopentane (C,H-subst.)	1.543	1.543	0.018	1.532	1.554	1 641	
	in cyclohexane (C,H-subst.)	1.535	1.535	0.016	1.525	1.545	2 814	
	cyclopropyl-C* (exocyclic)	1.518	1.518	0.019	1.505	1.531	366	7
	cyclobutyl-C* (exocyclic)	1.529	1.529	0.016	1.519	1.539	376	8
	cyclopentyl-C* (exocyclic)	1.540	1.541	0.017	1.527	1.549	956	
	cyclohexyl-C* (exocyclic)	1.539	1.538	0.016	1.529	1.549	2 682	
	in cyclobutene (any subst.)	1.573	1.574	0.017	1.566	1.586	25	8
	in cyclopentene (C,H-subst.)	1.541	1.539	0.015	1.532	1.549	208	
	in cyclohexene (C,H-subst.)	1.541	1.541	0.020	1.528	1.554	586	
	in oxirane (epoxide)	1.466	1.466	0.015	1.458	1.474	249	9
	in aziridine	1.480	1.481	0.021	1.465	1.496	67	9
	in oxetane	1.541	1.541	0.019	1.527	1.557	16	
	in azetidine	1.548	1.543	0.018	1.536	1.558	22	
	oxiranyl-C* (exocyclic)	1.509	1.507	0.018	1.497	1.519	333	9
	aziridinyl-C* (exocyclic)	1.512	1.512	0.018	1.496	1.526	13	9
C*sp*^{3}–C*sp*^{2}	CH_3–C=C	1.503	1.504	0.011	1.497	1.509	215	
	C#–CH_2–C=C	1.502	1.502	0.013	1.494	1.510	483	
	(C#)$_2$–CH–C=C	1.510	1.510	0.014	1.501	1.518	564	
	(C#)$_3$–C–C=C	1.522	1.522	0.016	1.511	1.533	193	
C*sp*^{3}–C*sp*^{2}	C*–C=C (overall)	1.507	1.507	0.015	1.499	1.517	1 456	5
	C*–C=C (endocyclic)							
	in cyclopropene	1.509	1.508	0.016	1.500	1.516	20	10
	in cyclobutene	1.513	1.512	0.018	1.500	1.525	50	8
	in cyclopentene	1.512	1.512	0.014	1.502	1.521	208	
	in cyclohexene	1.506	1.505	0.016	1.495	1.516	391	
	in cyclopentadiene	1.502	1.503	0.019	1.490	1.515	18	
	in cyclohexa-1,3-diene	1.504	1.504	0.017	1.491	1.517	56	
	C*–C=C (exocyclic):							
	cyclopropenyl-C*	1.478	1.475	0.012	1.470	1.485	7	10
	cyclobutenyl-C*	1.489	1.483	0.015	1.479	1.496	11	8

Bond	Substructure	d	m	σ	q_l	q_u	n	Note
	cyclopentenyl-C*	1.504	1.506	0.012	1.495	1.512	115	
	cyclohexenyl-C*	1.511	1.511	0.013	1.502	1.519	292	
	C*–CH=O in aldehydes	1.510	1.510	0.008	1.501	1.518	7	
	$(C^*)_2$–C=O							
	in ketones	1.511	1.511	0.015	1.501	1.521	952	11
	in cyclobutanone	1.529	1.530	0.016	1.514	1.545	18	
	in cyclopentanone	1.514	1.514	0.016	1.505	1.523	312	
	acyclic and 6+ rings	1.509	1.509	0.016	1.499	1.519	626	
	C*–COOH in carboxylic acids	1.502	1.502	0.014	1.495	1.510	176	
	C*–COO^- in carboxylate anions	1.520	1.521	0.011	1.516	1.528	57	
	C*–C(=O)(–OC*)							
	in acyclic esters	1.497	1.496	0.018	1.484	1.509	553	12
	in β-lactones	1.519	1.519	0.020	1.500	1.538	4	13
	in γ-lactones	1.512	1.512	0.015	1.501	1.521	110	12
	in δ-lactones	1.504	1.502	0.013	1.495	1.517	27	12
	cyclopropyl (C)–C=O in ketones, acids and esters	1.486	1.485	0.018	1.474	1.497	105	7
	C*–C(=O)(–NH_2) in acyclic amides	1.514	1.512	0.016	1.506	1.526	32	14
	C*–C(=O)(–NHC*) in acyclic amides	1.506	1.505	0.012	1.498	1.515	78	14
	C*–C(=O)[–N$(C^*)_2$] in acyclic amides	1.505	1.505	0.011	1.496	1.517	15	14
Csp^3–*Car*	CH_3–*Car*	1.506	1.507	0.011	1.501	1.513	454	
	C#–CH_2–*Car*	1.510	1.510	0.009	1.505	1.516	674	
	$(C\#)_2$–CH–*Car*	1.515	1.515	0.011	1.508	1.522	363	
	$(C\#)_3$–C–*Car*	1.527	1.530	0.016	1.517	1.539	308	
	C*–*Car* (overall)	1.513	1.513	0.014	1.505	1.521	1 813	
	cyclopropyl (C)–*Car*	1.490	1.490	0.015	1.479	1.503	90	7
Csp^3–Csp^1	C*–C≡C	1.466	1.465	0.010	1.460	1.469	21	15
	C#–C≡C	1.472	1.472	0.012	1.464	1.481	88	15
	C*–C≡N	1.470	1.469	0.013	1.463	1.479	106	7b
	cyclopropyl (C)–C≡N	1.444	1.447	0.010	1.436	1.451	38	7
Csp^2–Csp^2	C=C–C=C							
	(conjugated)	1.455	1.455	0.011	1.447	1.463	30	16,18
	(unconjugated)	1.478	1.476	0.012	1.470	1.479	8	17,18
	(overall)	1.460	1.460	0.015	1.450	1.470	38	
	C=C–C=C–C=C	1.443	1.445	0.013	1.431	1.454	29	18
	C=C–C=C (endocyclic in TCNQ)	1.432	1.433	0.012	1.424	1.441	280	19
	C=C–C(=O)(–C*)							
	(conjugated)	1.464	1.462	0.018	1.453	1.476	211	16,18
	(unconjugated)	1.484	1.486	0.017	1.475	1.497	14	17,18
	(overall)	1.465	1.462	0.018	1.453	1.478	226	
	C=C–C(=O)–C=C							
	in benzoquinone (C,H-subst. only)	1.478	1.476	0.011	1.469	1.488	28	
	in benzoquinone (any subst.)	1.478	1.478	0.031	1.464	1.498	172	
	non-quinonoid	1.456	1.455	0.012	1.447	1.464	28	
	C=C–COOH	1.475	1.476	0.015	1.461	1.488	22	
	C=C–COOC*	1.488	1.489	0.014	1.478	1.497	113	
	C=C–COO^-	1.502	1.499	0.017	1.488	1.510	11	
	HOOC–COOH	1.538	1.537	0.007	1.535	1.541	9	
	HOOC–COO^-	1.549	1.552	0.009	1.546	1.553	13	
	$^-$OOC–COO^-	1.564	1.559	0.022	1.554	1.568	9	
	formal Csp^2–Csp^2 single bond in selected non-fused heterocycles:							
	in 1*H*-pyrrole (C3–C4)	1.412	1.410	0.016	1.401	1.427	29	
	in furan (C3–C4)	1.423	1.423	0.016	1.412	1.433	62	
	in thiophene (C3–C4)	1.424	1.425	0.015	1.415	1.433	40	
	in pyrazole (C3–C4)	1.410	1.412	0.016	1.400	1.418	20	
	in isoxazole (C3–C4)	1.425	1.425	0.016	1.413	1.438	9	
	in furazan (C3–C4)	1.428	1.427	0.007	1.422	1.435	6	
	in furoxan (C3–C4)	1.417	1.417	0.006	1.412	1.422	14	
Csp^2–*Car*	C=C–*Car*							
	(conjugated)	1.470	1.470	0.015	1.463	1.480	37	16,18
Csp^2–*Car*		1.488	1.490	0.012	1.480	1.496	87	17,18
	(overall)	1.483	1.483	0.015	1.472	1.494	124	
	cyclopropenyl (C=C)–*Car*	1.447	1.448	0.006	1.441	1.452	8	10
	Car–C(=O)–C*	1.488	1.489	0.016	1.478	1.500	84	
	Car–C(=O)–*Car*	1.480	1.481	0.017	1.468	1.494	58	
	Car–COOH	1.484	1.485	0.014	1.474	1.491	75	
	Car–C(=O)(–OC*)	1.487	1.487	0.012	1.480	1.494	218	
	Car–COO^-	1.504	1.509	0.014	1.495	1.512	26	
	Car–C(=O)–NH_2	1.500	1.503	0.020	1.498	1.510	19	
	Car–C=N–C#							
	(conjugated)	1.476	1.478	0.014	1.466	1.486	27	16
	(unconjugated)	1.491	1.490	0.008	1.485	1.496	48	17
	(overall)	1.485	1.487	0.013	1.481	1.493	75	

BOND LENGTHS IN CRYSTALLINE ORGANIC COMPOUNDS (continued)

Bond	Substructure	*d*	*m*	σ	q_1	q_u	*n*	Note
	in indole (C3–C3a)	1.434	1.434	0.011	1.428	1.439	40	
Csp^2–Csp^1	C=C–C≡C	1.431	1.427	0.014	1.425	1.441	11	7b
	C=C–C≡N in TCNQ	1.427	1.427	0.010	1.420	1.433	280	19
Car–*Car*	in biphenyls (*ortho* subst. all H)	1.487	1.488	0.007	1.484	1.493	30	
	(≥1 non-H *ortho*-subst.)	1.490	1.491	0.010	1.486	1.495	212	
Car–Csp^1	*Car*–C≡C	1.434	1.436	0.006	1.430	1.437	37	
	Car–C≡N	1.443	1.444	0.008	1.436	1.448	31	
Csp^1–Csp^1	C≡C–C≡C	1.377	1.378	0.012	1.374	1.384	21	
Csp^2=Csp^2	C*–CH=CH_2	1.299	1.300	0.027	1.280	1.311	42	
	$(C^*)_2$–C=CH_2	1.321	1.321	0.013	1.313	1.328	77	
	C*–CH=CH–C*							
	(*cis*)	1.317	1.318	0.013	1.310	1.323	106	
	(*trans*)	1.312	1.311	0.011	1.304	1.320	19	
	(overall)	1.316	1.317	0.015	1.309	1.323	127	
	$(C^*)_2$–C=CH–C*	1.326	1.328	0.011	1.319	1.334	168	
	$(C^*)_2$–C=C–$(C^*)_2$	1.331	1.330	0.009	1.326	1.334	89	
	$(C^*,H)_2$–C=C–$(C^*,H)_2$ (overall)	1.322	1.323	0.014	1.315	1.331	493	5
	in cyclopropene (any subst.)	1.294	1.288	0.017	1.284	1.302	10	10
	in cyclobutene (any subst.)	1.335	1.335	0.019	1.324	1.347	25	8
	in cyclopentene (C,H-subst.)	1.323	1.324	0.013	1.314	1.331	104	
	in cyclohexene (C,H-subst.)	1.326	1.325	0.012	1.318	1.334	196	
	C=C=C (allenes, any subst.)	1.307	1.307	0.005	1.303	1.310	18	
	C=C–C=C (C,H subst., conjugated)	1.330	1.330	0.014	1.322	1.338	76	16
	C=C–C=C–C=C (C,H subst., conjugated)	1.345	1.345	0.012	1.337	1.350	58	16
	C=C-*Car* (C,H subst., conjugated)	1.339	1.340	0.011	1.334	1.346	124	16
	C=C in cyclopenta-1,3-diene (any subst.)	1.341	1.341	0.017	1.328	1.356	18	
	C=C in cyclohexa-1,3-diene (any subst.)	1.332	1.332	0.013	1.323	1.341	56	
	in C=C–C=O							
	(C,H subst., conjugated)	1.340	1.340	0.013	1.332	1.348	211	16,18
	(C,H subst., unconjugated)	1.331	1.330	0.008	1.326	1.339	14	17,18
	(C,H subst., overall)	1.340	1.339	0.013	1.332	1.348	226	
	in cyclohexa-2,5-dien-1-ones	1.329	1.327	0.011	1.321	1.335	28	
	in *p*-benzoquinones							
	(C*,H subst.)	1.333	1.337	0.011	1.325	1.338	14	
	(any subst.)	1.349	1.339	0.030	1.330	1.364	86	
	in TCNQ							
	(endocyclic)	1.352	1.353	0.010	1.345	1.358	142	19
	(exocyclic)	1.392	1.391	0.017	1.379	1.405	139	19
	C=C–OH in enol tautomers	1.362	1.360	0.020	1.349	1.370	54	
	in heterocycles (any subst.):							
	1*H*-pyrrole (C2–C3, C4–C5)	1.375	1.377	0.018	1.361	1.388	58	
	furan (C2–C3, C4–C5)	1.341	1.342	0.021	1.329	1.351	125	
	thiophene (C2–C3, C4–C5)	1.362	1.359	0.025	1.346	1.377	60	
	pyrazole (C4–C5)	1.369	1.372	0.019	1.362	1.383	20	
	imidazole (C4–C5)	1.360	1.361	0.014	1.352	1.367	44	
	isoxazole (C4–C5)	1.341	1.336	0.012	1.331	1.355	9	
	indole (C2–C3)	1.364	1.363	0.012	1.355	1.371	40	
Car≃*Car*	in phenyl rings with C*,H subst. only							
	H–C≃C–H	1.380	1.381	0.013	1.372	1.388	2 191	
	C*–C≃C–H	1.387	1.388	0.010	1.382	1.393	891	
	C*–C≃C–C*	1.397	1.397	0.009	1.392	1.403	182	
	C≃C (overall)	1.384	1.384	0.013	1.375	1.391	3 264	
	F–C≃C–F	1.372	1.374	0.011	1.366	1.380	84	4
	Cl–C≃C–Cl	1.388	1.389	0.014	1.380	1.398	152	4
	in naphthalene (D_{2h}, any subst.)							
	C1–C2	1.364	1.364	0.014	1.356	1.373	440	
	C2–C3	1.406	1.406	0.014	1.397	1.415	218	
	C1–C8a	1.420	1.419	0.012	1.412	1.426	440	
	C4a–C8a	1.422	1.424	0.011	1.417	1.429	109	
Car≃*Car*	in anthracene (D_{2h}, any subst.)							
	C1–C2	1.356	1.356	0.009	1.350	1.360	56	
	C2–C3	1.410	1.410	0.010	1.401	1.416	34	
	C1–C9a	1.430	1.430	0.006	1.426	1.434	56	
	C4a–C9a	1.435	1.436	0.007	1.429	1.440	34	
	C9–C9a	1.400	1.402	0.009	1.395	1.406	68	
	in pyridine (C,H subst.)	1.379	1.381	0.012	1.371	1.387	276	20
	(any subst.)	1.380	1.380	0.015	1.371	1.389	537	20
	in pyridinium cation							
	(N^+–H; C,H subst. on C)							
	C2–C3	1.373	1.375	0.012	1.368	1.380	30	
	C3–C4	1.379	1.380	0.011	1.371	1.388	30	
	(N^+–X; C,H subst. on C)							
	C2–C3	1.373	1.372	0.019	1.362	1.382	151	
	C3–C4	1.383	1.385	0.019	1.372	1.394	151	

Bond	Substructure	*d*	*m*	σ	q_1	q_u	*n*	Note
	in pyrazine (H subst. on C)	1.379	1.377	0.010	1.370	1.388	10	
	(any subst. on C)	1.405	1.405	0.024	1.388	1.420	60	
	in pyrimidine (C,H subst. on C)	1.387	1.389	0.018	1.379	1.400	28	
Csp^1≡*Csp*^{1}	X–C≡C–X	1.183	1.183	0.014	1.174	1.193	119	15
	C,H–C≡C–C,H	1.181	1.181	0.014	1.173	1.192	104	15
	in C≡C–C(*sp*^{2},*ar*)	1.189	1.193	0.010	1.181	1.195	38	15
	in C≡C–C≡C	1.192	1.192	0.010	1.187	1.197	42	15
	in CH≡C–C#	1.174	1.174	0.011	1.167	1.180	42	15
Csp^3–Cl	Omitting 1,2-dichlorides:							
	C–CH$_2$–Cl	1.790	1.790	0.007	1.783	1.795	13	4
	C$_2$–CH–Cl	1.803	1.802	0.003	1.800	1.807	8	4
	C$_3$–C–Cl	1.849	1.856	0.011	1.837	1.858	5	4
	X–CH$_2$–Cl (X = C,H,N,O)	1.790	1.791	0.011	1.783	1.797	37	4
	X$_2$–CH–Cl (X = C,H,N,O)	1.805	1.803	0.014	1.800	1.812	26	4
	X$_3$–C–Cl (X = C,H,N,O)	1.843	1.838	0.014	1.835	1.858	7	4
	X$_2$–C–Cl$_2$ (X = C,H,N,O)	1.779	1.776	0.015	1.769	1.790	18	4
	X–C–Cl$_3$ (X = C,H,N,O)	1.768	1.765	0.011	1.761	1.776	33	4
	Cl–CH(–C)–CH(–C)–Cl	1.793	1.793	0.013	1.786	1.800	66	4
	Cl–C(–C$_2$)–C(–C$_2$)–Cl	1.762	1.760	0.010	1.757	1.765	54	4
	cyclopropyl–Cl	1.755	1.756	0.011	1.749	1.763	64	
Csp^2–Cl	C=C–Cl (C,H,N,O subst. on C)	1.734	1.729	0.019	1.719	1.748	63	4
	C=C–Cl$_2$ (C,H,N,O subst. on C)	1.720	1.716	0.013	1.708	1.729	20	4
	Cl–C=C–Cl	1.713	1.711	0.011	1.705	1.720	80	4
Car–Cl	*Car*–Cl (mono-Cl + *m,p*-Cl$_2$)	1.739	1.741	0.010	1.734	1.745	340	4
	Car–Cl (*o*-Cl$_2$)	1.720	1.720	0.010	1.713	1.717	364	4
Csp^1–Cl	see HCLENE10 (1.634, 1.646)							
Csp^3–F	Omitting 1,2-difluorides							
	C–CH$_2$–F and C$_2$–CH–F	1.399	1.399	0.017	1.389	1.408	25	4
	C$_3$–C–F	1.428	1.431	0.009	1.421	1.435	11	4
	(C*,H)$_2$–C–F$_2$	1.349	1.347	0.012	1.342	1.356	58	4
	C*–C–F$_3$	1.336	1.334	0.007	1.330	1.344	12	4
	F–C*–C*–F	1.371	1.374	0.007	1.362	1.375	26	4
	X$_3$–C–F (X = C,H,N,O)	1.386	1.389	0.033	1.373	1.408	70	4
	X$_2$–C–F$_2$ (X = C,H,N,O)	1.351	1.349	0.013	1.342	1.356	58	4
	X–C–F$_3$ (X = C,H,N,O)	1.322	1.323	0.015	1.314	1.332	309	4
	F–C(–X)$_2$–C(–X)$_2$–F (X = C,H,N,O)	1.373	1.374	0.009	1.362	1.377	30	4
	F–C(–X)$_2$–NO$_2$ (X = any subst.)	1.320	1.319	0.009	1.312	1.327	18	
Csp^2–F	C=C–F (C,H,N,O subst. on C)	1.340	1.340	0.013	1.334	1.346	34	4
Car–F	*Car*–F (mono-F + *m,p*-F$_2$)	1.363	1.362	0.008	1.357	1.368	38	4
	Car–F (*o*-F$_2$)	1.340	1.340	0.009	1.336	1.344	167	4
Csp^3–H	C–C–H$_3$ (methyl)	1.059	1.061	0.030	1.039	1.083	83	21
	C$_2$–C–H$_2$ (primary)	1.092	1.095	0.013	1.088	1.099	100	21
	C$_3$–C–H (secondary)	1.099	1.097	0.004	1.095	1.103	14	21
	C$_{2,3}$–C–H (primary and secondary)	1.093	1.095	0.012	1.089	1.100	118	21
	X–C–H$_3$ (methyl)	1.066	1.074	0.028	1.049	1.087	160	21
	X$_2$–C–H$_2$ (primary)	1.092	1.095	0.012	1.088	1.099	230	21
	X$_3$–C–H (secondary)	1.099	1.099	0.007	1.095	1.103	117	21
	X$_{2,3}$–C–H (primary and secondary)	1.094	1.096	0.011	1.091	1.100	348	21
Csp^2–H	C–C=C–H	1.077	1.079	0.012	1.074	1.085	14	21
Car–H	*Car*–H	1.083	1.083	0.011	1.080	1.087	218	21
Csp^3–I	C*–I	2.162	2.159	0.015	2.149	2.179	15	4
Car–I	*Car*–I	2.095	2.095	0.015	2.089	2.104	51	4
Csp^3–N(4)	C*–NH$_3^+$	1.488	1.488	0.013	1.482	1.495	298	
	(C*)$_2$–NH$_2^+$	1.494	1.493	0.016	1.484	1.503	249	
	(C*)$_3$–NH$^+$	1.502	1.502	0.015	1.491	1.512	509	
	(C*)$_4$–N$^+$	1.510	1.509	0.020	1.496	1.523	319	
	C*–N$^+$ (overall)	1.499	1.498	0.018	1.488	1.510	1 370	
Csp^3–N(3)	C*–N$^+$ in N-subst. pyridinium	1.485	1.484	0.009	1.477	1.490	32	
	C*–NH$_2$ (N*sp*^{3}: pyramidal)	1.469	1.470	0.010	1.462	1.474	19	22
	(C*)$_2$–NH (N*sp*^{3}: pyramidal)	1.469	1.467	0.012	1.461	1.477	152	5,22
	(C*)$_3$–N (N*sp*^{3}: pyramidal)	1.469	1.468	0.014	1.460	1.476	1 042	5,22
	C*–N*sp*^{3} (overall)	1.469	1.468	0.014	1.460	1.476	1 201	
	Csp^3–N*sp*^{3}							
	in aziridine	1.472	1.471	0.016	1.464	1.482	134	
	in azetidine	1.484	1.481	0.018	1.472	1.495	21	
	in tetrahydropyrrole	1.475	1.473	0.016	1.464	1.483	66	
	in piperidine	1.473	1.473	0.013	1.460	1.479	240	
	Csp^3–N*sp*^{2} (N planar) in:							23
	acyclic amides C*–NH–C=O	1.454	1.451	0.011	1.446	1.461	78	14
	β-lactams C*–N(–X)–C=O (endo)	1.464	1.465	0.012	1.458	1.475	23	13
	γ-lactams							
	C*–NH–C=O (endo)	1.457	1.458	0.011	1.449	1.465	20	13
	C*–N(–C*)–C=O (endo)	1.462	1.461	0.010	1.453	1.466	15	13
	C*–N(–C*)–C=O (exo)	1.458	1.456	0.014	1.448	1.465	15	13

Bond	Substructure	*d*	*m*	σ	q_1	q_u	*n*	Note
	δ-lactams							
	C*–NH–C=O (endo)	1.478	1.472	0.016	1.467	1.491	6	14
	C*–N(–C*)–C=O (endo)	1.479	1.476	0.007	1.475	1.482	15	14
	C*–N(–C*)–C=O (exo)	1.468	1.471	0.009	1.462	1.477	15	14
	nitro compounds (1,2-dinitro omitted):							
	C–CH$_2$–NO$_2$	1.485	1.483	0.020	1.478	1.502	8	
	C$_2$–CH–NO$_2$	1.509	1.509	0.011	1.502	1.511	12	
	C$_3$–C–NO$_2$	1.533	1.533	0.013	1.530	1.539	17	
	C$_2$–C–(NO$_2$)$_2$	1.537	1.536	0.016	1.525	1.550	19	
	1,2-dinitro: NO$_2$–C*–C*–NO$_2$	1.552	1.550	0.023	1.536	1.572	32	
Csp^3–N(2)	C#–N=N	1.493	1.493	0.020	1.477	1.506	54	
	C*–N=C–*Car*	1.465	1.468	0.011	1.461	1.472	75	
Csp^2–N(3)	C=C–NH$_2$ Nsp^2 planar	1.336	1.344	0.017	1.317	1.348	10	23
	C=C–NH–C# Nsp^2 planar	1.339	1.340	0.016	1.327	1.351	17	23
	C=C–N–(C#)$_2$							
	Nsp^2 planar	1.355	1.358	0.014	1.341	1.363	22	23
	Nsp^3 pyramidal	1.416	1.418	0.018	1.397	1.432	18	22
	Csp^2–Nsp^2 (N planar) in:							23
	acyclic amides							
	NH$_2$–C=O	1.325	1.323	0.009	1.318	1.331	32	14
	C*–NH–C=O	1.334	1.333	0.011	1.326	1.343	78	14
	(C*)$_2$–N–C=O	1.346	1.342	0.011	1.339	1.356	5	14
	β-lactams C*–NH–C=O	1.385	1.388	0.019	1.374	1.396	23	13
	γ-lactams							
	C*–NH–C=O	1.331	1.331	0.011	1.326	1.337	20	13
	C*–N(–C*)–C=O	1.347	1.344	0.014	1.335	1.359	15	13
	δ-lactams							
	C*–NH–C=O	1.334	1.334	0.006	1.330	1.339	6	14
	C*–N(–C*)–C=O	1.352	1.353	0.010	1.344	1.356	15	14
	peptides C#–N(–X)–C(–C#)(=O)	1.333	1.334	0.013	1.326	1.340	380	24
	ureas							
	(NH$_2$)$_2$–C=O	1.334	1.334	0.008	1.329	1.339	48	25,26
	(C#–NH)$_2$–C=O	1.347	1.345	0.010	1.341	1.354	26	25
	[(C#)$_n$–N]$_2$–C=O	1.363	1.359	0.014	1.354	1.370	40	25,27
	thioureas	1.346	1.343	0.023	1.328	1.361	192	
	(X$_2$N)$_2$–C=S							
	imides							
	[C#–C(=O)]$_2$–NH	1.376	1.377	0.012	1.369	1.383	64	
	[C#–C(=O)]$_2$–N–C#	1.389	1.383	0.017	1.376	1.404	38	
	[Csp^2–C(=O)]$_2$–N–C#	1.396	1.396	0.010	1.389	1.403	46	
	[Csp^2–C(=O)]$_2$–N–Csp^2	1.409	1.406	0.020	1.391	1.419	28	
	guanidinium [C–(NH$_2$)$_3$]$^+$ (unsubst.)	1.321	1.320	0.008	1.314	1.327	39	
	(any subst.)	1.328	1.325	0.015	1.317	1.333	140	
	in heterocyclic systems (any subst.)							
	1*H*-pyrrole (N1–C2, N1–C5)	1.372	1.374	0.016	1.363	1.384	58	
	indole (N1–C2)	1.370	1.370	0.012	1.364	1.377	40	
	pyrazole (N1–C5)	1.357	1.359	0.012	1.347	1.365	20	
	imidazole (N1–C2)	1.349	1.349	0.018	1.338	1.358	44	
	imidazole (N1–C5)	1.370	1.370	0.010	1.365	1.377	44	
Csp^2–N(2)	in imidazole (N3–C4)	1.376	1.377	0.011	1.369	1.384	44	
Car–N(4)	*Car*–N$^+$–(C,H)$_3$	1.465	1.466	0.007	1.461	1.470	23	
Car–N(3)	*Car*–NH$_2$							
	(Nsp^2: planar)	1.355	1.360	0.020	1.340	1.372	33	23
	(Nsp^3: pyramidal)	1.394	1.396	0.011	1.385	1.403	25	22
	(overall)	1.375	1.377	0.025	1.363	1.394	98	28
Car–N(3)	*Car*–NH–C#							
	(Nsp^2: planar)	1.353	1.353	0.007	1.347	1.359	16	23
	(Nsp^3: pyramidal)	1.419	1.423	0.017	1.412	1.432	8	22
	(overall)	1.380	1.364	0.032	1.353	1.412	31	28
	Car–N–(C#)$_2$							
	(Nsp^2: planar)	1.371	1.370	0.016	1.363	1.382	41	23
	(Nsp^3: pyramidal)	1.426	1.425	0.011	1.421	1.431	22	22
	(overall)	1.390	1.385	0.030	1.366	1.420	69	28
	in indole (N1–C7a)	1.372	1.372	0.007	1.367	1.376	40	
	Car–NO$_2$	1.468	1.469	0.014	1.460	1.476	556	
Car–N(2)	*Car*–N=N	1.431	1.435	0.020	1.422	1.442	26	
Csp^2=N(3)	in furoxan ($^+$N2=C3)	1.316	1.316	0.009	1.311	1.324	14	
Csp^2=N(2)	*Car*–C=N–C#	1.279	1.279	0.008	1.275	1.285	75	
	(C,H)$_2$–C=N–OH in oximes	1.281	1.280	0.013	1.273	1.288	67	
	S–C=N–X	1.302	1.302	0.021	1.285	1.319	36	
	in pyrazole (N2=C3)	1.329	1.331	0.014	1.315	1.339	20	
	in imidazole (C2=N3)	1.313	1.314	0.011	1.307	1.319	44	
	in isoxazole (N2=C3)	1.314	1.315	0.009	1.305	1.320	9	

Bond	Substructure	*d*	*m*	σ	q_l	q_u	*n*	Note
	in furazan (N2=C3, C4=N5)	1.298	1.299	0.006	1.294	1.303	12	
	in furoxan (C4=N5)	1.304	1.306	0.008	1.300	1.308	14	
Car≃N(3)	C≃N^+–H (pyrimidinium)	1.335	1.334	0.015	1.325	1.342	30	
	C≃N^+–C* (pyrimidinium)	1.346	1.346	0.010	1.340	1.352	64	
	C≃N^+–O^- (pyrimidinium)	1.362	1.359	0.013	1.353	1.369	56	
Car≃N(2)	C≃N (pyridine)	1.337	1.338	0.012	1.330	1.344	269	
	C≃N (pyrazine)	1.336	1.335	0.022	1.319	1.347	120	
	C≃N≃C (pyrimidine)	1.339	1.338	0.015	1.333	1.342	28	
	N≃C≃N (pyrimidine)	1.333	1.335	0.013	1.326	1.337	28	
	C≃N (pyrimidine) (overall)	1.336	1.337	0.014	1.331	1.339	56	
	in any 6-membered N-containing aromatic ring:							
	H–C≃N≃C–H	1.334	1.334	0.014	1.327	1.341	146	
	H–C≃N≃C–C*	1.339	1.341	0.013	1.336	1.345	38	
	C*–C≃N≃C–C*	1.345	1.345	0.008	1.342	1.348	24	
	C≃N≃C (overall)	1.336	1.337	0.014	1.329	1.344	204	
Csp^1≡N(2)	X–S–N≡C^- (isothiocyanide)	1.144	1.147	0.006	1.140	1.148	6	
Csp^1≡N(1)	C*–C≡N	1.136	1.137	0.010	1.131	1.142	140	
	C=C–C≡N in TCNQ	1.144	1.144	0.008	1.139	1.149	284	19
	Car–C≡N	1.138	1.138	0.007	1.133	1.143	31	
	X–C≡N	1.144	1.141	0.012	1.138	1.151	10	
	$(S–C≡N)^-$	1.155	1.156	0.012	1.147	1.165	14	
Csp^3–O(2)	in alcohols							
	CH_3–OH	1.413	1.414	0.018	1.395	1.425	17	
	C–CH_2–OH	1.426	1.426	0.011	1.420	1.431	75	
	C_2–CH–OH	1.432	1.431	0.011	1.425	1.439	266	
	C_3–C–OH	1.440	1.440	0.012	1.432	1.449	106	
	C*–OH (overall)	1.432	1.431	0.013	1.424	1.441	464	
	in dialkyl ethers							29
	CH_3–O–C*	1.416	1.418	0.016	1.405	1.426	110	
	C–CH_2–O–C*	1.426	1.424	0.011	1.418	1.435	34	
	C_2–CH–O–C*	1.429	1.430	0.010	1.420	1.437	53	
	C_3–C–O–C*	1.452	1.450	0.011	1.445	1.458	39	
	C*–O–C* (overall)	1.426	1.425	0.019	1.414	1.437	236	5
	in aryl alkyl ethers							29
	CH_3–O–*Car*	1.424	1.424	0.012	1.417	1.431	616	
	C–CH_2–O–*Car*	1.431	1.430	0.013	1.422	1.438	188	
	C_2–CH–O–*Car*	1.447	1.446	0.020	1.435	1.466	58	
	C_3–C–O–*Car*	1.470	1.469	0.018	1.456	1.483	55	
	C*–O–*Car* (overall)	1.429	1.427	0.018	1.419	1.436	917	
	in alkyl esters of carboxylic acids							12,29
	CH_3–O–C(=O)–C*	1.448	1.449	0.010	1.442	1.455	200	
	C–CH_2–O–C(=O)–C*	1.452	1.453	0.009	1.445	1.458	32	
	C_2–CH–O–C(=O)–C*	1.460	1.460	0.010	1.454	1.465	78	
	C_3–C–O–C(=O)–C*	1.477	1.475	0.008	1.472	1.484	6	
	C*–O–C(=O)–C* (overall)	1.450	1.451	0.014	1.442	1.459	314	
	in alkyl esters of α,β-unsaturated acids:							
	C*–O–C(=O)–C=C (overall)	1.453	1.452	0.013	1.444	1.459	112	
	in alkyl esters of benzoic acid							
	C*–O–C(=O)–C(phenyl) (overall)	1.454	1.454	0.012	1.446	1.463	219	
	in ring systems							
	oxirane (epoxides) (any subst.)	1.446	1.446	0.014	1.438	1.456	498	9
	oxetane (any subst.)	1.463	1.460	0.015	1.451	1.474	16	
	tetrahydrofuran (C,H subst.)	1.442	1.441	0.017	1.430	1.451	154	
Csp^3–O(2)	tetrahydropyran (C,H subst.)	1.441	1.442	0.015	1.431	1.451	22	
	β-lactones: C*–O–C(=O)	1.492	1.494	0.010	1.481	1.501	4	16
	γ-lactones: C*–O–C(=O)	1.464	1.464	0.012	1.455	1.473	110	12
	δ-lactones: C*–O–C(=O)	1.461	1.464	0.017	1.452	1.473	27	12
	O–C–O system in *gem*-diols, and pyranose and furanose sugars:							30,31
	HO–C*–OH	1.397	1.401	0.012	1.388	1.405	18	
	C_5–O_5–C_1–O_1H in pyranoses							
	O_1 axial (α):							
	C_5–O_5	1.439	1.440	0.008	1.432	1.445	29	
	O_5–C_1	1.427	1.426	0.012	1.421	1.432	29	
	C_1–O_1	1.403	1.400	0.012	1.391	1.412	29	
	O_1 equatorial (β):							
	C_5–O_5	1.435	1.436	0.008	1.429	1.440	17	
	O_5–C_1	1.430	1.431	0.010	1.424	1.436	17	
	C_1–O_1	1.393	1.393	0.007	1.386	1.399	17	
	α + β (overall):							
	C_5–O_5	1.439	1.440	0.008	1.432	1.446	60	
	O_5–C_1	1.430	1.429	0.012	1.421	1.436	60	
	C_1–O_1	1.401	1.399	0.011	1.392	1.407	60	

Bond	Substructure	*d*	*m*	σ	q_1	q_u	*n*	Note
	C_4–O_4–C_1–O_1H in furanoses (overall values)							
	C_4–O_4	1.442	1.446	0.012	1.436	1.449	18	
	O_4–C_1	1.432	1.432	0.012	1.421	1.443	18	
	C_1–O_1	1.404	1.405	0.013	1.397	1.409	18	
	C_5–O_5–C_1–O_1–C* in pyranoses							
	O_1 axial (α):							
	C_5–O_5	1.439	1.438	0.010	1.433	1.446	67	
	O_5–C_1	1.417	1.417	0.009	1.410	1.424	67	
	C_1–O_1	1.409	1.409	0.014	1.401	1.417	67	
	O_1–C*	1.435	1.435	0.013	1.427	1.443	67	
	O_1 equatorial (β):							
	C_5–O_5	1.434	1.435	0.006	1.429	1.439	39	
	O_5–C_1	1.424	1.424	0.008	1.418	1.431	39	
	C_1–O_1	1.390	1.390	0.011	1.381	1.400	39	
	O_1–C*	1.437	1.438	0.013	1.428	1.445	39	
	α + β (overall):							
	C_5–O_5	1.436	1.436	0.009	1.431	1.442	126	
	O_5–C_1	1.419	1.419	0.011	1.412	1.426	126	
	C_1–O_1	1.402	1.403	0.016	1.391	1.413	126	
	O_1–C*	1.436	1.436	0.013	1.428	1.445	126	
	C_4–O_4–C_1–O_1–C* in furanoses (overall values)							
	C_4–O_4	1.443	1.445	0.013	1.429	1.453	23	
	O_4–C_1	1.421	1.418	0.012	1.413	1.431	23	
	C_1–O_1	1.410	1.409	0.014	1.401	1.420	23	
	O_1–C*	1.439	1.437	0.014	1.429	1.449	23	
	Miscellaneous:							
	C#–O–SiX_3	1.416	1.416	0.017	1.405	1.428	29	
	C*–O–SO_2–C	1.465	1.461	0.014	1.454	1.475	33	
Csp^2–O(2)	in enols: C=C–OH	1.333	1.331	0.017	1.324	1.342	53	
	in enol esters: C=C–O–C*	1.354	1.353	0.016	1.341	1.363	40	
	in acids:							
	C*–C(=O)–OH	1.308	1.311	0.019	1.298	1.320	174	
	C=C–C(=O)–OH	1.293	1.295	0.019	1.279	1.307	22	
	C*ar*-C(=O)–OH	1.305	1.311	0.020	1.291	1.317	75	
	in esters:							
	C*–C(=O)–O–C*	1.336	1.337	0.014	1.328	1.346	551	12,29
	C=C–C(=O)–O–C*	1.332	1.331	0.011	1.324	1.339	112	
	C*ar*–C(=O)–O–C*	1.337	1.335	0.013	1.329	1.344	219	12
	C*–C(=O)–O–C=C	1.362	1.359	0.018	1.351	1.374	26	
	C*–C(=O)–O–C=C	1.407	1.405	0.017	1.394	1.420	26	
	C*–C(=O)–O–C*ar*	1.360	1.359	0.011	1.355	1.367	40	12
	in anhydrides: O=C–O–C=O	1.386	1.386	0.011	1.379	1.393	70	
	in ring systems:							
	furan (O1–C2, O1–C5)	1.368	1.369	0.015	1.359	1.377	125	
	isoxazole (O1–C5)	1.354	1.354	0.010	1.345	1.360	9	
	β-lactones: C*–C(=O)–O–C*	1.359	1.359	0.013	1.348	1.371	4	13
	γ-lactones: C*–C(=O)–O–C*	1.350	1.349	0.012	1.342	1.359	110	12
	δ-lactones: C*–C(=O)–O–C*	1.339	1.339	0.016	1.332	1.347	27	12
C*ar*–O(2)	in phenols: C*ar*–OH	1.362	1.364	0.015	1.353	1.373	551	
	in aryl alkyl ethers: C*ar*–O–C*	1.370	1.370	0.011	1.363	1.377	920	29,32
C*ar*–O(2)	in diaryl ethers: C*ar*–O–C*ar*	1.384	1.381	0.014	1.375	1.391	132	
	in esters: C*ar*–O–C(=O)–C*	1.401	1.401	0.010	1.394	1.408	40	12
Csp^2=O(1)	in aldehydes and ketones:							
	C*–CH=O	1.192	1.192	0.005	1.188	1.197	7	
	$(C^*)_2$–C=O	1.210	1.210	0.008	1.206	1.215	474	5
	$(C\#)_2$–C=O							
	in cyclobutanones	1.198	1.198	0.007	1.194	1.204	12	
	in cyclopentanones	1.208	1.208	0.007	1.203	1.212	155	
	in cyclohexanones	1.211	1.211	0.009	1.207	1.216	312	
	C=C–C=O	1.222	1.222	0.010	1.216	1.229	225	
	$(C{=}C)_2$–C=O	1.233	1.229	0.010	1.226	1.242	28	
	C*ar*–C=O	1.221	1.218	0.014	1.212	1.229	85	
	$(Car)_2$–C=O	1.230	1.226	0.015	1.220	1.238	66	
	C=O in benzoquinones	1.222	1.220	0.013	1.211	1.231	86	
	delocalized double bonds in carboxylate anions:							
	H–C≃O_2^- (formate)	1.242	1.243	0.012	1.234	1.252	24	
	C*–C≃O_2^-	1.254	1.253	0.010	1.247	1.261	114	
	C=C–C≃O_2^-	1.250	1.248	0.017	1.238	1.261	52	
	C*ar*–C≃O_2^-	1.255	1.253	0.010	1.249	1.262	22	
	HOOC–C≃O_2^- (hydrogen oxalate)	1.243	1.247	0.015	1.232	1.256	26	
	$^-O_2$≃C–C≃O_2– (oxalate)	1.251	1.251	0.007	1.248	1.254	18	

Bond	Substructure	*d*	*m*	σ	q_l	q_u	*n*	Note
	in carboxylic acids (X–COOH)							
	C*–C(=**O**)–OH	1.214	1.214	0.019	1.203	1.224	175	
	C=C–C(=**O**)–OH	1.229	1.226	0.017	1.218	1.237	22	
	Car–C(=**O**)–OH	1.226	1.223	0.020	1.211	1.241	75	
	in esters:							
	C*–C(=**O**)–O–C*	1.196	1.196	0.010	1.190	1.202	551	12
	C=C–C(=**O**)–O–C*	1.199	1.198	0.009	1.193	1.203	113	
	Car–C(=**O**)–O–C*	1.202	1.201	0.009	1.196	1.207	218	12
	C*–C(=**O**)–O–C=C	1.190	1.190	0.014	1.184	1.198	26	
	C*–C(=**O**)–O–*Car*	1.187	1.188	0.011	1.181	1.195	40	12
	in anhydrides: **O**=C–O–C=**O**	1.187	1.187	0.010	1.184	1.193	70	
	in β-lactones: C*–C(=**O**)–O–C*	1.193	1.193	0.006	1.187	1.198	4	13
	γ-lactones: C*–C(=**O**)–O–C*	1.201	1.202	0.009	1.196	1.206	109	12
	δ-lactones: C*–C(=**O**)–O–C*	1.205	1.207	0.008	1.201	1.209	27	12
	in amides:							
	NH_2-C(–C*)=**O**	1.234	1.233	0.012	1.225	1.243	32	14
	(C*–)(C*,H–)N–C(–C*)=**O**	1.231	1.231	0.012	1.224	1.238	378	14
	β-lactams: C*–NH–C=**O**	1.198	1.200	0.012	1.193	1.204	23	13
	γ-lactams:							
	C*–NH–C=**O**	1.235	1.235	0.008	1.232	1.240	20	13
	C*–N(–C*)–C=**O**	1.225	1.226	0.011	1.217	1.233	15	13
	δ-lactams:							
	C*–NH–C=**O**	1.240	1.241	0.003	1.237	1.243	6	14
	C*–N(–C*)–C=**O**	1.233	1.233	0.007	1.229	1.239	15	14
	in ureas:							
	$(NH)_2)_2$–C=**O**	1.256	1.256	0.007	1.249	1.261	24	25,26
	$(C\#–NH)_2$–C=**O**	1.241	1.237	0.011	1.235	1.245	13	25
	$[(C\#)_n–N]_2$–C=**O**	1.230	1.230	0.007	1.224	1.234	20	25,27
Csp^3–P(4)	C_3–P^+–C*	1.800	1.802	0.015	1.790	1.812	35	33
	C_2–P(=O)–CH_3	1.791	1.790	0.006	1.786	1.795	10	
	C_2–P(=O)–CH_2–C	1.806	1.806	0.009	1.801	1.813	45	
	C_2–P(=O)–CH–C_2	1.821	1.821	0.009	1.815	1.828	15	
	C_2–P(=O)–C–C_3	1.841	1.842	0.008	1.835	1.847	14	
	C_2–P(=O)–C* (overall)	1.813	1.811	0.017	1.800	1.822	84	
Csp^3–P(3)	C_2–P–C*	1.855	1.857	0.019	1.840	1.870	23	
Car–P(4)	C_3–P^+–*Car*	1.793	1.792	0.011	1.786	1.800	276	
	C_2–P(=O)–*Car*	1.801	1.802	0.011	1.796	1.807	98	
	Ph_3–P=N^+=P–Ph_3	1.795	1.795	0.008	1.789	1.800	197	
Car–P(3)	C_2–P–*Car*	1.836	1.837	0.010	1.830	1.844	102	
	$(N\simeq)_2$P-*Car* (P≃N aromatic)	1.795	1.793	0.011	1.788	1.803	43	
Csp^3–S(4)	C*–SO_2–C (C* = CH_3 excluded)	1.786	1.782	0.018	1.774	1.797	75	
	C*–SO_2–C (overall)	1.779	1.778	0.020	1.764	1.790	94	
	C*–SO_2–O–X	1.745	1.744	0.009	1.738	1.754	7	34
	C*–SO_2–N–X_2	1.758	1.756	0.018	1.746	1.773	17	34
Csp^3–S(3)	C*–S(=O)–C (C* = CH_3 excluded)	1.818	1.814	0.024	1.802	1.829	69	
	C*–S(=O)–C (overall)	1.809	1.806	0.025	1.793	1.820	88	
	CH_3–S^+–X_2	1.786	1.787	0.007	1.779	1.792	21	
	C*–S^+–X_2 (C* = CH_3 excluded)	1.823	1.820	0.016	1.812	1.834	18	
	C*–S^+–X_2 (overall)	1.804	1.794	0.025	1.788	1.820	41	
Csp^3–S(2)	C*–SH	1.808	1.805	0.010	1.800	1.819	6	
	CH_3–S–C*	1.789	1.787	0.008	1.784	1.794	9	
Csp^3–S(2)	C-CH_2–S–C*	1.817	1.816	0.013	1.808	1.824	92	
	C_2–CH–S–C*	1.819	1.819	0.011	1.811	1.825	32	
	C_3–C–S–C*	1.856	1.860	0.011	1.854	1.863	26	
	C*–S–C* (overall)	1.819	1.817	0.019	1.809	1.827	242	
	in thiirane	1.834	1.835	0.025	1.810	1.858	4	9
	in thietane: see ZCMXSP (1.817, 1.844)							
	in tetrahydrothiophene	1.827	1.826	0.018	1.811	1.837	20	
	in tetrahydrothiopyran	1.823	1.821	0.014	1.812	1.832	24	
	C–CH_2–S–S–X	1.823	1.820	0.014	1.813	1.832	41	
	C_3–C–S–S–X	1.863	1.865	0.015	1.848	1.878	11	
	C*–S–S–X (overall)	1.833	1.828	0.022	1.818	1.848	59	
Csp^2–S(2)	C=C–S–C*	1.751	1.755	0.017	1.740	1.764	61	
	C=C–S–C=C (in tetrathiafulvalene)	1.741	1.741	0.011	1.733	1.750	88	
	C=C–S–C=C (in thiophene)	1.712	1.712	0.013	1.703	1.722	60	
	O=C–S–C#	1.762	1.759	0.018	1.747	1.778	20	
Car–S(4)	*Car*–SO_2–C	1.763	1.764	0.009	1.756	1.769	96	
	Car–SO_2–O–X	1.752	1.750	0.008	1.749	1.756	27	
	Car–SO_2–N–X_2	1.758	1.759	0.013	1.749	1.765	106	35
Car–S(3)	*Car*–S(=O)–C	1.790	1.790	0.010	1.783	1.798	41	
	Car–S^+–X_2	1.778	1.779	0.010	1.771	1.787	10	
Car–S(2)	*Car*–S–C*	1.773	1.774	0.009	1.765	1.779	44	
	Car–S–*Car*	1.768	1.767	0.010	1.762	1.774	158	
	Car–S–*Car* (in phenothiazine)	1.764	1.764	0.008	1.760	1.769	48	

Bond	Substructure	*d*	*m*	σ	q_1	q_u	*n*	Note
	Car–S–S–X	1.777	1.777	0.012	1.767	1.785	47	
Csp^1–S(2)	N≡C–S–X	1.679	1.683	0.026	1.645	1.698	10	
Csp^1–S(1)	$(N{\equiv}C{-}S)^-$	1.630	1.630	0.014	1.619	1.641	14	
Csp^2=S(1)	$(C^*)_2$–C=S: see IPMUDS (1.599)							
	$(Car)_2$–C=S: see CELDOM (1.611)							
	$(X)_2$–C=S (X = C,N,O,S)	1.671	1.675	0.024	1.656	1.689	245	
	X_2N–C(=S)–S–X	1.660	1.660	0.016	1.648	1.674	38	
	$(X_2N)_2$–C=S (thioureas)	1.681	1.684	0.020	1.669	1.693	96	
	N–C(≃S)$_2$	1.720	1.721	0.012	1.709	1.731	20	
Csp^3–Se	C#–Se	1.970	1.967	0.032	1.948	1.998	21	
Csp^2–Se(2)	C=C–Se–C=C (in tetraselenafulvalene)	1.893	1.895	0.013	1.882	1.902	32	
Car–Se(3)	Ph_3–Se^+	1.930	1.929	0.006	1.924	1.936	13	
Csp^3–Si(5)	C#–Si^-–X_4	1.874	1.876	0.015	1.859	1.884	9	
Csp^3–Si(4)	CH_3–Si–X_3	1.857	1.857	0.018	1.848	1.869	552	
	C*–Si–X_3 (C* = CH_3 excluded)	1.888	1.887	0.023	1.872	1.905	124	
	C*–Si–X_3 (overall)	1.863	1.861	0.024	1.850	1.875	681	
Car–Si(4)	*Car*–Si–X_3	1.868	1.868	0.014	1.857	1.878	178	
Csp^1–Si(4)	C≡C–Si–X_3	1.837	1.840	0.012	1.824	1.849	8	
Csp^3–Te	C#–Te	2.158	2.159	0.030	2.128	2.177	13	
Car–Te	*Car*–Te	2.116	2.115	0.020	2.104	2.130	72	
Csp^2=Te	see CEDCUJ (2.044)							
Cl–Cl	see PHASCL (2.306, 2.227)							
Cl–I	see CMBIDZ (2.563), HXPASC (2.541, 2.513), METAMM (2.552), BQUINI (2.416, 2.718)							
Cl–N	see BECTAE (1.743—1.757), BOGPOC (1.705)							
Cl–O(1)	in ClO_4^-	1.414	1.419	0.026	1.403	1.431	252	
Cl–P	(N≃)$_2$P–Cl (N≃P aromatic)	1.997	1.994	0.015	1.989	2.004	46	
	Cl–P (overall)	2.008	2.001	0.035	1.986	2.028	111	
Cl–S	Cl–S (overall)	2.072	2.079	0.023	2.047	2.091	6	
	see also longer bonds in CILSAR (2.283), BIHXIZ (2.357), CANLUY (2.749)							
Cl–Se	see BIRGUE10, BIRHAL10, CTCNSE (2.234—2.851)							
Cl–Si(4)	Cl–Si–X_3 (monochloro)	2.072	2.075	0.009	2.066	2.078	5	
	Cl_2–Si–X_2 and Cl_3–Si–X	2.020	2.012	0.015	2.007	2.036	5	
Cl–Te	Cl–Te in range 2.34—2.60	2.520	2.515	0.034	2.493	2.537	22	36
	see also longer bonds in BARRIV, BOJPUL, CETUTE, EPHTEA, OPNTEC10 (2.73—2.94)							
F–N(3)	F–N–C_2 and F_2–N–C	1.406	1.404	0.016	1.395	1.416	9	
F–P(6)	in hexafluorophosphate, PF_6^-	1.579	1.587	0.025	1.563	1.598	72	
F–P(3)	(N≃)$_2$P–F (N≃P aromatic)	1.495	1.497	0.016	1.481	1.510	10	
F–S	43 observations in range 1.409—1.770 in a wide variety of environments; F–S(6) in F_2–SO_2–C_2 (see FPSULF10, BETJOZ)	1.640	1.646	0.011	1.626	1.649	6	
	F–S(4) in F_2–S(=O)–N (see BUDTEZ)	1.527	1.528	0.004	1.524	1.530	24	37
F–Si(6)	in SiF_6^{2-}	1.694	1.701	0.013	1.677	1.703	6	
F–Si(5)	F–Si^-–X_4	1.636	1.639	0.035	1.602	1.657	10	
F–Si(4)	F–Si–X_3	1.588	1.587	0.014	1.581	1.599	24	
F–Te	see CUCPIZ (F–Te(6) = 1.942, 1.937), FPHTEL (F–Te(4) = 2.006)							
H–N(4)	X_3–N^+–H	1.033	1.036	0.022	1.026	1.045	87	21
H–N(3)	X_2–N–H	1.009	1.010	0.019	0.997	1.023	95	21
H–O(2)	in alcohols C*–O–H	0.967	0.969	0.010	0.959	0.974	63	21
	C#–O–H	0.967	0.970	0.010	0.959	0.974	73	21
	in acids O=C–O–H	1.015	1.017	0.017	1.001	1.031	16	21,38
I–I	in I_3^-	2.917	2.918	0.011	2.907	2.927	6	
I–N	see BZPRIB, CMBIDZ, HMTITI, HMTNTI, IFORAM, IODMAM (2.042—2.475)							
I–O	X–I–O (see BZPRIB, CAJMAB, IBZDAC11)	2.144	2.144	0.028	2.127	2.164	6	
	for IO_6^- see BOVMEE (1.829—1.912)							
I–P(3)	see CEHKAB (2.490—2.493)							†
I–S	see DTHIBR10 (2.687), ISUREA10 (2.629), BZTPPI (3.251)							
I–Te(4)	I–Te–X_3	2.926	2.928	0.026	2.902	2.944	8	
N(4)–N(3)	X_3–N^+–N^0–X_2 (N^0 planar)	1.414	1.414	0.005	1.412	1.418	13	
N(3)–N(3)	(C)(C,H)–N_a–N_b–(C)(C,H)							5,39
	N_a, N_b pyramidal	1.454	1.452	0.021	1.444	1.457	44	40
	N_a pyramidal, N_b planar	1.420	1.420	0.015	1.407	1.433	68	40
	N_a, N_b planar	1.401	1.401	0.018	1.384	1.418	40	40
	overall	1.425	1.425	0.027	1.407	1.443	139	
N(3)–N(2)	in pyrazole (N1–N2)	1.366	1.366	0.019	1.350	1.375	20	
	in pyridazinium ($N1^+$≃N2)	1.350	1.349	0.010	1.345	1.361	7	

Bond	Substructure	d	m	σ	q_1	q_u	n	Note
N(2)≃N(2)	N≃N (aromatic) in pyridazine							
	with C,H as *ortho* substituents	1.304	1.300	0.019	1.287	1.326	6	
	with N,Cl as *ortho* substituents	1.368	1.373	0.011	1.362	1.375	9	
N(2)=N(2)	C#–N=N–C#							
	cis	1.245	1.244	0.009	1.239	1.252	21	
	trans	1.222	1.222	0.006	1.218	1.227	6	
	(overall)	1.240	1.241	0.012	1.230	1.251	27	
	*C*ar–N=N–*C*ar	1.255	1.253	0.016	1.247	1.262	13	
	X–N=N=N (azides)	1.216	1.226	0.028	1.202	1.237	19	
N(2)=N(1)	X–N=N=N (azides)	1.124	1.128	0.015	1.114	1.137	19	
N(3)–O(2)	$(C,H)_2$–N–OH (Nsp^2: planar)	1.396	1.394	0.012	1.390	1.401	28	
	C_2–N–O–C							
	(Nsp^3: pyramidal)	1.463	1.465	0.012	1.457	1.468	22	
	(Nsp^2: planar)	1.397	1.394	0.011	1.388	1.409	12	
	in furoxan (N2–O1)	1.438	1.436	0.009	1.430	1.447	14	
N(3)–O(1)	$(C\simeq)_2N^+$–O^- in pyridine *N*-oxides	1.304	1.299	0.015	1.291	1.316	11	
	in furoxan ($^+$N2–O6$^-$)	1.234	1.234	0.008	1.228	1.240	14	
N(2)–O(2)	in oximes							
	$(C\#)_2$–C=N–OH	1.416	1.418	0.006	1.416	1.420	7	
	(H)(Csp^2)–C=N–OH	1.390	1.390	0.011	1.380	1.401	20	
	(C#)(Csp^2)–C=N–OH	1.402	1.403	0.010	1.393	1.410	18	
	$(Csp^2)_2$–C=N–OH	1.378	1.377	0.017	1.365	1.393	16	
	$(C,H)_2$–C=N–OH (overall)	1.394	1.395	0.018	1.379	1.408	67	
	in furazan (O1–N2, O1–N5)	1.385	1.383	0.013	1.378	1.392	12	
	in furoxan (O1–N5)	1.380	1.380	0.011	1.370	1.388	14	
	in isoxazole (O1–N2)	1.425	1.425	0.010	1.417	1.434	9	
N(3)=O(1)	in nitrate ions NO_3^-	1.239	1.240	0.020	1.227	1.251	105	
	in nitro groups							
	C*–NO_2	1.212	1.214	0.012	1.206	1.221	84	
	C#–NO_2	1.210	1.210	0.011	1.203	1.218	251	
	*C*ar– NO_2	1.217	1.218	0.011	1.211	1.215	1 116	
	C–NO_2 (overall)	1.218	1.219	0.013	1.210	1.226	1 733	
N(3)–P(4)	X_2–P(=X)–NX_2							
	Nsp^2: planar	1.652	1.651	0.024	1.634	1.670	205	
	Nsp^3: pyramidal	1.683	1.683	0.005	1.680	1.686	6	
	(overall)	1.662	1.662	0.029	1.639	1.682	358	
	subsets of this group are:							
	O_2–P(=S)–NX_2	1.628	1.624	0.015	1.615	1.634	9	
	C–P(=S)–$(NX_2)_2$	1.691	1.694	0.018	1.678	1.703	28	
	O–P(=S)–$(NX_2)_2$	1.652	1.654	0.014	1.642	1.664	28	
	P(=O)–$(NX_2)_3$	1.663	1.668	0.026	1.640	1.679	78	
N(3)–P(3)	–NX–P(–X)–NX–P(–X)– (P_2N_2 ring)	1.730	1.721	0.017	1.716	1.748	20	
	–NX–P(=S)–NX–P(=S)– (P_2N_2 ring)	1.697	1.697	0.015	1.690	1.703	44	
	in *P*-substituted phosphazenes:							
	$(N\simeq)_2$P–N (amino)	1.637	1.638	0.014	1.625	1.651	16	
	(aziridinyl)	1.672	1.674	0.010	1.665	1.676	15	
N(2)=P(4)	Ph_3–P=N^+=P–Ph_3	1.571	1.573	0.013	1.563	1.580	66	
N(2)=P(3)	Ph_3–P=N–C,S	1.599	1.597	0.018	1.580	1.615	7	
N(2)≃P(3)	N≃P aromatic							
	in phosphazenes	1.582	1.582	0.019	1.571	1.594	126	
	in P≃N≃S	1.604	1.606	0.009	1.594	1.612	36	
N(3)–S(4)	C–SO_2–NH_2	1.600	1.601	0.012	1.591	1.610	14	35
	C–SO_2–NH–C#	1.633	1.633	0.019	1.615	1.652	47	35
	C–SO_2–N–$C(\#)_2$	1.642	1.641	0.024	1.623	1.659	38	35
N(3)–S(2)	C–S–NX_2 Nsp^2: planar	1.710	1.707	0.019	1.698	1.722	22	23
	(for Nsp^3 pyramidal see MODIAZ: 1.765)							
	X–S–NX_2 Nsp^2: planar	1.707	1.705	0.012	1.699	1.715	30	23
N(2)–S(2)	C=N–S–X	1.656	1.663	0.027	1.632	1.677	36	
N(2)≃S(2)	N≃S aromatic in P≃N≃S	1.560	1.558	0.011	1.554	1.563	37	
N(2)=S(2)	N=S in N=S=N and N=S=S	1.541	1.546	0.022	1.521	1.558	37	
N(3)–Se	see COJCUZ (1.830), DSEMOR10 (1.846, 1.852), MORTRS10 (1.841)							
N(2)–Se	see SEBZQI (1.805), NAPSEZ10 (1.809, 1.820)							
N(2)=Se	see CISMUM (1.790, 1.791)							
N(3)–Si(5)	see DMESIP01, BOJLER, CASSAQ, CASYOK, CECXEN, CINTEY, CIPBUY, FMESIB, MNPSIL, PNPOSI (1.973—2.344)							
N(3)–Si(4)	X_3–Si–NX_2 (overall)	1.748	1.746	0.022	1.735	1.757	170	
	subsets of this group are:							
	X_3–Si–NHX	1.714	1.719	0.014	1.702	1.727	16	
	X_3–Si–NX–Si–X_3 acyclic	1.743	1.744	0.016	1.731	1.755	45	
	N–Si–N in 4-membered rings	1.742	1.742	0.009	1.735	1.748	53	
	N–Si–N in 5-membered rings	1.741	1.742	0.019	1.726	1.749	33	
N(2)–Si(4)	X_3–Si–N^-–Si–X_3	1.711	1.712	0.019	1.693	1.729	15	

Bond	Substructure	d	m	σ	q_l	q_u	n	Note
N–Te	see ACLTEP (2.402), BIBLAZ (1.980), CESSAU (2.023)							
O(2)–O(2)	C*–**O–O**–C*,H							
	τ(OO) = 70—85°	1.464	1.464	0.009	1.458	1.472	12	
	τ(OO) *ca.* 180°	1.482	1.480	0.005	1.478	1.486	5	
	overall	1.469	1.471	0.012	1.461	1.478	17	
	O=C–**O–O**–C=O see ACBZPO01 (1.446), CEYLUN (1.452), CIMHIP (1.454)							
	Si–**O–O**–Si	1.496	1.499	0.005	1.490	1.499	10	
O(2)–P(5)	X–**P**–$(\mathbf{O}X)_4$							41
	trigonal bipyramidal:							
	axial	1.689	1.685	0.024	1.675	1.712	20	
	equatorial	1.619	1.622	0.024	1.604	1.628	20	
	square pyramidal	1.662	1.661	0.020	1.649	1.673	28	
O(2)–P(4)	C–**O–P**$(\simeq O)_3^{2-}$	1.621	1.622	0.007	1.615	1.628	12	
	$(H–\mathbf{O})_2$–**P**$(\simeq O)_2^-$	1.560	1.561	0.009	1.555	1.566	16	
	$(C–\mathbf{O})_2$–**P**$(\simeq O)_2^-$	1.608	1.607	0.013	1.599	1.615	16	
	$(C\#–\mathbf{O})_3$–**P**=O	1.558	1.554	0.011	1.550	1.564	30	
	$(C$*ar*$–\mathbf{O})_3$–**P**=O	1.587	1.588	0.014	1.572	1.599	19	
	X–**O–P**(=O)–$(C,N)_2$	1.590	1.585	0.016	1.577	1.601	33	
	$(X–\mathbf{O})_2$–**P**(=O)–(C,N)	1.571	1.572	0.013	1.563	1.579	70	
O(2)–P(3)	$(N\simeq)_2$**P–O**–C (N≃P aromatic)	1.573	1.573	0.011	1.563	1.584	16	
O(1)=P(4)	C–O–**P**$(\simeq \mathbf{O})_3^{2-}$ (declocalized)	1.513	1.512	0.008	1.508	1.518	42	
	$(H–O)_2$–**P**$(\simeq \mathbf{O})_2^-$ (delocalized)	1.503	1.503	0.005	1.499	1.508	16	
	$(C–O)_2$–**P**$(\simeq \mathbf{O})_2^-$ (delocalized)	1.483	1.485	0.008	1.474	1.490	16	
	$(C–O)_3$–**P=O**	1.449	1.448	0.007	1.446	1.452	18	
	C_3–**P=O**	1.489	1.486	0.010	1.481	1.496	72	
	N_3–**P=O**	1.461	1.462	0.014	1.449	1.470	26	
	$(C)_2(N)$–**P=O**	1.487	1.489	0.007	1.479	1.493	5	
	$(C,N)_2(O)$–**P=O**	1.467	1.465	0.007	1.462	1.472	33	
	$(C,N)(O)_2$–**P=O**	1.457	1.458	0.009	1.454	1.462	35	
O(2)–S(4)	C–**O–S**O_2–C	1.577	1.576	0.015	1.566	1.584	41	
	C–**O–S**O_2–CH_3	1.569	1.569	0.013	1.556	1.582	7	
	C–**O–S**O_2–C*ar*	1.580	1.578	0.015	1.571	1.588	27	
O(1)=S(4)	C–$\mathbf{SO}_2$–C	1.436	1.437	0.010	1.431	1.442	316	42
	X–$\mathbf{SO}_2$–NX_2	1.428	1.428	0.010	1.422	1.434	326	
	C–$\mathbf{SO}_2$–N–$(C,H)_2$	1.430	1.430	0.009	1.425	1.435	206	
	C–$\mathbf{SO}_2$–O–C	1.423	1.423	0.008	1.418	1.428	82	
	in $\mathbf{SO}_4^{2-}$	1.472	1.473	0.013	1.463	1.481	104	
O(1)=S(3)	C–**S(=O)**–C	1.497	1.498	0.013	1.489	1.505	90	5
O–Se	see BAPPAJ, BIRGUE10, BIRHAL10, CXMSEO, DGLYSE, SPSEBU (1.597 for **O=Se** to 1.974 for **O–Se**)							
O(2)–Si(5)	$(X–\mathbf{O})_3$–**Si**–(N)(C)	1.663	1.658	0.023	1.650	1.665	21	
O(2)–Si(4)	X_3–**Si–O**–X (overall)	1.631	1.630	0.022	1.617	1.646	191	
O(2)–Si(4)	subsets of this group are:							
	X_3–**Si–O**–C#	1.645	1.647	0.012	1.634	1.652	29	
	X_3–**Si–O**–Si–X_3	1.622	1.625	0.014	1.614	1.631	70	
	X_3–**Si–O**–O–Si–X_3	1.680	1.676	0.008	1.673	1.688	10	
O(2)–Te(6)	$(X–\mathbf{O})_6$–**Te**	1.927	1.927	0.020	1.908	1.942	16	
O(2)–Te(4)	$(X–\mathbf{O})_2$–**Te**–X_2	2.133	2.136	0.054	2.078	2.177	12	
P(4)–P(4)	X_3–**P–P**–X_3	2.256	2.259	0.025	2.243	2.277	6	
P(4)–P(3)	see CECHEX (2.197), COZPIQ (2.249)							
P(3)–P(3)	X_2–**P–P**–X_2	2.214	2.210	0.022	2.200	2.224	41	
P(4)=P(4)	see BUTSUE (2.054)							
P(3)=P(3)	see BALXOB (2.034)							
P(4)=S(1)	C_3–**P=S**	1.954	1.952	0.005	1.950	1.957	13	
	$(N,O)_2(C)$–**P=S**	1.922	1.924	0.014	1.913	1.927	26	
	$(N,O)_3$–**P=S**	1.913	1.914	0.014	1.906	1.921	50	
P(4)=Se(1)	X_3–**P=Se**	2.093	2.099	0.019	2.075	2.108	12	
P(3)–Si(4)	X_2–**P–Si**–X_3: 3- and 4-rings excluded (see BOPFER, BOPFIV, CASTOF10, COZVIW: 2.201—2.317)	2.264	2.260	0.019	2.249	2.283	22	
P(4)=Te(1)	see MOPHTE (2.356), TTEBPZ (2.327)							
S(2)–S(2)	C–**S–S**–C							
	τ(SS) = 75—105°	2.031	2.029	0.015	2.021	2.038	46	
	τ(SS) = 0—20°	2.070	2.068	0.022	2.057	2.077	28	
	(overall)	2.048	2.045	0.026	2.028	2.068	99	
	in polysulphide chain–**S–S**–S–	2.051	2.050	0.022	2.037	2.065	126	
S(2)–S(1)	X–N=**S–S**	1.897	1.896	0.012	1.887	1.908	5	
S–Se(4)	see BUWZUO (2.264, 2.269)							
S–Se(2)	X–**Se–S** (any)	2.193	2.195	0.015	2.174	2.207	9	
S(2)–Si(4)	X_3–**Si–S**–X	2.145	2.138	0.020	2.130	2.158	19	

Bond	Substructure	d	m	σ	q_l	q_u	n	Note
S(2)–Te	X–S–Te (any)	2.405	2.406	0.022	2.383	2.424	10	
	X=S–Te (any)	2.682	2.686	0.035	2.673	2.694	28	
Se(2)–Se(2)	X–Se–Se–X	2.340	2.340	0.024	2.315	2.361	15	
Se(2)–Te(2)	see BAWFUA, BAWGAH (2.524—2.561)							†
Si(4)–Si(4)	X_3–Si–Si–X_3 3-membered rings excluded: see CIHRAM (2.511)	2.359	2.359	0.012	2.349	2.366	42	
Te–Te	see CAHJOK (2.751, 2.704)							

ppendix 1. (Footnotes to Table)

1. Sample dominated by B–CH_3. For longer bonds in B^-–CH_3 see LITMEB10 [B(4)–CH_3 = 1.621—1.644Å].
2. $p(\pi)$–$p(\pi)$ Bonding with Bsp^2 and Nsp^2 coplanar (τBN = 0 ± 15°) predominates. See G. Schmidt, R. Boese, and D. Bläser, *Z. Naturforsch.*, 1982, **37b**, 1230.
3. 84 observations range from 1.38 to 1.61 Å and individual values depend on substituents on B and O. For a discussion of borinic acid adducts see S. J. Rettig and J. Trotter, *Can. J. Chem.*, 1982, **60**, 2957.
4. See M. Kaftory in 'The Chemistry of Functional Groups. Supplement D: The Chemistry of Halides, Pseudohalides, and Azides' eds. S. Patai and Z. Rappoport, Wiley: New York, 1983, Part 2, ch. 24.
5. Bonds which are endocyclic or exocyclic to any 3- or 4-membered rings have been omitted from all averages in this section.
6. The overall average given here is for Csp^3–Csp^3 bonds which carry only C or H substituents. The value cited reflects the relative abundance of each 'substitution' group. The 'mean of means' for the 9 subgroups is 1.538 (σ = 0.022) Å.
7. See F. H. Allen, (*a*) *Acta Crystallogr.*, 1980, **B36**, 81; (*b*) 1981, **B37**, 890.
8. See F. H. Allen, *Acta Crystallogr.*, 1984, **B40**, 64.
9. See F. H. Allen, *Tetrahedron*, 1982, **38**, 2843.
10. See F. H. Allen, *Tetrahedron*, 1982, **38**, 645.
11. Cyclopropanones and cyclobutanones excluded.
12. See W. B. Schweizer and J. D. Dunitz, *Helv. Chim. Acta*, 1982, **65**, 1547.
13. See L. Norskov-Lauritsen, H.-B. Bürgi, P. Hoffmann, and H. R. Schmidt, *Helv. Chim. Acta*, 1985, **68**, 76.
14. See P. Chakrabarti and J. D. Dunitz, *Helv. Chim. Acta*, 1982, **65**, 1555.
15. See J. L. Hencher in 'The Chemistry of the C≡C Triple Bond,' ed. S. Patai, Wiley, New York, 1978, ch. 2.
16. Conjugated: torsion angle about central C–C single bond is 0 ± 20° (*cis*) or 180 ± 20° (*trans*).
7. Unconjugated: torsion angle about central C–C single bond is 20—160°.
8. Other conjugative substituents excluded.
9. TCNQ is tetracyanoquinodimethane.
0. No difference detected between C2≃C3 and C3≃C4 bonds.
1. Derived from neutron diffraction results only.
2. Nsp^3: pyramidal; mean valence angle at N is in range 108—114°.
3. Nsp^2: planar; mean valence angle at N is ⩾117.5°.
. Cyclic and acyclic peptides.
. See R. H. Blessing, *J. Am. Chem. Soc.*, 1983, **105**, 2776.
. See L. Lebioda, *Acta Crystallogr.*, 1980, **B36**, 271.
. n = 3 or 4, *i.e.* tri- or tetra-substituted ureas.
. Overall value also includes structures with mean valence angle at N in the range 115—118°.

See F. H. Allen and A. J. Kirby, *J. Am. Chem. Soc.*, 1984, **106**, 6197.
See A. J. Kirby, 'The Anomeric Effect and Related Stereoelectronic Effects at Oxygen,' Springer, Berlin, 1983.
See B. Fuchs, L. Schleifer, and E. Tartakovsky, *Nouv. J. Chim.*, 1984, **8**, 275.
See S. C. Nyburg and C. H. Faerman, *J. Mol. Struct.*, 1986, **140**, 347.
Sample dominated by P–CH_3 and P–CH_2–C.
Sample dominated by C* = methyl.
See A. Kalman, M. Czugler, and G. Argay, *Acta Crystallogr.*, 1981, **B37**, 868.
Bimodal distribution resolved into 22 'short' bonds and 5 longer outliers.
All 24 observations come from BUDTEZ.
'Long' O–H bonds in centrosymmetric O - - - H - - - O H-bonded dimers are excluded.
N–N bond length also dependent on torsion angle about N–N bond and on nature of substituent C atoms; these effects are ignored here.
N pyramidal has average angle at N in range 100—113.5°; N planar has average angle of ⩾117.5°.
See R. R. Holmes and J. A. Deiters, *J. Amer. Chem. Soc.*, 1977, **99**, 3318.
No detectable variation in S=O bond length with type of C-substituent.

ix 2.

orm references to individual CSD entries cited by reference code in the Table. A full list of CSD bibliographic entries is given in SUP 56701.

ACBZPO01	*J. Am. Chem. Soc.*, 1975, **97**, 6729.
ACLTEP	*J. Organomet. Chem.*, 1980, **184**, 417.
ASAZOC	*Dokl. Akad. Nauk SSSR*, 1979, **249**, 120.
BALXOB	*J. Am. Chem. Soc.*, 1981, **103**, 4587.
BAPPAJ	*Inorg. Chem.*, 1981, **20**, 3071.
BARRIV	*Acta Chem. Scand., Ser. A*, 1981, **35**, 443.
BAWFUA	*Cryst. Struct. Commun.*, 1981, **10**, 1345.
BAWGAH	*Cryst. Struct. Commun.*, 1981, **10**, 1353.
BECTAE	*J. Org. Chem.*, 1981, **46**, 5048, 1981.
BELNIP	*Z. Naturforsch., Teil B*, 1982, **37**, 299.
BEMLIO	*Chem. Ber.*, 1982, **115**, 1126.
BEPZEB	*Cryst. Struct. Commun.*, 1982, **11**, 175.
BETJOZ	*J. Am. Chem. Soc.*, 1982, **104**, 1683.
BETUTE10	*Acta Chem. Scand., Ser. A*, 1976, **30**, 719.
BIBLAZ	*Zh. Strukt. Khim.*, 1981, **22**, 118.
BICGEZ	*Z. Anorg. Allg. Chem.*, 1982, **486**, 90.
BIHXIZ	*J. Chem. Soc., Chem. Commun.*, 1982, 982.
BIRGUE10	*Z. Naturforsch., Teil B*, 1983, **38**, 20.
BIRHAL10	*Z. Naturforsch., Teil B*, 1982, **37**, 1410.
BIZJAV	*J. Organomet. Chem.*, 1982, **238**, C1.
BOGPOC	*Z. Naturforsch., Teil B*, 1982, **37**, 1402.
BOGSUL	*Z. Naturforsch., Teil* B, 1982, **37**, 1230.
BOJLER	*Z. Anorg. Allg. Chem.*, 1982, **493**, 53.
BOJPUL	*Acta Chem. Scand., Ser. A*, 1982, **36**, 829.
BOPFER	*Chem. Ber.*, 1983, **116**, 146.
BOPFIV	*Chem. Ber.*, 1983, **116**, 146.
BOVMEE	*Acta Crystallogr., Sect. B*, 1982, **38**, 1048.
BQUINI	*Acta Crystallogr., Sect. B*, 1979, **35**, 1930.

BTUPTE *Acta Chem. Scand., Ser. A*, 1975, **29**, 738.
BUDTEZ *Z. Naturforsch., Teil B*, 1983, **38**, 454.
BUPSIB10 *Z. Anorg. Allg. Chem.*, 1981, **474**, 31.
BUSHAY *Z. Naturforsch., Teil. B*, 1983, **38**, 692.
BUTHAZ10 *Inorg. Chem.*, 1984, **23**, 2582.
BUTSUE *J. Chem. Soc., Chem. Commun.*, 1983, 862.
BUWZUO *Acta Chem. Scand., Ser A*, 1983, **37**, 219.
BZPRIB *Z. Naturforsch., Teil B*, 1981, **36**, 922.
BZTPPI *Inorg. Chem.*, 1978, **17**, 894.
CAHJOK *Inorg. Chem.*, 1983, **22**, 1809.
CAJMAB *Chem. Z*, 1983, **107**, 169.
CANLUY *Tetrahedron Lett.*, 1983, **24**, 4337.
CASSAQ *J. Struct. Chem.*, 1983, **2**, 101.
CASTOF10 *Acta Crystallogr., Sect. C*, 1984, **40**, 1879.
CASYOK *J. Struct. Chem.*, 1983, **2**, 107.
CECHEX *Z. Anorg. Allg. Chem.*, 1984, **508**, 61.
CECXEN *J. Struct. Chem.*, 1983, **2**, 207.
CEDCUJ *J. Org. Chem.*, 1983, **48**, 5149.
CEHKAB *Z. Naturforsch., Teil B*, 1984, **39**, 139.
CELDOM *Acta Crystallogr., Sect. C*, 1984, **40**, 556.
CESSAU *Acta Crystallogr., Sect. C*, 1984, **40**, 653.
CETTAW *Chem. Ber.*, 1984, **117**, 1089.
CETUTE *Acta Chem. Scand., Ser A*, 1975, **29**, 763.
CEYLUN *Izv. Akad. Nauk SSSR, Ser. Khim.*, 1983, 2744.
CIFZUM *Acta Chem. Scand., Ser A*, 1984, **38**, 289.
CIHRAM *Angew. Chem., Int. Ed. Engl.*, 1984, **23**, 302.
CILRUK *J. Chem. Soc., Chem. Commun.*, 1984, 1023.
CILSAR *J. Chem. Soc., Chem. Commun.*, 1984, 1021.
CIMHIP *Acta Crystallogr., C*, 1984, **40**, 1458.
CINTEY *Dokl. Akad. Nauk SSSR*, 1984, **274**, 615.
CIPBUY *J. Struct. Chem.*, 1983, **2**, 281.
CISMUM *Z. Naturforsch., Teil B*, 1984, **39**, 485.
CISTED *Z. Anorg. Allg. Chem.*, 1984, **511**, 95.
CIWYIQ *Inorg. Chem.*, 1984, **23**, 1946.
CIYFOF *Inorg. Chem.*, 1984, **23**, 1790.
CMBIDZ *J. Org. Chem.*, 1979, **44**, 1447.
CODDEE *Z. Naturforsch., Teil B*, 1984, **39**, 1257.
CODDII *Z. Naturforsch., Teil B*, 1984, **39**, 1257.
COFVOI *Z. Naturforsch., Teil B*, 1984, **39**, 1027.
COJCUZ *Chem. Ber.*, 1984, **117**, 2686.
COSDIX *Z. Naturforsch., Teil B*, 1984, **39**, 1344.
COZPIQ *Chem. Ber.*, 1984, **117**, 2063.
COZVIW *Z. Anorg. Allg. Chem.*, 1984, **515**, 7.
CTCNSE *J. Am. Chem. Soc.*, 1980, **102**, 5430.
CUCPIZ *J. Am. Chem. Soc.*, 1984, **106**, 7529.
CUDLOC *J. Cryst. Spectrosc.*, 1985, **15**, 53.
CUDLUI *J. Cryst. Spectrosc.*, 1985, **15**, 53.
CUGBAH *Acta Crystallogr., Sect. C*, 1985, **41**, 476.
CXMSEO *Acta Crystallogr., Sect. B*, 1973, **29**, 595.
DGLYSE *Acta Crystallogr., Sect. B*, 1975, **31**, 1785.
DMESIP01 *Acta Crystallogr., Sect. C*, 1984, **40**, 895.
DSEMOR10 *J. Chem. Soc., Dalton Trans.*, 1980, 628.
DTHIBR10 *Inorg. Chem.*, 1971, **10**, 697.
EPHTEA *Inorg. Chem.*, 1980, **19**, 2487.
ESEARS *J. Chem. Soc. C*, 1971, 1511.
ETEARS *J. Chem. Soc. C*, 1971, 1511.
FMESIB *J. Organomet. Chem.*, 1980, **197**, 275.
FPHTEL *J. Chem. Soc., Dalton Trans.*, 1980, 2306.
FPSULF10 *J. Am. Chem. Soc.*, 1982, **104**, 1683.
HCLENE10 *Acta Crystallogr., Sect. B*, 1982, **38**, 3139.
HMTITI *Acta Crystallogr., Sect. B*, 1975, **31**, 1505.
HMTNTI *Z. Anorg. Allg. Chem.*, 1974, **409**, 237.
HXPASC *J. Chem. Soc., Dalton Trans.*, 1975, 1381.
IBZDAC11 *J. Chem. Soc., Dalton Trans.*, 1979, 854.
IFORAM *Monatsh. Chem.*, 1974, **105**, 621.
IODMAM *Acta Crystallogr., Sect. B*, 1977, **33**, 3209.
IPMUDS *Acta Crystallogr., Sect. B*, 1973, **29**, 2128.
ISUREA10 *Acta Crystallogr., Sect. B*, 1972, **28**, 643.
LITMEB10 *J. Am. Chem. Soc.*, 1975, **97**, 6401.
MESIAD *Z. Naturforsch., Teil B*, 1980, **35**, 789.
METAMM *Acta Crystallogr.*, 1964, **17**, 1336.
MNPSIL *J. Am. Chem. Soc.*, 1969, **91**, 4134.
MODIAZ *J. Heterocycl. Chem.*, 1980, **17**, 1217.
MOPHTE *Acta Chem. Scand., Ser. A*, 1980, **34**, 333.
MORTRS10 *J. Chem. Soc., Dalton Trans.*, 1980, 628.
NAPSEZ10 *J. Am. Chem. Soc.*, 1980, **102**, 5070.
NBBZAM *Z. Naturforsch., Teil B*, 1977, **32**, 1416.
OPIMAS *Aust. J. Chem.*, 1977, **30**, 2417.
OPNTEC10 *J. Chem. Soc., Dalton Trans.*, 1982, 251.
PHASCL *Acta Crystallogr., Sect. B*, 1981, **37**, 1357.
PHASOC01 *Aust. J. Chem.*, 1975, **28**, 15.
PNPOSI *J. Am. Chem. Soc.*, 1968, **90**, 5102.
SEBZQI *J. Chem. Soc., Chem. Commun.*, 1977, 325.
SPSEBU *Acta Chem. Scand., Ser. A*, 1979, **33**, 403.
TEACBR *Cryst. Struct. Commun.*, 1974, **3**, 753.
THINBR *J. Am. Chem. Soc.*, 1970, **92**, 4002.
TMPBTI *Acta Crystallogr., Sect. B*, 1975, **31**, 1116.
TPASSN *J. Chem. Soc., Dalton Trans.*, 1977, 514.
TPASTB *Cryst. Struct. Commun.*, 1976, **5**, 39.
TPHOSI *Z. Naturforsch., Teil B*, 1979, **34**, 1064.
TTEBPZ *Z. Naturforsch., Teil B*, 1979, **34**, 256.
ZCMXSP *Cryst. Struct. Commun.*, 1977, **6**, 93.

BOND LENGTHS IN ORGANOMETALLIC COMPOUNDS

This table summarizes the average values of interatomic distances of representative metal-ligand bonds. Sigma bonds between *d*- and *f*-block metals d the elements C, N, O, P, S, and As are included. The values are extracted from a much larger list in Reference 1. The tabulated values are the weighted means of reported measurements on compounds in each category. If four or more measurements are available, the standard deviation is en in parentheses. All values are in Ångstrom units (10^{-10} m).

The first part of the table covers metal-carbon bonds in different ligand categories, while the second part covers metal bonds to other elements. tands for any alkyl group; Me for a CH_3 group; C_6R_5 indicates an aryl group; and C(=O)R an acyl group. Metals are listed in atomic number order.

REFERENCE

)rpen, A. G., Brammer, L., Allen, F.H., Kennard, O., Watson, D. G., and Taylor, R., *J. Chem. Soc. Dalton Trans.*, 1989, S1-S83.

M	M-CH_3	M-CH_2R	M-CR=CR_2	M-C_6R_5	M-C(=O)R
Ti		2.167	2.215(0.042)	2.148	
V				2.114(0.012)	
Cr	2.168		2.035(0.009)	2.075(0.019)	
Mn	2.095(0.030)	2.176(0.024)	2.007	2.064(0.021)	2.044
Fe	2.074	2.091(0.030)	1.991(0.039)	2.031(0.062)	1.997(0.033)
Co	2.014(0.023)	2.039(0.032)	1.934(0.019)	1.974	1.990
Ni	2.029	1.964	1.892(0.017)	1.917(0.038)	1.850(0.059)
Cu				2.020	
Zn		1.964			
Zr	2.292(0.049)		2.257		
Nb	2.336	1.319			
Mo	2.254(0.065)	2.250(0.061)	2.204(0.049)	2.193(0.054)	2.109
Ru	2.179(0.045)	2.036(0.010)	2.063	2.092(0.057)	2.091
Rh	2.092(0.027)	2.100	2.040(0.054)	2.011(0.026)	1.995(0.031)
Pd		2.028	2.000(0.024)	1.981(0.032)	1.982(0.029)
Hf	2.275(0.049)		2.205		
Ta	2.217(0.035)	2.225(0.056)		2.199(0.073)	
W	2.189(0.039)	2.175	2.224		
Re	2.173(0.051)	2.290		2.027	2.190(0.027)
Os		2.221	2.052	2.090(0.032)	2.161
Ir	2.175		2.071(0.044)	2.070(0.038)	2.019
Pt	2.083(0.045)	2.062(0.031)	2.024(0.037)	2.049(0.046)	1.991(0.025)
Au	2.066(0.045)		2.042	2.059(0.024)	
Hg	2.072(0.026)	2.125		2.086(0.040)	
Th	2.567				

M	M-NH_3	M-OH_2	M-PMe_3	M-SR	M-AsR_3
Ti		2.066(0.052)		2.369	2.686
V		2.129(0.131)	2.510(0.010)	2.378(0.007)	
Cr	2.069(0.008)	1.997(0.070)	2.389(0.069)	2.362	2.460(0.040)
Mn		2.189(0.040)	2.455(0.164)	2.366(0.054)	2.400(0.013)
Fe		2.085(0.066)	2.246(0.042)	2.271(0.028)	2.352(0.043)
Co	1.965(0.021)	2.085(0.064)	2.217(0.043)	2.254(0.025)	2.323(0.021)
Ni	2.074(0.093)	2.079(0.038)	2.204(0.031)	2.187(0.007)	2.333(0.035)
Cu	1.987(0.017)	2.186(0.215)			2.367(0.016)
Zn	2.044	2.090(0.061)		2.295	
Y		2.398(0.068)			
Zr			2.692		
Nb		2.248(0.137)			2.741(0.008)
Mo	2.217	2.201(0.094)	2.462(0.046)	2.401(0.050)	2.582(0.036)
Ru	2.126(0.024)	2.074(0.051)	2.307(0.050)		2.446(0.031)
Rh	2.114(0.018)	2.190(0.096)	2.266(0.036)		2.416(0.039)
Pd	2.032	2.200	2.287(0.018)		2.386(0.052)
Ag		2.350			
Cd		2.318(0.065)		2.444	

M	M-NH$_3$	M-OH$_2$	M-PMe$_3$	M-SR	M-AsR$_3$
La		2.556(0.062)			
Ce		2.565(0.063)			
Pr		2.518(0.038)			
Nd		2.533(0.058)			
Sm		2.459(0.050)			
Eu		2.441(0.055)			
Gd		2.443(0.074)			
Tb		2.455			
Dy		2.409(0.074)			
Ho		2.407(0.069)			
Er		2.404(0.083)			
Yb		2.353(0.066)			
Lu		2.404(0.116)			
Ta			2.589(0.044)		
W		2.115(0.065)	2.485(0.039)		
Re	2.253	2.199(0.091)	2.369(0.065)		2.575(0.006)
Os	2.136	2.166	2.328(0.029)		
Ir	2.050(0.021)		2.323(0.028)	2.461	
Pt			2.295(0.036)	2.320(0.015)	2.366(0.058)
Au		2.157		2.293	
Hg		2.690(0.083)		2.402(0.065)	
Th		2.483(0.032)			
U		2.455(0.047)			

BOND LENGTHS AND ANGLES IN GAS-PHASE MOLECULES

This table is reprinted from *Kagaku Benran, 3rd Edition*, Vol. II, pp. 649—661 (1984), with permission of the publisher, Maruzen Company, LTD.
 Translation was carried out by Kozo Kuchitsu.

Internuclear distances and bond angles are represented in units of Å (1 Å = 10^{-10} m) and degrees, respectively. The same but inequivalent atoms e discriminated by subscripts a, b, etc. In some molecules ax for axial and eq for equatorial are also used. All measurements were made in the gas ase. The methods used are abbreviated as follows. UV: ultraviolet (including visible) spectroscopy; IR: infrared spectroscopy; R: Raman ectroscopy; MW: microwave spectroscopy; ED: electron diffraction; NMR: nuclear magnetic resonance; LMR: laser magnetic resonance; EPR: ctron paramagnetic resonance; MBE: molecular beam electric resonance. If two methods were used jointly for structure determination, they are ted together, as (ED, MW). If the numerical values listed refer to the equilibrium values, they are specified by r_e and θ_e. In other cases the listed values resent various average values in vibrational states; it is frequently the case that they represent the r_s structure derived from several isotopic species MW or the r_g structure (i.e., the average internuclear distances at thermal equilibrium) for ED. These internuclear distances for the same atom pair h different definitions may sometimes differ as much as 0.01 Å. Appropriate comments are made on the symmetry and conformation in the ilibrium structure.

In general, the numerical values listed in the following tables contain uncertainties in the last digits. However, for certain molecules such as diatomic lecules, with experimental uncertainties of the order of 10^{-5} Å or smaller, numerical values are listed to four decimal places.

REFERENCES

1. L. E. Sutton, ed., *Tables of Interatomic Distances and Configuration in Molecules and Ions*, The Chemical Society Special Publication, No. 11, 18, The Chemical Society (London) (1958, 1965).
2. K.-H. Hellwege, ed., *Landolt-Börnstein Numerical Data and Functional Relations in Science and Technology*, New Series, II/7, J. H. Callomon, E. Hirota, K. Kuchitsu, W. J. Lafferty, A. G. Maki, C. S. Pote, with assistance of I. Buck and B. Starck, *Structure Data of Free Polyatomic Molecules*, Springer-Verlag (1976).
3. K. P. Huber and G. Herzberg, *Molecular Spectra and Molecular Structure IV. Constants of Diatomic Molecules*, Van Nostrand Reinhold Co., London (1979).
4. B. Starck, *Microwave Catalogue and Supplements*.
5. B. Starck, *Electron Diffraction Catalogue and Supplements*.

STRUCTURES OF ELEMENTS AND INORGANIC COMPOUNDS

Compounds are Arranged in Alphabetical Order by their Chemical Formulas (Lengths in Å and Angles in Degrees)

pound	Structure				Method
	Ag—Br (r_e)	2.3931			MW
	Ag—Cl (r_e)	2.2808			MW
	Ag—F (r_e)	1.9832			MW
	Ag—H (r_e)	1.617			UV
	Ag—I (r_e)	2.5446			MW
	Ag—O (r_e)	2.0030			UV
	Al—Br (r_e)	2.295			UV
	Al—Cl (r_e)	2.1301			MW
	Al—F (r_e)	1.6544			MW
	Al—H (r_e)	1.6482			UV
	Al—I (r_e)	2.5371			MW
	Al—O (r_e)	1.6176			UV
	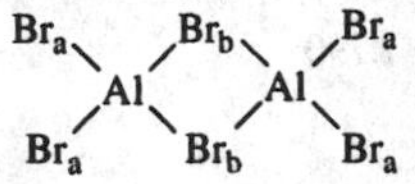		Al—Br_a Al—Br_b ∠Br_bAlBr_b ∠Br_aAlBr_a (D_{2h})	2.22 2.38 82 118	ED
	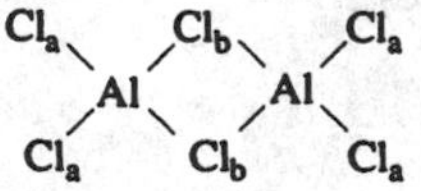		Al—Cl_a Al—Cl_b ∠Cl_bAlCl_b ∠Cl_aAlCl_a (D_{2h})	2.04 2.24 87 122	ED
	As—Br	2.324	∠BrAsBr	99.6	ED
	As—Cl	2.165	∠ClAsCl	98.6	ED, MW
	As—F	1.710	∠FAsF	95.9	ED, MW

Compound	Structure							Method
AsF_5	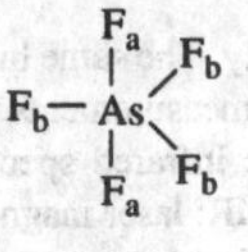		As—F_a	1.711	As—F_b (D_{3h})	1.656		
AsH_3	As—H (r_e)	1.511			∠HAsH (θ_e)	92.1		MW, IR
AsI_3	As—I	2.557			∠IAsI	100.2		ED
AuH	Au—H (r_e)	1.5237						UV
BBr_3	B—Br	1.893			(D_{3h})			ED
BCl_3	B—Cl	1.742			(D_{3h})			ED
BF	B—F (r_e)	1.2626						UV
BF_2H	B—H	1.189	B—F	1.311	∠FBF	118.3		MW
BF_2OH	B—F	1.32	B—O	1.34	O—H	0.941		MW
	∠FBF	118	∠FBO	123	∠BOH	114.1		
BF_3	B—F	1.313			(D_{3h})			ED, IR
BH	B—H (r_e)	1.2325						UV
BH_3PH_3	B—P	1.937	B—H	1.212	P—H	1.399		MW
	∠PBH	103.6	∠BPH	116.9	∠HBH	114.6		
	∠HPH	101.3	staggered form					
BI_3	B—I	2.118			(D_{3h})			ED
BN	B—N (r_e)	1.281						UV
BO	B—O (r_e)	1.2045						EPR
BO_2	B—O	1.265			linear			UV
BS	B—S	1.6091						UV
B_2H_6	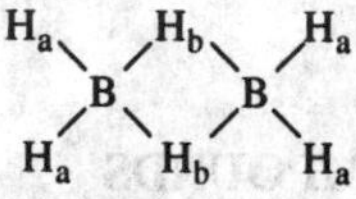				B—H_a	1.19		IR, ED
					B—H_b	1.33		
					B···B	1.77		
					∠H_aBH_a	122		
					∠H_bBH_b	97		
$B_3H_3O_3$	B—O	1.376			∠BOB≅∠OBO	120		ED
$B_3H_6N_3$	B—N	1.435	B—H	1.26	N—H	1.05		ED
	∠NBN	118	∠BNB	121	(C_2)			
BaH	Ba—H (r_e)	2.2318						UV
BaO	Ba—O (r_e)	1.9397						MW
BaS	Ba—S (r_e)	2.5074						MBE
BeF	Be—F (r_e)	1.3609						UV
BeH	Be—H (r_e)	1.3431						UV
BeO	Be—O (r_e)	1.3308						UV
BiBr	Bi—Br (r_e)	2.6095						MW
$BiBr_3$	Bi—Br	2.63			∠BrBiBr	90	(C_{3v})	ED
BiCl	Bi—Cl (r_e)	2.4716						MW
$BiCl_3$	Bi—Cl	2.423			∠ClBiCl	100	(C_{3v})	ED
BiF	Bi—F (r_e)	2.0516						MW
BiH	Bi—H (r_e)	1.805						UV
BiI	Bi—I (r_e)	2.8005						MW
BiO	Bi—O (r_e)	1.934						UV
BrCN	C—N (r_e)	1.157			C—Br (r_e)	1.790		IR
BrCl	Br—Cl (r_e)	2.1361						MW
BrF	Br—F (r_e)	1.7590						MW
BrF_3	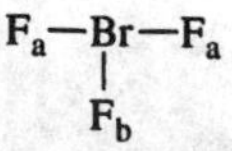		Br—F_a	1.810	Br—F_b	1.721		MW
			∠F_aBrF_b	86.2	(C_{2v})			
BrF_5	Br—F (average)	1.753						ED
	(Br—F_{eq}) − (Br—F_{ax}) = 0.069							
	∠$F_{ax}BrF_{eq}$	85.1			(C_{4v})			
BrO	Br—O (r_e)	1.7172						MW
Br_2	Br—Br (r_e)	2.2811						R
CBr_4	C—Br	1.935			(T_d)			ED
CCl	C—Cl	1.6512						UV

Compound	Structure						Method
ClF$_3$	C—Cl	1.752	C—F	1.325	∠FCF	108.6	ED, MW
Cl$_3$F	C—Cl	1.754	C—F	1.362	∠ClCCl (C$_{3v}$)	111	MW
Cl$_4$	C—Cl	1.767	(T$_d$)				ED
F	C—F (r_e)	1.2718					EPR
F$_3$I	C—I	2.138	C—F	1.330	∠FCF	108.1	ED, MW
F$_4$	C—F	1.323	(T$_d$)				ED
H	C—H (r_e)	1.1199					UV
I$_4$	C—I	2.15	(T$_d$)				ED
N	C—N (r_e)	1.1718					MW
O	C—O (r_e)	1.1283					MW
OBr$_2$	C—O	1.178	C—Br	1.923			ED, MW
	∠BrCBr	112.3					
OClF	C—F	1.334	C—O	1.173	C—Cl	1.725	ED, MW
	∠FCCl	108.8	∠ClCO	127.5			
OCl$_2$	C—O	1.179	C—Cl	1.742			ED, MW
	∠ClCCl	111.8					
F$_2$	C—F	1.3157	C—O	1.172			ED, MW
	∠FCF	107.71					
2	C—O (r_e)	1.1600					IR
	C—P (r_e)	1.562					UV
	C—S (r_e)	1.5349					MW
	C—S (r_e)	1.5526					IR
	C—C (r_e)	1.2425					UV
2	C—O	1.163	C—C	1.289			ED
	linear (large-amplitude bending vibration)						
	Ca—H (r_e)	2.002					UV
	Ca—O (r_e)	1.8221					UV
	Ca—S (r_e)	2.3178					UV
	Cd—H (r_e)	1.781					EPR
2	Cd—Br	2.35	linear				ED
2	Cd—Cl	2.24	linear				ED
	Cd—I	2.56	linear				ED
	C—Cl (r_e)	1.629	C—N (r_e)	1.160			MW
	Cl—F (r_e)	1.6283					MW
	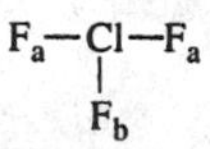		Cl—F$_a$	1.698	Cl—F$_b$	1.598	MW
			∠F$_a$ClF$_b$	87.5	(C$_{2v}$)		
	Cl—O (r_e)	1.5696					MW, UV
	O—Cl	1.690	O—H	0.975	∠HOCl	102.5	MW, IR
	Cl—O	1.470	∠OClO	117.38			MW
H)	O$_a$—Cl	1.407	O$_b$—Cl	1.639			ED
			∠O$_a$ClO$_a$	114.3	∠O$_a$ClO$_b$	104.1	
	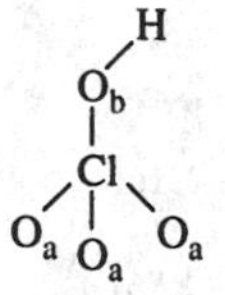						
	Cl—Cl (r_e)	1.9878					UV
	Cl—O	1.6959	∠ClOCl	110.89			MW
	Co—H (r_e)	1.542					UV
;	C—O	1.16	Cr—C	1.92			ED
	∠CrCO	180					
	Cr—O (r_e)	1.615					UV
	Cs—Br (r_e)	3.0723					MW
	Cs—Cl (r_e)	2.9063					MW
	Cs—F (r_e)	2.3454					MW
	Cs—H (r_e)	2.4938					UV
	Cs—I (r_e)	3.3152					MW

Compound	Structure						Method
CsOH	Cs—O (r_e)	2.395	O—H (r_e)	0.97			MW
CuBr	Cu—Br (r_e)	2.1734					MW
CuCl	Cu—Cl (r_e)	2.0512					MW
CuF	Cu—F (r_e)	1.7449					MW
CuH	Cu—H (r_e)	1.4626					UV
CuI	Cu—I (r_e)	2.3383					MW
FCN	C—F	1.262	C—N	1.159			MW
FOH	O—H	0.96	O—F	1.442	∠HOF	97.2	MW
F_2	F—F (r_e)	1.4119					R
$Fe(CO)_5$	Fe—C (average)	1.821					ED
	$(Fe—C)_{eq} - (Fe—C)_{ax}$	0.020					
	C—O (average)	1.153	(D_{3h})				
GaBr	Ga—Br (r_e)	2.3525					MW
GaCl	Ga—Cl (r_e)	2.2017					MW
GaF	Ga—F (r_e)	1.7744					MW
GaF_3	Ga—F	1.88	(D_{3h})				ED
GaI	Ga—I (r_e)	2.5747					MW
GaI_3	Ga—I	2.458	(D_{3h})				ED
GdI_3	Gd—I	2.841	∠IGdI	108	(C_{3v})		ED
$GeBrH_3$	Ge—H	1.526	Ge—Br	2.299			MW,
	∠HGeH	106.2					
$GeBr_4$	Ge—Br	2.272	(T_d)				ED
$GeClH_3$	Ge—H	1.537	Ge—Cl	2.150			IR, M
	∠HGeH	111.0					
$GeCl_2$	Ge—Cl	2.183	∠ClGeCl	100.3			ED
$GeCl_4$	Ge—Cl	2.113	(T_d)				ED
$GeFH_3$	Ge—H	1.522	Ge—F	1.732			MW,
	∠HGeH	113.0					
GeF_2	Ge—F (r_e)	1.7321	∠FGeF (θ_e)	97.17			MW
GeH	Ge—H (r_e)	1.5880					UV
GeH_4	Ge—H	1.5251	(T_d)				IR, R
GeO	Ge—O (r_e)	1.6246					MW
GeS	Ge—S (r_e)	2.0121					MW
GeSe	Ge—Se (r_e)	2.1346					MW
GeTe	Ge—Te (r_e)	2.3402					MW
Ge_2H_6	Ge—H	1.541	Ge—Ge	2.403			ED
	∠HGeH	106.4	∠GeGeH	112.5			
HBr	H—Br (r_e)	1.4145					MW
HCN	C—H (r_e)	1.0655	C—N (r_e)	1.1532			MW
			linear				
HCNO	H—C	1.027	C—N	1.161	N—O	1.207	MW
					linear		
HCl	H-Cl (r_e)	1.2746					MW
HF	H—F (r_e)	0.9169					MW
HI	H—I (r_e)	1.6090					MW
HNCO	N—H	0.986	N—C	1.209	C—O	1.166	MW
	∠HNC	128.0					
HNCS	N—H	0.989	N—C	1.216	C—S	1.561	MW
	∠HNC	135.0	∠NCS	180			
HNO	N—H	1.063	N—O	1.212	∠HNO	108.6	UV
HNO_2	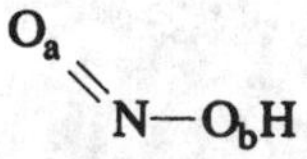		*s-trans* conformer		*s-cis* conformer		MW
			O_b—H	0.958	0.98		
			N—O_b	1.432	1.39		
			N—O_a	1.170	1.19		
			∠O_aNO_b	110.7	114		
			∠NO_bH	102.1	104		
HNO_3	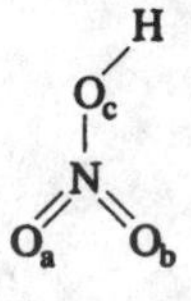		O_c—H	0.96	N—O_c	1.41	M
			N—O_a	1.20	N—O_b	1.21	
			∠HO_cN	102.2	∠O_cNO_a	113.9	
			∠O_cNO_b	115.9	planar		

Compound	Structure							Method
INSO	N—H	1.029	N—S	1.512	S—O	1.451		MW
	∠HNS	115.8	∠NSO	120.4				
			planar					
2	H—H (r_e)	0.7414						UV
$_2$O	O—H (r_e)	0.9575	∠HOH (θ_e)	104.51				MW, IR
$_2$O$_2$	O—O	1.475	∠OOH	94.8				IR
	dihedral angle of internal rotation		119.8	(C_2)				
$_2$S	H—S (r_e)	1.3356	∠HSH (θ_e)	92.12				MW, IR
SO$_4$	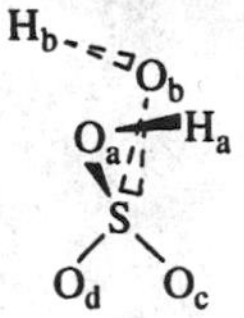		O—H	0.97	S—O$_a$	1.574		MW
			S—O$_c$	1.422	∠H$_a$O$_a$S	108.5		
			∠O$_a$SO$_b$	101.3	∠O$_c$SO$_d$	123.3		
			∠O$_a$SO$_c$	108.6	∠O$_a$SO$_d$	106.4		
			dihedral angle between the H$_a$O$_a$S and O$_a$SO$_c$ planes			20.8		
			dihedral angle between the H$_a$O$_a$S and O$_a$SO$_b$ planes			90.9		
			dihedral angle between the H$_a$SO$_b$ and O$_c$SO$_d$ planes			88.4	(C_2)	
S$_2$	S—S	2.055	S—H	1.327	∠SSH	91.3		ED, MW
	dihedral angle of internal rotation		90.6	(C_2)				
Cl$_4$	Hf—Cl	2.33	(T_d)					ED
Cl$_2$	Hg—Cl	2.252	linear					ED
H	Hg—H (r_e)	1.7404						UV
2	Hg—I	2.553	linear					ED
	I—Br (r_e)	2.4691						MW
	C—I	1.995	C—N	1.159				MW
	I—Cl (r_e)	2.3210						MW
	I—F (average)	1.860	(I—F)$_{eq}$ – (I—F)$_{ax}$	0.03				ED, MW
	∠F$_{ax}$IF$_{eq}$	82.1	(C_{4v})					
	I—O (r_e)	1.8676						MW
	I—I (r_e)	2.6663						R
	In—Br (r_e)	2.5432						MW
	In—Cl (r_e)	2.4012						MW
	In—F (r_e)	1.9854						MW
	In—H (r_e)	1.8376						UV
	In—I (r_e)	2.7537						MW
	Ir—F	1.830	(O_h)					ED
	K—Br (r_e)	2.8208						MW
	K—Cl (r_e)	2.6667						MW
	K—F (r_e)	2.1716						MW
	K—H (r_e)	2.244						UV
	K—I (r_e)	3.0478						MW
	O—H	0.91	K—O	2.212 linear				MW
	K—K (r_e)	3.9051						UV
	Kr—F	1.89	linear					ED
	Li—Br (r_e)	2.1704						MW
	Li—Cl (r_e)	2.0207						MW
	Li—F (r_e)	1.5639						MW
	Li—H (r_e)	1.5949						MW
	Li—I (r_e)	2.3919						MW
	Li—Li (r_e)	2.6729						UV
	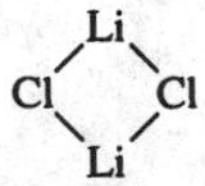		Li—Cl	2.23				ED
			Cl—Cl	3.61				
			∠ClLiCl	108				
	Lu—Cl	2.417	∠ClLuCl	112	(C_{3v})			ED
	Mg—F (r_e)	1.7500						UV
	Mg—H (r_e)	1.7297						UV
	Mg—O (r_e)	1.749						UV
	Mn—H (r_e)	1.7308						UV
6	Mo—C 2.063		C—O	1.145	(O_h)			ED
	Mo—Cl	2.279	Mo—O	1.658				ED
	∠ClMoCl	87.2	(C_{4v})					

Compound	Structure						Method
MoF_6	Mo—F	1.820	(O_h)				ED
$NClH_2$	N—H	1.017	N—Cl	1.748			MW, IR
	∠HNCl	103.7	∠HNH	107			
NCl_3	N—Cl	1.759	∠ClNCl	107.1			ED
NF_2	N—F	1.3528	∠FNF	103.18			MW
NH_2	N—H	1.024	∠HNH	103.3			UV
NH_2CN	N—H	1.00	N_a—C	1.35			MW
	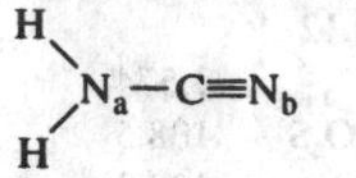		$C—N_b$	1.160	∠HNH	114	
			angle between the NH_2 plane and the N—C bond			142	
NH_2NO_2	N—N	1.427	N—H	1.005			MW
	∠HNH	115.2	∠ONO	130.1			
	dihedral angle between the NH_2 and NNO_2 planes			128.2			
NH_3	N—H (r_e)	1.012	∠HNH (θ_e)	106.7			IR
NH_4Cl	N—H	1.22	N—Cl	2.54	(C_{3v})		ED
NF_2CN	F_2N_b—C≡N_a		C—N_a	1.158	C—N_b	1.386	MW
	N_b—F	1.399	∠N_aCN_b	174			
	∠CN_bF	105.4	∠FN_bF	102.8			
NH	N—H (r_e)	1.0362					LMR
NH_2OH	N—H	1.02	N—O	1.453	O—H	0.962	MW
	∠HNH	107	∠HNO	103.3	∠NOH	101.4	
	The bisector of H—N—H angle is *trans* to the O—H bond						
NO	N—O (r_e)	1.1506					IR
NOCl	N—Cl	1.975	N—O	1.14	∠ONCl	113	MW
NOF	O—N	1.136	N—F	1.512	∠FNO	110.1	MW
NO_2	N—O	1.193	∠ONO	134.1			MW
NO_2Cl	N—Cl	1.840	N—O	1.202			MW
	∠ONO	130.6	(C_{2v})				
NO_2F	N—O	1.1798	N—F	1.467			MW
	∠ONO	136	(C_{2v})				
NS	N—S (r_e)	1.4940					IR
N_2	N—N (r_e)	1.0977					UV
N_2H_4	N—H	1.021	N—N	1.449			ED, M
	∠HNH	106.6 (assumed)	∠NNH_a	112			
	∠NNH_b	106	dihedral angle of internal rotation	91			
	H_a: the H atom closer to the C_2 axis, H_b: the H atom farther from the C_2 axis						
N_2O	N—N (r_e)	1.1284	N—O (r_e)	1.1841			MW,
N_2O_3	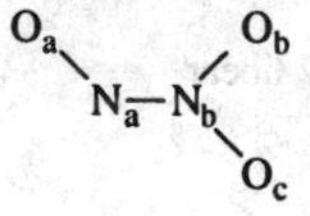		N_a—N_b	1.864	N_a—O_a	1.142	MW
			N_b—O_b	1.202	N_b—O_c	1.217	
			∠$O_aN_aN_b$	105.05			
			∠$N_aN_bO_b$	112.72			
			∠$N_aN_bO_c$	117.47			
N_2O_4	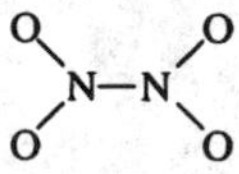		N—N	1.782	N—O	1.190	ED
			∠ONO	135.4	(D_{2h})		
NaBr	Na—Br (r_e)	2.5020					MW
NaCl	Na—Cl (r_e)	2.3609					MW
NaF	Na—F (r_e)	1.9260					MW
NaH	Na—H (r_e)	1.8873					UV
NaI	Na—I (r_e)	2.7115					MW
Na_2	Na—Na (r_e)	3.0789					UV
$NbCl_5$	Nb—Cl_{eq}	2.241	Nb—Cl_{ax}	2.338 (D_{3h})	ED		
NbO	Nb—O (r_e)	1.691					UV
$Ni(CO)_4$	Ni—C	1.838	C—O	1.141	(T_d)		ED
NiH	Ni—H (r_e)	1.476					UV
NpF_6	Np—F	1.981	(O_h)				ED
OCS	C—O (r_e)	1.1578	C—S (r_e)	1.5601			M

Compound	Structure							Method
)CSe	C—O	1.159			C—Se	1.709		MW
)F	O—F (r_e)	1.3579						LMR
$)F_2$	O—F (r_e)	1.4053			∠FOF (θ_e)	103.07	(C_{2v})	MW
$)(SiH_3)_2$	Si—H	1.486			Si—O	1.634		ED
	∠SiOSi	144.1						
2	O—O (r_e)	1.2074						MW
$_2F_2$	O—O	1.217			F—O	1.575		MW
	∠OOF	109.5			dihedral angle of internal rotation	87.5	(C_2)	
3	O—O (r_e)	1.2716			∠OOO (θ_e)	117.47	(C_{2v})	MW
sF_6	Os—F	1.831			(O_h)			ED
sO_4	Os—O	1.712			(T_d)			ED
Br_3	P—Br	2.220			∠BrPBr	101.0		ED
Cl_3	P—Cl	2.039			∠ClPCl	100.27		ED
Cl_5	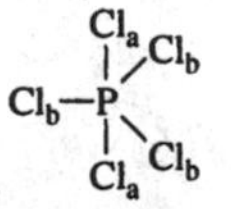		P—Cl_a	2.124	P—Cl_b	2.020	(D_{3h})	ED
	P—F (r_e)	1.5896						UV
	P—F	1.570			∠FPF	97.8		ED, MW
	P—F_{ax}	1.577	P—F_{eq}	1.534	(D_{3h})			ED
	P—H (r_e)	1.4223						LMR
	P—H	1.418			∠HPH	91.70		UV
	P—H	1.4200			∠HPH	93.345		MW
	N—P (r_e)	1.4909						MW
	O—P (r_e)	1.4759						UV
Cl_3	P—O	1.449			P—Cl	1.993		ED
	∠ClPCl	103.3						
	P—O	1.436	P—F	1.524	∠FPF	101.3		ED, MW
	P—P (r_e)	1.8931						UV
	P—F	1.587			P—P	2.281		ED
	∠PPF	95.4			∠FPF	99.1		
	The two PF_2 planes are *trans* to each other (the *gauche* conformer is less than 10%)							
	P—P	2.21			(T_d)			ED
	P—O	1.638	∠POP	126.4	(T_d)			ED
	Pb—H (r_e)	1.839						UV
	Pb—O (r_e)	1.9218						MW
	Pb—S (r_e)	2.2869						MW
	Pb—Se (r_e)	2.4022						MW
	Pb—Te (r_e)	2.5950						MW
	Pr—I	2.904	∠IPrI	113	(C_{3v})			ED
	Pt—O (r_e)	1.7273						UV
	Pu—F	1.971			(O_h)			ED
	Rb—Br (r_e)	2.9447						MW
	Rb—Cl (r_e)	2.7869						MW
	Rb—F (r_e)	2.2703						MW
	Rb—H (r_e)	2.367						UV
	Rb—I (r_e)	3.1768						MW
	Rb—O	2.301	O—H	0.957	linear			MW
	Re—O	1.702			Re—Cl	2.229		MW
	∠ClReO	109.4			(C_{3v})			
	Re—F	1.832			(O_h)			ED
	Ru—O	1.706			(T_d)			ED
	C—Se	1.693			C—S	1.553		MW
	C—S	1.557			C—Te	1.904		MW
	S—Cl	2.006	∠ClSCl	103.0	(C_{2v})			ED
	S—F (r_e)	1.6006						MW
	S—F	1.5921			∠FSF	98.20		MW
	S—F	1.561			(O_h)			ED

BOND LENGTHS AND ANGLES IN GAS-PHASE MOLECULES (continued)

Compound	Structure						Method
SO	S—O (r_e)	1.4811					MW
$SOCl_2$	S—O	1.44	S—Cl	2.072			MW
	∠ClSCl	97.2	∠OSCl	108.0			
SOF_2	S—O	1.420	S—F	1.583			ED
	∠OSF	106.2	∠FSF	92.2			
SOF_4	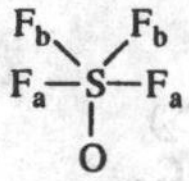		S—O	1.403	S—F_a	1.575	ED
			S—F_b	1.552	∠OSF_a	90.7	
			∠OSF_b	124.9	∠F_aSF_b	89.6	
			∠F_bSF_b	110.2	(C_{2v})		
SO_2	S—O (r_e)	1.4308	∠OSO (θ_e)	119.329			MW
SO_2Cl_2	S—O	1.404	S—Cl	2.011	∠OSO	123.5	ED
	∠ClSCl	100.0	(C_{2v})				
SO_2F_2	S—O	1.397	S—F	1.530	∠OSO	123	ED
	∠FSF	97	(C_{2v})				
SO_3	S—O	1.4198	(D_{3h})				IR
$S(SiH_3)_2$	Si—H	1.494	Si—S	2.136	∠SiSSi	97.4	ED
S_2	S—S (r_e)	1.8892					R
S_2Br_2	S—Br	2.24	S—S	1.98			ED
	∠SSBr	105	dihedral angle of internal rotation	83.5			
S_2Cl_2	S—Cl	2.057	S—S	1.931			ED
	∠SSCl	108.2	dihedral angle of internal rotation	84.1	(C_2)		
S_2O_2	S—O	1.458	S—S	2.025	∠OSS	112.8	MW
					planar *cis* form		
S_8	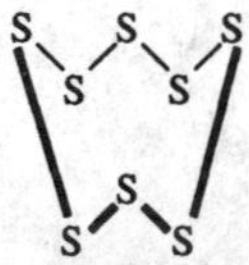		S—S	2.07			ED
			∠SSS	105			
			(D_{4d})				
$SbCl_3$	Sb—Cl	2.333	∠ClSbCl	97.2			ED
SbH_3	Sb—H	1.704	∠HSbH	91.6			MW
SeF	Se—F	1.742					MW
SeF_6	Se—F	1.69	(O_h)				ED
SeO	Se—O (r_e)	1.6393					MW
$SeOF_2$	Se—O	1.576	Se—F	1.730			MW
	∠OSeF	104.82	∠FSeF	92.22			
SeO_2	Se—O (r_e)	1.6076	∠OSeO (θ_e)	113.83			MW
SeO_3	Se—O	1.69	(D_{3h})				ED
Se_2	Se—Se (r_e)	2.1660					UV
Se_6	Se—Se	2.34	∠SeSeSe	102			ED
		six-membered ring with chair conformation					
$SiBrF_3$	Si—F	1.560	Si—Br	2.153			MW
	∠FSiBr	108.5	(C_{3v})				
$SiBrH_3$	Si—H	1.485	Si—Br	2.210			MW
	∠HSiBr	107.8	(C_{3v})				
$SiClH_3$	Si—H	1.482	Si—Cl	2.048			MW
	∠HSiCl	107.9	(C_{3v})				
$SiCl_4$	Si—Cl	2.019	(T_d)				ED
SiF	Si—F	1.6008					UV
$SiFH_3$	Si—H	1.484	Si—F	1.593			MW
	∠HSiH	110.63	(C_{3v})				
SiF_2	Si—F (r_e)	1.590	∠FSiF (θ_e)	100.8			MW
SiF_3H	Si—H (r_e)	1.4468	Si—F (r_e)	1.5624			MW
	∠HSiF (θ_e)	110.64					
SiF_4	Si—F	1.553	(T_d)				ED
SiH	Si—H (r_e)	1.5201					UV
SiH_3I	Si—H	1.485	Si—I	2.437			MW
	∠HSH	107.8					
SiH_4	Si—H	1.4798	(T_d)				IR
SiN	N—Si (r_e)	1.572					UV
SiO	Si—O (r_e)	1.5097					MW

Compound	Structure					Method
S	Si—S (r_e)	1.9293				MW
Se	Se—Si (r_e)	2.0583				MW
	Si—Si (r_e)	2.246				UV
Cl_6	Si—Si	2.32	Si—Cl	2.009		ED
	∠ClSiCl	109.7				
F_6	Si—Si	2.317	Si—F	1.564		ED
	∠FSiF	108.6				
H_6	Si—H	1.492	Si—Si	2.331		ED
	∠SiSiH	110.3	∠HSiH	108.6		
			staggered form (assumed)			
l_4	Sn—Cl	2.280	(T_d)			ED
	Sn—H (r_e)	1.7815				UV
$_4$	Sn—H	1.711	(T_d)			R, IR
	Sn—O	1.8325				MW
	S—Sn (r_e)	2.2090				MW
	Se—Sn (r_e)	2.3256				MW
	Sn—Te (r_e)	2.5228				MW
	Sr—H (r_e)	2.1455				UV
	Sr—O (r_e)	1.9198				MW
	S—Sr (r_e)	2.4405				UV
$_5$	Ta—Cl_{eq}	2.227	Ta—Cl_{ax}	2.369	(D_{3h})	ED
	Ta—O (r_e)	1.6875				UV
	Te—F	1.815	(O_h)			ED
	Te—Te (r_e)	2.5574				UV
	Th—Cl	2.58	(T_d)			ED
	Th—F	2.14	(T_d)			ED
	Tl—Br (r_e)	2.6182				MW
	Tl—Cl (r_e)	2.4848				MW
	Tl—F (r_e)	2.0844				MW
	Tl—H (r_e)	1.870				UV
	Tl—I (r_e)	2.8137				MW
	Ti—Br	2.339	(T_d)			ED
	Ti—Cl	2.170	(T_d)			ED
	Ti—O (r_e)	1.620				UV
	Ti—S (r_e)	2.0825				UV
	U—F	1.996	(O_h)			ED
	V—C	2.015	C—O	1.138		ED
		(O_h, involving dynamic Jahn-Teller effect)				
	V—O	1.570	V—Cl	2.142		ED, MW
	∠ClVCl	111.3				
	V—Cl	2.138	(T_d, involving dynamic Jahn-Teller effect)			ED
	V—F (average)	1.71				ED
	V—O (r_e)	1.5893				UV
	W—C	2.059	C—O	1.149	(O_h)	ED
	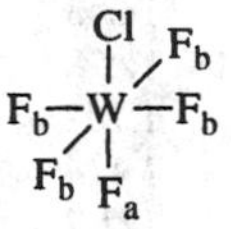		W—Cl	2.251		MW
			W—F (average)	1.836		
			∠F_aWF_b	88.7		
	W—O	1.666	W—F	1.847		ED
	∠FWF	86.2	(C_{4v})			
	W—F	1.832	(O_h)			ED
	Xe—F	1.977	linear			IR
	Xe—F	1.94	(D_{4h})			ED
	Xe—F	1.890	(large-amplitude bending vibration around the O_h structure)			ED
	Xe—O	1.736	(T_d)			ED
	Zn—H (r_e)	1.5949				UV
	Zr—Cl	2.32	(T_d)			ED
	Zr—F	1.902	(T_d)			ED
	Zr—O (r_e)	1.7116				UV

STRUCTURES OF ORGANIC MOLECULES

Compounds are Arranged in Alphabetical Order by Chemical Name; Cross References are Given for Common Synonyms (Lengths in Å and Angles in Degrees)

Compound	Structure				Method
Acetaldehyde	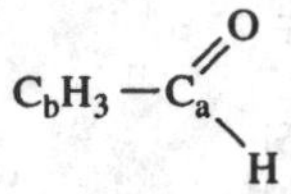		C_a—O	1.210	ED, MV
			C_b—H	1.107	
			C_a—H	1.128	
	C_a—C_b	1.515	∠C_bC_aO	124.1	
	∠HC_bH	109.8	∠C_bC_aH	115.3	
Acetamide CH_3CONH_2	C—O	1.220	C—N	1.380	ED
	C—C	1.519	N—H	1.022	
	C—H	1.124	∠NCO	122.0	
	∠CCN	115.1			
Acetic acid	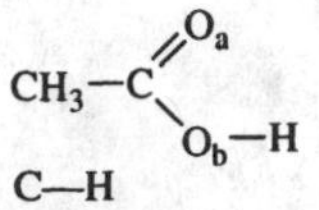		C—C	1.520	ED
			C—O_a	1.214	
			C—O_b	1.364	
	C—H	1.10	∠CCO_a	126.6	
	∠CCO_b	110.6			
Acetone $(CH_3)_2CO$	C—C	1.520	C—O	1.213	ED, M
	C—H	1.103	∠CCC	116.0	
	∠HCH	108.5	symmetry axis of each methyl group is tilted 2° from the C—C bond		
Acetonitrile CH_3CN	C—H	1.107	C—C	1.468	ED,
	C—N	1.159	∠CCH	109.7	
Acetonitrile oxide CH_3CNO	C—C	1.442	C—N	1.169	MW
	N—O	1.217	(C_{3v})		
Acetyl chloride CH_3COCl	C—H	1.105	C—O	1.187	ED,
	C—C	1.506	C—Cl	1.798	
	∠HCH	108.6	∠OCCl	121.2	
	∠CCCl	111.6			
Acetyl cyanide → Pyruvonitrile					
Acetylene HC≡CH	C—H (r_e)	1.060	C—C (r_e)	1.203	IR
Acrolein → Acrylaldehyde					
Acrylaldehyde	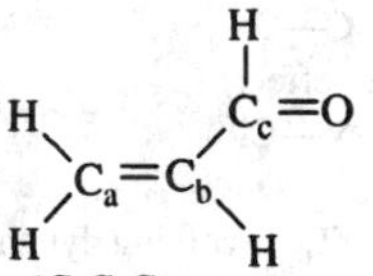		C_b—C_c	1.484	EI
			C_a—C_b	1.345	
			C_c—O	1.217	
			C_a—H	1.10	
			C_c—H	1.13	
	∠$C_aC_bC_c$	120.3	∠C_bC_cO	123.3	
	∠HC_cC_b	114	other CCH angles (average)	122	
	planar *s-trans* form				
Acrylonitrile	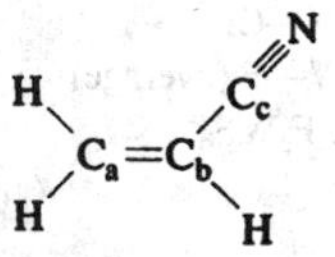		C_a—C_b	1.343	I
			C_b—C_c	1.438	
			C_c—N	1.167	
			C_a—H	1.114	
			∠$C_aC_bC_c$	121.7	
	∠C_bC_cN	178	∠HCC	120	
Acryloyl chloride	C—H	1.086 (assumed)	C_b—C_c	1.48	
	C_c—Cl	1.82	C_a—C_b	1.35	
	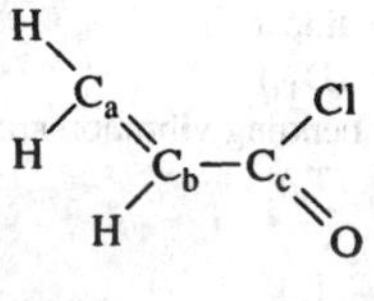		C_c—O	1.19	
			∠C_aC_bH	120 (assumed)	
			∠C_bC_aH	121.5 (assumed)	
			∠$C_aC_bC_c$	123	
			∠C_bC_cCl	116	
			∠C_bC_cO	127	

Compound	Structure				Method
llene CH_2=C=CH_2	C—C	1.3084	C—H	1.087	IR
	∠HCH	118.2			
llyl chloride	H_2C_a=C_bH—C_cH_2(Cl)		*cis* conformer C—Cl	1.811	MW
			∠CCCl	115.2	
			skew conformer C—Cl	1.809	
	∠CCCl	109.6	CCCCl dihedral angle of internal rotation	122.4	
iline $_5H_5NH_2$	C—C	1.392	C—N	1.431	MW
	N—H	0.998	∠HNH	113.9	
	dihedral angle between the NH_2 plane and the N—C bond			140.6	
etidine	CH_2—CH_2 / CH_2—NH (ring)		C—N	1.482	ED
			C—C	1.553	
			C—H	1.107	
	N—H	1.03	∠CNC	92.2	
	∠CCC	86.9	∠CCN	85.8	
	dihedral angle between the CCC and CNC planes			147	
idine	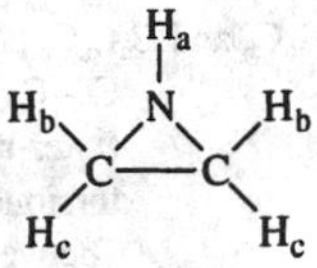		N—H	1.016	MW
			N—C	1.475	
			C—C	1.481	
			C—H	1.084	
			∠CNC	60.3	
			∠H_aNC	109.3	
	∠H_bCH_c	115.7	∠H_bCC	117.8	
	∠H_bCN	118.3	∠H_cCC	119.3	
	∠H_cCN	114.3			
ethane N=NCH$_3$	C—N	1.482	N—N	1.247	ED
	∠CNN	112.3	*trans* conformer		
ne $_5$	C—C	1.399	C—H	1.101	ED, IR
zoquinone	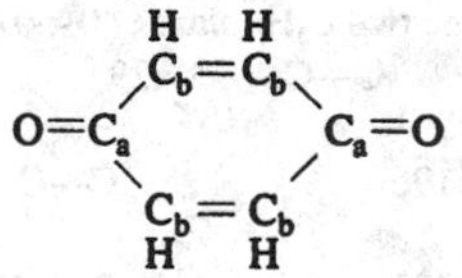		C_a—O	1.225	ED
			C_b—C_b	1.344	
			C_a—C_b	1.481	
			∠$C_bC_aC_b$	118.1	
l OCOCH$_3$	C—O	1.215	C—C (average)	1.524	ED
	C—H	1.108	∠CCO	119.5	
	∠CCC	116.2	*trans* conformer		
[1.1.0]butane	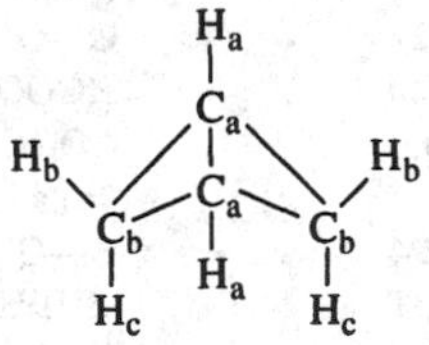		C_a—C_a	1.497	MW
			C_a—C_b	1.498	
			C_a—H_a	1.071	
			C_b—H_b, C_b—H_c	1.093	
			∠$H_bC_bH_c$	115.6	
	∠$C_bC_aH_a$	130.4	∠$C_aC_bC_a$	60.0	
	∠$C_aC_aH_a$	128.4	dihedral angle between the two $C_aC_aC_b$ planes	121.7	
2.2.1]hepta-2,5-diene	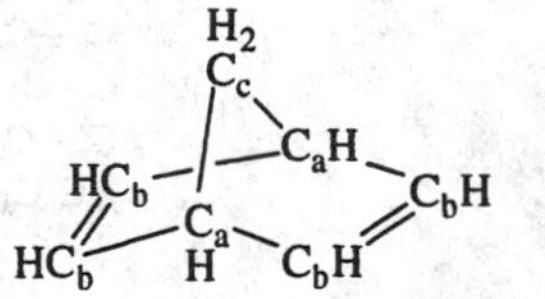		C_a—C_b	1.535	ED
			C_b—C_b	1.343	
			C_a—C_c	1.573	
			C—H	1.12	
			∠$C_aC_cC_a$	94	
	dihedral angle between the two $C_aC_bC_bC_a$ planes (C_{2v})			115.6	
.2.1]heptane	See the preceding molecule for the labels of the C atoms				ED
	C_a—C_b	1.54	C_b—C_b	1.56	
	C_a—C_c	1.56	C—C (average)	1.549	
	-$C_aC_cC_a$	93.1	dihedral angle between the two $C_aC_bC_bC_a$ planes	113.1	

Compound	Structure				Method
Bicyclo[2.2.0]hexa-2,5-diene	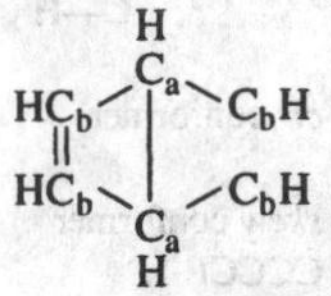		$C_b—C_b$	1.345	ED
			$C_a—C_a$	1.574	
			$C_a—C_b$	1.524	
	dihedral angle between the two $C_aC_bC_bC_a$ planes			117.3	
Bicyclo[2.2.2]octane	$HC_a(C_bH_2C_bH_2)_3C_aH$		$C_a—C_b$	1.54	ED
	$C_b—C_b$	1.55	$\angle C_aC_bC_b$	109.7	
	C—C (average)	1.542			
	large-amplitude torsional motion about the D_{3h} symmetry axis				
Bicyclo[1.1.1]pentane C_5H_8	C—C	1.557	∠CCC	74.2	ED
Bicyclo[2.1.0]pentane	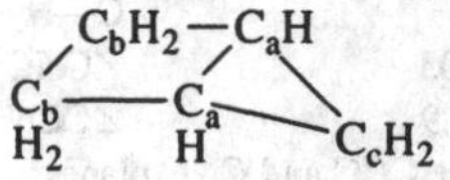		$C_a—C_a$	1.536	MW
			$C_b—C_b$	1.565	
			$C_a—C_b$	1.528	
			$C_a—C_c$	1.507	
	Dihedral angle between the $C_aC_aC_bC_b$ and $C_aC_aC_c$ planes			112.7	
Biphenyl	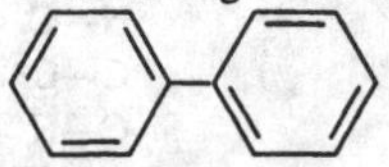		C—C (intra-ring)	1.396	ED
			(inter-ring)	1.49	
			torsional dihedral angle between the two rings	~40	
4,4′-Bipyridyl	C—C, C—N (intra-ring)		1.375		ED
	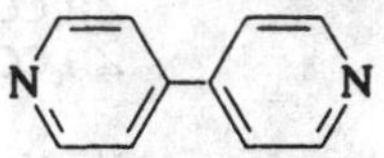		C—C (inter-ring)	1.465	
			torsional dihedral angle between the two rings	~37	
Bis (cyclopentadienyl) beryllium $(C_5H_5)_2Be$	Be—(cyclopentadienyl plane)		1.470, 1.92		ED
	C—C 1.423 (C_{5v}) (The Be atom has two equilibrium positions)				
Bis (cyclopentadienyl) iron → Ferrocene					
Bis (cyclopentadienyl) lead $(C_5H_5)_2Pb$	C—C	1.430	Pb—C	2.79	ED
	dihedral angle between the two C_5H_5 planes 40~50 (The two rings are not parallel.)				
Bis (cyclopentadienyl) manganese $(C_5H_5)_2Mn$	Mn—C 2.383		C—C 1.429	(D_{5h})	ED
Bis (cyclopentadienyl) nickel $(C_5H_5)_2Ni$	Ni—C	2.196	C—C	1.430 (D_{5h})	ED
Bis (cyclopentadienyl) ruthenium $(C_5H_5)_2Ru$	C—C	1.439	Ru—C	2.196	ED
Bis (cyclopentadienyl) tin $(C_5H_5)_2Sn$	C—C	1.431	Sn—C	2.71	EI
	C—H	1.14	(D_{5h})		
Bis (trifluoromethyl) peroxide CF_3OOCF_3	O—O	1.42	C—O	1.399	E
	C—F	1.320	∠COO	107	
	∠FCF	109.0	COOC dihedral angle of internal rotation	123	
Borine carbonyl BH_3CO	B—H	1.194	B—C	1.540	M
	C—O	1.131	∠HBH	113.9	
	∠BCO	180	(C_{3v})		
Bromobenzene	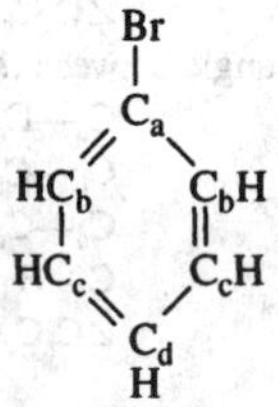		C—H	1.072	M
			$C_c—C_d$	1.401	
			$C_b—C_c$	1.375	
			C—Br	1.85	
			$C_a—C_b$	1.42	
			$\angle C_bC_aC_b$	117.4	
Bromoform $CHBr_3$	C—Br	1.924	C—H	1.11	
	∠BrCBr	111.7	(C_{3v})		
Bromoiodoacetylene IC≡CBr	C—I	1.972	C—C	1.206	
	C—Br	1.795			

Compound	Structure				Method
,3-Butadiene	$C_aH_2=C_bH-C_bH=C_aH_2$		C_b-C_b	1.467	ED
			C_a-C_b	1.349	
			C—H (average)	1.108	
			∠CCC	124.4	
	$\angle C_bC_aH$	120.9	*anti* conformer	(C_{2h})	
3-Butadiyne	$HC_a\equiv C_b-C_b\equiv C_aH$		C—H	1.09	ED
	C_a-C_b	1.218	C_b-C_b	1.384	
			linear		
utane	C—C	1.531	C—H	1.117	ED
$CH_3CH_2CH_2CH_3$	∠CCC	113.8	∠CCH	111.0	
	trans conformer	54%	dihedral angle for the *gauche* conformer	65	
Butanone → Ethyl methyl ketone					
tatriene	$H_2C_a=C_b=C_b=C_aH_2$		C—H	1.08	ED
	C_a-C_b	1.32	C_b-C_b	1.28 (D_{2h})	
utene	$C_aH_3-C_bH=C_bH-C_aH_3$		ED		
	C_a-C_b	*cis* conformer	1.506	*trans* conformer 1.508	
	C_b-C_b	1.346	1.347		
	$\angle C_aC_bC_b$	125.4	123.8		
uten-1-yne → Vinylacetylene					
Butyl chloride	C—H	1.102	C—C	1.528	ED, MW
$H_3)_3CCl$	C—Cl	1.828	∠CCCl	107.3	
	∠CCH	110.8	∠CCC	111.6	
Butyl cyanide → Pivalonitrile					
tyne	$C_aH_3-C_b\equiv C_b-C_aH_3$ C—H		1.116	ED	
	C_b-C_b	1.214	C_a-C_b	1.468	
	$\angle C_bC_aH$	110.7			
on C_2	C—C (r_e)	1.3119			UV
on C_3	C—C	1.277	linear		UV
on suboxide → Tricarbon dioxide					
on tetrabromide	C—Br	1.935	(T_d)		ED
4					
on tetrachloride	C—Cl	1.767	(T_d)		ED
n tetrafluoride	C—F	1.323	(T_d)		ED
n tetraiodide	C—I	2.15	(T_d)		ED
nyl cyanide	C—O	1.209	C—C	1.466	ED, MW
$CN)_2$	C—N	1.153	∠CCC	115	
	∠CCN	180			
acetylene	C—H	1.0550	C—C	1.2033	MW
CCl	C—Cl	1.6368			
benzene	C—C	1.400	C—Cl	1.737	ED
Cl	C—H	1.083	∠CC(Cl)C	121.7	
	∠CC(H)C	120			
bromoacetylene	Cl—C	1.636	C—C	1.206	ED
CBr	C—Br	1.784			
yanoacetylene	C—Cl	1.624	C—C	1.205	ED
CCN	C—CN	1.362	C—N	1.160	
thane → Ethyl chloride					
bethanol	C—O	1.413	C—C	1.519	ED
CH_2OH	C—Cl	1.801	C—H	1.093	
	O—H 1.033	∠CCCl 110.7	∠CCO	113.8	
	fraction of the *gauche* conformer at 37°C is 92 ~ 94%, dihedral angle of internal rotation		62.4		
hylene → Vinyl chloride					
rm	C—H	1.100	C—Cl	1.758	MW
	∠ClCCl	111.3	(C_{3v})		

Compound	Structure				Method
Chloroiodoacetylene	C—Cl	1.63	C—I	1.99	MW
ClC≡CI	C—C	1.209 (assumed)			
Chloromethane → Methyl chloride					
3-Chloropropene → Allyl chloride					
Cyanamide	N_a—C	1.346	C—N_b	1.160	MW
$H_2N_aCN_b$	N—H	1.00	∠HNH	114	
	dihedral angle between the NH_2 plane and the N—C bond			142	
Cyanoacetylene	C_b—H	1.058	C_a—C_b	1.205	MW
H—C_b≡C_a—C_c≡N	C_a—C_c	1.378	C_c—N	1.159	
Cyanocyclopropane	C—C (ring)	1.513	C—C_a	1.472	MW
$C_3H_5C_aN$	C—H	1.107	C_a—N	1.157	
	∠HCH	114.6	∠C_aCH	119.6	
Cyanogen	C—N	1.163	C—C	1.393	ED
$(CN)_2$				linear	
Cyclobutane	C—H	1.113	C—C	1.555	ED
$(CH_2)_4$	dihedral angle between the two CCC planes 145				
Cyclobutanone	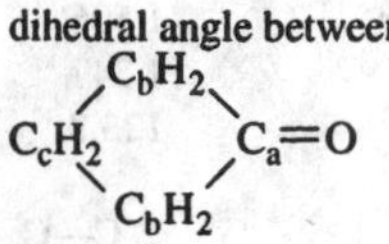		C_a—C_b	1.527	MW
			C_b—C_c	1.556	
			∠$C_bC_aC_b$	93.1	
			∠$C_aC_bC_c$	88.0	
Cyclobutene	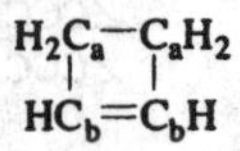 C_b—C_b 1.342; C_a—C_b 1.517; C_b—H 1.083		C_a—C_a 1.566; C_a—H 1.094		MW
	∠$C_aC_bC_b$ 94.2; ∠C_aC_aH 114.5; ∠HC_aH 109.2		∠C_bC_bH 133.5; ∠$C_aC_aC_b$ 85.8; dihedral angle between the CH_2 plane and the C_a—C_a bond 135.8		
Cyclohexane	C—C	1.536	C—H	1.119	ED
C_6H_{12}	∠CCC	111.3	chair form		
Cyclohexene	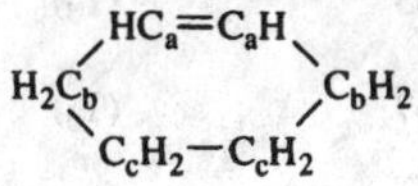		C_a—C_a	1.334	ED
			C_a—C_b	1.50	
			C_b—C_c	1.52	
			C_c—C_c	1.54	
	∠$C_aC_aC_b$	123.4	∠$C_aC_bC_c$	112.0	
	∠$C_bC_cC_c$	110.9	(C_2)	half-chair form	
Cyclooctatetraene	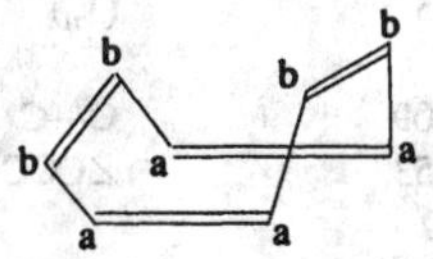		C_a—C_b	1.476	ED
			C—H	1.100	
			C_a—C_a,C_b—C_b	1.340	
			∠$C_bC_aC_a$, ∠$C_aC_bC_b$	126.1	
	dihedral angle between the $C_aC_aC_aC_a$ and $C_aC_bC_bC_a$ planes			136.9	
	tub form	(D_{2d})			
1,3-Cyclopentadiene	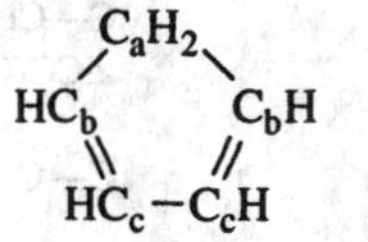		C_a—C_b	1.509	M
			C_b—C_c	1.342	
			C_c—C_c	1.469	
			∠$C_aC_bC_c$	109.3	
	∠$C_bC_cC_c$	109.4	∠$C_bC_aC_b$	102.8	
Cyclopentadienylindium	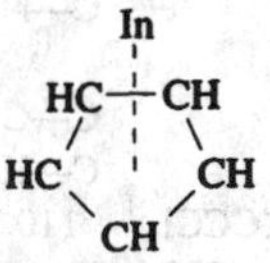		In—C	2.621	E
			C—C	1.426	
			(C_{5v})		
Cyclopentane	C—H	1.114	C—C	1.546	I
$(CH_2)_5$	∠CCH	111.7			
	(The out-of-plane vibration of the C atoms is essentially free pseudorotation; average value of the displacements of the C atoms from the molecular plane 0.43)				

Compound	Structure				Method
yclopentene	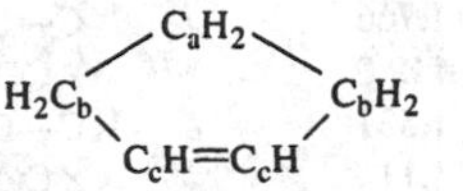		C_a—C_b	1.546	ED
			C_b—C_c	1.519	
			C_c—C_c	1.342	
			$\angle C_bC_aC_b$	104.0	
	$\angle C_bC_cC_c$	110.0	$\angle C_aC_bC_c$	103.0	
	dihedral angle between the $C_bC_aC_b$ and $C_bC_cC_cC_b$ planes			151.2	
'clopropane	C—C	1.512	C—H	1.083	R
$'H_2)_3$	∠HCH	114.0			
clopropanone	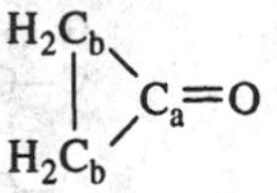 C—H	1.086	C_a—C_b	1.475	MW
	C_b—C_b	1.575	C_a—O	1.191	
	$\angle C_aC_bC_b$	57.7			
	$\angle HC_bH$	114	dihedral angle between the CH_2 plane and the C_b—C_b bond	151	
clopropene	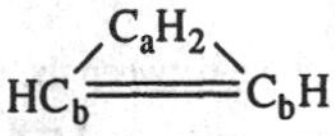		C_b—C_b	1.304	ED
			C_a—C_b	1.519	
			C_b—H	1.077	
	C_a—H	1.112	$\angle C_bC_bH$	133	
	$\angle HC_aH$	118			
alin	C—C (average)	1.530	C—H (average)	1.113	ED
H_{18}	∠CCC (average)	111.4			
er benzene → Bicyclo[2.2.0] a-2,5-diene					
etylene → 1,3-Butadiyne					
Diazabicyclo[2.2.2]octane	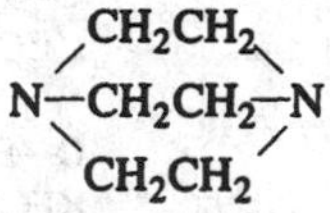		C—N	1.472	ED
			C—C	1.562	
			∠NCC	110.2	
			∠CNC	108.7	
	large-amplitude torsional motion about the D_{3h} symmetry axis				
iaza-1,3-butadiene → Formalde- azine					
ine	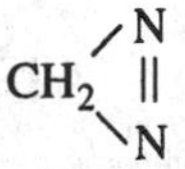		C—H	1.09	MW
			C—N	1.482	
			N—N	1.228	
			∠HCH	117	
acetonitrile	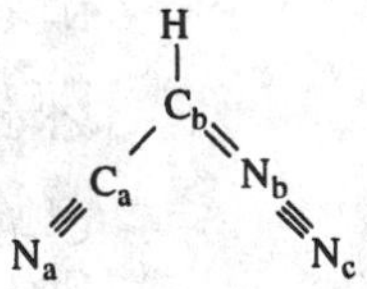		C_b—N_b	1.280	MW
			N_b—N_c	1.132	
			C_a—N_a	1.165	
			C—H	1.082	
			C_a—C_b	1.424	
	$\angle C_aC_bH$	117	$\angle C_aC_bN_b$	119.5	
ethane	C—H	1.075	C—N	1.32	MW, IR
2	N—N	1.12	∠HCH	126.0 linear	
romoethane	C—C	1.506	C—Br	1.950	ED
CH_2Br	C—H	1.108	∠CCBr	109.5	
	∠CCH 110	fraction of the *trans* conformer at 25°C		95%	
omethane	C—H	1.08	C—Br	1.924	ED
2	∠HCBr	109	∠BrCBr	113.2	
hlorobiphenyl	C—C	1.398	C—C inter-ring	1.495	ED
—C_6H_4Cl	C—Cl	1.732	C—H	1.10	
	∠CCCl	121.4	∠CCH	126	
	dihedral angle between the two aromatic rings 74 (defined to be 0 for that of the *cis* conformer)				
-Dichlorocyclohexane	C—H	1.102	C—Cl	1.810	ED
2	C—C	1.530	∠CCC	111.5	
	∠CCCl (*ee*)	108.6	∠CCCl (*aa*)	110.6	
	∠HCCl (*ee*)	111.5	∠HCCl (*aa*)	107.6	
	ee 49%	*aa* 51%	e: equatorial, a: axial		

Compound	Structure				Method
1,1-Dichloroethane $CHCl_2CH_3$	C—Cl	1.766	C—C	1.540	MW
	∠ClCCl	112.0	∠CCCl	111.0	
1,2-Dichloroethane CH_2ClCH_2Cl	C—C	1.531	C—Cl	1.790	ED
	C—H	1.11	∠CCCl	109.0	
	∠CCH	113			
	fraction of the *trans* conformer at room temperature 73%, that of the *gauche* conformer 27%				
1,1-Dichloroethylene $CH_2=CCl_2$	C—C	1.32 (assumed)	C—Cl	1.73	MW
	∠ClCC	123	(C_{2v})		
cis-1,2-Dichloroethylene CHCl=CHCl	C—Cl	1.718	C—C	1.354	ED
	∠ClCC	123.8			
Dichloromethane CH_2Cl_2	C—H (r_e)	1.087	C—Cl (r_e)	1.765	MW,
	∠HCH (θ_e)	111.5	∠ClCCl (θ_e)	112.0	
1,1-Difluoroethane CH_3CHF_2	C—C	1.498	C—H (average)	1.081	ED
	C—F	1.364	∠CCH (average)	111.0	
	∠CCF 110.7	dihedral angle between the two CCF planes		118.9	
1,2-Difluoroethane CH_2FCH_2F	C—F	1.389	C—C	1.503	ED
	C—H	1.103	∠CCF	110.3	
	∠CCH 111	dihedral angle of internal rotation		109	
	fraction of the *gauche* conformer at 22°C 94%				
1,1-Difluoroethane $CH_2=CF_2$	C—C	1.340	C—F	1.315	ED,
	C—H	1.091	∠CCF	124.7	
	∠CCH	119.0			
cis-1,2-Difluoroethylene CHF=CHF	C—C	1.33	C—F	1.342	ED
	C—H	1.099	∠CCF	122.0	
	∠CCH	124.1			
Difluoromethane CH_2F_2	C—H	1.093	C—F	1.357	M
	∠HCH	113.7	∠FCF	108.3	
Dimethoxymethane	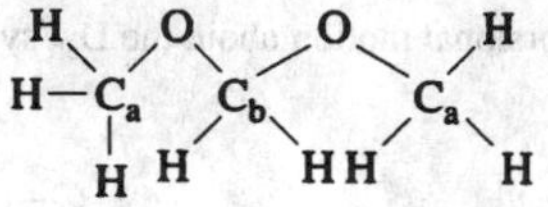		C_a—O	1.432	ED
			C_b—O	1.382	
			C—H (average)	1.108	
	∠COC	114.6	∠OCO	114.3	
	∠OCH	110.3			
Dimethylacetylene → 2-Butyne					
Dimethylamine $(CH)_2NH$	C—H	1.106	N—H	1.00	E
	C—N	1.455	∠CNC	111.8	
	∠CNH	107	∠NCH	112	
	∠HCH	107			
Dimethylberyllium $(CH_3)_2Be$	Be—C	1.698	C—H	1.127	E
	∠BeCH	113.9	CBeC linear		
Dimethylcadmium $(CH_3)_2Cd$	C—Cd	2.112	∠HCH	108.4	
Dimethyl carbonate	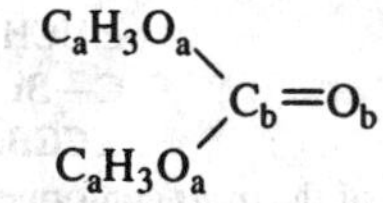		C_b—O_b	1.209	
			C_b—O_a	1.34	
			C_a—O_a	1.42	
	∠$O_aC_bO_a$	107	∠$C_bO_aC_a$	114.5	
Dimethylcyanamide $(C_aH_3)_2N_a$—C_b≡N_b	C_b—N_b	1.161	C_b—N_a	1.338	
	C_a—N_a	1.463	∠C_aNC_a	115.5	
	∠C_aNC_b	116.0			
1,2-Dimethyldiborane	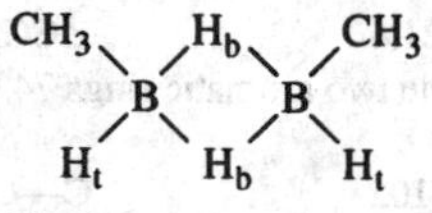		B—B	1.799	
			B—C	1.580	
			B—H_b	1.358 (*cis*), 1.365 (*trans*)	
	B—H_t	1.24			
	∠BBC	122.6 (*cis*), 121.8 (*trans*)			
Dimethyl diselenide $(CH_3)_2Se_2$	C—H	1.13	C—Se	1.95	
	Se—Se	2.326	∠CSeSe	98.9	
	∠HCSe 108	dihedral angle between the CSeSe and SeSeC planes		88	

Compound	Structure				Method
imethyl disulfide	C—S	1.816	S—S	2.029	ED
$CH_3)_2S_2$	C—H	1.105	∠SSC	103.2	
	∠SCH 111.3	CSSC dihedral angle of internal rotation		85	
S′-Dimethyl dithiocarbonate	$C_aH_3SC_bSC_aH_3$		C_b—O	1.206	ED
	‖		C_b—S	1.777	
	O		C_a—S	1.802	
	∠OCS	124.9	∠CSC	99.3	
			syn-syn conformer		
methyl ether	C—O	1.416	C—H	1.121	ED
$CH_3)_2O$	∠COC	112	∠HCH	108	
nethylglyoxal → Biacetyl					
′-Dimethylhydrazine	N—N	1.42	C—N	1.46	ED
H_3NH—$NHCH_3$	N—H	1.03	C—H	1.12	
	∠NNC 112	CNNC dihedral angle of internal rotation		90	
nethylmercury	C—Hg	2.083	C—H	1.160 (assumed)	ED
$H_3)_2Hg$	Hg···H	2.71			
ethylphosphine	C—P	1.848	P—H	1.419	MW
$H_3)_2PH$	∠CPC	99.7	∠CPH	97.0	
ethyl selenide	C—H	1.093	Se—C	1.943	MW
$I_3)_2Se$	∠CSeC	96.2	∠SeCH	108.7	
	∠HCH	110.3			
ethyl sulfide	C—S	1.807	C—H	1.116	ED, MW
$_3)_2S$	∠CSC	99.05	∠HCH	109.3	
thyl sulfone	C—H	1.114	S—O	1.435	ED
$_3)_2SO_2$	S—C	1.771	∠CSC	102	
	∠OSO	121			
thyl sulfoxide	C—H	1.081	C—S	1.799	MW
$_3)_2SO$	S—O	1.485	∠CSC	96.6	
	∠CSO	106.7	∠HCH	110.3	
	dihedral angle between the SCC plane and the S—O bond			115.5	
hylzinc	Zn—C	1.929	∠HCH	107.7	R
$)_2Zn$					
oxane	C—C	1.523	C—O	1.423	ED
I_2CH_2╲	C—H	1.112	∠COC	112.45	
O	∠CCO	109.2	chair form		
I_2CH_2╱					
l → Acetaldehyde					
	C—C	1.5351	C—H	1.0940	MW
	∠CCH	111.17	staggered conformation		
hiol	C_bH_3—C_aH_2—SH		C_a—H	1.090	MW
	C_b—H	1.093	C_a—C_b	1.530	
	C_a—S	1.829	S—H	1.350	
	$∠C_bC_aH$	109.6	$∠C_aC_bH$	109.7	
	$∠C_bC_aS$	108.3	$∠C_aSH$	96.4	
	$C_bH_3C_aH_2OH$		C—C	1.512	MW
	C—O	1.431	O—H	0.971	
	C_a—H	1.10	C_b—H	1.09	
	∠CCO	107.8	∠COH	105	
	$∠C_bC_aH$	111	$∠C_aC_bH$	110	
	staggered conformation				
oride	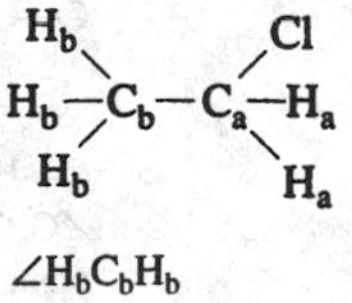		C—C	1.528	ED, MW
			C—Cl	1.802	
			C—H	1.103	
			C_a—H_a=C_b—H_b (assumed)		
			∠CCCl	110.7	
	$∠H_bC_bH_b$	109.8	$∠H_aC_aH_a$	109.2	
	$∠C_bC_aH_a$	110.6			
	C—H	1.087	C—C	1.339	MW
2	∠CCH	121.3			

Compound	Structure				Metho
Ethylenediamine $H_2NCH_2CH_2NH_2$	C—N	1.469	C—C	1.545	ED
	C—H	1.11	∠CCN	110.2	
	gauche conformer	dihedral angle between the NCC and CCN planes		64	
Ethylene dibromide → 1,2-Dibromoethane					
Ethylene dichloride → 1,2-Dichloroethane					
Ethyleneimine → Aziridine					
Ethylene oxide 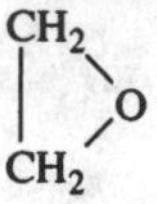	C—C	1.466	C—H	1.085	MW
	C—O	1.431	∠HCH	116.6	
	dihedral angle between the NH_2 plane and the N—C bond			158.0	
Ethylene sulfide → Thiirane					
Ethyl methyl ether $C_2H_5OCH_3$	C—O (average)	1.418	C—C	1.520	ED
	C—H (average)	1.118	∠COC	111.9	
	∠OCC	109.4	∠HCH	109.0	
	fraction of the *trans* conformer at 20°C		80%		
Ethyl methyl ketone	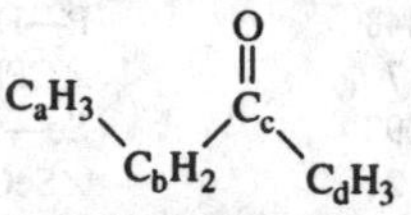		C—C (average)	1.518	ED
			C_c—O	1.219	
			C—H (average)	1.102	
			$∠C_aC_bC_c$	113.5	
	$∠C_bC_cO$, $∠C_dC_cO$	121.9	*trans* conformer	95%	
Ethyl methyl sulfide $C_2H_5SCH_3$	C—S (average)	1.813	C—C	1.536	ED
	C—H	1.111	∠CSC	97	
	∠SCC	114.0	∠HCH	110	
	fraction of the *gauche* conformer at 20°C		75%		
Ferrocene $(C_5H_5)_2Fe$	C—C	1.440	C—H	1.104	ED
	Fe—C	2.064	(D_{5h})		
Fluoroform CHF_3	C—H	1.098	C—F	1.332	M
	∠FCF	108.8	(C_{3v})		
Formaldehyde H_2CO	C—H	1.116	C—O	1.208	M
	∠HCH	116.5			
Formaldehyde azine	$H_2C{=}N{-}N{=}CH_2$	N—N	1.418		EI
	C—N	1.277	C—H	1.094	
	∠CNN	111.4	∠HCN	120.7	
	fraction of the *trans* conformer at −30°C		91%		
Formaldehyde dimethylacetal → Dimethoxy-methane					
Formaldoxime	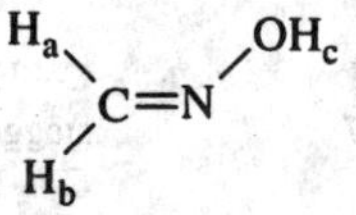		$C—H_a$	1.085	M
			$C—H_b$	1.086	
			C—N	1.276	
	N—O	1.408	$O—H_c$	0.956	
	$∠H_bCN$	115.6	∠CNO	110.2	
	$∠H_aCN$	121.8	$∠NOH_c$	102.7	
Formamide	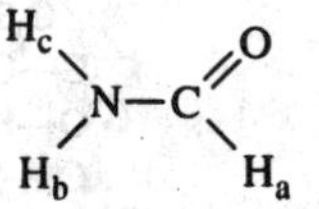		$C—H_a$	1.125	
			N—H	1.027	
			C—N	1.368	
	C—O	1.212	∠NCO	125.0	
	∠CNH (average)	119.2			
Formic acid	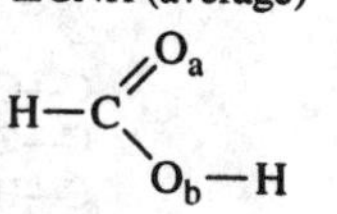		$C—O_a$	1.202	
			$C—O_b$	1.343	
	$O_b—H$	0.972	C—H	1.097	
	$∠HCO_a$	124.1	$∠O_aCO_b$	124.9	
	$∠CO_bH$	106.3	planar		

Compound	Structure				Method
Formic acid dimer	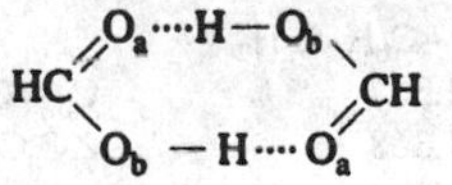		O_a···O_b	2.703	ED
			C—O_a	1.220	
			C—O_b	1.323	
			∠O_aCO_b	126.2	
			∠CO_aO_b	108.5	
Formyl radical	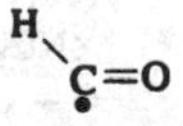 C—H	1.110	C—O	1.1712	MW
	∠HCO	127.43			
Fulvene	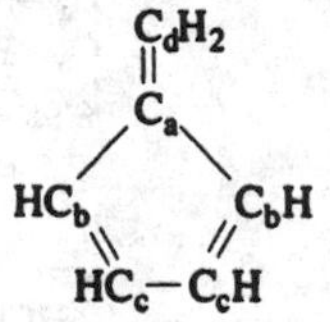		C_a—C_d	1.349	MW
			C_a—C_b	1.470	
			C_b—C_c	1.355	
			C_c—C_c	1.476	
			C_b—H	1.078	
			C_c—H	1.080	
	C_d—H	1.13	∠$C_bC_aC_b$	106.6	
	∠$C_aC_bC_c$	107.7	∠$C_bC_cC_c$	109	
	∠C_aC_bH	124.7	∠C_bC_cH	126.4	
	∠HC_dH	117			
2-Furaldehyde	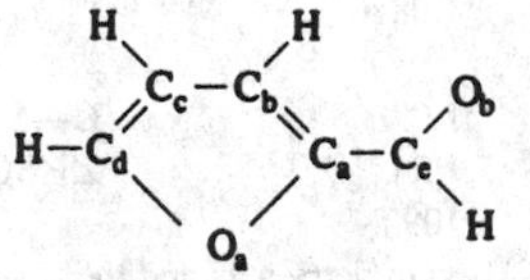		C_a—C_e	1.458	MW
			C_e—O_b	1.250	
			C_e—H	1.088	
			∠$C_eC_aC_b$	133.9	
	∠C_aC_eH	116.9	∠C_aC_eO	121.6	
	trans conformer (with respect to the O_a and O_b atoms)				
Furan	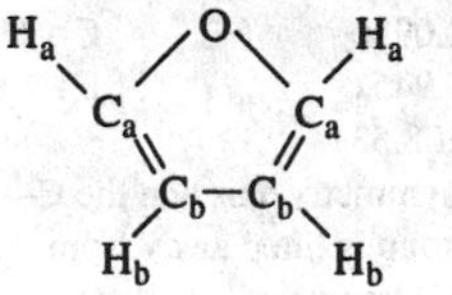		C_b—C_b	1.431	MW
			C_a—C_b	1.361	
			C_a—O	1.362	
			C_a—H_a	1.075	
			C_b—H_b	1.077	
	∠$C_aC_bC_b$	106.1	∠C_aOC_a	106.6	
	∠C_bC_aO	110.7	∠OC_aH_a	115.9	
	∠$C_bC_bH_b$	128.0			
Furfural → 2-Furaldehyde					
Glycolaldehyde	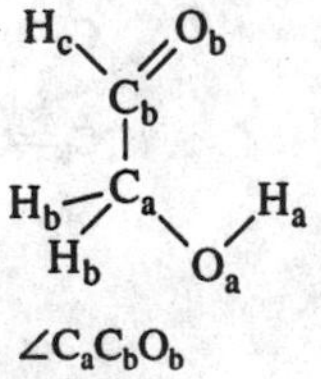		C_b—O_b	1.209	MW
			C_a—O_a	1.437	
			C_a—C_b	1.499	
			O_a—H_a	1.051	
			C_b—H_c	1.102	
			C_a—H_b	1.093	
	∠$C_aC_bO_b$	122.7	∠$C_aC_bH_c$	115.3	
	∠$C_bC_aO_a$	111.5	∠$C_aO_aH_a$	101.6	
	∠$C_bC_aH_b$	109.2	∠$H_bC_aH_b$	107.6	
	∠$H_bC_aO_a$	109.7			
Glyoxal	C—C	1.526	C—O	1.212	ED, UV
CHOCHO	C—H	1.132	∠CCO	121.2	
	∠HCO	112	*trans* conformer	(C_{2h} (assumed))	
Hexachloroethane	C—C	1.56	C—Cl	1.769	ED
Cl_3CCCl_3	∠CCCl	110.0			
2,4-Hexadiyne	C_aH_3—C_b≡C_c—C_c≡C_b—C_aH_3				ED
	C_a—C_b	1.450	C_b—C_c	1.208	
	C_c—C_c	1.377	C_a—H	1.09	
Hexafluoroethane	C—C	1.545	C—F	1.326	ED
F_3CCF_3	∠CCF	109.8	staggered conformation		
Hexafluoropropene	average value of the C=C and C—F distances			1.329	ED
CF_2=$CFCF_3$	C—C	1.513	∠CCC	127.8	
	∠FCC (CF_2)	124	∠FCC (CF)	120	
	∠FCC (CF_3)	110			

Compound	Structure				Method
1,3,5-Hexatriene	H_2C_a=C_bH—C_cH=C_cH—C_bH=C_aH_2				ED
	C_a—C_b	1.337	C_b—C_c	1.458	
	C_c—C_c	1.368	∠$C_aC_bC_c$	121.7	
	∠$C_bC_cC_c$	124.4			
Iminocyanide radical	N—H	1.034	N···N	2.470	UV
HṄCN	∠HNC	116.5	∠NCN	~180	
Iodocyanoacetylene	I—C_a	1.985	C_a—C_b	1.207	MW
I—C_a≡C_b—C_c≡N	C_b—C_c	1.370	C_c—N	1.160	
Isobutane	C_a—H	1.122	C_b—H	1.113	ED, MW
$(C_bH_3)_3C_aH$	C_a—C_b	1.535	∠$C_bC_aC_b$	110.8	
	∠C_aC_bH	111.4			
Isobutylene → 2-Methylpropene					
Ketene	C—C	1.317	C—O	1.161	MW
CH_2=C=O	C—H	1.080	∠HCH	123.0	
Malononitrile	C—H	1.091	C—C	1.480	MW
$C_aH_2(C_bN)_2$	C—N	1.147	∠CCC	110.4	
	∠HCH	108.4	∠CCN	176.6	
	(The two N atoms are bent away from each other in the plane of C_b—C_a—C_b)				
Methane	C—H (r_e)	1.0870	(T_d)		MW
CH_4					
Methanethiol	C—H	1.09	C—S	1.819	MW
CH_3SH	S—H	1.34	∠HSC	96.5	
	∠HCH	109.8			
	angle between the CH_3 symmetry axis and the C—S bond 2.2. (The axis of the CH_3 group is tilted away from the H atom with respect to the C—S bond.)				
Methanol	C—H	1.0936	C—O	1.4246	MW
CH_3OH	O—H	0.9451	∠HCH	108.63	
	∠COH	108.53			
	angle between the CH_3 symmetry axis and the C—O bond (The axis of the CH_3 group is tilted away from the H atom with respect to the C—O bond.)			3.27	
Methyl radical	C—H	1.08	planar		UV
·CH_3					
N-Methylacetamide	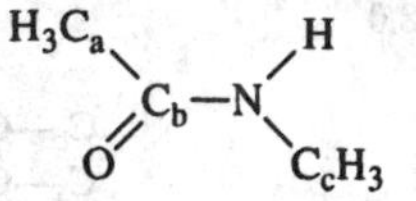		C_a—C_b	1.520	ED
			N—C_c	1.469	
			C—H	1.107	
	C_b—N	1.386	C_b—O	1.225	
	∠C_bNC_c	119.7			
	∠NC_bO	121.8			
	∠C_aC_bN	114.1			
Methylacetylene → Propyne					
Methylal → Dimethoxymethane					
Methylamine	N—H	1.010	C—N	1.471	MW
CH_3NH_2	C—H	1.099	∠NHN	107.1	
	∠HNC	110.3	∠HCH	108.0	
	dihedral angle between the CH_3 symmetry axis and the C—N bond (The axis of the CH_3 group is tilted away from the NH_2 group with respect to the C—N bond.)			2.9	
Methyl azide	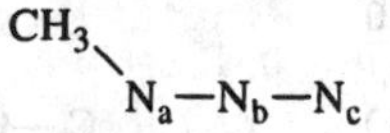		C—H	1.09	ED
			C—N_a	1.468	
			N_a—N_b	1.216	
	N_b—N_c	1.113	∠CN_aN_b	116.8	
	NNN linear				
Methyl bromide	C—H (r_e)	1.086	C—Br (r_e)	1.933	MW, IR
CH_3Br	∠HCH (θ_e)	111.2	(C_{3v})		
Methyl chloride	C—H	1.090	C—Cl	1.785	MW, IR
CH_3Cl	∠HCH	110.8			

Compound	Structure				Method
Methyldiazirine	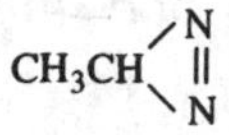 C—N	1.481	C—C	1.501	MW
	N—N	1.235	∠NCN	49.3	
	dihedral angle between the CNN plane and the C—C bond			122.3	
Methylene :CH_2	C—H	1.078	∠HCH	130	LMR
Methylenecyclopropane	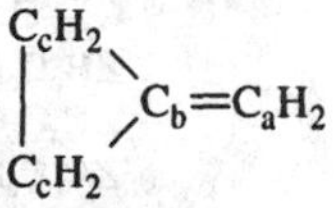		C_a—C_b	1.332	MW
			C_b—C_c	1.457	
			C_c—C_c	1.542	
	C_c—H	1.09	∠$C_cC_bC_c$	63.9	
	∠HC_aH	114.3	∠HC_cH	113.5	
	dihedral angle between the C_cH_2 plane and the C_c—C_c bond			150.8	
3-Methyleneoxetane	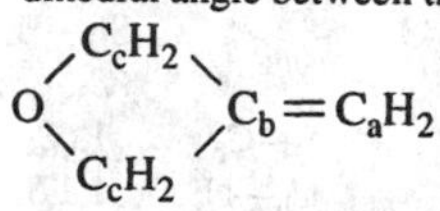		C_b—C_c	1.52	MW
			C_c—O	1.45	
			C_a—C_b	1.33	
	C—H	1.09 (assumed)	∠$C_cC_bC_c$	87	
	∠HC_cH	114 (assumed)	∠HC_aH	120 (assumed)	
Methyl fluoride CH_3F	C—H (r_e)	1.095			MW, IR
	C—F (r_e)	1.382	∠HCH (θ_e)	110.45 (C_{3v})	
Methyl formate	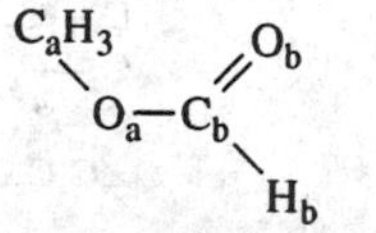		C_a—H	1.08	ED
			C_b—O_b	1.206	
			C—O (average)	1.393	
			C_b—H	1.101 (assumed)	
	∠$O_aC_bO_b$	127	∠COC	114	
	∠O_aC_aH	110			
Methylgermane CH_3GeH_3	C—H	1.083	Ge—H	1.529	MW
	C—Ge	1.945	∠HCH	108.4	
	∠HGeH	109.3			
Methyl hypochlorite CH_3OCl	C—H	1.103	O—Cl	1.674	MW
	O—C	1.389	∠HCH	109.6	
	∠COCl	112.8			
Methylidyne radical :ĊH	C—H (r_e)	1.1198			UV
Methylidyne phosphide HCP	H—C (r_e)	1.0692	C—P (r_e)	1.5398	MW
Methyl iodide CH_3I	C—H (r_e)	1.084	C—I (r_e)	2.132	MW, IR
	∠HCH (θ_e)	111.2	(C_{3v})		
Methyl isocyanide	C_aH_3—N≡C_b		C_a—H 1.102	C_a—N 1.424	MW
	N—C_b	1.166	∠NC_aH	109.12	
Methylketene	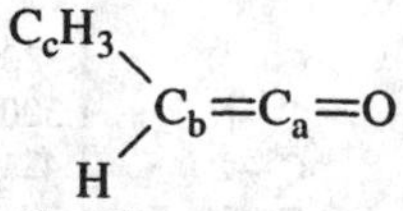		O—C_a	1.171	MW
			C_b—C_c	1.518	
			C_c—H	1.10	
	C_a—C_b	1.306	C_b—H	1.083	
	∠OC_aC_b	180.5	∠$C_aC_bC_c$	122.6	
	∠C_aC_bH	113.7	∠C_cC_bH	123.7	
	∠HCH	109.2			
Methylmercury chloride CH_3HgCl	Hg—Cl	2.282	C—H	1.15	MW, NMR
	Hg—C	1.99	(C_{3v})		
Methyl nitrate	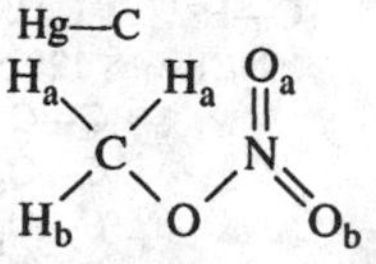		C—H_a	1.10	MW
			C—H_b	1.09	
			C—O	1.437	
			O—N	1.402	
	N—O_a	1.205	N—O_b	1.208	
	∠OCH_a	110	∠OCH_b	103	
	∠CON	112.7	∠ONO_a	118.1	
	∠ONO_b	112.4			

Compound	Structure						Method
Methylphosphine CH_3PH_2	C—P	1.858	C—H	1.094			ED
2-Methylpropane → Isobutane							
2-Methylpropene	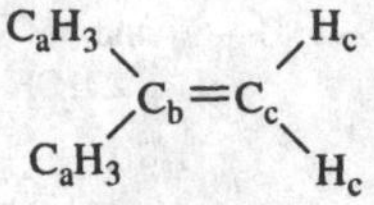		C_a—H	1.119			ED, MW
			C_c—H_c	1.10			
			C_a—C_b	1.508			
			C_b—C_c	1.342			
	$\angle HC_aC_b$ (average)	111.4	$\angle H_cC_cH_c$	118.5			
	$\angle C_aC_bC_a$	115.6	$\angle C_aC_bC_c$	122.2			
	$\angle HC_aH$	107.9	$\angle C_bC_cH$	121			
Methylsilane CH_3SiH_3	C—H	1.093	C—Si	1.867			MW
	Si—H	1.485	∠HCH	107.7			
	∠HSiH	108.3	(C_{3v})				
Methylstannane CH_3SnH_3	C—Sn	2.143	Sn—H	1.700			MW
			(C_{3v})				
Methyl thiocyanate	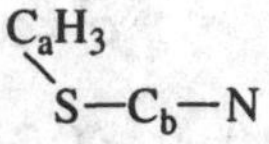		S—C_b	1.684	S—C_a	1.824	MW
			C_b—N	1.170	C—H	1.081	
	$\angle C_aSC_b$	99.0	∠HCH	110.6	∠HCS	108.3	
Naphthalene	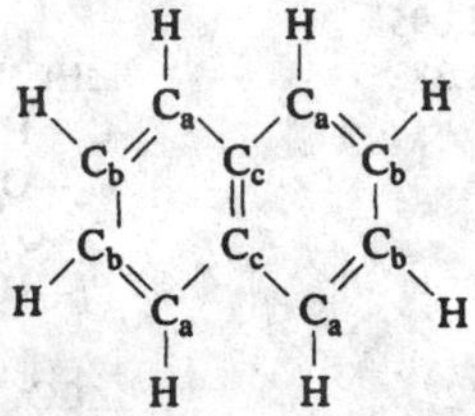		C_a—C_b	1.37			ED
			C_b—C_b	1.41			
			C_a—C_c	1.42			
			C_c—C_c	1.42			
			C—C (average)	1.40			
			$\angle C_aC_cC_c$	119.4			
Neopentane $C(CH_3)_4$	C—C	1.537	C—H	1.114			ED
	∠CCH	112					
Nickelocene → Bis (cyclopentadienyl) nickel							
Nitromethane CH_3NO_2	C—H	1.088 (assumed)	C—N	1.489			MW
	N—O	1.224	∠NCH	107			
	∠ONO	125.3					
N-Nitrosodimethylamine $(CH_3)_2NNO$	N—O	1.235	N—N	1.344			ED
	C—N	1.461	∠ONN	113.6			
	∠CNC	123.2	∠CNN	116.4			
Nitrosomethane CH_3NO	C—N	1.49	N—O	1.22			MW
	C—H	1.084	∠CNO	112.6			
	∠NCH	109.0					
Norbornane → Bicyclo[2.2.1]heptane							
Norbornadiene → Bicyclo[2.2.1]hepta-2,5-diene							
1,2,5-Oxadiazole	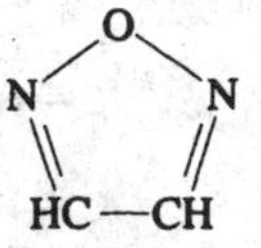		O—N	1.380	∠NON	110.4	MW
			C—N	1.300	∠ONC	105.8	
			C—C	1.421	∠CCN	109.0	
			C—H	1.076	∠CCH	130.2	
			∠NCH	120.9	planar		
1,3,4-Oxadiazole	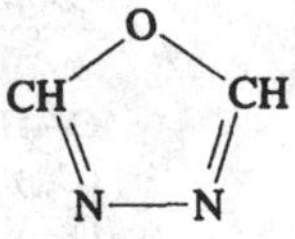		O—C	1.348	∠COC	102.0	MW
			C—N	1.297	∠OCN	113.4	
			N—N	1.399	∠CNN	105.6	
			C—H	1.075	∠OCH	118.1	
			∠NCH	128.5	planar		
Oxalic acid	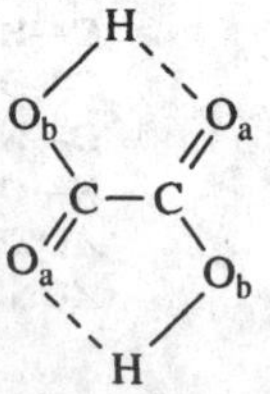		C—C	1.544			ED
			C—O_a	1.205			
			C—O_b	1.336			
			O_b—H	1.05			
			$\angle CCO_a$	123.1			
			$\angle O_aCO_b$	125.0			
			$\angle CO_bH$	104			

Compound	Structure				Method
Oxalyl dichloride	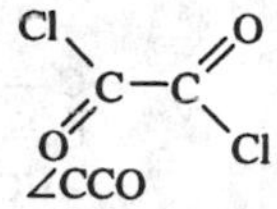				
			C—O	1.182	ED
			C—C	1.534	
			C—Cl	1.744	
	∠CCO	124.2	∠CCCl	111.7	
	fraction of the *trans* conformer at 0°C 68%, that of the *gauche* conformer 32%				
Oxetane	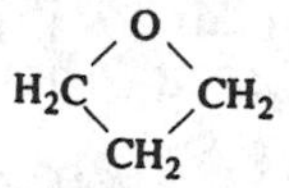				
			C—O	1.448	MW
			C—C	1.546	
			C—H (average)	1.090	
			∠COC	92	
	∠CCC	85	∠OCC	92	
	∠HCH (average)	109.9			
Oxirane → Ethylene oxide					
Phenol	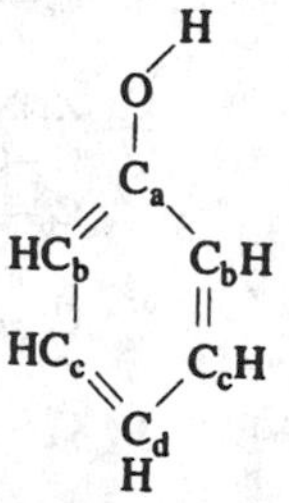				
			C—C (average)	1.397	MW
			C_b—H	1.084	
			C_c—H	1.076	
			C_d—H	1.082	
			C_a—O	1.364	
			O—H	0.956	
			∠COH	109.0	
Phosphirane	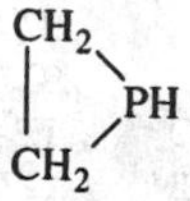				
	C—P 1.867	C—C 1.502	∠CPC 47.4	P—H 1.43; C—H 1.09; ∠HCH 114.4	MW
	∠HPC	95.2	∠CCH	118	
	dihedral angle between the PCC plane and the PH bond			95.7	
Piperazine	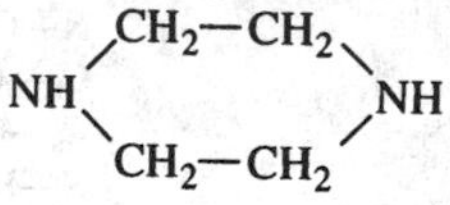				
			C—C	1.540	ED
			C—N	1.467	
			C—H	1.110	
	∠CNC	109.0	∠CCN	110.4	
			(C_{2h})		
Pivalonitrile $(C_cH_3)_3C_b—C_a≡N$	$C_a—C_b$	1.495	C_a—N	1.159	MW
	$C_b—C_c$	1.536	$∠C_cC_bC_c$	110.5	
Propadiene → Allene					
Propane C_3H_8	C—C	1.532	C—H	1.107	ED
	∠CCC	112	∠HCH	107	
Propenal → Acrylaldehyde					
Propene	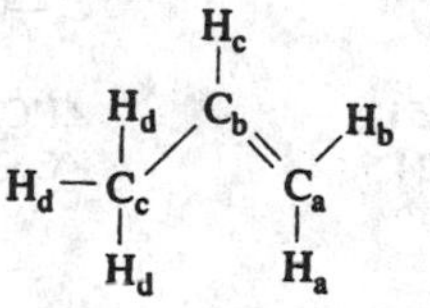				
			$C_a—H_a$	1.104	ED, MW
			$C_a—C_b$	1.341	
			$C_c—H_d$	1.117	
			$C_b—C_c$	1.506	
			$∠C_bC_aH_{a,b,c}$	121.3	
	$∠C_bC_cH_d$	110.7	$∠C_aC_bC_c$	124.3	
1-Propenyl chloride	$CH_3—C_bH{=}C_aH—Cl$	C_a—Cl	1.728	MW	
	$∠C_bC_aCl$	121.9	*trans* conformer		
Propiolaldehyde	$H_aC_a≡C_b—C_cH_cO$		$C_a—H_a$	1.085	ED, MW
	$C_a—C_b$	1.211	$C_b—C_c$	1.453	
	$C_c—H_c$	1.130	C_c—O	1.214	
	$∠C_bC_cO$	124.2	$∠C_bC_cH_c$	113.7	
	$∠C_aC_bC_c$	178.6	planar		
Propylene → Propene					
Propylene oxide	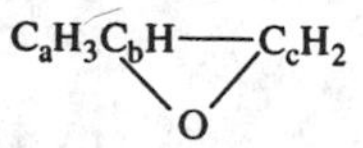				
			$C_a—C_b$	1.51	MW
			$∠C_aC_bC_c$	121.0	
			dihedral angle between the C_bC_cO plane and the C_aC_b bond	123.8	

Compound	Structure						Method
Propynal → Propiolaldehyde							
Propyne	$H_3C_c—C_b≡C_aH$		$C_c—H$	1.105			MW
	$C_c—C_b$	1.459	$C_b—C_a$	1.206			
	$C_a—H$	1.056	∠HC_cC_b	110.2			
Pyrazine			C—C	1.339	C—N	1.403	ED
			C—H	1.115	∠CCN	115.6	
			∠CCH	123.9			
Pyridazine	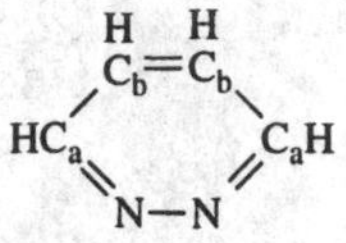		$N—C_a$	1.341			ED, MW
			$C_a—C_b$	1.393			
			N—N	1.330			
			$C_b—C_b$	1.375			
	∠NCC	123.7	∠NNC	119.3			
Pyridine	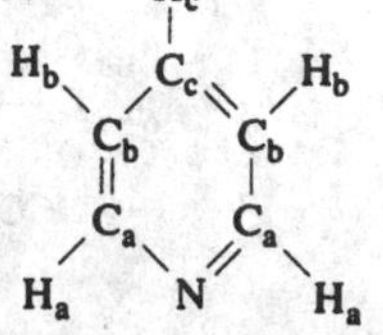		$N—C_a$	1.340			MW
			$C_b—C_c$	1.394			
			$C_b—H_b$	1.081			
			$C_a—C_b$	1.395			
			$C_a—H_a$	1.084			
			$C_c—H_c$	1.077			
	∠C_aNC_a	116.8	∠NC_aC_b	123.9			
	∠$C_aC_bC_c$	118.5	∠$C_bC_cC_b$	118.3			
	∠NC_aH_a	115.9	∠$C_cC_bH_b$	121.3			
Pyrimidine			N—C	1.340	C—C	1.393	ED
			∠NCN	127.6	∠CNC	115.5	
			(C_{2v} assumed)				
Pyrrole	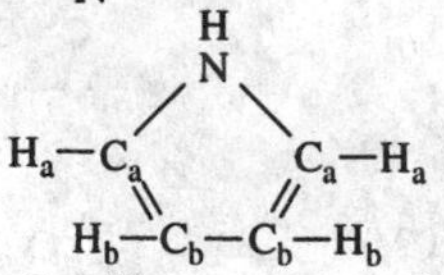		N—Ca	1.370			MW
			$C_b—C_b$	1.417			
			$C_a—C_b$	1.382			
			N—H	0.996			
			$C_a—H_a$	1.076			
	$C_b—H_b$	1.077	∠C_aNC_a	109.8			
	∠NC_aC_b	107.7	∠$C_aC_bC_b$	107.4			
	∠NC_aH_a	121.5	∠C_bC_bH	127.1			
Pyruvonitrile	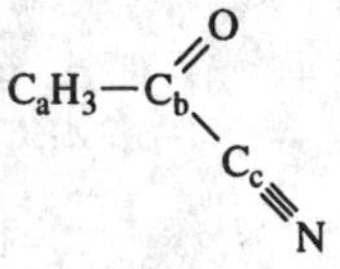		C—H	1.12			ED, MW
			C—N	1.17			
			C—O	1.208			
			$C_b—C_c$	1.477			
	$C_a—C_b$	1.518	∠HCH	109.2			
	∠C_aC_bO	124.5	∠$C_aC_bC_c$	114.2			
	∠CCN	179					
Ruthenocene → Bis (cyclopentadienyl) ruthenium							
Silacyclobutane	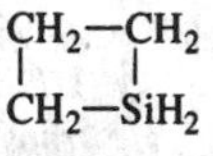		Si—C	1.892			ED
			C—C	1.600			
			Si—H	1.47			
	C—H	1.14	∠CSiC	80.7			
	∠SiCC	84.8	∠CCC	99.8			
	dihedral angle between the CCC and CSiC planes			146			
Spiropentane	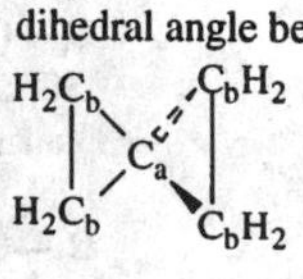		$C_b—C_b$	1.52			ED
			$C_a—C_b$	1.47			
			C—H	1.09			
			∠$C_bC_aC_b$	62			
	∠HCH	118	(D_{2d})				

Compound	Structure					Method
Succinonitrile	CH_2CN / CH_2CN	C—C 1.561	C—C(N) 1.465			ED
		C—N 1.161	C—H 1.09			
		∠CCC 110.4				
	fraction of the *anti* conformer at 170°C 74%,					
	dihedral angle of CCCC for the *gauche* conformer				75	
Tetrachloroethylene	C—Cl	1.718	C—C	1.354		ED
CCl_2=CCl_2	∠ClCCl	115.7				
Tetracyanoethylene	C—N	1.162	C—C	1.435		ED
$(CN)_2C$=$C(CN)_2$	C=C	1.357	∠CC=C	121.1		
Tetrafluoro-1,3-dithietane	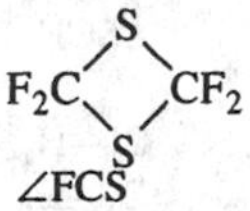		C—S	1.785		ED
			C—F	1.314		
			∠CSC	83.2		
	∠FCS	113.7	(D_{2h} assumed)			
Tetrafluoroethylene	C—C	1.31	C—F	1.319		ED
CF_2=CF_2	∠CCF	123.8	(D_{2h} assumed)			
Tetrahydrofuran 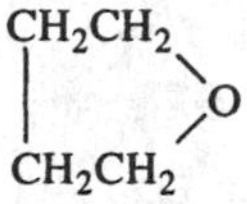	C—H	1.115	C—O	1.428		ED
	C—C	1.536				
	The skeletal bending vibration of the molecular plane is essentially free pseudorotation					
Tetrahydropyran	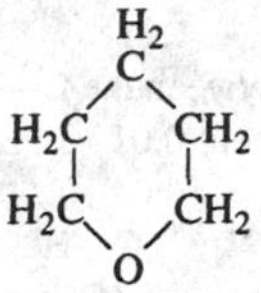		C—O	1.420		ED
			C—C	1.531		
			C—H	1.116		
			∠COC	111.5		
			∠OCC	111.8		
	∠CCC (C)	108	∠CCC (O)	111		
	chair form					
Tetrahydrothiophene 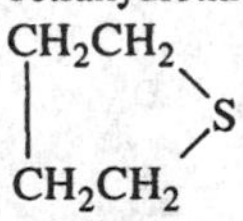	C—S	1.839	C—H	1.120		
	C—C	1.536	∠CSC	93.4		
	∠SCC	106.1	∠CCC	105.0		
Tetramethylgermane	Ge—C	1.945	C—H	1.12		ED
$(CH_3)_4Ge$	∠GeCH	108	(T_d excluding the H atoms)			
Tetramethyllead	Pb—C	2.238	(T_d excluding the H atoms)			ED
$(CH_3)_4Pb$						
Tetramethylsilane	C—H	1.115	C—Si	1.875		ED
$(CH_3)_4Si$	∠HCH	109.8	(T_d excluding the H atoms)			
Tetramethylstannane	C—Sn	2.144				ED
$(CH_3)_4Sn$	C—H	1.12	(T_d excluding the H atoms)			
1,2,5-Thiadiazole	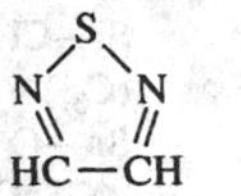	S—N 1.631	∠NSN	99.6		MW
		C—N 1.328	∠CCN	113.8		
		C—C 1.420	∠CCH	126.2		
		C—H 1.079	planar			
1,3,4-Thiadiazole	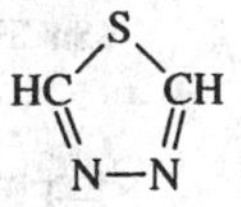	S—C 1.721	∠CSC	86.4		MW
		N—N 1.371	∠SCN	114.6		
		C—N 1.302	∠CCN	112.2		
		C—H 1.08	∠SCH	121.9		
		∠NCH 123.5	planar			
Thietane	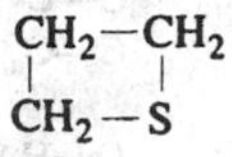		C—S	1.847		ED, MW
			C—C	1.549		
			C—H (average)	1.100		
	∠CSC	76.8	∠HCH (average)	112		
	dihedral angle between the CCC and CSC planes			154		
Thiirane	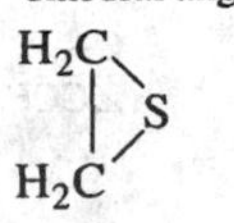	C—C 1.484	∠HCH	116		MW
		C—H 1.083	∠CSC	48.3		
		C—S 1.815	∠CCS	65.9		
		dihedral angle between the CH_2 plane and the C—C bond		152		

Compound	Structure				Method
Thioformaldehyde	C—S	1.611	C—H	1.093	MW
CH_2S	∠HCH	116.9			
Thioformamide	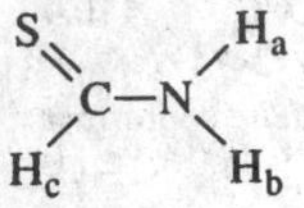		N—H_a	1.002	MW
			N—H_b	1.007	
			C—N	1.358	
	C—S	1.626	C—H_c	1.10	
	∠H_aNH_b	121.7	∠H_aNC	117.9	
	∠H_bNC	120.4	∠NCS	125.3	
	∠NCH_c	108	∠SCH_c	127	
Thiolane → Tetrahydrothiophene					
Thiophene	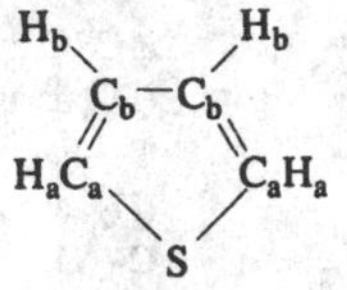		C_a—H_a	1.078	MW
			C_b—H_b	1.081	
			C_a—S	1.714	
			C_a—C_b	1.370	
			C_b—C_b	1.423	
			∠C_aSC_a	92.2	
	∠SC_aC_b	115.5	∠$C_aC_bC_b$	112.5	
	∠SC_aH_a	119.9	∠$C_bC_bH_b$	124.3	
Toluene	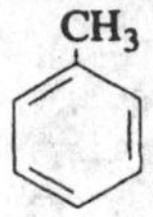 C—C (ring)	1.399	C—CH_3	1.524	ED
	C—H (average)	1.11			
	the difference between the C—H(CH_3) and C—H(ring): about 0.01				
1,1,1-Tribromoethane	C—Br	1.93	C—H	1.095 (assumed)	MW
CH_3CBr_3	C—C	1.51 (assumed)	∠CCBr	108	
	∠BrCBr	111	∠CCH	109.0 (assumed)	
Tribromomethane → Bromoform					
Tri-*tert*-butyl methane	C_a—C_b	1.611	C—H	1.111	ED
$HC_a[C_b(C_cH_3)_3]_3$	C_b—C_c	1.548	∠$C_aC_bC_c$	113.0	
Tricarbon dioxide	C—O	1.163	C—C	1.289	ED
OCCCO	linear (with a large-amplitude bending vibration)				
Trichloroacetonitrile	C—N	1.165	C—C	1.460	ED
CCl_3CN	C—Cl	1.763	∠ClCCl	110.0	
1,1,1-Trichloroethane	C—H	1.090	C—C	1.541	MW
CH_3CCl_3	C—Cl	1.771	∠HCH	110.0	
	∠CCH	108.9	∠ClCCl	109.4	
	∠CCCl	109.6			
Trichloro(methyl)germane	Ge—Cl	2.132	Ge—C	1.89	ED, MW
CH_3GeCl_3	C—H	1.103 (assumed)	∠ClGeCl	106.4	
	∠GeCH	110.5 (assumed)			
Trichloro(methyl)silane	C—Si	1.876	Si—Cl	2.021	MW
CH_3SiCl_3			(C_{3v})		
Trichloro(methyl)stannane	Sn—Cl	2.304	Sn—C	2.10	ED
CH_3SnCl_3	C—H	1.100	∠CSnCl	113.9	
	∠ClSnCl	104.7	∠SnCH	108	
Triethylenediamine → 1,4-Diazabicyclo [2.2.2]octane					
Trifluoroacetic acid	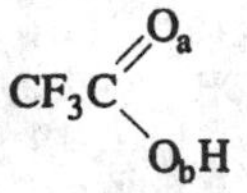		C—F	1.325	ED
			C—C	1.546	
			C—O_a	1.192	
	C—O_b	1.35	O—H	0.96 (assumed)	
	∠CCO_a	126.8	∠CCO_b	111.1	
	∠CCF	109.5			
1,1,1-Trifluoroethane	C—C	1.494	C—F	1.340	ED
CH_3CF_3	C—H	1.081	∠CCF	119.2	
	∠CCH	112			
Trifluoromethane → Fluoroform					

Compound	Structure				Method
1,1,1-Trifluoro-2,2,2-trichloroethane	C—C	1.54	C—F	1.33	MW
CF_3CCl_3	C—Cl	1.77	∠CCF	110	
	∠CCCl	109.6	staggered conformation		
Trimethylaluminium	C—H	1.113	Al—C	1.957	ED
$(CH_3)_3Al$	∠AlCH	111.7	∠CAlC	120	
Trimethylamine	C—N	1.458	C—H	1.100	ED
$(CH_3)_3N$	∠CNC	110.9	∠HCH	110	
Trimethylarsine	C—As	1.979	∠CAsC	98.8	ED
$(CH_3)_3As$	∠AsCH	111.4			
Trimethylbismuth	Bi—C	2.263	C—H	1.07	ED
$(CH_3)_3Bi$	∠CBiC	97.1			
Trimethylborane	C—B	1.578	C—H	1.114	ED
$(CH_3)_3B$	∠CBC	120.0	∠BCH	112.5	
Trimethyleneimine → Azetidine					
Trimethylphosphine	C—P	1.847	C—H	1.091	ED
$(CH_3)_3P$	∠CPC	98.6	∠PCH	110.7	
1,3,5-Trioxane	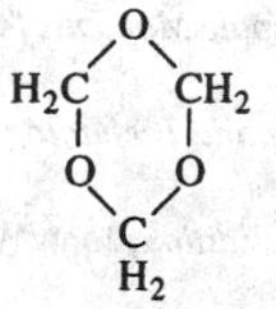		C—O	1.422	MW
			∠OCO	112.2	
			∠COC	110.3	
Triphenylamine	C—C	1.392	C—N	1.42	ED
$(C_6H_5)_3N$	∠CNC	116	(C_3)		
	torsional dihedral angle of the two phenyl rings 47° (defined to be 0 when the symmetry axis is contained in the phenyl planes)				
Tropone	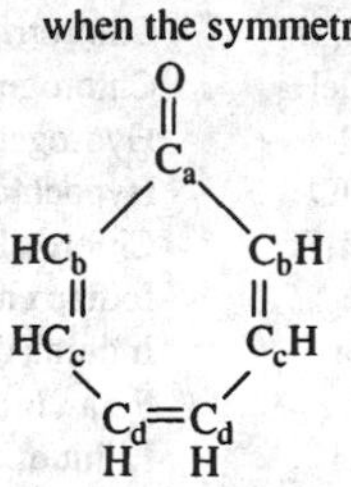		C_a—O	1.23	ED
			C_a—C_b	1.45	
			C_b—C_c	1.36	
			C_c—C_d	1.46	
			C_d—C_d	1.34	
			$∠C_bC_aC_b$	122	
			$∠C_aC_bC_c$	133	
	$∠C_bC_cC_d$	126	$∠C_cC_dC_d$	130	
				(C_{2v})	
Vinylacetylene	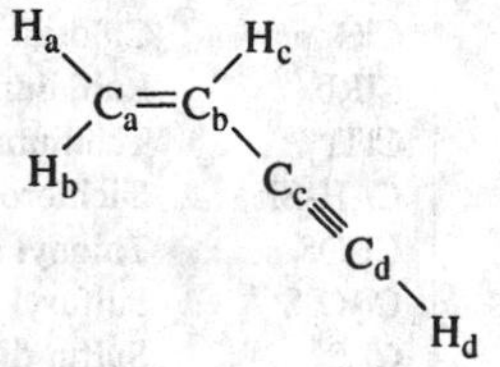		C_b—C_c	1.434	ED, MW
			C_a—C_b	1.344	
			C_c—C_d	1.215	
			C_a—H_a	1.11	
			C_d—H_d	1.09	
			$∠C_aC_bC_c$	123.1	
	$∠C_bC_cC_d$	178	$∠H_aC_aC_b$	119	
	$∠H_bC_aC_b$	122	$∠H_cC_bC_a$	122	
	$∠C_cC_dH_d$	182			
Vinyl chloride	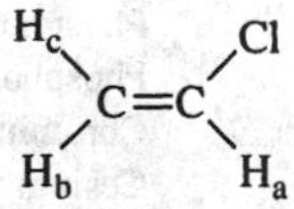		C—C	1.342	ED, MW
			C—Cl	1.730	
			C—H	1.09	
	∠CCCl	122.5	$∠CCH_a$	124	
	$∠CCH_b$	120	$∠CCH_c$	121.1	

DIPOLE MOMENTS

This table gives values of the electric dipole moment for about 800 molecules. When available, values determined by microwave spectroscopy, molecular beam electric resonance, and other high-resolution spectroscopic techniques were selected. Otherwise, the values come from measurements of the dielectric constant in the gas phase or, if these do not exist, in the liquid phase. Compounds are listed by molecular formula in Hill order; compounds not containing carbon are listed first, followed by compounds containing carbon.

The dipole moment μ is given in debye units (D). The conversion factor to SI units is $1\ D = 3.33564 \times 10^{-30}$ C m.

Dipole moments of individual conformers (rotational isomers) are given when they have been measured. The conformers are designated as *gauche, trans, axial*, etc. The meaning of these terms can be found in the references. In some cases an average value, obtained from measurements on the bulk gas, is also given. Other information on molecules that have been studied by spectroscopy, such as the components of the dipole moment in the molecular framework and the variation with vibrational state and isotopic species, is given in References 1 and 2.

When the accuracy of a value is explicitly stated (i.e., 1.234 ± 0.005), the stated uncertainty generally indicates two or three standard deviations. When no uncertainty is given, the value may be assumed to be precise to a few units in the last decimal place. However, if more than three decimal places are given, the exact interpretation of the final digits may require analysis of the vibrational averaging.

Values measured in the gas phase that are questionable because of undetermined error sources are indicated as approximate (≈). Values obtained by liquid phase measurements, which sometimes have large errors because of association effects, are enclosed in brackets, e.g., [1.8].

REFERENCES

1. Nelson, R. D., Lide, D. R., and Maryott, A. A., *Selected Values of Electric Dipole Moments for Molecules in the Gas Phase*, Natl. Stand. Ref. Data Ser. - Nat. Bur. Stnds. 10, 1967.
2. *Landolt-Börnstein, Numerical Data and Functional Relationships in Science and Technology, New Series*, II/6 (1974), II/14a (1982), II/14b (1983), II/19c (1992), Springer-Verlag, Heidelberg.
3. Riddick, J. A., Bunger, W. B., and Sakano, T. K., *Organic Solvents, Fourth Edition*, John Wiley & Sons, New York, 1986.

Mol. Form.	Name	μ/D
	Compounds not containing carbon	
AgBr	Silver(I) bromide	5.62 ± 0.03
AgCl	Silver(I) chloride	6.08 ± 0.06
AgF	Silver(I) fluoride	6.22 ± 0.30
AgI	Silver(I) iodide	4.55 ± 0.05
AlF	Aluminum monofluoride	1.53 ± 0.15
$AsCl_3$	Arsenic(III) chloride	1.59 ± 0.08
AsF_3	Arsenic(III) fluoride	2.59 ± 0.05
AsH_3	Arsine	0.217 ± 0.003
BClH2	Chloroborane	0.75 ± 0.05
BF	Fluoroborane(1)	≈0.5
BF_2H	Difluoroborane	0.971 ± 0.010
B_4H_{10}	Tetraborane	0.486 ± 0.002
B_5H_9	Pentaborane(9)	2.13 ± 0.04
B_6H_{10}	Hexaborane	2.50 ± 0.05
BaO	Barium oxide	7.954 ± 0.003
BaS	Barium sulfide	10.86 ± 0.02
BrCl	Bromine chloride	0.519 ± 0.004
BrF	Bromine fluoride	1.422 ± 0.016
BrF_3Si	Bromotrifluorosilane	0.83 ± 0.01
BrF_5	Bromine pentafluoride	1.51 ± 0.15
BrH	Hydrogen bromide	0.8272 ± 0.0003
BrH_3Si	Bromosilane	1.319
BrI	Iodine bromide	0.726 ± 0.003
BrK	Potassium bromide	10.628 ± 0.001
BrLi	Lithium bromide	7.268 ± 0.001
BrNO	Nitrosyl bromide	≈1.8
BrNa	Sodium bromide	9.1183 ± 0.0006
BrO	Bromine monoxide	1.76 ± 0.04
BrO_2	Bromine dioxide	2.8 ± 0.3
BrRb	Rubidium bromide	≈10.9
BrTl	Thallium(I) bromide	4.49 ± 0.05
CaCl	Calcium monochloride	≈3.6
ClCs	Cesium chloride	10.387 ± 0.004
ClF	Chlorine fluoride	0.888061
$ClFO_3$	Perchloryl fluoride	0.023 ± 0.001
ClF_3	Chlorine trifluoride	0.6 ± 0.10
ClF_3Si	Chlorotrifluorosilane	0.636 ± 0.004
$ClGeH_3$	Chlorogermane	2.13 ± 0.02
ClH	Hydrogen chloride	1.1086 ± 0.0003
ClHO	Hypochlorous acid	≈1.3
ClH_3Si	Chlorosilane	1.31 ± 0.01
ClI	Iodine chloride	1.24 ± 0.02
ClIn	Indium(I) chloride	3.79 ± 0.19
ClK	Potassium chloride	10.269 ± 0.001
ClLi	Lithium chloride	7.12887
$ClNO_2$	Nitryl chloride	0.53
ClNS	Thionitrosyl chloride	1.87 ± 0.02
ClNa	Sodium chloride	9.00117
ClO	Chlorine oxide	1.297 ± 0.001
ClRb	Rubidium chloride	10.510 ± 0.005
ClTl	Thallium(I) chloride	4.54299
Cl_2H_2Si	Dichlorosilane	1.17 ± 0.02
Cl_2OS	Thionyl chloride	1.45 ± 0.03
Cl_2O_2S	Sulfuryl chloride	1.81 ± 0.04
Cl_2S	Sulfur dichloride	0.36 ± 0.01
Cl_3FSi	Trichlorofluorosilane	0.49 ± 0.01
Cl_3HSi	Trichlorosilane	0.86 ± 0.01
Cl_3N	Nitrogen trichloride	0.39 ± 0.01
Cl_3OP	Phosphorus(V) oxychloride	2.54 ± 0.05
Cl_3P	Phosphorus(III) chloride	0.56 ± 0.02
CrO	Chromium monoxide	3.88 ± 0.13
CsF	Cesium fluoride	7.884 ± 0.001
CsNa	Cesium sodium	4.75 ± 0.20
CuF	Copper(I) fluoride	5.77 ± 0.29
CuO	Copper(II) oxide	4.5 ± 0.5
FGa	Gallium monofluoride	2.45 ± 0.05
$FGeH_3$	Fluorogermane	2.33 ± 0.12
FH	Hydrogen fluoride	1.826178
FHO	Hypofluorous acid	2.23 ± 0.11
FH_2N	Fluoramide	2.27 ± 0.18
FH_3Si	Fluorosilane	1.2969 ± 0.0006
FI	Iodine fluoride	1.948 ± 0.020

DIPOLE MOMENTS (continued)

Mol. Form.	Name	μ/D
FIn	Indium(I) fluoride	3.40 ± 0.07
FK	Potassium fluoride	8.585 ± 0.003
FLi	Lithium fluoride	6.3274 ± 0.0002
FNO	Nitrosyl fluoride	1.730 ± 0.003
FNO_2	Nitryl fluoride	0.466 ± 0.005
FNS	Thionitrosyl fluoride (NSF)	1.902 ± 0.012
FN_3	Fluorine azide	≈1.3
FNa	Sodium fluoride	8.156 ± 0.001
FO	Fluorine oxide	0.0043 ± 0.0004
FRb	Rubidium fluoride	8.5465 ± 0.0005
FS	Sulfur monofluoride	0.794 ± 0.02
FTl	Thallium(I) fluoride	4.2282 ± 0.0008
F_2Ge	Germanium(II) fluoride	2.61 ± 0.02
F_2HN	Difluoramine	1.92 ± 0.02
F_2H_2Si	Difluorosilane	1.55 ± 0.02
F_2N_2	*cis*-Difluorodiazine	0.16 ± 0.01
F_2O	Fluorine monoxide	0.308180
F_2OS	Thionyl fluoride	1.63 ± 0.01
F_2O_2	Fluorine dioxide	1.44 ± 0.07
F_2O_2S	Sulfuryl fluoride	1.12 ± 0.02
F_2S	Sulfur difluoride	1.05 ± 0.05
F_2Si	Difluorosilylene	1.23 ± 0.02
F_3HSi	Trifluorosilane	1.27 ± 0.03
$F_3H_3Si_2$	1,1,1-Trifluorodisilane	2.03 ± 0.10
F_3ISi	Trifluoroiodosilane	1.11 ± 0.03
F_3N	Nitrogen trifluoride	0.235 ± 0.004
F_3NO	Trifluoramine oxide	0.0390 ± 0.0004
F_3OP	Phosphorus(V) oxyfluoride	1.8685 ± 0.0001
F_3P	Phosphorus(III) fluoride	1.03 ± 0.01
F_3PS	Phosphorus(V) sulfide trifluoride	0.64 ± 0.02
F_4N_2	Tetrafluorohydrazine (*gauche*)	0.257 ± 0.002
F_4S	Sulfur tetrafluoride	0.632 ± 0.003
F_4Se	Selenium tetrafluoride	1.78 ± 0.09
F_5I	Iodine pentafluoride	2.18 ± 0.11
GeH_3N_3	Germylazide	2.579 ± 0.003
GeO	Germanium(II) oxide	3.2823 ± 0.0001
GeS	Germanium(II) sulfide	2.00 ± 0.06
GeSe	Germanium(II) selenide	1.65 ± 0.05
GeTe	Germanium(II) telluride	1.06 ± 0.07
HI	Hydrogen iodide	0.448 ± 0.001
HKO	Potassium hydroxide	7.415 ± 0.002
HLi	Lithium hydride	5.884 ± 0.001
HLiO	Lithium hydroxide	4.754 ± 0.002
HN	Imidogen	1.39 ± 0.07
HNO	Nitrosyl hydride	1.62 ± 0.03
HNO_2	Nitrous acid (*cis*)	1.423 ± 0.005
HNO_2	Nitrous acid (*trans*)	1.855 ± 0.016
HNO_3	Nitric acid	2.17 ± 0.02
HN_3	Hydrazoic acid	1.70 ± 0.09
HO	Hydroxyl	1.655 ± 0.001
HS	Mercapto	0.7580 ± 0.0001
H_2O	Water	1.8546 ± 0.0040
H_2O_2	Hydrogen peroxide	1.573 ± 0.001
H_2S	Hydrogen sulfide	0.97833
H_3N	Ammonia	1.4718 ± 0.0002
H_3NO	Hydroxylamine	0.59 ± 0.05
H_3P	Phosphine	0.5740 ± 0.0003
H_3Sb	Stibine	0.12 ± 0.05
H_4N_2	Hydrazine	1.75 ± 0.09
H_6OSi_2	Disiloxane	0.24 ± 0.02
IK	Potassium iodide	≈10.8
ILi	Lithium iodide	7.428 ± 0.001
INa	Sodium iodide	9.236 ± 0.003
IO	Iodine monoxide	2.45 ± 0.05
IRb	Rubidium iodide	≈11.5
ITl	Thallium(I) iodide	4.61 ± 0.07
KLi	Lithium potassium	3.45 ± 0.20
KNa	Potassium sodium	2.693 ± 0.014
LaO	Lanthanum monoxide	3.207 ± 0.011
LiNa	Lithium sodium	0.463 ± 0.002
LiO	Lithium monoxide	6.84 ± 0.03
LiRb	Lithium rubidium	4.0 ± 0.1
MgO	Magnesium oxide	6.2 ± 0.6
NO	Nitric oxide	0.15872
NO_2	Nitrogen dioxide	0.316 ± 0.010
NP	Phosphorus nitride	2.7470 ± 0.0001
NS	Nitrogen sulfide	1.81 ± 0.02
N_2O	Nitrous oxide	0.16083
N_2O_3	Nitrogen trioxide	2.122 ± 0.010
NaRb	Rubidium sodium	3.1 ± 0.3
OP	Phosphorus monoxide	1.88 ± 0.07
OPb	Lead(II) oxide	4.64 ± 0.50
OS	Sulfur monoxide	1.55 ± 0.02
OS_2	Sulfur oxide (SSO)	1.47 ± 0.03
OSi	Silicon monoxide	3.0982
OSn	Tin(II) oxide	4.32 ± 0.22
OSr	Strontium oxide	8.900 ± 0.003
OTi	Titanium(II) oxide	2.96 ± 0.05
OY	Yttrium monoxide	4.524 ± 0.007
OZr	Zirconium(II) oxide	2.55 ± 0.01
O_2S	Sulfur dioxide	1.63305
O_2Se	Selenium dioxide	2.62 ± 0.05
O_2Zr	Zirconium(IV) oxide	7.80 ± 0.02
O_3	Ozone	0.53373
PbS	Lead(II) sulfide	3.59 ± 0.18
SSi	Silicon monosulfide	1.73 ± 0.09
SSn	Tin(II) sulfide	3.18 ± 0.16

Compounds containing carbon

Mol. Form.	Name	μ/D
$CBrF_3$	Bromotrifluoromethane	0.65 ± 0.05
CBr_2F_2	Dibromodifluoromethane	0.66 ± 0.05
$CClF_3$	Chlorotrifluoromethane	0.50 ± 0.01
CClN	Cyanogen chloride	2.8331 ± 0.0002
CCl_2F_2	Dichlorodifluoromethane	0.51 ± 0.05
CCl_2O	Carbonyl chloride	1.17 ± 0.01
CCl_3F	Trichlorofluoromethane	0.46 ± 0.02
CF	Fluoromethylidyne	0.645 ± 0.005
CFN	Cyanogen fluoride	2.120 ± 0.001
CF_2	Difluoromethylene	0.47 ± 0.02
CF_2O	Carbonyl fluoride	0.95 ± 0.01
CF_3I	Trifluoroiodomethane	1.048 ± 0.003
CH	Methylidyne	≈1.46
CHBrClF	Bromochlorofluoromethane	1.5 ± 0.3
$CHBr_3$	Tribromomethane	0.99 ± 0.02
$CHClF_2$	Chlorodifluoromethane	1.42 ± 0.03
$CHCl_2F$	Dichlorofluoromethane	1.29 ± 0.03
$CHCl_3$	Trichloromethane	1.04 ± 0.02
CHFO	Formyl fluoride	2.081 ± 0.001
CHF_2N	Carboimidic difluoride	1.393 ± 0.001
CHF_3	Trifluoromethane	1.65150
CHN	Hydrogen cyanide	2.985188

Mol. Form.	Name	μ/D
CHN	Hydrogen isocyanide	3.05 ± 0.15
$CHNO$	Isocyanic acid (HNCO)	≈1.6
$CHNO$	Fulminic acid	3.09934
CH_2BrCl	Bromochloromethane	[1.66]
CH_2Br_2	Dibromomethane	1.43 ± 0.03
CH_2ClF	Chlorofluoromethane	1.82 ± 0.04
CH_2Cl_2	Dichloromethane	1.60 ± 0.03
CH_2F_2	Difluoromethane	1.9785 ± 0.02
CH_2I_2	Diiodomethane	[1.08]
CH_2N_2	Diazomethane	1.50 ± 0.01
CH_2N_2	Cyanamide	4.28 ± 0.10
CH_2N_4	1H-Tetrazole	2.19 ± 0.05
CH_2O	Formaldehyde	2.332 ± 0.002
CH_2O_2	Formic acid	1.425 ± 0.002
CH_2S	Thioformaldehyde	1.6491 ± 0.0004
CH_2Se	Selenoformaldehyde	1.41 ± 0.01
CH_3BCl_2	Dichloromethylborane	1.419 ± 0.013
CH_3BF_2	Difluoromethylborane	1.668 ± 0.003
CH_3BO	Borane carbonyl	1.698 ± 0.020
CH_3Br	Bromomethane	1.8203 ± 0.0004
CH_3Cl	Chloromethane	1.8963 ± 0.0002
CH_3Cl_3Si	Methyltrichlorosilane	1.91 ± 0.01
CH_3F	Fluoromethane	1.858 ± 0.002
CH_3F_2OP	Methylphosphonic difluoride	3.69 ± 0.26
CH_3F_2P	Methyldifluorophosphine	2.056 ± 0.006
CH_3F_3Si	Trifluoromethylsilane	2.3394 ± 0.0002
CH_3F_3Si	(Trifluoromethyl)silane	2.32 ± 0.02
CH_3I	Iodomethane	1.6406 ± 0.0004
CH_3NO	Formamide	3.73 ± 0.07
CH_3NO_2	Nitromethane	3.46 ± 0.02
CH_3N_3	Methyl azide	2.17 ± 0.04
CH_4O	Methanol	1.70 ± 0.02
CH_4O_2	Methylhydroperoxide	≈0.65
CH_4S	Methanethiol	1.52 ± 0.08
CH_5FSi	Fluoromethylsilane	1.700 ± 0.008
CH_5ISi	Iodomethylsilane	1.862 ± 0.005
CH_5N	Methylamine	1.31 ± 0.03
CH_6OSi	Methyl silyl ether	1.15 ± 0.02
CH_6Si	Methylsilane	0.73456
CH_8B_2	Methyldiborane(6)	0.566 ± 0.006
CIN	Cyanogen iodide	3.67 ± 0.02
CO	Carbon monoxide	0.10980
COS	Carbon oxysulfide	0.715189
$COSe$	Carbon oxyselenide	0.73 ± 0.02
CS	Carbon monosulfide	1.958 ± 0.005
CSe	Carbon monoselenide	1.99 ± 0.04
C_2BrF	Bromofluoroacetylene	0.448 ± 0.002
C_2ClF_3	Chlorotrifluoroethene	0.40 ± 0.10
C_2ClF_5	Chloropentafluoroethane	0.52 ± 0.05
$C_2Cl_2F_2$	1,1-Dichloro-2,2-difluoroethene	0.50
$C_2Cl_2F_4$	1,2-Dichloro-1,1,2,2-tetrafluoroethane	≈0.5
C_2F_3N	Trifluoroacetonitrile	1.262 ± 0.010
C_2F_3N	Trifluoroisocyanomethane	1.153 ± 0.010
C_2HBr	Bromoacetylene	0.22962
C_2HCl	Chloroacetylene	0.44408
C_2HCl_3	Trichloroethene	[0.8]
C_2HCl_5	Pentachloroethane	0.92 ± 0.05
C_2HF	Fluoroacetylene	0.7207 ± 0.0003
C_2HF_3	Trifluoroethene	1.32 ± 0.03
$C_2HF_3O_2$	Trifluoroacetic acid	2.28 ± 0.25
C_2HI	Iodoacetylene	0.02525
$C_2H_2Br_4$	1,1,2,2-Tetrabromoethane	[1.38]
$C_2H_2Cl_2$	1,1-Dichloroethene	1.34 ± 0.01
$C_2H_2Cl_2$	*cis*-1,2-Dichloroethene	1.90 ± 0.04
$C_2H_2Cl_2O$	Chloroacetyl chloride	2.23 ± 0.11
$C_2H_2Cl_4$	1,1,2,2-Tetrachloroethane	1.32 ± 0.07
$C_2H_2F_2$	1,1-Difluoroethene	1.3893 ± 0.0002
$C_2H_2F_2$	*cis*-1,2-Difluoroethene	2.42 ± 0.02
$C_2H_2F_4$	1,1,1,2-Tetrafluoroethane	1.80 ± 0.22
$C_2H_2N_2S$	1,2,5-Thiadiazole	1.579 ± 0.007
C_2H_2O	Ketene	1.42215
$C_2H_2O_2$	Glyoxal (*cis*)	4.8 ± 0.2
C_2H_3Br	Bromoethene	1.42 ± 0.03
C_2H_3Cl	Chloroethene	1.45 ± 0.03
$C_2H_3ClF_2$	1-Chloro-1,1-difluoroethane	2.14 ± 0.04
C_2H_3ClO	Acetyl chloride	2.72 ± 0.14
$C_2H_3Cl_3$	1,1,1-Trichloroethane	1.755 ± 0.015
$C_2H_3Cl_3$	1,1,2-Trichloroethane	[1.4]
C_2H_3F	Fluoroethene	1.468 ± 0.003
C_2H_3FO	Acetyl fluoride	2.96 ± 0.03
$C_2H_3F_3$	1,1,1-Trifluoroethane	2.3470 ± 0.005
C_2H_3HgN	Cyanomethylmercury	4.7 ± 0.1
C_2H_3I	Iodoethene	1.311 ± 0.005
C_2H_3N	Acetonitrile	3.92519
C_2H_3NO	Methyl cyanate	4.26 ± 0.18
C_2H_3NO	Methyl isocyanate	≈2.8
C_2H_3NS	Methyl isothiocyanate	3.453 ± 0.003
$C_2H_3N_3$	1H-1,2,4-Triazole	2.7 ± 0.1
C_2H_4BrCl	1-Bromo-2-chloroethane	[1.2]
$C_2H_4Br_2$	1,2-Dibromoethane	[1.19]
C_2H_4ClF	1-Chloro-1-fluoroethane	2.068 ± 0.014
$C_2H_4Cl_2$	1,1-Dichloroethane	2.06 ± 0.04
$C_2H_4Cl_2$	1,2-Dichloroethane	[1.83]
$C_2H_4F_2$	1,1-Difluoroethane	2.27 ± 0.05
$C_2H_4F_2$	1,2-Difluoroethane (*gauche*)	2.67 ± 0.13
C_2H_4O	Acetaldehyde	2.750 ± 0.006
C_2H_4O	Ethylene oxide	1.89 ± 0.01
$C_2H_4O_2$	Acetic acid	1.70 ± 0.03
$C_2H_4O_2$	Methyl formate	1.77 ± 0.04
$C_2H_4O_2$	Glycolaldehyde	2.73 ± 0.05
C_2H_5Br	Bromoethane	2.04 ± 0.02
C_2H_5Cl	Chloroethane	2.05 ± 0.02
C_2H_5ClO	2-Chloroethanol	1.78 ± 0.09
$C_2H_5Cl_3Si$	Trichloroethylsilane	[2.04]
C_2H_5F	Fluoroethane	1.937 ± 0.007
C_2H_5I	Iodoethane	1.976 ± 0.002
C_2H_5N	Ethyleneimine	1.90 ± 0.01
C_2H_5NO	Acetamide	3.68 ± 0.03
C_2H_5NO	*N*-Methylformamide	3.83 ± 0.08
$C_2H_5NO_2$	Nitroethane	3.23 ± 0.03
C_2H_6O	Ethanol (*gauche*)	1.68 ± 0.03
C_2H_6O	Ethanol (*trans*)	1.44 ± 0.03
C_2H_6O	Ethanol (*average*)	1.69 ± 0.03
C_2H_6O	Dimethyl ether	1.30 ± 0.01
C_2H_6OS	Dimethyl sulfoxide	3.96 ± 0.04
$C_2H_6O_2$	Ethylene glycol (*average*)	2.36 ± 0.10
C_2H_6S	Ethanethiol (*gauche*)	1.61 ± 0.08
C_2H_6S	Ethanethiol (*trans*)	1.58 ± 0.08
C_2H_6S	Dimethyl sulfide	1.554 ± 0.004
$C_2H_6S_2$	1,2-Ethanedithiol	2.03 ± 0.08
$C_2H_6S_2$	Dimethyl disulfide	[1.85]

Mol. Form.	Name	μ/D
C_2H_6Si	Vinylsilane	0.657 ± 0.002
C_2H_7N	Ethylamine (*gauche*)	1.210 ± 0.015
C_2H_7N	Ethylamine (*trans*)	1.304 ± 0.011
C_2H_7N	Ethylamine (*average*)	1.22 ± 0.10
C_2H_7N	Dimethylamine	1.01 ± 0.02
C_2H_7NO	Ethanolamine	[2.27]
$C_2H_8N_2$	1,2-Ethanediamine	1.99 ± 0.10
C_3HF_3	3,3,3-Trifluoro-1-propyne	2.317 ± 0.013
C_3HN	Cyanoacetylene	3.73172
$C_3H_2F_2$	3,3-Difluorocyclopropene	2.98 ± 0.02
C_3H_2O	2-Propynal	2.78 ± 0.02
$C_3H_3Cl_2F$	1,1-Dichloro-2-fluoropropene	2.43 ± 0.02
C_3H_3F	3-Fluoropropyne	1.73 ± 0.02
$C_3H_3F_3$	3,3,3-Trifluoropropene	2.45 ± 0.05
C_3H_3N	Acrylonitrile	3.92 ± 0.07
C_3H_3NO	Oxazole	1.503 ± 0.030
C_3H_3NO	Isoxazole	2.95 ± 0.04
C_3H_4	Propyne	0.784 ± 0.001
C_3H_4	Cyclopropene	0.454 ± 0.010
$C_3H_4F_2$	1,1-Difluoro-1-propene	0.889 ± 0.007
$C_3H_4N_2$	1*H*-Pyrazole	2.20 ± 0.01
$C_3H_4N_2$	Imidazole	3.8 ± 0.4
C_3H_4O	Propargyl alcohol	1.13 ± 0.06
C_3H_4O	Acrolein (*trans*)	3.117 ± 0.004
C_3H_4O	Acrolein (*cis*)	2.552 ± 0.003
C_3H_4O	Cyclopropanone	2.67 ± 0.13
$C_3H_4O_2$	Vinyl formate	1.49 ± 0.01
$C_3H_4O_2$	2-Oxetanone	4.18 ± 0.03
$C_3H_4O_2$	3-Oxetanone	0.887 ± 0.005
$C_3H_4O_3$	Ethylene carbonate	[4.9]
C_3H_5Br	2-Bromopropene	[1.51]
C_3H_5Br	3-Bromopropene	≈1.9
C_3H_5Cl	*cis*-1-Chloropropene	1.67 ± 0.08
C_3H_5Cl	*trans*-1-Chloropropene	1.97 ± 0.10
C_3H_5Cl	2-Chloropropene	1.647 ± 0.010
C_3H_5Cl	3-Chloropropene	1.94 ± 0.10
C_3H_5ClO	Epichlorohydrin	[1.8]
C_3H_5F	*cis*-1-Fluoropropene	1.46 ± 0.03
C_3H_5F	*trans*-1-Fluoropropene	≈1.9
C_3H_5F	2-Fluoropropene	1.61 ± 0.03
C_3H_5F	3-Fluoropropene (*gauche*)	1.939 ± 0.015
C_3H_5F	3-Fluoropropene (*cis*)	1.765 ± 0.014
C_3H_5N	Propanenitrile	4.05 ± 0.03
C_3H_5NO	Ethyl cyanate	4.72 ± 0.09
C_3H_5NO	3-Hydroxypropanenitrile (*gauche*)	3.17 ± 0.02
C_3H_6	Propene	0.366 ± 0.001
$C_3H_6Br_2$	1,2-Dibromopropane	[1.2]
$C_3H_6Cl_2$	1,2-Dichloropropane	[1.85]
$C_3H_6Cl_2$	1,3-Dichloropropane	2.08 ± 0.04
C_3H_6O	Acetone	2.88 ± 0.03
C_3H_6O	Propanal (*gauche*)	2.86 ± 0.01
C_3H_6O	Propanal (*cis*)	2.52 ± 0.05
C_3H_6O	Propanal (*average*)	2.72
C_3H_6O	Allyl alcohol (*gauche*)	1.55 ± 0.08
C_3H_6O	Allyl alcohol (*average*)	1.60 ± 0.08
C_3H_6O	Methyl vinyl ether	0.965 ± 0.002
C_3H_6O	Methyloxirane	2.01 ± 0.02
C_3H_6O	Oxetane	1.94 ± 0.01
$C_3H_6O_2$	Propanoic acid (*cis*)	1.46 ± 0.07
$C_3H_6O_2$	Propanoic acid (*average*)	1.75 ± 0.09
$C_3H_6O_2$	Ethyl formate (*gauche*)	1.81 ± 0.02
$C_3H_6O_2$	Ethyl formate (*trans*)	1.98 ± 0.02
$C_3H_6O_2$	Ethyl formate (*average*)	1.93
$C_3H_6O_2$	Methyl acetate	1.72 ± 0.09
$C_3H_6O_2$	1,3-Dioxolane	1.19 ± 0.06
$C_3H_6O_2S$	Thietane 1,1-dioxide	4.8 ± 0.1
$C_3H_6O_3$	1,3,5-Trioxane	2.08 ± 0.02
C_3H_6S	Thiacyclobutane	1.85 ± 0.09
C_3H_7Br	1-Bromopropane	2.18 ± 0.11
C_3H_7Br	2-Bromopropane	2.21 ± 0.11
C_3H_7Cl	1-Chloropropane (*gauche*)	2.02 ± 0.03
C_3H_7Cl	1-Chloropropane (*trans*)	1.95 ± 0.02
C_3H_7Cl	1-Chloropropane (*average*)	2.05 ± 0.04
C_3H_7Cl	2-Chloropropane	2.17 ± 0.11
C_3H_7F	1-Fluoropropane (*gauche*)	1.90 ± 0.10
C_3H_7F	1-Fluoropropane (*trans*)	2.05 ± 0.04
C_3H_7F	2-Fluoropropane	1.958 ± 0.001
C_3H_7I	1-Iodopropane	2.04 ± 0.10
C_3H_7I	2-Iodopropane	[1.95]
C_3H_7N	Allylamine	≈1.2
C_3H_7N	Cyclopropylamine	1.19 ± 0.01
C_3H_7N	Propyleneimine (*cis*)	1.77 ± 0.09
C_3H_7N	Propyleneimine (*trans*)	1.57 ± 0.03
C_3H_7NO	*N*,*N*-Dimethylformamide	3.82 ± 0.08
C_3H_7NO	*N*-Methylacetamide	[4.3]
$C_3H_7NO_2$	1-Nitropropane	3.66 ± 0.07
$C_3H_7NO_2$	2-Nitropropane	3.73 ± 0.07
C_3H_8	Propane	0.084 ± 0.001
C_3H_8O	1-Propanol (*gauche*)	1.58 ± 0.03
C_3H_8O	1-Propanol (*trans*)	1.55 ± 0.03
C_3H_8O	2-Propanol (*trans*)	1.58 ± 0.03
C_3H_8O	Ethyl methyl ether (*trans*)	1.17 ± 0.02
$C_3H_8O_2$	1,2-Propylene glycol	[2.25]
$C_3H_8O_2$	1,3-Propylene glycol	[2.55]
$C_3H_8O_2$	Ethylene glycol monomethyl ether (*gauche*)	2.36 ± 0.05
$C_3H_8O_2$	Dimethoxymethane	[0.74]
$C_3H_8O_3$	Glycerol	[2.56]
C_3H_8S	1-Propanethiol (*gauche*)	1.683 ± 0.010
C_3H_8S	1-Propanethiol (*trans*)	1.60 ± 0.08
C_3H_8S	2-Propanethiol (*gauche*)	1.53 ± 0.03
C_3H_8S	2-Propanethiol (*trans*)	1.61 ± 0.03
C_3H_8S	Ethyl methyl sulfide (*gauche*)	1.593 ± 0.004
C_3H_8S	Ethyl methyl sulfide (*trans*)	1.56 ± 0.03
C_3H_9N	Propylamine	1.17 ± 0.06
C_3H_9N	Isopropylamine	1.19 ± 0.06
C_3H_9N	Trimethylamine	0.612 ± 0.003
$C_3H_9O_4P$	Trimethyl phosphate	[3.18]
C_4H_4	1-Buten-3-yne	0.22 ± 0.02
C_4H_4	Methylenecyclopropene	1.90 ± 0.01
$C_4H_4N_2$	Succinonitrile	[3.7]
$C_4H_4N_2$	Pyrimidine	2.334 ± 0.010
$C_4H_4N_2$	Pyridazine	4.22 ± 0.02
C_4H_4O	Furan	0.66 ± 0.01
$C_4H_4O_2$	Diketene	3.53 ± 0.07
C_4H_4S	Thiophene	0.55 ± 0.01
C_4H_5N	2-Methylacrylonitrile	3.69 ± 0.18
C_4H_5N	Pyrrole	1.767 ± 0.001
C_4H_5N	Isocyanocyclopropane	4.03 ± 0.10
C_4H_5NO	2-Methyloxazole	1.37 ± 0.07
C_4H_5NO	4-Methyloxazole	1.08 ± 0.05

DIPOLE MOMENTS (continued)

Mol. Form.	Name	μ/D
C_4H_5NO	5-Methyloxazole	2.16 ± 0.04
C_4H_5NO	4-Methylisoxazole	3.583 ± 0.005
C_4H_6	1,2-Butadiene	0.403 ± 0.002
C_4H_6	1-Butyne	0.782 ± 0.004
C_4H_6	Cyclobutene	0.132 ± 0.001
C_4H_6O	Divinyl ether	0.78 ± 0.05
C_4H_6O	3-Methoxy-1,2-propadiene	0.963 ± 0.020
C_4H_6O	*trans*-2-Butenal	3.67 ± 0.07
C_4H_6O	2-Methylpropenal	2.68 ± 0.13
C_4H_6O	Cyclobutanone	2.89 ± 0.03
C_4H_6O	2,3-Dihydrofuran	1.32 ± 0.03
C_4H_6O	2,5-Dihydrofuran	1.63 ± 0.01
$C_4H_6O_2$	*trans*-Crotonic acid	[2.13]
$C_4H_6O_2$	Methacrylic acid	[1.65]
$C_4H_6O_2$	Vinyl acetate	[1.79]
$C_4H_6O_2$	Methyl acrylate	[1.77]
$C_4H_6O_2$	γ-Butyrolactone	4.27 ± 0.03
$C_4H_6O_2$	2,3-Dihydro-1,4-dioxin	0.939 ± 0.008
$C_4H_6O_2$	3,6-Dihydro-1,2-dioxin	2.329 ± 0.001
$C_4H_6O_3$	Acetic anhydride	≈2.8
$C_4H_6O_3$	Propylene carbonate	[4.9]
C_4H_6S	2,3-Dihydrothiophene	1.61 ± 0.20
C_4H_6S	2,5-Dihydrothiophene	1.75 ± 0.01
C_4H_7N	Butanenitrile (*gauche*)	3.91 ± 0.04
C_4H_7N	Butanenitrile (*anti*)	3.73 ± 0.06
C_4H_7N	2-Methylpropanenitrile	4.29 ± 0.09
C_4H_7N	2-Isocyanopropane	4.055 ± 0.001
C_4H_7NO	2-Pyrrolidone	[3.5]
C_4H_8	1-Butene (*cis*)	0.438 ± 0.007
C_4H_8	1-Butene (*skew*)	0.359 ± 0.011
C_4H_8	*cis*-2-Butene	0.253 ± 0.005
C_4H_8	Isobutene	0.503 ± 0.010
C_4H_8	Methylcyclopropane	0.139 ± 0.004
$C_4H_8Cl_2$	1,4-Dichlorobutane	2.22 ± 0.11
$C_4H_8Cl_2O$	Bis(2-chloroethyl) ether	[2.58]
C_4H_8O	*cis*-2-Buten-1-ol	1.96 ± 0.03
C_4H_8O	*trans*-2-Buten-1-ol	1.90 ± 0.02
C_4H_8O	2-Methyl-2-propenol (*skew*)	1.295 ± 0.022
C_4H_8O	Ethyl vinyl ether	[1.26]
C_4H_8O	1,2-Epoxybutane	1.891 ± 0.011
C_4H_8O	Butanal	2.72 ± 0.05
C_4H_8O	Isobutanal (*gauche*)	2.69 ± 0.01
C_4H_8O	Isobutanal (*trans*)	2.86 ± 0.01
C_4H_8O	2-Butanone	2.779 ± 0.015
C_4H_8O	Tetrahydrofuran	1.75 ± 0.04
C_4H_8OS	1,4-Oxathiane	0.295 ± 0.003
$C_4H_8O_2$	Butanoic acid	[1.65]
$C_4H_8O_2$	2-Methylpropanoic acid	[1.08]
$C_4H_8O_2$	Propyl formate	[1.89]
$C_4H_8O_2$	Ethyl acetate	1.78 ± 0.09
$C_4H_8O_2$	*cis*-2-Butene-1,4-diol	[2.48]
$C_4H_8O_2$	*trans*-2-Butene-1,4-diol	[2.45]
$C_4H_8O_2$	1,3-Dioxane	2.06 ± 0.04
$C_4H_8O_2S$	Sulfolane	[4.8]
C_4H_8S	3-Methylthietane	2.046 ± 0.009
C_4H_8S	Tetrahydrothiophene	[1.90]
$C_4H_8S_2$	1,3-Dithiane	2.14 ± 0.04
C_4H_9Br	1-Bromobutane	2.08 ± 0.10
C_4H_9Br	2-Bromobutane	2.23 ± 0.11
C_4H_9Br	2-Bromo-2-methylpropane	[2.17]
C_4H_9Cl	1-Chlorobutane	2.05 ± 0.04
C_4H_9Cl	2-Chlorobutane	2.04 ± 0.10
C_4H_9Cl	1-Chloro-2-methylpropane	2.00 ± 0.10
C_4H_9Cl	2-Chloro-2-methylpropane	2.13 ± 0.04
C_4H_9I	1-Iodobutane	[1.93]
C_4H_9I	2-Iodobutane	2.12 ± 0.11
C_4H_9I	1-Iodo-2-methylpropane	[1.87]
C_4H_9N	Pyrrolidine	[1.57]
C_4H_9NO	*N*-Methylpropanamide	3.61
C_4H_9NO	*N,N*-Dimethylacetamide	[3.7]
C_4H_9NO	Morpholine	1.55 ± 0.03
C_4H_{10}	Isobutane	0.132 ± 0.002
$C_4H_{10}O$	1-Butanol	1.66 ± 0.03
$C_4H_{10}O$	2-Butanol	[1.8]
$C_4H_{10}O$	2-Methyl-1-propanol	1.64 ± 0.08
$C_4H_{10}O$	2-Methyl-2-propanol	[1.66]
$C_4H_{10}O$	Diethyl ether	1.15 ± 0.02
$C_4H_{10}O$	Methyl propyl ether (*trans-trans*)	1.107 ± 0.013
$C_4H_{10}O$	Isopropyl methyl ether	1.247 ± 0.003
$C_4H_{10}O_2$	1,4-Butanediol	[2.58]
$C_4H_{10}O_2$	Ethylene glycol monoethyl ether	[2.08]
$C_4H_{10}O_3$	Diethylene glycol	[2.31]
$C_4H_{10}S$	1-Butanethiol	[1.53]
$C_4H_{10}S$	2-Methyl-2-propanethiol	1.66 ± 0.03
$C_4H_{10}S$	Diethyl sulfide	1.54 ± 0.08
$C_4H_{11}N$	Butylamine	≈1.0
$C_4H_{11}N$	*sec*-Butylamine	[1.28]
$C_4H_{11}N$	*tert*-Butylamine	[1.29]
$C_4H_{11}N$	Isobutylamine	[1.27]
$C_4H_{11}N$	Diethylamine	0.92 ± 0.05
$C_4H_{11}NO_2$	Diethanolamine	[2.8]
$C_4H_{13}N_3$	Diethylenetriamine	[1.89]
C_5F_5N	Perfluoropyridine	0.98 ± 0.08
C_5H_3NS	2-Thiophenecarbonitrile	4.59 ± 0.02
C_5H_3NS	3-Thiophenecarbonitrile	4.13 ± 0.02
C_5H_4	1,3-Pentadiyne	1.207 ± 0.001
C_5H_4ClN	4-Chloropyridine	0.756 ± 0.005
C_5H_4FN	3-Fluoropyridine	2.09 ± 0.26
C_5H_4O	2,4-Cyclopentadien-1-one	3.132 ± 0.007
C_5H_4OS	4H-Pyran-4-thione	3.95 ± 0.05
$C_5H_4O_2$	Furfural	[3.54]
$C_5H_4O_2$	4H-Pyran-4-one	3.79 ± 0.02
$C_5H_4S_2$	4H-Thiopyran-4-thione	3.9 ± 0.2
C_5H_5N	Pyridine	2.215 ± 0.010
C_5H_6	1,2,3-Pentatriene	0.51 ± 0.05
C_5H_6	1-Penten-3-yne	0.66 ± 0.02
C_5H_6	*cis*-3-Penten-1-yne	0.78 ± 0.02
C_5H_6	*trans*-3-Penten-1-yne	1.06 ± 0.05
C_5H_6	2-Methyl-1-buten-3-yne	0.513 ± 0.02
C_5H_6	1,3-Cyclopentadiene	0.419 ± 0.004
$C_5H_6N_2$	2-Methylpyrimidine	1.676 ± 0.010
$C_5H_6N_2$	5-Methylpyrimidine	2.881 ± 0.006
C_5H_6O	2-Methylfuran	0.65 ± 0.05
C_5H_6O	3-Methylfuran	1.03 ± 0.02
C_5H_6O	3-Cyclopenten-1-one	2.79 ± 0.03
$C_5H_6O_2$	5-Methyl-2(3H)-furanone	4.08 ± 0.02
$C_5H_6O_2$	Furfuryl alcohol	[1.92]
C_5H_6S	2-Methylthiophene	0.674 ± 0.005
C_5H_6S	3-Methylthiophene	0.914 ± 0.015
C_5H_7N	3-Methyl-2-butenenitrile	4.61 ± 0.13
C_5H_7N	Cyclobutanecarbonitrile	4.04 ± 0.04
$C_5H_7NO_2$	Ethyl cyanoacetate	[2.17]

DIPOLE MOMENTS (continued)

Mol. Form.	Name	μ/D
C_5H_8	*cis*-1,3-Pentadiene	0.500 ± 0.015
C_5H_8	*trans*-1,3-Pentadiene	0.585 ± 0.010
C_5H_8	2-Methyl-1,3-butadiene	0.25 ± 0.01
C_5H_8	1-Pentyne (*gauche*)	0.769 ± 0.028
C_5H_8	1-Pentyne (*trans*)	0.842 ± 0.010
C_5H_8	Cyclopentene	0.20 ± 0.02
C_5H_8	3,3-Dimethylcyclopropene	0.287 ± 0.003
C_5H_8O	Cyclopropyl methyl ketone	2.62 ± 0.25
C_5H_8O	Cyclopentanone	≈3.3
C_5H_8O	3,4-Dihydro-2H-pyran	1.400 ± 0.008
C_5H_8O	3,6-Dihydro-2H-pyran	1.283 ± 0.005
$C_5H_8O_2$	Ethyl acrylate	[1.96]
$C_5H_8O_2$	Methyl methacrylate	[1.67]
$C_5H_8O_2$	2,4-Pentanedione	[2.78]
$C_5H_8O_2$	Dihydro-3-methyl-2(3H)-furanone	4.56 ± 0.02
$C_5H_8O_2$	Dihydro-5-methyl-2(3H)-furanone	4.71 ± 0.05
$C_5H_8O_2$	Tetrahydro-4H-pyran-4-one	1.720 ± 0.003
C_5H_9N	Pentanenitrile	4.12 ± 0.08
C_5H_9N	2,2-Dimethylpropanenitrile	3.95 ± 0.04
C_5H_9N	1,2,5,6-Tetrahydropyridine	1.007 ± 0.003
C_5H_9NO	*N*-Methyl-2-pyrrolidone	[4.1]
C_5H_{10}	1-Pentene	≈0.5
C_5H_{10}	3-Methyl-1-butene (*gauche*)	0.398 ± 0.004
C_5H_{10}	3-Methyl-1-butene (*trans*)	0.320 ± 0.010
C_5H_{10}	1,1-Dimethylcyclopropane	0.142 ± 0.001
$C_5H_{10}O$	2,2-Dimethylpropanal	2.66 ± 0.05
$C_5H_{10}O$	2-Pentanone	[2.70]
$C_5H_{10}O$	3-Pentanone	[2.82]
$C_5H_{10}O$	Tetrahydropyran (*chair*)	1.58 ± 0.03
$C_5H_{10}O_2$	Pentanoic acid	[1.61]
$C_5H_{10}O_2$	3-Methylbutanoic acid	[0.63]
$C_5H_{10}O_2$	Butyl formate	[2.03]
$C_5H_{10}O_2$	Isobutyl formate	[1.88]
$C_5H_{10}O_2$	Propyl acetate	[1.78]
$C_5H_{10}O_2$	Ethyl propanoate	[1.74]
$C_5H_{10}O_2$	Tetrahydrofurfuryl alcohol	[2.1]
$C_5H_{10}O_3$	Diethyl carbonate	1.10 ± 0.06
$C_5H_{10}O_3$	Ethylene glycol monomethyl ether acetate	[2.13]
$C_5H_{10}O_3$	Ethyl lactate	[2.4]
$C_5H_{10}S$	Thiacyclohexane	1.781 ± 0.010
$C_5H_{11}Br$	1-Bromopentane	2.20 ± 0.11
$C_5H_{11}Cl$	1-Chloropentane	2.16 ± 0.11
$C_5H_{11}Cl$	1-Chloro-3-methylbutane	[1.92]
$C_5H_{11}N$	Piperidine (*equitorial*)	0.82 ± 0.02
$C_5H_{11}N$	Piperidine (*axial*)	1.19 ± 0.02
$C_5H_{11}N$	Piperidine (*average*)	[1.19]
$C_5H_{11}N$	*N*-Methylpyrrolidine	0.572 ± 0.003
C_5H_{12}	Isopentane	0.13 ± 0.05
$C_5H_{12}N_2O$	Tetramethylurea	[3.5]
$C_5H_{12}O$	1-Pentanol	[1.7]
$C_5H_{12}O$	2-Pentanol	[1.66]
$C_5H_{12}O$	3-Pentanol	[1.64]
$C_5H_{12}O$	2-Methyl-1-butanol	[1.88]
$C_5H_{12}O$	2-Methyl-2-butanol	[1.82]
$C_5H_{12}O_2$	1,5-Pentanediol	[2.5]
$C_5H_{12}O_3$	Diethylene glycol monomethyl ether	[1.6]
$C_6H_2F_4$	1,2,3,4-Tetrafluorobenzene	2.42 ± 0.05
$C_6H_2F_4$	1,2,3,5-Tetrafluorobenzene	1.46 ± 0.06
$C_6H_3F_3$	1,2,4-Trifluorobenzene	1.402 ± 0.009
$C_6H_4ClNO_2$	1-Chloro-2-nitrobenzene	4.64 ± 0.09
$C_6H_4ClNO_2$	1-Chloro-3-nitrobenzene	3.73 ± 0.07
$C_6H_4ClNO_2$	1-Chloro-4-nitrobenzene	2.83 ± 0.06
$C_6H_4Cl_2$	*o*-Dichlorobenzene	2.50 ± 0.05
$C_6H_4Cl_2$	*m*-Dichlorobenzene	1.72 ± 0.09
$C_6H_4FNO_2$	1-Fluoro-4-nitrobenzene	2.87 ± 0.06
$C_6H_4F_2$	*o*-Difluorobenzene	2.46 ± 0.05
$C_6H_4F_2$	*m*-Difluorobenzene	1.51 ± 0.02
$C_6H_4N_2$	2-Pyridinecarbonitrile	5.78 ± 0.11
$C_6H_4N_2$	3-Pyridinecarbonitrile	3.66 ± 0.11
$C_6H_4N_2$	4-Pyridinecarbonitrile	1.96 ± 0.03
$C_6H_4O_2$	3,5-Cyclohexadiene-1,2-dione	4.23 ± 0.02
C_6H_5Br	Bromobenzene	1.70 ± 0.03
C_6H_5Cl	Chlorobenzene	1.69 ± 0.03
C_6H_5ClO	*p*-Chlorophenol	2.11 ± 0.11
C_6H_5F	Fluorobenzene	1.60 ± 0.08
C_6H_5I	Iodobenzene	1.70 ± 0.09
C_6H_5NO	2-Pyridinecarboxaldehyde	3.56 ± 0.07
C_6H_5NO	3-Pyridinecarboxaldehyde	1.44
C_6H_5NO	4-Pyridinecarboxaldehyde	1.66
$C_6H_5NO_2$	Nitrobenzene	4.22 ± 0.08
C_6H_6	Fulvene	0.4236 ± 0.013
C_6H_6ClN	*o*-Chloroaniline	[1.77]
C_6H_6O	Phenol	1.224 ± 0.008
C_6H_6O	2-Vinylfuran	0.69 ± 0.07
$C_6H_6O_2$	*p*-Hydroquinone	2.38 ± 0.05
C_6H_6S	Benzenethiol	[1.23]
C_6H_7N	Aniline	1.13 ± 0.02
C_6H_7N	2-Methylpyridine	1.85 ± 0.04
C_6H_7N	3-Methylpyridine	[2.40]
C_6H_7N	4-Methylpyridine	2.70 ± 0.02
C_6H_8O	3-Methyl-2-cyclopenten-1-one	4.33 ± 0.002
$C_6H_8O_4$	Dimethyl maleate	[2.48]
C_6H_8Si	Phenylsilane	0.845 ± 0.012
C_6H_9F	1-Fluorocyclohexene	1.942 ± 0.010
C_6H_{10}	1-Hexyne	0.83 ± 0.05
C_6H_{10}	3,3-Dimethyl-1-butyne	0.661 ± 0.004
C_6H_{10}	Cyclohexene (*half-chair*)	0.332 ± 0.012
$C_6H_{10}F_2$	1,1-Difluorocyclohexane	2.556 ± 0.010
$C_6H_{10}O$	3-Methylcyclopentanone	3.14 ± 0.03
$C_6H_{10}O$	Cyclohexanone	3.246 ± 0.006
$C_6H_{10}O$	Mesityl oxide	[2.79]
$C_6H_{10}O_4$	Diethyl oxalate	[2.49]
$C_6H_{10}O_4$	Ethylene glycol diacetate	[2.34]
$C_6H_{11}Cl$	Chlorocyclohexane (*equitorial*)	2.44 ± 0.07
$C_6H_{11}Cl$	Chlorocyclohexane (*axial*)	1.91 ± 0.02
$C_6H_{11}F$	Fluorocyclohexane (*equitorial*)	2.11 ± 0.04
$C_6H_{11}F$	Fluorocyclohexane (*axial*)	1.81 ± 0.04
$C_6H_{11}N$	4-Methylpentanenitrile	[3.5]
$C_6H_{11}NO$	Caprolactam	[3.9]
$C_6H_{12}O$	Butyl vinyl ether	[1.25]
$C_6H_{12}O$	2-Hexanone	[2.66]
$C_6H_{12}O_2$	Hexanoic acid	[1.13]
$C_6H_{12}O_2$	Pentyl formate	1.90 ± 0.10
$C_6H_{12}O_2$	Butyl acetate	[1.87]
$C_6H_{12}O_2$	*sec*-Butyl acetate	[1.87]
$C_6H_{12}O_2$	Isobutyl acetate	[1.86]
$C_6H_{12}O_2$	Ethyl butanoate	[1.74]
$C_6H_{12}O_2$	Diacetone alcohol	[3.24]
$C_6H_{12}O_3$	Ethylene glycol monoethyl ether acetate	[2.25]

Mol. Form.	Name	μ/D
$C_6H_{12}O_3$	Paraldehyde	1.43 ± 0.07
$C_6H_{13}N$	Cyclohexylamine	[1.26]
$C_6H_{14}O$	Dipropyl ether	1.21 ± 0.06
$C_6H_{14}O$	Diisopropyl ether	1.13 ± 0.10
$C_6H_{14}O$	Butyl ethyl ether	[1.24]
$C_6H_{14}O_2$	2-Methyl-2,4-pentanediol	[2.9]
$C_6H_{14}O_2$	Ethylene glycol monobutyl ether	[2.08]
$C_6H_{14}O_2$	1,1-Diethoxyethane	[1.38]
$C_6H_{14}O_3$	Diethylene glycol monoethyl ether	[1.6]
$C_6H_{14}O_3$	Diethylene glycol dimethyl ether	[1.97]
$C_6H_{15}N$	Dipropylamine	[1.03]
$C_6H_{15}N$	Diisopropylamine	[1.15]
$C_6H_{15}N$	Triethylamine	0.66 ± 0.05
$C_6H_{15}NO_3$	Triethanolamine	[3.57]
$C_6H_{15}O_4P$	Triethyl phosphate	[3.12]
$C_6H_{18}N_3OP$	Hexamethylphosphoric triamide	[5.5]
$C_7H_5Cl_3$	(Trichloromethyl)benzene	[2.03]
$C_7H_5F_3$	(Trifluoromethyl)benzene	2.86 ± 0.06
C_7H_5N	Benzonitrile	4.18 ± 0.08
C_7H_5N	Isocyanobenzene	4.018 ± 0.003
$C_7H_6Cl_2$	2,4-Dichlorotoluene	[1.70]
$C_7H_6Cl_2$	3,4-Dichlorotoluene	[2.95]
$C_7H_6Cl_2$	(Dichloromethyl)benzene	[2.07]
C_7H_6O	2,4,6-Cycloheptatrien-1-one	4.1 ± 0.3
C_7H_6O	Benzaldehyde	[3.0]
$C_7H_6O_2$	Salicylaldehyde	[2.86]
C_7H_7Cl	*o*-Chlorotoluene	1.56 ± 0.08
C_7H_7Cl	*m*-Chlorotoluene	[1.82]
C_7H_7Cl	*p*-Chlorotoluene	2.21 ± 0.04
C_7H_7Cl	(Chloromethyl)benzene	[1.82]
C_7H_7F	*o*-Fluorotoluene	1.37 ± 0.07
C_7H_7F	*m*-Fluorotoluene	1.82 ± 0.04
C_7H_7F	*p*-Fluorotoluene	2.00 ± 0.10
$C_7H_7NO_3$	2-Nitroanisole	[5.0]
C_7H_8	Toluene	0.375 ± 0.010
C_7H_8	2,5-Norbornadiene	0.0587 ± 0.0001
C_7H_8O	*o*-Cresol	[1.45]
C_7H_8O	*m*-Cresol	[1.48]
C_7H_8O	*p*-Cresol	[1.48]
C_7H_8O	Benzyl alcohol	1.71 ± 0.09
C_7H_8O	Anisole	1.38 ± 0.07
C_7H_9N	*o*-Methylaniline	[1.60]
C_7H_9N	*m*-Methylaniline	[1.45]
C_7H_9N	*p*-Methylaniline	[1.52]
C_7H_9N	2,4-Dimethylpyridine	[2.30]
C_7H_9N	2,6-Dimethylpyridine	[1.66]
C_7H_{10}	1,3-Cycloheptadiene	0.740
C_7H_{12}	Methylenecyclohexane	0.62 ± 0.01
$C_7H_{12}O_4$	Diethyl malonate	[2.54]
$C_7H_{14}O$	2-Heptanone	[2.59]
$C_7H_{14}O$	3-Heptanone	[2.78]
$C_7H_{14}O$	2,4-Dimethyl-3-pentanone	[2.74]
$C_7H_{14}O$	*cis*-3-Methylcyclohexanol	[1.91]
$C_7H_{14}O$	*trans*-3-Methylcyclohexanol	[1.75]
$C_7H_{14}O_2$	Pentyl acetate	1.75 ± 0.10
$C_7H_{14}O_2$	Isopentyl acetate	[1.86]
$C_7H_{15}Br$	1-Bromoheptane	2.16 ± 0.11
$C_7H_{16}O$	2-Heptanol	[1.71]
$C_7H_{16}O$	3-Heptanol	[1.71]
C_8H_6	Phenylacetylene	0.656 ± 0.005
C_8H_7N	Benzeneacetonitrile	[3.5]
C_8H_8	Styrene	0.123 ± 0.003
C_8H_8O	Acetophenone	3.02 ± 0.06
$C_8H_8O_2$	Methyl benzoate	[1.94]
$C_8H_8O_3$	Methyl salicylate	[2.47]
C_8H_{10}	Ethylbenzene	0.59 ± 0.05
C_8H_{10}	*o*-Xylene	0.640 ± 0.005
$C_8H_{10}O$	2,4-Xylenol	[1.4]
$C_8H_{10}O$	2,5-Xylenol	[1.45]
$C_8H_{10}O$	2,6-Xylenol	[1.40]
$C_8H_{10}O$	3,4-Xylenol	[1.56]
$C_8H_{10}O$	3,5-Xylenol	[1.55]
$C_8H_{10}O$	Phenetole	1.45 ± 0.15
$C_8H_{10}O_2$	1,2-Dimethoxybenzene	[1.29]
$C_8H_{11}N$	*N,N*-Dimethylaniline	1.68 ± 0.17
$C_8H_{11}N$	2,4-Dimethylaniline	[1.40]
$C_8H_{11}N$	2,6-Dimethylaniline	[1.63]
$C_8H_{11}N$	2,4,6-Trimethylpyridine	[2.05]
$C_8H_{16}O$	2-Octanone	[2.70]
$C_8H_{16}O_2$	Octanoic acid	[1.15]
$C_8H_{16}O_2$	*sec*-Hexyl acetate	[1.9]
$C_8H_{16}O_2$	Isobutyl isobutanoate	[1.9]
$C_8H_{16}O_4$	Diethylene glycol monoethyl ether acetate	[1.8]
$C_8H_{17}Cl$	1-Chlorooctane	[2.00]
$C_8H_{18}O$	1-Octanol	[1.76]
$C_8H_{18}O$	2-Octanol	[1.71]
$C_8H_{18}O$	2-Ethyl-1-hexanol	[1.74]
$C_8H_{18}O$	Dibutyl ether	1.17 ± 0.06
$C_8H_{18}S$	Dibutyl sulfide	[1.61]
$C_8H_{19}N$	Dibutylamine	[0.98]
C_9H_7N	Quinoline	2.29 ± 0.11
C_9H_7N	Isoquinoline	2.73 ± 0.14
$C_9H_{10}O_2$	Ethyl benzoate	2.00 ± 0.10
$C_9H_{10}O_2$	Benzyl acetate	[1.22]
C_9H_{12}	Isopropylbenzene	≈0.79
$C_9H_{18}O$	2,6-Dimethyl-4-heptanone	[2.66]
$C_9H_{18}O_2$	Nonanoic acid	[0.79]
$C_{10}H_7Br$	1-Bromonaphthalene	[1.55]
$C_{10}H_7Cl$	1-Chloronaphthalene	[1.57]
$C_{10}H_8$	Azulene	0.80 ± 0.02
$C_{10}H_{14}$	*tert*-Butylbenzene	≈0.83
$C_{10}H_{16}O$	Camphor, (+)	[3.1]
$C_{10}H_{20}O_2$	2-Ethylhexyl acetate	[1.8]
$C_{10}H_{21}Br$	1-Bromodecane	[1.93]
$C_{10}H_{22}O$	Dipentyl ether	[1.20]
$C_{10}H_{22}O$	Diisopentyl ether	[1.23]
$C_{11}H_{12}O_2$	Ethyl *trans*-cinnamate	[1.84]
$C_{12}H_{10}$	Acenaphthene	≈0.85
$C_{12}H_{10}O$	Diphenyl ether	≈1.3
$C_{12}H_{27}BO_3$	Tributyl borate	[0.77]
$C_{12}H_{27}N$	Tributylamine	[0.78]
$C_{12}H_{27}O_4P$	Tributyl phosphate	[3.07]
$C_{14}H_{12}O_2$	Benzyl benzoate	[2.06]
$C_{16}H_{22}O_4$	Dibutyl phthalate	[2.82]
$C_{18}H_{34}O_2$	Oleic acid	[1.18]
$C_{18}H_{34}O_4$	Dibutyl sebacate	[2.48]
$C_{21}H_{21}O_4P$	Tri-*o*-cresyl phosphate	[2.87]
$C_{21}H_{21}O_4P$	Tri-*m*-cresyl phosphate	[3.05]
$C_{21}H_{21}O_4P$	Tri-*p*-cresyl phosphate	[3.18]
$C_{22}H_{44}O_2$	Butyl stearate	[1.88]
$C_{24}H_{38}O_4$	Bis(2-ethylhexyl) phthalate	[2.84]

STRENGTHS OF CHEMICAL BONDS*

J. Alistair Kerr and David W. Stocker

The strength of a chemical bond, $D°(R\text{-}X)$, often known as the bond dissociation energy, is defined as the standard enthalpy change of the reaction in which the bond is broken: $RX \rightarrow R + X$. It is given by the thermochemical equation, $D°(R\text{-}X) = \Delta_f H°(R) + \Delta_f H°(X) - \Delta_f H°(RX)$. Some authors list bond strengths at a temperature of absolute zero but here the values at 298 K are given because more thermodynamic data are available for this temperature. Bond strengths or bond dissociation energies are not equal to, and may differ considerably from, mean bond energies determined solely from thermochemical data on atoms and molecules.

Table 1
BOND STRENGTHS IN DIATOMIC MOLECULES

These have usually been measured spectroscopically or by mass spectrometric analysis of hot gases effusing a Knudsen cell. Excellent accounts of these and other methods are given in (i) *Dissociation Energies and Spectra of Diatomic Molecules*, by A. G. Gaydon, 3rd. ed., Chapman & Hall, London, 1968 and (ii) "Mass Spectrometric Determination of Bond Energies of High-Temperature Molecules", K. A. Gingerich, *Chimia*, 26, 619, 1972. The errors quoted in the table are those given in the original paper or review article. The references have been chosen primarily as a key to the literature. It should not be assumed that the author referred to was responsible for the value quoted, as the reference may be to a review article.

Bond strengths reported at a temperature of absolute zero, $D°_0$, have been converted to $D°_{298}$ by the use of enthalpy functions taken mainly from the JANAF Thermochemical Tables, Third Edition, *J. Phys. Chem. Ref. Data*, 14, Suppl. 1, 1985, wherever possible. For most bonds, however, this data is not available and the conversion has been made by the approximate relation:

$$D°_{298} = D°_0 + (3/2)RT$$

The list below does not include the increasing number of bond strengths of diatomic molecules now being calculated by *ab initio* methods. The Table has been arranged in an alphabetical order of the atoms.

Molecule	$D°_{298}$/kJ mol^{-1}	Ref.
Ag-Ag	160.3 ± 3.4	314
Ag-Al	183.7 ± 9.2	79
Ag-Au	202.9 ± 9.2	4
Ag-Bi	193 ± 42	246
Ag-Br	293 ± 29	120
Ag-Cl	314.2	184
Ag-Cu	174.1 ± 9.2	136
Ag-D	226.8	205
Ag-Dy	130 ± 19	203
Ag-Eu	129.7 ± 12.6	66
Ag-F	354.4 ± 16.3	120
Ag-Ga	180 ± 15	44
Ag-Ge	174.5 ± 20.9	135
Ag-H	215.1 ± 8	217
Ag-Ho	123.4 ± 16.7	62
Ag-I	234 ± 29	120
Ag-In	166.5 ± 4.9	13
Ag-Li	173.6 ± 6.3	276,303
Ag-Mn	100 ± 21	246
Ag-Na	138.1 ± 8.4	291,298
Ag-Nd	<209	221
Ag-O	220.1 ± 20.9	287
Ag-S	217.1	349
Ag-Se	202.5	349
Ag-Si	177.8 ± 10.0	320
Ag-Sn	136.0 ± 20.9	3
Ag-Te	195.8	349
Al-Al	133 ± 6	118
Al-Ar	5.182 ± 0.005	180
Al-As	202.9 ± 7.1	294,301
Al-Au	325.9 ± 6.3	124
Al-Br	429 ± 6	186
Al-Cl	511.3 ± 0.8	312
Al-Co	181.6 ± 0.2	22
Al-Cr	223.6 ± 0.6	19
Al-Cu	227.1 ± 1.2	21
Al-D	290.8	246
Al-F	663.6 ± 6.3	80
Al-H	284.9 ± 6.3	80
Al-I	369.9 ± 2.1	267
Al-Kr	6.047 ± 0.001	180
Al-Li	76.5	39
Al-N	297 ± 96	120
Al-Ni	225 ± 5	20
Al-O	511 ± 3	56,74
Al-P	216.7 ± 12.6	80
Al-Pd	254.4 ± 12.1	64
Al-S	373.6 ± 7.9	376
Al-Sb	216.3 ± 5.9	293
Al-Se	337.6 ± 10.0	376
Al-Si	229.3 ± 30.1	51
Al-Te	267.8 ± 10.0	376
Al-U	326 ± 29	123
Al-V	147.4 ± 1.0	22
Al-Xe	7.43 ± 0.69	43
Ar-Ar	4.73 ± 0.04	181
Ar-He	3.89	246
Ar-Hg	6.15	246
Ar-I	10.0	40
Ar-K	4.2	205
As-As	382.0 ± 10.5	247
As-Cl	448	80
As-D	270.3	205
As-F	410	205
As-Ga	209.6 ± 1.2	83
As-H	274.0 ± 2.9	29
As-I	296.6 ± 28.0	325
As-In	201	297
As-N	489 ± 2	310
As-O	481 ± 8	262
As-P	433.5 ± 12.6	137
As-S	379.5 ± 6.3	262
As-Sb	330.5 ± 5.4	94
As-Se	96	288
As-Tl	198.3 ± 14.6	300
At-At	~80	89
Au-Au	226.2 ± 0.5	210
Au-B	367.8 ± 10.5	145
Au-Ba	254.8 ± 10.0	135
Au-Be	285 ± 8	120
Au-Bi	297 ± 8.4	135
Au-Ca	243	135
Au-Ce	339 ± 21	135
Au-Cl	343 ± 9.6	120
Au-Co	222 ± 17	135
Au-Cr	213 ± 17	135
Au-Cs	255 ± 3.3	41
Au-Cu	228.0 ± 5.0	34,135
Au-D	318.4	205
Au-Dy	259 ± 21	203
Au-Eu	241.0 ± 10.5	66
Au-Fe	187.0 ± 16.7	220
Au-Ga	234 ± 38	135
Au-Ge	274.1 ± 5.0	135
Au-H	292.0 ± 8	217
Au-Ho	267.4 ± 16.7	62,250

* Revised to November 1998.

Table 1
BOND STRENGTHS IN DIATOMIC MOLECULES (continued)

Molecule	$D°_{298}$/kJ mol^{-1}	Ref.
Au-In	286.0 ± 5.7	13
Au-La	336.4 ± 20.9	127
Au-Li	284.5 ± 6.7	276
Au-Lu	332.2 ± 16.7	132
Au-Mg	243 ± 42	246
Au-Mn	185.4 ± 12.6	342
Au-Na	215.1 ± 12.6	298
Au-Nd	299.2 ± 20.9	127
Au-Ni	247 ± 16	352
Au-O	221.8 ± 20.9	287
Au-Pb	130 ± 42	246
Au-Pd	155 ± 21	135
Au-Pr	310 ± 21	135
Au-Rb	243 ± 2.9	41
Au-Rh	231.0 ± 29	66
Au-S	418 ± 25	131
Au-Sc	280.3 ± 16.7	128
Au-Se	243.1	349
Au-Si	305.4 ± 5.9	139
Au-Sn	254.8 ± 7.1	233
Au-Sr	264 ± 42	246
Au-Tb	289.5 ± 33.5	152,250
Au-Te	317.6	349
Au-U	318 ± 29	123
Au-V	240.6 ± 12.1	172
Au-Y	307.1 ± 8.4	177
B-B	297 ± 21	80
B-Br	396	32
B-C	448 ± 29	246
B-Ce	305 ± 21	246
B-Cl	511.3 ± 4	195
B-D	341.0 ± 6.3	246
B-F	757	257
B-H	340	302
B-I	220.5 ± 0.8	315
B-Ir	514.2 ± 17.2	381
B-La	339 ± 63	246
B-N	389 ± 21	80
B-O	808.8 ± 20.9	287
B-P	346.9 ± 16.7	147
B-Pd	329.3 ± 20.9	381
B-Pt	477.8 ± 16.7	268
B-Rh	475.7 ± 20.9	381
B-Ru	446.9 ± 20.9	381
B-S	580.7 ± 9.2	374
B-Sc	276 ± 63	246
B-Se	461.9 ± 14.6	374
B-Si	317 ± 7	390
B-Te	354.4 ± 20.1	374
B-Th	297	140
B-Ti	276 ± 63	246
B-U	322 ± 33	246
B-Y	293 ± 63	246
Ba-Br	362.8 ± 8.4	104,199,230
Ba-Cl	436.0 ± 8.4	197,199
Ba-D	≤193.7	205
Ba-F	587.0 ± 6.7	101,196
Ba-H	176 ± 14.6	120
Ba-I	320.8 ± 6.3	188
Ba-O	561.9 ± 13.4	287
Ba-Pd	221.8 ± 5.0	125
Ba-Rh	259.4 ± 25.1	125
Ba-S	400.0 ± 18.8	70
Be-Be	59	35,87
Be-Br	381 ± 84	246
Be-Cl	388.3 ± 9.2	108,191,382
Be-D	203.05	205
Be-F	577 ± 42	80,108
Be-H	200.0 ± 1.3	67
Be-O	434.7 ± 13.4	287
Be-S	372 ± 59	120
Bi-Bi	200.4 ± 7.5	307,324
Bi-Br	267.4 ± 4.2	76
Bi-Cl	301 ± 4	78
Bi-D	283.7	266
Bi-F	367 ± 13	400
Bi-Ga	159 ± 17	296
Bi-H	≤283.3	266
Bi-I	218.0 ± 4.6	77
Bi-In	153.6 ± 1.7	321
Bi-Li	154.0 ± 5.0	277,305
Bi-O	337.2 ± 12.6	287
Bi-P	280 ± 13	137
Bi-Pb	141.8 ± 14.6	324
Bi-S	315.5 ± 4.6	375
Bi-Sb	251 ± 4	244
Bi-Se	280.3 ± 5.9	375
Bi-Sn	210.0 ± 8.4	135
Bi-Te	232.2 ± 11.3	375
Bi-Tl	121 ± 13	84
Br-Br	192.807	1
Br-C	280 ± 21	120
Br-Ca	310.9 ± 9.2	319
Br-Cd	159 ± 96	120
Br-Cl	217.53 ± 0.29	59
Br-Co	331 ± 42	246
Br-Cr	328.0 ± 24.3	120
Br-Cs	389.1 ± 4	285,362
Br-Cu	331 ± 25	120
Br-D	370.74	205
Br-F	280 ± 12	211
Br-Fe	247 ± 96	120
Br-Ga	444 ± 17	80
Br-Ge	255 ± 29	120
Br-H	366.35	205
Br-Hg	72.8 ± 4	80
Br-I	179.1 ± 0.4	120,309
Br-In	414 ± 21	80
Br-K	379.9 ± 0.8	362,378
Br-Li	418.8 ± 4	362
Br-Mg	≤327.2	205
Br-Mn	314.2 ± 9.6	120
Br-N	276 ± 21	120
Br-Na	367.4 ± 0.8	362,378
Br-Ni	360 ± 13	120
Br-O	235.5 ± 2.4	46
Br-Pb	247 ± 38	120
Br-Rb	380.7 ± 4	362
Br-Sb	314 ± 59	120
Br-Sc	444 ± 63	246
Br-Se	297 ± 84	246
Br-Si	367.8 ± 10.0	106
Br-Sn	≥552	286
Br-Sr	333.0 ± 9.2	199
Br-Th	364	187
Br-Ti	439	246
Br-Tl	333.9 ± 1.7	28
Br-U	377.4 ± 6.3	260
Br-V	439 ± 42	246
Br-W	329.3	223
Br-Xe	5.94 ± 0.02	58
Br-Y	485 ± 84	246
Br-Zn	142 ± 29	246
C-C	610 ± 2.0	373
C-Ce	444 ± 13	236
C-Cl	397 ± 29	283
C-D	341.4	205
C-F	552	193
C-Ge	460 ± 21	120
C-H	338.4 ± 1.2	205
C-Hf	540 ± 25	357
C-I	209 ± 21	120
C-Ir	632 ± 4	171
C-La	462 ± 20	290
C-Mo	481 ± 15.9	167
C-N	748.0 ± 10	42
C-Nb	569 ± 13.0	167
C-O	1076.5 ± 0.4	80
C-Os	≥594	126
C-P	513.4 ± 8	350
C-Pt	598 ± 5.9	171,379
C-Rh	580.0 ± 3.8	332
C-Ru	616.2 ± 10.5	333
C-S	714.1 ± 1.2	71,354
C-Sc	≤444	148
C-Se	590.4 ± 5.9	343
C-Si	451.5	91,387
C-Tc	565 ± 29	322
C-Th	453 ± 17	166,357
C-Ti	423 ± 29	162,357
C-U	454.8 ± 15.1	165,169
C-V	427 ± 23.8	167
C-Y	418 ± 14	338
C-Zr	561 ± 25	357
Ca-Ca	~17	153
Ca-Cl	409 ± 9	269
Ca-D	≤169.9	205
Ca-F	527 ± 21	101,190
Ca-H	167.8	120
Ca-I	284.7 ± 8.4	188
Ca-Li	84.9 ± 8.4	397

Table 1
BOND STRENGTHS IN DIATOMIC MOLECULES (continued)

Molecule	$D°_{298}$/kJ mol^{-1}	Ref.	Molecule	$D°_{298}$/kJ mol^{-1}	Ref.	Molecule	$D°_{298}$/kJ mol^{-1}	Ref.
Ca-O	402.1 ± 16.7	287,326	Cl-Si	406	308,385	Cu-I	197 ± 21	120
Ca-S	337.6 ± 18.8	70,205	Cl-Sm	≥423	399	Cu-In	187.4 ± 7.9	13
Cd-Cd	7.36	251	Cl-Sn	414 ± 17	246	Cu-Li	192.9 ± 8.8	276
Cd-Cl	208.4	205	Cl-Sr	406 ± 13	197,199	Cu-Mn	158.6 ± 17	222
Cd-F	305 ± 21	31	Cl-Ta	544	23	Cu-Na	176.1 ± 16.7	299
Cd-H	69.0 ± 0.4	120	Cl-Th	489	261	Cu-Ni	202 ± 10	117
Cd-I	97.23	215	Cl-Ti	405.4 ± 10.5	192	Cu-O	269.0 ± 20.9	287
Cd-In	138	246	Cl-Tl	372.8 ± 2.1	28	Cu-S	276	349
Cd-O	235.6 ± 83.7	158,287	Cl-U	452 ± 8	259	Cu-Se	251	349
Cd-S	208.4 ± 20.9	158	Cl-V	477 ± 63	246	Cu-Si	221.3 ± 6.3	320
Cd-Se	127.6 ± 25.1	158	Cl-W	423 ± 42	246	Cu-Sn	169.5 ± 6.7	3,237
Cd-Te	100.0 ± 15.1	158	Cl-Xe	6.7	205	Cu-Tb	193 ± 19	203
Ce-Ce	245.2	127	Cl-Y	527 ± 84	246	Cu-Te	278.7	1
Ce-F	582 ± 42	246	Cl-Yb	~322	113	D-D	443.533	205
Ce-Ir	586	149	Cl-Zn	228.9 ± 19.7	73	D-F	576.6	205
Ce-N	519 ± 21	146	Cm-O	736	341	D-Ga	<272.8	253
Ce-O	795 ± 8	95	Co-Co	167 ± 25	218	D-Ge	≤322	205
Ce-Os	506 ± 33	126	Co-Cu	167 ± 17	135	D-H	439.433	205
Ce-Pd	322.2	63	Co-F	435 ± 63	246	D-Hg	42.05	205
Ce-Pt	556	149	Co-Ge	234 ± 21	135	D-In	246.0	205
Ce-Rh	548	149	Co-H	226 ± 42	369	D-Li	240.1892 ± 0.0046	207,360
Ce-Ru	531 ± 25	126	Co-I	285 ± 21	246	D-Lu	302	308
Ce-S	569	24	Co-Nb	267.0 ± 0.1	8	D-Mg	135.1	205
Ce-Se	494.5 ± 14.6	271	Co-O	384.5 ± 13.4	287	D-Ni	≤302.9	205
Ce-Te	189.4 ± 12.8	252	Co-S	331	349	D-Pt	≤350.2	205
Cl-Cl	242.580 ± 0.004	205	Co-Si	276 ± 17	380	D-S	351	205
Cl-Co	337.6 ± 6.7	194	Co-Ti	235.4 ± 0.1	353	D-Si	302.5	205
Cl-Cr	377.8 ± 6.7	194	Co-Y	253.7 ± 0.1	8	D-Sr	≥275.7	205
Cl-Cs	448 ± 8	285,363	Co-Zr	306.4 ± 0.1	8	D-Zn	88.7	205
Cl-Cu	377.8 ± 7.5	184	Cr-Cr	142.9 ± 5.4	201	Dy-F	531	406
Cl-D	436.47	205	Cr-Cu	155 ± 21	222	Dy-O	607 ± 17	95
Cl-Eu	~326	113	Cr-F	444.8 ± 19.7	227	Dy-S	414 ± 42	246
Cl-F	256.23	205,279	Cr-Ge	154 ± 7	202	Dy-Se	322 ± 42	246
Cl-Fe	329.7 ± 6.7	194	Cr-H	190.3 ± 7.0	53	Dy-Te	234 ± 42	246
Cl-Ga	481 ± 13	80	Cr-I	287.0 ± 24.3	120	Er-F	565 ± 17	406
Cl-Ge	~431	205	Cr-N	377.8 ± 18.8	152,355	Er-O	615 ± 13	95
Cl-H	431.62	205	Cr-O	461 ± 9	179	Er-S	418 ± 42	246
Cl-Hg	100 ± 8	120	Cr-Pb	105 ± 2	202	Er-Se	326 ± 42	246
Cl-I	211.3 ± 0.4	120	Cr-S	331	93	Er-Te	238 ± 42	246
Cl-In	439 ± 8	80	Cr-Sn	141 ± 3	202	Eu-Eu	33.5 ± 17	66
Cl-K	433.0 ± 8	363	Cs-Cs	43.919 ± 0.010	394	Eu-F	544	242
Cl-Li	469 ± 13	80	Cs-F	519 ± 8	285	Eu-Li	66.9 ± 2.9	275
Cl-Mg	327.6 ± 2.1	105,197,382	Cs-H	175.364	401	Eu-O	479 ± 10	95
Cl-Mn	338.5 ± 6.7	194	Cs-Hg	8	205	Eu-Rh	233.9 ± 33	66
Cl-N	333.9 ± 9.6	54	Cs-I	337.2 ± 2.1	285,361	Eu-S	362.3 ± 13.0	271,347
Cl-Na	412.1 ± 8	363	Cs-Na	63.2 ± 1.3	86	Eu-Se	301 ± 14.6	25,178,271
Cl-Ni	377.0 ± 6.7	194	Cs-O	295.8 ± 62.8	287	Eu-Te	243 ± 14.6	25,271
Cl-O	268.85 ± 0.10	2	Cs-Rb	49.57 ± 0.01	174	F-F	158.78	205
Cl-P	289 ± 42	246	Cu-Cu	176.52 ± 2.38	135,323	F-Ga	577 ± 14.6	270
Cl-Pb	301 ± 29	120	Cu-D	270.3	205	F-Gd	590.4 ± 27.2	405
Cl-Ra	343 ± 75	120	Cu-Dy	142 ± 21	203	F-Ge	485 ± 21	98
Cl-Rb	427.6 ± 8	363	Cu-F	413.4 ± 13	99	F-H	569.87 ± 0.06	402
Cl-S	277.0	224	Cu-Ga	215.9 ± 15.1	44	F-Hf	650 ± 15	16
Cl-Sb	360 ± 50	120	Cu-Ge	208.8 ± 21	273	F-Hg	~180	205
Cl-Sc	331	386	Cu-H	277.8	217,318	F-Ho	540	406
Cl-Se	322	246	Cu-Ho	142 ± 21	203	F-I	≤271.5	7,33,60,75

Table 1
BOND STRENGTHS IN DIATOMIC MOLECULES (continued)

Molecule	$D°_{298}$/kJ mol^{-1}	Ref.
F-In	506 ± 14.6	270
F-K	497.5 ± 2.5	17
F-La	598 ± 42	246
F-Li	577 ± 21	80
F-Lu	333.5	208
F-Mg	461.9 ± 5.0	101,196
F-Mn	423.4 ± 14.6	228
F-Mo	464.8	198
F-N	343	205
F-Na	519	205
F-Nd	545.2 ± 12.6	403
F-Ni	430 ± 20	85
F-O	219.54 ± 10	48
F-P	439 ± 96	120
F-Pb	356 ± 8	408
F-Pm	540 ± 42	246
F-Pr	582 ± 46	246
F-Pu	538.5 ± 29	229
F-Rb	494 ± 21	80
F-Ru	402	185
F-S	342.7 ± 5.0	30,200,231
F-Sb	439 ± 96	120
F-Sc	589.1 ± 13	404
F-Se	339 ± 42	246
F-Si	552.7 ± 2.1	109
F-Sm	565	242
F-Sn	466.5 ± 13	408
F-Sr	541.8 ± 6.7	101,196
F-Ta	573 ± 13	256
F-Tb	561 ± 42	246
F-Th	652	263
F-Ti	569 ± 33	407
F-Tl	445.2 ± 19.2	28
F-Tm	510	242
F-U	659.0 ± 10.5	157,258
F-V	590 ± 63	246
F-W	548 ± 63	246
F-Xe	15.77	317,368
F-Y	605.0 ± 20.9	404
F-Yb	≥521.3	18,113,399
F-Zn	368 ± 63	246
F-Zr	616 ± 15	16
Fe-Fe	75 ± 17	330
Fe-Ge	210.9 ± 29	219
Fe-H	180 ± 25	369
Fe-O	390.4 ± 17.2	287
Fe-S	322	93
Fe-Si	297 ± 25	380
Ga-Ga	112.1 ± 7	337
Ga-H	<274.1	253
Ga-I	339 ± 9.6	120
Ga-Li	133.1 ± 14.6	160
Ga-O	353.5 ± 41.8	287
Ga-P	229.7 ± 12.6	130
Ga-Sb	192.0 ± 12.6	295
Ga-Te	251 ± 25	377
Gd-O	719 ± 10	95
Gd-S	526.8 ± 10.5	116,345
Gd-Se	431 ± 14.6	25
Gd-Te	343 ± 14.6	25
Ge-Ge	263.6 ± 7.1	238
Ge-H	≤321.7	243
Ge-Ni	290.3 ± 10.9	335
Ge-O	659.4 ± 12.6	226,287
Ge-Pd	254.7 ± 10.5	334
Ge-S	534 ± 3	280
Ge-Sc	271.0 ± 11	234
Ge-Se	484.7 ± 1.7	282
Ge-Si	296.4 ± 8.6	391
Ge-Te	397 ± 3	282
Ge-Y	279.8 ± 11.4	235
H-H	435.990	205
H-Hg	39.844	205
H-I	298.407	205
H-In	243.1	205
H-K	174.576	206,401
H-Li	238.049 ± 0.004	393
H-Mg	126.4 ± 2.9	14,15,100
H-Mn	234 ± 29	120
H-N	≤339	205
H-Na	185.69 ± 0.25	274,316
H-Ni	252.3 ± 8	217
H-O	427.6	205
H-P	297	205
H-Pb	≤157	205
H-Pd	234 ± 25	369
H-Pt	≤335	205
H-Rb	167 ± 21	120
H-Rh	247 ± 21	369
H-Ru	234 ± 21	369
H-S	344.3 ± 12.1	212
H-Sc	~180	329
H-Se	314.47 ± 0.96	122
H-Si	≤299.2	205
H-Sn	264 ± 17	120
H-Sr	163 ± 8	120
H-Te	268 ± 2.1	119
H-Ti	204.6 ± 8.8	52
H-Tl	188 ± 8	120
H-V	208.7 ± 7.0	53
H-Yb	159 ± 38	120
H-Zn	85.8 ± 2.1	120
He-He	3.8	205
He-Hg	6.61	246
Hf-C	548 ± 63	246
Hf-N	536 ± 29	152,245
Hf-O	801.7 ± 13.4	287
Hg-Hg	8 ± 2	204
Hg-I	34.69 ± 0.96	388
Hg-K	8.24 ± 0.21	246
Hg-Li	13.8	205
Hg-Na	9.2	205,411
Hg-O	220.9 ± 33.1	158
Hg-Rb	8.4	205
Hg-S	217.1 ± 22.2	158
Hg-Se	144.3 ± 30.1	158
Hg-Te	≤142	246
Hg-Tl	4	183
Ho-Ho	84 ± 17	62
Ho-O	611 ± 17	95
Ho-S	428.4 ± 14.6	345
Ho-Se	335 ± 17	25
Ho-Te	259 ± 17	25
I-I	151.088	205,371
I-In	331	384
I-K	325.1 ± 0.8	361,378
I-Li	345.2 ± 4.2	361
I-Mg	~285	26
I-Mn	282.8 ± 9.6	120
I-N	159 ± 17	246
I-Na	304.2 ± 2.1	361,378
I-Ni	293 ± 21	120
I-O	230	47
I-Pb	193 ± 4	340
I-Rb	318.8 ± 2.1	361
I-Si	293	205
I-Sn	234 ± 42	246
I-Sr	269.9 ± 5.9	239
I-Te	192 ± 42	246
I-Ti	310 ± 42	246
I-Tl	272 ± 8	26
I-Zn	108.29	215
I-Zr	305	241
In-In	100 ± 8	246
In-Li	92.5 ± 14.6	160
In-O	<320.1	287
In-P	197.9 ± 8.4	291
In-S	289 ± 17	69
In-Sb	151.9 ± 10.5	81
In-Se	247 ± 17	69
In-Te	218 ± 17	69
Ir-La	577 ± 13	176
Ir-O	414.6 ± 42.3	287
Ir-Si	462.8 ± 20.9	381
Ir-Th	573	133
Ir-Ti	422 ± 13	289
Ir-Y	456.1 ± 16.7	177
K-K	54.63 ± 0.02	6,265
K-Kr	4.6	205
K-Li	82.0 ± 4.2	103,410
K-Na	65.994 ± 0.008	36,410
K-O	277.8 ± 20.9	287
K-Xe	5.0	205
Kr-Kr	5.23	50,205
Kr-O	<8	246
Kr-Xe	5.505 ± 0.002	9
La-La	247 ± 21	386
La-N	519 ± 42	246
La-O	799 ± 4	95
La-Pt	502 ± 21	272
La-Rh	527 ± 17	65

Table 1
BOND STRENGTHS IN DIATOMIC MOLECULES (continued)

Molecule	$D°_{298}$/kJ mol^{-1}	Ref.	Molecule	$D°_{298}$/kJ mol^{-1}	Ref.	Molecule	$D°_{298}$/kJ mol^{-1}	Ref.
La-S	573.2 ± 1.7	214,359	Nd-Se	385 ± 17	25,156,271	P-Tl	209 ± 13	293
La-Se	477 ± 17	25,271	Nd-Te	305 ± 17	25	P-U	297 ± 21	246
La-Te	381 ± 17	25,154	Ne-Ne	3.93	365	P-W	305 ± 4	150
La-Y	202.1	386	Ni-Ni	200.7 ± 0.2	304	Pb-Pb	86.6 ± 0.8	138,305
Li-Li	110.21 ± 4	383,398	Ni-O	382.0 ± 16.7	287	Pb-S	346.0 ± 1.7	375
Li-Mg	67.4 ± 6.3	396	Ni-Pt	273.7 ± 0.3	367	Pb-Sb	161.5 ± 10.5	409
Li-Na	87.181 ± 0.001	102,111	Ni-S	344.3	93	Pb-Se	302.9 ± 4	375
Li-O	333.5 ± 8.4	287	Ni-Si	318 ± 17	380	Pb-Te	251 ± 13	375
Li-Pb	78.7 ± 7.9	277	Ni-V	206.3 ± 0.1	353	Pd-Pd	100 ± 15	331
Li-S	312.5 ± 7.5	232	Ni-Y	283.9 ± 0.1	8	Pd-Si	261 ± 12	336
Li-Sb	172.8 ± 10.0	278	Ni-Zr	279.7 ± 0.1	8	Pd-Y	238 ± 17	313
Li-Sm	49.0 ± 4.2	275	Np-O	718.4 ± 41.8	287	Pm-S	423 ± 63	246
Li-Tm	69.0 ± 3.3	275	O-O	498.36 ± 0.17	38,205	Pm-Se	339 ± 63	246
Li-Yb	37.2 ± 2.9	275	O-Os	575	189	Pm-Te	255 ± 63	246
Lu-Lu	142 ± 33	246	O-P	599.1 ± 12.6	287	Po-Po	187.0	205
Lu-O	678 ± 8	95	O-Pa	788.3 ± 17.2	240	Pr-S	492.5 ± 4.6	112
Lu-Pt	402 ± 33	141	O-Pb	382.0 ± 12.6	287	Pr-Se	446.4 ± 23.0	155,271
Lu-S	507.1 ± 14.6	114,345	O-Pd	380.7 ± 83.7	287	Pr-Te	326 ± 42	246
Lu-Se	418 ± 17	25	O-Pm	674 ± 63	246	Pt-Pt	307 ± 2	366
Lu-Te	326 ± 17	25	O-Pr	753 ± 13	95	Pt-Si	501.2 ± 18.0	381
Mg-Mg	8.552 ± 0.004	264,397	O-Pt	391.6 ± 41.8	287	Pt-Th	552	133
Mg-O	363.2 ± 12.6	287	O-Pu	715.9 ± 33.9	287	Pt-Ti	397 ± 13	173
Mg-S	234	70	O-Rb	255 ± 84	37	Pt-Y	474.0 ± 12.1	170
Mn-Mn	25.9	221	O-Re	626.8 ± 83.7	287	Rb-Rb	48.898 ± 0.005	5
Mn-O	402.9 ± 41.8	225,287	O-Rh	405.0 ± 41.8	287	Rh-Rh	285.3 ± 0.05	255
Mn-S	301 ± 17	395	O-Ru	528.4 ± 41.8	287	Rh-Sc	443.9 ± 10.5	175
Mn-Se	239.3 ± 9.2	351	O-S	517.90 ± 0.05	57	Rh-Si	395.0 ± 18.0	381
Mo-Mo	406 ± 21	168	O-Sb	434.3 ± 41.8	287	Rh-Th	515 ± 21	129
Mo-Nb	456 ± 25	163	O-Sc	681.6 ± 11.3	287	Rh-Ti	390.8 ± 14.6	61
Mo-O	560.2 ± 20.9	287	O-Se	464.8 ± 21.3	287,344	Rh-U	519 ± 17	129
N-N	945.33 ± 0.59	205	O-Si	799.6 ± 13.4	287	Rh-V	364 ± 29	135
N-O	630.57 ± 0.13	205	O-Sm	565 ± 13	95	Rh-Y	445.2 ± 10.5	175
N-P	617.1 ± 20.9	72,151	O-Sn	531.8 ± 12.6	287	Ru-Si	397.1 ± 20.9	381
N-Pu	473 ± 63	246	O-Sr	426.3 ± 6.3	327	Ru-Th	591.6 ± 42	134
N-S	464 ± 21	246	O-Ta	799.1 ± 12.6	287	Ru-V	414 ± 29	135
N-Sb	301 ± 50	120	O-Tb	711 ± 13	95	S-S	425.30	205
N-Sc	469 ± 84	246	O-Te	376.1 ± 20.9	287	S-Sb	378.7	110
N-Se	370 ± 11	254	O-Th	878.6 ± 12.1	287	S-Sc	477 ± 13	359,372
N-Si	470 ± 15	311	O-Ti	672.4 ± 9.2	287	S-Se	371.1 ± 6.7	90
N-Ta	611 ± 84	246	O-Tm	502 ± 13	95	S-Si	623	205
N-Th	577.4 ± 33.1	144,152	O-U	759.4 ± 13.4	287	S-Sm	389	112
N-Ti	476.1 ± 33.1	152,356	O-V	626.8 ± 18.8	12,287	S-Sn	464 ± 3.3	88
N-U	531.4 ± 2.1	142	O-W	672.0 ± 41.8	287	S-Sr	339	45
N-V	477.4 ± 17.2	107,152	O-Xe	36.4	246	S-Tb	515 ± 42	246
N-Xe	23.0	182	O-Y	719.6 ± 11.3	209,287	S-Te	339 ± 21	88
N-Y	481 ± 63	246	O-Yb	397 ± 17	95	S-Ti	418 ± 3	96,292
N-Zr	564.8 ± 25.1	143,152	O-Zn	159 ± 4	55	S-Tm	368 ± 42	246
Na-Na	73.0813 ± 0.0001	213	O-Zr	776.1 ± 13.4	287	S-U	522.6 ± 9.6	359
Na-O	256.1 ± 16.7	287	P-P	489.5 ± 10.5	151	S-V	450	97,205
Na-Rb	63.25	392	P-Pt	≤416.7	348	S-Y	528.4 ± 10.5	358
Nb-Nb	510 ± 10.0	164	P-Rh	353.1 ± 17	348	S-Yb	167	246
Nb-Ni	271.9 ± 0.1	8	P-S	444 ± 8	92	S-Zn	205 ± 13	82,158
Nb-O	771.5 ± 25.1	287	P-Sb	356.9	249	S-Zr	575.3 ± 16.7	359
Nb-Ti	302.0 ± 0.1	254	P-Se	363.6 ± 10.0	92	Sb-Sb	299.2 ± 6.3	81,248
Nd-Nd	<163	246	P-Si	363.6	346	Sb-Te	277.4 ± 3.8	306,364
Nd-O	703 ± 13	95	P-Te	297.9 ± 10.0	92	Sb-Tl	126.8 ± 10.5	11,293
Nd-S	471.5	25	P-Th	550.2 ± 42	135	Sc-Sc	162.8 ± 21	136

Table 1
BOND STRENGTHS IN DIATOMIC MOLECULES (continued)

Molecule	$D°_{298}$/kJ mol^{-1}	Ref.
Sc-Se	385 ± 17	246
Sc-Si	228.7 ± 14	234
Sc-Te	289 ± 17	246
Se-Se	332.6 ± 0.4	90,375
Se-Si	538 ± 13	370
Se-Sm	331.0 ± 14.6	271
Se-Sn	401.2 ± 5.9	68
Se-Sr	~285	27
Se-Tb	423 ± 42	246
Se-Te	291.6 ± 4	88,90,159
Se-Ti	381 ± 42	246
Se-Tm	276 ± 42	246
Se-V	347 ± 21	246
Se-Y	435 ± 13	246
Se-Zn	170.7 ± 25.9	82,158
Si-Si	325 ± 7	328
Si-Te	452 ± 8	205,281
Si-Y	258.8 ± 17.3	235
Sm-Te	272.4 ± 14.6	271
Sn-Sn	187.1 ± 0.3	284
Sn-Te	359.8	205
Sr-Sr	15.5 ± 0.4	121
Tb-Tb	131.4 ± 25.1	250
Tb-Te	339 ± 42	246
Te-Te	257.6 ± 4.1	389
Te-Ti	289 ± 17	246
Te-Tm	276 ± 42	246
Te-Y	339 ± 13	246
Te-Zn	117.6 ± 18.0	158
Th-Th	≤289	140
Ti-Ti	141.4 ± 21	216
Ti-V	203.2 ± 0.1	353
Ti-Zr	214.3 ± 0.1	254
Tl-Tl	64.4 ± 17	10
U-U	222 ± 21	246
V-V	269.3 ± 0.1	353
V-Zr	260.6 ± 0.3	254
Xe-Xe	6.138 ± 0.001	49,115
Y-Y	159 ± 21	246
Yb-Yb	20.5 ± 17	161
Zn-Zn	29	339
Zr-Zr	298.2 ± 0.1	8

REFERENCES

1. Abbasov, A. S., Azizov, T. Kh., Alleva, N. A., Aliev, I. Ya, Mustafaev, F. M., and Mamedov, A. N., *Zh. Fiz. Khim.*, 50, 2172, 1976.
2. Abramowitz, S. and Chase, M. L., Pure Appl. Chem., 63, 1449, 1991.
3. Ackerman, M., Drowart, J., Stafford, F. E., and Verhaegen, G., *J. Chem. Phys.*, 36, 1557, 1962.
4. Ackerman, M., Stafford, F. E., and Drowart, J., *J. Chem. Phys.*, 33, 1784, 1960.
5. Amiot, C., *J. Chem. Phys.*, 93, 8591, 1990.
6. Amiot, C., *J. Mol. Spectrosc.*, 147, 370, 1991.
7. Appelman, E. H. and Clyne, M. A. A., *J. Chem. Soc. Faraday Trans. 1*, 71, 2072, 1975.
8. Arrington, C. A., Blume, T., Morse, M. D., Doverstal, M., and Sassenberg, U., J. Phys. Chem., 98, 1398, 1994.
9. Balakrishnan, A., Jones, W. J., Mahajan, C. G., and Stoicheff, B. P., *Chem. Phys. Lett.*, 155, 43, 1989.
10. Balducci, G. and Piacente, V., *J. Chem. Soc. Chem. Commun.*, 1287, 1980.
11. Balducci, G., Ferro, D., and Piacente, V., *High Temp. Sci.*, 14, 207, 1981.
12. Balducci, G., Gigli, G., and Guido, M., *J. Chem. Phys.,* 79, 5616, 1983.
13. Balducci, G., Nunzio, P. E., Gigli, G., and Guido, M., *J. Chem. Phys.,* 90, 406, 1989.
14. Balfour, W. J. and Cartwright, H. M., *Astron. Astrophys. Suppl. Ser.*, 26, 389, 1976.
15. Balfour, W. J. and Lingren, B., Can. J. Chem., 56, 767, 1978.
16. Barkovskii, N. V., Tsirel'nikov, V. I., Emel'yanov, A. M., and Khodeev, Yu. S., *Teplofiz. Vys. Temp.*, 29, 474, 1991.
17. Barrow, R. F. and Caunt, A. D., *Proc. R. Soc. London Ser. A,* 219, 120, 1953.
18. Barrow, R. F. and Chojnicki, A. H., *J. Chem. Soc. Faraday Trans. 2*, 71, 728, 1975.
19. Behm, J. M. and Morse, M. D., J. Chem. Phys., 101, 6500, 1994.
20. Behm, J. M., Arrington, C. A., and Morse, M. D., *J. Chem. Phys.*, 99, 6409, 1993.
21. Behm, J. M., Arrington, C. A., Langenberg, J. D., and Morse, M. D., *J. Chem. Phys.,* 99, 6394, 1993.
22. Behm, J. M., Brugh, D. J., and Morse, M. D., J. Chem. Phys., 101, 6487, 1994.
23. Behrens, R. G. and Feber, R. C., *J. Less-Common Met.*, 75, 281, 1980.
24. Bergman, C. and Gingerich, K. A., *J. Phys. Chem.,* 76, 2332, 1972.
25. Bergman, C., Coppens P., Drowart, J., and Smoes, S., *Trans. Faraday Soc.,* 66, 800, 1970.
26. Berkowitz, J. and Chupka, W. A., *J. Chem. Phys.*, 45, 1287, 1966.
27. Berkowitz, J. and Chupka, W. A., *J. Chem. Phys.*, 45, 4289, 1966.
28. Berkowitz, J. and Walter, T., *J. Chem. Phys.*, 49, 1184, 1968.
29. Berkowitz, J., *J. Chem. Phys.*, 89, 7065, 1988.
30. Berneike ,W., Kreuttle, U., and Neuert, H., *Chem. Phys. Lett.*, 76, 525, 1980.
31. Besenbruch, G., Kana'an, A. S., and Margrave, J. L., *J. Phys. Chem.*, 69, 3174, 1965.
32. Bharate, N. S., Bhartiya, J. B., and Behere, S. H., *Proc. Indian Natl. Sci. Acad. Part A*, 57, 419, 1991.
33. Birks, J. W., Gabelnick, S. D., and Johnston, H. S., *J. Mol. Spectrosc.*, 57, 23, 1975.
34. Bishea, G. A. and Morse, M. D., *Chem. Phys. Lett.*, 171, 430, 1990.
35. Bondybey, V. E., *Chem. Phys. Lett.,* 109, 436, 1984.
36. Breford, E. J. and Engelke, F., *J. Chem. Phys.,* 71, 1994, 1979.
37. Brewer, L. and Rosenblatt, G. M., *Adv. High Temp. Sci.*, 2, 1, 1969.
38. Brix, P. and Herzberg, G., *J. Chem. Phys.*, 21, 2240, 1953.
39. Brock, L. R., Pilgrim, J. S., and Duncan, M. A., Chem. Phys. Lett., 230, 93, 1994.
40. Burns, G., LeRoy, L. J., Morris, D. J., and Blake, J. A., *Proc. R. Soc. London Ser. A*, 316, 81, 1970.
41. Busse, V. B. and Weil, K. G., *Ber. Bunsenges. Phys. Chem.*, 85, 309, 1981.

Table 1

BOND STRENGTHS IN DIATOMIC MOLECULES (continued)

42. Calculated from $\Delta_f H°$(CN) = 441.4 ± 4.6 kJ mol^{-1} (Table 4), but see also Costes, M., Naulin, C., and Dorthe, G., *Astron. Astronphys.*, 232, 270, 1990.
43. Callender, C. L., Mitchell, S. A., and Hackett, P. A., *J. Chem. Phys.*, 90, 5252, 1989.
44. Carbonel, M., Bergman, C., and Laffite, M., *Colloq. Int. Cent. Nat. Rech. Sci.*, 201, 311, 1972.
45. Cater, E. D. and Johnson, E. W., *J. Chem. Phys.*, 47, 5353, 1967.
46. Chase, M. L., *J. Phys. Chem. Ref. Data,* 25, 1069, 1996.
47. Chase, M. L., *J. Phys. Chem. Ref. Data,* 25, 1297, 1996.
48. Chase, M. L., *J. Phys. Chem. Ref. Data,* 25, 551, 1996.
49. Chashchina, G. I. and Shreider, E. Ya., *Zh. Prikl. Spektrosk.*, 21, 696, 1974.
50. Chashchina, G. I. and Shreider, E. Ya., *Zh. Prikl. Spektrosk.*, 25, 163, 1976.
51. Chatillon, C., Allibert, M., and Pattoret, A., *C. R. Acad. Sci. Ser. C*, 280, 1505, 1975.
52. Chen, Y. M., Clemmer, D. E., and Armentrout, P. B., *J. Chem. Phys.*, 95, 1228, 1991.
53. Chen, Y.-M., Clemmer, D. E., and Armentrout, P. B., *J. Chem. Phys.*, 98, 4929, 1993.
54. Clarke, T. C. and Clyne, M. A. A., *Trans. Faraday Soc.*, 66, 877, 1970.
55. Clemmer, D. E., Daltaska, N. F., and Armentrout, P. B., *J. Chem. Phys.*, 95, 7263, 1991.
56. Clemmer, D. E., Weber, M. E., and Armentrout, P. B., *J. Phys. Chem.*, 96, 10888, 1992.
57. Clerbaux, C. and Colin, R., *J. Mol. Spectrosc.* 165, 334, 1994.
58. Clevenger, J. O. and Tellinghuisen, J., *J. Chem. Phys.*, 103, 9611, 1995.
59. Clyne, M. A. A. and McDermid, I. S., *Faraday Discuss. Chem. Soc.*, 67, 316, 1979.
60. Clyne, M. A. A. and McDermid, I. S., *J. Chem. Soc. Faraday Trans. 2*, 72, 2252, 1976.
61. Cocke, D. L. and Gingerich, K. A., *J. Chem. Phys.*, 60, 1958, 1974.
62. Cocke, D. L. and Gingerich, K. A., *J. Phys. Chem.*, 75, 3264, 1971.
63. Cocke, D. L. and Gingerich, K. A., *J. Phys. Chem.*, 76, 2332, 1972.
64. Cocke, D. L., Gingerich, K. A., and Chang, C. A., *J. Chem. Soc. Faraday Trans. 1*, 72, 268, 1976.
65. Cocke, D. L., Gingerich, K. A., and Kordis, J., *High Temp. Sci.*, 5, 474, 1973.
66. Cocke, D. L., Gingerich, K. A., and Kordis, J., *High Temp. Sci.*, 7, 61, 1975.
67. Colin, R. and De Greef, D., *Can. J. Phys.*, 53, 2142, 1975.
68. Colin, R. and Drowart, J., *Trans. Faraday Soc.*, 60, 673, 1964.
69. Colin, R. and Drowart, J., *Trans. Faraday Soc.*, 64, 2611, 1968.
70. Colin, R., Goldfinger, P., and Jeunehomme, M., *Trans. Faraday Soc.*, 60, 306, 1964.
71. Coppens, P., Reynaert, J. C., and Drowart, J., *J. Chem. Soc. Faraday Trans. 2*, 75, 292, 1979.
72. Coquart, B. and Prudhomme, J. C., *J. Mol. Spectrosc.*, 87, 75, 1981.
73. Corbett, J. D. and Lynde, R. A., *Inorg. Chem.,* 6, 2199, 1967.
74. Costes, M., Naulin, C., Dorthe, G., Vaucamps, C., Nouchi, G., *Faraday Discuss. Chem. Soc.*, 84, 75, 1987.
75. Coxon, J. A., *Chem. Phys. Lett.*, 33, 136, 1975.
76. Cubicciotti, D., *Inorg. Chem.*, 7, 208, 1968.
77. Cubicciotti, D., *Inorg. Chem.*, 7, 211, 1968.
79. Cubicciotti, D., *J. Phys. Chem.*, 71, 3066, 1967.
80. Cuthill, A. M., Fabian, D. J., and Shu-Shou-Shen, S., *J. Phys. Chem.*, 77, 2008, 1973.
81. De Maria, G., Drowart, J., and Inghram, M. G., *J. Chem. Phys.*, 31, 1076, 1959.
82. De Maria, G., Goldfinger, P., Malaspina, L., and Piacente, V., *Trans. Faraday Soc.*, 61, 2146, 1965.
83. De Maria, G., Malaspina, L., and Piacente, V., *J. Chem. Phys.*, 52, 1019, 1970.
84. De Maria, G., Malaspina, L., and Piacente, V., *J. Chem. Phys.*, 56, 1978, 1972.
85. Devore, T. C., McQuaid, M., and Gole, J. L., *High Temp. Sci.*, 30, 83, 1990.
86. Diemer, U., Weickenmeier, H., Wahl, M., and Demtroeder, W., *Chem. Phys. Lett.*, 104, 489, 1984.
87. Drowart, J. and Goldfinger, P., *Angew. Chem.*, 6, 581, 1967.
88. Drowart, J. and Goldfinger, P., *Q. Rev. (London),* 20, 545, 1966.
89. Drowart, J. and Honig, R. E., *J. Phys. Chem.*, 61, 980, 1957.
90. Drowart, J. and Smoes, S., *J. Chem. Soc. Faraday Trans. 2*, 73, 1755, 1977.
91. Drowart, J., De Maria, G., and Inghram, M. G., *J. Chem. Phys.*, 29, 1015, 1958.
92. Drowart, J., Myers, C. E., Szwarc, R.,Vander Auwera-Mahieu, A., and Uy, O. M., *High Temp. Sci.*, 5, 482, 1973.
93. Drowart, J., Pattoret, A., and Smoes, S., *Proc. Br. Ceramic Soc.*, No. 8, 67, 1967.
94. Drowart, J., Smoes, S., and Vander Auwera-Mahicu, A., *J. Chem. Thermodyn.*, 10, 453, 1978.
95. Dulick, M., Murad, E., and Barrow, R. F., *J. Chem. Phys.*, 85, 385, 1986.
96. Edwards, J. G., Franklin, H. F., and Gilles, P. W., *J. Chem. Phys.*, 54, 545, 1971.
97. Edwards, J. G., *J. Chem. Phys.*, 96, 866, 1992
98. Ehlert, T. C. and Margrave, J. L., *J. Chem. Phys.*, 41, 1066, 1964.
99. Ehlert, T. C. and Wang, J. S., *J. Phys. Chem.*, 81, 2069, 1977.
100. Ehlert, T. C., Hilmer, R. M., and Beauchamp, E. A., *J. Inorg. Nucl. Chem.*, 30, 3112, 1968.
101. Engelke, F., *Chem. Phys.*, 39, 279, 1979.
102. Engelke, F., Ennen, G., and Meiwes, K. H., *Chem. Phys.*, 66, 391, 1982.

Table 1
BOND STRENGTHS IN DIATOMIC MOLECULES (continued)

103. Engelke, F., Hage, H., and Sprick, U., *Chem. Phys.*, 88, 443, 1984.
104. Estler, C. and Zare, R. N., *Chem. Phys.*, 28, 253, 1978.
105. Farber, M. and Srivastava, R. D., *Chem. Phys. Lett.*, 42, 567, 1976.
106. Farber, M. and Srivastava, R. D., *High Temp. Sci.*, 12, 21, 1980.
107. Farber, M. and Srivastava, R. D., *J. Chem. Soc. Faraday Trans. 1*, 69, 390, 1973.
108. Farber, M. and Srivastava, R. D., *J. Chem. Soc. Faraday Trans. 1*, 70, 1581, 1974.
109. Farber, M. and Srivastava, R. D., *J. Chem. Soc. Faraday Trans. 1*, 74, 1089, 1978.
110. Faure, F. M., Mitchell, M. J., and Bartlett, R. W., *High Temp. Sci.*, 4, 181, 1972.
111. Fellows, C. E., *J. Chem. Phys.*, 94, 5855, 1991.
112. Fenochka, B. V. and Gorkienko, S. P., *Zh. Fiz. Khim*, 47, 2445, 1973.
113. Filippenko, N. V., Morozov, E. V., Giricheva, N. L., and Krasnev, K. S., *Izv. Vyssh. Ucheb. Zaved Khim.Technol.*, 15, 1416, 1972.
114. Franzen, H. and Hariharan, A. V., *J. Chem. Phys.*, 70, 4907, 1979.
115. Freeman, D. E., Yoshino, K., and Tanaka, Y., *J. Chem. Phys.*, 61, 4880, 1974.
116. Fries, J. A. and Cater, E. D., *J. Chem. Phys.*, 68, 3978, 1978.
117. Fu, Z. and Morse, M. D., *J. Chem. Phys.*, 90, 3417, 1989.
118. Fu, Z., Lemire, G. W., Bishea, G. A., and Morse, M. D., *J. Chem. Phys.*, 93, 8420, 1990.
119. Gal, J. F., Maria, P. C., and Decouzon, M., *Int. J. Mass Spectrom. Ion Processes*, 93, 87, 1989.
120. Gaydon, A. G., *Dissociation Energies and Spectra of Diatomic Molecules, 3rd ed.*, Chapman & Hall, London, 1968.
121. Gerber, G. and Moeller, R., *Contrib. Symp. At. Surf. Phys.*, 168, 1982.
122. Gibson, S. T., Greene, J. P., and Berkowitz, J., *J. Chem. Phys.*, 85, 4815, 1986.
123. Gingerich, K. A. and Blue, G. D., *J. Chem. Phys.*, 47, 5447, 1967.
124. Gingerich, K. A. and Blue, G. D., *J. Chem. Phys.*, 59, 186, 1973.
125. Gingerich, K. A. and Choudary, U. V., *J. Chem. Phys.*, 68, 3265, 1978.
126. Gingerich, K. A. and Cocke, D. L., *Inorg. Chim. Acta*, 28, L171, 1978.
127. Gingerich, K. A. and Finkbeiner, H. C., *J. Chem. Phys.*, 54, 2621, 1971.
128. Gingerich, K. A. and Finkbeiner, H. C., *Proc. 9th Rare Earth Res. Conf.*, 2, 795, 1971.
129. Gingerich, K. A. and Gupta, S. K., *J. Chem. Phys.*, 69, 505, 1978.
130. Gingerich, K. A. and Piacente, V., *J. Chem. Phys.*, 54, 2498, 1971.
131. Gingerich, K. A., *Chem. Commun.*, 580, 1970.
132. Gingerich, K. A., *Chem. Phys. Lett.*, 13, 262, 1972.
133. Gingerich, K. A., *Chem. Phys. Lett.*, 23, 270, 1973.
134. Gingerich, K. A., *Chem. Phys. Lett.*, 25, 523, 1974.
135. Gingerich, K. A., *Chem. Soc. Faraday, Symp.*, No.14, 109, 1980.
136. Gingerich, K. A., *Chimia*, 26, 619, 1972.
137. Gingerich, K. A., Cocke, D. L., and Kordis, J., *J. Phys. Chem.*, 78, 603, 1974.
138. Gingerich, K. A., Cocke, D. L., and Miller, F., *J. Chem. Phys.*, 64, 4027, 1976.
139. Gingerich, K. A., Haque, R., and Kingcade, J. E., *Thermochim. Acta*, 30, 61, 1979.
140. Gingerich, K. A., *High Temp. Sci.*, 1, 258, 1969.
141. Gingerich, K. A., *High Temp. Sci.*, 3, 415, 1971.
142. Gingerich, K. A., *J. Chem. Phys.*, 47, 2192, 1967.
143. Gingerich, K. A., *J. Chem. Phys.*, 49, 14, 1968.
144. Gingerich, K. A., *J. Chem. Phys.*, 49, 19, 1968.
145. Gingerich, K. A., *J. Chem. Phys.*, 54, 2646, 1971.
146. Gingerich, K. A., *J. Chem. Phys.*, 54, 3720, 1971.
147. Gingerich, K. A., *J. Chem. Phys.*, 56, 4239, 1972.
148. Gingerich, K. A., *J. Chem. Phys.*, 74, 6407, 1981.
149. Gingerich, K. A., *J. Chem. Soc. Faraday Trans. 2*, 70, 471, 1974.
150. Gingerich, K. A., *J. Phys. Chem.*, 68, 768, 1964.
151. Gingerich, K. A., *J. Phys. Chem.*, 73, 2734, 1969.
152. Gingerich, K. A., *NBS Spec. Publ. (U. S.)*, 561, 289, 1979.
153. Gondal, M. A., Khan, M. A., and Rais, M. H., Chem. Phys. Lett., 243, 94, 1995.
154. Gordienko, S. P. and Fenochka, B. V., *Izv. Akad. Nauk. SSSR Neorg. Mater.*, 18, 1811, 1982.
155. Gordienko, S. P., Fenochka, B. V., Viksman, G. Sh., Klockkova, L. A., and Mikhlina, T. M., *Izv. Akad. Nauk. SSSR Neorg. Mater.*, 18, 18, 1982.
156. Gordienko, S. P., *Izv. Akad. Nauk. SSSR Neorg. Mater.*, 20, 1472, 1984.
157. Gorokhov, L. N., Smirnov, V. K., and Khodeev, Yu. S., *Zh. Fiz. Khim.*, 58, 1603, 1984.
158. Grade, M. and Hirschwald, W., *Ber. Bunsenges. Phys. Chem.*, 86, 899, 1982.
159. Grade, M., Wienecke, J., Rosinger, W., and Hirschwald, W., *Ber. Bunsenges. Phys. Chem.*, 87, 355, 1983.
160. Guggi, D. J., Neubert, A., and Zmbov, K. F., *Conf. Int. Thermodyn. Chim. [C. R.] 4th*, 3, 124, 1975.
161. Guido, M. and Balducci, G., *J. Chem. Phys.*, 57, 5611, 1972.
162. Gupta, S. K. and Gingerich, K. A., *High Temp. - High Pressures*, 12, 273, 1980.

Table 1
BOND STRENGTHS IN DIATOMIC MOLECULES (continued)

163. Gupta, S. K. and Gingerich, K. A., *J. Chem. Phys.*, 69, 4318, 1978.
164. Gupta, S. K. and Gingerich, K. A., *J. Chem. Phys.*, 70, 5350, 1979.
165. Gupta, S. K. and Gingerich, K. A., *J. Chem. Phys.*, 71, 3072, 1979.
166. Gupta, S. K. and Gingerich, K. A., *J. Chem. Phys.*, 72, 2795, 1980.
167. Gupta, S. K. and Gingerich, K. A., *J. Chem. Phys.*, 74, 3584, 1981.
168. Gupta, S. K., Atkins, R. M., and Gingerich, K. A., *Inorg. Chem.*, 17, 3211, 1978.
169. Gupta, S. K., Kingcade, J. E., and Gingerich, K. A., *Adv. Mass Spectrom.*, 8A, 445, 1980.
170. Gupta, S. K., Nappi, B. M., and Gingerich, K. A., *Inorg. Chem.*, 20, 966, 1981.
171. Gupta, S. K., Nappi, B. M., and Gingerich, K. A., *J. Phys. Chem.*, 85, 971, 1981.
172. Gupta, S. K., Pelino, M., and Gingerich, K. A., *J. Chem. Phys.*, 70, 2044, 1979.
173. Gupta, S. K., Pelino, M., and Gingerich, K. A., *J. Phys. Chem.*, 83, 2335, 1979.
174. Gustavsson, T., Amiot, C., Verges, J., *Mol. Phys.*, 64, 279, 1988.
175. Haque, R. and Gingerich, K. A., *J. Chem. Thermodyn.*, 12, 439, 1980.
176. Haque, R., Pelino, M., and Gingerich, K. A., *J. Chem. Phys.*, 71, 2929, 1979.
177. Haque, R., Pelino, M., and Gingerich, K. A., *J. Chem. Phys.*, 73, 4045, 1980.
178. Hariharan, A. V. and Eick, H. A., *J. Chem. Thermodyn.*, 6, 373, 1974.
179. Hedgecock, I. M., Naulin, C., and Costes, M., Chem. Phys., 207, 379, 1996.
180. Heidecke, S. A., Fu, Z., Colt, J. R., and Morse, M. D., J. Chem. Phys., 97, 1692, 1992.
181. Herman, P. R., La Rocque, P. E., and Stoicheff, B., *J. Chem. Phys.*, 89, 4535, 1988.
182. Herman, R. and Herman, L., *J. Phys. Radium.*, 24, 73, 1963.
183. Herzberg, G., *Molecular Spectra and Molecular Structure. I. Spectra of Diatomic Molecules*, 2nd ed.,Van Nostrand, New York, 1950.
184. Hildenbrand, D. L. and Lau, K. H., *High Temp. Mater. Sci.*, 35, 11, 1996.
185. Hildenbrand, D. L. and Lau, K. H., *J. Chem. Phys.*, 89, 5825, 1988.
186. Hildenbrand, D. L. and Lau, K. H., *J. Chem. Phys.*, 91, 4909, 1989.
187. Hildenbrand, D. L. and Lau, K. H., *J. Chem. Phys.*, 93, 5983, 1990.
188. Hildenbrand, D. L. and Lau, K. H., *J. Chem. Phys.*, 96, 3830, 1992.
189. Hildenbrand, D. L. and Lau, K. H., *J. Phys. Chem.*, 96, 2325, 1992.
190. Hildenbrand, D. L. and Murad, E., *J. Chem. Phys.*, 44, 1524, 1966.
191. Hildenbrand, D. L. and Theard, L. P., *J. Chem. Phys.*, 50, 5350, 1969.
192. Hildenbrand, D. L., *High Temp. Mater. Sci.*, 35, 151, 1996.
193. Hildenbrand, D. L., *Chem. Phys. Lett.*, 32, 523, 1975.
194. Hildenbrand, D. L., *J. Chem. Phys.*, 103, 2634, 1995.
195. Hildenbrand, D. L., *J. Chem. Phys.*, 105, 10507, 1996.
196. Hildenbrand, D. L., *J. Chem. Phys.*, 48, 3657, 1968.
197. Hildenbrand, D. L., *J. Chem. Phys.*, 52, 5751, 1970.
198. Hildenbrand, D. L., *J. Chem. Phys.*, 65, 614, 1976.
199. Hildenbrand, D. L., *J. Chem. Phys.*, 66, 3526, 1977.
200. Hildenbrand, D. L., *J. Phys. Chem.*, 77, 897, 1973.
201. Hilpert, K. and Ruthardt, K., *Ber. Bunsenges. Phys. Chem.*, 91, 724, 1987.
202. Hilpert, K. and Ruthardt, K., *Ber. Bunsenges. Phys. Chem.*, 93, 1070, 1989.
203. Hilpert, K., *Ber. Kernforschungsanlage Juelich,* JUEL-1744, 272, 1981.
204. Hilpert, K., *J. Chem. Phys.*, 77, 1425, 1982.
205. Huber, K. P. and Herzberg, G., *Molecular Spectra and Molecular Structure Constants of Diatomic Molecules*, Van Nostrand, New York, 1979.
206. Hussein, K., Effantin, C., D'Incan, J., Verges, J., and Barrow, R. F., *Chem. Phys. Lett.*, 124, 105, 1986.
207. Ihle, H. R. and Wu, C. H., *J. Chem. Phys.*, 63, 1605, 1975.
208. Ishwar, N. B., Varma, M. P., and Jha, B. L., *Acta Phys. Pol. A*, A61, 503, 1982.
209. Ishwar, N. B., Varma, M. P., and Jha, B. L., *Indian J. Pure Appl. Phys.*, 20, 992, 1982.
210. James, A. M., Kowalczyk, P., Simard, B., Pinegar, J. C., Morse, M. D., J. Mol. Spectrosc., 168, 248, 1994.
211. Jeyagopal, T., Rajavel, S. R. K., Ramakrishnan, M., and Rajamanickam, N., *Acta Phys. Hung.*, 68, 145, 1990.
212. Johns, J. W. C. and Ramsey, D. A., *Can. J. Phys.*, 39, 210, 1961.
213. Jones, K. M., Maleki, S., Bize, S., Lett, P. D., Williams, C. J., Richling, H., Knöckel, H., Tiemann, E., Wang, H., Gould, P. L., and Stwalley, W. C., Phys. Rev., A, 54, R1006, 1996.
214. Jones, R. W. and Gole, J. L., *Chem. Phys.*, 20, 311, 1977.
215. Jordan, K. J., Lipson, R. H., McDonald, N. A., and Le Roy, R. J., *J. Phys. Chem.*, 96, 4778, 1992.
216. Kant, A. and Lin, S.-S., *J. Chem. Phys.*, 51, 1644, 1969.
217. Kant, A. and Moon, K. A., *High Temp. Sci.*, 11, 55, 1979.
218. Kant, A. and Strauss, B. H., *J. Chem. Phys.*, 41, 3806, 1964.
219. Kant, A. and Strauss, B., *J. Chem. Phys.*, 49, 3579, 1968.
220. Kant, A., *J. Chem. Phys.*, 49, 5144, 1968.
221. Kant, A., Lin, S.-S, and Strauss, B., *J. Chem. Phys.*, 49, 1983, 1968.

Table 1
BOND STRENGTHS IN DIATOMIC MOLECULES (continued)

222. Kant, A., Strauss, B., and Lin, S.-S., *J. Chem. Phys.*, 52, 2384, 1970.
223. Kaposi, O., *Magy. Kem. Foly.*, 83, 356, 1977.
224. Kaufel, R., Vahl, G., Nunkwitz, R., and Baumgaertel, H., *Z. Anorg. Allg. Chem.*, 481, 207, 1981.
225. Kazenas ,E., Tagirov, V. K., and Zviadadze, G. N., *Izv. Akad. Nauk. SSSR Met.*, 58, 1984.
226. Kazenas, E. K., Bol'shikh, M. A., and Petrov, A. A., *Izvestiya Rossiiskoi Akademii Nauk. Metally,* (3), 29, 1996.
227. Kent, R. A. and Margrave, J. L., *J. Am. Chem. Soc.*, 87, 3582, 1965.
228. Kent, R. A., Ehlert, T. C., and Margrave, J. L., *J. Am. Chem. Soc.*, 86, 5090, 1964.
229. Kent, R. A., *J. Am. Chem. Soc.*, 90, 5657, 1968.
230. Khitrov, A. N., Ryabova, V. G., and Gurvich, L. V., *Teplofiz. Vys. Tempo.*, 11, 1126, 1973.
231. Kiang, T. and Zare, R. N., *J. Am. Chem. Soc.*, 102, 4024, 1980.
232. Kimura, H., Asano, M., and Kubo, K., *J. Nucl. Mater.*, 97, 259, 1981.
233. Kingcade, J. E. Jr. and Gingerich, K. A., *J. Chem. Phys.*, 84, 3432, 1986.
234. Kingcade, J. E. Jr. and Gingerich, K. A., *J. Chem. Soc. Faraday Trans. 2*, 85, 195, 1989.
235. Kingcade, J. E. Jr., and Gingerich, K. A., *J. Chem. Phys.*, 84, 4574, 1986.
236. Kingcade, J. E., Cocke, D. L., and Gingerich, K. A., *High Temp. Sci.*, 16, 89, 1983,
237. Kingcade, J. E., Dufner, D. C., Gupta, S. K., and Gingerich, K. A., *High Temp. Sci.*, 10, 213, 1978.
238. Kingcade, J. E., Nagarathna, H. M., Shim, I., and Gingerich, K. A., *J. Phys. Chem.*, 90, 2830, 1986.
239. Kleinschmidt, P. D. and Hildenbrand, D. L., *J. Chem. Phys.*, 68, 2819, 1978.
240. Kleinschmidt, P. D. and Ward, J. W., *J. Less-Common. Met.*, 121, 61, 1986.
241. Kleinschmidt, P. D., Cubicciotti, D., and Hildenbrand, D. L., *J. Electrochem. Soc.*, 125, 1543, 1978; *Proc. Electrochem. Soc.*, 78, 217, 1978.
242. Kleinschmidt, P. D., Lau, K. H., and Hildenbrand, D. L., *J. Chem. Phys.*, 74, 653, 1981.
243. Klynning, L. and Lindgren, B., *Arkiv. Fysik.*, 32, 575, 1966.
244. Kohl, F. J. and Carlson, K. D., *J. Am. Chem. Soc.*, 90, 4814, 1968.
245. Kohl, F. J. and Stearns, C. A., *J. Phys. Chem.*, 78, 273, 1974.
246. Kondratiev, V. N., *Bond Dissociation Energies, Ionization Potentials and Electron Affinities*, Mauka Publishing House, Moscow, 1974.
247. Kordis, J. and Gingerich, K. A., *J. Chem. Eng. Data,* 18, 135, 1973.
248. Kordis, J. and Gingerich, K. A., *J. Chem. Phys.*, 58, 5141, 1973.
249. Kordis, J. and Gingerich, K. A., *J. Phys. Chem.*, 76, 2336, 1972.
250. Kordis, J., Gingerich, K. A., and Seyse, R. J., *J. Chem. Phys.*, 61, 5114, 1974.
251. Kowalski, A., Czaikowski, M., and Breckenridge, W. H., *Chem. Phys. Lett.*, 119, 368, 1985.
252. Koyama, T. and Yamawaki, M., *J. Nucl. Mater.*, 152, 30, 1988.
253. Kronekvist, M., Lagerqvist, A., and Neuhaus, H., *J. Mol. Spectrosc.*, 39, 516, 1971.
254. Langenberg, J. D. and Morse, M. D., *Chem. Phys. Lett.*, 239, 25, 1995.
255. Langenberg, J. D. and Morse, M. D., *J. Chem. Phys.*, 100, 2331, 1998.
256. Lau, K. H. and Hildenbrand, D. L., *J. Chem. Phys.*, 71, 1572, 1979.
257. Lau, K. H. and Hildenbrand, D. L., *J. Chem. Phys.*, 72, 4928, 1980.
258. Lau, K. H. and Hildenbrand, D. L., *J. Chem. Phys.*, 76, 2646, 1982.
259. Lau, K. H. and Hildenbrand, D. L., *J. Chem. Phys.*, 80, 1312, 1984.
260. Lau, K. H. and Hildenbrand, D. L., *J. Chem. Phys.*, 86, 2949, 1987.
261. Lau, K. H. and Hildenbrand, D. L., *J. Chem. Phys.*, 92, 6124, 1990.
262. Lau, K. H., Brittain, R. D., and Hildenbrand, D. L., *Chem. Phys. Lett.*, 81, 227, 1981; *J. Phys. Chem.*, 86, 4429, 1982.
263. Lau, K. H., Brittain, R. D., and Hildenbrand, D. L., *J. Chem. Phys.*, 90, 1158, 1989.
264. Li, K. C. and Stwalley, W. C., *J. Chem. Phys.*, 59, 4423, 1973.
265. Li, L., Lyyra, A. M., Luh, W. T., and Stwalley, W. C., *J. Chem. Phys.*, 93, 8452, 1990.
266. Lindgren, B. and Nilsson, Ch., *J. Mol. Spectrosc.*, 55, 407, 1975.
267. Martin, E. and Barrow, R. F., *Phys. Scr.*, 17, 501, 1978.
268. McIntyre, N. S., Vander Auwera-Mahieu, A., and Drowart, J., *Trans. Faraday Soc.*, 64, 3006, 1968.
269. Menendez, M., Garay, M., Verdasco, E., and Gonzalez, U. A., *J. Chem. Phys.*, 99, 2760, 1993.
270. Murad, E., Hildenbrand, D. L., and Main, R. P., *J. Chem. Phys.*, 45, 263, 1966.
271. Nagai, S., Shinmei, M., and Yokokawa, T., *J. Inorg. Nucl. Chem.*, 36, 1904, 1974.
272. Nappi, B. M. and Gingerich, K. A., *Inorg. Chem.*, 20, 522, 1981.
273. Neckel, A. and Sodeck, G., *Monatsch. Chem.*, 103, 367, 1972.
274. Nedelec, O. and Giroud, M., *J. Chem. Phys.*, 79, 2121, 1983.
275. Neubert, A. and Zmbov, K. F., *Chem. Phys.*, 76, 469, 1983.
276. Neubert, A. and Zmbov, K. F., J. *Chem. Soc. Faraday Trans. 1*, 70, 2219, 1974.
277. Neubert, A., Ihle, H. R., and Gingerich, K. A., *J. Chem. Phys.*, 73, 1406, 1980.
278. Neubert, A., Zmbov, K. F., Gingerich, K. A., and Ihle, H. R., *J. Chem. Phys.*, 77, 5218, 1982.
279. Nordine, P. C., *J. Chem. Phys.*, 61, 224, 1974.
280. O'Hare, P. A. G. and Curtiss, L. A., *J. Chem. Thermodyn.*, 27, 643, 1995.
281. O'Hare, P. A. G., *J. Phys. Chem. Ref. Data*, 22, 1455, 1993.

Table 1
BOND STRENGTHS IN DIATOMIC MOLECULES (continued)

282. O'Hare, P. A. G., Zywocinski, A., and Curtiss, L. A., *J. Chem. Thermodyn.*, 28, 459, 1996.
283. Ovcharenko, I. E., Ya, Kuzyankow, Y., and Tatevaskii, V. M., *Opt. Spectrosk.*, 19, 528, 1965.
284. Pak, K., Cai, M. F., Dzugan, T. P., and Bondybey, V. E., *Faraday Discuss. Chem. Soc.*, 86, 153, 1988.
285. Parks, E. K. and Wexler, S., *J. Phys. Chem.*, 88, 4492, 1984.
286. Parr, T. P., Behrens, R., Freedman, A., and Heron, R. R., *Chem. Phys. Lett.*, 56, 71, 1978.
287. Pedley, J. B. and Marshall, E. M., *J. Phys. Chem. Ref. Data*, 12, 967, 1984.
288. Pelevin, O. V., Mil'vidskii, M. G., Belyaev, A. I., and Khotin, B. A., *Izv. Akad. Nauk. SSSR Neorg. Mater.*, 2, 924, 1966.
289. Pelino, M. and Gingerich, K. A., *J. Chem. Phys.*, 90, 1286, 1989.
290. Pelino, M., and Gingerich, K. A., *J. Chem. Phys.*, 93, 1581, 1989.
291. Pelino, M., Piacente, V., and Ascenzo, G., *Thermochim. Acta*, 31, 383, 1979.
292. Pelino, M., Viswanadham, P., and Edwards, J. G., *J. Phys. Chem.*, 83, 2964, 1979.
293. Piacente, V. and Balducci, G., *Adv. Mass Spectrom.*, 7A, 626, 1978.
294. Piacente, V. and Balducci, G., *Dyn. Mass Spectrom.*, 4, 295, 1976.
295. Piacente, V. and Balducci, G., *High Temp. Sci.*, 6, 254, 1974.
296. Piacente, V. and Desideri, A., *J. Chem. Phys.*, 57, 2213, 1972.
297. Piacente, V. and Gigli, R., *J. Chem. Phys.*, 77, 4790, 1982.
298. Piacente, V. and Gingerich, K. A., *High Temp. Sci.*, 9, 189, 1977.
299. Piacente, V. and Gingerich, K. A., *Z. Naturforsch. Teil A*, 28, 316, 1973.
300. Piacente, V. and Malaspina, L., *J. Chem. Phys.*, 56, 1780, 1972.
301. Piacente, V., *J. Chem. Phys.*, 70, 5911, 1979.
302. Pianalto, F. S., O'Brien, L. C., Keller, P. C., and Bernath, P. F., *J. Mol. Spectrosc.*, 129, 348, 1988.
303. Pilgrim, J. C. and Duncan, M. A., *Chem. Phys. Lett.*, 232, 335, 1995.
304. Pinegar, J. C., Langenberg, J. D., Arrington, C.A., Spain, E. M., and Morse, M. D., *J. Chem. Phys.*, 102, 666, 1995.
305. Pitzer, K. S., *J. Chem. Phys.*, 74, 3078, 1981.
306. Porter, R. F. and Spencer, C. W. J., *J. Chem. Phys.*, 32, 943, 1960.
307. Prasad, R., Venugopal, V., and Sood, D. D., *J. Chem. Thermodyn.*, 9, 593, 1977.
308. Rajamanickam, N., Dhuvaragaikannan, N., and Raja Mohamed, K., *Acta Physica Hungarica*, 74, 385, 1994.
309. Rajamanickam, N., Palaniselvam, K., Rajavel, S. R. K., Rajesh, M., and Sureshkumar, G., *Acta Phys. Hung.*, 70, 141, 1991.
310. Rajamanickam, N., Senthilkumar, R. N., Ganesan, S., Gopalakrishnan, N., Rajkumer, J., Jegadesan, V., and Dandapani, C., *Acta Phys. Hung.*, 70, 71, 1991.
311. Rajamanickan, N., *Acta Ciencia Indica Phys.*, 14, 18, 1988.
312. Ram, R. S., Rai, S. B., Ram, R. S., Upadhya, K. N., J. *Chim. Phys. Phys-Chim. Biol.*, 76, 560, 1979.
313. Ramakrishnan, E. S., Shim, I., and Gingerich, K. A., *J. Chem. Soc. Faraday Trans. 2*, 80, 395, 1984.
314. Ran, Q., Schmude, R. W., Gingerich, K. A., Wilhite, D. W., and Kincade, J. E., *J. Phys. Chem.*, 97, 8535, 1993.
315. Rao, P. S. and Rao, T. V. R., *J. Quant. Spectrosc. Radiat. Transfer*, 27, 207, 1982.
316. Rao, S. P. and Rao, T. V. R., *Acta Ciencia Indica Phys.*, 7, 58, 1981.
617. Rao, T. V. R., Reddy, R. R., and Rao, P. S., *Indian J. Pure Appl. Phys.*, 19, 1219, 1981.
318. Rao, V. M., Rao, M. L. P., and Rao, P. T., *J. Quant. Spectrosc. Radiat. Transfer*, 25, 547, 1981.
319. Reddy, R. R., Reddy, A. S. R., and Rao, T. V. R., *Acta Phys. Slovaca*, 36, 273, 1986.
320. Riekert, G., Lamparter, P., and Steeb, S., *Z. Metallkd.*, 72, 765, 1981.
321. Riekert, G., Rainer-Harbach, G., Lamparter, P., and Steeb, S., *Z. Metallkd.*, 76, 406, 1981.
322. Rinehart, G. H. and Behrens, R. G., *J. Phys. Chem.*, 83, 2052, 1979.
323. Rohlfing, E. A. and Valentine, J. J., *J. Chem. Phys.*, 84, 6560, 1986.
324. Rovner, L., Drowart, A., and Drowart, J., *Trans. Faraday Soc.*, 63, 2910, 1967.
325. Rusin, A. D., Zhukov, E., Agamirova, L. M., and Kalinnikov, V. T., *Zh. Neorg. Khim.*, 24, 1457, 1979.
326. Samoilova, I. O. and Kazenas, E. K., *Izvestiya Rossiiskoi Akademii Nauk. Metally*, (1), 33, 1995.
327. Samoilova, I. O. and Kazenas, E. K., *Izvestiya Rossiiskoi Akademii Nauk. Metally*, (3), 36, 1994.
328. Schumde, R. W. Jr., Ran, Q., Gingerich, K. A., and Kingcade, J. E. Jr., *J. Chem. Phys.*, 102,2574, 1995.
329. Scott, P. R. and Richards, W. G., *J. Phys. B*, 7,1679, 1974.
330. Shim, I. and Gingerich, K. A., *J. Chem. Phys.*, 77, 2490, 1982.
331. Shim, I. and Gingerich, K. A., *J. Chem. Phys.*, 80, 5107, 1984.
332. Shim, I., and Gingerich, K. A., *J. Chem. Phys.*, 81, 5937, 1984.
333. Shim, I., Finkbeiner, H. C., and Gingerich, K. A., *J. Phys. Chem.*, 91, 3171, 1987.
334. Shim, I., Kingcade, J. E. Jr., and Gingerich, K. A., *J. Chem. Phys.*, 85, 6629, 1986.
335. Shim, I., Kingcade, J. E. Jr., and Gingerich, K. A., *J. Chem. Phys.*, 89, 3104, 1988.
336. Shim, I., Kingcade, J. E. Jr., and Gingerich, K. A., *Z. Phys. D - Atoms Molecules and Clusters*, 7, 261, 1987.
337. Shim, I., Mandix, K., and Gingerich, K. A., *J. Phys. Chem.*, 95, 5435, 1991.
338. Shim, I., Pelino, M., and Gingerich, K. A., *J. Chem. Phys.*, 97, 9240, 1992.
339. Siegel, B., *Q. Rev. (London)*, 19, 77, 1965.
340. Simons, J. W., Oldenberg, R. C., and Baughaim, S. L., *J. Phys. Chem.*, 91, 3840, 1987.

Table 1
BOND STRENGTHS IN DIATOMIC MOLECULES (continued)

341. Smith, P. K. and Peterson, D. E., *J .Chem. Phys.*, 52, 4963, 1970.
342. Smoes, S. and Drowart, J., *Chem. Commun.*, p.534, 1968.
343. Smoes, S. and Drowart, J., *J. Chem. Soc. Faraday Trans.* 2, 73, 1746, 1977.
344. Smoes, S. and Drowart, J., *J. Chem. Soc. Faraday Trans.* 2, 80, 1171, 1984.
345. Smoes, S., Coppens, P., Bergman, C., and Drowart, J., *Trans. Faraday Soc.*, 65, 682, 1969.
346. Smoes, S., Depiere, D., and Drowart, J., *Rev. Int. Hautes Temp. Refractaires Paris*, 9, 171, 1972.
347. Smoes, S., Drowart, J., Welter, J. M., *J. Chem. Thermodyn.*, 9, 275, 1977; *Adv. Mass Spectrom.*, 7A, 622, 1978.
348. Smoes, S., Huguet, R., and Drowart, J., *Z. Naturforsch. Teil A*, 26, 1934, 1971.
349. Smoes, S., Mandy, F., Vander Auwera-Mahieu, A., and Drowart, J., *Bull. Soc. Chim. Belg.*, 81, 45, 1972.
350. Smoes, S., Myers, C. E., and Drowart, J., *Chem. Phys. Lett.*, 8, 10, 1971.
351. Smoes, S., Pattje, W. R., and Drowart, J., *High Temp. Sci.*, 10, 109, 1978.
352. Spain, E. M. and Morse, M. D., *J. Chem. Phys.*, 97, 4605, 1992.
353. Spain, E. M. and Morse, M. D., *J. Phys. Chem.*, 96, 2479, 1992.
354. Sreedhara Murthy, N., *Indian J. Phys.*, 62B, 92, 1988.
355. Srivastara, R. D. and Farber, M., *High Temp. Sci.*, 5, 489, 1973.
356. Stearns, C. A. and Kohl, F. J., *High Temp. Sci.*, 2, 146, 1970.
357. Stearns, C. A. and Kohl, F. J., *High Temp. Sci.*, 6, 284, 1974.
358. Steiger, R. A. and Cater, E. D., *High Temp. Sci.*, 7, 204, 1975.
359. Steiger, R. P. and Cater, E. D., *High Temp. Sci.*, 7, 288, 1975.
360. Stwalley, W. C., Way, K. R., and Velasco, R., *J. Chem. Phys.*, 60, 3611, 1974.
361. Su, T.-M. R. and Riley, S. J., *J. Chem. Phys.*, 71, 3194, 1979.
362. Su, T.-M. R. and Riley, S. J., *J. Chem. Phys.*, 72, 1614, 1980.
363. Su, T.-M. R. and Riley, S. J., *J. Chem. Phys.*, 72, 6632, 1980.
364. Sullivan, C. L., Zehe, M. J., and Carlson, K. D., *High Temp. Sci.*, 6, 80, 1974.
365. Tanaka, Y., Yushina, K., and Freeman, D. E., *J. Chem. Phys.*, 59, 564, 1973.
366. Taylor, S., Lemire, G. W., Hamrick, Y. M., Fu, Z., and Morse, M. D., *J. Chem. Phys.*, 89, 5517, 1988.
367. Taylor, S., Spain, E. M., and Morse, M. D., *J. Chem. Phys.*, 92, 2698, 1990.
368. Tellinghuisen, J., Tisone, G. C., Hoffmann, J. M., and Hays, A. K., *J. Chem. Phys.*, 64, 4796, 1976.
369. Tolbert, M. A. and Beauchamp, J. L., *J. Phys. Chem.*, 90, 5015, 1986.
370. Tomaszkiewicz, I., Susman, S., Volin, K. J., and O'Hare, P. A. G., *J. Chem. Thermodyn.*, 26, 1081, 1994.
371. Tromp, J. W., LeRoy, R. J., Gerstenkorn, S., and Luc, P., *J. Mol. Spectrosc.*, 100, 82, 1983.
372. Tuenge, R. T., Laabs, F., and Franzen, H. F., *J. Chem. Phys.*, 65, 2400, 1976.
373. Urdahl, R. S., Bao, Y., and Jackson, W. M., *Chem. Phys. Lett.*, 178, 425, 1991.
374. Uy, O. M. and Drowart, J., *High Temp. Sci.*, 2, 293, 1970.
375. Uy, O. M. and Drowart, J., *Trans. Faraday Soc.*, 65, 3221, 1969.
376. Uy, O. M. and Drowart, J., *Trans. Faraday Soc.*, 67, 1293, 1971.
377. Uy, O. M., Muenow, D. W., Ficalora, P, J., and Margrave, J. L., *Trans. Faraday Soc.*, 64, 2998, 1968.
378. Van Veen, N. J. A., DeVries, M., and DeVries, A. E., *Chem. Phys. Lett.*, 64, 213, 1979.
379. Vander Auwera-Mahieu, A. and Drowart, J., *Chem. Phys. Lett.*, 1, 311, 1967.
380. Vander Auwera-Mahieu, A., McIntyre, N. S., and Drowart, J., *Chem. Phys. Lett.*, 4, 198, 1969.
381. Vander Auwera-Mahieu, A., Peeters, R., McIntyre, N. S., and Drowart, J., *Trans. Faraday Soc.*, 66, 809, 1970.
382. Varma, M. P., Ishwar, N. B., and Jha, B. L., *Indian J. Pure Appl. Phys.*, 20, 828, 1982.
383. Velasco, R., Ottinger, C., and Zare, R. N., *J. Chem. Phys.*, 51, 5522, 1969.
384. Vempati, S. N. and Jones, W. E., *J. Mol. Spectrosc.*, 127, 232, 1988.
385. Venkataramanaiah, M. and Lakshman, S. V. J., *J. Quant. Spectrosc. Radiat. Transfer*, 26, 11, 1981.
386. Verhaegen, G., Smoes, S., and Drowart, J., *J. Chem. Phys.*, 40, 239, 1964.
387. Verhaegen, G., Stafford, F. E., and Drowart, J., *J. Chem. Phys.*, 40, 1622, 1964.
388. Viswanathan, K. S. and Tellinghuisen, J., *J. Mol. Spectrosc.*, 98, 185, 1983.
389. Viswanathan, R., Baba, M. S., Raj, D. D. A., Balasubramanian, R., Narasimhan, T. S. L., and Mathews, C. K., *Spectrochim. Acta*, Part B, 49B, 243, 1994.
390. Viswanathan, R., Schmude, R. W., and Gingerich, K. A., *J. Phys. Chem.*, 100, 10784, 1996.
391. Viswanathan, R., Schumde, R. W., and Gingerich, K. A., *J. Chem. Thermodyn.*, 27, 763, 1995.
392. Wang, Y. C., Kajitani, M., Kasahara, S., Bata, M., Ishikawa, K., and Kato, H., *J. Chem. Phys.*, 95, 6229, 1991.
393. Way, K. R. and Stwalley, W. C., *J. Chem. Phys.*, 59, 5298, 1973.
394. Weickenmeier, W., Diemer, U., Wahl, M., Raab, M., Demtroeder, W., and Mueller, W., *J. Phys.Chem.*, 82, 5354, 1985.
395. Wiedemeier, H. and Gilles, P. W., *J. Chem. Phys.*, 42, 2765, 1965.
396. Wu, C. H. and Ihle, H. R., *Adv. Mass Spectrom.*, 8A, 374, 1980.
397. Wu, C. H., Ihle, H. R., and Gingerich, K. A., *Int. J. Mass Spectrom. Ion Phys.*, 47, 235, 1983.
398. Wu, C. H., *J. Chem. Phys.*, 65, 3181, 1976; 65, 2040, 1976.

Table 1
BOND STRENGTHS IN DIATOMIC MOLECULES (continued)

399. Yokozeki, A. and Menzinger, M., *Chem. Phys.*, 14, 427, 1976.
400. Yoo, R. K., Ruscic, B., and Berkowitz, J., *Chem. Phys.*, 166, 215, 1992.
401. Zemke, W. T. and Stwalley, W. C., *Chem. Phys. Lett.*, 143, 84, 1988.
402. Zemke, W. T., Stwalley, W. C., Langhoff, S. R., Valderrama, G. L., and Berry, M. J., *J. Chem. Phys.*, 95, 7846, 1991.
403. Zmbov, K. F. and Margrave, J. L., *J. Chem. Phys.*, 45, 3167, 1966.
404. Zmbov, K. F. and Margrave, J. L., *J. Chem. Phys.*, 47, 3122, 1967.
405. Zmbov, K. F. and Margrave, J. L., J. *Inorg. Nucl. Chem.*, 29, 59, 1967.
406. Zmbov, K. F. and Margrave, J. L., *J. Phys. Chem.*, 70, 3379, 1966.
407. Zmbov, K. F. and Margrave, J. L., *J. Phys. Chem.*, 71, 2893, 1967.
408. Zmbov, K. F., Hastie, J. W., and Margrave, J. L., *Trans. Faraday Soc.*, 64, 861, 1968.
409. Zmbov, K. F., Neubert, A., and Ihle, H. R., *Z. Naturforsch., A*, 36A, 914, 1981.
410. Zmbov, K. F., Wu, C. H., and Ihle, H. R., *J. Chem. Phys.*, 67, 4603, 1977.
411. Zollweg, R. J., *Contrib. Pap. Int. Conf. Phenom. Ioniz. Gases.* 11th, 402, 1973.

Table 2
ENTHALPY OF FORMATION OF GASEOUS ATOMS FROM ELEMENTS IN THEIR STANDARD STATES

For elements that are diatomic gases in their standard states these are readily obtained from the bond strength. For elements that are crystalline in their standard states they are derived from vapor pressure data.

Atom	$\Delta_f H°_{298}$/kJ mol^{-1}	Ref.	Atom	$\Delta_f H°_{298}$/kJ mol^{-1}	Ref.	Atom	$\Delta_f H°_{298}$/kJ mol^{-1}	Ref.
			Hf	619 ± 4	1	Re	774 ± 6.3	1
Ag	284.9 ± 0.8	2	Hg	61.38 ± 0.04	2	Rh	556 ± 4	1
Al	330.0 ± 4.0	2	I	106.76 ± 0.04	2	Ru	650.6 ± 6.3	1
As	302.5 ± 13	1	In	243 ± 4	1	S	277.17 ± 0.15	2
Au	368.2 ± 2.1	1	Ir	669 ± 4	1	Sb	264.4 ± 2.5	1
B	565 ± 5	2	K	89.0 ± 0.8	2	Sc	377.8 ± 4	1
Ba	177.8 ± 4	1	Li	159.3 ± 1.0	2	Se	227.2 ± 4	1
Be	324 ± 5	2	Mg	147.1 ± 0.8	2	Si	450 ± 8	2
Bi	209.6 ± 2.1	1	Mn	283.3 ± 4	1	Sn	301.2 ± 1.5	2
Br	111.87 ± 0.12	2	Mo	658.1 ± 2.1	1	Sr	163.6 ± 2.1	1
C	716.68 ± 0.45	2	N	472.68 ± 0.40	2	Ta	782.0 ± 2.5	1
Ca	177.8 ± 0.8	2	Na	107.5 ± 0.7	2	Te	196.6 ± 2.1	1
Cd	111.80 ± 0.20	2	Nb	721.3 ± 4	1	Th	602 ± 6	2
Ce	423 ± 13	1	Ni	430.1 ± 2.1	1	Ti	473 ± 3	2
Cl	121.301 ± 0.008	2	O	249.18 ± 0.10	2	Tl	182.21 ± 0.4	1
Co	428.4 ± 4	1	Os	787 ± 6.3	1	U	533 ± 8	2
Cr	397 ± 4	1	P	316.5 ± 1.0	2	V	514.2 ± 1.3	1
Cs	76.5 ± 1.0	2	Pb	195.2 ± 0.8	2	W	849.8 ± 4	1
Cu	337.4 ± 1.2	2	Pd	376.6 ± 2.1	1	Y	424.7 ± 2.1	1
Er	317.1 ± 4	1	Pt	565.7 ± 1.3	1	Yb	152.09 ± 0.8	1
F	79.38 ± 0.30	2	Pu	364.4 ± 17	1	Zn	130.40 ± 0.40	2
Ge	372 ± 3	2	Rb	80.9 ± 0.8	2	Zr	608.8 ± 4	1
H	217.998 ± 0.006	2						

REFERENCES

1. Brewer, L. and Rosenblatt, G. M., *Adv. High Temp. Chem.*, 2, 1, 1969.
2. Cox, J. D., Wagman, D. D., and Medvedev, V. A., Eds., *CODATA Key Values for Thermodynamics*, Hemisphere Publishing Corporation, New York, 1989.

Table 3
BOND STRENGTHS IN POLYATOMIC MOLECULES

The values below refer to a temperature of 298 K and have mostly been determined by kinetic methods (see (i) S. W. Benson, *J. Chem. Educ.*, 42, 502, 1965, (ii) J. A. Kerr, *Chem. Rev.*, 66, 465, 1966 and (iii) D. F. McMillen and D. M. Golden, *Ann. Rev. Phys. Chem.*, 33, 493, 1982, for a full description of the methods). An increasing number of bond strengths are being determined from gas-phase acidity cycles and from photoionization mass spectrometry (see J. Berkowitz, G. B. Ellison and D. Gutman, *J. Phys. Chem.*, 98, 2744, 1994).

Bond strengths in polyatomic molecules are notoriously difficult to measure accurately since the mechanisms of the kinetic systems involved in many of the measurements are seldom straightforward. Thus much controversy has taken place in the literature over the past 15 years concerning C-H bond strengths in simple alkanes, for which we recommend data based largely on kinetic studies involving time-resolved flow tube experiments with mass spectrometric determination of reactant radical concentrations (see Berkowitz, J., Ellison, G. B., and Gutman, D., *J. Phys. Chem.*, 98, 2744, 1994.). These alkane bond strengths and the enthalpies of formation of the corresponding radicals are significantly larger than values derived from experiments in very low pressure reactors (see Dobis, O. and Benson, S. W., *J. Phys. Chem.*, 101, 6030, 1997; and Benson, S. W. and Dobis, O., *J. Phys. Chem.*, 102, 5175, 1998). Other examples illustrating the difficulties involved are concerned with the C-H bond strengths in ethene and methanol or the corresponding enthalpies of formation of the vinyl and hydroxymethyl radicals and changes to the recommendations could well arise.

Some of the bond strengths have been calculated from the enthalpies of formation of the species involved according to the equations:

$$D°(R–X) = \Delta_f H°(R) + \Delta_f H°(X) - \Delta_f H°(RX)$$
$$D°(R–R) = 2\,\Delta_f H°(R) - \Delta_f H°(RR)$$

The enthalpies of formation of the atoms and radicals are taken from Tables 2 and 4 and for the molecules from the appropriate References following Table 3.

An attempt has been made to list all the important values obtained by methods that are considered to be valid. The references are intended to serve as a guide to the literature.

Bond	$D°_{298}$/kJ mol^{-1}	Ref.
H-CH	424.0 ± 4.2	1
H-CH_2	462.0 ± 4.0	1
H-CH_3	438.9 ± 0.4	16
H-CCH	556.1 ± 2.9	4,37
H-$CHCH_2$	465.3 ± 3.4	37,81
H-C_2H_5	423.0 ± 1.6	16
H-Cycloprop-2-en-1-yl	379.1 ± 17	63
H-CH_2CCH	374.0 ± 8	63
H-CH_2CHCH_2	361.9 ± 8.8	16,35
H-Cyclopropyl	444.8 ± 1.3	63
H-n-C_3H_7	423.3 ± 2.1	82
H-i-C_3H_7	409.1 ± 2.0	82
H-CH_2CCCH_3	364.8 ± 8	63
H-$CH(CH_3)CCH$	347.7 ± 9.2	63
H-Cyclobutyl	403.8 ± 4	63
H-Cyclopropylmethyl	407.5 ± 6.7	63
H-$CH(CH_3)CHCH_2$	345.2 ± 5.4	63
H-$CH_2CHCHCH_3$	358.2 ± 6.3	63
H-$CH_2C(CH_3)CH_2$	358.2 ± 4	91,95
H-n-C_4H_9	425.4 ± 2.1	82
H-i-C_4H_9	425.2 ± 2.1	82
H-s-C_4H_9	411.2 ± 2.0	82
H-t-C_4H_9	404.3 ± 1.3	82
H-Cyclopenta-1,3-dien-5-yl	346.7	1,12
H-Spiropentyl	413.4 ± 4	63
H-Cyclopent-1-en-3-yl	344.3 ± 4	63
H-$CH_2CHCHCHCH_2$	347 ± 13	63
H-$CH(C_2H_3)_2$	319.7	63,92
H-$CH(CH_3)CCCH_3$	365.3 ± 11.3	63
H-$C(CH_3)_2CCH$	338.9 ± 9.6	63
H-$C(CH_3)_2CHCH_2$	323.0 ± 6.3	63
H-Cyclopentyl	403.5 ± 2.5	22,74
H-$CH_2C(CH_3)_3$	418 ± 8	63
H-$C(CH_3)_2C_2H_5$	404.0 ± 6.3	1,74,89
H-C_6H_5	473.1 ± 3.0	46
H-Cyclohexa-1,3-dien-5-yl	305 ± 21	63
H-Cyclohexa-1,4-dien-3-yl	305.4 ± 8.4	45
H-Cyclohexyl	399.6 ± 4	63
H-$C(CH_3)_2CCCH_3$	344.3 ± 11.3	63
H-$CH_2C(CH_3)C(CH_3)_2$	326.4 ± 4.6	63
H-$C(CH_3)_2C(CH_3)CH_2$	319.2 ± 4.6	63
H-$CH_2C_6H_5$	375.7 ± 1.7	35
H-Cyclohepta-1,3,5-trien-7-yl	305.4 ± 8	63
H-Norbornyl	404.6 ± 10.5	63
H-Cycloheptyl	387.0 ± 4	63
H-$CH(CH_3)C_6H_5$	357.3 ± 6.3	63
H-Inden-1-yl	351 ± 13	63
H-$C(CH_3)_2C_6H_5$	353.1 ± 6.3	63
H-1-Naphthylmethyl	356.1 ± 6.3	63
H-$CH(C_6H_5)_2$	340.6	80
H-9,10-Dihydroanthracen-9-yl	315.1 ± 6.3	63
H-$C(CH_3)(C_6H_5)_2$	339 ± 8	63
H-9-Anthracenylmethyl	342.3 ± 6.3	63
H-9-Phenanthrenylmethyl	356.1 ± 6.3	63
H-CN	527.6 ± 1.7	16
H-CH_2CN	392.9 ± 8.4	16
H-CH_2NC	380.7 ± 8.8	16
H-$CH(CH_3)CN$	376.1 ± 9.6	63
H-$C(CH_3)_2CN$	361.9 ± 8.4	63
H-CH_2NH_2	390.4 ± 8.4	63
H-CH_2NHCH_3	364 ± 8	63
H-$CH_2N(CH_3)_2$	351 ± 8	63
H-CHO	368.5 ± 1.0	23
H-CHCO	440.6 ± 8.8	16
H-$COCH_3$	373.8 ± 1.5	67
H-$COCHCH_2$	364.4 ± 4.2	63
H-COC_2H_5	371.3	1,12
H-COC_6H_5	363.6 ± 4	63

Table 3
BOND STRENGTHS IN POLYATOMIC MOLECULES (continued)

Bond	$D°_{298}$/kJ mol^{-1}	Ref.	Bond	$D°_{298}$/kJ mol^{-1}	Ref.
H-$COCF_3$	380.7 ± 8	63	H-$N(CH_3)C_6H_5$	366.1 ± 8	63
H-CH_2CHO	394.6 ± 9.2	16	H-NO	195.35 ± 0.25	32
H-CH_2COCH_3	411.3 ± 7.5	63	H-NO_2	327.6 ± 2.1	63
H-$CH(CH_3)COCH_3$	386.2 ± 5.9	63	H-NF_2	316.7 ± 10.5	63
H-CH_2OCH_3	402.2	1,12	H-$NHNH_2$	366.1	44
H-$CH(CH_3)OC_2H_5$	383.7 ± 1.7	55	H-NH_3	385 ± 21	63
H-Tetrahydrofuran-2-yl	385 ± 4	63	H-OH	498 ± 4	63
H-2-Furylmethyl	361.9 ± 8	63	H-OCH_3	436.0 ± 3.8	16
H-CH_2OH	401.8 ± 1.5	50	H-OC_2H_5	437.7 ± 3.4	16
H-$CH(CH_3)OH$	401.4	1,12	H-$OC(CH_3)_3$	439.7 ± 4	63
H-$CH(OH)CHCH_2$	341.4 ± 7.5	63	H-$OCH_2C(CH_3)_3$	428.0 ± 6.3	63
H-$C(CH_3)_2OH$	381 ± 4	63	H-OC_6H_5	361.9 ± 8	63
H-$CH_2OCOC_6H_5$	419.2 ± 5.4	63	H-O_2H	369.0 ± 4.2	88
H-$COOCH_3$	387.9 ± 4	63	H-O_2CH_3	370.3 ± 2.1	56
H-CH_2F	423.8 ± 4	77	H-$O_2C(CH_3)_3$	374.0 ± 0.8	47
H-CHF_2	431.8 ± 4	77	H-$OCOCH_3$	442.7 ± 8	63
H-CF_3	449.5	3,61	H-$OCOC_2H_5$	445.2 ± 8	63
H-CHFCl	421.7 ± 5.4	97	H-OCO-n-C_3H_7	443.1 ± 8	63
H-CF_2Cl	421.3 ± 8.3	64	H-ONO	327.6 ± 2.1	15
H-$CHFCl_2$	413.8 ± 5.0	97	H-ONO_2	423.4 ± 2.1	15
H-CH_2Cl	419.0 ± 2.3	84	H-SiH	351	63
H-$CHCl_2$	402.5 ± 2.7	84	H-SiH_2	268	63
H-CH_2CH_2Cl	423.1 ± 2.4	85	H-SiH_3	384.1 ± 2.0	83
H-$CH(CH_3)Cl$	406.6 ± 1.5	84	H-SiH_2CH_3	374.9	99
H-$C(CH_3)Cl_2$	390.6 ± 1.5	84	H-$SiH(CH_3)_2$	374.0	99
H-CCl_3	392.5 ± 2.5	48	H-$Si(CH_3)_3$	377.8	63,99
H-CH_2Br	425.1 ± 4.2	97	D-$Si(CH_3)_3$	389 ± 7.1	36
H-$CHBr_2$	417.2 ± 7.5	97	H-$SiH_2C_6H_5$	369.0	63,99
H-CBr_3	401.7 ± 6.7	63	H-SiF_3	418.8	63,99
H-CH_2I	431 ± 8	63	H-$SiCl_3$	382.0	63,99
H-CHI_2	431 ± 8	63	H-Si_2H_5	361.1	63
H-$CHCF_2$	448 ± 8	90	H-$Si(CH_3)_2Si(CH_3)_3$	356.9 ± 8.4	45
H-CFCHF	448 ± 8	90	H-$Si(Si(CH_3)_3)_3$	330.5 ± 8.4	45
H-$CFCF_2$	452 ± 8	90	H-PH_2	351.0 ± 2.1	16
H-CH_2CF_3	446.4 ± 4.6	63	H-SH	381.6 ± 2.9	65
H-CF_2CH_3	416.3 ± 10.5	63	H-SCH_3	365.3 ± 2.5	65
H-C_2F_5	429.7 ± 2.1	63	H-SC_6H_5	348.5 ± 8	63
H-CFCFCl	444 ± 8	90	H-SO	172.8	100
H-$CHClCF_3$	425.9 ± 6.3	63	H-GeH_3	349.0 ± 8	16
H-CClCFCl	439 ± 8	90	H-GeH_2I	331 ± 8	68
H-$CClCH_2$	>433.5	81	H-$Ge(CH_3)_3$	339 ± 8	34
H-CClCHCl	435 ± 8	90	H-AsH_2	319.2 ± 0.8	16
H-CCl_2CHCl_2	393 ± 8	63	H-SeH	334.93 ± 0.75	16
H-C_2Cl_5	393.5 ± 6.0	66	H-Sn(n-$C_4H_9)_3$	308.4 ± 8.4	19
H-$CClBrCF_3$	404.2 ± 6.3	63	H-SbH_2	288.3 ± 2.1	16
H-n-C_3F_7	435 ± 8	63	H-TeH	277.0 ± 5.0	16
H-i-C_3F_7	433.5 ± 2.5	38	HC≡CH	965 ± 8	1,24,74
H-$CHClCHCH_2$	370.7 ± 5.9	63	H_2C=CH_2	728.3 ± 6	1,74
H-C_6F_5	476.6	63	CH_3-CH_3	376.0 ± 2.1	1,74,86
H-$CH_2Si(CH_3)_3$	415.1 ± 4	99	CH_3-CH_2CCH	318.0 ± 8	63
H-CSH	399.6 ± 5.0	16	CH_3-CH_2CCCH_3	308.4 ± 6.3	63
H-CH_2SH	392.9 ± 8.4	16	CH_3-$CH(CH_3)CCH$	305.4	63
H-CH_2SCH_3	384.9 ± 5.9	49	CH_3-$C(CH_3)CCH_2$	320.1 ± 9.2	63
H-NH_2	452.7 ± 1.3	16	CH_3-$CH_2CHCHCH_3$	305.0 ± 3.3	63
H-$NHCH_3$	418.4 ± 10.5	63	CH_3-$CH_2C(CH_3)CH_2$	301.2 ± 3.3	91
H-$N(CH_3)_2$	382.8 ± 8	63	CH_3-$CH(CH_3)CCCH_3$	320.9 ± 6.3	63
H-NHC_6H_5	368.2 ± 8	63	CH_3-$C(CH_3)_2CCH$	295.8 ± 6.3	63

Table 3
BOND STRENGTHS IN POLYATOMIC MOLECULES (continued)

Bond	D°_{298}/kJ mol^{-1}	Ref.
n-C_3H_7-CH_2CCH	306.3 ± 6.3	63
CH_3-$C(CH_3)_2CHCH_2$	284.9 ± 6.3	63
n-C_3H_7-CH_2CHCH_2	295.8	96
CH_3-$C(CH_3)_2CCCH_3$	303.3 ± 6.3	63
$CHCCH_2$-s-C_4H_9	300.0 ± 6.3	63
CH_3-$CH_2C_6H_5$	332.2 ± 4	63
CH_3-$CH(CH_3)C_6H_5$	312.1 ± 6.3	63
C_2H_5-$CH_2C_6H_5$	294.1 ± 4	63
CH_3-1-Naphthylmethyl	305.0 ± 6.3	63
CH_3-$C(CH_3)_2C_6H_5$	308.4 ± 6.3	63
$CHCCH_2$-$CH_2C_6H_5$	256.9 ± 8	63
n-C_3H_7-$CH_2C_6H_5$	292.9 ± 4	63
CH_3-9-Anthracenylmethyl	282.8 ± 6.3	63
CH_3-9-Phenanthrenylmethyl	305.0 ± 6.3	63
CH_3-$CH(C_6H_5)_2$	301 ± 8	63
CH_3-$C(CH_3)(C_6H_5)_2$	289 ± 8	63
CH_3-CN	509.6 ± 8	63
C_2H-CN	602 ± 4	70
C_2H_5-CH_2NH_2	332.2 ± 8	63
CH_3-CH_2CN	336.4 ± 4	93
C_2H_5-CH_2CN	321.7 ± 7.1	63
CH_3-$CH(CH_3)CN$	329.7 ± 8	63
C_2H_5-CH_2CN	321.7 ± 7.1	63
CH_3-$C(CH_3)_2CN$	312.5 ± 6.7	63
CH_3-$C(CH_3)(CN)C_6H_5$	250.6	63
$C_6H_5CH_2$-CH_2NH_2	284.5 ± 8	63
$C_6H_5CH_2$-C_5H_4N	362.8	80
CN-CN	536 ± 4	30
CH_3-2-Furylmethyl	314 ± 8	63
CH_3-COC_6H_5	355.6 ± 9.2	102
$C_6H_5CH_2$-$COCH_2C_6H_5$	273.6 ± 8	63
CH_3CO-$COCH_3$	282.0 ± 9.6	63
$C_6H_5CH_2$-COOH	280	63
C_6H_5CO-COC_6H_5	277.8	63
$(C_6H_5)_2CH$-COOH	248.5 ± 13	63
CF_3-COC_6H_5	308.8 ± 8	63
CF_2=CF_2	319.2 ± 13	103
CH_2F-CH_2F	368 ± 8	51
CH_3-CF_3	423.4 ± 4.6	79
CF_3-CF_3	413.0 ± 10.5	63
C_6F_5-C_6F_5	487.9 ± 24.7	78
CH_3-BF_2	~473	63
C_6H_5-BCl_2	~510	63
CH_2CHCH_2-$Si(CH_3)_3$	293	63
s-C_4H_9-$Si(CH_3)_3$	414	63
CH_3-NHC_6H_5	298.7 ± 8	63
$C_6H_5CH_2$-NH_2	297.5 ± 4	63
CH_3-$N(CH_3)C_6H_5$	296.2 ± 8	63
$C_6H_5CH_2$-$NHCH_3$	287.4 ± 8	63
$C_6H_5CH_2$-$N(CH_3)_2$	259.8 ± 8	63
CH_2=N_2	<175	58
CH_3-N_2CH_3	219.7	13
C_2H_5-$N_2C_2H_5$	209.2	13
i-C_3H_7-N_2i-C_3H_7	198.7	13
n-C_4H_9-N_2n-C_4H_9	209.2	13
i-C_4H_9-N_2i-C_4H_9	205.0	13
s-C_4H_9-N_2s-C_4H_9	195.4	13
t-C_4H_9-N_2t-C_4H_9	182.0	13
$C_6H_5CH_2$-$N_2CH_2C_6H_5$	157.3	13
CF_3-N_2CF_3	231.0	13
CH_3-NO	167.4 ± 3.3	63
i-C_3H_7-NO	152.7 ± 13	63
t-C_4H_9-NO	165.3 ± 6.3	63
C_6H_5-NO	212.5 ± 4	63
NC-NO	120.5 ± 10.5	43
CF_3-NO	179.1 ± 8	63
C_6F_5-NO	208.4 ± 4	63
CCl_3-NO	134 ± 13	63
t-C_4H_9-NOt-C_4H_9	121	21
CH_3-NO_2	254.4	63
$CH_2C(CH_3)$-NO_2	245.2	63
i-C_3H_7-NO_2	246.9	63
t-C_4H_9-NO_2	244.8	63
C_6H_5-NO_2	298.3 ± 4	63
$C(NO_2)_3$-NO_2	169.5 ± 4	63
CH_3-$OC(CH_3)CH_2$	277.4	101
CH_3-OC_6H_5	238 ± 8	73
CH_3-$OCH_2C_6H_5$	280.3	26
C_2H_5-OC_6H_5	264 ± 6.3	63
CH_2CHCH_2-OC_6H_5	208.4 ± 8	63
O=CO	532.2 ± 0.4	29
CH_3-O_2	137.0 ± 3.8	53
C_2H_5-O_2	148.4 ± 8.4	53
CH_2CHCH_2-O_2	76.2 ± 2.1	62
i-C_3H_7-O_2	155.4 ± 9.6	53
t-C_4H_9-O_2	152.8 ± 7.4	53
$C_6H_5CH_2$-O_2CCH_3	280 ± 8	63
$C_6H_5CH_2$-$O_2CC_6H_5$	289	13
CH_3-O_2SCH_3	279.5	63
CH_2CHCH_2-O_2SCH_3	207.5	63
$C_6H_5CH_2$-O_2SCH_3	221.3	63
CF_3-O_2CF_3	361.5	10
CH_2Cl-O_2	122.4 ± 10.5	53
$CHCl_2$-O_2	108.2 ± 8.2	53
CCl_3-O_2	92.0 ± 6.4	53
CH_3CHCl-O_2	131.2 ± 1.8	53
CH_3CCl_2-O_2	112.2 ± 2.2	53
$(CH_3)_2CCl$-O_2	136.0 ± 3.8	53
CH_3-SH	312.5 ± 4.2	65
t-C_4H_9-SH	286.2 ± 6.3	63
C_6H_5-SH	361.9 ± 8	63
CH_3-SCH_3	307.9 ± 3.3	65
CH_3-SC_6H_5	290.4 ± 8	63
$C_6H_5CH_2$-SCH_3	256.9 ± 8	63
S-CS	430.5 ± 13	63
F-CH_3	472	1,61
F-CN	469.9 ± 5.0	63
F-COF	535 ± 12	18
F-CHFCl	465.3 ± 9.6	97
F-CF_2Cl	490 ± 25	41
F-$CFCl_2$	462.3 ± 10.0	97
F-CF_2CH_3	522.2 ± 8	63

Table 3
BOND STRENGTHS IN POLYATOMIC MOLECULES (continued)

Bond	$D°_{298}$/kJ mol^{-1}	Ref.
F-C_2F_5	530.5 ± 7.5	63
Cl-CN	421.7 ± 5.0	63
Cl-COC_6H_5	310 ± 13	63
Cl-CSCl	265.3 ± 2.1	71
Cl-CF_3	360.2 ± 3.3	27
Cl-CHFCl	354.4 ± 11.7	97
Cl-CF_2Cl	346.0 ± 13.4	97
Cl-$CFCl_2$	305 ± 8	40
Cl-CH_2Cl	350.2 ± 0.8	97
Cl-$CHCl_2$	338.5 ± 4.2	97
Cl-CCl_3	305.9 ± 7.5	63
Cl-C_2F_5	346.0 ± 7.1	27
Cl-CF_2CF_2Cl	326 ± 8	63
Cl-$SiCl_3$	464	99
Br-CH_3	292.9 ± 5.0	39
Br-C_6H_5	336.8 ± 8	63
Br-CN	367.4 ± 5.0	63
Br-CH_2COCH_3	261.5	101
Br-COC_6H_5	268.6	13
Br-CHF_2	289 ± 8	63
Br-CF_3	296.2 ± 1.3	3
Br-CF_2CH_3	287.0 ± 5.4	76
Br-C_2F_5	287.4 ± 6.3	63
Br-n-C_3F_7	278.2 ± 10.5	63
Br-i-C_3F_7	274.1 ± 4.6	63
Br-$CH_2C_6F_5$	225 ± 6	54
Br-$CHClCF_3$	274.9 ± 6.3	63
Br-CCl_3	231.4 ± 4	63
Br-$CClBrCF_3$	251.0 ± 6.3	63
Br-CH_2Br	296.7 ± 1.3	97
Br-$CHBr_2$	292.0 ± 8	97
Br-CBr_3	235.1 ± 7.5	63
Br-NO_2	82.0 ± 7.1	57
Br-NF_2	≤222	25
I-$CHCH_2$	259.0 ± 4.2	20
I-n-C_4H_9	205.0 ± 4	63
I-Norbornyl	261.5 ± 10.5	69
I-CN	305 ± 4	30
I-CF_3	227.2 ± 1.3	3
I-CF_2CH_3	218.0 ± 4.2	63
I-CH_2CF_3	235.6 ± 4	63
I-C_2F_5	218.8 ± 2.9	2
I-n-C_3F_7	208.4 ± 4.2	63
I-i-C_3F_7	215.1 ± 2.9	2
I-n-C_4F_9	205.0 ± 4.2	72
I-$C(CF_3)_3$	206	33
I-C_6H_5	273.6 ± 8	63
I-C_6F_5	277.0	63
C_5H_5-FeC_5H_5	381 ± 13	59
CH_3-$ZnCH_3$	285 ± 17	63
C_2H_5-ZnC_2H_5	238 ± 17	63
CH_3-$Ga(CH_3)_2$	264 ± 17	63
C_2H_5-$Ga(C_2H_5)_2$	209 ± 17	63
CH_3-$Ge(CH_3)_3$	347 ± 17	63
CH_3-$As(CH_3)_2$	280 ± 17	63
CH_3-$CdCH_3$	251 ± 17	63
CH_3-$In(CH_3)_2$	205 ± 17	63
CH_3-$Sn(CH_3)_3$	297 ± 17	63
C_2H_5-$Sn(C_2H_5)_3$	264 ± 17	63
CH_3-$Sb(CH_3)_2$	255 ± 17	63
C_2H_5-$Sb(C_2H_5)_2$	243 ± 17	63
CH_3-$HgCH_3$	255 ± 17	63
C_2H_5-HgC_2H_5	205 ± 17	63
CH_3-$Tl(CH_3)_2$	167 ± 17	63
CH_3-$Pb(CH_3)_3$	238 ± 17	63
C_2H_5-$Pb(C_2H_5)_3$	230 ± 17	63
CH_3-$Bi(CH_3)_2$	218 ± 17	63
CO-$Cr(CO)_5$	155 ± 8	60
CO-$Fe(CO)_4$	172 ± 8	60
CO-$Mo(CO)_5$	167 ± 8	60
CO-$W(CO)_5$	192 ± 8	60
BH_3-BH_3	146	13
NH_2-NH_2	275.3	63
NH_2-$NHCH_3$	268.2 ± 8	63
NH_2-$N(CH_3)_2$	246.9 ± 8	63
NH_2-NHC_6H_5	218.8 ± 8	63
ON-NO_2	40.6 ± 2.1	63
O_2N-NO_2	56.9	63
NF_2-NF_2	88 ± 4	63
O-N_2	167	1,14
O-NO	305	1,14
O-NO_2	208.7 ± 1.1	31
HO-NO	206.3	63
HO-NO_2	206.7	63
HO_2-NO_2	96 ± 8	63
CH_3O-NO	174.9 ± 3.8	9,11
C_2H_5O-NO	175.7 ± 5.4	8,11
CH_3COO_2-NO_2	118.8 ± 3.0	17
n-C_3H_7O-NO	167.8 ± 7.5	11
i-C_3H_7O-NO	171.5 ± 5.4	7,11
n-C_4H_9O-NO	177.8 ± 6.3	11
i-C_4H_9O-NO	175.7 ± 6.3	11
s-C_4H_9O-NO	173.6 ± 3.3	5,11
t-C_4H_9O-NO	171.1 ± 3.3	6,11
HO-$NCHCH_3$	207.9	13
Cl-NF_2	~134	1,75
I-NO	77.8 ± 0.4	42
I-NO_2	76.6 ± 4	98
HO-OH	213 ± 4	63
HO-$OCH_2C(CH_3)_3$	193.7 ± 7.9	63
CH_3O-OCH_3	157.3 ± 8	63
C_2H_5O-OC_2H_5	158.6 ± 4	63
n-C_3H_7O-On-C_3H_7	155.2 ± 4	63
i-C_3H_7O-Oi-C_3H_7	157.7 ± 4	63
s-C_4H_9O-Os-C_4H_9	152.3 ± 4	63
t-C_4H_9O-Ot-C_4H_9	159.0 ± 4	63
$C_2H_5C(CH_3)_2O$-$OC(CH_3)_2C_2H_5$	164.4 ± 4	63
$(CH_3)_3CCH_2O$-$OCH_2C(CH_3)_3$	152.3 ± 4	63
CF_3O-OCF_3	193.3	63
$(CF_3)_3CO$-$OC(CF_3)_3$	148.5 ± 4.6	63
t-C_4H_9O-$OSi(CH_3)_3$	197	63
SF_5O-OSF_5	155.6	63

Table 3
BOND STRENGTHS IN POLYATOMIC MOLECULES (continued)

Bond	$D°_{298}$/kJ mol^{-1}	Ref.	Bond	$D°_{298}$/kJ mol^{-1}	Ref.
t-C_4H_9O-$OGe(C_2H_5)_3$	192	63	O=PF_3	544 ± 21	52
t-C_4H_9O-$OSn(C_2H_5)_3$	192	63	O=PCl_3	510 ± 21	52
$FClO_2$-O	244.3	13	O=PBr_3	498 ± 21	52
CF_3O-O_2CF_3	126.8 ± 8	63	HO-$Si(CH_3)_3$	536	63
SF_5O-O_2SF_5	126.8	63	HS-SH	276 ± 8	63
CH_3CO_2-O_2CCH_3	127.2 ± 8	63	CH_3S-SCH_3	272.8 ± 3.8	65
$C_2H_5CO_2$-$O_2CC_2H_5$	127.2 ± 8	63	F-SF_5	420 ± 10	94
n-$C_3H_7CO_2$-O_2Cn-C_3H_7	127.2 ± 8	63	I-SH	206.7 ± 8	63
O-SO	552 ± 8	29	I-SO	180	63
F-OCF_3	182.0 ± 2.1	28	I-SCH_3	206.3 ± 7.1	87
HO-Cl	251 ± 13	52	I-$Si(CH_3)_3$	322	99
O-ClO	247 ± 13	29	H_3Si-SiH_3	310	63,99
HO-Br	234 ± 13	52	$(CH_3)_3Si$-$Si(CH_3)_3$	336.8	63,99
HO-I	234 ± 13	52	$(C_6H_5)_3Si$-$Si(C_6H_5)_3$	368 ± 29	63,99

REFERENCES

1. A value calculated from one of the thermochemical equations above, taking enthalpy data fro atoms from Table 2 and for radicals from Table 4.
2. Ahonkhai, S. I. and Whittle, E., *Int. J. Chem. Kinet.,* 16, 543, 1984.
3. Asher, R. L. and Ruscic, B., *J. Chem. Phys.,* 106, 210, 1997.
4. Baldwin, D. P., Buntine, M. A., and Chandler, D.W., *J. Chem. Phys.*, 93, 6578, 1990.
5. Batt, L. and McCulloch, R. D., *Int. J. Chem. Kinet.*, 8, 911, 1976.
6. Batt, L. and Milne, R. T. , *Int. J. Chem. Kinet.*, 8, 59, 1976.
7. Batt, L. and Milne, R. T. , *Int. J. Chem. Kinet.*, 9, 141, 1977.
8. Batt, L. and Milne, R. T. , *Int. J. Chem. Kinet.*, 9, 549, 1977.
9. Batt, L. and Milne, R. T., and McCulloch, R. D., *Int. J. Chem. Kinet.*, 9, 567, 1977.
10. Batt, L. and Walsh, R., *Int. J. Chem. Kinet.*, 15, 605, 1983.
11. Batt, L., Christie, K., Milne, R. T., and Summers, A. J., *Int. J. Chem. Kinet.*, 6, 877, 1974.
12. Baulch, D. L., Bowman, C. T., Cobos, C. J., Cox, R. A., Just, Th., Kerr, J. A., Pilling, M. J., Stocker, D., Troe, J., Tsang, W., Walker, R. W., and Warnatz, J., *J. Phys. Chem. Ref. Data*, in press.
13. Benson, S. W. and O'Neal, H. E., *Kinetic data on Gas Phase Unimolecular Reactions*, National Bureau of Standards, NSRDS-NBS, Washington, D. C., 21, 1970.
14. Benson, S. W., *J. Chem. Educ.*, 42, 502, 1965.
15. Benson, S. W., *Thermochemical Kinetics, 2nd ed.*, John Wiley & Sons, New York, 1976.
16. Berkowitz, J., Ellison, G. B., and Gutman, D., *J. Phys. Chem.*, 98, 2744, 1994.
17. Bridier, I., Caralp, F., Loirat, H., Lesclaux, R., Veyret, B., Becker, K. H., Reimer, A., and Zabel, F., *J. Phys. Chem.*, 95, 3594, 1991.
18. Buckley, T. J., Johnson, R. D., Huie, R. E., Zhang, Z., Kuo, S. C., and Klemm, B., *J. Phys. Chem.*, 99, 4879, 1995
19. Burkey, T. J., Majewski, M., and Griller, D., *J. Am. Chem. Soc.,* 108, 2218, 1986.
20. Cao, J. R., Zhang, J. M., Zhong, X., Huang, Y. H., Fang, W. Q., Wu, X. J., and Zhu, Q. H., *Chem. Phys.*, 138, 377, 1989.
21. Carmichael, P. J., Gowenlock, B. G., and Johnson, C. A. F., *J. Chem. Soc. Perkin Trans. 2,* 1853, 1973.
22. Castelhano, A. L. and Griller, D., *J. Am. Chem. Soc.,* 104, 3655, 1982.
23. Chuang, M.-C., Foltz, M. F., and Moore, C. B., *J. Chem. Phys.,* 87, 3855, 1987.
24. Chupka, W. A. and Liftshitz, C., *J. Chem. Phys.,* 48, 1109, 1968.
25. Clyne, M. A. A. and Connor, J., *J. Chem. Soc. Faraday Trans. 1*, 68, 1220, 1972.
26. Colussi, A. J., Zabel, F., and Benson, S. W., *Int. J. Chem. Kinet.*, 9, 161, 1977.
27. Coomber, J. W. and Whittle, E., *Trans. Faraday Soc.*, 63, 2656, 1967.
28. Czarnarski, J., Castellano, E., and Schumacher, H. J., *Chem. Comm.*, p. 1255, 1968.
29. Darwent, D. deB., *Bond Dissociation Energies in Simple Molecules*, National Bureau of Standards, NSRDS-NBS, 31 Washington, D. C., 1970.
30. Davis, D. D. and Okabe, H., *J. Chem. Phys.*, 49, 5526, 1968.
31. Davis, H. F., Kim, B., Johnston, H. S., and Lee, Y. T., *J. Chem. Phys.*, 97, 2172, 1993.
32. Dixon, R. N., *J. Chem. Phys.,* 104, 6905, 1996.
33. Dobychin, S. L., Mashendzhinov, V. I., Mishin, V. I., Semenov, V. N., and Shpak, V. S., *Doklady Akademii Nauk SSSR*, 312, 1166 1991.
34. Doncaster, A. M. and Walsh, R., *J. Phys. Chem.*, 83, 578, 1979.
35. Ellison, G. B., Davico, G. E., Bierbaum, V. M., and DePuy, C. H., *Int. J. Mass Spectrom. Ion Proc.*, 156, 109, 1997.

Table 3
BOND STRENGTHS IN POLYATOMIC MOLECULES (continued)

36. Ellul, E., Potzinger, P., Reimann, B., and Camilleri, P., *Ber. Bunsenges. Phys. Chem.,* 85, 407, 1981.
37. Ervin, K. M., Gronert, S., Barlow, S. E., Gilles, M. K., Harrison, A. G., Bierbaum, V. M., DePuy, C. H., Lineberger, W. C., and Ellison, G. B., *J. Am. Chem. Soc.*, 112, 5750, 1990.
38. Evans, B. S., Weeks, I., and Whittle, E., *J. Chem. Soc. Faraday Trans. 1*, 79, 1471, 1983.
39. Ferguson, K. C., Okafo, E. N., and Whittle, E., *J. Chem. Soc. Faraday Trans. 1*, 69, 295, 1973.
40. Foon, R. and Tait, K. B., *J. Chem. Soc. Faraday Trans. 1*, 68, 104, 1972.
41. Foon, R. and Tait, K. B., *J. Chem. Soc. Faraday Trans. 1*, 68, 1121, 1972.
42. Forte, E., Hippler, H., and van den Bergh, H., *Int. J. Chem. Kinet.*, 13, 1227, 1981.
43. Gowenlock, B. G., Jonhson, C. A. F., Keary, C. M., and Pfaf, J., *J. Chem. Soc. Perkin Trans. 2,* 71, 351, 1975.
44. Grela, M. A. and Colussi, A. J., *Int. J. Chem. Kinet.*, 20, 713, 1988.
45. Griller, D. and Wayner, D. D. M., *Pure Appl. Chem.*, 61, 717, 1989.
46. Heckmann, E., Hippler, H., and Troe, J., *26th Symp. (Int.) Combust.,* Combustion Institute, Pittsburgh, Pennsylvania, pp 543, 1996.
47. Heneghan, S. P. and Benson, S. W., *Int. J. Chem. Kinet.*, 15, 815, 1983.
48. Hudgens, J. W., Johnson, R. D., Timonen, R. S., Seetula, J. A., and Gutman, D., *J. Phys. Chem.*, 95, 4400, 1991.
49. Jefferson, A., Nicovich, J. M., and Wine, P. H., *J. Phys. Chem.*, 98, 7128, 1994.
50. Johnson, R. D. and Hudgens, J. W., *J. Phys Chem.,* 100, 19874, 1996.
51. Kerr, J. A. and Timlin, D. M., *Int. J. Chem. Kinet.*, 3, 427, 1971.
52. Kerr, J. A., *Chem. Rev.*, 66, 465, 1966.
53. Knyazev, V. D. and Slagle, I. R., *J. Phys. Chem.,* 102, 1770, 1998.
54. Kominar, R. J., Krech, M. J., and Price, S. J. W., *Can. J. Chem.,* 58, 1906, 1980.
55. Kondo, O. and Benson, S. W., *Int. J. Chem. Kinet.*, 16, 949, 1984.
56. Kondo, O. and Benson, S. W., *J. Phys. Chem.*, 88, 6675, 1984.
57. Kreutter, K. D., Nicovich, J. M., and Wine, P. H., *J. Phys. Chem.*, 95, 4020, 1991.
58. Laufer, A. H. and Okabe, H., *J. Am. Chem. Soc.*, 93, 4137,1971.
59. Lewis, K. E. and Smith, G. P., *J. Am. Chem. Soc.*, 106, 4650, 1984.
60. Lewis, K. E., Golden, D. M., and Smith, G. P., *J. Am. Chem. Soc.,* 106, 3905, 1984.
61. Lias, S. G., Bartmess, J. E., Liebman, J. F., Holmes, J. L., Levin, R. D., and Mallard, W. G., *J. Phys. Chem. Ref. Data 17*, Suppl. 1, 1988.
62. Lightfoot, P. D., Cox, R. A., Crowley, J. N., Destriau, M., Hayman G. D., Jenkin, M. E., Mootrgat, G. K., and Zabel, F., *Atmos. Environ.*, 26A, 1805, 1992.
63. McMillen, D. F. and Golden, D. M., *Ann. Rev. Phys. Chem.,* 33, 493, 1982.
64. Miyokawa, K. and Tschuikow-Roux, E., *J. Phys. Chem.*, 96, 7328, 1992.
65. Nicovich, J. M., Kreutter, K. D., van Dijk, C. A., and Wine, P. H., *J. Phys. Chem.*, 96, 2518, 1992.
66. Nicovich, J. M., Wang, S., McKee, M. L., and Wine, P. H., *J. Phys. Chem.,* 100, 680, 1996.
67. Niiranen, J. T., Gutman, D., and Krasnoperov, L. N., *J. Phys. Chem.,* 96, 5881, 1992.
68. Noble, P. N. and Walsh, R., *Int. J. Chem. Kinet.*, 15, 561, 1983.
69. O'Neal, H. E., Bagg, J. W., and Richardson, W. H., *Int. J. Chem. Kinet.*, 2, 493, 1970.
70. Okabe, H. and Dibeler, V. H., *J. Chem. Phys.*, 59, 2430, 1973.
71. Okabe, H., *J. Chem. Phys.,* 66, 2058, 1977.
72. Okafo, E. N. and Whittle, E., *Int. J. Chem. Kinet.*, 7, 287, 1975.
73. Paul, S. and Back, M. H., *Can J. Chem.*, 53, 3330, 1975.
74. Pedley, J. B. and Rylance, J., "Sussex - N.P.L. Computer Analysed Thermochemical Data; Organic and Organometallic Compounds", University of Sussex, 1977.
75. Petry, R. C., *J. Am. Chem. Soc.*, 89, 4600, 1967.
76. Pickard, J. M. and Rodgers, A. S., *Int. J. Chem. Kinet.,* 9, 759, 1977.
77. Pickard, J. M. and Rodgers, A. S., *Int. J. Chem. Kinet.,* 15, 569, 1983.
78. Price, S. J. W. and Sapiano, H. J., *Can. J. Chem.*, 57, 1468, 1979.
79. Rogers, A. S. and Ford, W. G. F., *Int. J. Chem. Kinet.*, 5, 965, 1973.
80. Rossi, M., McMillen, D. F., and Golden, D. M., *J. Phys. Chem.*, 88, 5031, 1984.
81. Russell, J. J., Senkan, S. M., Seetula, J. A., and Gutman, D., *J. Phys. Chem.*, 93, 5184, 1989.
82. Seetula, J. A., and Slagle, I. R., *J. Chem. Soc. Faraday Trans.,* 93, 1709, 1997.
83. Seetula, J. A., Feng, Y., Gutman, D., Seakins, P. W., and Pilling, M. J., *J. Phys. Chem.,* 95, 1658, 1991.
84. Seetula, J. A., *J. Chem. Soc. Faraday Trans.,* 92, 3069, 1996.
85. Seetula, J. A., *J. Chem. Soc. Faraday Trans.,* 94, 891, 1998.
86. Seetula, J. A., Russell, J. J., and Gutman, D., *J. Am. Chem. Soc.,* 112, 1347, 1990.
87. Shum, L. G. S. and Benson, S. W., *Int. J. Chem.,* 15, 433, 1983.
88. Shum, L. G. S. and Benson, S. W., *J. Phys. Chem.*, 87, 3479, 1983.
89. Stein, S. E., SRD Thermochemical Database, 25. N.I.S.T. Structures and Properties Database and Estimation Program, U.S. Department of Commerce, 1992.

Table 3
BOND STRENGTHS IN POLYATOMIC MOLECULES (continued)

90. Steinkruger, F. J. and Rowland, F. S., *J. Phys. Chem.*, 85, 136, 1981.
91. Trenwith, A. B. and Wrigley, S. P., *J. Chem. Soc. Faraday Trans. 1*, 73, 817, 1977.
92. Trenwith, A. B., *J. Chem. Soc. Faraday Trans. 1*, 78, 3131, 1982.
93. Trenwith, A. B., *J. Chem. Soc. Faraday Trans. 1*, 79, 2755, 1983.
94. Tsang, W. and Herron, J. T., *J. Chem. Phys.*, 96, 4272, 1992.
95. Tsang, W., *Int. J. Chem. Kinet.*, 5, 929, 1973.
96. Tsang, W., *Int. J. Chem. Kinet.*, 10, 1119, 1978.
97. Tschuikow-Roux, E. and Paddison, S., *Int. J. Chem. Kinet.*, 19, 15, 1987.
98. van den Bergh, H. and Troe, J., *J. Chem. Phys.*, 64, 736, 1976.
99. Walsh, R., *Acc. Chem. Res.*, 14, 246, 1981.
100. White, J. N. and Gardiner, W. C., *Chem. Phys. Lett.*, 58, 470, 1978.
101. Zabel, F., Benson, S. W., and Golden, D. M., *Int. J. Chem. Kinet.*, 10, 295, 1978.
102. Zhao, H.-Q., Cheung, Y.-S., Liao, C.-L., Liao, C.-X., Ng, C. Y., and Li, W.-K., *J. Chem. Phys.*, 107, 7230, 1997.
103. Zmbov, K. F., Uy, O. M., and Margrave, J. L., *J. Am. Chem. Soc.*, 90, 5090, 1968.

Table 4
ENTHALPIES OF FORMATION OF FREE RADICALS

The enthalpies of formation of the free radicals are related to the corresponding bond strengths by the equations

$$D°(\text{R-X}) = \Delta_f H°(\text{R}) + \Delta_f H°(\text{X}) - \Delta_f H°(\text{RX})$$

or

$$D°(\text{R-R}) = 2\Delta_f H°(\text{R}) - \Delta_f H°(\text{RR})$$

For an excellent review of the methods of determining the enthalpies of formation of free radicals the reader is referred to "Thermochemistry of Free Radicals" by H. E. O'Neal and S. W. Benson in *Free Radicals*, Kochi, J. K., Ed., John Wiley & Sons, New York, 1973, 275 and the article by J. Berkowitz, G. B. Ellison, and D. Gutman, *J. Phys. Chem.*, 98, 2744, 1994.

Radical	$\Delta_f H°_{298}$/kJ mol^{-1}	Ref.
CH	596.4 ± 1.2	6
CH_2(triplet)	390.4 ± 4	15,34
CH_2(singlet)	428.3 ± 4	15
CH_3	146.4 ± 0.4	10
CH≡C	566.1 ± 2.9	5,27
CH_2=CH	300.0 ± 3.4	27,64
C_2H_5	120.9 ± 1.6	10
Cycloprop-2-en-1-yl	439.7 ± 17.2	48
CH≡CCH_2	340.6 ± 8.4	48
CH_2=$CHCH_2$	170.7 ± 8.8	10,26
CH_3CH=CH	262.7	77
Cyclopropyl	279.9 ± 1.1	48
n-C_3H_7	100.8 ± 2.1	65
i-C_3H_7	86.6 ± 2.0	65
CH_3C≡CCH_2	293.7	48
CH_2=$CHCHCH_3$	125.5 ± 6.3	48
CH≡$CCHCH_3$	295.0 ± 9.2	48
Cyclobutyl	214.2 ± 4.2	48
Cyclopropylmethyl	213.8 ± 6.7	48
CH_2=$C(CH_3)CH_2$	127.2 ± 5.4	48
CH_3CH=$CHCH_2$	125.5 ± 6.3	48
n-C_4H_9	80.9 ± 2.2	65
i-C_4H_9	72.7 ± 2.2	65
s-C_4H_9	66.7 ± 2.1	65
t-C_4H_9	51.8 ± 1.3	65
Cyclopenta-1,3-dien-5-yl	263.0	8
Spiropentyl	380.7 ± 4	48
Cyclopent-1-en-3-yl	160.7 ± 4	48
CH_2=CHCH=$CHCH_2$	205 ± 13	48
$(C_2H_3)_2CH$	205 ± 13	48
CH_3C≡$CCHCH_3$	272.8 ± 9.6	48
CH≡$CC(CH_3)_2$	257.3 ± 8.4	48
CH_2=$CHC(CH_3)_2$	77.4 ± 6.3	48
Cyclopentyl	107.1 ± 2.5	17
$(CH_3)_3CCH_2$	36.4 ± 8	48
$C_2H_5C(CH_3)_2$	32.2 ± 6.3	71
C_6H_5	338 ± 3	35
Cyclohexa-1,3-dien-5-yl	197 ± 21	48
Cyclohexyl	58.2 ± 4	48
CH_3C≡$CC(CH_3)_2$	221.8 ± 9.6	48
$(CH_3)_2C$=$C(CH_3)CH_2$	39.8 ± 6.3	48
CH_2=$C(CH_3)C(CH_3)_2$	37.7 ± 6.3	48
$C_6H_5CH_2$	208.0 ± 2.5	26
Cyclohepta-1,3,5-trien-7-yl	271.1 ± 8	48
$CH_3CH_2CH_2C(CH_3)_2$	3.4 ± 8.4	69
Norbornyl	136.4 ± 10.5	48
Cycloheptyl	51.1 ± 4	48
$C_6H_5CHCH_3$	169.0	48
$C_6H_5C(CH_3)_2$	134.7	48
1-Naphthylmethyl	252.7	48
$(C_6H_5)_2CH$	289	63
9,10-Dihydroanthracen-9-yl	256.9 ± 6.3	48

Table 4
ENTHALPIES OF FORMATION OF FREE RADICALS (continued)

Radical	$\Delta_f H^\circ_{298}$/kJ mol^{-1}	Ref.
9-Anthracenylmethyl	337.6	48
9-Phenanthrenylmethyl	311.3	48
CH_2CN	243.1 ± 11.3	10
CH_2NC	326.4 ± 11.3	10
CH_3CHCN	209.2 ± 9.6	48
$(CH_3)_2CCN$	166.5 ± 8.4	48
$C_6H_5C(CH_3)CN$	248.5	48
CH_2NH_2	149.4 ± 8	48
CH_3NHCH_2	126 ± 8	48
$(CH_3)_2NCH_2$	109 ± 8	48
CN	441.4 ± 4.6	10
CHN_2	494.42	30
CHO	43.1	9,22
CHCO	175.3 ± 8.4	10
CH_3CO	-10.0 ± 1.2	56
CH_2=CHCO	72.4	48
C_2H_5CO	-32.3	8
C_6H_5CO	123.0 ± 9.6	78
CH_2CHO	10.5 ± 9.2	10
CH_3COCH_2	-23.9 ± 10.9	48
$CH_3COCHCH_3$	-70.3 ± 7.1	48
CH_3OCH_2	-0.1	8
$C_2H_5OCHCH_3$	-84.5	42
Tetrahydrofuran-2-yl	-18.0 ± 6.3	48
CH_2OH	-17.8 ± 1.3	39
CH_2CH_2OH	-36.0	31
CH=CHOH	113.0	31
CH_3CHOH	-51.6	8
CH_2=CHCHOH	0.0	48
$(CH_3)_2COH$	-111.3 ± 4.6	48
COOH	-217 ± 10	32
$COOCH_3$	-169.0 ± 4	48
$C_6H_5COOCH_2$	-69.9 ± 8.4	48
CF	261.5 ± 4.6	3
CHF	143.1 ± 12.6	61
CH_2F	-31.8 ± 8.4	59
FCO	-152.1 ± 12	14
CHF_2	-238.9 ± 4	59
CF_2	-184.1 ± 8.4	61
CF_3	-466.1 ± 3.8	3
CHCl	336.4 ± 11.7	61
CH_2Cl	117.3 ± 3.1	67
CFCl	31.0 ± 13.4	61
CHFCl	-60.7 ± 10.0	73
CF_2Cl	-279.1 ± 8.3	49
ClCO	-21.8	55
$CHCl_2$	89.0 ± 3.0	67
$CFCl_2$	-89.1 ± 10.0	73
CCl_2	230.1 ± 8.4	61
CCl_3	71.1 ± 2.5	37
CH_2Br	169.0 ± 4.2	73
$CHBr_2$	188.3 ± 9.2	73
CBr_3	207.1 ± 8	73
CH_2I	230.1 ± 6.7	48
CHI_2	333.9 ± 9.2	48
CH_3CF_2	-302.5 ± 8	48
CF_3CH_2	-517.1 ± 5.0	48
C_2F_5	-892.9 ± 4	48
CCl=CH_2	>251	64
CH_3CHCl	76.5 ± 1.6	67
CH_3CCl_2	48.4 ± 7.6	40,67
CH_2CH_2Cl	93.0 ± 3.4	68
$CHCl_2CCl_2$	23.4 ± 8	48
CF_2ClCF_2	-686 ± 17	48
C_2Cl_5	33.5 ± 5.4	54
C_6F_5	-547.7 ± 8	48
$(CH_3)_3SiCH_2$	-34.7	48
CS	278.5 ± 3.8	62
HCS	300.4 ± 8.4	10
CH_2SH	151.9 ± 8.4	10
CH_3SCH_2	136.8 ± 5.9	38
NH	352.3 ± 9.6	60
NH_2	188.7 ± 1.3	10
HNO	112.9	25
NF_2	34 ± 4	48
N_2H_3	243.5	33
N_3	469 ± 21	48
CH_2NH	104.6 ± 13	33
CH_3NH	177.4 ± 8	48
$(CH_3)_2N$	145.2 ± 8	48
C_6H_5NH	237.2 ± 8	48
$C_6H_5NCH_3$	233.5 ± 8	48
NCO	127.0	13,79
CNO	407.01	41
CH_3N_2	215.5 ± 7.5	2
$C_2H_5N_2$	187.4 ± 10.5	2
i-$C_3H_7N_2$	158.6 ± 9.2	2
OH	39.3	48
CH_3O	17.2 ± 3.8	10
C_2H_5O	-15.5 ± 3.4	10
n-C_3H_7O	-41.4	48
i-C_3H_7O	-52.3	48
n-C_4H_9O	-62.8	48
s-C_4H_9O	-69.5 ± 3.3	48
t-C_4H_9O	-90.8	48
C_6H_5O	47.7	48
CF_3O	-655.6	7
FO	109 ± 10	21
ClO	101.63 ± 0.1	1
BrO	125.8 ± 2.4	19
IO	126 ± 18	20
HO_2	14.6	70
CH_3O_2	9.0 ± 5.1	40
$C_2H_5O_2$	-27.4 ± 9.9	40
CH_2=$CHCH_2O_2$	87.9 ± 5.5	43
i-$C_3H_7O_2$	-68.8 ± 11.3	16,40
t-$C_4H_9O_2$	-101.0 ± 9.2	40
$HOCH_2O_2$	-162.1 ± 2.1	43
CF_3O_2	-614.0	43
CF_2ClO_2	-406.5	43
$CFCl_2O_2$	-213.7	43
CH_2ClO_2	-5.1 ± 13.6	40

Table 4
ENTHALPIES OF FORMATION OF FREE RADICALS (continued)

Radical	$\Delta_f H°_{298}$/kJ mol^{-1}	Ref.	Radical	$\Delta_f H°_{298}$/kJ mol^{-1}	Ref.
$CHCl_2O_2$	-19.2 ± 11.2	40	SiF_3	-1025	48,76
CCl_3O_2	-20.9 ± 8.9	40	SiCl	195.8	48,76
CH_3CHClO_2	-54.7 ± 3.4	40	$SiCl_2$	-163.6	48,76
$CH_3CCl_2O_2$	-63.8 ± 9.8	40	$SiCl_3$	-318	48,76
CH_3CO_2	-207.5 ± 4	48	SiH_3SiH	269.9 ± 14.6	74
CH_3COO_2	-172 ± 20	12	Si_2H_5	223.0	48,76
$C_2H_5CO_2$	-228.5 ± 4	48	PH_2	138.5 ± 2.5	10
n-$C_3H_7CO_2$	-249.4 ± 4	48	HS	143.0 ± 2.8	53
FO_2	26.1	45,58	CH_3S	124.6 ± 1.8	53
ClO_2	97.5	4,52	C_6H_5S	229.7 ± 8	48
OClO	95.6	28,51	SO	5.0	18
NO_3	73.7 ± 1.4	24	HSO	-4	44
sym-ClO_3	232.6	23	HSO_2	-222	11
SiH	377	48,76	CH_3SO_2	-239.3	57
SiH_2	269.0 ± 1.3	29,47	$HOSO_2$	-385	46
SiH_3	200.5 ± 2.5	66	SO_3	-395.7	75
CH_3Si	310	76	SF_4	-746 ± 12	72
CH_3SiH	213.0 ± 14.6	74	SF_5	-879.9 ± 20	72
CH_3SiH_2	152.7	48,76	CH_3S_2	68.6 ± 8	36
$(CH_3)_2Si$	109	76	$C_2H_5S_2$	43.5 ± 8	36
$(CH_3)_2SiH$	59.8	48,76	i-$C_3H_7S_2$	13.8 ± 8	36
$(CH_3)_3Si$	-3.3	48,76	t-$C_4H_9S_2$	-19.3 ± 8	36
$C_6H_5Si(CH_3)_2$	163	57	$HOCS_2$	110.5	50
$(C_6H_5)_2SiCH_3$	326	57	GeH_3	222 ± 8	10
$(C_6H_5)_3Si$	486.2	57	AsH_2	167.8 ± 1.3	10
SiF	-19.3	48,76	HSe	144.8 ± 2.1	10
SiF_2	-587.9	48,76	SbH_2	215.5 ± 2.5	10
			HTe	158.6 ± 5.0	10

REFERENCES

1. Abramowitz, S. and Chase, M. W., *Pure Appl. Chem.*, 63, 1448, 1991.
2. Acs, G. and Peter, A., *Int. J. Chem. Kinet.*, 19, 929, 1987.
3. Asher, R. L. and Ruscic, B., *J. Chem. Phys.*, 106, 210, 1997.
4. Baer, S., Hippler, H., Rabu, R., Siefke, M., Seitzinger, N., and Troe, J., *J. Chem. Phys.*, 95, 6463, 1991.
5. Baldwin, D. P., Buntine, M. A., and Chandler, D. W., *J. Chem. Phys.*, 93, 6578, 1990.
6. Based on $D°$(C-H), see Table 1 and $\Delta_f H°$(C) and $D_f H°$(H), see Table 2.
7. Batt, L. and Walsh, R., Int. J. Chem. Kinet., 14, 933, 1982.
8. Baulch, D. L., Bowman, C. T., Cobos, C. J., Cox, R. A., Just, Th., Kerr, J. A., Pilling, M. J., Stocker, D., Troe, J., Tsang, W., Walker, R. W., and Warnatz, J., *J. Phys. Chem. Ref. Data,* in press.
9. Becerra,R., Carpenter, I. W., and Walsh, R., *J. Phys. Chem.,* A 101, 4185, 1997.
10. Berkowitz, J., Ellison, G. B., and Gutman, D., *J. Phys. Chem.,* 98, 2744, 1994.
11. Boyd, R. J., Gupta, A., Langler, R. F., Lownie, S. P., and Pincock, J. A., *Can. J. Chem.*, 58, 331, 1980.
12. Bridier, I., Caralp, F., Loirat, H., Lesclaux, R., Veyret, B., Becker, K. H., Reimer, A., and Zabel, F., *J. Phys. Chem.*, 95, 3594, 1991.
13. Brown, S. S., Berghout, H. L., and Crim, F. F., *J. Chem. Phys.,* 105, 8103, 1996.
14. Buckley, T. J., Johnson, R. D., Huie, R. E., Zhang, Z., Kuo, S. C., and Klemm, B., *J. Phys. Chem.*, 99, 4879, 1995.
15. Bunker, P. R. and Sears, T. J., *J. Chem. Phys.,* 83, 4866, 1985.
16. Calculated taking $\Delta_f H°$(i-C_3H_7) = 86.6 kJ mol^{-1}.
17. Castelhano, A. L. and Griller, D., *J. Am. Chem. Soc.*, 104, 3655, 1982; value adjusted to $\Delta_f H°(CH_3)$ = 146 kJ mol^{-1} and error limits assigned here.
18. Chase, M. W. Jr., Davies, C. A., Downey, J.R. Jr., Frurip, D. J., McDonald, R. A., and Syverud, A. N., *J. Phys. Chem. Ref. Data,* 14, Suppl. 1, 1985.
19. Chase, M. W., *J. Phys. Chem. Ref. Data,* 25, 1069, 1996.
20. Chase, M. W., *J. Phys. Chem. Ref. Data,* 25, 1297, 1996.
21. Chase, M. W., *J. Phys. Chem. Ref. Data,* 25, 551, 1996.
22. Chuang, M.-C., Foltz, M. F., and Moore, C. B., *J. Chem. Phys.*, 87, 3855, 1987.
23. Colussi, A. J., *J. Phys. Chem.*, 94, 8922, 1990.
24. Davis, H. F., Kim, B., Johnston, H. S., and Lee, Y. T., *J. Phys. Chem.,* 97, 2172, 1993.

Table 4
ENTHALPIES OF FORMATION OF FREE RADICALS (continued)

25. Dixon, R. N., *J. Chem. Phys.*, 104, 6905, 1996.
26. Ellison, G. B., Davico, G. E., Bierbaum, V. M., and DePuy, C. H., *Int. J. Mass Spectrom. Ion Proc.*, 156, 109, 1996.
27. Ervin, K. M., Gronert, S., Barlow, S. E., Gilles, M. K., Harrison, A. G., Bierbaum, V. M., DePuy, C. H., Lineberger, W. C., and Ellison, G. B., *J. Am. Chem. Soc.*, 112, 5750, 1990.
28. Flesch, R., E. Rühl, K. Hottmann, and H. Baumgartel, *J. Phys. Chem.*, 97, 837 (1993).
29. Frey, H. M., Walsh, R., and Watts, I. M., *J. Chem. Soc. Chem. Comm.*, 1189, 1986.
30. Fulle, D. and Hippler, H., *J. Chem. Phys.*, 105, 5423, 1996.
31. Fulle, D., Hamann, H. F., Hippler, H., and Jansch, C. P., *Ber. Bunsenges. Phys. Chem.*, 101, 1433, 1997.
32. Fulle, D., Hamann, H. F., Hippler, H., and Troe, *J. Chem. Phys.*, 105, 983, 1996.
33. Grela, M. A. and Colussi, A. J., *Int. J. Chem. Kinet.*, 20, 713, 1988.
34. Gurvich, I.V., Veyts, I.V., Alcock, C.B., *Thermodynamic Properties of Individual Substances, 4th ed.*, Hemisphere, New York, 1991, Vol. 2.
35. Heckmann, E., Hippler, H., and Troe, J., *26th Symp. (Int.) Combust.*, The Combustion Institute, Pittsburgh, Pennsylvania, pp 543, 1996.
36. Howari, J. A. Griller, D., and Lossing, F. P., *J. Am. Chem. Soc.*, 108, 3273, 1986.
37. Hudgens, J. W., Johnson, R. D., Timonen, R. S., Seetula, J. A., and Gutman, D., *J. Phys. Chem.*, 95, 4400, 1991.
38. Jefferson, A., Nicovich, J. M., and Wine, P. H., *J. Phys. Chem.*, 98, 7128, 1994.
39. Johnson, R. D. and Hudgens, J. W., *J. Phys. Chem.*, 100, 19874, 1996.
40. Knyazev, V. D., and Slagle, I. R., *J. Phys. Chem.*, 102, 1770, 1998.
41. Koch, W. and Frenking, G., *J. Phys. Chem.*, 91, 49, 1987.
42. Kondo, O. and Benson, S. W., *Int. J. Chem. Kinet.*, 16, 949, 1984.
43. Lightfoot, P. D., Cox, R. A., Crowley, J. N., Destriau, M., Hayman, G.D., Jenkin, M. E., Moortgat, G. K., and Zabel, F., *Atmos Environ.*, 26A, 1805, 1992.
44. Lovejoy, E. R., Wang, N. S., and Howard, C. J., *J. Phys. Chem.*, 91, 5749, 1987.
45. Lyman, J. L. and Holland, R., *J. Phys. Chem.*, 92, 7232, 1988.
46. Margitan, J. J., *J. Phys. Chem.*, 88, 3314, 1984.
47. Martin, J. G., Ring, M. A., and O'Neal, H. E., *Int. J. Chem. Kinet.*, 19, 715, 1987.
48. McMillen, D. F. and Golden, D. M., *Ann. Rev. Phys. Chem.*, 33, 493, 1982.
49. Miyokawa, K. and Tschuikow-Roux, E., *J. Phys. Chem.*, 96, 7328, 1992.
50. Murrells, T. P., Lovejoy, E.R., and Ravishankara, A. R., *J. Phys. Chem.*, 80, 4065, 1984.
51. Nickolaisen, S. L., R. R. Friedl, and S. P. Sander, *J. Phys. Chem.*, 98, 155 (1994).
52. Nicovich, J. M., Kreutter, K. D., Shockelford, C. J., and Wine, P. H., *Chem. Phys. Lett.*, 179, 367, 1991.
53. Nicovich, J. M., Kreutter, K. D., van Dijk, C. A., and Wine, P. H., *J. Phys. Chem.*, 96, 2518, 1992.
54. Nicovich, J. M., Wang, S., McKee, M. L., and Wine, P. H., *J. Phys. Chem.*, 100, 680, 1996.
55. Nicovich, J.M., Kreutter, K. D., and Wine, P.H., *J. Chem. Phys.*, 92, 3539 (1990).
56. Niiranen, J. T., Gutman, D., and Krasnoperov, L. N., *J. Phys. Chem.*, 96, 5881, 1992.
57. O'Neal, H. E. and Benson, S. W., in *Free Radicals*, Kochi, J. K., Ed., John Wiley & Sons, New York, 1973, 275.
58. Pagsberg, P., Ratajczak, E., Sillesen, A., and Jodkowski, J. T., *Chem. Phys. Lett.*, 141, 88, 1987.
59. Pickard, J. M. and Rodgers, A. S., *Int. J. Chem. Kinet.*, 15, 569, 1983.
60. Piper, L. G., *J. Chem. Phys.*, 70, 3417, 1979.
61. Poutsma, J. C., Paulino, J. A., and Squires, R. R., *J. Phys. Chem. A*, 101, 5327, 1997.
62. Prinslow, D. A. and Armentrout, P. B., *J. Chem. Phys.*, 94, 3563, 1991.
63. Rossi, M., McMillen, D. F., and Golden, D. M., *J. Phys. Chem.*, 88, 5031, 1984.
64. Russell, J. R., Senkan, S. M., Seetula, J. A., and Gutman, D., *J. Phys. Chem.*, 93, 5184, 1989.
65. Seetula, J. A. and Slagle I. R., *J. Chem. Soc. Faraday Trans.*, 93, 1709, 1997.
66. Seetula, J. A., Feng, Y., Gutman, D., Seakins, P. W., and Pilling, M. J., *J. Phys. Chem.*, 95, 1658, 1991.
67. Seetula, J. A., *J. Chem. Soc. Faraday Trans.*, 92, 3069, 1996.
68. Seetula, J. A., *J. Chem. Soc. Faraday Trans.*, 94, 891, 1998.
69. Seres, L., Gorgenyi, M., and Farkas, J., *Int. J. Chem. Kinet.*, 15, 1133, 1983.
70. Shum, L. G. S. and Benson, S. W., *J. Phys. Chem.*, 87, 3479, 1983.
71. Stein, S. E., SRD Thermochemical Database, 25. N.I.S.T. Structures and Properties Database and Estimation Program, U.S. Department of Commerce, 1992.
72. Tsang, W. and Herron, J. T., *J. Chem. Phys.*, 96, 4272, 1992.
73. Tschuikow-Roux, E. and Paddison, S., *Int. J. Chem. Kinet.*, 19, 15, 1987.
74. Vanderwielen, A. J., Ring, M. A., and O'Neal, H. E., *J. Am. Chem. Soc.*, 97, 993, 1975.
75. Wagman, D. D., Evans, W. H., Parker, V. B., Schumm, R. H., Halow, I., Bailey, S. M., Churney, K. L., and Nuttall, R.L., *J. Phys. Chem. Ref. Data*, 11, Suppl. 2, 1978.
76. Walsh, R., *Acc. Chem. Res.*, 14, 246, 1981.
77. Wu, C. H. and Kern, R. D., *J. Phys. Chem.*, 91, 6291, 1987.
78. Zhao, H.-Q., Cheung, Y.-S., Liao, C.-L., Liao, C.-X., Ng, C. Y., and Li, W.-K., *J. Chem. Phys.*, 107, 7230, 1997.
79. Zyrianov, M., Droz-Georget, T., Sanov, A., and Reisler, H., *J. Chem. Phys.*, 105, 8111, 1996.

Table 5
BOND STRENGTHS OF SOME ORGANIC MOLECULES

Bond strengths at 298 K expressed in kJ/mol^{-1} for some organic molecules of the general formula R-X are presented below. Some are experimental values taken from the preceding tables; the remainder are calculated from the enthalpies of formation of atoms (Table 2) and of radicals (Table 4), and the enthalpies of formation of the parent compounds from sources indicated by the references below. The table also includes bond strengths for the inorganic molecules, hydrogen, the hydrogen halides, water and ammonia.

	H	F	Cl	Br	I	OH	NH_2	CH_3O	CH_3	CH_3CO	NO	CF_3	CCl_3
H	435.990	569.87	431.62	366.35	298.407	498	453	436	439	374	195	450	393
CH_3	439	472	350[e]	293	239[e]	387[e]	358[e]	348[e]	377[e]	354[e]	167	423	362[e]
C_2H_5	423	463[d]	354[e]	295[e]	236[e]	395[e]	357[e]	355[e]	371[e]	349[e]	—	—	—
i-C_3H_7	409	460[e]	353[e]	298[e]	234[e]	399[e]	359[e]	356[e]	367[e]	339[e]	153	—	—
t-C_4H_9	404	—	355[e]	296[e]	231[e]	404[e]	362[e]	353[e]	366[e]	332[e]	165	—	—
C_6H_5	473	533[e]	407[e]	346[f]	280[e]	474[e]	439[e]	423[e]	434[e]	415[e]	213	—	—
$C_6H_5CH_2$	376	—	310[e]	256[f]	215[f]	348[e]	302[e]	—	332	297[e]	—	—	—
CCl_3	393	419[e]	288[e]	224[e]	167[e]	—	—	—	362[e]	—	134	335[b]	286[e]
CF_3	450	547[e]	362[e]	294[e]	227	—	—	—	423	—	179	413	335[b]
C_2F_5	430	531[e]	347[e]	283[e]	219	—	—	—	—	—	—	—	—
CH_3CO	374	512[e]	354[f]	292[e]	223[e]	462[e]	417[e]	421[e]	354[e]	282	—	—	—
CN	528	470	422	367	305	—	—	—	514[e]	—	121	—	—
C_6F_5	473	487[e]	383[e]	—	277[a]	499[e]	—	—	441[e]	—	208[a]	—	—

REFERENCES

a Choo, K. Y., Mendenhall, G. D., Golden, D.M., and Benson, S. W., *Int. J. Chem. Kinet.*, 6, 813, 1974.

b Kolesov, V. P. and Papina, T. S., *Russ. Chem. Rev.*, 52, 425, 1983.

c Kudchadker, S. A. and Kudchadker, A. P., *J. Phys. Chem. Ref. Data*, 1, 1285, 1978.

d Lias, S. G., Bartmess, J. E., Liebman, J. F., Holmes, J. L., Levin, R. D., and Mallard, W. G., *J. Phys. Chem. Ref. Data*, 17, Suppl. No. 1, 1988.

e Lide, D. R., Ed., *Handbook of Chemistry and Physics, 80th Edition,* CRC Press, Boca Raton, FL, 1999.

f Pedley, J. B. and Rylance, J., *Sussex.N.P.L. Computer Analysed Thermochemical Data: Organic and Organometallic Compounds*, University of Sussex, 1977.

ELECTRONEGATIVITY

Electronegativity is a parameter originally introduced by Pauling which describes, on a relative basis, the tendency of an atom in a molecule to attract bonding electrons. While electronegativity is not a precisely defined molecular property, the electronegativity difference between two atoms provides a useful measure of the polarity and ionic character of the bond between them. This table gives the electronegativity *X*, on the Pauling scale, for the most common oxidation state. Other scales are described in the references.

REFERENCES

1. Pauling, L., *The Nature of the Chemical Bond, Third Edition*, Cornell University Press, Ithaca, New York, 1960.
2. Allen, L.C., *J. Am. Chem. Soc.*, 111, 9003, 1989.
3. Allred, A.L., *J. Inorg. Nucl. Chem.*, 17, 215, 1961.

Z	Symbol	*X*
1	H	2.20
2	He	—
3	Li	0.98
4	Be	1.57
5	B	2.04
6	C	2.55
7	N	3.04
8	O	3.44
9	F	3.98
10	Ne	—
11	Na	0.93
12	Mg	1.31
13	Al	1.61
14	Si	1.90
15	P	2.19
16	S	2.58
17	Cl	3.16
18	Ar	—
19	K	0.82
20	Ca	1.00
21	Sc	1.36
22	Ti	1.54
23	V	1.63
24	Cr	1.66
25	Mn	1.55
26	Fe	1.83
27	Co	1.88
28	Ni	1.91
29	Cu	1.90
30	Zn	1.65
31	Ga	1.81
32	Ge	2.01
33	As	2.18
34	Se	2.55
35	Br	2.96
36	Kr	—
37	Rb	0.82
38	Sr	0.95
39	Y	1.22
40	Zr	1.33
41	Nb	1.6
42	Mo	2.16
43	Tc	2.10
44	Ru	2.2
45	Rh	2.28
46	Pd	2.20
47	Ag	1.93
48	Cd	1.69
49	In	1.78
50	Sn	1.96
51	Sb	2.05
52	Te	2.1
53	I	2.66
54	Xe	2.60
55	Cs	0.79
56	Ba	0.89
57	La	1.10
58	Ce	1.12
59	Pr	1.13
60	Nd	1.14
61	Pm	—
62	Sm	1.17
63	Eu	—
64	Gd	1.20
65	Tb	—
66	Dy	1.22
67	Ho	1.23
68	Er	1.24
69	Tm	1.25
70	Yb	—
71	Lu	1.0
72	Hf	1.3
73	Ta	1.5
74	W	1.7
75	Re	1.9
76	Os	2.2
77	Ir	2.2
78	Pt	2.2
79	Au	2.4
80	Hg	1.9
81	Tl	1.8
82	Pb	1.8
83	Bi	1.9
84	Po	2.0
85	At	2.2
86	Rn	—
87	Fr	0.7
88	Ra	0.9
89	Ac	1.1
90	Th	1.3
91	Pa	1.5
92	U	1.7
93	Np	1.3
94	Pu	1.3

FORCE CONSTANTS FOR BOND STRETCHING

Representative force constants (f) for stretching of chemical bonds are listed in this table. Except where noted, all force constants are derived from values of the harmonic vibrational frequencies ω_e. Values derived from the observed vibrational fundamentals ν, which are noted by a, are lower than the harmonic force constants, typically by 2 to 3% in the case of heavy atoms (often by 5 to 10% if one of the atoms is hydrogen). Values are given in the SI unit newton per centimeter (N/cm), which is identical to the commonly used cgs unit mdyn/Å.

REFERENCES

1. Huber, K. P., and Herzberg, G., *Molecular Spectra and Molecular Structure. IV. Constants of Diatomic Molecules*, Van Nostrand Reinhold, New York, 1979.
2. Shimanouchi, T., *The Molecular Force Field*, in Eyring, H., Henderson, D., and Yost, W., Eds., *Physical Chemistry: An Advanced Treatise*, Vol. IV, Academic Press, New York, 1970.
3. Tasumi, M., and Nakata, M., *Pure and Appl. Chem.*, 57, 121—147, 1985.

Bond	Molecule	f/(N/cm)	Note
H-H	H_2	5.75	
Be-H	BeH	2.27	
B-H	BH	3.05	
C-H	CH	4.48	
	CH_4	5.44	b
	C_2H_6	4.83	a,b,c
	CH_3CN	5.33	b
	CH_3Cl	5.02	a,b,c
	$CCl_2{=}CH_2$	5.57	b
	HCN	6.22	
N-H	NH	5.97	
O-H	OH	7.80	
	H_2O	8.45	
P-H	PH	3.22	
S-H	SH	4.23	
	H_2S	4.28	
F-H	HF	9.66	
Cl-H	HCl	5.16	
Br-H	HBr	4.12	
I-H	HI	3.14	
Li-H	LiH	1.03	
Na-H	NaH	0.78	
K-H	KH	0.56	
Rb-H	RbH	0.52	
Cs-H	CsH	0.47	
C-C	C_2	12.16	
	$CCl_2{=}CH_2$	8.43	
	C_2H_6	4.50	a,c
	CH_3CN	5.16	
C-F	CF	7.42	
	CH_3F	5.71	a,c
C-Cl	CCl	3.95	
	CH_3Cl	3.44	a,c
	$CCl_2{=}CH_2$	4.02	b
C-Br	CH_3Br	2.89	a,c
C-I	CH_3I	2.34	a,c
C-O	CO	19.02	
	CO_2	16.00	
	OCS	16.14	
	CH_3OH	5.42	a,c
C-S	CS	8.49	
	CS_2	7.88	
	OCS	7.44	
C-N	CN	16.29	
	HCN	18.78	
	CH_3CN	18.33	
	CH_3NH_2	5.12	a,c
C-P	CP	7.83	
Si-Si	Si_2	2.15	
Si-O	SiO	9.24	
Si-F	SiF	4.90	
Si-Cl	SiCl	2.63	
N-N	N_2	22.95	
	N_2O	18.72	
N-O	NO	15.95	
	N_2O	11.70	
P-P	P_2	5.56	
P-O	PO	9.45	
O-O	O_2	11.77	
	O_3	5.74	a
S-O	SO	8.30	
	SO_2	10.33	a
S-S	S_2	4.96	
F-F	F_2	4.70	
Cl-F	ClF	4.48	
Br-F	BrF	4.06	
Cl-Cl	Cl_2	3.23	
Br-Cl	BrCl	2.82	
Br-Br	Br_2	2.46	
I-I	I_2	1.72	
Li-Li	Li_2	0.26	
Li-Na	LiNa	0.21	
Na-Na	Na_2	0.17	
Li-F	LiF	2.50	
Li-Cl	LiCl	1.43	
Li-Br	LiBr	1.20	
Li-I	LiI	0.97	
Na-F	NaF	1.76	
Na-Cl	NaCl	1.09	
Na-Br	NaBr	0.94	
Na-I	NaI	0.76	
Be-O	BeO	7.51	
Mg-O	MgO	3.48	
Ca-O	CaO	3.61	

[a] Derived from fundamental frequency, without anharmonicity correction.
[b] Average of symmetric and antisymmetric (or degenerate) modes.
[c] Calculated from Local Symmetry Force Field (see Reference 2).

FUNDAMENTAL VIBRATIONAL FREQUENCIES OF SMALL MOLECULES

This table lists the fundamental vibrational frequencies of selected three-, four-, and five-atom molecules. Both stable molecules and transient free radicals are included. The data have been taken from evaluated sources. In general, the selected values are based on gas-phase infrared, Raman, or ultraviolet spectra; when these were not available, liquid-phase or matrix-isolation spectra were used.

Molecules are grouped by structural type. Within each group, related molecules appear together for convenient comparison.

The vibrational modes are described by their approximate character in terms of stretching, bending, deformation, etc. However, it should be emphasized that most such descriptions are only approximate, and that the true normal mode usually involves a mixture of motions. Abbreviations are:

sym.	symmetric
antisym.	antisymmetric
str.	stretch
deform.	deformation
scis.	scissors
rock.	rocking
deg.	degenerate

In the case of free radicals, strong interactions may exist between the electronic and bending vibrational motions. Details can be found in References 3 and 4. The references should be consulted for information on the accuracy of the data and for data on other molecules not listed here.

All fundamental frequencies (more precisely, wavenumbers) are given in units of cm^{-1}.

XY_2 Molecules
Point groups $D_{\infty h}$(linear) and C_{2v}(bent)

Molecule	Structure	Sym. str.	Bend	Antisym. str.
CO_2	Linear	1333	667	2349
CS_2	Linear	658	397	1535
C_3	Linear	1224	63	2040
CNC	Linear		321	1453
NCN	Linear	1197	423	1476
BO_2	Linear	1056	447	1278
BS_2	Linear	510	120	1015
KrF_2	Linear	449	233	590
XeF_2	Linear	515	213	555
$XeCl_2$	Linear	316		481
H_2O	Bent	3657	1595	3756
D_2O	Bent	2671	1178	2788
F_2O	Bent	928	461	831
Cl_2O	Bent	639	296	686
O_3	Bent	1103	701	1042
H_2S	Bent	2615	1183	2626
D_2S	Bent	1896	855	1999
SF_2	Bent	838	357	813
SCl_2	Bent	525	208	535
SO_2	Bent	1151	518	1362
H_2Se	Bent	2345	1034	2358
D_2Se	Bent	1630	745	1696
NH_2	Bent	3219	1497	3301
NO_2	Bent	1318	750	1618
NF_2	Bent	1075	573	942
ClO_2	Bent	945	445	1111
CH_2	Bent		963	
CD_2	Bent		752	
CF_2	Bent	1225	667	1114
CCl_2	Bent	721	333	748
CBr_2	Bent	595	196	641
SiH_2	Bent	2032	990	2022
SiD_2	Bent	1472	729	1468
SiF_2	Bent	855	345	870

Molecule	Structure	Sym. str.	Bend	Antisym. str.
$SiCl_2$	Bent	515		505
$SiBr_2$	Bent	403		400
GeH_2	Bent	1887	920	1864
$GeCl_2$	Bent	399	159	374
SnF_2	Bent	593	197	571
$SnCl_2$	Bent	352	120	334
$SnBr_2$	Bent	244	80	231
PbF_2	Bent	531	165	507
$PbCl_2$	Bent	314	99	299
ClF_2	Bent	500		576

XYZ Molecules
Point Groups $C_{\infty v}$ (linear) and C_s(bent)

Molecule	Structure	XY str.	Bend	YZ str.
HCN	Linear	3311	712	2097
DCN	Linear	2630	569	1925
FCN	Linear	1077	451	2323
ClCN	Linear	744	378	2216
BrCN	Linear	575	342	2198
ICN	Linear	486	305	2188
CCN	Linear	1060	230	1917
CCO	Linear	1063	379	1967
HCO	Bent	2485	1081	1868
HCC	Linear	3612		1848
OCS	Linear	2062	520	859
NCO	Linear	1270	535	1921
NNO	Linear	2224	589	1285
HNB	Linear	3675		2035
HNC	Linear	3653		2032
HNSi	Linear	3583	523	1198
HBO	Linear		754	1817
FBO	Linear		500	2075
ClBO	Linear	676	404	1958
BrBO	Linear	535	374	1937
FNO	Bent	766	520	1844
ClNO	Bent	596	332	1800
BrNO	Bent	542	266	1799
HNF	Bent		1419	1000
HNO	Bent	2684	1501	1565
HPO	Bent	2095	983	1179
HOF	Bent	3537	886	1393
HOCl	Bent	3609	1242	725
HOO	Bent	3436	1392	1098
FOO	Bent	579	376	1490
ClOO	Bent	407	373	1443
BrOO	Bent			1487
HSO	Bent		1063	1009
NSF	Bent	1372	366	640
NSCl	Bent	1325	273	414
HCF	Bent		1407	1181
HCCl	Bent		1201	815
HSiF	Bent	1913	860	834
HSiCl	Bent		808	522
HSiBr	Bent	1548	774	408

FUNDAMENTAL VIBRATIONAL FREQUENCIES OF SMALL MOLECULES (continued)

Symmetric XY_3 Molecules
Point Groups D_{3h} (planar) and C_{3v} (pyramidal)

Molecule	Structure	Sym. str.	Sym. deform.	Deg. str.	Deg. deform.
NH_3	Pyram.	3337	950	3444	1627
ND_3	Pyram.	2420	748	2564	1191
PH_3	Pyram.	2323	992	2328	1118
AsH_3	Pyram.	2116	906	2123	1003
SbH_3	Pyram.	1891	782	1894	831
NF_3	Pyram.	1032	647	907	492
PF_3	Pyram.	892	487	860	344
AsF_3	Pyram.	741	337	702	262
PCl_3	Pyram.	504	252	482	198
PI_3	Pyram.	303	111	325	79
AsI_3	Pyram.	219	94	224	71
$AlCl_3$	Pyram.	375	183	595	150
SO_3	Planar	1065	498	1391	530
BF_3	Planar	888	691	1449	480
BH_3	Planar		1125	2808	1640
CH_3	Planar		606	3161	1396
CD_3	Planar		453	2369	1029
CF_3	Pyram.	1090	701	1260	510
SiF_3	Pyram.	830	427	937	290

Linear XYYX Molecules
Point Group $D_{\infty h}$

Molecule	Sym. XY str.	Antisym. XY str.	YY str.	Bend	Bend
C_2H_2	3374	3289	1974	612	730
C_2D_2	2701	2439	1762	505	537
C_2N_2	2330	2158	851	507	233

Planar X_2YZ Molecules
Point Group C_{2v}

Molecule	Sym.XY str.	YZ str.	YX_2 scis.	Antisym. XY str.	YX_2 rock	YX_2 wag
H_2CO	2783	1746	1500	2843	1249	1167
D_2CO	2056	1700	1106	2160	990	938
F_2CO	965	1928	584	1249	626	774
Cl_2CO	567	1827	285	849	440	580
O_2NF	1310	822	568	1792	560	742
O_2NCl	1286	793	370	1685	408	652

Tetrahedral XY_4 Molecules
Point Group T_d

Molecule	Sym. str.	Deg. deform.(e)	Deg. str.(f)	Deg. deform.(f)
CH_4	2917	1534	3019	1306
CD_4	2109	1092	2259	996
CF_4	909	435	1281	632
CCl_4	459	217	776	314

Molecule	Sym. str.	Deg. deform.(e)	Deg. str.(f)	Deg. deform.(f)
CBr_4	267	122	672	182
CI_4	178	90	555	125
SiH_4	2187	975	2191	914
SiD_4	1558	700	1597	681
SiF_4	800	268	1032	389
$SiCl_4$	424	150	621	221
GeH_4	2106	931	2114	819
GeD_4	1504	665	1522	596
$GeCl_4$	396	134	453	172
$SnCl_4$	366	104	403	134
$TiCl_4$	389	114	498	136
$ZrCl_4$	377	98	418	113
$HfCl_4$	382	102	390	112
RuO_4	885	322	921	336
OsO_4	965	333	960	329

REFERENCES

1. T. Shimanouchi, Tables of Molecular Vibrational Frequencies, Consolidated Volume I, Natl. Stand. Ref. Data Ser. Natl. Bur. Stand. (U.S.), 39, 1972.
2. T. Shimanouchi, Tables of Molecular Vibrational Frequencies, Consolidated Volume II, *J. Phys. Chem. Ref. Data*, 6, 993, 1977.
3. G. Herzberg, *Electronic Spectra and Electronic Structure of Polyatomic Molecules*, D. Van Nostrand Co., Princeton, 1966.
4. M. E. Jacox, Ground state vibrational energy levels of polyatomic transient molecules, *J. Phys. Chem. Ref. Data*, 13, 945, 1984.

SPECTROSCOPIC CONSTANTS OF DIATOMIC MOLECULES

This table lists the leading spectroscopic constants and equilibrium internuclear distance r_e in the ground electronic state for selected diatomic molecules. The constants are those describing the vibrational and rotational energy through the expressions:

$$E_{vib}/hc = \omega_e(v+1/2) - \omega_e x_e(v+1/2)^2 + \cdots$$

$$E_{rot}/hc = B_v J(J+1) - D_v[J(J+1)]^2 + \cdots$$

where

$$B_v = B_e - \alpha_e(v+1/2) + \cdots$$

$$D_v = D_e + \cdots$$

Here v and J are the vibrational and rotational quantum numbers, respectively, h is Planck's constant, and c is the speed of light. In this customary formulation the constants ω_e, B_e, etc. have dimensions of inverse length; in this table they are given in units of cm^{-1}.

Users should note that higher order terms in the above energy expressions are required for very precise calculations; constants for many of these terms can be found in the references. Also, if the ground electronic state is not $^1\Sigma$, additional terms are needed to account for the interaction between electronic and pure rotational angular momentum. For some molecules in the table the data have been analyzed in terms of the Dunham series expansion:

$$E/hc = \Sigma_{lm} Y_{lm}(v+1/2)^l J^m (J+1)^m$$

In such cases it has been assumed that $Y_{10} = \omega_e$, $Y_{01} = B_e$, etc., although in the highest approximations these identities are not precisely correct. Some of the values of r_e in the table have been corrected for breakdown of the Born-Oppenheimer approximation, which can affect the last decimal place. Because of differences in the method of data analysis and limitations in the model, care should be taken in comparing r_e values for different molecules to a precision beyond 0.001 Å.

Molecules are listed in alphabetical order by formula as written in the most common form. In most cases this form places the more electropositive element first, but there are exceptions such as OH, NH, CH, etc.

* Indicates a value for the interval between $v = 0$ and $v = 1$ states instead of a value of ω_e.

REFERENCES

1. Huber, K. P., and Herzberg, G., *Molecular Spectra and Molecular Structure IV. Constants of Diatomic Molecules*, Van Nostrand Reinhold, New York, 1979.
2. Lovas, F. J., and Tiemann, E., *J. Phys. Chem. Ref. Data*, 3, 609, 1974.
3. *Landolt-Börnstein, Numerical Data and Functional Relationships in Science and Technology, New Series*, II/6 (1974), II/14a (1982), II/14b (1983), II/19a (1992), II/19d-1 (1995), *Molecular Constants*, Springer-Verlag, Heidelberg.

Molecule	State	ω_e cm^{-1}	$\omega_e x_e$ cm^{-1}	B_e cm^{-1}	α_e cm^{-1}	D_e $10^{-6}cm^{-1}$	r_e Å
$^{107}Ag^{79}Br$	$^1\Sigma^+$	249.57	0.63	0.064833	0.0002361	0.0175	2.39311
$^{107}Ag^{35}Cl$	$^1\Sigma^+$	343.49	1.17	0.12298388	0.00059541	0.06305	2.28079
$^{107}Ag^{19}F$	$^1\Sigma^+$	513.45	2.59	0.2657020	0.0019206	0.284	1.98318
$^{107}Ag^1H$	$^1\Sigma^+$	1759.9	34.06	6.449	0.201	344	1.618
$^{107}Ag^2H$	$^1\Sigma^+$	1250.70	17.17	3.2572	0.0722	85.9	1.6180
$^{107}Ag^{127}I$	$^1\Sigma^+$	206.50	0.46	0.04486821	0.0001414	0.00847	2.54463
$^{107}Ag^{16}O$	$^2\Pi_{1/2}$	490.2	3.1	0.3020	0.0025	0.45	2.003
$^{27}Al_2$	$^3\Sigma_g^-$	350.01	2.02	0.2054	0.0012	0.31	2.466
$^{27}Al^{79}Br$	$^1\Sigma^+$	378.0	1.28	0.15919713	0.00086045	0.11285	2.29481
$^{27}Al^{35}Cl$	$^1\Sigma^+$	481.30	1.95	0.24393012	0.00161113	0.2503	2.13011
$^{27}Al^{19}F$	$^1\Sigma^+$	802.3	4.77	0.5524798	0.0049841	1.0464	1.65437
$^{27}Al^1H$	$^1\Sigma^+$	1682.56	29.09	6.3907	0.1858	356.5	1.6478
$^{27}Al^2H$	$^1\Sigma^+$	1211.95	15.14	3.3186	0.0697	97	1.6463
$^{27}Al^{127}I$	$^1\Sigma^+$	316.1	1.0	0.11769985	0.00055859		2.53710
$^{27}Al^{16}O$	$^2\Sigma^+$	979.23	6.97	0.6414	0.0058	1.08	1.6179
$^{27}Al^{32}S$	$^2\Sigma^+$	617.1	3.33	0.2799	0.0018	0.22	2.029
$^{75}As_2$	$^1\Sigma_g^+$	429.55	1.12	0.10179	0.000333		2.1026
$^{75}As^1H$	$^3\Sigma^-$	2130*		7.3067	0.2117	327	1.52315
$^{75}As^2H$	$^3\Sigma^-$	1484*		3.6688		90	1.5306
$^{75}As^{14}N$	$^1\Sigma^+$	1068.54	5.41	0.54551	0.003366	0.53	1.6184
$^{75}As^{16}O$	$^2\Pi_{1/2}$	967.08	4.85	0.48482	0.003299	0.49	1.6236

SPECTROSCOPIC CONSTANTS OF DIATOMIC MOLECULES (continued)

Molecule	State	ω_e cm^{-1}	$\omega_e x_e$ cm^{-1}	B_e cm^{-1}	α_e cm^{-1}	D_e 10^{-6}cm^{-1}	r_e Å
$^{197}Au_2$	$^1\Sigma_g^+$	190.9	0.42	0.028013	0.0000723	0.00250	2.4719
$^{197}Au^1H$	$^1\Sigma^+$	2305.01	43.12	7.2401	0.2136	279	1.5239
$^{197}Au^2H$	$^1\Sigma^+$	1634.98	21.65	3.6415	0.07614	70.9	1.5238
$^{11}B_2$	$^3\Sigma_g^-$	1051.3	9.35	1.212	0.014		1.590
$^{11}B^{79}Br$	$^1\Sigma^+$	684.31	3.52	0.4894	0.0035	1.00	1.888
$^{11}B^{35}Cl$	$^1\Sigma^+$	840.29	5.49	0.684282	0.006812	1.84	1.71528
$^{11}B^{19}F$	$^1\Sigma^+$	1402.1	11.8	1.516950	0.019056	7.105	1.26267
$^{11}B^1H$	$^1\Sigma^+$	2366.9	49.40	12.021	0.412	1242	1.2324
$^{11}B^2H$	$^1\Sigma^+$	1703.3	28	6.54	0.17	400	1.2324
$^{11}B^{14}N$	$^3\Pi$	1514.6	12.3	1.666	0.025	8.1	1.281
$^{11}B^{16}O$	$^2\Sigma^+$	1885.69	11.81	1.7820	0.0166	6.32	1.2045
$^{11}B^{32}S$	$^2\Sigma^+$	1180.17	6.31	0.7949	0.0061	1.40	1.6092
$^{138}Ba^{79}Br$	$^2\Sigma^+$	193.77	0.41	0.0415082	0.0001219	0.00762	2.84449
$^{138}Ba^{35}Cl$	$^2\Sigma^+$	279.92	0.82	0.08396717	0.00033429	0.03022	2.68276
$^{138}Ba^{19}F$	$^2\Sigma^+$	468.9	1.79	0.2159	0.0012	0.175	2.163
$^{138}Ba^1H$	$^2\Sigma^+$	1168.31	14.50	3.38285	0.06599	112.67	2.23175
$^{138}Ba^2H$	$^2\Sigma^+$	829.77	7.32	1.7071	0.02363	28.77	2.2304
$^{138}Ba^{127}I$	$^2\Sigma^+$	152.14	0.27	0.02680587	0.00006634	0.00333	3.08476
$^{138}Ba^{16}O$	$^1\Sigma^+$	669.76	2.03	0.3126140	0.0013921	0.2724	1.93969
$^{138}Ba^{32}S$	$^1\Sigma^+$	379.42	0.88	0.10331	0.0003188	0.0306	2.5074
$^9Be^{19}F$	$^2\Sigma^+$	1247.36	9.12	1.4889	0.0176	8.28	1.3610
$^9Be^1H$	$^2\Sigma^+$	2060.78	36.31	10.3164	0.3030	1022.1	1.3426
$^9Be^2H$	$^2\Sigma^+$	1530.32	20.71	5.6872	0.1225	313.8	1.3419
$^9Be^{16}O$	$^1\Sigma^+$	1487.32	11.83	1.6510	0.0190	8.20	1.3309
$^9Be^{32}S$	$^1\Sigma^+$	997.94	6.14	0.79059	0.00664	2.00	1.7415
$^{209}Bi_2$	$^1\Sigma_g^+$	172.71	0.34	0.022781	0.000055	0.00150	2.6596
$^{209}Bi^1H$	$^3\Sigma^-$	1635.73	31.6	5.137	0.148	183	1.805
$^{209}Bi^2H$	$^3\Sigma^-$	1173.32	16.1	2.592	0.054	50.6	1.804
$^{79}Br_2$	$^1\Sigma_g^+$	325.32	1.08	0.082107	0.0003187	0.02092	2.2811
$^{79}Br^{35}Cl$	$^1\Sigma^+$	444.28	1.84	0.152470	0.000770	0.07183	2.13607
$^{79}Br^{19}F$	$^1\Sigma^+$	670.75	4.05	0.35584	0.00261	0.401	1.75894
$^{79}Br^{16}O$	$^2\Pi_{3/2}$	779	6.8	0.429598	0.003639	0.523	1.717
$^{12}C_2$	$^1\Sigma_g^+$	1854.71	13.34	1.8198	0.0177	6.92	1.2425
$^{12}C^{35}Cl$	$^2\Pi_{1/2}$	866.72*	6.2	0.6936	0.00672	1.9	1.6450
$^{12}C^{19}F$	$^2\Pi_{1/2}$	1308.1	11.10	1.4172	0.0184	6.5	1.2718
$^{12}C^1H$	$^2\Pi_{1/2}$	2858.5	63.0	14.457	0.534	1450	1.1199
$^{12}C^2H$	$^2\Pi_{1/2}$	2099.8	34.02	7.806	0.208	420	1.1190
$^{12}C^{14}N$	$^2\Sigma^+$	2068.59	13.09	1.8997830	0.0173717	6.4034	1.17181
$^{12}C^{16}O$	$^1\Sigma^+$	2169.81	13.29	1.93128075	0.01750390	6.1216	1.12823
$^{12}C^{31}P$	$^2\Sigma^+$	1239.67	6.86	0.7986	0.00597	1.33	1.562
$^{12}C^{32}S$	$^1\Sigma^+$	1285.15	6.50	0.8200434	0.0059182	1.336	1.53482
$^{12}C^{80}Se$	$^1\Sigma^+$	1035.36	4.86	0.5750	0.00379	0.71	1.67609
$^{40}Ca^{35}Cl$	$^2\Sigma^+$	367.53	1.31	0.1522302	0.0007990	0.1029	2.43676
$^{40}Ca^{19}F$	$^2\Sigma^+$	581.1	2.74	0.339	0.0026	0.45	1.967
$^{40}Ca^1H$	$^2\Sigma^+$	1298.34	19.10	4.2766	0.0970	183.7	2.0025
$^{40}Ca^2H$	$^2\Sigma^+$	910*		2.1769	0.035	47.9	2.002
$^{40}Ca^{127}I$	$^2\Sigma^+$	238.70	0.63	0.0693263	0.0002634	0.0234	2.82859
$^{40}Ca^{16}O$	$^1\Sigma^+$	732.03	4.83	0.444441	0.003282	0.6541	1.8221
$^{40}Ca^{32}S$	$^1\Sigma^+$	462.23	1.78	0.1766757	0.0008270	0.1032	2.31775
$^{114}Cd^1H$	$^2\Sigma^+$	1337.1*		5.323		314	1.781
$^{114}Cd^2H$	$^2\Sigma^+$			2.704		76	1.775
$^{35}Cl_2$	$^1\Sigma_g^+$	559.7	2.68	0.2440	0.0015	0.186	1.988
$^{35}Cl^{19}F$	$^1\Sigma^+$	786.15	6.16	0.516479	0.004358	0.88	1.62831
$^{35}Cl^{16}O$	$^2\Pi_{3/2}$	853.8	5.5	0.62345	0.0058	1.33	1.56963
$^{52}Cr^1H$	$^6\Sigma^+$	1581*	32	6.220	0.179	347	1.656
$^{52}Cr^2H$	$^6\Sigma^+$	1182*		3.14		88.8	1.664
$^{52}Cr^{16}O$	$^5\Pi$	898.4	6.8	0.5231	0.0070		1.615
$^{133}Cs_2$	$^1\Sigma_g^+$	42.02	0.08	0.0127	0.0000264	0.00464	4.47

Molecule	State	ω_e cm^{-1}	$\omega_e x_e$ cm^{-1}	B_e cm^{-1}	α_e cm^{-1}	D_e $10^{-6}cm^{-1}$	r_e Å
$^{133}Cs^{79}Br$	$^1\Sigma^+$	149.66	0.37	0.03606925	0.00012401	0.00838	3.07225
$^{133}Cs^{35}Cl$	$^1\Sigma^+$	214.17	0.73	0.07209149	0.00033756	0.03268	2.90627
$^{133}Cs^{19}F$	$^1\Sigma^+$	352.56	1.62	0.18436969	0.0011756	0.20168	1.34535
$^{133}Cs^1H$	$^1\Sigma^+$	891.0	12.9	2.7099	0.0579	113	2.4938
$^{133}Cs^2H$	$^1\Sigma^+$	619.1*		1.354		20	2.505
$^{133}Cs^{127}I$	$^1\Sigma^+$	119.18	0.25	0.02362736	0.00006826	0.00371	3.31519
$^{133}Cs^{16}O$	$^2\Sigma^+$	357.5*		0.223073	0.001303	0.348	2.3007
$^{63}Cu_2$	$^1\Sigma_g^+$	264.55	1.02	0.10874	0.000614	0.0716	2.2197
$^{63}Cu^{79}Br$	$^1\Sigma^+$	314.8	0.96	0.10192625	0.00045214	0.04274	2.17344
$^{65}Cu^{35}Cl$	$^1\Sigma^+$	415.29	1.58	0.17628802	0.00099647	0.12706	2.05118
$^{63}Cu^{19}F$	$^1\Sigma^+$	622.7	3.95	0.3794029	0.0032298	0.563	1.74493
$^{63}Cu^1H$	$^1\Sigma^+$	1941.26	37.51	7.9441	0.2563	520	1.46263
$^{63}Cu^2H$	$^1\Sigma^+$	1384.14	18.97	4.0381	0.0917	136.2	1.4626
$^{63}Cu^{127}I$	$^1\Sigma^+$	264.5	0.60	0.07328742	0.00028390	0.02244	2.33832
$^{63}Cu^{16}O$	$^2\Pi_{3/2}$	640.17	4.43	0.44454	0.00456	0.85	1.7244
$^{63}Cu^{32}S$	$^2\Pi_{3/2}$	415.0	1.75	0.1891		0.18	2.051
$^{19}F_2$	$^1\Sigma_g^+$	916.64	11.24	0.89019	0.013847	3.3	1.41193
$^{56}Fe^{16}O$	$^5\Delta$	965*		0.650		0.72	1.444
$^{69}Ga^{81}Br$	$^1\Sigma^+$	263.0	0.81	0.081839	0.0003207	0.032	2.35248
$^{69}Ga^{35}Cl$	$^1\Sigma^+$	365.67	1.25	0.1499046	0.0007936	0.1008	2.20169
$^{69}Ga^{19}F$	$^1\Sigma^+$	622.2	3.2	0.3595161	0.0028642	0.50	1.77437
$^{69}Ga^1H$	$^1\Sigma^+$	1604.52	28.77	6.137	0.181	342	1.663
$^{69}Ga^2H$	$^1\Sigma^+$			3.083	0.06	84	1.663
$^{69}Ga^{127}I$	$^1\Sigma^+$	216.38	0.47	0.0569359	0.0001897	0.015770	2.57464
$^{69}Ga^{16}O$	$^2\Sigma$	767.5	6.24	0.4271		0.37	1.744
$^{74}Ge^{79}Br$	$^2\Pi_{1/2}$	295	0.7				
$^{74}Ge^{35}Cl$	$^2\Pi_{1/2}$	407.6	1.36				
$^{72}Ge^1H$	$^2\Pi_{1/2}$	1833.77	37	6.726	0.192	326	1.5880
$^{72}Ge^2H$	$^2\Pi_{1/2}$	1320.09	19	3.415	0.070	83.2	1.5874
$^{74}Ge^{16}O$	$^1\Sigma^+$	986.49	4.47	0.4856981	0.0030787	0.4709	1.62464
$^{74}Ge^{32}S$	$^1\Sigma^+$	575.8	1.80	0.18656576	0.00074910	0.07883	2.01209
$^{74}Ge^{80}Se$	$^1\Sigma^+$	408.7	1.36	0.09634051	0.00028904	0.02207	2.13463
$^{74}Ge^{130}Te$	$^1\Sigma^+$	323.9	0.75	0.06533821	0.00017246	0.012	2.34017
$^1H_2$	$^1\Sigma_g^+$	4401.21	121.34	60.853	3.062	47100	0.74144
$^2H_2$	$^1\Sigma_g^+$	3115.50	61.82	30.444	1.0786	11410	0.74152
$^3H_2$	$^1\Sigma_g^+$	2546.5	41.23	20.335	0.5887		0.74142
$^1H^{81}Br$	$^1\Sigma^+$	2648.97	45.22	8.46488	0.23328	345.8	1.41444
$^2H^{81}Br$	$^1\Sigma^+$	1884.75	22.72	4.245596	0.084	88.32	1.4145
$^1H^{35}Cl$	$^1\Sigma^+$	2990.95	52.82	10.59342	0.30718	531.94	1.27455
$^2H^{35}Cl$	$^1\Sigma^+$	2145.16	27.18	5.448796	0.113292	140	1.27458
$^1H^{19}F$	$^1\Sigma^+$	4138.32	89.88	20.9557	0.798	2151	0.91681
$^2H^{19}F$	$^1\Sigma^+$	2998.19	45.76	11.0102	0.3017	594	0.91694
$^1H^{127}I$	$^1\Sigma^+$	2309.01	39.64	6.4263650	0.1689	206.9	1.60916
$^{202}Hg^1H$	$^2\Sigma^+$	1203.24*		5.3888		395.3	1.7662
$^{202}Hg^2H$	$^2\Sigma^+$	896.12*		2.739		91	1.757
$^{127}I_2$	$^1\Sigma_g^+$	214.50	0.61	0.03737	0.000114	0.0043	2.666
$^{127}I^{79}Br$	$^1\Sigma^+$	268.64	0.81	0.0568325	0.0001969	0.0102	2.46899
$^{127}I^{35}Cl$	$^1\Sigma^+$	384.29	1.50	0.1141587	0.0005354	0.0403	2.32088
$^{127}I^{19}F$	$^1\Sigma^+$	610.24	3.12	0.2797111	0.0018738	0.2356	1.90976
$^{127}I^{16}O$	$^2\Pi_{3/2}$	681.5	4.3	0.34026	0.00270	0.36	1.8676
$^{115}In^{81}Br$	$^1\Sigma^+$	221.0	0.65	0.05489468	0.00018672	0.01350	2.54315
$^{115}In^{35}Cl$	$^1\Sigma^+$	317.39	1.03	0.1090583	0.0005177	0.0515	2.40117
$^{115}In^{19}F$	$^1\Sigma^+$	535.4	2.6	0.2623241	0.0018798	0.252	1.98540
$^{115}In^1H$	$^1\Sigma^+$	1476.0	25.61	4.995	0.143	223	1.8380
$^{115}In^2H$	$^1\Sigma^+$	1048.2	12.4	2.523	0.051	58	1.837
$^{115}In^{127}I$	$^1\Sigma^+$	177.08	0.34	0.03686702	0.00010411	0.00639	2.75364
$^{39}K_2$	$^1\Sigma_g^+$	92.02	0.28	0.056743	0.000165	0.0863	3.9051
$^{39}K^{79}Br$	$^1\Sigma^+$	213	0.80	0.08122109	0.00040481	0.04462	2.82078

SPECTROSCOPIC CONSTANTS OF DIATOMIC MOLECULES (continued)

Molecule	State	ω_e cm^{-1}	$\omega_e x_e$ cm^{-1}	B_e cm^{-1}	α_e cm^{-1}	D_e 10^{-6}cm^{-1}	r_e Å
$^{39}K^{35}Cl$	$^1\Sigma^+$	281	1.30	0.1286348	0.0007899	0.1087	2.66665
$^{39}K^{19}F$	$^1\Sigma^+$	426.26	2.45	0.27993741	0.00233492	0.4829	2.17146
$^{39}K^1H$	$^1\Sigma^+$	983.6	14.3	3.416400	0.085313	163.55	2.243
$^{39}K^2H$	$^1\Sigma^+$	707	7.7	1.754	0.0318	50	2.240
$^{39}K^{127}I$	$^1\Sigma^+$	186.53	0.57	0.06087473	0.00026776	0.02593	3.04784
$^{139}La^{16}O$	$^2\Sigma^+$	812.8	2.22	0.35252001	0.00142365	0.2626	1.82591
$^7Li_2$	$^1\Sigma_g^+$	351.43	2.61	0.67264	0.00704	9.87	2.6729
$^7Li^{79}Br$	$^1\Sigma^+$	563.2	3.5	0.555399	0.005644	2.159	2.17043
$^7Li^{35}Cl$	$^1\Sigma^+$	642.95	4.47	0.7065225	0.0080102	3.409	2.02067
$^7Li^{19}F$	$^1\Sigma^+$	910.57	8.21	1.3452583	0.0202887	11.745	1.56386
$^7Li^1H$	$^1\Sigma^+$	1405.65	23.20	7.51373	0.21665	862	1.59490
$^7Li^2H$	$^1\Sigma^+$	1054.80	12.94	4.23310	0.09155	276	1.5941
$^7Li^{127}I$	$^1\Sigma^+$	496.85	2.85	0.4431766	0.0040862	1.4104	2.39192
$^7Li^{16}O$	$^2\Pi$	814.62	7.78	1.212830	0.017899	0.1079	1.68822
$^{24}Mg_2$	$^1\Sigma_g^+$	51.12	1.64	0.09287	0.00378	1.22	3.891
$^{24}Mg^{35}Cl$	$^2\Sigma^+$	462.12*	2.1	0.2456154	0.0016204	0.2723	2.19639
$^{24}Mg^{19}F$	$^2\Sigma^+$	711.69*	4.9	0.51922	0.00470	1.080	1.7500
$^{24}Mg^1H$	$^2\Sigma^+$	1495.20	31.89	5.8257	0.1859	344	1.7297
$^{24}Mg^2H$	$^2\Sigma^+$	1077.9	16.1	3.0306	0.06289	92	1.7302
$^{24}Mg^{16}O$	$^1\Sigma^+$	784.78	5.26	0.57470436	0.00532377	1.2328	1.74838
$^{55}Mn^1H$	$^7\Sigma$	1548.0	28.8	5.6841	0.1570	303.9	1.7311
$^{55}Mn^2H$	$^7\Sigma$	1103	13.9	2.8957	0.051	79.5	1.7310
$^{14}N_2$	$^1\Sigma_g^+$	2358.57	14.32	1.99824	0.017318	5.76	1.09769
$^{14}N^{79}Br$	$^3\Sigma^-$	691.75	4.72	0.444	0.0040		1.79
$^{14}N^{35}Cl$	$^3\Sigma^-$	827.96	5.30	0.649770	0.006414	1.598	1.61071
$^{14}N^{19}F$	$^3\Sigma^-$	1141.37	8.99	1.2057	0.01492	5.39	1.3170
$^{14}N^1H$	$^3\Sigma^-$	3282.3	78.4	16.6993	0.6490	1709.7	1.0362
$^{14}N^2H$	$^3\Sigma^-$	2398	42	8.7913	0.2531	490.4	1.0361
$^{14}N^{16}O$	$^2\Pi_{1/2}$	1904.20	14.07	1.67195	0.0171	0.5	1.15077
$^{14}N^{32}S$	$^2\Pi_{1/2}$	1218.7	7.28	0.769602	0.0064	1.2	1.4940
$^{23}Na_2$	$^1\Sigma_g^+$	159.13	0.72	0.154707	0.008736	0.581	3.0789
$^{23}Na^{79}Br$	$^1\Sigma^+$	302	1.5	0.1512533	0.0009410	0.1554	2.50204
$^{23}Na^{35}Cl$	$^1\Sigma^+$	366	2.05	0.2180631	0.0016248	0.3120	2.36080
$^{23}Na^{19}F$	$^1\Sigma^+$	535.66	3.57	0.4369011	0.0045580	1.163	1.92595
$^{23}Na^1H$	$^1\Sigma^+$	1172.2	19.72	4.9033634	0.1370919	343.40	1.88654
$^{23}Na^2H$	$^1\Sigma^+$	826.1*		2.557089	0.051600	93.46	1.88654
$^{23}Na^{127}I$	$^1\Sigma^+$	258	1.1	0.1178056	0.0006478	0.0973	2.71145
$^{23}Na^{16}O$	$^2\Pi$	492.3		0.424630	0.004506	1.2638	2.05155
$^{93}Nb^{16}O$	$^4\Sigma^-$	989.0	3.8	0.4321	0.0021	0.22	1.691
$^{58}Ni^1H$	$^2\Delta_{5/2}$	1926.6	38	7.700	0.23	481	1.476
$^{58}Ni^2H$	$^2\Delta_{5/2}$	1390.1	19	3.992	0.092	130	1.465
$^{16}O_2$	$^3\Sigma_g^-$	1580.19	11.98	1.44563	0.0159	4.839	1.20752
$^{16}O^1H$	$^2\Pi_{3/2}$	3737.76	84.88	18.911	0.7242	1938	0.96966
$^{16}O^2H$	$^2\Pi_{3/2}$	2720.24	44.05	10.021	0.276	537.4	0.9698
$^{31}P_2$	$^1\Sigma_g^+$	780.77	2.84	0.30362	0.00149	0.188	1.8934
$^{31}P^{35}Cl$	$^3\Sigma^-$	551.38	2.23	0.2528748	0.0015119	0.2124	2.01461
$^{31}P^{19}F$	$^3\Sigma^-$	846.75	4.49	0.5665	0.00456		1.58938
$^{31}P^1H$	$^3\Sigma^-$	2365.2	44.5	8.5371	0.2514	436	1.42140
$^{31}P^2H$	$^3\Sigma^-$	1699.2	23.0	4.4081	0.0928	116	1.4220
$^{31}P^{14}N$	$^1\Sigma^+$	1337.24	6.98	0.7864854	0.0055364	1.091	1.49087
$^{31}P^{16}O$	$^2\Pi_{1/2}$	1233.34	6.56	0.7337	0.0055	1.3	1.4759
$^{208}Pb_2$		110.5	0.35				
$^{208}Pb^{79}Br$	$^2\Pi_{1/2}$	207.5	0.50				
$^{208}Pb^{35}Cl$	$^2\Pi_{1/2}$	303.9	0.88				
$^{208}Pb^{19}F$	$^2\Pi_{1/2}$	502.73	2.28	0.22875	0.001473	0.183	2.0575
$^{208}Pb^1H$	$^2\Pi_{1/2}$	1564.1	29.75	4.971	0.144	201	1.839
$^{208}Pb^{16}O$	$^1\Sigma^+$	720.96	3.52	0.30730373	0.00190977	0.2138	1.92181
$^{208}Pb^{32}S$	$^1\Sigma^+$	429.17	1.26	0.11632307	0.00043510	0.03418	2.28678

Molecule	State	ω_e cm^{-1}	$\omega_e x_e$ cm^{-1}	B_e cm^{-1}	α_e cm^{-1}	D_e 10^{-6}cm^{-1}	r_e Å
$^{208}Pb^{80}Se$	$^1\Sigma^+$	277.6	0.51	0.05059953	0.00012993	0.0070	2.40218
$^{208}Pb^{130}Te$	$^1\Sigma^+$	212.0	0.43	0.03130774	0.00006743	0.0027	2.59492
$^{195}Pt^{12}C$	$^1\Sigma^+$	1051.13	4.86	0.53044	0.003273	0.546	1.6767
$^{195}Pt^1H$	$^2\Delta_{5/2}$	2294.68*	46	7.1963	0.1996	261	1.52852
$^{195}Pt^2H$	$^2\Delta_{5/2}$	1644.3*	23	3.640	0.071	66	1.524
$^{85}Rb^{79}Br$	$^1\Sigma^+$	169.46	0.46	0.04752798	0.00018596	0.01496	2.94474
$^{85}Rb^{35}Cl$	$^1\Sigma^+$	228	0.92	0.0876404	0.0004537	0.04947	2.78673
$^{85}Rb^{19}F$	$^1\Sigma^+$	376	1.9	0.2106640	0.0015228	0.2684	2.27033
$^{85}Rb^1H$	$^1\Sigma^+$	936.9	14.21	3.020	0.072	123	2.367
$^{85}Rb^{127}I$	$^1\Sigma^+$	138.51	0.33	0.03283293	0.00010946	0.00738	3.17688
$^{85}Rb^{16}O$	$^2\Sigma^+$	388.4*		0.246481	0.002174	0.397	2.25420
$^{32}S_2$	$^3\Sigma_g^-$	725.65	2.84	0.2955	0.001570	0.19	1.8892
$^{32}S^{19}F$	$^2\Pi_{3/2}$			0.552174			1.60058
$^{32}S^1H$	$^2\Pi_{3/2}$	2711.6	59.9	9.5995	0.2785	480.6	1.34066
$^{32}S^2H$	$^2\Pi_{3/2}$	1885	31	4.95130	0.10308	130	1.34049
$^{32}S^{16}O$	$^3\Sigma^-$	1149.2	5.6	0.7208171	0.005737	1.134	1.48109
$^{121}Sb^{35}Cl$	$^3\Sigma^-$	374.7	0.6				
$^{121}Sb^{19}F$	$^3\Sigma^-$	605.0	2.6	0.2792	0.0020	0.23	1.918
$^{121}Sb^1H$	$^3\Sigma^-$			5.684		240	1.723
$^{121}Sb^2H$	$^3\Sigma^-$			2.8782		45	1.7194
$^{121}Sb^{14}N$	$^1\Sigma^+$	942.0	5.6				
$^{121}Sb^{16}O$	$^2\Pi_{1/2}$	816	4.2	0.3580	0.0022	0.270	1.826
$^{45}Sc^{19}F$	$^1\Sigma^+$	735.6	3.8	0.3950	0.00266		1.788
$^{80}Se_2$	$^3\Sigma_g^-$	385.30	0.96	0.08992	0.000288	0.024	2.166
$^{80}Se^1H$	$^2\Pi_{3/2}$	2400*		8.02	0.23	330	1.48
$^{80}Se^2H$	$^2\Pi_{3/2}$	1708*		3.94			1.48
$^{80}Se^{16}O$	$^3\Sigma^-$	914.69	4.52	0.4655	0.00323	0.5	1.648
$^{28}Si_2$	$^3\Sigma_g^-$	510.98	2.02	0.2390	0.0014	0.21	2.246
$^{28}Si^{35}Cl$	$^2\Pi_{1/2}$	535.60	2.17	0.2561	0.0016	0.25	2.058
$^{28}Si^{19}F$	$^2\Pi_{1/2}$	857.19	4.73	0.5812	0.00494	1.07	1.6011
$^{28}Si^1H$	$^2\Pi_{1/2}$	2041.80	35.51	7.4996	0.2190	397	1.5201
$^{28}Si^2H$	$^2\Pi_{1/2}$	1469.32	18.23	3.8840	0.0781	105.4	1.5199
$^{28}Si^{14}N$	$^2\Sigma^+$	1151.4	6.47	0.7311	0.00565	1.2	1.572
$^{28}Si^{16}O$	$^1\Sigma^+$	1241.54	5.97	0.7267521	0.0050379	0.9923	1.50975
$^{28}Si^{32}S$	$^1\Sigma^+$	749.64	2.58	0.30352788	0.00147308	0.201	1.92926
$^{28}Si^{80}Se$	$^1\Sigma^+$	580.0	1.78	0.1920117	0.0007767	0.0842	2.05832
$^{120}Sn^{79}Br$	$^2\Pi_{1/2}$	247.2	0.6				
$^{120}Sn^{35}Cl$	$^2\Pi_{1/2}$	351.1	1.06	0.1117	0.0004		2.361
$^{118}Sn^{19}F$	$^2\Pi_{1/2}$	577.6	2.69	0.2727	0.0014	0.26	1.944
$^{120}Sn^1H$	$^2\Pi_{1/2}$			5.31488		207.5	1.78146
$^{120}Sn^2H$	$^2\Pi_{1/2}$	1188.0*		2.6950	0.049	53.4	1.7770
$^{120}Sn^{127}I$	$^2\Pi_{1/2}$	199.0	0.6				
$^{120}Sn^{16}O$	$^1\Sigma^+$	822.13	3.72	0.35571998	0.00214432	0.26638	1.83251
$^{120}Sn^{32}S$	$^1\Sigma^+$	487.26	1.36	0.13686139	0.00050563	0.0424	2.20898
$^{120}Sn^{80}Se$	$^1\Sigma^+$	331.2	0.74	0.0649978	0.0001705	0.011	2.32557
$^{120}Sn^{130}Te$	$^1\Sigma^+$	259.5	0.50	0.04247917	0.00009543	0.0055	2.52280
$^{88}Sr^{79}Br$	$^2\Sigma^+$	216.60	0.52	0.0541847	0.0001827	0.01356	2.73522
$^{88}Sr^{35}Cl$	$^2\Sigma^+$	302.3	0.95				
$^{88}Sr^{19}F$	$^2\Sigma^+$	502.4	2.3	0.2505346	0.0015513	0.2498	2.07537
$^{88}Sr^1H$	$^2\Sigma^+$	1206.2	17.0	3.6751	0.0814	135	2.1456
$^{88}Sr^2H$	$^2\Sigma^+$	841	8.6	1.8609	0.0292	34.7	2.1449
$^{88}Sr^{127}I$	$^2\Sigma^+$	173.77	0.35	0.0367097	0.0001060	0.00655	2.94364
$^{88}Sr^{16}O$	$^1\Sigma^+$	653.5	3.96	0.33798	0.00219	0.36	1.91983
$^{181}Ta^{16}O$	$^2\Delta_{3/2}$	1028.69	3.51	0.40284	0.00182	0.2450	1.68746
$^{130}Te_2$	$^3\Sigma_g^-$	247.07	0.51	0.039681	0.000106	0.0044	2.5574
$^{130}Te^1H$	$^2\Pi_{3/2}$			5.56			1.74
$^{130}Te^{16}O$	0^+	797.11	4.00	0.3554	0.00237	0.27	1.825
$^{232}Th^{16}O$	$^1\Sigma^+$	895.77	2.39	0.332644	0.001302	0.1833	1.84032

Molecule	State	ω_e cm⁻¹	$\omega_e x_e$ cm⁻¹	B_e cm⁻¹	α_e cm⁻¹	D_e 10^{-6}cm⁻¹	r_e Å
$^{48}Ti^{16}O$	$^3\Delta_1$	1009.02	4.50	0.53541	0.00301	0.603	1.6202
$^{205}Tl^{81}Br$	$^1\Sigma^+$	192.10	0.39	0.0423899	0.0001276	0.0083	1.61817
$^{205}Tl^{35}Cl$	$^1\Sigma^+$	284.71	0.86	0.09139702	0.00039784	0.0377	2.48483
$^{205}Tl^{19}F$	$^1\Sigma^+$	476.86	2.24	0.22315014	0.00150380	0.1955	2.08439
$^{205}Tl^{1}H$	$^1\Sigma^+$	1390.7	22.7	4.806	0.154	254	1.870
$^{205}Tl^{2}H$	$^1\Sigma^+$	987.7	12.04	2.419	0.057	60	1.869
$^{205}Tl^{127}I$	$^1\Sigma^+$	150*		0.0271676	0.0000664	0.0036	2.81361
$^{51}V^{16}O$	$^4\Sigma^-$	1011.3	4.86	0.54825	0.00352	0.6	1.5893
$^{89}Y^{35}Cl$	$^1\Sigma$	380.7	1.3	0.1160	0.0003	0.09	2.41
$^{89}Y^{19}F$	$^1\Sigma^+$	631.29	2.50	0.29042	0.00163	0.237	1.9257
$^{89}Y^{16}O$	$^2\Sigma^+$	861.0	2.9	0.3881	0.0018	0.32	1.790
$^{174}Yb^{1}H$	$^2\Sigma^+$	1249.54	21.06	3.9931	0.0957	161.8	2.0526
$^{174}Yb^{2}H$	$^2\Sigma^+$	886.6	10.57	2.01162	0.03425	41.60	2.0516
$^{64}Zn^{35}Cl$	$^2\Sigma$	390.5	1.6				
$^{64}Zn^{19}F$	$^2\Sigma$	628	3.5				
$^{64}Zn^{1}H$	$^2\Sigma^+$	1607.6	55.14	6.6794	0.2500	466	1.5949
$^{64}Zn^{2}H$	$^2\Sigma^+$	1072	28	3.350		124	1.6054
$^{64}Zn^{127}I$	$^2\Sigma$	223.4	0.6				
$^{90}Zr^{16}O$	$^1\Sigma^+$	969.8	4.9	0.42263	0.0023	0.319	1.7116

INFRARED CORRELATION CHARTS

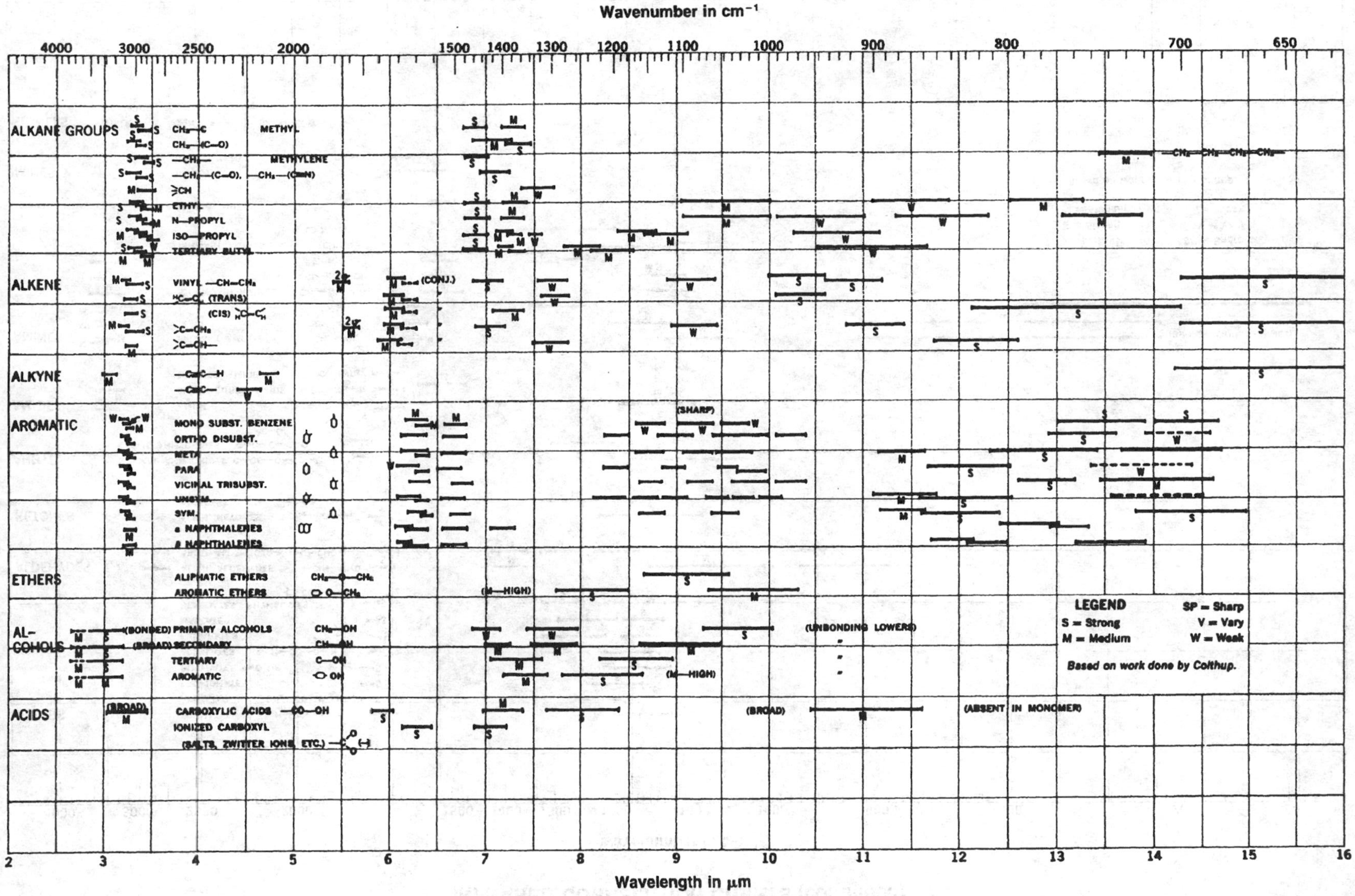

INFRARED CORRELATION CHARTS (continued)

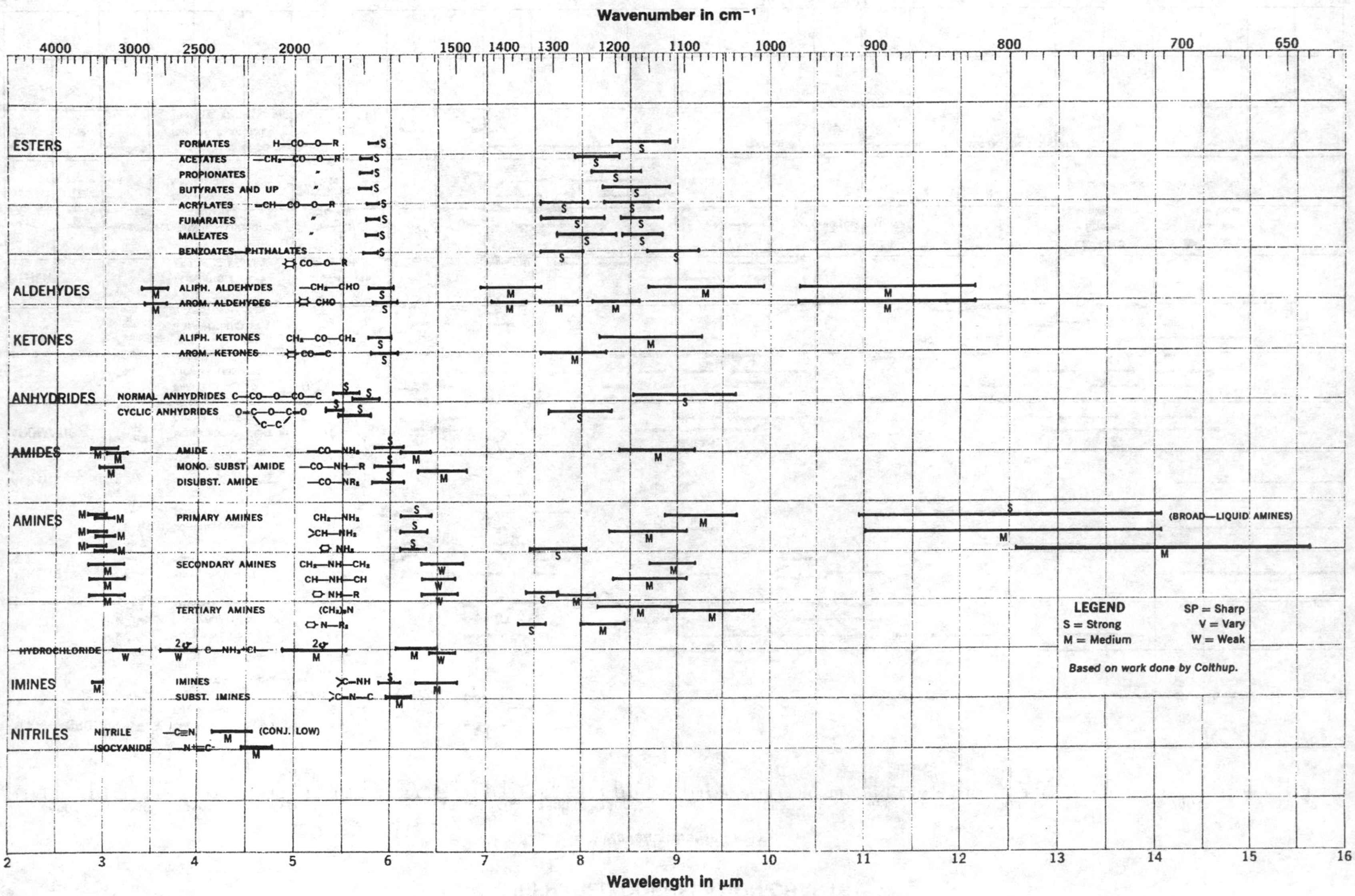

INFRARED CORRELATION CHARTS (continued)

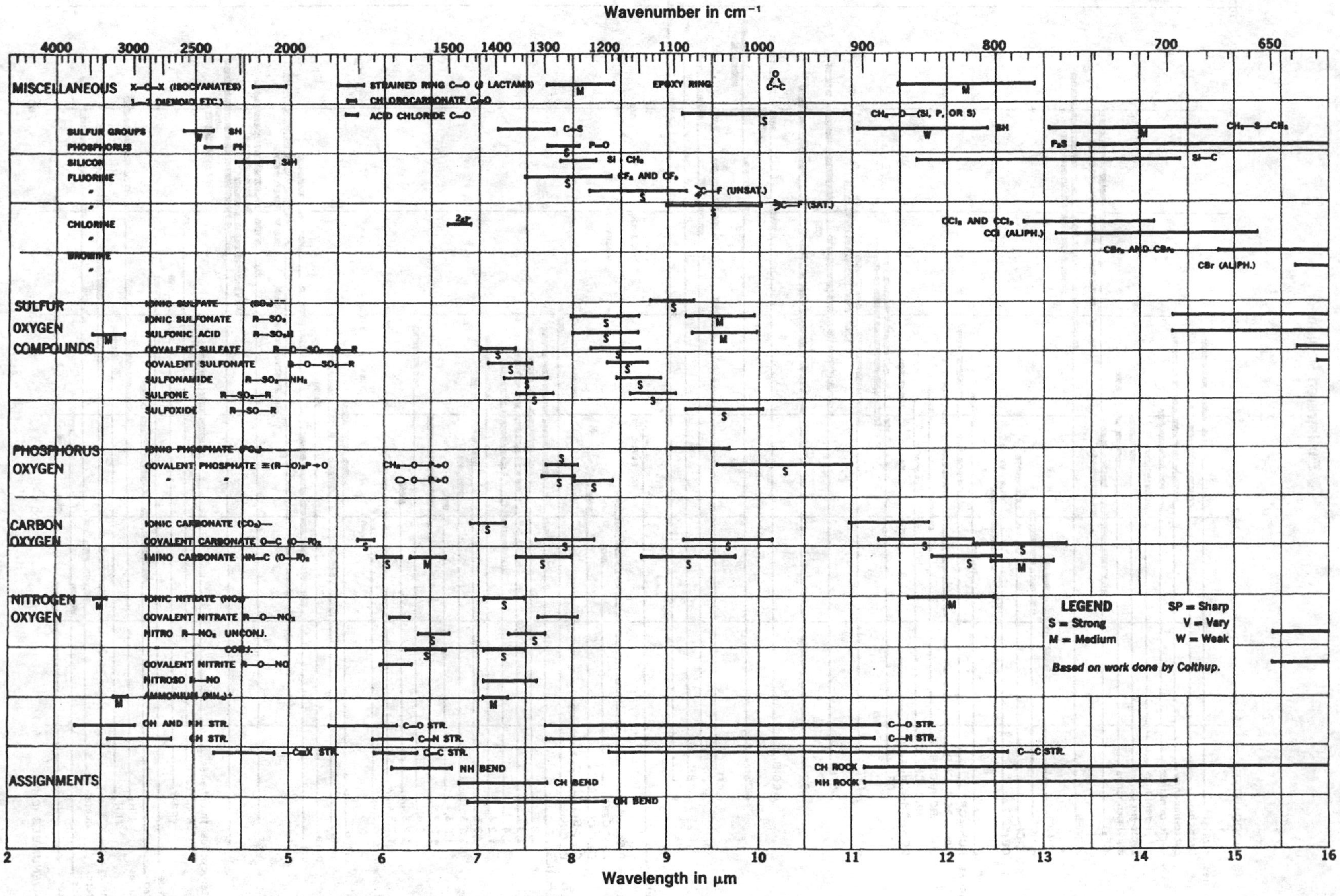

INFRARED CORRELATION CHARTS (continued)

Far Infrared Region

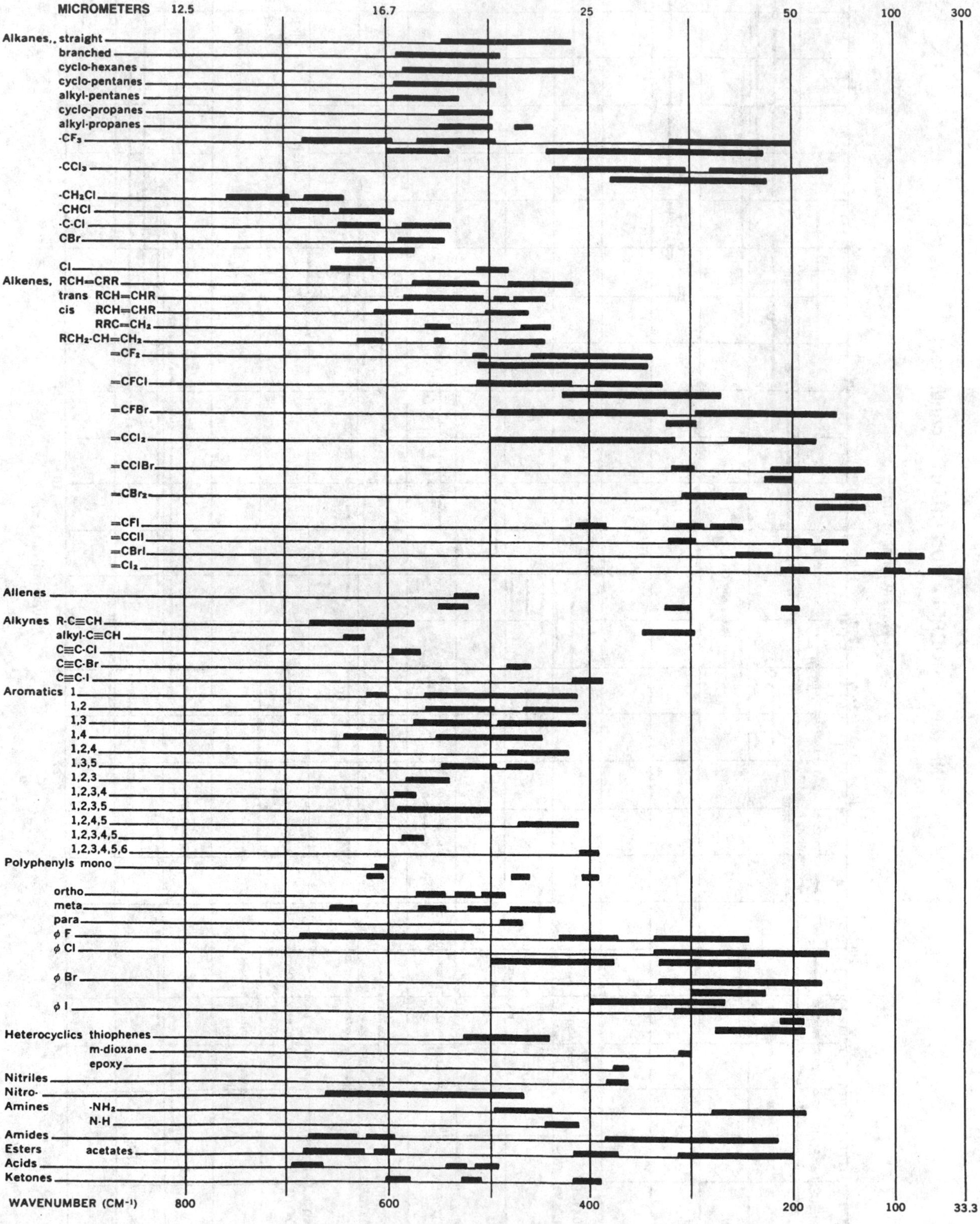

INFRARED CORRELATION CHARTS (continued)

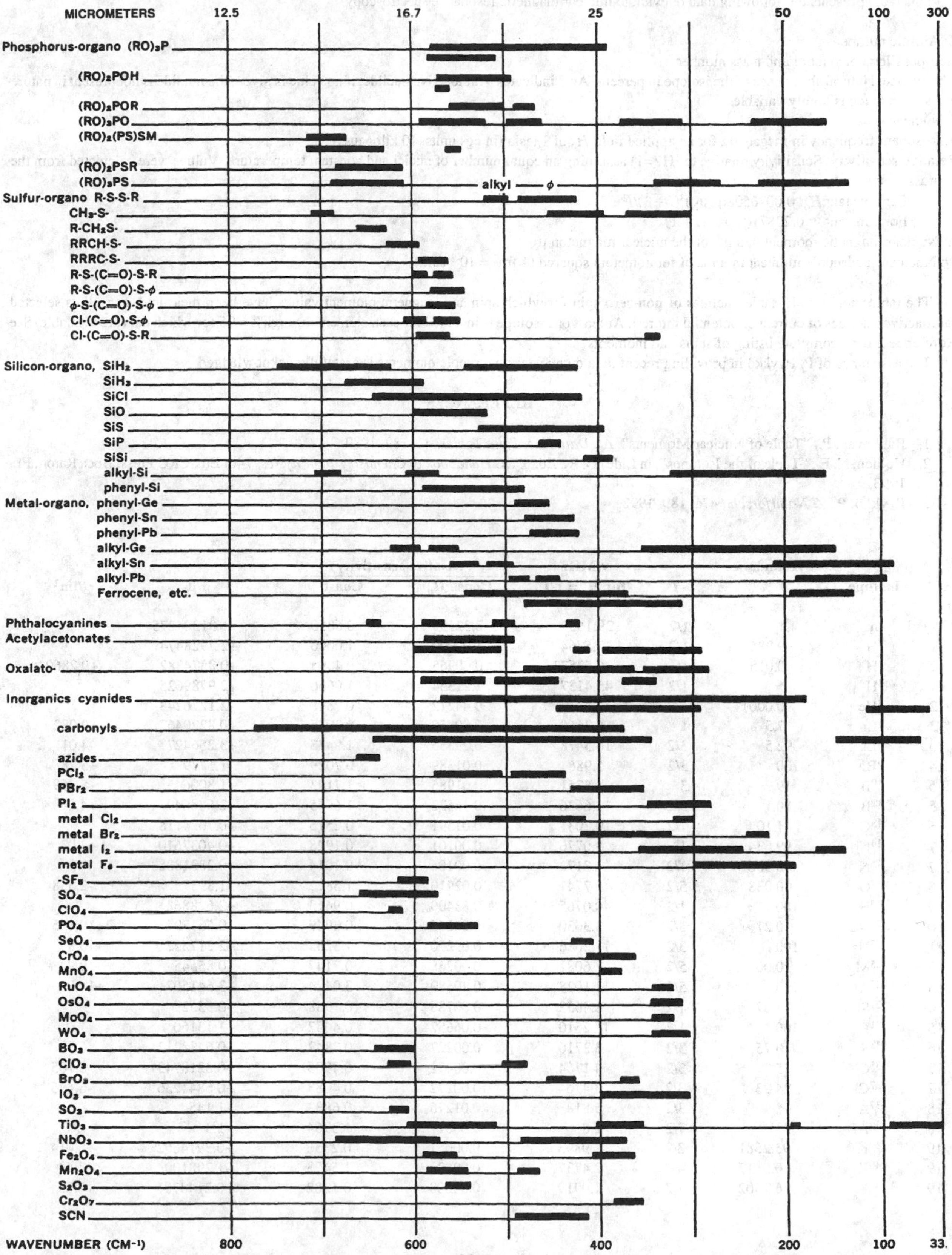

NUCLEAR SPINS, MOMENTS, AND OTHER DATA RELATED TO NMR SPECTROSCOPY

This table presents the following data relevant to nuclear magnetic resonance spectroscopy:

Z: Atomic number
Isotope: Element symbol and mass number
Abundance: Natural abundance of the isotope in percent. An * indicates a radioactive nuclide; if no value is given, the nuclide is not present in nature or its abundance is highly variable.
I: Nuclear spin
ν: Resonant frequency in megahertz for an applied field H_o of 1 tesla (in cgs units, 10 kilogauss)
Relative sensitivity: Sensitivity relative to ^{1}H (=1) assuming an equal number of nuclei and constant temperature. Values were calculated from the expressions:

For constant H_o: $0.0076508(\mu/\mu_N)^3(I+1)/I^2$
For constant ν: $0.23871(\mu/\mu_N)\ (I+1)$

μ: Nuclear magnetic moment in units of the nuclear magneton μ_N
Q: Nuclear quadrupole moment in units of femtometers squared (1 fm^2 = 10^{-2} barn)

The table includes all stable nuclides of non-zero spin for which spin and magnetic moment values have been measured, as well as selected radioactive nuclides of current or potential interest. At least one isotope is included for each element through Z = 95 for which data are available. See Reference 2 for a complete listing of spins and moments.

The assistance of P. Pyykkö in providing recent data on nuclear quadrupole moments is gratefully acknowledged.

REFERENCES

1. Raghavan, P., "Table of Nuclear Moments", *At. Data Nuc. Data Tables,* 42, 189, 1989.
2. Holden, N. E., "Table of the Isotopes", in Lide, D. R., Ed., *CRC Handbook of Chemistry and Physics,* 74th Ed., CRC Press, Boca Raton, FL, 1993.
3. Pyykkö, P., *Z. Naturforsch.*, 47a, 189, 1992.

Z	Isotope	Abundance %	I	ν/MHz for H_o = 1 T	Relative sensitivity Const. H_o	Relative sensitivity Const. ν	μ/μ$_N$	Q/fm^2
1	^{1}n	*	1/2	29.1639	0.32139	0.6850	-1.91304275	
1	^{1}H	99.985	1/2	42.5764	1.00000	1.0000	+2.7928474	
1	^{2}H	0.015	1	6.53573	0.00965	0.4093	+0.8574382	+0.2860
1	^{3}H	*	1/2	45.4137	1.21354	1.0666	+2.9789625	
2	^{3}He	0.0001	1/2	32.4352	0.44212	0.7618	-2.1276248	
3	^{6}Li	7.5	1	6.2660	0.00850	0.3925	+0.8220467	-0.082
3	^{7}Li	92.5	3/2	16.5478	0.29355	1.9433	+3.256427	-4.01
4	^{9}Be	100	3/2	5.986	0.01389	0.7029	-1.1779	+5.288
5	^{10}B	19.9	3	4.5751	0.01985	1.7193	+1.800645	+8.459
5	^{11}B	80.1	3/2	13.6626	0.16522	1.6045	+2.688649	+4.059
6	^{13}C	1.10	1/2	10.7081	0.01591	0.2515	+0.7024118	
7	^{14}N	99.634	1	3.0776	0.00101	0.1928	+0.4037610	+2.02
7	^{15}N	0.366	1/2	4.3172	0.00104	0.1014	-0.2831888	
8	^{17}O	0.038	5/2	5.7741	0.02910	1.5822	-1.89379	-2.558
9	^{19}F	100	1/2	40.0765	0.83400	0.9413	+2.628868	
10	^{21}Ne	0.27	3/2	3.3630	0.00246	0.3949	-0.661797	+10.155
11	^{23}Na	100	3/2	11.2686	0.09270	1.3233	+2.217522	+10.89
12	^{25}Mg	10.00	5/2	2.6082	0.00268	0.7147	-0.85545	+19.94
13	^{27}Al	100	5/2	11.1028	0.20689	3.0424	+3.641507	+14.03
14	^{29}Si	4.67	1/2	8.4653	0.00786	0.1988	-0.55529	
15	^{31}P	100	1/2	17.2510	0.06652	0.4052	+1.13160	
16	^{33}S	0.75	3/2	3.2716	0.00227	0.3842	+0.6438212	-6.78
17	^{35}Cl	75.77	3/2	4.1764	0.00472	0.4905	+0.8218743	-8.165
17	^{37}Cl	24.23	3/2	3.4764	0.00272	0.4083	+0.6841236	-6.435
18	^{37}Ar	*	3/2	5.818	0.01276	0.6833	+1.145	
18	^{39}Ar	*	7/2	2.8	0.00617	1.3964	-1.3	
19	^{39}K	93.2581	3/2	1.9893	0.00051	0.2336	+0.3914662	+6.01
19	^{40}K	0.0117	4	2.4737	0.00523	1.5493	-1.298100	-7.49
19	^{41}K	6.7302	3/2	1.0919	0.00008	0.1282	+0.2148701	+7.33

NUCLEAR SPINS, MOMENTS, AND OTHER DATA RELATED TO NMR SPECTROSCOPY (continued)

Z	Isotope	Abundance %	I	ν/MHz for $H_0 = 1$ T	Relative sensitivity Const. H_0	Const. ν	μ/μ_N	Q/fm^2
20	^{43}Ca	0.135	7/2	2.8688	0.00642	1.4150	-1.31726	-4.08
21	^{45}Sc	100	7/2	10.3588	0.30244	5.1093	+4.756487	-22
22	^{47}Ti	7.3	5/2	2.4040	0.00210	0.6587	-0.78848	+29
22	^{49}Ti	5.5	7/2	2.4047	0.00378	1.1861	-1.10417	+24
23	^{50}V	0.250	6	4.2504	0.05571	5.5904	+3.345689	+21
23	^{51}V	99.750	7/2	11.2130	0.38360	5.5306	+5.1487057	-5.2
24	^{53}Cr	9.501	3/2	2.4114	0.00091	0.2832	-0.47454	-15
25	^{55}Mn	100	5/2	10.5760	0.17881	2.8980	+3.46872	+33
26	^{57}Fe	2.1	1/2	1.3815	0.00003	0.0324	+0.0906230	
27	^{59}Co	100	7/2	10.077	0.27841	4.9702	+4.627	+42
28	^{61}Ni	1.140	3/2	3.8113	0.00359	0.4476	-0.75002	+16.2
29	^{63}Cu	69.17	3/2	11.2979	0.09342	1.3268	+2.22329	-22.0
29	^{65}Cu	30.83	3/2	12.1027	0.11484	1.4213	+2.38167	-20.4
30	^{67}Zn	4.1	5/2	2.6693	0.00287	0.7314	+0.875479	+15.0
31	^{69}Ga	60.108	3/2	10.2475	0.06971	1.2034	+2.01659	+17.0
31	^{71}Ga	39.892	3/2	13.0204	0.14300	1.5291	+2.56227	+10.0
32	^{73}Ge	7.73	9/2	1.4897	0.00141	1.1546	-0.8794677	-17.3
33	^{75}As	100	3/2	7.3148	0.02536	0.8590	+1.439475	+31.4
34	^{77}Se	7.63	1/2	8.1566	0.00703	0.1916	+0.535042	
35	^{79}Br	50.69	3/2	10.7039	0.07945	1.2570	+2.106400	+33.1
35	^{81}Br	49.31	3/2	11.5381	0.09951	1.3550	+2.270562	+27.6
36	^{83}Kr	11.5	9/2	1.6442	0.00190	1.2744	-0.970669	+25.3
37	^{85}Rb	72.165	5/2	4.1253	0.01061	1.1304	+1.35303	+27.4
37	^{87}Rb	27.835	3/2	13.9807	0.17703	1.6418	+2.75124	+13.2
38	^{87}Sr	7.00	9/2	1.8524	0.00272	1.4358	-1.093603	+33.5
39	^{89}Y	100	1/2	2.0949	0.00012	0.0492	-0.1374154	
40	^{91}Zr	11.22	5/2	3.9747	0.00949	1.0891	-1.30362	-20.6
41	^{93}Nb	100	9/2	10.4520	0.48821	8.1011	+6.1705	-32
42	^{95}Mo	15.92	5/2	2.7874	0.00327	0.7638	-0.9142	-2.2
42	^{97}Mo	9.55	5/2	2.8462	0.00349	0.7799	-0.9335	+25.5
43	^{99}Tc	*	9/2	9.63	0.38174	7.4633	+5.6847	-12.9
44	^{99}Ru	12.7	5/2	1.9553	0.00113	0.5358	-0.6413	+7.9
44	^{101}Ru	17.0	5/2	2.192	0.00159	0.6005	-0.7188	+45.7
45	^{103}Rh	100	1/2	1.3476	0.00003	0.0317	-0.08840	
46	^{105}Pd	22.33	5/2	1.957	0.00113	0.5364	-0.642	+66.0
47	^{107}Ag	51.839	1/2	1.7330	0.00007	0.0407	-0.1136796	
47	^{109}Ag	48.161	1/2	1.9924	0.00010	0.0468	-0.1306906	
48	^{111}Cd	12.80	1/2	9.0689	0.00966	0.2130	-0.5948861	
48	^{113}Cd	12.22	1/2	9.4868	0.01106	0.2228	-0.6223009	
49	^{113}In	4.3	9/2	9.3652	0.35121	7.2588	+5.5289	+79.9
49	^{115}In	95.7	9/2	9.3854	0.35348	7.2744	+5.5408	+81
50	^{115}Sn	0.34	1/2	14.0074	0.03561	0.3290	-0.91883	
50	^{117}Sn	7.68	1/2	15.2606	0.04605	0.3584	-1.00104	
50	^{119}Sn	8.59	1/2	15.9656	0.05273	0.3750	-1.04728	
51	^{121}Sb	57.36	5/2	10.2549	0.16302	2.8100	+3.3634	-36
51	^{123}Sb	42.64	7/2	5.5530	0.04659	2.7389	+2.5498	-49
52	^{123}Te	0.908	1/2	11.2346	0.01837	0.2639	-0.7369478	
52	^{125}Te	7.139	1/2	13.5451	0.03220	0.3181	-0.8885051	
53	^{127}I	100	5/2	8.5776	0.09540	2.3504	+2.813273	-78.9
54	^{129}Xe	26.4	1/2	11.8601	0.02162	0.2786	-0.7779763	
54	^{131}Xe	21.2	3/2	3.5158	0.00282	0.4129	+0.6918619	-12.0
55	^{133}Cs	100	7/2	5.6232	0.04838	2.7735	+2.582025	-0.37
56	^{135}Ba	6.592	3/2	4.2581	0.00500	0.5001	+0.837943	+16.0
56	^{137}Ba	11.23	3/2	4.7633	0.00700	0.5594	+0.937365	+24.5
57	^{138}La	* 0.0902	5	5.6614	0.09404	5.3188	+3.713646	+45
57	^{139}La	99.9098	7/2	6.0610	0.06058	2.9895	+2.7830455	+20
58	^{137}Ce	*	3/2	4.62	0.00641	0.5431	0.91	

Z	Isotope	Abundance %	I	ν/MHz for H_o = 1 T	Relative sensitivity Const. H_o	Const. ν	μ/μ_N	Q/fm²
58	^{139}Ce	*	3/2	4.62	0.00641	0.5431	0.91	
58	^{141}Ce	*	7/2	2.37	0.00364	1.1708	1.09	
59	^{141}Pr	100	5/2	13.0355	0.33483	3.5720	+4.2754	-5.9
60	^{143}Nd	12.18	7/2	2.319	0.00339	1.1440	-1.065	-63
60	^{145}Nd	8.30	7/2	1.429	0.00079	0.7047	-0.656	-33
61	^{143}Pm	*	5/2	11.6	0.23510	3.1748	3.8	
61	^{147}Pm	*	7/2	5.7	0.04940	2.7928	+2.6	+70
62	^{147}Sm	15.0	7/2	1.7747	0.00152	0.8753	-0.8149	-26
62	^{149}Sm	13.8	7/2	1.4631	0.00085	0.7216	-0.6718	+9.4
63	^{151}Eu	47.8	5/2	10.5854	0.17929	2.9006	+3.4718	+90.3
63	^{153}Eu	52.2	5/2	4.6744	0.01544	1.2809	+1.5331	+241
64	^{155}Gd	14.80	3/2	1.317	0.00015	0.1546	-0.2591	+127
64	^{157}Gd	15.65	3/2	1.727	0.00033	0.2028	-0.3399	+135
65	^{159}Tb	100	3/2	10.23	0.06945	1.2019	+2.014	+143.2
66	^{161}Dy	18.9	5/2	1.4653	0.00048	0.4015	-0.4806	+247
66	^{163}Dy	24.9	5/2	2.0507	0.00130	0.5619	+0.6726	+265
67	^{165}Ho	100	7/2	9.0881	0.20423	4.4825	+4.173	+358
68	^{167}Er	22.95	7/2	1.2281	0.00050	0.6057	-0.5639	+357
69	^{169}Tm	100	1/2	3.531	0.00057	0.0829	-0.2316	
70	^{171}Yb	14.3	1/2	7.5259	0.00552	0.1768	+0.49367	
70	^{173}Yb	16.12	5/2	2.0730	0.00135	0.5680	-0.67989	+280
71	^{175}Lu	97.41	7/2	4.8624	0.03128	2.3983	+2.2327	+349
71	^{176}Lu	* 2.59	7	3.451	0.03975	6.0516	+3.169	+497
72	^{177}Hf	18.606	7/2	1.7281	0.00140	0.8524	+0.7935	+336
72	^{179}Hf	13.629	9/2	1.0856	0.00055	0.8414	-0.6409	+379
73	^{180}Ta	0.012	9	4.04	0.10251	11.3862	4.77	
73	^{181}Ta	99.988	7/2	5.1625	0.03744	2.5463	+2.3705	+317
74	^{183}W	14.3	1/2	1.7956	0.00008	0.0422	+0.1177847	
75	^{185}Re	37.40	5/2	9.717	0.13870	2.6627	+3.1871	+218
75	^{187}Re	*62.60	5/2	9.817	0.14300	2.6900	+3.2197	+207
76	^{187}Os	1.6	1/2	0.9856	0.00001	0.0231	+0.06465189	
76	^{189}Os	16.1	3/2	3.3535	0.00244	0.3938	+0.659933	+85.6
77	^{191}Ir	37.3	3/2	0.766	0.00003	0.0899	+0.1507	+81.6
77	^{193}Ir	62.7	3/2	0.832	0.00004	0.0977	+0.1637	+75.1
78	^{195}Pt	33.8	1/2	9.2920	0.01039	0.2182	+0.60952	
79	^{197}Au	100	3/2	0.7406	0.00003	0.0870	+0.145746	+54.7
80	^{199}Hg	16.87	1/2	7.7121	0.00594	0.1811	+0.5058855	
80	^{201}Hg	13.18	3/2	2.8468	0.00149	0.3343	-0.5602257	+38.6
81	^{203}Tl	29.524	1/2	24.7310	0.19598	0.5809	+1.6222579	
81	^{205}Tl	70.476	1/2	24.9742	0.20182	0.5866	+1.6382146	
82	^{207}Pb	22.1	1/2	9.0338	0.00955	0.2122	+0.59258	
83	^{209}Bi	100	9/2	6.9628	0.14433	5.3967	+4.1106	-50
84	^{209}Po	*	1/2	11.7	0.02096	0.2757	+0.77	
86	^{211}Rn	*	1/2	9.16	0.00997	0.2152	+0.601	
87	^{223}Fr	*	3/2	5.94	0.01362	0.6982	+1.17	+117
88	^{223}Ra	*	3/2	1.3746	0.00017	0.1614	+0.2705	+119
88	^{225}Ra	*	1/2	11.187	0.01814	0.2627	-0.7338	
89	^{227}Ac	*	3/2	5.6	0.01131	0.6564	+1.1	+170
90	^{229}Th	*	5/2	1.40	0.00042	0.3843	+0.46	+430
91	^{231}Pa	*	3/2	10.2	0.06903	1.1995	2.01	-172
92	^{235}U	* 0.7200	7/2	0.83	0.00015	0.4082	-0.38	+493.6
93	^{237}Np	*	5/2	9.57	0.13264	2.6234	+3.14	+388.6
94	^{239}Pu	*	1/2	3.09	0.00038	0.0727	+0.203	
95	^{243}Am	*	5/2	4.91	0.01788	1.3451	+1.61	+421

PROTON NMR CHEMICAL SHIFTS FOR CHARACTERISTIC ORGANIC STRUCTURES

The chart below summarizes the range of chemical shifts for protons in several classes or organic compounds and substituent groups. The chemical shifts δ are given in parts per million relative to tetramethylsilane.

REFERENCE

Mohacsi, E., *J. Chem. Edu.*, 41, 38, 1964 (with permission).

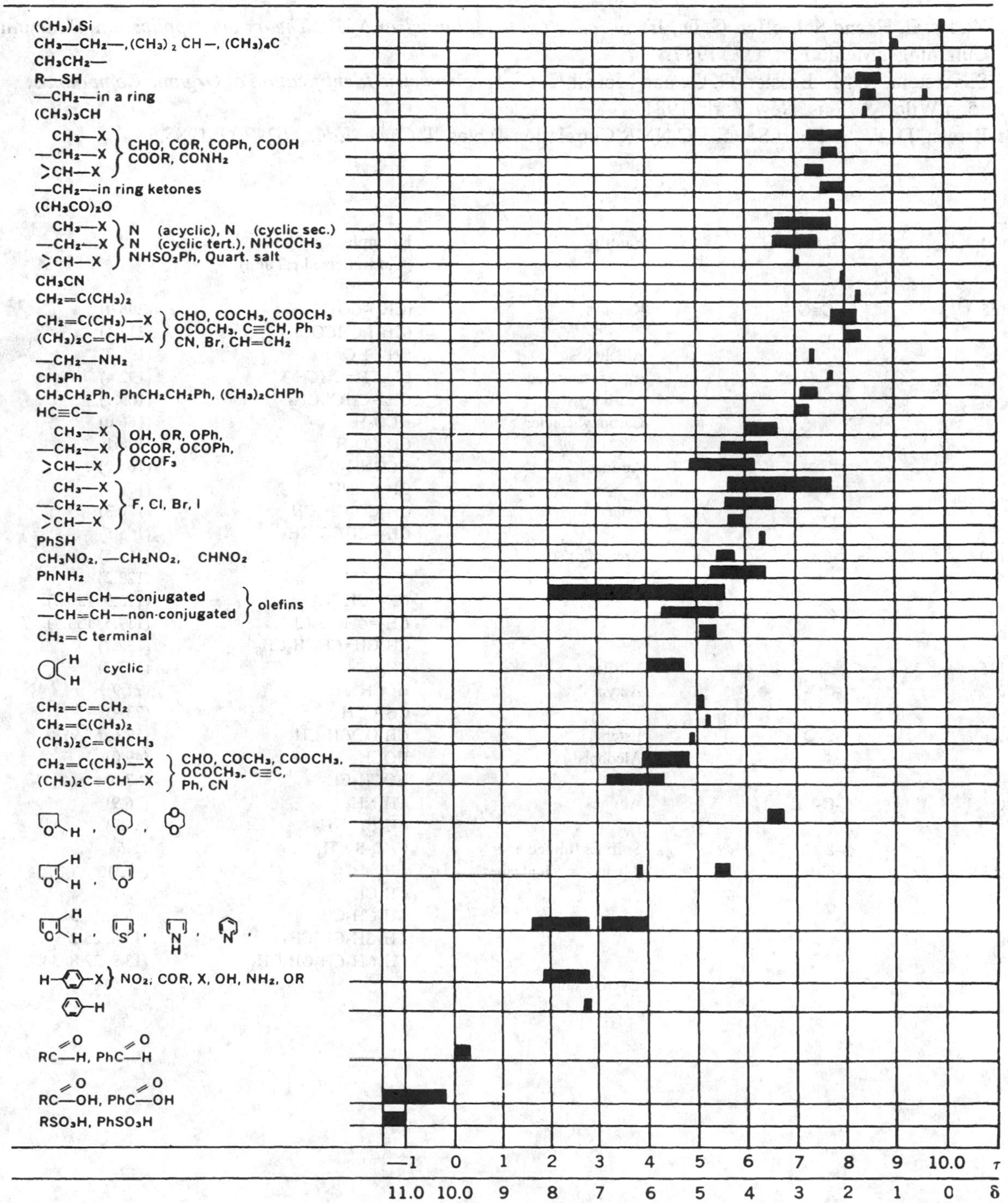

^{13}C-NMR ABSORPTIONS OF MAJOR FUNCTIONAL GROUPS

The table below lists the range of ^{13}C chemical shifts δ in parts per million relative to tetramethylsilane, in descending order, for various functional groups. Examples of simple compounds for each family are given to illustrate the correlations. The shifts for the carbons of interest, which are italicized, are given in parentheses; when two or more values appear, they refer to the sequence of italicized carbon atoms from left to right in the formula.

REFERENCES

1. Yoder, C. H. and Schaeffer, C. D., Jr., *Introduction to Multinuclear NMR: Theory and Application*, Benjamin/ Cummings, Menlo Part, CA, 1987.
2. Silverstein, R. M., Bassler, G. C., and Morrill, T. C., *Spectrometric Identification of Organic Compounds*, John Wiley & Sons, New York, 1981.
3. Brown, D. W., A Short Set of ^{13}C NMR Correlation Tables, *J. Chem. Educ.*, 62, 209, 1985.

δ (ppm)	Group	Family	Example (δ of italicized carbon)	
220-165	>C=O	Ketones	$(CH_3)_2CO$	(206.0)
			$(CH_3)_2CHCOCH_3$	(212.1)
		Aldehydes	CH_3CHO	(199.7)
		α,β-Unsaturated	$CH_3CH{=}CHCHO$	(192.4)
		carbonyls	$CH_2{=}CHCOCH_3$	(169.9)
		Carboxylic acids	HCO_2H	(166.0)
			CH_3CO_2H	(178.1)
		Amides	$HCONH_2$	(165.0)
			CH_3CONH_2	(172.7)
		Esters	$CH_3CO_2CH_2CH_3$	(170.3)
			$CH_2{=}CHCO_2CH_3$	(165.5)
140-120	>C=C<	Aromatic	C_6H_6	(128.5)
		Alkenes	$CH_2{=}CH_2$	(123.2)
			$CH_2{=}CHCH_3$	(115.9, 136.2)
			$CH_2{=}CHCH_2Cl$	(117.5, 133.7)
			$CH_3CH{=}CHCH_2CH_3$	(132.7)
125-115	-CN	Nitriles	CH_3-CN	(117.7)
80-70	-CC-	Alkynes	$HCCH$	(71.9)
			CH_3CCH_3	(73.9)
70-45	-C-O	Esters	$CH_3OOCH_2CH_3$	(57.6, 67.9)
		Alcohols	$HOCH_3$	(49.0)
			$HOCH_2CH_3$	(57.0)
40-20	$-C-NH_2$	Amines	CH_3NH_2	(26.9)
			$CH_3CH_2NH_2$	(35.9)
30-15	$-S-CH_3$	Sulfides (thioethers)	C_6H_5-S-CH_3	15.6
30-(-2.3)	-C-H	Alkanes, cycloalkanes	CH_4	(-2.3)
			CH_3CH_3	(5.7)
			$CH_3CH_2CH_3$	(15.8, 16.3)
			$CH_3CH_2CH_2CH_3$	(13.4, 25.2)
			$CH_3CH_2CH_2CH_2CH_3$	(13.9, 22.8, 34.7)
		Cyclohexane		(26.9)

Section 10
Atomic, Molecular, and Optical Physics

Line Spectra of the Elements **10-1**
NIST Atomic Transition Probability Tables **10-88**
Electron Affinities **10-147**
Atomic and Molecular Polarizabilities **10-160**
Ionization Potentials of Atoms and Atomic Ions **10-175**
Ionization Energies of Gas-Phase Molecules **10-178**
X-Ray Atomic Energy Levels **10-196**
Electron Binding Energies of the Elements **10-200**
Natural Width of X-Ray Lines **10-206**
Photon Attenuation Coefficients **10-207**
Classification of Electromagnetic Radiation **10-212**
Black Body Radiation **10-214**
Characteristics of Infrared Detectors **10-216**
Refractive Index and Transmittance of Representative Glasses **10-217**
Index of Refraction of Water **10-218**
Index of Refraction of Liquids for Calibration Purposes **10-219**
Index of Refraction of Air **10-220**
Characteristics of Laser Sources **10-221**
Infrared Laser Frequencies **10-228**
Infrared and Far-Infrared Absorption Frequency Standards **10-236**

LINE SPECTRA OF THE ELEMENTS

Joseph Reader and Charles H. Corliss

The original tables from which this table was derived were prepared under the auspices of the Committee on Line Spectra of the Elements of the National Academy of Sciences-National Research Council. The table contains the outstanding spectral lines of neutral (I) and singly ionized (II) atoms of the elements from hydrogen through plutonium (Z = 1-94); selected strong lines from doubly ionized (III), triply ionized (IV), and quadruply ionized (V) atoms are also included. Listed are lines that appear in emission from the vacuum ultraviolet to the far infrared. These lines were selected from much larger lists in such a way as to include the stronger observed lines in each spectral region. A more extensive list may be found in Reference 1.

The data were compiled by the following contributors.

J. G. Conway — Lawrence Berkeley Laboratory
C. H. Corliss — National Bureau of Standards
R. D. Cowan — Los Alamos Scientific Laboratory
C. R. Cowley — University of Michigan
Henry M. and Hannah Crosswhite — Argonne National Laboratory
S. P. Davis — University of California, Berkeley
V. Kaufman — National Bureau of Standards
R. L. Kelly — Naval Postgraduate School
J. F. Kielkopf — University of Louisville
W. C. Martin — National Bureau of Standards
T. K. McCubbin — Pennsylvania State University
L. J. Radziemski — Los Alamos Scientific Laboratory
J. Reader — National Bureau of Standards
C. J. Sansonetti — National Bureau of Standards
G. V. Shalimoff — Lawrence Berkeley Laboratory
R. W. Stanley — Purdue University
J. O. Stoner, Jr. — University of Arizona
H. H. Stroke — New York University
D. R. Wood — Wright State University
E. F. Worden — Lawrence Livermore Laboratory
J. J. Wynne — International Business Machines Corporation
R. Zalubas — National Bureau of Standards

All wavelengths are given in Ångstrom units (10^{-10} m). Below 2000 Å the wavelengths are in vacuum; above 2000 Å the wavelengths are in air. Wavelengths given to three decimal places have an uncertainty of less than 0.001 Å and are therefore suitable for calibration purposes. In the air region, the elements used most commonly for calibration are Ne, Ar, Kr, Fe, Th, and Hg; in the vacuum region, the most common are C, N, O, Si, Cu.

All data refer to natural isotopic abundance of the elements except that Kr I and Kr II lines below 11,000 Å given to three decimal places are for ^{86}Kr. A separate table for ^{198}Hg contains accurately known wavelengths that are frequently used for calibration.

A large number of the lines for neutral and singly ionized atoms were extracted from the National Bureau of Standards (NBS) *Tables of Spectral-Line Intensities* (Reference 2). The intensities of these lines represent quantitative estimates of relative line strengths that take account of varying detection sensitivity at different wavelengths. They are on a linear scale.For nearly all of the other lines the intensities represent qualitative estimates of the relative strengths of lines not greatly separated in wavelength. Because different observers frequently use different scales for their intensity estimates, these intensities are useful only as a rough indication of the appearance of a spectrum. In some cases the intensity scale is not intended to be linear. In the first and second spectra the intensities of the lines of the singly ionized atom (II) relative to those of the neutral atom (I) should be used with caution, inasmuch as the concentration of ions in a light source depends greatly on the excitation conditions.

Descriptive symbols that follow the wavelength have the following meanings:

c — complex
d — line consists of two unresolved lines
h — hazy
l — shaded to longer wavelengths
s — shaded to shorter wavelengths
p — perturbed by a close line
r — easily reversed
w — wide

The table is arranged alphabetically by element name (not symbol); for each element the lines are listed by wavelength. References to the sources of data for each element are given at the end of the table, starting on page **10**-82.

GENERAL REFERENCES

1. Reader, J., Corliss, C. H., Wiese, W. L., and Martin, G. A., *Tables of Line Spectra of the Elements, Part 1. Wavelengths and Intensities*, Nat. Stand. Ref. Data Sys.- Nat. Bur. Standards (U.S.), No. 68, 1980.
2. Meggers, W. F., Corliss, C. H., and Scribner, B. F., *Tables of Spectral Line Intensities, Part 1. Arranged by Elements*, Nat. Bur. Stand. (U.S.), Monograph 145, 1975.
3. Fuhr, J. R., Martin, W. C., Musgrove, A., Sugar, J., and Wiese, W. L., "NIST Atomic Spectroscopic Database" ver. 1.1, January 1996. *NIST Physical Reference Data*, National Institute of Standards and Technology, Gaithersburg, MD. Available at the WWW address: http://physics.nist.gov/PhysRefData/contents.html

Intensity	Wavelength/Å	
Actinium		
Ac Z = 89		
2000 h	2952.55	III
2000 h	3392.78	III
3000	3487.59	III
2000 s	3863.12	II
3000 s	4088.44	II
3000 s	4168.40	II
100	4179.98	I
20	4183.12	I
20	4194.40	I
20 l	4384.53	I
20	4396.71	I
2000 h	4413.09	III
20	4462.73	I
3000 h	4569.87	III
1000	5910.85	II
20	6359.86	I
20 l	6691.27	I
Aluminum		
Al Z = 13		
900	125.53	V
800	126.07	V
800	130.41	V
1000	130.85	V
900	131.00	V
900	131.44	V
800	160.07	IV
1000	278.69	V
900	281.39	V
70	486.884	III
30	486.912	III
250	511.138	III
150	511.191	III
500	560.317	III
200	560.433	III
100	670.068	III
200	671.118	III
500	695.829	III
400	696.217	III
200	725.683	III
300	726.915	III
400	855.034	III
500	856.746	III
400	892.024	III
50	893.887	III
450	893.897	III
800	1042.17	IV
50	1191.812	II
900	1237.19	IV
900	1257.62	IV
800	1264.18	IV
1000	1272.76	IV
150	1350.18	II
800	1384.13	III
800	1447.51	IV
800	1494.79	IV
1000	1526.14	V
800	1537.54	IV
800	1539.830	II
1000	1557.25	IV
100	1569.385	II
900	1582.04	IV
800	1584.46	IV
125	1596.059	II
700	1605.766	III
100	1611.814	III
800	1611.874	III
150	1625.627	II
800	1639.06	IV
100	1644.235	II
100	1644.809	II
1000	1670.787	II
100	1686.250	II
800	1719.440	II
500	1721.244	II
900	1721.271	II
500	1724.952	II
900	1724.984	II
350	1760.104	II
300	1761.975	II
290	1763.00	I
500	1763.869	II
700	1763.952	II
450	1765.64	I
300	1765.815	II
450	1766.38	I
400	1767.731	II
450	1769.14	I
1000	1818.56	IV
600	1828.588	II
400	1832.837	II
250	1834.808	II
1000	1854.716	III
300	1855.929	II
700	1858.026	II
120	1859.980	II
1000	1862.311	II
600	1862.790	III
200	1929.978	II
150	1931.048	II
200	1932.377	II
400	1934.503	II
150	1934.713	II
300	1935.840	III
200	1935.949	III
150	1936.907	II
220	1939.261	II
700	1990.531	II
150	2016.052	II
150	2016.234	II
100	2016.368	II
200	2074.008	II
700	2094.264	II
150	2094.744	II
300	2094.791	II
100	2095.104	II
200	2095.141	II
400	2269.10	I
120	2269.22	I
140	2321.56	I
460	2367.05	I
110	2367.61	I
110	2368.11	I
180	2369.30	I
140	2370.22	I
160	2372.07	I
850	2373.12	I
170	2373.35	I
110	2373.57	I
240	2567.98	I
480	2575.10	I
110	2637.70	II
150	2652.48	I
200	2660.39	I
160	2669.17	II
650	2816.19	II
150	3041.28	II
360	3050.07	I
450	3057.14	I
150	3074.64	II
4500 r	3082.153	I
7200 r	3092.710	I
1800 r	3092.839	I
150	3428.92	II
150	3443.64	I
900	3492.23	IV
800	3508.46	IV
450	3586.56	II
360	3587.07	II
290	3587.45	II
870	3601.63	III
220	3651.06	II
110	3651.10	II
150	3654.98	II
290	3655.00	II
450	3900.68	II
4500 r	3944.006	I
9000 r	3961.520	I
110	3995.86	II
290	4226.81	II
870	4529.19	III
150	4585.82	II
110	4588.19	II
550	4666.80	II
110	4898.76	II
110	4902.77	II
150	5280.21	II
290	5283.77	II
150	5285.85	II
110	5312.32	II
220	5316.07	II
150	5371.84	II
180	5557.06	I
110	5557.95	I
450	5593.23	II
1200	5696.60	III
1000	5722.73	III
110	5853.62	II
220	5971.94	II
290	6001.76	II
220	6001.88	II
450	6006.42	II
150	6061.11	II
290	6068.43	II
110	6068.53	II
450	6073.23	II
110	6181.57	II
150	6181.68	II
290	6182.28	II
220	6182.45	II
450 h	6183.42	II
450	6201.52	II
360	6201.70	II
290	6226.18	II
360	6231.78	II
450	6243.36	II
450	6335.74	II
360	6696.02	I
230	6698.67	I
110	7361.57	I
140	7362.30	I
230	7835.31	I
290	7836.13	I
110	8075.35	I
290	8640.70	II
360	8772.87	I
450	8773.90	I
110	8828.91	I
180	8841.28	I
140	8923.56	I
150	9290.65	II
110	9290.75	II
150	10076.29	II
110	10768.36	I
140	10782.04	I
110	10872.98	I
230	10891.73	I
450	11253.19	I
570	11254.88	I
570	13123.41	I
450	13150.76	I
230	16718.96	I
300	16750.56	I
140	16763.36	I
300	21093.04	I
360	21163.75	I
Antimony		
Sb Z = 51		
15	722.86	III
15	732.33	III
	861.5	IV
4	876.84	II
4	921.07	II
6	983.57	II
15	999.62	III
6	1001.13	II
6	1009.43	II
40	1011.94	III
6	1052.21	II
8	1056.27	II
8	1057.32	II
40	1065.90	III
6	1073.81	II
30	1075.82	III
	1087.6	IV
8	1104.32	V
30	1151.49	III
40	1157.74	III
	1199.1	IV
50	1205.20	III
50	1210.64	III
12	1226.00	V
6	1230.30	II
8	1274.98	II
20	1306.69	III
8	1327.40	II
6	1358.04	II
8	1384.70	II
20	1404.18	III
6	1407.83	II
8	1436.49	II
20 r	1486.57	I
40 h	1491.36	I

Intensity	Wavelength/Å	
	1499.2	IV
12	1505.70	V
50 r	1512.57	I
12	1524.47	V
120 r	1532.74	I
80 r	1535.06	I
6	1565.51	II
8	1576.11	II
7	1581.36	II
80 r	1599.96	I
10	1606.98	II
200 w	1612.8	I
100 w	1623.3	I
20	1657.04	II
100 w	1662.6	I
15	1673.89	III
15	1711.84	III
80 r	1716.93	I
150 r	1717.45	I
150 r	1723.43	I
15	1725.33	III
100 r	1736.19	I
100 h	1765.76	I
100 r	1780.87	I
100 r	1788.24	I
150	1800.18	I
50 r	1810.50	I
80 r	1814.20	I
100	1829.50	I
50 r	1868.17	I
300 r	1871.15	I
150 r	1882.56	I
100	1927.08	I
200 r	1950.39	I
60 r	2029.49	I
70 r	2039.77	I
150 r	2049.57	I
1000 r	2068.33	I
100	2079.56	I
50 r	2098.41	I
80 r	2118.48	I
100 r	2127.39	I
50 r	2137.05	I
100 r	2139.69	I
10	2141.80	II
50 r	2141.83	I
100 r	2144.86	I
1500 r	2175.81	I
250 r	2179.19	I
200 r	2201.32	I
300 r	2208.45	I
150 r	2220.73	I
100	2221.98	I
120 r	2224.93	I
300 r	2262.51	I
120	2288.98	I
150 r	2293.44	I
300 r	2306.46	I
2500 r	2311.47	I
150	2315.89	I
400 h	2373.67	I
300 h	2383.64	I
100	2395.22	I
150	2422.13	I
250	2426.35	I
400 r	2445.51	I
400	2478.32	I
150	2480.44	I
100	2510.54	I
2000 r	2528.52	I
15	2528.54	II
10	2567.75	II
150	2574.06	I
15	2590.13	III
1500 r	2598.05	I
500 r	2598.09	I
300 r	2612.31	I
12	2617.17	III
200 r	2652.60	I
20	2669.39	III
300 r	2670.64	I
200 r	2682.76	I
120	2692.25	I
150 r	2718.90	I
400 r	2769.95	I
1000 r	2877.92	I
15	2980.96	II
500 r	3029.83	I
600 r	3232.52	I
20	3241.28	II
700 r	3267.51	I
15	3498.46	II
25	3637.80	II
250	3637.83	I
20	3722.78	II
200 r	3722.79	I
20	3850.22	II
200	4033.55	I
20	4033.56	II
20	4133.63	II
15	4140.54	II
15	4195.17	II
20	4219.07	II
20	4314.32	II
15	4514.50	II
30	4596.90	II
20	4599.09	II
15	4604.77	II
30	4647.32	II
20	4675.74	II
40	4711.26	II
20	4757.81	II
20	4765.36	II
30	4784.03	II
20	4802.01	II
20	4832.82	II
20	4877.24	II
15	4947.40	II
15	5044.56	II
20	5238.94	II
20	5354.24	II
40 h	5556.10	I
100 l	5632.02	I
30	5639.75	II
60 h	5830.34	I
100	6005.21	II
20	6053.41	II
30	6079.80	II
50	6130.04	II
20	6154.94	II
20	6611.49	I
30	6647.44	II
30 h	7648.28	I
80	7844.44	I
200	7924.65	I
60	8411.69	I
150	8572.64	I
100	8619.55	I
400	9518.68	I
400	9949.14	I
200	10078.49	I
300	10261.01	I
200	10585.60	I
1000	10677.41	I
800	10741.94	I
80	10794.11	I
600	10839.73	I
200	10868.58	I
400	10879.55	I
300	11012.79	I
150	11266.23	I
5	12116.06	I

Argon
Ar $Z = 18$

Intensity	Wavelength/Å	
3	336.56	V
3	337.56	V
6	338.00	V
2	338.43	V
2	339.01	V
3	339.89	V
3	350.88	V
4	396.87	IV
4	398.55	IV
2	436.67	V
5	446.00	V
8	446.95	V
4	447.53	V
18	449.06	V
4	449.49	V
3	458.12	V
2	458.98	V
6 p	461.23	V
3	462.42	V
7	463.94	V
30	487.227	II
50	490.650	II
30	490.701	II
30	519.327	II
3	522.09	V
5	524.19	V
6	527.69	V
30	542.912	II
200	543.203	II
70	547.461	II
2	554.50	V
70	556.817	II
5	558.48	V
70	573.362	II
30	576.736	II
70	580.263	II
30	583.437	II
70	597.700	II
30	602.858	II
30	612.372	II
6	623.77	IV
3	635.12	V
500	661.867	II
30	664.562	II
200	666.011	II
1000	670.946	II
3000	671.851	II
70	676.242	II
30	677.952	II
30	679.218	II
200	679.401	II
10	683.28	IV
7	688.39	IV
12 p	689.01	IV
6	699.41	IV
8	700.28	IV
3	705.35	V
5	709.20	V
4	715.60	V
3	715.65	V
200	718.090	II
3000	723.361	II
2	725.11	V
500	725.548	II
70	730.930	II
200	740.269	II
200	744.925	II
70	745.322	II
4	754.20	IV
5	761.47	IV
12	769.15	III
5	800.57	IV
10	801.09	IV
10	801.41	IV
5	801.91	IV
20	802.859	I
100	806.471	I
60	806.869	I
30	807.218	I
40	807.653	I
50	809.927	I
120	816.232	I
70	816.464	I
80	820.124	I
4	822.16	V
120	825.346	I
120	826.365	I
5	827.05	V
3	827.35	V
150	834.392	I
4 p	834.88	V
100	835.002	I
2	836.13	V
15	840.03	IV
100	842.805	I
20	843.77	IV
25	850.60	IV
180	866.800	I
150	869.754	I
10	871.10	III
9	875.53	III
180 r	876.058	I
12	878.73	III
8	879.62	III
180 r	879.947	I
9	883.18	III
10	887.40	III
150	894.310	I
5	900.36	IV
9	901.17	IV

Intensity	Wavelength/Å	
1000	919.781	II
1000	932.054	II
1000 r	1048.220	I
500 r	1066.660	I
7	1669.67	III
7	1673.42	III
7	1675.48	III
9	1914.40	III
7	1915.56	III
10	2125.16	III
15	2133.87	III
10	2138.59	III
10	2148.73	III
15	2166.19	III
10	2168.26	III
20	2170.23	III
25	2177.22	III
8	2184.06	III
10	2188.22	III
15	2192.06	III
7	2248.73	III
10	2279.10	III
7	2281.22	III
7	2282.21	III
12	2293.03	III
4	2299.72	IV
10	2300.85	III
15	2302.17	III
9	2317.00	III
15	2317.47	III
12	2318.04	III
10	2319.13	III
10	2319.37	III
9	2345.17	III
7	2351.67	III
9	2360.26	III
10	2395.63	III
12	2399.15	III
10	2413.20	III
7	2415.61	III
10	2418.82	III
5	2420.456	II
12	2423.52	III
12	2423.93	III
7	2443.69	III
8	2447.71	IV
8	2472.95	III
7	2476.10	III
12	2488.86	III
12	2513.28	IV
10	2516.789	II
6	2518.40	IV
9	2525.69	IV
10	2534.709	II
15	2562.087	II
12	2562.17	IV
10	2568.07	IV
7	2569.53	IV
12	2599.47	IV
10	2608.06	IV
7	2608.44	IV
12	2615.68	IV
6	2619.98	IV
12	2621.36	IV
12	2624.92	IV
7	2631.90	III
15	2640.34	IV
10	2654.63	III
8	2674.02	III
9	2678.38	III
9	2682.63	IV
10	2724.84	III
14	2757.92	IV
7	2762.23	III
10	2776.26	IV
12	2784.47	IV
14	2788.96	IV
7	2797.11	IV
16	2809.44	IV
10	2830.25	IV
7	2842.88	III
8	2855.29	III
6	2874.40	IV
9	2884.12	III
25	2891.612	II
12	2913.00	IV
11	2926.33	IV
200	2942.893	II
100	2979.050	II
10	3010.02	III
12	3024.05	III
50	3033.508	II
6	3037.98	IV
12	3054.82	III
10	3064.77	III
8	3077.40	IV
10	3078.15	III
50	3093.402	II
7	3110.41	III
7	3127.90	III
8	3200.37	I
20	3243.689	II
25	3285.85	III
25	3293.640	II
20	3301.88	III
20	3307.228	II
15	3311.25	III
7	3319.34	I
7	3323.59	III
25	3336.13	III
20	3344.72	III
25	3350.924	II
15	3358.49	III
7	3361.28	III
7	3373.47	I
25	3376.436	II
25	3388.531	II
15	3391.85	III
7	3393.73	I
7	3417.49	III
9	3424.25	III
8	3438.04	III
7	3461.07	I
9	3471.32	III
70	3476.747	II
20	3478.232	II
20	3480.55	III
50	3491.244	II
100	3491.536	II
12	3499.67	III
15	3503.58	III
70	3509.778	II
8	3511.12	III
70	3514.388	II
70	3545.596	II
70	3545.845	II
7	3554.306	I
100	3559.508	II
100	3561.030	II
70	3576.616	II
25	3581.608	II
50	3582.355	II
70	3588.441	II
7	3606.522	I
25	3622.138	II
20	3639.833	II
35	3718.206	II
70	3729.309	II
50	3737.889	II
150	3765.270	II
50	3766.119	II
20	3770.369	I
20	3770.520	II
25	3780.840	II
20	3795.37	III
25	3803.172	II
50	3809.456	II
7	3834.679	I
70	3850.581	II
10	3858.32	III
35	3868.528	II
7	3907.84	III
35	3925.719	II
50	3928.623	II
25	3932.547	II
70	3946.097	II
7	3947.505	I
35	3948.979	I
8	3960.53	III
20	3979.356	II
35	3994.792	II
50	4013.857	II
6	4023.60	III
50	4033.809	II
20	4035.460	II
150	4042.894	II
50	4044.418	I
100	4052.921	II
200	4072.005	II
70	4072.385	II
25	4076.628	II
35	4079.574	II
25	4082.387	II
150	4103.912	II
300	4131.724	II
5	4146.70	III
35	4156.086	II
400	4158.590	I
50	4164.180	I
35	4179.297	II
50	4181.884	I
100	4190.713	I
50	4191.029	I
200	4198.317	I
400	4200.674	I
25	4218.665	II
25	4222.637	II
25	4226.988	II
100	4228.158	II
100	4237.220	II
25	4251.185	I
200	4259.362	I
100	4266.286	I
70	4266.527	II
150	4272.169	I
550	4277.528	II
20	4282.898	II
100	4300.101	I
25	4300.650	II
70	4309.239	II
200	4331.200	II
50	4332.030	II
100	4333.561	I
50	4335.338	I
25	4345.168	I
800	4348.064	II
50	4352.205	II
25	4362.066	II
50	4367.832	II
200	4370.753	II
70	4371.329	II
50	4375.954	II
150	4379.667	II
50	4385.057	II
70	4400.097	II
200	4400.986	II
400	4426.001	II
150	4430.189	II
50	4430.996	II
50	4433.838	II
20	4439.461	II
35	4448.879	II
100	4474.759	II
200	4481.811	II
100	4510.733	I
20	4522.323	I
20	4530.552	II
400	4545.052	II
20	4564.405	II
400	4579.350	II
400	4589.898	II
15	4596.097	I
550	4609.567	II
7	4628.441	I
35	4637.233	II
400	4657.901	II
15	4702.316	I
20	4721.591	II
550	4726.868	II
50	4732.053	II
300	4735.906	II
800	4764.865	II
550	4806.020	II
150	4847.810	II
50	4865.910	II
800	4879.864	II
70	4889.042	II
20	4904.752	II
35	4933.209	II
200	4965.080	II
50	5009.334	II
70	5017.163	II
70	5062.037	II
20	5090.495	II

Intensity	Wavelength/Å	
100	5141.783	II
70	5145.308	II
5	5151.391	I
15	5162.285	I
25	5165.773	II
20	5187.746	I
20	5216.814	II
7	5221.271	I
5	5421.352	I
10	5451.652	I
25	5495.874	I
5	5506.113	I
25	5558.702	I
10	5572.541	I
35	5606.733	I
20	5650.704	I
10	5739.520	I
5	5834.263	I
10	5860.310	I
15	5882.624	I
25	5888.584	I
50	5912.085	I
15	5928.813	I
5	5942.669	I
7	5987.302	I
5	5998.999	I
5	6025.150	I
70	6032.127	I
35	6043.223	I
10	6052.723	I
20	6059.372	I
7	6098.803	I
10	6105.635	I
100	6114.923	II
10	6145.441	I
7	6170.174	I
150	6172.278	II
10	6173.096	I
10	6212.503	I
5	6215.938	I
25	6243.120	II
7	6296.872	I
15	6307.657	I
7	6369.575	I
20	6384.717	I
70	6416.307	I
25	6483.082	II
15	6538.112	I
15	6604.853	I
25	6638.221	II
20	6639.740	II
50	6643.698	II
5	6660.676	I
5	6664.051	I
25	6666.359	II
100	6677.282	I
35	6684.293	II
150	6752.834	I
5	6756.163	I
15	6766.612	I
20	6861.269	II
150	6871.289	I
5	6879.582	I
10	6888.174	I
50	6937.664	I
7	6951.478	I
7	6960.250	I
10000	6965.431	I
150	7030.251	I
10000	7067.218	I
100	7068.736	I
25	7107.478	I
25	7125.820	I
1000	7147.042	I
15	7158.839	I
70	7206.980	I
15	7265.172	I
7	7270.664	I
2000	7272.936	I
35	7311.716	I
25	7316.005	I
5	7350.814	I
70	7353.293	I
200	7372.118	I
20	7380.426	II
10000	7383.980	I
20	7392.980	I
15	7412.337	I
10	7425.294	I
25	7435.368	I
10	7436.297	I
20000	7503.869	I
15000	7514.652	I
25000	7635.106	I
15000	7723.761	I
10000	7724.207	I
10	7891.075	I
20000	7948.176	I
20000	8006.157	I
25000	8014.786	I
7	8053.308	I
20000	8103.693	I
35000	8115.311	I
10000	8264.522	I
20	8392.27	I
15000	8408.210	I
20000	8424.648	I
15000	8521.442	I
7	8605.776	I
4500	8667.944	I
20	8771.860	II
180	8849.91	I
20	9075.394	I
35000	9122.967	I
550	9194.638	I
15000	9224.499	I
400	9291.531	I
1600	9354.220	I
25000	9657.786	I
4500	9784.503	I
180	10052.06	I
30	10332.72	I
100	10467.177	II
1600	10470.054	I
13	10478.034	I
180	10506.50	I
200	10673.565	I
11	10681.773	I
7	10683.034	II
30	10733.87	I
30	10759.16	I
7	10812.896	II
11	11078.869	I
30	11106.46	I
12	11441.832	I
400	11488.109	I
200	11668.710	I
12	11719.488	I
200	12112.326	I
50	12139.738	I
50	12343.393	I
200	12402.827	I
200	12439.321	I
100	12456.12	I
200	12487.663	I
150	12702.281	I
30	12733.418	I
12	12746.232	I
200	12802.739	I
50	12933.195	I
500	12956.659	I
200	13008.264	I
200	13213.99	I
200	13228.107	I
100	13230.90	I
500	13272.64	I
1000	13313.210	I
1000	13367.111	I
30	13499.41	I
1000	13504.191	I
11	13573.617	I
30	13599.333	I
400	13622.659	I
200	13678.550	I
1000	13718.577	I
10	13825.715	I
10	13907.478	I
200	14093.640	I
100	15046.50	I
25	15172.69	I
10	15329.34	I
30	15989.49	I
30	16519.86	I
500	16940.58	I
12	18427.76	I
50	20616.23	I
30	20986.11	I
20	23133.20	I
20	23966.52	I

Arsenic

As Z = 33

Intensity	Wavelength/Å	
510	871.7	III
325	889.0	III
325	927.5	III
325	937.2	III
325	953.6	III
325	963.8	III
250	987.7	V
340	1021.96	II
250	1029.5	V
340	1082.35	II
500	1139.40	II
615	1149.31	II
555	1181.51	II
555	1189.87	II
615	1196.38	II
615	1196.56	II
340	1207.44	II
800	1211.17	II
800	1218.10	II
340	1223.15	II
760	1241.31	II
965	1243.08	II
870	1245.67	II
800	1258.58	II
965	1263.77	II
800	1266.34	II
800	1267.59	II
715	1280.99	II
715	1287.54	II
715	1305.70	II
340	1307.74	II
760	1333.15	II
965	1341.55	II
760	1355.93	II
965	1369.77	II
800	1373.65	II
1000	1375.07	II
760	1375.78	II
800	1394.64	II
800	1400.31	II
500	1448.59	II
500	1558.88	II
500	1570.99	II
100 r	1593.60	I
500	1660.55	II
340	1860.34	II
1000 r	1890.42	I
500	1912.94	II
800 r	1937.59	I
585 r	1972.62	I
170 r	1990.35	I
100 r	1991.13	I
100 r	1995.43	I
230 r	2003.34	I
100 r	2009.19	I
200	2263.2	IV
350 r	2288.12	I
200	2301.0	IV
350 r	2349.84	I
100 r	2370.77	I
135 r	2381.18	I
250	2417.5	IV
250	2454.0	IV
170 r	2456.53	I
200	2461.4	IV
340	2602.00	II
170 r	2780.22	I
300	2830.359	II
300	2831.164	II
100 r	2860.44	I
300	2884.406	II
80	2926.3	III
615	2959.572	II
300	3003.819	II
300	3116.516	II
340	3842.60	II
325	3922.6	III
715	4190.082	II
615	4197.40	II
615	4242.982	II
500	4315.657	II
500	4323.867	II
500	4336.64	II

Intensity	Wavelength/Å	
500	4352.145	II
425	4352.864	II
375	4371.17	II
615	4427.106	II
615	4431.562	II
715	4458.469	II
340	4461.075	II
715	4466.348	II
500	4474.46	II
800	4494.230	II
850	4507.659	II
615	4539.74	II
715	4543.483	II
615	4602.427	II
340	4629.787	II
340	4707.586	II
340	4730.67	II
340	4888.557	II
340	5105.58	II
500	5107.55	II
425	5231.38	II
500	5331.23	II
340	5497.727	II
425	5558.09	II
425	5651.32	II
425	6110.07	II
500	6170.27	II
300	6511.74	II
300	7092.27	II
300	7102.72	II
340	7990.53	II
300	8174.51	II
200	9300.61	I
230	9597.95	I
290	9626.70	I
230	9833.76	I
170	9915.71	I
290	9923.05	I
290	10024.04	I
170	10614.07	I
Astatine		
At Z = 85		
8	2162.25	I
10	2244.01	I
Barium		
Ba Z = 56		
14	555.48	III
14	587.57	III
18	647.27	III
300	719.86	V
150	721.85	V
1000	766.87	V
40000	794.89	IV
300	877.41	V
50000	923.74	IV
200	946.26	V
200	1486.72	II
400	1504.01	II
300	1554.38	II
200	1572.73	II
	1573.92	II
	1630.40	II
100	1674.51	II
400	1694.37	II
	1697.16	II
	1761.75	II
	1771.03	II
	1786.93	II
100	1904.15	II
500	1924.70	II
	1985.60	II
300	1999.54	II
10	2001.30	III
	2009.20	II
400	2023.95	II
	2052.68	II
	2054.57	II
500	2214.7	II
800	2245.61	II
1000	2254.73	II
1400	2304.24	II
60	2331.10	III
2000	2335.27	II
190	2347.58	II
40	2512.28	III
40	2523.83	III
60	2528.51	II
50	2559.54	III
8 h	2596.64	I
100	2634.78	II
40	2681.89	III
8	2702.63	I
18	2771.36	II
15	2785.28	I
100 r	3071.58	I
40	3079.14	III
10 h	3108.21	I
8	3132.60	I
8 h	3135.72	I
10	3137.70	I
10	3155.34	I
10	3155.67	I
12	3158.05	I
12 h	3158.54	I
25	3165.60	I
15 h	3173.69	I
30	3183.16	I
15	3183.96	I
10	3193.91	I
25 h	3203.70	I
30	3221.63	I
40	3222.19	I
50	3261.96	I
60 r	3262.34	I
40	3281.50	I
15	3281.77	I
50	3322.80	I
80 h	3356.80	I
50	3368.18	III
60 r	3377.08	I
20	3377.39	I
70 r	3420.32	I
25	3421.01	I
30 h	3421.48	I
40	3463.74	I
200 r	3501.11	I
80 h	3524.97	I
30 h	3531.35	I
80 h	3544.66	I
20 h	3547.68	I
100	3552.45	II
200	3567.73	II
100	3576.28	II
30	3577.62	I
80 h	3579.67	I
200	3596.57	II
40	3630.64	I
40 h	3636.83	I
20 h	3688.47	I
400	3735.75	II
200	3816.69	II
200	3842.80	II
100	3854.76	II
20	3889.33	I
1400 l	3891.78	II
20	3892.65	I
40	3909.91	I
500	3914.73	II
25	3926.85	III
50	3935.72	I
20	3937.87	I
200	3939.67	II
500	3949.51	II
25	3993.06	III
80	3993.40	I
30	3995.66	I
300	4036.26	II
200	4083.77	II
30 h	4084.86	I
1500 h	4130.66	II
20	4132.43	I
200	4166.00	II
500	4216.04	II
800	4267.95	II
100	4283.10	I
300	4287.80	II
200	4297.60	II
800	4309.32	II
20 h	4323.00	I
600	4325.73	II
200	4326.74	II
300	4329.62	II
80	4350.33	I
60	4402.54	I
400	4405.23	II
40	4431.89	I
60 h	4488.98	I
50 h	4493.64	I
40	4505.92	I
200	4509.63	II
60 h	4523.17	I
130	4524.93	II
65000	4554.03	II
40	4573.85	I
80	4579.64	I
30	4599.75	I
20 h	4619.92	I
25 h	4628.33	I
300	4644.10	II
30	4673.62	I
35	4691.62	I
20	4700.43	I
800	4708.94	II
40	4726.44	I
800	4843.46	II
300	4847.14	II
200	4850.84	II
30 h	4877.65	I
400	4899.97	II
15	4902.90	I
20000	4934.09	II
8	4947.35	I
1000	4957.15	II
300	4997.81	II
1000	5013.00	II
20 h	5159.94	I
20	5267.03	I
800	5361.35	II
1000	5391.60	II
200	5421.05	II
100	5424.55	I
200	5428.79	II
300	5480.30	II
200	5519.05	I
1000 r	5535.48	I
20 h	5620.40	I
10	5680.18	I
400	5777.62	I
800	5784.18	II
100	5800.23	I
20	5805.69	I
150	5826.28	I
2800	5853.68	II
15	5907.64	I
100	5971.70	I
800	5981.25	II
100	5997.09	I
300	5999.85	II
100	6019.47	I
200	6063.12	I
300	6110.78	I
400	6135.83	II
20000	6141.72	II
150	6341.68	I
500	6378.91	II
10	6383.76	III
90	6450.85	I
150	6482.91	I
12000	6496.90	II
300	6498.76	I
150	6527.31	I
3000	6595.33	I
150	6654.10	I
1500	6675.27	I
1800	6693.84	I
1000	6769.62	II
600	6865.69	I
300 h	6867.85	I
1000	6874.09	II
6000	7059.94	I
2400 hs	7120.33	I
600	7195.24	I
600 hl	7228.84	I
3000	7280.30	I
1200	7392.41	I
300	7417.53	I
900 hl	7459.78	I
600	7488.08	I
450 hl	7636.90	I
600 hl	7642.91	I
1800	7672.09	I
1200	7780.48	I
180 h	7839.57	I
1500	7905.75	I

Intensity	Wavelength/Å	
600	7911.34	I
900 h	8210.24	I
8	8308.69	III
1800 h	8559.97	I
100	8710.74	II
100	8737.71	II
300 h	8799.76	I
300	8860.98	I
450	8914.99	I
300	9219.69	I
300	9308.08	I
300 h	9324.58	I
1500	9370.06	I
300	9455.92	I
8	9521.76	III
450	9589.37	I
900	9608.88	I
300 h	9645.72	I
1500 hl	9830.37	I
900	10001.08	I
600	10032.10	I
1200 h	10233.23	I
300	10471.26	I
120 hl	10791.25	I
180 hl	11012.69	I
150 h	11114.42	I
240	11303.04	I
120 h	11697.45	I
120	13207.30	I
120	13810.50	I
120	14077.90	I
120	15000.40	I
120	20712.00	I
150	25515.70	I
150	29223.90	I
	Beryllium	
	Be $Z = 4$	
	58.13	IV
	58.57	IV
	59.32	IV
	60.74	IV
	64.06	IV
	75.93	IV
1 h	76.10	III
2	76.48	III
3	78.53	III
4	78.66	III
1 h	78.92	III
5	81.89	III
10	82.38	III
	82.58	II
20	83.20	III
	83.66	II
30	84.76	III
50	88.31	III
	89.16	I
	89.80	II
	90.04	II
	90.21	I
	90.67	I
	91.06	II
	91.36	II
	91.74	II
	92.19	I
	92.61	II
	93.14	II
	93.42	II
	93.93	II
	94.78	II
	95.76	II
	96.29	I
	97.24	I
	97.44	I
	97.86	I
	97.97	I
	98.12	I
	98.37	I
	98.66	I
	98.94	I
	99.19	I
100	100.25	III
	100.86	I
	101.20	I
	102.13	I
	102.49	II
	104.40	II
	104.67	I
	105.80	I
	107.26	I
	107.38	I
3	509.99	III
2	549.31	III
6	582.08	III
4	661.32	III
8	675.59	III
	714.0	II
4	725.59	III
5	725.71	II
5	743.58	II
7	746.23	III
2	767.75	III
8	775.37	II
20	842.06	II
	865.3	II
2	925.25	II
10	943.56	II
10	973.27	II
	981.4	II
	1020.1	II
8	1026.93	II
5	1036.32	II
15	1048.23	II
1	1114.69	III
20	1143.03	II
	1155.9	II
60	1197.19	II
2	1213.12	III
1	1214.32	III
2	1362.25	III
1	1401.52	III
10	1421.26	III
5	1422.86	III
	1426.12	I
1	1435.17	III
2	1440.77	III
	1491.76	I
20	1512.30	II
60	1512.43	II
100	1661.49	I
2 h	1754.69	III
15	1776.12	II
20	1776.34	II
	1907.	I
	1909.0	II
	1912.	I
3	1917.03	III
	1919.	I
5	1929.67	I
10	1943.68	I
60 h	1954.97	III
	1956.	I
50	1964.59	I
5	1985.13	I
	1997.95	I
	1997.98	I
60	1998.01	I
	2033.25	I
	2033.28	I
	2033.38	I
50	2055.90	I
100	2056.01	I
75 h	2076.94	III
60 h	2080.38	III
25	2118.56	III
15 h	2122.27	III
10	2125.57	I
20	2125.68	I
15 h	2127.20	III
5	2137.25	III
25	2145.	I
55	2174.99	I
55	2175.10	I
5	2191.57	III
	2273.5	II
	2324.6	II
	2337.0	I
950	2348.61	I
20	2350.66	I
60	2350.71	I
200	2350.83	I
2	2413.34	II
16	2413.46	II
20	2453.84	II
	2480.6	I
35	2494.54	I
35	2494.58	I
100	2494.73	I
16	2507.43	II
5	2617.99	II
20	2618.13	II
100	2650.45	I
60	2650.55	I
200	2650.62	I
60	2650.69	I
100	2650.76	I
5	2697.46	II
20	2697.58	II
20	2728.88	II
30	2738.05	I
	2764.2	II
20	2898.13	I
10	2898.19	I
20	2898.25	I
30	2986.06	I
10	2986.42	I
60	3019.33	I
30	3019.49	I
30	3019.53	I
20	3019.60	I
10	3046.52	II
30	3046.69	II
	3090.3	I
10	3110.81	I
10	3110.92	I
20	3110.99	I
	3120.	I
480	3130.42	II
320	3131.07	II
	3136.	I
	3150.	I
	3160.6	I
	3163.	I
	3168.	I
	3180.7	II
	3187.	I
20	3193.81	I
20	3197.10	II
30	3197.15	II
20	3208.60	I
	3220.	I
60	3229.63	I
2	3233.52	II
10	3241.62	II
30	3241.83	II
15	3269.02	I
100	3274.58	II
30	3274.67	II
30	3282.91	I
30	3321.01	I
30	3321.09	I
220	3321.34	I
20	3345.43	I
60	3367.63	I
	3405.6	II
5	3451.37	I
300	3455.18	I
20	3476.56	I
300	3515.54	I
10	3555.	I
100	3720.36	III
	3720.92	III
	3722.98	III
100	3736.30	I
700	3813.45	I
40	3865.13	I
80	3865.42	I
1	3865.51	I
6	3865.72	I
100	3866.03	I
90 h	4249.14	III
100	4253.05	I
60	4253.76	I
300	4360.66	II
500	4360.99	II
400	4407.94	I
2	4485.52	III
100 h	4487.30	III
1	4495.09	III
140 h	4497.8	III
	4526.6	I
	4548.	I
12	4572.66	I
700	4673.33	II
1000	4673.42	II
6	4709.37	I

Intensity	Wavelength/Å		Intensity	Wavelength/Å		Intensity	Wavelength/Å		Intensity	Wavelength/Å	
200	4828.16	II	30	12098.18	II	35	1538.06	II	2800	2989.03	I
40	4849.16	I	100	14643.92	I	20	1563.67	II	700	2993.34	I
2 h	4858.22	II	60	14644.75	I	40	1573.70	II	100	3012.	IV
80	5087.75	I	200	16157.72	I	60	1591.79	II	2400	3024.64	I
8	5218.12	II	80	17855.38	I	25	1601.58	II	60	3034.87	I
20	5218.33	II	120	17856.63	I	60 h	1606.40	III	100	3042.	IV
3	5255.86	II	100	18143.54	I	40	1609.70	II	9000 c	3067.72	I
64	5270.28	II	160	31775.05	I	40	1611.38	II	140	3076.66	I
500	5270.81	II	200	31778.70	I	20	1652.81	II	35	3115.0	III
20	5403.04	II	**Bismuth**			20	1749.29	II	100	3239.	IV
20	5410.21	II	**Bi Z = 83**			80	1777.11	II	550 c	3397.21	I
	5558.	I	6	420.7	IV	60	1787.47	II	10	3430.83	II
140 h	6142.01	III	6	431.2	IV	70	1791.93	II	12	3431.23	II
10	6229.11	I	2	488.39	V	70	1823.80	II	40 h	3451.0	III
16	6279.43	II	3	563.62	V	100	1902.41	II	40	3473.8	III
30	6279.73	II	5	670.76	III	9000	1954.53	I	35	3485.5	III
30	6473.54	I	6	686.88	V	7000	1960.13	I	500 c	3510.85	I
60	6547.89	II	5	730.71	V	25	1989.35	II	380 c	3596.11	I
60	6558.36	II	10	738.17	V	7000	2021.21	I	45	3613.4	III
30	6564.52	I	4	775.16	III	9000	2061.70	I	100	3643.	IV
2 h	6636.44	II	6	790.5	IV	45 h	2068.9	II	12	3654.2	II
1	6756.72	II	6	790.6	IV	4600	2110.26	I	100	3682.	IV
2	6757.13	II	8	792.5	IV	2500	2133.63	I	50	3695.32	III
30	6786.56	I	10	820.3	IV	15	2143.40	II	50	3695.68	III
1 h	6884.22	I	9	822.9	IV	15	2143.46	II	100	3734.	IV
6 h	6884.44	I	12	824.9	IV	60	2186.9	II	70 h	3792.5	II
100	6982.75	I	15 d	864.45	V	40 h	2214.0	II	12	3811.1	II
6 h	7154.40	I	15	872.6	IV	360	2228.25	I	20	3815.8	II
40 h	7154.65	I	12	923.9	IV	1700	2230.61	I	10	3845.8	II
100	7209.13	I	15	943.3	IV	340	2276.58	I	30	3863.9	II
3	7401.20	II	25	1039.99	III	100	2311.	IV	100	3868.	IV
2	7401.43	II	50 h	1045.76	III	100	2326.	IV	40 h	4079.1	II
10	7551.90	I	30	1051.81	III	16	2368.12	II	10	4097.2	II
10 h	7618.68	I	15	1058.88	II	12	2368.25	II	140	4121.53	I
20 h	7618.88	I	20	1085.47	II	100	2376.	IV	140	4121.86	I
60	8090.06	I	10	1099.20	II	190	2400.88	I	75 h	4259.4	II
5 h	8158.99	I	24	1103.4	IV	75 h	2414.6	III	25	4272.0	II
10 h	8159.24	I	20	1139.01	III	10	2501.0	II	70 h	4301.7	II
4	8254.07	I	50	1224.64	III	25	2515.69	I	12 h	4339.8	II
10 h	8287.07	I	10	1225.43	II	70	2524.49	I	25 h	4340.5	II
30	8547.36	I	15	1232.78	II	20 h	2544.5	II	12 h	4379.4	II
60	8547.67	I	10	1241.05	II	700	2627.91	I	25 h	4476.8	II
300	8801.37	I	10	1265.35	II	100	2629.	IV	60 h	4705.3	II
6	8882.18	I	15	1283.73	II	100	2677.	IV	600 c	4722.52	I
40	9190.45	I	10	1306.18	II	12	2693.0	II	30	4730.3	II
20 h	9243.92	I	60	1317.0	IV	280 c	2696.76	I	20	4749.7	II
1 h	9343.89	II	20	1325.46	II	20	2713.3	II	40 h	4797.4	III
40	9392.74	I	40	1326.84	III	140 d	2730.50	I	12	4908.2	II
2	9476.43	II	20	1329.47	II	100	2767.	IV	10	4916.6	II
16	9477.03	II	60	1346.12	III	100	2772.	IV	12	4969.7	II
20	9847.32	I	20	1350.07	II	360	2780.52	I	20	4993.6	II
10 h	9895.63	I	25	1372.61	II	100	2786.	IV	45 h	5079.3	III
20 h	9895.96	I	15	1376.02	II	15	2803.42	II	10	5091.6	II
80	9939.78	I	20	1393.92	II	11	2803.70	II	50 h	5124.3	II
16	10095.52	II	35	1423.33	III	12	2805.3	II	60 h	5144.3	II
20	10095.73	II	35	1423.52	III	140 c	2809.62	I	20	5201.5	II
60	10119.92	II	45	1436.83	II	100	2842.	IV	75 h	5209.2	II
80	10331.03	I	25	1447.94	II	80 h	2855.6	III	40 h	5270.3	II
30	11066.46	I	50	1455.11	II	4000	2897.98	I	10	5397.8	II
	11173.	II	60 h	1461.00	III	100	2924.	IV	10 c	5552.35	I
1	11173.73	II	25	1462.14	II	100	2933.	IV	3	5599.41	I
120	11496.39	I	35	1486.93	II	100	2936.	IV	20	5655.2	II
2 h	11625.16	II	20	1502.50	II	15	2936.7	II	40 h	5719.2	II
	11659.	II	40	1520.57	II	3200	2938.30	I	6	5742.55	I
2	11660.25	II	40	1533.17	II	20	2950.4	II	12	5818.3	II
100	12095.36	II	30	1536.77	II	12	2963.4	II	20	5860.2	II

Intensity	Wavelength/Å	
20	5973.0	II
15	6059.1	II
15	6128.0	II
6	6134.82	I
3	6475.73	I
3	6476.24	I
15	6497.7	II
10	6577.2	II
40 h	6600.2	II
50 h	6808.6	II
4 h	6991.12	I
12	7033.	II
2	7036.15	I
10 h	7381.	II
2	7502.33	I
10 h	7637.	II
10	7750.	II
3	7838.70	I
2	7840.33	I
20	7965.	II
40	8008.	III
12 h	8050.	II
50	8070.	III
15	8328.	II
15	8388.	II
30	8532.	II
2	8544.54	I
1	8579.74	I
25	8653.	II
2	8754.88	I
3	8761.54	I
25	8863.	II
2	8907.81	I
2000 d	9657.04	I
40	9827.78	I
20	10104.5	I
15	10138.8	I
20	10300.6	I
20	10536.19	I
50	11072.44	I
1500 d	11710.37	I
40	11999.49	I
200	12165.08	I
200	12690.04	I
100	12817.8	I
200	14330.5	I
50	16001.5	I
60	22551.6	I
Boron		
B $Z = 5$		
	41.00	V
30	48.59	V
10	52.68	IV
30	60.31	IV
	194.37	V
	262.37	V
160	344.0	IV
450	385.0	IV
40	411.80	III
285	418.7	IV
20	510.77	III
40	510.85	III
	512.53	V
150	518.24	III
75	518.27	III
110	677.00	III
160	677.14	III
40	693.95	II
40	731.36	II
40	731.44	II
	749.74	V
40	758.48	III
70	758.67	III
110	882.54	II
110	882.68	II
40	984.67	II
110	1081.88	II
110	1082.07	II
70	1112.2	IV
450	1168.9	IV
70	1170.9	IV
110	1230.16	II
220	1362.46	II
70	1600.46	I
120	1600.73	I
160	1623.58	II
110	1623.77	II
220	1624.02	II
70	1624.16	II
160	1624.34	II
100	1663.04	I
150	1666.87	I
200	1667.29	I
150	1817.86	I
200	1818.37	I
300	1825.91	I
300	1826.41	I
110	1842.81	II
20	1953.83	III
550	2065.78	III
250	2066.38	I
250	2066.65	I
100	2066.93	I
300	2067.19	I
450	2067.23	III
160	2077.09	III
500	2088.91	I
500	2089.57	I
70	2220.30	II
40	2234.09	III
70	2234.59	III
40	2323.03	II
40	2328.67	II
40	2393.20	II
220	2395.05	II
40	2459.69	II
40	2459.90	II
1000	2496.77	I
1000	2497.73	I
70	2524.7	IV
160	2530.3	IV
450	2821.68	IV
70	2824.57	IV
285	2825.85	IV
160	2918.08	II
110	3032.26	II
70	3179.33	II
110	3323.18	II
110	3323.60	II
450	3451.29	II
285	4121.93	II
110	4194.79	II
40	4242.98	III
70	4243.61	III
110	4472.10	II
110	4472.85	II
220	4487.05	III
360	4497.73	III
70	4784.21	II
110	4940.38	II
110	6080.44	II
70	6285.47	II
70	7030.20	II
40	7031.90	II
110	7835.25	III
70	7841.41	III
20	8667.22	I
70	8668.57	I
800	11660.04	I
570	11662.47	I
125	15629.08	I
200	16240.38	I
250	16244.67	I
235	18994.33	I
Bromine		
Br $Z = 35$		
700	379.73	IV
700	400.37	IV
800	482.11	V
900	531.97	V
1000	545.43	IV
1000	547.90	V
1000	559.76	IV
1000	569.19	IV
1000	576.59	IV
1000	585.10	IV
1000	586.71	IV
1000	597.51	IV
1000	600.09	IV
1000	601.27	IV
1000	607.03	IV
1000	617.85	IV
1000	619.87	IV
1000	630.14	IV
1000	642.23	IV
1000	661.53	IV
1000	683.51	IV
1000	697.72	IV
1000	715.39	IV
1000	731.00	IV
1000	800.12	IV
700	812.95	V
1000	813.66	IV
1000	850.81	V
1000	889.23	II
1000	948.97	II
1000	1015.54	II
1000	1049.00	II
1000	1069.15	V
900	1112.13	V
1000	1143.56	V
1000	1189.28	I
1000	1189.50	I
1000	1210.73	I
1000	1221.13	I
1000	1223.24	I
1200	1224.41	I
1200	1226.90	I
7500	1232.43	I
1200	1243.90	I
1500	1251.66	I
1000	1255.80	I
1500	1259.20	I
1200	1261.66	I
1200	1266.20	I
1000	1279.48	I
1000	1286.26	I
3000	1309.91	I
3000	1316.74	I
1000	1317.37	I
2000	1317.70	I
12000	1384.60	I
3000	1449.90	I
50000	1488.45	I
30000	1531.74	I
25000	1540.65	I
30000	1574.84	I
20000	1576.39	I
25000	1582.31	I
75000	1633.40	I
1000	2133.79	IV
1000	2145.02	IV
1000	2257.21	IV
1000	2272.73	IV
1000	2307.40	IV
1000	2408.16	IV
1000	2411.58	IV
700	2491.14	IV
1000	2581.19	IV
600	2661.40	IV
1000	2842.88	IV
1100 h	2907.71	IV
500 h	2972.26	II
500	3041.18	IV
500	3074.42	III
500	3349.64	III
500	3380.56	IV
500	3540.16	III
500	3562.43	III
1200	3815.65	I
1500	3992.36	I
1000	4223.89	II
2000	4365.14	I
1000	4365.60	II
1500	4425.14	I
10000	4441.74	I
10000	4472.61	I
20000	4477.72	I
1000	4490.42	I
3000	4513.44	I
15000	4525.59	I
3000	4575.74	I
2500	4614.58	I
2500	4752.28	I
4000	4780.31	I
1600	4785.19	I
4000	4979.76	I
1200	5395.48	I
1200	5466.22	I
1800	5852.08	I
1600	5940.48	I
2400	6122.14	I
40000	6148.60	I
2000	6177.39	I

Intensity	Wavelength/Å	
1500	6335.48	I
60000	6350.73	I
2500	6410.32	I
1800	6483.56	I
1000	6514.62	I
20000	6544.57	I
1500	6548.09	I
50000 c	6559.80	I
1000	6571.31	I
1800	6579.14	I
20000	6582.17	I
1500	6620.47	I
50000 c	6631.62	I
20000	6682.28	I
10000	6692.13	I
8000	6728.28	I
2000	6760.06	I
2000	6779.48	I
2200	6786.74	I
6500	6790.04	I
1600 c	6791.48	I
1800	6861.15	I
10000	7005.19	I
2000	7260.45	I
10000	7348.51	I
40000	7512.96	I
1600	7591.61	I
1800	7595.07	I
2000	7616.41	I
30000	7803.02	I
1200	7827.23	I
2500 s	7881.45	I
2500	7881.57	I
2500	7925.81	I
30000 c	7938.68	I
3000	7947.94	I
3000	7950.18	I
8000	7978.44	I
10000	7978.57	I
30000	7989.94	I
2000	8026.35	I
2500	8026.54	I
30000	8131.52	I
1000 c	8152.65	I
10000	8153.75	I
25000	8154.00	I
5000	8246.86	I
15000	8264.96	I
75000 c	8272.44	I
20000	8334.70	I
10000	8343.70	I
1200	8384.04	I
40000	8446.55	I
4000	8477.45	I
1500	8513.38	I
1000	8557.73	I
1000	8566.28	I
20000	8638.66	I
4000	8698.53	I
10000 c	8793.47	I
15000	8819.96	I
25000	8825.22	I
4000	8888.98	I
30000	8897.62	I
6000	8932.40	I
1800	8949.39	I
9000	8964.00	I
30000	9166.06	I
15000	9173.63	I
20000	9178.16	I
40000	9265.42	I
15000	9320.86	I
6000	9793.48	I
10000	9896.40	I
3000	10140.08	I
6000	10237.74	I
1000	10299.62	I
1500	10377.65	I
30000	10457.96	I
1000	10742.14	I
3000	10755.92	I
1700	13217.17	I
1800	14354.57	I
1250	14888.70	I
1800	16731.19	I
1200	18568.31	I
3500	19733.62	I
1000	20281.73	I
1000	20624.67	I
1200	21787.24	I
4000	22865.65	I
1000	23513.15	I
500	28346.50	I
500	30380.85	I
600	31630.13	I
150	38345.75	I
120	39964.36	I

Cadmium
Cd Z = 48

Intensity	Wavelength/Å	
50	427.01	IV
50	447.85	IV
60	480.90	IV
70	493.00	IV
70	495.13	IV
70	498.14	IV
70	498.53	IV
80	504.09	IV
70	504.20	IV
70	504.50	IV
80	506.31	IV
60	508.01	IV
50	508.95	IV
70	509.55	IV
70	511.40	IV
80	513.00	IV
70	514.50	IV
60	519.42	IV
80	524.41	IV
70	524.47	IV
70	525.10	IV
60	525.19	IV
70	527.07	IV
80	531.09	IV
80	531.51	IV
70	534.29	IV
70	536.77	IV
60	540.90	IV
70	541.74	IV
80	542.60	IV
80	546.55	IV
60	553.06	IV
80	554.05	IV
60	567.01	IV
150	1118.16	IV
100	1164.65	IV
100	1183.40	IV
100	1256.00	II
150	1296.43	II
100	1326.50	II
60	1370.48	IV
150	1370.91	II
60	1418.89	IV
200	1514.26	II
50	1545.17	III
200	1571.58	II
100	1668.60	II
50	1702.47	II
40	1707.16	III
40	1722.95	III
50	1724.41	II
40	1747.67	III
40	1773.06	III
100	1785.84	II
75	1793.40	III
40	1823.41	III
100	1827.70	II
50	1844.66	III
40	1851.13	III
40	1855.85	III
200	1856.67	III
150	1874.08	III
300	1922.23	II
100	1943.54	II
40	1965.54	II
30	1986.89	II
200	1995.43	II
100	2007.49	II
50	2032.45	II
75	2036.23	II
40	2039.83	III
50	2045.61	III
75	2087.91	III
150	2096.00	II
50	2111.60	III
1000 r	2144.41	II
50	2155.06	II
100	2187.79	II
1000	2194.56	II
1000	2265.02	II
1500 r	2288.022	I
1000	2312.77	II
200	2321.07	II
40	2376.82	II
50	2418.69	II
50	2469.73	II
40	2487.93	II
40	2495.58	II
50	2509.11	II
30	2516.22	II
25 h	2525.196	I
50	2544.613	I
50	2551.98	II
25	2553.465	I
3	2565.789	I
500	2572.93	II
50	2580.106	I
30	2592.026	I
25 h	2602.048	I
50	2628.979	I
40	2632.190	I
75	2639.420	I
40	2659.23	II
50 h	2660.325	I
25	2668.20	II
50	2672.62	II
100	2677.540	I
25	2677.748	I
50	2707.00	II
75	2712.505	I
50	2733.820	I
1000	2748.54	II
100 h	2763.894	I
50 h	2764.230	I
50	2774.958	I
30	2823.19	II
200	2836.900	I
25	2856.46	II
100	2868.180	I
200 r	2880.767	I
50 r	2881.224	I
200	2914.67	II
50	2927.87	II
200	2929.27	II
1000 r	2980.620	I
200 r	2981.362	I
50	2981.845	I
50	3030.60	II
150	3080.822	I
25	3081.48	II
30	3082.593	I
100	3092.34	II
200	3133.167	I
50	3146.79	II
150	3250.33	II
300	3252.524	I
300	3261.055	I
50	3343.21	II
50	3385.49	II
30	3388.88	II
800	3403.652	I
50	3417.49	II
50	3442.42	II
100	3464.43	II
1000	3466.200	I
800	3467.655	I
25	3483.08	II
150	3495.44	II
25	3499.952	I
100	3524.11	II
100	3535.69	II
1000	3610.508	I
800	3612.873	I
60	3614.453	I
20	3649.558	I
10	3981.926	I
100	4029.12	II
200	4134.77	II
50	4141.49	II
100	4285.08	II
8	4306.672	I
100	4412.41	II
3	4412.989	I
1000	4415.63	II
30	4440.45	II

Intensity	Wavelength/Å	
8	4662.352	I
200	4678.149	I
30	4744.69	II
300	4799.912	I
50	4881.72	II
50	5025.50	II
1000 h	5085.822	I
6	5154.660	I
100	5268.01	II
100	5271.60	II
1000	5337.48	II
1000	5378.13	II
200	5381.89	II
40	5843.30	II
50	5880.22	II
300	6099.142	I
100	6111.49	I
100	6325.166	I
30	6330.013	I
400	6354.72	II
500	6359.98	II
2000	6438.470	I
400	6464.94	II
25	6567.65	II
500	6725.78	II
100	6759.19	II
30	6778.116	I
50	7237.01	II
100	7284.38	II
1000	7345.670	I
50	8066.99	II
5	8200.309	I
20	9289.	I
15	11652.	I
35	14487.	I
80	15708.	I
55 d	19120.	I
25	24371.	I
35	25448.	I
	Calcium	
	Ca *Z* = 20	
250	190.46	V
250	196.97	V
300	199.55	V
250	200.51	V
265	257.98	V
400	267.77	V
300	270.31	V
400	280.99	V
300	284.98	V
450 c	286.96	V
500	322.17	V
300	323.22	V
300	330.94	V
300	334.55	V
250 c	342.45	IV
250	343.93	IV
450	352.92	V
250	377.18	V
200	387.08	V
750	425.00	V
600	434.57	IV
250	437.77	IV
750	443.82	IV
500	450.57	IV
500	558.60	V
400	637.93	V
300	643.12	V
400	646.57	V
750	656.00	IV
300	656.76	V
500	669.70	IV
24	1341.89	II
12	1342.54	II
20	1433.75	II
20	1545.29	III
60	1649.86	II
20	1807.34	II
40	1814.50	II
40	1838.01	II
60	1840.06	II
20	1843.09	II
40	1850.69	II
17	2123.03	III
16	2152.43	III
16	2687.76	III
19	2881.78	III
21	2899.79	III
19	2924.33	III
20	2988.63	III
10	3006.86	I
15	3028.59	III
3	3055.32	I
19	3119.67	III
170	3158.87	II
180	3179.33	II
150	3181.28	II
20	3316.51	II
12	3361.92	I
19	3372.67	III
20	3461.87	II
13	3487.60	I
18	3537.77	III
20	3644.41	I
30	3683.70	II
40	3694.11	II
170	3706.03	II
180	3736.90	II
20	3755.67	II
30	3758.39	II
230	3933.66	II
220	3968.47	II
50	4097.10	II
60	4109.82	II
30	4110.28	II
40	4206.18	II
50	4220.07	II
50	4226.73	I
24	4283.01	I
22	4289.36	I
22	4298.99	I
25	4302.53	I
20	4302.81	III
23	4307.74	I
22	4318.65	I
20	4355.08	I
19	4399.59	III
25	4425.44	I
26	4434.96	I
25	4435.69	I
30	4454.78	I
28	4455.89	I
20	4456.61	I
20	4472.04	II
20	4489.18	II
19	4499.88	III
23	4526.94	I
22	4578.55	I
23	4581.40	I
23	4581.47	I
24	4585.87	I
24	4585.96	I
20	4685.27	I
30	4716.74	II
40	4721.03	II
40	4799.97	II
25	4878.13	I
70	5001.48	II
80	5019.97	II
40	5021.14	II
23	5041.62	I
25	5188.85	I
22	5261.71	I
23	5262.24	I
22	5264.24	I
24	5265.56	I
25	5270.27	I
60	5285.27	II
70	5307.22	II
50	5339.19	II
27	5349.47	I
23	5512.98	I
25	5581.97	I
27	5588.76	I
24	5590.12	I
26	5594.47	I
25	5598.49	I
24	5601.29	I
24	5602.85	I
30	5857.45	I
27	6102.72	I
29	6122.22	I
22	6161.29	I
30	6162.17	I
22	6163.76	I
24	6166.44	I
26	6169.06	I
28	6169.56	I
35	6439.07	I
30	6449.81	I
22	6455.60	I
80	6456.87	II
34	6462.57	I
29	6471.66	I
32	6493.78	I
28	6499.65	I
23	6572.78	I
30	6717.69	I
33	7148.15	I
31	7202.19	I
33	7326.15	I
30	7575.81	II
60	7581.11	II
80	7601.30	II
20	7602.32	II
40	7820.78	II
60	7843.38	II
20	8017.50	II
20	8020.50	II
70	8133.05	II
100	8201.72	II
110	8248.80	II
70	8254.73	II
130	8498.02	II
170	8542.09	II
160	8662.14	II
100	8912.07	II
110	8927.36	II
110	9213.90	II
90	9312.00	II
100	9319.56	II
110	9320.65	II
25	9416.97	I
100	9567.97	II
110	9599.24	II
80	9601.82	II
80	9854.74	II
110	9890.63	II
90	9931.39	II
100	10223.04	II
20	10343.81	I
20	11838.99	II
25	12816.04	I
24	12823.86	I
25	12909.10	I
30	13033.57	I
21	13086.44	I
24	13134.95	I
20	16150.77	I
22	16157.36	I
21	16197.04	I
20	18925.47	I
24	18970.14	I
30	19046.14	I
48	19309.20	I
49	19452.99	I
47	19505.72	I
50	19776.79	I
35	19853.10	I
34	19862.22	I
23	19917.19	I
24	19933.70	I
25	22624.93	I
30	22651.23	I
	Carbon	
	C *Z* = 6	
110	34.973	V
450	40.268	V
110	227.19	V
250	244.91	IV
160	248.66	V
160	248.74	V
200	289.14	IV
250	289.23	IV
570	312.42	IV
500	312.46	IV
250	371.69	III
250	371.75	III
150	371.78	III
650	384.03	IV
700	384.18	IV
500	386.203	III
400	419.52	IV
500	419.71	IV

Intensity	Wavelength/Å	
200	450.734	III
400	459.46	III
500	459.52	III
570	459.63	III
250	511.522	III
250	535.288	III
300	538.080	III
350	538.149	III
400	538.312	III
350	574.281	III
9	595.022	II
30	687.053	II
50	687.345	II
10	858.092	II
20	858.559	II
30	903.624	II
60	903.962	II
150	904.142	II
30	904.480	II
800	977.03	III
9	1009.86	II
10	1010.08	II
10	1010.37	II
80	1036.337	II
150	1037.018	II
150	1157.910	I
150	1158.019	I
150	1158.035	I
370	1174.93	III
350	1175.26	III
330	1175.59	III
500	1175.71	III
350	1175.99	III
370	1176.37	III
150	1188.992	I
150	1189.447	I
200	1189.631	I
300	1193.009	I
300	1193.031	I
300	1193.240	I
300	1193.264	I
100	1193.393	I
150	1193.649	I
150	1193.679	I
100	1194.064	I
100	1194.488	I
100	1261.552	I
250	1277.245	I
250	1277.282	I
300	1277.513	I
300	1277.550	I
200	1280.333	I
100	1311.363	I
9	1323.951	II
120	1329.578	I
120	1329.600	I
150	1334.532	II
300	1335.708	II
100	1354.288	I
150	1355.84	I
120	1364.164	I
100	1459.032	I
200	1463.336	I
120	1467.402	I
150	1481.764	I
1000	1548.202	IV
900	1550.774	IV
150	1560.310	I
400	1560.683	I
400	1560.708	I
100	1561.341	I
400	1561.438	I
150	1656.266	I
120	1656.928	I
300	1657.008	I
120	1657.380	I
120	1657.907	I
150	1658.122	I
500	1751.823	I
1000	1930.905	I
250	2162.94	III
40	2270.91	V
5	2277.25	V
20	2277.92	V
800	2296.87	III
800	2478.56	I
250	2509.12	II
350	2512.06	II
200 l	2524.41	IV
300 s	2529.98	IV
250 h	2574.83	II
150	2697.75	III
110 l	2724.85	III
150 l	2725.30	III
150 l	2725.90	III
350 l	2741.28	II
250	2746.49	II
1000	2836.71	II
800	2837.60	II
200	2982.11	III
800 h	2992.62	II
350	3876.19	II
350	3876.41	II
350	3876.66	II
570	3918.98	II
800	3920.69	II
150	4056.06	III
200	4067.94	III
250	4068.91	III
250	4070.26	III
250	4074.52	II
350 l	4075.85	II
150	4162.86	III
250 h	4186.90	III
800	4267.00	II
1000	4267.26	II
200	4325.56	III
600	4647.42	III
520	4650.25	III
375	4651.47	III
200 w	4658.30	IV
200	4665.86	III
200	4771.75	I
200	4932.05	I
5	4943.88	V
5	4944.56	V
200	5052.17	I
350	5132.94	II
350	5133.28	II
350	5143.49	II
570	5145.16	II
400	5151.09	II
300	5380.34	I
250	5648.07	II
350	5662.47	II
450	5695.92	III
250	5801.33	IV
200	5811.98	IV
150	5826.42	III
570	5889.77	II
350	5891.59	II
200	6001.13	I
250	6006.03	I
110	6007.18	I
150	6010.68	I
300	6013.22	I
250	6014.84	I
800	6578.05	II
570	6582.88	II
200	6587.61	I
150	6744.38	III
250	6783.90	II
150 h	7037.25	III
250	7113.18	I
250	7115.19	I
250	7115.63	II
200	7116.99	I
350	7119.90	II
800	7231.32	II
1000	7236.42	II
150	7612.65	III
90 w	7726.2	IV
200	7860.89	I
200	8058.62	I
300 h	8196.48	III
150	8332.99	III
520	8335.15	I
300	8500.32	III
250	9061.43	I
200	9062.47	I
200	9078.28	I
250	9088.51	I
450	9094.83	I
300	9111.80	I
800	9405.73	I
150	9603.03	I
250	9620.80	I
300	9658.44	I
200	10683.08	I
300	10691.25	I
12	11619.29	I
23	11628.83	I
13	11658.85	I
47	11659.68	I
24	11669.63	I
85	11748.22	I
142	11753.32	I
114	11754.76	I
11	11777.54	I
17	11892.91	I
30	11895.75	I
26	12614.10	I
20	13502.27	I
38	14399.65	I
16	14403.25	I
61	14420.12	I
12	14429.03	I
13	14442.24	I
12	16559.66	I
50	16890.38	I
10	17338.56	I
11	17448.60	I
13	18139.80	I
23	19721.99	I

Cerium

Ce $Z = 58$

Intensity	Wavelength/Å	
300	399.36	V
200	482.96	V
40	741.79	IV
30	754.60	IV
75	1332.16	IV
75	1372.72	IV
100	2000.42	IV
100	2009.94	IV
10000	2318.64	III
10000	2372.34	III
10000	2380.12	III
10000	2431.45	III
15000	2439.80	III
10000	2454.32	III
10000	2469.95	III
10000	2483.82	III
10000	2497.50	III
20000	2531.99	III
10000	2603.59	III
340	2651.01	II
270	2830.90	II
250	2874.14	II
10000	2923.81	III
10000	2931.54	III
400	2976.91	II
10000	3022.75	III
50000	3031.58	III
95000	3055.59	III
20000	3056.56	III
40000	3057.23	III
20000	3057.58	III
680	3063.01	II
40000	3085.10	III
20000	3106.98	III
30000	3110.53	III
30000	3121.56	III
20000	3141.29	III
20000	3143.96	III
20000	3147.06	III
710	3194.83	II
990	3201.71	II
710	3218.94	II
880	3221.17	II
710	3227.11	II
20000	3228.57	III
710	3234.16	II
990	3272.25	II
20000	3353.29	III
10000	3395.77	III
30000	3427.36	III
40000	3443.63	III
30000	3454.39	III
40000	3459.39	III
60000	3470.92	III
710	3485.05	II
50000	3497.81	III
60000	3504.64	III
770	3539.08	II

Intensity	Wavelength/Å	Spectrum
50000	3544.07	III
1200	3560.80	II
1000	3577.45	II
1800	3655.85	II
880	3660.64	II
880	3667.98	II
1000	3709.29	II
1000	3709.93	II
1400	3716.37	II
800	3728.42	II
860	3786.63	II
2500	3801.52	II
800	3803.09	II
1000	3808.11	II
1100	3838.54	II
860	3848.59	II
860	3853.15	II
1200	3854.18	II
1200	3854.31	II
1100	3878.36	II
1500	3882.45	II
1000	3889.98	II
770	3907.29	II
980	3912.44	II
770	3918.28	II
770	3931.09	II
770	3940.34	II
2000	3942.15	II
2700	3942.75	II
770	3943.89	II
3100	3952.54	II
980	3956.28	II
770	3960.91	II
770	3967.05	II
770	3978.65	II
770	3984.68	II
700	3992.39	II
910	3993.82	II
2800	3999.24	II
910	4003.77	II
2700	4012.39	II
910	4014.90	II
840	4024.49	II
840	4028.41	II
840	4031.34	II
2100	4040.76	II
910	4042.58	II
700	4053.51	II
1100	4071.81	II
1800	4073.48	II
1500	4075.71	II
1500	4075.85	II
910	4083.23	II
770	4118.14	II
980	4123.87	II
980	4127.37	II
2700	4133.80	II
2000	4137.65	II
770	4142.40	II
980	4149.94	II
1400	4151.97	II
1300	4165.61	II
3500	4186.60	II
840	4198.72	II
910	4202.94	II
1500	4222.60	II
770	4227.75	II
980	4239.92	II
1100	4248.68	II
2000	4289.94	II
1500	4296.67	II
770	4300.33	II
770	4306.72	II
980	4337.77	II
700	4349.79	II
910	4364.66	II
910	4382.17	II
700	4386.84	II
1700	4391.66	II
980	4418.78	II
770	4449.34	II
2400	4460.21	II
1400	4471.24	II
700	4479.36	II
700	4483.90	II
840	4486.91	II
770	4523.08	II
840	4527.35	II
840	4528.47	II
840	4539.75	II
2100	4562.36	II
1100	4572.28	II
840	4593.93	II
1700	4628.16	II
310	4737.28	II
470	5079.68	II
280	5159.69	I
280	5161.48	I
370	5187.46	II
260	5223.46	I
260	5245.92	I
340	5274.23	II
450	5353.53	II
300	5393.40	II
280	5409.23	II
260	5512.08	II
300	5696.99	I
370	5699.23	I
240	5719.03	I
230	5940.86	I
55	6001.90	I
55	6005.86	I
55	6006.82	I
75	6013.42	I
110	6024.20	I
10000	6032.54	III
110	6043.39	II
55	6047.40	I
10000	6060.91	III
45	6098.34	II
45	6123.67	I
35	6143.36	II
35	6186.17	I
35	6208.98	I
35	6228.94	I
23	6232.45	II
28	6237.45	I
45	6272.05	II
35	6295.58	I
28	6299.51	II
23	6300.21	I
35	6310.01	I
35	6343.95	II
35	6371.11	II
28	6386.84	I
23	6393.02	II
35	6430.07	I
23	6436.40	I
35	6458.03	I
28	6467.39	I
35	6473.72	I
23	6513.59	II
45	6555.65	I
23	6579.10	I
22	6612.06	I
30	6628.93	I
22	6652.72	II
26	6700.66	I
35	6704.27	I
30	6774.28	II
35	6775.59	I
30	6924.81	I
30	6986.02	I
35	7061.75	II
35	7086.35	II
22	7238.36	II
25	7252.75	I
25	7329.91	I
25	7397.77	I
25	7616.11	II
25	7689.17	II
22	7844.94	II
22	7857.54	II
30	8025.56	II
25	8772.14	II
30	8891.20	II

Cesium

Cs $Z = 55$

Intensity	Wavelength/Å	Spectrum
10000	614.01	III
2000	638.17	III
2500	666.25	III
5000	691.60	III
3500	703.89	III
15000	718.14	II
20000	721.79	III
20000	722.20	III
5000	731.56	III
12000	740.29	III
15000	808.76	II
15000	813.84	II
7500	830.39	III
35000	901.27	II
15000	920.35	III
40000	926.66	II
25000 c	1054.79	III
17 c	1673.99	III
12	1705.25	III
10	1801.83	III
20 c	1822.40	III
11	1823.93	III
12	1824.70	III
12	1841.80	III
25	1915.50	III
25 c	1923.29	III
12	1961.33	III
17	1996.56	III
710	2035.11	III
120	2056.43	III
330	2076.43	III
540	2077.30	III
410	2088.68	III
210	2101.63	III
200	2141.47	III
1000	2316.88	III
230	2325.95	III
390	2340.49	III
1600	2455.81	III
1600	2477.57	III
890	2485.45	III
410	2495.07	III
1400	2525.67	III
430	2573.05	III
16000	2596.86	III
390	2610.12	III
6200	2630.51	III
370	2700.32	III
710	2701.20	III
390	2776.44	III
270	2810.87	III
630	2845.70	III
3100	2859.32	III
200	2893.85	III
180	2921.13	III
3200	2976.86	III
210	3001.28	III
1700	3066.59	III
1100 c	3149.36	III
1400	3152.36	III
8400	3268.32	III
1300	3315.51	III
550	3340.60	III
430	3344.02	III
1200	3349.46	III
400	3463.45	III
580	3476.83	III
480	3559.82	III
7200	3597.45	III
1300	3608.31	III
2300	3618.19	III
300 c	3641.34	III
520	3651.08	III
4800	3661.40	III
640	3699.50	III
430	3837.46	III
2100 c	3876.15	I
2900	3888.37	III
600 c	3888.61	I
2700	3925.60	III
680 c	4001.70	III
3100	4006.55	III
420	4006.78	III
520	4043.42	III
14000	4264.70	II
18000 w	4277.13	II
370	4403.86	III
1200	4410.22	III
940	4425.68	III
530	4471.48	III
12000	4501.55	II
1200	4506.72	III
590	4522.86	III
20000	4526.74	II
1000 c	4555.28	I
460 c	4593.17	I

Intensity	Wavelength/Å	
99900	4603.79	II
420 h	4620.61	III
210	4665.52	III
25000	4830.19	II
140	4851.59	III
19000	4870.04	II
37000	4952.85	II
370	5035.72	III
27000	5043.80	II
75000	5227.04	II
29000	5249.38	II
11000	5274.05	II
10000 c	5349.13	II
22000	5370.99	II
230	5380.79	III
60 c	5465.94	I
37	5502.88	I
39000	5563.02	II
100	5635.21	I
210 c	5664.02	I
27	5745.72	I
24000	5831.14	II
59 c	5838.83	I
300	5845.14	I
51000	5925.63	II
140	5950.14	III
110	5979.97	III
640 c	6010.49	I
86	6034.09	I
150	6043.99	III
870	6079.86	III
9800	6128.61	II
330	6150.42	III
1000	6213.10	I
170	6217.60	I
450	6242.96	III
320 c	6354.55	I
510	6456.33	III
8300	6495.53	II
10000 w	6536.44	II
490	6586.51	I
97	6628.66	I
8800	6646.57	II
3300 c	6723.28	I
9600	6724.47	II
400	6753.12	III
200	6824.65	I
300	6870.45	I
37000	6955.50	II
4800	6973.30	I
16000	6979.67	II
980	6983.49	I
13000 w	7149.54	II
1900 c	7219.60	III
790	7228.53	I
130	7279.90	I
1100	7279.96	I
2600 c	7608.90	I
3300	7943.88	I
22000	7997.44	II
3500	8015.73	I
510	8078.94	I
4500	8079.04	I
59000 c	8521.13	I
15000 c	8761.41	I
61000 c	8943.47	I
18000	9172.32	I
5200	9208.53	I
19000	10024.36	I
4800	10123.41	I
26000	10123.60	I
2900	13424.31	I
38000 c	13588.29	I
8400	13602.56	I
5700	13758.81	I
55000 c	14694.91	I
820	16535.63	I
1500	17012.32	I
760	20138.47	I
880	22811.86	I
1100	23037.98	I
3900	23344.47	I
4400	24251.21	I
850	24374.96	I
890 d	25763.51	I
500	25764.73	I
680 c	29310.06	I
2800	30103.27	I
610 c	30953.06	I
1100	34900.13	I
190	36131.00	I
2 c	39177.28	I
2 d	39421.25	I
1	39424.11	I

Chlorine

Cl *Z* = 17

Intensity	Wavelength/Å	
500	392.43	V
800	486.17	IV
800	534.73	IV
700	535.67	IV
600	536.15	IV
900	537.61	IV
500	538.03	V
600	538.12	IV
800	542.23	V
600	542.30	V
1000	545.11	V
600	546.33	V
1000	547.63	V
500	549.22	IV
700	552.02	IV
600	553.30	IV
700	554.62	IV
600	556.23	III
700	556.61	III
700	557.12	III
350	559.305	II
700	561.53	III
700	561.68	III
700	561.74	III
400	571.904	II
800	574.406	II
500	601.50	IV
500	604.59	IV
500	606.35	III
700	618.057	II
600	619.982	II
800	620.298	II
700	626.735	II
800	635.881	II
1000	636.626	II
1000	650.894	II
1000	659.811	II
1300	661.841	II
2000	663.074	II
1500	682.053	II
1500	687.656	II
1500	693.594	II
2000	725.271	II
2500	728.951	II
2000	777.562	II
5000	787.580	II
5000	788.740	II
5000	793.342	II
500	834.84	IV
500	834.97	IV
6000	839.297	II
8000	839.599	II
600	840.93	IV
5000	851.691	II
2000	888.026	II
2000	893.549	II
2000	961.499	II
500	973.21	IV
600	977.56	IV
40	978.284	I
700	984.95	IV
25	998.372	I
25	998.432	I
75	1002.346	I
500	1005.28	III
600	1008.78	III
150	1013.664	I
700	1015.02	III
90	1025.553	I
6000	1063.831	II
3000	1067.945	II
9000	1071.036	II
6000	1071.767	II
5000	1075.230	II
5000	1079.080	II
200	1084.667	I
200	1085.171	I
250	1085.304	I
400	1088.06	I
350	1090.271	I
250	1090.982	I
250	1092.437	I
400	1094.769	I
350	1095.148	I
350	1095.662	I
400	1095.797	I
250	1096.810	I
300	1097.369	I
200	1098.068	I
200	1099.523	I
500	1107.528	I
800	1139.214	II
800	1167.148	I
3000	1179.293	I
1200	1188.774	I
900	1201.353	I
3000	1335.726	I
10000	1347.240	I
5000	1351.657	I
12000	1363.447	I
2500	1373.116	I
20000	1379.528	I
25000	1389.693	I
20000	1389.957	I
12000	1396.527	I
500	1441.470	II
500	1528.569	II
500	1542.942	II
500	1558.144	II
500	1565.050	II
600	1822.50	III
500	1828.40	III
500	1857.488	II
500	1901.61	III
500	1983.61	III
450 h	1997.370	II
450	2032.116	II
350 h	2088.583	II
350 h	2091.458	II
700	2253.07	III
500	2268.95	III
500	2278.34	III
700	2283.93	III
600	2323.50	III
500	2336.45	III
600	2340.64	III
600	2359.67	III
600	2370.37	III
700	2416.42	III
600	2447.14	III
600	2448.58	III
500	2486.91	III
500	2532.48	III
600	2580.67	III
500	2603.59	III
500	2632.67	III
500	2633.18	III
600	2665.54	III
700	2710.37	III
500	2724.03	IV
500	2751.23	IV
700	2782.47	IV
600	2965.56	III
500	3063.13	IV
600	3076.68	IV
600	3104.46	III
800	3139.34	III
900	3191.45	III
700	3289.80	III
700	3320.57	III
800	3329.06	III
900	3340.42	III
800	3392.89	III
800	3393.45	III
900	3530.03	III
800	3560.68	III
900	3602.10	III
800	3612.85	III
700	3622.69	III
700	3656.95	III
700	3670.28	III
700	3682.05	III
600	3705.45	III
600	3707.34	III
800	3720.45	III
800	3748.81	III
500	3779.35	III
10000	3850.99	II

Intensity	Wavelength/Å	
25000	3860.83	II
500	3925.87	III
700	3991.50	III
600	4018.50	III
600	4059.07	III
500	4104.23	III
500	4106.83	III
10000 h	4132.50	II
500	4608.21	III
40	4623.938	I
50	4654.040	I
80	4661.208	I
45	4691.523	I
40	4721.255	I
45	4740.729	I
13000	4781.32	II
99000	4794.55	II
29000	4810.06	II
16000	4819.47	II
81000	4896.77	II
47000	4904.78	II
26000	4917.73	II
10000	4995.48	II
26000	5078.26	II
30	5099.789	I
56000	5217.94	II
23000	5221.36	II
15000	5392.12	II
99000	5423.23	II
10000	5423.51	II
19000	5443.37	II
10000	5444.21	II
40	5532.162	I
50 d	5796.305	I
45	5799.914	I
30	5856.742	I
50	6019.812	I
200	6140.245	I
160	6194.757	I
150	6434.833	I
300	6932.903	I
300	6981.886	I
600	7086.814	I
7500	7256.62	I
5000	7414.11	I
550	7462.370	I
550	7489.47	I
700	7492.118	I
11000	7547.072	I
2300	7672.42	I
450	7702.828	I
7000	7717.581	I
10000	7744.97	I
2200	7769.16	I
650	7771.09	I
2200	7821.36	I
1700	7830.75	I
3000	7878.22	I
2300	7899.31	I
1800	7915.08	I
3000	7924.645	I
2100	7933.89	I
1700	7935.012	I
650	7952.52	I
1500	7974.72	I
1300	7976.97	I
600	7980.60	I
2900	7997.85	I
2200	8015.61	I
1100	8023.33	I
400	8051.07	I
1700	8084.51	I
2200	8085.56	I
3000	8086.67	I
1300	8087.73	I
2500	8194.42	I
2200	8199.13	I
2200	8200.21	I
800	8203.78	I
18000	8212.04	I
3000	8220.45	I
20000	8221.74	I
18000	8333.31	I
99900	8375.94	I
400	8406.199	I
15000	8428.25	I
2200	8467.34	I
2200	8550.44	I
20000	8575.24	I
750	8578.02	I
75000	8585.97	I
450	8628.54	I
300	8641.71	I
3500	8686.26	I
2200	8912.92	I
3000	8948.06	I
2000	9038.982	I
2500	9045.43	I
1000	9069.656	I
2000	9073.17	I
7500	9121.15	I
3000	9191.731	I
500	9197.596	I
4000	9288.86	I
1500	9393.862	I
3500	9452.10	I
500	9486.964	I
1000	9584.801	I
3500	9592.22	I
250	9632.509	I
1000	9702.439	I
250	9744.426	I
200	9807.057	I
400	9875.970	I
331	10392.549	I
300	11123.05	I
269	11409.69	I
1000	11436.33	I
350	13243.8	I
310	13296.0	I
550	13346.8	I
525	13821.7	I
294	14931.7	I
269	15108.0	I
381	15465.1	I
1094	15520.3	I
1487	15730.1	I
2780	15869.7	I
277	15883.3	I
342	15928.9	I
735	15960.0	I
283	15970.5	I
259	16198.5	I
717	19755.3	I
100	24470.0	I
	39716.0	I
	40085.5	I
	40089.5	I
	40532.2	I

Chromium

Cr Z = 24

Intensity	Wavelength/Å	
100	438.62	V
100	464.02	V
100	620.66	IV
100	629.26	IV
80	630.30	IV
100	666.55	IV
100	693.92	IV
60	1030.47	III
100	1033.69	III
100	1036.03	III
80	1055.89	IV
80	1068.41	III
100	1116.48	V
150	1121.07	V
150	1127.63	V
100	1263.50	V
100	1417.42	IV
150	1465.86	V
150	1497.97	V
170	1519.03	V
220	1579.70	V
170	1591.72	V
150	1603.19	V
120	1672.66	IV
120	1758.51	IV
140	1802.72	IV
130	1812.41	IV
200	1837.44	V
140	1873.89	IV
140	1967.18	IV
120	1972.07	IV
19000	2055.52	II
14000	2061.49	II
8900	2065.42	II
200	2226.72	III
200	2235.91	III
150	2237.59	III
150	2244.10	III
150	2284.44	III
150	2324.88	III
130	2383.33	I
140	2408.62	I
170	2496.31	I
110	2502.53	I
190	2504.31	I
110	2516.92	I
390	2519.52	I
190	2527.12	I
160	2549.54	I
130	2560.69	I
150	2571.74	I
100	2577.65	I
380	2591.85	I
250	2653.59	II
250	2658.59	II
320	2663.42	II
440	2666.02	II
280	2668.71	II
350	2671.81	II
280	2672.83	II
1800	2677.16	II
320	2678.79	II
230	2687.09	II
280	2691.04	II
180	2698.41	II
180	2698.69	II
110	2701.99	I
140	2712.31	II
170	2722.75	II
420 h	2726.51	I
280 h	2731.91	I
170 h	2736.47	I
250	2743.64	II
110 h	2748.29	I
330	2748.98	II
390	2750.73	II
280	2751.87	II
110 h	2752.88	I
150	2757.10	I
350	2757.72	II
750	2762.59	II
750	2766.54	II
250 h	2769.92	I
610	2780.70	I
180	2822.37	II
180	2830.47	II
2500	2835.63	II
110	2840.02	II
1700	2843.25	II
1200	2849.84	II
120	2851.36	II
880	2855.68	II
610	2858.91	II
440	2860.93	II
790	2862.57	II
750	2865.11	II
610	2866.74	II
480	2867.65	II
210	2870.44	II
110	2871.63	I
160	2873.48	II
320	2875.99	II
230	2876.24	II
180	2877.98	II
120	2879.27	I
170	2887.00	I
700	2889.29	I
370	2893.25	I
190	2894.17	I
210	2896.75	I
180	2905.49	I
260	2909.05	I
260	2910.90	I
250	2911.14	I
480	2967.64	I
480	2971.11	I
210	2971.91	II
480	2975.48	I
190	2979.74	II
350	2980.79	I
110	2985.32	II
480	2985.85	I
1500	2986.00	I

Intensity	Wavelength/Å	
2100	2986.47	I
660	2988.65	I
160	2989.19	II
480	2991.89	I
230	2994.07	I
300	2995.10	I
700	2996.58	I
210	2998.79	I
1100	3000.89	I
750	3005.06	I
140	3013.03	I
710	3013.71	I
710	3014.76	I
1400	3014.92	I
710	3015.19	I
2800	3017.57	I
430	3018.50	I
240	3018.82	I
430	3020.67	I
2800	3021.56	I
1100	3024.35	I
170	3029.16	I
710	3030.24	I
140	3031.35	I
390	3034.19	I
550	3037.04	I
550	3040.85	I
110	3050.14	II
710	3053.88	I
240	3118.65	II
430	3120.37	II
470	3124.94	II
120	3128.70	II
590	3132.06	II
140	3136.68	II
140	3147.23	II
100	3155.15	I
100	3163.76	I
240	3180.70	II
220	3197.08	II
170	3209.18	II
140	3217.40	II
120	3245.54	I
130	3251.84	I
130	3257.82	I
130	3339.80	II
110	3342.59	II
170	3358.50	II
160	3360.30	II
430	3368.05	II
140	3382.68	II
170	3403.32	II
360	3408.76	II
210	3421.21	II
270	3422.74	II
140	3433.31	II
270	3433.60	I
160	3436.19	I
140	3441.44	I
170	3445.62	I
170	3447.43	I
190	3453.33	I
130	3455.60	I
100	3460.43	I
120	3550.64	I
130	3566.16	I
130	3573.64	I
330 h	3574.80	I
19000	3578.69	I
160 h	3584.33	I
130	3585.30	II
17000	3593.49	I
350	3601.67	I
13000	3605.33	I
130	3632.84	I
350	3636.59	I
630	3639.80	I
220	3641.83	I
220	3649.00	I
170	3653.91	I
220	3656.26	I
130	3663.21	I
120	3685.55	I
130	3686.80	I
130	3687.25	I
130	3730.81	I
150	3732.03	I
480	3743.58	I
570	3743.88	I
340	3749.00	I
230	3757.66	I
260	3768.24	I
130	3791.38	I
130	3792.14	I
120	3793.29	I
130	3793.88	I
140	3797.13	I
200	3797.72	I
530	3804.80	I
110	3806.83	I
110	3807.93	I
180	3815.43	I
180	3819.56	I
130	3826.42	I
130	3830.03	I
380	3841.28	I
190	3848.98	I
140	3849.36	I
290	3850.04	I
140	3852.22	I
190	3854.22	I
110	3855.29	I
140	3855.57	I
260	3857.63	I
660	3883.29	I
570	3885.22	I
380	3886.79	I
260	3894.04	I
360	3902.92	I
960	3908.76	I
120 hd	3911.82	I
120	3915.84	I
190	3916.24	I
1900	3919.16	I
600	3921.02	I
600	3928.64	I
410	3941.49	I
1900	3963.69	I
120	3969.06	I
1600	3969.75	I
1600	3976.66	I
960	3983.91	I
190	3984.34	I
160	3989.99	I
960	3991.12	I
160	3991.67	I
190	3992.84	I
160	4001.44	I
120	4012.47	II
120	4026.17	I
190	4039.10	I
160	4048.78	I
120	4058.77	I
140	4126.52	I
120	4153.82	I
140	4163.62	I
170	4174.80	I
170	4179.26	I
110	4209.37	I
20000	4254.35	I
110	4263.14	I
16000	4274.80	I
10000	4289.72	I
780	4337.57	I
1100	4339.45	I
380	4339.72	I
1900	4344.51	I
380	4351.05	I
2300	4351.77	I
570	4359.63	I
530	4371.28	I
110	4374.16	I
530	4384.98	I
110	4458.54	I
660	4496.86	I
380	4526.47	I
380	4530.74	I
240	4535.72	I
240	4540.50	I
240	4540.72	I
140	4544.62	I
600	4545.96	I
120	4565.51	I
120	4571.68	I
360	4580.06	I
360	4591.39	I
480	4600.75	I
240	4613.37	I
600	4616.14	I
550	4626.19	I
1600	4646.17	I
570	4651.28	I
840	4652.16	I
240 d	4698.46	I
190	4708.04	I
240	4718.43	I
120	4730.71	I
140	4737.35	I
340	4756.11	I
190	4789.32	I
120	4801.03	I
110	4829.38	I
140	4870.80	I
130	4887.01	I
260	4922.27	I
110	4936.33	I
70	4942.50	I
110	4954.81	I
60	5013.32	I
70	5166.23	I
70	5184.59	I
70	5192.00	I
85	5196.44	I
5300	5204.52	I
8400	5206.04	I
11000	5208.44	I
85	5224.94	I
290	5247.56	I
530	5264.15	I
180	5265.72	I
95 h	5275.17	I
70 h	5276.03	I
340	5296.69	I
70 h	5297.36	I
660	5298.27	I
85	5300.75	I
340 h	5328.34	I
70 h	5329.17	I
780	5345.81	I
380	5348.32	I
40	5400.61	I
1400	5409.79	I
24	5628.64	I
7	5642.36	I
24	5664.04	I
24	5694.73	I
40	5698.33	I
24	5702.31	I
24	5712.78	I
24 h	5783.11	I
30 h	5783.93	I
24 h	5785.00	I
19 h	5785.82	I
60 h	5787.99	I
180 h	5791.00	I
35	6330.10	I
22	6362.87	I
19	6661.08	I
21 h	6883.03	I
27 h	6924.13	I
30 h	6978.48	I
85	7355.90	I
130	7400.21	I
150	7462.31	I
40	8947.15	I
19	8976.83	I

Cobalt
Co Z = 27

Intensity	Wavelength/Å	
20	355.52	V
18	355.88	V
12	356.06	V
66	609.16	IV
70	609.21	IV
64	609.28	IV
10	1018.36	V
10	1021.14	V
15	1231.73	V
50	1277.01	V
80	1299.58	II
80	1306.95	II
50	1345.67	V
1000	1696.01	III
800	1697.99	III
1000	1707.35	III

Line Spectra of the Elements (continued): Cobalt

Intensity	Wavelength/Å		Intensity	Wavelength/Å		Intensity	Wavelength/Å		Intensity	Wavelength/Å	
5000	1760.35	III	200	2291.98	II	200	2546.74	II	8800	3443.64	I
5000	1773.57	III	300 d	2293.38	II	340	2548.34	I	50	3446.39	II
2000	1780.05	III	300	2301.40	II	310	2553.37	I	4100	3449.17	I
3000	1782.97	III	800 d	2307.85	II	310	2555.07	I	2100	3449.44	I
1000	1787.08	III	2600	2309.02	I	300	2559.41	II	21000	3453.50	I
1000	1789.07	III	500	2311.60	II	200	2560.03	II	1000	3455.23	I
1000	1823.08	III	500	2314.05	II	960	2562.15	I	5100	3462.80	I
2000	1830.09	III	300	2314.96	II	500	2564.04	II	5100	3465.80	I
2000	1831.44	III	200 p	2317.06	II	1100	2567.35	I	8000	3474.02	I
5000	1835.00	III	2400	2323.14	I	960	2574.35	I	1900	3483.41	I
1500	1842.34	I	300 p	2324.31	II	800	2580.32	II	4800	3489.40	I
1800	1847.89	I	200 d	2326.11	II	300 d	2582.22	II	2400	3495.69	I
1800	1852.71	I	500	2326.47	II	500	2587.22	II	50	3501.72	II
2400	1855.05	I	1400	2335.99	I	500	2587.52	II	9600	3502.28	I
2000	1863.83	III	1600	2338.67	I	200	2588.91	II	7000	3506.32	I
1500	1878.28	I	200	2347.39	II	100 p	2605.71	II	50	3507.77	II
1800	1936.58	I	1600	2352.85	I	100	2612.50	II	2900	3509.84	I
1500	1946.79	I	200 d	2353.41	II	100	2614.36	II	1400	3510.43	I
1500	1951.90	I	2000	2353.42	I	100 p	2628.77	II	4800	3512.64	I
1800	1954.22	I	500	2363.80	II	100	2632.26	II	3800	3513.48	I
1800	1955.17	I	400	2378.62	II	100	2636.07	II	4800	3518.35	I
1500	1958.55	I	1400	2380.48	I	310	2646.42	I	1300	3520.08	I
1500	1961.59	I	200	2381.76	II	770	2648.64	I	2700	3521.57	I
1500 h	1968.69	I	300 p	2383.45	II	100	2653.72	II	3800	3523.43	I
1500 h	1968.93	I	1400	2384.86	I	100	2663.53	II	60	3523.51	II
3000	1970.71	I	200	2386.36	II	200	2666.73	II	6400	3526.85	I
1800 h	1971.16	I	500	2388.92	II	100	2675.85	II	2700	3529.03	I
1800 h	1972.52	I	200	2397.38	II	100	2684.42	II	7300	3529.81	I
1500	1973.85	I	1100 d	2402.06	I	100	2702.02	II	1900	3533.36	I
1800	1976.97	I	200 p	2404.16	II	200	2706.62	II	50	3545.03	II
2400 h	1980.89	I	5300	2407.25	I	200	2707.35	II	1100	3560.89	I
1500	1989.80	I	5300	2411.62	I	190	2715.99	I	80	3561.07	II
1800	1990.34	I	1600	2412.76	I	100	2727.78	II	8800	3569.38	I
1500 l	1998.49	I	4800	2414.46	I	80	2734.54	II	50	3574.95	II
1500	2002.32	I	4800	2415.30	I	190	2745.10	I	1600	3574.96	I
900	2008.04	I	300	2417.65	II	100	2753.22	II	60	3575.32	II
50	2011.51	II	4100	2424.93	I	190	2764.19	I	2500	3575.36	I
1200 h	2014.58	I	3300	2432.21	I	100	2766.70	II	60	3577.96	II
900	2016.17	I	2900	2436.66	I	100	2774.97	II	1000	3585.16	I
50	2022.35	II	2400	2439.05	I	100	2791.00	II	6700	3587.19	I
50	2027.04	II	200	2442.63	II	100	2793.73	II	1900	3594.87	I
900	2031.96	I	200 d	2446.03	II	150	2815.56	I	1600	3602.08	I
1500	2039.95	I	200 p	2447.69	II	80	2835.63	II	100	3621.21	II
1200	2041.11	I	200	2450.00	II	80	2847.35	II	1000	3627.81	I
50	2065.54	II	200	2464.20	II	80	2871.22	II	80	3643.61	II
1500 h	2077.76	I	200	2486.44	II	190	2886.44	I	60	3681.35	II
900	2085.67	I	200	2498.82	II	100	2918.38	II	1100	3745.50	I
900	2087.55	I	570	2504.52	I	100	2930.24	II	1400	3842.05	I
900	2089.35	I	500	2506.46	II	100	2954.73	II	6900	3845.47	I
900	2093.40	I	360	2506.88	I	690	2987.16	I	5500	3873.12	I
900	2094.86	I	200	2511.16	II	690	2989.59	I	2800	3873.96	I
900	2095.77	I	860	2517.87	I	60	3022.59	II	7900	3894.08	I
1200	2097.51	I	500	2519.82	II	3100	3044.00	I	1500	3935.97	I
1500	2104.73	I	4300	2521.36	I	1700	3061.82	I	80 h	3963.10	II
1500	2106.80	I	200 h	2524.65	II	80	3387.70	II	6000	3995.31	I
900	2108.98	I	300	2524.97	II	1100	3388.17	I	970	3997.91	I
900 s	2117.68	I	500	2528.62	II	2200	3395.38	I	350	4020.90	I
900	2137.78	I	2900	2528.97	I	11000	3405.12	I	370	4045.39	I
900	2138.97	I	200 p	2530.09	II	4500	3409.18	I	350	4066.37	I
900	2163.03	I	720	2530.13	I	6700	3412.34	I	830	4092.39	I
1100	2174.60	I	860	2532.18	I	2200	3412.63	I	550	4110.54	I
200	2193.60	II	200 d	2533.82	II	2700	3417.16	I	2800	4118.77	I
200	2256.73	II	2900	2535.96	I	50	3423.84	II	4400	4121.32	I
150	2260.00	II	860	2536.49	I	2500	3431.58	I	90	4190.71	I
200	2283.52	II	300	2541.94	II	4500	3433.04	I	90	4469.56	I
1000	2286.15	II	1700	2544.25	I	1600	3442.93	I	690	4530.96	I

Intensity	Wavelength/Å	
90	4549.66	I
140	4565.59	I
190	4581.60	I
120	4629.38	I
85	4663.41	I
110	4792.86	I
100	4840.27	I
150	4867.88	I
80 h	4964.18	II
50	5212.71	I
50	5230.22	I
50	5247.93	I
50	5342.71	I
50	5352.05	I
Copper		
Cu Z = 29		
80	685.141	II
100	709.313	II
100	718.179	II
150	724.489	II
200	735.520	II
250	736.032	II
80	779.295	II
100	797.455	II
150	810.998	II
200	813.883	II
300	826.996	II
150	848.808	II
250	851.303	II
250	858.487	II
400	861.994	II
400	865.390	II
250	869.336	II
150	873.263	II
200	876.723	II
250	877.012	II
200	877.555	II
500	878.699	II
100	884.133	II
250	885.847	II
600	886.943	II
600	890.567	II
500	892.414	II
800	893.678	II
400	894.227	II
600	896.759	II
400	896.976	II
600	901.073	II
400	906.113	II
800	914.213	II
600	922.019	II
500	924.239	II
400	935.232	II
600	935.898	II
600	943.335	II
600	945.525	II
500	945.965	II
200	954.383	II
250	956.290	II
400	958.154	II
200	960.414	II
250	968.042	II
200	974.759	II
250	977.567	II
100	987.657	II
250	992.953	II
300	1004.055	II
300	1008.569	II
300	1008.728	II
300	1010.269	II
250	1012.597	II
500	1018.707	II
500	1027.831	II
250	1028.328	II
200	1030.263	II
600	1036.470	II
600	1039.348	II
600	1039.582	II
800	1044.519	II
800	1044.744	II
500	1049.755	II
600	1054.690	II
400	1055.797	II
600	1056.955	II
400	1058.799	II
600	1059.096	II
600	1060.634	II
600	1063.005	II
200	1065.782	II
200	1066.134	II
500	1069.195	II
300	1073.745	II
200	1088.395	II
300	1094.402	II
250	1097.053	II
150	1119.947	II
200	1142.640	II
300	1144.856	II
100	1250.048	II
150	1265.506	II
300	1275.572	II
150	1282.455	II
150	1287.468	II
150	1298.395	II
300	1308.297	II
300	1314.337	II
100	1320.686	II
100	1326.395	II
150	1350.594	II
250	1351.837	II
150	1355.305	II
300	1358.773	II
200	1359.009	II
200	1362.600	II
250	1367.951	II
200	1371.840	II
300 r	1376.79	III
200 r	1377.49	III
100	1393.128	II
100	1398.642	II
150	1402.777	II
150	1407.169	II
100	1414.898	II
250	1418.426	II
250	1421.759	II
200	1427.829	II
400	1430.243	II
250	1434.904	II
150	1436.236	II
150	1442.139	II
200	1445.984	II
200	1449.058	II
250	1450.304	II
200	1452.294	II
300	1458.002	II
250	1459.412	II
200	1463.752	II
400	1463.838	II
200	1466.070	II
400	1470.697	II
200	1472.395	II
250	1473.978	II
200	1474.935	II
150	1476.059	II
300 r	1481.23	III
200	1481.544	II
200	1485.328	II
750	1488.831	II
300	1492.834	II
250	1493.366	II
250	1495.430	II
350	1496.687	II
150	1503.368	II
250	1504.757	II
200	1505.388	II
300	1508.632	II
350	1510.506	II
200	1512.465	II
200	1513.366	II
500	1514.492	II
200	1517.631	II
500	1519.492	II
600	1519.837	II
200	1520.540	II
200	1524.860	II
150	1525.764	II
500	1531.856	II
300	1532.131	II
250	1533.986	II
250	1535.002	II
500	1537.559	II
200	1540.239	II
300	1540.389	II
300	1540.588	II
750	1541.703	II
400	1544.677	II
100	1547.958	II
300	1550.653	II
300	1551.389	II
500	1552.646	II
250	1553.896	II
400	1555.134	II
500	1555.703	II
300	1558.345	II
400	1565.924	II
400	1566.415	II
100	1569.416	II
300	1579.492	II
300	1580.626	II
400	1581.995	II
500	1583.682	II
400	1590.165	II
600	1593.556	II
500 r	1593.75	III
400	1598.402	II
400	1602.388	II
200	1604.848	II
300	1605.281	II
400	1606.834	II
250	1608.639	II
150	1610.296	II
200	1617.915	II
600	1621.426	II
400	1622.428	II
250	1630.268	II
100	1636.605	II
1000 r	1642.21	III
250	1649.458	II
30 r	1655.32	I
200	1656.322	II
200	1660.001	II
300	1663.002	II
100	1672.776	II
30	1688.09	I
30	1691.08	I
30 r	1703.84	I
50 r	1713.36	I
150	1717.721	II
50 r	1725.66	I
100	1736.551	II
50 r	1741.57	I
150	1753.281	II
200 r	1774.82	I
100 r	1825.35	I
250	1929.751	II
250	1944.597	II
100	1946.493	II
200	1957.518	II
150	1970.495	II
150	1977.027	II
500	1979.956	II
300	1989.855	II
250	1999.698	II
270	2035.854	II
250	2037.127	II
350	2043.802	II
300	2054.980	II
100	2078.663	II
110	2098.398	II
320	2104.797	II
300	2112.100	II
320	2117.310	II
350	2122.980	II
350	2126.044	II
420	2134.341	II
900	2135.981	II
400	2148.984	II
150	2161.320	II
1300 r	2165.09	I
250	2174.982	II
1600 r	2178.94	I
700	2179.410	II
1700 r	2181.72	I
700	2189.630	II
900	2192.268	II
400	2195.683	II
1700 r	2199.58	I
1300 r	2199.75	I
100	2200.509	II
200	2209.806	II
750	2210.268	II
1600 r	2214.58	I
250	2215.106	II
1000 r	2215.65	I

Line Spectra of the Elements (continued): Copper

Intensity	Wavelength/Å		Intensity	Wavelength/Å		Intensity	Wavelength/Å		Intensity	Wavelength/Å	
750	2218.108	II	10000 r	3273.96	I	100	4758.433	II	550	6305.972	II
2100 r	2225.70	I	1400 h	3282.72	I	400	4812.948	II	400	6312.492	II
150	2226.780	II	400	3290.418	II	120	4851.262	II	120	6326.466	II
1600 r	2227.78	I	1500 h	3290.54	I	300	4854.988	II	400	6373.268	II
350	2228.868	II	110	3300.881	II	100	4873.304	II	750	6377.840	II
2500 r	2230.08	I	250	3301.229	II	150	4901.427	II	400	6403.384	II
1100 r	2238.45	I	2500 h	3307.95	I	1000	4909.734	II	850	6423.884	II
900	2242.618	II	200	3316.276	II	500	4918.376	II	200	6442.965	II
2300 r	2244.26	I	1500	3337.84	I	200	4926.424	II	750	6448.559	II
1000	2247.002	II	150	3338.648	II	900	4931.698	II	170	6466.246	II
1300 r	2260.53	I	200	3365.648	II	120	4943.026	II	950	6470.168	II
2200 r	2263.08	I	450	3370.454	II	700	4953.724	II	750	6481.437	II
150	2263.786	II	300	3374.952	II	500	4985.506	II	400	6484.421	II
200	2276.258	II	200	3380.712	II	400	5006.801	II	220	6517.317	II
100	2286.645	II	100	3384.945	II	350	5009.851	II	400	6530.083	II
2500 r	2293.84	I	1250 h	3483.76	I	400	5012.620	II	120	6551.286	II
170	2294.368	II	1250	3524.23	I	350	5021.279	II	200	6577.080	II
1000	2303.12	I	2000	3530.38	I	200	5039.016	II	750	6624.292	II
150	2369.890	II	1400	3599.13	I	300	5047.348	II	800	6641.396	II
2500 r	2392.63	I	1400	3602.03	I	900	5051.793	II	450	6660.962	II
120	2403.337	II	1000	3686.555	II	400	5058.910	II	100	6770.362	II
1500	2406.66	I	150	3786.270	II	500	5065.459	II	300	6806.216	II
1000 r	2441.64	I	170	3797.849	II	450	5067.094	II	400	6809.647	II
100	2485.792	II	100	3818.879	II	350	5072.302	II	320	6823.202	II
2000 r	2492.15	I	140	3826.921	II	450	5088.277	II	250	6844.157	II
150	2506.273	II	160	3864.137	II	420	5093.816	II	320	6868.791	II
120	2526.593	II	280	3884.131	II	350	5100.067	II	270	6872.231	II
300	2544.805	II	150	3892.924	II	1500	5105.54	I	270	6879.404	II
100	2571.756	II	170	3903.177	II	250	5124.476	II	220	6937.553	II
150	2590.529	II	140	3920.654	II	2000	5153.24	I	150	6952.871	II
200	2600.270	II	120	3933.268	II	100	5158.093	II	150	6977.572	II
2500 r	2618.37	I	120	3987.024	II	100	5183.367	II	200	7022.860	II
200	2666.291	II	150	3993.302	II	2500	5218.20	I	300	7194.896	II
750	2689.300	II	140	4003.476	II	100	5269.991	II	400	7326.008	II
700	2700.962	II	1250	4022.63	I	100	5276.525	II	300	7331.694	II
650	2703.184	II	100	4032.647	II	1650	5292.52	I	250	7382.277	II
700	2713.508	II	600	4043.484	II	100	5368.383	II	1000	7404.354	II
650	2718.778	II	500	4043.751	II	1500	5700.24	I	270	7434.156	II
300	2721.677	II	2000	4062.64	I	1500	5782.13	I	500	7562.015	II
120	2737.342	II	120	4068.106	II	150	5805.989	II	700	7652.333	II
270	2745.271	II	500	4131.363	II	100	5833.515	II	1000	7664.648	II
2500 r	2766.37	I	200	4143.017	II	200	5897.971	II	150	7681.788	II
800	2769.669	II	300	4153.623	II	120	5937.577	II	450	7744.097	II
200	2791.795	II	500	4161.140	II	400	5941.196	II	800	7778.738	II
170	2799.528	II	370	4164.284	II	100	5993.260	II	750	7805.184	II
100	2810.804	II	400	4171.851	II	650	6000.120	II	1500	7807.659	II
1250 r	2824.37	I	500	4179.512	II	100	6023.264	II	1000	7825.654	II
350	2837.368	II	500	4211.866	II	250	6072.218	II	350	7860.577	II
100	2857.748	II	320	4230.449	II	150	6080.343	II	300	7890.567	II
600	2877.100	II	200	4255.635	II	150	6099.990	II	700	7902.553	II
270	2884.196	II	950	4275.11	I	160	6107.412	II	1500	7933.13	I
2500 r	2961.16	I	300	4279.962	II	300	6114.493	II	400	7944.438	II
100	2986.335	II	500	4292.470	II	600	6150.384	II	400	7972.033	II
2000	2997.36	I	400	4365.370	II	750	6154.222	II	1200	7988.163	II
2000	3010.84	I	100	4444.831	II	500	6172.037	II	2000	8092.63	I
2500	3036.10	I	400	4506.002	II	550	6186.884	II	500	8277.560	II
2500	3063.41	I	150	4516.049	II	400	6188.676	II	800	8283.160	II
1400	3073.80	I	150	4541.032	II	300	6198.092	II	250	8503.396	II
1500	3093.99	I	500	4555.920	II	470	6204.261	II	750	8511.061	II
1250	3099.93	I	100	4596.906	II	450	6208.457	II	200	8609.134	II
2000	3108.60	I	120	4649.271	II	750	6216.939	II	500	9813.213	II
1400 h	3126.11	I	2000	4651.12	I	700	6219.844	II	250	9827.978	II
1500	3194.10	I	120	4661.363	II	500	6261.848	II	200	9830.798	II
1400	3208.23	I	320	4671.702	II	1000	6273.349	II	600	9861.280	II
1500 h	3243.16	I	300	4673.577	II	350	6288.696	II	600	9864.137	II
10000 r	3247.54	I	450	4681.994	II	900	6301.009	II	200	9883.969	II

Intensity	Wavelength/Å	
550	9916.419	II
500	9917.954	II
550	9925.594	II
450	9938.998	II
500	9960.354	II
450	10006.588	II
550	10022.969	II
550	10038.093	II
650	10054.938	II
450	10080.354	II
Dysprosium Dy Z = 66		
260	2356.91	II
240	2410.01	II
260	2439.84	II
220	2585.30	I
440	2634.80	II
220	2755.75	II
300	2816.39	II
390	2913.95	II
610	3038.28	II
830	3135.38	II
500	3141.14	II
1200	3156.52	II
670	3162.83	II
1000	3169.99	II
470	3215.19	II
830	3216.63	II
490	3235.89	II
490	3245.12	II
1200	3251.27	II
890	3280.09	II
490	3282.77	II
1100	3308.88	II
780	3316.32	II
1000	3319.88	II
780	3341.00	II
510	3353.58	II
510	3368.11	II
5300	3385.02	II
610	3388.85	II
3800	3393.57	II
1300	3396.16	II
5300	3407.80	II
1300	3413.78	II
530	3414.82	II
780	3419.63	II
530	3425.06	II
1900	3434.37	II
560	3440.93	II
1300	3441.45	II
3800	3445.57	II
830	3446.99	II
2700	3454.32	II
1300	3456.56	II
4400	3460.97	II
720	3468.43	II
560	3471.14	II
560 d	3471.53	II
1300	3477.07	II
4400	3494.49	II
560	3496.34	II
830	3498.71	II
830	3504.53	II
830	3505.45	II
1300	3506.81	II
560	3517.26	II
4400	3523.98	II
22000	3531.70	II
4400	3534.96	II
5500	3536.02	II
4400	3538.52	II
1700	3542.33	II
1400	3546.83	II
4400	3550.22	II
2200	3551.62	II
440 h	3558.23	II
440	3559.30	II
2200	3563.15	II
560	3563.69	II
780	3573.83	II
1400	3574.15	II
4400	3576.24	II
1700	3576.87	II
830	3577.98	II
440	3580.04	II
3300	3585.06	II
1400	3585.78	II
560	3586.11	II
1100	3591.41	II
560	3591.81	II
560	3592.11	II
1800	3595.04	II
560	3600.38	II
1800	3606.12	II
440	3618.51	II
560	3620.16	II
470	3624.27	II
1100	3629.42	II
4000	3630.24	II
440	3632.78	II
1100	3640.25	II
11000	3645.40	II
1000	3648.78	II
700	3664.62	II
990	3672.30	II
420	3672.70	II
1400	3674.08	II
2200	3676.59	II
640	3678.51	I
820	3684.85	I
1300	3685.78	I
4700	3694.81	II
990	3698.21	II
540	3701.63	II
440	3707.57	II
440	3708.22	II
420	3710.07	II
1600	3724.45	II
930	3739.34	I
1200	3747.82	II
1400	3753.51	II
1400	3753.75	II
1200	3757.05	I
4700	3757.37	II
640	3767.63	I
640	3773.05	I
420	3781.47	I
3300	3786.18	II
1600	3788.44	II
700	3791.87	II
510	3804.14	II
580	3806.27	II
470	3812.27	I
470	3813.67	II
1400	3816.76	II
700	3825.68	II
2300	3836.50	II
1400	3841.31	II
420	3846.34	II
420	3847.02	I
1200	3853.03	II
420	3858.40	I
560	3868.45	II
1600	3868.81	I
820	3869.86	II
7000	3872.11	II
1200	3873.99	II
470	3879.11	II
5800	3898.53	II
540	3914.87	II
540	3915.59	II
540 d	3917.29	I
420	3927.86	I
540	3930.14	I
2100	3931.52	II
10000	3944.68	II
800	3957.79	II
14000	3968.39	II
2700	3978.57	II
1400	3981.92	II
1600	3983.65	II
800	3984.21	II
540	3991.32	II
1600	3996.69	II
8000	4000.45	II
420	4005.84	II
540	4011.29	II
540	4013.82	I
540	4014.70	II
420	4027.78	II
520 d	4028.32	II
520	4032.47	II
420	4033.65	II
420	4036.32	II
12000	4045.97	I
1600	4050.56	II
520	4055.14	II
2500	4073.12	II
7400	4077.96	II
3900	4103.30	II
860	4103.87	I
1500	4111.34	II
490	4124.63	II
990	4129.42	II
1200	4143.10	II
990	4146.06	I
5700	4167.97	I
930	4183.72	I
12000	4186.82	I
2200	4191.64	I
6800	4194.84	I
800	4198.02	I
680	4201.30	I
680	4202.24	I
16000	4211.72	I
1800	4213.18	I
3700	4215.16	I
4400	4218.09	I
4400	4221.11	I
2700	4225.16	I
1000	4308.63	II
540	4409.38	II
740	4449.70	II
420	4577.78	I
2100	4589.36	I
990	4612.26	I
170	4731.84	II
120 h	4775.79	I
480	4957.34	II
70	5022.12	I
160	5042.63	I
95	5070.68	I
120	5077.67	I
80	5090.38	II
80	5110.32	I
130 h	5120.04	I
190	5139.60	II
110	5169.69	II
80	5185.30	I
290	5192.86	II
95	5197.66	II
70	5259.88	I
130	5260.56	I
65	5267.11	I
55	5282.07	I
160	5301.58	I
65	5340.30	I
85	5389.58	II
80	5419.13	I
70	5423.32	I
95	5451.11	I
65	5547.27	I
100	5639.50	I
55 h	5645.99	I
80	5652.01	I
70 h	5718.46	I
55	5745.53	I
55 h	5868.11	II
70	5945.80	I
120	5974.49	I
140	5988.56	I
140	6088.26	I
100	6168.43	I
270	6259.09	I
160	6579.37	I
75	6667.86	I
180	6835.42	I
80	6852.96	I
65	6899.32	II
55	7426.86	II
55	7543.73	I
80	7662.36	I
100	8201.57	II
45	8791.39	II
Erbium Er Z = 68		
600	2277.65	III
290	2586.73	II
490	2670.26	II
500	2739.27	III
610	2755.63	II
1000	2904.47	II
1500	2910.36	II

Intensity	Wavelength/Å	
1500	2964.52	II
1200	3002.41	II
1000	3055.10	III
1000	3070.40	III
610	3073.34	II
720	3082.08	II
610	3084.02	II
770	3122.72	II
1500	3166.25	III
870	3181.92	II
870	3220.73	II
610	3223.31	II
2300	3230.58	II
2700	3264.78	II
720	3279.33	II
720	3280.22	II
2000	3301.23	III
2300	3312.42	II
770	3323.19	II
770	3332.70	II
1300	3346.04	II
1400	3364.08	II
1400 d	3368.02	II
7700	3372.71	II
970	3374.17	II
1700	3385.08	II
2300	3392.00	II
770	3441.13	II
970	3471.71	II
610	3479.41	II
970	3485.85	II
6700	3499.10	II
610	3502.78	I
610	3524.91	II
820	3549.84	II
1500	3558.02	I
1000	3559.90	II
920	3570.75	II
1000	3580.52	II
610	3590.76	I
610	3599.50	II
1000	3599.83	II
3100	3616.56	II
720	3628.04	I
1000	3633.54	II
1600	3638.68	I
900	3645.94	II
7900	3692.65	II
1300	3729.52	II
900	3742.64	II
900	3747.43	I
1800	3786.84	II
1600	3810.33	I
4000	3816.78	III
3600	3830.48	II
680	3855.90	I
7500	3862.85	I
1500	3880.61	II
1200	3882.89	II
4200	3892.68	I
5200	3896.23	II
11000	3906.31	II
3200	3937.01	I
2100	3938.63	II
3200	3944.42	I
2700	3973.04	I
3200	3973.58	I
1400	3974.72	II
810	3977.02	I
1100	3982.33	I
810	3987.66	I
14000	4007.96	I
1100	4012.58	I
3000	4020.51	I
1000	4046.96	I
940	4055.47	II
690	4059.78	II
3500	4087.63	I
1100	4098.10	I
6900	4151.11	I
1000	4190.70	I
1400	4218.43	I
690	4286.56	I
40000	4290.06	III
20000	4386.86	III
810	4409.34	I
1000	4606.61	I
570	4675.62	II
15000	4735.56	III
2000	4783.12	III
250	5007.25	I
200	5035.94	I
210	5042.05	II
120	5124.56	I
130	5127.41	II
120	5131.53	I
130	5133.83	II
170	5164.77	II
130	5172.78	I
160	5188.90	II
150	5206.52	I
140	5255.93	II
80	5272.91	I
90	5348.06	I
60	5414.63	II
180	5456.62	I
90	5468.32	I
80	5485.97	II
80	5593.46	I
60	5611.82	I
70	5622.01	I
80	5626.53	II
90	5640.36	I
70	5664.95	I
70	5719.55	I
100	5739.19	I
290	5762.80	I
70	5784.66	I
70	5800.79	I
430	5826.79	I
100	5850.07	I
120	5855.31	I
140	5872.35	I
120	5881.14	I
8000	5903.30	III
70	6022.56	I
70	6061.25	I
60	6076.45	II
360	6221.02	I
55	6262.56	I
60	6268.87	I
130	6308.77	I
55	6326.13	I
55	6492.35	I
60	6583.48	I
70	6601.11	I
70	6759.87	I
35	6790.92	I
70	6848.10	I
55	6865.13	I
55	7459.55	I
120	7469.51	I
35	7680.01	I
35	7797.47	I
35	7921.85	I
30	7937.84	I
35	8312.82	I
55	8409.90	I
9	8866.84	II
Europium		
Eu $Z = 63$		
30	2124.69	III
200	2350.51	III
4000	2375.46	III
100 d	2435.14	III
1000	2444.38	III
4000	2445.99	III
2000	2513.76	III
200	2522.14	III
160	2564.17	II
110	2568.17	II
230	2577.14	II
1000	2638.77	II
380	2641.27	II
640	2668.34	II
110	2673.42	II
250	2678.29	II
250	2685.66	II
550	2692.03	II
700	2701.14	II
800	2701.90	II
240	2705.28	II
180	2709.99	I
700	2716.98	II
4200	2727.78	II
160	2740.62	II
120	2744.26	II
480	2781.89	II
1900	2802.84	II
220	2811.75	II
3400	2813.94	II
550	2816.18	II
2000	2820.78	II
400 cw	2828.72	II
260	2859.67	II
280	2862.57	II
200	2892.54	I
140	2893.03	I
360	2893.83	I
3200	2906.68	II
160	2908.99	I
850	2925.04	II
200 cw	2952.68	II
260	2960.21	II
300	2991.33	II
100 c	3023.93	III
200 c	3026.79	III
320 cw	3054.94	II
120	3058.98	I
220	3077.36	II
120	3097.45	II
320	3106.18	I
950	3111.43	I
120	3130.73	II
50 c	3171.00	III
50 c	3183.78	III
420	3210.57	I
1000	3212.81	I
420	3213.75	I
150	3272.77	II
210	3277.78	II
150	3301.95	II
140	3308.02	II
140	3313.33	II
950	3334.33	I
110	3350.40	I
140	3369.06	II
190	3391.99	II
280	3396.58	II
150	3425.02	II
150	3441.00	II
130	3461.38	II
470 cw	3521.09	II
150	3542.15	II
180	3552.52	II
150	3603.20	II
6400	3688.42	II
20000 cw	3724.94	II
350	3741.31	II
260	3761.12	II
39000 cw	3819.67	II
140	3844.23	II
190	3865.57	I
150	3884.75	I
28000 cw	3907.10	II
32000 cw	3930.48	II
30000 cw	3971.96	II
180	4011.69	II
150	4017.58	II
120	4039.19	I
120	4085.38	II
33000 cw	4129.70	II
60000 cw	4205.05	II
150	4298.73	I
240	4355.09	II
14000 cw	4435.56	II
3000	4522.57	II
11000	4594.03	I
9800	4627.22	I
8300	4661.88	I
110	4867.62	I
150	4907.18	I
180	4911.40	I
180	5013.17	I
170	5022.91	I
110	5029.54	I
170	5114.37	I
170	5129.10	I
210	5133.52	I
270	5160.07	I
210	5166.70	I
200	5199.85	I
110	5200.96	I
120	5206.44	I

Intensity	Wavelength/Å	
750	5215.10	I
300	5223.49	I
120	5239.24	I
200	5266.40	I
390	5271.96	I
110	5272.48	I
150	5282.82	I
120	5291.26	I
120	5294.64	I
540	5357.61	I
120	5361.61	I
110	5376.94	I
120	5392.94	I
450	5402.77	I
380	5451.51	I
260	5452.94	I
120	5488.65	I
120	5510.52	I
200	5547.44	I
150	5570.33	I
200	5577.14	I
120	5580.03	I
210	5645.80	I
330	5765.20	I
180	5783.69	I
170	5818.74	II
600 cw	5830.98	I
330	5966.07	II
480 cw	5967.10	I
170	5972.75	I
240	5992.83	I
110	6012.56	I
420	6018.15	I
170	6029.00	I
420	6049.51	II
140	6057.36	I
240	6083.84	I
240	6099.35	I
120	6118.78	I
330	6173.05	II
110	6178.76	I
260 cw	6188.13	I
140	6195.07	I
240	6262.25	I
170	6299.77	I
230	6303.41	II
120 cw	6350.04	I
120 cw	6400.93	I
180	6410.04	I
140	6411.32	I
830	6437.64	II
120	6457.96	I
1400	6645.11	II
50	6666.35	III
140	6802.72	I
360	6864.54	I
120	7040.20	I
330	7077.10	II
570	7194.81	II
570	7217.55	II
540	7301.17	II
720	7370.22	II
300	7426.57	II
160	7583.91	I
60 cw	7742.57	I
70	7746.19	I
35	7887.99	I
24 cw	8209.80	I
21 cw	8642.67	I
18	8870.30	I
Fluorine		
F Z = 9		
50	148.00	V
50	163.56	V
90	165.98	V
100	166.18	V
50	186.84	V
60	190.57	V
70	190.84	V
50	196.39	IV
60	196.45	IV
70	200.09	IV
80	201.16	IV
90	208.25	IV
90	240.08	IV
100	251.03	IV
140	419.65	IV
150	420.05	IV
160	420.73	IV
100	429.51	III
110	430.15	III
150	430.76	IV
90	464.29	III
120	465.98	V
130	490.57	IV
160	491.00	IV
50	497.38	IV
60	497.83	IV
70	498.80	IV
90	506.16	V
100	508.08	V
120	508.39	III
60	514.08	V
90	525.29	V
100	526.30	V
120	567.69	III
110	567.75	III
140	570.64	IV
140	571.30	IV
150	571.39	IV
160	572.66	IV
90	605.67	II
100	606.80	II
90	630.20	III
100	647.77	V
110	647.87	V
130	654.03	V
120	656.12	III
130	656.87	III
110	657.23	V
140	657.33	V
140	658.33	III
140	676.12	IV
130	677.15	IV
150	677.22	IV
130	678.99	IV
160	679.21	IV
60	757.04	V
150	806.96	I
125	809.60	I
500	951.87	I
1000	954.83	I
750	955.55	I
500	958.52	I
20	972.40	I
350	973.90	I
100	976.22	I
40	976.51	I
100	977.75	I
60	1082.31	V
70	1088.39	V
80	1219.03	III
80	1266.87	III
90	1267.71	III
70	1297.54	III
70	1359.92	III
110	1498.93	III
120	1502.01	III
110	1504.18	III
140	1504.79	III
130	1506.30	III
110	1506.77	III
100	1553.02	III
110	1557.59	III
100	1563.73	III
100	1565.54	III
100	1623.40	III
100	1650.76	III
130	1670.39	III
140	1677.40	III
100	1716.99	III
120	1770.09	III
150	1770.67	III
110	1772.93	III
140	1773.36	III
160	1791.65	III
110	1803.03	III
170	1805.90	III
110	1839.30	III
120	1839.97	III
110	1840.14	III
100	2027.44	III
120	2030.32	III
120	2217.17	III
50	2298.29	IV
40	2451.58	IV
120	2452.07	III
50	2456.92	IV
130	2464.85	III
130	2470.29	III
120	2478.73	III
150	2484.37	III
120	2542.77	III
120	2580.04	III
130	2583.81	III
120	2593.23	III
130	2595.53	III
140	2599.28	III
130	2625.01	III
140	2629.70	III
120	2656.44	III
130	2755.55	III
160	2759.63	III
120	2788.15	III
160	2811.45	III
40	2820.74	IV
50	2826.13	IV
140	2833.99	III
150	2835.63	III
150	2860.33	III
120	2862.86	III
140	2887.58	III
150	2889.45	III
120	2905.30	III
140	2913.29	III
160	2916.34	III
140	2932.49	III
140	2994.28	III
120 h	2997.21	III
130	2997.53	III
120	2999.47	III
130	3039.25	III
120	3039.75	III
160	3042.80	III
150	3049.14	III
140	3113.62	III
160	3115.70	III
180	3121.54	III
140	3124.79	III
140	3134.23	III
140	3146.99	III
180	3174.17	III
170	3174.76	III
120	3214.00	III
200	3501.45	II
200	3501.57	II
200	3502.96	II
6	3594.10	I
12	3668.17	I
270	3847.09	II
260	3849.99	II
250	3851.67	II
5	3898.48	I
8	3930.69	I
5	3934.26	I
5	3948.56	I
240	4024.73	II
220	4025.01	II
230	4025.49	II
200	4103.51	II
200	4246.23	II
200	4299.17	II
140 h	4420.30	III
120 h	4427.35	III
120 h	4432.32	III
140 h	4479.99	III
6	4960.65	I
150	5012.54	III
160	5110.99	III
15	5230.41	I
12	5279.01	I
18	5540.52	I
12	5552.43	I
10	5577.33	I
20	5624.06	I
12	5626.93	I
15	5659.15	I
40	5667.53	I
90	5671.67	I
18	5689.14	I
25	5700.82	I
25	5707.31	I
140	5753.17	III
120	5761.20	III

Intensity	Wavelength/Å	
12	5950.15	I
25	5959.19	I
70	5965.28	I
50	5994.43	I
150	6015.83	I
80	6038.04	I
900	6047.54	I
100	6080.11	I
150	6091.82	III
140	6125.50	III
800	6149.76	I
400	6210.87	I
130	6233.57	III
13000	6239.65	I
10000	6348.51	I
140	6363.05	III
8000	6413.65	I
450	6569.69	I
300	6580.39	I
400	6650.41	I
1800	6690.48	I
400	6708.28	I
7000	6773.98	I
1500	6795.53	I
9000	6834.26	I
50000	6856.03	I
8000	6870.22	I
15000	6902.48	I
6000	6909.82	I
4000	6966.35	I
45000	7037.47	I
30000	7127.89	I
15000	7202.36	I
1000	7309.03	I
15000	7311.02	I
700	7314.30	I
5000	7331.96	I
120	7336.77	III
130	7354.94	III
10000	7398.69	I
4000	7425.65	I
2200	7482.72	I
2500	7489.16	I
900	7514.92	I
5000	7552.24	I
5000	7573.38	I
7000	7607.17	I
18000	7754.70	I
15000	7800.21	I
300	7879.18	I
500	7898.59	I
350	7936.31	I
300	7956.32	I
80	8016.01	II
1000	8040.93	I
900	8075.52	I
350	8077.52	I
350	8126.56	I
600	8129.26	I
300	8159.51	I
600	8179.34	I
300	8191.24	I
350	8208.63	I
2500	8214.73	I
3000	8230.77	I
500	8232.19	I
1500	8274.62	I
2000	8298.58	I
600	8302.40	I
900	8807.58	I
1000	8900.92	I
300	8912.78	I
350	9025.49	I
400	9042.10	I
350	9178.68	I
200	9433.67	I
25	9505.30	I
12	9662.04	I
25	9734.34	I
15	9822.11	I
12	9902.65	I
80 h	10047.98	II
15	10285.45	I
20	10862.31	I

Francium
Fr Z = 87

Intensity	Wavelength/Å	
	7177.	I

Gadolinium
Gd Z = 64

Intensity	Wavelength/Å	
1200	1007.24	IV
1200	1063.84	IV
1600	1476.98	IV
1500	1705.03	IV
1600	1706.01	IV
2000	1736.24	IV
1500	1815.32	IV
2200	1975.24	III
3400	2018.07	III
2800	2359.31	III
1400	2397.87	IV
2800	2697.39	III
2800	2703.28	III
2700	2727.89	III
9000	2904.73	III
9500	2955.53	III
1200	2999.04	II
2100	3010.13	II
1900	3027.60	II
2100	3032.84	II
1600	3034.05	II
2100	3081.99	II
3500	3100.50	II
930	3145.00	II
980	3156.53	II
980	3161.37	II
4000	3176.66	III
1400	3331.38	II
1100	3336.18	II
5400	3350.47	II
4300	3358.62	II
5400	3362.23	II
1100	3392.53	II
1100 d	3407.56	II
6900	3422.47	II
1700	3439.21	II
2700	3439.99	II
1400	3450.38	II
1100	3451.23	II
2700	3463.98	II
1700	3467.27	II
1700	3468.99	II
1400	3473.22	II
2200	3481.28	II
1700	3481.80	II
1700	3494.40	II
1400	3505.51	II
1100	3512.50	II
4300	3545.80	II
3900	3549.36	II
1400	3557.05	II
5400	3584.96	II
1100	3592.71	II
1100	3604.87	I
6100	3646.19	II
3900	3654.62	II
3100	3656.15	II
1400	3662.26	II
2700	3664.60	II
2000	3671.20	II
2000	3684.13	I
3100	3687.74	II
2000	3697.73	II
1300	3699.73	II
2700	3712.70	II
2000	3713.57	I
1400	3716.36	II
2000	3717.48	I
1800 d	3719.45	II
1500	3730.84	II
4500	3743.47	II
1400	3758.31	II
8700	3768.39	II
1400	3770.69	II
2900	3783.05	I
5100	3796.37	II
3700	3813.97	II
3300	3850.69	II
5100	3850.97	II
4300	3852.45	II
1600	3866.99	I
1500	3894.70	II
2200	3916.51	II
1200	3934.79	I
1400	3945.54	I
1200	3957.67	II
1100	4023.14	I
1100	4028.15	I
1400	4037.33	II
1600	4045.01	I
1300	4049.43	II
2200	4049.86	II
2600	4053.64	I
2600	4058.22	I
1900	4063.39	II
1300	4078.44	II
2800	4078.70	I
1500	4085.56	II
1100	4092.71	I
2600	4098.61	II
2200	4130.37	II
1100	4132.28	II
2400	4175.54	I
2400	4184.25	II
2200	4190.78	I
1300	4212.00	II
4800	4225.85	I
1700	4251.73	II
1600	4262.09	I
1100	4306.34	I
1800	4313.84	I
2600 d	4325.57	II
1900	4327.12	I
1000	4344.30	II
2200	4346.46	I
1400	4401.86	I
1400	4422.41	I
1100	4430.63	I
1100	4519.66	I
910	4537.81	I
520	4614.50	I
700	4694.33	I
410	4743.65	I
470	4767.24	I
300	4784.62	I
320	4821.69	I
280	4934.12	I
750	5015.04	I
75	5039.09	I
5000	5091.70	III
130	5098.38	II
910	5103.45	I
180	5108.91	II
120	5125.56	II
860	5155.84	I
190	5176.28	II
410	5197.77	I
280	5219.40	I
130	5233.93	I
320	5251.18	I
120	5252.14	II
140	5255.80	I
280	5283.08	I
280	5301.67	I
220	5302.76	I
280	5307.30	I
130	5321.50	I
280	5321.78	I
110	5327.32	I
170	5333.30	I
300	5343.00	I
200	5348.67	I
300	5350.38	I
240	5353.26	I
3000	5365.96	III
150	5370.63	I
4000	5553.30	III
3000	5587.88	III
190	5617.91	I
110	5632.25	I
260	5643.24	I
3000	5658.98	III
390	5696.22	I
120	5733.86	II
240	5791.38	I
220	5851.63	I
280	5856.22	I
110	5904.56	I
170	5911.45	II
85	5930.29	I
85	5936.84	I
65	5937.71	I
110 h	5988.02	I
430	6114.07	I
75	6305.15	II

Intensity	Wavelength/Å	
40	6331.35	I
40	6380.95	II
40 h	6538.15	I
55	6564.78	I
50	6634.36	II
35	6681.23	II
85	6730.73	I
50	6752.67	II
26	6786.33	II
100	6828.25	I
100	6916.57	I
50	6985.89	II
75	6991.92	I
60	6996.76	II
45	7006.16	II
35	7122.57	I
170	7168.37	I
28	7189.57	II
28	7262.66	I
35	7441.85	I
40	7464.36	I
55	7562.97	I
80	7733.50	I
35	7846.35	II
35	7856.93	I
25	7930.25	II
18	8146.15	I
21	8668.63	I
21 h	8832.06	II
14 h	8849.14	I
18 h	8867.31	I
5000	14332.88	III

Gallium

Ga Z = 31

Intensity	Wavelength/Å	
14	294.53	IV
61	295.67	IV
30	298.44	V
30	300.01	V
30	302.86	V
41	304.99	IV
30	307.03	V
30	308.26	V
30	311.79	V
30	313.68	V
40	319.41	V
40	322.31	V
50	322.99	V
30	323.10	V
40	324.25	V
40	324.95	V
40	326.14	V
30	326.77	V
30	328.65	V
25	423.18	IV
16	439.92	IV
50	620.00	III
40	622.01	III
90	806.51	III
90	817.30	III
50	828.70	III
20	878.17	V
40	973.21	V
40	989.75	V
90	1014.47	V
90	1019.71	V
120	1050.48	V
80	1054.56	V
90	1058.12	V
80	1066.69	V
60	1069.60	V
80	1073.77	V
90	1078.83	V
110	1079.60	V
60	1080.99	V
80	1085.00	III
250	1085.01	V
80	1087.37	V
90	1091.71	V
100	1094.36	V
80	1095.10	V
160	1102.83	V
140	1103.03	V
60	1105.61	III
75	1105.62	V
70	1106.17	V
80	1118.34	V
120	1126.40	V
130	1128.10	V
120	1128.53	V
100	1129.94	V
130	1136.07	V
67	1137.06	IV
130	1150.23	V
90	1150.27	III
70	1156.10	IV
120	1156.51	V
35	1157.74	V
70	1163.60	IV
25	1169.40	V
75	1170.58	IV
48	1171.71	IV
40	1178.95	V
68	1185.23	IV
40	1186.06	IV
73	1190.89	IV
73	1193.02	IV
75	1195.02	IV
69	1201.54	IV
72	1206.89	IV
80	1213.17	V
63	1216.15	IV
50	1228.03	IV
60	1236.38	IV
60	1238.59	IV
75	1245.53	IV
83	1258.77	IV
81	1264.66	IV
82	1267.15	IV
90	1267.16	III
81	1279.24	IV
15	1283.64	V
80	1285.33	IV
80	1293.46	III
60	1295.36	III
82	1295.86	IV
83	1299.46	IV
82	1303.53	IV
80	1309.68	IV
80	1314.82	IV
60	1323.15	III
85	1338.09	IV
77	1347.03	IV
76	1351.06	IV
70	1353.92	III
74	1364.63	IV
60	1395.54	IV
77	1402.55	IV
70	1405.32	IV
73	1465.87	IV
90	1495.07	III
50	1534.46	III
10	1813.98	II
15	1845.30	II
20	2091.34	II
90	2417.70	III
90	2423.98	III
15	2424.36	III
10	2632.66	I
10	2665.05	I
20	2700.47	II
15	2780.15	II
50	3521.77	III
80	3581.19	III
100	3589.34	III
10	3731.10	III
10	3806.60	III
10	4032.99	I
10	4172.04	I
15	4251.16	II
10 h	4254.04	II
10	4255.77	II
40	4262.00	II
100	4380.69	III
150	4381.76	III
100	4863.00	III
150	4993.78	III
10	5808.28	III
20	5848.25	III
15	5993.51	III
10	6334.2	II
2000	6396.56	I
1000	6413.44	I
10 h	7251.4	I
20 h	7403.0	I
30 h	7464.0	I
10 h	7620.5	I
50 h	7734.77	I
100 h	7800.01	I
15 h	8002.55	I
20 h	8074.25	I
100 h	8311.86	I
200 h	8386.49	I
200 h	9492.92	I
200 h	9493.12	I
300 h	9589.36	I
100 h	10905.95	I
400	11949.12	I
200	12109.78	I
60	14996.64	I
60	22016.81	I
70	22568.71	I

Germanium

Ge Z = 32

Intensity	Wavelength/Å	
700	294.51	V
1000	295.64	V
200	304.98	V
20	621.52	V
50	724.21	V
60	746.88	V
60	760.05	V
10	862.234	II
15	875.493	II
15	905.977	II
20	920.554	II
300	971.35	V
300	990.66	V
50	999.101	II
300	1004.38	V
100	1016.638	II
900	1045.71	V
700	1072.66	V
100	1075.072	II
300	1085.51	II
40	1088.45	III
800	1089.49	V
200	1098.71	II
500	1106.74	II
1000	1116.94	V
500	1120.46	II
700	1163.39	V
200	1164.27	II
500	1181.19	II
500	1181.65	II
200	1188.73	II
20	1188.99	IV
300	1191.26	II
700	1222.30	V
20	1229.81	IV
500	1237.059	II
500	1261.905	II
100	1264.710	II
200	1401.24	II
200	1538.091	II
500	1576.855	II
75	1581.070	II
100	1602.486	II
3 r	1615.57	I
2 r	1624.130	I
2 r	1630.173	I
3 r	1636.31	I
4 r	1639.730	I
2	1647.531	I
200	1649.194	II
2	1651.528	I
4 r	1651.955	I
3	1661.345	I
4 r	1663.539	I
10 h	1665.275	I
4	1667.802	I
3 r	1670.608	I
100 r	1691.090	I
200 r	1716.784	I
100 h	1739.102	I
100	1742.195	I
50	1746.065	I
200	1750.043	I
100	1758.279	I
100 h	1764.185	I
100 h	1765.284	I
50 h	1766.433	I
200	1774.176	I
200	1785.046	I
100 h	1793.071	I
75 h	1801.432	I

Intensity	Wavelength/Å	
200 h	1841.328	I
200 h	1842.410	I
100 h	1844.410	I
100 h	1845.872	I
100 h	1846.958	I
200	1853.134	I
500 r	1860.086	I
100	1865.052	I
300 r	1874.256	I
100	1895.197	I
500 r	1904.702	I
50 h	1908.434	I
30	1912.409	I
300 r	1917.592	I
100 h	1923.467	I
500 r	1929.826	I
10 h	1934.048	I
100 r	1937.483	I
500	1938.008	II
100 r	1938.300	I
500	1938.891	II
30 s	1944.116	I
200	1944.731	I
200	1955.115	I
500	1962.013	I
30 h	1963.373	I
30	1965.383	I
200	1970.880	I
200	1979.274	II
300 h	1987.849	I
300	1988.267	I
500 r	1998.887	I
200	2011.29	I
1700	2019.068	I
2400 r	2041.712	I
1600 r	2043.770	I
420	2054.461	I
220 h	2057.238	I
750 r	2065.215	I
2600 r	2068.656	I
420	2086.021	I
2000 r	2094.258	I
25	2104.45	III
240	2105.824	I
95 h	2124.744	I
50 h	2186.451	I
340 r	2198.714	I
15	2220.375	I
18	2256.001	I
18	2314.201	I
24	2327.918	I
15	2359.233	I
20	2379.144	I
10	2389.472	I
15	2397.885	I
130	2417.367	I
30	2436.412	I
30	2488.25	IV
90	2497.962	I
500	2500.54	II
70	2533.230	I
20	2542.44	IV
3	2556.298	I
28	2589.188	I
500	2592.534	I
8	2644.184	I
1200	2651.172	I
500	2651.568	I
500	2691.341	I
850	2709.624	I
400	2729.78	II
40	2740.426	I
650	2754.588	I
70	2793.925	I
80	2829.008	I
1000	2831.843	II
1000	2845.527	II
750	3039.067	I
600	3067.021	I
20	3124.816	I
35	3211.86	III
40	3255.05	III
110	3269.489	I
40	3434.03	III
300	3499.21	II
60	3554.19	IV
50	3676.65	IV
200	4178.96	III
70	4226.562	I
200	4260.85	III
150	4291.71	III
10	4685.829	I
1000	4741.806	II
1000	4814.608	II
50	4824.097	II
100	5131.752	II
200	5178.648	II
3	5194.583	I
6	5265.892	I
6	5513.263	I
8	5564.741	I
8	5607.010	I
6	5616.135	I
7	5621.426	I
6	5664.226	I
5	5664.842	I
9	5691.954	I
6	5701.776	I
5	5717.877	I
6	5801.029	I
9	5802.093	I
1000	5893.389	II
500	6021.041	II
75	6283.452	II
100	6336.377	II
100	6484.181	II
6	6557.488	I
50	7049.369	II
30	7145.390	II
5	7353.334	I
7	7384.208	I
10	7833.575	I
10	8031.039	I
6	8044.165	I
10	8256.013	I
10	8482.21	I
9	8700.60	I
5	9068.785	I
5	9095.957	I
6	9398.868	I
20	9474.993	II
20	9475.645	II
4	9492.559	I
7	9625.664	I
5	10039.436	I
4	10200.952	I
10	10382.427	I
10	10404.913	I
8	10734.068	I
8	10947.416	I
10	11125.130	I
230	11252.83	I
600	11714.76	I
1300	12069.20	I
1050	12391.58	I
235	13107.61	I
470	14822.38	I
150	16759.79	I
135	17214.34	I
70	18811.86	I
62	19279.24	I
28	20673.64	I

Gold

Au Z = 79

Intensity	Wavelength/Å	
100	843.44	III
100	845.14	III
200	945.10	III
100	1040.63	III
80	1044.49	III
80	1046.81	III
100 h	1239.96	III
100	1278.51	III
100	1314.84	III
100 h	1328.37	I
200	1336.72	III
180	1341.68	III
100	1348.89	III
150	1350.32	III
150	1355.61	III
150	1356.13	III
50	1363.98	I
500	1365.40	III
200	1367.17	III
60	1368.62	I
70	1374.82	I
50	1375.76	I
180	1377.73	III
150	1378.69	III
150	1379.98	III
125	1380.53	III
200	1381.36	III
80	1382.75	I
50	1385.33	I
300	1385.79	III
100	1389.41	III
180	1391.46	III
60	1392.27	I
180	1396.00	III
50	1402.12	I
100	1402.91	III
70	1407.38	I
100	1408.45	I
225	1409.50	III
250	1413.80	III
100	1414.27	III
100	1417.09	III
125	1417.39	III
150	1427.42	III
300	1428.93	III
80	1429.19	I
250	1430.06	III
275	1433.37	III
50	1435.79	I
250	1435.81	III
300	1439.12	III
200	1441.21	III
150	1446.37	III
250	1448.42	III
250	1454.95	III
100	1464.72	III
150	1471.28	III
100	1474.73	III
150	1481.10	III
100	1481.76	I
300	1487.15	III
250	1487.91	III
200	1489.47	III
250	1500.37	III
200	1502.47	III
200	1503.74	III
100	1542.00	III
100	1548.50	III
200	1567.54	III
200	1574.85	III
200	1579.44	III
150	1584.10	III
200	1587.16	I
200	1589.56	III
150	1593.41	III
70	1598.24	I
200	1600.51	III
250	1617.16	III
100	1617.78	III
500	1621.93	III
100	1624.34	I
300 d	1629.13	III
250	1638.88	III
50	1639.90	I
100	1644.17	III
150	1646.67	I
250	1652.74	III
250	1664.77	III
100	1665.76	I
100	1668.11	III
125	1673.93	III
1000	1693.94	III
150	1697.09	III
200	1698.98	III
200	1699.34	I
200	1700.00	III
200	1702.25	III
100	1707.53	III
250	1710.16	III
200	1715.69	III
100	1716.71	III
300	1717.83	III
500	1727.31	III
100 d	1733.17	III
300	1738.48	III
150	1744.39	III
500	1746.10	III
500	1756.92	III
500	1761.95	III
300	1767.42	III

Intensity	Wavelength/Å	
100	1774.42	III
800	1775.17	III
200	1776.40	III
100	1780.57	III
60	1783.22	II
300	1786.11	III
150	1792.65	III
500	1793.76	III
200	1801.98	III
400	1805.24	III
100	1809.81	III
400	1821.17	III
400	1844.89	III
150	1848.83	III
500	1861.80	III
150	1871.92	III
100	1879.83	I
150	1918.28	III
100	1932.04	III
100	1935.42	III
200	1948.79	III
100	1958.47	III
400	1989.63	III
150	1996.85	III
11000	2012.00	I
2600	2021.38	I
150	2082.09	II
300	2083.09	III
60	2110.68	II
100	2159.08	III
80	2167.33	III
200	2172.20	III
100	2184.11	III
500	2188.97	III
70	2248.56	II
80	2263.62	II
300	2322.27	III
180	2352.65	I
100	2382.40	III
120	2387.75	I
150	2402.71	III
150	2405.12	III
2600	2427.95	I
60	2533.52	II
250	2641.48	I
3400	2675.95	I
1100	2748.25	I
100	2780.82	I
1000	2802.04	II
300	2819.79	II
100	2822.55	II
100 h	2825.44	II
300	2837.85	II
100	2846.92	II
100	2856.74	II
300	2883.45	I
300	2891.96	I
100	2893.25	II
100	2907.04	II
300	2913.52	II
300	2918.24	II
100	2954.22	II
100 h	2990.27	II
300	2994.80	II
320	3029.20	I
300	3065.42	I
100	3122.50	II
1600	3122.78	I
100	3194.72	I
100	3227.99	III
300	3230.63	I
300	3308.30	I
300	3309.64	I
100	3309.86	III
100	3320.12	I
100	3355.15	I
100 h	3391.31	I
100	3395.40	I
100	3467.21	I
300	3557.36	I
300	3586.73	I
100	3611.57	I
100 h	3631.31	I
300	3637.90	I
100 h	3645.02	I
100	3650.74	I
100	3709.62	I
100	3796.01	I
100	3874.73	I
100 h	3892.26	I
400	3897.86	I
300	3909.38	I
100	3927.69	I
400	4040.93	I
700	4065.07	I
100	4084.10	I
100	4241.80	I
200	4315.11	I
120 h	4437.27	I
250	4488.25	I
900 h	4607.51	I
100 h	4620.56	I
500	4792.58	I
100	4811.60	I
100	5147.44	I
300	5230.26	I
100 h	5261.76	I
100	5655.77	I
100 h	5721.36	I
300	5837.37	I
100 h	5862.93	I
300 h	5956.96	I
600	6278.17	I
100	6562.68	I
600	7510.73	I
10	8145.06	I
10	9254.28	I

Hafnium
Hf Z = 72

Intensity	Wavelength/Å	
220	545.41	V
180	600.00	V
200	618.27	IV
200	644.54	IV
400	647.39	IV
600	665.65	IV
200	673.49	IV
270	867.25	V
180	875.88	V
180	885.58	V
180	896.14	V
180	901.54	V
180	919.10	V
270	921.67	V
245	951.62	V
180	960.12	V
180	964.74	V
160	971.51	V
160	1092.76	V
160	1201.76	V
270	1232.03	V
200	1233.59	V
160	1237.42	V
160	1239.53	V
160	1244.46	V
440	1396.66	V
270	1400.09	V
160	1401.70	V
370	1407.17	V
370	1408.38	V
270	1412.28	V
270	1413.51	V
160	1421.96	V
220	1422.53	V
370	1433.43	V
370	1437.27	V
500	1437.73	V
370	1445.40	V
270	1457.91	V
100	1717.21	IV
270	1719.32	V
550	1729.08	V
750	1731.83	V
750	1733.96	V
440	1741.74	V
1000	1749.11	V
1000	1750.19	V
500	1760.89	V
370	1765.62	V
270	1774.02	V
6200	2012.78	II
8500	2028.18	II
300	2070.94	III
1200	2096.18	II
200 h	2099.30	III
200 h	2110.31	III
200	2155.66	III
200	2183.50	III
200	2195.44	III
540	2210.82	II
200	2234.59	III
320	2254.01	II
160	2255.15	II
250	2266.83	II
620	2277.16	II
200 h	2313.44	III
230	2321.14	II
580	2322.47	II
300	2323.25	II
300	2324.89	II
200	2332.97	II
300	2336.47	III
200	2337.33	II
230	2343.32	II
320	2347.44	II
540	2351.22	II
250	2380.30	II
250	2383.540	III
170	2393.18	II
450	2393.36	II
670	2393.83	II
540	2405.42	II
370	2410.14	II
320	2417.69	II
390	2447.25	II
450	2460.49	II
400	2461.74	III
430	2464.19	II
210	2469.18	II
2000	2495.16	III
290	2496.99	II
580	2512.69	II
580	2513.03	II
1000	2515.16	III
890	2516.88	II
340	2531.19	II
200	2537.33	II
320	2551.40	II
400 h	2560.74	III
250	2563.61	II
300 h	2567.46	III
890	2571.67	II
320	2573.90	II
320	2576.82	II
300	2578.14	II
320	2582.54	II
390	2606.37	II
450	2607.03	II
230	2613.60	II
450	2622.74	II
160	2637.00	I
1100	2638.71	II
1100	2641.41	II
160	2642.75	I
670	2647.29	II
160	2657.84	II
210	2661.88	II
290	2683.35	II
200	2687.22	III
670	2705.61	I
210	2712.42	II
250	2718.59	I
710	2738.76	II
200	2743.64	I
360	2751.81	II
500	2753.60	III
450	2761.63	I
160	2766.96	I
170	2773.02	I
980	2773.36	II
180	2774.02	II
390	2779.37	I
230	2808.00	II
230	2813.86	II
170	2814.48	II
230	2817.68	I
200	2819.74	I
1200	2820.22	II
490	2822.68	II
180	2833.28	I
410	2845.83	I
270	2849.21	II
270	2850.96	I
180	2851.21	II
180	2860.56	I

Intensity	Wavelength/Å	
760	2861.01	II
760	2861.70	II
2100	2866.37	I
210	2887.14	I
800	2889.62	I
1800	2898.26	I
1200	2904.41	I
890	2904.75	I
2000	2916.48	I
580	2918.58	I
320	2919.59	II
180	2924.62	I
490	2929.63	II
450	2929.90	I
710	2937.80	II
2000	2940.77	I
160	2944.71	I
1200	2950.68	I
1100	2954.20	I
540	2958.02	I
1400	2964.88	I
620	2966.93	I
710	2968.81	II
890	2975.88	II
1100	2980.81	I
210	2982.72	I
170	3000.10	II
800	3005.56	I
1100	3012.90	II
540	3016.78	I
1100	3016.94	II
980	3018.31	I
1200	3020.53	I
410	3031.16	II
710	3050.76	I
1100	3057.02	I
850	3067.41	I
2100	3072.88	I
170	3074.10	I
250	3074.79	I
430	3080.84	I
200	3096.76	I
340	3101.40	II
710	3109.12	II
710	3131.81	I
850	3134.72	II
170	3139.65	II
220	3145.32	II
220	3148.41	I
450	3156.63	I
270	3159.82	I
710	3162.61	II
450	3168.39	I
890	3172.94	I
450	3176.86	II
220	3181.01	I
360	3193.53	II
670	3194.19	II
200	3196.93	I
310	3206.11	I
180	3210.98	I
180	3217.30	II
180	3220.61	II
360	3247.66	I
220	3249.53	I
890	3253.70	II
270	3255.28	II
180	3273.66	II
200 h	3279.67	III
270	3279.98	II
160	3291.05	I
210	3306.12	I
340	3310.27	I
670	3312.86	I
180	3317.99	II
890	3332.73	I
370	3352.06	II
230	3358.91	I
180	3360.06	I
180	3378.93	I
230	3384.70	II
170	3386.21	I
800	3389.83	II
230	3392.81	I
230	3394.59	II
230	3397.26	I
230	3397.60	I
2300	3399.80	II
170	3400.21	I
180	3402.51	I
230	3410.17	II
230	3417.34	I
410	3419.18	I
200	3428.37	II
250	3438.24	II
710	3472.40	I
200	3478.99	II
480	3479.28	II
250	3495.75	II
250	3497.16	I
980	3497.49	I
1200	3505.23	II
980	3523.02	I
980	3535.54	II
760	3536.62	I
180	3548.81	I
540	3552.70	II
1300	3561.66	II
270	3567.36	I
1100	3569.04	II
210	3597.42	II
540	3599.87	I
800	3616.89	I
320	3630.87	II
800	3644.36	II
320	3649.10	I
200	3651.84	I
220	3665.35	II
200	3672.27	I
480	3675.74	I
2200	3682.24	I
280	3696.51	I
240	3699.72	II
340	3701.15	II
1000	3717.80	I
650	3719.28	II
160	3729.10	I
460	3733.79	I
160	3737.88	II
400	3746.80	I
170	3766.92	II
200	3768.25	I
1400	3777.64	I
1400	3785.46	I
650	3793.37	II
850 d	3800.38	I
320	3811.78	I
1300	3820.73	I
280	3830.02	I
800	3849.18	I
600	3858.31	I
230	3860.91	I
200	3872.55	II
160	3877.10	II
380	3880.82	II
200	3882.52	I
200	3889.23	I
200	3889.33	I
620	3899.94	I
620	3918.09	II
200	3923.90	II
320	3931.38	I
410	3951.83	I
160	3968.01	I
200	3973.48	I
180	4032.27	I
230	4062.84	I
180	4083.35	I
540	4093.16	II
1100	4174.34	I
160	4206.58	II
190	4209.70	I
170	4228.08	I
170	4232.44	II
170	4260.98	I
200	4263.39	I
170	4272.85	II
320	4294.79	I
160	4330.27	I
180	4336.66	II
250	4356.33	I
180	4370.97	II
160	4417.91	I
200	4438.04	I
250	4565.94	I
500 d	4598.80	I
230	4620.86	I
210	4655.19	I
120	4699.01	I
160	4782.74	I
310	4800.50	I
130	4859.24	I
120	4975.25	I
95	5018.20	I
95	5047.45	I
75	5170.18	I
230	5181.86	I
110	5243.99	I
120	5294.87	I
110	5354.73	I
110	5373.86	I
75	5452.92	I
230	5550.60	I
230	5552.12	I
95	5613.27	I
160	5719.18	I
95	6098.67	I
95	6185.13	I
85	6789.27	I
160	6818.94	I
160	7063.83	I
570	7131.81	I
650	7237.10	I
410	7240.87	I
360	7624.40	I
110	7740.17	I
310	7845.35	I
130	7920.71	I
250	7994.73	I
130	8204.58	I
150	8546.48	I
160	8640.06	I
40	8711.24	I
65	9004.73	I

Helium

He $Z = 2$

Intensity	Wavelength/Å	
15	231.454	II
20	232.584	II
30	234.347	II
50	237.331	II
100	243.027	II
300	256.317	II
1000	303.780	II
500	303.786	II
10	320.293	I
2	505.500	I
3	505.684	I
4	505.912	I
5	506.200	I
7	506.570	I
10	507.058	I
15	507.718	I
20	508.643	I
25	509.998	I
35	512.098	I
50	515.616	I
100	522.213	I
400	537.030	I
1000	584.334	I
50	591.412	I
5	958.70	II
6	972.11	II
8	992.36	II
15	1025.27	II
30	1084.94	II
35	1215.09	II
50	1215.17	II
120	1640.34	II
180	1640.47	II
7	2385.40	II
9	2511.20	II
50	2577.6	I
1	2723.19	I
12	2733.30	II
2	2763.80	I
10	2818.2	I
4	2829.08	I
10	2945.11	I
40	3013.7	I
20	3187.74	I
3	3202.96	II
15	3203.10	II
1	3354.55	I
2	3447.59	I

Intensity	Wavelength/Å	
1	3587.27	I
3	3613.64	I
2	3634.23	I
3	3705.00	I
1	3732.86	I
10	3819.607	I
1	3819.76	I
500	3888.65	I
20	3964.729	I
1	4009.27	I
50	4026.191	I
5	4026.36	I
12	4120.82	I
2	4120.99	I
3	4143.76	I
10	4387.929	I
3	4437.55	I
200	4471.479	I
25	4471.68	I
6	4685.4	II
30	4685.7	II
30	4713.146	I
4	4713.38	I
20	4921.931	I
100	5015.678	I
10	5047.74	I
5	5411.52	II
500	5875.62	I
100	5875.97	I
8	6560.10	II
100	6678.15	I
3	6867.48	I
200	7065.19	I
30	7065.71	I
50	7281.35	I
1	7816.15	I
2	8361.69	I
2	9063.27	I
2	9210.34	I
10	9463.61	I
4	9516.60	I
3	9526.17	I
1	9529.27	I
1	9603.42	I
3	9702.60	I
6	10027.73	I
2	10031.16	I
15	10123.6	II
1	10138.50	I
10	10311.23	I
2	10311.54	I
3	10667.65	I
300	10829.09	I
1000	10830.25	I
2000	10830.34	I
9	10913.05	I
3	10917.10	I
4	11626.4	II
30	11969.12	I
20	12527.52	I
50	12784.99	I
20	12790.57	I
7	12845.96	I
10	12968.45	I
2	12984.89	I
12	15083.64	I
200	17002.47	I
1	18555.55	I
6	18636.8	II
500	18685.34	I
200	18697.23	I
100	19089.38	I
20	19543.08	I
1000	20581.30	I
80	21120.07	I
10	21121.43	I
20	21132.03	I
3	30908.5	II
4	40478.90	I

Holmium
Ho Z = 67

Intensity	Wavelength/Å	
170	2502.91	II
170	2533.80	I
190	2605.86	II
270	2750.35	II
270	2769.89	II
300	2824.20	II
270 c	2831.69	II
270	2849.10	II
360	2880.26	II
460	2880.98	II
570 c	2909.41	II
410 c	2979.63	II
410	2987.64	II
480 c	3049.38	II
410 c	3054.00	II
500 c	3057.45	II
500 c	3082.34	II
910	3084.36	II
430 c	3086.54	II
760	3118.50	II
580 c	3166.62	II
810	3173.78	II
810 c	3181.50	II
980 c	3281.97	II
630 c	3337.23	II
980 c	3343.58	II
8100 c	3398.98	II
810 c	3410.26	II
1400 c	3414.90	II
5400	3416.46	II
1200	3421.63	II
2000 c	3425.34	II
2000 c	3428.13	II
3200	3453.14	II
16000 c	3456.00	II
1600	3461.97	II
810 c	3473.91	II
5400 c	3474.26	II
6300	3484.84	II
2500 c	3494.76	II
810 c	3498.88	II
810	3510.73	I
4100 c	3515.59	II
410 c	3519.94	II
630	3540.76	II
1600	3546.05	II
1100 c	3556.78	II
410	3560.15	II
410 c	3573.24	II
630 c	3574.80	II
810	3579.12	I
410	3580.75	II
410	3581.83	II
630 c	3592.23	II
1100 cw	3598.77	II
540 c	3600.95	II
410	3618.43	I
430 c	3626.69	II
490	3627.25	II
430 c	3631.76	II
430 c	3638.30	II
1600 c	3662.29	I
1400	3667.97	I
720	3679.19	I
670	3679.70	I
720	3682.65	I
580	3690.65	I
410	3700.04	I
490 c	3702.35	II
430	3712.88	I
450	3720.72	I
1100	3731.40	I
810	3736.35	I
3200 cw	3748.17	II
8900 c	3796.75	II
8900 c	3810.73	II
410 c	3835.35	II
1300 cw	3837.51	II
410 c	3842.05	II
1100	3843.86	II
490 c	3846.73	II
1800 c	3854.07	II
2700 c	3861.68	II
3000 c	3888.96	II
13000 c	3891.02	II
1300 cw	3905.68	II
580	3955.73	I
490	3959.68	I
2700	4040.81	I
5400 c	4045.44	II
8100	4053.93	I
1700	4065.09	II
720	4068.05	I
8900	4103.84	I
2900	4108.62	I
1500	4120.20	I
1300	4125.65	I
4300	4127.16	I
1500	4136.22	I
980 cw	4152.61	II
8100	4163.03	I
2500	4173.23	I
540	4194.35	I
2000	4227.04	I
1300 cw	4254.43	I
490	4264.05	I
1300	4350.73	I
300	4477.64	II
290	4629.10	II
80	4701.69	II
130	4709.84	II
130 c	4717.52	I
290	4742.04	II
100 c	4757.01	I
65	4782.92	I
290	4939.01	I
250 c	4967.21	II
220	4979.97	I
90	4995.05	I
130	5042.37	I
80	5093.07	I
140	5127.81	I
130	5142.59	II
110	5143.22	II
160	5149.59	II
90 c	5167.88	I
130 c	5182.11	I
90	5190.11	II
65	5251.82	I
90	5301.25	I
80	5330.11	I
90	5359.99	I
100	5407.08	I
70	5566.52	I
65	5627.60	I
140	5659.58	I
140 c	5691.47	I
140 c	5696.57	I
140 c	5860.28	I
70 c	5882.99	I
70	5921.76	I
70 cw	5948.03	I
70	5972.76	I
90	5973.52	I
230 c	5982.90	I
120	6081.79	I
70	6208.65	I
70 c	6305.36	I
70	6550.97	I
260	6604.94	I
120	6628.99	I
55 cw	6694.32	I
55 c	6785.43	I
40 cw	6939.49	I
45 cw	6950.39	I
140	7555.09	I
40 c	7628.42	I
50 c	7693.15	I
60 cw	7815.48	I
60	7894.64	I
50	8512.94	I
40	8670.19	I
90	8915.98	II

Hydrogen
H Z = 1

Intensity	Wavelength/Å	
15	926.226	I
20	930.748	I
30	937.803	I
50	949.743	I
100	972.537	I
300	1025.722	I
1000	1215.668	I
500	1215.674	I
5	3835.384	I
6	3889.049	I
8	3970.072	I
15	4101.74	I
30	4340.47	I
80	4861.33	I
120	6562.72	I
180	6562.852	I
5	9545.97	I
7	10049.4	I

Line Spectra of the Elements (continued): Hydrogen—Iodine

Intensity	Wavelength/Å	
12	10938.1	I
20	12818.1	I
40	18751.0	I
5	21655.3	I
8	26251.5	I
15	40511.6	I
4	46525.1	I
6	74578.	I
Indium		
In Z = 49		
17	378.61	V
17	386.21	V
14	388.91	V
25	393.89	V
25	400.57	V
25	402.39	V
622	472.71	IV
689	479.39	IV
709	498.62	IV
10	882.24	III
10	890.84	III
10	915.87	III
85	954.67	IV
87	973.50	IV
86	991.60	IV
89	1024.68	IV
85	1024.79	IV
88	1031.45	IV
82	1031.98	IV
80	1054.43	IV
84	1063.03	IV
83	1068.25	IV
82	1069.82	IV
86	1077.64	IV
90	1082.10	IV
83	1086.33	IV
82	1096.81	IV
84	1097.18	IV
85	1116.10	IV
80	1124.06	IV
90	1131.46	IV
85	1144.43	IV
80	1145.41	IV
89	1146.62	IV
83	1154.11	IV
84	1154.60	IV
90	1157.71	IV
90	1157.82	IV
85	1159.78	IV
88	1176.50	IV
85	1191.58	IV
83	1204.87	IV
90	1206.55	IV
88	1221.50	IV
85	1221.90	IV
85	1233.58	IV
87	1235.84	IV
90	1373.20	IV
88	1398.77	IV
81	1412.09	IV
100	1625.42	III
100	1748.83	III
30	1842.41	III
40	1850.30	III
30	2154.08	III
50	2205.28	II
40	2281.64	II
100 c	2306.05	II
90 d	2313.21	II
70 d	2327.95	II
80 h	2334.57	II
110 d	2382.63	II
70 h	2427.20	II
100	2447.90	II
110 d	2488.62	II
90	2488.95	II
80	2498.59	II
100	2499.60	II
90 d	2500.99	II
110 d	2512.31	II
100	2521.37	I
100	2527.41	III
160 d	2554.44	II
1100	2560.15	I
200	2601.76	I
90 d	2654.70	II
100 d	2662.63	II
140 d	2668.65	II
140 d	2674.56	II
80	2683.12	II
1600	2710.26	I
300	2713.94	I
80	2726.15	III
130 d	2749.75	II
700	2753.88	I
90 d	2818.97	II
180 c	2836.92	I
80	2865.68	II
120 d	2890.18	II
1100	2932.63	I
100	2941.05	II
100	2982.80	III
110 c	2999.40	II
100	3008.08	III
8000	3039.36	I
110 d	3099.80	II
180 c	3101.8	II
130 c	3138.60	II
80 c	3142.75	II
130 d	3146.70	II
150	3155.77	II
100 c	3158.40	II
90 c	3176.30	II
90 d	3198.11	II
13000	3256.09	I
3000	3258.56	I
90 c	3338.50	II
100 c	3404.28	II
110 d	3438.40	II
180 c	3693.91	II
95 c	3708.13	II
380 w	3716.14	II
120 c	3718.30	II
160 c	3718.72	II
160 c	3723.40	II
170 w	3795.21	II
230 c	3799.21	II
250 c	3834.65	II
200 c	3842.18	II
100	3852.82	III
100	3889.78	II
100 c	3902.07	II
250 w	3962.35	II
120 c	4004.66	II
140 d	4013.92	II
100	4023.77	III
150	4032.32	III
410 w	4056.94	II
100	4071.57	III
100	4072.93	III
17000	4101.76	I
140 c	4205.14	II
100 d	4213.04	II
110 d	4219.66	II
100	4252.68	III
150 d	4372.87	II
150 c	4500.78	II
18000	4511.31	I
110 c	4549.01	II
140 c	4570.85	II
180 w	4578.02	II
180 w	4578.40	II
140 c	4616.08	II
170 c	4617.17	II
250 c	4620.14	II
150 w	4620.70	II
170 c	4627.30	II
140 c	4637.04	II
380 c	4638.16	II
220 c	4644.58	II
360 c	4655.62	II
320 w	4656.74	II
190 c	4681.11	II
450 w	4684.8	II
90 d	4907.06	II
150 c	4973.77	II
80 h	5109.36	II
100 w	5115.14	II
140 c	5117.40	II
270 c	5120.80	II
200 w	5121.75	II
80 d	5129.85	II
240 c	5175.42	II
140 c	5184.44	II
200	5248.77	III
150 c	5309.45	II
80	5411.41	II
140 c	5418.45	II
220 w	5436.70	II
130 c	5497.50	II
140 c	5507.08	II
320 c	5513.00	II
250 w	5523.28	II
130 c	5536.50	II
190 w	5555.45	II
240 c	5576.90	II
200 w	5636.70	II
100	5645.15	III
160 c	5708.50	II
100 c	5721.80	II
100	5819.50	III
210 c	5853.15	II
490 w	5903.4	II
260 w	5915.4	II
120 c	5918.78	II
130 c	6062.9	II
250 c	6095.95	II
210 c	6108.66	II
180 w	6115.9	II
230 w	6128.7	II
240 w	6129.4	II
320 w	6132.1	II
150 c	6140.0	II
90	6143.23	II
140 c	6148.10	II
190 w	6149.5	II
80	6161.15	II
180 w	6162.45	II
200	6197.72	III
100 c	6224.28	II
280 w	6228.3	II
140 w	6231.1	II
270 w	6304.8	II
290 w	6362.3	II
300 w	6469.0	II
210 c	6541.20	II
190 c	6751.88	II
180 c	6765.9	II
100 c	6783.72	II
320 w	6891.5	II
380 w	7182.9	II
180 c	7255.0	II
210 c	7276.5	II
180 c	7303.4	II
320 c	7350.6	II
100 c	7632.7	II
100 c	7682.9	II
210 c	7740.7	II
100 c	7776.96	II
180 c	7789.0	II
90 c	7840.9	II
240 c	8227.0	II
100 w	8813.5	II
80 c	8832.6	II
120 c	9197.7	II
120 c	9202.0	II
220 w	9213.0	II
160 d	9241.1	II
100	9977.86	I
200	10257.03	I
100 h	10744.31	I
20	11334.72	I
20	11731.48	I
10	12912.59	I
9	13429.96	I
7	14719.08	I
6	22291.06	I
7	23879.13	I
Iodine		
I Z = 53		
30	363.78	V
36	380.74	V
45	565.53	V
50	607.57	V
6	612.46	IV
6	666.81	III
8	705.11	III
7	784.64	III
7	784.80	III
8	795.52	III
7	919.28	IV
10000	1034.66	II
8	1094.20	III
10000	1139.80	II

Intensity	Wavelength/Å	
10000	1160.56	II
20000	1166.48	II
10000	1178.65	II
15000	1187.34	II
10000	1190.85	II
200	1218.41	I
20000	1220.89	II
600	1224.05	I
600	1224.08	I
500	1228.89	I
20000	1234.06	II
600	1251.34	I
8	1252.35	III
2500	1259.15	I
3000	1259.51	I
800	1261.27	I
600	1267.57	I
600	1267.60	I
1500	1275.26	I
3000	1289.40	I
10000	1300.34	I
3000	1302.98	I
3000	1313.95	I
3000	1317.54	I
2000	1330.19	I
20000	1336.52	II
5000	1355.10	I
3000	1357.97	I
5000	1360.97	I
3000	1361.11	I
2500	1367.71	I
2500	1368.22	I
4000	1383.23	I
3000	1390.75	I
2000	1392.90	I
2000	1400.01	I
8000	1425.49	I
5000	1446.26	I
5000	1453.18	I
5000	1457.39	I
5000	1457.47	I
10000	1457.98	I
2500	1458.79	I
4000	1459.15	I
2500	1465.83	I
1000	1485.92	I
5000	1492.89	I
5000	1507.04	I
5000	1514.68	I
15000	1518.05	I
2500	1526.45	I
5000	1593.58	I
5000	1617.60	I
2500	1640.78	I
15000	1702.07	I
12000	1782.76	I
5000	1799.09	I
75000	1830.38	I
15000	1844.45	I
2000	2061.63	I
7	2361.13	IV
6	2372.45	IV
7	2376.46	IV
8	2387.11	IV
9	2426.10	IV
8	2475.35	IV
8	2519.74	IV
8	2545.67	IV
7	2545.71	III
8	2652.23	IV
5000	3078.75	II
200	4102.23	I
200	4129.21	I
100 d	4134.15	I
500 d	4321.84	I
250	4763.31	I
1000	4862.32	I
200	4916.94	I
10000	5119.29	I
3000 c	5161.20	II
1000	5234.57	I
3000 c	5245.71	II
10000	5338.22	II
5000 c	5345.15	II
600 c	5427.06	I
3000	5435.83	II
2000 c	5464.62	II
10000	5625.69	II
2000 c	5690.91	II
4000 c	5710.53	II
1000 d	5764.33	I
2000	5894.03	I
5000	5950.25	II
300	5984.86	I
2000 d	6024.08	I
2000 c	6074.98	II
1000	6082.43	I
2000 c	6127.49	II
800	6191.88	I
500	6213.10	I
800	6244.48	I
1000	6293.98	I
500	6313.13	I
800	6330.37	I
400	6333.50	I
2000	6337.85	I
1000	6339.44	I
500	6359.16	I
1000	6566.49	I
2000	6583.75	I
1000	6585.27	I
5000	6619.66	I
500	6661.11	I
500 c	6697.29	I
400	6732.03	I
4000	6812.57	II
500	6989.78	I
500	7120.05	I
1200	7122.05	I
2000	7142.06	I
1000	7164.79	I
400 d	7191.66	I
700	7227.30	I
1000	7236.78	I
500	7237.84	I
5000	7402.06	I
1000	7410.50	I
500	7416.48	I
5000	7468.99	I
500 c	7490.52	I
2000	7554.18	I
500 d	7556.65	I
2000 c	7700.20	I
600	7897.98	I
500	7969.48	I
1000	8003.63	I
99000	8043.74	I
300 d	8065.70	I
1000	8090.76	I
800 c	8169.38	I
500 d	8222.57	I
4000	8240.05	I
10000 c	8393.30	I
1000	8486.11	I
1500 c	8664.95	I
500 c	8700.80	I
250 d	8748.22	I
1000	8853.24	I
2000	8853.80	I
3000	8857.50	I
1000 d	8898.50	I
400	8964.69	I
400	8993.13	I
5000	9022.40	I
15000	9058.33	I
1000	9098.86	I
12000	9113.91	I
600	9128.03	I
600	9227.74	I
1000	9335.05	I
4000	9426.71	I
3000	9427.15	I
2000	9598.22	I
2000	9649.61	I
3000 d	9653.06	I
5000	9731.73	I
500	10003.05	I
750	10131.16	I
1000	10238.82	I
400	10375.20	I
400	10391.74	I
5000	10466.54	I
400	11236.56	I
350	11558.46	I
320	11778.34	I
450	11996.86	I
300	12033.69	I
150	12304.58	I
60	13149.16	I
140	13958.27	I
200	14287.02	I
100	14460.00	I
225	15032.57	I
105	15528.65	I
150	16037.33	I
15	18275.71	I
20	18348.52	I
15	18982.41	I
35	19070.17	I
110	19105.12	I
50	19370.02	I
10	20648.69	I
220	22183.03	I
150	22226.53	I
30	22309.21	I
32	24420.82	I
12	27365.42	I
9	27573.05	I
10	30361.93	I
8	30383.88	I
10	34295.73	I
9	34513.11	I
3	40228.54	I
2	41633.80	I

Iridium

Ir Z = 77

Intensity	Wavelength/Å	
9900	2010.65	I
8700	2022.35	I
15000	2033.57	I
6200	2052.22	I
5000	2060.64	I
3700	2083.22	I
3100	2085.74	I
17000	2088.82	I
14000	2092.63	I
2700	2112.68	I
1800	2119.54	I
2000	2125.44	I
4500	2126.81	II
2000	2127.52	I
4500	2127.94	I
3700	2148.22	I
2500	2150.54	I
3500	2152.68	II
2900	2155.81	I
7900	2158.05	I
2100	2162.88	I
5800	2169.42	II
4500	2175.24	I
2700	2178.17	I
1600	2187.43	II
1100	2190.38	II
740	2191.64	I
910	2208.09	II
1300	2220.37	I
790	2221.07	II
2500	2242.68	II
620	2245.76	II
2100	2253.38	I
2100	2255.10	I
1400	2255.81	I
350	2258.51	I
1400	2258.86	I
830	2264.61	I
1100	2266.33	I
1000	2268.90	I
660	2280.00	I
950	2281.02	II
660	2281.91	I
330	2284.60	I
330	2295.08	I
790	2298.05	I
460	2299.53	I
910	2300.50	I
2700	2304.22	I
410	2305.47	I
210	2307.27	I
910	2308.93	I
460	2315.38	I
410	2321.45	I
410	2321.58	I
210	2327.98	I
540	2333.30	I
740	2333.84	I

Intensity	Wavelength/Å	
580	2334.50	I
1600	2343.18	I
740	2343.61	I
580	2355.00	I
230	2357.53	II
410	2358.16	I
500	2360.73	I
2500	2363.04	I
370	2368.04	II
3500	2372.77	I
290	2375.09	II
250	2377.28	I
250	2377.98	I
500	2379.38	I
540	2381.62	I
210	2383.17	I
1300	2386.89	I
2500	2390.62	I
2700	2391.18	I
230	2407.59	I
290	2409.37	I
290	2410.17	I
290	2410.73	I
540	2413.31	I
370	2415.86	I
620	2418.11	I
210	2424.89	I
370	2424.99	I
290	2425.66	I
540	2427.61	I
540	2431.24	I
1300	2431.94	I
270	2435.14	I
250	2445.34	I
250	2447.76	I
910	2452.81	I
1300	2455.61	I
230	2455.87	I
210	2457.03	I
210	2457.23	I
870	2467.30	I
3300	2475.12	I
210	2478.11	I
2100	2481.18	I
620	2493.08	I
210	2496.27	I
250	2502.63	I
4100	2502.98	I
210	2513.71	I
990	2533.13	I
1100	2534.46	I
580	2537.22	I
580	2542.02	I
7900	2543.97	I
790	2546.03	I
210	2551.40	I
210	2555.35	I
910	2564.18	I
210	2569.88	I
230	2572.70	I
740	2577.26	I
740	2592.06	I
740	2599.04	I
700	2608.25	I
1800	2611.30	I
210	2614.98	I
330	2617.78	I
210	2619.88	I
250	2625.32	I
700	2634.17	I
250	2639.42	I
3500	2639.71	I
210	2644.19	I
1800	2661.98	I
350	2662.63	I
2700	2664.79	I
520	2669.91	I
520	2671.84	I
330	2673.61	I
270	2692.34	I
3000	2694.23	I
330	2772.46	I
250	2775.55	I
520	2781.29	I
330	2785.22	I
540	2797.35	I
1600	2797.70	I
380	2798.18	I
410	2800.82	I
680	2823.18	I
1200	2824.45	I
820	2836.40	I
1100	2839.16	I
820	2840.22	I
3800	2849.72	I
380	2875.60	I
380	2875.98	I
270	2877.68	I
820	2882.64	I
650	2897.15	I
260	2901.95	I
260	2904.80	I
200	2907.24	I
440	2916.36	I
230	2918.57	I
4400	2924.79	I
1200	2934.64	I
880	2936.68	I
250	2938.47	I
2700	2943.15	I
230	2946.97	I
200	2949.76	I
1200	2951.22	I
200	2974.95	I
440	2980.65	I
300	2996.08	I
220	3002.25	I
600	3003.63	I
270	3017.31	I
380	3029.36	I
330	3039.26	I
300	3047.16	I
300	3049.44	I
300	3057.28	I
1600	3068.89	I
320	3083.22	I
240	3086.44	I
390	3088.04	I
510	3100.29	I
510	3100.45	I
340	3120.76	I
200	3121.78	I
3400	3133.32	I
490	3168.88	I
370	3177.58	I
370	3198.92	I
610	3212.12	I
370	3219.51	I
5100	3220.78	I
300	3229.28	I
470	3241.52	I
200	3262.01	I
390	3266.44	I
200	3322.60	I
560	3368.48	I
660	3437.02	I
410	3448.97	I
3200	3513.64	I
220	3515.95	I
410	3522.03	I
320	3558.99	I
1200	3573.72	I
320	3594.39	I
220	3609.77	I
660	3628.67	I
220	3636.20	I
300	3661.71	I
300	3664.62	I
320	3674.98	I
200	3687.08	I
200	3731.36	II
530	3747.20	I
3100	3800.12	I
230	3817.24	I
480	3902.51	I
480	3915.38	I
400	3934.84	I
590	3976.31	I
460	3992.12	I
350	4033.76	I
370	4069.92	I
150	4070.68	I
100	4092.61	I
140	4115.78	I
90	4172.56	I
260	4268.10	I
220	4311.50	I
160	4399.47	I
65	4403.78	I
110	4426.27	I
55	4478.48	I
55	4545.68	I
30	4548.48	I
35	4568.09	I
75	4616.39	I
26	4656.18	I
50	4728.86	I
26	4756.46	I
65	4778.16	I
30	4795.67	I
50	4938.09	I
26	4970.48	I
25	4999.74	I
25	5002.74	I
30	5014.98	I
30	5123.66	I
35	5364.32	I
75	5449.50	I
45	5625.55	I
35	5894.06	I
20	6110.67	I
12	6288.28	I
10	6686.08	I
6	7834.32	I

Iron

Fe Z = 26

Intensity	Wavelength/Å	
350	386.16	V
350	386.88	V
400	387.20	V
400	387.50	V
400	387.76	V
400	387.78	V
400	395.90	V
400	404.62	V
400	405.50	V
800	407.42	V
600	407.44	V
400	407.49	V
500	407.75	V
400	409.71	V
400	410.20	V
600	411.55	V
700	417.39	V
700	418.04	V
500	418.47	V
700	421.06	V
500	421.78	V
500	422.31	V
500	426.06	V
500	426.11	V
350	426.97	V
17	525.69	IV
15	526.29	IV
13	526.63	IV
14	536.61	IV
15	537.10	IV
13	537.26	IV
14	537.79	IV
13	537.94	IV
13	552.14	IV
14	607.53	IV
13	608.80	IV
10	813.38	III
10	844.28	III
10 p	861.83	III
10	891.17	III
10	950.33	III
10	981.37	III
10 w	983.88	III
12	1055.27	II
15	1068.36	II
15	1071.60	II
15	1096.89	II
12	1099.12	II
18	1112.09	II
12	1121.99	II
12	1122.86	II
12	1128.07	II
12	1130.43	II
15	1133.41	II
12	1133.68	II
12	1138.64	II
12	1142.33	II
12	1143.23	II

Intensity	Wavelength/Å	
18	1144.95	II
12	1147.41	II
15	1148.29	II
12	1151.16	II
12	1267.44	II
12	1272.00	II
400	1317.86	V
400	1323.27	V
400	1330.40	V
400	1359.01	V
600	1361.82	V
700	1373.59	V
600	1373.67	V
500	1376.34	V
500	1378.56	V
800	1387.94	V
400	1397.97	V
600	1400.24	V
800	1402.39	V
400	1406.67	V
500	1406.82	V
400	1407.25	V
600	1409.45	V
400	1415.20	V
600	1420.46	V
800	1430.57	V
13	1431.43	IV
800	1440.53	V
400	1442.22	V
800	1446.62	V
700	1448.85	V
400	1449.93	V
700	1456.16	V
500	1459.83	V
400	1460.73	V
500	1462.63	V
700	1464.68	V
500	1465.38	V
400	1466.65	V
500	1469.00	V
500	1479.47	V
10 h	1505.17	III
13	1526.60	IV
13	1530.26	IV
14	1532.63	IV
13	1532.91	IV
15	1533.86	IV
13	1533.95	IV
14	1536.58	IV
10 h	1538.63	III
13	1542.16	IV
14	1542.70	IV
12 h	1550.20	III
13	1566.26	IV
14	1568.27	IV
13	1591.51	IV
13	1592.05	IV
13	1598.01	IV
13	1600.58	IV
10 h	1601.21	III
13	1601.67	IV
13	1603.18	IV
13	1603.73	IV
13	1604.88	IV
13	1605.68	IV
15	1605.97	IV
13	1606.98	IV
17	1609.10	IV
14	1609.83	IV
13	1610.47	IV
13	1611.20	IV
13	1613.64	IV
15	1614.02	IV
13	1614.64	IV
13	1615.00	IV
16	1616.68	IV
14	1617.68	IV
12	1618.47	II
14	1619.02	IV
13	1621.16	IV
14	1621.57	IV
13	1623.38	IV
13	1623.53	IV
15	1626.47	IV
14	1626.90	IV
13	1628.54	IV
13	1630.18	IV
17	1631.08	IV
15	1631.12	II
14	1632.40	IV
13	1634.01	IV
18	1635.40	II
15	1636.32	II
15	1639.40	II
14	1639.40	IV
16	1640.04	IV
14	1640.16	IV
12	1641.76	II
15	1641.87	IV
15	1647.09	IV
12	1647.16	II
15	1651.58	IV
15	1652.90	IV
13	1653.41	IV
13	1656.11	IV
15	1656.65	IV
14	1660.10	IV
13	1662.32	IV
13	1662.52	IV
13	1663.54	IV
13	1668.09	IV
12	1670.74	II
14	1671.04	IV
13	1673.68	IV
14	1675.66	IV
13	1681.36	IV
15	1687.69	IV
15	1698.88	IV
12	1702.04	II
13	1709.81	IV
15	1711.41	IV
14	1712.76	IV
14	1717.90	IV
14	1718.16	IV
14	1719.46	IV
14	1722.71	IV
14	1724.06	IV
16	1725.63	IV
13	1761.08	IV
12	1761.38	II
20	1785.26	II
20	1786.74	II
18	1788.07	II
13	1792.10	IV
13	1796.93	IV
13	1827.98	IV
10	1869.83	III
12	1877.99	III
10	1882.05	III
12	1886.76	III
13	1890.67	III
11	1893.98	III
20	1895.46	III
10 s	1907.58	III
19	1914.06	III
15	1915.08	III
15	1922.79	III
10 p	1926.01	III
18	1926.30	III
15	1930.39	III
14	1931.51	III
30	1934.538	I
25	1937.269	I
14	1937.34	III
10 l	1938.90	III
14 s	1943.48	III
12	1945.34	III
50	1946.988	I
10	1950.33	III
12	1951.01	III
25	1951.571	I
30	1952.59	I
11	1952.65	III
30	1953.005	I
13	1953.32	III
10	1953.49	III
10 w	1954.22	III
60	1957.823	I
11	1958.58	III
60	1960.144	I
13	1960.32	III
30	1961.25	I
50	1962.111	I
12	1963.11	II
15	1987.50	III
14	1991.61	III
13	1994.07	III
12	1995.56	III
12	1996.42	III
10	2061.55	III
12	2068.24	III
14	2078.99	III
100	2084.122	I
10	2084.35	III
12	2090.14	III
15	2097.48	III
12	2097.69	III
12	2103.80	III
10	2107.32	III
15	2151.78	III
12	2157.71	III
50	2157.794	I
12	2158.47	III
10	2161.27	III
40	2166.773	I
12	2166.95	III
12	2171.04	III
15	2174.66	III
300	2178.118	I
12	2180.41	III
250	2186.486	I
60	2186.892	I
120	2187.195	I
250	2191.839	I
150	2196.043	I
80	2200.390	I
80	2200.724	I
15	2208.41	II
10 p	2208.85	III
20	2213.65	II
12	2218.26	II
20	2220.38	II
10	2221.83	III
10	2229.27	III
10	2232.43	III
10	2232.69	III
10	2235.91	III
10	2238.16	III
12 p	2241.54	III
50	2250.790	I
60	2251.874	I
300	2259.511	I
12	2261.59	III
60	2264.389	I
80	2267.085	I
10	2267.42	III
80	2267.469	I
50	2270.862	I
150	2272.070	I
150	2276.026	I
80	2279.937	I
150	2284.086	I
150	2287.250	I
300	2292.524	I
10	2293.06	III
15	2295.86	III
200	2297.787	I
600	2298.169	I
80	2299.220	I
300	2300.142	I
50	2301.684	I
100	2303.424	I
150	2303.581	I
120	2308.999	I
150	2313.104	I
10 p	2317.70	III
10	2319.22	III
200	2320.358	I
10 p	2321.71	III
10	2326.95	III
100	2327.40	II
100	2331.31	II
300	2332.80	II
10 p	2336.77	III
200	2338.01	II
10	2338.96	III
600	2343.49	II
80	2343.96	II
150	2344.28	II
200	2348.11	II
250	2348.30	II
200	2359.12	II
150	2360.00	II
120	2360.29	II

Intensity	Wavelength/Å	
200	2364.83	II
80	2365.76	II
80	2368.59	II
80	2369.456	I
80	2369.95	II
120	2371.430	I
300	2373.624	I
150	2373.74	II
120	2374.518	I
120	2376.43	II
80	2379.27	II
120	2380.76	II
150	2381.835	I
1000	2382.04	II
300	2388.63	II
200	2389.973	I
1000	2395.62	II
300	2399.24	II
800	2404.88	II
250	2406.66	II
80	2406.97	II
300	2410.52	II
200	2411.07	II
150	2413.31	II
80	2417.87	II
60	2420.396	I
60	2423.089	I
150	2424.14	II
120	2428.36	II
120	2430.08	II
80	2432.26	II
60	2438.182	I
150	2439.30	II
150	2439.74	I
100	2442.57	I
250	2443.872	I
100	2444.51	II
50	2445.212	I
100	2445.57	II
60	2447.709	I
100	2453.476	I
1500	2457.598	I
150	2458.78	II
80	2461.28	II
100	2461.86	II
100	2462.181	I
1500	2462.647	I
50	2463.730	I
800	2465.149	I
60	2467.732	I
600	2468.879	I
80	2470.67	II
80	2470.965	I
800	2472.336	I
1000	2472.895	I
200	2473.16	I
600	2474.814	I
60	2476.657	I
120	2479.480	I
1200	2479.776	I
100	2480.16	II
80	2482.12	II
100	2482.66	II
10000	2483.271	I
300	2483.533	I
1000	2484.185	I
50	2485.990	I
800	2486.373	I
100	2486.691	I
100	2487.066	I
120	2487.370	I
4000	2488.143	I
100	2488.945	I
80	2489.48	II
1000	2489.750	I
50	2489.913	I
3000	2490.644	I
100	2490.71	II
2000	2491.155	I
100	2491.40	II
100	2493.18	II
500	2493.26	II
60	2494.000	I
50	2494.251	I
100	2495.87	I
600	2496.533	I
150	2498.90	I
1000	2501.132	I
50	2501.693	I
80	2506.09	II
500	2507.900	I
50	2508.753	I
1000	2510.835	I
120	2511.76	II
80	2512.275	I
400	2512.365	I
80	2516.570	I
300	2517.661	I
800	2518.102	I
150	2519.629	I
50	2522.480	I
4000	2522.849	I
200	2523.66	I
500	2524.293	I
100	2525.02	I
200	2525.39	II
300	2526.29	II
2000	2527.435	I
800	2529.135	I
250	2529.55	II
150	2529.836	I
200	2530.687	I
120	2533.63	II
100	2534.42	II
120	2535.49	II
400	2535.607	I
200	2536.792	I
200	2536.80	II
100	2538.80	II
100	2538.91	II
150	2538.99	II
50	2539.357	I
200	2540.66	II
600	2540.972	I
80	2541.10	II
300	2542.10	I
250	2543.92	I
150	2544.70	I
800	2545.978	I
80	2546.67	II
100	2548.74	II
80	2549.08	II
80	2549.39	II
600	2549.613	I
400	2562.53	II
200	2563.48	II
150	2574.36	II
300	2576.691	I
100	2582.58	II
1500	2584.54	I
650	2585.88	II
90	2591.54	II
90	2593.73	II
650	2598.37	II
2000	2599.40	II
300	2599.57	I
60	2605.657	I
300	2606.51	II
800	2606.827	I
650	2607.09	II
600	2611.87	II
320	2613.82	II
320	2617.62	II
250	2618.018	I
90	2620.41	II
400	2623.53	I
200	2625.67	II
150	2628.29	II
250	2631.05	II
250	2631.32	II
100	2632.237	I
300	2635.809	I
50	2641.646	I
200	2643.998	I
300	2666.812	I
60	2666.965	I
600	2679.062	I
500	2684.75	II
400	2689.212	I
10 h	2695.13	III
200	2699.106	I
80	2706.012	I
400	2706.582	I
60	2708.571	I
200	2711.655	I
80	2714.41	II
50	2716.257	I
50	2717.786	I
250	2718.436	I
4000	2719.027	I
100	2719.420	I
50	2720.197	I
1500	2720.903	I
400	2723.578	I
150	2724.953	I
50	2726.235	I
80	2727.54	II
200	2728.020	I
50	2728.820	I
80	2728.90	II
1000	2733.581	I
60	2734.005	I
50	2734.268	I
500	2735.475	I
50	2735.612	I
500	2737.310	I
120	2737.83	I
400	2739.55	II
250	2742.254	I
800	2742.405	I
200	2743.20	II
150	2743.565	I
200	2744.068	I
80	2744.527	I
300	2746.48	II
100	2749.32	II
500	2749.48	II
1200	2750.140	I
80	2753.29	II
150	2754.032	I
100	2754.426	I
800	2755.73	II
250	2756.328	I
100	2757.316	I
120	2761.780	I
150	2761.81	II
150	2762.026	I
120	2762.772	I
120	2763.109	I
80	2766.910	I
250	2767.522	I
300	2772.07	I
600	2778.220	I
3000	2788.10	I
200	2797.78	I
400	2804.521	I
1500	2806.98	I
10 p	2813.24	III
2500	2813.287	I
300	2823.276	I
600	2825.56	I
50	2825.687	I
120	2828.808	I
1500	2832.436	I
120	2835.950	I
200	2838.119	I
200	2843.631	I
1000	2843.977	I
100	2845.594	I
800	2851.797	I
50	2869.307	I
50	2872.334	I
80	2874.172	I
50	2894.504	I
12	2904.43	III
10	2907.50	III
12	2907.70	III
120	2912.157	I
120	2929.007	I
1200	2936.903	I
60	2941.343	I
1000	2947.876	I
600	2953.940	I
250	2957.364	I
150	2965.254	I
1500	2966.898	I
120	2969.36	I
800	2970.099	I
1200	2973.132	I
500	2973.235	I
600	2981.445	I
1000	2983.570	I
1000	2994.427	I
250	2994.502	I

…/Å	Wavelength/Å	
	2999.512	I
	3000.451	I
	3000.948	I
	3001.62	III
	3001.655	I
	3007.28	III
	3007.282	I
	3008.14	I
	3009.569	I
	3013.17	III
[illegible]	3017.627	I
…0	3018.983	I
…00	3020.491	I
…00	3020.639	I
…500	3021.073	I
500	3024.032	I
150	3025.638	I
500	3025.842	I
80	3030.148	I
60	3031.214	I
60	3034.484	I
800	3037.389	I
80	3041.637	I
800	3047.604	I
600	3057.446	I
1000	3059.086	I
250	3067.244	I
120	3075.719	I
120	3091.577	I
80	3098.189	I
100	3099.895	I
100	3099.968	I
60	3100.303	I
100	3100.665	I
10 p	3136.43	III
10	3174.09	III
80	3175.445	I
10	3175.99	III
10	3178.01	III
150	3184.895	I
250	3191.659	I
500	3193.226	I
800	3193.299	I
200	3196.928	I
80	3199.500	I
50	3205.398	I
100	3211.88	I
200	3214.011	I
200	3214.396	I
60	3215.938	I
50	3217.377	I
80	3219.583	I
60	3219.766	I
300	3222.045	I
600	3225.78	I
80	3227.796	I
50	3233.967	I
120	3234.613	I
300	3236.222	I
100	3239.433	I
80	3244.187	I
80	3246.005	I
80	3265.046	I
50	3265.617	I
13	3266.88	III
50	3271.000	I

Intensity	Wavelength/Å	
11	3276.08	III
150	3286.75	I
10	3288.81	III
120	3305.97	I
200	3306.343	I
400	3355.227	I
80	3355.517	I
60	3369.546	I
120	3370.783	I
50	3378.678	I
50	3380.110	I
60	3383.978	I
50	3392.304	I
150	3392.651	I
150	3399.333	I
80	3404.353	I
500	3407.458	I
250	3413.131	I
60	3424.284	I
500	3427.119	I
60	3428.748	I
6000	3440.606	I
2500	3440.989	I
1000	3443.876	I
200	3445.149	I
1200	3465.860	I
2000	3475.450	I
500	3476.702	I
2500	3490.574	I
500	3497.840	I
10	3501.76	III
250	3513.817	I
300	3521.261	I
400	3526.040	I
100	3526.166	I
60	3526.237	I
60	3526.381	I
60	3526.467	I
100	3533.199	I
200	3536.556	I
300	3541.083	I
250	3542.075	I
80	3553.739	I
400	3554.925	I
200	3556.878	I
400	3558.515	I
1000	3565.379	I
1200	3570.097	I
800	3570.25	I
120	3571.996	I
100	3573.393	I
60	3573.829	I
60	3573.888	I
4000	3581.19	I
150	3582.199	I
150	3584.660	I
120	3584.929	I
300	3585.319	I
150	3585.705	I
10	3586.04	III
200	3586.103	I
400	3586.984	I
100	3594.633	I
11	3600.94	III
150	3603.204	I
11	3603.88	III

Intensity	Wavelength/Å	
200	3605.454	I
500	3606.680	I
1500	3608.859	I
250	3610.16	I
60	3612.068	I
150	3617.788	I
1500	3618.768	I
200	3621.462	I
150	3622.004	I
150	3623.19	I
100	3631.096	I
1200	3631.463	I
60	3632.041	I
100	3638.298	I
200	3640.389	I
80	3643.717	I
1500	3647.842	I
250	3649.506	I
80	3650.279	I
200	3651.467	I
120	3670.024	I
150	3670.089	I
100	3676.311	I
150	3677.629	I
1500	3679.913	I
200	3682.242	I
120	3683.054	I
150	3684.107	I
120	3685.998	I
500	3687.456	I
120	3689.477	I
150	3694.008	I
120	3695.051	I
150	3701.086	I
80	3704.462	I
1200	3705.566	I
60	3707.041	I
150	3707.821	I
300	3707.919	I
600	3709.246	I
120	3716.442	I
8000	3719.935	I
1500	3722.563	I
120	3724.377	I
60	3725.491	I
60	3727.093	I
500	3727.619	I
150	3732.396	I
1200	3733.317	I
5000	3734.864	I
120	3735.324	I
6000	3737.131	I
100	3738.306	I
400	3743.362	I
6000	3745.561	I
1200	3745.899	I
3000	3748.262	I
80	3748.964	I
3000	3749.485	I
1500	3758.232	I
400	3760.05	I
1500	3763.788	I
400	3765.54	I
600	3767.191	I
60	3776.452	I
250	3785.95	I

Intensity	Wavelength/Å	
100	3786.68	I
250	3787.880	I
250	3790.092	I
150	3794.34	I
400	3795.002	I
120	3797.518	I
250	3798.511	I
400	3799.547	I
200	3805.345	I
80	3806.696	I
600	3812.964	I
60	3813.059	I
1500	3815.840	I
2500	3820.425	I
150	3821.179	I
80	3824.306	I
2500	3824.444	I
1500	3825.880	I
1200	3827.823	I
1000	3834.222	I
120	3839.257	I
500	3840.437	I
800	3841.047	I
120	3843.256	I
80	3846.800	I
200	3849.96	I
120	3850.817	I
2500	3856.372	I
150	3859.212	I
10000	3859.911	I
150	3865.523	I
60	3867.215	I
250	3872.501	I
150	3873.761	I
250	3878.018	I
2000	3878.573	I
4000	3886.282	I
200	3887.048	I
300	3888.513	I
800	3895.656	I
1200	3899.707	I
400	3902.945	I
250	3906.479	I
80	3916.731	I
600	3920.258	I
1200	3922.911	I
1200	3927.920	I
2000	3930.296	I
60	3948.774	I
60	3949.953	I
50	3951.164	I
50	3952.601	I
16	3954.33	III
60	3956.454	I
250	3956.68	I
60	3966.614	I
11	3968.72	III
100	3969.257	I
80	3977.741	I
10 w	3979.42	III
40	3981.771	I
50	3983.956	I
60	3994.114	I
200	3997.392	I
40	3998.053	I
400	4005.241	I

Intensity	Wavelength/Å	
60	4009.713	I
100	4021.867	I
10	4035.42	III
50	4040.638	I
4000	4045.813	I
11	4053.11	III
1500	4063.594	I
50	4066.975	I
50	4067.977	I
1200	4071.737	I
40	4076.629	I
12	4081.00	III
40	4100.737	I
40	4107.489	I
150	4118.544	I
10	4120.90	III
11	4122.02	III
11	4122.78	III
40	4127.608	I
400	4132.058	I
80	4134.676	I
40	4136.997	I
15	4137.76	III
13	4139.35	III
200	4143.415	I
800	4143.869	I
40	4153.898	I
50	4154.500	I
60	4156.799	I
18	4164.73	III
13	4166.84	III
50	4172.744	I
13	4174.26	III
60	4174.912	I
50	4175.635	I
50	4177.593	I
120	4181.754	I
50	4184.891	I
120	4187.038	I
120	4187.795	I
80	4191.430	I
40	4195.329	I
150	4198.304	I
40	4199.095	I
300	4202.029	I
40	4203.984	I
80	4206.696	I
80	4210.343	I
400	4216.183	I
100	4219.360	I
50	4222.212	I
11	4222.27	III
50	4225.956	I
200	4227.423	I
100	4233.602	I
13	4235.56	III
250	4235.936	I
50	4238.809	I
12	4243.75	III
50	4247.425	I
200	4250.118	I
300	4250.787	I
40	4258.315	I
800	4260.473	I
250	4271.153	I
1200	4271.759	I
12 h	4273.40	III
12	4279.72	III
1200	4282.402	I
14 h	4286.16	III
80	4291.462	I
16 h	4296.85	III
250	4299.234	I
18 h	4304.78	III
1200	4307.901	I
20 h	4310.36	III
150	4315.084	I
1500	4325.761	I
80	4352.734	I
80	4369.771	I
11 h	4372.31	III
14 h	4372.53	III
18 h	4372.81	III
800	4375.929	I
3000	4383.544	I
1200	4404.750	I
300	4415.122	I
12	4419.60	III
600	4427.299	I
400	4461.652	I
120	4466.551	I
80	4476.017	I
80	4482.169	I
200	4482.252	I
50	4489.739	I
50	4528.613	I
30	4647.433	I
30	4736.771	I
50	4859.741	I
120	4871.317	I
60	4872.136	I
30	4878.208	I
100	4890.754	I
250	4891.492	I
30	4903.309	I
150	4918.992	I
500	4920.502	I
1500	4957.597	I
80	5001.862	I
30	5005.711	I
100	5006.117	I
60	5012.067	I
30	5014.941	I
150	5041.755	I
30	5049.819	I
30	5051.634	I
25	5074.748	I
150	5110.357	I
40	5139.251	I
100	5139.462	I
25	5151.910	I
12	5156.12	III
80	5166.281	I
2500	5167.487	I
80	5168.897	I
500	5171.595	I
50	5191.454	I
80	5192.343	I
200	5194.941	I
10	5199.08	III
30	5204.582	I
25	5215.179	I
150	5216.274	I
60	5226.862	I
1000	5227.150	I
250	5232.939	I
10	5235.66	III
18	5243.31	III
13 l	5260.34	III
100	5266.555	I
1200	5269.537	I
800	5270.357	I
14	5272.98	III
15	5276.48	III
30	5281.789	I
16	5282.30	III
60	5283.621	I
12	5284.83	III
11	5298.12	III
12	5299.93	III
25	5302.299	I
14 w	5302.60	III
10	5306.76	III
10	5322.74	III
150	5324.178	I
800	5328.038	I
300	5328.531	I
100	5332.899	I
80	5339.928	I
500	5341.023	I
11	5346.88	III
12	5353.77	III
12	5363.76	III
10	5368.06	III
400	5371.489	I
11 l	5375.47	III
40	5393.167	I
300	5397.127	I
250	5405.774	I
250	5429.695	I
100	5434.523	I
200	5446.871	I
120	5455.609	I
25	5497.516	I
20	5501.464	I
30	5506.778	I
30	5569.618	I
60	5572.841	I
120	5586.755	I
200	5615.644	I
20	5624.541	I
50	5662.515	I
11	5719.88	III
10	5756.38	III
20	5762.990	I
18	5833.93	III
10	5854.62	III
30	5862.353	I
15	5891.91	III
30	5914.114	I
10 p	5920.13	III
18 p	5929.69	III
10	5952.31	III
14	5953.62	III
12	5979.32	III
30	5986.956	I
12 h	5989.08	III
18	5999.54	III
16	6032.59	III
13	6036.56	III
11	6048.72	III
11	6054.18	III
40	6065.482	I
30	6102.159	I
40	6136.614	I
40	6137.694	I
40	6191.558	I
30	6213.429	I
30	6219.279	I
40	6230.726	I
20	6246.317	I
80	6247.56	II
30	6252.554	I
20	6393.602	I
30	6399.999	I
20	6411.647	I
20	6421.349	I
30	6430.844	I
200	6456.38	II
60	6494.981	I
20	6546.239	I
20	6592.913	I
40	6677.989	I
25	7164.443	I
80	7187.313	I
30	7207.381	I
30	7445.746	I
40	7495.059	I
60	7511.045	I
80	7937.131	I
60	7945.984	I
80	7998.939	I
60	8046.047	I
50	8085.176	I
150	8220.41	I
120	8327.053	I
20	8331.908	I
120	8387.770	I
30	8468.404	I
15	8514.069	I
60	8661.898	I
150	8688.621	I
52	11422.32	I
87	11439.12	I
91	11593.59	I
255	11607.57	I
160	11638.26	I
230	11689.98	I
160	11783.26	I
580	11882.84	I
225	11884.08	I
1030	11973.05	I
96	14400.56	I
72	14512.23	I
50	14555.06	I
40	14826.43	I
94	15294.58	I
41	15769.42	I
105	18856.65	I

Krypton

Kr $Z = 36$

Intensity	Wavelength/Å	
30	467.35	III
150	472.16	V
100	484.39	V

Intensity	Wavelength/Å		Intensity	Wavelength/Å		Intensity	Wavelength/Å		Intensity	Wavelength/Å	
250	496.25	V	50	837.66	III	6	2774.70	IV	40 h	3868.70	III
120	500.77	V	22	842.04	IV	3	2829.60	IV	150 h	3875.44	II
200	507.20	V	100	844.06	II	100	2833.00	II	150	3906.177	II
30	540.86	III	50	854.73	III	3	2836.08	IV	200	3920.081	II
60	548.04	V	60	862.58	III	30	2841.00	III	5	3934.29	IV
30	565.64	III	60	864.82	II	30	2851.16	III	100	3994.840	II
30	569.16	III	60	868.87	II	5	2853.0	IV	100 h	3997.793	II
30	571.98	III	40	870.84	III	3	2859.3	IV	300	4057.037	II
30	579.83	III	50	876.08	III	50	2870.61	III	300	4065.128	II
30	585.14	III	200	884.14	II	100	2892.18	III	50	4067.37	III
30	585.96	III	1000	886.30	II	30	2909.17	III	500	4088.337	II
30	593.70	III	400	891.01	II	50	2952.56	III	250	4098.729	II
30	594.10	III	75	897.81	III	60	2992.22	III	100	4109.248	II
30	596.41	III	200	911.39	II	50	3022.30	III	40	4131.33	III
40	600.17	III	2000	917.43	II	80	3024.45	III	250	4145.122	II
30	603.67	III	50	945.44	I	50	3046.93	III	40	4154.46	III
50	605.86	III	50	946.54	I	30	3056.72	III	150	4250.580	II
35	606.47	III	20	951.06	I	60	3063.13	III	1000	4273.969	I
50	611.12	III	50	953.40	I	40	3097.16	III	100	4282.967	I
35	616.72	III	50	963.37	I	60	3112.25	III	600	4292.923	II
40	621.45	III	2000	964.97	II	30	3120.61	III	200	4300.49	II
45	622.80	III	50	987.29	III	100	3124.39	III	500 h	4317.81	II
50	625.02	III	100	1001.06	I	60	3141.35	III	400	4318.551	I
30	625.76	III	100	1003.55	I	3	3142.01	IV	1000	4319.579	I
45	628.59	III	100	1030.02	I	100	3189.11	III	150 h	4322.98	II
50	630.04	III	30	1158.74	III	80	3191.21	III	100	4351.359	I
35	633.09	III	200	1164.87	I	6	3224.99	IV	3000	4355.477	II
120	637.87	V	650	1235.84	I	40	3239.52	III	500	4362.641	I
50	639.98	III	6	1638.82	III	40	3240.44	III	200	4369.69	II
60	646.41	III	6	1914.09	III	300	3245.69	III	800	4376.121	I
50	651.20	III	3	2237.34	IV	3	3261.70	IV	300 h	4386.54	II
50	659.72	III	6	2291.26	IV	150	3264.81	III	200	4399.965	I
30	664.86	III	3	2329.3	IV	100	3268.48	III	100	4425.189	I
40	672.34	III	4	2336.75	IV	30	3271.65	III	500	4431.685	II
35	672.85	III	4	2348.27	IV	30	3285.89	III	600	4436.812	II
35	676.57	III	3	2358.5	IV	30	3304.75	III	600	4453.917	I
35	680.13	III	3	2388.05	IV	50	3311.47	III	800	4463.689	I
35	683.68	III	40	2393.94	III	200	3325.75	III	800	4475.014	II
45	686.25	III	4	2416.9	IV	60	3330.76	III	400 h	4489.88	II
45	687.98	III	3	2428.04	IV	50	3342.48	III	600	4502.353	I
	690.86	V	5	2442.68	IV	100	3351.93	III	400 h	4523.14	II
	691.75	V	4	2451.7	IV	40	3374.96	III	200 h	4556.61	II
45	691.93	III	6	2459.74	IV	100	3439.46	III	800	4577.209	II
50	695.61	III	100 h	2464.77	II	70	3474.65	III	300	4582.978	II
30	698.05	III	5	2474.06	IV	100	3488.59	III	150 h	4592.80	II
50	708.36	III	60	2492.48	II	200	3507.42	III	500	4615.292	II
600	708.85	V	40	2494.01	III	100	3564.23	III	1000	4619.166	II
50	714.00	III	4	2517.0	IV	100 h	3607.88	II	800	4633.885	II
100 p	722.04	III	5	2518.02	IV	200	3631.889	II	2000	4658.876	II
60	729.40	II	6	2519.38	IV	30	3641.34	III	500	4680.406	II
30	746.70	III	5	2524.5	IV	250	3653.928	II	100	4691.301	II
200	761.18	II	5	2546.0	IV	80	3665.324	I	200	4694.360	II
100	763.98	II	6	2547.0	IV	150	3669.01	II	3000	4739.002	II
60	766.20	II	4	2558.08	IV	100	3679.559	I	300	4762.435	II
200	771.03	II	30	2563.25	III	80	3686.182	II	1000	4765.744	II
60 p	773.69	II	3	2586.9	IV	30	3690.65	III	300	4811.76	II
200	782.10	II	5	2606.17	IV	300 h	3718.02	II	300	4825.18	II
100	783.72	II	10	2609.5	IV	200	3718.595	II	800	4832.077	II
60	785.97	III	8	2615.3	IV	150	3721.350	II	700	4846.612	II
	793.44	IV	7	2621.11	IV	200	3741.638	II	150	4857.20	II
	794.11	IV	60	2639.76	III	150	3744.80	II	300	4945.59	II
7	805.76	IV	30	2680.32	III	80	3754.245	II	20 h	5016.45	III
	810.70	V	40	2681.19	III	500	3778.089	II	200	5022.40	II
18	816.82	IV	80 h	2712.40	II	500	3783.095	II	250	5086.52	II
60	818.15	II	3	2730.55	IV	3	3809.30	IV	400 h	5125.73	II
60	830.38	II	8	2748.18	IV	5	3860.58	IV	500	5208.32	II

Intensity	Wavelength/Å	
200	5308.66	II
500	5333.41	II
200	5468.17	II
10	5501.43	III
500	5562.224	I
2000	5570.288	I
80	5580.386	I
100	5649.561	I
400	5681.89	II
200 h	5690.35	II
100	5832.855	I
3000	5870.914	I
200	5992.22	II
60	5993.849	I
10 h	6037.17	III
60	6056.125	I
10 h	6078.38	III
10	6310.22	III
300	6420.18	II
100	6421.026	I
200	6456.288	I
150	6570.07	II
60	6699.228	I
100	6904.678	I
250	7213.13	II
100	7224.104	I
80	7287.258	I
400	7289.78	II
400	7407.02	II
60	7425.541	I
200	7435.78	II
100	7486.862	I
300	7524.46	II
1000	7587.411	I
2000	7601.544	I
150	7641.16	II
1000	7685.244	I
1200	7694.538	I
250	7735.69	II
150	7746.827	I
800	7854.821	I
200	7913.423	I
180	7928.597	I
200	7933.22	II
120	7973.62	II
100	7982.401	I
1500	8059.503	I
4000	8104.364	I
6000	8112.899	I
60	8132.967	I
3000	8190.054	I
200	8202.72	II
80	8218.365	I
3000	8263.240	I
100	8272.353	I
1500	8281.050	I
5000	8298.107	I
100	8412.430	I
3000	8508.870	I
150	8764.110	I
6000	8776.748	I
2000	8928.692	I
500	9238.48	II
500 hl	9293.82	II
200 h	9320.99	II
300	9361.95	II
100	9362.082	I
200 h	9402.82	II
200 h	9470.93	II
500	9577.52	II
500 h	9605.80	II
400 h	9619.61	II
200	9663.34	II
200 h	9711.60	II
2000	9751.758	I
500	9803.14	II
500	9856.314	I
1000	10221.46	II
100	11187.108	I
200	11257.711	I
150	11259.126	I
500	11457.481	I
150	11792.425	I
1500	11819.377	I
600	11997.105	I
160	12077.224	I
100	12861.892	I
1100	13177.412	I
1000	13622.415	I
2400	13634.220	I
800	13658.394	I
200	13711.036	I
600	13738.851	I
150	13974.027	I
550	14045.657	I
140	14104.298	I
180	14402.22	I
2000	14426.793	I
100	14517.84	I
1600	14734.436	I
550	14762.672	I
450	14765.472	I
400	14961.894	I
120	15005.307	I
140	15209.526	I
1700	15239.615	I
130	15326.480	I
1500	15334.958	I
700	15372.037	I
200	15474.026	I
180	15681.02	I
120	15820.09	I
200	16726.513	I
2000	16785.128	I
1000	16853.488	I
2400	16890.441	I
1600	16896.753	I
1800	16935.806	I
600	17098.771	I
700	17367.606	I
120	17404.443	I
150	17616.854	I
650	17842.737	I
700	18002.229	I
2600	18167.315	I
100	18399.786	I
150	18580.896	I
300	18696.294	I
170	18785.460	I
200	18797.703	I
140	20209.878	I
300	20423.964	I
140	20446.971	I
600	21165.471	I
1800	21902.513	I
120	22485.775	I
180	23340.416	I
120	24260.506	I
180	24292.221	I
600	25233.820	I
180	28610.55	I
1000	28655.72	I
150	28769.71	I
140	28822.49	I
300	29236.69	I
300	30663.54	I
300	30979.16	I
500	39300.6	I
1100	39486.52	I
220	39557.25	I
100	39572.60	I
1400	39588.4	I
1100	39589.6	I
500	39954.8	I
300	39966.6	I
1300	40306.1	I
250	40685.16	I

Lanthanum

La Z = 57

Intensity	Wavelength/Å	
100	344.12	IV
400	390.72	V
1000	432.11	V
2500	435.28	V
10000	463.14	IV
5000	482.16	V
7000	498.08	V
15000	499.54	IV
10000	503.58	V
12000	526.76	V
10000	531.07	V
15000	533.23	V
8000	547.44	V
40000	552.02	IV
5000	600.24	V
30000	631.26	IV
400	796.99	III
2000	870.40	III
1000	882.34	III
400	942.86	III
50000	1081.61	III
95000	1099.73	III
2000	1255.63	III
10000	1349.18	III
25000	1368.04	IV
20000	1463.47	IV
15000	1507.87	IV
10000	1523.79	III
4000	1808.66	IV
5000	1902.97	IV
4000 c	2197.45	IV
770	2256.76	II
25000 w	2417.58	IV
50000	2532.75	IV
45000	2582.05	IV
95000 w	2597.50	IV
70000 w	2662.75	IV
420	2808.39	II
50000 w	2848.30	IV
30000 c	2962.58	IV
70000 w	3009.51	IV
90000 c	3056.68	IV
1000	3171.63	III
1500	3171.74	III
510	3245.13	II
550	3265.67	II
800	3303.11	II
1500	3337.49	II
870	3344.56	II
1500	3380.91	II
320	3628.83	II
1000	3645.42	II
550	3713.54	II
2400	3759.08	II
3700	3790.83	II
3900	3794.78	II
600	3840.72	II
1600	3849.02	II
3400	3871.64	II
1700	3886.37	II
1300	3916.05	II
1100	3921.54	II
2200	3929.22	II
9000	3949.10	II
4400	3988.52	II
3600	3995.75	II
2800	4031.69	II
3000	4042.91	II
850	4067.39	II
2800	4077.35	II
5500	4086.72	II
4400	4123.23	II
550	4141.74	II
1100	4151.97	II
1500	4196.55	II
1600	4238.38	II
480	4269.50	II
600	4286.97	II
600	4296.05	II
440	4322.51	II
4600	4333.74	II
550	4354.40	II
2000	4429.90	II
850	4522.37	II
420	4526.12	II
400	4558.46	II
400	4574.88	II
410	4613.39	II
410	4619.88	II
540	4655.50	II
360	4662.51	II
230	4692.50	II
230	4728.42	II
500	4740.28	II
390	4743.09	II
320	4748.73	II
320	4860.91	II
850	4899.92	II
1000	4920.98	II
1000	4921.79	II
370	4949.77	I
340	4970.39	II
370	4986.83	II
720	4999.47	II
210	5050.57	I

Intensity	Wavelength/Å	
470	5114.56	II
470	5122.99	II
450	5145.42	I
290	5158.69	I
580	5177.31	I
850	5183.42	II
260	5188.22	II
720	5211.86	I
520	5234.27	I
340	5253.46	I
370	5271.19	I
370	5301.98	II
180	5303.55	II
500	5455.15	I
470	5501.34	I
240	5648.25	I
180	5740.66	I
370	5769.34	I
320	5789.24	I
450	5791.34	I
140	5821.99	I
320	5930.62	I
720	6249.93	I
260 d	6262.30	II
450	6394.23	I
250	6455.99	I
180	6709.50	I
110	7045.96	I
160	7066.23	II
50	7161.25	I
110 w	7282.34	II
110 w	7334.18	I
75 cw	7483.50	II
50	7498.83	I
85	7539.23	I
40	7964.83	I
75	8086.05	I
85	8324.69	I
95	8346.53	I
65	8545.44	I
300	8583.45	III
40	8674.43	I
35	8825.82	I
120	9184.38	III
100	9212.63	III
140	10284.79	III

Lead

Pb Z = 82

Intensity	Wavelength/Å	
10	496.38	IV
12	499.94	IV
14	529.78	IV
20	570.16	IV
10	648.50	IV
20	703.73	V
12	749.46	V
10	752.52	V
10	761.09	IV
18	767.45	V
18	769.49	V
14	771.42	V
14	782.79	V
15	797.02	V
18	802.07	IV
12	802.82	IV
18	809.63	V
10	812.59	IV
10	827.41	IV
12	832.60	IV
12	845.94	IV
18	857.64	IV
16	862.33	IV
20	863.97	V
14	870.44	IV
6	873.71	II
12	879.96	IV
18	883.90	V
14	884.96	IV
14	884.99	IV
14	888.37	V
8	889.68	II
16	890.72	IV
14	894.40	V
12	896.08	V
12	908.51	IV
14	915.71	V
12	917.90	IV
12	918.09	V
12	920.28	V
12	920.66	V
10	922.12	IV
12	922.49	IV
10	927.64	IV
14	932.20	IV
12	954.35	V
10	967.23	II
10	986.71	II
10	995.89	II
10	1016.61	II
14	1028.61	IV
20	1032.05	IV
16	1041.24	IV
18	1044.14	IV
12	1048.9	III
10	1049.82	II
10	1050.77	II
10	1051.26	V
15	1056.53	IV
10	1060.66	II
12	1072.09	IV
18	1080.81	IV
20	1084.17	IV
10	1088.86	V
10	1103.94	II
10	1108.43	II
10	1109.84	II
20	1116.08	IV
10	1119.57	II
10	1121.36	II
10	1133.14	II
18	1137.84	IV
14	1144.93	IV
12	1157.88	V
14	1185.43	V
20	1189.95	IV
10	1203.63	II
10	1231.20	II
11	1233.50	V
10	1291.10	IV
20	1313.05	IV
10	1331.65	II
10	1335.20	II
12	1343.06	IV
10	1348.37	II
16	1388.94	IV
18	1400.26	IV
10	1404.34	IV
10	1433.96	II
10	1512.42	II
14	1535.71	IV
20	1553.1	III
10	1671.53	II
10	1682.15	II
20	1726.75	II
10	1796.670	II
10	1822.050	II
10	1904.77	I
7	1921.471	II
12	1959.34	IV
16	1973.16	IV
10	1998.83	V
5 r	2022.02	I
10	2042.58	IV
12	2049.34	IV
8 r	2053.28	I
12	2079.22	IV
6	2111.758	I
10	2115.066	I
15	2154.01	IV
500 r	2170.00	I
7	2175.580	I
12	2177.46	IV
7	2187.888	I
8	2189.603	I
10	2203.534	II
20	2237.425	I
20	2246.86	I
25	2246.89	I
20	2259.01	V
150	2332.418	I
16	2359.53	IV
180	2388.797	I
550 r	2393.792	I
140	2399.597	I
320 r	2401.940	I
320 r	2411.734	I
16	2417.61	IV
15	2424.81	V
150 r	2443.829	I
160 r	2446.181	I
130 r	2476.378	I
80 r	2577.260	I
500 r	2613.655	I
900 r	2614.175	I
160	2628.262	I
4	2634.256	II
10	2657.094	I
700	2663.154	I
10	2697.541	I
25000 r	2801.995	I
100	2822.58	I
14000 r	2823.189	I
35000 r	2833.053	I
6	2840.557	II
14000 r	2873.311	I
3	2914.442	II
15	2966.460	I
15	2972.991	I
15	2980.157	I
4	2986.876	II
10	3043.85	III
150	3118.894	I
10	3137.81	III
10	3176.50	III
600	3220.528	I
100	3229.613	I
400	3240.186	I
200	3262.355	I
35000	3572.729	I
50000 r	3639.568	I
20000	3671.491	I
70000 r	3683.462	I
10	3713.982	II
25000	3739.935	I
12	3854.08	III
15000	4019.632	I
95000	4057.807	I
14000	4062.136	I
10	4157.814	I
10000	4168.033	I
8	4272.66	III
200	4340.413	I
10	4496.15	IV
6	4499.34	III
16	4534.60	IV
7	4571.21	III
10	4579.051	II
6	4761.12	III
1000	5005.416	I
100	5006.572	I
50	5089.484	I
10	5107.242	I
2000	5201.437	I
10	5372.099	II
40	5692.346	I
200	5895.624	I
2000	6001.862	I
500	6011.667	I
500	6059.356	I
40	6081.409	II
50	6110.520	I
100	6235.266	I
50 c	6660.20	II
20000	7228.965	I
10	7346.676	I
20	7809.259	I
5	7896.737	I
10	8168.001	I
6	8191.886	I
5	8217.711	I
40	8272.690	I
20	8409.384	I
10	8478.492	I
5	8722.810	I
10	8857.457	I
8	9293.476	I
15	9438.05	I
15	9604.297	I
15	9674.351	I
200	10290.458	I
100	10498.965	I
50	10649.249	I
15	10886.688	I
40	10969.53	I
	13512.6	I

Intensity	Wavelength/Å	
	14743.0	I
	15349.6	I
	39039.4	I
	Lithium	
	Li Z = 3	
	102.9	III
	103.4	III
	104.1	III
	105.5	III
	108.0	III
	113.9	III
	125.5	II
	135.0	III
	136.5	II
	140.5	II
	167.21	II
	168.74	II
	171.58	II
	178.02	II
	199.28	II
	207.5	II
	456.	II
	483.	II
	540.	II
	540.0	III
	729.	II
	729.1	III
	800.	II
	820.	II
	861.	II
	905.5	II
	917.5	II
	936.	II
	945.	II
	965.	II
	972.	II
	988.	II
	1018.	II
	1032.	II
	1036.	II
	1093.	II
	1103.	II
	1109.	II
	1116.	II
	1132.1	II
	1141.	II
	1166.4	II
	1198.09	II
	1215.	II
	1238.	II
	1253.8	II
	1420.89	II
	1424.	II
3	1492.93	II
5	1492.97	II
1	1493.04	II
	1555.	II
3	1653.08	II
5	1653.13	II
1	1653.21	II
	1681.66	II
	1755.33	II
	2009.	II
	2039.	I
	2068.	II
	2131.	II
	2164.	II
	2173.4	I
	2183.	II
	2214.	II
	2222.	II
	2237.	II
h	2249.21	II
	2286.82	II
	2302.57	II
	2303.33	II
	2304.59	I
	2304.92	I
	2305.36	I
	2305.83	I
	2306.29	I
	2306.82	I
	2307.44	I
	2308.97	I
	2309.88	I
	2310.94	I
	2312.11	I
	2313.49	I
	2315.08	I
	2316.95	I
	2319.18	I
	2321.88	I
	2325.11	I
	2329.02	I
	2329.84	II
	2333.94	I
3	2336.88	II
5	2336.91	II
2	2337.00	II
	2340.15	I
	2348.22	I
	2358.93	I
	2373.54	I
	2381.54	II
	2383.20	II
1	2394.39	I
	2402.33	II
	2410.84	II
3	2425.43	I
	2429.81	II
	2460.2	I
10	2475.06	I
	2506.94	II
	2508.78	II
	2518.	I
	2539.49	II
24	2551.7	II
	2559.	II
15	2562.31	I
	2605.08	II
2	2657.29	II
3	2657.30	II
	2674.46	II
	2728.24	II
5	2728.29	II
2	2728.32	II
3	2730.47	II
1	2730.55	II
5	2741.20	I
	2766.99	II
	2790.31	II
	2801.	I
	2846.	I
	2868.	I
	2895.	I
2	2934.02	II
2	2934.07	II
5	2934.12	II
1	2934.25	II
	2968.	I
3	3029.12	II
3	3029.14	II
	3144.	I
3	3155.31	II
4	3155.33	II
1	3196.26	II
9	3196.33	II
4	3196.36	II
5	3199.33	II
2	3199.43	II
17	3232.66	I
	3249.87	II
	3306.28	II
	3488.	I
	3579.8	I
	3618.	I
	3662.	I
	3684.32	II
1	3714.00	II
5	3714.16	II
6 d	3714.27	II
8	3714.29	II
7 d	3714.40	II
10	3714.41	II
1	3714.51	II
	3714.58	II
3	3718.7	I
6	3794.72	I
20	3915.30	I
20	3915.35	I
10	3985.48	I
10	3985.54	I
40	4132.56	I
40	4132.62	I
	4196.	I
20	4273.07	I
20	4273.13	I
5	4325.42	II
5	4325.47	II
1	4325.54	II
	4516.45	II
13	4602.83	I
13	4602.89	I
	4607.34	II
	4671.51	II
6	4671.65	II
2	4671.70	II
3	4678.06	II
1	4678.29	II
	4760.	I
	4763.	II
	4788.36	II
	4843.0	II
4	4881.32	II
4	4881.39	II
1	4881.49	II
8	4971.66	I
8	4971.75	I
	5037.92	II
	5271.	I
	5315.	I
	5395.	I
	5440.	I
600 c	5483.55	II
600 c	5485.65	II
320	6103.54	I
320	6103.65	I
3600	6707.76	I
3600	6707.91	I
48	8126.23	I
48	8126.45	I
	8517.37	II
	9581.42	II
	10120.	II
	12232.	I
	12782.	I
	13566.	I
	17552.	I
	18697.	I
	19290.	I
	24467.	I
	40475.	I
	Lutetium	
	Lu Z = 71	
100	563.72	V
500	810.73	III
2000	832.28	III
100	861.92	V
400	876.80	IV
100	880.32	V
100	891.81	V
100	914.72	V
400	1001.18	III
100	1272.42	IV
800	1333.79	IV
400	1429.38	IV
200	1441.76	V
200	1453.35	V
200	1468.99	V
400	1472.12	V
200	1473.71	V
200	1485.58	V
400	1511.26	IV
600	1772.57	IV
100 c	1786.25	V
1000	1854.57	III
1500	2065.35	III
1500 c	2070.56	III
600 c	2086.47	IV
1000 c	2104.41	IV
1000 c	2108.31	IV
1700 h	2195.54	II
1000	2236.14	III
2000	2236.22	III
95	2276.94	II
190	2297.41	II
1300	2392.19	II
120	2399.14	II
80	2419.21	II
130	2459.64	II
370	2536.95	II
930	2571.23	II
1700	2578.79	II
4500 c	2603.35	III

Intensity	Wavelength/Å	
1800	2613.40	II
18000	2615.42	II
1800	2619.26	II
2700	2657.80	II
570 h	2685.08	I
4200	2701.71	II
180 d	2719.09	I
480 h	2728.95	I
3600	2754.17	II
750 h	2765.74	I
2000	2772.55	III
2700	2796.63	II
270 c	2834.35	II
330 h	2845.13	I
3000	2847.51	II
570 h	2885.14	I
6300	2894.84	II
4500	2900.30	II
300	2903.05	I
9000	2911.39	II
270 h	2949.73	I
1200	2951.69	II
4200	2963.32	II
2400	2969.82	II
1800	2989.27	I
3000	3020.54	II
2100	3056.72	II
1000	3057.86	III
7500	3077.60	II
390	3080.11	I
5100 h	3081.47	I
3000	3118.43	I
2400	3171.36	I
260	3191.80	II
1400	3198.12	II
4800	3254.31	II
3800	3278.97	I
7600	3281.74	I
6200	3312.11	I
7600	3359.56	I
6200	3376.50	I
950	3385.50	I
160 h	3391.55	I
1400	3396.82	I
4100	3397.07	II
4800	3472.48	II
8300 c	3507.39	II
1600	3508.42	I
4800	3554.43	II
4800	3567.84	I
340	3596.34	I
800	3623.99	II
680	3636.25	I
2600	3647.77	I
110	3756.70	I
110	3756.79	I
150	3800.67	I
2700	3841.18	I
530	3876.65	II
50	3918.86	I
480	3968.46	I
670	4054.45	I
310	4122.49	I
3100	4124.73	I
150 c	4131.79	I
460	4154.08	I
1600	4184.25	II
150	4277.50	I
250	4281.03	I
330 d	4295.97	I
150	4309.57	I
190 c	4430.48	I
190	4450.81	I
3300	4518.57	I
100 h	4648.21	I
1000	4658.02	I
85 h	4659.03	I
150	4785.42	II
85	4815.05	I
460	4904.88	I
180	4942.34	I
800	4994.13	II
800	5001.14	I
140	5134.05	I
2700	5135.09	I
170	5196.61	I
500	5402.57	I
140 c	5421.90	I
100	5437.88	I
2100	5476.69	II
550	5736.55	I
80	5800.59	I
690 cw	5983.9	II
140	5997.13	I
1400	6004.52	I
440	6055.03	I
150	6159.94	II
600	6198.13	III
160	6199.66	II
2100	6221.87	II
80	6235.36	II
160	6242.34	II
70 h	6345.35	I
1100	6463.12	II
29	6477.67	I
55 c	6523.18	I
35 cw	6611.28	II
23 c	6677.14	I
30 c	6793.77	I
45	6917.31	I
23	7031.24	I
45	7125.84	II
14 ch	7237.98	I
11 c	7441.52	I
9 c	8178.16	I
17	8382.08	I
35	8459.19	II
10 d	8478.50	I
29 c	8508.08	I
35 c	8610.98	I

Magnesium
Mg $Z = 12$

Intensity	Wavelength/Å	
400	146.95	IV
20	186.51	III
20	187.20	III
10	188.53	III
100	231.73	III
80	234.26	III
35	276.58	V
4000	320.99	IV
3000	323.31	IV
30	353.09	V
150	857.29	IV
50	919.03	IV
250	1037.41	IV
300	1210.99	IV
300	1342.19	IV
800	1346.57	IV
300	1346.68	IV
600	1352.05	IV
900	1384.46	IV
500	1385.77	IV
800	1387.53	IV
300	1404.68	IV
1000	1409.36	IV
500	1437.53	IV
1000	1437.64	IV
300	1447.42	IV
300	1459.54	IV
400	1459.62	IV
400	1481.51	IV
350	1490.45	IV
300	1495.50	IV
300	1607.11	IV
5	1668.43	I
500	1683.02	IV
10	1683.41	I
400	1698.81	IV
15	1707.06	I
40	1734.84	II
50	1737.62	II
20	1747.80	I
40	1750.65	II
50	1753.46	II
30	1827.93	I
300	1844.17	IV
9	2025.82	I
25	2064.90	III
20	2091.96	III
20	2177.70	III
3	2329.58	II
20	2395.15	III
6	2449.57	II
1	2557.23	I
1	2560.94	I
1	2562.26	I
1	2564.94	I
1	2570.91	I
1	2572.25	I
2	2574.94	I
1	2577.89	I
1	2580.59	I
1	2584.22	I
2	2585.56	I
3	2588.28	I
1	2591.89	I
1	2593.23	I
2	2595.97	I
2	2602.50	I
4	2603.85	I
5	2606.62	I
1	2613.36	I
2	2614.73	I
3	2617.51	I
3	2628.66	I
6	2630.05	I
8	2632.87	I
2	2644.80	I
3	2646.21	I
4	2649.06	I
8	2660.76	II
8	2660.82	II
6	2668.12	I
8	2669.55	I
10	2672.46	I
3	2693.72	I
5	2695.18	I
6	2698.14	I
8	2731.99	I
10	2733.49	I
12	2736.53	I
5	2765.22	I
7	2768.34	I
38	2776.69	I
32	2778.27	I
90	2779.83	I
8	2781.29	I
32	2781.42	I
36	2782.97	I
1000	2795.53	II
600	2802.70	II
3	2809.76	I
2	2811.11	I
1	2811.78	I
12	2846.72	I
12	2846.75	I
14	2848.34	I
14	2848.42	I
16	2851.65	I
16	2851.66	I
6000	2852.13	I
2	2902.92	I
4	2906.36	I
3	2915.45	I
10	2936.74	I
12	2938.47	I
2	2942.00	I
13	2942.00	I
20	3091.08	I
22	3092.99	I
14	3096.90	I
9	3104.71	II
8	3104.81	II
6	3168.98	II
6	3172.71	II
7	3175.78	II
2	3197.62	I
17	3329.93	I
6	3332.15	I
9	3336.68	I
7	3535.04	II
8	3538.86	II
7	3549.52	II
8	3553.37	II
140	3829.30	I
300	3832.30	I
500	3838.29	I
8	3848.24	II
7	3850.40	II
3	3878.31	I
3	3895.57	I
4	3903.86	I
6	3938.40	I
8	3986.75	I

Intensity	Wavelength/Å	
10	4057.50	I
15	4167.27	I
20	4351.91	I
9	4384.64	II
10	4390.59	II
8	4428.00	II
9	4433.99	II
14	4481.16	II
13	4481.33	II
28	4571.10	I
10	4730.03	I
7	4851.10	II
75	5167.33	I
220	5172.68	I
400	5183.61	I
8	5264.21	II
7	5264.37	II
9	5401.54	II
6	5528.41	I
30	5711.09	I
10	6318.72	I
9	6319.24	I
7	6319.49	I
10	6346.74	II
9	6346.96	II
11	6545.97	II
7	6781.45	II
8	6787.85	II
7	6812.86	II
8	6819.27	II
10	7193.17	I
10	7291.06	I
12	7387.69	I
20	7657.60	I
19	7659.15	I
17	7659.90	I
15	7691.55	I
12	7877.05	II
13	7896.37	II
10	8098.72	I
9	8115.22	II
8	8120.43	II
10	8209.84	I
20	8213.03	I
10	8213.99	II
11	8234.64	II
10	8310.26	I
15	8346.12	I
10	8710.18	I
12	8712.69	I
13	8717.83	I
10	8734.99	II
17	8736.02	I
11	8745.66	II
14	8806.76	I
10	8824.32	II
11	8835.08	II
20	8923.57	I
10	8997.16	I
14	9218.25	II
13	9244.27	II
12	9246.50	I
30	9255.78	I
10	9327.54	II
10	9340.54	II
25	9414.96	I
17	9429.81	I
19	9432.76	I
20	9438.78	I
12	9631.89	II
11	9632.43	II
15	9953.20	I
15	9983.20	I
17	9986.47	I
18	9993.21	I
14	10092.16	II
35	10811.08	I
11	10914.23	II
10	10951.78	II
25	10953.32	I
27	10957.30	I
28	10965.45	I
15	11032.10	I
14	11033.66	I
45	11828.18	I
30	12083.66	I
28	14877.62	I
35	15024.99	I
30	15040.24	I
25	15047.70	I
10	15765.84	I
30	17108.66	I
5	26392.90	I

Manganese
Mn Z = 25

Intensity	Wavelength/Å	
600	410.30	V
600	410.60	V
600	415.62	V
650	415.98	V
600	428.59	V
600	435.67	V
1000	441.72	V
850	442.49	V
60	579.79	IV
60	581.44	IV
60	581.65	IV
60	585.21	IV
90	1242.25	IV
90	1244.50	IV
95	1251.93	IV
95	1257.28	IV
90	1264.41	IV
500	1283.58	III
400	1287.59	III
300	1291.62	III
1000	1360.72	III
800	1365.20	III
500 h	1609.17	III
1000	1614.14	III
2000	1620.60	III
500	1633.80	III
80	1667.00	IV
80	1698.30	IV
20	1726.47	II
30	1732.70	II
50	1733.55	II
40	1734.49	II
30	1737.93	II
20	1740.16	II
20	1742.00	II
85	1742.10	IV
85	1766.27	IV
80	1795.65	IV
80	1795.79	IV
30	1853.27	II
20	1857.92	II
50	1902.95	II
20	1907.84	II
75	1910.25	IV
30	1911.41	II
20 d	1914.68	II
100	1915.10	II
20	1918.64	II
30	1919.64	II
80	1921.25	II
20	1923.07	II
20	1923.34	II
30	1925.52	II
50	1926.59	II
30	1931.40	II
500	1941.28	III
800	1943.21	III
20	1945.15	II
20	1947.93	II
20	1950.14	II
500	1952.36	III
1000	1952.52	III
30	1953.23	II
20 d	1954.81	II
30	1959.25	II
20	1969.24	II
500	1978.95	III
30	1994.23	II
9700	1996.06	I
14000	1999.51	I
18000	2003.85	I
1000 w	2027.83	III
500 w	2028.14	III
50	2037.31	II
40	2037.64	II
40	2039.97	II
500	2049.68	III
500	2066.38	III
1000	2069.02	III
30	2076.21	II
900	2077.38	III
800	2084.23	III
600	2090.05	III
1500	2092.16	I
500	2094.78	III
20	2097.46	II
500	2097.93	III
500	2099.97	III
20	2102.50	II
1700	2109.58	I
30	2113.96	II
1000	2169.78	III
700	2174.15	III
900	2176.87	III
800	2181.86	III
800	2184.87	III
290	2208.81	I
540	2213.85	I
900	2220.55	III
770	2221.84	I
1000	2227.42	III
20	2373.36	II
20	2427.38	II
50	2427.72	II
30	2427.94	II
30	2437.37	II
20	2437.84	II
30	2452.49	II
50	2499.00	II
30	2507.60	II
20	2516.60	II
30	2516.74	II
20	2521.66	II
20	2530.72	II
20	2531.80	II
50	2532.78	II
50	2533.33	II
30	2534.10	II
80	2534.22	II
100	2535.66	II
30	2535.98	II
100	2537.92	II
50	2541.11	II
80	2542.92	II
50	2543.45	II
100	2548.75	II
50	2551.85	II
30	2553.27	II
75	2556.57	II
30	2556.89	II
95	2558.59	II
30	2559.41	II
150	2563.65	II
30	2565.22	II
580	2572.76	I
480	2575.51	I
12000	2576.10	II
550	2584.31	I
30	2588.97	II
45	2589.71	II
250	2592.94	I
6200	2593.73	II
250	2595.76	I
95	2598.90	II
30	2602.72	II
45	2603.72	II
4300	2605.69	II
190	2610.20	II
500	2618.14	II
140	2622.90	I
150	2624.04	I
40	2624.80	II
200	2625.58	II
190	2632.35	II
130	2638.17	II
80	2639.84	II
27	2650.99	II
60	2655.91	II
30	2666.77	II
30	2667.03	II
110	2672.59	II
55	2673.37	II
55	2674.43	II
45	2680.34	II
30	2680.68	II
30	2681.25	II
55	2684.55	II
55	2685.94	II
110	2688.25	II

Intensity	Wavelength/Å	
27	2693.19	II
55	2695.36	II
27	2698.97	II
85	2701.00	II
50	2701.17	II
160	2701.70	II
100	2703.98	II
130	2705.74	II
80	2707.53	II
110	2708.45	II
45	2709.96	II
80	2710.33	II
110	2711.58	II
30	2716.80	II
30	2717.53	II
30	2719.01	II
50	2719.74	II
30	2722.10	II
30	2724.46	II
55	2728.61	II
6200	2794.82	I
5100	2798.27	I
220	2799.84	I
3700	2801.06	I
110	2809.11	I
60	2815.02	II
30	2816.33	II
60	2870.08	II
30	2872.94	II
80	2879.49	II
70	2886.68	II
160	2889.58	II
55	2892.39	II
50	2898.70	II
80	2900.16	II
140 h	2914.60	I
190 h	2925.57	I
1100	2933.06	II
1500	2939.30	II
250 h	2940.39	I
1900	2949.20	II
30	3019.92	II
55	3031.06	II
30	3035.35	II
330	3044.57	I
120	3045.59	I
200	3047.04	I
30	3050.65	II
250	3054.36	I
140	3062.12	I
170	3066.02	I
170	3070.27	I
160	3073.13	I
140 h	3178.50	I
220	3212.88	I
1000	3228.09	I
300	3230.72	I
850	3236.78	I
330	3243.78	I
650	3248.52	I
100	3251.14	I
310	3252.95	I
310	3256.14	I
220	3258.41	I
180	3260.23	I
180	3264.71	I
200	3330.78	II
720	3441.99	II
50	3460.03	II
360	3460.33	II
360 h	3474.04	II
	3474.13	II
290	3482.91	II
180	3488.68	II
140	3495.84	II
50	3496.81	II
100	3497.54	II
360	3531.85	I
1100	3532.12	I
1300	3547.80	I
1100	3548.03	I
390	3548.20	I
2200	3569.49	I
720	3569.80	I
1400	3577.88	I
720	3586.54	I
290	3595.12	I
150	3601.72	III
420	3607.54	I
420	3608.49	I
360	3610.30	I
290	3619.28	I
220	3623.79	I
140	3629.74	I
100	3660.40	I
280	3693.67	I
180	3696.57	I
210	3706.08	I
130	3718.93	I
130	3731.93	I
260	3790.22	I
110	3800.55	I
3200	3806.72	I
700	3809.59	I
2100	3823.51	I
390	3823.89	I
200	3829.68	I
480	3833.86	I
1300	3834.36	I
350	3839.78	I
670	3841.08	I
350	3843.98	I
120	3926.47	I
130	3982.58	I
150	3985.24	I
190	3986.83	I
150	3987.10	I
1500	4018.10	I
150	4026.44	I
27000	4030.76	I
19000	4033.07	I
11000	4034.49	I
1500	4035.73	I
5600	4041.36	I
210 d	4045.13	I
1100	4048.76	I
150	4055.21	I
1900	4055.54	I
210	4057.95	I
1100	4058.93	I
150	4059.39	I
730	4061.74	I
730	4063.53	I
290	4070.28	I
730	4079.24	I
730	4079.42	I
1100	4082.94	I
1100	4083.63	I
200	4110.90	I
150	4131.12	I
120	4135.04	I
150	4176.60	I
120	4189.99	I
370	4235.14	I
510	4235.29	I
190	4239.72	I
290	4257.66	I
290	4265.92	I
270	4281.10	I
50	4323.63	II
350	4414.88	I
210	4436.35	I
800	4451.59	I
160	4453.00	I
130	4455.01	I
160	4455.32	I
110	4455.82	I
210	4457.55	I
270	4458.26	I
150	4461.08	I
510	4462.02	I
290	4464.68	I
200	4470.14	I
130	4472.79	I
170	4490.08	I
240	4498.90	I
240	4502.22	I
160	4709.72	I
180	4727.48	I
130	4739.11	I
1000	4754.04	I
180	4761.53	I
750	4762.38	I
300	4765.86	I
500	4766.43	I
940	4783.42	I
1000	4823.52	I
19	5004.91	I
30	5074.79	I
200	5079.20	III
150	5100.03	III
60	5117.94	I
50	5150.89	I
50	5196.59	I
85	5255.32	I
160	5341.06	I
19	5349.88	I
95	5377.63	I
95	5394.67	I
50	5399.49	I
95	5407.42	I
35	5413.69	I
85	5420.36	I
35	5432.55	I
150	5454.07	III
12	5457.47	I
60	5470.64	I
200	5474.68	III
40	5481.40	I
30	5505.87	I
50	5516.77	I
40	5537.76	I
21	5551.98	I
200	5946.65	III
140	6013.50	I
200	6016.64	I
290	6021.80	I
200	6231.21	III
17	6440.97	I
24	6491.71	I
14 h	6942.52	I
12	6989.96	I
14	7069.84	I
12	7184.25	I
24 h	7283.82	I
35 h	7302.89	I
50	7326.51	I
12	7680.20	I
12 h	8672.06	I
12 h	8701.05	I
17 h	8703.76	I
30 h	8740.93	I
Mercury 198	**Hg Z = 80**	
80	1250.564	I
8	1259.242	I
100	1268.825	I
5	1307.751	I
20	1402.619	I
10	1435.503	I
1000	1849.492	I
60	2262.210	II
20	2302.065	I
20	2345.440	I
100	2378.325	I
20	2380.004	I
40	2399.349	I
20	2399.729	I
20	2446.900	I
15	2464.064	I
40	2481.999	I
30	2482.713	I
40	2483.821	I
90	2534.769	I
15000	2536.506	I
25	2563.861	I
25	2576.290	I
250	2652.043	I
400	2653.683	I
100	2655.130	I
50	2698.831	I
80	2752.783	I
20	2759.710	I
40	2803.471	I
30	2804.438	I
750	2847.675	II
50	2856.939	I
150	2893.598	I
150	2916.227	II
60	2925.413	I
1200	2967.283	I
300	3021.500	I
120	3023.476	I
30	3025.608	I

Intensity	Wavelength/Å	
50	3027.490	I
400	3125.670	I
320	3131.551	I
320	3131.842	I
80	3341.481	I
2800	3650.157	I
300	3654.839	I
80	3662.883	I
240	3663.281	I
30	3701.432	I
35	3704.170	I
30	3801.660	I
20	3901.867	I
60	3906.372	I
200	3983.839	II
1800	4046.572	I
150	4077.838	I
40	4108.057	I
250	4339.224	I
400	4347.496	I
4000	4358.337	I
80	4916.068	I
1100	5460.753	I
160	5675.922	I
240	5769.598	I
280	5790.663	I
20	6072.713	I
30	6234.402	I
160	6716.429	I
250	6907.461	I
240	11287.407	I

Mercury
Hg Z = 80

Intensity	Wavelength/Å	
3	621.44	III
2	679.68	III
2	878.59	III
1	886.48	III
400	893.08	II
300	915.83	II
150	923.39	II
200	940.80	II
100	962.74	II
50	969.13	II
1	988.89	III
2	1009.29	III
5	1068.03	III
800	1099.26	II
2	1161.95	III
80	1250.58	I
8	1259.24	I
100	1268.82	I
5	1307.75	I
300	1307.93	II
400	1321.71	II
400	1331.74	II
80	1350.07	II
200	1361.27	II
20	1402.62	I
200	1414.43	II
10	1435.51	I
15	1619.46	II
120	1623.95	II
20	1628.25	II
150	1649.94	II
50	1653.64	II
200	1672.41	II
9	1681.40	III
100	1702.73	II
100	1707.40	II
120	1727.18	II
250	1732.14	II
15	1759.75	III
20	1775.68	I
40	1783.70	II
30	1796.22	II
200	1796.90	II
60	1798.74	II
30	1803.89	II
40	1808.29	II
400	1820.34	II
5	1832.74	I
1000	1849.50	I
160	1869.23	II
300	1870.55	II
200	1875.54	II
1	1894.77	III
20	1900.28	II
30	1927.60	II
300	1942.27	II
100	1972.94	II
200	1973.89	II
150	1987.98	II
90	2026.97	II
90	2052.93	II
70	2148.00	II
5	2247.55	I
60	2262.23	II
20	2302.06	I
7	2314.15	III
15	2323.20	I
5	2340.57	I
20	2345.43	I
20	2352.48	I
100	2378.32	I
20	2380.00	I
4	2380.55	III
40	2399.38	I
20	2399.73	I
10	2400.49	I
60	2407.35	II
50	2414.13	II
8	2431.65	III
5	2441.06	I
20	2446.90	I
15	2464.06	I
5	2480.56	III
40	2482.00	I
30	2482.72	I
40	2483.82	I
7	2484.50	III
90	2534.77	I
15000	2536.52	I
25	2563.86	I
25	2576.29	I
5	2578.91	I
2	2612.92	III
4	2617.97	III
15	2625.19	I
5	2639.78	I
250	2652.04	I
400	2653.69	I
100	2655.13	I
3	2670.49	III
5	2674.91	I
50	2698.83	I
50	2699.38	I
80	2705.36	II
70	2724.43	III
80	2752.78	I
20	2759.71	I
6	2769.22	III
40	2803.46	I
30	2804.43	I
2	2805.34	I
2	2806.77	I
150	2814.93	II
3	2844.76	III
750	2847.68	II
50	2856.94	I
150	2893.60	I
150	2916.27	II
60	2925.41	I
150	2935.94	II
400	2947.08	II
1200	2967.28	I
300	3021.50	I
120	3023.47	I
30	3025.61	I
50	3027.49	I
15	3090.05	III
400	3125.67	I
320	3131.55	I
320	3131.84	I
400	3208.20	II
400	3264.06	II
5	3283.02	III
12	3312.28	III
80	3341.48	I
100	3385.25	II
8	3389.01	III
5	3450.77	III
400	3451.69	II
3	3500.35	III
4	3538.88	III
200	3549.42	II
5	3557.24	III
2800	3650.15	I
300	3654.84	I
80	3662.88	I
240	3663.28	I
30	3701.44	I
35	3704.17	I
30	3801.66	I
15	3803.51	III
100	3806.38	II
20	3901.87	I
60	3906.37	I
100	3918.92	II
200	3983.96	II
1800	4046.56	I
150	4077.83	I
40	4108.05	I
70	4122.07	III
10	4140.34	III
100	4216.74	III
250	4339.22	I
400	4347.49	I
4000	4358.33	I
100	4398.62	II
15	4470.58	III
12	4552.84	III
90	4660.28	II
50	4797.01	III
80	4855.72	II
10	4869.85	III
5	4883.00	I
5	4889.91	I
80	4916.07	I
5	4970.37	I
80	4973.57	III
5	4980.64	I
20	5102.70	I
40	5120.64	I
100	5128.45	II
20	5137.94	I
30	5210.82	III
20	5290.74	I
5	5316.78	I
60	5354.05	I
30	5384.63	I
1100	5460.74	I
30	5549.63	I
160	5675.86	I
6	5695.71	III
240	5769.60	I
100	5789.66	I
280	5790.66	I
140	5803.78	I
60	5859.25	I
60	5871.73	II
20	5871.98	I
20	6072.72	I
1000	6149.50	II
25	6220.35	III
30	6234.40	I
35	6418.98	III
40	6501.38	III
80	6521.13	II
10	6584.26	III
6	6610.12	III
30	6709.29	III
160	6716.43	I
250	6907.52	I
250	7081.90	I
200	7091.86	I
40	7346.37	II
100	7485.87	II
12	7517.46	III
20	7728.82	I
7	7808.10	III
100	7944.66	II
25	7946.75	III
50	7984.51	III
5	8151.64	III
2000	10139.75	I
240	11287.40	I
120	13209.95	I
140	13426.57	I
60	13468.38	I
80	13505.58	I
500	13570.21	I
450	13673.51	I
200	13950.55	I
500	15295.82	I

Intensity	Wavelength/Å	
100	16881.48	I
400	16920.16	I
300	16942.00	I
500	17072.79	I
400	17109.93	I
20	17116.75	I
20	17198.67	I
20	17213.20	I
70	17329.41	I
30	17436.18	I
50	18130.38	I
40	19700.17	I
	22493.28	I
250	23253.07	I
	32148.06	I
	36303.03	I

Molybdenum
Mo Z = 42

Intensity	Wavelength/Å	
50	867.92	IV
100	884.19	IV
60	886.05	IV
50	891.74	IV
100	1169.33	III
100	1254.93	III
100	1258.52	III
100	1262.21	III
100	1263.74	III
100	1274.37	III
100	1276.40	III
200	1277.40	III
200	1277.58	III
200	1278.40	III
150	1281.90	III
150	1283.60	III
100	1854.73	III
80	1926.26	IV
100	1929.24	IV
80	1971.06	IV
70	2010.92	IV
19000	2015.11	II
40000	2020.30	II
21000	2038.44	II
17000	2045.98	II
50	2060.38	IV
4800	2081.68	II
2400	2089.52	II
2200	2092.50	II
4000	2093.11	II
2700	2100.84	II
1500	2104.29	II
1400	2108.02	II
100	2184.37	III
100	2211.02	III
400	2269.69	II
150	2269.71	III
200	2294.97	III
160	2304.25	II
160	2306.97	II
150	2330.93	III
110	2332.12	II
190	2341.59	II
100	2359.76	III
110	2389.20	II
140	2403.61	II
120	2413.01	II
85	2498.28	II
200	2506.19	III
440	2538.46	II
330	2542.67	II
80	2558.88	II
85	2564.34	II
250	2593.70	II
250	2602.80	II
400	2616.78	I
440	2629.85	I
330	2636.67	II
720	2638.76	II
410	2640.99	I
600	2644.35	II
370	2646.49	II
640	2649.46	I
480	2653.35	II
560 h	2655.03	I
640	2660.58	II
720	2672.84	II
250	2673.27	II
1000	2679.85	I
95	2681.36	II
640	2683.23	II
880	2684.14	II
560	2687.99	II
480	2701.42	II
190	2713.51	II
290	2717.35	II
85	2726.97	II
140	2729.68	II
80	2730.20	II
330	2732.88	II
160	2736.96	II
80 h	2737.88	II
290	2746.30	II
110	2756.07	II
220	2763.62	II
240	2769.76	II
160	2773.78	II
190	2774.39	II
1700	2775.40	II
65	2777.86	II
880	2780.04	II
400	2784.99	II
100	2807.74	III
400	2807.76	II
1700	2816.15	II
220	2817.44	II
80	2827.74	II
80	2834.39	II
80	2835.33	II
160	2842.15	II
1700	2848.23	II
370	2853.23	II
370	2863.81	II
220	2866.69	II
1700	2871.51	II
85	2872.88	II
220	2879.05	II
65	2888.15	II
1300	2890.99	II
95	2891.28	II
190	2892.81	II
950	2894.45	II
140	2897.63	II
70	2900.80	II
290	2903.07	II
80	2907.12	II
600	2909.12	II
1100	2911.92	II
120	2918.83	II
1300	2923.39	II
140	2924.32	II
1100	2930.50	II
800	2934.30	II
95	2940.10	II
110	2941.22	II
150	2944.82	II
140	2946.69	II
95	2947.28	II
125	2947.32	III
95	2955.84	II
240	2956.06	II
70	2956.90	II
95	2960.24	II
250	2963.79	II
210	2965.27	II
70	2971.91	II
250	2972.61	II
80	2975.40	II
95	2992.84	II
95	3027.77	II
100	3060.78	II
800	3064.28	I
250	3065.04	II
800	3074.37	I
85	3077.66	II
800	3085.62	I
270	3087.62	II
190	3092.07	II
560	3094.66	I
560	3101.34	I
1400	3112.12	I
290	3122.00	II
14000	3132.59	I
110	3138.72	II
220	3152.82	II
55	3155.64	II
6000	3158.16	I
8700	3170.35	I
95	3172.03	II
160	3172.74	II
120 d	3187.59	II
7600	3193.97	I
880	3205.88	I
3000	3208.83	I
560	3215.07	I
880	3228.22	I
600	3229.79	I
1100	3233.14	I
950	3237.08	I
65	3240.71	II
950	3256.21	I
480	3264.40	I
800	3270.90	I
200	3271.69	III
1100	3289.02	I
950	3290.82	I
190	3292.31	II
100	3313.62	II
190	3320.90	II
640	3323.95	I
1300	3344.75	I
95	3346.40	II
1600	3358.12	I
950	3363.78	I
950	3379.97	I
1900	3384.62	I
130	3395.36	II
640	3404.34	I
1300	3405.94	I
640	3437.22	I
130	3446.08	II
3200	3447.12	I
640	3449.07	I
950	3456.39	I
640	3460.78	I
800	3504.41	I
560	3508.12	I
480	3521.41	I
640	3537.28	I
520	3558.10	I
400	3563.14	I
1400	3581.89	I
1400	3624.46	I
1000	3635.43	I
400	3657.35	I
540	3664.81	I
590	3672.82	I
1300	3680.60	I
65	3688.31	II
180	3692.64	II
1400	3694.94	I
500	3727.69	I
80	3744.37	II
29000	3798.25	I
520	3826.70	I
940	3828.87	I
1700	3833.75	I
29000	3864.11	I
580	3869.08	I
580	3886.82	I
19000	3902.96	I
65	3941.48	II
1400	4062.08	I
2300	4069.88	I
1300	4081.44	I
940	4084.38	I
730	4107.47	I
630	4120.10	I
2900	4143.55	I
480	4185.82	I
2500	4188.32	I
1500	4232.59	I
890	4276.91	I
1200	4277.24	I
1400	4288.64	I
680	4292.13	I
890	4293.21	I
840	4326.14	I
1900	4381.64	I
2500	4411.57	I
990	4434.95	I
480	4457.36	I
630	4474.56	I
400	4536.80	I
460	4626.47	I
640	4707.26	I

Intensity	Wavelength/Å	
700	4731.44	I
770	4760.19	I
410	4819.25	I
410	4830.51	I
180	5014.60	I
80	5029.00	I
65	5030.78	I
100	5047.71	I
50	5055.00	I
200	5059.88	I
100	5080.02	I
100	5096.65	I
130	5097.52	I
130	5109.71	I
80	5114.97	I
150	5145.38	I
110	5147.39	I
80	5163.19	I
100	5167.76	I
160 d	5171.08	I
230 h	5172.94	I
160 h	5174.18	I
110	5200.17	I
50	5200.74	I
50	5211.86	I
80	5219.40	I
65	5231.06	I
100	5234.26	I
460 h	5238.20	I
230 h	5240.88	I
110 h	5242.81	I
100	5245.51	I
150	5259.04	I
65	5261.14	I
65	5279.65	I
210	5280.86	I
55	5292.08	I
55	5295.47	I
55	5313.89	I
80	5354.88	I
65	5356.48	I
560 hl	5360.56	I
110 hl	5364.28	I
65	5394.52	I
50	5400.47	I
55	5435.68	I
65	5437.75	I
50	5501.54	I
7800	5506.49	I
5200	5533.05	I
50	5543.12	I
55	5556.28	I
2500	5570.45	I
100	5610.93	I
330	5632.47	I
50	5634.86	I
230	5650.13	I
55	5674.47	I
460	5689.14	I
80	5705.72	I
210	5722.74	I
620	5751.40	I
520	5791.85	I
55 h	5849.73	I
50 h	5851.52	I
520	5858.27	I
50	5869.33	I
820	5888.33	I
50 h	5893.38	I
160 h	5928.88	I
35	6025.49	I
1300	6030.66	I
40	6101.87	I
40	6357.22	I
35	6401.07	I
100	6424.37	I
230	6619.13	I
50	6650.38	I
110	6733.98	I
50	6746.27	I
35	6753.97	I
40	6838.88	I
35	6914.01	I
110	7109.87	I
150	7242.50	I
40	7245.85	I
40	7391.36	I
140	7485.74	I
27	7720.77	I
40 h	8328.44	I
45 h	8389.32	I
45 h	8483.39	I
Neodymium		
Nd *Z* = 60		
75	2764.98	I
80	2993.20	II
95	3007.97	II
95	3014.19	II
95	3018.35	II
140	3056.71	II
130	3069.73	II
160	3075.38	II
240	3092.92	II
260	3115.18	II
290	3133.60	II
220	3134.90	II
170	3141.46	II
170	3142.44	II
150	3203.47	II
220	3259.24	II
220	3265.12	II
320	3275.22	II
290	3285.10	II
410	3328.28	II
320	3353.59	II
410	3560.75	II
470	3587.51	II
370	3615.82	II
410	3653.15	II
470	3662.26	II
540	3665.18	II
540	3672.36	II
380	3673.54	II
1200	3685.80	II
440	3687.30	II
410	3689.69	II
410	3697.56	II
470	3713.70	II
640 d	3714.73	II
470	3715.68	II
410	3718.54	II
410	3721.35	II
780	3723.50	II
410	3724.87	II
710	3728.13	II
470	3730.58	II
1000 d	3735.54	II
440	3737.10	II
1000	3738.06	II
580	3752.49	II
510	3757.82	II
930	3758.95	II
930	3763.47	II
510	3769.65	II
1400	3775.50	II
710	3779.47	II
580	3780.40	II
510	3781.32	II
2400	3784.25	II
370	3801.12	II
1200	3803.47	II
2500	3805.36	II
470	3807.23	II
540	3808.77	II
440	3809.06	II
580	3810.49	II
710	3814.73	II
410	3822.47	II
1200	3826.42	II
540	3828.85	II
440	3829.16	II
510	3830.47	II
740	3836.54	II
1700	3838.98	II
410 d	3841.82	II
1700 d	3848.24	II
1500	3848.52	II
470	3850.22	II
2400 d	3851.66	II
3700 d	3863.33	II
850	3869.07	II
470	3875.87	II
1100	3878.58	II
1000	3879.55	II
780	3880.38	II
1200	3880.78	II
540	3887.87	II
1300	3889.93	II
1300	3890.58	II
1300	3890.94	II
580	3891.51	II
470	3892.06	II
810	3894.63	II
440	3897.63	II
2000	3900.21	II
1300	3901.84	II
1700	3905.89	II
510	3907.84	II
2000	3911.16	II
850	3912.23	II
440	3915.13	II
610	3915.95	II
1100	3920.96	II
510	3927.10	II
610	3934.82	II
410	3936.11	II
510	3938.86	II
2000	3941.51	II
2000	3951.16	II
810	3952.20	II
590	3958.00	II
510	3962.21	II
1400	3963.12	II
1100	3973.30	II
740	3973.69	II
740	3976.85	II
740	3979.49	II
470	3986.25	II
1400	3990.10	II
1000	3991.74	II
1100	3994.68	II
410	4000.50	II
540	4004.02	II
410	4007.43	II
3700	4012.25	II
540	4012.70	II
1000	4020.87	II
1000	4021.34	II
1000	4021.78	II
1200	4023.00	II
410	4030.47	II
1200	4031.82	II
3000	4040.80	II
410	4043.59	II
410	4048.81	II
850	4051.15	II
850	4059.96	II
4700	4061.09	II
1100	4069.28	II
710	4075.12	II
470	4075.28	II
470	4080.23	II
1400	4109.08	II
2500	4109.46	II
510	4110.48	II
410	4123.88	II
470	4133.36	II
510	4135.33	II
3000	4156.08	II
510	4156.26	II
410	4168.00	II
810	4175.61	II
2400	4177.32	II
640	4179.59	II
470	4205.60	II
470	4211.29	II
440	4227.73	II
1300	4232.38	II
2000	4247.38	II
850	4252.44	II
410	4261.84	II
470	4282.44	II
710	4284.52	II
5400	4303.58	II
470	4314.52	II
1100	4325.76	II
510	4327.93	II
540	4338.70	II
680	4351.29	II
850	4358.17	II
470 d	4374.93	II
710	4385.66	II
540	4400.83	II
510	4411.06	II

Intensity	Wavelength/Å	
580	4446.39	II
1400	4451.57	II
740	4462.99	II
410	4501.82	II
250	4516.36	II
340	4541.27	II
340	4542.61	II
340	4563.22	II
300	4621.94	I
510	4634.24	I
340	4641.10	I
250	4645.77	II
300	4649.67	I
310	4683.45	I
470	4706.54	II
240	4719.02	I
240	4811.34	II
350	4825.48	II
280	4859.02	II
350	4883.81	I
220	4890.70	II
240	4891.07	I
280	4896.93	I
210	4901.84	I
330	4920.68	II
470	4924.53	I
260	4944.83	I
290	4954.78	I
290	4959.13	II
250	4989.94	II
360	5076.59	II
360	5092.80	II
360	5107.59	II
340	5123.79	II
680	5130.60	II
500	5191.45	II
630	5192.62	II
330	5200.12	II
310	5212.37	II
450	5234.20	II
250	5239.79	II
720	5249.59	II
360	5255.51	II
590	5273.43	II
680	5293.17	II
220	5311.46	II
500	5319.82	II
290	5361.47	II
160	5431.53	II
240	5594.43	II
220	5620.54	I
140 d	5675.97	I
220	5688.53	II
130	5702.24	II
160	5708.28	II
100	5729.29	I
160	5804.02	II
80	5811.57	II
70	5825.87	II
80	5842.39	II
55	5858.91	I
45	6007.67	I
45	6034.24	II
55	6066.03	I
45	6178.59	I
45	6223.39	I
55	6310.49	I
65	6385.20	II
45	6630.14	I
45	6650.57	II
40	6740.11	II
40	6900.43	II
35	7037.30	II
40	7066.89	II
29	7129.35	II
24	7189.42	II
20	7192.01	II
15	7236.54	II
12	7316.81	II
10	7406.62	II
10	7418.18	II
12	7511.16	II
17	7513.73	II
12	7528.99	II
10	7538.26	II
12	7696.56	II
10	7750.95	II
10	7808.47	II
12	7863.04	II
12	7917.01	II
12	7958.95	I
12	7965.73	II
15	7982.09	II
12	7982.68	II
12	8000.76	II
10	8120.93	II
12	8122.07	II
12	8141.75	II
12	8143.27	II
10	8231.52	II
10	8307.72	II
12	8346.36	II
17	8839.10	II

Neon

Ne $Z = 10$

Intensity	Wavelength/Å	
66	119.01	V
200	122.52	V
66	125.12	V
45	131.99	V
50	132.04	V
150	140.76	V
150	140.79	V
100	142.44	V
100	142.50	V
150	142.72	V
100	143.27	V
150	143.34	V
150	147.13	V
66	151.23	V
120	151.42	V
15	151.82	IV
15	152.23	IV
45	154.50	V
15	158.65	IV
15	158.82	IV
100	164.02	V
100	164.14	V
80	172.62	IV
500	173.93	V
80	177.16	IV
150	186.58	IV
100	194.28	IV
100	208.48	IV
100	208.73	IV
80	208.90	IV
150	212.56	IV
140	223.24	IV
120	223.60	IV
140	234.32	IV
120	234.70	IV
20	251.14	III
20	251.56	III
20	251.73	III
40	267.06	III
40	267.52	III
20	267.71	III
40	283.18	III
160	283.21	III
110	283.69	III
40	283.89	III
220	301.12	III
220	313.05	III
220	313.68	III
40	313.95	III
90	352.956	I
60	354.962	I
50	357.83	IV
400	357.96	V
500	358.47	V
200	358.72	IV
500	359.38	V
90	361.433	II
60	362.455	II
1000	365.59	V
220	379.31	III
125	387.14	IV
100	388.22	IV
150	405.854	II
120	407.138	II
800	416.20	V
150	421.61	IV
200	445.040	II
300	446.256	II
250	446.590	II
180	447.815	II
150	454.654	II
200	455.274	II
10	456.275	II
120	456.348	II
90	456.896	II
1000	460.728	II
500	462.391	II
140	469.77	IV
200	469.82	IV
180	469.87	IV
140	469.92	IV
250	480.41	V
150	481.28	V
250	481.36	V
500	482.99	V
285	488.10	III
220	488.87	III
450	489.50	III
70	489.64	III
220	490.31	III
360	491.05	III
120	521.74	IV
140	521.82	IV
80	541.13	IV
100	542.07	IV
150	543.89	IV
400	568.42	V
250	569.76	V
500	569.83	V
250	572.11	V
800	572.34	V
35	587.213	I
35	589.179	I
35	589.911	I
70	591.830	I
100	595.920	I
75	598.706	I
35	598.891	I
70	600.036	I
170	602.726	I
170	615.628	I
170	618.672	I
120	619.102	I
200	626.823	I
200	629.739	I
1000	735.896	I
400	743.720	I
60	993.88	I
70	1068.65	I
90	1131.72	I
100	1131.85	II
90	1229.83	I
20	1255.03	III
110	1255.68	III
160	1257.19	III
90	1418.38	I
90	1428.58	I
90	1436.09	I
120	1681.68	II
180	1688.36	II
100	1888.11	II
100	1889.71	II
200	1907.49	II
500	1916.08	II
300	1930.03	II
200	1938.83	II
100 c	1945.46	II
80	2007.01	II
65	2018.44	IV
110	2022.19	IV
80	2025.56	II
150	2085.47	II
200	2086.96	III
300	2089.43	III
240	2092.44	III
400	2095.54	III
180	2096.11	II
120	2096.25	II
200	2161.22	III
300	2163.77	III
200	2180.89	III
30	2203.88	IV
200	2209.35	III
200	2211.85	III
240	2213.76	III
300	2216.07	III
10	2220.81	IV
75	2227.42	V
110	2232.41	V

Line Spectra of the Elements (continued): Neon

Intensity	Wavelength/Å	
65	2245.48	V
250	2258.02	IV
65	2259.57	V
175	2262.08	IV
240	2263.21	III
65	2263.39	V
110	2264.54	IV
200	2264.91	III
250	2265.71	V
550	2285.79	IV
30	2293.14	IV
250	2293.49	IV
250	2350.84	IV
450	2352.52	IV
700	2357.96	IV
250	2362.68	IV
250	2363.28	IV
110	2365.49	IV
350	2372.16	IV
65	2384.20	IV
350	2384.95	IV
300	2412.73	III
240	2412.94	III
200	2413.78	III
200	2473.40	III
80 p	2562.12	II
90 w	2567.12	II
800	2590.04	III
600	2593.60	III
400	2595.68	III
300	2610.03	III
240	2613.41	III
200	2615.87	III
80	2623.11	II
80	2629.89	II
90 w	2636.07	II
80	2638.29	II
200	2638.70	III
200	2641.07	III
80	2644.10	II
600	2677.90	III
500	2678.64	III
80	2762.92	II
90	2792.02	II
80	2794.22	II
100	2809.48	II
80	2906.59	II
80	2906.82	II
90	2910.06	II
90	2910.41	II
80	2911.14	II
80	2915.12	II
80	2925.62	II
80 w	2932.10	II
80	2940.65	II
90	2946.04	II
150	2955.72	II
150	2963.24	II
150	2967.18	II
100	2973.10	II
15	2974.72	I
100	2979.46	II
12	2982.67	I
150	3001.67	II
120 p	3017.31	II
300	3027.02	II
300	3028.86	II
100	3030.79	II
120	3034.46	II
100	3035.92	II
100	3037.72	II
100	3039.59	II
100	3044.09	II
100	3045.56	II
120	3047.56	II
100	3054.34	II
100	3054.68	II
100	3059.11	II
100	3062.49	II
100	3063.30	II
100	3070.89	II
100	3071.53	II
100	3075.73	II
120	3088.17	II
100	3092.09	II
120	3092.90	II
100	3094.01	II
100	3095.10	II
100	3097.13	II
100	3117.98	II
120	3118.16	II
10	3126.199	I
300	3141.33	II
100	3143.72	II
100 p	3148.68	II
100	3164.43	II
100	3165.65	II
100	3188.74	II
120	3194.58	II
500	3198.59	II
60	3208.96	II
120	3209.36	II
120	3213.74	II
150	3214.33	II
150	3218.19	II
120	3224.82	II
120	3229.57	II
200	3230.07	II
120	3230.42	II
120	3232.02	II
150	3232.37	II
100	3243.40	II
100	3244.10	II
100	3248.34	II
100	3250.36	II
150	3297.73	II
150	3309.74	II
300	3319.72	II
1000	3323.74	II
150	3327.15	II
100	3329.16	II
200	3334.84	II
150	3344.40	II
300	3345.45	II
150	3345.83	II
200	3355.02	II
120	3357.82	II
200	3360.60	II
120	3362.16	II
100	3362.71	II
120	3367.22	II
12	3369.808	I
40	3369.908	I
100	3371.80	II
500	3378.22	II
150	3388.42	II
120	3388.94	II
300	3392.80	II
100	3404.82	II
120	3406.95	II
100	3413.15	II
120	3416.91	II
120	3417.69	II
50	3417.904	I
15	3418.006	I
120	3428.69	II
60	3447.703	I
50	3454.195	I
100	3456.61	II
100	3459.32	II
25	3460.524	I
30	3464.339	I
30	3466.579	I
60	3472.571	I
150	3479.52	II
200	3480.72	II
200	3481.93	II
25	3498.064	I
30	3501.216	I
25	3515.191	I
150	3520.472	I
120	3542.85	II
120	3557.80	II
100	3561.20	II
250	3568.50	II
100	3574.18	II
200	3574.61	II
50	3593.526	I
30	3593.640	I
15	3600.169	I
20	3633.665	I
150	3643.93	II
200	3664.07	II
20	3682.243	I
12	3685.736	I
200	3694.21	II
10	3701.225	I
150	3709.62	II
250	3713.08	II
250	3727.11	II
800	3766.26	II
1000	3777.13	II
100	3818.43	II
120	3829.75	II
150	4219.74	II
100	4233.85	II
120	4250.65	II
120	4369.86	II
70	4379.40	II
150	4379.55	II
100	4385.06	II
200	4391.99	II
150	4397.99	II
150	4409.30	II
100	4413.22	II
100	4421.39	II
100 p	4428.52	II
100 p	4428.63	II
150 p	4430.90	II
150 p	4430.94	II
120	4457.05	II
100	4522.72	II
10	4537.754	I
10	4540.380	I
100	4569.06	II
15	4704.395	I
12	4708.862	I
10	4710.067	I
10	4712.066	I
15	4715.347	I
10	4752.732	I
12	4788.927	I
10	4790.22	I
10	4827.344	I
10	4884.917	I
4	5005.159	I
10	5037.751	I
10	5144.938	I
25	5330.778	I
20	5341.094	I
8	5343.283	I
60	5400.562	I
5	5562.766	I
10	5656.659	I
5	5719.225	I
12	5748.298	I
80	5764.419	I
12	5804.450	I
40	5820.156	I
500	5852.488	I
100	5872.828	I
100	5881.895	I
60	5902.462	I
60	5906.429	I
100	5944.834	I
100	5965.471	I
100	5974.627	I
120	5975.534	I
80	5987.907	I
100	6029.997	I
100	6074.338	I
80	6096.163	I
60	6128.450	I
100	6143.063	I
120	6163.594	I
250	6182.146	I
150	6217.281	I
150	6266.495	I
60	6304.789	I
100	6334.428	I
120	6382.992	I
200	6402.246	I
150	6506.528	I
60	6532.882	I
150	6598.953	I
70	6652.093	I
90	6678.276	I
20	6717.043	I
100	6929.467	I
90	7024.050	I
100	7032.413	I
50	7051.292	I
80	7059.107	I
100	7173.938	I

Intensity	Wavelength/Å	
150	7213.20	II
150	7235.19	II
100	7245.167	I
150	7343.94	II
40	7472.439	I
90	7488.871	I
100	7492.10	II
150	7522.82	II
80	7535.774	I
60	7544.044	I
100	7724.628	I
120	7740.74	II
300	7839.055	I
120	7926.20	II
400	7927.118	I
700	7936.996	I
2000	7943.181	I
2000	8082.458	I
100	8084.34	II
1000	8118.549	I
600	8128.911	I
3000	8136.406	I
2500	8259.379	I
100	8264.81	II
2500	8266.077	I
800	8267.117	I
6000	8300.326	I
100	8315.00	II
1500	8365.749	I
100	8372.11	II
8000	8377.606	I
1000	8417.159	I
4000	8418.427	I
1500	8463.358	I
800	8484.444	I
5000	8495.360	I
600	8544.696	I
1000	8571.352	I
4000	8591.259	I
6000	8634.647	I
3000	8647.041	I
15000	8654.383	I
4000	8655.522	I
100	8668.26	II
5000	8679.492	I
5000	8681.921	I
2000	8704.112	I
4000	8771.656	I
12000	8780.621	I
10000	8783.753	I
500	8830.907	I
7000	8853.867	I
1000	8865.306	I
1000	8865.755	I
3000	8919.501	I
2000	8988.57	I
100	9079.46	II
6000	9148.67	I
6000	9201.76	I
4000	9220.06	I
2000	9221.58	I
2000	9226.69	I
1000	9275.52	I
200	9287.56	II
6000	9300.85	I
1500	9310.58	I
3000	9313.97	I
6000	9326.51	I
2000	9373.31	I
5000	9425.38	I
3000	9459.21	I
5000	9486.68	I
5000	9534.16	I
3000	9547.40	I
120	9577.01	II
1000	9665.42	I
100	9808.86	II
800	10295.42	I
2000	10562.41	I
1500	10798.07	I
2000	10844.48	I
3000	11143.020	I
3500	11177.528	I
1600	11390.434	I
1100	11409.134	I
3000	11522.746	I
1500	11525.020	I
950	11536.344	I
500	11601.537	I
1200	11614.081	I
300	11688.002	I
2000	11766.792	I
1500	11789.044	I
500	11789.889	I
1000	11984.912	I
3000	12066.334	I
800	12459.389	I
1000	12689.201	I
1100	12912.014	I
700	13219.241	I
800	15230.714	I
400	17161.930	I
400	18035.80	I
1000	18083.21	I
350	18221.11	I
250	18227.02	I
2500	18276.68	I
2000	18282.62	I
1200	18303.97	I
250	18359.12	I
1200	18384.85	I
2000	18389.95	I
1000	18402.84	I
1200	18422.39	I
300	18458.65	I
400	18475.79	I
900	18591.55	I
1600	18597.70	I
350	18618.96	I
550	18625.16	I
1200	21041.295	I
750	21708.145	I
300	22247.35	I
350	22428.13	I
2250	22530.40	I
400	22661.81	I
600	23100.51	I
1000	23260.30	I
1050	23373.00	I
850	23565.36	I
3500	23636.52	I
300	23701.64	I
1100	23709.2	I
1800	23951.42	I
600	23956.46	I
1000	23978.12	I
200	24098.54	I
500	24161.42	I
600	24249.64	I
1500	24365.05	I
800	24371.60	I
400	24447.85	I
700	24459.4	I
300	24776.46	I
550	24928.88	I
250	25161.69	I
650	25524.37	I
125	28386.21	I
150	30200.	I
250	33173.09	I
450	33352.35	I
1300	33901.	I
2200	33912.10	I
600	34131.31	I
100	34471.44	I
120	35834.78	I

Neptunium
Np $Z = 93$

Intensity	Wavelength/Å	
300	3481.93	I
300 h	3501.50	I
300 l	3986.89	I
300 s	5044.66	I
300 l	5601.70	I
300 l	5652.75	I
300 l	5784.39	I
300 l	5878.04	I
300 s	6011.22	I
300	6056.09	I
300 s	6073.90	I
300 s	6080.05	I
300 l	6120.49	I
300	6188.59	I
300 l	6200.00	I
300 s	6215.90	I
300 s	6317.84	I
300 l	6341.38	I
300 l	6566.11	I
300 l	6720.68	I
300 s	6751.32	I
300 s	6795.21	I
300 l	6802.62	I
300 l	6805.81	I
300 s	6816.44	I
300 l	6865.45	I
300 s	6907.13	I
300 h	6912.91	I
1000 s	6930.31	I
300 l	6963.63	I
3000 s	6972.09	I
300	7014.02	I
300 l	7018.91	I
300 s	7039.14	I
300 s	7080.01	I
300 l	7174.83	I
300 l	7184.93	I
300 l	7284.28	I
300 l	7292.29	I
300 l	7332.52	I
300 s	7370.60	I
300 l	7381.03	I
300 l	7381.65	I
300 l	7402.70	I
300 s	7512.22	I
300 l	7515.15	I
300 l	7546.05	I
300 l	7624.83	I
300	7626.85	I
300 s	7681.01	I
300 s	7685.25	I
1000 l	7735.14	I
300 l	7761.61	I
1000 l	7765.75	I
300 s	7776.07	I
300	7787.46	I
1000 l	7791.38	I
300 l	7851.44	I
300 l	7887.88	I
300 l	7901.71	I
300 l	7975.98	I
300 h	8080.32	I
300 s	8124.59	I
300	8155.11	I
300 l	8167.42	I
300 l	8183.06	I
300 l	8188.61	I
300 l	8247.82	I
300 l	8287.11	I
300 s	8287.75	I
300 l	8306.22	I
300 s	8313.66	I
1000 l	8339.12	I
300	8356.79	I
300 l	8367.11	I
3000	8372.88	I
3000	8529.96	I
1000 s	8696.23	I
1000 s	8906.02	I
1000	8942.70	I
1000 s	9004.75	I
1000 l	9006.31	I
10000 l	9016.18	I
3000 l	9141.30	I
3000 s	9379.33	I
3000 l	9468.66	I
3000 s	9679.13	I
3000 l	9930.55	I
10000 l	10091.99	I
10000 s	10817.45	I
10000 l	11695.15	I
10000 l	11776.64	I
10000 s	12148.18	I
10000 s	12377.42	I
10000 l	12407.99	I
10000 l	13834.33	I

Nickel
Ni $Z = 28$

Intensity	Wavelength/Å	
55	315.24	V
56	315.71	V
72	354.18	V
76	354.42	V
68	354.49	V
500	630.71	III
500	676.94	III
300	713.33	III

Intensity	Wavelength/Å	
300	713.38	III
500	718.48	III
300	722.09	III
500	729.82	III
400	731.70	III
300	732.16	III
300	747.99	III
300	750.05	III
300	757.80	III
400	770.22	III
500	778.81	III
300	788.04	III
500	811.57	III
500	826.14	III
500	842.14	III
400	845.24	III
300	847.43	III
300	860.64	III
300	862.88	III
300	863.22	III
300	867.51	III
300	973.79	III
400	979.59	III
500	1317.22	II
76	1398.19	IV
74	1411.45	IV
70	1438.82	IV
73	1449.01	IV
76	1452.22	IV
73	1482.25	IV
72	1489.83	IV
75	1525.31	IV
74	1527.68	IV
74	1527.80	IV
76	1534.71	IV
73	1537.25	IV
75	1543.41	IV
74	1546.23	IV
300	1604.54	III
300	1652.87	III
400	1687.90	III
1000	1692.51	III
800	1709.90	III
650	1715.30	III
500	1719.46	III
400	1722.28	III
500	1738.25	III
300	1739.78	III
1000	1741.55	II
300	1741.96	III
550	1747.01	III
300	1752.43	III
400	1753.01	III
800	1764.69	III
500	1767.94	III
2000	1769.64	III
400	1776.07	III
300	1807.24	III
300	1819.28	III
800	1823.06	III
400	1830.01	III
650	1847.28	III
800	1854.15	III
300	1858.75	III
1000	2165.55	II
2000	2169.10	II
2000	2174.67	II
1500	2175.15	II
2500	2185.50	II
3000	2192.09	II
5000	2205.55	II
4000	2206.72	II
6000	2216.48	II
1000	2264.46	II
2000	2270.21	II
1600	2289.98	I
630	2300.78	I
1000	2303.00	II
2000	2310.96	I
1700	2312.34	I
1400	2313.66	I
1400	2313.98	I
1000	2316.04	II
1400	2317.16	I
2600	2320.03	I
1900	2321.38	I
1400	2325.79	I
940	2329.96	I
1200	2345.54	I
400	2347.52	I
1000	2375.42	II
240	2386.58	I
1000	2394.52	II
2000	2416.13	II
240	2419.31	I
160	2472.06	I
150	2798.65	I
250	2821.29	I
500	2943.91	I
570	2981.65	I
500	2992.60	I
1000	2994.46	I
4000	3002.49	I
2200	3003.63	I
3700	3012.00	I
1700	3037.94	I
3500	3050.82	I
1500	3054.32	I
1900	3057.64	I
500	3064.62	I
2600	3101.55	I
1300	3101.88	I
2900	3134.11	I
1100	3232.96	I
600	3243.06	I
660	3315.66	I
2000	3331.88	II
2900	3369.57	I
3300	3380.57	I
1300	3391.05	I
3300	3392.99	I
8200	3414.76	I
1600	3423.71	I
2600	3433.56	I
990	3437.28	I
4800	3446.26	I
1300	3452.89	I
5000	3458.47	I
5000	3461.65	I
1600	3472.54	I
550	3483.77	I
5500	3492.96	I
660	3500.85	I
2600	3510.34	I
6600	3515.05	I
660	3519.77	I
8200	3524.54	I
5000	3566.37	I
990	3571.87	I
1300	3597.70	I
1300	3610.46	I
530	3612.74	I
6600	3619.39	I
200	3664.10	I
130	3669.24	I
180	3670.43	I
260	3674.15	I
160	3688.42	I
80	3693.93	I
120	3722.48	I
150	3736.81	I
60	3739.23	I
600	3775.57	I
700	3783.53	I
700	3807.14	I
110	3831.69	I
1200	3858.30	I
110	3973.56	I
110	4401.55	I
85	4459.04	I
55	4470.48	I
65	4605.00	I
75	4648.66	I
110	4714.42	I
45	4786.54	I
45	4855.41	I
40	4904.41	I
45	4980.16	I
45	4984.13	I
50	5017.59	I
100	5035.37	I
100	5080.52	I
65	5081.11	I
40 h	5146.48	I
40 h	5155.76	I
180	5476.91	I
23	5709.56	I
16	5754.68	I
10	5857.76	I
10	5892.88	I
10	6108.12	I
10	6176.81	I
10	6191.18	I
13	6256.36	I
16	6643.64	I
22	6767.77	I
10	6914.56	I
26	7122.20	I
16	7393.60	I
16	7409.35	I
23	7422.28	I
13	7522.76	I
19	7555.60	I
23	7617.00	I
16	7714.32	I
19	7727.61	I
19	7748.89	I
10	7788.94	I
13	7797.59	I
1000	8096.75	II
700	8121.48	II
9	8862.55	I
500 w	9900.92	II
Niobium		
Nb Z = 41		
80	464.55	V
80	468.32	V
80	763.77	V
80	774.02	V
60	993.54	IV
400	1005.72	IV
500	1007.05	IV
500	1010.19	IV
100	1116.08	IV
150	1120.02	IV
100	1258.87	V
60	1314.56	III
80	1445.43	III
80	1445.98	III
80	1447.09	III
100	1456.68	III
80	1484.73	III
100	1495.94	III
80	1498.02	III
80	1499.45	III
100	1501.99	III
60	1502.30	IV
80	1513.81	III
60	1524.36	IV
100	1524.91	III
100	1590.21	III
80	1598.86	III
80	1604.72	III
80	1639.51	III
100	1682.77	III
100	1705.44	III
100	1707.14	III
100	1758.33	V
100	1877.34	V
100	1892.92	III
60	1922.41	IV
100	1938.84	III
60	1978.22	IV
3300	2029.32	II
65	2032.53	IV
3000	2032.99	II
2000	2109.42	II
1700	2125.21	II
1100	2126.54	II
80 h	2130.24	III
1500	2131.18	II
80	2273.92	III
100	2275.23	III
80	2279.36	III
100	2281.51	III
80	2284.40	III
100	2290.36	III
370	2295.68	II
280	2302.08	II
100	2313.30	III
100	2338.09	III
80	2344.12	III
90	2349.21	III
80	2355.54	III

Intensity	Wavelength/Å		Intensity	Wavelength/Å		Intensity	Wavelength/Å		Intensity	Wavelength/Å	
100	2362.06	III	470	2716.62	II	390	3215.60	II	2700	3726.24	I
80	2362.50	III	470	2721.98	II	800	3225.48	II	2700	3739.80	I
80	2365.70	III	310	2733.26	II	140	3229.56	II	670	3740.73	II
100	2372.73	III	110	2737.09	II	400	3236.40	II	1700	3742.39	I
170	2376.40	II	240	2768.13	II	200	3247.47	II	530	3763.49	I
110	2387.09	II	310	2773.20	I	120	3248.94	II	350	3765.08	I
100	2387.41	III	270	2780.24	II	320	3254.07	II	530	3771.85	I
140	2387.52	II	110	2793.05	II	230	3260.56	II	870	3781.01	I
80	2388.23	III	190	2827.08	II	160	3263.37	II	1700	3787.06	I
45	2388.27	II	250	2841.15	II	200	3283.46	II	1300	3790.15	I
160	2398.48	II	280	2842.65	II	160	3292.02	II	3500	3791.21	I
80	2404.89	III	160	2846.28	II	320	3296.01	I	2700	3798.12	I
55	2405.34	II	240	2861.09	II	400	3312.60	I	2700	3802.92	I
55	2405.85	II	100	2865.61	II	120	3319.58	II	670	3803.88	I
140	2412.46	II	500	2868.52	II	130	3341.60	II	530	3804.74	I
100	2413.94	III	800	2875.39	II	1300	3341.97	I	670	3810.49	I
160	2416.99	II	270	2876.95	II	1300	3343.71	I	530	3811.03	I
140	2418.69	II	530	2877.03	II	1700	3349.06	I	530	3815.51	I
100	2421.91	III	100	2880.72	II	420	3349.52	I	210	3818.86	II
75	2433.80	II	570	2883.18	II	340	3354.74	I	670	3824.88	I
40	2435.95	II	280	2888.83	II	1700	3358.42	I	350	3835.18	I
45	2437.42	II	470	2897.81	II	130	3365.58	II	350	3863.38	I
40	2442.14	II	400	2899.24	II	340	3366.96	I	530	3877.56	I
28	2442.68	II	470	2908.24	II	130	3369.16	II	870	3878.82	I
65	2451.87	II	670	2910.59	II	350	3374.92	I	670	3883.14	I
65	2453.95	II	470	2911.74	II	170	3386.24	II	1100	3885.44	I
100	2456.99	III	1100	2927.81	II	350	3392.34	I	670	3885.68	I
55	2458.09	II	110	2931.47	II	230	3408.68	II	580	3891.30	I
65	2462.89	I	870	2941.54	II	180	3409.19	II	670	3914.70	I
80	2468.72	III	110 h	2945.88	II	230	3412.94	II	530	3920.20	I
80	2475.87	III	110	2946.12	II	230	3425.42	II	670	3937.44	I
110	2477.38	II	110	2946.90	II	230	3426.57	II	520	3943.67	I
65	2478.29	II	1100	2950.88	II	180	3432.70	II	910 d	3966.09	I
65	2479.94	II	400	2972.57	II	180	3440.59	II	1100	4032.52	I
35	2483.88	II	320	2974.10	II	200	3479.56	II	16000 c	4058.94	I
100	2499.73	III	210	2977.68	II	100	3484.05	II	350	4060.79	I
110	2511.00	II	200	2982.11	II	500	3498.63	I	12000	4079.73	I
110	2521.40	II	330	2990.26	II	460	3507.96	I	440	4100.40	I
390	2544.80	II	470	2994.73	II	200	3510.26	II	6700	4100.92	I
100	2545.64	III	80	3001.84	III	200	3515.42	II	310	4116.90	I
110	2551.38	II	140	3024.74	II	200	3517.67	II	5300	4123.81	I
130	2556.94	II	350	3028.44	II	2000	3535.30	I	670	4129.43	I
80	2557.94	III	300	3032.77	II	1300	3537.48	I	770	4129.93	I
130	2562.41	II	100	3044.76	II	250	3540.96	II	2300	4137.10	I
110	2571.33	II	100	3055.52	II	500	3544.02	I	440	4139.44	I
390	2583.99	II	220	3064.53	II	300	3550.45	I	2700	4139.71	I
390	2590.94	II	110	3069.68	II	1000	3554.66	I	350	4143.21	I
80	2598.86	III	100	3070.90	II	630	3563.50	I	870	4150.12	I
80	2633.17	III	110	3071.56	II	630	3563.62	I	4400	4152.58	I
200	2642.24	II	100	3073.24	II	1500	3575.85	I	870	4163.47	I
320	2646.26	II	400	3076.87	II	5000	3580.27	I	4400	4163.66	I
330	2647.50	I	110	3080.35	II	500	3584.97	I	4000	4164.66	I
330	2654.45	I	1800	3094.18	II	750	3589.11	I	3500	4168.13	I
310	2656.08	II	140	3099.19	II	500	3589.36	I	310	4184.44	I
80	2657.99	III	270	3127.53	II	500	3593.97	I	1200	4190.88	I
110	2665.25	II	1500	3130.79	II	500	3602.56	I	870	4192.07	I
110	2666.59	II	80	3142.26	III	300	3619.51	II	870	4195.09	I
110	2667.30	II	390	3145.40	II	420	3649.85	I	1300	4195.66	I
400	2671.93	II	1200	3163.40	II	400	3651.19	II	310	4198.51	I
200	2673.57	II	150	3175.78	II	200	3659.61	II	350	4201.52	I
200	2675.94	II	390	3180.29	II	630	3660.37	I	870	4205.31	I
160	2691.77	II	300	3191.10	II	900	3664.70	I	350	4214.73	I
1000	2697.06	II	150	3191.43	II	1500	3697.85	I	420	4217.94	I
320	2698.86	II	1000	3194.98	II	330	3711.34	I	420	4229.15	I
320	2702.20	II	120	3203.35	II	3300	3713.01	I	770	4262.05	I
150	2702.52	II	300	3206.34	II	480	3716.99	I	420	4266.02	I

Intensity	Wavelength/Å	
400	4286.99	I
580	4299.60	I
580	4300.99	I
390	4311.27	I
350	4326.33	I
390	4331.37	I
330	4410.21	I
150	4503.04	I
530	4523.41	I
480	4546.82	I
370	4564.53	I
720	4573.08	I
480	4581.62	I
1200	4606.77	I
170	4616.17	I
450	4630.11	I
450	4648.95	I
450	4663.83	I
340	4666.24	I
240	4667.22	I
580	4672.09	I
530	4675.37	I
320	4685.14	I
130 c	4706.14	I
260	4708.29	I
150	4713.50	I
220 c	4749.70	I
130 c	4967.78	I
190	4988.97	I
230	5017.75	I
150	5026.36	I
210	5039.04	I
170	5058.01	I
130	5065.25	I
750	5078.96	I
420	5095.30	I
170	5100.16	I
170	5120.30	I
210	5134.75	I
250	5160.33	I
250	5164.38	I
230	5180.31	I
190	5189.20	I
170	5193.08	I
150	5195.84	I
150	5232.81	I
150 d	5251.62	I
270	5271.53	I
130 c	5276.20	I
250	5318.60	I
460	5344.17	I
340	5350.74	I
110	5437.27	I
85	5551.35	I
170	5642.11	I
130	5664.71	I
170	5665.63	I
130	5729.19	I
110	5760.34	I
110	5819.43	I
130 d	5838.64	I
190 cw	5900.62	I
150	5983.22	I
75	6221.96	I
85 c	6430.46	I
65	6544.61	I
210 cw	6660.84	I
150 cw	6677.33	I
130 c	6723.62	I
85	6828.11	I
85	6990.32	I
190 c	7046.81	I
130	7159.43	I
190 cw	7372.50	I
65	7515.93	I
170 c	7574.58	I
75 c	7726.68	I
35	7885.31	I
40	8135.20	I
29 cw	8320.93	I
29	8346.08	I
35	8905.78	I

Nitrogen
N Z = 7

Intensity	Wavelength/Å	
400	181.75	IV
52	186.069	V
62	186.153	V
400	191.7	IV
400	192.9	IV
500	196.87	IV
500	197.23	IV
500	202.60	IV
500	205.94	IV
500	205.97	IV
500	206.03	IV
90	209.303	V
500	217.20	IV
500 d	217.90	IV
500 d	223.4	IV
800 w	225.12	IV
800	225.21	IV
600 w	234.12	IV
600 w	234.20	IV
600 w	234.25	IV
550	236.07	IV
500	237.99	IV
500 w	238.7	IV
600	238.80	IV
500 w	239.62	IV
900	247.20	IV
90	247.561	V
120	247.706	V
500 w	248.43	IV
500 w	248.46	IV
500 w	248.48	IV
500	257.95	III
650	258.50	III
700	259.19	III
800	260.09	III
600	260.45	IV
800	261.28	III
500	262.91	III
500	265.23	III
500	265.27	III
150	266.196	V
200	266.379	V
500	268.70	III
650	270.99	IV
250	283.42	IV
300	283.48	IV
350	283.58	IV
600	285.56	IV
600 w	297.7	IV
700	297.82	IV
650	300.32	IV
90	303.123	IV
500	303.28	IV
150	314.715	III
200	314.850	III
90	314.877	III
150	315.053	IV
120	322.503	IV
150	322.570	IV
200	322.724	IV
120	323.175	IV
600	323.26	III
300	335.050	IV
500	338.35	III
500	340.20	III
500 w	351.93	IV
500	351.98	III
700	353.06	IV
120	362.833	III
150	362.881	III
150	362.946	III
90	362.985	III
300	374.204	III
350	374.441	III
500	387.48	III
500	420.77	IV
250	451.869	III
300	452.226	III
650	463.74	IV
285	644.634	II
360	644.837	II
450	645.178	II
140	647.50	I
360	660.286	II
170	671.016	II
285	671.386	II
150	671.630	II
160	671.773	II
170	672.001	II
500	684.996	III
570	685.513	III
650	685.816	III
500	686.335	III
350	692.70	I
90	713.518	V
150	713.860	V
285	746.984	II
150	748.195	V
200	748.291	V
500	763.336	III
570	764.359	III
570	765.148	IV
250	771.544	III
300	771.901	III
330	772.385	III
200	772.891	III
150	772.975	III
650	775.965	II
90	885.67	I
90	909.697	I
80	910.278	I
40	910.645	I
450	915.612	II
450	915.962	II
550	916.012	II
650	916.701	II
520	921.992	IV
500	922.519	IV
480	923.057	IV
520	924.283	IV
90	953.415	I
100	953.655	I
130	953.970	I
1000	955.335	IV
130	963.990	I
115	964.626	I
70	965.041	I
650	979.842	III
700	979.919	III
900	989.790	III
700	991.514	III
1000	991.579	III
150 w	1036.16	IV
90	1067.614	I
60	1068.612	I
90	1078.71	IV
450	1083.990	II
600	1084.580	II
430	1085.546	II
650	1085.701	II
175	1097.237	I
115	1098.095	I
115	1098.260	I
105	1100.360	I
40	1100.465	I
90	1101.291	I
360	1134.165	I
385	1134.415	I
410	1134.980	I
105	1143.65	I
130	1163.884	I
60	1164.206	I
105	1164.325	I
270	1167.448	I
105	1168.334	I
60	1168.417	I
195	1168.536	I
230	1176.510	I
105	1176.630	I
195	1177.695	I
500	1183.031	III
570	1184.550	III
90	1188.01	IV
410	1199.550	I
385	1200.223	I
360	1200.710	I
175	1225.026	I
160	1225.37	I
130	1228.41	I
160	1228.79	I
1000	1238.821	V
900	1242.804	V
360	1243.179	I
315	1243.306	I
290	1310.540	I
250	1310.95	I
230	1319.00	I
315	1319.68	I
115	1326.57	I
115	1327.92	I

Intensity	Wavelength/Å		Intensity	Wavelength/Å		Intensity	Wavelength/Å		Intensity	Wavelength/Å	
150	1387.371	III	160	2885.27	II	160	4950.23	I	160	7406.12	I
360	1411.94	I	90 l	2974.52	V	350	4963.98	I	265	7406.24	I
700	1492.625	I	150 w	2980.78	V	285	4987.37	II	685	7423.64	I
490	1492.820	I	250 w	2981.31	V	450	4994.36	II	785	7442.29	I
640	1494.675	I	60 w	2998.43	V	650	5001.48	II	900	7468.31	I
90	1549.336	V	220	3006.83	II	360	5002.70	II	185	7608.80	I
200 l	1616.33	V	90	3078.25	IV	870	5005.15	II	60 w	7618.46	V
350 l	1619.69	V	120	3367.34	III	550	5007.32	II	450	7762.24	II
1000	1718.55	IV	360	3437.15	II	450	5010.62	II	400	8184.87	I
250	1729.945	III	90	3463.37	IV	360	5016.39	II	400	8188.02	I
775	1742.729	I	570	3478.71	IV	360	5025.66	II	250	8200.36	I
700	1745.252	I	500	3482.99	IV	550	5045.10	II	300	8210.72	I
570	1747.848	III	400	3484.96	IV	185	5281.20	I	570	8216.34	I
350	1751.218	III	90	3747.54	IV	140	5292.68	I	400	8223.14	I
650	1751.657	III	90	3754.67	III	90	5314.35	III	400	8242.39	I
150	1804.486	III	120	3771.05	III	200	5320.82	III	550	8438.74	II
200	1805.669	III	285	3838.37	II	150	5327.18	III	500	8567.74	I
150	1846.42	III	360	3919.00	II	450	5495.67	II	570	8594.00	I
90 w	1860.37	V	90	3938.52	III	285	5535.36	II	650	8629.24	I
350	1885.06	III	450	3955.85	II	650	5666.63	II	500	8655.89	I
400	1885.22	III	1000	3995.00	II	550	5676.02	II	220	8676.08	II
200	1907.99	III	150	3998.63	III	870	5679.56	II	700	8680.28	I
150	1919.55	III	200	4003.58	III	450	5686.21	II	650	8683.40	I
150	1919.77	III	360	4035.08	II	450	5710.77	II	500	8686.15	I
300	1920.65	III	550	4041.31	II	285	5747.30	II	110	8687.43	II
150	1920.84	III	360	4043.53	II	700	5752.50	I	110 h	8699.00	II
200	1921.30	III	150	4057.76	IV	240	5764.75	I	500	8703.25	I
200	2064.01	III	250	4097.33	III	265	5829.54	I	160 h	8710.54	II
250	2064.42	III	140	4099.94	I	235	5854.04	I	570	8711.70	I
120	2068.68	III	200	4103.43	III	360	5927.81	II	500	8718.83	I
90	2071.09	III	185	4109.95	I	550	5931.78	II	250	8728.89	I
90	2080.34	IV	285	4176.16	II	285	5940.24	II	200	8747.36	I
160	2095.53	II	120	4195.76	III	650	5941.65	II	500	9386.80	I
70	2096.20	II	150	4200.10	III	285	5952.39	II	570	9392.79	I
110	2096.86	II	285	4227.74	II	160	5999.43	I	250	9460.68	I
90	2117.59	III	285	4236.91	II	210	6008.47	I	200	9863.33	I
90	2121.50	III	220	4237.05	II	285	6167.76	II	160 h	9865.41	II
110	2130.18	II	450	4241.78	II	360	6379.62	II	110 h	9868.21	II
160	2142.78	II	90	4332.91	III	150	6380.77	IV	160 h	9887.39	II
90	2147.31	III	120	4345.68	III	185	6411.65	I	220 h	9891.09	II
200	2188.20	III	300	4379.11	III	210	6420.64	I	160 h	9961.86	II
150	2188.38	III	285	4432.74	II	210	6423.02	I	220 h	9969.34	II
160	2206.09	II	650	4447.03	II	210	6428.32	I	285 h	10023.27	II
160	2286.69	II	90	4510.91	III	185	6437.68	I	220 h	10035.45	II
110	2288.44	II	120	4514.86	III	235	6440.94	I	220 h	10065.15	II
220	2316.49	II	360	4530.41	II	90	6454.11	III	160 h	10070.12	II
160	2316.69	II	550	4601.48	II	185	6457.90	I	250	10105.13	I
285	2317.05	II	350	4603.73	V	120	6467.02	III	300	10108.89	I
90 w	2318.09	IV	90	4606.33	IV	300	6468.44	I	350	10112.48	I
160	2461.27	II	450	4607.16	II	265	6481.71	I	400	10114.64	I
150	2477.69	IV	360	4613.87	II	750	6482.05	II	110 h	10126.27	II
110	2496.83	II	250	4619.98	V	360	6482.70	I	250	10539.57	I
70	2496.97	II	450	4621.39	II	300	6483.75	I	200	12074.51	I
110	2520.22	II	870	4630.54	II	325	6484.80	I	380	12186.82	I
160	2520.79	II	90	4634.14	III	160	6491.22	I	225	12288.97	I
220	2522.23	II	120	4640.64	III	210	6499.54	I	290	12328.76	I
110	2590.94	II	550	4643.08	II	185	6506.31	I	310	12381.65	I
250	2645.65	IV	285	4788.13	II	750	6610.56	II	180	12438.40	I
300	2646.18	IV	450	4803.29	II	185	6622.54	I	510	12461.25	I
350	2646.96	IV	180	4847.38	I	185	6636.94	I	920	12469.62	I
250 w	2682.18	III	90	4858.82	III	235	6644.96	I	500	13429.61	I
90	2689.20	III	150	4867.15	III	185	6646.50	I	840	13581.33	I
160	2709.84	II	285	4895.11	II	235	6653.46	I	180	13587.73	I
110	2799.22	II	160	4914.94	I	210	6656.51	I	180	13602.27	I
110	2823.64	II	210	4935.12	I	185	6722.62	I	290	13624.18	I
60 l	2859.16	V	200 w	4944.56	V	210	7398.64	I	250	14757.07	I

Intensity	Wavelength/Å	
100	14868.87	I
160	14966.60	I
180	15582.27	I
120 s	17516.58	I
100 l	17584.86	I
100	17878.26	I
Osmium		
Os Z = 76		
9600	2001.45	I
13000	2003.73	I
17000	2010.15	I
29000	2018.14	I
14000	2022.76	I
14000	2028.23	I
18000	2034.44	I
26000	2045.36	I
8600	2058.69	I
13000	2061.69	I
7800	2067.21	II
4200	2070.67	II
7200	2076.95	I
14000	2079.97	I
2900	2082.54	I
2900	2089.03	I
2900	2089.21	I
6000	2097.60	I
5300	2100.63	I
2100	2117.66	I
4800	2117.96	I
5300	2137.11	I
2600	2154.59	I
1300	2157.84	I
1200	2158.53	I
3100	2166.90	I
1100	2167.75	I
2100	2171.65	I
1100	2234.61	I
1300	2252.15	I
2000	2255.85	II
1400	2264.60	I
1400	2282.26	II
500	2367.35	II
2600	2377.03	I
1700	2387.29	I
1100	2395.88	I
200	2423.07	II
1400	2424.97	I
110	2454.91	II
1800	2461.42	I
110	2468.90	II
530	2486.24	II
4500	2488.55	I
2600	2498.41	I
2400	2513.25	I
780	2538.00	II
1000	2542.51	I
1000	2590.76	I
1800	2613.06	I
3800	2637.13	I
1900	2644.11	I
1900	2658.60	I
2100	2689.82	I
3000	2714.64	I
1300	2720.04	I
960	2770.71	I
2800	2806.91	I
5100	2838.63	I
2300	2844.40	I
1500	2850.76	I
1500	2860.96	I
9600	2909.06	I
2100	2912.33	I
2100	2919.79	I
1100 h	2948.23	I
1400	2949.53	I
4400	3018.04	I
1100	3030.70	I
2900	3040.90	I
120	3042.74	II
8600	3058.66	I
1100	3077.72	I
3100	3156.25	I
180	3173.93	II
150	3213.31	II
1900	3232.06	I
3100	3262.29	I
3100	3267.94	I
1200	3290.26	I
7600	3301.56	I
960	3336.15	I
960	3370.59	I
620	3387.84	I
620	3401.86	I
620	3504.66	I
1200	3528.60	I
1200	3560.86	I
620	3598.11	I
95	3604.48	II
480	3670.89	I
3700	3752.52	I
2100	3782.20	I
730	3876.77	I
1000	3963.63	I
730	3977.23	I
960	4066.69	I
1200	4112.02	I
2500	4135.78	I
1200	4173.23	I
1200	4211.86	I
4900	4260.85	I
560	4293.95	I
560	4311.40	I
4900	4420.47	I
540	4550.41	I
670	4793.99	I
55	5031.83	I
45	5039.12	I
35	5072.88	I
35	5074.77	I
35	5079.09	I
90	5103.50	I
55	5110.81	I
140	5149.74	I
40	5193.52	I
270	5202.63	I
35	5203.23	I
45	5255.82	I
55	5265.15	I
40	5298.78	I
110	5376.79	I
120	5416.34	I
45	5416.69	I
28	5417.51	I
55	5443.31	I
22	5446.93	I
22	5457.30	I
28	5470.00	I
22	5509.33	I
270	5523.53	I
22	5546.82	I
80	5584.44	I
35	5620.08	I
22	5642.56	I
28	5645.25	I
28	5680.88	I
170	5721.93	I
22	5765.05	I
170	5780.82	I
40	5800.60	I
110	5857.76	I
28	5860.64	I
65	5996.00	I
35	6227.70	I
22	6269.41	I
22	6403.15	I
27	6729.56	I
22	7145.54	I
26	7602.95	I
7	8041.29	I
Oxygen		
O Z = 8		
80	124.616	V
110	135.523	V
80	138.109	V
110	139.029	V
80	151.447	V
110	151.477	V
150	151.546	V
80	164.574	V
110	164.657	V
80	164.709	V
80	166.235	V
150	167.99	V
110	170.219	V
450	172.169	V
250	185.745	V
375	192.751	V
450	192.799	V
520	192.906	V
80	193.003	V
200	194.593	V
150	195.86	IV
200	196.01	IV
80	202.161	V
80	202.224	V
80	202.283	V
80	202.334	V
150	202.393	V
110	203.78	V
150	203.82	V
100	203.85	V
200	203.89	V
100	203.94	V
110	207.18	IV
150	207.24	IV
300	207.794	V
150	215.040	V
200	215.103	V
250	215.245	V
250	216.018	V
520	220.352	V
80	227.372	V
80	227.469	V
150	227.511	V
80	227.549	V
80	227.634	V
80	227.689	V
150	231.823	V
140	233.46	IV
150	233.50	IV
110	233.52	IV
200	233.56	IV
110	233.60	IV
90	238.36	IV
180	238.57	IV
110	248.459	V
110	252.56	IV
110	252.95	IV
150	253.08	IV
300	260.39	IV
250	260.56	IV
80 d	264.34	III
110	264.48	III
110	266.97	III
150	266.98	III
150	267.03	III
150	277.38	III
300	279.63	IV
375	279.94	IV
110	285.71	IV
150	285.84	IV
110	286.448	V
80	295.62	III
110	295.66	III
120	295.72	III
150	303.41	III
150	303.46	III
140	303.52	III
160	303.62	III
160	303.69	III
250	303.80	III
200	305.60	III
250	305.66	III
190	305.70	III
300	305.77	III
190	305.84	III
200	306.62	IV
150	306.88	IV
450	320.979	III
300	328.45	III
250	328.74	III
300	345.31	III
110	355.14	III
90	355.33	III
80	355.47	III
200	359.02	III
190	359.22	III
150	359.38	III
210	373.80	III
200	374.00	III
300	374.08	III
190	374.16	III
200	374.33	III
210	374.44	III

Line Spectra of the Elements (continued): Oxygen

Intensity	Wavelength/Å	Intensity	Wavelength/Å	Intensity	Wavelength/Å	Intensity	Wavelength/Å
450	395.558 III	80	805.810 I	30 d	2283.42 II	160	3305.15 II
300	434.98 III	240	832.762 II	30 d	2284.89 II	160	3306.60 II
800	507.391 III	600	832.927 III	110	2293.32 II	80	3312.30 III
900	507.683 III	450	833.332 II	200	2300.35 II	110	3340.74 III
1000	508.182 III	780	833.742 III	30 d	2313.05 II	230	3348.08 IV
1000	525.795 III	600	834.467 II	30 d	2316.12 II	270	3349.11 IV
250	537.83 II	600	835.096 III	30 d	2316.79 II	160	3354.27 IV
300	538.26 II	800	835.292 III	50 d	2319.68 II	200	3375.40 IV
220	539.09 II	40	877.879 I	30 d	2322.15 II	220	3377.20 II
200	539.55 II	130	921.296 IV	30 d	2339.31 II	130	3378.06 IV
150	539.85 II	160	921.366 IV	200 d	2390.44 III	360	3381.20 IV
700	553.330 IV	80	922.008 I	80	2394.33 III	360	3385.52 IV
775	554.075 IV	200	923.367 IV	110	2411.60 II	285	3390.25 II
850	554.514 IV	130	923.433 IV	80	2422.84 III	270	3396.79 IV
700	555.261 IV	90	935.193 I	80	2425.55 II	360	3403.52 IV
700	597.818 III	40	948.686 I	250	2433.56 II	220	3407.38 II
1000	599.598 III	90	971.738 I	80 d	2436.06 II	230	3409.66 IV
580	608.398 IV	40	976.448 I	80 d	2438.83 III	160	3409.84 II
110	609.70 III	160	988.773 I	80	2444.26 II	410	3411.69 IV
640	609.829 IV	40	990.204 I	300	2445.55 II	230	3413.64 IV
160	610.04 III	250	1025.762 I	200	2449.372 IV	80	3444.10 III
200	610.75 III	90	1027.431 I	200	2450.040 IV	80	3455.12 III
100	610.85 III	160	1039.230 I	200	2454.99 III	285	3470.81 II
270	616.952 IV	60	1040.942 I	200	2493.44 IV	200	3489.83 IV
150	617.005 IV	40	1152.152 I	200	2493.77 IV	160	3492.24 IV
200	617.036 IV	900	1302.168 I	200	2507.73 IV	230	3560.39 IV
520	624.617 IV	600	1304.858 I	230	2509.19 IV	270	3563.33 IV
580	625.130 IV	300	1306.029 I	200	2517.2 IV	80	3698.70 III
640	625.852 IV	200	1338.612 IV	200	2558.06 III	80	3702.75 III
1000	629.730 V	130	1342.992 IV	80	2687.53 III	80	3703.37 III
150	644.148 II	230	1343.512 IV	110	2695.49 III	110	3707.24 III
200	672.95 II	640	1371.292 V	300	2733.34 II	220	3712.75 II
150	673.77 II	160	1476.89 III	110	2747.46 II	110	3715.08 III
230	681.272 V	160 w	1506.72 V	1000	2781.01 V	315 w	3725.93 IV
70	685.544 I	285	1590.01 III	920	2786.99 V	285	3727.33 II
800	702.332 III	160	1591.33 III	775	2789.85 V	360	3729.03 IV
800	702.822 III	315 w	1643.68 V	160	2836.26 IV	410	3736.85 IV
900	702.899 III	160	1707.996 V	160	2921.45 IV	160	3739.92 II
1000	703.850 III	220	1760.12 III	200	2941.33 V	110	3744.00 III
900	718.484 II	110	1760.42 III	210	2941.65 V	230	3744.89 IV
600	718.562 II	220	1763.22 III	80	2959.68 III	360	3749.49 II
70	744.794 I	220	1764.48 III	265	2972.29 I	150	3754.67 III
700	758.678 V	750	1767.78 III	250	2983.78 III	80	3757.21 III
640	759.441 V	550	1768.24 III	80	3017.63 III	250	3759.87 III
580	760.228 V	360	1771.67 III	80	3023.45 III	110	3791.26 III
775	760.445 V	110	1773.00 III	80	3043.02 III	160	3803.14 II
640	761.128 V	110	1773.85 III	200	3047.13 III	120	3823.41 I
700	762.003 V	220	1779.16 III	110	3059.30 III	450	3911.96 II
70	770.793 I	160	1781.03 III	460	3063.42 IV	160	3919.29 II
90	771.056 I	160	1784.85 III	410	3071.61 IV	185	3947.29 I
520	774.518 V	220	1789.66 III	80	3121.71 III	160	3947.48 I
70	775.321 I	110	1848.26 III	160	3122.62 II	140	3947.59 I
200	779.734 IV	110	1856.62 III	220	3129.44 II	220	3954.37 II
315	779.821 IV	285	1872.78 III	110	3132.86 III	100	3954.61 I
360	779.912 IV	285	1872.87 III	450	3134.82 II	200	3961.59 III
200	779.997 IV	285	1874.94 III	285	3138.44 II	450	3973.26 II
640	787.711 IV	160	1920.04 III	160	3144.66 V	220	3982.20 II
520	790.109 IV	110	1920.75 III	160	3209.66 IV	160	4069.90 II
700	790.199 IV	110	1921.52 III	80	3238.57 III	285	4072.16 II
70	791.973 I	220	1923.49 III	200	3260.98 III	450	4075.87 II
300	796.66 II	110	1923.82 III	300	3265.46 III	80 d	4083.91 II
200	802.200 IV	110	1926.94 III	80	3267.31 III	50 d	4087.14 II
160	802.255 IV	360	2013.27 III	220	3270.98 II	150 d	4089.27 II
90	804.267 I	160	2026.96 III	220	3273.52 II	110	4097.24 II
70	804.848 I	220	2045.67 III	220	3277.69 II	220	4105.00 II
70	805.295 I	160	2052.74 III	360	3287.59 II	285	4119.22 II

Intensity	Wavelength/Å	
100	4123.99	V
160	4132.81	II
50	4146.06	II
220	4153.30	II
285	4185.46	II
450	4189.79	II
80	4233.27	I
50 d	4253.74	II
50 d	4253.98	II
50 d	4275.47	II
50 d	4303.78	II
285	4317.14	II
160	4336.86	II
220	4345.56	II
285	4349.43	II
220	4366.90	II
100	4368.25	I
220	4395.95	II
450	4414.91	II
285	4416.98	II
160	4448.21	II
160	4452.38	II
50	4465.45	II
50 d	4466.28	II
50	4467.83	II
50	4469.41	II
360	4590.97	II
285	4596.17	II
80 d	4609.39	II
160	4638.85	II
360	4641.81	II
450	4649.14	II
160	4650.84	II
360	4661.64	II
285	4676.23	II
220	4699.21	II
285	4705.36	II
160	4924.60	II
230 w	4930.27	V
220	4943.06	II
135	5329.10	I
160	5329.68	I
190	5330.74	I
90	5435.18	I
110	5435.78	I
135	5436.86	I
120	5577.34	I
110	5592.37	III
130	5597.91	V
160	5958.39	I
190	5958.58	I
80	5995.28	I
160	6046.23	I
190	6046.44	I
110	6046.49	I
100	6106.27	I
400	6155.98	I
450	6156.77	I
490	6158.18	I
80	6256.83	I
100	6261.55	I
100	6366.34	I
100	6374.32	I
320	6453.60	I
360	6454.44	I
400	6455.98	I
130	6500.24	V
80	6604.91	I
100	6653.83	I
360	7001.92	I
450	7002.23	I
210	7156.70	I
400	7254.15	I
450	7254.45	I
320	7254.53	I
210	7476.44	I
100	7477.24	I
120	7479.08	I
120	7480.67	I
100	7706.75	I
870	7771.94	I
810	7774.17	I
750	7775.39	I
80	7886.27	I
100	7943.15	I
100	7947.17	I
235	7947.55	I
210	7950.80	I
185	7952.16	I
110	7981.94	I
135	7982.40	I
190	7986.98	I
135	7987.33	I
250	7995.07	I
400	8221.82	I
265	8227.65	I
265	8230.02	I
325	8233.00	I
120	8235.35	I
120	8426.16	I
810	8446.25	I
1000	8446.36	I
935	8446.76	I
325	8820.43	I
160 d	9057.01	I
120	9118.29	I
80	9134.71	I
80	9150.14	I
80	9151.48	I
235	9156.01	I
450	9260.81	I
490	9260.84	I
450	9260.94	I
400	9262.58	I
540	9262.67	I
590	9262.77	I
490	9265.94	I
640	9266.01	I
185	9399.19	I
120	9481.16	I
120 d	9482.88	I
235	9487.43	I
140	9492.71	I
265	9497.97	I
160	9499.30	I
235	9505.59	I
210	9521.96	I
120	9523.36	I
120	9523.96	I
100	9528.28	I
100	9622.13	I
120	9625.29	I
160	9677.38	I
80	9694.66	I
65	9694.91	I
235	9741.50	I
235	9760.65	I
120	9909.05	I
140	9936.98	I
120	9940.41	I
160	9995.31	I
120 d	10421.18	I
590	11286.34	I
640	11286.91	I
490	11287.02	I
490	11287.32	I
490	11295.10	I
540	11297.68	I
590	11302.38	I
265	11358.69	I
490	12464.02	I
450	12570.04	I
120	12990.77	I
160	13076.91	I
700	13163.89	I
750	13164.85	I
640	13165.11	I
160	16212.06	I
120	17966.70	I
590	18021.21	I
120	18041.48	I
120	18042.19	I
120	18046.23	I
140	18229.23	I
540	18243.63	I
140	26173.56	I

Palladium
Pd Z = 46

Intensity	Wavelength/Å	
200	705.49	III
200	727.72	III
500	763.06	III
500	766.42	III
2000	781.02	III
500	794.08	III
500	797.52	III
500	800.03	III
500	800.10	III
500	803.67	III
500	825.35	III
500	840.58	III
500	856.47	III
500	864.04	III
500	880.59	III
500	888.84	III
1000	889.29	III
300	1596.89	III
500	1741.62	III
4000	1782.55	III
400	1843.49	III
1500	1851.59	III
2000	1852.27	III
1000	1859.21	III
1500	1874.63	III
2000	1885.83	III
1000	1887.40	III
1500	1891.34	III
4000	1914.62	III
1000	1930.33	III
2000	1941.64	III
800	2002.16	III
1000	2004.47	III
500	2055.11	III
500	2149.82	III
500	2177.55	III
500	2177.63	III
100 r	2231.59	II
200 r	2296.53	II
100	2426.87	II
100	2430.94	II
100	2433.11	II
100	2435.32	II
150	2446.17	II
1100	2447.91	I
100	2457.29	II
150	2469.29	II
100	2471.18	II
1700	2476.42	I
250	2486.52	II
300	2488.92	II
200	2498.81	II
150	2505.73	II
150	2551.84	II
150	2565.51	II
100	2569.56	II
150	2658.75	II
1900	2763.09	I
150 h	2776.85	II
100 h	2787.92	II
200	2854.59	II
100 h	2871.37	II
100 h	2878.01	II
520	2922.49	I
650	3002.65	I
1500	3027.91	I
1100	3065.31	I
2600	3114.04	I
11000	3242.70	I
2700	3251.64	I
3500	3258.78	I
3600	3302.13	I
5000	3373.00	I
24000	3404.58	I
13000	3421.24	I
5000	3433.45	I
6400	3441.40	I
7700	3460.77	I
10000	3481.15	I
2000	3489.77	I
12000	3516.94	I
12000	3553.08	I
4500	3571.16	I
20000	3609.55	I
20000	3634.70	I
5500	3690.34	I
1400	3718.91	I
1500	3799.19	I
1500	3832.29	I
2200	3894.20	I
1500	3958.64	I
290	4087.34	I
2500	4212.95	I
180	4473.59	I

Intensity	Wavelength/Å	
160	5163.84	I
120	5295.63	I
55	5542.80	I
75	5670.07	I
55 h	5695.09	I
65	6784.52	I
75	7368.12	I
120	7764.03	I
45	7915.80	I
55	8132.82	I
45	8300.83	I
65	8761.35	I
Phosphorus		
P Z = 15		
250	328.78	V
150	359.899	IV
500	388.318	IV
250	389.50	V
300	390.70	V
300	445.158	IV
375	475.60	V
120	498.180	III
520	542.57	V
600	544.92	V
200	569.853	III
200	581.831	III
350	629.008	IV
400	629.914	IV
500	631.779	IV
450	673.90	V
10	810.24	II
650	823.179	IV
700	824.730	IV
800	827.932	IV
300	847.669	III
350	855.624	III
500	859.652	III
10	865.44	II
450	865.45	V
600	871.39	V
700	877.476	IV
300	913.971	III
300	917.120	III
350	918.665	III
1000	950.655	IV
250	1003.598	III
570	1025.563	IV
500	1028.096	IV
570	1030.517	IV
500	1033.111	IV
500	1035.517	IV
900	1117.98	V
570	1118.551	IV
700	1128.01	V
20	1249.82	II
20	1301.87	II
20	1304.47	II
15	1304.68	II
35	1305.48	II
60	1310.70	II
500	1334.808	III
650	1344.327	III
300	1344.845	III
500	1366.695	IV
15	1372.033	I
400	1372.674	IV
15	1373.500	I
10	1374.732	I
15	1377.080	I
15	1377.937	I
25	1379.429	I
25	1381.469	I
15	1381.637	I
500	1484.507	IV
400	1487.788	IV
350	1502.228	III
80	1532.51	II
120	1535.90	II
450	1610.50	V
150	1618.632	III
200	1618.907	III
140	1671.070	I
100	1671.510	I
180	1671.680	I
140	1672.035	I
140	1672.474	I
600	1674.591	I
600	1679.695	I
140	1685.976	I
100	1694.028	I
100	1694.486	I
100	1706.376	I
100	1707.553	I
600	1774.951	I
500	1782.838	I
400	1787.656	I
140	1834.801	I
140	1847.165	I
100	1849.820	I
140	1851.194	I
100	1852.069	I
500	1858.886	I
400	1859.393	I
140	1864.348	I
650	1888.523	IV
180	1905.481	I
140	1906.403	I
280	1907.665	I
280	2023.489	I
180	2024.516	I
400	2032.432	I
400	2033.477	I
400	2135.465	I
400	2136.182	I
400	2149.145	I
280	2152.940	I
500	2154.080	I
180	2235.732	I
450	2440.93	V
250	2478.256	IV
750	2533.976	I
950	2535.603	I
750	2553.262	I
500	2554.915	I
250	2605.506	IV
300	2632.713	III
400	2644.295	IV
400	2728.770	IV
500	2739.309	IV
250	2739.872	IV
450	2978.55	V
700	3175.09	V
520	3204.04	V
300	3219.307	III
400	3233.602	III
650	3347.736	IV
570	3364.467	IV
400	3371.122	IV
300	3957.641	III
350	3978.307	III
400	4059.312	III
300	4080.084	III
500	4222.195	III
350	4246.720	III
400	4420.71	II
250	4479.776	III
250	4540.288	IV
250	4541.112	IV
500	4588.04	II
500	4589.86	II
600	4602.08	II
300	4626.70	II
300	4658.31	II
500	4943.53	II
300	4954.39	II
300	4969.71	II
100	5079.381	I
100	5098.221	I
100	5100.974	I
140	5109.628	I
140	5154.844	I
180	5162.290	I
300	5253.52	II
140	5293.539	I
400	5296.13	II
250	5316.07	II
300	5344.75	II
180	5345.851	I
100	5364.631	I
250	5378.20	II
300	5386.88	II
400	5425.91	II
100	5428.094	I
400	5450.74	II
140	5458.305	I
180	5477.672	I
140	5477.860	I
140	5478.267	I
100	5514.774	I
100	5516.997	I
250	5588.34	II
500	6024.18	II
400	6034.04	II
500	6043.12	II
250	6055.50	II
150	6083.409	III
350	6087.82	II
180	6097.690	I
350	6165.59	II
500	6199.024	I
180	6210.499	I
140	6375.681	I
100	6388.579	I
250	6435.32	II
600	6459.99	II
600	6503.46	II
600	6507.97	II
100	6717.411	I
150	6992.690	III
100	7102.200	I
100	7158.367	I
180	7165.465	I
180	7175.102	I
180	7176.660	I
200	7443.657	IV
250	7845.63	II
100	8046.801	I
150	8113.528	III
140	8278.058	I
100	8367.856	I
140	8531.475	I
140	8613.835	I
180	8637.578	I
400	8741.529	I
100	8872.174	I
180	9175.819	I
950	9193.85	I
600	9278.88	I
1250	9304.94	I
500	9323.50	I
950	9435.069	I
950	9441.86	I
600	9452.83	I
1250	9493.56	I
1700	9525.73	I
1500	9545.18	I
280	9556.81	I
1700	9563.439	I
280	9593.50	I
750	9609.04	I
400	9638.939	I
500	9676.24	I
180	9706.533	I
1500	9734.750	I
280	9736.680	I
1500	9750.77	I
600	9790.21	I
1700	9796.85	I
280	9834.80	I
400	9903.68	I
280	9976.67	I
229	10084.27	I
458	10511.58	I
962	10529.52	I
1235	10581.57	I
415	10596.90	I
435	10681.40	I
265	10813.13	I
764	11183.23	I
402	11186.75	I
479	14241.64	I
256	14307.83	I
714	15711.52	I
228	15962.53	I
296	16254.77	I
203	16292.97	I
1627	16482.92	I
588	16590.07	I
225	16613.05	I
221	16738.68	I
419	16803.39	I
471	17112.48	I
289	17286.91	I
299	17423.67	I

Intensity	Wavelength/Å	
287	23844.97	I
311	29097.16	I
Platinum		
Pt Z = 78		
30	1621.66	II
30	1723.13	II
30	1751.70	II
50 r	1777.09	II
30	1781.86	II
30	1879.09	II
40	1883.05	II
50	1889.52	II
50	1911.70	II
30	1929.25	II
30	1929.68	II
30	1939.80	II
30	1949.90	II
30	1983.74	II
40	2014.93	II
3200	2030.63	I
4400	2032.41	I
100	2036.46	II
40	2041.57	II
5500	2049.37	I
1500	2067.50	I
3000	2084.59	I
1000	2103.33	I
30	2115.57	II
950	2128.61	I
30	2130.69	II
1900	2144.23	I
100	2144.24	II
600	2165.17	I
1500	2174.67	I
30	2190.32	II
400	2202.22	I
50 h	2202.58	II
320	2222.61	I
50 h	2233.11	II
30 h	2240.99	II
100	2245.52	II
150	2249.30	I
30	2251.52	II
30 h	2251.92	II
190	2268.84	I
30 h	2271.72	II
280	2274.38	I
50 h	2287.50	II
30	2288.20	II
150	2289.27	I
150	2292.40	I
240	2308.04	I
50	2310.96	II
90	2315.50	I
220	2318.29	I
100	2326.10	I
170	2340.18	I
280	2357.10	I
180	2368.28	I
50	2377.28	II
130	2383.64	I
40	2386.81	I
120	2389.53	I
35	2396.17	I
70	2401.87	I
200	2403.09	I

Intensity	Wavelength/Å	
100	2418.06	I
50	2424.87	II
80	2428.04	I
50	2428.20	I
25	2429.10	I
180	2436.69	I
650	2440.06	I
60	2450.97	I
440	2467.44	I
35	2471.01	I
1000	2487.17	I
25	2488.74	II
200	2490.12	I
160	2495.82	I
240	2498.50	I
50	2505.93	I
120	2508.50	I
50	2514.07	I
60	2515.03	I
240	2515.58	I
140	2524.30	I
40	2529.41	I
50	2536.49	I
160	2539.20	I
18	2549.46	I
50	2552.25	I
50	2596.00	I
70	2603.14	I
30	2616.76	II
50	2619.57	I
30	2625.34	II
1100	2628.03	I
130	2639.35	I
1000	2646.89	I
500	2650.86	I
20	2658.17	I
2800	2659.45	I
40	2674.57	I
440	2677.15	I
200	2698.43	I
2000	2702.40	I
1600	2705.89	I
60	2713.13	I
1300	2719.04	I
130	2729.92	I
1800	2733.96	I
70	2738.48	I
70	2747.61	I
80	2753.86	I
200	2754.92	I
30	2769.84	I
500	2771.67	I
40	2773.24	I
20	2774.00	I
50	2774.77	II
50	2793.27	I
100	2794.21	II
40 h	2799.98	II
140	2803.24	I
10	2808.51	I
50	2818.25	I
30 h	2822.27	II
1400	2830.30	I
70	2834.71	I
16	2853.11	I
80 h	2860.68	II

Intensity	Wavelength/Å	
40 h	2865.05	II
40 h	2875.85	II
100 h	2877.52	II
25	2888.20	I
25	2893.22	I
600	2893.86	I
300	2897.87	I
60	2905.90	I
120	2912.26	I
120	2913.54	I
70	2919.34	I
30	2921.38	I
1700	2929.79	I
30	2942.76	I
30	2944.75	I
25	2959.10	I
60	2960.75	I
1800	2997.97	I
35	3001.17	II
220	3002.27	I
30	3017.88	I
30 h	3031.22	II
130	3036.45	I
800	3042.64	I
3200	3064.71	I
30	3071.94	I
130	3100.04	I
320	3139.39	I
140	3156.56	I
120	3200.71	I
320	3204.04	I
30	3230.29	I
20	3233.42	I
20	3250.36	I
40	3251.98	I
160	3255.92	I
25	3268.42	I
25	3281.97	I
120	3290.22	I
500	3301.86	I
60	3315.05	I
35	3323.80	I
340	3408.13	I
35	3427.93	I
60	3483.43	I
160	3485.27	I
120	3628.11	I
70	3638.79	I
70	3643.17	I
50	3663.10	I
80	3671.99	I
80	3674.04	I
35	3699.91	I
18	3706.53	I
80	3818.69	I
40	3900.73	I
110	3922.96	I
35	3948.40	I
100	3966.36	I
20	3996.57	I
110	4118.69	I
80	4164.56	I
40	4192.43	I
18	4327.06	I
18	4391.83	I
80	4442.55	I

Intensity	Wavelength/Å	
14	4445.55	I
25	4498.76	I
12	4520.90	I
35	4552.42	I
12	4879.53	I
14	5044.04	I
30	5059.48	I
35	5227.66	I
40	5301.02	I
12	5368.99	I
12	5390.79	I
14	5475.77	I
14	5478.50	I
6	5763.57	I
20	5840.12	I
8	5844.84	I
6	6026.04	I
7	6318.37	I
8	6326.58	I
9	6523.45	I
10	6710.42	I
20	6760.02	I
60	6842.60	I
20	7113.73	I
10	8224.74	I
Plutonium		
Pu Z = 94		
10000	2806.11	II
10000	2950.06	II
10000	3000.31	II
10000	3200.23	II
10000	3418.88	II
10000	3805.93	I
10000	4097.12	I
10000	4170.95	I
10000	4367.41	I
10000	5590.54	I
10000	7068.90	I
10000	8691.94	I
3000	9533.07	I
3000	12144.46	I
3000	16897.38	I
Polonium		
Po Z = 84		
1500 w	2450.08	I
1500 w	2558.01	I
2500 w	3003.21	I
1200	4170.52	I
800	4493.21	I
500	8618.26	I
Potassium		
K Z = 19		
100	214.35	V
150	271.82	IV
100	273.06	IV
150	282.35	V
150	293.33	V
300	294.84	V
200	296.17	V
200	297.06	V
200	300.25	V
200	300.50	V
200	311.24	V
250	312.77	V
200	315.18	V
250	327.38	V

Intensity	Wavelength/Å	
25	330.68	III
300	340.46	IV
150	340.74	IV
30	341.92	III
15	348.00	III
200	349.50	V
300	354.93	IV
150	356.26	IV
300	359.73	IV
200	359.91	IV
250	362.08	IV
150	362.15	IV
150	363.02	IV
500	372.15	V
200	372.46	V
200	372.77	V
300	375.96	IV
300	375.96	V
250	377.76	V
30	379.12	III
300	379.12	V
300	379.88	IV
25	380.48	III
250	380.48	IV
200	381.70	IV
30	382.23	III
300	382.23	IV
150	382.49	IV
200	382.65	IV
300	382.91	IV
250	384.10	IV
200	386.61	IV
300	387.80	V
250	388.92	IV
250	389.07	IV
250	389.07	V
250	390.11	V
250	390.42	IV
300	390.57	IV
200	391.46	IV
200	392.47	IV
500	393.14	IV
250	395.40	V
200	398.36	V
15	398.63	III
200	398.88	V
200	399.75	V
400	400.21	IV
20	402.10	III
300	402.91	IV
250	403.97	IV
150	404.41	IV
30	406.48	III
250	408.08	IV
40	408.96	III
50	413.79	III
30	414.87	III
250	415.05	V
200	415.79	V
30	416.00	III
150	417.28	IV
30	417.54	III
30	418.62	III
400	422.18	V
300	425.16	V
500	425.59	V
75	434.72	III
50	435.68	III
250	438.02	V
25	441.81	II
200	442.30	IV
300	443.57	IV
75	444.34	III
200	445.61	IV
250	446.83	IV
75	448.60	III
750	448.60	IV
200	449.71	V
200	452.90	V
250	455.67	V
400	456.33	IV
400	456.33	V
75	466.79	III
100	470.09	III
75	471.57	III
45	474.92	III
10	476.03	II
40	479.18	III
10	482.11	III
10	482.41	III
200	482.71	V
200	483.75	V
30	495.14	II
75	497.10	III
10	514.94	III
50	520.61	III
250	523.00	IV
25	523.79	III
200	526.45	IV
150	527.62	IV
40	529.80	III
15	539.71	III
15	546.12	III
750	580.32	V
250	585.51	V
500	586.32	V
30	600.77	II
250	602.27	V
400	603.43	V
25	607.93	II
30	612.62	II
250	638.67	V
750	646.19	IV
300	687.50	V
20	708.84	III
300	720.43	V
400	724.42	V
600	731.86	V
500	737.14	IV
500	741.95	IV
500	745.26	IV
400	746.35	IV
300	749.99	IV
150	754.19	IV
400	754.67	IV
20	765.31	III
30	765.64	III
150	770.29	V
150	771.46	V
35	778.53	III
20	872.31	III
10	873.86	III
15	874.04	III
6	2550.02	III
5	2635.11	III
5	2689.90	III
5	2938.45	III
5	2986.20	III
6	2992.42	III
6	3052.07	III
5	3056.84	III
5	3062.18	II
4	3101.79	I
3	3102.04	I
7	3217.16	I
6	3217.62	I
11	3446.37	I
10	3447.38	I
3	3648.84	I
4	3648.98	I
18	4044.14	I
17	4047.21	I
10	4641.88	I
11	4642.37	I
4	4740.91	I
6	4744.35	I
5	4753.93	I
7	4757.39	I
5	4786.49	I
7	4791.05	I
6	4799.75	I
8	4804.35	I
7	4849.86	I
8	4856.09	I
8	4863.48	I
9	4869.76	I
8	4942.02	I
9	4950.82	I
9	4956.15	I
10	4965.03	I
10	5084.23	I
11	5097.17	I
11	5099.20	I
12	5112.25	I
12	5323.28	I
13	5339.69	I
12	5342.97	I
14	5359.57	I
16	5782.38	I
17	5801.75	I
15	5812.15	I
17	5831.89	I
8	6120.27	II
7	6307.29	II
19	6911.08	I
12	6936.28	I
20	6938.77	I
7	6964.18	I
12	6964.67	I
25	7664.90	I
24	7698.96	I
5	7955.37	I
4	7956.83	I
7	8078.11	I
6	8079.62	I
9	8250.18	I
8	8251.74	I
3	8390.22	I
11	8503.45	I
10	8505.11	I
4	8763.96	I
3	8767.05	I
13	8902.19	I
12	8904.02	I
5	8923.31	I
4	8925.44	I
7	9347.24	I
3	9349.25	I
6	9351.59	I
15	9595.70	I
14	9597.83	I
6	9949.67	I
5	9954.14	I
9	10479.63	I
5	10482.15	I
8	10487.11	I
17	11019.87	I
16	11022.67	I
17	11690.21	I
16	11769.62	I
17	11772.83	I
	12522.11	I
	13377.86	I
	13397.09	I
	15163.08	I
	15168.40	I
	40158.37	I

Praseodymium
Pr $Z = 59$

Intensity	Wavelength/Å	
7000	865.90	V
5000	869.17	V
2000	1228.59	IV
5000	1293.22	IV
5000	1295.28	IV
5000	1321.36	IV
5000	1333.57	IV
5000	1354.66	IV
2000	1360.64	IV
2000	1365.77	IV
5000	1374.41	IV
5000	1435.56	IV
2000	1520.98	IV
5000	1574.55	IV
5000	1575.10	IV
3000	1578.38	IV
2000	1622.30	IV
10000	1884.87	IV
2000	2083.23	IV
3300	2246.20	V
2000 c	2378.98	IV
40 h	2598.04	II
100 h	2707.37	II
60	2760.35	II
270	3168.24	II
200 d	3195.99	II
190	3219.48	II
200	3584.21	II
250	3645.66	II
250	3646.30	II
370	3668.83	II
290	3714.05	II
410	3739.18	II
680	3761.87	II
680	3800.30	II

Intensity	Wavelength/Å	
390	3811.84	II
1300 h	3816.02	II
680	3818.28	II
310	3821.80	II
960	3830.72	II
480	3840.99	II
580	3846.59	II
1200	3850.79	II
720 c	3851.55	II
960	3852.80	II
480 c	3865.45	II
480	3876.19	II
1700 c	3877.18	II
680	3880.47	II
440 c	3885.19	II
440 c	3889.34	II
770 c	3908.05	II
630	3912.90	II
310	3913.55	II
1300 c	3918.85	II
420	3919.63	II
960	3925.47	II
480	3927.46	II
370	3929.29	II
370	3935.82	II
730 c	3947.63	II
900 c	3949.43	II
900 c	3953.51	II
380	3956.75	II
470	3962.45	II
560	3964.26	II
1600 c	3964.81	II
560 c	3966.57	II
500	3971.16	II
320	3971.67	II
620 c	3972.14	II
320	3974.85	II
1300 c	3989.68	II
340	3992.16	II
1600	3994.79	II
560 c	3997.04	II
320	3999.12	II
620 c	4000.17	II
730	4004.70	II
1900	4008.69	II
620	4010.60	II
730	4015.39	II
620	4020.96	II
470	4022.71	II
360	4025.54	II
360 c	4029.72	II
730 c	4031.75	II
960	4033.83	II
730	4038.45	II
470	4039.34	II
1300	4044.81	II
340	4047.08	II
450	4051.13	II
2200	4054.88	II
2200	4056.54	II
450	4058.80	II
3400	4062.81	II
500 c	4079.77	II
500 c	4080.98	II
790	4081.85	II
500	4083.34	II
560	4096.82	II
380	4098.40	II
2900 c	4100.72	II
1700 c	4118.46	II
340	4130.77	II
1500 c	4141.22	II
2700	4143.11	II
1700 c	4164.16	II
620	4171.82	II
730	4172.25	II
5200	4179.39	II
2500	4189.48	II
560 c	4191.60	II
2500 c	4206.72	II
500	4208.32	II
320	4211.86	II
320	4217.81	II
3800	4222.93	II
3800	4225.35	II
320	4233.11	II
320 c	4236.15	II
960	4241.01	II
340	4243.51	II
840 c	4247.63	II
500	4254.40	II
320	4269.09	II
790 c	4272.27	II
470 c	4280.07	II
790 c	4282.42	II
450 c	4298.98	II
1500	4305.76	II
1300	4333.97	II
360	4338.70	II
620 cw	4344.30	II
470 c	4347.49	II
340	4350.40	II
450	4354.91	II
410 c	4359.79	II
1200	4368.33	II
320	4371.62	II
430	4405.83	II
1700	4408.82	II
410	4413.77	II
1200 c	4429.13	II
730	4449.83	II
960	4468.66	II
1100	4496.46	II
790	4510.15	II
340 c	4534.15	II
340	4535.92	II
270 c	4628.74	II
270 c	4672.09	II
290	4695.77	I
250	4736.69	I
200	4924.60	I
320	4939.74	I
380	4951.37	I
270	5034.41	II
320	5045.52	I
360	5110.38	II
560	5110.76	II
410	5129.52	II
620	5173.90	II
360	5206.55	II
360	5219.05	II
560	5220.11	II
680	5259.73	II
340 c	5292.02	II
340	5292.62	II
430	5322.76	II
65	5509.15	II
150	5535.17	II
110	5623.05	II
90	5624.45	II
90	5756.17	II
90	5779.28	I
160 d	5815.17	II
90	5823.72	II
90	5859.68	II
160	5939.90	II
7000 w	5956.05	III
90	5956.60	II
110	5967.82	II
90	6006.33	II
150	6017.80	II
150	6025.72	II
140	6055.13	I
65	6087.52	II
9000 w	6090.02	III
65	6114.38	II
65	6148.23	I
5000	6160.24	III
190	6161.18	II
270	6165.94	II
45	6244.35	II
110	6281.28	II
55 c	6359.03	I
55	6411.23	I
45	6429.63	II
45	6431.84	II
45	6486.55	I
45	6566.77	II
55	6616.67	I
75	6656.83	II
55	6673.41	II
75	6673.78	II
35 c	6747.09	I
55 cw	6798.60	I
35 cw	6827.60	II
7000	6910.14	III
40	7021.51	II
5000	7030.39	III
4500	7076.62	III
20	7114.55	I
24	7227.70	II
16	7407.56	II
20 c	7451.74	II
14	7541.02	II
20	7645.66	II
16	7721.84	I
14	7871.67	I
14	8067.44	I
10 cw	8122.78	II
11	8141.10	I
5000 w	8602.74	III
10	8714.59	II

Promethium
Pm Z = 61

Intensity	Wavelength/Å	
1000	3892.15	II
1000	3910.26	II
1000	3919.10	II
1000	3957.74	II
1000 r	3998.96	II
1000	4417.96	II
900 r	4728.36	I
900	6100.21	I
1000 d	6520.45	I

Protactinium
Pa Z = 91

Intensity	Wavelength/Å	
3000	2599.16	II
3000	2699.22	II
3000	2822.79	II
3000 h	2871.42	II
3000 h	2891.14	II
3000 l	3011.10	II
3000 s	3033.59	II
3000 l	3071.24	II
3000 l	3093.23	II
3000 l	3126.23	II
3000 l	3146.28	II
3000 l	3170.89	II
3000 l	3171.54	II
3000 l	3240.58	II
3000	3274.46	II
3000 l	3332.69	II
3000 s	3346.66	II
3000 l	3452.82	II
3000	3504.97	I
3000 s	3530.65	II
3000	3570.56	I
3000	3571.82	I
3000	3618.07	I
10000	3636.52	I
3000	3702.74	I
3000	3752.67	I
3000	3873.35	I
3000	3931.83	I
3000 s	3952.62	II
10000 l	3957.85	II
3000 s	3970.07	II
3000	3981.82	I
10000	3982.23	I
3000 l	4012.96	II
3000 s	4018.21	II
3000	4030.16	II
3000 s	4046.93	II
10000 s	4056.20	II
10000 s	4070.40	II
3000 l	4176.18	II
10000 l	4217.23	II
10000 s	4248.08	II
3000 s	4291.34	II
3000 s	4601.43	II
3000 l	6035.78	I
3000	6162.56	I
3000 l	6358.61	I
3000	6379.25	I
3000 l	6438.97	I
3000 h	6792.75	I
10000	6945.72	I
3000	6960.09	I
3000 h	6961.78	I
3000 s	6992.73	I
3000	7076.27	I
3000 h	7100.94	I
10000 s	7114.89	I
3000 h	7171.55	I
3000	7227.13	I

Intensity	Wavelength/Å	
3000	7318.79	I
10000 l	7368.25	I
3000 h	7471.89	I
10000 h	7493.15	I
3000 h	7558.26	I
10000 h	7608.20	I
10000	7626.79	I
10000 s	7635.18	I
10000	7669.34	I
3000	7679.20	I
10000 h	7749.19	I
3000	7872.95	I
3000 l	7945.56	I
10000	8039.34	I
10000 h	8099.84	I
10000	8199.04	I
10000	8271.87	I
3000 s	8358.98	I
3000 s	8369.60	I
3000 h	8441.04	I
10000 h	8532.66	I
10000 s	8572.96	I
3000 h	8639.91	I
3000 h	8653.51	I
10000	8735.27	I
3000	10923.32	I
10000	11791.73	I
10000	14344.76	I
3000	18478.61	I
Radium		
Ra *Z* = 88		
100	3649.55	II
200	3814.42	II
100	4340.64	II
100	4682.28	II
100	4825.91	I
50	5660.81	I
50	7141.21	I
50	8019.70	II
Radon		
Rn *Z* = 86		
100	4349.60	I
200	7055.42	I
100	7268.11	I
300	7450.00	I
100	7809.82	I
100	8099.51	I
100	8270.96	I
100	8600.07	I
Rhenium		
Re *Z* = 75		
25000	2003.53	I
16000	2017.87	I
27000	2049.08	I
10000	2085.59	I
9800	2097.12	I
3400	2139.04	II
3700	2156.67	I
4900	2167.94	I
3400	2176.21	I
4200 c	2214.26	II
5200 c	2275.25	II
2900	2287.51	I
2700	2294.49	I
390	2298.09	II
610	2302.99	I
680	2306.54	I
800	2322.49	I
300	2328.66	I
860	2344.78	I
230	2349.39	I
680	2352.07	I
250	2356.50	I
1200	2365.90	I
570	2367.68	I
520	2369.27	I
220	2370.76	II
320	2375.07	I
370	2379.77	I
340	2388.57	I
230	2393.65	I
320	2394.37	I
320	2396.79	I
210 d	2400.72	I
210	2401.68	I
1500	2405.06	I
740	2405.60	I
320	2406.70	I
270	2410.37	I
1200	2419.81	I
300	2421.73	I
300	2421.88	I
2500	2428.58	I
490	2431.54	I
420	2432.18	I
340 c	2441.47	I
230	2442.51	I
250	2444.94	I
610	2446.98	I
610	2449.71	I
390	2461.20	I
800 c	2461.84	II
1200	2483.92	I
390	2485.81	I
980	2487.33	I
370	2496.04	I
370	2501.72	I
570	2502.35	II
230	2504.60	II
270	2505.94	I
1800 c	2508.99	I
570	2520.01	I
540	2521.50	I
370	2534.80	I
570	2540.51	I
740 d	2544.74	I
370	2545.48	I
300	2552.02	I
370	2554.63	II
1000	2556.51	I
250	2559.08	I
340	2564.19	I
540	2568.64	II
370	2571.81	II
380	2586.79	I
290	2599.86	I
290	2603.89	I
660	2608.50	II
610 d	2611.54	I
310	2635.83	II
550	2636.64	I
270	2642.75	I
270	2649.05	I
660	2651.90	I
400	2654.12	I
220	2663.63	I
940	2674.34	I
220	2688.53	I
1300	2715.47	I
220	2732.21	I
610	2733.04	II
220	2758.00	I
210	2763.79	I
310	2767.74	I
220	2768.85	I
220	2769.32	I
350	2770.42	I
550	2783.57	I
220	2791.29	I
220	2814.68	I
880	2819.95	I
310	2834.08	I
220	2843.00	I
270	2850.98	I
240	2867.19	I
2900	2887.68	I
490	2896.01	I
830 c	2902.48	I
210	2905.58	I
550	2909.82	I
830 c	2927.42	I
270	2930.61	I
440	2943.14	I
270	2962.27	I
720	2965.11	I
1500	2965.76	I
310	2976.29	I
210	2978.15	I
220	2980.82	I
220	2982.19	I
220	2988.47	I
1800	2992.36	I
5500	2999.60	I
350	3001.14	I
220	3004.14	I
500	3016.02	I
300	3016.49	I
380	3030.45	I
240	3047.25	I
1600	3067.40	I
320	3069.94	I
260	3071.16	I
550	3082.43	I
340	3088.76	I
700	3100.67	I
700	3108.81	I
340	3110.86	I
340 c	3118.19	I
340	3121.36	I
420	3128.94	I
260	3134.02	I
250	3141.38	I
440	3151.64	I
330	3153.79	I
360 c	3158.31	I
220	3164.52	I
700	3168.37	I
220	3174.61	I
440	3177.71	I
260	3178.61	I
600	3182.87	I
1100	3184.76	I
1100	3185.57	I
260	3190.78	I
260	3192.36	I
220	3198.58	I
1100 c	3204.25	I
380	3235.94	I
600	3258.85	I
600	3259.55	I
300	3268.89	I
280	3296.70	I
280	3296.99	I
280	3301.60	I
240	3302.23	I
320	3303.21	II
280	3303.75	I
240	3313.95	I
600	3322.48	I
2000	3338.18	I
1600	3342.24	I
810	3344.32	I
320	3346.20	I
240 d	3356.33	I
240	3377.74	I
320	3379.06	II
320	3379.70	I
240	3389.43	I
4000	3399.30	I
650	3404.72	I
650	3405.89	I
240	3408.67	I
320	3409.83	I
320	3417.77	I
810	3419.41	I
8000	3424.62	I
400	3426.19	I
300	3427.61	I
320	3437.71	I
400	3449.37	I
16000 c	3451.88	I
240	3453.50	I
55000 c	3460.46	I
40000 c	3464.73	I
400	3467.96	I
240	3476.44	I
400	3480.38	I
320	3480.85	I
240	3482.23	I
560	3503.06	I
320	3516.65	I
320	3517.33	I
320	3537.46	I
240	3549.89	I
240	3570.26	I
360	3579.12	I
810 c	3580.15	II
650	3580.97	I
810	3583.02	I
320	3617.08	I
810	3637.84	I
440	3651.97	I
320	3670.53	I
860 c	3689.50	I

Intensity	Wavelength/Å		Intensity	Wavelength/Å		Intensity	Wavelength/Å		Intensity	Wavelength/Å	
1500 c	3691.48	I	800	1880.66	III	160	2728.94	I	880 d	3538.14	I
520	3703.24	I	500	1884.91	III	100	2771.51	I	280	3541.91	I
240	3709.93	I	500	1887.36	III	130	2783.03	I	1200	3543.95	I
360 c	3717.28	I	700	1888.62	III	150	2826.43	I	1800	3549.54	I
4000	3725.76	I	800	1901.32	III	180	2826.68	I	1200	3570.18	I
240 c	3735.01	I	500	1910.16	III	280	2862.94	I	4700	3583.10	I
810	3735.31	I	600	1919.37	III	110	2878.66	I	4700	3596.19	I
910	3740.10	I	500	1927.07	III	140	2882.37	I	5900	3597.15	I
300 cw	3745.44	I	700	1931.79	III	160	2907.21	I	3100	3612.47	I
700	3787.52	I	500	1954.25	III	65	2910.17	II	1800	3626.59	I
240	3869.94	I	500	1994.26	III	180	2924.02	I	8200	3657.99	I
240	3875.26	I	800	2013.71	III	130	2929.11	I	1300	3666.22	I
240	3876.86	I	500	2017.47	III	130	2931.94	I	560	3681.04	I
380 c	3917.27	I	500	2028.53	III	230	2968.66	I	1900	3690.70	I
550	3929.85	I	800	2036.72	III	160	2977.68	I	9400	3692.36	I
280	3961.04	I	600	2037.61	III	450	2986.20	I	940	3695.52	I
350 c	3962.48	I	1000	2040.18	III	110	3004.46	I	280	3698.26	I
220	4033.31	I	3000	2048.67	III	50	3006.43	III	380	3698.60	I
240	4081.43	I	2000	2064.11	III	130	3023.91	I	7600	3700.91	I
240 c	4110.89	I	800	2076.84	III	50	3052.44	III	940	3713.02	I
240 cw	4133.42	I	1000	2118.53	III	180	3083.96	I	650	3735.28	I
1800	4136.45	I	1000	2118.63	III	140	3121.76	I	420	3737.27	I
700	4144.36	I	1000	2139.44	III	240	3123.70	I	420	3744.17	I
220	4182.90	I	1000	2152.23	III	130	3155.78	I	1200	3748.22	I
220	4183.06	I	3000	2158.17	III	140	3189.05	I	240	3754.12	I
650	4221.08	I	3000	2163.19	III	470	3191.19	I	380	3754.27	I
3600 c	4227.46	I	3000	2167.33	III	190	3197.13	I	490	3755.58	I
260 c	4257.60	I	150	2276.21	II	520	3263.14	I	1000	3760.40	I
380	4358.69	I	140	2288.57	I	520	3271.61	I	2300	3765.08	I
360 cw	4394.38	I	110	2309.82	I	2300	3280.55	I	490	3769.97	I
2600	4513.31	I	350	2322.58	I	2300	3283.57	I	380	3778.13	I
260	4516.64	I	140	2326.47	I	280	3289.14	I	1000	3788.47	I
500	4522.73	I	190	2334.77	II	210	3294.28	I	1300	3792.18	I
2200 cw	4889.14	I	300	2361.92	I	260	3300.46	I	3800	3793.22	I
220	4923.90	I	110	2368.34	I	50	3310.69	III	4900	3799.31	I
1300	5270.95	I	270	2382.89	I	4200	3323.09	I	760	3805.92	I
1600 cw	5275.56	I	230	2383.40	I	330	3338.54	I	1300	3806.76	I
100	5667.88	I	270	2386.14	II	280	3360.80	I	470	3815.01	I
110 c	5752.93	I	80	2415.84	II	420	3368.38	I	760	3816.47	I
110 cw	5776.83	I	130	2427.68	I	1100	3372.25	I	1300	3818.19	I
550	5834.31	I	230	2429.52	I	110	3377.14	I	3800	3822.26	I
200	6307.70	I	110	2437.90	I	110	3385.78	I	2300	3828.48	I
200	6321.90	I	330	2440.34	I	5600	3396.82	I	2000	3833.89	I
100 cw	6605.19	I	90	2461.04	II	820	3399.70	I	5900	3856.52	I
180 c	6813.41	I	130	2473.09	I	160	3406.55	I	490	3870.01	I
260	6829.90	I	150	2487.47	I	820	3412.27	I	380	3877.34	I
50 cw	7640.94	I	100	2490.77	II	330	3421.22	I	120	3913.51	I
65 cw	7912.94	I	130	2502.46	I	120 d	3424.38	I	240	3922.19	I
Rhodium			300	2504.29	II	8200	3434.89	I	2000	3934.23	I
Rh Z = 45			150	2505.67	I	1400	3440.53	I	590	3942.72	I
50	813.44	III	350	2509.70	I	120	3447.74	I	3800	3958.86	I
80	882.51	III	300	2511.03	II	120	3450.29	I	45	3964.54	II
100	925.75	III	200	2515.75	I	400	3455.22	I	380	3975.31	I
150	937.28	III	130	2520.53	II	180	3457.07	I	240	3984.40	I
500	991.62	III	110	2537.04	II	220	3457.93	I	240	3995.61	I
400	992.48	III	350	2545.70	I	5900	3462.04	I	380	3996.15	I
500 d	1009.60	III	550	2555.36	I	100	3469.62	I	120	4023.14	I
200	1012.22	III	150	2622.58	I	4700	3470.66	I	560	4082.78	I
200	1015.17	III	230	2625.88	I	120	3472.25	I	140	4097.52	I
200	1073.87	III	100	2630.42	I	4700	3474.78	I	120	4119.68	I
150	1784.24	III	110	2647.28	I	2100	3478.91	I	1100	4121.68	I
200	1784.94	III	400	2652.66	I	110	3494.44	I	1500	4128.87	I
150	1796.50	III	100	2680.63	I	1200	3498.73	I	2100	4135.27	I
200	1816.03	III	400	2703.73	I	5900	3502.52	I	240	4154.37	I
1000	1832.05	III	100	2715.31	II	2800	3507.32	I	330	4196.50	I
500	1859.85	III	180	2718.54	I	8800	3528.02	I	3300	4211.14	I

Intensity	Wavelength/Å	
820	4288.71	I
4200	4374.80	I
130	4569.00	I
150	4675.03	I
70	4745.11	I
70	5090.63	I
60	5155.54	I
60	5175.97	I
95	5193.14	I
130	5354.40	I
95	5390.44	I
160	5599.42	I
40	5686.38	I
29	5792.66	I
40	5806.91	I
35	5831.58	I
130	5983.60	I
35	6102.72	I
14	6199.99	I
16	6253.72	I
29	6319.53	I
40	6752.35	I
13	6827.33	I
11	6857.68	I
20	6879.94	I
65	6965.67	I
16	6979.15	I
16	7001.58	I
18	7101.64	I
15	7104.45	I
18	7268.18	I
35	7270.82	I
18 h	7442.39	I
12	7475.74	I
12	7495.24	I
11	7557.67	I
29	7791.61	I
55	7824.91	I
21	8029.91	I
29	8045.36	I
15	8136.20	I
8	8425.59	I

Rubidium

Rb ***Z*** **= 37**

Intensity	Wavelength/Å	
30	465.85	III
40	481.118	II
500	482.83	III
500	489.66	III
600	493.48	III
90	497.430	II
20	508.434	II
150	513.266	II
300	530.173	II
75	533.801	II
1200	535.86	III
40	542.887	II
200	555.036	II
1200	556.19	III
1500	566.71	III
1000	572.82	III
1500	576.65	III
2500	579.63	III
1500	581.26	III
2500	589.419	II
1000	594.94	III
1300	595.88	III
1200	598.49	III
1500	643.878	II
25	663.76	IV
3000	697.049	II
6000	711.187	II
25	716.24	IV
50	740.85	IV
10000	741.456	II
5000	769.04	III
25	776.89	IV
2500	815.28	III
15	850.18	IV
1000	1604.12	II
5000	1760.50	II
2000	2068.92	II
10000	2075.95	II
30000	2143.83	II
10000	2217.08	II
5000	2291.71	II
50000	2472.20	II
1000	2631.75	III
2000	2956.07	III
500	3086.84	III
500	3111.36	III
5000 c	3148.90	II
25	3157.54	I
50	3227.98	I
500	3286.41	III
60	3348.72	I
75	3350.82	I
100	3587.05	I
40	3591.57	I
5000	3600.60	II
10000	3600.64	II
25000	3940.51	II
1000	4201.80	I
500	4215.53	I
90000	4244.40	II
15000	4273.14	II
20000	4571.77	II
10000	4648.57	II
30000	4775.95	II
2	5087.987	I
2	5132.471	I
10	5150.134	I
10000	5152.08	II
1	5165.023	I
2	5165.142	I
15	5195.278	I
2	5233.968	I
20	5260.034	I
1	5260.228	I
3	5322.380	I
40	5362.601	I
4	5390.568	I
75	5431.532	I
3	5431.830	I
6	5578.788	I
40	5647.774	I
20	5653.750	I
60	5724.121	I
3	5724.614	I
75	6070.755	I
30 c	6159.626	I
75 c	6206.309	I
120 c	6298.325	I
5	6299.224	I
10000	6458.33	II
5000	6560.81	II
100 l	7279.997	I
150	7408.173	I
200 l	7618.933	I
300	7757.651	I
60	7759.436	I
90000 c	7800.27	I
45000 c	7947.60	I
40 l	8271.41	I
30	8271.71	I
40 l	8868.512	I
30	8868.852	I
30 l	9522.65	I
20 l	9540.18	I
2000 c	9689.05	II
35 l	10075.282	I
30 l	10075.708	I
100	13235.17	I
20	13442.81	I
30	13443.57	I
75	13665.01	I
1000	14752.41	I
800	15288.43	I
150	15289.48	I
20	22529.65	I
4	27314.31	I

Ruthenium

Ru ***Z*** **= 44**

Intensity	Wavelength/Å	
250	850.09	III
200	850.30	III
250	919.74	III
500	940.09	III
500	966.54	III
750	974.14	III
900	979.43	III
500	981.35	III
900	986.84	III
900	994.56	III
300	1001.65	III
500	1009.13	III
900	1009.87	III
500	1014.68	III
800	1190.51	III
500	1200.07	III
500	1207.17	III
500	1209.77	III
300	1211.31	III
500	1941.35	III
500	2009.28	III
2400	2076.43	I
2600	2083.77	I
2400	2090.89	I
690	2255.52	I
780	2272.09	I
780	2279.57	I
480	2317.80	I
120	2334.96	II
190 h	2342.85	II
310	2351.33	I
170	2357.91	II
780	2402.72	II
150	2407.92	II
180	2455.53	II
150	2456.44	II
370	2456.57	II
280	2478.93	II
140	2498.42	II
140	2498.57	II
260	2507.01	II
110	2513.32	II
110	2517.32	II
150	2535.59	II
550	2549.58	I
370	2609.06	I
830	2612.07	I
460	2642.96	I
330	2651.84	I
400	2659.62	I
330	2661.61	II
690	2678.76	II
330	2692.06	II
200	2712.41	II
690	2719.52	I
140	2725.47	II
310	2734.35	II
1800	2735.72	I
100	2778.38	II
110	2787.83	II
350	2810.03	I
1700	2810.55	I
350	2818.36	I
400	2829.16	I
640	2854.07	I
420	2861.41	I
550	2866.64	I
1800	2874.98	I
740	2886.54	I
370	2908.88	I
1100	2916.26	I
180	2945.67	II
370	2949.50	I
550	2965.16	I
170	2965.55	II
140	2976.59	II
550	2976.92	I
1400	2988.95	I
460	2994.96	I
440	3006.59	I
330	3017.24	I
310	3020.88	I
390	3064.84	I
330	3096.57	I
830	3099.28	I
740	3100.84	I
490	3294.11	I
370	3301.59	I
930	3339.55	I
3100	3417.35	I
4900	3428.31	I
6400	3436.74	I
8300	3498.94	I
640	3514.49	I
790	3539.37	I
690	3570.59	I
6400	3589.22	I
6900	3593.02	I
6400	3596.18	I
1300	3599.76	I
3100	3634.93	I
6200	3661.35	I

Intensity	Wavelength/Å	
830	3663.37	I
650	3669.49	I
550	3726.10	I
8700	3726.93	I
11000	3728.03	I
7100	3730.43	I
3500	3742.28	I
870	3742.78	I
2800	3745.59	I
760	3753.54	I
870	3755.93	I
1200	3759.84	I
600	3761.51	I
600	3767.35	I
1500	3777.59	I
600	3782.74	I
3900	3786.06	I
6000	3790.51	I
760	3798.05	I
7600	3798.90	I
7600	3799.35	I
600	3812.72	I
760	3817.27	I
760	3819.03	I
650	3822.09	I
550	3824.93	I
760	3831.80	I
930	3839.70	I
760	3850.43	I
1300	3857.55	I
650	3862.69	I
1300	3867.84	I
650	3892.21	I
760	3909.08	I
1500	3923.47	I
3300	3925.92	I
600	3931.76	I
760	3945.57	I
600	3978.44	I
600	3979.42	I
870	3984.86	I
1500	4022.16	I
600	4023.83	I
1400	4051.40	I
710	4054.05	I
760	4068.37	I
980	4076.73	I
6000	4080.60	I
930	4097.79	I
1900	4112.74	I
2000	4144.16	I
650	4145.74	I
870	4167.51	I
550	4197.58	I
550	4198.88	I
7600	4199.90	I
1500	4206.02	I
5400	4212.06	I
760	4214.44	I
930	4217.27	I
550	4230.31	I
760	4241.05	I
760	4243.06	I
760	4284.33	I
550	4295.93	I
3700	4297.71	I
930	4307.60	I
550	4319.87	I
550	4342.07	I
710	4354.13	I
870	4361.21	I
2400	4372.21	I
870	4385.39	I
1300	4385.65	I
1700	4390.44	I
1600	4410.03	I
1100	4460.04	I
5400	4554.51	I
1700	4584.44	I
720	4647.61	I
1400	4709.48	I
500	4757.84	I
550	4869.15	I
160	5011.23	I
450	5057.33	I
120	5076.32	I
200	5093.83	I
530	5136.55	I
170	5142.76	I
250	5147.24	I
110	5151.07	I
500	5155.14	I
920	5171.03	I
180	5195.02	I
130	5284.08	I
260	5309.27	I
110	5335.93	I
130	5361.77	I
110 h	5401.04	I
80	5484.32	I
130	5510.71	I
90	5559.75	I
290	5636.24	I
180	5699.05	I
65	5814.98	I
55	5919.34	I
80	5921.45	I
21 h	5973.38	I
16	5988.67	I
35	5993.65	I
18	6116.77	I
26	6199.42	I
26	6225.20	I
18	6295.22	I
16	6390.23	I
26 h	6444.84	I
21	6663.14	I
55	6690.00	I
21	6766.95	I
30	6775.02	I
21	6824.17	I
26	6911.48	I
110	6923.23	I
26	6982.01	I
26	7027.98	I
35	7238.92	I
16	7393.93	I
18	7468.91	I
26	7485.79	I
70	7499.75	I
26	7559.61	I
18	7621.50	I
18	7722.87	I
22	7791.86	I
30	7847.80	I
80	7881.49	I
18	8264.96	I
22	8710.84	I

Samarium
Sm $Z = 62$

Intensity	Wavelength/Å	
150	2789.38	II
410	3152.52	II
720	3183.92	II
600	3211.73	II
530	3216.85	II
600	3218.61	II
720	3230.56	II
720	3236.64	II
720	3239.66	II
720	3250.37	II
850	3254.38	II
1700	3306.39	II
1200	3321.18	II
1200	3365.86	II
1200	3382.40	II
4200	3568.27	II
4200	3592.60	II
1700	3604.28	II
3400	3609.49	II
1700	3621.23	II
3400	3634.29	II
2200	3661.36	II
2200	3670.84	II
1100	3693.99	II
1600	3728.47	II
2100	3731.26	II
1600	3735.98	II
2900	3739.12	II
1200	3743.87	II
930	3745.46	I
800	3756.41	I
1200	3757.53	II
1900	3760.69	II
1100	3764.37	II
370 d	3773.33	I
1100	3778.14	II
1500	3788.12	II
1600	3793.97	II
1600	3797.73	II
1600	3826.20	II
1100	3831.50	II
560	3834.48	I
1600	3843.50	II
530	3853.30	I
2700	3854.21	II
480	3854.56	I
400	3858.74	I
3700	3885.29	II
1600	3896.98	II
1300	3903.42	II
2500	3922.40	II
1900	3928.28	II
1300	3941.87	II
470	3951.89	I
1500	3963.00	II
1500	3971.40	II
620	3974.66	I
1000	3976.43	II
1500	3990.00	II
1400	4064.58	II
1000	4092.27	II
1900	4118.55	II
1200	4152.21	II
1000	4188.13	II
1100	4203.05	II
1000	4225.33	II
1200	4236.74	II
2100	4256.39	II
1300	4262.68	II
1200	4279.68	II
2200	4280.79	II
710	4282.21	I
470	4282.83	I
1600	4296.74	I
1900	4318.94	II
470	4319.53	I
1800	4329.02	II
440	4330.02	I
1300	4334.15	II
880	4336.14	I
1100	4347.80	II
440	4362.91	I
530	4380.42	I
1600	4390.86	II
410	4401.17	I
470	4419.33	I
1500	4420.53	II
2900	4424.34	II
470	4429.66	I
1600	4433.88	II
1800	4434.32	II
530	4441.81	I
440	4442.28	I
710	4445.15	I
1300	4452.73	II
1200	4454.63	II
1000	4458.52	II
2200	4467.34	II
810	4470.89	I
370	4499.11	I
440	4581.73	I
380	4649.49	I
470 d	4670.75	I
1100	4674.60	II
370	4688.73	I
530	4704.40	II
730	4716.10	I
770	4728.42	I
470	4745.68	II
730	4760.27	I
580	4783.10	I
350	4785.86	I
970	4841.70	I
730	4883.97	I
630	4910.40	I
350	4913.25	II
430	4918.99	I
400	5044.28	I
540	5071.20	I
510	5117.16	I
350	5122.14	I
360	5155.03	II
470	5175.42	I
250	5200.59	I

Intensity	Wavelength/Å	
260	5251.92	I
400	5271.40	I
250	5282.91	I
220	5453.00	I
230	5493.72	I
230	5516.09	I
140	5550.40	I
140	5659.86	I
120	5696.73	I
85	5706.20	I
70	5773.77	I
60	5778.33	I
70 d	5786.98	II
60	5788.38	I
60	5800.52	I
65	5802.84	I
65	5867.79	I
50	5874.21	I
50	5898.96	I
65	5965.71	II
50	6045.00	I
50	6070.06	I
45	6084.12	I
45 h	6159.56	I
45	6256.54	I
100	6267.28	II
140	6569.31	II
110	6589.72	II
50	6671.51	I
120 d	6731.84	II
95	6794.20	II
120	6860.93	I
120	6955.29	II
90	7020.44	II
90	7039.22	II
90	7042.24	II
90	7051.52	II
90	7082.37	II
26	7088.30	I
30	7095.50	I
30	7104.54	I
26	7115.96	I
85 d	7149.60	II
23	7213.82	I
60	7240.90	II
26	7347.30	I
30	7444.56	I
26	7445.41	I
13	7470.76	I
45	7645.09	II
12	7645.82	I
40 w	7835.08	II
16	7895.96	I
90	7928.14	II
40	8048.70	II
16	8065.16	I
45	8068.46	II
40 w	8305.79	II
19	8383.71	I
45 w	8485.99	II
45 w	8708.43	II
95	8913.66	II
Scandium		
Sc Z = 21		
350	180.14	V
500	243.87	V
500	252.85	V
500	253.73	V
900	283.91	V
800	284.45	V
600	288.29	V
900	289.59	V
15	289.85	IV
1000 d	291.93	V
800	293.25	V
15	296.31	IV
15	299.04	IV
700	300.00	V
1000	573.36	V
600	587.94	V
10	785.12	IV
25	1168.61	III
15	1550.80	IV
180	1603.06	III
150	1610.19	III
160	2010.42	III
12	2118.97	IV
11	2185.43	IV
11	2205.46	IV
14	2222.22	IV
11	2271.33	IV
110	2438.62	I
560	2545.22	II
2900	2552.37	II
560	2555.82	II
2300	2560.25	II
1100	2563.21	II
11	2586.93	IV
120	2692.78	I
350	2699.07	III
360	2706.77	I
210	2707.95	I
580	2711.35	I
230	2734.05	III
340	2965.86	I
1200	2974.01	I
1400	2980.75	I
340	2988.95	I
2200	3015.36	I
2700	3019.34	I
360	3030.76	I
120 h	3056.31	I
130	3065.11	II
990	3251.32	II
1500	3255.69	I
4400	3269.91	I
5500	3273.63	I
110 d	3343.28	II
270	3352.05	II
9900	3353.73	II
2000	3359.68	II
1700	3361.27	II
1700	3361.94	II
4000	3368.95	II
6600	3372.15	II
130	3418.51	I
200	3429.21	I
200	3429.48	I
270	3431.36	I
530	3435.56	I
270	3457.45	I
180	3462.19	I
130 d	3469.65	I
110	3471.13	I
200	3498.91	I
2700	3535.73	II
6600	3558.55	II
6100	3567.70	II
13000	3572.53	II
9900	3576.35	II
7700	3580.94	II
4000	3589.64	II
4000	3590.48	II
28000	3613.84	II
110	3617.43	I
20000	3630.75	II
13000	3642.79	II
6600	3645.31	II
110	3646.90	I
5300	3651.80	II
110	3664.25	II
290	3666.54	II
75 h	3717.10	I
270	3833.07	II
610	3843.03	II
20000	3907.49	I
23000	3911.81	I
4400	3933.38	I
5500	3996.61	I
530	4014.49	II
20000	4020.40	I
20000	4023.69	I
220	4030.67	I
140	4031.39	I
220	4043.80	I
200	4046.48	I
2700	4047.79	I
120	4049.95	I
5500	4054.55	I
220	4056.59	I
100	4068.66	III
160 h	4074.97	I
160	4078.57	I
6100	4082.40	I
200	4086.67	I
400	4087.16	I
440 h	4133.00	I
530 h	4140.30	I
720	4152.36	I
1100 h	4165.19	I
110 h	4218.26	I
110 h	4219.73	I
180	4231.93	I
200	4233.61	I
400	4238.05	I
15000	4246.83	II
290	4294.77	II
350	4305.71	II
4200	4314.09	II
3300	4320.74	II
2400	4325.01	II
180	4354.61	II
110	4358.64	I
2000	4374.46	II
130	4384.81	II
1100	4400.37	II
880	4415.56	II
120 h	4557.24	I
160 h	4573.99	I
350	4670.40	II
120	4706.97	I
120	4709.34	I
200	4728.77	I
490	4729.23	I
590	4734.10	I
690	4737.65	I
790	4741.02	I
1200	4743.81	I
200	4753.16	I
220	4779.35	I
170	4839.44	I
90	4909.76	I
90	4922.84	I
90	4934.25	I
170	4954.06	I
120	4973.66	I
150	4980.37	I
140	4991.92	I
530	5031.02	II
250	5064.32	I
530	5070.23	I
250	5075.81	I
2100	5081.56	I
1200	5083.72	I
1100	5085.55	I
750	5086.95	I
390	5087.14	I
270	5089.89	I
390	5096.73	I
620	5099.23	I
370	5101.12	I
180	5109.06	I
150	5112.86	I
320	5116.69	I
390	5210.52	I
280	5219.67	I
350	5239.82	II
280	5258.33	I
210	5285.76	I
120	5341.05	I
350	5349.30	I
120	5349.71	I
210	5355.75	I
530	5356.10	I
270	5375.35	I
370	5392.08	I
270	5446.20	I
120	5451.34	I
750	5481.99	I
530	5484.62	I
570	5514.22	I
660	5520.50	I
660	5526.82	II
70	5564.86	I
110	5591.33	I
80	5640.98	II
250	5657.88	II
1500	5671.81	I
1200	5686.84	I
1100	5700.21	I
10	5706.82	IV
190	5708.61	I
880	5711.75	I
230	5717.28	I

Intensity	Wavelength/Å	
180	5724.08	I
14	5771.63	IV
620	6210.68	I
320	6239.78	I
120	6245.63	II
110	6249.96	I
80	6256.01	III
250	6258.96	I
750	6305.67	I
60	6378.82	I
90	6413.35	I
60	6604.60	II
65	6737.87	I
50	6819.52	I
50	6835.03	I
90	7449.16	III
55 h	7741.17	I
30	7800.44	I
19 h	8761.40	I
50	8829.78	III
30 h	8834.35	I
400	22051.86	I
150	22065.05	I
Selenium		
Se $Z = 34$		
360	613.0	V
360	652.7	IV
450	670.1	IV
360	724.3	III
450	746.4	IV
450	759.1	V
360	808.7	V
360	830.3	V
360	832.7	II
450	839.5	V
360	843.0	III
360	845.8	V
360	912.9	II
360	959.6	IV
360	974.8	III
450	996.7	IV
360	1013.4	II
360	1014.0	II
450	1033.6	II
450	1049.6	II
360	1057.4	II
360	1094.7	V
360	1099.1	III
450	1119.2	III
360	1141.9	II
450	1192.3	II
450	1227.6	V
285	1291.0	II
285	1308.9	II
285	1314.4	IV
120	1435.3	I
120	1435.8	I
150	1449.2	I
150	1500.9	I
250	1530.4	I
150	1531.3	I
200	1531.8	I
150	1575.3	I
150	1577.6	I
150	1577.9	I
150	1579.5	I
200	1580.0	I
150	1587.5	I
150	1593.2	I
250	1606.5	I
200	1617.4	I
150	1621.2	I
150	1643.4	I
250	1671.2	I
250	1675.3	I
250	1690.7	I
250	1793.3	I
300	1795.3	I
300	1855.2	I
250	1858.8	I
400	1898.6	I
350	1913.8	I
300	1919.2	I
500	1960.9	I
500	2039.8	I
285	2057.5	III
500	2074.8	I
285	2136.6	IV
500	2164.2	I
600	2413.5	I
300	2548.0	I
360	2665.5	IV
285	2724.3	IV
285	2767.2	III
220	2773.8	III
160	2951.6	IV
450	3387.2	III
450	3413.9	III
450	3457.8	III
450	3637.6	III
450	3738.7	III
450	3800.9	III
450	4169.1	III
360	4175.3	II
450	4180.9	II
285	4382.9	II
285	4446.0	II
220	4449.2	II
285	4467.6	II
500	4730.8	I
400	4739.0	I
300	4742.2	I
285	4840.6	II
360	4845.0	II
450	5227.5	II
360	5305.4	II
100	5365.5	I
120	5369.9	I
110	5374.1	I
285	5522.4	II
285	5566.9	II
285	5866.3	II
450	6056.0	II
285	6303.8	III
200	6325.6	I
360	6444.2	II
285	6490.5	II
285	6535.0	II
150	6831.3	I
120	6990.690	I
100	6991.792	I
200	7010.809	I
150	7013.875	I
300	7062.065	I
200	7575.1	I
250	7583.4	I
150	7592.2	I
300	8001.0	I
200	8036.4	I
150	8093.2	I
150	8094.7	I
180	8149.3	I
150	8152.0	I
200	8157.7	I
180	8163.1	I
150	8182.9	I
150	8440.47	I
150	8450.38	I
150	8742.33	I
300	8918.86	I
200	9001.97	I
200	9038.61	I
100	9432.50	I
200	10217.25	I
377	10307.45	I
900	10327.26	I
640	10386.36	I
275	11946.87	I
170	11952.64	I
205	11972.93	I
315	14817.93	I
410	14917.47	I
500	15151.44	I
320	15471.00	I
265	15520.97	I
395	15618.40	I
360	16659.44	I
505	16813.78	I
205	16866.54	I
235	21374.24	I
680	21442.56	I
415	21473.48	I
270	21716.36	I
240	21730.60	I
150	23388.85	I
265	24148.18	I
375	24385.99	I
255	25017.51	I
510	25127.43	I
Silicon		
Si $Z = 14$		
10	85.18	V
15	96.44	V
10	97.14	V
20	117.86	V
20	118.97	V
4	457.82	IV
8	566.61	III
8	653.33	III
7	815.05	IV
8	818.13	IV
9	823.41	III
40 h	845.77	II
100	889.72	II
200	892.00	II
9	967.95	III
100	989.87	II
200	992.68	II
10	993.52	III
13	994.79	III
16	997.39	III
50	1023.69	II
8	1066.63	IV
14	1108.37	III
16	1109.97	III
18	1113.23	III
8	1122.49	IV
10	1128.34	IV
100	1190.42	II
200	1193.28	II
250	1194.50	II
100	1197.39	II
30	1206.51	III
30	1206.53	III
9	1207.52	III
10	1210.46	III
50	1226.81	II
100	1227.60	II
150	1228.75	II
200	1229.39	II
100	1246.74	II
150	1248.43	II
100	1250.09	II
150	1250.43	II
200	1251.16	II
40	1256.49	I
50	1258.80	I
1000	1260.42	II
2000	1264.73	II
200	1265.02	II
17	1294.54	III
14	1296.73	III
15	1298.89	III
18	1298.96	III
14	1301.15	III
16	1303.32	III
100	1304.37	II
50 h	1305.59	II
200	1309.27	II
13	1312.59	III
100	1346.87	II
100	1348.54	II
150	1350.06	II
100	1352.64	II
100	1353.72	II
15	1393.76	IV
12	1402.77	IV
13	1417.24	III
90 h	1485.02	II
100 h	1485.51	II
12	1500.24	III
10	1501.19	III
9	1501.87	III
100 h	1509.10	II
50 h	1512.07	II
60 p	1516.91	II
500	1526.72	II
1000	1533.45	II
150	1594.55	I
100	1622.87	I
300	1629.43	I
200	1629.92	I
100	1667.62	I
100	1668.52	I

Intensity	Wavelength/Å	
100	1672.59	I
200	1675.20	I
200	1696.20	I
200	1697.94	I
100 h	1770.92	I
100 h	1776.83	I
100 h	1799.12	I
150	1808.00	II
500 h	1814.07	I
200	1816.92	II
200	1836.51	I
200	1841.44	I
9	1842.55	III
200	1843.77	I
300	1845.51	I
400	1847.47	I
200	1848.14	I
500	1850.67	I
200	1852.46	I
500 h	1874.84	I
200	1881.85	I
200	1887.70	I
200 h	1893.25	I
1000 h	1901.33	I
100 h	1902.46	II
50 h	1910.62	II
50	1941.67	II
100	1949.56	II
100	1954.97	I
50	2058.65	II
50	2059.01	II
200	2072.02	II
200	2072.70	II
100	2124.12	I
50 h	2136.56	II
110	2207.98	I
115	2210.89	I
110	2211.74	I
120	2216.67	I
120	2218.06	I
10	2296.87	III
10	2308.19	III
100 h	2356.30	II
30 h	2357.18	II
50 h	2357.97	II
300	2435.15	I
11	2449.48	III
425	2506.90	I
375	2514.32	I
500	2516.113	I
7	2517.51	IV
350	2519.202	I
425	2524.108	I
450	2528.509	I
110	2532.381	I
25	2541.82	III
10	2546.09	III
14	2559.21	III
30	2563.679	I
85	2568.641	I
45	2577.151	I
190	2631.282	I
11	2640.79	III
14	2655.51	III
9	2817.11	III
1000	2881.579	I
300	2904.28	II
500	2905.69	II
55	2970.355	I
150	2987.645	I
50	3006.739	I
75	3020.004	I
100 h	3030.00	II
9	3040.93	III
100 h	3043.69	II
50 h	3048.30	II
150 h	3053.18	II
25	3086.24	III
20	3093.42	III
16	3096.83	III
9	3165.71	IV
16	3185.13	III
13	3186.02	III
150	3188.97	II
150	3193.09	II
100	3195.41	II
14	3196.50	III
200	3199.51	II
100 h	3203.87	II
200 h	3210.03	II
15	3210.55	III
75	3214.66	II
12	3230.50	III
14	3233.95	III
15	3241.62	III
12	3258.66	III
10	3276.26	III
300	3333.14	II
500	3339.82	II
15	3486.91	III
9	3525.94	III
20	3590.47	III
8	3762.44	IV
20 c	3791.41	III
25	3796.11	III
30	3806.54	III
100 h	3853.66	II
500 h	3856.02	II
200 h	3862.60	II
300	3905.523	I
20	3924.47	III
10	4088.85	IV
70	4102.936	I
9	4116.10	IV
300 h	4128.07	II
500 h	4130.89	II
100 h	4190.72	II
50	4198.13	II
9	4338.50	III
30	4552.62	III
25	4567.82	III
20	4574.76	III
100	4621.42	II
150	4621.72	II
9 h	4631.24	IV
10 h	4654.32	IV
9	4683.02	III
16	4716.65	III
50	4782.991	I
35	4792.212	I
80	4792.324	I
15	4813.33	III
16	4819.72	III
18	4828.97	III
30	4947.607	I
40	5006.061	I
1000	5041.03	II
1000	5055.98	II
10 h	5091.42	III
100	5181.90	II
100 h	5185.25	II
200 h	5192.86	II
500 h	5202.41	II
100 h	5405.34	II
100 h	5438.62	II
100 h	5456.45	II
500 h	5466.43	II
500 h	5466.87	II
100 h	5469.21	II
200 h	5496.45	II
35	5517.535	I
100 h	5540.74	II
150 h	5576.66	II
30	5622.221	I
100 h	5632.97	II
200 h	5639.48	II
90	5645.611	I
150 h	5660.66	II
80	5665.554	I
1000 h	5669.56	II
120	5684.484	I
300 h	5688.81	II
100	5690.425	I
90	5701.105	I
200 h	5701.37	II
100 h	5706.37	II
160	5708.397	I
20	5739.73	III
45	5747.667	I
45	5753.625	I
45	5754.220	I
45	5762.977	I
70	5772.145	I
70	5780.384	I
90	5793.071	I
100	5797.859	I
150 h	5800.47	II
200	5806.74	II
50	5846.13	II
300 h	5868.40	II
40	5873.764	I
10 h	5898.79	III
150	5915.22	II
200	5948.545	I
500	5957.56	II
500	5978.93	II
90	6125.021	I
85	6131.574	I
90	6131.850	I
100	6142.487	I
100	6145.015	I
160	6155.134	I
160	6237.320	I
40	6238.287	I
125	6243.813	I
125	6244.468	I
180	6254.188	I
45	6331.954	I
1000	6347.10	II
1000	6371.36	II
45	6526.609	I
45	6527.199	I
45	6555.462	I
50 h	6660.52	II
100	6671.88	II
7	6701.21	IV
50 h	6717.04	II
100	6721.853	I
50	6829.82	II
30	6848.568	I
80	6976.523	I
180	7003.567	I
180	7005.883	I
90	7017.646	I
250	7034.903	I
6 h	7047.94	IV
200	7165.545	I
100	7226.206	I
100	7235.326	I
180	7250.625	I
160	7275.294	I
400	7289.173	I
375	7405.774	I
200	7409.082	I
275	7415.946	I
425	7423.497	I
9 h	7466.32	III
12 h	7612.36	III
100	7680.267	I
6 h	7723.82	IV
30	7800.008	I
400	7848.80	II
500	7849.72	II
30	7849.967	I
90	7918.386	I
120	7932.349	I
140	7944.001	I
35	7970.306	I
35	8035.619	I
70	8093.241	I
9 h	8102.86	III
11 h	8103.45	III
35	8230.642	I
9 h	8262.57	III
40	8443.982	I
40	8501.547	I
60	8502.221	I
40	8536.165	I
120	8556.780	I
50	8648.462	I
40	8728.011	I
75	8742.451	I
100	8752.009	I
35	8790.389	I
100	9412.72	II
100	9413.506	I
30	10371.269	I
120	10585.141	I
120	10603.431	I
120	10660.975	I
30	10694.251	I
30	10727.408	I
60	10749.384	I
30	10784.550	I

Intensity	Wavelength/Å	
80	10786.856	I
140	10827.091	I
60	10843.854	I
130	10869.541	I
30	10882.802	I
30	10885.336	I
80	10979.308	I
30	10982.061	I
80	11017.965	I
370	11984.19	I
220	11991.57	I
440	12031.51	I
190	15888.39	I
95	16060.03	I
110	19722.50	I
	Silver	
	Ag $Z = 47$	
25	730.83	II
30	752.80	II
400	799.41	III
15	1005.32	II
10	1065.49	II
12	1072.23	II
250	1074.22	II
150	1107.03	II
150	1112.46	II
60	1195.83	II
50	1223.33	II
50	1240.80	II
50	1246.87	II
55	1256.81	II
55	1257.55	II
50	1266.63	II
70	1273.67	II
65	1297.51	II
85	1311.20	II
55	1313.81	II
50	1314.61	II
60	1323.84	II
60	1342.09	II
50	1342.57	II
70	1346.62	II
50	1353.54	II
150	1364.50	II
100	1396.00	II
100	1410.93	II
90	1419.72	II
95	1432.60	II
100	1464.72	II
50	1466.23	II
50 r	1507.37	I
100 r	1515.63	I
50 r	1548.58	I
100	1555.16	II
100	1644.50	II
60	1651.52	I
50	1652.10	I
700	1656.18	III
120	1682.82	II
500	1693.51	III
10	1708.11	I
50	1709.27	I
125	1736.44	II
750	1751.03	III
10 h	1766.14	I
75	1790.37	II
600	1917.08	III
700	1957.62	III
100	1967.38	II
600	1975.92	III
500	1977.03	III
600	2000.24	III
150	2015.96	II
150	2033.98	II
200	2061.17	I
100	2069.85	I
80 r	2113.82	II
60	2145.60	II
600	2161.89	III
50	2186.76	II
60	2229.53	II
100 r	2246.43	II
500	2246.51	III
75 r	2248.74	II
75	2280.03	II
30 h	2309.56	I
700	2310.04	III
70 r	2317.05	II
80 r	2320.29	II
70 r	2324.68	II
80 r	2331.40	II
70	2357.92	II
50 h	2375.02	I
75	2411.41	II
90 r	2413.23	II
100 r	2437.81	II
80	2447.93	II
80	2473.84	II
60	2506.63	II
50 h	2575.63	I
60	2660.49	II
60	2721.77	I
75	2767.54	II
100 h	2824.39	I
30 h	3130.02	I
90	3180.70	II
100	3267.35	II
55000 r	3280.68	I
28000 r	3382.89	I
30	3469.16	I
70	3475.82	II
80	3495.28	II
50	3542.61	I
50 h	3624.68	I
75	3682.46	II
30	3682.50	I
80	3683.34	II
50 h	3709.20	I
200	3810.94	I
50	3811.78	I
100 h	3840.74	I
50 h	3907.41	I
50	3909.31	II
50 h	3914.40	I
70	3920.10	II
60	3949.43	II
100 h	3981.58	I
70	3985.19	II
100 h	4055.48	I
80	4085.91	II
100	4185.48	II
90 h	4210.96	I
100	4212.82	I
50	4311.07	I
50 h	4476.04	I
30 h	4615.69	I
80	4620.04	II
50	4620.46	II
60 h	4668.48	I
30 h	4677.60	I
100	4788.40	II
30 h	4847.82	I
100	4874.10	I
80	5027.35	II
1000	5209.08	I
1000	5465.50	I
100	5471.55	I
100	5667.34	I
10 h	6268.50	I
320	7687.78	I
25	8005.4	II
500	8273.52	I
25	8403.8	II
30 h	8645.70	I
10 h	8704.85	I
12	8747.6	II
15	9000.9	II
10	12551.0	I
60	16819.5	I
20	17416.7	I
15	18307.9	I
15	18382.3	I
	Sodium	
	Na $Z = 11$	
7	142.232	IV
8	146.064	IV
9	150.298	IV
8	150.687	IV
8	155.510	IV
8	156.537	IV
12	162.448	IV
10	163.190	IV
12	168.411	IV
10	168.546	IV
5	183.95	III
10	190.445	IV
10	199.772	IV
8	202.49	III
8	202.76	III
8 p	203.06	III
8	203.28	III
8	203.33	III
15	229.87	III
50 c	250.52	III
30	251.37	III
25	266.90	III
70	267.65	III
50	267.87	III
50	268.63	III
20 p	272.08	III
20	272.45	III
10	319.644	IV
300	372.08	II
350	376.38	II
100	378.14	III
70	380.10	III
12	408.684	IV
10	409.614	IV
15	410.372	IV
10	411.334	IV
13	412.242	IV
11	1582.18	IV
11 d	1583.98	IV
12	1584.14	IV
12 d	1587.05	IV
11	1615.92	IV
12	1618.57	IV
11	1655.47	IV
15 c	1701.97	IV
20 d	1887.47	III
12	1960.76	IV
11	1965.08	IV
12 d	2106.33	IV
30	2230.33	III
16	2232.19	III
20 h	2246.70	III
300	2315.65	II
18	2386.99	III
17	2394.03	III
300	2420.99	II
300	2424.73	II
25	2459.31	III
18	2468.85	III
20	2474.73	III
1000	2493.15	II
25	2497.03	III
17	2510.26	III
20	2543.84	I
10	2543.87	I
70	2593.87	I
35	2593.92	I
850	2611.81	II
850	2661.00	II
1000	2671.83	II
200	2680.34	I
100	2680.43	I
1000	2841.72	II
400	2852.81	I
200	2853.01	I
2	2893.62	I
1100	2904.92	II
1100	2917.52	II
1100	2919.05	II
1200	2919.85	II
1300	2920.95	II
1000	2923.49	II
1200	2951.24	II
1100	2952.40	II
1000	2977.13	II
1100	2979.66	II
1100	2980.63	II
1300	2984.19	II
1700	3124.42	II
2500	3135.48	II
1700	3137.86	II
2000	3149.28	II
2000	3163.74	II
1000	3179.06	II
1700	3189.79	II
1600	3212.19	II
1500	3257.96	II
1700	3285.60	II
1700	3301.35	II
1200	3302.37	I

Intensity	Wavelength/Å	
600	3302.98	I
1500	3304.96	II
1000	3318.04	II
50	3426.86	I
1500	3533.05	II
1200	3631.27	II
6	4238.99	I
10	4242.08	I
1	4249.41	I
2	4252.52	I
15	4273.64	I
20	4276.79	I
2	4287.84	I
3	4291.01	I
30	4321.40	I
40	4324.62	I
3	4341.49	I
5	4344.74	I
40	4390.03	I
60	4393.34	I
5	4419.88	I
8	4423.25	I
60	4494.18	I
100	4497.66	I
10	4541.63	I
15	4545.19	I
120	4664.811	I
200	4668.560	I
20	4747.941	I
30	4751.822	I
200	4978.541	I
400	4982.813	I
40	5148.838	I
80	5153.402	I
280	5682.633	I
70	5688.193	I
560	5688.205	I
80000	5889.950	I
40000	5895.924	I
120	6154.225	I
240	6160.747	I
130	6530.70	II
130	6544.04	II
130	6545.75	II
20	7373.23	I
10	7373.49	I
50	7809.78	I
25	7810.24	I
4400	8183.256	I
800	8194.790	I
8800	8194.824	I
100	8649.92	I
60	8650.89	I
25	8942.96	I
40	9153.88	I
60	9465.94	I
80	9961.28	I
20	10566.00	I
60	10572.28	I
200	10746.44	I
80	10749.29	I
120	10834.87	I
35	11190.19	I
50	11197.21	I
400	11381.45	I
1000	11403.78	I

Intensity	Wavelength/Å	
400	12679.17	I
60	14767.48	I
100	14779.73	I
60	16373.85	I
100	16388.85	I
400	18465.25	I
50	22056.44	I
25	22083.67	I
60	23348.41	I
100	23379.13	I

Strontium
Sr Z = 38

Intensity	Wavelength/Å	
15	298.12	IV
15	300.12	IV
125	330.67	III
500	351.62	III
75	358.80	III
250	363.49	III
150	371.21	III
20	378.53	IV
75	392.44	IV
50	393.00	IV
50	396.22	IV
1000	437.24	III
1875	491.79	III
1250	507.04	III
3750	514.38	III
10	517.28	V
2500	562.75	III
25	578.01	V
30	624.93	V
25	642.23	V
50	649.21	V
25	660.94	V
200	664.43	IV
35	686.23	V
100	710.35	IV
12	747.82	V
50	1025.23	III
35	1125.49	III
50	1236.23	III
1400	2152.84	II
1400	2165.96	II
100	2273.71	III
100	2340.13	III
50	2346.97	IV
160	2428.10	I
100	2486.52	III
40	2555.60	IV
40	2571.04	IV
100	3002.61	III
200	3012.32	III
10	3019.29	IV
100	3021.73	III
50	3061.43	III
50	3182.61	III
100	3235.39	III
400	3351.25	I
650	3380.71	II
50	3430.76	III
950	3464.46	II
600	3969.26	I
1300	4030.38	I
46000	4077.71	II
32000	4215.52	II
9	4298.57	IV

Intensity	Wavelength/Å	
340	4305.45	II
65000	4607.33	I
9	4685.08	IV
3200	4722.28	I
2200	4741.92	I
1400	4784.32	I
4800	4811.88	I
3600	4832.08	I
3000	4872.49	I
2000	4876.32	I
1000	4891.98	I
8000	4962.26	I
1300	4967.94	I
800 h	5156.07	I
1400	5222.20	I
2000	5225.11	I
2000	5229.27	I
2800	5238.55	I
4800	5256.90	I
40	5257.71	III
40	5443.48	III
1500	5450.84	I
7000	5480.84	I
3500	5504.17	I
2600	5521.83	I
2000	5534.81	I
2000	5540.05	I
1000	6380.75	I
900 h	6386.50	I
600 h	6388.24	I
9000	6408.47	I
5500	6504.00	I
1000	6546.79	I
1700	6550.26	I
3000	6617.26	I
1800	6791.05	I
4800	6878.38	I
1200	6892.59	I
5500	7070.10	I
2500	7309.41	I
500	7621.50	I
400 h	7673.06	I
200 h	8422.80	I
120	8505.69	II
200	8688.91	II
100	9294.10	I
400 h	9448.95	I
600	9596.00	I
300	9624.70	I
100	9638.10	I
100 h	9647.70	II
300	10036.66	II
1000	10327.31	II
200	10914.88	II
700	11241.25	I
100	12014.76	II
60	12445.90	II
40	12495.00	I
75	12974.70	II
100	13123.80	II
50	17447.40	I
230	20261.40	I
120	20700.70	I
30	26023.60	I

Sulfur
S Z = 16

Intensity	Wavelength/Å	
5	437.4	V
5	438.2	V
5	439.6	V
20	519.3	IV
20	520.1	IV
40	520.8	IV
20	522.0	IV
20	522.5	IV
20	551.2	IV
40	652.5	IV
40	653.0	IV
70	653.6	IV
40	654.0	IV
70	655.6	IV
20	655.9	IV
110	657.3	IV
40	658.3	V
70	659.8	V
40	660.9	IV
160	661.4	IV
110	663.2	V
40	663.7	IV
40	664.8	IV
70	666.1	IV
20	678.1	V
40	680.3	V
110	680.9	V
40	681.6	V
20	693.5	V
70	729.5	III
110	732.42	III
70	735.2	III
70	738.5	III
110	744.9	IV
110	748.4	IV
110	750.2	IV
110	753.8	IV
285	786.5	V
70	789.0	III
70	796.7	III
70	800.5	IV
70	804.0	IV
70	809.7	IV
110	816.0	IV
70	824.9	III
70	836.3	III
160	849.2	V
110	852.2	V
220	854.8	V
110	857.9	V
110	860.5	V
40	906.9	II
40	910.5	II
40	912.7	II
40	937.4	II
40	937.7	II
160	1062.7	IV
160	1073.0	IV
70	1073.5	IV
285	1077.1	III
40	1102.3	II
70	1194.0	III
70	1201.0	III
40	1234.1	II

Intensity	Wavelength/Å	
40	1250.5	II
110	1253.8	II
110	1259.5	II
275	1270.782	I
250	1277.216	I
280	1295.653	I
275	1302.337	I
235	1302.863	I
235	1303.110	I
245	1303.430	I
260	1305.883	I
265	1310.194	I
355	1316.542	I
290	1316.618	I
375	1323.515	I
355	1326.643	I
775	1381.552	I
710	1385.510	I
960	1388.435	I
640	1389.154	I
775	1392.588	I
1000	1396.112	I
300	1409.337	I
510	1425.030	I
425	1433.280	I
300	1436.968	I
300	1448.229	I
425	1472.972	I
550	1473.995	I
300	1474.380	I
355	1481.665	I
485	1483.039	I
300	1483.233	I
330	1485.622	I
390	1487.150	I
20	1624.0	IV
20	1629.2	IV
680	1666.688	I
640	1687.530	I
710	1807.311	I
680	1820.343	I
640	1826.245	I
710	1900.286	I
550	1914.698	I
20	2387.0	IV
40	2398.9	IV
110	2460.5	III
110	2489.6	III
160	2496.2	III
160	2499.1	III
220	2508.2	III
70	2636.9	III
220	2665.4	III
110	2691.8	III
110	2702.8	III
220	2718.9	III
110	2721.4	III
220	2726.8	III
220	2731.1	III
110	2741.0	III
285	2756.9	III
110	2775.2	III
160	2785.5	III
110	2863.5	III
160	2904.3	III
160	2986.0	III
110	3097.5	IV
110	3497.3	III
160	3632.0	III
110	3709.4	III
160	3717.8	III
160	3838.3	III
285	3867.6	I
285	3902.0	I
160	3928.6	III
360	3933.3	II
450	4120.8	I
280	4142.3	II
360	4145.1	II
450	4153.1	II
450	4162.7	II
360	4253.6	III
450	4694.1	I
285	4695.4	I
160	4696.2	I
280	4716.2	II
450	4815.5	II
360	4924.1	II
450	4925.3	II
285	4993.5	I
360	5428.6	II
650	5432.8	II
1000	5453.8	II
1000	5473.6	II
1000	5509.7	II
280	5564.9	II
1000	5606.1	II
450	5640.0	II
450	5640.3	II
280	5647.0	II
650	5659.9	II
450	5664.7	II
160	5706.1	I
450	5819.2	II
450	6052.7	I
280	6286.4	II
450	6287.1	II
450	6305.5	II
450	6312.7	II
280	6384.9	II
280	6397.3	II
280	6398.0	II
360	6413.7	II
160	6743.6	I
285	6748.8	I
450	6757.2	I
450	7579.0	I
450	7629.8	I
285	7686.1	I
450	7696.7	I
1000	7924.0	I
160	7928.8	I
285	7930.3	I
450	7931.7	I
450	7967.4	I
450	7967.4	II
450	8314.7	I
450	8314.7	II
450	8585.6	I
285	8680.5	I
450	8694.7	I
360	8874.5	I
110	8882.5	I
220	8884.2	I
160	9035.9	I
450	9212.9	I
450	9228.1	I
450	9237.5	I
285	9413.5	I
285	9421.9	I
285	9437.1	I
650	9649.9	I
450	9672.3	I
450	9680.8	I
450	9693.7	I
285	9697.3	I
285	9739.7	I
285	9932.3	I
285	9949.8	I
285	9958.9	I
285	10455.5	I
285	10459.5	I

Tantalum
Ta $Z = 73$

Intensity	Wavelength/Å	
60	493.07	V
1000	890.87	V
500	947.30	V
67	999.34	IV
79	1116.10	IV
78	1136.17	IV
85	1175.51	IV
80	1189.28	IV
80	1192.67	IV
85	1213.09	IV
500	1213.42	V
85	1215.53	IV
90	1223.73	IV
88	1238.12	IV
95	1240.06	IV
87	1258.34	IV
94	1264.91	IV
98	1272.42	IV
94	1275.48	IV
86	1275.94	IV
92	1308.51	IV
87	1315.58	IV
92	1332.38	IV
86	1343.30	IV
92	1365.88	IV
5000	1392.56	V
91	1398.78	IV
93	1413.40	IV
91	1454.32	IV
92	1464.41	IV
93	1469.82	IV
90	1495.25	IV
95	1514.19	IV
85	1607.70	IV
7000	1709.10	V
85	1712.16	IV
85	1716.13	IV
85	2055.75	IV
1100	2140.13	II
1500	2146.87	II
1200	2182.71	II
1100	2193.88	II
1500	2196.03	II
90	2199.58	IV
1500	2199.67	II
90	2207.64	IV
1400 d	2210.03	II
1400	2239.48	II
1200	2250.76	II
840	2261.42	II
990	2262.30	II
990	2272.59	II
790	2285.25	II
600	2286.59	II
990	2289.16	II
440	2302.24	II
440	2302.93	II
440	2312.60	II
420	2315.46	II
690	2331.98	II
550	2332.19	II
250	2357.30	I
260	2361.09	I
600	2364.24	II
320	2371.58	I
1400	2387.06	II
2400	2400.63	II
320	2416.89	II
360	2427.64	I
360	2429.71	II
480	2432.70	II
380	2470.90	II
600	2474.62	I
500	2484.95	I
600	2488.70	II
500	2490.46	I
600	2504.45	I
600	2507.45	I
1200 d	2526.35	I
600	2532.12	II
1200	2559.43	I
460	2562.10	I
600	2577.37	II
600	2603.49	II
1400	2608.63	I
1200	2635.58	II
860	2636.90	I
2400	2647.47	I
2600	2653.27	I
1900	2656.61	I
1500	2661.34	I
770	2675.90	II
1500	2685.17	II
470	2694.52	II
1000	2698.30	I
1200	2710.13	I
2600	2714.67	I
470	2727.44	II
1200	2748.78	I
860	2749.83	I
410	2752.49	II
1000	2758.31	I
430	2761.68	II
770	2775.88	I
680	2796.34	I
680	2797.76	II
510	2806.58	I
640	2844.25	I
560	2848.52	I
1500	2850.49	I

Intensity	Wavelength/Å	
1900	2850.98	I
360	2861.98	I
470	2871.42	I
380	2880.02	I
770	2891.84	I
560	2902.05	I
310	2915.49	I
410	2925.19	I
310	2932.70	I
1700	2933.55	I
470	2940.06	I
1200	2940.22	I
510	2951.92	I
340	2953.56	I
1500	2963.32	I
770	2965.13	II
770	2965.54	I
340	2969.47	I
430	2975.56	I
1800	3012.54	II
290 d	3027.48	I
530	3049.56	I
530	3069.24	I
360	3077.24	I
560	3103.25	I
380	3124.97	I
380	3130.58	I
270	3132.64	I
320	3170.29	I
270	3173.59	I
600	3180.95	I
300	3223.83	I
1100	3311.16	I
680	3318.84	I
330 d	3330.99	II
640	3371.54	I
360	3385.05	I
450	3406.94	I
490	3480.52	I
380	3497.85	I
490	3511.04	I
750	3607.41	I
980	3626.62	I
500	3642.06	I
210	3918.51	I
210	3970.10	I
210	3996.17	I
410	4061.40	I
310	4067.91	I
300	4205.88	I
360 c	4510.98	I
340	4574.31	I
260	4619.51	I
450	4681.88	I
200	5037.37	I
100	5067.87	I
110	5115.84	I
100	5141.62	I
100	5143.69	I
330	5156.56	I
110	5212.74	I
110 d	5218.45	I
140	5341.05	I
200	5402.51	I
130	5419.19	I
90	5518.91	I
150	5645.91	I
130	5664.90	I
130	5776.77	I
90	5780.71	I
130	5811.10	I
240	5877.36	I
130	5882.30	I
90	5901.91	I
90	5918.95	I
130	5939.76	I
240	5944.02	I
190 c	5997.23	I
100	6020.72	I
250	6045.39	I
100	6047.25	I
100	6101.58	I
65	6144.56	I
130	6154.50	I
150	6256.68	I
150	6268.70	I
150	6309.58	I
75	6325.08	I
65	6341.17	I
75	6356.16	I
65	6360.84	I
90	6389.45	I
65	6428.60	I
250	6430.79	I
200	6450.36	I
380	6485.37	I
65	6505.52	I
100	6514.39	I
100	6516.10	I
100	6574.84	I
110	6611.95	I
75	6621.30	I
100	6673.73	I
180	6675.53	I
75 c	6740.73	I
75	6771.74	I
160 c	6813.25	I
210	6866.23	I
180	6875.27	I
150	6902.10	I
140	6927.38	I
140	6928.54	I
65	6951.26	I
180	6966.13	I
110 d	6995.22	I
75	7006.96	I
150	7148.63	I
110	7172.90	I
140	7301.74	I
160	7346.41	I
140 c	7352.86	I
100	7356.96	I
90 cw	7369.09	I
160	7407.89	I
100	7882.37	I
75	8026.50	I
75	8281.62	I

Technetium
Tc Z = 43

Intensity	Wavelength/Å	
10000 c	3636.07	I
20000 c	4031.63	I
15000	4095.67	I
20000	4262.27	I
30000	4297.06	I
20000	4853.59	I

Tellurium
Te Z = 52

Intensity	Wavelength/Å	
8	802.28	II
8	1059.51	II
8	1077.66	II
10	1161.42	II
10	1174.34	II
12	1175.79	II
9	1208.54	II
9	1220.98	II
9	1253.62	II
9	1270.52	II
10	1324.92	II
9	1363.24	II
8	1366.73	II
10	1374.80	II
10	1608.41	II
10	1613.15	II
5	1655.4	I
5	1688.5	I
6	1700.0	I
5	1708.0	I
10	1822.4	I
26000	2002.02	I
6500	2081.16	I
18000	2142.81	I
3200	2147.25	I
500	2259.02	I
1200	2383.26	I
1500	2385.78	I
50	2438.69	II
120	2530.72	I
100	2649.66	II
80	2661.10	II
110	2677.13	I
100	2858.29	II
150	2895.41	II
70	2967.29	II
70	3047.00	II
100	3175.14	I
60	3256.80	II
60	3329.22	II
150	3406.79	II
50	3442.25	II
50	3521.11	II
50	3552.19	II
100	3611.78	II
50	3617.57	II
50	4006.52	II
70	4127.32	II
100	4169.77	II
80	4225.73	II
100	4261.11	II
60	4273.43	II
80	4285.85	II
150	4364.00	II
75	4385.10	II
170	4478.63	II
80	4537.07	II
100	4557.78	II
70	4630.62	II
100	4641.12	II
180	4654.37	II
200	4686.91	II
100	4696.38	II
100	4706.53	II
100	4766.05	II
100	4784.87	II
100	4827.14	II
150	4831.28	II
150	4842.90	II
130	4865.12	II
200	4866.24	II
8	5083.0	I
50	5449.84	II
50	5487.95	II
150	5576.35	II
150	5649.26	II
100	5666.20	II
200	5708.12	II
150	5755.85	II
100	5974.68	II
50	6367.13	II
10 h	6790.0	I
20 h	6837.6	I
20 h	6854.7	I
15 h	7191.1	I
20 h	7263.5	I
12	7460.98	II
15	7468.75	II
15	7921.69	II
15	7943.14	II
10	7950.34	II
30 h	8061.4	I
10	8122.44	II
20	8186.44	II
15	8273.53	II
15	8672.95	II
10	8733.81	II
205	8758.18	I
81	9004.37	I
5660	9722.74	I
532	9868.92	I
689	9956.30	I
325	9977.13	I
5950	10051.41	I
4097	10091.01	I
381	10118.08	I
397	10300.56	I
745	10493.57	I
1880	10918.34	I
10200	11089.56	I
508	11163.74	I
6620	11487.23	I
1580	13247.75	I
1050	14513.51	I
1480	15452.45	I
2430	15546.23	I
3760	16403.90	I
1960	17303.54	I
2780	18291.59	I
1020	21043.73	I
464	21602.50	I
74	22555.29	I
38	26539.17	I

Terbium
Tb Z = 65

Intensity	Wavelength/Å	
1000	1259.40	IV
1000	1327.67	IV

Intensity	Wavelength/Å	
1000	1373.86	IV
5000	1595.39	IV
2000	1633.19	IV
2000	2027.79	IV
1000	2089.98	IV
1000	2332.54	IV
110	2584.61	II
110	2608.57	II
130	2628.69	II
140	2669.29	II
190	2704.07	II
270	2769.53	II
320	2897.44	II
250	2956.21	II
230	3010.59	II
230	3016.18	II
460	3053.55	II
460	3070.05	II
670	3078.86	II
480	3082.36	II
480	3089.58	II
480	3102.96	II
440	3139.64	II
480	3187.26	II
480	3199.56	II
1100	3218.93	II
1200	3219.98	II
480	3252.32	II
760	3280.31	II
760	3281.40	II
1000	3285.04	II
1500	3293.07	II
3800	3324.40	II
760	3349.42	II
760	3364.93	II
810	3454.06	II
810 d	3472.79	II
810	3500.84	II
5700	3509.17	II
1300	3523.66	II
1100	3540.24	II
810	3543.89	II
3200	3561.74	II
810	3567.35	II
4200	3568.52	II
1600	3568.98	II
1100	3579.20	II
710	3585.03	II
810	3596.38	II
1600	3600.44	II
810	3625.54	II
2300	3650.40	II
810	3654.88	II
2000	3658.88	II
3800	3676.35	II
810	3682.26	II
450	3693.50	I
450	3700.12	I
4700	3702.86	II
2400	3703.92	II
1000 d	3711.76	II
650	3745.04	I
870	3747.17	II
870	3747.34	II
1100	3755.24	II
650	3759.35	I
1700	3765.14	I
2100	3776.49	II
600	3783.53	I
410	3789.92	I
760 d	3806.85	II
1500	3830.26	I
540	3833.42	I
920 d	3842.50	II
3700	3848.73	II
3500 w	3874.17	II
480	3888.22	I
490	3894.64	I
2400	3899.20	II
1600	3901.33	I
480	3908.06	I
650	3915.43	I
760	3925.45	II
810 d	3939.52	II
2200 d	3976.84	II
1800	3981.87	II
970	4002.59	II
1900	4005.47	II
760	4012.75	II
870	4032.28	I
2100	4033.03	II
430	4054.12	I
410	4060.37	I
1300	4061.58	I
650	4105.37	I
1100	4144.41	II
350	4158.53	I
390	4196.74	I
650	4203.74	I
600	4206.49	I
480	4215.09	I
480	4232.82	I
650	4266.34	I
760 cw	4278.52	II
450	4310.42	I
2200	4318.83	I
600	4322.23	I
600	4325.83	II
3000	4326.43	I
600	4332.12	I
870	4336.43	I
600	4337.64	I
1700	4338.41	I
700	4340.62	I
870	4356.81	I
330	4382.45	I
300	4388.23	I
260	4390.91	I
350	4423.10	I
240	4436.12	I
240	4448.04	I
430	4493.07	I
75	4514.31	II
110	4549.07	I
110	4550.45	I
110	4556.46	I
110	4563.69	II
210	4578.69	II
65	4584.84	II
65	4591.56	II
75 d	4626.32	II
95	4626.94	II
65	4632.07	I
65 h	4636.59	I
85	4641.00	II
210	4641.98	II
260 cw	4645.31	II
80	4647.23	I
80	4662.79	I
80	4676.90	I
70 c	4681.87	I
80	4688.63	II
80	4693.11	II
200	4702.41	II
110	4707.94	II
80	4739.93	I
70	4747.80	I
410 cw	4752.53	II
180	4786.78	I
100	4813.77	I
80	4875.57	II
80	4881.15	II
95	4915.90	I
65	4931.79	I
85	4993.82	II
110	5078.25	I
75	5089.12	II
85	5186.13	I
120	5228.12	I
75	5248.71	I
75 w	5262.11	II
75	5281.05	I
65	5304.72	I
110	5319.23	I
65 w	5337.90	I
160	5354.88	I
75	5369.72	I
75	5375.98	I
50	5424.10	II
55	5459.81	I
55	5509.61	I
50	5514.54	I
65	5524.12	I
85 c	5747.58	I
75	5795.64	I
75	5803.13	II
65	5815.36	I
65	5851.07	I
65	5870.62	I
65 c	5920.78	I
75	5967.34	II
35	6331.68	II
35 cw	6518.68	I
35	6581.82	I
90	6677.94	II
40 cw	6702.61	I
130	6794.58	II
55	6896.37	II
45 h	6899.95	I
40	6901.98	I
65	7204.28	I
40	7257.73	I
45	7348.88	II
45	7496.12	I
27 h	7582.03	II
45	7590.24	I
65	7596.44	I
30	7627.81	I
30	7737.63	I
30	7855.79	II
27	7927.90	II
30	8025.42	II
30	8085.06	II
65	8194.82	II
95	8212.57	I
40	8450.06	II
30 h	8511.80	I
45	8583.45	II
30	8603.40	I
65	8765.74	II

Thallium

Tl $Z = 81$

Intensity	Wavelength/Å	
10	570.49	IV
5 r	670.87	II
15 r	696.30	II
5 r	709.23	II
10 r	817.18	II
5 r	836.34	II
8 r	1018.85	II
30	1028.69	IV
20	1034.73	IV
20	1036.61	IV
10 r	1049.73	II
8 r	1050.30	II
5 r	1074.97	II
30	1079.68	IV
10 r	1130.17	II
15 r	1162.55	II
10 r	1167.43	II
10 r	1183.41	II
12 r	1194.84	II
5 r	1246.00	II
10	1266.33	III
15 r	1307.50	II
8 r	1310.20	II
25 r	1321.71	II
8 r	1330.40	II
10 r	1373.52	II
10	1477.14	III
8 r	1489.65	I
10 r	1499.30	II
10 r	1507.82	II
15 r	1561.58	II
10 r	1568.57	II
7 r	1593.26	II
5 h	1616.	I
5	1685.40	I
10 r	1792.76	II
12 r	1814.85	II
25 r	1908.64	II
100 r	2007.56	I
100 r	2210.71	I
30	2298.04	II
140	2315.90	I
900 h	2379.69	I
20	2530.86	II
700	2580.14	I
420	2709.23	I
4400 d	2767.87	I
10	2849.80	II
2800	2918.32	I
20	3091.56	II
15	3185.51	II
15	3186.56	II

Intensity	Wavelength/Å		Intensity	Wavelength/Å		Intensity	Wavelength/Å		Intensity	Wavelength/Å	
15	3187.74	II	510	3119.526	II	310	3747.539	I	30	7045.795	II
1200	3229.75	I	510	3122.963	II	650	3752.569	II	30	7084.171	I
15	3291.01	II	480	3125.507	II	180	3770.056	I	30	7168.896	I
15	3369.15	II	100	3136.216	I	590	3803.075	I	40	7191.132	II
9	3456.34	III	420	3139.306	II	450	3828.384	I	35	7208.006	I
20000	3519.24	I	420	3142.835	II	840	3839.746	II	50	7525.508	II
5000	3529.43	I	420	3175.726	II	450	3863.405	II	30	7647.380	I
8	3540.08	II	1100	3180.193	II	210	3875.374	I	30	8330.451	I
9	3560.68	II	770	3188.233	II	340	3895.419	I	40	8967.641	I
12000 w	3775.72	I	560	3221.292	II	590	3929.669	II	20	9833.42	I
10	3832.30	II	560	3229.009	II	200	3932.911	I	20	10726.93	I
10	3887.15	II	480	3235.84	II	390	3967.392	I	20	10942.24	II
7	4109.85	III	590	3238.116	II	200	3972.155	I	30	11230.259	I
6	4269.81	III	910	3256.274	II	150	3980.089	I	20	11984.67	II
20	4274.98	II	180	3257.366	I	530	3994.549	II	20	17208.22	II
40	4306.80	II	910	3262.668	II	220	4008.210	I	15	18811.88	I
20	4737.05	II	620	3287.789	II	220	4009.056	I	10	22264.35	II
15	4981.35	II	910	3291.739	II	280	4012.495	I	**Thulium**		
25	5078.54	II	620	3292.520	II	4200	4019.129	II	**Tm**	***Z* = 69**	
25	5152.14	II	240	3301.650	I	250	4030.842	I	5000	2185.94	III
18000	5350.46	I	480	3304.238	I	250	4036.047	I	360	2284.79	II
15 d	5384.85	II	510	3321.450	II	250	4063.407	I	20000	2296.21	III
10	5410.97	II	840	3325.120	II	700	4086.520	II	5000	2305.03	III
25	5949.48	II	250	3330.476	I	700	4094.747	II	20000	2311.16	III
10	6179.98	II	620	3334.604	II	150	4100.341	I	5000	2312.72	III
10	6378.32	II	620	3337.870	II	840	4108.421	II	5000	2326.19	III
16 h	6549.84	I	310	3348.768	I	240	4112.754	I	6000	2328.50	III
10	6966.5	II	980	3351.228	II	280	4115.758	I	6000	2329.29	III
10	7815.80	I	620	3358.602	II	1100	4116.713	II	3000	2331.80	III
20	8373.6	I	250	3374.974	I	200	4127.411	I	3000	2357.05	III
10	8474.27	I	1300	3392.035	II	200	4134.067	I	4000	2406.63	III
10	8664.1	II	200	3396.727	I	450	4149.986	II	450	2409.02	II
20	9130.	II	250	3398.544	I	620	4178.060	II	450	2426.17	II
20	9130.5	I	200	3405.558	I	620	4208.890	II	770	2480.13	II
40	9509.4	I	250	3413.012	I	110	4253.538	I	30000	2489.44	III
20	9930.4	I	390	3421.210	I	110	4260.333	I	2000	2504.71	III
30	10011.9	I	270	3423.989	I	480	4277.313	II	1300	2509.08	II
40	10488.80	I	980	3433.998	II	700	4282.042	II	3000	2519.78	III
1000	11512.82	I	770	3435.976	II	130	4337.277	I	130	2527.02	I
150	12736.4	I	1300	3469.920	II	1300	4381.860	II	10000	2552.46	III
700	13013.2	I	170	3471.218	I	1100	4391.110	II	360	2552.76	I
Thorium			200	3486.552	I	110	4498.940	I	540	2561.65	II
Th	***Z* = 90**		670	3539.587	II	280	4510.527	II	430	2588.27	II
150	1707.37	IV	180	3544.018	I	90	4723.438	I	170 h	2596.49	I
200	1959.02	IV	170	3549.595	I	50	4840.843	I	810	2607.06	II
200	2002.34	IV	200	3555.013	I	280	4863.163	II	730	2624.33	II
200	2413.50	III	530	3559.451	II	260	5017.255	II	5000	2682.32	III
200	2427.94	III	200	3576.557	I	110	5067.974	I	2000	2707.03	III
200	2431.68	III	270	3592.780	I	120	5148.211	II	3000	2719.47	III
200	2441.24	III	270	3598.120	I	95	5216.596	II	540	2721.19	II
500	2565.593	II	980	3609.445	II	110	5231.160	I	3000	2724.44	III
480	2692.415	II	200	3612.427	I	95	5247.654	II	4000	2727.56	III
520	2747.156	II	480	3615.133	II	60	5343.581	I	680	2794.60	II
410	2752.166	II	270	3635.943	I	60	5587.026	I	730	2797.27	II
800	2832.315	II	210	3642.248	I	95	5707.103	II	2000	2806.77	III
1200	2837.295	II	170	3649.735	I	70	5760.551	I	580	2827.92	II
100	2848.084	I	220	3663.202	I	85	5989.044	II	200	2854.17	I
550	2870.406	II	280	3669.968	I	60	6169.822	I	1600	2869.23	II
100	2936.086	I	700	3675.567	II	50	6182.622	I	1000	2947.72	III
100	2943.729	I	150	3682.486	I	50	6274.116	II	490	2973.22	I
420	3049.092	II	170	3692.566	I	50	6274.117	II	1000	2998.28	III
450	3067.729	II	180	3698.105	I	50	6355.911	II	1500	3015.30	II
670	3078.828	II	340	3706.767	I	60	6457.283	I	360	3081.12	I
480	3080.217	II	590	3719.435	I	50	6462.614	I	7400	3131.26	II
510	3108.296	II	770	3721.825	II	50 h	6531.342	I	2300	3133.89	II
100	3116.263	I	1300	3741.183	II	55	6989.656	I	1900	3151.04	II

Intensity	Wavelength/Å	
1500	3157.34	II
450	3172.65	I
2300	3172.83	II
1200	3236.81	II
1600	3240.23	II
2300	3241.54	II
320	3246.96	I
1900	3258.05	II
1600	3266.64	II
1200	3267.40	II
1100	3276.81	II
1200	3283.40	II
1200	3285.61	II
2300	3291.00	II
2000	3302.46	II
1200	3309.80	II
230	3349.99	I
4000	3362.61	II
1700	3397.50	II
850	3410.05	I
340	3412.59	I
340	3416.59	I
6400	3425.08	II
340	3429.33	I
4900	3441.50	II
4900	3453.66	II
8500	3462.20	II
210	3467.51	I
340	3476.69	I
340	3480.98	I
420	3487.38	I
340	3499.95	I
250	3517.60	I
1700	3535.52	II
420	3537.91	I
210	3555.82	I
340	3560.92	I
420	3563.88	I
1300	3566.47	II
420	3567.36	I
280	3586.07	I
2100	3608.77	II
1000	3629.09	III
380	3638.41	I
1100	3668.09	II
4800	3700.26	II
3800	3701.36	II
7700	3717.91	I
2400	3734.12	II
5000	3744.06	I
1700	3751.81	I
6000	3761.33	II
4800	3761.91	II
7100	3795.75	II
770	3798.54	I
600	3807.72	I
290	3826.39	I
1300	3838.20	II
290	3840.87	I
8900	3848.02	II
6800	3883.13	I
1800	3883.44	II
5400	3887.35	I
440	3896.62	I
3500	3916.48	I
1500	3949.27	I
1500	3958.10	II
1800	3996.52	II
220	4024.23	I
380	4044.47	I
10000	4094.19	I
9500	4105.84	I
1100	4138.33	I
8800	4187.62	I
6000	4203.73	I
380	4222.67	I
3000	4242.15	II
270	4271.71	I
150	4298.36	I
2700	4359.93	I
1400	4386.43	I
200	4394.42	I
140	4396.50	I
120	4454.03	I
540	4481.26	II
150	4519.60	I
260	4522.57	II
110	4548.60	I
270	4599.02	I
300	4615.94	II
80	4626.33	II
95	4626.56	II
110	4634.26	II
120	4655.09	I
160	4681.92	I
120	4691.11	I
110	4724.26	I
680	4733.34	I
70	4759.90	I
80	4831.20	II
140	4957.18	I
160	5009.77	II
160	5034.22	II
150	5060.90	I
95	5113.97	I
80	5213.38	I
650	5307.12	I
80	5346.49	II
270	5631.41	I
520	5675.84	I
40	5684.76	II
35	5709.97	II
190	5764.29	I
35	5838.76	II
240	5895.63	I
140	5971.26	I
200	6460.26	I
95	6604.96	I
110	6779.77	I
120	6844.26	I
80	6845.76	I
10	6937.37	I
10	7017.90	I
12	7034.34	I
10	7106.14	I
17	7272.62	I
14	7310.51	I
14	7432.18	I
75	7481.08	I
75	7490.20	I
140	7558.33	I
80	7731.53	I
40	7856.08	I
55	7927.51	I
110	7930.84	I
95	8017.90	I
27	8472.01	II

Tin
Sn Z = 50

Intensity	Wavelength/Å	
7	169.47	II
150	361.01	V
100	753.01	III
200	910.92	III
500	956.25	IV
7	985.13	II
500	1019.72	IV
1000	1044.49	IV
1000	1073.41	IV
200	1089.35	V
8	1108.19	II
1000	1119.34	IV
1000	1139.29	III
1000	1158.33	III
200	1160.74	V
10	1161.43	II
1000	1184.25	III
2000	1210.52	III
9	1219.07	II
13	1223.70	II
11	1243.00	II
2000	1251.38	V
1000	1259.92	III
20	1290.86	II
200	1294.36	V
1000	1305.97	III
1000	1314.55	IV
20	1316.59	II
1000	1327.34	III
1000	1347.65	III
1000	1386.74	III
25	1400.52	II
1000	1437.52	IV
20	1475.15	II
9	1489.22	II
1000	1570.36	III
10 r	1737.21	I
15 r	1751.46	I
20 r	1764.98	I
30 r	1790.75	I
80 r	1804.60	I
15	1811.34	II
500	1811.71	III
40 r	1815.74	I
120 r	1823.00	I
9	1831.89	II
50 r	1848.75	I
200 r	1860.32	I
80	1886.05	I
100	1891.40	I
12	1899.91	II
50	1909.30	I
80	1925.31	I
500	1941.86	III
150	1952.15	I
50 h	1977.6	I
80	1984.20	I
50	2040.66	I
50	2054.03	I
70	2058.31	I
80	2068.58	I
100	2072.89	I
100	2073.08	I
200	2096.39	I
100	2100.93	I
100 r	2113.93	I
50	2121.26	I
40 r	2148.73	I
20 r	2151.43	I
30	2151.54	II
80	2171.32	I
150 r	2194.49	I
300 r	2199.34	I
400 r	2209.65	I
80 r	2231.72	I
400 r	2246.05	I
60	2251.17	I
400 r	2268.91	I
200 r	2286.68	I
600 r	2317.23	I
300 r	2334.80	I
1000 r	2354.84	I
22	2368.33	II
100	2408.15	I
800 r	2421.70	I
1000 r	2429.49	I
15	2448.98	II
300	2483.39	I
13	2483.48	II
10	2486.99	II
200	2495.70	I
400	2546.55	I
500 r	2571.58	I
200	2594.42	I
200 r	2661.24	I
700 r	2706.51	I
150	2779.81	I
1400 r	2839.99	I
1000 r	2863.32	I
700 r	3009.14	I
850 r	3034.12	I
12	3047.50	II
550 r	3175.05	I
550 r	3262.34	I
50	3283.21	II
110	3330.62	I
60	3351.97	II
10	3472.46	II
11	3575.45	II
280 r	3801.02	I
10	5332.36	II
20	5561.95	II
25	5588.92	II
500	5631.71	I
15	5799.18	II
50	5925.44	I
100	5970.30	I
150	6037.70	I
250	6069.00	I
100	6073.46	I
400	6149.71	I
200	6154.60	I
150	6171.50	I
100	6310.78	I
70	6453.50	II

Intensity	Wavelength/Å	
25	6844.05	II
20	7191.40	II
10	7387.79	II
13	7741.80	II
100	7754.97	I
100 h	8030.5	I
200	8114.09	I
80	8357.04	I
300	8422.72	I
400	8552.60	I
50 h	8681.7	I
50 h	9410.86	I
80 h	9415.37	I
150	9616.40	I
50	9741.1	I
100 h	9742.8	I
300 h	9805.38	I
500	9850.52	I
54	10894.00	I
70	11191.85	I
56	11277.66	I
200	11454.59	I
200	11616.26	I
258	11739.78	I
96	11825.18	I
106	11835.82	I
254	11932.99	I
48	12009.50	I
111	12313.24	I
42	12530.87	I
42	12536.5	I
89	12888.5	I
187	12981.7	I
187	13018.5	I
68	13081.5	I
378	13460.2	I
144	13608.2	I
40	20861.7	I
4	24738.2	I

Titanium
Ti Z = 22

Intensity	Wavelength/Å	
17	252.96	V
15	498.26	V
14	502.08	V
13	526.57	V
18	779.07	IV
20	1298.66	III
20	1298.97	III
23	1455.19	III
20	1467.34	IV
11	1717.40	V
10	1841.49	V
20	2067.56	IV
18	2103.16	IV
180	2273.28	I
190	2279.96	I
190	2305.67	I
22	2413.99	III
25	2516.05	III
360	2525.60	II
24	2527.84	III
210	2529.85	I
190	2531.25	II
190	2534.62	II
130	2535.87	II
23	2540.06	III
24	2563.44	III
23	2565.42	III
22	2567.56	III
270	2599.92	I
340	2605.15	I
510	2611.28	I
300	2619.94	I
640	2641.10	I
800	2644.26	I
950	2646.64	I
250	2742.32	I
250	2802.50	I
190	2841.94	II
180	2877.44	II
280	2884.11	II
450	2912.08	I
340	2928.34	I
1100	2942.00	I
1300	2948.26	I
1600	2956.13	I
22	2984.75	III
1300 d	3066.22	II
1100	3072.97	II
1600	3075.22	II
2300	3078.64	II
3600	3088.02	II
720	3119.72	I
500	3161.20	II
780	3161.77	II
1000	3162.57	II
1600	3168.52	II
2400	3186.45	I
1000	3190.87	II
3100	3191.99	I
3800	3199.92	I
780	3202.54	II
1100	3217.06	II
1300	3222.84	II
6600	3234.52	II
5200	3236.57	II
4100	3239.04	II
2600	3241.99	II
1200	3248.60	II
1200	3252.91	II
1200	3254.25	II
1200	3261.60	II
840	3314.42	I
2900	3322.94	II
2100	3329.46	II
1800	3335.20	II
1100	3340.34	II
5700	3341.88	I
4300	3349.04	II
12000	3349.41	II
4100	3354.64	I
7200	3361.21	II
1100	3370.44	I
4300	3371.45	I
5700	3372.80	II
2900 d	3377.48	I
1400	3380.28	II
5700	3383.76	II
1400	3385.95	I
1400	3387.84	II
1100	3394.58	II
890	3444.31	II
600	3461.50	II
600	3477.18	II
480	3491.05	II
890	3504.89	II
600	3510.84	II
17	3576.44	IV
600	3610.16	I
4800	3635.46	I
6600	3642.68	I
7200	3653.50	I
600	3671.67	I
3100	3685.20	II
600	3689.91	I
2900	3729.82	I
3300	3741.06	I
330	3741.64	II
5200	3752.86	I
3300	3759.30	II
2900	3761.32	II
840	3786.04	I
500	3882.89	I
530	3900.54	II
2600	3904.78	I
500	3913.46	II
500	3914.34	I
15	3915.47	III
1100	3924.53	I
890	3929.88	I
1100	3947.78	I
4500	3948.67	I
4500	3956.34	I
5200	3958.21	I
950	3962.85	I
950	3964.27	I
4800	3981.76	I
570	3982.48	I
5700	3989.76	I
7800	3998.64	I
950	4008.93	I
1200	4024.57	I
840	4078.47	I
890	4286.01	I
840	4287.40	I
950	4289.07	I
840	4290.94	I
840	4295.76	I
2000	4298.66	I
200	4300.05	II
2900	4300.56	I
4100	4301.09	I
6000	4305.92	I
1200	4314.80	I
330	4395.04	II
890	4427.10	I
230	4443.80	II
840	4449.15	I
550	4450.90	I
840	4453.32	I
950	4455.33	I
1100	4457.43	I
240	4468.50	II
530	4481.26	I
780	4512.74	I
1000	4518.03	I
1000	4522.80	I
780	4527.31	I
6000	4533.24	I
240	4533.97	II
3600	4534.78	I
2400	4535.58	I
1200	4535.92	I
1200	4536.05	I
720	4544.69	I
950	4548.77	I
240	4549.63	II
15	4549.84	III
950	4552.46	I
720	4555.49	I
240	4571.98	II
15 d	4572.20	III
950	4617.27	I
480	4623.09	I
720	4656.47	I
840	4667.59	I
950	4681.92	I
470	4840.87	I
400	4885.08	I
380	4899.91	I
5800	4981.73	I
4600	4991.07	I
4000	4999.51	I
3600	5007.21	I
3200 d	5014.19	I
840	5020.03	I
840	5022.87	I
1200	5035.91	I
840	5036.47	I
740	5038.40	I
1200	5039.95	I
1400	5064.66	I
1100	5173.75	I
1300	5192.98	I
1400	5210.39	I
17	5278.12	III
20	5398.93	IV
340	5512.53	I
270	5514.35	I
320	5514.54	I
250	5644.14	I
130	5675.44	I
95	5689.47	I
95	5715.13	I
85	5739.51	I
400	5866.46	I
230	5899.32	I
120	5918.55	I
150	5922.12	I
120	5941.76	I
300	5953.17	I
200	5965.84	I
270	5978.56	I
340	5999.04	I
110	6064.63	I
120	6085.23	I
120	6091.17	I
120	6126.22	I
17	6246.65	IV
380	6258.10	I
380	6258.70	I
300	6261.10	I
55	6546.28	I
65	6554.23	I

Intensity	Wavelength/Å	
75	6556.07	I
18	6621.58	III
18	6667.99	III
80	6743.12	I
20	7072.64	III
18	7084.57	III
260	7209.44	I
130	7244.86	I
130	7251.72	I
120	7344.72	I
90	7357.74	I
60	7364.11	I
60	7978.88	I
55	8024.84	I
75	8364.24	I
100	8377.85	I
100	8382.54	I
75	8396.87	I
120	8412.36	I
170	8426.52	I
490	8434.94	I
240	8435.70	I
20	8466.87	III
90	8675.39	I

Tungsten
W Z = 74

Intensity	Wavelength/Å	
5800	2001.71	II
13000	2008.07	II
5100	2009.98	II
4100	2010.23	II
4100	2014.23	II
7300	2026.08	II
15000	2029.98	II
5300	2049.63	II
9700	2079.11	II
6100	2094.75	II
2100	2118.87	II
2400	2121.59	II
1500	2166.32	II
1300	2204.48	II
460	2249.80	I
510	2277.58	I
530 d	2294.49	I
340	2309.02	I
440	2313.17	I
460	2321.63	I
390 d	2326.56	I
320	2354.61	I
580	2360.44	I
850	2363.07	I
510	2374.47	I
670	2384.82	I
730	2397.09	II
560	2397.73	I
560	2397.98	I
1700 d	2405.58	I
610	2413.68	I
870	2424.21	I
1800	2435.96	I
580	2444.06	I
780	2451.48	II
870	2452.00	I
630	2454.98	I
780	2455.51	I
780	2456.53	I
1100	2459.30	I
1400	2466.85	I
480	2472.51	I
1200	2474.15	I
870	2480.13	I
1500	2481.44	I
480 d	2482.10	I
580	2484.74	I
390	2487.50	I
390	2489.23	II
630	2495.26	I
680	2504.70	I
75	2510.47	II
310	2520.46	I
780	2521.32	I
270	2522.04	II
780	2523.41	I
430	2527.76	I
780	2533.64	I
1200	2547.14	I
780	2550.38	I
2700	2551.35	I
730	2561.97	I
870	2580.49	I
390	2584.39	I
390	2589.17	II
370	2601.96	I
680	2606.39	I
370	2608.32	I
970	2613.08	I
480	2613.82	I
400	2620.25	I
400	2622.21	I
400	2625.22	I
400	2632.48	I
400	2632.70	I
810	2633.13	I
400 d	2638.62	I
650	2646.18	I
400	2646.73	I
1600	2656.54	I
810	2662.84	I
810	2671.47	I
650	2677.28	I
2100	2681.42	I
650	2695.67	I
650	2699.59	I
400	2700.01	I
400	2706.58	I
400	2708.59	I
400 d	2708.80	I
400	2715.50	I
2100	2718.91	I
2600	2724.35	I
400	2725.03	I
650	2748.84	I
400	2762.34	I
400	2764.27	II
400	2769.74	I
810	2770.88	I
810	2774.00	I
810	2774.48	I
810	2792.70	I
400	2799.93	I
810	2818.06	I
1600	2831.38	I
810	2833.63	I
810	2848.02	I
1500	2896.44	I
690	2935.00	I
2400	2944.40	I
2400	2946.99	I
730 d	2979.71	I
360	3013.79	I
520	3016.47	I
770	3017.44	I
210	3024.93	I
310 d	3026.67	I
440 d	3041.73	I
270	3043.80	I
440	3046.44	I
810	3049.69	I
180	3073.28	I
180 d	3084.83	I
370	3093.50	I
240	3107.23	I
240	3108.02	I
230	3117.57	I
260	3120.18	I
290	3163.42	I
320	3176.60	I
190	3181.82	I
390	3191.57	I
390	3198.84	I
520	3207.25	I
1000	3215.56	I
190	3232.49	I
210	3254.36	I
210	3259.66	I
210 d	3266.62	I
730	3300.82	I
440	3311.38	I
440	3326.20	I
440	3331.69	I
390	3373.75	I
230	3429.59	I
240	3443.00	I
400	3495.24	I
650	3545.22	I
240	3570.65	I
1900	3617.52	I
650	3682.08	I
400	3683.30	I
570	3688.06	I
810	3707.92	I
510	3757.92	I
680	3760.13	I
1000	3768.45	I
340	3773.71	I
1000	3780.77	I
290	3809.22	I
190	3810.38	I
260	3810.79	I
1400	3817.48	I
1100	3835.06	I
730	3846.22	I
1800	3867.99	I
730	3881.41	I
8600	4008.75	I
540	4015.22	I
910	4045.59	I
730	4069.95	I
5000	4074.36	I
1000	4102.70	I
540	4137.46	I
450	4171.17	I
220	4207.05	I
250	4219.37	I
540	4244.36	I
1400	4269.38	I
4100	4294.61	I
2200	4302.11	I
200	4378.48	I
180	4384.85	I
200	4408.28	I
640	4484.19	I
170	4588.73	I
640	4659.87	I
640	4680.51	I
790	4843.81	I
380	4886.90	I
220	4982.59	I
820	5053.28	I
770	5224.66	I
220	5514.68	I
65	5648.37	I
55	5735.09	I
45	5804.85	I
40	5902.64	I
55	5947.57	I
55	5965.86	I
55	6012.78	I
40	6021.52	I
45	6292.02	I
35	6404.21	I
40	6445.12	I
17	6611.62	I
13	6678.42	I
15	6693.08	I
13	6984.27	I
15	7140.52	I
9	7162.64	I
11	7200.16	I
10	7278.24	I
15	7285.81	I
15	7296.55	I
10	7509.00	I
17	7569.92	I
17	7614.15	I
13	7688.97	I
11	7784.15	I
22	8017.19	I
22	8055.64	I
13	8123.82	I
10	8338.08	I
27	8585.11	I
10	8594.42	I
13	8865.53	I

Uranium
U Z = 92

Intensity	Wavelength/Å	
440	2565.41	II
610	2635.53	II
830	2793.94	II
870	2802.56	II
630	2807.05	II
630	2817.96	II
870	2821.12	II
680	2828.90	II
920	2832.06	II

Intensity	Wavelength/Å	
970	2865.68	II
1200	2889.62	II
780	2906.80	II
780	2908.28	II
580	2931.41	II
530 p	2940.37	II
1300	2941.92	II
830	2943.90	II
580	2956.06	II
580	2967.94	II
580	2971.06	II
530	2984.61	II
630	3022.21	II
630	3031.99	II
580	3050.20	II
630	3057.91	II
630	3062.54	II
580	3072.78	II
580	3093.01	II
580	3102.39	II
970	3111.62	II
530	3119.35	II
680	3124.95	II
530	3139.61	II
680	3149.24	II
530	3153.11	II
730	3229.50	II
680	3232.16	II
730	3291.33	II
1100	3305.89	II
730	3390.38	I
580	3424.56	II
580	3435.49	I
630	3466.30	I
680	3482.49	II
1600	3489.37	I
530	3496.41	II
630	3500.08	I
780	3507.34	I
1600	3514.61	I
630	3533.57	II
530	3540.47	II
1200	3550.82	II
680	3555.32	I
1200	3561.80	I
2300	3566.59	I
530	3569.08	I
630	3578.72	II
3200	3584.88	I
840	3638.20	I
2800	3670.07	II
1100	3701.52	II
600	3738.04	II
680	3746.42	II
950	3748.68	II
600	3751.17	I
1900	3782.84	II
570	3793.10	II
1900	3811.99	I
750	3826.51	II
2000	3831.46	II
1200	3839.63	I
2400	3854.64	II
4900	3859.57	II
1900	3865.92	II
1500	3871.03	I
1000	3881.45	II
2200	3890.36	II
2000	3932.02	II
1200	3943.82	I
1200	3985.79	II
1000	4042.75	I
1600	4050.04	II
880	4062.54	II
2200	4090.13	II
810	4116.10	II
880	4153.97	I
1400	4171.59	II
1000	4241.67	II
600	4472.33	II
620	4543.63	II
170	4689.07	II
150	4756.81	I
110	5008.21	II
170	5027.38	I
80	5160.32	II
70	5280.38	I
80	5475.70	II
70	5480.26	II
70	5481.20	II
160	5492.95	II
70	5780.59	I
70	5798.53	II
230	5915.39	I
100	5976.32	I
90	6077.29	I
55	6372.46	I
90	6395.42	I
110	6449.16	I
90	6826.92	I
45	7533.93	I
50	7881.94	I
35	8445.39	I
75	8607.95	I
30	8757.76	I
100	10554.93	I
75	11167.84	I
100	11384.13	I
100	11859.42	I
100	11908.83	I
100	12250.46	I
100	13185.16	I
75	13306.23	I
100	13961.58	I
75	18634.43	I
75	21910.22	I

Vanadium

V $Z = 23$

Intensity	Wavelength/Å	
20	225.46	V
20	251.66	V
20	286.84	V
35	483.01	V
50	633.94	III
200	677.34	IV
500	684.37	IV
400	737.85	IV
100	864.27	III
500	1006.46	III
500	1149.94	III
100	1426.65	IV
1000	1643.03	III
1000	1650.14	III
100	1680.20	V
1000	1694.78	III
1000	1760.07	III
1000	1788.26	III
1000	1794.60	III
1000	1812.19	III
300	1861.56	IV
500	1939.06	IV
400	1951.43	IV
500	1997.72	IV
2100	2092.44	I
500	2268.30	IV
1000	2292.86	III
2500	2330.42	III
2500	2371.06	III
1000	2382.46	III
240	2507.78	I
410	2526.22	I
210	2527.90	II
80 h	2570.72	IV
230	2574.02	I
250	2593.05	III
250	2595.10	III
80 h	2645.54	IV
180	2661.42	I
1100	2687.96	II
680	2700.94	II
530	2706.17	II
640	2715.69	II
180	2731.35	I
240	2864.36	I
900	2891.64	II
900	2892.66	II
1400	2893.32	II
900	2906.46	II
2400	2908.82	II
710	2923.62	I
2400	2924.02	II
1700	2924.64	II
900	2941.37	II
1100	2944.57	II
410	2962.77	I
600	2968.38	II
1200	3056.33	I
1400	3060.46	I
2400	3066.38	I
3800	3093.11	II
3000	3102.30	II
2600	3110.71	II
2000	3118.38	II
1500	3125.28	II
3200	3183.41	I
5300	3183.98	I
3800	3185.40	I
410	3187.71	II
530	3188.51	II
750	3190.68	II
1100	3267.70	II
900	3271.12	II
750	3276.12	II
80 h	3514.25	IV
560	3517.30	II
560	3533.68	I
560	3545.20	II
560	3556.80	II
560	3589.76	II
490	3592.02	II
560	3592.53	I
100	3679.86	III
1300	3688.07	I
1000	3690.28	I
1500	3692.22	I
1000	3695.86	I
3800	3703.58	I
1800	3704.70	I
320	3715.47	II
250	3727.34	II
280	3732.76	II
520	3790.32	I
1100	3794.96	I
570	3799.91	I
570	3803.47	I
1000	3813.49	I
1300	3818.24	I
1700	3828.56	I
2600	3840.75	I
1200	3855.37	I
3000	3855.84	I
1300	3864.86	I
1500	3875.08	I
700	3890.18	I
2400	3902.25	I
700	3909.89	I
540	3990.57	I
430	3998.73	I
170	4005.71	II
1100	4090.58	I
1800	4092.69	I
890	4095.49	I
2800	4099.80	I
590	4102.16	I
2800	4105.17	I
2300	4109.79	I
8900	4111.78	I
4300	4115.18	I
1800	4116.47	I
2000	4123.57	I
3100	4128.07	I
3100	4132.02	I
2300	4134.49	I
20	4200.32	V
360	4232.46	I
560	4268.64	I
460	4271.55	I
460	4276.96	I
430	4284.06	I
460	4330.02	I
510	4332.82	I
760	4341.01	I
1000	4352.87	I
12000	4379.24	I
7000	4384.72	I
4800	4389.97	I
3600	4395.23	I
1400	4400.58	I
2300	4406.64	I
2800	4407.64	I
3600	4408.20	I
4600	4408.51	I
640	4416.47	I
640	4421.57	I
640	4437.84	I

Intensity	Wavelength/Å	
830	4441.68	I
640	4444.21	I
610	4452.01	I
1000	4459.76	I
2000	4460.29	I
610	4462.36	I
510	4577.17	I
640	4580.40	I
830	4586.36	I
1300	4594.11	I
230	4619.77	I
100	4635.18	I
130	4646.40	I
160	4670.49	I
130	4776.36	I
110	4786.51	I
130	4796.92	I
130	4807.53	I
130	4827.45	I
150	4831.64	I
120	4832.43	I
320	4851.48	I
480	4864.74	I
620	4875.48	I
740	4881.56	I
110	5128.53	I
110	5138.42	I
110	5192.99	I
110	5194.83	I
110	5234.07	I
110	5240.87	I
100	5401.93	I
140	5415.26	I
140	5584.50	I
100	5592.42	I
200	5624.60	I
400	5627.64	I
110	5657.44	I
110	5668.36	I
310	5670.85	I
1200	5698.52	I
920	5703.56	I
570	5706.98	I
850	5727.03	I
230	5731.25	I
230	5737.06	I
450	6039.73	I
480	6081.44	I
1300	6090.22	I
600	6119.52	I
450	6199.19	I
450	6216.37	I
430	6230.74	I
710	6243.10	I
280	6251.82	I
130	6268.82	I
170	6274.65	I
200	6285.16	I
200	6292.83	I
170	6296.49	I
110	6531.43	I
65 c	6753.00	I
50 c	6766.49	I
40	6784.98	I
40	7338.92	I
35	7356.54	I
29 c	8027.39	I
120 w	8116.80	I
70 c	8161.07	I
60 c	8919.85	I
Xenon		
Xe Z = 54		
8	657.8	III
8	660.1	III
9	673.8	III
9	674.0	III
9	676.6	III
10	694.0	III
20	698.5	III
12	705.1	III
10	721.2	III
15	731.0	III
10	733.3	III
350	740.41	II
15	742.6	III
10	756.0	III
10	761.5	III
10	769.1	III
25	779.1	III
15	792.9	III
12	796.1	III
15	802.0	III
350	803.07	II
25	823.2	III
30	824.9	III
25	853.0	III
600	880.80	II
350	885.54	II
15	889.3	III
20	894.0	III
20	896.0	III
600	925.87	II
250	935.40	II
10	965.5	III
800	972.77	II
700	976.68	II
35	1003.4	III
35	1017.7	III
500	1032.44	II
700	1037.68	II
1100	1041.31	II
10	1047.8	III
1000	1048.27	II
1200	1051.92	II
12	1066.4	III
2000	1074.48	II
600	1083.86	II
1200	1100.43	II
30	1130.3	III
600	1158.47	II
250	1169.63	II
800 p	1183.05	II
250	1192.04	I
25	1232.1	III
600	1244.76	II
250	1250.20	I
1000	1295.59	I
600	1469.61	I
80	2668.98	III
100	2717.33	III
30	2814.45	III
40	2815.91	III
30	2827.45	III
40	2847.65	III
30	2862.40	III
200	2864.73	II
80 w	2871.10	III
60 w	2871.24	III
30	2871.7	III
150 h	2895.22	II
30	2896.62	III
50	2906.6	III
40	2911.89	III
80 w	2912.36	III
40	2940.2	III
60	2945.2	III
40	2947.5	III
40	2948.1	III
80 w	2970.47	III
400	2979.32	II
40	2992.87	III
30	3004.25	III
100 h	3017.43	II
100	3023.81	III
40	3083.5	III
50	3091.1	III
30	3106.46	III
300	3128.87	II
100 w	3138.3	III
80 c	3150.82	III
40	3185.2	III
100	3242.86	III
80	3268.98	III
30	3287.82	III
80 w	3301.55	III
40	3331.6	III
30	3358.0	III
200 h	3366.72	II
80	3384.12	III
2	3400.07	I
2	3418.37	I
2	3420.00	I
3	3442.66	I
60	3444.2	III
70	3454.2	III
100 w	3458.7	III
100 h	3461.26	II
40	3468.22	III
4	3469.81	I
4	3472.36	I
5	3506.74	I
80	3522.83	III
50	3542.3	III
10	3549.86	I
50	3552.1	III
10	3554.04	I
40	3561.4	III
100	3579.7	III
80	3583.6	III
100 w	3595.4	III
100	3606.06	III
40	3607.0	III
15	3610.32	I
8	3613.06	I
100 w	3615.9	III
40	3623.1	III
600	3624.08	III
6	3633.06	I
10	3669.91	I
50	3676.67	III
40	3685.90	I
40	3693.49	I
40	3776.3	III
300	3781.02	III
100	3841.5	III
200	3877.8	III
60	3880.5	III
100 l	3907.91	II
500	3922.55	III
300	3950.59	III
100	4037.59	II
200	4050.07	III
200 l	4057.46	II
60	4060.4	III
100 h	4098.89	II
100	4109.1	III
100	4145.7	III
200 l	4158.04	II
1000 h	4180.10	II
500 h	4193.15	II
300 h	4208.48	II
100 h	4209.47	II
300 h	4213.72	II
100	4215.60	II
300 h	4223.00	II
400 h	4238.25	II
500 h	4245.38	II
100 l	4251.57	II
30	4285.9	III
500 h	4296.40	II
500 h	4310.51	II
1000 l	4330.52	II
200 h	4369.20	II
100 l	4373.78	II
500 h	4393.20	II
500 l	4395.77	II
200 l	4406.88	II
150 l	4416.07	II
50	4434.2	III
500 h	4448.13	II
100 w	4462.1	III
1000 h	4462.19	II
500 l	4480.86	II
100 l	4521.86	II
100 w	4569.1	III
100 w	4570.1	III
100 w	4641.4	III
30	4673.7	III
60	4683.57	III
30	4723.60	III
600	4734.152	I
100 w	4757.3	III
150	4792.619	I
500	4807.02	I
400	4829.71	I
300	4843.29	I
40	4869.5	III
500	4916.51	I
500	4923.152	I
200 l	4971.71	II
400	4972.71	II
300	4988.77	II
100 l	4991.17	II
200	5028.280	I

Intensity	Wavelength/Å	
200	5044.92	II
1000	5080.62	II
300	5122.42	II
100	5125.70	II
100	5178.82	II
300	5188.04	II
400	5191.37	II
100	5192.10	II
60	5239.0	III
500	5260.44	II
500	5261.95	II
2000	5292.22	II
300	5309.27	II
1000	5313.87	II
2000	5339.33	II
200	5363.20	II
30	5367.1	III
200	5368.07	II
500	5372.39	II
100	5392.80	I
50	5401.0	III
3000	5419.15	II
800	5438.96	II
300	5445.45	II
200	5450.45	II
400	5460.39	II
1000	5472.61	II
100 l	5494.86	II
40	5524.4	III
200	5525.53	II
600	5531.07	II
100	5566.62	I
300	5616.67	II
300	5659.38	II
600	5667.56	II
150	5670.91	II
100	5695.75	I
200	5699.61	II
200	5716.10	II
500	5726.91	II
500	5751.03	II
300	5758.65	II
300	5776.39	II
100	5815.96	II
300	5823.89	I
150	5824.80	I
100	5875.02	I
300	5893.29	II
100	5894.99	I
200	5905.13	II
100	5934.17	I
500	5945.53	II
300	5971.13	II
2000	5976.46	II
200	6008.92	II
1000	6036.20	II
2000	6051.15	II
600	6093.50	II
1500	6097.59	II
400	6101.43	II
100	6115.08	II
100	6146.45	II
150	6178.30	I
120	6179.66	I
300	6182.42	I
500	6194.07	II
100	6198.26	I
60	6205.97	III
100	6220.02	II
25	6221.7	III
60	6238.2	III
60	6259.05	III
500	6270.82	II
400	6277.54	II
100	6284.41	II
100	6286.01	I
250	6300.86	II
500	6318.06	I
400	6343.96	II
600	6356.35	II
200	6375.28	II
100	6397.99	II
300	6469.70	I
150	6472.84	I
120	6487.76	I
100	6498.72	I
200 h	6504.18	I
300	6512.83	II
200	6528.65	II
100	6533.16	I
1000	6595.01	II
100	6595.56	I
400	6597.25	II
100	6598.84	II
150	6668.92	I
300	6694.32	II
200	6728.01	I
150	6788.71	II
100	6790.37	II
1000	6805.74	II
200	6827.32	I
100	6872.11	I
300	6882.16	I
80	6910.22	II
100	6925.53	I
800 h	6942.11	II
100	6976.18	I
2000	6990.88	II
150	7082.15	II
500	7119.60	I
50 s	7147.50	II
200	7149.03	II
500	7164.83	II
100	7284.34	II
200	7301.80	II
200	7339.30	II
100	7386.00	I
150	7393.79	I
300	7548.45	II
200	7584.68	I
80	7618.57	II
500	7642.02	I
100	7643.91	I
200	7670.66	II
60	7787.04	II
100	7802.65	I
100	7881.32	I
300	7887.40	I
500	7967.34	I
100	8029.67	I
200	8057.26	I
150	8061.34	I
100	8101.98	I
150 h	8151.80	II
100	8171.02	I
700	8206.34	I
10000	8231.635	I
500	8266.52	I
7000	8280.116	I
2000	8346.82	I
100	8347.24	II
2000	8409.19	I
50 h	8515.19	II
200	8576.01	I
50 h	8604.23	II
250	8648.54	I
100	8692.20	I
200	8696.86	I
50 h	8716.19	II
300	8739.39	I
100	8758.20	I
5000	8819.41	I
300	8862.32	I
200	8908.73	I
200	8930.83	I
1000	8952.25	I
100	8981.05	I
200	8987.57	I
400	9045.45	I
500	9162.65	I
100	9167.52	I
100	9374.76	I
200	9513.38	I
50 h	9591.35	II
150	9685.32	I
50 l	9698.68	II
100	9718.16	I
2000	9799.70	I
3000	9923.19	I
100	10838.37	I
90	11742.01	I
375	12235.24	I
100	12257.76	I
300	12590.20	I
2500	12623.391	I
250	13544.15	I
2000	13657.055	I
1250	14142.444	I
800	14240.96	I
375	14364.99	I
140	14660.81	I
3000	14732.806	I
100	15099.72	I
2500	15418.394	I
150	15557.13	I
250	15979.54	I
100	16039.90	I
1000	16053.28	I
125	16554.49	I
1500	16728.15	I
1500	17325.77	I
350	18788.13	I
150	20187.19	I
3000	20262.242	I
250	21470.09	I
1250	23193.33	I
110	23279.54	I
1800	24824.71	I
175	25145.84	I
2000	26269.08	I
2500	26510.86	I
250	28381.54	I
750	28582.25	I
300	29384.41	I
150	29448.06	I
100	29649.58	I
100	29813.62	I
600	30253.14	I
1500	30475.46	I
100	30504.12	I
500	30794.18	I
6000	31069.23	I
125	31336.01	I
550	31607.91	I
100	32293.08	I
1800	32739.26	I
3500	33666.69	I
150	34014.67	I
450	34335.27	I
170	34744.00	I
5000	35070.25	I
110	35246.92	I
250	36209.21	I
150	36231.74	I
450	36508.36	I
850	36788.83	I
140	38685.98	I
175	38737.82	I
270	38939.60	I
120	39955.14	I

Ytterbium
Yb Z = 70

Intensity	Wavelength/Å	
1000	1050.24	IV
1000	1054.46	IV
5000	1134.43	IV
900	1316.04	IV
800	1326.36	IV
900	1350.26	IV
80	1561.42	III
80 h	1765.21	III
800	1791.06	IV
100	1863.32	III
800	1873.91	III
500	1898.25	III
500	1998.82	III
900	2116.65	IV
2500	2116.67	II
800	2123.32	IV
3000	2126.74	II
800	2139.99	IV
20000	2144.77	IV
15000	2154.18	IV
370	2161.60	II
850	2185.71	II
640	2224.46	II
300	2240.11	III
300	2305.32	III
140	2320.81	II
170	2390.74	II
460	2464.50	I
140	2512.06	II
270	2538.67	II
2000	2567.61	III
1000	2579.57	III

Intensity	Wavelength/Å	
800	2599.14	III
600	2621.11	III
1000	2642.56	III
1000	2651.74	III
700	2652.25	III
990	2653.75	II
200	2665.04	II
2000	2666.13	III
2000	2666.99	III
390	2671.96	I
390	2672.66	II
170	2718.35	II
230	2748.66	II
1300	2750.48	II
170	2776.28	II
600	2795.60	III
1000	2803.43	III
600	2816.92	III
1000	2818.72	III
140	2821.15	II
190	2830.99	II
230 h	2847.18	II
360	2851.13	II
430	2859.80	II
140	2861.21	II
200	2867.06	II
45	2873.49	I
200	2888.04	II
3600	2891.38	II
600	2898.30	III
1000	2906.31	III
170	2914.21	II
140	2915.28	II
280	2919.35	II
35	2934.36	I
140	2945.91	II
2000	2970.56	II
200	2983.99	II
170	2994.80	II
800	2998.00	III
310	3005.77	II
160	3017.56	II
160	3026.67	II
2000	3029.49	III
920	3031.11	II
3000	3092.50	III
28	3100.74	I
170	3107.90	II
190	3117.81	II
4000	3126.01	III
1000	3138.58	III
230	3140.94	II
28	3162.29	I
800	3191.35	III
390	3192.88	II
240	3201.16	II
2000	3228.58	III
35	3239.58	I
18000	3289.37	II
130	3305.25	I
140	3305.73	II
80	3319.41	I
2000	3325.51	III
240	3337.17	II
280 d	3342.93	II
240	3375.48	II
2000	3384.01	III
140	3387.50	I
50	3412.45	I
140	3418.39	I
360	3426.04	I
240	3431.11	I
85	3452.40	I
500	3454.08	II
190 d	3458.29	II
360	3460.27	I
2400	3464.37	I
500	3476.30	II
500	3478.84	II
50	3517.00	I
230	3520.29	II
35	3559.03	I
200	3560.33	II
170	3560.70	II
360	3585.47	II
200	3619.80	II
240	3637.76	II
70	3648.15	I
90	3655.73	I
240	3669.69	II
140	3675.08	II
32000	3694.19	II
70	3700.58	I
400	3711.91	III
180	3734.69	I
550	3770.10	I
80	3774.32	I
60 h	3791.74	I
170	3839.91	I
340	3872.85	I
340	3900.85	I
140	3911.27	I
500	3931.23	III
32000	3987.99	I
930	3990.88	I
50	4007.36	I
2000	4028.14	III
70	4052.28	I
440	4089.68	I
470	4149.07	I
120	4174.56	I
340	4180.81	II
300	4213.64	III
150 d	4218.56	II
120	4231.97	I
70	4277.74	I
120	4305.97	I
60 h	4393.69	I
60 h	4430.21	I
440	4439.19	I
85 h	4482.42	I
100	4517.58	III
85 h	4563.95	I
640	4576.21	I
200	4582.36	I
70	4589.21	I
140	4590.83	I
40	4684.27	I
190	4726.08	II
170 h	4781.87	I
170	4786.61	II
35	4816.43	I
40	4837.46	I
40 h	4894.60	I
27	4912.36	I
710	4935.50	I
140	4966.90	I
30	5067.80	I
70	5069.14	I
220	5074.34	I
50	5076.74	I
60	5196.08	I
85	5211.60	I
100	5244.11	I
150 h	5277.04	I
170	5335.15	II
30 h	5351.29	I
150	5352.95	II
30	5363.66	I
40	5449.27	II
60	5481.92	I
40	5505.49	I
17	5524.54	I
85 h	5539.05	I
2400	5556.47	I
60	5651.98	II
220	5719.99	I
27	5771.66	II
35	5833.99	II
35	5837.14	II
27	5854.51	I
17	5989.33	I
40	5991.51	II
60	6152.57	II
60	6274.78	II
200	6328.52	III
35 h	6400.35	I
35 h	6417.91	I
340	6489.06	I
180	6667.82	I
25	6727.61	II
690	6799.60	I
9 h	7244.41	I
8 h	7305.22	I
10 h	7313.05	I
16 h	7350.04	I
25	7448.28	I
30 h	7527.46	I
750	7699.48	I
100	7971.46	III
70 h	8922.56	II
200	10110.60	III
100	10830.36	III

Yttrium

Y Z = 39

Intensity	Wavelength/Å	
150	264.64	IV
150	273.03	IV
900	333.09	V
500	333.80	V
400	335.14	V
500	336.62	V
500	339.02	V
500	344.59	V
900	355.86	IV
300	370.42	IV
300	372.05	V
400	379.96	V
500	386.82	IV
300	403.45	V
300	420.74	V
600	425.03	IV
300	473.10	IV
4000	584.98	V
2000	630.97	V
5000	805.20	III
7000	809.92	III
15000	989.21	III
25000	996.37	III
5000	1314.51	III
4000	1334.04	III
4000	2068.98	III
10000	2127.98	III
16000	2191.16	III
350	2243.06	II
10000	2284.34	III
10000	2327.31	III
50	2354.20	I
50000	2367.23	III
40000	2414.64	III
560	2422.20	II
60	2694.21	I
95	2723.00	I
70	2742.53	I
140	2760.10	I
90000	2817.04	III
45	2822.56	I
70	2854.43	II
95	2886.48	I
160	2919.05	I
99000	2946.01	III
390	2948.40	I
350	2964.96	I
480	2974.59	I
750	2984.26	I
140	2996.94	I
130	3021.73	I
190	3045.37	I
95	3095.88	II
110	3173.06	II
220	3179.41	II
70	3191.31	I
2300	3195.62	II
2200	3200.27	II
2200	3203.32	II
3900	3216.69	II
6200	3242.28	II
4700	3327.89	II
85	3388.59	I
85	3412.47	I
170	3485.73	I
1700	3496.09	II
3900	3549.01	II
130	3551.80	I
540	3552.69	I
170	3558.76	I
190	3571.43	I
260	3576.05	I
3300	3584.52	II
300	3587.75	I
100	3589.69	I
2800	3592.92	I
10000	3600.73	II
6200	3601.92	II
7800	3611.05	II

Intensity	Wavelength/Å	
4300	3620.94	I
1900	3628.71	II
7800	3633.12	II
3000	3664.61	II
170	3692.53	I
13000	3710.30	II
1200	3747.55	II
10000	3774.33	II
1400	3776.56	II
7400	3788.70	II
1300	3818.35	II
4000	3832.88	II
80	3876.82	I
480	3878.28	II
4400	3950.36	II
3600	3982.60	II
940	4039.83	I
2400	4047.64	I
9400	4077.38	I
2000	4083.71	I
9900	4102.38	I
8900	4128.31	I
7500	4142.85	I
100 h	4157.63	I
2400	4167.52	I
2000	4174.14	I
8000	4177.54	II
160	4217.80	I
280 h	4220.63	I
600	4235.73	II
2200	4235.94	I
300	4251.20	I
360 h	4302.30	I
2800	4309.63	II
110	4330.78	I
440 h	4348.79	I
120	4357.73	I
800	4358.73	II
120	4366.03	I
12000	4374.94	II
150 h	4375.61	I
100	4387.74	I
1800	4398.02	II
890	4422.59	II
100	4443.66	I
130	4446.63	I
170	4475.72	I
180	4476.96	I
160	4477.45	I
110	4487.28	I
300	4487.47	I
500	4505.95	I
890	4527.25	I
440	4527.80	I
100	4544.32	I
100	4559.37	I
130	4596.55	I
95	4604.80	I
2000	4643.70	I
200 h	4658.32	I
2000	4674.84	I
180	4696.81	I
170	4728.53	I
160	4752.79	I
410	4760.98	I
120	4781.04	I
170	4786.89	I
180	4799.30	I
140	4819.64	I
120	4822.13	I
770	4839.87	I
550	4845.68	I
410	4852.69	I
120	4854.25	I
890	4854.87	II
330	4859.84	I
1900	4883.69	II
95	4893.44	I
1100	4900.12	II
100	4906.11	I
150	4921.87	I
120	4974.30	I
100	5006.97	I
75	5070.21	I
75	5072.19	I
1100	5087.42	II
180	5135.20	I
960	5200.41	II
1500	5205.72	II
10000	5238.10	III
180	5240.81	I
75	5380.62	I
220	5402.78	II
90	5424.37	I
190	5438.24	I
710	5466.46	I
100	5468.47	I
240	5497.41	II
300	5503.45	I
250	5509.90	II
120	5521.63	I
740	5527.54	I
120	5544.50	I
180	5577.42	I
620	5581.87	I
120	5606.33	I
560	5630.13	I
120	5644.69	I
120	5648.47	I
740	5662.94	II
90	5675.27	I
160	5706.73	I
90	5743.85	I
75	5765.64	I
100	5781.69	II
120	6009.19	I
120	6023.41	I
120	6135.04	I
150	6138.43	I
1200	6191.73	I
300	6222.59	I
1000	6435.00	I
90	6538.60	I
70	6557.39	I
95	6613.75	II
40	6650.61	I
150	6687.58	I
70	6700.71	I
190	6793.71	I
21	6815.16	I
45	6845.24	I
29	6887.22	I
24 h	6950.31	I
24	6979.88	I
29	7052.94	I
35	7191.66	I
35	7264.17	II
50	7346.46	I
29	7450.30	II
9000	7558.71	III
35	7563.13	I
29	7855.52	I
110	7881.90	II
10000	7991.43	III
24	8344.43	I
10000	8796.21	III
95	8800.62	I
19 h	8835.85	II

Zinc

Zn Z = 30

Intensity	Wavelength/Å	
200	425.90	IV
200	428.54	IV
200	430.59	IV
1000	677.63	III
750	677.96	III
200	713.90	III
60	1193.23	II
50	1239.12	IV
50	1249.69	IV
500	1265.74	IV
500	1306.66	IV
200	1456.72	III
200	1459.98	IV
300	1499.42	III
300	1500.42	III
300	1505.92	III
300	1515.85	III
300	1552.30	III
90	1572.99	II
200	1629.19	III
200	1639.33	III
200	1673.05	III
80 d	1735.61	II
100	1767.69	III
100	1797.64	II
100 d	1811.05	II
100 d	1833.57	II
100	1864.12	II
100	1866.08	II
100	1872.13	II
100 d	1918.96	II
100 d	1929.67	II
100	1969.40	II
100	1982.11	II
100	1986.99	II
500	2025.48	II
500	2062.00	II
200	2064.23	II
120	2079.08	I
300	2099.94	II
200	2102.18	II
800 r	2138.56	I
1000	2501.99	II
150	2515.81	I
1000	2557.95	II
300	2582.49	I
200	2608.56	I
300	2608.64	I
200	2670.53	I
300	2684.16	I
300	2712.49	I
200	2756.45	I
300	2770.86	I
300	2770.98	I
400	2800.87	I
100	2801.06	I
200	3035.78	I
200	3072.06	I
300	3196.31	II
500 r	3282.33	I
800	3302.58	I
700 r	3302.94	I
800	3345.02	I
500	3345.57	I
50	3883.34	I
300	4680.14	I
400	4722.15	I
400	4810.53	I
800	4911.62	II
500	4924.03	II
200	5181.98	I
500	5894.33	II
500	6021.18	II
500	6102.49	II
500	6214.61	II
1000 h	6362.34	I
300	7588.5	II
300	7732.5	II
100	11054.25	I
100	13053.63	I
100	13150.59	I
100	14038.70	I
20	16483.45	I
20	16491.98	I
20	16505.23	I
10	24375.02	I

Zirconium

Zr Z = 40

Intensity	Wavelength/Å	
500	304.01	V
60	480.66	IV
60	497.23	IV
60	500.22	IV
600	628.66	IV
500	633.56	IV
50	690.39	III
2000	740.61	V
10000	800.00	V
10000	806.89	V
10000	812.05	V
3000	841.40	V
300	863.65	IV
500	864.59	IV
9000	1183.97	IV
9000	1201.77	IV
10000	1219.86	IV
500	1303.93	V
500 p	1323.81	V
1000	1469.47	IV
10000	1546.17	IV
10000	1598.95	IV
5000	1607.95	IV
100	1612.38	III
700	1725.02	V
200	1790.19	III

Intensity	Wavelength/Å	
150	1793.56	III
125	1798.13	III
600	1860.86	V
200	1940.25	III
600	2028.54	V
125	2070.43	III
200	2086.78	III
10000	2091.49	IV
10000	2092.36	IV
600	2132.42	V
10000	2163.68	IV
100	2175.80	III
100	2191.15	III
10000	2286.67	IV
100	2301.60	III
90	2539.65	I
570	2567.64	II
1600	2568.87	II
2100	2571.39	II
250	2620.56	III
200	2643.79	III
150	2664.26	III
1800	2678.63	II
90	2687.75	I
750	2700.13	II
1300	2722.61	II
800	2726.49	II
1400	2734.86	II
1100	2742.56	II
660	2745.86	II
660	2752.21	II
530	2758.81	II
620	2814.90	I
390	2818.74	II
530	2825.56	II
710	2837.23	I
660	2844.58	II
350	2848.52	I
350	2851.97	II
340	2869.81	II
490	2875.98	I
300	2915.99	II
270	2918.24	II
320	2926.99	II
320	2948.94	II
320	2955.78	II
320	2960.87	I
320	2962.68	II
320	2968.96	II
320	2978.05	II
820	2985.39	I
320	3003.74	II
820	3011.75	I
350	3020.47	II
500	3028.04	II
880	3029.52	I
350 d	3036.39	II
690	3054.84	II
690	3106.58	II
350	3120.74	I
500	3129.18	II
500	3129.76	II
350	3132.07	I
690	3138.68	II
540	3164.31	II
880	3165.97	II
880	3182.86	II
540	3191.21	I
540	3212.01	I
760	3214.19	II
630	3231.69	II
630	3234.12	I
760	3241.05	II
1000	3273.05	II
1300	3279.26	II
880	3284.71	II
540	3305.15	II
880	3306.28	II
380	3322.99	II
380	3326.80	II
380	3334.25	II
760	3340.56	II
380	3344.79	II
760	3356.09	II
540	3357.26	II
380	3374.73	II
570	3387.87	II
760	3388.30	II
5700	3391.98	II
570	3393.12	II
570	3404.83	II
760	3410.25	II
380	3414.66	I
1000	3430.53	II
4700	3438.23	II
600	3447.36	I
410	3457.56	II
820	3463.02	II
600	3471.19	I
1200	3479.39	II
1300	3481.15	II
4100	3496.21	II
820	3505.67	II
1000	3509.32	I
2000	3519.60	I
440	3525.81	II
440	3533.22	I
630	3542.62	II
1800	3547.68	I
630	3550.46	I
1800	3551.95	II
2100	3556.60	II
1100	3566.10	I
2100	3572.47	II
1100	3575.79	I
1300	3576.85	II
880	3586.29	I
3500	3601.19	I
690	3611.89	II
1100	3613.10	II
1100	3614.77	II
1100	3623.86	I
1100	3663.65	I
390	3671.27	II
800	3674.72	II
390	3697.46	II
960	3698.17	II
720	3709.26	II
560	3745.98	II
880	3751.60	II
480	3764.39	I
480	3766.72	I
340	3766.82	II
720	3780.54	I
560	3791.40	I
560	3822.41	I
2200	3835.96	I
1300	3836.76	II
550	3843.02	II
550	3847.01	I
550	3849.25	I
2900	3863.87	I
770	3864.34	I
990	3877.60	I
1500	3885.42	I
2900	3890.32	I
2000	3891.38	I
610	3921.79	I
1200	3929.53	I
940	3958.22	II
490	3966.66	I
990	3968.26	I
660	3973.50	I
770	3991.13	II
770	3998.97	II
400	4023.98	I
770	4024.92	I
990	4027.20	I
400	4029.68	II
490	4030.04	I
400	4035.89	I
610	4043.58	I
490	4044.56	I
400	4045.61	II
610	4048.67	II
770	4055.03	I
600	4055.71	I
1500	4064.16	I
2000	4072.70	I
240	4078.31	I
2000	4081.22	I
400	4121.46	I
1200	4149.20	II
400	4161.21	II
400	4166.36	I
660	4187.56	I
400	4194.76	I
610	4199.09	I
610	4201.46	I
610	4208.98	II
400	4213.86	I
2000	4227.76	I
2000	4239.31	I
770	4240.34	I
770	4241.20	I
1200	4241.69	I
550	4282.20	I
550	4294.79	I
550	4341.13	I
1000	4347.89	I
290	4359.74	II
310	4360.81	I
350	4366.45	I
550	4507.12	I
610	4535.75	I
490	4542.22	I
490	4575.52	I
350	4602.57	I
700	4633.98	I
2300	4687.80	I
510	4688.45	I
1900	4710.08	I
1400	4739.48	I
870	4772.31	I
700	4815.63	I
250	5046.58	I
360	5064.91	I
470	5078.25	I
300	5155.45	I
200	5158.00	I
100	5191.60	II
270	5385.14	I
160	5664.51	I
160	5797.74	I
340	5879.80	I
170	6045.85	I
170	6121.91	I
680	6127.44	I
340	6134.55	I
440	6143.20	I
300	6313.02	I
150	6953.84	I
150	6990.84	I
540	7097.70	I
280	7102.91	I
170	7103.72	I
590	7169.09	I
160	7944.61	I
160	8005.27	I
150	8063.09	I
790	8070.08	I
390	8132.99	I
280	8212.53	I

SOURCES OF DATA FOR EACH ELEMENT

Numbers following the element name refer to the references on the following pages.

Actinium: 193
Aluminum: 6,8,81,89,127,144,146,227,228,282
Americium: 92
Antimony: 164,167,194,386,406
Argon: 190,203,204,219,367,368,372,373,374,375,414,421
Arsenic: 163,168,197,244,280
Astatine: 188
Barium: 1,78,111,252,259,277,279
Berkelium: 53,339
Beryllium: 15,44,73,102,115,134,135,171,175,198,335
Bismuth: 1,357,358,359,360,361
Boron: 66,69,74,94,104,171,221,222
Bromine: 42,122,124,139,142,240.243,246,248,249,250,316
Cadmium: 44,285,296,353,399
Calcium: 16,25,70,150,270
Californium: 52,331
Carbon: 22,66,211
Cerium: 1,136,166,261,305
Cesium: 78,82,154,155,200,201,259,263,325
Chlorine: 11,28,30,31,85,233,238,239
Chromium: 1,379,380,412
Cobalt: 1,100,125,159,236,276,291
Copper: 199,273,290,295,324
Curium: 51,332
Dysprosium: 1
Einsteinium: 333
Erbium: 1,301
Europium: 1,312
Fluorine: 68,169,224,225,226
Francium: 408
Gadolinium: 1,46,137,151,152
Gallium: 2,19,62,132,140,141,143,195,281
Germanium: 5,119,293,340,341,342
Gold: 38,72,234,393,395
Hafnium: 1,369,404,410,425
Helium: 16,94,173,183,317
Holmium: 1
Hydrogen: 214
Indium: 1,132,348,349,350,351,352,353,435,436
Iodine: 20,21,58,84,124,153,161,176,184
Iridium: 1
Iron: 56,63,71,101,105,138,174,278,381,382
Krypton: 61,121,123,147,208,232,366,390,409,417,421
Lanthanum: 1,78,79,220,309
Lead: 54,64,106,256,274,297,283,329,330
Lithium: 3,15,17,18,37,44,112,284,321,335
Lutetium: 1,148,310,401
Magnesium: 4,7,49,83,103,128,129,177,217,269,315,335
Manganese: 1,126,385,405,433
Mercury (198): 43,50,69,145,229,242
Mercury (Natural): 34,45,90,117,133,189,235,304,327,328,343
Molybdenum: 1,383,420
Neodymium: 1
Neon: 56,58,69,118,150,230,364,365,371,388,389,400,402,413,430
Neptunium: 93
Nickel: 1,294,415,416,422
Niobium: 1,392,407,431
Nitrogen: 66,107,108,212,213,318
Osmium: 1
Oxygen: 23,24,36,66,69,209,210,215
Palladium: 1,287,424
Phosphorus: 179,180,182,336
Platinum: 1,288
Plutonium: 91
Polonium: 47,48
Potassium: 32,59,60,75,76,86,150,160,172,268,314,322
Praseodymium: 1,149,306,308,337,338
Promethium: 196,260
Protactinium: 96
Radium: 253,254
Radon: 251
Rhenium: 1
Rhodium: 1,396
Rubidium: 12,109,130,241,257,258,262,264
Ruthenium: 1,423
Samarium: 1
Scandium: 1,88,150,298,323
Selenium: 9,80,181,216,245,247,275
Silicon: 87,170,237,292,319,320
Silver: 13,99,255,286,289,363,387,398
Sodium: 178,205,206,207,268,299,334
Strontium: 1,109,110,218,231,265,279,313
Sulfur: 29,144,202,209,210,266
Tantalum: 1,411,426
Technetium: 35
Tellurium: 1,344,345,346,347
Terbium: 1,302
Thallium: 1,195,348,354,355,356
Thorium: 1,97,98,156,157,165,434
Thulium: 1,307
Tin: 187,191,399,423
Titanium: 1,378,427,428
Tungsten: 1
Uranium: 1,303
Vanadium: 1,394,397,432
Xenon: 33,116,118,120,232,384,391,429
Ytterbium: 1,40,192,311
Yttrium: 1,77,265,419
Zinc: 39,55,113,131,185,186,370,376,377
Zirconium: 1,362,403,418

REFERENCES

1. Meggers, W. F., Corliss, C. H., and Scribner, B. F., *Natl. Bur. Stand. (U.S.) Monogr.*, 145, Washington, D.C., 1975.
2. Aksenov, V. P. and Ryabtsev, A. N., *Opt. Spectrosc.*, 37, 860, 1970.
3. Andersen, N., Bickel, W. S., Carriveau, G. W., Jensen, K., and Veje, E., *Phys. Scr.*, 4, 113, 1971.
4. Andersson, E. and Johannesson, G. A., *Phys. Scr.*, 3, 203, 1971.
5. Andrew, K. L. and Meissner, K. W., *J. Opt. Soc. Am.*, 49, 146, 1959.
6. Artru, M. C. and Brillet, W. U. L., *J. Opt. Soc. Am.*, 64, 1063, 1974.
7. Artru, M. C. and Kaufman, V., *J. Opt. Soc. Am.*, 62, 949, 1972.
8. Artru, M. C. and Kaufman, V., *J. Opt. Soc. Am.*, 65, 594, 1975.
9. Badami, J. S. and Rao, K. R., *Proc. R. Soc. London*, 140(A), 387, 1933.
10. Baird, K. M. and Smith, D. S., *J. Opt. Soc. Am.*, 48, 300, 1958.
11. Bashkin, S. and Martinson, I., *J. Opt. Soc. Am.*, 61, 1686, 1971.
12. Beacham, J. R., Ph.D. thesis, Purdue University, 1970.
13. Benschop, H., Joshi, Y. N., and van Kleef, T. A. M., *Can. J. Phys.*, 53, 700, 1975.
14. Berry, H. G., Bromander, J., and Buchta, R., *Phys. Scr.*, 1, 181, 1970.
15. Berry, H. G., Bromander, J., Martinson, I., and Buchta, R., *Phys. Scr.*, 3, 63, 1971.
16. Berry, H. G., Desesquelles, J., and Dufay, M., *Phys. Rev. Sect. A.*, 6, 600, 1972.
17. Berry, H. G., Desesquelles, J., and Dufay, M., *Nucl. Instrum. Methods*, 110, 43, 1973.
18. Berry, H. G., Pinnington, E. H., and Subtil, J. L., *J. Opt. Soc. Am.*, 62, 767, 1972.
19. Bidelman, W. P. and Corliss, C. H., *Astrophys. J.*, 135, 968, 1962.
20. Bloch, L. and Bloch, E., *Ann. Phys.* (Paris), 10(11), 141, 1929.
21. Bloch, L., Bloch, E., and Felici, N., *J. Phys. Radium*, 8, 355, 1937.
22. Bockasten, K., *Ark. Fys.*, 9, 457, 1955.
23. Bockasten, K., Hallin, R., Johansson, K. B., and Tsui, P., *Phys. Lett.* (Netherlands), 8, 181, 1964.
24. Bockasten, K. and Johansson, K. B., *Ark. Fys.*, 38, 563, 1969.
25. Borgstrom, A., *Ark. Fys.*, 38, 243, 1968.
26. Borgstrom, A., *Phys. Scr.*, 3, 157, 1971.
27. Bowen, I. S., *Phys. Rev.*, 29, 231, 1927.
28. Bowen, I. S., *Phys. Rev.*, 31, 34, 1928.
29. Bowen, I. S., *Phys. Rev.*, 39, 8, 1932.
30. Bowen, I. S., *Phys. Rev.*, 45, 401, 1934.
31. Bowen, I. S., *Phys. Rev.*, 46, 377, 1934.
32. Bowen, I. S., *Phys. Rev.*, 46, 791, 1934.
33. Boyce, J. C., *Phys. Rev.*, 49, 730, 1936.
34. Boyce, J. C. and Robinson, H. A., *J. Opt. Soc. Am.*, 26, 133, 1936.
35. Bozman, W. R., Meggers, W. F., and Corliss, C. H., *J. Res. Natl. Bur. Stand. Sect. A*, 71, 547, 1967.
36. Bromander, J., *Ark. Fys.*, 40, 257, 1969.
37. Bromander, J. and Buchta, R., *Phys. Scr.*, 1, 184, 1970.
38. Brown, C. M. and Ginter, M. L., *J. Opt. Soc. Am.*, 68, 243, 1978.
39. Brown, C. M., Tilford, S. G., and Ginter, M. L., *J. Opt. Soc. Am.*, 65, 1404, 1975.
40. Bryant, B. W., *Johns Hopkins Spectroscopic Report* No. 21, 1961.
41. Buchet, J. P., Buchet-Poulizac, M. C., Berry, H. G., and Drake, G. W. F., *Phys. Rev. Sect. A*, 7, 922, 1973.
42. Budhiraja, C. J. and Joshi, Y. N., *Can. J. Phys.*, 49, 391, 1971.
43. Burns, K. and Adams, K. B., *J. Opt. Soc. Am.*, 42, 56, 1952.
44. Burns, K. and Adams, K. B., *J. Opt. Soc. Am.*, 46, 94, 1956.
45. Burns, K., Adams, K. B., and Longwell, J., *J. Opt. Soc. Am.*, 40, 339, 1950.
46. Callahan, W. R., Ph.D. thesis, Johns Hopkins University, 1962.
47. Charles, G. W., *J. Opt. Soc. Am.*, 56, 1292, 1966.
48. Charles, G. W., Hunt, D. J., Pish, G., and Timma, D. L., *J. Opt. Soc. Am.*, 45, 869, 1955.
49. Codling, K., *Proc. Phys. Soc.*, 77, 797, 1961.
50. Comite Consulatif Pour La Definition du Metre, *J. Phys. Chem. Ref. Data*, 3, 852, 1974.
51. Conway, J. G., Blaise, J., and Verges, J., *Spectrochim. Acta Part B*, 31, 31, 1976.
52. Conway, J. G., Worden, E. F., Blaise, J., and Verges, J., *Spectrochim. Acta Part B*, 32, 97, 1977.
53. Conway, J. G., Worden, E. F., Blaise, J., Camus, P., and Verges, J., *Spectrochim. Acta Part B*, 32, 101, 1977.
54. Crooker, A. M., *Can. J. Res. Sect. A,* 14, 115, 1936.
55. Crooker, A. M. and Dick, K. A., *Can. J. Phys.*, 46, 1241, 1968.
56. Crosswhite, H. M., *J. Res. Natl. Bur. Stand. Sect. A*, 79, 17, 1975.
58. Crosswhite, H. M. and Dieke, G. H., *American Institute of Physics Handbook*, Section 7, 1972.
59. de Bruin, T. L., *Z. Phys.*, 38, 94, 1926.
60. de Bruin, T. L., *Z. Phys.*, 53, 658, 1929.
61. de Bruin, T. L., Humphreys, C. J., and Meggers, W. F., *J. Res. Natl. Bur. Stand.*, 11, 409, 1933.
62. Dick, K. A., *J. Opt. Soc. Am.*, 64, 702, 1973.
63. Dobbie, J. C., *Ann. Solar Phys. Observ.* (Cambridge), 5, 1, 1938.
64. Earls, L. T. and Sawyer, R. A., *Phys. Rev.*, 47, 115, 1935.
65. Edlen, B., *Z. Phys.*, 85, 85, 1933.
66. Edlen, B., *Nova Acta Reglae Soc. Sci. Ups.*, (IV) 9, No. 6, 1934.
67. Edlen, B., *Z. Phys.*, 93, 726, 1935.
68. Edlen, B., *Z. Phys.*, 94, 47, 1935.
69. Edlen, B., *Rep. Prog. Phys.*, 26, 181, 1963.
70. Edlen, B. and Risberg, P., *Ark. Fys.*, 10, 553, 1956.
71. Edlen, B. and Swings, P., *Astrophys. J.*, 95, 532, 1942.
72. Ehrhardt, J. C. and Davis, S. P., *J. Opt. Soc. Am.*, 61, 1342, 1971.
73. Eidelsberg, M., *J. Phys. B*, 5, 1031, 1972.
74. Eidelsberg, M., *J. Phys. B*, 7, 1476, 1974.
75. Ekberg, J. O. and Svensson, L. A., *Phys. Scr.*, 2, 283, 1970.
76. Ekefors, E., *Z. Phys.*, 71, 53, 1931.
77. Epstein, G. L. and Reader, J., *J. Opt. Soc. Am.*, 65, 310, 1975.
78. Epstein, G. L. and Reader, J., *J. Opt. Soc. Am.*, 66, 590, 1976.
79. Epstein, G. L. and Reader, J., unpublished.
80. Eriksson, K. B. S., *Phys. Lett. A.*, 41, 97, 1972.
81. Eriksson, K. B. S. and Isberg, H. B. S., *Ark. Fys.*, 23, 527, 1963.
82. Eriksson, K. B. S. and Wenaker, I., *Phys. Scr.*, 1, 21, 1970.
83. Esteva, J. M. and Mehlman, G., *Astrophys. J.*, 193, 747, 1974.
84. Even-Zohar, M. and Fraenkel, B. S., *J. Phys. B*, 5, 1596, 1972.
85. Fawcett, B. C., *J. Phys. B*, 3, 1732, 1970.
86. Fawcett, B. C., Culham Laboratory Report ARU-R4, 1971.
87. Ferner, E., *Ark. Mat. Astron. Fys.*, 28(A), 4, 1941.
88. Fischer, R. A., Knopf, W. C., and Kinney, F. E., *Astrophys. J.*, 130, 683, 1959.

89. Fowler, A., *Report on Series in Line Spectra*, Fleetway Press, London, 1922.
90. Fowles, G. R., *J. Opt. Soc. Am.*, 44, 760, 1954.
91. Fred, M., *Argonne Natl. Lab.*, unpublished, 1977.
92. Fred, M. and Tomkins, F. S., *J. Opt. Soc. Am.*, 47, 1076, 1957.
93. Fred, M., Tomkins, F. S., Blaise, J. E., Camus, P., and Verges, J., Argonne National Laboratory Report No. 76-68, 1976.
94. Garcia, J. D. and Mack, J. E., *J. Opt. Soc. Am.*, 55, 654, 1965.
96. Giacchetti, A., *Argonne Natl. Lab.*, unpublished, 1975.
97. Giacchetti, A., Blaise, J., Corliss, C. H., and Zalubas, R., *J. Res. Natl. Bur. Stand. Sect. A*, 78, 247, 1974.
98. Giacchetti, A., Stanley, R. W., and Zalubas, R., *J. Opt. Soc. Am.*, 69, 474, 1970.
99. Gilbert, W. P., *Phys. Rev.*, 47, 847, 1935.
100. Gilroy, H. T., *Phys. Rev.*, 38, 2217, 1931.
101. Glad, S., *Ark. Fys.*, 10, 291, 1956.
102. Goldsmith, S., *J. Phys. B*, 2, 1075, 1969.
103. Goorvitch, D., Mehlmam-Balloffet, G., and Valero, F. P. J., *J. Opt. Soc. Am.*, 60, 1458, 1970.
104. Goorvitch, D. and Valero, F. P. J., *Astrophys. J.*, 171, 643, 1972.
105. Green, L. C., *Phys. Rev.*, 55, 1209, 1939.
106. Gutman, F., *Diss. Abstr. Int. B*, 31, 363, 1970.
107. Hallin, R., *Ark. Fys.*, 31, 511, 1966.
108. Hallin, R., *Ark. Fys.*, 32, 201, 1966.
109. Hansen, J. E. and Persson, W., *J. Opt. Soc. Am.*, 64, 696, 1974.
110. Hansen, J. E. and Persson, W., *Phys. Scr.*, 13, 166, 1976.
111. Hellintin, P., *Phys. Scr.*, 13, 155, 1976.
112. Herzberg, G. and Moore, H. R., *Can. J. Phys.*, 37, 1293, 1959.
113. Hetzler, C. W., Boreman, R. W., and Burns, K., *Phys. Rev.*, 48, 656, 1935.
114. Holmstrom, J. E. and Johansson, L., *Ark. Fys.*, 40, 133, 1969.
115. Hontzeas, S., Martinson, I., Erman, P., and Buchta, R., *Nucl. Instrum. Methods*, 110, 51, 1973.
116. Humphreys, C. J., *J. Res. Natl. Bur. Stand.*, 22, 19, 1939.
117. Humphreys, C. J., *J. Opt. Soc. Am.*, 43, 1027, 1953.
118. Humphreys, C. J., *J. Phys. Chem. Ref. Data*, 2, 519, 1973.
119. Humphreys, C. J. and Andrew, K. L., *J. Opt. Soc. Am.*, 54, 1134, 1964.
120. Humphreys, C. J. and Meggers, W. F., *J. Res. Natl. Bur. Stand.*, 10, 139, 1933.
121. Humphreys, C. J. and Paul, E., Jr., *J. Opt. Soc. Am.*, 60, 200, 1970.
122. Humphreys, C. J. and Paul, E., Jr., *J. Opt. Soc. Am.*, 62, 432, 1972.
123. Humphreys, C. J., Paul, E., Jr., Cowan, R. D., and Andrew, K. L., *J. Opt. Soc. Am.*, 57, 855, 1967.
124. Humphreys, C. J., Paul, E., Jr., and Minnhagen, L., *J. Opt. Soc. Am.*, 61, 110, 1971.
125. Iglesias, L., Inst. of Optics, Madrid, unpublished, 1977.
126. Iglesias, L. and Velasco, R., *Publ. Inst. Opt. Madrid*, No. 23, 1964.
127. Isberg, B., *Ark. Fys.*, 35, 551, 1967.
128. Johannesson, G. A., Lundstrom, T., and Minnhagen, L., *Phys. Scr.*, 6, 129, 1972.
129. Johannesson, G. A. and Lundstrom, T., *Phys. Scr.*, 8, 53, 1973.
130. Johansson, I. *Ark. Fys.*, 20, 135, 1961.
131. Johansson, I. and Contreras, R., *Ark. Fys.*, 37, 513, 1968.
132. Johansson, I. and Litzen, U., *Ark. Fys.*, 34, 573, 1967.
133. Johansson, I. and Svensson, K. F., *Ark. Fys.*, 16, 353, 1960.
134. Johansson, L., *Ark. Fys.*, 20, 489, 1961.
135. Johansson, L., *Ark. Fys.*, 23, 119, 1963.
136. Johansson, S. and Litzen, U., *Phys. Scr.*, 6, 139, 1972.
137. Johansson, S. and Litzen, U., *Phys. Scr.*, 8, 43, 1973.
138. Johansson, S. and Litzen, U., *Phys. Scr.*, 10, 121, 1974.
139. Joshi, Y. N., St. Francis Xavier Univ., Nova Scotia, unpublished.
140. Joshi, Y. N., Bhatia, K. S., and Jones, W. E., *Sci. Light Tokyo*, 21, 113, 1972.
141. Joshi, Y. N., Bhatia, K. S., and Jones, W. E., *Spectrochim. Acta Part B*, 28, 149, 1973.
142. Joshi, Y. N. and Budhiraja, C. J., *Can. J. Phys.*, 49, 670, 1971.
143. Joshi, Y. N. and van Kleef, T. A. M., *Can. J. Phys.*, 52, 1891, 1974.
144. Kaufman, V., *Natl. Bur. Stand.*, unpublished.
145. Kaufman, V., *J, Opt. Soc. Am.*, 52, 866, 1962.
146. Kaufman, V., Artru, M. C., and Brillet, W. U. L., *J. Opt. Soc. Am.*, 64, 197, 1974.
147. Kaufman, V. and Humphreys, C. J., *J. Opt. Soc. Am.*, 59, 1614, 1969.
148. Kaufman, V. and Sugar, J., *J. Opt. Soc. Am.*, 61, 1693, 1971.
149. Kaufman, V. and Sugar, J., *J. Res. Natl. Bur. Stand. Sect. A*, 71, 583, 1967.
150. Kelly, R. L. and Palumbo, L. J., *Naval Research Laboratory Report 7599*, Washington, DC., 1973.
151. Kielkopf, J. F., *Univ. of Louisville*, unpublished. 1975.
152. Kielkopf, J. F., *Univ. of Louisville*, unpublished, 1976.
153. Kiess, C. C. and Corliss, C. H., *J. Res. Natl. Bur. Stand. Sect. A*, 63, 1, 1959.
154. Kleiman, H., *J. Opt. Soc. Am.*, 52, 441, 1962.
155. Eriksson, K. B., Johansson, I., and Norlen, G., *Ark. Fys.*, 28, 233, 1964.
156. Klinkenberg, P. F. A., *Physica*, 15, 774, 1949.
157. Klinkenberg, P. F. A., *Physica*, 16, 618, 1950.
158. Krishnamurty, S. G., *Proc. Phys. Soc. London*, 48, 277, 1936.
159. Kruger, P. G. and Gilroy, H. T., *Phys. Rev.*, 48, 720, 1935.
160. Kruger, P. G. and Pattin, H. S., *Phys. Rev.*, 52, 621, 1937.
161. Lacroute, P., *Ann. Phys.* (Paris), 3, 5, 1935.
162. Lang, R. J., *Phys. Rev.*, 30, 762, 1927.
163. Lang, R. J., *Phys. Rev.*, 32, 737, 1928.
164. Lang, R. J., *Phys. Rev.*, 35, 445, 1930.
165. Lang, R. J., *Can. J. Res. Sect. A*, 14, 43, 1936.
166. Lang, R. J., *Can. J. Res. Sect. A*, 14, 127, 1936.
167. Lang, R. J. and Vestine, E. H., *Phys. Rev.*, 42, 233, 1932.
168. Li, H. and Andrew, K. L., *J. Opt. Soc. Am.*, 61, 96, 1971.
169. Liden, K., *Ark. Fys.*, 1, 229, 1949.
170. Litzen, U., *Ark. Fys.*, 28, 239, 1965.
171. Litzen, U., *Phys. Scr.*, 1, 251, 1970.
172. Litzen, U., *Phys. Scr.*, 1, 253, 1970.
173. Litzen, U., *Phys. Scr.*, 2, 103, 1970.
174. Litzen, U. and Verges, I., *Phys. Scr.*, 13, 240, 1976.
175. Lofstrand, B., *Phys. Scr.*, 8, 57, 1973.
176. Luc-Koenig, E., Morillon, C., and Verges, J., *Phys. Scr.*, 12, 199, 1975.
177. Lundstrom, T., *Phys. Scr.*, 7, 62, 1973.
178. Lundstrom, T. and Minnhagen, L., *Phys. Scr.*, 5, 243, 1972.
179. Magnusson, C. E. and Zetterberg, P. O., *Phys. Scr.*, 10, 177, 1974.
180. Magnusson, C. E., and Zetterberg, P. O., *Phys. Scr.*, 15, 237, 1977.
181. Martin, D. C., *Phys. Rev.*, 48. 938, 1935.
182. Svendenius, N., *Phys. Scr.*, 22, 240, 1980.
183. Martin, W. C., *J. Res. Natl. Bur. Stand. Sect. A*, 64, 19, 1960.
184. Martin, W. C. and Corliss, C. H., *J. Res. Natl. Bur. Stand. Sect. A*, 64, 443, 1960.
185. Martin, W. C. and Kaufman, V., *J. Res. Natl. Bur. Stand. Sect. A*, 74, 11, 1970.
186. Martin, W. C. and Kaufman, V., *J. Opt. Soc. Am.*, 60, 1096, 1970.

187. McCormick, W. W. and Sawyer, R. A., *Phys. Rev.*, 54, 71, 1938.
188. McLaughlin, R., *J. Opt. Soc. Am.*, 54, 965, 1964.
189. McLennan, J. C., McLay, A. B., and Crawford, M. F., *Proc. R. Soc. London Ser. A*, 134, 41, 1931.
190. Meissner, K. W., *Z. Phys.*, 39, 172, 1926.
191. Meggers, W. F., *J. Res. Natl. Bur. Stand.*, 24, 153, 1940.
192. Meggers, W. F. and Corliss, C. H., *J. Res. Natl. Bur. Stand. Sect. A*, 70, 63, 1966.
193. Meggers, W. F., Fred, M., and Tomkins, F. S., *J. Res. Natl. Bur. Stand.*, 58, 297, 1957.
194. Meggers, W. F. and Humphreys, C. J., *J. Res. Natl. Bur. Stand.*, 28, 463, 1942.
195. Meggers, W. F. and Murphy, R. J., *J. Res. Natl. Bur. Stand.*, 48, 334, 1952.
196. Meggers, W. F., Scribner, B. F., and Bozman, W. R., *J. Res. Natl. Bur. Stand.*, 46, 85, 1951.
197. Meggers, W. F., Shenstone, A. G., and Moore, C. E., *J. Res. Natl. Bur. Stand.*, 45, 346, 1950.
198. Mehlman, G. and Esteva, J. M., *Astrophys. J.*, 188, 191, 1974.
199. Meinders, E., *Physica*, 84(C), 117, 1976.
200. Sansonetti, C. J., Dissertation, Purdue University, 1981.
201. Sansonetti, C. J., *Natl. Bur. Stand. (U.S.)*, unpublished.
202. Millikan, R. A. and Bowen, I. S., *Phys. Rev.*, 25, 600, 1925.
203. Minnhagen, L., *J. Opt. Soc. Am.*, 61, 1257, 1925.
204. Minnhagen, L., *J. Opt. Soc. Am.*, 63, 1185, 1973.
205. Minnhagen, L., *Phys. Scr.*, 11, 38, 1975.
206. Minnhagen, L., *J. Opt. Soc. Am.*, 66, 659, 1976.
207. Minnhagen, L. and Nietsche, H., *Phys. Scr.*, 5, 237, 1972.
208. Minnhagen. L., Strihed, H., and Petersson, B., *Ark. Fys.*, 39, 471, 1969.
209. Moore, C. E., *Natl. Bur. Stand. (U.S.) Circ.*, 488, 1950.
210. Moore, C. E., *Revised Multiplet Table*, Princeton University Observatory No. 20, 1945.
211. Moore, C. E., National Standard Reference Data Series - National Bureau of Standards 3, Sect. 3, 1970.
212. Moore, C. E., National Standard Reference Data Series - National Bureau of Standards 3, Sect. 4, 1971.
213. Moore, C. E., National Standard Reference Data Series - National Bureau of Standards 3, Sect. 5, 1975.
214. Moore, C. E., National Standard Reference Data Series - National Bureau of Standards 3, Sect. 6, 1972.
215. Moore, C. E., National Standard Reference Data Series - National Bureau of Standards 3, Sect. 7, 1975.
216. Morillon, C. and Verges, J., *Phys. Scr.*, 10, 227, 1974.
217. Newsom, G. H., *Astrophys. J.*, 166, 243, 1971.
218. Newsom, G. H., O'Connor, S., and Learner, R. C. M., *J. Phys. B*, 6, 2162, 1973.
219. Norlen, G., *Phys. Scr.*, 8, 249, 1973.
220. Odabasi, H., *J. Opt. Soc. Am.*, 57, 1459, 1967.
221. Olme, A., *Ark. Fys.*, 40, 35, 1969.
222. Olme, A., *Phys. Scr.*, 1, 256, 1970.
223. Johansson, S., and Litzen, U., *J. Opt. Soc. Am.*, 61, 1427, 1971.
224. Palenius, H. P., *Ark. Fys.*, 39, 15, 1969.
225. Palenius, H. P., *Phys. Scr.*, 1, 113, 1970.
226. Palenius, H. P., *Univ. of Lund, Sweden*, unpublished.
227. Paschen, F., *Ann. Phys.*, Series 5, 12, 509, 1932.
228. Paschen, F. and Ritschl, R., *Ann. Phys.*, Series 5, 18, 867, 1933.
229. Peck, E. R., Khanna, B. N., and Anderholm, N. C., *J. Opt. Soc. Am.*, 52, 53, 1962.
230. Persson, W., *Phys. Scr.*, 3, 133, 1971.
231. Persson, W. and Valind, S., *Phys. Scr.*, 5, 187, 1972.
232. Petersson, B., *Ark. Fys.*, 27, 317, 1964.
233. Phillips, L. W. and Parker, W. L., *Phys. Rev.*, 60, 301, 1941.
234. Platt, J. R. and Sawyer, R. A., Phys. Rev., 60, 866, 1941.
235. Plyer, E. K., Blaine, L. R., and Tidwell, E., *J. Res. Natl. Bur. Stand.*, 55, 279, 1955.
236. Poppe, R., van Kleef, T. A. M., and Raassen, A. J. J., *Physica*, 77, 165, 1974.
237. Radziemski, L. J., Jr. and Andrew, K. L., *J. Opt. Soc. Am.*, 55, 474, 1965.
238. Radziemski, L. J., Jr. and Kaufman, V., *J. Opt. Soc. Am.*, 59, 424, 1969.
239. Radziemski, L. J., Jr. and Kaufman, V., *J. Opt. Soc. Am.*, 64, 366, 1974.
240. Ramanadham, R. and Rao, K. R., *Indian J. Phys.*, 18, 317, 1944.
241. Ramb, R., *Ann. Phys.*, 10, 311, 1931.
242. Rank, D. H., Bennett, J. M., and Bennett, H. E., *J. Opt. Soc. Am.*, 40, 477, 1950.
243. Rao, A. S. and Krishnamurty, S. G., *Proc. Phys. Soc. London*, 46, 531, 1943.
244. Rao, K. R., *Proc. R. Soc. London, Ser. A*, 134, 604, 1932.
245. Rao, K. R. and Badami, J. S., *Proc. R. Soc. London Ser. A*, 131, 154, 1931.
246. Rao, K. R. and Krishnamurty, S. G., *Proc. R. Soc. London Ser. A*, 161, 38, 1937.
247. Rao, K. R. and Murti, S. G. K., *Proc. R. Soc. London Ser. A*, 145, 681, 1934.
248. Rao, Y. B., *Indian J. Phys.*, 32, 497, 1958.
249. Rao, Y. B., *Indian J. Phys.*, 33, 546, 1959.
250. Rao, Y. B., *Indian J. Phys.*, 35, 386, 1961.
251. Rasmussen, E., *Z. Phys.*, 80, 726, 1933.
252. Rasmussen, E., *Z. Phys.*, 83, 404, 1933.
253. Rasmussen, E., *Z. Phys.*, 86, 24, 1934.
254. Rasmussen, E., *Z. Phys.*, 87, 607, 1934.
255. Rasmussen, E., *Phys. Rev.*, 57, 840, 1940.
256. Rau, A. S. and Narayan, A. L., *Z. Phys.*, 59, 687, 1930.
257. Reader, J., *J. Opt. Soc. Am.*, 65, 286, 1975.
258. Reader, J., *J. Opt. Soc. Am.*, 65, 988, 1975.
259. Reader, J., *J. Opt Soc. Am.*, 73, 349, 1983.
260. Reader, J. and Davis, S., *J. Res. Natl. Bur. Stand. Sect. A*, 71, 587, 1967, and unpublished.
261. Reader, J. and Ekberg, J. O., *J. Opt. Soc. Am.*, 62, 464, 1972.
262. Reader, J. and Epstein, G. L., *J. Opt. Soc. Am.*, 62, 1467, 1972.
263. Reader, J. and Epstein, G. L., *J. Opt. Soc. Am.*, 65, 638, 1975.
264. Reader, J. and Epstein, G. L., *Natl. Bur. Stand.*, unpublished.
265. Reader, J., Epstein, G. L., and Ekberg, J. O., *J. Opt. Soc. Am.*, 62, 273, 1972.
266. Kaufman, V., *Phys. Scr.*, 26, 439, 1982.
267. Ricard, R., Givord, M., and George, F., *C. R. Acad. Sci. Paris*, 205, 1229, 1937.
268. Risberg, P., *Ark. Fys.*, 10, 583, 1956.
269. Risberg, G., *Ark. Fys.*, 28, 381, 1965.
270. Risberg, G., *Ark. Fys.*, 37, 231, 1968.
271. Robinson, H. A., *Phys. Rev.*, 49, 297, 1936.
272. Robinson, H. A., *Phys. Rev.*, 50, 99, 1936.
273. Ross, C. B., Jr., Doctoral dissertation, Purdue University, 1969.
274. Ross, C. B., Wood, D. R., and Scholl, P. S., *J. Opt. Soc. Am.*, 66, 36, 1976.
275. Ruedy, J. E. and Gibbs, R. C., *Phys. Rev.*, 46, 880, 1934.
276. Russell, H. N., King, R. B., and Moore, C. E., *Phys. Rev.*, 58, 407, 1940.
277. Russell, H. N. and Moore, C. E., *J. Res. Natl. Bur. Stand.*, 55, 299, 1955.
278. Russell, H. N., Moore, C. E., and Weeks, D. W., *Trans. Am. Philos. Soc.*, 34(2), 111, 1944.

279. Saunders, F., Schneider, E., and Buckingham, E., *Proc. Natl. Acad. Sci.*, 20, 291, 1934.
280. Sawyer, R. A. and Humphreys, C. J., *Phys. Rev.*, 32, 583, 1928.
281. Sawyer, R. A. and Lang, R. J., *Phys. Rev.*, 34, 712, 1929.
282. Sawyer, R. A. and Paschen, F., *Ann. Phys.*, 84(4),1, 1927.
283. Scholl, P. S., M.S. thesis, Wright State Univ., 1975.
284. Schurmann, D., *Z. Phys.*, 17, 4, 1975.
285. Seguier, J., *C. R. Acad. Sci. Paris*, 256, 1703, 1963.
286. Shenstone, A. G., *Phys. Rev.*, 31, 317, 1928.
287. Shenstone, A. G., *Phys. Rev.*, 32, 30, 1928.
288. Shenstone, A. G., *Trans. R. Soc. London*, 237(A), 57, 1938.
289. Shenstone, A. G., *Phys. Rev.*, 57, 894, 1940.
290. Shenstone, A. G. *Philos. Trans. R. Soc. London Ser. A*, 241, 297, 1948.
291. Shenstone, A. G., *Can. J. Phys.*, 38, 677, 1960.
292. Shenstone, A. G., *Proc. R. Soc. London*, 261(A), 153, 1961.
293. Shenstone, A. G., *Proc. R. Soc. London*, 276(A), 293, 1963.
294. Shenstone, A. G., *J. Res. Natl. Bur. Stand. Sect. A*, 74, 801, 1970.
295. Shenstone, A. G., *J. Res. Natl. Bur. Stand. Sect. A*, 79, 497, 1975.
296. Shenstone, A. G. and Pittenger, J. T., *J. Opt. Soc. Am.*, 39, 219, 1949.
297. Smith, S., *Phys. Rev.*, 36, 1, 1930.
298. Smitt, R., *Phys. Scr.*, 8, 292, 1973.
299. Soderqvist, J., *Ark. Mat. Astronom. Fys.*, 32(A), 1, 1946.
300. Sommer, L. A., *Ann. Phys.*, 75, 163, 1924.
301. Spector, N., *J. Opt. Soc. Am.*, 63, 358, 1973.
302. Spector, N. and Sugar, J., *J. Opt. Soc. Am.*, 66, 436, 1976.
303. Steinhaus, D. W., Radziemski, L. J., Jr., and Blaise, J., *Los Alamos Sci. Lab.*, unpublished, 1975.
304. Subbaraya, T. S., *Z. Phys.*, 78, 541, 1932.
305. Sugar, J., *J. Opt. Soc. Am.*, 55, 33, 1965.
306. Sugar, J., *J. Res. Natl. Bur. Stand. Sect. A*, 73, 333, 1969.
307. Sugar, J., *J. Opt. Soc. Am.*, 60, 454, 1970.
308. Sugar, J., *J. Res. Natl. Bur. Stand. Sect. A*, 78, 555, 1974.
309. Sugar, J. and Kaufman, V., *J. Opt. Soc. Am.*, 55, 1283, 1965.
310. Sugar, J. and Kaufman, V., *J. Opt. Soc. Am.*, 62, 562, 1972.
311. Sugar, J., Kaufman, V., and Spector, N., *J. Res. Natl. Bur. Stand., Sect. A*, 83, 233, 1978.
312. Sugar, J. and Spector, N., *J. Opt. Soc. Am.*, 64, 1484, 1974.
313. Sullivan, F. J. *Univ. Pittsburgh Bull.*, 35, 1, 1938.
314. Svensson, L. A. and Ekberg, J. O., *Ark. Fys.*, 37, 65, 1968.
315. Swensson, J. W. and Risberg, G., *Ark. Fys.*, 31, 237, 1966.
316. Tech, J. L., *J. Res. Natl. Bur. Stand. Sect. A*, 67, 505, 1963.
317. Tech, J. L. and Ward, J. F., *Phys. Rev. Lett.*, 27, 367, 1971.
318. Tilford, S. G., *J. Opt. Soc. Am.*, 53, 1051, 1963.
319. Toresson, Y. G., *Ark. Fys.*, 17, 179, 1960.
320. Toresson, Y. G., *Ark. Fys.*, 18, 389, 1960.
321. Toresson, Y. G. and Edlen, B., *Ark. Fys.*, 23, 117, 1963.
322. Tsien, W. Z., *Chin. J. Phys.*, Peiping, 3, 117, 1939.
323. van Deurzen, C. H. H., Conway, J., and Davis, S. P., *J. Opt. Soc. Am.*, 63, 158, 1973.
324. van Kleef, T. A. M., Raassen, A. J. J., and Joshi, Y. N., *Physica*, 84(C), 401, 1976.
325. Sansonetti, C. J., Andrew, K. L., and Verges, J., *J. Opt. Soc. Am.*, 71, 423, 1981.
326. Wheatley, M. A. and Sawyer, R. A., *Phys. Rev.*, 61, 591, 1942.
327. Wilkinson, P. G., *J. Opt. Soc. Am.*, 45, 862, 1955.
328. Wilkinson, P. G. and Andrew, K. L., *J. Opt. Soc. Am.*, 53, 710, 1963.
329. Wood, D. and Andrew, K. L., *J. Opt. Soc. Am.*, 58, 818, 1968.
330. Wood, D. R., Ron, C. B., Scholl, P. S., and Hoke, M., *J. Opt. Soc. Am.*, 64, 1159, 1974.
331. Worden, E. F. and Conway, J. G., *Lawrence Livermore Lab.*, unpublished, 1977.
332. Worden, E. F., Hulet, E. K., Gutmacher, R. G., Conway, J. G., *At. Data Nucl. Data Tables*, 18, 459, 1976.
333. Worden, E. F., Lougheed, R. W., Gutmacher, R. G., and Conway, J. G., *J. Opt. Soc. Am.*, 64, 77, 1974.
334. Wu, C. M., Ph.D. thesis, University of British Columbia, 1971.
335. Zaidel, A. N., Prokofev, V. K., Raiskii, S. M., Slavnyi, V. A., and Schreider, E. Y., *Tables of Spectral Lines*, 3rd ed., Plenum, New York, 1970.
336. Zetterberg, P. O. and Magnusson, C. E., *Phys. Scr.*, 15, 189, 1977.
337. Sugar, J., *J. Opt. Soc. Am.*, 55, 1058, 1965.
338. Sugar, J., *J. Opt. Soc. Am.*, 61, 727, 1971.
339. Worden, E. F., and Conway, J. G., *At. Data Nucl. Data Tables*, 22, 329, 1978.
340. Kaufman, V. and Edlen, B., *J. Phys. Chem. Ref. Data*, 3, 825, 1974.
341. Lang, R. J., *Phys. Rev.*, 34, 697, 1929.
342. Ryabtsev, A. N., *Opt. Spectros.*, 39, 455, 1975.
343. Foster, E. W., *Proc. R. Soc. London*, 200(A), 429, 1950.
344. Morillon, C. and Verges, J., *Phys. Scr.*, 12, 129, 1975.
345. Ruedy, J. E., *Phys. Rev.*, 41, 588, 1932.
346. McLennan, J. C., McLay, A. B., and McLeod, J. H., *Philos. Mag.*, 4 ,486, 1927.
347. Handrup, M. B. and Mack, J. E., *Physica*, 30, 1245, 1964.
348. Clearman, H. E., *J. Opt. Soc. Am.*, 42, 373, 1952.
349. Paschen, F., *Ann. Physik*, 424, 148, 1938.
350. Paschen, F. and Campbell, J. S., *Ann. Phys.*, 31(5), 29, 1938.
351. Nodwell, R., *Univ. of British Columbia, Vancouver*, unpublished, 1955.
352. Gibbs, R. C. and White, H. E., *Phys. Rev.*, 31, 776, 1928.
353. Green, M., *Phys. Rev.*, 60, 117, 1941.
354. Ellis, C. B. and Sawyer, R. A., *Phys. Rev.*, 49, 145, 1936.
355. McLennan, J. C., McLay, A. B., and Crawford, M. F., *Proc. R. Soc. London Ser. A*, 125, 50, 1929.
356. Mack, J. E. and Fromer, M., *Phys. Rev.*, 48, 346, 1935.
357. Humphreys, C. J. and Paul, E., U.S. Nav. Ord. Lab., Navord Rep. 4589, 25, 1956.
358. Walters, F. M., *Sci. Pap. Bur. Stand.*, 17, 161, 1921.
359. Crawford, M. F. and McLay, A. B., *Proc. R. Soc. London Ser. A*, 143, 540, 1934.
360. McLay, A. D. and Crawford, M. F., Phys. Rev., 44, 986, 1933.
361. Schoepfle, G. K., *Phys. Rev.*, 47, 232, 1935.
362. Acquista, N., and Reader, J., *J. Opt. Soc. Am.*, 70, 789, 1980.
363. Benschop, H., Joshi, Y. N., and van Kleef, T. A. M., *Can. J. Phys.*, 53, 498, 1975.
364. Bockasten, K., Hallin, R., and Hughes, T. P., *Proc. Phys. Soc.*, 81, 522, 1963.
365. Boyce, J. C., *Phys. Rev.*, 46, 378, 1934.
366. Boyce, J. C., *Phys. Rev.*, 47, 718, 1935.
367. Boyce, J. C., *Phys. Rev.*, 48, 396, 1935.
368. Boyce, J. C., *Phys. Rev.*, 49, 351, 1936.
369. Corliss, C. H. and Meggers, W. F., *J. Res. Natl. Bur. Stand.*, 61, 269, 1958.
370. Crooker, A. M. and Dick, K. A., *Can. J. Phys.*, 42, 766, 1964.
371. De Bruin, T. L., *Z. Physik*, 77, 505, 1932.
372. De Bruin, T. L., *Proc. Roy. Acad. Amsterdam*, 36, 727, 1933.
373. De Bruin, T. L., *Zeeman Verhandelingen*, (The Hague), 1935, p. 415.
374. De Bruin, T. L., *Physica*, 3, 809, 1936.
375. De Bruin, T. L., *Proc. Roy. Acad. Amsterdam*, 40, 339, 1937.
376. Dick, K. A., *Can. J. Phys.*, 46, 1291, 1968.
377. Dick, K. A., unpublished, 1978.

378. Edlen, B. and Swensson, J. W., *Phys. Scr.*, 12, 21, 1975.
379. Ekberg, J. O., *Phys. Scr.*, 7, 55, 1973.
380. Ekberg, J. O., *Phys. Scr.*, 7, 59, 1973.
381. Ekberg, J. O., *Phys. Scr.*, 12, 42, 1975.
382. Ekberg, J. O. and Edlen, B., *Phys. Scr.*, 18, 107, 1978.
383. Eliason, A. Y., *Phys. Rev.*, 43, 745, 1933.
384. Gallardo, M., Massone, C. A., Tagliaferri, A. A., Garavaglia, M., and Persson, W., *Phys. Scr.*, 19, 538, 1979.
385. Garcia-Riquelme, O., *Optica Pura Y Aplicada*, 1, 53, 1968.
386. Gibbs, R. C., Vieweg, A. M., and Gartlein, C. W., *Phys. Rev.*, 34, 406, 1929.
387. Gilbert, W. P., *Phys. Rev.*, 48, 338, 1935.
388. Goldsmith, S. and Kaufman, A. S., *Proc. Phys. Soc.*, 81, 544, 1963.
389. Hermansdorfer, H., *J. Opt. Soc. Am.*, 62, 1149, 1972.
390. Humphreys, C. J., *Phys. Rev.*, 47, 712, 1935.
391. Humphreys, C. J., *J. Res. Natl. Bur. Stand.*, 16, 639, 1936.
392. Iglesias, L., *J. Opt. Soc. Am.*, 45, 856, 1955.
393. Iglesias, L., *J. Res. Natl. Bur. Stand.*, 64A, 481, 1960.
394. Iglesias, L., *Anales Fisica Y Quimica*, 58A, 191, 1962.
395. Iglesias, L., *J. Res. Natl. Bur. Stand.*, 70A, 465, 1966.
396. Iglesias, L., *Can. J. Phys.*, 44, 895, 1966.
397. Iglesias, L., *J. Res. Natl. Bur. Stand.*, 72A, 295, 1968.
398. Joshi, Y. N., *Can. Spectrosc.*, 15, 96, 1970.
399. Joshi, Y. N. and van Kleef, T. A. M., *Can. J. Phys.*, 55, 714, 1977.
400. Kaufman, A. S., Hughes, T. P., and Williams, R. V., *Proc. Phys. Soc.*, 76, 17, 1960.
401. Kaufman, V. and Sugar, J., *J. Opt. Soc. Am.*, 68, 1529, 1978.
402. Keussler, V., *Z. Physik*, 85, 1, 1933.
403. Kiess, C. C., *J. Res. Natl. Bur. Stand.*, 56, 167, 1956.
404. Klinkenberg, P. F. A., van Kleef, T. A. M., and Noorman, P. E., *Physica*, 27, 1177, 1961.
405. Kovalev, V. I., Romanos, A. A., and Ryabtsev, A. N., *Opt. Spectrosc.*, 43, 10, 1977.
406. Lang, R. J., *Proc. Natl. Acad. Sci.*, 13, 341, 1927.
407. Lang, R. J., *Zeeman Verhandelingen*, (The Hague), 44, 1935.
408. Liberman, S., et al., *C. R. Acad. Sci.* (Paris), 286, 253, 1978.
409. Livingston, A. E., *J. Phys.*, B9, L215, 1976.
410. Meijer, F. G., *Physica*, 72, 431, 1974.
411. Meijer, F. G. and Metsch, B. C., *Physica*, 94C, 259, 1978.
412. Moore, F. L., thesis, Princeton, 1949.
413. Paul, F. W. and Polster, H. D., *Phys. Rev.*, 59, 424, 1941.
414. Phillips, L. W. and Parker, W. L., *Phys. Rev.*, 60, 301, 1941.
415. Poppe, R., *Physica*, 81C, 351, 1976.
416. Raassen, A. J. J., van Kleef, T. A. M., and Metsch, B. C., *Physica*, 84C, 133, 1976.
417. Rao, A. B. and Krishnamurty, S. G., *Proc. Phys. Soc. (London)*, 51, 772, 1939.
418. Reader, J. and Acquista, N., *J. Opt. Soc. Am.*, 69, 239, 1979.
419. Reader, J. and Epstein, G. L., *J. Opt. Soc. Am.*, 62, 619, 1972.
420. Rico, F. R., *Anales, Real Soc. Esp. Fis. Quim.*, 61, 103, 1965.
421. Schonheit, E., *Optik*, 23, 409, 1966.
422. Shenstone, A. G., *J. Opt. Soc. Am.*, 44, 749, 1954.
423. Shenstone, A. G., unpublished, 1958.
424. Shenstone, A. G., *J. Res. Natl. Bur. Stand.*, 67A, 87, 1963.
425. Sugar, J. and Kaufman, V., *J. Opt. Soc. Am.*, 64, 1656, 1974.
426. Sugar, J. and Kaufman, V., *Phys. Rev.*, C12, 1336, 1975.
427. Svensson, L. A., *Phys. Scr.*, 13, 235, 1976.
428. Swensson, J. W. and Edlen, B., *Phys. Scr.*, 9, 335, 1974.
429. Tagliaferri, A. A., Gallego Lluesma, E., Garavaglia, M., Gallardo, M., and Massone, C. A., *Optica Pura Y Aplica*, 7, 89.
430. Tilford, S. G. and Giddings, L. E., *Astrophys. J.*, 141, 1222, 1965.
431. Trawick, M. W., *Phys. Rev.*, 46, 63, 1934.
432. Van Deurzen, C. H. H., *J. Opt. Soc. Am.*, 67, 476, 1977.
433. Yarosewick, S. L. and Moore, F. L., *J. Opt. Soc. Am.*, 57, 1381, 1967.
434. Zalubas, R., unpublished, 1979.
435. Bhatia, K. S., Jones, W. E., and Crooker, A. M., *Can. J. Phys.*, 50, 2421, 1972.
436. van Kleef, T. A. M. and Joshi, Y. N., *Phys. Scr.*, 24, 557, 1981.

NIST ATOMIC TRANSITION PROBABILITY TABLES

J.R. Fuhr and W.L. Wiese

These tables substantially update and enlarge our earlier tables in this *Handbook.* The new tables contain critically evaluated atomic transition probabilities for about 9000 selected lines of all elements for which reliable data are available on an absolute scale. The material is largely for neutral and singly ionized spectra, but also includes a number of prominent lines of more highly charged ions of important elements.

Many of the data are obtained from comprehensive compilations of the Data Center on Atomic Transition Probabilities at the National Institute of Standards and Technology (formerly the National Bureau of Standards). Specifically, data have been taken from three recent comprehensive critical compilations on C, N and O,[1] on Sc through Mn,[2] and Fe through Ni.[3] Material from earlier compilations for the elements H through Ne[4] and Na through Ca[5] was supplemented by more recent material taken directly from the original literature. For the highly charged ions, some of the data were derived from studies of the systematic behavior of transition probabilities.[6-8] Most of the original literature is cited in the above tables and in recent bibliographies[9,10]; for lack of space, individual literature references are not cited here.

The wavelength range for the neutral species is normally the visible spectrum or shorter wavelengths; only the very prominent near infrared lines are included. For the higher ions, most of the strong lines are located in the far UV. The tabulation is limited to electric dipole — including intercombination — lines and comprises essentially the fairly strong transitions with estimated uncertainties of 50% or less. With the exception of hydrogen, helium, and the alkalis, most transitions are between states with low principal quantum numbers.

The transition probability, A, is given in units of 10^8 s^{-1} and is listed to as many digits as is consistent with the indicated accuracy. The power of 10 is indicated by the E notation (i.e., E-02 means 10^{-2}). Generally, the estimated uncertainties of the A-values are ±25 to 50% for two-digit numbers, ±10 to 25% for three-digit numbers and ±1% or better for four- and five-digit numbers.

Each transition is identified by the wavelength, λ, in angstroms; and the statistical weights, g_i and g_k, of the lower (i) and upper (k) states [the product g_kA (or g_if) is needed for many applications]. Whenever the wavelengths of individual lines within a multiplet are extremely close, only an average wavelength for the multiplet as well as the multiplet A-value are given, and this is indicated by an asterisk (*) to the left of the wavelength. This also has been done when the transition probability for an entire multiplet has been taken from the literature and values for individual lines cannot be determined because of insufficient knowledge of the coupling of electrons. The wavelength data have been taken either from recent compilations or from the original literature cited in bibliographies published by the Atomic Energy Levels Data Center[11,12] at the National Institute of Standards and Technology. Wavelength values are consistent with those given in the table "Line Spectra of the Elements", which appears elsewhere in this *Handbook.*

The transition probabilities for hydrogen and hydrogen-like ions are known precisely. Because of the hydrogen degeneracy, a "transition" is actually the sum of all fine-structure transitions between the principal quantum numbers listed in the transition column; therefore, the special hydrogen table which appears below gives weighted average A-values.

In addition to the transition probability A, the atomic oscillator strength f and the line strength S are often used in the literature. The conversion factors between these quantities are (for electric-dipole transitions):

$$g_if = 1.499 \times 10^{-8}\,\lambda^2\,g_k\,A = 303.8\,\lambda^{-1}\,S$$

where λ is in angstroms, A is in 10^8 s^{-1}, and S is in atomic units, which are $a_0{}^2e^2 = 7.188 \times 10^{-59}$ m^2 C^2.

After the special table for hydrogen, the tables for other elements appear in alphabetical sequence by element name (not symbol). Within each element, the tables are ordered by increasing ionization stage (e.g., Al I, Al II, etc.).

REFERENCES

1. Wiese, W.L., Fuhr, J.R., and Deters, T.M., *Atomic Transition Probabilities of Carbon, Nitrogen and Oxygen, J. Phys. Chem. Ref. Data, Monograph 7*, 1996.
2. Martin, G.A., Fuhr, J.R., and Wiese, W.L., *Atomic Transition Probabilities—Scandium through Manganese, J. Phys. Chem. Ref. Data, 17, Suppl.* 3, 1988.
3. Fuhr, J.R., Martin, G.A., and Wiese, W.L., *Atomic Transition Probabilities—Iron through Nickel, J. Phys. Chem. Ref. Data, 17, Suppl.* 4, 1988.
4. Wiese, W.L., Smith, M.W., and Glennon, B.M., *Atomic Transition Probabilities (H through Ne—A Critical Data Compilation)*, National Standard Reference Data Series, National Bureau of Standards 4, Vol. I, U.S. Government Printing Office, Washington, D.C., 1966.
5. Wiese, W.L., Smith, M.W., and Miles, B.M., *Atomic Transition Probabilities (Na through Ca—A Critical Data Compilation)*, National Standard Reference Data Series, National Bureau of Standards 22, Vol. II, U. S. Government Printing Office, Washington, D.C., 1969.
6. Wiese, W.L. and Weiss, A.W., *Phys. Rev.*, 175, 50, 1968.
7. Smith, M.W. and Wiese,M.L., *Astrophys. J., Suppl. Ser.*, 23, No. 196, 103, 1971.
8. Martin, G.A., and Wiese,W.L., *J. Phys. Chem. Ref. Data,* 5, 537, 1976.
9. Fuhr, J.R., Miller, B.J., and Martin, G.A., *Bibliography on Atomic Transition Probabilities (1914 through October 1997)*, National Bureau of Standards Special Publication 505, 1978; Miller, B.J., Fuhr, J.R., and Martin, G.A., *Bibliography on Atomic Transition Probabilities (November 1977 through February 1980)*, National Bureau of Standards Special Publication 505, Supplement 1, 1980.
10. Wiese, W.L., Reports on Astronomy, *Trans. Int. Astron. Union*, 18A, 116—123, 1982; 19A, 122—138, 1985.; 20A, 117—123, 1988, Reidel, D., Ed., Kluwer, Dordrecht, Holland.
11. Moore, C.E., *Bibliography on the Analyses of Optical Atomic Spectra*, National Bureau of Standards Special Publication 306—Section 1, 1968; Sections 2—4, 1969.

12. Hagan, L. and Martin, W.C., *Bibliography on Atomic Energy Levels and Spectra (July 1968 through June 1971)*, National Bureau of Standards Special Publication 363, 1972; Hagan, L., *Bibliography on Atomic Energy Levels and Spectra (July 1971 through June 1975)*, National Bureau of Standards Special Publication 363, Supplement 1, 1977; Zalubas, R. and Albright, A., *Bibliography on Atomic Energy Levels and Spectra (July 1975 through June 1979)*, National Bureau of Standards Special Publication 363, Supplement 2, 1980; Musgrove, A. and Zalubas, R., *Bibliography on Atomic Energy Levels and Spectra (July 1979 through December 1983)*, National Bureau of Standards Special Publication 363, Supplement 3, 1985.
13. Younger, S.M. and Weiss, A., *J. Res. Natl. Bur. Stand.*, 79A, 629, 1975.

Transition Probabilities for Allowed Lines of Hydrogen

λ Å	Weights g_i	Weights g_k	A 10^8 s^{-1}
Hydrogen			
H I			
912.768	2	1800	5.167E-06
912.839	2	1682	6.122E-06
912.918	2	1568	7.297E-06
913.006	2	1458	8.753E-06
913.104	2	1352	1.057E-05
913.215	2	1250	1.286E-05
913.339	2	1152	1.578E-05
913.480	2	1058	1.952E-05
913.641	2	968	2.438E-05
913.826	2	882	3.077E-05
914.039	2	800	3.928E-05
914.286	2	722	5.077E-05
914.576	2	648	6.654E-05
914.919	2	578	8.858E-05
915.329	2	512	1.200E-04
915.824	2	450	1.657E-04
916.429	2	392	2.341E-04
917.181	2	338	3.393E-04
918.129	2	288	5.066E-04
919.351	2	242	7.834E-04
920.963	2	200	1.263E-03
923.150	2	162	2.143E-03
926.226	2	128	3.869E-03
930.748	2	98	7.568E-03
937.803	2	72	1.644E-02
949.743	2	50	4.125E-02
972.537	2	32	1.278E-01
1025.72	2	18	5.575E-01
1215.67	2	8	4.699E+00
3662.26	8	1800	2.847E-06
3663.40	8	1682	3.374E-06
3664.68	8	1568	4.022E-06
3666.10	8	1458	4.826E-06
3667.68	8	1352	5.830E-06
3669.46	8	1250	7.096E-06
3671.48	8	1152	8.707E-06
3673.76	8	1058	1.078E-05
3676.36	8	968	1.347E-05
3679.35	8	882	1.700E-05
3682.81	8	800	2.172E-05
3686.83	8	722	2.809E-05
3691.55	8	648	3.685E-05
3697.15	8	578	4.910E-05
3703.85	8	512	6.658E-05
3711.97	8	450	9.210E-05
3721.94	8	392	1.303E-04
3734.37	8	338	1.893E-04
3750.15	8	288	2.834E-04
3770.63	8	242	4.397E-04
3797.90	8	200	7.122E-04
3835.38	8	162	1.216E-03
3889.05	8	128	2.215E-03
3970.07	8	98	4.389E-03
4101.73	8	72	9.732E-03
4340.46	8	50	2.530E-02
4861.32	8	32	8.419E-02
6562.80	8	18	4.410E-01
8392.40	18	800	1.517E-05
8413.32	18	722	1.964E-05
8437.96	18	648	2.580E-05
8467.26	18	578	3.444E-05
8502.49	18	512	4.680E-05
8545.39	18	450	6.490E-05
8598.40	18	392	9.211E-05
8665.02	18	338	1.343E-04
8750.48	18	288	2.021E-04
8862.79	18	242	3.156E-04
9014.91	18	200	5.156E-04
9229.02	18	162	8.905E-04
9545.97	18	128	1.651E-03
10049.4	18	98	3.358E-03
10938.1	18	72	7.783E-03
12818.1	18	50	2.201E-02
16407.2	32	288	1.620E-04
16806.5	32	242	2.556E-04
17362.1	32	200	4.235E-04
18174.1	32	162	7.459E-04
18751.0	18	32	8.986E-02
19445.6	32	128	1.424E-03
21655.3	32	98	3.041E-03
26251.5	32	72	7.711E-03
27575	50	288	1.402E-04
28722	50	242	2.246E-04
30384	50	200	3.800E-04
32961	50	162	6.908E-04
37395	50	128	1.388E-03
40511.5	32	50	2.699E-02
43753	72	288	1.288E-04
46525	50	98	3.253E-03
46712	72	242	2.110E-04
51273	72	200	3.688E-04
59066	72	162	7.065E-04
74578	50	72	1.025E-02
75004	72	128	1.561E-03
123680	72	98	4.561E-03

For hydrogen-like ions of nuclear charge Z, the following scaling laws hold:

$$A_Z = Z^4 A_{Hydrogen};\ f_Z = f_H;\ S_Z = Z^{-2} S_H$$
$$\text{(For wavelengths, } \lambda_Z = Z^{-2} \lambda_H\text{)}$$

For very highly charged hydrogen-like ions, starting at about $Z>25$, relativistic corrections[13] must be applied.

Transition Probabilities for Other Elements

λ Å	Weights g_i	Weights g_k	A $10^8\ s^{-1}$
Aluminum			
Al I			
2263.5	2	4	6.6E-01
2269.1	4	6	7.9E-01
2269.2	4	4	1.3E-01
2367.1	2	4	7.2E-01
2373.1	4	6	8.6E-01
2373.4	4	4	1.4E-01
2568.0	2	4	2.3E-01
2575.1	4	6	2.8E-01
2575.4	4	4	4.4E-02
2652.5	2	2	1.33E-01
2660.4	4	2	2.64E-01
3082.2	2	4	6.3E-01
3092.7	4	6	7.4E-01
3092.8	4	4	1.2E-01
3944.0	2	2	4.93E-01
3961.5	4	2	9.8E-01
6696.0	2	4	1.69E-02
6698.7	2	2	1.69E-02
7835.3	4	6	5.7E-02
7836.1	6	8	6.2E-02
Al II			
1047.9	1	3	3.6E-01
1048.6	3	5	4.8E-01
1539.8	3	5	8.8E+00
1670.8	1	3	1.46E+01
1719.4	1	3	6.79E+00
1764.0	5	5	9.8E+00
1772.8	1	3	9.5E+00
1777.0	5	7	1.7E+01
*1819.0	15	15	5.6E+00
1855.9	1	3	8.32E-01
1858.0	3	3	2.48E+00
1862.3	5	3	4.12E+00
1931.0	3	1	1.08E+01
1990.5	3	5	1.47E+01
2816.2	3	1	3.83E+00
4663.1	5	3	5.3E-01
6226.2	1	3	6.2E-01
6231.8	3	5	8.4E-01
6243.4	5	7	1.1E+00
6335.7	5	3	1.4E-01
6823.4	3	3	3.4E-01
6837.1	5	3	5.7E-01
6920.3	3	1	9.6E-01
7042.1	3	5	5.9E-01
7056.7	3	3	5.8E-01
7471.4	5	7	9.4E-01
Al III			
*560.36	2	6	4.0E-01
695.83	2	4	7.4E-01
696.22	2	2	7.2E-01
*1352.8	10	14	4.40E+00
1379.7	2	2	4.59E+00
1384.1	4	2	9.1E+00
1605.8	2	4	1.22E+01
1611.8	4	4	2.42E+00
1611.9	4	6	1.45E+01
1854.7	2	4	5.40E+00
1862.8	2	2	5.33E+00
*1935.9	10	14	1.22E+01
3601.6	6	4	1.34E+00
3601.9	4	4	1.49E-01
3612.4	4	2	1.5E+00
Al X			
39.925	1	3	2.22E+03
51.979	1	3	4.8E+03
55.227	1	3	5.2E+03
55.272	3	5	7.2E+03
55.376	5	7	9.5E+03
59.107	3	5	4.6E+03
332.78	1	3	5.6E+01
394.83	3	1	8.3E+01
395.36	3	5	1.2E+01
397.76	1	3	1.7E+01
400.43	3	3	1.3E+01
401.12	5	5	3.6E+01
403.55	3	1	4.9E+01
406.31	5	3	1.9E+01
670.06	3	5	9.8E+00
2535	1	3	3.8E-01
Al XI			
*36.675	2	6	1.5E+03
39.091	2	4	2.6E+03
39.180	4	6	3.1E+03
39.530	2	2	1.8E+02
39.623	4	2	3.7E+02
48.298	2	4	3.09E+03
48.338	2	2	3.08E+03
52.299	2	4	8.1E+03
52.446	4	6	9.6E+03
52.458	4	4	1.6E+03
54.217	2	2	4.8E+02
54.388	4	2	9.6E+02
*99.083	2	6	2.2E+02
103.6	2	4	4.2E+02
103.8	4	6	5.0E+02
*141.6	2	6	4.07E+02
150.31	2	4	8.5E+02
150.61	4	6	9.9E+02
157.0	2	2	1.3E+02
157.4	4	2	2.6E+02
*205.0	2	6	6.3E+01
*308.6	2	6	9.9E+01
*341.3	6	2	1.3E+02
550.05	2	4	8.55E+00
568.12	2	2	7.73E+00
1997	2	4	1.07E+00
2069	2	2	9.7E-01
*4761	2	6	2.55E-01
5172	2	4	3.95E-02
5551	4	6	3.85E-02
5687	4	4	6.0E-03
Argon			
Ar I			
1048.22	1	3	5.36E+00
1066.66	1	3	1.29E+00
3406.18	3	1	3.9E-03
3461.08	3	5	6.7E-04
3554.30	5	5	2.7E-03
3563.29	1	3	1.2E-03
3567.66	5	7	1.1E-03
3572.30	3	1	5.1E-03
3606.52	3	1	7.6E-03
3632.68	3	5	6.6E-04
3634.46	3	3	1.3E-03
3643.12	3	5	2.4E-04
3649.83	3	1	8.0E-03
3659.53	3	3	4.4E-04
3670.67	3	5	3.1E-04
3675.23	3	3	4.9E-04
3770.37	1	3	7.0E-04
3834.68	3	1	7.5E-03
3894.66	3	3	5.7E-04
3947.50	5	5	5.6E-04
3948.98	5	3	4.55E-03
4044.42	3	5	3.33E-03
4045.96	3	3	4.1E-04
4054.53	3	3	2.7E-04
4158.59	5	5	1.40E-02
4164.18	5	3	2.88E-03
4181.88	1	3	5.61E-03
4190.71	5	5	2.80E-03
4191.03	1	3	5.39E-03
4198.32	3	1	2.57E-02
4200.67	5	7	9.67E-03
4251.18	5	3	1.11E-03
4259.36	3	1	3.98E-02
4266.29	3	5	3.12E-03
4272.17	3	3	7.97E-03
4300.10	3	5	3.77E-03
4333.56	3	5	5.68E-03
4335.34	3	3	3.87E-03
4345.17	3	3	2.97E-03
4363.79	3	3	1.2E-04
4424.00	1	3	7.3E-05
4510.73	3	1	1.18E-02
4522.32	1	3	8.98E-04
4544.75	3	3	8.3E-04
4554.32	3	5	3.8E-04
4584.96	3	5	1.6E-03
4586.61	3	3	2.3E-03
4587.21	3	1	4.9E-03
4589.29	3	5	6.2E-05
4596.10	3	3	9.47E-04

λ Å	Weights g_i	Weights g_k	A 10^8 s^{-1}
4628.44	3	5	3.83E-04
4642.15	3	5	9.6E-04
4647.49	3	3	1.2E-03
4702.32	3	3	1.09E-03
4746.82	3	1	3.6E-03
4752.94	3	3	4.5E-03
4768.68	3	5	8.6E-03
4798.74	7	9	8.8E-04
4835.97	7	9	9.3E-04
4836.70	3	5	1.02E-03
4876.26	3	5	7.8E-03
4886.29	7	9	1.2E-03
4887.95	3	3	1.3E-02
4894.69	3	1	1.8E-02
4921.04	5	7	5.9E-04
4937.72	7	5	3.6E-04
4956.75	7	9	1.8E-03
4989.95	5	7	1.1E-03
5032.03	7	5	8.2E-04
5048.81	3	5	4.6E-03
5054.18	3	3	4.5E-03
5056.53	3	1	5.7E-03
5060.08	7	9	3.7E-03
5070.99	5	3	2.6E-03
5073.08	3	5	5.9E-04
5078.03	7	7	4.7E-04
5087.09	5	7	1.6E-03
5104.74	3	5	8.7E-04
5118.21	5	7	2.7E-03
5127.80	5	5	3.3E-04
5151.39	3	1	2.39E-02
5152.30	3	5	1.1E-03
5162.29	3	3	1.90E-02
5177.54	7	5	2.4E-03
5192.72	7	7	1.2E-04
5194.02	3	1	7.8E-03
5210.49	7	7	1.1E-03
5214.77	5	3	2.1E-03
5216.28	5	3	1.3E-03
5221.27	7	9	8.8E-03
5241.09	5	5	1.3E-03
5246.24	5	7	1.2E-03
5249.20	5	5	7.9E-04
5252.79	5	7	5.4E-03
5254.47	3	5	3.6E-03
5286.07	5	7	9.6E-04
5290.00	5	3	9.0E-04
5309.52	5	5	1.2E-03
5317.73	5	7	2.6E-03
5373.50	3	5	2.7E-03
5393.27	5	5	9.6E-04
5410.48	5	7	2.0E-03
5421.35	7	5	6.0E-03
5439.99	3	3	1.9E-03
5442.24	7	7	9.3E-04
5451.65	3	5	4.7E-03
5457.42	5	3	3.6E-03
5459.65	7	7	3.8E-04
5467.16	5	5	7.6E-04
5473.46	5	3	2.0E-03
5490.12	5	5	8.5E-04
5492.09	3	1	5.6E-03
5495.87	7	9	1.69E-02
5506.11	5	7	3.6E-03
5524.96	7	7	1.7E-03
5528.97	1	3	1.2E-03
5534.49	5	3	2.7E-03
5540.87	7	5	4.1E-04
5552.77	3	3	7.9E-04
5558.70	3	5	1.42E-02
5559.66	3	5	2.2E-03
5572.54	5	7	6.6E-03
5574.22	3	5	4.6E-04
5581.87	7	5	5.6E-04
5588.72	5	5	1.5E-03
5597.48	5	7	4.2E-03
5606.73	3	3	2.20E-02
5618.01	3	3	2.1E-03
5620.92	3	1	3.6E-03
5623.78	5	5	1.4E-03
5635.58	3	5	9.6E-04
5637.33	1	3	9.1E-04
5639.12	1	3	2.1E-03
5641.39	3	5	8.7E-04
5648.69	5	3	1.2E-03
5650.70	3	1	3.20E-02
5659.13	5	5	2.6E-03
5681.90	5	7	2.0E-03
5683.73	5	5	2.0E-03
5700.87	5	7	5.9E-03
5712.51	1	3	8.7E-04
5739.52	3	5	8.7E-03
5772.11	5	7	2.0E-03
5773.99	5	5	1.1E-03
5783.54	3	5	8.1E-04
5789.48	5	5	4.6E-04
5790.40	5	3	3.4E-04
5802.08	5	3	4.2E-03
5843.77	3	5	3.3E-04
5882.62	3	1	1.23E-02
5888.58	7	5	1.29E-02
5916.58	5	3	5.9E-04
5927.11	7	7	3.7E-04
5928.81	5	3	1.1E-02
5940.86	1	3	1.2E-03
5942.67	5	5	1.8E-03
5943.89	7	5	3.6E-04
5949.26	3	3	1.5E-03
5964.48	1	3	7.7E-04
5968.32	3	3	1.8E-03
5971.60	3	1	1.1E-02
5981.90	5	7	1.2E-04
5987.30	7	7	1.2E-03
5988.13	3	5	6.1E-04
5994.66	3	5	2.6E-04
5999.00	5	5	1.4E-03
6005.73	5	3	1.4E-03
6013.68	7	5	1.4E-03
6025.15	5	3	9.0E-03
6043.22	5	7	1.47E-02
6052.73	3	5	1.9E-03
6064.76	5	7	5.8E-04
6081.25	3	3	7.5E-04
6085.86	3	3	9.0E-05
6090.79	1	3	3.0E-03
6098.81	3	3	5.2E-03
6101.16	3	3	3.3E-03
6104.58	3	1	3.4E-03
6105.64	3	5	1.21E-02
6113.46	3	5	4.7E-04
6119.66	3	3	5.1E-04
6121.86	3	5	1.3E-04
6127.42	5	3	1.1E-03
6128.73	3	5	8.6E-04
6145.44	5	7	7.6E-03
6155.24	5	3	5.1E-03
6165.12	5	5	9.89E-04
6170.17	5	5	5.0E-03
6173.10	3	5	6.7E-03
6179.41	5	3	6.6E-04
6212.50	5	7	3.9E-03
6215.94	5	5	5.7E-03
6230.93	5	5	1.2E-04
6243.40	3	1	1.3E-03
6244.73	3	5	2.0E-04
6248.41	3	5	6.8E-04
6278.65	5	7	2.0E-04
6296.87	3	5	9.0E-03
6307.66	5	5	6.0E-03
6309.14	3	3	7.6E-04
6364.89	3	1	5.6E-03
6369.58	5	3	4.2E-03
6384.72	3	3	4.21E-03
6416.31	3	5	1.16E-02
6431.56	5	3	5.1E-04
6466.55	1	3	1.5E-03
6481.14	1	3	9.4E-04
6513.85	3	3	5.4E-04
6538.11	7	7	1.1E-03
6596.12	7	5	2.3E-04
6598.68	5	5	3.6E-04
6604.02	7	5	2.8E-03
6604.85	5	7	1.3E-04
6632.09	3	3	5.3E-04
6656.88	3	3	3.1E-04
6660.68	3	1	7.8E-03
6664.05	5	5	1.5E-03
6677.28	3	1	2.36E-03
6684.73	3	5	3.9E-04
6698.47	3	3	2.5E-04
6698.88	5	3	1.6E-03
6719.22	1	3	2.4E-03
6722.88	5	7	3.2E-04
6752.84	3	5	1.93E-02
6754.37	3	3	2.1E-03
6756.10	5	5	3.6E-03
6766.61	5	3	4.0E-03

λ Å	Weights g_i	Weights g_k	A 10^8 s^{-1}
6779.93	1	3	1.21E-03
6818.29	3	1	2.0E-03
6827.25	5	3	2.4E-03
6851.88	3	5	6.7E-04
6871.29	3	3	2.78E-02
6879.59	3	5	1.8E-03
6887.10	5	7	1.3E-03
6888.17	3	5	2.5E-03
6925.01	3	3	1.2E-03
6937.67	3	1	3.08E-02
6951.46	5	5	2.2E-03
6960.23	5	5	2.4E-03
6965.43	5	3	6.39E-02
6992.17	3	1	7.5E-03
7030.25	7	5	2.67E-02
7067.22	5	5	3.80E-02
7068.73	5	3	2.0E-02
7086.70	1	3	1.5E-03
7107.48	5	5	4.5E-03
7125.83	3	3	6.0E-03
7147.04	5	3	6.25E-03
7158.83	3	1	2.1E-02
7162.57	1	3	5.8E-04
7206.98	5	3	2.48E-02
7229.93	5	5	6.6E-04
7265.17	3	3	1.7E-03
7270.66	7	7	1.1E-03
7272.93	3	3	1.83E-02
7285.44	5	3	1.2E-03
7311.72	3	3	1.7E-02
7316.01	3	3	9.6E-03
7350.78	3	1	1.2E-02
7353.32	5	7	9.6E-03
7372.12	7	9	1.9E-02
7383.98	3	5	8.47E-02
7392.97	5	3	7.2E-03
7412.33	3	5	3.9E-03
7422.26	3	5	6.6E-04
7425.29	5	7	3.1E-03
7435.33	5	5	9.0E-03
7436.25	7	5	2.7E-03
7471.17	3	3	2.2E-04
7484.24	3	5	3.4E-03
7503.84	3	1	4.45E-01
7510.42	5	5	4.5E-03
7514.65	3	1	4.02E-01
7618.33	3	5	2.9E-03
7628.86	3	5	2.9E-03
7635.11	5	5	2.45E-01
7670.04	5	3	2.8E-03
7704.81	5	7	6.3E-04
7723.76	5	3	5.18E-02
7724.21	1	3	1.17E-01
7798.55	3	5	8.7E-04
7868.20	1	3	3.50E-03
7891.08	5	5	9.5E-03
7916.45	3	3	1.2E-03
7948.18	1	3	1.86E-01
8006.16	3	5	4.90E-02
8014.79	5	5	9.28E-02
8037.23	1	3	3.59E-03
8046.13	3	1	1.12E-02
8053.31	5	3	8.6E-03
8066.60	5	5	1.4E-03
8103.69	3	3	2.5E-01
8115.31	5	7	3.31E-01
8264.52	3	3	1.53E-01
8384.73	5	7	2.4E-03
8408.21	3	5	2.23E-01
8424.65	3	5	2.15E-01
8490.30	3	5	9.6E-04
8521.44	3	3	1.39E-01
8605.78	5	5	1.04E-02
8620.46	1	3	9.2E-03
8667.94	1	3	2.43E-02
8761.69	3	5	9.5E-03
8784.61	3	1	2.4E-03
8799.08	5	3	4.6E-03
8962.19	3	3	1.6E-03
9075.42	3	1	1.2E-02
9122.97	5	3	1.89E-01
9194.64	3	3	1.76E-02
9224.50	3	5	5.03E-02
9291.53	3	1	3.26E-02
9354.22	3	3	1.06E-02
9657.78	3	3	5.43E-02
9784.50	3	5	1.47E-02
10470.05	1	3	9.8E-03
10478.0	3	3	2.44E-02
10950.7	5	3	3.96E-03
11078.9	5	5	8.3E-03
11393.7	3	1	2.22E-02
11441.8	5	3	1.39E-02
11467.5	3	5	3.69E-03
11488.11	3	3	1.9E-03
11668.7	5	5	3.76E-02
11719.5	5	3	9.52E-03
12026.6	1	3	4.2E-03
12112.2	7	7	3.1E-02
12139.8	3	3	4.5E-02
12343.7	5	7	2.0E-02
12402.9	3	3	1.1E-01
12439.2	3	5	4.9E-02
12456.1	5	3	8.9E-02
12487.6	7	5	1.1E-01
12554.4	7	5	1.2E-03
12702.4	3	3	7.1E-02
12733.6	5	5	1.1E-02
12746.3	3	3	2.0E-02
12802.7	5	5	5.7E-02
12933.3	3	1	1.0E-01
12956.6	3	3	7.4E-02
13008.5	5	3	8.9E-02
13214.7	3	1	8.1E-02
13273.1	5	7	1.5E-01
13313.4	3	5	1.3E-01
13504.0	5	7	1.1E-01
13599.2	5	5	2.2E-02
13622.4	3	5	7.3E-02
13678.5	3	5	6.2E-02
14093.6	1	3	4.3E-02
14739.1	5	7	8.8E-04
15046.4	1	3	5.2E-02
15172.3	1	3	1.3E-02
15329.6	5	5	1.2E-03
15555.5	5	7	9.8E-05
15734.9	5	3	2.9E-04
15816.8	5	3	8.7E-04
15989.3	1	3	1.9E-02
16122.7	5	3	3.9E-04
16180.0	5	5	1.2E-03
16264.1	3	3	3.0E-04
16520.1	3	5	2.6E-03
16739.8	3	5	3.1E-03
16940.4	5	5	2.5E-02
20317.0	1	3	1.6E-03
20616.5	5	5	3.9E-03
20812.0	5	7	7.6E-04
21332.2	3	3	3.2E-04
21534.9	3	5	1.1E-03
22039.2	3	1	1.2E-03
22077.4	5	3	1.4E-03
23133.4	3	3	1.7E-03
23844.8	9	7	1.1E-02
23967.5	3	1	3.6E-03
Ar II			
2317.7	6	4	1.4E-01
2891.6	4	2	1.82E-01
2942.9	4	4	5.3E-01
2979.1	2	2	4.16E-01
3033.5	2	4	9.9E-02
3139.0	6	6	5.2E-01
3169.7	4	6	4.9E-01
3181.0	6	4	3.7E-01
3212.5	4	4	5.2E-02
3221.6	6	6	1.8E-02
3226.0	4	4	2.1E-02
3243.7	4	2	1.06E+00
3249.8	2	4	6.3E-01
3263.6	2	4	1.55E-01
3281.7	2	2	4.2E-01
3430.4	6	8	6.2E-02
3454.1	6	4	3.14E-01
3466.3	8	6	3.0E-02
3476.7	6	6	1.25E+00
3491.2	4	4	1.79E+00
3491.5	6	8	2.31E+00
3509.8	2	2	2.55E+00
3514.4	4	6	1.36E+00
3520.0	6	6	5.2E-01
3521.3	8	8	2.27E-01
3535.3	2	4	5.7E-01
3548.5	4	4	8.7E-01
3550.0	6	6	2.6E-02
3556.9	2	2	5.0E-02
3559.5	6	8	2.88E+00

λ Å	Weights g_i	g_k	A 10^8 s^{-1}
3565.0	2	4	5.5E-01
3576.6	6	8	2.75E+00
3581.6	2	4	1.76E+00
3582.4	4	6	2.53E+00
3588.4	8	10	3.03E+00
3605.9	4	6	4.4E-02
3656.0	6	6	7.6E-02
3682.5	4	2	1.7E-02
3709.9	4	4	4.7E-02
3717.2	6	8	5.2E-02
3729.3	6	4	4.80E-01
3746.9	4	6	2.1E-02
3763.5	8	6	1.78E-01
3766.1	4	4	7.4E-02
3777.5	2	2	1.1E-02
3780.8	8	8	7.7E-01
3786.4	8	6	1.5E-02
3799.4	6	4	1.7E-01
3808.6	6	6	1.0E-02
3826.8	6	6	2.81E-01
3841.5	4	2	2.69E-01
3844.7	6	8	4.8E-02
3845.4	6	4	1.6E-02
3850.6	4	4	3.87E-01
3868.5	4	6	1.4E+00
3872.1	4	4	1.5E-01
3875.3	4	2	8.2E-02
3880.3	2	2	2.32E-01
3891.4	2	2	4.3E-02
3892.0	6	4	6.3E-02
3900.6	4	6	7.2E-02
3911.6	2	4	7.7E-02
3914.8	4	4	3.7E-02
3928.6	2	4	2.44E-01
3931.2	2	4	2.0E-02
3932.5	4	4	9.3E-01
3944.3	8	6	4.1E-02
3952.7	4	4	2.08E-01
3958.4	6	4	3.8E-02
3968.4	6	6	4.8E-02
3979.4	4	2	9.8E-01
3988.2	6	6	4.1E-02
3992.1	4	6	1.6E-02
4013.9	8	8	1.05E-01
4031.4	4	2	7.5E-02
4035.5	4	6	4.4E-02
4038.8	6	8	1.2E-02
4042.9	4	4	4.06E-01
4045.7	4	4	1.6E-02
4052.9	2	4	6.7E-01
4065.1	4	4	1.1E-02
4072.0	6	6	5.8E-01
4079.6	6	4	1.19E-01
4082.4	6	6	2.9E-02
4112.8	4	4	1.1E-02
4128.6	8	6	1.4E-02
4131.7	4	2	8.5E-01
4178.4	6	4	1.2E-02
4202.0	2	4	2.1E-02
4228.2	4	6	1.31E-01
4237.2	4	4	1.12E-01
4266.5	6	6	1.64E-01
4277.5	6	4	8.0E-01
4282.9	4	2	1.32E-01
4300.6	6	6	5.7E-02
4331.2	4	4	5.74E-01
4332.0	4	2	1.92E-01
4348.1	6	8	1.17E+00
4352.2	2	2	2.12E-01
4362.1	4	6	5.5E-02
4370.8	4	4	6.6E-01
4371.3	6	4	2.21E-01
4376.0	4	2	2.05E-01
4379.7	2	2	1.00E+00
4383.8	4	4	1.1E-02
4400.1	4	4	1.60E-01
4401.0	8	6	3.04E-01
4412.9	6	8	6.1E-02
4420.9	2	4	3.1E-02
4426.0	4	6	8.17E-01
4430.2	2	4	5.69E-01
4431.0	6	6	1.09E-01
4460.6	4	6	1.5E-02
4474.8	4	2	2.90E-01
4481.8	6	6	4.55E-01
4491.0	6	4	4.6E-02
4530.5	6	4	2.1E-02
4545.1	4	4	4.71E-01
4579.4	2	2	8.0E-01
4589.9	4	6	6.64E-01
4598.8	4	4	6.7E-02
4609.6	6	8	7.89E-01
4637.2	6	6	7.1E-02
4657.9	4	2	8.92E-01
4726.9	4	4	5.88E-01
4732.1	6	4	6.7E-02
4735.9	6	4	5.80E-01
4764.9	2	4	6.4E-01
4806.0	6	6	7.80E-01
4847.8	4	2	8.49E-01
4879.9	4	6	8.23E-01
4889.0	2	2	1.9E-01
4904.8	6	8	3.7E-02
4933.2	4	4	1.44E-01
4965.1	2	4	3.94E-01
4972.2	2	2	9.7E-02
5009.3	4	6	1.51E-01
5017.2	4	6	2.07E-01
5017.6	4	4	1.1E-02
5062.0	2	4	2.23E-01
5141.8	6	8	8.1E-02
5145.3	4	6	1.06E-01
5176.2	6	6	1.7E-02
6103.5	2	2	1.7E-02
6114.9	10	8	2.00E-01
6138.7	6	4	1.2E-02
6172.3	8	6	2.00E-01
6243.1	8	6	3.0E-02
6483.1	4	2	1.06E-01
6638.2	6	4	1.37E-01
6639.7	4	2	1.69E-01
6643.7	10	8	1.47E-01
6666.4	2	2	8.8E-02
6684.3	8	6	1.07E-01
6756.6	4	4	2.0E-02
6863.5	6	6	2.5E-02
7233.5	2	4	3.7E-02
7380.4	4	4	5.6E-02
7589.3	6	4	1.07E-01
Ar III			
769.15	5	3	6.0E+00
871.10	5	3	1.59E+00
875.53	3	1	3.74E+00
878.73	5	5	2.79E+00
879.62	3	3	9.2E-01
883.18	1	3	1.22E+00
887.40	3	5	9.0E-01
3024.1	5	7	2.6E+00
3027.2	5	5	6.4E-01
3054.8	3	5	1.9E+00
3064.8	3	3	1.0E+00
3078.2	1	3	1.4E+00
3285.9	5	7	2.0E+00
3301.9	5	5	2.0E+00
3311.3	5	3	2.0E+00
3336.1	7	9	2.0E+00
3344.7	5	7	1.8E+00
3352.1	7	7	2.2E-01
3358.5	3	5	1.6E+00
3361.3	5	5	3.0E-01
3472.6	5	7	2.0E-01
3480.6	7	7	1.6E+00
3499.7	3	3	1.3E+00
3500.6	3	5	2.6E-01
3502.7	5	3	4.3E-01
3503.6	5	5	1.2E+00
3511.7	7	5	2.6E-01
Ar IV			
840.03	4	2	2.73E+00
843.77	4	4	2.70E+00
850.60	4	6	2.63E+00
Ar VI			
292.15	2	2	6.9E+01
294.05	4	2	1.36E+02
Ar VII			
*250.41	9	3	2.78E+02
*477.54	9	15	9.92E+01
585.75	1	3	7.83E+01
*637.30	9	9	6.7E+01
Ar VIII			
158.92	2	4	1.1E+02
159.18	2	2	1.11E+02

λ Å	Weights g_i	g_k	A 10^8 s^{-1}
229.44	2	2	1.12E+02
230.88	4	2	2.21E+02
337.09	4	4	1.2E+01
337.26	6	4	1.0E+02
338.22	4	2	1.1E+02
519.43	2	4	6.3E+01
526.46	4	6	7.2E+01
526.87	4	4	1.2E+01
700.24	2	4	2.55E+01
713.81	2	2	2.4E+01
Ar IX			
48.739	1	3	1.69E+03
Ar XIII			
162.96	5	3	3.4E+02
*163.08	9	3	5.3E+02
184.90	5	5	1.66E+02
186.38	1	3	8.8E+01
*207.89	9	9	9.5E+01
*245.10	9	15	3.7E+01
Ar XIV			
180.29	2	4	4.5E+01
183.41	2	2	1.69E+02
187.95	4	4	1.97E+02
191.35	4	2	7.5E+01
194.39	2	2	4.6E+01
203.35	4	2	7.8E+01
Ar XV			
25.05	1	3	1.7E+04
221.10	1	3	9.55E+01
*265.3	9	9	8.1E+01
Ar XVI			
*23.52	2	6	1.43E+04
*24.96	6	10	4.4E+04
353.88	2	4	1.5E+01
389.11	2	2	1.1E+01
1268	2	4	1.9E+00
1401	2	2	1.4E+00
2975	2	4	9.0E-02
3514	4	6	6.5E-02
Arsenic			
As I			
1890.4	4	6	2.0E+00
1937.6	4	4	2.0E+00
1972.6	4	2	2.0E+00
2288.1	6	4	2.8E+00
2344.0	2	4	3.5E-01
2349.8	4	2	3.1E+00
2369.7	4	4	6.0E-01
2370.8	4	6	4.2E-01
2456.5	6	4	7.2E-02
2492.9	4	2	1.2E-01
2745.0	2	4	2.6E-01
2780.2	4	4	7.8E-01
2860.4	2	2	5.5E-01
2898.7	4	2	9.9E-02
Barium			
Ba I			
2409.2	1	3	8.6E-04
2414.1	1	3	1.5E-03
2420.1	1	3	2.3E-03
2427.4	1	3	5.6E-03
2432.5	1	3	7.2E-03
2438.8	1	3	1.4E-03
2444.6	1	3	4.5E-03
2452.4	1	3	8.1E-04
2473.2	1	3	4.6E-03
2500.2	1	3	1.5E-02
2543.2	1	3	4.1E-02
2596.6	1	3	1.2E-01
2646.5	1	3	1.1E-02
2702.6	1	3	2.5E-02
2739.2	1	3	9.1E-03
2785.3	1	3	2.8E-02
3071.6	1	3	4.1E-01
3501.1	1	3	1.9E-01
3889.3	1	3	8.8E-03
3909.9	3	5	4.9E-01
3935.7	5	7	4.7E-01
3937.9	5	5	1.1E-01
3993.4	7	9	5.5E-01
3995.7	7	7	8.8E-02
4132.4	1	3	7.1E-03
4239.6	5	3	2.4E-01
4242.6	3	5	5.6E-02
4264.4	1	3	1.5E-01
4283.1	5	7	6.4E-01
4323.0	3	5	1.5E-01
4325.2	5	7	7.1E-02
4332.9	3	3	1.5E-01
4350.3	3	5	6.0E-01
4402.5	3	5	2.7E-01
4406.8	5	5	1.0E-01
4431.9	1	3	1.2E+00
4467.1	5	7	6.6E-02
4489.0	5	7	4.2E-01
4493.6	5	5	3.6E-01
4505.9	3	3	1.1E+00
4523.2	5	5	9.6E-01
4573.9	3	1	1.21E+00
4579.6	5	5	7.0E-01
4591.8	5	5	1.6E-02
4599.7	3	1	4.07E-01
4605.0	3	1	7.7E-02
4619.9	1	3	9.3E-02
4628.3	5	3	6.0E-02
4673.6	7	5	6.5E-02
4691.6	5	3	1.6E+00
4700.4	3	3	2.4E-01
4726.4	5	3	4.6E-01
5519.1	3	5	5.0E-01
5535.5	1	3	1.19E+00
5777.6	5	7	6.5E-01
5784.0	3	5	2.1E-01
5800.2	5	5	9.9E-02
5805.7	7	7	1.1E-02
5826.3	5	3	5.6E-01
5907.6	3	5	1.5E-02
5971.7	5	5	1.8E-01
5997.1	3	3	2.7E-01
6019.5	3	1	1.4E+00
6063.1	5	3	5.7E-01
6083.4	3	1	1.1E-01
6110.8	7	5	5.5E-01
6129.2	3	1	6.0E-02
6341.7	5	7	1.9E-01
6450.9	3	5	1.1E-01
6482.9	5	7	4.4E-01
6498.8	7	7	8.6E-01
6527.3	5	5	5.9E-01
6595.3	3	3	3.9E-01
6675.3	5	3	1.9E-01
6693.8	7	5	2.8E-01
6865.7	5	5	2.3E-02
7059.9	7	9	7.1E-01
7120.3	3	5	2.1E-01
7195.2	1	3	2.4E-01
7280.3	5	7	5.3E-01
7392.4	3	3	5.0E-01
7417.5	7	5	2.5E-02
7488.1	7	7	1.0E-01
7528.2	5	5	2.7E-02
7672.1	3	5	3.1E-01
7780.5	5	5	1.3E-01
7905.8	5	3	6.3E-01
7911.3	1	3	2.98E-03
8147.7	5	5	6.3E-02
9645.6	7	5	1.1E-01
9704.3	3	1	1.6E-01
9821.5	3	1	5.5E-02
10370.3	3	5	1.3E-02
10649.1	5	5	2.7E-02
11075.7	3	3	3.6E-05
11303.1	5	3	1.2E-03
11373.8	3	1	1.3E-01
14158.4	9	7	2.0E-03
14723.2	3	5	8.6E-03
14999.9	5	3	2.8E-03
17123.7	7	7	3.3E-03
17187.1	3	1	2.7E-02
20563.9	5	7	2.6E-03
Ba II			
1413.4	6	8	1.7E-02
1417.1	4	6	3.8E-02
1444.9	4	6	8.1E-02
1461.5	6	8	8.7E-02
1487.0	4	6	1.4E-01
1503.9	6	8	1.5E-01
1554.4	4	6	2.6E-01
1572.7	6	8	2.4E-01

λ Å	Weights g_i	Weights g_k	A 10^8 s^{-1}
1573.9	6	6	1.6E-02
1630.4	2	2	1.7E-02
1674.5	4	6	2.2E-01
1694.4	6	8	2.1E-01
1697.2	6	6	1.7E-02
1761.8	4	4	3.9E-03
1771.0	4	2	3.4E-02
1786.9	6	4	4.4E-02
1892.7	2	4	9.0E-02
1904.2	4	6	1.1E-02
1906.8	2	2	5.1E-02
1924.7	6	8	3.1E-02
1954.2	4	6	1.3E-01
1955.1	4	4	1.8E-02
1970.2	4	2	6.7E-02
1985.6	2	4	2.5E-01
1999.5	2	4	1.0E-01
2009.2	2	2	8.6E-02
2052.7	4	6	2.0E-01
2054.6	4	4	2.9E-02
2080.0	4	2	1.0E-01
2153.9	2	4	5.3E-01
2200.9	2	2	2.0E-01
2232.8	4	6	2.9E-01
2235.4	4	4	4.4E-02
2286.0	4	2	1.3E-01
2528.5	2	4	7.1E-01
2634.8	4	6	7.6E-01
2641.4	4	4	1.2E-01
2647.3	2	2	2.0E-01
2771.4	4	2	4.0E-01
3816.7	4	6	2.3E-03
3842.8	6	8	2.2E-03
3891.8	2	4	1.67E+00
4024.1	6	4	5.3E-03
4057.5	8	6	1.2E-02
4130.7	4	6	1.80E+00
4166.0	4	4	3.7E-01
4216.0	2	4	5.8E-02
4287.8	2	2	2.4E-02
4325.7	4	6	5.9E-02
4329.6	4	4	8.8E-03
4405.2	4	2	3.9E-02
4470.7	6	4	1.4E-02
4509.6	8	6	1.2E-02
4524.9	2	2	7.2E-01
4554.0	2	4	1.17E+00
4708.9	2	4	9.7E-02
4843.5	4	6	9.3E-02
4847.1	2	2	4.1E-02
4850.8	4	4	1.4E-02
4900.0	4	2	7.75E-01
4934.1	2	2	9.55E-01
4997.8	4	2	6.1E-02
5185.0	2	4	1.8E-02
5361.4	4	6	4.8E-02
5391.6	6	8	5.2E-02
5413.6	6	6	8.4E-04
5421.1	6	6	1.9E-03
5428.8	6	4	2.3E-02
5480.3	8	6	1.8E-02
5784.2	2	4	2.0E-01
5853.7	4	4	4.8E-02
5981.3	4	6	1.6E-01
5999.9	4	4	2.6E-02
6135.8	2	2	8.5E-02
6141.7	6	4	3.7E-01
6363.2	6	4	2.9E-03
6372.9	4	4	6.7E-04
6378.9	4	2	9.9E-02
6457.7	6	4	3.0E-03
6496.9	4	2	3.32E-01
7556.8	6	4	1.6E-03
7678.2	8	6	6.6E-04
8710.7	6	8	8.0E-01
8737.7	4	6	9.3E-01

Beryllium

Be I

λ Å	Weights g_i	Weights g_k	A 10^8 s^{-1}
1491.8	1	3	1.3E-02
1661.5	1	3	2.0E-01
2348.6	1	3	5.55E+00
*2494.7	9	15	1.6E+00
*2650.6	9	9	4.24E+00
4572.7	3	5	7.9E-01

Be II

λ Å	Weights g_i	Weights g_k	A 10^8 s^{-1}
1197.1	2	2	4.7E-01
1197.2	4	2	9.4E-01
1512.3	2	4	9.2E+00
1512.4	4	6	1.1E+01
1776.1	2	2	1.4E+00
1776.3	4	2	2.9E+00
*2453.8	2	6	1.42E-01
3046.5	2	4	4.8E-01
3046.7	4	6	5.9E-01
3130.4	2	4	1.14E+00
3131.1	2	2	1.15E+00
3241.6	2	2	1.41E-01
3241.8	4	2	2.8E-01
3274.6	2	4	1.9E-01
3274.7	2	2	1.9E-01
4360.7	2	4	9.2E-01
4361.0	4	6	1.1E+00
*5255.9	2	6	2.56E-02
5270.3	2	2	3.30E-01
5270.8	4	2	6.6E-01
6279.4	2	4	1.2E-01
6279.7	4	6	1.43E-01
6756.7	2	2	5.1E-02
6757.1	4	2	1.02E-01
7401.2	2	4	3.0E-02
7401.4	2	2	3.0E-02

Bismuth

Bi I

λ Å	Weights g_i	Weights g_k	A 10^8 s^{-1}
1954.5	4	6	1.2E+00
2021.2	4	4	6.0E-02
2061.7	4	6	9.9E-01
2110.3	4	2	9.1E-01
2177.3	4	2	2.6E-02
2228.3	4	4	8.9E-01
2230.6	4	6	2.6E+00
2276.6	4	4	2.5E-01
2515.7	4	6	4.3E-02
2627.9	4	4	4.7E-01
2696.8	4	6	6.4E-02
2780.5	4	2	3.09E-01
2798.7	6	6	3.6E-02
2898.0	4	2	1.53E+00
2938.3	6	4	1.23E+00
2989.0	4	4	5.5E-01
2993.3	4	6	1.6E-01
3024.6	6	6	8.8E-01
3067.7	4	2	2.07E+00
3076.7	4	4	3.5E-02
3397.2	6	4	1.81E-01
3402.9	6	6	1.6E-02
3510.9	6	4	6.8E-02
3596.1	2	4	1.98E-01
3888.2	2	2	6.9E-02
4121.5	2	2	1.64E-01
4308.5	2	4	1.6E-02
4493.0	2	4	1.5E-02
4722.5	4	2	1.17E-01
6134.8	4	4	1.8E-02

Boron

B I

λ Å	Weights g_i	Weights g_k	A 10^8 s^{-1}
1378.6	2	4	3.50E+00
1378.9	2	2	1.40E+01
1378.9	4	4	1.75E+01
1379.2	4	2	7.0E+00
1465.5	2	4	3.34E+00
1465.7	4	4	6.7E+00
1465.8	6	4	1.00E+01
1825.9	2	4	1.76E+00
1826.4	4	6	2.11E+00
2088.9	2	4	2.8E-01
2089.6	4	6	3.3E-01
2496.8	2	2	8.64E-01
2497.7	4	2	1.73E+00

Bromine

Br I

λ Å	Weights g_i	Weights g_k	A 10^8 s^{-1}
1488.5	4	4	1.2E+00
1540.7	4	4	1.4E+00
1574.8	2	4	2.0E-01
1576.4	4	6	2.1E-02
1633.4	2	4	8.1E-02
4365.1	2	4	7.5E-03
4425.1	4	2	4.2E-03
4441.7	6	4	7.5E-03
4472.6	4	4	9.3E-03
4477.7	6	8	1.3E-02
4513.4	6	4	2.8E-03

λ Å	Weights g_i	g_k	A 10^8 s^{-1}
4525.6	6	6	7.2E-03
4575.7	4	4	1.6E-02
4614.6	4	6	5.4E-03
4979.8	4	4	2.6E-03
5245.1	2	4	3.1E-03
5345.4	2	4	7.6E-03
7348.5	4	6	1.2E-01
7513.0	6	4	1.2E-01
7803.0	2	4	5.3E-02
7938.7	6	6	1.9E-01
8131.5	2	4	3.8E-02
8343.7	2	2	2.2E-01
8446.6	4	4	1.2E-01
8638.7	6	4	9.7E-02
Br II			
4704.9	5	7	1.1E+00
4785.5	5	5	9.4E-01
4816.7	5	3	1.1E+00
Cadmium			
Cd I			
2288.0	1	3	5.3E+00
2836.9	1	3	2.8E-01
2880.8	3	5	4.2E-01
2881.2	3	3	2.4E-01
2980.6	5	7	5.9E-01
2981.4	5	5	1.5E-01
3261.1	1	3	4.06E-03
3403.7	1	3	7.7E-01
3466.2	3	5	1.2E+00
3467.7	3	3	6.7E-01
3610.5	5	7	1.3E+00
3612.9	5	5	3.5E-01
4140.5	3	5	4.7E-02
4662.4	3	5	5.5E-02
4678.1	1	3	1.3E-01
4799.9	3	3	4.1E-01
5085.8	5	3	5.6E-01
6438.5	3	5	5.9E-01
Cd II			
2144.4	2	4	2.8E+00
2265.0	2	2	3.0E+00
2572.9	2	2	1.7E+00
2748.5	4	2	2.8E+00
4415.6	4	6	1.4E-02
Calcium			
Ca I			
2275.5	1	3	3.01E-01
2995.0	1	3	3.67E-01
2997.3	3	5	2.41E-01
2999.6	3	3	2.79E-01
3000.9	3	1	1.58E+00
3006.9	5	5	7.5E-01
3009.2	5	3	4.30E-01
3344.5	1	3	1.51E-01
3350.2	3	5	1.78E-01

λ Å	Weights g_i	g_k	A 10^8 s^{-1}
3361.9	5	7	2.23E-01
3624.1	1	3	2.12E-01
3630.8	3	5	2.97E-01
3631.0	3	3	1.53E-01
3644.4	5	7	3.55E-01
3644.8	5	5	9.4E-02
3870.5	3	5	7.2E-02
3957.1	3	3	9.8E-02
3973.7	5	3	1.75E-01
4092.6	3	5	1.1E-01
4094.9	5	7	1.2E-01
4098.5	7	9	1.3E-01
4108.5	5	7	9.0E-01
4226.7	1	3	2.18E+00
4283.0	3	5	4.34E-01
4289.4	1	3	6.0E-01
4299.0	3	3	4.66E-01
4302.5	5	5	1.36E+00
4307.7	3	1	1.99E+00
4318.7	5	3	7.4E-01
4355.1	5	7	1.9E-01
4425.4	1	3	4.98E-01
4435.0	3	5	6.7E-01
4435.7	3	3	3.42E-01
4454.8	5	7	8.7E-01
4455.9	5	5	2.0E-01
4526.9	5	3	4.1E-01
4578.6	3	5	1.76E-01
4581.4	5	7	2.09E-01
4585.9	7	9	2.29E-01
4685.3	3	5	8.0E-02
4878.1	5	7	1.88E-01
5041.6	5	3	3.3E-01
5188.9	3	5	4.0E-01
5261.7	3	3	1.5E-01
5262.2	3	1	6.0E-01
5264.2	5	5	9.1E-02
5265.6	5	3	4.4E-01
5270.3	7	5	5.0E-01
5582.0	5	7	6.0E-02
5588.8	7	7	4.9E-01
5590.1	3	5	8.3E-02
5594.5	5	5	3.8E-01
5598.5	3	3	4.3E-01
5601.3	7	5	8.6E-02
5602.9	5	3	1.4E-01
5857.5	3	5	6.6E-01
6102.7	1	3	9.6E-02
6122.2	3	3	2.87E-01
6161.3	5	5	3.3E-02
6162.2	5	3	3.54E-01
6163.8	3	3	5.6E-02
6166.4	3	1	2.2E-01
6169.1	5	3	1.7E-01
6169.6	7	5	1.9E-01
6439.1	7	9	5.3E-01
6449.8	3	5	9.0E-02
6462.6	5	7	4.7E-01
6471.7	7	7	5.9E-02

λ Å	Weights g_i	g_k	A 10^8 s^{-1}
6493.8	3	5	4.4E-01
6499.7	5	5	8.1E-02
Ca II			
1341.9	2	4	1.5E-02
1342.5	2	2	1.5E-02
1649.9	2	4	3.2E-03
1652.0	2	2	3.1E-03
1673.9	2	4	2.24E-01
1680.1	4	6	2.65E-01
1680.1	4	4	4.41E-02
1807.3	2	4	3.54E-01
1814.5	4	6	4.2E-01
1814.7	4	4	7.0E-02
1843.1	2	2	1.6E-01
1850.7	4	2	3.08E-01
2103.2	2	4	8.2E-01
2112.8	4	6	9.7E-01
2113.2	4	4	1.6E-01
2197.8	2	2	3.1E-01
2208.6	4	2	6.2E-01
3158.9	2	4	3.1E+00
3179.3	4	6	3.6E+00
3181.3	4	4	5.8E-01
3706.0	2	2	8.8E-01
3736.9	4	2	1.7E+00
3933.7	2	4	1.47E+00
3968.5	2	2	1.4E+00
Ca III			
357.97	1	3	8.8E+02
439.69	1	3	1.9E-01
490.55	1	3	1.6E-02
Ca V			
558.60	5	3	2.2E+01
637.93	5	3	3.9E+00
643.12	3	1	9.1E+00
646.57	5	5	6.9E+00
647.88	3	3	2.3E+00
651.55	1	3	2.9E+00
656.76	3	5	2.1E+00
Ca VII			
550.20	5	5	1.8E+01
624.39	1	3	3.3E+00
630.54	3	5	4.5E+00
630.79	3	3	2.2E+00
639.15	5	7	5.7E+00
640.41	5	5	1.3E+00
Ca VIII			
182.71	2	2	1.6E+02
184.16	4	2	3.2E+02
Ca IX			
163.23	5	3	3.76E+02
371.89	1	3	8.8E+01
373.81	3	5	1.16E+02

λ Å	Weights g_i	g_k	A 10^8 s^{-1}
378.08	5	7	1.5E+02
395.03	3	5	2.2E+02
466.24	1	3	1.12E+02
498.01	3	5	2.49E+01
506.18	5	5	7.2E+01
515.57	5	3	3.75E+01
Ca X			
110.96	2	4	2.9E+02
111.20	2	2	2.92E+02
151.84	2	2	2.3E+02
153.02	4	2	4.5E+02
206.57	4	4	2.9E+01
206.75	6	4	2.6E+02
207.39	4	2	2.8E+02
411.70	2	4	8.3E+01
419.75	4	6	9.5E+01
420.47	4	4	1.6E+01
557.76	2	4	3.50E+01
574.01	2	2	3.2E+01
Ca XI			
30.448	1	3	6.2E+03
30.867	1	3	4.9E+04
35.212	1	3	2.0E+03
Ca XII			
140.05	4	2	3.7E+02
147.27	2	2	1.6E+02
Ca XV			
141.69	5	3	4.08E+02
*142.23	9	3	6.3E+02
161.00	5	5	1.9E+02
Ca XVII			
19.558	1	3	3.8E+04
21.198	3	5	4.9E+04
192.82	1	3	1.21E+02
218.82	3	5	2.76E+01
223.02	1	3	3.44E+01
228.72	3	3	2.37E+01
232.83	5	5	6.5E+01
244.06	5	3	3.28E+01
Ca XVIII			
*18.71	2	6	2.31E+04
*19.74	6	10	7.0E+04
302.19	2	4	2.0E+01
344.76	2	2	1.3E+01
Carbon			
C I			
945.191	1	3	3.79E+00
945.338	3	3	1.14E+01
945.579	5	3	1.89E+01
1193.24	5	7	1.22E+00
1260.74	1	3	5.32E-01
1260.93	3	1	1.70E+00
1261.00	3	3	4.42E-01
1261.12	3	5	3.71E-01
1261.43	5	3	7.06E-01
1261.55	5	5	1.27E+00
1274.11	5	7	1.03E-02
1277.25	1	3	1.27E+00
1277.28	3	5	1.73E+00
1277.51	3	3	9.12E-01
1277.55	5	7	2.31E+00
1277.72	5	5	6.35E-01
1277.95	5	3	5.56E-02
1279.23	5	7	1.10E-01
1279.89	3	5	3.08E-01
1280.14	1	3	3.11E-01
1280.33	5	5	5.77E-01
1280.40	3	3	1.73E-01
1280.60	3	1	8.22E-01
1280.85	5	3	3.33E-01
1328.83	1	3	7.95E-01
1329.09	3	1	2.41E+00
1329.58	5	5	1.79E+00
1329.60	5	3	1.00E+00
1355.84	5	7	1.04E+00
1364.16	5	5	1.57E-01
1431.60	5	7	2.11E+00
1432.10	5	5	2.01E+00
1432.53	5	3	2.11E+00
1459.03	5	3	4.76E-01
1463.34	5	7	1.88E+00
1467.40	5	3	5.49E-01
1468.41	5	3	3.90E-02
1470.09	5	7	1.37E-02
1472.23	5	3	8.01E-03
1481.76	5	5	3.92E-01
1560.31	1	3	6.57E-01
1561.34	5	5	2.94E-01
1561.44	5	7	1.18E+00
1656.27	3	5	8.58E-01
1656.93	1	3	1.13E+00
1657.01	5	5	2.52E+00
1657.38	3	3	8.64E-01
1657.91	3	1	3.43E+00
1658.12	5	3	1.44E+00
1751.83	1	3	9.07E-01
1763.91	1	3	3.59E-02
1765.37	1	3	1.04E-02
1930.90	5	3	3.51E+00
2478.56	1	3	3.40E-01
2902.23	1	3	4.32E-03
2903.27	3	3	1.29E-02
2905.00	5	3	2.15E-02
4371.37	3	3	1.27E-02
4762.31	1	3	3.37E-03
4762.53	3	5	2.72E-03
4766.67	3	3	2.36E-03
4770.03	3	1	1.07E-02
4771.74	5	5	7.97E-03
4775.90	5	3	4.84E-03
4812.92	1	3	4.03E-04
4817.37	3	3	8.76E-04
4826.80	5	3	6.28E-04
4932.05	3	1	6.02E-02
5023.84	7	9	1.81E-03
5039.06	7	9	4.73E-03
5041.48	3	5	5.25E-03
5041.79	5	7	3.28E-03
5052.17	3	5	2.60E-02
5380.34	3	3	1.86E-02
5545.05	3	3	3.04E-03
5668.94	3	3	2.35E-02
5793.12	7	5	3.44E-03
5794.47	5	5	6.44E-04
5800.23	3	3	1.04E-03
5800.60	5	3	3.04E-03
5805.20	3	1	4.12E-03
6001.12	5	5	3.22E-03
6006.02	7	5	1.79E-02
6007.18	3	3	5.34E-03
6010.68	3	1	2.13E-02
6013.17	7	5	1.79E-02
6013.21	7	9	4.35E-03
6014.83	5	3	1.60E-02
6016.45	5	7	3.86E-03
6587.61	3	3	5.09E-02
6655.52	3	3	5.03E-03
6828.12	3	5	9.89E-03
7111.47	3	5	2.17E-02
7113.18	7	9	2.47E-02
7115.17	5	7	2.19E-02
7115.18	3	1	4.43E-02
7116.99	7	5	3.26E-02
7119.66	5	3	3.12E-02
7860.88	5	5	1.53E-02
8058.62	5	5	1.09E-02
8335.15	3	1	3.51E-01
9061.44	3	5	7.31E-02
9062.49	1	3	9.48E-02
9078.29	3	3	7.07E-02
9088.51	3	1	3.00E-01
9094.83	5	5	2.28E-01
9111.81	5	3	1.35E-01
9405.73	3	5	2.91E-01
9603.03	1	3	3.06E-02
9620.78	3	3	8.62E-02
9658.43	5	3	1.25E-01
C II			
687.345	4	6	2.84E+01
858.092	2	2	1.18E+00
858.559	4	2	2.35E+00
903.623	2	4	6.85E+00
903.962	2	2	2.74E+01
904.142	4	4	3.42E+01
904.480	4	2	1.37E+01
1009.86	2	4	5.71E+00
1010.08	4	4	1.14E+01
1010.37	6	4	1.71E+01
1036.34	2	2	7.61E+00

λ Å	Weights g_i	g_k	A 10^8 s^{-1}
1037.02	4	2	1.52E+01
1323.91	4	4	4.33E+00
1323.95	6	6	4.49E+00
1334.53	2	4	2.37E+00
1335.71	4	6	2.84E+00
2091.14	2	4	1.00E-01
2091.19	4	6	1.69E-01
2091.65	6	8	2.41E-01
2093.16	6	6	7.20E-02
2173.85	2	4	2.31E-01
2174.17	2	2	2.31E-01
2509.13	2	4	4.53E-01
2511.74	4	4	9.04E-02
2512.06	4	6	5.42E-01
2727.31	2	4	6.63E-02
2728.72	4	4	3.31E-01
2729.21	2	2	2.65E-01
2730.63	4	2	1.32E-01
5132.95	2	4	3.89E-01
5133.28	4	6	2.80E-01
5137.26	2	2	1.55E-01
5139.17	4	4	1.24E-01
5143.49	4	2	7.73E-01
5145.16	6	6	6.49E-01
5151.08	6	4	4.16E-01
5640.55	2	4	9.89E-02
5648.07	4	4	1.97E-01
5662.46	6	4	2.93E-01
5818.31	2	2	3.38E-02
5822.98	2	4	3.38E-03
5823.18	4	2	3.38E-02
5827.85	4	4	2.16E-02
5836.37	6	4	4.22E-02
5843.62	6	6	1.20E-02
5856.06	8	6	5.31E-02
6095.29	2	4	4.20E-01
6098.51	4	6	5.03E-01
6102.56	4	4	8.37E-02
6578.05	2	4	3.63E-01
6582.88	2	2	3.62E-01
6724.56	2	4	3.17E-02
6727.07	2	2	6.34E-02
6727.26	4	6	2.96E-02
6731.07	4	4	5.06E-02
6733.58	4	2	6.32E-02
6734.00	6	8	1.80E-02
6738.61	6	6	7.23E-02
6742.43	6	4	4.41E-02
6750.54	8	8	1.08E-01
6755.16	8	6	2.38E-02
6779.94	4	6	2.56E-01
6780.59	2	4	1.52E-01
6783.91	6	8	3.65E-01
6787.21	2	2	3.04E-01
6791.47	4	4	1.94E-01
6798.10	4	2	6.04E-02
6800.69	6	6	1.09E-01
6812.28	6	4	1.80E-02

λ Å	Weights g_i	g_k	A 10^8 s^{-1}
7046.25	4	2	3.20E-01
7053.09	4	4	3.19E-01
7063.68	4	6	3.17E-01
7112.48	2	4	2.94E-01
7113.04	4	6	3.15E-01
7115.63	6	8	3.60E-01
7119.76	4	4	1.17E-01
7119.91	8	10	4.19E-01
7125.72	6	6	1.02E-01
7132.47	6	4	8.33E-03
7134.10	8	8	5.93E-02
7231.33	2	4	3.52E-01
7236.42	4	6	4.22E-01
7237.17	4	4	7.03E-02
8028.85	2	2	1.71E-02
8037.73	2	4	4.26E-02
8039.40	4	2	8.51E-02
8048.31	4	4	1.36E-02
8062.10	4	6	3.04E-02
8062.80	6	4	4.56E-02
8076.64	6	6	7.05E-02
9238.30	4	6	3.34E-02
9251.01	2	4	2.77E-02
9863.06	2	4	5.56E-02
9870.78	4	6	9.31E-02
9882.68	6	8	1.33E-01

C III

λ Å	Weights g_i	g_k	A 10^8 s^{-1}
310.170	1	3	6.56E+00
386.203	1	3	3.46E+01
459.466	1	3	5.91E+01
459.514	3	5	7.97E+01
459.627	5	7	1.06E+02
574.281	3	5	6.24E+01
977.020	1	3	1.767E+01
1174.93	3	5	3.293E+00
1175.26	1	3	4.385E+00
1175.59	3	3	3.287E+00
1175.71	5	5	9.856E+00
1175.99	3	1	1.313E+01
1176.37	5	3	5.468E+00
1247.38	3	1	2.082E+01
2296.87	3	5	1.376E+00
2849.05	3	1	1.95E-01
3703.70	3	3	5.90E-01
4325.56	3	5	1.24E-01
4647.42	3	5	7.26E-01
4650.25	3	3	7.25E-01
4651.02	3	5	2.28E-01
4651.47	3	1	7.24E-01
4652.05	1	3	3.04E-01
4659.06	3	3	2.27E-01
4663.64	3	1	9.05E-01
4665.86	5	5	6.78E-01
4673.95	5	3	3.75E-01
5244.66	1	3	5.30E-02
5253.58	3	3	1.58E-01
5272.52	5	3	2.61E-01
5695.92	3	5	4.27E-01
5858.34	3	1	1.34E-01
5863.25	3	3	3.35E-02
5871.68	5	3	1.00E-01
5880.56	5	5	1.99E-02
5894.07	7	5	1.11E-01
6727.48	1	3	1.12E-01
6731.04	3	5	1.50E-01
6742.15	3	3	8.32E-02
6744.39	5	7	1.99E-01
6762.17	5	5	4.95E-02
6773.39	5	3	5.47E-03
6851.18	3	5	7.60E-03
6853.68	5	7	5.64E-03
6857.24	3	3	3.79E-02
6862.69	5	5	3.51E-02
6868.78	5	3	1.26E-02
6872.04	7	7	4.46E-02
6881.10	7	5	7.80E-03
7353.88	5	3	3.09E-02
7707.43	3	5	1.30E-01
7771.76	3	1	1.77E-01
7780.41	3	3	1.76E-01
7796.00	3	5	1.75E-01
8500.32	1	3	1.01E-01
9593.32	3	3	5.32E-03
9651.47	5	5	1.57E-02
9696.48	5	7	7.53E-03
9696.54	3	5	7.12E-03
9699.57	7	9	8.47E-03
9701.10	1	3	4.40E-02
9705.41	3	5	5.93E-02
9706.44	3	3	3.29E-02
9715.09	5	7	7.88E-02
9717.75	5	5	1.97E-02
9718.79	5	3	2.19E-03

C IV

λ Å	Weights g_i	g_k	A 10^8 s^{-1}
*312.43	2	6	4.63E+01
*384.13	6	10	1.76E+02
1548.19	2	4	2.65E+00
1550.77	2	2	2.64E+00
5801.31	2	4	3.17E-01
5811.97	2	2	3.16E-01

C V

λ Å	Weights g_i	g_k	A 10^8 s^{-1}
34.9728	1	3	2.554E+03
40.2678	1	3	8.873E+03
*227.19	3	9	1.363E+02
247.315	1	3	1.278E+02
*248.71	9	15	4.247E+02
*260.19	9	3	6.680E+01
267.267	3	5	3.947E+02
*2273.9	3	9	5.646E-01
3526.66	1	3	1.663E-01
8420.72	3	5	6.898E-02
*8433.2	3	9	6.868E-02
8448.12	3	1	6.832E-02

λ Å	Weights g_i	g_k	A $10^8\ s^{-1}$
8449.19	3	3	6.829E-02
Cesium			
Cs I			
3203.5	2	4	7.6E-06
3205.3	2	4	7.9E-06
3207.5	2	4	8.5E-06
3210.0	2	4	9.4E-06
3212.8	2	4	1.19E-05
3216.2	2	4	1.49E-05
3220.1	2	4	1.7E-05
3220.2	2	2	1.07E-07
3224.8	2	4	2.0E-05
3225.0	2	2	1.43E-07
3230.5	2	4	2.5E-05
3230.7	2	2	1.97E-07
3237.4	2	4	2.8E-05
3237.6	2	2	2.63E-07
3245.9	2	4	3.45E-05
3246.2	2	2	3.7E-07
3256.7	2	4	4.25E-05
3257.1	2	2	7.0E-07
3270.5	2	4	5.6E-05
3271.0	2	2	9.8E-07
3288.6	2	4	1.0E-04
3289.3	2	2	2.7E-06
3313.1	2	4	1.6E-04
3314.0	2	2	5.2E-06
3347.5	2	4	2.2E-04
3348.8	2	2	1.1E-05
3397.9	2	4	4.0E-04
3400.0	2	2	2.4E-05
3476.8	2	4	6.6E-04
3480.0	2	2	6.6E-05
3611.4	2	4	1.5E-03
3617.3	2	2	2.5E-04
3876.1	2	4	3.8E-03
3888.6	2	2	9.7E-04
4555.3	2	4	1.88E-02
4593.2	2	2	8.0E-03
8521.1	2	4	3.276E-01
8943.5	2	2	2.87E-01
Chlorine			
Cl I			
1188.8	4	6	2.33E+00
1188.8	4	4	2.71E-01
1201.4	2	4	2.39E+00
1335.7	4	2	1.74E+00
1347.2	4	4	4.19E+00
1351.7	2	2	3.23E+00
1363.4	2	4	7.5E-01
4323.3	4	4	1.1E-02
4363.3	4	6	6.8E-03
4379.9	4	4	1.4E-02
4389.8	6	8	1.4E-02
4526.2	4	4	5.1E-02
4601.0	2	2	4.2E-02
4661.2	2	4	1.2E-02

λ Å	Weights g_i	g_k	A $10^8\ s^{-1}$
7256.6	6	4	1.5E-01
7414.1	6	4	4.7E-02
7547.1	4	4	1.2E-01
7717.6	4	4	3.0E-02
7745.0	2	4	6.3E-02
7769.2	6	6	6.0E-02
7821.4	6	8	9.8E-02
7830.8	4	4	9.7E-02
7878.2	6	6	1.8E-02
7899.3	4	6	5.1E-02
7924.6	2	4	2.1E-02
7935.0	6	8	3.9E-02
7997.9	4	4	2.1E-02
Cl II			
3329.1	5	7	1.5E+00
3522.1	7	7	1.4E+00
3798.8	5	7	1.6E+00
3805.2	7	9	1.8E+00
3809.5	3	5	1.5E+00
3851.0	5	7	1.8E+00
3851.4	5	5	1.6E+00
3854.7	3	5	2.2E+00
3861.9	5	7	2.4E+00
3868.6	7	9	2.7E+00
3913.9	9	9	8.2E-01
3990.2	5	7	8.4E-01
4132.5	5	5	1.6E+00
4276.5	9	7	7.6E-01
4768.7	3	5	7.7E-01
4781.3	5	7	1.0E+00
4794.6	5	7	1.04E+00
4810.1	5	5	9.9E-01
4819.5	5	3	1.00E+00
4904.8	5	7	8.1E-01
4917.7	3	5	7.5E-01
5078.3	7	7	7.7E-01
5219.1	3	9	8.6E-01
5392.1	5	7	1.0E+00
Cl III			
2298.5	4	4	4.2E+00
2340.6	6	6	4.2E+00
2370.4	8	6	2.8E+00
2531.8	2	4	4.4E+00
2532.5	4	6	5.3E+00
2577.1	4	6	4.3E+00
2580.7	6	8	4.7E+00
2601.2	2	4	4.6E+00
2603.6	4	6	5.0E+00
2609.5	6	8	5.7E+00
2617.0	8	10	6.6E+00
2661.6	4	6	3.4E+00
2665.5	6	8	4.8E+00
2691.5	4	4	3.5E+00
2710.4	4	6	3.5E+00
3340.4	6	6	1.5E+00
3392.9	4	4	1.9E+00
3393.5	6	6	1.9E+00

λ Å	Weights g_i	g_k	A $10^8\ s^{-1}$
3530.0	6	8	1.8E+00
3560.7	4	6	1.7E+00
3602.1	6	8	1.7E+00
3612.9	4	6	1.2E+00
3720.5	4	6	1.7E+00
Chromium			
Cr I			
1999.95	9	9	1.4E+00
2383.30	9	11	4.1E-01
2389.21	3	5	2.3E-01
2408.60	9	7	6.7E-01
2408.72	7	5	2.9E-01
2492.57	3	5	4.5E-01
2495.08	3	3	2.7E-01
2496.30	5	7	5.6E-01
2502.55	7	9	2.2E-01
2504.31	7	9	4.5E-01
2508.11	5	5	2.1E-01
2508.97	5	3	3.8E-01
2527.11	9	9	5.3E-01
2549.55	3	3	4.8E-01
2560.70	5	5	4.3E-01
2571.74	7	5	6.4E-01
2577.66	7	7	2.6E-01
2591.84	9	7	6.5E-01
2620.48	5	3	1.9E-01
2673.64	3	3	1.8E-01
2701.99	9	11	2.1E-01
2726.50	5	7	7.5E-01
2731.90	5	5	7.8E-01
2736.46	5	3	7.5E-01
2752.85	3	3	8.7E-01
2757.09	5	5	6.8E-01
2761.74	5	3	6.8E-01
2764.36	7	7	3.7E-01
2769.90	7	5	1.1E+00
2780.70	9	7	1.4E+00
2879.27	5	7	2.1E-01
2887.00	3	5	2.7E-01
2889.22	9	9	6.6E-01
2893.25	7	7	5.2E-01
2894.17	1	3	3.3E-01
2896.76	5	5	3.0E-01
2905.48	3	1	1.3E+00
2909.05	5	3	6.8E-01
2910.89	7	5	3.4E-01
2911.15	9	7	2.6E-01
2967.64	7	9	3.9E-01
2971.10	5	7	7.1E-01
2975.48	3	5	8.9E-01
2980.78	1	3	5.10E-01
2988.64	5	7	5.2E-01
2991.88	3	1	3.0E+00
2994.06	5	5	2.5E-01
2995.09	5	5	4.3E-01
2996.57	5	3	2.0E+00
2998.78	5	3	4.07E-01
3000.88	7	5	1.6E+00

λ Å	Weights g_i	Weights g_k	A 10^8 s^{-1}
3005.06	9	7	9.2E-01
3013.72	3	5	8.3E-01
3015.20	1	3	1.63E+00
3020.67	3	3	1.5E+00
3021.58	9	11	2.91E+00
3024.36	5	5	1.27E+00
3029.17	5	3	3.8E-01
3030.25	7	7	1.1E+00
3031.35	5	3	3.1E-01
3034.19	7	7	3.5E-01
3037.05	9	9	5.4E-01
3040.84	7	5	7.4E-01
3053.87	9	7	7.97E-01
3148.44	9	11	5.6E-01
3155.16	11	13	5.7E-01
3163.76	13	15	6.0E-01
3237.73	9	9	1.3E+00
3238.09	11	11	2.0E-01
3578.68	7	9	1.48E+00
3593.48	7	7	1.50E+00
3605.32	7	5	1.62E+00
3639.80	13	11	1.8E+00
3743.89	13	13	7.61E-01
3757.66	7	7	4.13E-01
3768.24	5	5	5.10E-01
3804.80	9	9	6.9E-01
3963.69	13	15	1.3E+00
3969.75	11	13	1.2E+00
3983.90	7	9	1.05E+00
3991.12	5	7	1.07E+00
4001.44	9	11	6.8E-01
4039.10	15	15	6.7E-01
4048.78	13	13	6.4E-01
4058.78	11	11	6.7E-01
4065.71	9	11	3.5E-01
4165.52	11	13	7.5E-01
4204.48	13	11	3.1E-01
4254.33	7	9	3.15E-01
4263.15	15	17	6.4E-01
4274.81	7	7	3.07E-01
4275.98	11	11	2.2E-01
4280.42	13	15	4.7E-01
4289.73	7	5	3.16E-01
4291.97	7	5	2.4E-01
4297.75	11	13	4.9E-01
4298.05	9	9	2.6E-01
4300.52	9	7	1.9E-01
4301.19	11	9	2.6E-01
4302.78	11	11	2.5E-01
4319.66	5	3	1.8E-01
4337.25	5	7	2.0E-01
4373.65	9	9	2.8E-01
4376.80	13	13	3.2E-01
4413.86	7	5	2.7E-01
4422.70	5	5	2.7E-01
4424.29	9	7	2.1E-01
4429.93	3	3	2.4E-01
4432.16	1	3	1.8E-01
4432.77	15	15	4.9E-01
4443.72	3	1	4.5E-01
4482.88	3	3	3.0E-01
4490.55	9	7	3.9E-01
4492.31	5	3	4.47E-01
4495.28	9	7	2.0E-01
4500.29	7	7	2.1E-01
4506.84	13	11	2.7E-01
4540.72	11	11	3.14E-01
4564.17	11	13	5.1E-01
4595.60	13	13	4.7E-01
4622.47	7	7	4.1E-01
4663.33	3	3	2.0E-01
4665.90	3	3	3.0E-01
4689.38	7	5	2.3E-01
4698.46	9	7	2.2E-01
4708.02	11	9	4.31E-01
4718.43	13	11	3.4E-01
4730.69	7	5	3.83E-01
4737.33	9	7	3.38E-01
4741.09	3	5	2.2E-01
4752.07	13	13	6.2E-01
4756.09	11	9	4.0E-01
4792.49	7	5	2.6E-01
4801.02	9	7	3.06E-01
4816.13	9	9	1.8E-01
4870.79	7	9	3.5E-01
4887.01	9	11	3.2E-01
4922.28	11	13	4.0E-01
4966.80	3	1	3.0E-01
5204.51	5	3	5.09E-01
5206.02	5	5	5.14E-01
5208.42	5	7	5.06E-01
5243.38	5	3	2.19E-01
5297.37	7	9	3.88E-01
5297.99	7	7	3.0E-01
5328.36	9	11	6.2E-01
5329.17	9	9	2.25E-01
5783.11	3	3	2.1E-01
5783.89	5	5	2.02E-01
5787.97	5	7	2.35E-01
Cr II			
2653.57	4	6	3.5E-01
2658.59	2	4	5.8E-01
2666.02	6	8	5.9E-01
2668.71	4	2	1.4E+00
2671.80	6	4	1.0E+00
2672.83	8	6	5.5E-01
2744.97	4	6	8.5E-01
2787.61	6	6	1.5E+00
2822.38	14	16	2.3E+00
2835.63	10	12	2.0E+00
2840.01	10	12	2.7E+00
2843.24	8	10	6.4E-01
2849.83	6	8	9.2E-01
2851.35	8	10	2.2E+00
2856.77	4	6	4.3E-01
2857.40	6	8	2.8E-01
2860.92	2	4	6.9E-01
2862.57	8	8	6.3E-01
2866.72	4	4	1.2E+00
2867.09	4	4	1.1E+00
2867.65	2	2	1.1E+00
2870.43	6	6	1.3E+00
2873.81	4	2	8.8E-01
2880.86	6	4	7.9E-01
2898.53	10	12	1.2E+00
2921.81	8	10	9.0E-01
2930.83	2	4	1.1E+00
2935.12	6	8	1.8E+00
2953.34	2	2	1.8E+00
2966.03	10	8	5.4E-01
2971.90	14	14	2.0E+00
2979.73	12	12	1.8E+00
2985.32	10	10	2.2E+00
2989.18	8	8	2.2E+00
3118.64	2	4	1.7E+00
3120.36	4	6	1.5E+00
3122.59	12	12	4.4E-01
3128.69	4	4	8.1E-01
3136.68	6	6	6.4E-01
4588.22	8	6	1.2E-01
Cr V			
434.306	9	9	1.5E+01
436.351	9	7	2.4E+01
436.601	7	5	2.1E+01
437.420	7	7	1.4E+01
437.655	5	5	1.3E+01
441.056	5	3	2.3E+01
456.357	1	3	9.5E+00
456.637	3	1	3.3E+01
456.743	3	3	9.1E+00
457.028	5	5	2.7E+01
457.504	5	3	1.2E+01
464.015	9	7	3.6E+01
469.634	5	5	2.3E+01
1106.25	7	9	1.2E+01
1121.07	7	9	2.1E+01
1127.63	9	11	3.5E+01
1465.86	5	3	1.1E+01
1481.65	3	1	1.0E+01
1519.03	5	7	9.5E+00
1579.70	7	9	8.6E+00
Cr VI			
161.687	6	6	1.7E+02
168.088	4	6	2.0E+02
201.007	4	4	2.5E+03
201.224	4	6	1.8E+02
201.388	6	4	2.7E+02
201.606	6	6	2.6E+03
202.442	6	4	1.0E+03
202.739	4	2	1.2E+03
226.241	6	8	7.2E+02

λ Å	Weights g_i	Weights g_k	A 10^8 s^{-1}
227.202	4	6	6.6E+02
Cr X			
216.72	6	8	9.0E+02
223.86	4	2	7.7E+02
224.74	4	4	7.6E+02
226.24	4	6	7.3E+02
227.42	4	4	5.2E+02
227.50	4	6	1.8E+01
228.63	6	4	8.1E+01
228.71	6	6	4.5E+02
231.21	2	4	1.2E+02
232.96	4	4	4.4E+02
242.20	2	4	5.0E+01
244.19	4	6	5.8E+01
395.984	4	4	2.4E+01
398.150	6	6	2.1E+01
Cr XI			
214.31	5	7	1.4E+01
226.45	5	7	6.0E+02
232	3	1	4.1E+02
235.53	5	7	5.5E+02
240.76	1	3	4.8E+02
250.28	5	7	1.0E+01
366.491	3	3	1.2E+01
366.942	3	1	3.0E+01
374.927	5	5	2.3E+01
422.083	3	5	1.0E+01
Cr XII			
216	4	6	2.4E+02
218	6	8	2.4E+02
239	2	2	1.6E+02
244.70	2	4	3.0E+02
247	4	2	2.4E+02
247	2	2	3.3E+02
248	6	8	1.4E+02
250	6	8	3.5E+02
250	6	6	2.2E+02
251.52	4	6	3.4E+02
252	4	6	2.0E+02
256	2	2	1.5E+02
259	2	4	3.2E+02
269	2	2	2.1E+02
300.32	2	2	1.4E+02
305.81	4	4	2.76E+02
309	4	2	2.7E+02
309	6	6	1.6E+02
311.55	4	2	1.6E+02
324	4	6	2.2E+02
327	6	8	2.2E+02
332.06	6	4	1.4E+02
Cr XIII			
49.59	1	3	9.9E+02
67.01	1	3	1.67E+03
228	5	7	1.8E+02
267.73	5	7	1.9E+02
270	3	1	1.7E+02
276.4	5	7	2.2E+02
277	1	3	2.1E+02
279.32	3	5	3.5E+02
286	3	1	4.6E+02
328.29	1	3	1.86E+02
345	7	9	1.74E+02
Cr XIV			
*38.036	2	6	2.47E+02
39.796	2	4	3.05E+02
40.018	4	6	3.6E+02
40.782	2	4	3.9E+02
40.800	2	2	3.9E+02
41.556	2	4	4.5E+02
41.788	4	6	5.3E+02
44.597	2	4	7.1E+02
44.869	4	6	8.3E+02
46.125	4	2	3.1E+02
46.468	2	4	6.6E+02
46.527	2	2	6.7E+02
48.300	4	6	5.9E+02
48.338	6	8	6.3E+02
50.821	2	4	1.2E+03
51.172	4	6	1.4E+03
51.180	4	4	2.3E+02
52.321	4	6	1.0E+03
52.363	6	8	1.1E+03
53.760	2	2	3.0E+02
54.164	4	2	5.9E+02
60.699	4	6	2.05E+03
60.756	6	8	2.19E+03
63.324	2	4	1.07E+03
63.539	2	2	1.13E+03
68.594	2	4	1.98E+03
69.213	4	6	2.31E+03
69.247	4	4	3.8E+02
86.060	4	6	5.3E+03
86.169	6	8	5.9E+03
86.185	6	6	3.9E+02
101.05	6	4	4.4E+02
101.42	4	2	4.83E+02
104.4	4	6	3.0E+02
104.5	6	8	3.1E+02
109.8	2	4	2.3E+02
110.4	4	6	2.8E+02
118.3	4	2	2.1E+02
125.2	4	6	5.0E+02
125.3	6	8	5.4E+02
148.5	2	4	2.18E+02
149.1	2	2	2.1E+02
157.1	2	4	3.3E+02
158.4	4	6	3.7E+02
187.02	4	6	9.3E+02
187.30	6	8	9.6E+02
189.1	2	2	2.13E+02
191.0	4	2	4.11E+02
222.9	4	2	2.2E+02
346.3	4	6	2.4E+02
346.5	6	8	2.5E+02
Cr XV			
18.497	1	3	1.62E+05
18.782	1	3	2.8E+04
19.015	1	3	6.3E+02
20.863	1	3	6.0E+03
21.153	1	3	5.6E+03
102	3	3	1.6E+02
102.18	5	3	7.0E+02
103	3	1	3.8E+02
105	7	5	5.3E+02
111.27	3	3	1.7E+02
Cr XVI			
17.073	4	6	1.2E+04
17.242	2	4	8.6E+04
17.299	4	4	2.5E+04
17.372	4	4	1.4E+05
17.438	4	2	1.1E+05
17.514	2	4	1.1E+05
17.587	2	4	2.0E+04
17.656	2	2	2.0E+04
19.442	4	2	9.9E+03
19.714	2	2	1.1E+04
Cr XVII			
16.31	5	3	9.6E+03
16.32	5	7	3.2E+04
16.37	3	1	9.7E+04
16.44	5	7	1.3E+05
16.59	3	1	5.7E+04
16.65	5	5	1.1E+04
16.66	1	3	1.8E+05
16.68	5	7	6.8E+04
16.80	5	7	4.4E+04
16.97	1	3	2.63E+04
16.97	3	3	1.5E+04
17.968	5	3	8.6E+03
18.336	5	3	1.7E+04
18.336	5	5	1.6E+04
18.389	1	3	9.2E+03
Cr XVIII			
95.77	4	2	3.08E+02
102.32	4	4	1.54E+02
104.98	6	4	8.7E+02
106.84	4	2	3.4E+02
110.41	4	2	7.9E+02
112.27	4	2	4.24E+02
119.62	2	2	3.2E+02
123.87	6	4	3.9E+02
125.51	4	4	3.4E+02
128.10	6	6	2.8E+02
136.52	4	2	1.66E+02
139.87	4	4	1.49E+02
140.82	4	2	2.66E+02
155.46	2	2	2.84E+02
157.40	4	4	2.83E+02

λ Å	Weights g_i	g_k	A 10^8 s^{-1}
Cr XIX			
14.73	3	3	7.1E+04
14.80	1	3	1.3E+05
14.81	5	3	3.4E+04
14.84	5	7	1.3E+05
109.64	3	3	2.46E+02
110.37	5	3	6.0E+02
113.97	5	3	5.5E+02
118.31	3	1	3.29E+02
118.67	5	3	2.1E+02
118.83	3	3	1.35E+02
126.30	1	3	1.56E+02
126.33	5	5	4.35E+02
130.99	7	5	2.9E+02
134.89	3	1	1.98E+02
138.15	3	1	1.75E+02
138.45	5	5	1.71E+02
140.92	5	3	1.38E+02
143.57	3	1	7.2E+02
163.94	5	5	3.1E+02
179.18	3	1	1.45E+02
Cr XX			
14.13	2	4	1.1E+05
14.26	4	6	1.3E+05
128.42	4	4	3.8E+02
131.31	6	4	1.27E+02
133.82	2	4	8.3E+01
135.26	4	2	2.41E+02
140.75	4	4	1.35E+02
148.99	6	4	1.75E+02
156.00	2	4	8.4E+01
167.97	6	6	1.12E+02
180.85	4	4	1.6E+02
Cr XXI			
12.97	3	1	4.8E+04
12.98	5	5	3.9E+04
13.02	3	5	3.8E+04
13.02	5	7	3.9E+04
13.08	1	3	5.2E+04
13.22	3	1	4.6E+04
13.34	3	5	5.2E+04
13.49	1	3	9.0E+04
13.53	3	3	6.6E+04
13.55	3	5	1.2E+05
13.65	5	7	1.5E+05
13.66	3	1	1.2E+05
13.67	5	5	3.9E+04
13.68	3	3	8.2E+04
13.75	5	3	4.5E+04
13.75	5	5	9.5E+04
13.76	1	3	1.51E+05
13.78	5	7	1.7E+05
13.84	5	7	2.59E+05
13.87	3	5	8.5E+04
13.92	3	5	8.5E+04
13.93	5	7	4.2E+04
13.95	5	5	3.8E+04
14.04	3	5	1.2E+05
14.24	1	3	1.41E+05
Cr XXII			
2.190	4	2	1.7E+06
2.191	2	2	2.5E+06
2.198	4	4	4.5E+06
2.199	2	4	2.3E+06
2.202	4	6	1.6E+06
2.203	4	2	1.3E+06
13.149	2	4	1.29E+05
13.292	4	6	1.54E+05
Cr XXIII			
1.7632	1	3	3.68E+05
1.8557	1	3	8.97E+05
2.095	3	1	3.5E+06
2.101	1	3	2.0E+06
2.101	5	5	7.9E+05
2.102	3	5	2.1E+06
2.103	3	5	1.2E+06
2.104	1	3	1.4E+06
2.105	3	3	9.6E+05
2.106	3	3	2.0E+06
2.107	5	5	2.3E+06
2.107	3	5	3.3E+06
2.109	5	3	1.7E+06
2.113	3	5	5.9E+05
2.119	3	1	2.7E+05
2.129	3	1	5.1E+05
2.1818	1	3	3.37E+06
2.1923	1	3	2.34E+05
Cobalt			
Co I			
2287.80	8	8	8.6E-01
2295.22	10	8	2.2E-01
2309.03	10	10	5.6E-01
2323.13	8	8	5.0E-01
2325.53	6	8	1.1E-01
2335.98	6	6	5.1E-01
2338.66	4	4	7.7E-01
2353.36	8	10	1.5E-01
2355.48	6	8	1.3E-01
2358.18	4	6	1.4E-01
2365.06	10	10	1.3E-01
2371.85	6	8	7.3E-02
2384.86	10	8	2.4E-01
2392.03	6	6	4.0E-01
2402.06	8	6	5.1E-01
2407.25	10	12	3.6E+00
2412.76	4	6	6.5E-01
2414.46	6	8	3.4E+00
2415.29	4	6	3.6E+00
2424.93	10	10	3.2E+00
2432.21	8	8	2.6E+00
2436.66	6	6	2.6E+00
2439.04	4	4	2.7E+00
2460.80	4	6	1.2E-01
2467.69	6	8	7.0E-02
2470.27	10	12	1.5E-01
2476.64	10	8	2.2E-01
2504.52	10	8	1.8E-01
2511.02	10	10	9.2E-01
2521.36	10	8	3.0E+00
2528.97	8	6	2.8E+00
2530.13	6	6	7.1E-02
2535.96	6	4	1.9E+00
2536.50	8	8	3.0E-01
2544.25	4	2	3.0E+00
2562.12	4	4	3.9E-01
2567.34	6	6	3.0E-01
2574.35	8	8	1.7E-01
2685.34	6	8	5.5E-02
3017.55	8	6	6.9E-02
3044.00	10	10	1.9E-01
3048.89	6	4	7.5E-02
3061.82	8	8	1.6E-01
3072.34	6	6	1.5E-01
3086.78	4	4	1.9E-01
3354.37	8	6	1.1E-01
3367.11	10	8	6.0E-02
3385.22	8	6	1.1E-01
3388.16	6	4	2.4E-01
3395.37	6	8	2.9E-01
3405.12	10	10	1.0E+00
3409.17	8	8	4.2E-01
3412.34	8	10	6.1E-01
3412.63	10	8	1.2E-01
3414.74	4	4	8.8E-02
3417.15	6	6	3.2E-01
3431.58	8	6	1.1E-01
3433.05	4	4	1.0E+00
3442.92	6	4	1.2E-01
3443.64	8	8	6.9E-01
3449.17	6	6	7.6E-01
3449.44	10	10	1.8E-01
3453.51	10	12	1.1E+00
3455.24	4	2	1.9E-01
3462.80	4	6	7.9E-01
3465.79	10	12	9.2E-02
3474.02	6	8	5.6E-01
3483.41	8	10	5.5E-02
3489.40	8	6	1.3E+00
3491.32	4	4	5.0E-02
3495.68	4	6	4.9E-01
3502.28	10	8	8.0E-01
3502.63	6	6	5.2E-02
3506.32	8	6	8.2E-01
3509.84	6	8	3.2E-01
3512.64	6	4	1.0E+00
3513.48	8	10	7.8E-02
3518.34	6	4	1.6E+00
3521.58	10	8	1.8E-01
3523.42	4	2	9.8E-01
3526.85	10	10	1.3E-01
3529.03	6	8	8.8E-02
3529.82	8	10	4.6E-01

λ Å	Weights g_i	g_k	A 10^8 s^{-1}
3533.36	4	6	9.1E-02
3560.89	4	4	2.3E-01
3564.95	6	8	7.0E-02
3569.37	8	8	1.6E+00
3574.97	6	6	1.5E-01
3575.36	8	8	9.6E-02
3585.15	8	8	7.1E-02
3587.19	6	6	1.4E+00
3594.87	6	6	9.2E-02
3602.08	4	4	1.0E-01
3704.06	6	8	1.2E-01
3745.49	8	8	7.5E-02
3842.05	8	6	1.3E-01
3845.47	8	10	4.6E-01
3861.16	6	4	1.4E-01
3873.12	10	8	1.2E-01
3873.95	8	6	1.0E-01
3881.87	6	4	8.2E-02
3894.07	6	8	6.9E-01
3894.98	4	2	8.8E-02
3935.96	8	10	6.2E-02
3995.31	8	10	2.5E-01
3997.90	6	8	7.0E-02
4092.39	8	8	5.7E-02
4110.53	6	6	5.5E-02
4118.77	6	8	1.6E-01
4121.32	8	10	1.9E-01
5146.75	8	8	1.5E-01
5212.70	10	10	1.9E-01
5265.79	6	8	5.0E-02
5280.63	10	8	2.8E-01
5352.05	12	10	2.7E-01
5477.09	6	8	6.8E-02
5483.96	8	10	7.3E-02
6082.43	10	10	5.4E-02
6455.00	8	10	9.0E-02
7838.12	8	10	5.4E-02
8093.93	12	10	2.0E-01
8372.79	10	10	8.7E-02
Co II			
2286.15	11	13	3.3E+00
2307.85	9	11	2.6E+00
2311.61	7	9	2.8E+00
2314.05	5	7	2.8E+00
2314.97	3	5	2.7E+00
2330.36	5	3	1.32E+00
2344.28	3	3	1.5E+00
2353.41	7	7	1.9E+00
2363.80	9	9	2.1E+00
2378.62	11	9	1.9E+00
2383.45	9	7	1.8E+00
2388.92	11	11	2.8E+00
2389.54	5	3	1.5E+00
2404.17	3	3	1.5E+00
2417.66	9	9	8.5E-01

λ Å	Weights g_i	g_k	A 10^8 s^{-1}
Copper			
Cu I			
*2024.3	2	6	9.8E-02
2165.1	2	4	5.1E-01
2178.9	2	4	9.13E-01
2181.7	2	2	1.0E+00
2225.7	2	2	4.6E-01
2244.3	2	4	1.19E-02
2441.6	2	2	2.0E-02
2492.2	2	4	3.11E-02
2618.4	6	4	3.07E-01
2766.4	4	4	9.6E-02
2824.4	6	6	7.8E-02
2961.2	6	8	3.76E-02
3063.4	4	4	1.55E-02
3194.1	4	4	1.55E-02
3247.5	2	4	1.39E+00
3274.0	2	2	1.37E+00
3337.8	6	8	3.8E-03
4022.6	2	4	1.90E-01
4062.6	4	6	2.10E-01
4249.0	2	2	1.95E-01
4275.1	6	8	3.45E-01
4480.4	2	2	3.0E-02
4509.4	4	2	2.75E-01
4530.8	4	2	8.4E-02
4539.7	6	4	2.12E-01
4587.0	8	6	3.20E-01
4651.1	10	8	3.80E-01
4704.6	8	8	5.5E-02
5105.5	6	4	2.0E-02
5153.2	2	4	6.0E-01
5218.2	4	6	7.5E-01
5220.1	4	4	1.50E-01
5292.5	8	8	1.09E-01
5700.2	4	4	2.4E-03
5782.1	4	2	1.65E-02
Cu II			
2489.7	5	5	1.5E-02
2544.8	9	7	1.1E+00
2689.3	7	7	4.1E-01
2701.0	5	5	6.7E-01
2703.2	3	3	1.2E+00
2713.5	5	5	6.8E-01
2769.7	7	7	6.1E-01
Dysprosium			
Dy I			
2862.7	17	15	6.5E-02
2964.6	17	17	6.5E-02
3147.7	15	17	1.1E-01
3263.2	15	13	1.4E-01
3511.0	15	13	3.1E-01
3571.4	15	13	2.0E-01
3757.1	17	19	3.0E+00

λ Å	Weights g_i	g_k	A 10^8 s^{-1}
3868.8	17	17	3.1E+00
3967.5	17	19	8.7E-01
4046.0	17	15	1.5E+00
4103.9	13	11	1.7E+00
4186.8	17	17	1.32E+00
4194.8	17	17	7.2E-01
4211.7	17	19	2.08E+00
4218.1	15	15	1.85E+00
4221.1	15	17	1.52E+00
4225.2	13	15	4.5E+00
4268.3	15	15	3.6E-02
4276.7	13	13	7.3E-01
4292.0	15	15	5.8E-02
4577.8	17	19	2.2E-02
4589.4	17	15	1.3E-01
4612.3	17	15	8.2E-02
5077.7	17	17	5.7E-03
5301.6	17	15	1.1E-02
5547.3	17	17	2.7E-03
5639.5	17	19	4.7E-03
5974.5	17	17	4.0E-03
5988.6	17	15	5.3E-03
6010.8	15	15	2.6E-02
6088.3	15	13	3.5E-02
6168.4	15	17	2.5E-02
6259.1	17	19	8.5E-03
6579.4	17	15	7.5E-03
Erbium			
Er I			
3862.9	13	13	2.5E+00
4008.0	13	15	2.6E+00
4151.1	13	11	1.8E+00
Europium			
Eu I			
2372.9	8	6	1.9E-01
2375.3	8	8	2.0E-01
2379.7	8	10	2.0E-01
2619.3	8	10	7.0E-03
2643.8	8	8	6.6E-03
2659.4	8	10	1.2E-02
2682.6	8	6	1.2E-02
2710.0	8	10	1.4E-01
2724.0	8	8	1.2E-01
2731.4	8	8	3.1E-02
2732.6	8	6	3.7E-02
2735.3	8	10	4.7E-02
2738.6	8	10	1.3E-02
2743.3	8	6	1.1E-01
2745.6	8	6	5.0E-02
2747.8	8	8	5.2E-02
2772.9	8	6	1.0E-02
2878.9	8	10	2.8E-02
2892.5	8	8	1.0E-01
2893.0	8	6	1.0E-01

λ Å	Weights g_i	g_k	A 10^8 s^{-1}
2909.0	8	10	6.9E-02
2958.9	8	6	1.6E-02
3059.0	8	8	3.8E-02
3067.0	8	10	9.1E-03
3106.2	8	10	5.5E-02
3111.4	8	10	3.0E-01
3168.3	8	10	6.9E-02
3185.5	8	10	5.8E-03
3210.6	8	8	1.1E-01
3212.8	8	8	2.9E-01
3213.8	8	6	1.8E-01
3235.1	8	10	1.0E-02
3241.4	8	8	2.3E-02
3246.0	8	6	1.4E-02
3247.6	8	8	2.3E-02
3322.3	8	6	3.5E-02
3334.3	8	6	3.4E-01
3350.4	8	10	1.5E-02
3353.7	8	8	5.8E-03
3457.1	8	8	8.4E-03
3467.9	8	8	1.0E-02
3589.3	8	6	6.9E-03
4594.0	8	10	1.4E+00
4627.2	8	8	1.3E+00
4661.9	8	6	1.3E+00
5645.8	8	6	5.4E-03
5765.2	8	8	1.1E-02
6018.2	8	10	8.5E-03
6291.3	8	6	1.8E-03
6864.5	8	10	5.8E-03
7106.5	8	8	2.6E-03
Fluorine			
F I			
806.96	4	6	3.3E+00
809.60	2	4	2.8E+00
951.87	4	2	2.6E+00
954.83	4	4	5.77E+00
955.55	2	2	5.1E+00
958.52	2	4	1.3E+00
6239.7	6	4	2.5E-01
6348.5	4	4	1.8E-01
6413.7	2	4	1.1E-01
6708.3	6	4	1.4E-02
6774.0	6	6	1.0E-01
6795.5	4	2	5.2E-02
6834.3	4	4	2.1E-01
6856.0	6	8	4.94E-01
6870.2	2	2	3.8E-01
6902.5	4	6	3.2E-01
6909.8	2	4	2.2E-01
6966.4	4	2	1.1E-01
7037.5	4	4	3.0E-01
7127.9	2	2	3.8E-01
7309.0	6	8	4.7E-01
7311.0	4	2	3.9E-01
7314.3	4	6	4.8E-01
7332.0	6	4	3.1E-01
7398.7	6	6	2.85E-01
7425.7	4	2	3.4E-01
7482.7	4	4	5.6E-02
7489.2	2	2	1.1E-01
7514.9	2	2	5.2E-02
7552.2	4	6	7.8E-02
7573.4	2	4	1.0E-01
7607.2	4	4	7.0E-02
7754.7	4	6	3.82E-01
7800.2	2	4	2.1E-01
Gallium			
Ga I			
2195.4	2	2	1.9E-02
2199.7	4	2	3.3E-02
2214.4	4	6	1.2E-02
2235.9	4	2	4.3E-02
2255.0	2	2	3.1E-02
2259.2	4	6	3.1E-02
2294.2	2	4	7.0E-02
2297.9	4	2	5.8E-02
2338.2	4	6	9.8E-02
2371.3	2	2	5.7E-02
2418.7	4	2	1.0E-01
2450.1	2	4	2.8E-01
2500.2	4	6	3.4E-01
2659.9	2	2	1.2E-01
2719.7	4	2	2.3E-01
2874.2	2	4	1.2E+00
2943.6	4	6	1.4E+00
2944.2	4	4	2.7E-01
4033.0	2	2	4.9E-01
4172.0	4	2	9.2E-01
Ga II			
829.60	1	3	2.2E-01
1414.4	1	3	1.88E+01
Germanium			
Ge I			
1944.7	3	1	7.0E-01
1955.1	3	3	2.8E-01
1988.3	5	3	2.5E-01
1998.9	5	5	5.5E-01
2041.7	1	3	1.1E+00
2065.2	3	3	8.5E-01
2068.7	3	5	1.2E+00
2086.0	3	5	4.0E-01
2094.3	5	7	9.7E-01
2105.8	5	5	1.7E-01
2256.0	5	5	3.2E-02
2417.4	5	5	9.6E-01
2498.0	1	3	1.3E-01
2533.2	3	3	1.0E-01
2589.2	5	3	5.1E-02
2592.5	3	5	7.1E-01
2651.2	5	5	2.0E+00
2651.6	1	3	8.5E-01
2691.3	3	3	6.1E-01
2709.6	3	1	2.8E+00
2754.6	5	3	1.1E+00
3039.1	5	3	2.8E+00
3124.8	5	5	3.1E-02
3269.5	5	3	2.9E-01
4226.6	1	3	2.1E-01
4685.8	1	3	9.5E-02
Ge II			
999.10	2	4	1.9E+00
1016.6	4	6	2.1E+00
1017.1	4	4	3.5E-01
1055.0	2	2	6.9E-01
1075.1	4	2	1.3E+00
1237.1	2	4	1.9E+01
1261.9	4	6	2.2E+01
1264.7	4	4	3.5E+00
1602.5	2	2	3.4E+00
1649.2	4	2	6.5E+00
4741.8	2	4	4.6E-01
4814.6	4	6	5.1E-01
4824.1	4	4	8.6E-02
5131.8	4	6	1.9E+00
5178.5	6	6	1.3E-01
5178.6	6	8	2.0E+00
5893.4	2	4	9.2E-01
6021.0	2	2	8.4E-01
6336.4	2	2	4.4E-01
6484.2	4	2	8.5E-01
Gold			
Au I			
2427.95	2	4	1.99E+00
2675.95	2	2	1.64E+00
3122.78	6	4	1.90E-01
6278.30	4	2	3.4E-02
Helium			
He I			
510.00	1	3	4.6224E-01
512.10	1	3	7.3174E-01
515.62	1	3	1.2582E+00
522.21	1	3	2.4356E+00
537.03	1	3	5.6634E+00
584.33	1	3	1.7989E+01
*2677.1	3	9	4.4174E-03
*2696.1	3	9	6.0234E-03
*2723.2	3	9	8.4996E-03
*2763.8	3	9	1.2508E-02
*2829.1	3	9	1.9389E-02
*2945.1	3	9	3.2006E-02
*3187.7	3	9	5.6361E-02
3231.3	1	3	5.1015E-03
3258.3	1	3	6.9627E-03
3296.8	1	3	9.8432E-03
3354.6	1	3	1.4537E-02
3447.6	1	3	2.2691E-02
*3554.4	9	15	7.5971E-03
*3563.0	9	3	4.8362E-03
*3587.3	9	15	1.8107E-02

λ Å	Weights g_i	Weights g_k	A 10^8 s^{-1}
3613.6	1	3	3.8022E-02
*3634.2	9	15	2.6062E-02
*3652.0	9	3	9.7444E-03
*3705.0	9	15	3.9528E-02
*3819.6	9	15	6.4351E-02
3833.6	3	5	9.6470E-03
*3867.5	9	3	2.4465E-02
3871.8	3	5	1.3386E-02
*3888.7	3	9	9.4746E-02
3926.5	3	5	1.9371E-02
3935.9	3	1	7.4475E-03
3964.7	1	3	6.9507E-02
4009.3	3	5	2.9612E-02
4024.0	3	1	1.1281E-02
*4026.2	9	15	1.1600E-01
*4120.8	9	3	4.4529E-02
4143.8	3	5	4.8812E-02
4169.0	3	1	1.8298E-02
4387.9	3	5	8.9889E-02
4437.6	3	1	3.2689E-02
*4471.5	9	15	2.4578E-01
*4713.2	9	3	9.5209E-02
4921.9	3	5	1.9863E-01
5015.7	1	3	1.3372E-01
5047.7	3	1	6.7712E-02
*5875.7	9	15	7.0703E-01
6678.2	3	5	6.3705E-01
*7065.2	9	3	2.7853E-01
7281.4	3	1	1.8299E-01
*8361.7	3	9	3.8126E-03
*9463.6	3	9	5.6868E-03
9603.4	1	3	5.8286E-03
*9702.6	9	3	8.6511E-03
*10311	9	15	1.9945E-02
*10668	9	3	1.4471E-02
*10830	3	9	1.0216E-01
*10913	15	21	1.9801E-02
10917	5	7	1.6083E-02
*10997	15	9	1.4253E-03
11013	1	3	9.2496E-03
11045	3	5	1.8457E-02
11226	3	1	1.1168E-02
*11969	9	15	3.4781E-02
*12528	3	9	7.0932E-03
12756	5	3	1.2754E-03
*12785	15	21	4.1339E-02
12791	5	7	3.2475E-02
*12846	9	3	2.7317E-02
12968	3	5	3.3615E-02
*12985	15	9	2.7292E-03
Indium			
In I			
2560.2	2	4	4.0E-01
2710.3	4	6	4.0E-01
3039.4	2	4	1.3E+00
3256.1	4	6	1.3E+00
4101.8	2	2	5.6E-01
4511.3	4	2	1.02E+00

λ Å	Weights g_i	Weights g_k	A 10^8 s^{-1}
In II			
2941.1	3	1	1.4E+00
Iodine			
I I			
1782.8	4	4	2.71E+00
1830.4	4	6	1.6E-01
Iridium			
Ir I			
2475.12	10	10	2.1E-01
2502.98	10	12	3.2E-01
2639.71	10	10	4.7E-01
2661.98	10	10	2.5E-01
2664.79	10	8	4.0E-01
2694.23	10	12	4.8E-01
2849.72	10	10	2.2E-01
2853.31	10	10	2.0E-03
2882.64	10	8	7.2E-02
2924.79	10	12	1.42E-01
2934.64	8	10	2.0E-01
2951.22	10	8	2.8E-02
3003.63	8	10	5.9E-02
3168.88	8	10	5.47E-02
3220.78	10	8	2.4E-01
3558.99	6	8	1.5E-02
3573.72	8	10	5.4E-02
3617.21	6	8	2.0E-02
3628.67	8	8	2.8E-02
3661.71	8	10	4.0E-02
3734.77	8	8	2.7E-02
4033.76	8	10	2.7E-02
4069.92	6	8	3.6E-02
4913.35	12	12	3.3E-02
4939.24	10	12	2.5E-03
Iron			
Fe I			
1934.54	9	7	2.5E-01
1937.27	9	7	2.2E-01
1940.66	7	5	2.6E-01
2084.12	9	7	3.7E-01
2102.35	7	7	8.8E-02
2112.97	1	3	1.9E-01
2132.02	9	9	7.6E-02
2145.19	7	7	5.7E-02
2153.01	5	5	6.9E-02
2161.58	3	5	5.0E-02
2166.77	9	7	2.7E+00
2171.30	5	7	5.1E-02
2173.21	3	5	8.3E-02
2176.84	1	3	1.0E-01
2191.20	1	3	7.3E-02
2191.84	5	5	1.2E+00
2196.04	3	3	1.2E+00
2200.72	3	5	2.8E-01
2259.51	9	11	7.0E-02
2267.08	7	5	7.1E-02
2272.07	7	9	3.8E-02

λ Å	Weights g_i	Weights g_k	A 10^8 s^{-1}
2276.03	9	7	1.7E-01
2277.11	7	5	3.7E+01
2287.25	5	3	3.4E-01
2292.52	7	9	4.3E-02
2294.41	3	1	6.1E-01
2300.14	5	7	8.0E-02
2301.68	1	3	1.3E-01
2303.42	1	3	9.4E-02
2303.58	3	5	7.6E-02
2309.00	3	5	1.5E-01
2313.10	5	7	1.4E-01
2320.36	7	9	1.2E-01
2371.43	5	5	5.2E-02
2373.62	7	7	6.7E-02
2374.52	1	3	2.9E-01
2381.83	3	5	5.4E-02
2389.97	5	7	5.0E-02
2462.18	7	5	1.5E-01
2462.65	9	9	5.8E-01
2479.78	5	5	1.8E+00
2483.27	9	11	4.9E+00
2488.14	7	9	4.7E+00
2490.64	5	7	3.8E+00
2491.15	3	5	3.0E+00
2501.13	9	7	6.8E-01
2510.83	7	5	1.3E+00
2518.10	5	3	1.9E+00
2522.85	9	9	2.9E+00
2524.29	3	1	3.4E+00
2527.43	7	7	1.9E+00
2529.13	5	5	9.8E-01
2535.61	1	3	9.7E-01
2540.97	3	5	9.2E-01
2545.98	5	7	6.7E-01
2549.61	7	9	3.6E-01
2584.54	11	13	4.6E-01
2606.83	9	11	4.2E-01
2618.02	7	7	4.0E-01
2623.53	7	9	3.3E-01
2656.15	13	15	2.8E-01
2669.49	11	13	1.7E-01
2679.06	11	11	1.9E-01
2719.03	9	7	1.4E+00
2720.90	7	5	1.1E+00
2723.58	5	3	6.4E-01
2733.58	11	9	8.6E-01
2735.48	9	7	6.2E-01
2737.31	3	3	8.5E-01
2742.41	5	5	6.3E-01
2744.07	1	3	3.5E-01
2750.14	7	7	3.9E-01
2756.33	3	5	2.0E-01
2788.10	11	13	6.3E-01
2894.50	5	5	6.2E-01
2899.42	5	3	5.9E-01
2920.69	5	5	5.2E-02
2923.29	11	11	1.6E+00
2925.36	7	9	1.8E-01
2929.01	7	5	7.3E-02

λ Å	Weights g_i	g_k	A 10^8 s^{-1}
2936.90	9	9	1.3E-01
2941.34	5	3	5.6E-02
2947.88	7	7	2.0E-01
2953.94	5	5	1.89E-01
2954.65	5	7	1.0E-01
2957.36	3	3	1.77E-01
2965.25	1	3	1.16E-01
2966.90	9	11	2.72E-01
2969.36	3	1	3.66E-02
2973.13	5	7	1.35E-01
2973.24	7	9	1.83E-01
2980.53	7	7	2.2E-01
2981.45	7	5	6.54E-02
2983.57	9	7	2.80E-01
2987.29	9	7	6.6E-02
2990.39	9	11	3.9E-01
2994.43	7	5	4.4E-01
2996.39	3	5	1.6E-01
2999.51	11	11	2.3E-01
3000.95	5	3	6.42E-01
3008.14	3	1	1.07E+00
3009.09	13	11	6.7E-02
3009.57	9	9	1.7E-01
3011.48	7	9	4.7E-01
3015.92	11	9	5.9E-02
3016.18	5	3	8.5E-02
3017.63	3	3	6.82E-02
3018.98	7	7	1.3E-01
3021.07	7	7	4.56E-01
3024.03	3	5	4.88E-02
3025.84	1	3	3.48E-01
3026.46	5	5	1.1E-01
3031.63	3	3	1.5E-01
3037.39	3	5	3.2E-01
3042.02	3	5	4.9E-02
3042.66	5	7	5.7E-02
3047.60	5	7	2.84E-01
3053.07	3	5	1.5E-01
3057.45	11	9	4.4E-01
3059.09	7	9	1.7E-01
3067.24	9	7	3.4E-01
3068.17	5	3	9.8E-02
3075.72	7	5	2.9E-01
3083.74	5	3	3.0E-01
3091.58	3	1	5.4E-01
3098.19	11	11	1.1E-01
3100.67	7	7	1.4E-01
3119.49	11	9	8.2E-02
3120.43	9	7	8.9E-02
3156.27	7	7	5.4E-01
3160.66	9	9	1.9E-01
3161.95	11	13	1.2E-01
3166.44	9	7	1.14E-01
3168.85	5	7	5.7E-02
3175.45	11	11	1.3E-01
3176.36	5	3	9.2E-02
3196.93	9	11	9.0E-01
3199.53	9	9	2.6E-01
3205.40	3	3	1.2E+00
3215.94	5	5	8.0E-01
3217.38	11	9	2.2E-01
3219.58	7	9	6.2E-01
3222.07	11	11	3.3E-01
3225.79	11	13	8.8E-01
3227.80	9	7	1.4E+00
3228.25	5	3	4.5E-01
3229.99	9	11	4.5E-01
3230.21	5	5	1.9E-01
3230.96	7	5	3.9E-01
3233.05	13	15	5.4E-01
3233.97	9	9	2.0E-01
3246.96	5	3	9.9E-02
3248.20	7	7	2.2E-01
3253.60	7	9	1.8E-01
3254.36	11	13	5.1E-01
3257.59	7	5	1.4E-01
3265.62	7	5	3.8E-01
3268.23	3	3	5.9E-02
3271.00	5	3	6.6E-01
3280.26	9	11	5.4E-01
3282.89	3	5	3.0E-01
3284.59	5	5	5.4E-02
3290.99	3	5	6.0E-02
3292.02	7	9	6.1E-01
3292.59	3	3	2.6E-01
3298.13	3	5	8.1E-02
3305.97	5	7	4.7E-01
3306.36	3	5	6.1E-01
3307.23	13	13	2.0E-01
3314.74	5	7	6.9E-01
3322.47	9	11	6.2E-02
3323.74	5	5	3.0E-01
3328.87	11	11	2.7E-01
3337.66	11	9	5.7E-02
3347.93	5	5	4.0E-02
3354.06	1	3	7.7E-02
3355.23	9	9	3.2E-01
3369.55	9	9	2.4E-01
3370.78	11	11	3.3E-01
3380.11	7	7	2.4E-01
3383.98	7	7	9.3E-02
3392.65	7	7	2.6E-01
3394.58	5	3	9.9E-02
3399.33	5	5	3.8E-01
3402.26	13	13	2.8E-01
3406.44	3	5	3.0E-01
3407.46	7	9	5.8E-01
3410.17	3	5	4.7E-01
3411.35	9	9	5.5E-02
3413.13	5	7	3.6E-01
3417.84	3	3	5.1E-01
3418.51	3	1	1.3E+00
3424.28	7	7	2.0E-01
3425.01	9	7	2.8E-01
3427.12	7	9	5.5E-01
3428.19	5	5	2.1E-01
3428.75	7	5	2.7E-01
3440.99	7	5	8.4E-02
3442.36	5	5	4.55E-02
3443.88	5	3	6.2E-02
3445.15	5	7	2.8E-01
3447.28	5	5	9.1E-02
3450.33	3	3	2.0E-01
3476.70	1	3	5.4E-02
3477.85	3	1	4.2E-02
3485.34	5	3	1.4E-01
3495.29	9	7	9.46E-02
3497.10	7	7	1.4E-01
3505.07	5	3	9.9E-02
3506.50	5	5	7.1E-02
3508.49	9	11	5.7E-02
3510.44	1	3	4.4E-02
3516.56	7	5	3.7E-02
3521.84	3	5	9.6E-02
3523.31	5	3	7.6E-02
3524.08	7	5	7.5E-02
3524.24	5	7	4.2E-02
3527.79	9	9	2.0E-01
3529.82	3	3	7.6E-01
3536.56	5	7	7.8E-01
3537.73	5	3	1.1E-01
3537.90	11	11	8.4E-02
3540.12	7	9	1.2E-01
3541.08	9	11	6.2E-01
3542.08	7	9	7.4E-01
3543.67	3	5	1.8E-01
3548.02	5	3	9.7E-02
3552.11	3	5	4.5E-02
3552.83	5	5	1.5E-01
3553.74	11	9	8.1E-01
3556.88	9	11	4.4E-01
3559.50	3	3	1.9E-01
3560.70	7	9	6.5E-02
3565.38	7	9	3.8E-01
3567.03	5	7	6.5E-02
3568.42	5	3	5.3E-02
3568.82	7	9	5.6E-02
3570.10	9	11	6.77E-01
3572.00	11	11	2.4E-01
3573.39	5	7	7.5E-02
3576.76	11	9	9.6E-02
3578.38	1	3	6.3E-02
3581.19	11	13	1.02E+00
3582.20	13	11	2.5E-01
3583.33	1	3	2.3E-01
3585.32	7	7	1.3E-01
3585.71	9	9	3.75E-02
3586.98	5	5	1.6E-01
3591.48	1	3	6.0E-02
3592.67	7	5	4.0E-02
3594.63	9	9	2.7E-01
3595.30	5	5	5.4E-02
3597.02	5	3	1.7E-01
3599.62	11	9	1.8E-01
3603.20	11	11	2.6E-01
3603.82	3	3	1.7E-01
3605.45	9	9	6.4E-01

λ Å	Weights g_i	Weights g_k	A 10^8 s^{-1}
3606.68	11	13	8.2E-01
3608.86	3	5	8.14E-01
3610.16	13	13	4.8E-01
3610.70	5	3	7.1E-02
3612.07	11	13	7.5E-02
3613.45	7	7	6.7E-02
3615.19	3	3	5.8E-02
3617.79	5	7	6.5E-01
3618.77	5	7	7.3E-01
3621.46	9	11	5.1E-01
3622.00	7	7	5.1E-01
3623.19	13	13	7.4E-02
3624.06	5	3	5.4E-02
3630.35	9	7	7.6E-02
3631.46	7	9	5.17E-01
3632.04	3	5	4.8E-01
3632.55	11	9	5.2E-02
3635.19	5	3	1.4E-01
3637.86	9	9	5.5E-02
3638.30	7	9	2.6E-01
3640.39	9	11	3.8E-01
3644.80	7	5	7.8E-02
3645.82	1	3	5.7E-01
3647.84	9	11	2.92E-01
3649.51	11	9	4.2E-01
3650.03	7	7	9.9E-02
3651.47	7	9	6.2E-01
3655.46	5	5	1.0E-01
3659.52	9	9	5.8E-02
3667.25	9	7	1.4E-01
3669.15	9	7	7.4E-02
3669.52	9	7	3.0E-01
3670.09	11	13	7.6E-02
3674.77	5	3	6.7E-02
3676.31	9	11	4.63E-02
3677.31	5	7	3.1E-01
3677.63	7	5	8.0E-01
3678.86	3	5	4.1E-02
3682.24	5	5	1.7E+00
3684.11	9	7	3.4E-01
3686.00	9	11	2.6E-01
3686.26	3	1	1.2E-01
3687.46	11	9	8.01E-02
3688.48	7	9	6.9E-02
3690.73	11	11	2.7E-01
3694.01	5	7	6.8E-01
3697.43	7	7	2.1E-01
3698.60	5	7	3.8E-02
3699.15	5	7	4.5E-02
3701.09	7	9	4.8E-01
3702.03	3	1	3.5E-01
3703.69	9	11	5.3E-02
3703.82	1	3	1.2E-01
3704.46	11	9	1.3E-01
3709.25	9	7	1.56E-01
3711.41	3	5	7.3E-02
3718.41	7	7	5.3E-02
3719.93	9	11	1.62E-01
3722.56	5	5	4.97E-02
3724.38	5	7	1.3E-01
3726.93	5	5	4.6E-01
3727.09	9	7	2.0E-01
3727.62	7	5	2.25E-01
3730.39	9	11	1.3E-01
3730.95	5	7	3.8E-02
3732.40	5	5	2.8E-01
3733.32	3	3	6.2E-02
3734.86	11	11	9.02E-01
3735.32	9	9	2.4E-01
3737.13	7	9	1.42E-01
3738.31	11	13	3.8E-01
3740.24	7	9	1.4E-01
3742.62	9	9	1.0E-01
3743.36	5	3	2.60E-01
3744.10	5	3	3.6E-01
3745.56	5	7	1.15E-01
3745.90	1	3	7.33E-02
3746.93	7	7	2.2E-01
3748.26	3	5	9.15E-02
3749.48	9	9	7.64E-01
3753.61	7	5	9.3E-02
3756.94	11	11	2.4E-01
3757.45	5	3	1.2E-01
3758.23	7	7	6.34E-01
3760.05	13	15	4.47E-02
3760.53	3	5	4.8E-02
3763.79	5	5	5.44E-01
3765.54	13	15	9.8E-01
3766.67	5	3	9.7E-02
3767.19	3	3	6.40E-01
3768.03	3	1	8.4E-02
3774.82	3	3	4.7E-02
3778.51	7	5	1.2E-01
3781.94	5	7	3.7E-02
3785.95	11	13	4.2E-02
3786.19	5	5	1.2E-01
3787.16	5	5	1.0E-01
3787.88	3	5	1.29E-01
3789.82	9	7	3.9E-02
3791.73	5	3	6.3E-02
3793.87	3	3	7.4E-02
3794.34	9	11	3.8E-02
3795.00	5	7	1.15E-01
3799.55	7	9	7.32E-02
3801.68	5	7	6.6E-02
3802.00	11	13	3.5E-02
3802.28	5	5	5.0E-02
3804.01	11	9	4.7E-02
3805.35	9	11	9.8E-01
3806.22	3	3	2.3E-01
3806.70	11	11	5.4E-01
3807.54	3	5	8.0E-02
3808.73	9	9	3.54E-02
3810.76	5	3	2.0E-01
3813.88	13	11	8.7E-02
3815.84	9	7	1.3E+00
3817.64	11	11	8.3E-02
3819.50	7	5	4.6E-02
3820.43	11	9	6.68E-01
3821.18	11	13	7.0E-01
3821.83	5	5	7.8E-02
3825.88	9	7	5.98E-01
3827.82	7	5	1.05E+00
3833.31	9	9	4.69E-02
3834.22	7	5	4.53E-01
3836.33	5	5	3.7E-01
3839.26	9	9	2.8E-01
3839.61	3	5	3.9E-01
3840.44	5	3	4.70E-01
3841.05	5	3	1.3E+00
3843.26	9	7	4.7E-01
3845.17	3	3	6.8E-02
3845.69	5	7	4.9E-02
3846.00	9	7	4.3E-02
3846.41	11	9	1.9E-01
3846.80	7	7	6.6E-01
3849.96	3	1	6.06E-01
3856.37	7	5	4.64E-02
3859.21	13	11	8.5E-02
3859.91	9	9	9.70E-02
3865.52	3	3	1.55E-01
3867.22	5	5	3.4E-01
3871.75	11	11	6.7E-02
3872.50	5	5	1.05E-01
3873.76	11	9	8.0E-02
3878.02	7	7	7.72E-02
3878.57	5	3	6.6E-02
3883.28	7	7	1.6E-01
3884.36	11	9	3.5E-02
3885.51	3	5	5.8E-02
3886.28	7	7	5.30E-02
3887.05	9	9	3.52E-02
3888.51	5	5	2.6E-01
3888.82	5	3	2.7E-01
3891.93	3	3	4.0E-01
3893.39	11	11	1.3E-01
3895.66	3	1	9.40E-02
3900.52	7	7	7.5E-02
3902.95	7	7	2.14E-01
3903.90	9	9	9.6E-02
3906.75	5	7	6.7E-02
3907.93	7	5	6.7E-02
3909.66	3	5	5.3E-02
3909.83	3	3	6.5E-02
3914.27	3	3	5.4E-02
3916.73	13	11	1.2E-01
3919.07	9	9	3.9E-02
3925.20	1	3	5.7E-02
3931.12	5	7	4.5E-02
3941.28	5	5	8.4E-02
3942.44	3	5	9.0E-02
3946.99	9	11	4.4E-02
3948.77	11	9	2.2E-01
3949.14	3	3	3.9E-02
3949.95	7	5	5.9E-02
3951.16	3	5	3.6E-01
3952.60	11	11	4.1E-02

λ Å	Weights g_i	Weights g_k	A 10^8 s^{-1}
3953.15	7	9	3.7E-02
3955.34	3	3	1.4E-01
3955.96	3	3	5.7E-02
3956.45	13	11	2.1E-01
3957.02	5	7	1.6E-01
3960.28	5	7	4.2E-02
3963.10	3	5	1.7E-01
3967.42	9	7	2.3E-01
3967.96	7	9	6.3E-02
3969.26	9	7	2.3E-01
3970.39	3	1	3.5E-01
3971.32	11	9	5.7E-02
3973.65	5	7	6.6E-02
3976.61	3	5	1.8E-01
3977.74	5	5	7.0E-02
3981.77	9	9	3.9E-02
3983.96	9	7	7.6E-02
3985.39	5	5	6.7E-02
3989.86	5	7	5.0E-02
3996.97	9	9	6.7E-02
3997.39	9	11	1.5E-01
3998.05	11	9	6.6E-02
4003.76	3	3	7.1E-02
4005.24	7	5	2.04E-01
4006.31	11	9	4.7E-02
4007.27	7	5	4.2E-02
4009.71	3	5	5.2E-02
4014.53	11	11	2.4E-01
4017.15	9	11	4.5E-02
4021.87	7	9	1.0E-01
4024.72	7	9	8.9E-02
4031.96	3	5	7.1E-02
4040.64	5	7	4.4E-02
4044.61	5	3	1.1E-01
4045.81	9	9	8.63E-01
4054.87	5	3	1.6E-01
4058.22	9	7	4.9E-02
4059.73	5	3	8.1E-02
4062.44	3	3	2.2E-01
4063.59	7	7	6.8E-01
4065.40	3	1	1.9E-01
4067.98	9	9	1.7E-01
4070.77	7	5	1.3E-01
4071.74	5	5	7.65E-01
4073.76	5	3	1.6E-01
4074.79	9	9	4.8E-02
4076.63	9	9	1.9E-01
4078.35	5	3	4.2E-02
4079.18	5	5	5.1E-02
4079.84	1	3	6.3E-02
4080.21	3	1	2.4E-01
4082.44	3	3	3.8E-02
4084.49	11	9	1.1E-01
4085.00	3	5	4.2E-02
4085.30	7	7	1.1E-01
4085.98	7	5	5.0E-02
4088.57	5	3	3.9E-02
4098.18	7	7	6.8E-02
4107.49	5	3	2.5E-01
4109.07	1	3	4.5E-02
4109.80	3	3	1.6E-01
4112.96	11	13	1.4E-01
4114.45	5	5	4.7E-02
4118.54	11	13	5.8E-01
4126.18	11	11	3.9E-02
4127.61	1	3	1.3E-01
4132.06	5	7	1.2E-01
4132.90	3	5	9.4E-02
4134.68	5	7	1.8E-01
4137.00	3	5	2.2E-01
4137.42	5	7	6.1E-02
4142.63	3	5	7.4E-02
4143.87	7	9	1.5E-01
4149.37	11	13	3.6E-02
4150.25	3	3	7.1E-02
4153.90	7	9	2.3E-01
4154.80	9	11	1.5E-01
4156.80	5	5	1.9E-01
4158.79	3	5	1.6E-01
4170.90	5	5	6.1E-02
4172.12	7	5	9.7E-02
4175.64	3	5	1.6E-01
4181.75	5	7	3.6E-01
4182.38	5	5	4.9E-02
4184.89	5	5	1.1E-01
4187.04	7	5	2.15E-01
4187.79	9	7	1.52E-01
4191.68	1	3	4.8E-02
4196.21	7	7	9.8E-02
4198.30	11	9	8.03E-02
4198.64	5	5	1.3E-01
4199.09	9	11	6.1E-01
4200.09	7	7	4.0E-02
4200.92	7	9	4.2E-02
4202.03	9	9	8.22E-02
4203.67	7	9	8.6E-02
4203.94	13	13	1.3E-01
4205.54	5	5	3.6E-02
4207.13	5	3	4.3E-02
4210.34	3	3	1.7E-01
4213.65	3	1	1.9E-01
4217.55	3	5	2.3E-01
4219.36	11	13	3.8E-01
4220.34	3	1	1.9E-01
4222.21	7	7	5.77E-02
4224.17	9	11	1.3E-01
4224.51	3	5	7.1E-02
4225.45	5	7	1.7E-01
4226.42	3	3	3.7E-02
4233.60	3	5	1.85E-01
4235.94	9	9	1.88E-01
4238.81	7	9	2.2E-01
4240.37	5	3	5.7E-02
4245.26	1	3	8.3E-02
4246.08	7	5	5.7E-02
4247.43	9	11	2.0E-01
4248.22	3	5	3.5E-02
4250.12	5	7	2.08E-01
4250.79	7	7	1.0E-01
4260.47	11	11	3.2E-01
4267.83	1	3	9.4E-02
4268.75	5	3	4.2E-02
4271.15	7	9	1.82E-01
4271.76	9	11	2.28E-01
4282.40	7	5	1.1E-01
4300.83	5	5	4.7E-02
4305.45	5	3	6.0E-02
4307.90	7	9	3.4E-01
4315.08	5	5	7.7E-02
4325.76	5	7	5.0E-01
4327.09	5	5	7.8E-02
4352.73	3	5	3.9E-02
4369.77	9	9	7.2E-02
4383.54	9	11	5.00E-01
4387.89	3	3	3.9E-02
4388.41	7	7	1.3E-01
4401.29	7	7	5.9E-02
4404.75	7	9	2.75E-01
4415.12	5	7	1.19E-01
4422.57	3	3	8.8E-02
4430.61	3	1	7.45E-02
4433.22	5	3	2.3E-01
4438.34	3	1	7.9E-02
4442.34	5	5	3.76E-02
4443.19	1	3	1.1E-01
4446.83	3	3	5.3E-02
4447.72	3	3	5.11E-02
4454.38	5	5	3.8E-02
4455.03	9	7	3.9E-02
4466.55	5	7	1.2E-01
4469.37	5	7	2.6E-01
4481.61	3	3	4.2E-02
4484.22	7	9	7.0E-02
4485.67	3	3	1.1E-01
4528.61	7	9	5.44E-02
4533.13	3	1	3.7E-02
4547.85	5	7	7.6E-02
4619.29	7	5	4.7E-02
4669.17	5	3	4.0E-02
4673.16	5	7	4.6E-02
4678.85	7	9	7.4E-02
4704.95	3	1	8.1E-02
4736.77	9	11	4.9E-02
4789.65	5	5	7.2E-02
4859.74	5	3	1.3E-01
4871.32	7	5	2.2E-01
4872.14	3	3	2.4E-01
4878.21	1	3	9.1E-02
4890.75	5	5	2.1E-01
4891.49	9	7	2.9E-01
4892.87	3	3	4.8E-02
4903.31	3	5	4.7E-02
4917.23	5	3	6.1E-02
4918.01	1	3	4.0E-02
4918.99	7	7	1.7E-01
4920.50	11	9	3.5E-01
4930.31	3	3	4.1E-02

λ Å	Weights g_i	g_k	A 10^8 s^{-1}
4969.92	3	3	1.8E-01
4973.10	3	3	1.0E-01
4978.60	5	3	1.1E-01
4988.95	7	7	4.9E-02
4991.27	5	7	8.2E-02
5001.86	9	7	3.9E-01
5004.04	5	3	3.5E-02
5014.94	7	5	3.0E-01
5022.24	5	3	2.6E-01
5074.75	9	11	1.5E-01
5090.78	7	5	2.0E-01
5109.65	3	5	5.4E-02
5121.64	5	5	7.9E-02
5125.11	9	7	2.6E-01
5133.69	11	13	2.7E-01
5137.38	11	9	1.1E-01
5159.06	5	3	1.3E-01
5162.27	11	11	2.4E-01
5184.26	5	7	3.5E-02
5208.59	7	5	5.2E-02
5232.94	9	11	1.4E-01
5263.30	5	5	5.2E-02
5266.55	7	9	8.6E-02
5283.62	7	7	8.0E-02
5302.30	3	5	6.3E-02
5324.18	9	9	1.5E-01
5339.93	5	7	7.0E-02
5353.39	9	7	4.8E-02
5364.87	5	7	5.5E-01
5367.47	7	9	5.8E-01
5369.96	9	11	4.7E-01
5373.71	7	9	3.5E-02
5383.37	11	13	5.6E-01
5389.48	7	7	1.3E-01
5398.29	5	5	9.8E-02
5400.50	9	9	1.8E-01
5410.91	7	9	4.8E-01
5415.20	11	13	5.6E-01
5424.07	13	15	5.0E-01
5432.95	5	5	4.1E-02
5445.04	11	11	2.0E-01
5463.27	9	9	3.2E-01
5466.39	9	7	7.5E-02
5473.90	7	7	5.5E-02
5480.87	3	1	1.2E-01
5487.74	7	5	8.6E-02
5554.89	9	9	8.7E-02
5569.62	5	3	2.1E-01
5572.84	7	5	2.1E-01
5576.09	3	1	2.1E-01
5586.76	9	7	1.9E-01
5598.30	5	5	1.8E-01
5615.64	11	9	1.7E-01
5624.54	5	5	5.3E-02
5633.97	11	13	8.7E-02
5638.27	9	7	4.0E-02
5650.01	3	5	5.0E-02
5655.18	7	9	5.3E-02
5658.82	7	7	3.6E-02
5679.02	5	7	3.6E-02
5686.53	9	11	4.4E-02
5691.51	3	1	6.2E-02
5705.99	7	9	6.7E-02
5717.85	1	3	5.0E-02
5753.12	3	5	7.0E-02
5762.99	5	7	1.0E-01
5816.36	9	11	3.7E-02
5905.67	5	3	1.2E-01
5927.80	5	3	5.1E-02
5930.17	5	7	1.6E-01
6020.17	7	9	1.1E-01
6024.07	9	11	1.3E-01
6055.99	7	9	7.0E-02
6170.49	5	5	1.3E-01
6336.84	3	3	4.9E-02
6338.90	5	3	4.8E-02
6400.00	7	9	5.5E-02
6411.65	5	7	3.5E-02
6419.98	7	7	1.3E-01
6469.21	3	3	9.0E-02
6495.78	3	3	6.0E-02
6496.46	5	5	8.5E-02
6569.23	7	9	6.5E-02
6633.76	7	7	3.6E-02
6733.16	3	1	3.9E-02
6841.35	5	7	3.6E-02
7130.94	3	5	4.3E-02
Fe II			
1144.94	10	12	4.8E+00
1635.40	8	6	2.4E+00
1641.76	6	4	1.8E+00
1647.16	6	6	5.2E-01
2208.41	10	10	1.8E+00
2213.66	14	14	4.4E-01
2218.27	8	10	1.9E+00
2327.40	6	4	5.9E-01
2331.31	10	8	2.9E-01
2332.80	8	6	1.5E+00
2338.01	4	4	1.1E+00
2343.49	10	8	1.7E+00
2343.96	8	6	2.9E-01
2344.28	2	4	8.2E-01
2348.11	10	8	5.1E-01
2348.30	6	6	1.2E+00
2351.67	6	6	1.7E+00
2352.31	2	4	4.2E+00
2353.68	8	8	1.3E+00
2354.89	6	4	2.4E-01
2360.00	10	10	2.4E-01
2360.29	8	6	5.9E-01
2362.02	8	8	1.3E-01
2363.86	8	10	5.1E+00
2364.83	8	8	6.1E-01
2365.77	6	6	2.1E+00
2366.59	6	6	9.9E-02
2368.60	6	4	5.9E-01
2369.95	10	12	5.7E+00
2370.50	4	4	1.4E-01
2373.74	10	10	3.3E-01
2375.19	4	2	9.8E-01
2379.27	8	8	1.5E-01
2380.76	6	8	3.1E-01
2382.04	10	12	3.8E+00
2382.90	12	14	2.2E-01
2383.25	6	6	3.4E-01
2384.39	4	4	2.3E-01
2388.37	10	12	2.2E-01
2388.63	8	8	1.0E+00
2390.10	14	16	5.5E+00
2390.77	6	6	9.3E-01
2395.42	6	4	3.3E-01
2395.62	8	10	2.5E+00
2399.24	6	6	1.4E+00
2400.06	12	14	5.2E+00
2401.29	6	8	2.5E+00
2404.43	4	2	7.1E-01
2404.89	6	8	1.7E+00
2406.66	4	4	1.6E+00
2410.52	4	6	1.5E+00
2411.07	2	2	2.4E+00
2413.31	2	4	1.1E+00
2416.45	8	10	1.6E+00
2418.44	6	8	1.6E+00
2423.21	4	6	1.4E+00
2428.36	8	10	2.7E+00
2432.87	14	14	3.2E+00
2434.06	8	6	7.0E-01
2434.24	8	10	2.0E+00
2434.73	12	12	3.2E+00
2439.30	12	14	2.8E+00
2445.11	12	12	1.9E+00
2445.80	4	6	1.5E+00
2446.47	12	14	2.9E-01
2447.20	6	6	1.2E+00
2453.98	8	10	7.3E-01
2455.71	8	8	1.0E+00
2458.78	10	12	2.7E+00
2458.97	6	4	2.0E+00
2460.44	10	12	5.3E+00
2461.28	6	8	2.6E+00
2461.86	8	10	2.6E+00
2466.52	2	4	2.1E+00
2469.51	8	6	2.8E+00
2472.61	8	10	3.7E+00
2475.12	4	6	3.9E+00
2475.54	6	8	3.5E+00
2481.05	12	12	1.9E-01
2484.44	8	8	2.3E+00
2492.34	10	12	1.6E-01
2493.26	14	16	3.4E+00
2501.31	2	2	1.4E+00
2503.87	10	10	2.4E+00
2508.34	8	10	2.7E+00
2533.63	12	12	1.3E+00
2534.42	8	8	1.2E+00
2535.36	6	4	3.3E+00

NIST ATOMIC TRANSITION PROBABILITY TABLES (continued)

λ Å	Weights g_i	g_k	A $10^8\ s^{-1}$
2535.49	10	8	5.4E-01
2536.67	12	12	4.0E-01
2537.14	10	10	1.4E+00
2538.20	14	12	1.2E+00
2538.50	8	6	3.3E-01
2538.80	12	10	8.2E-01
2538.91	10	8	7.8E-01
2538.99	14	12	1.2E+00
2540.52	2	2	1.5E+00
2541.10	8	6	7.3E-01
2541.84	8	6	7.7E-01
2542.73	2	2	1.9E+00
2543.38	10	12	4.4E-01
2543.43	6	4	7.1E-01
2544.97	4	6	4.0E-01
2545.22	8	10	3.3E-01
2545.44	8	10	1.4E-01
2546.67	8	8	6.2E-01
2547.34	8	8	2.0E-01
2548.33	4	6	2.0E-01
2548.59	10	10	1.9E-01
2548.74	4	2	1.7E+00
2548.92	12	10	4.8E-01
2549.08	10	8	1.5E+00
2549.40	4	4	1.3E+00
2549.46	6	6	8.0E-01
2549.77	8	6	2.5E-01
2550.03	10	10	1.2E+00
2550.15	8	10	4.0E-01
2550.68	12	12	8.9E-01
2551.21	10	8	3.2E-01
2555.07	6	8	1.8E-01
2555.45	4	6	2.5E-01
2557.51	10	8	1.3E-01
2559.77	6	8	2.4E-01
2559.92	6	8	2.4E-01
2560.28	4	4	1.5E+00
2562.09	4	2	1.5E+00
2562.54	8	6	1.5E+00
2563.48	6	4	1.3E+00
2566.22	8	10	2.5E+00
2566.40	8	6	2.1E+00
2566.91	4	2	1.1E+00
2568.41	2	4	4.4E-01
2569.78	2	4	1.2E+00
2570.53	6	8	1.2E+00
2570.85	8	6	1.7E+00
2573.21	8	10	1.4E-01
2574.36	6	4	1.6E+00
2576.86	10	12	1.1E+00
2577.92	2	2	1.3E+00
2582.41	6	8	2.4E-01
2582.58	4	4	7.7E-01
2585.63	10	10	3.6E-01
2585.88	10	8	8.1E-01
2587.95	8	10	1.4E+00
2588.18	2	2	1.6E-01
2590.55	4	6	9.1E-02
2591.54	6	6	5.1E-01
2592.78	14	16	2.1E+00
2593.72	2	4	1.3E-01
2594.96	8	8	1.0E-01
2598.37	8	6	1.3E+00
2599.40	10	10	2.2E+00
2604.05	8	8	1.1E-01
2605.04	6	8	2.1E+00
2605.34	4	4	1.6E+00
2605.42	6	6	2.6E-01
2605.90	4	2	1.2E+00
2606.51	6	6	1.8E+00
2607.09	6	4	1.7E+00
2609.13	8	10	3.0E-01
2609.87	8	8	1.8E-01
2611.87	8	8	1.1E+00
2613.82	4	2	2.0E+00
2617.62	6	6	4.4E-01
2619.07	10	10	2.7E-01
2620.17	6	6	1.3E-01
2620.70	8	8	3.3E-01
2621.67	2	2	4.9E-01
2623.11	14	14	1.1E-01
2623.73	6	6	2.2E-01
2625.49	12	14	2.2E+00
2625.67	8	10	3.4E-01
2626.50	4	6	3.4E-01
2628.29	2	4	8.6E-01
2629.59	6	8	6.2E-01
2630.07	4	6	5.7E-01
2631.05	4	6	7.7E-01
2631.32	6	8	6.0E-01
2631.61	10	12	5.3E-01
2633.20	6	4	1.7E+00
2636.69	4	4	1.2E-01
2637.50	6	6	5.2E-01
2637.64	2	4	8.3E-01
2639.56	2	2	1.1E+00
2642.01	6	6	3.6E-01
2649.47	6	8	1.8E+00
2650.48	6	8	1.6E+00
2654.63	4	4	7.7E-01
2658.25	8	8	3.2E-01
2662.56	2	2	9.6E-01
2664.66	8	10	1.5E+00
2666.64	6	8	1.7E+00
2667.22	4	6	9.2E-01
2669.93	2	4	4.7E-01
2671.40	2	4	5.6E-01
2682.51	8	10	7.0E-01
2683.00	4	6	6.4E-01
2684.75	8	10	1.4E+00
2692.60	10	12	1.2E+00
2697.33	4	4	2.7E-01
2697.46	4	2	1.8E+00
2699.20	4	4	6.6E-01
2703.99	8	8	1.2E+00
2707.13	4	6	8.5E-01
2709.05	4	6	3.7E-01
2711.84	12	14	3.8E-01
2712.39	10	12	1.3E-01
2714.41	8	6	5.5E-01
2716.22	6	6	1.1E+00
2716.56	14	12	1.6E+00
2717.87	16	14	1.4E+00
2718.64	10	8	1.3E+00
2719.30	6	8	3.7E-01
2722.06	8	8	1.1E-01
2722.74	6	8	7.8E-01
2724.88	6	6	9.7E-01
2727.38	12	10	3.2E-01
2727.54	6	4	8.5E-01
2728.91	8	10	8.8E-02
2730.73	4	4	2.5E-01
2732.94	8	6	7.8E-01
2739.55	8	8	1.9E+00
2741.40	6	6	1.7E-01
2743.20	2	4	1.8E+00
2746.48	4	6	1.9E+00
2746.98	6	6	1.6E+00
2749.18	4	4	1.1E+00
2749.32	6	8	2.1E+00
2749.49	2	2	1.1E+00
2753.29	10	12	1.2E+00
2754.91	8	6	8.4E-01
2755.73	8	10	2.1E+00
2761.81	2	4	1.1E-01
2762.34	6	6	3.7E-01
2763.66	14	12	1.3E+00
2765.13	10	8	1.2E+00
2767.50	12	14	1.9E+00
2769.36	12	14	1.6E-01
2774.69	2	4	2.4E-01
2776.91	8	8	3.0E-01
2779.30	10	8	7.6E-01
2779.91	2	4	2.3E-01
2780.04	2	2	2.9E-01
2783.69	12	10	7.0E-01
2785.19	12	10	1.0E+00
2787.24	8	6	1.3E-01
2793.89	10	12	9.6E-02
2796.63	10	10	1.0E-01
2799.29	10	8	1.1E-01
2809.78	8	8	1.6E-01
2817.09	6	4	2.1E-01
2831.56	4	6	5.8E-01
2833.09	6	6	2.7E-01
2835.71	4	6	3.1E-01
2838.22	4	2	4.2E-01
2839.51	10	8	9.9E-01
2839.80	8	10	4.1E-01
2840.65	2	4	5.3E-01
2840.76	10	12	1.1E-01
2844.96	2	2	4.5E-01
2847.77	4	4	3.3E-01
2848.11	6	6	7.0E-01
2848.32	6	4	1.1E+00
2855.69	8	10	1.0E-01
2856.38	6	8	2.7E-01

λ Å	Weights g_i	g_k	A 10^8 s^{-1}
2856.91	8	8	8.7E-01
2857.17	6	8	9.5E-02
2872.39	10	8	1.5E-01
2873.40	8	10	3.4E-01
2875.35	8	10	9.5E-02
2883.71	12	14	1.0E-01
2884.77	6	8	1.4E-01
2895.22	8	10	8.0E-02
2897.27	6	4	1.4E-01
2944.40	4	2	4.6E-01
2947.66	6	4	2.0E-01
2949.18	10	8	2.0E-01
2959.84	8	6	1.6E-01
2964.63	2	2	9.3E-02
2969.93	8	6	1.8E-01
2982.06	4	6	2.1E-01
2984.82	6	6	3.6E-01
2985.55	2	4	1.8E-01
2997.30	6	8	8.3E-02
3002.65	4	6	1.4E-01
3036.96	6	6	1.6E-01
3048.99	4	4	2.8E-01
3062.23	12	10	1.2E-01
3071.12	2	4	1.9E-01
3076.44	4	6	2.8E-01
3077.17	14	12	1.1E-01
3078.68	6	8	4.2E-01
3135.36	6	6	8.4E-02
3154.20	10	10	1.5E-01
3167.86	8	8	1.3E-01
3177.54	8	8	8.1E-02
3179.50	6	8	9.9E-02
5247.95	4	6	1.7E+00
5506.20	12	14	1.4E+00
5961.71	10	12	7.7E-01
Fe III			
1843.4	9	7	4.8E+00
1844.3	7	5	4.9E+00
1846.9	5	3	5.5E+00
1854.38	3	1	5.7E+00
1865.20	7	7	6.1E+00
1893.98	11	9	5.5E+00
1896.80	13	11	5.0E+00
1904.3	5	5	5.7E+00
1907.58	15	13	5.3E+00
1915.08	13	15	6.0E+00
1922.79	11	13	5.5E+00
1930.39	9	11	5.1E+00
1931.51	9	11	5.3E+00
1937.35	7	9	5.1E+00
1943.48	5	7	5.0E+00
1950.33	13	15	5.5E+00
1951.01	11	11	5.3E+00
1952.65	9	9	4.9E+00
1953.32	7	7	5.1E+00
1987.50	13	13	4.9E+00
Fe VII			
150.807	5	7	1.3E+03
150.852	7	9	1.3E+03
151.023	9	11	1.6E+03
151.046	7	7	2.2E+02
151.145	9	9	2.1E+02
151.432	5	7	2.2E+02
151.512	5	5	5.3E+02
151.675	7	7	3.9E+02
151.782	9	9	2.4E+02
154.307	3	1	8.9E+02
154.335	5	7	1.2E+03
154.363	3	3	4.2E+02
154.565	5	3	3.5E+02
154.650	5	5	8.8E+02
154.848	1	3	7.7E+02
154.921	3	5	9.7E+02
154.941	3	3	2.4E+02
154.949	5	7	1.0E+03
155.994	9	11	1.8E+03
158.481	9	9	2.3E+02
165.087	1	3	6.9E+02
165.919	7	5	2.8E+03
166.365	9	7	2.9E+03
173.441	9	9	3.6E+03
176.744	9	9	2.7E+03
176.928	7	7	2.4E+03
177.172	5	5	1.5E+03
235.221	5	3	1.7E+02
240.053	3	1	1.3E+02
243.379	9	7	2.1E+02
Fe VIII			
112.472	4	4	3.6E+02
112.486	6	6	4.3E+02
116.196	4	6	4.5E+02
117.197	6	8	3.8E+02
167.486	4	4	3.0E+03
168.172	6	6	3.1E+03
168.545	6	4	2.0E+03
168.929	4	2	2.1E+03
185.213	6	8	1.0E+03
186.601	4	6	9.4E+02
Fe X			
76.822	2	2	1.8E+03
77.865	4	6	1.6E+03
100.026	8	10	2.6E+03
101.733	6	8	1.8E+03
101.846	4	6	1.7E+03
102.095	10	12	2.9E+03
102.192	10	12	2.9E+03
102.829	4	6	2.1E+03
103.319	6	8	2.6E+03
103.724	6	8	1.7E+03
104.638	8	10	2.1E+03
174.534	4	6	1.8E+03
175.266	2	4	1.72E+03
Fe XI			
72.166	5	7	2.9E+03
72.310	5	5	1.5E+03
72.635	5	7	1.6E+03
91.394	5	7	2.6E+03
91.472	7	9	2.5E+03
91.63	3	5	2.3E+03
91.63	7	9	3.4E+03
91.63	5	7	2.8E+03
91.733	9	11	4.1E+03
92.81	9	11	3.7E+03
92.87	11	13	3.9E+03
93.433	9	11	3.2E+03
179.762	5	7	1.67E+03
Fe XII			
65.905	4	4	2.0E+03
66.526	6	8	1.7E+03
66.960	4	6	1.6E+03
67.164	4	2	1.1E+03
67.821	4	6	1.4E+03
68.382	2	4	1.7E+03
80.541	6	6	8.7E+02
81.943	6	4	1.4E+03
82.226	4	2	1.9E+03
84.48	4	6	4.5E+03
84.48	8	10	4.9E+03
84.52	10	12	5.2E+03
84.52	6	8	4.0E+03
84.85	6	8	2.3E+03
85.14	8	10	3.4E+03
85.477	10	12	4.6E+03
186.880	6	8	1.0E+03
192.394	4	2	9.0E+02
193.509	4	4	9.1E+02
195.119	4	6	8.6E+02
Fe XIII			
62.353	1	3	2.0E+03
62.46	5	7	1.2E+03
62.699	3	5	2.3E+03
63.188	5	7	3.9E+03
64.139	1	3	2.1E+03
74.845	5	5	1.0E+03
75.892	5	3	7.7E+02
76.117	5	3	2.1E+03
78.452	9	11	6.3E+03
84.270	7	9	5.5E+03
107.384	7	5	1.8E+03
Fe XIV			
58.963	2	4	2.7E+03
59.579	4	6	3.1E+03
69.176	4	6	5.6E+02
69.386	2	4	7.6E+02

λ Å	Weights g_i	Weights g_k	A 10^8 s^{-1}
69.66	2	2	8.9E+02
69.66	6	6	1.3E+03
70.251	6	4	8.1E+02
70.613	4	2	1.7E+03
72.80	10	12	7.9E+03
76.022	4	6	6.6E+03
76.152	6	8	7.0E+03
91.009	6	4	5.1E+02
91.273	4	2	5.6E+02
188	4	6	2.7E+02
190	6	8	2.8E+02
207	2	2	2.1E+02
211.316	2	4	3.6E+02
213	4	2	2.8E+02
214	2	2	4.0E+02
216	6	8	1.7E+02
217	6	8	4.0E+02
217	6	6	2.6E+02
219	2	4	4.8E+02
219	4	6	2.4E+02
219.123	4	6	3.9E+02
220	4	4	3.2E+02
221	4	6	5.9E+02
226	2	4	3.9E+02
234	2	2	2.8E+02
264.787	4	4	3.38E+02
265	4	4	1.5E+02
266	6	4	1.7E+02
268	6	6	2.1E+02
268	4	2	3.3E+02
270.524	4	2	2.1E+02
274.203	2	2	1.8E+02
280	4	6	2.8E+02
283	6	8	2.7E+02
288.45	6	4	1.6E+02
Fe XV			
38.95	1	3	1.69E+03
52.911	1	3	2.94E+03
59.404	3	5	3.4E+03
63.959	5	7	1.6E+03
65.370	1	3	3.2E+02
65.612	3	3	9.8E+02
66.238	5	3	1.6E+03
68.860	9	11	9.2E+03
69.7	3	1	1.9E+03
69.942	3	5	7.4E+03
69.989	5	7	7.9E+03
70.052	7	9	8.8E+03
70.224	1	3	4.13E+03
70.53	7	5	2.6E+02
70.59	7	7	1.7E+03
73.199	7	9	8.8E+03
73.473	5	7	6.2E+03
233.857	5	7	2.2E+02
235	1	3	2.5E+02
243	1	3	2.4E+02
243	5	7	2.3E+02
243.790	3	5	4.2E+02
248	3	1	5.4E+02
284.160	1	3	2.28E+02
Fe XVI			
31.041	2	4	5.2E+02
31.242	4	6	6.1E+02
32.166	2	4	6.8E+02
32.192	2	2	6.7E+02
32.433	2	4	7.7E+02
32.652	4	6	9.1E+02
34.857	2	4	1.23E+03
35.106	4	6	1.44E+03
35.333	4	6	6.4E+02
35.368	6	8	6.8E+02
36.01	4	2	5.0E+02
36.749	2	4	1.1E+03
36.803	2	2	1.2E+03
37.096	4	6	1.0E+03
37.138	6	8	1.07E+03
39.827	2	4	2.1E+03
40.153	4	6	2.5E+03
40.161	4	4	4.1E+02
40.199	4	6	1.7E+03
40.245	6	8	1.8E+03
41.91	2	2	4.72E+02
42.30	4	2	9.2E+02
46.661	4	6	3.46E+03
46.718	6	8	3.7E+03
50.350	2	4	1.86E+03
50.555	2	2	1.98E+03
54.142	2	4	3.41E+03
54.728	4	6	4.16E+03
54.769	4	4	6.97E+02
62.879	2	2	1.05E+03
63.719	4	2	2.18E+03
66.263	4	6	9.39E+03
66.368	6	8	1.00E+04
66.392	6	6	6.69E+02
76.502	6	4	6.7E+02
76.796	4	2	7.72E+02
80.192	4	6	5.2E+02
80.270	6	8	5.4E+02
85.587	2	4	4.0E+02
86.133	4	6	4.8E+02
96.256	4	6	8.7E+02
96.348	6	8	9.3E+02
117.2	2	4	3.93E+02
117.7	2	2	3.9E+02
123.4	2	4	5.9E+02
124.5	4	6	7.0E+02
144.06	4	6	1.6E+03
144.25	6	8	1.6E+03
148	4	2	6.5E+02
266.7	4	6	3.9E+02
267.0	6	8	4.3E+02
Fe XVII			
11.023	1	3	2.1E+04
12.123	1	3	8.0E+04
12.264	1	3	5.9E+04
12.526	1	3	3.0E+03
12.681	1	3	3.5E+03
13.823	1	3	3.3E+04
13.891	1	3	3.4E+03
15.015	1	3	2.28E+05
15.262	1	3	6.0E+04
16.777	1	3	8.29E+03
17.054	1	3	9.33E+03
41.37	9	11	4.8E+03
49.427	3	3	4.0E+03
50.26	7	9	6.0E+03
58.76	9	11	1.2E+04
Fe XIX			
13.413	5	3	1.3E+04
13.426	5	7	4.8E+04
13.47	3	1	1.5E+05
13.520	5	7	2.0E+05
13.56	3	5	1.0E+04
13.68	3	1	8.0E+04
13.69	5	7	2.3E+04
13.700	1	3	2.7E+05
13.71	5	5	2.2E+04
13.738	5	7	1.0E+04
13.796	5	7	7.0E+04
13.83	5	5	1.4E+04
13.934	1	3	4.51E+04
13.961	3	3	2.0E+04
14.668	5	7	1.1E+04
14.671	5	3	1.1E+04
14.929	3	3	1.2E+04
14.966	5	3	2.5E+04
14.995	5	5	2.2E+04
15.015	1	3	1.4E+04
16.668	3	1	1.1E+04
Fe XX			
12.67	6	6	1.0E+04
12.69	4	6	1.2E+04
12.73	4	2	4.0E+04
12.77	4	4	2.1E+05
12.78	4	2	6.9E+04
12.78	2	4	1.4E+05
12.79	6	4	1.7E+04
12.82	4	4	1.1E+05
12.88	6	4	2.7E+04
12.89	4	4	4.4E+04
12.90	4	2	6.2E+03
12.90	4	6	1.4E+05
12.92	2	4	1.7E+04
12.93	4	6	1.6E+05
12.93	2	2	1.2E+04
12.98	2	2	6.7E+04
12.99	6	6	5.1E+04
13.00	6	4	1.1E+04
13.01	2	4	3.0E+04
13.03	4	2	8.6E+04
13.07	6	4	8.2E+03

λ Å	Weights g_i	Weights g_k	A $10^8\ s^{-1}$
13.13	2	4	8.9E+04
13.24	4	4	1.2E+04
13.28	4	4	6.1E+03
13.70	4	6	1.1E+04
13.71	2	2	9.9E+03
13.78	4	4	1.0E+04
13.79	6	6	1.2E+04
13.83	4	2	9.8E+03
13.90	4	2	1.2E+04
13.98	6	4	1.6E+04
13.99	4	2	2.2E+04
14.05	4	4	1.7E+04
14.23	2	2	6.3E+03
Fe XXI			
8.53	3	1	1.8E+04
8.53	3	5	6.1E+03
8.53	3	3	1.5E+04
8.56	5	7	2.0E+04
8.56	1	3	2.1E+04
8.56	5	3	6.5E+03
8.64	5	7	1.5E+04
8.65	5	7	3.9E+04
8.66	5	5	4.4E+03
8.74	1	3	2.5E+04
9.42	3	1	4.3E+04
9.42	3	3	3.3E+04
9.44	3	5	1.7E+04
9.45	1	3	5.2E+04
9.46	5	3	1.5E+04
9.47	5	7	4.9E+04
9.47	5	5	6.1E+03
9.52	3	3	8.1E+03
9.58	5	5	5.2E+03
9.59	5	5	1.0E+04
9.67	1	3	5.7E+04
9.68	5	7	4.0E+03
9.74	5	3	5.3E+03
12.02	1	3	1.3E+04
12.13	3	3	1.8E+04
12.18	5	7	2.2E+04
12.19	5	3	6.4E+03
12.21	3	1	1.5E+05
12.21	3	3	1.2E+05
12.25	1	3	2.1E+05
12.28	5	3	5.2E+04
12.30	5	7	2.1E+05
12.36	3	3	3.6E+04
12.37	5	7	3.1E+05
12.38	5	3	6.9E+03
12.47	5	7	3.8E+04
12.47	5	3	1.3E+04
12.49	5	7	1.3E+04
12.53	5	5	1.5E+04
12.57	1	3	7.2E+04
12.73	5	5	8.2E+03
12.95	3	5	6.2E+03
13.00	1	3	7.2E+03
13.03	5	5	1.3E+04
13.14	3	1	2.0E+04
13.41	1	3	7.3E+03
Fe XXII			
9.002	4	6	5.5E+04
9.006	6	8	5.7E+04
9.006	6	6	5.3E+04
9.163	4	6	6.9E+04
9.183	6	8	8.3E+04
9.241	4	6	5.1E+04
11.748	4	4	1.2E+05
11.748	4	6	1.6E+05
11.748	4	2	1.8E+05
11.763	2	4	1.6E+05
11.789	2	2	2.6E+05
11.789	6	8	1.2E+05
11.797	2	4	1.7E+05
11.823	6	4	7.9E+04
11.837	6	8	2.3E+05
11.837	6	6	1.7E+05
11.886	4	6	1.3E+05
11.898	2	4	8.2E+04
11.922	4	6	1.8E+05
11.976	6	8	5.9E+04
12.027	2	4	6.9E+04
12.045	6	8	2.4E+05
12.045	4	4	9.7E+04
12.053	4	6	6.1E+04
12.077	2	4	1.0E+05
12.077	4	6	2.4E+05
12.095	6	6	7.8E+04
12.193	2	4	7.2E+04
12.193	4	6	9.9E+04
12.325	2	2	1.5E+05
Fe XXIII			
7.733	5	7	3.0E+04
7.849	5	7	4.9E+04
8.307	1	3	4.8E+04
8.529	1	3	4.3E+04
8.550	3	5	6.0E+04
8.552	3	3	3.2E+04
8.614	5	7	7.7E+04
8.664	3	3	4.4E+04
8.669	5	7	6.1E+04
8.672	1	3	6.8E+04
8.752	5	7	1.2E+05
8.764	5	7	4.6E+04
8.814	3	5	6.2E+04
10.902	5	5	5.3E+04
10.910	3	1	6.7E+04
10.927	5	7	6.0E+04
10.934	3	5	5.4E+04
10.979	1	3	7.9E+04
11.018	1	3	4.9E+04
11.086	3	1	6.5E+04
11.165	3	5	6.7E+04
11.255	3	3	3.7E+04
11.298	1	3	1.3E+05
11.325	3	5	1.7E+05
11.338	3	3	9.3E+04
11.429	3	1	1.7E+05
11.433	3	3	1.2E+05
11.441	5	7	2.2E+05
11.445	5	5	5.6E+04
11.485	3	5	1.40E+05
11.491	5	3	5.9E+04
11.519	5	5	1.16E+05
11.520	1	3	2.16E+05
11.524	5	7	2.3E+05
11.593	5	7	3.58E+05
11.613	3	5	1.0E+05
11.615	3	3	4.4E+04
11.691	5	7	7.7E+04
11.698	5	5	7.3E+04
11.737	3	5	1.8E+05
11.898	1	3	2.03E+05
Fe XXIV			
1.8523	2	2	1.0E+05
1.8552	2	4	4.82E+06
1.8563	4	2	2.43E+06
1.8572	2	2	3.06E+06
1.858	2	4	1.2E+05
1.8614	4	4	6.24E+06
1.8626	2	4	3.16E+06
1.8627	2	2	5.47E+06
1.8637	2	2	1.91E+06
1.8655	4	6	2.14E+06
1.8672	4	2	1.63E+06
1.8678	4	4	3.5E+05
1.8721	4	6	3.2E+05
1.8721	2	2	2.0E+05
1.8730	2	4	1.5E+05
1.8739	4	4	8.3E+04
1.891	2	2	9.7E+04
1.897	4	2	9.8E+04
8.231	2	4	6.10E+04
8.316	4	6	7.07E+04
10.619	2	4	7.28E+04
10.663	2	2	7.51E+04
11.030	2	4	1.84E+05
11.171	4	6	2.18E+05
Fe XXV			
1.4607	1	3	2.54E+05
1.4945	1	3	5.05E+05
1.5730	1	3	1.24E+06
1.5749	1	3	1.5E+05
1.778	3	3	8.7E+04
1.782	3	1	4.69E+06
1.787	1	3	2.57E+06
1.787	5	5	1.19E+06
1.788	3	5	2.68E+06
1.788	3	5	1.63E+06
1.789	1	3	1.78E+06
1.790	3	3	1.23E+06
1.791	3	5	4.10E+06

λ Å	Weights g_i	g_k	A 10^8 s^{-1}
1.791	3	3	2.59E+06
1.792	3	1	4.92E+06
1.792	5	5	2.81E+06
1.793	3	1	2.67E+06
1.794	5	3	2.22E+06
1.797	3	5	8.8E+05
1.798	3	3	1.0E+05
1.800	1	3	8.6E+04
1.802	3	1	4.1E+05
1.810	3	1	5.9E+05
1.8502	1	3	4.57E+06
1.8593	1	3	4.42E+05
10.038	3	3	8.08E+04
Krypton			
Kr I			
1164.9	1	3	3.16E+00
1235.8	1	3	3.12E+00
4274.0	5	5	2.6E-02
4351.4	3	1	3.2E-02
4362.6	5	3	8.4E-03
4376.1	3	1	5.6E-02
4400.0	3	5	2.0E-02
4410.4	3	3	4.4E-03
4425.2	3	3	9.7E-03
4453.9	3	5	7.8E-03
4463.7	3	3	2.3E-02
4502.4	3	5	9.2E-03
5562.2	5	5	2.8E-03
5570.3	5	3	2.1E-02
5649.6	1	3	3.7E-03
5870.9	3	5	1.8E-02
6904.7	3	5	1.3E-02
7224.1	3	5	1.4E-02
7587.4	3	1	5.1E-01
7601.5	5	5	3.1E-01
7685.2	3	1	4.9E-01
7694.5	5	3	5.6E-02
7854.8	1	3	2.3E-01
8059.5	1	3	1.9E-01
8104.4	5	5	1.3E-01
8112.9	5	7	3.6E-01
8190.1	3	5	1.1E-01
8263.2	3	5	3.5E-01
8281.1	3	3	1.9E-01
8298.1	3	3	3.2E-01
8508.9	3	3	2.4E-01
8776.7	3	5	2.7E-01
8928.7	5	3	3.7E-01
Kr II			
4250.6	4	4	1.2E-01
4292.9	4	4	9.6E-01
4355.5	6	8	1.0E+00
4431.7	2	2	1.8E+00
4436.8	2	4	6.6E-01
4577.2	6	8	9.6E-01
4583.0	6	4	7.6E-01
4615.3	4	4	5.4E-01
4619.2	4	6	8.1E-01
4633.9	4	6	7.1E-01
4658.9	6	4	6.5E-01
4739.0	6	6	7.6E-01
4762.4	2	4	4.2E-01
4765.7	4	6	6.7E-01
4811.8	2	4	1.7E-01
4825.2	2	4	1.9E-01
4832.1	4	2	7.3E-01
5208.3	4	4	1.4E-01
5308.7	4	6	2.4E-02
7407.0	6	6	7.0E-02
Lead			
Pb I			
2022.0	1	3	5.2E-02
2053.3	1	3	1.2E-01
2170.0	1	3	1.5E+00
2401.9	3	3	1.9E-01
2446.2	3	3	2.5E-01
2476.4	3	5	2.8E-01
2577.3	5	3	5.0E-01
2613.7	3	3	2.7E-01
2614.2	3	5	1.9E+00
2628.3	5	3	3.1E-02
2657.1	3	5	9.8E-04
2663.2	5	5	7.1E-01
2802.0	5	7	1.6E+00
2823.2	5	5	2.6E-01
2833.1	1	3	5.8E-01
2873.3	5	5	3.7E-01
3572.7	5	3	9.9E-01
3639.6	3	3	3.4E-01
3671.5	5	3	4.4E-01
3683.5	3	1	1.5E+00
3739.9	5	5	7.3E-01
4019.6	5	7	3.5E-02
4057.8	5	3	8.9E-01
4062.1	5	3	9.2E-01
4168.0	5	5	1.2E-02
5005.4	1	3	2.7E-01
5201.4	1	3	1.9E-01
7229.0	5	3	8.9E-03
Lithium			
Li I			
*2741.2	2	6	1.3E-02
*3232.7	2	6	1.17E-02
*4602.9	6	10	2.23E-01
*6103.6	6	10	6.860E-01
*6707.8	2	6	3.691E-01
Lutetium			
Lu I			
3376.5	4	4	2.23E+00
3567.8	4	6	5.9E-01
3620.3	6	4	1.1E-02
3841.2	6	6	2.5E-01
4518.6	4	4	2.1E-01
Magnesium			
Mg I			
2025.8	1	3	8.4E-01
*2779.8	9	9	5.2E+00
*2850.0	9	15	2.3E-01
2852.1	1	3	4.95E+00
*3094.9	9	15	5.2E-01
3329.9	1	3	3.3E-02
3332.2	3	3	9.7E-02
3336.7	5	3	1.6E-01
*3835.3	9	15	1.68E+00
4703.0	3	5	2.55E-01
5167.3	1	3	1.16E-01
5172.7	3	3	3.46E-01
5183.6	5	3	5.75E-01
5528.4	3	5	1.99E-01
Mg II			
1239.9	2	4	1.4E-02
1240.4	2	2	1.4E-02
*2660.8	10	14	3.8E-01
2790.8	2	4	4.0E+00
2795.5	2	4	2.6E+00
2797.9	4	4	7.9E-01
2798.1	4	6	4.8E+00
2802.7	2	2	2.6E+00
2928.8	2	2	1.2E+00
2936.5	4	2	2.3E+00
*3104.8	10	14	8.1E-01
3848.2	6	4	2.8E-02
3848.3	4	4	3.0E-03
3850.4	4	2	3.0E-02
*4481.2	10	14	2.23E+00
9218.3	2	4	3.6E-01
9244.3	2	2	3.6E-01
Mg IV			
320.99	4	2	1.2E+02
323.31	2	2	5.9E+01
1219.0	6	6	5.9E+00
1375.5	4	4	4.5E+00
1459.6	6	4	4.6E+00
1495.5	4	6	6.4E+00
1510.7	4	4	6.7E+00
1683.0	6	8	5.8E+00
1698.8	4	6	3.9E+00
1893.9	6	6	2.8E+00
Mg VI			
*269.92	10	6	3.1E+02
*292.53	6	6	9.0E+01
*314.64	6	2	1.8E+02
*349.15	10	10	6.1E+01
*387.94	6	10	1.3E+01
399.29	4	2	2.8E+01
400.68	4	4	2.8E+01
403.32	4	6	2.7E+01

λ Å	Weights g_i	Weights g_k	A 10^8 s^{-1}
Mg VII			
277.01	3	3	9.5E+01
278.41	5	3	1.5E+02
280.74	5	3	2.0E+02
319.02	5	5	8.9E+01
*366.42	9	9	4.4E+01
*433.04	9	15	1.6E+01
1334.3	5	5	5.3E+00
1410.0	5	5	2.57E+00
1487.0	3	5	3.02E+00
1487.9	5	7	3.66E+00
Mg VIII			
*74.976	6	10	4.3E+03
315.02	4	4	1.2E+02
*342.29	10	6	6.3E+01
353.86	4	4	3.89E+01
356.00	6	4	5.7E+01
*428.52	10	10	3.24E+01
*434.62	6	10	1.6E+01
*489.33	6	6	3.9E+01
*686.92	6	10	9.4E+00
Mg IX			
62.751	1	3	2.87E+03
*67.189	9	15	6.20E+03
*71.965	9	3	1.22E+03
72.312	3	5	4.43E+03
77.737	3	1	3.92E+02
368.07	1	3	5.27E+01
438.69	3	1	7.9E+01
*443.74	9	9	4.19E+01
749.55	3	5	8.2E+00
1639.8	3	5	2.1E+00
2814.2	1	3	3.35E-01
Mg X			
57.876	2	4	2.09E+03
57.920	2	2	2.09E+03
63.152	2	4	5.6E+03
63.295	4	6	6.7E+03
609.79	2	4	7.53E+00
624.94	2	2	7.01E+00
2212.5	2	4	9.64E-01
2278.7	2	2	8.82E-01
5918.7	2	4	3.20E-02
6229.6	4	6	3.30E-02
Mg XI			
7.310	1	3	1.15E+04
7.473	1	3	2.27E+04
7.850	1	3	5.50E+04
9.169	1	3	1.97E+05
Manganese			
Mn I			
2794.82	6	8	3.7E+00
2798.27	6	6	3.6E+00
2801.08	6	4	3.7E+00
3007.65	6	8	1.8E-01
3011.38	8	10	3.1E-01
3016.45	10	12	2.9E-01
3043.36	8	8	5.9E-01
3044.57	10	8	5.7E-01
3045.59	10	10	6.7E-01
3045.80	8	10	1.7E-01
3047.03	12	12	6.1E-01
3054.36	8	6	4.6E-01
3070.27	6	6	1.9E-01
3073.18	4	4	3.7E-01
3082.71	14	14	2.9E-01
3110.68	6	8	2.7E-01
3113.80	12	10	2.6E-01
3118.10	4	6	1.7E-01
3122.88	10	10	1.9E-01
3126.85	8	6	2.3E-01
3132.28	10	10	2.1E-01
3132.79	8	8	2.7E-01
3175.58	8	10	1.8E-01
3201.11	4	6	2.2E-01
3228.09	10	12	6.4E-01
3230.23	10	12	1.9E-01
3230.72	8	8	3.5E-01
3240.88	6	4	2.2E-01
3243.78	6	6	5.3E-01
3251.13	4	2	2.3E-01
3252.95	4	4	1.8E-01
3256.14	4	6	5.0E-01
3258.41	2	2	9.7E-01
3260.24	2	4	3.8E-01
3267.79	14	14	3.5E-01
3268.72	6	8	3.3E-01
3270.35	12	12	2.6E-01
3273.02	10	10	2.7E-01
3298.23	6	4	2.8E-01
3303.28	4	4	1.9E-01
3463.66	8	8	3.2E-01
3470.01	6	8	2.4E-01
3511.83	12	12	2.7E-01
3535.30	10	10	1.7E-01
3559.81	6	6	2.1E-01
3577.87	10	8	9.4E-01
3595.11	6	4	1.8E-01
3601.27	12	10	2.3E-01
3607.53	8	8	2.3E-01
3608.49	6	6	3.6E-01
3610.30	4	4	4.2E-01
3635.70	10	8	2.1E-01
3660.40	12	14	9.1E-01
3675.67	6	8	2.2E-01
3676.96	10	12	7.3E-01
3680.15	12	10	1.9E-01
3682.09	8	10	7.6E-01
3684.87	6	8	2.6E-01
3706.08	12	14	1.4E+00
3718.92	10	12	9.6E-01
3731.94	8	10	1.0E+00
3771.44	14	14	1.9E-01
3773.86	12	12	2.5E-01
3800.55	6	8	2.7E-01
3806.72	10	12	5.9E-01
3823.51	8	10	5.21E-01
3823.89	6	6	2.31E-01
3833.87	4	4	3.14E-01
3834.37	6	8	4.29E-01
3839.78	2	2	4.64E-01
3841.07	4	6	3.3E-01
3843.99	2	4	2.11E-01
3889.46	12	14	3.1E-01
3898.37	6	8	1.7E-01
3899.34	4	6	2.4E-01
3924.08	2	4	9.4E-01
3926.48	6	8	5.4E-01
3951.98	2	2	3.1E-01
3952.84	6	6	4.1E-01
3975.88	2	4	1.8E-01
3982.16	4	2	3.5E-01
3982.58	6	4	2.3E-01
3982.90	6	4	5.5E-01
3991.60	2	2	2.1E-01
4011.91	8	8	2.3E-01
4018.11	10	8	2.54E-01
4030.76	6	8	1.7E-01
4033.07	6	6	1.65E-01
4034.49	6	4	1.58E-01
4041.36	10	10	7.87E-01
4048.75	6	4	7.5E-01
4052.48	6	8	3.8E-01
4055.55	8	8	4.31E-01
4058.94	4	2	7.25E-01
4061.74	8	6	1.9E-01
4063.53	6	6	1.69E-01
4065.08	12	14	2.5E-01
4066.24	10	8	2.2E-01
4070.28	2	2	2.3E-01
4079.42	2	4	3.8E-01
4082.95	4	6	2.95E-01
4083.63	6	8	2.8E-01
4089.94	8	10	1.7E-01
4105.37	10	8	1.7E-01
4135.03	12	12	3.0E-01
4141.06	10	10	2.6E-01
4148.80	8	8	2.3E-01
4176.61	14	12	2.4E-01
4189.99	12	10	2.0E-01
4201.78	10	8	2.3E-01
4235.30	8	6	9.17E-01
4239.74	4	2	3.9E-01
4257.67	2	2	3.7E-01
4265.93	4	4	4.92E-01
4281.10	6	6	2.3E-01
4411.87	12	10	2.6E-01
4414.89	8	6	2.93E-01
4419.77	10	8	2.1E-01
4436.36	6	4	4.37E-01
4451.58	8	8	7.98E-01
4453.01	4	2	5.44E-01

λ	Weights		A
Å	g_i	g_k	10^8 s^{-1}
4455.82	4	6	1.7E-01
4457.04	6	4	2.34E-01
4457.55	6	6	4.27E-01
4458.26	6	8	4.62E-01
4461.09	8	8	1.7E-01
4462.03	8	10	7.00E-01
4464.68	6	6	4.39E-01
4470.14	4	4	3.00E-01
4472.79	2	2	4.35E-01
4479.40	8	10	3.4E-01
4490.08	2	4	2.49E-01
4498.90	4	6	2.49E-01
4502.22	6	8	1.86E-01
4605.37	10	12	3.6E-01
4626.54	12	14	3.6E-01
4709.71	8	8	1.72E-01
4727.46	6	6	1.7E-01
4739.11	4	4	2.40E-01
4754.05	6	8	3.03E-01
4761.53	2	4	5.35E-01
4762.38	8	10	7.83E-01
4765.86	4	6	4.1E-01
4766.43	6	8	4.6E-01
4783.43	8	8	4.01E-01
4823.53	10	8	4.99E-01
6013.48	4	6	1.72E-01
6021.79	8	6	3.32E-01
Mn II			
2593.72	7	7	2.6E+00
2605.68	7	5	2.7E+00
2933.05	5	3	2.0E+00
2939.31	5	5	1.9E+00
2949.20	5	7	1.9E+00
3441.99	9	7	4.3E-01
3460.32	7	5	3.2E-01
3474.13	5	3	1.5E-01
3482.90	5	5	2.0E-01
3488.68	3	3	2.5E-01
Mn VI			
307.999	9	9	3.7E+01
309.440	9	7	5.7E+01
309.579	7	5	4.4E+01
310.058	7	7	3.4E+01
310.182	5	5	2.8E+01
311.748	5	3	5.7E+01
320.598	3	5	1.5E+01
320.681	1	3	2.2E+01
320.874	3	1	7.8E+01
320.979	3	3	2.2E+01
321.176	5	5	6.0E+01
321.541	5	3	2.7E+01
325.146	9	7	1.3E+02
328.431	5	5	4.4E+01
328.558	3	5	1.2E+01
329.043	1	3	1.1E+01
1236.23	5	3	1.3E+01
1255.77	3	1	1.2E+01
1285.10	5	7	1.1E+01
1333.87	7	9	1.0E+01
Mercury			
Hg I			
2536.52	1	3	8.00E-02
2652.04	3	5	3.88E-01
2655.13	3	5	1.1E-01
2752.78	1	3	6.10E-02
2856.94	3	1	1.1E-02
2893.60	3	3	1.6E-01
2925.4	5	3	7.7E-02
2967.3	1	3	4.5E-01
3021.50	5	7	5.09E-01
3023.48	5	5	9.4E-02
3027.49	5	5	2.0E-02
3125.66	3	5	6.56E-01
3341.48	5	3	1.68E-01
3650.15	5	7	1.3E+00
3654.83	5	5	1.8E-01
4046.56	1	3	2.1E-01
4077.81	3	1	4.0E-02
4108.1	3	1	3.0E-02
4339.22	3	5	2.88E-02
4347.50	3	5	8.4E-02
4358.34	3	3	5.57E-01
4916.07	3	1	5.8E-02
5025.64	3	3	2.7E-04
5460.75	5	3	4.87E-01
5769.59	3	5	2.36E-01
6234.4	1	3	5.3E-03
6716.4	1	3	4.3E-03
6907.5	3	5	2.8E-02
7728.8	1	3	9.7E-03
10139.79	3	1	2.71E-01
Molybdenum			
Mo I			
2616.79	3	5	7.34E-01
2621.06	7	7	1.16E-01
2628.96	3	3	2.81E-01
2629.85	5	7	7.75E-01
2631.50	1	3	2.54E-01
2638.30	5	5	7.57E-01
2640.98	7	5	1.20E+00
2644.36	5	7	1.96E-01
2649.46	7	9	9.84E-01
2655.02	9	7	4.08E-01
2658.11	7	7	6.43E-01
2665.09	7	9	1.32E-01
2679.85	9	11	1.31E+00
2684.16	9	9	4.18E-01
2706.11	3	5	2.03E-01
2710.74	3	3	1.57E-01
2725.15	3	5	2.79E-01
2728.71	3	3	1.26E-01
2733.39	5	7	2.95E-01
2743.71	1	3	2.47E-01
2745.38	13	11	1.29E-01
2751.47	7	9	2.54E-01
2756.26	5	3	1.18E-01
2761.53	9	11	2.06E-01
2763.02	3	1	4.44E-01
2766.25	3	5	1.17E-01
2787.83	9	7	2.85E-01
2792.96	5	3	1.53E-01
2798.02	7	5	1.22E-01
2801.47	5	7	1.24E-01
2825.68	5	7	2.53E-01
2826.75	7	7	4.23E-01
2876.54	9	9	2.84E-01
2886.60	11	11	4.74E-01
2906.06	3	3	8.04E-01
2913.52	5	3	1.38E-01
2915.38	5	3	7.31E-01
2918.84	5	3	3.79E-01
2930.39	1	3	1.91E-01
2936.50	11	11	2.33E-01
2945.43	7	7	3.66E-01
2945.66	3	3	4.08E-01
2946.01	5	5	1.68E-01
2951.45	9	9	1.43E-01
2959.48	9	11	1.75E-01
2972.96	5	3	2.69E-01
2977.27	9	7	3.28E-01
2978.28	7	5	1.50E-01
2983.04	1	3	2.82E-01
2987.92	3	5	8.43E-01
2988.23	5	7	4.28E-01
2988.68	7	9	1.61E-01
2989.80	9	7	9.27E-01
3000.24	9	9	1.40E-01
3000.44	5	5	1.25E-01
3000.85	5	7	2.58E-01
3001.43	5	5	2.31E-01
3007.71	7	5	1.90E-01
3013.39	7	5	6.06E-01
3016.78	9	9	2.75E-01
3025.00	5	5	8.49E-01
3036.31	3	5	5.81E-01
3041.70	13	11	5.94E-01
3046.80	13	11	1.63E-01
3047.31	11	9	5.01E-01
3055.32	9	7	4.29E-01
3057.56	7	5	2.64E-01
3061.59	7	5	4.41E-01
3064.27	13	13	8.46E-01
3065.04	13	13	3.08E-01
3069.51	5	5	1.52E-01
3069.96	11	11	2.72E-01
3070.89	9	11	1.87E-01
3074.37	11	11	1.42E+00
3079.88	9	11	9.55E-01
3080.40	7	9	3.61E-01
3081.16	3	5	2.35E-01
3085.62	9	9	1.63E+00
3089.13	11	9	1.53E-01
3089.71	5	7	2.34E-01

λ Å	Weights g_i	Weights g_k	A 10^8 s^{-1}
3094.66	7	7	1.63E+00
3099.92	9	7	1.45E-01
3100.88	7	9	1.20E+00
3101.34	5	5	1.92E+00
3106.34	7	5	2.21E-01
3117.54	13	13	1.89E-01
3123.03	3	3	2.81E-01
3125.03	5	3	1.98E-01
3132.59	7	9	1.79E+00
3135.90	9	11	3.68E-01
3136.75	9	11	1.57E-01
3142.75	3	5	4.10E-01
3147.35	13	11	2.41E-01
3155.19	7	7	2.75E-01
3158.17	7	7	4.63E-01
3170.34	7	7	1.37E+00
3171.38	5	7	2.03E-01
3175.59	13	11	8.40E-01
3179.78	11	13	2.33E-01
3183.03	11	9	3.98E-01
3184.58	7	5	2.77E-01
3185.10	7	7	2.54E-01
3185.71	5	3	6.10E-01
3188.10	7	9	3.45E-01
3188.41	5	7	4.40E-01
3192.79	9	11	1.88E-01
3193.98	7	5	1.53E+00
3194.88	9	11	1.75E-01
3195.96	9	7	4.10E-01
3197.18	1	3	1.47E-01
3198.85	15	13	7.22E-01
3200.89	3	5	1.82E-01
3205.22	1	3	4.27E-01
3205.43	9	11	2.55E-01
3205.89	9	9	5.35E-01
3208.84	7	5	2.77E-01
3210.97	7	5	6.94E-01
3214.44	9	7	2.01E-01
3215.07	3	5	4.20E-01
3216.78	15	13	2.10E-01
3221.73	3	1	1.41E+00
3228.21	5	7	3.85E-01
3229.79	9	11	1.44E-01
3233.14	13	13	6.33E-01
3237.06	7	9	2.95E-01
3244.47	5	3	2.80E-01
3247.61	5	5	1.71E-01
3249.93	5	3	1.87E-01
3251.65	3	5	3.05E-01
3256.21	5	3	6.89E-01
3256.72	3	3	1.31E-01
3259.16	11	13	1.62E-01
3262.63	7	9	3.62E-01
3264.40	11	9	5.42E-01
3265.14	5	7	2.60E-01
3266.16	9	11	1.95E-01
3270.90	7	7	3.59E-01
3276.07	11	9	1.18E-01
3285.03	1	3	1.41E-01
3285.35	9	7	4.49E-01
3287.38	5	5	1.38E-01
3289.01	9	9	5.08E-01
3290.82	7	5	5.44E-01
3305.56	5	3	1.74E-01
3305.91	7	9	3.06E-01
3307.13	7	9	1.25E-01
3312.33	7	5	1.62E-01
3323.95	9	7	2.82E-01
3325.13	5	3	2.26E-01
3325.67	5	5	1.72E-01
3327.30	1	3	2.88E-01
3336.56	9	9	1.64E-01
3340.16	5	3	1.20E-01
3344.73	3	5	6.04E-01
3346.83	11	11	1.13E-01
3347.00	3	3	2.72E-01
3358.12	5	7	7.59E-01
3361.37	9	9	1.38E-01
3363.78	5	7	2.74E-01
3363.87	5	7	1.39E-01
3373.81	3	3	2.03E-01
3375.22	7	7	1.38E-01
3375.65	7	9	1.56E-01
3378.19	3	1	1.88E-01
3378.46	13	13	3.75E-01
3379.96	5	5	4.11E-01
3382.48	3	3	2.66E-01
3384.61	7	9	7.32E-01
3385.87	9	11	3.30E-01
3389.79	5	7	1.85E-01
3392.17	9	9	1.97E-01
3393.65	11	11	2.08E-01
3404.33	7	7	2.10E-01
3413.37	11	11	1.25E-01
3415.27	9	9	1.83E-01
3415.61	7	9	1.29E-01
3416.14	9	11	2.45E-01
3418.52	5	3	1.41E-01
3419.69	7	7	1.15E-01
3420.04	5	5	3.28E-01
3422.31	9	9	2.52E-01
3425.13	11	11	2.29E-01
3427.90	11	13	4.09E-01
3434.79	7	7	1.75E-01
3435.45	15	15	1.50E+00
3437.21	11	9	8.06E-01
3438.87	1	3	2.34E-01
3441.87	5	3	1.34E-01
3442.66	3	3	2.94E-01
3445.03	7	9	1.53E-01
3445.26	7	5	2.96E-01
3445.80	9	9	1.14E-01
3447.12	9	11	8.75E-01
3447.29	5	3	1.79E-01
3449.07	7	9	1.52E-01
3449.85	5	7	1.65E-01
3452.60	7	7	2.48E-01
3456.15	5	5	3.60E-01
3456.52	3	3	2.96E-01
3460.22	5	3	2.77E-01
3460.78	9	7	6.03E-01
3465.84	3	1	9.99E-01
3466.19	9	7	2.11E-01
3466.96	7	7	1.52E-01
3467.85	5	7	2.63E-01
3469.22	5	3	6.96E-01
3469.63	13	15	1.51E-01
3470.92	3	5	2.91E-01
3475.03	3	3	4.68E-01
3479.42	7	5	2.26E-01
3483.67	7	7	1.13E-01
3483.83	7	5	1.41E-01
3489.43	7	7	3.27E-01
3504.41	7	9	8.06E-01
3505.31	7	9	2.25E-01
3508.11	9	9	1.59E-01
3510.77	13	13	4.75E-01
3517.55	11	11	5.41E-01
3518.21	3	3	3.64E-01
3521.38	9	9	1.39E-01
3521.41	9	11	6.06E-01
3524.65	5	3	3.10E-01
3524.98	7	9	2.25E-01
3538.92	11	11	2.24E-01
3540.57	5	3	4.46E-01
3542.17	7	5	4.93E-01
3552.71	9	7	3.64E-01
3555.64	3	3	3.46E-01
3558.09	5	7	5.43E-01
3563.75	1	3	1.53E-01
3566.05	9	9	2.67E-01
3566.74	7	7	1.43E-01
3570.64	15	15	7.18E-01
3573.88	3	5	3.58E-01
3580.54	13	11	5.49E-01
3581.88	11	13	3.81E-01
3584.25	3	3	1.73E-01
3585.57	7	5	3.95E-01
3588.95	7	7	1.18E-01
3590.74	7	9	2.23E-01
3595.55	5	5	2.32E-01
3598.88	13	11	5.67E-01
3600.73	9	9	2.07E-01
3601.88	7	9	1.15E-01
3602.94	5	7	2.96E-01
3604.07	9	7	3.25E-01
3610.61	5	3	1.78E-01
3611.99	7	7	1.16E-01
3615.16	7	9	1.96E-01
3623.22	11	9	5.58E-01
3624.46	9	11	5.27E-01
3624.62	5	7	1.37E-01
3638.20	5	3	3.51E-01
3638.21	5	3	3.33E-01
3640.62	7	5	1.94E-01
3647.84	7	7	2.11E-01
3648.70	7	5	1.15E-01

λ Å	Weights g_i	g_k	A 10^8 s^{-1}
3654.58	3	3	1.80E-01
3657.36	5	7	2.03E-01
3658.13	9	9	1.86E-01
3659.36	7	9	6.70E-01
3660.92	3	5	1.34E-01
3662.15	7	9	1.45E-01
3662.99	11	11	3.48E-01
3663.27	7	5	2.30E-01
3664.81	11	13	9.54E-01
3664.88	1	3	1.92E-01
3669.34	9	7	2.16E-01
3672.81	9	11	1.95E-01
3672.82	9	9	1.13E-01
3676.23	3	1	5.22E-01
3680.68	11	11	2.96E-01
3681.72	9	7	1.68E-01
3683.01	3	5	1.20E-01
3687.96	5	7	2.12E-01
3688.97	11	9	3.26E-01
3690.59	11	9	2.07E-01
3694.94	5	7	6.36E-01
3696.04	11	11	3.59E-01
3698.07	7	5	1.48E-01
3708.55	7	9	1.28E-01
3715.75	9	7	2.38E-01
3718.48	5	7	1.34E-01
3720.25	7	9	2.86E-01
3725.55	7	7	1.60E-01
3727.68	9	11	1.51E-01
3728.30	7	5	1.55E-01
3728.50	7	9	2.20E-01
3733.02	7	7	1.45E-01
3733.41	13	13	2.80E-01
3735.62	11	11	1.66E-01
3742.28	7	7	1.56E-01
3747.19	5	7	3.07E-01
3748.48	9	11	3.95E-01
3755.10	3	5	1.41E-01
3755.16	9	9	2.48E-01
3758.52	9	9	1.22E-01
3759.60	9	7	1.82E-01
3760.88	9	9	2.16E-01
3768.73	9	9	2.88E-01
3769.99	7	9	2.46E-01
3777.72	13	11	1.66E-01
3788.25	7	9	2.87E-01
3794.43	9	9	1.22E-01
3797.47	7	5	1.48E-01
3798.25	7	9	6.90E-01
3801.84	9	7	3.16E-01
3805.99	5	5	2.44E-01
3819.78	9	11	1.47E-01
3824.78	5	7	1.40E-01
3827.15	7	7	1.94E-01
3828.88	7	7	1.35E-01
3830.81	5	5	1.83E-01
3831.07	7	9	1.20E-01
3832.11	9	9	3.05E-01
3833.75	9	9	1.70E-01

λ Å	Weights g_i	g_k	A 10^8 s^{-1}
3834.64	3	5	1.20E-01
3846.18	7	7	1.26E-01
3847.25	3	1	2.41E-01
3848.30	9	9	1.26E-01
3851.99	11	9	1.78E-01
3864.10	7	7	6.24E-01
3866.69	3	5	1.74E-01
3867.67	5	3	2.22E-01
3869.08	5	3	1.35E-01
3874.15	7	5	1.67E-01
3902.95	7	5	6.17E-01
3909.54	9	7	1.13E-01
3911.94	5	5	1.15E-01
3915.43	5	5	1.40E-01
3916.43	5	3	1.78E-01
3919.55	11	13	2.24E-01
3955.48	13	11	1.71E-01
3973.76	11	13	4.39E-01
3977.90	9	7	1.35E-01
3980.20	5	3	2.70E-01
3991.85	11	9	1.29E-01
4010.13	5	3	4.38E-01
4021.01	9	11	2.65E-01
4051.18	13	11	1.36E-01
4062.08	11	9	1.96E-01
4069.88	13	11	3.25E-01
4076.19	9	9	1.16E-01
4084.37	9	7	1.94E-01
4102.15	5	3	1.22E-01
4107.46	7	5	2.02E-01
4120.09	13	15	6.05E-01
4131.92	9	11	1.56E-01
4148.98	9	11	1.56E-01
4157.40	13	11	2.17E-01
4157.90	9	11	1.60E-01
4185.82	11	13	3.82E-01
4188.32	11	13	3.32E-01
4194.56	11	11	2.70E-01
4232.59	9	11	3.17E-01
4240.83	5	5	1.68E-01
4246.02	11	13	2.00E-01
4251.88	13	11	1.76E-01
4254.95	7	9	2.01E-01
4269.28	11	11	1.36E-01
4276.91	7	9	2.85E-01
4277.24	9	11	1.35E-01
4317.92	15	15	1.28E-01
4325.80	3	3	1.84E-01
4326.14	5	7	2.56E-01
4340.74	5	7	1.23E-01
4381.63	13	13	2.93E-01
4382.41	11	13	3.83E-01
4409.94	13	13	1.38E-01
4411.69	11	11	2.63E-01
4434.95	9	9	2.51E-01
4446.42	11	11	1.90E-01
4457.35	7	7	1.28E-01
4474.57	5	5	2.10E-01
4491.65	11	11	2.09E-01

λ Å	Weights g_i	g_k	A 10^8 s^{-1}
4536.80	13	15	5.03E-01
4598.23	1	3	1.47E-01
4624.23	9	9	1.32E-01
4633.08	3	5	2.35E-01
4649.06	3	1	1.25E-01
4652.24	5	7	1.55E-01
4686.08	3	3	1.72E-01
4688.21	13	15	1.54E-01
4707.25	7	9	3.63E-01
4718.86	5	5	2.17E-01
4723.05	9	9	1.23E-01
4731.44	9	11	4.49E-01
4758.50	11	9	3.01E-01
4760.18	11	13	4.67E-01
4764.11	9	7	2.16E-01
4811.05	13	11	4.36E-01
4819.25	11	9	2.71E-01
4830.51	9	7	4.07E-01
4858.39	13	11	1.24E-01
4868.02	7	5	3.11E-01
5037.18	9	7	1.14E-01
5044.36	7	5	1.31E-01
5047.70	3	1	2.61E-01
5163.18	9	11	2.03E-01
5171.06	5	7	1.84E-01
5172.94	5	5	4.11E-01
5174.18	5	3	5.83E-01
5191.45	7	9	1.62E-01
5238.21	7	9	3.74E-01
5240.87	7	7	3.89E-01
5242.80	7	5	2.01E-01
5261.53	5	7	1.13E-01
5280.85	5	5	1.28E-01
5355.52	9	9	1.21E-01
5356.46	11	11	2.11E-01
5360.51	9	11	6.19E-01
5364.28	9	9	2.26E-01
5460.50	5	3	3.46E-01
5493.76	7	5	2.13E-01
5506.49	5	7	3.61E-01
5533.03	5	5	3.72E-01
5570.44	5	3	3.30E-01
5849.71	3	3	3.02E-01
5851.50	3	5	1.55E-01
5893.36	5	5	2.60E-01
5895.93	5	7	3.12E-01
5926.37	7	7	1.63E-01
5928.88	7	9	5.32E-01
7154.11	9	9	3.45E-01
Neodymium			
Nd II			
3780.4	16	18	1.4E-01
3805.4	14	16	6.9E-01
3807.2	10	12	4.9E-02
3863.3	8	10	1.5E-01
3941.5	10	10	6.1E-01
3951.2	12	12	6.0E-01
3973.3	18	18	6.3E-01

λ Å	Weights g_i	g_k	A 10^8 s^{-1}
3979.5	10	12	2.7E-01
3990.1	16	16	5.2E-01
4012.3	18	20	5.5E-01
4061.1	16	18	4.4E-01
4106.6	14	16	6.8E-02
4109.5	14	16	3.7E-01
4133.4	14	12	1.5E-01
4156.1	12	14	3.4E-01
4205.6	18	16	1.8E-01
4284.5	18	18	8.5E-02
4303.6	8	10	4.7E-01
4325.8	16	16	1.6E-01
4358.2	14	14	1.5E-01
4382.7	12	10	4.0E-02
4400.8	10	10	6.8E-02
4451.6	12	14	2.5E-01
4456.4	16	18	6.4E-02
4463.0	14	16	1.8E-01
4958.1	12	10	1.2E-02
5130.6	22	20	1.6E-01
5192.6	20	18	1.7E-01
5249.6	18	16	1.8E-01
5276.9	12	10	1.2E-01
5293.2	16	14	1.2E-01
5302.3	20	18	1.1E-01
5311.5	14	12	1.1E-01
5319.8	12	10	1.6E-01
5357.0	18	16	1.8E-01
5371.9	20	20	5.1E-02
5485.7	18	18	5.7E-02
5594.4	16	16	7.0E-02
5620.6	18	18	1.3E-01
5688.5	14	14	5.9E-02
5718.1	16	16	8.7E-02
5726.8	10	10	5.6E-02
5740.9	12	12	7.2E-02
5804.0	10	10	4.6E-02
5865.1	16	18	1.3E-02
6051.9	12	10	1.1E-02
Neon			
Ne I			
615.63	1	3	3.8E-01
618.67	1	3	9.3E-01
619.10	1	3	3.3E-01
626.82	1	3	7.4E-01
629.74	1	3	4.8E-01
735.90	1	3	6.11E+00
743.72	1	3	4.86E-01
3369.8	5	5	1.0E-03
3369.9	5	3	7.6E-03
3375.6	5	3	2.2E-03
3417.9	3	5	9.2E-03
3418.0	3	3	2.2E-03
3423.9	3	3	1.0E-03
3447.7	5	5	2.1E-02
3450.8	5	3	4.9E-03
3454.2	3	1	3.7E-02
3460.5	1	3	7.0E-03
3464.3	5	5	6.7E-03
3466.6	1	3	1.3E-02
3472.6	5	7	1.7E-02
3498.1	3	5	5.1E-03
3501.2	3	3	1.2E-02
3510.7	5	3	2.2E-03
3515.2	3	5	6.9E-03
3520.5	3	1	9.3E-02
3593.5	3	5	9.9E-03
3593.6	3	3	6.6E-03
3600.2	3	3	4.3E-03
3633.7	3	1	1.1E-02
3682.2	3	5	1.6E-03
3685.7	3	3	3.9E-03
3701.2	3	5	2.2E-03
4536.3	3	3	5.0E-03
4702.5	3	3	2.1E-03
4708.9	3	3	4.2E-02
4955.4	3	3	3.3E-03
5113.7	3	3	1.0E-02
5120.5	3	3	5.6E-03
5154.4	3	3	1.9E-02
5191.3	3	3	1.3E-02
5326.4	3	3	6.8E-03
5333.3	3	3	5.3E-03
5341.1	3	3	1.1E-01
5400.6	3	1	9.0E-03
5418.6	3	3	5.2E-03
5433.7	3	3	2.83E-03
5652.6	3	3	8.9E-03
5662.5	3	3	6.9E-03
5852.5	3	1	6.82E-01
5868.4	3	3	1.4E-02
5881.9	5	3	1.15E-01
5913.6	3	3	4.8E-02
5939.3	5	3	2.00E-03
5944.8	5	5	1.13E-01
5961.6	3	3	3.3E-02
5975.5	5	3	3.51E-02
6030.0	3	3	5.61E-02
6046.1	3	3	2.26E-03
6074.3	3	1	6.03E-01
6096.2	3	5	1.81E-01
6118.0	5	3	6.09E-03
6128.5	3	3	6.7E-03
6143.1	5	5	2.82E-01
6150.3	3	3	1.5E-02
6163.6	1	3	1.46E-01
6217.3	5	3	6.37E-02
6266.5	1	3	2.49E-01
6273.0	3	3	9.7E-03
6293.7	3	3	6.39E-03
6304.8	3	5	4.16E-02
6328.2	5	3	3.39E-02
6330.9	3	3	2.3E-02
6334.4	5	5	1.61E-01
6351.9	1	3	3.45E-03
6383.0	3	3	3.21E-01
6401.1	3	3	1.39E-02
6402.2	5	7	5.14E-01
6506.5	3	5	3.00E-01
6532.9	1	3	1.08E-01
6599.0	3	3	2.32E-01
6602.9	3	3	5.9E-03
6652.1	3	1	2.9E-03
6678.3	3	5	2.33E-01
6717.0	3	3	2.17E-01
6721.1	3	3	4.9E-04
6929.5	3	5	1.74E-01
7024.1	3	3	1.89E-02
7032.4	5	3	2.53E-01
7051.3	3	3	3.0E-02
7059.1	3	5	6.8E-02
7173.9	3	5	2.87E-02
7245.2	3	3	9.35E-02
7304.8	1	3	2.55E-03
7438.9	1	3	2.31E-02
7472.4	3	3	4.0E-02
7535.8	3	3	4.3E-01
7937.0	5	5	7.8E-03
8082.5	3	3	1.2E-03
8118.5	3	3	4.9E-02
8128.9	3	5	7.2E-03
8259.4	5	5	2.03E-02
8571.4	3	3	5.5E-02
8582.9	3	5	1.00E-02
8647.0	5	5	3.91E-02
8681.9	3	3	2.1E-01
8767.5	3	3	1.1E-03
8771.7	3	3	1.6E-01
8783.8	3	5	3.13E-01
8865.3	3	3	9.4E-03
9201.8	3	3	9.1E-02
9433.0	3	3	1.1E-03
9486.7	3	3	2.5E-02
9534.2	3	3	6.3E-02
10621	3	3	2.4E-03
11409	3	3	4.2E-02
11525	3	3	8.4E-02
11767	3	3	6.9E-02
12459	3	3	1.5E-02
Ne II			
*357.03	6	10	3.8E+01
*361.77	6	2	1.6E+01
*406.28	6	10	1.8E+01
*446.37	6	6	4.07E+01
460.73	4	2	4.7E+01
462.39	2	2	2.3E+01
1907.5	4	2	2.8E-01
1916.1	4	4	6.9E-01
1930.0	2	2	5.7E-01
1938.8	2	4	1.3E-01
2858.0	6	6	7.9E-01
2870.0	6	6	1.7E-01
2873.0	6	4	3.8E-01
2876.3	4	6	7.8E-01
2876.5	6	4	3.3E-01

λ Å	Weights g_i	Weights g_k	A 10^8 s^{-1}
2878.1	2	2	6.9E-02
2888.4	4	6	7.0E-02
2891.5	4	4	6.1E-02
2897.0	6	8	5.2E-02
2906.8	2	4	5.5E-01
2910.1	4	2	1.7E+00
2910.4	2	4	5.9E-01
2916.2	6	4	9.6E-02
2925.6	2	2	5.6E-01
2933.7	6	6	6.9E-02
2955.7	6	4	1.2E+00
3001.7	4	4	8.7E-01
3017.3	6	4	3.5E-01
3027.0	6	6	1.4E+00
3028.7	4	2	8.5E-01
3028.9	2	4	4.7E-01
3034.5	6	8	3.1E+00
3037.7	4	4	2.1E+00
3045.6	2	2	2.5E+00
3047.6	4	6	1.8E+00
3054.7	2	4	9.4E-01
3092.9	6	6	1.3E+00
3097.1	8	8	1.3E+00
3118.0	8	6	4.2E-02
3134.1	6	4	2.6E-01
3140.4	8	6	2.4E-01
3151.1	6	6	4.8E-02
3154.8	8	6	1.8E-02
3164.4	8	8	1.6E-01
3165.7	6	6	1.2E-01
3173.6	6	4	4.5E-02
3176.1	4	6	6.0E-02
3187.6	4	6	1.4E-02
3188.7	6	6	3.9E-01
3190.9	4	6	1.5E-01
3194.6	4	4	5.2E-01
3198.6	6	8	1.7E+00
3198.9	4	4	2.3E-01
3209.0	8	8	1.6E-01
3209.4	2	4	6.0E-01
3213.7	2	4	1.7E+00
3214.3	4	6	2.2E+00
3218.2	8	10	3.6E+00
3224.8	6	8	3.5E+00
3229.5	8	8	1.3E-01
3229.6	8	10	3.6E+00
3230.1	6	6	1.8E+00
3230.4	4	6	1.4E-01
3232.0	6	4	2.7E-01
3232.4	4	4	1.6E+00
3243.4	6	6	2.3E-01
3244.1	6	8	1.5E+00
3248.1	4	4	2.4E-01
3255.4	6	4	3.8E-02
3263.4	2	4	3.9E-01
3269.9	4	6	5.1E-01
3270.8	6	4	5.7E-02
3297.7	6	6	4.3E-01
3309.7	4	2	3.1E-01
3310.5	4	4	6.9E-02
3311.3	4	2	2.6E-01
3314.7	6	6	4.4E-02
3319.7	4	2	1.6E+00
3320.2	8	6	2.1E-01
3323.7	4	4	1.6E+00
3327.2	4	4	9.1E-01
3329.2	8	8	8.8E-01
3330.7	6	6	3.9E-02
3334.8	6	8	1.8E+00
3336.1	4	6	1.1E+00
3344.4	2	2	1.5E+00
3345.5	6	4	1.4E+00
3345.8	4	4	2.2E-01
3353.6	4	2	1.2E-01
3355.0	4	6	1.3E+00
3356.3	6	6	2.0E-01
3357.8	6	6	5.0E-01
3360.3	2	4	8.6E-01
3360.6	2	4	8.2E-01
3362.9	4	2	3.5E-01
3371.8	4	6	2.2E-01
3374.1	4	4	3.0E-01
3378.2	2	2	1.7E+00
3379.3	2	2	3.0E-01
3386.2	4	6	5.5E-02
3388.4	4	6	2.2E+00
3390.6	2	4	7.7E-02
3392.8	2	4	4.4E-01
3404.8	4	6	1.9E+00
3407.0	6	8	2.3E+00
3411.4	4	2	6.1E-01
3413.2	4	4	1.8E+00
3414.9	4	6	1.8E-02
3416.9	6	6	6.4E-01
3417.7	6	8	1.6E+00
3438.9	2	2	1.4E+00
3440.7	2	4	3.5E-01
3453.1	4	4	4.6E-01
3454.8	4	4	1.6E+00
3456.6	2	4	9.6E-01
3457.1	4	6	9.9E-02
3459.3	6	6	1.6E+00
3475.2	4	4	1.2E-02
3477.6	4	6	4.3E-01
3481.9	4	2	1.4E+00
3503.6	2	2	2.0E+00
3522.7	4	2	2.3E-02
3538.0	4	2	7.6E-01
3539.9	4	4	3.6E-02
3542.2	6	4	6.0E-01
3542.9	4	6	1.2E+00
3546.2	2	4	6.3E-02
3551.6	2	4	3.7E-02
3557.8	2	2	1.9E-01
3561.2	4	6	2.1E-01
3565.8	4	4	6.2E-01
3568.5	6	8	1.4E+00
3571.2	4	4	6.3E-01
3574.2	6	6	1.0E-01
3574.6	4	6	1.3E+00
3590.4	4	6	3.6E-02
3594.2	4	2	1.3E+00
3612.3	2	4	2.6E-01
3628.0	4	4	6.0E-01
3632.7	4	4	1.3E-01
3643.9	4	4	3.2E-01
3644.9	2	4	9.9E-01
3659.9	4	6	6.7E-02
3664.1	6	4	7.0E-01
3679.8	4	2	3.2E-01
3694.2	6	6	1.0E+00
3697.1	2	2	2.8E-01
3701.8	4	6	2.7E-01
3709.6	4	2	1.1E+00
3713.1	4	6	1.3E+00
3721.8	4	6	2.0E-01
3726.9	4	4	1.2E-01
3727.1	2	4	9.8E-01
3734.9	4	4	1.9E-01
3744.6	2	4	2.6E-01
3751.2	2	2	1.8E-01
3753.8	4	6	4.5E-01
3766.3	4	6	2.9E-01
3777.1	2	4	4.2E-01
3800.0	4	4	3.7E-01
3818.4	2	4	6.1E-01
3829.8	4	6	8.4E-01
3942.3	4	6	1.0E-02
Ne V			
*142.61	9	9	6.7E+02
*143.32	9	15	1.2E+03
147.13	5	7	1.5E+03
151.23	5	5	3.38E+02
154.50	1	3	7.0E+02
*167.69	9	9	1.5E+02
*358.93	9	3	2.1E+02
365.59	5	3	1.35E+02
*482.15	9	9	3.01E+01
*571.04	9	15	1.0E+01
2259.6	3	5	1.9E+00
2265.7	5	7	2.4E+00
Ne VII			
97.502	1	3	1.07E+03
*115.46	9	3	4.8E+02
116.69	3	5	1.6E+03
127.66	3	1	1.9E+02
465.22	1	3	4.09E+01
558.61	3	5	8.11E+00
559.95	1	3	1.07E+01
561.38	3	3	7.99E+00
561.73	5	5	2.39E+01
562.99	3	1	3.17E+01

λ	Weights		A
Å	g_i	g_k	$10^8\ s^{-1}$
564.53	5	3	1.31E+01
Ne VIII			
*88.09	2	6	8.4E+02
*98.208	6	10	2.77E+03
770.41	2	4	5.90E+00
780.32	2	2	5.69E+00
2820.7	2	4	7.20E-01
2860.1	2	2	6.88E-01
Nickel			
Ni I			
1963.85	7	7	1.1E-01
1976.87	7	9	1.1E+00
1981.61	5	5	1.3E-01
1990.25	5	7	8.3E-01
2007.01	5	5	1.7E-01
2007.69	7	7	9.0E-02
2014.25	3	5	9.3E-01
2025.40	7	5	2.3E-01
2026.62	9	7	2.4E-01
2047.35	7	5	1.8E-01
2052.04	9	9	9.7E-02
2055.50	5	3	3.3E-01
2059.92	7	5	2.1E-01
2060.20	5	3	2.3E-01
2064.39	3	1	4.0E-01
2069.52	5	5	1.1E-01
2085.57	5	5	2.6E+00
2089.09	7	5	9.7E-02
2095.13	5	7	1.1E-01
2114.43	5	5	9.7E-02
2121.40	7	5	2.8E-01
2124.80	5	3	3.8E-01
2147.80	5	3	4.7E-01
2157.83	5	3	4.1E-01
2158.31	7	5	6.9E-01
2161.04	5	5	1.3E-01
2173.54	5	3	1.5E-01
2174.48	3	1	8.9E-01
2182.38	7	5	1.3E-01
2183.91	5	5	1.2E-01
2190.22	5	5	3.0E-01
2197.35	3	3	7.8E-01
2201.59	5	3	7.3E-01
2221.94	5	3	2.2E-01
2244.46	5	5	3.8E-01
2253.57	7	7	1.9E-01
2254.81	9	9	9.6E-02
2258.15	7	5	1.7E-01
2259.56	5	3	2.0E-01
2261.42	9	7	9.1E-02
2287.32	3	5	1.8E-01
2289.98	9	7	2.1E+00
2293.11	5	5	3.8E-01
2300.77	7	7	7.5E-01
2302.97	3	3	4.5E-01
2307.35	5	7	1.6E-01
2312.34	7	7	5.5E+00
2313.98	5	5	5.0E+00
2317.16	7	5	3.8E+00
2320.03	9	11	6.9E+00
2321.38	5	7	5.6E+00
2324.65	7	9	1.8E-01
2325.79	7	9	3.5E+00
2329.96	5	3	5.3E+00
2345.54	9	7	2.2E+00
2346.63	7	5	5.5E-01
2347.51	9	9	2.2E-01
2348.73	7	7	2.2E-01
2419.31	7	5	2.0E-01
2943.91	7	5	1.1E-01
2981.65	5	3	2.8E-01
3002.48	7	7	8.0E-01
3003.62	5	5	6.9E-01
3012.00	5	5	1.3E+00
3037.93	7	7	2.8E-01
3050.82	7	9	6.0E-01
3054.31	5	5	4.0E-01
3057.64	3	3	1.0E+00
3064.62	5	7	1.1E-01
3101.56	5	7	6.3E-01
3101.88	5	7	4.9E-01
3134.11	3	5	7.3E-01
3225.02	5	3	9.3E-02
3369.56	9	7	1.8E-01
3380.57	5	3	1.3E+00
3392.98	7	7	2.4E-01
3414.76	7	9	5.5E-01
3423.71	3	3	3.3E-01
3433.56	7	7	1.7E-01
3446.26	5	5	4.4E-01
3452.88	5	7	9.8E-02
3458.46	3	5	6.1E-01
3461.66	7	9	2.7E-01
3472.55	5	7	1.2E-01
3483.77	5	3	1.4E-01
3492.96	5	3	9.8E-01
3510.33	3	1	1.2E+00
3515.05	5	7	4.2E-01
3524.54	7	5	1.0E+00
3566.37	5	5	5.6E-01
3597.70	3	3	1.4E-01
3619.39	5	7	6.6E-01
4027.67	5	7	1.3E-01
4295.88	9	7	1.7E-01
4401.54	9	11	3.8E-01
4462.46	3	5	1.7E-01
4470.48	5	7	1.9E-01
4600.37	5	3	2.6E-01
4604.99	9	7	2.3E-01
4606.23	5	3	1.0E-01
4648.66	11	9	2.4E-01
4686.22	5	5	1.4E-01
4701.54	9	9	1.4E-01
4714.42	13	11	4.6E-01
4715.78	7	7	2.0E-01
4732.47	7	9	9.3E-02
4752.43	3	3	2.0E-01
4756.52	9	9	1.5E-01
4786.54	11	11	1.8E-01
4812.00	3	1	9.5E-02
4829.03	5	7	1.9E-01
4831.18	9	7	1.6E-01
4838.64	9	7	2.2E-01
4855.41	5	5	5.7E-01
4904.41	5	3	6.2E-01
4912.03	3	3	1.5E-01
4913.97	1	3	2.2E-01
4918.36	9	7	2.3E-01
4935.83	7	5	2.4E-01
4937.34	9	9	1.2E-01
4953.20	5	5	1.2E-01
4980.17	9	11	1.9E-01
5000.34	7	7	1.4E-01
5012.46	7	7	1.1E-01
5017.58	11	11	2.0E-01
5035.37	7	9	5.7E-01
5042.20	3	5	1.4E-01
5048.85	7	7	1.6E-01
5080.53	9	11	3.2E-01
5081.11	7	9	5.7E-01
5082.35	3	3	2.5E-01
5084.08	7	9	3.1E-01
5099.95	7	7	2.9E-01
5115.40	11	9	2.2E-01
5129.37	7	5	1.2E-01
5155.14	5	5	1.1E-01
5155.76	5	7	2.9E-01
5176.57	5	5	1.8E-01
5371.33	7	7	1.6E-01
5476.91	1	3	9.5E-02
5637.12	3	3	1.1E-01
5664.02	5	7	1.1E-01
5695.00	3	3	1.7E-01
6086.29	3	5	1.1E-01
6175.42	3	3	1.7E-01
7122.24	5	7	2.1E-01
7381.94	9	11	9.7E-02
7422.30	7	5	1.8E-01
7727.66	7	7	1.1E-01
Ni II			
2165.55	10	10	2.4E+00
2169.10	8	8	1.58E+00
2174.67	8	10	1.43E+00
2175.15	6	6	1.77E+00
2184.61	4	4	2.90E+00
2201.41	4	6	1.3E+00
2206.72	6	8	1.66E+00
2216.48	10	12	3.4E+00
2220.40	6	8	2.3E+00
2222.96	10	10	9.8E-01
2224.86	8	8	1.55E+00
2226.33	6	6	1.3E+00
2253.85	4	6	1.98E+00
2264.46	6	8	1.43E+00

λ Å	Weights g_i	Weights g_k	A 10^8 s^{-1}
2270.21	8	10	1.56E+00
2278.77	8	6	2.8E+00
2287.09	6	4	2.8E+00
2296.55	8	8	1.98E+00
2297.14	6	4	2.70E+00
2297.49	4	2	3.0E+00
2298.27	6	6	2.8E+00
2303.00	8	6	2.9E+00
2316.04	10	8	2.88E+00
2334.58	8	8	8.0E-01
2375.42	6	8	6.6E-01
2394.52	8	10	1.70E+00
2416.13	6	8	2.1E+00
2437.89	8	10	5.4E-01
2510.87	8	10	5.8E-01
Ni III			
1692.51	11	13	7.9E+00
1709.90	9	11	6.3E+00
1719.46	5	7	6.0E+00
1722.28	3	5	5.9E+00
1724.52	3	1	6.7E+00
1741.96	9	7	5.7E+00
1752.43	7	5	5.5E+00
1760.56	5	3	6.5E+00
1769.64	11	11	6.2E+00
1823.06	9	9	5.6E+00
Ni XIV			
164.13	6	8	1.2E+03
168	2	4	2.4E+02
168.12	4	2	8.5E+02
169.69	4	4	9.8E+02
170.50	4	4	7.1E+02
171.37	4	6	9.4E+02
172.16	6	6	4.7E+02
172.80	6	4	1.4E+02
177.28	4	4	5.6E+02
178	2	4	8.9E+01
181	4	6	7.4E+01
182.14	4	2	1.5E+02
196	4	2	3.8E+01
288.894	4	4	4.6E+01
292.399	6	6	3.6E+01
Ni XV			
50.249	5	7	6.8E+03
60.890	9	11	1.0E+04
64.635	7	9	9.6E+03
163.64	5	7	5.6E+01
173.73	5	7	7.6E+02
175	3	1	5.7E+02
179.28	5	7	7.5E+02
181	1	3	6.8E+02
269	3	1	5.3E+01
278.386	5	5	4.3E+01
Ni XVI			
166	4	6	3.1E+02
168	6	8	3.2E+02
182	2	2	2.5E+02
185.23	2	4	4.2E+02
187	4	6	1.2E+02
187	4	2	3.3E+02
188	2	2	4.7E+02
190	6	8	2.0E+02
192	6	8	4.54E+02
192	6	6	3.1E+02
194	4	6	2.8E+02
194	2	4	5.5E+02
194	2	2	1.1E+02
194	4	4	3.5E+02
194.04	4	6	4.6E+02
195.27	4	4	9.5E+01
196	4	6	6.7E+02
197	4	6	1.5E+02
197	4	2	1.2E+02
199	2	4	4.9E+02
206	2	2	3.7E+02
217	4	4	1.1E+02
218.391	2	4	9.5E+01
223.119	2	2	1.3E+02
231	4	4	1.6E+02
232.475	4	4	4.07E+02
233	6	4	2.4E+02
235	4	2	3.8E+02
235	6	6	2.5E+02
236	4	4	1.2E+02
237.875	4	2	2.6E+02
238	6	4	1.3E+02
239.550	2	2	2.6E+02
245	4	4	1.4E+02
245	4	6	3.2E+02
249	6	8	3.3E+02
249	6	4	1.2E+02
250	4	2	1.6E+02
254	6	4	1.8E+02
Ni XVII			
30.919	1	3	2.77E+03
42.855	1	3	4.75E+03
54.451	9	11	1.5E+04
55.361	1	3	6.7E+03
57.348	7	9	1.4E+04
197.39	1	3	1.6E+02
199.87	3	5	2.1E+02
204	3	3	1.8E+02
205	3	1	2.4E+02
206	1	3	3.0E+02
207.50	5	7	2.5E+02
215.89	3	5	4.8E+02
216	1	3	2.7E+02
217	5	7	2.4E+02
227	5	5	1.6E+02
249.180	1	3	2.75E+02
281.50	3	1	2.1E+02
282	3	1	2.4E+02
284	5	3	1.5E+02
292	5	7	2.2E+02
Ni XVIII			
24.881	2	4	8.6E+02
25.070	4	6	9.9E+02
26.02	2	4	1.26E+03
26.020	2	4	1.1E+03
26.046	2	2	1.1E+03
26.218	4	6	1.5E+03
27.98	4	6	1.0E+03
27.982	2	4	2.0E+03
28.018	6	8	1.1E+03
28.220	4	6	2.33E+03
29.383	4	6	1.58E+03
29.422	6	8	1.69E+03
29.779	2	4	1.9E+03
29.829	2	2	1.9E+03
31.845	4	6	2.7E+03
31.890	6	8	3.0E+03
32.034	2	4	3.4E+03
32.340	4	6	4.0E+03
36.990	4	6	5.5E+03
37.049	6	8	5.9E+03
41.015	2	4	2.97E+03
41.218	2	2	3.2E+03
43.814	2	4	5.5E+03
44.365	4	6	6.8E+03
44.405	4	4	1.14E+03
52.615	4	6	1.5E+04
52.720	6	8	1.6E+04
52.745	6	6	1.06E+03
59.950	6	4	9.6E+02
60.212	4	2	1.1E+03
63.512	4	6	7.9E+02
63.589	6	8	8.5E+02
69.075	4	6	8.0E+02
76.254	4	6	1.38E+03
76.359	6	8	1.47E+03
99.275	2	4	1.0E+03
100.4	4	6	1.2E+03
114.46	4	6	2.5E+03
114.74	6	8	2.7E+03
Ni XIX			
9.140	1	3	3.1E+04
9.153	1	3	5.2E+03
9.977	1	3	1.1E+05
10.110	1	3	9.4E+04
10.283	1	3	4.7E+03
10.433	1	3	5.1E+03
11.539	1	3	4.8E+04
11.599	1	3	6.3E+03
12.435	1	3	3.66E+05
12.656	1	3	1.0E+05
13.779	1	3	1.23E+04
14.043	1	3	1.31E+04
40.7	3	3	6.4E+03
40.7	3	1	8.4E+03
41.132	7	9	9.4E+03

λ Å	Weights g_i	g_k	A 10^8 s^{-1}
Ni XXI			
11.13	3	3	1.7E+04
11.23	5	3	1.7E+04
11.239	5	7	5.7E+04
11.28	3	1	2.2E+05
11.318	5	7	2.8E+05
11.48	3	1	1.1E+05
11.48	1	3	4.0E+05
11.517	5	7	1.4E+05
11.539	5	7	1.2E+05
11.67	1	3	8.0E+04
11.72	3	3	2.3E+04
12.454	5	3	3.3E+04
12.472	3	3	1.8E+04
12.502	5	5	2.8E+04
Ni XXII			
72.52	4	2	2.84E+02
84.06	6	4	1.2E+03
84.24	4	2	5.6E+02
85.86	4	2	4.9E+02
88.00	4	2	1.2E+03
95.95	2	2	4.4E+02
98.16	4	4	5.2E+02
98.58	4	4	2.45E+02
100.60	6	6	3.9E+02
101.31	6	4	4.83E+02
103.31	4	2	2.66E+02
106.04	4	4	2.36E+02
106.16	4	2	5.1E+02
124.31	2	2	3.7E+02
126.32	4	4	3.3E+02
Ni XXIII			
87.66	3	3	2.8E+02
88.11	5	3	8.3E+02
90.49	3	3	1.77E+02
90.96	5	3	2.5E+02
91.83	5	3	7.5E+02
92.32	3	1	4.39E+02
100.42	1	3	2.1E+02
102.08	5	5	5.3E+02
103.23	3	3	2.4E+02
103.67	5	5	1.78E+02
104.70	3	1	2.94E+02
106.02	5	5	2.87E+02
108.27	7	5	3.32E+02
111.23	3	1	2.26E+02
111.78	5	3	2.19E+02
111.86	1	3	1.7E+02
112.55	3	1	1.0E+03
128.87	5	5	4.02E+02
133.54	3	3	1.86E+02
137.55	3	1	2.53E+02
Ni XXIV			
101.13	6	4	1.63E+02
102.11	4	4	5.4E+02
103.43	2	4	1.3E+02
103.53	4	2	4.17E+02
104.64	2	2	4.7E+02
106.68	4	2	3.67E+02
113.14	4	4	1.65E+02
118.52	2	4	1.5E+02
122.72	6	4	2.17E+02
134.73	6	6	1.44E+02
135.47	4	4	8.0E+01
137.01	4	4	2.6E+02
138.80	4	6	7.2E+01
153.47	2	2	1.27E+02
159.69	2	4	8.9E+01
Ni XXV			
9.30	3	1	9.3E+04
9.31	5	7	8.2E+04
9.32	3	5	7.8E+04
9.34	1	3	1.1E+05
9.42	3	1	9.0E+04
9.49	3	5	8.9E+04
9.60	1	3	1.8E+05
9.63	3	5	2.4E+05
9.64	3	3	1.3E+05
9.71	3	1	2.3E+05
9.71	3	3	1.8E+05
9.74	5	7	3.0E+05
9.75	3	5	1.3E+05
9.76	1	3	3.03E+05
9.76	5	3	7.5E+04
9.78	5	7	2.9E+05
9.86	5	7	4.8E+05
9.87	3	5	2.03E+05
9.92	5	5	1.3E+05
9.94	5	7	1.29E+05
9.97	3	5	2.5E+05
10.08	1	3	2.80E+05
Ni XXVI			
1.5930	4	2	3.4E+06
1.5935	2	2	4.0E+06
1.5973	4	4	8.1E+06
1.5977	2	4	4.4E+06
1.5982	2	2	7.3E+06
1.5996	2	2	2.7E+06
1.6005	4	6	2.7E+06
1.6036	4	2	2.1E+06
9.390	2	4	2.59E+05
9.535	4	6	2.96E+05
Ni XXVII			
1.2534	1	3	3.35E+05
1.2824	1	3	6.38E+05
1.3500	1	3	1.63E+06
1.3516	1	3	2.4E+05
1.531	3	3	2.0E+05
1.534	3	1	6.9E+06
1.537	5	5	2.3E+06
1.537	1	3	3.7E+06
1.538	3	5	3.9E+06
1.539	1	3	2.6E+06
1.539	3	5	2.6E+06
1.540	3	3	1.7E+06
1.541	3	5	5.5E+06
1.542	3	3	3.6E+06
1.542	5	5	3.5E+06
1.544	5	3	3.2E+06
1.546	3	5	1.6E+06
1.547	3	3	2.1E+05
1.549	1	3	2.0E+05
1.551	3	1	8.2E+05
1.558	3	1	6.5E+05
1.5883	1	3	6.02E+06
1.5963	1	3	7.70E+05
Nitrogen			
N I			
1163.88	6	6	7.52E-01
1164.00	4	6	1.27E-02
1164.21	6	4	5.17E-02
1164.32	4	4	6.94E-01
1167.45	6	8	1.29E+00
1168.42	6	6	4.24E-02
1168.54	4	6	1.24E+00
1176.51	6	4	9.22E-01
1176.63	4	4	1.02E-01
1177.69	4	2	1.02E+00
1199.55	4	6	4.01E+00
1200.22	4	4	3.99E+00
1200.71	4	2	3.98E+00
1310.54	4	6	8.42E-01
1316.29	4	6	1.42E-02
1492.63	6	4	3.13E+00
1492.82	4	4	3.51E-01
1494.68	4	2	3.72E+00
3822.03	2	2	3.70E-02
3830.43	4	4	4.67E-02
3834.22	4	2	1.89E-02
4099.94	2	4	3.48E-02
4109.95	4	6	3.90E-02
4113.97	4	4	6.62E-03
4137.64	2	4	2.80E-03
4143.43	4	4	6.09E-03
4151.48	6	4	1.01E-02
4249.87	4	2	2.59E-02
4264.00	6	4	2.26E-02
4356.29	6	8	5.10E-02
4385.54	2	2	8.84E-03
4392.41	4	2	1.76E-02
4435.43	2	4	7.51E-03
4442.45	4	4	3.81E-02
4669.89	4	4	7.49E-03
4914.94	2	2	8.08E-03
4935.12	4	2	1.76E-02
5199.84	2	2	1.87E-02
5201.61	2	4	1.87E-02
5281.20	6	6	2.45E-03
5344.05	6	6	6.10E-04
5356.62	4	6	1.41E-03

λ Å	Weights g_i	g_k	A 10^8 s^{-1}
5367.01	4	4	1.07E-03
5372.61	2	4	8.34E-04
5378.27	2	2	1.66E-03
6606.18	4	6	8.87E-04
6622.54	6	6	7.93E-03
6626.99	2	4	2.20E-03
6636.94	4	4	1.40E-02
6644.96	8	6	3.49E-02
6646.50	2	2	2.18E-02
6653.46	6	4	2.74E-02
6656.51	4	2	2.17E-02
6926.67	4	6	7.75E-03
6945.18	6	6	1.83E-02
6951.60	2	4	1.03E-02
6960.50	4	4	4.67E-03
6973.07	2	2	3.83E-03
6979.18	6	4	9.83E-03
6982.03	4	2	2.04E-02
7423.64	2	4	5.95E-02
7442.30	4	4	1.24E-01
7468.31	6	4	1.93E-01
7898.98	6	4	2.82E-01
7899.28	4	4	3.28E-02
7915.42	4	2	3.13E-01
8184.86	4	6	8.58E-02
8188.01	2	4	1.27E-01
8200.36	2	2	4.95E-02
8210.72	4	4	4.84E-02
8216.34	6	6	2.23E-01
8223.13	4	2	2.64E-01
8242.39	6	4	1.36E-01
8567.74	2	4	4.92E-02
8594.00	2	2	2.09E-01
8629.24	4	4	2.66E-01
8655.88	4	2	1.05E-01
8680.28	6	8	2.46E-01
8683.40	4	6	1.80E-01
8686.15	2	4	1.09E-01
8703.25	2	2	2.10E-01
8711.70	4	4	1.28E-01
8718.84	6	6	6.75E-02
8728.90	4	2	3.76E-02
8747.37	6	4	1.04E-02
9028.92	2	2	3.02E-01
9045.88	6	8	2.80E-01
9049.49	6	6	1.88E-02
9049.89	4	6	2.60E-01
9060.48	2	4	2.95E-01
9187.45	6	6	2.44E-01
9187.86	4	6	1.76E-02
9207.59	6	4	2.70E-02
9208.00	4	4	2.33E-01
9386.81	2	4	2.24E-01
9392.79	4	6	2.63E-01
9460.68	4	4	3.98E-02
9776.90	2	4	1.18E-02
9786.78	4	6	1.13E-02
9788.29	2	2	2.99E-02
9798.56	4	4	2.75E-02

λ Å	Weights g_i	g_k	A 10^8 s^{-1}
9810.01	4	2	5.30E-02
9814.02	6	8	6.56E-03
9822.75	6	6	4.95E-02
9834.61	6	4	4.50E-02
9863.33	8	8	9.62E-02
9872.15	8	6	2.97E-02
9883.38	2	2	2.93E-02
9905.52	4	2	3.11E-03
9909.22	2	4	7.58E-03
9931.47	4	4	3.64E-02
9947.07	6	8	1.08E-02
9965.75	4	6	7.60E-03
9968.51	6	4	4.50E-03
9980.42	4	6	8.10E-03
9997.73	8	8	9.20E-03
N II			
474.891	5	5	9.66E+00
475.647	1	3	1.17E+01
475.698	3	5	1.58E+01
475.757	3	3	8.75E+00
475.803	5	7	2.10E+01
475.884	5	5	5.25E+00
508.697	5	5	1.91E+00
510.758	5	7	1.87E+01
513.849	5	5	1.24E+01
529.355	1	3	7.23E+00
529.413	3	1	2.43E+01
529.491	3	3	6.75E+00
529.637	3	5	4.92E+00
529.722	5	3	1.03E+01
529.867	5	5	1.94E+01
533.511	1	3	2.39E+01
533.581	3	5	3.20E+01
533.650	3	3	1.66E+01
533.729	5	7	4.13E+01
533.815	5	5	9.19E+00
547.818	5	3	2.16E+00
559.762	1	3	1.14E+01
574.650	5	7	3.60E+01
582.156	5	5	2.85E+01
635.197	1	3	2.33E+01
644.634	1	3	1.21E+01
644.837	3	3	3.64E+01
645.178	5	3	6.07E+01
660.286	5	3	3.69E+01
671.016	3	5	2.47E+00
671.386	5	5	7.40E+00
671.411	1	3	3.04E+00
671.630	3	3	2.27E+00
671.773	3	1	9.85E+00
672.001	5	3	3.87E+00
745.841	1	3	1.25E+01
746.984	5	3	3.85E+01
748.369	5	3	3.83E+00
775.965	5	5	3.08E+01
915.612	1	3	4.38E+00
915.962	3	1	1.32E+01
1083.99	1	3	2.18E+00

λ Å	Weights g_i	g_k	A 10^8 s^{-1}
1085.55	5	5	9.47E-01
1085.70	5	7	3.87E+00
3408.13	3	1	2.19E-01
3437.14	3	1	2.07E+00
3593.60	3	5	1.21E-01
3609.10	3	3	1.41E-01
3615.86	3	1	1.53E-01
3829.80	3	5	2.42E-01
3838.37	5	5	6.98E-01
3842.19	1	3	3.06E-01
3847.40	3	3	2.22E-01
3855.10	3	1	8.82E-01
3856.06	5	3	3.71E-01
3919.00	3	3	6.76E-01
3955.85	3	5	1.31E-01
3995.00	3	5	1.35E+00
4114.33	3	3	1.42E-03
4124.08	3	5	3.20E-01
4133.67	5	5	5.30E-01
4145.77	7	5	7.36E-01
4374.99	3	5	5.55E-03
4447.03	3	5	1.14E+00
4459.94	3	1	1.12E-01
4465.53	3	3	2.36E-02
4477.68	5	3	8.85E-02
4488.09	5	5	1.30E-02
4507.56	7	5	1.00E-01
4564.76	3	5	1.41E-02
4601.48	3	5	2.35E-01
4607.15	1	3	3.26E-01
4613.87	3	3	2.26E-01
4621.39	3	1	9.55E-01
4630.54	5	5	7.72E-01
4643.09	5	3	4.51E-01
4654.53	3	5	2.43E-02
4667.21	3	3	2.99E-02
4674.91	3	1	1.05E-01
4694.27	1	3	1.23E-01
4695.90	3	5	1.29E-01
4697.64	3	3	3.06E-02
4698.55	3	1	3.67E-01
4700.03	5	7	1.05E-01
4702.50	5	5	9.15E-02
4704.25	5	3	2.13E-01
4706.40	7	9	6.09E-02
4709.58	7	7	1.82E-01
4712.07	7	5	1.46E-01
4718.38	9	9	3.02E-01
4721.58	9	7	7.75E-02
4774.24	3	5	3.24E-02
4779.72	3	3	2.52E-01
4781.19	5	7	2.05E-02
4788.14	5	5	2.52E-01
4793.65	5	3	7.77E-02
4803.29	7	7	3.18E-01
4810.30	7	5	4.75E-02
4860.17	3	5	1.61E-02
4987.38	3	1	7.48E-01
4991.24	3	5	3.54E-01

λ Å	Weights g_i	Weights g_k	A 10^8 s^{-1}
4994.36	5	7	2.62E-01
4994.37	3	3	7.60E-01
4997.22	3	3	1.96E-01
5001.13	3	5	9.76E-01
5001.47	5	7	1.05E+00
5002.70	1	3	8.45E-02
5005.15	7	9	1.16E+00
5005.30	5	5	6.51E-02
5007.33	3	5	7.89E-01
5010.62	3	3	2.19E-01
5011.31	5	3	5.84E-01
5012.04	7	7	5.19E-01
5016.38	5	5	1.62E-01
5023.05	7	5	3.61E-01
5025.66	7	7	1.07E-01
5040.71	7	5	3.78E-03
5045.10	5	3	3.42E-01
5073.59	3	3	2.59E-02
5168.05	3	5	3.06E-01
5170.16	3	3	6.54E-01
5171.27	3	1	8.71E-01
5171.47	5	7	5.81E-01
5172.34	3	5	6.01E-01
5172.97	1	3	5.01E-01
5173.39	5	7	7.36E-01
5174.46	5	5	5.07E-01
5175.89	7	9	8.93E-01
5176.57	5	3	2.17E-01
5177.06	3	3	5.00E-01
5179.34	7	9	8.67E-01
5179.52	9	11	1.07E+00
5180.36	5	5	4.28E-01
5183.20	7	7	2.88E-01
5184.96	7	7	3.20E-01
5185.09	5	3	7.11E-02
5186.21	7	5	5.76E-02
5190.38	9	9	1.77E-01
5191.96	7	5	4.25E-02
5199.50	9	7	1.51E-02
5313.42	3	3	1.41E-01
5320.20	5	3	4.20E-01
5320.96	3	5	2.52E-01
5327.76	5	5	4.65E-02
5338.73	5	7	1.85E-01
5340.21	7	5	2.59E-01
5351.23	7	7	3.67E-01
5383.72	3	5	3.31E-03
5452.07	1	3	8.89E-02
5454.22	3	1	3.34E-01
5462.58	3	3	1.00E-01
5478.09	3	5	4.75E-02
5480.05	5	3	1.30E-01
5495.65	5	5	2.40E-01
5526.23	3	5	2.13E-01
5530.24	5	7	4.04E-01
5535.35	7	9	6.04E-01
5535.38	3	3	4.53E-01
5540.06	3	1	6.03E-01
5543.47	5	5	3.51E-01
5551.92	7	7	2.00E-01
5552.68	5	3	1.50E-01
5565.26	7	5	3.97E-02
5666.63	3	5	3.74E-01
5676.02	1	3	2.96E-01
5679.56	5	7	5.25E-01
5686.21	3	3	1.94E-01
5710.77	5	5	1.24E-01
5730.66	5	3	1.34E-02
5747.30	3	5	3.40E-02
5767.45	3	3	2.44E-02
5893.15	5	7	2.88E-01
5897.25	3	5	2.16E-01
5899.83	1	3	1.60E-01
5927.81	1	3	3.22E-01
5931.78	3	5	4.27E-01
5940.24	3	3	2.26E-01
5941.65	5	7	5.54E-01
5952.39	5	5	1.27E-01
5960.91	5	3	1.34E-02
6065.00	3	5	2.21E-03
6284.32	5	3	7.74E-02
6379.62	3	3	6.11E-02
6482.05	3	3	3.01E-01
6610.56	5	7	6.34E-01
6857.03	5	3	2.53E-01
6869.58	5	5	2.51E-01
6887.83	5	7	2.49E-01
7762.24	5	5	8.74E-02
8438.74	1	3	2.24E-01
8831.75	1	3	8.42E-03
8855.30	3	3	2.51E-02
8893.29	5	3	4.12E-02
N III			
374.198	2	4	9.89E+01
451.871	2	2	1.03E+01
452.227	4	2	2.05E+01
684.998	2	4	9.63E+00
685.515	2	2	3.83E+01
685.817	4	4	4.54E+01
686.336	4	2	1.95E+01
763.334	2	2	9.58E+00
764.351	4	2	1.85E+01
771.545	2	4	8.19E+00
771.901	4	4	1.64E+01
772.384	6	4	2.45E+01
772.889	6	4	2.09E+01
772.955	4	2	2.34E+01
979.832	4	4	8.84E+00
979.905	6	6	9.21E+00
989.799	2	4	4.18E+00
991.511	4	4	8.17E-01
991.577	4	6	4.97E+00
1747.85	2	4	1.28E+00
1751.22	4	4	2.48E-01
1751.66	4	6	1.51E+00
2972.55	2	2	6.67E-01
2977.33	4	2	3.32E-01
2978.84	2	4	1.66E-01
2983.64	4	4	8.24E-01
3342.76	2	2	3.80E-01
3353.98	2	4	7.66E-01
3354.32	4	6	5.51E-01
3355.46	4	2	7.51E-01
3358.78	2	2	3.05E-01
3360.98	4	4	2.44E-01
3365.80	4	2	1.52E+00
3367.36	6	6	1.27E+00
3374.07	6	4	8.13E-01
3745.95	2	4	1.90E-01
3752.63	2	2	6.67E-02
3754.69	4	4	3.78E-01
3762.60	4	4	4.24E-02
3771.03	6	4	5.59E-01
3771.36	6	4	8.28E-02
3792.97	8	6	1.03E-01
3934.50	2	4	7.49E-01
3938.51	4	6	8.96E-01
3942.88	4	4	1.49E-01
4097.36	2	4	8.70E-01
4103.39	2	2	8.67E-01
4195.74	2	4	9.37E-01
4200.07	4	6	1.12E+00
4215.77	4	4	1.85E-01
4318.78	2	4	5.40E-02
4321.22	2	2	1.08E-01
4321.39	4	6	5.03E-02
4325.43	4	4	8.60E-02
4327.69	6	8	3.06E-02
4327.88	4	2	1.07E-01
4332.95	6	6	1.23E-01
4337.01	6	4	7.47E-02
4345.81	8	8	1.82E-01
4351.11	8	6	4.01E-02
4510.88	2	4	2.84E-01
4510.96	4	6	4.77E-01
4514.85	6	8	6.80E-01
4518.14	2	2	5.65E-01
4523.56	4	4	3.61E-01
4530.86	4	2	1.12E-01
4534.58	6	6	2.01E-01
4547.30	6	4	3.33E-02
4634.13	2	4	6.36E-01
4640.64	4	6	7.60E-01
4641.85	4	4	1.26E-01
4858.70	2	4	4.35E-01
4858.98	4	6	4.66E-01
4861.27	6	8	5.32E-01
4867.12	4	4	1.73E-01
4867.17	8	10	6.18E-01
4873.60	6	6	1.50E-01
4881.78	6	4	1.22E-02
4884.14	8	8	8.71E-02
4896.58	8	6	5.86E-03
5260.86	2	2	2.80E-02
5270.57	2	4	6.95E-02
5272.68	4	2	1.39E-01

λ Å	Weights g_i	g_k	A $10^8\ s^{-1}$
5282.43	4	4	2.21E-02
5297.75	4	6	4.93E-02
5298.95	6	4	7.38E-02
5314.36	6	6	1.14E-01
5320.87	6	8	5.68E-01
5327.19	4	6	5.29E-01
5352.46	6	6	3.72E-02
6365.84	2	2	2.18E-01
6394.75	2	4	2.15E-01
6445.34	2	4	8.89E-02
6450.79	2	2	1.77E-01
6454.08	4	6	1.49E-01
6463.09	4	4	1.13E-01
6467.02	6	8	2.11E-01
6468.57	4	2	3.52E-02
6478.76	6	6	6.31E-02
6487.84	6	4	1.05E-02
7371.51	4	4	3.53E-02
7404.54	6	6	3.61E-02
8307.51	2	4	1.65E-02
8344.95	2	2	6.52E-02
8386.39	4	4	8.03E-02
8424.56	4	2	3.17E-02

N IV

λ Å	Weights g_i	g_k	A $10^8\ s^{-1}$
247.205	1	3	1.19E+02
*283.52	9	15	3.05E+02
*322.64	9	3	8.99E+01
335.047	3	5	1.845E+02
387.356	3	1	2.55E+01
765.147	1	3	2.320E+01
*923.16	9	9	1.759E+01
955.334	3	1	2.919E+01
1718.55	3	5	2.321E+00
2649.88	3	3	1.07E+00
3052.20	1	3	1.33E-01
3059.60	3	3	3.95E-01
3075.19	5	3	6.48E-01
3443.61	3	5	3.46E-01
3445.22	1	3	4.60E-01
3454.65	3	3	3.42E-01
3461.36	3	1	1.36E+00
3463.36	5	5	1.02E+00
3474.53	5	3	5.61E-01
3478.72	3	5	1.06E+00
*3480.8	3	9	1.06E+00
3483.00	3	3	1.06E+00
3484.93	3	1	1.06E+00
3689.94	3	1	9.10E-02
3694.14	3	3	2.27E-02
3707.39	5	3	6.73E-02
3714.43	5	5	1.34E-02
3735.43	7	5	7.37E-02
3747.54	3	5	9.92E-01
4057.76	3	5	6.62E-01
4740.26	3	5	1.53E-02
4747.96	3	3	7.60E-02
4752.49	5	7	1.13E-02
4762.09	5	5	6.99E-02
4769.86	5	3	2.50E-02
4786.92	7	7	8.79E-02
4796.66	7	5	1.53E-02
5200.41	3	5	2.67E-01
5204.28	5	7	3.55E-01
5205.15	1	3	1.97E-01
5226.70	3	3	1.46E-01
5245.60	5	5	8.66E-02
5272.35	5	3	9.48E-03
5288.25	5	3	3.22E-02
5736.93	3	5	1.84E-01
5776.31	1	3	1.85E-02
5784.76	3	1	5.51E-02
5795.09	3	3	1.37E-02
5812.31	3	5	1.36E-02
5826.43	5	3	2.25E-02
5843.84	5	5	4.01E-02
6380.75	1	3	1.42E-01
7103.24	1	3	6.28E-02
7109.35	3	5	8.46E-02
7111.28	3	3	4.70E-02
*7116.8	9	15	1.12E-01
7122.98	5	7	1.12E-01
7127.25	5	5	2.80E-02
7127.25	5	3	3.11E-03
9165.07	3	5	4.23E-02
9182.16	5	7	4.45E-02
9222.99	7	9	4.95E-02
9247.04	5	5	7.66E-03
9311.55	7	7	5.36E-03

N V

λ Å	Weights g_i	g_k	A $10^8\ s^{-1}$
*209.29	2	6	1.21E+02
*247.66	6	10	4.26E+02
1238.82	2	4	3.40E+00
1242.80	2	2	3.37E+00
4603.74	2	4	4.14E-01
4619.97	2	2	4.10E-01

N VI

λ Å	Weights g_i	g_k	A $10^8\ s^{-1}$
24.8980	1	3	5.158E+03
28.7870	1	3	1.809E+04
*161.220	3	9	2.859E+02
173.275	1	3	2.697E+02
*173.93	9	15	8.756E+02
185.192	3	5	8.205E+02
*1901	3	9	6.780E-01
2896.4	1	3	2.079E-01
*6991.1	3	9	8.384E-02
9622.0	1	3	3.276E-02

Oxygen

O I

λ Å	Weights g_i	g_k	A $10^8\ s^{-1}$
791.973	5	5	4.94E+00
792.938	1	3	2.19E+00
792.967	3	5	1.64E+00
877.798	5	3	2.85E+00
877.879	5	5	5.12E+00
922.008	5	7	1.23E+00
935.193	5	5	1.33E+00
1028.16	1	3	4.22E-01
1152.15	5	5	5.28E+00
1217.65	1	3	2.06E+00
1302.17	5	3	3.41E+00
1304.86	3	3	2.03E+00
1306.03	1	3	6.76E-01
3823.41	7	7	6.63E-03
3823.87	5	3	1.87E-03
3824.35	5	5	5.19E-03
3825.02	3	3	5.59E-03
3825.19	5	7	8.31E-04
3855.01	5	5	1.63E-02
3947.29	5	7	4.91E-03
3947.48	5	5	4.88E-03
3947.59	5	3	4.87E-03
3951.93	3	1	3.10E-03
3952.98	5	3	1.29E-03
3953.00	1	3	1.03E-03
3954.52	3	5	7.73E-04
3954.61	5	5	2.32E-03
3997.95	5	3	2.41E-02
4217.09	3	1	5.44E-03
4222.77	5	3	2.26E-03
4222.82	1	3	1.81E-03
4233.27	5	5	4.04E-03
4368.19	3	1	7.56E-03
4368.24	3	5	7.59E-03
4967.38	3	5	4.43E-03
4967.88	5	7	8.44E-03
4968.79	7	9	1.27E-02
5019.29	5	5	7.13E-03
5020.22	7	5	9.98E-03
5329.11	3	5	9.48E-03
5329.69	5	7	1.81E-02
5330.74	7	9	2.71E-02
5435.18	3	5	7.74E-03
5435.77	5	5	1.29E-02
5436.86	7	5	1.80E-02
5512.60	3	5	2.69E-03
5512.77	5	7	3.58E-03
5554.83	3	3	5.83E-03
5555.00	5	3	9.71E-03
5958.39	3	5	6.80E-03
5958.58	5	7	9.06E-03
6046.23	3	3	1.05E-02
6046.44	5	3	1.75E-02
6046.49	1	3	3.50E-03
6155.99	3	5	2.67E-02
6156.78	5	7	5.08E-02
6158.19	7	9	7.62E-02
6324.84	7	5	3.76E-05
6453.60	3	5	1.65E-02
6454.44	5	5	2.75E-02
6455.98	7	5	3.85E-02
6726.28	5	5	1.18E-05
6726.54	5	3	6.44E-06
7001.92	3	5	2.65E-02
7002.23	5	7	3.53E-02

λ Å	Weights g_i	Weights g_k	A 10^8 s^{-1}
7254.15	3	3	2.24E-02
7254.45	5	3	3.73E-02
7254.53	1	3	7.45E-03
7771.94	5	7	3.69E-01
7774.17	5	5	3.69E-01
7775.39	5	3	3.69E-01
7981.94	3	3	2.33E-04
7982.40	1	3	3.09E-04
7986.98	3	5	4.19E-04
7987.33	5	5	1.41E-04
7995.07	5	7	5.63E-04
8221.82	7	7	2.89E-01
8227.65	5	3	8.13E-02
8230.00	5	5	2.26E-01
8233.00	3	3	2.43E-01
8235.35	3	5	4.86E-02
8446.25	3	1	3.22E-01
8446.36	3	5	3.22E-01
8446.76	3	3	3.22E-01
8820.42	5	7	2.93E-01
9260.81	3	1	4.46E-01
9260.85	3	3	3.34E-01
9260.94	3	5	1.56E-01
9262.58	5	3	1.11E-01
9262.67	5	5	2.60E-01
9262.78	5	7	2.97E-01
9265.83	7	5	2.97E-02
9265.93	7	7	1.48E-01
9266.01	7	9	4.45E-01
9482.89	5	3	2.34E-01
9622.11	5	3	5.22E-04
9622.16	3	3	1.57E-03
9625.26	7	5	3.25E-04
9625.30	7	7	1.85E-03
9694.66	5	7	4.54E-04
9694.91	5	5	4.54E-04
9695.06	5	3	4.54E-04

O II

λ Å	Weights g_i	Weights g_k	A 10^8 s^{-1}
429.918	4	2	4.25E+01
430.041	4	4	4.13E+01
430.176	4	6	4.36E+01
483.760	4	2	2.05E+01
483.980	6	4	1.80E+01
484.027	4	4	3.22E+00
485.087	6	8	2.60E+01
485.470	6	6	1.20E+00
485.518	4	6	1.93E+01
2290.85	2	4	7.41E-02
2293.30	2	2	3.25E-01
2300.33	4	4	4.17E-01
2302.81	4	2	1.67E-01
2365.14	4	2	1.52E-01
2375.72	6	4	1.35E-01
2406.38	6	4	1.85E-01
2407.48	4	4	2.25E-01
2411.60	4	2	2.05E-01
2411.64	2	2	1.10E-01
2415.13	4	2	2.20E-01
2418.46	6	6	2.30E-01
2425.57	6	6	1.77E-01
2433.54	2	4	4.21E-01
2436.06	4	4	1.69E-01
2444.25	4	4	7.56E-02
2445.53	4	6	4.98E-01
2517.96	4	6	7.72E-02
2523.21	2	2	9.63E-02
2526.87	4	4	1.20E-01
2530.28	6	8	8.16E-02
2571.46	2	4	1.15E-01
2575.28	4	6	1.37E-01
3134.73	8	6	1.23E+00
3273.43	8	6	9.99E-01
3377.15	2	2	1.27E+00
3390.21	2	4	1.22E+00
3407.28	6	6	1.02E+00
3712.74	2	4	2.84E-01
3727.32	4	4	5.81E-01
3749.48	6	4	9.31E-01
3833.07	6	8	1.02E-02
3842.81	2	4	7.45E-02
3843.58	4	6	3.55E-02
3847.89	2	2	1.95E-01
3850.80	4	6	6.00E-03
3851.03	4	4	1.59E-01
3851.47	8	8	2.72E-02
3856.13	4	2	2.28E-01
3857.16	6	6	6.59E-02
3863.50	6	8	6.49E-02
3864.13	2	2	9.12E-02
3864.43	6	6	2.15E-01
3864.67	6	4	1.80E-01
3874.09	2	4	3.26E-02
3875.80	8	6	3.38E-02
3882.19	8	8	5.50E-01
3882.45	4	4	8.94E-02
3883.14	8	6	1.13E-01
3893.52	4	6	1.89E-02
3907.45	6	6	8.64E-02
3911.96	6	4	1.09E+00
3912.12	4	4	1.41E-01
3919.27	4	2	1.22E+00
3945.04	2	4	2.05E-01
3954.36	2	2	8.57E-01
3973.26	4	4	1.04E+00
3982.71	4	2	4.27E-01
4069.62	2	4	1.52E+00
4069.88	4	6	1.53E+00
4072.15	6	8	1.98E+00
4075.86	8	10	2.11E+00
4078.84	4	4	5.52E-01
4084.65	6	8	7.28E-02
4085.11	6	6	4.55E-01
4092.93	8	8	2.65E-01
4094.14	6	4	4.70E-02
4096.53	4	6	1.73E-01
4097.22	2	4	3.62E-01
4103.00	2	2	5.09E-01
4104.72	4	6	3.14E-01
4104.99	4	4	9.14E-01
4106.02	8	6	1.70E-02
4109.84	6	6	1.21E-02
4110.19	6	4	2.54E-01
4110.79	4	2	7.70E-01
4112.02	6	6	1.81E-01
4113.83	8	6	2.41E-01
4119.22	6	8	1.33E+00
4120.28	6	6	2.15E-01
4120.55	6	4	2.60E-01
4121.46	2	2	5.60E-01
4129.32	4	2	1.79E-01
4132.80	2	4	9.13E-01
4140.70	4	4	4.09E-02
4153.30	4	6	7.91E-01
4156.53	6	4	2.11E-01
4169.22	6	6	2.71E-01
4185.44	6	8	1.91E+00
4189.58	8	8	7.06E-02
4189.79	8	10	1.98E+00
4192.51	6	4	3.21E-01
4196.27	4	4	3.56E-02
4196.70	4	2	3.56E-01
4317.14	2	4	3.70E-01
4319.63	4	6	2.55E-01
4319.87	2	2	5.62E-01
4325.76	2	2	1.47E-01
4327.46	6	6	6.76E-01
4327.85	6	4	7.24E-02
4328.59	4	2	1.12E+00
4331.47	4	6	4.82E-02
4331.86	4	4	6.50E-01
4336.86	4	4	1.57E-01
4345.56	4	2	8.31E-01
4347.22	6	4	1.19E-01
4347.41	4	4	9.32E-01
4349.43	6	6	6.91E-01
4351.26	6	6	9.89E-01
4351.46	4	6	5.82E-02
4359.40	4	6	1.44E-02
4366.89	6	4	3.98E-01
4369.27	4	4	3.57E-01
4395.93	6	6	3.91E-01
4405.98	6	4	4.30E-02
4414.90	4	6	8.34E-01
4416.97	2	4	7.13E-01
4443.01	6	6	5.05E-01
4443.52	6	8	1.89E-02
4447.68	8	6	2.52E-02
4448.19	8	8	5.10E-01
4452.38	4	4	1.37E-01
4466.24	2	4	9.00E-01
4467.46	2	2	9.00E-01
4563.18	4	4	7.18E-03
4590.97	6	8	8.85E-01
4595.96	6	6	4.87E-02
4596.18	4	6	8.34E-01
4638.86	2	4	3.71E-01

λ Å	Weights g_i	g_k	A 10^8 s^{-1}
4641.81	4	6	5.96E-01
4649.13	6	8	7.81E-01
4650.84	2	2	6.86E-01
4661.63	4	4	4.10E-01
4673.73	4	2	1.35E-01
4676.23	6	6	2.05E-01
4690.89	2	4	1.86E-01
4691.42	2	2	7.43E-01
4696.35	6	4	3.25E-02
4698.44	6	6	6.59E-02
4699.01	6	8	9.88E-01
4699.22	4	6	9.36E-01
4701.18	4	4	9.23E-01
4701.71	4	2	3.69E-01
4703.16	4	6	9.20E-01
4705.35	6	8	1.10E+00
4710.01	4	6	2.98E-01
4741.70	6	6	4.71E-02
4751.28	6	8	6.39E-02
4752.69	6	6	1.45E-02
4844.92	4	6	1.02E-02
4856.39	4	6	5.58E-02
4856.76	4	4	1.00E-01
4860.97	2	4	4.70E-01
4864.88	4	2	8.07E-02
4871.52	4	6	5.60E-01
4872.02	4	4	9.34E-02
4890.86	4	2	4.80E-01
4906.83	4	4	4.54E-01
4924.53	4	6	5.43E-01
4941.07	2	4	5.87E-01
4943.01	4	6	7.78E-01
4955.71	4	4	1.82E-01
5159.94	2	2	3.29E-01
5175.90	4	2	1.49E-01
5190.50	2	4	1.26E-01
5206.65	4	4	3.58E-01
5583.22	2	4	2.17E-02
5611.07	2	2	2.14E-02
6627.37	4	4	1.73E-01
6641.03	2	2	9.88E-02
6666.66	4	2	6.78E-02
6677.87	2	4	3.37E-02
6717.75	2	2	1.33E-01
6721.39	4	2	1.81E-01
6810.48	6	8	1.64E-03
6844.10	4	6	2.97E-03
6846.80	8	8	3.17E-02
6869.48	6	6	5.35E-02
6884.88	4	4	6.12E-02
6895.10	10	8	2.72E-01
6906.44	8	6	2.48E-01
6907.87	4	2	3.03E-01
6910.56	6	4	2.43E-01

O III

λ Å	Weights g_i	g_k	A 10^8 s^{-1}
263.694	1	3	3.32E+01
263.727	3	5	4.48E+01
263.773	3	3	2.49E+01
263.817	5	7	5.97E+01
263.861	5	5	1.49E+01
277.386	5	7	9.43E+01
279.788	5	5	4.25E+01
295.942	1	3	5.56E+01
303.413	1	3	4.29E+01
303.461	3	1	1.29E+02
303.517	3	3	3.21E+01
303.622	3	5	3.21E+01
303.695	5	3	5.34E+01
303.800	5	5	9.61E+01
305.596	1	3	1.20E+02
305.656	3	5	1.62E+02
305.702	3	3	9.01E+01
305.767	5	7	2.16E+02
305.836	5	5	5.40E+01
320.978	5	7	2.17E+02
328.448	5	5	1.04E+02
345.312	1	3	1.35E+02
374.073	5	5	2.85E+01
395.557	5	3	2.80E+01
507.388	1	3	1.61E+01
507.680	3	3	4.82E+01
508.178	5	3	8.04E+01
525.794	5	3	9.60E+01
597.814	1	3	1.49E+01
599.590	5	5	5.41E+01
702.337	1	3	6.06E+00
702.838	3	1	1.83E+01
832.929	1	3	3.41E+00
835.092	5	5	1.44E+00
835.289	5	7	5.99E+00
1679.03	3	5	6.57E-01
1686.73	3	3	6.48E-01
1760.41	3	5	8.38E-01
1764.46	5	5	2.50E+00
1766.63	1	3	1.11E+00
1772.28	3	1	3.29E+00
1772.97	5	3	1.37E+00
2390.43	3	3	1.62E+00
2454.97	3	1	3.43E+00
2665.68	3	5	6.75E-01
2674.58	5	5	1.11E+00
2683.66	3	1	1.85E+00
2686.15	7	5	1.54E+00
2687.55	3	3	1.84E+00
2695.48	3	5	1.82E+00
2794.14	3	1	1.82E-01
2798.93	3	3	4.52E-02
2809.66	5	3	1.34E-01
2818.70	5	5	2.66E-02
2836.31	7	5	1.46E-01
2959.69	3	5	1.83E+00
2983.78	3	5	2.15E+00
2992.08	3	5	9.32E-02
2996.48	3	3	4.64E-01
2997.69	5	7	6.88E-02
3004.34	5	5	4.27E-01
3008.78	5	3	1.53E-01
3017.62	7	7	5.38E-01
3023.43	3	5	4.79E-01
3024.36	7	5	9.39E-02
3024.54	1	3	6.16E-01
3035.41	3	3	4.59E-01
3042.07	3	1	1.94E+00
3047.10	5	5	1.49E+00
3059.28	5	3	8.72E-01
3064.98	1	3	2.17E-01
3068.13	3	1	6.49E-01
3068.26	3	3	5.41E-02
3068.67	3	5	2.27E-01
3074.14	5	7	1.84E-01
3074.72	5	3	3.76E-01
3075.13	5	5	1.61E-01
3075.95	7	9	1.07E-01
3083.65	7	7	3.20E-01
3084.64	7	5	2.55E-01
3088.04	9	9	5.30E-01
3095.79	9	7	1.35E-01
3115.67	3	1	1.39E+00
3121.63	3	3	1.38E+00
3132.79	3	5	1.37E+00
3198.18	3	5	9.57E-02
3201.14	3	3	4.77E-01
3202.51	5	7	7.08E-02
3207.61	5	5	4.40E-01
3210.58	5	3	1.58E-01
3216.07	7	7	5.58E-01
3221.21	7	5	9.75E-02
3260.86	5	7	1.68E+00
3265.33	7	9	1.88E+00
3267.20	3	5	1.58E+00
3281.83	5	5	2.89E-01
3284.45	7	7	2.06E-01
3299.39	1	3	1.64E-01
3312.33	3	3	4.60E-01
3326.06	3	3	2.65E-01
3330.30	3	5	6.81E-01
3330.32	3	5	4.76E-01
3332.41	5	3	7.92E-01
3332.93	5	7	5.04E-01
3336.67	3	3	3.76E-01
3336.69	5	5	8.77E-02
3340.76	5	3	6.57E-01
3344.20	5	5	1.25E-01
3344.51	5	7	3.48E-01
3347.98	7	5	4.86E-01
3350.62	5	3	1.12E+00
3350.92	7	7	9.91E-01
3355.86	7	7	6.89E-01
3362.31	7	5	6.87E-01
3376.61	3	1	1.49E+00
3376.76	3	3	1.12E+00
3377.26	3	5	5.20E-01
3382.61	5	7	9.86E-01
3383.31	5	3	3.70E-01
3383.81	5	5	8.62E-01
3384.90	7	9	1.48E+00

λ Å	Weights g_i	g_k	A $10^8\ s^{-1}$
3394.22	7	7	4.88E-01
3395.43	7	5	9.75E-02
3406.88	1	3	1.93E-01
3408.13	3	1	5.79E-01
3415.26	3	3	1.44E-01
3428.63	3	5	1.42E-01
3430.57	5	3	2.37E-01
3444.05	5	5	4.21E-01
3446.68	3	5	9.71E-01
3447.15	1	3	8.09E-01
3447.97	5	7	1.19E+00
3450.91	7	9	1.44E+00
3451.30	3	3	8.06E-01
3454.84	5	5	6.89E-01
3454.99	9	11	1.72E+00
3459.48	5	3	1.14E-01
3459.94	7	7	5.14E-01
3466.13	9	9	2.84E-01
3466.85	7	5	6.82E-02
3475.24	9	7	2.42E-02
3520.94	1	3	1.50E-01
3531.22	3	1	4.45E-01
3533.38	3	3	1.11E-01
3534.90	3	5	1.11E-01
3555.24	5	3	1.82E-01
3556.78	5	5	3.26E-01
3695.38	3	5	4.01E-01
3698.72	5	7	7.62E-01
3703.36	7	9	1.14E+00
3704.75	3	3	8.53E-01
3707.27	3	5	7.34E-01
3709.54	3	1	1.13E+00
3712.49	5	5	6.59E-01
3714.03	3	3	4.06E-01
3715.09	5	7	9.73E-01
3720.89	7	7	3.74E-01
3721.95	5	3	2.80E-01
3725.31	5	5	2.41E-01
3728.51	5	7	1.29E+00
3728.84	7	9	1.45E+00
3729.80	3	5	1.22E+00
3732.13	5	3	2.67E-02
3734.83	7	5	7.40E-02
3742.63	5	5	2.24E-01
3746.90	7	7	1.59E-01
3754.70	3	5	7.53E-01
3757.23	1	3	5.56E-01
3759.88	5	7	9.79E-01
3774.03	3	3	3.91E-01
3791.28	5	5	2.24E-01
3810.98	5	3	2.37E-02
3816.75	5	3	9.63E-02
3961.57	5	7	1.25E+00
4072.64	1	3	3.37E-01
4073.98	3	5	4.54E-01
4081.02	5	7	6.02E-01
4089.30	3	3	2.49E-01
4103.07	5	5	1.48E-01
4118.60	5	3	1.63E-02
4440.09	5	3	4.42E-01
4447.69	5	5	4.40E-01
4461.61	5	7	4.36E-01
4524.22	3	1	3.38E-01
4532.78	5	3	1.40E-01
4535.29	3	3	8.40E-02
4555.39	5	5	2.49E-01
4557.91	3	5	8.27E-02
5268.30	1	3	3.50E-01
5508.24	5	5	1.06E-01
5592.25	3	3	3.27E-01
O IV			
238.360	2	4	2.96E+02
238.570	4	6	3.54E+02
238.579	4	4	5.90E+01
279.631	2	2	2.68E+01
279.933	4	2	5.34E+01
553.329	2	4	1.22E+01
554.076	2	2	4.86E+01
554.513	4	4	6.06E+01
555.263	4	2	2.41E+01
608.397	2	2	1.21E+01
609.829	4	2	2.40E+01
616.952	6	4	2.60E+01
617.005	4	4	2.89E+00
617.036	4	2	2.89E+01
624.619	2	4	1.07E+01
625.127	4	4	2.13E+01
625.853	6	4	3.19E+01
779.736	6	4	1.46E+00
779.820	4	4	1.31E+01
779.912	6	6	1.36E+01
779.997	4	6	9.70E-01
787.710	2	4	5.95E+00
790.112	4	4	1.18E+00
790.199	4	6	7.08E+00
921.296	2	4	2.21E+00
921.365	2	2	8.83E+00
923.367	4	4	1.10E+01
923.436	4	2	4.39E+00
1338.61	2	4	2.17E+00
1342.99	4	4	4.29E-01
1343.51	4	6	2.57E+00
2120.58	2	2	1.05E+00
2132.64	4	4	1.29E+00
2493.39	2	4	1.18E+00
2493.75	4	6	8.48E-01
2493.99	2	2	6.09E-01
2499.27	2	2	4.68E-01
2501.81	4	4	3.73E-01
2507.73	4	2	2.32E+00
2509.22	6	6	1.94E+00
2510.58	4	2	1.19E+00
2517.37	6	4	1.24E+00
2781.22	2	2	1.03E-01
2803.57	6	4	1.26E-01
2805.87	2	4	2.90E-01
2812.50	6	6	3.58E-02
2816.53	4	4	5.74E-01
2829.17	8	6	1.56E-01
2836.27	6	4	8.43E-01
2916.31	2	4	1.06E+00
2921.46	4	6	1.27E+00
2926.18	4	4	2.11E-01
3063.43	2	4	1.30E+00
3071.60	2	2	1.29E+00
3177.89	2	4	7.59E-02
3180.77	2	2	1.51E-01
3180.99	4	6	7.06E-02
3185.74	4	4	1.21E-01
3188.22	6	8	4.28E-02
3188.64	4	2	1.50E-01
3194.78	6	6	1.71E-01
3199.58	6	4	1.04E-01
3209.65	8	8	2.53E-01
3216.31	8	6	5.56E-02
3348.06	2	4	8.51E-01
3349.11	4	6	1.02E+00
3354.27	4	2	7.71E-01
3362.55	4	4	7.65E-01
3375.40	4	6	7.56E-01
3378.02	4	4	1.66E-01
3381.21	4	6	7.19E-01
3381.30	2	4	4.28E-01
3385.52	6	8	1.02E+00
3390.19	2	2	8.49E-01
3396.80	4	4	5.40E-01
3405.77	4	2	1.67E-01
3409.70	6	6	3.00E-01
3411.30	4	4	1.69E-01
3411.69	4	6	1.02E+00
3425.55	6	4	4.94E-02
3489.89	4	6	7.29E-01
3492.21	2	4	6.06E-01
3493.43	4	4	1.21E-01
3560.39	4	6	1.03E+00
3563.33	6	8	1.10E+00
3593.08	6	6	7.15E-02
3725.89	2	4	5.61E-01
3725.94	4	6	6.01E-01
3729.03	6	8	6.86E-01
3736.68	4	4	2.23E-01
3736.85	8	10	7.95E-01
3744.89	6	6	1.92E-01
3758.39	8	8	1.11E-01
3930.68	2	2	3.80E-02
3942.06	2	4	9.42E-02
3945.31	4	2	1.88E-01
3956.77	4	4	2.98E-02
3974.58	4	6	6.62E-02
3977.09	6	4	9.91E-02
3995.08	6	6	1.52E-01
4687.03	2	4	2.79E-01
4772.60	2	4	1.23E-01
4779.10	2	2	2.45E-01
4783.42	4	6	2.06E-01
4794.18	4	4	1.56E-01

λ Å	Weights g_i	g_k	A 10^8 s^{-1}
4798.27	6	8	2.91E-01
4813.15	6	6	8.65E-02
5305.51	4	4	6.10E-02
5362.51	6	6	6.12E-02
6876.49	2	4	1.88E-02
6931.60	2	2	7.35E-02
7004.11	4	4	8.90E-02
7061.30	4	2	3.48E-02

O V

λ Å	Weights g_i	g_k	A 10^8 s^{-1}
172.169	1	3	2.94E+02
*192.85	9	15	6.90E+02
*215.17	9	3	1.83E+02
220.353	3	5	4.292E+02
248.460	3	1	5.59E+01
629.732	1	3	2.872E+01
758.677	3	5	5.547E+00
759.442	1	3	7.373E+00
760.227	3	3	5.514E+00
760.446	5	5	1.652E+01
761.128	3	1	2.197E+01
762.004	5	3	9.125E+00
774.518	3	1	3.804E+01
1371.30	3	5	3.336E+00
2729.31	3	5	4.52E-01
2731.45	1	3	5.90E-01
2743.61	3	3	4.38E-01
2752.23	3	1	1.82E+00
2755.13	5	5	1.37E+00
2769.69	5	3	7.88E-01
2781.01	3	5	1.40E+00
*2784.0	3	9	1.40E+00
2786.99	3	3	1.39E+00
2789.85	3	1	1.38E+00
3058.68	3	5	1.39E+00
3144.66	3	5	8.86E-01
3219.24	3	1	1.54E-01
3222.29	1	3	1.16E-01
3227.54	3	3	3.38E-02
3239.21	3	3	3.28E-01
3248.28	5	3	1.18E-01
3263.54	5	5	1.86E-02
3275.64	5	3	4.76E-01
3297.62	7	5	1.30E-01
3690.17	3	5	1.97E-02
3698.36	3	3	1.03E-01
3702.72	5	7	1.41E-02
3717.31	5	5	9.63E-02
3725.63	5	3	2.91E-02
3746.64	7	7	1.18E-01
3761.58	7	5	1.61E-02
4119.37	3	5	3.66E-01
4120.49	3	1	3.33E-01
4123.96	5	7	4.81E-01
4125.49	1	3	2.70E-01
4134.11	3	3	3.34E-01
4153.27	3	3	1.92E-01
4158.86	3	5	3.39E-01
4178.46	5	5	1.12E-01
4213.35	5	3	1.19E-02
4522.66	5	3	1.02E-02
4554.53	3	5	2.41E-01
5114.06	1	3	1.80E-01
5339.94	1	3	1.85E-02
5349.74	3	1	7.04E-02
5372.71	3	3	1.42E-02
5414.59	3	5	9.29E-03
5428.38	5	3	2.68E-02
5471.12	5	5	4.86E-02
5571.81	1	3	8.33E-02
5580.12	3	5	1.11E-01
5583.23	3	3	6.20E-02
*5589.9	9	15	1.49E-01
5597.89	5	7	1.48E-01
5604.27	5	5	3.68E-02
5607.41	5	3	4.08E-03
6330.05	5	7	1.21E-01
6460.12	3	5	9.37E-02
6466.14	5	7	1.01E-01
6500.24	7	9	1.11E-01
6543.77	5	5	1.64E-02
6601.28	7	7	1.14E-02
6764.72	1	3	4.37E-02
6789.62	3	5	5.79E-02
6817.40	3	3	3.00E-02
6828.95	5	7	7.35E-02
6878.76	5	5	1.65E-02

O VI

λ Å	Weights g_i	g_k	A 10^8 s^{-1}
*150.10	2	6	2.62E+02
*173.03	6	10	8.78E+02
1031.91	2	4	4.16E+00
1037.61	2	2	4.09E+00
3811.35	2	4	5.14E-01
3834.24	2	2	5.05E-01

O VII

λ Å	Weights g_i	g_k	A 10^8 s^{-1}
18.6270	1	3	9.365E+03
21.6020	1	3	3.309E+04
*120.33	3	9	5.334E+02
128.411	1	3	8.982E+02
*128.46	9	15	1.615E+03
135.820	3	5	1.523E+03
*1630.3	3	9	7.935E-01
2448.98	1	3	2.514E-01
*5933.1	3	9	1.002E-01
8241.76	1	3	3.864E-02

Phosphorus

P I

λ Å	Weights g_i	g_k	A 10^8 s^{-1}
1671.7	4	2	3.9E-01
1674.6	4	4	4.0E-01
1679.7	4	6	3.9E-01
1775.0	4	6	2.17E+00
1782.9	4	4	2.14E+00
1787.7	4	2	2.13E+00
2135.5	4	4	2.11E-01
2136.2	6	4	2.83E+00
2149.1	4	2	3.18E+00
2152.9	2	4	4.85E-01
2154.1	4	4	1.73E-01
2154.1	4	6	5.8E-01
2534.0	2	4	2.00E-01
2535.6	4	4	9.5E-01
2553.3	2	2	7.1E-01
2554.9	4	2	3.00E-01

P II

λ Å	Weights g_i	g_k	A 10^8 s^{-1}
1301.9	1	3	5.0E-01
1304.5	3	1	1.5E+00
1304.7	3	3	3.7E-01
1305.5	3	5	3.8E-01
1309.9	5	3	6.2E-01
1310.7	5	5	1.1E+00
4475.3	5	7	1.3E+00
4499.2	5	7	1.4E+00
4530.8	3	5	1.0E+00
4554.8	3	5	9.6E-01
4588.0	5	7	1.7E+00
4589.9	3	5	1.6E+00
4602.1	7	9	1.9E+00
4943.5	7	5	6.3E-01
5253.5	3	5	1.0E+00
5425.9	5	5	6.9E-01
6024.2	3	5	5.1E-01
6043.1	5	7	6.8E-01

P III

λ Å	Weights g_i	g_k	A 10^8 s^{-1}
1334.8	2	4	5.5E-01
1344.3	4	6	6.4E-01
1344.8	4	4	1.1E-01
4057.4	4	4	1.0E-01
4059.3	6	4	9.0E-01
4080.1	4	2	9.9E-01

Potassium

K I

λ Å	Weights g_i	g_k	A 10^8 s^{-1}
4044.1	2	4	1.24E-02
4047.2	2	2	1.24E-02
5084.2	2	2	3.50E-03
5099.2	4	2	7.0E-03
5323.3	2	2	6.3E-03
5339.7	4	2	1.26E-02
5343.0	2	4	4.0E-03
5359.6	4	6	4.6E-03
5782.4	2	2	1.23E-02
5801.8	4	2	2.46E-02
5812.2	2	4	2.8E-03
5831.9	4	6	3.2E-03
6911.1	2	2	2.72E-02
6938.8	4	2	5.4E-02
7664.9	2	4	3.87E-01
7699.0	2	2	3.82E-01

K II

λ Å	Weights g_i	g_k	A 10^8 s^{-1}
607.93	1	3	1.3E-02

λ Å	Weights g_i	Weights g_k	A 10^8 s^{-1}
K III			
2550.0	6	4	2.0E+00
2635.1	4	4	1.2E+00
2992.4	6	8	2.5E+00
3052.1	4	6	1.7E+00
3202.0	4	4	1.8E+00
3289.1	4	6	2.0E+00
3322.4	6	6	1.3E+00
3421.8	2	4	1.5E+00
K XVI			
206.27	1	3	9.4E+01
K XVII			
22.020	2	4	4.7E+04
22.163	4	6	5.6E+04
22.18	4	4	9.3E+03
22.60	2	2	2.5E+03
22.76	4	2	4.7E+03
Praseodymium			
Pr II			
3997.0	15	15	1.87E-01
4062.8	13	15	1.00E+00
4100.7	17	19	8.4E-01
4143.1	15	17	5.8E-01
4179.4	13	15	5.2E-01
4222.9	11	13	3.91E-01
4241.0	17	15	2.30E-01
4359.8	15	15	1.1E-01
4405.8	17	17	9.0E-02
4429.3	15	15	2.28E-01
4449.8	13	13	1.24E-01
4468.7	11	13	1.54E-01
4510.2	13	15	1.16E-01
4534.2	15	17	4.9E-02
4734.2	15	13	2.5E-02
4879.1	15	15	1.8E-02
4886.0	15	15	1.3E-02
4912.6	17	15	5.7E-02
5034.4	19	19	1.1E-01
5110.8	21	19	2.78E-01
5135.1	17	17	1.25E-01
5173.9	19	17	3.18E-01
5219.1	15	15	9.5E-02
5220.1	17	15	2.35E-01
5251.7	15	13	1.1E-02
5259.7	15	13	2.24E-01
5292.6	13	13	9.3E-02
5810.6	17	19	2.3E-02
5879.3	15	15	7.6E-02
6200.8	15	17	1.8E-02
6278.7	13	15	2.6E-02
6398.0	11	13	1.9E-02
Rhodium			
Rh I			
3083.96	8	6	4.8E-02
3114.91	6	4	4.45E-02
3121.76	6	6	1.1E-01
3123.70	10	8	4.6E-02
3137.71	4	6	3.3E-02
3189.05	6	6	3.03E-01
3197.13	6	4	4.35E-02
3263.14	6	6	1.3E-01
3271.61	6	4	2.0E-01
3280.55	8	8	2.36E-01
3283.57	6	8	4.4E-01
3289.14	4	4	1.0E-01
3323.09	8	10	6.3E-01
3331.09	4	2	5.40E-02
3338.54	8	6	3.5E-02
3360.80	4	4	1.2E-01
3368.38	6	4	1.1E-01
3396.82	10	10	6.5E-01
3399.70	6	8	1.2E-01
3462.04	6	6	6.2E-01
3470.66	4	4	8.5E-01
3478.91	6	6	3.32E-01
3484.04	6	8	9.3E-03
3498.73	4	6	2.12E-01
3502.52	10	10	4.3E-01
3507.32	6	8	3.4E-01
3528.02	8	8	8.5E-01
3543.95	4	4	4.65E-01
3549.54	6	6	2.22E-01
3570.18	4	6	1.82E-01
3583.10	8	10	2.6E-01
3596.19	6	4	5.5E-01
3597.15	6	8	5.9E-01
3612.47	4	2	8.90E-01
3620.46	6	4	8.5E-02
3654.87	8	8	6.0E-02
3657.99	8	6	8.8E-01
3666.22	6	8	8.4E-02
3690.70	6	4	3.23E-01
3692.36	10	8	9.1E-01
3700.91	8	10	3.9E-01
3713.02	4	4	8.3E-02
3788.47	4	6	1.4E-01
3793.22	8	6	4.2E-01
3799.31	8	8	5.5E-01
3806.76	6	6	6.2E-02
3818.19	6	4	5.8E-01
3822.26	6	6	8.5E-01
3828.48	6	6	6.2E-01
3833.89	6	4	5.8E-01
3856.52	8	10	5.9E-01
3872.39	4	6	6.7E-03
3877.34	8	6	3.7E-02
3913.51	8	8	2.5E-03
3922.19	4	2	6.25E-02
3934.23	8	8	1.58E-01
3942.72	4	2	7.15E-01
3958.86	6	8	5.5E-01
3984.40	4	4	1.1E-01
3995.61	4	6	4.7E-02
4053.44	2	2	2.8E-02
4056.34	6	4	9.5E-03
4082.78	6	4	1.4E-01
4097.52	2	4	7.0E-02
4121.68	6	6	9.8E-02
4128.87	6	8	1.73E-01
4135.27	8	8	1.0E-01
4196.50	6	8	3.9E-02
4211.14	8	10	1.62E-01
4244.44	4	4	6.5E-03
4278.60	4	6	9.2E-03
4288.71	6	8	6.1E-02
4373.04	2	4	1.8E-02
4374.80	8	10	1.64E-01
4379.92	6	6	2.48E-02
4492.47	6	6	4.5E-03
4528.72	6	8	1.35E-02
4548.73	4	6	5.5E-03
4551.64	4	4	4.00E-02
4565.19	4	4	1.1E-02
4569.00	6	8	1.0E-02
4608.12	2	2	2.1E-02
4675.03	8	8	6.4E-03
4721.00	6	4	3.43E-03
4745.11	6	6	5.2E-03
4755.58	4	4	6.0E-03
4842.43	6	8	1.6E-03
4963.71	2	2	3.0E-02
4977.75	4	4	9.8E-03
4979.18	4	6	1.0E-02
5090.63	6	6	5.0E-03
5120.69	6	8	3.1E-03
5130.76	4	4	4.35E-03
5155.54	2	4	9.8E-03
5184.19	6	8	1.6E-03
5212.73	4	2	5.95E-03
5292.14	10	10	3.7E-03
5390.44	4	6	9.5E-03
5424.72	4	4	5.0E-03
5599.42	6	8	1.3E-02
5983.60	10	10	2.1E-02
Rubidium			
Rb I			
3022.5	2	4	4.13E-05
3032.0	2	4	4.93E-05
3044.2	2	4	8.2E-05
3060.2	2	4	1.05E-04
3082.0	2	4	1.49E-04
3112.6	2	4	2.5E-04
3113.1	2	2	1.3E-04
3157.5	2	4	3.38E-04
3158.3	2	2	2.0E-04
3228.0	2	4	6.4E-04
3229.2	2	2	3.8E-04
3348.7	2	4	1.37E-03
3350.8	2	2	8.9E-04
3587.1	2	4	3.97E-03
3591.6	2	2	2.9E-03
4201.8	2	4	1.8E-02

λ Å	Weights g_i	Weights g_k	A $10^8\ s^{-1}$
4215.5	2	2	1.5E-02
7800.3	2	4	3.70E-01
7947.6	2	2	3.40E-01
Scandium			
Sc I			
2116.7	4	4	2.0E-01
2120.4	6	6	2.0E-01
2262.3	4	4	5.8E-02
2266.6	4	2	4.8E-01
2270.9	6	4	4.6E-01
2280.8	4	6	2.8E-01
2289.6	6	6	4.1E-02
2311.29	4	6	4.1E-02
2315.69	4	4	2.5E-01
2320.32	6	6	2.4E-01
2324.75	6	4	4.1E-02
2328.19	4	6	4.6E-02
2334.67	4	2	1.7E-01
2346.03	6	4	1.3E-01
2429.19	4	4	2.8E-01
2438.63	6	6	2.1E-01
2468.40	4	2	4.9E-02
2692.78	4	2	1.61E-01
2699.02	4	6	2.4E-02
2706.74	4	4	3.1E-01
2707.93	6	4	1.49E-01
2711.34	6	6	3.2E-01
2965.88	4	6	7.5E-02
2974.01	4	4	5.5E-01
2980.76	6	6	5.4E-01
2988.97	6	4	6.9E-02
3015.37	4	6	7.8E-01
3019.35	6	8	8.7E-01
3030.76	6	6	1.00E-01
3255.68	4	4	3.2E-01
3269.90	4	2	3.13E+00
3273.63	6	4	2.81E+00
3907.48	4	6	1.66E+00
3911.81	6	8	1.79E+00
3933.38	6	6	1.62E-01
3996.60	4	6	1.65E-01
4020.39	4	4	1.63E+00
4023.22	4	4	3.0E-01
4023.68	6	6	1.65E+00
4031.38	6	6	2.9E-01
4036.86	6	4	7.9E-02
4043.80	8	8	3.11E-01
4047.80	6	4	1.54E-01
4051.83	8	6	7.7E-02
4054.54	4	2	1.67E-01
4067.00	6	8	1.91E-01
4067.63	10	8	4.1E-02
4074.96	4	6	3.7E-01
4078.56	2	4	4.3E-01
4080.57	4	4	6.6E-02
4082.39	6	4	2.73E-01
4086.66	6	8	3.7E-01
4087.47	4	6	1.12E-01
4093.12	4	4	1.23E-01
4094.86	6	6	1.44E-01
4098.36	8	8	8.7E-02
4132.98	4	6	1.19E+00
4140.27	6	8	1.17E+00
4147.38	6	6	1.74E-01
4161.85	8	8	1.77E-01
4171.53	6	4	1.36E-01
4186.42	6	8	8.4E-02
4187.61	8	6	1.28E-01
4193.53	4	6	6.1E-02
4204.52	6	8	3.5E-02
4205.20	10	8	1.12E-01
4212.32	4	6	1.58E-01
4212.48	6	6	8.6E-02
4216.08	2	4	2.36E-01
4218.23	4	4	2.26E-01
4225.54	6	8	9.5E-02
4225.69	4	6	7.6E-02
4231.64	4	4	1.31E-01
4233.59	6	6	4.0E-01
4238.05	8	8	7.1E-01
4239.55	6	4	2.27E-01
4246.14	8	6	1.15E-01
4542.55	6	4	1.28E-01
4544.67	8	6	1.33E-01
4706.94	4	6	2.81E-01
4709.31	6	8	4.0E-01
4711.72	2	4	1.81E-01
4714.30	4	4	2.14E-01
4719.31	6	6	1.04E-01
4728.77	8	8	1.16E-01
4729.20	4	4	2.20E-01
4729.24	6	6	1.93E-01
4734.11	4	2	1.10E+00
4737.65	6	4	8.8E-01
4741.02	8	6	9.1E-01
4743.82	10	8	9.8E-01
4973.67	4	2	8.4E-01
4980.36	6	4	5.6E-01
4983.43	4	4	2.58E-01
4991.91	6	6	3.8E-01
4995.00	4	6	5.9E-02
5018.41	6	4	2.09E-01
5021.52	4	4	2.30E-01
5064.31	8	10	7.3E-02
5066.38	6	6	3.6E-02
5070.17	6	8	1.16E-01
5072.71	2	4	2.0E-02
5075.82	4	6	1.15E-01
5080.22	4	4	4.1E-02
5081.56	10	10	7.6E-01
5083.72	8	8	6.2E-01
5085.55	6	6	5.7E-01
5086.94	4	4	6.6E-01
5096.72	6	4	1.69E-01
5099.27	4	6	1.50E-01
5101.12	10	8	8.8E-02
5331.79	4	4	1.11E-01
5339.43	6	6	1.06E-01
5341.07	4	2	3.8E-01
5349.34	6	4	5.9E-01
5350.28	8	8	6.8E-02
5355.79	6	4	3.0E-01
5356.10	8	6	5.7E-01
5375.37	8	6	3.4E-01
5392.06	10	8	4.2E-01
5416.16	4	6	4.4E-02
5416.41	6	6	2.0E-02
5425.55	6	8	4.5E-02
5429.42	2	4	9.0E-02
5432.98	4	4	5.4E-02
5433.25	6	4	9.7E-02
5438.28	4	6	3.4E-02
5439.04	2	2	1.74E-01
5442.62	4	2	2.15E-01
5446.20	8	8	2.8E-01
5451.37	6	6	1.50E-01
5455.24	4	4	6.6E-02
5464.95	4	2	3.2E-02
5468.40	6	4	9.7E-02
5472.19	8	6	9.7E-02
5482.01	8	8	5.2E-01
5484.63	6	6	5.2E-01
5514.23	6	8	4.1E-01
5520.52	8	10	4.3E-01
5526.10	4	4	7.1E-02
5541.07	6	6	5.5E-02
5631.04	2	4	3.0E-02
5671.83	10	12	5.4E-01
5686.86	8	10	4.9E-01
5700.19	6	8	4.6E-01
5708.64	10	10	4.7E-02
5711.79	4	6	4.5E-01
5717.31	8	8	7.5E-02
5724.13	6	6	7.4E-02
5988.43	6	6	6.6E-02
6026.16	4	4	7.2E-02
6146.20	6	8	4.2E-02
6198.43	4	6	3.5E-02
6249.96	6	8	3.2E-02
6262.22	4	6	8.4E-02
6280.16	2	4	4.0E-02
6284.16	6	6	3.9E-02
6284.73	4	4	7.1E-02
6293.02	2	2	1.04E-01
7741.16	10	10	3.8E-02
7800.42	8	8	5.1E-02
Sc II			
1880.6	5	3	5.0E+00
2064.3	7	5	2.2E+00
2068.0	5	3	2.0E+00
2273.1	1	3	7.7E+00
2545.20	5	5	4.0E-01
2552.35	7	5	2.21E+00
2555.79	3	3	6.9E-01
2560.23	5	3	2.01E+00

λ Å	Weights g_i	Weights g_k	A $10^8\ s^{-1}$
2563.19	3	1	2.70E+00
2611.19	5	5	2.2E+00
2667.70	3	5	1.5E+00
2746.36	3	1	3.9E+00
2782.31	5	5	1.3E+00
2789.15	7	7	1.3E+00
2801.31	9	9	1.3E+00
2819.49	3	5	2.3E+00
2822.12	5	7	2.5E+00
2826.64	7	9	2.8E+00
2870.85	5	3	1.1E+00
2912.98	5	3	1.1E+00
2979.68	3	5	1.2E+00
2988.92	5	7	2.9E+00
3039.92	7	9	3.5E+00
3045.73	5	7	3.68E+00
3052.92	7	9	3.92E+00
3060.54	7	7	3.0E-01
3065.12	9	11	4.00E+00
3075.36	9	9	2.5E-01
3128.27	3	3	1.9E+00
3133.07	5	5	1.8E+00
3139.72	7	7	2.1E+00
3190.98	3	3	1.1E+00
3199.33	5	3	1.9E+00
3312.72	5	7	1.2E+00
3320.40	5	3	1.2E+00
3343.23	9	7	1.1E+00
3353.72	5	7	1.51E+00
3359.67	5	5	2.16E-01
3361.26	3	3	3.4E-01
3361.93	3	1	1.17E+00
3368.94	5	3	8.3E-01
3372.15	7	5	9.9E-01
3379.16	3	3	2.5E+00
3535.71	5	3	6.1E-01
3558.53	5	7	3.0E-01
3567.70	3	5	3.5E-01
3572.53	7	7	1.38E+00
3576.34	5	5	1.06E+00
3580.93	3	3	1.23E+00
3589.63	5	3	4.6E-01
3590.47	7	5	2.9E-01
3613.83	7	9	1.48E+00
3630.74	5	7	1.20E+00
3642.78	3	5	1.13E+00
3645.31	7	7	2.74E-01
3651.80	5	5	3.0E-01
3859.59	7	5	1.1E+00
4246.82	5	5	1.29E+00
4314.08	9	7	4.1E-01
4320.75	7	5	4.0E-01
4325.00	5	3	4.3E-01
4374.46	9	9	1.48E-01
4400.39	7	7	1.43E-01
4415.54	5	5	1.47E-01
4670.41	5	7	1.16E-01
5031.01	5	3	3.5E-01
5239.81	1	3	1.39E-01
5526.79	9	7	3.3E-01
5657.91	5	5	1.04E-01
5669.06	3	1	1.31E-01
Silicon			
Si I			
1977.6	1	3	1.8E-01
1979.2	3	1	5.1E-01
1980.6	3	3	1.3E-01
1983.2	3	5	1.4E-01
1986.4	5	3	2.1E-01
1989.0	5	5	4.1E-01
2208.0	1	3	3.11E-01
2210.9	3	5	4.16E-01
2211.7	3	3	2.32E-01
2216.7	5	7	5.5E-01
2218.1	5	5	1.38E-01
2506.9	3	5	4.66E-01
2514.3	1	3	6.1E-01
2516.1	5	5	1.21E+00
2519.2	3	3	4.56E-01
2524.1	3	1	1.81E+00
2528.5	5	3	7.7E-01
2532.4	1	3	2.6E-01
2631.3	1	3	9.7E-01
2881.6	5	3	1.89E+00
3905.5	1	3	1.18E-01
4738.8	3	3	1.0E-02
4783.0	5	3	1.7E-02
4792.3	5	5	1.7E-02
4818.1	5	7	1.1E-02
4821.2	3	5	8.0E-03
4947.6	3	1	4.2E-02
5006.1	3	5	2.8E-02
5622.2	3	3	1.6E-02
5690.4	3	3	1.2E-02
5708.4	5	5	1.4E-02
5754.2	5	3	1.5E-02
5772.1	3	1	3.6E-02
5948.5	3	5	2.2E-02
7226.2	3	5	7.9E-03
7405.8	3	5	3.7E-02
7409.1	5	7	2.3E-02
7680.3	3	5	4.6E-02
7918.4	3	5	5.2E-02
7932.3	5	7	5.1E-02
7944.0	7	9	5.8E-02
7970.3	5	5	7.1E-03
Si II			
989.87	2	4	6.7E+00
992.68	4	6	8.0E+00
1020.7	2	2	1.3E+00
1190.4	2	4	6.9E+00
1193.3	2	2	2.8E+01
1194.5	4	4	3.6E+01
1197.4	4	2	1.4E+01
1248.4	4	4	1.3E+01
1251.2	6	4	1.9E+01
1260.4	2	4	2.0E+01
1264.7	4	6	2.3E+01
1304.4	2	2	3.6E+00
1309.3	4	2	7.0E+00
1526.7	2	2	3.73E+00
1533.5	4	2	7.4E+00
1808.0	2	4	3.7E-02
2904.3	4	6	6.7E-01
2905.7	6	8	7.1E-01
3210.0	4	6	4.6E-01
4128.1	4	6	1.32E+00
4130.9	6	8	1.42E+00
5041.0	2	4	9.8E-01
5056.0	4	6	1.2E+00
5957.6	2	2	4.2E-01
5978.9	4	2	8.1E-01
6347.1	2	4	7.0E-01
6371.4	2	2	6.9E-01
7848.8	4	6	3.9E-01
7849.7	6	8	4.2E-01
Si III			
883.40	5	7	6.3E+01
994.79	3	3	7.89E+00
997.39	5	3	1.31E+01
1141.6	3	5	3.0E+01
1144.3	5	7	3.9E+01
1161.6	5	5	1.6E+01
1206.5	1	3	2.59E+01
1206.5	3	5	4.89E+01
1207.5	5	5	1.9E+01
1294.5	3	5	5.42E+00
1296.7	1	3	7.19E+00
1298.9	3	3	5.36E+00
1299.0	5	5	1.61E+01
1301.2	3	1	2.13E+01
1303.3	5	3	8.85E+00
1328.8	1	3	2.7E+01
1417.2	3	1	2.60E+01
1435.8	5	7	2.1E+01
1589.0	5	3	1.1E+01
1778.7	7	9	4.4E+00
1783.1	5	7	3.8E+00
3241.6	5	3	2.3E+00
*3486.9	15	21	1.8E+00
3590.5	3	5	3.9E+00
4552.6	3	5	1.26E+00
4554.0	5	3	7.6E-01
4567.8	3	3	1.25E+00
4683.0	5	5	9.5E 01
4716.7	5	7	2.8E+00
5451.5	3	5	6.0E-01
5473.1	5	7	7.9E-01
5716.3	9	7	1.9E-01
5739.7	1	3	4.7E-01
7462.6	5	3	4.9E-01
7466.3	7	5	5.4E-01
7612.4	3	5	1.1E+00

λ Å	Weights g_i	g_k	A 10^8 s^{-1}
Si IV			
457.82	2	4	3.6E+00
458.16	2	2	3.6E+00
515.12	2	2	4.1E+00
516.35	4	2	8.2E+00
*560.50	6	10	1.0E+00
*749.94	10	14	1.45E+01
815.05	2	2	1.23E+01
818.13	4	2	2.44E+01
*860.74	10	6	1.8E+00
*1066.6	10	14	3.91E+01
1122.5	2	4	2.05E+01
1128.3	4	4	4.03E+00
1128.3	4	6	2.42E+01
1393.8	2	4	7.73E+00
1402.8	2	2	7.58E+00
*1724.1	10	6	5.5E+00
Si V			
96.439	1	3	4.8E+02
97.143	1	3	2.0E+03
117.86	1	3	3.0E+02
Si VI			
246.00	4	2	1.7E+02
249.12	2	2	8.5E+01
Si VII			
217.83	5	3	4.3E+02
272.64	5	3	5.1E+01
274.18	3	1	1.2E+02
275.35	5	5	8.9E+01
275.67	3	3	3.0E+01
276.84	1	3	3.9E+01
278.45	3	5	2.9E+01
Si VIII			
214.76	4	2	4.1E+02
216.92	6	4	3.6E+02
232.86	2	2	8.0E+01
235.56	4	4	9.7E+01
250.45	2	2	7.7E+01
250.79	4	2	1.6E+02
314.31	4	2	5.2E+01
316.20	4	4	5.0E+01
319.83	4	6	4.9E+01
Si IX			
223.73	1	3	4.2E+01
225.03	3	3	1.2E+02
227.01	5	3	2.0E+02
227.30	5	3	2.3E+02
258.10	5	5	1.04E+02
*294.37	9	9	5.9E+01
*347.36	9	15	2.2E+01
Si X			
253.77	2	4	2.9E+01
256.57	2	2	1.1E+02
258.35	4	4	1.4E+02
261.05	4	2	5.4E+01
272.00	2	2	3.0E+01
277.26	4	2	5.7E+01
287.08	2	4	2.6E+01
289.19	4	4	5.0E+01
292.22	6	4	7.3E+01
*347.73	10	10	4.3E+01
*353.09	6	10	2.1E+01
Si XI			
43.763	1	3	6.11E+03
*49.116	9	3	2.45E+03
49.222	3	5	8.9E+03
52.296	3	1	7.6E+02
303.30	1	3	6.42E+01
358.29	3	1	1.03E+02
358.63	3	5	1.38E+01
361.41	1	3	1.80E+01
364.50	3	3	1.32E+01
365.42	5	5	3.90E+01
368.28	3	1	5.1E+01
371.48	5	3	2.07E+01
604.14	3	5	1.12E+01
2300.8	1	3	4.34E-01
Si XII			
*40.924	2	6	4.42E+03
*44.118	6	10	1.4E+04
499.43	2	4	9.56E+00
520.72	2	2	8.47E+00
1862	2	4	1.15E+00
1949	2	2	1.0E+00
4620	2	4	4.6E-02
4942	4	6	4.5E-02
Silver			
Ag I			
2061.2	2	4	3.1E-02
2069.9	2	2	1.5E-02
3280.7	2	4	1.4E+00
3382.9	2	2	1.3E+00
5209.1	2	4	7.5E-01
5465.5	4	6	8.6E-01
5471.6	4	4	1.4E-01
Sodium			
Na I			
3302.4	2	4	2.81E-02
3303.0	2	2	2.81E-02
4390.0	2	4	7.7E-03
4393.3	4	4	1.6E-03
4393.3	4	6	9.2E-03
4494.2	2	4	1.2E-02
4497.7	4	6	1.4E-02
4497.7	4	4	2.4E-03
4664.8	2	4	2.33E-02
4668.6	4	4	4.1E-03
4668.6	4	6	2.5E-02
4747.9	2	2	6.3E-03
4751.8	4	2	1.27E-02
4978.5	2	4	4.1E-02
4982.8	4	4	8.2E-03
4982.8	4	6	4.89E-02
5148.8	2	2	1.17E-02
5153.4	4	2	2.33E-02
5682.6	2	4	1.03E-01
5688.2	4	6	1.2E-01
5688.2	4	4	2.1E-02
5890.0	2	4	6.11E-01
5895.9	2	2	6.10E-01
6154.2	2	2	2.6E-02
6160.8	4	2	5.2E-02
8183.3	2	4	4.53E-01
8194.8	4	6	5.4E-01
8194.8	4	4	9.0E-02
11381	2	2	8.9E-02
11404	4	2	1.76E-01
Na II			
300.15	1	3	3.0E+01
301.44	1	3	4.9E+01
372.08	1	3	3.4E+01
Na III			
378.14	4	2	7.7E+01
380.10	2	2	3.7E+01
1991.0	4	6	8.3E+00
2004.2	2	4	4.6E+00
2011.9	6	8	8.4E+00
2151.5	2	4	4.4E+00
2174.5	4	6	5.3E+00
2230.3	6	8	3.7E+00
2232.2	4	4	3.3E+00
2246.7	4	6	2.4E+00
2459.3	4	6	3.0E+00
2468.9	2	4	2.4E+00
2497.0	6	6	1.7E+00
Na V			
*307.89	10	6	2.0E+02
*333.46	6	6	5.6E+01
*369.01	10	6	1.2E+02
*400.72	10	10	5.0E+01
*445.14	6	10	7.1E+00
459.90	4	2	2.3E+01
461.05	4	4	2.3E+01
463.26	4	6	2.2E+01
510.10	2	2	5.6E+01
511.19	4	4	6.8E+01
Na VI			
313.75	5	3	1.3E+02
361.25	5	5	7.7E+01
*416.53	9	9	3.7E+01
*492.80	9	15	1.3E+01
1550.6	5	5	4.35E+00
1567.8	5	3	2.68E+00

λ Å	Weights g_i	Weights g_k	A $10^8 s^{-1}$
1608.5	3	1	2.6E+00
1649.4	5	5	2.05E+00
1741.5	3	5	2.59E+00
1747.5	5	7	3.1E+00
Na VII			
*94.409	6	10	2.7E+03
*105.27	6	2	4.5E+02
353.29	4	4	1.0E+02
381.30	4	2	4.0E+01
397.49	4	4	3.5E+01
399.18	6	4	5.2E+01
*483.28	10	10	2.9E+01
486.74	2	4	1.1E+01
491.95	4	6	1.3E+01
555.80	4	4	2.3E+01
777.83	4	6	6.8E+00
Na VIII			
*83.34	9	15	3.94E+03
*89.88	9	3	8.09E+02
90.536	3	5	2.86E+03
411.15	1	3	4.42E+01
1239.4	3	3	3.02E+00
1802.7	3	1	2.70E+00
1867.7	3	5	2.01E+00
2059.1	3	5	1.80E+00
2558.2	5	3	2.26E-02
2772.0	3	5	4.19E-01
3021.0	5	7	4.90E-01
3108.9	1	3	2.58E-01
3182.3	1	3	2.92E-01
Na IX			
70.615	2	4	1.35E+03
70.653	2	2	1.35E+03
77.764	2	4	3.6E+03
77.911	4	6	4.3E+03
681.72	2	4	6.63E+00
694.17	2	2	6.30E+00
2487.7	2	4	8.32E-01
2535.8	2	2	7.89E-01
6841.8	2	4	2.59E-02
7103.4	4	6	2.78E-02
Strontium			
Sr I			
2206.2	1	3	6.6E-03
2211.3	1	3	8.5E-03
2217.8	1	3	1.2E-02
2226.3	1	3	1.6E-02
2237.7	1	3	2.3E-02
2253.3	1	3	3.7E-02
2275.3	1	3	6.7E-02
2307.3	1	3	1.2E-01
2354.3	1	3	1.8E-01
2428.1	1	3	1.7E-01
2569.5	1	3	5.3E-02
2931.8	1	3	1.9E-02

λ Å	Weights g_i	Weights g_k	A $10^8 s^{-1}$
4607.3	1	3	2.01E+00
Sr II			
2018.7	2	2	1.2E-01
2051.9	4	2	2.4E-01
2282.0	2	4	8.3E-01
2322.4	4	6	9.1E-01
2324.5	4	4	1.5E-01
2423.5	2	2	2.4E-01
2471.6	4	2	4.8E-01
3464.5	4	6	3.1E+00
3474.9	4	4	5.1E-01
4077.7	2	4	1.42E+00
4161.8	2	2	6.5E-01
4215.5	2	2	1.27E+00
4305.5	4	2	1.4E+00
4414.8	4	6	1.1E-01
4417.5	4	4	1.8E-02
4585.9	4	2	7.0E-02
5303.1	2	4	1.9E-01
5379.1	4	6	2.2E-01
5385.5	4	4	3.7E-02
5723.7	2	2	7.1E-02
5819.0	4	2	1.4E-01
8688.9	4	6	5.5E-01
8719.6	4	4	9.7E-02
Sulfur			
S I			
1295.7	5	5	4.9E+00
1296.2	5	3	2.7E+00
1302.3	3	5	1.8E+00
1302.9	3	3	1.6E+00
1303.1	3	1	6.6E+00
1303.4	5	3	1.9E+00
1305.9	1	3	2.4E+00
1401.5	5	3	9.1E-01
1409.3	3	3	5.0E-01
1412.9	1	3	1.6E-01
1425.0	5	7	4.5E+00
1425.2	5	5	1.2E+00
1433.3	3	5	3.3E+00
1433.3	3	3	1.9E+00
1437.0	1	3	2.4E+00
1448.2	5	3	7.3E+00
1473.0	5	7	4.2E-01
1474.0	5	7	1.6E+00
1474.4	5	5	5.0E-01
1474.6	5	3	6.2E-02
1481.7	3	5	1.7E-01
1483.0	3	5	1.2E+00
1483.2	3	3	7.5E-01
1487.2	1	3	8.7E-01
1666.7	5	5	6.3E+00
1687.5	1	3	9.4E-01
1782.3	1	3	1.9E+00
1807.3	5	3	3.8E+00
1820.3	3	3	2.2E+00
1826.2	1	3	7.2E-01

λ Å	Weights g_i	Weights g_k	A $10^8 s^{-1}$
4694.1	5	7	6.7E-03
4695.4	5	5	6.7E-03
4696.2	5	3	6.5E-03
6403.6	3	5	5.7E-03
6408.1	5	5	9.5E-03
6415.5	7	5	1.3E-02
*6751.2	15	25	7.9E-02
7679.6	3	5	1.2E-02
7686.1	5	5	2.0E-02
7696.7	7	5	2.8E-02
S II			
1124.4	2	4	1.0E+00
1125.0	4	4	4.6E+00
1131.0	2	2	3.5E+00
1131.6	4	2	1.4E+00
1250.5	4	2	4.6E-01
1253.8	4	4	4.2E-01
1259.5	4	6	3.4E-01
4463.6	8	6	5.3E-01
4483.4	6	4	3.1E-01
4486.7	4	2	6.6E-01
4524.7	4	4	9.3E-02
4525.0	6	4	1.2E+00
4552.4	4	2	1.2E+00
4656.7	2	4	9.0E-02
4716.2	4	4	2.9E-01
4815.5	6	4	8.8E-01
4885.6	2	4	1.7E-01
4917.2	2	2	6.6E-01
4924.1	4	6	2.2E-01
4925.3	2	4	2.4E-01
4942.5	2	2	1.5E-01
4991.9	4	4	1.5E-01
5009.5	4	2	7.0E-01
5014.0	4	4	8.4E-01
5027.2	4	2	2.6E-01
5032.4	6	6	8.1E-01
5047.3	4	2	3.6E-01
5103.3	6	4	5.0E-01
5142.3	2	2	1.9E-01
5201.0	4	4	7.5E-01
5201.3	6	4	6.5E-02
5212.6	4	6	9.8E-02
5212.6	6	6	8.5E-01
5320.7	6	8	9.2E-01
5345.7	4	6	8.8E-01
5345.7	6	6	1.1E-01
5428.6	2	4	4.2E-01
5432.8	4	6	6.8E-01
5453.8	6	8	8.5E-01
5473.6	2	2	7.3E-01
5509.7	4	4	4.0E-01
5526.2	8	8	8.1E-02
5536.8	4	6	6.6E-02
5556.0	4	2	1.1E-01
5564.9	6	6	1.7E-01
5578.8	6	6	1.1E-01
5606.1	10	8	5.4E-01

λ Å	Weights g_i	g_k	A $10^8\ s^{-1}$
5616.6	4	4	1.2E-01
5640.0	4	6	6.6E-01
5645.6	6	4	1.8E-02
5647.0	2	4	5.7E-01
5659.9	6	4	4.6E-01
5664.7	4	2	5.8E-01
5819.2	4	4	8.5E-02
6305.5	8	6	1.8E-01
6312.7	6	4	3.0E-01
S III			
2496.2	7	5	2.5E+00
2508.2	5	3	2.3E+00
2636.9	3	5	4.5E-01
2665.4	5	5	1.4E+00
2680.5	1	3	6.2E-01
2691.8	3	3	4.6E-01
2702.8	3	1	1.9E+00
2718.9	3	3	1.2E+00
2721.4	5	3	7.7E-01
2726.8	3	5	6.0E-01
2731.1	5	5	1.1E+00
2756.9	7	7	1.4E+00
2785.5	3	3	6.1E-01
2856.0	5	7	5.1E+00
2863.5	7	9	5.7E+00
2872.0	3	5	4.7E+00
2950.2	3	5	3.0E+00
2964.8	5	7	4.0E+00
3662.0	3	3	6.4E-01
3717.8	5	3	1.0E+00
3778.9	3	5	4.4E-01
3831.8	1	3	5.6E-01
3837.8	3	3	4.2E-01
3838.3	5	5	1.3E+00
3860.6	3	1	1.6E+00
3899.1	5	3	6.7E-01
4253.6	5	7	1.2E+00
4285.0	3	5	9.0E-01
S IV			
551.17	2	2	2.06E+01
554.07	4	2	4.08E+01
3097.5	2	4	2.6E+00
3117.7	2	2	2.5E+00
S V			
437.37	1	3	1.12E+01
438.19	3	3	3.33E+01
439.65	5	3	5.5E+01
*661.52	9	15	6.44E+01
*679.01	9	15	8.6E+01
*690.75	9	9	5.0E+01
786.48	1	3	5.25E+01
*854.85	9	9	4.18E+01
S VI			
248.99	2	4	3.1E+01
249.27	2	2	3.1E+01
388.94	2	2	4.5E+01
390.86	4	2	8.8E+01
706.48	2	4	4.17E+01
712.68	4	6	4.85E+01
712.84	4	4	8.1E+00
933.38	2	4	1.7E+01
944.52	2	2	1.6E+01
S VII			
60.161	1	3	9.46E+03
60.804	1	3	5.1E+02
72.029	1	3	8.61E+02
S VIII			
198.55	4	2	2.5E+02
202.61	2	2	1.2E+02
S XI			
*189.90	9	3	4.3E+02
190.37	5	3	2.8E+02
215.95	5	5	1.4E+02
217.63	1	3	7.2E+01
239.81	1	3	2.6E+01
242.57	3	5	1.9E+01
242.82	3	3	1.9E+01
246.90	5	5	5.4E+01
247.12	5	3	3.0E+01
*288.49	9	15	2.9E+01
S XII			
212.14	2	4	3.7E+01
215.18	2	2	1.4E+02
218.20	4	4	1.7E+02
221.44	4	2	6.4E+01
227.50	2	2	3.7E+01
234.48	4	2	6.8E+01
S XIII			
32.236	1	3	1.09E+04
37.600	3	1	1.3E+03
256.66	1	3	8.7E+01
299.89	3	5	1.78E+01
303.37	1	3	2.28E+01
307.36	3	3	1.64E+01
308.91	5	5	4.82E+01
312.68	3	1	6.3E+01
316.84	5	3	2.50E+01
500.42	3	5	1.43E+01
S XIV			
*30.434	2	6	8.28E+03
*32.517	6	10	2.6E+04
417.67	2	4	1.2E+01
445.71	2	2	1.0E+01
1550	2	4	1.4E+00
1663	2	2	1.2E+00
3967	2	4	5.4E-02
4153	4	6	5.7E-02
Tantalum			
Ta I			
3127.9	4	6	5.7E-03
3168.3	4	4	6.0E-03
3170.3	8	10	8.5E-02
3205.5	6	8	5.6E-03
3260.2	4	4	5.8E-03
3337.8	6	6	1.3E-02
3383.9	6	4	5.3E-03
3406.9	4	6	6.8E-02
3419.7	8	8	1.91E-02
3463.8	4	6	2.62E-02
3484.6	4	4	8.5E-03
3488.8	6	4	7.3E-03
3497.9	6	8	4.9E-02
3505.0	8	6	2.72E-02
3553.4	4	6	3.3E-03
3607.4	6	8	4.6E-02
3625.2	10	8	1.0E-02
3626.6	8	10	7.1E-02
3642.1	10	12	5.5E-02
3657.5	6	6	4.3E-03
3731.0	4	6	5.3E-03
3754.5	8	8	6.5E-03
3784.3	4	6	4.3E-02
3792.1	4	4	9.0E-03
3826.9	6	6	5.2E-03
3836.6	8	10	4.0E-03
3848.1	10	8	1.30E-02
3858.6	10	10	2.5E-03
3918.5	4	2	2.5E-02
3922.8	4	4	3.98E-02
3996.2	2	4	3.35E-02
3999.3	4	4	1.8E-02
4003.7	10	8	3.1E-03
4006.8	6	8	7.6E-03
4026.9	4	4	3.60E-02
4029.9	10	10	2.8E-02
4030.7	8	10	2.3E-03
4040.9	10	12	7.3E-03
4061.4	2	4	6.5E-02
4064.6	4	4	3.83E-02
4067.2	6	4	6.8E-03
4067.9	6	8	8.4E-03
4097.2	10	10	2.1E-03
4105.0	6	4	1.1E-02
4136.2	8	6	1.82E-02
4147.9	10	8	1.79E-02
4175.2	6	8	2.8E-02
4205.9	8	10	8.9E-03
4303.0	6	6	2.08E-02
4378.8	8	6	4.8E-03
4386.1	4	6	1.0E-02
4402.5	6	6	2.28E-02
4415.7	2	4	2.53E-02
4441.0	4	6	7.5E-03

λ Å	Weights g_i	Weights g_k	A 10^8 s^{-1}
4441.7	10	8	9.0E-03
4473.5	6	8	1.36E-02
4511.0	10	12	1.56E-02
4511.5	10	8	3.6E-03
4514.2	10	10	3.1E-03
4521.1	10	10	2.3E-03
4530.9	4	6	2.42E-02
4547.2	4	6	5.3E-03
4553.7	6	8	9.5E-03
4565.9	8	8	2.5E-02
4574.3	4	4	1.2E-02
4580.7	8	10	2.1E-03
4619.5	6	4	5.3E-02
4633.1	4	4	1.2E-02
4669.1	6	4	2.85E-02
4681.9	6	6	1.5E-02
4684.9	10	8	2.8E-03
4685.3	6	8	3.4E-03
4691.9	2	4	4.08E-02
4706.1	6	6	1.4E-02
4740.2	4	4	5.0E-02
4758.0	4	6	7.5E-03
4769.0	8	8	2.8E-02
4780.9	10	8	2.16E-02
4812.8	4	4	1.2E-02
4825.4	6	6	2.63E-02
4832.2	4	4	1.7E-02
4852.2	4	4	1.7E-02
4884.0	6	8	1.1E-02
4904.6	12	10	1.95E-02
4920.9	8	10	2.1E-03
4921.3	2	4	1.2E-02
4926.0	4	4	1.5E-02
4936.4	8	6	4.5E-02
4969.7	4	4	1.0E-02
5012.5	4	4	1.9E-02
5037.4	10	8	4.4E-02
5043.3	6	4	2.73E-02
5067.9	8	6	2.92E-02
5069.9	10	12	1.7E-03
5082.3	10	12	1.9E-03
5087.4	6	4	1.5E-02
5090.7	8	6	9.5E-03
5095.3	6	6	5.0E-03
5136.5	2	2	4.5E-02
5141.6	4	2	1.2E-02
5143.7	6	4	1.7E-02
5147.6	6	4	9.0E-03
5161.8	4	6	6.3E-03
5218.7	8	6	8.2E-03
5235.4	6	6	4.7E-03
5295.0	6	6	7.5E-03
5336.1	6	8	5.5E-03
5349.6	6	4	2.2E-02
5354.7	4	4	6.5E-03
5396.0	6	8	2.5E-03
5402.5	4	2	1.4E-02
5435.3	4	6	1.1E-02
5499.4	10	10	6.1E-03
5518.9	8	10	3.8E-02
5620.7	8	10	6.0E-03
5640.2	6	8	4.9E-03
5645.9	6	8	1.43E-02
5699.2	6	6	4.2E-03
5767.9	6	8	2.6E-03
5780.7	4	6	3.3E-03
5811.1	8	6	5.7E-03
5849.7	10	8	2.8E-03
5877.4	10	12	2.3E-02
5939.8	2	4	1.6E-02
5944.0	4	6	2.13E-02
5997.2	10	10	2.4E-02
6020.7	2	4	1.0E-02
6045.4	6	8	2.6E-02
6047.3	8	10	9.0E-03
6249.8	6	6	3.5E-03
6258.7	6	8	3.3E-03
6309.6	4	6	1.83E-02
6360.8	6	8	4.6E-03
6428.6	6	6	6.0E-03
6430.8	8	8	2.9E-02
6450.4	8	10	2.2E-02
6485.4	10	10	5.8E-02
6514.4	6	4	2.2E-02
6516.1	6	8	1.25E-02
6612.0	6	4	1.9E-02
6673.7	2	4	9.0E-03
6771.7	4	4	5.8E-03
6866.2	8	6	2.58E-02
6927.4	10	12	1.01E-02
6928.5	10	8	1.69E-02
6951.3	10	10	3.7E-03
6953.9	6	8	8.3E-03
6966.1	8	8	1.2E-02
6969.5	10	10	2.9E-03
7407.9	6	4	2.0E-02

Thallium

Tl I

λ Å	Weights g_i	Weights g_k	A 10^8 s^{-1}
2104.6	2	4	4.0E-02
2118.9	2	2	2.0E-02
2129.3	2	4	5.8E-02
2151.9	2	2	3.1E-02
2168.6	2	4	9.8E-02
2237.8	2	4	1.9E-01
2316.0	2	2	7.8E-02
2379.7	2	4	4.4E-01
2507.9	4	2	1.1E-02
2538.2	4	2	1.6E-02
2580.1	2	2	1.8E-01
2609.0	4	6	1.0E-01
2609.8	4	4	1.9E-02
2665.6	4	2	5.7E-02
2709.2	4	6	1.7E-01
2710.7	4	4	3.7E-02
2767.9	2	4	1.26E+00
2826.2	4	2	8.0E-02
2918.3	4	6	4.2E-01
2921.5	4	4	7.6E-02
3229.8	4	2	1.73E-01
3519.2	4	6	1.24E+00
3529.4	4	4	2.20E-01
3775.7	2	2	6.25E-01
5350.5	4	2	7.05E-01

Thulium

Tm I

λ Å	Weights g_i	Weights g_k	A 10^8 s^{-1}
2513.8	8	10	6.9E-02
2527.0	8	8	1.7E-01
2596.5	8	10	1.6E-01
2601.1	8	6	1.7E-01
2622.5	8	10	6.1E-02
2841.1	6	6	2.0E-01
2854.2	8	6	2.7E-01
2914.8	8	8	7.7E-02
2933.0	8	6	1.0E-01
2973.2	8	8	2.3E-01
3046.9	8	8	1.8E-01
3081.1	8	8	1.9E-01
3122.5	6	6	5.2E-01
3142.4	6	6	8.8E-02
3172.7	8	8	1.8E-01
3233.7	8	10	5.1E-02
3247.0	6	8	3.0E-01
3251.8	6	4	5.2E-01
3380.7	6	8	2.0E-01
3406.0	6	8	1.5E-01
3410.1	8	10	1.0E-01
3416.6	8	8	5.7E-02
3418.6	6	6	1.1E-01
3563.9	8	6	9.8E-02
3567.4	8	10	4.2E-02
3744.1	8	8	9.5E-01
3751.8	8	10	1.9E-01
3798.5	6	4	1.2E+00
3807.7	6	6	3.9E-01
3883.1	8	6	1.0E+00
3887.4	8	8	3.8E-01
3916.5	6	8	1.5E+00
3949.3	6	6	1.0E+00
4022.6	6	8	4.0E-02
4044.5	6	4	2.9E-01
4094.2	8	6	9.0E-01
4105.8	8	10	6.0E-01
4138.3	6	4	7.0E-01
4158.6	6	8	5.5E-02
4187.6	8	8	6.1E-01
4203.7	8	10	2.5E-01
4222.7	6	8	1.5E-01
4271.7	6	6	1.1E-01
4359.9	8	6	1.3E-01
4386.4	8	8	4.2E-02
4394.4	6	4	1.1E-01
4643.1	6	6	3.4E-02
4681.9	6	8	3.9E-02
4691.1	6	6	3.9E-02
5307.1	8	10	2.3E-02

λ Å	Weights g_i	g_k	A $10^8\ s^{-1}$
5658.3	6	8	1.0E-02
5675.8	8	10	1.3E-02
5760.2	6	6	1.3E-02
Tin			
Sn I			
2073.1	1	3	3.6E-02
2199.3	3	5	2.9E-01
2209.7	5	5	5.6E-01
2246.1	1	3	1.6E+00
2268.9	5	7	1.2E+00
2286.7	5	5	3.1E-01
2317.2	5	7	2.0E+00
2334.8	3	3	6.6E-01
2354.8	3	5	1.7E+00
2380.7	3	5	3.1E-02
2408.2	5	3	1.8E-01
2421.7	5	7	2.5E+00
2429.5	5	7	1.5E+00
2433.5	5	3	8.0E-03
2455.2	5	5	1.1E-02
2476.4	5	3	1.1E-02
2483.4	5	5	2.1E-01
2491.8	1	3	1.7E-01
2495.7	5	5	6.2E-01
2523.9	5	3	7.4E-02
2546.6	1	3	2.1E-01
2558.0	1	3	3.4E-01
2571.6	5	7	4.5E-01
2594.4	5	5	3.0E-01
2636.9	1	3	1.1E-01
2661.2	3	3	1.1E-01
2706.5	3	5	6.6E-01
2761.8	5	5	3.7E-03
2779.8	5	7	1.8E-01
2785.0	5	3	1.4E-01
2788.0	1	3	1.4E-01
2812.6	1	3	2.3E-01
2813.6	5	5	1.2E-01
2840.0	5	5	1.7E+00
2850.6	5	5	3.3E-01
2863.3	1	3	5.4E-01
2913.5	1	3	8.3E-01
3009.1	3	3	3.8E-01
3032.8	1	3	6.2E-01
3034.1	3	1	2.0E+00
3141.8	1	3	1.9E-01
3175.1	5	3	1.0E+00
3218.7	1	3	4.7E-02
3223.6	5	5	1.2E-03
3262.3	5	3	2.7E+00
3330.6	5	5	2.0E-01
3655.8	1	3	4.1E-02
3801.0	5	3	2.8E-01
4524.7	1	3	2.6E-01
5631.7	1	3	2.4E-02
5970.3	5	3	9.6E-02
6037.7	5	5	5.0E-02
6069.0	1	3	4.6E-02
6073.5	3	1	6.3E-02
6171.5	3	3	4.9E-02
Sn II			
2368.3	4	2	4.4E-03
2449.0	4	6	3.7E-01
2487.0	6	8	5.5E-01
3283.2	4	6	1.0E+00
3352.0	6	8	1.0E+00
3472.5	2	4	1.6E-01
3575.5	4	6	1.3E-01
5332.4	2	4	8.6E-01
5562.0	4	6	1.2E+00
5588.9	4	6	8.5E-01
5596.2	4	4	1.5E-01
5797.2	6	6	2.8E-01
5799.2	6	8	8.1E-01
6453.5	2	4	1.2E+00
6761.5	2	2	3.2E-01
6844.1	2	2	6.6E-01
Titanium			
Ti I			
2276.75	7	5	1.3E+00
2280.00	9	7	9.4E-01
2299.86	5	5	6.9E-01
2302.75	7	7	5.7E-01
2305.69	9	9	5.2E-01
2424.26	9	9	1.7E-01
2520.54	5	3	3.8E-01
2529.87	7	5	3.8E-01
2541.92	9	7	4.3E-01
2599.91	5	5	6.7E-01
2605.16	7	7	6.4E-01
2611.29	9	9	6.4E-01
2611.47	7	5	3.3E-01
2619.94	9	7	2.1E-01
2631.55	7	7	1.7E-01
2632.42	5	5	2.7E-01
2641.12	5	3	1.8E+00
2644.28	7	5	1.4E+00
2646.65	9	7	1.5E+00
2733.27	5	5	1.9E+00
2735.30	3	1	4.1E+00
2912.07	5	7	1.3E+00
2942.00	5	5	1.0E+00
2948.26	7	7	9.3E-01
2956.13	9	9	9.7E-01
2956.80	7	5	1.8E-01
3186.45	5	7	8.0E-01
3191.99	7	9	8.5E-01
3199.92	9	11	9.4E-01
3341.88	5	7	6.5E-01
3354.63	7	9	6.9E-01
3370.44	5	3	7.6E-01
3371.45	9	11	7.2E-01
3377.58	7	5	6.9E-01
3385.94	9	7	5.0E-01
3635.46	5	7	8.04E-01
3642.68	7	9	7.74E-01
3653.50	9	11	7.54E-01
3724.57	9	9	9.1E-01
3725.16	5	3	7.3E-01
3729.81	5	5	4.27E-01
3741.06	7	7	4.17E-01
3752.86	9	9	5.04E-01
3786.04	5	3	1.4E+00
3948.67	5	3	4.85E-01
3956.34	7	5	3.00E-01
3958.21	9	7	4.05E-01
3981.76	5	5	3.76E-01
3989.76	7	7	3.79E-01
3998.64	9	9	4.08E-01
4013.24	7	5	2.0E-01
4055.01	1	3	2.8E-01
4060.26	3	5	2.4E-01
4064.20	3	3	2.4E-01
4065.09	3	1	7.0E-01
4186.12	9	9	2.10E-01
4266.23	5	5	3.1E-01
4284.99	5	5	3.2E-01
4289.07	5	5	3.0E-01
4290.93	3	3	4.5E-01
4295.75	3	1	1.3E+00
4393.93	9	11	3.3E-01
4417.27	11	9	3.6E-01
4449.14	11	11	9.7E-01
4450.90	9	9	9.6E-01
4453.31	5	5	5.98E-01
4453.71	7	7	4.7E-01
4455.32	7	7	4.8E-01
4457.43	9	9	5.6E-01
4465.81	5	7	3.28E-01
4481.26	7	7	5.7E-01
4496.15	7	5	4.4E-01
4518.02	7	9	1.72E-01
4522.80	5	7	1.9E-01
4527.31	3	5	2.2E-01
4533.24	11	11	8.83E-01
4534.78	9	9	6.87E-01
4544.69	5	3	3.3E-01
4548.76	7	5	2.85E-01
4552.45	9	7	2.1E-01
4563.43	9	11	2.1E-01
4617.27	7	9	8.51E-01
4623.10	5	7	5.74E-01
4639.94	3	3	6.64E-01
4640.43	3	1	5.0E-01
4645.19	3	1	8.57E-01
4650.02	5	3	2.6E-01
4742.79	9	9	5.3E-01
4758.12	11	11	7.13E-01
4759.27	13	13	7.40E-01
4778.26	9	9	2.0E-01
4805.42	5	7	5.8E-01
4840.87	5	5	1.76E-01
4856.01	13	15	5.2E-01
4885.08	11	13	4.90E-01

λ Å	Weights g_i	Weights g_k	A 10^8 s^{-1}
4913.62	7	9	4.44E-01
4928.34	3	5	6.2E-01
4981.73	11	13	6.60E-01
4989.14	7	5	3.25E-01
4991.07	9	11	5.84E-01
4999.50	7	9	5.27E-01
5000.99	9	7	3.52E-01
5007.21	5	7	4.92E-01
5014.28	3	5	6.8E-01
5036.47	7	9	3.94E-01
5038.40	5	7	3.87E-01
5062.11	5	3	2.98E-01
5210.39	9	9	3.57E-02
5222.69	3	3	1.95E-01
5224.30	11	11	3.6E-01
5259.98	5	7	2.3E-01
5351.07	7	7	3.4E-01
5503.90	11	9	2.6E-01
5774.04	9	11	5.5E-01
5785.98	11	13	6.1E-01
5804.27	13	15	6.8E-01
6098.66	9	7	2.5E-01
6220.46	9	7	1.8E-01
Ti II			
2440.91	4	4	5.1E-01
2451.18	6	6	4.5E-01
2525.59	10	8	5.6E-01
2531.28	8	6	4.9E-01
2534.63	6	4	5.4E-01
2535.89	4	2	6.8E-01
2555.99	6	8	3.2E-01
2635.44	4	4	1.9E+00
2638.56	6	6	1.7E+00
2642.02	8	8	1.9E+00
2645.86	10	10	2.7E+00
2746.54	6	8	2.6E+00
2751.59	8	10	3.7E+00
2752.68	8	10	1.1E+00
2757.62	6	8	7.2E-01
2758.35	4	6	9.9E-01
2758.79	2	4	4.4E-01
2764.28	4	4	7.4E-01
2804.82	6	8	4.6E+00
2810.30	8	10	5.1E+00
2817.83	10	12	3.8E+00
2819.87	8	8	6.5E-01
2821.26	6	8	7.9E-01
2827.12	8	10	1.0E+00
2828.06	12	14	1.1E+00
2828.64	6	6	1.2E+00
2828.83	10	10	9.1E-01
2834.02	10	12	7.9E-01
2836.47	8	8	1.2E+00
2839.64	12	12	8.3E-01
2845.93	10	10	1.2E+00
2851.11	2	4	4.1E-01
2856.10	12	12	1.5E+00
2862.33	4	6	4.0E-01
2877.47	8	8	5.7E-01
2884.13	10	10	5.2E-01
2910.65	8	8	4.6E-01
2926.64	10	8	8.9E-01
2931.10	6	6	3.2E+00
2936.02	4	6	2.7E+00
2938.57	6	8	2.4E+00
2941.90	8	10	1.8E+00
2942.97	8	8	1.1E+00
2945.30	10	12	2.7E+00
2952.00	8	8	3.0E-01
2954.59	10	12	4.0E+00
2958.80	8	10	4.0E+00
2979.06	4	6	1.2E+00
2990.06	6	8	5.6E-01
3017.17	12	12	3.6E-01
3022.64	10	10	1.2E+00
3023.67	8	8	1.0E+00
3029.76	10	10	3.5E-01
3056.75	2	4	3.2E-01
3058.08	6	6	5.0E-01
3066.34	4	4	3.3E-01
3071.25	6	4	3.6E-01
3072.99	4	2	1.6E+00
3075.23	6	4	1.13E+00
3078.65	8	6	1.09E+00
3081.52	10	8	1.1E+00
3088.04	10	8	1.25E+00
3089.44	8	6	1.3E+00
3097.20	4	6	4.4E-01
3103.81	10	8	1.1E+00
3105.10	2	4	6.3E-01
3106.26	6	6	7.8E-01
3117.67	4	2	1.1E+00
3119.83	6	4	5.9E-01
3127.86	6	6	1.6E+00
3128.50	8	8	1.1E+00
3161.23	4	2	5.9E-01
3161.80	6	4	4.6E-01
3162.59	8	6	3.9E-01
3168.55	10	8	4.1E-01
3181.73	6	8	4.6E-01
3182.54	4	6	4.3E-01
3189.49	4	4	9.2E-01
3190.91	6	8	1.3E+00
3202.56	4	6	1.1E+00
3224.25	12	10	7.0E-01
3228.62	4	2	2.0E+00
3232.29	8	6	6.0E-01
3231.51	10	10	1.38E+00
3236.13	4	4	7.0E-01
3236.58	8	8	1.11E+00
3239.04	6	6	9.87E-01
3239.66	6	4	9.4E-01
3241.99	4	4	1.16E+00
3251.91	6	4	3.38E-01
3252.92	8	6	3.9E-01
3272.07	2	4	3.2E-01
3278.28	4	4	9.6E-01
3278.91	6	4	1.0E+00
3282.32	2	2	1.6E+00
3287.66	8	10	1.4E+00
3315.32	2	4	3.8E-01
3321.70	4	4	7.2E-01
3322.94	10	10	3.96E-01
3329.46	8	8	3.25E-01
3332.11	6	4	1.1E+00
3340.34	4	4	3.6E-01
3361.23	8	10	1.1E+00
3372.80	6	8	1.11E+00
3383.77	4	6	1.09E+00
3452.49	2	2	7.7E-01
3456.40	4	4	8.2E-01
3465.56	4	2	4.1E-01
3483.63	10	8	9.7E-01
3492.37	8	6	9.8E-01
3504.90	10	10	8.2E-01
3510.86	8	8	9.3E-01
3520.27	2	4	4.8E-01
3535.41	4	6	5.5E-01
3641.33	4	2	4.9E-01
3706.23	4	4	3.1E-01
3741.64	6	6	6.2E-01
3757.70	4	4	4.1E-01
3759.30	8	8	9.4E-01
3761.33	6	6	9.9E-01
4911.18	6	4	3.2E-01
Ti III			
865.79	5	3	6.6E+01
1002.37	5	5	7.6E+00
1004.67	7	5	4.3E+01
1005.80	3	3	1.3E+01
1007.16	5	3	3.8E+01
1008.12	3	1	5.1E+01
1286.37	9	9	2.0E+00
1289.30	7	7	2.2E+00
1291.62	5	5	2.4E+00
1293.23	9	7	1.0E+00
1298.97	7	5	4.9E+00
1327.59	5	3	3.2E+00
1420.44	1	3	1.2E+00
1421.63	3	1	4.0E+00
1422.41	5	5	3.0E+00
1424.14	5	3	1.6E+00
1455.19	9	7	6.4E+00
1498.70	5	5	2.8E+00
2007.36	3	3	3.4E+00
2007.60	1	3	1.2E+00
2010.80	5	3	5.4E+00
2097.30	5	7	3.3E+00
2099.86	3	5	2.5E+00
2104.86	3	3	1.1E+00
2105.09	1	3	1.7E+00
2199.22	3	3	5.7E+00
2237.77	7	7	2.4E+00
2331.35	3	1	4.3E+00
2331.66	3	3	1.2E+00

λ Å	Weights g_i	Weights g_k	A $10^8\ s^{-1}$
2339.00	5	3	3.0E+00
2346.79	7	5	3.3E+00
2374.99	5	3	4.0E+00
2413.99	5	7	3.8E+00
2516.05	7	9	3.4E+00
2567.56	3	3	2.3E+00
2984.75	5	5	1.9E+00
3066.51	3	3	2.5E+00
3228.89	3	3	1.5E+00
3278.31	7	9	3.4E+00
3320.94	3	5	2.8E+00
3340.20	7	9	3.7E+00
3346.18	9	11	3.7E+00
3354.71	11	13	4.4E+00
3397.24	3	1	1.8E+00
3404.46	3	3	1.8E+00
3417.62	3	5	1.9E+00
3915.47	9	11	2.1E+00
4119.14	5	5	9.9E-01
4213.26	9	11	2.2E+00
4215.53	9	11	2.2E+00
4247.62	11	13	1.1E+00
4248.54	5	7	2.3E+00
4250.09	3	5	9.5E-01
4259.01	11	13	9.4E-01
4269.84	9	11	1.7E+00
4285.61	13	15	3.0E+00
4288.66	11	13	1.1E+00
4296.70	11	13	1.6E+00
4319.56	9	11	1.1E+00
4343.25	3	1	1.0E+00
4378.94	3	5	1.6E+00
4433.91	11	13	1.8E+00
4440.66	1	3	1.2E+00
4533.26	3	5	1.5E+00
4576.53	9	7	1.3E+00
4628.07	3	1	1.5E+00
4652.86	7	9	2.6E+00
4874.00	5	7	1.5E+00
4914.32	3	3	1.1E+00
4971.19	9	11	2.1E+00
5083.80	5	3	9.7E-01
5278.33	3	3	9.4E-01
7506.87	11	13	1.1E+00
Ti IV			
423.49	4	6	4.9E+01
424.16	6	8	5.3E+01
433.63	4	2	5.5E+00
433.76	6	4	5.0E+00
729.36	4	2	5.7E+00
1183.64	2	2	6.9E+00
1195.21	4	2	1.4E+01
1451.74	2	4	1.8E+01
1467.34	4	6	2.1E+01
2067.56	2	4	5.1E+00
2103.16	2	2	5.0E+00
2541.79	4	6	6.9E+00
2546.88	6	8	7.4E+00
2862.60	4	2	4.1E+00
3576.44	4	6	4.6E+00
Ti VIII			
249	6	4	1.0E+01
258.610	6	8	7.5E+02
269.533	4	6	6.0E+02
272.037	4	4	4.3E+02
272.843	6	4	6.2E+01
276.701	2	4	9.3E+01
277.813	4	4	3.8E+02
289.375	2	4	3.6E+01
478.971	4	4	1.7E+01
480.376	6	6	1.5E+01
Ti IX			
267.941	5	7	5.1E+02
278.713	5	7	4.7E+02
281.446	3	1	3.2E+02
285.128	1	3	4.1E+02
433.567	1	3	6.9E+00
439.513	3	3	7.5E+00
439.745	3	1	2.1E+01
447.484	5	5	1.6E+01
447.701	5	3	6.5E+00
507.174	3	5	6.5E+00
516.215	5	7	6.9E+00
Ti X			
253	4	6	2.1E+02
254	6	8	2.3E+02
281	2	2	1.1E+02
289.579	2	4	2.5E+02
290.294	4	6	1.1E+02
291	4	2	1.8E+02
291	2	2	2.3E+02
292	6	8	1.1E+02
293.684	6	8	2.97E+02
293.798	6	6	1.7E+02
295.584	4	6	2.9E+02
296	4	6	1.4E+02
297	4	6	9.9E+01
298	4	6	4.3E+02
302	2	2	1.6E+02
305	2	4	2.5E+02
317	2	2	1.5E+02
355.815	2	2	1.3E+02
360.133	4	4	2.19E+02
363	4	2	2.1E+02
363	6	6	1.3E+02
365.628	4	2	1.2E+02
382	4	6	1.8E+02
385	6	8	1.8E+02
389.99	6	4	1.1E+02
Ti XI			
65.403	1	3	5.1E+02
87.725	1	3	8.5E+02
266	5	7	1.8E+02
308.250	3	5	1.3E+02
313.229	5	7	1.6E+02
318	3	1	1.4E+02
322.75	5	7	1.99E+02
323	1	3	1.8E+02
327.192	3	5	2.9E+02
332	3	1	3.25E+02
386.140	1	3	1.48E+02
408	7	9	1.37E+02
425.74	3	1	1.2E+02
446.69	3	1	1.2E+02
453	5	7	1.3E+02
Ti XII			
52.896	2	4	1.61E+02
53.140	4	6	1.9E+02
53.433	2	4	2.1E+02
53.457	2	2	2.1E+02
55.181	2	4	2.4E+02
55.443	4	6	2.81E+02
59.133	2	4	3.72E+02
59.435	4	6	4.41E+02
60.701	2	4	3.4E+02
60.762	2	2	3.5E+02
61.286	4	2	1.8E+02
62.433	4	6	2.08E+02
62.470	6	8	2.22E+02
65.540	4	6	3.2E+02
65.577	6	8	3.5E+02
67.171	2	4	6.2E+02
67.555	4	6	7.2E+02
70.986	4	6	5.7E+02
71.031	6	8	6.1E+02
71.545	2	2	1.8E+02
71.987	4	2	3.48E+02
82.121	2	4	5.9E+02
82.307	4	6	1.13E+03
82.344	2	2	5.8E+02
82.368	6	8	1.2E+03
89.844	2	4	9.9E+02
90.512	4	6	1.16E+03
90.547	4	4	1.9E+02
116.497	4	6	3.0E+03
116.597	6	8	3.2E+03
116.62	6	6	2.1E+02
139.884	6	4	2.6E+02
140.361	4	2	2.9E+02
141.6	4	6	1.7E+02
141.7	6	8	1.7E+02
169.7	4	6	2.8E+02
169.8	6	8	2.9E+02
207.2	2	4	1.5E+02
208.5	4	6	1.8E+02
252.8	4	6	4.8E+02
253.1	6	8	5.2E+02
257.5	4	2	2.4E+02
Ti XIII			
23.356	1	3	1.02E+05

λ Å	Weights g_i	g_k	A 10^8 s^{-1}
23.698	1	3	1.2E+04
23.991	1	3	3.4E+02
26.641	1	3	4.06E+03
26.960	1	3	3.06E+03
117.1	3	3	1.3E+02
117.3	3	1	2.8E+02
120.2	5	3	5.4E+02
120.2	7	5	4.4E+02
128.7	3	3	1.2E+02

Ti XIV

λ Å	Weights g_i	g_k	A 10^8 s^{-1}
21.341	4	6	9.8E+03
21.522	2	4	4.5E+04
21.657	4	4	1.3E+04
21.733	4	4	8.8E+04
21.82	4	2	6.4E+04
21.883	2	4	7.0E+04
21.958	2	4	1.2E+04
22.05	2	2	1.4E+04
24.592	4	2	6.1E+03
24.891	2	2	7.5E+03

Ti XV

λ Å	Weights g_i	g_k	A 10^8 s^{-1}
20.19	5	7	6.9E+03
20.234	5	7	1.9E+04
20.234	3	3	4.9E+04
20.246	1	3	4.2E+04
20.250	5	3	6.5E+03
20.29	3	3	1.1E+04
20.30	1	3	3.4E+04
20.30	1	1	5.8E+04
20.313	5	3	7.5E+04
20.418	5	7	8.0E+04
20.538	3	3	3.8E+04
20.54	3	1	4.1E+04
20.551	1	3	1.3E+04
20.689	5	7	4.3E+04
20.698	1	3	1.1E+05
20.771	5	3	1.1E+04
20.897	5	7	2.85E+04
20.928	5	5	8.4E+03
21.065	3	3	1.1E+04
21.079	1	3	1.58E+04
21.102	3	5	1.3E+04
22.482	5	3	6.4E+03
22.936	5	5	1.1E+04
22.966	5	3	1.1E+04
23.034	1	3	6.3E+03

Ti XVI

λ Å	Weights g_i	g_k	A 10^8 s^{-1}
110.561	4	2	3.36E+02
116.198	4	4	1.45E+02
118.215	6	4	7.4E+02
121.382	4	2	2.4E+02
124.805	4	2	6.1E+02
129.075	4	2	3.81E+02
134.724	2	2	2.6E+02
138.800	6	4	3.5E+02
143.459	4	4	2.8E+02
145.665	6	6	2.3E+02
157.812	4	2	1.32E+02
161.168	4	4	1.2E+02
163.610	4	2	1.92E+02
169.740	4	6	1.0E+02
176.267	2	2	2.45E+02
178.240	4	4	2.52E+02

Ti XVII

λ Å	Weights g_i	g_k	A 10^8 s^{-1}
18.05	3	3	4.5E+04
18.13	5	3	2.4E+04
18.13	1	3	8.1E+04
18.176	5	7	9.2E+04
123.654	3	3	2.3E+02
124.553	5	3	5.2E+02
127.782	5	3	4.6E+02
135.202	3	1	2.93E+02
136.160	5	3	1.95E+02
136.393	3	3	1.14E+02
141.948	5	5	3.87E+02
142.589	1	3	1.35E+02
144.405	5	5	9.4E+01
146.067	7	5	2.6E+02
154.133	3	1	1.63E+02
156.54	3	1	1.44E+02
158.469	5	5	1.4E+02
159.62	5	3	1.03E+02
163.049	3	1	6.2E+02
186.863	5	5	2.66E+02
207.73	3	1	1.07E+02

Ti XVIII

λ Å	Weights g_i	g_k	A 10^8 s^{-1}
17.22	2	4	7.3E+04
17.365	4	6	8.6E+04
17.39	4	4	1.4E+04
133.852	2	4	5.2E+01
144.759	4	4	3.2E+02
150.15	6	4	1.15E+02
153.15	4	2	1.97E+02
153.23	2	4	6.7E+01
159.00	4	4	1.16E+02
166.225	6	4	1.54E+02
179.902	2	4	6.3E+01
189.663	6	6	9.6E+01
191.23	4	4	6.6E+01
197.838	4	6	4.56E+01
208.07	4	4	1.2E+02

Ti XIX

λ Å	Weights g_i	g_k	A 10^8 s^{-1}
15.67	3	1	3.3E+04
15.68	5	5	2.7E+04
15.74	5	7	2.7E+04
15.75	3	5	2.4E+04
15.83	1	3	3.2E+04
15.86	1	3	2.9E+04
16.02	3	1	3.1E+04
16.18	3	5	3.8E+04
16.41	1	3	6.1E+04
16.43	3	5	8.2E+04
16.46	3	3	4.4E+04
16.51	5	7	1.0E+05
16.55	5	5	2.7E+04
16.61	3	1	8.0E+04
16.64	3	3	5.3E+04
16.69	1	3	1.02E+05
16.71	3	5	7.3E+04
16.72	5	3	3.3E+04
16.72	5	5	7.3E+04
16.74	5	7	1.2E+05
16.77	3	3	2.6E+04
16.80	5	7	1.81E+05
16.85	3	5	4.4E+04
17.08	3	5	8.3E+04
17.36	1	3	9.5E+04

Ti XX

λ Å	Weights g_i	g_k	A 10^8 s^{-1}
2.629	2	4	4.9E+04
2.6295	4	4	3.2E+06
2.631	2	2	6.1E+05
2.6319	2	4	1.5E+06
2.632	2	2	2.7E+06
2.6355	4	6	1.2E+06
8.621	4	2	1.1E+06
9.788	4	6	5.26E+03
10.046	2	4	7.29E+03
10.109	4	6	8.6E+03
*10.278	2	6	8.4E+03
10.620	2	4	1.34E+04
10.690	4	6	1.58E+04
*11.452	2	6	1.7E+04
11.872	2	4	2.8E+04
11.958	4	6	3.4E+04
11.958	4	4	5.6E+03
15.211	2	4	3.50E+04
15.253	2	2	3.58E+04
15.907	2	4	8.84E+04
16.049	4	6	1.05E+05
16.067	4	4	1.8E+04
31.586	4	6	5.49E+03
45.650	2	4	9.6E+03
45.996	4	6	1.1E+04

Ti XXI

λ Å	Weights g_i	g_k	A 10^8 s^{-1}
2.0633	1	3	1.32E+05
2.1108	1	3	2.60E+05
2.2211	1	3	6.35E+05
2.497	3	1	2.4E+06
2.505	5	5	3.5E+05
2.505	1	3	1.4E+06
2.507	3	5	1.4E+06
2.508	3	5	7.9E+05
2.510	3	3	6.9E+05
2.510	1	3	9.6E+05
2.511	3	3	1.4E+06
2.512	5	5	1.8E+06
2.512	3	1	1.4E+06
2.513	3	1	2.7E+06
2.513	3	5	2.4E+06

λ Å	Weights g_i	g_k	A 10^8 s^{-1}
2.514	5	3	1.2E+06
2.520	3	5	2.6E+05
2.527	3	1	1.2E+05
2.539	3	1	4.1E+05
2.6102	1	3	2.40E+06
2.6227	1	3	1.12E+05

Tungsten

W I

λ Å	Weights g_i	g_k	A 10^8 s^{-1}
2879.4	1	3	2.4E-01
2911.0	1	3	7.7E-02
2923.5	7	9	1.54E-02
2935.0	3	5	1.5E-01
3013.8	7	9	6.4E-02
3016.5	9	11	9.27E-02
3017.4	7	9	1.21E-01
3024.9	3	3	1.4E-01
3046.4	3	5	5.8E-02
3049.7	7	5	1.7E-01
3064.9	5	7	1.1E-02
3084.9	5	5	1.3E-02
3093.5	7	9	4.4E-02
3107.2	5	7	2.33E-02
3108.0	7	9	1.58E-02
3145.5	9	9	4.8E-03
3170.2	7	5	6.0E-03
3176.6	3	5	2.12E-02
3183.5	7	7	2.64E-03
3184.4	5	3	2.3E-02
3191.6	1	3	3.2E-02
3198.8	7	9	4.6E-02
3207.3	7	9	3.0E-02
3208.3	5	5	4.4E-02
3215.6	9	11	2.1E-01
3221.9	5	7	1.61E-02
3223.1	5	3	3.53E-03
3232.5	9	9	2.4E-02
3235.1	7	5	2.68E-03
3259.7	7	7	1.3E-02
3300.8	7	9	8.1E-02
3311.4	7	5	5.6E-02
3363.3	9	7	6.6E-03
3371.0	7	5	1.0E-02
3371.4	3	3	6.7E-03
3386.1	7	7	2.64E-03
3413.0	7	9	9.7E-03
3459.5	9	9	2.04E-03
3510.0	7	9	5.2E-03
3545.2	1	3	3.2E-02
3570.6	5	3	6.7E-03
3606.1	3	5	9.6E-03
3617.5	7	7	1.1E-01
3631.9	3	5	1.3E-02
3675.6	9	11	1.20E-02
3682.1	9	11	2.0E-02
3707.9	7	7	2.9E-02
3757.9	7	9	1.38E-02
3760.1	5	7	1.99E-02
3768.5	3	3	3.47E-02
3780.8	7	5	4.2E-02
3809.2	7	5	9.0E-03
3817.5	7	7	3.1E-02
3829.1	3	3	3.83E-03
3835.1	5	5	5.2E-02
3846.3	3	5	2.14E-02
3847.5	1	3	8.3E-03
3864.3	5	5	5.6E-03
3868.0	7	9	4.6E-02
3881.4	7	7	3.6E-02
3968.5	1	3	5.07E-03
3975.5	9	11	4.1E-03
4001.4	9	9	5.6E-03
4008.8	7	9	1.63E-01
4019.3	5	3	6.7E-03
4028.8	1	3	2.0E-02
4045.6	7	5	2.88E-02
4055.2	7	9	1.79E-03
4070.0	7	5	3.60E-02
4070.6	3	5	5.6E-03
4074.4	7	7	1.0E-01
4088.3	5	3	4.13E-03
4102.7	9	7	4.9E-02
4115.6	11	11	4.8E-03
4137.5	5	7	8.4E-03
4171.2	7	9	8.6E-03
4203.8	9	7	4.9E-03
4219.4	9	7	6.1E-03
4244.4	9	11	1.38E-02
4269.4	7	5	3.04E-02
4283.8	9	7	1.69E-03
4294.6	7	5	1.2E-01
4302.1	7	7	3.6E-02
4355.2	9	9	5.1E-03
4361.8	9	7	1.64E-03
4378.5	7	5	3.48E-03
4458.1	3	5	4.2E-03
4466.3	7	5	1.5E-02
4472.5	13	11	1.55E-03
4484.2	3	5	5.6E-03
4492.3	9	11	3.6E-03
4495.3	11	11	3.3E-03
4504.8	9	7	7.0E-03
4552.5	9	9	1.42E-03
4586.8	1	3	4.20E-03
4592.6	7	9	3.4E-03
4609.9	7	9	1.42E-02
4613.3	9	9	2.9E-03
4634.8	9	9	8.8E-03
4659.9	1	3	1.0E-02
4680.5	7	7	1.4E-02
4720.4	3	5	3.22E-03
4729.6	7	5	7.8E-03
4752.6	3	3	5.20E-03
4757.5	7	5	2.72E-03
4757.8	11	9	4.1E-03
4788.4	9	11	2.6E-03
4843.8	5	5	1.9E-02
4886.9	9	11	8.1E-03
4924.6	13	11	1.75E-03
4931.6	7	5	1.0E-02
4948.6	9	11	1.36E-03
4972.6	9	11	3.9E-03
4982.6	1	3	4.17E-03
4986.9	11	9	6.3E-03
5006.2	9	7	1.2E-02
5015.3	7	9	5.4E-03
5040.4	3	5	5.2E-03
5053.3	3	3	1.9E-02
5071.5	13	11	3.4E-03
5117.6	11	11	1.61E-03
5124.2	5	5	4.0E-03
5141.2	7	9	1.12E-03
5224.7	7	5	1.2E-02
5243.0	9	7	1.1E-02
5254.5	7	5	3.86E-03
5268.6	9	9	1.4E-03
5500.5	11	9	6.9E-03
5514.7	5	3	7.3E-03
5537.7	9	11	2.2E-03
5617.1	7	7	1.47E-03
5631.9	9	7	1.43E-03
5660.7	13	11	6.8E-03
5675.4	5	5	2.20E-03
5796.5	9	7	2.21E-03
5891.6	7	7	1.47E-03
5947.6	5	7	2.40E-03
5965.9	7	5	1.0E-02
6021.5	5	3	8.7E-03
6081.4	5	3	4.7E-03
6203.5	7	7	3.0E-03
6285.9	7	5	6.6E-03
6292.0	3	5	2.26E-03
6303.2	9	9	1.84E-03
6404.2	5	7	1.50E-03
6439.7	9	9	1.29E-03
6445.1	7	5	6.4E-03
6532.4	3	5	4.6E-03
6538.1	11	9	2.7E-03
6563.2	5	5	2.04E-03
6814.9	9	9	1.46E-03
7285.8	13	11	1.47E-03
7569.9	5	3	3.73E-03
7664.9	5	3	3.80E-03
8017.2	5	7	1.6E-03
8358.7	5	7	1.89E-03
9381.4	9	7	1.53E-03

Uranium

U I

λ Å	Weights g_i	g_k	A 10^8 s^{-1}
3553.0	13	13	2.0E-02
3553.0	9	7	1.4E-02
3553.4	15	13	2.2E-02
3554.5	11	9	8.4E-03
3554.9	15	17	7.9E-03
3555.3	13	15	2.7E-02
3555.8	13	11	4.1E-03
3556.9	13	11	7.5E-03

λ Å	Weights g_i	Weights g_k	A 10^8 s^{-1}
3557.8	13	13	2.9E-02
3558.0	11	13	1.6E-02
3558.6	9	7	3.9E-02
3559.4	7	9	1.5E-02
3560.3	9	7	6.4E-02
3561.4	15	13	5.5E-02
3561.5	9	9	2.5E-02
3561.8	13	11	5.7E-02
3563.7	13	13	2.9E-02
3563.8	7	7	1.1E-02
3565.0	13	11	2.9E-02
3566.0	13	15	1.7E-02
3566.6	11	11	2.4E-01
3568.8	13	13	3.8E-02
3569.1	17	15	1.1E-01
3569.4	9	9	1.5E-02
3570.1	13	11	1.3E-02
3570.2	11	9	5.3E-03
3570.6	13	15	2.7E-02
3570.7	15	15	1.2E-02
3571.2	11	11	6.3E-03
3571.6	17	15	1.3E-01
3572.9	13	15	1.5E-02
3573.9	13	11	4.0E-02
3574.1	13	15	3.5E-02
3574.8	13	15	1.9E-02
3577.1	17	15	4.3E-02
3577.5	15	13	7.8E-03
3577.8	11	11	8.3E-03
3577.9	13	13	2.3E-02
3578.3	13	11	2.0E-02
3580.0	9	9	1.2E-02
3580.2	11	9	2.9E-02
3580.4	11	13	7.5E-03
3580.9	13	13	2.1E-02
3582.6	13	13	2.9E-02
3584.6	7	5	2.4E-02
3584.9	13	15	1.8E-01
3585.4	11	11	1.9E-02
3585.8	11	9	2.8E-02
3587.8	9	11	1.3E-02
3588.3	7	9	1.8E-02
3589.7	11	13	2.1E-02
3589.8	15	13	5.9E-02
3590.7	9	7	2.2E-02
3591.7	11	9	5.3E-02
3593.0	11	11	1.4E-02
3593.2	13	15	4.2E-02
3593.7	11	11	7.2E-02
Vanadium			
V I			
3043.12	6	8	2.3E-01
3050.39	10	8	5.3E-01
3053.65	4	4	1.3E+00
3056.33	6	6	1.3E+00
3060.46	8	8	1.4E+00
3066.37	10	10	2.1E+00
3066.53	6	4	3.2E-01
3075.93	4	6	2.8E-01
3080.33	2	4	2.7E-01
3083.54	6	8	2.5E-01
3087.06	2	2	9.2E-01
3088.11	4	6	4.9E-01
3089.13	4	4	5.3E-01
3093.79	6	6	4.1E-01
3094.69	2	4	4.3E-01
3112.92	4	2	5.0E-01
3183.41	6	8	2.4E+00
3183.96	8	10	2.5E+00
3183.98	4	6	2.4E+00
3185.38	10	12	2.7E+00
3198.01	6	6	3.9E-01
3202.39	8	8	4.0E-01
3205.58	8	10	1.3E+00
3207.41	10	10	2.6E-01
3212.43	10	12	1.4E+00
3218.87	8	6	3.5E-01
3233.19	10	8	3.2E-01
3273.03	8	8	2.7E-01
3284.36	10	10	2.8E-01
3309.18	4	4	3.2E-01
3329.85	6	4	7.7E-01
3356.35	4	6	3.1E-01
3365.55	2	4	4.8E-01
3376.05	4	4	3.2E-01
3377.39	4	2	9.0E-01
3377.62	6	6	6.0E-01
3397.58	6	4	2.3E-01
3400.39	8	8	2.5E-01
3529.73	4	6	4.1E-01
3533.68	6	8	5.2E-01
3533.76	2	4	3.7E-01
3543.49	2	2	6.7E-01
3545.33	4	4	3.7E-01
3553.27	6	6	2.2E-01
3555.14	4	2	2.6E-01
3663.60	4	6	3.1E+00
3667.74	6	8	2.7E+00
3672.41	12	12	9.2E-01
3673.41	8	10	2.7E+00
3676.70	14	14	1.3E+00
3680.12	10	12	2.2E+00
3686.26	10	12	2.3E-01
3687.50	12	14	2.9E+00
3688.07	8	8	3.5E-01
3690.28	2	4	4.5E-01
3692.22	6	6	5.4E-01
3695.34	14	16	2.8E+00
3695.86	4	4	6.6E-01
3703.57	10	8	9.2E-01
3704.70	8	6	6.6E-01
3705.04	6	4	3.6E-01
3706.03	10	10	5.2E-01
3708.71	12	12	4.4E-01
3790.46	10	8	2.3E-01
3794.96	10	10	2.3E-01
3806.79	10	10	2.5E-01
3818.24	4	2	6.73E-01
3828.56	6	4	5.33E-01
3840.75	8	6	5.48E-01
3855.36	4	4	3.30E-01
3855.85	10	8	5.78E-01
3863.86	8	6	3.1E-01
3864.86	6	6	2.70E-01
3871.07	10	8	2.8E-01
3875.07	8	8	2.36E-01
3902.26	10	10	2.68E-01
3921.86	4	2	2.7E-01
3922.43	6	6	2.6E-01
3930.02	10	10	3.3E-01
3934.01	8	8	6.2E-01
3992.80	12	10	1.2E+00
3998.73	14	12	1.0E+00
4050.96	10	10	1.4E+00
4051.35	12	12	1.3E+00
4090.57	8	10	8.5E-01
4092.68	8	10	2.30E-01
4095.48	6	8	7.2E-01
4099.78	6	8	4.10E-01
4102.15	4	6	7.1E-01
4104.77	10	8	2.1E+00
4105.16	4	6	4.9E-01
4109.78	2	4	5.00E-01
4111.78	10	10	1.01E+00
4115.18	8	8	5.80E-01
4116.47	6	6	3.2E-01
4116.59	2	2	2.90E-01
4123.50	4	2	1.00E+00
4128.06	6	4	7.70E-01
4131.99	8	6	5.5E-01
4134.49	10	8	2.90E-01
4232.46	10	10	9.8E-01
4232.95	8	8	7.7E-01
4268.64	14	14	1.2E+00
4271.55	12	12	9.6E-01
4276.95	10	10	9.4E-01
4284.05	8	8	1.2E+00
4291.82	12	14	8.8E-01
4296.10	10	12	7.7E-01
4297.67	8	10	7.0E-01
4298.03	6	8	7.8E-01
4379.23	10	12	1.1E+00
4384.71	8	10	1.1E+00
4389.98	6	8	6.9E-01
4395.22	4	6	5.5E-01
4400.57	2	4	3.4E-01
4406.64	10	10	2.2E-01
4407.63	8	8	4.4E-01
4408.20	6	6	6.0E-01
4416.47	4	2	2.6E-01
4452.01	14	16	9.2E-01
4457.75	10	12	2.7E-01
4460.33	10	8	3.0E-01
4462.36	12	14	7.6E-01
4468.00	8	10	2.3E-01
4469.71	10	12	6.2E-01

λ Å	Weights g_i	Weights g_k	A 10^8 s^{-1}
4474.04	10	8	4.7E-01
4496.06	8	6	4.0E-01
4514.18	6	4	3.3E-01
4524.21	12	10	3.0E-01
4525.17	4	2	4.1E-01
4529.58	10	8	2.4E-01
4545.40	10	12	7.6E-01
4560.72	8	10	7.0E-01
4571.79	6	8	6.0E-01
4578.73	4	6	6.8E-01
4706.16	6	4	2.4E-01
4757.47	4	2	7.6E-01
4766.62	6	4	5.6E-01
4776.36	8	6	5.1E-01
4786.50	10	8	4.7E-01
4796.92	12	10	4.8E-01
4807.52	14	12	5.8E-01
5193.00	12	12	4.0E-01
5195.39	8	8	2.3E-01
5234.08	10	10	4.9E-01
5240.87	12	12	4.3E-01
5415.25	12	14	3.1E-01
5487.91	12	10	2.9E-01
5507.75	10	8	3.5E-01
6090.21	8	6	2.60E-01

V II

λ Å	Weights g_i	Weights g_k	A 10^8 s^{-1}
2527.90	13	13	6.1E-01
2528.47	9	9	5.2E-01
2528.83	11	11	5.3E-01
2554.04	9	9	5.4E-01
2589.10	9	9	7.7E-01
2640.86	5	7	1.2E+00
2677.80	3	5	3.4E-01
2679.33	7	7	3.4E-01
2683.09	1	3	3.4E-01
2687.96	9	9	7.6E-01
2689.88	3	1	9.2E-01
2690.25	7	5	3.4E-01
2690.79	5	3	5.2E-01
2700.94	9	11	3.5E-01
2706.17	7	9	3.4E-01
2734.22	9	7	6.2E-01
2753.41	13	11	4.2E-01
2784.20	9	9	1.3E+00
2787.91	7	9	5.0E-01
2825.86	9	7	1.2E+00
2843.82	7	5	9.9E-01
2847.57	9	7	4.6E-01
2854.34	11	9	5.0E-01
2862.31	11	11	3.6E-01
2868.11	5	3	2.1E+00
2869.13	13	11	4.8E-01
2882.49	5	5	4.2E-01
2884.78	3	3	5.6E-01
2889.61	3	1	1.9E+00
2891.64	5	3	1.4E+00
2892.43	9	9	3.6E-01
2892.65	7	5	1.3E+00
2893.31	9	7	1.2E+00
2903.07	3	5	3.4E-01
2906.45	7	7	7.8E-01
2908.81	11	9	1.6E+00
2910.01	5	5	1.1E+00
2910.38	3	3	1.2E+00
2911.05	7	9	3.7E-01
2912.46	11	9	5.0E-01
2915.88	9	7	4.9E-01
2924.02	11	11	1.7E+00
2924.63	9	9	1.2E+00
2930.80	7	7	5.8E-01
2941.37	11	9	3.5E-01
2944.57	9	7	7.6E-01
2948.08	9	11	4.0E-01
2952.07	7	5	7.2E-01
2955.58	7	9	3.3E-01
2968.37	7	9	7.0E-01
2972.26	5	7	5.2E-01
2973.98	9	11	3.5E-01
2985.18	7	9	4.4E-01
3001.20	7	7	7.5E-01
3014.82	5	3	8.9E-01
3016.78	7	5	5.0E-01
3020.21	9	7	5.0E-01
3048.21	11	13	7.0E-01
3063.25	9	11	1.0E+00
3100.94	7	7	5.8E-01
3113.56	11	11	5.0E-01
3122.89	11	13	7.6E-01
3134.93	13	13	5.9E-01
3136.50	11	11	5.3E-01
3139.73	9	9	5.2E-01
3151.32	3	5	4.4E-01
3190.69	9	9	3.3E-01
3250.78	11	9	5.2E-01
3251.87	5	7	3.5E-01
3271.12	7	9	6.9E-01
3276.12	9	11	5.2E-01
3279.84	9	11	5.8E-01
3287.71	5	7	7.5E-01
3337.85	5	7	5.3E-01
3517.30	9	7	3.8E-01
3530.77	5	3	4.5E-01
3545.19	7	5	4.3E-01
3556.80	9	7	5.1E-01
3592.01	7	5	4.4E-01
3618.92	3	5	3.3E-01

V III

λ Å	Weights g_i	Weights g_k	A 10^8 s^{-1}
2318.06	8	10	4.6E+00
2323.82	6	8	3.8E+00
2330.42	10	10	3.2E+00
2331.75	8	8	2.5E+00
2334.21	6	6	2.2E+00
2337.13	4	4	2.7E+00
2343.10	6	8	3.6E+00
2358.73	6	8	4.2E+00
2366.31	8	10	4.2E+00
2371.06	10	12	5.2E+00
2373.06	4	6	2.9E+00
2382.46	8	10	5.0E+00
2393.58	6	8	4.3E+00
2404.18	4	6	2.5E+00
2516.14	10	10	3.7E+00
2521.55	8	8	3.5E+00
2548.21	6	4	2.0E+00
2554.22	8	6	1.2E+00
2593.05	6	6	2.8E+00
2595.10	8	8	2.8E+00

V IV

λ Å	Weights g_i	Weights g_k	A 10^8 s^{-1}
677.345	9	9	6.7E+00
680.632	9	7	1.2E+01
681.145	7	5	1.1E+01
682.455	7	7	6.5E+00
682.923	5	5	6.9E+00
684.450	7	5	7.7E+00
691.530	5	3	1.1E+01
723.537	3	1	1.5E+01
724.068	5	5	1.1E+01
724.809	5	3	5.6E+00
737.854	9	7	2.4E+01
750.110	5	5	1.0E+01
884.146	1	3	4.7E+00
1071.05	5	5	6.1E+00
1110.72	3	3	5.0E+00
1112.20	7	7	6.3E+00
1112.44	5	5	5.0E+00
1127.84	7	5	8.9E+00
1131.26	9	7	9.4E+00
1194.46	7	5	1.0E+01
1226.52	5	5	1.5E+01
1243.72	3	1	9.4E+00
1247.07	5	3	4.7E+00
1272.97	3	1	2.7E+01
1304.17	3	5	1.5E+01
1305.42	5	7	7.0E+00
1308.06	7	9	7.9E+00
1309.50	5	5	8.7E+00
1312.72	7	7	8.6E+00
1317.57	5	7	8.7E+00
1321.92	7	9	9.9E+00
1326.81	3	5	4.0E+00
1329.29	5	5	1.5E+01
1329.97	3	3	4.8E+00
1330.36	1	3	6.0E+00
1331.67	3	1	1.7E+01
1332.46	5	3	7.5E+00
1334.49	9	9	8.3E+00
1355.13	7	9	2.5E+01
1356.53	5	3	4.9E+00
1395.00	5	7	1.4E+01
1400.42	5	7	7.5E+00
1403.62	7	9	8.4E+00
1412.69	3	3	1.1E+01
1414.41	5	7	1.2E+01
1414.84	5	5	4.6E+00

λ Å	Weights g_i	Weights g_k	A 10^8 s^{-1}
1418.53	7	7	5.2E+00
1419.58	7	9	1.3E+01
1423.72	3	5	7.1E+00
1426.65	9	11	2.2E+01
1429.11	5	5	5.0E+00
1434.84	7	7	5.4E+00
1451.04	3	3	7.0E+00
1454.00	5	3	1.1E+01
1520.14	5	7	7.2E+00
1522.49	3	5	5.5E+00
1601.92	3	3	1.2E+01
1611.88	7	7	5.2E+00
1806.18	5	3	7.3E+00
1809.85	3	1	7.2E+00
1817.68	5	3	4.8E+00
1825.84	7	5	5.3E+00
1861.56	5	7	6.6E+00
1939.07	7	9	5.8E+00
1951.43	5	7	5.0E+00
1963.10	3	5	4.8E+00
1997.72	7	7	4.7E+00
2084.43	5	5	4.0E+00
2120.05	7	9	8.1E+00
2141.20	3	5	7.0E+00
2146.83	7	9	6.6E+00
2149.85	5	7	5.1E+00
2151.09	7	9	4.3E+00
2155.34	11	13	1.2E+01
2446.80	9	11	5.3E+00
2570.72	9	11	7.6E+00
3284.56	7	9	5.3E+00
3496.42	7	9	4.4E+00
3514.25	9	11	4.7E+00
Xenon			
Xe I			
1043.8	1	3	5.9E-01
1047.1	1	3	1.3E+00
1050.1	1	3	8.5E-02
1056.1	1	3	2.45E+00
1061.2	1	3	1.9E-01
1068.2	1	3	3.99E+00
1085.4	1	3	4.10E-01
1099.7	1	3	4.34E-01
1110.7	1	3	1.5E+00
1129.3	1	3	4.4E-02
1170.4	1	3	1.6E+00
1192.0	1	3	6.2E+00
1250.2	1	3	1.4E-01
1295.6	1	3	2.46E+00
1469.6	1	3	2.81E+00
4501.0	5	3	6.2E-03
4524.7	5	5	2.1E-03
4624.3	5	5	7.2E-03
4671.2	5	7	1.0E-02
4807.0	3	1	2.4E-02
7119.6	7	9	6.6E-02
7967.3	1	3	3.0E-03
8409.2	5	3	1.0E-02
Xe II			
4180.1	4	4	2.2E+00
4330.5	6	8	1.4E+00
4414.8	6	6	1.0E+00
4603.0	4	4	8.2E-01
4844.3	6	8	1.1E+00
4876.5	6	8	6.3E-01
5260.4	2	4	2.2E-01
5262.0	4	4	8.5E-01
5292.2	6	6	8.9E-01
5372.4	4	2	7.1E-01
5419.2	4	6	6.2E-01
5439.0	4	2	7.4E-01
5472.6	8	8	9.9E-02
5531.1	8	6	8.8E-02
5719.6	4	6	6.1E-02
5976.5	4	4	2.8E-01
6036.2	6	6	7.5E-02
6051.2	8	6	1.7E-01
6097.6	6	4	2.6E-01
6270.8	4	6	1.8E-01
6277.5	4	6	3.6E-02
6805.7	8	6	6.1E-02
6990.9	10	8	2.7E-01
Ytterbium			
Yb I			
2464.5	1	3	9.1E-01
2672.0	1	3	1.18E-01
3464.4	1	3	6.2E-01
3988.0	1	3	1.76E+00
5556.5	1	3	1.14E-02
Yb II			
3289.4	2	4	1.8E+00
3694.2	2	2	1.4E+00
Yttrium			
Y I			
2948.41	4	4	3.5E-01
2974.59	4	6	3.5E-01
2984.25	6	8	4.8E-01
2995.26	6	4	5.1E-02
2996.94	4	6	8.4E-02
3005.26	4	4	4.8E-02
3022.28	6	6	6.6E-02
3045.36	6	6	1.07E-01
3053.95	6	4	1.9E-03
3155.65	4	6	2.7E-03
3172.84	4	4	9.9E-03
3185.96	6	8	1.2E-03
3209.38	6	6	3.0E-03
3227.16	6	4	1.10E-03
3484.05	4	6	1.2E-02
3549.66	6	6	1.0E-03
3552.69	4	4	2.3E-01
4077.36	4	6	1.1E+00
4083.71	4	4	2.5E-01
4102.36	6	8	1.3E+00
4128.30	6	6	1.6E+00
4142.84	4	4	1.6E+00
4167.51	6	6	2.38E-01
4235.93	6	4	3.0E-01
4352.40	4	4	6.7E-03
4379.33	6	4	7.83E-01
4385.47	4	4	6.9E-02
4394.01	8	8	1.9E-02
4409.70	4	6	2.7E-03
4417.43	10	8	3.2E-02
4437.34	6	6	8.64E-02
4443.65	10	8	1.1E-01
4459.01	4	6	1.8E-02
4476.95	8	6	2.8E-01
4491.74	10	10	2.3E-02
4514.01	4	6	3.34E-01
4527.78	8	6	8.33E-01
4534.09	6	8	4.4E-02
4544.31	6	6	4.10E-01
4559.36	2	4	4.0E-01
4581.33	6	4	1.5E-01
4613.00	6	4	1.8E-01
4643.70	4	6	1.8E-01
4653.78	4	6	1.6E-01
4674.85	6	8	1.3E-01
4725.84	4	4	1.5E-01
4762.96	6	4	4.2E-02
4780.16	2	4	8.9E-02
4781.03	8	10	1.0E-01
4799.30	6	8	1.6E-01
4804.31	6	4	2.6E-01
4804.80	4	4	3.84E-01
4821.63	6	6	1.0E-01
4845.67	8	8	6.8E-01
4852.68	6	6	6.2E-01
4856.71	6	6	2.0E-01
4859.84	4	4	7.26E-01
4893.44	6	4	2.2E-01
4900.08	8	6	2.0E-01
4906.11	10	8	1.2E-01
4950.01	8	6	2.0E-02
4963.49	4	4	1.4E-02
4981.97	4	6	4.7E-03
5004.44	6	4	1.2E-02
5205.01	4	4	8.4E-03
5258.47	6	6	2.9E-03
5271.82	8	6	1.1E-02
5380.63	6	4	3.2E-01
5381.24	4	4	9.9E-03
5388.39	6	8	1.1E-02
5390.81	8	6	2.9E-02
5401.88	6	8	6.0E-03
5424.36	6	4	3.47E-01
5466.24	4	4	1.0E-01
5466.47	10	12	6.3E-01
5469.10	4	6	3.6E-03
5513.65	6	6	2.39E-01
5519.88	4	6	1.2E-02
5526.43	6	4	3.9E-03

λ Å	Weights g_i	g_k	A 10^8 s^{-1}
5527.56	8	10	5.4E-01
5541.63	8	8	5.2E-02
5551.00	4	4	6.9E-02
5573.03	6	4	1.8E-02
5594.12	6	8	5.0E-02
5606.34	10	10	5.84E-02
5619.96	6	4	2.0E-02
5630.14	4	6	4.9E-01
5641.78	2	4	1.9E-02
5675.27	6	6	9.3E-02
5675.64	4	6	4.3E-02
5693.63	4	4	1.1E-01
5714.94	8	6	2.0E-02
5729.25	6	6	2.2E-03
5732.09	6	6	7.5E-02
5740.22	8	6	4.0E-02
5757.59	4	6	7.6E-03
5788.36	4	4	9.4E-03
5844.13	6	4	5.6E-03
5879.93	4	2	8.5E-02
5902.91	6	8	4.0E-02
6087.94	6	4	1.1E-01
6191.72	4	4	4.7E-02
6222.58	4	6	5.9E-03
6402.01	6	4	2.7E-03
6435.02	6	6	4.0E-02
6437.17	10	8	4.8E-02
6538.57	10	10	1.5E-01
6622.48	8	6	4.5E-03
6815.15	2	4	7.18E-02
7009.89	2	4	4.4E-02
7035.15	4	4	6.3E-02
Y II			
3112.03	1	3	1.3E-02
3179.42	3	5	3.8E-02
3195.62	3	3	8.23E-01
3200.27	5	5	4.8E-01
3203.32	3	1	2.77E+00
3216.69	5	3	2.0E+00
3242.28	7	5	2.0E+00
3448.81	5	5	4.1E-02
3467.88	5	3	2.7E-02
3496.08	1	3	3.49E-01
3549.01	5	7	3.97E-01
3584.51	3	5	4.02E-01
3600.74	7	7	1.4E+00
3601.91	3	3	1.13E+00
3611.04	5	5	1.04E+00
3628.70	5	3	3.3E-01
3664.62	7	5	3.7E-01
3710.29	7	9	1.5E+00
3747.55	3	3	1.9E-01
3774.34	5	7	1.1E+00
3776.56	5	3	2.42E-01
3788.70	3	5	8.1E-01
3818.34	5	5	9.70E-02
3832.90	7	7	3.0E-01
3878.29	7	5	2.9E-02
3930.66	5	5	2.1E-02
3950.36	3	5	2.80E-01
3951.59	5	3	1.5E-02
3982.60	5	5	2.7E-01
4124.91	5	7	1.8E-02
4177.54	5	5	5.27E-01
4199.27	3	5	5.36E-03
4204.69	1	3	2.20E-02
4235.73	5	5	2.3E-02
4309.62	7	5	1.29E-01
4358.73	3	3	5.55E-02
4374.95	5	5	9.97E-01
4398.01	5	3	1.16E-01
4422.59	3	1	1.83E-01
4682.33	5	5	1.9E-02
4786.58	7	7	2.1E-02
4823.31	5	5	4.3E-02
4854.87	5	3	3.9E-01
4881.44	5	3	1.5E-03
4883.69	9	7	4.7E-01
4900.11	7	5	4.51E-01
4982.13	7	9	1.5E-02
5087.42	9	9	2.0E-01
5119.11	5	7	1.6E-02
5200.41	5	5	1.3E-01
5205.73	7	7	1.6E-01
5289.82	7	5	6.7E-03
5320.78	9	7	3.9E-03
5473.39	3	5	4.3E-02
5480.73	1	3	7.62E-02
5497.41	5	5	1.2E-01
5509.90	5	5	4.24E-02
5544.61	3	1	1.8E-01
5546.01	5	3	5.8E-02
5728.89	5	5	3.0E-02
6613.74	5	7	1.7E-02
6832.48	5	5	3.3E-03
7264.16	5	3	1.3E-02
Zinc			
Zn I			
748.29	1	3	6.0E-02
765.60	1	3	7.6E-02
792.05	1	3	5.7E-02
793.85	1	3	1.8E-01
809.92	1	3	2.6E-01
1109.1	1	3	3.05E-01
2138.6	1	3	7.09E+00
3075.9	1	3	3.29E-04
3282.3	1	3	9.0E-01
3302.6	3	5	1.2E+00
3302.9	3	3	6.7E-01
3345.0	5	7	1.7E+00
3345.6	5	5	4.0E-01
3345.9	5	3	4.5E-02
6362.3	3	5	4.74E-01
11054	3	1	2.43E-01
Zn II			
2025.5	2	4	3.3E+00
2064.2	2	4	4.6E+00
2099.9	4	6	5.6E+00
2102.2	4	4	9.3E-01
4911.6	4	6	1.6E+00

ELECTRON AFFINITIES

Thomas M. Miller

Electron affinity is defined as the energy difference between the lowest (ground) state of the neutral and the lowest state of the corresponding negative ion. The accuracy of electron affinity measurements has been greatly improved since the advent of laser photodetachment experiments with negative ions. Electron affinities can be determined with optical precision, though a detailed understanding of atomic and molecular states and splittings is required to specify the photodetachment threshold corresponding to the electron affinity.

Atomic and molecular electron affinities are discussed in two excellent articles reviewing photodetachment studies which appear in *Gas Phase Ion Chemistry*, Vol. 3, Bowers, M. T., Ed., Academic Press, Orlando, 1984: Chapter 21 by Drzaic, P. S., Marks, J., and Brauman, J. I., "Electron Photodetachment from Gas Phase Negative Ions", p. 167, and Chapter 22 by Mead, R. D., Stevens, A. E., and Lineberger, W. C., "Photodetachment in Negative Ion Beams", p. 213. Persons interested in photodetachment details should consult these articles and the critical review of Hotop, H., and Lineberger, W. C., *J. Phys. Chem. Ref. Data*, 14, 731, 1985. For simplicity in the tables below, any electron affinity which was discussed in the articles by Drzaic *et al.* or Hotop and Lineberger is referenced to these sources, where original references are given. A great many additional electron affinities have been provided here by G. B. Ellison, W. C. Lineberger, H. Hotop, D. G. Leopold, and K. H. Bowen. Electron affinities for the lanthanides and actinides have not been measured, but theoretical estimates have been made by Bratch, S. G., *Chem. Phys. Lett.*, 98, 113, 1983, and Bratch, S. G., and Lagowski, J. J., *Chem. Phys. Lett.*, 107, 136, 1984. The development of cluster-ion photodetachment apparatuses has brought an explosion of electron affinity estimates for atomic and molecular clusters. [See Arnold, S. T., Eaton, J. G., Patel-Mistra, D., Sarkas, H. W., and Bowen, K. H., in *Ion and Cluster Ion Spectroscopy and Structure*, Maier, J. P., Ed., Elsevier Science, New York, 1989, p. 417.] The policy in this tabulation is to list the electron affinities for the atoms, diatoms, and triatoms, if adiabatic electron affinities have been determined, but to refer the reader to original sources for higher-order clusters. Additional data on molecular electron affinities may be found in Lias, S. G., Bartmess, J. E., Liebman, J. F., Holmes, J. L., Levin, R. D., and Mallard, W. G., *Gas Phase Ion and Neutral Thermochemistry*, *J. Phys. Chem. Ref. Data*, 17, (Supplement No. 1), 1988.

For the present tabulation the value $e/hc = 8065.5410 \pm 0.0024$ cm^{-1} eV^{-1} (Cohen, E. R., and Taylor, B. N., *Rev. Mod. Phys.*, 59, 1121, 1987) has been used to convert electron affinities from the units used in spectroscopic work, cm^{-1}, into eV for these tables. The uncertainties in the electron affinities given in the tables do not include the 0.3 ppm uncertainty in e/hc; only in a few cases, *e.g.*, atomic oxygen, is this additional uncertainty significant.

Abbreviations used in the tables: calc = calculated value; PT = photodetachment threshold using a lamp as a light source; LPT = laser photodetachment threshold; LPES = laser photoelectron spectroscopy; DA = dissociative attachment; e-scat = electron scattering; CT = charge transfer; and CD = collisional detachment.

Table 1
Atomic Electron Affinities

Atomic number	Atom	Electron affinity in eV	Uncertainty in eV	Method	Ref.
1	D	0.754593	0.000074	LPT	89
	H	0.754195	0.000019	LPT	89
		0.754209	0.000003	calc	1
2	He	not stable	—	calc	1
3	Li	0.6180	0.0005	LPT	1
4	Be	not stable	—	calc	1
5	B	0.277	0.010	LPES	1
6	C	1.2629	0.0003	LPT	1
7	N	not stable	—	DA	1
8	O	1.4611103	0.0000007	LPT	4
9	F	3.401190	0.000004	LPT	74
10	Ne	not stable	—	calc	1
11	Na	0.547926	0.000025	LPT	1
12	Mg	not stable	—	e-scat	1
13	Al	0.441	0.010	LPES	1
14	Si	1.385	0.005	LPES	1
15	P	0.7465	0.0003	LPT	1
16	S	2.077104	0.000001	LPT	1
17	Cl	3.61269	0.00006	LPT	52
18	Ar	not stable	—	calc	1
19	K	0.50147	0.00010	LPT	1
20	Ca	0.02455	0.00010	LPT	44
21	Sc	0.188	0.020	LPES	1
22	Ti	0.079	0.014	LPES	1
23	V	0.525	0.012	LPES	1
24	Cr	0.666	0.012	LPES	1
25	Mn	not stable	—	calc	1

Table 1
Atomic Electron Affinities (continued)

Atomic number	Atom	Electron affinity in eV	Uncertainty in eV	Method	Ref.
26	Fe	0.151	0.003	LPES	27
27	Co	0.662	0.003	LPES	27
28	Ni	1.156	0.010	LPES	1
29	Cu	1.235	0.005	LPES	37
30	Zn	not stable	—	e-scat	1
31	Ga	0.3	0.15	PT	1
32	Ge	1.233	0.003	LPES	28
33	As	0.81	0.03	PT	1
34	Se	2.020670	0.000025	LPT	1
35	Br	3.363590	0.000003	LPT	74
36	Kr	not stable	—	calc	1
37	Rb	0.48592	0.00002	LPT	1
38	Sr	0.048	0.006	LPT	122
39	Y	0.307	0.012	LPES	1
40	Zr	0.426	0.014	LPES	1
41	Nb	0.893	0.025	LPES	1
42	Mo	0.748	0.002	LPES	127
43	Tc	0.55	0.20	calc	1
44	Ru	1.05	0.15	calc	1
45	Rh	1.137	0.008	LPES	1
46	Pd	0.562	0.005	LPES	116
47	Ag	1.302	0.007	LPES	1
48	Cd	not stable	—	e-scat	1
49	In	0.3	0.2	PT	1
50	Sn	1.112	0.004	LPES	28
51	Sb	1.046	0.005	LPES	108
52	Te	1.9708	0.0003	LPT	1
53	I	3.059038	0.000010	LPT	92
54	Xe	not stable	—	calc	1
55	Cs	0.471626	0.000025	LPT	1
56	Ba	0.15	—	calc	57
57	La	0.5	0.3	calc	1
72	Hf	≈0	—	calc	1
73	Ta	0.322	0.012	LPES	1
74	W	0.815	0.002	LPES	37
75	Re	0.15	0.15	calc	1
76	Os	1.1	0.2	calc	1
77	Ir	1.5638	0.0005	LPT	141
78	Pt	2.128	0.002	LPT	1
79	Au	2.30863	0.00003	LPT	1
80	Hg	not stable	—	e-scat	1
81	Tl	0.2	0.2	PT	1
82	Pb	0.364	0.008	LPES	1
83	Bi	0.946	0.010	LPES	1
84	Po	1.9	0.3	calc	1
85	At	2.8	0.2	calc	1
86	Rn	not stable	—	calc	1
87	Fr	0.46	—	calc	82
118	ekaradon	0.056	—	calc	140

Table 2
Electron Affinities for Diatomic Molecules

Molecule	Electron affinity in eV	Uncertainty in eV	Method	Ref.
Ag_2	1.023	0.007	LPES	37
Al_2	1.10	0.15	LPES	68
AlO	2.60	0.02	LPES	143
AlS	2.60	0.03	LPES	129
As_2	0	—	PT	2
AsH	1.0	0.1	PT	2
Au_2	1.938	0.007	LPES	37
BO	3.12	0.09	PT	6
BeH	0.7	0.1	PT	2
Bi_2	1.271	0.008	LPES	119
Br_2	2.55	0.10	CT	2
BrO	2.353	0.006	LPES	88
C_2	3.269	0.006	LPES	87
CH	1.238	0.008	LPES	2
CN	3.862	0.004	LPES	111
CS	0.205	0.021	LPES	2
CaH	0.93	0.05	PT	2
Cl_2	2.38	0.10	CT	2
ClO	2.275	0.006	LPES	88
Co_2	1.110	0.008	LPES	27
CoD	0.680	0.010	LPES	29
CoH	0.671	0.010	LPES	29
Cr_2	0.505	0.005	LPES	114
CrD	0.568	0.010	LPES	29
CrH	0.563	0.010	LPES	29
CrO	1.221	0.006	LPES	5
Cs_2	0.469	0.015	LPES	104
CsCl	0.455	0.010	LPES	30
CsO	0.273	0.012	LPES	133
Cu_2	0.836	0.006	LPES	37
CuO	1.777	0.006	LPES	118
F_2	3.08	0.10	CT	2
FO	2.272	0.006	LPES	88
Fe_2	0.902	0.008	LPES	27
FeD	0.932	0.015	LPES	9
FeH	0.934	0.011	LPES	9
FeO	1.493	0.005	LPES	45
Ge_2	2.035	0.001	LPES	123
I_2	2.55	0.05	CT	2
IBr	2.55	0.10	CT	2
IO	2.378	0.006	LPES	88
InP	1.95	0.05	LPES	137
K_2	0.497	0.012	LPES	104
KBr	0.642	0.010	LPES	30
KCl	0.582	0.010	LPES	30
KCs	0.471	0.020	LPES	104
KI	0.728	0.010	LPES	30
KRb	0.486	0.020	LPES	104
LiCl	0.593	0.010	LPES	30
LiD	0.337	0.012	LPES	102
LiH	0.342	0.012	LPES	102
MgCl	1.589	0.011	LPES	31
MgH	1.05	0.06	PT	2
MgI	1.899	0.018	LPES	31
MnD	0.866	0.010	LPES	9
MnH	0.869	0.010	LPES	9
MoO	1.290	0.006	LPES	127
NH	0.370	0.004	LPT	32
NO	0.026	0.005	LPES	73
NS	1.194	0.011	LPES	2
Na_2	0.430	0.015	LPES	104
NaBr	0.788	0.010	LPES	30
NaCl	0.727	0.010	LPES	30
NaF	0.520	0.010	LPES	30
NaI	0.865	0.010	LPES	30
NaK	0.465	0.030	LPES	104
Ni_2	0.926	0.010	LPES	112
NiCu	0.889	0.010	LPES	128
NiAg	0.979	0.010	LPES	128
NiD	0.477	0.007	LPES	29
NiH	0.481	0.007	LPES	29
NiO	1.470	0.003	LPES	146
O_2	0.451	0.007	LPES	73
OD	1.825534	0.000037	LPT	142
OH	1.8276542	0.0000037	LPT	142
P_2	0.589	0.025	LPES	42
PH	1.028	0.010	LPES	2
PO	1.092	0.010	LPES	2
Pb_2	1.366	0.010	LPES	117
PbO	0.722	0.006	LPES	105
Pd_2	1.685	0.008	LPES	112
Pt_2	1.898	0.008	LPES	112
PtN	1.240	0.010	LPES	46
Rb_2	0.498	0.015	LPES	104
RbCl	0.544	0.010	LPES	30
RbCs	0.478	0.020	LPES	104
Re_2	1.571	0.008	LPES	33
S_2	1.670	0.015	LPES	53
SD	2.315	0.002	LPES	10
SF	2.285	0.006	LPES	93
SH	2.314344	0.000004	LPT	47
SO	1.125	0.005	LPES	84
Sb_2	1.282	0.008	LPES	108
Se_2	1.94	0.07	LPES	38
SeH	2.212520	0.000025	LPT	48
SeO	1.456	0.020	LPES	41
Si_2	2.201	0.010	LPES	100
SiH	1.277	0.009	LPES	2
Sn_2	1.962	0.010	LPES	117
SnPb	1.569	0.008	LPES	117
Te_2	1.92	0.07	LPES	38
TeH	2.102	0.015	LPES	39
TeO	1.697	0.022	LPES	40
ZnH	≤0.95	—	PT	2

Table 3
Electron Affinities for Triatomic Molecules

Molecule	Electron affinity in eV	Uncertainty in eV	Method	Ref.
Ag_3	2.32	0.05	LPES	37
Al_3	1.4	0.15	LPES	68
AlO_2	4.23	0.02	LPES	143
Al_2S	0.80	0.12	LPES	129
AsH_2	1.27	0.03	PT	2
Au_3	3.7	0.3	LPES	37
BO_2	3.57	0.13	CT	6
BO_2	4.3	0.2	CT	98
Bi_3	1.60	0.03	LPES	119
C_3	1.981	0.020	LPES	11
CCl_2	1.591	0.010	LPES	95
CD_2	0.645	0.006	LPES	12
CDF	0.535	0.005	LPES	95
CF_2	0.165	0.010	LPES	95
CH_2	0.652	0.006	LPES	12
CHBr	1.454	0.005	LPES	95
CHCl	1.210	0.005	LPES	95
CHF	0.542	0.005	LPES	95
CHI	1.42	0.17	LPES	95
C_2H	2.969	0.006	LPES	87
C_2O	1.848	0.027	LPES	13
COS	0.46	0.20	CD	2
CS_2	0.895	0.020	LPES	11
C_2Ti	1.542	0.020	LPES	147
CoD_2	1.465	0.013	LPES	34
CoH_2	1.450	0.014	LPES	34
CrH_2	>2.5	—	LPES	34
Cr_2D	1.464	0.005	LPES	107
Cr_2H	1.474	0.005	LPES	107
CrO_2	2.413	0.008	LPES	144
Cs_3	0.864	0.030	LPES	18
Cu_3	2.11	0.05	LPES	37
DCO	0.301	0.005	LPES	35
DNO	0.330	0.015	LPES	14
DO_2	1.089	0.017	LPES	15
DS_2	1.912	0.015	LPES	53
HS_2	1.907	0.015	LPES	53
Fe_3	1.47	0.08	LPES	149
FeCO	1.157	0.005	LPES	103
FeD_2	1.038	0.013	LPES	34
FeH_2	1.049	0.014	LPES	34
FeO_2	2.358	0.030	LPES	130
Ge_3	2.23	0.01	LPES	123
GeH_2	1.097	0.015	LPES	28
HCO	0.313	0.005	LPES	35
HCl_2	4.896	0.005	LPES	69
HNO	0.338	0.015	LPES	14
HO_2	1.078	0.017	LPES	15
InP_2	1.61	0.05	LPES	137
In_2P	2.36	0.05	LPES	137
K_3	0.956	0.050	LPES	18
MnD_2	0.465	0.014	LPES	34
MnH_2	0.444	0.016	LPES	34
N_3	2.70	0.12	PT	2
NCN	2.484	0.006	LPES	154
NCND	2.622	0.005	LPES	154
NCNH	2.622	0.005	LPES	154
NCO	3.609	0.005	LPES	111
NCS	3.537	0.005	LPES	111
NH_2	0.771	0.005	LPES	58
N_2O	0.22	0.10	CT	59
NO_2	2.273	0.005	LPES	63
(NO)R	R=Ar,Kr,Xe	—	LPES	90
Na_3	1.019	0.060	LPES	18
Ni_3	1.41	0.05	LPES	55
NiCO	0.804	0.012	LPES	2
NiD_2	1.926	0.007	LPES	34
NiH_2	1.934	0.008	LPES	34
O_3	2.1028	0.0025	LPT	2
O_2Ar	0.52	0.02	LPES	75
OClO	2.140	0.008	LPES	88
OIO	2.577	0.008	LPES	88
PH_2	1.271	0.010	LPES	2
PO_2	3.42	0.01	LPES	124
Pt_3	1.87	0.02	LPES	55
Pd_3	<1.5	0.1	LPES	55
Rb_3	0.920	0.030	LPES	18
S_3	2.093	0.025	LPES	16
SO_2	1.107	0.008	LPES	16
S_2O	1.877	0.008	LPES	16
Sb_3	1.85	0.03	LPES	108
SeO_2	1.823	0.050	LPES	38
SiH_2	1.124	0.020	LPES	2
VO_2	2.3	0.2	CT	101

Table 4
Electron Affinities for Larger Polyatomic Molecules

Molecule	Electron Affinity in eV	Uncertainty in eV	Method	Ref.	Name
Ag_n	n = 1-10	—	LPES	37	
Al_n	n = 3-32	—	LPES	68	
Al_nO_m	n=1,2	m=1-5	LPES	143	
Al_3O	1.00	0.15	LPES	68	
Al_nS_m	n = 1-5	m = 1-3	LPES	129	
$Ar(H_2O)_n$	n = 2,6,7	—	LPES	77	

Table 4
Electron Affinities for Larger Polyatomic Molecules (continued)

Molecule	Electron Affinity in eV	Uncertainty in eV	Method	Ref.	Name
Au_n	n = 2-5	—	LPES	37	
AuF_6	7.5	estimate	CT	98	
BD_3	0.027	0.014	LPES	62	
BH_3	0.038	0.015	LPES	62	
Bi_4	1.05	0.010	LPES	119	
$Br(CO_2)$	3.582	0.017	LPES	131	
C_n	n = 2-84	—	LPES	70	
$(CO_2)_n$	n = 1,2	—	LPES	75	
$(CS)_n$	n = 2	—	LPES	75	
$(CS_2)_n$	n = 1,2	—	LPES	75	
CF_3Br	0.91	0.2	CD	2	
CF_3I	1.57	0.2	CD	2	
CDO_2	3.510	0.015	LPES	109	
CHO_2	3.498	0.015	LPES	109	
CH_2S	0.465	0.023	LPES	53	
CH_3	0.08	0.03	LPES	2	
CH_3I	0.2	0.1	CT	2	
CH_3NO_2	0.48	0.10	CT	61	
CH_3Si	0.852	0.010	LPES	97	CH_3-Si
CH_3Si	2.010	0.010	LPES	97	$CH_2 = SiH$
CD_3O	1.552	0.022	LPES	2	
CH_3O	1.570	0.022	LPES	2	
CD_3S	1.856	0.006	LPT	2	
CH_3S	1.861	0.004	LPT	2	
CD_3S_2	1.748	0.022	LPES	53	
CH_3S_2	1.757	0.022	LPES	53	
CH_3SiH_2	1.19	0.04	LPT	65	
CFO_2	4.277	0.030	LPES	131	
CO_3	2.69	0.14	LPES	2	
$CO_3(H_2O)$	2.1	0.2	PT	2	
C_2F_2	2.255	0.006	LPES	106	difluorovinylidene
C_2DO	2.350	0.020	LPES	13	
C_2HO	2.350	0.020	LPES	13	
C_2D_2	0.492	0.006	LPES	83	vinylidene-d_2
C_2HD	0.489	0.006	LPES	83	vinylidene-d_1
C_2HF	1.718	0.006	LPES	106	monofluorovinylidene
C_2H_2	0.490	0.006	LPES	83	vinylidene
C_2H_2FO	2.22	0.09	PT	2	acetyl fluoride enolate
C_2D_2N	1.538	0.012	LPES	21	cyanomethyl-d_2 radical
C_2D_2N	1.070	0.024	LPES	21	isocyanomethyl-d_2 radical
C_2H_2N	1.543	0.014	LPES	21	cyanomethyl radical
C_2H_2N	1.059	0.024	LPES	21	isocyanomethyl radical
C_2H_3	0.667	0.024	LPES	90	vinyl
C_2D_3O	1.81897	0.00012	LPT	22	acetaldehyde-d_3 enolate
C_2H_3O	1.82476	0.00012	LPT	22	acetaldehyde enolate
C_2H_5N	0.56	0.01	PT	2	ethyl nitrine
C_2D_5O	1.702	0.033	LPES	23	ethoxide-d_5
C_2H_5O	1.726	0.033	LPES	23	ethoxide
C_2H_5S	1.953	0.006	LPT	2	ethyl sulfide
C_2H_5S	0.868	0.051	LPES	53	CH_3SCH_2
$C_2H_7O_2$	2.26	0.08	PT	50	MeOHOMe
C_3Fe	1.69	0.08	LPES	132	
C_3H	1.858	0.023	LPES	11	
C_3HFe	1.58	0.06	LPES	132	
C_3H_2	1.794	0.008	LPES	153	
$C_3H_2F_3O$	2.625	0.010	LPT	113	1,1,1-trifluoroacetone enolate

Table 4
Electron Affinities for Larger Polyatomic Molecules (continued)

Molecule	Electron Affinity in eV	Uncertainty in eV	Method	Ref.	Name
C_3H_3	0.893	0.025	LPES	24	propargyl radical
C_3H_2D	0.88	0.15	LPES	24	propargyl-d_1 radical
C_3D_2H	0.907	0.023	LPES	24	propargyl-d_2 radical
C_3H_3N	1.247	0.012	LPES	21	CH_3CH-CN
C_3D_5	0.464	0.006	LPES	138	allyl-d_5
C_3H_5	0.481	0.008	LPES	138	allyl
C_3H_4D	0.373	0.019	LPES	25	allyl-d_1
C_3H_5O	1.758	0.019	LPT	113	acetone enolate
C_3H_5O	1.621	0.006	LPT	113	propionaldehyde enolate
$C_3H_5O_2$	1.80	0.06	PT	2	methyl acetate enolate
C_3H_7O	1.789	0.033	LPES	23	*n*-propyl oxide
C_3H_7O	1.839	0.029	LPES	23	isopropyl oxide
C_3H_7S	2.00	0.02	PT	2	*n*-propyl sulfide
C_3H_7S	2.02	0.02	PT	2	isopropyl sulfide
C_3O	1.34	0.15	LPES	11	
C_3O_2	0.85	0.15	LPES	11	
C_3Ti	1.561	0.015	LPES	147	
$C_4F_4O_3$	0.5	0.2	CD	2	tetrafluorosuccinic anhydride
C_4Fe	≤2.2	0.2	LPES	132	
C_4HFe	1.67	0.06	LPES	132	
$C_4H_2O_3$	1.44	0.10	CT	61	maleic anhydride
C_4D_4	0.909	0.015	LPES	125	vinylvinylidene-d_4
C_4H_4	0.914	0.015	LPES	125	vinylvinylidene
C_4H_4N	2.39	0.13	PT	2	pyrrolate
C_4H_5O	1.801	0.008	LPT	113	cyclobutanone enolate
C_4H_6	0.431	0.006	LPES	135	trimethylenemethane
$C_4H_6O_2$	0.69	0.10	CT	61	2,3-butanedione
C_4H_6D	0.493	0.008	LPES	138	2-methylallyl-d_7
C_4H_7	0.505	0.006	LPES	138	2-methylallyl
C_4H_7O	1.67	0.05	PT	2	butyraldehyde enolate
C_4H_5DO	1.67	0.05	PT	2	2-butanone-3-d_1 enolate
$C_4H_5D_2O$	1.75	0.06	PT	2	2-butanone-3,3-d_2 enolate
C_4H_9O	1.912	0.054	LPES	23	*t*-butoxyl
C_4H_9S	2.03	0.02	PT	2	*n*-butyl sulfide
C_4H_9S	2.07	0.02	PT	2	*t*-butyl sulfide
C_4O	2.05	0.15	LPES	11	
C_4O_2	2.0	0.2	LPES	11	
C_4Ti	1.494	0.020	LPES	147	
C_5	2.853	0.001	LPT	99	
C_5F_5N	0.68	0.11	CT	67	pentafluoropyridine
$C_5F_6O_3$	1.5	0.2	CD	2	hexafluoroglutaric anhydride
C_5D_5	1.790	0.008	LPES	11	cyclopentadienyl-d_5
C_5H_5	1.804	0.007	LPES	11	cyclopentadienyl
C_5H_7	0.91	0.03	PT	2	pentadienyl
C_5H_7O	1.598	0.007	LPT	113	cyclopentanone enolate
C_5H_9O	1.69	0.05	PT	2	3-penanone enolate
$C_5H_{11}O$	1.93	0.05	LPT	2	neopentoxyl
$C_5H_{11}S$	2.09	0.02	PT	2	*n*-pentyl sulfide
C_5O_2	1.2	0.2	LPES	11	
C_5Ti	1.748	0.050	LPES	147	
C_6	4.180	0.001	LPT	8	
$C_6Br_4O_2$	2.44	0.20	CT	2	tetrabromo-BQ
$C_6Cl_4O_2$	2.78	0.10	CT	61	tetrachloro-BQ
$C_6F_4O_2$	2.70	0.10	CT	61	tetrafluoro-BQ
C_6F_5Br	1.15	0.11	CT	67	pentafluorobromobenzene

Table 4
Electron Affinities for Larger Polyatomic Molecules (continued)

Molecule	Electron Affinity in eV	Uncertainty in eV	Method	Ref.	Name
C_6F_5Cl	0.82	0.11	CT	67	pentafluorochlorobenzene
C_6F_5I	1.41	0.11	CT	67	pentafluoroiodobenzene
$C_6F_5NO_2$	1.52	0.11	CT	67	pentafluoro-NB
C_6F_6	0.52	0.10	CT	51	hexafluorobenzene
C_6F_{10}	>1.4	0.3	CT	2	perfluorocyclohexene
$C_6H_2Cl_2O_2$	2.48	0.10	CT	61	2,6-dichloro-BQ
$C_6H_3F_2NO_2$	1.17	0.10	CT	61	2,4-difluoro-NB
C_6D_4	0.551	0.010	LPES	36	*o*-benzyne-d_4
C_6H_4	0.560	0.010	LPES	36	*o*-benzyne
$C_6H_4BrNO_2$	1.16	0.10	CT	61	*o*-bromo-NB
$C_6H_4BrNO_2$	1.32	0.10	CT	61	*m*-bromo-NB
$C_6H_4BrNO_2$	1.29	0.10	CT	61	*p*-bromo-NB
$C_6H_4ClNO_2$	1.14	0.10	CT	61	*o*-chloro-NB
$C_6H_4ClNO_2$	1.28	0.10	CT	61	*m*-chloro-NB
$C_6H_4ClNO_2$	1.26	0.10	CT	61	*p*-chloro-NB
C_6H_4ClO	≤2.58	0.08	PT	2	*o*-chloroperoxide
$C_6H_4FNO_2$	1.07	0.10	CT	61	*o*-fluoro-NB
$C_6H_4FNO_2$	1.23	0.10	CT	61	*m*-fluoro-NB
$C_6H_4FNO_2$	1.12	0.10	CT	61	*p*-fluoro-NB
$C_6H_4N_2O_4$	1.65	0.10	CT	61	*o*-diNB
$C_6H_4N_2O_4$	1.65	0.10	CT	61	*m*-diNB
$C_6H_4N_2O_4$	2.00	0.10	CT	61	*p*-diNB
$C_6H_4O_2$	1.91	0.10	CT	61	1,4-benzoquinone (BQ)
C_6D_5	1.092	0.020	LPES	26	phenyl-d_5
C_6D_5N	1.44	0.02	LPES	96	phenylnitrene-d_5
C_6H_5	1.096	0.006	LPES	26	phenyl
C_6H_5N	1.429	0.011	LPT	115	phenylnitrene
$C_6H_5NO_2$	1.01	0.10	CT	61	nitrobenzene (NB)
C_6H_5O	2.253	0.006	LPES	26	phenoxyl
C_6H_5S	≤2.47	0.06	PT	2	thiophenoxide
C_6H_5NH	1.70	0.03	PT	2	anilide
C_6H_7	<1.67	0.04	PT	2	methylchylopentadienyl
C_6H_8Si	1.435	0.004	LPT	65	$C_6H_5SiH_3$
C_6H_9O	1.526	0.010	LPT	113	cyclohexanone enolate
C_6H_{10}	0.645	0.015	LPES	126	*t*-butyl vinylidene
$C_6H_{11}O$	1.755	+0.05/-0.005	LPT	113	pinacolone enolate
$C_6H_{11}O$	1.82	0.06	PT	2	3,3-dimethylbutananl enolate
C_6N_4	2.3	0.3	PT	2	TCNE
C_7F_5N	1.11	0.11	CT	67	pentafluorobenzonitrile
C_7F_8	0.86	0.11	CT	67	octafluorotoluene
C_7F_{14}	1.08	0.10	CT	61	perfluoromethylcyclohexane
C_7HF_5O	1.10	0.11	CT	67	pentafluorobenzaldehyde
$C_7H_3N_3O_4$	2.16	0.10	CT	61	3,5-$(NO_2)_2$-benzonitrile
$C_7H_4F_3NO_2$	1.41	0.10	CT	61	*m*-trifluoromethyl-NB
$C_7H_4N_2O_2$	1.61	0.10	CT	61	*o*-cyano-NB
$C_7H_4N_2O_2$	1.56	0.10	CT	61	*m*-cyno-NB
$C_7H_4N_2O_2$	1.72	0.10	CT	61	*p*-cyano-NB
C_7H_6FO	2.218	0.010	LPT	2	*m*-fluoroacetophenone enolate
C_7H_6FO	2.176	0.010	LPT	2	*p*-fluoroacetophenone enolate
$C_7H_6FeO_3$	0.990	0.10	CT	120	η^4-1,3-butadiene-$Fe(CO)_3$
$C_7H_6N_2O_4$	1.77	0.05	PT	60	3,4-dintrotoluene
$C_7H_6N_2O_4$	1.77	0.05	PT	60	2,3-dinitrotoluene
$C_7H_6N_2O_4$	1.60	0.05	PT	60	2,4-dinitrotoluene
$C_7H_6N_2O_4$	1.55	0.05	PT	60	2,6-dinitrotoluene
C_7H_7	0.912	0.006	LPES	26	benzyl

Table 4
Electron Affinities for Larger Polyatomic Molecules (continued)

Molecule	Electron Affinity in eV	Uncertainty in eV	Method	Ref.	Name
C_7H_7	0.868	0.006	LPES	136	1-quadricyclanide
C_7H_7	0.962	0.006	LPES	136	2-quadricyclanide
C_7H_7	1.286	0.006	LPES	136	norbornadienide
C_7H_7	0.39	0.04	LPES	136	cycloheptatrienide
C_7H_7	3.046	0.006	LPES	136	1-(1,6-heptadiynide)
C_7H_7	≥1.140	0.006	LPES	136	3-(1,6-heptadiynide)
$C_7H_7NO_2$	0.92	0.10	CT	61	*o*-methyl-NB
$C_7H_7NO_2$	0.99	0.10	CT	61	*m*-methyl-NB
$C_7H_7NO_2$	0.95	0.10	CT	61	*p*-methyl-NB
$C_7H_7NO_3$	1.04	0.10	CT	61	*m*-OCH_3-NB
$C_7H_7NO_3$	0.91	0.10	CT	61	*p*-OCH_3-NB
C_7H_7O	≤2.36	0.06	PT	2	*o*-methyl phenoxide
C_7H_7O	2.14	0.02	PT	50	benzyloxide
$C_7H_7O_2$	1.85	0.10	CT	61	*o*-CH_3-BQ
C_7H_8FO	<3.05	0.06	PT	50	$PhCH_2OHF$
C_7H_9	1.27	0.03	PT	2	heptatrienyl
C_7H_9O	1.61	0.05	PT	2	2-norbornanone enolate
C_7H_9Si	1.33	0.04	LPT	65	$C_6H_5(CH_3)SiH$
$C_7H_{11}O$	1.598	0.007	LPT	113	cycloheptanone enolate
$C_7H_{11}O$	1.49	0.04	PT	2	2,5-dimethylcyclopentanone enolate
$C_7H_{13}O$	1.72	0.06	PT	2	4-heptanone enolate
$C_7H_{13}O$	1.46	0.04	PT	2	di-isopropyl ketone enolate
$C_8F_{14}N_2$	1.89	0.10	CT	51	1,4-$(CN)_2C_6F_4$
$C_8H_3F_5O$	0.88	0.11	CT	67	pentafluoroacetophenone
$C_8H_3F_6NO_2$	1.79	0.10	CT	61	3,5-$(CF_3)_2$-NB
$C_8H_4O_3$	1.21	0.10	CT	61	phthalic anhydride
C_8H_6	1.044	0.008	LPES	148	
C_8H_7	1.091	0.008	LPES	134	
C_8H_7O	2.057	0.010	PT	2	acetophenone enolate
C_8H_7O	2.10	0.08	LPT	2	phenylacetaldehyde enolate
C_8H_8	0.55	0.02	CT	134	cycooctatetraene
C_8H_8	0.65	0.04	LPES	134	cycooctatetraene
C_8H_8	0.919	0.008	LPES	139	*m*-xylylene
$C_8H_9NO_2$	1.21	0.05	PT	60	3,5-dimethyl-NB
$C_8H_9NO_2$	2.61	0.05	PT	60	2,6-dimethyl-NB
$C_8H_9NO_2$	0.86	0.10	CT	61	2,3-dimethyl-NB
$C_8H_{13}O$	1.63	0.06	PT	2	cyclooctanone enolate
$C_9H_8FeO_3$	0.76	0.10	CT	120	η^4-1,3-cyclohexadiene-$Fe(CO)_3$
C_9H_9O	2.030	0.010	LPT	2	*m*-methylacetophenone enolate
C_9H_9SiN	1.43	0.10	PT	2	trimethylsilylnitrene
$C_9H_{11}NO_2$	0.70	0.10	CT	61	2,4,6-trimethyl-NB
$C_9H_{15}O$	1.69	0.06	PT	2	cyclononanone enolate
$C_{10}H_4Cl_2O_2$	2.19	0.10	CT	61	2,3-dichloro-1,4-naphthoquinone
$C_{10}H_6N_2O_4$	1.78	0.10	CT	61	1,3-dinitronaphthalene
$C_{10}H_6N_2O_4$	1.77	0.10	CT	61	1,5-dinitronaphthalene
$C_{10}H_6O_2$	1.81	0.10	CT	61	1,4-naphthoquinone
$C_{10}H_7NO_2$	1.23	0.10	CT	61	1-nitronaphthalene
$C_{10}H_7NO_2$	1.18	0.10	CT	61	2-nitronaphthalene
$C_{10}H_8$	0.69	0.10	CT	61	azulene
$C_{10}H_8CrO_3$	0.93	0.10	CT	120	η^4-1,3,5-cycloheptatriene $Cr(CO)_3$
$C_{10}H_8FeO_3$	0.98	0.10	CT	120	η^4-1,3,5-cycloheptatriene-$Fe(CO)_3$
$C_{10}H_{17}O$	1.83	0.06	PT	2	cyclodecanone enolate
$C_{11}H_8FeO_3$	1.29	0.10	CT	120	η^4-1,3-butadiene-$Fe(CO)_3$
$C_{12}F_{10}$	0.82	0.11	CT	67	decafluorobiphenyl
$C_{12}H_4N_4$	2.8	0.3	CD	2	TCNQ

Table 4
Electron Affinities for Larger Polyatomic Molecules (continued)

Molecule	Electron Affinity in eV	Uncertainty in eV	Method	Ref.	Name
$C_{12}H_9$	1.07	0.10	PT	2	perinaphthenyl
$C_{12}H_{15}O$	2.032	0.010	LPT	2	*t*-butylacetophenone enolate
$C_{12}H_{21}O$	1.90	0.07	PT	2	cyclododecanone enolate
$C_{13}F_{10}O$	1.52	0.11	CT	67	decafluorobenzophenone
$C_{13}H_9FO$	0.64	0.10	CT	61	4-fluorobenzophenone
$C_{13}H_{10}O$	0.62	0.10	CT	61	benzophenone
$C_{14}H_9NO_2$	1.43	0.10	CT	61	9-nitroanthracene
$C_{14}H_{10}$	0.57	0.10	CT	66	anthracene
$C_{18}H_{12}$	1.04	0.10	CT	66	tetracene
$C_{20}H_{12}$	0.79	0.10	CT	66	benz[*a*]pyrene
$C_{20}H_{12}$	0.97	0.10	CT	66	perylene
$C_{22}H_{14}$	1.35	0.10	CT	66	pentacene
CeF_4	3.8	0.4	CT	98	
$Cl(CO_2)$	3.907	0.010	LPES	131	
CoF_4	6.4	0.3	CT	98	
$Cr(CO)_3$	1.349	0.006	LPES	94	
CrO_3	3.6	0.2	CT	98	
Cu_n	$n = 1\text{-}41$	—	LPES	37	
Fe_n	n=3-24	—	LPES	149	
$Fe(CO)_2$	1.22	0.02	LPES	2	
$Fe(CO)_3$	1.8	0.2	LPES	2	
$Fe(CO)_4$	2.4	0.3	LPES	2	
FeF_3	3.6	0.1	CT	98	
FeF_4	6.0	estimate	CT	98	
Fe_nO_m	n=1-4	m=1-6	LPES	152	
Ge_n	$n = 3\text{-}15$	—	LPES	71	
Ge_xAs_y	$n = 5\text{-}30$	$n = x + y$	LPES	72	
GeH_3	≤1.74	0.04	PT	2	
$H(NH_3)_n$	$n = 1,2$	—	LPES	76	
HNO_3	0.57	0.15	CD	2	
$(H_2O)_n$	$n = 2\text{-}19$	—	LPES	77	
$I(CO_2)$	3.225	0.001	LPES	131	
In_xP_y	$n = 2\text{-}8$	$n = x + y$	LPES	137	
IrF_4	4.7	0.3	CT	98	
IrF_6	6.5	0.4	CT	98	
K_n	$n = 2\text{-}7$	—	LPES	18	
MnF_4	5.5	0.2	CT	98	
$Mo(CO)_3$	1.337	0.006	LPES	94	
MoF_5	3.5	0.2	CT	98	
MoF_6	3.8	0.2	CT	98	
MoO_3	2.9	0.2	CT	98	
$(NH_3)_n$	$n = 41\text{-}1100$	—	LPES	77	
$NH_2(NH_3)_n$	$n = 1,2$	—	LPES	78	
$NO(H_2O)_n$	$n = 1,2$	—	LPES	75	
NO_3	3.937	0.014	LPES	85	
$NO(N_2O)_n$	$n = 1,2$	—	LPES	79	
$(NO)_2$	≥2.1	—	LPES	75	
$(N_2O)_n$	$n = 1,2$	—	LPES	81	
Na_n	$n = 2\text{-}5$	—	LPES	18	
$(NaF)_n$	$n = 1\text{-}7,12$	—	LPES	64	
$Na(NaF)_n$	$n = 5,7\text{-}12$	—	LPES	64	
$Ni(CO)_2$	0.643	0.014	LPES	2	
$Ni(CO)_3$	1.077	0.013	LPES	2	
NiO_2	3.05	0.01	LPES	145	ONiO
NiO_2	0.82	0.03	LPES	145	$Ni(O_2)$

Table 4
Electron Affinities for Larger Polyatomic Molecules (continued)

Molecule	Electron Affinity in eV	Uncertainty in eV	Method	Ref.	Name
$OH(H_2O)$	<2.95	0.15	PT	2	
OsF_4	3.9	0.3	CT	98	
OsF_6	6.0	0.3	CT	98	
PBr_3	1.59	0.15	CD	2	
PBr_2Cl	1.63	0.20	CD	2	
PCl_2Br	1.52	0.20	CD	2	
PCl_3	0.82	0.10	CD	2	
PF_5	0.75	0.15	CT	121	
PO_3	4.5	0.5	CT	98	
$POCl_2$	3.83	0.25	CD	2	
$POCl_3$	1.41	0.20	CD	2	
PtF_4	5.5	0.3	CT	98	
PtF_6	7.0	0.4	CT	98	
ReF_6	4.7	estimate	CT	98	
RhF_4	5.4	0.3	CT	98	
RuF_4	4.8	0.3	CT	98	
RuF_5	5.2	0.4	CT	98	
RuF_6	7.5	0.3	CT	98	
SF_4	1.5	0.2	CT	91	
SF_6	1.05	0.10	CT	56	
SO_3	≥1.70	0.15	CD	2	
$(SO_2)_2$	0.6	0.2	LPES	80	
SeF_6	2.9	0.2	CD	2	
Si_4	2.17	0.01	LPES	110	
Si_n	$n = 3\text{-}20$	—	LPES	71	
SiD_3	1.386	0.022	LPES	43	
SiF_3	≤2.95	0.10	PT	17	
SiH_3	1.406	0.014	LPES	43	
TeF_6	3.34	0.17	CD	2	
Ti_n	n=3-65	—	LPES	151	
UF_5	3.7	0.2	CT	98	
UF_6	5.1	0.2	CT	98	
UO_3	<2.1	—	CT	98	
V_n	n=3-65	—	LPES	150	
VF_4	3.5	0.2	CT	98	
V_4O_{10}	4.2	0.6	CT	101	
$W(CO)_3$	1.859	0.006	LPES	94	
WF_5	1.25	0.3	CD	18	
WF_6	3.36	+0.04/-0.20	CT	19	
WO_3	3.33	+0.04/-0.15	LPT	86	
WO_3	3.9	0.2	CT	98	

REFERENCES

1. Hotop, H., and Lineberger, W. C., *J. Phys. Chem. Ref. Data*, 14, 731, 1985.
2. Drzaic, P. S., Marks, J., and Brauman, J. I., in *Gas Phase Ion Chemistry*, Vol. 3, Bowers, M. T., Ed., Academic Press, Orlando, 1984, p. 167.
3. Schulz, P. A., Mead, R. D., Jones, P. L., and Lineberger, W. C., *J. Chem. Phys.*, 77, 1153, 1982.
4. Neumark, D. M., Lykke, K. R., Anderson, T., and Lineberger, W. C., *Phys. Rev. A*, 32, 1890, 1985.
5. Wenthold, P. G., Gunion, R. F., and Lineberger, W. C., *Chem. Phys. Lett.*, 258, 101, 1996.
6. Srivastava, R. D., Uy, O. M., and Farber, M., *Trans. Faraday Soc.*, 67, 2941, 1971.
7. Klein, R., McGinnis, R. P., and Leone, S. R., *Chem. Phys. Lett.*, 100, 475, 1983.
8. Arnold, C. C., Zhao, Y., Kitsopoulos, T. N., and Neumark, D. M., *J. Chem. Phys.*, 97, 6121, 1992; linear C_6^- to linear C_6.
9. Stevens, A. E., Fiegerle, C. S., and Lineberger, W. C., *J. Chem. Phys.*, 78, 5420, 1983.
10. Breyer, F., Frey, P., and Hotop, H., *Z. Phys.*, A 300, 7, 1981.
11. Oakes, J. M., and Ellison, G. B., *Tetrahedron*, 42, 6263, 1986.
12. Leopold, D. G., Murray, K. K., Miller, A. E. S., and Lineberger, W. C., *J. Chem. Phys.*, 83, 4849, 1985.

13. Oakes, J. M., Jones, M.E., Bierbaum, V. M., and Ellison, G. B., *J. Chem. Phys.*, 87, 4810, 1983.
14. Ellis, H. B., Jr. and Ellison, G. B., *J. Chem. Phys.*, 78, 6541, 1983.
15. Oakes, J. M., Harding, L. B., and Ellison, G. B., *J. Chem. Phys.*, 83, 5400, 1985.
16. Nimlos, M. E., and Ellison, G. B., *J. Chem. Phys.*, 90, 2574, 1986.
17. Richardson, L. M., Stephenson, L. M., and Brauman, J. I., *Chem. Phys. Lett.*, 30, 17, 1975.
18. McHugh, K. M., Eaton, J. G., Lee, G. H., Sarkas, H. W., Kidder, L. H., Snodgrass, J. T., Manaa, M. R., and Bowen, K. H., *J. Chem. Phys.*, 91, 3792, 1989. See also Ref. 104.
19. Viggiano, A. A., Paulson, J. F., Dale, F., Henchman, M., Adams, N. G., and Smith, D., *J. Phys. Chem.*, 89, 2264, 1985.
20. Burnett, S. M., Stevens, A. E., Fiegerle, C. S., and Lineberger, W. C., *Chem. Phys. Lett.*, 100, 124, 1983.
21. Moran, S., Ellis, H. B., DeFrees, D. J., McLean, A. D., and Ellison, G. B., *J. Am. Chem. Soc.*, 109, 5996, 1987; Moran, S., Ellis, H. B., DeFrees, D. J., McLean, A. D., Paulson, S. E., and Ellison, G. B., *J. Am. Chem. Soc.*, 109, 6004, 1987; see also Lykke, K. R., Neumark, D. M., Andersen, T., Trapa, V. J., and Lineberger, W. C., *J. Chem. Phys.*, 87, 6842, 1987.
22. Mead, R. D., Lykke, K. R., Lineberger, W. C., Marks, J., and Brauman, J. I., *J. Chem. Phys.*, 81, 4883, 1984; Lykke, K. R., Mead, R. D., and Lineberger, W. C., *Phys. Rev. Lett.*, 52, 2221, 1984. The EAs are 14717.7 ± 1.0 cm^{-1} for actaldehyde enolate and 14671.0 ± 1.0 cm^{-1} for acetaldehyde-d_3 enolate.
23. Ellison, G. B., Engelking, P. C., and Lineberger, W. C., *J. Chem. Phys.*, 86, 4873, 1982.
24. Oakes, J. M., and Ellison, G. B., *J. Am. Chem. Soc.*, 105, 2969, 1983.
25. Ellison, G. B., and Oakes, J. M., *J. Am. Chem. Soc.*, 106, 7734, 1984. EA(allyl) and EA(allyl-d_5) are 0.119 and 0.083 eV too low, respectively, in this work, according to Reference 138. Therefore, EA(allyl-d_1) is likely too low by a similar amount.
26. Gunion, R. F., Gilles, M. K., Polak, M. L., and Lineberger, W. C., *Int. J. Mass Spectrom. Ion Processes*, 117, 601, 1992; also, Miller, A. E. S., and Lineberger, W. C., unpublished. See also Reference 136.
27. Leopold, D. G., and Lineberger, W. C., *J. Chem. Phys.*, 85, 51, 1986.
28. Miller, T. M., Miller, A. E. S., and Lineberger, W. C., *Phys. Rev. A*, 33, 3558, 1986.
29. Miller, A. E. S., Fiegerle, C. S., and Lineberger, W. C., *J. Chem. Phys.*, 87, 1549, 1987.
30. Miller, T. M., Leopold, D. G., Murray, K. K., and Lineberger, W. C., *J. Chem. Phys.*,85, 2368, 1986.
31. Miller, T. M., and Lineberger, W. C., *Chem. Phys. Lett.*, 146, 364, 1988.
32. Neumark, D. M., Lykke, K. R., Andersen, T., and Lineberger, W. C., *J. Chem. Phys.*, 83, 4364, 1985.
33. Leopold, D. G., Miller, T. M., and Lineberger, W. C., *J. Am. Chem. Soc.*, 108, 178, 1986.
34. Miller, A. E. S., Fiegerle, C. S., and Lineberger, W. C., *J. Chem. Phys.*, 84, 4127, 1986.
35. Murray, K. K., Miller, T. M., Leopold, D. G., and Lineberger, W. C., *J. Chem. Phys.*, 84, 2520, 1986.
36. Leopold, D. G., Miller, A. E. S., and Lineberger, W. C., *J. Am. Chem. Soc.*, 108, 1379, 1986.
37. Ho, J., Ervin, K. M., and Lineberger, W. C., *J. Chem. Phys.*, 93, 6987, 1990; Leopold, D. G., Ho, J., and Lineberger, W. C., *J. Chem. Phys.*, 86, 1715, 1987; Pettiette, C. L., Yang, S. H., Craycraft, M. J., Conceicao, J., Laaksonen, R. T., Cheshnovsky, O., and Smalley, R. E., *J. Chem. Phys.*, 88, 5377, 1988.
38. Snodgrass, J. T., Coe, J. V., McHugh, K. M., Friedhoff, C. B., and Bowen, K. H., *J. Phys. Chem.*, 93, 1249, 1989.
39. Friedhoff, C. B., Snodgrass, J. T., Coe, J. V., McHugh, K. M., and Bowen, K. H., *J. Chem. Phys.*,84, 1051, 1986.
40. Friedhoff, C. B., Coe, J. V., Snodgrass, J. T., McHugh, K. M., and Bowen, K. H., *Chem. Phys. Lett.*, 124, 268, 1986.
41. Coe, J. V., Snodgrass, J. T., Friedhoff, C. B., McHugh, K. M., and Bowen, K. H., *J. Chem. Phys.*, 84, 619, 1986.
42. Snodgrass, J. T., Coe, J. V., Friedhoff, C. B., McHugh, K. M., and Bowen, K. H., *Chem. Phys. Lett.*, 122, 352, 1985.
43. Nimlos, M. R., and Ellison, G. B., *J. Am. Chem. Soc.*, 108, 6522, 1986.
44. Petrunin, V., Andersen, H., Balling, P., and Andersen, T., *Phys. Rev. Lett.*, 76, 744, 1996
45. Andersen, T., Lykke, K. R., Neumark, D. M., and Lineberger, W. C., *J. Chem. Phys.*, 86, 1858, 1987.
46. Murray, K. K., Lykke, K. R., and Lineberger, W. C., *Phys. Rev. A*, 36, 699, 1987.
47. Mansour, N. B., and Larson, D. J., *Abstracts of the XV Int. Conf. on the Phys. of Electronic and Atomic Collisions*, p. 70, 1987.
48. Stonemann, R. C., and Larson, D. J., *Phys. Rev. A*, 35, 2928, 1987.
49. Nimlos, M. R., Harding, L. B., and Ellison, G. B., *J. Chem. Phys.*, 87, 5116, 1987.
50. Moylan, C. R., Dodd, J. A., Han, C.-C., and Braumann, J. I., *J. Chem. Phys.*, 86, 5350, 1987.
51. Chowdhury, S., Grimsrud, E. P., Heinis, T., and Kebarle, P., *J. Am. Chem. Soc.*, 108, 3630, 1986.
52. Trainham, R., Fletcher, G. D., and Larson, D. J., *J. Phys. B: At. Mol. Phys.*, 20, L777, 1987.
53. Moran, S., and Ellison, G. B., *J. Phys. Chem.*, 92, 1794, 1988.
54. Murray, K. K., Leopold, D. G., Miller, T. M., and Lineberger, W. C., *J. Chem. Phys.*, 89, 5442, 1988.
55. Ervin, K. M., Ho, J., and Lineberger, W. C., *J. Chem. Phys.*, 89, 4514, 1988.
56. Grimsrud, E. P., Chowdhury, S., and Kebarle, P., *J. Chem. Phys.*, 85, 4989, 1985.
57. Fischer, C. F., *Phys. Rev. A*, 39, 963, 1989.
58. Wickham-Jones, C. T., Ervin, K. M., Ellision, G. B., and Lineberger, W. C., *J. Chem. Phys.*, 91, 2762, 1989.
59. Hopper, D. G., Wahl, A. C., Wu, R. L. C., and Tiernan, T. O., *J. Chem. Phys.*, 65, 5474, 1976.
60. Mock, R. S., and Grimsrud, E. P., *J. Am. Chem. Soc.*, 111, 2861, 1989.
61. Chowdhury, S., Heinis, T., Grimsrud, E. P., and Kebarle, P., *J. Phys. Chem.*, 90, 2747, 1986. The uncertainty and other results are quoted in Ref. 60.
62. Wickham-Jones, C. T., Moran, S., and Ellison, G. B., *J. Chem. Phys.*, 90, 795, 1989.
63. Ervin, K. M., Ho, J., and Lineberger, W. C., *J. Phys. Chem.*, 92, 5405, 1988.
64. Miller, T. M., and Lineberger, W. C., *Int. J. Mass Spectrom. Ion Processes*, 102, 239, 1990.
65. Wetzel, D. M., Salomon, K. E., Berger, S., and Brauman, J. I., *J. Am. Chem. Soc.*, 111, 3835, 1989.
66. Crocker, L., Wang, T., and Kebarle, P., *J. Am. Chem. Soc.*, 115, 7818, 1993.
67. Dillow, G. W., and Kebarle, P., *J. Am. Chem. Soc.*, 111, 5592, 1989.
68. Gantefor, G., Gausa, M., Meiwes-Broer, K. H., and Lutz, H. O., *Z. Phys. D*, 9, 253, 1988; Taylor, K. J., Petteitte, C. L., Craycraft, M. J., Chesnovsky, O., and Smalley, R. E., *Chem. Phys. Lett.*, 152, 347, 1988.
69. Metz, R. B., Kitsopoulos, T., Weaver, A., and Neumark, D. M., *J. Chem. Phys.*, 88, 1463, 1988.
70. Yang, S., Pettiette, C. L., Conceicao, J., Cheshnovsky, O., and Smalley, R. E., *Chem. Phys. Lett.*, 139, 233, 1987; Yang, S., Taylor, K. J., Craycraft, M. J., Conceicao, J., Pettiette, C. L., Cheshnovsky, O., and Smalley, R. E., *Chem. Phys. Lett.*, 144, 431, 1988; Arnold, D. W., Bradforth, S. E., Kitsopoulos, T. N., and Neumark, D. M., *J. Chem. Phys.*, 95, 5479, 1991.
71. Cheshnovsky, O., Yang, S., Pettiette, C. L., Craycraft, M. J., Liu, Y., and Smalley, R. E., *Chem. Phys. Lett.*, 138, 119, 1987.

73. Travers, M. J., Cowles, D. C., and Ellison, G. B., *Chem. Phys. Lett.*, 164, 449, 1989.
74. Blondel, C., Cacciani, P., Delsart, C., and Trainham, R., *Phys. Rev. A*, 40, 3698, 1989.
75. Bowen, K. H., and Eaton, J. G., in *The Structure of Small Molecules and Ions*, Naaman, R., and Vager, Z., Eds., Plenum, New York, 1988, pp. 147-169; Arnold, S. T., Eaton, J. G., Patel-Mistra, D., Sarkas, H. W., and Bowen, K. H., in *Ion and Cluster Ion Spectroscopy and Structure*, Maier, J. P., Ed., Elsevier Science, New York, 1989, p. 417.
76. Snodgrass, J. T., Coe, J. V., Friedhoff, C. B., McHugh, K. M., and Bowen, K. H., *Faraday Disc. Chem. Soc.*, 88, 1988.
77. Lee, G. H., Arnold, S. T., Eaton, J. G., Sarkas, H. W., Bowen, K. H., Ludewigt, C., and Haberland, H., *Z. Phys. D - Atoms, Mol. and Clusters*, 20, 9, 1991; Coe, J. V., Lee, G. H., Eaton, J. G., Arnold, S. T., Sarkas, H. W., Bowen, K. H., Ludewigt, C., Haberland, H., and Worsnop, D. R., *J. Chem. Phys.*, 92, 3980, 1990.
78. Snodgrass, J. T., Coe, J. V., Friedhoff, C. B., McHugh, K. M., and Bowen, K. H., to be published, quoted in Ref. 75.
79. Coe, J. V., Snodgrass, J. T., Friedhoff, C. B., McHugh, K. M., and Bowen, K. H., *J. Chem. Phys.*, 87, 4302, 1987.
80. Friedhoff, C. B., Snodgrass, J. T., and Bowen, K. H., to be published, quoted in Ref. 75.
81. Coe, J. V., Snodgrass, J. T., Friedhoff, C. B., McHugh, K. M., and Bowen, K. H., *Chem. Phys. Lett.*, 124, 274, 1986.
82. Greene, C. H., *Phys. Rev. A*, 42, 1405, 1990.
83. Ervin, K. M., Ho, J., and Lineberger, W. C., *J. Chem. Phys.*, 91, 5974, 1989.
84. Polak, M. L., Fiala, B. L., Ervin, K. M., and Lineberger, W. C., *J. Chem. Phys.*, 94, 6924, 1991.
85. Weaver, A., Arnold, D. W., Bradforth, S. E., Neumark, D. M., *J. Chem. Phys.*, 94, 1740, 1991.
86. Walter, C. W., Devynck, P., Hertzler, C. F., Bae, Y. K., Smith, G. P., and Peterson, J. R., *Bull. Am. Phys. Soc.*, 35, 1163, 1990.
87. Ervin, K. M., and Lineberger, W. C., *J. Phys. Chem.*, 95, 1167, 1991.
88. Gilles, M. K., Polak, M. L., and Lineberger, W. C., *J. Chem. Phys.*, 96, 8012, 1992.
89. Lykke, K. R., Murray, K. K., and Lineberger, W. C., *Phys. Rev. A*, 43, 6104, 1991.
90. Ervin, K. M., Gronert, S., Barlow, S. E., Gilles, M. K., Harrison, A. G., Bierbaum, V. M., DePuy, C. H., Lineberger, W. C., and Ellison, G. B., *J. Am. Chem. Soc.*, 112, 5750, 1990.
91. Viggiano, A. A., Miller, T. M., Miller, A. E. S., Morris, R. A., Van Doren, J. M., and Paulson, J. F., *Int. J. Mass Spectrom. Ion Processes*, 109, 327, 1991.
92. Hanstorp, D., and Gustafsson, M., *J. Phys. B: At. Mol. Opt. Phys.*, 25, 1773, 1992.
93. Polak, M. L., Gilles, M. K., and Lineberger, W. C., *J. Chem. Phys.*, 96, 7191, 1992.
94. Bengali, A. A., Casey, S. M., Cheng, C.-L., Dick, J. P., Fenn, P. T., Villalta, P. W., and Leopold, D. G., *J. Am. Chem. Soc.*, 114, 5257, 1992.
95. Gilles, M. K., Ervin, K. M., Ho, J., and Lineberger, W. C., *J. Phys. Chem.*, 96, 1130, 1992.
96. Travers, M. J., Cowles, D. C., Clifford, E. P., and Ellison, G. B., *J. Am. Chem. Soc.*, 114, 8699, 1992.
97. Bengali, A. A., and Leopold, D. G., *J. Am. Chem. Soc.*, 114, 9192, 1992.
98. Rudnyi, E. B., Kaibicheva, E. A., and Sidorov, L. N., *Rapid Comm. in Mass Spectrom.*, 6, 356, 1992; Sidorov, L. N., *High Temp. Sci.*, 29, 153, 1990.
99. Kitsopoulos, T. N., Chick, C. J., Zhao, Y., and Neumark, D. M., *J. Chem. Phys.*, 95, 5479, 1991.
100. Arnold, C. C., Kitsopoulos, T. N., and Neumark, D. M., *J. Chem. Phys.*, 99, 766, 1993.
101. Rudnyi, E. B., Kaibicheva, E. A., and Sidorov, L. N., *J. Chem. Thermodynamics*, 25, 929, 1993.
102. Sarkas, H. W., Hendricks, J. H., Arnold, S. T., and Bowen, K. H., *J. Chem. Phys.* 100, 1884, 1994.
103. Villalta, P. W., and Leopold, D. G., *J. Chem. Phys.* 98, 7730, 1993.
104. Eaton, J. G., Sarkas, H. W., Arnold, S. T., McHugh, K. M., and Bowen, K. H., *Chem. Phys. Lett.*, 193, 141, 1992. See also Ref. 18.
105. Polak, M. L., Gilles, M. K., Gunion, R. F., and Lineberger, W. C., *Chem. Phys. Lett.*, 210, 55, 1993.
106. Gilles, M. K., Lineberger, W. C., and Ervin, K. M., *J. Am. Chem. Soc.*, 115, 1031, 1993.
107. Casey, S. M., and Leopold, D. G., *Chem. Phys. Lett.*, 201, 205, 1993.
108. Polak, M. L., Gerber, G., Ho, J., and Lineberger, W. C., *J. Chem. Phys.*, 97, 8990, 1992.
109. Kim, E. H., Bradforth, S. E., Arnold, D. W., Metz, R. B., and Neumark, D. M., *J. Chem. Phys.*, 103, 7801, 1995.
110. Arnold, C. C., and Neumark, D. M., *J. Chem. Phys.*, 99, 3353, 1993.
111. Bradforth, S. E., Kim, E. H., Arnold, D. W., and Neumark, D. M., *J. Chem. Phys.*, 98, 800, 1993.
112. Ho, J., Polak, M. L., Ervin, K. M., and Lineberger, W. C., *J. Chem. Phys.*, 99, 8542, 1993.
113. Brinkman, E. A., Berger, S., Marks, J., and Brauman, J. I., *J. Chem. Phys.*, 99, 7586, 1993.
114. Casey, S. M., and Leopold, D. G., *J. Phys. Chem.*, 97, 816, 1993.
115. McDonald, R. N., and Davidson, S. J., *J. Am. Chem. Soc.*, 115, 10857, 1993.
116. Ho, J., Ervin, K. M., Polak, M. L., Gilles, M. K., and Lineberger, W. C., *J. Chem. Phys.*, 95, 4845, 1991.
117. Ho, J., Polak, M. L., and Lineberger, W. C., *J. Chem. Phys.*, 96, 144, 1992.
118. Polak, M. L., Gilles, M. K., Ho, J., and Lineberger, W. C., *J. Phys. Chem.*, 95, 3460, 1991.
119. Polak, M. L., Ho, J., Gerber, G., and Lineberger, W. C., *J. Chem. Phys.*, 95, 3053, 1991.
120. Sharpe, P., and Kebarle, P., *J. Am. Chem. Soc.*, 115, 782, 1993.
121. Miller, T. M., Miller, A. E. S., Viggiano, A. A., Morris, R. A., and Paulson, J. F., *J. Chem. Phys.*, 100, 7200, 1994.
122. Berkovits, D., Boaretto, E., Gehlberg, S., Heber, O., and Paul, M., *Phys. Rev. Lett.*, 75, 414, 1995.
123. Arnold, C. C., Xu, C., Burton, G. R., and Neumark, D. M., *J. Chem. Phys.*, 102, 6982, 1995. Burton, G. R., Xu, C., Arnold, C. C., and Neumark, D. M., *J. Chem. Phys.*, in press.
124. Xu, C., de Beer, E., and Neumark, D. M., *J. Chem. Phys.*, 104, 2749, 1996.
125. Gunion, R. F., Koppel, H., Leach, G. W., and Lineberger, W. C., *J. Chem. Phys.*, 103, 1250, 1995.
126. Gunion, R. F., and Lineberger, W. C., *J. Phys. Chem.*, 100, 4395, 1996.
127. Gunion, R. F., Dixon-Warren, St. J., Lineberger, W. C., and Morse, M. D., *J. Chem. Phys.*, in press.
128. Dixon-Warren, St. J., Gunion, R. F., and Lineberger, W. C., *J. Chem. Phys.*, 104, 4902, 1996.
129. Nakajima, A., Zhang, N., Kawamata, H., Hayase, T., Nakao, K., and Kaya, K., *Chem. Phys. Lett.*, 241, 295, 1995; Nakajima, A., Taguwa, T., Nakao, K., Hoshino, K., Iwata, S., and Kaya, K., *J. Chem. Phys.*, 102, 660, 1995.
130. Fan, J., and Wang, L.-S., *J. Chem. Phys.*, 102, 8714, 1995.
131. Arnold, D. W., Bradforth, S. E., Kim, E. H., and Neumark, D. M., *J. Chem. Phys.*, 102, 3493, 1995; Zhao, Y., Arnold, C. C., and Neumark, D. M., *J. Chem. Soc. Faraday Trans. 2*, 89, 1449, 1992.
132. Fan, J., Lou, L., and Wang, L.-S., *J. Chem. Phys.*, 102, 2701, 1995.

133. Sarkas, H. W., Hendricks, J. H., Arnold, S. T., Slager, V. L., and Bowen, K. H., *J. Chem. Phys.*, 100, 3358, 1994.
134. Kato, S., Lee, H. S., Gareyev, R., Wenthold, P. G., Lineberger, W. C., DePuy, C. H., and Bierbaum, V. M., *J. Am. Chem. Soc.*, 119, 7863, 1997.
135. Wenthold, P. G., Hu, J., Squires, R. R., and Lineberger, W. C., *J. Am. Chem. Soc.*, 118, 475, 1996.
136. Gunion, R. F., Karney, W., Wenthold, P. G., Borden, W. T., and Lineberger, W. C., *J. Am. Chem. Soc.*, 118, 5074, 1996. The numbers in the abstract for 1,6-heptadiyne were misprinted. EA(cycloheptatrienide) quoted here derives from the LPES data combined with other thermochemical data in Reference 136.
137. Xu, C., de Beer, E., Arnold, D. W., Arnold, C. C., and Neumark, D. M., *J. Chem. Phys.*, 101, 5406, 1996.
138. Wenthold, P. G., Polak, M. L., and Lineberger, W. C., *J. Phys. Chem.*, 100, 6920, 1996.
139. Wenthold, P. G., Kim, J. B., and Lineberger, W. C., *J. Am. Chem. Soc.*, 119, 1354, 1997.
140. Eliav, E., Kaldor, U., Ishikawa, Y., and Pyykkoe, *Phys. Rev. Lett.*, 77, 5350, 1996.
141. Davies, B. J., Ingram, C. W., Larson, D. J., and Ljungblad, U., *J. Chem. Phys.* 106, 5783, 1997.
142. Smith, J. R., Kim, J. B., and Lineberger, W. C., *Phys. Rev. A.*, 55, 2036, 1997. Schulz, P.A., Mead, R. D., Jones, P. L., and Lineberger, W. C., *J. Chem. Phys.*, 77, 1153, 1982. See also Rudmin, J. D., Ratliff, L. P., Yukich, J. N., and Larson, D. J., *J. Phys. B: At. Mol. Opt. Phys.*, 29, L881, 1996.
143. Desai, S. R., Wu, H., Rohlfing, C. M., and Wantg, L.-S., *Int. J. Chem. Phys.*, 106, 1309, 1997.
144. Wenthold, P. G., Joans, K.-L., and Lineberger, W. C., *J. Chem. Phys.*, 106, 9961, 1997.
145. Wu, H. and Wang, L.-S., *J. Chem. Phys.*, 107, 16, 1997.
146. Moravec, V. D. and Jarrold, C. C., *J. Chem. Phys.*, in press.
147. Wang, X.-B., Ding, C.-F., and Wang, L.-S., *J. Phys. Chem. A*, 101, 7699, 1997.
148. Wenthold, P. G. and Lineberger, W. C., *J. Am. Chem. Soc.*, 19, 7772, 1997.
149. Wang, L.-S., Cheng, H.-S., and Fan, J., *J. Chem. Phys.*, 102, 9480, 1995.
150. Wu, H., Desai, S. R., and Wang, L.-S., *Phys. Rev. Lett.*, 77, 2436, 1996.
151. Wu, H., Desai, S. R., and Wang, L.-S., *Phys. Rev. Lett.*, 76, 212, 1996.
152. Wang, L.-S., Wu, H., and Desai, S. R., *Phys. Rev. Lett.*, 76, 4853, 1996.
153. Robinson, M. S., Polak, M. L., Bierbaum, V. M., DePuy, C. H., and Lineberger, W. C., *J. Am. Chem. Soc.*, 117, 6766, 1995.
154. Clifford, E. P., Wenthold, P. G., Lineberger, W. C., Petersson, G. A., and Ellison, G. B., *J. Phys. Chem.*, 101, 4338, 1997.
155. Wenthold, P. G., Kim, J. B., Jonas, K., and Lineberger, W. C., *J. Phys. Chem.*, in press, 1997.

ATOMIC AND MOLECULAR POLARIZABILITIES

Thomas M. Miller

The *polarizability* of an atom or molecule describes the response of the electron cloud to an external field. The atomic or molecular energy shift ΔW due to an external electric field E is proportional to E^2 for external fields which are weak compared to the internal electric fields between the nucleus and electron cloud. The *electric dipole polarizability* α is the constant of proportionality defined by $\Delta W = -\alpha E^2/2$. The induced electric dipole moment is αE. *Hyperpolarizabilities*, coefficients of higher powers of E, are less often required. Technically, the polarizability is a tensor quantity but for spherically symmetric charge distributions reduces to a single number. In any case, an *average polarizability* is usually adequate in calculations. Frequency-dependent or *dynamic polarizabilities* are needed for electric fields which vary in time, except for frequencies which are much lower than electron orbital frequencies, where *static polarizabilities* suffice.

Polarizabilities for atoms and molecules in excited states are found to be larger than for ground states and may be positive or negative. Molecular polarizabilities are very slightly temperature dependent since the size of the molecule depends on its rovibrational state. Only in the case of dihydrogen has this effect been studied enough to warrant consideration in Table 3.

Polarizabilities are normally expressed in cgs units of cm^3. Ground state polarizabilities are in the range of 10^{-24} cm^3 = 1 Å^3 and hence are often given in Å^3 units. Theorists tend to use atomic units of a_o^3 where a_o is the Bohr radius. The conversion is $\alpha(\text{cm}^3) = 0.148184 \times 10^{-24} \times \alpha(a_o^3)$. Polarizabilities are only recently encountered in SI units, C·m^2/V = J/(V/m)^2. The conversion from cgs units to SI units is $\alpha(\text{C·m}^2/\text{V}) = 4\pi\varepsilon_o \times 10^{-6} \times \alpha(\text{cm}^3)$, where ε_o is the permittivity of free space in SI units and the factor 10^{-6} simply converts cm^3 into m^3. Thus, $\alpha(\text{C·m}^2/\text{V}) = 1.11265 \times 10^{-16} \times \alpha(\text{cm}^3)$. Persons measuring excited state polarizabilities by optical methods tend to use units of MHz/(V/cm)^2, where the energy shift, ΔW, is expressed in frequency units with a factor of h understood. The polarizability is $-2\,\Delta W/E^2$. The conversion into cgs units is $\alpha(\text{cm}^3) = 5.95531 \times 10^{-16} \times \alpha[\text{MHz/(V/cm)}^2]$.

The polarizability appears in many formulas for low-energy processes involving the valence electrons of atoms or molecules. These formulas are given below in cgs units: the polarizability α is in cm^3; masses m or μ are in grams; energies are in ergs; and electric charges are in esu, where $e = 4.8032 \times 10^{-10}$ esu. The symbol $\alpha(\nu)$ denotes a frequency (ν) dependent polarizability, where $\alpha(\nu)$ reduces to the static polarizability α for $\nu = 0$. For further information and references, see Miller, T. M., and Bederson, B., *Advances in Atomic and Molecular Physics*, 13, 1, 1977. Details on polarizability-related interactions, especially in regard to hyperpolarizabilities and nonlinear optical phenomena, are given by Bogaard, M. P., and Orr, B. J., in *Physical Chemistry*, Series Two, Vol. 2, *Molecular Structure and Properties*, Buckingham, A. D., Ed., Butterworths, London, 1975, pp. 149-194. A tabulation of tensor and hyperpolarizabilities is included. The gas number density, n, in Table 1 is usually taken to be that of 1 atm at 0°C in reporting experimental data.

Table 1
Formulas Involving Polarizability

Description	Formula	Remarks
Lorentz-Lorenz relation	$\alpha(\nu) = \dfrac{3}{4\pi n}\left[\dfrac{\eta^2(\nu)-1}{\eta^2(\nu)+2}\right]$	For a gas of atoms or nonpolar molecules; the index of refraction is $\eta(\nu)$
Refraction by polar molecules	$\alpha(\nu) + \dfrac{d^2}{3kT} = \dfrac{3}{4\pi n}\left[\dfrac{\eta^2(\nu)-1}{\eta^2(\nu)+2}\right]$	The dipole moment is d, in esu·cm (= 10^{-18} D)
Dielectric constant (dimensionless)	$\kappa(\nu) = 1 + 4\pi n\ \alpha(\nu)$	From the Lorentz-Lorenz relation for the usual case of $\kappa(\upsilon) \approx 1$
Index of refraction (dimensionless)	$\eta(\nu) = 1 + 2\pi n\ \alpha(\nu)$	From $\eta^2(\nu) = \kappa(\nu)$
Diamagnetic susceptibility	$\chi_m = e^2 (a_o N\alpha)^{1/2} / 4m_e c^2$	From the approximation that the static polarizability is given by the variational formula $\alpha = (4/9a_o)\Sigma(N_i r_i^2)^2$; N is the number of electrons, m_e is the electron mass; a crude approximation is $\chi_m = (E_i/4m_e c^2)\alpha$, where E_i is the ionization energy
Long-range electron- or ion-molecule interaction energy	$V(r) = -e^2\alpha / 2r^4$	The target molecule polarizability is α
Ion mobility in a gas	$\kappa = 13.87/(\alpha\mu)^{1/2}\ \text{cm}^2/\text{V}\cdot\text{s}$	This one formula is not in cgs units. Enter α in Å^3 or 10^{-24} cm^3 units and the reduced mass μ of the ion-molecule pair in amu. Classical limit; pure polarization potential

Table 1
Formulas Involving Polarizability (continued)

Description	Formula	Remarks
Langevin capture cross section	$\sigma(v_o) = (2\pi e/v_o)(\alpha/\mu)^{1/2}$	The relative velocity of approach for an ion-molecule pair is v_o; the target molecular polarizability is α and the reduced mass of the ion-molecule pair is μ
Langevin reaction rate coefficient	$k = 2\pi e(\alpha/\mu)^{1/2}$	Collisional rate coefficient for an ion-molecule reaction
Rate coefficient for polar molecules	$k_d = 2\pi e\left[(\alpha/\mu)^{1/2} + cd(2/\mu\pi kT)^{1/2}\right]$	The dipole moment of the neutral is *d* in esu·cm; the number c is a "locking factor" that depends on α and *d*, and is between 0 and 1
Modified effective range cross section for electron-neutral scattering	$\sigma(k) = 4\pi A^2 + 32\pi^4\mu e^2\alpha Ak/3h^2 + \mathrm{L}$	Here, *k* is the electron momentum divided by *h*/2π, where *h* is Planck's constant; *A* is called the "scattering length"; the reduced mass is μ
van der Waals constant between two systems A, B	$C_6 = \frac{3}{2}\left[\frac{\alpha^A\alpha^B E^A E^B}{E^A + E^B}\right]$	For the interaction potential term $V_6(r) = -C_6r^6$; $E^{A,B}$ represents average dipole transition energies and $\alpha^{A,B}$ the respective polarizabilities of A, B
Dipole-quadrupole constant between two systems A, B	$C_8 = \frac{15}{4}\left[\frac{\alpha^A\alpha_q{}^B E^A E_q{}^B}{E^A + E_q{}^B}\right] + \frac{15}{4}\left[\frac{\alpha_q{}^A\alpha^B E_q{}^A E^B}{E_q{}^A + E^B}\right]$	For the interaction potential term $V_8(r) = -C_8r^8$; $E_q{}^{A,B}$ represents average quadrupole transition energies and $\alpha_q{}^{A,B}$ are the respective quadrupole polarizabilities of A, B
van der Waals constant between an atom and a surface	$C_3 = \frac{\alpha g E^A E^S}{8(E^A + E^S)}$	For an interaction potential $V_3(r) = -C_3r^3$; $E^{A,S}$ are characteristic energies of the atom and surface; $g = 1$ for a free-electron metal and $g = (\varepsilon_\infty - 1)/(\varepsilon_\infty + 1)$ for an ionic crystal
Relation between α(ν) and oscillator strengths	$\alpha(\nu) = \frac{e^2h^2}{4\pi^2 m_e}\Sigma\frac{f_k}{E_k^2 - (h\nu)^2}$	Here, f_k is the oscillator strength from the ground state to an excited state *k*, with excitation energy E_k. This formula is often used to estimate static polarizabilities (ν = 0)
Dynamic polarizability	$\alpha(\nu) = \frac{\alpha E_r{}^2}{E_r{}^2 - (h\nu)^2}$	Approximate variation of the frequency-dependent polarizability α(ν) from ν = 0 up to the first dipole-allowed electronic transition, of energy E_r; the static dipole polarizability is α(0); infrared contributions ignored
Rayleigh scattering cross section	$\alpha(\nu) = \frac{8\pi}{9c^4}(2\pi\nu)^4 \times \left[3\alpha^2(\nu) + 2\gamma^2(\nu)/3\right]$	The photon frequency is ν; the polarizability anisotropy (the difference between polarizabilities parallel and perpendicular to the molecular axis) is γ(ν)
Verdet constant	$V(\nu) = \frac{\nu n}{2m_ec^2}\left[\frac{d\alpha(\nu)}{d\nu}\right]$	Defined from θ = *V*(ν)*B*, where θ is the angle of rotation of linearly polarized light through a medium of number density *n*, per unit length, for a longitudinal magnetic field strength *B* (Faraday effect)

Table 2
Static Average Electric Dipole Polarizabiilities for Ground State Atoms (in Units of 10^{-24} cm^3)

Atomic number	Atom	Polarizability	Estimated accuracy (%)	Method	Ref.
1	H	0.666793	"exact"	calc	MB77
2	He	0.204956	"exact"	calc	MB77
		0.2050	0.1	index/diel	NB65/OC67
3	Li	24.3	2	beam	MB77
4	Be	5.60	2	calc	MB77
5	B	3.03	2	calc	MB77
6	C	1.76	2	calc	MB77
7	N	1.10	2	calc/index	MB77
8	O	0.802	2	calc/index	MB77
9	F	0.557	2	calc	MB77
10	Ne	0.3956	0.1	diel	OC67
11	Na	24.08	0.4	interferom	ESCHP94
12	Mg	10.6	2	calc	MB77
13	Al	6.8	4.4	beam	MMD90
14	Si	5.38	2	calc	MB77
15	P	3.63	2	calc	MB77
16	S	2.90	2	calc	MB77
17	Cl	2.18	2	calc	MB77
18	Ar	1.6411	0.05	index/diel	NB65/OC67
19	K	43.4	2	beam	MB77
20	Ca	22.8	2	calc	MB77
		25.0	8	beam	MB77
21	Sc	17.8	25	calc	D84
22	Ti	14.6	25	calc	D84
23	V	12.4	25	calc	D84
24	Cr	11.6	25	calc	D84
25	Mn	9.4	25	calc	D84
26	Fe	8.4	25	calc	D84
27	Co	7.5	25	calc	D84
28	Ni	6.8	25	calc	D84
29	Cu	6.1	25	calc	D84
		7.31	25	calc	G84
30	Zn	7.1	25	calc	MB77
		5.6	25	calc	D84
31	Ga	8.12	2	calc	MB77
32	Ge	6.07	2	calc	MB77
33	As	4.31	2	calc	MB77
34	Se	3.77	2	calc	MB77
35	Br	3.05	2	calc	MB77
36	Kr	2.4844	0.05	diel	OC67
37	Rb	47.3	2	beam	MB77
38	Sr	27.6	8	beam	MB77
39	Y	22.7	25	calc	D84
40	Zr	17.9	25	calc	D84
41	Nb	15.7	25	calc	D84
42	Mo	12.8	25	calc	D84
43	Tc	11.4	25	calc	D84
44	Ru	9.6	25	calc	D84
45	Rh	8.6	25	calc	D84
46	Pd	4.8	25	calc	D84
47	Ag	7.2	25	calc	D84
		8.56	25	calc	G84
48	Cd	7.2	25	calc	D84
49	In	10.2	12	beam	GMBSJ84
		9.1	25	calc	D84

Table 2
Static Average Electric Dipole Polarizabiilities for Ground State Atoms (in Units of 10^{-24} cm^3) (continued)

Atomic number	Atom	Polarizability	Estimated accuracy (%)	Method	Ref.
50	Sn	7.7	25	calc	D84
51	Sb	6.6	25	calc	D84
52	Te	5.5	25	calc	D84
53	I	5.35	25	index	A56
		4.7	25	calc	D84
54	Xe	4.044	0.5	diel	MB77
55	Cs	59.6	2	beam	MB77
56	Ba	39.7	8	beam	MB77
57	La	31.1	25	calc	D84
58	Ce	29.6	25	calc	D84
59	Pr	28.2	25	calc	D84
60	Nd	31.4	25	calc	D84
61	Pm	30.1	25	calc	D84
62	Sm	28.8	25	calc	D84
63	Eu	27.7	25	calc	D84
64	Gd	23.5	25	calc	D84
65	Tb	25.5	25	calc	D84
66	Dy	24.5	25	calc	D84
67	Ho	23.6	25	calc	D84
68	Er	22.7	25	calc	D84
69	Tm	21.8	25	calc	D84
70	Yb	21.0	25	calc	D84
71	Lu	21.9	25	calc	D84
72	Hf	16.2	25	calc	D84
73	Ta	13.1	25	calc	D84
74	W	11.1	25	calc	D84
75	Re	9.7	25	calc	D84
76	Os	8.5	25	calc	D84
77	Ir	7.6	25	calc	D84
78	Pt	6.5	25	calc	D84
79	Au	5.8	25	calc	D84
		6.48	25	calc	G84
80	Hg	5.7	25	calc	D84
		5.1	15	diel	MB77
81	Tl	7.6	15	beam	NYU84
		7.5	25	calc	D84
82	Pb	6.8	25	calc	D84
83	Bi	7.4	25	calc	D84
84	Po	6.8	25	calc	D84
85	At	6.0	25	calc	D84
86	Rn	5.3	25	calc	D84
87	Fr	48.7	25	calc	D84
88	Ra	38.3	25	calc	D84
89	Ac	32.1	25	calc	D84
90	Th	32.1	25	calc	D84
91	Pa	25.4	25	calc	D84
92	U	24.9	6	beam	KB94
93	Np	24.8	25	calc	D84
94	Pu	24.5	25	calc	D84
95	Am	23.3	25	calc	D84
96	Cm	23.0	25	calc	D84
97	Bk	22.7	25	calc	D84
98	Cf	20.5	25	calc	D84
99	Es	19.7	25	calc	D84

Table 2
Static Average Electric Dipole Polarizabiilities for Ground State Atoms (in Units of 10^{-24} cm^3) (continued)

Atomic number	Atom	Polarizability	Estimated accuracy (%)	Method	Ref.
100	Fm	23.8	25	calc	D84
101	Md	18.2	25	calc	D84
102	No	17.5	25	calc	D84

Note: calc = calculated value; beam = atomic beam deflection technique; interferom = atomic beam interference; index = determination based on the measured index of refraction; diel = determination based on the measured dielectric constant.

REFERENCES

A56. Atoji, M., *J. Chem. Phys.*, 25, 174, 1956. Semiempirical method based on molecular polarizabilities and atomic radii.

D84. Doolen, G. D., Los Alamos National Laboratory, unpublished. A relativistic linear response method was used. The method is that described by Zangwill, A., and Soven, P., *Phys. Rev. A*, 21, 1561, 1980. Adjustments of less than 10% across the periodic table have been made to these results to bring them into agreement with accurate experimental values where avaialble, for the purpose of presenting "recommended" polarizabilities in Table 2.

ESCHP94. Ekstrom, C. R., Schmiedmayer, J., Chapman, M. S., Hammond, T. D., and Pritchard, D. E., *Phys. Rev. Lett.*, in press, 1994.

G84. Gollisch, H., *J. Phys. B: Atom. Mol. Phys.*, 17, 1463, 1984. Other results and useful references are contained in this paper.

GMBSJ84. Guella, T. P., Miller, T. M., Bederson, B., Stockdale, J. A. D., and Jaduszliwer, B., *Phys. Rev. A*, 29, 2977, 1984.

KB94. Kadar-Kallen, M. A., and Bonin, K. D., *Phys. Rev. Lett.*, 72, 828, 1994.

MB77. Miller, T. M., and Bederson, B., *Adv. At. Mol. Phys.*, 13, 1, 1977. For simplicity, any value in Table 2 which has not changed since this 1977 review is referenced as MB77. Persons interested in original references and further details should consult MB77.

MMD90. Milani, P., Moullet, I., and de Heer, W. A., *Phys. Rev. A*, 42, 5150, 1990.

NB65. Newell, A. C., and Baird, R. D., *J. Appl. Phys.*, 36, 3751, 1965.

NYU84. Preliminary value from the New York University group. See GMBSJ84.

OC67. Orcutt, R. H., and Cole, R. H., *J. Chem. Phys.*, 46, 697, 1967; see also the later references from this group, given following the tables.

Table 3
Average Electric Dipole Polarizabilities for Ground State Diatomic Molecules (in Units of 10^{-24} cm^3)

Molecule	Polarizability	Ref.	Molecule	Polarizability	Ref.
Al_2	19	23	HI	5.44	3
BH	3.32*	1		5.35	2
Br_2	7.02	2	HgCl	7.4*	9
CO	1.95	3	ICl	12.3	2
Cl_2	4.61	3	K_2	77	22
Cs_2	104	22		72	21
CsK	89	22	Li_2	34	22
D_2 (v = 0,J = 0)	0.7921*	5	LiCl	3.46*	10
D_2 (293 K)	0.7954	6	LiF	10.8*	11
DCl	2.84	2	LiH	3.84*	12
F_2	1.38*	7		3.68*	13
H_2 (v = 0,J = 0)	0.8023*	5		3.88*	14
H_2 (293 K)	0.8045*	5	N_2	1.7403	6,8
H_2 (293 K)	0.8042	6	NO	1.70	2
H_2 (322 K)	0.8059	8	Na_2	40	22
HBr	3.61	3		38	21
HCl	2.63	3	NaK	51	22
	2.77	2	NaLi	40	4
HD (v = 0,J = 0)	0.7976*	5	O_2	1.5812	6
HF	0.80	27	Rb_2	79	22

Table 4
Average Electric Dipole Polarizabilities for Ground State Triatomic Molecules (in Units of 10^{-24} cm^3)

Molecule	Polarizability	Ref.
BeH_2	4.34*	14
CO_2	2.911	8
CS_2	8.74	3
	8.86	2
D_2O	1.26	2
H_2O	1.45	2
H_2S	3.78	3
	3.95	2
HCN	2.59	3
	2.46	2
$HgBr_2$	14.5	2
$HgCl_2$	11.6	2
HgI_2	19.1	2
N_2O	3.03	8
NO_2	3.02	2†
Na_3	70	21
O_3	3.21	2
OCS	5.71	2
	5.2	15
SO_2	3.72	3
	4.28	2

Table 5
Average Electric Dipole Polarizabilities for Ground State Inorganic Polyatomic Molecules (Larger than Triatomic) (in Units of 10^{-24} cm^3)

Molecule	Polarizability	Ref.
$AsCl_3$	14.9	2
AsN_3	5.75	2
BCl_3	9.38	20
BF_3	3.31	2
$(BN_3)_2$	5.73	2
$(BH_2N)_3$	8.0	2†
ClF_3	6.32	2
$(CsBr)_2$	54.5	16
$(CsCl)_2$	42.4	16
$(CsF)_2$	28.4	16
$(CsI)_2$	51.8	16
$GeCl_4$	15.1	2
GeH_3Cl	6.7	2†
$(HgCl)_2$	14.7	9
K_n	$n = 2,5,7\text{-}9,11,20$	21
$(KBr)_2$	42.0	16
$(KCl)_2$	32.1	16
$(KF)_2$	21.0	16
$(KI)_2$	36.3	16
$(LiBr)_2$	18.9	16
$(LiCl)_2$	13.1	16
$(LiF)_2$	6.9	16
$(LiI)_2$	23.4	16
ND_3	1.70	2
NF_3	3.62	2
NH_3	2.81	20
	2.10	2
	2.26	3
$(NO_2)_2$	6.69	2
Na_n	$n = 1\text{-}40$	21
$(NaBr)_2$	26.8	16
$(NaCl)_2$	23.4	16
$(NaF)_2$	20.7	16
$(NaI)_2$	26.9	16
OsO_4	8.17	2
PCl_3	12.8	2
PF_5	6.10	2
PH_3	4.84	2
$(RbBr)_2$	48.2	16
$(RbCl)_2$	43.2	16
$(RbF)_2$	40.7	16
$(RbI)_2$	46.3	16
SF_6	6.54	8
$(SF_5)_2$	13.2	2
SO_3	4.84	2
SO_2Cl_2	10.5	2
SeF_6	7.33	2
SiF_4	5.45	2
SiH_4	5.44	2
$(SiH_3)_2$	11.1	2
$SiHCl_3$	10.7	2
SiH_2Cl_2	8.92	2
SiH_3Cl	7.02	2
$SnBr_4$	22.0	2
$SnCl_4$	18.0	2
	13.8	15
SnI_4	32.3	2
TeF_6	9.00	2
$TiCl_4$	16.4	2
UF_6	12.5	2

Table 6
Average Electric Dipole Polarizabilities for Ground State Hydrocarbon Molecules (in Units of 10^{-24} cm^3)

Molecule	Name	Polarizability	Ref.
CH_4	methane	2.593	8
C_2H_2	acetylene	3.33	3
		3.93	2
C_2H_4	ethylene	4.252	8
C_2H_6	ethane	4.47	3
		4.43	2
C_3H_4	propyne	6.18	2
C_3H_6	propene	6.26	2
	cyclopropane	5.66	2
C_3H_8	propane	6.29	3
		6.37	2
C_4H_6	1-butyne	7.41	2†
	1,3-butadiene	8.64	2
C_4H_8	1-butene	7.97	2
		8.52	2
	trans-2-butene	8.49	2
	2-methylpropene	8.29	2
C_4H_{10}	*n*-butane	8.20	2
	isobutane	8.14	27
C_5H_6	1,3-cyclopentadiene	8.64	2
C_5H_8	1-pentyne	9.12	2
	trans-1,3-pentadiene	10.0	2
	isoprene	9.99	2
C_5H_{10}	cyclopentane	9.15	18
	1-pentene	9.65	27
	2-pentene	9.84	27
C_5H_{12}	pentane	9.99	2
	neopentane	10.20	18
C_6H_6	benzene	10.0	25
		10.32	3
		10.74	2
C_6H_{10}	1-hexyne	10.9	2†
	2-ethyl-1,3-butadiene	11.8	2†
	3-methyl-1,3-pentadiene	11.8	2†
	2-methyl-1,3-pentadiene	12.1	2†
	2,3-dimethyl-1,3-butadiene	11.8	2†
	cyclohexene	10.7	2†
C_6H_{12}	cyclohexane	11.0	18
		10.87	15
	1-hexene	11.7	27
C_6H_{14}	*n*-hexane	11.9	2
C_7H_8	toluene	11.8	25
		12.26	15
		12.3	2
C_7H_{12}	1-heptyne	12.8	2†
C_7H_{14}	methylcyclohexane	13.1	2
	1-heptene	13.5	27
C_7H_{16}	*n*-heptane	13.7	2
C_8H_8	styrene	15.0	2
		14.4	27
C_8H_{10}	ethylbenzene	14.2	2
	o-xylene	14.9	2
		14.1	15
	p-xylene	13.7	25
		14.2	15
		14.9	2
	m-xylene	14.2	15
C_8H_{16}	ethylcyclohexane	15.9	2
C_8H_{18}	*n*-octane	15.9	2
	3-methylheptane	15.4	27
	2,2,4-trimethylpentane	15.4	27
C_9H_{10}	α-methylstyrene	16.05	27
C_9H_{12}	isopropylbenzene	16.0	2–
	mesitylene	15.5	25
		16.1	27
C_9H_{18}	isopropylcyclohexane	17.2	2
C_9H_{20}	nonane	17.4	27
$C_{10}H_8$	napthalene	16.5	17
		17.5	27
$C_{10}H_{14}$	durene	17.3	25
	t-butylbenzene	17.2	25
		17.8	2†
$C_{10}H_{20}$	*t*-butylcyclohexane	19.8	2
$C_{10}H_{22}$	decane	19.1	27
$C_{11}H_{10}$	α-methylnaphthalene	19.35	27
	β-methylnaphthalene	19.52	27
$C_{11}H_{14}$	α,β,β-trimethylstyrene	19.64	27
$C_{11}H_{16}$	pentamethylbenzene	19.1	25
$C_{11}H_{24}$	undecane	21.0	27
$C_{12}H_{10}$	acenaphthene	20.6	27
$C_{12}H_{12}$	α-ethylnaphthalene	21.19	27
	β-ethylnaphthalene	21.36	27
$C_{12}H_{18}$	hexamethylbenzene	20.9	25
$C_{12}H_{26}$	dodecane	22.8	27
$C_{13}H_{10}$	fluorene	21.7	27
$C_{14}H_{10}$	anthracene	25.4	17
		25.9	27
	phenanthrene	36.8*	17
		24.7	27
$C_{14}H_{22}$	*p*-di-*t*-butylbenzene	24.5	25
$C_{16}H_{10}$	pyrene	28.2	27
$C_{17}H_{12}$	2,3-benzfluorene	30.2	27
$C_{18}H_{12}$	napthacene	32.3	27
	1,2-benzanthracene	32.9	27
	chrysene	33.1	27
	triphenylene	31.1	27
$C_{18}H_{30}$	1,3,5-tri-*t*-butylbenzene	31.8	25
$C_{24}H_{12}$	coronene		

Table 7
Average Electric Dipole Polarizabilities for Ground State Organic Halides (in Units of 10^{-24} cm^3)

Molecule	Name	Polarizability	Ref.
CBr_2F_2	dibromodifluoromethane	9.0	2†
$CClF_3$	chlorotrifluoromethane	5.72	20
		5.59	2
CCl_2F_2	dichlorodifluoromethane	7.93	20
		7.81	2
CCl_2O	phosgene	7.29	2
CCl_2S	thiophosgene	10.2	2
CCl_3F	trichlorofluoromethane	9.47	2
CCl_3NO_2	trichloronitromethane	10.8	2†
CCl_4	carbon tetrachloride	11.2	2
		10.5	3
CF_4	carbon tetrafluoride	3.838	8
CF_2O	carbonylfluoride	1.88*	17
$CHBr_3$	bromoform	11.8	17
$CHBrF_2$	bromodifluoromethane	5.7	2†
$CHClF_2$	chlorodifluoromethane	6.38	20
		5.91	2
$CHCl_2F$	dichlorofluoromethane	6.82	2
$CHCl_3$	chloroform	9.5	8
		8.23	27
CHF_3	fluoroform	3.52	20
		3.57	8
$CHFO$	fluoroformaldehyde	1.76*	17
CHI_3	iodoform	18.0	17
CH_2Br_2	dibromomethane	9.32	2
		8.62	27
CH_2ClNO_2	chloronitromethane	6.9	2†
CH_2Cl_2	dichloromethane	6.48	3
		7.93	2
CH_2I_2	diiodomethane	12.90	27
CH_3Br	bromomethane	5.87	20
		6.03	2
		5.55	15
CH_3Cl	chloromethane	5.35	20
		4.72	8
CH_3F	fluoromethane	2.97	8
CH_3I	iodomethane	7.97	2
C_2ClF_5	chloropentafluoroethane	6.3	2†
$C_2Cl_2F_4$	1,2-dichlorotetrafluoroethane	8.5	2†
C_2Cl_3N	trichloroacetonitrile	6.10	18
C_2F_6	hexafluoroethane	6.82	2
C_2HBr	bromoacetylene	7.39	2
C_2HCl	chloroacetylene	6.07	2
C_2HCl_5	pentachloroethane	14.0	2
$C_2H_2Cl_2$	1,1-dichloroethylene	7.83	27
	trans-1,2-dichloroethylene	8.15	27
	cis-1,2-dichloroethylene	8.03	27
$C_2H_2Cl_2F_2$	1,1-dichloro-2,2-difluoroethane	8.4	2†
$C_2H_2Cl_2O$	chloroacetyl chloride	8.92	2
$C_2H_2Cl_3F$	1,2,2-trichloro-1-fluoroethane	10.2	2†
$C_2H_2Cl_4$	1,1,2,2-tetrachloroethane	12.1	2†
C_2H_2ClN	chloroacetonitrile	6.10	18
$C_2H_2F_2$	1,1-difluoroethylene	5.01	20
C_2H_3Br	bromoethylene	7.59	2
C_2H_3Cl	chloroethylene	6.41	2

Table 7
Average Electric Dipole Polarizabilities for Ground State Organic Halides (in Units of 10^{-24} cm^3) (continued)

Molecule	Name	Polarizability	Ref.
$C_2H_3ClF_2$	1-chloro-1,1-difluoroethane	8.05	2
C_2H_3ClO	acetyl chloride	6.62	2
$C_2H_3ClO_2$	ethyl chloroformate	7.1	2†
$C_2H_3Cl_3$	1,1,1-trichloroethane	10.7	2
$C_2H_3F_3$	1,1,1-trifluoroethane	4.4	2†
C_2H_3I	iodoethylene	9.3	2†
C_2H_4BrCl	1-bromo-2-chloroethane	9.5	2†
$C_2H_4Br_2$	1,2-dibromomethane	10.7	2†
C_2H_4ClF	1-chloro-2-fluoroethane	6.5	2†
$C_2H_4ClNO_2$	1-chloro-1-nitroethane	10.9	2
$C_2H_4Cl_2$	1,1-dichloroethane	8.64	2
	1,2-dichloroethane	8.0	2†
C_2H_5Br	bromoethane	8.05	2
		7.28	27
C_2H_5Cl	chloroethane	7.27	20
		8.29	2
		6.4	15
C_2H_5ClO	2-chloroethanol	7.1	2†
		6.88	27
	chloromethyl methyl ether	7.1	2†
C_2H_5F	fluoroethane	4.96	2
C_2H_5I	iodoethane	10.0	2
$C_3H_4Cl_2$	dichloropropene	10.1	2†
C_3H_5Cl	chloropropene	8.3	2
C_3H_5ClO	chloroacetone	8.4	2†
$C_3H_3ClO_2$	ethyl chloroformate	9.0	2†
$C_3H_6ClNO_2$	1-chloro-1-nitropropane	10.4	2†
$C_3H_6Cl_2$	dichloropropane	10.9	2†
C_3H_7Br	1-bromopropane	9.4	2†
		9.07	27
	2-bromopropane	9.6	2†
C_3H_7Cl	chloropropane	10.0	2
C_3H_7ClO	β-chloroethyl methyl ether	8.71	27
	2-chloro-1-propanol	8.89	27
	3-chloro-1-propanol	8.84	27
C_3H_7I	1-iodopropane	11.5	2†
C_4H_5Cl	4-chloro-1,2-butadiene	10.0	2†
C_4H_7Cl	1-chloro-2-methylpropene	10.8	2
$C_4H_7ClO_2$	2-chlorobutyric acid	10.7	27
	3-chlorobutyric acid	10.7	27
	4-chlorobutyric acid	10.6	27
$C_4H_8Cl_2$	1,4-dichlorobutane	12.0	2†
C_4H_9Br	bromobutane	13.9	2
		10.86	27
C_4H_9Cl	1-chlorobutane	11.3	2
	1-chloro-2-methylpropane	11.1	2
	2-chloro-2-methylpropane	12.5	2†
	2-chlorobutane	12.4	2
C_4H_9ClO	β-chloroethyl ethyl ether	10.56	27
	2-chloro-1-butanol	10.70	27
	3-chloro-1-butanol	10.38	27
C_4H_9I	1-iodobutane	13.3	2†
		12.65	27
$C_5H_9ClO_2$	2-chlorobutanoate acid	12.33	27
	3-chlorobutanoate acid	12.31	27

Table 7
Average Electric Dipole Polarizabilities for Ground State Organic Halides (in Units of 10^{-24} cm^3) (continued)

Molecule	Name	Polarizability	Ref.
	4-chlorobutanoate acid	12.27	27
	2-chloropentanoic acid	12.69	27
	3-chloropentanoic acid	12.57	27
	4-chloropentanoic acid	12.53	27
$C_5H_{11}Br$	1-bromopentane	13.1	2†
$C_5H_{11}Cl$	1-chloropentane	12.0	2†
$C_5H_{11}F$	fluoropentane	9.95	27
C_6F_6	hexafluorobenzene	9.58	27
C_6HF_5	pentafluorobenzene	9.63	27
$C_6H_2Cl_2O_2$	2,5-dichloro-1,4-benzoquinone	18.4	2
$C_6H_2F_4$	1,2,3,4-tetrafluorobenzene	9.69	27
	1,2,4,5-tetrafluorobenzene	9.69	27
$C_6H_3F_3$	1,3,5-trifluorobenzene	9.74	27
C_6H_4BrF	*p*-bromofluorobenzene	13.4	2†
$C_6H_4ClNO_2$	chloronitrobenzene	14.6	2†
$C_6H_4Cl_2$	*o*-dichlorobenzene	14.17	27
	m-dichlorobenzene	14.23	27
	p-dichlorobenzene	14.20	27
C_6H_4FI	*p*-fluoroiodobenzene	15.5	2†
$C_6H_4FNO_2$	*p*-fluoronitrobenzene	12.8	2†
$C_6H_4F_2$	*o*-difluorobenzene	9.80	27
	m-difluorobenzene	10.3	2†
	p-difluorobenzene	9.80	27
C_6H_5Br	bromobenzene	14.7	2
		13.62	27
C_6H_5Cl	chlorobenzene	14.1	2
		12.3	15
C_6H_5ClO	chlorophenol	13.0	2†
C_6H_5F	fluorobenzene	10.3	2
C_6H_5I	iodobenzene	15.5	2†
$C_6H_{11}ClO_2$	2-chlorobutanoate acid	14.16	27
	3-chlorobutanoate acid	14.13	27
	4-chlorobutanoate acid	14.11	27
$C_6H_{13}Br$	bromohexane	14.44	27
$C_6H_{13}F$	fluorohexane	11.80	27
C_7H_7Br	*p*-bromotoluene	14.80	27
C_7H_7Cl	*p*-chlorotoluene	13.70	27
C_7H_7F	*p*-fluorotoluene	11.70	27
C_7H_7I	*p*-iodotoluene	17.10	27
$C_7H_{15}Br$	1-bromoheptane	16.8	2†
		16.23	27
$C_7H_{15}F$	fluoroheptane	13.66	27
$C_8H_{17}Br$	bromooctane	18.02	27
$C_8H_{17}F$	fluorooctane	15.46	27
$C_9H_{19}Br$	bromononane	19.81	27
$C_9H_{19}F$	fluorononane	17.34	27
$C_{10}F_8$	octafluoronaphthalene	17.64	27
$C_{10}H_7Br$	α-bromonaphthalene	20.34	27
$C_{10}H_7Cl$	α-chloronaphthalene	19.30	27
	β-chloronaphthalene	19.58	27
$C_{10}H_7I$	α-iodonaphthalene	22.41	27
	β-iodonaphthalene	22.95	27
$C_{10}H_{21}Br$	bromodecane	21.60	27
$C_{10}H_{21}F$	fluorodecane	19.18	27

Table 7
Average Electric Dipole Polarizabilities for Ground State Organic Halides (in Units of 10^{-24} cm^3) (continued)

Molecule	Name	Polarizability	Ref.
$C_{11}H_{23}F$	fluoroundecane	21.00	27
$C_{12}H_{25}Br$	bromododecane	25.18	27
$C_{12}H_{25}F$	fluorododecane	22.83	27
$C_{12}H_8Br_2O$	4,4′–dibromodiphenyl ether	27.8	$2^{\dagger}$
$C_{12}H_9BrO$	4-bromodiphenyl ether	24.2	$2^{\dagger}$
$C_{13}H_{11}BrO$	*p*-bromophenyl-*p*-tolyl ether	26.6	$2^{\dagger}$
$C_{14}H_9Br$	9-bromoanthracene	28.32	27
$C_{14}H_9Cl$	9-chloroanthracene	27.35	27
$C_{14}H_9F$	fluoranthracene	28.34	27
$C_{14}H_{29}F$	fluorotetradecane	26.57	27
$C_{16}H_{33}Br$	bromohexadecane	32.34	27
$C_{18}H_{37}Br$	bromooctadecane	35.92	27

Table 8
Static Average Electric Dipole Polarizabilities for Other Ground State Organic Molecules (in Units of 10^{-24} cm^3)

Molecule	Name	Polarizability	Ref.
CN_4O_8	tetranitromethane	15.3	2
CH_2O	formaldehyde	2.8	$2^{\dagger}$
		2.45	18
CH_2O_2	formic acid	3.4	$2^{\dagger}$
CH_3NO	formamide	4.2	$2^{\dagger}$
		4.08	18
CH_3NO_2	nitromethane	7.37	2
CH_4O	methanol	3.29	2
		3.23	15
		3.32	18
CH_5N	methyl amine	4.7	2
		4.01	19
C_2N_2	cyanogen	7.99	2
C_2H_2O	ketene	4.4	$2^{\dagger}$
C_2H_3N	acetonitrile	4.40	$2^{\dagger}$
		4.48	18
C_2H_4O	acetaldehyde	4.6	$2^{\dagger}$
		4.59	18
	ethylene oxide	4.43	18
$C_2H_4O_2$	acetic acid	5.1	$2^{\dagger}$
	methyl formate	5.05	27
$C_2H_4O_4$	formic acid dimer	12.7	2
C_2H_5NO	acetamide	5.67	18
	N-methyl formamide	5.91	18
$C_2H_5NO_2$	nitroethane	9.63	2
	ethyl nitrite	7.0	15
C_2H_6O	ethanol	5.41	2
		5.11	18
	methyl ether	5.29	20
		5.84	2
		5.16	15
$C_2H_6O_2$	ethylene glycol	5.7	$2^{\dagger}$
		5.61	27
$C_2H_6O_2S$	dimethyl sulfone	7.3	$2^{\dagger}$

Table 8
Static Average Electric Dipole Polarizabilities for Other Ground State Organic Molecules (in Units of 10^{-24} cm^3) (continued)

Molecule	Name	Polarizability	Ref.
C_2H_6S	ethanethiol	7.41	2
C_2H_7N	ethyl amine	7.10	2
	dimethyl amine	6.37	2
$C_2H_8N_2$	ethylene diamine	7.2	2†
$C_3H_2N_2$	malononitrile	5.79	18
C_3H_3N	acrylonitrile	8.05	2
$C_3H_4N_2$	pyrazole	7.23	27
C_3H_4O	propenal	6.38	2†
C_3H_5N	propionitrile	6.70	2
		6.24	18
C_3H_6O	acetone	6.33	15
		6.4	2†
		6.39	18
	allyl alcohol	7.65	2
	propionaldehyde	6.50	2
$C_3H_6O_2$	propionic acid	6.9	2†
	ethyl formate	8.01	2
		6.88	27
	methyl acetate	6.94	2
		6.81	27
$C_3H_6O_3$	dimethyl carbonate	7.7	2†
C_3H_7NO	*N*-methyl acetamide	7.82	18
	N,N-dimethyl formamide	7.81	18
$C_3H_7NO_2$	nitropropane	8.5	2†
C_3H_8O	2-propanol	7.61	2
		6.97	18
	1-propanol	6.74	2
	ethyl methyl ether	7.93	2
$C_3H_8O_2$	dimethoxymethane	7.7	2†
	monomethylether ethylene glycol	7.44	27
C_3H_9N	*n*-propylamine	7.70	27
		9.20	2
	isopropylamine	7.77	27
	trimethylamine	8.15	2
$C_4H_2N_2$	fumaronitrile	11.8	2
$C_4H_4N_2$	succinonitrile	8.1	2†
	pyrimidene	8.53*	17
	pyridazine	9.27*	17
$C_4H_4O_2$	diketene	8.0	2†
C_4H_4S	thiophene	9.67	2
C_4H_5N	methacrylonitrile	8.0	2†
	trans-crotononitrile	8.2	2†
$C_4H_6N_2$	*N*-methylpyrazole	8.99	27
C_4H_6O	crotonaldehyde	8.5	2†
	methacrylaldehyde	8.3	2†
$C_4H_6O_2$	biacetyl	8.2	2†
$C_4H_6O_3$	acetic anhydride	8.9	2†
C_4H_6S	divinyl sulfide	10.9	2†
C_4H_7N	butyronitrile	8.4	2†
	isobutyronitrile	8.05	18
C_4H_8O	butanal	8.2	2†
	methyl ethyl ketone	8.13	15
	trans-2,3-epoxy butane	8.22*	17
$C_4H_8O_2$	ethyl acetate	9.7	2
		8.62	27

Table 8
Static Average Electric Dipole Polarizabilities for Other Ground State Organic Molecules (in Units of 10^{-24} cm^3) (continued)

Molecule	Name	Polarizability	Ref.
	1,4-dioxane	10.0	2
	p-dioxane	8.60	18
	2-methyl-1,3-dioxolane	9.44	15
	butyric acid	8.38	27
	methyl propionate	8.97	27
$C_4H_9NO_2$	1-nitrobutane	10.4	2†
	2-methyl-2-nitropropane	10.3	2†
$C_4H_{10}O$	ethyl ether	10.2	2
		8.73	15
	1-butanol	8.88	2
	2-methylpropanol	8.92	2
	n-propyl methyl ether	8.86	27
$C_4H_{10}O_2$	monoethylether ethylene glycol	9.28	27
$C_4H_{10}S$	ethyl sulfide	10.8	2
$C_4H_{11}N$	*n*-butylamine	13.5	2
	diethylamine	10.2	2
		9.61	27
C_5H_5N	pyridine	9.5	15
		9.18	27
	4-cyano-1,2-butadiene	10.5	2†
$C_5H_8N_2$	1,5-dimethylpyrazole	10.72	27
$C_5H_8O_2$	acetyl acetone	10.5	2†
C_5H_9N	valeronitrile	10.4	2
	22-DMPN	9.59	18
$C_5H_{10}O$	diethyl ketone	9.93	15
	methyl propyl ketone	9.93	15
$C_5H_{10}O_2$	ethyl propionate	10.41	27
	methyl butanoate	10.41	27
$C_5H_{10}O_3$	diethyl carbonate	11.3	2
$C_5H_{12}O$	ethyl propyl ether	10.68	27
$C_5H_{12}O_4$	tetramethyl orthocarbonate	13.0	2†
$C_6H_4N_2O_4$	*p*-dinitrobenzene	18.4	2
$C_6H_4O_2$	*p*-benzoquinone	14.5	2
$C_6H_5NO_2$	nitrobenzene	14.7	2
		12.92	15
C_6H_6O	phenol	11.1	2†
		9.94*	17
C_6H_7N	aniline	12.1	2†
$C_6H_8N_2$	phenylenediamine	13.8	2†
	phenylhydrazine	12.91	27
$C_6H_{10}N_2$	1-ethyl-5-methylpyrazole	12.50	27
$C_6H_{10}O_3$	ethyl acetoacetate	12.9	2†
$C_6H_{12}N_2$	dimethylketazine	15.6	2
$C_6H_{12}O$	cyclohexanol	11.56	18
$C_6H_{12}O_2$	amyl formate	14.2	2
$C_6H_{12}O_3$	paraldehyde	17.9	2
$C_6H_{14}O$	propyl ether	12.8	2
		12.5	15
$C_6H_{14}O_2$	1,1-diethoxyethane	13.2	2†
	1,2-diethoxyethane	11.3	2†
$C_6H_{15}N$	triethylamine	13.1	2
		13.38	27
	di-*n*-propylamine	13.29	27
$C_7H_4N_2O_2$	*p*-cyanonitrobenzene	19.0	2
C_7H_5N	benzonitrile	12.5	2†

Table 8
Static Average Electric Dipole Polarizabilities for Other Ground State Organic Molecules (in Units of 10^{-24} cm^3) (continued)

Molecule	Name	Polarizability	Ref.
$C_7H_7NO_3$	nitroanisole	15.7	2†
C_7H_7O	anisole	13.1	2†
C_7H_9NO	*o*-anisidine	14.2	2†
$C_7H_{10}N_2$	1,1-methylphenylhydrazine	14.81	27
$C_7H_{14}O$	cyclohexyl methyl ether	13.4	2†
	2,4-dimethyl-3-pentanone	13.5	15
$C_7H_{14}O_2$	pentyl acetate	14.9	2
$C_8H_4N_2$	*p*-dicyanobenzene	19.2	2
$C_8H_6N_2$	quinoxaline	15.13	27
C_8H_8O	acetophenone	15.0	2
$C_8H_8O_2$	2,5-dimethyl-1,4-benzoquinone	18.8	2
$C_8H_{10}O$	phenetole	14.9	2
$C_8H_{11}N$	*N*-dimethylaniline	16.2	2†
$C_8H_{12}N_2$	1,1-ethylphenylhydrazine	16.62	27
$C_8H_{12}O_2$	ethyl sorbate	17.2	2†
	tetramethylcyclobutane-1,3-dione	18.6	2
$C_8H_{14}O_4$	diethyl succinate	16.8	2†
$C_8H_{18}O$	*n*-butyl ether	17.2	2
C_9H_7N	quinoline	15.70	27
	isoquinoline	16.43	27
$C_9H_{10}O_2$	ethyl benzoate	16.9	2†
$C_9H_{21}N$	tri-*n*-propylamine	18.87	27
$C_{10}H_9N$	α-naphthylamine	19.50	27
	β-naphthylamine	19.73	27
	1-methylquinoline	18.65	27
	1-methylisoquinoline	18.28	27
$C_{10}H_{10}Fe$	ferrocene	17.1	26
$C_{10}H_{10}N_2$	2,3-dimethylquinoxaline	18.70	27
$C_{10}H_{14}BeO_4$	beryllium acetylacetonate	34.1	2
$C_{11}H_8O$	1-naphthaldehyde	19.75	27
	2-naphthaldehyde	20.06	27
$C_{14}H_8O_2$	anthraquinone	24.46	27
$C_{12}H_8N_2$	phenazine	23.43	27
$C_{12}H_9NO_3$	4-nitrodiphenyl ether	24.7	2†
$C_{14}H_{14}O$	di-*p*-tolyl ether	24.9	2†
$C_{15}H_{21}AlO_6$	aluminum acetylacetonate	51.9	2
$C_{15}H_{21}CrO_6$	chromium acetylacetonate	53.7	2
$C_{15}H_{21}FeO_6$	ferric acetylacetonate	58.1	2
$C_{20}H_{28}O_8Th$	thorium acetylacetonate	79.0	2
C_{60}	buckminsterfullerene	~80	24

Note: All polarizabilities in the tables are experimental values except those values marked by an asterisk (*), which indicates a calculated result. The experimental polarizabilities are mostly determined by measurements of a dielectric constant or refractive index which are quite accurate (0.5% or better). However, one should treat many of the results with several percent of caution because of the age of the data and because some of the results refer to optical frequencies rather than static. Comments given with the references are intended to allow one to judge the degree of caution required. Interested persons should consult these references. In many cases, the reference given is to a theoretical paper in which the experimental results are quoted. These papers, noted in the References, contain valuable information on polarizability calculations and experimental data which often includes the tensor components of the polarizability.

REFERENCES

1. McCullough, E. A., Jr., *J. Chem. Phys.*, 63, 5050, 1975. This calculation is for the parallel component, not the average polarizability.
2. Maryott, A. A., and Buckley, F., U. S. National Bureau of Standards Circular No. 537, 1953. A tabulation of dipole moments, dielectric constants, and molar refractions measured between 1910 and 1952, and used here to determine polarizabilities if no more recent result exists. The polarizability is $3/(4\pi N_o)$ times the molar polarization or molar refraction, where N_o is Avogadro's number. The value $3/(4\pi N_o) = 0.3964308 \times 10^{-24}$ cm^3 was used for this conversion. A dagger (†) following the reference number in the tables indicates that the polarizability was derived from the molar refraction and hence may not include some low-frequency contributions to the static polarizability; these "static" polarizabilities are therefore low by 1 to 30%.
3. Hirschfelder, J. O., Curtis, C. F., and Bird, R. B., *Molecular Theory of Gases and Liquids*, Wiley, New York, 1954, p. 950. Fundamental information on molecular polarizabilities.
4. Miller, T. M., and Bederson, B., *Adv. At. Mol. Phys.*, 13, 1, 1977. Review emphasizing atomic polarizabilities and measurement techniques. The data quoted in Table 3 are accurate to 8 to 12%.
5. Kolos, W., and Wolniewicz, L., *J. Chem. Phys.*, 46, 1426, 1967. Highly accurate molecular hydrogen calculations.
6. Newell, A. C., and Baird, R. C., *J. Appl. Phys.*, 36, 3751, 1965. Highly accurate refractive index measurements at 47.7 GHz (essentially static).
7. Jao, T. C., Beebe, N. H. F., Person, W. B., and Sabin, J. R., *Chem. Phys. Lett.*, 26, 47, 1974. Tensor polarizabilities, derivatives, and other results are reported.
8. Orcutt, R. H., and Cole, R. H., *J. Chem. Phys.*, 46, 697, 1967 (He, Ne, Ar, Kr, H_2, N_2); Sutter, H., and Cole, R. H., *J. Chem. Phys.*, 52, 132, 1970 (CF_3H, CFH_3, $CClF_3$, $CClH_3$); Bose, T. K., and Cole, R. H., *J. Chem. Phys.*, 52, 140, 1970 (CO_2), and 54, 3829, 1971 (C_2H_2, C_2H_4); Nelson, R. D., and Cole, R. H., *J. Chem. Phys.*, 54, 4033, 1971 (SF_6, $CClF_3$); Bose, T. K., Sochanski, J. S., and Cole, R. H., *J. Chem. Phys.*, 57, 3592, 1972 (CH_4, CF_4); Kirouac, S., and Bose, T. K., *J. Chem. Phys.*, 59, 3043, 1973 (N_2O), and 64, 1580, 1976 (He). Highly accurate dielectric constant measurements. These modern data give the most accurate polarizabilities available. A criticism of the interpretation of these data in the case of polar molecules is given in Ref. 20, p. 2905.
9. Huestis, D. L., Technical Report #MP 78-25, SRI International (project PYU 6158), Menlo Park, CA 94025. Molar refractions for mercury-chlorine compounds are analyzed.
10. Bounds, D. G., Clarke, J. H. R., and Hinchliffe, A., *Chem. Phys. Lett.*, 45, 367, 1977. Theoretical tensor polarizability for LiCl.
11. Kolker, H. J., and Karplus, M., *J. Chem. Phys.*, 39, 2011, 1963. Theoretical.
12. Cutschick, V. P., and McKoy, V., *J. Chem. Phys.*, 58, 2397, 1973. Theoretical tensor polarizabilities.
13. Gready, J. E., Bacskay, G. B., and Hush, N. S., *Chem. Phys.*, 22, 141, 1977, and 23, 9, 1977. Theoretical.
14. Amos, A. T., and Yoffe, J. A., *J. Chem. Phys.*, 63, 4723, 1975. Theoretical.
15. Stuart, H. A., *Landolt-Börnstein Zahlenwerte and Funktionen*, Vol. 1, Part 3, Eucken, A., and Hellwege, K. H., Eds., Springer-Verlag, Berlin, 1951, p. 511. Tabulation of molecular polarizabilities. Two misprints in the chemical symbols have been corrected.
16. Guella, T., Miller, T. M., Stockdale, J. A. D., Bederson, B., and Vuskovic, L., *J. Chem. Phys.*, 94, 6857, 1991. Beam measurements with accuracies between 12-24%.
17. Marchese, F. T., and Jaffé, *Theoret. Chim. Acta (Berlin)*, 45, 241, 1977. Theoretical and experimental tensor polarizabilities are tabulated in this paper.
18. Applequist, J., Carl, J. R., and Fung, K.-K., *J. Am. Chem. Soc.*, 94, 2952, 1972. Excellent reference on the calculation of molecular polarizabilities, including extensive tables of tensor polarizabilities, both theoretical and experimental, at 589.3 nm wavelength.
19. Bridge, N. J., and Buckingham, A. D., *Proc. Roy. Soc. (London)*, A295, 334, 1966. Measured tensor polarizabilities at 633 nm wavelength.
20. Barnes, A. N. M., Turner, D. J., and Sutton, L. E., *Trans. Faraday Soc.*, 67, 2902, 1971. Dielectric constants yielding polarizabilities accurate from 0.3-8%.
21. Knight, W. D., Clemenger, K., de Heer, W. A., and Saunders, W. A., *Phys. Rev. B*, 31, 2539, 1985. These data probably correspond to a very low internal temperature.
22. Tarnovsky, V., Bunimovicz, M., Vuskovic, L., Stumpf, B., and Bederson, B., *J. Chem. Phys.*, 98, 3894, 1993. These data correspond to internal temperatures 480-948 K.
23. Milani, P., Moullet, I., and de Heer, W. A., *Phys. Rev. A*, 42, 5150, 1990. Beam measurements accurate to 11%.
24. Bonin, K. D., and Kadar-Kallen, M. A., *Int. J. Mod. Phys.*, 8, 3313, 1994. Review article.
25. Aroney, M. J., and Pratten, S. J., *J. Chem. Soc., Faraday Trans. 1*, 80, 1201, 1984. Uncertainties in the range 1-3%.
26. Le Fevre, R. J. W., Murthy, D. S. N., and Saxby, J. D., *Aust. J. Chem.*, 24, 1057, 1971. Kerr effect.
27. No, K. T., Cho, K. H., Jhon, M. S., and Scheraga, H. A., *J. Am. Chem. Soc.*, 115, 2005, 1993. Theoretical; these results are quoted in numerous valuable papers on calculated polarizabilities, e.g., Miller, K. J., and Savchik, J. A., *J. Am. Chem. Soc.*, 101, 7206, 1979.

IONIZATION POTENTIALS OF ATOMS AND ATOMIC IONS

The ionization potentials of neutral and partially ionized atoms are listed in this table. Data were obtained from the compilations cited below, supplemented by results from the recent research literature. All values have been corrected to the currently recommended value of the conversion factor from wave number to energy, namely 1 eV = 8065.541 cm^{-1} (Reference 5). Values are given in eV.

Following the traditional spectroscopic notation, columns are headed I, II, III, etc. up to XXX, where I indicates the neutral atom, II the singly ionized atom, III the doubly ionized atom, etc. The first section of the table includes spectra I to VIII of all the elements; subsequent sections cover higher spectra (ionization stages) for those elements for which data are available.

REFERENCES

1. Moore, C. E., *Ionization Potentials and Ionization Limits Derived from the Analysis of Optical Spectra*, Natl. Stand. Ref. Data Ser. — Natl. Bur. Stand. (U.S.) No. 34, 1970.
2. Martin, W. C., Zalubas, R., and Hagan, L., *Atomic Energy Levels — The Rare Earth Elements*, Natl. Stand. Ref. Data Ser. — Natl. Bur. Stand. (U.S.), No. 60, 1978.
3. Sugar, J. and Corliss, C., *Atomic Energy Levels of the Iron Period Elements: Potassium through Nickel, J. Phys. Chem. Ref. Data*, Vol.14, Suppl. 2, 1985.
4. References to papers in *J. Phys. Chem. Ref. Data*, in the period 1973—91 covering other elements may be found in the cumulative index to that journal.
5. Cohen, E. R. and Taylor, B. N., *J. Phys. Chem. Ref. Data*, 17, 1795, 1988.
6. Martin, W.C., and Wiese, W.L., in *Atomic, Molecular, and Optical Physics Handbook*, Drake, G.W.F., Ed., AIP Press, New York, 1996.

Neutral Atoms to +7 Ions

Z	Element	I	II	III	IV	V	VI	VII	VIII
1	H	13.59844							
2	He	24.58741	54.41778						
3	Li	5.39172	75.64018	122.45429					
4	Be	9.3227	18.21116	153.89661	217.71865				
5	B	8.29803	25.15484	37.93064	259.37521	340.22580			
6	C	11.26030	24.38332	47.8878	64.4939	392.087	489.99334		
7	N	14.53414	29.6013	47.44924	77.4735	97.8902	552.0718	667.046	
8	O	13.61806	35.11730	54.9355	77.41353	113.8990	138.1197	739.29	871.4101
9	F	17.42282	34.97082	62.7084	87.1398	114.2428	157.1651	185.186	953.9112
10	Ne	21.5646	40.96328	63.45	97.12	126.21	157.93	207.2759	239.0989
11	Na	5.13908	47.2864	71.6200	98.91	138.40	172.18	208.50	264.25
12	Mg	7.64624	15.03528	80.1437	109.2655	141.27	186.76	225.02	265.96
13	Al	5.98577	18.82856	28.44765	119.992	153.825	190.49	241.76	284.66
14	Si	8.15169	16.34585	33.49302	45.14181	166.767	205.27	246.5	303.54
15	P	10.48669	19.7694	30.2027	51.4439	65.0251	220.421	263.57	309.60
16	S	10.36001	23.3379	34.79	47.222	72.5945	88.0530	280.948	328.75
17	Cl	12.96764	23.814	39.61	53.4652	67.8	97.03	114.1958	348.28
18	Ar	15.75962	27.62967	40.74	59.81	75.02	91.009	124.323	143.460
19	K	4.34066	31.63	45.806	60.91	82.66	99.4	117.56	154.88
20	Ca	6.11316	11.87172	50.9131	67.27	84.50	108.78	127.2	147.24
21	Sc	6.5615	12.79967	24.75666	73.4894	91.65	110.68	138.0	158.1
22	Ti	6.8281	13.5755	27.4917	43.2672	99.30	119.53	140.8	170.4
23	V	6.7463	14.66	29.311	46.709	65.2817	128.13	150.6	173.4
24	Cr	6.7665	16.4857	30.96	49.16	69.46	90.6349	160.18	184.7
25	Mn	7.43402	15.63999	33.668	51.2	72.4	95.6	119.203	194.5
26	Fe	7.9024	16.1878	30.652	54.8	75.0	99.1	124.98	151.06
27	Co	7.8810	17.083	33.50	51.3	79.5	102.0	128.9	157.8
28	Ni	7.6398	18.16884	35.19	54.9	76.06	108	133	162
29	Cu	7.72638	20.29240	36.841	57.38	79.8	103	139	166
30	Zn	9.3942	17.96440	39.723	59.4	82.6	108	134	174
31	Ga	5.99930	20.5142	30.71	64				
32	Ge	7.8994	15.93462	34.2241	45.7131	93.5			
33	As	9.7886	18.633	28.351	50.13	62.63	127.6		
34	Se	9.75238	21.19	30.8204	42.9450	68.3	81.7	155.4	
35	Br	11.81381	21.8	36	47.3	59.7	88.6	103.0	192.8
36	Kr	13.99961	24.35985	36.950	52.5	64.7	78.5	111.0	125.802
37	Rb	4.17713	27.285	40	52.6	71.0	84.4	99.2	136
38	Sr	5.6949	11.03013	42.89	57	71.6	90.8	106	122.3
39	Y	6.2171	12.24	20.52	60.597	77.0	93.0	116	129
40	Zr	6.63390	13.13	22.99	34.34	80.348			
41	Nb	6.75885	14.32	25.04	38.3	50.55	102.057	125	

Z	Element	I	II	III	IV	V	VI	VII	VIII
42	Mo	7.09243	16.16	27.13	46.4	54.49	68.8276	125.664	143.6
43	Tc	7.28	15.26	29.54					
44	Ru	7.36050	16.76	28.47					
45	Rh	7.45890	18.08	31.06					
46	Pd	8.3369	19.43	32.93					
47	Ag	7.5763	21.49	34.83					
48	Cd	8.9938	16.90832	37.48					
49	In	5.78636	18.8698	28.03	54				
50	Sn	7.3439	14.63225	30.50260	40.73502	72.28			
51	Sb	8.6084	16.53051	25.3	44.2	56	108		
52	Te	9.0096	18.6	27.96	37.41	58.75	70.7	137	
53	I	10.45126	19.1313	33					
54	Xe	12.1298	21.20979	32.1230					
55	Cs	3.89390	23.15745						
56	Ba	5.21170	10.00390						
57	La	5.5770	11.060	19.1773	49.95	61.6			
58	Ce	5.5387	10.85	20.198	36.758	65.55	77.6		
59	Pr	5.464	10.55	21.624	38.98	57.53			
60	Nd	5.5250	10.73	22.1	40.41				
61	Pm	5.58	10.90	22.3	41.1				
62	Sm	5.6436	11.07	23.4	41.4				
63	Eu	5.6704	11.241	24.92	42.7				
64	Gd	6.1501	12.09	20.63	44.0				
65	Tb	5.8638	11.52	21.91	39.79				
66	Dy	5.9389	11.67	22.8	41.47				
67	Ho	6.0215	11.80	22.84	42.5				
68	Er	6.1077	11.93	22.74	42.7				
69	Tm	6.18431	12.05	23.68	42.7				
70	Yb	6.25416	12.1761	25.05	43.56				
71	Lu	5.4259	13.9	20.9594	45.25	66.8			
72	Hf	6.82507	14.9	23.3	33.33				
73	Ta	7.5496							
74	W	7.8640							
75	Re	7.8335							
76	Os	8.4382							
77	Ir	8.9670							
78	Pt	8.9587	18.563						
79	Au	9.2255	20.5						
80	Hg	10.43750	18.756	34.2					
81	Tl	6.1082	20.428	29.83					
82	Pb	7.41666	15.0322	31.9373	42.32	68.8			
83	Bi	7.2856	16.69	25.56	45.3	56.0	88.3		
84	Po	8.41671							
85	At								
86	Rn	10.74850							
87	Fr	4.0727							
88	Ra	5.2784	10.14716						
89	Ac	5.17	12.1						
90	Th	6.3067	11.5	20.0	28.8				
91	Pa	5.89							
92	U	6.19405							
93	Np	6.2657							
94	Pu	6.0262							
95	Am	5.9738							
96	Cm	6.02							
97	Bk	6.23							
98	Cf	6.30							
99	Es	6.42							
100	Fm	6.50							
101	Md	6.58							
102	No	6.65							

IONIZATION POTENTIALS OF ATOMS AND ATOMIC IONS (continued)

+8 Ions to +15 Ions

Z	Element	IX	X	XI	XII	XIII	XIV	XV	XVI
9	F	1103.1176							
10	Ne	1195.8286	1362.1995						
11	Na	299.864	1465.121	1648.702					
12	Mg	328.06	367.50	1761.805	1962.6650				
13	Al	330.13	398.75	442.00	2085.98	2304.1410			
14	Si	351.12	401.37	476.36	523.42	2437.63	2673.182		
15	P	372.13	424.4	479.46	560.8	611.74	2816.91	3069.842	
16	S	379.55	447.5	504.8	564.44	652.2	707.01	3223.78	3494.1892
17	Cl	400.06	455.63	529.28	591.99	656.71	749.76	809.40	3658.521
18	Ar	422.45	478.69	538.96	618.26	686.10	755.74	854.77	918.03
19	K	175.8174	503.8	564.7	629.4	714.6	786.6	861.1	968
20	Ca	188.54	211.275	591.9	657.2	726.6	817.6	894.5	974
21	Sc	180.03	225.18	249.798	687.36	756.7	830.8	927.5	1009
22	Ti	192.1	215.92	265.07	291.500	787.84	863.1	941.9	1044
23	V	205.8	230.5	255.7	308.1	336.277	896.0	976	1060
24	Cr	209.3	244.4	270.8	298.0	354.8	384.168	1010.6	1097
25	Mn	221.8	248.3	286.0	314.4	343.6	403.0	435.163	1134.7
26	Fe	233.6	262.1	290.2	330.8	361.0	392.2	457	489.256
27	Co	186.13	275.4	305	336	379	411	444	511.96
28	Ni	193	224.6	321.0	352	384	430	464	499
29	Cu	199	232	265.3	369	401	435	484	520
30	Zn	203	238	274	310.8	419.7	454	490	542
36	Kr	230.85	268.2	308	350	391	447	492	541
37	Rb	150	277.1						
38	Sr	162	177	324.1					
39	Y	146.2	191	206	374.0				
42	Mo	164.12	186.4	209.3	230.28	279.1	302.60	544.0	570

+16 Ions to +23 Ions

Z	Element	XVII	XVIII	XIX	XX	XXI	XXII	XXIII	XXIV
17	Cl	3946.2960							
18	Ar	4120.8857	4426.2296						
19	K	1033.4	4610.8	4934.046					
20	Ca	1087	1157.8	5128.8	5469.864				
21	Sc	1094	1213	1287.97	5674.8	6033.712			
22	Ti	1131	1221	1346	1425.4	6249.0	6625.82		
23	V	1168	1260	1355	1486	1569.6	6851.3	7246.12	
24	Cr	1185	1299	1396	1496	1634	1721.4	7481.7	7894.81
25	Mn	1224	1317	1437	1539	1644	1788	1879.9	8140.6
26	Fe	1266	1358	1456	1582	1689	1799	1950	2023
27	Co	546.58	1397.2	1504.6	1603	1735	1846	1962	2119
28	Ni	571.08	607.06	1541	1648	1756	1894	2011	2131
29	Cu	557	633	670.588	1697	1804	1916	2060	2182
30	Zn	579	619	698	738	1856			
36	Kr	592	641	786	833	884	937	998	1051
42	Mo	636	702	767	833	902	968	1020	1082

+24 Ions to +29 Ions

Z	Element	XXV	XXVI	XXVII	XXVIII	XXIX	XXX
25	Mn	8571.94					
26	Fe	8828	9277.69				
27	Co	2219.0	9544.1	10012.12			
28	Ni	2295	2399.2	10288.8	10775.40		
29	Cu	2308	2478	2587.5	11062.38	11567.617	
36	Kr	1151	1205.3	2928	3070	3227	3381
42	Mo	1263	1323	1387	1449	1535	1601

IONIZATION ENERGIES OF GAS-PHASE MOLECULES

Sharon G. Lias

This table presents values for the first ionization energies (IP) of approximately 1000 molecules and atoms. Substances are listed by molecular formula in the modified Hill order (see introduction). Values enclosed in parentheses are considered not to be well established. Data appearing in the 1988 reference, were updated in 1996 for inclusion in the database of ionization energies available at the Internet site of the Standard Reference Data program of the National Institute of Standards and Technology (http://webbook.nist.gov). The list appearing here includes these updates.

The list also includes values for enthalpies of formation of the ions at 298 K, $\Delta_f H_{ion}$, given according to the ion convention used by mass spectrometrists; to convert these values to the electron convention used by thermodynamicists, add 6 kJ/mol. Details on the calculation of $\Delta_f H_{ion}$ as well as data for a much larger number of molecules, may be found in the reference and on the Internet site.

REFERENCE

Lias, S.G., Bartmess, J.E., Liebman, J.F., Holmes, J.L., Levin, R.D., and Mallard, W.G., *Gas-Phase Ion and Neutral Thermochemistry, J. Phys. Chem. Ref. Data*, Vol. 17, Suppl. No. 1, 1988.

Mol. Form.	Name	IP/eV	$\Delta_f H_{ion}$ kJ/mol
	Substances not containing carbon		
Ac	Actinium	5.17	905
Ag	Silver	7.57624	1016
AgCl	Silver(I) chloride	(≤ 10.08)	≤ 1065
AgF	Silver(I) fluoride	(11.0 ± 0.3)	1071
Al	Aluminum	5.98577	905
AlBr	Aluminum monobromide	(9.3)	913
$AlBr_3$	Aluminum tribromide	(10.4)	593
AlCl	Aluminum monochloride	9.4	855
$AlCl_3$	Aluminum trichloride	(12.01)	573
AlF	Aluminum monofluoride	9.73 ± 0.01	673
AlF_3	Aluminum trifluoride	≤ 15.45	≤ 282
AlI	Aluminum monoiodide	9.3 ± 0.3	965
AlI_3	Aluminum triiodide	(9.1)	673
Am	Americium	5.9738 ± 0.0002	860
Ar	Argon	15.75962	1521
As	Arsenic	9.8152	1250
$AsCl_3$	Arsenic(III) chloride	(10.55 ± 0.025)	754
AsF_3	Arsenic(III) fluoride	(12.84 ± 0.05)	452
AsH_3	Arsine	(9.89)	1021
Au	Gold	9.22567	1254
B	Boron	8.29803	1363
BBr_3	Boron tribromide	(10.51)	809
BCl_3	Boron trichloride	11.60 ± 0.02	718
BF	Fluoroborane	11.12 ± 0.01	957
	Difluoroborane	(9.4)	317
BF_3	Boron trifluoride	15.7 ± 0.3	365
BH	Boron monohydride	(9.77)	1385
BH_3	Borane	12.026 ± 0.024	1261
BI_3	Boron triiodide	(9.25 ± 0.03)	964
BO_2	Boron dioxide	(13.5 ± 0.3)	1001
B_2H_6	Diborane	11.38 ± 0.05	1134
B_2O_3	Boron oxide	13.5 ± 0.15	460
B_4H_{10}	Tetraborane	10.76 ± 0.04	1105
B_5H_9	Pentaborane(9)	9.90 ± 0.04	1028
B_6H_{10}	Hexaborane	(9.0)	965
Ba	Barium	5.21170	683
BaO	Barium oxide	6.91 ± 0.06	543
Be	Beryllium	9.32263	1224
BeO	Beryllium oxide	(10.1 ± 0.4)	1111
Bi	Bismuth	7.2855	908
$BiCl_3$	Bismuth trichloride	(10.4)	736
Bk	Berkelium	6.23	911

Mol. Form.	Name	IP/eV	$\Delta_f H_{ion}$ kJ/mol
Br	Bromine (atomic)	11.81381	1252
BrCl	Bromine chloride	11.01	1079
BrF	Bromine fluoride	11.86	1086
BrF_5	Bromine pentafluoride	13.172 ± 0.002	840
BrH	Hydrogen bromide	11.66 ± 0.03	1087
BrH_3Si	Bromosilane	10.6	943
BrI	Iodine bromide	9.790 ± 0.004	986
BrK	Potassium bromide	7.85 ± 0.1	578
BrLi	Lithium bromide	(8.7)	685
BrNO	Nitrosyl bromide	10.17 ± 0.03	1065
BrNa	Sodium bromide	8.31 ± 0.1	660
BrO	Bromine monoxide	10.46 ± 0.02	1135
BrRb	Rubidium bromide	7.94 ± 0.03	583
BrTl	Thallium(I) bromide	9.14 ± 0.02	844
Br_2	Bromine	10.516 ± 0.005	1046
Br_2Hg	Mercury(II) bromide	10.560 ± 0.003	935
Br_2Sn	Tin(II) bromide	9.0	839
Br_3Ga	Gallium(III) bromide	10.40	711
Br_3P	Phosphorus(III) bromide	9.7	798
Br_4Hf	Hafnium(IV) bromide	(10.9)	366
Br_4Sn	Tin(IV) bromide	10.6	709
Br_4Ti	Titanium(IV) bromide	10.3	375
Br_4Zr	Zirconium(IV) bromide	(10.7)	388
Ca	Calcium	6.11316	768
CaCl	Calcium monochloride	5.86 ± 0.07	462
CaO	Calcium oxide	6.66 ± 0.18	668
Cd	Cadmium	8.99367	980
Ce	Cerium	5.5387	957
Cf	Californium	6.30	805
Cl	Chlorine (atomic)	12.96764	1373
ClCs	Cesium chloride	(7.84 ± 0.05)	510
ClF	Chlorine fluoride	12.66 ± 0.01	1171
$ClFO_3$	Perchloryl fluoride	(12.945 ± 0.005)	1224
ClF_2	Chlorine difluoride	(12.77 ± 0.05)	1128
ClF_3	Chlorine trifluoride	(12.65 ± 0.05)	1057
ClF_5S	Sulfur chloride pentafluoride	(12.335 ± 0.005)	144
ClH	Hydrogen chloride	12.749 ± 0.009	1137
ClHO	Hypochlorous acid	(11.12 ± 0.01)	993
ClH_3Si	Chlorosilane	11.4	899
ClI	Iodine chloride	10.088 ± 0.01	991
ClIn	Indium(I) chloride	(9.51)	842
ClK	Potassium chloride	(8.0 ± 0.4)	557
ClLi	Lithium chloride	9.57	727
ClNO	Nitrosyl chloride	10.87 ± 0.01	1099
$ClNO_2$	Nitryl chloride	(11.84)	1155
ClNa	Sodium chloride	8.92 ± 0.06	681
ClO	Chlorine monoxide	10.95	1159
ClO_2	Chlorine dioxide	10.33 ± 0.02	1093
ClRb	Rubidium chloride	(8.50 ± 0.03)	590
ClTl	Thallium(I) chloride	9.70 ± 0.03	869
Cl_2	Chlorine	11.480 ± 0.005	1108
Cl_2CrO_2	Chromyl chloride	11.6	580
Cl_2Ge	Germanium(II) chloride	(10.20 ± 0.05)	813
Cl_2H_2Si	Dichlorosilane	11.4	765
Cl_2Hg	Mercury(II) chloride	11.380 ± 0.003	952
Cl_2O	Chlorine oxide	10.94	1135
Cl_2OS	Thionyl chloride	10.96	844
Cl_2O_2S	Sulfuryl chloride	12.05	807
Cl_2Pb	Lead(II) chloride	(10.2)	791
Cl_2S	Sulfur dichloride	9.45 ± 0.03	895

Mol. Form.	Name	IP/eV	$\Delta_f H_{ion}$ kJ/mol
Cl_2Si	Dichlorosilylene	(10.93 ± 0.10)	887
Cl_2Sn	Tin(II) chloride	(10.0)	760
Cl_3Ga	Gallium(III) chloride	11.52	664
Cl_3HSi	Trichlorosilane	(11.7)	648
Cl_3N	Nitrogen trichloride	(10.12 ± 0.1)	1244
Cl_3OP	Phosphorus(V) oxychloride	11.36 ± 0.02	540
Cl_3OV	Vanadyl trichloride	(11.6)	425
Cl_3P	Phosphorus(III) chloride	9.91	668
Cl_3PS	Phosphorus(V) sulfide trichloride	9.71 ± 0.03	573
Cl_3Sb	Antimony(III) chloride	(≤ 10.7)	s719
Cl_4Ge	Germanium(IV) chloride	11.68 ± 0.05	629
Cl_4Hf	Hafnium(IV) chloride	(11.7)	246
Cl_4Si	Tetrachlorosilane	11.79 ± 0.01	527
Cl_4Sn	Tin(IV) chloride	11.7 ± 0.2	656
Cl_4Ti	Titanium(IV) chloride	(11.5)	349
Cl_4V	Vanadium(IV) chloride	(9.2)	361
Cl_4Zr	Zirconium(IV) chloride	(11.2)	210
Cl_5Mo	Molybdenum(V) chloride	(8.7)	392
Cl_5Nb	Niobium(V) chloride	(10.97)	356
Cl_5P	Phosphorus(V) chloride	(10.2)	608
Cl_5Ta	Tantalum(V) chloride	(11.08)	303
Cl_6W	Tungsten(VI) chloride	(9.5)	348
Cm	Curium	6.02	966
Co	Cobalt	7.8810	1187
Cr	Chromium	6.76664	1050
Cs	Cesium	3.89390	452
CsF	Cesium fluoride	(8.80 ± 0.10)	489
CsNa	Cesium sodium	(4.05 ± 0.04)	535
Cu	Copper	7.72638	1084
CuF	Copper(I) fluoride	10.15 ± 0.02	984
Dy	Dysprosium	5.9389	862
Er	Erbium	6.1078	907
Es	Einsteinium	6.42	753
Eu	Europium	5.6704	723
F	Fluorine (atomic)	17.42282	1761
FGa	Gallium monofluoride	(9.6 ± 0.5)	700
FH	Hydrogen fluoride	16.044 ± 0.003	1276
FHO	Hypofluorous acid	12.71 ± 0.01	1130
FH_3Si	Fluorosilane	11.7	752
FI	Iodine fluoride	10.54 ± 0.01	922
FIn	Indium monofluoride	(9.6 ± 0.5)	740
FNO	Nitrosyl fluoride	12.63 ± 0.03	1152
FNO_2	Nitryl fluoride	(13.09)	1154
FNS	Thionitrosyl fluoride (NSF)	11.51 ± 0.04	1090
FO	Fluorine monoxide	12.78 ± 0.03	1342
FO_2	Fluorine superoxide (FOO)	(12.6 ± 0.2)	1228
FS	Sulfur fluoride	10.09	986
FTl	Thallium(I) fluoride	10.52	835
F_2	Fluorine	15.697 ± 0.003	1515
F_2Ge	Germanium(II) fluoride	(≤ 11.65)	551
F_2HN	Difluoramine	(11.53 ± 0.08)	1046
F_2H_2Si	Difluorosilane	(12.2)	386
F_2Mg	Magnesium fluoride	(13.6 ± 0.3)	588
F_2N	Difluoroamidogen	11.628 ± 0.01	1155
F_2N_2	*trans*-Difluorodiazine	(12.8)	1315
F_2O	Fluorine monoxide	13.11 ± 0.01	1290
F_2OS	Thionyl fluoride	12.25	688
F_2O_2S	Sulfuryl fluoride	13.04 ± 0.01	501
F_2Pb	Lead(II) fluoride	(11.5)	679
F_2S	Sulfur difluoride	(10.08)	676

IONIZATION ENERGIES OF GAS-PHASE MOLECULES (continued)

Mol. Form.	Name	IP/eV	$\Delta_f H_{ion}$ kJ/mol
F_2Si	Difluorosilylene	10.78 ± 0.05	450
F_2Sn	Tin(II) fluoride	(11.1)	586
F_2Xe	Xenon difluoride	12.35 ± 0.01	1083
F_3HSi	Trifluorosilane	(14.0)	150
F_3N	Nitrogen trifluoride	13.00 ± 0.02	1125
F_3NO	Trifluoramine oxide	13.31 ± 0.06	1121
F_3OP	Phosphorus(V) oxyfluoride	12.76 ± 0.01	-24
F_3P	Phosphorus(III) fluoride	11.60 ± 0.05	161
F_3PS	Phosphorus(V) sulfide trifluoride	≤ 11.05 ± 0.035	≤ 58
F_3Si	Trifluorosilyl	(9.99)	- 32
F_4Ge	Germanium(IV) fluoride	(15.5)	307
F_4N_2	Tetrafluorohydrazine	11.94 ± 0.03	1119
F_4S	Sulfur tetrafluoride	12.0 ± 0.3	399
F_4Si	Tetrafluorosilane	15.24 ± 0.14	-144
F_4Xe	Xenon tetrafluoride	12.65 ± 0.1	1016
F_5I	Iodine pentafluoride	12.943 ± 0.005	408
F_5P	Phosphorus(V) fluoride	(15.1)	-137
F_5S	Sulfur pentafluoride	9.60 ± 0.05	10
F_6Mo	Molybdenum(VI) fluoride	(14.5 ± 0.1)	-159
F_6S	Sulfur hexafluoride	15.32 ± 0.02	258
F_6U	Uranium(VI) fluoride	14.00 ± 0.10	-796
Fe	Iron	7.9024	1177
Fm	Fermium	6.50	627
Ga	Gallium	5.99930	851
GaI_3	Gallium(III) iodide	9.40	765
Gd	Gadolinium	6.1500	991
Ge	Germanium	7.900	1139
GeH_4	Germane	≤ 10.53	≤ 1108
GeI_4	Germanium(IV) iodide	(9.42)	850
GeO	Germanium(II) oxide	11.25 ± 0.01	1044
GeS	Germanium(II) sulfide	(9.98)	1055
H	Hydrogen (atomic)	13.59844	1530
HI	Hydrogen iodide	10.386 ± 0.001	1028
HLi	Lithium hydride	7.7	882
HN	Imidogen	≤ 13.49 ± 0.01	1678
HNO	Nitrosyl hydride	(10.1)	1075
HNO_2	Nitrous acid	≤ 11.3	≤ 1011
HNO_3	Nitric acid	11.95 ± 0.01	1019
HN_3	Hydrazoic acid	10.72 ± 0.025	1328
HO	Hydroxyl	13.0170 ± 0.0002	1294
HO_2	Hydroperoxy	11.35 ± 0.01	1106
HS	Mercapto	10.4219 ± 0.0004	1145
H_2	Hydrogen	15.42593 ± 0.00005	1488
H_2N	Amidogen	11.14 .01	1264
H_2O	Water	12.6206 ± 0.0020	976
H_2O_2	Hydrogen peroxide	10.58 ± 0.04	885
H_2S	Hydrogen sulfide	10.457 ± 0.012	989
H_2Se	Hydrogen selenide	9.892 ± 0.005	984
H_2Si	Silylene	8.244 ± 0.025	1084
H_3N	Ammonia	10.070 ± 0.020	925
H_3NO	Hydroxylamine	(10.00)	923
H_3P	Phosphine	9.869 ± 0.002	958
H_3Sb	Stibine	9.54 ± 0.03	1067
H_4N_2	Hydrazine	8.1 ± 0.15	877
H_4Si	Silane	11.00 ± 0.02	1095
H_4Sn	Stannane	(10.75)	1200
H_6Si_2	Disilane	9.74 ± 0.02	1019
H_8Si_3	Trisilane	(9.2)	1009
He	Helium	24.58741	2372
Hf	Hafnium	6.82507 ± 0.00004	1278

Mol. Form.	Name	IP/eV	$\Delta_f H_{ion}$ kJ/mol
Hg	Mercury	10.43750	1069
HgI_2	Mercury(II) iodide	9.5088 ± 0.0022	900
Ho	Holmium	6.0216	882
I	Iodine (atomic)	10.45126	1115
IK	Potassium iodide	(7.21 ± 0.3)	570
ILi	Lithium iodide	(7.5)	633
INa	Sodium iodide	7.64 ± 0.02	659
ITl	Thallium(I) iodide	8.47 ± 0.02	826
I_2	Iodine	9.3074 ± 0.0002	960
I_4Ti	Titanium(IV) iodide	(9.1)	602
I_4Zr	Zirconium(IV) iodide	(9.3)	500
In	Indium	5.78636	802
Ir	Iridium	9.1	1543
K	Potassium	4.34066	508
KLi	Lithium potassium	4.57 ± 0.04	512
KNa	Potassium sodium	4.41636 ± 0.00017	561
K_2	Dipotassium	4.0637 ± 0.0002	519
Kr	Krypton	13.99961	1351
La	Lanthanum	5.5770	969
Li	Lithium	5.39172	680
LiNa	Lithium sodium	5.05 ± 0.04	571
LiO	Lithium monoxide	(8.44)	894
LiRb	Lithium rubidium	4.3 ± 0.1	486
Li_2	Dilithium	5.1127 ± 0.0003	709
Lu	Lutetium	5.42585	950
Md	Mendelevium	6.58	635
Mg	Magnesium	7.64624	885
MgO	Magnesium oxide	(8.76 ± 0.22)	901
Mn	Manganese	7.43402	998
Mo	Molybdenum	7.09243	1343
N	Nitrogen (atomic)	14.53414	1875
NO	Nitric oxide	9.26438 ± 0.00005	985
NO_2	Nitrogen dioxide	9.586 ± 0.002	958
NP	Phosphorus nitride	11.84 ± 0.04	1247
NS	Nitrogen sulfide	8.87 ± 0.01	1119
N_2	Nitrogen	15.5808	1503
N_2O	Nitrous oxide	12.886	1325
N_2O_4	Nitrogen tetroxide	(10.8)	1050
N_2O_5	Nitrogen pentoxide	(11.9)	1161
Na	Sodium	5.13908	603
NaRb	Rubidium sodium	4.32 ± 0.04	480
Na_2	Disodium	4.894 ± 0.003	614
Nb	Niobium	6.75885	1384
Nd	Neodymium	5.5250	859
Ne	Neon	21.56454	2081
Ni	Nickel	7.6398	1167
No	Nobelium	6.65	642
Np	Neptunium	6.2657 ± 0.0003	1069
O	Oxygen (atomic)	13.61806	1563
OPb	Lead(II) oxide	9.08 ± 0.10	939
OS	Sulfur monoxide	10.294 ± 0.004	998
OS_2	Sulfur oxide (SSO)	10.584 ± 0.005	971
OSi	Silicon monoxide	11.49 ± 0.20	1008
OSn	Tin(II) oxide	9.60 ± 0.02	944
OSr	Strontium oxide	6.6 ± 0.2	623
O_2	Oxygen	12.0697 ± 0.0002	1165
O_2S	Sulfur dioxide	12.349 ± 0.001	894
O_2Th	Thorium(IV) oxide	(8.7 ± 0.15)	342
O_2Ti	Titanium(IV) oxide	(9.54 ± 0.1)	623
O_2U	Uranium(IV) oxide	(5.4 ± 0.1)	57

Mol. Form.	Name	IP/eV	$\Delta_f H_{ion}$ kJ/mol
O_3	Ozone	12.43	1342
O_3S	Sulfur trioxide	12.82 ± 0.03	841
O_3U	Uranium(VI) oxide	(10.5 ± 0.5)	214
O_4Os	Osmium(VIII) oxide	(12.32)	850
O_4Ru	Ruthenium(VIII) oxide	12.15 ± 0.03	988
O_7Re_2	Rhenium(VII) oxide	(12.7 ± 0.2)	125
Os	Osmium	8.7	1630
P	Phosphorus	10.48669	1328
P_2	Diphosphorus	10.53	1160
Pa	Protactinium	5.89	1133
Pb	Lead	7.41666	911
PbS	Lead(II) sulfide	(8.5 ± 0.5)	954
Pd	Palladium	8.3367	1181
Pm	Promethium	5.55	536
Pr	Praseodymium	5.464	883
Pt	Platinum	9.0	1433
Pu	Plutonium	6.025	926
Ra	Radium	5.27892	668
Rb	Rubidium	4.17713	484
Re	Rhenium	7.88	1530
Rh	Rhodium	7.45890	1276
Rn	Radon	10.74850	1037
Ru	Ruthenium	7.36050	1355
S	Sulfur	10.36001	1277
SSn	Tin(II) sulfide	(8.8)	966
S_2	Disulfur	9.356 ± 0.002	1031
Sb	Antimony	8.64	1096
Sc	Scandium	6.56144	1010
Se	Selenium	9.75238	1168
Si	Silicon	8.15169	1238
Sm	Samarium	5.6437	751
Sn	Tin	7.34381	1011
Sr	Strontium	5.69484	713
Ta	Tantalum	7.89	1544
Tb	Terbium	5.8639	955
Tc	Technetium	7.28	1380
Te	Tellurium	9.0096	1066
Th	Thorium	6.308 ± 0.003	1207
Ti	Titanium	6.8282	1127
Tl	Thallium	6.10829	771
Tm	Thulium	6.18431	827
U	Uranium	6.19405	1129
V	Vanadium	6.746 ± 0.002	1166
W	Tungsten	7.98	1621
Xe	Xenon	12.12987	1170
Y	Yttrium	6.217	1022
Yb	Ytterbium	6.25416	754
Zn	Zinc	9.39405	1037
Zr	Zirconium	6.63390	1251
	Substances containing carbon		
C	Carbon	11.26030	1803
$CBrClF_2$	Bromochlorodifluoromethane	(11.21)	642
$CBrCl_3$	Bromotrichloromethane	(10.6)	980
$CBrF_3$	Bromotrifluoromethane	(11.40)	451
CBr_2F_2	Dibromodifluoromethane	11.03 ± 0.04	683
CBr_4	Tetrabromomethane	(10.31 ± 0.02)	1079
CCl	Chloromethylidyne	(8.9 ± 0.2)	1244
$CClF_3$	Chlorotrifluoromethane	12.6 ± 0.2	505

Mol. Form.	Name	IP/eV	$\Delta_f H_{ion}$ kJ/mol
CClN	Cyanogen chloride	12.34 ± 0.01	1329
CCl_2	Dichloromethylene	(9.27)	1058
CCl_2F_2	Dichlorodifluoromethane	12.05 ± 0.24	685
CCl_2O	Carbonyl chloride	(11.5)	888
CCl_3F	Trichlorofluoromethane	11.77 ± 0.02	868
CCl_4	Tetrachloromethane	11.47 ± 0.01	1010
CF	Fluoromethylidyne	9.11 ± 0.01	1134
CFN	Cyanogen fluoride	13.34 ± 0.02	1325
CF_2	Difluoromethylene	11.44 ± 0.03	899
CF_2O	Carbonyl fluoride	13.035 ± 0.030	617
CF_3	Trifluoromethyl	8.7 ± 0.2	379
CF_3I	Trifluoroiodomethane	10.23	397
CH	Methylidyne	10.64 ± 0.01	1622
$CHBrCl_2$	Bromodichloromethane	10.6	973
$CHBr_2Cl$	Chlorodibromomethane	10.59 ± 0.01	1030
$CHBr_3$	Tribromomethane	10.48 ± 0.02	1035
CHCl	Chloromethylene	9.84	1247
$CHClF_2$	Chlorodifluoromethane	(12.2)	693
$CHCl_2F$	Dichlorofluoromethane	(11.5)	829
$CHCl_3$	Trichloromethane	11.37 ± 0.02	992
CHF	Fluoromethylene	10.06 ± 0.05	1121
CHF_3	Trifluoromethane	(13.86)	643
CHI_3	Triiodomethane	9.25 ± 0.02	1010
CHN	Hydrogen cyanide	13.60 ± 0.01	1447
CHN	Hydrogen isocyanide	(12.5 ± 0.1)	1407
CHNO	Isocyanic acid	11.595 ± 0.005	1016
CHNO	Fulminic acid	(10.83)	1263
CHO	Oxomethyl (HCO)	(8.55)	826
CH_2	Methylene	10.396 ± .003	1392
CH_2BrCl	Bromochloromethane	10.77 ± 0.01	1085
CH_2Br_2	Dibromomethane	(10.50 ± 0.02)	1013
CH_2ClF	Chlorofluoromethane	11.71 ± 0.01	870
CH_2Cl_2	Dichloromethane	11.32 ± .01	996
CH_2F_2	Difluoromethane	12.71	774
CH_2I_2	Diiodomethane	9.46 ± 0.02	1030
CH_2N_2	Diazomethane	8.999 ± 0.001	1098
CH_2N_2	Cyanamide	(10.4)	1137
CH_2O	Formaldehyde	10.88 ± 0.01	941
CH_2O_2	Formic acid	11.33 ± 0.01	715
CH_3	Methyl	9.843 ± 0.002	1095
CH_3BO	Borane carbonyl	11.14 ± 0.02	962
CH_3Br	Bromomethane	10.541 ± 0.003	979
CH_3Cl	Chloromethane	11.22 ± 0.01	1001
CH_3Cl_3Si	Methyltrichlorosilane	(11.36 ± 0.03)	548
CH_3F	Fluoromethane	12.47 ± 0.02	956
CH_3I	Iodomethane	9.538	936
CH_3NO	Formamide	10.16 ± 0.06	796
CH_3NO_2	Nitromethane	11.08 ± 0.07	994
CH_3N_3	Methyl azide	9.81 ± 0.02	1227
CH_3O	Methoxy	(10.72)	1050
CH_4	Methane	12.61 ± 0.01	1143
CH_4N_2O	Urea	9.7	690
CH_4O	Methanol	10.85 ± 0.01	845
CH_4S	Methanethiol	9.44 ± 0.005	888
CH_5N	Methylamine	(8.80)	826
CH_6N_2	Methylhydrazine	7.7 ± 0.15	835
CH_6Si	Methylsilane	(10.7)	1003
CN	Cyanide	13.5984	1748
CNO	Cyanate	11.76 ± 0.01	1290
CO	Carbon monoxide	14.014 ± 0.0003	1242

Mol. Form.	Name	IP/eV	$\Delta_f H_{ion}$ kJ/mol
COS	Carbon oxysulfide	11.18 ± 0.01	936
$COSe$	Carbon oxyselenide	10.36 ± 0.01	929
CO_2	Carbon dioxide	13.773 ± 0.002	935
CS	Carbon sulfide	11.33 ± 0.01	1361
CS_2	Carbon disulfide	10.0685 ± 0.0020	1089
C_2	Dicarbon	(11.4 ± 0.3)	2000
$C_2Br_2F_4$	1,2-Dibromotetrafluoroethane	(11.1)	280
C_2ClF_3	Chlorotrifluoroethylene	9.81 ± 0.03	373
C_2ClF_5	Chloropentafluoroethane	(12.6)	99
C_2Cl_2	Dichloroacetylene	9.9	1165
$C_2Cl_2F_4$	1,2-Dichlorotetrafluoroethane	12.2	252
$C_2Cl_3F_3$	1,1,1-Trichlorotrifluoroethane	11.5	386
$C_2Cl_3F_3$	1,1,2-Trichlorotrifluoroethane	11.99 ± 0.02	429
C_2Cl_4	Tetrachloroethylene	9.326 ± 0.001	887
$C_2Cl_4F_2$	1,1,2,2-Tetrachloro-1,2-difluoroethane	(11.3)	563
C_2Cl_4O	Trichloroacetyl chloride	(11.0)	827
C_2Cl_6	Hexachloroethane	(11.1)	920
C_2F_3N	Trifluoroacetonitrile	13.93 ± 0.07	845
C_2F_4	Tetrafluoroethylene	10.12 ± 0.02	315
C_2F_6	Hexafluoroethane	(13.6)	-30
C_2H	Ethynyl	(11.61 ± 0.07)	1685
C_2HBr	Bromoacetylene	10.31 ± 0.02	1242
$C_2HBrClF_3$	2-Bromo-2-chloro-1,1,1-trifluoroethane	(11.0)	363
C_2HCl	Chloroacetylene	10.58 ± 0.02	1276
C_2HClF_2	1-Chloro-2,2-difluoroethylene	9.80 ± 0.04	628
C_2HCl_3	Trichloroethylene	9.46 ± 0.02	894
C_2HCl_3O	Dichloroacetyl chloride	(10.9)	809
C_2HCl_5	Pentachloroethane	(11.0)	919
C_2HF	Fluoroacetylene	11.26	1195
C_2HF_3	Trifluoroethylene	10.14	489
$C_2HF_3O_2$	Trifluoroacetic acid	11.46	75
C_2H_2	Acetylene	11.400 ± 0.002	1328
$C_2H_2Cl_2$	1,1-Dichloroethylene	9.81 ± 0.04	949
$C_2H_2Cl_2$	*cis*-1,2-Dichloroethylene	9.66 ± 0.01	936
$C_2H_2Cl_2$	*trans*-1,2-Dichloroethylene	9.64 ± 0.02	934
$C_2H_2Cl_2O$	Chloroacetyl chloride	(≤ 10.3)	815
$C_2H_2Cl_4$	1,1,1,2-Tetrachloroethane	(11.1)	920
$C_2H_2Cl_4$	1,1,2,2-Tetrachloroethane	(≤ 11.62)	≤ 971
$C_2H_2F_2$	1,1-Difluoroethylene	10.29 ± 0.01	650
$C_2H_2F_2$	*cis*-1,2-Difluoroethylene	10.23 ± 0.02	690
C_2H_2O	Ketene	9.617 ± 0.003	880
$C_2H_2O_2$	Glyoxal	10.2	773
$C_2H_2S_2$	Thiirene	8.61	892
C_2H_3Br	Bromoethylene	9.83 ± 0.02	1028
C_2H_3Cl	Chloroethylene	9.99 ± 0.02	985
$C_2H_3ClF_2$	1-Chloro-1,1-difluoroethane	(11.98)	626
C_2H_3ClO	Acetyl chloride	10.82 ± 0.04	801
C_2H_3ClO	Chloroacetaldehyde	(10.48)	815
$C_2H_3ClO_2$	Chloroacetic acid	(10.7)	597
$C_2H_3Cl_3$	1,1,1-Trichloroethane	(11.0)	917
$C_2H_3Cl_3$	1,1,2-Trichloroethane	(11.0)	911
C_2H_3F	Fluoroethylene	10.36 ± 0.01	861
C_2H_3FO	Acetyl fluoride	(11.5)	667
$C_2H_3F_3$	1,1,1-Trifluoroethane	13.3 ± 0.5	536
C_2H_3N	Acetonitrile	12.20 ± 0.01	1253
C_2H_3NO	Methylisocyanate	(10.67)	900
C_2H_4	Ethylene	10.5138 ± 0.0006	1067
$C_2H_4Br_2$	1,2-Dibromoethane	10.35 ± 0.04	961
$C_2H_4Cl_2$	1,1-Dichloroethane	11.04 ± 0.02	935
$C_2H_4Cl_2$	1,2-Dichloroethane	11.04 ± 0.02	931

Mol. Form.	Name	IP/eV	$\Delta_f H_{ion}$ kJ/mol
$C_2H_4F_2$	1,1-Difluoroethane	(11.87)	643
C_2H_4O	Acetaldehyde	10.229 ± 0.0007	821
C_2H_4O	Ethylene oxide	10.56 ± 0.01	966
$C_2H_4O_2$	Acetic acid	10.65 ± 0.02	595
$C_2H_4O_2$	Methyl formate	10.835 ± 0.005	690
C_2H_5Br	Bromoethane	10.29 ± 0.01	931
C_2H_5Cl	Chloroethane	10.98 ± 0.02	947
C_2H_5ClO	2-Chloroethanol	(10.5)	756
C_2H_5F	Fluoroethane	(11.78)	873
C_2H_5I	Iodoethane	9.3492 ± 0.0006	893
C_2H_5N	Ethyleneimine	(9.5 ± 0.3)	1044
C_2H_5NO	Acetamide	9.65 ± 0.03	693
C_2H_5NO	*N*-Methylformamide	9.83 ± 0.04	760
$C_2H_5NO_2$	Nitroethane	10.88 ± 0.05	948
C_2H_6	Ethane	11.56 ± 0.02	1031
$C_2H_6Cl_2Si$	Dichlorodimethylsilane	(10.7)	576
C_2H_6O	Ethanol	10.43 ± 0.05	772
C_2H_6O	Dimethyl ether	10.025 ± 0.025	783
C_2H_6OS	Dimethyl sulfoxide	9.10 ± 0.03	727
$C_2H_6O_2$	Ethylene glycol	10.16	593
C_2H_6S	Ethanethiol	9.31 ± 0.03	851
C_2H_6S	Dimethyl sulfide	8.69 ± 0.02	801
$C_2H_6S_2$	Dimethyl disulfide	(7.4 ± 0.3)	690
C_2H_7N	Ethylamine	8.86 ± 0.02	808
C_2H_7N	Dimethylamine	8.24 ± 0.08	777
C_2H_7NO	Ethanolamine	8.96	664
$C_2H_8N_2$	1,2-Ethanediamine	(8.6)	812
$C_2H_8N_2$	1,1-Dimethylhydrazine	7.29 ± 0.05	787
C_2N_2	Cyanogen	13.37 ± 0.01	1597
C_3F_6	Perfluoropropene	10.60 ± 0.03	-103
C_3F_6O	Perfluoroacetone	(11.57 ± 0.13)	-282
C_3F_8	Perfluoropropane	(13.38)	-491
C_3HN	Cyanoacetylene	11.64 ± 0.01	1475
C_3H_2O	2-Propynal	(10.7 ± 0.1)	1145
$C_3H_3F_3$	3,3,3-Trifluoropropene	(10.9)	437
C_3H_3N	2-Propenenitrile	10.91 ± 0.01	1237
C_3H_3NO	Oxazole	(9.9)	940
C_3H_3NO	Isoxazole	(9.93)	1038
C_3H_4	Allene	9.692 ± 0.004	1126
C_3H_4	Propyne	10.37 ± 0.01	1187
C_3H_4	Cyclopropene	9.67 ± 0.01	1209
$C_3H_4N_2$	Imidazole	(8.81)	997
C_3H_4O	Propargyl alcohol	10.49 ± 0.02	1060
C_3H_4O	Acrolein	10.103 ± 0.006	900
C_3H_4O	Cyclopropanone	(9.1 ± 0.1)	895
$C_3H_4O_2$	Propenoic acid	10.60	701
$C_3H_4O_2$	2-Oxetanone	(9.70 ± 0.01)	653
C_3H_5Br	3-Bromopropene	(9.96)	1008
C_3H_5Cl	3-Chloropropene	10.04 ± 0.01	965
C_3H_5ClO	Epichlorohydrin	(10.64)	919
$C_3H_5ClO_2$	Methyl chloroacetate	(10.3)	575
C_3H_5F	3-Fluoropropene	(10.11)	821
C_3H_5N	Propanenitrile	11.84 ± 0.02	1194
C_3H_5NO	Acrylamide	(9.5)	720
C_3H_6	Propene	9.73 ± 0.02	959
C_3H_6	Cyclopropane	9.86	1005
$C_3H_6Br_2$	1,2-Dibromopropane	(10.1)	903
$C_3H_6Br_2$	1,3-Dibromopropane	(≤ 10.2)	≤ 919
$C_3H_6Cl_2$	1,2-Dichloropropane	10.8 ± 0.1	886
$C_3H_6Cl_2$	1,3-Dichloropropane	10.89 ± 0.04	892

IONIZATION ENERGIES OF GAS-PHASE MOLECULES (continued)

Mol. Form.	Name	IP/eV	$\Delta_f H_{ion}$ kJ/mol
C_3H_6O	Allyl alcohol	9.67±0.05	808
C_3H_6O	Methyl vinyl ether	8.95±0.01	763
C_3H_6O	Propanal	9.96±0.01	772
C_3H_6O	Acetone	9.703±0.006	719
C_3H_6O	Methyloxirane	(10.22)	892
C_3H_6O	Oxetane	9.65±0.01	851
$C_3H_6O_2$	Propanoic acid	10.525±0.003	568
$C_3H_6O_2$	Ethyl formate	10.61±0.01	639
$C_3H_6O_2$	Methyl acetate	10.25±0.02	579
$C_3H_6O_2$	1,3-Dioxolane	(9.9)	658
$C_3H_6O_3$	1,3,5-Trioxane	(10.3)	528
C_3H_7Br	1-Bromopropane	10.18±0.01	898
C_3H_7Br	2-Bromopropane	10.10±0.03	877
C_3H_7Cl	1-Chloropropane	10.81±0.01	911
C_3H_7Cl	2-Chloropropane	10.79±0.02	896
C_3H_7F	1-Fluoropropane	(11.3)	806
C_3H_7F	2-Fluoropropane	(11.08)	776
C_3H_7I	1-Iodopropane	9.25±0.01	860
C_3H_7I	2-Iodopropane	9.19±0.02	845
C_3H_7N	Allylamine	(8.76)	891
C_3H_7N	Cyclopropylamine	(8.8)	926
C_3H_7N	Propyleneimine	(9.0)	960
C_3H_7NO	*N,N*-Dimethylformamide	(9.12)	688
$C_3H_7NO_2$	1-Nitropropane	(10.81)	919
$C_3H_7NO_2$	2-Nitropropane	(10.71)	894
C_3H_8	Propane	10.95±0.05	952
C_3H_8O	1-Propanol	10.18±0.06	727
C_3H_8O	2-Propanol	10.17±0.02	709
C_3H_8O	Ethyl methyl ether	9.72±0.07	722
$C_3H_8O_2$	Dimethoxymethane	9.7	588
C_3H_8S	1-Propanethiol	9.20±0.01	819
C_3H_8S	2-Propanethiol	9.145±0.005	806
C_3H_8S	Ethyl methyl sulfide	(8.55)	765
$C_3H_9BO_3$	Trimethyl borate	(10.0)	65
C_3H_9ClSi	Trimethylchlorosilane	(10.15)	624
C_3H_9N	Propylamine	(8.78)	777
C_3H_9N	Isopropylamine	(8.72)	758
C_3H_9N	Trimethylamine	7.82±0.06	731
C_3H_9NO	3-Amino-1-propanol	(9.0)	651
$C_4H_2O_3$	Maleic anhydride	(10.8)	645
C_4H_4	1-Buten-3-yne	9.58±0.02	1230
$C_4H_4N_2$	Succinonitrile	(12.1±0.25)	1377
$C_4H_4N_2$	Pyrimidine	9.23	1087
$C_4H_4N_2$	Pyridazine	8.67±0.03	1112
C_4H_4O	Furan	8.883±0.003	822
$C_4H_4O_2$	Diketene	(9.6±0.02)	736
$C_4H_4O_3$	Succinic anhydride	(10.6)	500
$C_4H_4O_4$	Fumaric acid	(10.7)	355
C_4H_4S	Thiophene	8.86±0.02	970
C_4H_5N	Methylacrylonitrile	10.34	1127
C_4H_5N	Pyrrole	8.207±0.005	900
C_4H_5N	Cyclopropanecarbonitrile	(10.25)	1173
C_4H_6	1,2-Butadiene	(9.03)	1034
C_4H_6	1,3-Butadiene	9.082±0.004	986
C_4H_6	1-Butyne	10.19±0.02	1148
C_4H_6	2-Butyne	9.59±0.03	1071
C_4H_6	Cyclobutene	9.43±0.02	1067
C_4H_6O	Divinyl ether	(8.7)	827
C_4H_6O	*trans*-2-Butenal	9.73±0.01	835
C_4H_6O	2-Methylpropenal	(9.92)	834

Mol. Form.	Name	IP/eV	$\Delta_f H_{ion}$ kJ/mol
C_4H_6O	Cyclobutanone	(9.35	815
$C_4H_6O_2$	*cis*-Crotonic acid	(10.08)	625
$C_4H_6O_2$	*trans*-Crotonic acid	(9.9)	604
$C_4H_6O_2$	Methacrylic acid	(10.15)	611
$C_4H_6O_2$	Vinyl acetate	9.19 ± 0.05	572
$C_4H_6O_2$	Methyl acrylate	(9.9)	641
$C_4H_6O_3$	Acetic anhydride	(10.0)	398
$C_4H_6O_4$	Dimethyl oxalate	(10.0)	287
C_4H_6S	2,5-Dihydrothiophene	(8.4)	898
C_4H_7N	Butanenitrile	(11.2)	1110
C_4H_7N	2-Methylpropanenitrile	(11.3)	1115
C_4H_7NO	2-Pyrrolidone	(9.2)	674
C_4H_8	1-Butene	9.55 ± 0.06	921
C_4H_8	*cis*-2-Butene	9.11 ± 0.01	871
C_4H_8	*trans*-2-Butene	9.10 ± 0.01	866
C_4H_8	Isobutene	9.239 ± 0.003	875
C_4H_8	Cyclobutane	(9.82 ± 0.05)	976
C_4H_8	Methylcyclopropane	(9.46)	936
$C_4H_8Br_2$	1,4-Dibromobutane	(10.15)	879
C_4H_8O	Ethyl vinyl ether	(8.98)	709
C_4H_8O	1,2-Epoxybutane	(≤ 10.15)	862
C_4H_8O	Butanal	9.84 ± 0.02	742
C_4H_8O	Isobutanal	9.71 ± 0.01	721
C_4H_8O	2-Butanone	9.52 ± 0.04	678
C_4H_8O	Tetrahydrofuran	9.38 ± 0.05	721
$C_4H_8O_2$	Butanoic acid	10.17 ± 0.05	509
$C_4H_8O_2$	2-Methylpropanoic acid	10.33 ± 0.03	516
$C_4H_8O_2$	Propyl formate	10.52 ± 0.02	555
$C_4H_8O_2$	Ethyl acetate	10.01 ± 0.05	522
$C_4H_8O_2$	Methyl propanoate	10.15 ± 0.03	548
$C_4H_8O_2$	1,3-Dioxane	9.8	607
$C_4H_8O_2$	1,4-Dioxane	9.19 ± 0.01	571
$C_4H_8O_2S$	Sulfolane	(9.8)	577
C_4H_8S	Tetrahydrothiophene	8.38	774
C_4H_9Br	1-Bromobutane	(10.12)	869
C_4H_9Br	2-Bromobutane	10.01 ± 0.02	845
C_4H_9Br	1-Bromo-2-methylpropane	10.09 ± 0.02	861
C_4H_9Br	2-Bromo-2-methylpropane	9.92 ± 0.03	823
C_4H_9Cl	1-Chlorobutane	10.67 ± 0.03	875
C_4H_9Cl	2-Chlorobutane	10.53	857
C_4H_9Cl	1-Chloro-2-methylpropane	10.73 ± 0.07	877
C_4H_9Cl	2-Chloro-2-methylpropane	(10.61)	842
C_4H_9I	1-Iodobutane	9.23 ± 0.01	840
C_4H_9I	2-Iodobutane	9.10 ± 0.02	815
C_4H_9I	1-Iodo-2-methylpropane	9.19 ± 0.01	824
C_4H_9I	2-Iodo-2-methylpropane	(9.02)	798
C_4H_9N	Pyrrolidine	(8.0)	769
C_4H_9NO	*N,N*-Dimethylacetamide	8.81 ± 0.03	616
C_4H_9NO	Morpholine	(8.2)	841
C_4H_{10}	Butane	10.53 ± 0.10	890
C_4H_{10}	Isobutane	(10.57)	886
$C_4H_{10}O$	1-Butanol	9.99 ± 0.05	689
$C_4H_{10}O$	2-Butanol	9.88 ± 0.03	658
$C_4H_{10}O$	2-Methyl-1-propanol	10.02 ± 0.04	683
$C_4H_{10}O$	2-Methyl-2-propanol	9.90 ± 0.02	642
$C_4H_{10}O$	Diethyl ether	9.51 ± 0.03	666
$C_4H_{10}O$	Methyl propyl ether	9.41 ± 0.07	670
$C_4H_{10}O$	Isopropyl methyl ether	9.45 ± 0.04	661
$C_4H_{10}O_2$	Ethylene glycol monoethyl ether	(9.6)	529
$C_4H_{10}O_2$	Ethylene glycol dimethyl ether	(9.3)	558

Mol. Form.	Name	IP/eV	$\Delta_f H_{ion}$ kJ/mol
$C_4H_{10}S$	1-Butanethiol	9.14±0.01	794
$C_4H_{10}S$	2-Butanethiol	(9.10)	781
$C_4H_{10}S$	2-Methyl-1-propanethiol	(9.12)	783
$C_4H_{10}S$	2-Methyl-2-propanethiol	(9.03)	762
$C_4H_{10}S$	Diethyl sulfide	(8.43)	730
$C_4H_{10}S$	Methyl propyl sulfide	(8.8)	767
$C_4H_{10}S$	Isopropyl methyl sulfide	(8.7)	749
$C_4H_{10}S_2$	Diethyl disulfide	(8.27)	724
$C_4H_{11}N$	Butylamine	8.7±0.1	748
$C_4H_{11}N$	*sec*-Butylamine	8.46±0.1	711
$C_4H_{11}N$	*tert*-Butylamine	8.46±0.1	695
$C_4H_{11}N$	Isobutylamine	8.50±0.1	721
$C_4H_{11}N$	Diethylamine	7.85±0.1	684
$C_4H_{12}Si$	Tetramethylsilane	9.80±0.04	713
$C_4H_{12}Sn$	Tetramethylstannane	8.89±0.05	837
C_4NiO_4	Nickel carbonyl	8.27±0.04	200
$C_5H_4O_2$	Furfural	9.22±0.01	739
C_5H_5N	Pyridine	9.25	1031
C_5H_6	1-Penten-3-yne	9.00±0.01	1119
C_5H_6	*cis*-3-Penten-1-yne	9.14±0.04	1137
C_5H_6	*trans*-3-Penten-1-yne	9.05±0.01	1128
C_5H_6	2-Methyl-1-buten-3-yne	9.25±0.02	1152
C_5H_6	1,3-Cyclopentadiene	8.55±0.02	955
C_5H_6O	2-Methylfuran	8.38±0.02	729
C_5H_6O	3-Methylfuran	(8.64)	763
C_5H_6S	2-Methylthiophene	(8.14)	867
C_5H_6S	3-Methylthiophene	(8.40)	893
C_5H_8	*cis*-1,3-Pentadiene	8.63±0.03	914
C_5H_8	*trans*-1,3-Pentadiene	8.59±0.02	905
C_5H_8	1,4-Pentadiene	9.60±0.02	1032
C_5H_8	2-Methyl-1,3-butadiene	8.84±0.01	928
C_5H_8	1-Pentyne	10.10±0.01	1119
C_5H_8	Cyclopentene	9.01±0.01	905
C_5H_8	Spiropentane	(9.26)	1078
C_5H_8O	Cyclopropyl methyl ketone	(≤ 9.46)	796
C_5H_8O	Cyclopentanone	9.26±0.01	701
C_5H_8O	3,4-Dihydro-2H-pyran	8.35±0.01	681
$C_5H_8O_2$	Ethyl acrylate	(≤ 10.3)	617
$C_5H_8O_2$	Methyl methacrylate	(9.7)	589
$C_5H_8O_2$	2,4-Pentanedione	8.85±0.01	469
C_5H_9NO	*N*-Methyl-2-pyrrolidone	(≤ 9.17)	≤ 676
C_5H_{10}	1-Pentene	9.51±0.01	896
C_5H_{10}	*cis*-2-Pentene	9.01±0.03	843
C_5H_{10}	*trans*-2-Pentene	9.04±0.01	841
C_5H_{10}	2-Methyl-1-butene	9.12±0.01	844
C_5H_{10}	3-Methyl-1-butene	9.52±0.01	891
C_5H_{10}	2-Methyl-2-butene	8.69±0.01	796
C_5H_{10}	Cyclopentane	(10.33±0.15)	918
$C_5H_{10}O$	2,2-Dimethylpropanal	9.51±0.01	675
$C_5H_{10}O$	Cyclopentanol	(9.72)	695
$C_5H_{10}O$	Pentanal	9.74±0.04	709
$C_5H_{10}O$	2-Pentanone	9.38±0.01	646
$C_5H_{10}O$	3-Pentanone	9.31±0.01	640
$C_5H_{10}O$	3-Methyl-2-butanone	9.30±0.01	635
$C_5H_{10}O$	Tetrahydropyran	9.25±0.01	670
$C_5H_{10}O_2$	Pentanoic acid	(≤ 10.53)	≤ 527
$C_5H_{10}O_2$	3-Methylbutanoic acid	(≤ 10.51)	≤ 499
$C_5H_{10}O_2$	Butyl formate	10.52±0.02	584
$C_5H_{10}O_2$	Propyl acetate	(≤ 9.92)	501
$C_5H_{10}O_2$	Isopropyl acetate	9.99±0.03	482

Mol. Form.	Name	IP/eV	$\Delta_f H_{ion}$ kJ/mol
$C_5H_{10}O_2$	Ethyl propanoate	(10.00)	500
$C_5H_{10}O_2$	Methyl butanoate	(10.07)	520
$C_5H_{10}S$	Thiacyclohexane	(8.2)	728
$C_5H_{11}Br$	1-Bromopentane	10.10 ± 0.01	846
$C_5H_{11}I$	1-Iodopentane	9.20 ± 0.01	817
$C_5H_{11}N$	Piperidine	8.03 ± 0.11	726
$C_5H_{11}N$	*N*-Methylpyrrolidine	≤ 8.41 ± 0.02	≤ 809
C_5H_{12}	Pentane	10.28 ± 0.10	845
C_5H_{12}	Isopentane	10.32 ± 0.05	843
C_5H_{12}	Neopentane	(≤ 10.2)	≤ 818
$C_5H_{12}O$	1-Pentanol	(10.00)	668
$C_5H_{12}O$	2-Pentanol	(9.78)	630
$C_5H_{12}O$	3-Pentanol	9.78	628
$C_5H_{12}O$	2-Methyl-1-butanol	(9.86)	649
$C_5H_{12}O$	2-Methyl-2-butanol	(9.8)	615
$C_5H_{12}O$	3-Methyl-2-butanol	(9.88 ± 0.13)	637
$C_5H_{12}O$	Butyl methyl ether	(9.4 ± 0.1)	648
$C_5H_{12}O$	Methyl *tert*-butyl ether	(9.24)	608
$C_5H_{12}O$	Ethyl propyl ether	(9.45)	640
$C_5H_{12}S$	*tert*-Butyl methyl sulfide	(8.38)	687
$C_5H_{12}S$	Ethyl propyl sulfide	(8.50)	716
$C_5H_{12}S$	Ethyl isopropyl sulfide	(8.35)	689
C_6BrF_5	Bromopentafluorobenzene	(9.67)	222
C_6ClF_5	Chloropentafluorobenzene	(9.72)	126
C_6Cl_6	Hexachlorobenzene	(8.98)	822
C_6F_6	Hexafluorobenzene	9.89 ± 0.04	8
C_6F_{12}	Perfluorocyclohexane	(13.2)	-1095
C_6HF_5	Pentafluorobenzene	(9.63)	122
C_6HF_5O	Pentafluorophenol	(9.20)	-71
$C_6H_2F_4$	1,2,3,4-Tetrafluorobenzene	(9.53)	284
$C_6H_2F_4$	1,2,3,5-Tetrafluorobenzene	(9.53)	263
$C_6H_2F_4$	1,2,4,5-Tetrafluorobenzene	(9.35)	254
$C_6H_3Cl_3$	1,2,4-Trichlorobenzene	(9.04)	880
$C_6H_3Cl_3$	1,3,5-Trichlorobenzene	9.32 ± 0.02	899
$C_6H_4ClNO_2$	1-Chloro-3-nitrobenzene	(9.92 ± 0.1)	995
$C_6H_4ClNO_2$	1-Chloro-4-nitrobenzene	(9.96 ± 0.1)	999
$C_6H_4Cl_2$	*o*-Dichlorobenzene	9.06 ± 0.02	907
$C_6H_4Cl_2$	*m*-Dichlorobenzene	9.10 ± 0.02	906
$C_6H_4Cl_2$	*p*-Dichlorobenzene	8.92 ± 0.02	885
$C_6H_4FNO_2$	1-Fluoro-4-nitrobenzene	(9.90)	826
$C_6H_4F_2$	*o*-Difluorobenzene	9.29 ± 0.01	602
$C_6H_4F_2$	*m*-Difluorobenzene	9.33 ± 0.01	591
$C_6H_4F_2$	*p*-Difluorobenzene	9.1589 ± 0.0003	577
$C_6H_4O_2$	*p*-Benzoquinone	10.01 ± 0.06	844
C_6H_5Br	Bromobenzene	9.00 ± 0.02	971
C_6H_5Cl	Chlorobenzene	9.07 ± 0.02	930
C_6H_5ClO	*m*-Chlorophenol	8.655 ± 0.001	680
C_6H_5ClO	*p*-Chlorophenol	(≤ 8.69)	≤ 692
C_6H_5F	Fluorobenzene	9.20 ± 0.01	772
C_6H_5I	Iodobenzene	8.685	1003
$C_6H_5NO_2$	Nitrobenzene	9.86 ± 0.02	1019
$C_6H_5NO_3$	*o*-Nitrophenol	(9.1)	782
$C_6H_5NO_3$	*m*-Nitrophenol	(9.0)	755
$C_6H_5NO_3$	*p*-Nitrophenol	(9.1)	761
C_6H_6	Benzene	9.24378 ± 0.00007	975
C_6H_6	Fulvene	(8.36)	1031
C_6H_6ClN	*o*-Chloroaniline	(8.50)	883
C_6H_6ClN	*m*-Chloroaniline	(8.09)	835
C_6H_6ClN	*p*-Chloroaniline	(≤ 8.18)	≤ 844
$C_6H_6N_2O_2$	*o*-Nitroaniline	(8.27)	861

Mol. Form.	Name	IP/eV	$\Delta_f H_{ion}$ kJ/mol
$C_6H_6N_2O_2$	*m*-Nitroaniline	(8.31)	865
$C_6H_6N_2O_2$	*p*-Nitroaniline	(8.34)	859
C_6H_6O	Phenol	8.49 ± 0.02	723
$C_6H_6O_2$	*p*-Hydroquinone	7.94 ± 0.01	503
C_6H_6S	Benzenethiol	(8.32)	915
C_6H_7N	Aniline	7.720 ± 0.002	832
C_6H_7N	2-Methylpyridine	(9.02)	970
C_6H_7N	3-Methylpyridine	(9.04)	979
C_6H_7N	4-Methylpyridine	(9.04)	976
$C_6H_8N_2$	*o*-Phenylenediamine	(7.2)	787
$C_6H_8N_2$	*m*-Phenylenediamine	(7.14)	777
$C_6H_8N_2$	*p*-Phenylenediamine	(6.87 ± 0.05)	759
C_6H_{10}	1,5-Hexadiene	9.27 ± 0.05	978
C_6H_{10}	1-Hexyne	10.03 ± 0.05	1089
C_6H_{10}	3,3-Dimethyl-1-butyne	9.90 ± 0.04	1060
C_6H_{10}	Cyclohexene	8.945 ± 0.01	859
$C_6H_{10}O$	Cyclohexanone	9.14 ± 0.01	656
$C_6H_{10}O$	Mesityl oxide	9.10 ± 0.01	694
$C_6H_{10}O_4$	Diethyl oxalate	(9.8)	205
$C_6H_{11}NO$	Caprolactam	(9.07 ± 0.02)	629
C_6H_{12}	1-Hexene	9.44 ± 0.04	869
C_6H_{12}	*cis*-2-Hexene	(8.97 ± 0.01)	818
C_6H_{12}	*trans*-2-Hexene	(8.97 ± 0.01)	814
C_6H_{12}	2-Methyl-1-pentene	(9.08 ± 0.01)	817
C_6H_{12}	4-Methyl-1-pentene	9.45 ± 0.01	862
C_6H_{12}	2-Methyl-2-pentene	(8.58)	761
C_6H_{12}	4-Methyl-*cis*-2-pentene	8.98 ± 0.01	809
C_6H_{12}	4-Methyl-*trans*-2-pentene	(8.97 ± 0.01)	804
C_6H_{12}	2-Ethyl-1-butene	(9.06 ± 0.02)	818
C_6H_{12}	2,3-Dimethyl-1-butene	(9.07 ± 0.01)	812
C_6H_{12}	2,3-Dimethyl-2-butene	8.27 ± 0.01	729
C_6H_{12}	Cyclohexane	9.86 ± 0.03	828
C_6H_{12}	Methylcyclopentane	(9.85)	845
$C_6H_{12}O$	Hexanal	9.72 ± 0.05	691
$C_6H_{12}O$	2-Hexanone	9.3 ± 0.1	626
$C_6H_{12}O$	3-Hexanone	9.12 ± 0.02	600
$C_6H_{12}O$	3-Methyl-2-pentanone	9.21 ± 0.01	600
$C_6H_{12}O$	4-Methyl-2-pentanone	9.30 ± 0.01	609
$C_6H_{12}O$	2-Methyl-3-pentanone	9.10 ± 0.01	592
$C_6H_{12}O$	3,3-Dimethyl-2-butanone	9.12 ± 0.02	589
$C_6H_{12}O$	Cyclohexanol	(9.75)	651
$C_6H_{12}O_2$	Hexanoic acid	≤ 10.12	≤ 463
$C_6H_{12}O_2$	Butyl acetate	(9.92 ± .05)	471
$C_6H_{12}O_2$	*sec*-Butyl acetate	9.90	453
$C_6H_{12}O_2$	Methyl 2,2-dimethylpropanoate	(9.90 ± 0.04)	466
$C_6H_{13}I$	1-Iodohexane	9.179	794
$C_6H_{13}N$	Cyclohexylamine	(8.86)	750
C_6H_{14}	Hexane	10.13	810
C_6H_{14}	2-Methylpentane	(10.12)	802
C_6H_{14}	3-Methylpentane	(10.08)	801
C_6H_{14}	2,2-Dimethylbutane	(10.06)	787
C_6H_{14}	2,3-Dimethylbutane	(10.02)	791
$C_6H_{14}O$	1-Hexanol	(9.89)	639
$C_6H_{14}O$	2-Hexanol	(9.80 ± 0.03)	611
$C_6H_{14}O$	3-Hexanol	(9.63 ± 0.03)	599
$C_6H_{14}O$	Dipropyl ether	(9.27)	602
$C_6H_{14}O$	Diisopropyl ether	9.20 ± 0.05	569
$C_6H_{14}O$	Butyl ethyl ether	(9.36)	610
$C_6H_{14}O$	Methyl pentyl ether	(≤ 9.67)	≤ 657
$C_6H_{14}O_2$	1,1-Diethoxyethane	(9.2)	434

Mol. Form.	Name	IP/eV	$\Delta_f H_{ion}$ kJ/mol
$C_6H_{14}O_3$	Diethylene glycol dimethyl ether	≤ 9.8	≤ 448
$C_6H_{14}S$	Dipropyl sulfide	8.30 ± 0.02	676
$C_6H_{14}S$	Diisopropyl sulfide	(8.2 ± 0.2)	649
$C_6H_{15}N$	Hexylamine	(8.63 ± 0.05)	699
$C_6H_{15}N$	Dipropylamine	(7.84 ± 0.02)	641
$C_6H_{15}N$	Diisopropylamine	(7.73 ± 0.03)	602
$C_6H_{15}N$	Triethylamine	(7.50 ± 0.02)	631
$C_6H_{15}NO_3$	Triethanolamine	(7.9)	206
$C_7H_3F_5$	2,3,4,5,6-Pentafluorotoluene	(9.4)	64
C_7H_5ClO	Benzoyl chloride	(9.53)	815
$C_7H_5Cl_3$	(Trichloromethyl)benzene	(≤ 9.60)	≤ 914
$C_7H_5F_3$	(Trifluoromethyl)benzene	9.685 ± 0.005	335
C_7H_5N	Benzonitrile	9.70 ± 0.01	1154
C_7H_6O	Benzaldehyde	9.49 ± 0.02	878
$C_7H_6O_2$	Benzoic acid	(9.3)	604
C_7H_7Br	*p*-Bromotoluene	8.67 ± 0.02	908
C_7H_7Cl	*o*-Chlorotoluene	(8.7 ± 0.1)	856
C_7H_7Cl	*m*-Chlorotoluene	(8.83)	869
C_7H_7Cl	*p*-Chlorotoluene	(8.69)	855
C_7H_7Cl	(Chloromethyl)benzene	9.10 ± 0.02	897
C_7H_7F	*o*-Fluorotoluene	8.91 ± 0.01	709
C_7H_7F	*m*-Fluorotoluene	8.91 ± 0.01	709
C_7H_7F	*p*-Fluorotoluene	8.79 ± 0.01	701
C_7H_7NO	Benzamide	(9.25)	792
$C_7H_7NO_2$	*o*-Nitrotoluene	9.24	946
$C_7H_7NO_2$	*m*-Nitrotoluene	9.45 ± 0.1	941
$C_7H_7NO_2$	*p*-Nitrotoluene	9.46 ± 0.05	942
C_7H_8	Toluene	8.8276 ± 0.0006	901
C_7H_8O	*o*-Cresol	(8.24)	670
C_7H_8O	*m*-Cresol	8.29 ± 0.07	668
C_7H_8O	*p*-Cresol	(8.3)	675
C_7H_8O	Benzyl alcohol	(8.3)	701
C_7H_8O	Anisole	8.22 ± 0.03	725
C_7H_9N	Benzylamine	(8.64)	917
C_7H_9N	*o*-Methylaniline	(7.44 ± 0.02)	772
C_7H_9N	*m*-Methylaniline	(7.50 ± 0.02)	778
C_7H_9N	*p*-Methylaniline	(7.24 ± 0.02)	753
C_7H_9N	*N*-Methylaniline	7.34 ± 0.04	792
C_7H_9N	2,3-Dimethylpyridine	(8.85 ± 0.02)	922
C_7H_9N	2,4-Dimethylpyridine	(8.85 ± 0.03)	918
C_7H_9N	2,5-Dimethylpyridine	(≤ 8.80 ± 0.05)	≤ 916
C_7H_9N	2,6-Dimethylpyridine	8.86 ± 0.03	913
C_7H_9N	3,4-Dimethylpyridine	(≤ 9.15)	≤ 953
C_7H_9N	3,5-Dimethylpyridine	(≤ 9.25)	≤ 965
$C_7H_{10}O$	Dicyclopropyl ketone	(9.1)	1041
C_7H_{14}	1-Heptene	9.34 ± 0.10	839
C_7H_{14}	*trans*-3-Heptene	(8.92)	790
C_7H_{14}	Cycloheptane	9.97	844
C_7H_{14}	Methylcyclohexane	9.64	775
C_7H_{14}	*cis*-1,2-Dimethylcyclopentane	(9.92 ± 0.05)	828
C_7H_{14}	*trans*-1,2-Dimethylcyclopentane	9.7 ± 0.2	799
$C_7H_{14}O$	1-Heptanal	(9.65)	668
$C_7H_{14}O$	2-Heptanone	9.28 ± 0.10	594
$C_7H_{14}O$	3-Heptanone	9.18 ± 0.08	589
$C_7H_{14}O$	4-Heptanone	9.10 ± 0.06	577
$C_7H_{14}O$	5-Methyl-2-hexanone	(9.28)	586
$C_7H_{14}O$	2,4-Dimethyl-3-pentanone	8.95 ± 0.01	552
$C_7H_{14}O$	1-Methylcyclohexanol	(9.8 ± 0.2)	586
C_7H_{16}	Heptane	9.93 ± 0.10	771
$C_7H_{16}O$	1-Heptanol	(9.84)	614

Mol. Form.	Name	IP/eV	$\Delta_f H_{ion}$ kJ/mol
$C_7H_{16}O$	2-Heptanol	(9.70)	580
$C_7H_{16}O$	3-Heptanol	(9.68)	578
$C_7H_{16}O$	4-Heptanol	(9.61)	572
$C_7H_{16}O$	Ethyl pentyl ether	(≤ 9.49)	≤ 602
$C_8H_4O_3$	Phthalic anhydride	(10.1)	603
$C_8H_6O_4$	Isophthalic acid	(9.98)	268
$C_8H_6O_4$	Terephthalic acid	(9.86)	232
C_8H_7N	2-Methylbenzonitrile	(≤ 9.38)	1085
C_8H_7N	3-Methylbenzonitrile	(≤ 9.34)	1085
C_8H_7N	4-Methylbenzonitrile	9.32 ± 0.02	1083
C_8H_7N	Indole	7.7602 ± 0.0006	908
C_8H_8	Styrene	8.464 ± 0.001	964
C_8H_8O	*p*-Tolualdehyde	(9.33)	825
C_8H_8O	Acetophenone	9.29 ± 0.03	810
$C_8H_8O_2$	*o*-Toluic acid	(9.1)	558
$C_8H_8O_2$	*m*-Toluic acid	(9.43)	579
$C_8H_8O_2$	*p*-Toluic acid	(9.23)	560
$C_8H_8O_2$	Benzeneacetic acid	(8.26)	479
$C_8H_8O_2$	Methyl benzoate	9.32 ± 0.03	611
C_8H_{10}	Ethylbenzene	8.77 ± 0.01	876
C_8H_{10}	*o*-Xylene	8.56 ± 0.01	844
C_8H_{10}	*m*-Xylene	8.56 ± 0.01	843
C_8H_{10}	*p*-Xylene	8.44 ± 0.01	832
$C_8H_{10}O$	*p*-Ethylphenol	(7.84)	613
$C_8H_{10}O$	2,3-Xylenol	(8.26)	640
$C_8H_{10}O$	2,4-Xylenol	(8.0)	609
$C_8H_{10}O$	2,6-Xylenol	(8.05)	615
$C_8H_{10}O$	3,4-Xylenol	(8.09)	624
$C_8H_{10}O$	Phenetole	(8.13)	683
$C_8H_{11}N$	2,4,6-Trimethylpyridine	(≤ 8.9)	≤ 880
$C_8H_{11}N$	*N*-Ethylaniline	(≤ 7.67)	≤ 794
$C_8H_{11}N$	*N,N*-Dimethylaniline	7.12 ± 0.02	787
C_8H_{14}	1-Octyne	(9.95 ± 0.02)	1040
C_8H_{14}	2-Octyne	9.31 ± 0.01	961
C_8H_{14}	3-Octyne	9.22 ± 0.01	952
C_8H_{14}	4-Octyne	9.20 ± 0.01	946
C_8H_{16}	1-Octene	9.43 ± 0.01	829
C_8H_{16}	Cyclooctane	9.75 ± 0.05	816
C_8H_{16}	Ethylcyclohexane	(9.54)	748
C_8H_{16}	1,1-Dimethylcyclohexane	(9.42)	728
C_8H_{16}	*cis*-1,2-Dimethylcyclohexane	(<9.78)	772
C_8H_{16}	*trans*-1,2-Dimethylcyclohexane	9.41	728
C_8H_{16}	*cis*-1,3-Dimethylcyclohexane	(<9.98)	778
C_8H_{16}	*trans*-1,3-Dimethylcyclohexane	9.53	743
C_8H_{16}	*cis*-1,4-Dimethylcyclohexane	(<9.93)	782
C_8H_{16}	*trans*-1,4-Dimethylcyclohexane	(9.56)	738
C_8H_{16}	Propylcyclopentane	(9.34)	753
$C_8H_{16}O$	2,2,4-Trimethyl-3-pentanone	(8.80)	511
C_8H_{18}	Octane	9.80 ± 0.10	737
C_8H_{18}	2-Methylheptane	(9.84)	734
C_8H_{18}	2,2,4-Trimethylpentane	(9.86)	713
C_8H_{18}	2,2,3,3-Tetramethylbutane	9.8	720
$C_8H_{18}O$	Dibutyl ether	(9.28)	s 560
$C_8H_{18}O$	Di-*sec*-butyl ether	(9.11)	511
$C_8H_{18}O$	Di-*tert*-butyl ether	8.88 ± 0.07	493
$C_8H_{18}S$	Dibutyl sulfide	(8.2)	624
$C_8H_{18}S$	Di-*tert*-butyl sulfide	(8.0)	583
$C_8H_{18}S$	Diisobutyl sulfide	(8.34)	625
$C_8H_{19}N$	Dibutylamine	(7.69)	586
$C_8H_{19}N$	Diisobutylamine	(7.8)	574

Mol. Form.	Name	IP/eV	$\Delta_f H_{ion}$ kJ/mol
$C_8H_{20}Si$	Tetraethylsilane	(8.9)	595
C_9H_7N	Quinoline	8.62±0.01	1041
C_9H_7N	Isoquinoline	8.53±0.03	1032
C_9H_8	Indene	8.14±0.01	949
C_9H_{10}	*o*-Methylstyrene	(8.20)	908
C_9H_{10}	*m*-Methylstyrene	(8.15)	899
C_9H_{10}	*p*-Methylstyrene	(8.1)	895
C_9H_{10}	Cyclopropylbenzene	(8.35)	956
C_9H_{10}	Indan	(8.3)	864
$C_9H_{10}O_2$	Ethyl benzoate	(8.9)	537
C_9H_{12}	Propylbenzene	8.713±0.010	848
C_9H_{12}	Isopropylbenzene	8.73±0.01	847
C_9H_{12}	1,2,3-Trimethylbenzene	8.42±0.02	803
C_9H_{12}	1,2,4-Trimethylbenzene	8.27±0.01	784
C_9H_{12}	1,3,5-Trimethylbenzene	8.41±0.01	796
$C_9H_{13}N$	*N,N*-Dimethyl-*o*-toluidine	7.40±0.02	814
$C_9H_{14}O$	Isophorone	(≤ 9.07)	≤ 670
C_9H_{18}	Butylcyclopentane	(9.95)	793
C_9H_{18}	Propylcyclohexane	(9.46)	720
C_9H_{18}	Isopropylcyclohexane	(9.33)	704
$C_9H_{18}O$	2-Nonanone	(9.16)	545
$C_9H_{18}O$	5-Nonanone	(9.07)	530
$C_9H_{18}O$	2,6-Dimethyl-4-heptanone	9.01±0.06	512
C_9H_{20}	Nonane	9.71±0.10	709
$C_{10}F_8$	Perfluoronaphthalene	8.85	-368
$C_{10}H_7Br$	1-Bromonaphthalene	8.08±0.03	955
$C_{10}H_7Cl$	1-Chloronaphthalene	(8.13)	906
$C_{10}H_8$	Naphthalene	8.1442±0.0009	936
$C_{10}H_8$	Azulene	7.38±0.05	1001
$C_{10}H_8O$	1-Naphthol	7.76±0.03	719
$C_{10}H_8O$	2-Naphthol	7.87±0.06	729
$C_{10}H_{10}O_4$	Dimethyl phthalate	(9.64±0.07)	277
$C_{10}H_{12}$	1,2,3,4-Tetrahydronaphthalene	8.46±0.02	841
$C_{10}H_{14}$	Butylbenzene	8.69±0.02	826
$C_{10}H_{14}$	*sec*-Butylbenzene	8.68±0.02	820
$C_{10}H_{14}$	*tert*-Butylbenzene	8.68±0.05	816
$C_{10}H_{14}$	Isobutylbenzene	8.69±0.02	817
$C_{10}H_{14}$	*p*-Cymene	(8.29)	771
$C_{10}H_{14}$	*o*-Diethylbenzene	(≤ 8.51)	≤ 804
$C_{10}H_{14}$	*m*-Diethylbenzene	(8.49)	798
$C_{10}H_{14}$	*p*-Diethylbenzene	(8.40)	790
$C_{10}H_{14}$	1,2,4,5-Tetramethylbenzene	8.04±0.02	730
$C_{10}H_{14}O$	*p-tert*-Butylphenol	(7.8)	552
$C_{10}H_{16}$	α-Pinene	(8.07)	808
$C_{10}H_{16}O$	Camphor	(8.76)	577
$C_{10}H_{18}$	*cis*-Decahydronaphthalene	9.36±0.04	734
$C_{10}H_{18}$	*trans*-Decahydronaphthalene	9.34±0.04	720
$C_{10}H_{20}$	1-Decene	9.42±0.05	786
$C_{10}H_{20}$	Butylcyclohexane	(9.41)	695
$C_{10}H_{22}$	Decane	(9.65)	682
$C_{11}H_{10}$	1-Methylnaphthalene	7.97±0.03	882
$C_{11}H_{10}$	2-Methylnaphthalene	7.91±0.08	877
$C_{11}H_{16}$	*p-tert*-Butyltoluene	(8.12)	730
$C_{11}H_{24}$	Undecane	(9.56)	650
$C_{11}H_{24}$	2-Methyldecane	(9.7)	658
$C_{12}H_8$	Acenaphthylene	(8.22)	1053
$C_{12}H_9N$	Carbazole	(7.57)	961
$C_{12}H_{10}$	Acenaphthene	7.75±0.07	903
$C_{12}H_{10}$	Biphenyl	8.23±0.10	977
$C_{12}H_{10}N_2O$	*trans*-Azoxybenzene	(8.1)	1123

Mol. Form.	Name	IP/eV	$\Delta_f H_{ion}$ kJ/mol
$C_{12}H_{10}O$	Diphenyl ether	(8.09)	766
$C_{12}H_{11}N$	Diphenylamine	7.16 ± 0.04	908
$C_{12}H_{18}$	5,7-Dodecadiyne	(8.67)	1079
$C_{12}H_{18}$	Hexamethylbenzene	7.85 ± 0.01	670
$C_{12}H_{22}$	Cyclohexylcyclohexane	(9.41)	690
$C_{12}H_{27}N$	Tributylamine	(7.4)	492
$C_{13}H_{10}$	9H-Fluorene	7.91 ± 0.02	952
$C_{13}H_{10}O$	Benzophenone	9.08 ± 0.05	926
$C_{13}H_{12}$	Diphenylmethane	(8.55)	963
$C_{14}H_{10}$	Anthracene	7.439 ± 0.006	948
$C_{14}H_{10}$	Phenanthrene	7.8914 ± 0.0006	966
$C_{14}H_{10}$	Diphenylacetylene	7.94 ± 0.03	1168
$C_{14}H_{12}$	*cis*-Stilbene	(7.80)	1005
$C_{14}H_{12}$	*trans*-Stilbene	7.656 ± 0.001	973
$C_{14}H_{14}$	1,2-Diphenylethane	8.9 ± 0.1	1002
$C_{16}H_{10}$	Fluoranthene	7.9 ± 0.1	1052
$C_{16}H_{10}$	Pyrene	7.4256 ± 0.0006	935
$C_{18}H_{12}$	Chrysene	7.60 ± 0.01	1017
$C_{18}H_{14}$	*o*-Terphenyl	(7.99)	1056
$C_{18}H_{14}$	*m*-Terphenyl	(8.01)	1057
$C_{18}H_{14}$	*p*-Terphenyl	7.80 ± 0.03	1037
$C_{20}H_{12}$	Perylene	6.960 ± 0.001	981
$C_{24}H_{12}$	Coronene	7.29 ± 0.01	1026

X-RAY ATOMIC ENERGY LEVELS

The energy levels in this tables are the values recommended by Bearden and Burr on the basis of a thorough review of the literature on x-ray wavelengths and related data. All values are in electron volts (eV). Values in parentheses are interpolated, and an asterisk * indicates a level which is not resolved from the level above it. See Reference 1 for uncertainties in the levels and a complete description of how the recommended values were obtained.

REFERENCES

1. Bearden, J. A., and Burr, A. F., *Rev. Mod. Phys.*, 39, 125, 1967; also published as *X-Ray Wavelengths and X-Ray Atomic Energy Levels*, Natl. Stand. Ref. Data Sys.- Natl. Bur. Standards (U.S.), No. 14, 1967.
2. Gray, D. E., Editor, *American Institute of Physics Handbook, Third Edition*, pp. 7-158 to 7-167, McGraw-Hill, New York, 1972.

Level	^{1}H	^{2}He	^{3}Li	^{4}Be	^{5}B	^{6}C	^{7}N	^{8}O
K	13.59811	24.58678	54.75	111.0	188.0	283.8	401.6	532.0
L_I								23.7
$L_{II,III}$					4.7	6.4	9.2	7.1

Level	^{9}F	^{10}Ne	^{11}Na	^{12}Mg	^{13}Al	^{14}Si	^{15}P	^{16}S
K	685.4	866.9	1072.1	1305.0	1559.6	1838.9	2145.5	2472.0
L_I	(31)	(45)	63.3	89.4	117.7	148.7	189.3	229.2
$L_{II,III}$	8.6	18.3	31.1	51.4	73.1	99.2	132.2	164.8

Level	^{17}Cl	^{18}Ar	^{19}K	^{20}Ca	^{21}Sc	^{22}Ti	^{23}V	^{24}Cr
K	2822.4	3202.9	3607.4	4038.1	4492.8	4966.4	5465.1	5989.2
L_I	270.2	320	377.1	437.8	500.4	563.7	628.2	694.6
L_{II}	201.6	247.3	296.3	350.0	406.7	461.5	520.5	583.7
L_{III}	200.0	245.2	293.6	346.4	402.2	455.5	512.9	574.5
M_I	17.5	25.3	33.9	43.7	53.8	60.3	66.5	74.1
$M_{II,III}$	6.8	12.4	17.8	25.4	32.3	34.6	37.8	42.5
$M_{IV,V}$					6.6	3.7	2.2	2.3

Level	^{25}Mn	^{26}Fe	^{27}Co	^{28}Ni	^{29}Cu	^{30}Zn	^{31}Ga	^{32}Ge
K	6539.0	7112.0	7708.9	8332.8	8978.9	9658.6	10367.1	11103.1
L_I	769.0	846.1	925.6	1008.1	1096.1	1193.6	1297.7	1414.3
L_{II}	651.4	721.1	793.6	871.9	951.0	1042.8	1142.3	1247.8
L_{III}	640.3	708.1	778.6	854.7	931.1	1019.7	1115.4	1216.7
M_I	83.9	92.9	100.7	111.8	119.8	135.9	158.1	180.0
M_{II}	48.6	54.0	59.5	68.1	73.6	86.6	106.8	127.9
M_{III}	48.6*	54.0*	59.5*	68.1*	73.6*	86.6*	102.9	120.8
$M_{IV,V}$	3.3	3.6	2.9	3.6	1.6	8.1	17.4	28.7

Level	^{33}As	^{34}Se	^{35}Br	^{36}Kr	^{37}Rb	^{38}Sr	^{39}Y	^{40}Zr
K	11866.7	12657.8	13473.7	14325.6	15199.7	16104.6	17038.4	17997.6
L_I	1526.5	1653.9	1782.0	1921.0	2065.1	2216.3	2372.5	2531.6
L_{II}	1358.6	1476.2	1596.0	1727.2	1863.9	2006.8	2155.5	2306.7
L_{III}	1323.1	1435.8	1549.9	1674.9	1804.4	1939.6	2080.0	2222.3
M_I	203.5	231.5	256.5		322.1	357.5	393.6	430.3
M_{II}	146.4	168.2	189.3	222.7	247.4	279.8	312.4	344.2
M_{III}	140.5	161.9	181.5	213.8	238.5	269.1	300.3	330.5
M_{IV}	41.2	56.7	70.1	88.9	111.8	135.0	159.6	182.4
M_V	41.2*	56.7*	69.0	88.9*	110.3	133.1	157.4	180.0
N_I			27.3	24.0	29.3	37.7	45.4	51.3
N_{II}	2.5	5.6	5.2	10.6	14.8	19.9	25.6	28.7
N_{III}	2.5*	5.6*	4.6	10.6*	14.0	19.9*	25.6*	28.7*

X-RAY ATOMIC ENERGY LEVELS (continued)

Level	^{41}Nb	^{42}Mo	^{43}Tc	^{44}Ru	^{45}Rh	^{46}Pd	^{47}Ag	^{48}Cd
K	18985.6	19999.5	21044.0	22117.2	23219.9	24350.3	25514.0	26711.2
L_I	2697.7	2865.5	3042.5	3224.0	3411.9	3604.3	3805.8	4018.0
L_{II}	2464.7	2625.1	2793.2	2966.9	3146.1	3330.3	3523.7	3727.0
L_{III}	2370.5	2520.2	2676.9	2837.9	3003.8	3173.3	3351.1	3537.5
M_I	468.4	504.6		585.0	627.1	669.9	717.5	770.2
M_{II}	378.4	409.7	444.9	482.8	521.0	559.1	602.4	650.7
M_{III}	363.0	392.3	425.0	460.6	496.2	531.5	571.4	616.5
M_{IV}	207.4	230.3	256.4	283.6	311.7	340.0	372.8	410.5
M_V	204.6	227.0	252.9	279.4	307.0	334.7	366.7	403.7
N_I	58.1	61.8		74.9	81.0	86.4	95.2	107.6
N_{II}	33.9	34.8	38.9	43.1	47.9	51.1	62.6	66.9
N_{III}	33.9*	34.8*	38.9*	43.1*	47.9*	51.1*	55.9	66.9*
$N_{IV,V}$	3.2	1.8		2.0	2.5	1.5	3.3	9.3

Level	^{49}In	^{50}Sn	^{51}Sb	^{52}Te	^{53}I	^{54}Xe	^{55}Cs	^{56}Ba
K	27939.9	29200.1	30491.2	31813.8	33169.4	34561.4	35984.6	37440.6
L_I	4237.5	4464.7	4698.3	4939.2	5188.1	5452.8	5714.3	5988.8
L_{II}	3938.0	4156.1	4380.4	4612.0	4852.1	5103.7	5359.4	5623.6
L_{III}	3730.1	3928.8	4132.2	4341.4	4557.1	4782.2	5011.9	5247.0
M_I	825.6	883.8	943.7	1006.0	1072.1		1217.1	1292.8
M_{II}	702.2	756.4	811.9	869.7	930.5	999.0	1065.0	1136.7
M_{III}	664.3	714.4	765.6	818.7	874.6	937.0	997.6	1062.2
M_{IV}	450.8	493.3	536.9	582.5	631.3		739.5	796.1
M_V	443.1	484.8	527.5	572.1	619.4	672.3	725.5	780.7
N_I	121.9	136.5	152.0	168.3	186.4		230.8	253.0
N_{II}	77.4	88.6	98.4	110.2	122.7	146.7	172.3	191.8
N_{III}	77.4*	88.6*	98.4*	110.2*	122.7*	146.7*	161.6	179.7
N_{IV}	16.2	23.9	31.4	39.8	49.6		78.8	92.5
N_V	16.2*	23.9*	31.4*	39.8*	49.6*		76.5	89.9
O_I	0.1	0.9	6.7	11.6	13.6		22.7	39.1
O_{II}	0.8	1.1	2.1	2.3	3.3		13.1	16.6
O_{III}	0.8*	1.1*	2.1*	2.3*	3.3*		11.4	14.6

Level	^{57}La	^{58}Ce	^{59}Pr	^{60}Nd	^{61}Pm	^{62}Sm	^{63}Eu	^{64}Gd
K	38924.6	40443.0	41990.6	43568.9	45184.0	46834.2	48519.0	50239.1
L_I	6266.3	6548.8	6834.8	7126.0	7427.9	7736.8	8052.0	8375.6
L_{II}	5890.6	6164.2	6440.4	6721.5	7012.8	7311.8	7617.1	7930.3
L_{III}	5482.7	5723.4	5964.3	6207.9	6459.3	6716.2	6976.9	7242.8
M_I	1361.3	1434.6	1511.0	1575.3		1722.8	1800.0	1880.8
M_{II}	1204.4	1272.8	1337.4	1402.8	1471.4	1540.7	1613.9	1688.3
M_{III}	1123.4	1185.4	1242.2	1297.4	1356.9	1419.8	1480.6	1544.0
M_{IV}	848.5	901.3	951.1	999.9	1051.5	1106.0	1160.6	1217.2
M_V	831.7	883.3	931.0	977.7	1026.9	1080.2	1130.9	1185.2
N_I	270.4	289.6	304.5	315.2		345.7	360.2	375.8
N_{II}	205.8	223.3	236.3	243.3	242	265.6	283.9	288.5
N_{III}	191.4	207.2	217.6	224.6	242*	247.4	256.6	270.9
$N_{IV,V}$	98.9	110.0	113.2	117.5	120.4	129.0	133.2	140.5
$N_{VI,VII}$		0.1	2.0	1.5		5.5	0.0	0.1
O_I	32.3	37.8	37.4	37.5		37.4	31.8	36.1
$O_{II,III}$	14.4	19.8	22.3	21.1		21.3	22.0	20.3

Level	^{65}Tb	^{66}Dy	^{67}Ho	^{68}Er	^{69}Tm	^{70}Yb	^{71}Lu	^{72}Hf
K	51995.7	53788.5	55617.7	57485.5	59389.6	61332.3	63313.8	65350.8
L_I	8708.0	9045.8	9394.2	9751.3	10115.7	10486.4	10870.4	11270.7
L_{II}	8251.6	8580.6	8917.8	9264.3	9616.9	9978.2	10348.6	10739.4
L_{III}	7514.0	7790.1	8071.1	8357.9	8648.0	8943.6	9244.1	9560.7
M_I	1967.5	2046.8	2128.3	2206.5	2306.8	2398.1	2491.2	2600.9

Level	^{65}Tb	^{66}Dy	^{67}Ho	^{68}Er	^{69}Tm	^{70}Yb	^{71}Lu	^{72}Hf
M_{II}	1767.7	1841.8	1922.8	2005.8	2089.8	2173.0	2263.5	2365.4
M_{III}	1611.3	1675.6	1741.2	1811.8	1884.5	1949.8	2023.6	2107.6
M_{IV}	1275.0	1332.5	1391.5	1453.3	1514.6	1576.3	1639.4	1716.4
M_{V}	1241.2	1294.9	1351.4	1409.3	1467.7	1527.8	1588.5	1661.7
N_{I}	397.9	416.3	435.7	449.1	471.7	487.2	506.2	538.1
N_{II}	310.2	331.8	343.5	366.2	385.9	396.7	410.1	437.0
N_{III}	385.0	292.9	306.6	320.0	336.6	343.5	359.3	380.4
N_{IV}	147.0	154.2	161.0	176.7	179.6	198.1	204.8	223.8
N_{V}	147.0*	154.2*	161.0*	167.6	179.6*	184.9	195.0	213.7
$N_{VI,VII}$	2.6	4.2	3.7	4.3	5.3	6.3	6.9	17.1
O_{I}	39.0	62.9	51.2	59.8	53.2	54.1	56.8	64.9
O_{II}	25.4	26.3	20.3	29.4	32.3	23.4	28.0	38.1
O_{III}	25.4*	26.3*	20.3*	29.4*	32.3*	23.4*	28.0*	30.6

Level	^{73}Ta	^{74}W	^{75}Re	^{76}Os	^{77}Ir	^{78}Pt	^{79}Au	^{80}Hg
K	67416.4	69525.0	71676.4	73870.8	76111.0	78394.8	80724.9	83102.3
L_{I}	11681.5	12099.8	12526.7	12968.0	13418.5	13879.9	14352.8	14839.3
L_{II}	11136.1	11544.0	11958.7	12385.0	12824.1	13272.6	13733.6	14208.7
L_{III}	9881.1	10206.8	10535.3	10870.9	11215.2	11563.7	11918.7	12283.9
M_{I}	2708.0	2819.6	2931.7	3048.5	3173.7	3296.0	3424.9	3561.6
M_{II}	2468.7	2574.9	2681.6	2792.2	2908.7	3026.5	3147.8	3278.5
M_{III}	2194.0	2281.0	2367.3	2457.2	2550.7	2645.4	2743.0	2847.1
M_{IV}	1793.2	1871.6	1948.9	2030.8	2116.1	2201.9	2291.1	2384.9
M_{V}	1735.1	1809.2	1882.9	1960.1	2040.4	2121.6	2205.7	2294.9
N_{I}	565.5	595.0	625.0	654.3	690.1	722.0	758.8	800.3
N_{II}	464.8	491.6	517.9	546.5	577.1	609.2	643.7	676.9
N_{III}	404.5	425.3	444.4	468.2	494.3	519.0	545.4	571.0
N_{IV}	241.3	258.8	273.7	289.4	311.4	330.8	352.0	378.3
N_{V}	229.3	245.4	260.2	272.8	294.9	313.3	333.9	359.8
N_{VI}	25.0	36.5	40.6	46.3	63.4	74.3	86.4	102.2
N_{VII}	25.0*	33.6	40.6*	46.3*	60.5	71.1	82.8	98.5
O_{I}	71.1	77.1	82.8	83.7	95.2	101.7	107.8	120.3
O_{II}	44.9	46.8	45.6	58.0	63.0	65.3	71.7	80.5
O_{III}	36.4	35.6	34.6	45.4	50.5	51.7	53.7	57.6
$O_{IV,V}$	5.7	6.1	3.5		3.8	2.2	2.5	6.4

Level	^{81}Tl	^{82}Pb	^{83}Bi	^{84}Po	^{85}At	^{86}Rn	^{87}Fr	^{88}Ra
K	85530.4	88004.5	90525.9	93105.0	95729.9	98404	101137	103921.9
L_{I}	15346.7	15860.8	16387.5	16939.3	17493	18049	18639	19236.7
L_{II}	14697.9	15200.0	15711.1	16244.3	16784.7	17337.1	17906.5	18484.3
L_{III}	12657.5	13035.2	13418.6	13813.8	14213.5	14619.4	15031.2	15444.4
M_{I}	3704.1	3850.7	3999.1	4149.4	(4317)	(4482)	(4652)	4822.0
M_{II}	3415.7	3554.2	3696.3	3854.1	4008	4159	4327	4489.5
M_{III}	2956.6	3066.4	3176.9	3301.9	3426	3538	3663	3791.8
M_{IV}	2485.1	2585.6	2687.6	2798.0	2908.7	3021.5	3136.2	3248.4
M_{V}	2389.3	2484.0	2579.6	2683.0	2786.7	2892.4	2999.9	3104.9
N_{I}	845.5	893.6	938.2	995.3	(1042)	(1097)	(1153)	1208.4
N_{II}	721.3	763.9	805.3	851	886	929	980	1057.6
N_{III}	609.0	644.5	678.9	705	740	768	810	879.1
N_{IV}	406.6	435.2	463.6	500.2	533.2	566.6	603.3	635.9
N_{V}	386.2	412.9	440.0	473.4			577	602.7
N_{VI}	122.8	142.9	161.9					298.9
N_{VII}	118.5	138.1	157.4					298.9*
O_{I}	136.3	147.3	159.3					254.4
O_{II}	99.6	104.8	116.8					200.4
O_{III}	75.4	86.0	92.8					152.8

Level	^{81}Tl	^{82}Pb	^{83}Bi	^{84}Po	^{85}At	^{86}Rn	^{87}Fr	^{88}Ra
O_{IV}	15.3	21.8	26.5	31.4				67.2
O_{V}	13.1	19.2	24.4	31.4*				67.2*
P_{I}		3.1						43.5
$P_{II,III}$		0.7	2.7					18.8

Level	^{89}Ac	^{90}Th	^{91}Pa	^{92}U	^{93}Np	^{94}Pu	^{95}Am	^{96}Cm
K	106755.3	109650.9	112601.4	115606.1	118678	121818	125027	128220
L_{I}	19840	20472.1	21104.6	21757.4	22426.8	23097.2	23772.9	24460
L_{II}	19083.2	19693.2	20313.7	20947.6	21600.5	22266.2	22944.0	23779
L_{III}	15871.0	16300.3	16733.1	17166.3	17610.0	18056.8	18504.1	18930
M_{I}	(5002)	5182.3	5366.9*	5548.0	5723.2	5932.9	6120.5	6288
M_{II}	4656	4830.4	5000.9	5182.2	5366.2	5541.2	5710.2	5895
M_{III}	3909	4046.1	4173.8	4303.4	4434.7	4556.6	4667.0	4797
M_{IV}	3370.2	3490.8	3611.2	3727.6	3850.3	3972.6	4092.1	4227
M_{V}	3219.0	3332.0	3441.8	3551.7	3665.8	3778.1	3886.9	3971
N_{I}	(1269)	1329.5	1387.1	1440.8	1500.7	1558.6	1617.1	1643
N_{II}	1080	1168.2	1224.3	1272.6	1327.7	1372.1	1411.8	1440
N_{III}	890	967.3	1006.7	1044.9	1086.8	1114.8	(1135.7)	1154
N_{IV}	674.9	714.1	743.4	780.4	815.9	848.9	878.7	
N_{V}		676.4	708.2	737.7	770.3	801.4	827.6	
N_{VI}		344.4	371.2	391.3	415.0	445.8		
N_{VII}		335.2	359.5	380.9	404.4	432.4		
O_{I}		290.2	309.6	323.7		351.9		385
O_{II}		229.4	222.9	259.3	283.4	274.1		
O_{III}		181.8	222.9*	195.1	206.1	206.5		
O_{IV}		94.3	94.1	105.0	109.3	116.0	115.8	
O_{V}		87.9	94.1*	96.3	101.3	105.4	103.3	
P_{I}		59.5		70.7				
P_{II}		49.0		42.3				
P_{III}		43.0		32.3				

Level	^{97}Bk	^{98}Cf	^{99}Es	^{100}Fm	^{101}Md	^{102}No	^{103}Lr
K	131590	135960	139490	143090	146780	150540	154380
L_{I}	25275	26110	26900	27700	28530	29380	30240
L_{II}	24385	25250	26020	26810	27610	28440	29280
L_{III}	19452	19930	20410	20900	21390	21880	22360
M_{I}	6556	6754	6977	7205	7441	7675	7900
M_{II}	6147	6359	6574	6793	7019	7245	7460
M_{III}	4977	5109	5252	5397	5546	5688	5710
M_{IV}	4366	4497	4630	4766	4903	5037	5150
M_{V}	4132	4253	4374	4498	4622	4741	4860
N_{I}	1755	1799	1868	1937	2010	2078	2140
N_{II}	1554	1616	1680	1747	1814	1876	1930
N_{III}	1235	1279	1321	1366	1410	1448	1480
O_{I}	398	419	435	454	472	484	490

ELECTRON BINDING ENERGIES OF THE ELEMENTS

Gwyn P. Williams

This table gives the binding energies in electron volts (eV) for selected electronic levels of the elements. For metallic elements the binding energy is referred to the Fermi level; for semiconductors, to the valence band maximum; and for gases and insulators, to the vacuum level. The atomic number is listed after the element name.

REFERENCES

1. Fluggle and Martensson, J. *Elect. Spect.*, 21, 275, 1980.
2. Cardona, M. and Ley, L., *Photoemission from Solids*, Springer Verlag, Heidelberg, 1978.
3. Bearden, J. A. and Burr, A. F., *Rev. Mod. Phys.*, 39, 125, 1967.

Actinium (89)

K	1s	106755
L I	2s	19840
L II	$2p_{1/2}$	19083
L III	$2p_{3/2}$	15871
M I	3s	5002
M II	$3p_{1/2}$	4656
M III	$3p_{3/2}$	3909
M IV	$3d_{3/2}$	3370
M V	$3d_{5/2}$	3219
N I	4s	1269[a]
N II	$4p_{1/2}$	1080[a]
N III	$4p_{3/2}$	890[a]
N IV	$4d_{3/2}$	675[a]
N V	$4d_{5/2}$	639[a]
N VI	$4f_{5/2}$	319[a]
N VII	$4f_{7/2}$	319[a]
O I	5s	272[a]
O II	$5p_{1/2}$	215[a]
O III	$5p_{3/2}$	167[a]
O IV	$5d_{3/2}$	80[a]
O V	$5d_{5/2}$	80[a]
P I	6s	—
P II	$6p_{1/2}$	—
P III	$6p_{3/2}$	—

Aluminum (13)

K	1s	1559.0
L I	2s	117.8[a]
L II	$2p_{1/2}$	72.9[a]
L III	$2p_{3/2}$	72.5[a]

Antimony (51)

K	1s	30419
L I	2s	4698
L II	$2p_{1/2}$	4380
L III	$2p_{3/2}$	4132
M I	3s	946[b]
M II	$3p_{1/2}$	812.7[b]
M III	$3p_{3/2}$	766.4[b]
M IV	$3d_{3/2}$	537.5[b]
M V	$3d_{5/2}$	528.2[b]
N I	4s	153.2[b]
N II	$4p_{1/2}$	95.6[b,c]
N III	$4p_{3/2}$	95.6[b]
N IV	$4d_{3/2}$	33.3[b]
N V	$4d_{5/2}$	32.1[b]

Argon (18)

K	1s	3205.9[a]
L I	2s	326.3[a]
L II	$2p_{1/2}$	250.6[a]
L III	$2p_{3/2}$	248.4[a]
M I	3s	29.3[a]
M II	$3p_{1/2}$	15.9[a]
M III	$3p_{3/2}$	15.7[a]

Arsenic (33)

K	1s	11867
L I	2s	1527.0[a,d]
L II	$2p_{1/2}$	1359.1[a,d]
L III	$2p_{3/2}$	1323.6[a,d]
M I	3s	204.7[a]
M II	$3p_{1/2}$	146.2[a]
M III	$3p_{3/2}$	141.2[a]
M IV	$3d_{3/2}$	41.7[a]
M V	$3d_{5/2}$	41.7[a]

Astatine (85)

K	1s	95730
L I	2s	17493
L II	$2p_{1/2}$	16785
L III	$2p_{3/2}$	14214
M I	3s	4317
M II	$3p_{1/2}$	4008
M III	$3p_{3/2}$	3426
M IV	$3d_{3/2}$	2909
M V	$3d_{5/2}$	2787
N I	4s	1042[a]
N II	$4p_{1/2}$	886[a]
N III	$4p_{3/2}$	740[a]
N IV	$4d_{3/2}$	533[a]
N V	$4d_{5/2}$	507[a]
N VI	$4f_{5/2}$	210[a]
N VII	$4f_{7/2}$	210[a]
O I	5s	195[a]
O II	$5p_{1/2}$	148[a]
O III	$5p_{3/2}$	115[a]
O IV	$5d_{3/2}$	40[a]
O V	$5d_{5/2}$	40[a]

Barium (56)

K	1s	37441
L I	2s	5989
L II	$2p_{1/2}$	5624
L III	$2p_{3/2}$	5247
M I	3s	1293[a,d]
M II	$3p_{1/2}$	1137[a,d]
M III	$3p_{3/2}$	1063[a,d]
M IV	$3d_{3/2}$	795.7[a]
M V	$3d_{5/2}$	780.5[a]
N I	4s	253.5[b]
N II	$4p_{1/2}$	192
N III	$4p_{3/2}$	178.6[b]
N IV	$4d_{3/2}$	92.6[b]
N V	$4d_{5/2}$	89.9[b]
N VI	$4f_{5/2}$	—
N VII	$4f_{7/2}$	—
O I	5s	30.3[b]
O II	$5p_{1/2}$	17.0[b]
O III	$5p_{3/2}$	14.8[b]

Beryllium (4)

K	1s	111.5[a]

Bismuth (83)

K	1s	90526
L I	2s	16388
L II	$2p_{1/2}$	15711
L III	$2p_{3/2}$	13419
M I	3s	3999
M II	$3p_{1/2}$	3696
M III	$3p_{3/2}$	3177
M IV	$3d_{3/2}$	2688
M V	$3d_{5/2}$	2580
N I	4s	939[b]
N II	$4p_{1/2}$	805.2[b]
N III	$4p_{3/2}$	678.8[b]
N IV	$4d_{3/2}$	464.0[b]
N V	$4d_{5/2}$	440.1[b]
N VI	$4f_{5/2}$	162.3[b]
N VII	$4f_{7/2}$	157.0[b]
O I	5s	159.3[a,d]
O II	$5p_{1/2}$	119.0[b]
O III	$5p_{3/2}$	92.6[b]
O IV	$5d_{3/2}$	26.9[b]
O V	$5d_{5/2}$	23.8[b]

Boron (5)

K	1s	188[a]

Bromine (35)

K	1s	13474
L I	2s	1782[a]
L II	$2p_{1/2}$	1596[a]
L III	$2p_{3/2}$	1550[a]
M I	3s	257[a]
M II	$3p_{1/2}$	189[a]
M III	$3p_{3/2}$	182[a]
M IV	$3d_{3/2}$	70[a]
M V	$3d_{5/2}$	69[a]

Cadmium (48)

K	1s	26711
L I	2s	4018
L II	$2p_{1/2}$	3727
L III	$2p_{3/2}$	3538
M I	3s	772.0[b]
M II	$3p_{1/2}$	652.6[b]
M III	$3p_{3/2}$	618.4[b]
M IV	$3d_{3/2}$	411.9[b]
M V	$3d_{5/2}$	405.2[b]
N I	4s	109.8[b]
N II	$4p_{1/2}$	63.9[b,c]
N III	$4p_{3/2}$	63.9[b,c]
N IV	$4d_{3/2}$	11.7[b]
N V	$4d_{5/2}$	10.7[b]

Calcium (20)

K	1s	4038.5[a]
L I	2s	438.4[b]
L II	$2p_{1/2}$	349.7[b]
L III	$2p_{3/2}$	346.2[b]
M I	3s	44.3[b]
M II	$3p_{1/2}$	25.4[b]
M III	$3p_{3/2}$	25.4[b]

Carbon (6)

K	1s	284.2[a]

Cerium (58)

K	1s	40443
L I	2s	6548
L II	$2p_{1/2}$	6164
L III	$2p_{3/2}$	5723
M I	3s	1436[a,d]
M II	$3p_{1/2}$	1274[a,d]
M III	$3p_{3/2}$	1187[a,d]
M IV	$3d_{3/2}$	902.4[a]
M V	$3d_{5/2}$	883.8[a]

N I	4s	291.0[a]
N II	$4p_{1/2}$	223.3
N III	$4p_{3/2}$	206.5[a]
N IV	$4d_{3/2}$	109[a]
N V	$4d_{5/2}$	—
N VI	$4f_{5/2}$	0.1
N VII	$4f_{7/2}$	0.1
O I	5s	37.8
O II	$5p_{1/2}$	19.8[a]
O III	$5p_{3/2}$	17.0[a]

Cesium (55)

K	1s	35985
L I	2s	5714
L II	$2p_{1/2}$	5359
L III	$2p_{3/2}$	5012
M I	3s	1211[a,d]
M II	$3p_{1/2}$	1071[a]
M III	$3p_{3/2}$	1003[a]
M IV	$3d_{3/2}$	740.5[a]
M V	$3d_{5/2}$	726.6[a]
N I	4s	232.3[a]
N II	$4p_{1/2}$	172.4[a]
N III	$4p_{3/2}$	161.3[a]
N IV	$4d_{3/2}$	79.8[a]
N V	$4d_{5/2}$	77.5[a]
N VI	$4f_{5/2}$	—
N VII	$4f_{7/2}$	—
O I	5s	22.7
O II	$5p_{1/2}$	14.2[a]
O III	$5p_{3/2}$	12.1[a]

Chlorine (17)

K	1s	2822.0
L I	2s	270[a]
L II	$2p_{1/2}$	202[a]
L III	$2p_{3/2}$	200[a]

Chromium(24)

K	1s	5989
L I	2s	696.0[b]
L II	$2p_{1/2}$	583.8[b]
L III	$2p_{3/2}$	574.1[b]
M I	3s	74.1[b]
M II	$3p_{1/2}$	42.2[b]
M III	$3p_{3/2}$	42.2[b]

Cobalt (27)

K	1s	7709
L I	2s	925.1[b]
L II	$2p_{1/2}$	793.2[b]
L III	$2p_{3/2}$	778.1[b]
M I	3s	101.0[b]
M II	$3p_{1/2}$	58.9[b]
M III	$3p_{3/2}$	58.9[b]

Copper (29)

K	1s	8979
L I	2s	1096.7[b]
L II	$2p_{1/2}$	952.3[b]
L III	$2p_{3/2}$	932.5[b]
M I	3s	122.5[b]
M II	$3p_{1/2}$	77.3[b]
M III	$3p_{3/2}$	75.1[b]

Dysprosium (66)

K	1s	53789
L I	2s	9046
L II	$2p_{1/2}$	8581
L III	$2p_{3/2}$	7790
M I	3s	2047
M II	$3p_{1/2}$	1842
M III	$3p_{3/2}$	1676
M IV	$3d_{3/2}$	1333
M V	$3d_{5/2}$	1292[a]
N I	4s	414.2[a]
N II	$4p_{1/2}$	333.5[a]
N III	$4p_{3/2}$	293.2[a]
N IV	$4d_{3/2}$	153.6[a]
N V	$4d_{5/2}$	153.6[a]
N VI	$4f_{5/2}$	8.0[a]
N VII	$4f_{7/2}$	4.3[a]
O I	5s	49.9[a]
O II	$5p_{1/2}$	26.3
O III	$5p_{3/2}$	26.3

Erbium (68)

K	1s	57486
L I	2s	9751
L II	$2p_{1/2}$	9264
L III	$2p_{3/2}$	8358
M I	3s	2206
M II	$3p_{1/2}$	2006
M III	$3p_{3/2}$	1812
M IV	$3d_{3/2}$	1453
M V	$3d_{5/2}$	1409
N I	4s	449.8[a]
N II	$4p_{1/2}$	366.2
N III	$4p_{3/2}$	320.2[a]
N IV	$4d_{3/2}$	167.6[a]
N V	$4d_{5/2}$	167.6[a]
N VI	$4f_{5/2}$	—
N VII	$4f_{7/2}$	4.7[a]
O I	5s	50.6[a]
O II	$5p_{1/2}$	31.4[a]
O III	$5p_{3/2}$	24.7[a]

Europium (63)

K	1s	48519
L I	2s	8052
L II	$2p_{1/2}$	7617
L III	$2p_{3/2}$	6977
M I	3s	1800
M II	$3p_{1/2}$	1614
M III	$3p_{3/2}$	1481
M IV	$3d_{3/2}$	1158.6[a]
M V	$3d_{5/2}$	1127.5[a]
N I	4s	360
N II	$4p_{1/2}$	284
N III	$4p_{3/2}$	257
N IV	$4d_{3/2}$	133
N V	$4d_{5/2}$	1227[a]
N VI	$4f_{5/2}$	0
N VII	$4f_{7/2}$	0
O I	5s	32
O II	$5p_{1/2}$	22
O III	$5p_{3/2}$	22

Fluorine (9)

K	1s	696.7[a]

Francium (87)

K	1s	101137
L I	2s	18639
L II	$2p_{1/2}$	17907
L III	$2p_{3/2}$	15031
M I	3s	4652
M II	$3p_{1/2}$	4327
M III	$3p_{3/2}$	3663
M IV	$3d_{3/2}$	3136
M V	$3d_{5/2}$	3000
N I	4s	1153[a]
N II	$4p_{1/2}$	980[a]
N III	$4p_{3/2}$	810[a]
N IV	$4d_{3/2}$	603[a]
N V	$4d_{5/2}$	577[a]
N VI	$4f_{5/2}$	268[a]
N VII	$4f_{7/2}$	268[a]
O I	5s	234[a]
O II	$5p_{1/2}$	182[a]
O III	$5p_{3/2}$	140[a]
O IV	$5d_{3/2}$	58[a]
O V	$5d_{5/2}$	58[a]
P I	6s	34
P II	$6p_{1/2}$	15
P III	$6p_{3/2}$	15

Gadolinium (64)

K	1s	50239
L I	2s	8376
L II	$2p_{1/2}$	7930
L III	$2p_{3/2}$	7243
M I	3s	1881
M II	$3p_{1/2}$	1688
M III	$3p_{3/2}$	1544
M IV	$3d_{3/2}$	1221.9[a]
M V	$3d_{5/2}$	1189.6[a]
N I	4s	378.6[a]
N II	$4p_{1/2}$	286
N III	$4p_{3/2}$	271
N IV	$4d_{3/2}$	—
N V	$4d_{5/2}$	142.6[a]
N VI	$4f_{5/2}$	8.6[a]
N VII	$4f_{7/2}$	8.6[a]
O I	5s	36
O II	$5p_{1/2}$	20
O III	$5p_{3/2}$	20

Gallium (31)

K	1s	10367
L I	2s	1299.0[a,d]
L II	$2p_{1/2}$	1143.2[b]
L III	$2p_{3/2}$	1116.4[b]
M I	3s	159.5[b]
M II	$3p_{1/2}$	103.5[b]
M III	$3p_{3/2}$	100.0[b]
M IV	$3d_{3/2}$	18.7[b]
M V	$3d_{5/2}$	18.7[b]

Germanium (32)

K	1s	11103
L I	2s	1414.6[a,d]
L II	$2p_{1/2}$	1248.1[a,d]
L III	$2p_{3/2}$	1217.0[a,d]
M I	3s	180.1[a]
M II	$3p_{1/2}$	124.9[a]
M III	$3p_{3/2}$	120.8[a]
M IV	$3d_{3/2}$	29.8[a]
M V	$3d_{5/2}$	29.2[a]

Gold (79)

K	1s	80725
L I	2s	14353
L II	$2p_{1/2}$	13734
L III	$2p_{3/2}$	11919
M I	3s	3425
M II	$3p_{1/2}$	3148
M III	$3p_{3/2}$	2743
M IV	$3d_{3/2}$	2291
M V	$3d_{5/2}$	2206
N I	4s	762.1[b]
N II	$4p_{1/2}$	642.7[b]
N III	$4p_{3/2}$	546.3[b]
N IV	$4d_{3/2}$	353.2[b]
N V	$4d_{5/2}$	335.1[b]
N VI	$4f_{5/2}$	87.6[b]
N VII	$4f_{7/2}$	83.9[b]
O I	5s	107.2[a,d]
O II	$5p_{1/2}$	74.2[b]
O III	$5p_{3/2}$	57.2[b]

Hafnium (72)

K	1s	65351
L I	2s	11271
L II	$2p_{1/2}$	10739
L III	$2p_{3/2}$	9561
M I	3s	2601
M II	$3p_{1/2}$	2365
M III	$3p_{3/2}$	2107
M IV	$3d_{3/2}$	1176
M V	$3d_{5/2}$	1662
N I	4s	538[a]
N II	$4p_{1/2}$	438.2[b]
N III	$4p_{3/2}$	380.7[b]
N IV	$4d_{3/2}$	220.0[b]
N V	$4d_{5/2}$	211.5[b]
N VI	$4f_{5/2}$	15.9[b]

N VII	$4f_{7/2}$	14.2[b]
O I	5s	64.2[b]
O II	$5p_{1/2}$	38[a]
O III	$5p_{3/2}$	29.9[b]

Helium (2)

K	1s	24.6[a]

Holmium (67)

K	1s	55618
L I	2s	9394
L II	$2p_{1/2}$	8918
L III	$2p_{3/2}$	8071
M I	3s	2128
M II	$3p_{1/2}$	1923
M III	$3p_{3/2}$	1741
M IV	$3d_{3/2}$	1392
M V	$3d_{5/2}$	1351
N I	4s	432.4[a]
N II	$4p_{1/2}$	343.5
N III	$4p_{3/2}$	308.2[a]
N IV	$4d_{3/2}$	160[a]
N V	$4d_{5/2}$	160[a]
N VI	$4f_{5/2}$	8.6[a]
N VII	$4f_{7/2}$	5.2[a]
O I	5s	49.3[a]
O II	$5p_{1/2}$	30.8[a]
O III	$5p_{3/2}$	24.1[a]

Hydrogen (1)

K	1s	13.6

Indium (49)

K	1s	27940
L I	2s	4238
L II	$2p_{1/2}$	3938
L III	$2p_{3/2}$	3730
M I	3s	827.2[b]
M II	$3p_{1/2}$	703.2[b]
M III	$3p_{3/2}$	665.3[b]
M IV	$3d_{3/2}$	451.4[b]
M V	$3d_{5/2}$	443.9[b]
N I	4s	122.9[b]
N II	$4p_{1/2}$	73.5[b,c]
N III	$4p_{3/2}$	73.5[b,c]
N IV	$4d_{3/2}$	17.7[b]
N V	$4d_{5/2}$	16.9[b]

Iodine (53)

K	1s	33169
L I	2s	5188
L II	$2p_{1/2}$	4852
L III	$2p_{3/2}$	4557
M I	3s	1072[a]
M II	$3p_{1/2}$	931[a]
M III	$3p_{3/2}$	875[a]
M IV	$3d_{3/2}$	631[a]
M V	$3d_{5/2}$	620[a]
N I	4s	186[a]
N II	$4p_{1/2}$	123[a]
N III	$4p_{3/2}$	123[a]
N IV	$4d_{3/2}$	50[a]
N V	$4d_{5/2}$	50[a]

Iridium (77)

K	1s	76111
L I	2s	13419
L II	$2p_{1/2}$	12824
L III	$2p_{3/2}$	11215
M I	3s	3174
M II	$3p_{1/2}$	2909
M III	$3p_{3/2}$	2551
M IV	$3d_{3/2}$	2116
M V	$3d_{5/2}$	2040
N I	4s	691.1[b]
N II	$4p_{1/2}$	577.8[b]
N III	$4p_{3/2}$	495.8[b]
N IV	$4d_{3/2}$	311.9[b]
N V	$4d_{5/2}$	296.3[b]
N VI	$4f_{5/2}$	63.8[b]
N VII	$4f_{7/2}$	60.8[b]
O I	52	95.2[a,d]
O II	$5p_{1/2}$	63.0[a,d]
O III	$5p_{3/2}$	48.0[b]

Iron (26)

K	1s	7112
L I	2s	844.6[b]
L II	$2p_{1/2}$	719.9[b]
L III	$2p_{3/2}$	706.8[b]
M I	3s	91.3[b]
M II	$3p_{1/2}$	52.7[b]
M III	$3p_{3/2}$	52.7[b]

Krypton (36)

K	1s	14326
L I	2s	1921
L II	$2p_{1/2}$	1730.9[a]
L III	$2p_{3/2}$	1678.4[a]
M I	3s	292.8[a]
M II	$3p_{1/2}$	222.2[a]
M III	$3p_{3/2}$	214.4[a]
M IV	$3d_{3/2}$	95.0[a]
M V	$3d_{5/2}$	93.8[a]
N I	4s	27.5[a]
N II	$4p_{1/2}$	14.1[a]
N III	$4p_{3/2}$	14.1[a]

Lanthanum (57)

K	1s	38925
L I	2s	6266
L II	$2p_{1/2}$	5891
L III	$2p_{3/2}$	5483
M I	3s	1362[a,d]
M II	$3p_{1/2}$	1209[a,d]
M III	$3p_{3/2}$	1128[a,d]
M IV	$3d_{3/2}$	853[a]
M V	$3d_{5/2}$	836[a]
N I	4s	247.7[a]
N II	$4p_{1/2}$	205.8
N III	$4p_{3/2}$	196.0[a]
N IV	$4d_{3/2}$	105.3[a]
N V	$4d_{5/2}$	102.5[a]
N VI	$4f_{5/2}$	—
N VII	$4f_{7/2}$	—
O I	5s	34.3[a]
O II	$5p_{1/2}$	19.3[a]
O III	$5p_{3/2}$	16.8[a]

Lead (82)

K	1s	88005
L I	2s	15861
L II	$2p_{1/2}$	15200
L III	$2p_{3/2}$	13055
M I	3s	3851
M II	$3p_{1/2}$	3554
M III	$3p_{3/2}$	3066
M IV	$3d_{3/2}$	2586
M V	$3d_{5/2}$	2484
N I	4s	891.8[b]
N II	$4p_{1/2}$	761.9[b]
N III	$4p_{3/2}$	643.5[b]
N IV	$4d_{3/2}$	434.3[b]
N V	$4d_{5/2}$	412.2[b]
N VI	$4f_{5/2}$	141.7[b]
N VII	$4f_{7/2}$	136.9[b]
O I	5s	147[a,d]
O II	$5p_{1/2}$	106.4[b]
O III	$5p_{3/2}$	83.3[b]
O IV	$5d_{3/2}$	20.7[b]
O V	$5d_{5/2}$	18.1[b]

Lithium (3)

K	1s	54.7[a]

Lutetium

K	1s	63314
L I	2s	10870
L II	$2p_{1/2}$	10349
L III	$2p_{3/2}$	9244
M I	3s	2491
M II	$3p_{1/2}$	2264
M III	$3p_{3/2}$	2024
M IV	$3d_{3/2}$	1639
M V	$3d_{5/2}$	1589
N I	4s	506.8[a]
N II	$4p_{1/2}$	412.4[a]
N III	$4p_{3/2}$	359.2[a]
N IV	$4d_{3/2}$	206.1[a]
N V	$4d_{5/2}$	196.3[a]
N VI	$4f_{5/2}$	8.9[a]
N VII	$4f_{7/2}$	7.5[a]
O I	5s	57.3[a]
O II	$5p_{1/2}$	33.6[a]
O III	$5p_{3/2}$	26.7[a]

Magnesium (12)

K	1s	1303.0[b]
L I	2s	88.6[a]
L II	$2p_{1/2}$	49.6[b]
L III	$2p_{3/2}$	49.2[a]

Manganese (25)

K	1s	6539
L I	2s	769.1[b]
L II	$2p_{1/2}$	649.9[b]
L III	$2p_{3/2}$	638.7[b]
M I	3s	82.3[b]
M II	$3p_{1/2}$	47.2[b]
M III	$3p_{3/2}$	47.2[b]

Mercury (80)

K	1s	83102
L I	2s	14839
L II	$2p_{1/2}$	14209
L III	$2p_{3/2}$	12284
M I	3s	3562
M II	$3p_{1/2}$	3279
M III	$3p_{3/2}$	2847
M IV	$3d_{3/2}$	2385
M V	$3d_{5/2}$	2295
N I	4s	802.2[b]
N II	$4p_{1/2}$	680.2[b]
N III	$4p_{3/2}$	576.6[b]
N IV	$4d_{3/2}$	378.2[b]
N V	$4d_{5/2}$	358.8[b]
N VI	$4f_{5/2}$	104.0[b]
N VII	$4f_{7/2}$	99.9[b]
O I	5s	127[b]
O II	$5p_{3/2}$	83.1[b]
O III	$5p_{3/2}$	64.5[b]
O IV	$5d_{3/2}$	9.6[b]
O V	$5d_{5/2}$	7.8[b]

Molybdenum (42)

K	1s	20000
L I	2s	2866
L II	$2p_{1/2}$	2625
L III	$2p_{3/2}$	2520
M I	3s	506.3[b]
M II	$3p_{1/2}$	411.6[b]
M III	$3p_{3/2}$	394.0[b]
M IV	$3d_{3/2}$	231.1[b]
M V	$3d_{5/2}$	227.9[b]
N I	4s	63.2[b]
N II	$4p_{1/2}$	37.6[b]
N III	$4p_{3/2}$	35.5[b]

Neodymium (60)

K	1s	43569
L I	2s	7126
L II	$2p_{1/2}$	6722
L III	$2p_{3/2}$	6208
M I	3s	1575
M II	$3p_{1/2}$	1403
M III	$3p_{3/2}$	1297
M IV	$3d_{3/2}$	1003.3[a]
M V	$3d_{5/2}$	980.4[a]
N I	4s	319.2[a]
N II	$4p_{1/2}$	243.3
N III	$4p_{3/2}$	224.6
N IV	$4d_{3/2}$	120.5[a]

N V	$4d_{5/2}$	120.5[a]
N VI	$4f_{5/2}$	1.5
N VII	$4f_{7/2}$	1.5
O I	5s	37.5
O II	$5p_{1/2}$	21.1
O III	$5p_{3/2}$	21.1

Neon (10)

K	1s	870.2[a]
L I	2s	48.5[a]
L II	$2p_{1/2}$	21.7[a]
L III	$2p_{3/2}$	21.6[a]

Nickel (28)

K	1s	8333
L I	2s	1008.6[b]
L II	$2p_{1/2}$	870.0[b]
L III	$2p_{3/2}$	852.7[b]
M I	3s	110.8[b]
M II	$3p_{1/2}$	68.0[b]
M III	$3p_{3/2}$	66.2[b]

Niobium (41)

K	1s	18986
L I	2s	2698
L II	$2p_{1/2}$	2465
L III	$2p_{3/2}$	2371
M I	3s	466.6[b]
M II	$3p_{1/2}$	376.1[b]
M III	$3p_{3/2}$	360.6[b]
M IV	$3d_{3/2}$	205.0[b]
M V	$3d_{5/2}$	202.3[b]
N I	4s	56.4[b]
N II	$4p_{1/2}$	32.6[b]
N III	$4p_{3/2}$	30.8[b]

Nitrogen (7)

K	1s	409.9[a]
L I	2s	37.3[a]

Osmium (76)

K	1s	73871
L I	2s	12968
L II	$2p_{1/2}$	12385
L III	$2p_{3/2}$	10871
M I	3s	3049
M II	$3p_{1/2}$	2792
M III	$3p_{3/2}$	2457
M IV	$3d_{3/2}$	2031
M V	$3d_{5/2}$	1960
N I	4s	658.2[b]
N II	$4p_{1/2}$	549.1[b]
N III	$4p_{3/2}$	470.7[b]
N IV	$4d_{3/2}$	293.1[b]
N V	$4d_{5/2}$	278.5[b]
N VI	$4f_{5/2}$	53.4[b]
N VII	$4f_{7/2}$	50.7[b]
O I	5s	84[a]
O II	$5p_{1/2}$	58[a]
O III	$5p_{3/2}$	44.5[b]

Oxygen (8)

K	1s	543.1[a]
L I	2s	41.6[a]

Palladium (46)

K	1s	24350
L I	2s	3604
L II	$2p_{1/2}$	3330
L III	$2p_{3/2}$	3173
M I	3s	671.6[b]
M II	$3p_{1/2}$	559.9[b]
M III	$3p_{3/2}$	532.3[b]
M IV	$3d_{3/2}$	340.5[b]
M V	$3d_{5/2}$	335.2[b]
N I	4s	87.1[a,d]
N II	$4p_{1/2}$	55.7[b,c]
N III	$4p_{3/2}$	50.9[b,c]

Phosphorus (15)

K	1s	2145.5
L I	2s	189[a]
L II	$2p_{1/2}$	136[a]
L III	$2p_{3/2}$	135[a]

Platinum (78)

K	1s	78395
L I	2s	13880
L II	$2p_{1/2}$	13273
L III	$2p_{3/2}$	11564
M I	3s	3296
M II	$3p_{1/2}$	3027
M III	$3p_{3/2}$	2645
M IV	$3d_{3/2}$	2202
M V	$3d_{5/2}$	2122
N I	4s	725.4[b]
N II	$4p_{1/2}$	609.1[b]
N III	$4p_{3/2}$	519.4[b]
N IV	$4d_{3/2}$	331.6[b]
N V	$4d_{5/2}$	314.6[b]
N VI	$4f_{5/2}$	74.5[b]
N VII	$4f_{7/2}$	71.2[b]
O I	5s	101.7[a,d]
O II	$5p_{1/2}$	65.3[a,b]
O III	$5p_{3/2}$	51.7[b]

Polonium (84)

K	1s	93105
L I	2s	16939
L II	$2p_{1/2}$	16244
L III	$2p_{3/2}$	13814
M I	3s	4149
M II	$3p_{1/2}$	3854
M III	$3p_{3/2}$	3302
M IV	$3d_{3/2}$	2798
M V	$3d_{5/2}$	2683
N I	4s	995[a]

N II	$4p_{1/2}$	851[a]
N III	$4p_{3/2}$	705[a]
N IV	$4d_{3/2}$	500[a]
N V	$4d_{5/2}$	473[a]
N VI	$4f_{5/2}$	184[a]
N VII	$4f_{7/2}$	184[a]
O I	5s	177[a]
O II	$5p_{1/2}$	132[a]
O III	$5p_{3/2}$	104[a]
O IV	$5d_{3/2}$	31[a]
O V	$5d_{5/2}$	31[a]

Potassium (19)

K	1s	3608.4[a]
L I	2s	378.6[a]
L II	$2p_{1/2}$	297.3[a]
L III	$2p_{3/2}$	294.6[a]
M I	3s	34.8[a]
M II	$3p_{1/2}$	18.3[a]
M III	$3p_{3/2}$	18.3[a]

Praseodymium (59)

K	1s	41991
L I	2s	6835
L II	$2p_{1/2}$	6440
L III	$2p_{3/2}$	5964
M I	3s	1511
M II	$3p_{1/2}$	1337
M III	$3p_{3/2}$	1242
M IV	$3d_{3/2}$	948.3[a]
M V	$3d_{5/2}$	928.8[a]
N I	4s	304.5
N II	$4p_{1/2}$	236.3
N III	$4p_{3/2}$	217.6
N IV	$4d_{3/2}$	115.1[a]
N V	$4d_{5/2}$	115.1[a]
N VI	$4f_{5/2}$	2.0
N VII	$4f_{7/2}$	2.0
O I	5s	37.4
O II	$5p_{1/2}$	22.3
O III	$5p_{3/2}$	22.3

Promethium (61)

K	1s	45184
L I	2s	7428
L II	$2p_{1/2}$	7013
L III	$2p_{3/2}$	6459
M I	3s	—
M II	$3p_{1/2}$	1471.4
M III	$3p_{3/2}$	1357
M IV	$3d_{3/2}$	1052
M V	$3d_{5/2}$	1027
N I	4s	—
N II	$4p_{1/2}$	242
N III	$4p_{3/2}$	242
N IV	$4d_{3/2}$	120
N V	$4d_{5/2}$	120

Protactinium (91)

K	1s	112601
L I	2s	21105
L II	$2p_{1/2}$	20314
L III	$2p_{3/2}$	16733
M I	3s	5367
M II	$3p_{1/2}$	5001
M III	$3p_{3/2}$	4174
M IV	$3d_{3/2}$	3611
M V	$3d_{5/2}$	3442
N I	4s	1387[a]
N II	$4p_{1/2}$	1224[a]
N III	$4p_{3/2}$	1007[a]
N IV	$4d_{3/2}$	743[a]
N V	$4d_{5/2}$	708[a]
N VI	$4f_{5/2}$	371[a]
N VII	$4f_{7/2}$	360[a]
O I	5s	310[a]
O II	$5p_{1/2}$	232[a]
O III	$5p_{3/2}$	232[a]
O IV	$5d_{3/2}$	94[a]
O V	$5d_{5/2}$	94[a]
P I	6s	—
P II	$6p_{1/2}$	—
P III	$6p_{3/2}$	—

Radium (88)

K	1s	103922
L I	2s	19237
L II	$2p_{1/2}$	18484
L III	$2p_{3/2}$	15444
M I	3s	4822
M II	$3p_{1/2}$	4490
M III	$3p_{3/2}$	3792
M IV	$3d_{3/2}$	3248
M V	$3d_{5/2}$	3105
N I	4s	1208[a]
N II	$4p_{1/2}$	1058
N III	$4p_{3/2}$	879[a]
N IV	$4d_{3/2}$	636[a]
N V	$4d_{5/2}$	603[a]
N VI	$4f_{5/2}$	299[a]
N VII	$4f_{7/2}$	299[a]
O I	5s	254[a]
O II	$5p_{1/2}$	200[a]
O III	$5p_{3/2}$	153[a]
O IV	$5d_{3/2}$	68[a]
O V	$5d_{5/2}$	68[a]
P I	6s	44
P II	$6p_{1/2}$	19
P III	$6p_{3/2}$	19

Radon (86)

K	1s	98404
L I	2s	18049
L II	$2p_{1/2}$	17337
L III	$2p_{3/2}$	14619
M I	3s	4482
M II	$3p_{1/2}$	4159
M III	$3p_{3/2}$	3538
M IV	$3d_{3/2}$	3022
M V	$3d_{5/2}$	2892

N I	4s	1097[a]
N II	$4p_{1/2}$	929[a]
N III	$4p_{3/2}$	768[a]
N IV	$4d_{3/2}$	567[a]
N V	$4d_{5/2}$	541[a]
N VI	$4f_{5/2}$	238[a]
N VII	$4f_{7/2}$	238[a]
O I	5s	214[a]
O II	$5p_{1/2}$	164[a]
O III	$5p_{3/2}$	127[a]
O IV	$5d_{3/2}$	48[a]
O V	$5d_{5/2}$	48[a]
P I	6s	26

Rhenium (75)

K	1s	71676
L I	2s	12527
L II	$2p_{1/2}$	11959
L III	$2p_{3/2}$	10535
M I	3s	2932
M II	$3p_{1/2}$	2682
M III	$3p_{3/2}$	2367
M IV	$3d_{3/2}$	1949
M V	$3d_{5/2}$	1883
N I	4s	625.4[b]
N II	$4p_{1/2}$	518.7[b]
N III	$4p_{3/2}$	446.8[b]
N IV	$4d_{3/2}$	273.9[b]
N V	$4d_{5/2}$	260.5[b]
N VI	$4f_{5/2}$	42.9[a]
N VII	$4f_{7/2}$	40.5[a]
O I	5s	83[b]
O II	$5p_{1/2}$	45.6[b]
O III	$5p_{3/2}$	34.6[a,d]

Rhodium (45)

K	1s	23220
L I	2s	3412
L II	$2p_{1/2}$	3146
L III	$2p_{3/2}$	3004
M I	3s	628.1[b]
M II	$3p_{1/2}$	521.3[b]
M III	$3p_{3/2}$	496.5[b]
M IV	$3d_{3/2}$	311.9[b]
M V	$3d_{5/2}$	307.2[b]
N I	4s	81.4[a,d]
N II	$4p_{1/2}$	50.5[b]
N III	$4p_{3/2}$	47.3[b]

Rubidium (37)

K	1s	15200
L I	2s	2065
L II	$2p_{1/2}$	1864
L III	$2p_{3/2}$	1804
M I	3s	326.7[a]
M II	$3p_{1/2}$	248.7[a]
M III	$3p_{3/2}$	239.1[a]
M IV	$3d_{3/2}$	113.0[a]
M V	$3d_{5/2}$	112[a]
N I	4s	30.5[a]
N II	$4p_{1/2}$	16.3[a]
N III	$4p_{3/2}$	15.3[a]

Ruthenium (44)

K	1s	22117
L I	2s	3224
L II	$2p_{1/2}$	2967
L III	$2p_{3/2}$	2838
M I	3s	586.2[b]
M II	$3p_{1/2}$	483.3[b]
M III	$3p_{3/2}$	461.5[b]
M IV	$3d_{3/2}$	284.2[b]
M V	$3d_{5/2}$	280.0[b]
N I	4s	75.0[b]
N II	$4p_{1/2}$	46.5[b]
N III	$4p_{3/2}$	43.2[b]

Samarium (62)

K	1s	46834
L I	2s	7737
L II	$2p_{1/2}$	7312
L III	$2p_{3/2}$	6716
M I	3s	1723
M II	$3p_{1/2}$	1541
M III	$3p_{3/2}$	1419.8
M IV	$3d_{3/2}$	1110.9[a]
M V	$3d_{5/2}$	1083.4[a]
N I	4s	347.2[a]
N II	$4p_{1/2}$	265.6
N III	$4p_{3/2}$	247.4
N IV	$4d_{3/2}$	129.0
N V	$4d_{5/2}$	129.0
N VI	$4f_{5/2}$	5.2
N VII	$4f_{7/2}$	5.2
O I	5s	37.4
O II	$5p_{1/2}$	21.3
O III	$5p_{3/2}$	21.3

Scandium (21)

K	1s	4492
L I	2s	498.0[a]
L II	$2p_{1/2}$	403.6[a]
L III	$2p_{3/2}$	389.7[a]
M I	3s	51.1[a]
M II	$3p_{1/2}$	28.3[a]
M III	$3p_{3/2}$	28.3[a]

Selenium (34)

K	1s	12658
L I	2s	1652.0[a,d]
L II	$2p_{1/2}$	1474.3[a,d]
L III	$2p_{3/2}$	1433.9[a,d]
M I	3s	229.6[a]
M II	$3p_{1/2}$	166.5[a]
M III	$3p_{3/2}$	160.7[a]
M IV	$3d_{3/2}$	55.5[a]
M V	$3d_{5/2}$	54.6[a]

Silicon (14)

K	1s	1839
L I	2s	149.7[a,d]
L II	$2p_{1/2}$	99.8[a]
L III	$2p_{3/2}$	99.2[a]

Silver (47)

K	1s	25514
L I	2s	3806
L II	$2p_{1/2}$	3524
L III	$2p_{3/2}$	3351
M I	3s	719.0[b]
M II	$3p_{1/2}$	603.8[b]
M III	$3p_{3/2}$	573.0[b]
M IV	$3d_{3/2}$	374.0[b]
M V	$3d_{5/2}$	368.0[b]
N I	4s	97.0[b]
N II	$4p_{1/2}$	63.7[b]
N III	$4p_{3/2}$	58.3[b]

Sodium (11)

K	1s	1070.8[b]
L I	2s	63.5[b]
L II	$2p_{1/2}$	30.4[b]
L III	$2p_{3/2}$	30.5[a]

Strontium (38)

K	1s	16105
L I	2s	2216
L II	$2p_{1/2}$	2007
L III	$2p_{3/2}$	1940
M I	3s	358.7[b]
M II	$3p_{1/2}$	280.3[b]
M III	$3p_{3/2}$	270.0[b]
M IV	$3d_{3/2}$	136.0[b]
M V	$3d_{5/2}$	134.2[b]
N I	4s	38.9[b]
N II	$4p_{1/2}$	21.6[b]
N III	$4p_{3/2}$	20.1[b]

Sulfur (16)

K	1s	2472
L I	2s	230.9[a,d]
L II	$2p_{1/2}$	163.6[a]
L III	$2p_{3/2}$	162.5[a]

Tantalum (73)

K	1s	67416
L I	2s	11682
L II	$2p_{1/2}$	11136
L III	$2p_{3/2}$	9881
M I	3s	2708
M II	$3p_{1/2}$	2469
M III	$3p_{3/2}$	2194
M IV	$3d_{3/2}$	1793
M V	$3d_{5/2}$	1735
N I	4s	563.4[b]
N II	$4p_{1/2}$	463.4[b]
N III	$4p_{3/2}$	400.9[b]
N IV	$4d_{3/2}$	237.9[b]
N V	$4d_{5/2}$	226.4[b]
N VI	$4f_{5/2}$	23.5[b]
N VII	$4f_{7/2}$	21.6[b]
O I	5s	69.7[b]
O II	$5p_{1/2}$	42.2[a]
O III	$5p_{3/2}$	32.7[b]

Technetium (43)

K	1s	21044
L I	2s	3043
L II	$2p_{1/2}$	2793
L III	$2p_{3/2}$	2677
M I	3s	586.1[a]
M II	$3p_{1/2}$	447.6[a]
M III	$3p_{3/2}$	417.7[a]
M IV	$3d_{3/2}$	257.6[a]
M V	$3d_{5/2}$	253.9[a]
N I	4s	69.5[a]
N II	$4p_{1/2}$	42.3[a]
N III	$4p_{3/2}$	39.9[a]

Tellurium (52)

K	1s	31814
L I	2s	4939
L II	$2p_{1/2}$	4612
L III	$2p_{3/2}$	4341
M I	3s	1006[b]
M II	$3p_{1/2}$	870.8[b]
M III	$3p_{3/2}$	820.0[b]
M IV	$3d_{3/2}$	583.4[b]
M V	$3d_{5/2}$	573.0[b]
N I	4s	169.4[b]
N II	$4p_{1/2}$	103.3[b,c]
N III	$4p_{3/2}$	103.3[b,c]
N IV	$4d_{3/2}$	41.9[b]
N V	$4d_{5/2}$	40.4[b]

Terbium (65)

K	1s	51996
L I	2s	8708
L II	$2p_{1/2}$	8252
L III	$2p_{3/2}$	7514
M I	3s	1968
M II	$3p_{1/2}$	1768
M III	$3p_{3/2}$	1611
M IV	$3d_{3/2}$	1267.9[a]
M V	$3d_{5/2}$	1241.1[a]
N I	4s	396.0[a]
N II	$4p_{1/2}$	322.4[a]
N III	$4p_{3/2}$	284.1[a]
N IV	$4d_{3/2}$	150.5[a]
N V	$4d_{5/2}$	150.5[a]
N VI	$4f_{5/2}$	7.7[a]
N VII	$4f_{7/2}$	2.4[a]
O I	5s	45.6[a]
O II	$5p_{1/2}$	28.7[a]
O III	$5p_{3/2}$	22.6[a]

ELECTRON BINDING ENERGIES OF THE ELEMENTS (continued)

Thallium (81)

K	1s	85530
L I	2s	15347
L II	$2p_{1/2}$	14698
L III	$2p_{3/2}$	12658
M I	3s	3704
M II	$3p_{1/2}$	3416
M III	$3p_{3/2}$	2957
M IV	$3d_{3/2}$	2485
M V	$3d_{5/2}$	2389
N I	4s	846.2[b]
N II	$4p_{1/2}$	720.5[b]
N III	$4p_{3/2}$	609.5[b]
N IV	$4d_{3/2}$	405.7[b]
N V	$4d_{5/2}$	385.0[b]
N VI	$4f_{5/2}$	122.2[b]
N VII	$4f_{7/2}$	117.8[b]
O I	5s	136[a,d]
O II	$5p_{1/2}$	94.6[b]
O III	$5p_{3/2}$	73.5[b]
O IV	$5d_{3/2}$	14.7[b]
O V	$5d_{5/2}$	12.5[b]

Thorium (90)

K	1s	109651
L I	2s	20472
L II	$2p_{1/2}$	19693
L III	$2p_{3/2}$	16300
M I	3s	5182
M II	$3p_{1/2}$	4830
M III	$3p_{3/2}$	4046
M IV	$3d_{3/2}$	3491
M V	$3d_{5/2}$	3332
N I	4s	1330[a]
N II	$4p_{1/2}$	1168[a]
N III	$4p_{3/2}$	966.4[b]
N IV	$4d_{3/2}$	712.1[b]
N V	$4d_{5/2}$	675.2[b]
N VI	$4f_{5/2}$	342.4[b]
N VII	$4f_{7/2}$	333.1[b]
O I	5s	290[a,c]
O II	$5p_{1/2}$	229[a,c]
O III	$5p_{3/2}$	182[a,c]
O IV	$5d_{3/2}$	92.5[b]
O V	$5d_{5/2}$	85.4[b]
P I	6s	41.4[b]
P II	$6p_{1/2}$	24.5[b]
P III	$6p_{3/2}$	16.6[b]

Thulium (69)

K	1s	59390
L I	2s	10116
L II	$2p_{1/2}$	9617
L III	$2p_{3/2}$	8648
M I	3s	2307
M II	$3p_{1/2}$	2090
M III	$3p_{3/2}$	1885
M IV	$3d_{3/2}$	1515
M V	$3d_{5/2}$	1468
N I	4s	470.9[a]
N II	$4p_{1/2}$	385.9[a]
N III	$4p_{3/2}$	332.6[a]
N IV	$4d_{3/2}$	175.5[a]
N V	$4d_{5/2}$	175.5[a]
N VI	$4f_{5/2}$	—
N VII	$4f_{7/2}$	4.6
O I	5s	54.7[a]
O II	$5p_{1/2}$	31.8[a]
O III	$5p_{3/2}$	25.0[a]

Tin (50)

K	1s	29200
L I	2s	4465
L II	$2p_{1/2}$	4156
L III	$2p_{3/2}$	3929
M I	3s	884.7[b]
M II	$3p_{1/2}$	756.5[b]
M III	$3p_{3/2}$	714.6[b]
M IV	$3d_{3/2}$	493.2[b]
M V	$3d_{5/2}$	484.9[b]
N I	4s	137.1[b]
N II	$4p_{1/2}$	83.6[b,c]
N III	$4p_{3/2}$	83.6[b,c]
N IV	$4d_{3/2}$	24.9[b]
N V	$4d_{5/2}$	23.9[b]

Titanium (22)

K	1s	4966
L I	2s	560.9[b]
L II	$2p_{1/2}$	460.2[b]
L III	$2p_{3/2}$	453.8[b]
M I	3s	58.7[b]
M II	$3p_{1/2}$	32.6[b]
M III	$3p_{3/2}$	32.6[b]

Tungsten (74)

K	1s	69525
L I	2s	12100
L II	$2p_{1/2}$	11544
L III	$2p_{3/2}$	10207
M I	3s	2820
M II	$3p_{1/2}$	2575
M III	$3p_{3/2}$	2281
M IV	$3d_{3/2}$	1949
M V	$3d_{5/2}$	1809
N I	4s	594.1[b]
N II	$4p_{1/2}$	490.4[b]
N III	$4p_{3/2}$	423.6[b]
N IV	$4d_{3/2}$	255.9[b]
N V	$4d_{5/2}$	243.5[b]
N VI	$4f_{5/2}$	33.6[a]
N VII	$4f_{7/2}$	31.4[b]
O I	5s	75.6[b]
O II	$5p_{1/2}$	453[a,d]
O III	$5p_{3/2}$	36.8[b]

Uranium (92)

K	1s	115606
L I	2s	21757
L II	$2p_{1/2}$	20948
L III	$2p_{3/2}$	17166
M I	3s	5548
M II	$3p_{1/2}$	5182
M III	$3p_{3/2}$	4303
M IV	$3d_{3/2}$	3728
M V	$3d_{5/2}$	3552
N I	4s	1439[a,d]
N II	$4p_{1/2}$	1271[a,d]
N III	$4p_{3/2}$	1043[b]
N IV	$4d_{3/2}$	778.3[b]
N V	$4d_{5/2}$	736.2[b]
N VI	$4f_{5/2}$	388.2[a]
N VII	$4f_{7/2}$	377.4[b]
O I	5s	321[a,c,d]
O II	$5p_{1/2}$	257[a,c,d]
O III	$5p_{3/2}$	192[a,c,d]
O IV	$5d_{3/2}$	102.8[b]
O V	$5d_{5/2}$	94.2[b]
P I	6s	43.9[b]
P II	$6p_{1/2}$	26.8[b]
P III	$6p_{3/2}$	16.8[b]

Vanadium (23)

K	1s	5465
L I	2s	626.7[b]
L II	$2p_{1/2}$	519.8[b]
L III	$2p_{3/2}$	521.1[b]
M I	3s	66.3[b]
M II	$3p_{1/2}$	37.2[b]
M III	$3p_{3/2}$	37.2[b]

Xenon (54)

K	1s	34561
L I	2s	5453
L II	$2p_{1/2}$	5107
L III	$2p_{3/2}$	4786
M I	3s	1148.7[a]
M II	$3p_{1/2}$	1002.1[a]
M III	$3p_{3/2}$	940.6[a]
M IV	$3d_{3/2}$	689.0[a]
M V	$3d_{5/2}$	676.4[a]
N I	4s	213.2[a]
N II	$4p_{1/2}$	146.7
N III	$4p_{3/2}$	145.5[a]
N IV	$4d_{3/2}$	69.5[a]
N V	$4d_{5/2}$	67.5[a]
N VI	$4f_{5/2}$	—
N VII	$4f_{7/2}$	—
O I	5s	23.3[a]
O II	$5p_{1/2}$	13.4[a]
O III	$5p_{3/2}$	12.1[a]

Ytterbium (70)

K	1s	61332
L I	2s	10486
L II	$2p_{1/2}$	9978
L III	$2p_{3/2}$	8944
M I	3s	2398
M II	$3p_{1/2}$	2173
M III	$3p_{3/2}$	1950
M IV	$3d_{3/2}$	1576
M V	$3d_{5/2}$	1528
N I	4s	480.5[a]
N II	$4p_{1/2}$	388.7[a]
N III	$4p_{3/2}$	339.7[a]
N IV	$4d_{3/2}$	191.2[a]
N V	$4d_{5/2}$	182.4[a]
N VI	$4f_{5/2}$	2.5[a]
N VII	$4f_{7/2}$	1.3[a]
O I	5s	52.0[a]
O II	$5p_{1/2}$	30.3[a]
O III	$5p_{3/2}$	24.1[a]

Yttrium (39)

K	1s	17038
L I	2s	2373
L II	$2p_{1/2}$	2156
L III	$2p_{3/2}$	2080
M I	3s	392.0[a,d]
M II	$3p_{1/2}$	310.6[a]
M III	$3p_{3/2}$	298.8[a]
M IV	$3d_{3/2}$	157.7[b]
M V	$3d_{5/2}$	155.8[b]
N I	4s	43.8[a]
N II	$4p_{1/2}$	24.4[a]
N III	$4p_{3/2}$	23.1[a]

Zinc (30)

K	1s	9659
L I	2s	1196.2[a]
L II	$2p_{1/2}$	1044.9[a]
L III	$2p_{3/2}$	1021.8[a]
M I	3s	139.8[a]
M II	$3p_{1/2}$	91.4[a]
M III	$3p_{3/2}$	88.6[a]
M IV	$3d_{3/2}$	10.2[a]
M V	$3d_{5/2}$	10.1[a]

Zirconium (40)

K	1s	17998
L I	2s	2532
L II	$2p_{1/2}$	2307
L III	$2p_{3/2}$	2223
M I	3s	430.3[b]
M II	$3p_{1/2}$	343.5[b]
M III	$3p_{3/2}$	329.8[b]
M IV	$3d_{3/2}$	181.1[b]
M V	$3d_{5/2}$	178.8[b]
N I	4s	50.6[b]
N II	$4p_{1/2}$	28.5[b]
N III	$4p_{3/2}$	27.1[b]

[a] Reference 1.
[b] Reference 2 (remaining values from Reference 3).
[c] One-particle approximation not valid.
[d] Derived using energy differences from Reference 3.

NATURAL WIDTH OF X-RAY LINES

Natural widths of K X-ray lines in eV:

Element	$K\alpha_1$	$K\alpha_2$	$K\beta_1$	$K\beta_3$
Ca	1.00	0.98		
Ti	1.45	2.13		
Cr	2.05	2.64		
Fe	2.45	3.20		
Ni	3.00	3.70		
Zn	3.40	3.96		
Ge	3.75	4.18		
Se	4.10	4.43		
Kr	4.23	4.62		
Sr	5.17	4.97		
Zr	5.70	5.25		
Mo	6.82	6.80		
Ru	7.41	7.96		
Pd	8.80	9.20		
Cd	9.80	10.40		
Sn	11.20	12.40	11.80	11.00
Te	12.80	14.20	13.30	13.10
Xe	14.20	15.10	15.30	14.50
Ba	16.10	16.80	18.15	16.70
Ce	18.60	19.50	20.60	18.60
Nd	21.50	21.50	23.25	21.33
Sm	26.00	24.70	25.65	24.65
Gd	29.50	28.00	29.37	28.00
Dy	33.90	32.20	32.73	32.00
Er	35.00	35.50	36.20	35.70
Yb	38.80	40.60	41.43	41.15
Hf	42.70	44.30	46.00	46.10
W	46.80	48.00	51.83	51.50
Os	49.00	49.40	55.90	55.95
Pt	54.10	54.30	59.98	62.13
Hg	64.75	68.20	65.75	68.95
Pb	67.10	72.30	72.20	73.80
Po	73.20	75.10	78.60	80.10
Rn	80.00	81.50	85.50	86.50
Ra	87.00	88.20	94.20	95.50
Th	94.70	95.00	99.70	101.00
U	103.00	104.30	105.00	107.30

From Salem, S. I. and Lee, P. L., *At. Data Nucl. Data Tables,* 18, 233, 1976.

Natural widths of L X-ray lines in eV:

Element	$L\alpha_1$	$L\alpha_2$	$L\beta_1$	$L\beta_2$	$L\beta_3$	$L\beta_4$	$L\gamma_1$
Zr	1.68	1.52	1.87	5.13	5.50	5.60	3.34
Mo	1.86	1.80	2.03	5.30	5.90	5.78	3.76
Ru	2.03	1.98	2.18	5.45	6.35	5.96	4.15
Pd	2.21	2.16	2.36	5.63	6.80	6.18	4.50
Gd	2.43	2.40	2.54	5.82	7.23	6.28	4.83
Sn	2.62	2.62	2.75	6.10	7.70	6.60	5.23
Tc	2.88	2.88	2.96	6.25	8.22	6.82	5.60
Xe	3.15	3.15	3.20	6.43	8.70	7.15	5.95
Ba	3.39	3.45	3.45	6.70	9.20	7.42	6.35
Ce	3.70	3.78	3.73	6.86	9.70	7.82	6.75
Nd	3.93	4.08	4.00	7.18	10.30	8.15	7.16
Sm	4.13	4.50	4.33	7.42	10.80	8.60	7.50
Cd	4.46	4.90	4.63	7.70	11.20	9.08	7.83
Dy	4.81	5.35	5.03	7.90	11.50	9.60	8.30
Er	5.17	5.73	5.45	8.28	11.85	10.03	8.75
Yb	5.40	6.22	5.90	8.58	12.20	11.00	9.20
Hf	5.83	6.70	6.36	8.92	12.40	12.80	9.63
W	6.50	7.20	6.90	9.06	13.10	14.60	10.20
Os	7.04	7.70	7.42	9.60	14.60	16.50	10.65
Pt	7.60	8.28	8.00	9.95	16.10	18.00	11.20
Hg	8.10	8.80	8.70	10.40	17.40	19.70	11.80
Pb	8.82	9.35	9.35	10.75	18.65	21.30	12.30
Po	9.50	9.95	10.10	11.25	19.90	22.70	13.05
Rn	10.03	10.50	10.65	11.65	21.00	24.00	13.55
Ra	11.00	11.20	11.60	12.20	22.00	25.20	14.30
Th	11.90	11.80	12.40	12.80	22.85	26.35	15.00
U	12.40	12.40	13.50	13.30	23.70	27.50	15.70
Pu	13.20	13.00	14.10	13.90	24.10	28.30	16.40
Cm	14.80	13.60	15.70	14.60	25.00	29.40	17.10

PHOTON ATTENUATION COEFFICIENTS

Martin J. Berger and John H. Hubbell

This table gives mass attenuation coefficients for photons for all elements at energies between 1 keV (soft x-rays) and 1 GeV (hard gamma rays). The mass attenuation coefficient μ describes the attenuation of radiation as it passes through matter by the relation

$$I(x)/I_o = e^{-\mu\rho x}$$

where I_o is the initial intensity, $I(x)$ the intensity after path length x, and ρ is the mass density of the element in question. To a high approximation the mass attenuation coefficient is additive for the elements present, independent of the way in which they are bound in chemical compounds.

The power of ten is indicated beside each number in the table; i.e., 7.41 + 03 means 7.41×10^3. A vertical line between two columns indicates that an absorption edge lies between those energy values. The various edges are labeled at the bottom of the table.

The attenuation coefficients were calculated with the computer program XCOM (Reference 1), which uses a cross-section database compiled at the Photon and Charged Particle Data Center at the National Institute of Standards and Technology. Their accuracy has been confirmed at all energies by extensive comparisons with experimental attenuation coefficients. Such comparisons for X-ray energies up to 100 keV can be found in Reference 2.

REFERENCES

1. Berger, M. J. and Hubbell, J. H., National Bureau of Standards Report NBSIR-87-3597, 1987.
2. Saloman, E. B., Hubbell, J. H., and Scofield, J. H., *Atomic Data and Nuclear Data Tables*, 38, 1, 1988.

Mass attenuation coefficient, cm²/g

	Atomic no.	Photon energy, MeV 0.001	0.002	0.005	0.01	0.02	0.05	0.1	0.2	0.5
H	1	7.21 + 00	1.06 + 00	4.19-01	3.85-01	3.69-01	3.36-01	2.94-01	2.43-01	1.73-01
He	2	6.08 + 01	6.86 + 00	5.77-01	2.48-01	1.96-01	1.70-01	1.49-01	1.22-01	8.71-02
Li	3	2.34 + 02	2.71 + 01	1.62 + 00	3.40-01	1.86-01	1.49-01	1.29-01	1.06-01	7.53-02
Be	4	6.04 + 02	7.47 + 01	4.37 + 00	6.47-01	2.25-01	1.55-01	1.33-01	1.09-01	7.74-02
B	5	1.23 + 03	1.60 + 02	9.68 + 00	1.25 + 00	3.01-01	1.66-01	1.39-01	1.14-01	8.07-02
C	6	2.21 + 03	3.03 + 02	1.91 + 01	2.37 + 00	4.42-01	1.87-01	1.51-01	1.23-01	8.72-02
N	7	3.31 + 03	4.77 + 02	3.14 + 01	3.88 + 00	6.18-01	1.98-01	1.53-01	1.23-01	8.72-02
O	8	4.59 + 03	6.95 + 02	4.79 + 01	5.95 + 00	8.65-01	2.13-01	1.55-01	1.24-01	8.73-02
F	9	5.65 + 03	9.05 + 02	6.51 + 01	8.21 + 00	1.13 + 00	2.21-01	1.50-01	1.18-01	8.27-02
Ne	10	7.41 + 03	1.24 + 03	9.34 + 01	1.20 + 01	1.61 + 00	2.58-01	1.60-01	1.24-01	8.66-02
Na	11	6.54 + 02	1.52 + 03	1.19 + 02	1.56 + 01	2.06 + 00	2.80-01	1.59-01	1.20-01	8.37-02
Mg	12	9.22 + 02	1.93 + 03	1.58 + 02	2.11 + 01	2.76 + 00	3.29-01	1.69-01	1.24-01	8.65-02
Al	13	1.19 + 03	2.26 + 03	1.93 + 02	2.62 + 01	3.44 + 00	3.68-01	1.70-01	1.22-01	8.44-02
Si	14	1.57 + 03	2.78 + 03	2.45 + 02	3.39 + 01	4.46 + 00	4.38-01	1.84-01	1.28-01	8.75-02
P	15	1.91 + 03	3.02 + 02	2.86 + 02	4.04 + 01	5.35 + 00	4.92-01	1.87-01	1.25-01	8.51-02
S	16	2.43 + 03	3.85 + 02	3.49 + 02	5.01 + 01	6.71 + 00	5.85-01	2.02-01	1.30-01	8.78-02
Cl	17	2.83 + 03	4.52 + 02	3.90 + 02	5.73 + 01	7.74 + 00	6.48-01	2.05-01	1.27-01	8.45-02
Ar	18	3.18 + 03	5.12 + 02	4.23 + 02	6.32 + 01	8.63 + 00	7.01-01	2.04-01	1.20-01	7.96-02
K	19	4.06 + 03	6.59 + 02	5.19 + 02	7.91 + 01	1.09 + 01	8.68-01	2.34-01	1.32-01	8.60-02
Ca	20	4.87 + 03	8.00 + 02	6.03 + 02	9.34 + 01	1.31 + 01	1.02 + 00	2.57-01	1.38-01	8.85-02
Sc	21	5.24 + 03	8.70 + 02	6.31 + 02	9.95 + 01	1.41 + 01	1.09 + 00	2.58-01	1.31-01	8.31-02
Ti	22	5.87 + 03	9.86 + 02	6.84 + 02	1.11 + 02	1.59 + 01	1.21 + 00	2.72-01	1.31-01	8.19-02
V	23	6.50 + 03	1.11 + 03	9.29 + 01	1.22 + 02	1.77 + 01	1.35 + 00	2.88-01	1.32-01	8.07-02
Cr	24	7.40 + 03	1.28 + 03	1.08 + 02	1.39 + 02	2.04 + 01	1.55 + 00	3.17-01	1.38-01	8.28-02
Mn	25	8.09 + 03	1.42 + 03	1.21 + 02	1.51 + 02	2.25 + 01	1.71 + 00	3.37-01	1.39-01	8.19-02
Fe	26	9.09 + 03	1.63 + 03	1.40 + 02	1.71 + 02	2.57 + 01	1.96 + 00	3.72-01	1.46-01	8.41-02
Co	27	9.80 + 03	1.78 + 03	1.54 + 02	1.84 + 02	2.80 + 01	2.14 + 00	3.95-01	1.48-01	8.32-02
Ni	28	9.86 + 03	2.05 + 03	1.79 + 02	2.09 + 02	3.22 + 01	2.47 + 00	4.44-01	1.58-01	8.70-02
Cu	29	1.06 + 04	2.15 + 03	1.90 + 02	2.16 + 02	3.38 + 01	2.61 + 00	4.58-01	1.56-01	8.36-02
Zn	30	1.55 + 03	2.37 + 03	2.12 + 02	2.33 + 02	3.72 + 01	2.89 + 00	4.97-01	1.62-01	8.45-02
Ga	31	1.70 + 03	2.52 + 03	2.27 + 02	3.42 + 01	3.93 + 01	3.08 + 00	5.20-01	1.62-01	8.24-02
Ge	32	1.89 + 03	2.71 + 03	2.47 + 02	3.74 + 01	4.22 + 01	3.34 + 00	5.55-01	1.66-01	8.21-02
As	33	2.12 + 03	2.93 + 03	2.71 + 02	4.12 + 01	4.56 + 01	3.63 + 00	5.97-01	1.72-01	8.26-02
Se	34	2.32 + 03	3.10 + 03	2.90 + 02	4.41 + 01	4.82 + 01	3.86 + 00	6.28-01	1.74-01	8.13-02
Br	35	2.62 + 03	3.41 + 03	3.21 + 02	4.91 + 01	5.27 + 01	4.26 + 00	6.86-01	1.84-01	8.33-02
Kr	36	2.85 + 03	3.60 + 03	3.43 + 02	5.26 + 01	5.55 + 01	4.52 + 00	7.22-01	1.87-01	8.23-02
Rb	37	3.17 + 03	3.41 + 03	3.74 + 02	5.77 + 01	5.98 + 01	4.92 + 00	7.80-01	1.96-01	8.36-02
Sr	38	3.49 + 03	2.59 + 03	4.06 + 02	6.27 + 01	6.39 + 01	5.31 + 00	8.37-01	2.04-01	8.44-02
Y	39	3.86 + 03	7.42 + 02	4.42 + 02	6.87 + 01	6.86 + 01	5.76 + 00	9.05-01	2.15-01	8.61-02
Zr	40	4.21 + 03	8.12 + 02	4.76 + 02	7.42 + 01	7.24 + 01	6.17 + 00	9.66-01	2.24-01	8.69-02
Nb	41	4.60 + 03	8.89 + 02	5.13 + 02	8.04 + 01	7.71 + 01	6.64 + 00	1.04 + 00	2.34-01	8.83-02
Mo	42	4.94 + 03	9.60 + 02	5.45 + 02	8.58 + 01	1.31 + 01	7.04 + 00	1.10 + 00	2.42-01	8.85-02
Tc	43	5.36 + 03	1.04 + 03	5.84 + 02	9.23 + 01	1.41 + 01	7.52 + 00	1.17 + 00	2.53-01	8.97-02
Ru	44	5.72 + 03	1.12 + 03	6.17 + 02	9.80 + 01	1.50 + 01	7.92 + 00	1.23 + 00	2.62-01	8.99-02
Rh	45	6.17 + 03	1.21 + 03	6.59 + 02	1.05 + 02	1.61 + 01	8.45 + 00	1.31 + 00	2.74-01	9.13-02
Pd	46	6.54 + 03	1.29 + 03	6.91 + 02	1.11 + 02	1.70 + 01	8.85 + 00	1.38 + 00	2.83-01	9.13-02
Ag	47	7.04 + 03	1.40 + 03	7.39 + 02	1.19 + 02	1.84 + 01	9.45 + 00	1.47 + 00	2.97-01	9.32-02
Cd	48	7.35 + 03	1.47 + 03	7.69 + 02	1.24 + 02	1.92 + 01	9.78 + 00	1.52 + 00	3.04-01	9.25-02
In	49	7.81 + 03	1.58 + 03	8.13 + 02	1.32 + 02	2.04 + 01	1.03 + 01	1.61 + 00	3.17-01	9.37-02
Sn	50	8.16 + 03	1.66 + 03	8.47 + 02	1.38 + 02	2.15 + 01	1.07 + 01	1.68 + 00	3.26-01	9.37-02

L_3 L_1 L_2 K EDGE

PHOTON ATTENUATION COEFFICIENTS (continued)

Mass attenuation coefficient, cm²/g

	Atomic no.	1.0	2.0	5.0	10.0	20.0	50.0	100.0	500.0	1000.0
		Photon energy, MeV								
H	1	1.26-01	8.77-02	5.05-02	3.25-02	2.15-02	1.42-02	1.19-02	1.14-02	1.16-02
He	2	6.36-02	4.42-02	2.58-02	1.70-02	1.18-02	8.61-03	7.78-03	7.79-03	7.95-03
Li	3	5.50-02	3.83-02	2.26-02	1.53-02	1.11-02	8.68-03	8.21-03	8.61-03	8.87-03
Be	4	5.65-02	3.94-02	2.35-02	1.63-02	1.23-02	1.02-02	9.94-03	1.08-02	1.12-02
B	5	5.89-02	4.11-02	2.48-02	1.76-02	1.37-02	1.19-02	1.19-02	1.32-02	1.37-02
C	6	6.36-02	4.44-02	2.71-02	1.96-02	1.58-02	1.43-02	1.46-02	1.64-02	1.70-02
N	7	6.36-02	4.45-02	2.74-02	2.02-02	1.67-02	1.57-02	1.63-02	1.85-02	1.92-02
O	8	6.37-02	4.46-02	2.78-02	2.09-02	1.77-02	1.71-02	1.79-02	2.06-02	2.13-02
F	9	6.04-02	4.23-02	2.66-02	2.04-02	1.77-02	1.75-02	1.86-02	2.14-02	2.21-02
Ne	10	6.32-02	4.43-02	2.82-02	2.20-02	1.95-02	1.96-02	2.11-02	2.43-02	2.51-02
Na	11	6.10-02	4.28-02	2.75-02	2.18-02	1.97-02	2.03-02	2.19-02	2.53-02	2.62-02
Mg	12	6.30-02	4.43-02	2.87-02	2.31-02	2.13-02	2.23-02	2.42-02	2.81-02	2.90-02
Al	13	6.15-02	4.32-02	2.84-02	2.32-02	2.17-02	2.31-02	2.52-02	2.93-02	3.03-02
Si	14	6.36-02	4.48-02	2.97-02	2.46-02	2.34-02	2.52-02	2.76-02	3.23-02	3.34-02
P	15	6.18-02	4.36-02	2.91-02	2.45-02	2.36-02	2.58-02	2.84-02	3.33-02	3.45-02
S	16	6.37-02	4.50-02	3.04-02	2.59-02	2.53-02	2.79-02	3.08-02	3.62-02	3.75-02
Cl	17	6.13-02	4.33-02	2.95-02	2.55-02	2.52-02	2.81-02	3.11-02	3.67-02	3.80-02
Ar	18	5.76-02	4.07-02	2.80-02	2.45-02	2.45-02	2.76-02	3.07-02	3.62-02	3.75-02
K	19	6.22-02	4.40-02	3.05-02	2.70-02	2.74-02	3.11-02	3.46-02	4.09-02	4.24-02
Ca	20	6.39-02	4.52-02	3.17-02	2.84-02	2.90-02	3.32-02	3.71-02	4.40-02	4.56-02
Sc	21	5.98-02	4.24-02	3.00-02	2.72-02	2.80-02	3.23-02	3.62-02	4.30-02	4.45-02
Ti	22	5.89-02	4.18-02	2.98-02	2.73-02	2.84-02	3.30-02	3.71-02	4.40-02	4.56-02
V	23	5.79-02	4.11-02	2.96-02	2.74-02	2.88-02	3.36-02	3.78-02	4.49-02	4.65-02
Cr	24	5.93-02	4.21-02	3.06-02	2.86-02	3.03-02	3.56-02	4.01-02	4.76-02	4.93-02
Mn	25	5.85-02	4.16-02	3.04-02	2.87-02	3.07-02	3.63-02	4.09-02	4.86-02	5.04-02
Fe	26	5.99-02	4.26-02	3.15-02	2.99-02	3.22-02	3.83-02	4.33-02	5.15-02	5.33-02
Co	27	5.91-02	4.20-02	3.13-02	3.00-02	3.26-02	3.88-02	4.40-02	5.23-02	5.41-02
Ni	28	6.16-02	4.39-02	3.29-02	3.18-02	3.48-02	4.17-02	4.73-02	5.61-02	5.81-02
Cu	29	5.90-02	4.20-02	3.18-02	3.10-02	3.41-02	4.10-02	4.66-02	5.53-02	5.72-02
Zn	30	5.94-02	4.24-02	3.22-02	3.18-02	3.51-02	4.24-02	4.82-02	5.72-02	5.91-02
Ga	31	5.77-02	4.11-02	3.16-02	3.13-02	3.48-02	4.22-02	4.80-02	5.70-02	5.89-02
Ge	32	5.73-02	4.09-02	3.16-02	3.16-02	3.53-02	4.30-02	4.89-02	5.80-02	6.00-02
As	33	5.73-02	4.09-02	3.19-02	3.21-02	3.60-02	4.40-02	5.01-02	5.95-02	6.15-02
Se	34	5.62-02	4.01-02	3.14-02	3.19-02	3.60-02	4.41-02	5.03-02	5.97-02	6.17-02
Br	35	5.73-02	4.09-02	3.23-02	3.29-02	3.74-02	4.60-02	5.24-02	6.22-02	6.43-02
Kr	36	5.63-02	4.02-02	3.20-02	3.28-02	3.74-02	4.61-02	5.26-02	6.25-02	6.46-02
Rb	37	5.69-02	4.06-02	3.25-02	3.36-02	3.85-02	4.75-02	5.43-02	6.45-02	6.67-02
Sr	38	5.71-02	4.08-02	3.29-02	3.41-02	3.93-02	4.87-02	5.56-02	6.61-02	6.83-02
Y	39	5.80-02	4.14-02	3.35-02	3.50-02	4.05-02	5.03-02	5.75-02	6.83-02	7.06-02
Zr	40	5.81-02	4.15-02	3.38-02	3.55-02	4.12-02	5.13-02	5.87-02	6.98-02	7.22-02
Nb	41	5.87-02	4.18-02	3.44-02	3.63-02	4.22-02	5.27-02	6.03-02	7.17-02	7.42-02
Mo	42	5.84-02	4.16-02	3.44-02	3.65-02	4.26-02	5.33-02	6.10-02	7.26-02	7.51-02
Tc	43	5.88-02	4.19-02	3.48-02	3.71-02	4.35-02	5.45-02	6.24-02	7.43-02	7.68-02
Ru	44	5.85-02	4.16-02	3.48-02	3.73-02	4.39-02	5.50-02	6.30-02	7.51-02	7.77-02
Rh	45	5.89-02	4.20-02	3.53-02	3.80-02	4.48-02	5.63-02	6.45-02	7.69-02	7.94-02
Pd	46	5.85-02	4.16-02	3.52-02	3.80-02	4.50-02	5.66-02	6.49-02	7.73-02	8.00-02
Ag	47	5.92-02	4.21-02	3.58-02	3.88-02	4.61-02	5.81-02	6.67-02	7.93-02	8.20-02
Cd	48	5.83-02	4.14-02	3.54-02	3.85-02	4.59-02	5.79-02	6.64-02	7.91-02	8.18-02
In	49	5.85-02	4.15-02	3.56-02	3.90-02	4.65-02	5.88-02	6.75-02	8.04-02	8.32-02
Sn	50	5.80-02	4.11-02	3.55-02	3.90-02	4.66-02	5.90-02	6.78-02	8.07-02	8.35-02

PHOTON ATTENUATION COEFFICIENTS (continued)

Mass attenuation coefficients, cm²/g

	Atomic no.	Photon energy, MeV 0.001	0.002	0.005	0.01	0.02	0.05	0.1	0.2	0.5
Sb	51	8.58 + 03	1.77 + 03	8.85 + 02	1.46 + 02	2.27 + 01	1.12 + 01	1.76 + 00	3.38-01	9.45-02
Te	52	8.43 + 03	1.83 + 03	9.01 + 02	1.50 + 02	2.34 + 01	1.14 + 01	1.80 + 00	3.43-01	9.33-02
I	53	9.10 + 03	2.00 + 03	8.43 + 02	1.63 + 02	2.54 + 01	1.23 + 01	1.94 + 00	3.66-01	9.70-02
Xe	54	9.41 + 03	2.09 + 03	6.39 + 02	1.69 + 02	2.65 + 01	1.27 + 01	2.01 + 00	3.76-01	9.70-02
Cs	55	9.37 + 03	2.23 + 03	2.30 + 02	1.79 + 02	2.82 + 01	1.34 + 01	2.12 + 00	3.94-01	9.91-02
ZB	56	8.54 + 03	2.32 + 03	2.41 + 02	1.86 + 02	2.94 + 01	1.38 + 01	2.20 + 00	4.05-01	9.92-02
La	57	9.09 + 03	2.46 + 03	2.58 + 02	1.97 + 02	3.12 + 01	1.45 + 01	2.32 + 00	4.24-01	1.01-01
Ce	58	9.71 + 03	2.61 + 03	2.74 + 02	2.08 + 02	3.31 + 01	1.52 + 01	2.45 + 00	4.45-01	1.04-01
Pr	59	1.06 + 04	2.77 + 03	2.92 + 02	2.21 + 02	3.53 + 01	1.60 + 01	2.59 + 00	4.69-01	1.07-01
Nd	60	6.63 + 03	2.88 + 03	3.06 + 02	2.30 + 02	3.68 + 01	1.65 + 01	2.69 + 00	4.84-01	1.08-01
Pm	61	2.06 + 03	3.05 + 03	3.26 + 02	2.44 + 02	3.92 + 01	1.73 + 01	2.84 + 00	5.10-01	1.12-01
Sm	62	2.11 + 03	3.12 + 03	3.36 + 02	2.50 + 02	4.03 + 01	1.77 + 01	2.90 + 00	5.19-01	1.11-01
Eu	63	2.22 + 03	3.28 + 03	3.54 + 02	2.63 + 02	4.24 + 01	1.85 + 01	3.04 + 00	5.43-01	1.14-01
Gd	64	2.29 + 03	3.36 + 03	3.65 + 02	2.69 + 02	4.36 + 01	3.86 + 00	3.11 + 00	5.54-01	1.14-01
Tb	65	2.40 + 03	3.51 + 03	3.84 + 02	2.82 + 02	4.59 + 01	4.06 + 00	3.25 + 00	5.77-01	1.17-01
Dy	66	2.49 + 03	3.47 + 03	3.99 + 02	2.90 + 02	4.76 + 01	4.23 + 00	3.36 + 00	5.95-01	1.18-01
Ho	67	2.62 + 03	3.59 + 03	4.17 + 02	3.01 + 02	4.98 + 01	4.43 + 00	3.49 + 00	6.18-01	1.20-01
Er	68	2.75 + 03	3.52 + 03	4.36 + 02	3.13 + 02	5.20 + 01	4.63 + 00	3.63 + 00	6.41-01	1.23-01
Tm	69	2.90 + 03	3.69 + 03	4.57 + 02	2.83 + 02	5.45 + 01	4.87 + 00	3.78 + 00	6.68-01	1.26-01
Yb	70	3.02 + 03	3.80 + 03	4.72 + 02	2.94 + 02	5.63 + 01	5.04 + 00	3.88 + 00	6.86-01	1.27-01
Lu	71	3.19 + 03	3.45 + 03	4.94 + 02	2.21 + 02	5.88 + 01	5.28 + 00	4.03 + 00	7.13-01	1.30-01
Hf	72	3.34 + 03	3.60 + 03	5.11 + 02	2.30 + 02	6.09 + 01	5.48 + 00	4.15 + 00	7.34-01	1.32-01
Ta	73	3.51 + 03	3.77 + 03	5.33 + 02	2.38 + 02	6.33 + 01	5.72 + 00	4.30 + 00	7.60-01	1.35-01
W	74	3.68 + 03	3.92 + 03	5.53 + 02	9.69 + 01	6.57 + 01	5.95 + 00	4.44 + 00	7.84-01	1.38-01
Re	75	3.87 + 03	3.77 + 03	5.76 + 02	1.01 + 02	6.84 + 01	6.21 + 00	4.59 + 00	8.12-01	1.41-01
Os	76	4.03 + 03	2.22 + 03	5.93 + 02	1.04 + 02	7.04 + 01	6.41 + 00	4.70 + 00	8.33-01	1.43-01
Ir	77	4.24 + 03	1.03 + 03	6.18 + 02	1.09 + 02	7.32 + 01	6.69 + 00	4.86 + 00	8.63-01	1.46-01
Pt	78	4.43 + 03	1.08 + 03	6.40 + 02	1.13 + 02	7.57 + 01	6.95 + 00	4.99 + 00	8.90-01	1.49-01
Au	79	4.65 + 03	1.14 + 03	6.66 + 02	1.18 + 02	7.88 + 01	7.26 + 00	5.16 + 00	9.22-01	1.53-01
Hg	80	4.83 + 03	1.18 + 03	6.87 + 02	1.22 + 02	8.12 + 01	7.50 + 00	5.28 + 00	9.46-01	1.56-01
Tl	81	5.01 + 03	1.23 + 03	7.07 + 02	1.26 + 02	8.36 + 01	7.75 + 00	5.40 + 00	9.69-01	1.58-01
Pb	82	5.21 + 03	1.29 + 03	7.30 + 02	1.31 + 02	8.64 + 01	8.04 + 00	5.55 + 00	9.99-01	1.61-01
Bi	83	5.44 + 03	1.35 + 03	7.58 + 02	1.36 + 02	8.95 + 01	8.38 + 00	5.74 + 00	1.03 + 00	1.66-01
Po	84	5.72 + 03	1.42 + 03	7.93 + 02	1.43 + 02	9.35 + 01	8.80 + 00	5.99 + 00	1.08 + 00	1.71-01
At	85	5.87 + 03	1.49 + 03	8.25 + 02	1.49 + 02	9.70 + 01	9.19 + 00	6.17 + 00	1.12 + 00	1.77-01
Rn	86	5.83 + 03	1.49 + 03	8.16 + 02	1.48 + 02	9.56 + 01	9.12 + 00	6.09 + 00	1.10 + 00	1.73-01
Fr	87	6.08 + 03	1.56 + 03	8.49 + 02	1.54 + 02	9.93 + 01	9.52 + 00	1.66 + 00	1.14 + 00	1.78-01
Ra	88	6.20 + 03	1.62 + 03	8.74 + 02	1.59 + 02	1.02 + 02	9.85 + 00	1.71 + 00	1.17 + 00	1.82-01
Ac	89	6.47 + 03	1.70 + 03	8.69 + 02	1.65 + 02	1.06 + 02	1.03 + 01	1.79 + 00	1.21 + 00	1.87-01
Th	90	6.61 + 03	1.74 + 03	8.88 + 02	1.69 + 02	9.37 + 01	1.05 + 01	1.83 + 00	1.23 + 00	1.90-01
Pa	91	6.53 + 03	1.83 + 03	8.76 + 02	1.77 + 02	7.03 + 01	1.10 + 01	1.92 + 00	1.29 + 00	1.97-01
U	92	6.63 + 03	1.86 + 03	8.89 + 02	1.79 + 02	7.11 + 01	1.12 + 01	1.95 + 00	1.30 + 00	1.98-01
Np	93	6.95 + 03	1.96 + 03	9.32 + 02	1.87 + 02	7.45 + 01	1.18 + 01	2.05 + 00	1.35 + 00	2.05-01
Pu	94	7.19 + 03	2.04 + 03	9.65 + 02	1.94 + 02	7.71 + 01	1.22 + 01	2.13 + 00	1.39 + 00	2.10-01
Am	95	7.37 + 03	2.10 + 03	9.90 + 02	1.98 + 02	7.93 + 01	1.25 + 01	2.19 + 00	1.42 + 00	2.14-01
Cm	96	7.54 + 03	2.15 + 03	1.02 + 03	2.03 + 02	8.14 + 01	1.28 + 01	2.25 + 00	1.44 + 00	2.18-01
Bk	97	7.84 + 03	2.25 + 03	1.06 + 03	2.10 + 02	8.39 + 01	1.34 + 01	2.35 + 00	1.50 + 00	2.25-01
Cf	98	7.89 + 03	2.31 + 03	9.27 + 02	2.15 + 02	8.58 + 01	1.37 + 01	2.41 + 00	1.52 + 00	2.29-01
Es	99	7.79 + 03	2.40 + 03	9.59 + 02	2.22 + 02	4.01 + 01	1.42 + 01	2.51 + 00	1.57 + 00	2.36-01
Fm	100	7.13 + 03	2.46 + 03	9.77 + 02	2.26 + 02	4.09 + 01	1.45 + 01	2.57 + 00	1.59 + 00	2.39-01

Edge labels: N_5, N_4, N_3, N_2, N_1, M_5, M_4, M_3, M_2, M_1, L_3, L_2, L_1, K EDGE

PHOTON ATTENUATION COEFFICIENTS (continued)

Mass attenuation coefficients, cm²/g

	Atomic no.	Photon Energy, MeV 1.0	2.0	5.0	10.0	20.0	50.0	100.0	500.0	1000.0
Sb	51	5.80-02	4.10-02	3.56-02	3.92-02	4.70-02	5.96-02	6.85-02	8.16-02	8.44-02
Te	52	5.67-02	4.01-02	3.49-02	3.86-02	4.64-02	5.89-02	6.77-02	8.07-02	8.35-02
I	53	5.84-02	4.12-02	3.61-02	4.00-02	4.82-02	6.13-02	7.04-02	8.40-02	8.69-02
Xe	54	5.78-02	4.08-02	3.58-02	3.99-02	4.82-02	6.12-02	7.04-02	8.40-02	8.69-02
Cs	55	5.85-02	4.12-02	3.64-02	4.06-02	4.91-02	6.25-02	7.19-02	8.58-02	8.88-02
ZB	56	5.80-02	4.08-02	3.61-02	4.04-02	4.90-02	6.25-02	7.19-02	8.58-02	8.88-02
La	57	5.88-02	4.12-02	3.66-02	4.11-02	5.00-02	6.37-02	7.34-02	8.76-02	9.06-02
Ce	58	5.96-02	4.18-02	3.73-02	4.19-02	5.10-02	6.52-02	7.50-02	8.96-02	9.27-02
Pr	59	6.07-02	4.24-02	3.80-02	4.29-02	5.23-02	6.68-02	7.69-02	9.19-02	9.50-02
Nd	60	6.07-02	4.24-02	3.81-02	4.30-02	5.26-02	6.72-02	7.74-02	9.25-02	9.56-02
Pm	61	6.19-02	4.31-02	3.88-02	4.40-02	5.38-02	6.89-02	7.94-02	9.48-02	9.81-02
Sm	62	6.11-02	4.24-02	3.83-02	4.35-02	5.34-02	6.84-02	7.88-02	9.41-02	9.73-02
Eu	63	6.19-02	4.28-02	3.88-02	4.42-02	5.42-02	6.96-02	8.02-02	9.57-02	9.90-02
Gd	64	6.12-02	4.23-02	3.84-02	4.38-02	5.38-02	6.91-02	7.97-02	9.51-02	9.83-02
Tb	65	6.20-02	4.27-02	3.89-02	4.45-02	5.47-02	7.03-02	8.11-02	9.67-02	1.00-01
Dy	66	6.20-02	4.26-02	3.90-02	4.46-02	5.49-02	7.06-02	8.15-02	9.72-02	1.00-01
Ho	67	6.26-02	4.29-02	3.93-02	4.50-02	5.55-02	7.14-02	8.24-02	9.83-02	1.02-01
Er	68	6.32-02	4.32-02	3.96-02	4.55-02	5.61-02	7.23-02	8.34-02	9.95-02	1.03-01
Tm	69	6.40-02	4.36-02	4.01-02	4.61-02	5.70-02	7.35-02	8.48-02	1.01-01	1.04-01
Yb	70	6.40-02	4.35-02	4.00-02	4.61-02	5.70-02	7.35-02	8.49-02	1.01-01	1.04-01
Lu	71	6.48-02	4.39-02	4.05-02	4.66-02	5.77-02	7.45-02	8.60-02	1.02-01	1.06-01
Hf	72	6.50-02	4.39-02	4.05-02	4.68-02	5.80-02	7.48-02	8.64-02	1.03-01	1.06-01
Ta	73	6.57-02	4.41-02	4.08-02	4.72-02	5.85-02	7.56-02	8.73-02	1.04-01	1.07-01
W	74	6.62-02	4.43-02	4.10-02	4.75-02	5.89-02	7.62-02	8.80-02	1.05-01	1.08-01
Re	75	6.69-02	4.46-02	4.14-02	4.79-02	5.95-02	7.70-02	8.89-02	1.06-01	1.09-01
Os	76	6.71-02	4.46-02	4.13-02	4.79-02	5.96-02	7.71-02	8.90-02	1.06-01	1.10-01
Ir	77	6.79-02	4.50-02	4.17-02	4.84-02	6.02-02	7.80-02	9.01-02	1.07-01	1.11-01
Pt	78	6.86-02	4.52-02	4.20-02	4.87-02	6.06-02	7.86-02	9.08-02	1.08-01	1.12-01
Au	79	6.95-02	4.57-02	4.24-02	4.93-02	6.14-02	7.95-02	9.19-02	1.09-01	1.13-01
Hg	80	6.99-02	4.57-02	4.25-02	4.94-02	6.15-02	7.98-02	9.22-02	1.10-01	1.13-01
Tl	81	7.03-02	4.58-02	4.25-02	4.94-02	6.16-02	8.00-02	9.24-02	1.10-01	1.14-01
Pb	82	7.10-02	4.61-02	4.27-02	4.97-02	6.21-02	8.06-02	9.31-02	1.11-01	1.15-01
Bi	83	7.21-02	4.66-02	4.32-02	5.03-02	6.28-02	8.15-02	9.42-02	1.12-01	1.16-01
Po	84	7.39-02	4.75-02	4.40-02	5.12-02	6.40-02	8.32-02	9.61-02	1.15-01	1.18-01
At	85	7.54-02	4.82-02	4.46-02	5.20-02	6.49-02	8.44-02	9.76-02	1.16-01	1.20-01
Rn	86	7.30-02	4.65-02	4.30-02	5.01-02	6.26-02	8.14-02	9.42-02	1.12-01	1.16-01
Fr	87	7.45-02	4.72-02	4.36-02	5.08-02	6.35-02	8.26-02	9.56-02	1.14-01	1.18-01
Ra	88	7.53-02	4.75-02	4.38-02	5.10-02	6.38-02	8.31-02	9.61-02	1.15-01	1.19-01
Ac	89	7.69-02	4.82-02	4.44-02	5.17-02	6.47-02	8.43-02	9.75-02	1.16-01	1.20-01
Th	90	7.71-02	4.81-02	4.42-02	5.15-02	6.45-02	8.40-02	9.72-02	1.16-01	1.20-01
Pa	91	7.94-02	4.93-02	4.52-02	5.26-02	6.59-02	8.60-02	9.95-02	1.19-01	1.23-01
U	92	7.90-02	4.88-02	4.46-02	5.19-02	6.51-02	8.49-02	9.83-02	1.17-01	1.21-01
Np	93	8.13-02	4.99-02	4.56-02	5.30-02	6.65-02	8.68-02	1.01-01	1.20-01	1.24-01
Pu	94	8.26-02	5.05-02	4.60-02	5.34-02	6.71-02	8.76-02	1.01-01	1.21-01	1.25-01
Am	95	8.33-02	5.06-02	4.60-02	5.34-02	6.70-02	8.77-02	1.02-01	1.21-01	1.25-01
Cm	96	8.41-02	5.08-02	4.60-02	5.34-02	6.70-02	8.77-02	1.02-01	1.21-01	1.26-01
Bk	97	8.62-02	5.18-02	4.68-02	5.42-02	6.81-02	8.92-02	1.03-01	1.24-01	1.28-01
Cf	98	8.70-02	5.20-02	4.68-02	5.42-02	6.81-02	8.92-02	1.04-01	1.24-01	1.28-01
Es	99	8.89-02	5.28-02	4.74-02	5.48-02	6.89-02	9.04-02	1.05-01	1.25-01	1.29-01
Fm	100	8.94-02	5.28-02	4.72-02	5.45-02	6.86-02	9.00-02	1.05-01	1.25-01	1.29-01

CLASSIFICATION OF ELECTROMAGNETIC RADIATION

Hans Dolezalek

Basic Conversions: $c = \lambda\nu = \nu/k$; $\nu = c/\lambda = ck$; $\lambda = c/\nu = 1/k$; $k = \nu/c = 1/\lambda$

c = speed of light = 2.99792458×10^8 m/s

Frequency (ν)	Wavelength (λ)	Wave number (k)	Names of bands	Approximate photon energies
3×10^0 — 3×10^1 Hz 3 — 30 Hz	10^8 — 10^7 m 100 — 10 Mm	10^{-8} — 10^{-7} m^{-1} 10 — 100 Gm^{-1}	ELF-(ELF 1), ITU band no. 1	
3×10^1 — 3×10^2 Hz 30 — 300 Hz	10^7 — 10^6 m 10 — 1 Mm	10^{-7} — 10^{-6} m^{-1} 100 Gm^{-1} — 1 Mm^{-1}	SLF-(ELF 2), ITU band no. 2, megameter waves	
3×10^2 — 3×10^3 Hz 300 Hz — 3 kHz	10^6 — 10^5 m 1 Mm — 100 km	10^{-6} — 10^{-5} m^{-1} 1 — 10 Mm^{-1}	ULF-(ELF 3), ITU band no. 3	
3×10^3 — 3×10^4 Hz 3 — 30 kHz	10^5 — 10^4 m 100 — 10 km	10^{-5} — 10^{-4} m^{-1} 10 — 100 Mm^{-1}	VLF, ITU band no. 4, myriameter waves	
3×10^4 — 3×10^5 Hz 30 — 300 kHz	10^4 — 10^3 m 10 — 1 km	10^{-4} — 10^{-3} m^{-1} 100 Mm^{-1} — 1 km^{-1}	LF, ITU band no. 5, kilometer waves	
3×10^5 — 3×10^6 Hz 300 kHz — 3 MHz	10^3 — 10^2 m 1 km — 100 m	10^{-3} — 10^{-2} m^{-1} 1 — 10 km^{-1}	MF, ITU band no. 6, hectometer waves	
3×10^6 — 3×10^7 Hz 3 — 30 MHz	10^2 — 10^1 m 100 — 10 m	10^{-2} — 10^{-1} m^{-1} 10 — 100 km^{-1}	HF, ITU band no. 7, decameter waves	
3×10^7 — 3×10^8 Hz 30 — 300 MHz	10^1 — 10^0 m 10 — 1 m	10^{-1} — 10^0 m^{-1} 100 km^{-1} — 1 m^{-1}	VHF, ITU band no. 8, meter waves	
3×10^8 — 3×10^9 Hz 300 MHz — 3 GHz	10^0 — 10^{-1} m 1 m — 100 mm	10^0 — 10^1 m^{-1} 1 — 10 m^{-1}	UHF, ITU band no. 9, decimeter waves[a]	
3×10^9 — 3×10^{10} Hz 3 — 30 GHz	10^{-1} — 10^{-2} m 100 — 10 mm	10^1 — 10^2 m^{-1} 10 — 100 m^{-1}	SHF, ITU band no. 10, centimeter waves[a]	
3×10^{10} — 3×10^{11} Hz 30 — 300 GHz	10^{-2} — 10^{-3} m 10 — 1 mm	10^2 — 10^3 m^{-1} 100 m^{-1} — 1 mm^{-1} (1 — 10 cm^{-1})	EHF, ITU band no. 11, millimeter waves	
3×10^{11} — 3×10^{12} Hz 300 GHz — 3 THz	10^{-3} — 10^{-4} m 1 mm — 100 μm	10^3 — 10^4 m^{-1} 1 — 10 mm^{-1} (10 — 100 cm^{-1})	Part of micrometer waves, includes part of far or thermal infrared; ITU band no. 12	
3×10^{12} — 3×10^{13} Hz 3 — 30 THz	10^{-4} — 10^{-5} m 100 — 10 μm	10^4 — 10^5 m^{-1} 10 — 100 mm^{-1} (100 — 1000 cm^{-1})	Part of micrometer waves includes part of far (thermal) infrared	
3×10^{13} — 3×10^{14} Hz 30 — 300 THz	10^{-5} — 10^{-6} m 10 — 1 μm (100,000 — 10,000 Å)	10^5 — 10^6 m^{-1} 100 mm^{-1} — 1 μm^{-1}	Part of μm waves, part of infrared	(1.6—16) × 10^{-20} joule {0.1 — 1 eV}
3×10^{14} — 3×10^{15} Hz 300 THz — 3 PHz	10^{-6} — 10^{-7} m 1 μm — 100 nm (10,000 — 1000 Å)	10^6 — 10^7 m^{-1} 1 — 10 μm^{-1}	Near infrared, visible, near ultraviolet	(1.6—16) × 10^{-19} joule {1 — 10 eV}
3×10^{15} — 3×10^{16} Hz 3 — 30 PHz	10^{-7} — 10^{-8} m 100 — 10 nm (1000 — 100 Å)	10^7 — 10^8 m^{-1} 10 — 100 μm^{-1}	Part of "vacuum" - ultraviolet	(1.6—16) × 10^{-18} joule {10 — 100 eV}
3×10^{16} — 3×10^{17} Hz 30 — 300 PHz	10^{-8} — 10^{-9} m 10 — 1 nm (100 — 10 Å)	10^8 — 10^9 m^{-1} 100 μm^{-1} — 1 nm^{-1}	Part of soft X-rays	(1.6—16) × 10^{-17} joule {100 — 1000 eV)}
3×10^{17} — 3×10^{18} Hz 300 PHz — 3 EHz	10^{-9} — 10^{-10} m 1 nm — 100 pm (10 — 1 Å)	10^9 — 10^{10} m^{-1} 1 — 10 nm^{-1}	Part of soft X-rays	(1.6—16) × 10^{-16} joule {1 — 10 keV}
3×10^{18} — 3×10^{19} Hz 3 — 30 EHz	10^{-10} — 10^{-11} m 100 — 10 pm (1 — 0.1 Å)	10^{10} — 10^{11} m^{-1} 10 — 100 nm^{-1}	Hard X-rays and part of soft γ-rays	(1.6—16) × 10^{-15} joule {10 — 100 keV}
3×10^{19} — 3×10^{20} Hz 30 — 300 EHz	10^{-11} — 10^{-12} m 10 — 1 pm (0.1 — 0.01 Å)	10^{11} — 10^{12} m^{-1} 100 nm^{-1} — 1 pm^{-1}	Part of soft and part of hard γ-rays (limit at 510 keV)	(1.6—16) × 10^{-14} joule {100 keV — 1 MeV}
3×10^{20} — 3×10^{21} Hz 300 — 3000 EHz	10^{-12} — 10^{-13} m 1 pm — 100 fm (0.01 — 0.001 Å)	10^{12} — 10^{13} m^{-1} 1 — 10 pm^{-1}	Part of hard γ-rays and part of "cosmic" γ-rays	(1.6—16) × 10^{-13} joule {1 — 10 MeV}
3×10^{21} — 3×10^{22} Hz 3000 — 30,000 EHz	10^{-13} — 10^{-14} m 100 — 10 fm (0.001 — 0.0001 Å)	10^{13} — 10^{14} m^{-1} 10 — 100 pm^{-1}	γ-rays produced by cosmic rays	(1.6—16) × 10^{-12} joule {10 — 100 MeV}

Note: Abbreviations used in this table: Å—ångstrom (1 Å = 10^{-10} m); EHz—exahertz (10^{18} hertz); EHF—extremely high frequency; ELF—extremely low frequency; eV—electron volt (1 eV = 1.60219×10^{-19} joule); PHz—petahertz (10^{15} hertz); fm—femtometer (10^{-15} m); GHz—gigahertz (10^{9} hertz); Gm—gigameter (10^{9} m); HF—high frequency; Hz—hertz (s^{-1}); ITU—International Telecommunications Union; keV—kiloelectron volt (10^{3} eV); km—kilometer (10^{3} m); LF—low frequency; m—meter; MeV—megaelectron volt (10^{6} eV); MF—medium frequency; MHz—megahertz (10^{6} hertz); Mm—megameter (10^{6} meter); mm—millimeter (10^{-3} meter); µm—micrometer (10^{-6} meter); nm—nanometer (10^{-9} meter); pm—picometer (10^{-12} meter); SHF—super high frequency; SLF—super low frequency; THz—terahertz; UHF—ultra high frequency; ULF—ultra low frequency; VHF—very high frequency; VLF—very low frequency.

[a] Also called "microwaves"; not to be confused with "micrometer waves".

LETTER DESIGNATIONS OF MICROWAVE BANDS

Frequency (GHz)	Wavelength (cm)	Wavenumber (cm^{-1})	Band
1—2	30—15	0.033—0.067	L-Band
1—4	15—7.5	0.067—0.133	S-Band
4—8	7.5—3.7	0.133—0.267	C-Band
8—12	3.7—2.5	0.267—0.4	X-Band
12—18	2.5—1.7	0.4—0.6	Ku-Band
18—27	1.7—1.1	0.6—0.9	K-Band
27—40	1.1—0.75	0.9—1.33	Ka-Band

BLACK BODY RADIATION

The total power radiated from an ideal black body and the wavelength corresponding to maximum power are given here as a function of absolute temperature. Constants used in the calculation are taken from the table "Fundamental Physical Constants" in Section 1. The radiated power in a band $\Delta\lambda$ at λ_{max} may be calculated from:

$$P_{max} = 0.657548\ (\Delta\lambda/\lambda_{max})\ P_{tot}$$

T/K	P_{tot}	λ_{max}/μm	T/K	P_{tot}	λ_{max}/μm	T/K	P_{tot}	λ_{max}/μm
50	0.354 W/m²	57.955	740	17.004	3.916	1520	302.689	1.906
100	5.671	28.978	750	17.942	3.864	1540	318.937	1.882
150	28.707	19.318	760	18.918	3.813	1560	335.831	1.858
200	90.728	14.489	770	19.934	3.763	1580	353.387	1.834
250	221.504	11.591	780	20.989	3.715	1600	371.623	1.811
273	314.973	10.614	790	22.087	3.668	1620	390.555	1.789
280	348.541	10.349	800	23.226	3.622	1640	410.202	1.767
290	401.064	9.992	810	24.410	3.577	1660	430.581	1.746
300	459.311	9.659	820	25.638	3.534	1680	451.710	1.725
310	523.684	9.348	830	26.911	3.491	1700	473.607	1.705
320	594.596	9.055	840	28.232	3.450	1720	496.290	1.685
330	672.478	8.781	850	29.600	3.409	1740	519.779	1.665
340	757.771	8.523	860	31.018	3.369	1760	544.093	1.646
350	850.931	8.279	870	32.486	3.331	1780	569.249	1.628
360	952.428	8.049	880	34.006	3.293	1800	595.267	1.610
370	1.063 kW/m²	7.832	890	35.578	3.256	1820	622.168	1.592
380	1.182	7.626	900	37.204	3.220	1840	649.970	1.575
390	1.312	7.430	910	38.886	3.184	1860	678.694	1.558
400	1.452	7.244	920	40.623	3.150	1880	708.359	1.541
410	1.602	7.068	930	42.418	3.116	1900	738.987	1.525
420	1.764	6.899	940	44.272	3.083	1920	770.597	1.509
430	1.939	6.739	950	46.187	3.050	1940	803.210	1.494
440	2.125	6.586	960	48.162	3.018	1960	836.848	1.478
450	2.325	6.439	970	50.201	2.987	1980	871.531	1.464
460	2.539	6.299	980	52.303	2.957	2000	907.282	1.449
470	2.767	6.165	990	54.471	2.927	2020	944.121	1.435
480	3.010	6.037	1000	56.705	2.898	2040	982.071	1.420
490	3.269	5.914	1020	61.379	2.841	2060	1.021 MW/m²	1.407
500	3.544	5.796	1040	66.337	2.786	2080	1.061	1.393
510	3.836	5.682	1060	71.589	2.734	2100	1.103	1.380
520	4.146	5.573	1080	77.147	2.683	2120	1.145	1.367
530	4.474	5.467	1100	83.022	2.634	2140	1.189	1.354
540	4.822	5.366	1120	89.227	2.587	2160	1.234	1.342
550	5.189	5.269	1140	95.773	2.542	2180	1.281	1.329
560	5.577	5.175	1160	102.672	2.498	2200	1.328	1.317
570	5.986	5.084	1180	109.939	2.456	2220	1.377	1.305
580	6.417	4.996	1200	117.584	2.415	2240	1.428	1.294
590	6.871	4.911	1220	125.621	2.375	2260	1.479	1.282
600	7.349	4.830	1240	134.063	2.337	2280	1.532	1.271
610	7.851	4.750	1260	142.924	2.300	2300	1.587	1.260
620	8.379	4.674	1280	152.217	2.264	2320	1.643	1.249
630	8.933	4.600	1300	161.955	2.229	2340	1.700	1.238
640	9.514	4.528	1320	172.154	2.195	2360	1.759	1.228
650	10.122	4.458	1340	182.827	2.163	2380	1.819	1.218
660	10.760	4.391	1360	193.989	2.131	2400	1.881	1.207
670	11.427	4.325	1380	205.655	2.100	2420	1.945	1.197
680	12.124	4.261	1400	217.838	2.070	2440	2.010	1.188
690	12.853	4.200	1420	230.556	2.041	2460	2.077	1.178
700	13.615	4.140	1440	243.822	2.012	2480	2.145	1.168
710	14.410	4.081	1460	257.652	1.985	2500	2.215	1.159
720	15.239	4.025	1480	272.063	1.958	2550	2.398	1.136
730	16.103	3.970	1500	287.070	1.932	2600	2.591	1.115

T/K	P_{tot}	λ_{max}/μm	T/K	P_{tot}	λ_{max}/μm	T/K	P_{tot}	λ_{max}/μm
2650	2.796	1.093	3600	9.524	0.805	5100	38.362	0.568
2700	3.014	1.073	3650	10.065	0.794	5200	41.461	0.557
2750	3.243	1.054	3700	10.627	0.783	5300	44.743	0.547
2800	3.485	1.035	3750	11.214	0.773	5400	48.217	0.537
2850	3.741	1.017	3800	11.824	0.763	5500	51.889	0.527
2900	4.011	0.999	3850	12.458	0.753	5600	55.767	0.517
2950	4.294	0.982	3900	13.118	0.743	5700	59.858	0.508
3000	4.593	0.966	3950	13.804	0.734	5800	64.170	0.500
3050	4.907	0.950	4000	14.517	0.724	5900	68.712	0.491
3100	5.237	0.935	4100	16.024	0.707	6000	73.490	0.483
3150	5.583	0.920	4200	17.645	0.690	6500	101.222	0.446
3200	5.946	0.906	4300	19.386	0.674	7000	136.149	0.414
3250	6.326	0.892	4400	21.254	0.659	7500	179.418	0.386
3300	6.725	0.878	4500	23.253	0.644	8000	232.264	0.362
3350	7.142	0.865	4600	25.389	0.630	8500	296.004	0.341
3400	7.578	0.852	4700	27.670	0.617	9000	372.042	0.322
3450	8.033	0.840	4800	30.101	0.604	9500	461.867	0.305
3500	8.509	0.828	4900	32.689	0.591	10000	567.051	0.290
3550	9.006	0.816	5000	35.441	0.580			

The curves below show, for various temperatures, the fraction of radiant power as a function of wavelength. The function plotted is $P_\lambda/\Delta\lambda\ P_{tot}$, where P_λ is the power at wavelength λ in a small interval $\Delta\lambda$ (in μm), and P_{tot} is the total power.

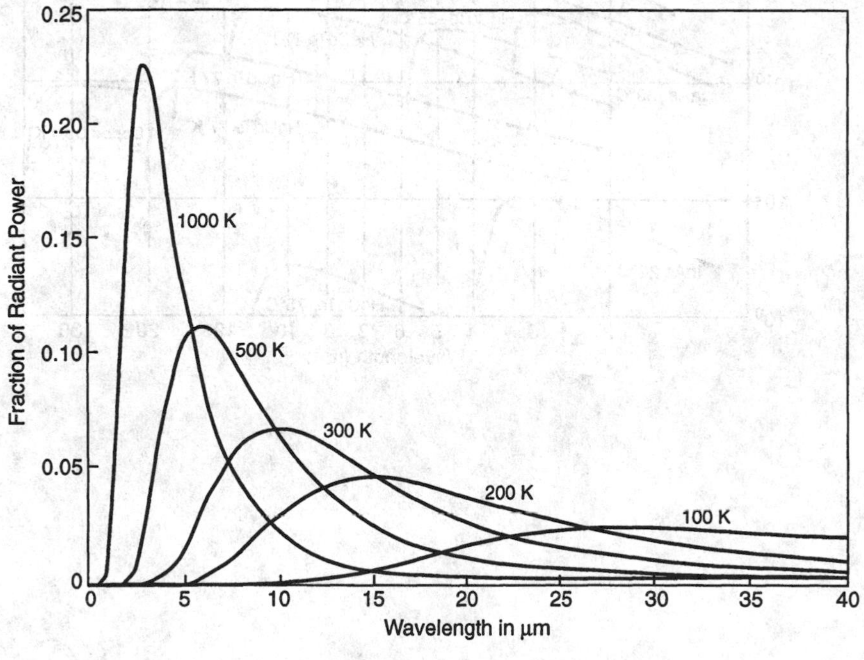

CHARACTERISTICS OF INFRARED DETECTORS

This graph summarizes the wavelength response of some semiconductors used as detectors for infrared radiation. The quantity $D^*(\lambda)$ is the signal to noise ratio for an incident radiant power density of 1 W/cm^2 and a bandwidth of 1 Hz (60° field of view). The Ge, InAs, and InSb detectors are photovoltaics, while the HgCdTe series are photoconductive devices. The cutoff wavelength of the latter can be varied by adjusting the relative amounts of Hg, Cd, and Te (three examples are shown at 77 K). The graph also shows the theoretical background limited sensitivity for ideal detectors which introduce no intrinsic noise.

REFERENCE

Infrared Detectors 1995, EG&G Judson, Montgomeryville, PA.

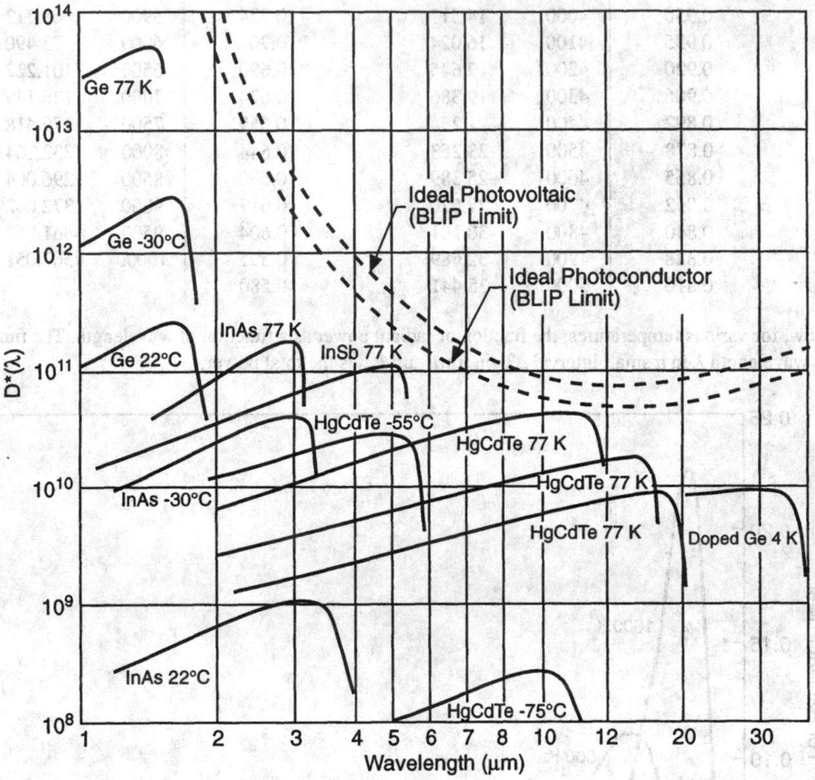

REFRACTIVE INDEX AND TRANSMITTANCE OF REPRESENTATIVE GLASSES

Typical values of the index of refraction and internal transmittance (fraction of light transmitted through a one centimeter thickness) are tabulated here for selected types of glasses, as well as for synthetic fused (vitreous) silica. Nominal compositions are given in the first part of the table. The second part gives the index of refraction, relative to air, and the internal transmittance for representative samples of each glass at wavelengths in the infrared, visible, and near-ultraviolet regions. It should be emphasized that wide variation of these parameters may be found among subtypes of each glass. More detailed data may be found in Reference 3.

Assuming that the Lambert-Beer Law is followed, the transmittance of a glass plate of thickness *d* (in centimeters) can be obtained by raising the transmittance value in the table to the power *d*.

REFERENCES

1. Weber, M.J., *CRC Handbook of Laser Science and Technology*, Vol. IV, Part 2, CRC Press, Boca Raton, FL ,1988.
2. Gray, D.E., Ed., *American Institute of Physics Handbook, Third Edition*, McGraw Hill, New York, 1972.
3. *Schott Optical Glass*, Schott Glass Technologies, Inc., 400 York Ave., Duryea, PA 18642.
4. Kaye, G.W.C., and Laby, T.H., *Tables of Physical and Chemical Constants, 15th Edition*, Longman, London, 1986.

Type	Name	Composition in percent by mass									
		SiO_2	B_2O_3	Al_2O_3	Na_2O	K_2O	CaO	BaO	ZnO	PbO	P_2O_5
PK	Phosphate crown		3	10		12	5				70
PSK	Dense phosphate crown		3	5			4	28			60
BK	Borosilicate crown	70	10		8	8	1	3			
K	Crown	74			9	11	6				
ZK	Zinc crown	71			17				12		
BaK	Barium crown	60	3		3	10		19	5		
SK	Dense crown	39	15	5				41			
KF	Crown flint	67		2	16				3	12	
BaLF	Barium light flint	51			6	5		20	14	4	
SSK	Extra dense crown	35	10	5				42	8		
LLF	Extra light flint	63			5	8				24	
BaF	Barium flint	46				8		16	8	22	
LF	Light flint	53			5	8				34	
F	Flint	47			2	7				44	
BaSF	Dense barium flint	43			1	7		11	5	33	
SF	Dense flint	33				5				62	
KzFS	Short flint										
SiO_2	Fused silica	100									

Type	Index of refraction				Transmittance of 1 cm plate			
	1.060 μm	546.1 nm	365.0 nm	312.6 nm	1.060 μm	546.1 nm	365.0 nm	310 nm
PK	1.51519	1.52736	1.54503	1.5574	0.997	0.998	0.987	0.46
PSK	1.54154	1.55440	1.57342	1.5868	0.996	0.998	0.984	0.46
BK	1.50669	1.51872	1.53627	1.5486	0.999	0.998	0.987	0.35
K	1.50091	1.51314	1.53189	1.5454	0.998	0.998	0.988	0.40
ZK	1.52220	1.53534	1.55588	1.5708	0.996	0.998	0.976	0.27
BaK	1.55695	1.57124	1.59407	1.6108	0.998	0.997	0.986	0.28
SK	1.59490	1.60994	1.63398		0.998	0.998	0.959	0.28
KF	1.50586	1.51978	1.54251	1.5600	0.998	0.996	0.989	0.49
BaLF	1.57579	1.59166	1.61804		0.996	0.998	0.933	0.010
SSK	1.60402	1.61993	1.64595		0.999	0.998	0.915	0.010
LaK	1.69710	1.71616	1.74573		0.999	0.998	0.882	0.17
LLF	1.52775	1.54344	1.57038		0.998	0.997	0.990	0.32
BaF	1.56873	1.58565	1.61524		0.999	0.997	0.992	0.004
LF	1.56594	1.58482	1.61926		0.999	0.998	0.981	0.008
F	1.58636	1.60718	1.64606		0.997	0.998	0.959	
BaSF	1.60889	1.62987	1.66926		0.999	0.998	0.857	
SF	1.71350	1.74620	1.8145		0.998	0.997	0.650	
KzFS	1.59680	1.61639	1.64849	1.6739		0.998	0.672	0.012
SiO_2	1.44968	1.46008	1.47435[a]	1.53430[b]				

[a] At 366.3 nm.
[b] At 213.9 nm.

INDEX OF REFRACTION OF WATER

This table gives the index of refraction of liquid water at atmospheric pressure, relative to a vacuum, at several temperatures and wavelengths. It is generated from the formulation in Reference 1, which covers a wide range of temperature, pressure, and wavelength.The wavelengths listed here correspond to prominent lines of cadmium (226.50 and 361.05 nm), potassium (404.41 nm), sodium (589.00 nm), Ne (632.80 nm, from a helium - neon laser), and mercury (1.01398 μm).

REFERENCES

1. Schiebener, P., Straub, J., Levelt Sengers, J.M.H., and Gallagher, J.S., *J. Phys. Chem. Ref. Data*, 19, 677 (1990); 19, 1617, 1990.
2. Marsh, K.N., Editor, *Recommended Reference Materials for the Realization of Physicochemical Properties*, Blackwell Scientific Publications, Oxford, 1987.

T/°C	226.50 nm	361.05 nm	404.41 nm	589.00 nm	632.80 nm	1.01398 μm
0	1.39450	1.34896	1.34415	1.33432	1.33306	1.32612
10	1.39422	1.34870	1.34389	1.33408	1.33282	1.32591
20	1.39336	1.34795	1.34315	1.33336	1.33211	1.32524
30	1.39208	1.34682	1.34205	1.33230	1.33105	1.32424
40	1.39046	1.34540	1.34065	1.33095	1.32972	1.32296
50	1.38854	1.34373	1.33901	1.32937	1.32814	1.32145
60	1.38636	1.34184	1.33714	1.32757	1.32636	1.31974
70	1.38395	1.33974	1.33508	1.32559	1.32438	1.31784
80	1.38132	1.33746	1.33284	1.32342	1.32223	1.31576
90	1.37849	1.33501	1.33042	1.32109	1.31991	1.31353
100	1.37547	1.33239	1.32784	1.31861	1.31744	1.31114

INDEX OF REFRACTION OF LIQUIDS FOR CALIBRATION PURPOSES

This table gives the index of refraction of six liquids which are available in highly pure form and whose index of refraction has been accurately measured as a function of wavelength and temperature. They are therefore useful for calibration of refractometers. The estimated uncertainty in the values is:

2,2,4-Trimethylpentane	±0.00003
Hexadecane	±0.00008
trans-Bicyclo[4.0.0]decane	±0.00008
1-Methylnaphthalene	±0.00008
Toluene	±0.00003
Methylcyclohexane	±0.00003

Full details are given in the references. This table is reprinted from Reference 1 by permission of the International Union of Pure and Applied Chemistry.

REFERENCES

1. Marsh, K. N., Editor, *Recommended Reference Materials for the Realization of Physicochemical Properties,* Blackwell Scientific Publications, Oxford, 1987.
2. Tilton, L. W., J. *Opt. Soc. Am.*, 32, 71, 1941.

λ	2,2,4-Trimethylpentane			Hexadecane		
nm	20°C	25°C	30°C	20°C	25°C	30°C
667.81	1.38916	1.38670	1.38424	1.43204	1.43001	1.42798
656.28	1.38945	1.38698	1.38452	1.43235	1.43032	1.42829
589.26	1.39145	1.38898	1.38650	1.43453	1.43250	1.43047
546.07	1.39316	1.39068	1.38820	1.43640	1.43436	1.43232
501.57	1.39544	1.39294	1.39044	1.43888	1.43684	1.43480
486.13	1.39639	1.39389	1.39138	1.43993	1.43788	1.43583
435.83	1.40029	1.39776	1.39523	1.44419	1.44213	1.44007

λ	*trans*-Bicyclo[4.4.0]decane			1-Methylnaphthalene		
nm	20°C	25°C	30°C	20°C	25°C	30°C
667.81	1.46654	1.46438	1.46222	1.60828	1.60592	1.60360
656.28	1.46688	1.46472	1.46256	1.60940	1.60703	1.60471
589.26	1.46932	1.46715	1.46498	1.61755	1.61512	1.61278
546.07	1.47141	1.46923	1.46705	1.62488	1.62240	1.62005
501.57	1.47420	1.47200	1.46980	1.63513	1.63259	1.63022
486.13	1.47535	1.47315	1.47095	1.63958	1.63701	1.63463
435.83	1.48011	1.47789	1.47567		1.65627	1.65386

λ	Toluene			Methylcyclohexane		
nm	20°C	25°C	30°C	20°C	25°C	30°C
667.81	1.49180	1.48903	1.48619	1.42064	1.41812	1.41560
656.28	1.49243	1.48966	1.48682	1.42094	1.41842	1.41591
589.26	1.49693	1.49413	1.49126	1.42312	1.42058	1.41806
546.07	1.50086	1.49803	1.49514	1.42497	1.42243	1.41989
501.57	1.50620	1.50334	1.50041	1.42744	1.42488	1.42233
486.13	1.50847	1.50559	1.50265	1.42847	1.42590	1.42334
435.83	1.51800	1.51506	1.51206	1.43269	1.43010	1.42752

INDEX OF REFRACTION OF AIR

This is a table of the index of refraction n of dry air at 15°C and a pressure of 101.325 kPa and containing 0.03% by volume of carbon dioxide ("standard air"). The index of refraction is defined by $n = \lambda_{vac}/\lambda_{air}$, where λ is the wavelength of the radiation. The index is calculated from the expression:

$$(n-1)\times 10^8 = 8342.13 + 2406030(130-\sigma^2)^{-1} + 15997(38.9-\sigma^2)^{-1}$$

where $\sigma = 1/\lambda_{vac}$ and λ_{vac} has units of µm. The equation is valid for λ_{vac} from 200 nm to 2 µm.
The table also gives the correction $(n-1)\lambda_{air}$ which must be added to the wavelength in air to obtain λ_{vac}.
If the air is at a temperature t in °C and a pressure p in pascals, a value of $(n-1)$ from this table should be multiplied by

$$\frac{p\left[1+p(61.3-t)\times 10^{-10}\right]}{96095.4(1+0.003661t)}$$

REFERENCE

Edlen, B., *Metrologia*, 2, 71, 1966.

λ_{vac}	$(n-1)\times 10^8$	$\lambda_{vac}-\lambda_{air}$	λ_{vac}	$(n-1)\times 10^8$	$\lambda_{vac}-\lambda_{air}$	λ_{vac}	$(n-1)\times 10^8$	$\lambda_{vac}-\lambda_{air}$
200 nm	32408	0.0648 nm	540	27803	0.1501	880	27461	0.2416
210	31746	0.0666	550	27782	0.1528	890	27457	0.2443
220	31224	0.0687	560	27763	0.1554	900	27452	0.2470
230	30799	0.0708	570	27745	0.1581	910	27448	0.2497
240	30445	0.0730	580	27728	0.1608	920	27444	0.2524
250	30146	0.0753	590	27712	0.1635	930	27440	0.2551
260	29890	0.0777	600	27697	0.1661	940	27436	0.2578
270	29669	0.0801	610	27682	0.1688	950	27432	0.2605
280	29475	0.0825	620	27669	0.1715	960	27429	0.2632
290	29306	0.0850	630	27656	0.1742	970	27425	0.2660
300	29155	0.0874	640	27643	0.1769	980	27422	0.2687
310	29022	0.0899	650	27631	0.1796	990 nm	27419	0.2714 nm
320	28902	0.0925	660	27620	0.1822			
330	28795	0.0950	670	27609	0.1849	1.00 µm	27416	0.000274 µm
340	28698	0.0975	680	27599	0.1876	1.05	27401	0.000288
350	28611	0.1001	690	27589	0.1903	1.10	27389	0.000301
360	28531	0.1027	700	27579	0.1930	1.15	27378	0.000315
370	28458	0.1053	710	27570	0.1957	1.20	27368	0.000328
380	28392	0.1079	720	27562	0.1984	1.25	27360	0.000342
390	28331	0.1105	730	27553	0.2011	1.30	27352	0.000355
400	28275	0.1131	740	27545	0.2038	1.35	27346	0.000369
410	28223	0.1157	750	27538	0.2065	1.40	27340	0.000383
420	28175	0.1183	760	27530	0.2092	1.45	27334	0.000396
430	28131	0.1209	770	27523	0.2119	1.50	27330	0.000410
440	28090	0.1236	780	27516	0.2146	1.55	27325	0.000423
450	28052	0.1262	790	27510	0.2173	1.60	27321	0.000437
460	28016	0.1288	800	27504	0.2200	1.65	27318	0.000451
470	27983	0.1315	810	27498	0.2227	1.70	27314	0.000464
480	27952	0.1341	820	27492	0.2254	1.75	27311	0.000478
490	27923	0.1368	830	27486	0.2281	1.80	27309	0.000491
500	27896	0.1394	840	27481	0.2308	1.85	27306	0.000505
510	27870	0.1421	850	27476	0.2335	1.90	27304	0.000519
520	27846	0.1448	860	27471	0.2362	1.95	27302	0.000532
530	27824	0.1474	870	27466	0.2389	2.00 µm	27300	0.000546 µm

CHARACTERISTICS OF LASER SOURCES

William F. Krupke

Light Amplification by Stimulated Emission of Radiation was first demonstrated by Maiman in 1960, the result of a population inversion produced between energy levels of chromium ions in a ruby crystal when irradiated with a xenon flashlamp. Since then population inversions and coherent emission have been generated in literally thousands of substances (neutral and ionized gases, liquids, and solids) using a variety of incoherent excitation techniques (optical pumping, electrical discharges, gas-dynamic flow, electron-beams, chemical reactions, nuclear decay).

The extrema of laser output parameters which have been demonstrated to date and the laser media used are summarized in Table 1. Note that the extreme power and energy parameters listed in this table were attained with laser *systems* rather than with simple laser oscillators.

Laser sources are commonly classified in terms of the state-of-matter of the active medium: gas, liquid, and solid. Each of these classes is further subdivided into one or more types as shown in Table 2. A well-known representative example of each type of laser is also given in Table 2 together with its nominal operation wavelength and the methods by which it is pumped.

The various lasers together cover a wide spectral range from the far ultraviolet to the far infrared. The particular wavelength of emission (usually a narrow line) is presented for some six dozen lasers in Figures 1A and 1B.

By suitably designing the excitation source and/or by controlling the laser resonator structure, laser systems can provide continuous or pulsed radiation as shown in Table 3.

Besides the method of excitation and the temporal behavior of a laser, there are many other parameters that characterize its operation and efficiency, as shown in Tables 4 and 5.

Although many lasers only emit in one or more narrow spectral "lines", an increasing number of lasers can be tuned by changing the composition or the pressure of the medium, or by varying the wavelength of the pump bands. The spectral regions in which these tunable lasers operate are presented in Figure 2.

REFERENCE

Krupke, W. F., in *Handbook of Laser Science and Technology,* Vol. I, Weber, M. J., Ed., CRC Press, Boca Raton, FL, 1986.

TABLE 1
Extrema of Output Parameters of Laser Devices or Systems

Parameter	Value	Laser medium
Peak power	1×10^{14} W (collimated)	Nd:glass
Peak power density	10^{18} W/cm^2 (focused)	Nd:glass
Pulse energy	$>10^5$ J	CO_2, Nd:glass
Average power	10^5 W	CO_2
Pulse duration	3×10^{-15} s continuous wave (cw)	Rh6G dye; various gases, liquids, solids
Wavelength	60 nm ↔ 385 μm	Many required
Efficiency (nonlaser pumped)	70%	CO
Beam quality	Diffraction limited	Various gases, liquids, solids
Spectral linewidth	20 Hz (for 10^{-1} s)	Neon-helium
Spatial coherence	10 m	Ruby

TABLE 2
Classes, Types, and Representative Examples of Laser Sources

Class	Type (characteristic)	Representative example	Nominal operating wavelength (nm)	Method(s) of excitation
Gas	Atom, neutral (electronic transition)	Neon-Helium (Ne-He)	633	Glow dischage
	Atom, ionic (electronic transition)	Argon (Ar^+)	488	Arc discharge
	Molecule, neutral (electronic transition)	Krypton fluoride (KrF)	248	Glow discharge; e-beam
	Molecule, neutral (vibrational transition)	Carbon dioxide (CO_2)	10600	Glow discharge; gas-dynamic flow
	Molecule, neutral (rotational transition)	Methyl fluoride (CH_3F)	496000	Laser pumping
	Molecule, ionic (electronic transition)	Nitrogen ion (N_2^+)	420	E-beam
Liquid	Organic solvent (dye-chromophore)	Rhodamine dye (Rh6G)	580–610	Flashlamp; laser pumping
	Organic solvent (rare earth chelate)	Europium:TTF	612	Flashlamp
	Inorganic solvent (trivalent rare earth ion)	Neodymium:$POCl_4$	1060	Flashlamp
Solid	Insulator, crystal (impurity)	Neodymium:YAG	1064	Flashlamp, arc lamp
	Insulator, crystal (stoichiometric)	Neodymium:UP(NdP_5O_{14})	1052	Flashlamp
	Insulator, crystal (color center)	F_2^-:LiF	1120	Laser pumping
	Insulator, amorphous (impurity)	Neodymium:glass	1061	Flashlamp
	Semiconductor (p-n junction)	GaAs	820	Injection current
	Semiconductor (electron-hole plasma)	GaAs	890	E-beam, laser pumping

Table 3
Temporal Characteristics of Lasers and Laser Systems

Form	Technique	Pulse width range (s)
Continuous wave	Excitation is continuous; resonator Q is held constant at some moderate value	∞
Pulsed	Excitation is pulsed; resonator Q is held constant at some moderate value	10^{-8}–10^{-3}
Q-Switched	Excitation is continuous or pulsed; resonator Q is switched from a very low value to a moderate value	10^{-8}–10^{-6}
Cavity dumped	Excitation is continuous or pulsed; resonator Q is switched from a very high value to a low value	10^{-7}–10^{-5}
Mode locked	Excitation is continuous or pulsed; phase or loss of the resonator modes is modulated at a rate related to the resonator transit time	10^{-12}–10^{-9}

Table 4
Properties and Performance of Some Continuous Wave (CW) Lasers

Parameter	Unit	Gas			Liquid	Solid	
		Neon helium	Argon ion	Carbon dioxide	Rhodamine 6G dye	Nd:YAG	GaAs
Excitation method		DC discharge	DC discharge	DC discharge	Ar^+ laser pump	Krypton arc lamp	DC injection
Gain medium composition		Neon:helium	Argon	CO_2:N_2:He	Rh 6G:H_2O	Nd:YAG	p:n:GaAs
Gain medium density	Torr	0.1:1.0	0.4	0.4:0.8:5.0			
	ions/cm^3				2(18):2(22)	1.5(20):2(22)	2(19):3(18):3(22)
Wavelength	nm	633	488	10600	590	1064	810
Laser cross-section	cm^{-2}	3(-13)	1.6(-12)	1.5(-16)	1.8(-16)	7(-19)	~6(-15)
Radiative lifetime (upper level)	s	~1(-7)	7.5(-9)	4(-3)	6.5(-9)	2.6(-4)	~1(-9)
Decay lifetime (upper level)	s	~1(-7)	~5.0(-9)	~4(-3)	6.0(-9)	2.3(-4)	~1(-9)
Gain bandwidth	nm	2(-3)	5(-3)	1.6(-2)	80	0.5	10
Type, gain saturation		Inhomogeneous	Inhomogeneous	Homogeneous	Homogeneous	Homogeneous	Homogeneous
Homogeneous saturation flux	W cm^{-2}			~20	3(5)	2.3(3)	~2(4)
Decay lifetime (lower level)	s	~1(-8)	~4(-10)	~5(-6)[b]	<1(-12)	<1(-7)	<1(-12)
Inversion density	cm^{-3}	~1(9)	2(10)	2(15)	2(16)	6(16)	1(16)
Small signal gain coefficient	cm^{-1}	~1(-3)	~3(-2)	1(-2)	4	5(-2)	40
Pump power density	W cm^{-3}	3	900	0.15	1(6)	150	7(7)
Output power density	W cm^{-3}	2.6(-3)	~1	2(-2)	3(5)	95	5(6)
Laser size (diameter:length)	cm:cm	0.5:100	0.3:100	5.0:600	1(-3):0.3	0.6:10	5(-4):7(-3);2(-2)[a]
Excitation current/voltage	A/V	3(-2):2(3)	30:300	0.1:1.5(4)		90:125	1.0/1.7
Excitation current density	A cm^{-2}	0.15	600	6(-3)		140	4.5(3)
Excitation power	W	60	9(3)	1.5(3)	4	1.1(4)	1.7
Output power	W	0.06	10	240	0.3	300	0.12
Efficiency	%	0.1	0.1	13	7	2.6	7

[a] Junction thickness:width:length.
[b] Pressure dependent.

Table 5
Properties and Performance of Some Pulsed Lasers

Parameter	Unit	Gas: Carbon dixoxide		Gas: Krypton fluoride		Liquid: Rhodamine 6G	Solid: Nd:YAG	Solid: Nd:glass
Excitation method		TEA-discharge	E-beam/sust.	Glow discharge	E-beam	Xenon flashlamp	Xenon flashlamp	Xenon flashlamp
Gain medium composition		CO_2:N_2:He	CO_2:N_2:He	He:Kr:F_2	Ar:Kr:F_2	Rh6G:alcohol	Nd:YAG	Nd:Glass
Gain medium density	torr	100:50:600	240:240:320	1070:70:3	1235:52:3			-
	ions/cm^3					1(18):1.5(22)	1.5(20):1(22)	3(20):2(22)
Wavelength	nm	10600	10600	249	249	590	1064	1061
Laser cross-section	cm^{-2}	2(-18)	2(-18)	2(-16)	2(-16)	1.8(-16)	7(-19)	2.8(-20)
Radiative lifetime (upper level)	s	4(-3)	4(-3)	7(-9)	7(-9)	6.5(-9)	2.6(-4)	4.1(-4)
Decay lifetime (upper level)	s	~1(-4)	5(-5)	2(-9)	3(-9)	6.0(-9)	2.3(-4)	3.7(-4)
Gain bandwidth	nm	1	1	2	2	80	0.5	26
Homogeneous saturation fluence	J/cm^2	0.2	0.2	4(-3)	4(-3)	2(-3)	0.6	~5
Decay lifetime (lower level)	s	5(-8)[a]	1(-8)[a]	<1(-12)	<1(-12)	<1(-12)	<1(-7)	<1(-8)
Inversion density	cm^{-3}	3(17)	6(17)	4(14)	2(14)	2(16)	4(17)	3(18)
Small signal gain coefficient	cm^{-1}	2(-2)	4(-2)	8–92)	4(-2)	4	0.3	8(-2)
Medium excitation energy density	J/cm^3	0.1	0.36	0.15	0.13	2.8	0.15	0.6
Output energy density	J/cm^3	2(-2)	1.8(-2)	1.5(-3)	1.2(-2)	0.85	5(-2)	2(-2)
Laser dimensions	cm:cm:cm	4.5:4.5:87	10:10:100	1.5:4.5:100	8.5:10:100	1.2ϕ25	0.6ϕ7.5	0.6ϕ8.3
Excitation current/voltage	A/V	6(4)/3.3(3)	2.4(4)/4(4)	2.5(4)/1.5(5)	1.2(4)/2.5(5)	2(5)/2.5(4)		
Excitation current density	A cm^2	8.5	22	170	11.5	2.6(3)		
Excitation peak power	W	2(8)	9(8)	4(9)	3(9)	5.4(9)	4(4)	9(4)
Output pulse energy	J	35	180	1	102	32	0.1	1.0
Output pulse length	s	1(-6)	4(-6)	2.5(-8)	6(-7)	3.2(-6)	2(-8)	1(-4)
Output pulse power	W	3.5(7)	4(7)	4(7)	2(8)	1(7)	5(6)	1(4)
Efficiency	%	17	5	1	10[b]	0.2	1.5	3.7

[a] Pressure dependent.

[b] Intrinsic efficiency ≡ energy output/energy deposited in gas.

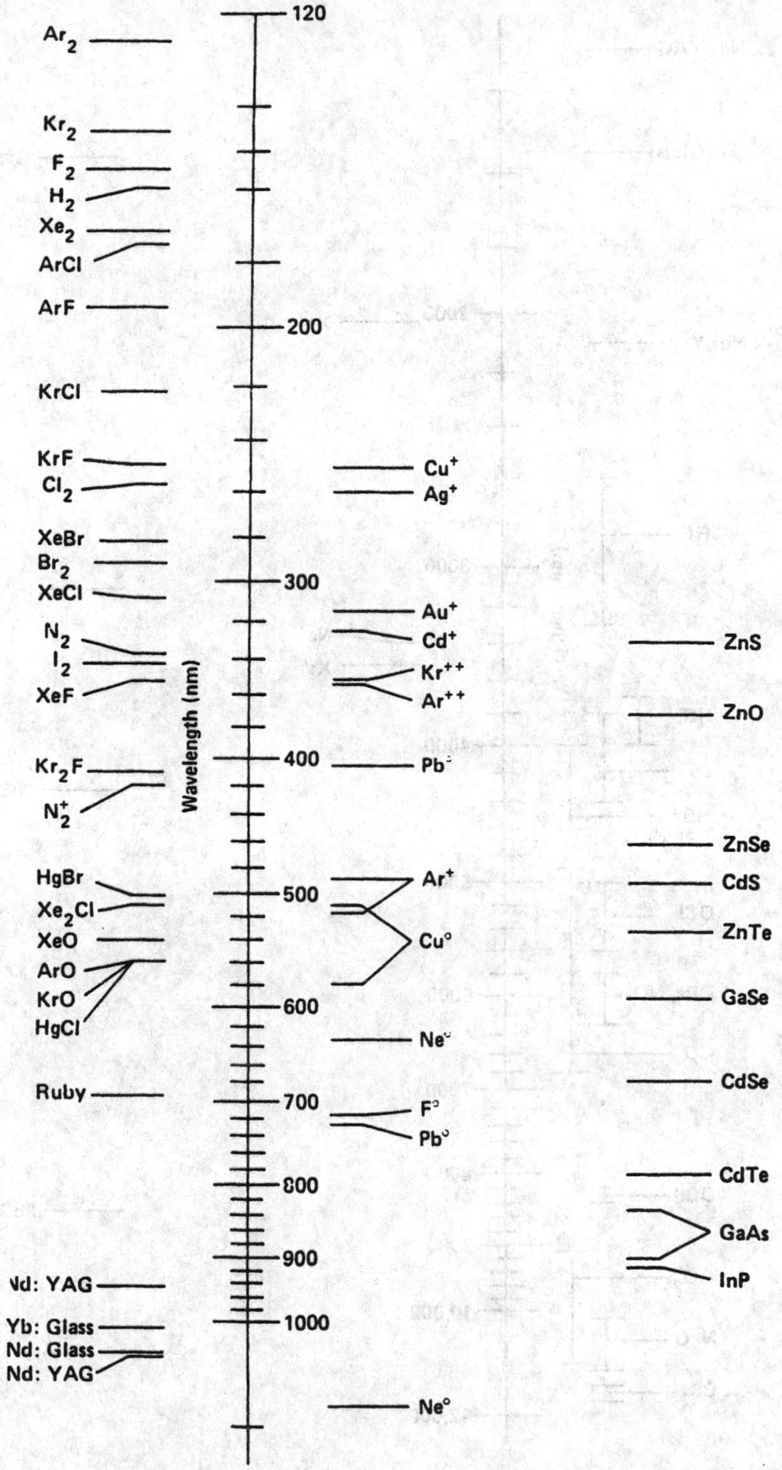

FIGURE 1A. Wavelengths of lasers operating in the 120 to 1200 nm spectral region.

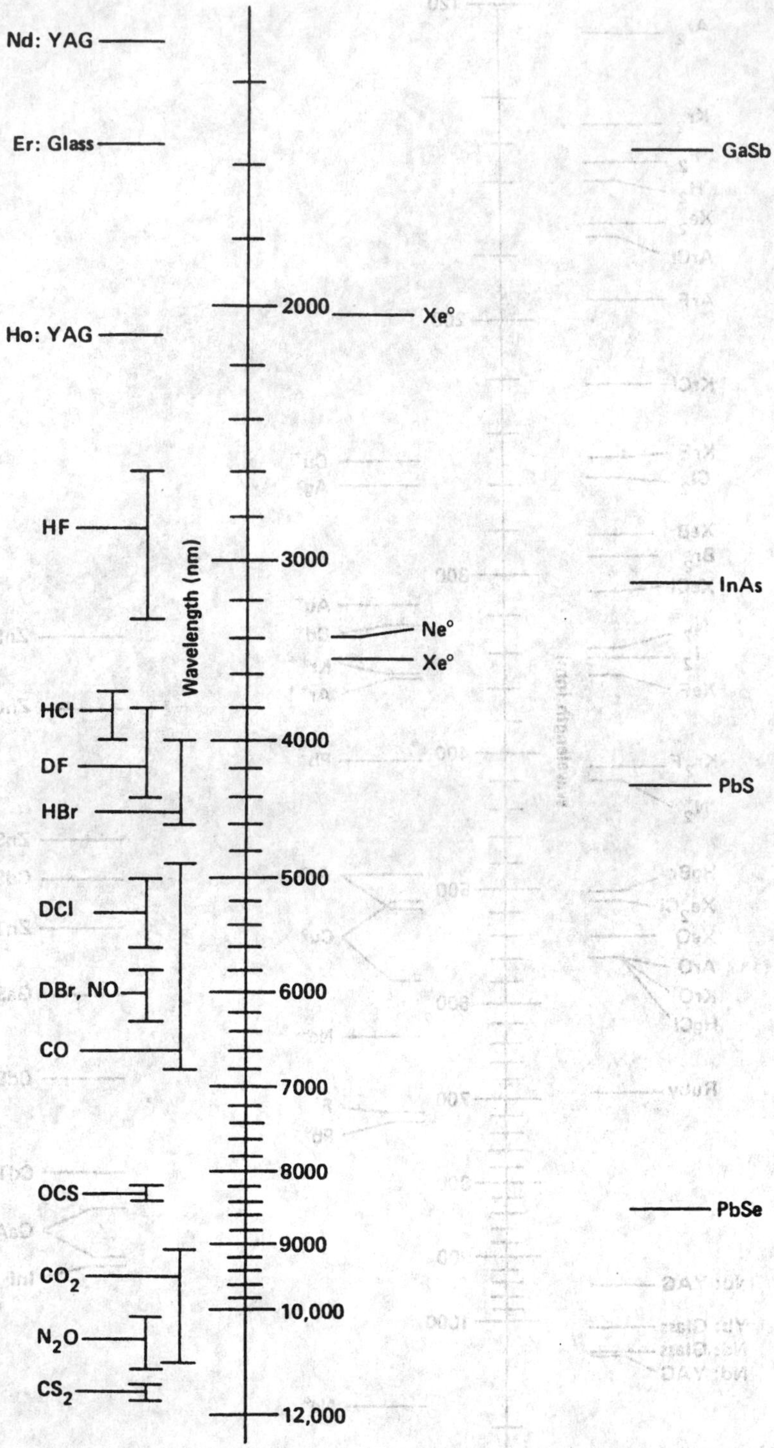

FIGURE 1B. Wavelength of lasers operating in the 1300 to 12,000 nm spectral region.

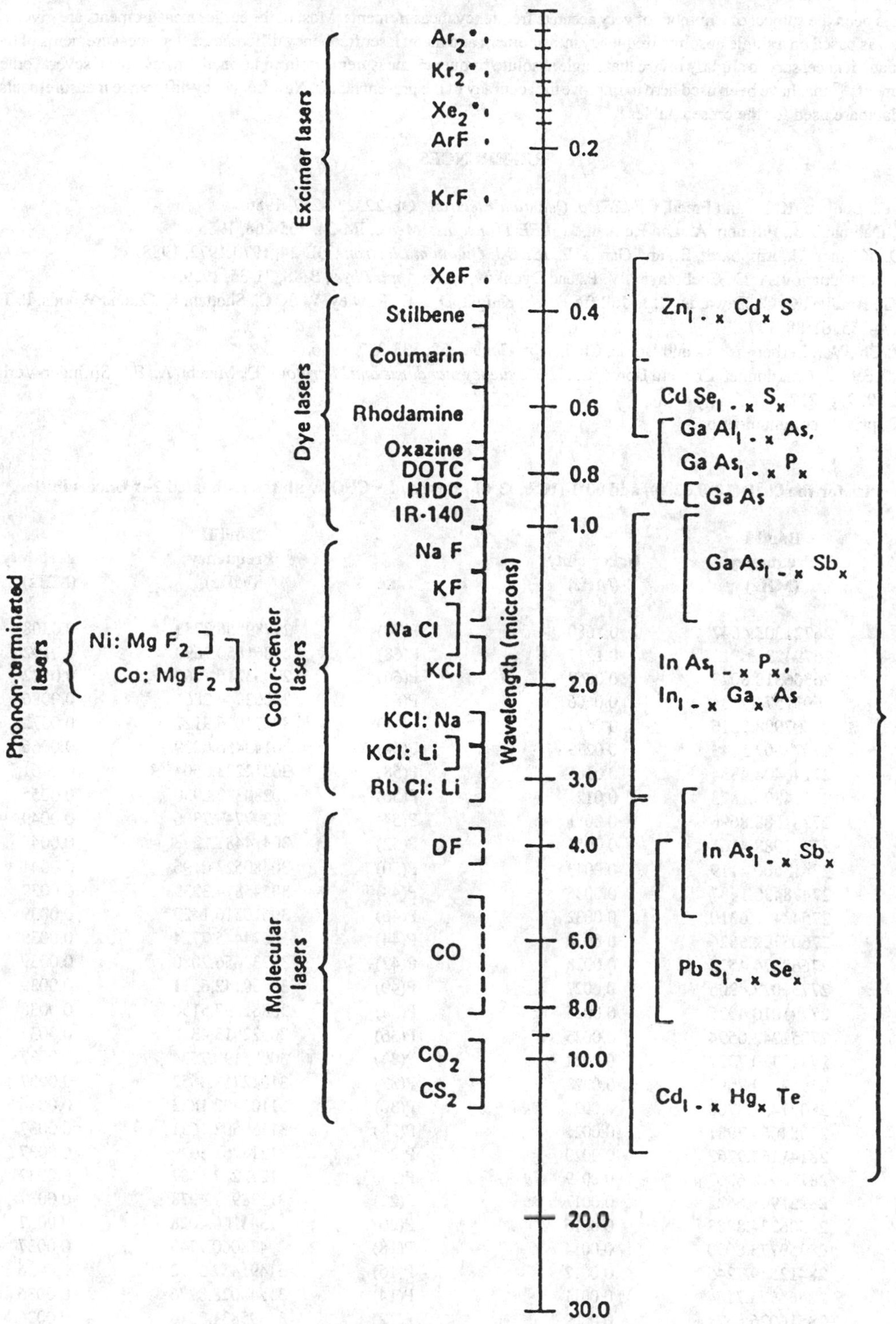

FIGURE 2. Spectral tuning ranges of various types of tunable lasers.

INFRARED LASER FREQUENCIES

Arthur Maki

The CO_2 laser has been the subject of a number of very accurate frequency measurements. Most of the earlier measurements are given by Bradley et al.[1] That analysis was based on a single absolute frequency measurement and many laser frequency differences. New measurements of the methane frequency[2-4] have made it necessary to slightly revise that single absolute frequency measurement. In addition, there have been several other absolute frequency measurements[5-7] that have been used here to improve the accuracy of the present tables. New frequency difference measurements have also been added to the database used for the present tables.[8]

REFERENCES

1. Bradley, L. C., Soohoo, K. L., and Freed, C., *IEEE J. Quantum Electron.*, QE-22, 234-267, 1986.
2. Clairon, A., Dahmani, B., Filimon, A., and Rutman, J., *IEEE Trans. Inst. Meas.*, IM-34, 265-268, 1985.
3. Weiss, C. O., Kramer, G., Lipphardt, B., and Garcia, E., *IEEE J. Quantum Electron.*, QE-24, 1970-1972, 1988.
4. Bagayev, S. N., Baklanov, A. E., Chebotayev, V. P., and Dychkov, A. S., *Appl. Phys.*, B-48, 31-35, 1989.
5. Blaney, T. G., Bradley, C. C., Edwards, G. J., Jolliffe, B. W., Knight, D. J. E., Rowley, W. R. C., Shotten, K. C., and Woods, P. T., *Proc. R. Soc. Lond.*, A-355, 61-88, 1977.
6. Chardonnet, Ch., Van Lerberghe, A., and Bordé, Ch. J., *Opt. Comm.*, 58, 333-337, 1986.
7. Clairon, A., Acef, O., Chardonnet, Ch., and Bordé, Ch. J., *Frequency Standards and Metrology*, De Marchi, A., Ed., Springer-Verlag, Berlin, Heidelberg, 1989, p. 212.
8. Evenson, K., private communication.

Frequencies for the 00^01-$(10^00,02^00)_I$ and 00^01-$(10^00,02^00)_{II}$ Bands of $^{12}C^{16}O_2$ with the Estimated 2-σ Uncertainties

Line	Band I Frequency (MHz)	Uncertainty (MHz)	Line	Band II Frequency (MHz)	Uncertainty (MHz)
P(70)	26721305.4647	0.1680	P(70)	29789856.3783	0.0308
P(68)	26794232.6712	0.1217	P(68)	29861850.7690	0.0192
P(66)	26866318.8073	0.0867	P(66)	29933216.1760	0.0122
P(64)	26937571.7234	0.0606	P(64)	30003944.2861	0.0086
P(62)	27007998.9216	0.0415	P(62)	30074026.9127	0.0072
P(60)	27077607.5643	0.0279	P(60)	30143456.0039	0.0066
P(58)	27146404.4834	0.0185	P(58)	30212223.6504	0.0061
P(56)	27214396.1873	0.0121	P(56)	30280322.0930	0.0055
P(54)	27281588.8696	0.0081	P(54)	30347743.7306	0.0049
P(52)	27347988.4161	0.0057	P(52)	30414481.1273	0.0044
P(50)	27413600.4119	0.0043	P(50)	30480527.0196	0.0041
P(48)	27478430.1487	0.0036	P(48)	30545874.3239	0.0039
P(46)	27542482.6310	0.0032	P(46)	30610516.1429	0.0039
P(44)	27605762.5826	0.0030	P(44)	30674445.7724	0.0039
P(42)	27668274.4525	0.0028	P(42)	30737656.7080	0.0039
P(40)	27730022.4206	0.0027	P(40)	30800142.6511	0.0039
P(38)	27791010.4036	0.0026	P(38)	30861897.5150	0.0038
P(36)	27851242.0594	0.0025	P(36)	30922915.4310	0.0037
P(34)	27910720.7927	0.0024	P(34)	30983190.7534	0.0037
P(32)	27969449.7593	0.0023	P(32)	31042718.0652	0.0037
P(30)	28027431.8708	0.0022	P(30)	31101492.1833	0.0036
P(28)	28084669.7981	0.0021	P(28)	31159508.1631	0.0037
P(26)	28141165.9762	0.0020	P(26)	31216761.3029	0.0037
P(24)	28196922.6067	0.0019	P(24)	31273247.1487	0.0037
P(22)	28251941.6622	0.0017	P(22)	31328961.4978	0.0037
P(20)	28306224.8888	0.0016	P(20)	31383900.4028	0.0037
P(18)	28359773.8090	0.0014	P(18)	31438060.1749	0.0037
P(16)	28412589.7245	0.0012	P(16)	31491437.3872	0.0036
P(14)	28464673.7184	0.0011	P(14)	31544028.8776	0.0036
P(12)	28516026.6574	0.0009	P(12)	31595831.7516	0.0036
P(10)	28566649.1935	0.0008	P(10)	31646843.3843	0.0035
P(8)	28616541.7661	0.0008	P(8)	31697061.4225	0.0035
P(6)	28665704.6027	0.0008	P(6)	31746483.7868	0.0035

Line	Band I Frequency (MHz)	Uncertainty (MHz)	Line	Band II Frequency (MHz)	Uncertainty (MHz)
P(4)	28714137.7205	0.0008	P(4)	31795108.6724	0.0035
P(2)	28761840.9272	0.0008	P(2)	31842934.5511	0.0035
R(0)	28832026.2198	0.0008	R(0)	31913172.5691	0.0035
R(2)	28877902.4382	0.0007	R(2)	31958996.0621	0.0034
R(4)	28923046.4303	0.0006	R(4)	32004017.3822	0.0034
R(6)	28967457.0657	0.0005	R(6)	32048236.2498	0.0034
R(8)	29011133.0054	0.0003	R(8)	32091652.6619	0.0034
R(10)	29054072.7010	0.0001	R(10)	32134266.8917	0.0034
R(12)	29096274.3935	0.0003	R(12)	32176079.4878	0.0034
R(14)	29137736.1129	0.0005	R(14)	32217091.2721	0.0035
R(16)	29178455.6759	0.0007	R(16)	32257303.3386	0.0036
R(18)	29218430.6852	0.0009	R(18)	32296717.0510	0.0037
R(20)	29257658.5269	0.0010	R(20)	32335334.0408	0.0038
R(22)	29296136.3689	0.0011	R(22)	32373156.2044	0.0039
R(24)	29333861.1583	0.0012	R(24)	32410185.7003	0.0041
R(26)	29370829.6191	0.0011	R(26)	32446424.9459	0.0042
R(28)	29407038.2491	0.0011	R(28)	32481876.6140	0.0042
R(30)	29442483.3168	0.0011	R(30)	32516543.6293	0.0042
R(32)	29477160.8582	0.0012	R(32)	32550429.1641	0.0042
R(34)	29511066.6733	0.0013	R(34)	32583536.6340	0.0042
R(36)	29544196.3221	0.0015	R(36)	32615869.6937	0.0041
R(38)	29576545.1205	0.0017	R(38)	32647432.2320	0.0040
R(40)	29608108.1360	0.0019	R(40)	32678228.3665	0.0039
R(42)	29638880.1831	0.0022	R(42)	32708262.4386	0.0038
R(44)	29668855.8183	0.0024	R(44)	32737539.0081	0.0039
R(46)	29698029.3350	0.0027	R(46)	32766062.8469	0.0041
R(48)	29726394.7582	0.0032	R(48)	32793838.9334	0.0045
R(50)	29753945.8385	0.0037	R(50)	32820872.4463	0.0055
R(52)	29780676.0464	0.0042	R(52)	32847168.7576	0.0071
R(54)	29806578.5659	0.0047	R(54)	32872733.4269	0.0099
R(56)	29831646.2878	0.0052	R(56)	32897572.1935	0.0141
R(58)	29855871.8032	0.0058	R(58)	32921690.9701	0.0202
R(60)	29879247.3960	0.0074	R(60)	32945095.8355	0.0288
R(62)	29901765.0357	0.0113	R(62)	32967793.0268	0.0407
R(64)	29923416.3695	0.0186	R(64)	32989788.9322	0.0567
R(66)	29944192.7145	0.0302	R(66)	33011090.0831	0.0780
R(68)	29964085.0488	0.0475	R(68)	33031703.1467	0.1060
R(70)	29983084.0036	0.0720	R(70)	33051634.9172	0.1423

Frequencies for the 00°1-(10°0,02°0)$_I$ and 00°1-(10°0,02°0)$_{II}$ Bands of $^{13}C^{16}O_2$ with the Estimated 2-σ Uncertainties

Line	Band I Frequency (MHz)	Uncertainty (MHz)	Line	Band II Frequency (MHz)	Uncertainty (MHz)
P(66)	25523832.1808	0.7836	P(66)	28512082.5283	1.2894
P(64)	25590013.4703	0.5415	P(64)	28585121.9396	0.9194
P(62)	25655543.6502	0.3629	P(62)	28657449.4180	0.6420
P(60)	25720428.2487	0.2339	P(60)	28729056.6374	0.4373
P(58)	25784672.4840	0.1430	P(58)	28799935.4147	0.2897
P(56)	25848281.2771	0.0810	P(56)	28870077.7187	0.1853
P(54)	25911259.2627	0.0405	P(54)	28939475.6771	0.1135
P(52)	25973610.8005	0.0157	P(52)	29008121.5846	0.0659
P(50)	26035339.9857	0.0045	P(50)	29076007.9109	0.0357
P(48)	26096450.6582	0.0079	P(48)	29143127.3077	0.0180
P(46)	26156946.4123	0.0101	P(46)	29209472.6164	0.0090

Line	Band I Frequency (MHz)	Uncertainty (MHz)	Line	Band II Frequency (MHz)	Uncertainty (MHz)
P(44)	26216830.6053	0.0101	P(44)	29275036.8754	0.0058
P(42)	26276106.3655	0.0090	P(42)	29339813.3270	0.0050
P(40)	26334776.6003	0.0077	P(40)	29403795.4243	0.0044
P(38)	26392844.0030	0.0068	P(38)	29466976.8383	0.0037
P(36)	26450311.0599	0.0063	P(36)	29529351.4635	0.0032
P(34)	26507180.0565	0.0061	P(34)	29590913.4252	0.0029
P(32)	26563453.0836	0.0060	P(32)	29651657.0844	0.0028
P(30)	26619132.0428	0.0058	P(30)	29711577.0447	0.0028
P(28)	26674218.6515	0.0055	P(28)	29770668.1566	0.0031
P(26)	26728714.4479	0.0054	P(26)	29828925.5239	0.0035
P(24)	26782620.7952	0.0054	P(24)	29886344.5074	0.0041
P(22)	26835938.8858	0.0054	P(22)	29942920.7308	0.0046
P(20)	26888669.7451	0.0055	P(20)	29998650.0838	0.0051
P(18)	26940814.2347	0.0055	P(18)	30053528.7271	0.0054
P(16)	26992373.0555	0.0055	P(16)	30107553.0955	0.0055
P(14)	27043346.7508	0.0054	P(14)	30160719.9016	0.0055
P(12)	27093735.7083	0.0052	P(12)	30213026.1388	0.0054
P(10)	27143540.1624	0.0051	P(10)	30264469.0839	0.0054
P(8)	27192760.1962	0.0049	P(8)	30315046.2994	0.0054
P(6)	27241395.7431	0.0048	P(6)	30364755.6359	0.0055
P(4)	27289446.5880	0.0047	P(4)	30413595.2335	0.0056
P(2)	27336912.3682	0.0046	P(2)	30461563.5231	0.0057
R(0)	27407012.8882	0.0045	P(0)	30531879.5415	0.0057
R(2)	27453013.4589	0.0043	P(2)	30577664.6138	0.0056
R(4)	27498426.5430	0.0040	P(4)	30622575.1885	0.0054
R(6)	27543251.1200	0.0037	P(6)	30666611.0128	0.0051
R(8)	27587486.0225	0.0034	P(8)	30709772.1257	0.0047
R(10)	27631129.9356	0.0031	P(10)	30752058.8571	0.0045
R(12)	27674181.3963	0.0029	P(12)	30793471.8269	0.0044
R(14)	27716638.7917	0.0029	P(14)	30834011.9425	0.0043
R(16)	27758500.3577	0.0029	P(16)	30873680.3976	0.0044
R(18)	27799764.1770	0.0029	P(18)	30912478.6694	0.0044
R(20)	27840428.1773	0.0030	P(20)	30950408.5159	0.0044
R(22)	27880490.1283	0.0029	P(22)	30987471.9732	0.0043
R(24)	27919947.6395	0.0029	P(24)	31023671.3517	0.0042
R(26)	27958798.1567	0.0028	P(26)	31059009.2327	0.0042
R(28)	27997038.9591	0.0028	P(28)	31093488.4642	0.0042
R(30)	28034667.1551	0.0027	P(30)	31127112.1569	0.0043
R(32)	28071679.6785	0.0027	P(32)	31159883.6793	0.0045
R(34)	28108073.2842	0.0026	P(34)	31191806.6529	0.0046
R(36)	28143844.5432	0.0026	P(36)	31222884.9469	0.0048
R(38)	28178989.8377	0.0026	P(38)	31253122.6730	0.0053
R(40)	28213505.3554	0.0028	P(40)	31282524.1795	0.0061
R(42)	28247387.0838	0.0033	P(42)	31311094.0452	0.0077
R(44)	28280630.8035	0.0046	P(44)	31338837.0736	0.0108
R(46)	28313232.0818	0.0083	P(46)	31365758.2858	0.0173
R(48)	28345186.2652	0.0161	P(48)	31391862.9147	0.0295
R(50)	28376488.4720	0.0301	P(50)	31417156.3972	0.0505
R(52)	28407133.5839	0.0531	P(52)	31441644.3679	0.0845
R(54)	28437116.2372	0.0887	P(54)	31465332.6516	0.1366
R(56)	28466430.8141	0.1419	P(56)	31488227.2557	0.2138
R(58)	28495071.4324	0.2188	P(58)	31510334.3631	0.3247
R(60)	28523031.9357	0.3271	P(60)	31531660.3243	0.4800
R(62)	28550305.8819	0.4763	P(62)	31552211.6497	0.6932
R(64)	28576886.5323	0.6781	P(64)	31571995.0017	0.9805
R(66)	28602766.8393	0.9467	P(66)	31591017.1868	1.3619

INFRARED LASER FREQUENCIES (continued)

Frequencies for the 00°1-(10°0,02°0)$_I$ and 00°1-(10°0,02°0)$_{II}$ Bands of $^{12}C^{18}O_2$ with the Estimated 2-σ Uncertainties

Line	Band I Frequency (MHz)	Uncertainty (MHz)	Line	Band II Frequency (MHz)	Uncertainty (MHz)
P(70)	27045326.3119	0.4540	P(70)	30695237.5856	0.0858
P(68)	27114914.0922	0.3324	P(68)	30755520.2231	0.0570
P(66)	27183635.7945	0.2392	P(66)	30815311.4928	0.0364
P(64)	27251496.4118	0.1688	P(64)	30874607.2084	0.0223
P(62)	27318500.7361	0.1165	P(62)	30933403.2309	0.0131
P(60)	27384653.3618	0.0783	P(60)	30991695.4724	0.0075
P(58)	27449958.6881	0.0510	P(58)	31049479.9009	0.0049
P(56)	27514420.9224	0.0319	P(56)	31106752.5446	0.0041
P(54)	27578044.0828	0.0191	P(54)	31163509.4964	0.0040
P(52)	27640832.0010	0.0108	P(52)	31219746.9183	0.0040
P(50)	27702788.3248	0.0059	P(50)	31275461.0455	0.0039
P(48)	27763916.5206	0.0035	P(48)	31330648.1908	0.0039
P(46)	27824219.8762	0.0028	P(46)	31385304.7490	0.0039
P(44)	27883701.5029	0.0026	P(44)	31439427.2006	0.0039
P(42)	27942364.3379	0.0025	P(42)	31493012.1163	0.0038
P(40)	28000211.1464	0.0024	P(40)	31546056.1605	0.0038
P(38)	28057244.5242	0.0022	P(38)	31598556.0954	0.0037
P(36)	28113466.8992	0.0021	P(36)	31650508.7847	0.0037
P(34)	28168880.5335	0.0020	P(34)	31701911.1970	0.0037
P(32)	28223487.5256	0.0019	P(32)	31752760.4093	0.0037
P(30)	28277289.8118	0.0017	P(30)	31803053.6105	0.0037
P(28)	28330289.1679	0.0016	P(28)	31852788.1043	0.0038
P(26)	28382487.2111	0.0015	P(26)	31901961.3125	0.0038
P(24)	28433885.4012	0.0013	P(24)	31950570.7773	0.0038
P(22)	28484485.0420	0.0012	P(22)	31998614.1649	0.0038
P(20)	28534287.2828	0.0011	P(20)	32046089.2669	0.0037
P(18)	28583293.1193	0.0010	P(18)	32092994.0036	0.0037
P(16)	28631503.3952	0.0010	P(16)	32139326.4254	0.0036
P(14)	28678918.8025	0.0009	P(14)	32185084.7154	0.0036
P(12)	28725539.8830	0.0010	P(12)	32230267.1907	0.0036
P(10)	28771367.0288	0.0010	P(10)	32274872.3041	0.0037
P(8)	28816400.4829	0.0010	P(8)	32318898.6455	0.0038
P(6)	28860640.3403	0.0011	P(6)	32362344.9434	0.0039
P(4)	28904086.5477	0.0011	P(4)	32405210.0652	0.0041
P(2)	28946738.9048	0.0011	P(2)	32447493.0185	0.0041
R(0)	29009228.1702	0.0010	P(0)	32509824.0580	0.0042
R(2)	29049894.0586	0.0010	P(2)	32550648.1723	0.0042
R(4)	29089764.2368	0.0009	P(4)	32590887.7542	0.0042
R(6)	29128837.8426	0.0008	P(6)	32630542.4457	0.0041
R(8)	29167113.8668	0.0008	P(8)	32669612.0295	0.0041
R(10)	29204591.1529	0.0009	P(10)	32708096.4282	0.0040
R(12)	29241268.3964	0.0010	P(12)	32745995.7040	0.0040
R(14)	29277144.1444	0.0011	P(14)	32783310.0573	0.0040
R(16)	29312216.7955	0.0012	P(16)	32820039.8258	0.0040
R(18)	29346484.5984	0.0012	P(18)	32856185.4827	0.0040
R(20)	29379945.6517	0.0013	P(20)	32891747.6358	0.0040
R(22)	29412597.9024	0.0013	P(22)	32926727.0254	0.0040
R(24)	29444439.1458	0.0013	P(24)	32961124.5220	0.0040
R(26)	29475467.0236	0.0014	P(26)	32994941.1249	0.0040
R(28)	29505679.0230	0.0015	P(28)	33028177.9594	0.0040
R(30)	29535072.4755	0.0016	P(30)	33060836.2743	0.0040
R(32)	29563644.5557	0.0018	P(32)	33092917.4394	0.0041
R(34)	29591392.2794	0.0020	P(34)	33124422.9429	0.0043
R(36)	29618312.5023	0.0023	P(36)	33155354.3878	0.0046
R(38)	29644401.9182	0.0028	P(38)	33185713.4894	0.0049

Line	Band I Frequency (MHz)	Uncertainty (MHz)	Line	Band II Frequency (MHz)	Uncertainty (MHz)
R(40)	29669657.0575	0.0036	P(40)	33215502.0716	0.0056
R(42)	29694074.2853	0.0053	P(42)	33244722.0637	0.0068
R(44)	29717649.7992	0.0082	P(44)	33273375.4969	0.0092
R(46)	29740379.6276	0.0128	P(46)	33301464.5003	0.0134
R(48)	29762259.6274	0.0200	P(48)	33328991.2976	0.0199
R(50)	29783285.4820	0.0307	P(50)	33355958.2027	0.0294
R(52)	29803452.6988	0.0461	P(52)	33382367.6161	0.0427
R(54)	29822756.6072	0.0681	P(54)	33408222.0209	0.0607
R(56)	29841192.3558	0.0985	P(56)	33433523.9780	0.0848
R(58)	29858754.9100	0.1401	P(58)	33458276.1228	0.1165
R(60)	29875439.0495	0.1960	P(60)	33482481.1601	0.1576
R(62)	29891239.3658	0.2702	P(62)	33506141.8605	0.2104
R(64)	29906150.2589	0.3673	P(64)	33529261.0556	0.2775
R(66)	29920165.9352	0.4930	P(66)	33551841.6335	0.3621
R(68)	29933280.4042	0.6540	P(68)	33573886.5352	0.4679
R(70)	29945487.4756	0.8581	P(70)	33595398.7493	0.5992

Frequencies for the 00°1-(10°0,02°0)$_I$ and 00°1-(10°0,02°0)$_{II}$ Bands of $^{13}C^{18}O_2$ with the Estimated 2-σ Uncertainties

Line	Band I Frequency (MHz)	Uncertainty (MHz)	Line	Band II Frequency (MHz)	Uncertainty (MHz)
P(70)	25967863.7652	1.1146	P(70)	28960476.2278	0.4069
P(68)	26033448.2798	0.8152	P(68)	29022326.9578	0.2861
P(66)	26098273.9159	0.5860	P(66)	29083661.3546	0.1961
P(64)	26162346.4813	0.4129	P(64)	29144473.5795	0.1303
P(62)	26225671.5466	0.2844	P(62)	29204757.8761	0.0833
P(60)	26288254.4494	0.1906	P(60)	29264508.5768	0.0507
P(58)	26350100.2984	0.1237	P(58)	29323720.1086	0.0290
P(56)	26411213.9778	0.0772	P(56)	29382386.9988	0.0152
P(54)	26471600.1504	0.0459	P(54)	29440503.8809	0.0073
P(52)	26531263.2618	0.0258	P(52)	29498065.4997	0.0038
P(50)	26590207.5442	0.0138	P(50)	29555066.7172	0.0032
P(48)	26648437.0195	0.0077	P(48)	29611502.5178	0.0031
P(46)	26705955.5026	0.0057	P(46)	29667368.0132	0.0031
P(44)	26762766.6051	0.0055	P(44)	29722658.4475	0.0034
P(42)	26818873.7378	0.0055	P(42)	29777369.2022	0.0039
P(40)	26874280.1143	0.0056	P(40)	29831495.8006	0.0044
P(38)	26928988.7531	0.0056	P(38)	29885033.9125	0.0049
P(36)	26983002.4809	0.0056	P(36)	29937979.3584	0.0053
P(34)	27036323.9351	0.0055	P(34)	29990328.1139	0.0054
P(32)	27088955.5657	0.0054	P(32)	30042076.3132	0.0055
P(30)	27140899.6384	0.0051	P(30)	30093220.2534	0.0055
P(28)	27192158.2363	0.0049	P(28)	30143756.3978	0.0054
P(26)	27242733.2620	0.0047	P(26)	30193681.3793	0.0053
P(24)	27292626.4396	0.0044	P(24)	30242992.0038	0.0052
P(22)	27341839.3165	0.0042	P(22)	30291685.2529	0.0051
P(20)	27390373.2651	0.0040	P(20)	30339758.2870	0.0049
P(18)	27438229.4843	0.0037	P(18)	30387208.4477	0.0048
P(16)	27485409.0008	0.0035	P(16)	30434033.2603	0.0046
P(14)	27531912.6704	0.0033	P(14)	30480230.4356	0.0045
P(12)	27577741.1795	0.0031	P(12)	30525797.8725	0.0044
P(10)	27622895.0455	0.0031	P(10)	30570733.6593	0.0043
P(8)	27667374.6182	0.0031	P(8)	30615036.0750	0.0043
P(6)	27711180.0803	0.0033	P(6)	30658703.5912	0.0044
P(4)	27754311.4480	0.0034	P(4)	30701734.8727	0.0045

Line	Band I Frequency (MHz)	Uncertainty (MHz)	Line	Band II Frequency (MHz)	Uncertainty (MHz)
P(2)	27796768.5718	0.0036	P(2)	30744128.7785	0.0045
R(0)	27859189.3155	0.0036	P(0)	30806522.5414	0.0045
R(2)	27899959.0889	0.0035	P(2)	30847319.2956	0.0044
R(4)	27940052.7921	0.0033	P(4)	30887476.2168	0.0043
R(6)	27979469.5315	0.0031	P(6)	30926993.0424	0.0042
R(8)	28018208.2478	0.0028	P(8)	30965869.7046	0.0041
R(10)	28056267.7161	0.0026	P(10)	31004106.3298	0.0040
R(12)	28093646.5448	0.0025	P(12)	31041703.2379	0.0040
R(14)	28130343.1757	0.0025	P(14)	31078660.9408	0.0040
R(16)	28166355.8825	0.0025	P(16)	31114980.1420	0.0040
R(18)	28201682.7706	0.0025	P(18)	31150661.7340	0.0041
R(20)	28236321.7757	0.0025	P(20)	31185706.7976	0.0042
R(22)	28270270.6628	0.0024	P(22)	31220116.5992	0.0043
R(24)	28303527.0249	0.0024	P(24)	31253892.5891	0.0043
R(26)	28336088.2817	0.0023	P(26)	31287036.3991	0.0044
R(28)	28367951.6781	0.0024	P(28)	31319549.8396	0.0043
R(30)	28399114.2823	0.0025	P(30)	31351434.8973	0.0043
R(32)	28429572.9843	0.0026	P(32)	31382693.7318	0.0042
R(34)	28459324.4940	0.0028	P(34)	31413328.6728	0.0042
R(36)	28488365.3390	0.0029	P(36)	31443342.2165	0.0041
R(38)	28516691.8625	0.0029	P(38)	31472737.0219	0.0040
R(40)	28544300.2211	0.0031	P(40)	31501515.9074	0.0039
R(42)	28571186.3823	0.0032	P(42)	31529681.8467	0.0040
R(44)	28597346.1222	0.0032	P(44)	31557237.9646	0.0042
R(46)	28622775.0223	0.0038	P(46)	31584187.5329	0.0046
R(48)	28647468.4672	0.0071	P(48)	31610533.9656	0.0057
R(50)	28671421.6417	0.0148	P(50)	31636280.8146	0.0088
R(52)	28694629.5272	0.0286	P(52)	31661431.7650	0.0151
R(54)	28717086.8993	0.0510	P(54)	31685990.6298	0.0261
R(56)	28738788.3239	0.0852	P(56)	31709961.3449	0.0434
R(58)	28759728.1540	0.1355	P(58)	31733347.9642	0.0693
R(60)	28779900.5263	0.2075	P(60)	31756154.6537	0.1068
R(62)	28799299.3572	0.3078	P(62)	31778385.6867	0.1594
R(64)	28817918.3393	0.4447	P(64)	31800045.4375	0.2317
R(66)	28835750.9374	0.6283	P(66)	31821138.3761	0.3291
R(68)	28852790.3843	0.8707	P(68)	31841669.0622	0.4581
R(70)	28869029.6768	1.1863	P(70)	31861642.1394	0.6268

Frequencies for the $01^{1e}1$-$(11^{1e}0,03^{1e}0)_I$ and $01^{1e}1$-$(11^{1e}0,03^{1e}0)_{II}$ Bands of $^{12}C^{16}O_2$ with the Estimated 2-σ Uncertainties

Line	Band I Frequency (MHz)	Uncertainty (MHz)	Line	Band II Frequency (MHz)	Uncertainty (MHz)
P(59)	26125213.2723	1.6633	P(59)	30427055.2899	0.1962
P(57)	26191576.6703	1.0880	P(57)	30494640.3229	0.1332
P(55)	26257240.7898	0.6844	P(55)	30561557.5929	0.0865
P(53)	26322208.2302	0.4094	P(53)	30627802.0344	0.0530
P(51)	26386481.4313	0.2286	P(51)	30693368.7014	0.0306
P(49)	26450062.6783	0.1155	P(49)	30758252.7710	0.0175
P(47)	26512954.1076	0.0498	P(47)	30822449.5469	0.0123
P(45)	26575157.7109	0.0191	P(45)	30885954.4624	0.0114
P(43)	26636675.3402	0.0160	P(43)	30948763.0834	0.0109
P(41)	26697508.7115	0.0182	P(41)	31010871.1119	0.0100
P(39)	26757659.4084	0.0177	P(39)	31072274.3882	0.0091
P(37)	26817128.8857	0.0160	P(37)	31132968.8940	0.0091
P(35)	26875918.4726	0.0144	P(35)	31192950.7549	0.0102

Line	Band I Frequency (MHz)	Uncertainty (MHz)	Line	Band II Frequency (MHz)	Uncertainty (MHz)
P(33)	26934029.3751	0.0131	P(33)	31252216.2430	0.0118
P(31)	26991462.6787	0.0119	P(31)	31310761.7788	0.0134
P(29)	27048219.3509	0.0106	P(29)	31368583.9339	0.0147
P(27)	27104300.2431	0.0096	P(27)	31425679.4328	0.0155
P(25)	27159706.0925	0.0093	P(25)	31482045.1550	0.0157
P(23)	27214437.5237	0.0097	P(23)	31537678.1367	0.0154
P(21)	27268495.0505	0.0104	P(21)	31592575.5725	0.0147
P(19)	27321879.0769	0.0108	P(19)	31646734.8172	0.0137
P(17)	27374589.8987	0.0108	P(17)	31700153.3868	0.0127
P(15)	27426627.7040	0.0104	P(15)	31752828.9602	0.0119
P(13)	27477992.5747	0.0098	P(13)	31804759.3803	0.0113
P(11)	27528684.4867	0.0096	P(11)	31855942.6551	0.0113
P(9)	27578703.3113	0.0101	P(9)	31906376.9582	0.0116
P(7)	27628048.8151	0.0113	P(7)	31956060.6304	0.0122
P(5)	27676720.6609	0.0127	P(5)	32004992.1796	0.0129
P(3)	27724718.4080	0.0141	P(3)	32053170.2819	0.0136
R(1)	27841759.7696	0.0152	P(1)	32170312.0391	0.0149
R(3)	27887393.2105	0.0146	P(3)	32215845.0845	0.0151
R(5)	27932349.2934	0.0135	P(5)	32260620.8121	0.0152
R(7)	27976627.0108	0.0124	P(7)	32304638.8261	0.0152
R(9)	28020225.2521	0.0115	P(9)	32347898.8990	0.0150
R(11)	28063142.8031	0.0110	P(11)	32390400.9714	0.0148
R(13)	28105378.3457	0.0109	P(13)	32432145.1513	0.0145
R(15)	28146930.4576	0.0109	P(15)	32473131.7137	0.0142
R(17)	28187797.6116	0.0107	P(17)	32513361.0997	0.0140
R(19)	28227978.1750	0.0103	P(19)	32552833.9153	0.0140
R(21)	28267470.4088	0.0099	P(21)	32591550.9309	0.0141
R(23)	28306272.4666	0.0099	P(23)	32629513.0796	0.0143
R(25)	28344382.3939	0.0107	P(25)	32666721.4564	0.0144
R(27)	28381798.1267	0.0122	P(27)	32703177.3164	0.0142
R(29)	28418517.4902	0.0141	P(29)	32738882.0732	0.0136
R(31)	28454538.1976	0.0165	P(31)	32773837.2976	0.0136
R(33)	28489857.8477	0.0213	P(33)	32808044.7156	0.0174
R(35)	28524473.9240	0.0312	P(35)	32841506.2063	0.0279
R(37)	28558383.7917	0.0486	P(37)	32874223.8000	0.0462
R(39)	28591584.6963	0.0754	P(39)	32906199.6761	0.0735
R(41)	28624073.7602	0.1131	P(41)	32937436.1606	0.1114
R(43)	28655847.9806	0.1644	P(43)	32967935.7238	0.1624
R(45)	28686904.2261	0.2328	P(45)	32997700.9775	0.2292
R(47)	28717239.2334	0.3239	P(47)	33026734.6728	0.3151
R(49)	28746849.6038	0.4465	P(49)	33055039.6965	0.4238
R(51)	28775731.7988	0.6142	P(51)	33082619.0689	0.5595
R(53)	28803882.1361	0.8465	P(53)	33109475.9403	0.7272

Frequencies for the $01^{1f}1$-$(11^{1f}0,03^{1f}0)_I$ and $01^{1f}1$-$(11^{1f}0,03^{1f}0)_{II}$ Bands of $^{12}C^{16}O_2$ with the Estimated 2-σ Uncertainties

Line	Band I Frequency (MHz)	Uncertainty (MHz)	Line	Band II Frequency (MHz)	Uncertainty (MHz)
P(60)	26051570.0104	4.4521	P(60)	30355115.0204	0.2752
P(58)	26120964.4932	3.0629	P(58)	30425283.5969	0.1926
P(56)	26189552.8496	2.0516	P(56)	30494732.8293	0.1301
P(54)	26257339.6006	1.3305	P(54)	30563455.6325	0.0840
P(52)	26324329.0344	0.8289	P(52)	30631445.1076	0.0512
P(50)	26390525.2136	0.4901	P(50)	30698694.5456	0.0292
P(48)	26455931.9824	0.2698	P(48)	30765197.4310	0.0163

INFRARED LASER FREQUENCIES (continued)

Line	Band I Frequency (MHz)	Uncertainty (MHz)	Line	Band II Frequency (MHz)	Uncertainty (MHz)
P(46)	26520552.9722	0.1334	P(46)	30830947.4444	0.0111
P(44)	26584391.6075	0.0551	P(44)	30895938.4662	0.0104
P(42)	26647451.1105	0.0181	P(42)	30960164.5794	0.0105
P(40)	26709734.5057	0.0151	P(40)	31023620.0723	0.0105
P(38)	26771244.6242	0.0174	P(38)	31086299.4415	0.0107
P(36)	26831984.1067	0.0157	P(36)	31148197.3941	0.0114
P(34)	26891955.4069	0.0126	P(34)	31209308.8510	0.0126
P(32)	26951160.7945	0.0105	P(32)	31269628.9481	0.0138
P(30)	27009602.3576	0.0096	P(30)	31329153.0395	0.0147
P(28)	27067282.0045	0.0092	P(28)	31387876.6994	0.0151
P(26)	27124201.4662	0.0090	P(26)	31445795.7236	0.0149
P(24)	27180362.2977	0.0089	P(24)	31502906.1318	0.0141
P(22)	27235765.8792	0.0090	P(22)	31559204.1695	0.0128
P(20)	27290413.4182	0.0093	P(20)	31614686.3091	0.0113
P(18)	27344305.9494	0.0096	P(18)	31669349.2515	0.0098
P(16)	27397444.3368	0.0097	P(16)	31723189.9280	0.0086
P(14)	27449829.2733	0.0096	P(14)	31776205.5007	0.0081
P(12)	27501461.2824	0.0096	P(12)	31828393.3642	0.0085
P(10)	27552340.7179	0.0101	P(10)	31879751.1463	0.0095
P(8)	27602467.7649	0.0111	P(8)	31930276.7092	0.0107
P(6)	27651842.4399	0.0125	P(6)	31979968.1497	0.0120
P(4)	27700464.5912	0.0139	P(4)	32028823.8002	0.0131
P(2)	27748333.8988	0.0148	P(2)	32076842.2290	0.0139
R(2)	27864709.8633	0.0146	P(2)	32193218.1935	0.0150
R(4)	27909939.2762	0.0135	P(4)	32238298.4853	0.0151
R(6)	27954412.3294	0.0122	P(6)	32282538.0393	0.0153
R(8)	27998127.7801	0.0112	P(8)	32325936.7244	0.0153
R(10)	28041084.2173	0.0107	P(10)	32368494.6458	0.0154
R(12)	28083280.0620	0.0108	P(12)	32410212.1438	0.0155
R(14)	28124713.5668	0.0110	P(14)	32451089.7941	0.0155
R(16)	28165382.8151	0.0112	P(16)	32491128.4063	0.0154
R(18)	28205285.7213	0.0111	P(18)	32530329.0234	0.0152
R(20)	28244420.0302	0.0110	P(20)	32568692.9211	0.0149
R(22)	28282783.3158	0.0114	P(22)	32606221.6061	0.0147
R(24)	28320372.9812	0.0129	P(24)	32642916.8154	0.0144
R(26)	28357186.2574	0.0149	P(26)	32678780.5147	0.0140
R(28)	28393220.2023	0.0168	P(28)	32713814.8971	0.0133
R(30)	28428471.6994	0.0175	P(30)	32748022.3813	0.0120
R(32)	28462937.4565	0.0165	P(32)	32781405.6101	0.0106
R(34)	28496614.0042	0.0142	P(34)	32813967.4482	0.0122
R(36)	28529497.6934	0.0163	P(36)	32845710.9809	0.0212
R(38)	28561584.6939	0.0309	P(38)	32876639.5111	0.0385
R(40)	28592870.9914	0.0593	P(40)	32906756.5580	0.0646
R(42)	28623352.3850	0.1054	P(42)	32936065.8540	0.1015
R(44)	28653024.4839	0.1779	P(44)	32964571.3426	0.1518
R(46)	28681882.7038	0.2907	P(46)	32992277.1760	0.2184
R(48)	28709922.2632	0.4635	P(48)	33019187.7118	0.3048
R(50)	28737138.1785	0.7231	P(50)	33045307.5105	0.4152

INFRARED AND FAR-INFRARED ABSORPTION FREQUENCY STANDARDS

Arthur Maki

Aside from the CO_2 laser transitions, the absorption spectrum of CO has been more accurately and thoroughly measured than any other spectrum. A bibliography of earlier measurements on CO is given by Maki and Wells,[1] and the present tables were calculated from the measurements referred to in that work. In addition, some new and very accurate frequency measurements[2,3] have been made and were incorporated in the present tables. The frequencies of the rotational transitions of HF and HCl were calculated from constants obtained from fitting the measurements of Evenson et al.[4,5] and Jennings and Wells.[6]

A new report on infrared wavenumber standards from the International Union of Pure and Applied Chemistry, Commission on Molecular Structure and Spectroscopy, may be found in Reference 7.

REFERENCES

1. Maki, A. G. and Wells, J. S., *Wavenumber Calibration Tables from Heterodyne Frequency Measurements*, NIST Special Publication 821, U.S. Dept. of Commerce, Washington, D.C., 1991.
2. Evenson, K. and Stroh, F., private communication.
3. George, T., Urban, W., and co-workers, private communication.
4. Jennings, D. A., Evenson, K. M., Zink, L. R., Demuynck, C., Destombes, J. L., Lemoine, B., and Johns, J. W. C., *J. Mol. Spectrosc.*, 122, 477-480, 1987.
5. Nolt, I. G., Radostitz, J. V., DiLonardo, G., Evenson, K. M., Jennings, D. A., Leopold, K. R., Vanek, M. D., Zink, L. R., Hinz, A., and Chance, K. V., *J. Mol. Spectrosc.*, 125, 274-287, 1987.
6. Jennings, D. A. and Wells, J. S., *J. Mol. Spectrosc.*, 130, 267-268, 1988.
7. *High Resolution Wavenumber Standards for the Infrared, Pure Appl. Chemistry,* 68, 193, 1996.

Wavenumbers for the ν = 1-0 Band of CO

Wavenumber (unc)* cm^{-1}	Transition	Wavenumber (unc) cm^{-1}	Transition
		2147.081132(01)	R(0)
2139.426071(01)	P(1)	2150.856006(01)	R(1)
2135.546178(01)	P(2)	2154.595581(01)	R(2)
2131.631574(01)	P(3)	2158.299710(01)	R(3)
2127.682404(01)	P(4)	2161.968245(01)	R(4)
2123.698816(01)	P(5)	2165.601041(01)	R(5)
2119.680957(01)	P(6)	2169.197949(01)	R(6)
2115.628973(01)	P(7)	2172.758824(01)	R(7)
2111.543012(01)	P(8)	2176.283519(01)	R(8)
2107.423221(01)	P(9)	2179.771887(01)	R(9)
2103.269746(01)	P(10)	2183.223782(01)	R(10)
2099.082734(01)	P(11)	2186.639057(01)	R(11)
2094.862333(01)	P(12)	2190.017565(01)	R(12)
2090.608688(01)	P(13)	2193.359161(01)	R(13)
2086.321947(01)	P(14)	2196.663698(01)	R(14)
2082.002256(01)	P(15)	2199.931030(01)	R(15)
2077.649762(01)	P(16)	2203.161010(01)	R(16)
2073.264612(01)	P(17)	2206.353492(01)	R(17)
2068.846952(01)	P(18)	2209.508331(02)	R(18)
2064.396929(01)	P(19)	2212.625379(02)	R(19)
2059.914688(02)	P(20)	2215.704492(02)	R(20)
2055.400377(02)	P(21)	2218.745522(02)	R(21)
2050.854140(02)	P(22)	2221.748326(03)	R(22)
2046.276126(03)	P(23)	2224.712755(03)	R(23)
2041.666479(03)	P(24)	2227.638666(03)	R(24)
2037.025345(03)	P(25)	2230.525912(04)	R(25)
2032.352870(04)	P(26)	2233.374349(04)	R(26)
2027.649200(04)	P(27)	2236.183829(04)	R(27)
2022.914480(04)	P(28)	2238.954210(05)	R(28)
2018.148857(05)	P(29)	2241.685344(05)	R(29)
2013.352474(05)	P(30)	2244.377088(06)	R(30)
2008.525477(06)	P(31)	2247.029296(07)	R(31)
2003.668012(06)	P(32)	2249.641824(08)	R(32)
1998.780224(07)	P(33)	2252.214527(10)	R(33)
1993.862257(09)	P(34)	2254.747262(14)	R(34)

Wavenumber (unc)* cm⁻¹	Transition	Wavenumber (unc) cm⁻¹	Transition
1988.914257(11)	P(35)	2257.239883(18)	R(35)
1983.936367(14)	P(36)	2259.692248(24)	R(36)
1978.928733(18)	P(37)	2262.104213(33)	R(37)
1973.891500(25)	P(38)	2264.475634(45)	R(38)
1968.824811(34)	P(39)	2266.806368(61)	R(39)
1963.728813(46)	P(40)	2269.096273(81)	R(40)
1958.603648(61)	P(41)	2271.345206(106)	R(41)
1953.449462(82)	P(42)	2273.553027(139)	R(42)

* The uncertainty in the last digits (twice the standard error) is given in parentheses.

Wavenumbers for the v = 2-0 Band of CO

Wavenumber (unc)* cm⁻¹	Transition	Wavenumber (unc) cm⁻¹	Transition
		4263.837198(02)	R(0)
4256.217140(02)	P(1)	4267.542066(02)	R(1)
4252.302244(02)	P(2)	4271.176630(02)	R(2)
4248.317633(02)	P(3)	4274.740746(02)	R(3)
4244.263453(02)	P(4)	4278.234264(02)	R(4)
4240.139852(02)	P(5)	4281.657039(02)	R(5)
4235.946975(02)	P(6)	4285.008924(02)	R(6)
4231.684972(02)	P(7)	4288.289772(02)	R(7)
4227.353987(02)	P(8)	4291.499437(02)	R(8)
4222.954169(02)	P(9)	4294.637773(02)	R(9)
4218.485665(02)	P(10)	4297.704631(02)	R(10)
4213.948620(02)	P(11)	4300.699868(02)	R(11)
4209.343182(02)	P(12)	4303.623334(02)	R(12)
4204.669499(02)	P(13)	4306.474886(02)	R(13)
4199.927716(02)	P(14)	4309.254375(02)	R(14)
4195.117980(02)	P(15)	4311.961657(02)	R(15)
4190.240439(02)	P(16)	4314.596584(02)	R(16)
4185.295239(02)	P(17)	4317.159011(02)	R(17)
4180.282526(02)	P(18)	4319.648791(02)	R(18)
4175.202447(02)	P(19)	4322.065779(03)	R(19)
4170.055149(03)	P(20)	4324.409829(03)	R(20)
4164.840777(03)	P(21)	4326.680794(03)	R(21)
4159.559478(03)	P(22)	4328.878530(03)	R(22)
4154.211398(03)	P(23)	4331.002889(04)	R(23)
4148.796683(04)	P(24)	4333.053728(04)	R(24)
4143.315479(04)	P(25)	4335.030899(05)	R(25)
4137.767932(04)	P(26)	4336.934259(06)	R(26)
4132.154187(05)	P(27)	4338.763661(07)	R(27)
4126.474391(06)	P(28)	4340.518961(09)	R(28)
4120.728689(07)	P(29)	4342.200014(11)	R(29)
4114.917226(09)	P(30)	4343.806675(16)	R(30)
4109.040148(12)	P(31)	4345.338799(21)	R(31)
4103.097600(16)	P(32)	4346.796243(29)	R(32)
4097.089728(21)	P(33)	4348.178862(40)	R(33)
4091.016676(29)	P(34)	4349.486513(54)	R(34)
4084.878591(40)	P(35)	4350.719052(73)	R(35)
4078.675618(54)	P(36)	4351.876336(96)	R(36)
4072.407901(73)	P(37)	4352.958224(127)	R(37)
4066.075588(97)	P(38)	4353.964572(166)	R(38)
4059.678822(127)	P(39)	4354.895240(214)	R(39)

* The uncertainty in the last digits (twice the standard error) is given in parentheses.

Wavenumbers for the v = 3-0 Band of CO

Wavenumber (unc)* cm^{-1}	Transition	Wavenumber (unc) cm^{-1}	Transition
		6354.179057(13)	R(0)
6346.594000(13)	P(1)	6357.813923(13)	R(1)
6342.644103(13)	P(2)	6361.343487(13)	R(2)
6338.589491(13)	P(3)	6364.767599(13)	R(3)
6334.430309(13)	P(4)	6368.086115(13)	R(4)
6330.166705(13)	P(5)	6371.298887(13)	R(5)
6325.798826(13)	P(6)	6374.405768(12)	R(6)
6321.326819(13)	P(7)	6377.406611(12)	R(7)
6316.750831(12)	P(8)	6380.301271(12)	R(8)
6312.071008(12)	P(9)	6383.089600(12)	R(9)
6307.287498(12)	P(10)	6385.771452(12)	R(10)
6302.400447(12)	P(11)	6388.346680(13)	R(11)
6297.410003(12)	P(12)	6390.815139(13)	R(12)
6292.316311(13)	P(13)	6393.176681(13)	R(13)
6287.119520(13)	P(14)	6395.431160(13)	R(14)
6281.819775(13)	P(15)	6397.578430(13)	R(15)
6276.417224(13)	P(16)	6399.618344(13)	R(16)
6270.912012(13)	P(17)	6401.550757(13)	R(17)
6265.304287(13)	P(18)	6403.375523(13)	R(18)
6259.594194(13)	P(19)	6405.092495(14)	R(19)
6253.781880(13)	P(20)	6406.701527(14)	R(20)
6247.867492(14)	P(21)	6408.202474(14)	R(21)
6241.851176(14)	P(22)	6409.595189(15)	R(22)
6235.733077(14)	P(23)	6410.879527(15)	R(23)
6229.513342(15)	P(24)	6412.055343(16)	R(24)
6223.192117(15)	P(25)	6413.122491(17)	R(25)
6216.769547(16)	P(26)	6414.080825(19)	R(26)
6210.245778(17)	P(27)	6414.930201(23)	R(27)
6203.620957(19)	P(28)	6415.670474(28)	R(28)
6196.895229(23)	P(29)	6416.301500(37)	R(29)
6190.068739(28)	P(30)	6416.823133(50)	R(30)
6183.141633(37)	P(31)	6417.235231(67)	R(31)
6176.114058(50)	P(32)	6417.537649(90)	R(32)
6168.986159(67)	P(33)		
6161.758082(90)	P(34)		

* The uncertainty in the last digits (twice the standard error) is given in parentheses.

Frequencies and Wavenumbers for the Rotational Lines of CO

Frequency MHz	Uncertainty* MHz	J'	J''	Wavenumber cm^{-1}	Uncertainty* cm^{-1}
115271.2029	0.0004	1	0	3.84503345	0.00000001
230538.0016	0.0008	2	1	7.68991999	0.00000003
345795.9923	0.0012	3	2	11.53451273	0.00000004
461040.7712	0.0016	4	3	15.37866477	0.00000005
576267.9350	0.0019	5	4	19.22222923	0.00000006
691473.0809	0.0021	6	5	23.06505926	0.00000007
806651.8065	0.0023	7	6	26.90700800	0.00000008
921799.7104	0.0025	8	7	30.74792863	0.00000008
1036912.3919	0.0027	9	8	34.58767438	0.00000009
1151985.4515	0.0029	10	9	38.42609848	0.00000010
1267014.4906	0.0031	11	10	42.26305422	0.00000010
1381995.1119	0.0034	12	11	46.09839491	0.00000011
1496922.9195	0.0038	13	12	49.93197392	0.00000013

Frequency MHz	Uncertainty* MHz	J'	J''	Wavenumber cm^{-1}	Uncertainty* cm^{-1}
1611793.5189	0.0042	14	13	53.76364468	0.00000014
1726602.5173	0.0047	15	14	57.59326065	0.00000016
1841345.5237	0.0052	16	15	61.42067535	0.00000017
1956018.1486	0.0057	17	16	65.24574239	0.00000019
2070616.0050	0.0061	18	17	69.06831542	0.00000020
2185134.7075	0.0065	19	18	72.88824816	0.00000022
2299569.8733	0.0069	20	19	76.70539441	0.00000023
2413917.1217	0.0071	21	20	80.51960806	0.00000024
2528172.0747	0.0073	22	21	84.33074306	0.00000024
2642330.3567	0.0074	23	22	88.13865346	0.00000025
2756387.5949	0.0075	24	23	91.94319341	0.00000025
2870339.4194	0.0077	25	24	95.74421713	0.00000026
2984181.4631	0.0080	26	25	99.54157896	0.00000027
3097909.3621	0.0085	27	26	103.33513334	0.00000028
3211518.7558	0.0090	28	27	107.12473480	0.00000030
3325005.2869	0.0096	29	28	110.91023800	0.00000032
3438364.6013	0.0102	30	29	114.69149772	0.00000034
3551592.3489	0.0107	31	30	118.46836884	0.00000036
3664684.1829	0.0111	32	31	122.24070637	0.00000037
3777635.7608	0.0118	33	32	126.00836545	0.00000039
3890442.7435	0.0137	34	33	129.77120137	0.00000046
4003100.7965	0.0179	35	34	133.52906952	0.00000060
4115605.5892	0.0254	36	35	137.28182546	0.00000085
4227952.7954	0.0370	37	36	141.02932487	0.00000123
4340138.0932	0.0531	38	37	144.77142361	0.00000177
4452157.1657	0.0746	39	38	148.50797766	0.00000249
4564005.7001	0.1025	40	39	152.23884318	0.00000342

* The uncertainty given is twice the standard error.

Frequencies and Wavenumbers for the Rotational Lines of HF

Frequency MHz	Uncertainty* MHz	J'	J''	Wavenumber cm^{-1}	Uncertainty* cm^{-1}
1232476.21	0.12	1	0	41.110981	0.000004
2463428.09	0.19	2	1	82.171116	0.000006
3691334.81	0.25	3	2	123.129676	0.000008
4914682.58	0.51	4	3	163.936165	0.000017
6131968.11	1.10	5	4	204.540439	0.000037
7341702.00	2.00	6	5	244.892818	0.000067
8542412.1	3.21	7	6	284.944197	0.000107
9732646.8	4.72	8	7	324.646153	0.000157
10910978.2	6.51	9	8	363.951056	0.000217
12076004.8	8.55	10	9	402.81216	0.000285
13226355.2	10.81	11	10	441.18372	0.000361
14360689.8	13.25	12	11	479.02105	0.00044
15477704.4	15.86	13	12	516.28065	0.00053
16576131.8	18.61	14	13	552.92024	0.00062
17654744.4	21.48	15	14	588.89888	0.00072
18712356.5	24.44	16	15	624.17703	0.00082
19747825.6	27.43	17	16	658.71656	0.00092
20760054.3	30.32	18	17	692.4809	0.00101
21747991.7	32.91	19	18	725.4349	0.00110
22710634.7	34.94	20	19	757.5452	0.00117
23647028.7	36.08	21	20	788.7800	0.00120
24556268.8	35.93	22	21	819.1090	0.00120
25437499.9	34.12	23	22	848.5037	0.00114

Frequency MHz	Uncertainty* MHz	*J′*	*J″*	Wavenumber cm^{-1}	Uncertainty* cm^{-1}
26289917.4	30.32	24	23	876.9373	0.00101
27112767.2	24.41	25	24	904.38457	0.00081
27905345.6	16.88	26	25	930.82214	0.00056
28666999.3	10.80	27	26	956.22817	0.00036
29397124.8	14.65	28	27	980.58253	0.00049
30095168.2	24.62	29	28	1003.86676	0.00082
30760624.2	33.36	30	29	1026.0640	0.00111
31393035.7	36.17	31	30	1047.1590	0.00121

* The uncertainty given is twice the standard error.

Frequencies and Wavenumbers for the Rotational Lines of $H^{35}Cl$

Frequency MHz	Uncertainty* MHz	*J′*	*J″*	Wavemumber cm^{-1}	Uncertainty* cm^{-1}
1876226.517	0.065	3	2	62.584180	0.000002
2499864.439	0.066	4	3	83.386502	0.000002
3121986.563	0.064	5	4	104.138262	0.000002
3742216.601	0.076	6	5	124.826909	0.000003
4360180.042	0.098	7	6	145.439951	0.000003
4975504.51	0.11	8	7	165.964966	0.000004
5587820.10	0.12	9	8	186.389615	0.000004
6196759.76	0.22	10	9	206.701656	0.000007
6801959.63	0.50	11	10	226.888951	0.000017
7403059.41	1.02	12	11	246.939481	0.000034
7999702.7	1.8	13	12	266.841359	0.000062
8591537.3	3.1	14	13	286.582837	0.000103
9178215.8	4.8	15	14	306.152324	0.000161

* The uncertainty given is twice the standard error.

Frequencies and Wavenumbers for the Rotational Lines of $H^{37}Cl$

Frequency MHz	Uncertainty* MHz	*J′*	*J″*	Wavenumber cm^{-1}	Uncertainty* cm^{-1}
1873410.72	0.05	3	2	62.490255	0.000002
2496115.33	0.05	4	3	83.261445	0.000002
3117308.69	0.05	5	4	103.982225	0.000002
3736615.64	0.06	6	5	124.640082	0.000002
4353662.84	0.08	7	6	145.222561	0.000003
4968079.04	0.09	8	7	165.717279	0.000003
5579495.53	0.10	9	8	186.111938	0.000003
6187546.42	0.19	10	9	206.394332	0.000006
6791869.04	0.45	11	10	226.552365	0.000015
7392104.3	0.9	12	11	246.574057	0.000030
7987896.9	1.6	13	12	266.447561	0.000054
8578896.1	2.7	14	13	286.161170	0.000089

* The uncertainty given is twice the standard error.

Section 11
Nuclear and Particle Physics

Summary Tables of Particle Properties 11-1
Table of the Isotopes 11-50
Neutron Scattering and Absorption Properties 11-199
Cosmic Radiation 11-216

SUMMARY TABLES OF PARTICLE PROPERTIES

Extracted from the Particle Listings of the

Review of Particle Physics

Published in Eur. Jour. Phys. **C3**, 1 (1998)

Available at `http://pdg.lbl.gov`

Particle Data Group Authors:

C. Caso, G. Conforto, A. Gurtu, M. Aguilar-Benitez, C. Amsler, R.M. Barnett, P.R. Burchat, C.D. Carone, O. Dahl, M. Doser, S. Eidelman, J.L. Feng, M. Goodman, C. Grab, D.E. Groom, K. Hagiwara, K.G. Hayes, J.J. Hernández, K. Hikasa, K. Honscheid, F. James, M.L. Mangano, A.V. Manohar, K. Mönig, H. Murayama, K. Nakamura, K.A. Olive, A. Piepke, M. Roos, R.H. Schindler, R.E. Shrock, M. Tanabashi, N.A. Törnqvist, T.G. Trippe, P. Vogel, C.G. Wohl, R.L. Workman, W.-M. Yao

Technical Associates: B. Armstrong, J.L. Casas Serradilla, B.B. Filimonov, P.S. Gee, S.B. Lugovsky, S. Mankov, F. Nicholson

Other Authors who have made substantial contributions to reviews since the 1994 edition:

K.S. Babu, D. Besson, O. Biebel, R.N. Cahn, R.L. Crawford, R.H. Dalitz, T. Damour, K. Desler, R.J. Donahue, D.A. Edwards, J. Erler, V.V. Ezhela, A. Fassò, W. Fetscher, D. Froidevaux, T.K. Gaisser, L. Garren, S. Geer, H.-J. Gerber, F.J. Gilman, H.E. Haber, C. Hagmann, I. Hinchliffe, C.J. Hogan, G. Höhler, J.D. Jackson, K.F. Johnson, D. Karlen, B. Kayser, K. Kleinknecht, I.G. Knowles, C. Kolda, P. Kreitz, P. Langacker, R. Landua, L. Littenberg, D.M. Manley, J. March-Russell, T. Nakada, H. Quinn, G. Raffelt, B. Renk, M.T. Ronan, L.J. Rosenberg, M. Schmitt, D.N. Schramm, D. Scott, T. Sjöstrand, G.F. Smoot, S. Spanier, M. Srednicki, T. Stanev, M. Suzuki, N.P. Tkachenko, G. Valencia, K. van Bibber, R. Voss, L. Wolfenstein, S. Youssef

(Approximate closing date for data: January 1, 1998)

GAUGE AND HIGGS BOSONS

γ

$I(J^{PC}) = 0,1(1^{--})$

Mass $m < 2 \times 10^{-16}$ eV
Charge $q < 5 \times 10^{-30}$ e
Mean life τ = Stable

g or gluon

$I(J^P) = 0(1^-)$

Mass $m = 0$ [a]
SU(3) color octet

W

$J = 1$

Charge $= \pm 1$ e
Mass $m = 80.41 \pm 0.10$ GeV
$m_Z - m_W = 10.78 \pm 0.10$ GeV
$m_{W^+} - m_{W^-} = -0.2 \pm 0.6$ GeV
Full width $\Gamma = 2.06 \pm 0.06$ GeV

W^- modes are charge conjugates of the modes below.

W^+ DECAY MODES		Fraction (Γ_i/Γ)	Confidence level	p (MeV/c)
$\ell^+\nu$	[b]	(10.74±0.33) %		–
$e^+\nu$		(10.9 ±0.4) %		40205
$\mu^+\nu$		(10.2 ±0.5) %		40205
$\tau^+\nu$		(11.3 ±0.8) %		40185
hadrons		(67.8 ±1.0) %		–
$\pi^+\gamma$		< 2.2 × 10^{-4}	95%	40205

Z

$J = 1$

Charge = 0
Mass $m = 91.187 \pm 0.007$ GeV [c]
Full width $\Gamma = 2.490 \pm 0.007$ GeV
$\Gamma(\ell^+\ell^-) = 83.83 \pm 0.27$ MeV [b]
$\Gamma(\text{invisible}) = 498.3 \pm 4.2$ MeV [d]
$\Gamma(\text{hadrons}) = 1740.7 \pm 5.9$ MeV
$\Gamma(\mu^+\mu^-)/\Gamma(e^+e^-) = 1.000 \pm 0.005$
$\Gamma(\tau^+\tau^-)/\Gamma(e^+e^-) = 0.998 \pm 0.005$ [e]

Average charged multiplicity

$\langle N_{charged} \rangle = 21.00 \pm 0.13$

Couplings to leptons

$g_V^\ell = -0.0377 \pm 0.0007$
$g_A^\ell = -0.5008 \pm 0.0008$
$g^{\nu_e} = 0.53 \pm 0.09$
$g^{\nu_\mu} = 0.502 \pm 0.017$

Asymmetry parameters [f]

$A_e = 0.1519 \pm 0.0034$
$A_\mu = 0.102 \pm 0.034$
$A_\tau = 0.143 \pm 0.008$
$A_c = 0.59 \pm 0.19$
$A_b = 0.89 \pm 0.11$

Charge asymmetry (%) at *Z* pole

$A_{FB}^{(0\ell)} = 1.59 \pm 0.18$
$A_{FB}^{(0u)} = 4.0 \pm 7.3$
$A_{FB}^{(0s)} = 9.9 \pm 3.1$ (S = 1.2)
$A_{FB}^{(0c)} = 7.32 \pm 0.58$
$A_{FB}^{(0b)} = 10.02 \pm 0.28$

Z DECAY MODES		Fraction (Γ_i/Γ)	Confidence level	p (MeV/c)
e^+e^-		(3.366±0.008) %		45594
$\mu^+\mu^-$		(3.367±0.013) %		45593
$\tau^+\tau^-$		(3.360±0.015) %		45559
$\ell^+\ell^-$	[b]	(3.366±0.006) %		–
invisible		(20.01 ±0.16) %		–
hadrons		(69.90 ±0.15) %		–
$(u\overline{u}+c\overline{c})/2$		(10.1 ±1.1) %		–
$(d\overline{d}+s\overline{s}+b\overline{b})/3$		(16.6 ±0.6) %		–
$c\overline{c}$		(12.4 ±0.6) %		–
$b\overline{b}$		(15.16 ±0.09) %		–
ggg		< 1.1 %	95%	–
$\pi^0\gamma$		< 5.2 × 10^{-5}	95%	45593
$\eta\gamma$		< 5.1 × 10^{-5}	95%	45592
$\omega\gamma$		< 6.5 × 10^{-4}	95%	45590
$\eta'(958)\gamma$		< 4.2 × 10^{-5}	95%	45588
$\gamma\gamma$		< 5.2 × 10^{-5}	95%	45594
$\gamma\gamma\gamma$		< 1.0 × 10^{-5}	95%	45594
$\pi^\pm W^\mp$	[g]	< 7 × 10^{-5}	95%	10139
$\rho^\pm W^\mp$	[g]	< 8.3 × 10^{-5}	95%	10114
$J/\psi(1S)$X		(3.66 ±0.23) × 10^{-3}		–
$\psi(2S)$X		(1.60 ±0.29) × 10^{-3}		–
$\chi_{c1}(1P)$X		(2.9 ±0.7) × 10^{-3}		–
$\chi_{c2}(1P)$X		< 3.2 × 10^{-3}	90%	–
$\Upsilon(1S)$ X + $\Upsilon(2S)$ X + $\Upsilon(3S)$ X		(1.0 ±0.5) × 10^{-4}		–
$\Upsilon(1S)$X		< 5.5 × 10^{-5}	95%	–
$\Upsilon(2S)$X		< 1.39 × 10^{-4}	95%	–
$\Upsilon(3S)$X		< 9.4 × 10^{-5}	95%	–
$(D^0/\overline{D}^0)$ X		(20.7 ±2.0) %		–
$D^\pm$ X		(12.2 ±1.7) %		–
$D^*(2010)^\pm$X	[g]	(11.4 ±1.3) %		–
B_s^0X		seen		–
anomalous γ+ hadrons	[h]	< 3.2 × 10^{-3}	95%	–
$e^+e^-\gamma$	[h]	< 5.2 × 10^{-4}	95%	45594
$\mu^+\mu^-\gamma$	[h]	< 5.6 × 10^{-4}	95%	45593
$\tau^+\tau^-\gamma$	[h]	< 7.3 × 10^{-4}	95%	45559
$\ell^+\ell^-\gamma\gamma$	[i]	< 6.8 × 10^{-6}	95%	–
$q\overline{q}\gamma\gamma$	[i]	< 5.5 × 10^{-6}	95%	–

$\nu\overline{\nu}\gamma\gamma$		[i] <	3.1	$\times 10^{-6}$	95%	45594
$e^{\pm}\mu^{\mp}$	LF	[g] <	1.7	$\times 10^{-6}$	95%	45593
$e^{\pm}\tau^{\mp}$	LF	[g] <	9.8	$\times 10^{-6}$	95%	45576
$\mu^{\pm}\tau^{\mp}$	LF	[g] <	1.2	$\times 10^{-5}$	95%	45576

Higgs Bosons — H^0 and $H^{\pm}$, Searches for

H^0 Mass $m >$ 77.5 GeV, CL = 95%

H_1^0 in Supersymmetric Models ($m_{H_1^0} < m_{H_2^0}$)

Mass $m >$ 62.5 GeV, CL = 95%

A^0 Pseudoscalar Higgs Boson in Supersymmetric Models [j]

Mass $m >$ 62.5 GeV, CL = 95% $\tan\beta > 1$

$H^{\pm}$ Mass $m >$ 54.5 GeV, CL = 95%

See the Particle Listings for a Note giving details of Higgs Bosons.

Heavy Bosons Other Than Higgs Bosons, Searches for

Additional W Bosons

W_R — right-handed W
Mass $m >$ 549 GeV
(assuming light right-handed neutrino)
W' with standard couplings decaying to $e\nu$, $\mu\nu$
Mass $m >$ 720 GeV, CL = 95%

Additional Z Bosons

Z'_{SM} with standard couplings
Mass $m >$ 690 GeV, CL = 95% ($p\overline{p}$ direct search)
Mass $m >$ 779 GeV, CL = 95% (electroweak fit)
Z_{LR} of $SU(2)_L \times SU(2)_R \times U(1)$
(with $g_L = g_R$)
Mass $m >$ 630 GeV, CL = 95% ($p\overline{p}$ direct search)
Mass $m >$ 389 GeV, CL = 95% (electroweak fit)
Z_χ of $SO(10) \to SU(5) \times U(1)_\chi$
(coupling constant derived from G.U.T.)
Mass $m >$ 595 GeV, CL = 95% ($p\overline{p}$ direct search)
Mass $m >$ 321 GeV, CL = 95% (electroweak fit)
Z_ψ of $E_6 \to SO(10) \times U(1)_\psi$
(coupling constant derived from G.U.T.)
Mass $m >$ 590 GeV, CL = 95% ($p\overline{p}$ direct search)
Mass $m >$ 160 GeV, CL = 95% (electroweak fit)
Z_η of $E_6 \to SU(3) \times SU(2) \times U(1) \times U(1)_\eta$
(coupling constant derived from G.U.T.);
charges are $Q_\eta = \sqrt{3/8}Q_\chi - \sqrt{5/8}Q_\psi$)
Mass $m >$ 620 GeV, CL = 95% ($p\overline{p}$ direct search)
Mass $m >$ 182 GeV, CL = 95% (electroweak fit)

Scalar Leptoquarks

Mass $m >$ 225 GeV, CL = 95% (1st generation, pair prod.)
Mass $m >$ 237 GeV, CL = 95% (1st gener., single prod.)
Mass $m >$ 119 GeV, CL = 95% (2nd gener., pair prod.)
Mass $m >$ 73 GeV, CL = 95% (2nd gener., single prod.)
Mass $m >$ 99 GeV, CL = 95% (3rd gener., pair prod.)
(See the Particle Listings for assumptions on leptoquark quantum numbers and branching fractions.)

Axions (A^0) and Other Very Light Bosons, Searches for

The standard Peccei-Quinn axion is ruled out. Variants with reduced couplings or much smaller masses are constrained by various data. The Particle Listings in the full *Review* contain a Note discussing axion searches.

The best limit for the half-life of neutrinoless double beta decay with Majoron emission is $> 7.2 \times 10^{24}$ years (CL = 90%).

NOTES

In this Summary Table:

When a quantity has "(S = ...)" to its right, the error on the quantity has been enlarged by the "scale factor" S, defined as $S = \sqrt{\chi^2/(N-1)}$, where N is the number of measurements used in calculating the quantity. We do this when S > 1, which often indicates that the measurements are inconsistent. When S > 1.25, we also show in the Particle Listings an ideogram of the measurements. For more about S, see the Introduction.

A decay momentum p is given for each decay mode. For a 2-body decay, p is the momentum of each decay product in the rest frame of the decaying particle. For a 3-or-more-body decay, p is the largest momentum any of the products can have in this frame.

[a] Theoretical value. A mass as large as a few MeV may not be precluded.

[b] ℓ indicates each type of lepton (e, μ, and τ), not sum over them.

[c] The Z-boson mass listed here corresponds to a Breit-Wigner resonance parameter. It lies approximately 34 MeV above the real part of the position of the pole (in the energy-squared plane) in the Z-boson propagator.

[d] This partial width takes into account Z decays into $\nu\overline{\nu}$ and any other possible undetected modes.

[e] This ratio has not been corrected for the τ mass.

[f] Here $A \equiv 2g_V g_A/(g_V^2+g_A^2)$.

[g] The value is for the sum of the charge states of particle/antiparticle states indicated.

[h] See the Z Particle Listings for the γ energy range used in this measurement.

[i] For $m_{\gamma\gamma} = (60 \pm 5)$ GeV.

[j] The limits assume no invisible decays.

LEPTONS

e

$J = \frac{1}{2}$

Mass $m = 0.51099907 \pm 0.00000015$ MeV [a]
$= (5.485799111 \pm 0.000000012) \times 10^{-4}$ u
$(m_{e^+} - m_{e^-})/m < 4 \times 10^{-8}$, CL = 90%
$|q_{e^+} + q_{e^-}|/e < 4 \times 10^{-8}$
Magnetic moment $\mu = 1.001159652193 \pm 0.000000000010\ \mu_B$
$(g_{e^+} - g_{e^-}) / g_{\text{average}} = (-0.5 \pm 2.1) \times 10^{-12}$
Electric dipole moment $d = (0.18 \pm 0.16) \times 10^{-26}$ $e$cm
Mean life $\tau > 4.3 \times 10^{23}$ yr, CL = 68% [b]

μ

$J = \frac{1}{2}$

Mass $m = 105.658389 \pm 0.000034$ MeV [c]
$= 0.113428913 \pm 0.000000017$ u
Mean life $\tau = (2.19703 \pm 0.00004) \times 10^{-6}$ s
$\tau_{\mu^+}/\tau_{\mu^-} = 1.00002 \pm 0.00008$
$c\tau = 658.654$ m
Magnetic moment $\mu = 1.0011659230 \pm 0.0000000084\ e\hbar/2m_\mu$
$(g_{\mu^+} - g_{\mu^-}) / g_{\text{average}} = (-2.6 \pm 1.6) \times 10^{-8}$
Electric dipole moment $d = (3.7 \pm 3.4) \times 10^{-19}$ $e$cm

Decay parameters [d]

$\rho = 0.7518 \pm 0.0026$
$\eta = -0.007 \pm 0.013$
$\delta = 0.749 \pm 0.004$
$\xi P_\mu = 1.003 \pm 0.008$ [e]
$\xi P_\mu \delta/\rho > 0.99682$, CL = 90% [e]
$\xi' = 1.00 \pm 0.04$
$\xi'' = 0.7 \pm 0.4$
$\alpha/A = (0 \pm 4) \times 10^{-3}$
$\alpha'/A = (0 \pm 4) \times 10^{-3}$
$\beta/A = (4 \pm 6) \times 10^{-3}$
$\beta'/A = (2 \pm 6) \times 10^{-3}$
$\overline{\eta} = 0.02 \pm 0.08$

μ^+ modes are charge conjugates of the modes below.

μ^- DECAY MODES		Fraction (Γ_i/Γ)		Confidence level	p (MeV/c)
$e^- \overline{\nu}_e \nu_\mu$		$\approx 100\%$			53
$e^- \overline{\nu}_e \nu_\mu \gamma$		[f]	(1.4 ± 0.4) %		53
$e^- \overline{\nu}_e \nu_\mu e^+ e^-$		[g]	$(3.4 \pm 0.4) \times 10^{-5}$		53
Lepton Family number (*LF*) violating modes					
$e^- \nu_e \overline{\nu}_\mu$	*LF*	[h] < 1.2	%	90%	53
$e^- \gamma$	*LF*	< 4.9	$\times 10^{-11}$	90%	53
$e^- e^+ e^-$	*LF*	< 1.0	$\times 10^{-12}$	90%	53
$e^- 2\gamma$	*LF*	< 7.2	$\times 10^{-11}$	90%	53

τ

$J = \frac{1}{2}$

Mass $m = 1777.05^{+0.29}_{-0.26}$ MeV
Mean life $\tau = (290.0 \pm 1.2) \times 10^{-15}$ s
$c\tau = 86.93$ μm
Magnetic moment anomaly > -0.052 and < 0.058, CL = 95%
Electric dipole moment $d > -3.1$ and $< 3.1 \times 10^{-16}$ $e$cm, CL = 95%

Weak dipole moment

$\text{Re}(d^w_\tau) < 0.56 \times 10^{-17}$ $e$cm, CL = 95%
$\text{Im}(d^w_\tau) < 1.5 \times 10^{-17}$ $e$cm, CL = 95%

Weak anomalous magnetic dipole moment

$\text{Re}(\alpha^w_\tau) < 4.5 \times 10^{-3}$, CL = 90%
$\text{Im}(\alpha^w_\tau) < 9.9 \times 10^{-3}$, CL = 90%

Decay parameters

See the τ Particle Listings for a note concerning τ-decay parameters.

$\rho^\tau(e \text{ or } \mu) = 0.748 \pm 0.010$
$\rho^\tau(e) = 0.745 \pm 0.012$
$\rho^\tau(\mu) = 0.741 \pm 0.030$
$\xi^\tau(e \text{ or } \mu) = 1.01 \pm 0.04$
$\xi^\tau(e) = 0.98 \pm 0.05$
$\xi^\tau(\mu) = 1.07 \pm 0.08$
$\eta^\tau(e \text{ or } \mu) = 0.01 \pm 0.07$
$\eta^\tau(\mu) = -0.10 \pm 0.18$
$(\delta\xi)^\tau(e \text{ or } \mu) = 0.749 \pm 0.026$
$(\delta\xi)^\tau(e) = 0.733 \pm 0.033$
$(\delta\xi)^\tau(\mu) = 0.78 \pm 0.05$
$\xi^\tau(\pi) = 0.99 \pm 0.05$
$\xi^\tau(\rho) = 0.996 \pm 0.010$
$\xi^\tau(a_1) = 1.02 \pm 0.04$
$\xi^\tau(\text{all hadronic modes}) = 0.997 \pm 0.009$

τ^+ modes are charge conjugates of the modes below. "$h^\pm$" stands for $\pi^\pm$ or $K^\pm$. "ℓ" stands for e or μ. "Neutral" means neutral hadron whose decay products include γ's and/or π^0's.

τ^- DECAY MODES		Fraction (Γ_i/Γ)	Scale factor/ Confidence level	p (MeV/c)
Modes with one charged particle				
particle$^-$ ≥ 0 neutrals $\geq 0 K^0_L \nu_\tau$ ("1-prong")		(84.71 ± 0.13) %	S=1.2	–
particle$^-$ ≥ 0 neutrals $\geq 0 K^0 \nu_\tau$		(85.30 ± 0.13) %	S=1.2	–
$\mu^- \overline{\nu}_\mu \nu_\tau$	[l]	(17.37 ± 0.09) %		885
$\mu^- \overline{\nu}_\mu \nu_\tau \gamma$	[g]	$(3.0 \pm 0.6) \times 10^{-3}$		–
$e^- \overline{\nu}_e \nu_\tau$	[l]	(17.81 ± 0.07) %		889
$h^- \geq 0$ neutrals $\geq 0 K^0_L \nu_\tau$		(49.52 ± 0.16) %	S=1.2	–
$h^- \geq 0 K^0_L \nu_\tau$		(12.32 ± 0.12) %	S=1.5	–
$h^- \nu_\tau$		(11.79 ± 0.12) %	S=1.5	–
$\pi^- \nu_\tau$	[l]	(11.08 ± 0.13) %	S=1.4	883
$K^- \nu_\tau$	[l]	$(7.1 \pm 0.5) \times 10^{-3}$		820
$h^- \geq 1$ neutrals ν_τ		(36.91 ± 0.17) %	S=1.2	–
$h^- \pi^0 \nu_\tau$		(25.84 ± 0.14) %	S=1.1	–
$\pi^- \pi^0 \nu_\tau$	[i]	(25.32 ± 0.15) %	S=1.1	878
$\pi^- \pi^0$ non-$\rho(770) \nu_\tau$		$(3.0 \pm 3.2) \times 10^{-3}$		878
$K^- \pi^0 \nu_\tau$	[i]	$(5.2 \pm 0.5) \times 10^{-3}$		814
$h^- \geq 2\pi^0 \nu_\tau$		(10.79 ± 0.16) %	S=1.2	–
$h^- 2\pi^0 \nu_\tau$		(9.39 ± 0.14) %	S=1.2	–
$h^- 2\pi^0 \nu_\tau$ (ex.K^0)		(9.23 ± 0.14) %	S=1.2	–
$\pi^- 2\pi^0 \nu_\tau$ (ex.K^0)	[i]	(9.15 ± 0.15) %	S=1.2	862
$K^- 2\pi^0 \nu_\tau$ (ex.K^0)	[l]	$(8.0 \pm 2.7) \times 10^{-4}$		796
$h^- \geq 3\pi^0 \nu_\tau$		(1.40 ± 0.11) %	S=1.1	–
$h^- 3\pi^0 \nu_\tau$		(1.23 ± 0.10) %	S=1.1	–
$\pi^- 3\pi^0 \nu_\tau$ (ex.K^0)	[l]	(1.11 ± 0.14) %		836
$K^- 3\pi^0 \nu_\tau$ (ex.K^0)	[i]	$(4.3\ ^{+10.0}_{-\ 2.9}) \times 10^{-4}$		766
$h^- 4\pi^0 \nu_\tau$ (ex.K^0)		$(1.7 \pm 0.6) \times 10^{-3}$		–
$h^- 4\pi^0 \nu_\tau$ (ex.K^0,η)	[l]	$(1.1 \pm 0.6) \times 10^{-3}$		–
$K^- \geq 0\pi^0 \geq 0 K^0 \nu_\tau$		(1.66 ± 0.10) %		–
$K^- \geq 1$ (π^0 or K^0) ν_τ		$(9.5 \pm 1.0) \times 10^{-3}$		–
Modes with K^0's				
K^0(particles)$^- \nu_\tau$		(1.66 ± 0.09) %	S=1.4	–
$h^- \overline{K}^0 \geq 0$ neutrals $\geq 0 K^0_L \nu_\tau$		(1.62 ± 0.09) %	S=1.4	–
$h^- \overline{K}^0 \nu_\tau$		$(9.9 \pm 0.8) \times 10^{-3}$	S=1.5	–
$\pi^- \overline{K}^0 \nu_\tau$	[l]	$(8.3 \pm 0.8) \times 10^{-3}$	S=1.4	812
$\pi^- \overline{K}^0$ (non-$K^*(892)^-$) ν_τ		$< 1.7 \times 10^{-3}$	CL=95%	812
$K^- K^0 \nu_\tau$	[l]	$(1.59 \pm 0.24) \times 10^{-3}$		737
$h^- \overline{K}^0 \pi^0 \nu_\tau$		$(5.5 \pm 0.5) \times 10^{-3}$		–
$\pi^- \overline{K}^0 \pi^0 \nu_\tau$	[l]	$(3.9 \pm 0.5) \times 10^{-3}$		794
$\overline{K}^0 \rho^- \nu_\tau$		$(1.9 \pm 0.7) \times 10^{-3}$		–
$K^- K^0 \pi^0 \nu_\tau$	[l]	$(1.51 \pm 0.29) \times 10^{-3}$		685
$\pi^- \overline{K}^0 \pi^0 \pi^0 \nu_\tau$		$(6 \pm 4) \times 10^{-4}$		–
$K^- K^0 \pi^0 \pi^0 \nu_\tau$		$< 3.9 \times 10^{-4}$	CL=95%	–
$\pi^- K^0 \overline{K}^0 \nu_\tau$	[i]	$(1.21 \pm 0.21) \times 10^{-3}$	S=1.2	682

$\pi^- K^0_S K^0_S \nu_\tau$		$(3.0 \pm 0.5) \times 10^{-4}$	S=1.2	–
$\pi^- K^0_S K^0_L \nu_\tau$		$(6.0 \pm 1.0) \times 10^{-4}$	S=1.2	–
$\pi^- K^0_S K^0_S \pi^0 \nu_\tau$		$< 2.0 \times 10^{-4}$	CL=95%	–
$\pi^- K^0_S K^0_L \pi^0 \nu_\tau$		$(3.1 \pm 1.2) \times 10^{-4}$		–
$K^- K^0 \geq 0$ neutrals ν_τ		$(3.1 \pm 0.4) \times 10^{-3}$		–
$K^0 h^+ h^- h^- \geq 0$ neutrals ν_τ		$< 1.7 \times 10^{-3}$	CL=95%	–
$K^0 h^+ h^- h^- \nu_\tau$		$(2.3 \pm 2.0) \times 10^{-4}$		–
Modes with three charged particles				
$h^- h^- h^+ \geq 0$neut. ν_τ ("3-prong")		(15.18 ± 0.13) %	S=1.2	–
$h^- h^- h^+ \geq 0$ neutrals ν_τ (ex. $K^0_S \to \pi^+\pi^-$)		(14.60 ± 0.13) %	S=1.2	–
$\pi^- \pi^+ \pi^- \geq 0$ neutrals ν_τ		(14.60 ± 0.14) %		–
$h^- h^- h^+ \nu_\tau$		(9.96 ± 0.10) %	S=1.1	–
$h^- h^- h^+ \nu_\tau$ (ex.K^0)		(9.62 ± 0.10) %	S=1.1	–
$h^- h^- h^+ \nu_\tau$ (ex.K^0,ω)		(9.57 ± 0.10) %	S=1.1	–
$\pi^- \pi^+ \pi^- \nu_\tau$		(9.56 ± 0.11) %	S=1.1	–
$\pi^- \pi^+ \pi^- \nu_\tau$ (ex.K^0)		(9.52 ± 0.11) %	S=1.1	–
$\pi^- \pi^+ \pi^- \nu_\tau$ (ex.K^0,ω)	[l]	(9.23 ± 0.11) %	S=1.1	–
$h^- h^- h^+ \geq 1$ neutrals ν_τ		(5.18 ± 0.11) %	S=1.2	–
$h^- h^- h^+ \geq 1$ neutrals ν_τ (ex. $K^0_S \to \pi^+\pi^-$)		(4.98 ± 0.11) %	S=1.2	–
$h^- h^- h^+ \pi^0 \nu_\tau$		(4.50 ± 0.09) %	S=1.1	–
$h^- h^- h^+ \pi^0 \nu_\tau$ (ex.K^0)		(4.31 ± 0.09) %	S=1.1	–
$h^- h^- h^+ \pi^0 \nu_\tau$ (ex. K^0, ω)		(2.59 ± 0.09) %		–
$\pi^- \pi^+ \pi^- \pi^0 \nu_\tau$		(4.35 ± 0.10) %		–
$\pi^- \pi^+ \pi^- \pi^0 \nu_\tau$ (ex.K^0)		(4.22 ± 0.10) %		–
$\pi^- \pi^+ \pi^- \pi^0 \nu_\tau$ (ex.K^0,ω)	[l]	(2.49 ± 0.10) %		–
$h^- (\rho\pi)^0 \nu_\tau$		(2.88 ± 0.35) %		–
$(a_1(1260) h)^- \nu_\tau$		< 2.0 %	CL=95%	–
$h^- \rho \pi^0 \nu_\tau$		(1.35 ± 0.20) %		–
$h^- \rho^+ h^- \nu_\tau$		$(4.5 \pm 2.2) \times 10^{-3}$		–
$h^- \rho^- h^+ \nu_\tau$		(1.17 ± 0.23) %		–
$h^- h^- h^+ 2\pi^0 \nu_\tau$		$(5.4 \pm 0.4) \times 10^{-3}$		–
$h^- h^- h^+ 2\pi^0 \nu_\tau$ (ex.K^0)		$(5.3 \pm 0.4) \times 10^{-3}$		–
$h^- h^- h^+ 2\pi^0 \nu_\tau$ (ex.K^0,ω,η)	[l]	$(1.1 \pm 0.4) \times 10^{-3}$		–
$h^- h^- h^+ \geq 3\pi^0 \nu_\tau$	[l]	$(1.4 {}^{+0.9}_{-0.7}) \times 10^{-3}$	S=1.5	–
$h^- h^- h^+ 3\pi^0 \nu_\tau$		$(2.9 \pm 0.8) \times 10^{-4}$		–
$K^- h^+ h^- \geq 0$ neutrals ν_τ		$(5.4 \pm 0.7) \times 10^{-3}$	S=1.1	–
$K^- \pi^+ \pi^- \geq 0$ neutrals ν_τ		$(3.1 \pm 0.6) \times 10^{-3}$	S=1.1	–
$K^- \pi^+ \pi^- \nu_\tau$		$(2.3 \pm 0.4) \times 10^{-3}$		–
$K^- \pi^+ \pi^- \nu_\tau$ (ex.K^0)	[l]	$(1.8 \pm 0.5) \times 10^{-3}$		–
$K^- \pi^+ \pi^- \pi^0 \nu_\tau$		$(8 \pm 4) \times 10^{-4}$		–
$K^- \pi^+ \pi^- \pi^0 \nu_\tau$ (ex.K^0)	[l]	$(2.4 {}^{+4.3}_{-1.6}) \times 10^{-4}$		–
$K^- \pi^+ K^- \geq 0$ neut. ν_τ		$< 9 \times 10^{-4}$	CL=95%	–
$K^- K^+ \pi^- \geq 0$ neut. ν_τ		$(2.3 \pm 0.4) \times 10^{-3}$		–
$K^- K^+ \pi^- \nu_\tau$	[l]	$(1.61 \pm 0.26) \times 10^{-3}$		685
$K^- K^+ \pi^- \pi^0 \nu_\tau$	[l]	$(6.9 \pm 3.0) \times 10^{-4}$		–
$K^- K^+ K^- \geq 0$ neut. ν_τ		$< 2.1 \times 10^{-3}$	CL=95%	–
$K^- K^+ K^- \nu_\tau$		$< 1.9 \times 10^{-4}$	CL=90%	–
$\pi^- K^+ \pi^- \geq 0$ neut. ν_τ		$< 2.5 \times 10^{-3}$	CL=95%	–
$e^- e^- e^+ \overline{\nu}_e \nu_\tau$		$(2.8 \pm 1.5) \times 10^{-5}$		889
$\mu^- e^- e^+ \overline{\nu}_\mu \nu_\tau$		$< 3.6 \times 10^{-5}$	CL=90%	885
Modes with five charged particles				
$3h^- 2h^+ \geq 0$ neutrals ν_τ (ex. $K^0_S \to \pi^-\pi^+$) ("5-prong")		$(9.7 \pm 0.7) \times 10^{-4}$		–
$3h^- 2h^+ \nu_\tau$ (ex.K^0)	[l]	$(7.5 \pm 0.7) \times 10^{-4}$		–
$3h^- 2h^+ \pi^0 \nu_\tau$ (ex.K^0)	[l]	$(2.2 \pm 0.5) \times 10^{-4}$		–
$3h^- 2h^+ 2\pi^0 \nu_\tau$		$< 1.1 \times 10^{-4}$	CL=90%	–
Miscellaneous other allowed modes				
$(5\pi)^- \nu_\tau$		$(7.4 \pm 0.7) \times 10^{-3}$		–
$4h^- 3h^+ \geq 0$ neutrals ν_τ ("7-prong")		$< 2.4 \times 10^{-6}$	CL=90%	–
$K^*(892)^- \geq 0(h^0 \neq K^0_S) \nu_\tau$		(1.94 ± 0.31) %		–
$K^*(892)^- \geq 0$ neutrals ν_τ		(1.33 ± 0.13) %		–
$K^*(892)^- \nu_\tau$		(1.28 ± 0.08) %		665
$K^*(892)^0 K^- \geq 0$ neutrals ν_τ		$(3.2 \pm 1.4) \times 10^{-3}$		–
$K^*(892)^0 K^- \nu_\tau$		$(2.1 \pm 0.4) \times 10^{-3}$		539
$\overline{K}^*(892)^0 \pi^- \geq 0$ neutrals ν_τ		$(3.8 \pm 1.7) \times 10^{-3}$		–
$\overline{K}^*(892)^0 \pi^- \nu_\tau$		$(2.2 \pm 0.5) \times 10^{-3}$		653
$(\overline{K}^*(892)\pi)^- \nu_\tau \to \pi^- \overline{K}^0 \pi^0 \nu_\tau$		$(1.1 \pm 0.5) \times 10^{-3}$		–
$K_1(1270)^- \nu_\tau$		$(4 \pm 4) \times 10^{-3}$		433
$K_1(1400)^- \nu_\tau$		$(8 \pm 4) \times 10^{-3}$		335
$K^*_2(1430)^- \nu_\tau$		$< 3 \times 10^{-3}$	CL=95%	317
$\eta \pi^- \nu_\tau$		$< 1.4 \times 10^{-4}$	CL=95%	798
$\eta \pi^- \pi^0 \nu_\tau$	[l]	$(1.74 \pm 0.24) \times 10^{-3}$		778
$\eta \pi^- \pi^0 \pi^0 \nu_\tau$		$(1.4 \pm 0.7) \times 10^{-4}$		746
$\eta K^- \nu_\tau$		$(2.7 \pm 0.6) \times 10^{-4}$		720
$\eta \pi^+ \pi^- \pi^- \geq 0$ neutrals ν_τ		$< 3 \times 10^{-3}$	CL=90%	–
$\eta \pi^- \pi^+ \pi^- \nu_\tau$		$(3.4 \pm 0.8) \times 10^{-4}$		–
$\eta a_1(1260)^- \nu_\tau \to \eta \pi^- \rho^0 \nu_\tau$		$< 3.9 \times 10^{-4}$	CL=90%	–
$\eta \eta \pi^- \nu_\tau$		$< 1.1 \times 10^{-4}$	CL=95%	637
$\eta \eta \pi^- \pi^0 \nu_\tau$		$< 2.0 \times 10^{-4}$	CL=95%	559
$\eta'(958) \pi^- \nu_\tau$		$< 7.4 \times 10^{-5}$	CL=90%	–
$\eta'(958) \pi^- \pi^0 \nu_\tau$		$< 8.0 \times 10^{-5}$	CL=90%	–
$\phi \pi^- \nu_\tau$		$< 2.0 \times 10^{-4}$	CL=90%	585
$\phi K^- \nu_\tau$		$< 6.7 \times 10^{-5}$	CL=90%	–
$f_1(1285) \pi^- \nu_\tau$		$(5.8 \pm 2.3) \times 10^{-4}$		–
$f_1(1285) \pi^- \nu_\tau \to \eta \pi^- \pi^+ \pi^- \nu_\tau$		$(1.9 \pm 0.7) \times 10^{-4}$		–
$h^- \omega \geq 0$ neutrals ν_τ		(2.36 ± 0.08) %		–
$h^- \omega \nu_\tau$	[l]	(1.93 ± 0.06) %		–
$h^- \omega \pi^0 \nu_\tau$	[l]	$(4.3 \pm 0.5) \times 10^{-3}$		–
$h^- \omega 2\pi^0 \nu_\tau$		$(1.9 \pm 0.8) \times 10^{-4}$		–

Lepton Family number (*LF*), Lepton number (*L*), or Baryon number (*B*) violating modes
(In the modes below, ℓ means a sum over *e* and μ modes)

L means lepton number violation (*e.g.* $\tau^- \to e^+\pi^-\pi^-$). Following common usage, *LF* means lepton family violation *and not* lepton number violation (*e.g.* $\tau^- \to e^-\pi^+\pi^-$). *B* means baryon number violation.

$e^- \gamma$	*LF*	$< 2.7 \times 10^{-6}$	CL=90%	888
$\mu^- \gamma$	*LF*	$< 3.0 \times 10^{-6}$	CL=90%	885
$e^- \pi^0$	*LF*	$< 3.7 \times 10^{-6}$	CL=90%	883
$\mu^- \pi^0$	*LF*	$< 4.0 \times 10^{-6}$	CL=90%	880
$e^- K^0$	*LF*	$< 1.3 \times 10^{-3}$	CL=90%	819
$\mu^- K^0$	*LF*	$< 1.0 \times 10^{-3}$	CL=90%	815
$e^- \eta$	*LF*	$< 8.2 \times 10^{-6}$	CL=90%	804
$\mu^- \eta$	*LF*	$< 9.6 \times 10^{-6}$	CL=90%	800
$e^- \rho^0$	*LF*	$< 2.0 \times 10^{-6}$	CL=90%	722
$\mu^- \rho^0$	*LF*	$< 6.3 \times 10^{-6}$	CL=90%	718
$e^- K^*(892)^0$	*LF*	$< 5.1 \times 10^{-6}$	CL=90%	663
$\mu^- K^*(892)^0$	*LF*	$< 7.5 \times 10^{-6}$	CL=90%	657
$e^- \overline{K}^*(892)^0$	*LF*	$< 7.4 \times 10^{-6}$	CL=90%	663
$\mu^- \overline{K}^*(892)^0$	*LF*	$< 7.5 \times 10^{-6}$	CL=90%	657
$e^- \phi$	*LF*	$< 6.9 \times 10^{-6}$	CL=90%	596
$\mu^- \phi$	*LF*	$< 7.0 \times 10^{-6}$	CL=90%	590
$\pi^- \gamma$	*L*	$< 2.8 \times 10^{-4}$	CL=90%	883
$\pi^- \pi^0$	*L*	$< 3.7 \times 10^{-4}$	CL=90%	878
$e^- e^+ e^-$	*LF*	$< 2.9 \times 10^{-6}$	CL=90%	888
$e^- \mu^+ \mu^-$	*LF*	$< 1.8 \times 10^{-6}$	CL=90%	882
$e^+ \mu^- \mu^-$	*LF*	$< 1.5 \times 10^{-6}$	CL=90%	882
$\mu^- e^+ e^-$	*LF*	$< 1.7 \times 10^{-6}$	CL=90%	885
$\mu^+ e^- e^-$	*LF*	$< 1.5 \times 10^{-6}$	CL=90%	885
$\mu^- \mu^+ \mu^-$	*LF*	$< 1.9 \times 10^{-6}$	CL=90%	873
$e^- \pi^+ \pi^-$	*LF*	$< 2.2 \times 10^{-6}$	CL=90%	877
$e^+ \pi^- \pi^-$	*L*	$< 1.9 \times 10^{-6}$	CL=90%	877
$\mu^- \pi^+ \pi^-$	*LF*	$< 8.2 \times 10^{-6}$	CL=90%	866
$\mu^+ \pi^- \pi^-$	*L*	$< 3.4 \times 10^{-6}$	CL=90%	866
$e^- \pi^+ K^-$	*LF*	$< 6.4 \times 10^{-6}$	CL=90%	814
$e^- \pi^- K^+$	*LF*	$< 3.8 \times 10^{-6}$	CL=90%	814
$e^+ \pi^- K^-$	*L*	$< 2.1 \times 10^{-6}$	CL=90%	814
$e^- K^+ K^-$	*LF*	$< 6.0 \times 10^{-6}$	CL=90%	739
$e^+ K^- K^-$	*L*	$< 3.8 \times 10^{-6}$	CL=90%	739
$\mu^- \pi^+ K^-$	*LF*	$< 7.5 \times 10^{-6}$	CL=90%	800
$\mu^- \pi^- K^+$	*LF*	$< 7.4 \times 10^{-6}$	CL=90%	800
$\mu^+ \pi^- K^-$	*L*	$< 7.0 \times 10^{-6}$	CL=90%	800
$\mu^- K^+ K^-$	*LF*	$< 1.5 \times 10^{-5}$	CL=90%	699
$\mu^+ K^- K^-$	*L*	$< 6.0 \times 10^{-6}$	CL=90%	699
$e^- \pi^0 \pi^0$	*LF*	$< 6.5 \times 10^{-6}$	CL=90%	878
$\mu^- \pi^0 \pi^0$	*LF*	$< 1.4 \times 10^{-5}$	CL=90%	867
$e^- \eta \eta$	*LF*	$< 3.5 \times 10^{-5}$	CL=90%	700
$\mu^- \eta \eta$	*LF*	$< 6.0 \times 10^{-5}$	CL=90%	654
$e^- \pi^0 \eta$	*LF*	$< 2.4 \times 10^{-5}$	CL=90%	798
$\mu^- \pi^0 \eta$	*LF*	$< 2.2 \times 10^{-5}$	CL=90%	784
$\overline{p} \gamma$	*L,B*	$< 2.9 \times 10^{-4}$	CL=90%	641
$\overline{p} \pi^0$	*L,B*	$< 6.6 \times 10^{-4}$	CL=90%	632
$\overline{p} \eta$	*L,B*	$< 1.30 \times 10^{-3}$	CL=90%	476
e^- light boson	*LF*	$< 2.7 \times 10^{-3}$	CL=95%	–
μ^- light boson	*LF*	$< 5 \times 10^{-3}$	CL=95%	–

Heavy Charged Lepton Searches

$L^{\pm}$ – charged lepton

Mass $m >$ 80.2 GeV, CL = 95% $m_{\nu} \approx 0$

$L^{\pm}$ – stable charged heavy lepton

Mass $m >$ 84.2 GeV, CL = 95%

Neutrinos

See the Particle Listings for a Note "Neutrino Mass" giving details of neutrinos, masses, mixing, and the status of experimental searches.

ν_e

$J = \frac{1}{2}$

Mass m: Unexplained effects have resulted in significantly negative m^2 in the new, precise tritium beta decay experiments. It is felt that a real neutrino mass as large as 10–15 eV would cause observable spectral distortions even in the presence of the end-point count excesses.

Mean life/mass, τ/m_{ν_e} $>$ 7×10^9 s/eV (solar)

Mean life/mass, τ/m_{ν_e} $>$ 300 s/eV, CL = 90% (reactor)

Magnetic moment μ $<$ 1.8×10^{-10} μ_B, CL = 90%

ν_μ

$J = \frac{1}{2}$

Mass $m <$ 0.17 MeV, CL = 90%

Mean life/mass, τ/m_{ν_μ} $>$ 15.4 s/eV, CL = 90%

Magnetic moment μ $<$ 7.4×10^{-10} μ_B, CL = 90%

ν_τ

$J = \frac{1}{2}$

Mass $m <$ 18.2 MeV, CL = 95%

Magnetic moment μ $<$ 5.4×10^{-7} μ_B, CL = 90%

Electric dipole moment $d <$ 5.2×10^{-17} $e$cm, CL = 95%

Number of Light Neutrino Types

(including ν_e, ν_μ, and ν_τ)

Number $N = 2.994 \pm 0.012$ (Standard Model fits to LEP data)

Number $N = 3.07 \pm 0.12$ (Direct measurement of invisible Z width)

Massive Neutrinos and Lepton Mixing, Searches for

For excited leptons, see Compositeness Limits below.

See the Particle Listings for a Note "Neutrino Mass" giving details of neutrinos, masses, mixing, and the status of experimental searches.

While no direct, uncontested evidence for massive neutrinos or lepton mixing has been obtained, suggestive evidence has come from solar neutrino observations, from anomalies in the relative fractions of ν_e and ν_μ observed in energetic cosmic-ray air showers, and possibly from a $\overline{\nu}_e$ appearance experiment at Los Alamos. Sample limits are:

Stable Neutral Heavy Lepton Mass Limits

Mass $m >$ 45.0 GeV, CL = 95% (Dirac)

Mass $m >$ 39.5 GeV, CL = 95% (Majorana)

Neutral Heavy Lepton Mass Limits

Mass $m >$ 69.0 GeV, CL = 95% (Dirac ν_L coupling to e, μ, τ with $|U_{\ell j}|^2 > 10^{-12}$)

Mass $m >$ 58.2 GeV, CL = 95% (Majorana ν_L coupling to e, μ, τ with $|U_{\ell j}|^2 > 10^{-12}$)

Solar Neutrinos

Detectors using gallium ($E_\nu \gtrsim 0.2$ MeV), chlorine ($E_\nu \gtrsim 0.8$ MeV), and Čerenkov effect in water ($E_\nu \gtrsim 7$ MeV) measure significantly lower neutrino rates than are predicted from solar models. The deficit in the solar neutrino flux compared with solar model calculations could be explained by oscillations with $\Delta m^2 \leq 10^{-5}$ eV^2 causing the disappearance of ν_e.

Atmospheric Neutrinos

Underground detectors observing neutrinos produced by cosmic rays in the atmosphere have measured a ν_μ/ν_e ratio much less than expected and also a deficiency of upward going ν_μ compared to downward. This could be explained by oscillations leading to the disappearance of ν_μ with $\Delta m^2 \approx 10^{-3}$ to 10^{-2} eV^2.

ν oscillation: $\overline{\nu}_e \not\to \overline{\nu}_e$ (θ = mixing angle)

Δm^2 $<$ 9×10^{-4} eV^2, CL = 90% (if $\sin^2 2\theta = 1$)

$\sin^2 2\theta$ $<$ 0.02, CL = 90% (if $\Delta(m^2)$ is large)

ν oscillation: ν_μ ($\overline{\nu}_\mu$) $\to$ ν_e ($\overline{\nu}_e$) (any combination)

Δm^2 $<$ 0.075 eV^2, CL = 90% (if $\sin^2 2\theta = 1$)

$\sin^2 2\theta$ $<$ 1.8×10^{-3}, CL = 90% (if $\Delta(m^2)$ is large)

NOTES

In this Summary Table:

When a quantity has "(S = ...)" to its right, the error on the quantity has been enlarged by the "scale factor" S, defined as $S = \sqrt{\chi^2/(N-1)}$, where N is the number of measurements used in calculating the quantity. We do this when S $>$ 1, which often indicates that the measurements are inconsistent. When S $>$ 1.25, we also show in the Particle Listings an ideogram of the measurements. For more about S, see the Introduction.

A decay momentum p is given for each decay mode. For a 2-body decay, p is the momentum of each decay product in the rest frame of the decaying particle. For a 3-or-more-body decay, p is the largest momentum any of the products can have in this frame.

[a] The uncertainty in the electron mass in unified atomic mass units (u) is ten times smaller than that given by the 1986 CODATA adjustment, quoted in the Table of Physical Constants (Section 1). The conversion to MeV via the factor 931.49432(28) MeV/u is more uncertain because of the electron charge uncertainty. Our value in MeV differs slightly from the 1986 CODATA result.

[b] This is the best "electron disappearance" limit. The best limit for the mode $e^- \to \nu\gamma$ is $> 2.35 \times 10^{25}$ yr (CL=68%).

[c] The muon mass is most precisely known in u (unified atomic mass units). The conversion factor to MeV via the factor 931.49432(28) MeV/u is more uncertain because of the electron charge uncertainty.

[d] See the "Note on Muon Decay Parameters" in the μ Particle Listings for definitions and details.

[e] P_μ is the longitudinal polarization of the muon from pion decay. In standard $V-A$ theory, $P_\mu = 1$ and $\rho = \delta = 3/4$.

[f] This only includes events with the γ energy $>$ 10 MeV. Since the $e^-\overline{\nu}_e\nu_\mu$ and $e^-\overline{\nu}_e\nu_\mu\gamma$ modes cannot be clearly separated, we regard the latter mode as a subset of the former.

[g] See the μ Particle Listings for the energy limits used in this measurement.

[h] A test of additive vs. multiplicative lepton family number conservation.

[i] Basis mode for the τ.

QUARKS

The u-, d-, and s-quark masses are estimates of so-called "current-quark masses," in a mass-independent subtraction scheme such as $\overline{\text{MS}}$ at a scale $\mu \approx 2$ GeV. The c- and b-quark masses are estimated from charmonium, bottomonium, D, and B masses. They are the "running" masses in the $\overline{\text{MS}}$ scheme. These can be different from the heavy quark masses obtained in potential models.

u

$I(J^P) = \frac{1}{2}(\frac{1}{2}^+)$

Mass m = 1.5 to 5 MeV [a] Charge = $\frac{2}{3}$ e $I_z = +\frac{1}{2}$

m_u/m_d = 0.20 to 0.70

d

$I(J^P) = \frac{1}{2}(\frac{1}{2}^+)$

Mass m = 3 to 9 MeV [a] Charge = $-\frac{1}{3}$ e $I_z = -\frac{1}{2}$

m_s/m_d = 17 to 25

$\overline{m} = (m_u + m_d)/2$ = 2 to 6 MeV

s

$I(J^P) = 0(\frac{1}{2}^+)$

Mass m = 60 to 170 MeV [a] Charge = $-\frac{1}{3}$ e Strangeness = -1

$(m_s - (m_u + m_d)/2)/(m_d - m_u)$ = 34 to 51

c

$I(J^P) = 0(\frac{1}{2}^+)$

Mass m = 1.1 to 1.4 GeV Charge = $\frac{2}{3}$ e Charm = +1

b

$I(J^P) = 0(\frac{1}{2}^+)$

Mass m = 4.1 to 4.4 GeV Charge = $-\frac{1}{3}$ e Bottom = -1

t

$I(J^P) = 0(\frac{1}{2}^+)$

Charge = $\frac{2}{3}$ e Top = +1

Mass m = 173.8 ± 5.2 GeV (direct observation of top events)

Mass m = 170 ± 7 (+14) GeV (Standard Model electroweak fit, assuming $M_H = M_Z$. Number in parentheses is shift from changing M_H to 300 GeV.

b' (4th Generation) Quark, Searches for

Mass $m >$ 128 GeV, CL = 95% ($p\overline{p}$, charged current decays)

Mass $m >$ 46.0 GeV, CL = 95% ($e^+ e^-$, all decays)

Free Quark Searches

All searches since 1977 have had negative results.

NOTES

[a] The ratios m_u/m_d and m_s/m_d are extracted from pion and kaon masses using chiral symmetry. The estimates of u and d masses are not without controversy and remain under active investigation. Within the literature there are even suggestions that the u quark could be essentially massless. The s-quark mass is estimated from SU(3) splittings in hadron masses.

LIGHT UNFLAVORED MESONS ($S = C = B = 0$)

For $I = 1$ (π, b, ρ, a): $u\overline{d}$, $(u\overline{u}-d\overline{d})/\sqrt{2}$, $d\overline{u}$;
for $I = 0$ (η, η', h, h', ω, ϕ, f, f'): $c_1(u\overline{u} + d\overline{d}) + c_2(s\overline{s})$

$\pi^{\pm}$

$I^G(J^P) = 1^-(0^-)$

Mass $m = 139.56995 \pm 0.00035$ MeV
Mean life $\tau = (2.6033 \pm 0.0005) \times 10^{-8}$ s (S = 1.2)
$c\tau = 7.8045$ m

$\pi^{\pm} \to \ell^{\pm}\nu\gamma$ form factors [a]

$F_V = 0.017 \pm 0.008$
$F_A = 0.0116 \pm 0.0016$ (S = 1.3)
$R = 0.059^{+0.009}_{-0.008}$

π^- modes are charge conjugates of the modes below.

π^+ DECAY MODES		Fraction (Γ_i/Γ)	Confidence level	p (MeV/c)
$\mu^+\nu_\mu$	[b]	(99.98770±0.00004) %		30
$\mu^+\nu_\mu\gamma$	[c]	(1.24 ±0.25) × 10^{-4}		30
$e^+\nu_e$	[b]	(1.230 ±0.004) × 10^{-4}		70
$e^+\nu_e\gamma$	[c]	(1.61 ±0.23) × 10^{-7}		70
$e^+\nu_e\pi^0$		(1.025 ±0.034) × 10^{-8}		4
$e^+\nu_e e^+ e^-$		(3.2 ±0.5) × 10^{-9}		70
$e^+\nu_e\nu\overline{\nu}$		< 5 × 10^{-6}	90%	70
Lepton Family number (*LF*) or Lepton number (*L*) violating modes				
$\mu^+\overline{\nu}_e$	*L* [d]	< 1.5 × 10^{-3}	90%	30
$\mu^+\nu_e$	*LF* [d]	< 8.0 × 10^{-3}	90%	30
$\mu^- e^+ e^+ \nu$	*LF*	< 1.6 × 10^{-6}	90%	30

π^0

$I^G(J^{PC}) = 1^-(0^{-+})$

Mass $m = 134.9764 \pm 0.0006$ MeV
$m_{\pi^\pm} - m_{\pi^0} = 4.5936 \pm 0.0005$ MeV
Mean life $\tau = (8.4 \pm 0.6) \times 10^{-17}$ s (S = 3.0)
$c\tau = 25.1$ nm

π^0 DECAY MODES		Fraction (Γ_i/Γ)	Scale factor/ Confidence level	p (MeV/c)
2γ		(98.798±0.032) %	S=1.1	67
$e^+e^-\gamma$		(1.198±0.032) %	S=1.1	67
γ positronium		(1.82 ±0.29) × 10^{-9}		67
$e^+e^+e^-e^-$		(3.14 ±0.30) × 10^{-5}		67
e^+e^-		(7.5 ±2.0) × 10^{-8}		67
4γ		< 2 × 10^{-8}	CL=90%	67
$\nu\overline{\nu}$	[e]	< 8.3 × 10^{-7}	CL=90%	67
$\nu_e\overline{\nu}_e$		< 1.7 × 10^{-6}	CL=90%	67
$\nu_\mu\overline{\nu}_\mu$		< 3.1 × 10^{-6}	CL=90%	67
$\nu_\tau\overline{\nu}_\tau$		< 2.1 × 10^{-6}	CL=90%	67
Charge conjugation (*C*) or Lepton Family number (*LF*) violating modes				
3γ	*C*	< 3.1 × 10^{-8}	CL=90%	67
$\mu^+e^- + e^-\mu^+$	*LF*	< 1.72 × 10^{-8}	CL=90%	26

η

$I^G(J^{PC}) = 0^+(0^{-+})$

Mass $m = 547.30 \pm 0.12$ MeV
Full width $\Gamma = 1.18 \pm 0.11$ keV [f] (S = 1.8)

C-nonconserving decay parameters

$\pi^+\pi^-\pi^0$ Left-right asymmetry = $(0.09 \pm 0.17) \times 10^{-2}$
$\pi^+\pi^-\pi^0$ Sextant asymmetry = $(0.18 \pm 0.16) \times 10^{-2}$
$\pi^+\pi^-\pi^0$ Quadrant asymmetry = $(-0.17 \pm 0.17) \times 10^{-2}$
$\pi^+\pi^-\gamma$ Left-right asymmetry = $(0.9 \pm 0.4) \times 10^{-2}$
$\pi^+\pi^-\gamma$ β (*D*-wave) = 0.05 ± 0.06 (S = 1.5)

Dalitz plot parameter

$\pi^0\pi^0\pi^0$ $\alpha = -0.039 \pm 0.015$

η DECAY MODES		Fraction (Γ_i/Γ)	Scale factor/ Confidence level	p (MeV/c)
Neutral modes				
neutral modes		(71.5 ±0.6) %	S=1.4	–
2γ	[f]	(39.21±0.34) %	S=1.4	274
$3\pi^0$		(32.2 ±0.4) %	S=1.3	178
$\pi^0 2\gamma$		(7.1 ±1.4) × 10^{-4}		257
other neutral modes		< 2.8 %	CL=90%	–
Charged modes				
charged modes		(28.5 ±0.6) %	S=1.4	–
$\pi^+\pi^-\pi^0$		(23.1 ±0.5) %	S=1.4	173
$\pi^+\pi^-\gamma$		(4.77±0.13) %	S=1.3	235
$e^+e^-\gamma$		(4.9 ±1.1) × 10^{-3}		274
$\mu^+\mu^-\gamma$		(3.1 ±0.4) × 10^{-4}		252
e^+e^-		< 7.7 × 10^{-5}	CL=90%	274
$\mu^+\mu^-$		(5.8 ±0.8) × 10^{-6}		252
$\pi^+\pi^-e^+e^-$		(1.3 $^{+1.2}_{-0.8}$) × 10^{-3}		235
$\pi^+\pi^-2\gamma$		< 2.1 × 10^{-3}		235
$\pi^+\pi^-\pi^0\gamma$		< 6 × 10^{-4}	CL=90%	173
$\pi^0\mu^+\mu^-\gamma$		< 3 × 10^{-6}	CL=90%	210
Charge conjugation (*C*), Parity (*P*), Charge conjugation × Parity (*CP*), or Lepton Family number (*LF*) violating modes				
$\pi^+\pi^-$	*P,CP*	< 9 × 10^{-4}	CL=90%	235
3γ	*C*	< 5 × 10^{-4}	CL=95%	274
$\pi^0e^+e^-$	*C* [g]	< 4 × 10^{-5}	CL=90%	257
$\pi^0\mu^+\mu^-$	*C* [g]	< 5 × 10^{-6}	CL=90%	210
$\mu^+e^- + \mu^-e^+$	*LF*	< 6 × 10^{-6}	CL=90%	263

$f_0(400\text{–}1200)$ [h] or σ

$I^G(J^{PC}) = 0^+(0^{++})$

Mass $m = (400\text{–}1200)$ MeV
Full width $\Gamma = (600\text{–}1000)$ MeV

$f_0(400\text{–}1200)$ DECAY MODES	Fraction (Γ_i/Γ)	p (MeV/c)
$\pi\pi$	dominant	–
$\gamma\gamma$	seen	–

ρ(770) [l]

$I^G(J^{PC}) = 1^+(1^{--})$

Mass $m = 770.0 \pm 0.8$ MeV (S = 1.8)
Full width $\Gamma = 150.7 \pm 1.1$ MeV
$\Gamma_{ee} = 6.77 \pm 0.32$ keV

ρ(770) DECAY MODES	Fraction (Γ_i/Γ)	Scale factor/ Confidence level	p (MeV/c)
$\pi\pi$	~ 100 %		358
ρ(770)± decays			
$\pi^\pm\gamma$	(4.5 ±0.5) × 10^{-4}	S=2.2	372
$\pi^\pm\eta$	< 6 × 10^{-3}	CL=84%	146
$\pi^\pm\pi^+\pi^-\pi^0$	< 2.0 × 10^{-3}	CL=84%	249
ρ(770)⁰ decays			
$\pi^+\pi^-\gamma$	(9.9 ±1.6) × 10^{-3}		358
$\pi^0\gamma$	(6.8 ±1.7) × 10^{-4}		372
$\eta\gamma$	(2.4 $^{+0.8}_{-0.9}$) × 10^{-4}	S=1.6	189
$\mu^+\mu^-$	[j] (4.60±0.28) × 10^{-5}		369
e^+e^-	[j] (4.49±0.22) × 10^{-5}		384
$\pi^+\pi^-\pi^0$	< 1.2 × 10^{-4}	CL=90%	319
$\pi^+\pi^-\pi^+\pi^-$	< 2 × 10^{-4}	CL=90%	246
$\pi^+\pi^-\pi^0\pi^0$	< 4 × 10^{-5}	CL=90%	252

ω(782)

$I^G(J^{PC}) = 0^-(1^{--})$

Mass $m = 781.94 \pm 0.12$ MeV (S = 1.5)
Full width $\Gamma = 8.41 \pm 0.09$ MeV
$\Gamma_{ee} = 0.60 \pm 0.02$ keV

ω(782) DECAY MODES	Fraction (Γ_i/Γ)	Scale factor/ Confidence level	p (MeV/c)
$\pi^+\pi^-\pi^0$	(88.8 ±0.7) %		327
$\pi^0\gamma$	(8.5 ±0.5) %		379
$\pi^+\pi^-$	(2.21±0.30) %		365
neutrals (excluding $\pi^0\gamma$)	(5.3 $^{+8.7}_{-3.5}$) × 10^{-3}		–
$\eta\gamma$	(6.5 ±1.0) × 10^{-4}		199
$\pi^0e^+e^-$	(5.9 ±1.9) × 10^{-4}		379
$\pi^0\mu^+\mu^-$	(9.6 ±2.3) × 10^{-5}		349
e^+e^-	(7.07±0.19) × 10^{-5}	S=1.1	391
$\pi^+\pi^-\pi^0\pi^0$	< 2 %	CL=90%	261
$\pi^+\pi^-\gamma$	< 3.6 × 10^{-3}	CL=95%	365
$\pi^+\pi^-\pi^+\pi^-$	< 1 × 10^{-3}	CL=90%	256
$\pi^0\pi^0\gamma$	(7.2 ±2.5) × 10^{-5}		367
$\mu^+\mu^-$	< 1.8 × 10^{-4}	CL=90%	376
3γ	< 1.9 × 10^{-4}	CL=95%	391
Charge conjugation (C) violating modes			
$\eta\pi^0$	C < 1 × 10^{-3}	CL=90%	162
$3\pi^0$	C < 3 × 10^{-4}	CL=90%	329

η′(958)

$I^G(J^{PC}) = 0^+(0^{-+})$

Mass $m = 957.78 \pm 0.14$ MeV
Full width $\Gamma = 0.203 \pm 0.016$ MeV (S = 1.3)

η′(958) DECAY MODES	Fraction (Γ_i/Γ)	Scale factor/ Confidence level	p (MeV/c)
$\pi^+\pi^-\eta$	(43.8 ±1.5) %	S=1.1	232
$\rho^0\gamma$ (including non-resonant $\pi^+\pi^-\gamma$)	(30.2 ±1.3) %	S=1.1	169
$\pi^0\pi^0\eta$	(20.7 ±1.3) %	S=1.2	239
$\omega\gamma$	(3.01±0.30) %		160
$\gamma\gamma$	(2.11±0.13) %	S=1.2	479
$3\pi^0$	(1.54±0.26) × 10^{-3}		430
$\mu^+\mu^-\gamma$	(1.03±0.26) × 10^{-4}		467
$\pi^+\pi^-\pi^0$	< 5 %	CL=90%	427
$\pi^0\rho^0$	< 4 %	CL=90%	118
$\pi^+\pi^+\pi^-\pi^-$	< 1 %	CL=90%	372
$\pi^+\pi^+\pi^-\pi^-$ neutrals	< 1 %	CL=95%	–
$\pi^+\pi^+\pi^-\pi^-\pi^0$	< 1 %	CL=90%	298
6π	< 1 %	CL=90%	189
$\pi^+\pi^-e^+e^-$	< 6 × 10^{-3}	CL=90%	458
$\pi^0\gamma\gamma$	< 8 × 10^{-4}	CL=90%	469
$4\pi^0$	< 5 × 10^{-4}	CL=90%	379
e^+e^-	< 2.1 × 10^{-7}	CL=90%	479
Charge conjugation (C) or Parity (P) violating modes			
$\pi^+\pi^-$	P,CP < 2 %	CL=90%	458
$\pi^0\pi^0$	P,CP < 9 × 10^{-4}	CL=90%	459
$\pi^0e^+e^-$	C [g] < 1.3 %	CL=90%	469
ηe^+e^-	C [g] < 1.1 %	CL=90%	322
3γ	C < 1.0 × 10^{-4}	CL=90%	479
$\mu^+\mu^-\pi^0$	C [g] < 6.0 × 10^{-5}	CL=90%	445
$\mu^+\mu^-\eta$	C [g] < 1.5 × 10^{-5}	CL=90%	274

f_0(980) [k]

$I^G(J^{PC}) = 0^+(0^{++})$

Mass $m = 980 \pm 10$ MeV
Full width Γ = 40 to 100 MeV

f_0(980) DECAY MODES	Fraction (Γ_i/Γ)	Confidence level	p (MeV/c)
$\pi\pi$	dominant		470
$K\overline{K}$	seen		–
$\gamma\gamma$	(1.19±0.33) × 10^{-5}		490
e^+e^-	< 3 × 10^{-7}	90%	490

a_0(980) [k]

$I^G(J^{PC}) = 1^-(0^{++})$

Mass $m = 983.4 \pm 0.9$ MeV
Full width Γ = 50 to 100 MeV

a_0(980) DECAY MODES	Fraction (Γ_i/Γ)	p (MeV/c)
$\eta\pi$	dominant	321
$K\overline{K}$	seen	–
$\gamma\gamma$	seen	492

φ(1020)

$I^G(J^{PC}) = 0^-(1^{--})$

Mass $m = 1019.413 \pm 0.008$ MeV
Full width $\Gamma = 4.43 \pm 0.05$ MeV

φ(1020) DECAY MODES	Fraction (Γ_i/Γ)	Scale factor/ Confidence level	p (MeV/c)
K^+K^-	(49.1 ±0.8) %	S=1.3	127
$K^0_LK^0_S$	(34.1 ±0.6) %	S=1.2	110
$\rho\pi + \pi^+\pi^-\pi^0$	(15.5 ±0.7) %	S=1.5	–
$\eta\gamma$	(1.26±0.06) %	S=1.1	363
$\pi^0\gamma$	(1.31±0.13) × 10^{-3}		501
e^+e^-	(2.99±0.08) × 10^{-4}	S=1.2	510
$\mu^+\mu^-$	(2.5 ±0.4) × 10^{-4}		499
ηe^+e^-	(1.3 $^{+0.8}_{-0.6}$) × 10^{-4}		363
$\pi^+\pi^-$	(8 $^{+5}_{-4}$) × 10^{-5}	S=1.5	490
$\omega\gamma$	< 5 %	CL=84%	210
$\rho\gamma$	< 7 × 10^{-4}	CL=90%	219
$\pi^+\pi^-\gamma$	< 3 × 10^{-5}	CL=90%	490
$f_0(980)\gamma$	< 1 × 10^{-4}	CL=90%	39
$\pi^0\pi^0\gamma$	< 1 × 10^{-3}	CL=90%	492
$\pi^+\pi^-\pi^+\pi^-$	< 8.7 × 10^{-4}	CL=90%	410
$\pi^+\pi^+\pi^-\pi^-\pi^0$	< 1.5 × 10^{-4}	CL=95%	341
$\pi^0e^+e^-$	< 1.2 × 10^{-4}	CL=90%	501
$\pi^0\eta\gamma$	< 2.5 × 10^{-3}	CL=90%	346
$a_0(980)\gamma$	< 5 × 10^{-3}	CL=90%	36
$\eta'(958)\gamma$	(1.2 $^{+0.7}_{-0.5}$) × 10^{-4}		–
$\mu^+\mu^-\gamma$	(2.3 ±1.0) × 10^{-5}		–

h_1(1170)

$I^G(J^{PC}) = 0^-(1^{+-})$

Mass $m = 1170 \pm 20$ MeV
Full width $\Gamma = 360 \pm 40$ MeV

h_1(1170) DECAY MODES	Fraction (Γ_i/Γ)	p (MeV/c)
$\rho\pi$	seen	310

$b_1(1235)$ $\quad I^G(J^{PC}) = 1^+(1^{+-})$

Mass $m = 1229.5 \pm 3.2$ MeV $\quad (S = 1.6)$
Full width $\Gamma = 142 \pm 9$ MeV $\quad (S = 1.2)$

$b_1(1235)$ DECAY MODES	Fraction (Γ_i/Γ)	Confidence level	p (MeV/c)
$\omega\pi$	dominant		348
[D/S amplitude ratio $= 0.29 \pm 0.04$]			
$\pi^{\pm}\gamma$	$(1.6\pm0.4) \times 10^{-3}$		608
$\eta\rho$	seen		–
$\pi^+\pi^+\pi^-\pi^0$	< 50 %	84%	536
$(K\overline{K})^{\pm}\pi^0$	< 8 %	90%	248
$K^0_S K^0_L \pi^{\pm}$	< 6 %	90%	238
$K^0_S K^0_S \pi^{\pm}$	< 2 %	90%	238
$\phi\pi$	< 1.5 %	84%	146

$a_1(1260)$ [l] $\quad I^G(J^{PC}) = 1^-(1^{++})$

Mass $m = 1230 \pm 40$ MeV [m]
Full width $\Gamma = 250$ to 600 MeV

$a_1(1260)$ DECAY MODES	Fraction (Γ_i/Γ)	p (MeV/c)
$\rho\pi$	dominant	356
[D/S amplitude ratio $= -0.100 \pm 0.028$]		
$\pi\gamma$	seen	607
$\pi(\pi\pi)_{S\text{-wave}}$	possibly seen	575

$f_2(1270)$ $\quad I^G(J^{PC}) = 0^+(2^{++})$

Mass $m = 1275.0 \pm 1.2$ MeV
Full width $\Gamma = 185.5^{+3.8}_{-2.7}$ MeV $\quad (S = 1.5)$

$f_2(1270)$ DECAY MODES	Fraction (Γ_i/Γ)	Scale factor/ Confidence level	p (MeV/c)
$\pi\pi$	$(84.6\ ^{+2.5}_{-1.3})$ %	S=1.3	622
$\pi^+\pi^-2\pi^0$	$(7.2\ ^{+1.5}_{-2.7})$ %	S=1.3	562
$K\overline{K}$	(4.6 ± 0.4) %	S=2.8	403
$2\pi^+2\pi^-$	(2.8 ± 0.4) %	S=1.2	559
$\eta\eta$	$(4.5 \pm 1.0) \times 10^{-3}$	S=2.4	327
$4\pi^0$	$(3.0 \pm 1.0) \times 10^{-3}$		564
$\gamma\gamma$	$(1.32^{+0.17}_{-0.16}) \times 10^{-5}$		637
$\eta\pi\pi$	$< 8 \times 10^{-3}$	CL=95%	475
$K^0K^-\pi^+$ + c.c.	$< 3.4 \times 10^{-3}$	CL=95%	293
e^+e^-	$< 9 \times 10^{-9}$	CL=90%	637

$f_1(1285)$ $\quad I^G(J^{PC}) = 0^+(1^{++})$

Mass $m = 1281.9 \pm 0.6$ MeV $\quad (S = 1.7)$
Full width $\Gamma = 24.0 \pm 1.2$ MeV $\quad (S = 1.4)$

$(4\pi = \rho(\pi\pi)_{P\,wave})$

$f_1(1285)$ DECAY MODES	Fraction (Γ_i/Γ)	Scale factor/ Confidence level	p (MeV/c)
4π	(35 ± 4) %	S=1.6	563
$\quad \pi^0\pi^0\pi^+\pi^-$	(23.5 ± 3.0) %	S=1.6	566
$\quad 2\pi^+2\pi^-$	(11.7 ± 1.5) %	S=1.6	563
$\quad\quad \rho^0\pi^+\pi^-$	(11.7 ± 1.5) %	S=1.6	340
$4\pi^0$	$< 7 \times 10^{-4}$	CL=90%	568
$\eta\pi\pi$	(50 ± 18) %		479
$\quad a_0(980)\pi$ [ignoring $a_0(980) \to K\overline{K}$]	(34 ± 8) %	S=1.2	234
$\quad \eta\pi\pi$ [excluding $a_0(980)\pi$]	(15 ± 7) %	S=1.1	–
$K\overline{K}\pi$	(9.0 ± 1.2) %	S=1.5	308
$\quad K\overline{K}^*(892)$	not seen		–
$\gamma\rho^0$	(5.4 ± 1.2) %	S=2.3	410
$\phi\gamma$	$(7.9 \pm 3.0) \times 10^{-4}$		236

$\eta(1295)$ $\quad I^G(J^{PC}) = 0^+(0^{-+})$

Mass $m = 1297.0 \pm 2.8$ MeV
Full width $\Gamma = 53 \pm 6$ MeV

$\eta(1295)$ DECAY MODES	Fraction (Γ_i/Γ)	p (MeV/c)
$\eta\pi^+\pi^-$	seen	488
$a_0(980)\pi$	seen	245
$\eta\pi^0\pi^0$	seen	–
$\eta(\pi\pi)_{S\text{-wave}}$	seen	–

$\pi(1300)$ $\quad I^G(J^{PC}) = 1^-(0^{-+})$

Mass $m = 1300 \pm 100$ MeV [m]
Full width $\Gamma = 200$ to 600 MeV

$\pi(1300)$ DECAY MODES	Fraction (Γ_i/Γ)	p (MeV/c)
$\rho\pi$	seen	406
$\pi(\pi\pi)_{S\text{-wave}}$	seen	–

$a_2(1320)$ $\quad I^G(J^{PC}) = 1^-(2^{++})$

Mass $m = 1318.1 \pm 0.6$ MeV $\quad (S = 1.1)$
Full width $\Gamma = 107 \pm 5$ MeV [m] $\quad$ ($K^{\pm}K^0_S$ and $\eta\pi$ modes)

$a_2(1320)$ DECAY MODES	Fraction (Γ_i/Γ)	Scale factor/ Confidence level	p (MeV/c)
$\rho\pi$	(70.1 ± 2.7) %	S=1.2	419
$\eta\pi$	(14.5 ± 1.2) %		535
$\omega\pi\pi$	(10.6 ± 3.2) %	S=1.3	362
$K\overline{K}$	(4.9 ± 0.8) %		437
$\eta'(958)\pi$	$(5.3 \pm 0.9) \times 10^{-3}$		287
$\pi^{\pm}\gamma$	$(2.8 \pm 0.6) \times 10^{-3}$		652
$\gamma\gamma$	$(9.4 \pm 0.7) \times 10^{-6}$		659
$\pi^+\pi^-\pi^-$	< 8 %	CL=90%	621
e^+e^-	$< 2.3 \times 10^{-7}$	CL=90%	659

$f_0(1370)$ [k] $\quad I^G(J^{PC}) = 0^+(0^{++})$

Mass $m = 1200$ to 1500 MeV
Full width $\Gamma = 200$ to 500 MeV

$f_0(1370)$ DECAY MODES	Fraction (Γ_i/Γ)	p (MeV/c)
$\pi\pi$	seen	–
4π	seen	–
$\quad 4\pi^0$	seen	–
$\quad 2\pi^+2\pi^-$	seen	–
$\quad \pi^+\pi^-2\pi^0$	seen	–
$\quad 2(\pi\pi)_{S\text{-wave}}$	seen	–
$\eta\eta$	seen	–
$K\overline{K}$	seen	–
$\gamma\gamma$	seen	–
e^+e^-	not seen	–

$f_1(1420)$ [n] $\quad I^G(J^{PC}) = 0^+(1^{++})$

Mass $m = 1426.2 \pm 1.2$ MeV $\quad (S = 1.3)$
Full width $\Gamma = 55.0 \pm 3.0$ MeV

$f_1(1420)$ DECAY MODES	Fraction (Γ_i/Γ)	p (MeV/c)
$K\overline{K}\pi$	dominant	439
$\quad K\overline{K}^*(892)$ + c.c.	dominant	155
$\eta\pi\pi$	possibly seen	571

$\omega(1420)$ [o] $\quad I^G(J^{PC}) = 0^-(1^{--})$

Mass $m = 1419 \pm 31$ MeV
Full width $\Gamma = 174 \pm 60$ MeV

$\omega(1420)$ DECAY MODES	Fraction (Γ_i/Γ)	p (MeV/c)
$\rho\pi$	dominant	488

$\eta(1440)$ [p]

$I^G(J^{PC}) = 0^+(0^{-+})$

Mass m = 1400 - 1470 MeV [m]
Full width Γ = 50 - 80 MeV [m]

$\eta(1440)$ DECAY MODES	Fraction (Γ_i/Γ)	p (MeV/c)
$K\overline{K}\pi$	seen	–
$K\overline{K}^*(892)$+ c.c.	seen	–
$\eta\pi\pi$	seen	–
$a_0(980)\pi$	seen	–
$\eta(\pi\pi)_{S\text{-wave}}$	seen	–
4π	seen	–

$a_0(1450)$

$I^G(J^{PC}) = 1^-(0^{++})$

Mass m = 1474 ± 19 MeV
Full width Γ = 265 ± 13 MeV

$a_0(1450)$ DECAY MODES	Fraction (Γ_i/Γ)	p (MeV/c)
$\pi\eta$	seen	613
$\pi\eta'(958)$	seen	392
$K\overline{K}$	seen	530

$\rho(1450)$ [q]

$I^G(J^{PC}) = 1^+(1^{--})$

Mass m = 1465 ± 25 MeV [m]
Full width Γ = 310 ± 60 MeV [m]

$\rho(1450)$ DECAY MODES	Fraction (Γ_i/Γ)	Confidence level	p (MeV/c)
$\pi\pi$	seen		719
4π	seen		665
$\omega\pi$	<2.0 %	95%	512
e^+e^-	seen		732
$\eta\rho$	<4 %		317
$\phi\pi$	<1 %		358
$K\overline{K}$	$<1.6 \times 10^{-3}$	95%	541

$f_0(1500)$ [r]

$I^G(J^{PC}) = 0^+(0^{++})$

Mass m = 1500 ± 10 MeV (S = 1.3)
Full width Γ = 112 ± 10 MeV

$f_0(1500)$ DECAY MODES	Fraction (Γ_i/Γ)	p (MeV/c)
$\eta\eta'(958)$	seen	–
$\eta\eta$	seen	513
4π	seen	–
$4\pi^0$	seen	690
$2\pi^+2\pi^-$	seen	686
2π	seen	–
$\pi^+\pi^-$	seen	737
$2\pi^0$	seen	738
$K\overline{K}$	seen	563

$f_2'(1525)$

$I^G(J^{PC}) = 0^+(2^{++})$

Mass m = 1525 ± 5 MeV [m]
Full width Γ = 76 ± 10 MeV [m]

$f_2'(1525)$ DECAY MODES	Fraction (Γ_i/Γ)	p (MeV/c)
$K\overline{K}$	(88.8 ±3.1) %	581
$\eta\eta$	(10.3 ±3.1) %	531
$\pi\pi$	(8.2 ±1.5) $\times 10^{-3}$	750
$\gamma\gamma$	(1.32±0.21) $\times 10^{-6}$	763

$\omega(1600)$ [s]

$I^G(J^{PC}) = 0^-(1^{--})$

Mass m = 1649 ± 24 MeV (S = 2.3)
Full width Γ = 220 ± 35 MeV (S = 1.6)

$\omega(1600)$ DECAY MODES	Fraction (Γ_i/Γ)	p (MeV/c)
$\rho\pi$	seen	637
$\omega\pi\pi$	seen	601
e^+e^-	seen	824

$\omega_3(1670)$

$I^G(J^{PC}) = 0^-(3^{--})$

Mass m = 1667 ± 4 MeV
Full width Γ = 168 ± 10 MeV [m]

$\omega_3(1670)$ DECAY MODES	Fraction (Γ_i/Γ)	p (MeV/c)
$\rho\pi$	seen	647
$\omega\pi\pi$	seen	614
$b_1(1235)\pi$	possibly seen	359

$\pi_2(1670)$

$I^G(J^{PC}) = 1^-(2^{-+})$

Mass m = 1670 ± 20 MeV [m]
Full width Γ = 258 ± 18 MeV [m] (S = 1.7)

$\pi_2(1670)$ DECAY MODES	Fraction (Γ_i/Γ)	p (MeV/c)
3π	(95.8±1.4) %	806
$f_2(1270)\pi$	(56.2±3.2) %	325
$\rho\pi$	(31 ±4) %	649
$f_0(1370)\pi$	(8.7±3.4) %	–
$K\overline{K}^*(892)$+ c.c.	(4.2±1.4) %	453

$\phi(1680)$

$I^G(J^{PC}) = 0^-(1^{--})$

Mass m = 1680 ± 20 MeV [m]
Full width Γ = 150 ± 50 MeV [m]

$\phi(1680)$ DECAY MODES	Fraction (Γ_i/Γ)	p (MeV/c)
$K\overline{K}^*(892)$+ c.c.	dominant	463
$K_S^0 K\pi$	seen	620
$K\overline{K}$	seen	681
e^+e^-	seen	840
$\omega\pi\pi$	not seen	622

$\rho_3(1690)$

$I^G(J^{PC}) = 1^+(3^{--})$

J^P from the 2π and $K\overline{K}$ modes.
Mass m = 1691 ± 5 MeV [m]
Full width Γ = 160 ± 10 MeV [m] (S = 1.5)

$\rho_3(1690)$ DECAY MODES	Fraction (Γ_i/Γ)	Scale factor	p (MeV/c)
4π	(71.1 ± 1.9) %		788
$\pi^\pm\pi^+\pi^-\pi^0$	(67 ±22) %		788
$\omega\pi$	(16 ± 6) %		656
$\pi\pi$	(23.6 ± 1.3) %		834
$K\overline{K}\pi$	(3.8 ± 1.2) %		628
$K\overline{K}$	(1.58± 0.26) %	1.2	686
$\eta\pi^+\pi^-$	seen		728

$\rho(1700)$ [q]

$I^G(J^{PC}) = 1^+(1^{--})$

Mass $m = 1700 \pm 20$ MeV [m] ($\eta\rho^0$ and $\pi^+\pi^-$ modes)
Full width $\Gamma = 240 \pm 60$ MeV [m] ($\eta\rho^0$ and $\pi^+\pi^-$ modes)

$\rho(1700)$ DECAY MODES	Fraction (Γ_i/Γ)	p (MeV/c)
$\rho\pi\pi$	dominant	640
$2(\pi^+\pi^-)$	large	792
$\rho^0\pi^+\pi^-$	large	640
$\rho^\pm\pi^\mp\pi^0$	large	642
$\pi^+\pi^-$	seen	838
$\pi^-\pi^0$	seen	839
$K\overline{K}^*(892)$+ c.c.	seen	479
$\eta\rho$	seen	533
$K\overline{K}$	seen	692
e^+e^-	seen	850
$\pi^0\omega$	seen	662

$f_J(1710)$ [t]

$I^G(J^{PC}) = 0^+(\text{even}^{++})$

Mass $m = 1712 \pm 5$ MeV (S = 1.1)
Full width $\Gamma = 133 \pm 14$ MeV (S = 1.2)

$f_J(1710)$ DECAY MODES	Fraction (Γ_i/Γ)	p (MeV/c)
$K\overline{K}$	seen	690
$\eta\eta$	seen	648
$\pi\pi$	seen	837

$\pi(1800)$

$I^G(J^{PC}) = 1^-(0^{-+})$

Mass $m = 1801 \pm 13$ MeV (S = 1.9)
Full width $\Gamma = 210 \pm 15$ MeV

$\pi(1800)$ DECAY MODES	Fraction (Γ_i/Γ)	p (MeV/c)
$\pi^+\pi^-\pi^-$	seen	–
$f_0(980)\pi^-$	seen	623
$f_0(1370)\pi^-$	seen	–
$\rho\pi^-$	not seen	728
$\eta\eta\pi^-$	seen	–
$a_0(980)\eta$	seen	459
$f_0(1500)\pi^-$	seen	240
$\eta\eta'(958)\pi^-$	seen	–
$K_0^*(1430)K^-$	seen	–
$K^*(892)K^-$	not seen	560

$\phi_3(1850)$

$I^G(J^{PC}) = 0^-(3^{--})$

Mass $m = 1854 \pm 7$ MeV
Full width $\Gamma = 87^{+28}_{-23}$ MeV (S = 1.2)

$\phi_3(1850)$ DECAY MODES	Fraction (Γ_i/Γ)	p (MeV/c)
$K\overline{K}$	seen	785
$K\overline{K}^*(892)$+ c.c.	seen	602

$f_2(2010)$

$I^G(J^{PC}) = 0^+(2^{++})$

Seen by one group only.
Mass $m = 2011^{+60}_{-80}$ MeV
Full width $\Gamma = 202 \pm 60$ MeV

$f_2(2010)$ DECAY MODES	Fraction (Γ_i/Γ)	p (MeV/c)
$\phi\phi$	seen	–

$a_4(2040)$

$I^G(J^{PC}) = 1^-(4^{++})$

Mass $m = 2020 \pm 16$ MeV
Full width $\Gamma = 387 \pm 70$ MeV

$a_4(2040)$ DECAY MODES	Fraction (Γ_i/Γ)	p (MeV/c)
$K\overline{K}$	seen	892
$\pi^+\pi^-\pi^0$	seen	–
$\eta\pi^0$	seen	941

$f_4(2050)$

$I^G(J^{PC}) = 0^+(4^{++})$

Mass $m = 2044 \pm 11$ MeV (S = 1.4)
Full width $\Gamma = 208 \pm 13$ MeV (S = 1.2)

$f_4(2050)$ DECAY MODES	Fraction (Γ_i/Γ)	p (MeV/c)
$\omega\omega$	(26 ±6) %	658
$\pi\pi$	(17.0±1.5) %	1012
$K\overline{K}$	$(6.8^{+3.4}_{-1.8}) \times 10^{-3}$	895
$\eta\eta$	$(2.1\pm0.8) \times 10^{-3}$	863
$4\pi^0$	< 1.2 %	977

$f_2(2300)$

$I^G(J^{PC}) = 0^+(2^{++})$

Mass $m = 2297 \pm 28$ MeV
Full width $\Gamma = 149 \pm 40$ MeV

$f_2(2300)$ DECAY MODES	Fraction (Γ_i/Γ)	p (MeV/c)
$\phi\phi$	seen	529

$f_2(2340)$

$I^G(J^{PC}) = 0^+(2^{++})$

Mass $m = 2339 \pm 60$ MeV
Full width $\Gamma = 319^{+80}_{-70}$ MeV

$f_2(2340)$ DECAY MODES	Fraction (Γ_i/Γ)	p (MeV/c)
$\phi\phi$	seen	573

STRANGE MESONS
$(S = \pm 1,\ C = B = 0)$

$K^+ = u\overline{s}$, $K^0 = d\overline{s}$, $\overline{K}^0 = \overline{d}s$, $K^- = \overline{u}s$, similarly for K^*'s

$K^\pm$

$I(J^P) = \frac{1}{2}(0^-)$

Mass $m = 493.677 \pm 0.016$ MeV [u] (S = 2.8)
Mean life $\tau = (1.2386 \pm 0.0024) \times 10^{-8}$ s (S = 2.0)
$c\tau = 3.713$ m

Slope parameter g [v]

(See Particle Listings for quadratic coefficients)

$K^+ \rightarrow \pi^+\pi^+\pi^- = -0.2154 \pm 0.0035$ (S = 1.4)
$K^- \rightarrow \pi^-\pi^-\pi^+ = -0.217 \pm 0.007$ (S = 2.5)
$K^\pm \rightarrow \pi^\pm\pi^0\pi^0 = 0.594 \pm 0.019$ (S = 1.3)

$K^\pm$ decay form factors [a,w]

K^+_{e3} $\lambda_+ = 0.0286 \pm 0.0022$
$K^+_{\mu 3}$ $\lambda_+ = 0.032 \pm 0.008$ (S = 1.6)
$K^+_{\mu 3}$ $\lambda_0 = 0.006 \pm 0.007$ (S = 1.6)
K^+_{e3} $|f_S/f_+| = 0.084 \pm 0.023$ (S = 1.2)
K^+_{e3} $|f_T/f_+| = 0.38 \pm 0.11$ (S = 1.1)
$K^+_{\mu 3}$ $|f_T/f_+| = 0.02 \pm 0.12$
$K^+ \rightarrow e^+\nu_e\gamma$ $|F_A + F_V| = 0.148 \pm 0.010$
$K^+ \rightarrow \mu^+\nu_\mu\gamma$ $|F_A + F_V| < 0.23$, CL = 90%
$K^+ \rightarrow e^+\nu_e\gamma$ $|F_A - F_V| < 0.49$
$K^+ \rightarrow \mu^+\nu_\mu\gamma$ $|F_A - F_V| = -2.2$ to 0.3

K^- modes are charge conjugates of the modes below.

K^+ DECAY MODES		Fraction (Γ_i/Γ)	Scale factor/ Confidence level	p (MeV/c)
$\mu^+\nu_\mu$		(63.51 ± 0.18) %	S=1.3	236
$e^+\nu_e$		$(1.55 \pm 0.07) \times 10^{-5}$		247
$\pi^+\pi^0$		(21.16 ± 0.14) %	S=1.1	205
$\pi^+\pi^+\pi^-$		(5.59 ± 0.05) %	S=1.8	125
$\pi^+\pi^0\pi^0$		(1.73 ± 0.04) %	S=1.2	133
$\pi^0\mu^+\nu_\mu$		(3.18 ± 0.08) %	S=1.5	215
Called $K^+_{\mu 3}$.				
$\pi^0 e^+\nu_e$		(4.82 ± 0.06) %	S=1.3	228
Called K^+_{e3}.				
$\pi^0\pi^0 e^+\nu_e$		$(2.1 \pm 0.4) \times 10^{-5}$		206
$\pi^+\pi^- e^+\nu_e$		$(3.91 \pm 0.17) \times 10^{-5}$		203
$\pi^+\pi^-\mu^+\nu_\mu$		$(1.4 \pm 0.9) \times 10^{-5}$		151
$\pi^0\pi^0\pi^0 e^+\nu_e$		$< 3.5 \times 10^{-6}$	CL=90%	135
$\pi^+\gamma\gamma$	[x]	$(1.10 \pm 0.32) \times 10^{-6}$		227
$\pi^+ 3\gamma$	[x]	$< 1.0 \times 10^{-4}$	CL=90%	227
$\mu^+\nu_\mu\nu\overline{\nu}$		$< 6.0 \times 10^{-6}$	CL=90%	236
$e^+\nu_e\nu\overline{\nu}$		$< 6 \times 10^{-5}$	CL=90%	247
$\mu^+\nu_\mu e^+ e^-$		$(1.3 \pm 0.4) \times 10^{-7}$		236
$e^+\nu_e e^+ e^-$		$(3.0\ ^{+3.0}_{-1.5}) \times 10^{-8}$		247
$\mu^+\nu_\mu\mu^+\mu^-$		$< 4.1 \times 10^{-7}$	CL=90%	185
$\mu^+\nu_\mu\gamma$	[x,y]	$(5.50 \pm 0.28) \times 10^{-3}$		236
$\pi^+\pi^0\gamma$	[x,y]	$(2.75 \pm 0.15) \times 10^{-4}$		205
$\pi^+\pi^0\gamma$(DE)	[x,z]	$(1.8 \pm 0.4) \times 10^{-5}$		205
$\pi^+\pi^+\pi^-\gamma$	[x,y]	$(1.04 \pm 0.31) \times 10^{-4}$		125
$\pi^+\pi^0\pi^0\gamma$	[x,y]	$(7.5\ ^{+5.5}_{-3.0}) \times 10^{-6}$		133
$\pi^0\mu^+\nu_\mu\gamma$	[x,y]	$< 6.1 \times 10^{-5}$	CL=90%	215
$\pi^0 e^+\nu_e\gamma$	[x,y]	$(2.62 \pm 0.20) \times 10^{-4}$		228
$\pi^0 e^+\nu_e\gamma$(SD)	[aa]	$< 5.3 \times 10^{-5}$	CL=90%	228
$\pi^0\pi^0 e^+\nu_e\gamma$		$< 5 \times 10^{-6}$	CL=90%	206

Lepton Family number (*LF*), Lepton number (*L*), $\Delta S = \Delta Q$ (*SQ*) violating modes, or $\Delta S = 1$ weak neutral current (*S1*) modes

Mode			Fraction	CL	p (MeV/c)
$\pi^+\pi^+ e^-\overline{\nu}_e$	*SQ*		$< 1.2 \times 10^{-8}$	CL=90%	203
$\pi^+\pi^+\mu^-\overline{\nu}_\mu$	*SQ*		$< 3.0 \times 10^{-6}$	CL=95%	151
$\pi^+ e^+ e^-$	*S1*		$(2.74 \pm 0.23) \times 10^{-7}$		227
$\pi^+\mu^+\mu^-$	*S1*		$(5.0 \pm 1.0) \times 10^{-8}$		172
$\pi^+\nu\overline{\nu}$	*S1*		$(4.2\ ^{+9.7}_{-3.5}) \times 10^{-10}$		227
$\mu^-\nu e^+ e^+$	*LF*		$< 2.0 \times 10^{-8}$	CL=90%	236
$\mu^+\nu_e$	*LF*	[d]	$< 4 \times 10^{-3}$	CL=90%	236
$\pi^+\mu^+ e^-$	*LF*		$< 2.1 \times 10^{-10}$	CL=90%	214
$\pi^+\mu^- e^+$	*LF*		$< 7 \times 10^{-9}$	CL=90%	214
$\pi^-\mu^+ e^+$	*L*		$< 7 \times 10^{-9}$	CL=90%	214
$\pi^- e^+ e^+$	*L*		$< 1.0 \times 10^{-8}$	CL=90%	227
$\pi^-\mu^+\mu^+$	*L*	[d]	$< 1.5 \times 10^{-4}$	CL=90%	172
$\mu^+\overline{\nu}_e$	*L*	[d]	$< 3.3 \times 10^{-3}$	CL=90%	236
$\pi^0 e^+\overline{\nu}_e$	*L*		$< 3 \times 10^{-3}$	CL=90%	228

K^0

$I(J^P) = \frac{1}{2}(0^-)$

50% K_S, 50% K_L

Mass $m = 497.672 \pm 0.031$ MeV
$m_{K^0} - m_{K^\pm} = 3.995 \pm 0.034$ MeV (S = 1.1)
$|m_{K^0} - m_{\overline{K}^0}| / m_{\text{average}} < 10^{-18}$ [bb]

K^0_S

$I(J^P) = \frac{1}{2}(0^-)$

Mean life $\tau = (0.8934 \pm 0.0008) \times 10^{-10}$ s
$c\tau = 2.6762$ cm

***CP*-violation parameters** [cc]

$\text{Im}(\eta_{+-0}) = -0.002 \pm 0.008$
$\text{Im}(\eta_{000})^2 < 0.1$, CL = 90%

K^0_S DECAY MODES		Fraction (Γ_i/Γ)	Scale factor/ Confidence level	p (MeV/c)
$\pi^+\pi^-$		(68.61 ± 0.28) %	S=1.2	206
$\pi^0\pi^0$		(31.39 ± 0.28) %	S=1.2	209
$\pi^+\pi^-\gamma$	[y,dd]	$(1.78 \pm 0.05) \times 10^{-3}$		206
$\gamma\gamma$		$(2.4 \pm 0.9) \times 10^{-6}$		249
$\pi^+\pi^-\pi^0$		$(3.4\ ^{+1.1}_{-0.9}) \times 10^{-7}$		133
$3\pi^0$		$< 3.7 \times 10^{-5}$	CL=90%	139
$\pi^\pm e^\mp\nu$	[ee]	$(6.70 \pm 0.07) \times 10^{-4}$	S=1.1	229
$\pi^\pm\mu^\mp\nu$	[ee]	$(4.69 \pm 0.06) \times 10^{-4}$	S=1.1	216

$\Delta S = 1$ weak neutral current (*S1*) modes

Mode		Fraction	CL	p (MeV/c)
$\mu^+\mu^-$	*S1*	$< 3.2 \times 10^{-7}$	CL=90%	225
e^+e^-	*S1*	$< 1.4 \times 10^{-7}$	CL=90%	249
$\pi^0 e^+ e^-$	*S1*	$< 1.1 \times 10^{-6}$	CL=90%	231

K^0_L

$I(J^P) = \frac{1}{2}(0^-)$

$m_{K_L} - m_{K_S} = (0.5301 \pm 0.0014) \times 10^{10}\ \hbar\ \text{s}^{-1}$
$= (3.489 \pm 0.009) \times 10^{-12}$ MeV
Mean life $\tau = (5.17 \pm 0.04) \times 10^{-8}$ s (S = 1.1)
$c\tau = 15.51$ m

Slope parameter g [v]

(See Particle Listings for quadratic coefficients)

$K^0_L \rightarrow \pi^+\pi^-\pi^0 = 0.670 \pm 0.014$ (S = 1.6)

K_L decay form factors [w]

K^0_{e3} $\lambda_+ = 0.0300 \pm 0.0016$ (S = 1.2)
$K^0_{\mu 3}$ $\lambda_+ = 0.034 \pm 0.005$ (S = 2.3)
$K^0_{\mu 3}$ $\lambda_0 = 0.025 \pm 0.006$ (S = 2.3)
K^0_{e3} $|f_S/f_+| < 0.04$, CL = 68%
K^0_{e3} $|f_T/f_+| < 0.23$, CL = 68%
$K^0_{\mu 3}$ $|f_T/f_+| = 0.12 \pm 0.12$
$K_L \rightarrow e^+e^-\gamma$: $\alpha_{K^*} = -0.28 \pm 0.08$

CP-violation parameters [cc]

$\delta = (0.327 \pm 0.012)\%$

$|\eta_{00}| = (2.275 \pm 0.019) \times 10^{-3}$ (S = 1.1)

$|\eta_{+-}| = (2.285 \pm 0.019) \times 10^{-3}$

$|\eta_{00}/\eta_{+-}| = 0.9956 \pm 0.0023$ [ff] (S = 1.8)

$\epsilon'/\epsilon = (1.5 \pm 0.8) \times 10^{-3}$ [ff] (S = 1.8)

$\phi_{+-} = (43.5 \pm 0.6)°$

$\phi_{00} = (43.4 \pm 1.0)°$

$\phi_{00} - \phi_{+-} = (-0.1 \pm 0.8)°$

j for $K^0_L \to \pi^+\pi^-\pi^0 = 0.0011 \pm 0.0008$

$|\eta_{+-\gamma}| = (2.35 \pm 0.07) \times 10^{-3}$

$\phi_{+-\gamma} = (44 \pm 4)°$

$|\epsilon'_{+-\gamma}|/\epsilon < 0.3$, CL = 90%

$\Delta S = -\Delta Q$ in $K^0_{\ell 3}$ decay

Re $x = 0.006 \pm 0.018$ (S = 1.3)

Im $x = -0.003 \pm 0.026$ (S = 1.2)

CPT-violation parameters

Re $\Delta = 0.018 \pm 0.020$

Im $\Delta = 0.02 \pm 0.04$

K^0_L DECAY MODES		Fraction (Γ_i/Γ)	Scale factor/ Confidence level	p (MeV/c)
$3\pi^0$		(21.12 ± 0.27) %	S=1.1	139
$\pi^+\pi^-\pi^0$		(12.56 ± 0.20) %	S=1.7	133
$\pi^\pm\mu^\mp\nu$ Called $K^0_{\mu 3}$.	[gg]	(27.17 ± 0.25) %	S=1.1	216
$\pi^\pm e^\mp\nu_e$ Called K^0_{e3}.	[gg]	(38.78 ± 0.27) %	S=1.1	229
2γ		$(5.92 \pm 0.15) \times 10^{-4}$		249
3γ		$< 2.4 \times 10^{-7}$	CL=90%	249
$\pi^0 2\gamma$	[hh]	$(1.70 \pm 0.28) \times 10^{-6}$		231
$\pi^0\pi^\pm e^\mp\nu$	[gg]	$(5.18 \pm 0.29) \times 10^{-5}$		207
$(\pi\mu\,\text{atom})\nu$		$(1.06 \pm 0.11) \times 10^{-7}$		–
$\pi^\pm e^\mp\nu_e\gamma$	[y,gg,hh]	$(3.62\ ^{+0.26}_{-0.21}) \times 10^{-3}$		229
$\pi^+\pi^-\gamma$	[y,hh]	$(4.61 \pm 0.14) \times 10^{-5}$		206
$\pi^0\pi^0\gamma$		$< 5.6 \times 10^{-6}$		209

Charge conjugation × Parity (CP, CPV) or Lepton Family number (LF) violating modes, or $\Delta S = 1$ weak neutral current (S1) modes

$\pi^+\pi^-$	CPV		$(2.067 \pm 0.035) \times 10^{-3}$	S=1.1	206
$\pi^0\pi^0$	CPV		$(9.36 \pm 0.20) \times 10^{-4}$		209
$\mu^+\mu^-$	S1		$(7.2 \pm 0.5) \times 10^{-9}$	S=1.4	225
$\mu^+\mu^-\gamma$	S1		$(3.25 \pm 0.28) \times 10^{-7}$		225
e^+e^-	S1		$< 4.1 \times 10^{-11}$	CL=90%	249
$e^+e^-\gamma$	S1		$(9.1 \pm 0.5) \times 10^{-6}$		249
$e^+e^-\gamma\gamma$	S1	[hh]	$(6.5 \pm 1.2) \times 10^{-7}$		249
$\pi^+\pi^-e^+e^-$	S1	[hh]	$< 4.6 \times 10^{-7}$	CL=90%	206
$\mu^+\mu^-e^+e^-$	S1		$(2.9\ ^{+6.7}_{-2.4}) \times 10^{-9}$		225
$e^+e^-e^+e^-$	S1		$(4.1 \pm 0.8) \times 10^{-8}$	S=1.2	249
$\pi^0\mu^+\mu^-$	CP,S1	[ii]	$< 5.1 \times 10^{-9}$	CL=90%	177
$\pi^0e^+e^-$	CP,S1	[ii]	$< 4.3 \times 10^{-9}$	CL=90%	231
$\pi^0\nu\overline{\nu}$	CP,S1	[jj]	$< 5.8 \times 10^{-5}$	CL=90%	231
$e^\pm\mu^\mp$	LF	[gg]	$< 3.3 \times 10^{-11}$	CL=90%	238
$e^\pm e^\pm\mu^\mp\mu^\mp$	LF	[gg]	$< 6.1 \times 10^{-9}$	CL=90%	–

$K^*(892)$

$I(J^P) = \frac{1}{2}(1^-)$

$K^*(892)^\pm$ mass $m = 891.66 \pm 0.26$ MeV

$K^*(892)^0$ mass $m = 896.10 \pm 0.28$ MeV (S = 1.4)

$K^*(892)^\pm$ full width $\Gamma = 50.8 \pm 0.9$ MeV

$K^*(892)^0$ full width $\Gamma = 50.5 \pm 0.6$ MeV (S = 1.1)

$K^*(892)$ DECAY MODES	Fraction (Γ_i/Γ)	Confidence level	p (MeV/c)
$K\pi$	~ 100 %		291
$K^0\gamma$	$(2.30 \pm 0.20) \times 10^{-3}$		310
$K^\pm\gamma$	$(9.9 \pm 0.9) \times 10^{-4}$		309
$K\pi\pi$	$< 7 \times 10^{-4}$	95%	224

$K_1(1270)$

$I(J^P) = \frac{1}{2}(1^+)$

Mass $m = 1273 \pm 7$ MeV [m]

Full width $\Gamma = 90 \pm 20$ MeV [m]

$K_1(1270)$ DECAY MODES	Fraction (Γ_i/Γ)	p (MeV/c)
$K\rho$	(42 ± 6) %	76
$K^*_0(1430)\pi$	(28 ± 4) %	–
$K^*(892)\pi$	(16 ± 5) %	301
$K\omega$	(11.0 ± 2.0) %	–
$Kf_0(1370)$	(3.0 ± 2.0) %	–

$K_1(1400)$

$I(J^P) = \frac{1}{2}(1^+)$

Mass $m = 1402 \pm 7$ MeV

Full width $\Gamma = 174 \pm 13$ MeV (S = 1.6)

$K_1(1400)$ DECAY MODES	Fraction (Γ_i/Γ)	p (MeV/c)
$K^*(892)\pi$	(94 ± 6) %	401
$K\rho$	(3.0 ± 3.0) %	298
$Kf_0(1370)$	(2.0 ± 2.0) %	–
$K\omega$	(1.0 ± 1.0) %	285
$K^*_0(1430)\pi$	not seen	–

$K^*(1410)$

$I(J^P) = \frac{1}{2}(1^-)$

Mass $m = 1414 \pm 15$ MeV (S = 1.3)

Full width $\Gamma = 232 \pm 21$ MeV (S = 1.1)

$K^*(1410)$ DECAY MODES	Fraction (Γ_i/Γ)	Confidence level	p (MeV/c)
$K^*(892)\pi$	> 40 %	95%	408
$K\pi$	(6.6 ± 1.3) %		611
$K\rho$	< 7 %	95%	309

$K^*_0(1430)$ [kk]

$I(J^P) = \frac{1}{2}(0^+)$

Mass $m = 1429 \pm 6$ MeV

Full width $\Gamma = 287 \pm 23$ MeV

$K^*_0(1430)$ DECAY MODES	Fraction (Γ_i/Γ)	p (MeV/c)
$K\pi$	(93 ± 10) %	621

$K^*_2(1430)$

$I(J^P) = \frac{1}{2}(2^+)$

$K^*_2(1430)^\pm$ mass $m = 1425.6 \pm 1.5$ MeV (S = 1.1)

$K^*_2(1430)^0$ mass $m = 1432.4 \pm 1.3$ MeV

$K^*_2(1430)^\pm$ full width $\Gamma = 98.5 \pm 2.7$ MeV (S = 1.1)

$K^*_2(1430)^0$ full width $\Gamma = 109 \pm 5$ MeV (S = 1.9)

$K^*_2(1430)$ DECAY MODES	Fraction (Γ_i/Γ)	Scale factor/ Confidence level	p (MeV/c)
$K\pi$	(49.9 ± 1.2) %		622
$K^*(892)\pi$	(24.7 ± 1.5) %		423
$K^*(892)\pi\pi$	(13.4 ± 2.2) %		375
$K\rho$	(8.7 ± 0.8) %	S=1.2	331
$K\omega$	(2.9 ± 0.8) %		319
$K^+\gamma$	$(2.4 \pm 0.5) \times 10^{-3}$	S=1.1	627
$K\eta$	$(1.5\ ^{+3.4}_{-1.0}) \times 10^{-3}$	S=1.3	492
$K\omega\pi$	$< 7.2 \times 10^{-4}$	CL=95%	110
$K^0\gamma$	$< 9 \times 10^{-4}$	CL=90%	631

$K^*(1680)$

$I(J^P) = \frac{1}{2}(1^-)$

Mass $m = 1717 \pm 27$ MeV (S = 1.4)
Full width $\Gamma = 322 \pm 110$ MeV (S = 4.2)

$K^*(1680)$ DECAY MODES	Fraction (Γ_i/Γ)	p (MeV/c)
$K\pi$	(38.7 ± 2.5) %	779
$K\rho$	$(31.4^{+4.7}_{-2.1})$ %	571
$K^*(892)\pi$	$(29.9^{+2.2}_{-4.7})$ %	615

$K_2(1770)$ [ll]

$I(J^P) = \frac{1}{2}(2^-)$

Mass $m = 1773 \pm 8$ MeV
Full width $\Gamma = 186 \pm 14$ MeV

$K_2(1770)$ DECAY MODES	Fraction (Γ_i/Γ)	p (MeV/c)
$K\pi\pi$		–
$K_2^*(1430)\pi$	dominant	287
$K^*(892)\pi$	seen	653
$Kf_2(1270)$	seen	–
$K\phi$	seen	441
$K\omega$	seen	608

$K_3^*(1780)$

$I(J^P) = \frac{1}{2}(3^-)$

Mass $m = 1776 \pm 7$ MeV (S = 1.1)
Full width $\Gamma = 159 \pm 21$ MeV (S = 1.3)

$K_3^*(1780)$ DECAY MODES	Fraction (Γ_i/Γ)	Confidence level	p (MeV/c)
$K\rho$	(31 ± 9) %		612
$K^*(892)\pi$	(20 ± 5) %		651
$K\pi$	(18.8± 1.0) %		810
$K\eta$	(30 ±13) %		715
$K_2^*(1430)\pi$	< 16 %	95%	284

$K_2(1820)$ [mm]

$I(J^P) = \frac{1}{2}(2^-)$

Mass $m = 1816 \pm 13$ MeV
Full width $\Gamma = 276 \pm 35$ MeV

$K_2(1820)$ DECAY MODES	Fraction (Γ_i/Γ)	p (MeV/c)
$K_2^*(1430)\pi$	seen	325
$K^*(892)\pi$	seen	680
$Kf_2(1270)$	seen	186
$K\omega$	seen	638

$K_4^*(2045)$

$I(J^P) = \frac{1}{2}(4^+)$

Mass $m = 2045 \pm 9$ MeV (S = 1.1)
Full width $\Gamma = 198 \pm 30$ MeV

$K_4^*(2045)$ DECAY MODES	Fraction (Γ_i/Γ)	p (MeV/c)
$K\pi$	(9.9 ± 1.2) %	958
$K^*(892)\pi\pi$	(9 ±5) %	800
$K^*(892)\pi\pi\pi$	(7 ±5) %	764
$\rho K\pi$	(5.7 ± 3.2) %	742
$\omega K\pi$	(5.0 ± 3.0) %	736
$\phi K\pi$	(2.8 ± 1.4) %	591
$\phi K^*(892)$	(1.4 ± 0.7) %	363

CHARMED MESONS
$(C = \pm1)$

$D^+ = c\overline{d}$, $D^0 = c\overline{u}$, $\overline{D}^0 = \overline{c}u$, $D^- = \overline{c}d$, similarly for D^*'s

$D^\pm$

$I(J^P) = \frac{1}{2}(0^-)$

Mass $m = 1869.3 \pm 0.5$ MeV (S = 1.1)
Mean life $\tau = (1.057 \pm 0.015) \times 10^{-12}$ s
$c\tau = 317$ μm

CP-violation decay-rate asymmetries

$A_{CP}(K^+K^-\pi^\pm) = -0.017 \pm 0.027$
$A_{CP}(K^\pm K^{*0}) = -0.02 \pm 0.05$
$A_{CP}(\phi\pi^\pm) = -0.014 \pm 0.033$
$A_{CP}(\pi^+\pi^-\pi^\pm) = -0.02 \pm 0.04$

$D^+ \to \overline{K}^*(892)^0 \ell^+ \nu_\ell$ form factors

$r_2 = 0.72 \pm 0.09$
$r_v = 1.85 \pm 0.12$
$\Gamma_L/\Gamma_T = 1.23 \pm 0.13$
$\Gamma_+/\Gamma_- = 0.16 \pm 0.04$

D^- modes are charge conjugates of the modes below.

D^+ DECAY MODES		Fraction (Γ_i/Γ)	Scale factor/ Confidence level	p (MeV/c)
Inclusive modes				
e^+ anything		(17.2 ±1.9) %		–
K^- anything		(24.2 ±2.8) %	S=1.4	–
$\overline{K}^0$ anything + K^0 anything		(59 ±7) %		–
K^+ anything		(5.8 ±1.4) %		–
η anything	[nn]	< 13 %	CL=90%	–
Leptonic and semileptonic modes				
$\mu^+\nu_\mu$		< 7.2 $\times 10^{-4}$	CL=90%	932
$\overline{K}^0\ell^+\nu_\ell$	[oo]	(6.8 ±0.8) %		868
$\overline{K}^0 e^+\nu_e$		(6.7 ±0.9) %		868
$\overline{K}^0\mu^+\nu_\mu$		($7.0\ ^{+3.0}_{-2.0}$) %		865
$K^-\pi^+e^+\nu_e$		($4.1\ ^{+0.9}_{-0.7}$) %		863
$\overline{K}^*(892)^0 e^+\nu_e$ × B($\overline{K}^{*0} \to K^-\pi^+$)		(3.2 ±0.33) %		720
$K^-\pi^+e^+\nu_e$ nonresonant		< 7 $\times 10^{-3}$	CL=90%	863
$K^-\pi^+\mu^+\nu_\mu$		(3.2 ±0.4) %	S=1.1	851
$\overline{K}^*(892)\mu^+\nu_\mu$ × B($\overline{K}^{*0} \to K^-\pi^+$)		(2.9 ±0.4) %		715
$K^-\pi^+\mu^+\nu_\mu$ nonresonant		(2.7 ±1.1) $\times 10^{-3}$		851
$(\overline{K}^*(892)\pi)^0 e^+\nu_e$		< 1.2 %	CL=90%	714
$(\overline{K}\pi\pi)^0 e^+\nu_e$ non-$\overline{K}^*(892)$		< 9 $\times 10^{-3}$	CL=90%	846
$K^-\pi^+\pi^0\mu^+\nu_\mu$		< 1.4 $\times 10^{-3}$	CL=90%	825
$\pi^0\ell^+\nu_\ell$	[pp]	(3.1 ±1.5) $\times 10^{-3}$		930
Fractions of some of the following modes with resonances have already appeared above as submodes of particular charged-particle modes.				
$\overline{K}^*(892)^0\ell^+\nu_\ell$	[oo]	(4.7 ±0.4) %		720
$\overline{K}^*(892)^0 e^+\nu_e$		(4.8 ±0.5) %		720
$\overline{K}^*(892)^0\mu^+\nu_\mu$		(4.4 ±0.6) %	S=1.1	715
$\rho^0 e^+\nu_e$		(2.2 ±0.8) $\times 10^{-3}$		776
$\rho^0\mu^+\nu_\mu$		(2.7 ±0.7) $\times 10^{-3}$		772
$\phi e^+\nu_e$		< 2.09 %	CL=90%	657
$\phi\mu^+\nu_\mu$		< 3.72 %	CL=90%	651
$\eta\ell^+\nu_\ell$		< 5 $\times 10^{-3}$	CL=90%	–
$\eta'(958)\mu^+\nu_\mu$		< 9 $\times 10^{-3}$	CL=90%	684
Hadronic modes with a $\overline{K}$ or $\overline{K}K\overline{K}$				
$\overline{K}^0\pi^+$		(2.89±0.26) %	S=1.1	862
$K^-\pi^+\pi^+$	[qq]	(9.0 ±0.6) %		845
$\overline{K}^*(892)^0\pi^+$ × B($\overline{K}^{*0} \to K^-\pi^+$)		(1.27±0.13) %		712
$\overline{K}_0^*(1430)^0\pi^+$ × B($\overline{K}_0^*(1430)^0 \to K^-\pi^+$)		(2.3 ±0.3) %		368
$\overline{K}^*(1680)^0\pi^+$ × B($\overline{K}^*(1680)^0 \to K^-\pi^+$)		(3.7 ±0.8) $\times 10^{-3}$		65
$K^-\pi^+\pi^+$ nonresonant		(8.5 ±0.8) %		845
$\overline{K}^0\pi^+\pi^0$	[qq]	(9.7 ±3.0) %	S=1.1	845

$\overline{K}^0 \rho^+$		(6.6 ±2.5) %		680
$\overline{K}^*(892)^0 \pi^+$ × B($\overline{K}^{*0} \to \overline{K}^0 \pi^0$)		(6.3 ±0.4) × 10^{-3}		712
$\overline{K}^0 \pi^+ \pi^0$ nonresonant		(1.3 ±1.1) %		845
$K^- \pi^+ \pi^+ \pi^0$	[qq]	(6.4 ±1.1) %		816
$\overline{K}^*(892)^0 \rho^+$ total × B($\overline{K}^{*0} \to K^- \pi^+$)		(1.4 ±0.9) %		423
$\overline{K}_1(1400)^0 \pi^+$ × B($\overline{K}_1(1400)^0 \to K^- \pi^+ \pi^0$)		(2.2 ±0.6) %		390
$K^- \rho^+ \pi^+$ total		(3.1 ±1.1) %		616
$K^- \rho^+ \pi^+$ 3-body		(1.1 ±0.4) %		616
$\overline{K}^*(892)^0 \pi^+ \pi^0$ total × B($\overline{K}^{*0} \to K^- \pi^+$)		(4.5 ±0.9) %		687
$\overline{K}^*(892)^0 \pi^+ \pi^0$ 3-body × B($\overline{K}^{*0} \to K^- \pi^+$)		(2.8 ±0.9) %		687
$K^*(892)^- \pi^+ \pi^+$ 3-body × B($K^{*-} \to K^- \pi^0$)		(7 ±3) × 10^{-3}		688
$K^- \pi^+ \pi^+ \pi^0$ nonresonant	[rr]	(1.2 ±0.6) %		816
$\overline{K}^0 \pi^+ \pi^+ \pi^-$	[qq]	(7.0 ±0.9) %		814
$\overline{K}^0 a_1(1260)^+$ × B($a_1(1260)^+ \to \pi^+ \pi^+ \pi^-$)		(4.0 ±0.9) %		328
$\overline{K}_1(1400)^0 \pi^+$ × B($\overline{K}_1(1400)^0 \to \overline{K}^0 \pi^+ \pi^-$)		(2.2 ±0.6) %		390
$K^*(892)^- \pi^+ \pi^+$ 3-body × B($K^{*-} \to \overline{K}^0 \pi^-$)		(1.4 ±0.6) %		688
$\overline{K}^0 \rho^0 \pi^+$ total		(4.2 ±0.9) %		614
$\overline{K}^0 \rho^0 \pi^+$ 3-body		(5 ±5) × 10^{-3}		614
$\overline{K}^0 \pi^+ \pi^+ \pi^-$ nonresonant		(8 ±4) × 10^{-3}		814
$K^- \pi^+ \pi^+ \pi^+ \pi^-$	[qq]	(7.2 ±1.0) × 10^{-3}		772
$\overline{K}^*(892)^0 \pi^+ \pi^+ \pi^-$ × B($\overline{K}^{*0} \to K^- \pi^+$)		(5.4 ±2.3) × 10^{-3}		642
$\overline{K}^*(892)^0 \rho^0 \pi^+$ × B($\overline{K}^{*0} \to K^- \pi^+$)		(1.9 $^{+1.1}_{-1.0}$) × 10^{-3}		242
$\overline{K}^*(892)^0 \pi^+ \pi^+ \pi^-$ no-ρ × B($\overline{K}^{*0} \to K^- \pi^+$)		(2.9 ±1.1) × 10^{-3}		642
$K^- \rho^0 \pi^+ \pi^+$		(3.1 ±0.9) × 10^{-3}		529
$K^- \pi^+ \pi^+ \pi^+ \pi^-$ nonresonant		< 2.3 × 10^{-3}	CL=90%	772
$K^- \pi^+ \pi^+ \pi^0 \pi^0$		(2.2 $^{+5.0}_{-0.9}$) %		775
$\overline{K}^0 \pi^+ \pi^+ \pi^- \pi^0$		(5.4 $^{+3.0}_{-1.4}$) %		773
$\overline{K}^0 \pi^+ \pi^+ \pi^+ \pi^- \pi^-$		(8 ±7) × 10^{-4}		714
$K^- \pi^+ \pi^+ \pi^+ \pi^- \pi^0$		(2.0 ±1.8) × 10^{-3}		718
$\overline{K}^0 \overline{K}^0 K^+$		(1.8 ±0.8) %		545

Fractions of some of the following modes with resonances have already appeared above as submodes of particular charged-particle modes.

$\overline{K}^0 \rho^+$		(6.6 ±2.5) %		680
$\overline{K}^0 a_1(1260)^+$		(8.0 ±1.7) %		328
$\overline{K}^0 a_2(1320)^+$		< 3 × 10^{-3}	CL=90%	199
$\overline{K}^*(892)^0 \pi^+$		(1.90±0.19) %		712
$\overline{K}^*(892)^0 \rho^+$ total	[rr]	(2.1 ±1.3) %		423
$\overline{K}^*(892)^0 \rho^+$ *S*-wave	[rr]	(1.6 ±1.6) %		423
$\overline{K}^*(892)^0 \rho^+$ *P*-wave		< 1 × 10^{-3}	CL=90%	423
$\overline{K}^*(892)^0 \rho^+$ *D*-wave		(10 ±7) × 10^{-3}		423
$\overline{K}^*(892)^0 \rho^+$ *D*-wave longitudinal		< 7 × 10^{-3}	CL=90%	423
$\overline{K}_1(1270)^0 \pi^+$		< 7 × 10^{-3}	CL=90%	487
$\overline{K}_1(1400)^0 \pi^+$		(4.9 ±1.2) %		390
$\overline{K}^*(1410)^0 \pi^+$		< 7 × 10^{-3}	CL=90%	382
$\overline{K}^*_0(1430)^0 \pi^+$		(3.7 ±0.4) %		368
$\overline{K}^*(1680)^0 \pi^+$		(1.43±0.30) %		65
$\overline{K}^*(892)^0 \pi^+ \pi^0$ total		(6.7 ±1.4) %		687
$\overline{K}^*(892)^0 \pi^+ \pi^0$ 3-body	[rr]	(4.2 ±1.4) %		687
$K^*(892)^- \pi^+ \pi^+$ 3-body		(2.0 ±0.9) %		688
$K^- \rho^+ \pi^+$ total		(3.1 ±1.1) %		616
$K^- \rho^+ \pi^+$ 3-body		(1.1 ±0.4) %		616
$\overline{K}^0 \rho^0 \pi^+$ total		(4.2 ±0.9) %	CL=90%	614
$\overline{K}^0 \rho^0 \pi^+$ 3-body		(5 ±5) × 10^{-3}		614
$\overline{K}^0 f_0(980) \pi^+$		< 5 × 10^{-3}	CL=90%	461
$\overline{K}^*(892)^0 \pi^+ \pi^+ \pi^-$		(8.1 ±3.4) × 10^{-3}	S=1.7	642
$\overline{K}^*(892)^0 \rho^0 \pi^+$		(2.9 $^{+1.7}_{-1.5}$) × 10^{-3}	S=1.8	242
$\overline{K}^*(892)^0 \pi^+ \pi^+ \pi^-$ no-ρ		(4.3 ±1.7) × 10^{-3}		642
$K^- \rho^0 \pi^+ \pi^+$		(3.1 ±0.9) × 10^{-3}		529

Pionic modes

$\pi^+ \pi^0$		(2.5 ±0.7) × 10^{-3}		925
$\pi^+ \pi^+ \pi^-$		(3.6 ±0.4) × 10^{-3}		908
$\rho^0 \pi^+$		(1.05±0.31) × 10^{-3}		769
$\pi^+ \pi^+ \pi^-$ nonresonant		(2.2 ±0.4) × 10^{-3}		908
$\pi^+ \pi^+ \pi^- \pi^0$		(1.9 $^{+1.5}_{-1.2}$) %		882
$\eta \pi^+$ × B($\eta \to \pi^+ \pi^- \pi^0$)		(1.7 ±0.6) × 10^{-3}		848
$\omega \pi^+$ × B($\omega \to \pi^+ \pi^- \pi^0$)		< 6 × 10^{-3}	CL=90%	764
$\pi^+ \pi^+ \pi^+ \pi^- \pi^-$		(2.1 ±0.4) × 10^{-3}		845
$\pi^+ \pi^+ \pi^+ \pi^- \pi^- \pi^0$		(2.9 $^{+2.9}_{-2.0}$) × 10^{-3}		799

Fractions of some of the following modes with resonances have already appeared above as submodes of particular charged-particle modes.

$\eta \pi^+$		(7.5 ±2.5) × 10^{-3}		848
$\rho^0 \pi^+$		(1.05±0.31) × 10^{-3}		769
$\omega \pi^+$		< 7 × 10^{-3}	CL=90%	764
$\eta \rho^+$		< 1.2 %	CL=90%	658
$\eta'(958) \pi^+$		< 9 × 10^{-3}	CL=90%	680
$\eta'(958) \rho^+$		< 1.5 %	CL=90%	355

Hadronic modes with a $K\overline{K}$ pair

$K^+ \overline{K}^0$		(7.4 ±1.0) × 10^{-3}		792
$K^+ K^- \pi^+$	[qq]	(8.8 ±0.8) × 10^{-3}		744
$\phi \pi^+$ × B($\phi \to K^+ K^-$)		(3.0 ±0.3) × 10^{-3}		647
$K^+ \overline{K}^*(892)^0$ × B($\overline{K}^{*0} \to K^- \pi^+$)		(2.8 ±0.4) × 10^{-3}		610
$K^+ K^- \pi^+$ nonresonant		(4.5 ±0.9) × 10^{-3}		744
$K^0 \overline{K}^0 \pi^+$		—		741
$K^*(892)^+ \overline{K}^0$ × B($K^{*+} \to K^0 \pi^+$)		(2.1 ±1.0) %		611
$K^+ K^- \pi^+ \pi^0$		—		682
$\phi \pi^+ \pi^0$ × B($\phi \to K^+ K^-$)		(1.1 ±0.5) %		619
$\phi \rho^+$ × B($\phi \to K^+ K^-$)		< 7 × 10^{-3}	CL=90%	268
$K^+ K^- \pi^+ \pi^0$ non-ϕ		(1.5 $^{+0.7}_{-0.6}$) %		682
$K^+ \overline{K}^0 \pi^+ \pi^-$		< 2 %	CL=90%	678
$K^0 K^- \pi^+ \pi^+$		(1.0 ±0.6) %		678
$K^*(892)^+ \overline{K}^*(892)^0$ × $B^2(K^{*+} \to K^0 \pi^+)$		(1.2 ±0.5) %		273
$K^0 K^- \pi^+ \pi^+$ non-$K^{*+} \overline{K}^{*0}$		< 7.9 × 10^{-3}	CL=90%	678
$K^+ K^- \pi^+ \pi^+ \pi^-$		—		600
$\phi \pi^+ \pi^+ \pi^-$ × B($\phi \to K^+ K^-$)		< 1 × 10^{-3}	CL=90%	565
$K^+ K^- \pi^+ \pi^+ \pi^-$ nonresonant		< 3 %	CL=90%	600

Fractions of the following modes with resonances have already appeared above as submodes of particular charged-particle modes.

$\phi \pi^+$		(6.1 ±0.6) × 10^{-3}		647
$\phi \pi^+ \pi^0$		(2.3 ±1.0) %		619
$\phi \rho^+$		< 1.4 %	CL=90%	268
$\phi \pi^+ \pi^+ \pi^-$		< 2 × 10^{-3}	CL=90%	565
$K^+ \overline{K}^*(892)^0$		(4.2 ±0.5) × 10^{-3}		610
$K^*(892)^+ \overline{K}^0$		(3.2 ±1.5) %		611
$K^*(892)^+ \overline{K}^*(892)^0$		(2.6 ±1.1) %		273

Doubly Cabibbo suppressed (*DC*) modes, $\Delta C = 1$ weak neutral current (*C1*) modes, or Lepton Family number (*LF*) or Lepton number (*L*) violating modes

$K^+ \pi^+ \pi^-$	*DC*	(6.8 ±1.5) × 10^{-4}		845
$K^+ \rho^0$	*DC*	(2.5 ±1.2) × 10^{-4}		681
$K^*(892)^0 \pi^+$	*DC*	(3.6 ±1.6) × 10^{-4}		712
$K^+ \pi^+ \pi^-$ nonresonant	*DC*	(2.4 ±1.2) × 10^{-4}		845
$K^+ K^+ K^-$	*DC*	< 1.4 × 10^{-4}	CL=90%	550
ϕK^+	*DC*	< 1.3 × 10^{-4}	CL=90%	527
$\pi^+ e^+ e^-$	*C1*	< 6.6 × 10^{-5}	CL=90%	929
$\pi^+ \mu^+ \mu^-$	*C1*	< 1.8 × 10^{-5}	CL=90%	917
$\rho^+ \mu^+ \mu^-$	*C1*	< 5.6 × 10^{-4}	CL=90%	759
$K^+ e^+ e^-$	[ss]	< 2.0 × 10^{-4}	CL=90%	869
$K^+ \mu^+ \mu^-$	[ss]	< 9.7 × 10^{-5}	CL=90%	856
$\pi^+ e^+ \mu^-$	*LF*	< 1.1 × 10^{-4}	CL=90%	926
$\pi^+ e^- \mu^+$	*LF*	< 1.3 × 10^{-4}	CL=90%	926
$K^+ e^+ \mu^-$	*LF*	< 1.3 × 10^{-4}	CL=90%	866

$K^+ e^- \mu^+$	*LF*	< 1.2	$\times 10^{-4}$	CL=90%	866
$\pi^- e^+ e^+$	*L*	< 1.1	$\times 10^{-4}$	CL=90%	929
$\pi^- \mu^+ \mu^+$	*L*	< 8.7	$\times 10^{-5}$	CL=90%	917
$\pi^- e^+ \mu^+$	*L*	< 1.1	$\times 10^{-4}$	CL=90%	926
$\rho^- \mu^+ \mu^+$	*L*	< 5.6	$\times 10^{-4}$	CL=90%	759
$K^- e^+ e^+$	*L*	< 1.2	$\times 10^{-4}$	CL=90%	869
$K^- \mu^+ \mu^+$	*L*	< 1.2	$\times 10^{-4}$	CL=90%	856
$K^- e^+ \mu^+$	*L*	< 1.3	$\times 10^{-4}$	CL=90%	866
$K^*(892)^- \mu^+ \mu^+$	*L*	< 8.5	$\times 10^{-4}$	CL=90%	703

D^0

$I(J^P) = \frac{1}{2}(0^-)$

Mass $m = 1864.6 \pm 0.5$ MeV (S = 1.1)

$m_{D^\pm} - m_{D^0} = 4.76 \pm 0.10$ MeV (S = 1.1)

Mean life $\tau = (0.415 \pm 0.004) \times 10^{-12}$ s

$c\tau = 124.4\ \mu$m

$\left|m_{D_1^0} - m_{D_2^0}\right| < 24 \times 10^{10}\ \hbar\ s^{-1}$, CL = 90% [tt]

$\left|\Gamma_{D_1^0} - \Gamma_{D_2^0}\right| / \Gamma_{D^0} < 0.20$, CL = 90% [tt]

$\Gamma(K^+ \ell^- \overline{\nu}_\ell \text{ (via } \overline{D}^0)) / \Gamma(K^- \ell^+ \nu_\ell) < 0.005$, CL = 90%

$\frac{\Gamma(K^+\pi^- \text{ or } K^+\pi^-\pi^+\pi^- \text{ (via } \overline{D}^0))}{\Gamma(K^-\pi^+ \text{ or } K^-\pi^+\pi^+\pi^-)} < 0.0085$ (or < 0.0037), CL = 90% [uu]

CP-violation decay-rate asymmetries

$A_{CP}(K^+K^-) = 0.026 \pm 0.035$

$A_{CP}(\pi^+\pi^-) = -0.05 \pm 0.08$

$A_{CP}(K_S^0 \phi) = -0.03 \pm 0.09$

$A_{CP}(K_S^0 \pi^0) = -0.018 \pm 0.030$

$\overline{D}^0$ modes are charge conjugates of the modes below.

D^0 DECAY MODES		Fraction (Γ_i/Γ)	Scale factor/ Confidence level	p (MeV/c)
Inclusive modes				
e^+ anything		(6.75±0.29) %		–
μ^+ anything		(6.6 ±0.8) %		–
K^- anything		(53 ±4) %	S=1.3	–
$\overline{K}^0$ anything + K^0 anything		(42 ±5) %		–
K^+ anything		($3.4^{+0.6}_{-0.4}$) %		–
η anything	[nn]	< 13 %	CL=90%	–
Semileptonic modes				
$K^- \ell^+ \nu_\ell$	[oo]	(3.50±0.17) %	S=1.3	867
$K^- e^+ \nu_e$		(3.66±0.18) %		867
$K^- \mu^+ \nu_\mu$		(3.23±0.17) %		863
$K^- \pi^0 e^+ \nu_e$		($1.6^{+1.3}_{-0.5}$) %		861
$\overline{K}^0 \pi^- e^+ \nu_e$		($2.8^{+1.7}_{-0.9}$) %		860
$\overline{K}^*(892)^- e^+ \nu_e$ $\times$ B($K^{*-} \to \overline{K}^0 \pi^-$)		(1.35±0.22) %		719
$K^*(892)^- \ell^+ \nu_\ell$		—		–
$\overline{K}^*(892)^0 \pi^- e^+ \nu_e$		—		708
$K^- \pi^+ \pi^- \mu^+ \nu_\mu$		< 1.2 $\times 10^{-3}$	CL=90%	821
$(\overline{K}^*(892)\pi)^- \mu^+ \nu_\mu$		< 1.4 $\times 10^{-3}$	CL=90%	693
$\pi^- e^+ \nu_e$		(3.7 ±0.6) $\times 10^{-3}$		927

A fraction of the following resonance mode has already appeared above as a submode of a charged-particle mode.

$K^*(892)^- e^+ \nu_e$		(2.02±0.33) %		719
Hadronic modes with a $\overline{K}$ or $\overline{K}K\overline{K}$				
$K^- \pi^+$		(3.85±0.09) %		861
$\overline{K}^0 \pi^0$		(2.12±0.21) %	S=1.1	860
$\overline{K}^0 \pi^+ \pi^-$	[qq]	(5.4 ±0.4) %	S=1.2	842
$\overline{K}^0 \rho^0$		(1.21±0.17) %		676
$\overline{K}^0 f_0(980)$ $\times$ B($f_0 \to \pi^+\pi^-$)		(3.0 ±0.8) $\times 10^{-3}$		549
$\overline{K}^0 f_2(1270)$ $\times$ B($f_2 \to \pi^+\pi^-$)		(2.4 ±0.9) $\times 10^{-3}$		263
$\overline{K}^0 f_0(1370)$ $\times$ B($f_0 \to \pi^+\pi^-$)		(4.3 ±1.3) $\times 10^{-3}$		–
$K^*(892)^- \pi^+$ $\times$ B($K^{*-} \to \overline{K}^0 \pi^-$)		(3.4 ±0.3) %		711
$K_0^*(1430)^- \pi^+$ $\times$ B($K_0^*(1430)^- \to \overline{K}^0 \pi^-$)		(6.4 ±1.6) $\times 10^{-3}$		364
$\overline{K}^0 \pi^+ \pi^-$ nonresonant		(1.47±0.24) %		842
$K^- \pi^+ \pi^0$	[qq]	(13.9 ±0.9) %	S=1.3	844
$K^- \rho^+$		(10.8 ±1.0) %		678
$K^*(892)^- \pi^+$ $\times$ B($K^{*-} \to K^- \pi^0$)		(1.7 ±0.2) %		711
$\overline{K}^*(892)^0 \pi^0$ $\times$ B($\overline{K}^{*0} \to K^- \pi^+$)		(2.1 ±0.3) %		709
$K^- \pi^+ \pi^0$ nonresonant		(6.9 ±2.5) $\times 10^{-3}$		844
$\overline{K}^0 \pi^0 \pi^0$		—		843
$\overline{K}^*(892)^0 \pi^0$ $\times$ B($\overline{K}^{*0} \to \overline{K}^0 \pi^0$)		(1.1 ±0.2) %		709
$\overline{K}^0 \pi^0 \pi^0$ nonresonant		(7.9 ±2.1) $\times 10^{-3}$		843
$K^- \pi^+ \pi^+ \pi^-$	[qq]	(7.6 ±0.4) %	S=1.1	812
$K^- \pi^+ \rho^0$ total		(6.3 ±0.4) %		612
$K^- \pi^+ \rho^0$ 3-body		(4.8 ±2.1) $\times 10^{-3}$		612
$\overline{K}^*(892)^0 \rho^0$ $\times$ B($\overline{K}^{*0} \to K^- \pi^+$)		(9.8 ±2.2) $\times 10^{-3}$		418
$K^- a_1(1260)^+$ $\times$ B($a_1(1260)^+ \to \pi^+\pi^+\pi^-$)		(3.6 ±0.6) %		327
$\overline{K}^*(892)^0 \pi^+ \pi^-$ total $\times$ B($\overline{K}^{*0} \to K^- \pi^+$)		(1.5 ±0.4) %		683
$\overline{K}^*(892)^0 \pi^+ \pi^-$ 3-body $\times$ B($\overline{K}^{*0} \to K^- \pi^+$)		(9.5 ±2.1) $\times 10^{-3}$		683
$K_1(1270)^- \pi^+$ $\times$ B($K_1(1270)^- \to K^- \pi^+ \pi^-$)	[rr]	(3.6 ±1.0) $\times 10^{-3}$		483
$K^- \pi^+ \pi^+ \pi^-$ nonresonant		(1.76±0.25) %		812
$\overline{K}^0 \pi^+ \pi^- \pi^0$	[qq]	(10.0 ±1.2) %		812
$\overline{K}^0 \eta$ $\times$ B($\eta \to \pi^+\pi^-\pi^0$)		(1.6 ±0.3) $\times 10^{-3}$		772
$\overline{K}^0 \omega$ $\times$ B($\omega \to \pi^+\pi^-\pi^0$)		(1.9 ±0.4) %		670
$K^*(892)^- \rho^+$ $\times$ B($K^{*-} \to \overline{K}^0 \pi^-$)		(4.1 ±1.6) %		422
$\overline{K}^*(892)^0 \rho^0$ $\times$ B($\overline{K}^{*0} \to \overline{K}^0 \pi^0$)		(4.9 ±1.1) $\times 10^{-3}$		418
$K_1(1270)^- \pi^+$ $\times$ B($K_1(1270)^- \to \overline{K}^0 \pi^- \pi^0$)	[rr]	(5.1 ±1.4) $\times 10^{-3}$		483
$\overline{K}^*(892)^0 \pi^+ \pi^-$ 3-body $\times$ B($\overline{K}^{*0} \to \overline{K}^0 \pi^0$)		(4.8 ±1.1) $\times 10^{-3}$		683
$\overline{K}^0 \pi^+ \pi^- \pi^0$ nonresonant		(2.1 ±2.1) %		812
$K^- \pi^+ \pi^0 \pi^0$		(15 ±5) %		815
$K^- \pi^+ \pi^+ \pi^- \pi^0$		(4.1 ±0.4) %		771
$\overline{K}^*(892)^0 \pi^+ \pi^- \pi^0$ $\times$ B($\overline{K}^{*0} \to K^- \pi^+$)		(1.2 ±0.6) %		641
$\overline{K}^*(892)^0 \eta$ $\times$ B($\overline{K}^{*0} \to K^- \pi^+$) $\times$ B($\eta \to \pi^+\pi^-\pi^0$)		(2.9 ±0.8) $\times 10^{-3}$		580
$K^- \pi^+ \omega$ $\times$ B($\omega \to \pi^+\pi^-\pi^0$)		(2.7 ±0.5) %		605
$\overline{K}^*(892)^0 \omega$ $\times$ B($\overline{K}^{*0} \to K^- \pi^+$) $\times$ B($\omega \to \pi^+\pi^-\pi^0$)		(7 ±3) $\times 10^{-3}$		406
$\overline{K}^0 \pi^+ \pi^+ \pi^- \pi^-$		(5.8 ±1.6) $\times 10^{-3}$		768
$\overline{K}^0 \pi^+ \pi^- \pi^0 \pi^0 (\pi^0)$		($10.6^{+7.3}_{-3.0}$) %		771
$\overline{K}^0 K^+ K^-$		(9.4 ±1.0) $\times 10^{-3}$		544
$\overline{K}^0 \phi$ $\times$ B($\phi \to K^+ K^-$)		(4.3 ±0.5) $\times 10^{-3}$		520
$\overline{K}^0 K^+ K^-$ non-ϕ		(5.1 ±0.8) $\times 10^{-3}$		544
$K_S^0 K_S^0 K_S^0$		(8.4 ±1.5) $\times 10^{-4}$		538
$K^+ K^- K^- \pi^+$		(2.1 ±0.5) $\times 10^{-4}$		434
$K^+ K^- \overline{K}^0 \pi^0$		($7.2^{+4.8}_{-3.5}$) $\times 10^{-3}$		435

Fractions of many of the following modes with resonances have already appeared above as submodes of particular charged-particle modes. (Modes for which there are only upper limits and $\overline{K}^*(892)\rho$ submodes only appear below.)

$\overline{K}^0 \eta$		(7.1 ±1.0) $\times 10^{-3}$		772
$\overline{K}^0 \rho^0$		(1.21±0.17) %		676
$K^- \rho^+$		(10.8 ±1.0) %	S=1.2	678
$\overline{K}^0 \omega$		(2.1 ±0.4) %		670
$\overline{K}^0 \eta'(958)$		(1.72±0.26) %		565
$\overline{K}^0 f_0(980)$		(5.7 ±1.6) $\times 10^{-3}$		549
$\overline{K}^0 \phi$		(8.6 ±1.0) $\times 10^{-3}$		520
$K^- a_1(1260)^+$		(7.3 ±1.1) %		327
$\overline{K}^0 a_1(1260)^0$		< 1.9 %	CL=90%	322
$\overline{K}^0 f_2(1270)$		(4.2 ±1.5) $\times 10^{-3}$		263
$K^- a_2(1320)^+$		< 2 $\times 10^{-3}$	CL=90%	197
$\overline{K}^0 f_0(1370)$		(7.0 ±2.1) $\times 10^{-3}$		–
$K^*(892)^- \pi^+$		(5.1 ±0.4) %	S=1.2	711
$\overline{K}^*(892)^0 \pi^0$		(3.2 ±0.4) %		709
$\overline{K}^*(892)^0 \pi^+ \pi^-$ total		(2.3 ±0.5) %		683
$\overline{K}^*(892)^0 \pi^+ \pi^-$ 3-body		(1.43±0.32) %		683

Mode	Fraction (Γ_i/Γ)	Scale factor/Confidence level	p (MeV/c)
$K^-\pi^+\rho^0$ total	(6.3 ±0.4) %		612
$K^-\pi^+\rho^0$ 3-body	(4.8 ±2.1) × 10^{-3}		612
$\overline{K}^*(892)^0\rho^0$	(1.47±0.33) %		418
$\overline{K}^*(892)^0\rho^0$ transverse	(1.5 ±0.5) %		418
$\overline{K}^*(892)^0\rho^0$ S-wave	(2.8 ±0.6) %		418
$\overline{K}^*(892)^0\rho^0$ S-wave long.	< 3 × 10^{-3}	CL=90%	418
$\overline{K}^*(892)^0\rho^0$ P-wave	< 3 × 10^{-3}	CL=90%	418
$\overline{K}^*(892)^0\rho^0$ D-wave	(1.9 ±0.6) %		418
$K^*(892)^-\rho^+$	(6.1 ±2.4) %		422
$K^*(892)^-\rho^+$ longitudinal	(2.9 ±1.2) %		422
$K^*(892)^-\rho^+$ transverse	(3.2 ±1.8) %		422
$K^*(892)^-\rho^+$ P-wave	< 1.5 %	CL=90%	422
$K^-\pi^+f_0(980)$	< 1.1 %	CL=90%	459
$\overline{K}^*(892)^0 f_0(980)$	< 7 × 10^{-3}	CL=90%	–
$K_1(1270)^-\pi^+$ [rr]	(1.06±0.29) %		483
$K_1(1400)^-\pi^+$	< 1.2 %	CL=90%	386
$\overline{K}_1(1400)^0\pi^0$	< 3.7 %	CL=90%	387
$K^*(1410)^-\pi^+$	< 1.2 %	CL=90%	378
$K_0^*(1430)^-\pi^+$	(1.04±0.26) %		364
$K_2^*(1430)^-\pi^+$	< 8 × 10^{-3}	CL=90%	367
$\overline{K}_2^*(1430)^0\pi^0$	< 4 × 10^{-3}	CL=90%	363
$\overline{K}^*(892)^0\pi^+\pi^-\pi^0$	(1.8 ±0.9) %		641
$\overline{K}^*(892)^0\eta$	(1.9 ±0.5) %		580
$K^-\pi^+\omega$	(3.0 ±0.6) %		605
$\overline{K}^*(892)^0\omega$	(1.1 ±0.5) %		406
$K^-\pi^+\eta'(958)$	(7.0 ±1.8) × 10^{-3}		479
$\overline{K}^*(892)^0\eta'(958)$	< 1.1 × 10^{-3}	CL=90%	99
Pionic modes			
$\pi^+\pi^-$	(1.53±0.09) × 10^{-3}		922
$\pi^0\pi^0$	(8.5 ±2.2) × 10^{-4}		922
$\pi^+\pi^-\pi^0$	(1.6 ±1.1) %	S=2.7	907
$\pi^+\pi^+\pi^-\pi^-$	(7.4 ±0.6) × 10^{-3}		879
$\pi^+\pi^+\pi^-\pi^-\pi^0$	(1.9 ±0.4) %		844
$\pi^+\pi^+\pi^+\pi^-\pi^-\pi^-$	(4.0 ±3.0) × 10^{-4}		795
Hadronic modes with a $K\overline{K}$ pair			
K^+K^-	(4.27±0.16) × 10^{-3}		791
$K^0\overline{K}^0$	(6.5 ±1.8) × 10^{-4}	S=1.2	788
$K^0K^-\pi^+$	(6.4 ±1.0) × 10^{-3}	S=1.1	739
$\overline{K}^*(892)^0K^0$ × B($\overline{K}^{*0}\to K^-\pi^+$)	< 1.1 × 10^{-3}	CL=90%	605
$K^*(892)^+K^-$ × B($K^{*+}\to K^0\pi^+$)	(2.3 ±0.5) × 10^{-3}		610
$K^0K^-\pi^+$ nonresonant	(2.3 ±2.3) × 10^{-3}		739
$\overline{K}^0K^+\pi^-$	(5.0 ±1.0) × 10^{-3}		739
$K^*(892)^0\overline{K}^0$ × B($K^{*0}\to K^+\pi^-$)	< 5 × 10^{-4}	CL=90%	605
$K^*(892)^-K^+$ × B($K^{*-}\to \overline{K}^0\pi^-$)	(1.2 ±0.7) × 10^{-3}		610
$\overline{K}^0K^+\pi^-$ nonresonant	(3.9 $^{+2.3}_{-1.9}$) × 10^{-3}		739
$K^+K^-\pi^0$	(1.3 ±0.4) × 10^{-3}		742
$K_S^0K_S^0\pi^0$	< 5.9 × 10^{-4}		739
$K^+K^-\pi^+\pi^-$ [vv]	(2.52±0.24) × 10^{-3}		676
$\phi\pi^+\pi^-$ × B($\phi\to K^+K^-$)	(5.3 ±1.4) × 10^{-4}		614
$\phi\rho^0$ × B($\phi\to K^+K^-$)	(3.0 ±1.6) × 10^{-4}		260
$K^+K^-\rho^0$ 3-body	(9.1 ±2.3) × 10^{-4}		309
$K^*(892)^0K^-\pi^+$+c.c. × B($K^{*0}\to K^+\pi^-$) [ww]	< 5 × 10^{-4}		528
$K^*(892)^0\overline{K}^*(892)^0$ × $B^2(K^{*0}\to K^+\pi^-)$	(6 ±2) × 10^{-4}		257
$K^+K^-\pi^+\pi^-$ non-ϕ	—		676
$K^+K^-\pi^+\pi^-$ nonresonant	< 8 × 10^{-4}	CL=90%	676
$K^0\overline{K}^0\pi^+\pi^-$	(6.9 ±2.7) × 10^{-3}		673
$K^+K^-\pi^+\pi^-\pi^0$	(3.1 ±2.0) × 10^{-3}		600

Fractions of most of the following modes with resonances have already appeared above as submodes of particular charged-particle modes.

Mode	Fraction (Γ_i/Γ)	Scale factor/Confidence level	p (MeV/c)
$\overline{K}^*(892)^0K^0$	< 1.6 × 10^{-3}	CL=90%	605
$K^*(892)^+K^-$	(3.5 ±0.8) × 10^{-3}		610
$K^*(892)^0\overline{K}^0$	< 8 × 10^{-4}	CL=90%	605
$K^*(892)^-K^+$	(1.8 ±1.0) × 10^{-3}		610
$\phi\pi^0$	< 1.4 × 10^{-3}	CL=90%	644
$\phi\eta$	< 2.8 × 10^{-3}	CL=90%	489
$\phi\omega$	< 2.1 × 10^{-3}	CL=90%	239
$\phi\pi^+\pi^-$	(1.08±0.29) × 10^{-3}		614
$\phi\rho^0$	(6 ±3) × 10^{-4}		260
$\phi\pi^+\pi^-$ 3-body	(7 ±5) × 10^{-4}		614
$K^*(892)^0K^-\pi^+$ + c.c. [ww]	< 8 × 10^{-4}	CL=90%	–
$K^*(892)^0\overline{K}^*(892)^0$	(1.4 ±0.5) × 10^{-3}		257

Doubly Cabibbo suppressed (*DC*) modes, $\Delta C = 2$ forbidden via mixing (*C2M*) modes, $\Delta C = 1$ weak neutral current (*C1*) modes, or Lepton Family number (*LF*) violating modes

Mode		Fraction (Γ_i/Γ)	Scale factor/Confidence level	p (MeV/c)
$K^+\ell^-\overline{\nu}_\ell$ (via $\overline{D}^0$)	C2M	< 1.7 × 10^{-4}	CL=90%	–
$K^+\pi^-$ or $K^+\pi^-\pi^+\pi^-$ (via $\overline{D}^0$)	C2M	< 1.0 × 10^{-3}	CL=90%	–
$K^+\pi^-$	DC	(2.8 ±0.9) × 10^{-4}		861
$K^+\pi^-$ (via $\overline{D}^0$)		< 1.9 × 10^{-4}	CL=90%	861
$K^+\pi^-\pi^+\pi^-$	DC	(1.9 ±2.7) × 10^{-4}		812
$K^+\pi^-\pi^+\pi^-$ (via $\overline{D}^0$)		< 4 × 10^{-4}	CL=90%	812
μ^- anything (via $\overline{D}^0$)		< 4 × 10^{-4}	CL=90%	–
e^+e^-	C1	< 1.3 × 10^{-5}	CL=90%	932
$\mu^+\mu^-$	C1	< 4.1 × 10^{-6}	CL=90%	926
$\pi^0e^+e^-$	C1	< 4.5 × 10^{-5}	CL=90%	927
$\pi^0\mu^+\mu^-$	C1	< 1.8 × 10^{-4}	CL=90%	915
ηe^+e^-	C1	< 1.1 × 10^{-4}	CL=90%	852
$\eta\mu^+\mu^-$	C1	< 5.3 × 10^{-4}	CL=90%	838
$\rho^0e^+e^-$	C1	< 1.0 × 10^{-4}	CL=90%	773
$\rho^0\mu^+\mu^-$	C1	< 2.3 × 10^{-4}	CL=90%	756
ωe^+e^-	C1	< 1.8 × 10^{-4}	CL=90%	768
$\omega\mu^+\mu^-$	C1	< 8.3 × 10^{-4}	CL=90%	751
ϕe^+e^-	C1	< 5.2 × 10^{-5}	CL=90%	654
$\phi\mu^+\mu^-$	C1	< 4.1 × 10^{-4}	CL=90%	631
$\overline{K}^0e^+e^-$	[ss]	< 1.1 × 10^{-4}	CL=90%	866
$\overline{K}^0\mu^+\mu^-$	[ss]	< 2.6 × 10^{-4}	CL=90%	852
$\overline{K}^*(892)^0e^+e^-$	[ss]	< 1.4 × 10^{-4}	CL=90%	717
$\overline{K}^*(892)^0\mu^+\mu^-$	[ss]	< 1.18 × 10^{-3}	CL=90%	698
$\pi^+\pi^-\pi^0\mu^+\mu^-$	C1	< 8.1 × 10^{-4}	CL=90%	863
$\mu^\pm e^\mp$	LF [gg]	< 1.9 × 10^{-5}	CL=90%	929
$\pi^0e^\pm\mu^\mp$	LF [gg]	< 8.6 × 10^{-5}	CL=90%	924
$\eta e^\pm\mu^\mp$	LF [gg]	< 1.0 × 10^{-4}	CL=90%	848
$\rho^0e^\pm\mu^\mp$	LF [gg]	< 4.9 × 10^{-5}	CL=90%	769
$\omega e^\pm\mu^\mp$	LF [gg]	< 1.2 × 10^{-4}	CL=90%	764
$\phi e^\pm\mu^\mp$	LF [gg]	< 3.4 × 10^{-5}	CL=90%	648
$\overline{K}^0e^\pm\mu^\mp$	LF [gg]	< 1.0 × 10^{-4}	CL=90%	862
$\overline{K}^*(892)^0e^\pm\mu^\mp$	LF [gg]	< 1.0 × 10^{-4}	CL=90%	712

$D^*(2007)^0$

$I(J^P) = \frac{1}{2}(1^-)$
I, J, P need confirmation.

Mass $m = 2006.7 \pm 0.5$ MeV (S = 1.1)
$m_{D^{*0}} - m_{D^0} = 142.12 \pm 0.07$ MeV
Full width $\Gamma < 2.1$ MeV, CL = 90%

$\overline{D}^*(2007)^0$ modes are charge conjugates of modes below.

$D^*(2007)^0$ DECAY MODES	Fraction (Γ_i/Γ)	p (MeV/c)
$D^0\pi^0$	(61.9±2.9) %	43
$D^0\gamma$	(38.1±2.9) %	137

$D^*(2010)^\pm$

$I(J^P) = \frac{1}{2}(1^-)$
I, J, P need confirmation.

Mass $m = 2010.0 \pm 0.5$ MeV (S = 1.1)
$m_{D^*(2010)^+} - m_{D^+} = 140.64 \pm 0.10$ MeV (S = 1.1)
$m_{D^*(2010)^+} - m_{D^0} = 145.397 \pm 0.030$ MeV
Full width $\Gamma < 0.131$ MeV, CL = 90%

$D^*(2010)^-$ modes are charge conjugates of the modes below.

$D^*(2010)^\pm$ DECAY MODES	Fraction (Γ_i/Γ)	p (MeV/c)
$D^0\pi^+$	(68.3±1.4) %	39
$D^+\pi^0$	(30.6±2.5) %	38
$D^+\gamma$	(1.1$^{+2.1}_{-0.7}$) %	136

$D_1(2420)^0$

$I(J^P) = \frac{1}{2}(1^+)$
I, J, P need confirmation.

Mass $m = 2422.2 \pm 1.8$ MeV (S = 1.2)
Full width $\Gamma = 18.9^{+4.6}_{-3.5}$ MeV

$\overline{D}_1(2420)^0$ modes are charge conjugates of modes below.

$D_1(2420)^0$ DECAY MODES	Fraction (Γ_i/Γ)	p (MeV/c)
$D^*(2010)^+\pi^-$	seen	355
$D^+\pi^-$	not seen	474

$D_2^*(2460)^0$

$I(J^P) = \frac{1}{2}(2^+)$

$J^P = 2^+$ assignment strongly favored (ALBRECHT 89B).

Mass $m = 2458.9 \pm 2.0$ MeV (S = 1.2)
Full width $\Gamma = 23 \pm 5$ MeV

$\overline{D}_2^*(2460)^0$ modes are charge conjugates of modes below.

$D_2^*(2460)^0$ DECAY MODES	Fraction (Γ_i/Γ)	p (MeV/c)
$D^+\pi^-$	seen	503
$D^*(2010)^+\pi^-$	seen	387

$D_2^*(2460)^\pm$

$I(J^P) = \frac{1}{2}(2^+)$

$J^P = 2^+$ assignment strongly favored (ALBRECHT 89B).

Mass $m = 2459 \pm 4$ MeV (S = 1.7)
$m_{D_2^*(2460)^\pm} - m_{D_2^*(2460)^0} = 0.9 \pm 3.3$ MeV (S = 1.1)
Full width $\Gamma = 25^{+8}_{-7}$ MeV

$D_2^*(2460)^-$ modes are charge conjugates of modes below.

$D_2^*(2460)^\pm$ DECAY MODES	Fraction (Γ_i/Γ)	p (MeV/c)
$D^0\pi^+$	seen	508
$D^{*0}\pi^+$	seen	390

CHARMED, STRANGE MESONS ($C = S = \pm1$)

$D_s^+ = c\overline{s}$, $D_s^- = \overline{c}s$, similarly for D_s^*'s

$D_s^\pm$ was $F^\pm$

$I(J^P) = 0(0^-)$

Mass $m = 1968.5 \pm 0.6$ MeV (S = 1.1)
$m_{D_s^\pm} - m_{D^\pm} = 99.2 \pm 0.5$ MeV (S = 1.1)
Mean life $\tau = (0.467 \pm 0.017) \times 10^{-12}$ s
$c\tau = 140\ \mu$m

D_s^+ form factors

$r_2 = 1.6 \pm 0.4$
$r_v = 1.5 \pm 0.5$
$\Gamma_L/\Gamma_T = 0.72 \pm 0.18$

Branching fractions for modes with a resonance in the final state include all the decay modes of the resonance. D_s^- modes are charge conjugates of the modes below.

D_s^+ DECAY MODES		Fraction (Γ_i/Γ)	Scale factor/ Confidence level	p (MeV/c)
Inclusive modes				
K^- anything		$(13\ ^{+14}_{-12})$ %		–
$\overline{K}^0$ anything + K^0 anything		(39 ± 28) %		–
K^+ anything		$(20\ ^{+18}_{-14})$ %		–
non-$K\overline{K}$ anything		(64 ± 17) %		–
e^+ anything		$(8\ ^{+6}_{-5})$ %		–
ϕ anything		$(18\ ^{+15}_{-10})$ %		–
Leptonic and semileptonic modes				
$\mu^+\nu_\mu$		$(4.0\ ^{+2.2}_{-2.0}) \times 10^{-3}$	S=1.4	981
$\tau^+\nu_\tau$		(7 ± 4) %		182
$\phi\ell^+\nu_\ell$	[xx]	(2.0 ± 0.5) %		–
$\eta\ell^+\nu_\ell + \eta'(958)\ell^+\nu_\ell$	[xx]	(3.4 ± 1.0) %		–
$\eta\ell^+\nu_\ell$		(2.5 ± 0.7) %		–
$\eta'(958)\ell^+\nu_\ell$		$(8.8 \pm 3.4) \times 10^{-3}$		–
Hadronic modes with a $K\overline{K}$ pair (including from a ϕ)				
$K^+\overline{K}^0$		(3.6 ± 1.1) %		850
$K^+K^-\pi^+$	[qq]	(4.4 ± 1.2) %	S=1.1	805
$\phi\pi^+$	[yy]	(3.6 ± 0.9) %		712
$K^+\overline{K}^*(892)^0$	[yy]	(3.3 ± 0.9) %		682
$f_0(980)\pi^+$	[yy]	(1.8 ± 0.8) %	S=1.3	732
$K^+\overline{K}_0^*(1430)^0$	[yy]	$(7 \pm 4) \times 10^{-3}$		186
$f_J(1710)\pi^+ \to K^+K^-\pi^+$	[zz]	$(1.5 \pm 1.9) \times 10^{-3}$		204
$K^+K^-\pi^+$ nonresonant		$(9 \pm 4) \times 10^{-3}$		805
$K^0\overline{K}^0\pi^+$		—		802
$K^*(892)^+\overline{K}^0$	[yy]	(4.3 ± 1.4) %		683
$K^+K^-\pi^+\pi^0$		—		748
$\phi\pi^+\pi^0$	[yy]	(9 ± 5) %		687
$\phi\rho^+$	[yy]	(6.7 ± 2.3) %		407
$\phi\pi^+\pi^0$ 3-body	[yy]	< 2.6 %	CL=90%	687
$K^+K^-\pi^+\pi^0$ non-ϕ		< 9 %	CL=90%	748
$K^+\overline{K}^0\pi^+\pi^-$		< 2.8 %	CL=90%	744
$K^0K^-\pi^+\pi^+$		(4.3 ± 1.5) %		744
$K^*(892)^+\overline{K}^*(892)^0$	[yy]	(5.8 ± 2.5) %		412
$K^0K^-\pi^+\pi^+$ non-$K^{*+}\overline{K}^{*0}$		< 2.9 %	CL=90%	744
$K^+K^-\pi^+\pi^+\pi^-$		$(8.3 \pm 3.3) \times 10^{-3}$		673
$\phi\pi^+\pi^+\pi^-$	[yy]	(1.18 ± 0.35) %		640
$K^+K^-\pi^+\pi^+\pi^-$ non-ϕ		$(3.0\ ^{+3.0}_{-2.0}) \times 10^{-3}$		673
Hadronic modes without K's				
$\pi^+\pi^+\pi^-$		(1.0 ± 0.4) %	S=1.2	959
$\rho^0\pi^+$		$< 8 \times 10^{-4}$	CL=90%	827
$f_0(980)\pi^+$	[yy]	(1.8 ± 0.8) %	S=1.7	732
$f_2(1270)\pi^+$	[yy]	$(2.3 \pm 1.3) \times 10^{-3}$		559
$f_0(1500)\pi^+ \to \pi^+\pi^-\pi^+$	[aaa]	$(2.8 \pm 1.6) \times 10^{-3}$		391
$\pi^+\pi^+\pi^-$ nonresonant		$< 2.8 \times 10^{-3}$	CL=90%	959
$\pi^+\pi^+\pi^-\pi^0$		< 12 %	CL=90%	935
$\eta\pi^+$	[yy]	(2.0 ± 0.6) %		902

$\omega\pi^+$	[yy]	$(3.1 \pm 1.4) \times 10^{-3}$		822
$\pi^+\pi^+\pi^+\pi^-\pi^-$		$(6.9 \pm 3.0) \times 10^{-3}$		899
$\pi^+\pi^+\pi^-\pi^0\pi^0$		—		902
$\eta\rho^+$	[yy]	(10.3 ± 3.2) %		727
$\eta\pi^+\pi^0$ 3-body	[yy]	< 3.0 %	CL=90%	886
$\pi^+\pi^+\pi^+\pi^-\pi^-\pi^0$		(4.9 ± 3.2) %		856
$\eta'(958)\pi^+$	[yy]	(4.9 ± 1.8) %		743
$\pi^+\pi^+\pi^+\pi^-\pi^-\pi^0\pi^0$		—		803
$\eta'(958)\rho^+$	[yy]	(12 ± 4) %		470
$\eta'(958)\pi^+\pi^0$ 3-body	[yy]	< 3.1 %	CL=90%	720
Modes with one or three K's				
$K^0\pi^+$		$< 8 \times 10^{-3}$	CL=90%	916
$K^+\pi^+\pi^-$		(1.0 ± 0.4) %		900
$K^+\rho^0$		$< 2.9 \times 10^{-3}$	CL=90%	747
$K^*(892)^0\pi^+$	[yy]	$(6.5 \pm 2.8) \times 10^{-3}$		773
$K^+K^+K^-$		$< 6 \times 10^{-4}$	CL=90%	628
ϕK^+	[yy]	$< 5 \times 10^{-4}$	CL=90%	607
$\Delta C = 1$ weak neutral current ($C1$) modes, or Lepton number (L) violating modes				
$\pi^+\mu^+\mu^-$	[ss]	$< 4.3 \times 10^{-4}$	CL=90%	968
$K^+\mu^+\mu^-$	$C1$	$< 5.9 \times 10^{-4}$	CL=90%	909
$K^*(892)^+\mu^+\mu^-$	$C1$	$< 1.4 \times 10^{-3}$	CL=90%	765
$\pi^-\mu^+\mu^+$	L	$< 4.3 \times 10^{-4}$	CL=90%	968
$K^-\mu^+\mu^+$	L	$< 5.9 \times 10^{-4}$	CL=90%	909
$K^*(892)^-\mu^+\mu^+$	L	$< 1.4 \times 10^{-3}$	CL=90%	765

$D_s^{*\pm}$

$I(J^P) = 0(?^?)$

J^P is natural, width and decay modes consistent with 1^-.

Mass $m = 2112.4 \pm 0.7$ MeV (S = 1.1)

$m_{D_s^{*\pm}} - m_{D_s^\pm} = 143.8 \pm 0.4$ MeV

Full width $\Gamma < 1.9$ MeV, CL = 90%

D_s^{*-} modes are charge conjugates of the modes below.

D_s^{*+} DECAY MODES	Fraction (Γ_i/Γ)	p (MeV/c)
$D_s^+\gamma$	(94.2 ± 2.5) %	139
$D_s^+\pi^0$	(5.8 ± 2.5) %	48

$D_{s1}(2536)^\pm$

$I(J^P) = 0(1^+)$

J, P need confirmation.

Mass $m = 2535.35 \pm 0.34 \pm 0.5$ MeV

Full width $\Gamma < 2.3$ MeV, CL = 90%

$D_{s1}(2536)^-$ modes are charge conjugates of the modes below.

$D_{s1}(2536)^+$ DECAY MODES	Fraction (Γ_i/Γ)	p (MeV/c)
$D^*(2010)^+K^0$	seen	150
$D^*(2007)^0K^+$	seen	169
D^+K^0	not seen	382
D^0K^+	not seen	392
$D_s^{*+}\gamma$	possibly seen	389

$D_{sJ}(2573)^\pm$

$I(J^P) = 0(?^?)$

J^P is natural, width and decay modes consistent with 2^+.

Mass $m = 2573.5 \pm 1.7$ MeV

Full width $\Gamma = 15^{+5}_{-4}$ MeV

$D_{sJ}(2573)^-$ modes are charge conjugates of the modes below.

$D_{sJ}(2573)^+$ DECAY MODES	Fraction (Γ_i/Γ)	p (MeV/c)
D^0K^+	seen	436
$D^*(2007)^0K^+$	not seen	245

BOTTOM MESONS
($B = \pm1$)

$B^+ = u\bar{b}$, $B^0 = d\bar{b}$, $\overline{B}^0 = \bar{d}\,b$, $B^- = \bar{u}\,b$, similarly for B^*'s

B-particle organization

Many measurements of B decays involve admixtures of B hadrons. Previously we arbitrarily included such admixtures in the $B^\pm$ section, but because of their importance we have created two new sections: "$B^\pm/B^0$ Admixture" for $\Upsilon(4S)$ results and "$B^\pm/B^0/B_s^0/b$-baryon Admixture" for results at higher energies. Most inclusive decay branching fractions are found in the Admixture sections. B^0-$\overline{B}^0$ mixing data are found in the B^0 section, while B_s^0-$\overline{B}_s^0$ mixing data and B-$\overline{B}$ mixing data for a B^0/B_s^0 admixture are found in the B_s^0 section. CP-violation data are found in the B^0 section. b-baryons are found near the end of the Baryon section.

The organization of the B sections is now as follows, where bullets indicate particle sections and brackets indicate reviews.

[Production and Decay of b-flavored Hadrons]

[Semileptonic Decays of B Mesons]

- $B^\pm$
  - mass, mean life
  - branching fractions
- B^0
  - mass, mean life
  - branching fractions
  - polarization in B^0 decay
  - B^0-$\overline{B}^0$ mixing
  - [B^0-$\overline{B}^0$ Mixing and CP Violation in B Decay]
  - CP violation
- $B^\pm$ B^0 Admixtures
  - branching fractions
- $B^\pm/B^0/B_s^0/b$-baryon Admixtures
  - mean life
  - production fractions
  - branching fractions
- B^*
  - mass
- B_s^0
  - mass, mean life
  - branching fractions
  - polarization in B_s^0 decay
  - B_s^0-$\overline{B}_s^0$ mixing
  - B-$\overline{B}$ mixing (admixture of B^0, B_s^0)

At end of Baryon Listings:

- Λ_b
  - mass, mean life
  - branching fractions
- b-baryon Admixture
  - mean life
  - branching fractions

$B^{\pm}$

$I(J^P) = \frac{1}{2}(0^-)$

I, J, P need confirmation. Quantum numbers shown are quark-model predictions.

Mass $m_{B^{\pm}} = 5278.9 \pm 1.8$ MeV
Mean life $\tau_{B^{\pm}} = (1.65 \pm 0.04) \times 10^{-12}$ s
$c\tau = 495$ μm

B^- modes are charge conjugates of the modes below. Modes which do not identify the charge state of the B are listed in the $B^{\pm}/B^0$ ADMIXTURE section.

The branching fractions listed below assume 50% $B^0\overline{B}^0$ and 50% B^+B^- production at the $\Upsilon(4S)$. We have attempted to bring older measurements up to date by rescaling their assumed $\Upsilon(4S)$ production ratio to 50:50 and their assumed D, D_s, D^*, and ψ branching ratios to current values whenever this would affect our averages and best limits significantly.

Indentation is used to indicate a subchannel of a previous reaction. All resonant subchannels have been corrected for resonance branching fractions to the final state so the sum of the subchannel branching fractions can exceed that of the final state.

B^+ DECAY MODES	Fraction (Γ_i/Γ)	Scale factor/ Confidence level	p (MeV/c)
Semileptonic and leptonic modes			
$\ell^+ \nu_\ell$ anything	[pp] (10.3 ± 0.9) %		–
$\overline{D}^0 \ell^+ \nu_\ell$	[pp] (1.86 ± 0.33) %		–
$\overline{D}^*(2007)^0 \ell^+ \nu_\ell$	[pp] (5.3 ± 0.8) %		–
$\pi^0 e^+ \nu_e$	$< 2.2 \times 10^{-3}$	CL=90%	2638
$\omega \ell^+ \nu_\ell$	[pp] $< 2.1 \times 10^{-4}$	CL=90%	–
$\rho^0 \ell^+ \nu_\ell$	[pp] $< 2.1 \times 10^{-4}$	CL=90%	–
$e^+ \nu_e$	$< 1.5 \times 10^{-5}$	CL=90%	2639
$\mu^+ \nu_\mu$	$< 2.1 \times 10^{-5}$	CL=90%	2638
$\tau^+ \nu_\tau$	$< 5.7 \times 10^{-4}$	CL=90%	2340
$e^+ \nu_e \gamma$	$< 2.0 \times 10^{-4}$	CL=90%	–
$\mu^+ \nu_\mu \gamma$	$< 5.2 \times 10^{-5}$	CL=90%	–
***D*, *D**, or *D_s* modes**			
$\overline{D}^0 \pi^+$	(5.3 ± 0.5) $\times 10^{-3}$		2308
$\overline{D}^0 \rho^+$	(1.34 ± 0.18) %		2238
$\overline{D}^0 \pi^+ \pi^+ \pi^-$	(1.1 ± 0.4) %		2289
$\overline{D}^0 \pi^+ \pi^+ \pi^-$ nonresonant	(5 ± 4) $\times 10^{-3}$		2289
$\overline{D}^0 \pi^+ \rho^0$	(4.2 ± 3.0) $\times 10^{-3}$		2209
$\overline{D}^0 a_1(1260)^+$	(5 ± 4) $\times 10^{-3}$		2123
$D^*(2010)^- \pi^+ \pi^+$	(2.1 ± 0.6) $\times 10^{-3}$		2247
$D^- \pi^+ \pi^+$	$< 1.4 \times 10^{-3}$	CL=90%	2299
$\overline{D}^*(2007)^0 \pi^+$	(4.6 ± 0.4) $\times 10^{-3}$		2256
$D^*(2010)^+ \pi^0$	$< 1.7 \times 10^{-4}$	CL=90%	2254
$\overline{D}^*(2007)^0 \rho^+$	(1.55 ± 0.31) %		2183
$\overline{D}^*(2007)^0 \pi^+ \pi^+ \pi^-$	(9.4 ± 2.6) $\times 10^{-3}$		2236
$\overline{D}^*(2007)^0 a_1(1260)^+$	(1.9 ± 0.5) %		2062
$D^*(2010)^- \pi^+ \pi^+ \pi^0$	(1.5 ± 0.7) %		2235
$D^*(2010)^- \pi^+ \pi^+ \pi^+ \pi^-$	< 1 %	CL=90%	2217
$\overline{D}_1^*(2420)^0 \pi^+$	(1.5 ± 0.6) $\times 10^{-3}$	S=1.3	2081
$\overline{D}_1^*(2420)^0 \rho^+$	$< 1.4 \times 10^{-3}$	CL=90%	1997
$\overline{D}_2^*(2460)^0 \pi^+$	$< 1.3 \times 10^{-3}$	CL=90%	2064
$\overline{D}_2^*(2460)^0 \rho^+$	$< 4.7 \times 10^{-3}$	CL=90%	1979
$\overline{D}^0 D_s^+$	(1.3 ± 0.4) %		1815
$\overline{D}^0 D_s^{*+}$	(9 ± 4) $\times 10^{-3}$		1734
$\overline{D}^*(2007)^0 D_s^+$	(1.2 ± 0.5) %		1737
$\overline{D}^*(2007)^0 D_s^{*+}$	(2.7 ± 1.0) %		1650
$D_s^+ \pi^0$	$< 2.0 \times 10^{-4}$	CL=90%	2270
$D_s^{*+} \pi^0$	$< 3.3 \times 10^{-4}$	CL=90%	2214
$D_s^+ \eta$	$< 5 \times 10^{-4}$	CL=90%	2235
$D_s^{*+} \eta$	$< 8 \times 10^{-4}$	CL=90%	2177
$D_s^+ \rho^0$	$< 4 \times 10^{-4}$	CL=90%	2198
$D_s^{*+} \rho^0$	$< 5 \times 10^{-4}$	CL=90%	2139
$D_s^+ \omega$	$< 5 \times 10^{-4}$	CL=90%	2195
$D_s^{*+} \omega$	$< 7 \times 10^{-4}$	CL=90%	2136
$D_s^+ a_1(1260)^0$	$< 2.2 \times 10^{-3}$	CL=90%	2079
$D_s^{*+} a_1(1260)^0$	$< 1.6 \times 10^{-3}$	CL=90%	2014
$D_s^+ \phi$	$< 3.2 \times 10^{-4}$	CL=90%	2141
$D_s^{*+} \phi$	$< 4 \times 10^{-4}$	CL=90%	2079
$D_s^+ \overline{K}^0$	$< 1.1 \times 10^{-3}$	CL=90%	2241
$D_s^{*+} \overline{K}^0$	$< 1.1 \times 10^{-3}$	CL=90%	2184
$D_s^+ \overline{K}^*(892)^0$	$< 5 \times 10^{-4}$	CL=90%	2171
$D_s^{*+} \overline{K}^*(892)^0$	$< 4 \times 10^{-4}$	CL=90%	2110
$D_s^- \pi^+ K^+$	$< 8 \times 10^{-4}$	CL=90%	2222
$D_s^{*-} \pi^+ K^+$	$< 1.2 \times 10^{-3}$	CL=90%	2164
$D_s^- \pi^+ K^*(892)^+$	$< 6 \times 10^{-3}$	CL=90%	2137
$D_s^{*-} \pi^+ K^*(892)^+$	$< 8 \times 10^{-3}$	CL=90%	2075
Charmonium modes			
$J/\psi(1S) K^+$	(9.9 ± 1.0) $\times 10^{-4}$		1683
$J/\psi(1S) K^+ \pi^+ \pi^-$	(1.4 ± 0.6) $\times 10^{-3}$		1612
$J/\psi(1S) K^*(892)^+$	(1.47 ± 0.27) $\times 10^{-3}$		1571
$J/\psi(1S) \pi^+$	(5.0 ± 1.5) $\times 10^{-5}$		1727
$J/\psi(1S) \rho^+$	$< 7.7 \times 10^{-4}$	CL=90%	1613
$J/\psi(1S) a_1(1260)^+$	$< 1.2 \times 10^{-3}$	CL=90%	1414
$\psi(2S) K^+$	(6.9 ± 3.1) $\times 10^{-4}$	S=1.3	1284
$\psi(2S) K^*(892)^+$	$< 3.0 \times 10^{-3}$	CL=90%	1115
$\psi(2S) K^+ \pi^+ \pi^-$	(1.9 ± 1.2) $\times 10^{-3}$		909
$\chi_{c1}(1P) K^+$	(1.0 ± 0.4) $\times 10^{-3}$		1411
$\chi_{c1}(1P) K^*(892)^+$	$< 2.1 \times 10^{-3}$	CL=90%	1265
***K* or *K** modes**			
$K^0 \pi^+$	(2.3 ± 1.1) $\times 10^{-5}$		2614
$K^+ \pi^0$	$< 1.6 \times 10^{-5}$	CL=90%	2615
$\eta' K^+$	(6.5 ± 1.7) $\times 10^{-5}$		2528
$\eta' K^*(892)^+$	$< 1.3 \times 10^{-4}$	CL=90%	2472
ηK^+	$< 1.4 \times 10^{-5}$	CL=90%	2587
$\eta K^*(892)^+$	$< 3.0 \times 10^{-5}$	CL=90%	2534
$K^*(892)^0 \pi^+$	$< 4.1 \times 10^{-5}$	CL=90%	2561
$K^*(892)^+ \pi^0$	$< 9.9 \times 10^{-5}$	CL=90%	2562
$K^+ \pi^- \pi^+$ nonresonant	$< 2.8 \times 10^{-5}$	CL=90%	2609
$K^- \pi^+ \pi^+$ nonresonant	$< 5.6 \times 10^{-5}$	CL=90%	–
$K_1(1400)^0 \pi^+$	$< 2.6 \times 10^{-3}$	CL=90%	2451
$K_2^*(1430)^0 \pi^+$	$< 6.8 \times 10^{-4}$	CL=90%	2443
$K^+ \rho^0$	$< 1.9 \times 10^{-5}$	CL=90%	2559
$K^0 \rho^+$	$< 4.8 \times 10^{-5}$	CL=90%	2559
$K^*(892)^+ \pi^+ \pi^-$	$< 1.1 \times 10^{-3}$	CL=90%	2556
$K^*(892)^+ \rho^0$	$< 9.0 \times 10^{-4}$	CL=90%	2505
$K_1(1400)^+ \rho^0$	$< 7.8 \times 10^{-4}$	CL=90%	2389
$K_2^*(1430)^+ \rho^0$	$< 1.5 \times 10^{-3}$	CL=90%	2382
$K^+ \overline{K}^0$	$< 2.1 \times 10^{-5}$	CL=90%	2592
$K^+ K^- \pi^+$ nonresonant	$< 7.5 \times 10^{-5}$	CL=90%	–
$K^+ K^- K^+$	$< 2.0 \times 10^{-4}$	CL=90%	2522
$K^+ \phi$	$< 1.2 \times 10^{-5}$	CL=90%	2516
$K^+ K^- K^+$ nonresonant	$< 3.8 \times 10^{-5}$	CL=90%	2516
$K^*(892)^+ K^+ K^-$	$< 1.6 \times 10^{-3}$	CL=90%	2466
$K^*(892)^+ \phi$	$< 7.0 \times 10^{-5}$	CL=90%	2460
$K_1(1400)^+ \phi$	$< 1.1 \times 10^{-3}$	CL=90%	2339
$K_2^*(1430)^+ \phi$	$< 3.4 \times 10^{-3}$	CL=90%	2332
$K^+ f_0(980)$	$< 8 \times 10^{-5}$	CL=90%	2524
$K^*(892)^+ \gamma$	(5.7 ± 3.3) $\times 10^{-5}$		2564
$K_1(1270)^+ \gamma$	$< 7.3 \times 10^{-3}$	CL=90%	2486
$K_1(1400)^+ \gamma$	$< 2.2 \times 10^{-3}$	CL=90%	2453
$K_2^*(1430)^+ \gamma$	$< 1.4 \times 10^{-3}$	CL=90%	2447
$K^*(1680)^+ \gamma$	$< 1.9 \times 10^{-3}$	CL=90%	2361
$K_3^*(1780)^+ \gamma$	$< 5.5 \times 10^{-3}$	CL=90%	2343
$K_4^*(2045)^+ \gamma$	$< 9.9 \times 10^{-3}$	CL=90%	2243
Light unflavored meson modes			
$\pi^+ \pi^0$	$< 2.0 \times 10^{-5}$	CL=90%	2636
$\pi^+ \pi^+ \pi^-$	$< 1.3 \times 10^{-4}$	CL=90%	2630
$\rho^0 \pi^+$	$< 4.3 \times 10^{-5}$	CL=90%	2582
$\pi^+ f_0(980)$	$< 1.4 \times 10^{-4}$	CL=90%	2547
$\pi^+ f_2(1270)$	$< 2.4 \times 10^{-4}$	CL=90%	2483
$\pi^+ \pi^- \pi^+$ nonresonant	$< 4.1 \times 10^{-5}$	CL=90%	–
$\pi^+ \pi^0 \pi^0$	$< 8.9 \times 10^{-4}$	CL=90%	2631
$\rho^+ \pi^0$	$< 7.7 \times 10^{-5}$	CL=90%	2582
$\pi^+ \pi^- \pi^+ \pi^0$	$< 4.0 \times 10^{-3}$	CL=90%	2621
$\rho^+ \rho^0$	$< 1.0 \times 10^{-3}$	CL=90%	2525
$a_1(1260)^+ \pi^0$	$< 1.7 \times 10^{-3}$	CL=90%	2494
$a_1(1260)^0 \pi^+$	$< 9.0 \times 10^{-4}$	CL=90%	2494
$\omega \pi^+$	$< 4.0 \times 10^{-4}$	CL=90%	2580

$\eta\pi^+$		< 1.5	$\times 10^{-5}$	CL=90%	2609
$\eta'\pi^+$		< 3.1	$\times 10^{-5}$	CL=90%	2550
$\eta'\rho^+$		< 4.7	$\times 10^{-5}$	CL=90%	2493
$\eta\rho^+$		< 3.2	$\times 10^{-5}$	CL=90%	2554
$\pi^+\pi^+\pi^+\pi^-\pi^-$		< 8.6	$\times 10^{-4}$	CL=90%	2608
$\rho^0 a_1(1260)^+$		< 6.2	$\times 10^{-4}$	CL=90%	2434
$\rho^0 a_2(1320)^+$		< 7.2	$\times 10^{-4}$	CL=90%	2411
$\pi^+\pi^+\pi^+\pi^-\pi^-\pi^0$		< 6.3	$\times 10^{-3}$	CL=90%	2592
$a_1(1260)^+ a_1(1260)^0$		< 1.3	%	CL=90%	2335
Baryon modes					
$p\overline{p}\pi^+$		< 1.6	$\times 10^{-4}$	CL=90%	2439
$p\overline{p}\pi^+$ nonresonant		< 5.3	$\times 10^{-5}$	CL=90%	–
$p\overline{p}\pi^+\pi^+\pi^-$		< 5.2	$\times 10^{-4}$	CL=90%	2369
$p\overline{p}K^+$ nonresonant		< 8.9	$\times 10^{-5}$	CL=90%	–
$p\overline{\Lambda}$		< 6	$\times 10^{-5}$	CL=90%	2430
$p\overline{\Lambda}\pi^+\pi^-$		< 2.0	$\times 10^{-4}$	CL=90%	2367
$\overline{\Delta}^0 p$		< 3.8	$\times 10^{-4}$	CL=90%	2402
$\Delta^{++}\overline{p}$		< 1.5	$\times 10^{-4}$	CL=90%	2402
$\Lambda_c^- p\pi^+$		(6.2 ±2.7)	$\times 10^{-4}$		–
$\Lambda_c^- p\pi^+\pi^0$		< 3.12	$\times 10^{-3}$	CL=90%	–
$\Lambda_c^- p\pi^+\pi^+\pi^-$		< 1.46	$\times 10^{-3}$	CL=90%	–
$\Lambda_c^- p\pi^+\pi^+\pi^-\pi^0$		< 1.34	%	CL=90%	–
Lepton Family number (*LF*) or Lepton number (*L*) violating modes, or $\Delta B = 1$ weak neutral current (*B1*) modes					
$\pi^+e^+e^-$	*B1*	< 3.9	$\times 10^{-3}$	CL=90%	2638
$\pi^+\mu^+\mu^-$	*B1*	< 9.1	$\times 10^{-3}$	CL=90%	2633
$K^+e^+e^-$	*B1*	< 6	$\times 10^{-5}$	CL=90%	2616
$K^+\mu^+\mu^-$	*B1*	< 1.0	$\times 10^{-5}$	CL=90%	2612
$K^*(892)^+e^+e^-$	*B1*	< 6.9	$\times 10^{-4}$	CL=90%	2564
$K^*(892)^+\mu^+\mu^-$	*B1*	< 1.2	$\times 10^{-3}$	CL=90%	2560
$\pi^+e^+\mu^-$	*LF*	< 6.4	$\times 10^{-3}$	CL=90%	2637
$\pi^+e^-\mu^+$	*LF*	< 6.4	$\times 10^{-3}$	CL=90%	2637
$K^+e^+\mu^-$	*LF*	< 6.4	$\times 10^{-3}$	CL=90%	2615
$K^+e^-\mu^+$	*LF*	< 6.4	$\times 10^{-3}$	CL=90%	2615
$\pi^-e^+e^+$	*L*	< 3.9	$\times 10^{-3}$	CL=90%	2638
$\pi^-\mu^+\mu^+$	*L*	< 9.1	$\times 10^{-3}$	CL=90%	2633
$\pi^-e^+\mu^+$	*LF*	< 6.4	$\times 10^{-3}$	CL=90%	2637
$K^-e^+e^+$	*L*	< 3.9	$\times 10^{-3}$	CL=90%	2616
$K^-\mu^+\mu^+$	*L*	< 9.1	$\times 10^{-3}$	CL=90%	2612
$K^-e^+\mu^+$	*LF*	< 6.4	$\times 10^{-3}$	CL=90%	2615

B^0

$I(J^P) = \frac{1}{2}(0^-)$

I, *J*, *P* need confirmation. Quantum numbers shown are quark-model predictions.

Mass $m_{B^0} = 5279.2 \pm 1.8$ MeV

$m_{B^0} - m_{B^\pm} = 0.35 \pm 0.29$ MeV $\quad (S = 1.1)$

Mean life $\tau_{B^0} = (1.56 \pm 0.04) \times 10^{-12}$ s

$c\tau = 468\ \mu$m

$\tau_{B^+}/\tau_{B^0} = 1.02 \pm 0.04$ (average of direct and inferred)

$\tau_{B^+}/\tau_{B^0} = 1.04 \pm 0.04$ (direct measurements)

$\tau_{B^+}/\tau_{B^0} = 0.95^{+0.15}_{-0.12}$ (inferred from branching fractions)

B^0-$\overline{B}^0$ mixing parameters

$\chi_d = 0.172 \pm 0.010$

$\Delta m_{B^0} = m_{B^0_H} - m_{B^0_L} = (0.464 \pm 0.018) \times 10^{12}\ \hbar\ s^{-1}$

$x_d = \Delta m_{B^0}/\Gamma_{B^0} = 0.723 \pm 0.032$

***CP* violation parameters**

$|\mathrm{Re}(\epsilon_{B^0})| = 0.002 \pm 0.008$

$\overline{B}^0$ modes are charge conjugates of the modes below. Reactions indicate the weak decay vertex and do not include mixing. Modes which do not identify the charge state of the *B* are listed in the $B^\pm/B^0$ ADMIXTURE section.

The branching fractions listed below assume 50% $B^0\overline{B}^0$ and 50% B^+B^- production at the $\Upsilon(4S)$. We have attempted to bring older measurements up to date by rescaling their assumed $\Upsilon(4S)$ production ratio to 50:50 and their assumed D, D_s, D^*, and ψ branching ratios to current values whenever this would affect our averages and best limits significantly.

Indentation is used to indicate a subchannel of a previous reaction. All resonant subchannels have been corrected for resonance branching fractions to the final state so the sum of the subchannel branching fractions can exceed that of the final state.

B^0 DECAY MODES	Fraction (Γ_i/Γ)	Scale factor/ Confidence level	*p* (MeV/c)
$\ell^+\nu_\ell$ anything	[*pp*] (10.5 ± 0.8) %		–
$D^-\ell^+\nu_\ell$	[*pp*] (2.00± 0.25) %		–
$D^*(2010)^-\ell^+\nu_\ell$	[*pp*] (4.60± 0.27) %		–
$\rho^-\ell^+\nu_\ell$	[*pp*] ($2.5^{+0.8}_{-1.0}$) $\times 10^{-4}$		–
$\pi^-\ell^+\nu_\ell$	(1.8 ± 0.6) $\times 10^{-4}$		–
Inclusive modes			
K^+ anything	(78 ±80) %		–
***D*, *D**, or D_s modes**			
$D^-\pi^+$	(3.0 ± 0.4) $\times 10^{-3}$		2306
$D^-\rho^+$	(7.9 ± 1.4) $\times 10^{-3}$		2236
$\overline{D}^0\pi^+\pi^-$	< 1.6 $\times 10^{-3}$	CL=90%	2301
$D^*(2010)^-\pi^+$	(2.76± 0.21) $\times 10^{-3}$		2254
$D^-\pi^+\pi^+\pi^-$	(8.0 ± 2.5) $\times 10^{-3}$		2287
$(D^-\pi^+\pi^+\pi^-)$ nonresonant	(3.9 ± 1.9) $\times 10^{-3}$		2287
$D^-\pi^+\rho^0$	(1.1 ± 1.0) $\times 10^{-3}$		2207
$D^-a_1(1260)^+$	(6.0 ± 3.3) $\times 10^{-3}$		2121
$D^*(2010)^-\pi^+\pi^0$	(1.5 ± 0.5) %		2247
$D^*(2010)^-\rho^+$	(6.7 ± 3.3) $\times 10^{-3}$		2181
$D^*(2010)^-\pi^+\pi^+\pi^-$	(7.6 ± 1.7) $\times 10^{-3}$	S=1.3	2235
$(D^*(2010)^-\pi^+\pi^+\pi^-)$ non-resonant	(0.0 ± 2.5) $\times 10^{-3}$		2235
$D^*(2010)^-\pi^+\rho^0$	(5.7 ± 3.1) $\times 10^{-3}$		2151
$D^*(2010)^-a_1(1260)^+$	(1.30± 0.27) %		2061
$D^*(2010)^-\pi^+\pi^+\pi^-\pi^0$	(3.4 ± 1.8) %		2218
$\overline{D}_2^*(2460)^-\pi^+$	< 2.2 $\times 10^{-3}$	CL=90%	2064
$\overline{D}_2^*(2460)^-\rho^+$	< 4.9 $\times 10^{-3}$	CL=90%	1979
$D^-D_s^+$	(8.0 ± 3.0) $\times 10^{-3}$		1812
$D^*(2010)^-D_s^+$	(9.6 ± 3.4) $\times 10^{-3}$		1735
$D^-D_s^{*+}$	(1.0 ± 0.5) %		1731
$D^*(2010)^-D_s^{*+}$	(2.0 ± 0.7) %		1649
$D_s^+\pi^-$	< 2.8 $\times 10^{-4}$	CL=90%	2270
$D_s^{*+}\pi^-$	< 5 $\times 10^{-4}$	CL=90%	2214
$D_s^+\rho^-$	< 7 $\times 10^{-4}$	CL=90%	2198
$D_s^{*+}\rho^-$	< 8 $\times 10^{-4}$	CL=90%	2139
$D_s^+a_1(1260)^-$	< 2.6 $\times 10^{-3}$	CL=90%	2079
$D_s^{*+}a_1(1260)^-$	< 2.2 $\times 10^{-3}$	CL=90%	2014
$D_s^-K^+$	< 2.4 $\times 10^{-4}$	CL=90%	2242
$D_s^{*-}K^+$	< 1.7 $\times 10^{-4}$	CL=90%	2185
$D_s^-K^*(892)^+$	< 9.9 $\times 10^{-4}$	CL=90%	2172
$D_s^{*-}K^*(892)^+$	< 1.1 $\times 10^{-3}$	CL=90%	2112
$D_s^-\pi^+K^0$	< 5 $\times 10^{-3}$	CL=90%	2221
$D_s^{*-}\pi^+K^0$	< 3.1 $\times 10^{-3}$	CL=90%	2164
$D_s^-\pi^+K^*(892)^0$	< 4 $\times 10^{-3}$	CL=90%	2136
$D_s^{*-}\pi^+K^*(892)^0$	< 2.0 $\times 10^{-3}$	CL=90%	2074
$\overline{D}^0\pi^0$	< 1.2 $\times 10^{-4}$	CL=90%	2308
$\overline{D}^0\rho^0$	< 3.9 $\times 10^{-4}$	CL=90%	2238
$\overline{D}^0\eta$	< 1.3 $\times 10^{-4}$	CL=90%	2274
$\overline{D}^0\eta'$	< 9.4 $\times 10^{-4}$	CL=90%	2198
$\overline{D}^0\omega$	< 5.1 $\times 10^{-4}$	CL=90%	2235
$\overline{D}^*(2007)^0\pi^0$	< 4.4 $\times 10^{-4}$	CL=90%	2256
$\overline{D}^*(2007)^0\rho^0$	< 5.6 $\times 10^{-4}$	CL=90%	2183
$\overline{D}^*(2007)^0\eta$	< 2.6 $\times 10^{-4}$	CL=90%	2220
$\overline{D}^*(2007)^0\eta'$	< 1.4 $\times 10^{-3}$	CL=90%	2141
$\overline{D}^*(2007)^0\omega$	< 7.4 $\times 10^{-4}$	CL=90%	2180
$D^*(2010)^+D^*(2010)^-$	< 2.2 $\times 10^{-3}$	CL=90%	1711
$D^*(2010)^+D^-$	< 1.8 $\times 10^{-3}$	CL=90%	1790
$D^+D^*(2010)^-$	< 1.2 $\times 10^{-3}$	CL=90%	1790
Charmonium modes			
$J/\psi(1S)K^0$	(8.9 ± 1.2) $\times 10^{-4}$		1683
$J/\psi(1S)K^+\pi^-$	(1.1 ± 0.6) $\times 10^{-3}$		1652
$J/\psi(1S)K^*(892)^0$	(1.35± 0.18) $\times 10^{-3}$		1570
$J/\psi(1S)\pi^0$	< 5.8 $\times 10^{-5}$	CL=90%	1728
$J/\psi(1S)\eta$	< 1.2 $\times 10^{-3}$	CL=90%	1672
$J/\psi(1S)\rho^0$	< 2.5 $\times 10^{-4}$	CL=90%	1614

Mode		Fraction	Confidence level	p (MeV/c)
$J/\psi(1S)\omega$		$< 2.7 \times 10^{-4}$	CL=90%	1609
$\psi(2S)K^0$		$< 8 \times 10^{-4}$	CL=90%	1283
$\psi(2S)K^+\pi^-$		$< 1 \times 10^{-3}$	CL=90%	1238
$\psi(2S)K^*(892)^0$		$(1.4 \pm 0.9) \times 10^{-3}$		1113
$\chi_{c1}(1P)K^0$		$< 2.7 \times 10^{-3}$	CL=90%	1411
$\chi_{c1}(1P)K^*(892)^0$		$< 2.1 \times 10^{-3}$	CL=90%	1263
K or K^* modes				
$K^+\pi^-$		$(1.5 ^{+0.5}_{-0.4}) \times 10^{-5}$		2615
$K^0\pi^0$		$< 4.1 \times 10^{-5}$	CL=90%	2614
$\eta' K^0$		$(4.7 ^{+2.8}_{-2.2}) \times 10^{-5}$		2528
$\eta' K^*(892)^0$		$< 3.9 \times 10^{-5}$	CL=90%	2472
$\eta K^*(892)^0$		$< 3.0 \times 10^{-5}$	CL=90%	2534
ηK^0		$< 3.3 \times 10^{-5}$	CL=90%	2593
K^+K^-		$< 4.3 \times 10^{-6}$	CL=90%	2593
$K^0\overline{K}^0$		$< 1.7 \times 10^{-5}$	CL=90%	2592
$K^+\rho^-$		$< 3.5 \times 10^{-5}$	CL=90%	2559
$K^0\rho^0$		$< 3.9 \times 10^{-5}$	CL=90%	2559
$K^0 f_0(980)$		$< 3.6 \times 10^{-4}$	CL=90%	2523
$K^*(892)^+\pi^-$		$< 7.2 \times 10^{-5}$	CL=90%	2562
$K^*(892)^0\pi^0$		$< 2.8 \times 10^{-5}$	CL=90%	2562
$K_2^*(1430)^+\pi^-$		$< 2.6 \times 10^{-3}$	CL=90%	2445
$K^0K^+K^-$		$< 1.3 \times 10^{-3}$	CL=90%	2522
$K^0\phi$		$< 8.8 \times 10^{-5}$	CL=90%	2516
$K^-\pi^+\pi^+\pi^-$	[bbb]	$< 2.3 \times 10^{-4}$	CL=90%	2600
$K^*(892)^0\pi^+\pi^-$		$< 1.4 \times 10^{-3}$	CL=90%	2556
$K^*(892)^0\rho^0$		$< 4.6 \times 10^{-4}$	CL=90%	2504
$K^*(892)^0 f_0(980)$		$< 1.7 \times 10^{-4}$	CL=90%	2467
$K_1(1400)^+\pi^-$		$< 1.1 \times 10^{-3}$	CL=90%	2451
$K^- a_1(1260)^+$	[bbb]	$< 2.3 \times 10^{-4}$	CL=90%	2471
$K^*(892)^0K^+K^-$		$< 6.1 \times 10^{-4}$	CL=90%	2466
$K^*(892)^0\phi$		$< 4.3 \times 10^{-5}$	CL=90%	2459
$K_1(1400)^0\rho^0$		$< 3.0 \times 10^{-3}$	CL=90%	2389
$K_1(1400)^0\phi$		$< 5.0 \times 10^{-3}$	CL=90%	2339
$K_2^*(1430)^0\rho^0$		$< 1.1 \times 10^{-3}$	CL=90%	2380
$K_2^*(1430)^0\phi$		$< 1.4 \times 10^{-3}$	CL=90%	2330
$K^*(892)^0\gamma$		$(4.0 \pm 1.9) \times 10^{-5}$		2564
$K_1(1270)^0\gamma$		$< 7.0 \times 10^{-3}$	CL=90%	2486
$K_1(1400)^0\gamma$		$< 4.3 \times 10^{-3}$	CL=90%	2453
$K_2^*(1430)^0\gamma$		$< 4.0 \times 10^{-4}$	CL=90%	2445
$K^*(1680)^0\gamma$		$< 2.0 \times 10^{-3}$	CL=90%	2361
$K_3^*(1780)^0\gamma$		< 1.0 %	CL=90%	2343
$K_4^*(2045)^0\gamma$		$< 4.3 \times 10^{-3}$	CL=90%	2244
$\phi\phi$		$< 3.9 \times 10^{-5}$	CL=90%	2435
Light unflavored meson modes				
$\pi^+\pi^-$		$< 1.5 \times 10^{-5}$	CL=90%	2636
$\pi^0\pi^0$		$< 9.3 \times 10^{-6}$	CL=90%	2636
$\eta\pi^0$		$< 8 \times 10^{-6}$	CL=90%	2609
$\eta\eta$		$< 1.8 \times 10^{-5}$	CL=90%	2582
$\eta'\pi^0$		$< 1.1 \times 10^{-5}$	CL=90%	2551
$\eta'\eta'$		$< 4.7 \times 10^{-5}$	CL=90%	2460
$\eta'\eta$		$< 2.7 \times 10^{-5}$	CL=90%	2522
$\eta'\rho^0$		$< 2.3 \times 10^{-5}$	CL=90%	2493
$\eta\rho^0$		$< 1.3 \times 10^{-5}$	CL=90%	2554
$\pi^+\pi^-\pi^0$		$< 7.2 \times 10^{-4}$	CL=90%	2631
$\rho^0\pi^0$		$< 2.4 \times 10^{-5}$	CL=90%	2582
$\rho^\mp\pi^\pm$	[gg]	$< 8.8 \times 10^{-5}$	CL=90%	2582
$\pi^+\pi^-\pi^+\pi^-$		$< 2.3 \times 10^{-4}$	CL=90%	2621
$\rho^0\rho^0$		$< 2.8 \times 10^{-4}$	CL=90%	2525
$a_1(1260)^\mp\pi^\pm$	[gg]	$< 4.9 \times 10^{-4}$	CL=90%	2494
$a_2(1320)^\mp\pi^\pm$	[gg]	$< 3.0 \times 10^{-4}$	CL=90%	2473
$\pi^+\pi^-\pi^0\pi^0$		$< 3.1 \times 10^{-3}$	CL=90%	2622
$\rho^+\rho^-$		$< 2.2 \times 10^{-3}$	CL=90%	2525
$a_1(1260)^0\pi^0$		$< 1.1 \times 10^{-3}$	CL=90%	2494
$\omega\pi^0$		$< 4.6 \times 10^{-4}$	CL=90%	2580
$\pi^+\pi^+\pi^-\pi^-\pi^0$		$< 9.0 \times 10^{-3}$	CL=90%	2609
$a_1(1260)^+\rho^-$		$< 3.4 \times 10^{-3}$	CL=90%	2434
$a_1(1260)^0\rho^0$		$< 2.4 \times 10^{-3}$	CL=90%	2434
$\pi^+\pi^+\pi^+\pi^-\pi^-\pi^-$		$< 3.0 \times 10^{-3}$	CL=90%	2592
$a_1(1260)^+a_1(1260)^-$		$< 2.8 \times 10^{-3}$	CL=90%	2336
$\pi^+\pi^+\pi^+\pi^-\pi^-\pi^-\pi^0$		< 1.1 %	CL=90%	2572
Baryon modes				
$p\overline{p}$		$< 1.8 \times 10^{-5}$	CL=90%	2467
$p\overline{p}\pi^+\pi^-$		$< 2.5 \times 10^{-4}$	CL=90%	2406
$p\overline{\Lambda}\pi^-$		$< 1.8 \times 10^{-4}$	CL=90%	2401
$\Delta^0\overline{\Delta}^0$		$< 1.5 \times 10^{-3}$	CL=90%	2334
$\Delta^{++}\Delta^{--}$		$< 1.1 \times 10^{-4}$	CL=90%	2334
$\overline{\Sigma}_c^{--}\Delta^{++}$		$< 1.0 \times 10^{-3}$	CL=90%	1839
$\Lambda_c^- p\pi^+\pi^-$		$(1.3 \pm 0.6) \times 10^{-3}$		–
$\Lambda_c^- p$		$< 2.1 \times 10^{-4}$	CL=90%	2021
$\Lambda_c^- p\pi^0$		$< 5.9 \times 10^{-4}$	CL=90%	–
$\Lambda_c^- p\pi^+\pi^-\pi^0$		$< 5.07 \times 10^{-3}$	CL=90%	–
$\Lambda_c^- p\pi^+\pi^-\pi^+\pi^-$		$< 2.74 \times 10^{-3}$	CL=90%	–
Lepton Family number (LF) violating modes, or $\Delta B = 1$ weak neutral current ($B1$) modes				
$\gamma\gamma$	$B1$	$< 3.9 \times 10^{-5}$	CL=90%	2640
e^+e^-	$B1$	$< 5.9 \times 10^{-6}$	CL=90%	2640
$\mu^+\mu^-$	$B1$	$< 6.8 \times 10^{-7}$	CL=90%	2637
$K^0e^+e^-$	$B1$	$< 3.0 \times 10^{-4}$	CL=90%	2616
$K^0\mu^+\mu^-$	$B1$	$< 3.6 \times 10^{-4}$	CL=90%	2612
$K^*(892)^0e^+e^-$	$B1$	$< 2.9 \times 10^{-4}$	CL=90%	2564
$K^*(892)^0\mu^+\mu^-$	$B1$	$< 2.3 \times 10^{-5}$	CL=90%	2559
$K^*(892)^0\nu\overline{\nu}$	$B1$	$< 1.0 \times 10^{-3}$	CL=90%	2244
$e^\pm\mu^\mp$	LF [gg]	$< 5.9 \times 10^{-6}$	CL=90%	2639
$e^\pm\tau^\mp$	LF [gg]	$< 5.3 \times 10^{-4}$	CL=90%	2341
$\mu^\pm\tau^\mp$	LF [gg]	$< 8.3 \times 10^{-4}$	CL=90%	2339

$B^\pm/B^0$ ADMIXTURE

The branching fraction measurements are for an admixture of B mesons at the $\Upsilon(4S)$. The values quoted assume that $B(\Upsilon(4S) \to B\overline{B}) = 100\%$.

For inclusive branching fractions, *e.g.*, $B \to D^\pm$ anything, the treatment of multiple D's in the final state must be defined. One possibility would be to count the number of events with one-or-more D's and divide by the total number of B's. Another possibility would be to count the total number of D's and divide by the total number of B's, which is the definition of average multiplicity. The two definitions are identical when only one of the specified particles is allowed in the final state. Even though the "one-or-more" definition seems sensible, for practical reasons inclusive branching fractions are almost always measured using the multiplicity definition. For heavy final state particles, authors call their results inclusive branching fractions while for light particles some authors call their results multiplicities. In the B sections, we list all results as inclusive branching fractions, adopting a multiplicity definition. This means that inclusive branching fractions can exceed 100% and that inclusive partial widths can exceed total widths, just as inclusive cross sections can exceed total cross sections.

$\overline{B}$ modes are charge conjugates of the modes below. Reactions indicate the weak decay vertex and do not include mixing.

B DECAY MODES		Fraction (Γ_i/Γ)	Scale factor/ Confidence level	p (MeV/c)
Semileptonic and leptonic modes				
$B \to e^+\nu_e$ anything	[ccc]	(10.41 ± 0.29) %	S=1.2	–
$B \to \overline{p}e^+\nu_e$ anything		$< 1.6 \times 10^{-3}$	CL=90%	–
$B \to \mu^+\nu_\mu$ anything	[ccc]	(10.3 ± 0.5) %		–
$B \to \ell^+\nu_\ell$ anything	[pp,ccc]	(10.45 ± 0.21) %		–
$B \to D^-\ell^+\nu_\ell$ anything	[pp]	(2.7 ± 0.8) %		–
$B \to \overline{D}^0\ell^+\nu_\ell$ anything	[pp]	(7.0 ± 1.4) %		–
$B \to \overline{D}^{**}\ell^+\nu_\ell$	[pp,ddd]	(2.7 ± 0.7) %		–
$B \to \overline{D}_1(2420)\ell^+\nu_\ell$ anything		$(7.4 \pm 1.6) \times 10^{-3}$		–
$B \to D\pi\ell^+\nu_\ell$ anything + $D^*\pi\ell^+\nu_\ell$ anything		(2.3 ± 0.4) %		–
$B \to \overline{D}_2^*(2460)\ell^+\nu_\ell$ anything		$< 6.5 \times 10^{-3}$	CL=95%	–
$B \to D^{*-}\pi^+\ell^+\nu_\ell$ anything		(1.00 ± 0.34) %		–
$B \to D_s^-\ell^+\nu_\ell$ anything	[pp]	$< 9 \times 10^{-3}$	CL=90%	–
$B \to D_s^-\ell^+\nu_\ell K^+$ anything	[pp]	$< 6 \times 10^{-3}$	CL=90%	–
$B \to D_s^-\ell^+\nu_\ell K^0$ anything	[pp]	$< 9 \times 10^{-3}$	CL=90%	–
$B \to K^+\ell^+\nu_\ell$ anything	[pp]	(6.0 ± 0.5) %		–
$B \to K^-\ell^+\nu_\ell$ anything	[pp]	$(10 \pm 4) \times 10^{-3}$		–
$B \to K^0/\overline{K}^0\ell^+\nu_\ell$ anything	[pp]	(4.4 ± 0.5) %		–

Mode		Fraction		
D, D^*, or D_s modes				
$B \to D^{\pm}$ anything		(24.1 ± 1.9) %		–
$B \to D^0/\overline{D}^0$ anything		(63.1 ± 2.9) %	S=1.1	–
$B \to D^*(2010)^{\pm}$ anything		(22.7 ± 1.6) %		–
$B \to D^*(2007)^0$ anything		(26.0 ± 2.7) %		–
$B \to D_s^{\pm}$ anything	[gg]	(10.0 ± 2.5) %		–
$b \to c\overline{c}s$		(22 ± 4) %		–
$B \to D_s D$, $D_s^* D$, $D_s D^*$, or $D_s^* D^*$	[gg]	(4.9 ± 1.3) %		–
$B \to D^*(2010)\gamma$		$< 1.1 \times 10^{-3}$	CL=90%	–
$B \to D_s^+\pi^-$, $D_s^{*+}\pi^-$, $D_s^+\rho^-$, $D_s^{*+}\rho^-$, $D_s^+\pi^0$, $D_s^{*+}\pi^0$, $D_s^+\eta$, $D_s^{*+}\eta$, $D_s^+\rho^0$, $D_s^{*+}\rho^0$, $D_s^+\omega$, $D_s^{*+}\omega$	[gg]	$< 5 \times 10^{-4}$	CL=90%	–
$B \to D_{s1}(2536)^+$ anything		$< 9.5 \times 10^{-3}$	CL=90%	–
Charmonium modes				
$B \to J/\psi(1S)$ anything		(1.13 ± 0.06) %		–
$B \to J/\psi(1S)$ (direct) anything		$(8.0 \pm 0.8) \times 10^{-3}$		–
$B \to \psi(2S)$ anything		$(3.5 \pm 0.5) \times 10^{-3}$		–
$B \to \chi_{c1}(1P)$ anything		$(4.2 \pm 0.7) \times 10^{-3}$		–
$B \to \chi_{c1}(1P)$ (direct) anything		$(3.7 \pm 0.7) \times 10^{-3}$		–
$B \to \chi_{c2}(1P)$ anything		$< 3.8 \times 10^{-3}$	CL=90%	–
$B \to \eta_c(1S)$ anything		$< 9 \times 10^{-3}$	CL=90%	–
K or K^* modes				
$B \to K^{\pm}$ anything	[gg]	(78.9 ± 2.5) %		–
$B \to K^+$ anything		(66 ± 5) %		–
$B \to K^-$ anything		(13 ± 4) %		–
$B \to K^0/\overline{K}^0$ anything	[gg]	(64 ± 4) %		–
$B \to K^*(892)^{\pm}$ anything		(18 ± 6) %		–
$B \to K^*(892)^0/\overline{K}^*(892)^0$ anything	[gg]	(14.6 ± 2.6) %		–
$B \to K_1(1400)\gamma$		$< 4.1 \times 10^{-4}$	CL=90%	–
$B \to K_2^*(1430)\gamma$		$< 8.3 \times 10^{-4}$	CL=90%	–
$B \to K_2(1770)\gamma$		$< 1.2 \times 10^{-3}$	CL=90%	–
$B \to K_3^*(1780)\gamma$		$< 3.0 \times 10^{-3}$	CL=90%	–
$B \to K_4^*(2045)\gamma$		$< 1.0 \times 10^{-3}$	CL=90%	–
$B \to \overline{b} \to \overline{s}\gamma$		$(2.3 \pm 0.7) \times 10^{-4}$		–
$B \to \overline{b} \to \overline{s}$ gluon		< 6.8 %	CL=90%	–
Light unflavored meson modes				
$B \to \pi^{\pm}$ anything	[gg,eee]	(359 ± 7) %		–
$B \to \eta$ anything		(17.6 ± 1.6) %		–
$B \to \rho^0$ anything		(21 ± 5) %		–
$B \to \omega$ anything		< 81 %	CL=90%	–
$B \to \phi$ anything		(3.5 ± 0.7) %	S=1.8	–
Baryon modes				
$B \to \Lambda_c^{\pm}$ anything		(6.4 ± 1.1) %		–
$B \to \Lambda_c^- e^+$ anything		$< 3.2 \times 10^{-3}$	CL=90%	–
$B \to \Lambda_c^- p$ anything		(3.6 ± 0.7) %		–
$B \to \Lambda_c^- p e^+ \nu_e$		$< 1.5 \times 10^{-3}$	CL=90%	–
$B \to \overline{\Sigma}_c^{--}$ anything		$(4.2 \pm 2.4) \times 10^{-3}$		–
$B \to \overline{\Sigma}_c^{-}$ anything		$< 9.6 \times 10^{-3}$	CL=90%	–
$B \to \overline{\Sigma}_c^{0}$ anything		$(4.6 \pm 2.4) \times 10^{-3}$		–
$B \to \overline{\Sigma}_c^{0} N$ ($N = p$ or n)		$< 1.5 \times 10^{-3}$	CL=90%	–
$B \to \Xi_c^0$ anything $\times$ B($\Xi_c^0 \to \Xi^-\pi^+$)		$(1.4 \pm 0.5) \times 10^{-4}$		–
$B \to \Xi_c^+$ anything $\times$ B($\Xi_c^+ \to \Xi^-\pi^+\pi^+$)		$(4.5\ ^{+1.3}_{-1.2}) \times 10^{-4}$		–
$B \to p/\overline{p}$ anything	[gg]	(8.0 ± 0.4) %		–
$B \to p/\overline{p}$(direct) anything	[gg]	(5.5 ± 0.5) %		–
$B \to \Lambda/\overline{\Lambda}$ anything	[gg]	(4.0 ± 0.5) %		–
$B \to \Xi^-/\overline{\Xi}^+$ anything	[gg]	$(2.7 \pm 0.6) \times 10^{-3}$		–
$B \to$ baryons anything		(6.8 ± 0.6) %		–
$B \to p\overline{p}$ anything		(2.47 ± 0.23) %		–
$B \to \Lambda\overline{p}/\overline{\Lambda}p$ anything	[gg]	(2.5 ± 0.4) %		–
$B \to \Lambda\overline{\Lambda}$ anything		$< 5 \times 10^{-3}$	CL=90%	–

Lepton Family number (LF) violating modes or $\Delta B = 1$ weak neutral current ($B1$) modes

$B \to e^+e^-s$	$B1$	$< 5.7 \times 10^{-5}$	CL=90%	–
$B \to \mu^+\mu^-s$	$B1$	$< 5.8 \times 10^{-5}$	CL=90%	–
$B \to e^{\pm}\mu^{\mp}s$	LF	$< 2.2 \times 10^{-5}$	CL=90%	–

$B^{\pm}/B^0/B_s^0/b$-baryon ADMIXTURE

These measurements are for an admixture of bottom particles at high energy (LEP, Tevatron, $Sp\overline{p}S$).

Mean life $\tau = (1.564 \pm 0.014) \times 10^{-12}$ s

Mean life $\tau = (1.72 \pm 0.10) \times 10^{-12}$ s Charged b-hadron admixture

Mean life $\tau = (1.58 \pm 0.14) \times 10^{-12}$ s Neutral b-hadron admixture

$\tau_{\text{charged } b\text{-hadron}}/\tau_{\text{neutral } b\text{-hadron}} = 1.09 \pm 0.13$

The branching fraction measurements are for an admixture of B mesons and baryons at energies above the $\Upsilon(4S)$. Only the highest energy results (LEP, Tevatron, $Sp\overline{p}S$) are used in the branching fraction averages. The production fractions give our best current estimate of the admixture at LEP.

For inclusive branching fractions, *e.g.*, $B \to D^{\pm}$ anything, the treatment of multiple D's in the final state must be defined. One possibility would be to count the number of events with one-or-more D's and divide by the total number of B's. Another possibility would be to count the total number of D's and divide by the total number of B's, which is the definition of average multiplicity. The two definitions are identical when only one of the specified particles is allowed in the final state. Even though the "one-or-more" definition seems sensible, for practical reasons inclusive branching fractions are almost always measured using the multiplicity definition. For heavy final state particles, authors call their results inclusive branching fractions while for light particles some authors call their results multiplicities. In the B sections, we list all results as inclusive branching fractions, adopting a multiplicity definition. This means that inclusive branching fractions can exceed 100% and that inclusive partial widths can exceed total widths, just as inclusive cross sections can exceed total cross sections.

The modes below are listed for a $\overline{b}$ initial state. b modes are their charge conjugates. Reactions indicate the weak decay vertex and do not include mixing.

PRODUCTION FRACTIONS

The production fractions for weakly decaying b-hadrons at the Z have been calculated from the best values of mean lives, mixing parameters, and branching fractions in this edition by the LEP B Oscillation Working Group as described in the note "Production and Decay of b-Flavored Hadrons" in the $B^{\pm}$ Particle Listings. Values assume

$$B(\overline{b} \to B^+) = B(\overline{b} \to B^0)$$
$$B(\overline{b} \to B^+) + B(\overline{b} \to B^0) + B(\overline{b} \to B_s^0) + B(b \to \Lambda_b) = 100\ \%.$$

The notation for production fractions varies in the literature (f_{B^0}, $f(b \to \overline{B}^0)$, Br($b \to \overline{B}^0$)). We use our own branching fraction notation here, B($\overline{b} \to B^0$).

$\overline{b}$ DECAY MODES		Fraction (Γ_i/Γ)	Confidence level	p (MeV/c)
B^+		$(39.7\ ^{+1.8}_{-2.2})$ %		–
B^0		$(39.7\ ^{+1.8}_{-2.2})$ %		–
B_s^0		$(10.5\ ^{+1.8}_{-1.7})$ %		–
Λ_b		$(10.1\ ^{+3.9}_{-3.1})$ %		–

DECAY MODES

Semileptonic and leptonic modes

ν anything		(23.1 ± 1.5) %		–
$\ell^+\nu_\ell$ anything	[pp,ccc]	(10.99 ± 0.23) %		–
$e^+\nu_e$ anything	[ccc]	(10.9 ± 0.5) %		–
$\mu^+\nu_\mu$ anything	[ccc]	(10.8 ± 0.5) %		–
$D^-\ell^+\nu_\ell$ anything	[pp]	(2.02 ± 0.29) %		–
$\overline{D}^0\ell^+\nu_\ell$ anything	[pp]	(6.3 ± 0.6) %		–
$D^{*-}\ell^+\nu_\ell$ anything	[pp]	(2.76 ± 0.29) %		–
$\overline{D}_j^0\ell^+\nu_\ell$ anything	[pp,fff]	seen		–
$D_j^-\ell^+\nu_\ell$ anything	[pp,fff]	seen		–
$\overline{D}_2^*(2460)^0\ell^+\nu_\ell$ anything		seen		–
$D_2^*(2460)^-\ell^+\nu_\ell$ anything		seen		–
$\tau^+\nu_\tau$ anything		(2.6 ± 0.4) %		–
$\overline{c} \to \ell^-\overline{\nu}_\ell$ anything	[pp]	(7.8 ± 0.6) %		–

Charmed meson and baryon modes				
$\overline{D}^0$ anything		(60.1 ± 3.2) %		–
D^- anything		(23.7 ± 2.3) %		–
$\overline{D}_s$ anything		(18 ± 5) %		–
Λ_c anything		(9.7 ± 2.9) %		–
$\overline{c}/c$ anything	[eee]	(117 ± 4) %		–
Charmonium modes				
$J/\psi(1S)$ anything		(1.16 ± 0.10) %		–
$\psi(2S)$ anything		(4.8 ± 2.4) $\times 10^{-3}$		–
$\chi_{c1}(1P)$ anything		(1.8 ± 0.5) %		–
K or K^* modes				
$\overline{s}\gamma$		< 5.4 $\times 10^{-4}$	90%	–
$K^\pm$ anything		(88 ± 19) %		–
K^0_S anything		(29.0 ± 2.9) %		–
Pion modes				
π^0 anything	[eee]	(278 ± 60) %		–
Baryon modes				
$p/\overline{p}$ anything		(14 ± 6) %		–
Other modes				
charged anything	[eee]	(497 ± 7) %		–
hadron$^+$ hadron$^-$		($1.7^{+1.0}_{-0.7}$) $\times 10^{-5}$		–
charmless		(7 ± 21) $\times 10^{-3}$		–
Baryon modes				
$\Lambda/\overline{\Lambda}$ anything		(5.9 ± 0.6) %		–
$\Delta B = 1$ weak neutral current ($B1$) modes				
$\mu^+\mu^-$ anything	$B1$	< 3.2 $\times 10^{-4}$	90%	–

B^*

$I(J^P) = \frac{1}{2}(1^-)$

I, *J*, *P* need confirmation. Quantum numbers shown are quark-model predictions.

Mass $m_{B^*} = 5324.9 \pm 1.8$ MeV

$m_{B^*} - m_B = 45.78 \pm 0.35$ MeV

B^* DECAY MODES	Fraction (Γ_i/Γ)	p (MeV/c)
$B\gamma$	dominant	46

BOTTOM, STRANGE MESONS ($B = \pm1, S = \mp1$)

$B^0_s = s\overline{b}$, $\overline{B}^0_s = \overline{s}b$, similarly for B^*_s's

B^0_s

$I(J^P) = 0(0^-)$

I, *J*, *P* need confirmation. Quantum numbers shown are quark-model predictions.

Mass $m_{B^0_s} = 5369.3 \pm 2.0$ MeV

Mean life $\tau = (1.54 \pm 0.07) \times 10^{-12}$ s

$c\tau = 462\ \mu$m

B^0_s-$\overline{B}^0_s$ mixing parameters

χ_B at high energy $= f_d\chi_d + f_s\chi_s = 0.118 \pm 0.006$

$\Delta m_{B^0_s} = m_{B^0_{sH}} - m_{B^0_{sL}} > 9.1 \times 10^{12}\ \hbar\ s^{-1}$, CL = 95%

$x_s = \Delta m_{B^0_s}/\Gamma_{B^0_s} > 14.0$, CL = 95%

$\chi_s > 0.4975$, CL = 95%

These branching fractions all scale with B($\overline{b} \to B^0_s$), the LEP B^0_s production fraction. The first four were evaluated using B($\overline{b} \to B^0_s$) = $(10.5^{+1.8}_{-1.7})$% and the rest assume B($\overline{b} \to B^0_s$) = 12%.

The branching fraction B($B^0_s \to D^-_s \ell^+ \nu_\ell$ anything) is not a pure measurement since the measured product branching fraction B($\overline{b} \to B^0_s$) × B($B^0_s \to D^-_s \ell^+ \nu_\ell$ anything) was used to determine B($\overline{b} \to B^0_s$), as described in the note on "Production and Decay of b-Flavored Hadrons."

B^0_s DECAY MODES		Fraction (Γ_i/Γ)	Confidence level	p (MeV/c)
D^-_s anything		(92 ± 33) %		–
$D^-_s \ell^+ \nu_\ell$ anything	[ggg]	(8.1 ± 2.5) %		–
$D^-_s \pi^+$		< 13 %		2321
$J/\psi(1S)\phi$		(9.3 ± 3.3) $\times 10^{-4}$		1590
$J/\psi(1S)\pi^0$		< 1.2 $\times 10^{-3}$	90%	1788
$J/\psi(1S)\eta$		< 3.8 $\times 10^{-3}$	90%	1735
$\psi(2S)\phi$		seen		1122
$\pi^+\pi^-$		< 1.7 $\times 10^{-4}$	90%	1122
$\pi^0\pi^0$		< 2.1 $\times 10^{-4}$	90%	2861
$\eta\pi^0$		< 1.0 $\times 10^{-3}$	90%	2655
$\eta\eta$		< 1.5 $\times 10^{-3}$	90%	2628
π^+K^-		< 2.1 $\times 10^{-4}$	90%	2660
K^+K^-		< 5.9 $\times 10^{-5}$	90%	2639
$p\overline{p}$		< 5.9 $\times 10^{-5}$	90%	2515
$\gamma\gamma$		< 1.48 $\times 10^{-4}$	90%	2685
$\phi\gamma$		< 7 $\times 10^{-4}$	90%	2588
Lepton Family number (LF) violating modes or $\Delta B = 1$ weak neutral current ($B1$) modes				
$\mu^+\mu^-$	$B1$	< 2.0 $\times 10^{-6}$	90%	2682
e^+e^-	$B1$	< 5.4 $\times 10^{-5}$	90%	2864
$e^\pm\mu^\mp$	LF [gg]	< 4.1 $\times 10^{-5}$	90%	2864
$\phi\nu\overline{\nu}$	$B1$	< 5.4 $\times 10^{-3}$	90%	–

$c\bar{c}$ MESONS

$\eta_c(1S)$

$I^G(J^{PC}) = 0^+(0^{-+})$

Mass $m = 2979.8 \pm 2.1$ MeV (S = 2.1)
Full width $\Gamma = 13.2^{+3.8}_{-3.2}$ MeV

$\eta_c(1S)$ DECAY MODES	Fraction (Γ_i/Γ)	Confidence level	p (MeV/c)
Decays involving hadronic resonances			
$\eta'(958)\pi\pi$	(4.1 ± 1.7) %		1319
$\rho\rho$	(2.6 ± 0.9) %		1275
$K^*(892)^0 K^- \pi^+$ + c.c.	(2.0 ± 0.7) %		1273
$K^*(892)\overline{K}^*(892)$	$(8.5 \pm 3.1) \times 10^{-3}$		1193
$\phi\phi$	$(7.1 \pm 2.8) \times 10^{-3}$		1086
$a_0(980)\pi$	< 2 %	90%	1323
$a_2(1320)\pi$	< 2 %	90%	1193
$K^*(892)\overline{K}$ + c.c.	< 1.28 %	90%	1307
$f_2(1270)\eta$	< 1.1 %	90%	1142
$\omega\omega$	$< 3.1 \times 10^{-3}$	90%	1268
Decays into stable hadrons			
$K\overline{K}\pi$	(5.5 ± 1.7) %		1378
$\eta\pi\pi$	(4.9 ± 1.8) %		1425
$\pi^+\pi^- K^+ K^-$	$(2.0\ ^{+0.7}_{-0.6})$ %		1342
$2(K^+K^-)$	(2.1 ± 1.2) %		1053
$2(\pi^+\pi^-)$	(1.2 ± 0.4) %		1457
$p\overline{p}$	$(1.2 \pm 0.4) \times 10^{-3}$		1157
$K\overline{K}\eta$	< 3.1 %	90%	1262
$\pi^+\pi^- p\overline{p}$	< 1.2 %	90%	1023
$\Lambda\overline{\Lambda}$	$< 2 \times 10^{-3}$	90%	987
Radiative decays			
$\gamma\gamma$	$(3.0 \pm 1.2) \times 10^{-4}$		1489

$J/\psi(1S)$

$I^G(J^{PC}) = 0^-(1^{--})$

Mass $m = 3096.88 \pm 0.04$ MeV
Full width $\Gamma = 87 \pm 5$ keV
$\Gamma_{ee} = 5.26 \pm 0.37$ keV (Assuming $\Gamma_{ee} = \Gamma_{\mu\mu}$)

$J/\psi(1S)$ DECAY MODES		Fraction (Γ_i/Γ)	Scale factor/ Confidence level	p (MeV/c)
hadrons		(87.7 ± 0.5) %		–
virtual $\gamma \to$ hadrons		(17.0 ± 2.0) %		–
e^+e^-		(6.02 ± 0.19) %		1548
$\mu^+\mu^-$		(6.01 ± 0.19) %		1545
Decays involving hadronic resonances				
$\rho\pi$		(1.27 ± 0.09) %		1449
$\rho^0\pi^0$		$(4.2 \pm 0.5) \times 10^{-3}$		1449
$a_2(1320)\rho$		(1.09 ± 0.22) %		1125
$\omega\pi^+\pi^+\pi^-\pi^-$		$(8.5 \pm 3.4) \times 10^{-3}$		1392
$\omega\pi^+\pi^-$		$(7.2 \pm 1.0) \times 10^{-3}$		1435
$\omega f_2(1270)$		$(4.3 \pm 0.6) \times 10^{-3}$		1143
$K^*(892)^0\overline{K}_2^*(1430)^0$ + c.c.		$(6.7 \pm 2.6) \times 10^{-3}$		1005
$\omega K^*(892)\overline{K}$ + c.c.		$(5.3 \pm 2.0) \times 10^{-3}$		1098
$K^+\overline{K}^*(892)^-$ + c.c.		$(5.0 \pm 0.4) \times 10^{-3}$		1373
$K^0\overline{K}^*(892)^0$ + c.c.		$(4.2 \pm 0.4) \times 10^{-3}$		1371
$\omega\pi^0\pi^0$		$(3.4 \pm 0.8) \times 10^{-3}$		1436
$b_1(1235)^\pm\pi^\mp$	[gg]	$(3.0 \pm 0.5) \times 10^{-3}$		1299
$\omega K^\pm K_S^0\pi^\mp$	[gg]	$(3.0 \pm 0.7) \times 10^{-3}$		1210
$b_1(1235)^0\pi^0$		$(2.3 \pm 0.6) \times 10^{-3}$		1299
$\phi K^*(892)\overline{K}$ + c.c.		$(2.04 \pm 0.28) \times 10^{-3}$		969
$\omega K\overline{K}$		$(1.9 \pm 0.4) \times 10^{-3}$		1268
$\omega f_J(1710) \to \omega K\overline{K}$		$(4.8 \pm 1.1) \times 10^{-4}$		878
$\phi 2(\pi^+\pi^-)$		$(1.60 \pm 0.32) \times 10^{-3}$		1318
$\Delta(1232)^{++}\overline{p}\pi^-$		$(1.6 \pm 0.5) \times 10^{-3}$		1030
$\omega\eta$		$(1.58 \pm 0.16) \times 10^{-3}$		1391
$\phi K\overline{K}$		$(1.48 \pm 0.22) \times 10^{-3}$		1179
$\phi f_J(1710) \to \phi K\overline{K}$		$(3.6 \pm 0.6) \times 10^{-4}$		875
$p\overline{p}\omega$		$(1.30 \pm 0.25) \times 10^{-3}$	S=1.3	769
$\Delta(1232)^{++}\overline{\Delta}(1232)^{--}$		$(1.10 \pm 0.29) \times 10^{-3}$		938
$\Sigma(1385)^-\overline{\Sigma}(1385)^+$ (or c.c.)	[gg]	$(1.03 \pm 0.13) \times 10^{-3}$		692
$p\overline{p}\eta'(958)$		$(9 \pm 4) \times 10^{-4}$	S=1.7	596
$\phi f_2'(1525)$		$(8 \pm 4) \times 10^{-4}$	S=2.7	871
$\phi\pi^+\pi^-$		$(8.0 \pm 1.2) \times 10^{-4}$		1365
$\phi K^\pm K_S^0\pi^\mp$	[gg]	$(7.2 \pm 0.9) \times 10^{-4}$		1114
$\omega f_1(1420)$		$(6.8 \pm 2.4) \times 10^{-4}$		1062
$\phi\eta$		$(6.5 \pm 0.7) \times 10^{-4}$		1320
$\Xi(1530)^-\overline{\Xi}^+$		$(5.9 \pm 1.5) \times 10^{-4}$		597
$pK^-\overline{\Sigma}(1385)^0$		$(5.1 \pm 3.2) \times 10^{-4}$		645
$\omega\pi^0$		$(4.2 \pm 0.6) \times 10^{-4}$	S=1.4	1447
$\phi\eta'(958)$		$(3.3 \pm 0.4) \times 10^{-4}$		1192
$\phi f_0(980)$		$(3.2 \pm 0.9) \times 10^{-4}$	S=1.9	1182
$\Xi(1530)^0\overline{\Xi}^0$		$(3.2 \pm 1.4) \times 10^{-4}$		608
$\Sigma(1385)^-\overline{\Sigma}^+$ (or c.c.)	[gg]	$(3.1 \pm 0.5) \times 10^{-4}$		857
$\phi f_1(1285)$		$(2.6 \pm 0.5) \times 10^{-4}$	S=1.1	1032
$\rho\eta$		$(1.93 \pm 0.23) \times 10^{-4}$		1398
$\omega\eta'(958)$		$(1.67 \pm 0.25) \times 10^{-4}$		1279
$\omega f_0(980)$		$(1.4 \pm 0.5) \times 10^{-4}$		1271
$\rho\eta'(958)$		$(1.05 \pm 0.18) \times 10^{-4}$		1283
$p\overline{p}\phi$		$(4.5 \pm 1.5) \times 10^{-5}$		527
$a_2(1320)^\pm\pi^\mp$	[gg]	$< 4.3 \times 10^{-3}$	CL=90%	1263
$K\overline{K}_2^*(1430)$ + c.c.		$< 4.0 \times 10^{-3}$	CL=90%	1159
$K_2^*(1430)^0\overline{K}_2^*(1430)^0$		$< 2.9 \times 10^{-3}$	CL=90%	588
$K^*(892)^0\overline{K}^*(892)^0$		$< 5 \times 10^{-4}$	CL=90%	1263
$\phi f_2(1270)$		$< 3.7 \times 10^{-4}$	CL=90%	1036
$p\overline{p}\rho$		$< 3.1 \times 10^{-4}$	CL=90%	779
$\phi\eta(1440) \to \phi\eta\pi\pi$		$< 2.5 \times 10^{-4}$	CL=90%	946
$\omega f_2'(1525)$		$< 2.2 \times 10^{-4}$	CL=90%	1003
$\Sigma(1385)^0\overline{\Lambda}$		$< 2 \times 10^{-4}$	CL=90%	911
$\Delta(1232)^+\overline{p}$		$< 1 \times 10^{-4}$	CL=90%	1100
$\Sigma^0\overline{\Lambda}$		$< 9 \times 10^{-5}$	CL=90%	1032
$\phi\pi^0$		$< 6.8 \times 10^{-6}$	CL=90%	1377
Decays into stable hadrons				
$2(\pi^+\pi^-)\pi^0$		(3.37 ± 0.26) %		1496
$3(\pi^+\pi^-)\pi^0$		(2.9 ± 0.6) %		1433
$\pi^+\pi^-\pi^0$		(1.50 ± 0.20) %		1533
$\pi^+\pi^-\pi^0 K^+K^-$		(1.20 ± 0.30) %		1368
$4(\pi^+\pi^-)\pi^0$		$(9.0 \pm 3.0) \times 10^{-3}$		1345
$\pi^+\pi^- K^+K^-$		$(7.2 \pm 2.3) \times 10^{-3}$		1407
$K\overline{K}\pi$		$(6.1 \pm 1.0) \times 10^{-3}$		1440
$p\overline{p}\pi^+\pi^-$		$(6.0 \pm 0.5) \times 10^{-3}$	S=1.3	1107
$2(\pi^+\pi^-)$		$(4.0 \pm 1.0) \times 10^{-3}$		1517
$3(\pi^+\pi^-)$		$(4.0 \pm 2.0) \times 10^{-3}$		1466
$n\overline{n}\pi^+\pi^-$		$(4 \pm 4) \times 10^{-3}$		1106
$\Sigma^0\overline{\Sigma}^0$		$(1.27 \pm 0.17) \times 10^{-3}$		992
$2(\pi^+\pi^-)K^+K^-$		$(3.1 \pm 1.3) \times 10^{-3}$		1320
$p\overline{p}\pi^+\pi^-\pi^0$	[hhh]	$(2.3 \pm 0.9) \times 10^{-3}$	S=1.9	1033
$p\overline{p}$		$(2.14 \pm 0.10) \times 10^{-3}$		1232
$p\overline{p}\eta$		$(2.09 \pm 0.18) \times 10^{-3}$		948
$p\overline{n}\pi^-$		$(2.00 \pm 0.10) \times 10^{-3}$		1174
$n\overline{n}$		$(1.9 \pm 0.5) \times 10^{-3}$		1231
$\Xi\overline{\Xi}$		$(1.8 \pm 0.4) \times 10^{-3}$	S=1.8	818
$\Lambda\overline{\Lambda}$		$(1.35 \pm 0.14) \times 10^{-3}$	S=1.2	1074
$p\overline{p}\pi^0$		$(1.09 \pm 0.09) \times 10^{-3}$		1176
$\Lambda\overline{\Sigma}^-\pi^+$ (or c.c.)	[gg]	$(1.06 \pm 0.12) \times 10^{-3}$		945
$pK^-\overline{\Lambda}$		$(8.9 \pm 1.6) \times 10^{-4}$		876
$2(K^+K^-)$		$(7.0 \pm 3.0) \times 10^{-4}$		1131
$pK^-\overline{\Sigma}^0$		$(2.9 \pm 0.8) \times 10^{-4}$		820
K^+K^-		$(2.37 \pm 0.31) \times 10^{-4}$		1468
$\Lambda\overline{\Lambda}\pi^0$		$(2.2 \pm 0.7) \times 10^{-4}$		998
$\pi^+\pi^-$		$(1.47 \pm 0.23) \times 10^{-4}$		1542
$K_S^0 K_L^0$		$(1.08 \pm 0.14) \times 10^{-4}$		1466
$\Lambda\overline{\Sigma}$ + c.c.		$< 1.5 \times 10^{-4}$	CL=90%	1032
$K_S^0 K_S^0$		$< 5.2 \times 10^{-6}$	CL=90%	1466
Radiative decays				
$\gamma\eta_c(1S)$		(1.3 ± 0.4) %		116
$\gamma\pi^+\pi^- 2\pi^0$		$(8.3 \pm 3.1) \times 10^{-3}$		1518
$\gamma\eta\pi\pi$		$(6.1 \pm 1.0) \times 10^{-3}$		1487
$\gamma\eta(1440) \to \gamma K\overline{K}\pi$	[p]	$(9.1 \pm 1.8) \times 10^{-4}$		1223
$\gamma\eta(1440) \to \gamma\gamma\rho^0$		$(6.4 \pm 1.4) \times 10^{-5}$		1223
$\gamma\eta(1440) \to \gamma\eta\pi^+\pi^-$		$(3.4\ +0.7) \times 10^{-4}$		–
$\gamma\rho\rho$		$(4.5 \pm 0.8) \times 10^{-3}$		1343
$\gamma\eta'(958)$		$(4.31 \pm 0.30) \times 10^{-3}$		1400
$\gamma 2\pi^+ 2\pi^-$		$(2.8 \pm 0.5) \times 10^{-3}$	S=1.9	1517
$\gamma f_4(2050)$		$(2.7 \pm 0.7) \times 10^{-3}$		874
$\gamma\omega\omega$		$(1.59 \pm 0.33) \times 10^{-3}$		1337
$\gamma\eta(1440) \to \gamma\rho^0\rho^0$		$(1.7 \pm 0.4) \times 10^{-3}$	S=1.3	1223
$\gamma f_2(1270)$		$(1.38 \pm 0.14) \times 10^{-3}$		1286
$\gamma f_J(1710) \to \gamma K\overline{K}$		$(8.5\ ^{+1.2}_{-0.9}) \times 10^{-4}$	S=1.2	1075

$\gamma\eta$	$(8.6 \pm 0.8) \times 10^{-4}$		1500
$\gamma f_1(1420) \to \gamma K\overline{K}\pi$	$(8.3 \pm 1.5) \times 10^{-4}$		1220
$\gamma f_1(1285)$	$(6.5 \pm 1.0) \times 10^{-4}$		1283
$\gamma f_2'(1525)$	$(4.7 \ {}^{+0.7}_{-0.5}) \times 10^{-4}$		1173
$\gamma\phi\phi$	$(4.0 \pm 1.2) \times 10^{-4}$	S=2.1	1166
$\gamma p\overline{p}$	$(3.8 \pm 1.0) \times 10^{-4}$		1232
$\gamma\eta(2225)$	$(2.9 \pm 0.6) \times 10^{-4}$		834
$\gamma\eta(1760) \to \gamma\rho^0\rho^0$	$(1.3 \pm 0.9) \times 10^{-4}$		1048
$\gamma\pi^0$	$(3.9 \pm 1.3) \times 10^{-5}$		1546
$\gamma p\overline{p}\pi^+\pi^-$	$< 7.9 \times 10^{-4}$	CL=90%	1107
$\gamma\gamma$	$< 5 \times 10^{-4}$	CL=90%	1548
$\gamma\Lambda\overline{\Lambda}$	$< 1.3 \times 10^{-4}$	CL=90%	1074
3γ	$< 5.5 \times 10^{-5}$	CL=90%	1548
$\gamma f_J(2220)$	$> 2.50 \times 10^{-3}$	CL=99.9%	–
$\gamma f_0(1500)$	$(5.7 \pm 0.8) \times 10^{-4}$		1184
γe^+e^-	$(8.8 \pm 1.4) \times 10^{-3}$		–

$\chi_{c0}(1P)$

$I^G(J^{PC}) = 0^+(0^{++})$

Mass $m = 3417.3 \pm 2.8$ MeV
Full width $\Gamma = 14 \pm 5$ MeV

$\chi_{c0}(1P)$ DECAY MODES	Fraction (Γ_i/Γ)	Confidence level	p (MeV/c)
Hadronic decays			
$2(\pi^+\pi^-)$	(3.7 ± 0.7) %		1679
$\pi^+\pi^- K^+K^-$	(3.0 ± 0.7) %		1580
$\rho^0\pi^+\pi^-$	(1.6 ± 0.5) %		1608
$3(\pi^+\pi^-)$	(1.5 ± 0.5) %		1633
$K^+\overline{K}^*(892)^0\pi^-$ + c.c.	(1.2 ± 0.4) %		1522
$\pi^+\pi^-$	$(7.5 \pm 2.1) \times 10^{-3}$		1702
K^+K^-	$(7.1 \pm 2.4) \times 10^{-3}$		1635
$\pi^+\pi^- p\overline{p}$	$(5.0 \pm 2.0) \times 10^{-3}$		1320
$p\overline{p}$	$< 9.0 \times 10^{-4}$	90%	1427
Radiative decays			
$\gamma J/\psi(1S)$	$(6.6 \pm 1.8) \times 10^{-3}$		303
$\gamma\gamma$	$< 5 \times 10^{-4}$	95%	1708

$\chi_{c1}(1P)$

$I^G(J^{PC}) = 0^+(1^{++})$

Mass $m = 3510.53 \pm 0.12$ MeV
Full width $\Gamma = 0.88 \pm 0.14$ MeV

$\chi_{c1}(1P)$ DECAY MODES	Fraction (Γ_i/Γ)	p (MeV/c)
Hadronic decays		
$3(\pi^+\pi^-)$	(2.2 ± 0.8) %	1683
$2(\pi^+\pi^-)$	(1.6 ± 0.5) %	1727
$\pi^+\pi^- K^+K^-$	$(9 \pm 4) \times 10^{-3}$	1632
$\rho^0\pi^+\pi^-$	$(3.9 \pm 3.5) \times 10^{-3}$	1659
$K^+\overline{K}^*(892)^0\pi^-$ + c.c.	$(3.2 \pm 2.1) \times 10^{-3}$	1576
$\pi^+\pi^- p\overline{p}$	$(1.4 \pm 0.9) \times 10^{-3}$	1381
$p\overline{p}$	$(8.6 \pm 1.2) \times 10^{-5}$	1483
$\pi^+\pi^-$ + K^+K^-	$< 2.1 \times 10^{-3}$	–
Radiative decays		
$\gamma J/\psi(1S)$	(27.3 ± 1.6) %	389

$\chi_{c2}(1P)$

$I^G(J^{PC}) = 0^+(2^{++})$

Mass $m = 3556.17 \pm 0.13$ MeV
Full width $\Gamma = 2.00 \pm 0.18$ MeV

$\chi_{c2}(1P)$ DECAY MODES	Fraction (Γ_i/Γ)	Confidence level	p (MeV/c)
Hadronic decays			
$2(\pi^+\pi^-)$	(2.2 ± 0.5) %		1751
$\pi^+\pi^- K^+K^-$	(1.9 ± 0.5) %		1656
$3(\pi^+\pi^-)$	(1.2 ± 0.8) %		1707
$\rho^0\pi^+\pi^-$	$(7 \pm 4) \times 10^{-3}$		1683
$K^+\overline{K}^*(892)^0\pi^-$ + c.c.	$(4.8 \pm 2.8) \times 10^{-3}$		1601
$\pi^+\pi^- p\overline{p}$	$(3.3 \pm 1.3) \times 10^{-3}$		1410
$\pi^+\pi^-$	$(1.9 \pm 1.0) \times 10^{-3}$		1773
K^+K^-	$(1.5 \pm 1.1) \times 10^{-3}$		1708
$p\overline{p}$	$(10.0 \pm 1.0) \times 10^{-5}$		1510
$J/\psi(1S)\pi^+\pi^-\pi^0$	< 1.5 %	90%	185
Radiative decays			
$\gamma J/\psi(1S)$	(13.5 ± 1.1) %		430
$\gamma\gamma$	$(1.6 \pm 0.5) \times 10^{-4}$		1778

$\psi(2S)$

$I^G(J^{PC}) = 0^-(1^{--})$

Mass $m = 3686.00 \pm 0.09$ MeV
Full width $\Gamma = 277 \pm 31$ keV (S = 1.1)
$\Gamma_{ee} = 2.14 \pm 0.21$ keV (Assuming $\Gamma_{ee} = \Gamma_{\mu\mu}$)

$\psi(2S)$ DECAY MODES	Fraction (Γ_i/Γ)	Scale factor/ Confidence level	p (MeV/c)
hadrons	(98.10 ± 0.30) %		–
virtual$\gamma \to$ hadrons	(2.9 ± 0.4) %		–
e^+e^-	$(8.5 \pm 0.7) \times 10^{-3}$		1843
$\mu^+\mu^-$	$(7.7 \pm 1.7) \times 10^{-3}$		1840
Decays into $J/\psi(1S)$ and anything			
$J/\psi(1S)$anything	(54.2 ± 3.0) %		–
$J/\psi(1S)$neutrals	(22.8 ± 1.7) %		–
$J/\psi(1S)\pi^+\pi^-$	(30.2 ± 1.9) %		477
$J/\psi(1S)\pi^0\pi^0$	(17.9 ± 1.8) %		481
$J/\psi(1S)\eta$	(2.7 ± 0.4) %	S=1.7	200
$J/\psi(1S)\pi^0$	$(9.7 \pm 2.1) \times 10^{-4}$		527
$J/\psi(1S)\mu^+\mu^-$	$(10.0 \pm 3.3) \times 10^{-3}$		–
Hadronic decays			
$3(\pi^+\pi^-)\pi^0$	$(3.5 \pm 1.6) \times 10^{-3}$		1746
$2(\pi^+\pi^-)\pi^0$	$(3.0 \pm 0.8) \times 10^{-3}$		1799
$\pi^+\pi^- K^+K^-$	$(1.6 \pm 0.4) \times 10^{-3}$		1726
$\pi^+\pi^- p\overline{p}$	$(8.0 \pm 2.0) \times 10^{-4}$		1491
$K^+\overline{K}^*(892)^0\pi^-$ + c.c.	$(6.7 \pm 2.5) \times 10^{-4}$		1673
$2(\pi^+\pi^-)$	$(4.5 \pm 1.0) \times 10^{-4}$		1817
$\rho^0\pi^+\pi^-$	$(4.2 \pm 1.5) \times 10^{-4}$		1751
$\overline{p}p$	$(1.9 \pm 0.5) \times 10^{-4}$		1586
$3(\pi^+\pi^-)$	$(1.5 \pm 1.0) \times 10^{-4}$		1774
$\overline{p}p\pi^0$	$(1.4 \pm 0.5) \times 10^{-4}$		1543
K^+K^-	$(1.0 \pm 0.7) \times 10^{-4}$		1776
$\pi^+\pi^-\pi^0$	$(9 \pm 5) \times 10^{-5}$		1830
$\rho\pi$	$< 8.3 \times 10^{-5}$	CL=90%	1760
$\pi^+\pi^-$	$(8 \pm 5) \times 10^{-5}$		1838
$\Lambda\overline{\Lambda}$	$< 4 \times 10^{-4}$	CL=90%	1467
$\Xi^-\overline{\Xi}^+$	$< 2 \times 10^{-4}$	CL=90%	1285
$K^+K^-\pi^0$	$< 2.96 \times 10^{-5}$	CL=90%	1754
$K^+\overline{K}^*(892)^-$ + c.c.	$< 5.4 \times 10^{-5}$	CL=90%	1698
Radiative decays			
$\gamma\chi_{c0}(1P)$	(9.3 ± 0.9) %		261
$\gamma\chi_{c1}(1P)$	(8.7 ± 0.8) %		171
$\gamma\chi_{c2}(1P)$	(7.8 ± 0.8) %		127
$\gamma\eta_c(1S)$	$(2.8 \pm 0.6) \times 10^{-3}$		639
$\gamma\eta'(958)$	$< 1.1 \times 10^{-3}$	CL=90%	1719
$\gamma\gamma$	$< 1.6 \times 10^{-4}$	CL=90%	1843
$\gamma\eta(1440) \to \gamma K\overline{K}\pi$	$< 1.2 \times 10^{-4}$	CL=90%	1569

$\psi(3770)$

$I^G(J^{PC}) = ?^?(1^{--})$

Mass $m = 3769.9 \pm 2.5$ MeV (S = 1.8)
Full width $\Gamma = 23.6 \pm 2.7$ MeV (S = 1.1)
$\Gamma_{ee} = 0.26 \pm 0.04$ keV (S = 1.2)

$\psi(3770)$ DECAY MODES	Fraction (Γ_i/Γ)	Scale factor	p (MeV/c)
$D\overline{D}$	dominant		242
e^+e^-	$(1.12 \pm 0.17) \times 10^{-5}$	1.2	1885

$\psi(4040)$ [iii]

$I^G(J^{PC}) = ?^?(1^{--})$

Mass $m = 4040 \pm 10$ MeV
Full width $\Gamma = 52 \pm 10$ MeV
$\Gamma_{ee} = 0.75 \pm 0.15$ keV

$\psi(4040)$ DECAY MODES	Fraction (Γ_i/Γ)	p (MeV/c)
e^+e^-	$(1.4 \pm 0.4) \times 10^{-5}$	2020
$D^0\overline{D}^0$	seen	777
$D^*(2007)^0\overline{D}^0$ + c.c.	seen	578
$D^*(2007)^0\overline{D}^*(2007)^0$	seen	232

$\psi(4160)$ [///]

$I^G(J^{PC}) = ?^?(1^{--})$

Mass $m = 4159 \pm 20$ MeV
Full width $\Gamma = 78 \pm 20$ MeV
$\Gamma_{ee} = 0.77 \pm 0.23$ keV

$\psi(4160)$ DECAY MODES	Fraction (Γ_i/Γ)	p (MeV/c)
e^+e^-	$(10\pm4) \times 10^{-6}$	2079

$\psi(4415)$ [///]

$I^G(J^{PC}) = ?^?(1^{--})$

Mass $m = 4415 \pm 6$ MeV
Full width $\Gamma = 43 \pm 15$ MeV (S = 1.8)
$\Gamma_{ee} = 0.47 \pm 0.10$ keV

$\psi(4415)$ DECAY MODES	Fraction (Γ_i/Γ)	p (MeV/c)
hadrons	dominant	–
e^+e^-	$(1.1\pm0.4) \times 10^{-5}$	2207

$b\overline{b}$ MESONS

$\Upsilon(1S)$

$I^G(J^{PC}) = 0^-(1^{--})$

Mass $m = 9460.37 \pm 0.21$ MeV (S = 2.7)
Full width $\Gamma = 52.5 \pm 1.8$ keV
$\Gamma_{ee} = 1.32 \pm 0.05$ keV

$\Upsilon(1S)$ DECAY MODES	Fraction (Γ_i/Γ)	Scale factor/ Confidence level	p (MeV/c)
$\tau^+\tau^-$	$(2.67^{+0.14}_{-0.16})$ %		4384
e^+e^-	(2.52 ± 0.17) %		4730
$\mu^+\mu^-$	(2.48 ± 0.07) %	S=1.1	4729
Hadronic decays			
$J/\psi(1S)$anything	$(1.1\ \pm0.4\ ) \times 10^{-3}$		4223
$\rho\pi$	$< 2 \times 10^{-4}$	CL=90%	4698
$\pi^+\pi^-$	$< 5 \times 10^{-4}$	CL=90%	4728
K^+K^-	$< 5 \times 10^{-4}$	CL=90%	4704
$p\overline{p}$	$< 5 \times 10^{-4}$	CL=90%	4636
Radiative decays			
$\gamma 2h^+2h^-$	$(7.0\ \pm1.5\ ) \times 10^{-4}$		4720
$\gamma 3h^+3h^-$	$(5.4\ \pm2.0\ ) \times 10^{-4}$		4703
$\gamma 4h^+4h^-$	$(7.4\ \pm3.5\ ) \times 10^{-4}$		4679
$\gamma\pi^+\pi^-K^+K^-$	$(2.9\ \pm0.9\ ) \times 10^{-4}$		4686
$\gamma 2\pi^+2\pi^-$	$(2.5\ \pm0.9\ ) \times 10^{-4}$		4720
$\gamma 3\pi^+3\pi^-$	$(2.5\ \pm1.2\ ) \times 10^{-4}$		4703
$\gamma 2\pi^+2\pi^-K^+K^-$	$(2.4\ \pm1.2\ ) \times 10^{-4}$		4658
$\gamma\pi^+\pi^-p\overline{p}$	$(1.5\ \pm0.6\ ) \times 10^{-4}$		4604
$\gamma 2\pi^+2\pi^-p\overline{p}$	$(4\ \ \pm6\ \ ) \times 10^{-5}$		4563
$\gamma 2K^+2K^-$	$(2.0\ \pm2.0\ ) \times 10^{-5}$		4601
$\gamma\eta'(958)$	$< 1.3 \times 10^{-3}$	CL=90%	4682
$\gamma\eta$	$< 3.5 \times 10^{-4}$	CL=90%	4714
$\gamma f'_2(1525)$	$< 1.4 \times 10^{-4}$	CL=90%	4607
$\gamma f_2(1270)$	$< 1.3 \times 10^{-4}$	CL=90%	4644
$\gamma\eta(1440)$	$< 8.2 \times 10^{-5}$	CL=90%	4624
$\gamma f_J(1710) \to \gamma K\overline{K}$	$< 2.6 \times 10^{-4}$	CL=90%	4576
$\gamma f_0(2200) \to \gamma K^+K^-$	$< 2 \times 10^{-4}$	CL=90%	4475
$\gamma f_J(2220) \to \gamma K^+K^-$	$< 1.5 \times 10^{-5}$	CL=90%	4469
$\gamma\eta(2225) \to \gamma\phi\phi$	$< 3 \times 10^{-3}$	CL=90%	4469
γX	$< 3 \times 10^{-5}$	CL=90%	–
X = pseudoscalar with $m <$ 7.2 GeV)			
$\gamma X\overline{X}$	$< 1 \times 10^{-3}$	CL=90%	–
$X\overline{X}$ = vectors with $m <$ 3.1 GeV)			

$\chi_{b0}(1P)$ [///]

$I^G(J^{PC}) = 0^+(0^{++})$
J needs confirmation.

Mass $m = 9859.8 \pm 1.3$ MeV

$\chi_{b0}(1P)$ DECAY MODES	Fraction (Γ_i/Γ)	Confidence level	p (MeV/c)
$\gamma\Upsilon(1S)$	<6 %	90%	391

$\chi_{b1}(1P)$ [///]

$I^G(J^{PC}) = 0^+(1^{++})$
J needs confirmation.

Mass $m = 9891.9 \pm 0.7$ MeV

$\chi_{b1}(1P)$ DECAY MODES	Fraction (Γ_i/Γ)	p (MeV/c)
$\gamma\Upsilon(1S)$	(35 ± 8) %	422

$\chi_{b2}(1P)$ [///]

$I^G(J^{PC}) = 0^+(2^{++})$
J needs confirmation.

Mass $m = 9913.2 \pm 0.6$ MeV

$\chi_{b2}(1P)$ DECAY MODES	Fraction (Γ_i/Γ)	p (MeV/c)
$\gamma\Upsilon(1S)$	(22 ± 4) %	443

$\Upsilon(2S)$

$I^G(J^{PC}) = 0^-(1^{--})$

Mass $m = 10.02330 \pm 0.00031$ GeV
Full width $\Gamma = 44 \pm 7$ keV
$\Gamma_{ee} = 0.520 \pm 0.032$ keV

$\Upsilon(2S)$ DECAY MODES	Fraction (Γ_i/Γ)	Confidence level	p (MeV/c)
$\Upsilon(1S)\pi^+\pi^-$	$(18.5\ \pm0.8\ )$ %		475
$\Upsilon(1S)\pi^0\pi^0$	$(\ 8.8\ \pm1.1\ )$ %		480
$\tau^+\tau^-$	$(\ 1.7\ \pm1.6\ )$ %		4686
$\mu^+\mu^-$	$(\ 1.31\pm0.21)$ %		5011
e^+e^-	$(\ 1.18\pm0.20)$ %		5012
$\Upsilon(1S)\pi^0$	$< 8 \times 10^{-3}$	90%	531
$\Upsilon(1S)\eta$	$< 2 \times 10^{-3}$	90%	127
$J/\psi(1S)$anything	$< 6 \times 10^{-3}$	90%	4533
Radiative decays			
$\gamma\chi_{b1}(1P)$	$(\ 6.7\ \pm0.9\ )$ %		131
$\gamma\chi_{b2}(1P)$	$(\ 6.6\ \pm0.9\ )$ %		110
$\gamma\chi_{b0}(1P)$	$(\ 4.3\ \pm1.0\ )$ %		162
$\gamma f_J(1710)$	$< 5.9 \times 10^{-4}$	90%	4866
$\gamma f'_2(1525)$	$< 5.3 \times 10^{-4}$	90%	4896
$\gamma f_2(1270)$	$< 2.41 \times 10^{-4}$	90%	4931

$\chi_{b0}(2P)$ [///]

$I^G(J^{PC}) = 0^+(0^{++})$
J needs confirmation.

Mass $m = 10.2321 \pm 0.0006$ GeV

$\chi_{b0}(2P)$ DECAY MODES	Fraction (Γ_i/Γ)	p (MeV/c)
$\gamma\Upsilon(2S)$	(4.6 ± 2.1) %	210
$\gamma\Upsilon(1S)$	$(9\ \ \pm6\ \ ) \times 10^{-3}$	746

$\chi_{b1}(2P)$ [///]

$I^G(J^{PC}) = 0^+(1^{++})$
J needs confirmation.

Mass $m = 10.2552 \pm 0.0005$ GeV
$m_{\chi_{b1}(2P)} - m_{\chi_{b0}(2P)} = 23.5 \pm 1.0$ MeV

$\chi_{b1}(2P)$ DECAY MODES	Fraction (Γ_i/Γ)	Scale factor	p (MeV/c)
$\gamma\Upsilon(2S)$	$(21\ \ \pm4\ \ )$ %	1.5	229
$\gamma\Upsilon(1S)$	$(\ 8.5\pm1.3)$ %	1.3	764

$\chi_{b2}(2P)$ [///]

$I^G(J^{PC}) = 0^+(2^{++})$
J needs confirmation.

Mass $m = 10.2685 \pm 0.0004$ GeV
$m_{\chi_{b2}(2P)} - m_{\chi_{b1}(2P)} = 13.5 \pm 0.6$ MeV

$\chi_{b2}(2P)$ DECAY MODES	Fraction (Γ_i/Γ)	p (MeV/c)
$\gamma\Upsilon(2S)$	(16.2 ± 2.4) %	242
$\gamma\Upsilon(1S)$	$(\ 7.1\pm1.0)$ %	776

$\Upsilon(3S)$

$I^G(J^{PC}) = 0^-(1^{--})$

Mass $m = 10.3553 \pm 0.0005$ GeV
Full width $\Gamma = 26.3 \pm 3.5$ keV

$\Upsilon(3S)$ DECAY MODES	Fraction (Γ_i/Γ)	Scale factor/ Confidence level	p (MeV/c)
$\Upsilon(2S)$anything	(10.6 ±0.8) %		296
$\Upsilon(2S)\pi^+\pi^-$	(2.8 ±0.6) %	S=2.2	177
$\Upsilon(2S)\pi^0\pi^0$	(2.00±0.32) %		190
$\Upsilon(2S)\gamma\gamma$	(5.0 ±0.7) %		327
$\Upsilon(1S)\pi^+\pi^-$	(4.48±0.21) %		814
$\Upsilon(1S)\pi^0\pi^0$	(2.06±0.28) %		816
$\Upsilon(1S)\eta$	< 2.2 × 10^{-3}	CL=90%	–
$\mu^+\mu^-$	(1.81±0.17) %		5177
e^+e^-	seen		5177
Radiative decays			
$\gamma\chi_{b2}(2P)$	(11.4 ±0.8) %	S=1.3	87
$\gamma\chi_{b1}(2P)$	(11.3 ±0.6) %		100
$\gamma\chi_{b0}(2P)$	(5.4 ±0.6) %	S=1.1	123

$\Upsilon(4S)$ or $\Upsilon(10580)$

$I^G(J^{PC}) = ?^?(1^{--})$

Mass $m = 10.5800 \pm 0.0035$ GeV
Full width $\Gamma = 10 \pm 4$ MeV
$\Gamma_{ee} = 0.248 \pm 0.031$ keV (S = 1.3)

$\Upsilon(4S)$ DECAY MODES	Fraction (Γ_i/Γ)	Confidence level	p (MeV/c)
$B\overline{B}$	> 96 %	95%	–
non-$B\overline{B}$	< 4 %	95%	–
e^+e^-	(2.8±0.7) × 10^{-5}		5290
$J/\psi(3097)$anything	(2.2±0.7) × 10^{-3}		–
D^{*+}anything + c.c.	< 7.4 %	90%	5099
ϕanything	< 2.3 × 10^{-3}	90%	5240
$\Upsilon(1S)$anything	< 4 × 10^{-3}	90%	1053

$\Upsilon(10860)$

$I^G(J^{PC}) = ?^?(1^{--})$

Mass $m = 10.865 \pm 0.008$ GeV (S = 1.1)
Full width $\Gamma = 110 \pm 13$ MeV
$\Gamma_{ee} = 0.31 \pm 0.07$ keV (S = 1.3)

$\Upsilon(10860)$ DECAY MODES	Fraction (Γ_i/Γ)	p (MeV/c)
e^+e^-	(2.8±0.7) × 10^{-6}	5432

$\Upsilon(11020)$

$I^G(J^{PC}) = ?^?(1^{--})$

Mass $m = 11.019 \pm 0.008$ GeV
Full width $\Gamma = 79 \pm 16$ MeV
$\Gamma_{ee} = 0.130 \pm 0.030$ keV

$\Upsilon(11020)$ DECAY MODES	Fraction (Γ_i/Γ)	p (MeV/c)
e^+e^-	(1.6±0.5) × 10^{-6}	5509

NOTES

In this Summary Table:

When a quantity has "(S = ...)" to its right, the error on the quantity has been enlarged by the "scale factor" S, defined as $S = \sqrt{\chi^2/(N-1)}$, where N is the number of measurements used in calculating the quantity. We do this when S > 1, which often indicates that the measurements are inconsistent. When S > 1.25, we also show in the Particle Listings an ideogram of the measurements. For more about S, see the Introduction.

A decay momentum p is given for each decay mode. For a 2-body decay, p is the momentum of each decay product in the rest frame of the decaying particle. For a 3-or-more-body decay, p is the largest momentum any of the products can have in this frame.

[a] See the "Note on $\pi^\pm \to \ell^\pm\nu\gamma$ and $K^\pm \to \ell^\pm\nu\gamma$ Form Factors" in the $\pi^\pm$ Particle Listings for definitions and details.

[b] Measurements of $\Gamma(e^+\nu_e)/\Gamma(\mu^+\nu_\mu)$ always include decays with γ's, and measurements of $\Gamma(e^+\nu_e\gamma)$ and $\Gamma(\mu^+\nu_\mu\gamma)$ never include low-energy γ's. Therefore, since no clean separation is possible, we consider the modes with γ's to be subreactions of the modes without them, and let $[\Gamma(e^+\nu_e) + \Gamma(\mu^+\nu_\mu)]/\Gamma_{total} = 100\%$.

[c] See the $\pi^\pm$ Particle Listings for the energy limits used in this measurement; low-energy γ's are not included.

[d] Derived from an analysis of neutrino-oscillation experiments.

[e] Astrophysical and cosmological arguments give limits of order 10^{-13}; see the π^0 Particle Listings.

[f] See the "Note on the Decay Width $\Gamma(\eta \to \gamma\gamma)$" in our 1994 edition, Phys. Rev. **D50**, 1 August 1994, Part I, p. 1451.

[g] C parity forbids this to occur as a single-photon process.

[h] See the "Note on scalar mesons" in the $f_0(1370)$ Particle Listings . The interpretation of this entry as a particle is controversial.

[i] See the "Note on $\rho(770)$" in the $\rho(770)$ Particle Listings .

[j] The e^+e^- branching fraction is from $e^+e^- \to \pi^+\pi^-$ experiments only. The $\omega\rho$ interference is then due to $\omega\rho$ mixing only, and is expected to be small. If $e\mu$ universality holds, $\Gamma(\rho^0 \to \mu^+\mu^-) = \Gamma(\rho^0 \to e^+e^-) \times 0.99785$.

[k] See the "Note on scalar mesons" in the $f_0(1370)$ Particle Listings .

[l] See the "Note on $a_1(1260)$" in the $a_1(1260)$ Particle Listings .

[m] This is only an educated guess; the error given is larger than the error on the average of the published values. See the Particle Listings for details.

[n] See the "Note on the $f_1(1420)$" in the $\eta(1440)$ Particle Listings.

[o] See also the $\omega(1600)$ Particle Listings.

[p] See the "Note on the $\eta(1440)$" in the $\eta(1440)$ Particle Listings.

[q] See the "Note on the $\rho(1450)$ and the $\rho(1700)$" in the $\rho(1700)$ Particle Listings.

[r] See the "Note on non-$q\overline{q}$ mesons" in the Particle Listings (see the index for the page number).

[s] See also the $\omega(1420)$ Particle Listings.

[t] See the "Note on $f_J(1710)$" in the $f_J(1710)$ Particle Listings .

[u] See the note in the $K^\pm$ Particle Listings.

[v] The definition of the slope parameter g of the $K \to 3\pi$ Dalitz plot is as follows (see also "Note on Dalitz Plot Parameters for $K \to 3\pi$ Decays" in the $K^\pm$ Particle Listings):

$$|M|^2 = 1 + g(s_3 - s_0)/m_{\pi^+}^2 + \cdots .$$

[*w*] For more details and definitions of parameters see the Particle Listings.

[*x*] See the $K^{\pm}$ Particle Listings for the energy limits used in this measurement.

[*y*] Most of this radiative mode, the low-momentum γ part, is also included in the parent mode listed without γ's.

[*z*] Direct-emission branching fraction.

[*aa*] Structure-dependent part.

[*bb*] Derived from measured values of ϕ_{+-}, ϕ_{00}, $|\eta|$, $|m_{K^0_L} - m_{K^0_S}|$, and $\tau_{K^0_S}$, as described in the introduction to "Tests of Conservation Laws."

[*cc*] The *CP*-violation parameters are defined as follows (see also "Note on *CP* Violation in $K_S \to 3\pi$" and "Note on *CP* Violation in K^0_L Decay" in the Particle Listings):

$$\eta_{+-} = |\eta_{+-}|e^{i\phi_{+-}} = \frac{A(K^0_L \to \pi^+\pi^-)}{A(K^0_S \to \pi^+\pi^-)} = \epsilon + \epsilon'$$

$$\eta_{00} = |\eta_{00}|e^{i\phi_{00}} = \frac{A(K^0_L \to \pi^0\pi^0)}{A(K^0_S \to \pi^0\pi^0)} = \epsilon - 2\epsilon'$$

$$\delta = \frac{\Gamma(K^0_L \to \pi^-\ell^+\nu) - \Gamma(K^0_L \to \pi^+\ell^-\nu)}{\Gamma(K^0_L \to \pi^-\ell^+\nu) + \Gamma(K^0_L \to \pi^+\ell^-\nu)},$$

$$\mathrm{Im}(\eta_{+-0})^2 = \frac{\Gamma(K^0_S \to \pi^+\pi^-\pi^0)^{CP\ \mathrm{viol.}}}{\Gamma(K^0_L \to \pi^+\pi^-\pi^0)},$$

$$\mathrm{Im}(\eta_{000})^2 = \frac{\Gamma(K^0_S \to \pi^0\pi^0\pi^0)}{\Gamma(K^0_L \to \pi^0\pi^0\pi^0)}.$$

where for the last two relations *CPT* is assumed valid, *i.e.*, $\mathrm{Re}(\eta_{+-0}) \simeq 0$ and $\mathrm{Re}(\eta_{000}) \simeq 0$.

[*dd*] See the K^0_S Particle Listings for the energy limits used in this measurement.

[*ee*] Calculated from K^0_L semileptonic rates and the K^0_S lifetime assuming $\Delta S = \Delta Q$.

[*ff*] ϵ'/ϵ is derived from $|\eta_{00}/\eta_{+-}|$ measurements using theoretical input on phases.

[*gg*] The value is for the sum of the charge states of particle/antiparticle states indicated.

[*hh*] See the K^0_L Particle Listings for the energy limits used in this measurement.

[*ii*] Allowed by higher-order electroweak interactions.

[*jj*] Violates *CP* in leading order. Test of direct *CP* violation since the indirect *CP*-violating and *CP*-conserving contributions are expected to be suppressed.

[*kk*] See the "Note on $f_0(1370)$" in the $f_0(1370)$ Particle Listings and in the 1994 edition.

[*ll*] See the note in the $L(1770)$ Particle Listings in Reviews of Modern Physics **56** No. 2 Pt. II (1984), p. S200. See also the "Note on $K_2(1770)$ and the $K_2(1820)$" in the $K_2(1770)$ Particle Listings .

[*mm*] See the "Note on $K_2(1770)$ and the $K_2(1820)$" in the $K_2(1770)$ Particle Listings .

[*nn*] This is a weighted average of $D^{\pm}$ (44%) and D^0 (56%) branching fractions. See "D^+ and $D^0 \to$ (η anything) / (total D^+ and D^0)" under "D^+ Branching Ratios" in the Particle Listings.

[*oo*] This value averages the e^+ and μ^+ branching fractions, after making a small phase-space adjustment to the μ^+ fraction to be able to use it as an e^+ fraction; hence our ℓ^+ here is really an e^+.

[*pp*] An ℓ indicates an e or a μ mode, not a sum over these modes.

[*qq*] The branching fraction for this mode may differ from the sum of the submodes that contribute to it, due to interference effects. See the relevant papers in the Particle Listings.

[*rr*] The two experiments measuring this fraction are in serious disagreement. See the Particle Listings.

[*ss*] This mode is not a useful test for a ΔC=1 weak neutral current because both quarks must change flavor in this decay.

[*tt*] The D^0_1-D^0_2 limits are inferred from the D^0-$\overline{D}^0$ mixing ratio $\Gamma(K^+\ell^-\overline{\nu}_\ell(\text{via } \overline{D}^0)) \,/\, \Gamma(K^-\ell^+\nu_\ell)$.

[*uu*] The larger limit (from E791) allows interference between the doubly Cabibbo-suppressed and mixing amplitudes; the smaller limit (from E691) doesn't. See the papers for details.

[*vv*] The experiments on the division of this charge mode amongst its submodes disagree, and the submode branching fractions here add up to considerably more than the charged-mode fraction.

[*ww*] However, these upper limits are in serious disagreement with values obtained in another experiment.

[*xx*] For now, we average together measurements of the $X e^+ \nu_e$ and $X \mu^+ \nu_\mu$ branching fractions. This is the *average*, not the *sum*.

[*yy*] This branching fraction includes all the decay modes of the final-state resonance.

[*zz*] This value includes only K^+K^- decays of the $f_J(1710)$, because branching fractions of this resonance are not known.

[*aaa*] This value includes only $\pi^+\pi^-$ decays of the $f_0(1500)$, because branching fractions of this resonance are not known.

[*bbb*] B^0 and B^0_s contributions not separated. Limit is on weighted average of the two decay rates.

[*ccc*] These values are model dependent. See 'Note on Semileptonic Decays' in the B^+ Particle Listings.

[*ddd*] D^{**} stands for the sum of the $D(1\,^1P_1)$, $D(1\,^3P_0)$, $D(1\,^3P_1)$, $D(1\,^3P_2)$, $D(2\,^1S_0)$, and $D(2\,^1S_1)$ resonances.

[*eee*] Inclusive branching fractions have a multiplicity definition and can be greater than 100%.

[*fff*] D_j represents an unresolved mixture of pseudoscalar and tensor D^{**} (P-wave) states.

[*ggg*] Not a pure measurement. See note at head of B^0_s Decay Modes.

[*hhh*] Includes $p\overline{p}\pi^+\pi^-\gamma$ and excludes $p\overline{p}\eta$, $p\overline{p}\omega$, $p\overline{p}\eta'$.

[*iii*] J^{PC} known by production in e^+e^- via single photon annihilation. I^G is not known; interpretation of this state as a single resonance is unclear because of the expectation of substantial threshold effects in this energy region.

[*jjj*] Spectroscopic labeling for these states is theoretical, pending experimental information.

Meson Summary Table

See also the table of suggested $q\bar{q}$ quark-model assignments in the Quark Model section.

- • Indicates particles that appear in the preceding Meson Summary Table. We do not regard the other entries as being established.
- † Indicates that the value of J given is preferred, but needs confirmation.

LIGHT UNFLAVORED ($S = C = B = 0$)

	$I^G(J^{PC})$
• $\pi^\pm$	$1^-(0^-)$
• π^0	$1^-(0^{-+})$
• η	$0^+(0^{-+})$
• $f_0(400\text{–}1200)$	$0^+(0^{++})$
• $\rho(770)$	$1^+(1^{--})$
• $\omega(782)$	$0^-(1^{--})$
• $\eta'(958)$	$0^+(0^{-+})$
• $f_0(980)$	$0^+(0^{++})$
• $a_0(980)$	$1^-(0^{++})$
• $\phi(1020)$	$0^-(1^{--})$
• $h_1(1170)$	$0^-(1^{+-})$
• $b_1(1235)$	$1^+(1^{+-})$
• $a_1(1260)$	$1^-(1^{++})$
• $f_2(1270)$	$0^+(2^{++})$
• $f_1(1285)$	$0^+(1^{++})$
• $\eta(1295)$	$0^+(0^{-+})$
• $\pi(1300)$	$1^-(0^{-+})$
• $a_2(1320)$	$1^-(2^{++})$
• $f_0(1370)$	$0^+(0^{++})$
$h_1(1380)$	$?^-(1^{+-})$
$\hat{\rho}(1405)$	$1^-(1^{-+})$
• $f_1(1420)$	$0^+(1^{++})$
• $\omega(1420)$	$0^-(1^{--})$
$f_2(1430)$	$0^+(2^{++})$
• $\eta(1440)$	$0^+(0^{-+})$
• $a_0(1450)$	$1^-(0^{++})$
• $\rho(1450)$	$1^+(1^{--})$
• $f_0(1500)$	$0^+(0^{++})$
$f_1(1510)$	$0^+(1^{++})$
• $f_2'(1525)$	$0^+(2^{++})$
$f_2(1565)$	$0^+(2^{++})$
• $\omega(1600)$	$0^-(1^{--})$
$X(1600)$	$2^+(2^{++})$
$f_2(1640)$	$0^+(2^{++})$
$\eta_2(1645)$	$0^+(2^{-+})$
$X(1650)$	$0^+(?^{?-})$
• $\omega_3(1670)$	$0^-(3^{--})$
• $\pi_2(1670)$	$1^-(2^{-+})$
• $\phi(1680)$	$0^-(1^{--})$
• $\rho_3(1690)$	$1^+(3^{--})$
• $\rho(1700)$	$1^+(1^{--})$
• $f_J(1710)$	$0^+(\text{even}^{++})$
$\eta(1760)$	$0^+(0^{-+})$
$X(1775)$	$1^-(?^{-+})$
• $\pi(1800)$	$1^-(0^{-+})$
$f_2(1810)$	$0^+(2^{++})$
• $\phi_3(1850)$	$0^-(3^{--})$
$\eta_2(1870)$	$0^+(2^{-+})$
$X(1910)$	$0^+(?^{?+})$
$f_2(1950)$	$0^+(2^{++})$
$X(2000)$	$1^-(?^{?+})$
• $f_2(2010)$	$0^+(2^{++})$
$f_0(2020)$	$0^+(0^{++})$
• $a_4(2040)$	$1^-(4^{++})$
• $f_4(2050)$	$0^+(4^{++})$
$f_0(2060)$	$0^+(0^{++})$
$\pi_2(2100)$	$1^-(2^{-+})$
$f_2(2150)$	$0^+(2^{++})$
$\rho(2150)$	$1^+(1^{--})$
$f_0(2200)$	$0^+(0^{++})$
$f_J(2220)$	$0^+(2^{++}$ or $4^{++})$
$\eta(2225)$	$0^+(0^{-+})$
$\rho_3(2250)$	$1^+(3^{--})$
• $f_2(2300)$	$0^+(2^{++})$
$f_4(2300)$	$0^+(4^{++})$
• $f_2(2340)$	$0^+(2^{++})$
$\rho_5(2350)$	$1^+(5^{--})$
$a_6(2450)$	$1^-(6^{++})$
$f_6(2510)$	$0^+(6^{++})$
$X(3250)$	$?^?(?^{??})$

OTHER LIGHT UNFLAVORED ($S = C = B = 0$)

$e^+e^-(1100\text{–}2200)$	$?^?(1^{--})$
$\overline{N}N(1100\text{–}3600)$	
$X(1900\text{–}3600)$	

STRANGE ($S = \pm1$, $C = B = 0$)

	$I(J^P)$
• $K^\pm$	$1/2(0^-)$
• K^0	$1/2(0^-)$
• K^0_S	$1/2(0^-)$
• K^0_L	$1/2(0^-)$
• $K^*(892)$	$1/2(1^-)$
• $K_1(1270)$	$1/2(1^+)$
• $K_1(1400)$	$1/2(1^+)$
• $K^*(1410)$	$1/2(1^-)$
• $K^*_0(1430)$	$1/2(0^+)$
• $K^*_2(1430)$	$1/2(2^+)$
$K(1460)$	$1/2(0^-)$
$K_2(1580)$	$1/2(2^-)$
$K_1(1650)$	$1/2(1^+)$
• $K^*(1680)$	$1/2(1^-)$
• $K_2(1770)$	$1/2(2^-)$
• $K^*_3(1780)$	$1/2(3^-)$
• $K_2(1820)$	$1/2(2^-)$
$K(1830)$	$1/2(0^-)$
$K^*_0(1950)$	$1/2(0^+)$
$K^*_2(1980)$	$1/2(2^+)$
• $K^*_4(2045)$	$1/2(4^+)$
$K_2(2250)$	$1/2(2^-)$
$K_3(2320)$	$1/2(3^+)$
$K^*_5(2380)$	$1/2(5^-)$
$K_4(2500)$	$1/2(4^-)$
$K(3100)$	$?^?(?^{??})$

CHARMED ($C = \pm1$)

	$I(J^P)$
• $D^\pm$	$1/2(0^-)$
• D^0	$1/2(0^-)$
• $D^*(2007)^0$	$1/2(1^-)$
• $D^*(2010)^\pm$	$1/2(1^-)$
• $D_1(2420)^0$	$1/2(1^+)$
$D_1(2420)^\pm$	$1/2(?^?)$
• $D^*_2(2460)^0$	$1/2(2^+)$
• $D^*_2(2460)^+$	$1/2(2^+)$

CHARMED, STRANGE ($C = S = \pm1$)

	$I(J^P)$
• $D^\pm_s$	$0(0^-)$
• $D^{*\pm}_s$	$0(?^?)$
• $D_{s1}(2536)^\pm$	$0(1^+)$
• $D_{sJ}(2573)^\pm$	$0(?^?)$

BOTTOM ($B = \pm1$)

	$I(J^P)$
• $B^\pm$	$1/2(0^-)$
• B^0	$1/2(0^-)$
• B^*	$1/2(1^-)$
$B^*_J(5732)$	$?(?^?)$

BOTTOM, STRANGE ($B = \pm1$, $S = \mp1$)

	$I^G(J^{PC})$
• B^0_s	$0(0^-)$
B^*_s	$0(1^-)$
$B^*_{sJ}(5850)$	$?(?^?)$

BOTTOM, CHARMED ($B = C = \pm1$)

$B^\pm_c$	$0(0^-)$

$c\bar{c}$

	$I^G(J^{PC})$
• $\eta_c(1S)$	$0^+(0^{-+})$
• $J/\psi(1S)$	$0^-(1^{--})$
• $\chi_{c0}(1P)$	$0^+(0^{++})$
• $\chi_{c1}(1P)$	$0^+(1^{++})$
$h_c(1P)$	$?^?(?^{??})$
• $\chi_{c2}(1P)$	$0^+(2^{++})$
$\eta_c(2S)$	$?^?(?^{?+})$
• $\psi(2S)$	$0^-(1^{--})$
• $\psi(3770)$	$?^?(1^{--})$
• $\psi(4040)$	$?^?(1^{--})$
• $\psi(4160)$	$?^?(1^{--})$
• $\psi(4415)$	$?^?(1^{--})$

$b\bar{b}$

	$I^G(J^{PC})$
• $\Upsilon(1S)$	$0^-(1^{--})$
• $\chi_{b0}(1P)$	$0^+(0^{++})$
• $\chi_{b1}(1P)$	$0^+(1^{++})$
• $\chi_{b2}(1P)$	$0^+(2^{++})$
• $\Upsilon(2S)$	$0^-(1^{--})$
• $\chi_{b0}(2P)$	$0^+(0^{++})$
• $\chi_{b1}(2P)$	$0^+(1^{++})$
• $\chi_{b2}(2P)$	$0^+(2^{++})$
• $\Upsilon(3S)$	$0^-(1^{--})$
• $\Upsilon(4S)$	$?^?(1^{--})$
• $\Upsilon(10860)$	$?^?(1^{--})$
• $\Upsilon(11020)$	$?^?(1^{--})$

NON-$q\bar{q}$ CANDIDATES

Non-$q\bar{q}$ Candidates

This short table gives the name, the quantum numbers (where known), and the status of baryons in the Review. Only the baryons with 3- or 4-star status are included in the main Baryon Summary Table. Due to insufficient data or uncertain interpretation, the other entries in the short table are not established as baryons. The names with masses are of baryons that decay strongly. See our 1986 edition (Physics Letters **170B**) for listings of evidence for Z baryons (KN resonances).

p	P_{11}	****	$\Delta(1232)$	P_{33}	****	Λ	P_{01}	****	Σ^+	P_{11}	****	Ξ^0	P_{11}	****
n	P_{11}	****	$\Delta(1600)$	P_{33}	***	$\Lambda(1405)$	S_{01}	****	Σ^0	P_{11}	****	Ξ^-	P_{11}	****
$N(1440)$	P_{11}	****	$\Delta(1620)$	S_{31}	****	$\Lambda(1520)$	D_{03}	****	Σ^-	P_{11}	****	$\Xi(1530)$	P_{13}	****
$N(1520)$	D_{13}	****	$\Delta(1700)$	D_{33}	****	$\Lambda(1600)$	P_{01}	***	$\Sigma(1385)$	P_{13}	****	$\Xi(1620)$		*
$N(1535)$	S_{11}	****	$\Delta(1750)$	P_{31}	*	$\Lambda(1670)$	S_{01}	****	$\Sigma(1480)$		*	$\Xi(1690)$		***
$N(1650)$	S_{11}	****	$\Delta(1900)$	S_{31}	**	$\Lambda(1690)$	D_{03}	****	$\Sigma(1560)$		**	$\Xi(1820)$	D_{13}	***
$N(1675)$	D_{15}	****	$\Delta(1905)$	F_{35}	****	$\Lambda(1800)$	S_{01}	***	$\Sigma(1580)$	D_{13}	**	$\Xi(1950)$		***
$N(1680)$	F_{15}	****	$\Delta(1910)$	P_{31}	****	$\Lambda(1810)$	P_{01}	***	$\Sigma(1620)$	S_{11}	**	$\Xi(2030)$		***
$N(1700)$	D_{13}	***	$\Delta(1920)$	P_{33}	***	$\Lambda(1820)$	F_{05}	****	$\Sigma(1660)$	P_{11}	***	$\Xi(2120)$		*
$N(1710)$	P_{11}	***	$\Delta(1930)$	D_{35}	***	$\Lambda(1830)$	D_{05}	****	$\Sigma(1670)$	D_{13}	****	$\Xi(2250)$		**
$N(1720)$	P_{13}	****	$\Delta(1940)$	D_{33}	*	$\Lambda(1890)$	P_{03}	****	$\Sigma(1690)$		**	$\Xi(2370)$		**
$N(1900)$	P_{13}	**	$\Delta(1950)$	F_{37}	****	$\Lambda(2000)$		*	$\Sigma(1750)$	S_{11}	***	$\Xi(2500)$		*
$N(1990)$	F_{17}	**	$\Delta(2000)$	F_{35}	**	$\Lambda(2020)$	F_{07}	*	$\Sigma(1770)$	P_{11}	*			
$N(2000)$	F_{15}	**	$\Delta(2150)$	S_{31}	*	$\Lambda(2100)$	G_{07}	****	$\Sigma(1775)$	D_{15}	****	Ω^-		****
$N(2080)$	D_{13}	**	$\Delta(2200)$	G_{37}	*	$\Lambda(2110)$	F_{05}	***	$\Sigma(1840)$	P_{13}	*	$\Omega(2250)^-$		***
$N(2090)$	S_{11}	*	$\Delta(2300)$	H_{39}	**	$\Lambda(2325)$	D_{03}	*	$\Sigma(1880)$	P_{11}	**	$\Omega(2380)^-$		**
$N(2100)$	P_{11}	*	$\Delta(2350)$	D_{35}	*	$\Lambda(2350)$	H_{09}	***	$\Sigma(1915)$	F_{15}	****	$\Omega(2470)^-$		**
$N(2190)$	G_{17}	****	$\Delta(2390)$	F_{37}	*	$\Lambda(2585)$		**	$\Sigma(1940)$	D_{13}	***			
$N(2200)$	D_{15}	**	$\Delta(2400)$	G_{39}	**				$\Sigma(2000)$	S_{11}	*	Λ_c^+		****
$N(2220)$	H_{19}	****	$\Delta(2420)$	$H_{3,11}$	****				$\Sigma(2030)$	F_{17}	****	$\Lambda_c(2593)^+$		***
$N(2250)$	G_{19}	****	$\Delta(2750)$	$I_{3,13}$	**				$\Sigma(2070)$	F_{15}	*	$\Lambda_c(2625)^+$		***
$N(2600)$	$I_{1,11}$	***	$\Delta(2950)$	$K_{3,15}$	**				$\Sigma(2080)$	P_{13}	**	$\Sigma_c(2455)$		****
$N(2700)$	$K_{1,13}$	**							$\Sigma(2100)$	G_{17}	*	$\Sigma_c(2520)$		***
									$\Sigma(2250)$		***	Ξ_c^+		***
									$\Sigma(2455)$		**	Ξ_c^0		***
									$\Sigma(2620)$		**	$\Xi_c(2645)$		***
									$\Sigma(3000)$		*	Ω_c^0		***
									$\Sigma(3170)$		*			
												Λ_b^0		***
												Ξ_b^0, Ξ_b^-		*

**** Existence is certain, and properties are at least fairly well explored.

*** Existence ranges from very likely to certain, but further confirmation is desirable and/or quantum numbers, branching fractions, etc. are not well determined.

** Evidence of existence is only fair.

* Evidence of existence is poor.

N BARYONS ($S = 0$, $I = 1/2$)

p, $N^+ = uud$; n, $N^0 = udd$

p

$I(J^P) = \frac{1}{2}(\frac{1}{2}^+)$

Mass $m = 938.27231 \pm 0.00028$ MeV [a]
$= 1.007276470 \pm 0.000000012$ u
$|\frac{q_{\bar{p}}}{m_{\bar{p}}}|/(\frac{q_p}{m_p}) = 1.0000000015 \pm 0.0000000011$
$|q_p + q_{\bar{p}}|/e < 2 \times 10^{-5}$
$|q_p + q_e|/e < 1.0 \times 10^{-21}$ [b]
Magnetic moment $\mu = 2.79284739 \pm 0.00000006\ \mu_N$
Electric dipole moment $d = (-4 \pm 6) \times 10^{-23}$ $e$cm
Electric polarizability $\overline{\alpha} = (12.1 \pm 0.9) \times 10^{-4}$ fm^3
Magnetic polarizability $\overline{\beta} = (2.1 \pm 0.9) \times 10^{-4}$ fm^3
Mean life $\tau > 1.6 \times 10^{25}$ years (independent of mode)
$> 10^{31}$ to 5×10^{32} years [c] (mode dependent)

Below, for N decays, p and n distinguish proton and neutron partial lifetimes. See also the "Note on Nucleon Decay" in our 1994 edition (Phys. Rev. **D50**, 1673) for a short review.

The "partial mean life" limits tabulated here are the limits on τ/B_j, where τ is the total mean life and B_j is the branching fraction for the mode in question.

p DECAY MODES	Partial mean life (10^{30} years)	Confidence level	p (MeV/c)
Antilepton + meson			
$N \to e^+\pi$	> 130 (n), > 550 (p)	90%	459
$N \to \mu^+\pi$	> 100 (n), > 270 (p)	90%	453
$N \to \nu\pi$	> 100 (n), > 25 (p)	90%	459
$p \to e^+\eta$	> 140	90%	309
$p \to \mu^+\eta$	> 69	90%	296
$n \to \nu\eta$	> 54	90%	310
$N \to e^+\rho$	> 58 (n), > 75 (p)	90%	153
$N \to \mu^+\rho$	> 23 (n), > 110 (p)	90%	119
$N \to \nu\rho$	> 19 (n), > 27 (p)	90%	153
$p \to e^+\omega$	> 45	90%	142
$p \to \mu^+\omega$	> 57	90%	104
$n \to \nu\omega$	> 43	90%	144
$N \to e^+K$	> 1.3 (n), > 150 (p)	90%	337
$p \to e^+K^0_S$	> 76	90%	337
$p \to e^+K^0_L$	> 44	90%	337
$N \to \mu^+K$	> 1.1 (n), > 120 (p)	90%	326
$p \to \mu^+K^0_S$	> 64	90%	326
$p \to \mu^+K^0_L$	> 44	90%	326
$N \to \nu K$	> 86 (n), > 100 (p)	90%	339
$p \to e^+K^*(892)^0$	> 52	90%	45
$N \to \nu K^*(892)$	> 22 (n), > 20 (p)	90%	45
Antilepton + mesons			
$p \to e^+\pi^+\pi^-$	> 21	90%	448
$p \to e^+\pi^0\pi^0$	> 38	90%	449
$n \to e^+\pi^-\pi^0$	> 32	90%	449
$p \to \mu^+\pi^+\pi^-$	> 17	90%	425
$p \to \mu^+\pi^0\pi^0$	> 33	90%	427
$n \to \mu^+\pi^-\pi^0$	> 33	90%	427
$n \to e^+K^0\pi^-$	> 18	90%	319
Lepton + meson			
$n \to e^-\pi^+$	> 65	90%	459
$n \to \mu^-\pi^+$	> 49	90%	453
$n \to e^-\rho^+$	> 62	90%	154
$n \to \mu^-\rho^+$	> 7	90%	120
$n \to e^-K^+$	> 32	90%	340
$n \to \mu^-K^+$	> 57	90%	330
Lepton + mesons			
$p \to e^-\pi^+\pi^+$	> 30	90%	448
$n \to e^-\pi^+\pi^0$	> 29	90%	449
$p \to \mu^-\pi^+\pi^+$	> 17	90%	425
$n \to \mu^-\pi^+\pi^0$	> 34	90%	427
$p \to e^-\pi^+K^+$	> 20	90%	320
$p \to \mu^-\pi^+K^+$	> 5	90%	279
Antilepton + photon(s)			
$p \to e^+\gamma$	> 460	90%	469
$p \to \mu^+\gamma$	> 380	90%	463
$n \to \nu\gamma$	> 24	90%	470
$p \to e^+\gamma\gamma$	> 100	90%	469
Three (or more) leptons			
$p \to e^+e^+e^-$	> 510	90%	469
$p \to e^+\mu^+\mu^-$	> 81	90%	457
$p \to e^+\nu\nu$	> 11	90%	469
$n \to e^+e^-\nu$	> 74	90%	470
$n \to \mu^+e^-\nu$	> 47	90%	464
$n \to \mu^+\mu^-\nu$	> 42	90%	458
$p \to \mu^+e^+e^-$	> 91	90%	464
$p \to \mu^+\mu^+\mu^-$	> 190	90%	439
$p \to \mu^+\nu\nu$	> 21	90%	463
$p \to e^-\mu^+\mu^+$	> 6	90%	457
$n \to 3\nu$	> 0.0005	90%	470
Inclusive modes			
$N \to e^+$ anything	> 0.6 (n, p)	90%	–
$N \to \mu^+$ anything	> 12 (n, p)	90%	–
$N \to e^+\pi^0$anything	> 0.6 (n, p)	90%	–
$\Delta B = 2$ dinucleon modes			
The following are lifetime limits per iron nucleus.			
$pp \to \pi^+\pi^+$	> 0.7	90%	–
$pn \to \pi^+\pi^0$	> 2	90%	–
$nn \to \pi^+\pi^-$	> 0.7	90%	–
$nn \to \pi^0\pi^0$	> 3.4	90%	–
$pp \to e^+e^+$	> 5.8	90%	–
$pp \to e^+\mu^+$	> 3.6	90%	–
$pp \to \mu^+\mu^+$	> 1.7	90%	–
$pn \to e^+\overline{\nu}$	> 2.8	90%	–
$pn \to \mu^+\overline{\nu}$	> 1.6	90%	–
$nn \to \nu_e\overline{\nu}_e$	> 0.000012	90%	–
$nn \to \nu_\mu\overline{\nu}_\mu$	> 0.000006	90%	–

$\overline{p}$ DECAY MODES

$\overline{p}$ DECAY MODES	Partial mean life (years)	Confidence level	p (MeV/c)
$\overline{p} \to e^-\gamma$	> 1848	95%	469
$\overline{p} \to e^-\pi^0$	> 554	95%	459
$\overline{p} \to e^-\eta$	> 171	95%	309
$\overline{p} \to e^-K^0_S$	> 29	95%	337
$\overline{p} \to e^-K^0_L$	> 9	95%	337

n

$I(J^P) = \frac{1}{2}(\frac{1}{2}^+)$

Mass $m = 939.56563 \pm 0.00028$ MeV [a]
$= 1.008664904 \pm 0.000000014$ u
$m_n - m_p = 1.293318 \pm 0.000009$ MeV
$= 0.001388434 \pm 0.000000009$ u
Mean life $\tau = 886.7 \pm 1.9$ s ($S = 1.2$)
$c\tau = 2.658 \times 10^8$ km
Magnetic moment $\mu = -1.9130428 \pm 0.0000005\ \mu_N$
Electric dipole moment $d < 0.97 \times 10^{-25}$ $e$cm, CL = 90%
Electric polarizability $\alpha = (0.98^{+0.19}_{-0.23}) \times 10^{-3}$ fm^3 ($S = 1.1$)
Charge $q = (-0.4 \pm 1.1) \times 10^{-21}$ e
Mean $n\overline{n}$-oscillation time $> 1.2 \times 10^8$ s, CL = 90% [d] (bound n)
$> 0.86 \times 10^8$ s, CL = 90% (free n)

Decay parameters [e]

$pe^-\overline{\nu}_e$ $g_A/g_V = -1.2670 \pm 0.0035$ ($S = 1.9$)
" $A = -0.1162 \pm 0.0013$ ($S = 1.8$)
" $B = 0.990 \pm 0.008$
" $a = -0.102 \pm 0.005$
" $\phi_{AV} = (180.07 \pm 0.18)^\circ$ [f]
" $D = (-0.5 \pm 1.4) \times 10^{-3}$

n DECAY MODES	Fraction (Γ_i/Γ)	Confidence level	p (MeV/c)
$pe^-\overline{\nu}_e$	100 %		1.19
Charge conservation (Q) violating mode			
$p\nu_e\overline{\nu}_e$	Q < 8×10^{-27}	68%	1.29

$N(1440)\ P_{11}$

$I(J^P) = \frac{1}{2}(\frac{1}{2}^+)$

Breit-Wigner mass = 1430 to 1470 (≈ 1440) MeV
Breit-Wigner full width = 250 to 450 (≈ 350) MeV
$p_{beam} = 0.61$ GeV/c $4\pi\lambda\!\!\!^{-2} = 31.0$ mb
Re(pole position) = 1345 to 1385 (≈ 1365) MeV
−2Im(pole position) = 160 to 260 (≈ 210) MeV

N(1440) DECAY MODES	Fraction (Γ_i/Γ)	p (MeV/c)
$N\pi$	60–70 %	397
$N\pi\pi$	30–40 %	342
$\Delta\pi$	20–30 %	143
$N\rho$	<8 %	†
$N(\pi\pi)^{I=0}_{S\text{-wave}}$	5–10 %	–
$p\gamma$	0.035–0.048 %	414
$p\gamma$, helicity=1/2	0.035–0.048 %	414
$n\gamma$	0.009–0.032 %	413
$n\gamma$, helicity=1/2	0.009–0.032 %	413

$N(1520)\ D_{13}$

$I(J^P) = \frac{1}{2}(\frac{3}{2}^-)$

Breit-Wigner mass = 1515 to 1530 (≈ 1520) MeV
Breit-Wigner full width = 110 to 135 (≈ 120) MeV
$p_{beam} = 0.74$ GeV/c $4\pi\lambda\!\!\!^{-2} = 23.5$ mb
Re(pole position) = 1505 to 1515 (≈ 1510) MeV
−2Im(pole position) = 110 to 120 (≈ 115) MeV

N(1520) DECAY MODES	Fraction (Γ_i/Γ)	p (MeV/c)
$N\pi$	50–60 %	456
$N\pi\pi$	40–50 %	410
$\Delta\pi$	15–25 %	228
$N\rho$	15–25 %	†
$N(\pi\pi)^{I=0}_{S\text{-wave}}$	<8 %	–
$p\gamma$	0.46–0.56 %	470
$p\gamma$, helicity=1/2	0.001–0.034 %	470
$p\gamma$, helicity=3/2	0.44–0.53 %	470
$n\gamma$	0.30–0.53 %	470
$n\gamma$, helicity=1/2	0.04–0.10 %	470
$n\gamma$, helicity=3/2	0.25–0.45 %	470

$N(1535)\ S_{11}$

$I(J^P) = \frac{1}{2}(\frac{1}{2}^-)$

Breit-Wigner mass = 1520 to 1555 (≈ 1535) MeV
Breit-Wigner full width = 100 to 250 (≈ 150) MeV
$p_{beam} = 0.76$ GeV/c $4\pi\lambda\!\!\!^{-2} = 22.5$ mb
Re(pole position) = 1495 to 1515 (≈ 1505) MeV
−2Im(pole position) = 90 to 250 (≈ 170) MeV

N(1535) DECAY MODES	Fraction (Γ_i/Γ)	p (MeV/c)
$N\pi$	35–55 %	467
$N\eta$	30–55 %	182
$N\pi\pi$	1–10 %	422
$\Delta\pi$	<1 %	242
$N\rho$	<4 %	†
$N(\pi\pi)^{I=0}_{S\text{-wave}}$	<3 %	–
$N(1440)\pi$	<7 %	†
$p\gamma$	0.15–0.35 %	481
$p\gamma$, helicity=1/2	0.15–0.35 %	481
$n\gamma$	0.004–0.29 %	480
$n\gamma$, helicity=1/2	0.004–0.29 %	480

$N(1650)\ S_{11}$

$I(J^P) = \frac{1}{2}(\frac{1}{2}^-)$

Breit-Wigner mass = 1640 to 1680 (≈ 1650) MeV
Breit-Wigner full width = 145 to 190 (≈ 150) MeV
$p_{beam} = 0.96$ GeV/c $4\pi\lambda\!\!\!^{-2} = 16.4$ mb
Re(pole position) = 1640 to 1680 (≈ 1660) MeV
−2Im(pole position) = 150 to 170 (≈ 160) MeV

N(1650) DECAY MODES	Fraction (Γ_i/Γ)	p (MeV/c)
$N\pi$	55–90 %	547
$N\eta$	3–10 %	346
ΛK	3–11 %	161
$N\pi\pi$	10–20 %	511
$\Delta\pi$	1–7 %	344
$N\rho$	4–12 %	†
$N(\pi\pi)^{I=0}_{S\text{-wave}}$	<4 %	–
$N(1440)\pi$	<5 %	147
$p\gamma$	0.04–0.18 %	558
$p\gamma$, helicity=1/2	0.04–0.18 %	558
$n\gamma$	0.003–0.17 %	557
$n\gamma$, helicity=1/2	0.003–0.17 %	557

$N(1675)\ D_{15}$

$I(J^P) = \frac{1}{2}(\frac{5}{2}^-)$

Breit-Wigner mass = 1670 to 1685 (≈ 1675) MeV
Breit-Wigner full width = 140 to 180 (≈ 150) MeV
$p_{beam} = 1.01$ GeV/c $4\pi\lambda\!\!\!^{-2} = 15.4$ mb
Re(pole position) = 1655 to 1665 (≈ 1660) MeV
−2Im(pole position) = 125 to 155 (≈ 140) MeV

N(1675) DECAY MODES	Fraction (Γ_i/Γ)	p (MeV/c)
$N\pi$	40–50 %	563
ΛK	<1 %	209
$N\pi\pi$	50–60 %	529
$\Delta\pi$	50–60 %	364
$N\rho$	< 1–3 %	†
$p\gamma$	0.004–0.023 %	575
$p\gamma$, helicity=1/2	0.0–0.015 %	575
$p\gamma$, helicity=3/2	0.0–0.011 %	575
$n\gamma$	0.02–0.12 %	574
$n\gamma$, helicity=1/2	0.006–0.046 %	574
$n\gamma$, helicity=3/2	0.01–0.08 %	574

$N(1680)\ F_{15}$

$I(J^P) = \frac{1}{2}(\frac{5}{2}^+)$

Breit-Wigner mass = 1675 to 1690 (≈ 1680) MeV
Breit-Wigner full width = 120 to 140 (≈ 130) MeV
$p_{beam} = 1.01$ GeV/c $4\pi\lambda\!\!\!^{-2} = 15.2$ mb
Re(pole position) = 1665 to 1675 (≈ 1670) MeV
−2Im(pole position) = 105 to 135 (≈ 120) MeV

N(1680) DECAY MODES	Fraction (Γ_i/Γ)	p (MeV/c)
$N\pi$	60–70 %	567
$N\pi\pi$	30–40 %	532
$\Delta\pi$	5–15 %	369
$N\rho$	3–15 %	†
$N(\pi\pi)^{I=0}_{S\text{-wave}}$	5–20 %	–
$p\gamma$	0.21–0.32 %	578
$p\gamma$, helicity=1/2	0.001–0.011 %	578
$p\gamma$, helicity=3/2	0.20–0.32 %	578
$n\gamma$	0.021–0.046 %	577
$n\gamma$, helicity=1/2	0.004–0.029 %	577
$n\gamma$, helicity=3/2	0.01–0.024 %	577

$N(1700)\ D_{13}$

$I(J^P) = \frac{1}{2}(\frac{3}{2}^-)$

Breit-Wigner mass = 1650 to 1750 (≈ 1700) MeV
Breit-Wigner full width = 50 to 150 (≈ 100) MeV
p_{beam} = 1.05 GeV/c $4\pi\lambda\!\!\!^-{}^2$ = 14.5 mb
Re(pole position) = 1630 to 1730 (≈ 1680) MeV
−2Im(pole position) = 50 to 150 (≈ 100) MeV

N(1700) DECAY MODES	Fraction (Γ_i/Γ)	p (MeV/c)
$N\pi$	5–15 %	580
ΛK	<3 %	250
$N\pi\pi$	85–95 %	547
$N\rho$	<35 %	†
$p\gamma$	0.01–0.05 %	591
$p\gamma$, helicity=1/2	0.0–0.024 %	591
$p\gamma$, helicity=3/2	0.002–0.026 %	591
$n\gamma$	0.01–0.13 %	590
$n\gamma$, helicity=1/2	0.0–0.09 %	590
$n\gamma$, helicity=3/2	0.01–0.05 %	590

$N(1710)\ P_{11}$

$I(J^P) = \frac{1}{2}(\frac{1}{2}^+)$

Breit-Wigner mass = 1680 to 1740 (≈ 1710) MeV
Breit-Wigner full width = 50 to 250 (≈ 100) MeV
p_{beam} = 1.07 GeV/c $4\pi\lambda\!\!\!^-{}^2$ = 14.2 mb
Re(pole position) = 1670 to 1770 (≈ 1720) MeV
−2Im(pole position) = 80 to 380 (≈ 230) MeV

N(1710) DECAY MODES	Fraction (Γ_i/Γ)	p (MeV/c)
$N\pi$	10–20 %	587
ΛK	5–25 %	264
$N\pi\pi$	40–90 %	554
$\Delta\pi$	15–40 %	393
$N\rho$	5–25 %	48
$N(\pi\pi)^{I=0}_{S\text{-wave}}$	10–40 %	–
$p\gamma$	0.002–0.05%	598
$p\gamma$, helicity=1/2	0.002–0.05%	598
$n\gamma$	0.0–0.02%	597
$n\gamma$, helicity=1/2	0.0–0.02%	597

$N(1720)\ P_{13}$

$I(J^P) = \frac{1}{2}(\frac{3}{2}^+)$

Breit-Wigner mass = 1650 to 1750 (≈ 1720) MeV
Breit-Wigner full width = 100 to 200 (≈ 150) MeV
p_{beam} = 1.09 GeV/c $4\pi\lambda\!\!\!^-{}^2$ = 13.9 mb
Re(pole position) = 1650 to 1750 (≈ 1700) MeV
−2Im(pole position) = 110 to 390 (≈ 250) MeV

N(1720) DECAY MODES	Fraction (Γ_i/Γ)	p (MeV/c)
$N\pi$	10–20 %	594
ΛK	1–15 %	278
$N\pi\pi$	>70 %	561
$N\rho$	70–85 %	104
$p\gamma$	0.003–0.10 %	604
$p\gamma$, helicity=1/2	0.003–0.08 %	604
$p\gamma$, helicity=3/2	0.001–0.03 %	604
$n\gamma$	0.002–0.39 %	603
$n\gamma$, helicity=1/2	0.0–0.002 %	603
$n\gamma$, helicity=3/2	0.001–0.39 %	603

$N(2190)\ G_{17}$

$I(J^P) = \frac{1}{2}(\frac{7}{2}^-)$

Breit-Wigner mass = 2100 to 2200 (≈ 2190) MeV
Breit-Wigner full width = 350 to 550 (≈ 450) MeV
p_{beam} = 2.07 GeV/c $4\pi\lambda\!\!\!^-{}^2$ = 6.21 mb
Re(pole position) = 1950 to 2150 (≈ 2050) MeV
−2Im(pole position) = 350 to 550 (≈ 450) MeV

N(2190) DECAY MODES	Fraction (Γ_i/Γ)	p (MeV/c)
$N\pi$	10–20 %	888

$N(2220)\ H_{19}$

$I(J^P) = \frac{1}{2}(\frac{9}{2}^+)$

Breit-Wigner mass = 2180 to 2310 (≈ 2220) MeV
Breit-Wigner full width = 320 to 550 (≈ 400) MeV
p_{beam} = 2.14 GeV/c $4\pi\lambda\!\!\!^-{}^2$ = 5.97 mb
Re(pole position) = 2100 to 2240 (≈ 2170) MeV
−2Im(pole position) = 370 to 570 (≈ 470) MeV

N(2220) DECAY MODES	Fraction (Γ_i/Γ)	p (MeV/c)
$N\pi$	10–20 %	905

$N(2250)\ G_{19}$

$I(J^P) = \frac{1}{2}(\frac{9}{2}^-)$

Breit-Wigner mass = 2170 to 2310 (≈ 2250) MeV
Breit-Wigner full width = 290 to 470 (≈ 400) MeV
p_{beam} = 2.21 GeV/c $4\pi\lambda\!\!\!^-{}^2$ = 5.74 mb
Re(pole position) = 2080 to 2200 (≈ 2140) MeV
−2Im(pole position) = 280 to 680 (≈ 480) MeV

N(2250) DECAY MODES	Fraction (Γ_i/Γ)	p (MeV/c)
$N\pi$	5–15 %	923

$N(2600)\ I_{1,11}$

$I(J^P) = \frac{1}{2}(\frac{11}{2}^-)$

Breit-Wigner mass = 2550 to 2750 (≈ 2600) MeV
Breit-Wigner full width = 500 to 800 (≈ 650) MeV
p_{beam} = 3.12 GeV/c $4\pi\lambda\!\!\!^-{}^2$ = 3.86 mb

N(2600) DECAY MODES	Fraction (Γ_i/Γ)	p (MeV/c)
$N\pi$	5–10 %	1126

Δ BARYONS
($S = 0$, $I = 3/2$)

$\Delta^{++} = uuu$, $\Delta^{+} = uud$, $\Delta^{0} = udd$, $\Delta^{-} = ddd$

$\Delta(1232)\ P_{33}$

$I(J^P) = \frac{3}{2}(\frac{3}{2}^+)$

Breit-Wigner mass (mixed charges) = 1230 to 1234 (≈ 1232) MeV
Breit-Wigner full width (mixed charges) = 115 to 125 (≈ 120) MeV
p_{beam} = 0.30 GeV/c $4\pi\lambda\!\!\!^-{}^2$ = 94.8 mb
Re(pole position) = 1209 to 1211 (≈ 1210) MeV
−2Im(pole position) = 98 to 102 (≈ 100) MeV

Δ(1232) DECAY MODES	Fraction (Γ_i/Γ)	p (MeV/c)
$N\pi$	>99 %	227
$N\gamma$	0.52–0.60 %	259
$N\gamma$, helicity=1/2	0.11–0.13 %	259
$N\gamma$, helicity=3/2	0.41–0.47 %	259

$\Delta(1600)\ P_{33}$

$I(J^P) = \frac{3}{2}(\frac{3}{2}^+)$

Breit-Wigner mass = 1550 to 1700 (≈ 1600) MeV
Breit-Wigner full width = 250 to 450 (≈ 350) MeV
p_{beam} = 0.87 GeV/c $4\pi\lambda\!\!\!^-{}^2$ = 18.6 mb
Re(pole position) = 1500 to 1700 (≈ 1600) MeV
−2Im(pole position) = 200 to 400 (≈ 300) MeV

Δ(1600) DECAY MODES	Fraction (Γ_i/Γ)	p (MeV/c)
$N\pi$	10–25 %	512
$N\pi\pi$	75–90 %	473
$\Delta\pi$	40–70 %	301
$N\rho$	<25 %	†
$N(1440)\pi$	10–35 %	74
$N\gamma$	0.001–0.02 %	525
$N\gamma$, helicity=1/2	0.0–0.02 %	525
$N\gamma$, helicity=3/2	0.001–0.005 %	525

$\Delta(1620)\ S_{31}$ $\quad I(J^P) = \frac{3}{2}(\frac{1}{2}^-)$

Breit-Wigner mass = 1615 to 1675 (≈ 1620) MeV
Breit-Wigner full width = 120 to 180 (≈ 150) MeV
$p_{beam} = 0.91$ GeV/c $\quad 4\pi\lambda\!\!\!^{-2} = 17.7$ mb
Re(pole position) = 1580 to 1620 (≈ 1600) MeV
−2Im(pole position) = 100 to 130 (≈ 115) MeV

Δ(1620) DECAY MODES	Fraction (Γ_i/Γ)	p (MeV/c)
$N\pi$	20–30 %	526
$N\pi\pi$	70–80 %	488
$\Delta\pi$	30–60 %	318
$N\rho$	7–25 %	†
$N\gamma$	0.004–0.044 %	538
$N\gamma$, helicity=1/2	0.004–0.044 %	538

$\Delta(1700)\ D_{33}$ $\quad I(J^P) = \frac{3}{2}(\frac{3}{2}^-)$

Breit-Wigner mass = 1670 to 1770 (≈ 1700) MeV
Breit-Wigner full width = 200 to 400 (≈ 300) MeV
$p_{beam} = 1.05$ GeV/c $\quad 4\pi\lambda\!\!\!^{-2} = 14.5$ mb
Re(pole position) = 1620 to 1700 (≈ 1660) MeV
−2Im(pole position) = 150 to 250 (≈ 200) MeV

Δ(1700) DECAY MODES	Fraction (Γ_i/Γ)	p (MeV/c)
$N\pi$	10–20 %	580
$N\pi\pi$	80–90 %	547
$\Delta\pi$	30–60 %	385
$N\rho$	30–55 %	†
$N\gamma$	0.12–0.26 %	591
$N\gamma$, helicity=1/2	0.08–0.16 %	591
$N\gamma$, helicity=3/2	0.025–0.12 %	591

$\Delta(1905)\ F_{35}$ $\quad I(J^P) = \frac{3}{2}(\frac{5}{2}^+)$

Breit-Wigner mass = 1870 to 1920 (≈ 1905) MeV
Breit-Wigner full width = 280 to 440 (≈ 350) MeV
$p_{beam} = 1.45$ GeV/c $\quad 4\pi\lambda\!\!\!^{-2} = 9.62$ mb
Re(pole position) = 1800 to 1860 (≈ 1830) MeV
−2Im(pole position) = 230 to 330 (≈ 280) MeV

Δ(1905) DECAY MODES	Fraction (Γ_i/Γ)	p (MeV/c)
$N\pi$	5–15 %	713
$N\pi\pi$	85–95 %	687
$\Delta\pi$	<25 %	542
$N\rho$	>60 %	421
$N\gamma$	0.01–0.03 %	721
$N\gamma$, helicity=1/2	0.0–0.1 %	721
$N\gamma$, helicity=3/2	0.004–0.03 %	721

$\Delta(1910)\ P_{31}$ $\quad I(J^P) = \frac{3}{2}(\frac{1}{2}^+)$

Breit-Wigner mass = 1870 to 1920 (≈ 1910) MeV
Breit-Wigner full width = 190 to 270 (≈ 250) MeV
$p_{beam} = 1.46$ GeV/c $\quad 4\pi\lambda\!\!\!^{-2} = 9.54$ mb
Re(pole position) = 1830 to 1880 (≈ 1855) MeV
−2Im(pole position) = 200 to 500 (≈ 350) MeV

Δ(1910) DECAY MODES	Fraction (Γ_i/Γ)	p (MeV/c)
$N\pi$	15–30 %	716
$N\gamma$	0.0–0.2 %	725
$N\gamma$, helicity=1/2	0.0–0.2 %	725

$\Delta(1920)\ P_{33}$ $\quad I(J^P) = \frac{3}{2}(\frac{3}{2}^+)$

Breit-Wigner mass = 1900 to 1970 (≈ 1920) MeV
Breit-Wigner full width = 150 to 300 (≈ 200) MeV
$p_{beam} = 1.48$ GeV/c $\quad 4\pi\lambda\!\!\!^{-2} = 9.37$ mb
Re(pole position) = 1850 to 1950 (≈ 1900) MeV
−2Im(pole position) = 200 to 400 (≈ 300) MeV

Δ(1920) DECAY MODES	Fraction (Γ_i/Γ)	p (MeV/c)
$N\pi$	5–20 %	722

$\Delta(1930)\ D_{35}$ $\quad I(J^P) = \frac{3}{2}(\frac{5}{2}^-)$

Breit-Wigner mass = 1920 to 1970 (≈ 1930) MeV
Breit-Wigner full width = 250 to 450 (≈ 350) MeV
$p_{beam} = 1.50$ GeV/c $\quad 4\pi\lambda\!\!\!^{-2} = 9.21$ mb
Re(pole position) = 1840 to 1940 (≈ 1890) MeV
−2Im(pole position) = 200 to 300 (≈ 250) MeV

Δ(1930) DECAY MODES	Fraction (Γ_i/Γ)	p (MeV/c)
$N\pi$	10–20 %	729
$N\gamma$	0.0–0.02 %	737
$N\gamma$, helicity=1/2	0.0–0.01 %	737
$N\gamma$, helicity=3/2	0.0–0.01 %	737

$\Delta(1950)\ F_{37}$ $\quad I(J^P) = \frac{3}{2}(\frac{7}{2}^+)$

Breit-Wigner mass = 1940 to 1960 (≈ 1950) MeV
Breit-Wigner full width = 290 to 350 (≈ 300) MeV
$p_{beam} = 1.54$ GeV/c $\quad 4\pi\lambda\!\!\!^{-2} = 8.91$ mb
Re(pole position) = 1880 to 1890 (≈ 1885) MeV
−2Im(pole position) = 210 to 270 (≈ 240) MeV

Δ(1950) DECAY MODES	Fraction (Γ_i/Γ)	p (MeV/c)
$N\pi$	35–40 %	741
$N\pi\pi$		716
$\Delta\pi$	20–30 %	574
$N\rho$	<10 %	469
$N\gamma$	0.08–0.13 %	749
$N\gamma$, helicity=1/2	0.03–0.055 %	749
$N\gamma$, helicity=3/2	0.05–0.075 %	749

$\Delta(2420)\ H_{3,11}$ $\quad I(J^P) = \frac{3}{2}(\frac{11}{2}^+)$

Breit-Wigner mass = 2300 to 2500 (≈ 2420) MeV
Breit-Wigner full width = 300 to 500 (≈ 400) MeV
$p_{beam} = 2.64$ GeV/c $\quad 4\pi\lambda\!\!\!^{-2} = 4.68$ mb
Re(pole position) = 2260 to 2400 (≈ 2330) MeV
−2Im(pole position) = 350 to 750 (≈ 550) MeV

Δ(2420) DECAY MODES	Fraction (Γ_i/Γ)	p (MeV/c)
$N\pi$	5–15 %	1023

Λ BARYONS
$(S = -1, I = 0)$

$\Lambda^0 = uds$

Λ

$I(J^P) = 0(\frac{1}{2}^+)$

Mass $m = 1115.683 \pm 0.006$ MeV
Mean life $\tau = (2.632 \pm 0.020) \times 10^{-10}$ s (S = 1.6)
$c\tau = 7.89$ cm
Magnetic moment $\mu = -0.613 \pm 0.004\ \mu_N$
Electric dipole moment $d < 1.5 \times 10^{-16}$ ecm, CL = 95%

Decay parameters

$p\pi^-$	$\alpha_- = 0.642 \pm 0.013$
"	$\phi_- = (-6.5 \pm 3.5)°$
"	$\gamma_- = 0.76$ [g]
"	$\Delta_- = (8 \pm 4)°$ [g]
$n\pi^0$	$\alpha_0 = +0.65 \pm 0.05$
$pe^-\overline{\nu}_e$	$g_A/g_V = -0.718 \pm 0.015$ [e]

Λ DECAY MODES	Fraction (Γ_i/Γ)	p (MeV/c)
$p\pi^-$	(63.9 ±0.5) %	101
$n\pi^0$	(35.8 ±0.5) %	104
$n\gamma$	(1.75±0.15) × 10^{-3}	162
$p\pi^-\gamma$	[h] (8.4 ±1.4) × 10^{-4}	101
$pe^-\overline{\nu}_e$	(8.32±0.14) × 10^{-4}	163
$p\mu^-\overline{\nu}_\mu$	(1.57±0.35) × 10^{-4}	131

Λ(1405) S_{01}

$I(J^P) = 0(\frac{1}{2}^-)$

Mass $m = 1407 \pm 4$ MeV
Full width $\Gamma = 50.0 \pm 2.0$ MeV
Below $\overline{K}N$ threshold

Λ(1405) DECAY MODES	Fraction (Γ_i/Γ)	p (MeV/c)
$\Sigma\pi$	100 %	152

Λ(1520) D_{03}

$I(J^P) = 0(\frac{3}{2}^-)$

Mass $m = 1519.5 \pm 1.0$ MeV [i]
Full width $\Gamma = 15.6 \pm 1.0$ MeV [i]
$p_{beam} = 0.39$ GeV/c $4\pi\lambda\!\!\!^{-2} = 82.8$ mb

Λ(1520) DECAY MODES	Fraction (Γ_i/Γ)	p (MeV/c)
$N\overline{K}$	45 ± 1%	244
$\Sigma\pi$	42 ± 1%	267
$\Lambda\pi\pi$	10 ± 1%	252
$\Sigma\pi\pi$	0.9 ± 0.1%	152
$\Lambda\gamma$	0.8 ± 0.2%	351

Λ(1600) P_{01}

$I(J^P) = 0(\frac{1}{2}^+)$

Mass $m = 1560$ to 1700 (≈ 1600) MeV
Full width $\Gamma = 50$ to 250 (≈ 150) MeV
$p_{beam} = 0.58$ GeV/c $4\pi\lambda\!\!\!^{-2} = 41.6$ mb

Λ(1600) DECAY MODES	Fraction (Γ_i/Γ)	p (MeV/c)
$N\overline{K}$	15–30 %	343
$\Sigma\pi$	10–60 %	336

Λ(1670) S_{01}

$I(J^P) = 0(\frac{1}{2}^-)$

Mass $m = 1660$ to 1680 (≈ 1670) MeV
Full width $\Gamma = 25$ to 50 (≈ 35) MeV
$p_{beam} = 0.74$ GeV/c $4\pi\lambda\!\!\!^{-2} = 28.5$ mb

Λ(1670) DECAY MODES	Fraction (Γ_i/Γ)	p (MeV/c)
$N\overline{K}$	15–25 %	414
$\Sigma\pi$	20–60 %	393
$\Lambda\eta$	15–35 %	64

Λ(1690) D_{03}

$I(J^P) = 0(\frac{3}{2}^-)$

Mass $m = 1685$ to 1695 (≈ 1690) MeV
Full width $\Gamma = 50$ to 70 (≈ 60) MeV
$p_{beam} = 0.78$ GeV/c $4\pi\lambda\!\!\!^{-2} = 26.1$ mb

Λ(1690) DECAY MODES	Fraction (Γ_i/Γ)	p (MeV/c)
$N\overline{K}$	20–30 %	433
$\Sigma\pi$	20–40 %	409
$\Lambda\pi\pi$	~ 25 %	415
$\Sigma\pi\pi$	~ 20 %	350

Λ(1800) S_{01}

$I(J^P) = 0(\frac{1}{2}^-)$

Mass $m = 1720$ to 1850 (≈ 1800) MeV
Full width $\Gamma = 200$ to 400 (≈ 300) MeV
$p_{beam} = 1.01$ GeV/c $4\pi\lambda\!\!\!^{-2} = 17.5$ mb

Λ(1800) DECAY MODES	Fraction (Γ_i/Γ)	p (MeV/c)
$N\overline{K}$	25–40 %	528
$\Sigma\pi$	seen	493
$\Sigma(1385)\pi$	seen	345
$N\overline{K}^*(892)$	seen	†

Λ(1810) P_{01}

$I(J^P) = 0(\frac{1}{2}^+)$

Mass $m = 1750$ to 1850 (≈ 1810) MeV
Full width $\Gamma = 50$ to 250 (≈ 150) MeV
$p_{beam} = 1.04$ GeV/c $4\pi\lambda\!\!\!^{-2} = 17.0$ mb

Λ(1810) DECAY MODES	Fraction (Γ_i/Γ)	p (MeV/c)
$N\overline{K}$	20–50 %	537
$\Sigma\pi$	10–40 %	501
$\Sigma(1385)\pi$	seen	356
$N\overline{K}^*(892)$	30–60 %	†

Λ(1820) F_{05}

$I(J^P) = 0(\frac{5}{2}^+)$

Mass $m = 1815$ to 1825 (≈ 1820) MeV
Full width $\Gamma = 70$ to 90 (≈ 80) MeV
$p_{beam} = 1.06$ GeV/c $4\pi\lambda\!\!\!^{-2} = 16.5$ mb

Λ(1820) DECAY MODES	Fraction (Γ_i/Γ)	p (MeV/c)
$N\overline{K}$	55–65 %	545
$\Sigma\pi$	8–14 %	508
$\Sigma(1385)\pi$	5–10 %	362

Λ(1830) D_{05}

$I(J^P) = 0(\frac{5}{2}^-)$

Mass $m = 1810$ to 1830 (≈ 1830) MeV
Full width $\Gamma = 60$ to 110 (≈ 95) MeV
$p_{beam} = 1.08$ GeV/c $4\pi\lambda\!\!\!^{-2} = 16.0$ mb

Λ(1830) DECAY MODES	Fraction (Γ_i/Γ)	p (MeV/c)
$N\overline{K}$	3–10 %	553
$\Sigma\pi$	35–75 %	515
$\Sigma(1385)\pi$	>15 %	371

$\Lambda(1890)\ P_{03}$

$I(J^P) = 0(\frac{3}{2}^+)$

Mass m = 1850 to 1910 ($\approx$ 1890) MeV
Full width Γ = 60 to 200 ($\approx$ 100) MeV
p_{beam} = 1.21 GeV/c $4\pi\lambda\!\!\!^{-2}$ = 13.6 mb

Λ(1890) DECAY MODES	Fraction (Γ_i/Γ)	p (MeV/c)
$N\overline{K}$	20–35 %	599
$\Sigma\pi$	3–10 %	559
$\Sigma(1385)\pi$	seen	420
$N\overline{K}^*(892)$	seen	233

$\Lambda(2100)\ G_{07}$

$I(J^P) = 0(\frac{7}{2}^-)$

Mass m = 2090 to 2110 ($\approx$ 2100) MeV
Full width Γ = 100 to 250 ($\approx$ 200) MeV
p_{beam} = 1.68 GeV/c $4\pi\lambda\!\!\!^{-2}$ = 8.68 mb

Λ(2100) DECAY MODES	Fraction (Γ_i/Γ)	p (MeV/c)
$N\overline{K}$	25–35 %	751
$\Sigma\pi$	~ 5 %	704
$\Lambda\eta$	<3 %	617
ΞK	<3 %	483
$\Lambda\omega$	<8 %	443
$N\overline{K}^*(892)$	10–20 %	514

$\Lambda(2110)\ F_{05}$

$I(J^P) = 0(\frac{5}{2}^+)$

Mass m = 2090 to 2140 ($\approx$ 2110) MeV
Full width Γ = 150 to 250 ($\approx$ 200) MeV
p_{beam} = 1.70 GeV/c $4\pi\lambda\!\!\!^{-2}$ = 8.53 mb

Λ(2110) DECAY MODES	Fraction (Γ_i/Γ)	p (MeV/c)
$N\overline{K}$	5–25 %	757
$\Sigma\pi$	10–40 %	711
$\Lambda\omega$	seen	455
$\Sigma(1385)\pi$	seen	589
$N\overline{K}^*(892)$	10–60 %	524

$\Lambda(2350)\ H_{09}$

$I(J^P) = 0(\frac{9}{2}^+)$

Mass m = 2340 to 2370 ($\approx$ 2350) MeV
Full width Γ = 100 to 250 ($\approx$ 150) MeV
p_{beam} = 2.29 GeV/c $4\pi\lambda\!\!\!^{-2}$ = 5.85 mb

Λ(2350) DECAY MODES	Fraction (Γ_i/Γ)	p (MeV/c)
$N\overline{K}$	~ 12 %	915
$\Sigma\pi$	~ 10 %	867

Σ BARYONS
($S = -1$, $I = 1$)

$\Sigma^+ = uus$, $\Sigma^0 = uds$, $\Sigma^- = dds$

Σ^+

$I(J^P) = 1(\frac{1}{2}^+)$

Mass m = 1189.37 $\pm$ 0.07 MeV (S = 2.2)
Mean life τ = (0.799 $\pm$ 0.004) $\times 10^{-10}$ s
$c\tau$ = 2.396 cm
Magnetic moment μ = 2.458 $\pm$ 0.010 μ_N (S = 2.1)
$\Gamma(\Sigma^+ \to n\ell^+\nu)/\Gamma(\Sigma^- \to n\ell^-\overline{\nu})$ < 0.043

Decay parameters

$p\pi^0$	$\alpha_0 = -0.980^{+0.017}_{-0.015}$
"	$\phi_0 = (36 \pm 34)^\circ$
"	$\gamma_0 = 0.16$ [g]
"	$\Delta_0 = (187 \pm 6)^\circ$ [g]
$n\pi^+$	$\alpha_+ = 0.068 \pm 0.013$
"	$\phi_+ = (167 \pm 20)^\circ$ (S = 1.1)
"	$\gamma_+ = -0.97$ [g]
"	$\Delta_+ = (-73^{+133}_{-10})^\circ$ [g]
$p\gamma$	$\alpha_\gamma = -0.76 \pm 0.08$

Σ^+ DECAY MODES		Fraction (Γ_i/Γ)	Confidence level	p (MeV/c)
$p\pi^0$		(51.57$\pm$0.30) %		189
$n\pi^+$		(48.31$\pm$0.30) %		185
$p\gamma$		(1.23$\pm$0.05) $\times 10^{-3}$		225
$n\pi^+\gamma$	[h]	(4.5 $\pm$0.5) $\times 10^{-4}$		185
$\Lambda e^+\nu_e$		(2.0 $\pm$0.5) $\times 10^{-5}$		71
$\Delta S = \Delta Q$ (SQ) violating modes or $\Delta S = 1$ weak neutral current ($S1$) modes				
$ne^+\nu_e$	SQ	< 5 $\times 10^{-6}$	90%	224
$n\mu^+\nu_\mu$	SQ	< 3.0 $\times 10^{-5}$	90%	202
pe^+e^-	$S1$	< 7 $\times 10^{-6}$		225

Σ^0

$I(J^P) = 1(\frac{1}{2}^+)$

Mass m = 1192.642 $\pm$ 0.024 MeV
$m_{\Sigma^-} - m_{\Sigma^0}$ = 4.807 $\pm$ 0.035 MeV (S = 1.1)
$m_{\Sigma^0} - m_\Lambda$ = 76.959 $\pm$ 0.023 MeV
Mean life τ = (7.4 $\pm$ 0.7) $\times 10^{-20}$ s
$c\tau$ = 2.22 $\times 10^{-11}$ m
Transition magnetic moment $|\mu_{\Sigma\Lambda}|$ = 1.61 $\pm$ 0.08 μ_N

Σ^0 DECAY MODES		Fraction (Γ_i/Γ)	Confidence level	p (MeV/c)
$\Lambda\gamma$		100 %		74
$\Lambda\gamma\gamma$		< 3 %	90%	74
Λe^+e^-	[i]	5×10^{-3}		74

Σ^-

$I(J^P) = 1(\frac{1}{2}^+)$

Mass $m = 1197.449 \pm 0.030$ MeV (S = 1.2)
$m_{\Sigma^-} - m_{\Sigma^+} = 8.08 \pm 0.08$ MeV (S = 1.9)
$m_{\Sigma^-} - m_{\Lambda} = 81.766 \pm 0.030$ MeV (S = 1.2)
Mean life $\tau = (1.479 \pm 0.011) \times 10^{-10}$ s (S = 1.3)
$c\tau = 4.434$ cm
Magnetic moment $\mu = -1.160 \pm 0.025\ \mu_N$ (S = 1.7)

Decay parameters

$n\pi^-$	$\alpha_- = -0.068 \pm 0.008$
"	$\phi_- = (10 \pm 15)°$
"	$\gamma_- = 0.98$ [g]
"	$\Delta_- = (249^{+12}_{-120})°$ [g]
$ne^-\overline{\nu}_e$	$g_A/g_V = 0.340 \pm 0.017$ [e]
"	$f_2(0)/f_1(0) = 0.97 \pm 0.14$
"	$D = 0.11 \pm 0.10$
$\Lambda e^-\overline{\nu}_e$	$g_V/g_A = 0.01 \pm 0.10$ [e] (S = 1.5)
"	$g_{WM}/g_A = 2.4 \pm 1.7$ [e]

Σ^- DECAY MODES	Fraction (Γ_i/Γ)	p (MeV/c)
$n\pi^-$	(99.848±0.005) %	193
$n\pi^-\gamma$	[h] (4.6 ±0.6) $\times 10^{-4}$	193
$ne^-\overline{\nu}_e$	(1.017±0.034) $\times 10^{-3}$	230
$n\mu^-\overline{\nu}_\mu$	(4.5 ±0.4) $\times 10^{-4}$	210
$\Lambda e^-\overline{\nu}_e$	(5.73 ±0.27) $\times 10^{-5}$	79

$\Sigma(1385)\ P_{13}$

$I(J^P) = 1(\frac{3}{2}^+)$

$\Sigma(1385)^+$ mass $m = 1382.8 \pm 0.4$ MeV (S = 2.0)
$\Sigma(1385)^0$ mass $m = 1383.7 \pm 1.0$ MeV (S = 1.4)
$\Sigma(1385)^-$ mass $m = 1387.2 \pm 0.5$ MeV (S = 2.2)
$\Sigma(1385)^+$ full width $\Gamma = 35.8 \pm 0.8$ MeV
$\Sigma(1385)^0$ full width $\Gamma = 36 \pm 5$ MeV
$\Sigma(1385)^-$ full width $\Gamma = 39.4 \pm 2.1$ MeV (S = 1.7)
Below $\overline{K}N$ threshold

$\Sigma(1385)$ DECAY MODES	Fraction (Γ_i/Γ)	p (MeV/c)
$\Lambda\pi$	88±2 %	208
$\Sigma\pi$	12±2 %	127

$\Sigma(1660)\ P_{11}$

$I(J^P) = 1(\frac{1}{2}^+)$

Mass $m = 1630$ to 1690 (≈ 1660) MeV
Full width $\Gamma = 40$ to 200 (≈ 100) MeV
$p_{beam} = 0.72$ GeV/c $4\pi\lambda\!\!\!^-{}^2 = 29.9$ mb

$\Sigma(1660)$ DECAY MODES	Fraction (Γ_i/Γ)	p (MeV/c)
$N\overline{K}$	10–30 %	405
$\Lambda\pi$	seen	439
$\Sigma\pi$	seen	385

$\Sigma(1670)\ D_{13}$

$I(J^P) = 1(\frac{3}{2}^-)$

Mass $m = 1665$ to 1685 (≈ 1670) MeV
Full width $\Gamma = 40$ to 80 (≈ 60) MeV
$p_{beam} = 0.74$ GeV/c $4\pi\lambda\!\!\!^-{}^2 = 28.5$ mb

$\Sigma(1670)$ DECAY MODES	Fraction (Γ_i/Γ)	p (MeV/c)
$N\overline{K}$	7–13 %	414
$\Lambda\pi$	5–15 %	447
$\Sigma\pi$	30–60 %	393

$\Sigma(1750)\ S_{11}$

$I(J^P) = 1(\frac{1}{2}^-)$

Mass $m = 1730$ to 1800 (≈ 1750) MeV
Full width $\Gamma = 60$ to 160 (≈ 90) MeV
$p_{beam} = 0.91$ GeV/c $4\pi\lambda\!\!\!^-{}^2 = 20.7$ mb

$\Sigma(1750)$ DECAY MODES	Fraction (Γ_i/Γ)	p (MeV/c)
$N\overline{K}$	10–40 %	486
$\Lambda\pi$	seen	507
$\Sigma\pi$	<8 %	455
$\Sigma\eta$	15–55 %	81

$\Sigma(1775)\ D_{15}$

$I(J^P) = 1(\frac{5}{2}^-)$

Mass $m = 1770$ to 1780 (≈ 1775) MeV
Full width $\Gamma = 105$ to 135 (≈ 120) MeV
$p_{beam} = 0.96$ GeV/c $4\pi\lambda\!\!\!^-{}^2 = 19.0$ mb

$\Sigma(1775)$ DECAY MODES	Fraction (Γ_i/Γ)	p (MeV/c)
$N\overline{K}$	37–43%	508
$\Lambda\pi$	14–20%	525
$\Sigma\pi$	2–5%	474
$\Sigma(1385)\pi$	8–12%	324
$\Lambda(1520)\pi$	17–23%	198

$\Sigma(1915)\ F_{15}$

$I(J^P) = 1(\frac{5}{2}^+)$

Mass $m = 1900$ to 1935 (≈ 1915) MeV
Full width $\Gamma = 80$ to 160 (≈ 120) MeV
$p_{beam} = 1.26$ GeV/c $4\pi\lambda\!\!\!^-{}^2 = 12.8$ mb

$\Sigma(1915)$ DECAY MODES	Fraction (Γ_i/Γ)	p (MeV/c)
$N\overline{K}$	5–15 %	618
$\Lambda\pi$	seen	622
$\Sigma\pi$	seen	577
$\Sigma(1385)\pi$	<5 %	440

$\Sigma(1940)\ D_{13}$

$I(J^P) = 1(\frac{3}{2}^-)$

Mass $m = 1900$ to 1950 (≈ 1940) MeV
Full width $\Gamma = 150$ to 300 (≈ 220) MeV
$p_{beam} = 1.32$ GeV/c $4\pi\lambda\!\!\!^-{}^2 = 12.1$ mb

$\Sigma(1940)$ DECAY MODES	Fraction (Γ_i/Γ)	p (MeV/c)
$N\overline{K}$	<20 %	637
$\Lambda\pi$	seen	639
$\Sigma\pi$	seen	594
$\Sigma(1385)\pi$	seen	460
$\Lambda(1520)\pi$	seen	354
$\Delta(1232)\overline{K}$	seen	410
$N\overline{K}^*(892)$	seen	320

$\Sigma(2030)\ F_{17}$

$I(J^P) = 1(\frac{7}{2}^+)$

Mass $m = 2025$ to 2040 (≈ 2030) MeV
Full width $\Gamma = 150$ to 200 (≈ 180) MeV
$p_{beam} = 1.52$ GeV/c $4\pi\lambda\!\!\!^-{}^2 = 9.93$ mb

$\Sigma(2030)$ DECAY MODES	Fraction (Γ_i/Γ)	p (MeV/c)
$N\overline{K}$	17–23 %	702
$\Lambda\pi$	17–23 %	700
$\Sigma\pi$	5–10 %	657
ΞK	<2 %	412
$\Sigma(1385)\pi$	5–15 %	529
$\Lambda(1520)\pi$	10–20 %	430
$\Delta(1232)\overline{K}$	10–20 %	498
$N\overline{K}^*(892)$	<5 %	438

$\Sigma(2250)$

$I(J^P) = 1(?^?)$

Mass m = 2210 to 2280 (≈ 2250) MeV
Full width Γ = 60 to 150 (≈ 100) MeV
$p_{beam} = 2.04$ GeV/c $\quad 4\pi\lambda\!\!\!^{-2} = 6.76$ mb

$\Sigma(2250)$ DECAY MODES	Fraction (Γ_i/Γ)	p (MeV/c)
$N\overline{K}$	<10 %	851
$\Lambda\pi$	seen	842
$\Sigma\pi$	seen	803

Ξ BARYONS ($S=-2$, $I=1/2$)

$\Xi^0 = uss$, $\Xi^- = dss$

Ξ^0

$I(J^P) = \frac{1}{2}(\frac{1}{2}^+)$

P is not yet measured; + is the quark model prediction.

Mass $m = 1314.9 \pm 0.6$ MeV
$m_{\Xi^-} - m_{\Xi^0} = 6.4 \pm 0.6$ MeV
Mean life $\tau = (2.90 \pm 0.09) \times 10^{-10}$ s
$c\tau = 8.71$ cm
Magnetic moment $\mu = -1.250 \pm 0.014$ μ_N

Decay parameters

$\Lambda\pi^0$	$\alpha = -0.411 \pm 0.022$ (S = 2.1)
"	$\phi = (21 \pm 12)^\circ$
"	$\gamma = 0.85$ [g]
"	$\Delta = (218^{+12}_{-19})^\circ$ [g]
$\Lambda\gamma$	$\alpha = 0.4 \pm 0.4$
$\Sigma^0\gamma$	$\alpha = 0.20 \pm 0.32$

Ξ^0 DECAY MODES		Fraction (Γ_i/Γ)	Confidence level	p (MeV/c)
$\Lambda\pi^0$		(99.54 ± 0.05) %		135
$\Lambda\gamma$		$(1.06 \pm 0.16) \times 10^{-3}$		184
$\Sigma^0\gamma$		$(3.5 \pm 0.4) \times 10^{-3}$		117
$\Sigma^+ e^- \overline{\nu}_e$		$< 1.1 \times 10^{-3}$	90%	120
$\Sigma^+ \mu^- \overline{\nu}_\mu$		$< 1.1 \times 10^{-3}$	90%	64
$\Delta S = \Delta Q$ (SQ) violating modes or $\Delta S = 2$ forbidden ($S2$) modes				
$\Sigma^- e^+ \nu_e$	SQ	$< 9 \times 10^{-4}$	90%	112
$\Sigma^- \mu^+ \nu_\mu$	SQ	$< 9 \times 10^{-4}$	90%	49
$p\pi^-$	$S2$	$< 4 \times 10^{-5}$	90%	299
$p e^- \overline{\nu}_e$	$S2$	$< 1.3 \times 10^{-3}$		323
$p \mu^- \overline{\nu}_\mu$	$S2$	$< 1.3 \times 10^{-3}$		309

Ξ^-

$I(J^P) = \frac{1}{2}(\frac{1}{2}^+)$

P is not yet measured; + is the quark model prediction.

Mass $m = 1321.32 \pm 0.13$ MeV
Mean life $\tau = (1.639 \pm 0.015) \times 10^{-10}$ s
$c\tau = 4.91$ cm
Magnetic moment $\mu = -0.6507 \pm 0.0025$ μ_N

Decay parameters

$\Lambda\pi^-$	$\alpha = -0.456 \pm 0.014$ (S = 1.8)
"	$\phi = (4 \pm 4)^\circ$
"	$\gamma = 0.89$ [g]
"	$\Delta = (188 \pm 8)^\circ$ [g]
$\Lambda e^- \overline{\nu}_e$	$g_A/g_V = -0.25 \pm 0.05$ [e]

Ξ^- DECAY MODES		Fraction (Γ_i/Γ)	Confidence level	p (MeV/c)
$\Lambda\pi^-$		(99.887 ± 0.035) %		139
$\Sigma^-\gamma$		$(1.27 \pm 0.23) \times 10^{-4}$		118
$\Lambda e^- \overline{\nu}_e$		$(5.63 \pm 0.31) \times 10^{-4}$		190
$\Lambda \mu^- \overline{\nu}_\mu$		$(3.5 {}^{+3.5}_{-2.2}) \times 10^{-4}$		163
$\Sigma^0 e^- \overline{\nu}_e$		$(8.7 \pm 1.7) \times 10^{-5}$		122
$\Sigma^0 \mu^- \overline{\nu}_\mu$		$< 8 \times 10^{-4}$	90%	70
$\Xi^0 e^- \overline{\nu}_e$		$< 2.3 \times 10^{-3}$	90%	6
$\Delta S = 2$ forbidden ($S2$) modes				
$n\pi^-$	$S2$	$< 1.9 \times 10^{-5}$	90%	303
$n e^- \overline{\nu}_e$	$S2$	$< 3.2 \times 10^{-3}$	90%	327
$n \mu^- \overline{\nu}_\mu$	$S2$	< 1.5 %	90%	314
$p\pi^-\pi^-$	$S2$	$< 4 \times 10^{-4}$	90%	223
$p\pi^- e^- \overline{\nu}_e$	$S2$	$< 4 \times 10^{-4}$	90%	304
$p\pi^- \mu^- \overline{\nu}_\mu$	$S2$	$< 4 \times 10^{-4}$	90%	250
$p\mu^-\mu^-$	L	$< 4 \times 10^{-4}$	90%	272

$\Xi(1530)$ P_{13}

$I(J^P) = \frac{1}{2}(\frac{3}{2}^+)$

$\Xi(1530)^0$ mass $m = 1531.80 \pm 0.32$ MeV (S = 1.3)
$\Xi(1530)^-$ mass $m = 1535.0 \pm 0.6$ MeV
$\Xi(1530)^0$ full width $\Gamma = 9.1 \pm 0.5$ MeV
$\Xi(1530)^-$ full width $\Gamma = 9.9^{+1.7}_{-1.9}$ MeV

$\Xi(1530)$ DECAY MODES	Fraction (Γ_i/Γ)	Confidence level	p (MeV/c)
$\Xi\pi$	100 %		152
$\Xi\gamma$	<4 %	90%	200

$\Xi(1690)$

$I(J^P) = \frac{1}{2}(?^?)$

Mass $m = 1690 \pm 10$ MeV [l]
Full width $\Gamma < 50$ MeV

$\Xi(1690)$ DECAY MODES	Fraction (Γ_i/Γ)	p (MeV/c)
$\Lambda\overline{K}$	seen	240
$\Sigma\overline{K}$	seen	51
$\Xi^-\pi^+\pi^-$	possibly seen	214

$\Xi(1820)$ D_{13}

$I(J^P) = \frac{1}{2}(\frac{3}{2}^-)$

Mass $m = 1823 \pm 5$ MeV [l]
Full width $\Gamma = 24^{+15}_{-10}$ MeV [l]

$\Xi(1820)$ DECAY MODES	Fraction (Γ_i/Γ)	p (MeV/c)
$\Lambda\overline{K}$	large	400
$\Sigma\overline{K}$	small	320
$\Xi\pi$	small	413
$\Xi(1530)\pi$	small	234

$\Xi(1950)$

$I(J^P) = \frac{1}{2}(?^?)$

Mass $m = 1950 \pm 15$ MeV [l]
Full width $\Gamma = 60 \pm 20$ MeV [l]

$\Xi(1950)$ DECAY MODES	Fraction (Γ_i/Γ)	p (MeV/c)
$\Lambda\overline{K}$	seen	522
$\Sigma\overline{K}$	possibly seen	460
$\Xi\pi$	seen	518

$\Xi(2030)$

$I(J^P) = \frac{1}{2}(\geq \frac{5}{2}^?)$

Mass $m = 2025 \pm 5$ MeV [l]
Full width $\Gamma = 20^{+15}_{-5}$ MeV [l]

$\Xi(2030)$ DECAY MODES	Fraction (Γ_i/Γ)	p (MeV/c)
$\Lambda\overline{K}$	~ 20 %	589
$\Sigma\overline{K}$	~ 80 %	533
$\Xi\pi$	small	573
$\Xi(1530)\pi$	small	421
$\Lambda\overline{K}\pi$	small	501
$\Sigma\overline{K}\pi$	small	430

Ω BARYONS
$(S=-3, I=0)$

$\Omega^- = sss$

Ω^-

$I(J^P) = 0(\frac{3}{2}^+)$

J^P is not yet measured; $\frac{3}{2}^+$ is the quark model prediction.

Mass $m = 1672.45 \pm 0.29$ MeV
Mean life $\tau = (0.822 \pm 0.012) \times 10^{-10}$ s
$c\tau = 2.46$ cm
Magnetic moment $\mu = -2.02 \pm 0.05\ \mu_N$

Decay parameters

ΛK^-	$\alpha = -0.026 \pm 0.026$
$\Xi^0 \pi^-$	$\alpha = 0.09 \pm 0.14$
$\Xi^- \pi^0$	$\alpha = 0.05 \pm 0.21$

Ω^- DECAY MODES		Fraction (Γ_i/Γ)	Confidence level	p (MeV/c)
ΛK^-		(67.8 ± 0.7) %		211
$\Xi^0 \pi^-$		(23.6 ± 0.7) %		294
$\Xi^- \pi^0$		(8.6 ± 0.4) %		290
$\Xi^- \pi^+ \pi^-$		$(4.3^{+3.4}_{-1.3}) \times 10^{-4}$		190
$\Xi(1530)^0 \pi^-$		$(6.4^{+5.1}_{-2.0}) \times 10^{-4}$		17
$\Xi^0 e^- \overline{\nu}_e$		$(5.6 \pm 2.8) \times 10^{-3}$		319
$\Xi^- \gamma$		$< 4.6 \times 10^{-4}$	90%	314
$\Delta S = 2$ forbidden ($S2$) modes				
$\Lambda \pi^-$	$S2$	$< 1.9 \times 10^{-4}$	90%	449

$\Omega(2250)^-$

$I(J^P) = 0(?^?)$

Mass $m = 2252 \pm 9$ MeV
Full width $\Gamma = 55 \pm 18$ MeV

$\Omega(2250)^-$ DECAY MODES	Fraction (Γ_i/Γ)	p (MeV/c)
$\Xi^- \pi^+ K^-$	seen	531
$\Xi(1530)^0 K^-$	seen	437

CHARMED BARYONS
$(C=+1)$

$\Lambda_c^+ = udc$, $\Sigma_c^{++} = uuc$, $\Sigma_c^+ = udc$, $\Sigma_c^0 = ddc$,
$\Xi_c^+ = usc$, $\Xi_c^0 = dsc$, $\Omega_c^0 = ssc$

Λ_c^+

$I(J^P) = 0(\frac{1}{2}^+)$

J not confirmed; $\frac{1}{2}$ is the quark model prediction.

Mass $m = 2284.9 \pm 0.6$ MeV
Mean life $\tau = (0.206 \pm 0.012) \times 10^{-12}$ s
$c\tau = 61.8\ \mu$m

Decay asymmetry parameters

$\Lambda \pi^+$	$\alpha = -0.98 \pm 0.19$
$\Sigma^+ \pi^0$	$\alpha = -0.45 \pm 0.32$
$\Lambda \ell^+ \nu_\ell$	$\alpha = -0.82^{+0.11}_{-0.07}$

Nearly all branching fractions of the Λ_c^+ are measured relative to the $pK^-\pi^+$ mode, but there are no model-independent measurements of this branching fraction. We explain how we arrive at our value of B($\Lambda_c^+ \to pK^-\pi^+$) in a Note at the beginning of the branching-ratio measurements, in the Listings. When this branching fraction is eventually well determined, all the other branching fractions will slide up or down proportionally as the true value differs from the value we use here.

Λ_c^+ DECAY MODES		Fraction (Γ_i/Γ)	Scale factor/ Confidence level	p (MeV/c)
Hadronic modes with a p and one $\overline{K}$				
$p\overline{K}^0$		(2.5 ± 0.7) %		872
$pK^-\pi^+$	[k]	(5.0 ± 1.3) %		822
$p\overline{K}^*(892)^0$	[l]	(1.8 ± 0.6) %		681
$\Delta(1232)^{++} K^-$		$(8 \pm 5) \times 10^{-3}$		709
$\Lambda(1520)\pi^+$	[l]	$(4.5^{+2.5}_{-2.1}) \times 10^{-3}$		626
$pK^-\pi^+$ nonresonant		(2.8 ± 0.9) %		822
$p\overline{K}^0\eta$		(1.3 ± 0.4) %		567
$p\overline{K}^0\pi^+\pi^-$		(2.4 ± 1.1) %		753
$pK^-\pi^+\pi^0$		seen		758
$pK^*(892)^-\pi^+$	[l]	(1.1 ± 0.6) %		579
$p(K^-\pi^+)_{\text{nonresonant}}\pi^0$		(3.6 ± 1.2) %		758
$\Delta(1232)\overline{K}^*(892)$		seen		416
$pK^-\pi^+\pi^+\pi^-$		$(1.1 \pm 0.8) \times 10^{-3}$		670
$pK^-\pi^+\pi^0\pi^0$		$(8 \pm 4) \times 10^{-3}$		676
$pK^-\pi^+\pi^0\pi^0\pi^0$		$(5.0 \pm 3.4) \times 10^{-3}$		573
Hadronic modes with a p and zero or two K's				
$p\pi^+\pi^-$		$(3.5 \pm 2.0) \times 10^{-3}$		926
$pf_0(980)$	[l]	$(2.8 \pm 1.9) \times 10^{-3}$		621
$p\pi^+\pi^+\pi^-\pi^-$		$(1.8 \pm 1.2) \times 10^{-3}$		851
pK^+K^-		$(2.3 \pm 0.9) \times 10^{-3}$		615
$p\phi$	[l]	$(1.2 \pm 0.5) \times 10^{-3}$		589
Hadronic modes with a hyperon				
$\Lambda\pi^+$		$(9.0 \pm 2.8) \times 10^{-3}$		863
$\Lambda\pi^+\pi^0$		(3.6 ± 1.3) %		843
$\Lambda\rho^+$		< 5 %	CL=95%	638
$\Lambda\pi^+\pi^+\pi^-$		(3.3 ± 1.0) %		806
$\Lambda\pi^+\eta$		(1.7 ± 0.6) %		690
$\Sigma(1385)^+\eta$	[l]	$(8.5 \pm 3.3) \times 10^{-3}$		569
$\Lambda K^+\overline{K}^0$		$(6.0 \pm 2.1) \times 10^{-3}$		441
$\Sigma^0\pi^+$		$(9.9 \pm 3.2) \times 10^{-3}$		824
$\Sigma^+\pi^0$		(1.00 ± 0.34) %		826
$\Sigma^+\eta$		$(5.5 \pm 2.3) \times 10^{-3}$		712
$\Sigma^+\pi^+\pi^-$		(3.4 ± 1.0) %		803
$\Sigma^+\rho^0$		< 1.4 %	CL=95%	578
$\Sigma^-\pi^+\pi^+$		(1.8 ± 0.8) %		798
$\Sigma^0\pi^+\pi^0$		(1.8 ± 0.8) %		802
$\Sigma^0\pi^+\pi^+\pi^-$		(1.1 ± 0.4) %		762
$\Sigma^+\pi^+\pi^-\pi^0$		—		766
$\Sigma^+\omega$	[l]	(2.7 ± 1.0) %		568
$\Sigma^+\pi^+\pi^+\pi^-\pi^-$		$(3.0^{+4.1}_{-2.1}) \times 10^{-3}$		707
$\Sigma^+K^+K^-$		$(3.5 \pm 1.2) \times 10^{-3}$		346
$\Sigma^+\phi$	[l]	$(3.5 \pm 1.7) \times 10^{-3}$		292
$\Sigma^+K^+\pi^-$		$(7^{+6}_{-4}) \times 10^{-3}$		668
Ξ^0K^+		$(3.9 \pm 1.4) \times 10^{-3}$		652
$\Xi^-K^+\pi^+$		$(4.9 \pm 1.7) \times 10^{-3}$		564
$\Xi(1530)^0K^+$	[l]	$(2.6 \pm 1.0) \times 10^{-3}$		471
Semileptonic modes				
$\Lambda\ell^+\nu_\ell$	[m]	(2.0 ± 0.6) %		–
$\Lambda e^+\nu_e$		(2.1 ± 0.6) %		–
$\Lambda\mu^+\nu_\mu$		(2.0 ± 0.7) %		–
e^+ anything		(4.5 ± 1.7) %		–
pe^+ anything		(1.8 ± 0.9) %		–
Λe^+ anything		—		–
$\Lambda\mu^+$ anything		—		–
$\Lambda\ell^+\nu_\ell$ anything		—		–
Inclusive modes				
p anything		(50 ± 16) %		–
p anything (no Λ)		(12 ± 19) %		–
p hadrons		—		–
n anything		(50 ± 16) %		–
n anything (no Λ)		(29 ± 17) %		–
Λ anything		(35 ± 11) %	S=1.4	–
$\Sigma^\pm$ anything	[n]	(10 ± 5) %		–

$\Delta C = 1$ weak neutral current ($C1$) modes, or Lepton number (L) violating modes

$p\mu^+\mu^-$	$C1$	< 3.4	$\times 10^{-4}$	CL=90%	936
$\Sigma^-\mu^+\mu^+$	L	< 7.0	$\times 10^{-4}$	CL=90%	811

$\Lambda_c(2593)^+$

$I(J^P) = 0(\frac{1}{2}^-)$

The spin-parity follows from the fact that $\Sigma_c(2455)\pi$ decays, with little available phase space, are dominant.

Mass $m = 2593.9 \pm 0.8$ MeV

$m - m_{\Lambda_c^+} = 308.9 \pm 0.6$ MeV (S = 1.1)

Full width $\Gamma = 3.6^{+2.0}_{-1.3}$ MeV

$\Lambda_c^+\pi\pi$ and its submode $\Sigma_c(2455)\pi$ — the latter just barely — are the only strong decays allowed to an excited Λ_c^+ having this mass; and the $\Lambda_c^+\pi^+\pi^-$ mode seems to be largely via $\Sigma_c^{++}\pi^-$ or $\Sigma_c^0\pi^+$.

$\Lambda_c(2593)^+$ DECAY MODES	Fraction (Γ_i/Γ)	p (MeV/c)
$\Lambda_c^+\pi^+\pi^-$	[o] ≈ 67 %	124
$\Sigma_c(2455)^{++}\pi^-$	24 ± 7 %	17
$\Sigma_c(2455)^0\pi^+$	24 ± 7 %	23
$\Lambda_c^+\pi^+\pi^-$ 3-body	18 ± 10 %	124
$\Lambda_c^+\pi^0$	not seen	261
$\Lambda_c^+\gamma$	not seen	290

$\Lambda_c(2625)^+$

$I(J^P) = 0(?^?)$

J^P is expected to be $3/2^-$.

Mass $m = 2626.6 \pm 0.8$ MeV (S = 1.2)

$m - m_{\Lambda_c^+} = 341.7 \pm 0.6$ MeV (S = 1.6)

Full width $\Gamma < 1.9$ MeV, CL = 90%

$\Lambda_c^+\pi\pi$ and its submode $\Sigma(2455)\pi$ are the only strong decays allowed to an excited Λ_c^+ having this mass.

$\Lambda_c(2625)^+$ DECAY MODES	Fraction (Γ_i/Γ)	p (MeV/c)
$\Lambda_c^+\pi^+\pi^-$	seen	184
$\Sigma_c(2455)^{++}\pi^-$	small	100
$\Sigma_c(2455)^0\pi^+$	small	101
$\Lambda_c^+\pi^+\pi^-$ 3-body	large	184
$\Lambda_c^+\pi^0$	not seen	293
$\Lambda_c^+\gamma$	not seen	319

$\Sigma_c(2455)$

$I(J^P) = 1(\frac{1}{2}^+)$

J^P not confirmed; $\frac{1}{2}^+$ is the quark model prediction.

$\Sigma_c(2455)^{++}$ mass $m = 2452.8 \pm 0.6$ MeV

$\Sigma_c(2455)^+$ mass $m = 2453.6 \pm 0.9$ MeV

$\Sigma_c(2455)^0$ mass $m = 2452.2 \pm 0.6$ MeV

$m_{\Sigma_c^{++}} - m_{\Lambda_c^+} = 167.87 \pm 0.19$ MeV

$m_{\Sigma_c^+} - m_{\Lambda_c^+} = 168.7 \pm 0.6$ MeV

$m_{\Sigma_c^0} - m_{\Lambda_c^+} = 167.30 \pm 0.20$ MeV

$m_{\Sigma_c^{++}} - m_{\Sigma_c^0} = 0.57 \pm 0.23$ MeV

$m_{\Sigma_c^+} - m_{\Sigma_c^0} = 1.4 \pm 0.6$ MeV

$\Lambda_c^+\pi$ is the only strong decay allowed to a Σ_c having this mass.

$\Sigma_c(2455)$ DECAY MODES	Fraction (Γ_i/Γ)	p (MeV/c)
$\Lambda_c^+\pi$	≈ 100 %	90

$\Sigma_c(2520)$

$I(J^P) = 1(?^?)$

$\Sigma_c(2520)^{++}$ mass $m = 2519.4 \pm 1.5$ MeV

$\Sigma_c(2520)^0$ mass $m = 2517.5 \pm 1.4$ MeV

$m_{\Sigma_c(2520)^{++}} - m_{\Lambda_c^+} = 234.5 \pm 1.4$ MeV

$m_{\Sigma_c(2520)^0} - m_{\Lambda_c^+} = 232.6 \pm 1.3$ MeV

$m_{\Sigma_c(2520)^{++}} - m_{\Sigma_c(2520)^0} = 1.9 \pm 1.9$ MeV

$\Sigma_c(2520)^{++}$ full width $\Gamma = 18 \pm 5$ MeV

$\Sigma_c(2520)^0$ full width $\Gamma = 13 \pm 5$ MeV

Ξ_c^+

$I(J^P) = \frac{1}{2}(\frac{1}{2}^+)$

$I(J^P)$ not confirmed; $\frac{1}{2}(\frac{1}{2}^+)$ is the quark model prediction.

Mass $m = 2465.6 \pm 1.4$ MeV

Mean life $\tau = (0.35^{+0.07}_{-0.04}) \times 10^{-12}$ s

$c\tau = 106$ μm

Ξ_c^+ DECAY MODES	Fraction (Γ_i/Γ)	p (MeV/c)
$\Lambda K^-\pi^+\pi^+$	seen	784
$\Lambda\overline{K}^*(892)^0\pi^+$	not seen	601
$\Sigma(1385)^+K^-\pi^+$	not seen	676
$\Sigma^+K^-\pi^+$	seen	808
$\Sigma^+\overline{K}^*(892)^0$	seen	653
$\Sigma^0K^-\pi^+\pi^+$	seen	733
$\Xi^0\pi^+$	seen	875
$\Xi^-\pi^+\pi^+$	seen	850
$\Xi(1530)^0\pi^+$	not seen	748
$\Xi^0\pi^+\pi^0$	seen	854
$\Xi^0\pi^+\pi^+\pi^-$	seen	817
$\Xi^0e^+\nu_e$	seen	882

Ξ_c^0

$I(J^P) = \frac{1}{2}(\frac{1}{2}^+)$

$I(J^P)$ not confirmed; $\frac{1}{2}(\frac{1}{2}^+)$ is the quark model prediction.

Mass $m = 2470.3 \pm 1.8$ MeV (S = 1.3)

$m_{\Xi_c^0} - m_{\Xi_c^+} = 4.7 \pm 2.1$ MeV (S = 1.2)

Mean life $\tau = (0.098^{+0.023}_{-0.015}) \times 10^{-12}$ s

$c\tau = 29$ μm

Ξ_c^0 DECAY MODES	Fraction (Γ_i/Γ)	p (MeV/c)
$\Lambda\overline{K}^0$	seen	864
$\Xi^-\pi^+$	seen	875
$\Xi^-\pi^+\pi^+\pi^-$	seen	816
$pK^-\overline{K}^*(892)^0$	seen	406
Ω^-K^+	seen	522
$\Xi^-e^+\nu_e$	seen	882
$\Xi^-\ell^+$ anything	seen	–

$\Xi_c(2645)$

$I(J^P) = ?(?^?)$

$\Xi_c(2645)^+$ mass $m = 2644.6 \pm 2.1$ MeV (S = 1.2)

$\Xi_c(2645)^0$ mass $m = 2643.8 \pm 1.8$ MeV

$m_{\Xi_c(2645)^+} - m_{\Xi_c^0} = 174.3 \pm 1.1$ MeV

$m_{\Xi_c(2645)^0} - m_{\Xi_c^+} = 178.2 \pm 1.1$ MeV

$\Xi_c(2645)^+$ full width $\Gamma < 3.1$ MeV, CL = 90%

$\Xi_c(2645)^0$ full width $\Gamma < 5.5$ MeV, CL = 90%

$\Xi_c\pi$ is the only strong decay allowed to a Ξ_c resonance having this mass.

$\Xi_c(2645)$ DECAY MODES	Fraction (Γ_i/Γ)	p (MeV/c)
$\Xi_c^0\pi^+$	seen	101
$\Xi_c^+\pi^-$	seen	107

Ω_c^0 $I(J^P) = 0(\frac{1}{2}^+)$

$I(J^P)$ not confirmed; $0(\frac{1}{2}^+)$ is the quark model prediction.

Mass $m = 2704 \pm 4$ MeV (S = 1.8)
Mean life $\tau = (0.064 \pm 0.020) \times 10^{-12}$ s
$c\tau = 19\ \mu$m

Ω_c^0 DECAY MODES	Fraction (Γ_i/Γ)	p (MeV/c)
$\Sigma^+ K^- K^- \pi^+$	seen	697
$\Xi^- K^- \pi^+ \pi^+$	seen	838
$\Omega^- \pi^+$	seen	827
$\Omega^- \pi^- \pi^+ \pi^+$	seen	759

BOTTOM BARYONS
$(B = -1)$

$\Lambda_b^0 = udb$, $\Xi_b^0 = usb$, $\Xi_b^- = dsb$

Λ_b^0 $I(J^P) = 0(\frac{1}{2}^+)$

$I(J^P)$ not yet measured; $0(\frac{1}{2}^+)$ is the quark model prediction.

Mass $m = 5624 \pm 9$ MeV (S = 1.8)
Mean life $\tau = (1.24 \pm 0.08) \times 10^{-12}$ s
$c\tau = 372\ \mu$m

These branching fractions are actually an average over weakly decaying b-baryons weighted by their production rates in Z decay (or high-energy $p\overline{p}$), branching ratios, and detection efficiencies. They scale with the LEP Λ_b production fraction B($b \to \Lambda_b$) and are evaluated for our value B($b \to \Lambda_b$) = $(10.1^{+3.9}_{-3.1})$%.

The branching fractions B(b-baryon $\to \Lambda \ell^- \overline{\nu}_\ell$anything) and B($\Lambda_b^0 \to \Lambda_c^+ \ell^- \overline{\nu}_\ell$anything) are not pure measurements because the underlying measured products of these with B($b \to \Lambda_b$) were used to determine B($b \to \Lambda_b$), as described in the note "Production and Decay of b-Flavored Hadrons."

Λ_b^0 DECAY MODES		Fraction (Γ_i/Γ)	Confidence level	p (MeV/c)
$J/\psi(1S)\Lambda$		$(4.7\pm2.8) \times 10^{-4}$		1744
$\Lambda_c^+ \pi^-$		seen		2345
$\Lambda_c^+ a_1(1260)^-$		seen		2156
$\Lambda_c^+ \ell^- \overline{\nu}_\ell$anything	[p]	$(9.0^{+3.1}_{-3.8})$ %		–
$p\pi^-$		$< 5.0 \times 10^{-5}$	90%	2732
pK^-		$< 5.0 \times 10^{-5}$	90%	2711

b-baryon ADMIXTURE (Λ_b, Ξ_b, Σ_b, Ω_b)

Mean life $\tau = (1.20 \pm 0.07) \times 10^{-12}$ s

These branching fractions are actually an average over weakly decaying b-baryons weighted by their production rates in Z decay (or high-energy $p\overline{p}$), branching ratios, and detection efficiencies. They scale with the LEP Λ_b production fraction B($b \to \Lambda_b$) and are evaluated for our value B($b \to \Lambda_b$) = $(10.1^{+3.9}_{-3.1})$%.

The branching fractions B(b-baryon $\to \Lambda \ell^- \overline{\nu}_\ell$anything) and B($\Lambda_b^0 \to \Lambda_c^+ \ell^- \overline{\nu}_\ell$anything) are not pure measurements because the underlying measured products of these with B($b \to \Lambda_b$) were used to determine B($b \to \Lambda_b$), as described in the note "Production and Decay of b-Flavored Hadrons."

b-baryon ADMIXTURE (Λ_b,Ξ_b,Σ_b,Ω_b)	Fraction (Γ_i/Γ)	p (MeV/c)
$p\mu^- \overline{\nu}$anything	(4.9 ± 2.4) %	–
$\Lambda \ell^- \overline{\nu}_\ell$anything	$(3.1^{+1.0}_{-1.2})$ %	–
$\Lambda/\overline{\Lambda}$anything	$(35\ ^{+12}_{-14})$ %	–
$\Xi^- \ell^- \overline{\nu}_\ell$anything	$(5.5^{+2.0}_{-2.4}) \times 10^{-3}$	–

NOTES

This Summary Table only includes established baryons. The Particle Listings include evidence for other baryons. The masses, widths, and branching fractions for the resonances in this Table are Breit-Wigner parameters, but pole positions are also given for most of the N and Δ resonances.

For most of the resonances, the parameters come from various partial-wave analyses of more or less the same sets of data, and it is not appropriate to treat the results of the analyses as independent or to average them together. Furthermore, the systematic errors on the results are not well understood. Thus, we usually only give ranges for the parameters. We then also give a best guess for the mass (as part of the name of the resonance) and for the width. The *Note on N and Δ Resonances* and the *Note on Λ and Σ Resonances* in the Particle Listings review the partial-wave analyses.

When a quantity has "(S = ...)" to its right, the error on the quantity has been enlarged by the "scale factor" S, defined as $S = \sqrt{\chi^2/(N-1)}$, where N is the number of measurements used in calculating the quantity. We do this when S > 1, which often indicates that the measurements are inconsistent. When S > 1.25, we also show in the Particle Listings an ideogram of the measurements. For more about S, see the Introduction.

A decay momentum p is given for each decay mode. For a 2-body decay, p is the momentum of each decay product in the rest frame of the decaying particle. For a 3-or-more-body decay, p is the largest momentum any of the products can have in this frame. For any resonance, the *nominal* mass is used in calculating p. A dagger ("†") in this column indicates that the mode is forbidden when the nominal masses of resonances are used, but is in fact allowed due to the nonzero widths of the resonances.

[a] The masses of the p and n are most precisely known in u (unified atomic mass units). The conversion factor to MeV, 1 u = 931.49432 ± 0.00028 MeV, is less well known than are the masses in u.

[b] The limit is from neutrality-of-matter experiments; it assumes $q_n = q_p + q_e$. See also the charge of the neutron.

[c] The first limit is geochemical and independent of decay mode. The second entry, a range of limits, assumes the dominant decay modes are among those investigated. For antiprotons the best limit, inferred from the observation of cosmic ray $\overline{p}$'s is $\tau_{\overline{p}} > 10^7$ yr, the cosmic-ray storage time, but this limit depends on a number of assumptions. The best direct observation of stored antiprotons gives $\tau_{\overline{p}}/B(\overline{p} \to e^-\gamma) > 1848$ yr.

[d] There is some controversy about whether nuclear physics and model dependence complicate the analysis for bound neutrons (from which the best limit comes). The second limit here is from reactor experiments with free neutrons.

[e] The parameters g_A, g_V, and g_{WM} for semileptonic modes are defined by $\overline{B}_f[\gamma_\lambda(g_V + g_A\gamma_5) + i(g_{WM}/m_{B_i})\,\sigma_{\lambda\nu}\,q^\nu]B_i$, and ϕ_{AV} is defined by $g_A/g_V = |g_A/g_V|e^{i\phi_{AV}}$. See the "Note on Baryon Decay Parameters" in the neutron Particle Listings.

[f] Time-reversal invariance requires this to be 0° or 180°.

[g] The decay parameters γ and Δ are calculated from α and ϕ using

$$\gamma = \sqrt{1-\alpha^2}\,\cos\phi\,, \qquad \tan\Delta = -\frac{1}{\alpha}\,\sqrt{1-\alpha^2}\,\sin\phi\,.$$

See the "Note on Baryon Decay Parameters" in the neutron Particle Listings.

[h] See the Particle Listings for the pion momentum range used in this measurement.

[i] The error given here is only an educated guess. It is larger than the error on the weighted average of the published values.

[j] A theoretical value using QED.

[k] See the "Note on Λ_c^+ Branching Fractions" in the Branching Fractions of the Λ_c^+ Particle Listings.

[l] This branching fraction includes all the decay modes of the final-state resonance.

[m] An ℓ indicates an e or a μ mode, not a sum over these modes.

[n] The value is for the sum of the charge states of particle/antiparticle states indicated.

[o] Assuming isospin conservation, so that the other third is $\Lambda_c^+\pi^0\pi^0$.

[p] Not a pure measurement. See note at head of Λ_b^0 Decay Modes.

MONOPOLES, SUPERSYMMETRY, COMPOSITENESS, etc., SEARCHES FOR

Magnetic Monopole Searches

Isolated supermassive monopole candidate events have not been confirmed. The most sensitive experiments obtain negative results.

Best cosmic-ray supermassive monopole flux limit:

$< 1.0 \times 10^{-15}\ \mathrm{cm}^{-2}\mathrm{sr}^{-1}\mathrm{s}^{-1}$ for $1.1 \times 10^{-4} < \beta < 0.1$

Supersymmetric Particle Searches

Limits are based on the Minimal Supersymmetric Standard Model.

Assumptions include: 1) $\widetilde{\chi}_1^0$ (or $\widetilde{\gamma}$) is lightest supersymmetric particle; 2) R-parity is conserved; 3) All scalar quarks (except $\widetilde{t}_L$ and $\widetilde{t}_R$) are degenerate in mass, and $m_{\widetilde{q}_R} = m_{\widetilde{q}_L}$. 4) Limits for selectrons and smuons refer to the $\widetilde{\ell}_R$ states.

See the Particle Listings for a Note giving details of supersymmetry.

$\widetilde{\chi}_i^0$ — neutralinos (mixtures of $\widetilde{\gamma}$, $\widetilde{Z}^0$, and $\widetilde{H}_i^0$)

Mass $m_{\widetilde{\chi}_1^0}$ > 10.9 GeV, CL = 95%

Mass $m_{\widetilde{\chi}_2^0}$ > 45.3 GeV, CL = 95% [$\tan\beta > 1$]

Mass $m_{\widetilde{\chi}_3^0}$ > 75.8 GeV, CL = 95% [$\tan\beta > 1$]

Mass $m_{\widetilde{\chi}_4^0}$ > 127 GeV, CL = 95% [$\tan\beta > 3$]

$\widetilde{\chi}_i^\pm$ — charginos (mixtures of $\widetilde{W}^\pm$ and $\widetilde{H}_i^\pm$)

Mass $m_{\widetilde{\chi}_1^\pm}$ > 65.7 GeV, CL = 95% [$m_{\widetilde{\chi}_1^\pm} - m_{\widetilde{\chi}_1^0} \geq 2$ GeV]

Mass $m_{\widetilde{\chi}_2^\pm}$ > 99 GeV, CL = 95% [GUT relations assumed]

$\widetilde{\nu}$ — scalar neutrino (sneutrino)

Mass m > 37.1 GeV, CL = 95% [one flavor]

Mass m > 43.1 GeV, CL = 95% [three degenerate flavors]

$\widetilde{e}$ — scalar electron (selectron)

Mass m > 58 GeV, CL = 95% [$m_{\widetilde{e}_R} - m_{\widetilde{\chi}_1^0} \geq 4$ GeV]

$\widetilde{\mu}$ — scalar muon (smuon)

Mass m > 55.6 GeV, CL = 95% [$m_{\widetilde{\mu}_R} - m_{\widetilde{\chi}_1^0} \geq 4$ GeV]

$\widetilde{\tau}$ — scalar tau (stau)

Mass m > 45 GeV, CL = 95% [if $m_{\widetilde{\chi}_1^0} < 38$ GeV]

$\widetilde{q}$ — scalar quark (squark)

These limits include the effects of cascade decays, evaluated assuming a fixed value of the parameters μ and $\tan\beta$. The limits are weakly sensitive to these parameters over much of parameter space. Limits assume GUT relations between gaugino masses and the gauge coupling; in particular that for $|\mu|$ not small, $m_{\widetilde{\chi}_1^0} \approx m_{\widetilde{g}}/6$.

Mass m > 176 GeV, CL = 95% [any $m_{\widetilde{g}} < 300$ GeV, $\mu = -250$ GeV, $\tan\beta = 2$]

Mass m > 224 GeV, CL = 95% [$m_{\widetilde{g}} \leq m_{\widetilde{q}}$, $\mu = -400$ GeV, $\tan\beta = 4$]

$\widetilde{g}$ — gluino

There is some controversy on whether gluinos in a low-mass window ($1 \lesssim m_{\widetilde{g}} \lesssim 5$ GeV) are excluded or not. See the Supersymmetry Listings for details.

The limits summarised here refere to the high-mass region ($m_{\widetilde{g}} \gtrsim 5$ GeV), and include the effects of cascade decays, evaluated assuming a fixed value of the parameters μ and $\tan\beta$. The limits are weakly sensitive to these parameters over much of parameter space. Limits assume GUT relations between gaugino masses and the gauge coupling; in particular that for $|\mu|$ not small, $m_{\widetilde{\chi}_1^0} \approx m_{\widetilde{g}}/6$.

Mass m > 173 GeV, CL = 95% [any $m_{\widetilde{q}}$, $\mu = -200$ GeV, $\tan\beta = 2$]

Mass m > 212 GeV, CL = 95% [$m_{\widetilde{g}} \geq m_{\widetilde{q}}$, $\mu = -250$ GeV, $\tan\beta = 2$]

Quark and Lepton Compositeness, Searches for

Scale Limits Λ for Contact Interactions (the lowest dimensional interactions with four fermions)

If the Lagrangian has the form

$$\pm \frac{g^2}{2\Lambda^2} \overline{\psi}_L \gamma_\mu \psi_L \overline{\psi}_L \gamma^\mu \psi_L$$

(with $g^2/4\pi$ set equal to 1), then we define $\Lambda \equiv \Lambda^\pm_{LL}$. For the full definitions and for other forms, see the Note in the Listings on Searches for Quark and Lepton Compositeness in the full *Review* and the original literature.

$\Lambda^+_{LL}(eeee)$ > 2.4 TeV, CL = 95%

$\Lambda^-_{LL}(eeee)$ > 3.6 TeV, CL = 95%

$\Lambda^+_{LL}(ee\mu\mu)$ > 2.6 TeV, CL = 95%

$\Lambda^-_{LL}(ee\mu\mu)$ > 2.9 TeV, CL = 95%

$\Lambda^+_{LL}(ee\tau\tau)$ > 1.9 TeV, CL = 95%

$\Lambda^-_{LL}(ee\tau\tau)$ > 3.0 TeV, CL = 95%

$\Lambda^+_{LL}(\ell\ell\ell\ell)$ > 3.5 TeV, CL = 95%

$\Lambda^-_{LL}(\ell\ell\ell\ell)$ > 3.8 TeV, CL = 95%

$\Lambda^+_{LL}(eeqq)$ > 2.5 TeV, CL = 95%

$\Lambda^-_{LL}(eeqq)$ > 3.7 TeV, CL = 95%

$\Lambda^+_{LL}(eebb)$ > 3.1 TeV, CL = 95%

$\Lambda^-_{LL}(eebb)$ > 2.9 TeV, CL = 95%

$\Lambda^+_{LL}(\mu\mu qq)$ > 2.9 TeV, CL = 95%

$\Lambda^-_{LL}(\mu\mu qq)$ > 4.2 TeV, CL = 95%

$\Lambda^\pm_{LR}(\nu_\mu \nu_e \mu e)$ > 3.1 TeV, CL = 90%

$\Lambda^\pm_{LL}(qqqq)$ > 1.6 TeV, CL = 95%

Excited Leptons

The limits from $\ell^{*+}\ell^{*-}$ do not depend on λ (where λ is the $\ell\ell^*$ transition coupling). The λ-dependent limits assume chiral coupling, except for the third limit for e^* which is for nonchiral coupling. For chiral coupling, this limit corresponds to $\lambda_\gamma = \sqrt{2}$.

$e^{*\pm}$ — excited electron

Mass m > 85.0 GeV, CL = 95% (from $e^{*+}e^{*-}$)

Mass m > 91 GeV, CL = 95% (if $\lambda_Z > 1$)

Mass m > 194 GeV, CL = 95% (if $\lambda_\gamma = 1$)

$\mu^{*\pm}$ — excited muon

Mass m > 85.3 GeV, CL = 95% (from $\mu^{*+}\mu^{*-}$)

Mass m > 91 GeV, CL = 95% (if $\lambda_Z > 1$)

$\tau^{*\pm}$ — excited tau

Mass m > 84.6 GeV, CL = 95% (from $\tau^{*+}\tau^{*-}$)

Mass m > 90 GeV, CL = 95% (if $\lambda_Z > 0.18$)

ν^* — excited neutrino

Mass m > 84.9 GeV, CL = 95% (from $\nu^*\overline{\nu}^*$)

Mass m > 91 GeV, CL = 95% (if $\lambda_Z > 1$)

Mass m = none 40–96 GeV, CL = 95% (from $ep \to \nu^* X$)

q^* — excited quark

Mass m > 45.6 GeV, CL = 95% (from $q^*\overline{q}^*$)

Mass m > 88 GeV, CL = 95% (if $\lambda_Z > 1$)

Mass m > 570 GeV, CL = 95% ($p\overline{p} \to q^* X$)

Color Sextet and Octet Particles

Color Sextet Quarks (q_6)

Mass m > 84 GeV, CL = 95% (Stable q_6)

Color Octet Charged Leptons (ℓ_8)

Mass m > 86 GeV, CL = 95% (Stable ℓ_8)

Color Octet Neutrinos (ν_8)

Mass m > 110 GeV, CL = 90% ($\nu_8 \to \nu g$)

TESTS OF CONSERVATION LAWS

Revised by L. Wolfenstein and T.G. Trippe, May 1998.

In keeping with the current interest in tests of conservation laws, we collect together a Table of experimental limits on all weak and electromagnetic decays, mass differences, and moments, and on a few reactions, whose observation would violate conservation laws. The Table is given only in the full *Review of Particle Physics*, not in the Particle Physics Booklet. For the benefit of Booklet readers, we include the best limits from the Table in the following text. Limits in this text are for CL=90% unless otherwise specified. The Table is in two parts: "Discrete Space-Time Symmetries," *i.e.*, C, P, T, CP, and CPT; and "Number Conservation Laws," *i.e.*, lepton, baryon, hadronic flavor, and charge conservation. The references for these data can be found in the the Particle Listings in the *Review*. A discussion of these tests follows.

CPT INVARIANCE

General principles of relativistic field theory require invariance under the combined transformation CPT. The simplest tests of CPT invariance are the equality of the masses and lifetimes of a particle and its antiparticle. The best test comes from the limit on the mass difference between K^0 and $\overline{K}^0$. Any such difference contributes to the CP-violating parameter ϵ. Assuming CPT invariance, ϕ_ϵ, the phase of ϵ should be very close to 44°. (See the "Note on CP Violation in K^0_L Decay" in the Particle Listings.) In contrast, if the entire source of CP violation in K^0 decays were a $K^0 - \overline{K}^0$ mass difference, ϕ_ϵ would be 44° + 90°. Assuming that there is no other source of CPT violation than this mass difference, it is possible to deduce that [1]

$$m_{\overline{K}^0} - m_{K^0} \approx \frac{2(m_{K^0_L} - m_{K^0_S})\,|\eta|\,(\frac{2}{3}\phi_{+-} + \frac{1}{3}\phi_{00} - \phi_0)}{\sin\phi_0} ,$$

where $\phi_0 = 43.5°$ with an uncertainty of less than 0.1°. Using our best values of the CP-violation parameters, we get $|(m_{\overline{K}^0} - m_{K^0})/m_{K^0}| \leq 10^{-18}$. Limits can also be placed on specific CPT-violating decay amplitudes. Given the small value of $(1 - |\eta_{00}/\eta_{+-}|)$, the value of $\phi_{00} - \phi_{+-}$ provides a measure of CPT violation in $K^0_L \to 2\pi$ decay. Results from CERN [1] and Fermilab [2] indicate no CPT-violating effect.

CP AND T INVARIANCE

Given CPT invariance, CP violation and T violation are equivalent. So far the only evidence for CP or T violation comes from the measurements of η_{+-}, η_{00}, and the semileptonic decay charge asymmetry for K_L, *e.g.*, $|\eta_{+-}| = |A(K^0_L \to \pi^+\pi^-)/A(K^0_S \to \pi^+\pi^-)| = (2.285 \pm 0.019) \times 10^{-3}$ and $[\Gamma(K^0_L \to \pi^-e^+\nu) - \Gamma(K^0_L \to \pi^+e^-\overline{\nu})]/[\text{sum}] = (0.333 \pm 0.014)\%$. Other searches for CP or T violation divide into (a) those that involve weak interactions or parity violation, and (b) those that involve processes otherwise allowed by the strong or electromagnetic interactions. In class (a) the most sensitive are probably the searches for an electric dipole moment of the neutron, measured to be $< 1.0 \times 10^{-25}$ e cm, and the electron $(-0.18 \pm 0.16) \times 10^{-26}$ e cm. A nonzero value requires both P and T violation. Class (b) includes the search for C violation in η decay, believed to be an electromagnetic process, *e.g.*, as measured by $\Gamma(\eta \to \mu^+\mu^-\pi^0)/\Gamma(\eta \to \text{all}) < 5 \times 10^{-6}$, and searches for T violation in a number of nuclear and electromagnetic reactions.

CONSERVATION OF LEPTON NUMBERS

Present experimental evidence and the standard electroweak theory are consistent with the absolute conservation of three separate lepton numbers: electron number L_e, muon number L_μ, and tau number L_τ. Searches for violations are of the following types:

a) $\Delta L = 2$ for one type of lepton. The best limit comes from the search for neutrinoless double beta decay $(Z, A) \to (Z + 2, A) + e^- + e^-$. The best laboratory limit is $t_{1/2} > 1.1 \times 10^{25}$ yr (CL=90%) for ^{76}Ge.

b) Conversion of one lepton type to another. For purely leptonic processes, the best limits are on $\mu \to e\gamma$ and $\mu \to 3e$, measured as $\Gamma(\mu \to e\gamma)/\Gamma(\mu \to \text{all}) < 5 \times 10^{-11}$ and $\Gamma(\mu \to 3e)/\Gamma(\mu \to \text{all}) < 1.0 \times 10^{-12}$. For semileptonic processes, the best limit comes from the coherent conversion process in a muonic atom, $\mu^- + (Z, A) \to e^- + (Z, A)$, measured as $\Gamma(\mu^-\text{Ti} \to e^-\text{Ti})/\Gamma(\mu^-\text{Ti} \to \text{all}) < 4 \times 10^{-12}$. Of special interest is the case in which the hadronic flavor also changes, as in $K_L \to e\mu$ and $K^+ \to \pi^+e^-\mu^+$, measured as $\Gamma(K_L \to e\mu)/\Gamma(K_L \to \text{all}) < 3.3 \times 10^{-11}$ and $\Gamma(K^+ \to \pi^+e^-\mu^+)/\Gamma(K^+ \to \text{all}) < 2.1 \times 10^{-10}$. Limits on the conversion of τ into e or μ are found in τ decay and are much less stringent than those for $\mu \to e$ conversion, *e.g.*, $\Gamma(\tau \to \mu\gamma)/\Gamma(\tau \to \text{all}) < 3.0 \times 10^{-6}$ and $\Gamma(\tau \to e\gamma)/\Gamma(\tau \to \text{all}) < 2.7 \times 10^{-6}$.

c) Conversion of one type of lepton into another type of antilepton. The case most studied is $\mu^- + (Z, A) \to e^+ + (Z - 2, A)$, the strongest limit being $\Gamma(\mu^-\text{Ti} \to e^+\text{Ca})/\Gamma(\mu^-\text{Ti} \to \text{all}) < 9 \times 10^{-11}$.

d) Relation to neutrino mass. If neutrinos have mass, then it is expected even in the standard electroweak theory that the lepton numbers are not separately conserved, as a consequence of lepton mixing analogous to Cabibbo quark mixing. However, in this case lepton-number-violating processes such as $\mu \to e\gamma$ are expected to have extremely small probability. For small neutrino masses, the lepton-number violation would be observed first in neutrino oscillations, which have been the subject of extensive experimental searches. For example, searches for $\overline{\nu}_e$ disappearance, which we label as $\overline{\nu}_e \not\to \overline{\nu}_e$, give measured limits $\Delta(m^2) < 9 \times 10^{-4}$ eV^2 for $\sin^2(2\theta) = 1$, and $\sin^2(2\theta) < 0.02$ for large $\Delta(m^2)$, where θ is the neutrino mixing angle. Possible evidence for mixing has come from two sources. The deficit in the solar neutrino flux compared with solar model calculations could be explained by oscillations with $\Delta(m^2) \leq 10^{-5}$ eV^2 causing the disappearance of ν_e. In addition underground detectors observing neutrinos produced by cosmic rays in the atmosphere have measured a ν_μ/ν_e ratio much less than expected and also a deficiency of upward going ν_μ compared to downward. This could be explained by oscillations leading to the disappearance of ν_μ with $\Delta(m^2)$ of the order 10^{-2}–10^{-3} eV^2.

CONSERVATION OF HADRONIC FLAVORS

In strong and electromagnetic interactions, hadronic flavor is conserved, *i.e.* the conversion of a quark of one flavor (d, u, s, c, b, t) into a quark of another flavor is forbidden. In the Standard Model, the weak interactions violate these conservation laws in a manner described by the Cabibbo-Kobayashi-Maskawa mixing (see the section "Cabibbo-Kobayashi-Maskawa Mixing Matrix"). The way in which these conservation laws are violated is tested as follows:

a) $\Delta S = \Delta Q$ rule. In the semileptonic decay of strange particles, the strangeness change equals the change in charge of the hadrons. Tests come from limits on decay rates such as

$\Gamma(\Sigma^+ \to ne^+\nu)/\Gamma(\Sigma^+ \to \text{all}) < 5 \times 10^{-6}$, and from a detailed analysis of $K_L \to \pi e\nu$, which yields the parameter x, measured to be $(\text{Re}\,x, \text{Im}\,x) = (0.006 \pm 0.018, -0.003 \pm 0.026)$. Corresponding rules are $\Delta C = \Delta Q$ and $\Delta B = \Delta Q$.

b) Change of flavor by two units. In the Standard Model this occurs only in second-order weak interactions. The classic example is $\Delta S = 2$ via $K^0 - \overline{K}^0$ mixing, which is directly measured by $m(K_S) - m(K_L) = (3.489 \pm 0.009) \times 10^{-12}$ MeV. There is now evidence for $B^0 - \overline{B}^0$ mixing ($\Delta B = 2$), with the corresponding mass difference between the eigenstates $(m_{B^0_H} - m_{B^0_L}) = (0.723 \pm 0.032)\Gamma_{B^0} = (3.05 \pm 0.12) \times 10^{-10}$ MeV, and for B^0_s–$\overline{B}^0_s$ mixing, with $(m_{B^0_{sH}} - m_{B^0_{sL}}) > 14\Gamma_{B^0_s}$ or $> 6 \times 10^{-9}$ MeV (CL=95%). No evidence exists for $D^0 - \overline{D}^0$ mixing, which is expected to be much smaller in the Standard Model.

c) Flavor-changing neutral currents. In the Standard Model the neutral-current interactions do not change flavor. The low rate $\Gamma(K_L \to \mu^+\mu^-)/\Gamma(K_L \to \text{all}) = (7.2 \pm 0.5) \times 10^{-9}$ puts limits on such interactions; the nonzero value for this rate is attributed to a combination of the weak and electromagnetic interactions. The best test should come from $K^+ \to \pi^+\nu\overline{\nu}$, which occurs in the Standard Model only as a second-order weak process with a branching fraction of (1 to 8)$\times 10^{-10}$. Observation of one event has been reported [4], yielding $\Gamma(K^+ \to \pi^+\nu\overline{\nu})/\Gamma(K^+ \to \text{all}) = (4.2^{+9.7}_{-3.5}) \times 10^{-10}$. Limits for charm-changing or bottom-changing neutral currents are much less stringent: $\Gamma(D^0 \to \mu^+\mu^-)/\Gamma(D^0 \to \text{all}) < 4 \times 10^{-6}$ and $\Gamma(B^0 \to \mu^+\mu^-)/\Gamma(B^0 \to \text{all}) < 7 \times 10^{-7}$. One cannot isolate flavor-changing neutral current (FCNC) effects in non leptonic decays. For example, the FCNC transition $s \to d + (\overline{u} + u)$ is equivalent to the charged-current transition $s \to u + (\overline{u} + d)$. Tests for FCNC are therefore limited to hadron decays into lepton pairs. Such decays are expected only in second-order in the electroweak coupling in the Standard Model.

References

1. R. Carosi *et al.*, Phys. Lett. **B237**, 303 (1990).
2. M. Karlsson *et al.*, Phys. Rev. Lett. **64**, 2976 (1990); L.K. Gibbons *et al.*, Phys. Rev. Lett. **70**, 1199 (1993).
3. B. Schwingenheuer *et al.*, Phys. Rev. Lett. **74**, 4376 (1995).
4. S. Adler *et al.*, Phys. Rev. Lett. **79**, 2204 (1997).

TESTS OF DISCRETE SPACE-TIME SYMMETRIES

CHARGE CONJUGATION (*C*) INVARIANCE

$\Gamma(\pi^0 \to 3\gamma)/\Gamma_{total}$		$<3.1 \times 10^{-8}$, CL = 90%
η *C*-nonconserving decay parameters		
$\pi^+\pi^-\pi^0$ left-right asymmetry parameter		$(0.09 \pm 0.17) \times 10^{-2}$
$\pi^+\pi^-\pi^0$ sextant asymmetry parameter		$(0.18 \pm 0.16) \times 10^{-2}$
$\pi^+\pi^-\pi^0$ quadrant asymmetry parameter		$(-0.17 \pm 0.17) \times 10^{-2}$
$\pi^+\pi^-\gamma$ left-right asymmetry parameter		$(0.9 \pm 0.4) \times 10^{-2}$
$\pi^+\pi^-\gamma$ parameter β (*D*-wave)		0.05 ± 0.06 (S = 1.5)
$\Gamma(\eta \to 3\gamma)/\Gamma_{total}$		$<5 \times 10^{-4}$, CL = 95%
$\Gamma(\eta \to \pi^0 e^+ e^-)/\Gamma_{total}$	[a]	$<4 \times 10^{-5}$, CL = 90%
$\Gamma(\eta \to \pi^0 \mu^+ \mu^-)/\Gamma_{total}$	[a]	$<5 \times 10^{-6}$, CL = 90%
$\Gamma(\omega(782) \to \eta\pi^0)/\Gamma_{total}$		$<1 \times 10^{-3}$, CL = 90%
$\Gamma(\omega(782) \to 3\pi^0)/\Gamma_{total}$		$<3 \times 10^{-4}$, CL = 90%
$\Gamma(\eta'(958) \to \pi^0 e^+ e^-)/\Gamma_{total}$	[a]	$<1.3 \times 10^{-2}$, CL = 90%
$\Gamma(\eta'(958) \to \eta e^+ e^-)/\Gamma_{total}$	[a]	$<1.1 \times 10^{-2}$, CL = 90%
$\Gamma(\eta'(958) \to 3\gamma)/\Gamma_{total}$		$<1.0 \times 10^{-4}$, CL = 90%
$\Gamma(\eta'(958) \to \mu^+\mu^-\pi^0)/\Gamma_{total}$	[a]	$<6.0 \times 10^{-5}$, CL = 90%
$\Gamma(\eta'(958) \to \mu^+\mu^-\eta)/\Gamma_{total}$	[a]	$<1.5 \times 10^{-5}$, CL = 90%

Limits are given at the 90% confidence level, while errors are given as ±1 standard deviation.

PARITY (*P*) INVARIANCE

e electric dipole moment	$(0.18 \pm 0.16) \times 10^{-26}$ *e*cm
μ electric dipole moment	$(3.7 \pm 3.4) \times 10^{-19}$ *e*cm
τ electric dipole moment (d_τ)	> -3.1 and $< 3.1 \times 10^{-16}$ *e*cm, CL = 95%
$\Gamma(\eta \to \pi^+\pi^-)/\Gamma_{total}$	$<9 \times 10^{-4}$, CL = 90%
$\Gamma(\eta'(958) \to \pi^+\pi^-)/\Gamma_{total}$	$<2 \times 10^{-2}$, CL = 90%
$\Gamma(\eta'(958) \to \pi^0\pi^0)/\Gamma_{total}$	$<9 \times 10^{-4}$, CL = 90%
p electric dipole moment	$(-4 \pm 6) \times 10^{-23}$ *e*cm
n electric dipole moment	$<0.97 \times 10^{-25}$ *e*cm, CL = 90%
Λ electric dipole moment	$<1.5 \times 10^{-16}$ *e*cm, CL = 95%

TIME REVERSAL (*T*) INVARIANCE

Limits on *e*, μ, τ, *p*, *n*, and Λ electric dipole moments under Parity Invariance above are also tests of Time Reversal Invariance.

μ decay parameters		
transverse e^+ polarization normal to plane of μ spin, e^+ momentum		0.007 ± 0.023
α'/A		$(0 \pm 4) \times 10^{-3}$
β'/A		$(2 \pm 6) \times 10^{-3}$
τ electric dipole moment (d_τ)		> -3.1 and $< 3.1 \times 10^{-16}$ *e*cm, CL = 95%
Im(ξ) in $K^{\pm}_{\mu 3}$ decay (from transverse μ pol.)		-0.017 ± 0.025
Im(ξ) in $K^{0}_{\mu 3}$ decay (from transverse μ pol.)		-0.007 ± 0.026
$n \to pe^-\nu$ decay parameters		
ϕ_{AV}, phase of g_A relative to g_V	[b]	$(180.07 \pm 0.18)^\circ$
triple correlation coefficient *D*		$(-0.5 \pm 1.4) \times 10^{-3}$
triple correlation coefficient *D* for $\Sigma^- \to ne^-\overline{\nu}_e$		0.11 ± 0.10

CP INVARIANCE

Re(d^W_τ)		$<0.56 \times 10^{-17}$ *e*cm, CL = 95%
Im(d^W_τ)		$<1.5 \times 10^{-17}$ *e*cm, CL = 95%
$\Gamma(\eta \to \pi^+\pi^-)/\Gamma_{total}$		$<9 \times 10^{-4}$, CL = 90%
$\Gamma(\eta'(958) \to \pi^+\pi^-)/\Gamma_{total}$		$<2 \times 10^{-2}$, CL = 90%
$\Gamma(\eta'(958) \to \pi^0\pi^0)/\Gamma_{total}$		$<9 \times 10^{-4}$, CL = 90%
$K^\pm \to \pi^\pm\pi^+\pi^-$ rate difference/average		(0.07 ± 0.12)%
$K^\pm \to \pi^\pm\pi^0\pi^0$ rate difference/average		(0.0 ± 0.6)%
$K^\pm \to \pi^\pm\pi^0\gamma$ rate difference/average		(0.9 ± 3.3)%
$(g_{\tau+} - g_{\tau-}) / (g_{\tau+} + g_{\tau-})$ for $K^\pm \to \pi^\pm\pi^+\pi^-$		(-0.7 ± 0.5)%
CP-violation parameters in K^0_S decay		
$\text{Im}(\eta_{+-0}) = \text{Im}(A(K^0_S \to \pi^+\pi^-\pi^0$, *CP*-violating) / $A(K^0_L \to \pi^+\pi^-\pi^0))$		-0.002 ± 0.008
$\text{Im}(\eta_{000})^2 = \Gamma(K^0_S \to 3\pi^0) / \Gamma(K^0_L \to 3\pi^0)$		<0.1, CL = 90%
charge asymmetry *j* for $K^0_L \to \pi^+\pi^-\pi^0$		0.0011 ± 0.0008
$\lvert\epsilon'_{+-\gamma}\rvert/\epsilon$ for $K^0_L \to \pi^+\pi^-\gamma$		<0.3, CL = 90%
$\Gamma(K^0_L \to \pi^0\mu^+\mu^-)/\Gamma_{total}$	[c]	$<5.1 \times 10^{-9}$, CL = 90%
$\Gamma(K^0_L \to \pi^0 e^+ e^-)/\Gamma_{total}$	[c]	$<4.3 \times 10^{-9}$, CL = 90%
$\Gamma(K^0_L \to \pi^0\nu\overline{\nu})/\Gamma_{total}$	[d]	$<5.8 \times 10^{-5}$, CL = 90%
$A_{CP}(K^+K^-\pi^\pm)$ in $D^\pm \to K^+K^-\pi^\pm$		-0.017 ± 0.027
$A_{CP}(K^\pm K^{*0})$ in $D^+ \to K^+\overline{K}^{*0}$ and $D^- \to K^-K^{*0}$		-0.02 ± 0.05
$A_{CP}(\phi\pi^\pm)$ in $D^\pm \to \phi\pi^\pm$		-0.014 ± 0.033
$A_{CP}(\pi^+\pi^-\pi^\pm)$ in $D^\pm \to \pi^+\pi^-\pi^\pm$		-0.02 ± 0.04
$A_{CP}(K^+K^-)$ in $D^0, \overline{D}^0 \to K^+K^-$		0.026 ± 0.035
$A_{CP}(\pi^+\pi^-)$ in $D^0, \overline{D}^0 \to \pi^+\pi^-$		-0.05 ± 0.08
$A_{CP}(K^0_S\phi)$ in $D^0, \overline{D}^0 \to K^0_S\phi$		-0.03 ± 0.09
$A_{CP}(K^0_S\pi^0)$ in $D^0, \overline{D}^0 \to K^0_S\pi^0$		-0.018 ± 0.030
$\lvert\text{Re}(\epsilon_{B^0})\rvert$		0.002 ± 0.008
$\lvert\alpha_-(\Lambda) + \alpha_+(\Lambda)\rvert / \lvert\alpha_-(\Lambda) - \alpha_+(\Lambda)\rvert$		-0.03 ± 0.06

Tests of Conservation Laws

CP VIOLATION OBSERVED

K^0_L branching ratios		
charge asymmetry in $K^0_{\ell 3}$ decays		
$\delta(\mu) = [\Gamma(\pi^-\mu^+\nu_\mu) - \Gamma(\pi^+\mu^-\overline{\nu}_\mu)]/\text{sum}$		$(0.304 \pm 0.025)\%$
$\delta(e) = [\Gamma(\pi^-e^+\nu_e) - \Gamma(\pi^+e^-\overline{\nu}_e)]/\text{sum}$		$(0.333 \pm 0.014)\%$
parameters for $K^0_L \to 2\pi$ decay		
$\lvert\eta_{00}\rvert = \lvert A(K^0_L \to 2\pi^0) / A(K^0_S \to 2\pi^0)\rvert$		$(2.275 \pm 0.019) \times 10^{-3}$ (S = 1.1)
$\lvert\eta_{+-}\rvert = \lvert A(K^0_L \to \pi^+\pi^-) / A(K^0_S \to \pi^+\pi^-)\rvert$		$(2.285 \pm 0.019) \times 10^{-3}$
$\epsilon'/\epsilon \approx \text{Re}(\epsilon'/\epsilon) = (1 - \lvert\eta_{00}/\eta_{+-}\rvert)/3$	[e]	$(1.5 \pm 0.8) \times 10^{-3}$ (S = 1.8)
ϕ_{+-}, phase of η_{+-}		$(43.5 \pm 0.6)^\circ$
ϕ_{00}, phase of η_{00}		$(43.4 \pm 1.0)^\circ$
parameters for $K^0_L \to \pi^+\pi^-\gamma$ decay		
$\lvert\eta_{+-\gamma}\rvert = \lvert A(K^0_L \to \pi^+\pi^-\gamma, CP \text{ violating})/A(K^0_S \to \pi^+\pi^-\gamma)\rvert$		$(2.35 \pm 0.07) \times 10^{-3}$
$\phi_{+-\gamma}$ = phase of $\eta_{+-\gamma}$		$(44 \pm 4)^\circ$
$\Gamma(K^0_L \to \pi^+\pi^-)/\Gamma_{\text{total}}$		$(2.067 \pm 0.035) \times 10^{-3}$ (S = 1.1)
$\Gamma(K^0_L \to \pi^0\pi^0)/\Gamma_{\text{total}}$		$(9.36 \pm 0.20) \times 10^{-4}$

CPT INVARIANCE

$(m_{W^+} - m_{W^-}) / m_{\text{average}}$		-0.002 ± 0.007
$(m_{e^+} - m_{e^-}) / m_{\text{average}}$		$<4 \times 10^{-8}$, CL = 90%
$\lvert q_{e^+} + q_{e^-}\rvert/e$		$<2 \times 10^{-18}$
$(g_{e^+} - g_{e^-}) / g_{\text{average}}$		$(-0.5 \pm 2.1) \times 10^{-12}$
$(\tau_{\mu^+} - \tau_{\mu^-}) / \tau_{\text{average}}$		$(2 \pm 8) \times 10^{-5}$
$(g_{\mu^+} - g_{\mu^-}) / g_{\text{average}}$		$(-2.6 \pm 1.6) \times 10^{-8}$
$(m_{\pi^+} - m_{\pi^-}) / m_{\text{average}}$		$(2 \pm 5) \times 10^{-4}$
$(\tau_{\pi^+} - \tau_{\pi^-}) / \tau_{\text{average}}$		$(6 \pm 7) \times 10^{-4}$
$(m_{K^+} - m_{K^-}) / m_{\text{average}}$		$(-0.6 \pm 1.8) \times 10^{-4}$
$(\tau_{K^+} - \tau_{K^-}) / \tau_{\text{average}}$		$(0.11 \pm 0.09)\%$ (S = 1.2)
$K^\pm \to \mu^\pm\nu_\mu$ rate difference/average		$(-0.5 \pm 0.4)\%$
$K^\pm \to \pi^\pm\pi^0$ rate difference/average	[f]	$(0.8 \pm 1.2)\%$
$\lvert m_{K^0} - m_{\overline{K}^0}\rvert / m_{\text{average}}$	[g]	$<10^{-18}$
phase difference $\phi_{00} - \phi_{+-}$		$(-0.1 \pm 0.8)^\circ$
CPT-violation parameters in K^0 decay		
real part of Δ		0.018 ± 0.020
imaginary part of Δ		0.02 ± 0.04
$(\lvert\frac{q_{\overline{p}}}{m_{\overline{p}}}\rvert - \frac{q_p}{m_p})/\lvert\frac{q}{m}\rvert_{\text{average}}$		$(1.5 \pm 1.1) \times 10^{-9}$
$\lvert q_p + q_{\overline{p}}\rvert/e$		$<2 \times 10^{-5}$
$(\mu_p + \mu_{\overline{p}}) / \lvert\mu\rvert_{\text{average}}$		$(-2.6 \pm 2.9) \times 10^{-3}$
$(m_n - m_{\overline{n}}) / m_{\text{average}}$		$(9 \pm 5) \times 10^{-5}$
$(m_\Lambda - m_{\overline{\Lambda}}) / m_\Lambda$		$(-1.0 \pm 0.9) \times 10^{-5}$
$(\tau_\Lambda - \tau_{\overline{\Lambda}}) / \tau_{\text{average}}$		0.04 ± 0.09
$(\mu_{\Sigma^+} + \mu_{\overline{\Sigma}^-}) / \lvert\mu\rvert_{\text{average}}$		0.014 ± 0.015
$(m_{\Xi^-} - m_{\overline{\Xi}^+}) / m_{\text{average}}$		$(1.1 \pm 2.7) \times 10^{-4}$
$(\tau_{\Xi^-} - \tau_{\overline{\Xi}^+}) / \tau_{\text{average}}$		0.02 ± 0.18
$(m_{\Omega^-} - m_{\overline{\Omega}^+}) / m_{\text{average}}$		$(0 \pm 5) \times 10^{-4}$

TESTS OF NUMBER CONSERVATION LAWS

LEPTON FAMILY NUMBER

Lepton family number conservation means separate conservation of each of L_e, L_μ, L_τ.

$\Gamma(Z \to e^\pm\mu^\mp)/\Gamma_{\text{total}}$	[h]	$<1.7 \times 10^{-6}$, CL = 95%
$\Gamma(Z \to e^\pm\tau^\mp)/\Gamma_{\text{total}}$	[h]	$<9.8 \times 10^{-6}$, CL = 95%
$\Gamma(Z \to \mu^\pm\tau^\mp)/\Gamma_{\text{total}}$	[h]	$<1.2 \times 10^{-5}$, CL = 95%
limit on $\mu^- \to e^-$ conversion		
$\sigma(\mu^-\,{}^{32}\text{S} \to e^-\,{}^{32}\text{S}) / \sigma(\mu^-\,{}^{32}\text{S} \to \nu_\mu\,{}^{32}\text{P}^*)$		$<7 \times 10^{-11}$, CL = 90%
$\sigma(\mu^-\,\text{Ti} \to e^-\,\text{Ti}) / \sigma(\mu^-\,\text{Ti} \to \text{capture})$		$<4.3 \times 10^{-12}$, CL = 90%
$\sigma(\mu^-\,\text{Pb} \to e^-\,\text{Pb}) / \sigma(\mu^-\,\text{Pb} \to \text{capture})$		$<4.6 \times 10^{-11}$, CL = 90%
limit on muonium → antimuonium conversion $R_g = G_C / G_F$		<0.018, CL = 90%
$\Gamma(\mu^- \to e^-\nu_e\overline{\nu}_\mu)/\Gamma_{\text{total}}$	[i]	$<1.2 \times 10^{-2}$, CL = 90%
$\Gamma(\mu^- \to e^-\gamma)/\Gamma_{\text{total}}$		$<4.9 \times 10^{-11}$, CL = 90%
$\Gamma(\mu^- \to e^-e^+e^-)/\Gamma_{\text{total}}$		$<1.0 \times 10^{-12}$, CL = 90%
$\Gamma(\mu^- \to e^-2\gamma)/\Gamma_{\text{total}}$		$<7.2 \times 10^{-11}$, CL = 90%
$\Gamma(\tau^- \to e^-\gamma)/\Gamma_{\text{total}}$		$<2.7 \times 10^{-6}$, CL = 90%
$\Gamma(\tau^- \to \mu^-\gamma)/\Gamma_{\text{total}}$		$<3.0 \times 10^{-6}$, CL = 90%
$\Gamma(\tau^- \to e^-\pi^0)/\Gamma_{\text{total}}$		$<3.7 \times 10^{-6}$, CL = 90%
$\Gamma(\tau^- \to \mu^-\pi^0)/\Gamma_{\text{total}}$		$<4.0 \times 10^{-6}$, CL = 90%
$\Gamma(\tau^- \to e^-K^0)/\Gamma_{\text{total}}$		$<1.3 \times 10^{-3}$, CL = 90%
$\Gamma(\tau^- \to \mu^-K^0)/\Gamma_{\text{total}}$		$<1.0 \times 10^{-3}$, CL = 90%
$\Gamma(\tau^- \to e^-\eta)/\Gamma_{\text{total}}$		$<8.2 \times 10^{-6}$, CL = 90%
$\Gamma(\tau^- \to \mu^-\eta)/\Gamma_{\text{total}}$		$<9.6 \times 10^{-6}$, CL = 90%
$\Gamma(\tau^- \to e^-\rho^0)/\Gamma_{\text{total}}$		$<2.0 \times 10^{-6}$, CL = 90%
$\Gamma(\tau^- \to \mu^-\rho^0)/\Gamma_{\text{total}}$		$<6.3 \times 10^{-6}$, CL = 90%
$\Gamma(\tau^- \to e^-K^*(892)^0)/\Gamma_{\text{total}}$		$<5.1 \times 10^{-6}$, CL = 90%
$\Gamma(\tau^- \to \mu^-K^*(892)^0)/\Gamma_{\text{total}}$		$<7.5 \times 10^{-6}$, CL = 90%
$\Gamma(\tau^- \to e^-\overline{K}^*(892)^0)/\Gamma_{\text{total}}$		$<7.4 \times 10^{-6}$, CL = 90%
$\Gamma(\tau^- \to \mu^-\overline{K}^*(892)^0)/\Gamma_{\text{total}}$		$<7.5 \times 10^{-6}$, CL = 90%
$\Gamma(\tau^- \to e^-\phi)/\Gamma_{\text{total}}$		$<6.9 \times 10^{-6}$, CL = 90%
$\Gamma(\tau^- \to \mu^-\phi)/\Gamma_{\text{total}}$		$<7.0 \times 10^{-6}$, CL = 90%
$\Gamma(\tau^- \to e^-e^+e^-)/\Gamma_{\text{total}}$		$<2.9 \times 10^{-6}$, CL = 90%
$\Gamma(\tau^- \to e^-\mu^+\mu^-)/\Gamma_{\text{total}}$		$<1.8 \times 10^{-6}$, CL = 90%
$\Gamma(\tau^- \to e^+\mu^-\mu^-)/\Gamma_{\text{total}}$		$<1.5 \times 10^{-6}$, CL = 90%
$\Gamma(\tau^- \to \mu^-e^+e^-)/\Gamma_{\text{total}}$		$<1.7 \times 10^{-6}$, CL = 90%
$\Gamma(\tau^- \to \mu^+e^-e^-)/\Gamma_{\text{total}}$		$<1.5 \times 10^{-6}$, CL = 90%
$\Gamma(\tau^- \to \mu^-\mu^+\mu^-)/\Gamma_{\text{total}}$		$<1.9 \times 10^{-6}$, CL = 90%
$\Gamma(\tau^- \to e^-\pi^+\pi^-)/\Gamma_{\text{total}}$		$<2.2 \times 10^{-6}$, CL = 90%
$\Gamma(\tau^- \to \mu^-\pi^+\pi^-)/\Gamma_{\text{total}}$		$<8.2 \times 10^{-6}$, CL = 90%
$\Gamma(\tau^- \to e^-\pi^+K^-)/\Gamma_{\text{total}}$		$<6.4 \times 10^{-6}$, CL = 90%
$\Gamma(\tau^- \to e^-\pi^-K^+)/\Gamma_{\text{total}}$		$<3.8 \times 10^{-6}$, CL = 90%
$\Gamma(\tau^- \to e^-K^+K^-)/\Gamma_{\text{total}}$		$<6.0 \times 10^{-6}$, CL = 90%
$\Gamma(\tau^- \to \mu^-\pi^+K^-)/\Gamma_{\text{total}}$		$<7.5 \times 10^{-6}$, CL = 90%
$\Gamma(\tau^- \to \mu^-\pi^-K^+)/\Gamma_{\text{total}}$		$<7.4 \times 10^{-6}$, CL = 90%
$\Gamma(\tau^- \to \mu^-K^+K^-)/\Gamma_{\text{total}}$		$<1.5 \times 10^{-5}$, CL = 90%
$\Gamma(\tau^- \to e^-\pi^0\pi^0)/\Gamma_{\text{total}}$		$<6.5 \times 10^{-6}$, CL = 90%
$\Gamma(\tau^- \to \mu^-\pi^0\pi^0)/\Gamma_{\text{total}}$		$<1.4 \times 10^{-5}$, CL = 90%
$\Gamma(\tau^- \to e^-\eta\eta)/\Gamma_{\text{total}}$		$<3.5 \times 10^{-5}$, CL = 90%
$\Gamma(\tau^- \to \mu^-\eta\eta)/\Gamma_{\text{total}}$		$<6.0 \times 10^{-5}$, CL = 90%
$\Gamma(\tau^- \to e^-\pi^0\eta)/\Gamma_{\text{total}}$		$<2.4 \times 10^{-5}$, CL = 90%
$\Gamma(\tau^- \to \mu^-\pi^0\eta)/\Gamma_{\text{total}}$		$<2.2 \times 10^{-5}$, CL = 90%
$\Gamma(\tau^- \to e^-\text{light boson})/\Gamma_{\text{total}}$		$<2.7 \times 10^{-3}$, CL = 95%
$\Gamma(\tau^- \to \mu^-\text{light boson})/\Gamma_{\text{total}}$		$<5 \times 10^{-3}$, CL = 95%

ν oscillations. (For other lepton mixing effects in particle decays, see the Particle Listings.)

$\overline{\nu}_e \not\to \overline{\nu}_e$	
$\Delta(m^2)$ for $\sin^2(2\theta) = 1$	$<9 \times 10^{-4}$ eV^2, CL = 90%
$\sin^2(2\theta)$ for "Large" $\Delta(m^2)$	<0.02, CL = 90%
$\nu_e \to \nu_\tau$	
$\Delta(m^2)$ for $\sin^2(2\theta) = 1$	<9 eV^2, CL = 90%
$\sin^2(2\theta)$ for "Large" $\Delta(m^2)$	<0.25, CL = 90%
$\overline{\nu}_e \to \overline{\nu}_\tau$	
$\sin^2(2\theta)$ for "Large" $\Delta(m^2)$	<0.7, CL = 90%
$\nu_\mu \to \nu_e$	

Limits are given at the 90% confidence level, while errors are given as ±1 standard deviation.

$\Delta(m^2)$ for $\sin^2(2\theta)=1$		<0.09 eV^2, CL = 90%
$\sin^2(2\theta)$ for "Large" $\Delta(m^2)$		$<3.0\times10^{-3}$, CL = 90%
$\overline{\nu}_\mu \to \overline{\nu}_e$		
$\Delta(m^2)$ for $\sin^2(2\theta)=1$		<0.14 eV^2, CL = 90%
$\sin^2(2\theta)$ for "Large" $\Delta(m^2)$		<0.004, CL = 95%
$\nu_\mu(\overline{\nu}_\mu) \to \nu_e(\overline{\nu}_e)$		
$\Delta(m^2)$ for $\sin^2(2\theta)=1$		<0.075 eV^2, CL = 90%
$\sin^2(2\theta)$ for "Large" $\Delta(m^2)$		$<1.8\times10^{-3}$, CL = 90%
$\nu_\mu \to \nu_\tau$		
$\Delta(m^2)$ for $\sin^2(2\theta)=1$		<0.9 eV^2, CL = 90%
$\sin^2(2\theta)$ for "Large" $\Delta(m^2)$		<0.004, CL = 90%
$\overline{\nu}_\mu \to \overline{\nu}_\tau$		
$\Delta(m^2)$ for $\sin^2(2\theta)=1$		<2.2 eV^2, CL = 90%
$\sin^2(2\theta)$ for "Large" $\Delta(m^2)$		$<4.4\times10^{-2}$, CL = 90%
$\nu_\mu(\overline{\nu}_\mu) \to \nu_\tau(\overline{\nu}_\tau)$		
$\Delta(m^2)$ for $\sin^2(2\theta)=1$		<1.5 eV^2, CL = 90%
$\sin^2(2\theta)$ for "Large" $\Delta(m^2)$		$<8\times10^{-3}$, CL = 90%
$\nu_e \not\to \nu_e$		
$\Delta(m^2)$ for $\sin^2(2\theta)=1$		<0.17 eV^2, CL = 90%
$\sin^2(2\theta)$ for "Large" $\Delta(m^2)$		$<7\times10^{-2}$, CL = 90%
$\nu_\mu \not\to \nu_\mu$		
$\Delta(m^2)$ for $\sin^2(2\theta)=1$		<0.23 or >1500 eV^2
$\sin^2(2\theta)$ for $\Delta(m^2)=100$eV^2	[j]	<0.02, CL = 90%
$\overline{\nu}_\mu \not\to \overline{\nu}_\mu$		
$\Delta(m^2)$ for $\sin^2(2\theta)=1$		<7 or >1200 eV^2
$\sin^2(2\theta)$ for 190 eV^2 $<\Delta(m^2)<$ 320 eV^2	[k]	<0.02, CL = 90%
$\Gamma(\pi^+\to\mu^+\nu_e)/\Gamma_{total}$	[l]	$<8.0\times10^{-3}$, CL = 90%
$\Gamma(\pi^+\to\mu^-e^+e^+\nu)/\Gamma_{total}$		$<1.6\times10^{-6}$, CL = 90%
$\Gamma(\pi^0\to\mu^+e^-+e^-\mu^+)/\Gamma_{total}$		$<1.72\times10^{-8}$, CL = 90%
$\Gamma(\eta\to\mu^+e^-+\mu^-e^+)/\Gamma_{total}$		$<6\times10^{-6}$, CL = 90%
$\Gamma(K^+\to\mu^-\nu e^+e^+)/\Gamma_{total}$		$<2.0\times10^{-8}$, CL = 90%
$\Gamma(K^+\to\mu^+\nu_e)/\Gamma_{total}$	[l]	$<4\times10^{-3}$, CL = 90%
$\Gamma(K^+\to\pi^+\mu^+e^-)/\Gamma_{total}$		$<2.1\times10^{-10}$, CL = 90%
$\Gamma(K^+\to\pi^+\mu^-e^+)/\Gamma_{total}$		$<7\times10^{-9}$, CL = 90%
$\Gamma(K^0_L\to e^\pm\mu^\mp)/\Gamma_{total}$	[h]	$<3.3\times10^{-11}$, CL = 90%
$\Gamma(K^0_L\to e^\pm e^\pm\mu^\mp\mu^\mp)/\Gamma_{total}$	[h]	$<6.1\times10^{-9}$, CL = 90%
$\Gamma(D^+\to\pi^+e^+\mu^-)/\Gamma_{total}$		$<1.1\times10^{-4}$, CL = 90%
$\Gamma(D^+\to\pi^+e^-\mu^+)/\Gamma_{total}$		$<1.3\times10^{-4}$, CL = 90%
$\Gamma(D^+\to K^+e^+\mu^-)/\Gamma_{total}$		$<1.3\times10^{-4}$, CL = 90%
$\Gamma(D^+\to K^+e^-\mu^+)/\Gamma_{total}$		$<1.2\times10^{-4}$, CL = 90%
$\Gamma(D^0\to\mu^\pm e^\mp)/\Gamma_{total}$	[h]	$<1.9\times10^{-5}$, CL = 90%
$\Gamma(D^0\to\pi^0e^\pm\mu^\mp)/\Gamma_{total}$	[h]	$<8.6\times10^{-5}$, CL = 90%
$\Gamma(D^0\to\eta e^\pm\mu^\mp)/\Gamma_{total}$	[h]	$<1.0\times10^{-4}$, CL = 90%
$\Gamma(D^0\to\rho^0e^\pm\mu^\mp)/\Gamma_{total}$	[h]	$<4.9\times10^{-5}$, CL = 90%
$\Gamma(D^0\to\omega e^\pm\mu^\mp)/\Gamma_{total}$	[h]	$<1.2\times10^{-4}$, CL = 90%
$\Gamma(D^0\to\phi e^\pm\mu^\mp)/\Gamma_{total}$	[h]	$<3.4\times10^{-5}$, CL = 90%
$\Gamma(D^0\to\overline{K}^0e^\pm\mu^\mp)/\Gamma_{total}$	[h]	$<1.0\times10^{-4}$, CL = 90%
$\Gamma(D^0\to\overline{K}^*(892)^0e^\pm\mu^\mp)/\Gamma_{total}$	[h]	$<1.0\times10^{-4}$, CL = 90%
$\Gamma(B^+\to\pi^+e^+\mu^-)/\Gamma_{total}$		$<6.4\times10^{-3}$, CL = 90%
$\Gamma(B^+\to\pi^+e^-\mu^+)/\Gamma_{total}$		$<6.4\times10^{-3}$, CL = 90%
$\Gamma(B^+\to K^+e^+\mu^-)/\Gamma_{total}$		$<6.4\times10^{-3}$, CL = 90%
$\Gamma(B^+\to K^+e^-\mu^+)/\Gamma_{total}$		$<6.4\times10^{-3}$, CL = 90%
$\Gamma(B^+\to\pi^-e^+\mu^+)/\Gamma_{total}$		$<6.4\times10^{-3}$, CL = 90%
$\Gamma(B^+\to K^-e^+\mu^+)/\Gamma_{total}$		$<6.4\times10^{-3}$, CL = 90%
$\Gamma(B^0\to e^\pm\mu^\mp)/\Gamma_{total}$	[h]	$<5.9\times10^{-6}$, CL = 90%
$\Gamma(B^0\to e^\pm\tau^\mp)/\Gamma_{total}$	[h]	$<5.3\times10^{-4}$, CL = 90%
$\Gamma(B^0\to\mu^\pm\tau^\mp)/\Gamma_{total}$	[h]	$<8.3\times10^{-4}$, CL = 90%
$\Gamma(B\to e^\pm\mu^\mp s)/\Gamma_{total}$		$<2.2\times10^{-5}$, CL = 90%
$\Gamma(B^0_s\to e^\pm\mu^\mp)/\Gamma_{total}$	[h]	$<4.1\times10^{-5}$, CL = 90%

TOTAL LEPTON NUMBER

Violation of total lepton number conservation also implies violation of lepton family number conservation.

limit on $\mu^-\to e^+$ conversion		
$\sigma(\mu^-\,{}^{32}S\to e^+\,{}^{32}Si^*)$ / $\sigma(\mu^-\,{}^{32}S\to\nu_\mu\,{}^{32}P^*)$		$<9\times10^{-10}$, CL = 90%
$\sigma(\mu^-\,{}^{127}I\to e^+\,{}^{127}Sb^*)$ / $\sigma(\mu^-\,{}^{127}I\to$ anything)		$<3\times10^{-10}$, CL = 90%
$\sigma(\mu^-\,Ti\to e^+\,Ca)$ / $\sigma(\mu^-\,Ti\to$ capture)		$<8.9\times10^{-11}$, CL = 90%
$\Gamma(\tau^-\to\pi^-\gamma)/\Gamma_{total}$		$<2.8\times10^{-4}$, CL = 90%
$\Gamma(\tau^-\to\pi^-\pi^0)/\Gamma_{total}$		$<3.7\times10^{-4}$, CL = 90%
$\Gamma(\tau^-\to e^+\pi^-\pi^-)/\Gamma_{total}$		$<1.9\times10^{-6}$, CL = 90%
$\Gamma(\tau^-\to\mu^+\pi^-\pi^-)/\Gamma_{total}$		$<3.4\times10^{-6}$, CL = 90%
$\Gamma(\tau^-\to e^+\pi^-K^-)/\Gamma_{total}$		$<2.1\times10^{-6}$, CL = 90%
$\Gamma(\tau^-\to e^+K^-K^-)/\Gamma_{total}$		$<3.8\times10^{-6}$, CL = 90%
$\Gamma(\tau^-\to\mu^+\pi^-K^-)/\Gamma_{total}$		$<7.0\times10^{-6}$, CL = 90%
$\Gamma(\tau^-\to\mu^+K^-K^-)/\Gamma_{total}$		$<6.0\times10^{-6}$, CL = 90%
$\Gamma(\tau^-\to\overline{p}\gamma)/\Gamma_{total}$		$<2.9\times10^{-4}$, CL = 90%
$\Gamma(\tau^-\to\overline{p}\pi^0)/\Gamma_{total}$		$<6.6\times10^{-4}$, CL = 90%
$\Gamma(\tau^-\to\overline{p}\eta)/\Gamma_{total}$		$<1.30\times10^{-3}$, CL = 90%
$\nu_e\to(\overline{\nu}_e)_L$		
$\alpha\Delta(m^2)$ for $\sin^2(2\theta)=1$		<0.14 eV^2, CL = 90%
$\alpha^2\sin^2(2\theta)$ for "Large" $\Delta(m^2)$		<0.032, CL = 90%
$\nu_\mu\to(\overline{\nu}_e)_L$		
$\alpha\Delta(m^2)$ for $\sin^2(2\theta)=1$		<0.16 eV^2, CL = 90%
$\alpha^2\sin^2(2\theta)$ for "Large" $\Delta(m^2)$		<0.001, CL = 90%
$\Gamma(\pi^+\to\mu^+\overline{\nu}_e)/\Gamma_{total}$	[l]	$<1.5\times10^{-3}$, CL = 90%
$\Gamma(K^+\to\pi^-\mu^+e^+)/\Gamma_{total}$		$<7\times10^{-9}$, CL = 90%
$\Gamma(K^+\to\pi^-e^+e^+)/\Gamma_{total}$		$<1.0\times10^{-8}$, CL = 90%
$\Gamma(K^+\to\pi^-\mu^+\mu^+)/\Gamma_{total}$	[l]	$<1.5\times10^{-4}$, CL = 90%
$\Gamma(K^+\to\mu^+\overline{\nu}_e)/\Gamma_{total}$	[l]	$<3.3\times10^{-3}$, CL = 90%
$\Gamma(K^+\to\pi^0e^+\overline{\nu}_e)/\Gamma_{total}$		$<3\times10^{-3}$, CL = 90%
$\Gamma(D^+\to\pi^-e^+e^+)/\Gamma_{total}$		$<1.1\times10^{-4}$, CL = 90%
$\Gamma(D^+\to\pi^-\mu^+\mu^+)/\Gamma_{total}$		$<8.7\times10^{-5}$, CL = 90%
$\Gamma(D^+\to\pi^-e^+\mu^+)/\Gamma_{total}$		$<1.1\times10^{-4}$, CL = 90%
$\Gamma(D^+\to\rho^-\mu^+\mu^+)/\Gamma_{total}$		$<5.6\times10^{-4}$, CL = 90%
$\Gamma(D^+\to K^-e^+e^+)/\Gamma_{total}$		$<1.2\times10^{-4}$, CL = 90%
$\Gamma(D^+\to K^-\mu^+\mu^+)/\Gamma_{total}$		$<1.2\times10^{-4}$, CL = 90%
$\Gamma(D^+\to K^-e^+\mu^+)/\Gamma_{total}$		$<1.3\times10^{-4}$, CL = 90%
$\Gamma(D^+\to K^*(892)^-\mu^+\mu^+)/\Gamma_{total}$		$<8.5\times10^{-4}$, CL = 90%
$\Gamma(D^+_s\to\pi^-\mu^+\mu^+)/\Gamma_{total}$		$<4.3\times10^{-4}$, CL = 90%
$\Gamma(D^+_s\to K^-\mu^+\mu^+)/\Gamma_{total}$		$<5.9\times10^{-4}$, CL = 90%
$\Gamma(D^+_s\to K^*(892)^-\mu^+\mu^+)/\Gamma_{total}$		$<1.4\times10^{-3}$, CL = 90%
$\Gamma(B^+\to\pi^-e^+e^+)/\Gamma_{total}$		$<3.9\times10^{-3}$, CL = 90%
$\Gamma(B^+\to\pi^-\mu^+\mu^+)/\Gamma_{total}$		$<9.1\times10^{-3}$, CL = 90%
$\Gamma(B^+\to K^-e^+e^+)/\Gamma_{total}$		$<3.9\times10^{-3}$, CL = 90%
$\Gamma(B^+\to K^-\mu^+\mu^+)/\Gamma_{total}$		$<9.1\times10^{-3}$, CL = 90%
$\Gamma(\Xi^-\to p\mu^-\mu^-)/\Gamma_{total}$		$<4\times10^{-4}$, CL = 90%
$\Gamma(\Lambda^+_c\to\Sigma^-\mu^+\mu^+)/\Gamma_{total}$		$<7.0\times10^{-4}$, CL = 90%

BARYON NUMBER

$\Gamma(\tau^-\to\overline{p}\gamma)/\Gamma_{total}$		$<2.9\times10^{-4}$, CL = 90%
$\Gamma(\tau^-\to\overline{p}\pi^0)/\Gamma_{total}$		$<6.6\times10^{-4}$, CL = 90%
$\Gamma(\tau^-\to\overline{p}\eta)/\Gamma_{total}$		$<1.30\times10^{-3}$, CL = 90%
p mean life		$>1.6\times10^{25}$ years

A few examples of proton or bound neutron decay follow. For limits on many other nucleon decay channels, see the Baryon Summary Table.

$\tau(N\to e^+\pi)$		>130 (n), >550 (p) $\times10^{30}$ years, CL = 90%
$\tau(N\to\mu^+\pi)$		>100 (n), >270 (p) $\times10^{30}$ years, CL = 90%
$\tau(N\to e^+K)$		>1.3 (n), >150 (p) $\times10^{30}$ years, CL = 90%
$\tau(N\to\mu^+K)$		>1.1 (n), >120 (p) $\times10^{30}$ years, CL = 90%
limit on $n\overline{n}$ oscillations (bound n)	[m]	$>1.2\times10^{8}$ s, CL = 90%
limit on $n\overline{n}$ oscillations (free n)		$>0.86\times10^{8}$ s, CL = 90%

Limits are given at the 90% confidence level, while errors are given as ±1 standard deviation.

Tests of Conservation Laws

ELECTRIC CHARGE (Q)

e mean life / branching fraction	[n]	$>4.3 \times 10^{23}$ yr, CL = 68%
$\Gamma(n \rightarrow p \nu_e \overline{\nu}_e)/\Gamma_{total}$		$<8 \times 10^{-27}$, CL = 68%

$\Delta S = \Delta Q$ RULE

Allowed in second-order weak interactions.

$\Gamma(K^+ \rightarrow \pi^+\pi^+e^-\overline{\nu}_e)/\Gamma_{total}$		$<1.2 \times 10^{-8}$, CL = 90%
$\Gamma(K^+ \rightarrow \pi^+\pi^+\mu^-\overline{\nu}_\mu)/\Gamma_{total}$		$<3.0 \times 10^{-6}$, CL = 95%
$x = A(\overline{K}^0 \rightarrow \pi^-\ell^+\nu)/A(K^0 \rightarrow \pi^-\ell^+\nu) = A(\Delta S{=}-\Delta Q)/A(\Delta S{=}\Delta Q)$		
real part of x		0.006 ± 0.018 (S = 1.3)
imaginary part of x		-0.003 ± 0.026 (S = 1.2)
$\Gamma(\Sigma^+ \rightarrow n\ell^+\nu)/\Gamma(\Sigma^- \rightarrow n\ell^-\overline{\nu})$		<0.043
$\Gamma(\Sigma^+ \rightarrow ne^+\nu_e)/\Gamma_{total}$		$<5 \times 10^{-6}$, CL = 90%
$\Gamma(\Sigma^+ \rightarrow n\mu^+\nu_\mu)/\Gamma_{total}$		$<3.0 \times 10^{-5}$, CL = 90%
$\Gamma(\Xi^0 \rightarrow \Sigma^-e^+\nu_e)/\Gamma_{total}$		$<9 \times 10^{-4}$, CL = 90%
$\Gamma(\Xi^0 \rightarrow \Sigma^-\mu^+\nu_\mu)/\Gamma_{total}$		$<9 \times 10^{-4}$, CL = 90%

$\Delta S = 2$ FORBIDDEN

Allowed in second-order weak interactions.

$\Gamma(\Xi^0 \rightarrow p\pi^-)/\Gamma_{total}$		$<4 \times 10^{-5}$, CL = 90%
$\Gamma(\Xi^0 \rightarrow pe^-\overline{\nu}_e)/\Gamma_{total}$		$<1.3 \times 10^{-3}$
$\Gamma(\Xi^0 \rightarrow p\mu^-\overline{\nu}_\mu)/\Gamma_{total}$		$<1.3 \times 10^{-3}$
$\Gamma(\Xi^- \rightarrow n\pi^-)/\Gamma_{total}$		$<1.9 \times 10^{-5}$, CL = 90%
$\Gamma(\Xi^- \rightarrow ne^-\overline{\nu}_e)/\Gamma_{total}$		$<3.2 \times 10^{-3}$, CL = 90%
$\Gamma(\Xi^- \rightarrow n\mu^-\overline{\nu}_\mu)/\Gamma_{total}$		$<1.5 \times 10^{-2}$, CL = 90%
$\Gamma(\Xi^- \rightarrow p\pi^-\pi^-)/\Gamma_{total}$		$<4 \times 10^{-4}$, CL = 90%
$\Gamma(\Xi^- \rightarrow p\pi^-e^-\overline{\nu}_e)/\Gamma_{total}$		$<4 \times 10^{-4}$, CL = 90%
$\Gamma(\Xi^- \rightarrow p\pi^-\mu^-\overline{\nu}_\mu)/\Gamma_{total}$		$<4 \times 10^{-4}$, CL = 90%
$\Gamma(\Omega^- \rightarrow \Lambda\pi^-)/\Gamma_{total}$		$<1.9 \times 10^{-4}$, CL = 90%

$\Delta S = 2$ VIA MIXING

Allowed in second-order weak interactions, e.g. mixing.

$m_{K^0_L} - m_{K^0_S}$		$(0.5301 \pm 0.0014) \times 10^{10}\ \hbar\ s^{-1}$
$m_{K^0_L} - m_{K^0_S}$		$(3.489 \pm 0.009) \times 10^{-12}$ MeV

$\Delta C = 2$ VIA MIXING

Allowed in second-order weak interactions, e.g. mixing.

$\vert m_{D^0_1} - m_{D^0_2}\vert$	[o]	$<24 \times 10^{10}\ \hbar\ s^{-1}$, CL = 90%
$\vert\Gamma_{D^0_1} - \Gamma_{D^0_2}\vert/\Gamma_{D^0}$ mean life difference/average	[o]	<0.20, CL = 90%
$\Gamma(K^+\ell^-\overline{\nu}_\ell(\text{via } \overline{D}^0))/\Gamma(K^-\ell^+\nu_\ell)$		<0.005, CL = 90%
$\Gamma(K^+\pi^- \text{ or } K^+\pi^-\pi^+\pi^- (\text{via } \overline{D}^0))/\Gamma(K^-\pi^+ \text{ or } K^-\pi^+\pi^+\pi^-)$	[p]	< 0.0085 (or < 0.0037), CL = 90%
$\Gamma(D^0 \rightarrow K^+\ell^-\overline{\nu}_\ell(\text{via } \overline{D}^0))/\Gamma_{total}$		$<1.7 \times 10^{-4}$, CL = 90%
$\Gamma(D^0 \rightarrow K^+\pi^- \text{ or } K^+\pi^-\pi^+\pi^- (\text{via } \overline{D}^0))/\Gamma_{total}$		$<1.0 \times 10^{-3}$, CL = 90%

$\Delta B = 2$ VIA MIXING

Allowed in second-order weak interactions, e.g. mixing.

x_d		0.172 ± 0.010
$\Delta m_{B^0} = m_{B^0_H} - m_{B^0_L}$		$(0.464 \pm 0.018) \times 10^{12}\ \hbar\ s^{-1}$
$x_d = \Delta m_{B^0}/\Gamma_{B^0}$		0.723 ± 0.032
χ_B at high energy		0.118 ± 0.006
$\Delta m_{B^0_s} = m_{B^0_{sH}} - m_{B^0_{sL}}$		$>9.1 \times 10^{12}\ \hbar\ s^{-1}$, CL = 95%
$x_s = \Delta m_{B^0_s}/\Gamma_{B^0_s}$		>14.0, CL = 95%
χ_s		>0.4975, CL = 95%

$\Delta S = 1$ WEAK NEUTRAL CURRENT FORBIDDEN

Allowed by higher-order electroweak interactions.

$\Gamma(K^+ \rightarrow \pi^+e^+e^-)/\Gamma_{total}$		$(2.74 \pm 0.23) \times 10^{-7}$
$\Gamma(K^+ \rightarrow \pi^+\mu^+\mu^-)/\Gamma_{total}$		$(5.0 \pm 1.0) \times 10^{-8}$
$\Gamma(K^+ \rightarrow \pi^+\nu\overline{\nu})/\Gamma_{total}$		$(4.2^{+9.7}_{-3.5}) \times 10^{-10}$
$\Gamma(K^0_S \rightarrow \mu^+\mu^-)/\Gamma_{total}$		$<3.2 \times 10^{-7}$, CL = 90%
$\Gamma(K^0_S \rightarrow e^+e^-)/\Gamma_{total}$		$<1.4 \times 10^{-7}$, CL = 90%
$\Gamma(K^0_S \rightarrow \pi^0e^+e^-)/\Gamma_{total}$		$<1.1 \times 10^{-6}$, CL = 90%
$\Gamma(K^0_L \rightarrow \mu^+\mu^-)/\Gamma_{total}$		$(7.2 \pm 0.5) \times 10^{-9}$ (S = 1.4)
$\Gamma(K^0_L \rightarrow \mu^+\mu^-\gamma)/\Gamma_{total}$		$(3.25 \pm 0.28) \times 10^{-7}$
$\Gamma(K^0_L \rightarrow e^+e^-)/\Gamma_{total}$		$<4.1 \times 10^{-11}$, CL = 90%
$\Gamma(K^0_L \rightarrow e^+e^-\gamma)/\Gamma_{total}$		$(9.1 \pm 0.5) \times 10^{-6}$
$\Gamma(K^0_L \rightarrow e^+e^-\gamma\gamma)/\Gamma_{total}$	[q]	$(6.5 \pm 1.2) \times 10^{-7}$
$\Gamma(K^0_L \rightarrow \pi^+\pi^-e^+e^-)/\Gamma_{total}$	[q]	$<4.6 \times 10^{-7}$, CL = 90%
$\Gamma(K^0_L \rightarrow \mu^+\mu^-e^+e^-)/\Gamma_{total}$		$(2.9^{+6.7}_{-2.4}) \times 10^{-9}$
$\Gamma(K^0_L \rightarrow e^+e^-e^+e^-)/\Gamma_{total}$		$(4.1 \pm 0.8) \times 10^{-8}$ (S = 1.2)
$\Gamma(K^0_L \rightarrow \pi^0\mu^+\mu^-)/\Gamma_{total}$		$<5.1 \times 10^{-9}$, CL = 90%
$\Gamma(K^0_L \rightarrow \pi^0e^+e^-)/\Gamma_{total}$		$<4.3 \times 10^{-9}$, CL = 90%
$\Gamma(K^0_L \rightarrow \pi^0\nu\overline{\nu})/\Gamma_{total}$		$<5.8 \times 10^{-5}$, CL = 90%
$\Gamma(\Sigma^+ \rightarrow pe^+e^-)/\Gamma_{total}$		$<7 \times 10^{-6}$

$\Delta C = 1$ WEAK NEUTRAL CURRENT FORBIDDEN

Allowed by higher-order electroweak interactions.

$\Gamma(D^+ \rightarrow \pi^+e^+e^-)/\Gamma_{total}$		$<6.6 \times 10^{-5}$, CL = 90%
$\Gamma(D^+ \rightarrow \pi^+\mu^+\mu^-)/\Gamma_{total}$		$<1.8 \times 10^{-5}$, CL = 90%
$\Gamma(D^+ \rightarrow \rho^+\mu^+\mu^-)/\Gamma_{total}$		$<5.6 \times 10^{-4}$, CL = 90%
$\Gamma(D^0 \rightarrow e^+e^-)/\Gamma_{total}$		$<1.3 \times 10^{-5}$, CL = 90%
$\Gamma(D^0 \rightarrow \mu^+\mu^-)/\Gamma_{total}$		$<4.1 \times 10^{-6}$, CL = 90%
$\Gamma(D^0 \rightarrow \pi^0e^+e^-)/\Gamma_{total}$		$<4.5 \times 10^{-5}$, CL = 90%
$\Gamma(D^0 \rightarrow \pi^0\mu^+\mu^-)/\Gamma_{total}$		$<1.8 \times 10^{-4}$, CL = 90%
$\Gamma(D^0 \rightarrow \eta e^+e^-)/\Gamma_{total}$		$<1.1 \times 10^{-4}$, CL = 90%
$\Gamma(D^0 \rightarrow \eta\mu^+\mu^-)/\Gamma_{total}$		$<5.3 \times 10^{-4}$, CL = 90%
$\Gamma(D^0 \rightarrow \rho^0e^+e^-)/\Gamma_{total}$		$<1.0 \times 10^{-4}$, CL = 90%
$\Gamma(D^0 \rightarrow \rho^0\mu^+\mu^-)/\Gamma_{total}$		$<2.3 \times 10^{-4}$, CL = 90%
$\Gamma(D^0 \rightarrow \omega e^+e^-)/\Gamma_{total}$		$<1.8 \times 10^{-4}$, CL = 90%
$\Gamma(D^0 \rightarrow \omega\mu^+\mu^-)/\Gamma_{total}$		$<8.3 \times 10^{-4}$, CL = 90%
$\Gamma(D^0 \rightarrow \phi e^+e^-)/\Gamma_{total}$		$<5.2 \times 10^{-5}$, CL = 90%
$\Gamma(D^0 \rightarrow \phi\mu^+\mu^-)/\Gamma_{total}$		$<4.1 \times 10^{-4}$, CL = 90%
$\Gamma(D^0 \rightarrow \pi^+\pi^-\pi^0\mu^+\mu^-)/\Gamma_{total}$		$<8.1 \times 10^{-4}$, CL = 90%
$\Gamma(D^+_s \rightarrow K^+\mu^+\mu^-)/\Gamma_{total}$		$<5.9 \times 10^{-4}$, CL = 90%
$\Gamma(D^+_s \rightarrow K^*(892)^+\mu^+\mu^-)/\Gamma_{total}$		$<1.4 \times 10^{-3}$, CL = 90%
$\Gamma(\Lambda^+_c \rightarrow p\mu^+\mu^-)/\Gamma_{total}$		$<3.4 \times 10^{-4}$, CL = 90%

Limits are given at the 90% confidence level, while errors are given as ±1 standard deviation.

ΔB = 1 WEAK NEUTRAL CURRENT FORBIDDEN

Allowed by higher-order electroweak interactions.

$\Gamma(B^+ \to \pi^+ e^+ e^-)/\Gamma_{total}$	$<3.9 \times 10^{-3}$, CL = 90%
$\Gamma(B^+ \to \pi^+ \mu^+ \mu^-)/\Gamma_{total}$	$<9.1 \times 10^{-3}$, CL = 90%
$\Gamma(B^+ \to K^+ e^+ e^-)/\Gamma_{total}$	$<6 \times 10^{-5}$, CL = 90%
$\Gamma(B^+ \to K^+ \mu^+ \mu^-)/\Gamma_{total}$	$<1.0 \times 10^{-5}$, CL = 90%
$\Gamma(B^+ \to K^*(892)^+ e^+ e^-)/\Gamma_{total}$	$<6.9 \times 10^{-4}$, CL = 90%
$\Gamma(B^+ \to K^*(892)^+ \mu^+ \mu^-)/\Gamma_{total}$	$<1.2 \times 10^{-3}$, CL = 90%
$\Gamma(B^0 \to \gamma\gamma)/\Gamma_{total}$	$<3.9 \times 10^{-5}$, CL = 90%
$\Gamma(B^0 \to e^+ e^-)/\Gamma_{total}$	$<5.9 \times 10^{-6}$, CL = 90%
$\Gamma(B^0 \to \mu^+ \mu^-)/\Gamma_{total}$	$<6.8 \times 10^{-7}$, CL = 90%
$\Gamma(B^0 \to K^0 e^+ e^-)/\Gamma_{total}$	$<3.0 \times 10^{-4}$, CL = 90%
$\Gamma(B^0 \to K^0 \mu^+ \mu^-)/\Gamma_{total}$	$<3.6 \times 10^{-4}$, CL = 90%
$\Gamma(B^0 \to K^*(892)^0 e^+ e^-)/\Gamma_{total}$	$<2.9 \times 10^{-4}$, CL = 90%
$\Gamma(B^0 \to K^*(892)^0 \mu^+ \mu^-)/\Gamma_{total}$	$<2.3 \times 10^{-5}$, CL = 90%
$\Gamma(B^0 \to K^*(892)^0 \nu\overline{\nu})/\Gamma_{total}$	$<1.0 \times 10^{-3}$, CL = 90%
$\Gamma(B \to e^+ e^- s)/\Gamma_{total}$	$<5.7 \times 10^{-5}$, CL = 90%
$\Gamma(B \to \mu^+ \mu^- s)/\Gamma_{total}$	$<5.8 \times 10^{-5}$, CL = 90%
$\Gamma(\overline{b} \to \mu^+ \mu^- \text{anything})/\Gamma_{total}$	$<3.2 \times 10^{-4}$, CL = 90%
$\Gamma(B_s^0 \to \mu^+ \mu^-)/\Gamma_{total}$	$<2.0 \times 10^{-6}$, CL = 90%
$\Gamma(B_s^0 \to e^+ e^-)/\Gamma_{total}$	$<5.4 \times 10^{-5}$, CL = 90%
$\Gamma(B_s^0 \to \phi\nu\overline{\nu})/\Gamma_{total}$	$<5.4 \times 10^{-3}$, CL = 90%

Limits are given at the 90% confidence level, while errors are given as ±1 standard deviation.

NOTES

In this Summary Table:

When a quantity has "(S = ...)" to its right, the error on the quantity has been enlarged by the "scale factor" S, defined as $S = \sqrt{\chi^2/(N-1)}$, where N is the number of measurements used in calculating the quantity. We do this when S > 1, which often indicates that the measurements are inconsistent. When S > 1.25, we also show in the Particle Listings an ideogram of the measurements. For more about S, see the Introduction.

[a] *C* parity forbids this to occur as a single-photon process.

[b] Time-reversal invariance requires this to be 0° or 180°.

[c] Allowed by higher-order electroweak interactions.

[d] Violates *CP* in leading order. Test of direct *CP* violation since the indirect *CP*-violating and *CP*-conserving contributions are expected to be suppressed.

[e] ϵ'/ϵ is derived from $|\eta_{00}/\eta_{+-}|$ measurements using theoretical input on phases.

[f] Neglecting photon channels. See, *e.g.*, A. Pais and S.B. Treiman, Phys. Rev. **D12**, 2744 (1975).

[g] Derived from measured values of ϕ_{+-}, ϕ_{00}, $|\eta|$, $|m_{K_L^0} - m_{K_S^0}|$, and $\tau_{K_S^0}$, as described in the introduction to "Tests of Conservation Laws."

[h] The value is for the sum of the charge states of particle/antiparticle states indicated.

[i] A test of additive vs. multiplicative lepton family number conservation.

[j] $\Delta(m^2) = 100$ eV^2.

[k] 190 eV^2 $< \Delta(m^2) <$ 320 eV^2.

[l] Derived from an analysis of neutrino-oscillation experiments.

[m] There is some controversy about whether nuclear physics and model dependence complicate the analysis for bound neutrons (from which the best limit comes). The second limit here is from reactor experiments with free neutrons.

[n] This is the best "electron disappearance" limit. The best limit for the mode $e^- \to \nu\gamma$ is $> 2.35 \times 10^{25}$ yr (CL=68%).

[o] The D_1^0-D_2^0 limits are inferred from the D^0-$\overline{D}^0$ mixing ratio $\Gamma(K^+\ell^-\overline{\nu}_\ell(\text{via } \overline{D}^0))$ / $\Gamma(K^-\ell^+\nu_\ell)$.

[p] The larger limit (from E791) allows interference between the doubly Cabibbo-suppressed and mixing amplitudes; the smaller limit (from E691) doesn't. See the papers for details.

[q] See the K_L^0 Particle Listings for the energy limits used in this measurement.

TABLE OF THE ISOTOPES

Norman E. Holden

This table presents an evaluated set of values for the experimental quantities which characterize the decay of radioactive nuclides. A list of the major references used in this evaluation is given below. When uncertainties are not listed, they are assumed to be five or less in the last digit quoted. If they exceed five in the last digit, the value is prefaced by an approximate sign. The effective literature cutoff date for data in this edition of the Table is December, 2000.

Table Layout

Column No.	Column Title	Description
1	Isotope or Element	For elements, the atomic number and chemical symbol are listed. For nuclides, the mass number and chemical symbol are listed. Isomers are indicated by the addition of m, m1, or m2.
2	Isotopic Abundance	in atom percent.
3	Atomic Mass or Atomic Weight	Atomic mass relative to $^{12}C = 12$. Atomic weight is given on the same scale.
4	Half-life	Half-life in decimal notation. μs = microseconds; ms = milliseconds; s = seconds; m = minutes; h = hours; d = days; and y = years.
5	Decay Mode/Energy	Decay modes are α = alpha particle emission; β^- = negative beta emission; β^+ = positron emission; EC = orbital electron capture; IT = isomeric transition from upper to lower isomeric state; n = neutron emission; SF = spontaneous fission. Total disintegration energy in MeV units.
6	Particle Energy/Intensity	End point energies of beta transitions and discrete energies of alpha particles in MeV and their intensities in percent.
7	Spin and Parity	Nuclear spin or angular momentum of the nuclides in units of $h/2\pi$; parity is positive or negative.
8	Magnetic Dipole Moment	Magnetic dipole moments in nuclear magneton units.
9	Electric Quadrupole Moment	Electric quadrupole moments in barn units (10^{-24} cm^2).
10	Gamma Ray Energy/Intensity	Gamma ray energies in MeV and intensities in percent. Ann. rad. refers to the 511.006 keV photons emitted in the annihilation of positrons in matter.

General Nuclear Data References

The following references represent the major sources of the nuclear data presented, along with subsequent published journals and reports:

1. G. Audi, O. Bersillon, J. Blachot, A.H. Wapstra, *The Nubase Evaluation of Nuclear and Decay Properties,* Nuclear Physics A624, 1 (1997).
2. International Commission on Atomic Weights, *Atomic Weights of the Elements - 1999,* Pure & Applied Chemistry 73, to be published (2000).
3. J.R. Parrington, H.D. Knox, S. Breneman, E.M. Baum, F. Feiner, *Chart of the Nuclides, 15th Edition,* Knolls Atomic Power Lab. (1996).
4. N.E. Holden, *Total and Spontaneous Fission Half-lives for Uranium, Plutonium, Americium and Curium Nuclides,* Pure & Applied Chemistry 61, 1483 (1989).
5. N.E. Holden, *Half-lives of Selected Nuclides,* Pure & Applied Chemistry 62, 941 (1990).
6. N.E. Holden, *Review of Thermal Neutron Cross Sections and Isotopic Composition of the Elements,* BNL-NCS-42224 (March 1989).
7. P. Raghavan, *Table of Nuclear Moments,* Atomic Data Nuclear Data Tables 42, 189 (1989).
8. E. Brown, R. Firestone, *Radioactivity Handbook,* Wiley Interscience Press (1986).
9. J.K. Tuli, *Nuclear Wallet Cards,* Brookhaven National Laboratory (Jan. 2000).
10. N.E. Holden, D.C. Hoffman, *Spontaneous Fission Half-lives for Ground State Nuclides,* Pure & Applied Chemistry 72, 1525 (2000).
11. N. Stone, Table of New Nuclear Moments, private communication, www.nndc.bnl.gov/nndc/stone_moments/moments.html (Dec. 2000).

*This research was carried out under the auspices of the US Department of Energy Contract No. DE-AC02-98CH10886.

TABLE OF ISOTOPES (CONTINUED)

Elem. or Isot.	Natural Abundance (%)	Atomic Mass or Weight	Half-Life	Decay Mode/Energy (/MeV)	Particle Energy /Intensity (MeV/%)	Spin (h/2π)	Nuclear Magnetic Mom. (nm)	Elect. Quadr. Mom. (b)	γ-ray/Energy Intensity (MeV/%)
${}_{0}n$		1.008664924	614. s	β^-/0.78235	0.782/100.	1/2+	-1.913043		
${}_{1}H$		1.00794(7)							
${}^{1}H$	99.985(1)	1.007825032				1/2+	+2.79285		
${}^{2}H$	0.015(1)	2.014101778				1+	+0.85744	+2.86 mb	
${}^{3}H$		3.016049268	12.33 y	β^-/0.01859	0.01860/100.	1/2+	+2.97896		
${}^{4}H$		4.0278	1.9×10^{-22} s	n/	/100	2-			
${}^{5}H$		5.040	$8.\times10^{-23}$ s	n/	/100				
${}^{6}H$		6.0449	$3.\times10^{-22}$ s						
${}_{2}He$		4.002602(2)							
${}^{3}He$	1.37×10^{-4}	3.016029309				1/2+	-2.12762		
${}^{4}He$	≈100.	4.002603250				0+			
${}^{5}He$		5.01222	7.6×10^{-22} s	n, α		3/2-			
${}^{6}He$		6.018888	0.807 s	β^-/3.508,d	3.510/100.	0+			
${}^{7}He$		7.02803	$3.\times10^{-21}$ s	n		(3/2)-			
${}^{8}He$		8.03392	0.119 s	β^-/10.65, t	13/88.	0+			0.9807/84.
				n/	/12.				0.4776/5.
${}^{9}He$		9.0438	$7.\times10^{-21}$ s	n	/100	(1/2–)			
${}^{10}He$		10.0524	$3.\times10^{-21}$ s	2n	/100	0+			
${}_{3}Li$		6.941(2)							
${}^{4}Li$		4.0272	$9.\times10^{-23}$ s	p/	/100	2-			
${}^{5}Li$		5.01254	$\approx3.\times10^{-22}$ s	p/		3/2-			
${}^{6}Li$	7.5(2)	6.0151223				1+	+0.82205	-0.8 mb	
${}^{7}Li$	92.5(2)	7.0160041				3/2-	+3.25644	-0.041	
${}^{8}Li$		8.022486	0.84 s	β^-/16.004	12.5/100.	2+	+1.6536	+0.032	
				α/	α(1.6)				
${}^{9}Li$		9.026789	0.178 s	β^-/13.606	13.5/75.	3/2-	3.439	-0.027	
				β^-/	11/25.				
${}^{10}Li$		10.03590	$4.\times10^{-22}$ s	β^-/20.84.					
${}^{11}Li$		11.04380	8.4 ms	β^-/20.6		3/2(-)	3.668	-0.031	3.367/35.
				n,2n,3n,α	n//106.				(0.22-2.81)
${}^{12}Li$		12.054	<0.01 μs						
${}_{4}Be$		9.012182(3)							
${}^{5}Be$		5.041							
${}^{6}Be$		6.01973	5.0×10^{-21} s	2p,α		0+			
${}^{7}Be$		7.0169293	53.28 d	EC/0.8618		3/2-			0.4776/10.4
${}^{8}Be$		8.00530509	$\approx7.\times10^{-17}$ s	2α/0.046		0+			
${}^{9}Be$	100.	9.0121822				3/2-	-1.1776	+0.0529	
${}^{10}Be$		10.0135338	1.52×10^{6} y	β^-/0.5559	0.555/100.	0+			
${}^{11}Be$		11.02166	13.8 s	β^-,β^-α/11.51	11.48/61.	1/2+			2.125/35.5
									(0.478-7.97)
${}^{12}Be$		12.02692	24. ms	β^-,(n)/11.71	n//0.5	0+			(0.95 - 4.4)
${}^{13}Be$		13.0361	$\approx 3.\times10^{-21}$ s						
${}^{14}Be$		14.0428	4.3 ms	β^-,(n)/16.2	n//100.	0+			
${}_{5}B$		10.811(7)							
${}^{7}B$		7.0299	$4.\times10^{-22}$ s	p					
${}^{8}B$		8.024607	0.770 s	β^+, 2α/17.979	13.7(β^+)/93.	2+	1.0355	0.068	ann.rad.
${}^{9}B$		9.013329	8×10^{-19} s	p2α/		3/2-			
${}^{10}B$	19.9(2)	10.0129371				3+	+1.8006	+0.085	
${}^{11}B$	80.1(2)	11.0093055				3/2-	+2.6886	+0.0406	
${}^{12}B$		12.014352	0.0202 s	β^-/13.369		1+	+1.0027	0.0132	4.438/1.3
				β^- α/1.6/					3.215/0.00065
${}^{13}B$		13.017780	0.0174 s	β^-/13.437	13.4	3/2-	+3.17778	0.037	3.68/7.6
				β^- n/0.25/	2.43(n)/0.09				
					3.55(n)/0.16				
${}^{14}B$		14.02540	14. ms	β^-/20.64		2-	1.185	0.0298	6.094/90.
${}^{15}B$		15.03110	10.4 ms	β^-,(n)/19.09		(3/2-)	2.66	0.038	
${}^{16}B$		16.0398	$<1.9\times10^{-10}$ s						
${}^{17}B$		17.0469	5.1 ms	β^-,(n)/22.7			2.54		
${}^{18}B$		18.056	<0.026 μs						

Elem. or Isot.	Natural Abundance (%)	Atomic Mass or Weight	Half-Life	Decay Mode/Energy (/MeV)	Particle Energy /Intensity (MeV/%)	Spin (h/2π)	Nuclear Magnetic Mom. (nm)	Elect. Quadr. Mom. (b)	γ-ray/Energy Intensity (MeV/%)
^{19}B		19.0637	3.3 ms	β$^-$,(n)/26.5	n//125.				
$_6$C		12.0107(8)							
^{8}C		8.03768	2.0x10^{-21} s	p		0+			
^{9}C		9.031040	127. ms	β$^+$,p, 2α/16.498		(3/2-)	-1.391		ann.rad.
^{10}C		10.0168532	19.3 s	β$^+$/3.648	1.865	0+			ann.rad.
									0.71829/100.
^{11}C		11.011433	20.3 m	β$^+$,EC/1.982	0.9608/99.	3/2-	-0.964	0.0333	ann.rad.
^{12}C	98.93(8)	12.000000000				0+			
^{13}C	1.07(8)	13.003354838				1/2-	+0.70241		
^{14}C		14.003241991	5715. y	β$^-$/0.15648	0.1565/100.	0+			
^{15}C		15.010599	2.45 s	β$^-$/9.772	4.51/68.	1/2+	1.32		5.298/68.
					9.82/32.				(7.30-9.05)
^{16}C		16.014701	0.75 s	β$^-$,n/8.012		0+			
^{17}C		17.02258	0.19 s	β$^-$,n/13.17					1.375
									1.849
									1.906
^{18}C		18.02676	0.09 s	β$^-$,n/11.81		0+			
^{19}C		19.0353	0.05 s	n					
^{20}C		20.0403	0.01 s			0+			
^{21}C		21.0493	<0.03 μs						
^{22}C		22.056	9 ms	β$^-$,n	n//99.	0+			
$_7$N		14.0067(2)							
^{10}N		10.0426							
^{11}N		11.0268	5.x10^{-22} s						
^{12}N		12.018613	11.00 ms	β$^+$,β$^+$α/17.338	16.38/95.	1+	+0.457	+10. mb	ann.rad.
									4.438/2.
^{13}N		13.0057386	9.97 m	β$^+$/2.2204	1.190/100.	1/2-	0.3222		
^{14}N	99.632(7)	14.003074007				1+	+0.40376	+0.0200	
^{15}N	0.368(7)	15.00010897				1/2-	-0.28319		
^{16}N		16.006100	7.13 s	β$^-$/10.419	4.27/68.	2-			6.129/68.8
					10.44/26.				7.115/4.7
				β$^-$, α	1.85/.0012				(0.99-8.87)
^{17}N		17.00845	4.17 s	β$^-$,β$^-$ n/8.68	3.7/100.	1/2-	0.352		0.871/3.
				0.4-1.7n/95.					2.1842/0.3
				β$^-$ α/	8.0, 8.2				
^{18}N		18.01408	0.62 s	β$^-$/13.90	9.4/100.	1-	0.328	0.012	0.822/61.
									1.65/60.5
									1.982/98.
									(0.535-7.13)
^{19}N		19.01703	0.32 s	β$^-$/12.53					(0.096-3.14)
^{20}N		20.02337	0.14 s	β$^-$/17.97					
^{21}N		21.0271	0.08 s						
^{22}N		22.0344	0.02 s						
^{23}N		23.0405	15 ms	β$^-$,n	n//80.				
^{24}N		24.050	<0.052 μs						
$_8$O		15.9994(3)							
^{12}O		12.03440	≈1.x10^{-21} s	2p					
^{13}O		13.02481	8.9 ms	β$^+$,p/17.77	1.56 (p)/	(3/2-)	1.389	0.026	ann.rad.
									4.438/0.56
^{14}O		14.0085953	70.60 s	β$^+$/5.1430	1.81/99.	0+			ann.rad.
									2.312/99.4
^{15}O		15.0030655	122.2 s	β$^+$/2.754	1.723/100.	1/2-	0.7195		ann.rad.
^{16}O	99.757(16)	15.994914622				0+			
^{17}O	0.038(1)	16.9991315				5/2+	-1.8938	-0.026	
^{18}O	0.205(14)	17.999160				0+			
^{19}O		19.003579	26.9 s	β$^-$/4.820	3.25/60.	5/2+	1.5320	3.7 mb	0.197/95.9
					4.60/40.				1.3569/50.4
									(0.11-4.18)
^{20}O		20.004076	13.5 s	β$^-$/3.814		0+			1.057/100.

TABLE OF ISOTOPES (CONTINUED)

Elem. or Isot.	Natural Abundance (%)	Atomic Mass or Weight	Half-Life	Decay Mode/Energy (/MeV)	Particle Energy /Intensity (MeV/%)	Spin (h/2π)	Nuclear Magnetic Mom. (nm)	Elect. Quadr. Mom. (b)	γ-ray/Energy Intensity (MeV/%)
^{21}O		21.00866	3.4 s	β^-/8.11					(0.28-4.6)
^{22}O		22.00997	2.2 s	β^-/6.5					(0.64-1.86)
^{23}O		23.0157	0.08 s						
^{24}O		24.0204	≈65 ms	β^-,n	n//18.				1.83/28
									0.52/14.
									1.31/12.
^{25}O		25.029	<0.05 μs						
^{26}O		26.038	<0.04 μs						
$_{9}F$		18.9984032(5)							
^{14}F		14.036							
^{15}F		15.0180	5.x10^{-22} s	p		(1/2+)			
^{16}F		16.01147	≈1.x10^{-20} s	p		0-			
^{17}F		17.0020952	64.5 s	β^+/2.761	1.75/	5/2+	+4.721	0.058	ann.rad.
^{18}F		18.000938	1.830 h	β^+,EC/1.656	0.635/97.	1+			ann.rad.
^{19}F	100.	18.9984032				1/2+	+2.62887	0.072	
^{20}F		19.9999813	11.00 s	β^-/7.0245	5.398/100.	2+	+2.0934	0.042	1.634/100.
									3.33/0.009
^{21}F		20.999949	4.16 s	β^-/5.684	3.7/8.	5/2+	3.9		0.3507/90.
					5.0/63.				1.395/15.
					5.4/29.				(1.746-4.684)
^{22}F		22.00300	4.23 s	β^-/10.82	3.48/15.	4+			1.2746/100.
					4.67/7.				2.0826/82.
					5.50/62.				(0.82-4.37)
^{23}F		23.00357	2.2 s	β^-/8.5		5/2+			1.701/48.
									2.129/34.
									(0.493-3.83)
^{24}F		24.0081	0.3 s	β^-/13.5					1.9816/
^{25}F		25.0121	≈50 ms	β^-,(n)	n//14.				1.70/39.
									(0.57-2.19)
^{26}F		26.0196	10 ms	β^-,(n)	n//11.				2.02/67.
									1.67/19.
^{27}F		27.0269	5. ms	β^-,(n)	n//90.				
^{29}F		29.043	3. ms	β^-,(n)	n//100.				
^{31}F									
$_{10}Ne$		20.1797(6)							
^{16}Ne		16.02575	4.x10^{-21} s	2p		0+			
^{17}Ne		17.01770	109. ms	β^+,p/14.53	1.4-10.6/	1/2-			ann.rad./
									0.495
^{18}Ne		18.005697	1.67 s	β^+/4.446	3.416/92.	0+			ann.rad./
									1.0413/7.8
									(0.658-1.70)
^{19}Ne		19.001880	17.22 s	β^+/3.238	2.24/99.	1/2+	-1.885		ann.rad./
									(0.11-1.55)
^{20}Ne	90.48(3)	19.992440176				0+			
^{21}Ne	0.27(1)	20.99384674				3/2+	-0.66180	+0.103	
^{22}Ne	9.25(3)	21.9913855				0+		-0.19	
^{23}Ne		22.9944673	37.2 s	β^-/4.376	3.95/32.	5/2+	-1.08		0.440/33.
					4.39/67.				(1.64-2.98)
^{24}Ne		23.99362	3.38 m	β^-/2.47	1.10/8.	0+			0.4723/100.
					1.90/92.				0.874/7.9
^{25}Ne		24.99779	0.61 s	β^-/7.30	6.3/	1/2+			0.0895/96.
					7.3/				(0.98-3.69)
^{26}Ne		26.00046	197 ms	β^-/7.3		0+			0.233/
^{27}Ne		27.0076	32 ms	β^-, n/12.7		(3/2+)			
^{28}Ne		28.0121	18. ms	β^-, n/12.3	n//11.	0+			2.06/19.
									0.86/3.
^{29}Ne		29.0194	15. ms	β^-,(n)/15.4	n//27.	(3/2+)			2.92/54.
									(0.22-1.18)
^{30}Ne		30.024	7. ms	β^-, (n)	n//9.	0+			

Elem. or Isot.	Natural Abundance (%)	Atomic Mass or Weight	Half-Life	Decay Mode/Energy (/MeV)	Particle Energy /Intensity (MeV/%)	Spin ($h/2\pi$)	Nuclear Magnetic Mom. (nm)	Elect. Quadr. Mom. (b)	γ-ray/Energy Intensity (MeV/%)
^{31}Ne		31.033	>0.26 μs						
^{32}Ne		32.040	>0.20 μs			0+			
$_{11}$Na		22.989770(2)							
^{18}Na		18.0272							
^{19}Na		19.01388	0.03 s	β^+,p/11.18					
^{20}Na		20.00735	0.446 s	β^+/13.89		2+	+0.3694		ann.rad./
				α	2.15/				1.634/79.
^{21}Na		20.997655	22.48 s	β^+/3.547	2.50/95.	3/2+	+2.3863	+0.05	ann.rad./
									0.351/5.
^{22}Na		21.9944366	2.605 y	β^+/90/2.842	0.545/90.	3+	+1.746		ann.rad./
				EC/10/					1.2745/99.9
^{23}Na	100.	22.9897697				3/2+	+2.21752	+0.104	
^{24m}Na			20.2 ms	I.T.,β^-		1+			0.4723/100.
^{24}Na		23.9909633	14.96 h	β^-/5.5158	1.389/>99.	4+	+1.690		1.3686/100.
									2.754/100.
									(0.997-4.238)
^{25}Na		24.989954	59.3 s	β^-/3.835	2.6/7.	5/2+	+3.683	-0.10	0.3897/12.7
					3.15/25.				0.5850/13.
					4.0/65.				0.9747/14.9
									(0.836-2.80)
^{26}Na		25.99259	1.07 s	β^-/9.31		3+	+2.851	-0.08	1.809/98.9
^{27}Na		26.99401	0.290 s	β^-/9.01	7.95/	5/2+	+3.90	0.24	0.9847/87.4
				β^-,n/					1.698/11.9
^{28}Na		27.9989	31. ms	β^-/14.0	12.3/	1+	+2.43	-0.02	1.473/37.
				β^-,n/					2.389/18.6
^{29}Na		29.0028	44. ms	β^-,n/13.3	11.5/	3/2+	+2.45	-1.3	2.560/36.
									(1.04-3.99)
^{30}Na		30.0092	50. ms	β^-/17.5		2	+2.08		1.483/46.
^{31}Na		31.0136	17.2 ms	/15.9		(3/2-)	+2.31		1.483/14.
				β^-,n					(0.05-3.54)
^{32}Na		32.0197	13.5 ms	β^-/19.1					0.886/60.
^{33}Na		33.027	8.1 ms	β^-/20.					0.886/16.
^{34}Na		34.035	5. ms	β^-/24.					0.886/60.
^{35}Na		35.044	1.5 ms	β^-/24					
$_{12}$Mg		24.3050(6)							
^{20}Mg		20.01886	96. ms	β^+,p/10.73		0+			
^{21}Mg		21.01171	122. ms	β^+,p/13.10		5/2+			0.332/51.
^{22}Mg		21.999574	3.86 s	β^+/4.786	3.05/	0+			0.0729/60.
									0.5820/100.
									(1.28-1.93)
^{23}Mg		22.994125	11.32 s	β^+/4.057	3.09/92.	3/2+	0.536	1.25	0.440/8.2
^{24}Mg	78.99(4)	23.9850419				0+			
^{25}Mg	10.00(1)	24.9858370				5/2+	-0.85545	+0.199	
^{26}Mg	11.01(3)	25.9825930				0+			
^{27}Mg		26.9843407	9.45 m	β^-/2.6103	1.59/41.	1/2+			0.17068/0.9
					1.75/58.				0.84376/72.
					2.65/0.3				1.01443/28.
^{28}Mg		27.983877	20.9 h	β^-/1.832	0.459/95.	0+			0.0306/95.
									0.4006/36.
									0.9418/36.
									1.342/54.
^{29}Mg		28.98855	1.3 s	β^-/7.55	5.4/	3/2+			0.960/15.
									1.398/16.
									2.224/36.
^{30}Mg		29.9905	0.32 s	β^-/7.0		0+			0.224/85.
^{31}Mg		30.9966	0.24 s	β^-/11.7		(3/2+)			1.61/26.
^{32}Mg		31.9992	0.12 s	β^-/10.3		0+			2.765/25.
^{33}Mg		33.0056	0.09 s	β^-/13.7					1.848/
^{34}Mg		34.0091	0.02 s	β^-/11.3		0+			

Elem. or Isot.	Natural Abundance (%)	Atomic Mass or Weight	Half-Life	Decay Mode/Energy (/MeV)	Particle Energy /Intensity (MeV/%)	Spin (h/2π)	Nuclear Magnetic Mom. (nm)	Elect. Quadr. Mom. (b)	γ-ray/Energy Intensity (MeV/%)
^{35}Mg		35.0175	0.07 s			(7/2-)			
^{36}Mg		36.022	>0.2 μs			0+			
^{37}Mg		37.031	>0.26 μs			(7/2-)			
^{38}Mg						0+			
$_{13}$Al		26.981538(2)							
^{21}Al		21.028	<0.035 μs						
^{22}Al		22.0195	59. ms	β^+/18.6		4+			ann.rad./
				β^+,p,2p,α/					
^{23m}Al			≈0.35 s	β^+,p/0.17					0.554
									0.839
^{23}Al		23.00727	0.47 s	β^+/12.24					ann.rad./
				β^+,p/					
^{24m}Al			0.129 s	I.T./0.4259					
				β^+	13.3	1+			1.3686/5.3
^{24}Al		23.999941	2.07 s	β^+/13.878,p	3.40/48.	4+			1.078(2)/16.
					4.42/41.				1.368(2)/96.
					6.80/3.				2.753(2)/43.
					8.74/8.				4.315(3)/15.
									5.392(3)/20.
									7.0662(2)/41.
^{25}Al		24.990429	7.17 s	β^+/4.277	3.27/	5/2+	3.646		ann.rad./
									1.6115(2)/100.
									0.975(2)/5.
^{26m}Al			6.345 s	β^+ /	3.2/	0+			ann.rad./
^{26}Al		25.9868917	7.1x10^5 y	β^+/82/4.0042	1.16/	5+	+2.804	+0.17	ann.rad./
				EC/18/					1.8087/99.8
^{27}Al	100.	26.9815384				5/2+	+3.64151	+0.140	
^{28}Al		27.9819102	2.25 m	β^-/4.6422	2.865/100.	3+	3.24	0.18	1.7778(6)/100.
^{29}Al		28.980445	6.5 m	β^-/3.680	1.4/30.	5/2+			1.2732(8)/89.
					2.5/70.				2.0282(8)/4.
									2.4262(8)/7.
^{30}Al		29.98296	3.68 s	β^-/8.56	5.05/	3+			1.26313(3)/35.
									2.23525(5)/65.
^{31}Al		30.98395	0.64 s	β^-/8.00	6.25/				0.75223(3)/18.
									1.69473(3)/59.
									2.31664(4)/73.
^{32}Al		31.9881	33. ms	β^-/13.0		1+			
^{33}Al		32.9909	41. ms						
^{34}Al		33.9969	≈42. ms	β^-/17.1					
^{35}Al		34.9999	30 ms						
^{36}Al		36.0064	0.09 s						
^{37}Al		37.010	>1 μs						
^{38}Al		38.0169	>0.2 μs						
^{39}Al		39.022	>0.2 μs						
^{40}Al									
^{41}Al									
$_{14}$Si		28.0855(3)							
^{22}Si		22.0345	29. ms	β^+,p	1.99/20	0+			
^{23}Si		23.0255	40.7 ms	β^+,p/5.9	1.32,2.40,2.83				
^{24}Si		24.01155	0.14 s	β^+,p/10.81	1.51,4.09,1.73	0+			ann.rad./
					1.13-4.38				
^{25}Si		25.00411	221 ms	β^+,p/12.74		5/2+			ann.rad./
^{26}Si		25.992330	2.23 s	β^+/5.066	3.282/	0+			ann.rad./
									0.8294(8)/22.
^{27}Si		26.9867048	4.14 s	β^+/4.8118	3.85/100.	5/2+	-0.8554		ann.rad./
									2.211(5)/0.2
^{28}Si	92.22(2)	27.97692653				0+			
^{29}Si	4.69(1)	28.97649472				1/2+	-0.5553		
^{30}Si	3.09(1)	29.97377022				0+			

Elem. or Isot.	Natural Abundance (%)	Atomic Mass or Weight	Half-Life	Decay Mode/Energy (/MeV)	Particle Energy /Intensity (MeV/%)	Spin ($h/2\pi$)	Nuclear Magnetic Mom. (nm)	Elect. Quadr. Mom. (b)	γ-ray/Energy Intensity (MeV/%)
^{31}Si		30.9753633	2.62 h	β^-/1.4920	1.471/99.9	3/2+			1.2662(5)/0.05
^{32}Si		31.974148	1.6x10^2 y	β^-/0.224	0.213/100.	0+			
^{33}Si		32.97800	6.1 s	β^-/5.85	3.92	(3/2+)	1.21		1.4313(5)/13.
									1.8477/100.
									2.538(2)/10.
^{34}Si		33.97858	2.8 s	β^-/4.60	3.09/	0+			0.42907(5)/60.
									1.17852(2)/64.
									1.60756(5)/36.
^{35}Si		34.98458	0.9 s	β^-/10.50					
^{36}Si		35.9867	0.5 s	β^-/7.9		0+			
^{37}Si		36.9930	≈0.09 s						
^{38}Si		37.9960	>1 μs			0+			
^{39}Si		39.0023	>1 μs						
^{40}Si		40.0058	>0.2 μs			0+			
^{41}Si		41.013	>0.2 μs						
^{42}Si		42.016	>0.2 μs			0+			
$_{15}$P		30.973761(2)							
^{24}P		24.0344							
^{25}P		25.0203	<0.03 μs						
^{26}P		26.0118	≈20. ms	β^+,p/18.1		3+			
^{27}P		26.99919	0.3 s	β^+,p/11.63		1/2+			
^{28}P		27.992312	270. ms	β^+/14.332	3.94/13.	3+			ann.rad./
					5.25/13.				1.779(2)/98.
					6.96/16.				2.839(2)/2.8
					8.8/7.				3.040(2)/3.2
					11.49/52.				4.498(2)/12.
									7.537(2)/9.
^{29}P		28.981801	4.14 s	β^+/4.9431	3.945/98.	1/2+	1.2349		ann.rad./
									1.273/1.32
									2.426/0.39
^{30}P		29.9783138	2.50 m	β^+/4.2323	3.245/99.9	1+			ann.rad./
									2.230(3)/0.07
^{31}P	100.	30.9737615				1/2+	+1.13160		
^{32}P		31.9739071	14.28 d	β^-/1.7106	1.710/100.	1+	-0.2524		
^{33}P		32.971725	25.3 d	β^-/0.249	0.249/100.	1/2+			
^{34}P		33.973636	12.4 s	β^-/5.374	3.2/15.	1+			1.78-4.1/
					5.1/85.				2.127(5)/15.
^{35}P		34.973314	47. s	β^-/3.989	2.34/100.	1/2+			1.572(1)/100.
^{36}P		35.97826	5.7 s	β^-/10.41					0.902/77.
									3.291/100.
^{37}P		36.97961	2.3 s	β^-/7.90					0.6462/
									1.5829/
^{38}P		37.9845	0.6 s	β^-/12.4					1.2923/
									2.224/
^{39}P		38.9864	≈0.16 s						
^{40}P		39.9911	≈0.26 s						
^{41}P		40.9948	0.12 s						
^{42}P		42.0001	0.11 s						
^{43}P		43.0033	33. ms	β^-,(n)/					
^{44}P		44.010	>0.2 μs						
^{45}P		45.015	>0.2 μs						
^{46}P		46.024	>0.2 μs						
$_{16}$S		32.065(5)							
^{26}S		26.0278	≈ 10 ms			0+			
^{27}S		27.0188	21. ms	β^+, 2p/18.3					
^{28}S			0.13 s			0+			
^{29}S		28.99661	0.188 s	β^+/13.79		5/2+			ann.rad./
				β^+,p/					
^{30}S		29.984903	1.18 s	β^+/6.138	4.42/78.	0+			ann.rad./

Elem. or Isot.	Natural Abundance (%)	Atomic Mass or Weight	Half-Life	Decay Mode/Energy (/MeV)	Particle Energy /Intensity (MeV/%)	Spin (h/2π)	Nuclear Magnetic Mom. (nm)	Elect. Quadr. Mom. (b)	γ-ray/Energy Intensity (MeV/%)
					5.08/20.				0.678/79.
^{31}S		30.979555	2.56 s	β^+/5.396	4.39/99.	1/2+	0.48793		ann.rad./
									1.2662(5)/1.2
^{32}S	94.93(31)	31.9720707				0+			
^{33}S	0.76(2)	32.9714585				3/2+	+0.64382	-0.68	
^{34}S	4.29(28)	33.9678668				0+			
^{35}S		34.9690321	87.2 d	β^-/0.1672	0.1674/100.	3/2+	+1.00	+0.047	
^{36}S	0.02(1)	35.9670809				0+			
^{37}S		36.9711257	5.05 m	β^-/4.8653	1.64/94.	7/2-			0.9083(4)/0.06
					4.75/5.6				3.1033(2)/94.2
^{38}S		37.97116	2.84 h	β^-/2.94	1.00/	0+			0.1962(4)/0.2
									1.9421(3)/84.
^{39}S		38.97514	11.5 s	β^-/6.64					1.301/52.
									1.697/44.
^{40}S		39.9755	9. s	β^-/4.7		0+			
^{41}S		40.9800	≈2.6 s						
^{42}S		41.9815	≈0.56 s	β,(n)/		0+			
^{43}S		42.987	0.22 s						
^{44}S		43.9883	0.12 s	β^-, n/9.		0+			
^{45}S		44.9948	0.08 s	β^-,n/					
^{46}S		45.9996	>0.2 μs			0+			
^{47}S		47.008	>0.2 μs						
^{48}S		48.013	>0.2 μs			0+			
^{49}S		49.022	<0.2 μs						
$_{17}$Cl		35.453(2)							
^{28}Cl		28.0285							
^{29}Cl		29.0141	<0.02 μs						
^{30}Cl		30.0048	<0.03 μs						
^{31}Cl		30.99242	0.15 s	β^+,p/11.98	1.52	3/2+			ann.rad./
^{32}Cl		31.98569	297. ms	β^+/12.69	4.75/25.	1+	1.11		ann.rad./
					6.18/10.				1.548(2)/3.5
					7.48/14.				2.2305(1)/92.
					9.47/50.				2.4638(1)/4.
					11.6/1.				2.885(1)/1.
									4.770(1)/20.
^{33}Cl		32.977452	2.511 s	β^+/5.583	4.51/98.	3/2+	+0.752		ann.rad./
									0.8409/0.52
									1.966/0.45
									2.866/0.44
^{34m}Cl			32.2 m	β^+/	1.35/24.	3+			ann.rad./
					2.47/28.				
				I.T./					0.1457(8)/42.
									2.1276(5)/42.
^{34}Cl		33.9737620	1.528 s	β^+/5.4922	4.50/100.	0+			ann.rad./
^{35}Cl	75.78(4)	34.96885271				3/2+	+0.82187	-0.0825	
^{36}Cl		35.9683069	3.01x10^5 y	β^-/0.7086	0.7093/98.	0+	+1.28547	-0.018	
				β^+,EC/1.1421	0.115/0.002				ann.rad./
^{37}Cl	24.22(4)	36.96590260				3/2+	+0.68412	-0.0649	
^{38m}Cl			0.715 s	I.T./		5-			0.6714/100
^{38}Cl		37.0680106	37.2 m	β^-/4.9168	1.11/31.	2-	2.05		1.64216(1)/31
					2.77/11.				2.16760(2)/42.
					4.91/58.				
^{39}Cl		38.968008	55.6 m	β^-/3.442	1.91/85.	3/2+			0.25026(1)/47.
					2.18/8.				1.26720(5)/54.
					3.45/7.				0.986-1.517
^{40}Cl		39.97042	1.38 m	β^-/7.48		2-			0.6431(3)/6.
									1.4608(1)/77.
									2.8402(2)/17.
^{41}Cl		40.9707	34. s	β^-/5.7	3.8/				(0.167-1.359)

TABLE OF ISOTOPES (CONTINUED)

Elem. or Isot.	Natural Abundance (%)	Atomic Mass or Weight	Half-Life	Decay Mode/Energy (/MeV)	Particle Energy /Intensity (MeV/%)	Spin (h/2π)	Nuclear Magnetic Mom. (nm)	Elect. Quadr. Mom. (b)	γ-ray/Energy Intensity (MeV/%)
^{42}Cl		41.9732	6.8 s	β^-/9.4					
^{43}Cl		42.9742	3.3 s	β^-/8.0					
^{44}Cl		43.9785	≈0.43 s	β^-,n/12.3					
^{45}Cl		44.980	0.40 s	β^-, n/11.					
^{46}Cl		45.984	0.22 s	β^-, n/14.9					
^{47}Cl		46.988	>0.2 μs	β^-, n/15.					
^{48}Cl		47.995	>0.2 μs						
^{49}Cl		48.9999	>0.17 s						
^{50}Cl		50.008							
^{51}Cl		51.014	>0.2 μs						
$_{18}$Ar		39.948(1)							
^{30}Ar		30.0216	<0.02 μs			0+			
^{31}Ar		31.0121	≈14.1 ms	β^+/18.4	p/2.08/100.	5/2			
				β^+, 2p/<10^{-4}	p/1.42/37				
				β^+, 3p/<10^{-3}	p/0.45-11.67				
^{32}Ar		31.99766	98. ms	β^+,p/11.2		0+			ann.rad./
^{33}Ar		32.98993	174. ms	β^+/11.62	3.12/	1/2+	-0.72		ann.rad./
				β^+,p/					0.810(2)/48.
^{34}Ar		33.980270	0.844 s	β^+/6.061	5.0/95.	0+			ann.rad./
									0.6658(1)/2.5
									3.1290(1)/1.3
^{35}Ar		34.975257	1.77 s	β^+/5.965	4.94/93.	3/2+	+0.633	-0.08	ann.rad./
									1.2185(5)/1.22
									1.763(1)/0.25
									2.964(1)/0.2
^{36}Ar	0.3365(30)	35.9675463				0+			
^{37}Ar		36.9667759	35.0 d	EC/.813		3/2+	+1.15	+0.076	
^{38}Ar	0.0632(5)	37.9627322				0+			
^{39}Ar		38.964313	268. y	β^-/0.565	0.565/100.	7/2-	-1.59	-0.12	
^{40}Ar	99.6003(30)	39.962383123				0+			
^{41}Ar		40.964501	1.82 h	β^-/2.492	1.198/	7/2-			1.29364(5)/99.
									1.6770(3)/0.05
^{42}Ar		41.96305	33. y	β^-/0.60	0.60/100.	0+			
^{43}Ar		42.9657	5.4 m	β^-/4.6					0.4791(2)/10.
									0.7380(1)/43.
									0.9752(1)/100.
									1.4400(3)/39.
^{44}Ar		43.96537	11.87 m	β^-/3.55		0+			0.182-1.866
^{45}Ar		44.96809	21.5 s	β^-/6.9		7/2-			0.0610/25.
									1.020/35.
									3.707/34.
^{46}Ar		45.96809	8.4 s	β^-/5.70		0+			1.944/
^{47}Ar		46.9722	≈0.7 s	β^-					
^{48}Ar		47.9751							
^{49}Ar		48.9822	>0.17 μs	β^-					
^{50}Ar		49.986	>0.17 μs	β^-					
^{51}Ar		50.993	>0.2 μs	β^-					
^{52}Ar		51.998	10 ms	β^-					
^{53}Ar		52.994		β^-					
$_{19}$K		39.0983(1)							
^{32}K		32.0219							
^{33}K		33.0073	<0.025 μs						
^{34}K		33.9984	<0.04 μs						
^{35}K		34.98801	0.19 s	β^+/11.88		3/2+			ann.rad./
				β^+,p/					1.751/14.
									2.5698/26.
									2.9827/51.
^{36}K		35.98129	0.342 s	β^+/12.81	5.3/42.	2+	+0.548		ann.rad./
					9.9/44.				1.97044(5)/82.

TABLE OF ISOTOPES (CONTINUED)

Elem. or Isot.	Natural Abundance (%)	Atomic Mass or Weight	Half-Life	Decay Mode/Energy (/MeV)	Particle Energy /Intensity (MeV/%)	Spin (h/2π)	Nuclear Magnetic Mom. (nm)	Elect. Quadr. Mom. (b)	γ-ray/Energy Intensity (MeV/%)
									2.20783(5)/30.
									2.43343(2)/32.
^{37}K		36.9733769	1.23 s	β^+/6.149	5.13/	3/2+	+0.2032		ann.rad./
									2.7944(8)/2.
									3.602(2)/0.05
^{38m}K			0.924 s	β^+/6.742	5.02/100.	0+			ann.rad./
^{38}K		37.969080	7.63 m	β^+/5.913	2.60/99.8	3+	+1.37		ann.rad./
									2.1675(3)/99.8
									3.9356(5)/0.2
^{39}K	93.2581(44)	38.9637069				3/2+	+0.39146	+0.049	
^{40}K	0.0117(1)	39.9639987	1.26x10^9 y	β^-/1.3111	1.312/89.	4-	-1.29810	-0.061	ann.rad./
				β^+,EC/1.505	1.50/10.7				1.4608/10.5
^{41}K	6.7302(44)	40.9618260				3/2+	+0.21487	+0.060	
^{42}K		41.9624031	12.36 h	β^-/3.525	1.97/19.	2-	-1.1425		0.31260(2)/0.3
					3.523/81.				1.5246(3)/18.
^{43}K		42.96072	22.3 h	β^-/1.82	0.465/8.	3/2+	+0.163		0.2211(2)/4.
					0.825/87.				0.3729(2)/88.
					1.24/3.5				0.3971(2)/11.
					1.814/1.3				0.6178(2)/81.
^{44}K		43.96156	22.1 m	β^-/5.66	5.66/34.	2-	-0.856		0.36821/2.2
									1.15700(1)/58.
									2.15079(2)/22.
^{45}K		44.96070	17.8 m	β^-/4.20	1.1/23.	3/2+	+0.173		0.1743(5)/80.
					2.1/69.				1.2607(8)/7.
					4.0/8.				1.7056(6)/69.
									2.3542(5)/14.
^{46}K		45.96198	1.8 m	β^-/7.72	6.3/	2-	-1.05		1.347(1)/91.
									3.700(5)/28.
^{47}K		46.96168	17.5 s	β^-/6.64	4.1/99.	1/2+	+1.93		0.56474(3)/15.
					6.0/1.				0.58575(3)/85.
									2.0131/100
^{48}K		47.96551	6.8 s	β^-/12.09	5.0/	(2-)			0.67122(1)/4.
									0.6723(5)/20.
									0.78016(1)/32.
									3.83153(7)/80.
^{49}K		48.9675	1.26 s	β^-/11.0					2.025/
									2.252/
^{50}K		49.9728	0.472 s	β^-/14.2					
^{51}K		50.9764	0.365 s	β^-/					
^{52}K		51.983	0.105 s	β^-					
^{53}K		52.987	30. ms	β^-		3/2+			
^{54}K		53.994	10. ms	β^-					
$_{20}$Ca		40.078(4)							
^{34}Ca		34.0141	<0.035 μs						
^{35}Ca		35.0048	25.7 ms	β^+,p/15.6	p/1.43/49				
					1.9-8.8				
^{36}Ca		35.99309	0.10 s	β^+,(p)/10.99	2.52				ann.rad./
				β^+,n/					
^{37}Ca		36.98587	0.18 s	β^+/11.64	3.103	3/2+			ann.rad./
				β^+,n/					1.369
^{38}Ca		37.976319	0.44 s	β^+/6.74		0+			ann.rad./
									1.5677(5)/25.
									3.210(2)/1.
^{39}Ca		38.970718	0.861 s	β^+/6.531	5.49/100.	3/2+	1.02168		ann.rad./
^{40}Ca	96.941(156)	39.9625912				0+			
^{41}Ca		40.9622783	1.02x10^5 y	EC/0.4214		7/2-	-1.5948	-0.08	
^{42}Ca	0.647(23)	41.9586183				0+			
^{43}Ca	0.135(10)	42.9587668				7/2-	-1.3173	-0.05	
^{44}Ca	2.086(110)	43.955481				0+			

Elem. or Isot.	Natural Abundance (%)	Atomic Mass or Weight	Half-Life	Decay Mode/Energy (/MeV)	Particle Energy /Intensity (MeV/%)	Spin (h/2π)	Nuclear Magnetic Mom. (nm)	Elect. Quadr. Mom. (b)	γ-ray/Energy Intensity (MeV/%)
^{45}Ca		44.956186	162.7 d	β^-/0.257	0.257/100.	7/2-	-1.327	+0.05	
^{46}Ca	0.004(3)	45.953693	>0.4×10^{16} y	$\beta^-\beta^-$		0+			
^{47}Ca		46.954546	4.536 d	β^-/1.992	0.684/84.	7/2-	-1.38	+0.02	1.297/75
					1.98/16.				(0.041-1.88)
^{48}Ca	0.187(21)	47.952533	4.3×10^{19} y	$\beta^-\beta^-$		0+			
^{49}Ca		48.955673	8.72 m	β^-/5.262	0.89/7.	3/2-			3.0844(1)/92.
					1.95/92.				4.0719(1)/7.
^{50}Ca		49.95752	14. s	β^-/4.97	3.12/	0+			0.2569/98.
									(0.0715 -1.59)
^{51}Ca		50.9615	10. s	β^-/7.3		(3/2-)			
^{52}Ca		51.9651	4.6 s	β^-/8.0					
^{53}Ca		52.9701	0.09 s	β^-/10.9					
^{54}Ca		53.975							
^{55}Ca		54.981							
^{56}Ca		55.986							
$_{21}$Sc		44.955910(8)							
^{36}Sc		36.0149							
^{37}Sc		37.0030							
^{38}Sc		37.9947	<0.3 μs						
^{39}Sc		38.98479	<0.3 μs	p					
^{40}Sc		39.977964	0.182 s	β^+/14.320	5.73/50.	4-			ann.rad./
					7.53/15.				0.752/41.
					8.76/15.				3.732/99.5
					9.58/20.				(1.12-3.92)
^{41}Sc		40.9692513	0.596 s	β^+/6.4953	5.61/100.	7/2-	+5.431	-0.156	ann.rad./
^{42m}Sc			61.6 s	β^+/	2.82/	7+			ann.rad./
									0.4375(5)/100.
									1.2270(5)/100.
									1.5245(5)/100.
^{42}Sc		41.9655168	0.682 s	β^+/6.4259	5.32/100.	0+			ann.rad./
^{43}Sc		42.961151	3.89 h	β^+,EC/2.221	0.82/22.	7/2-	+4.62	-0.26	ann.rad./
					1.22/78.				0.3729(1)/22.
^{44m}Sc			58.2 h	I.T./0.27		6+	+3.88		0.27124(1)/87.
				EC/3.926					(1.00-1.16)
^{44}Sc		43.959403	3.93 h	β^+, EC/3.653	1.47/	2+	+2.56	+0.10	ann.rad./
									1.157/100
^{45}Sc	100.	44.955910				7/2-	+4.75649	-0.220	
^{46m}Sc			18.7 s	I.T./0.14253		1-			0.14253(2)/62.
^{46}Sc		45.955170	83.81 d	β^-/2.367	0.357/100.	4+	+3.03	+0.12	0.8893/100
									1.121/100
^{47}Sc		46.952408	3.349 d	β^-/0.600	0.439/69.	7/2-	+5.34	-0.22	0.15938(1)/68.
					0.601/31.				
^{48}Sc		47.95224	43.7 h	β^-/3.99	0.655/	6+			0.9835/100
									1.03750(1)/97.
									1.3121/100
^{49}Sc		48.950024	57.3 m	β^-/2.006	2.00/99.9.	7/2-			1.7619(3)/0.05
^{50}Sc		49.95219	1.71 m	β^-/6.89	3.05/76.	(5+)			0.5235(1)/88.
					3.60/24.				1.1210(1)/100.
									1.5537(2)/100.
^{51}Sc		50.95360	12.4 s	β^-/6.51	4.4/	7/2-			1.4373(4)/52.
					5.0/				0.718-2.144
^{52}Sc		51.9566	8.2 s	β^-/9.0		(3+)			
^{53}Sc		52.9592	> 3. ms	β^-/8.1					
^{54m}Sc			≈ 7 μs			(5+)			0.110/IT
^{54}Sc		53.9630	0.23 s	β^-/11.6					0.100/50
									1.70/40
									0.50/40
^{55}Sc		54.967	0.12 s	β^-/13					
^{56}Sc		55.973							

Elem. or Isot.	Natural Abundance (%)	Atomic Mass or Weight	Half-Life	Decay Mode/Energy (/MeV)	Particle Energy /Intensity (MeV/%)	Spin ($h/2\pi$)	Nuclear Magnetic Mom. (nm)	Elect. Quadr. Mom. (b)	γ-ray/Energy Intensity (MeV/%)
^{57}Sc		56.977							
^{58}Sc		57.983							
$_{22}$Ti		47.867(1)							
^{38}Ti		38.0098	<0.12 μs						
^{39}Ti		39.0013	28. ms	β^+/15.4					
^{40}Ti		39.9905	52. ms	β^+/11.7	p/2.17/28				
				β^+,p	3.73/23				
					1.7/22				
					0.242-5.74				
^{41}Ti		40.98313	80. ms	β^+,p/12.93	p/4.73/107	3/2+			ann.rad./
					3.10/67				
					3.75/39				
					0.744-6.73				
^{42}Ti		41.97303	0.20 s	β^+/7.000	6.0/				ann.rad./
									0.6107(5)/56.
^{43}Ti		42.96852	0.50 s	β^+/6.87	5.80/	7/2-	0.85		ann.rad./
^{44}Ti		43.959690	60. y	EC/0.268		0+			0.06787/91
									0.07832/97
^{45}Ti		44.958124	3.078 h	β^+/86/2.062	1.04	7/2-	0.095	0.015	ann.rad./
				EC/14/					(0.36-1.66)
^{46}Ti	8.25(3)	45.952630				0+			
^{47}Ti	7.44(2)	46.951764				5/2-	-0.78848	+0.30	
^{48}Ti	73.72(3)	47.947947				0+			
^{49}Ti	5.41(2)	48.947871				7/2-	-1.10417	+0.24	
^{50}Ti	5.18(2)	49.944792				0+			
^{51}Ti		50.946616	5.76 m	β^-/2.471	1.50/92.	3/2-			0.3197(2)/93.
					2.13/				0.6094-0.9291
^{52}Ti		51.94690	1.7 m	β^-/1.97	1.8/100.	0+			0.0170(5)/100.
									0.1245/100
^{53}Ti		52.9497	33. s	β^-/5.0	(2.2-3)/	3/2-			0.1008(1)/20.
									0.1276(1)/45.
									0.2284(1)/39.
									1.6755(5)/45.
									(1.72-2.8)/
^{54}Ti		53.9509	1.5 s	β^-/4.3					
^{55}Ti		54.9551	0.32 s	β^-/7.4					
^{56}Ti		55.9580	0.19 s	β^-/7.0					
^{57}Ti		56.963	0.06 s	β^-/11.					
^{58}Ti		57.966	≈47 ms						
^{59}Ti		58.972	0.06 s						
^{60}Ti		59.976	>0.15 μs						
^{61}Ti		60.982	>0.15 μs						
$_{23}$V		50.9415(1)							
^{40}V		40.0111							
^{41}V		40.9997							
^{42}V		41.9912	<0.055 μs						
^{43}V		42.9807	>0.8 s	β^+/11.3					
^{44}V		43.9744	0.09 s	β^+,α/13.7					ann.rad./
^{45}V		44.96578	0.54 s	β^+/7.13		7/2-			
^{46}V		45.960200	0.4223 s	β^+/7.051	6.03/100.	0+			ann.rad./
^{47}V		46.954907	32.6 m	β^+,EC/2.928	1.90/99.+	3/2-			ann.rad./
									1.7949(8)/0.19
									(0.2-2.16)
^{48}V		47.952254	15.98 d	β^+/4.012	0.698/50.	4+	2.01		ann.rad./
									0.9835/100
									(1.3-2.4)
^{49}V		48.948517	337. d	EC/0.602		7/2-	4.47		
^{50}V	0.250(4)	49.947163	>1.4x10^{17}y	EC, β^-		6+	+3.34569	+0.21	
^{51}V	99.750(4)	50.943964				7/2-	+5.148706	-0.04	

Elem. or Isot.	Natural Abundance (%)	Atomic Mass or Weight	Half-Life	Decay Mode/Energy (/MeV)	Particle Energy /Intensity (MeV/%)	Spin (h/2π)	Nuclear Magnetic Mom. (nm)	Elect. Quadr. Mom. (b)	γ-ray/Energy Intensity (MeV/%)
^{52}V		51.944780	3.76 m	β^-/3.976	2.47/	3+			1.4341(1)/100.
^{53}V		52.944342	1.56 m	β^-/3.436	2.52/	7/2-			1.0060(5)/90.
									1.2891(3)/10.
^{54m}V			0.9 μs			(5+)			0.108/IT
^{54}V		53.94644	49.8 s	β^-/7.04	1.00/5.	3+			0.8348/97.
					2.00/12.				0.9887/80.
					2.95/45.				2.259/46.
					5.20/11.				(0.56-3.38)
^{55}V		54.9472	6.5 s	β^-/6.0	6.0/	(7/2-)			0.5177/73.
									(0.224-1.21)
^{56}V		55.9504	0.23 s	β^-/9.1					0.70/50.
									0.34/40.
									1.00/30.
^{57}V		56.9524	0.33 s	β^-/8.1					0.30/60.
									0.60/30.
									0.80/30.
^{58}V		57.9567	0.20 s	β^-/11.6					
^{59}V		58.9593	0.13 s	β^-/9.9					0.90/80.
^{60}V		59.965	0.20 s	β^-/14.					0.102-0.208
^{61}V		60.967	0.04 s						0.646
^{62}V		61.973	≈ 65 ms						
^{63}V		62.977	>0.15 μs						
^{64}V			>0.15 μs						
$_{24}$Cr		51.9961(6)							
^{42}Cr		42.0064	>0.35 μs						
^{43}Cr		42.9977	21. ms						
^{44}Cr		43.9855	53. ms	β^+,(p)/10.3	p/0.95-3.1				
^{45}Cr		44.9792	0.05 s	β^+,p/12.5		7/2-			ann.rad./
^{46}Cr		45.96836	0.3 s	β^+/7.60					ann.rad./
^{47}Cr		46.96291	0.51 s	β^+/7.45		3/2-			ann.rad./
^{48}Cr		47.95404	21.6 h	EC/1.66					ann.rad./
									0.116(2)/95.
									0.305(10)/100.
^{49}Cr		48.951341	42.3 m	β^+,EC/2.631	1.39/	5/2-	0.476		ann.rad./
					1.45/				0.09064(1)/51.
					1.54/				0.15293(1)/27.
									(0.062-1.6)
^{50}Cr	4.345(13)	49.946050				0+			
^{51}Cr		50.944772	27.70 d	EC/0.7527		7/2-	-0.934		0.3201/10.2
^{52}Cr	83.789(18)	51.940512				0+			
^{53}Cr	9.501(17)	52.940653				3/2-	-0.47454	-0.15	
^{54}Cr	2.365(7)	53.938885				0+			
^{55}Cr		54.940844	3.497 m	β^-/2.603	2.5/	3/2-			1.5282(2)/0.04
									(0.13-2.37)
^{56}Cr		55.94065	5.9 m	β^-/1.62	1.50/100.	0+			0.026(2)/100.
									0.083(3)/100.
^{57}Cr		56.9438	21. s	β^-/5.1	3.3/	3/2-	0.0834		0.850/8.
					3.5/				(0.083-2.62)
^{58}Cr		57.9443	7.0 s	β^-/4.0					(0.131-0.683)
^{59m}Cr			0.10 ms			(9/2+)			0.208/IT
									0.193
									0.102
^{59}Cr		58.9487	1.0 s	β^-/7.7					1.236
^{60}Cr		59.9497	0.6 s	β^-/6.0					
^{61}Cr		60.9541	0.26 s	β^-/8.8					0.354-1.860
^{62}Cr		61.9558	0.19 s	β^-/7.3					0.285
^{63}Cr		62.962	0.11 s						
^{64}Cr		63.964	0.04 s						
^{65}Cr		64.970	>0.15 μs						

Elem. or Isot.	Natural Abundance (%)	Atomic Mass or Weight	Half-Life	Decay Mode/Energy (/MeV)	Particle Energy /Intensity (MeV/%)	Spin (h/2π)	Nuclear Magnetic Mom. (nm)	Elect. Quadr. Mom. (b)	γ-ray/Energy Intensity (MeV/%)
^{66}Cr			>0.15 μs						
^{67}Cr									
$_{25}$Mn		54.938049(9)							
^{44}Mn		44.0069	<0.105 μs						
^{45}Mn		44.9945	<0.07 μs						
^{46}Mn		45.9867	≈41. ms	β^+/17.1					
^{47}Mn		46.9761	≈0.1 s	β^+/12.3					
^{48}Mn		47.9689	0.15 s	β^+/13.5	5.79/58.	4+			
					4.43/10.				
^{49}Mn		48.95962	0.38 s	β^+/7.72	6.69/	5/2-			ann.rad./
^{50m}Mn			1.74 m	β^+/7.887	3.54/	5+			ann.rad./
									1.0980/94.
									0.783/91.
									(0.66-3.11)
^{50}Mn		49.954244	0.283 s	β^+/7.6330	6.61/	0+			ann.rad./
^{51}Mn		50.948215	46.2 m	β^+,EC/3.208	2.2/	5/2-	3.568	0.4	ann.rad./
									0.7491(1)/0.26
									(1.148-1.164)
^{52m}Mn			21.1 m	β^+/98/5.09	2.631/	2+	0.0076		ann.rad./
				I.T./2/0.378					0.3778 (I.T.)
									1.43406(1)/98.
									(0.7-4.8)
^{52}Mn		51.945570	5.591 d	β^+/4.712	0.575/	6+	+3.063	+0.5	ann.rad./
				EC/					0.74421(1)/90.
									1.4341/100
^{53}Mn		52.941294	3.7x10^6 y	EC/0.5970		7/2-	5.024		
^{54}Mn		53.940363	312.1 d	EC/1.377		3+	+3.282	+0.33	0.8340/100
^{55}Mn	100.	54.938049				5/2-	+3.4687	+0.32	
^{56}Mn		55.938909	2.579 h	β^-/3.6954	0.718/18.	3+	+3.2266		0.84675/99
					1.028/34.				1.81072(4)/27.
									2.113/14.5
^{57}Mn		56.938287	1.45 m	β^-/2.691		5/2-			
^{58}Mn		57.93999	65 s	β^-/6.25	3.8/	3+			0.45916(2)/20.
					5.1/				0.81076(1)/82.
									1.32309(5)/53.
^{59}Mn		58.94045	4.6 s	β^-/5.19	4.5/				0.471/
									0.531-0.726
^{60m}Mn			1.77 s	β^-/IT	5.7/	3+			0.824/
^{60}Mn		59.9433	50. s	β^-/8.6		0+			1.969/
^{61}Mn		60.9446	0.67 s	β^-/7.4		(5/2)-			
^{62}Mn		61.9480	0.67 s	β^-/10.4		(3+)			0.877/
									0.942-1.299
^{63}Mn		62.9498	0.28 s	β^-/8.8					0.356,0.450
^{64m}Mn			> 0.1 ms						0.135/IT
^{64}Mn		63.9537	87 ms	β^-/11.8					0.746
^{65}Mn		64.9561	0.09 s	β^-/10.					0.366
^{66}Mn		65.961	66 ms						0.471
^{67}Mn		66.964	42 ms						
^{68}Mn			28 ms						
^{69}Mn			14 ms						
$_{26}$Fe		55.845(2)							
^{45}Fe		45.0146	>0.35 μs						
^{46}Fe		46.0008	≈0.02 s	β^+/13.1					
^{47}Fe		46.9929	≈0.03 s	β^+/15.6					
^{48}Fe		47.9806	≈ 44. ms	β^+/11.2					
^{49}Fe		48.9763	70. ms	β^+/13.0		(7/2-)			ann.rad./
^{50}Fe		49.9630	0.15 s	β^+/8.2					0.651
^{51}Fe		50.95683	0.31 s	β^+/8.02		(5/2-)			ann.rad./
^{52m}Fe			46. s	β^+/4.4		(12+)			ann.rad./

TABLE OF ISOTOPES (CONTINUED)

Elem. or Isot.	Natural Abundance (%)	Atomic Mass or Weight	Half-Life	Decay Mode/Energy (/MeV)	Particle Energy /Intensity (MeV/%)	Spin (h/2π)	Nuclear Magnetic Mom. (nm)	Elect. Quadr. Mom. (b)	γ-ray/Energy Intensity (MeV/%)
									(0.622-2.286)/
^{52}Fe		51.94812	8.28 h	β^+/57/2.37	0.804/	0+			ann.rad./
				EC/43/					0.16868(1)/99.
				I.T./					0.377 (I.T.)/
^{53m}Fe			2.6 m	I.T./3.0407		19/2-			0.7011(1)/99.
									1.0115(1)/87.
									1.3281(1)/87.
									2.3396(1)/13.
^{53}Fe		52.945312	8.51 m	β^+/3.743	2.40/42.	7/2-			ann.rad./
					2.80/57.				0.3779(1)/42.
									(1.2 - 3.2)
^{54}Fe	5.845(35)	53.939615				0+			
^{55}Fe		54.938298	2.73 y	EC/0.2314		3/2-			
^{56}Fe	91.754(36)	55.934942				0+			
^{57}Fe	2.119(10)	56.935398				1/2-	+0.0906	0.16	
^{58}Fe	0.282(4)	57.933280				0+			
^{59}Fe		58.934880	44.51 d	β^-/1.565	0.273/48.	3/2-	- 0.336		1.099/57
					0.475/51.				1.292/43.
									(0.14-1.48)
^{60}Fe		59.934077	1.5x10^6 y	β^-/0.237	0.184/100.	0+			0.0586/100
^{61m}Fe			0.25 μs			(9/2+)			0.654/IT
									0.207
^{61}Fe		60.93675	6.0 m	β^-/3.98	2.5/13.				1.205/44.
					2.63/54.				1.028/43.
					2.80/31.				(0.12-3.37)
^{62}Fe		61.93677	68. s	β^-/2.53	2.5/100.	0+			0.5061(1)/100.
^{63}Fe		62.9404	6. s	β^-/6.3		5/2-			0.995/
									(1.365-1.427)
^{64}Fe		63.9411	2.0 s	β^-/4.9					
^{65m}Fe			0.4 μs			(5/2-)			0.364/IT
^{65}Fe		64.9449	1.3 s	β^-/7.9					
^{66}Fe		65.9460	0.44 s	β^-/5.7					0.471-1.425
^{67m}Fe			≈0.04 ms			(5/2-)			0.367/IT
^{67}Fe		66.9500	0.48 s	β^-/8.8					0.189
^{68}Fe		67.953	0.15 s	β^-/≈7.6					
^{69}Fe		68.958	0.17 s						
^{70}Fe			>0.15 μs						
^{71}Fe			>0.15 μs						
^{72}Fe			>0.15 μs						
$_{27}$Co		58.933200(9)							
^{48}Co		48.0018							
^{49}Co		48.990	<0.035 μs						
^{50}Co		49.9812	44. ms	β^+/17.0	2.03-2.79				
^{51}Co		50.9705	>0.2 μs	β^+/12.8					
^{52}Co		51.9632	0.12 s	β^+/14.0					0.849-1.942
^{53m}Co			0.25 s	β^+,p/		19/2-			ann.rad./
^{53}Co		52.95423	0.26 s	β^+/8.30		7/2-			ann.rad./
^{54m}Co			1.46 m	β^+/8.44	4.25/100.	7+			ann.rad./
									0.411(1)/99.
									1.130(1)/100.
									1.408(1)/100.
^{54}Co		53.948464	0.1932 s	β^+/8.2430	7.34/100.	0+			ann.rad./
^{55}Co		54.942003	17.53 h	β^+/3.4513	0.53/	7/2-	+4.822		ann.rad./
				EC/	1.03/				0.9312/75.
					1.50/				0.4772/20.
									(0.092-3.11)
^{56}Co		55.939844	77.3 d	β^+/4.566	1.459/18.	4+	3.85	+0.25	ann.rad./
				EC/					0.8468/99.9
									1.2383/68.

TABLE OF ISOTOPES (CONTINUED)

Elem. or Isot.	Natural Abundance (%)	Atomic Mass or Weight	Half-Life	Decay Mode/Energy (/MeV)	Particle Energy /Intensity (MeV/%)	Spin (h/2π)	Nuclear Magnetic Mom. (nm)	Elect. Quadr. Mom. (b)	γ-ray/Energy Intensity (MeV/%)
									(0.26-3.61)
^{57}Co		56.936296	271.8 d	EC/0.8361		7/2-	+4.72	+0.5	0.12206/86
									(0.014-0.706)
^{58m}Co			9.1 h	I.T./		5+			0.02489/0.035
^{58}Co		57.935757	70.88 d	β^+/2.307		2+	+4.04	+0.22	ann.rad./
				EC/					0.81076/99
^{59}Co	100.	58.933200				7/2-	+4.63	+0.41	
^{60m}Co			10.47 m	I.T./99.8/0.059		2+	+4.40	+0.3	0.0586/2.0
				β^-/0.2/1.56					
^{60}Co		59.933822	5.271 y	β^-/2.824	0.315/99.7	5+	+3.799	+0.44	1.1732/100
									1.3325/100
^{61}Co		60.932479	1.650 h	β^-/1.322	1.22/95.	7/2-			0.0674/86.
									0.842-0.909
^{62m}Co			13.9 m	β^-/	0.88/25.	5+			1.1635(3)/70.
					2.88/75.				1.1730(3)/98.
									2.0039(3)/19.
^{62}Co		61.93405	1.50 m	β^-/5.32	1.03/10.	2+			1.1292(3)/13.
					1.76/5.				1.1730(3)/83.
					2.9/20.				1.9851(1)/3.
					4.05/60.				2.3020(1)/19.
^{63}Co		62.93362	27.5 s	β^-/3.67	3.6/	7/2-			0.08713(1)/49.
									0.9817(3)/2.6
									0.156-2.17
^{64}Co		63.93581	0.30 s	β^-/7.31	7.0/	1+			
^{65}Co		64.93648	1.14 s	β^-/5.96		(7/2)-			
^{66m2}Co			>0.1 ms			(8-)			0.252/IT
									0.214
									0.175
^{66m1}Co			1.2 μs			(5+)			0.175/IT
^{66}Co		65.9398	0.25 s	β^-/10.0					(1.245-1.425)
^{67}Co		66.9406	0.43 s	β^-/8.4					0.694
^{68}Co		67.9444	0.19 s	β^-/11.7					
^{69}Co		68.9452	0.20 s	β^-/9.3					
^{70}Co		69.950	0.09 s	β^- 13.					
^{71}Co		70.952	0.21 s	β					
^{72}Co		71.956	0.09 s	β					
^{73}Co			>0.15 μs						
^{74}Co			>0.15 μs						
^{75}Co			>0.15 μs						
$_{28}$Ni		58.6934(2)							
^{49}Ni			>0.35 μs						
^{50}Ni		49.9959	>0.3 μs						
^{51}Ni		50.9877	>0.2 μs	β^+/16.0					
^{52}Ni		51.9757	38. ms	β^+/11.7					
^{53}Ni		52.9685	0.05 s	β^+,p/13.3		7/2-			ann.rad./
^{54}Ni		53.95791	0.14 s	β^+/8.80					0.937
^{55}Ni		54.95134	0.20 s	β^+/8.70	7.66/	7/2-			ann.rad./
^{56}Ni		55.94214	6.08 d	EC/2.14		0+			0.15838/99
				β^+/< 10^{-6}					0.81185(3)/87.
									0.2695-0.7500
^{57}Ni		56.939800	35.6 h	β^+/3.264	0.712/10.	3/2-	- 0.798		ann.rad./
				EC/	0.849/76.				1.3776/78.
									(0.127-3.177)
^{58}Ni	68.0769(89)		57.935348			0+			
^{59}Ni		58.934351	≈7.6x10^4 y	EC/		3/2-			
^{60}Ni	26.2231(77)		59.930790			0+			
^{61}Ni	1.1399(6)	60.931060				3/2-	-0.75002	+0.16	
^{62}Ni	3.6345(17)	61.928348				0+			
^{63}Ni		62.929673	100. y	β^-/0.066945	0.065/	1/2-			

TABLE OF ISOTOPES (CONTINUED)

Elem. or Isot.	Natural Abundance (%)	Atomic Mass or Weight	Half-Life	Decay Mode/Energy (/MeV)	Particle Energy /Intensity (MeV/%)	Spin (h/2π)	Nuclear Magnetic Mom. (nm)	Elect. Quadr. Mom. (b)	γ-ray/Energy Intensity (MeV/%)
^{64}Ni	0.9256(9)	63.927969				0+			
^{65}Ni		64.930088	2.517 h	β^-/2.137	0.65/30.	5/2-	0.69		0.36627(3)/5.
					1.020/11.				1.11553(4)/16.
					2.140/58.				1.48184(5)/23.
^{66}Ni		65.92912	54.6 h	β^-/0.23		0+			
^{67m}Ni			13.3 μs			9/2+			0.313/IT
									0.694
^{67}Ni		66.93157	21. s	β^-/3.56	3.8/	1/2-	+0.601		1.0722/100.
									1.6539/100.
									(0.10-1.98)
^{68m2}Ni			0.34 μs			0+			0.511
^{68m1}Ni			0.86 ms			(5-)			0.814/IT
									2.033
^{68}Ni		67.93185	29. s	β^-/2.06					
^{69m2}Ni			0.44 μs			(17/2)			0.148/IT
									0.593
									1.959
^{69m1}Ni			3.5 s						
^{69}Ni		68.9352	11. s	β^-/5.4					0.6807(3)/100.
									(0.207-1.213)
^{70m}Ni			0.21 μs			(8+)			0.183/IT
									0.448
									0.970
									1.259
^{70}Ni		69.9361	6.0 s	β^-/3.5					
^{71}Ni		70.9400	2.56 s	β^-/6.9					
^{72}Ni		71.9413	1.6 s	β^-/5.2					
^{73}Ni		72.946	0.84 s	β^-/9.					
^{74}Ni		73.948	1.1 s	β^-/7.					
^{75}Ni		74.953	≈ 0.47 s						
^{76}Ni		75.955	≈ 0.24 s						
^{77}Ni		76.961	>0.15 μs						
^{78}Ni		77.964	>0.15 μs						
$_{29}$Cu		63.546(3)							
^{52}Cu		51.9972							
^{53}Cu		52.9856	<0.3 μs						
^{54}Cu		53.9767	<0.075 μs						
^{55}Cu		54.9655	>0.2 μs	β^+/13.2					
^{56}Cu		55.9586	0.08 s	β^+/15.3					0.511/233
									2.700/100
									1.23-2.78
^{57}Cu		56.94922	196. ms	β^+/8.77		3/2-			0.77-3.01
^{58}Cu		57.944541	3.21 s	β^+/8.563	4.5/15.	1+			ann.rad./
				EC/	7.439/83.				0.0403(4)/5.
									1.4483(2)/11.
									1.4546(2)/16.
^{59}Cu		58.939504	1.36 m	β^+/4.800	1.9/	3/2-			ann.rad./
					3.75/				0.3393(1)/8.
									0.8780(1)/12.
									1.3015(1)/15.
									(0.4 - 2.6)
^{60}Cu		59.937368	23.7 m	β^+/6.127	2.00/69.	2+	+1.219		ann.rad./
				EC/	3.00/18.				1.3325/88.
					3.92/6.				1.7915/45.
									(0.12-5.048)
^{61}Cu		60.933462	3.35 h	β^+/2.237	0.56/3.	3/2-	+2.14		ann.rad./
					0.94/5.				0.2830/13.
					1.15/2.				0.6560/11.

Elem. or Isot.	Natural Abundance (%)	Atomic Mass or Weight	Half-Life	Decay Mode/Energy (/MeV)	Particle Energy /Intensity (MeV/%)	Spin (h/2π)	Nuclear Magnetic Mom. (nm)	Elect. Quadr. Mom. (b)	γ-ray/Energy Intensity (MeV/%)
					1.220/51.				(0.067-2.123)
^{62}Cu		61.932587	9.67 m	β^+/98/3.948	2.93/98.	1+	-0.380		ann.rad./
				EC/					1.17302(1)/0.6
									(0.87-3.37)
^{63}Cu	69.17(3)	62.929601				3/2-	+2.2233	-0.211	
^{64}Cu		63.929768	12.701 h	β^-/39/0.579	0.578/	1+	-0.217		ann.rad./
				β^+/19/1.6751	0.65/				1.3459(3)/0.6
				EC/41/					
^{65}Cu	30.83(3)	64.927794				3/2-	+2.3817	-0.195	
^{66}Cu		65.928873	5.09 m	β^-/2.642	1.65/6.	1+	-0.282		0.8330(1)/0.22
					2.7/94.				1.0392(2)/9.2
^{67}Cu		66.92775	2.580 d	β^-/0.58	0.395/56.	3/2-			0.09125(1)/7.
					0.484/23.				0.09325(1)/17.
					0.577/20.				0.18453(1)/47.
^{68m}Cu			3.79 m	I.T./86/		6-			0.0843(5)/70.
				β^-/14/1.8					0.1112(5)/18.
									0.5259(5)/74.
									(0.64-1.34)
^{68}Cu		67.92964	31. s	β^-/4.46	3.5/40.	1+			1.0774(5)/58.
					4.6/31.				1.2613(5)/17.
									(0.15-2.34)
^{69m}Cu			0.36 μs			(13/2+)			0.075/IT
									0.190/IT
									0.680
									1.871
^{69}Cu		68.92943	2.8 m	β^-/2.68	2.48/80.	3/2-	+2.84		0.5307(3)/3.
									0.8340(5)/6.
									1.0065(8)/10.
^{70m}Cu			47. s	β^-/	2.52/10.	5-			0.8848(2)/100.
									0.9017(2)/90.
									1.2517(5)/60.
									(0.39-3.06)
^{70}Cu		69.93241	5. s	β^-/6.60	5.42/54.	1+			0.8848(2)/54.
					6.09/46.				
^{71m}Cu			0.28 μs			(19/2)			0.133/IT
									0.494
									0.939
									1.189
^{71}Cu		70.93262	20. s	β^-/4.56		3/2-			0.490/
^{72m}Cu			1.76 μs			(4-)			0.051/IT
									0.082
									0.138
^{72}Cu		71.9357	6.6 s	β^-/8.2		(1+)			0.652/
^{73}Cu		72.9365	4.2 s	β^-/6.3	5.8/43				0.450/100
					6.25/42				0.307-1.559
^{74}Cu		73.9401	1.6 s	β^-/9.9					
^{75}Cu		74.9414	1.2 s	β^-/7.9					
^{76}Cu		75.9455	0.64 s	β^-/11.					
^{77}Cu		76.947	0.47 s	β^-/≈10.					
^{78}Cu		77.952	0.34 s	β^-/12					
^{79}Cu		78.954	0.19 s	β^-/11.					
^{80}Cu		79.962	>0.15 μs						
$_{30}$Zn		65.39(2)							
^{54}Zn		53.9929							
^{55}Zn		54.9840							
^{56}Zn		55.9724	0.04 s						
^{57}Zn		56.9649	0.04 s	β^+,p/14.6		(7/2-)			ann.rad./
^{58}Zn		57.9546	0.09 s	β^+					
^{59}Zn		58.94927	183. ms	β^+,p/9.09	8.1/	3/2-			ann.rad./

Elem. or Isot.	Natural Abundance (%)	Atomic Mass or Weight	Half-Life	Decay Mode/Energy (/MeV)	Particle Energy /Intensity (MeV/%)	Spin (h/2π)	Nuclear Magnetic Mom. (nm)	Elect. Quadr. Mom. (b)	γ-ray/Energy Intensity (MeV/%)
									(0.491-0.914)
^{60}Zn		59.94183	2.40 m	β^+/97/4.16		0+			ann.rad./
				EC/3/					0.669/47.
									(0.062-0.947)
^{61}Zn		60.93951	1.485 m	β^+/5.64	4.38/68.	3/2-			ann.rad./
									0.4748/17.
									(0.15-3.52)
^{62}Zn		61.93433	9.22 h	β^+/3/1.63	0.66/7.	0+			ann.rad./
				EC/93/					0.0408/25
									0.5967/26.
									(0.20-1.526)/
^{63}Zn		62.933215	38.5 m	β^+/93/3.367	1.02/	3/2-	-0.28164	+0.29	ann.rad./
				EC/7/	1.40/				0.66962(5)/8.4
					1.71/				0.96206(5)/6.6
					2.36/84.				(0.24-3.1)
^{64}Zn	48.63(60)	63.929146				0+			
^{65}Zn		64.929245	243.8 d	β^+/98/1.3514	0.325/	5/2-	+0.7690	-0.023	ann.rad./
				EC/1.5/					1.116/50.8
^{66}Zn	27.90(27)	65.926036				0+			
^{67}Zn	4.10(13)	66.927131				5/2-	+0.8755	+0.15	
^{68}Zn	18.75(51)	67.924847				0+			
^{69m}Zn			13.76 h	I.T./99+/0.439		9/2+			0.4390(2)/95.
^{69}Zn		68.926553	56. m	β^-/0.906	0.905/99.9	1/2-			0.318/
^{70}Zn	0.62(3)	69.925325				0+			
^{71m}Zn			3.97 h	β^-/	1.45/	9/2+			0.3864/93.
									0.4874/62.
									0.6203/57.
									(0.099-2.489)
^{71}Zn		70.92773	2.4 m	β^-/2.81		1/2-			0.5116(1)/30.
									0.9103(1)/7.5
									(0.12-2.29)
^{72}Zn		71.92686	46.5 h	β^-/0.46	0.25/14.	0+			0.0164(3)/8.
					0.30/86.				0.1447(1)/83.
									0.1915(2)/9.4
^{73m}Zn			6. s		I.T./0.196	(7/2+)			0.042
^{73}Zn		72.92978	24. s	β^-/4.29	4.7/	(1/2-)			0.216(1)/100.
									0.496-0.911
^{74}Zn		73.92946	1.60 m	β^-/2.3	2.1/				0.0565/
									0.1401/
									(0.05-0.35)
^{75}Zn		74.9329	10.2 s	β^-/6.0					0.229/
^{76}Zn		75.9334	5.7 s	β^-/4.2	3.6/				0.119/
^{77m}Zn			1.0 s	β^-/		(1/2-)			0.772
^{77}Zn		76.9371	2.1 s	β^-/7.3	4.8/				0.189/
^{78m}Zn			>0.03 ms						1.070
^{78}Zn		77.9386	1.5 s	β^-/6.4					0.225/
^{79}Zn		78.9421	1.0 s	β^-/8.6					0.702/
^{80}Zn		79.9444	0.54 s	β^-/7.3					0.713/
									0.2248/
^{81}Zn		80.9505	0.29 s	β^-/11.9					
^{82}Zn		81.9548	>0.15 μs						
^{83}Zn			>0.15 μs						
$_{31}$Ga		69.723(1)							
^{56}Ga		55.9949							
^{57}Ga		56.9829							
^{58}Ga		57.9742							
^{59}Ga		58.9634							
^{60}Ga		59.9571							
^{61}Ga		60.9492	0.15 s	β^+/9.0		3/2-			

Elem. or Isot.	Natural Abundance (%)	Atomic Mass or Weight	Half-Life	Decay Mode/Energy (/MeV)	Particle Energy /Intensity (MeV/%)	Spin (h/2π)	Nuclear Magnetic Mom. (nm)	Elect. Quadr. Mom. (b)	γ-ray/Energy Intensity (MeV/%)
^{62}Ga		61.94418	0.116 s	β^+/9.17	8.3/	0+			ann.rad./
				EC/					
^{63}Ga		62.9391	32. s	β^+/5.5	4.5/				ann.rad./
				EC/					0.6271(2)/10.
									0.6370(2)/11.
									1.0652(4)/45.
^{64m}Ga			0.022 ms						0.0429
^{64}Ga		63.936838	2.63 m	β^+/7.165	2.79/	0+			ann.rad./
					6.05/				0.80785(1)/14.
									0.99152(1)/43.
									1.38727(1)/12.
									3.3659(1)/13.
^{65}Ga		64.9394	15.2 m	β^+/86/3.255	0.82/10.	3/2-			ann.rad./
				EC/	1.39/19.				0.1151(2)/55.
					2.113/56.				0.1530(2)/96.
					2.237/15.				0.2069(2)/39.
									(0.06-2.4)
^{66}Ga		65.931592	9.5 h	β^+/56/5.175	0.74/1.	0+			ann.rad./
				EC/43/	1.84/54.				1.03935(8)/38.
					4.153/51.				2.7523(1)/23.
									(0.28-5.01)
^{67}Ga		66.928205	3.260 d	EC/1.001		3/2-	+1.8507	0.20	0.09332/37.
									0.18459/20.
									0.30024/17.
									(0.091-0.89)
^{68}Ga		67.927983	1.130 h	β^+/90/2.921	1.83/	1+	0.01175	0.028	ann.rad./
				EC/10/					1.0774(1)/3.
									(0.57-2.33)/
^{69}Ga	60.108(9)	68.925581				3/2-	+2.01659	+0.17	
^{70}Ga		69.926027	21.1 m	EC/0.2/0.655		1+			0.1755(5)/0.15
				β^-/99.8/1.656	1.65/99.				1.042(5)/0.48
^{71}Ga	39.892(9)	70.924707				3/2-	+2.56227	+0.11	
^{72}Ga		71.926372	14.10 h	β^-/4.001	0.64/40.	3-	-0.13224	+0.5	0.62986(5)/24.
					1.51/9.				2.2016(2)/26.
					2.52/8.				2.5077(2)/12.8
					3.15/11.				(0.11-3.3)/
^{73}Ga		72.92517	74.87 h	β^-/1.59		3/2-			0.05344(5)/10.
									0.29732(5)/47.
									(0.01-1.00)/
^{74m}Ga			10. s	I.T./		1+			0.0565(1)/75.
^{74}Ga		73.92694	8.1 m	β^-/5.4	2.6/	3-			0.5959/92.
									2.354/45.
									(0.23-3.99)
^{75}Ga		74.92650	2.10 m	β^-/3.39	3.3/	3/2-			0.2529/
									0.5746/
									(0.12-2.10)
^{76}Ga		75.9289	29. s	β^-/7.0		3-			0.5629/66.
									0.5455/26.
									(0.31-1.26)
^{77}Ga		76.9293	13.0 s	β^-/5.3	5.2/				0.469/
									0.459/
^{78}Ga		77.9317	5.09 s	β^-/8.2		3+			0.619/77.
									1.187/20.
^{79}Ga		78.9329	2.85 s	β^-/7.0	4.6/				0.465/
^{80}Ga		79.9366	1.68 s	β^-/10.4	10./				0.659/
^{81}Ga		80.9377	1.22 s	β^-/8.3	5.1/				0.217/
^{82}Ga		81.9432	0.599 s	β^-/12.6					1.348/
^{83}Ga		82.9469	0.308 s	β^-/≈ 11.5					

Elem. or Isot.	Natural Abundance (%)	Atomic Mass or Weight	Half-Life	Decay Mode/Energy (/MeV)	Particle Energy /Intensity (MeV/%)	Spin (h/2π)	Nuclear Magnetic Mom. (nm)	Elect. Quadr. Mom. (b)	γ-ray/Energy Intensity (MeV/%)
^{84}Ga		83.952	≈0.085 s	β⁻/14					
^{85}Ga			>0.15 μs						
^{86}Ga			>0.15 μs						
$_{32}$Ge		72.64(1)							
^{58}Ge		57.9910							
^{59}Ge		58.9817							
^{60}Ge		59.9702							
^{61}Ge		60.9638	0.04 s	β⁺/13.6					
^{63}Ge		62.9496	0.10 s	β⁻/9.8					
^{64}Ge		63.9416	1.06 m	β⁺/4.4	3.0/	0+			ann.rad./
				EC/					0.1282(2)/11.
				β⁺,p					0.4270(3)/37.
									0.6671(3)/17.
^{65}Ge		64.9394	31. s	β⁺/6.2	0.82/10.				ann.rad./
				EC/	1.39/19.				0.0620/27.
				EC,p	2.113/56.				0.6497/33.
					2.237/15.				0.8091/21.
									(0.19-3.28)
^{66}Ge		65.93385	2.26 h	β⁺/27/2.10		0+			ann.rad./
				EC/73/					0.0438/29.
									0.3819/28.
									(0.022-1.77)
^{67}Ge		66.932738	19.0 m	β⁺/96/4.225	1.6/	1/2-			ann.rad./
				EC/4/	2.3/				0.1670/84.
					3.15/				(0.25-3.73)
^{68}Ge		67.92810	270.8 d	EC/0.11		0+			Ga k x-ray/39.
^{69}Ge		68.927973	1.63 d	β⁺/36/2.2273	0.70/	5/2-	0.735	0.02	ann.rad./
				EC/64/	1.2/				0.574/13.
									1.1068/36.
									(0.2-2.04)
^{70}Ge	20.84(87)	69.924250				0+			
^{71m}Ge			20.4 ms		I.T./0.0234	9/2+			0.1749
^{71}Ge		70.924954	11.2 d	EC/0.229		1/2-	+0.547		
^{72}Ge	27.54(34)	71.922076				0+			
^{73}Ge	7.73(5)	72.923460				9/2+	-0.879467	-0.17	
^{74}Ge	36.28(73)	73.921178				0+			
^{75m}Ge			48. s	I.T./		7/2+			0.13968(3)/39.
^{75}Ge		74.922860	1.380 h	β⁻/1.177	1.19/	1/2-	+0.510		0.26461(5)/11.
									0.41931(5)/0.2
^{76}Ge	7.61(38)	75.921403	≈1×10^{21}y	β⁻β⁻		0+			
^{77m}Ge			53. s	I.T./20/		1/2-			1.605/0.22
				β⁻/80/2.861	2.9/				1.676/0.16
									0.195-1.482
^{77}Ge		76.923549	11.25 h	β⁻/2.702	0.71/23.	7/2+			0.2110/29.
					1.38/35.				0.2155/27.
					2.19/42.				0.2644/51.
									(0.15-2.35)
^{78}Ge		77.922853	1.45 h	β⁻/0.95	0.70/	0+			0.2773(5)/96.
									0.2939(5)/4.
^{79m}Ge			39. s	β⁻/IT		7/2+			
^{79}Ge		78.9254	19.1 s	β⁻/4.2	4.0/20.	1/2-			0.1096/21.
					4.3/80.				(0.10-2.59)
									0.5427(4)/15.
^{80}Ge		79.92545	29.5 s	β⁻/2.67	2.4/	0+			0.1104(4)/6.
									0.2656(4)/25.
^{81m}Ge			≈7.6 s	β⁻/	3.75/	1/2+			0.3362(4)/
									0.7935(4)/
^{81}Ge		80.9288	≈7.6 s	β⁻/6.2	3.44/	9/2+			0.1976(4)/21.
									0.3362(4)/100.

Elem. or Isot.	Natural Abundance (%)	Atomic Mass or Weight	Half-Life	Decay Mode/Energy (/MeV)	Particle Energy /Intensity (MeV/%)	Spin (h/2π)	Nuclear Magnetic Mom. (nm)	Elect. Quadr. Mom. (b)	γ-ray/Energy Intensity (MeV/%)
^{82}Ge		81.9296	4.6 s	β^-/4.7		0+			1.093/
^{83}Ge		82.9345	1.9 s	β^-/8.9					
^{84}Ge		83.9373	0.98 s	β^-/7.7					
^{85}Ge		84.943	0.54 s	β^-/10.					
^{86}Ge		85.946	>0.15 μs						
^{87}Ge			>0.15 μs						
^{88}Ge			>0.15 μs						
^{89}Ge			>0.15 μs						
$_{33}$As		74.92160(2)							
^{60}As		59.993							
^{61}As		60.981							
^{62}As		61.9732							
^{63}As		62.9637							
^{64}As		63.9576	>1.2 μs						
^{65}As		64.9495	0.19 s	β^+/9.4					
^{66m2}As			1.9 μs						
^{66m1}As			0.018 ms						
^{66}As		65.94410	95.8 ms	β^+/9.55					
^{67}As		66.9392	42. s	β^+/6.0	5.0/	5/2-			0.121/
				EC/					0.123/
									0.244/
^{68}As		67.9368	2.53 m	β^+/8.1		3+			ann.rad./
									0.652/32.
									0.762/33.
									1.016/77.
									(0.61-3.55)
^{69}As		68.93228	15.2 m	β^+/98/4.01	2.95/	5/2-	1.6		ann.rad./
				EC/2/					0.0868(5)/1.5
									0.1458(3)/2.4
^{70}As		69.93093	52.6 m	β^+/84/6.22	1.44/	4+	+2.1061	+0.09	ann.rad./
				EC/16/2.14					1.0395(7)/82.
				/2.89					(0.17-4.4)/
^{71}As		70.927114	2.72 d	β^+/32/2.013		5/2-	+1.6735	-0.02	ann.rad./
				EC/68/					0.1749(2)/84.
									1.0957(2)/4.2
^{72}As		71.926753	26.0 h	β^+/77/4.356	0.669/5.	2-	-2.1566	-0.08	ann.rad./
					1.884/12.				0.83395(5)/80.
					2.498/62.				1.0507(1)/9.6
					3.339/19.				(0.1-4.0)
^{73}As		72.923825	80.3 d	EC/0.341		3/2-			0.0133/0.1
									0.0534/10.5
									Se k x-ray/90.
^{74}As		73.923829	17.78 d	β^+/31/2.562	0.94/26.	2-	-1.597		ann.rad./
				EC/37/	1.53/3.				0.59588(1)/60.
				β^-/1.353	0.71/16.				0.6084(1)/0.6
					1.35/16.				0.6348(1)/15.
^{75m}As			0.017 s						
^{75}As	100.	74.921597				3/2-	+1.43947	+0.31	
^{76}As		75.922394	26.3 h	β^-/2.962	0.54/3.	2-	-0.903		0.5591(1)/45.
					1.785/8.				0.65703(5)/6.2
					2.410/36.				1.21602(1)/3.4
					2.97/51.				(0.3-2.67)
^{77}As		76.920648	38.8 h	β^-/0.683	0.70/98.	3/2-	+1.295		0.2391(2)/1.6
									0.2500(3)/0.4
									0.5208/0.43
^{78}As		77.92183	1.512 h	β^-/4.21	3.00/12.	2-			0.6136(3)/54.
					3.70/17.				0.6954(3)/18.
					4.42/37.				1.3088(3)/10.
^{79m}As			1.21 μs			9/2+			0.542/IT

TABLE OF ISOTOPES (CONTINUED)

Elem. or Isot.	Natural Abundance (%)	Atomic Mass or Weight	Half-Life	Decay Mode/Energy (/MeV)	Particle Energy /Intensity (MeV/%)	Spin (h/2π)	Nuclear Magnetic Mom. (nm)	Elect. Quadr. Mom. (b)	γ-ray/Energy Intensity (MeV/%)
									0.231
^{79}As		78.92095	9.0 m	β^-/2.28	1.80/95.	3/2-			0.0955(5)/16.
									0.3645(5)/1.9
^{80}As		79.92258	16. s	β^-/5.64	3.38/	1+			0.6662(2)/42.
									(2.5-3.0)
^{81}As		80.92213	33. s	β^-/3.856		3/2-			0.4676(2)/20.
									0.4911(2)/8.
^{82m}As			13.7 s	β^-/	3.6/	5-			0.6544(1)/72.
									0.8186(4)/27.
									1.7313(2)/27.
									1.8954(2)/38.
^{82}As		81.9246	19. s	β^-/7.4	7.2/80.	1+			0.6544(1)/15.
^{83}As		82.9250	13.4 s	β^-/5.5					0.7345/100.
									1.1131/34.
									2.0767/28.
^{84m}As			0.6 s	β^-					
^{84}As		83.9291	4. s	β^-, n/7.2		1-			0.6671(2)/21.
									1.4439(5)/49.
									(0.325-5.150)
^{85}As		84.9318	2.03 s	β^-, n/8.9		3/2-			0.667(1)/42.
									1.4551(2)/100.
^{86}As		85.9362	0.95 s	β^-, n/11.4					0.704/
^{87}As		86.9396	0.49 s	β^-, n/10.					0.704/
^{88}As		87.945	>0.15 μs						
^{89}As		88.949	>0.15 μs						
^{90}As			>0.15 μs						
^{91}As			>0.15 μs						
^{92}As			>0.15 μs						
$_{34}$Se		78.96(3)							
^{65}Se		64.965	0.011 s	β^+/60/14.					
				β^+, p	3.55/				
^{66}Se		65.9552							
^{67}Se		66.9501	0.06 s	β^+/10.2					ann.rad./
				β^+,(p)/					0.352
^{68}Se		67.9419	36. s	β^+/4.7					ann.rad./
									(0.050-0.426)
^{69}Se		68.93956	27.4 s	β^+/6.78	5.006/				ann.rad./
				EC/					0.0664(4)/27.
									0.0982(4)/63.
^{70}Se		69.9335	41.1 m	β^+/2.4		0+			ann.rad
									0.04951(5)/35.
									0.4262(2)/29.
^{71}Se		70.9319	4.7 m	β^+/4.4	3.4/36.	5/2-			ann.rad
				EC/					0.1472(3)/47.
									0.8309(3)/13.
									1.0960(3)/10.
^{72}Se		71.92711	8.5 d	EC/0.34		0+			0.0460(2)/57.
^{73m}Se			40. m	I.T./73/0.0257	0.85	3/2-			ann.rad.
				β^+/27/2.77	1.45/				0.0257(2)/27.
					1.70/				0.2538(1)/2.5
^{73}Se		72.92678	7.1 h	β^+/65/2.74	0.80/	9/2+	0.86		ann.rad
				EC/35/	1.32/95.				0.0670(1)/72.
					1.68/1.				0.3609(1)/97.
									(0.6-1.5)
^{74}Se	0.89(4)	73.922477				0+			
^{75}Se		74.922524	119.78 d	EC/0.864		5/2+	0.67	1.0	0.13600/55
									0.26465/58
									(0.024-0.821)
^{76}Se	9.37(29)	75.919214				0+			

TABLE OF ISOTOPES (CONTINUED)

Elem. or Isot.	Natural Abundance (%)	Atomic Mass or Weight	Half-Life	Decay Mode/Energy (/MeV)	Particle Energy /Intensity (MeV/%)	Spin (h/2π)	Nuclear Magnetic Mom. (nm)	Elect. Quadr. Mom. (b)	γ-ray/Energy Intensity (MeV/%)
^{77m}Se			17.4 s	I.T./		7/2+			0.1619(2)/52.
^{77}Se	7.63(16)	76.919915				1/2-	+0.53506		
^{78}Se	23.77(28)	77.917310				0+			
^{79m}Se			3.92 m	I.T./					0.09573(3)/9.5
^{79}Se		78.918500	1.1x10^{6} y	β$^-$/0.151		7/2+	-1.018	+0.8	
^{80}Se	49.61(41)	79.916522				0+			
^{81m}Se			57.3 m	I.T./99/0.1031		7/2+			0.1031(3)/9.7
									0.2602(2)/0.06
									0.2760/0.06
^{81}Se		80.917993	18.5 m	β$^-$/1.585	1.6/98.	1/2-			0.2759/0.85
									0.2901/0.75
									0.8283/0.32
^{82}Se	8.73(22)	81.916700	≈1×10^{20} y	β$^-$β$^-$		0+			
^{83m}Se			1.17 m	β$^-$/3.96	2.88/	1/2-			0.35666(6)/17.
					3.92/				0.9879(1)/15.
									1.0305(1)/21.
									2.0514(2)/11.
									(0.19-3.1)
^{83}Se		82.919119	22.3 m	β$^-$/3.668	0.93/	9/2+			0.22516(6)/33.
					1.51/				0.35666(6)/69.
									0.51004(8)/45.
									(0.21-2.42)
^{84}Se		83.91847	3.3 m	β$^-$/1.83	1.41/100.	0+			0.4088(5)/100.
^{85}Se		84.92225	32. s	β$^-$/6.18	5.9/	5/2+			0.3450(1)/22.
									0.6094(1)/41.
^{86}Se		85.92428	15. s	β$^-$/5.10		5/2+			2.0124(1)/24.
									2.4433(8)/100.
									2.6619(1)/49.
^{87}Se		86.92853	5.4 s	β$^-$/7.28					0.468(1)/100.
				n/					1.4979(1)/23.
^{88}Se		87.93143	1.5 s	β$^-$,n/6.85					0.5346/
^{89}Se		88.9360	0.41s	β$^-$,n/9.0					
^{90}Se		89.9394	>0.15 μs						
^{91}Se		90.945	0.27 s	β$^-$,n/8.					
^{92}Se		91.949	>0.15 μs						
^{93}Se			>0.15 μs						
^{94}Se			>0.15 μs						
$_{35}$Br		79.904(1)							
^{67}Br		66.9648							
^{68}Br		67.958	<1.2 μs						
^{69}Br		68.9502	<0.024 μs	β$^+$/9.6					
^{70}Br		69.9446	79. ms	β$^+$/10.0	/0.75				
^{71}Br		70.9392	21. s	β$^+$/6.9					
^{72}Br		71.9365	1.31 m	β$^+$/8.7		3	≈0.55		0.4547-1.3167
^{73}Br		72.9318	3.4 m	β$^+$/4.7	3.7/	3/2-			ann.rad
									0.065-0.700
^{74m}Br			46. m	β$^+$/	4.5/	4-	1.82		ann.rad
									0.6348
									0.7285
									(0.2 - 4.38)
^{74}Br		73.92989	25.4 m	β$^+$/6.91					ann.rad
									0.6341
									0.6348
									(0.2-4.7)
^{75}Br		74.92578	1.62 h	β$^+$/76/3.03		3/2-	+0.75		ann.rad
									0.28650
									(0.1-1.56)
^{76m}Br			1.4 s	I.T./5.05		4+			0.104548
									0.05711

TABLE OF ISOTOPES (CONTINUED)

Elem. or Isot.	Natural Abundance (%)	Atomic Mass or Weight	Half-Life	Decay Mode/Energy (/MeV)	Particle Energy /Intensity (MeV/%)	Spin (h/2π)	Nuclear Magnetic Mom. (nm)	Elect. Quadr. Mom. (b)	γ-ray/Energy Intensity (MeV/%)
^{76}Br		75.92454	16.0 h	β^+/57/4.96	1.9/	1-	0.54821	0.270	ann.rad
					3.68/				0.55911
									1.85368
									(0.4-4.6)
^{77m}Br			4.3 m	I.T./0.1059		9/2+			0.1059
^{77}Br		76.921380	2.376 d	EC/99/1.365		3/2-	0.973	+0.53	ann.rad.
									0.23898
									0.52069
									(0.08-1.2)
^{78}Br		77.921146	6.45 m	β^+/92/3.574	1.2/	1+	0.13		ann.rad.
				EC/8/	2.5/				0.61363
									(0.7-3.0)
^{79m}Br			4.86 s	I.T./0.207		9/2+			0.2072
^{79}Br	50.69(7)	78.918338				3/2-	+2.106400	+0.331	
^{80m}Br			4.42 h	I.T./0.04885		5-	+1.3177	+0.75	Br k x-ray
									0.03705/39.1
									0.04885/0.3
^{80}Br		79.918530	17.66 m	β^-/92/2.004	1.38 β^-/7.6	1+	0.5140	0.196	ann.rad.
				EC/5.7/1.8706	1.99 β^-/82				0.6169/6.7
				β^+/2.6/	0.85 β^+/2.8				(0.64-1.45)
^{81}Br	49.31(7)	80.916291				3/2-	+2.270562	+0.276	
^{82m}Br			6.1 m	I.T./98/0.046		2-			0.046/0.24
				β^-/2/3.139					(0.62-2.66)
^{82}Br		81.916805	1.471 d	β^-/3.093	0.444/	5-	+1.6270	0.751	0.5544/71
									0.61905/43
									0.77649/84
									(0.013-1.96)
^{83}Br		82.915181	2.40 h	β^-/0.972	0.395/1	3/2-			0.52964
					0.925/99				(0.12-0.68)
^{84m}Br			6.0 m	β^-/4.97	2.2/100	(6-)			0.4240/100
									0.8817/98
									1.4637/101
^{84}Br		83.91651	31.8 m	β^-/4.65	2.70/11	2-	2.		0.8816/41
					3.81/20				1.8976/13
					4.63/34				(0.23-4.12)
^{85}Br		84.91561	2.87 m	β^-/2.87	2.57	3/2-			0.80241/2.56
									0.92463/1.6
									(0.09-2.4)
^{86}Br		85.91880	55.5 s	β^-/7.63	3.3	(2-)			1.56460/64
					7.4				2.75106/21
									(0.5-6.8)
^{87}Br		86.92072	55.6 s	β^-/6.85	6.1/	3/2-			1.41983
				n/					1.4762
									(0.2-6.1)
^{88m}Br			5.1 μs						
^{88}Br		87.92407	16.3 s	β^-/8.96		1-			0.7649
				n/					0.7753
									0.8021
									(0.1-6.99)
^{89}Br		88.92640	4.35 s	β^-/8.16		3/2-			0.7753
				n/					1.0978
^{90}Br		89.9306	1.91 s	β^-/10.4	8.3/	2-			0.6555
				n/	9.8/				0.7071
									1.3626
^{91}Br		90.9339	0.54 s	β^-/90/9.80					0.263
				β^- n/10/					0.803
^{92}Br		91.9392	0.31 s	β^-/12.20					0.740

Elem. or Isot.	Natural Abundance (%)	Atomic Mass or Weight	Half-Life	Decay Mode/Energy (/MeV)	Particle Energy /Intensity (MeV/%)	Spin (h/2π)	Nuclear Magnetic Mom. (nm)	Elect. Quadr. Mom. (b)	γ-ray/Energy Intensity (MeV/%)
				β^- n/					
^{93}Br		92.9431	0.10 s	β^- n/11.1					
^{94}Br		93.9487	0.07 s	β^- n/					
^{95}Br			>0.15 μs						
^{96}Br			>0.15 μs						
^{97}Br			>0.15 μs						
$_{36}$Kr		83.80(1)							
^{69}Kr		68.9653	0.03 s	β^+,(p)	4.07/				
^{70}Kr		69.9560	>1.2 μs						
^{71}Kr		70.9505	100. ms	β^+,EC/10.1					(0.198-0.207)
^{72}Kr		71.9419	17. s	β^+/5.0		0+			ann.rad
				EC/					0.3100/29
									0.4150/36
									(0.12-0.58)
^{73}Kr		72.9389	28. s	β^+/6.7		5/2-			ann.rad.
				EC/					0.1781/66
				β^+,p/	/0.25				(0.06-0.86)
^{74}Kr		73.9333	11.5 m	β^+/3.1		0+			ann.rad.
				EC/					0.08970/31
									0.2030/20
									(0.010-1.06)
^{75}Kr		74.93104	4.3 m	β^+/4.90	3.2/	5/2+	- 0.531	+ 1.1	ann.rad.
				EC/					0.1325/68
									0.1547/21
									(0.02-1.7)
^{76}Kr		75.92595	14.8 h	EC/1.31		0+			Br k x-ray
									0.270/21
									0.3158/39
									(0.03-1.07)
^{77}Kr		76.92467	1.24 h	β^+/80/3.06		5/2+	- 0.583	+ 0.9	ann.rad.
				EC/20/	1.55/				0.1297/80
					1.70/				0.1465/38
					1.87/				(0.02-2.3)
^{78}Kr	0.35(1)	77.92039	>0.9×10^{20} y	$\beta^-\beta^-$		0+			
^{79m}Kr			53. s	I.T./0.1299		7/2+	- 0.786	+ 0.40	Kr x-ray
^{79}Kr		78.920083	1.455 d	β^+/7/1.626		1/2-	+ 0.536		ann.rad.
				EC/93/					0.2613/13
									0.39756/19
									0.6061/8
									(0.04-1.3)
^{80}Kr	2.28(6)	79.916379				0+			
^{81m}Kr			13.1 s	I.T./0.1904		1/2-	+ 0.586		0.1904
^{81}Kr		80.916593	2.1x10^5 y	EC/0.2807		7/2+	- 0.908	+ 0.63	Br k x-ray
									0.2760
^{82}Kr	11.58(14)	81.913485				0+			
^{83m}Kr			1.86 h	I.T./0.0416		1/2-	+ 0.591		Kr k x-ray
									0.00940
									0.03216
^{83}Kr	11.49(6)	82.914137				9/2+	-0.970699	+0.253	
^{84}Kr	57.00(4)	83.911508				0+			
^{85m}Kr			4.48 h	β^-/79/	0.83/79	1/2-	+ 0.633		0.30487
				I.T./21/0.305					0.15118
^{85}Kr		84.912530	10.73 y	β^-/0.687	0.15/0.4	9/2+	1.005	+0.43	0.51399
^{86}Kr	17.30(22)	85.910615				0+			
^{87}Kr		86.913359	1.27 h	β^-/3.887	1.33/8	5/2+	-1.023	- 0.30	0.40258/49.6
					3.49/43				2.5548/9.2
					3.89/30				(0.13-3.31)
^{88}Kr		87.91445	2.84 h	β^-/2.91		0+			0.19632/26.
									2.392/34.6

TABLE OF ISOTOPES (CONTINUED)

Elem. or Isot.	Natural Abundance (%)	Atomic Mass or Weight	Half-Life	Decay Mode/Energy (/MeV)	Particle Energy /Intensity (MeV/%)	Spin (h/2π)	Nuclear Magnetic Mom. (nm)	Elect. Quadr. Mom. (b)	γ-ray/Energy Intensity (MeV/%)
									(0.03-2.8)
^{89}Kr		88.91764	3.15 m	β^-/4.99	3.8/	5/2+	- 0.330	+ 0.16	0.19746
					4.6/				0.2209/19.9
					4.9/				0.5858/16.4
									1.4728/6.8
									(0.2-4.7)
^{90}Kr		89.91953	32.3 s	β^-/4.39	2.6/77	0+			0.12182/32.9
					2.8/6				0.5395/28.6
									1.1187/36.2
									(0.1 - 4.2)
^{91}Kr		90.9234	8.6 s	β^-/6.4	4.33/	5/2+	- 0.583	+ 0.30	0.10878/43.5
					4.59/				0.50658/19.
									(0.2-4.4)
^{92}Kr		91.92611	1.84 s	β^-/5.99					0.1424/66.
				n/					(0.14 - 3.7)
^{93}Kr		92.9312	1.29 s	β^-/8.6	7.1/	1/2+	- 0.413		0.1820
				n/					0.2534/42.
									0.32309/24.6
									(0.057-4.03)
^{94}Kr		93.9343	0.21 s	β^-/7.3					0.2196/67
									0.6293/100.
							- 0.410		(0.098-0.985)
^{95}Kr		94.9397	0.78 s	β^-/9.7					
^{96}Kr		95.9431	> 50 ms						
^{97}Kr		96.9486	< 0.1 s	β^-					
^{98}Kr			>0.15 μs						
^{99}Kr			>0.15 μs						
^{100}Kr			>0.15 μs						
$_{37}$Rb		85.4678(3)							
^{71}Rb		70.9653							
^{72}Rb		71.9591	<1.2 μs						
^{73}Rb		72.9504	<0.03 μs						
^{74}Rb		73.9445	65. ms	β^+/10.4					
^{75}Rb		74.93857	19. s	β^+/7.02	2.31/				ann. rad.
									0.179
^{76}Rb		75.93508	39. s	β^+/8.50	4.7/	1-	-0.372623	+0.4	ann.rad.
									0.4240/92.
									(0.064-1.68)
^{77}Rb		76.93041	3.8 m	β^+/5.34	3.86/	3/2-	+0.654468	+0.70	ann.rad.
									0.0665/59
									(0.04 - 2.82)
^{78m}Rb			5.7 m	I.T./0.1034		4-	+2.549	+0.81	ann.rad.
				β^+/	3.4				0.4553/81.
				EC/					(0.103-4.01)
^{78}Rb		77.92814	17.7 m	β^+/7.22		0+			ann.rad.
				EC/					0.4553/63.
									(0.42-5.57)
^{79}Rb		78.92400	23. m	β^+/84/3.65		5/2+	+0.3358	-0.10	ann.rad.
				EC/16/					0.68812/23.
									(0.017-3.02)
^{80}Rb		79.92252	34. s	β^+/5.72	4.1/22	1+	-0.0836	+0.35	ann.rad.
					4.7/74				0.6167/25.
^{81m}Rb			30.5 m	I.T./0.85	1.4	9/2+	+5.598	-0.74	ann.rad.
				β^+,EC/					(0.085-1.9)
^{81}Rb		80.91900	4.57 h	β^+/27/2.24	1.05/	3/2-	+2.060	+0.40	ann.rad./
				EC/73					0.19030/64.
									(0.05 - 1.9)
^{82m}Rb			6.47 h	β^+/26/	0.80/	5-	+1.5100	+1.0	ann.rad./
				EC/74/					0.5544/63.

Elem. or Isot.	Natural Abundance (%)	Atomic Mass or Weight	Half-Life	Decay Mode/Energy (/MeV)	Particle Energy /Intensity (MeV/%)	Spin (h/2π)	Nuclear Magnetic Mom. (nm)	Elect. Quadr. Mom. (b)	γ-ray/Energy Intensity (MeV/%)
									0.7765/85.
									(0.092 - 2.3)
^{82}Rb		81.91821	1.258 m	β^+/96/4.40	3.3/	1+	+0.554508	+0.19	ann.rad./
				EC/4/					0.7665/13.
									(0.47 - 3.96)
^{83}Rb		82.91511	86.2 d	EC/0.91		5/2-	+1.425	+0.20	Kr x-ray
									0.5205/46.
									(0.03-0.80)
^{84m}Rb			20.3 m	I.T./0.216		6-	+0.2129	+0.6	0.2163/34.
									0.2482/63.
									0.4645/32.
^{84}Rb		83.914387	32.9 d	β^+/22/2.681	0.780/11	2-	-1.32412	-0.015	ann.rad./
				EC/75/	1.658/11				0.8817/68.
				β^-/3/0.894	0.893/				(1.02-1.9)
^{85}Rb	72.17(2)	84.911792				5/2-	+1.353	+0.23	
^{86m}Rb			1.018 m	I.T./0.5560		6-	+1.815	+0.37	0.556/98.
^{86}Rb		85.911170	18.65 d	β^-/1.775	1.774/8.8	2-	-1.6920	+0.19	1.0768/8.8
^{87}Rb	27.83(2)	86.909186	4.88x10^{10}y	β^-/0.283	0.273/100	3/2-	+2.7512	+0.13	
^{88}Rb		87.911323	17.7 m	β^-/5.316	5.31	2-	0.508		0.8980/14.
									1.8360/21.
									(0.34-4.85)
^{89}Rb		88.91229	15.4 m	β^-/4.50	1.26/38	3/2-	+2.304	+0.14	1.032/58.
					1.9/5				1.248/42.
					2.2/34				2.1960/13
					4.49/18				(0.12-4.09)
^{90m}Rb			4.3 m	β^-/4.50	1.7/	4-	+1.616	+0.20	0.1069(IT)
					6.5/				0.8317/94
									(0.20-5.00)
^{90}Rb		89.91481	2.6 m	β^-/6.59	6.6	1-			0.8317/28.
									(0.31-5.60)
^{91}Rb		90.91649	58.0 s	β^-/5.861	5.9	3/2-	+2.182	+0.15	0.0936/34.
									(0.35-4.70)
^{92}Rb		91.91968	4.48 s	β^-/8.11	8.1/94	1-			0.8148/8.
									(0.1-6.1)
^{93}Rb		92.92195	5.85 s	β^-/7.46	7.4/	5/2	+1.410	+0.18	0.2134/4.8
				n/1					0.4326/12.5
									0.9861/4.9
									(0.16-5.41)
^{94}Rb		93.92643	2.71 s	β^-/10.31	9.5/	3	+1.498	+0.16	0.8369/87.
				n/10					1.5775/32.
									(0.12-6.35)
^{95}Rb		94.92929	0.377 s	β^-/9.30	8.6/	5/2	+1.334	+0.21	0.352/65.
				n/8					0.680/22.
									(0.20-2.27)
^{96m}Rb			1.7 μs						0.2999
									0.4612
									0.2400
									0.093-0.369
^{96}Rb		95.93427	0.199 s	β^-/11.76	10.8/	2+	+1.466	+0.25	0.815/76.
				n/13/					(0.20-5.42)
^{97}Rb		96.93733	0.169 s	β^-/10.42	10.0	3/2	+1.841	+0.58	0.167/100.
				n/27/					0.585/79.
									0.599/56.
									1.258/52.
									(0.14-2.08)
^{98}Rb		97.94174	0.107 s	β^-/12.34	0.144/				
				n/13					(0.07-3.68)
^{99}Rb		98.9453	59. ms	β^-/11.3					
^{100}Rb		99.9499	53. ms	β^-/13.5					

Elem. or Isot.	Natural Abundance (%)	Atomic Mass or Weight	Half-Life	Decay Mode/Energy (/MeV)	Particle Energy /Intensity (MeV/%)	Spin (h/2π)	Nuclear Magnetic Mom. (nm)	Elect. Quadr. Mom. (b)	γ-ray/Energy Intensity (MeV/%)
^{101}Rb		100.9532	0.03 s	β^-/11.8					
^{102}Rb		101.9592	0.09 s	β^-					
$_{36}$Sr		87.62(1)							
^{73}Sr		72.966	> 25 ms						
^{74}Sr		73.9563	>1.2 μs						
^{75}Sr		74.9499	≈ 0.07 s						
^{76}Sr		75.9416	8.9 s	β^+/6.1					
^{77}Sr		76.9378	9.0 s	β^+/6.9	5.6		-0.35	+1.4	0.147
^{78}Sr		77.93218	2.7 m	β^+/3.76					(0.047-0.793)
^{79}Sr		78.92971	2.1 m	β^+/5.32	4.1	3/2-	-0.474	+0.74	ann.rad./
									0.039/28.
									0.105/22.
									(0.135-0.612)
^{80}Sr		79.92453	1.77 h	β^+/1.87		0+			ann.rad./
									0.174/10.
									0.589/39.
									(0.24-0.55)
^{81}Sr		80.92322	22.3 m	β^+/87/3.93	2.43/	1/2-	+0.544		ann.rad./
				EC/13/	2.68/				0.148/31.
									0.1534/35
									(0.06-1.7)
^{82}Sr		81.91840	25.36 d	EC/0.18					Rb x-ray
^{83m}Sr			5.0 s	I.T./0.2591		1/2-	+0.582		0.2591/87.5
^{83}Sr		82.91756	1.350 d	β^+/24/2.28	0.465/	7/2+	-0.898	+0.79	ann.rad./
				EC/76/	0.803/				0.3816/12.
					1.227/				0.3816
									0.7627/30.
									(0.094-2.15)
^{84}Sr	0.56(1)	83.913426				0+			
^{85m}Sr			1.127 h	I.T./87/0.2387		1/2-	+0.601		0.2318/84.
				EC/13					(0.15-0.24)
^{85}Sr		84.912936	64.85 d	EC/1.065		9/2+	-1.001	+0.30	0.51399/99.3
^{86}Sr	9.86(1)	85.909265				0+			
^{87m}Sr			2.81 h	I.T./0.3884		1/2-	+0.63		0.3884(IT)
^{87}Sr	7.00(1)	86.908882				9/2+	-1.09360	+0.34	
^{88}Sr	82.58(1)	87.905617				0+			
^{89}Sr		88.907455	50.52 d	β^-/1.497	1.492/100	5/2+	-1.149	-0.3	0.9092
^{90}Sr		89.907738	29.1 y	β^-/0.546	0.546/100	0+			
^{91}Sr		90.91020	9.5 h	β^-/2.70	0.61/7	5/2+	-0.887	+0.044	0.5556/61.
					1.09/33				0.7498/24.
					1.36/29				1.0243/33.
					2.66/26				(0.12-2.4)
^{92}Sr		91.91098	2.71 h	β^-/1.91	0.55/96	0+			1.3831/90.
					1.5/3				(0.24-1.1)
^{93}Sr		92.91394	7.4 m	β^-/4.08	2.2/10	5/2+	-0.794	+0.26	0.5903/
					2.6/25				0.7104
					3.2/65				0.87573
									0.8883/
									(0.17-3.97)
^{94}Sr		93.91537	1.25 m	β^-/3.511	2.1/	0+			0.6219
					3.3/				0.7043
									0.7241
									0.8064
									1.4283
^{95}Sr		94.91931	25.1 s	β^-/6.08		1/2+	-0.5379		0.6859
					6.1/50				0.8269
									2.7173
									2.9332
^{96}Sr		95.92165	1.06 s	β^-/5.37	4.2/	0+			0.1222

Elem. or Isot.	Natural Abundance (%)	Atomic Mass or Weight	Half-Life	Decay Mode/Energy (/MeV)	Particle Energy /Intensity (MeV/%)	Spin (h/2π)	Nuclear Magnetic Mom. (nm)	Elect. Quadr. Mom. (b)	γ-ray/Energy Intensity (MeV/%)
									0.5305
									0.8094
									0.9318
^{97}Sr		96.92615	0.42 s	β^-/7.47	5.3	(1/2+)	-0.500		0.2164
									0.3071
									0.6522
									0.9538
									1.2580
									1.9050
^{98}Sr		97.92845	0.65 s	β^-/5.83	5.1				0.0365
									0.1190
									0.4286
									0.4447
									0.5636
^{99}Sr		98.9333	0.27 s	β^-/8.0			-0.26	0.8	
^{100}Sr		99.9354	0.201 s	β^-/7.1					
^{101}Sr		100.9405	0.115 s	β^-/9.5					
^{102}Sr		101.9430	68. ms	β^-/8.8					
^{103}Sr		102.9490	>0.15 μs						
^{104}Sr		103.952	>0.15 μs						
^{105}Sr			>0.15 μs						
$_{39}$Y		88.90585(2)							
^{77}Y		76.9496	> 0.5 μs						
^{78m}Y			0.06 s						
^{78}Y		77.9435	5.8 s	β^+/10.5		(5+)			0.279/100
									0.504/90
									0.713/40
^{79}Y		78.9374	15. s	β^+/7.1					(0.152-1.106)
^{80m}Y			4.7 s						0.2285
^{80}Y		79.9320	30. s	β^+/7.0	5.5	(4)			ann.rad./
					5.0/				0.3858/100
									0.5951/42
									0.756-1.396
^{81}Y		80.9291	1.21 m	β^+/5.5	3.7/				ann.rad./
					4.2/				0.428
									0.469
^{82}Y		81.9268	9.5 s	β^+/7.8	6.3/	1+			ann.rad./
									0.5736
									0.6017
									0.7375
^{83m}Y			2.85 m	β^+/95/4.6	2.9	1/2-			ann.rad./
				EC/5/					0.2591
									0.4218
									0.4945
^{83}Y		82.92235	7.1 m	β^+/4.47	3.3	9/2+			ann.rad./
				EC/					0.0355
									0.4899
									0.8821
									(0.03 - 3.4)
^{84m}Y			4.6 s	β^+/		1+			ann.rad./
				EC/					0.7930
^{84}Y		83.9203	40. m	β^+/6.4	1.64/47	5-			ann.rad./
				EC/	2.24/25				0.4628
					2.64/21				0.6606
					3.15/7				0.7931
									0.9744
									1.0398
									(0.2 - 3.3)
^{85m}Y			4.9 h	β^+/70/		9/2+	6.2		ann.rad./

Elem. or Isot.	Natural Abundance (%)	Atomic Mass or Weight	Half-Life	Decay Mode/Energy (/MeV)	Particle Energy /Intensity (MeV/%)	Spin (h/2π)	Nuclear Magnetic Mom. (nm)	Elect. Quadr. Mom. (b)	γ-ray/Energy Intensity (MeV/%)
				EC/30/					0.2317
									0.5356
									0.7673
									2.1238
									(0.1 - 3.1)
^{85}Y		84.91643	2.6 h	β^+/55/3.26	1.54/	1/2-			ann.rad./
				EC/45/					0.2317
									0.5045
									0.9140
									(0.07 - 1.4)
^{86m}Y			48. m	I.T./99/		8+	4.8		ann.rad./
				β^+/					0.0102(IT)
				EC/					0.2080
									(0.09 - 1.1)
^{86}Y		85.91489	14.74 h	β^+/5.24		4-	<0.6		ann.rad./
				EC/					0.3070
									0.6277
									1.0766
									1.1531
									1.9207
									(0.1 - 3.8)
^{87m}Y			13. h	I.T./98/		9/2+	6.1		0.3807
				β^+/0.7/	1.15/0.7				
				EC/					
^{87}Y		86.910880	3.35 d	EC/99+/1.862	0.78/	1/2-			0.3880
									0.4870
^{88}Y		87.909506	106.6 d	EC/99+/3.623	0.76/	4-			ann.rad./
				β^+/0.2/					0.89802
									1.83601
									2.73404
									3.2190
^{89m}Y			15.7 s	I.T./0.909		9/2+			0.9092(IT)
^{89}Y	100.	88.905849				1/2-	-0.13742		
^{90m}Y			3.24 h	I.T./99+/0.68204		7+	5.1		0.2025
				β^-/0.002/					0.4794
									0.6820
^{90}Y		89.907152	2.67 d	β^-/2.282	2.28/	2-	-1.630	-0.155	
^{91m}Y			49.7 m	I.T./0.555		9/2+	5.96		0.5556(IT)
^{91}Y		90.907301	58.5 d	β^-/1.544	1.545/	1/2-	0.1641		1.208
^{92}Y		91.90893	3.54 h	β^-/3.63	3.64/	2-			0.4485
									0.5611
									0.9345
									1.4054
									(0.4 - 3.3)
^{93m}Y			0.82 s	I.T./0.759		9/2+			0.1686(IT)
									0.5902
^{93}Y		92.90956	10.2 h	β^-/2.87	2.88/90	1/2-			0.2669
									0.9471
									1.9178
^{94m}Y			1.4 μs						0.4322
									0.7699
									1.2024
^{94}Y		93.91160	18.7 m	β^-/4.919	4.92/	2-			0.3816
									0.9188
									1.1389
									(0.3 - 4.1)
^{95}Y		94.91279	10.3 m	β^-/4.42		1/2-			0.4324
									0.9542
									2.1760

Elem. or Isot.	Natural Abundance (%)	Atomic Mass or Weight	Half-Life	Decay Mode/Energy (/MeV)	Particle Energy /Intensity (MeV/%)	Spin (h/2π)	Nuclear Magnetic Mom. (nm)	Elect. Quadr. Mom. (b)	γ-ray/Energy Intensity (MeV/%)
									3.5770
^{96m}Y			9.6 s	β^-/		(3+)			0.1467
									0.6174
									0.9150
									1.1071
									1.7507
^{96}Y		95.91588	6.2 s	β^-/7.09	7.12/	0-			1.594
^{97m}Y			1.21 s	β^-/7.4	4.8/	9/2+			0.1614
					6.0/				0.9700
									1.1030
^{97}Y		96.91813	3.76 s	β^-/6.69	6.7	1/2-			0.2969
									1.9960
									3.2876
									3.4013
^{98m}Y			2.1 s	β^-/9.8	5.5/	(4-)			0.2415
									0.6205
									0.6473
									1.2228
									1.8016
^{98}Y		97.92224	0.59 s	β^-/8.83	8.7/	1+			0.2131
									1.2228
									1.5907
									2.9413
									4.4501
^{99m}Y			0.011 ms						
^{99}Y		98.92463	1.47 s	β^-/7.57		1/2-			0.1218/43.8
				n	/2.5/				0.5362
									0.7242
									1.0130
^{100m}Y			0.94 s	β^-,n/		3+			
^{100}Y		99.9278	0.73 s	β^-,n/9.3	n/1.8/	1+			
^{101}Y		100.9303	0.43 s	β^-,n/8.6	n/1.5/	(5/2)			
^{102}Y		101.9336	0.36 s	β^-,n/9.9	n/4.0/				
^{103}Y		102.9369	0.23 s	β^-,n	n/8.3/				
^{104}Y		103.9414	0.18 s						
^{105}Y		104.9451	>0.15 μs						
^{106}Y		105.950	>0.15 μs						
^{107}Y			>0.15 μs						
^{108}Y			>0.15 μs						
$_{40}$Zr		91.224(2)							
^{79}Zr		78.949	0.06 s						
^{80}Zr		79.9406	4. s	β^+/8.0					0.290
									0.538
^{81}Zr		80.9368	5. s	β^+/7.2	6.1	(3/2-)			
^{82}Zr		81.9311	32. s	β^+/4.0	3.				ann.rad./
^{83m}Zr			7. s	β^+/7.0		(7/2+)			ann.rad./
^{83}Zr		82.9287	44. s	β^+/5.9	4.8	(1/2-)			ann.rad./
				EC					0.0556
									0.1050
									0.2560
									0.474
									1.525
^{84}Zr		83.9233	26. m	β^+/2.7		0+			ann.rad./
				EC/					0.0449
									0.1125
									0.3729
									0.667
^{85m}Zr			10.9 s	IT/0.2922		1/2-			ann.rad./
				β^+,EC/					0.2922(IT)

Elem. or Isot.	Natural Abundance (%)	Atomic Mass or Weight	Half-Life	Decay Mode/Energy (/MeV)	Particle Energy /Intensity (MeV/%)	Spin (h/2π)	Nuclear Magnetic Mom. (nm)	Elect. Quadr. Mom. (b)	γ-ray/Energy Intensity (MeV/%)
									0.4165
^{85}Zr		84.9215	7.9 m	β⁺ /4.7	3.1	7/2+			ann.rad./
				EC/					0.2663
									0.4163
									0.4543
^{86}Zr		85.91647	16.5 h	EC/1.47		0+			0.0280
									0.243
									0.612
^{87m}Zr			14.0 s	I.T./0.3362		1/2-			0.1352(IT)
									0.2010
^{87}Zr		86.91482	1.73 h	β⁺ /3.67	2.26	9/2+			ann.rad./
				EC/					0.3811
									1.228
^{88}Zr		87.91023	83.4 d	EC/0.67		0+			0.3929
^{89m}Zr			4.18 m	I.T./94/0.5877		1/2-			ann.rad./
				β⁺ /1.5/					0.5877(IT)
				EC/4.7/					1.507
^{89}Zr		88.908889	3.27 d	β⁺ /23/2.832	0.9/	9/2+	-1.07		ann.rad./
				EC/77/					0.9092
^{90m}Zr			0.809 s	I.T./		5-	6.3		0.1326
									2.1862
									2.3189(IT)
^{90}Zr	51.45(40)	89.904702				0+			
^{91}Zr	11.22(5)	90.905643				5/2+	-1.30362	-0.21	
^{92}Zr	17.15(8)	91.905039				0+			
^{93}Zr		92.906474	1.5x10^{6}y	β⁻ /0.091		5/2+			0.0304
^{94}Zr	17.38(28)	93.906314	> 10^{17} y	β⁻β⁻		0+			
^{95}Zr		94.908041	64.02 d	β⁻ /1.125	0.366/55	5/2+	1.13	+0.29	0.7242
					0.400/44				0.7567
^{96}Zr	2.80(9)	95.908275	>2×10^{19} y	β⁻β⁻		0+			
^{97}Zr		96.910950	16.8 h	β⁻ /2.658	1.91/	1/2-			0.7434
^{98}Zr		97.91276	30.7 s	β⁻ /2.26	2.2/100	0+			
^{99}Zr		98.91651	2.2 s	β⁻ /4.56	3.9/	1/2+			0.4692/55.2
					3.5/				0.5459/48
									0.028-1.321
^{100}Zr		99.91776	7.1 s	β⁻ /3.34		0+			0.4006
									0.5043
^{101}Zr		100.92114	2.1 s	β⁻ /5.49	6.2/	3/2-			0.1194
									0.2057
									0.2089
^{102}Zr		101.92298	2.9 s	β⁻ /4.61					
^{103}Zr		102.9266	1.3 s	β⁻ /7.0					
^{104}Zr		103.9288	1.2 s	β⁻ /5.9					
^{105}Zr		104.9331	≈1. s	β⁻ /8.5					
^{106}Zr		105.9359	>0.24 μs						
^{107}Zr		106.941	>0.24 μs						
^{108}Zr		107.944	>0.15 μs						
^{109}Zr			>0.15 μs						
^{110}Zr			>0.15 μs						
$_{41}$Nb		92.90638(2)							
^{81}Nb		80.949	< 0.08 μs						
^{82}Nb		81.9431	50 ms	β⁺/11.					
^{83}Nb		82.9367		β⁺/7.5					
^{84}Nb		83.9336	12. s	β⁺ ,EC/9.6		(3+)			
^{85}Nb		84.9279	2.3 m	β⁺ /6.0					
^{86m}Nb			56. s	β⁺					
^{86}Nb		85.9250	1.46 m	β⁺ /8.0					ann.rad./
									0.751
									1.003

Elem. or Isot.	Natural Abundance (%)	Atomic Mass or Weight	Half-Life	Decay Mode/Energy (/MeV)	Particle Energy /Intensity (MeV/%)	Spin (h/2π)	Nuclear Magnetic Mom. (nm)	Elect. Quadr. Mom. (b)	γ-ray/Energy Intensity (MeV/%)
^{87m}Nb			3.7 m	β^+ /		1/2-			ann.rad./
				EC/					0.1352
									0.2010
^{87}Nb		86.92036	2.6 m	β^+ 5.2/		(9/2+)			ann.rad./
				EC/					0.2010
									0.4706
									0.6165
									1.0665
									1.8842
^{88m}Nb			7.7 m	β^+ /		4-			ann.rad./
				EC/					0.2625
									0.3996
									1.0569
									1.0825
^{88}Nb		87.9183	14.3 m	β^+ /7.6	3.2/	8+			ann.rad./
				EC/					1.0570
									1.0828
									(0.07 - 2.5)
^{89m}Nb			2.0 h	β^+ /	3.3/	9/2+			0.5880/10(D)
				EC/					(0.17 - 4.0)
^{89}Nb		88.91349	1.10 h	β^+ /74/4.29	2.8/	1/2-	+6.216		ann.rad./
				EC/26/					0.5074
									0.5880
									0.7696
									1.2775
^{90m}Nb			18.8 s	I.T./0.1246		4-			0.002
									0.1225
^{90}Nb		89.911263	14.6 h	β53/6.111	0.86/5	8+	4.961		ann.rad./
				EC/47/	1.5/92				0.1412
									1.1292
									2.1862
									2.3189
									(0.1 - 3.3)
^{91m}Nb			62. d	I.T./97/		1/2-			0.1045(IT)
				EC/3/					1.2050
^{91}Nb		90.906989	$7x10^2$y	EC/1.253		9/2+			Mo k x-ray
^{92m}Nb			10.13 d	EC/99+/		2+	6.114		0.9126
									0.9345
									1.8475
^{92}Nb		91.907192	$3.7x10^7$y	EC/2.006		7+			0.5611
									0.9345
^{93m}Nb			16.1 y	I.T./0.0304		1/2-			Nb x-ray
									0.0304
^{93}Nb	100.	92.906376				9/2+	+6.1705	-0.32	
^{94m}Nb			6.26 m	I.T./99+/2.086		3+			Nb k x-ray
				β^- /0.5/					0.0409
									0.87109
^{94}Nb		93.907282	$2.4x10^4$y	β^- /2.045	0.47/	6+			0.70263
									0.87109
^{95m}Nb			3.61 d	I.T./97.5/	0.2357	1/2-			0.2040
				β^- /2.5/					0.2356
^{95}Nb		94.906834	34.97 d	β^- /0.926	0.160/	9/2+	6.141		0.76578
^{96}Nb		95.908099	23.4 h	β^- /3.187	0.5/10	6+	4.976		0.7782
					0.75/90				0.2191-1.498
^{97m}Nb			58.1 s	I.T./0.7434	0.734/98	1/2-			0.7434
^{97}Nb		96.908096	1.23 h	β^- /1.934	1.27/98	9/2+	6.15		0.4809
									0.6579
^{98m}Nb			51. m	β^- /4.67		5+			0.7874
									0.1726-1.89

Elem. or Isot.	Natural Abundance (%)	Atomic Mass or Weight	Half-Life	Decay Mode/Energy (/MeV)	Particle Energy /Intensity (MeV/%)	Spin (h/2π)	Nuclear Magnetic Mom. (nm)	Elect. Quadr. Mom. (b)	γ-ray/Energy Intensity (MeV/%)
^{98}Nb		97.91033	2.9 s	β^- /4.59	4.6/	1+			0.6451
									0.7874
									1.0243
^{99m}Nb			2.6 m	β^- /	3.2/	1/2-			0.0978/100
									(0.138-3.010)
^{99}Nb		98.91162	15.0 s	β^- /3.64	3.5/100	9/2+			0.0977
									0.1378/3.1
^{100m2}Nb			0.013 ms						
^{100m1}Nb			3.0 s	β^- /6.74	5.8				Nb k x-ray
									0.159
									0.6364
									1.0637
^{100}Nb		99.91418	1.5 s	β^- /6.25	6.2/				0.5354
					5.3/				0.6001-1.566
^{101}Nb		100.91525	7.1 s	β^- /4.57	4.3/				0.1105-0.810
^{102m}Nb			4.3 s	β^- /					
^{102}Nb		101.91804	1.3 s	β^- /7.21	7.2/				0.2960-2.184
^{103}Nb		102.91914	1.5 s	β^- /5.53	5.3/	5/2+			
^{104m}Nb			0.9 s	β^-,n/	n/0.06				
^{104}Nb		103.9225	4.8 s	β^-,n/8.1	n/0.05				
^{105}Nb		104.9239	3.0 s	β^-,n/6.5	n/1.7				
^{106}Nb		105.9282	1.0 s	β^-,n/9.3	n/4.5				
^{107}Nb		106.9303	0.30 s	β^-,n/7.9	n/6.0				
^{108}Nb		107.9350	0.19 s	β,n/	n/6.2				(0.193-0.590)
^{109}Nb		108.9376	0.19 s	β,n/	n/31				
^{110}Nb		109.943	0.17 s	β,n/	n/40				
^{111}Nb			>0.15 μs						
^{112}Nb			>0.15 μs						
^{113}Nb			>0.15 μs						
$_{42}$Mo		95.94(1)							
^{83}Mo		82.949							
^{84}Mo		83.9401	>0.15 μs	β^+/6.		(1/2-)			
^{85}Mo		84.9366	3.2 s	β^+/8.1					
^{86}Mo		85.9302	20. s	β^+/4.8					
^{87}Mo		86.9273	14. s	EC, β^+/6.5					(0.752-1.004)
^{88}Mo		87.92195	8.0 m	β^+ /3.4		0+	+0.5		ann.rad./
				EC					0.0800
									0.1399
									0.1707
^{89m}Mo			0.19 s	I.T./0.118		1/2-			0.118(IT)
									0.268
^{89}Mo		88.91948	2.2 m	β^+ /5.58		9/2+			ann.rad./
				EC/					0.659
									0.803
									1.155
									1.272
^{90}Mo		89.91394	5.7 h	β^+ /25/2.489 1.085/		0+			ann.rad./
				EC/75/					0.04274
									0.12237
									0.25734
^{91m}Mo			1.08 m	I.T./50/0.653		1/2-			ann.rad./
				β^+ ,EC/50/	2.5/				0.6529
					2.8/				1.2081
					4.0/				1.5080
									2.2407
^{91}Mo		90.91175	15.5 m	β^+ /94/4.43	3.44/94	9/2-			ann.rad./
				EC/6/					1.6373
									2.6321

TABLE OF ISOTOPES (CONTINUED)

Elem. or Isot.	Natural Abundance (%)	Atomic Mass or Weight	Half-Life	Decay Mode/Energy (/MeV)	Particle Energy /Intensity (MeV/%)	Spin (h/2π)	Nuclear Magnetic Mom. (nm)	Elect. Quadr. Mom. (b)	γ-ray/Energy Intensity (MeV/%)
									3.0286
									(0.1 - 4.2)
^{92}Mo	14.84(35)	91.906810				0+			
^{93m}Mo			6.9 h	I.T./99+/2.425		21/2+	+9.21		0.26306(IT)
									0.68461
									1.47711
^{93}Mo		92.906811	3.5x10^{3}y	EC/0.405		5/2+			0.0304
^{94}Mo	9.25(12)	93.905087				0+			
^{95}Mo	15.92(13)	94.905841				5/2+	-0.9142	-0.02	
^{96}Mo	16.68(2)	95.904678				0+			
^{97}Mo	9.55(8)	96.906020				5/2+	-0.9335	+0.26	
^{98}Mo	24.13(31)	97.905407				0+			
^{99}Mo		98.907711	2.7476 d	β^- /1.357	0.45/14	1/2+	0.375		0.144048
					0.84/2				0.18109
					1.21/84				0.36644
									0.73947
^{100}Mo	9.63(23)	99.90748	≈ 1×10^{19} y	$\beta^-\beta^-$		0+			
^{101}Mo		100.91035	14.6 m	β^- /2.82	2.23/	1/2+			0.0063
					0.7/				0.19193
									0.5909
									(0.0809-2.405)
^{102}Mo		101.91030	11.3 m	β^-/1.01	1.2/	0+			0.1493/89.
									0.2116/100.
									0.2243/32.
^{103}Mo		102.91320	1.13 m	β^-/3.8		3/2+			0.1028(2)/
									0.1440(2)
									0.2511(2)
^{104}Mo		103.91376	1.00 m	β^-/2.16		0+			0.0686(1)/100.
									0.4239(4)/21.
^{105}Mo		104.9170	36. s	β^-/4.95		3/2+			0.0642/
									0.0856/
									0.2495/
^{106}Mo		105.91814	8.4 s	β^-/3.52		0+			0.1894(2)/22.
									0.3644(2)/6.
									0.3723(2)/12.
^{107}Mo		106.9217	3.5 s	β^-/6.2					
^{108}Mo		107.9236	1.1 s	β^-/5.1					(0.028-0.636)
^{109}Mo		108.9278	0.5 s	β^-/7.2					
^{110}Mo		109.9297	0.30 s	β^-/5.7					Tc k x-ray
									0.142
									(0.039-0.599)
^{111}Mo		110.9345	>0.15 μs						
^{112}Mo		111.937	>0.15 μs						
^{113}Mo		112.942	>0.15 μs						
^{114}Mo			>0.15 μs						
^{115}Mo			>0.15 μs						
^{116}Mo			>0.15 μs						
^{117}Mo			>0.15 μs						
$_{43}$Tc									
^{85}Tc		84.949	< 0.1 ms						
^{86}Tc		85.9430	0.05 s	β^+/11.9					
^{87}Tc		86.9365	>0.15 μs	β^+/8.6					
^{88}Tc		87.9328	5.8 s	β^+/10.1					
^{89m}Tc			13. s						
^{89}Tc		88.9275	13. s	β^+/7.5					
^{90m}Tc			49.2 s	β^+	5.3/	6+			ann.rad./
									0.9479/
									1.0542/
^{90}Tc		89.9235	8.3 s	β^+/8.9	7.0/15	1+			ann.rad./

Elem. or Isot.	Natural Abundance (%)	Atomic Mass or Weight	Half-Life	Decay Mode/Energy (/MeV)	Particle Energy /Intensity (MeV/%)	Spin (h/2π)	Nuclear Magnetic Mom. (nm)	Elect. Quadr. Mom. (b)	γ-ray/Energy Intensity (MeV/%)
					7.9/95.				0.9479/
^{91m}Tc			3.3 m	β^+		1/2+			ann.rad./170.
				EC					0.8110(5)/5.
									1.6052(1)/7.8
									1.6339(1)/9.1
									1.9023(1)/6.
									2.4509(1)/13.5
^{91}Tc		90.9184	3.14 m	β^+/6.2	5.2	9/2+			ann.rad./200.
^{92}Tc		91.91526	4.4 m	β^+/7.87	4.1	8+			ann.rad./200.
				EC					0.0850/
									0.1475
									0.3293
									0.7731
									1.5096
^{93m}Tc			43. m	I.T./13		1/2-			0.3924(IT)
				EC/20					0.9437
									2.6445
^{93}Tc		92.910248	2.73 h	β^+/13/3.201	0.81	9/2+	6.26		ann.rad./
				EC/87/					1.3629
									1.4771
									1.5203
									(0.1 - 3.0)
^{94m}Tc			52. m	β^+/72/4.33		2+			ann.rad./
				EC/28/					0.8710
									1.8686
^{94}Tc		93.909655	4.88 h	β^+/11/4.256		7+	5.08		ann.rad./
				EC/89/					0.4491
									0.7026
									0.8496
									0.8710
^{95m}Tc			61. d	I.T./4/		1/2-			ann.rad./
				β^+/0.3	0.5/				0.0389(IT)
				EC/96	0.7/				0.2041
									0.5821
									0.5821
									0.8351
^{95}Tc		94.90766	20.0 h	EC/100/1.691		9/2+	5.89		0.7657
									1.0738
^{96m}Tc			52. m	I.T./90/		4+			0.0342(IT)
				β^+,EC/2/					0.7782
									1.2002
^{96}Tc		95.90787	4.3 d	EC/2.973		7+	+5.04		Mo k x-ray
									0.7782
									0.8125
									0.8498
									1.12168
^{97m}Tc			91. d	I.T./0.0965		1/2-			Tc k x-ray
				EC	/3.9				0.0965
^{97}Tc		96.906364	4.2x10^6 y	EC/100/0.320		9/2+			Mo k x-ray
^{98}Tc		97.907215	≈6.6x10^6 y	β^-/1.80	0.40/100	6+			0.65241
									0.74535
^{99m}Tc			6.01 h	I.T./100/0.142		1/2-			Tc k x-ray
									0.14049
									0.14261
^{99}Tc		98.906254	2.13x10^5 y	β^-/0.294	0.293/100	9/2+	+5.6847	-0.129	
^{100}Tc		99.907657	15.8 s	β^-/3.202	2.2/	1+			0.5396
				EC/1.8(10)$^{-3}$/0.17	2.9/				0.5908
					3.3				1.5122
									(0.3 - 2.6)

TABLE OF ISOTOPES (CONTINUED)

Elem. or Isot.	Natural Abundance (%)	Atomic Mass or Weight	Half-Life	Decay Mode/Energy (/MeV)	Particle Energy /Intensity (MeV/%)	Spin (h/2π)	Nuclear Magnetic Mom. (nm)	Elect. Quadr. Mom. (b)	γ-ray/Energy Intensity (MeV/%)
^{101}Tc		100.90731	14.2 m	β^-/1.61	1.32/	9/2+			0.1272
									0.1841
									0.3068
									0.5451
									(0.073-0.969)
^{102m}Tc			4.4 m	I.T./2/4.8	1.8/				0.4184
				β^-/98/					0.4752
									0.6281
									0.6302
									1.0464
									1.1033
									1.6163
									2.2447
^{102}Tc		101.90921	5.3 s	β^-/4.53	3.4/	1+			0.4686
					4.2				0.4751
					2.2/				1.1055
^{103}Tc		102.90918	54. s	β^-/2.66	2.0/	5/2+			0.1361
					2.2/				0.1743
									0.2104
									0.3464
									0.5629
									(0.13 - 1.0)
^{104m}Tc			0.005 ms						
^{104}Tc		103.91144	18.2 m	β^-/5.60	5.3/	(3+)			0.3483
									0.3580
									0.5305
									0.5351
									0.8844
									0.8931
									1.6768
									(0.3 - 3.7)
^{105}Tc		104.91166	7.6 m	β^-/3.6	3.4/	5/2+			0.1079
									0.1432
									0.3215
^{106}Tc		105.91436	36. s	β^-/6.55		2+			0.2703
									0.5222
									1.9694
									2.2393
									2.7893
^{107}Tc		106.9151	21.2 s	β^-/4.8					0.1027
									0.1063
									0.1770
									0.4587
^{108}Tc		107.9185	5.1 s	β^-/7.72		(3)			0.2422
									0.4656
									0.7078
									0.7326
									1.5835
^{109}Tc		108.9200	1.4 s	β^-/6.3	p/0.08				
^{110}Tc		109.9234	0.83 s	β^-/8.8	p/0.04				0.2407
^{111}Tc		110.9250	0.30 s	β^-.n/7.0	n/0.85				0.150/92.7
									0.063-1.435
^{112}Tc		111.9292	0.26 s	β,n	n/2.6				
^{113}Tc		112.931	0.15 s	β^-,n/8.	/2.1				0.0985/100
									0.0658-1.520
^{114}Tc		113.936	0.15 s	β^-,n	/1.3				
^{115}Tc		114.938	>0.15 μs						
^{116}Tc			>0.15 μs						
^{117}Tc			>0.15 μs						

TABLE OF ISOTOPES (CONTINUED)

Elem. or Isot.	Natural Abundance (%)	Atomic Mass or Weight	Half-Life	Decay Mode/Energy (/MeV)	Particle Energy /Intensity (MeV/%)	Spin (h/2π)	Nuclear Magnetic Mom. (nm)	Elect. Quadr. Mom. (b)	γ-ray/Energy Intensity (MeV/%)
^{118}Tc			>0.15 μs						
$_{44}$Ru		101.07(2)							
^{87}Ru		86.949	>1.5 μs						
^{88}Ru		87.9404	>0.15 μs			0^+			
^{89}Ru		88.936	1.2 s	β^+.p/8.					
^{90}Ru		89.9298	11. s	β^+/5.9		0^+			ann.rad./
									0.155 - 1.551
^{91}Ru		90.9264	9. s	β^+,EC/7.4		9/2+			ann.rad./
^{92}Ru		91.9201	3.7 m	β^+/53/4.5		0+			ann.rad./
				EC/47/					0.1346
									0.2138
									0.2593
^{93m}Ru			10.8 s	I.T./21/		1/2-			ann.rad./
				β^+,EC/79/	5.3/				0.7344
									1.1112
									1.3962
									2.0931
^{93}Ru		92.9171	1.0 m	β^+/6.3		9/2+			ann.rad./
				EC/					0.6807
									1.4349
									(0.5- 4.2)weak
^{94}Ru		93.91137	52. m	EC/100/1.59		0^+			0.3672
									0.5247
									0.8922
^{95}Ru		94.91042	1.64 h	EC/85/2.57	1.20/	5/2+	0.86		ann.rad./
				β^+/15/	0.91/				0.2904
									0.3364
									0.6268
^{96}Ru	5.52(20)	95.90760				0+			
^{97}Ru		96.90756	2.89 d	EC/1.12		5/2+	-0.78		Tc k x-ray
									0.2157
									0.3245
									0.4606
^{98}Ru	1.88(9)	97.90529				0+			
^{99}Ru	12.74(26)	98.905939				5/2+	-0.6413	+0.079	
^{100}Ru	12.60(19)	99.904219				0+			
^{101}Ru	17.05(7)	100.905582				5/2+	-0.7188	+0.46	
^{102}Ru	31.57(31)	101.904349				0+			
^{103}Ru		102.906323	39.27 d	β^-/0.763	0.223	3/2+	0.206	+0.62	0.05329
									0.29498
									0.4438
									0.49708
									0.55704
									0.61033
									(0.04 - 1.6)
^{104}Ru	18.66(44)	103.905430				0+			
^{105}Ru		104.907750	4.44 h	β^-/1.917	1.11/22	3/2+	-0.3		0.12968
					1.134/13				0.1491
					1.187/49				0.2629
									0.31664
									0.46943
									0.67634
									0.72420
									(0.1 - 1.8)
^{106}Ru		105.90733	1.020 y	β^-/0.0394	0.0394/100	0+			
^{107}Ru		106.9099	3.8 m	β^-/2.9	2.1/				0.1939
					3.2/				0.3741
									0.4625
									0.8488

Elem. or Isot.	Natural Abundance (%)	Atomic Mass or Weight	Half-Life	Decay Mode/Energy (/MeV)	Particle Energy /Intensity (MeV/%)	Spin (h/2π)	Nuclear Magnetic Mom. (nm)	Elect. Quadr. Mom. (b)	γ-ray/Energy Intensity (MeV/%)
^{108}Ru		107.9102	4.5 m	β^-/1.4	1.2/	0+			0.0923
									0.1651
									0.4339
									0.4975
									0.6189
^{109}Ru		108.91320	34.5 s	β^-/4.2					0.1164
									0.3584
^{110}Ru		109.9140	15. s	β^-/2.81					0.1121
									0.3737
									0.4397
									0.7967
^{111}Ru		110.9176	1.5 s	β^-/5.5					
^{112}Ru		111.9188	4.5 s	β^-/4.5					
^{113}Ru		112.9225	2.7 s	β^-/7.					
^{114}Ru		113.9239	0.57 s	β^-/6.1					0.127/24
									(0.053-0.180)
^{115}Ru		114.928	≈0.74 s		β^-/8.				
^{116}Ru		115.930	>0.15 μs						
^{117}Ru		116.935	>0.15 μs						
^{118}Ru		117.937	>0.15 μs						
^{119}Ru			>0.15 μs						
^{120}Ru			>0.15 μs						
$_{45}$Rh		102.90550(2)							
^{89}Rh		88.9494	>0.15 μs						
^{90}Rh		89.9429	>0.15 μs						
^{91}Rh		90.9366	>0.15 μs						
^{92}Rh		91.9320	>0.15 μs	β^+/11.1					
^{93}Rh		92.9257	>0.15 μs	β^+/8.1					(0.138-1.493)
^{94m}Rh			25.8 s	β^+/		8+			ann.rad./
									0.1264
									0.3117
									0.7562
									1.0752
									1.4307
^{94}Rh		93.9217	1.18 m	β^+/9.6	6.4/	3+			ann.rad./
									0.1461
									0.3117
									0.7562
									1.4307
^{95m}Rh			1.96 m	I.T./88/		1/2+			ann.rad./
				β^+,EC/12/					0.5433(IT)
									0.7837
^{95}Rh		94.9159	5.0 m	β^+/5.1	3.2	9/2+			ann.rad./
									0.2293
									0.4103
									0.6610
									0.9416
									1.3520
									(0.2 - 3.8)
^{96m}Rh			1.51 m	I.T./60/0.052		2+			ann.rad./
				β^+,EC/40/	4.70/				Tc,Ru x-rays
									0.8326
									1.0985
									1.6921
									(0.4 - 3.3)
^{96}Rh		95.91452	9.6 m	β^+/6.45	3.3/	5+			ann.rad./
				EC/					0.4299
									0.6315
									0.6853

TABLE OF ISOTOPES (CONTINUED)

Elem. or Isot.	Natural Abundance (%)	Atomic Mass or Weight	Half-Life	Decay Mode/Energy (/MeV)	Particle Energy /Intensity (MeV/%)	Spin (h/2π)	Nuclear Magnetic Mom. (nm)	Elect. Quadr. Mom. (b)	γ-ray/Energy Intensity (MeV/%)
									0.7418
									0.8326
									(0.2 - 3.4)
^{97m}Rh			46.m	I.T./5/	2.6/	1/2-			ann.rad./
				β^+,EC/95/					0.1886
									0.4215
									2.2452
^{97}Rh		96.91134	31.0m	β^+/3.52	2.1/	9/2+			ann.rad./
									0.1886
									0.3892
									0.4515
									0.8398
									0.8788
									(0.2 - 3.5)
^{98m}Rh			3.5 m	β^+ /		5+			ann.rad./
									0.6154
									0.6524
									0.7452
^{98}Rh		97.91072	8.7 m	β^+/90/5.06	3.4/	2+			ann.rad./
									0.6524
									0.7623
^{99m}Rh			4.7 h	β^+/8/	.74/	9/2+	5.67		ann.rad./
				EC/92/					0.2766/
									0.3408
									0.6178
									1.2612
^{99}Rh		98.90820	16. d	β^+/4/2.10	0.54/	1/2-			ann.rad./
				EC/97/	0.68/				0.0894/
									0.3530
									0.5277
									(0.1 - 2.0)
^{100m}Rh			4.7 m	I.T./99/		5+			ann.rad./
				β^+ /0.4/					0.0748/
									0.2647(IT)
^{100}Rh		99.90812	20.8 h	β^+ /3.63	2.62/	1-			0.4462
				EC/	2.07/				0.5396
									0.5882
									0.8225
									1.5534
									2.3761
^{101m}Rh			4.35 d	EC/92/		9/2+	+5.51		Rh k x-ray
				I.T./8/0.1573					0.1272/
									0.3069
									0.5451
^{101}Rh		100.90616	3.3 y	EC/0.54		1/2-			Ru k x-ray
									0.1272
									0.1980
									0.3252
^{102m}Rh			3.74 y	EC/2.323		6^+	4.04		0.4751
				IT/0.0419					0.6313
									0.6975
									0.7668
									1.0466
									1.1032
^{102}Rh		101.906842	207. d	EC/62			0.5		ann.rad./
				β^- /19/					0.4686
				β^+ /14/					0.4751
									0.5566
									0.6280

Elem. or Isot.	Natural Abundance (%)	Atomic Mass or Weight	Half-Life	Decay Mode/Energy (/MeV)	Particle Energy /Intensity (MeV/%)	Spin (h/2π)	Nuclear Magnetic Mom. (nm)	Elect. Quadr. Mom. (b)	γ-ray/Energy Intensity (MeV/%)
									1.1032
									(0.4 - 1.6)
^{103m}Rh			56.12 m	IT		7/2+	4.54		
^{103}Rh	100.	102.905504				1/2-	-0.0884		
^{104m}Rh			4.36 m	I.T./99+/		5+			Rh k x-ray
				β^-	1.3/				0.0514
									0.0971
									0.5558
^{104}Rh		103.906655	42.3 s	β^-/99+/2.441	1.88/2	1+			0.3581
				EC/0.4/1.141	2.44/98				0.5558
									1.2370
									(0.35 - 1.8)
^{105m}Rh			43. s	I.T./1.296		1/2-			Rh k x-ray
									0.1296
^{105}Rh		104.905692	35.4 h	β^- /0.567	0.247/30	7/2+	+4.45		0.2801
					0.567/70				0.3061
									0.3189
^{106m}Rh			2.18 h	β^- /	0.92/	6+			0.2217
									0.4510
									0.5119
									0.6162
									0.7173
									0.7484
									1.0458
									1.5277
^{106}Rh		105.90729	29.9 s	β^- /3.54	2.4/2	1+	+2.58		0.51186/
					3.0/12				0.61612
					3.54/79				0.62187
									(0.05 - 3.04)
^{107}Rh		106.90675	21.7 m	β^- /1.51	1.20/65	7/2+			0.2776
					1.5/17				0.3028
									0.3925
^{108m}Rh			6.0 m	β^- /	1.57/				0.4339
									0.4973
									0.6189
^{108}Rh		107.9087	17. s	β^- /4.5		1+			0.4046
									0.4339
									0.4973
									0.5811
									0.6146
									0.9014
									0.9471
^{109}Rh		108.90874	1.34 m	β^- /2.59	2.25/	7/2+			0.1134
									0.1780
									0.2914
									0.3254
									0.3268
									0.4261
									(0.1 - 1.6)
^{110m}Rh			20. s	β^- /	0.6/				0.3737
									0.4397
									0.7967
^{110}Rh		109.9110	3.1 s	β^- /5.4	5.5/	1+			0.3737
									0.4400
									0.5463
									0.6877
									0.8381
									0.9045
^{111}Rh		110.9117	11. s	β^- /3.7					0.275

Elem. or Isot.	Natural Abundance (%)	Atomic Mass or Weight	Half-Life	Decay Mode/Energy (/MeV)	Particle Energy /Intensity (MeV/%)	Spin (h/2π)	Nuclear Magnetic Mom. (nm)	Elect. Quadr. Mom. (b)	γ-ray/Energy Intensity (MeV/%)
^{112m}Rh			6.8 s	β^- /					
^{112}Rh		111.9140	3.5 s	β^- /6.2		1+			0.3489
^{113}Rh		112.9154	0.9 s	β^- /4.9					0.1285
^{114m}Rh			1.8 s	β^- /					
^{114}Rh		113.9173	1.8 s	β^- /6.5		1+			
^{115}Rh		114.9201	0.99 s	β^- /6.0					
^{116m}Rh			0.9 s	β^- /					0.3405
^{116}Rh		115.9228	0.7 s	β^- /8.0		1+			
^{117}Rh		116.925	0.44 s	β^-/7.					0.0346
									0.1317
^{118}Rh		117.929	>0.15 μs						
^{119}Rh		118.931	>0.15 μs						
^{120}Rh		119.936	>0.15 μs						
^{121}Rh		120.938	>0.15 μs						
^{122}Rh									
$_{46}$Pd		106.42(1)							
^{91}Pd		90.949	>1.5 μs						
^{92}Pd		91.9404	>0.15 μs						
^{93}Pd		92.9359	0.9 s	β^+,p					0.240/81
									0.382-0.864
^{94}Pd		93.9288	9. s	EC,β^+ /≈ 6.6					0.5582
									(0.0546-0.798)
^{95m}Pd		94.92684	13.4 s	EC,β^+ /10.2		21/2+			
^{95}Pd									
^{96}Pd		95.9182	2.03 m	EC,β^+ /3.5	1.15/				0.1248
									0.4995
^{97}Pd		96.9165	3.1 m	β^+ ,EC/4.8	3.5/	5/2+			ann.rad./
									0.2653
									0.4752
									0.7927
									(0.2 - 3.4)
^{98}Pd		97.91273	17.7 m	β^+ /1.87		0+			ann.rad./
				EC/					0.0677
									0.1125
									0.6630
									0.8379
^{99}Pd		98.91181	21.4 m	β^+ /49/3.37	2.18/	5/2+			ann.rad./
				EC/51/					0.1360
									0.2636
									0.6734
									(0.2 - 2.85)
^{100}Pd		99.90851	3.7 d	EC/0.36		0+			0.03271
									0.0748
									0.0840
^{101}Pd		100.90829	8.4 h	β^+ /5/1.980	0.776/	5/2+	-0.66		ann.rad./
				EC/95/					0.0244
									0.2963
									0.5904
^{102}Pd	1.02(1)	101.905607				0+			
^{103}Pd		102.906087	16.99 d	EC/0.543		5/2+			Rh k x-ray
									0.03975
									0.3575
									0.4971
^{104}Pd	11.14(8)	103.904034				0+			
^{105}Pd	22.33(8)	104.905083				5/2+	-0.642	+0.66	
^{106}Pd	27.33(3)	105.903484				0+			
^{107m}Pd			20.9 s	I.T./0.2149		11/2-			Pd k x-ray
									0.2149(IT)
^{107}Pd		106.90513	6.5x10^{6}y	β^- /0.033	0.03/	5/2+			

TABLE OF ISOTOPES (CONTINUED)

Elem. or Isot.	Natural Abundance (%)	Atomic Mass or Weight	Half-Life	Decay Mode/Energy (/MeV)	Particle Energy /Intensity (MeV/%)	Spin (h/2π)	Nuclear Magnetic Mom. (nm)	Elect. Quadr. Mom. (b)	γ-ray/Energy Intensity (MeV/%)
^{108}Pd	26.46(9)	107.903895				0+			
^{109m}Pd			4.75 m	I.T./0.1889		11/2-			Pd x-ray
									0.1889(IT)
^{109}Pd		108.905954	13.5 h	β^- /1.116	1.028	5/2+			0.0880
									(0.08 - 1.0)
^{110}Pd	11.72(9)	109.905153				0+			
^{111m}Pd			5.5 h	I.T./73/0.172		11/2-			0.0704
				β^- /27/	0.35				0.1722
					0.77				0.3912
									(0.1 - 1.97)
^{111}Pd		110.90764	23.4 m	β^- /2.19	2.2/95	5/2+			0.0598
									0.2454
									0.5800
									0.6504
									1.3885
									1.4590
^{112}Pd		111.90731	21.04 h	β^- /0.29	0.28/	0+			0.018
^{113m}Pd			1.48 m	β^- /		5/2+			0.0959
^{113}Pd		112.91015	1.64 m	β^- /3.34					0.0958
									0.4824
									0.6436
									0.7394
^{114}Pd		113.91037	2.48 m	β^- /1.45		0+			0.1266
									0.2320
									0.5582
									0.5760
^{115}Pd		114.9137	47. s	β^- /4.58					0.1255
									0.2554
									0.3428
^{116}Pd		115.9142	12.7 s	β^- /2.61					0.1015
									0.1147
									0.1778
^{117}Pd		116.9178	4.4 s	β^-/5.7					0.2473
									0.077-0.403
^{118}Pd		117.9189	2.4 s	β^-/4.1					0.1254
									0.028-0.596
^{119}Pd		118.9227	0.9 s	β^-/6.5					0.2566
									0.070-0.326
^{120}Pd		119.9240	0.5 s	β^-/5.0					0.1581
									0.053-0.595
^{121}Pd		120.9282	>0.24 μs						
^{122}Pd		121.9298	>0.24 μs						
^{123}Pd		122.934	>0.15 μs						
^{124}Pd									
$_{47}$Ag		107.8682(2)							
^{93}Ag									
^{94}Ag		93.9428	0.42 s	β^+, p/					
^{95}Ag		94.9355	2.0 s	β^+, p/					(0.539-2.025)
^{96}Ag		95.9307	5.1 s	β^+/11.6					ann.rad./
				EC/					0.1248
									0.4995
									(0.1066-1.416)
^{97}Ag		96.9240	19. s	β^+ /7.0					ann.rad./
				EC/					0.6862
									1.2941
									(0.352-3.294)
^{98}Ag		97.9218	47. s	β^+ /8.4		5+			ann.rad./
				EC/					0.5711
									0.6786

Elem. or Isot.	Natural Abundance (%)	Atomic Mass or Weight	Half-Life	Decay Mode/Energy (/MeV)	Particle Energy /Intensity (MeV/%)	Spin (h/2π)	Nuclear Magnetic Mom. (nm)	Elect. Quadr. Mom. (b)	γ-ray/Energy Intensity (MeV/%)
									0.8631
									(0.153-1.185)
^{99m}Ag			11. s	I.T./100/		1/2-			Ag k x-ray
									0.1636(IT)
									0.3426
^{99}Ag		98.9176	2.07 m	β^+ /87/5.4		9/2+			ann.rad./
				EC/13/					0.2199
									0.2645
									0.8056
									0.8323
									(0.2 - 3.5)
^{100m}Ag			2.3 m	β^+ /		2+			ann.rad./
				EC/					0.6657
									1.6941
^{100}Ag		99.9161	2.0m	β^+/7.1	4.7/	5+			ann.rad./
				EC/					0.2807
									0.4503
									0.6657
									0.7508
									0.7732
^{101m}Ag			3.1 s	I.T./0.23		1/2-			Ag k x-ray
									0.0981
									0.176(IT)
^{101}Ag		100.9128	11.1 m	β^+/69/4.2	2.7/	9/2+	5.7		ann.rad./
				EC/31/					0.2610
					2.18/				0.2747
					2.73/				0.3269
					3.38/				0.4392
									0.6673
									1.1739
									(0.2 - 3.1)
^{102m}Ag			7.8 m	β^+/38/	3.4	2+	+4.14		ann.rad./
				EC/13/					0.5567
				I.T./49/					0.9777
									1.8347
									2.0545
									2.1594
									3.2386
^{102}Ag		101.91197	13.0 m	β^+/78/5.92	2.26/	5+	4.6		ann.rad./
				EC/22/					0.5567
									0.7194
									0.8354
									1.2571
									1.5816
									1.7446
^{103m}Ag			5.7 s	I.T./0.134		1/2-			Ag k x-ray
									0.1344
^{103}Ag		102.90897	1.10 h	β^+/28/2.69	1.7	7/2+	+4.47		ann.rad./
				EC/72/	1.3				0.1187
									0.1482
^{104m}Ag			33. m	β^+/64/	2.71/	2+	+3.7		ann.rad./
				EC/36/					0.5558
				I.T./0.07/					0.7657
									(0.5 - 3.4)
^{104}Ag		103.90863	69. m	β^+/16/4.28	0.99/	5+	3.92		ann.rad./
				EC/84/					0.5558
									0.9259
									0.9416
									(0.18 - 2.27)

Elem. or Isot.	Natural Abundance (%)	Atomic Mass or Weight	Half-Life	Decay Mode/Energy (/MeV)	Particle Energy /Intensity (MeV/%)	Spin (h/2π)	Nuclear Magnetic Mom. (nm)	Elect. Quadr. Mom. (b)	γ-ray/Energy Intensity (MeV/%)
^{105m}Ag			7.2 m	I.T./98/0.0255		7/2+	+4.41		Ag x-ray
				EC/2/					0.3063
									0.3192
									(0.1 - 1.0)
^{105}Ag		104.90653	41.3 d	EC/1.35		1/2-	0.1014		0.0640
									0.2804
									0.3445
									0.4434
^{106m}Ag			8.4 d	EC/		6+	3.71	+1.1	Pd k x-ray
									0.4510
									0.5118
									0.7173
									1.0458
^{106}Ag		105.90667	24.0 m	β^+/59/2.965	/1.96	1+	+2.85		ann.rad./
				EC/41/					0.5119
^{107m}Ag			44.2 s	I.T./0.093		7/2+	+4.40	1.0	Ag x-ray
									0.0931
^{107}Ag	51.839(8)	106.905093				1/2-	-0.11357		
^{108m}Ag			418.y	EC/92/		6+	3.580	+1.3	Ag k x-ray
				I.T./8/0.079					Pd k x-ray
									0.43392
									0.61427
									0.72290
^{108}Ag		107.905954	2.39 m	β^-/97/1.65	1.02/1.7	1+	+2.6884		ann.rad./
				EC/2/	1.65/96				0.43392
				β^-/1/1.92	0.88/0.3				0.61885
									0.63298
^{109m}Ag			39.8 s	I.T./0.088		7/2+	+4.40	+1.0	Ag k x-ray
									0.0880
^{109}Ag	48.161(8)	108.904756				1/2-	-0.13069		
^{110m}Ag			249.8 d	β^-/99/	0.087	6+	+3.60	+1.4	0.65774
				I.T./1/0.1164	0.530				0.76393
									0.88467
									0.93748
									1.38427
									(0.447-1.56)
^{110}Ag		109.906111	24.6 s	β^-/2.892	2.22/5	1+	+2.7271	0.2	0.65774
					2.89/95				0.8154
									1.1257
^{111m}Ag			1.08 m	IT/99/0.0598	7/2+				Ag k x-ray
				β^-/1/					0.0598
									0.2454
^{111}Ag		110.905295	7.47 d	β^-/1.037	1.035/	1/2-	-0.146		0.2454
									0.3421
^{112}Ag		111.90701	3.13 h	β^-/3.96	3.94/	2-	0.0547		0.6067
					3.4				0.6174
									1.3877
									(0.4 - 2.9)
^{113m}Ag			1.14 m	I.T./80/0.043	7/2+				0.1422
				β^-/20/	1.5				0.2983
									0.3161
									0.3923
^{113}Ag		112.90657	5.3 h	β^-/2.02	2.01/	1/2-	0.159		0.2588
									0.2986
^{114}Ag		113.90881	4.6 s	β^-/5.08	4.9/	1+			0.5582
									0.5760
									1.9946
^{115m}Ag			18.7 s	β^-/		7/2+			0.1134
									0.1315

Elem. or Isot.	Natural Abundance (%)	Atomic Mass or Weight	Half-Life	Decay Mode/Energy (/MeV)	Particle Energy /Intensity (MeV/%)	Spin ($h/2\pi$)	Nuclear Magnetic Mom. (nm)	Elect. Quadr. Mom. (b)	γ-ray/Energy Intensity (MeV/%)
									0.2288
									0.3887
^{115}Ag		114.90876	20. m	β^-/3.10		1/2-			0.1316
									0.2128
									0.2291
									0.4727
									(0.13 - 2.49)
^{116m}Ag			10.5 s	I.T./2/	3.2/	5+			0.1027
				β^-/98/	2.9				0.2549
									0.5134
									0.7055
									1.0289
^{116}Ag		115.91137	2.68 m	β^-/6.16	5.3	2-			0.5134
									0.6993
									2.4779
^{117m}Ag			5.3 s	β^-/	3.2/	7/2+			0.1354
									0.2981
									0.3868
^{117}Ag		116.91171	1.22 m	β^-/4.18	2.3	1/2-			0.1354
									0.1571
									0.3377
^{118m}Ag			2.8 s	β^-/59/					0.1277
				I.T./41/0.1277					0.4878
									0.6771
									0.7709
									1.0586
^{118}Ag		117.9145	4.0 s	β^-/7.1					0.4878
									0.6771
									3.2259
^{119}Ag		118.9157	2.1 s	β^-/5.35		7/2+			0.0674
									0.3662
									0.3991
									0.6264
^{120m}Ag			0.32 s	β^-/					0.2030
				I.T./					0.5059
									0.6978
									0.8300
									0.9258
^{120}Ag		119.9188	1.23 s	β^-/8.2					0.5059
									0.6978
									0.8171
									1.3231
^{121}Ag		120.9198	0.78 s	β^-/6.4					0.1150
									0.3148
									0.3537
									0.3696
									0.5007
									1.5105
									(0.11 - 2.5)
^{122m}Ag			1. s	β^-/					
^{122}Ag		121.9233	0.44 s	β^-/9.2					
^{123}Ag		122.9249	0.31 s	β^-/7.4					
^{124}Ag		123.9285	0.22 s	β^-/10.1					
^{125}Ag		124.9305	0.17 s	β^-					
^{126}Ag		125.9345	0.11 s	β^-					
^{127}Ag		126.9369	0.11 s	β^-					
^{128}Ag			58 ms	β^-					
^{129}Ag			0.05 s	β^-,n					
$_{48}$Cd		112.411(8)							

Elem. or Isot.	Natural Abundance (%)	Atomic Mass or Weight	Half-Life	Decay Mode/Energy (/MeV)	Particle Energy /Intensity (MeV/%)	Spin (h/2π)	Nuclear Magnetic Mom. (nm)	Elect. Quadr. Mom. (b)	γ-ray/Energy Intensity (MeV/%)
^{96}Cd		95.9398							
^{97}Cd		96.9349	3. s	β⁺,(p)					
^{98}Cd		97.9276	9.2 s	β⁺/5.4					
				(p)	/0.025				
^{99}Cd		98.9250	16. s	β⁺,EC/6.9					ann.rad./
^{100}Cd		99.9203	1.1 m	β⁺,EC/3.9					ann.rad./
									(0.090-1.043)
^{101}Cd		100.9187	1.2 m	β⁺/83/5.5	4.5	5/2+			In k x-ray
				EC/17/					0.0985
									1.7225
									0.31 - 2.84)
^{102}Cd		101.91474	5.8 m	β⁺/27/2.59		0+			ann.rad./
				EC/73					0.0974
									0.4810
									1.0366
									1.3598
^{103}Cd		102.91342	7.5 m	β⁺/33/4.14		5/2+	-0.81	-0.8	ann.rad./
				EC/67/					Ag k x-ray
									1.0799
									1.4487
									1.4618
									(0.1 - 2.8)
^{104}Cd		103.90985	58. m	EC/1.14		0+			Ag k x-ray
									0.0835
									0.7093
^{105}Cd		104.90947	55.5 m	β⁺/26/2.739	1.69/	5/2+	-0.7393	+0.43	Ag k x-ray
				EC/74/					0.3469
									0.6072
									0.9618
									1.3025
									(0.25 - 2.4)
^{106}Cd	1.25(6)	105.90646	>2.6x10^{17} y	β⁺,EC		0+			
^{107}Cd		106.90661	6.52 h	EC/99+/1.417		5/2+	-0.615055	+0.68	Ag k x-ray
				β⁺/					0.0931
									0.8289
^{108}Cd	0.89(3)	107.90418				0+			
^{109}Cd		108.904985	462.0 d	EC/0.214		5/2+	-0.827846	+0.69	Ag k x-ray
									0.08804
^{110}Cd	12.49(18)	109.903006				0+			
^{111m}Cd			48.5 m	I.T./		11/2-			Cd k x-ray
									0.1508(IT)
									0.2454
^{111}Cd	12.80(12)	110.904182				1/2+	-0.594886		
^{112}Cd	24.13(21)	111.902758				0+			
^{113m}Cd			14.1 y	β⁻/99.9/0.59	0.59/99.9	11/2-	-1.087	-0.71	0.2637
^{113}Cd	12.22(12)	112.904401	7.7x10^{15}y	β⁻		1/2+	-0.622301		
^{114}Cd	28.73(42)	113.903359				0+			
^{115m}Cd			44.6 d	β⁻/1.629	0.68/1.6	11/2-	-1.042	-0.54	0.48450
					1.62/97				0.93381
									1.29064
^{115}Cd		114.905431	2.228 d	β⁻/1.446	0.593/42	1/2+	-0.648426		0.23141
					1.11/58				0.26085
									0.33624
									0.49227
									0.52780
^{116}Cd	7.49(18)	115.904756	2.3×10^{19} y	β⁻β⁻		0+			
^{117m}Cd			3.4 h	β⁻/2.66	0.72/	11/2-			0.1586
									0.5529

Elem. or Isot.	Natural Abundance (%)	Atomic Mass or Weight	Half-Life	Decay Mode/Energy (/MeV)	Particle Energy /Intensity (MeV/%)	Spin (h/2π)	Nuclear Magnetic Mom. (nm)	Elect. Quadr. Mom. (b)	γ-ray/Energy Intensity (MeV/%)
									0.37 - 2.42
^{117}Cd		116.907219	2.49 h	β^-/2.52	0.67/51	1/2+			0.2209
					2.2/10				0.2733
									0.3445
									1.3033
^{118}Cd		117.90692	50.3 m	β^-/0.52		0+			
^{119m}Cd			2.20 m	β^-/		11/2-			0.1056
									0.7208
									1.0250
									2.0213
^{119}Cd		118.90992	2.69 m	β^-/3.8	≈ 3.5/	1/2+			0.1340
									0.2929
									0.3429
^{120}Cd		119.90985	50.8 s	β^-/1.76	1.5/	0+			
^{121m}Cd			8. s	β^-/		11/2-			0.1008
									0.9878
									1.0209
									1.1815
									2.0594
^{121}Cd		120.9131	13.5 s	β^-/4.9		(3/2+)			0.2102
									0.3242
									0.3492
									1.0403
^{122}Cd		121.9135	5.3 s	β^-/3.0		0+			
^{123m}Cd			1.9 s	β^-/					
^{123}Cd		122.91770	2.09 s	β^-/6.12		3+			
^{124}Cd		123.9177	1.24 s	β^-/4.17		0+			0.0365
									0.0628
									0.1799
^{125m}Cd			0.66 s	β^-/					
^{125}Cd		124.92129	0.68 s	β^-/7.16		3/+			
^{126}Cd		125.9224	0.52 s	β^-/5.49		0+			0.2601
^{127}Cd		126.9264	0.4 s	β^-/8.5		3/+			
^{128}Cd		127.9278	0.28 s	β^-/7.1		0+			0.247
^{129}Cd		128.9323	0.27 s	β^-/5.9					0.281
^{130}Cd		129.9340	0.20 s	β^-/		0+			
^{131}Cd			68 ms	p/	/3.5				
^{132}Cd			0.10 ms	p/	/60				
$_{49}$In		114.818(3)							
^{98}In		97.9422	>1.5 μs						
^{99}In		98.9346	>0.15 μs	β^+/8.9					
^{100}In		99.9316	6. s	β^+,(p)/10.5					
^{101}In		100.9266	15. s	β^+/7.3					
^{102}In		101.9243	22. s	EC/8.9		(5)			0.1566
									0.7767
									(0.397-0.923)
^{103m}In			34. s						
^{103}In		102.91991	1.1 m	β^+,EC/6.05	4.2	9/2+			ann.rad./
				EC	/45				0.1879
									(0.157-3.98)
^{104m}In			16. s	IT/0.0935					
^{104}In		103.9183	1.84 m	β^+,EC/7.9	4.8	5+	+4.44	+0.7	ann.rad./
									0.6580
									0.8341
									0.8781
^{105m}In			43. s	I.T.		1/2-			In k x-ray
									0.6740
^{105}In		104.91467	5.1 m	β^+,EC/4.85	3.7	9/2+	+5.675	+0.83	0.1310
									0.2600

TABLE OF ISOTOPES (CONTINUED)

Elem. or Isot.	Natural Abundance (%)	Atomic Mass or Weight	Half-Life	Decay Mode/Energy (/MeV)	Particle Energy /Intensity (MeV/%)	Spin (h/2π)	Nuclear Magnetic Mom. (nm)	Elect. Quadr. Mom. (b)	γ-ray/Energy Intensity (MeV/%)
									0.6038
^{106m}In			5.3 m	β^+/85/	4.90	3+			ann.rad./
				EC/15/					0.6326
									0.8611
									1.7164
^{106}In		105.91346	6.2 m	β^+/65/6.52	2.6	7+	+4.92	+0.97	ann.rad./
				EC/35/					0.2259
									0.6327
									0.8611
									0.9978
									1.0091
^{107m}In			51. s	I.T./0.6786		1/2-			In k x-ray
									0.6785
^{107}In		106.91029	32.4 m	β^+/35/3.43	2.20/	9/2+	+5.59	+0.81	ann.rad./
				E.C/65/					Cd k x-ray
									0.2050
									0.3209
									0.5055
									(0.2 - 2.99)
^{108m}In			57. m	β^+/53/	1.3	6+	+4.94	+0.47	ann.rad./
				EC/47/					Cd k x-ray
									0.6329
									1.9863
									3.4522
^{108}In		107.90971	40. m	β^+/33/5.15	3.49/	3+	+4.56	+1.01	ann.rad./
				EC/67/					Cd k x-ray
									0.2429
									0.6331
									0.8756
^{109m}In			1.3 m	I.T./0.650		1/2-			In k x-ray
									0.6498
^{109}In		108.90715	4.2 h	β^+/8/2.02	0.79/	9/2+	+5.54	+0.84	ann.rad./
				EC/92/					Cd k x-ray
									0.2035
									0.6235
^{110m}In			4.9 h	EC/		7+	+4.72	+1.00	Cd k x-ray
									0.6577
									0.8847
									0.9375
									(0.1 - 1.98)
^{110}In		109.90717	1.15 h	β^+/62/3.88	2.22/	2+	+4.37	+0.35	ann.rad./
				EC/38/					Cd k x-ray
									0.6577
									(0.6 - 3.6)
^{111m}In			7.7 m	I.T./0.537		1/2-	+5.53		In k x-ray
									0.537
^{111}In		110.90511	2.8049 d	EC/0.866		9/2+	+5.50	+0.80	Cd k x-ray
									0.1712
									0.2453
^{112m}In			20.8 m	I.T./0.155		4+			In k x-ray
									0.1555
^{112}In		111.90553	14.4 m	β^+/22/2.586		1+	+2.82	+0.09	ann.rad./
				EC/34/					Cd k x-ray
				β^-/0.663					0.6171
^{113m}In			1.658 h	I.T./0.3917		1/2-	-0.210		In k x-ray
									0.3917
^{113}In	4.29(5)	112.904062				9/2+	+5.529	+0.80	
^{114m}In			49.51 d	I.T./97/0.190		5+	+4.65	+0.74	In k x-ray
				EC/3/					0.19027

Elem. or Isot.	Natural Abundance (%)	Atomic Mass or Weight	Half-Life	Decay Mode/Energy (/MeV)	Particle Energy /Intensity (MeV/%)	Spin (h/2π)	Nuclear Magnetic Mom. (nm)	Elect. Quadr. Mom. (b)	γ-ray/Energy Intensity (MeV/%)
^{114}In		113.904918	1.198 m	$β^-$/97/1.989					Cd k x-ray
				EC/3/1.453	1.984/	1+	+2.82		0.5584
									0.5727
									1.2998
^{115m}In			4.486 h	I.T./95/0.336		1/2-	-0.255		In k x-ray
				$β^-$/5/0.83					0.3362
									0.4974
^{115}In	95.71(5)	114.903879	4.4x10^{14} y	$β^-$/0.495		9/2+	+5.541	+0.81	
$^{116m2}In$			2.16 s	I.T./0.162		8-	+3.22	+0.31	In k x-ray
				EC	/0.023				0.1624
$^{116m1}In$			54.1 m	$β^-$/	1.0	5+	+4.43	+0.80	0.13792
									0.41688/27
									1.09723/58.5
									1.29349/85
^{116}In		115.905261	14.1 s	$β^-$/3.274	3.3/99	1+	2.788	0.11	0.46313
									1.2526
									1.29349
^{117m}In			1.94 h	$β^-$/53/1.769	1.77/	1/2-	-0.2517		In k x-ray
				I.T./47/					0.15855
									0.31531
									0.55294
^{117}In		116.90452	44. m	$β^-$/1.455	0.74/	9/2+	+5.52	+0.83	0.15855
									0.3966
									0.55294
$^{118m2}In$			8.5 s	I.T./98/		(8-)	+3.32	+0.44	In k x-ray
				$β^-$/2/					0.1382
$^{118m1}In$			4.40 m	$β^-$/	1.3	5+	+4.23	+0.80	0.2086
					2.0				0.6833
									1.2295
^{118}In		117.90636	5.0 s	$β^-$/4.42	4.2/	1+			0.5282
									1.1734
									1.2295
									2.0432
^{119m}In			17.9 m	$β^-$/97/	2.7/	1/2-	-0.32		0.3114
				I.T./3/0.311					0.7631
^{119}In		118.90585	2.3 m	$β^-$/2.36	1.6/	9/2+	+5.52	+0.85	0.0239
									0.6495
									0.7631
									1.2149
$^{120m2}In$			47 s	$β^-$/6.1		8-	+3.692	+0.53	1.171
									1.023
$^{120m1}In$			46. s	$β^-$/5.8	2.2/	5+	+4.30	+0.81	1.171
									1.023
^{120}In		119.90796	3.1 s	$β^-$/5.37	5.6/	(1+)			0.4146
					3.1/				0.5924
									0.8637
									1.0232
									1.1714
									(0.4 - 2.7)
^{121m}In			3.8 m	$β^-$/99/	3.7/	1/2-	-0.36		0.0601
				I.T./1/0.313					0.3136
									0.9256
									1.0412
									1.1022
									1.1204
^{121}In		120.90785	23. s	$β^-$/3.36	2.5	9/2+	+5.50	+0.81	0.2620
									0.6573
									0.9256
^{122m}In			10. s	$β^-$/	4.4/	8-	+3.78	+0.59	1.0014

Elem. or Isot.	Natural Abundance (%)	Atomic Mass or Weight	Half-Life	Decay Mode/Energy (/MeV)	Particle Energy /Intensity (MeV/%)	Spin (h/2π)	Nuclear Magnetic Mom. (nm)	Elect. Quadr. Mom. (b)	γ-ray/Energy Intensity (MeV/%)
									1.1403
^{122}In		121.91028	1.5 s	β^-/6.37	5.3/	(1+)			0.2391
									1.0014
									1.1403
									1.164
									1.1903
^{123m}In			47. s	β^-/	4.6/	(1/2-)	-0.40		0.1258
									1.170
									3.234
^{123}In		122.91044	6.0 s	β^-/4.39	3.3/	(9/2+)	+5.49	+0.76	0.6188
									1.0197
									1.1305
^{124m}In			3.4 s	β^-		8-	+3.89	+0.66	0.1029
									0.9699
									1.0729
									1.1316
^{124}In		123.91318	3.18 s	β^-/7.36	5/	3+	+4.04	+0.61	0.7070
									0.9978
									1.1316
									3.2142
									(0.3 - 4.6)
^{125m}In			12.2 s	β^-/	5.5/	1/2-	-0.43		0.1876
^{125}In		124.91360	2.33 s	β^-/5.42	4.1/	9/2+	+5.50	+0.71	0.4260
									1.0318
									1.3350
^{126m}In			1.53 s		4.9/	3+	+4.03	+0.49	0.9086
									0.9696
									1.1411
^{126}In		125.91646	1.63 s	β^-/8.21	4.2/	8-	+4.06		0.1118
									0.9086
									1.1411
^{127m}In			3.73 s	β^-/	6.4/	(1/2-)			0.2523
									3.074
^{127}In		126.91734	1.14 s	β^-/6.51	4.9/	(9/2+)	+5.52	+0.59	0.4680
									0.6461
									0.8051
									1.5977
^{128m}In			0.7 s	β^-/	5.4/	(8-)			1.8670
									1.9739
									(0.1205-2.12)
^{128}In		127.92017	0.80 s	β^-/8.98	5.0/	3+			0.9352
									1.1688
									3.5198
									4.2970
^{129m}In			1.23 s	β^-/98/	≈ 7.5/	1/2-			0.3153
				n/2/					0.9067
									1.2220
^{129}In		128.9217	0.63 s	β^-/7.66	5.5/	9/2+			0.2853
									0.7693
									1.8650
									2.1180
^{130m2}In			0.53 s	β^-/	8.8/	5+			0.0892
									0.7744
									1.2212
^{130m1}In			0.51 s	β^-/	6.1/	10-			0.0892
									0.1298
									0.7744
									1.2212
									1.9052

TABLE OF ISOTOPES (CONTINUED)

Elem. or Isot.	Natural Abundance (%)	Atomic Mass or Weight	Half-Life	Decay Mode/Energy (/MeV)	Particle Energy /Intensity (MeV/%)	Spin (h/2π)	Nuclear Magnetic Mom. (nm)	Elect. Quadr. Mom. (b)	γ-ray/Energy Intensity (MeV/%)
^{130}In		129.92486	0.29 s	β^-/10.25	10.0/	1-			
^{131m2}In			0.3 s	β^-/		(21/2+)			
^{131m1}In			0.35 s	β^-/		(1/2-)			
^{131}In		130.9268	0.28 s	β^-/9.18	6.4/	(9/2+)			0.3328
									2.433
^{132}In		131.9323	0.20 s	β^-/13.6	6.0/	(7-)			0.1320
					8.8/				0.2992
									0.3747
									4.0406
^{133}In		132.9383	0.18 s	β^-,(n)					
^{134}In		133.9447	0.14 s						(0.354-2.005)
^{135}In									
$_{50}$Sn		118.710(7)							
^{100}Sn		99.9394	≈ 0.9 s	β^+/7.3	3.4/				
^{101}Sn		100.9361	3. s	β^+/9.					
^{102}Sn		101.9243	≈ 5 s	β^+/5.8					
^{103}Sn		102.9281	7. s	β^+/7.7					
^{104}Sn		103.9232	21. s	β^+, EC/4.5					
^{105}Sn		104.9214	28. s	β^+/6.3					In-x-ray
									(0.2879-3.819)
^{106}Sn		105.91688	2.0 m	β^+/20/3.18					ann.rad./
				EC/80/					In k x-ray
									0.3865
									0.4772
^{107}Sn		106.9157	2.92 m	EC/5.0	1.2/				0.4218
				β^+/					0.6105
									0.6785
									1.0013
									1.1290
									1.542
^{108}Sn		107.91196	10.3 m	β^+/1/2.09	0.36/	0+			In k x-ray
				EC/99/					0.2724
									0.3965
									(0.105-1.68)
^{109}Sn		108.91129	18.0 m	β^+/9/3.85	1.52/	7/2+	-1.08	+0.3	ann.rad./
				EC/91/					In k x-ray
									0.6498
									1.0992
^{110}Sn		109.90785	4.1 h	EC/0.64		0+			In k x-ray
									0.283
^{111}Sn		110.90774	35. m	β^+/31/2.45	1.5/	7/2+	+0.61	+0.2	In k x-ray
				EC/69/					0.7620
									1.1530
									1.9147
^{112}Sn	0.97(1)	111.904822				0+			
^{113m}Sn			21.4 m	I.T./92/0.077		7/2+			Sn k x-ray
				EC/8/					In x-ray
									0.0774
^{113}Sn		112.905174	115.1 d	EC/1.036		1/2+	-0.879		In k x-ray
									0.25511
									0.39169
^{114}Sn	0.65(1)	113.902783				0+			
^{115}Sn	0.34(1)	114.903347				1/2+	-0.9188		
^{116}Sn	14.54(9)	115.901745				0+			
^{117m}Sn			13.60 d	I.T./0.3146		11/2-	-1.396	-0.4	Sn k x-ray
									0.15856
^{117}Sn	7.68(7)	116.902955				1/2+	-1.0010		
^{118}Sn	24.22(9)	117.901608				0+			
^{119m}Sn			293. d	I.T./0.0896		11/2-	-1.4	0.21	Sn k x-ray

Elem. or Isot.	Natural Abundance (%)	Atomic Mass or Weight	Half-Life	Decay Mode/Energy (/MeV)	Particle Energy /Intensity (MeV/%)	Spin (h/2π)	Nuclear Magnetic Mom. (nm)	Elect. Quadr. Mom. (b)	γ-ray/Energy Intensity (MeV/%)
									0.02387
^{119}Sn	8.59(4)	118.903311				1/2+	-1.0473		
^{120}Sn	32.59(9)	119.902199				0+			
^{121m}Sn			≈ 55. y	I.T./78/0.006		11/2-	-1.388	-0.14	Sn k x-ray
				β^-/22/	0.354/				0.03715
^{121}Sn		120.904239	1.128 d	β^-/0.388	0.383/100	3/2+	0.698	-0.02	
^{122}Sn	4.63(3)	121.903441				0+			
^{123m}Sn			40.1 m	β^-/1.428	1.26/99	3/2+			0.1603
									0.3814
^{123}Sn		122.905723	129.2 d	β^-/1.404	1.42/99.4	11/2-	-1.370	+0.03	0.1603
									1.0302
									1.0886
^{124}Sn	5.79(5)	123.905275				0+			
^{125m}Sn			9.51 m	β^-/2.387	2.03/98	3/2+			0.3321
									1.4040
^{125}Sn		124.907785	9.63 d	β^-/2.364	2.35/82	11/2-	-1.35	+0.1	1.0671
									(0.2-2.3)
^{126}Sn		125.90765	2.34x10^{5}y	β^-/0.38	0.25/100	0+			0.0643
									0.0876
									0.4148
									0.6663
									0.6950
^{127m}Sn			4.15 m	β^-/3.21	2.72/	3/2+			0.4909
									1.3480
									1.5640
^{127}Sn		126.91035	2.12 h	β^-/3.20	2.42/	11/2-			0.8231
					3.2/				1.0956
									(0.120-2.84)
^{128m}Sn			6.5 s	IT/0.091		(7-)			
^{128}Sn		127.91054	59.1 m	β^-/1.27	0.48/	0+			0.4823
					0.63/				0.5573
									0.6805
^{129m}Sn			6.9 m	β^-/		11/2-			1.1611
^{129}Sn		128.9134	2.4 m	β^-/4.0		3/2+			0.6456
^{130m}Sn			1.7 m	β^-/		(7-)			0.1449
									0.8992
^{130}Sn		129.91386	3.7 m	β^-/2.15	1.10/	0+			0.0700
									0.1925
									0.7798
^{131m}Sn			1.02m	β^-/	3.4/	11/2-			0.3043
									0.4500
									0.7985
									1.2260
									(0.08 - 3.21)
^{131}Sn		130.9169	39. s	β^-/4.69	3.8/	3/2+			see ^{131m} Sn
^{132}Sn		131.91775	40. s	β^-/3.12	1.8/				0.0855
									0.2467
									0.3402
									0.8985
^{133}Sn		132.9236	1.44 s	β^-/7.8	7.5/	7/2-			
^{134}Sn		133.9278	1.04 s	β^-/6.8					
^{135}Sn		134.9347	>0.15 μs						
^{136}Sn		135.9393	>0.15 μs						
^{137}Sn		136.946	>0.15 μs						
$_{51}$Sb		121.760(1)							
^{103}Sb		102.9401	>1.5 μs						
^{104}Sb		103.9363	0.5 s						
^{105}Sb		104.9315	1.1 s						

Elem. or Isot.	Natural Abundance (%)	Atomic Mass or Weight	Half-Life	Decay Mode/Energy (/MeV)	Particle Energy /Intensity (MeV/%)	Spin (h/2π)	Nuclear Magnetic Mom. (nm)	Elect. Quadr. Mom. (b)	γ-ray/Energy Intensity (MeV/%)
^{106}Sb		105.9288	0.6 s	β^+/10.5					
^{107}Sb		106.9242	≈ 4.6 s	β^+/7.9					(0.253-2.154)
^{108}Sb		107.9222	7.0 s	β^+/9.5					(0.151-1.280)
^{109}Sb		108.91814	17. s	β^+/6.38	4.42/	5/2+			0.6645/63
				EC/	4.67/				0.9254/100
					4.33/				1.0617/75
									0.247-1.495
^{110}Sb		109.9175	24. s	β^+/9.0	6.8/	3+			ann.rad./
				EC/					0.6365
									0.9847
									1.2117
									1.2433
^{111}Sb		110.91254	1.25 m	β^+/87/4.47	3.3/	5/2+			ann.rad./
				EC/13/					0.1002
									0.1545
									0.4891
									1.0326
^{112}Sb		111.91240	51.4 s	β^+/90/7.06	4.75/	3+			ann.rad./
				EC/10/					0.6700
									0.9909
									1.2571
									(0.3 - 3.6)
^{113}Sb		112.90937	6.7 m	β^+/65/3.91	2.42/	5/2+			ann.rad./
				EC/35/					Sn k x-ray
									0.3324
									0.4980
^{114}Sb		113.9091	3.49 m	β^+/78/5.9	3.4/	3+	1.7		ann.rad./
				EC/22/					Sn k x-ray
									0.8876
									1.2999
^{115}Sb		114.90660	32.1 m	β^+/67/3.03	1.51/	5/2+	+3.46	-0.4	ann.rad./
				EC/33/					Sn k x-ray
									0.4973
^{116m}Sb			1.00 h	β^+/78/	1.16/	8-	2.6		ann.rad./
				EC/22/					Sn k x-ray
									0.4073
									0.5429
									0.9725
									1.2935
									(0.0998-1.501)
^{116}Sb		115.90680	16. m	β^+/50/4.707	1.3/	3+	2.72		ann.rad./
				EC/50/	2.3/				Sn k x-ray
									0.93180
									1.29354
									(0.138-3.903)
^{117}Sb		116.90484	2.80 h	β^+/2/1.76	0.57/	5/2+	+3.4		Sn k x-ray
				EC/98/					0.1586
^{118m}Sb			5.00 h	EC/99/		8-	2.3		Sn k x-ray
									0.25368
									1.05069
									1.22964
^{118}Sb		117.905533	3.6 m	β^+/74/3.657	2.65/	1+	2.5		ann.rad./
				EC/26/					Sn k x-ray
									1.22964
^{119}Sb		118.90395	38.1 h	EC/0.59		5/2+	+3.45	-0.4	Sn k x-ray
									0.0239
^{120m}Sb			5.76 d	EC/		8-	2.34		Sn k x-ray
									0.0898
									0.19730

Elem. or Isot.	Natural Abundance (%)	Atomic Mass or Weight	Half-Life	Decay Mode/Energy (/MeV)	Particle Energy /Intensity (MeV/%)	Spin (h/2π)	Nuclear Magnetic Mom. (nm)	Elect. Quadr. Mom. (b)	γ-ray/Energy Intensity (MeV/%)
									1.02301
									1.17121
^{120}Sb		119.90508	15.89 m	β^+/41/2.68	1.72/	1+	+2.3		ann.rad./
				EC/59/					Sn k x-ray
									0.7038
									1.17121
^{121}Sb	57.21(5)	120.903822				5/2+	+3.363	-0.4	
^{122m}Sb			4.19 m	I.T./0.162		8-			Sb x-ray
									0.0614
									0.0761
^{122}Sb		121.90518	2.72 d	β^-/98/1.979	1.414/65	2-	-1.90	+0.9	0.56409
				β^+/2/1.620	1.980/26				0.69277
									1.14050
									1.2569
^{123}Sb	42.79(5)	122.904216				7/2+	+2.550	-0.5	
^{124m2}Sb			20.3 m	I.T./0.035		8-			
^{124m1}Sb			1.6 m	I.T./80/	1.2/	5+			0.4984
				β^-/20/	1.7/				0.6027
									0.6458
									1.1010
^{124}Sb		123.905938	60.20 d	β^-/2.905	0.61/52	3-	1.2	+1.9	0.60271/97.8
					2.301/23				0.64583/7.4
									0.72277/10.5
									1.69094/48.2
									(0.0274-2.808)
^{125}Sb		124.905247	2.758 y	β^-/0.767	0.13/30	7/2+	+2.63		0.0355
					0.302/45				0.17632
					0.62/13				0.38044
									0.42786
									0.46336
									0.60060
									0.63595
^{126m2}Sb			11. s	I.T./		3-			L x-ray
									0.0227
^{126m1}Sb			19.0 m	β^-/86/	1.9	5+			0.4148
				I.T./14/					0.6663
									0.6950
^{126}Sb		125.90725	12.4 d	β^-/3.67	1.9	8-	1.3		0.2786
									0.4148/83.3
									0.6663/99.7
									0.6950/99
									0.7205
^{127}Sb		126.906914	3.84 d	β^-/1.581	0.89/	7/2+	2.70		0.2524
					1.10/				0.2908
					1.50/				0.4121
									0.4370
									0.6857
									0.7837
^{128m}Sb			10.1 m	β^-/96/	2.6/	5+			0.3140
				I.T/4/					0.5941
									0.7432
									0.7539
^{128}Sb		127.90917	9.1 h	β^-/4.38	2.3/	8-	1.3		0.2148
									0.3141
									0.5265
									0.7433
									0.7540
^{129m}Sb			17.7 m	β^-/					0.4338
									0.6578

TABLE OF ISOTOPES (CONTINUED)

Elem. or Isot.	Natural Abundance (%)	Atomic Mass or Weight	Half-Life	Decay Mode/Energy (/MeV)	Particle Energy /Intensity (MeV/%)	Spin (h/2π)	Nuclear Magnetic Mom. (nm)	Elect. Quadr. Mom. (b)	γ-ray/Energy Intensity (MeV/%)
									0.7598
^{129}Sb		128.90915	4.40 h	β^-/2.38	0.65/	7/2-	2.82		0.0278
									0.1808
									0.3594
									0.4596
									0.5447
									0.8128
									0.9146
									1.0301
^{130m}Sb			6.5 m	β^-/2.6	2.12/				0.1023
									0.7934
									0.8394
^{130}Sb		129.91155	38.4 m	β^-/4.96	2.9/	8-			0.1823
									0.3309
									0.4680
									0.7394
									0.8394
^{131}Sb		130.9120	23.0 m	β^-/3.20	1.31/	7/2+			0.6423
					3.0/				0.6579
									0.9331
									0.9434
^{132m}Sb			2.8 m	β^-/	3.9/	4+			0.1034
									0.3538
									0.6968
									0.9739
									0.9896
^{132}Sb		131.91420	4.2 m	β^-/5.49		8-			0.1034
									0.1506
									0.6968
									0.9739
^{133}Sb		132.9152	2.5 m	β^-/4.00	1.20/	7/2+	3.00		0.4235
									0.6318
									0.8165
									1.0764
^{134m}Sb			10.4 s	β^-/	6.1	7-			
^{134}Sb		133.9206	0.8 s	β^-/8.4	8.4	0-			0.1152
									0.2970
									0.7063
									1.2791
^{135}Sb		134.9252	1.71 s	β^-/8.12		7/2+			1.127
									1.279
^{136}Sb		135.9301	0.82 s	β^-/9.3					
^{137}Sb		136.9353	>0.15 μs						
^{138}Sb		137.9410	>0.15 μs						
^{139}Sb		138.946	>0.15 μs						
$_{52}$Te		127.60(3)							
^{106}Te		105.9377	0.06 ms	α/4.32					
^{107}Te		106.9350	3.1 ms	α/ 70/	3.86(1)/				
				β^+,EC/10.1					
^{108}Te		107.9295	2.1 s	α/68/	3.314(4)/	0+			
				β^+,EC/32/6.8					
^{109}Te		108.9275	4.6 s	β^+ EC/96/8.7					
				α/4/	3.107(4)/				
^{110}Te		109.9224	19. s	β^+,EC/4.5		0+			ann.rad./
									0.2191
									0.6059
^{111}Te		110.9211	19.3 s	β^+,EC/8.0		(7/2+)			ann.rad./
									0.267
									0.322

Elem. or Isot.	Natural Abundance (%)	Atomic Mass or Weight	Half-Life	Decay Mode/Energy (/MeV)	Particle Energy /Intensity (MeV/%)	Spin (h/2π)	Nuclear Magnetic Mom. (nm)	Elect. Quadr. Mom. (b)	γ-ray/Energy Intensity (MeV/%)
									0.341
^{112}Te		111.9171	2.0 m	β^+,EC/4.3		0+			ann.rad./
									0.2962
									0.3727
									0.4187
^{113}Te		112.9154	1.7 s	β^+/85/5.7	4.5/	(7/2+)			ann.rad./
				EC/15/					Sb k x-ray
									0.8144
									1.0181
									1.1812
^{114}Te		113.9125	15. m	β^+/40/3.2		0+			ann.rad./
				EC/60/					Sb k x-ray
									0.0838
									0.0903
^{115m}Te			6.7 m	β^+/45/		(1/2+)			ann.rad./
				EC/55/					Sb k x-ray
									0.7236
									0.7704
^{115}Te		114.9116	5.8 m	β^+/45/4.6	2.7/	7/2+			ann.rad./
				EC/55/					Sb k x-ray
									0.7236
									1.3268
									1.3806
									(0.22 - 2.7)
^{116}Te		115.9084	2.49 h	EC/1.5		0+			Sb k x-ray
									0.0937
^{117}Te		116.90864	1.03 h	EC/75/3.54	1.78/	1/2+			ann.rad./
				β^+/25/					Sb k x-ray
									0.9197
									1.7164
									2.3000
^{118}Te		117.90583	6.00 d	EC/0.28		0+			Sb k x-ray
^{119m}Te			4.69 d	EC/		11/2-	0.89		Sb k x-ray
									0.15360
									0.2705
									1.21271
^{119}Te		118.90641	16.0 h	β^+/2/2.293	0.627/	1/2+	0.25		ann.rad.
				EC/98/					Sb k x-ray
									0.6440
									0.6998
^{120}Te	0.09(1)	119.90403				0+			
^{121m}Te			≈ 154. d	I.T.(89%)		11/2-	0.90		Te k x-ray
				EC(11%)					0.2122
^{121}Te		120.90494	16.8 d	EC/1.04		1/2+			Sb k x-ray
									0.5076
									0.5731
^{122}Te	2.55(12)	121.903056				0+			
^{123m}Te			119.7 d	I.T./0.247		11/2-	-0.93		Te k x-ray
									0.1590/84.1
^{123}Te	0.89(3)	122.904271	2.4×10^{19} y	EC/0.051		1/2+	-0.73695		
^{124}Te	4.74(14)	123.902819				0+			
^{125m}Te			58. d	I.T./0.145		11/2-	-0.99	-0.06	Te k x-ray
									0.0355
^{125}Te	7.07(15)	124.904424				1/2+	-0.8885		
^{126}Te	18.84(25)	125.903305				0+			
^{127m}Te			109. d	I.T./98/0.088		11/2-	-1.04		Te k x-ray
				β^-/2/0.77					0.0883
^{127}Te		126.905217	9.4 h	β^-/0.698	0.696/	3/2+	0.64		0.3603
^{128}Te	31.74(8)	127.904462	$>0.6 \times 10^{23}$ y	$\beta^-\beta^-$		0+			

Elem. or Isot.	Natural Abundance (%)	Atomic Mass or Weight	Half-Life	Decay Mode/Energy (/MeV)	Particle Energy /Intensity (MeV/%)	Spin (h/2π)	Nuclear Magnetic Mom. (nm)	Elect. Quadr. Mom. (b)	γ-ray/Energy Intensity (MeV/%)
^{129m}Te			33.6 d	I.T./63/0.105		11/2-	-1.09		Te k x-ray
				β^-/37/	1.60/				0.45984
									0.6959
^{129}Te		128.906596	1.16 h	β^-/1.498	0.99/9	3/2+	0.70	0.06	0.0278
					1.45/89				0.45984
									0.48728
^{130}Te	34.08(62)	129.906223	≈2×10^{21} y	$\beta^-\beta^-$		0+			
^{131m}Te			1.35 d	β^-/78/2.4	0.42/	11/2-	-1.04		0.0811
				I.T./22/0.18					0.1021
									0.14973
									0.77369
									0.79375
									0.85225
^{131}Te		130.908522	25.0 m	β^-/2.233	1.35/12	3/2+	0.70		0.14973
					1.69/22				0.45327
					2.14/60				0.49269
^{132}Te		131.90852	3.26 d	β^-/0.51	0.215	0+			0.049725
									0.11198
									0.22830
^{133m}Te			55.4 m	β^-/82/	2.4/30	11/2-			Te k x-ray
				I.T./18/0.334					0.0949
									0.1689
									0.3121
									0.3341
^{133}Te		132.9109	12.4 m	β^-/2.94	2.25/25	3/2+			0.3121
					2.65				0.4079
									1.3334
^{134}Te		133.9116	42. m	β^-/1.51	0.6/	0+			0.7672/29
					0.7/				0.0794-0.9255
^{135}Te		134.9165	19.0 s	β^-/6.0	5.4/				0.267
					6.0				0.603
									0.870
^{136}Te		135.92010	17.5 s	β^-/5.1	2.5/	0+			2.0779/25
									0.0873-3.235
^{137}Te		136.9253	2.5 s	β^-/98/6.9	6.8	7/2-			0.2436
				n/2/					
^{138}Te		137.9292	1.4 s	β^-/6.4					
^{139}Te		138.9347	>0.15 μs						
^{140}Te		139.9387	>0.15 μs						
^{141}Te		140.9444	>0.15 μs						
^{142}Te		141.949	>0.15 μs						
$_{53}$I		126.90447(3)							
^{108}I		107.9436	0.04 s	α/91/4.	3.95				
^{109}I		108.9382	0.11 ms	p					0.593/100
									0.717/63
									0.496-1.057
^{110}I		109.9346	0.65 s	β^+,EC/83/11.4					ann.rad./
				α/17/≈3.6	3.457(10)/				
				p/11/					
^{111}I		110.9303	2.5 s	β^+,E../8.5					ann.rad./
									0.2665
									0.3215
									0.3412
^{112}I		111.9280	3.4 s	β^+,EC/10.2					ann.rad./
									0.6889
									0.7869
^{113}I		112.9237	5.9 s	β^+,EC/7.6					ann.rad./
									0.4625/100
									0.6224/74

TABLE OF ISOTOPES (CONTINUED)

Elem. or Isot.	Natural Abundance (%)	Atomic Mass or Weight	Half-Life	Decay Mode/Energy (/MeV)	Particle Energy /Intensity (MeV/%)	Spin (h/2π)	Nuclear Magnetic Mom. (nm)	Elect. Quadr. Mom. (b)	γ-ray/Energy Intensity (MeV/%)
									0.0550-1.422
^{114}I		113.9219	2.1 s	β^+,EC/8.7					ann.rad./
									0.6826
									0.7088
^{115}I		114.9188	1.3 m	β^+,EC/6.7		5/2+			ann.rad./
									0.275
									0.284
									0.460
									0.709
^{116}I		115.9167	2.9 s	β^+/97/7.8	6.7/	1+			ann.rad./
				EC/3/					0.5402
									0.6789
^{117}I		116.9136	2.22 m	β^+,EC/4.7	3.2/	(5/2+)	3.1		ann.rad./
									0.2744
									0.3259
^{118m}I			8.5 m	β^+,EC/	4.9/	7-	4.2		ann.rad./
				I.T.					0.104
									0.5998
									0.6052
									0.6138
^{118}I		117.9134	14. m	β^+,EC/7.0		2-	2.0		ann.rad./
									0.5448
									0.6052
									1.3384
^{119}I		118.9102	19. m	β^+/54/3.5	2.4/	(5/2+)	+2.9		ann.rad./
				EC/46/					Te k x-ray
									0.2575
^{120m}I			53. m	β^+/80/	3.8		4.2		ann.rad.
				EC/20/					Te k x-ray
									0.4257
									0.5604
									0.6147
									1.3459
^{120}I		119.91005	1.35 h	β^+/56/5.62	4.03	2-	1.23		ann.rad./
				EC/	4.60				Te k x-ray
									0.5604
									0.6411
									1.5230
									(0.43 - 3.1)
^{121}I		120.90737	2.12 h	β^+/13/2.27	1.2/	5/2+	2.3		ann.rad./
				EC/87/					Te k x-ray
									0.2122
									(0.14 - 1.1)
^{122}I		121.90760	3.6 m	β^+/4.234	3.1/	1+	+0.94		ann.rad./
				EC/					Te k x-ray
									0.5641
^{123}I		122.905605	13.2 h	EC/1.242		5/2+	2.82		Te k x-ray
									0.1590
^{124}I		123.906211	4.18 d	β^+/23/3.160	1.54/	2-	1.44		ann.rad./
				EC/77/	2.14/				Te k x-ray
					0.75/				0.6027/62.9
									0.7228/10.3
									1.6910/11.2
									(0.31-1.73)
^{125}I		124.904624	59.4 d	EC/0.1861		5/2+	2.82	-0.89	Te k x-ray
									0.0355
^{126}I		125.905619	13.0 d	EC/		2-	1.44		ann.rad./
				β^+/2.155	1.13/				Te k x-ray
				β^-/1.258/47	0.87/				0.3887

Elem. or Isot.	Natural Abundance (%)	Atomic Mass or Weight	Half-Life	Decay Mode/Energy (/MeV)	Particle Energy /Intensity (MeV/%)	Spin ($h/2\pi$)	Nuclear Magnetic Mom. (nm)	Elect. Quadr. Mom. (b)	γ-ray/Energy Intensity (MeV/%)
					1.25/				0.6622
^{127}I	100.	126.904468				5/2+	+2.8133	-0.79	
^{128}I		127.905805	25.00 m	β^-/2.118	2.13/	1+			Te k x-ray
				EC/1.251					0.44287
									0.52658
^{129}I		128.904988	$1.7\text{x}10^7$ y	β^-/0.194	0.15/	7/2+	+2.621	-0.55	Xe k x-ray
									0.0396
^{130m}I			9.0 m	I.T./83/0.048		2+			I k x-ray
				β^-/17/					0.5361
^{130}I		129.906674	12.36 h	β^-/2.949	1.04/	5+	3.35		0.4180
					0.62				0.5361
									0.6685
									0.7395
^{131}I		130.906125	8.040 d	β^-/0.971	0.606/	7/2+	+2.742	-0.40	0.08017
									0.28431
									0.36446
									0.63699
^{132m}I			1.39 h	IT		8-			
^{132}I		131.90800	2.28 h	β^-/14/3.58	0.80/	4+	3.09	0.09	I k x-ray
				I.T./86/	1.03/				0.0980
					1.2/				0.5059
					1.6/				0.52264
					2.16/				0.63019
									0.6506
									0.66768
									0.77260
									0.95457
^{133m}I			9. s	I.T./1.63		19/2-			I kx-ray
									0.0730
									0.6474
									0.9126
^{133}I		132.90781	20.8 h	β^-/1.77	1.24/85	7/2+	+2.86	-0.27	0.51056
									0.52989
									0.87537
^{134m}I			3.7 m	I.T./98/0.316		8-			I k x-ray
				β^-/2/					0.0444
									0.2719
^{134}I		133.9099	52.6 m	β^-/4.05	1.2/	4+			0.1354
									0.84702
									0.88409
^{135}I		134.91005	6.57 h	β^-/2.63	0.9/	7/2+	2.94		0.2884
					1.3/				0.41768
									0.52658
									1.13156
									1.26046
^{136m}I			47. s	β^-/	4.7/	6-			0.1973
					5.2/				0.3468
									0.3701
									0.3814
									1.3130
									(0.16 - 2.36)
^{136}I		135.91466	1.39 m	β^-/6.93	4.3/	2-			0.3447
					5.6/				1.3130
									1.3211
									2.2896
									(0.3 - 6.1)
^{137}I		136.91787	24.5 s	β^-/5.88	5.0/	(7/2+)			0.6010
									1.2180
									1.2201

TABLE OF ISOTOPES (CONTINUED)

Elem. or Isot.	Natural Abundance (%)	Atomic Mass or Weight	Half-Life	Decay Mode/Energy (/MeV)	Particle Energy /Intensity (MeV/%)	Spin (h/2π)	Nuclear Magnetic Mom. (nm)	Elect. Quadr. Mom. (b)	γ-ray/Energy Intensity (MeV/%)
									1.3026
									1.5343
									(0.25 - 4.4)
^{138}I		137.9224	6.5 s	β^-/7.8	6.9/	2-			0.4836
					7.4/				0.5888
									0.8752
									(0.4 - 5.3)
^{139}I		138.92609	2.30 s	β^-/6.81					0.192
				n/					0.198
									0.273
									0.382
									0.386
									0.468
									0.683
									1.313
^{140}I		139.9310	0.86 s	β^-/8.8		(3)			0.372
				n/					0.377
									0.457
^{141}I		140.9351	0.45 s	β^-/7.8					
^{142}I		141.9402	≈ 0.2 s	β^-					
^{143}I		142.9441	>0.15 μs						
^{144}I		143.9496	>0.15 μs						
$_{54}$Xe		131.293(6)							
^{110}Xe		109.9445	0.2 s	β^+/9.2					
^{111m}Xe			0.9 s	EC,β^+					
^{111}Xe		110.9416	0.7 s	EC,β^+/10.6					
				α/	3.58(1)/				
^{112}Xe		111.9357	3. s	EC,β^+/7.2 α/0.8/					
^{113}Xe		112.9334	2.8 s	EC,β^+/9.1					
^{114}Xe		113.9281	10.0 s	β^+,EC/5.9		0+			ann.rad./
									0.1031
									0.1616
									0.3085
									0.6826
									0.7088
^{115}Xe		114.9270	18. s	β^+,EC/7.6		(5/2+)			ann.rad./
^{116}Xe		115.9214	56. s	β^+,EC/4.3	3.3/	0+			ann.rad./
									0.1042
									0.1916
									0.2477
									0.3107
									0.4127
^{117}Xe		116.9206	1.02 m	β^+,EC/6.5		(5/2+)	-0.594	+1.16	ann.rad./
									0.2214
									0.5190
									0.6389
									0.6613
^{118}Xe		117.917	≈ 4. m	β^+,EC/3.	2.7/	0+			ann.rad./
									0.0535
									0.0600
									0.1199
^{119}Xe		118.9156	5.8 m	β^+,EC/5.0	3.5/	7/2+	-0.654	+1.31	0.0873
									0.1000
									0.2318
									0.4615
^{120}Xe		119.91216	40. m	β^+,EC/97/1.96		0+			I k x-ray
				β^+/3/					0.0251
									0.0726
									0.1781

Elem. or Isot.	Natural Abundance (%)	Atomic Mass or Weight	Half-Life	Decay Mode/Energy (/MeV)	Particle Energy /Intensity (MeV/%)	Spin ($h/2\pi$)	Nuclear Magnetic Mom. (nm)	Elect. Quadr. Mom. (b)	γ-ray/Energy Intensity (MeV/%)
									(0.1 - 1.03)
^{121}Xe		120.91138	39. m	β^+/44/3.73	2.8/	5/2+	-0.701	+1.33	ann.rad./
				EC/56/					I k x-ray
									0.1328
									0.2527
									0.4452
									(0.1 - 3.1)
^{122}Xe		121.9086	20.1 h	EC/0.9		0+			I k x-ray
									0.3501
^{123}Xe		122.90848	2.00 h	β^+/23/2.68	1.51/	1/2+	-0.150		ann.rad./
				EC/77/					I k x-ray
									0.1489
									0.1781
									(0.1 - 2.1)
^{124}Xe	0.09(1)	123.905895	> 10^{17} y	$\beta^-\beta^-$					
^{125m}Xe			57. s	I.T./0.252		(9/2-)	-0.745	+0.42	Xe k x-ray
									0.1111
									0.141
^{125}Xe		124.906398	17.1 h	EC/1.653	0.47/	1/2+	-0.269		I k x-ray
									0.1884
									0.2434
^{126}Xe	0.09(1)	125.90427				0+			
^{127m}Xe			1.15 m	I.T./0.297		(9/2-)	-0.884	+0.69	Xe k x-ray
									0.1246
									0.1725
^{127}Xe		126.905179	36.4 d	EC/0.662		1/2+	-0.504		I k x-ray
									0.1721
									0.2029
									0.3750
^{128}Xe	1.92(3)	127.903531				0+			
^{129m}Xe			8.89 d	I.T./0.236		11/2-	-0.891	+0.64	Xe k x-ray
									0.0396
									0.1966
^{129}Xe	26.44(24)	128.904780				1/2+	-0.7780		
^{130}Xe	4.08(2)	129.903509				0+			
^{131m}Xe			11.9 d	I.T./0.164		11/2-	-0.9940	+0.73	Xe k x-ray
									0.16398
^{131}Xe	21.18(3)	130.905083				3/2+	+0.69186	-0.12	
^{132}Xe	26.89(6)	131.904155				0+			
^{133m}Xe			2.19 d	I.T./0.233		11/2-	-1.082	+0.77	Xe k x-ray
									0.23325
^{133}Xe		132.905906	5.243 d	β^-/0.427	0.346/99	3/2+	+0.813	+0.14	Cs k x-ray
									0.080998
									0.1606
^{134}Xe	10.44(10)	133.905395				0+			
^{135m}Xe			15.3 m	I.T./		11/2-	1.103	+0.62	Xe k x-ray
									0.52658
^{135}Xe		134.90721	9.10 h	β^-/1.15	0.91/	3/2+	0.903	+0.21	0.24975
									0.60807
^{136}Xe	8.87(16)	135.90722	>0.8×10^{21} y	$\beta^-\beta^-$		0+			
^{137}Xe		136.91156	3.82 m	β^-/4.17	4.1/	7/2-	-0.970	-0.49	0.45549
					3.6/				0.8489
									0.9822
									1.2732
									1.7834
									2.8498
^{138}Xe		137.91399	14.1 m	β^-/2.77	0.8/	0+			0.1538
					2.4/				0.2426
									0.2583

Elem. or Isot.	Natural Abundance (%)	Atomic Mass or Weight	Half-Life	Decay Mode/Energy (/MeV)	Particle Energy /Intensity (MeV/%)	Spin (h/2π)	Nuclear Magnetic Mom. (nm)	Elect. Quadr. Mom. (b)	γ-ray/Energy Intensity (MeV/%)
									0.4345
									1.76826
									2.0158
^{139}Xe		138.91879	39.7 s	β^-/5.06	4.5/		-0.304	+0.40	0.1750
					5.0/				0.2186
									0.2965
									(0.1 - 3.37)
^{140}Xe		139.9216	13.6 s	β^-/4.1	2.6	0+			0.0801
									0.6220
									0.8055
									1.4137
									(0.04 - 2.3)
^{141}Xe		140.9267	1.72 s	β^-/6.2	6.2/	5/2+	+0.010	-0.58	0.1187
									0.9095
									(0.05 - 2.55)
^{142}Xe		141.9297	1.22 s	β^-/5.0	3.7/	0+			0.0338
					4.2/				0.0729
									0.2038
									0.3091
									0.4145
									0.5382
									0.5718
									0.6181
									0.6448
^{143m}Xe			0.96 s	β^-					
^{143}Xe		142.9352	0.30 s	β^-/7.3			-0.460	+0.93	
^{144}Xe		143.9385	1.2 s	β^-/6.1					
^{145}Xe		144.9437	0.9 s	β^-,(n)					
^{146}Xe		145.9473	>0.15 μs						
^{147}Xe		146.9530	>0.15 μs						
$_{55}$Cs		132.90545(2)							
^{112}Cs		111.9503	0.5 ms	p	0.81				
^{113}Cs		112.9445	17. μs	p	0.96				
^{114}Cs		113.9408	0.58 s	β^+,EC/11.8		1+			ann.rad./
									0.6826
									0.7088
^{115}Cs		114.9359	≈ 1.4 s	β^+,EC/8.4					ann.rad./
^{116m}Cs			0.7 s	β ,EC/					ann.rad./
									0.3935
^{116}Cs		115.9330	3.8 s	β^+,EC/10.8					ann.rad./
									0.3935
									0.5243
									0.6151
									0.6223
^{117m}Cs			6.5 s	β^+,EC/					
^{117}Cs		116.9286	≈ 8.4 s	β^+,EC/7.5					ann.rad./
^{118m}Cs			17. s	β^+,EC/			5.		
^{118}Cs		117.92654	14. s	β^+,EC/9.		2	+3.88	+1.4	ann.rad./
									0.3372
									0.4727
									0.5865
									0.5906
^{119m}Cs			28. s			3/2	+0.84	+0.9	
^{119}Cs		118.92234	38. s	β^+,EC/6.3		9/2+	+5.5	+2.8	ann.rad./
									0.169
									0.176
									0.224
									0.257
^{120m}Cs			60. s	β^+,EC/					

TABLE OF ISOTOPES (CONTINUED)

Elem. or Isot.	Natural Abundance (%)	Atomic Mass or Weight	Half-Life	Decay Mode/Energy (/MeV)	Particle Energy /Intensity (MeV/%)	Spin ($h/2\pi$)	Nuclear Magnetic Mom. (nm)	Elect. Quadr. Mom. (b)	γ-ray/Energy Intensity (MeV/%)
^{120}Cs		119.92066	64. s	β^+,EC/7.92		2+	+3.87	+1.45	ann.rad./
									0.3224
									0.4735
									0.5534
									(0.3 - 3.28)
^{121m}Cs			2.0 m	I.T./60/		(9/2+)	+5.41	+2.7	ann.rad./
				β^+/40/	4.4				0.1794
									0.1961
^{121}Cs		120.91718	2.3 m	β^+,EC/5.40	4.38/	3/2+	+0.77	+0.84	ann.rad./
									0.1537
									(0.08 - 0.56)
^{122m2}Cs			4.4 m	β^+,EC		8-	+4.77	+3.3	ann.rad./
^{122m1}Cs			0.36 s	IT					
									0.3311
									0.4971
									0.6385
									(0.27 - 2.22)
^{122}Cs		121.91614	21. s	β^+,EC/7.1	5.8/	(1+)	-0.133	-0.19	ann.rad./
									0.3311
									0.5120
									0.8179
^{123m}Cs			1.6 s	I.T./		11/2-			Cs k x-ray
									0.0946
^{123}Cs		122.91299	5.87 m	β^+/75/4.20	3.0/	1/2+	+1.38		ann.rad./
				EC/25/					Xe k x-ray
									0.0974
									0.5964
^{124m}Cs			6.3 s	IT		7+			
^{124}Cs		123.91225	30. s	β^+/92/5.92	≈ 5.	1+	+0.673	-0.74	ann.rad./
				EC/8/					Xe k x-ray
									0.3539
									0.4925
									0.9418
^{125}Cs		124.90972	45. m	β^+/40/3.09	2.06/	1/2+	+1.41		ann.rad./
				EC/60/					Xe k x-ray
									0.112
									0.526
^{126}Cs		125.90945	1.64 m.	β^+/81/4.83	3.4	1+	+0.78	-0.68	ann.rad./
				EC/19/	3.7/				Xe k x-ray
									0.3886
									0.4912
									0.9252
^{127}Cs		126.90741	6.2 h	β^+/96/2.08	0.65/	1/2+	+1.46		Xe k x-ray
				EC/4/	1.06				0.1247
									0.4119
^{128}Cs		127.90775	3.62 m	β^+/68/3.930	2.44/	1+	+0.97	-0.57	ann.rad./
				EC/32/	2.88/				Xe k x-ray
									0.4429
^{129}Cs		128.90606	1.336 d	EC/1.195		1/2+	+1.49		Xe k x-ray
									0.3719
									0.4115
^{130m}Cs			3.5 m	IT,β^+,EC		5-	+0.629		+1.45
^{130}Cs		129.90671	29.21 m	β^+/55/2.98	1.98/	1+	+1.46	-0.06	ann.rad./
				EC/43/					Xe k x-ray
				β^-/1.6/0.37	0.44/1.6				0.5361
^{131}Cs		130.90546	9.69 d	EC/0.352		5/2+	+3.54	-0.58	Xe k x-ray
^{132}Cs		131.906430	6.48 d	EC/98/		2-	+2.22	+0.51	Xe k x-ray
				β^+/0.3/2.120					0.4646
				β^- /1.280					0.6302

TABLE OF ISOTOPES (CONTINUED)

Elem. or Isot.	Natural Abundance (%)	Atomic Mass or Weight	Half-Life	Decay Mode/Energy (/MeV)	Particle Energy /Intensity (MeV/%)	Spin (h/2π)	Nuclear Magnetic Mom. (nm)	Elect. Quadr. Mom. (b)	γ-ray/Energy Intensity (MeV/%)
									0.66769
^{133}Cs	100.	132.905447				7/2+	+2.582	-0.0037	
^{134m}Cs			2.91 h	I.T./0.139		8-	+1.098	+1.0	Cs k x-ray
									0.12749
^{134}Cs		133.906714	2.065 y	β^-/2.059	0.089/27	4+	+2.994	+0.39	0.56327
					0.658/70				0.56935
				EC/1.22					0.60473
									0.79584
^{135m}Cs			53. m	I.T./1.627		19/2-	+2.18	+0.9	0.7869
									0.8402
^{135}Cs		134.905972	2.3x10^6 y	β^-/0.269	0.205/100	7/2+	+2.732	+0.05	
^{136m}Cs			19. s	I.T./		8	+1.32	+0.7	
^{136}Cs		135.907307	13.16 d	β^-/2.548	0.341/	5+	+3.71	+0.2	0.06691
									0.34057
									0.81850
									1.04807
^{137}Cs		136.907085	30.2 y	β^-/1.176	0.514/95	7/2+	+2.84	+0.05	Ba k x-ray
									0.66164
^{138m}Cs			2.9 m	I.T./75/0.080		6-	+1.71	-0.40	Cs k x-ray
				β^-/25/	3.3				0.0799
									0.1917
									0.4628
									1.43579
^{138}Cs		137.91101	32.2 m	β^-/5.37	2.9/	3-	+0.700	+0.12	0.1381
									0.46269
									1.00969
									1.43579
									2.21788
^{139}Cs		138.913359	9.3 m	β^-/4.213	4.21	7/2+	+2.70	-0.07	0.6272
									1.2832
									(0.4 - 3.66)
^{140}Cs		139.91727	1.06 m	β^-/6.22	5.7/	1-	+0.13390	-0.11	0.5283
					6.21/				0.6023
									0.9084
									(0.41 - 3.94)
^{141}Cs		140.92005	24.9 s	β^-/5.26	5.20/	7/2+	+2.44	-0.4	Ba k x-ray
									0.0485
									0.5616
									0.5887
									1.1940
									(0.05 - 3.33)
^{142}Cs		141.92430	1.8 s	β^-/7.31	6.9/				0.3596
					7.28				0.9668
									1.1759
									1.3265
^{143}Cs		142.92732	1.78 s	β^-/6.24	6.1	(3/2+)	+0.87	+0.47	0.1955
									0.2324
									0.3064
									(0.17 - 1.98)
^{144}Cs		143.93203	1.01 s	β^-/8.47	8.46/	1	-0.546	+0.30	0.1993
					7.9/				0.5598
									0.6392
									0.7587
^{145}Cs		144.93541	0.59 s	β^-/7.89	7.4/	3/2+	+0.784	+0.6	0.1126
					7.9/				0.1755
									0.1990
^{146}Cs		145.94024	0.322 s	β^-,(n)/9.38	≈ 9.0	2-	-0.515	+0.22	
^{147}Cs		146.9439	0.227 s	β^-,(n)/9.3					
^{148}Cs		147.9490	0.15 s	β^-,(n)/10.5					

TABLE OF ISOTOPES (CONTINUED)

Elem. or Isot.	Natural Abundance (%)	Atomic Mass or Weight	Half-Life	Decay Mode/Energy (/MeV)	Particle Energy /Intensity (MeV/%)	Spin (h/2π)	Nuclear Magnetic Mom. (nm)	Elect. Quadr. Mom. (b)	γ-ray/Energy Intensity (MeV/%)
^{149}Cs		148.9527	> 50 ms						
^{150}Cs		149.9580	> 50 ms						
^{151}Cs		150.9620	> 50 ms						
$_{56}$Ba		137.327(7)							
^{114}Ba		113.9509	0.43 s	β$^+$,(p)	p/20				
^{115}Ba		114.948	0.45 s	β$^+$,(p)	p/<15				
^{116}Ba		115.9417	1.3 s	β$^+$,(p)	p/3				
^{117}Ba		116.9377	1.8 s	β$^+$,(p),EC/8.4	p/13	(3/2-)			(0.0457-0.364)
^{118}Ba		117.9466	5.2 s	β$^+$,					(0.040-0.156)
^{119}Ba		118.931	5.4 s	β$^+$,EC/8.					
^{120}Ba		119.9260	24. s	β$^+$,EC/5.0		0+			ann.rad./
									0.140
									(0.075-0.146)
^{121}Ba		120.9245	30. s	β$^+$,EC/6.8		5/2	+0.660	+1.8	ann.rad./
^{122}Ba		121.9203	2.0 m	β$^+$,EC/3.8		0+			ann.rad./
^{123}Ba		122.9189	2.7 m	β$^+$,EC/5.5			-0.68	+1.5	ann.rad./
									0.0306
									0.0927
									0.1161
									0.1235
^{124}Ba		123.91509	12. m	β$^+$,EC/2.65					ann.rad./
									0.1695
									0.1888
									1.2160
^{125m}Ba			8. m	β$^+$,EC/	4.5		0.174		
^{125}Ba		124.9146	3.5 m	β$^+$,EC/4.6	3.4	1/2+	+0.18		ann.rad./
									0.0550
									0.0776
									0.0854
									0.1409
^{126}Ba		125.91124	1.65 h	β$^+$/2/1.67		0+			Cs k x-ray
				EC/98/					0.2179
									0.2336
									0.2576
^{127m}Ba			1.9 s	IT		7/2-	-0.723	1.6	
^{127}Ba		126.9111	12.9 m	β$^+$/54/3.5		1/2+	+0.083		ann.rad./
				EC/46/					Cs k x-ray
									0.1148
									0.1808
									(0.07 - 2.5)
^{128}Ba		127.90831	2.43 d	EC/0.52		0+			Cs k x-ray
									0.27344
^{129m}Ba			2.17 h	EC/98/		7/2+	+0.93	+1.6	Cs k x-ray
				β$^+$/2/					0.1769
									0.1823
									0.2023
									1.4593
^{129}Ba		128.90868	2.2 h	β$^+$/20/2.43	1.42/	1/2+	-0.40		ann.rad./
				EC/80/					Cs k x-ray
									0.1291
									0.2143
									0.2208
^{130}Ba	0.106(1)	129.90631	>0.5×10^{15} y	β$^-$β$^-$		0+			
^{131m}Ba			14.6 m	I.T./0.187		9/2-	-0.87	+1.5	Ba k x-ray
									0.1085
^{131}Ba		130.90693	11.7 d	EC/1.37		1/2+	0.7081		Cs k x-ray
									0.12381/28.4
									0.21608/21.3
									0.49636/42.9

Elem. or Isot.	Natural Abundance (%)	Atomic Mass or Weight	Half-Life	Decay Mode/Energy (/MeV)	Particle Energy /Intensity (MeV/%)	Spin (h/2π)	Nuclear Magnetic Mom. (nm)	Elect. Quadr. Mom. (b)	γ-ray/Energy Intensity (MeV/%)
									(0.0549-1.171)
^{132}Ba	0.101(1)	131.905056				0+			
^{133m}Ba			1.621 d	I.T./0.288		11/2-	-0.91	+0.9	Ba k x-ray
									0.2761
^{133}Ba		132.906003	10.53 y	EC/0.517		1/2+	0.7717		Cs k x-ray
									0.08099
									0.35600
^{134}Ba	2.417(18)	133.904504				0+			
^{135m}Ba			1.20 d	I.T./0.2682		11/2-	-1.00	+1.0	Ba k x-ray
									0.2682
^{135}Ba	6.592(12)	134.905684				3/2+	+0.838	+0.16	
^{136m}Ba			0.308 s	I.T./2.0305		7-			Ba k x-ray
									0.8185
									1.0481
^{136}Ba	7.854(24)	135.904571				0+			
^{137m}Ba			2.552 m	I.T./0.6617		11/2-	-0.99	+0.8	Ba k x-ray
									0.66164
^{137}Ba	11.232(24)	136.905822				3/2+	+0.9374	+0.245	
^{138}Ba	71.698(42)	137.905242				0+			
^{139}Ba		138.908836	1.396 h	β^-/2.317	2.14/27	7/2-	-0.97	-0.57	0.16585
					2.27/72				1.2544
									1.42033
^{140}Ba		139.91060	12.75 d	β^-/1.05	0.48	0+			0.16268
					1.0/				0.30485
					1.02/				0.53727
^{141}Ba		140.91441	18.3 m	β^-/3.22	2.59/	3/2-	-0.34	+0.45	0.1903
					2.73/				0.2770
									0.3042
									(0.1 - 2.5)
^{142}Ba		141.91645	10.7 m	β^-/2.212	1.0/	0+			0.23152
					1.10/				0.25512
									0.3090
									1.2040
^{143}Ba		142.92061	14.3 s	β^-/4.24	4.2/	5/2+	+0.44	-0.88	0.1786
									0.21148
									0.7988
									(0.17 - 2.4)
^{144}Ba		143.92294	11.4 s	β^-/3.1	2.4/	0+			La k x-ray
					2.9/				0.10386
									0.1566
									0.1728
									0.3882
									0.43048
^{145}Ba		144.9269	4.0 s	β^-/4.9	4.9/	(5/2-)	-0.28	+1.22	La k x-ray
									0.0918
									0.09709
^{146}Ba		145.9302	2.20 s	β^-/4.12	3.9/	0+			0.0644
									0.2513
									0.3270
									0.3329
									0.3622
^{147}Ba		146.9340	0.892 s	β^-/5.75	5.5/				
^{148}Ba		147.9377	0.64 s	β^-,n/5.11					
^{149}Ba		148.9421	0.36 s	β^-,(n)/7.3					
^{150}Ba		149.9456	0.3 s						
^{151}Ba		150.9507	>0.15 μs						
^{152}Ba		151.9542							
^{153}Ba		151.9596							
$_{57}$La		138.9055(2)							

Elem. or Isot.	Natural Abundance (%)	Atomic Mass or Weight	Half-Life	Decay Mode/Energy (/MeV)	Particle Energy /Intensity (MeV/%)	Spin (h/2π)	Nuclear Magnetic Mom. (nm)	Elect. Quadr. Mom. (b)	γ-ray/Energy Intensity (MeV/%)
^{117}La		116.950							
^{118}La		117.946							
^{119}La		118.941							
^{120}La		119.938	2.8 s	EC,β^+/11.					
^{121}La		120.9330	5.3 s						
^{122}La		121.9307	9. s	EC,β^+/≈ 9.7					
^{123}La		122.9262	17. s	EC/7.					
^{124}La		123.9245	30. s	EC/≈ 8.8		(7+)			
^{125m}La			0.39 s						
^{125}La		124.9207	1.2 m	β^+,EC/5.6		11/2-			ann.rad./
									0.0436
									0.0676
^{126}La		125.9194	1.0 m	β^+,EC/7.6					ann.rad./
									0.2561
									0.340
									0.4555
									0.6214
^{127}La		126.9162	3.8 m	β^+,EC/4.7		3/2+			ann.rad./
									0.025
									0.0562
^{128}La		127.9155	5.0 m	β^+/80/6.7		(5-)			ann.rad./
				EC/20/					Ba k x-ray
									0.2841/87
									0.4793/54
									(0.315-2.212)
^{129m}La			0.56 s	IT		(11/2-)			
^{129}La		128.91267	11.6 m	β^+/58/3.72	2.42/	3/2+			ann.rad./
				EC/42/					Ba k x-ray
									0.1105
									0.2786
									(0.1 - 1.8)
^{130}La		129.9123	8.7 m	β^+/78/5.6		3+			ann.rad./
				EC/22/					Ba k x-ray
									0.3573/81
									0.5506/27
									(0.1965-1.989)
^{131}La		130.9101	59. m	β^+/76/3.0	1.42/	3/2+			ann.rad./
				EC/24/	1.94/				Ba k x-ray
									0.1085
									0.3658
									0.5263
^{132m}La			24. m	I.T./76/		6-			La k x-ray
				β^+,EC/24/					0.1352
									0.4645
^{132}La		131.91011	4.8 h	β^+/40/4.71	2.6/	2-			ann.rad./
				EC/60/	3.2				Ba k x-ray
					3.7/				0.4645
									0.5671
^{133}La		132.9084	3.91 h	β^+/4/2.2	1.2/	5/2+			Ba k x-ray
				EC/96/					0.2788
									0.2901
									0.3024
^{134}La		133.90849	6.5 m	β^+/63/3.71	2.67/	1+			ann.rad./
				EC/37/					Ba k x-ray
									0.6047
									(0.5 - 1.9)
^{135}La		134.90697	19.5 h	EC/1.20		5/2+			Ba k x-ray
									0.4805
^{136}La		135.9077	9.87 m	β^+/36/2.9	1.8/	1+			ann.rad./

TABLE OF ISOTOPES (CONTINUED)

Elem. or Isot.	Natural Abundance (%)	Atomic Mass or Weight	Half-Life	Decay Mode/Energy (/MeV)	Particle Energy /Intensity (MeV/%)	Spin (h/2π)	Nuclear Magnetic Mom. (nm)	Elect. Quadr. Mom. (b)	γ-ray/Energy Intensity (MeV/%)
				EC/64/					Ba k x-ray
									0.8185
^{137}La		136.90647	6x10^{4} y	EC/0.60		7/2+	+2.70	+0.2	0.2836
^{138}La	0.090(1)	137.907107	1.06x10^{11} y			5+	+3.7136	+0.4	1.4358/65
									0.7887/35
^{139}La	99.910(1)	138.906349				7/2+	+2.7830	+0.20	
^{140}La		139.909473	1.678 d	β^-/3.762	1.35	3-	+0.73	+0.09	
					1.24/				
					1.67/				
^{141}La		140.910958	3.90 h	β^-/2.502	2.43/	7/2+			
^{142}La		141.91408	1.54 h	β^-/4.505	2.11/	2-			
					2.98/				
					4.52/				
^{143}La		142.91606	14.1 m	β^-/3.43	3.3/	7/2-			
^{144}La		143.9196	40.7 s	β^-/5.5	4.1/				
^{145}La		144.9217	24. s	β^-/4.1	4.1/	3/2+			
^{146m}La			10.0 s	β^-/6.7	5.5/	(6)			
^{146}La		145.9258	6.3 s	β^-/6.6	6.2/	(2-)			
^{147}La		146.9278	4.02 s	β^-/5.0	4.6/				
^{148}La		147.9322	1.1 s	β^-/7.26		2-			
^{149}La		148.9342	1.10 s	β^-/5.5					
^{150}La		149.9386	0.51 s						x-ray
									(0.097-0.209)
^{151}La		150.9416	>0.15 μs						
^{152}La		151.946	>0.15 μs						
^{153}La		152.949	>0.15 μs						
^{154}La		153.954							
^{155}La		154.958							
$_{58}$Ce		140.116(1)							
^{119}Ce		118.953							
^{120}Ce		119.947							
^{121}Ce		120.944	1.1 s	β^+,p					
^{122}Ce		121.938							
^{123}Ce		122.936	3.8 s	β^+,EC/≈8.6					ann.rad./
^{124}Ce		123.931	6. s	EC/≈5.6					
^{125}Ce		124.929	9.6 s	β^+,EC/7.		(5/2+)			ann.rad./
^{126}Ce		125.9241	50. s	EC/4.					
^{127}Ce		126.9228	32. s	β^+,EC/6.1					ann.rad./
									(0.058-1.148)
^{128}Ce		127.9189	4.1 m	β^+,EC/3.2					ann.rad./
									(0.023-0.880)
^{129}Ce		128.9187	3.5 m	β^+,EC/5.6					ann.rad./
									(0.0675-1.015)
^{130}Ce		129.9147	26. m	β^+,EC/2.2		0+			ann.rad./
									La k x-ray
									(0.047-1.431)
^{131m}Ce			5. m	β^+,EC/					ann.rad./
									0.2304
									0.3955
									0.4213
^{131}Ce		130.9144	10 m	β^+,EC/4.0	2.8/				ann.rad./
									0.119
									0.169
									0.414
^{132}Ce		131.9115	3.5 h	EC/1.3		0+			La k x-ray
									0.1554
									0.1821
^{133m}Ce			1.6 h	β^+,EC/		1/2+			ann.rad./
									0.0769

Elem. or Isot.	Natural Abundance (%)	Atomic Mass or Weight	Half-Life	Decay Mode/Energy (/MeV)	Particle Energy /Intensity (MeV/%)	Spin (h/2π)	Nuclear Magnetic Mom. (nm)	Elect. Quadr. Mom. (b)	γ-ray/Energy Intensity (MeV/%)
									0.0973
									0.5577
^{133}Ce		132.9116	5.4 h	β^+/8/2.9	1.3/	9/2-			ann.rad./
				EC/92/					La k x-ray
									0.0584
									0.1308
									0.4722
									0.5104
^{134}Ce		133.9090	3.16 d	EC/0.5		0+			La k x-ray
									0.1304
									0.1623
									0.6047
^{135m}Ce			20. s	I.T./0.446		11/2-			Ce k x-ray
									0.0826
									0.1497
									0.2134
^{135}Ce		134.90915	17.7 h	β^+/1/2.026	0.8/	1/2+			La k x-ray
				EC/99/					0.0345
									0.2656
									0.3001
									0.6068
^{136}Ce	0.19(1)	135.90714				0+			
^{137m}Ce			1.43 d	I.T./99/0.254		11/2-	1.0		Ce k x-ray
				EC/0.8/					0.1693
									0.2543
^{137}Ce		136.90778	9.0 h	β^+/1.222		3/2+	0.96		La k x-ray
									0.4472
^{138}Ce	0.25(1)	137.90599				0+			
^{139m}Ce			56.4 s	I.T./0.7542		11/2-			Ce k x-ray
									0.7542
^{139}Ce		138.90665	137.6 d	EC/0.28		3/2+	1.06		La k x-ray
									0.16585
^{140}Ce	88.48(10)	139.905435				0+			
^{141}Ce		140.908272	32.50 d	β^-/0.581	0.436/69	7/2-	1.1		Pr k x-ray
					0.581/31				0.14544/48.0
^{142}Ce	11.08(10)	141.909241				0+			
^{143}Ce		142.912382	1.38 d	β^-/1.462	1.404/	3/2-	≈1.		Pr k x-ray
					1.110/47				0.0574
									0.2933
^{144}Ce		143.913643	284.6 d	β^-/0.319	0.185/20	0+			Pr k x-ray
					0.318/				0.0801
									0.1335
^{145}Ce		144.91723	3.00 m	β^-/2.54	1.7/24	3/2-			Pr k x-ray
					1.3				0.0627
									0.7245
^{146}Ce		145.9187	13.5 m	β^-/1.04	0.7/90	0+			Pr k x-ray
									0.0986
									0.2182
									0.3167
^{147}Ce		146.9225	56. s	β^-/3.29	3.3/				0.0930
									0.2687
^{148}Ce		147.9244	56. s	β^-/2.1	1.66/	0+			0.0904
									0.0985
									0.1212
									0.2918
^{149}Ce		148.9283	5.2 s	β^-/4.2					0.0577
									0.0864
									0.3800
^{150}Ce		149.9302	4.4 s	β^-/3.0					0.1099

Elem. or Isot.	Natural Abundance (%)	Atomic Mass or Weight	Half-Life	Decay Mode/Energy (/MeV)	Particle Energy /Intensity (MeV/%)	Spin (h/2π)	Nuclear Magnetic Mom. (nm)	Elect. Quadr. Mom. (b)	γ-ray/Energy Intensity (MeV/%)
^{151}Ce		150.9340	1.0 s	β^-/5.3					0.0526
^{152}Ce		151.9366	1.4 s	β^-/4.4					Pr k x-ray
									0.098
									0.115
^{153}Ce		152.9406	>0.15 μs						
^{154}Ce		153.943	>0.15 μs						
^{155}Ce		154.947	>0.15 μs						
^{156}Ce		155.951							
^{157}Ce		156.956							
$_{59}$Pr		140.90765(2)							
^{121}Pr		120.955	0.6 s						
^{122}Pr		121.952							
^{123}Pr		122.946							
^{124}Pr		123.943	1.2 s	β^+,EC/12.					ann.rad./
^{125}Pr		124.9378	≈ 3.3 s	β^+					ann.rad./
									0.1358
^{126}Pr		125.9353	3.1 s	β^+,EC/≈10.4					ann.rad./
									(0.170-0.985)
^{127}Pr		126.9308	4.2 s	β^+/≈7.5					ann.rad./
									(0.028-0.8949)
^{128}Pr		127.9288	3.0 s	β^+,EC/≈9.3					ann.rad./
									0.207/100
									0.400-1.373
^{129}Pr		128.9249	32 s	β^+,EC/5.8					ann.rad./
									(0.0395-1.865)
^{130}Pr		129.9234	40. s	β^+,EC/8.1					ann.rad./
^{131m}Pr			5.7 s						(0.06 - 0.16)
^{131}Pr		130.9201	1.7 m	β^+,EC/5.3				≈5.5	ann.rad./
									(0.059-0.980)
^{132}Pr		131.9191	1.6 m	β^+,EC/7.1					ann.rad./
									0.325
									0.496
									0.533
^{133}Pr		132.9162	6.5 m	β^+,EC/4.3		5/2+			ann.rad./
									0.074
									0.1343
									0.2419
									0.3156
									0.3308
									0.4650
^{134m}Pr			≈ 11. m	β^+,EC/					ann.rad./
									0.294
									0.460
									0.495
									0.632
^{134}Pr		133.9157	17. m	β^+,EC/6.2		2+			ann.rad./
									0.294
									0.495
^{135}Pr		134.9131	24. m	β^+,EC/3.7	2.5/	3/2+			ann.rad./
									0.0826
									0.2135
									0.2961
									0.5832
^{136}Pr		135.91265	13.1 m	β^+/57/5.13	2.98/	2+			ann.rad./
				EC/43					Ce k x-ray
									0.5398
									0.5522
^{137}Pr		136.91068	1.28 h	β^+/26/2.70	1.68/	5/2+			ann.rad./
				EC/74/					Ce k x-ray

TABLE OF ISOTOPES (CONTINUED)

Elem. or Isot.	Natural Abundance (%)	Atomic Mass or Weight	Half-Life	Decay Mode/Energy (/MeV)	Particle Energy /Intensity (MeV/%)	Spin ($h/2\pi$)	Nuclear Magnetic Mom. (nm)	Elect. Quadr. Mom. (b)	γ-ray/Energy Intensity (MeV/%)
									0.4339
									0.5140
									0.8367
									(0.16 - 1.8)
^{138m}Pr			2.1 h	β^+/24/	1.65/	7-			ann.rad./
				EC/76/					Ce k x-ray
									0.3027
									0.7887
									1.0378
									(0.07 - 2.0)
^{138}Pr		137.91075	1.45 m	β^+/75/4.44	3.42/	1+			ann.rad./
				EC/25/					Ce k x-ray
									0.7887
^{139}Pr		138.90893	4.41 h	β^+/8/2.129	1.09/	5/2+			ann.rad./
				EC/92/					Ce k x-ray
									0.2551
									1.3473
									1.6307
^{140}Pr		139.90907	3.39 m	β^+/51/3.39	2.37/	1+			ann.rad./
				EC/49/					Ce k x-ray
									0.3069
									1.5965
^{141}Pr	100.	140.907648				5/2+	+4.275	-0.08	
^{142m}Pr			14.6 m	I.T./0.004	c.e./	5-	2.2		
^{142}Pr		141.910041	19.12 h	β^-/2.162	0.58/4	2-	+0.234	+0.030	0.5088
				EC/0.744	2.16/96				1.57580
^{143}Pr		142.910813	13.57 d	β^-/0.934	0.933/	7/2+	+2.70	+0.8	0.7420
^{144m}Pr			7.2 m	IT/99+/0.059		3-			Pr k x-ray
				β^-/					0.0590
									0.6965
									0.8142
^{144}Pr		143.913301	17.28 m	β^-/2.998	0.807/1	0-			0.69649
					2.30/				1.48912
					2.996/98				2.18562
^{145}Pr		144.91451	5.98 h	β^-/1.81	1.80/97	7/2+			0.0725
									0.6758
									0.7483
^{146}Pr		145.9176	24.2 m	β^-/4.2	2.2/30	2-			0.4539/48
					3.7/10				1.5247
					4.2/40				
^{147}Pr		146.91898	13.4 m	β^-/2.69	1.5/	3/2+			0.3146/24.
					2.1/				0.5779/16
									0.6413/19.
^{148m}Pr			2.0 m	β^-/	4.0/	(4)			0.3016
					3.8/				0.4506
									0.6975
^{148}Pr		147.9222	2.27 m	β^-/4.9	4.8/	1-			0.3017
					4.5/				
^{149}Pr		148.92379	2.3 m	β^-/3.40	3.0	(5/2+)			0.1085
									0.1385
									0.1651
^{150}Pr		149.9270	6.2 s	β^-/5.7		1-			0.1302
					≈ 5.5				0.8044
									0.8527
^{151}Pr		150.9283	22.4 s	β^-/4.2					

Elem. or Isot.	Natural Abundance (%)	Atomic Mass or Weight	Half-Life	Decay Mode/Energy (/MeV)	Particle Energy /Intensity (MeV/%)	Spin (h/2π)	Nuclear Magnetic Mom. (nm)	Elect. Quadr. Mom. (b)	γ-ray/Energy Intensity (MeV/%)
^{152}Pr		151.9319	3.2 s	β^-/6.7		4+			0.0726
									0.164
									0.285
^{153}Pr		152.9339	4.3 s	β^-/5.5					
^{154}Pr		153.9381	2.3 s	β^-/7.9					
^{155}Pr		154.9400							
^{156}Pr		155.944							
^{157}Pr		156.947							
^{158}Pr		157.952							
^{159}Pr		158.955							
$_{60}$Nd		144.24(3)							
^{125}Nd			0.6 s	β^+,p					
^{126}Nd		125.943							
^{127}Nd		126.941	1.8 s	β^+,EC/9.		(5/2)			ann.rad./
^{128}Nd		127.935	4. s	β^+,EC/6.					ann.rad./
^{129}Nd		128.933	4.9 s	β^+,EC/8.		5/2(-)			ann.rad./
									(0.091-0.875)
^{130}Nd		129.929	28. s	β^+,EC/5.					ann.rad./
^{131}Nd		130.9271	0.5 m	β^+,EC/6.6					ann.rad./
									(0.09-0.36)
^{132}Nd		131.9231	1.5 m	β^+,EC/3.7					ann.rad./
									(0.099-0.567)
^{133}Nd		132.9222	1.2 m	β^+,EC/5.6					ann.rad./
									(0.06-0.37)
^{134}Nd		133.9187	≈ 8.5 m	β^+/17/2.8		0+			ann.rad./
				EC/83/					Pr k x-ray
									0.1631/58
									(0.09-1.00)
^{135m}Nd			5.5 m	β^+/					
^{135}Nd		134.9182	12. m	β^+/65/4.8		9/2-	-0.78	+2.0	ann.rad./
				EC/35/					Pr k x-ray
									0.0415/23.
									0.204/51.
									(0.11-1.8)
^{136}Nd		135.9150	50.6 m	EC/94/2.21	1.04/	0+			Pr kx-ray
				β^+/6/					0.0401/21.
									0.1091/35.
									(0.10-0.97)
^{137m}Nd			1.6 s	I.T./0.5196		11/2-			Nd k x-ray
									0.1084
									0.1775
									0.2337
^{137}Nd		136.9146	38. m	β^+/40/3.69	1.7/20	1/2+	-0.63		ann.rad./
				EC/60/	2.40/20				Pr k x-ray
									0.0755
									0.5806
^{138}Nd		137.9119	5.1 h	EC/1.1		0+			Pr k x-ray
									0.1995
									0.3258
^{139m}Nd			5.5 h	I.T./12/0.231	1.17/	11/2-			Nd k x-ray
				β^+/88/					Pr k x-ray
									0.1139/34.
									0.7382/30.
^{139}Nd		138.91192	30. m	β^+/25/2.79	1.77/	3/2+	0.91	+0.3	ann.rad./
				EC/75/					Pr k x-ray
									0.4050
^{140}Nd		139.90931	3.37 d	EC/0.22		0+			Pr k x-ray
^{141m}Nd			1.04 m	IT/99+/0.756		11/2-			Nd k x-ray
									0.7565

Elem. or Isot.	Natural Abundance (%)	Atomic Mass or Weight	Half-Life	Decay Mode/Energy (/MeV)	Particle Energy /Intensity (MeV/%)	Spin (h/2π)	Nuclear Magnetic Mom. (nm)	Elect. Quadr. Mom. (b)	γ-ray/Energy Intensity (MeV/%)
^{141}Nd		140.909605	2.49 h	EC/98/1.823	0.802/	3/2+	1.01	+0.3	Pr k x-ray
				β^+/2/					(0.15-1.7)
^{142}Nd	27.13(12)	141.907719				0+			
^{143}Nd	12.18(6)	142.909810				7/2-	-1.07	-0.60	
^{144}Nd	23.80(12)	143.910083	2.1x10^{15} y			0+			
^{145}Nd	8.30(6)	144.912569				7/2-	-0.66	-0.31	
^{146}Nd	17.19(9)	145.913113				0+			
^{147}Nd		146.916096	10.98 d	β^-/0.896	0.805/	5/2-	0.58	0.9	Pr k x-ray
									0.53102
									0.09111-0.686
^{148}Nd	5.76(3)	147.916889				0+			
^{149}Nd		148.920145	1.73 h	β^-/1.691	1.03/25	5/2-	0.35	1.3	Pr k x-ray
					1.13/26				0.1143/19.
					1.42/				0.2113/27.
									(0.06 - 1.6)
^{150}Nd	5.64(3)	149.920887	≈1×10^{19} y	$\beta^-\beta^-$		0+			
^{151}Nd		150.923825	12.4 m	β^-/2.442	1.2/	(3/2+)			Pm k x-ray
									0.1168
									0.2557
									1.1806
									(0.10 - 1.9)m
^{152}Nd		151.92468	11.4 m	β^-/1.1		0+			0.2785/29.
									0.2501/18.
									(0.016 - 0.66)
^{153}Nd		152.9280	28.9 s	β^-/3.6					0.418
^{154}Nd		153.9296	25.9 s	β^-/2.8					0.1519
									0.7998
^{155}Nd		154.9334	8.9 s	β^-/5.0					0.1807
^{156}Nd		155.9355	5.5 s	β^-/4.1					0.0848
^{157}Nd		156.9393							
^{158}Nd		157.942							
^{159}Nd		158.946							
^{160}Nd		159.949							
^{161}Nd		160.954							
$_{61}$Pm									
^{128}Pm		127.948	1.0 s	β^+,p					Ann.rad.
^{129}Pm		128.943							
^{130}Pm		129.940	2.5 s	β^+,EC/11.					0.1589
									0.326-1.062
^{131}Pm		130.936	≈ 6.3 s	β^+					0.185
									0.220
									0.146
^{132}Pm		131.934	6. s	β^+,EC/10.					ann.rad./
^{133}Pm		132.930	12. s	β^+,EC/≈ 7.0					ann.rad./
^{134}Pm		133.9282	24. s	β^+,EC/≈ 8.9		(5+)			ann.rad./
									0.294
									0.495
^{135}Pm		134.9247	0.8 m	β^+,EC/6.0		11/2-			(0.13-0.47)
^{136}Pm		135.9235	1.8 m	β^+/89/7.9		(3+)			ann.rad./
				EC/11/					Nd k x-ray
									0.3735
									0.6027
^{137}Pm		136.9206	2.4 m	β^+,EC/5.6		(11/2-)			ann.rad./
									0.1086
									0.1775
^{138m}Pm			3.2 m	β^+/50/≈ 7.0	3.9/	3+	3.		ann.rad./
				EC/50/					Nd k x-ray
									0.5209
									0.7290

Elem. or Isot.	Natural Abundance (%)	Atomic Mass or Weight	Half-Life	Decay Mode/Energy (/MeV)	Particle Energy /Intensity (MeV/%)	Spin (h/2π)	Nuclear Magnetic Mom. (nm)	Elect. Quadr. Mom. (b)	γ-ray/Energy Intensity (MeV/%)
^{138}Pm		137.9193	10. s	β^+/6.9	6.1/	1+			ann.rad./
^{139m}Pm			0.18 s	IT/		(11/2-)			0.1887
^{139}Pm		138.91678	4.14 m	β^+/68/4.52	3.52/	(5/2+)			ann.rad./
				EC/32/					Nd k x-ray
									0.4028
									(0.27 - 2.4)
^{140m}Pm			5.87 m	β^+/70/	3.2	7/2-			ann.rad./
				EC/30/					Nd k x-ray
									0.4199
									0.7738
									1.0283
^{140}Pm		139.91585	9.2 s	β^+/89/6.09	5.07/74	1+			ann.rad./
				EC/11/					Nd k x-ray
									0.7738
									1.4898
^{141}Pm		140.91359	20.9 m	β^+/52/3.72	2.71	5/2+			ann.rad./
				EC/48/					Nd k x-ray
									0.8862
									1.2233
^{142}Pm		141.91295	40.5 s	β^+/86/4.87	3.8/	1+			ann.rad./
				EC/20/					Nd k x-ray
									0.6414
									1.5758
^{143}Pm		142.910928	265. d	EC/1.041		5/2+	3.8		Nd k x-ray
				β^+/<6 x 10^{-6}/					0.7420
^{144}Pm		143.912586	360. d	EC/2.332		5-	1.7		Nd k x-ray
				β^+/7x10^{-6}/					0.6180
									0.6965
^{145}Pm		144.912745	17.7 y	EC/0.163		5/2+	+3.8	+0.2	Nd k x-ray
									0.0723
^{146}Pm		145.914693	5.53 y	EC/63/1.472		3-			Nd k x-ray
				β^-/37/1.542	0.795/				0.4538
									0.7362
									0.7474
^{147}Pm		146.915134	2.623 y	β^-/0.224	0.224/	7/2+	+2.6	+0.7	0.1213
									0.1974
^{148m}Pm			41.3 d	β^-/95/2.6	0.4/60	6-	1.8		0.5503/94.
				I.T./5/0.137	0.5/17				0.6300/89.
					0.7/21				0.7257/33
^{148}Pm		147.91747	5.37 d	β^-/2.47	1.02/	1-	+2.0	+0.2	0.5503
					2.47/				0.9149
									1.4651
^{149}Pm		148.918330	2.212 d	β^-/1.071	0.78/9	7/2+	3.3		0.2859
					1.072/90				0.5909
									0.8594
^{150}Pm		149.92098	2.68 h	β^-/3.45	1.6/	(1-)			0.3339/69.
					2.3/				1.1658/16.
					1.8/				1.3245/17.
									(0.25 - 2.9)
^{151}Pm		150.92120	1.183 d	β^-/1.187	0.84/	5/2+	+1.8	1.9	0.1677/8
									0.2751/7
									0.3401/22
^{152m2}Pm			15. m	β^-,I.T./		(>6)			(0.14-1.4)
^{152m1}Pm			7.5 m	β^-/		(4-)			0.1218
									0.2447
									0.3404
									1.0971
									1.4375
^{152}Pm		151.9235	4.1 m	β^-/3.5	3.5/20	1+			0.1218

Elem. or Isot.	Natural Abundance (%)	Atomic Mass or Weight	Half-Life	Decay Mode/Energy (/MeV)	Particle Energy /Intensity (MeV/%)	Spin (h/2π)	Nuclear Magnetic Mom. (nm)	Elect. Quadr. Mom. (b)	γ-ray/Energy Intensity (MeV/%)
					3.50/60				(0.12 - 2.1)
^{153}Pm		152.92414	5.4 m	β^-/1.90	1.7/	(5/2-)			0.0910
									0.1198
									0.1273
^{154m}Pm			2.7 m	β^-/	2.0/				0.0820
									0.1848
									1.4403
^{154}Pm		153.9266	1.7 m	β^-/4.1	1.9/				0.0820
									0.8396
									1.3940
									2.0589
									(0.08 - 2.8)
^{155}Pm		154.9280	48. s	β^-/3.2		(5/2-)			(0.05-0.78)
^{156}Pm		155.93106	26.7 s	β^-/5.16					
^{157}Pm		156.9332	10.9 s	β^-/4.6					
^{158}Pm		157.9367	5. s	β^-/6.3					
^{159}Pm		158.939	2 s						
^{160}Pm		159.943							
^{161}Pm		160.946							
^{162}Pm		161.950							
^{163}Pm		162.954							
$_{62}$Sm		150.36(3)							
^{129}Sm			≈ 0.55 s	β^+,p					
^{130}Sm		129.949							
^{131}Sm		130.946	1.2 s	β^+,EC/					ann.rad./
^{132}Sm		131.941	4.0 s	β^+					
^{133}Sm		132.939	2.9 s	β^+,EC/≈8.4		5/2+			ann.rad./
^{134}Sm		133.934	11. s	β^+,EC/5.		0+			ann.rad./
^{135}Sm		134.932	10. s	β^+ ,EC/7.		7/2+			ann.rad./
^{136}Sm		135.9283	42. s	β^+ ,EC/4.5		0+			ann.rad./
^{137}Sm		136.9271	45. s	β^+ ,EC/6.1					ann.rad./
^{138}Sm		137.9235	3.0 m	β^+ ,EC/3.9		0+			ann.rad./
									0.0536
									0.0747
^{139m}Sm			10. s	I.T./94/0.457		(11/2-)	1.1		Sm k x-ray
				β^+/6/	4.7				0.1118
									0.1553
									0.1901
									0.2673
^{139}Sm		138.9226	2.6 m	β^+/75/5.5	4.1/	1/2+	-0.53		Pm k x-ray
				EC/25/					0.3678
									0.4028
									(0.27 - 2.4)
^{140}Sm		139.9195	14.8 m	β^+,EC/3.4	1.9/	0+			ann.rad./
									Pm k x-ray
									0.1396
									0.2255
									(0.07 - 1.7)
^{141m}Sm			22.6 m	β^+/32/	1.6/	11/2-	-0.83	+1.6	ann.rad./
				EC/68/	2.19/				Pm k x-ray
				I.T./0.3/0.1758					0.1966
									0.4318
									0.7774
^{141}Sm		140.91847	10.2 m	β^+/52/4.54	3.2/	1/2+	-0.74		ann.rad./
				EC/48/					Pm k x-ray
									0.4382
^{142}Sm		141.91520	1.208 h	β^+/6/2.10	1.0/	0+			ann.rad./
				EC/94/					Pm k x-ray
^{143m}Sm			1.10 m	IT/99/0.7540		11/2-			Sm k x-ray

Elem. or Isot.	Natural Abundance (%)	Atomic Mass or Weight	Half-Life	Decay Mode/Energy (/MeV)	Particle Energy /Intensity (MeV/%)	Spin (h/2π)	Nuclear Magnetic Mom. (nm)	Elect. Quadr. Mom. (b)	γ-ray/Energy Intensity (MeV/%)
									0.7540
^{143}Sm		142.914624	8.83 m	β$^+$/46/3.443	2.47/	3/2+	• +1.01	+0.4	ann.rad./
				EC/54/					Pm k x-ray
									1.0565
^{144}Sm	3.1(1)	143.911996				0+			
^{145}Sm		144.913407	340. d	EC/0.617		7/2-	-1.12	-0.60	Pm k x-ray
									0.0613
									0.4924
^{146}Sm		145.913038	1.03x10^8 y	α/	2.50/	0+			
^{147}Sm	15.0(2)	146.914894	1.06x10^{11} y	α/	2.23/	7/2-	-0.815	-0.26	
^{148}Sm	11.3(1)	147.914818	7x10^{15} y	α/	1.96/	0+			
^{149}Sm	13.8(1)	148.917180	10^{16} y	α/		7/2-	-0.672	+0.075	
^{150}Sm	7.4(1)	149.917272				0+			
^{151}Sm		150.919929	90. y	β$^-$/0.0768	0.076/	5/2-	-0.363	+0.7	0.02154
^{152}Sm	26.7(2)	151.919729				0+			
^{153}Sm		152.922094	1.929 d	β$^-$/0.808	0.64/	3/2+	-0.0216	+1.3	Eu k x-ray
					0.69/				0.0697/4.7
									0.10318/29
									0.075-0.714
^{154}Sm	22.7(2)	153.922206				0+			
^{155}Sm		154.924636	22.2 m	β$^-$/1.627	1.52	3/2-		1.1	Eu k x-ray
									0.1043/75.
^{156}Sm		155.92553	9.4 h	β$^-$/0.72	0.43/	0+			0.0872
					0.71/				0.1657
									0.2038
^{157}Sm		156.9283	8.0 m	β$^-$/2.7	2.4/	3/2-			Eu k x-ray
									0.1964
									0.1978
									0.3942
^{158}Sm		157.9299	5.5 m	β$^-$/2.0		0+			0.1894/100.
									0.3636/82.
^{159}Sm		158.9332	11.3 s	β$^-$/3.8					0.1898
^{160}Sm		159.9353	9.6 s	β$^-$/3.6		0+			0.110
^{161}Sm		160.9388	≈4.8 s						0.264
^{162}Sm		161.941							
^{163}Sm		162.945							
^{164}Sm		163.948							
^{165}Sm		164.953							
$_{63}$Eu		151.964(1)							
^{131}Eu			≈26. ms	β$^+$,p	p/0.95				
^{132}Eu		131.954							
^{133}Eu		132.949							
^{134}Eu		133.946	0.5 s	EC,β$^+$					ann.rad./
^{135}Eu		134.942	1.5 s	EC,β$^+$/≈8.7					ann.rad./
^{136m}Eu			≈ 3.2 s			7+			0.255
^{136}Eu		135.940	≈ 3.9 s	EC,β$^+$/10.		1+			ann.rad./
^{137}Eu		136.935	11. s	EC/≈7.5		11/2-			ann.rad./
^{138}Eu		137.9335	12. s	EC,β$^+$/≈ 9.2		7+	5		ann.rad./
^{139}Eu		138.9298	18. s	EC,β$^+$/6.7			6		ann.rad./
^{140m}Eu			0.125 s	EC,β$^+$					ann.rad./
^{140}Eu		139.9285	1.51 s	EC,β$^+$/8.4		1-			ann.rad./
^{141m}Eu			3.0 s	β$^+$/58/		11/2-			ann.rad./
				EC/9/					Eu k x-ray
				I.T./33/0.0964					(0.09 - 1.6)
^{141}Eu		140.9244	40. s	β$^+$/81/5.6		5/2+	+3.49	+0.85	ann.rad./
				EC/15/					Sm k x-ray
									0.3845
									0.3940
^{142m}Eu			1.22 m	β$^+$/83/	4.8/	8-	+2.98	+1.4	ann.rad./

Elem. or Isot.	Natural Abundance (%)	Atomic Mass or Weight	Half-Life	Decay Mode/Energy (/MeV)	Particle Energy /Intensity (MeV/%)	Spin (h/2π)	Nuclear Magnetic Mom. (nm)	Elect. Quadr. Mom. (b)	γ-ray/Energy Intensity (MeV/%)
				EC/17/					Sm k x-ray
									0.5566
									0.7680
									1.0233
^{142}Eu		141.9231	2.4 s	β^-/94/7.4	7.0/	1+	+1.54	+0.12	ann.rad./
				EC/6/					0.7680
^{143}Eu		142.92017	2.62 m	β^+/72/5.17	4.1/	5/2+	+3.67	+0.51	ann.rad./
				EC/28/	5.1/				Sm k x-ray
									0.1107/7
									1.5368/3.
									1.9127/2.
^{144}Eu		143.91879	10.2 s	β^+/86/6.33	5.31/	1+	+1.89	+0.10	ann.rad./
				EC/13/					Sm k x-ray
									1.6601
^{145}Eu		144.916263	5.93 d	β^+/2/2.660	0.79/	5/2+	+4.00	+0.29	ann.rad./
				EC/98/1.71					Sm k x-ray
									0.6535
									0.8937
									1.6587
^{146}Eu		145.91720	4.57 d	β^+/5/3.88	1.47/	4-	+1.42	-0.18	ann.rad./
				EC/95/					Sm k x-ray
									0.6336
									0.6341
									0.7470
									(0.27 - 2.64)
^{147}Eu		146.916742	24.4 d	EC/99./1.722		5/2+	+3.72	+0.53	Sm k x-ray
				β^+/0.4/					0.12113
									0.19725
									0.6776
^{148}Eu		147.91815	54.5 d	EC/3.11	0.92	5-	+2.34	+0.35	Sm k x-ray
									0.5503/99.
									0.6299/71.
									(0.067-2.17)
^{149}Eu		148.91792	93.1 d	EC/0.692		5/2+	+3.57	+0.75	Sm k x-ray
									0.2770
									0.3275
^{150}Eu		149.91970	36. y	EC/2.26		5-	+2.71	+1.13	Sm k x-ray
									0.3340
									0.4394
									0.5843
									(0.25 - 1.8)
^{150m}Eu			12.8 h	β^-/92/	1.013/	0-			Sm k x-ray
				β^+/0.4/	1.24/				0.3339
				EC/8/					0.4065
^{151}Eu	47.8(15)	150.919846				5/2+	+3.472	+0.90	
^{152m2}Eu			1.60 h	I.T./0.1478		8-			Eu k x-ray
									0.0898
^{152m1}Eu			9.30 h	β^-/72/	1.85/	0-			Sm k x-ray
				EC/28/	0.89/				0.12178
									0.84153
									0.96334
^{152}Eu		151.921741	13.5 y	EC/72/1.874	0.69/	3-	-1.941	+2.71	Sm k x-ray
				β^-/28/1.818	1.47/				Gd k x-ray
									0.12178
									0.34427
									1.40802
									(0.252-1.528)
^{153}Eu	52.2(15)	152.921227				5/2+	+1.533	+2.41	
^{154m}Eu			46.1 m	I.T./≈ 0.16		8-			Eu k x-ray

Elem. or Isot.	Natural Abundance (%)	Atomic Mass or Weight	Half-Life	Decay Mode/Energy (/MeV)	Particle Energy /Intensity (MeV/%)	Spin (h/2π)	Nuclear Magnetic Mom. (nm)	Elect. Quadr. Mom. (b)	γ-ray/Energy Intensity (MeV/%)
									0.0682
									0.1009
^{154}Eu		153.922976	8.59 y	β^-/99.9/1.969	0.27/29	3-	-2.01	+2.8	Gd k x-ray
				EC/0.02/0.717	0.58/38				0.12299/40.
					0.84/17				0.72331/20.
					0.98/4				1.2745/36
					1.87/11				(0.059-1.90)
^{155}Eu		154.922890	4.76 y	β^-/0.252	0.15/	5/2+	+1.52	+2.4	Gd k x-ray
									0.0865/30
									0.1053/20
^{156}Eu		155.92475	15.2 d	β^-/2.451	0.30/11	1+	≈ 1.1		0.08899/9.
					0.49/30				0.64623/7.
					1.2/12				0.723441/6.
					2.45/31				0.8118/10.
^{157}Eu		156.92542	15.13 h	β^-/1.36	0.98/	(5/2+)	+1.50	+2.6	Gd k x-ray
					1.30/41				0.0639/100.
									0.3705/48.
									0.4107/76.
^{158}Eu		157.9278	45.9 m	β^-/3.5	2.5/	(1-)	+1.44	+0.7	0.0795
									0.8976
									0.9442
									0.9771
^{159}Eu		158.92909	18.1 m	β^-/2.51	2.4/	(5/2+)	+1.38	+2.7	0.0678
					2.57/				0.0786
									0.0957
^{160}Eu		159.9315	38. s	β^-/4.1	2.7/	(0-)			0.0753
					4.1/				0.1735
									0.4131
									0.5155
									0.8217
									0.9110
									0.9246
^{161}Eu		160.9337	27. s	β^-/3.7					0.0719
^{162}Eu		161.9370	11. s	β^-/5.6					
^{163}Eu		162.9392							
^{164}Eu		163.943							
^{165}Eu		164.946							
^{166}Eu		165.950							
^{167}Eu		166.953							
$_{64}$Gd		157.25(3)							
^{135}Gd			1.1 s	β^+					(0.163-0.360)
^{136}Gd		135.947							
^{137}Gd		136.945	7. s	EC,β^+/≈8.8					ann.rad./
^{138}Gd		137.9400	≈4.7 s	EC,β^+					0.0647
^{139m}Gd			≈4.8 s						0.1216
^{139}Gd		138.9381	5. s	EC,β^+/≈7.7					0.104-0.323
^{140}Gd		139.934	16. s	EC/4.8		0+			0.1748
^{141m}Gd			25. s	EC,β^+/		11/2-			ann.rad./
^{141}Gd		140.9322	21. s	β^+/7.3		0+			ann.rad./
^{142}Gd		141.0276	1.17 m	EC,β^+/4.2		1/2+			ann.rad./
^{143m}Gd			1.84 m	β^+/67/		11/2-			ann.rad./
				EC/33/					Eu k x-ray
				I.T./					0.1176
									0.2719
									0.5880
									0.6681
									0.7999
^{143}Gd		142.9266	39. s	β^+/82/6.0		1/2+			ann.rad./
				EC/18/					Eu k x-ray

TABLE OF ISOTOPES (CONTINUED)

Elem. or Isot.	Natural Abundance (%)	Atomic Mass or Weight	Half-Life	Decay Mode/Energy (/MeV)	Particle Energy /Intensity (MeV/%)	Spin (h/2π)	Nuclear Magnetic Mom. (nm)	Elect. Quadr. Mom. (b)	γ-ray/Energy Intensity (MeV/%)
									0.2048
									0.2588
^{144}Gd		143.9234	4.5 m	β^+/45/4.3	3.3/	0+			ann.rad./
				EC/55/					Eu k x-ray
									0.3332
^{145m}Gd			1.44 m	I.T./95/0.749		11/2-			0.0273
				β^+/4/5.7					0.3295
									0.3866
									0.7214
^{145}Gd		144.92169	23.4 m	β^+/33/5.05	2.5/	1/2+			ann.rad./
				EC/67/					Eu k x-ray
									1.7579
									1.8806
									(0.32 - 3.69)
^{146}Gd		145.91831	48.3 d	EC/99.9/1.03	0.35/	0+			Eu k x-ray
				β^+/0.2					0.1147
									0.1155
									0.1546
^{147}Gd		146.919090	1.588 d	EC/99.8/2.188	0.93/	7/2-	1.0		Eu k x-ray
				EC/0.2/					0.2293
									0.3699
									0.3960
									0.9289
									(0.1 - 1.8)
^{148}Gd		147.918111	75. y	α/3.27	3.1828/	0+			
^{149}Gd		148.919339	9.3 d	EC/1.32		7/2-	0.9		Eu k x-ray
									0.1496
									0.2985
									0.3465
^{150}Gd		149.91866	1.8x10^6 y	α/2.80	2.73/	0+			
^{151}Gd		150.920345	124. d	EC/0.464		7/2-	0.8		Eu k x-ray
									0.1536
									0.2432
^{152}Gd	0.20(1)	151.919789				0+			
^{153}Gd		152.921747	241.6 d	EC/0.485		3/2-	0.4		Eu k x-ray
									0.09743
									0.10318
^{154}Gd	2.18(3)	153.920862				0+			
^{155}Gd	14.80(5)	154.922619				3/2-	-2.59	+1.30	
^{156}Gd	20.47(4)	155.922120				0+			
^{157}Gd	15.65(3)	156.923957				3/2-	-3.40	+1.36	
^{158}Gd	24.84(12)	157.924101				0+			
^{159}Gd		158.926385	18.6 h	β^- 0.971	0.60/11	3/2-	-0.44		Tb k x-ray
					0.89/26				0.36351
					0.96/63				0.058-0.855
^{160}Gd	21.86(4)	159.927051				0+			
^{161}Gd		160.929666	3.66 m	β^-/1.956	1.56/85	5/2-			Tb k x-ray
									0.1023
									0.3149
									0.3609
^{162}Gd		161.930981	8.4 m	β^-/1.39	1.0/	0+			0.4030
									0.4421
^{163}Gd		162.9340	1.13 m	β^-/3.1					0.2868
									0.214
									1.685
^{164}Gd		163.9359	45. s	β^-/2.3					
^{165}Gd		164.9394	10 s	β^-					
^{166}Gd		165.942							
^{167}Gd		166.946							

TABLE OF ISOTOPES (CONTINUED)

Elem. or Isot.	Natural Abundance (%)	Atomic Mass or Weight	Half-Life	Decay Mode/Energy (/MeV)	Particle Energy /Intensity (MeV/%)	Spin (h/2π)	Nuclear Magnetic Mom. (nm)	Elect. Quadr. Mom. (b)	γ-ray/Energy Intensity (MeV/%)
^{168}Gd		167.948							
^{169}Gd		168.953							
$_{65}$Tb		158.92534(2)							
^{138}Tb									
^{139}Tb		138.948	1.6 s						0.109
									0.120
^{140}Tb		139.946	2.4 s	β^+,EC/11					0.329
									0.355-0.740
^{141}Tb		140.941	3.5 s	β^+,EC/≈ 8.3					
^{142m}Tb			0.30 s	β^+,EC/		4-			
^{142}Tb		141.939	0.60 s	β^+,EC/10.		0+			
^{143}Tb		142.9346	12. s	β^+,EC/7.4		11/2-			
^{144m}Tb			4.1 s	IT		5-			
^{144}Tb		143.9324	< 1.5 s	β^+,EC/8.4		1+			
^{145m}Tb			30. s	β^+,EC/≈ 6.6		11/2-			ann.rad./
									0.2577
									0.5370
									0.9876
^{145}Tb		144.9287		β^+,EC/6.5		1/2+			
^{146m}Tb			23. s	β^+/76/		(5-)			ann.rad./
				EC/24/					Gd k x-ray
									1.0789
									1.5795
^{146}Tb		145.9270	≈ 8. s	β^+/8.1		1+			
^{147m}Tb			1.8 m	β^+/35/		11/2-			ann.rad./
				EC/65/					Gd k x-ray
									1.3977
									1.7978
^{147}Tb		146.92404	1.6 h	β^+/42/4.61		5/2+	+1.70		ann.rad./
				EC/58/					Gd k x-ray
									0.6944
									1.1522
									(0.120-3.318)
^{148m}Tb			2.3 m	β^+/25/		9+			ann.rad./
				EC/75/					Gd k x-ray
									0.3945
									0.6319
									0.7845
									0.8824
^{148}Tb		147.92422	1.00 h	β^+,EC/5.69		2-	-1.75	-0.3	ann.rad./
									Gd k x-ray
									0.4888
									0.7845
									(0.14 - 3.8)
^{149m}Tb			4.16 m	EC/88/		11/2-			ann.rad./
				β^+/12/					Gd k x-ray
									0.1650
									0.7960
^{149}Tb		148.923243	4.13 h	β^+/4/3.636	1.8/	1/2+	+1.35		Gd k x-ray
				α/16/	3.07/				0.1650
									0.3522
									0.3886
									(0.1 - 3.2)
^{150m}Tb			6.0 m	β^+/17/					ann.rad./
				EC/83/					Gd k x-ray
									0.4384
									0.6380
									0.6504
									0.8275

Elem. or Isot.	Natural Abundance (%)	Atomic Mass or Weight	Half-Life	Decay Mode/Energy (/MeV)	Particle Energy /Intensity (MeV/%)	Spin (h/2π)	Nuclear Magnetic Mom. (nm)	Elect. Quadr. Mom. (b)	γ-ray/Energy Intensity (MeV/%)
^{150}Tb		149.92366	3.3 h	β+,EC/4.66		2-	-0.90		ann.rad./
									0.4963
									0.6380
									(0.3 - 4.29)
^{151m}Tb			25. s	I.T./95/		11/2-			0.0229
				β+,EC/7/					0.0495
									0.3797
									0.8305
^{151}Tb		150.923099	17.61 h	β+/1/2.565	0.70/	1/2+	+0.92		Gd k x-ray
				EC/99/					0.1083
									0.2517
									0.2870
									(0.1 - 1.8)
^{152m}Tb			4.3 m	I.T./79/0.5018		(8+)			Tb k x-ray
				EC/21/4.35					Gd k x-ray
									0.2833
									0.3443
									0.4111
^{152}Tb		151.92407	17.5 h	β+/20/3.99	2.5/	2-	-0.58	+0.3	ann.rad./
				EC/80/	2.8/				Gd k x-ray
									0.3443
									(0.2 - 2.88)
^{153}Tb		152.923433	2.34 d	EC/1.570		5/2+	+3.44	+1.1	Gd k x-ray
									0.2119
									(0.05 - 1.1)
^{154m2}Tb			23.1 h	EC/98/		(7-)	0.9		Gd k x-ray
				I.T./2/					0.1231
									0.2479
									0.3467
									1.4199
^{154m1}Tb			9. h	β+/78/		(3-)	1.7	+3.	Gd k x-ray
				I.T./22/					0.1231
									0.2479
									0.5401
									(0.12 - 2.57)
^{154}Tb		153.92469	21.5 h	EC/99/3.56	1.86/	0-			Gd k x-ray
				β+/1/	2.45				0.1231
									1.2744
									2.1872
									(0.12 - 3.14)
^{155}Tb		154.92350	5.3 d	EC/0.82		3/2+	+2.01	+1.41	Gd k x-ray
									0.08654
									0.10530
^{156m2}Tb			1.02 d	I.T./		(7-)			Tb k x-ray
									0.0496
^{156m1}Tb			5.3 h	I.T./0.0884		(0+)			Tb k x-ray
									0.0884
^{156}Tb		155.924744	5.3 d	EC/2.444		3-	≈1.7	+2.	Gd k x-ray
									0.08896
									0.19921
									0.53435
									1.22245
^{157}Tb		156.924021	1.1x10^2 y	EC/0.0601		3/2+	+2.01	+1.4	Gd k x-ray
									0.0545
^{158m}Tb			10.5 s	I.T./0.11		0-			Gd k x-ray
									0.0110
^{158}Tb		157.925410	1.8x10^2 y	EC/80/1.220		3-	+1.76	+2.7	Gd k x-ray
				β-/20/0.937					0.0795
									0.9442

TABLE OF ISOTOPES (CONTINUED)

Elem. or Isot.	Natural Abundance (%)	Atomic Mass or Weight	Half-Life	Decay Mode/Energy (/MeV)	Particle Energy /Intensity (MeV/%)	Spin ($h/2\pi$)	Nuclear Magnetic Mom. (nm)	Elect. Quadr. Mom. (b)	γ-ray/Energy Intensity (MeV/%)
									0.9621
^{159}Tb	100.	158.925343				3/2+	+2.014	+1.43	
^{160}Tb		159.927164	72.3 d	β^-/1.835	0.57/47	3-	+1.79	3.8	Dy k x-ray
					0.86/27				0.08678
									0.29857
									0.87936
									0.96615
^{161}Tb		160.927566	6.91 d	β^-/0.593	0.46/23	3/2+	2.2	+1.2	Dy k x-ray
					0.52/66				0.02565
					0.6/10				0.04892
									0.07458
^{162}Tb		161.92948	7.6 m	β^-/2.51	1.4	(1/2-)			Dy k x-ray
									0.2600
									0.8075
									0.8882
^{163}Tb		162.930644	19.5 m	β^-/1.785	0.80/	3/2+			Dy k x-ray
									0.3511
									0.3897
									0.4945
^{164}Tb		163.9334	3.0 m	β^-/3.9	1.7/	(5+)			Dy k x-ray
									0.1689
									0.2157
									0.6110
									0.6885
									0.7548
^{165}Tb		164.9349	2.1 m	β^-/3.0		3/2+			0.5389
									1.1785
									1.2920
									1.6648
^{166}Tb		165.9380	≈ 21 s	β^-/					
^{167}Tb		166.9401	19 s						0.057
									0.070
^{168}Tb		167.9436	8 s						0.075-0.227
^{169}Tb		168.946							
^{170}Tb		169.950							
^{171}Tb		170.953							
$_{66}$Dy		162.50(3)							
^{139}Dy			0.6 s	β^+,p					
^{140}Dy		139.954							
^{141}Dy		140.951	0.9 s	EC,β^+/9.					
^{142}Dy		141.946	2.3 s	EC,β^+/7.1					
^{143}Dy		142.9440	3.9 s	EC,β^+/≈ 8.8					
^{144}Dy		143.9391	9.1 s	EC,β^+/≈ 6.2					
^{145m}Dy		144.9365	14. s	EC,β^+		11/2-			
^{146m}Dy			0.15 s	I.T.		10+			
^{146}Dy		145.9325	30. s	EC,β^+/5.2					
^{147m}Dy			56. s	I.T./40/		(11/2-)	-0.66	+0.7	Dy k x-ray
				β^+,EC/60/					0.072
									0.6787
^{147}Dy		146.9309	75 s	EC,β^+/6.37		1/2+	-0.92		ann.rad./
									0.1007
									0.2534
									0.3653
^{148}Dy		147.92710	3.1 m	β^+/4/2.68	1.2/	0+			ann.rad./
				EC/96/					Tb k x-ray
									0.6202
^{149}Dy		148.92734	4.2 m	β^+,EC/3.81		(7/2-)	-0.12	-0.62	ann.rad./
									0.1008
									0.1063

TABLE OF ISOTOPES (CONTINUED)

Elem. or Isot.	Natural Abundance (%)	Atomic Mass or Weight	Half-Life	Decay Mode/Energy (/MeV)	Particle Energy /Intensity (MeV/%)	Spin (h/2π)	Nuclear Magnetic Mom. (nm)	Elect. Quadr. Mom. (b)	γ-ray/Energy Intensity (MeV/%)
									0.2534
									0.6536
									0.7894
									1.7765
									1.8062
^{150}Dy		149.92558	7.18 m	β^+,EC/67/1.79		0+			Tb k x-ray
				α/33/	4.233/				0.3967
^{151}Dy		150.926181	17. m	β^+/5/2.871		7/2-	-0.95	-0.30	Tb k x-ray
				EC/89/					0.1764
				α/6/	4.067/				0.3030
									0.3861
									0.5463
									(0.16 - 2.09)
^{152}Dy		151.92472	2.37 h	EC/0.60		0+			Tb k x-ray
				α/	3.63/				0.2569
^{153}Dy		152.925763	6.3 h	β^+/1/2.171	0.89/	(7/2-)	-0.78	≈-0.15	Tb k x-ray
				EC/99/					0.0807
				α/0.01/	3.46/				0.0997
									0.2137
									(0.08 - 1.66)
^{154}Dy		153.92442	$3.x10^6$ y	α/2.95	2.87/	0+			
^{155}Dy		154.92575	9.9 h	β^+/2/2.095	0.845/	3/2-	-0.385	+1.04	Tb k x-ray
				EC/98/					0.0655
									0.2269
^{156}Dy	0.06(1)	155.92428				0+			
^{157}Dy		156.92546	8.1 h	EC/1.34		3/2-	-0.301	+1.30	Tb k x-ray
									(0.0609-1.319)
^{158}Dy	0.10(1)	157.924405				0+			
^{159}Dy		158.925736	144. d	EC/0.366		3/2-	-0.354	+1.37	Tb k x-ray
									0.3262
^{160}Dy	2.34(6)	159.925194				0+			
^{161}Dy	18.9(2)	160.926930				5/2+	-0.480	+2.51	
^{162}Dy	25.5(2)	161.926795				0+			
^{163}Dy	24.9(2)	162.928728				5/2-	+0.673	+2.65	
^{164}Dy	28.2(2)	163.929171				0+			
^{165m}Dy			1.26 m	I.T./98/0.108		1/2-			Dy k x-ray
				β^-/2/					0.1082
									0.5155
^{165}Dy		164.931700	2.33 h	β^-/1.286	1.29/	7/2+	-0.52	-3.5	Ho k x-ray
									0.09468
^{166}Dy		165.932803	3.400 d	β^-/0.486	0.40/	0+			Ho k x-ray
									0.0282
									0.0825
^{167}Dy		166.9357	6.2 m	β^-/≈ 2.35	1.78	(1/2-)			Ho k x-ray
									0.2593
									0.3103
									0.5697
									(0.06 - 1.4)
^{168}Dy		167.9372	8.5 m	β^-/1.6		0+			Ho k x-ray
									0.1925
									0.4867
^{169}Dy		168.9403	≈ 39. s	β^-/3.2					
^{170}Dy		169.9427							
^{171}Dy		170.9465							
^{172}Dy		171.949							
^{173}Dy		172.953							
$_{67}$Ho		164.93032(2)							
^{140}Ho			6 ms	p/	p/1.09				
^{141m}Ho			8 μs	p/	p/1.23				

Elem. or Isot.	Natural Abundance (%)	Atomic Mass or Weight	Half-Life	Decay Mode/Energy (/MeV)	Particle Energy /Intensity (MeV/%)	Spin (h/2π)	Nuclear Magnetic Mom. (nm)	Elect. Quadr. Mom. (b)	γ-ray/Energy Intensity (MeV/%)
^{141}Ho			4.2 ms	β^+,p	p/1.71				
^{142}Ho		141.960							
^{143}Ho		142.955							
^{144}Ho		143.952	0.7 s	β^+,EC/12					
^{145}Ho		144.947	2.4 s	β^+					
^{146}Ho		145.9440	3.3 s	β^+,EC/10.7		(10+)			ann.rad./
^{147}Ho		146.9396	5.8 s	β^+,EC/8.2		11/2-			ann.rad./
^{148m}Ho			9. s	β^+,EC/		4-			ann.rad./
^{148}Ho		147.9372	2. s	β^+,EC/9.4		1+			ann.rad./
									0.6615
									1.6883
^{149m}Ho			21. s	β^+,EC/		11/2-			ann.rad./
									1.0733
									1.0911
^{149}Ho		148.93379	> 30. s	β^+,EC/6.01		1/2+			
^{150m}Ho			25. s	β^+,EC/		(9+)			ann.rad./
									0.3939
									0.5511
									0.6534
									0.8034
^{150}Ho		149.9326	1.3 m	β^+,EC/6.6					ann.rad./
									0.5913
									0.6534
									0.8034
^{151m}Ho			47. s	β^+,EC/87/					ann.rad./
				α/13	4.605/				0.2102
									0.4889
									0.6948
									0.7762
^{151}Ho		150.93169	35.2 s	β^+,EC/80/5.13					ann.rad./
				α/20/	4.519/				0.3522
									0.5274
									0.9676
									1.0471
^{152m}Ho			50. s	β^+,EC/90/		(9+)	+5.9	-1.	ann.rad./
				α/10/	4.453/				0.4929
									0.6138
									0.6474
									0.6835
^{152}Ho		151.93166	2.4 m	β^+,EC/88/6.47		(3+)	-1.02	+0.1	ann.rad./
				α/12/	4.387/				0.6140
									0.6476
^{153m}Ho			9.3 m	β^+,EC/99+/4.12		5/2	+1.19		ann.rad./
				α/	4.01/				0.0905
									0.1089
									0.1618
									0.2302
									0.2707
									0.3659
									0.4565
^{153}Ho		152.93020	2.0 m	β^+,EC/99+/4.13		11/2-	+6.8	-1.1	ann.rad./
				α/	3.91/				0.2958
									0.3346
									0.4381
									0.6383
^{154m}Ho			3.3 m	β^+,EC/		(8+)	5.7	-1.0	ann.rad./
									0.3346
									0.4124
									0.4771

Elem. or Isot.	Natural Abundance (%)	Atomic Mass or Weight	Half-Life	Decay Mode/Energy (/MeV)	Particle Energy /Intensity (MeV/%)	Spin (h/2π)	Nuclear Magnetic Mom. (nm)	Elect. Quadr. Mom. (b)	γ-ray/Energy Intensity (MeV/%)
^{154}Ho		153.93060	12. m	β^+,EC/5.75		1-	-0.64	+0.2	ann.rad./
									Dy k x-ray
									0.3346
									0.5700
									0.8734
^{155}Ho		154.92908	48. m	β^+/6/3.10		(5/2+)	+3.51	+1.5	ann.rad./
				EC/94/					Dy k x-ray
									0.0474
									0.1363
									0.3254
									(0.06 - 2.24)
^{156m}Ho			5.8 m	I.T./0.0352			+2.99	+2.3	ann.rad./
				β^+/25/	1.8/				Dy k x-ray
				EC/75/	2.9/				0.1378
									0.2666
									(0.28 - 2.9)
^{156}Ho		155.9290	56. m	β^+,EC/4.4		(5+)			ann.rad./
									0.1378
									0.2665
^{157}Ho		156.92819	12.6 m	β^+/5/2.54	1.18/	7/2-	+4.35	+3.0	ann.rad./
				EC/95/					Dy k x-ray
									0.2800
									0.3411
^{158m2}Ho			28. m	I.T./44/		2-	+2.44	+1.6	ann.rad./
				EC/56/					Dy k x-ray
									0.0989
									0.2182
^{158m1}Ho			21. m	β^+,EC/		(9+)			ann.rad./
									0.0981
									0.1664
									0.2182
									0.3205
									0.4062
									0.9774
									1.0532
									0.4846
^{158}Ho		157.92895	11.3 m	β^+/8/4.24	1.30/	5+	+3.77	+4.1	ann.rad./
				EC/92/					Dy k x-ray
									0.0989
									0.2182
									0.9488
^{159m}Ho			8.3 s	IT/0.206		1/2+			Ho k x-ray
									0.1660
									0.2059
^{159}Ho		158.927708	33.0 m	EC/1.838		7/2-	+4.28	+3.2	Dy k x-ray
									0.1210
									0.1320
									0.2529
									0.3096
									(0.06 - 1.2)
^{160m2}Ho			3. s			1+			
^{160m}Ho			5.0 h	IT/67/0.060		2-	+2.52	+1.8	0.0868
				EC/33/3.35					0.1970
									0.6464
									0.7281
									0.8791
									0.9619
									0.9658
^{160}Ho		159.92873	25.6 m	β^+,EC/3.29	0.57/	5+	+3.71	+4.0	See Ho[166m]

Elem. or Isot.	Natural Abundance (%)	Atomic Mass or Weight	Half-Life	Decay Mode/Energy (/MeV)	Particle Energy /Intensity (MeV/%)	Spin (h/2π)	Nuclear Magnetic Mom. (nm)	Elect. Quadr. Mom. (b)	γ-ray/Energy Intensity (MeV/%)
									0.7282
									0.8794
^{161m}Ho			6.8 s	IT/0.211					Ho k x-ray
									0.2112
^{161}Ho		160.927852	2.48 h	EC/0.859		7/2-	+4.25	+3.2	Dy k x-ray
									0.0256
									0.0592
									0.0774
									0.1031
^{162m}Ho			1.12 h	IT/61/		6-	+3.60	+4.	Dy k x-ray
				EC/39/					Ho k x-ray
									0.0807
									0.1850
									0.2828
									0.9372
									1.2200
^{162}Ho		161.929092	15. m	EC/96/0.295		1+			Dy k x-ray
				β^+/4/					0.0807
									1.3196
									1.3728
^{163m}Ho			1.09 s	I.T./0.298		(1/2+)			Ho k x-ray
									0.2798
^{163}Ho		162.928730	4.57x10^3 y	EC/0.00258		7/2-	+4.23	+3.6	Dy M x-rays
^{164m}Ho			38. m	I.T./0.140		(6-)			Ho k x-ray
									0.0373
									0.0566
									0.0940
^{164}Ho		163.930231	29. m	EC/58/0.987		1+			Dy k x-ray
				β^-/42/0.963					0.0734
									0.0914
^{165}Ho	100.	164.930319				7/2-	+4.17	+3.49	
^{166m}Ho			1.2x10^3 y	β^-/		7-	3.6	-3.	Er k x-ray
									0.18407
									0.71169
									0.81031
^{166}Ho		165.932281	1.117 d	β^-/1.855	1.776/48	0-			Er k x-ray
					1.855/51				0.08057
									1.37943
^{167}Ho		166.933127	3.1 h	β^-/1.007	0.31/43	(7/2-)			Er k x-ray
					0.61/21				0.0793
					0.96/15				0.0835
					0.97/15				0.2379
									0.3213
									0.3465
^{168m}Ho			2.2 m	I.T./					
^{168}Ho		167.93550	3.0 m	β^-/2.91	2.0/	3+			Er k x-ray
									0.7413
									0.8159
									0.8211
									(0.00 2.34)
^{169}Ho		168.93687	4.7 m	β^-/2.12	1.2/	(7/2-)			
					2.0/				0.1496
									0.7610
									0.7784
									0.7884
									0.8529
^{170m}Ho			43. s	β^-/		1+			0.0787
									0.8123
									1.8940

TABLE OF ISOTOPES (CONTINUED)

Elem. or Isot.	Natural Abundance (%)	Atomic Mass or Weight	Half-Life	Decay Mode/Energy (/MeV)	Particle Energy /Intensity (MeV/%)	Spin (h/2π)	Nuclear Magnetic Mom. (nm)	Elect. Quadr. Mom. (b)	γ-ray/Energy Intensity (MeV/%)
									1.9726
^{170}Ho		169.93962	2.8 m	β^-/3.87		6+			Er k x-ray
									0.1816
									0.2582
									0.8902
									0.9321
									0.9414
									1.1387
^{171}Ho		170.941	53 s	β^-/					
^{172}Ho		171.9448	25. s	β^-/					Er k x-ray
									(0.077-1.186)
^{173}Ho		172.947							
^{174}Ho		173.951							
^{175}Ho		174.954							
$_{68}$Er		167.259(3)							
^{144}Er		143.961							
^{145}Er		144.957	0.9 s	β^+					
^{146}Er		145.952	≈ 1.7 s	β^+					
^{147}Er		146.9494	2.5 s	E.C,β^+/≈ 9.1					
^{148}Er		147.9444	4.5 s	β^+,EC/6.8					
^{149m}Er			10. s	IT		11/2-			
^{149}Er		148.9425	10.7 s	ECβ^+/8.1		1/2+			
^{150}Er		149.9370	18. s	β^+/36/4.11		0+			ann.rad./
				EC/64/					Ho k x-ray
									0.4758
^{151}Er		150.9373	23. s	β^+,EC/5.2		7/2-			ann.rad./
^{152}Er		151.93500	10.2 s	β^+,EC/10/3.11		0+			ann.rad./
				α/90/	4.804/				
^{153}Er		152.93509	37.1 s	α/	4.674		-0.934	-0.42	0.351
				β^+,EC/47/4.56	4.35/				(0.0945-1.700)
^{154}Er		153.93278	3.7 m	β^+,EC/99+/2.03		0+			ann.rad./
				α/0.5/	4.166/				
^{155}Er		154.93321	5.3 m	β^+,EC/47/3.84		(7/2-)	-0.669	-0.27	ann.rad./
				EC/53/					Ho k x-ray
									0.1101
									0.2415
^{156}Er		155.9308	20. m	β^+,EC/1.7		0+			ann.rad./
									0.0298
									0.0352
									0.0522
									0.1336
^{157}Er		156.9319	25. m	β^+,EC/3.5		3/2-	-0.412	+0.92	ann.rad./
									0.117
									0.385
									1.320
									1.660
									1.820
									2.000
^{158}Er		157.93087	2.2 h	EC/99.5/1.78	0.74/	0+			Ho k x-ray
				β^+/0.5/					0.0719
									0.2486
									0.3868
^{159}Er		158.930681	36. m	β^+/7/2.769		3/2-	-0.304	+1.17	ann.rad./
				EC/93/					Ho k x-ray
									0.6245
									0.6493
									(0.07 - 2.5)
^{160}Er		159.92908	1.191 d	EC/0.33		0+			Ho k x-ray
									(0.05 - 0.96)

Elem. or Isot.	Natural Abundance (%)	Atomic Mass or Weight	Half-Life	Decay Mode/Energy (/MeV)	Particle Energy /Intensity (MeV/%)	Spin (h/2π)	Nuclear Magnetic Mom. (nm)	Elect. Quadr. Mom. (b)	γ-ray/Energy Intensity (MeV/%)
^{161}Er		160.93000	3.21 h	EC/2.00		3/2-	-0.37	+1.36	Ho k x-ray
									0.8265
									(0.07 - 1.74)
^{162}Er	0.14(1)	161.928775				0+			
^{163}Er		162.93003	1.25 h	EC/1.210		5/2-	+0.557	+2.55	Ho k x-ray
									0.4361
									0.4399
									1.1135
^{164}Er	1.61(2)	163.929197				0+			
^{165}Er		164.930723	10.36 h	EC/0.376		5/2-	+0.643	+2.71	Ho k x-ray
^{166}Er	33.6(2)	165.930290				0+			
^{167m}Er			2.27 s	I.T./0.208		1/2-			Er k x-ray
									0.2078
^{167}Er	22.95(15)	166.932046				7/2+	-0.5639	+3.57	
^{168}Er	26.8(2)	167.932368				0+			
^{169}Er		168.934588	9.40 d	β^-/0.351	0.35/≈ 100	1/2-	+0.485		Tm k x-ray
									0.1098
									0.1182
^{170}Er	14.9(2)	169.935461				0+			
^{171}Er		170.938026	7.52 h	β^-/1.491		5/2-	0.66	2.9	Tm k x-ray
									0.11160
									0.29591
									0.30832
									(0.08 - 1.4)
^{172}Er		171.939352	2.05 d	β^-/0.891	0.28/48				Tm k x-ray
					0.36/46				0.0597
									0.4073
									0.6101
^{173}Er		172.9424	1.4 m	β^-/2.6		(7/2-)			Tm k x-ray
									0.1928
									0.1992
									0.8952
^{174}Er		173.9441	3.1 m	β^-/1.8					Tm k x-ray
									(0.100-0.152)
^{175}Er		174.9479	1.2 m	β^-					(0.0765-1.168)
^{176}Er		175.9503							
^{177}Er		176.954							
$_{69}$Tm		168.93421(2)							
^{145}Tm			≈3.5 μs						
^{146m}Tm			0.21 s	β^+,p	p/1.118				
^{146}Tm		145.967	0.06 s	β^+/14.					
				p	1.119/				
					1.189/				
^{147m}Tm			0.4 ms	β^+,p	p/1.115				
^{147}Tm		146.961	0.56 s	EC, β^+/85/≈10.7					
				p/15/	1.052/				
^{148m}Tm		147.9573	0.7 s	β^+,EC/12.					ann.rad./
^{148}Tm									
^{149}Tm		148.9524	0.9 s	β^+,EC/≈9.2		11/2-			
^{150}Tm		149.9494	2.3 s	β^+,EC/~11.5		6			(0.1007 2.177)
^{151}Tm		150.9454	4. s	β^+,EC/7.5					ann.rad./
^{152m}Tm			8. s	β^+,EC/		9+			
^{152}Tm		151.9443	5. s	β^+,EC/8.8					ann.rad./
^{153}Tm		152.94203	1.6 s	β^+,EC/10/6.46					ann.rad./
				α/90/	5.109/				
^{154m}Tm			3.3 s	β^+,EC/15/	α/5.031/100				ann.rad./
				α/	4.84/0.24				0.4605-0.7960
^{154}Tm		153.9407	8.1 s	β^+,EC/56/7.4	α/4.956/100				ann.rad./
				α/44/	4.83/0.45				

TABLE OF ISOTOPES (CONTINUED)

Elem. or Isot.	Natural Abundance (%)	Atomic Mass or Weight	Half-Life	Decay Mode/Energy (/MeV)	Particle Energy /Intensity (MeV/%)	Spin (h/2π)	Nuclear Magnetic Mom. (nm)	Elect. Quadr. Mom. (b)	γ-ray/Energy Intensity (MeV/%)
^{155}Tm		154.93919	30. s	β^+,EC/5.58					0.0315
				α/	4.46/				0.0638
									0.0881
									0.2268
									0.5320
									0.6067
^{156m}Tm			19. s	α/	4.46/				
^{156}Tm		155.9389	1.40 m	β^+,EC/7.6		2-	+0.40	-0.5	ann.rad./
				α/	4.23/				0.3446
									0.4529
									0.5860
^{157}Tm		156.9367	3.6 m	β^+,EC/4.5	2.6	1/2	+0.48		ann.rad./
				α/	3.97/				0.1104
									0.3484
									0.3855
									0.4550
									(0.1 - 1.58)
^{158}Tm		157.9379	4.0 m	β^+,EC/74/6.5		(2-)	+0.04	+0.7	ann.rad./
				EC/26/					Er k x-ray
									0.1921
									0.3351
									0.6280
									1.1498
									(0.18 - 2.81)
^{159}Tm		158.9348	9.1 m	β^+/23/3.9		5/2+	+3.42	+1.9	ann.rad./
				EC/77/					Er k x-ray
									0.0591
									0.0848
									0.2713
									(0.05 - 1.27)
^{160m}Tm			1.24 m	IT		(5)			
^{160}Tm		159.9354	9.4 m	β^+/15/5.9		1-	+0.16	+0.58	ann.rad./
				EC/85/					Er k x-ray
									0.1264
									0.2642
									0.7285
									0.8544
									0.8614
									1.3685
^{161}Tm		160.9334	31. m	β^+,EC/3.2		7/2+	+2.40	+2.9	ann.rad./
									Er k x-ray
									0.0595
									0.0844
									1.6481
									(0.04 - 2.15)
^{162m}Tm			24. s	I.T./90/		5+			Tm k x-ray
				β^+,EC/10/					Er k x-ray
									0.0669
									0.8115
									0.9003
^{162}Tm		161.93394	21.7 m	β^+/8/4.81		1-	+0.07	+0.69	ann.rad./
				EC/92/					Er k x-ray
									0.1020
									0.7987
									(0.1 - 3.75)m
^{163}Tm		162.93265	1.81 h	EC/98/2.439		1/2+	-0.082		Er k x-ray
				β^+/1/					0.0692
									0.1043
									0.2414

Elem. or Isot.	Natural Abundance (%)	Atomic Mass or Weight	Half-Life	Decay Mode/Energy (/MeV)	Particle Energy /Intensity (MeV/%)	Spin (h/2π)	Nuclear Magnetic Mom. (nm)	Elect. Quadr. Mom. (b)	γ-ray/Energy Intensity (MeV/%)
^{164m}Tm			5.1 m	I.T./80/		6-			0.0914
				β+,EC/20/					0.1394
									0.2081
									0.2405
									0.3149
^{164}Tm		163.93345	2.0 m	β+/36/3.96	2.94/	1+	+2.38	+0.71	ann.rad./
				EC/64/					Er k x-ray
									0.0914
^{165}Tm		164.932433	1.253 d	EC/1.593		1/2+	-0.139		Er k x-ray
									0.0472
									0.0544
									0.29728
									0.80636
^{166}Tm		165.93355	7.70 h	EC/98/3.04		2+	+0.092	+2.14	Er k x-ray
				β+/2/					0.0806
									0.1844
									0.7789
									1.2734
									2.0524
^{167}Tm		166.932849	9.24 d	EC/0.748		1/2+	-0.197		Er k x-ray
									0.0571
									0.20778
^{168}Tm		167.934171	93.1 d	EC/1.679		3+	+0.23	+3.2	Er k x-ray
									0.19825
									0.4475
									0.81595
^{169}Tm	100	168.934211				1/2+	-0.232	-1.2	
^{170}Tm		169.935798	128.6 d	β⁻/99.8/0.968	0.883/24	1-	+0.247	+0.74	Yb k x-ray
				EC/0.2/0.314	0.968/76				0.08425
^{171}Tm		170.936426	1.92 y	β⁻/0.096	0.03/2	1/2+	-0.230		0.06674
					0.096/98				
^{172}Tm		171.93840	2.65 d	β⁻/1.88	1.79/36	2-			Yb k x-ray
					1.88/29				0.07879
									1.38722
									1.46601
									1.52982
									1.60861
^{173}Tm		172.93960	8.2 h	β⁻/1.298	0.80/21	1/2+			Yb k x-ray
					0.86/71				0.3988
									0.4613
^{174}Tm		173.94216	5.4 m	β⁻/3.08	0.70/14	(4-)			Yb k x-ray
					1.20/83				0.07664
									0.17669
									0.27332
									0.3666
									0.99205
									(0.08 - 1.6)
^{175}Tm		174.94383	15.2 m	β⁻/2.39	0.9/36	(1/2+)			Yb k x-ray
					1.9/23				0.36396
									0.51487
									0.94125
									0.98247
^{176}Tm		175.9471	1.9 m	β⁻/4.2	2.0/	(4+)			Yb k x-ray
					1.2/				0.1898
									0.3819
									1.0691
^{177}Tm		176.9490	1.4 m	β⁻					
^{178}Tm		177.9526							
^{179}Tm		178.9553							

TABLE OF ISOTOPES (CONTINUED)

Elem. or Isot.	Natural Abundance (%)	Atomic Mass or Weight	Half-Life	Decay Mode/Energy (/MeV)	Particle Energy /Intensity (MeV/%)	Spin (h/2π)	Nuclear Magnetic Mom. (nm)	Elect. Quadr. Mom. (b)	γ-ray/Energy Intensity (MeV/%)
$_{70}$Yb		173.04(3)							
^{148}Yb		147.967							
^{149}Yb		148.963							
^{150}Yb		149.958							
^{151}Yb		150.9545	1.6 s	β+/8.5					
^{152}Yb		151.9502	3.2 s	β+ EC/5.5					
^{153}Yb		152.9492	4. s	β+ EC/6.7					
^{154}Yb		153.9455	0.40 s	β+ EC/7/4.49					ann.rad./
				α/93/	5.32/				
^{155}Yb		154.9456	1.7 s	β+,EC/16/6.0			-0.8	-1.	ann.rad./
				α/84/	5.19/				
^{156}Yb		155.94277	26. s	β+,EC/21/3.57		0+			ann.rad./
				α/79/	4.69/				
^{157}Yb		156.9427	39. s	β+,EC/99+/5.5			-0.64		ann.rad./
				α/0.5/	4.69/				0.231
									(0.035-0.670)
^{158}Yb		157.93986	1.5 m	β+,EC/1.9		0+			ann.rad./
									0.0741
									0.2526
^{159}Yb		158.9402	1.4 m	EC,β+/5.1			-0.37	-.022	Tm k x-ray
									0.1661
									0.1772
									0.3297
									0.3903
^{160}Yb		159.9376	4.8 m	β+,EC/2.0		0+			ann.rad./
									0.1404
									0.1737
									0.2158
^{161}Yb		160.9375	4.2 m	β+,EC/3.9		3/2-	-0.33	+1.03	ann.rad./
									Tm k x-ray
									0.0782
									0.5999
									0.6315
^{162}Yb		161.9358	18.9 m	β+,EC/1.7		0+			ann.rad./
									Tm k x-ray
									0.1188
									0.1635
^{163}Yb		162.9363	11.1 m	β+/26/3.4	1.4/	3/2-	-0.37	+1.24	ann.rad./
									Tm k x-ray
									0.0636
									0.8603
									(0.06 - 1.9)
^{164}Yb		163.9345	1.26 h	EC/1.0		0+			Tm k x-ray
									0.0914
									0.6752
^{165}Yb		164.93540	9.9 m	β+/10/2.76	1.58/	(5/2-)	+0.48	+2.48	ann.rad./
				EC/90/					Tm k x-ray
									0.0801
									1.0903
^{166}Yb		165.93388	2.363 d	EC/0.30		0+			Tm k x-ray
									0.0828
									0.1844
									0.7789
									1.2734
									2.0524
^{167}Yb		166.934947	17.5 m	β+/0.5/1.954	0.639/	5/2-	+0.62	+2.70	Tm k x-ray
				EC/99.5/					0.06296
									0.10616
									0.11337

Elem. or Isot.	Natural Abundance (%)	Atomic Mass or Weight	Half-Life	Decay Mode/Energy (/MeV)	Particle Energy /Intensity (MeV/%)	Spin (h/2π)	Nuclear Magnetic Mom. (nm)	Elect. Quadr. Mom. (b)	γ-ray/Energy Intensity (MeV/%)
									0.17633
^{168}Yb	0.13(1)	167.933895				0+			
^{169m}Yb			46. s	I.T./0.0242		1/2-			Yb L x-ray
									0.0242
^{169}Yb		168.935187	32.03 d	EC/0.909		7/2+	-0.63	+3.5	Tm k x-ray
									0.1979/35.9
									0.0498-0.3078
^{170}Yb	3.05(6)	169.934759				0+			
^{171}Yb	14.3(2)	170.936323				1/2-	+0.49367		
^{172}Yb	21.9(3)	171.936378				0+			
^{173}Yb	16.12(21)	172.938207				5/2-	-0.67989	+2.80	
^{174}Yb	31.8(4)	173.938858				0+			
^{175}Yb		174.941273	4.19 d	β^-/0.470	0.466/73	7/2-	0.77		Lu k x-ray
					0.071/21				0.3963/13
					0.353/6.2				(0.114 - 0.28)
^{176m}Yb			11.4 s	I.T./1.051		(8-)			Yb k x-ray
									0.0961
									0.1901
									0.2929
									0.3897
^{176}Yb	12.7(2)	175.942569	10^{26} y	$\beta^-\beta^-$		0+			
^{177m}Yb			6.41 s	I.T./0.3315		1/2-			Yb k x-ray
									0.1131
									0.2084
^{177}Yb		176.945257	1.9 h	β^-/1.399	1.40	9/2+			Lu k x-ray
									0.1504
^{178}Yb		177.94664	1.23 h	β^-/0.65	0.25/	0+			0.1415
									0.3246
									0.3516
									0.3815
									0.6125
^{179}Yb		178.9499	8. m	β^-/2.4					
^{180}Yb		179.9523	2. m	β^-					0.1028-0.4423
^{181}Yb		180.9561							
$_{70}$Lu		174.967(1)							
^{150m}Lu			≈0.03 ms	p/1.295					
^{150}Lu		149.973	49. ms	p					
^{151m}Lu			16 μs	p/1.31					
^{151}Lu		150.967	0.08 s	p/1.231					
^{152}Lu		151.963	0.7 s						
^{153}Lu		152.959							
^{154}Lu		153.9571	1.0 s	β^+,EC/10.8					
^{155m}Lu			2.6 ms	α/7.41					
^{155}Lu		154.9542	0.07 s	EC/8.0					
				α/	5.66/90				
^{156m}Lu			0.20 s	α/	5.57/				
^{156}Lu		155.9529	≈ 0.5 s	β^+,EC/9.5					ann.rad./
				α/	5.45/				
^{157m}Lu			≈9.6 s	α	4.925/				
^{157}Lu		156.95010	4.8 s	β^+,EC/94/0.90					ann.rad./
				α/	5.00/				
^{158}Lu		157.94984	10.4 s	β^+,EC/99/8.0					ann.rad./
				α/	4.67/				0.3682
									0.4770
^{159}Lu		158.9467	12.3 s	β^+,EC/6.0					ann.rad./
									0.1505
									0.1875
									0.3693
^{160}Lu		159.94654	36.1 s	β^+,EC/7.3					ann.rad./

Elem. or Isot.	Natural Abundance (%)	Atomic Mass or Weight	Half-Life	Decay Mode/Energy (/MeV)	Particle Energy /Intensity (MeV/%)	Spin (h/2π)	Nuclear Magnetic Mom. (nm)	Elect. Quadr. Mom. (b)	γ-ray/Energy Intensity (MeV/%)
									0.2434
									0.3957
									0.5773
^{161}Lu		160.9432	1.2 m	β+,EC/5.3					ann.rad./
									0.0437
									0.0671
									0.1003
									0.1108
									0.1562
									0.2562
^{162m}Lu			≈ 1.5 m	EC/		4-			
^{162}Lu		161.9432	1.37 m	β+,EC/6.9		1-			ann.rad./
									0.1666
									0.6314
^{163}Lu		162.9412	4.1 m	β+,EC/4.6					ann.rad./
									0.0539
									0.0581
									0.1504
									0.1631
									0.3717
^{164}Lu		163.9412	3.14 m	β+,EC/6.3	1.6/				0.1238
					3.8/				0.2621
									0.7404
									0.8639
									0.8804
^{165}Lu		164.9396	10.7 m	β+,EC/3.9	2.06/	1/2+			ann.rad./
									0.1206
									0.1324
									0.1742
									0.2036
									(0.04 - 2.0)
^{166m2}Lu			2.1 m	β+/35/		(0-)			ann.rad./
				EC/65/					Yb k x-ray
									1.0673
									1.2566
									2.0986
^{166m1}Lu			1.4 m	β+,EC/58/		(3-)			ann.rad./
				I.T./42/0.0344					0.1024
									0.2281
									0.2861
									0.8119
									0.8301
^{166}Lu		165.9398	2.8 m	β+/25/5.5		(6-)			ann.rad./
				EC/75/					Yb k x-ray
									0.1024
									0.2281
									0.3375
									0.3679
^{167}Lu		166.9383	52. m	β+/2/3.1	2.1/	7/2+			Yb k x-ray
				EC/98/					0.0297
									0.2392
									(0.03 - 2.0)
^{168m}Lu			6.7 m	β+/12/		3+			ann.rad./
				EC/88/					Yb k x-ray
				IT/<0.8					0.1988/190
									0.8960/100
									0.9792/128
									0.018-2.65
^{168}Lu		167.9387	5.5 m	β+/6/4.5	1.2/	(6-)			ann.rad./

TABLE OF ISOTOPES (CONTINUED)

Elem. or Isot.	Natural Abundance (%)	Atomic Mass or Weight	Half-Life	Decay Mode/Energy (/MeV)	Particle Energy /Intensity (MeV/%)	Spin (h/2π)	Nuclear Magnetic Mom. (nm)	Elect. Quadr. Mom. (b)	γ-ray/Energy Intensity (MeV/%)
				EC/94/					Yb k x-ray
									0.1114
									0.1124
									0.2286
									0.3483
									1.4836
^{169m}Lu			2.7 m	I.T./0.0290		1/2-			Lu L x-ray
									0.0290
^{169}Lu		168.93765	1.419 d	EC/2.293	1.271/	7/2+	2.30	3.5	Yb k x-ray
									0.19121
									0.9606
									(0.08 - 2.1)
^{170m}Lu			0.7 s	I.T./0.0929		4-			Lu L x-ray
									0.04449
									0.0484
^{170}Lu		169.93847	2.01 d	EC/3.46	2.44/	0+			Yb k x-ray
									0.58711
									0.5908
									1.28029
									(0.1 - 3.38)
^{171m}Lu			1.31 m	I.T./0.0711		1/2-			Lu k x-ray
									0.07119
^{171}Lu		170.937910	8.24 d	EC/1.479	0.362/	7/2+	2.30	3.42	Yb k x-ray
									0.01939
									0.66744
									(0.02 - 1.3)
^{172m}Lu			3.7 m	I.T./0.0419		1-			Lu L x-rays
									0.04186
^{172}Lu		171.939082	6.70 d	EC/2.519		4-	2.90	3.80	Yb k x-ray
									0.18156
									1.09367
									(0.07 - 2.2)
^{173}Lu		172.938927	1.37 y	EC/0.671		7/2+	2.28	3.63	Yb k x-ray
									0.07860
									0.27198
^{174m}Lu			142. d	IT/99.3/	0.17086	6-	1.50		Lu k x-ray
				EC/0.7/					0.067055
^{174}Lu		173.940334	3.3 y	EC/1.374		1-	1.9		Yb k x-ray
									0.07664
									1.2419
^{175}Lu	97.41(2)	174.940768				7/2+	+2.2327	+3.49	
^{176m}Lu			3.66 h	β^-/1.315	1.229/	1-	+0.318	-1.47	Hf k x-ray
					1.317/				0.088372
^{176}Lu	2.59(2)	175.942683	3.8×10^{10} y	β^-/1.192		7-	+3.169	+4.92	Hf k x-ray
									0.20187
									0.30691
^{177m}Lu			160.7 d	IT/22/0.9702		23/2-	2.33	5.4	Lu k x-ray
				β^-/78					Hf k x-ray
									0.11295
									0.20030
									0.37850
									0.41853
^{177}Lu		176.943755	6.75 d	β^-/0.498	0.497/	7/2+	+2.239	+3.39	0.11295
									0.20836
^{178m}Lu			23.1 m	β^-/		(9-)			0.2166
									0.3317
^{178}Lu		177.945952	28.5 m	β^-/2.099	2.03/	1+			Hf k x-ray
									0.0932
									1.3099

Elem. or Isot.	Natural Abundance (%)	Atomic Mass or Weight	Half-Life	Decay Mode/Energy (/MeV)	Particle Energy /Intensity (MeV/%)	Spin (h/2π)	Nuclear Magnetic Mom. (nm)	Elect. Quadr. Mom. (b)	γ-ray/Energy Intensity (MeV/%)
									1.3408
									(0.09 - 1.7)
^{179}Lu		178.94732	4.6 h	β⁻/1.405	1.35/	7/2+			0.2143
									0.3377
^{180}Lu		179.9499	5.7 m	β⁻/3.1	1.49/				0.40795/50.
									(0.07-1.9)
^{181}Lu		180.9518	3.5 m	β⁻/2.5		(7/2+)			0.0458
									0.2059
									0.5749
^{182}Lu		181.9552	2.0 m	β⁻/≈ 4.1					0.0978
									0.7208
									0.8182
^{183}Lu		182.9576	58. s	β⁻/		7/2+			
^{184}Lu		183.9612	20 s	β⁻					
$_{72}$Hf		178.49(2)							
^{154}Hf		153.964	2. s	EC,β⁺/≈ 6.7					
^{155}Hf		154.963	0.9 s	EC,β⁺/8.					
^{156}Hf		155.9593	25. ms	α/					
^{157}Hf		156.9581	0.11 s	α/					
^{158}Hf		157.9539	2.9 s	EC/54/5.1		0+			
				α/46/	5.27/				
^{159}Hf		158.9538	5.6 s	β⁺,EC/88/6.9					ann.rad./
				α/12/	5.09/				
^{160}Hf		159.95063	≈ 12. s	β⁺,EC/97/4.9		0+			ann.rad./
				α/4.78					
^{161}Hf		160.9503	17. s	α/	4.60/				
^{162}Hf		161.94720	38. s	β⁺,EC/3.7		0+			ann.rad./
									0.1739
									0.1963
									0.4101
^{163}Hf		162.9471	40. s	β⁺,EC/5.5					ann.rad./
									0.0454
									0.0621
									0.0710
									0.6882
^{164}Hf		163.9536	2.8 m	EC,β⁺/3.0					
^{165}Hf		164.9445	1.32 m	EC/4.6		11/2-			
^{166}Hf		165.9423	6.8 m	EC/93/2.3					ann.rad./
				β⁺/7/					Lu k x-ray
									0.0788
^{167}Hf		166.9426	2.0 m	β⁺/40/4.0		(5/2-)			ann.rad./
				EC/60/					Lu k x-ray
									0.1754
									0.3152
^{168}Hf		167.9406	25.9 m	β⁺,EC/1.8		0+			ann.rad./
									(0.0144-1.311)
^{169}Hf		168.9412	3.25 m	EC/85/3.3		(5/2-)			ann.rad./
				β⁺/15/					Lu k x-ray
									0.3695
									0.4929
^{170}Hf		169.9397	16.0 h	EC/1.1		0+			Lu k x-ray
									0.0985
									0.1202
									0.1647
									0.5729
									0.6207
^{171m}Hf			30. s			(1/2-)	+0.53		
^{171}Hf		170.9405	12.2 h	EC,β⁺/2.4		7/2+	-0.67	+3.46	ann.rad./
									Lu k x-ray

Elem. or Isot.	Natural Abundance (%)	Atomic Mass or Weight	Half-Life	Decay Mode/Energy (/MeV)	Particle Energy /Intensity (MeV/%)	Spin (h/2π)	Nuclear Magnetic Mom. (nm)	Elect. Quadr. Mom. (b)	γ-ray/Energy Intensity (MeV/%)
									0.1221
									0.6620
									1.0714
^{172}Hf		171.93946	1.87 y	EC/0.35		0+			Lu k x-ray
									0.02399
									0.12582
									(0.0818-0.123)
^{173}Hf		172.9407	23.6 h	EC/1.6		1/2-			Lu k x-ray
									0.12367
									0.13963
									0.29697
									0.31124
									(0.1 - 2.1)
^{174}Hf	0.162(3)	173.940042	2.0x10^{15} y			0+			
^{175}Hf		174.941504	70. d	EC/0.686		5/2-	-0.60	+2.7	Lu k x-ray
									0.08936
									0.34340
^{176}Hf	5.206(5)	175.941403				0+			
^{177m2}Hf			51.4 m	I.T./2.740		37/2-			Hf k x-ray
									0.2140
									0.2951
									0.3115
									0.3267
^{177m1}Hf			1.1 s	I.T./		23/2+			Hf k x-ray
									0.20836
									0.22847
									0.37851
^{177}Hf	18.606(4)	176.943220				7/2-	+0.7935	+0.337	
^{178m2}Hf			31. y	I.T./		16+	+8.2	+6.0	Hf k x-ray
									0.32555
									0.42635
									(0.0889-0.5742)
^{178m1}Hf			4.0 s	I.T./		8-			Hf k x-ray
									0.21342
									0.32555
									0.42635
^{178}Hf	27.297(4)	177.943698				0+			
^{179m2}Hf			25.1 d	I.T./1.1057		25/2-	7.4		Hf k x-ray
									0.1227
									0.1461
									0.3626
									0.4537
^{179m1}Hf			18.7 s	I.T./0.375		1/2-			Hf k x-ray
									0.1607
									0.2141
^{179}Hf	13.629(6)	178.945815				9/2+	-0.641	+3.79	
^{180m}Hf			5.52 h	I.T./1.1416		8-	+9.	+4.6	Hf k x-ray
									0.2152
									0.3323
									0.4432
^{180}Hf	35.100(7)	179.946549				0+			
^{181}Hf		180.949099	42.4 d	β^-/1.027	0.408/	1/2-			Ta k x-ray
									0.13294
									0.48200
^{182m}Hf			62. m	β^-/54/1.60	0.49/43	8-			Hf k x-ray
				IT/46/1.173	0.95/10				0.0509
									0.2244
									0.3441
									0.4558

TABLE OF ISOTOPES (CONTINUED)

Elem. or Isot.	Natural Abundance (%)	Atomic Mass or Weight	Half-Life	Decay Mode/Energy (/MeV)	Particle Energy /Intensity (MeV/%)	Spin (h/2π)	Nuclear Magnetic Mom. (nm)	Elect. Quadr. Mom. (b)	γ-ray/Energy Intensity (MeV/%)
									0.5066
									0.9428
^{182}Hf		181.95055	9.x10^{6} y	β^{-}/0.37		0+			Ta k x-ray
									0.2704
^{183}Hf		182.95353	1.07 h	β^{-}/2.01	1.18/68	3/2-			Ta k x-ray
					1.54/25				0.0732
									0.4591
									0.7837
^{184}Hf		183.95545	4.1 h	β^{-}/1.34	0.74/38	0+			Ta k x-ray
					0.85/16				0.0414
					1.10/46				0.1391
									0.3449
^{185}Hf		184.9588	≈3.5m	β^{-}/					0.165
^{186}Hf		185.9609	≈2.1 m						0.738
$_{73}$Ta		180.9479(1)							
^{155}Ta			12 μs	p/1.77					
^{156}Ta		155.972	0.11 s	β^{+}/≈11.6					
				p/	1.02/≈100				
^{157}Ta		156.968	10 ms	α/	6.117				
				p/	0.927/3.4				
^{158}Ta		157.9664	37. ms	α/	6.05/100				
					5.97/100				
^{159}Ta		158.9629	0.6 s	β^{+},EC/20/8.5	α/5.52/34				ann.rad./
				α/80/	5.60/55				
^{160}Ta		159.9615	1.4 s	β^{+},EC/10.1					ann.rad./
				α	5.41/				
^{161}Ta		160.9584	2.9 s	β^{+},EC/7.5					ann.rad./
				α/	5.15				
^{162}Ta		161.9564	4. s	EC/8.6					
^{163}Ta		162.9544	10.6 s	EC/6.8					
^{164}Ta		163.9536	14.2 s	β^{+}/8.5		3+			ann.rad./
				α/	4.62/				0.2110
									0.3768
^{165}Ta		164.9508	31. s	ECβ^{+}/5.9					
^{166}Ta		165.9505	34. s	β^{+}/82/7.7					ann.rad./
				EC/18/					Hf k x-ray
									0.1587
									0.3117
									0.8101
^{167}Ta		166.9486	1.4 m	β^{+},EC/5.6					ann.rad./
^{168}Ta		167.9478	2.4 m	β^{+}/77/6.7		3+			ann.rad./
				EC/23/					Hf k x-ray
									0.1239
									0.2615
									0.7502
^{169}Ta		168.9459	4.9 m	β^{+},EC/4.4					ann.rad./
									0.0288
									0.1535
									0.1924
^{170}Ta		169.9461	6.8 m	β^{+}/70/6.0		(3+)			ann.rad./
				EC/35/					Hf k x-ray
									0.1008
									0.2212
^{171}Ta		170.9445	23.3 m	β^{+},EC/3.7		(5/2-)			0.0496
									0.5018
									0.5064
									(0.05 - 1.02)
^{172}Ta		171.9447	36.8 m	β^{+}/25/4.9		(3-)			ann.rad./
				EC/75/					Hf k x-ray

Elem. or Isot.	Natural Abundance (%)	Atomic Mass or Weight	Half-Life	Decay Mode/Energy (/MeV)	Particle Energy /Intensity (MeV/%)	Spin (h/2π)	Nuclear Magnetic Mom. (nm)	Elect. Quadr. Mom. (b)	γ-ray/Energy Intensity (MeV/%)
									0.21396
									1.10923
									(0.09 - 3.8)
^{173}Ta		172.9446	3.6 h	β^+/24/3.7		(5/2-)	1.70	-1.9	ann.rad./
				EC/76/					Hf k x-ray
									0.06972
									0.17219
									(0.06 - 2.7)
^{174}Ta		173.9442	1.12 h	β^+/27/3.8		(3+)			ann.rad./
				EC/73/					Hf k x-ray
									0.09089
									0.20638
									(0.09 - 3.64)
^{175}Ta		174.9437	10.5 h	EC/2.0		7/2+	2.27	+3.7	Hf k x-ray
									0.2077
									0.2671
									0.3487
^{176}Ta		175.9447	8.1 h	EC/3.1		1-			Hf k x-ray
									0.08837
									1.15735
^{177}Ta		176.944472	2.356 d	EC/1.166		7/2+	2.25		Hf k x-ray
									0.11295
									(0.07 - 1.06)
^{178m}Ta			2.4 h	EC/		(7-)			Hf k x-ray
									0.08886
									0.21342
									0.32555
									0.42635
^{178}Ta		177.9458	9.29 m	EC/99/1.9		1+	+2.74	+0.65	ann.rad./
				β^+/1/					Hf k x-ray
									0.09316
^{179}Ta		178.94593	1.8 y	EC/0.110		7/2+	2.29	3.37	Hf k x-ray
^{180m}Ta	0.012(2)		>1.2x10^{15} y			(9-)	4.82		
^{180}Ta		179.947466	8.15 h	EC/87/0.854		1+			Hf k x-ray
				β^-/13/0.708	0.61/3				W k x-ray
					0.71/10				0.09333
									0.10340
^{181}Ta	99.988(2)	180.947996				7/2+	+2.370	+3.3	
^{182m}Ta			15.8 m	I.T./0.5198		10-			Ta k x-ray
									0.14678
									0.17157
^{182}Ta		181.950152	114.43 d	β^-/1.814	0.25/30	3-	+3.02	+2.6	W k x-ray
					0.44/20				1.12127/100
					0.52/40				1.22138/79
									0.085-1.289
^{183}Ta		182.951373	5.1 d	β^-/1.070	0.45/5	7/2+	+2.36		W k x-ray
					0.62/91				0.0847
									0.0991
									0.1079
									0.2461
									0.3540
^{184}Ta		183.95401	8.7 h	β^-/2.87	1.11/15	(5-)			W k x-ray
					1.17/81				0.2528/44.
									0.4140/74.
									(0.09-1.4)
^{185}Ta		184.95556	49. m	β^-/1.99	1.21/5	(7/2+)			W k x-ray
					1.77/81				0.0697
									0.1739
									0.1776

Elem. or Isot.	Natural Abundance (%)	Atomic Mass or Weight	Half-Life	Decay Mode/Energy (/MeV)	Particle Energy /Intensity (MeV/%)	Spin (h/2π)	Nuclear Magnetic Mom. (nm)	Elect. Quadr. Mom. (b)	γ-ray/Energy Intensity (MeV/%)
^{186}Ta		185.9586	10.5 m	β^-/3.9	2.2/	(3-)			W k x-ray
									0.1979
									0.2149
									0.5106
									(0.09 - 1.5)
^{187}Ta		186.9604							
^{188}Ta		187.9637							
$_{74}$W		183.84(1)							
^{158m}W			0.14 ms	α	8.28(3)/				
^{158}W		157.974	1.3 ms	α/	6.433/96				
^{159}W		158.972	7. ms	α/					
^{160}W		159.9684	0.08 s	α/	5.92/	0+			
^{161}W		160.9671	0.41 s	β^+,EC/18/8.1					
				α/82/	5.78/				
^{162}W		161.9626	1.39 s	β^+,EC/54/5.8		0+			
				α/46/	5.54/				
^{163}W		162.9624	2.8 s	β^+,EC/59/7.5					
				α/41/	5.38/				
^{164}W		163.95890	6. s	β^+,EC/97/5.0		0+			ann.rad./
				α/3/	5.15/				
^{165}W		164.9583	5.1 s	β^+,EC/99/7.0					ann.rad./
				α/1/	4.91/				
^{166}W		165.95502	16. s	β^+,EC/99/4.2		0+			ann.rad./
				α/1/	4.74/				
^{167}W		166.9547	20. s	EC/5.6					
^{168}W		167.9519	53. s	EC/3.8					ann.rad./
				α/10^{-5}/	4.40(1)				Ta k x-ray
									0.1755
									(0.037-0.573)
^{169}W		168.9518	1.3 m	EC/5.4					ann.rad./
									Ta k x-ray
									0.123
									(0.097-0.699)
^{170}W		169.9485	2.4 m	EC/2.2					ann.rad./
									Ta k x-ray
									0.3162
									(0.060-0.144)
^{171}W		170.9494	2.4 m	EC/4.6					ann.rad./
									Ta k x-ray
									0.1842
									(0.052-0.479)
^{172}W		171.9474	6.6 m	β^+,EC/2.5					ann.rad./
									Ta k x-ray
									0.0389
									(0.034-0.674)
^{173}W		172.9489	6.3 m	EC/4.0					ann.rad./
									Ta k x-ray
									0.4576
									(0.035-0.623)
^{174}W		173.9462	35. m	EC/1.9		0+			ann.rad./
									Ta k x-ray
									0.3287
									0.4288
									(0.056-0.429)
^{175}W		174.9468	35. m	EC/2.9		1/2-			(0.015-0.27)
^{176}W		175.9456	2.5 h	β^+,EC/0.8		0+			0.03358
									0.06129
									0.09487
									0.10020

Elem. or Isot.	Natural Abundance (%)	Atomic Mass or Weight	Half-Life	Decay Mode/Energy (/MeV)	Particle Energy /Intensity (MeV/%)	Spin (h/2π)	Nuclear Magnetic Mom. (nm)	Elect. Quadr. Mom. (b)	γ-ray/Energy Intensity (MeV/%)
^{177}W		176.9466	2.21 h	EC/2.0		(1/2-)			Ta k x-ray
									0.15505
									0.18569
									0.42694
^{178}W		177.9459	21.6 d	EC/0.091		0+			Ta k x-ray
^{179m}W			6.4 m	IT/99.7/0.222		(1/2-)			W k x-ray
				EC/0.3/					0.2220
^{179}W		178.94707	38. m	EC/1.06		(7/2-)			Ta k x-ray
									0.0307
^{180}W	0.120(1)	179.946706	7.4x10^{16} y	α/		0+			
^{181}W		180.94820	121.2 d	EC/0.188		9/2+			Ta k x-ray
									0.13617
									0.15221
^{182}W	26.498(29)	181.948205	8.3x10^{18} y	α/		0+			
^{183m}W			5.15 s	I.T./		(11/2+)			W k x-ray
									0.0465
									0.0526
									0.0991
									0.1605
^{183}W	14.314(4)	182.950224	1.9x10^{18} y	α/		1/2-	+0.1177848		
^{184}W	30.642(8)	183.950932	4.0x10^{18} y	α/		0+			
^{185m}W			1.6 m	I.T./0.1974		11/2+			W k x-ray
									0.0659
									0.1315
									0.1737
^{185}W		184.953420	74.8 d	β$^-$/0.433	0.433/99.9	3/2-			0.12536
^{186}W	28.426(37)	185.954362	6.5x10^{18} y	α/		0+			
^{187}W		186.957158	23.9 h	β$^-$/1.311	0.624/66	3/2-	0.62		Re k x-ray
					1.315/16				0.68572/33
					0.081-1.18				0.134-0.773
^{188}W		187.958487	69.4 d	β$^-$/0.349	0.349/99	0+			0.0636
									0.2271
									0.2907
^{189}W		188.9619	11.5 m	β$^-$/2.5	1.4/	(3/2-)			(0.1262-1.466)
					2.5/				
^{190m}W			0.3 ms						
^{190}W		189.9632	30. m	β$^-$/1.3	0.95/	0+			Re k x-ray
									0.1576
									0.1621
$_{75}$Re		186.207(1)							
^{160}Re		159.981	0.7 ms	p/	1.261(6)/91				
				α/	6.54/				
^{161}Re		160.978	14 ms	α/	6.24				
				p	1.35				
^{162}Re		161.9757	0.10 s	α/	6.12/94				
					6.09/94				
^{163}Re		162.9721	0.26 s	β$^+$,EC/9.0	α/5.87/32				
				α/	5.92/66				
^{164}Re		163.9704	0.9 s	β$^+$,EC/10.7					
				α/	5.78/				
^{165}Re		164.9671	2. s	β$^+$,EC/87/8.1					
				α/	5.51/				
^{166}Re		165.9651	2.5 s	β$^+$,EC/9.4					
				α/	5.50/				
^{167m}Re			6.2 s	α, EC/					
^{167}Re		166.9626	3.4 s	β$^+$,EC/7.4					
				α/	5.015/				
^{168}Re		167.9616	4.4 s	β$^+$,EC/9.1					
				α/	4.833/				0.1117

Elem. or Isot.	Natural Abundance (%)	Atomic Mass or Weight	Half-Life	Decay Mode/Energy (/MeV)	Particle Energy /Intensity (MeV/%)	Spin (h/2π)	Nuclear Magnetic Mom. (nm)	Elect. Quadr. Mom. (b)	γ-ray/Energy Intensity (MeV/%)
^{169m}Re			8.1 s	α	4.70/				
					4.87/				
^{169}Re		168.9588	16. s						
^{170}Re		169.9582	9.2 s	β+, EC/9.0					0.1560
									0.3055
									0.4125
^{171}Re		170.9555	15.2 s	EC/≈ 5.7					
^{172m}Re			55. s	β+,EC/		(2)			ann.rad./
									0.1234
									0.2537
									0.3504
^{172}Re		171.9553	15. s	β+,EC/7.3					ann.rad./
									0.1234
									0.2537
^{173}Re		172.9531	2.0 m	EC/≈3.9					ann.rad./
^{174}Re		173.9521	2.4 m	β+,EC/5.6					ann.rad./
									0.1119
									0.2430
^{175}Re		174.9514	5.8 m	β+,EC/4.3					ann.rad./
^{176}Re		175.9516	5.3 m	β+,EC/5.6		(3+)			ann.rad./
									0.1089
									0.2406
^{177}Re		176.9503	14. m	EC/78/3.4		(5/2-)			ann.rad./
				β+/22/					W k x-ray
									0.0797
									0.0843
									0.1968
^{178}Re		177.9509	13.2 m	β+/11/4.7	3.3/	(3)			ann.rad./
				EC/89/					W k x-ray
									0.1059
									0.2373
									0.9391
^{179}Re		178.9500	19.7 m	EC/99/2.71	0.95/	(5/2+)	2.8		W k x-ray
				β+/1/					0.1199
									0.2900
									0.4154
									0.4302
									1.6803
^{180}Re		179.95079	2.45 m	EC/92/3.80	1.76/	1-	1.6		ann.rad./
				β+/8/					W k x-ray
									0.1036
									0.9028
									(0.07 - 2.2)
^{181}Re		180.95006	20. h	EC/1.74		5/2+	3.19		W k x-ray
									0.3607
									0.3655
									0.6390
^{182m}Re			12.7 h	EC/	0.55/	2+	3.3	+1.8	W k x-ray
					1.74/				0.0677
									1.1214
									1.2215
									(0.06 - 2.2)
^{182}Re		181.9512	2.67 d	EC/2.8		(7+)	2.8	+4.1	W k x-ray
									0.0678
									0.2293
									1.1213
									1.2214
^{183}Re		182.95082	70. d	EC/0.56		(5/2+)	+3.17	+2.3	W k x-ray
									0.16232

Elem. or Isot.	Natural Abundance (%)	Atomic Mass or Weight	Half-Life	Decay Mode/Energy (/MeV)	Particle Energy /Intensity (MeV/%)	Spin (h/2π)	Nuclear Magnetic Mom. (nm)	Elect. Quadr. Mom. (b)	γ-ray/Energy Intensity (MeV/%)
^{184m}Re			165. d	I.T./75/0.188		8+	+2.9		Re k x-ray
				EC/25/					0.1047
									0.2165
									0.92093
									(0.10 - 1.1)
^{184}Re		183.95252	38. d	EC/1.48		3-	+2.53	+2.8	W k x-ray
									0.79207
									0.90328
									(0.1 - 1.4)
^{185}Re	37.40(2)	184.952955				5/2+	+3.1871	+2.18	
^{186m}Re			2.0×10^5 y	I.T./0.150		8+			Re k x-ray
									0.0590
^{186}Re		185.954986	3.718 d	β^-/92/1.070	0.973/21	1-	+1.739	+0.62	W k x-ray
				EC/8/0.582	1.07/71				0.1227/0.6
									0.1372/9.5
									(0.63-0.77)
^{187}Re	62.60(2)	186.955751	4.4×10^{10} y	β^-/0.00266	0.0025/	5/2+	+3.2197		+2.07
^{188m}Re			18.6 m	I.T./0.172		(6-)			Re k x-ray
									0.0925
									0.1059
^{188}Re		187.958112	16.94 h	β^-/2.120	1.962/20	1-	+1.788	+0.57	Os k x-ray
					2.118/79				0.15502
									0.309-2.022
^{189}Re		188.959228	24. h	β^-/1.01	1.01/	(5/2+)			0.1471
									0.2167
									0.2194
									0.2451
^{190m}Re			3.0 h	β^-/51/		(6-)			Re k x-ray
				I.T./49/					0.1191
									0.2238
									0.6731
									(0.1 - 1.79)
^{190}Re		189.9618	3.0 m	β^-/3.2	1.8/	(2-)			Os k x-ray
									0.1867
									0.5580
									0.6051
^{191}Re		190.96312	9.7 m	β^-/2.05	1.8/				
^{192}Re		191.9660	16. s	β^-/4.2	≈ 2.5/				(0.2-0.75)
$_{76}$Os		190.23(3)							
^{162}Os		161.984	1.8 ms	α/	6.60				
^{163}Os		162.982	5.5 ms	α/	6.51				
^{164}Os		163.9779	0.04 s	α					
^{165}Os		164.9765	0.07 s	α					
^{166}Os		165.9718	0.18 s	β^+,EC/28/6.3	6.27/	0+			ann. rad./
				α/72/	5.98/				
^{167}Os		166.9714	0.7 s	β^+,EC/76/8.2					ann.rad./
				α/24/	5.84/				
^{168}Os		167.96775	2.2 s	β^+,EC/51/5.7		0+			ann. rad./
				α/49/					
^{169}Os		168.0671	3.3 s	β^+,EC/89/7.7	5.57/80				ann.rad./
				α/13/	5.51/12				
					5.54/8				
^{170}Os		169.96357	7.1 s	β^+,EC/5.0		0+			ann.rad./
				α/	5.40/				(0.162-0.216)
^{171}Os		170.9630	8.4 s	β^+,EC/98/7.1	α/5.24/93.5				ann.rad./
				α/19/	5.17/6.5				0.190-0.705
^{172}Os		171.9601	19. s	β^+,EC/99/4.5		0+			ann.rad./
				α/1.1/	5.10/				(0.063-1.120)
^{173}Os		172.9598	16. s	β^+,EC/6.3					ann.rad./

Elem. or Isot.	Natural Abundance (%)	Atomic Mass or Weight	Half-Life	Decay Mode/Energy (/MeV)	Particle Energy /Intensity (MeV/%)	Spin (h/2π)	Nuclear Magnetic Mom. (nm)	Elect. Quadr. Mom. (b)	γ-ray/Energy Intensity (MeV/%)
				α/0.4/	4.94/				0.142-0.299
^{174}Os		173.9563	44. s	β⁺,EC/3.9		0+			0.118
				α/0.02/	4.76/				0.138/ 0.001
									0.158
									0.325
^{175}Os		174.9570	1.4 m	β⁺,EC/5.3					0.125
									0.181
									0.248
^{176}Os		175.9550	3.6 m	β⁺,EC/3.2		0+			0.8155
									0.7758
									0.8573
									1.2093
									1.2909
^{177}Os		176.9551	2.8 m	β⁺,EC/4.5		(1/2-)			0.0848
									0.1958
									0.3002
									1.2686
^{178}Os		177.9534	5.0 m	β⁺,EC/2.3		0+			ann.rad./
									0.5946
									0.6850
									0.9687
									1.3311
^{179}Os		178.9539	7. m	β⁺,EC/3.7					ann.rad./
									0.0654
									0.2186
									0.5938
^{180}Os		179.9524	21.5 m	β⁺,EC/1.5		0+			Re k x-ray
									0.0202-0.7174
^{181m}Os			1.75 h	EC/		(1/2-)			ann.rad./
									0.0489
^{181}Os		180.9532	2.7 m	EC/2.9		(7/2-)			ann.rad./
									0.11794
									0.23868
									0.8267
									(0.07 - 2.64)
^{182}Os		181.95219	21.5 h	EC/0.9		0+			Re k x-ray
									0.1802
									0.5100
^{183m}Os			9.9 h	EC/84/		1/2-			Os k x-ray
				I.T./16/					Re k x-ray
									1.1020
									1.1080
^{183}Os		182.9531	13. h	EC/2.1		9/2+	-0.79	+3.1	Re k x-ray
									0.1144
									0.3818
^{184}Os	0.020(3)	183.952491				0+			
^{185}Os		184.954043	93.6 d	EC/1.013		1/2-			Re k x-ray
									0.6461
									0.8748
									0.8805
^{186}Os	1.58(10)	185.953838	2x10^{15} y	α/	≈ 2.75/	0+			
^{187}Os	1.6(1)	186.955748				1/2-	+0.0646519		
^{188}Os	13.3(2)	187.955836				0+			
^{189m}Os			5.8 h	I.T./0.0308		9/2-			Os L x-ray
									0.0308
^{189}Os	16.1(3)	188.958145				3/2+	+0.65993	+0.86	
^{190m}Os			9.9 m	I.T./1.705		10-	-0.6		Os k x-ray
									0.1867
									0.3611

Elem. or Isot.	Natural Abundance (%)	Atomic Mass or Weight	Half-Life	Decay Mode/Energy (/MeV)	Particle Energy /Intensity (MeV/%)	Spin (h/2π)	Nuclear Magnetic Mom. (nm)	Elect. Quadr. Mom. (b)	γ-ray/Energy Intensity (MeV/%)
									0.5026
									0.6161
^{190}Os	26.4(4)	189.958445				0+			
^{191m}Os			13.1 h	I.T./0.0744		3/2-			Os k x-ray
									0.0744
^{191}Os		190.960928	15.4 d	β^-/0.314	0.140/100	9/2-		+2.5	Ir k x-ray
									0.1294
^{192m}Os			6.0 s	I.T./2.0154		(10-)			Os k x-ray
									0.2058/65.9
									0.5692/70
									(0.201-1.000)
^{192}Os	41.0(3)	191.961479				0+			
^{193}Os		192.964148	30.5 h	β^-/1.141	1.04/20	3/2-	+0.730	+0.47	Ir k x-ray
									0.1389
									0.4605
^{194}Os		193.965179	6.0 y	β^-/0.097	0.054/33	0+			Ir L x-ray
					0.096/67				0.0429
^{195}Os		194.9681	6.5 m	β^-/2.0	2.0/				
^{196}Os		195.96962	34.9 m	β^-/1.16	0.84/	0+			0.1262/5
									0.4079/5.9
$_{77}$Ir		192.217(3)							
^{164}Ir									
^{165}Ir		164.9876	0.3 ms	p/87	1.71				
				α/13	6.72				
^{166m}Ir			15 ms	α/98.2	6.56				
				p/1.8	1.32				
^{166}Ir		165.9855	≈ 11 ms	α/93	6.56				
				p/6.9	1.15				
^{167m}Ir			30 ms	α/48,β^+	6.41/80				
				p/32	1.24/0.4				
^{167}Ir		166.9817	35. ms	α/80,β^+	6.35/48				
				p/0.4	1.04/32				
^{168}Ir		167.9799	0.16 s	α/82					
^{169m}Ir			0.3 s	α/	6.11/84				
^{169}Ir		168.9764	0.6 s	α/	6.00/50				
^{170}Ir		169.9743	1.0 s	α/	6.03/				
^{171}Ir		170.9718	1.5 s	α/	5.91/				
^{172}Ir		171.9706	2.1 s	α/	5.811/				0.228
									(0.379-0.475)
^{173}Ir		172.9677	3.0 s	α/	5.665/				0.0493
									(0.092-0.296)
^{174}Ir		173.9668	4. s	α/	5.478/				0.1587
									(0.276-1.33)
^{175}Ir		174.9641	≈ 4.5 s	α/	5.393/				0.1056
^{176}Ir		175.9635	8. s	EC, β^+/80					0.260
				α/3.2/	5.118/				(0.135-0.415)
^{177}Ir		176.9612	30. s	EC, β^+/5.7					0.184
				α/0.06/	5.011/				(0.062-0.194)
^{178}Ir		177.9601	12. s	β^+,EC/6.3					
									0.1320
									0.2667
									0.3633
^{179}Ir		178.9592	4. m	EC/4.9					0.0975
									(0.045-0.220)
^{180}Ir		179.9593	1.5 m	EC/6.4					0.2765
									((0.132-1.106)
^{181}Ir		180.9576	4.9 m	β^+,EC/4.1		(7/2+)			ann.rad./
									0.1076
									(0.0196-1.715)

Elem. or Isot.	Natural Abundance (%)	Atomic Mass or Weight	Half-Life	Decay Mode/Energy (/MeV)	Particle Energy /Intensity (MeV/%)	Spin (h/2π)	Nuclear Magnetic Mom. (nm)	Elect. Quadr. Mom. (b)	γ-ray/Energy Intensity (MeV/%)
^{182}Ir		181.9582	15. m	β+/44/5.6					ann.rad./
				EC/56/					Os k x-ray
									0.1273
									0.2370
^{183}Ir		182.9568	57. m	β+,EC/3.5					ann.rad./
									0.0877
									0.2285
									0.2824
^{184}Ir		183.9574	3.0 h	β+/12/4.6	2.3/	5-	0.70	+2.41	ann.rad./
				EC/88/	2.9/				Os k x-ray
									0.11968
									0.2640
									0.3904
^{185}Ir		184.9566	14. h	β+/3/2.4		(5/2-)	2.60	-2.1	ann.rad./
				EC/97/					Os k x-ray
									0.2543
									1.8288
^{186m}Ir			1.7 h	EC/		(2-)	0.64	+1.46	Os k x-ray
									0.1371
									0.7675
^{186}Ir		185.95795	15.7 h	EC/98/3.83		(5+)	3.9	-2.55	Os k x-ray
				β+/2/					0.1372
									0.2968
									0.4348
									(0.13 - 3.0)
^{187}Ir		186.95736	10.5 h	EC/1.50		3/2+		+0.94	Os k x-ray
									0.0743
									0.4009
									0.4271
									0.6109
									0.9128
^{188}Ir		187.95885	1.72 d	β+/2.81	1.13/	(2-)	0.30	+0.48	Os k x-ray
				EC/99+/	1.64/				0.1550
									0.4780
									0.6330
									2.2146
^{189}Ir		188.95872	13.2 d	EC/0.53		3/2+	0.13	+0.88	Os k x-ray
									0.2449
^{190m2}Ir			3.09 h	β+,EC/95/		(11-)			0.376
				I.T./5/					
^{190m1}Ir			1.12 h	I.T./0.0263		7+			Ir L x-ray
^{190}Ir		189.9606	11.8 d	EC/2.0		(4+)	0.04	+2.8	Os k x-ray
									0.1867
									0.4072
									0.5186
									0.5580
									0.6051
									(0.2 - 1.4)
^{191m}Ir			4.93 s	I.T./0.1714		11/2-	+0.603		Ir k x-ray
									0.1294
^{191}Ir	37.3(5)	190.960591				3/2+	+0.151	+0.82	
^{192m2}Ir			241. y	I.T./0.161		(9+)			Ir k x-ray
^{192m1}Ir			1.44 m	I.T./0.0580		(1+)			Ir L x-ray
									0.0580
									0.3165
^{192}Ir		191.962602	73.83 d	β-/1.460		(4-)	+1.92	+2.15	Pt k x-ray
									0.31649/83.
									0.46806/48.
^{193m}Ir			10.53 d	I.T./0.0802		11/2-			Ir L x-ray

Elem. or Isot.	Natural Abundance (%)	Atomic Mass or Weight	Half-Life	Decay Mode/Energy (/MeV)	Particle Energy /Intensity (MeV/%)	Spin (h/2π)	Nuclear Magnetic Mom. (nm)	Elect. Quadr. Mom. (b)	γ-ray/Energy Intensity (MeV/%)
									0.0803
^{193}Ir	62.7(5)	192.962923				3/2+	+0.164	+0.75	
^{194m}Ir			170. d	β^-/		11			Pt k x-ray
									0.3284
									0.4829
									0.5624
^{194}Ir		193.965075	19.3 h	β^-/2.247	1.92/9	1-	+0.39	+0.34	0.2935
					2.25/86				0.3284
									0.6451
									(0.1 - 2.2)
^{195m}Ir			3.9 h	β^-/	0.41/	(11/2-)			Pt k x-ray
					0.97/				0.3199/9.6
									0.3649/9.5
									0.4329/9.6
									0.6849/9.6
^{195}Ir		194.965976	2.8 h	β^-/1.120	1.0/80	(3/2+)			Pt k x-ray
					1.11/13				0.0989/9.7
^{196m}Ir			1.40 h	β^-/	1.16/				Pt k x-ray
									0.3557
									0.3935
									0.4471
									0.5214
									0.6473
^{196}Ir		195.96838	52.s	β^-/3.21	2.1/15	0-			0.3329
					3.2/80				0.3557
									0.7796
^{197m}Ir			8.9 m	β^-/		(11/2-)			0.3465
				I.T./					See Ir[197]
^{197}Ir		196.96964	5.8 m	β^-/2.16	1.5/	(3/2+)			0.0531
					2.0/				0.1351
									0.4306
									0.4697
^{198}Ir		197.9723	8. s	β^-/4.1					0.4074
									0.5070
^{199}Ir		198.97378							
$_{78}$Pt		195.078(2)							
^{166}Pt			0.3 ms	α/	7.11/				
^{167}Pt			0.7 ms	α/	6.99/				
^{168}Pt		167.9880	2.0 ms	α	6.83				0.582/69
									0.594/69
									0.725/62
^{169}Pt		168.9864	3. ms	α					
^{170}Pt		169.9816	14 ms	α	6.55				0.509/100
									0.662/86
									0.214-0.726
^{171}Pt		170.9811	0.03 s	α					
^{172}Pt		171.97730	0.10 s	α/	6.31/94	0+			
^{173}Pt		172.9765	0.34 s	β^+,EC/8.2					
				α/	6.20/				
^{174}Pt		173.97281	0.89 s	β^+,EC/17/5.6		0+			
				α/83/	6.040/				
^{175}Pt		174.9723	2.5 s	β^+,EC/65/7.6					0.0774
				α/35/	5.831/5				0.1354
					5.96/54				0.2128
					6.038/				
^{176}Pt		175.9690	6.3 s	β^+,EC/60/5.1		0+			ann.rad./
				α/40/	5.528/0.6				0.2277
					5.750/41				
^{177}Pt		176.9685	11. s	EC/91/6.8	5.53/				0.0908

TABLE OF ISOTOPES (CONTINUED)

Elem. or Isot.	Natural Abundance (%)	Atomic Mass or Weight	Half-Life	Decay Mode/Energy (/MeV)	Particle Energy /Intensity (MeV/%)	Spin (h/2π)	Nuclear Magnetic Mom. (nm)	Elect. Quadr. Mom. (b)	γ-ray/Energy Intensity (MeV/%)
				α/9/	5.485/3				
					5.525/6				
^{178}Pt		177.9649	21. s	EC/93/4.5		0+			
				α/7/	5.286/0.2				
					5.442/7				
^{179}Pt		178.9653	33. s	β^+,EC/5.7					
				α/	5.16/				
^{180}Pt		179.9632	52. s	β^+,EC/99.7/3.7		0+			
				α/0.3/	5.140/				
^{181}Pt		180.9632	51. s	β^+,EC/5.2					
^{182}Pt		181.9613	2.7 m	β^+,EC/2.9		0+			ann.rad./
									0.1360
									0.1460
									0.2100
^{183m}Pt			43. s	β^+,EC/		(7/2-)	1.0		ann.rad./
				I.T./					0.3132/26
									0.3164/59
									0.6296/100
									0.058-1.75
^{183}Pt		182.9617	7. m	β^+,EC/4.6			+0.51		ann.rad./
									0.119/100
									0.307/93
									0.260/90
									0.058-1.377
^{184}Pt		183.9599	17.3 m	β^+,EC/2.3					ann.rad./
									0.1549
									0.1919
									0.5484
^{185m}Pt			33. m	β^+,EC/		1/2-	+0.54		
^{185}Pt		184.9607	1.18 h	β^+,EC/3.8		(9/2+)	-0.80	+4.5	ann.rad./
									0.1353
									0.1974
									0.2296
									0.2551
^{186}Pt		185.95943	2.0 h	β^+,EC/1.38		0+			ann.rad./
									0.6115
									0.6892
^{187}Pt		186.9607	2.35 h	β^+,EC/3.1		3/2	-0.41	-1.2	ann.rad./
									Ir k x-ray
									0.1064
									0.1100
									0.2015
									0.2849
									0.7092
^{188}Pt		187.95940	10.2 d	EC/0.51		0+			Ir k x-ray
									0.1876
									0.1951
^{189}Pt		188.96083	10.9 h	β^+,EC/1.97		3/2-	0.43	-1.1	Ir k x-ray
									0.0943
									0.6076
									0.7214
									(0.09 - 1.47)
^{190}Pt	0.01(1)	189.95993	4.5x10^{11} y			0+			
^{191}Pt		190.961684	2.96 d	EC/1.02		(3/2-)	0.50	-0.9	Ir k x-ray
									0.3599
									0.4094
									0.5389
^{192}Pt	0.79(6)	191.961035				0+			
^{193m}Pt			4.33 d	I.T./0.1498		13/2+	-0.75		Pt k x-ray

Elem. or Isot.	Natural Abundance (%)	Atomic Mass or Weight	Half-Life	Decay Mode/Energy (/MeV)	Particle Energy /Intensity (MeV/%)	Spin (h/2π)	Nuclear Magnetic Mom. (nm)	Elect. Quadr. Mom. (b)	γ-ray/Energy Intensity (MeV/%)
									0.1355
^{193}Pt		192.962984	60. y	EC/0.0566		(1/2-)	+0.60		Ir k x-rays
^{194}Pt	32.9(6)	193.962663				0+			
^{195m}Pt			4.02 d	I.T./0.2952		13/2+	-0.61	+1.4	Pt k x-ray
									0.0989
^{195}Pt	33.8(6)	194.964774				1/2-	+0.6095		
^{196}Pt	25.3(6)	195.964934				0+			
^{197m}Pt			1.590 h	I.T./97/		13/2+			Pt k x-ray
				β^-/3/					0.0530
									0.3465
^{197}Pt		196.967323	18.3 h	β^-/0.719		1/2-	0.51		Au k x-ray
									0.1914
									0.2688
^{198}Pt	7.2(2)	197.967875				0+			
^{199m}Pt			13.6 s	I.T./0.424		13/2+			Pt k x-ray
									0.3919
^{199}Pt		198.970576	30.8 m	β^-/1.70	0.90/18	(5/2-)			0.3170/4.9
					1.14/14				0.49375/5.7
									0.5430/14.8
									(0.055-1.293)
^{200}Pt		199.97142	12.5 h	β^-/≈0.66		0+			Au k x-ray
									0.13590
									0.22747
									0.24371
^{201}Pt		200.9745	2.5 m	β^-/2.66		(5/2-)			0.070
									0.152
									0.222
									1.760
^{202}Pt		201.9757	1.8 d						0.440
$_{79}$Au		196.96655(2)							
^{171}Au		170.9918	1.0 ms	p/46	1.44/100				
				α/54	7.00				
^{172}Au		171.9901	4 ms	α/7.02	6.86				
^{173m}Au			12 ms	α/92	6.732				
^{173}Au		172.9864	0.020 s	α/94	6.672				
^{174}Au		173.9842	0.12 s	α					
^{175}Au		174.9817	0.20 s	α					
^{176}Au		175.9803	1.2 s	β^+,EC/10.5					
				α/	6.260/80				
					6.290/20				
^{177}Au		176.9772	1.2 s	α/	6.115/				
					6.150/				
^{178}Au		177.9760	2.6 s	α/	5.920/				
^{179}Au		178.9732	7.5 s	α/	5.85/				
^{180}Au		179.9724	8.1 s	EC/8.6	5.65				0.1522
				α/	5.61				0.2564
					5.50				0.5242
									0.6765
									0.8084
									0.8597
^{181}Au		180.9700	11.4 s	EC/97.5/6.3	5.482/				
				α/2.7/					
^{182}Au		181.9686	21. s	β^+,EC/6.9					ann.rad./
				α/0.13/					0.1549
									0.2649
									(0.13 - 1.4)
^{183}Au		182.9676	42. s	EC/5.5			+1.97		0.1630
				α/0.8/					0.2730
									0.3625

TABLE OF ISOTOPES (CONTINUED)

Elem. or Isot.	Natural Abundance (%)	Atomic Mass or Weight	Half-Life	Decay Mode/Energy (/MeV)	Particle Energy /Intensity (MeV/%)	Spin (h/2π)	Nuclear Magnetic Mom. (nm)	Elect. Quadr. Mom. (b)	γ-ray/Energy Intensity (MeV/%)
^{184m}Au			48 s			(2+)	+1.44	+1.9	0.069(IT)
^{184}Au		183.9675	21. s	EC,β^+/7.1		(5+)	+2.07	+4.7	
				α/0.013/					
^{185m}Au			6.8 m	β^+,EC/					
				I.T./0.145					
^{185}Au		184.9657	4.3 m	β^+,EC/4.71		(5/2-)	+2.17	-1.1	ann.rad./
				α/0.26/					
^{186m}Au			< 2. m	β^+,EC/					0.1915
^{186}Au		185.9659	10.7 m	β^+,EC/6.0		3-	-1.26	+3.1	ann.rad./
				α/8(10)$^{-4}$/					0.1915
									0.2988
^{187m}Au			2.3 s	IT		9/2-			
^{187}Au		186.9646	8.3 m	β^+,EC/3.60		1/2+	+0.54		ann.rad./
									0.9152
									1.2668
									1.3321
									1.4081
^{188}Au		187.9651	8.8 m	β^+,EC/5.3		(1-)	-0.07		ann.rad./
									0.2660
									0.3404
									0.6061
^{189m}Au			4.6 m	β^+,EC/		11/2-	+6.19		0.1667
^{189}Au		188.9642	28.7 m	EC/96/3.2		1/2+	+0.49		ann.rad./
				β^+/4/					Pt k x-ray
									0.4478
									0.7133
									0.8128
^{190}Au		189.96470	43. m	β^+/2/4.44		1-	-0.07		ann.rad./
				EC/98/					Pt k x-ray
									0.2958
									0.3018
									0.5977
^{191m}Au			0.9 s	I.T./0.2663		11/2-	6.6		Au k x-ray
									0.2414
									0.2526
^{191}Au		190.96365	3.2 h	EC/1.83		3/2+	+0.137	+0.72	Pt k x-ray
									0.5864/16
									(0.088-1.30)
^{192}Au		191.96481	4.9 h	β^+/5/3.52	2.19/	1-	-0.011	-0.23	ann.rad./
				EC/95/	2.49/				Pt k x-ray
									0.2959
									0.3165
^{193m}Au			3.9 s	I.T./0.2901		11/2-	6.2	+1.98	Au k x-ray
									0.2580
^{193}Au		192.96413	17.6 h	EC/1.07		3/2+	+0.140	+0.66	Pt k x-ray
									0.1862
									0.2556
^{194}Au		193.96534	1.64 d	β^+/3/2.49	1.49/	1-	+0.076	-0.24	ann.rad./
				EC/97/					Pt k x-ray
									0.2935
									0.3284/61
^{195m}Au			30.5 s	I.T./0.3186		11/2-	6.2	+1.9	Au k x-ray
									0.2617
^{195}Au		194.965017	186.12 d	EC/0.227		3/2+	+0.149	+0.61	Pt k x-ray
^{196m2}Au			9.7 h	I.T./0.5954		12-	5.7		Au k x-ray
									0.1478
									0.1883
^{196m1}Au			8.1 s	I.T./0.0846		8+			0.0847
^{196}Au		195.966551	6.18 d	EC/92/1.506		2-	+0.591	0.81	Pt k x-ray

Elem. or Isot.	Natural Abundance (%)	Atomic Mass or Weight	Half-Life	Decay Mode/Energy (/MeV)	Particle Energy /Intensity (MeV/%)	Spin (h/2π)	Nuclear Magnetic Mom. (nm)	Elect. Quadr. Mom. (b)	γ-ray/Energy Intensity (MeV/%)
^{197m}Au			7.8 s	I.T./0.4094		11/2-	+6.0	+1.7	Au k x-ray
				β^-/8/0.686					0.1302
									0.2790
^{197}Au	100.	196.966551				3/2+	+0.14575	+0.55	
^{198m}Au			2.30 d	I.T./0.812		(12-)			Au k x-ray
									0.0972
									0.1803
									0.2419
^{198}Au		197.968225	2.694 d	β^-/1.372	0.290/1	2-	+0.5934	+0.64	Hg k x-ray
					0.961/99				0.411794
^{199}Au		198.968748	3.14 d	β^-/0.453	0.25/22	3/2+	+0.2715	+0.51	Hg k x-ray
					0.292/72				0.15837
					0.462/6				0.20820
^{200m}Au			18.7 h	β^-/84/1.0	0.56/	12-	5.9		Au k x-ray
				I.T./16/					0.2559/71
									0.3680/77
									0.4978/73
									0.5793/72
									0.084-0.904)
^{200}Au		199.97072	48.4 m	β^-/2.24	0.7/15	1-			0.3679/19
					2.2/77				1.2254/10.6
									(0.077-1.570)
^{201}Au		200.97165	26. m	β^-/1.28	1.27/82	3/2+			(0.027-0.732)
^{202}Au		201.9738	29. s	β^-/3.0		(1-)			0.4396
^{203}Au		202.97515	1.0 m	β^-/2.14	≈ 1.9/	3/2+			(0.04-0.37)
^{204}Au		203.9783	40. s	β^-/4.5		(2-)			0.4366
									1.5113
^{205}Au		204.9796	31. s	β^-/					(0.38 - 1.33)
$_{80}$Hg		200.59(2)							
^{172}Hg			≈0.25 ms	α	7.35				
^{173}Hg			0.9 ms	α	7.21				
^{174}Hg			1.9 ms	α	7.07				
^{175}Hg		174.9912	0.02 s	α					
^{176}Hg		175.98733	21 ms	α	6.74/94				
^{177}Hg		176.9863	0.13 s	α					
^{178}Hg		177.98248	0.26 s	EC/50/6.1		0+			
				α/50/	6.43/				
^{179}Hg		178.9818	1.09 s	EC/8.0					
				α/	6.29/				
^{180}Hg		179.9783	2.6 s	EC/5.5		0+			0.1250
				α/	6.12/33				0.3005
					5.69/.03				0.3812
^{181}Hg		180.9778	3.6 s	β^+ EC/74/≈7.3		(1/2-)	+0.507		0.0663
				α/26/					0.0811
									0.0924
									0.1474
									0.1587
									0.2142
									0.2398
^{182}Hg		181.9739	10.8 s	β^+,EC/85/5.0		0+			0.1200
				α/15/	5.87/8.6				0.2168
					5.45/0.03				0.4126
^{183}Hg		182.9744	9. s	β^+,EC/77/6.3		1/2-	+0.524		0.0714
				α/	5.83/				0.0874
					5.91/				0.1538
^{184}Hg		183.9719	30.9 s	β^+,EC/99/4.1		0+			0.0915
				α/1/	5.54/1.3				0.1265
					5.07/2 x 10^{-3}				0.1560
									0.2362

Elem. or Isot.	Natural Abundance (%)	Atomic Mass or Weight	Half-Life	Decay Mode/Energy (/MeV)	Particle Energy /Intensity (MeV/%)	Spin (h/2π)	Nuclear Magnetic Mom. (nm)	Elect. Quadr. Mom. (b)	γ-ray/Energy Intensity (MeV/%)
^{185m}Hg			21. s	β⁺,EC,IT,α/	5.37/	13/2+	-1.02	+0.2	0.211
									0.292
^{185}Hg		184.9720	51. s	β⁺,EC/95/5.8		1/2-	+0.509		(0.02 - 0.55)
^{186}Hg		185.9695	1.4 m	β⁺,EC/3.3		0+			0.1119
				α	5.09/0.02				0.2518
^{187m}Hg			1.7 m	β⁺,EC/		13/2+	-1.04	+0.5	See Hg[187]
^{187}Hg		186.9698	2.4 m	β⁺,EC/4.9		3/2-	-0.594	-0.8	0.1034/32.
									0.2334/100.
									0.2403/33.
									0.27151/31.
									0.3763/38.
									0.5254/30.
									(0.10-2.18)
^{188}Hg		187.9676	3.2 m	β⁺,EC/2.3		0+			0.0988
				α	4.61				0.1148
									0.1424
									0.1900
^{189m}Hg			8.6 m	EC/		13/2+	-1.06	+0.7	0.0780
									0.3210
									0.4345
									0.5655
									(0.08 - 2.170)
^{189}Hg		188.9687	7.6 m	EC/4.2		3/2-	-0.6086	-0.8	0.2005
									0.2038
									0.2386
									0.2485
^{190}Hg		189.9663	20.0 m	EC/1.5		0+			0.1296
									0.1426
^{191m}Hg			51. m	β⁺/6/		13/2+	-1.07	+0.6	ann.rad./
				EC/94/					Au k x-ray
									0.2741
									0.4203
									0.5787
									(0.07 - 1.9)
^{191}Hg		190.9671	50. m	β⁺,EC/3.2		(3/2-)	-0.62	-0.8	0.1963
									0.2247
									0.2524
^{192}Hg		191.9653	5.0 h	EC/≈0.5		0+			Au k x-ray
									0.1572
									0.2748
									0.3065
^{193m}Hg			11.8 h	β⁺,EC/91/		13/2+	-1.05843	+0.92	Hg k x-ray
				I.T./9/0.2901					0.1866
									0.2580
									0.4076
									0.5733
									0.9324
									(0.1 - 1.96)
^{193}Hg		192.96664	3.8 h	EC,B⁺/2.34		3/2-	-0.6276	-0.7	0.1866
									0.2580
									0.8611
^{194}Hg		193.96538	520. y	EC/0.04		0+			Au L x-rays
^{195m}Hg			1.67 d	I.T./(54)/0.3186		13/2+	-1.04465	+1.1	Hg k x-ray
				EC/(46)/					Au k x-ray
									0.2617
									0.5603
									0.7798
^{195}Hg		194.96664	9.5 h	EC/1.51		1/2-	+0.541475		Au k x-ray
									0.0614

Elem. or Isot.	Natural Abundance (%)	Atomic Mass or Weight	Half-Life	Decay Mode/Energy (/MeV)	Particle Energy /Intensity (MeV/%)	Spin (h/2π)	Nuclear Magnetic Mom. (nm)	Elect. Quadr. Mom. (b)	γ-ray/Energy Intensity (MeV/%)
									0.7798
^{196}Hg	0.15(1)	195.965814				0+			
^{197m}Hg			23.8 h	I.T./(93)/0.2989		13/2+	-1.02768	+1.2	Hg k x-ray
									Au k x-ray
									0.13398
^{197}Hg		196.967195	2.672 d	EC/0.600		1/2-	+0.527374		Au k x-ray
									0.07735
^{198}Hg	9.97(8)	197.966752				0+			
^{199m}Hg			42.6 m	I.T./0.532		13/2+	-1.014703	+1.2	Hg k x-ray
									0.15841
^{199}Hg	16.87(10)	198.968262				1/2-	+0.505885		
^{200}Hg	23.10(16)	199.968309				0+			
^{201}Hg	13.18(8)	200.970285				3/2-	-0.560226	+0.39	
^{202}Hg	29.86(20)	201.970625				0+			
^{203}Hg		202.972857	46.61 d	β^-/0.492	0.213/100	5/2-	+0.8489	+0.34	Tl k x-ray
									0.279188
^{204}Hg	6.87(4)	203.973475				0+			
^{205}Hg		204.976056	5.2 m	β^-/1.531	1.33/4	1/2-	+0.6010		0.20378
									(0.2 - 1.4)
^{206}Hg		205.97750	8.2 m	β^-/1.31	0.935/34	0+			Tl k x-ray
					1.3/63				0.3052
									0.6502
^{207}Hg		206.9825	2.9 m	β^-/4.8		(9/2+)			
^{208}Hg		207.9859	0.7 h	β^-					0.474
$_{81}$Tl		204.3833(2)							
^{177m}Tl			0.23 ms	p/51					
				α/49					
^{177}Tl		176,9969	0.017 s	α/73					
				P/27					
^{178}Tl		177.9952	≈0.2 s						
^{179m}Tl			1.7 ms	α	/7.21/80				
				α	/7.10/20				
^{179}Tl		178.9917	0.2 s	α					
^{180}Tl		179.9912	1.5 s	α	6.28/30				
					6.36/30				
					6.21/18				
					6.56/15				
					6.47/7				
^{181m}Tl			1.4 ms	α	6.58/100				
^{181}Tl		180.9869	3.2 ms	α	6.19/100				
^{182}Tl		181.9856	3. s	β^+, EC/10.9					0.351
									(0.26 - 0.41)
^{183m}Tl			0.06 s	α		9/2-			
^{183}Tl		182.9826	5. s	β^+, EC/7.7		1/2+			0.208
^{184}Tl		183.9818	11. s	β^+, EC/(98)/9.2					0.2868
				α/(2)/	6.16/				0.3399
									0.3667
^{185m}Tl			1.8 s	I.T./0.453		(9/2-)			0.1688
				α/5.97	6.01				0.2840
^{185}Tl		184.0701	20. s	EC/β^+/6.6					
^{186m}Tl			4. s	I.T./0.374					0.3738
^{186}Tl		185.9776	28. s	β^+,EC/7.5					0.3567
									0.4026
									0.4053
^{187m}Tl			15.6 s	I.T./≈ 0.33		(9/2+)	+3.8	-2.4	0.2995
^{187}Tl		186.9762	50. s	β^+,EC/6.0		1/2+	1.6		
^{188m}Tl			1.18 m	β^+,EC/		(7+)			Hg k x-ray
									0.4129
									0.5043

Elem. or Isot.	Natural Abundance (%)	Atomic Mass or Weight	Half-Life	Decay Mode/Energy (/MeV)	Particle Energy /Intensity (MeV/%)	Spin (h/2π)	Nuclear Magnetic Mom. (nm)	Elect. Quadr. Mom. (b)	γ-ray/Energy Intensity (MeV/%)
									0.5921
^{188}Tl		187.9759	1.2 m	β⁺,EC/7.8		(2-)	+0.48	+0.13	See Tl[188m]
									0.4129
^{189m}Tl			1.4 m	β⁺,EC/		(9/2-)	+3.878	-2.29	0.2156
									0.2284
									0.3175
									0.4452
^{189}Tl		188.9743	2.3 m	β⁺,EC/5.2		(1/2+)			0.3337
									0.4510
									0.5223
									0.9422
^{190m}Tl			3.7 m	β⁺,EC/	4.2/	(7+)	+0.495	+0.29	0.1968
									0.4164
									0.7311
^{190}Tl		189.9738	2.6 m	β⁺,EC/7.0	5.7/	(2-)	+0.25	-0.33	0.4164
									0.6254
									0.6838
									1.0999
^{191m}Tl			5.2 m	β⁺,EC/(98)/		(9/2+)	+3.903	-2.3	0.2157
									0.2647
									0.3256
									0.3359
^{191}Tl		190.9723				(1/2)	1.59		
^{192m}Tl			10.8 m	β⁺,EC/		(7+)	+0.518	0.46	0.1740
									0.4228
									0.6348
									0.7863
									0.7455
^{192}Tl		191.972	9.6 m	β⁺,EC/6.4		(2-)	+0.20	-0.33	0.3975
									0.4228
									0.6908
^{193m}Tl			2.1 m	I.T./(75)/		(9/2-)	+3.948	-2.2	0.3650
^{193}Tl		192.9706	22. m	β⁺,EC/3.6		(1/2+)	+1.591		0.2077
									0.3244
									0.3440
									0.6761
									1.0447
									1.5793
^{194m}Tl			32.8 m	β⁺/(20)/≈0.30		(7+)	+0.540	+0.61	ann.rad./
				EC/(80)/					Hg k x-ray
									0.4282
									0.6363
									0.7490
^{194}Tl		193.9711	34. m	β⁺,EC/5.3		2-	0.140	-0.28	0.3955
									0.4282
									0.6363
^{195m}Tl			3.6 s	I.T./0.483		9/2-			Tl k x-ray
									0.0990
									0.3836
^{195}Tl		194.9697	1.16 h	EC/97/2.8		1/2+	+1.58		ann.rad./
				β⁺/(3)/					Hg k x-ray
									0.2422
									0.5635
									0.8845
									1.3639
									(0.13 - 2.5)
^{196m}Tl			1.41 h	β⁺,EC/95/4.9		(7+)	0.55	+0.76	0.0840
									0.4261
									0.6353

Elem. or Isot.	Natural Abundance (%)	Atomic Mass or Weight	Half-Life	Decay Mode/Energy (/MeV)	Particle Energy /Intensity (MeV/%)	Spin (h/2π)	Nuclear Magnetic Mom. (nm)	Elect. Quadr. Mom. (b)	γ-ray/Energy Intensity (MeV/%)
									0.6954
									(0.08 - 1.0)
^{196}Tl		195.9705	1.84 h	β^+/(15)/4.4		2-	+0.072	-0.18	ann.rad./
				EC/(85)/					Hg k x-ray
									0.4257
									0.6105
									(0.03 - 2.4)
^{197m}Tl			0.54 s	IT/53/0.608		9/2-			Tl k x-ray
				β^+,EC/47/					0.2262
									0.4118
									0.5872
									0.6367
^{197}Tl		196.96954	2.83 h	β^+/(1)/2.18		1/2+	+1.58		Hg k x-ray
				EC/(99)/					0.1522/8.2
									0.4258
^{198m}Tl			1.87 h	β^+,EC/(53)/		7+	+0.64		Hg k x-ray
				IT/47/0.5347					Tl k x-ray
									0.4118
									0.5872
									0.6367
^{198}Tl		197.9405	5.3 h	EC,β^+/(1)/3.5	1.4/	2-			Hg k x-ray
					2.1/				0.4118
					2.4/				0.6367
									0.6759
									(0.23 - 2.8)
^{199}Tl		198.9698	7.4 h	EC/1.4		1/2-	+1.60		Hg k x-ray
									0.2082
									0.2473
									0.4555
^{200}Tl		199.97095	1.087 d	EC/2.46	1.07/	2-	0.04		Hg k x-ray
					1.44/				0.36799
									1.2057
									(0.11 - 2.3)
^{201}Tl		200.97080	3.040 d	EC/0.48		1/2+	+1.605		Hg k x-ray
									0.13528
									0.16740/10.0
^{202}Tl		201.97209	12.23 d	EC/1.36		2-	0.06		Hg k x-ray
									0.43957
^{203}Tl	29.524(14)	202.972329				1/2+	+1.622258		
^{204}Tl		203.973848	3.78 y	β^-/97/0.7637	0.763/97	2-	0.09		Hg k x-ray
				EC/(3)/0.347					
^{205}Tl	70.476(14)	204.974412				1/2+	+1.638215		
^{206m}Tl			3.76 m	I.T./2.644		12-			Tl k x-ray
									0.2166
									0.2661
									0.4534
									0.6866
									1.0219
^{206}Tl		205.976095	4.20 m	β^-/1.533	1.53/99.9	0-			Pb k x-ray
									0.80313
^{207m}Tl			1.3 s	I.T./1.350		11/2-			Tl k x-ray
									0.3501
									1.0000
^{207}Tl		206.97741	4.77 m	β^-/1.423	1.43/99.8	1/2+	+1.88		0.89723
^{208}Tl		207.982004	3.053 m	β^-/5.001	1.28/23	(5+)	+0.29		Pb k x-ray
					1.52/22				0.27728
					1.796/51				0.51061
									0.58302
									2.61448

Elem. or Isot.	Natural Abundance (%)	Atomic Mass or Weight	Half-Life	Decay Mode/Energy (/MeV)	Particle Energy /Intensity (MeV/%)	Spin (h/2π)	Nuclear Magnetic Mom. (nm)	Elect. Quadr. Mom. (b)	γ-ray/Energy Intensity (MeV/%)
^{209}Tl		208.98535	2.16 m	β^-/3.98	1.8 /100	(1/2+)			Pb k x-ray
									1.5670/100
									0.4651/95
									(0.12 - 1.33)
^{210}Tl		209.99006	1.30 m	β^-/5.48	1.3/25	(5+)			Pb k x-ray
					1.9/56				0.081
									0.2981
									0.79788
$_{82}$Pb		207.2(1)							
^{178}Pb			$\approx$0.2 ms						
^{180}Pb			5 ms	α/	7.25				
^{181}Pb		180.9967	0.05 s	α/	7.07				
^{182}Pb		181.99268	55 ms	α	6.90				
^{183}Pb		182.9919	0.3 s	α/		1/2+			
^{184}Pb		183.9882	0.48 s	α/	6.63/	0+			
^{185}Pb		184.9876	4.1 s	α/	6.34/				
					6.40/				
					6.48/				
^{186}Pb		185.9835	5. s	β^+,EC/95/5.5		0+			
				α/(5)/	6.32/				
					6.34/<100				
					6.01/<0.2				
^{187m}Pb			15.2 s	β^+,EC/	5.99/	(1/2-)			0.0674
				α/12	6.19/				0.2080
									0.2755
									0.2995
									0.4487
									0.7477
^{187}Pb		186.9839	18.3 s	EC/7.2		13/2+			0.1930
				α/7	6.08/				0.3314
									0.3435
									0.3934
^{188}Pb		187.9811	23. s	EC/(78)/4.8		0+			0.1850
				α/(22)/	5.98/<10				0.7582
					5.61/<0.1				
^{189}Pb		188.9809	51. s	EC/6.1					
				α/	5.58/				
^{190}Pb		189.9782	1.2 m	β^+ (13)/4.1		0+			ann.rad./
				EC/(86)/					Tl k x-ray
				α/(0.9)/	5.58/				0.1415
									0.1512
									0.9422
^{191m}Pb			2.2 m	β^+,EC/		13/2+	-1.17	+0.085	ann.rad./
									0.3871
									0.6135
									0.7122
^{191}Pb		190.9782	1.3 m	β^+,EC/5.5					ann.rad./
									0.9368
^{192}Pb		191.9758	3.5 m	β^+,EC/$\approx$3.4		0+			ann.rad./
				α/.006/	5.11				0.1675
									0.6082
									1.1954
^{193m}Pb			5.8 m	β^+,EC/		13/2+	-1.15	+0.19	ann.rad./
									0.3650
									0.3922
^{193}Pb		192.9761	$\approx$ 2. m	EC/5.2		3/2			
^{194}Pb		193.9740	11. m	β^+,EC/2.7		0+			ann.rad./
				α	4.64				0.2036
^{195m}Pb			15. m	β^+/(8)/		13/2+	-1.132	+0.30	ann.rad./

Elem. or Isot.	Natural Abundance (%)	Atomic Mass or Weight	Half-Life	Decay Mode/Energy (/MeV)	Particle Energy /Intensity (MeV/%)	Spin (h/2π)	Nuclear Magnetic Mom. (nm)	Elect. Quadr. Mom. (b)	γ-ray/Energy Intensity (MeV/%)
				EC/(92)/					Tl k x-ray
									0.3836
									0.3942
									0.8784
^{195}Pb		194.976	≈ 15. m	β$^+$,EC/5.8					ann.rad./
									0.3836
									0.3937
									0.7776
^{196}Pb		195.9727	37. m	β$^+$,EC/2.1		0+			Tl k x-ray
									0.2531
									0.5021
^{197m}Pb			43. m	EC/79/		13/2+	-1.104	+0.38	Tl k x-ray
				β$^+$/2/					0.3079
				IT/19/0.3193					0.3877
									0.7743
									(0.2 - 2.2)
^{197}Pb		196.9734	≈ 8. m	EC/97/3.6		(3/2-)	-1.075	-0.08	Tl k x-ray
				β$^+$/3/					0.3755
									0.3858
									0.7611
^{198}Pb		197.9720	2.4 h	EC/1.4		0+			Tl k x-ray
									0.1734
									0.2903
									0.3654
^{199m}Pb			12.2 m	IT/93/0.4248		13/2+			Pb k x-ray
				β$^+$,EC/(7)/					0.4255
^{199}Pb		198.9729	1.5 h	EC/(99)/2.9		5/2-	-1.074	+0.08	Tl k x-ray
				β$^+$/(1)/					0.3534
									0.7202
									1.1350
									(0.22 - 2.4)
^{200}Pb		199.97182	21.5 h	EC/0.81		0+			Tl k x-ray
									0.14763
^{201m}Pb			1.02 m	I.T./0.6291		13/2+			Pb k x-ray
									0.6288
^{201}Pb		200.97285	9.33 h	EC/1.90		5/2-	+0.675	-0.009	Tl k x-ray
									0.33120
									0.36131
									(0.11 - 1.8)
^{202m}Pb			3.53 h	IT/90/2.170		9-	-0.228	+0.58	Pb k x-ray
				β$^+$/10/					Tl k x-ray
									0.42219
									0.78700
									0.96271
^{202}Pb		201.97214	5.3x10^4 y	EC/0.05		0+			Tl L x-ray
^{203m}Pb			6.2 s	I.T./0.8252		13/2+			Pb k x-ray
									0.8203
									0.8252
^{203}Pb		202.97338	2.1615 d	EC/0.98		5/2-	+0.686	+0.10	Tl k x-ray
									0.279188
^{204m}Pb			1.12 h	I.T./2.185		9-			Pb k x-ray
									0.37481
									0.89922
									0.91175
^{204}Pb	1.4(1)	203.973028				0+			
^{205}Pb		204.974467	1.51x10^7 y	EC/0.0512		5/2-	+0.712	+0.23	Tl L x-ray
^{206}Pb	24.1(1)	205.974449				0+			
^{207m}Pb			0.80 s	I.T./1.632		13/2+			Pb k x-ray
									0.56915

Elem. or Isot.	Natural Abundance (%)	Atomic Mass or Weight	Half-Life	Decay Mode/Energy (/MeV)	Particle Energy /Intensity (MeV/%)	Spin (h/2π)	Nuclear Magnetic Mom. (nm)	Elect. Quadr. Mom. (b)	γ-ray/Energy Intensity (MeV/%)
									1.06310
^{207}Pb	22.1(1)	206.975880				1/2-	+0.59258		
^{208}Pb	52.4(1)	207.976636	>2x10^{19} y	SF		0+			
^{209}Pb		208.981075	3.25 h	β$^-$/0.644	0.645/100	9/2+	-1.474	-0.3	
^{210}Pb		209.984174	22.6 y	β$^-$/0.0635	0.017/81	0+			
					0.061/19				
				α	3.72				
^{211}Pb		210.988732	36.1 m	β$^-$/1.37	0.57/5	(9/2+)	-1.404	+0.09	0.40486
					1.36/92				0.42700
									0.83186
									(0.09 - 1.27)
^{212}Pb		211.991887	10.64 h	β$^-$/0.574	0.28/83	0+			Bi k x-ray
					0.57/12				0.23858
^{213}Pb		212.9966	10.2 m	β$^-$/2.1					
^{214}Pb		213.999797	26.9 m	β$^-$/1.0	0.67/48	0+			Bi k x-ray
					0.73/42				0.24192
									0.29509
									0.35187
^{215}Pb			36 s						
$_{83}$Bi		208.98038(2)							
^{185}Bi		184.9977	0.04 ms	p/86	1.59				
				α/14					
^{186}Bi		185.9965	10 ms	α	7.16				
					7.26				
^{187m}Bi			≈ 8. ms	α/12					
^{187}Bi		186.9935	32. ms	α/7	7.00/88.3				
					7.61/8.0				
					7.37/3.7				
^{188}Bi		187.9922		α					
^{189m}Bi			7.0 ms	α	7.30				
^{189}Bi		188.9895	0.68 s	α					
^{190}Bi		189.9875	5. s	β$^+$,EC/(10)/8.7					
				α/(90)/	6.45/				
^{191m}Bi			0.12 ms	α	6.87				
^{191}Bi		190.9861	12. s	β$^+$,EC/(60)/7.3					
				α/(40)/	6.32/				
^{192}Bi		191.9854	40. s	β$^+$,EC/(80)/9.0					
				α/(20)/	6.06/				
^{193m}Bi			3.2 s	β$^+$,EC/		1/2+			
				α/	6.48/				
^{193}Bi		192.9837	1.11 m	β$^+$,EC/40/7.1		9/2+			
				α/(60)/	5.91/				
^{194}Bi		193.9828	1.8 m	β$^+$,EC/99.9/8.2		(10-)			0.1661
				α/0.1/					0.1740
									0.2802
									0.421
									0.5754
									0.9650
^{195m}Bi			1.45 m	β$^+$,EC/(94)/					
				α/(6)/	6.11/				
^{195}Bi		194.9811	2.9 m	β$^+$,EC/99.8/5.8		3/2-			
				α/(0.2)	5.45/				
^{196}Bi		195.9806	5. m	EC/≈7.4					0.1376
									0.3720
									0.6880
									1.0486
^{197}Bi		196.9789	5. m	β$^+$,EC/5.2		1/2+			
^{198m}Bi			7.7 s	I.T./0.2485		(10-)			0.2485
^{198}Bi		197.9790	11.8 m	β$^+$,EC/6.6		(7+)			0.0900

Elem. or Isot.	Natural Abundance (%)	Atomic Mass or Weight	Half-Life	Decay Mode/Energy (/MeV)	Particle Energy /Intensity (MeV/%)	Spin (h/2π)	Nuclear Magnetic Mom. (nm)	Elect. Quadr. Mom. (b)	γ-ray/Energy Intensity (MeV/%)
									0.1976
									0.5624
									1.0635
^{199m}Bi			24.7 m	β$^+$,EC/					ann.rad./
^{199}Bi		198.9776	27. m	β$^+$,EC/4.3		9/2-	4.6		0.7203
									0.8374
									0.8417
									0.9460
									1.0528
									1.3056
									(0.12 - 3.2)
^{200m}Bi			31. m	β$^+$,EC/		(2+)			0.2453
									0.4198
									0.4624
									1.0265
^{200}Bi		199.9781	36. m	EC/(90)/5.9		7+			ann.rad./
				β$^+$/(10)/					Pb k x-ray
									0.4198
									0.4623
									1.0265
^{201m}Bi			59.1 m	I.T./0.846		(1/2+)			Bi k x-ray
				β$^+$,EC/					0.8464
^{201}Bi		200.97697	1.8 h	EC/3.84		9/2-	4.8		Pb k x-ray
									0.6288
									0.9357
									1.0138
									(0.13 - 2.4)
^{202}Bi		201.97768	1.72 h	β$^+$/(3)/5.16		5+	+4.26	-0.72	ann.rad./
				EC/(97)/					Pb k x-ray
									0.57860
									0.92734
									(0.08 - 3.5)
^{203}Bi		202.97687	11.8 h	EC/99.8/3.25		9/2-	+4.02	-0.69	Pb k x-ray
				β$^+$/(0.2)/	1.35/				0.1865
									0.8203
									0.8969
									1.8475
									(0.1 - 2.9)
^{204}Bi		203.97779	11.2 h	EC/4.44		6+	+4.32	-0.43	Pb k x-ray
									0.37481
									0.89922
									0.98409
^{205}Bi		204.97737	15.31 d	EC/2.71		9/2-	+4.07	-0.59	Pb k x-ray
									0.70347
									1.76435
^{206}Bi		205.97848	6.243 d	EC/3.76		6+	+4.36	-0.39	Pb k x-ray
									0.51619
									0.80313
									0.88100
^{207}Bi		206.978456	35. y	EC/2.399		9/2-	4.08	-0.6	Pb k x-ray
									0.56915
									1.06310
^{208}Bi		207.979727	3.68x10^5 y	EC/2.880		5+	4.63	-0.64	Pb k x-ray
									2.61435
^{209}Bi	100.	208.980384				9/2-	+4.111	-0.37	
^{210m}Bi			3.0x10^6 y	α/	4.420(3)/0.29	9-	+2.73	-0.47	Tl k x-ray
					4.569(3)/3.9				0.2661
					4.584(3)/1.4				0.3052
					4.908(4)/39				0.6502

Elem. or Isot.	Natural Abundance (%)	Atomic Mass or Weight	Half-Life	Decay Mode/Energy (/MeV)	Particle Energy /Intensity (MeV/%)	Spin (h/2π)	Nuclear Magnetic Mom. (nm)	Elect. Quadr. Mom. (b)	γ-ray/Energy Intensity (MeV/%)
					4.946(3)/55				
^{210}Bi		209.984105	5.01 d	β^-/1.163	1.16/99	1-	-0.0445	+0.136	0.2661
									0.3.52
^{211}Bi		210.98726	2.14 m	α/(99.7)/	6.279/16	9/2-			Tl k x-ray
				β^-/(0.3)/0.58	6.623/84				0.3501
^{212m2}Bi			7. m	β^-/		(15-)			
^{212m1}Bi			25.0 m	α/(93)/	6.300/40	(9-)			0.120
				β^-/(7)/	6.340/53				0.233
									0.275
									0.404
									0.727
^{212}Bi		211.991271	1.009 h	β^-/(64)/2.254		(1-)	+0.32	+0.1	Tl k x-ray
				α/(36)/	6.051/25				Po k x-ray
					6.090/9.6				0.2881
									0.72725
									0.78551
									1.62066
^{213}Bi		212.99437	45.6 m	β^-/(98)/1.43	1.02/31	9/2-	+3.72	-0.60	Po k x-ray
				α/(2)/	1.42/66				0.4404
					5.549/0.16				(0.15 - 1.328)
					5.869/2.0				
									1.10006
^{214}Bi		213.99870	19.7 m	β^-/3.27					0.60931
									1.12027
									1.76449
									(0.19 - 3.2)
^{215}Bi		215.0018	7.7 m	β^-/2.3					0.2937
									(0.27 - 0.835)
^{216}Bi		216.0062	2.3 m	β^-/4.0					0.5498
									0.4192
^{217}Bi			97 s	β/					
$_{84}$Po									
^{188}Po			0.4 ms	α	7.92				
					7.35				
^{189}Po			5 ms	α	7.54				
					7.25				
					7.32				
^{190}Po		189.9951	2.4 ms	α/	7.53				
^{191m}Po			0.10 s	α	7.38				
^{191}Po		190.9947	22 ms	α/	7.33				
^{192}Po		191.9915	34. ms	α/8.5	7.17				
^{193m}Po			0.24 s	α/	7.00				
^{193}Po		192.9911	0.45 s	α/	6.95				
^{194}Po		193.9883	0.39 s	α/	6.84/93	0+			
					6.19/0.22				
^{195m}Po			1.9 s	α/	6.70/				
^{195}Po		194.9881	4.6 s	α/	6.61/				
^{196}Po		195.9855	5.8 s	α/(95)/	6.52/94	0+			
				β^+,EC/(5)/≈4.6	5.77/0.02				
^{197m}Po			25.8 s	α/(84)/	6.385(3)/55	13/2+			
				β^+,EC/(16)/					
^{197}Po		196.9856	53. s	α/(44)/	6.282(4)/76	(3/2-)			
				β^+,EC/(56)/6.2					
^{198}Po		197.9834	1.76 m	α/(70)/	6.18/57	0+			
				β^+,EC/(30)/4.0	5.27/7.6x10^{-4}				
^{199m}Po			4.2 m	β^+,EC/(51)/		13/2+	0.99		ann.rad./
				α/(39)/	6.059/24				0.2745
									0.4998
									1.0020

TABLE OF ISOTOPES (CONTINUED)

Elem. or Isot.	Natural Abundance (%)	Atomic Mass or Weight	Half-Life	Decay Mode/Energy (/MeV)	Particle Energy /Intensity (MeV/%)	Spin (h/2π)	Nuclear Magnetic Mom. (nm)	Elect. Quadr. Mom. (b)	γ-ray/Energy Intensity (MeV/%)
^{199}Po		198.985	5.2 m	β$^+$,EC/(88)/7.		(3/2-)			Bi k x-ray
				α/(12)/	5.952/7.5				0.1877
									0.3616
									1.0214
									1.0344
^{200}Po		199.9817	11.5 m	β$^+$,EC/85/3.4		0+			0.14748
				α/(15)/	5.863/11.1				0.32792
									0.6176
									0.6709
^{201m}Po			8.9 m	β$^+$,EC/(57)/		13/2+	1.00		Bi k x-ray
				IT/40/0.418					Po k x-ray
				α/(3)/	5.786/≈3.				0.2726
									0.4123
									0.4179
									0.9670
^{201}Po		200.9822	15.3 m	β$^+$,EC/98/4.9		3/2-	0.94		Bi k x-ray
				α/(2)/	5.683(3)/1.1				0.2056
									0.2250
									0.8483
									0.9048
^{202}Po		201.9807	45. m	β$^+$,EC/98/2.8		0+			0.0410
				α/(2)/	5.588/1.9				0.1656
									0.3158
									0.6884
^{203m}Po			1.2 m	IT/96/0.6414		13/2+			Bi k x-ray
				β$^-$ EC/(4)/					Po k x-ray
									0.6414
^{203}Po		202.9814	35. m	β$^+$,EC/4.2		5/2-	+0.74		0.17516
									0.21477
									0.89350
									0.90863
									1.09095
^{204}Po		203.98031	3.53 h	EC/2.34		0+			Bi k x-ray
				α	5.377/0.66				0.2702
									0.8844
									1.0162
									(0.11 - 1.9)
^{205}Po		204.98117	1.7 h	β$^+$,EC/3.53		5/2-	+0.76	+0.17	Bi k x-ray
									0.83681
									0.84983
									0.87241
									1.00124
									(0.12 - 2.7)
^{206}Po		205.98047	8.8 d	EC/(95)/1.85		0+			Bi k x-ray
				α/(5)/	5.223/5.5				0.28644
									0.31156
									0.51134
									0.80737
									1.03228
									(0.11 - 1.5)
^{207m}Po			2.8 s	I.T./1.383		19/2-			Po k x-ray
									0.2682
									0.30074
									0.81448
^{207}Po		206.98158	5.80 h	EC,β$^+$/2.91		5/2-	+0.79	+0.28	Bi k x-ray
									0.74263
									0.91176
									0.99225
^{208}Po		207.981231	2.898 y	α/5.213	4.233/0.0002	0+			

TABLE OF ISOTOPES (CONTINUED)

Elem. or Isot.	Natural Abundance (%)	Atomic Mass or Weight	Half-Life	Decay Mode/Energy (/MeV)	Particle Energy /Intensity (MeV/%)	Spin (h/2π)	Nuclear Magnetic Mom. (nm)	Elect. Quadr. Mom. (b)	γ-ray/Energy Intensity (MeV/%)
					5.1158/100				
^{209}Po		208.982415	102. y	α/4.976	4.624/0.56	1/2-	≈+0.77		0.26049
					4.879/99.2				0.8964
^{210}Po		209.982857	138.4 d	α/5.407	4.516/0.001	0+			0.80313
					5.304/100				
^{211m}Po			25.2 s	α/	7.273/91	25/2+			Pb k x-ray
					7.994/1.7				0.32808
					8.316/0.25				0.56915
					8.875/7.0				0.89723
									1.06310
^{211}Po		210.986637	0.516 s	α/7.594	6.570/0.54	9/2+			0.56915
					6.892/0.55				0.89723
					7.450/98.9				
^{212m}Po			45. s	α/	8.514/2.0	16+			
					9.086/1.0				
					11.650/97				
^{212}Po		211.988852	0.298 μs	α/8.953	8.784/100	0+			
^{213}Po		212.992843	3.7 μs	α/8.537	7.614/0.003	9/2+			
					8.375/100				
^{214}Po		213.995186	163.7 μs	α/7.833	6.904/0.01	0+			0.7995
					7.686/99.99				0.298
^{215}Po		214.999415	1.780 ms	α/7.526	6.950/0.02	(9/2+)			
					6.957/0.03				
					7.386/100				
^{216}Po		216.001905	0.145 s	α/6.906	5.895/0.002	0+			
					6.778/99.99				
^{217}Po		217.0064	< 10. s	α/6.662	6.539/				
^{218}Po		218.008965	3.04 m	α/6.114	5.181/1.00	0+			
$_{85}$At									
^{193}At		192.9998	40 ms	α/					
^{194}At		193.9990	40 ms	α/					
^{195m}At			0.39 s	α	6.96				
^{195}At		194.9965	140 ms	α/	7.11				
^{196m}At			8 μs						0.158
^{196}At		195.9957	0.39 s	α/	7.05/				
^{197m}At			4. s	α		(1/2+)			
^{197}At		196.9939	0.35 s	β^+,EC/7.8		(9/2-)			
				α/	6.96/				
^{198m}At			1.5 s	β^+,EC/(75)/					
				α/(25)/	6.85/86				
^{198}At		197.9928	5. s	α/	6.75/94				
^{199}At		198.9910	7.1 s	β^+,EC/8/5.6		9/2-			
				α/(92)/	6.64/				
^{200m}At			4.3 s	β^+,EC/(80)		10-			
				α/(20)/	6.536/12				
^{200}At		199.990	43. s	β^+,EC/65/≈8.0		5+			
				α/(35)/	6.412/44				
					6.465/57				
^{201}At		200.9885	1.48 s	β^+,EC/29/5.9		9/2-			
				α/(71)/6.474	6.344/				
^{202m}At			≤ 1.5 s	I.T./0.391					
^{202}At		201.9885	3.02 m	β^+,EC/88/7.2		5+			ann.rad./
				α/(12)/	6.135/7.7				0.4413
					6.225/4.3				0.5697
									0.6753
^{203}At		202.9868	7.4 m	β^+,EC/69/5.1		9/2-			0.1458
				α/(31)/6.210	6.088/				0.2459
									0.6414
									1.0020

Elem. or Isot.	Natural Abundance (%)	Atomic Mass or Weight	Half-Life	Decay Mode/Energy (/MeV)	Particle Energy /Intensity (MeV/%)	Spin (h/2π)	Nuclear Magnetic Mom. (nm)	Elect. Quadr. Mom. (b)	γ-ray/Energy Intensity (MeV/%)
									1.0340
^{204}At		203.9873	9.1 m	β^+,EC/95/6.5		(5+)			Po k x-ray
				α/(5)/	5.951/				0.3271
									0.4254
									0.5156
									0.6837
^{205}At		204.98604	26. m	β^+,EC/90/4.54		(9/2-)			Po k x-ray
				α/(10)/6.020	5.902/				0.1543
									0.6696
									0.7194
^{206}At		205.98660	29.4 m	β^+,EC/99/5.72		5+			Po k x-ray
				α/(1)/5.881	5.703/				0.20186
									0.39561
									0.47716
									0.70071
^{207}At		206.98578	1.81 h	β^+,EC/90/3.91		9/2-			Po k x-ray
				α/(10)/5.873	5.758/				0.16801
									0.58842
									0.81448
^{208}At		207.98657	1.63 h	β^+,EC/99/4.97		(6+)			Po k x-ray
				α/(1)/5.752	5.626/0.01				0.1770
					5.641/0.53				0.2060
									0.6601
									0.6852
									0.8450
									1.0281
^{209}At		208.98616	5.4 h	β^+,EC/96/3.49		(6+)			Po k x-ray
				α/(4)/5.757	5.647/4.1				0.10422
									0.54503
									0.78189
									0.79020
									(0.1 - 2.6)
^{210}At		209.98713	8.1 h	EC/99.8/3.98		5+			Po k x-ray
				α/(0.2)/5.632	5.361/0.05				0.24535
					5.442/0.05				0.52758
									1.18143
									1.43678
									1.48335
									(0.04 - 2.4)
^{211}At		210.987481	7.21 h	EC/(58)/0.787		9/2-			Po k x-ray
				α/(42)/5.980	5.211/0.004				0.66956
					5.868/42				0.6870
									0.74263
^{212m}At			0.119 s	α/	7.837/65	(9-)			
					7.897/33				
^{212}At		211.990735	0.314 s	α/7.828	7.058/0.4	(1-)			
					7.088/0.6				
					7.618/15				
					7.681/84				
^{213}At		212.992922	0.11 μs	α/9.254	9.080/	9/2-			
^{214m}At			0.76 μs	α/8.762		(9-)			
^{214}At		213.996357	0.56 μs	α/8.987	8.819/100	(1-)			
^{215}At		214.99864	0.10 ms	α/8.178	7.626/0.045	(9/2-)			0.40486
					8.023/99.9				
^{216}At		216.002408	0.30 ms	α/7.947	7.595/0.2	(1-)			
					7.697/2.1				
					7.800/97				
^{217}At		217.00471	32. ms	α/7.202	6.812/0.06	(9/2-)			0.2595
					7.067/99.9				0.3345

Elem. or Isot.	Natural Abundance (%)	Atomic Mass or Weight	Half-Life	Decay Mode/Energy (/MeV)	Particle Energy /Intensity (MeV/%)	Spin (h/2π)	Nuclear Magnetic Mom. (nm)	Elect. Quadr. Mom. (b)	γ-ray/Energy Intensity (MeV/%)
									0.5940
^{218}At		218.00868	1.6 s	α/6.883	6.654/6				
					6.695/90				
					6.748/4				
^{219}At		219.0113	50. s	α/6.390	6.275/				
^{220}At		220.0153	3.71 m	β⁻/3.7					(0.24-0.70)
^{221}At		221.0181	2.3 m	β					
^{222}At		222.0223	0.9 m	β					
^{223}At		223.0253	50. s	β					
$_{86}$Rn									
^{196}Rn		195.9977	≈ 3 ms	α/	7.49				
^{197m}Rn			0.02 s	α	7.36				
^{197}Rn		196.9983	0.07 s	α/	7.26				
^{198}Rn		197.9988	0.05 s	α					
^{199m}Rn			0.3 s	α		(13/2+)			
^{199}Rn		198.9983	0.62 s	α/		3/2-			
^{200}Rn		199.9957	1.06 s	α/(98)/	6.901/	0+			0.4329
				EC/(2)/5.					0.5043
^{201m}Rn			3.8 s	EC/(10)/		13/2+			
				α/(90)/	6.773/				
^{201}Rn		200.9955	7.0 s	α/(80)/	6.725/	(3/2-)			
				EC/(20)/	α/6.778				
^{202}Rn		201.9932	9.9 s	α/(12)/	6.641/	0+			0.5695
				EC/(88)/					0.2876-0.6255
^{203m}Rn			28. s	α/	6.551	13/2+	-0.96	+1.3	
^{203}Rn		202.9948	45. s	α/(66)/6.629	6.499/	0			
				EC/(34)/≈7.4					
^{204}Rn		203.9914	1.24 m	α/(68)/	6.420/	0+			
				EC/(32)/3.8					
^{205}Rn		204.9917	2.8 m	α/(23)/6.390	6.123(3)/0.02	(5/2-)	+0.80	+0.06	0.2652
				EC/(77)/5.2	6.262(3)/23				0.3553
									0.4648
									0.6205
									0.6753
									0.7300
^{206}Rn		205.9902	5.7 m	α/(68)/6.384	6.258(3)/	0+			0.06170
				EC/(32)/3.3					0.0968
									0.3245
									0.3862
									0.4822
									0.4973
									0.7728
^{207}Rn		206.9907	9.3 m	β⁺,EC/77/4.6		5/2-	+0.82	+0.22	At k x-ray
				α/(23)/6.252	5.995(4)/0.02				0.32947
					6.068(3)/0.15				0.34455
					6.126(3)/22.8				0.36767
									0.40267
									0.74723
									(0.18 - 1.4)
^{208}Rn		207.98963	24.3 m	α/(60)/6.260	5.469(2)/0.003	0+			
				EC/(40)/2.85	6.140(2)/60				
^{209}Rn		208.99038	29. m	β⁺/(83)/3.93	2.16/2.3	5/2-	+0.8388	+0.31	At k x-ray
				α/(17)/	5.887(3)/0.04				0.27933
					5.898(3)/0.02				0.33753
					6.039(2)/16.9				0.40841
									0.68942
									0.74594
									(0.18 - 3.2)
^{210}Rn		209.98968	2.4 h	α/(96)/6.157	5.351(2)/0.005	0+			At k x-ray

Elem. or Isot.	Natural Abundance (%)	Atomic Mass or Weight	Half-Life	Decay Mode/Energy (/MeV)	Particle Energy /Intensity (MeV/%)	Spin (h/2π)	Nuclear Magnetic Mom. (nm)	Elect. Quadr. Mom. (b)	γ-ray/Energy Intensity (MeV/%)
				EC/(4)/2.37	6.039(2)/96				0.19625
									0.45824
									0.57104
									0.64868
									(0.14 - 1.7)
^{211}Rn		210.99059	14.6 h	β$^+$,EC/74/2.89		1/2-	+0.60		At k x-ray
				α/(26)/5.964	5.619(1)/0.7				0.16877
					5.784(1)/16.4				0.25022
					5.851(1)/8.8				0.37049
									0.67412
									0.67839
									1.36298
									(0.11 - 2.7)
^{212}Rn		211.990689	24. m	α/6.385	5.587(4)/0.05	0+			
					6.260(4)/99.95				
^{213}Rn		212.99387	20 ms	α/8.243	7.552(8)/2	9/2+			0.540
					8.087(8)/98				
^{214}Rn		213.99535	0.27 μs	α/9.209	9.037(9)/	0+			
^{215}Rn		214.99873	2.3 μs	α/8.840	8.674(8)/	(9/2+)			
^{216}Rn		216.00026	45. μs	α					
^{217}Rn		217.003915	0.6 ms	α/7.885	7.500/0.1	9/2+			
					7.742(4)/100				
^{218}Rn		218.005586	35. ms	α/7.267	6.534(1)/0.16	0+			0.6093
					7.133(1)/99.8				0.6653
^{219}Rn		219.009475	3.96 s	α/6.946(1)	6.3130(5)/0.05	(5/2+)	-0.44	+0.93	Po k x-ray
					6.425(3)/7.5				0.13057
					6.5309(4)/0.12				0.27113
					6.5531(3)/12.2				0.40170
					6.8193(3)/81				(0.1 - 1.05)
^{220}Rn		220.011384	55.6 s	α/6.404	5.7486(5)/0.07	0+			
					6.2883(1)/99.9				
^{221}Rn		221.0156	25. m	α/(22)/6.148	5.778(3)/1.8	7/2+	-0.020	-0.38	Fr L x-ray
				β$^-$/(78)/1.2	5.788(3)/2.2				0.07384
					6.037(3)/18				0.08323
									0.0610
									0.18639
^{222}Rn		222.017570	3.823 d	α/5.590	4.987(1)/0.08	0+			0.510
					5.4897(3)/99.9				
^{223}Rn		223.0218	23. m	β$^-$/			-0.78	+0.80	
^{224}Rn		224.0241	1.8 h	β$^-$/		0+			0.1085
									0.2601
									0.2655
^{225}Rn		225.0284	4.5 m	β$^-$/		7/2	-0.70	+0.84	
^{226}Rn		226.0309	7.4 m	β$^-$/					
^{227}Rn		227.0354	2. s	β$^-$/					
^{228}Rn		228.0381	65. s	β$^-$/					
$_{87}$Fr									
^{199}Fr			12 s	α	7.66				
^{200}Fr		200.0065	≈ 20 ms	α	7.47				
^{201}Fr		201.0040	0.05 s	α/	7.36/	(9/2-)			
^{202}Fr		202.0033	0.34 s	α/7.590	7.237(8)/100				
^{203}Fr		203.0014	0.55 s	α/7.280	7.132(5)/	(9/2-)			
^{204}Fr		204.001	2.1 s	α/	7.03/96				
					6.97/90				
					7.01/74				
^{205}Fr		204.9987	3.9 s	α/7.050	6.914(5)/	(9/2-)			
^{206m}Fr			0.7 s	α/	6.93				0.531(IT)
^{206}Fr		205.9985	16.0 s	α/7.416	6.792(5)/84				
^{207}Fr		206.9969	14.8 s	α/6.900	6.766(5)/	9/2-	+3.9	-0.16	

TABLE OF ISOTOPES (CONTINUED)

Elem. or Isot.	Natural Abundance (%)	Atomic Mass or Weight	Half-Life	Decay Mode/Energy (/MeV)	Particle Energy /Intensity (MeV/%)	Spin (h/2π)	Nuclear Magnetic Mom. (nm)	Elect. Quadr. Mom. (b)	γ-ray/Energy Intensity (MeV/%)
^{208}Fr		207.99713	59.1 s	α/(77)/6.770	6.636(5)/	7+	-4.8	+0.004	
				EC/(23)/6.99					
^{209}Fr		208.99592	50.0 s	α/(89)/5.1	6.646(3)/	9/2-	+3.9	-0.24	0.7978
				EC/(11)/5.16					(0.1103-1.384)
^{210}Fr		209.99640	3.2 m	α/6.670	6.543(5)/	6+	+4.4	+0.19	0.2030
				EC/6.26					0.6438
									0.8175
									0.9008
^{211}Fr		210.99553	3.10 m	α/6.660	6.534(5)/	9/2-	+4.0	-0.19	0.220
				EC/4.61					0.2799
									0.5389
									0.9169
^{212}Fr		211.99618	20. m	EC/(57)/5.12	6.261(1)/16	(5+)	+4.6	-0.10	Rn x-ray
				α/(43)/6.529	6.335(1)/4				0.08107
					6.335(1)/4				0.08378
					6.343(1)/1.3				0.2277
					6.383(1)/10				1.1856
					6.406(1)/9.5				1.2748
					6.08-6.18				0.014-1.178
^{213}Fr		212.99617	34.6 s	α/6.905	8.476(4)/51	9/2-	+4.0	-0.14	
^{214m}Fr			3.4 ms	α/	8.547(4)/46	9-			
					6.775-8.046				
^{214}Fr		213.99895	5.1 ms	α/8.587	7.409(3)/0.3	(1-)			
					7.605(8)/1.0				
					7.940(3)/1.0				
					8.355(3)/4.7				
					8.427(3)/93				
^{215}Fr		215.00033	0.12 μs	α/9.537	9.360(8)/	(9/2-)			
^{216}Fr		216.00319	0.70 μs	α/9.175	9.005(10)/95				(0.045-0.160)
^{217}Fr		217.00462	0.016 ms	α/8.471	8.315(8)/	(9/2-)			
^{218m}Fr			22. ms	α					
^{218}Fr		218.00756	1. ms	α/8.014	7.384(10)/0.5	(1-)			
					7.542(15)/1.0				
					7.572(10)/5				
					7.732(10)/0.5				
					7.867(2)/93				
^{219}Fr		219.00924	21. ms	α/8.132	6.802(2)/0.25	(9/2-)			
					6.967(2)/0.6				
					7.146(2)/0.25				
					7.313(2)/99				
^{220}Fr		220.012313	27.4 s	α/6.800	6.582(1)/10	1+	-0.67	+0.47	0.0450
					6.630(2)/6				0.061
					6.641(1)/12				0.1060
					6.686(1)/61				0.1539
					6.39-6.58				0.1617
^{221}Fr		221.01425	4.8 m	α/6.457	5.9393(7)/0.17	(5/2-)	+1.58	-1.0	At k x-ray
					5.9797(7)/0.49				0.0995
					6.0751(7)/0.15				0.21798
					6.1270(7)/				0.4091
					6.2433(3)/1.3				
					6.3410(7)/83.4				
^{222}Fr		222.01754	14.3 m	β^-/2.03	1.78/	2-	+0.63	+0.51	
				α/5.850					
^{223}Fr		223.019731	22.0 m	β^-/1.149	1.17/65	(3/2+)	+1.17	+1.17	0.05014
									0.07972
									(0.13 - 0.9)
^{224}Fr		224.02323	3.0 m	β^-/2.82		1-	+0.40	+0.517	0.13150
									0.21575
									0.8367

Elem. or Isot.	Natural Abundance (%)	Atomic Mass or Weight	Half-Life	Decay Mode/Energy (/MeV)	Particle Energy /Intensity (MeV/%)	Spin (h/2π)	Nuclear Magnetic Mom. (nm)	Elect. Quadr. Mom. (b)	γ-ray/Energy Intensity (MeV/%)
									(0.1 - 2.21)
^{225}Fr		225.02561	3.9 m	β⁻/1.87		3/2	+1.07	+1.3	
^{226}Fr		226.0293	49. s	β⁻/3.6		1	+0.071	-1.35	0.18606
									0.25373
^{227}Fr		227.0318	2.48 m	β⁻/2.5		1/2	+1.50		
^{228}Fr		228.0357	39. s	β⁻/≈3.5		2-	-0.76	+2.4	
^{229}Fr		229.0384	50. s	β⁻/					
^{230}Fr		230.0425	19. s	β⁻/		(3)			
^{231}Fr		231.0454	17. s	β⁻/					
^{232}Fr		232.0500	5. s	β⁻/					
$_{88}$Ra									
^{202}Ra			≈ 3 ms	α	7.86				
^{203m}Ra			0.03 s	α	7.62				
^{203}Ra		203.0092	≈ 4 ms	α	7.58				
^{204}Ra		204.0065	0.06 s	α	7.48				
^{205m}Ra			≈ 0.17 s						
^{205}Ra		205.0062	0.22 s	α	7.34				
^{206}Ra		206.0038	0.4 s	α/7.416	7.272(5)/	0+			
^{207}Ra		207.0037	1.3 s	α/7.270	7.133(5)/				
^{208}Ra		208.0018	1.4 s	α/7.273	7.133(5)/	0+			
^{209}Ra		209.0019	4.6 s	α/7.150	7.008(5)/	5/2	+0.87	+0.40	
^{210}Ra		210.0005	3.7 s	α/7.610	7.020(5)/	0+			
^{211}Ra		211.0009	13. s	α/7.046	6.912(5)/	(5/2-)	+0.878	+0.48	
				EC/5.0					
^{212}Ra		211.99978	13.0 s	α/7.033	6.901(2)/	0+			
^{213m}Ra			2.1 ms	IT					
^{213}Ra		213.00034	2.7 m	EC/(20)/3.88		(1/2-)	+0.613		0.1024
				α/(80)/6.860	6.521(3)/4.8				0.11010
					6.622(3)/39				0.2125
					6.730(3)/36				
^{214}Ra		214.00009	2.46 s	α/7.272	7.14/99.8/	0+			0.642
					6.51/0.2				
^{215}Ra		215.00270	1.7 ms	α/8.864	7.883(6)/2.8	(9/2+)			0.773/100
					8.171(3)/1.4				0.852/74
					8.700(3)/95.9				0.055-1.048
^{216}Ra		216.00352	0.18 μs	α/9.526	9.349(8)/	0+			
^{217}Ra		217.00631	1.6 μs	α/9.161	8.992(8)/	9/2-			
^{218}Ra		218.00712	26. μs	α/8.547	8.390(8)/	0+			
^{219}Ra		219.01006	0.010 s	α/8.132	7.680(10)/65				
					7.982(9)/35				
^{220}Ra		220.01101	18. ms	α/7.593	7.39/5	0+			0.465
					7.45/95				
^{221}Ra		221.01391	29. s	α/6.879	6.254(10)/0.7	5/2	-0.180	+1.9	
					6.578(5)/3				
					6.585(3)/8				
					6.608(3)/35				
					6.669(3)/21				
					6.758(3)/31				
^{222}Ra		222.015361	36.2 s	α/5.590	6.237(2)/3.0	0+			0.324
					6.556(2)/97				0.1448-0.8402
^{223}Ra		223.018497	11.43 d	α/5.979	5.287(1)/0.15	(3/2+)	+0.271	+1.25	Rn k x-ray
					5.338(1)/0.13				0.12231
					5.365(1)/0.13				0.14418
					5.433(5)/2.3				0.15418
					5.502(1)/1.0				0.15859
					5.540(1)/9.2				0.26939
					5.607(3)/24				0.32388
					5.716(3)/52				0.33328
					5.747(1)/9				0.44494

Elem. or Isot.	Natural Abundance (%)	Atomic Mass or Weight	Half-Life	Decay Mode/Energy (/MeV)	Particle Energy /Intensity (MeV/%)	Spin (h/2π)	Nuclear Magnetic Mom. (nm)	Elect. Quadr. Mom. (b)	γ-ray/Energy Intensity (MeV/%)
					5.857(1)/0.32				(0.10 - 0.7)
					5.872(1)/0.85				
^{224}Ra		224.020202	3.66 d	α/5.789	5.034(10)/0.003	0+			Rn k x-ray
					5.047(1)/0.007				0.2407
					5.164(5)/0.007				0.4093
					5.449(2)/4.9				0.6501
					5.685(2)/95				
^{225}Ra		225.023603	14.9 d	β^-/0.36	0.32/100	(3/2+)	-0.734		Ac k x-ray
				α	5.01/2×10^{-5}				0.0434
					4.98×10^{-6}				
^{226}Ra		226.025402	1599. y	α/4.870	4.194(1)/0.001	0+			Rn k x-ray
			$>4\times10^{18}$ y	SI/4×10^{-14}	4.343(1)/0.006				0.1861
					4.601(1)/5.5				0.2624
					4.784(1)/94				
^{227}Ra		227.029170	42. m	β^-/1.325	1.03/	(3/2+)	-0.404	+1.5	Ac L x-ray
					1.30/				Ac k x-ray
									0.02739
^{228}Ra		228.031063	5.76 y	β^-/0.046	0.039/50	0+			0.0135
					0.014/30				(0.006-0.0306)
					0.026/20				
^{229}Ra		229.0348	4.0 m	β^-/1.76	1.76/	(3/2+)	+0.503	+3.1	0.0145-0.1715
^{230}Ra		230.03708	1.5 h	β^-/1.0	0.7/	0+			0.0631
									0.0720
									0.2028
									0.4698
									0.4787
^{231}Ra		231.0412	1.7 m	β^-					
^{232}Ra		232.0437	4. m	β^-					
^{233}Ra		233.0480	30. s	β^-					
^{234}Ra		234.051	≈ 30. s	β^-/					
$_{89}$Ac									
^{206m}Ac			0.04 s	α	7.79				
^{206}Ac			≈ 26 ms	α	7.75				
^{207}Ac		207.0121	27 ms	α/	7.69				
^{208m}Ac			≈25. ms	α/	7.72				
^{208}Ac		208.0115	≈0.1 s	α/	7.62				
^{209}Ac		209.0096	≈0.10 s	α/	7.58				
^{210}Ac		210.0093	0.34 s	α/7.610	7.462(8)/				
^{211}Ac		211.0076	0.20 s	α/7.620	7.480(8)/				
^{212}Ac		212.0078	0.9 s	α/7.520	7.379(8)/				
^{213}Ac		213.0066	0.73 s	α/7.500	7.364(8)/	(9/2-)			
^{214}Ac		214.0069	8.2 s	α/(86)/7.350	7.007(8)/3	(5+)			
				EC/(14)/6.34	7.082(5)/38				
					7.214(5)/45				
^{215}Ac		215.0065	0.17 s	α/7.750	7.60/99.2	(9/2-)			0.399
					7.21/0.46				0.582
					7.03/0.20				0.654
					6.96/0.14				
^{216m}Ac			0.44 ms	α/	8.198(8)/1.7	(9-)			
					8.283(8)/2.5				
					9.028(5)/49				
					9.106(5)/46				
^{216}Ac		216.00871	≈ 0.3 ms	α/9.241	8.990(2)/10	(1)			
					9.070(8)/90				
^{217m}Ac			0.7 μs	α/	10.540/100				
^{217}Ac		217.00933	0.07 μs	α/9.832	9.650(10)/100	9/2-			
^{218}Ac		218.01162	1.1 μs	α/9.380	9.205(15)/				
^{219}Ac		219.01241	0.012 ms	α/8.830	8.664(10)/	(9/2-)			
^{220}Ac		220.0148	26. ms	α/8.350	7.610(20)/23				

Elem. or Isot.	Natural Abundance (%)	Atomic Mass or Weight	Half-Life	Decay Mode/Energy (/MeV)	Particle Energy /Intensity (MeV/%)	Spin (h/2π)	Nuclear Magnetic Mom. (nm)	Elect. Quadr. Mom. (b)	γ-ray/Energy Intensity (MeV/%)
					4.680(20)/21				
					7.790(10)/13				
					7.850(10)/24				
					7.985(10)/4				
					8.005(10)/5				
					8.060(10)/6				
					8.195(10)/3				
^{221}Ac		221.01558	52. ms	α/7.790	7.170(10)/2				
					7.375(10)/10				
					7.440(15)/20				
					7.645(10)/70				
^{222m}Ac			63. s	α/(>89)/	6.710(20)/7				
				EC/(1)/	6.750(20)/13				
				I.T./(<10)/	6.810(20)/24				
					6.840(20)/9				
					6.890(20)/13				
					6.970(20)/7				
					7.000(20)/13				
^{222}Ac		222.01782	5. s	α/7.141	6.967(10)/6	1-			
					7.013(2)/94				
^{223}Ac		223.01913	2.1 m	α/(99)/6.783	6.131(2)/0.12	(5/2-)			0.0725
				EC/(1)/0.59	6.177(2)/0.94				0.0839
					6.293(1)/0.47				0.0927
					6.326(1)/0.3				0.0990
					6.332(2)/0.14				0.1917
					6.360(1)/0.22				0.2158
					6.397(1)/0.13				0.3588
					6.448(1)/0.2				0.4768
					6.473(1)/3.1				
					6.523(2)/0.6				
					6.528(1)/3.1				
					6.563(1)/13.6				
					6.582(3)/0.3				
					6.646(1)/44				
					6.661(1)/31				
^{224}Ac		224.021708	2.7 h	EC/(90)/1.403	5.841(1)/0.5	0-			Ra L kx-ray
				α/(10)/6.323	5.860(1)/0.75				Ra k x-ray
					5.875(1)/1.7				0.08426
					5.941(1)/4.4				0.13150
					6.000(1)/6.7				0.1571
					6.013(1)/1.4				0.21575
					6.056(1)/22				0.2619
					6.138(1)/26				(0.03 - 0.3)
					6.154(1)/1.0				
					6.204(1)/12				
					6.210(1)/20				
^{225}Ac		225.02322	10.0 d	α/5.935	5.286(1)/0.2	3/2			Fr k x-ray
					5.444(3)/0.1				0.9958
					5.554(1)/0.1				0.9982
					5.608(1)/1.1				0.1084
					5.636(1)/4.5				0.1116
					5.681(1)/1.4				0.1451
					5.722(1)/2.9				0.1539
					5.731(1)/10				0.15724
					5.791(1)/9				0.18799
					5.793(1)/18				0.19575
									0.2162
									0.21686
									(0.025 - 0.52)

Elem. or Isot.	Natural Abundance (%)	Atomic Mass or Weight	Half-Life	Decay Mode/Energy (/MeV)	Particle Energy /Intensity (MeV/%)	Spin (h/2π)	Nuclear Magnetic Mom. (nm)	Elect. Quadr. Mom. (b)	γ-ray/Energy Intensity (MeV/%)
^{226}Ac		226.026089	1.224 d	EC/(17)/0.640		(1-)			Ra k x-ray
				β^-/(83)/1.116					Th k x-ray
				α/(0.006)/5.51	5.399(5)/0.006				0.07218
									0.15816
									0.23034
^{227}Ac		227.027747	21.77 y	β^-/98.6/0.045	0.045/54	(3/2-)	+1.1	+1.7	0.0838/23.
				α/(1.4)/5.043	4.869(1)/0.09				0.0811/14.
					4.938(1)/0.52				0.2696/13.
					4.951(1)/0.65				(0.044 - 1.27)
^{228}Ac		228.031014	6.15 h	β^-/2.127	1.11/32	(3+)			Th L x-ray
					1.85/12				Th k x-ray
					2.18/11				0.12903
									0.33842
									0.91116
									0.96897
									(0.2 - 1.96)
^{229}Ac		229.03293	1.04 h	β^-/1.10	1.1/	(3/2+)			0.07450
									0.16451
									0.26188
									0.5085
									0.56916
^{230}Ac		230.0360	2.03 m	β^-/2.7	1.4/	1+			Th k x-ray
									0.45497
									0.50820
									(0.12 - 2.5)
^{231}Ac		231.0386	7.5 m	β^-/2.1	2.1/100	(1/2+)			0.14379
									0.18574
									0.22140
									0.28250
									0.3070
^{232}Ac		232.0420	2.0 m	β^-/3.7		(2-)			
^{233}Ac		233.0446	2.4 m	β^-/		(1/2+)			
^{234}Ac		234.0484	40. s	β^-/		(1+)			
$_{90}$Th		232.0381(1)							
^{209}Th			≈ 0.01 s	α	8.08				
^{210}Th		210.0150	≈ 12 ms	α	7.90				
^{211}Th		211.0149	0.04 s	α	7.79				
^{212}Th		212.0129	≈ 30. ms	α/	7.80/	0+			
^{213}Th		213.0130	0.14 s	α/7.840	7.692(10)/				
^{214}Th		214.0115	0.09 s	α/7.825	7.677(10)/	0+			
^{215}Th		215.0117	1.2 s	α/7.660	7.33(10)/8	(1/2-)			0.134
					7.395(8)/52				0.192
					7.524(8)/40				
^{216m}Th			0.14 ms	α	9.93				
^{216}Th		216.01105	28. ms	α/8.071	7.92/99.46	0+			0.628
					7.30/0.54				
^{217}Th		217.01306	0.25 ms	α/9.424	9.27/94.6				
					8.46/3.8				
					8.73/1.6				
^{218}Th		218.01327	0.11 μs	α/9.847	9.665(10)/	0+			
^{219}Th		219.01552	1.05 μs	α/9.510	9.340(20)/				
^{220}Th		220.01573	10. μs	α/8.953	8.790(20)/	0+			
^{221}Th		221.01817	1.7 ms	α/8.628	7.743(8)/6				
					8.146(5)/56				
					8.4272(5)/39				
^{222}Th		222.01845	2.8 ms	α/8.129	7.982(8)/9.7	0+			
					7.600(15)/3				
^{223}Th		223.02079	0.65 s	α/7.454	7.29(1)/41(5)				
					7.32(1)/29(5)				

Elem. or Isot.	Natural Abundance (%)	Atomic Mass or Weight	Half-Life	Decay Mode/Energy (/MeV)	Particle Energy /Intensity (MeV/%)	Spin (h/2π)	Nuclear Magnetic Mom. (nm)	Elect. Quadr. Mom. (b)	γ-ray/Energy Intensity (MeV/%)
					7.350(15)/20(5)				
					7.390(15)/10(4)				
^{224}Th		224.02146	1.05 s	α/7.305	6.768(5)/1.2				
					6.997(5)/19				
					7.170(5)/79				
^{225}Th		225.02394	8.72 m	EC/(10)/0.68		(3/2+)			
				α/(90)/6.920	6.441(2)/15				
					6.479(2)/43				
					6.501(3)/14				
					6.627(3)/3				
					6.650(5)/3				
					6.700(5)/2				
					6.743(3)/7				
					6.796(2)/9				
^{226}Th		226.024891	30.83 m	α/6.454	6.026(1)/0.2	0+			Ra k x-ray
					6.041(1)/0.19				0.1112
					6.098(1)/1.3				0.2421
					6.2283(4)/23				0.1310
					6.3375(4)/75				0.1733-0.9295
^{227}Th		227.027699	18.72 d	α/6.146		(3/2+)			Ra L x-ray
									Ra k x-ray
									0.05014
									0.23597
									0.25624
									(0.02 - 1.0)
^{228}Th		228.028731	1.913 y	α/5.520	5.1770(2)/0.18	0+			
					5.2114(1)/0.4				
					5.3405(1)/26.7				
					5.4233(1)/73				
^{229}Th		229.031754	7.9x10^{3} y	α/5.168	4.814/9.3	5/2+	+0.46	+4.	
					4.845(5)/56				
					4.9008(5)/10.2				
					4.689-5.077				
^{230}Th		230.033126	7.54x10^{4} y	α/4.771	4.4383(6)/0.03	0+			0.0677/0.46
					4.4798(6)/0.12				0.1439/0.078
			>2.x10^{18} y	SF/<4×10^{-12}	4.6211(6)/23.4				
					4.6876(6)/76.3				
^{231}Th		231.036296	1.063 d	β^{-}/0.390	0.138/22	5/2+			Pa L x-ray
					0.218/20				Pa k x-ray
					0.305/52				0.02564
									0.084203/
									(0.02 - 0.3)
^{232}Th	100.	232.038050	1.40x10^{10} y	α/4.081	3.830(10)/0.2	0+			0.0590
			1.2x10^{21} y	SF/1.1x10^{-9}	3.952(5)/23				0.124
					4.010(5)/77				
^{233}Th		233.041576	22.3 m	β^{-}/1.245	1.245/	1/2+			Pa L x-ray
									Pa k x-ray
									0.02938
									0.08653
									0.45930
									(0.02 - 1.2)
^{234}Th		234.043596	24.10 d	β^{-}/0.273	0.102/20	0+			Pa L x-ray
					0.198/72				0.06329/4.1
									0.09235/2.4
									0.09278/2.4
^{235}Th		235.04751	7.2 m	β^{-}/1.9					0.4162
									0.6594
									0.7272
									0.747

TABLE OF ISOTOPES (CONTINUED)

Elem. or Isot.	Natural Abundance (%)	Atomic Mass or Weight	Half-Life	Decay Mode/Energy (/MeV)	Particle Energy /Intensity (MeV/%)	Spin (h/2π)	Nuclear Magnetic Mom. (nm)	Elect. Quadr. Mom. (b)	γ-ray/Energy Intensity (MeV/%)
									0.9318
^{236}Th		236.0497	37.5 m	β^-/≈ 1.0					Pa k x-ray
									0.1107
^{237}Th		237.0539	5.0 m	β^-					
$_{91}$Pa		231.03588(2)							
^{212}Pa			≈ 5 ms	α	8.27				
^{213}Pa		213.0212	7 ms	α	8.24				
^{214}Pa		214.0207	17 ms	α	8.12				
^{215}Pa		215.0190	15. ms	α	8.08/100				
^{216}Pa		216.0190	0.19 s	α/	7.95/51				0.134
					7.82/45				
					7.79/4				
^{217m}Pa			1.5 ms	α/	10.16/80				
					9.55/17				
					9.69/3				
^{217}Pa		217.0183	3.4 ms	α/8.490	8.340(10)/100				
^{218}Pa		218.0200	0.12 ms	α/	9.54/31				0.092
					9.61/69				
^{219}Pa		219.0199	0.05 μs	α					
^{220}Pa		220.0219	0.8 μs	α					
^{221}Pa		221.0219	6. μs	α	9.08(3)				
^{222}Pa		222.0237	≈ 4.3 ms	α/8.700	8.180/50				
					8.330/20				
					8.540/30				
^{223}Pa		223.0240	≈ 6.5 ms	α/8.340	8.006(10)/55				
					8.196(10)/45				
^{224}Pa		224.0256	0.84 s	α/7.630	7.555(10)/75(3)				0.1945
					7.46(1)/25(3)				(0.028-0.412)
^{225}Pa		225.0261	1.8 s	α/7.380	7.195(10)/30				
					7.245(10)/70				
^{226}Pa		226.02792	1.8 m	α/(74)/6.987	6.728(10)/0.7				
				EC/(26)/2.83	6.823(10)/35				
					6.863(10)/39				
^{227}Pa		227.02879	38.3 m	α/(85)/6.582	6.357(4)/7	(5/2-)			0.0649
				EC/(15)/1.02	6.376(10)/2.2				0.0669
					6.401(4)/8				0.1100
					6.416(4)/13				
					6.423(10)/10				
					6.465(4)/43				
^{228}Pa		228.03100	22. h	EC/(98)/2.111		(3+)	+3.5		Th k x-ray
				α/(2)	5.779/0.23				0.409/100
					5.805/0.15				0.4631/222
					6.078/0.4				0.91116/242
					6.105/0.25				0.96464/120
					6.118/0.22				0.96897/149
									0.058-1.96
^{229}Pa		229.03209	1.5 d	EC/(99.8)/0.32		(5/2)			0.04244
				α/(0.2)/5.836	5.536(2)/0.02				(0.024 - 0.18)
					5.579(2)/0.09				
					5.668(2)/0.05				
^{230}Pa		230.034532	17.4 d	EC/(90)/1.310	0.51/	(2-)	2.0		Th L x-ray
				β^-/(10)/0.563					Th k x-ray
									0.4437
									0.45477
									0.89876
									0.91856
									0.95199
									(0.053-1.07)
^{231}Pa		231.035878	3.25x10^4 y	α/5.148	4.6781(5)/1.5	3/2-	2.01	-1.7	Ac L x-ray

Elem. or Isot.	Natural Abundance (%)	Atomic Mass or Weight	Half-Life	Decay Mode/Energy (/MeV)	Particle Energy /Intensity (MeV/%)	Spin (h/2π)	Nuclear Magnetic Mom. (nm)	Elect. Quadr. Mom. (b)	γ-ray/Energy Intensity (MeV/%)
					4.7102(5)/1.0				Ac k x-ray
			>2x10^{17} y	SF/<1.6x10^{-15}	4.7343(5)/8.4				0.01899
					4.8513(5)/1.4				0.027396
					4.9339(5)/3				0.03823
					4.9505(5)/22.8				0.04639
					4.9858(5)/1.4				0.25586
					5.0131(5)/25.4				0.26029
					5.0292(5)/20				0.28367
					5.0318(5)/2.5				0.30007
					5.0587(5)/11				0.30264
									0.33007
									(0.02 - 0.61)
^{232}Pa		232.03858	1.31 d	β^-/1.34		(2-)			U k x-ray
									0.10900
									0.15009
									0.89439
									0.96934
									(0.10 - 1.17)
^{233}Pa		233.040239	27.0 d	β^-/0.571	0.15/40	3/2-	+4.0	-3.0	U L x-ray
					0.256/60				U k x-ray
									0.30017
									0.31201
									0.34059
^{234m}Pa			1.17 m	β^-/99.9/2.29		(0-)			U k x-ray
				IT/0.13/					0.25818/0.07
									0.76641/0.32
									1.0009/0.85
									(0.06 - 1.96)
^{234}Pa		234.043303	6.69 h	β^-/2.197	0.51/	(4+)			U L x-ray
									U k x-ray
									0.1312/0.03
									0.5695/0.02
									0.9256/0.02
									(0.02 - 1.99)
^{235}Pa		235.04544	24.4 m	β^-/1.41	1.4/97	(3/2-)			0.0308-0.65893
^{236}Pa		236.0487	9.1 m	β^-/2.9	1.1/40	(1-)			U k x-ray
					2.0/50				0.64235
					3.1/10				0.68759
									1.7630
									(0.04 - 2.18)
^{237}Pa		237.0511	8.7 m	β^-/2.3	1.1/60	(1/2+)			0.4986
					1.6/30				0.5293
					2.3/10				0.5407
									0.8536
									0.8650
									(0.04 - 1.4)
^{238}Pa		238.0545	2.3 m	β^-/3.5	1.2/	(3-)			0.10350
					1.7/				0.1785
									0.4484
									0.6350
									0.6800
									1.01446
									(0.04 - 2.5)
^{239}Pa		239.0571	1.8 h						
$_{92}$U		238.02891(3)							
^{217}U			≈16 ms	α	8.005				
^{218}U		218.0235	≈0.002 s	α	8.63(3)/				
^{219}U		219.0249	0.04 ms	α	9.68(4)/				
^{222}U		222.0261	≈ 1 μs	α					

Elem. or Isot.	Natural Abundance (%)	Atomic Mass or Weight	Half-Life	Decay Mode/Energy (/MeV)	Particle Energy /Intensity (MeV/%)	Spin (h/2π)	Nuclear Magnetic Mom. (nm)	Elect. Quadr. Mom. (b)	γ-ray/Energy Intensity (MeV/%)
^{223}U		223.0277	0.02 s	α/	8.78(4)/				
^{224}U		224.02759	≈ 1. ms	α/	8.46/100				
^{225}U		225.02938	0.09 s	α/	7.89/58				
					7.83/37				
					7.62/5				
^{226}U		226.02933	0.5 s	α/7.560	7.55/82	0+			
					7.37/15				
					7.32/3				
^{227}U		227.03113	1.1 m	α/7.200	6.870/				
^{228}U		228.03137	9.1 m	α/6.803	6.404(6)/0.6	0+			0.095
					6.440(5)/0.7				0.152
					6.589(5)/29				0.187
					6.681(6)/70				0.246
^{229}U		229.03350	58. m	EC/(80)/1.31	6.223/3	(3/2+)			
				α/(20)/6.473	6.297(3)/11				
					6.332(3)/20				
					6.360(3)/64				
^{230}U		230.033927	20.8 d	α/5.992	5.5866(3)/0.01	0+			Th L x-ray
			>4x10^{10} y	SF/<10^{-10}	5.6624(3)/0.26				0.07218
					5.6663(3)/0.38				0.15421
					5.8178(3)/32				0.23034
					5.8887(3)/67				(0.081-0.8565)
^{231}U		231.03626	4.2 d	EC/0.36		(5/2-)			Pa L x-ray
				α/(10^{-3})	5.46/1.6 x 10^{-3}				Pa k x-ray
					5.47/1.4 x 10^{-3}				0.02564
					5.40/1. x 10^{-3}				0.08420
^{232}U		232.037146	70. y	α/5.414	4.9979(1)/0.003	0+			
			2.6x10^{15} y	SF/2.7x10^{-12}	5.1367(1)/0.3				
					5.2635(1)/31				
					5.3203(1)/69				
^{233}U		233.039627	1.592x10^{5} y	α/4.909	4.7830(8)/13.2	5/2+	+0.59	3.66	Th L x-ray
			>2.7x10^{17} y	SF/6x10^{-11}	4.8247(8)/84.4				0.04244
					4.510-4.804				0.09714
									(0.0252-1.119)
^{234}U	0.0055(5)	234.040945	2.455x10^{5} y	α/4.856	4.604(1)/0.24	0+			0.05323/0.156
			1.5x10^{16} y	SF/1.6x10^{-9}	4.7231(1)/27.5				0.12091
					4.776(1)/72.5				
^{235m}U			26. m	IT/0.0007		1/2+			
^{235}U	0.720(1)	235.043922	7.04x10^{8} y	α/4.6793	4.1525(9)/0.9	7/2-	-0.38	4.9	Th L x-ray
			1.0x10^{19} y	SF/7x10^{-9}	4.2157(9)/5.7				Th k x-ray
					4.3237(9)/4.6				0.10917
					4.3641(9)/11				0.14378
					4.370(4)/6				0.16338
					4.3952(9)/55				0.18574
					4.4144(9)/2.1				0.20213
					4.5025(9)/1.7				0.20533
					4.5558(9)/4.2				0.22140
					4.5970(9)/5.0				(0.03 - 0.79)
^{236}U		236.045561	2.342x10^{7} y	α/4.569	4.332(8)/0.26	0+			Th L x-ray
			2.5x10^{16} y	SF/9x10^{-8}	4.445(5)/26				0.04937
					4.494(3)/74				0.11275
^{237}U		237.048723	6.75 d	β^{-}/0.519	0.24/	1/2+			Np L x-ray
					0.25/				Np k x-ray
									0.05953
									0.20801
^{238}U	99.2745(15)	238.050784	4.47x10^{9} y	α	4.0395/0.23	0+			Th L x-ray
			8.2x10^{15}y	SF/5x10^{-5}	4.147(5)/23				0.04955/.06
					4.196(5)/77				0.1135/.01
^{239}U		239.054289	23.5 m	β^{-}/1.265	1.2/	5/2+			(0.522-0.681)

TABLE OF ISOTOPES (CONTINUED)

Elem. or Isot.	Natural Abundance (%)	Atomic Mass or Weight	Half-Life	Decay Mode/Energy (/MeV)	Particle Energy /Intensity (MeV/%)	Spin (h/2π)	Nuclear Magnetic Mom. (nm)	Elect. Quadr. Mom. (b)	γ-ray/Energy Intensity (MeV/%)
					1.3/				
^{240}U		240.056585	14.1 h	β^-/0.39	0.36/	0+			Np L x-ray
									0.04410
									0.05558
									0.06760
^{242}U		242.0629	16.8 m	β^-/$\approx$ 1.2					
$_{93}$Np									
^{225}Np		225.0339	> 2 μs						
^{226}Np		226.0351	0.03 s	α/	8.04(2)/				
^{227}Np		227.0350	0.51 s	α/	7.65(2)/				
					7.68(1)/				
^{228}Np		228.0362	61. s	EC/60(7)/					
				α/40(7)/,SF					
^{229}Np		229.0363	4.0 m	α/7.010	6.890(20)				
^{230}Np		230.0378	4.6 m	EC/97/3.6					
				α/3	6.660(20)				
^{231}Np		231.03823	48.8 m	EC/98 /1.8		5/2			0.2629
				α/2 /6.368	6.280/2				0.3475
									0.3703
^{232}Np		232.0400	14.7 m	EC/99/2.7		(4-)			U L x-ray
									U k x-ray
									0.3268
									0.81925
									0.86683
^{233}Np		233.0410	36.2 m	EC/1.2		(5/2+)			U L x-ray
									U k x-ray
									0.29887
									0.31201
^{234}Np		234.04289	4.4 d	β^+,EC/1.81	0.79/	(0+)			U L x-ray
									U k x-ray
									1.5272
									1.5587
									1.6022
^{235}Np		235.044055	1.085 y	EC/99.9/0.124		5/2+			U k x-ray
				α/0.001/5.191					
^{236m}Np			22.5 h	EC/52/		(1-)			U L x-ray
				β^-/48/					Pu L x-ray
									U k x-ray
									0.64235
									0.68759
^{236}Np		236.04657	1.55x10^5 y	EC/91/0.94		(6-)			U L x-ray
				β^-/9/0.49					U k x-ray
									0.10423
									0.16031
^{237}Np		237.048166	2.14x10^6 y	α/4.957	4.6395(5)/6.5	5/2+	+3.14	+3.89	Pa L x-ray
			1x10^{18} y	SF/2.1x10^{-10}	4.766(5)/9.7				Pa k x-ray
					4.7715(5)/22.7				0.029378/15
					4.7884(5)/47.8				0.08653/12
					4.558-4.873				(0.03-0.28)
^{238}Np		238.050940	2.117 d	β^-/1.292	1.2/	2+			Pu L x-ray
									Pu k x-ray
									0.98447/25.2
									1.02855/18.3
									(.044-1.026)
^{239}Np		239.052931	2.355 d	β^-/0.722	0.341/30	5/2+			Pu L x-ray
					0.438/48				Pu k x-ray
									0.10613
									0.228186/11
									0.27760/15

Elem. or Isot.	Natural Abundance (%)	Atomic Mass or Weight	Half-Life	Decay Mode/Energy (/MeV)	Particle Energy /Intensity (MeV/%)	Spin (h/2π)	Nuclear Magnetic Mom. (nm)	Elect. Quadr. Mom. (b)	γ-ray/Energy Intensity (MeV/%)
									(0.04-0.50)
^{240m}Np			7.22 m	β^-/99.9/	2.18/	(1+)			0.25143
				IT/0.1/					0.26333
									0.55454
									0.59735
^{240}Np		240.05617	1.032 h	β^-/2.20	0.89/	5+			0.1471/
									0.5664
									0.6008
^{241}Np		241.0583	13.9 m	β^-/1.3	1.3/	5/2+			0.1330/
									0.1740
									0.280
^{242m}Np			2.2 m	β^-/		(1+)			0.15910
									0.2651/
									0.78570
									0.9448/
^{242}Np		242.0616	5.5 m	β^-/2.7	2.7/	6+			0.6209
									0.73620
									0.78074
									1.47340
									(0.04-2.37)
^{243}Np		243.0643	1.9 m						
^{244}Np		244.0678	2.3 m						
$_{94}$Pu									
^{228}Pu		228.0387		α/	7.81(2)/				
^{229}Pu		229.0362	4.0 m	α/	7.46(3)/				
^{230}Pu		230.03964	4.6 m	α/	7.05/				
^{231}Pu		231.04126	8.6 m	EC/90					
				α/10	6.72				
^{232}Pu		232.04118	34. m	EC/>80/1.1		0+			
				α/<20/6.716	6.542(10)/38				
					6.600(10)/62				
^{233}Pu		233.04299	20.9 m	EC(99.9)/1.9					0.1503
				α/0.1/6.416	6.300(20)/0.1				0.1804
									0.2353
									0.5002
									0.5346/
									1.0352/
^{234}Pu		234.04331	8.8 h	EC/94/0.39		0+			
				α/6/6.310	6.035(3)/0.024				
					6.149(3)/1.9				
					6.200(3)/4.0				
^{235}Pu		235.0453	25.3 m	EC/99+/1.2		(5/2+)			
				α/0.003/5.957	5.850(20)/0.003				
^{236}Pu		236.046048	2.87 y	α/5.867	5.611/0.21	0+			0.0476/0.07
			$1.5x10^9$ y	SF/$1.9x10^{-7}$	5.7210/30.5				0.109/0.02
					5.7677(1)/69.3				(0.17 - 0.97)
^{237}Pu		237.048403	45.7 d	EC/99.9/0.220		7/2-			Np L x-ray
				α/0.003/5.747	5.334(4)/0.0015				Np k x-ray
					5.356(4)/0.0006				0.026344
					5.650(4)/0.0007				0.03319
									0.05954
									(0.03-0.5)
^{238}Pu		238.049553	87.7 y	α/5.593	5.3583(1)/0.10	0+			U k x-ray
			$4.75x10^{10}$ y	SF/$1.8x10^{-7}$	5.465(1)/28.3				0.04347
					5.4992(1)/71.6				(0.04-1.1)
^{239}Pu		239.052156	$2.410x10^4$ y	α/5.244	5.055/0.047	1/2+	+0.203		U k x-ray
			$8.x10^{15}$ y	SF/$3x10^{-10}$	5.076/0.078				0.05162
					5.106/11.9				0.05682
					5.144/17.1				0.12928

Elem. or Isot.	Natural Abundance (%)	Atomic Mass or Weight	Half-Life	Decay Mode/Energy (/MeV)	Particle Energy /Intensity (MeV/%)	Spin (h/2π)	Nuclear Magnetic Mom. (nm)	Elect. Quadr. Mom. (b)	γ-ray/Energy Intensity (MeV/%)
					5.157/70.8				0.37502
					(4.74 -5.03)				0.41369
^{240}Pu		240.053807	6.56x10^{3} y	α/5.255	5.0212(1)/0.07	0+			U L x-ray
			1.14x10^{11} y	SF/5.7x10^{-6}	5.1237(1)/26.4				0.04524
					5.1681(1)/73.5				0.10423
									(0.04-0.97)
^{241}Pu		241.056844	14.4 y	β^{-}/99+/0.0208	4.853(7)/3x10^{-4}	5/2+	-0.683	+6.	0.14854
				α/0.002/5.139	4.8966(7)/0.002				0.1600
			<6.x10^{16} y	SF/>2.4x10^{-14}					
^{242}Pu		242.058736	3.75x10^{5} y	α/4.983	4.7546(7)/0.098	0+			U L x-ray
			6.77x10^{10} y	SF/5.5x10^{-4}	4.8564(7)/22.4				0.04491
					4.9006(7)/78				0.10350
^{243}Pu		243.061996	4.956 h	β^{-}/0.582	0.49/21	7/2+			Am L x-ray
					0.58/60				0.0417
									0.0839
^{244}Pu		244.064197	8.00x10^{7} y	α/99.9/4.665	4.546(1)/19.4	0+			U L x-ray
			6.6x10^{10} y	SF/0.12	4.589(1)/80.5				0.0439
^{245}Pu		245.06774	10.5 h	β^{-}/1.21	0.93/57	(9/2-)			Am L x-ray
					1.21/11				Am k x-ray
									0.2804/
									0.30832
									0.32752
									0.56014
									(0.03-1.2)
^{246}Pu		246.07020	10.85 d	β^{-}/0.40	0.150/85	0+			Am L x-ray
					0.35/10				Am k x-ray
									0.04379
									0.22371
^{247}Pu		247.0741	2.3 d						
$_{95}$Am									
^{232}Am		232.0466	0.9 m	EC/≈ 5.0					
^{234}Am		234.0478	2.3 m	EC/4.2					
^{235}Am		235.0480	≈ 15 m	EC					Pu K x-ray
^{236}Am		236.0456	≈4.4 m						
^{237}Am		237.0503	1.22 h	EC/99.98/1.7		(5/2-)			Pu k x-ray
				α/0.02/6.20	6.042(5)/0.02				0.14559
									0.28026
									0.43845
^{238}Am		238.05198	1.63 h	EC/2.26		1+			Pu L x-ray
				α/0.0001/6.04	5.940/0.0001				Pu k x-ray
									0.91870
									0.96278
^{239}Am		239.053018	11.9 h	EC/99.99/0.803		5/2-			Pu L x-ray
				α/0.01/5.924	5.734(2)/0.001				Pu k x-ray
					5.776(2)/0.008				0.18172
									0.22818
									0.27760
^{240}Am		240.05529	2.12 d	EC/1.38		(3-)			Pu L x-ray
				α/5.592	5.378(1)/16x10^{-4}				Pu k x-ray
									0.88878
									0.98764
									(0.1-1.3)
^{241}Am		241.056822	432.7 y	α/5.637	5.2443(1)/0.002	5/2-	+1.58	+3.1	Np L x-ray
			1.2x10^{14} y	SF/3.6x10^{-10}	5.3221(1)/0.015				0.02634
					5.3884(1)/1.4				0.033192
					5.4431(1)/12.8				0.059536
					5.4857(1)/85.2				(0.03-1.128)
					5.5116(1)/0.20				
					5.5442(1)/0.34				

Elem. or Isot.	Natural Abundance (%)	Atomic Mass or Weight	Half-Life	Decay Mode/Energy (/MeV)	Particle Energy /Intensity (MeV/%)	Spin (h/2π)	Nuclear Magnetic Mom. (nm)	Elect. Quadr. Mom. (b)	γ-ray/Energy Intensity (MeV/%)
^{242m}Am			141. y	IT/99.5/0.048		5-	+1.0	+6.5	Am L x-ray
				α/0.5/5.62	5.141(4)/0.026				0.04863
			>3.x10^{12} y	SF/<4.7x10^{-9}	5.2070(2)/0.4				0.08648
									0.10944
									0.16304
^{242}Am		242.059542	16.02 h	β^-/83/0.665	0.63/46	1-	+0.388	-2.4	Pu L x-ray
				EC/17/0.750	0.67/37				Cm L x-ray
									Pu k x-ray
									0.0422
									0.04453
^{243}Am		243.061372	7.37x10^{3} y	α/5.438	5.1798(5)/1.1	5/2-	+1.5	+2.9	0.04354
			2.x10^{14} y	SF/3.7x10^{-9}	5.2343(5)/11				0.07467
					5.2766(5)/88				0.08657
					5.394(5)/0.12				0.11770
					5.3500(5)/0.16				0.14197
^{244m}Am			≈ 26. m	β^-/1.498		(1-)			0.0429
^{244}Am		244.064279	10.1 h	β^-/1.428					Am L x-ray
									Cm k x-ray
									0.7460
									0.9000
^{245}Am		245.066444	2.05 h	β^-/0.894	0.65/19	(5/2+)			Cm L x-ray
					0.90/77				Cm k x-ray
									0.25299
^{246m}Am			25.0 m	β^-/	1.3/79.	2-			Cm L x-ray
					1.60/14				Cm k x-ray
					2.1/7				0.27002
									0.79881
									1.06201
									1.07885
									(0.04-2.29)
^{246}Am		246.06977	39. m	β^-/2.38	1.2/	(7-)			Cm L x-ray
									Cm k x-ray
									0.1529
									0.2046
									0.6786
^{247}Am		247.0722	22. m	β^-/1.7					Cm L x-ray
									Cm k x-ray
									0.2267/
									0.2853/
$_{96}$Cm									
^{235}Cm		235.0516							
^{236}Cm		236.0514		EC/1.7					
^{237}Cm		237.0529		EC/2.5					
^{238}Cm		238.05302	2.4 h	EC/>90/0.97		0+			
				α/<10/6.632	6.520(50)/<10				
^{239}Cm		239.0548	≈ 3. h	EC/1.7					
									0.0407
									0.1466
									0.1874
^{240}Cm		240.055519	27. d	α/6.397	5.989/0.014	0+			
					6.147/0.05				
			1.9x10^{6} y	SF/3.9x10^{-6}	6.2478(6)/28.8				
					6.2906(6)/70.6				
^{241}Cm		241.057646	32.8 d	EC/99/0.768		1/2+			Am k x-ray
				α/1/6.184	5.8842(4)/0.12				0.13241
					5.9291(4)/0.18				0.16505
					5.9389(4)/0.69				0.18028
									0.43063
									0.47181

TABLE OF ISOTOPES (CONTINUED)

Elem. or Isot.	Natural Abundance (%)	Atomic Mass or Weight	Half-Life	Decay Mode/Energy (/MeV)	Particle Energy /Intensity (MeV/%)	Spin (h/2π)	Nuclear Magnetic Mom. (nm)	Elect. Quadr. Mom. (b)	γ-ray/Energy Intensity (MeV/%)
^{242}Cm		242.058828	162.8 d	α/6.216	5.9694(1)/0.035	0+			Pu L x-ray
					6.069(1)/25				0.04408
			7.0x10^{6} y	SF/6.4x10^{-6}	6.1129(1)/74				0.10189
									(0.04-1.2)
^{243}Cm		243.061381	29.1 y	α/6.167	5.6815(5)/0.2	5/2+	0.41		Pu L x-ray
					5.6856(5)/1.6				Pu k x-ray
			5.5x10^{11} y	SF/5.3x10^{-9}	5.7420(5)/10.6				0.10612
					5.7859(5)/73.3				0.20975
					5.9922(5)/6.5				0.22819
					6.0103(5)/1.0				0.27760
					6.0589(5)/5				0.28546
					6.0666(5)/1.5				0.33431
									(0.04-0.7)
^{244}Cm		244.062745	18.1 y	α/5.902	5.6656/0.02	0+			Pu L x-ray
					5.7528/23				0.04282
			1.32x10^{7} y	SF/1.4x10^{-4}	5.8050/77				0.09885
					5.515/0.004				0.15262
^{245}Cm		245.065485	8.48x10^{3} y	α/5.623	5.235(10)/0.3	7/2+	0.5		Pu L x-ray
					5.3038(10)/5.0				Pu k x-ray
			1.4x10^{12} y	SF/6.1x10^{-7}	5.3620(7)/93				0.04195
					5.4927(11)/0.8				0.13299
					5.5331(11)/0.6				0.13606
									0.17494
^{246}Cm		246.067217	4.76x10^{3} y	α/5.476	5.343(3)/21	0+			Pu L x-ray
			1.8x10^{7} y	SF/0.026	5.386(3)/79				0.04453
^{247}Cm		247.070346	1.56x10^{7} y	α/5.352	4.818(4)/4.7	9/2-	0.37		Pu k x-ray
					4.8690(20)/71	9/2-			0.2792
					4.941(4)/1.6				0.2886
					4.9820(20)/2.0				0.3471
					5.1436(20)/1.2				0.4035
					5.2104(20)/5.7				
					5.2659(20)/13.8				
^{248}Cm		248.072341	3.48x10^{5} y	α/99.92/5.162	4.931(5)/0.07	0+			
					5.0349(2)/16.5				
			4.15x10^{6} y	SF/8.38	5.0784(2)/(75)/1				
^{249}Cm		249.075946	64.15 m	β^{-}/0.900	0.9/	1/2+			Bk k x-ray
									0.56039
									0.63431
^{250}Cm		250.07835	≈ 9.7x10^{3} y		SF/85.8	0+			
				α/5.27					
^{251}Cm		251.08228	16.8 m	β^{-}/1.42	0.90/16	(1/2+)			0.3896/
									0.5299
									0.5425
^{252}Cm		252.0849	< 2 d						
$_{97}$Bk									
^{238}Bk		238.0583	2.4 m	EC/5.0					
^{239}Bk		239.0584							
^{240}Bk		240.0598	≈ 4.8 m						
^{242}Bk		242.0621	7.0 m	EC/3.0					
^{243}Bk		243.063001	4.5 h	EC/99.8/1.508	6.542(4)/0.03	(3/2-)			0.1466
				α/0.15/6.871	6.5738(2)/0.04				0.1874
					6.7180(22)/0.02				0.755
					6.7581(20)/0.02				0.840
									0.946
^{244}Bk		244.0652	4.4 h	EC/99.99/2.26		(4-)			0.1445
				α/0.01/6.778	6.625(4)/0.003				0.1876
					6.667(4)/0.003				0.2176
									0.9815
									0.9215/

Elem. or Isot.	Natural Abundance (%)	Atomic Mass or Weight	Half-Life	Decay Mode/Energy (/MeV)	Particle Energy /Intensity (MeV/%)	Spin (h/2π)	Nuclear Magnetic Mom. (nm)	Elect. Quadr. Mom. (b)	γ-ray/Energy Intensity (MeV/%)
^{245}Bk		245.066355	4.94 d	EC/99.9/0.810		3/2-			Cm L x-ray
				α/0.1/6.453	5.8851(5)/0.03				Cm k x-ray
					6.1176(9)/0.01				0.25299
					6.1467(5)/0.02				0.3809
					6.3087(5)/0.014				0.3851
					6.3492(5)/0.018				
^{246}Bk		246.0687	1.80 d	EC/1.35		(2-)			Cm L x-ray
									Cm k x-ray
									0.79881
									1.08142
^{247}Bk		247.07030	1.4x10^{3} y	α/5.889	5.465(5)/1.5	(3/2-)			0.04175
					5.501(5)/7				0.0839
					5.532(5)/45				0.268
					5.6535(20)/5.5				
					5.678(2)/13				
					5.712(2)/17				
					5.753(2)/4.3				
					5.794(2)/5.5				
^{248}Bk		248.07311	23.7 h	β^{-}/70/0.87	0.86/	(1-)			Cm L x-ray
				EC/30/0.72					Cf L x-ray
									Cm k x-ray
									Cf k x-ray
									0.5507
^{249}Bk		249.074980	320. d	β^{-}/0.125	0.125/100	7/2+	2.0		0.327/10^{-5}
				α/0.001/5.525	5.390(1)/0.0002				0.308/10^{-6}
			1.8x10^{9} y	SF/4.9x10^{-8}	5.4174(6)/0.001				
^{250}Bk		250.078309	3.217 h	β^{-}/1.780	0.74/	2-			Cf L x-ray
									Cf k x-ray
									0.98912
									1.03184
									(0.04-1.6)
^{251}Bk		251.08075	56. m	β^{-}/1.09		(3/2-)			0.02481
									0.1528
									0.1776
^{252}Bk		252.0843	1.8 m						
$_{98}$Cf									
^{237}Cf		237.0621	2.1 s	α,SF/10					
^{238}Cf		238.0614	21 ms	SF/					
^{239}Cf		239.0626	≈ 0.7 m	α					
^{240}Cf		240.0623	1.1 m	α/7.719	7.590(10)/	0+			
				SF/ ≈2.1					
^{241}Cf		241.0637	4. m	EC/3.3					
				α/7.60	7.335(5)/				
^{242}Cf		242.06369	3.5 m	α/7.509	7.351(6)/20	0+			
				SF/œ0.014	7.385(4)/80				
^{243}Cf		243.0654	11. m	EC/86/2.2	7.060(6)/20	(1/2+)			
				α/14/7.40	7.170/4				
^{244}Cf		244.065990	20. m	α/7.328	7.168(5)/25	0+			
					7.210(5)/75				
^{245}Cf		245.068038	44. m	α/36/7.255	7.15/91.7				Cm K x-ray
				EC/64/1.569	6.983/0.31				0.5709
					7.09/7				0.6014
					7.065/0.68				0.6163
^{246}Cf		246.068798	1.49 d	α/6.869	6.6156(10)/0.18	0+			Cm L x-ray
					6.7086(7)/21.8				0.04221
			1.8x10^{3} y	SF/2.3x10^{-4}	6.7501(7)/78.0				0.0945
									0.147
^{247}Cf		247.07099	3.11 h	EC/99.96/0.65		7/2+			Bk k x-ray

Elem. or Isot.	Natural Abundance (%)	Atomic Mass or Weight	Half-Life	Decay Mode/Energy (/MeV)	Particle Energy /Intensity (MeV/%)	Spin (h/2π)	Nuclear Magnetic Mom. (nm)	Elect. Quadr. Mom. (b)	γ-ray/Energy Intensity (MeV/%)
				α/0.04/6.55	6.301(5)/				0.2941
									0.4778
^{248}Cf		248.07218	334. d	α/6.369	6.220(5)/17	0+			
			3.2x10^{4} y	SF/0.0029	6.262(5)/83				
^{249}Cf		249.074846	351. y	α/6.295	5.7582(2)/3.7	9/2-			Cm L x-ray
					5.8119(2)/84				Cm k x-ray
			8.x10^{10} y	SF/4.4x10^{-7}	5.8488(2)/1.0				0.25299
					5.9029(2)/2.8				0.33351
					5.9451(2)/4.0				0.38832
					6.1401(2)/1.1				(0.0376-1.103)
					6.1940(2)/2.2				
^{250}Cf		250.076399	13.1 y	α/6.129	5.8913(4)/0.3	0+			Cm L x-ray
			1.7x10^{4} y	SF/0.077	5.9889(4)/15				0.04285
					6.0310(4)/84.5				
^{251}Cf		251.079579	9.0x10^{2} y	α/6.172	5.56448(7)/1.5	1/2+			
					5.632(1)/4.5				
					5.648(1)/3.5				
					5.6773(6)/35				
					5.762(3)/3.8				
					5.7937(7)/2.0				
					5.8124(8)/4.2				
					5.8514(6)/27				
					6.0140(7)/11.6				
					6.0744(7)/2.7				
^{252}Cf		252.081619	2.65 y	α/96.9/6.217	5.7977(1)/0.23	0+			Cm L x-ray
			86. Y	SF/3.1/	6.0756(4)/15.2				0.04339
					6.1184(4)/81.6				0.1002
^{253}Cf		253.08512	17.8 d	β^-/99.7/0.29	0.27/100	(7/2+)			
				α/0.3/6.126	5.921(5)/0.02				
^{254}Cf		254.08732	60.5 d	SF/99.7/		0+			
				α/0.3/5.930	5.792(5)/0.05				
					5.834(5)/0.26				
^{255}Cf		255.0910	1.4 h	β^-/0.7					
^{256}Cf		256.0934	12. m	SF					
$_{99}$Es									
^{241}Es		241.0687	≈ 8 s	α	8.11				
^{242}Es		242.0697	16 s	α	7.92				
^{243}Es		243.0696	21. s	α/>30/	7.89/>30				
				EC/<70/4.0					
^{244}Es		244.0709	37. s	EC/76/4.6					
				α/4/	7.57/4				
^{245}Es		245.0713	1.3 m	α/40/7.858	7.74				
				EC/60/3.1					
^{246}Es		246.0730	7.7 m	EC/90/3.9					
				α/10/	7.35				
^{247}Es		247.07365	4.8 m	EC/93/2.48					
				α/7/	7.32				
^{248}Es		248.0755	26. m	EC/99.7/3.1					
				α/0.3/	6.87				
^{249}Es		249.07640	1.70 h	EC/99.4/1.45		(7/2+)			0.3795
				α/0.6/	6.77				0.8132
^{250m}Es			2.2 h	EC/		(1-)			Cf L x-ray
				β^+					Cf k x-ray
									0.9891
									1.0319
^{250}Es		250.0787	8.6 h	EC/2.1		(6+)			Cf L x-ray
									Cf k x-ray
									0.30339
									0.34948

Elem. or Isot.	Natural Abundance (%)	Atomic Mass or Weight	Half-Life	Decay Mode/Energy (/MeV)	Particle Energy /Intensity (MeV/%)	Spin (h/2π)	Nuclear Magnetic Mom. (nm)	Elect. Quadr. Mom. (b)	γ-ray/Energy Intensity (MeV/%)
									0.82883
^{251}Es		251.07998	1.38 d	EC/99.5/0.38		(3/2-)			
				α/0.5/	6.462/0.05				
					6.492/0.4				
^{252}Es		252.08297	1.29 y	α/76/	6.632/61.0	(5-)			
				EC/24/1.26	6.562/10.3				
^{253}Es		253.084818	20.47 d	α/	6.633/89.8	7/2+	+4.10	7.	0.04180
			6.3x10^{5} y	SF/8.9x10^{-6}	6.5916/6.6				0.3892
^{254m}Es			1.64 d	β^{-}/99.6/	0.475	2+	2.9	3.7	Fm L x-ray
				α/0.3/6.67	6.382	2+			Fm k x-ray
			>10. Y	SF/0.045					0.6488
									0.6938
^{254}Es		254.088017	276. d	α/	6.429	(7+)			0.064
			>2.5x10^{7} y	SF/<3x10^{-6}					
^{255}Es		255.09027	40. d	β^{-}/92/0.29		(7/2+)			
				α/8/	6.26				
			2.6x10^{3} y	SF/0.0042	6.300				
^{256m}Es			7.6 h	β^{-}/		(8+)			0.218
									0.232
									0.862
^{256}Es		256.0936	25. m	β^{-}/1.7		(1+)			
^{257}Es		257.0960	7.7 d	β^{-}					
$_{100}$Fm									
^{242}Fm		242.0734	0.8 ms	SF/>96					
^{243}Fm		243.0745	0.2 s	α/	8.55				
				<SF/0.4					
^{244}Fm		244.0741	3.3 ms	SF/>97		0+			
^{245}Fm		245.0754	4. s	α/	8.15/				
				SF/<0.1					
^{246}Fm		246.07528	1.2 s	α/85/	8.24/	0+			
				SF/15/					
^{247m}Fm			9. s	α/	8.18/				
^{247}Fm		247.0768	35. s	α/8.20	7.87/70				
				EC/2.9	7.93/30				
^{248}Fm		248.07718	36. s	α/99.9/8.001	7.83/20	0+			
				SF/0.1/	7.87/80				
^{249}Fm		249.0790	3. m	EC/2.4		(7/2+)			
				α/	7.53				
^{250m}Fm			1.8 s	IT/					
				SF/<8x10^{-5}					
^{250}Fm		250.07951	30. m	α/	7.43/	0+			
				EC/0.8					
				SF/0.007					
^{251}Fm		251.08157	5.3 h	EC/98/1.47		(9/2-)			
				α/2/	6.833				
^{252}Fm		252.08246	1.058 d	α/7.154	6.998/15	0+			
				SF/0.0023	7.039/85				
^{253}Fm		253.085175	3.0 d	EC(88%)/0.333	6.676/	1/2+			Es k x-ray
				α/12/	6.943/				0.2719
^{254}Fm		254.086847	3.240 h	α/	7.150	0+			
				SF/0.059	7.192				
^{255}Fm		255.089955	20.1 h	α/	6.9635(5)/5.0	7/2+			
			1.0x10^{4} y	SF/2.3x10^{-5}	7.0225(5)/93.4				
^{256}Fm		256.09177	2.63 h	SF/91		0+			
				α/19	6.92/				
^{257}Fm		257.09510	100.5 d	α/99.79	6.519	(9/2+)			0.1794
				SF/0.21					0.2410
^{258}Fm		258.0971	0.37 ms	SF/					
^{259}Fm		259.1006	1.5 s	SF/					

Elem. or Isot.	Natural Abundance (%)	Atomic Mass or Weight	Half-Life	Decay Mode/Energy (/MeV)	Particle Energy /Intensity (MeV/%)	Spin (h/2π)	Nuclear Magnetic Mom. (nm)	Elect. Quadr. Mom. (b)	γ-ray/Energy Intensity (MeV/%)
^{260}Fm			≈4 ms	SF/					
${}_{101}$Md									
^{245m}Md			≈ 0.4 s	α	8.64,8.68				
^{245}Md		245.0810	0.9 ms	SF					
^{246}Md		246.0819	1.0 s	α	8.74				
					8.50-8.56				
^{247m}Md			≈0.2 s	SF/					
^{247}Md		247.0818	3. s	α	8.43				
^{248}Md		248.0828	7. s	EC/80/5.3	8.32/15				
				α/20/	8.36/5				
				SF/<0.05					
^{249}Md		249.0830	24. s	EC>/<80/3.7					
				α/>20/8.46	8.030(20)/				
^{250}Md		250.0845	50. s	EC/94/4.6	7.75/4				
				α/6/8.25	7.83/2				
^{251}Md		251.0849	4.0 m	EC/>94/3.1					
				α/<6/	7.55/				
^{252}Md		252.0866	2. m	EC/>50/3.9					
				α/<50/	7.73/				
^{253}Md		253.0873	≈6 m	EC/2.0					
^{254m}Md			30. m	EC/					
^{254}Md		254.0897	10. m	EC/2.7					
^{255}Md		255.09108	27. m	EC/92/1.04	α/7.33/93	(7/2-)			0.121/100
				α/8/	7.27/5				0.115/65
				SF/≤0.15	7.75/1				0.136/35
					7.71/1				0.141-0.453
^{256}Md		256.0941	1.30 h	EC/89/2.13	7.21/71				Fm k x-ray
				α/11/	7.14/22				0.121/409
				SF/<2.6	7.68/2.5				0.115/266
					7.25/2.5				0.136/143
					7.64/2.1				0.634/119
									0.141-1.374
^{257}Md		257.095535	5.5 h	EC/85/0.41	7.074	(7/2-)			Fm k x-ray
				α/15,SF/≤1	7.014				(0.181-0.389)
^{258m}Md			57. m	EC/		(1-)			Fm k x-ray
				SF/≤30					
^{258}Md		258.098427	51.5 d	α/7.40	6.718(2)/	(8-)			0.3678
				SF/œ0.003	6.763(4)/				0.057 - 0.448
^{259}Md		259.1005	1.64 h	SF/>98.7		7/2+			
				α/<1.3					
^{260}Md		260.104	≈ 27.8 d	SF/73-100					
${}_{102}$No									
^{250}No		250.0875	0.25 ms	SF/		0+			
^{251}No		251.0889	0.76 s	α/	8.62/96				
				SF/0.26	8.58/4				
^{252}No		252.08897	2.3 s	α/74/8.551	8.42	0+			
				SF/26/	8.37				
^{253}No		253.0907	1.7 m	α/	8.010(20)	(9/2-)			
				EC/3.2					
^{254m}No			0.28 s	I.T./					
				SF/≤.2					
^{254}No		254.09095	49. s	α/	8.09	0+			
				EC/1.1					
				SF/0.17					
^{255}No		255.09323	3.1 m	α/62/	8.12/	1/2+			0.187
				EC/38/2.01	7.93				
					8.08				
^{256}No		256.09428	2.9 s	α/	8.43	0+			
				SF/0.5					

Elem. or Isot.	Natural Abundance (%)	Atomic Mass or Weight	Half-Life	Decay Mode/Energy (/MeV)	Particle Energy /Intensity (MeV/%)	Spin (h/2π)	Nuclear Magnetic Mom. (nm)	Elect. Quadr. Mom. (b)	γ-ray/Energy Intensity (MeV/%)
^{257}No		257.09685	25. s	α/	8.22	(7/2+)			
				SF/<1.5	8.27				
					8.32				
^{258}No		258.0983	≈ 1.2 ms	SF/		0+			
^{259}No		259.1011	58. m	α/78/7.794	7.52	(9/2+)			
				EC/22/0.5	7.55				
				SF/<9.7					
^{260}No		260.103	0.11 s	SF/					
^{262}No		262.108	≈ 8. ms	SF/					
$_{103}$Lr									
^{251}Lr		251.0944	39 m	SF					
^{252}Lr		252.0953	≈0.36 s	α	9.02/73				
				SF/<1	8.97/27				
^{253m}Lr			≈0.57 s	α	8.79				
				SF/1					
^{253}Lr		253.0953	1.5 s	α/	8.72				
				SF/≈5					
^{254}Lr		254.0965	13. s	α/	8.45				
				EC/5.2					
				SF/<0.1					
^{255}Lr		255.0967	22. s	α/	8.37/60				
				EC/3.2	8.43/40				
				SF/<0.1					
^{256}Lr		256.0988	28. s	α/99.7/8.554	8.43/				
				EC/4.2	8.39				
				SF/<0.03					
^{257}Lr		257.0996	0.65 s	α/	8.80	7/2+			
				EC/2.5	8.80				
				SF/<0.03					
^{258}Lr		258.1019	3.9 s	α/	8.60/46				
				EC/3.4	8.62/25				
				SF/<5	8.56/20				
					8.65/9				
^{259}Lr		259.1030	6.1 s	α/80	8.44(1)				
				SF/20					
^{260}Lr		260.1056	3. m	α/	8.03				
^{261}Lr		261.1069	40. m	SF					
^{262}Lr		262.1097	3.6 h	EC/2.					
				SF/<10					
$_{104}$Rf									
^{253}Rf		253.1007	≈ 48. μs	SF					
				α/<10					
^{254}Rf		254.1002	23. μs	SF/>98.5					
				α/<1.5					
^{255}Rf		255.1015	1.6 s	α	8.72/55				
				SF/45	8.77/25				
					8.80/7				
					8.69/3.5				
					8.83/3.5				
					8.89/2.5				
					8.92/2.5				
^{256}Rf		256.10118	6.2 ms	SF/99.68					
				α/0.32	8.81				
^{257}Rf		257.1031	4.7 s	α/9.22	8.77				0.117
				EC/11	9.01				
				SF/<1.4	8.95				
					8.62				
^{258}Rf		258.1036	12. ms	SF/87					
				α/13					

Elem. or Isot.	Natural Abundance (%)	Atomic Mass or Weight	Half-Life	Decay Mode/Energy (/MeV)	Particle Energy /Intensity (MeV/%)	Spin (h/2π)	Nuclear Magnetic Mom. (nm)	Elect. Quadr. Mom. (b)	γ-ray/Energy Intensity (MeV/%)
^{259}Rf		259.1056	3.4 s.	α/9.09/93	8.77(2)/				
				SF/7	8.86/				
^{260}Rf		260.1064	20. ms	SF/					
^{261}Rf		261.1088	1.1 m	α/8.78,SF/<10	8.28/				
^{262}Rf		262.1099	2.1 s	SF/>99.2					
^{263}Rf		263.1125	10. m	SF,α					
$_{105}$Db									
^{255}Db		255.1074	≈ 1.5 s	α,					
				SF/≈20					
^{256}Db		256.1081	1.9 s	α/64	9.02/≈64				
				EC/35	8.89/≈12				
				SF/0.05	9.08/≈12				
					9.12/≈12				
^{257}Db		257.1079	1.5 s	α/	8.97/33				
				SF/<6	9.07/38				
					9.12/5.5				
					8.94/9				
					9.02/9				
					8.89/5.5				
^{258}Db		258.1094	4.2 s	α/	9.20/				
				E.C/5.3	9.16/				
				SF/<33					
^{259}Db		259.1097	≈ 1.2 s	SF/					
^{260}Db		260.1114	1.5 s	α/	9.05/				
				SF/<9.6	9.08/				
					9.13/				
^{261}Db		261.1121	1.8 s	α/	8.93/				
				SF/<18					
^{262}Db		262.1141	34. s	SF/<33					
				α/	8.45/				
					8.53/				
					8.67/				
^{263}Db		263.1151	≈0.45 m	SF/57/, α/43/	8.35/43				
$_{106}$Sg									
^{258}Sg		258.1132	≈ 2.9 ms	SF					
				α/<20					
^{259}Sg		259.1147	0.5 s	α/	9.62				
				SF/<20	9.35				
					9.03				
^{260}Sg		260.11444	4. ms	α/50	9.76				
				SF/50	9.72				
					9.81				
^{261}Sg		261.1162	0.3 s	α,SF/<10	9.56				
^{263}Sg		263.1183	0.8 s	α	9.06				
				SF/<30	9.25				
^{265}Sg		265.1211	≈7.4 s	α/>65	8.84/46				
				SF/<35	8.76/23				
					8.94/23				
					8.69/8				
^{266}Sg		266.1219	≈21. s	α/	8.77/66				
				SF/<82	8.52/33				
^{269}Sg									
$_{107}$Bh									
^{260}Bh		260.122		α					
^{261}Bh		261.1218	12. ms	α/,SF<10	10.40				
					10.10				
					10.03				
^{262m}Bh			8. ms	α/	10.37				
				SF/<12	10.24				

Elem. or Isot.	Natural Abundance (%)	Atomic Mass or Weight	Half-Life	Decay Mode/Energy (/MeV)	Particle Energy /Intensity (MeV/%)	Spin (h/2π)	Nuclear Magnetic Mom. (nm)	Elect. Quadr. Mom. (b)	γ-ray/Energy Intensity (MeV/%)
^{262}Bh		262.1230	0.10 s	α/	10.06				
				SF/<12	9.91				
					9.74				
^{264}Bh		264.1247	0.44 s	α/	9.48				
				SF/	9.62				
^{266}Bh		266.1270	≈1 s	9.29					
^{267}Bh		267.1277	≈17 s	8.83					
$_{108}$Hs									
^{263}Hs		263.1287		α/					
^{264}Hs		264.1284	≈ 0.08 ms	α/,SF/≈50	11.0				
^{265m}Hs			≈0.75 ms	α	10.57/63				
					10.73				
					10.52				
					10.34				
^{265}Hs		265.1300	2.0 ms	α/	10.30/90				
				SF/<1	10.43				
					10.37				
					10.25				
^{267}Hs		267.1371	33 ms	α/>88	9.88				
					9.83				
					9.75				
^{269}Hs		269.1341	9.3 s	α	9.23				
					9.17				
^{273}Hs			≈1.2 s	α	9.78				
					9.47				
^{277}Hs			≈11 m	SF					
$_{109}$Mt									
^{266m}Mt			≈1.2 ms	α	10.46-10.81				
^{266}Mt		266.1379	≈0.7 ms	α	10.48-11.31				
^{267}Mt		267.138	19 ms	α					
^{268}Mt		268.1388	0.07 s	α/>68	10.10,10.24				
$_{110}$110									
^{267}110		267.1440	≈ 3 μs	α/>32	11.6				
^{269}110		269.1451	0.17 ms	α/>75	11.11				
^{271m}110			≈1.1 ms	α	10.68				
					10.74				
^{271}110		271.1461	≈ 56 ms	α	10.71				
^{273m}110			0.076 ms	α	11.8				
^{273}110		273.1492	118 ms	α/	9.73				
^{277}110			≈3.0 ms	α	10.2				
^{280}110			≈7.5 s		SF/				
^{281}110			≈ 1 m	α	8.83				
$_{111}$111									
^{272}111		272.1535	≈ 1.5 ms	α/>68	10.82				
$_{112}$112									
^{277}112			≈ 0.24 ms	α	11.45				
					11.65				
^{281}112			≈ 0.89 ms	α	10.7				
^{283}112			≈ 1. m		sf/>0.7				
				α/<0.3					
^{284}112			≈9.8 s	α	9.17				
^{285}112			≈11. m	α	8.67				
$_{114}$114									
^{285}114			≈ 0.58 ms	α	11.3				
^{287}114			≈ 5.5 s	α	10.3				
^{288}114			≈1.9 s	α	9.84				
^{289}114			≈ 20. s	α	9.71				
$_{116}$116									
^{289}116			≈ 0.60 ms	α	11.6				

TABLE OF ISOTOPES (CONTINUED)

Elem. or Isot.	Natural Abundance (%)	Atomic Mass or Weight	Half-Life	Decay Mode/Energy (/MeV)	Particle Energy /Intensity (MeV/%)	Spin (h/2π)	Nuclear Magnetic Mom. (nm)	Elect. Quadr. Mom. (b)	γ-ray/Energy Intensity (MeV/%)
$_{118}$118									
^{293}118			≈ 0.12 ms	α	12.4				

NEUTRON SCATTERING AND ABSORPTION PROPERTIES*
(Revised 1996)

Norman E. Holden
Reactor Division
Brookhaven National Laboratory
Upton, New York 11973

This Table presents an evaluated set of values for the experimental quantities, which characterize the properties for scattering and absorption of neutrons. The neutron cross section is given for room temperature neutrons, 20.43° C, corresponds to a thermal neutron energy of 0.0253 electron volts (eV) or a neutron velocity of 2200 meters/second. The neutron resonance integral is defined over the energy range from 0.5 eV to 0.1 x 10^6 eV, or 0.1 MeV. Bound neutron scattering lengths and neutron cross sections averaged over a maxwellian spectrum at 30 keV for astrophysical applications are also presented. A list of the major references used is given below. The literature cutoff date is December 1996. Uncertainties are given in parentheses. Parentheses with two or more numbers indicate values to the excited state(s) and to the ground state of the product nucleus. Discussions with Frank Feiner and Harold Knox of KAPL on thermal cross sections and resonance integrals and the assistance of Felicia Kramer in assembling this file are gratefully acknowledged.

TABLE LAYOUT

Column No.	Column Title	Description
1	Isotope or Element	For elements, the atomic number and chemical symbol are listed. For nuclides, the mass number and chemical symbol are listed. Isomers are indicated by the addition of m, m1, or m2.
2	Isotopic Abundance	in atom percent.
3	Half-life	Half-life in decimal notation. μs = microseconds; ms = milliseconds; s = seconds; m = minute; h = hours; d = days; and y = years.
4	Thermal Neutron Cross Section	Cross sections for neutron capture reactions in barns (10^{-24} cm²) or in millibarns (mb). Proton, alpha production and fission reactions are designated by σ_p, σ_α, σ_f, respectively. Separate values are listed for isomeric production.
5	Neutron Resonance Integral	Resonance integrals for neutron capture reactions in barns (10^{24} cm²) or millibarns (mb). Proton, alpha production and fission reactions are designated by RI_p, RI_α, RI_f, respectively. Separate values are listed for isomeric production.
6	Neutron Scattering Length	Bound Coherent scattering length for neutron scattering reactions in units of femtometer (fm), which is equal to fermis (10^{-13} cm).
7	Maxwellian Averaged Cross Section	Astrophysical Cross Sections, averaged over a stellar neutron maxwellian spectrum characterized by a thermal energy of 30 keV, expressed in barns (10^{-24} cm²), millibarns (mb) or microbarns (μb).

GENERAL NUCLEAR DATA REFERENCES

The following references represent the major sources of the neutron data presented:

1. Mughabghab, S.F.; Divadeenam, M.; Holden, N.E.; Neutron Cross Sections, Vol. I *Neutron Resonance Parameters and Thermal Cross Sections* Part A Z=1-60. Academic Press Inc., New York, New York (1981); Mughabghab, S.F.; Part B, Z-61-100. Academic Press Inc., Orlando, Florida (1984).

2. Parrington, J.R.; Knox, H.D.; Breneman, S.L.; Baum, E.M.; Feiner, F.: *Chart of the Nuclides, 15th Edition*, Knolls Atomic Power Laboratory (Oct. 1996).

3. Holden, N.E.; *Fifty Years with Nuclear Fission* Conference, Wash., D.C., Gaithersburg, Md. April 26-29, 1989, p. 946. American Nuclear Society, LaGrange Park, Illinois (1989).

4. Holden, N.E.; *Review of Thermal Neutron Cross Sections and Isotopic Composition of the Elements*, BNL-NCS-42224 (March 1989) and Holden, N.E.; *Thermal Neutron Cross Sections for the 1991 Table of the Isotopes*, BNL-45255 (Feb. 1991).

5. Tuli, J.K.; *Nuclear Wallet Cards*, Brookhaven National Laboratory (July 1995).

6. Holden, N.E.; *Half-lives of Selected Nuclides*, Pure & Applied Chemistry 62, 941 (1990).

7. Koester, L; Rauch, H.; Seymann, E.; *Neutron Scattering Lengths: A Survey of Experimental Data and Methods*, Atomic Data Nuclear Data Tables 49, 65 (1991).

8. Sears, V.F.; *Neutron Scattering Lengths and Cross Sections,* Neutron News 3, (3), 26 (1992).

*Research carried out under auspices of the US Department of Energy, Contract No. DE-AC02-76CH00636

Elem. or Isot.	Natural Abundance (%)	Half-life	Thermal Neut. Cross Section (barns)	Resonance Integral (barns)	Coh. Scat. Length (fm)	σ(30KEV) Maxw. Avg. (barns)
$_1H$			0.332(2)	0.149(1)	- 3.739(1)	
^1H	99.985(1)		0.332(2)	0.149(1)	- 3.741(1)	0.25(2) mb*
^2H	0.015(1)		0.51(1)mb	0.23(2) mb	6.671(4)	
^3H		12.32 y	< 6. μb		4.79(3)	
$_2He$			< 0.05		3.26(3)	
^3He	1.37x10^{-6}		σ_p=5.33(1)x10^3 0.05(1) mb	RI$_p$=2.39(1)x10^3	5.74(7)	8.(1) μb*
^4He	≈ 100.				3.26(3)	
$_3Li$			71.(2)	32.(1)	- 1.90(2)	
^6Li	7.5(2)		σ_α=9.4(1)x10^2 39.(5) mb	RI$_\alpha$=422.(4) 17.(2) mb	2.0(1)	
^7Li	92.5(2)		45.(5) mb	20.(2) mb	- 2.22(2)	39(6) μb
$_4Be$			8.8(4) mb	3.9(2) mb	7.79(1)	
^7Be		53.28 d	σ_p=3.9(1)x10^4 σ_α ≈ 0.1	RI$_p$=1.75(5)x10^4		σ_p 16(4)
^9Be	100.		8.8(4) mb	3.9(2) mb	7.79(1)	
^{10}Be		1.52x10^6 y	<1. mb			
$_5B$			7.6(1)x10^2	343.(5)	5.30(4)	
^{10}B	19.9(2)		σ_α=38.4(1)x10^2 0.3(1) σ_p=7.(1) mb σ_t=8.(2) mb	RI$_\alpha$=17.3(1)x10^2 0.13(4)	- 0.1(3)	
^{11}B	80.1(2)		5.(3) mb	2(1) mb	6.65(4)	
$_6C$			3.5(1) mb	1.6(1) mb	6.646(1)	
^{12}C	98.89(1)		3.5(1) mb	1.6(1) mb	6.651(2)	16(1) μb*
^{13}C	1.11(1)		1.4(1) mb	1.7(2) mb	6.19(9)	0.2(1) mb
^{14}C		5715. y	<1. μb			1.5(4) μb*
$_7N$			1.9(1)	0.85(5)	9.36(2)	
^{14}N	99.634(9)		σ_p=1.83(7) 0.080(1)	RI$_p$=0.82(3) 0.034(1)	9.37(2)	σ_p 1.8(2) mb* 0.04(7) mb
^{15}N	0.366(9)		0.04(1) mb	0.11(3) mb	6.44(3)	4.9(5) μb*
$_8O$			0.29(1) mb	0.40(4) mb	5.805(4)	
^{16}O	99.76(1)		0.19(1) mb	0.36(4) mb	5.805(5)	34(4) μb
^{17}O	0.04		σ_α=0.24(1) 0.54(7) mb	0.11(1) 0.39(5) mb	5.8(2)	σ_α 3.9(5) mb*
^{18}O	0.20(1)		0.16(1) mb	0.81(4) mb	5.84(7)	7.2(6) μb*

Elem. or Isot.	Natural Abundance (%)	Half-life	Thermal Neut. Cross Section (barns)	Resonance Integral (barns)	Coh. Scat. Length (fm)	σ(30KEV) Maxw. Avg. (barns)
$_{9}$F			9.5(7) mb	21.(3) mb	5.65(1)	
^{19}F	100.		9.5(7) mb	21.(3) mb	5.65(1)	6.(1) mb
$_{10}$Ne			42(5) mb	19(3) mb	4.566(6)	
^{20}Ne	90.48(3)		39(5) mb	18(3) mb	4.631(6)	0.12(1) mb
^{21}Ne	0.27(1)		0.7(1),σ_α<1.5	0.31(5)	6.7(2)	≈ 1.6 mb
^{22}Ne	9.25(3)		51(5) mb	23(3) mb	3.87(1)	60.(5) μb*
$_{11}$Na			0.53(2)	0.32(2)	3.63(2)	2.3(2) mb
^{22}Na		2.605 y	σ_p=2.8(3)x10^4	< 2x10^5		
			σ_α=2.6(4)x10^2	1.2(2) x 10^2		
^{23}Na	100.		σ_m=0.43(3)	RI$_m$=0.30(6)	3.63(2)	
$_{12}$Mg			66(6) mb	38.(5) mb	5.375(4)	
^{24}Mg	78.99(3)		0.053(6)	32.(4) mb	5.7(2)	4.2(2) mb
^{25}Mg	10.00(1)		.20(1)	98.(15) mb	3.6(2)	6.5(3) mb
^{26}Mg	11.01(2)		.038(1)	25.(2) mb	4.9(2)	66.(3) μb
^{27}Mg		9.45 m	0.07(2)	0.03(1)		
$_{13}$Al			0.230(2)	0.17(1)	3.45(1)	
^{26}Al			σ_p ≈ 2.1			0.14(2)
^{27}Al	100.		0.230(2)	0.17(1)	3.45(1)	3.8(3) mb
$_{14}$Si			0.166(9)	0.12(2)	4.15(1)	
^{28}Si	92.21(2)		0.17(1)	0.11(2)	4.11(1)	1.8(2) mb
^{29}Si	4.69(2)		0.12(1)	0.08(2)	4.7(1)	7.9(9) mb
^{30}Si	3.10(1)		0.107(4)	0.62(6)	4.61(1)	6.4(6) mb
^{31}Si		2.62 h	73.(6) mb	33.(3) mb		
$_{15}$P			0.17(1)	0.08(1)	5.13(1)	
^{31}P	100.		0.17(1)	0.08(1)	5.13(1)	1.7(1) mb
$_{16}$S			0.54(2)	0.24(2)	2.847(1)	
^{32}S	95.02(9)		0.55(5)	0.25(2)	2.804(2)	4.7(2) mb
			σ_α < 0.5 mb			
^{33}S	0.75(4)		0.46(3)	0.21(2)	4.7(2)	7.4(15) mb
			σ_α = 0.12(1)	RI$_\alpha$ = 0.05(1)		
			σ_p = 2 mb			σ_α = 0.18(1)
^{34}S	4.21(8)		0.30(2)	0.13	3.48(3)	2.9(9) mb
^{36}S	0.02(1)		0.23(2)	0.26(3)		0.17(1) mb
$_{17}$Cl			33.6(3)	15.(2)	9.58(1)	
^{35}Cl	75.77(7)		43.7(4);σ_p=0.44(1)	20.(2);RI$_p$=0.2	11.7(1)	10.(3) mb
			σ_α ≈ 0.08 mb			σ_p 1.7(2) mb

Elem. or Isot.	Natural Abundance (%)	Half-life	Thermal Neut. Cross Section (barns)	Resonance Integral (barns)	Coh. Scat. Length (fm)	σ(30KEV) Maxw. Avg. (barns)
^{36}Cl		3.01x10^{5} y	σ_p=46.(2) mb	RI_p=0.02		σ_p 91.(8) mb
			<10.			
			σ_α = 0.59(7) mb			σ_α 0.9(2) mb
^{37}Cl	24.23(7)		(0.05 + 0.38)	(0.04+0.26)	3.1(1)	2.15(8) mb
$_{18}$Ar			0.66(3)	0.42(5)	1.91(1)	
^{36}Ar	0.337(3)		5(1);σ_p<1.5 mb	2.(1)	24.9(1)	
			σ_α 6(1) mb			
^{37}Ar		35.0 d	σ_α=2.0(3)x10^{3}	RI_α=900.		
			σ_p = 69(14)	RI_p=31.		
^{38}Ar	0.063(1)		0.8(2)	0.4(1)	3.5(35)	
^{39}Ar		268. y	6.(2)x10^{2}			
^{40}Ar	99.600(3)		0.64(3)	0.41(5)	1.83(1)	2.3(2)mb
^{41}Ar		1.82 h	0.5(1)	0.2(1)		
$_{19}$K			2.1(1)	1.0(1)	3.67(2)	
^{39}K	93.2581(44)		2.1(2)	0.9(1)	3.74(2)	11.8(4)mb
			σ_α = 4.3(5) mb			
			σ_p < 0.05 mb			
^{40}K	0.0117(1)	1.26x10^{9} y	30.(8)	13.(4)		σ_p .7(1)mb
			σ_p=4.4(4)	2.0(2)		σ_α 40(6)mb
			σ_α = 0.42(8)			
^{41}K	6.7302(44)		1.46(3)	1.4(2)	2.69(8)	22.0(7)mb
$_{20}$Ca			0.43(2)	0.23(2)	4.70(2)	
^{40}Ca	96.941(18)		0.41(3)	0.22(4)	4.80(2)	6.7(7)mb
			σ_α 0.13(4) mb			
^{41}Ca		1.02x10^{5} y	≈ 4.			
			σ_α = 0.19			
^{42}Ca	0.647(9)		0.65(10)	0.39(4)	3.4(1)	16.(2)mb
^{43}Ca	0.135(6)		6.(1)	3.9(2)	- 1.56(9)	51.(6)mb
^{44}Ca	2.086(12)		0.8(2)	0.56(1)	1.42(6)	9.(1)mb
^{45}Ca		162.7 d	≈ 15.			
^{46}Ca	0.004(3)		0.70(3)	0.9(1)	3.6(2)	5.7(5)mb
^{48}Ca	0.187(4)		1.1(1)	0.5(1)	0.39(9)	1.0(2)mb
$_{21}$Sc			27.2(2)	12.(1)	12.3(1)	
^{45}Sc	100.		(10.+17.)	(5.6+6.4)	12.3(1)	90.(6)mb
^{46}Sc		83.81 d	8.(1)	3.6(5)		
$_{22}$Ti			6.1(1)	2.8(2)	- 3.438(2)	
^{46}Ti	8.25(3)		0.6(2)	0.4(1)	4.93(6)	27.(3)mb
^{47}Ti	7.44(2)		1.6(2)	1.6(2)	3.63(1)	64.(8)mb
^{48}Ti	73.72(3)		7.9(9)	3.6(2)	- 6.09(2)	31.(5)mb
^{49}Ti	5.41(2)		1.9(5)	1.2(2)	1.04(5)	22.(2)mb
^{50}Ti	5.18(2)		0.179(3)	0.12(2)	6.18(8)	4.0(5)mb

Elem. or Isot.	Natural Abundance (%)	Half-life	Thermal Neut. Cross Section (barns)	Resonance Integral (barns)	Coh. Scat. Length (fm)	σ(30KEV) Maxw. Avg. (barns)
$_{23}$V			5.0(2)	2.8(1)	- 0.382(1)	
^{50}V	0.250(4)	> 1.4x10^{17} y	21(4)	50.(20)	7.6(6)	
			σ_p = 0.7(4) mb			
^{51}V	99.750(4)		4.9(1)	2.7(2)	- 0.402(2)	42.(3)mb
$_{24}$Cr			3.0(2)	1.7(1)	3.635(7)	
^{50}Cr	4.345(13)		15.(1)	8.(1)	- 4.5(1)	52.(1)mb
^{51}Cr		27.70 d	< 10.			
^{52}Cr	83.79(2)		0.8(1)	0.6(2)	4.91(2)	8.8(4)mb
^{53}Cr	9.50(2)		18.(2)	9(1)	- 4.2(1)	0.06(1)
^{54}Cr	2.365(7)		0.36(4)	0.25(5)	4.6(1)	7.(2)mb
$_{25}$Mn			13.3(1)	14.0(3)	- 3.75(2)	
^{53}Mn		3.7x10^{6} y	70.(10)	32.(5)		
^{54}Mn		312.2 d	< 10.			
^{55}Mn	100.		13.3(1)	14.0(3)	- 3.75(2)	40.(2)mb
$_{26}$Fe			2.6(1)	1.4(2)	9.45(2)	
^{54}Fe	5.85(4)		2.7(5)	1.3(2)	4.2(1)	29.(2)mb
			σ_α = 10 μb	RI$_\alpha$ = 1.1(1) mb		
^{55}Fe		2.73 y	13.(2)	6.(1)		
			σ_α = 0.03			
^{56}Fe	91.75(4)		2.6(2)	1.4(2)	9.93(3)	13.5(5)mb
^{57}Fe	2.12(1)		2.5(5)	1.6(2)	2.3(1)	35.(4)mb
^{58}Fe	0.28(1)		1.3(1)	1.3(2)	15.(7)	13.(2)mb
^{59}Fe		44.51 d	13.(3)	6.(1)		
$_{27}$Co			37.19(8)	74.(2)	2.49(2)	
^{58m}Co		9.1 h	1.4(1)x10^{5}	2.5(10)x10^{5}		
^{58}Co		70.88 d	1.9(2)x10^{3}	7.(1)x10^{3}		
^{59}Co	100.		(20.7+16.5)	(39.+35.)	2.49(2)	38.(4)mb
^{60m}Co		10.47 m	58.(3)	230.(50)		
^{60}Co		5.271 y	2.0(2)	4.3(10)		
^{28}Ni			4.5(2)	2.3(2)	10.3(1)	
^{58}Ni	68.077(9)		4.6(4)	2.3(2)	14.4(1)	42.(3)mb
			σ_α < 0.03 mb			
^{59}Ni		≈7.6x10^{4} y	σabs=92.(4)	RIabs=1.4(1)x10^{2}		
			σ_α = 14(2)			
			σ_p = 2(1)			
^{60}Ni	26.223(8)		2.9(3)	1.5(2)	2.8(1)	27.(2)mb
^{61}Ni	1.140(1)		2.5(5)	1.5(4)	7.60(6)	82.(1)mb
			σ_α = 0.03 mb			
^{62}Ni	3.634(2)		15.(1)	6.8(3)	- 8.7(2)	36.(4)mb
^{63}Ni		100. y	20(5)	9.(2)		26.(8)mb
^{64}Ni	0.926(1)		1.6(1)	1.(2)	- 0.37(7)	10.(1)mb
^{65}Ni		2.517 h	22.(2)	10.(1)		

Elem. or Isot.	Natural Abundance (%)	Half-life	Thermal Neut. Cross Section (barns)	Resonance Integral (barns)	Coh. Scat. Length (fm)	σ(30KEV) Maxw. Avg. (barns)
$_{29}$Cu			3.8(1)	4.1(4)	7.718(4)	
^{63}Cu	69.17(3)		4.5(2)	5.(1)	6.43(15)	0.09(1)
^{64}Cu		12.701 h	≈ 270.			
^{65}Cu	30.83(3)		2.17(3)	2.2(1)	10.61(19)	53(5)mb
^{66}Cu		5.09 m	1.4(1)x10^{2}	60.(20)		
$_{30}$Zn			1.1(2)	2.8(4)	5.680(5)	
^{64}Zn	48.6(6)		0.74(5) σ_p < 12 μb σ_α = 11(3) μb	1.4(3)	5.23(4)	59(5)mb
^{65}Zn		243.8 d	66.(8) σ_α = 2.0(2)	30.(4)		
^{66}Zn	27.9(3)		0.9(3) σ_α < 0.02 mb	1.8(2)	5.98(5)	35(3)mb
^{67}Zn	4.1(1)		6.9(1.4) σ_α = 0.4 mb	25.(5)	7.58(8)	0.15(2)
^{68}Zn	18.8(5)		(0.072 + 0.8) σ_α <0.02 mb	(0.2 + 2.9)	6.04(3)	19(2)mb
^{70}Zn	0.6(3)		(8.1+83.) mb	0.9(2)		
$_{31}$Ga			2.9(1)	22.(3)	7.288(2)	
^{69}Ga	60.11(3)		1.68(7)	16.(2)	7.88(4)	0.15(1)
^{71}Ga	39.89(3)		4.7(2);σ_m=0.15(5)	31.(3)	6.40(3)	0.12(1)
$_{32}$Ge			2.2(1)	6.(2)	8.19(2)	
^{70}Ge	21.23(4)		(0.3 + 2.7)	2.3(1)	10.0(1)	86(5)mb
^{72}Ge	27.66(3)		0.9(2)	0.8(3)	8.5(1)	0.06(2)
^{73}Ge	7.73(1)		15.(1)	66.(20)	5.02(4)	0.28(8)
^{74}Ge	35.94(2)		(0.14 + 0.28)	(0.4+0.5)	7.6(1)	54(6)mb
^{76}Ge	7.44(2)		(0.09 + 0.06)	(1.3+0.6)	8.2(15)	16(1)mb
$_{33}$As			4.0(4)	61.(5)	6.58(1)	
^{75}As	100.		4.0(4)	61.(5)	6.58(1)	0.46(2)
$_{34}$Se			12.(1)	14.(3)	7.970(9)	
^{74}Se	0.89(4)		50.(2)	520(50)	0.8(3)	
^{75}Se		119.78 d	3.3(10)x10^{2}			
^{76}Se	9.37(29)		(22. + 63.)	(9.+31.)	12.2(1)	164(8)mb
^{77}Se	7.63(16)		42.(4),σ_α=0.97(3)μb	30.(5)	8.25(8)	
^{78}Se	23.77(28)		σ_m=0.38(2)	RI$_m$=4.3(4)	8.24(9)	0.08
^{80}Se	49.61(41)		(0.06+0.35)	(0.36+1.0)	7.48(3)	44.(3)mb
^{82}Se	8.73(22)		(39.+ 5.2) mb	39.(4)mb	6.34(8)	
$_{35}$Br			6.8(2)	92.(8)	6.79(2)	
^{76}Br		16.0 h	224.(42)			
^{79}Br	50.69(7)		(2.5+8.3)	(36.+96)	6.79(7)	0.74(3)
^{81}Br	49.31(7)		(2.4+0.24)	51.(5)	6.78(7)	0.24(1)

Elem. or Isot.	Natural Abundance (%)	Half-life	Thermal Neut. Cross Section (barns)	Resonance Integral (barns)	Coh. Scat. Length (fm)	σ(30KEV) Maxw. Avg. (barns)
$_{36}$Kr			24.(1)	39.(6)	7.81(2)	
^{78}Kr	0.35(1)		(0.17+6.)	20.(1)		0.34(4)
^{80}Kr	2.28(6)		(4.6+7.)	57.(6)		0.24(1)
^{82}Kr	11.58(14)		(14.+7.)	130.(13)		79(6)mb
^{83}Kr	11.50(6)		183.(30)	183.(20)	0.24(2)	
^{84}Kr	56.99(4)		σ_m=0.09;σ_m+g=0.11	2.4(3)		(17+17)mb
^{85}Kr		10.73 y	1.7(2)	1.8(10)		0.07(2)
^{86}Kr	17.30(22)		3.(2)mb	≈1. mb	8.1(3)	3.5(3)mb
$_{37}$Rb			0.39(4)	6.(3)	7.08(2)	
^{84}Rb		32.9 d	σ_p=12.(2)			
^{85}Rb	72.17(2)		(0.06+0.38)	(0.7+7.)	7.0(1)	240(9)mb
^{86}Rb		18.65 d	<20.			
^{87}Rb	27.83(2)	4.88x10^{10} y	0.10(1)	2.3(4)	7.3(1)	18.0(5)mb
^{88}Rb		17.7 m	1.2(3)	0.5(1)		
$_{38}$Sr			1.2(1)	10.(1)	7.02(2)	
^{84}Sr	0.56(1)		(0.6+0.2)	(9.+1.)		
^{86}Sr	9.86(1)		σ_m=0.81(4)	RI$_m$=4.(1)	5.68(5)	(48+22)mb
^{87}Sr	7.00(1)		16.(3)	118.(30)	7.41(7)	97.(5)mb
^{88}Sr	82.58(1)		5.8(4)mb	0.07(3)	7.16(6)	6.1(2)mb
^{89}Sr		50.52 d	0.42(4)	0.2		
^{90}Sr		29.1 y	9.7(7)mb	4. mb		
$_{39}$Y			1.25(5)	1.0(1)	7.75(2)	
^{89}Y	100.		(0.001+1.25)	(0.006+1.0)	7.75(2)	19.(1)mb
^{90}Y		2.67 d	<6.5			
^{91}Y		58.5 d	1.4(3)	0.6(1)		
$_{40}$Zr			0.19(1);σ_α<0.1mb	0.95(9)	7.16(3)	
^{90}Zr	51.45(40)		≈0.014	0.2(1)	6.4(1)	21.(2)mb
^{91}Zr	11.22(5)		1.2(3)	5.(2)	8.8(1)	60.(8)mb
^{92}Zr	17.15(8)		0.2(1)	0.6(2)	7.5(2)	33.(4)mb
^{93}Zr		1.5x10^{6} y	<4.	16.(5)		0.10(1)
^{94}Zr	17.38(28)		0.049(6)	0.25(3)	8.3(2)	26.(1)mb
^{96}Zr	2.80(9)		0.020(3)	5.0(5)	5.5(1)	10.0(3)mb
$_{41}$Nb			1.11(1);σ_α<0.1mb	8.5(6)	7.14(3)	
^{93}Nb	100.		1.1;σ_m=0.86	(6.3+2.2)	7.14(3)	266.(5)mb
^{94}Nb		2.4x10^{4} y	σ_m+g=15.(1) σ_m=0.6(1)	126.(13)		
^{95}Nb		34.97 d	<7.	<200.		
$_{42}$Mo			2.5(1),σ_α<0.1mb	26.(5)	6.72(2)	
^{92}Mo	14.84(35)		0.06,σ_m=0.2μb	≈0.8	6.93(8)	
^{94}Mo	9.25(12)		0.02	≈0.8	6.82(7)	
^{95}Mo	15.92(13)		13.4(3),σ_α=30.(4)μb	109.(5)	6.93(6)	0.29(1)

Elem. or Isot.	Natural Abundance (%)	Half-life	Thermal Neut. Cross Section (barns)	Resonance Integral (barns)	Coh. Scat. Length (fm)	σ(30KEV) Maxw. Avg. (barns)
^{96}Mo	16.68(2)		0.5	17.(3)	6.22(6)	112.(5)mb
^{97}Mo	9.55(8)		2.5(2),σ_α=0.4(2)μb	14.(3)	7.26(8)	0.34(1)
^{98}Mo	24.13(31)		0.14(1)	7.2(7)	6.60(7)	99.(5)mb
^{100}Mo	9.63(23)		0.19(1)	3.6(3)	6.75(7)	
$_{43}$Tc						
^{98}Tc		4.2x10^6 y	σ_m = 0.9(2)			
^{99}Tc		2.13x10^5 y	23.(2)	4.0(4)x10^2	6.8(3)	0.78(4)
$_{44}$Ru			2.6 (1)	48.(5)	7.03(3)	
^{96}Ru	5.52(20)		0.23(4)	7.(2)		
^{98}Ru	1.86(9)		< 8.			
^{99}Ru	12.74(26)		4.(1)	195.(20)		
^{100}Ru	12.60(19)		5.8(6)	11.(2)		0.21(1)
^{101}Ru	17.05(7)		5.(1),σ_α<0.15μb	1.1(3)x10^2		1.00(4)
^{102}Ru	31.57(31)		1.2(1)	4.3(5)		0.19(1)
^{103}Ru		39.27 d	<20.	≈30.		
^{104}Ru	18.66(44)		0.49(2)	6.(2)		0.16(1)
^{105}Ru		4.44 h	0.29(3)	0.13(1)		
^{106}Ru		1.020 y	0.15(4)	2.0(6)		
$_{45}$Rh			145.(2)	1.2(1)x10^3	5.88(4)	
^{103}Rh	100.		(11.+ 134.)	(0.08+1.1)x10^3	5.88(4)	0.81(1)
^{104m}Rh		4.36 m	800.(100)			
^{104}Rh		42.3 s	40.(30)			
^{105}Rh		35.4 h	1.1(3)x10^4	1.7(4)x10^4		
$_{46}$Pd			7.(1)	82.(8)	5.91(6)	
^{102}Pd	1.02(1)		3.2(10)	10.(2)		
^{104}Pd	11.14(11)			16.(2)		
^{105}Pd	22.33(13)		22.(2),σ_α=0.5(2)μb	60.(20)	5.5(3)	1.20(6)
^{106}Pd	27.33(6)		(0.013+0.28)	(0.2+5.5)	6.4(4)	0.25(3)
^{107}Pd		6.5x10^6 y	1.8(2)	108.(4)		1.34(6)
^{108}Pd	26.46(15)		(0.19+8.5)	(2.+240.)	4.1(3)	0.20(2)
^{110}Pd	11.72(16)		(0.033+0.7)	(0.7+8.)		0.15(2)
$_{47}$Ag			62.(1)	767.(60)	5.922(7)	
^{107}Ag	51.839(8)		(0.37+35.)	(1.2+105.)	7.56(1)	0.80(2)
^{109}Ag	48.161(8)		(4.2 + 87.)	(0.7+14.1)x10^2	4.17(1)	0.78(2)
^{110m}Ag		249.8 d	82.(11)	20.(4))		
^{111}Ag		7.47 d	3.(2)	105.(20)		
$_{48}$Cd			2.52(5)x10^3	73.(8)	4.87(5)	
^{106}Cd	1.25(6)		0.20(3)	4.(1)		0.56(6)
^{108}Cd	0.89(3)		1.	14.(3)	5.4(1)	0.41(7)
^{109}Cd		462.0 d	≈180.,σ_α<0.05	6.7(12)x10^3		
^{110}Cd	12.49(18)		(0.06+11.)	(6.+34.)	5.9(1)	0.25(3)
^{111}Cd	12.80(12)		3.5(20)	51.(6)	6.5(1)	1.06(13)

Elem. or Isot.	Natural Abundance (%)	Half-life	Thermal Neut. Cross Section (barns)	Resonance Integral (barns)	Coh. Scat. Length (fm)	σ(30KEV) Maxw. Avg. (barns)
^{112}Cd	24.13(21)		(0.012+2.2)	15.	6.4(1)	0.23(3)
^{113}Cd	12.22(12)	9x10^{15} y	2.06(4)x10^{4},σ_α<1.μb	390.(40)	- 8.0(2)	0.73(8)
^{114}Cd	28.73(42)		(0.04+0.29)	16.(7)	7.5(1)	0.16(3)
^{116}Cd	7.49(18)		(26.+52.)mb	1.2	6.3(1)	0.09(1)
$_{49}$In			197.(4)	3.3(2)x10^{3}	4.07(2)	
^{113}In	4.29(5)		(3.1+5.0+3.9)	(220.+90)	5.39(6)	0.79(7)
^{115}In	95.71(5)	4.4x10^{14} y	(88.+73.+44.)	(1.5+1.2+0.7)x10^{3}	4.01(2)	0.74(7)
$_{50}$Sn			0.61(3)	8.(2)	6.225(2)	
^{112}Sn	0.97(1)		(0.15+0.40)	(8.+19.)		0.20(5)
^{113}Sn		115.1 d	≈ 9.	210.(50)		
^{114}Sn	0.66(1)		≈ 0.12	5.(1)	6.2(3)	134.(3) mb
^{115}Sn	0.34(1)		σ_α=0.06 mb	29.(6)		0.34(1)
^{116}Sn	14.54(9)		(0.006+0.14)	(0.5+11.)	5.93(5)	91.(2)mb
^{117}Sn	7.68(7)		1.1(1)	16.(5)	6.48(5)	319.(7)mb
^{118}Sn	24.22(9)		σ_m=4. mb	4.7(5)	6.07(5)	62.(1)mb
^{119}Sn	8.58(4)		2.(1)	2.9(5)	6.12(5)	0.25(5)
^{120}Sn	32.59(9)		(0.001+0.13)	1.2(3)	6.49(5)	(0.5+36)mb
^{122}Sn	4.63(3)		(0.15+0.001)	0.81(4)	5.74(5)	
^{124}Sn	5.79(5)		(0.13+0.005)	RI$_m$=8.0(2)	5.97(5)	
$_{51}$Sb			5.2(2)	169.(20)	5.57(3)	
^{121}Sb	57.21(5)		(0.4+5.6)	(13.+192.)	5.71(6)	0.53(2)
^{123}Sb	42.79(5)		(0.02+0.04+4.0)	(1.+119.)	5.38(7)	0.30(1)
^{124}Sb		60.20 d	17.(3)	≈8.		
$_{52}$Te			4.2(1)	47.(3)	5.80(3)	
^{120}Te	0.096(2)		(0.25+2.0)	≈1.	5.3(5)	
^{122}Te	2.603(4)		2.4(5)	80.(20)	3.8(2)	295.(3)mb
^{123}Te	0.908(2)	1.3x10^{13} y	370.(40),σ_α=0.05mb	4.5(3)x10^{3}	- 0.05	0.83(1)
^{124}Te	4.816(6)		(0.05+7.)	5.2(7)	8.0(1)	155.(2)mb
^{125}Te	7.139(6)		1.6(2)	21.(4)	5.02(8)	431.(4)mb
^{126}Te	18.952(11)		(0.12+0.8)	8.2(6)	5.56(7)	81.(1)mb
^{128}Te	31.687(11)		(0.016+0.20)	(0.08+1.5)	5.89(7)	
^{130}Te	33.799(10)	2.5x10^{21} y	(0.03+0.20)	(0.08+0.34)	6.02(7)	
$_{53}$I			6.15(10)	148.(7)	5.28(2)	
^{125}I		59.4 d	900.(100)	1.4(2)x10^{4}		
^{127}I	100.		6.15(10)	148.(7)	5.28(2)	0.64(3)
^{128}I		25.00 m	22.(4)	≈10.		
^{129}I		1.7x10^{7} y	(20.7+10.3)	36.(4)		0.44(2)
^{130}I		12.36 h	18.(3)	≈8.		
^{131}I		8.040 d	≈ 0.7	8.(4)		
$_{54}$Xe			25.(1)	263.(50)	4.92(3)	
^{124}Xe	0.09(1)		(28.+137.)	(0.6+3.0)x10^{3}		
^{125}Xe		17.1 h	σ_α≤<0.03			

Elem. or Isot.	Natural Abundance (%)	Half-life	Thermal Neut. Cross Section (barns)	Resonance Integral (barns)	Coh. Scat. Length (fm)	σ(30KEV) Maxw. Avg. (barns)
^{126}Xe	0.09(1)		(0.45+3.)	(8.+52.)		
^{127}Xe		36.4 d	$\sigma_\alpha \leq 0.01$			
^{128}Xe	1.92(3)		$\sigma_m=0.48$	$RI_m=38.(10)$		0.35(11)
^{129}Xe	26.44(24)		22.(5)	250.(50)		0.76(7)
^{130}Xe	4.08(2)		$\sigma_m=0.45$	$RI_m=16.(4)$		0.18(5)
^{131}Xe	21.18(3)		90.(10)	9.(1)x10^2		0.45(8)
^{132}Xe	26.89(6)		(0.05+0.4)	(0.9+3.7)		65.(5)mb
^{133}Xe		5.243 d	190.(90)			
^{134}Xe	10.44(10)		(0.003 + 0.26)	0.40(4)		20.(2)mb
^{135}Xe		9.10 h	2.65(11)x10^6	7.6(5)x10^3		
^{136}Xe	8.87(16)		0.26(2)	0.7(2)		0.9(1)mb
$_{55}$Cs			30.4(8)	422.(50)	5.42(2)	
^{132}Cs		6.48 d	$\sigma_\alpha <$ 0.15			
^{133}Cs	100.		(2.7+27.7)	(32.+390.)	5.42(2)	0.51(2)
^{134}Cs		2.065 y	140.(10)	54.(9)		
^{135}Cs		2.3x10^6 y	8.9(5)	90.(20)		
^{137}Cs		30.2 y	0.25(1)	0.36(7)		
$_{56}$Ba			1.3(2)	10.(2)	5.07(3)	
^{130}Ba	0.106(1)		(1.+8.)	(25.+200.)	- 3.6(6)	
^{132}Ba	0.101(1)		(0.84+9.7)	(4.7+24.)	7.8(3)	
^{133}Ba		10.53 y	4.(1)	85.(30)		
^{134}Ba	2.417(18)		(0.1+1.3)	(5.6+18.)	5.7(1)	0.18(6)
^{135}Ba	6.592(12)		(0.014+5.8)	(0.47+131.)	4.7(1)	0.30(6)
^{136}Ba	7.854(24)		(0.010+0.44)	(0.1+1.5)	4.91(8)	62.(2)mb
^{137}Ba	11.232(24)		5.(1)	4.(1)	6.8(1)	0.05(1)
^{138}Ba	71.698(42)		0.41(2)	0.4(1)	4.84(8)	4.1(2)mb
^{139}Ba		1.396 h	5.(1)	2.2(5)		
^{140}Ba		12.75 d	1.6(3)	14.(1)		
$_{57}$La			9.2(2)	12.(1)	8.24(4)	
^{138}La	0.090(1)	1.06x10^{11} y	57.(6)	4.1(9)x10^2		
^{139}La	99.910(1)		9.2(2)	12.(1)	8.24(4)	38.(3)mb
^{140}La		1.678 d	2.7(3)	69.(4)		
$_{58}$Ce			0.64(4)	0.71(6)	4.84(2)	
^{136}Ce	0.185(2)		(1.0+6.5)	58.(12)	5.80(9)	(28+272)mb
^{138}Ce	0.251(2)		(0.018+1.)	(1.5+5.2)	6.70(9)	179.(5)mb
^{140}Ce	88.450(18)		0.58(4)	0.50(5)	4.84(9)	11.0(4)mb
^{141}Ce		32.50 d	29.(3)	13.(2)		
^{142}Ce	11.114(17)		0.97(3)	1.3(3)	4.75(9)	28.(1)mb
^{143}Ce		1.38 d	6.1(7)	2.7(3)		
^{144}Ce		284.6 d	1.0(1)	2.6(3)		
$_{59}$Pr			11.5(4)	14.(3)	4.58(5)	
^{141}Pr	100.		(4.+7.5)	14.(3)	4.58(5)	0.12(2)
^{142}Pr		19.12 h	20.(3)	9.(1)		
^{143}Pr		13.57 d	90.(10)	190.(25)		

Elem. or Isot.	Natural Abundance (%)	Half-life	Thermal Neut. Cross Section (barns)	Resonance Integral (barns)	Coh. Scat. Length (fm)	σ(30KEV) Maxw. Avg. (barns)
$_{60}$Nd			51.(2)	49.(5)	7.69(5)	
^{142}Nd	27.2(5)		19.(1)	34.(11)	7.7(3)	46.(4)mb
^{143}Nd	12.2(2)		330.(10),σ_α=17.mb	128.(30)		0.24(1)
^{144}Nd	23.8(3)	2.1x10^{15} y	3.6(3)	3.9(5)	2.8(3)	0.11(1)
^{145}Nd	8.30(1)		47.(6),σ_α=12.μb	260.(40)		0.49(10)
^{146}Nd	17.2(3)		1.5(2)	3.0(4)	8.7(2)	0.16(4)
^{147}Nd		10.98 d	440.(150)	200.		
^{148}Nd	5.7(1)		2.4(1)	13.(2)	5.7(3)	0.19(4)
^{150}Nd	5.6(2)		1.0(1)	14.(2)	5.3(2)	
$_{61}$Pm						
^{146}Pm		5.53 y	8.4(1.7)x10^{3}			
^{147}Pm		2.6234 y	(84.+96.)	(1000.+1280.)	12.6(4)	
^{148m}Pm		41.3 d	10600.(800)			
^{148}Pm		5.37 d	≈ 10^{3}	2.6(2.4)x10^{3}		
^{149}Pm		2.212 d	1400.(200)			
^{151}Pm		1.183 d	≈ 150.			
$_{62}$Sm			5.6(1)x10^{3}	1.4(2)x10^{3}		
^{144}Sm	3.08(7)		1.6(1)	2.4(3)		
^{145}Sm		340. d	280.(20)	600.(90)		
^{147}Sm	14.99(17)	1.06x10^{11} y	56.(4),σ_α=0.6mb	710.(50)	14.(3)	0.97(1)
^{148}Sm	11.24(10)	7x10^{15} y	2.4(6)	27.(14)		241.(2)mb
^{149}Sm	13.82(7)	10^{16} y	40100.(600),σ_α=31.mb	3100.(500)		1.82(2)
^{150}Sm	7.38(1)		102.(5)	290.(30)	14.(3)	422.(4)mb
^{151}Sm		90. y	15200.(300)	3520.(60)		
^{152}Sm	26.74(14)		206.(15)	3000.(300)	- 5.0(6)	473.(4)mb
^{153}Sm		1.929 d	420.(180)			
^{154}Sm	22.75(27)		7.5(3)	32.(6)	9.(1)	
$_{63}$Eu			4570.(100)	3.8(5)x10^{3}	5.3(3)	
^{151}Eu	47.81(6)		(4.+3150.+6000.) σ_α=8.7(3)μb	(2000.+4000.)		4.4(2)
^{152m1}Eu		9.30 h	68000.(15000)	< 100000.		
^{152}Eu		13.5 y	11000.(2000)	1600.(200)		
^{153}Eu	52.19(6)		300.(20),σ_α<1.μb	1800.(400)	8.2(1)	3.2(3)
^{154}Eu		8.59 y	1500.(300)	1600.(200)		4.4(7)
^{155}Eu		4.76 y	3900.(200)	16000.(2000)		
$_{64}$Gd			48.8(6)x10^{3}	400.(10)	9.5(2)	
^{152}Gd	0.20(1)		700.(200),σ_α<7.mb	700.(200)		1.05(7)
^{153}Gd		241.6 d	20000.(10000),σ_α=0.03			
^{154}Gd	2.18(3)		(0.06+60.)	230.(50)		0.88(3)
^{155}Gd	14.80(12)		61.(1)x10^{3},σ_α=.08mb	1540.(100)		2.72(9)
^{156}Gd	20.47(9)		≈2.0	104.(15)	6.3(4)	0.64(6)
^{157}Gd	15.65(2)		2.54(3)x10^{5},σ_α<0.05	800.(100)		1.36(4)
^{158}Gd	24.84(7)		2.3(3)	73.(7)	9.(2)	0.22(2)
^{160}Gd	21.86(19)		1.5(7)	6.(1)	9.15(5)	
^{161}Gd		3.66 m	2.0(6)x10^{4}			

Elem. or Isot.	Natural Abundance (%)	Half-life	Thermal Neut. Cross Section (barns)	Resonance Integral (barns)	Coh. Scat. Length (fm)	σ(30KEV) Maxw. Avg. (barns)
$_{65}$Tb			23.2(5)	420.(50)	7.38(3)	
^{159}Tb	100.		23.2(5)	420.(50)	7.38(3)	1.75(10)
^{160}Tb		72.3 d	570.(110)			
$_{66}$Dy			9.5(2)x10^{2}	1.5(2)x10^{3}	16.9(3)	
^{156}Dy	0.06(1)		33.(3),σ_α<9mb	1000.(100)		
^{158}Dy	0.10(1)		43.(6),σ_α<6mb	120.(10)	6.1(5)	
^{159}Dy		144. d	8.(2)x10^{3}			
^{160}Dy	2.34(8)		60.(10),σ_α<0.3mb	1100.(200)	6.7(4)	0.77(4)
^{161}Dy	18.91(24)		600.(50),σ_α<1μb	1100.(100)	10.3(4)	2.01(6)
^{162}Dy	25.51(26)		170.(20)	2755.(300)	- 1.4(5)	0.47(5)
^{163}Dy	24.90(16)		120.(10),σ_α<20μb	1600.(400)	5.0(4)	1.14(4)
^{164}Dy	28.18(36)		(1.7+1.0)x10^{3}	(420.+100.)	49.4(2)	0.27(3)
^{165m}Dy		1.26 m	2.0(6)x10^{3}			
^{165}Dy		2.33 h	3.5(3)x10^{3}	2.2(3)x10^{4}		
$_{67}$Ho			61.(2)	670.(40)	8.01(8)	
^{165}Ho	100.		(3.1+58.),σ_α<20μb	(?+670.)	8.01(8)	1.32(7)
^{166}Ho		1.2x10^{3} y	9140.(650)	1140.(90)		
$_{68}$Er			169.(20)	730.(10)	7.79(2)	
^{162}Er	0.14(1)		19.(3),σ_α<11mb	480.(50)	8.8(2)	
^{164}Er	1.61(3)		13.(3),σ_α<1.2mb	105.(10)	8.2(2)	0.71(6)
^{166}Er	33.61(35)		(15.+5.),σ_α<70μb	96.(12)	10.6(2)	0.61(6)
^{167}Er	22.93(17)		700.(200),σ_α=3μb	2970.(70)	3.0(3)	1.5(2)
^{168}Er	26.78(26)		2.0(6),σ_α=0.09mb	37.(5)	7.4(4)	0.37(4)
^{170}Er	14.93(27)		6.(1)	26.(4)	9.6(5)	0.22(3)
^{171}Er		7.52 h	370.(40)	170.(20)		
$_{69}$Tm			106.(5)	1.5(2)x10^{3}	7.07(3)	
^{169}Tm	100		(8.+98.)	1.5(2)x10^{3}	7.07(3)	1.13(6)
^{170}Tm		128.6 d	100.(20)	460.(50)		
^{171}Tm		1.92 y	≈ 160.	118.(6)		
$_{70}$Yb				52.(10)	1.7(2)x10^{2}	12.43(3)
^{168}Yb	0.13(1)		2.4(2)x103,σ_α<0.1mb	2.0(5)x10^{4}	-4.07(2)	
^{169}Yb		32.03 d	3.6(3)x10^{3}	5200.(500)		
^{170}Yb	3.04(15)		12.(2),σ_α<10μb	320.(30)	6.8(1)	0.74(3)
^{171}Yb	14.3(6)		53.(5),σ_α<1.5μb	315.(30)	9.7(1)	1.41(5)
^{172}Yb	21.8(7)		≈ 1.3,σ_α<1μb	25.(3)	9.4(1)	0.41(3)
^{173}Yb	16.13(3)		16.(2),σ_α<1μb	380.(30)	9.56(7)	0.87(7)
^{174}Yb	31.8(9)		(46.+74.),σ_α<0.02mb	(13.+47.)	19.3(1)	0.18(2)
^{176}Yb	12.8(4)		3.1(2),σ_α<1μb	8.(2)	8.7(1)	
$_{71}$Lu			78.(7)	8.3(7)x10^{2}	7.21(3)	
^{175}Lu	97.41(2)		(16.+8.)	(550.+270.)	7.24(3)	1.18(4)
^{176}Lu	2.59(2)	3.8x10^{10} y	(2.+2100.)	(3.+1100.)	6.1(2)	1.54(6)

Elem. or Isot.	Natural Abundance (%)	Half-life	Thermal Neut. Cross Section (barns)	Resonance Integral (barns)	Coh. Scat. Length (fm)	σ(30KEV) Maxw. Avg. (barns)
^{177m}Lu		160.7 d	3.2(3)	1.4(2)		
^{177}Lu		6.75 d	1000.(300)			
$_{72}$Hf			106.(3)	19.7(5)x10^{2}	7.8(1)	
^{174}Hf	0.162(3)	2.0x10^{15} y	600.(50)	400.(50)	11.(1)	
^{176}Hf	5.206(5)		23.(4)	700.(100)	6.6(2)	0.46(2)
^{177}Hf	18.606(13)		(1.+375.),σ_α<20μb	7170.(200)		1.37(6)
^{178}Hf	27.297(4)		(53.+32.)	(340.+1500.)	5.9(2)	0.31(1)
^{179}Hf	13.629(6)		(0.43+46.)	(6.8+620.)	7.5(2)	0.99(3)
^{180}Hf	35.100(7)		13.(1),σ_α<13μb	32.(2)	13.2(3)	175.(5)mb
^{181}Hf		42.4 d	30.(25)			
$_{73}$Ta			20.(1)	650(20.)	6.91(7)	
^{180m}Ta	0.012(2)	> 1.2x10^{15} y	≈ 560.	1350.(100)		
^{181}Ta	99.998(2)		(0.012 + 20),σ_α<1μb	(0.4+650.)	6.91(7)	0.77(2)
^{182}Ta		114.43 d	8200.(600)	900.(90)		
$_{74}$W			18.(1)	3.6(3)x10^{2}	4.86(2)	
^{180}W	0.12(1)		≈ 4.	210.(30)		
^{182}W	26.50(16)		20.(1)	600.(90)	6.97(4)	274.(8)mb
^{183}W	14.31(4)		10.5(3)	340.(50)	6.53(4)	0.52(2)
^{184}W	30.64(2)		(0.002 + 2.0)	15.(2)	7.48(6)	0.23(1)
^{185}W		74.8 d	≈ 3.3	300.(50)		
^{186}W	28.43(19)		37.(2)	510.(50)	- 0.72(4)	176.(5)mb
^{187}W		23.9 h	70.(10)	2760.(550)		
$_{75}$Re			90.(4)	8.4(2)x10^{2}	9.2(3)	
^{185}Re	37.40(2)		(0.33+110.)	1700.(50)	9.0(3)	1.54(6)
^{187}Re	62.60(2)	4.4x10^{10} y	(2.+72.)	(9.+310.)	9.3(3)	1.16(6)
$_{76}$Os			17.(1)	1.5(1)x10^{2}	10.7(2)	
^{184}Os	0.020(3)		3.3(3)x10^{3},σ_α<10mb	1.4(1)x10^{3}		
^{186}Os	1.58(2)	2.x10^{15} y	≈ 80.,σ_α<0.1mb	280.(40)	12(2)	0.42(2)
^{187}Os	1.6(4)		2.(1)x10^{2},σ_α<0.1mb	500.(70)		0.87(3)
^{188}Os	13.3(1)		≈ 5.,σ_α<30μb	150.(20)	7.6(3)	0.40(2)
^{189}Os	16.1(1)		(0.00026+40),σ_α<10μb	(0.013+670.)	10.7(3)	1.17(5)
^{190}Os	26.4(2)		(9.+4.),σ_α<20μb	(22.+8.)	11.0(3)	0.30(5)
^{191}Os		15.4 d	3.8(6)x10^{2}			
^{192}Os	41.0(3)		3.(1),σ_α<10μb	6.(1)	11.5(4)	0.31(5)
^{193}Os		30.5 h	40.(10)			
$_{77}$Ir			4.2(1)x10^{2}	2.8(4)x10^{3}	10.6(3)	
^{191}Ir	37.3(2)		(660.+260.)	(1000.+4200.)		1.35(7)
^{192}Ir		73.83 d	1450.(250)	4800.(700)		
^{193}Ir	62.7(2)		(6.+110.)	1400.(200)		0.86(4)
^{194}Ir		19.3 h	1.5(3)x10^{3}			
$_{78}$Pt			10.(1)	1.3(1)x10^{2}	9.60(1)	

Elem. or Isot.	Natural Abundance (%)	Half-life	Thermal Neut. Cross Section (barns)	Resonance Integral (barns)	Coh. Scat. Length (fm)	σ(30KEV) Maxw. Avg. (barns)
^{190}Pt	0.014(1)	6.5x10^{11} y	1.5(1)x10^2,σ_α<8mb	70.(10)	9(1)	
^{192}Pt	0.782(7)		(2.0+6.),σ_α<0.2mb	115.(20)	9.9(5)	0.20(6)
^{194}Pt	32.97(10)		(0.1+1.1),σ_α< 5μb	(4.+?)	10.55(8)	0.37(8)
^{195}Pt	33.832(10)		28.(1),σ_α< 5μb	365.(50)	8.8(1)	0.9(2)
^{196}Pt	25.242(41)		(0.045+0.55)	7.(2)	9.89(8)	0.20(2)
^{198}Pt	7.16(6)		(0.027+3.6)	(0.5+56.)	7.8(1)	
^{199}Pt		30.8 m	≈ 16.			
$_{79}$Au			98.7(1)	1.55(3)x10^3	7.63(6)	
^{197}Au	100.		98.7(1)	1.55(3)x10^3	7.63(6)	582.(9)mb
^{198}Au		2.694 d	26.5(15)x10^3	≈ 4.x10^4		
^{199}Au		3.14 d	≈ 30.			
$_{80}$Hg			3.7(1)x10^2	87.(5)	12.69(2)	
^{196}Hg	0.15(1)		(105.+3000.)	(53.+410.)	30.(1)	
^{198}Hg	9.97(20)		(0.017+2.)	(1.7+70.)		0.17(2)
^{199}Hg	16.87(22)		2.1(2)x10^3	435(20)	16.9(4)	0.37(2)
^{200}Hg	23.10(19)		< 60.	2.1(5)		0.12(1)
^{201}Hg	13.18(9)		< 60.	30.(3)		0.26(1)
^{202}Hg	29.86(26)		4.9(5)	4.5(2)		74.(6)mb
^{204}Hg	6.87(15)		0.4(1)	0.8(2)		42.(4)mb
$_{81}$Tl			3.3(1)	12.5(8)	8.776(5)	
^{203}Tl	29.52(1)		11.(1),σ_α<0.3mb	41.(2)	7.0(2)	124.(8)mb
^{204}Tl		3.78 y	22.(2)	90.(20)		
^{205}Tl	70.48(1)		0.11(2)	0.6(2)	9.52(7)	54.(4)mb
$_{82}$Pb			0.172(2)	0.14(4)	9.402(2)	
^{204}Pb	1.4(1)		0.68(7)	2.0(2)	10.6(20)	90.(6)mb
^{205}Pb		1.51x10^7 y	≈ 5.			
^{206}Pb	24.1(1)		0.030(1)	0.10(1)	9.23(5)	14.(1)mb
^{207}Pb	22.1(1)		0.70(1)	0.38(1)	9.28(4)	8.4(4)mb
^{208}Pb	52.4(1)		0.49(3)mb,σ_α<8μb	2.0(2) mb	9.50(6)	0.36(2)mb
^{210}Pb		22.6 y	< 0.5			
$_{83}$Bi			0.034(1)	0.19(2)	8.532(2)	
^{209}Bi	100.		(11+23)mb,σ_α<0.3μb	0.19(2)	8.532(2)	3.1(3)mb
^{210m}Bi		3.0x10^6 y	54(4)mb	0.20(3)		
$_{84}$Po						
^{210}Po		138.38 d	σ_m<0.5mb,σ_α< 2.mb σ_g<30.mb,σ_f< 0.1			
$_{85}$At						
$_{86}$Rn						
^{220}Rn		55.6 s	<0.2			

Elem. or Isot.	Natural Abundance (%)	Half-life	Thermal Neut. Cross Section (barns)	Resonance Integral (barns)	Coh. Scat. Length (fm)	σ(30KEV) Maxw. Avg. (barns)
^{222}Rn		3.8235 d	0.74(5)			
$_{88}Ra$						
^{223}Ra		11.435d	$1.3(2)x10^2, \sigma_f<0.7$			
^{224}Ra		3.66 d	12.0(5)			
^{226}Ra		1599. y	$\approx$ 13., $\sigma_f<$ 7.μb	280.(50)	10.(1)	
^{228}Ra		5.76 y	36.(5), $\sigma_f<2.$			
$_{89}Ac$						
^{227}Ac		21.77 y	$8.8(7)x10^2, \sigma_f<.35$ mb	$1.5(4)x10^3$		
$_{90}Th$			7.4	85.(3)	10.31(3)	
^{227}Th		18.72 d	$\sigma_f=2.0(2)x10^2$			
^{228}Th		1.913 y	$1.2(2)x10^2, \sigma_f<0.3$	1014.(400)		
^{229}Th		$7.9x10^3$ y	$\approx$ 60. $\sigma_f=30(3)$	$1.0(2)x10^3$ $RI_f=466.(75)$		
^{230}Th		$7.54x10^4$ y	23.4(5) $\sigma_f<0.5$mb	$1.0(1)x10^3$		
^{232}Th	100.	$1.4x10^{10}$y	7.37(4) $\sigma_f=3.(1)$ μb $\sigma_\alpha<1.$ μb	85.(3)	10.31(3)	
^{233}Th		22.3 m	$1.5(1)x10^3$ $\sigma_f=15.(2)$	$4.(1)x10^2$		
^{234}Th		24.10 d	1.8(5) $\sigma_f<0.01$			
$_{91}Pa$						
^{230}Pa		17.4 d	$1.5(3)x10^3$			
^{231}Pa		$3.25x10^4$ y	$2.0(1)x10^2$ $\sigma_f=20.(1)$ mb	750.(80) $RI_f=0.05(1)$	9.1(3)	
^{232}Pa		1.31 d	$4.6(10)x10^2$ $\sigma_f=7.(1)x10^2$	300.(70)		
^{233}Pa		27.0 d	39.(2) $\sigma_m=20.(4)$ $\sigma_g=19.(3)$ $\sigma_f<$ 0.1	(460.+440)		
$_{92}U$			3.4(3); $\sigma_f=4.2(1)$	280.(20), $RI_f=2.0$	8.417(5)	
^{230}U		20.8 d	$\sigma_f\approx$ 25.			
^{231}U		4.2 d	$\sigma_f\approx$ 250.			
^{232}U		68.9 y	73.(2) $\sigma_f=74.(8)$	280.(15) $RI_f=350.(30)$		
^{233}U		$1.59x10^5$ y	47.(2) $\sigma_f=5.3(1)x10^2$ $\sigma_\alpha<0.2$ mb	137.(6) $RI_f=760.(17)$	10.1(2)	
^{234}U	0.0055(5)	$2.45x10^5$ y	96.(2) $\sigma_f<$ 5. mb	660.(70) $RI_f=6.5$	12.(4)	

Elem. or Isot.	Natural Abundance (%)	Half-life	Thermal Neut. Cross Section (barns)	Resonance Integral (barns)	Coh. Scat. Length (fm)	σ(30KEV) Maxw. Avg. (barns)
^{235}U	0.720(1)	7.04x10^{8} y	95.(5)	144.(6)	10.47(4)	
			σ_f=586.(2)	RI_f=275(5)		
			σ_α< 0.1 mb			
^{236}U		2.34x10^{7} y	5.1(3)	360.(15)		
			σ_f=0.04	RI_f=4.38(50)		
^{237}U		6.75 d	≈ 10^{2}	1200.(200)		
			σ_f< 0.35			
^{238}U	99.2745(15)	4.46x10^{9} y	2.7(1)	277.(3)	8.402(5)	
			σ_f≈3.μb	1.54(15) mb		
			σ_α=1.4(5) μb			
^{239}U		23.5 m	22.(2)			
			σ_f=15.(3)			
$_{93}$Np						
^{234}Np		4.4 d	σ_f=9.(3)x10^{2}			
^{235}Np		1.085 y	1.6(1)x10^{2}			
^{236m}Np		22.5 h	σ_f=2.7(2)x10^{3}	700.(400)		
^{236}Np		1.55x10^{5} y	σ_f=2.6(5)x10^{3}	1030.(100)		
^{237}Np		2.14x10^{6} y	160.(10)	8.(2)x10^{2}	10.6(1)	
			σ_f=20.(1)mb	RI_f=0.32		
^{238}Np		2.117 d	σ_f=2.1(1)x10^{3}	9.(1)x10^{2}		
^{239}Np		2.355 d	(32.+19.)			
			σ_f<1.			
$_{94}$Pu						
^{236}Pu		2.87 y	σ_f=1.6(3)x10^{2}	1000.(60)		
^{237}Pu		45.7 d	σ_f=2.3(3)x10^{3}			
^{238}Pu		87.74 y	5.1(2)x10^{2}	1.6(2)x10^{2}	14.1(5)	
			σ_f=17.(1)	RI_f=33.(5)		
^{239}Pu		2.411 x 10^{4}y	2.7(1)x10^{2}	2.0(2)x10^{2}	7.7(1)	
			σ_f=752.(3)	3.0(1)x10^{2}		
			σ_α≤0.3 mb			
^{240}Pu		6537. y	2.9(1)x10^{2}	8.4(3)x10^{3}	3.5(1)	
			σ_f≈44.mb	RI_f=2.4		
^{241}Pu		14.4 y	3.7(1)x10^{2},σ_α<0.2mb	1.6(1)x10^{2}		
			σ_f=1.01(1)x10^{3}	5.7(4)x10^{2}		
^{242}Pu		3.76 x 10^{5} y	19.(1)	1.1(1)x10^{3}	8.1(1)	
			σ_f<0.2	RI_f=0.23		
^{243}Pu		4.956 h	<100.			
			σ_f=2.0(2)x10^{2}			
^{244}Pu		8.2x10^{7} y	1.7(1)	41.(3)		
^{245}Pu		10.5 h	1.5(3)x10^{2}	220(40)		
$_{95}$Am						
^{241}Am		432.2 y	(0.5+5.7)x10^{2}	(2.0+12.3)x10^{2}		
			σ_f=3.1(2)	14.(1)		
^{242m}Am		141. y	1.7(4)x10^{3}	≈ 200.		
			σ_f=7.0(3)x10^{3}	RI_f=1.8(1)x10^{3}		
^{242}Am		16.02 h	σ_f=2.1(2)x10^{3}	RI_f=< 300.		
^{243}Am		7.37x10^{3}y	(75.+5.)	(17.1+1.0)x10^{2}	8.3(2)	
			σ_f=74.(4) mb	RI_f=0.056		

Elem. or Isot.	Natural Abundance (%)	Half-life	Thermal Neut. Cross Section (barns)	Resonance Integral (barns)	Coh. Scat. Length (fm)	σ(30KEV) Maxw. Avg. (barns)
^{244m}Am		≈ 26. m	σ_f=1.6(3)x10^3			
^{244}Am		10.1 h	σ_f=2.2(3)x10^3			
$_{96}$Cm						
^{242}Cm		162.8 d	≈ 20.	120.(50)		
			σ_f≈ 5.			
^{243}Cm		28.5 y	1.3(1)x10^2	214.(20)		
			σ_f=6.2(2)x10^2	RI$_f$=1.6(1)x10^3		
^{244}Cm		18.11 y	15.(1)	640.(50)	9.5(3)	
			σ_f=1.1(2)	RI$_f$=10.8(8)		
^{245}Cm		8.5x10^3 y	3.5(2)x10^2	110(10)		
			σ_f=2.1(1)x10^3	RI$_f$=8.(1)x10^2		
^{246}Cm		4.78x10^{3}y	1.2(2)	120.(10)	9.3(2)	
			σ_f=0.16(7)	13.(2)		
^{247}Cm		1.56x10^{7}y	60.(30)	5.(1)x10^2		
			σ_f=82.(5)	7.3(7)x10^2		
^{248}Cm		3.4x10^5 y	2.6(3)	270.(30)	7.7(2)	
			σ_f=0.36(7)	13.(2)		
^{249}Cm		64.15 m	≈1.6			
^{250}Cm		≈ 9.7x10^3 y	≈80.			
$_{97}$Bk						
^{249}Bk		320. d	7.(1)x10^2	9.(1)x10^2		
			σ_f≈0.1			
^{250}Bk		3.217 h	σ_f=1.0(2)x10^3			
$_{98}$Cf						
^{249}Cf		351. y	5.0(3)x10^2	7.7(4)x10^2		
			σ_f=1.7(1)x10^3	RI$_f$=2.1(3)x10^3		
^{250}Cf		13.1 y	2.0(2)x10^3	12.(2)x10^3		
			σ_f=110.(90)	RI$_f$=160.(40)		
^{251}Cf		9.0x10^2 y	2.9(2)x10^3	1.6(1)x10^3		
			σ_f=4.5(5)x10^3	RI$_f$=5.5(3)x10^3		
^{252}Cf		2.65 y	20.(2)	43.(3)		
			σ_f=32.(4)	RI$_f$=1.1(3)x10^2		
^{253}Cf		17.8 d	18.(2)	8.(1)		
			σ_f=1.3(2)x10^3			
^{254}Cf		60.5 d	4.5(10)	2.		
$_{99}$Es						
^{253}Es		20.47 d	(180.+5.8)	(37.5+1.1)x10^2		
^{254m}Es		1.64 d	σ_f=1.8(1)x10^3			
^{254}Es		276. d	28.(3)	18.(2)		
			σ_f=1.8(2)x10^3	RI$_f$=1.2(3)x10^3		
^{255}Es		40. d	≈ 55.			
$_{100}$Fm						
^{255}Fm		20.1 h	26.(3)	14.(2)		
			σ_f=3.3(2)x10^3			
^{257}Fm		100.5 d	σ_f=3.0(2)x10^3			

COSMIC RADIATION

A.G. Gregory and R.W. Clay

The Nature of Cosmic Rays

Primary cosmic radiation, in the form of high energy nuclear particles, electrons and photons from outside the solar system and from the Sun, continually bombards our atmosphere. Secondary radiation, resulting from the interaction of the primary cosmic rays with atmospheric gas, is present at sea-level and throughout the atmosphere.

The secondary radiation is collimated by absorption and scattering in the atmosphere and consists of a number of components associated with different particle species. High energy primary particles can produce large numbers of secondary particles forming an extensive air shower. Thus, a number of particles may then be detected simultaneously at sea-level.

Primary particle energies accessible in the vicinity of the earth range from ~10^8 eV to ~10^{20} eV. At the lower energies, the limit is determined by the inability of charged particles to traverse the heliosphere to us through the outward-moving solar wind. The upper energy limit is set by the practicality of building detectors to record particles with the extremely low fluxes found at those energies (J.G. Wilson, 1976; O.C. Allkofer, 1975a).

Primary Cosmic Rays

Primary Particle Energy Spectrum

Figure 1 shows the spectrum of primary particle energies. This includes all particle species. In differential form it is roughly a power law of intensity versus energy with an index of ~ –3. There appears to be a knee (a steepening) at a little above 10^{15} eV and an ankle (a flattening) above ~10^{18} eV. Figure 2 emphasizes the features in the spectrum at the highest energies through multiplying the flux with a strongly rising power law of energy. This figure should be used with caution as errors for the two axes are not now independent.

Data on the high energy cosmic ray spectrum are uncertain largely because of limited event statistics due to the very low flux which might best be measured in particles per square kilometer per century. The highest energy event recorded to 1995 had an energy of 3×10^{20} eV (D.J. Bird et al., 1993).

It is expected that the highest energy cosmic rays will interact with the 2.7 K cosmic microwave background through photoproduction or photodisintegration. These interactions will appreciably reduce the observed flux of cosmic rays with energies above 5×10^{19} eV if they travel further than ~150 million light years. This process is known as the Greisen-Zatsepin-Kuz'min (GZK) cut off (P. Sokolsky, 1989).

At energies below ~10^{13} eV, solar system magnetic fields and plasma can modulate the primary component and Figure 3 shows the extent of this modulation between solar maximum and minimum (E. Juliusson, 1975; J. Linsley, 1981).

Primary Particle Energy Density

If the above spectrum is corrected for solar effects, the energy density above a particle energy of 10^9 eV outside the solar system is found to be ~5×10^5 eV m^{-3}. As the threshold energy is increased, the energy density decreases rapidly, being 2×10^4 eV m^{-3} above 10^{12} eV and 10^2 eV m^{-3} above 10^{15} eV. The energy density at lower energies outside the heliosphere is unknown but may be substantially greater if the particle rest mass energy is included together with the kinetic energy (A. W. Wolfendale, 1979).

Primary Particle Isotropy

This is measured as an anisotropy $(I_{max} - I_{min})/(I_{max} + I_{min}) \times 100\%$, where I, the intensity (m^{-2}s^{-1}sr^{-1}), is usually measured with an angular resolution of a few degrees.

The measured anisotropy is small and energy dependent. It is roughly constant in amplitude at between 0.05 and 0.1% (with a phase of 0 to 6 hours in right ascension) for energies between 10^{11} eV and 10^{14} eV and appears to increase at higher energies roughly as $0.4 \times (\text{Energy(eV)}/10^{16})^{0.5}\%$ up to ~10^{18} eV. The latter rise may well be an artifact of the progressively more limited statistics as the flux drops rapidly with energy. It appears possible that a real anisotropy has been observed at the highest energies (above a few times 10^{19} eV) with a directional preference for the supergalactic plane (this plane reflects the directions of galaxies within about 100 million light years) (A.W. Wolfendale, 1979; R.W. Clay, 1987; T. Stanev et al., 1995).

Primary Particle Composition

The composition of low energy cosmic rays is close to universal nuclear abundances except where propagation effects are present. For example, Li, Be, and B which are spallation products, are over-abundant by about six orders of magnitude.

Composition at 10^{11} eV per nucleus

Charge	1	2	(3–5)	(6–8)	(10–14)	(16–24)	(26–28)	≥30
% Composition (10% uncertainty)	50	25	1	12	7	4	4	0.1

Measurements at higher energies indicate that there is an increase in the relative abundances of nuclei with charge greater than 6 at energies above 50 TeV/nucleus (K. Asakimori et al., 1993) (1 TeV = 10^{12} eV).

Cosmic ray composition at low energies is often quoted at a fixed energy per nucleon. When presented in this way, protons constitute roughly 90% of the flux, helium nuclei about 10% and the remainder sum to a total of about 1%.

Certain radioactive isotopic ratios show lifetime effects. The ratio of Be^{10}/B^9 abundances is used to measure an "age" of cosmic rays since Be^{10} is unstable with a half life of about 1.6×10^6 years. A ratio of 0.6 is expected in the absence of Be^{10} decay and a ratio of about 0.2 is found experimentally (E. Juliusson, 1975; P. Meyer, 1981).

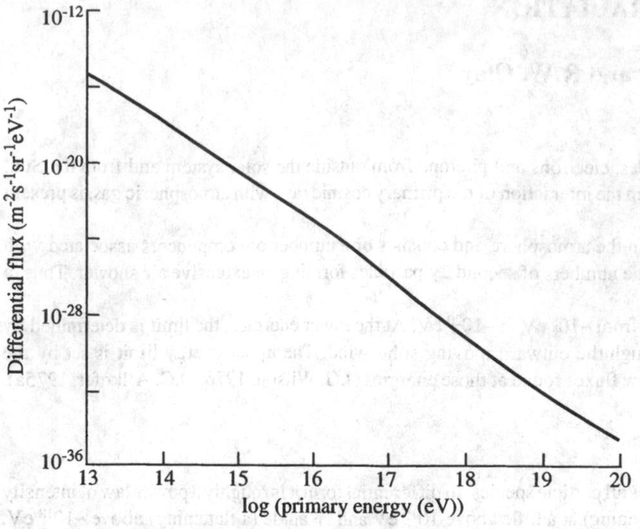

FIGURE 1. The energy spectrum of cosmic ray particles. This spectrum is of a differential form and can be converted to an integral spectrum by integration over all energies above a required threshold (E). Insofar as the spectrum approximates a power law of index –3, a simple conversion to the integral at an energy *E*/1.8 is obtained by multiplying the differential flux by the energy and dividing by 0.62.

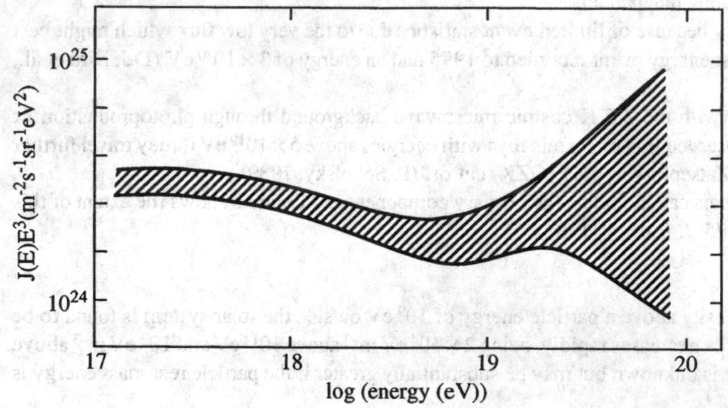

FIGURE 2. Energy spectrum at the highest energies. This spectrum (after Yoshida et al., 1995) has the differential spectrum multiplied by energy cubed. It is from a compilation of a number of measurements and indicates the good general agreement at the lower energies and a spread due to inadequate statistics at the highest energies.

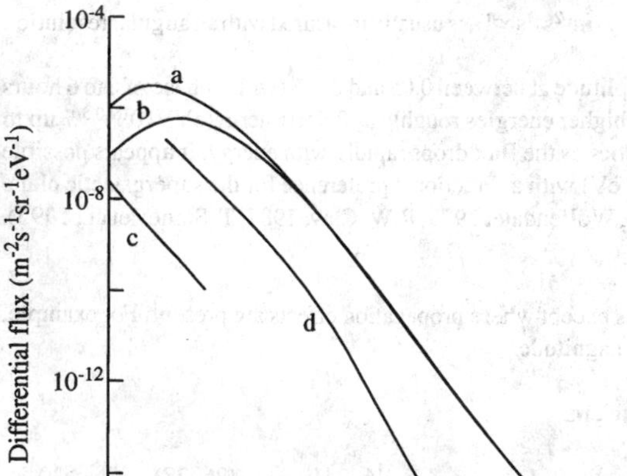

FIGURE 3. Energy spectrum of particles at lower energies. (a) Solar minimum proton energy spectrum. (b) Solar maximum proton energy spectrum. (c) Gamma-ray energy spectrum. (d) Local interstellar electron spectrum.

At higher energies, composition determinations are indirect and are rather contradictory and controversial. Experiments aim to differentiate between broad composition models. The measurement technique is based on studies of cosmic ray shower development. A rather direct technique for such studies is to use fluorescence observations of the shower development to determine the atmospheric depth of maximum development of the shower. Such observations suggest a heavy composition (large atomic number) at energies ~10^{17} eV which changes with increasing energy to a light composition (perhaps protonic) above ~10^{19} eV (T. K. Gaisser et al., 1993).

Primary Electrons

Primary electrons constitute about 1% of the cosmic ray beam. The positron to negative electron ratio is about 10% (J. M. Clem et al., 1995).

Antimatter in the Primary eam

The ratio of antiprotons to protons in the primary cosmic ray beam (at about 400 MeV) is about 10^{-5}. At about 10 GeV the ratio is about 10^{-3}. At the highest measured energies (10 TeV), the upper limit to the ratio is about 20% (S. Orito et al., 1995; M. Amenomori et al., 1995).

Primary Gamma-Rays

The flux of primary gamma-rays is low at high energies. At 1 GeV the ratio of gamma-rays to protons is about 10^{-6}. The arrival directions of these gamma-rays are strongly concentrated in the plane of the Milky Way although there is a diffuse, near isotropic background flux and some point sources have been detected.

Since the absorption cross section for gamma-rays above 100 MeV is approximately 20 mbarn/electron, less than 10% of gamma-rays reach mountain altitudes (A. W. Wolfendale, 1979; P. F. Michelson, 1994).

Sea Level Cosmic Radiation

The sea level cosmic ray dose is 300 millirad·yr^{-1} and the sea level ionization is 2.2×10^6 ion pairs $m^{-3}s^{-1}$. The sea level flux has a soft component, which can be absorbed in about 100 mm of lead (about 100 g·cm^{-2} of absorber) and a more penetrating (largely muon) hard component. The sea level radiation is largely produced in the atmosphere and is a secondary component from interactions of the primary particles. The steep primary energy spectrum means that most secondaries at sea level are from rather low energy primaries. Thus the secondary flux is dependent on the solar cycle and the geomagnetic latitude of the observer.

Absolute Flux of the Hard Component

Vertical Integral Intensity I(0) ~100 $m^{-2}s^{-1}sr^{-1}$
Angular dependence $I(\theta) \sim I(0)\cos^2(\theta)$
Integrated Intensity ~200 $m^{-2}s^{-1}$
(O.C. Allkofer, 1975b).

Flux of the Soft Component

In free air, the soft component comprises about one third of the total cosmic ray flux.

Latitude Effect

The geomagnetic field influences the trajectories of lower energy cosmic rays approaching the Earth. As a result, the background flux is reduced by about 7% at the geomagnetic equator. The effect decreases towards the poles and is negligible at latitudes above about 40°.

Flux of Protons

The proton component is strongly attenuated by the atmosphere with an attenuation length (reduction by a factor of e) of about 120 g·cm^{-2}. It constitutes about 1% of the total vertical sea level flux.

Absorption

The soft component is absorbed in about 100 g·cm^{-2} of matter. The hard component is absorbed much more slowly:

Absorption in lead, 6% per 100 g·cm^{-2}
Absorption in rock, 8.5% per 100 g·cm^{-2}
Absorption in water, 10% per 100 g·cm^{-2}
(Absorption for depths less than 100 g·pd cm^{-2} is given by K. Greisen, 1943.)

Altitude Dependence

The cosmic ray background in the atmosphere has a maximum intensity of about 15 times that at sea level at a depth of about 150 g·cm^{-2} (15 km altitude). At maximum intensity, the soft and hard components contribute roughly equally but the hard component is then attenuated more slowly (S. Hayakawa, 1969).

Cosmic Ray Showers

High energy cosmic rays produce particle cascades in the atmosphere which can be detected at sea level provided that their energy exceeds about 100 GeV (such low energy cascades may be detected by using the most sensitive atmospheric Cerenkov detectors). The primary particle progressively loses energy which is transferred through the production of successive generations of secondary particles to a cascade of hadrons, an electromagnetic shower component (both positively and negatively charged electrons and gamma-rays) and muons. The secondary particles are relativistic and all travel effectively at the speed of light. As a result, they reach sea level at approximately the same time but, due to Coulomb scattering (for the electrons) and production angles (for the pions producing the muons), are spread laterally into a disk-like shower front with a characteristic lateral width of several tens of meters and thickness (near the central shower core) of 2 to 3 m. The number of particles at sea level is roughly proportional to the primary particle energy:

Number of particles at sea level ~10^{-10} × energy (eV).

At altitudes below a few kilometers, the number of particles in a shower attenuates with an *attenuation length* of about 200 g·cm^{-2}.

i.e., particle number = original number × exp(–(depth increase)/200)

The above applies to an individual shower. The rate of observation of showers of a given size (particle number at the detector) at different depths of absorber attenuates with an *absorption length* of about 100 g·cm^{-2} (J.G. Wilson, 1976).

Atmospheric Background Light from Cosmic Rays

Cosmic ray particles produce Cerenkov light in the atmosphere and produce fluorescent light through the excitation of atmospheric molecules.

Cerenkov Light

High energy charged particles will cause the emission of Cerenkov light in air if their energies are above about 30 MeV (electrons). This threshold is pressure (and hence altitude) dependent. A typical Cerenkov light pulse (at sea level, 100 m from the central shower core) has a time spread of a few nanoseconds. Over this time, the photon flux between 430 and 530 nm would be ~10^{14} m^{-2}s^{-1} for a primary particle energy of 10^{16} eV. For comparison, the night sky background flux is ~6 × 10^{11} photons m^{-2}s^{-1}sr^{-1} in the same wavelength band (J.V. Jelley, 1967).

Fluorescence Light

Cosmic ray particles in the atmosphere excite atmospheric molecules which then emit fluorescence light. This is weak compared to the highly collimated Cerenkov component when viewed in the direction of the incident cosmic ray particle but is emitted isotropically. Typical pulse widths are longer than 50 ns and may be up to several microseconds for the total pulse from distant large showers (R.M. Baltrusaitis et al., 1985).

Effects of Cosmic Rays

Cerenkov Effects in Transparent Media

Background cosmic ray particles will produce Cerenkov light in transparent material with a photon yield between wavelengths λ_1 and λ_2

$$\sim (2\pi/137)\sin^2(\theta_c)\int_{\lambda_1}^{\lambda_2} d\lambda/\lambda^2 \text{ photons (unit length)}^{-1}$$

where θ_c (the Cerenkov angle) = $\cos^{-1}$ (1/refractive index).

This background light is known to affect light detectors, e.g., photomultipliers, and can be a major source of background noise (R.W. Clay and A.G. Gregory, 1977).

Effects on Electronic Components

If background cosmic ray particles pass through electronic components, they may deposit sufficient energy to affect the state of, e.g., a transistor flip-flop. This effect may be significant where reliability is of great importance or the background flux is high. For instance, it has been estimated that, in communication satellite operation, an error rate of about 2 × 10^{-3} per transistor per year may be found. Permanent damage may also result. A significant error rate may be found even at sea level in large electronic memories. This error rate is dependent on the sensitivity of the component devices to the deposition of electrons in their sensitive volumes (J.F. Ziegler, 1981).

Biophysical Significance

When cosmic rays interact with living tissue, they produce radiation damage. The amount of the damage depends on the total dose of radiation. At sea level, this dose is small compared with doses from other sources but both the quantity and quality of the radiation change rapidly with altitude. Approximate dose rates under various conditions are:

Dose rates (mrem·yr^{-1})
Sea level cosmic rays, 30
Cosmic rays at 10 km (subsonic jets), 2000
Cosmic rays at 18 km (supersonic transports), 10,000
(c.f., mean total sea level dose, 300)

Astronauts would be subject to radiation from galactic (0.05 rads per day) and solar (a few hundred rads per solar flare) cosmic rays as well as large fluxes of low energy radiation when passing through the Van Allen belts (about 0.3 rads per traverse).
Both astronauts and SST travellers would be subject to a small flux of low energy heavy nuclei stopping in the body. Such particles are capable of destroying cell nuclei and could be particularly harmful in the early stages of the development of an embryo. The rates of heavy nuclei stopping in tissue in supersonic transports and spacecraft are approximately as follows:

Stopping nuclei ((cm^3 tissue)$^{-1}$ hr^{-1})
Supersonic transport (16 km), 0.0005
Supersonic transport (20 km), 0.005
Spacecraft, 0.15
(O. C. Allkofer, 1975a; O. C. Allkofer et al., 1974).

Carbon Dating

Radiocarbon is produced in the atmosphere due to the action of cosmic ray slow neutrons. Solar cycle modulation of the very low energy cosmic rays causes an anticorrelation of the atmospheric ^{14}C activity with sunspot number with a mean amplitude of about 0.5%. In the long term, modulation of cosmic rays by a varying magnetic field may be important (A.A. Burchuladze et al., 1979).

Practical Uses of Cosmic Rays

There are few direct practical uses of cosmic rays. Their attenuation in water and snow have, however, enabled automatic monitors of water and snow depth to be constructed. A search for hidden cavities in pyramids has been carried out using a muon "telescope".

Other Effects

Stellar X-rays have been observed to affect the transmission times of radio signals between distant stations by altering the depth of the ionospheric reflecting layer. It has also been suggested that variations in ionization of the atmosphere due to solar modulation may have observable effects on climatic conditions.

REFERENCES

O.C. Allkofer, (1975a) *Introduction to Cosmic Radiation,* Verlag Karl Thiemig, Munchen, Germany.
O.O. Allkofer, (1975b) *J. Phys. G: Nucl. Phys.,* 1, L51.
O.C. Allkofer and W. Heinrich, (1974) *Health Phys.,* 27, 543.
M. Amenomori et al., (1995) Proc. 24th Int. Cosmic Ray Conf. Rome, 3, 85. Universita La Sapienza, Roma.
K. Asakimori et al., (1993) Proc. 23rd Int. Cosmic Ray Conf. Calgary, 2, 25, University of Calgary, Canada.
R.M. Baltrusaitis et al., (1985) *Nucl. Inst. Meth.,* A420, 410.
D.J. Bird et al., (1993) *Phys. Rev. Lett.,* 71, 3401.
A.A. Burchuladze, S.V. Pagava, P. Povinec, G. I. Togonidze, S. Usacev, (1979) Proc. 16th Int. Cosmic Ray Conf. Kyoto, 3, 201, Univ. of Tokyo, Japan.
R.W. Clay, (1987) *Aust. J. Phys.,* 40, 423.
R.W. Clay and A.G. Gregory, (1977) *J. Phys. A: Math. Gen.,* 10, 135.
J.M. Clem et al., (1995) Proc. 24th Int. Cosmic Ray Conf. Rome, 3, 5, Universita La Sapienza, Roma.
T.K. Gaisser et al., (1993) *Phys. Rev. D,* 47, 1919.
K. Greisen, (1943) *Phys. Rev.,* 63, 323.
S. Hayakawa, (1969) *Cosmic Ray Physics,* Wiley-Interscience, New York.
J.V. Jelley, (1967) *Prog. in Elementary Particle and Cosmic Ray Physics,* 9, 41.
E. Juliusson, (1975) Proc. 14th Int. Cosmic Ray Conf. Munich, 8, 2689, Max Planck Institute fur Extraterrestriche Physik, Munchen, Germany.
J. Linsley, (1981) *Origin of Cosmic Rays,* I.A.U. Symposium 94, 53, D. Reidel Publishing Co Dordrecht, Holland.
P. Meyer, (1981) *Origin of Cosmic Rays,* I.A.U. Symposium 94, 7, D. Reidel Publishing Co. Dordrecht, Holland.
P.F. Michelson (1994) in *Towards a Major Atmospheric Cerenkov Detector III,* 257, Ed. T. Kifune, Universal Academy Press Inc., Tokyo, Japan.
P. Sokolsky, (1989) *Introduction to Ultrahigh Energy Cosmic Ray Physics,* Addison Wesley Publishing Company.
T. Stanev et al., (1995) *Phys. Rev. Lett.,* 75, 3056.
S. Orito et al., (1995) Proc. 24th Int. Cosmic Ray Conf. Rome, 3, 76. Universita La Sapienza, Roma.
J.G. Wilson, (1976) *Cosmic Rays,* Wykeham Pub. (London) Lt., U.K.
A.W. Wolfendale, (1979) *Pramana,* 12, 631.
S.Yoshida et al., (1995) *Astroparticle Phys.,* 3, 105.
J.F. Ziegler, (1981) IEEE Trans. Electron Devices, ED-28, 560.

Section 12
Properties of Solids

Techniques for Materials Characterization 12-1
Symmetry of Crystals 12-7
Ionic Radii in Crystals 12-14
Polarizability of Atoms and Ions in Solids 12-17
Crystal Structures and Lattice Parameters of Allotropes of the Elements 12-19
Lattice Energies 12-22
The Madelung Constant and Crystal Lattice Energy 12-36
Elastic Constants of Single Crystals 12-37
Electrical Resistivity of Pure Metals 12-45
Electrical Resistivity of Selected Alloys 12-48
Permittivity (Dielectric Constant) of Inorganic Solids 12-52
Curie Temperature of Selected Ferroelectric Crystals 12-61
Properties of Antiferroelectric Crystals 12-62
Dielectric Constants of Glasses 12-63
Properties of Superconductors 12-64
High Temperature Superconductors 12-91
Organic Superconductors 12-94
Properties of Semiconductors 12-97
Diffusion Data for Semiconductors 12-108
Properties of Magnetic Materials 12-117
Organic Magnets 12-126
Electron Work Function of the Elements 12-130
Secondary Electron Emission 12-131
Optical Properties of Selected Elements 12-133
Optical Properties of Selected Inorganic and Organic Solids 12-157
Elasto-optic, Electro-optic, and Magneto-optic Constants 12-179
Nonlinear Optical Constants 12-196
Phase Diagrams 12-200
Heat Capacity of Selected Solids 12-218
Thermal and Physical Properties of Pure Metals 12-219
Thermal Conductivity of Metals and Semiconductors as a Function of Temperature 12-221
Thermal Conductivity of Alloys as a Function of Temperature 12-223
Thermal Conductivity of Crystalline Dielectrics 12-224
Thermal Conductivity of Ceramics and Other Insulating Materials 12-226
Thermal Conductivity of Glasses 12-228
Commercial Metals and Alloys 12-232
Hardness of Minerals and Ceramics 12-233

TECHNIQUES FOR MATERIALS CHARACTERIZATION

EXPERIMENTAL TECHNIQUES USED TO DETERMINE THE COMPOSITION, STRUCTURE, AND ENERGY STATES OF SOLIDS AND LIQUIDS

H.P.R.Frederikse

The many experimental methods, originally designed to study the chemical and physical behavior of solids and liquids, have grown into a new field known as Materials Characterization (or Materials Analysis). During the past 30 years a host of techniques aimed at the study of surfaces and thin films has been added to the many tools for the analysis of bulk samples. The field has benefitted particularly from the development of computers and microprocessors, which have vastly increased the speed and accuracy of the measuring devices and the recording of their output. Materials characterization was and is a very important tool in the search for new physical and chemical phenomena. It plays an essential role in new applications of solids and liquids in industry, communications, and medicine. Many of its techniques are used in quality control, in safety regulations, and in the fight against pollution.

In most Materials Characterization experiments the sample is subjected to some kind of radiation: electromagnetic, acoustic, thermal, or particles (electrons, ions, neutrons, etc.). The surface analysis techniques usually require a high vacuum. As a result of interactions between the solid (or liquid) and the incoming radiation a beam of a similar (or a different) nature will emerge from the sample. Measurement of the physical and/or chemical attributes of this emerging radiation will yield qualitative, and often quantitative, information about the composition and the properties of the material being probed.

The modern tendency of describing practically everything in this world by a combination of a few letters (acronyms) has also penetrated the field of Materials Characterization. The table below gives the meaning of the acronym for every technique listed, the form and size of the required sample (bulk, surface, film, liquid, powder, etc.), the nature of the incoming and of the emerging radiation, the depth and the lateral spatial resolution that can be probed, and the information obtained from the experiment. The last column lists one or two major references to the technique described.

	Technique	Sample	In	Out	Depth	Lateral resolution	Information obtained	Ref.
	OPTICAL AND MASS SPECTROSCOPIES FOR CHEMICAL ANALYSIS							
1.	AAS Atomic Absorption Spectroscopy	Atomize (flame, electro, thermal, etc.)	Light e.g., glow discharge	Absorption spectrum	—	—	Concentration of atomic species (quantitative, using standards)	1,2
2.	ICP-AES Induct. Coupled Plasma — Atomic Emission Spectroscopy	Atomize (flame, electro, thermal, ICP, etc.)	—	Emission spectrum	—	—	Concentration of atomic species (quantitative, using standards)	3
3.	Dynamic SIMS Dynamic Secondary Ion Mass Spectroscopy	Surface	Ion beam (1–20 keV)	Secondary ions; analysis with mass spectrometer	2 nm–1 μm (or deeper: ion milling)	0.50 nm	Elemental and isotopic analysis; depth profile (all elements); detection limits: ppb-ppm	4
4.	Static SIMS Static Secondary Ion Mass Spectroscopy	Surface	Ion beam (0.5–20 keV)	Secondary ions, analysis with mass spectrometer	0.1–0.5 nm	10 μm	Elemental analysis of surface layers; molecular analysis; detection limits: ppb-ppm	4
5.	SNMS Sputtered Neutral Mass Spectroscopy	Surface, bulk	Plasma discharge; noble gases: 0.5–20 keV	Sputtered atoms ionized by atoms or electrons; then mass analyzed	0.1–0.5 nm (or deeper: ion milling)	1 cm	Elemental analysis $Z \geq 3$; depth profile; detection limit: ppm	4,6
6.	SALI Surface Analysis by Laser Ionization	Surface	e-beam, ion-beam, or laser for sputtering	Sputtered atoms ionized by laser; then mass analyzed	0.1–0.5 nm up to 3 μm in milling mode	60 nm	Surface analysis; depth profiling	7
7.	LIMS Laser Ionization Mass Spectroscopy	Surface, bulk	u.v. laser (ns pulses)	Ionized species; analyzed with mass spectrometer	50–150 nm	5 μm–1 mm	Elemental (micro)analysis; detection limits: 1–100 ppm	8
8.	SSMS Spark Source Mass Spectroscopy	Sample in the form of two electrodes	High voltage R.F. spark produces ions	Ions — analyzed in mass spectrometer	1–5 μm	—	Survey of trace elements; detection limit: 0.01–0.05 ppm	9
9.	GDMS Glow Discharge Mass Spectroscopy	Sample forms the cathode for a D.C. glow discharge	Sputtered atoms ionized in plasma	Ions — analyzed in mass spectrometer	0.1–100 μm	3–4 mm	(Bulk) trace element analysis; detection limit: sub-ppb	9,10
10.	ICPMS Induct. Coupled Plasma Mass Spectroscopy	Liquid-dissolved sample carried by gas stream into R.F. induction coil	Ions produced in argon plasma	Ions — analyzed in quadrupole mass spectrometer	—	—	High sensitivity analysis of trace elements	11

TECHNIQUES FOR MATERIALS CHARACTERIZATION (continued)

Technique	Sample	In	Out	Depth	Lateral resolution	Information obtained	Ref.
PHOTONS — ABSORPTION, REFLECTION AND ELECTRON EMISSION							
11. IRS Infrared Spectroscopy	Thin crystal, glass, liquid	I.R. light (W-filament, globar, Hg-arc)	I.R. spectrum	—	—	Electronic transitions (mainly in semiconductors and superconductors); vibrational modes (in crystals and molecules)	12,13, 14
12. FTIR Fourier Transform I.R. Spectroscopy	Solid, liquid; transmission or reflection	White light (all frequencies)	Fourier Transform of spectrum (interferometer)	—	—	Spectra obtained at higher speed and resolution	15
13. ATR Attenuated Total Reflection	Surface or thin crystal	—	—	μm's	—	Atomic or molecular spectra of surfaces and films	16
14. (μ)-RS (Micro-) Raman Spectroscopy	Solid, liquid (1 μm–1 cm)	Laser beam, e.g., Ar-line, YAG-line	Raman spectra	0.5 μm	0.5 μm	Molecular and crystal vibrations	12,14, 17
15. CARS Coherent Anti-Stokes Raman Spectroscopy	Solid, liquid (50 μm–3 cm)	Pump beam (ω_0)+ probe beam (ω_s)	Anti-Stokes spectrum	—	—	High resolution Raman spectra	14
16. Ellipsometry	Transparent films, crystals, adsorbed layers	Polarized light	Change in polarization	0.05 nm–5 μm	25 μm (or sample thickness)	Refractive index *and* absorption	18,19
17. UPS Ultraviolet Photo-electron Spectroscopy	Surfaces, adsorbed layers	u.v. light, 10–100 eV; 200 eV (synchrotron)	Electrons	0.2–10 nm	0.1–10 nm	Energies of electronic states of surfaces and free molecules	20,21
18. PSD Photon Stimulated Desorption	Surfaces with adsorbed species	Far u.v. light E > 10 eV	Ions — analyzed with mass spectrometer	0.1–2 nm	—	Structure and desorption kinetics of adsorbed atoms and molecules	22
X-RAYS							
19. XRD X-Ray Diffraction	Single crystals, powders films	X-rays: λ = 0.05–0.2 nm (6–17 keV)	Diffracted X-ray beam	1–1000 μm	0.1–10 mm	Identification of crystallographic structures; all elements (low Z difficult)	23,24
20. XRF/EDS X-Ray Fluorescence/Energy Dispersive Spectroscopy	Thin films, single layer	Prim. X-ray beam λ = 0.02–0.1 nm 12–80 keV	Fluorescent X-rays	1–100 μm	10 mm	Elemental analysis; all elements except H, He, Li — (EDS also used in XRD, SEM, TEM and EPMA)	25,26
21. EXAFS Extended X-Ray Absorption Fine Structure	Films, foils	High intensity X-rays (synchrotron)	Spectrum near absorption edge	nm–μm	—	Local atomic structure: order/disorder in vicinity of absorbing atom	27
22. XPS/ESCA X-Ray Photo-electron Spectroscopy/Electron Spect. for Chemical Analysis	Surfaces, thin films (≈20 atomic layers)	Soft X-rays (1–20 keV)	Core electrons; valence electrons	0.5–10 nm	5 nm–50 μm	(Quantitative) identification of all elements in surface layer or film	28,29
ELECTRONS							
23. CL Cathode Luminescence	Insulators, semiconductors	Electrons 5–50 keV	Photons 0.1–5 eV	1 nm–2 μn	1 or 2 μm	Energy levels of impurities and point defects	30
24. APS Appearance Potential Spectroscopy	Surface (≈20 atomic layers)	Electrons (energy scan) 50–2000 eV	X-rays to pinpoint electron energy threshold	—	—	Identification of surface species	21, see also C

TECHNIQUES FOR MATERIALS CHARACTERIZATION (continued)

Technique	Sample	In	Out	Depth	Lateral resolution	Information obtained	Ref.
25. AES Auger Electron Spectroscopy	Thin films, surfaces	Electrons 3–10 keV	Auger electrons 20–2000 eV	0.3–3 nm	≈30 nm	Elemental composition of surface (except H, He); detection limit 0.1–1%	28,29
26. EELS Electron Energy Loss Spectroscopy	Very thin samples (<200 nm)	Electrons (100–400 keV)	(Retarded) electrons; minus 1–1000 eV	<200 nm	1–100 nm	Local elemental concentration; electronic structure, chem. bonding; interatomic distances	31
27. EXELFS Extended Electron Energy Loss Fine Structure	Thin films	Electrons (100–400 keV)	Electrons energies 0–30 eV above edge	<200 nm	1–100 nm	Density of states of valence electrons (above Fermi level)	27,32
28. ESD Electron Stimulated Desorption	Adsorbed species	Electrons E > 10 eV	Ions — analyzed with mass spectrometer	—	—	Structure and desorption properties of adsorbed atoms and molecules	22
29. ESDIAD ESD-Ion Angular Distribution	(See ESD)	(See ESD)	Directional dependence of emitted ions	—	—	Geometries of adsorbed species (atoms or molecules)	22
30. EPMA Electron Probe (X-Ray) Micro Analysis	Solid conductors and insulators <1 cm thick	Electrons 5–30 keV	Characteristic X-ray 0.1–15 keV	100 nm–5 μm	1 μm	Elemental analysis, $Z \le 4$, major, minor and trace amounts	33,34
31. LEED Low Energy Electron Diffraction	Surface	Mono-energetic electron beam 10–1000 eV	Diffracted electrons	0.4–2 nm	<5 μm	Crystallographic structure of surface; resolution: 0.01 nm	35
32. RHEED Reflection High Energy Electron Diffraction	Surface	Electron beam at grazing angle 5–50 keV	Reflected electrons	0.2–10 nm	<5 μm	Surface symmetry	36,37
33. SEM Scanning Electron Microscopy	Bulk, films (conducting)	High energy electrons usually ~30 keV	Secondary and backscattered electrons	1 nm–5 μm	1–20 nm	Surface image, defect structure; resolution 5–15 nm; magnification 300,000×	33,34
34. (S)TEM (Scanning) Transmission Electron Microscopy	Thin specimen — <200 nm	High energy electrons typically 300 keV	Transmitted and diffracted electrons	(Sample thickness)	2–20 nm	(Defect) structure of cryst. solids; microchemistry; high resol.: 0.2 nm	33
35. FEM Field Emission Microscopy	Metals, alloys (sharp point)	—	Electron emission (with appl. electric field — 50 kV)	≈0.5 nm	10–100 nm	Surface image, crystallographic structure	34
36. STM Scanning Tunneling Microscopy	Polished or cleaved surface (conducting)	Tunneling current controls distance between sample and very sharp tip		1–5 nm	2–10 nm	Atomic-scale relief map of surface; resolution: vert. 0.002 nm, hor. 0.2 nm	39
37. SPM Scanned Probe Microscopy	Very flat surface	Any field: e.g. mechan. vibration recorded with laser probe; same with magnetic, electric or thermal field		1–100 nm	1–100 nm	Surface-magnetic field, surface-thermal conductivity, etc.	39a
38. AFM Atomic Force Microscopy	Very flat surface	Similar to STM; force measured with cantilever spring		0.5–5 nm	0.2–130 nm	Surface topography with atomic resolution; interatomic force	40
			IONS AND NEUTRONS				
39. ISS (or LEIS) Ion Scattering Spectroscopy (Low Energy Ion Scattering)	Surface	Ion beam He^+ or Ne^+ <3 keV	Sputtered ions (energy analysis)	0.1–0.5 nm	1–100 μm	Elemental analysis (better for low Z) detection limits: 0.01–1%	41
40. FIM Field Ion Microscopy	Surface: metals, alloys; very sharp tip	(He gas above sample)	He ions + high electric field produce image	≈0.1 nm	0.1–2 nm	Atomic structure of surface	34,42
41. RBS Rutherford Back Scattering	Solids, thin films	Mono-energetic ions (H^+ or He^{++}) 0.5–3 MeV	Backscattered ions	10 nm–1 μm	1 mm	Element identification (Li to U) detection limit: 0.01–1%	46

TECHNIQUES FOR MATERIALS CHARACTERIZATION (continued)

Technique	Sample	In	Out	Depth	Lateral resolution	Information obtained	Ref.
42. NRA Nuclear Reaction Analysis	Solids, thin films	Mono-energetic ions (Li, Be, B, etc.) 200 keV–6 MeV	Protons, deuterons ^3He, α-particles, γ-rays	0.1–5 µm	10 µm–10 mm	Element identification (all) detection limit: 10^{-12}–10^{-2}	47
43. PIXE Particle Induced X-ray Emission	Thin films, surface layers	High energy ions (H^+ or He^{++})	Characteristic X-rays	<10 µm	1 µm–2 mm	Trace impurities: $Z > 3$ detection limit: 0.1–100 ppm (depending on sample thickness)	48
44. INS Ion Neutralization Spectroscopy	Surface	He-ions (≈5 eV)	Electrons	—	—	Energies of valence electrons	49
45. NAA Neutron Activation Analysis	Bulk, >0.5 g	Thermal neutrons	Characteristic γ-rays, (≈1 MeV)	Bulk	—	Trace concentrations (of isotopes) of elements: trans. metals, Pt-group; detection limit: 10^8–10^{14} atoms/cm^3	43
46. N(P)D Neutron (Powder) Diffraction	Crystalline solids	Thermal neutrons E ≈0.0025 eV	Diffracted neutrons	Bulk	—	Crystallographic structure; porosity, particle size	44
47. SANS Small Angle Neutron Scattering	Inhomogeneous solids; powders; porous samples	Thermal neutrons $2\theta = 10^{-2}$–10^{-4}	Scattered neutrons	1–25 mm	—	Average size of inhomogeneities; range: 1 nm–1 mm	45
			ACOUSTIC				
48. SLAM Scanning Laser Acoustic Microscopy	Bulk, film	Acoustic wave produced by laser 1 MHz–1 GHz	Reflected acoustic wave	µm–cm	0.1–20 mm	Defect structure; thickness measurement	50
			THERMAL				
49. DTA Differential Thermal Analysis	Specimen and reference sample	Uniform heating	Temperature difference	Bulk	—	Phase transitions, crystallization	51
50. DSC Differential Scanning Calorimetry	Specimen and ref. sample	Controlled heating	Measure heat required for equal temperature	Bulk	—	Phase transitions, crystallization; activation energies	51
51. TGA Thermo Gravimetric Analysis	Bulk, 1–100 g	Controlled heating	Weight as function of temperature (and time)	Bulk	—	Decomposition, non-stoichiometry, kinetics of reaction	52
			RESONANCE				
52. EPR (ESR) Electron Paramagnetic (Spin) Resonance	Paramagnetic solids or liquids	Microwave radiation in magnetic field 3–300 GHz; 1–100 kG	Microwave absorption (at resonance)	Bulk	—	Local environment of paramagnetic ion; concentration of paramagnetic, species; detection limit: 10^{11} spins/cm^3	53,54
53. ECR Electron Cyclotron Resonance	Semiconductors, metals; free electrons (low temperature)	Microwave radiation in magnetic field 10–30 GHz; 5–10 kG	Microwave absorption (at resonance)	Bulk	—	Electronic energy bands, effective masses	55
54. Mössbauer Effect	Source and absorber	Mono-energetic γ-rays: 5–100 keV	Mössbauer spectrum (Doppler shifted (lines)	50 m	1 cm	Interaction between nucleus and its environment (local electric, magnetic fields; bonds; valency; diffusion, etc.)	56
55. NMR (MRI) Nuclear Magnetic Resonance (Magnetic Resonance Imaging)	Solids, liquids	R.F. radiation + magnetic field; e.g. for protons: 60 MHz, 14 kG	R.F. absorption	<1 cm	1 cm	Quant. analysis; local magnetic environment; diffusion; imaging	58
56. ENDOR Electron Nuclear Double Resonance	Solids, liquids	R.F. + microwave radiation in magn. field.	Microwave absorption	—	—	Hyperfine interaction → local atomic structure	54

TECHNIQUES FOR MATERIALS CHARACTERIZATION (continued)

Technique	Sample	In	Out	Depth	Lateral resolution	Information obtained	Ref.
57. NQR Nuclear Quadrupole Resonance	Solids	R.F. radiation 0.5–1000 MHz	R.F. absorption	—	—	Asymmetry of the charge distribution at the nucleus	55,59
OTHER							
58. BET Brunauer-Emmett-Teller	(Large) surface area 1–20 m^2/g	Adsorbed gas (e.g., N_2 at low temp.) as function of pressure (monolayer coverage)		—	—	Surface area measurement	60

REFERENCES

General References

A. Wachtman, J. B., *Characterization of Materials*, Butterworth-Heinemann, Boston, 1993.
B. Brundle, C. R., Evans, C. A., and Wilson, S., Eds., *Encyclopedia of Materials*, Butterworth-Heinemann, Boston, 1992.
C. Woodruff, D. P. and Delchar, T. A., *Modern Techniques of Surface Science*, Cambridge University Press, Cambridge, 1986.
D. *Metals Handbook,* 9th Edition, Vol. 10, Materials Characterization, Whan, R. E., Coordinator, American Society for Metals, Metals Park, OH, 1986.

Specific References

1. Slavin, M., *Atomic Absorption Spectroscopy*, 2nd Edition, John Wiley & Sons, New York, 1978.
2. Schrenk, W. G., *Analytical Atomic Spectroscopy*, Plenum Press, New York, 1975.
3. Dean, J. A. and Rains, T. E., *Flame Emission and Atomic Absorption Spectroscopy*, Vols. 1—3, Marcel Dekker, New York, 1969.
4. Benninghoven, A., Rudenauer, F. G., and Werner, H. W., *Secondary Ion Mass Spectroscopy,* John Wiley & Sons, New York, 1987.
5. Bird, J. R. and Williams, J. S., Eds., in *Ion Beams for Materials Analysis*, Academic Press, New York, 1989, pp. 515—537.
6. Smith, G. C., *Quantitative Surface Analysis for Materials Science*, The Institute of Metals, London, 1991.
7. Becker, E. H., in *Ion Spectroscopies for Surface Analysis*, Czanderna, A. W. and Hercules, D. M., Eds., Plenum Press, New York, 1991, p. 273.
8. Simons, D. S., *Int. J. Mass Spectrometry and Ion Processes*, 55, 15, 1983.
9. White, F. A. and Wood, G. M., *Mass Spectrometry: Applications in Science and Engineering*, John Wiley & Sons, New York, 1986.
10. Harrison, W. W. and Bentz, B. L., *Prog. Anal. Spectrometry,* 11, 53, 1988.
11. Bowmans, P. W. J. M., *Inductively Coupled Plasma Emission Spectroscopy*, Parts I and II, John Wiley & Sons, New York, 1987.
12. Brame, Jr., E. G. and Grasselli, J., *Infrared and Raman Spectroscopy*, Practical Spectroscopy Series, Vol. I, Marcel Dekker, New York, 1976.
13. Hollas, J. M., *Modern Spectroscopy*, John Wiley & Sons, New York, 1987.
14. Turrell, G., *Infrared and Raman Spectroscopy of Crystals*, Academic Press, New York and London, 1972.
15. Griffith, P. R. and Haseth, J. A., *Fourier Transform Infrared Spectroscopy*, John Wiley & Sons, New York, 1986.
16. Barnowski, M. K., *Fundamentals of Optical Fiber Communications*, Academic Press, New York, 1976.
17. Long, D. A., *Raman Spectroscopy*, McGraw-Hill, New York, 1977.
18. Azzam, R. M. A., *Ellipsometry and Polarized Light*, Elsevier-North Holland, Amsterdam, 1977.
19. Hecht, E., *Optics,* 2nd Edition, Addison-Wesley, Reading MA, 1987.
20. Brundle, C. R., in *Molecular Spectroscopy*, West, A. R., Ed., Heyden, London, 1976.
21. Park, R. L., in *Experimental Methods in Catalytic Research*, Vol. III, Anderson, R. B. and Dawson, P. T., Academic Press, New York, 1976, pp. 1—39.
22. Madey, T. E. and Stockbauer, R., in *Solid State Physics: Surfaces,* Vol. 22 of Methods of Experimental Physics, Park, R.L. and Lagally, M. G., Eds., Academic Press, New York, 1985.
23. Cullity, B. D., *Elements of X-Ray Diffraction*, 2nd Edition, Addison-Wesley, Reading, MA, 1978.
24. Schwartz, L. H. and Cohen, J. B., *Diffraction from Materials*, Springer Verlag, Berlin, 1987.

25. deBoer, D. K. G., in *Advances in X-Ray Analysis*, Vol. 34, Barrett, C. S. et. al., Eds., Plenum Press, New York, 1991.
26. Birks, L. S., *X-Ray Spectrochemical Analysis*, 2nd Edition, John Wiley & Sons, New York, 1969.
27. Bonnelle, C. and Mande, C., *Advances in X-Ray Spectroscopy*, Pergamon Press, Oxford, 1982.
28. *Practical Surface Analysis by Auger and X-Ray Photo-Electric Spectroscopy*, Briggs, D. and Seah, M. P., Eds., John Wiley & Sons, New York, 1983.
29. Powell, C. J. and Seah, M. P., *J. Vac. Sci. Technol. A*, Vol. 8, 735, 1990.
30. Yacobi, G. G. and Holt, D. B., *Cathodeluminescence Microscopy of Inorganic Solids*, Plenum Press, New York, 1990.
31. Egerton, R. F., *Electron Energy Loss Spectroscopy in the Electron Microscope*, Plenum Press, New York, 1986.
32. Disko, M. M., Krivanek, O. L., and Rez, P., *Phys. Rev.*, B25, 4252, 1982.
33. Goldstein, J. I., et. al., *Scanning Electron Microscopy and X-Ray Microanalysis*, 2nd Edition, Plenum Press, New York, 1986.
34. Murr, L. E., *Electron and Ion Microscopy and Microanalysis*, Marcel Dekker, New York, 1982.
35. Armstrong, R. A., in *Experimental Methods in Catalytic Research*, Vol. III, Anderson, R. B., and Dawson, P. T.,Eds., Academic Press, New York, 1976.
36. Dobson, P. J. et. al., *Vacuum*, 33, 593, 1983.
37. Rymer, T. B., *Electron Diffraction*, Methuen, London, 1970.
38. Reimer, L., *Transmission Election Microscopy*, Springer-Verlag, Berlin, 1984.
39. *Scanning Tunneling Microscopy and Related Methods*, Behm, R. J., Garcia, N., and Rohrer, H., Kluwer, Eds., Academic Publishers, Norwell, MA, 1990.
39a. Wikramasinghe, H.K., *Scientific American*, Vol. 261, No. 4, pp. 98—105, Oct. 1989.
40. Rugar, D. and Hansma, P., *Physics Today*, 43(10), pp. 23—30, 1990.
41. Feldman, C. C. and Mayer, J. W., *Fundamentals of Surface and Thin Film Analysis*, North-Holland, Amsterdam, 1986.
42. Muller, E. W. and Tsong, T. T., *Field Ion Microscopy*, Elsevier, Amsterdam, 1969.
43. Amiel, S., *Nondestructive Activation Analysis*, Elsevier, Amsterdam, 1981.
44. Bacon, G. E., *Neutron Diffraction*, 3rd Edition, Clarendon Press, Oxford, 1975.
45. Neutron Scattering, Part A., in *Methods of Experimental Physics*, Vol. 23, Skold, K. and Price, D. L., Eds., Academic Press, New York, 1986.
46. Chu, W. K., Mayer, J. W., and Nicolet, M. A., *Backscattering Spectroscopy*, Academic Press, New York, 1987.
47. Rickey, F. A., in *High Energy and Heavy Ion Beams in Materials Analysis*, Tesmer, J. R., et. al., Eds., MRS, 1990, pp. 3—26.
48. Johansson, S. A. E. and Campbell, J. L., *PIXE: A Novel Technique for Elemental Analysis*, John Wiley & Sons, New York, 1988.
49. Hagstrum, H. D., in *Inelastic Ion-Surface Collisions*, Tolk, N. H. et. al., Eds., Academic Press, New York, 1977, pp. 1—46.
50. Nikoonahad, M., in *Research Techniques in Nondestructive Testing*, Vol. VI, Sharpe, R.S., Ed., Academic Press, New York, 1984, pp. 217—257.
51. Gallagher, P. K., *Characterization of Materials by Thermoanalytical Techniques*, MRS - Bulletin, Vol. 13, No. 7, pp. 23—27, 1988.
52. Earnest, C. M., *Compositional Analysis by Thermogravimetry*, ASTM Special Technical Publication 997, 1988.
53. Poole, C. P., *Electron Spin Resonance — A Comprehensive Treatise on Experimental Techniques*, 2nd Edition, John Wiley & Sons, New York, 1983.
54. Atherton, N. M., *Principles of Electron Spin Resonance*, Ellis Horwood Ltd., Chichester, U.K., 1993.
55. Kittel, C., *Introduction to Solid State Physics*, 6th Edition, John Wiley & Sons, New York, 1986, p. 196.
56. Gibb, T. C., *Principles of Mössbauer Spectroscopy*, Chapman & Hall, London, 1976.
57. Slichter, C. P., *Principles of Magnetic Resonance*, 3rd Edition, Springer-Verlag, Berlin, 1990.
58. *NMR Spectroscopy Techniques*, Dybrowski, C. and Lichter, R. L., Eds., Marcel Dekker, New York, 1987.
59. Das, T. P. and Hahn, E. L., *Nuclear Quadrupole Resonance Spectroscopy*, Academic Press, New York, 1958.
60. Somorjai, G. A., *Principles of Surface Chemistry*, Prentice-Hall, Englewood Cliffs, NJ, 1972, p. 216.

SYMMETRY OF CRYSTALS

L. I. Berger

The ability of a body to coincide with itself in its different positions regarding a coordinate system is called its symmetry. This property reveals itself in iteration of the parts of the body in space. The iteration may be done by reflection in mirror planes, rotation about certain axes, inversions and translations. These actions are called the symmetry operations. The planes, axes, points, etc., are known as symmetry elements. Essentially, mirror reflection is the only truly primitive symmetry operation. All other operations may be done by a sequence of reflections in certain mirror planes. Hence, the mirror plane is the only true basic symmetry element. But for clarity, it is convenient to use the other symmetry operations, and accordingly, the other aforementioned symmetry elements. The symmetry elements and operations are presented in Table 1.

The entire set of symmetry elements of a body is called its symmetry class. There are thirty-two symmetry classes that describe all crystals which have ever been noted in mineralogy or been synthesized (more than 150,000). The denominations and symbols of the symmetry classes are presented in Table 2.

There are several known approaches to classification of individual crystals in accordance with their symmetry and crystallochemistry. The particles which form a crystal are distributed in certain points in space. These points are separated by certain distances (translations) equal to each other in any chosen direction in the crystal. Crystal lattice is a diagram that describes the location of particles (individual or groups) in a crystal. The lattice parameters are three non-coplanar translations that form the crystal lattice. Three basic translations form the unit cell of a crystal. August Bravais (1848) has shown that all possible crystal lattice structures belong to one or another of fourteen lattice types (Bravais lattices). The Bravais lattices, both primitive and non-primitive, are the contents of Table 3.

Among the three-dimensional figures, there is a group of polyhedrons that are called regular, which have all faces of the same shape and all edges of the same size (regular polygons). It has been shown that there are only five regular polyhedrons. Because of their importance in crystallography and solid state physics, a brief description of these polyhedrons is included in Table 4.

The systematic description of crystal structures is presented primarily in the well known *Structurbericht*. The classification of crystals by the Structurbericht does not reflect their crystal class, the Bravais lattice, but is based on the crystallochemical type. This makes it inconvenient to use the Structurbericht categories for comparison of some individual crystals. Thus, there have been several attempts to provide a more convenient classification of crystals. Table 5 presents a compilation of different classifications which allows the reader to correlate the Structurbericht type with the international and Schoenflies point and space groups and with Pearson's symbols, based on the Bravais lattice and chemical composition of the class prototype. The information included in Table 5 has been chosen as an introduction to a more detailed crystallophysical and crystallochemical description of solids.

TABLE 1
Symmetry Operations and Elements

	Symmetry element			Presentation on the stereographic projection	
		Symbol			
Symmetry operation	**Name**	**International (Hermann-Mauguin)**	**Schoenflies**	**Parallel**	**Perpendicular**
Reflection in a plane	Plane	m	C_s		
Rotation by angle $\alpha = 360°/n$ about an axis	Axis	n = 1, 2, 3, 4 or 6	C_n		
		n = 2	C_2		
		n = 3	C_3		
		n = 4	C_4		
		n = 6	C_6		
Rotation about an axis and inversion in a symmetry center lying on the axis	Inversion (improper) axis	$\bar{n} = \bar{3}, \bar{4}, \bar{6}$	C_{ni}		
		$\bar{n} = \bar{3}$	C_{3i}		
		$\bar{n} = \bar{4}$	C_{4i}		

TABLE 1
Symmetry Operations and Elements (continued)

Symmetry operation	Symmetry element: Name	Symbol: International (Hermann-Mauguin)	Symbol: Schoenflies	Presentation on the stereographic projection: Parallel	Presentation on the stereographic projection: Perpendicular
		$\bar{n} = \bar{6}$	C_{6i}		
Inversion in a point	Center	$\bar{1}$	C_i	● ○	×
Parallel translation	Translation vector $\vec{a}, \vec{b}, \vec{c}$				
Reflection in a plane and translation parallel to the plane	Glide–plane	a, b, c, n, d			
Rotation about an axis and translation parallel to the axis	Screw axis	n_m (m = 1, 2, .., n – 1)			
Rotation about an axis and reflection in a plane perpendicular to the axis	Rotatory-reflection axis	$\tilde{n}$ $\tilde{n} = \tilde{1}, \tilde{2}, \tilde{3}, \tilde{4}, \tilde{6}$	S_n		

TABLE 2
The Thirty-Two Symmetry Classes

Crystal symbol	Primitive Int	Primitive Sch	Central Int	Central Sch	Planal Int	Planal Sch	Axial Int	Axial Sch	Plane-axial Int	Plane-axial Sch	Inversion primitive Int	Inversion primitive Sch	Inversion-planal Int	Inversion-planal Sch
Triclinic	1	C_1	$\bar{1}$	C_i										
Monoclinic					m	C_s	2	C_2	2/m	C_{2h}				
Orthorhombic					mm2	C_{2v}	222	D_2	mmm	D_{2h}				
Trigonal	3	C_3	$\bar{3}$	C_{3i}	3m	C_{3v}	32	D_3	$\bar{3}$m	C_{3d}				
Tetragonal	4	C_4	4/m	C_{4h}	4mm	C_{4v}	422	D_4	4/mmm	D_{4h}	$\bar{4}$	S_4	$\bar{4}$2m	D_{2d}
Hexagonal	6	C_6	6/m	C_{6h}	6mm	C_{6v}	622	D_6	6/mmm	D_{6h}	$\bar{6}$	C_{3h}	$\bar{6}$m2	D_{3h}
Cubic	23	T	m3	T_h	$\bar{4}$3m	T_d	432	O	m3m	O_h				

Class name[a] and its symbol — International (Int) and Schoenflies (Sch)

[a] Per Fedorov Institute of Crystallography, USSR Academy of Sciences, nomenclature.

TABLE 3
The Fourteen Possible Space Lattices (Bravais Lattices)

Crystal system	Metric category of the system	No. of different lattices in the system	Lattice type[a] (marked by +)					No. of identi-points per unit cell	Characteristic parameters (marked by +)						Description of characteristic parameters $a \subset X$, $b \subset Y$, $c \subset Z$; $\alpha \equiv (b,c)$, $\beta \equiv (a,c)$, $\gamma \equiv (a,b)$	Symmetry of the lattice	
			P	C	I	F	R		a	b	c	α	β	γ		Int	Sch
Triclinic	Trimetric	1	+					1	+	+	+	+	+	+	$a \neq b \neq c, \alpha \neq \beta \neq \gamma$	1	C_i
Monoclinic	Trimetric	2	+	+				1 or 2	+	+	+		+		$a \neq b \neq c, \alpha = \gamma = 90° \neq \beta$	2/m	C_{2h}
Orthorhombic	Trimetric	4	+	+	+	+		1, 2 or 4	+	+	+				$a \neq b \neq c, \alpha = \beta = \gamma = 90°$	mmm	D_{2h}
Trigonal (rhombohedral)	Dimetric	1					+	1	+			+			$a = b = c, 120° > \alpha = \beta = \gamma \neq 90°$	3m	D_{3d}
Tetragonal	Dimetric	2	+		+			1 or 2	+		+				$a = b \neq c, \alpha = \beta = \gamma = 90°$	4/mmm	D_{4h}
Hexagonal	Dimetric	1	+					1	+		+				$a = b \neq c, \alpha = \beta = 90°, \gamma = 120°$	6/mmm	D_{6h}
Isometric (cubic)	Monometric	3	+		+	+		1, 2 or 4	+						$a = b = c, \alpha = \beta = \gamma = 90°$	m3m	O_h

[a] Designations of the space-lattice types: P — primitive, C — side-centered (base-centered), I — body-centered, F — face-centered, R — rhombohedral.

TABLE 4
The Five Possible Regular Polyhedrons

Polyhedron	Symmetry (Schoenflies) Class	Symmetry (Schoenflies) Elements	Form of faces	Number of[a] Faces (F)	Number of[a] Edges (E)	Number of[a] Vertices (V)
Tetrahedron	T	$4C_3 3C_2$	Equilateral triangle	4	6	4
Cube (hexahedron)	O	$3C_4 4C_3 6C_2$	Square	6	12	8
Octahedron	O	$3C_4 4C_3 6C_2$	Equilateral triangle	8	12	6
Pentagonal dodecahedron	J	$6C_5 10C_3 15C_2$	Regular pentagon	12	30	20
Icosahedron	J	$6C_5 10C_3 15C_2$	Equilateral triangle	20	30	12

[a] Per formula by Leonhard Euler: F + V – E = 2

TABLE 5
Classification of Crystals

Strukturbericht symbol	Structure name	Symmetry group International	Symmetry group Schoenflies	Pearson symbol[a]	Standard ASTM E157-82a symbol[b]
1	2	3	4	5	6
A1	Cu	Fm3m	O^4_h	cF4	F
A2	W	Im3m	O^9_h	cI2	B
A3	Mg	$P6_3$/mmc	D^4_{6h}	hP2	H
A4	C	Fd3m	O^7_h	cF8	F
A5	Sn	If_1/amd	D^{19}_{4h}	tI4	U
A6	In	I4/mmm	D^{17}_{4h}	tI2	U
A7	As	R$\bar{3}$m	D^5_{3d}	hR2	R
A8	Se	$P3_121$ or $P3_221$	D^4_3 (D^6_3)	hP3	H
A10	Hg	R$\bar{3}$m	D^5_{3d}	hR1	R
A11	Ga	Cmca	D^{18}_{2h}	oC8	Q
A12	α-Mn	I$\bar{4}$3m	T^3_d	cI58	B
A13	β-Mn	$P4_132$	O^7	cP20	C
A15	OW_3	Pm3n	O^3_h	cP8	C
A20	α-U	Cmcm	D^{17}_{2h}	oC4	Q
B1	ClNa	Fm3m	O^5_h	cF8	F
B2	ClCs	Pm3m	O^1_h	cP2	C
B3	SZn	F$\bar{4}$3m	T^2_d	cF8	F
B4	SZn	$P6_3$mc	C^4_{6v}	hP4	H
$B8_1$	AsNi	$P6_3$/mmc	D^4_{6h}	hP4	H
$B8_2$	$InNi_2$	$P6_3$/mmc	D^4_{6h}	hP6	H
B9	HgS	$P3_121$ or $P3_221$	D^4_3 or D^6_3	hP6	H
B10	OPb	P4/nmm	D^7_{4h}	tP4	T
B11	γ-CuTi	P4/nmm	D^7_{4h}	tP4	T
B13	NiS	R$\bar{3}$m	D^5_{3d}	hR6	R
B16	GeS	Pnma	D^{16}_{2h}	oP8	O
B17	PtS	$P4_2$/mmc	D^9_{4h}	tP4	T
B18	CuS	$P6_3$/mmc	D^4_{6h}	hP12	H
B19	AuCd	Pmma	D^5_{2h}	oP4	O
B20	FeSi	$P2_13$	T^4	cP8	C
B27	BFe	Pnma	D^{16}_{2h}	oP8	O
B31	MnP	Pnma	D^{16}_{2h}	oP8	O
B32	NaTl	Fd3m	O^7_h	cF16	F
B34	Pds	$P4_2$/m	C^2_{4h}	tP16	T

TABLE 5
Classification of Crystals (continued)

Strukturbericht symbol	Structure name	Symmetry group International	Symmetry group Schoenflies	Pearson symbol[a]	Standard ASTM E157-82a symbol[b]
1	2	3	4	5	6
B35	CoSn	P6/mmm	D^1_{6h}	hP6	H
B37	SeTl	I4/mcm	D^{18}_{4h}	tI16	U
B_e	CdSb	Pbca	D^{15}_{2h}	oP16	O
B_f (B33)	ξ-BCr	Cmcm	D^{17}_{2h}	oC8	Q
B_g	BMo	$I4_1$/amd	D^{19}_{4h}	tI4	U
B_h	CW	P6m2	D^1_{3h}	hP2	H
B_i	γ-CMo (AsTi)	$P6_3$/mmc	D^4_{6h}	hP8	H
C1	CaF_2	Fm3m	O^5_h	cF12	F
$C1_b$	AgAsMg	F$\bar{4}$3m	T^2_d	cF12	F
C2	FeS_2	Pa3	T^6_h	cP12	C
C3	Cu_2O	Pn3m	O^4_h	cP6	C
C4	O_2Ti	$P4_2$/mnm	D^{14}_{4h}	tP6	T
C6	CdI_2	P$\bar{3}$m1	D^3_{3d}	hP3	H
C7	MoS_2	$P6_3$/mmc	D^4_{6h}	hP6	H
$C11_a$	C_2Ca	I4/mmm	D^{17}_{4h}	tI6	U
$C11_b$	$MoSi_2$	I4/mmm	D^{17}_{4h}	tI6	U
C12	$CaSi_2$	R$\bar{3}$m	D^5_{3d}	hR6	R
C14	$MgZn_2$	$P6_3$/mmc	D^4_{6h}	hP12	H
C15	Cu_2Mg	Fd3m	O^7_h	cF24	F
$C15_b$	$AuBe_5$	F$\bar{4}$3m or F23	T^2_d or T^2	cF24	F
C16	Al_2Cu	I4/mcm	D^{18}_{4h}	tI12	U
C18	FeS_2	Pnnm	D^{12}_{2h}	oP6	O
C19	$CdCl_2$	R$\bar{3}$m	D^5_{3d}	hR3	R
C22	Fe_2P	P$\bar{2}$6m	D^1_{3h}	hP9	H
C23	Cl_2Pb	Pnma	D^{16}_{2h}	oP12	O
C32	AlB_2	P6/mmm	D^1_{6h}	hP3	H
C33	Bi_2STe_2	R$\bar{3}$m	D^5_{3d}	hR5	R
C34	$AuTe_2$	C2/m (P2/m)	C^3_{2h} (C^1_{2h})	mC6	N
C36	$MgNi_2$	$P6_3$/mmc	D^4_{6h}	hP24	H
C38	Cu_2Sb	P4/nmm	D^7_{4h}	tP6	T
C40	$CrSi_2$	$P6_222$	D^4_6	hP9	H
C42	SiS_2	Ibam	D^{26}_{2h}	oI12	P
C44	GeS_2	Fdd2	C^{19}_{2v}	oF72	S
C46	$AuTe_2$	Pma2	C^4_{2v}	oP24	O
C49	Si_2Zr	Cmcm	D^{17}_{2h}	oC12	Q
C54	Si_2Ti	Fddd	D^{24}_{2h}	oF24	S
C_c	Si_2Th	$I4_1$/amd	D^{19}_{4h}	tI12	U
C_e	$CoGe_2$	Aba2	C^{17}_{2v}	oC23	Q
DO_2	As_3Co	Im3	T^5_h	cI32	B
DO_3	BiF_3	Fm3m	O^5_h	cF16	F
DO_9	O_3Re	Pm3m	O^1_h	cP4	C
DO_{11}	CFe_3	Pnma	D^{16}_{2h}	oP16	O
DO_{18}	$AsNa_3$	$P6_3$/mmc	D^4_{6h}	hP8	H
DO_{19}	Ni_3Sn	$P6_3$/mmc	D^4_{6h}	hP8	H
DO_{20}	Al_3Ni	Pnma	D^{16}_{2h}	oP16	O
DO_{21}	Cu_3P	P$\bar{3}$c1	D^4_{3d}	hP24	H
DO_{22}	Cu_3P	I4/mmm	D^{17}_{4h}	tI8	U
DO_{23}	Al_3Zr	I4/mmm	D^{17}_{4h}	tI16	U
DO_{24}	Ni_3Ti	$P6_3$/mmc	D^4_{6h}	hP16	H
DO_c	SiU_3	I4/mcm	D^{18}_{4h}	tI16	U
DO_e	Ni_3P	I$\bar{4}$	S^2_4	tI32	U
$D1_3$	Al_4Ba	I4/mmm	D^{17}_{4h}	tI10	U
$D1_a$	$MoNi_4$	I4/m	C^5_{4h}	tI10	U

TABLE 5
Classification of Crystals (continued)

Strukturbericht symbol	Structure name	Symmetry group: International	Symmetry group: Schoenflies	Pearson symbol[a]	Standard ASTM E157-82a symbol[b]
1	2	3	4	5	6
$D1_b$	Al_4U	Imma	D^{28}_{2h}	oI20	P
$D1_c$	$PtSn_4$	Aba2	C^{17}_{2v}	oC20	Q
$D1_e$	B_4Th	P4/mbm	D^5_{4h}	tP20	T
$D1_f$	BMn_4	Fddd	D^{24}_{2h}	oF40	S
$D2_1$	B_6Ca	Pm3m	O^1_h	cP7	C
$D2_3$	$NaZn_{13}$	Fm3m	O^5_h	cF112	F
$D2_b$	$Mn_{12}Th$	I4/mmm	D^{17}_{4h}	tI26	U
$D2_c$	MnU_6	I4/mcm	D^{18}_{4h}	tI28	U
$D2_d$	$CaCu_5$	P6/mmm	D^1_{6h}	hP6	H
$D2_f$	$B_{12}U$	Fm3m	O^5_h	cF52	F
$D2_h$	Al_6Mn	Cmcm	D^{17}_{2h}	oC28	Q
$D5_1$	α-Al_2O_3	$R\bar{3}c$	D^6_{3d}	hR10	R
$D5_2$	La_2O_3	$P\bar{3}m1$	D^3_{3d}	hP5	H
$D5_3$	Mn_2O_3	Ia3	T^7_h	cI80	B
$D5_8$	S_3Sb_2	Pnma	D^{16}_{2h}	oP20	O
$D5_9$	P_2Zn_3	$P4_2/mmc$	D^9_{4h}	tP40	T
$D5_{10}$	C_2C_3	Pnma	D^{16}_{2h}	oP20	O
$D5_{13}$	Al_3Ni_2	$P\bar{3}m1$	D^3_{3d}	hP5	H
$D5_a$	Si_2U_3	P4/mbm	D^5_{4h}	tP10	T
$D5_c$	C_3Pu_2	$I\bar{4}3d$	T^6_d	cI40	B
$D7_1$	Al_4C_3	$R\bar{3}m$	D^5_{3d}	hR7	R
$D7_3$	P_4Th_3	$I\bar{4}3d$	T^6_d	cI28	B
$D7_b$	B_4Ta_3	Immm	D^{25}_{2h}	oI14	P
$D8_1$	Fe_3Zn_{10}	Im3m	O^9_h	cI52	B
$D8_2$	Cu_5Zn_8	$I\bar{4}3m$	T^3_d	cI52	B
$D8_3$	Al_4Cu_9	$P\bar{4}3m$	T^1_d	cP52	C
$D8_4$	C_6Cr23	Fm3m	O^5_h	cF116	F
$D8_5$	Fe_7W_6	$R\bar{3}m$	D^5_{3d}	hR13	R
$D8_6$	$Cu_{15}Si_4$	$I\bar{4}3m$	T^3_d	cI76	B
$D8_8$	Mn_5Si_3	$P6_3/mcm$	D^3_{6h}	hP16	H
$D8_9$	Co_9S_8	Fm3m	O^5_h	cF68	F
$D8_{10}$	Al_8Cr_5	R3m	C^5_{3v}	hR26	R
$D8_{11}$	Al_5Co_2	$P6_3/mcm$	D^3_{6h}	hP28	H
$D8_a$	$Mn_{23}Th_6$	Fm3m	O^5_h	cF116	F
$D8_b$	σ-phase of Cr-Fe	$p\bar{4}_2/mnm$	D^{14}_{4h}	tP30	T
$D8_e$	$(Al,Zn)_{49}Mg_{32}$	Im3	T^5_h	cI162	B
$D8_f$	Ge_7Ir_3	Im3m	O^9_h	cI40	B
$D8_h$	B_5W_2	$P6_3/mmc$	D^4_{6h}	hP14	H
$D8_i$	B_5Mo_2	$R\bar{3}m$	D^5_{3d}	hR7	R
$D8_l$	B_3Cr_5	I4/mcm	D^{18}_{4h}	tI32	U
$D8_m$	Si_3W_5	I4/mcm	D^{18}_{4h}	tI32	U
$D10_1$	C_3Cr_7	P31c	C^4_{3v}	hP80	H
$D10_2$	Fe_3Th_7	$P6_3mc$	C^4_{6v}	hP20	H
$E0_1$	ClFPb	P4/nmm	D^7_{4h}	tP6	T
$E1_1$	$CuFeS_2$	$I\bar{4}2d$	D^{12}_{2d}	tI16	U
$E2_1$	CaO_3Ti	Pm3m	O^1_h	cP5	C
$E2_4$	S_3Sn_2	Pnma	D^{16}_{2h}	oP20	O
E3	Al_2CdS_4	$I\bar{4}$	S^2_4	tI14	U
$E9_3$	$SiFe_3W_3$	Fd3m	O^7_h	cF112	F
$E9_a$	Al_7Cu_2Fe	P4/mnc	D^6_{4h}	tP40	T
$E9_b$	$AlLi_3N_2$	Ia3	T^7_h	cI96	B
$F0_1$	NiSSb	$P2_13$	T^4	cP12	C
$F5_1$	$CrNaS_2$	$R\bar{3}m$ or R32	D^5_{3d} or D^7_3	hR4	R
$F5_6$	CuS_2Sb	Pnma	D^{16}_{2h}	oP16	O

TABLE 5
Classification of Crystals (continued)

Strukturbericht symbol 1	Structure name 2	Symmetry group International 3	Symmetry group Schoenflies 4	Pearson symbol[a] 5	Standard ASTM E157-82a symbol[b] 6
$H1_1$	Al_2MgO_4	Fd3m	O^7_h	cF56	F
$H2_4$	Cu_3S_4V	$P\bar{4}3m$	T^1_d	cP8	C
$H2_5$	$AsCu_3S_4$	$Pmn2_1$	C^7_{2v}	oP16	O
$L1_0$	AuCu	P4/mmm	D^1_{4h}	tP4	T
$L1_2$	$AlCu_3$	Pm3m	O^1_h	cP4	C
$L2_1$	$AlCu_2Mn$	Fm3m	O^5_h	cF16	F
$L2_2$	Sb_2Tl_7	Im3m	O^9_h	cI54	B
$L'2_b$	H_2Th	I4/mmm	D^{17}_{4h}	tI6	U
L'3	Fe_2N	$P6_3/mmc$	D^4_{6h}	hP3	H
$L6_0$	$CuTi_3$	P4/mmm	D^1_{4h}	tP4	T

[a] The first letter denotes the crystal system: triclinic (a), monoclinic (m), orthorhombic (o), tetragonal (t), hexagonal (h) and cubic (c). Trigonal (rhombohedral) system is presented by combination hR. The second letter of Pearson's symbol denotes lattice type: primitive (P), edge-(base-) centered (C), body-centered (I) or face-centered (F). The following number denotes amount of atoms in the crystal unit cell.

[b] Standard ASTM E157-82a has the Bravais lattices designations as following: C — primitive cubic; B — body-centered cubic; F — face-centered cubic; T — primitive tetragonal; U — body-centered tetragonal; R — rhombohedral; H — hexagonal; O — primitive orthorhombic; P — body-centered orthorhombic; Q — base-centered orthorhombic; S — face-centered orthorhombic; M — primitive monoclinic; N — centered monoclinic; A — triclinic.

REFERENCES

1. A. Schoenflies, *Kristallsysteme und Kristallstructur*, Leipzig, 1891.
2. E. S. Fedorow, Zusammenstellung der kristallographischen Resultate, *Zs. Krist.*, 20, 1892.
3. P. Groth, *Elemente der physikalischen und chemischen Krystallographie*, R. Oldenbourg, München/Berlin, 1921.
4. N. V. Belov, *Class Method of Deriving Space Groups of Symmetry*, Trudy Instituta Kristallodraffi imeni Fedorova (Transactions of the Fedorov Inst. of Crystallography), 5, 25, 1951, in Russian.
5. W. B. Pearson, *Handbook of Lattice Spacings and Structures of Metals and Alloys*, Vol. 1, Pergamon Press, 1958; Vol. 2, 1967.
6. Ch. Kittel, *Introduction to Solid State Physics*, John Wiley & Sons, 1956.
7. G. S. Zhdanov, *Fizika Tverdogo Tela (Solid State Physics)*, Moscow University Press, 1962, in Russian.
8. M. J. Buerger, *Elementary Crystallography*, John Wiley & Sons, 1963.
9. F. D. Bloss, *Crystallography & Crystal Chemistry*, Holt, Rinehart & Winston, 1971.
10. T. Janssen, *Crystallographic Groups*, North-Holland/American Elsevier, 1973.
11. M. P. Shaskolskaya, *Kristallografiya (Crystallography)*, Vysshaya Shkola, Moscow, 1976, in Russian.
12. T. Hahn, Ed., Internat. *Tables for Crystallography*, Vol. A, D. Reidel Publishing, Boston, 1983.
13. Crystal Data. Determinative Tables, Volumes 1—6, 1966—1983, JCPDS-Intern Centre for Diffraction Data and U.S. Dept. of Commerce.
14. R. W. G. Wyckoff, *Crystal Structures*, 2nd ed., Volumes 1–6, Interscience, New York, 1963.
15. C.J. Bradley and A.P. Cracknell, *The Mathematical Theory of Symmetry in Solids*, Clarendon Press, Oxford, 1972.
16. International Tables for Crystallography. Volume A, *Space–Group Symmetry*, T. Hahn, Ed., 1989; Volume B, *Reciprocal Space*, U. Schmueli, Ed.; Volume C, *Mathematical, Physical and Chemical Tables*, A. J. C. Wilson, Ed., Kluwer Academic Publishers, Dordrecht, 1989.
17. G. R. Desiraju, *Crystal Engineering: The Design of Organic Solids*, Elsevier, Amsterdam, 1989.
18. M. Senechal, *Crystalline Symmetries: An Informal Mathematical Introduction*, Adam Hilger Publ., Bristol, 1990.
19. C. Hammond, *Introduction to Crystallography*, Oxford University Press, 1990.
20. N.W. Alcock, *Bonding and Structure: Structural Principles in Inorganic and Organic Chemistry*, Ellis Norwood Publ., 1990.
21. T. C. W. Mak and G. D. Zhou. *Crystallography in Modern Chemistry: A Resource Book of Crystal Structures*, Wiley–Interscience, New York, 1992.
22. S. C. Abrahams, K. Mirsky, and R. M. Nielson, *Acta Cryst*, B52, 806 (1996); B52, 1057 (1996).
23. C. Marcos, A. Panalague, D. B. Morciras, S. Garcia–Granda and M. R. Dias. *Acta Cryst*, B52, 899 (1996).

Crystallographic Computing

24. A. C. Larson, *Crystallographic Computing*, Manksgaard, Copenhagen, 1970.
25. G. M. Sheldrick, SHELXS86. Crystallographic Computing 3, Clarendon Press, Oxford, 1986; SHELXL93. Program for the Refinement of Crystal Structures, University of Göttingen Press, 1993.
26. Inorganic Crystal Structure Database, CD–ROM. Sci. Inf. Service. E-mail: SISI@Delphi.com.

IONIC RADII IN CRYSTALS

Ionic radii are a useful tool for predicting and visualizing crystal structures. This table lists a set of ionic radii R_i in Å units for the most common coordination numbers *CN* of positive and negative ions. The values are based on experimental crystal structure determinations, supplemented by empirical relationships, and theoretical calculations. The notation sq after the coordination number indicates a square configuration, while py indicates pyramidal.

The advice of Howard T. Evans and Marvin J. Weber in preparing this table is appreciated.

REFERENCES

1. Shannon, R. D., *Acta Crystallogr.* A32, 751, 1976.
2. Jia, Y.Q., *J. Solid State Chem.* 95, 184, 1991.

Ion	*CN*	R_i/Å
Anions		
F^{-1}	6	1.33
Cl^{-1}	6	1.81
Br^{-1}	6	1.96
I^{-1}	6	2.20
OH^{-1}	4	1.35
	6	1.37
O^{-2}	2	1.21
	6	1.40
	8	1.42
S^{-2}	6	1.84
Se^{-2}	6	1.98
Te^{-2}	6	2.21
Cations		
Ac^{+3}	6	1.12
Ag^{+1}	4	1.00
	6	1.15
	8	1.28
Ag^{+2}	4sq	0.79
	6	0.94
Al^{+3}	4	0.39
	5	0.48
	6	0.54
Am^{+3}	6	0.98
	8	1.09
Am^{+4}	6	0.85
	8	0.95
As^{+3}	6	0.58
As^{+5}	4	0.34
	6	0.46
Au^{+1}	6	1.37
Au^{+3}	4sq	0.64
	6	0.85
Ba^{+2}	6	1.35
	8	1.42
	12	1.61
Be^{+2}	4	0.27
	6	0.45
Bi^{+3}	5	0.96
	6	1.03
	8	1.17
Bi^{+5}	6	0.76
Bk^{+3}	6	0.96
Bk^{+4}	6	0.83
	8	0.93
Br^{+5}	3py	0.31
Br^{+7}	4	0.25
	6	0.39
C^{+4}	4	0.15
	6	0.16
Ca^{+2}	6	1.00
	8	1.12
	10	1.23
	12	1.34
Cd^{+2}	4	0.78
	6	0.95
	8	1.10
	12	1.31
Ce^{+3}	6	1.01
	8	1.14
	10	1.25
	12	1.34
Ce^{+4}	6	0.87
	8	0.97
	10	1.07
	12	1.14
Cf^{+3}	6	0.95
Cf^{+4}	6	0.82
	8	0.92
Cl^{+5}	3py	0.12
Cl^{+7}	4	0.08
Cm^{+3}	6	0.97
Cm^{+4}	6	0.85
	8	0.95
Co^{+2}	4	0.56
	6	0.65
	8	0.90
Co^{+3}	6	0.55
Cr^{+2}	6	0.73
Cr^{+3}	6	0.62
Cr^{+4}	4	0.41
	6	0.55
Cr^{+6}	4	0.26
	6	0.44
Cs^{+1}	6	1.67
	8	1.74
	10	1.81
	12	1.88
Cu^{+1}	2	0.46
	4	0.60
	6	0.77
Cu^{+2}	4sq	0.57
	6	0.73
Dy^{+2}	6	1.07
	8	1.19
Dy^{+3}	6	0.91
	8	1.03
Er^{+3}	6	0.89
	8	1.00
Eu^{+2}	6	1.17
	8	1.25
	10	1.35
Eu^{+3}	6	0.95
	8	1.07
F^{+7}	6	0.08
Fe^{+2}	4	0.63
	6	0.61
	8	0.92
Fe^{+3}	4	0.49
	6	0.55
	8	0.78
Fr^{+1}	6	1.80
Ga^{+3}	4	0.47
	6	0.62
Gd^{+3}	6	0.94
	8	1.05
Ge^{+2}	6	0.73
Ge^{+4}	4	0.39
	6	0.53
Hf^{+4}	4	0.58
	6	0.71
	8	0.83
Hg^{+1}	6	1.19
Hg^{+2}	2	0.69
	4	0.96
	6	1.02
	8	1.14
I^{+5}	3py	0.44
	6	0.95
I^{+7}	4	0.42
	6	0.53
In^{+3}	4	0.62
	6	0.80
Ir^{+3}	6	0.68
Ir^{+4}	6	0.63
Ir^{+5}	6	0.57
K^{+1}	4	1.37
	6	1.38
	8	1.51
	12	1.64
La^{+3}	6	1.03
	8	1.16
	10	1.27
	12	1.36
Li^{+1}	4	0.59
	6	0.76

IONIC RADII IN CRYSTALS (continued)

Ion	*CN*	R_i/Å
	8	0.92
Lu^{+3}	6	0.86
	8	0.97
Mg^{+2}	4	0.57
	6	0.72
	8	0.89
Mn^{+2}	4	0.66
	6	0.83
	8	0.96
Mn^{+3}	6	0.58
Mn^{+4}	4	0.39
	6	0.53
Mn^{+5}	4	0.33
Mn^{+6}	4	0.26
Mn^{+7}	4	0.25
Mo^{+3}	6	0.69
Mo^{+4}	6	0.65
Mo^{+5}	4	0.46
	6	0.61
Mo^{+6}	4	0.41
	6	0.59
	7	0.73
N^{+3}	6	0.16
N^{+5}	6	0.13
Na^{+1}	4	0.99
	6	1.02
	8	1.18
	9	1.24
	12	1.39
Nb^{+3}	6	0.72
	8	0.79
Nb^{+4}	6	0.68
Nb^{+5}	4	0.48
	6	0.64
	8	0.74
Nd^{+3}	6	0.98
	8	1.12
	9	1.16
	12	1.27
Ni^{+2}	4sq	0.49
	6	0.69
Ni^{+3}	6	0.56
Np^{+3}	6	1.01
Np^{+4}	6	0.87
Np^{+5}	6	0.75
Np^{+6}	6	0.72
Os^{+4}	6	0.63
Os^{+5}	6	0.58
Os^{+6}	6	0.55
Os^{+8}	4	0.39
P^{+5}	4	0.17
	6	0.38
Pa^{+3}	6	1.04
Pa^{+4}	6	0.90
Pa^{+5}	6	0.78
Pb^{+2}	6	1.19
	8	1.29
	10	1.40
	12	1.49
Pb^{+4}	4	0.65

Ion	*CN*	R_i/Å
	6	0.78
	8	0.94
Pd^{+2}	4sq	0.64
	6	0.86
Pd^{+3}	6	0.76
Pd^{+4}	6	0.62
Pm^{+3}	6	0.97
	8	1.09
Po^{+4}	6	0.97
Pr^{+3}	6	0.99
	8	1.13
Pr^{+4}	6	0.85
	8	0.96
Pt^{+2}	4sq	0.60
	6	0.80
Pt^{+4}	6	0.63
Pu^{+3}	6	1.00
Pu^{+4}	6	0.86
Pu^{+5}	6	0.74
Pu^{+6}	6	0.71
Ra^{+2}	8	1.48
	12	1.70
Rb^{+1}	6	1.52
	8	1.61
	10	1.66
	12	1.72
Re^{+4}	6	0.63
Re^{+5}	6	0.58
Re^{+6}	6	0.55
Re^{+7}	4	0.38
	6	0.53
Rh^{+3}	6	0.67
Rh^{+4}	6	0.60
Rh^{+5}	6	0.55
Ru^{+3}	6	0.68
Ru^{+4}	6	0.62
Ru^{+5}	6	0.57
Ru^{+7}	4	0.38
Ru^{+8}	4	0.36
S^{+4}	6	0.37
S^{+6}	4	0.12
	6	0.29
Sb^{+3}	4py	0.76
	6	0.76
Sb^{+5}	6	0.60
Sc^{+3}	6	0.75
	8	0.87
Se^{+4}	6	0.50
Se^{+6}	4	0.28
	6	0.42
Si^{+4}	4	0.26
	6	0.40
Sm^{+2}	6	1.19
	8	1.27
Sm^{+3}	6	0.96
	8	1.08
	12	1.24
Sn^{+4}	4	0.55
	6	0.69
	8	0.81

Ion	*CN*	R_i/Å
Sr^{+2}	6	1.18
	8	1.26
	10	1.36
	12	1.44
Ta^{+3}	6	0.72
Ta^{+4}	6	0.68
Ta^{+5}	6	0.64
Tb^{+3}	6	0.92
	8	1.04
Tb^{+4}	6	0.76
	8	0.88
Tc^{+4}	6	0.65
Te^{+4}	4	0.66
	6	0.97
Te^{+6}	4	0.43
	6	0.56
Th^{+4}	6	0.94
	8	1.05
	10	1.13
	12	1.21
Ti^{+2}	6	0.86
Ti^{+3}	6	0.67
Ti^{+4}	4	0.42
	6	0.61
	8	0.74
Tl^{+1}	6	1.50
	8	1.59
	12	1.70
Tl^{+3}	4	0.75
	6	0.89
	8	0.98
Tm^{+2}	6	1.01
	7	1.09
Tm^{+3}	6	0.88
	8	0.99
U^{+3}	6	1.03
U^{+4}	6	0.89
	8	1.00
	12	1.17
U^{+5}	6	0.76
U^{+6}	2	0.45
	4	0.52
	6	0.73
	8	0.86
V^{+2}	6	0.79
V^{+3}	6	0.64
V^{+4}	5	0.53
	6	0.58
	8	0.72
V^{+5}	4	0.36
	5	0.46
	6	0.54
W^{+4}	6	0.66
W^{+5}	6	0.62
W^{+6}	4	0.42
	5	0.51
	6	0.60
Y^{+3}	6	0.90
	8	1.02
	9	1.08

Ion	CN	R_i/Å	Ion	CN	R_i/Å	Ion	CN	R_i/Å
Yb^{+2}	6	1.02	Zn^{+2}	4	0.60		6	0.72
	8	1.14		6	0.74		8	0.84
Yb^{+3}	8	0.99		8	0.90		9	0.89
	9	1.04	Zr^{+4}	4	0.59			

POLARIZABILITIES OF ATOMS AND IONS IN SOLIDS

H. P. R. Frederikse

The polarization of a solid dielectric medium, ***P***, is defined as the dipole moment per unit volume averaged over the volume of a crystal cell. A component of ***P*** can be expanded as a function of the electric field ***E***:

$$P_i = \sum_j a_j E_j + \sum_{jk} b_{jk} E_j E_k$$

For relatively small electric fields in isotropic substances $\boldsymbol{P} = \chi_e \boldsymbol{E}$, where χ_e is the electric susceptibility. If the medium is made up of N atoms (or ions) per unit volume, the polarization is $\boldsymbol{P} = N\boldsymbol{p}_m$ where $\boldsymbol{p}_m$ is the average dipole moment per atom. The polarizability α can be defined as $\boldsymbol{p}_m = \alpha \boldsymbol{E}_0$, where $\boldsymbol{E}_0$ is the local field at the position of the atom. Using the Lorentz method to calculate the local field one finds:

$$\boldsymbol{P} = N\alpha(\boldsymbol{E} + 4\pi\boldsymbol{P}) = \chi_c \boldsymbol{E}$$

Together with the definition of the dielectric constant (relative permittivity), $\varepsilon = 1 + 4\pi\chi_e$, this leads to:

$$\alpha = \frac{3}{4\pi N}\left(\frac{\varepsilon - 1}{\varepsilon + 2}\right)$$

This expression is known as the Clausius-Mossotti equation.

The total polarization associated with atoms, ions, or molecules is due to three different sources:

1. Electronic polarization arises because the center of the local electronic charge cloud around the nucleus is displaced under the action of the field: $P_e = N\alpha_e E_0$ where α_e is the *electronic polarizability.*
2. Ionic polarization occurs in ionic materials because the electric field displaces cations and anions in opposite directions: $P_i = N\alpha_i E_0$, where α_i is the *ionic polarizability.*
3. Orientational polarization can occur in substances composed of molecules that have permanent electric dipoles. The alignment of these dipoles depends on temperature and leads to an *orientational polarizability* per molecule: $\alpha_{or} = p^2/3kT$, where p is the permanent dipole moment per molecule, k is the Boltzmann constant, and T is the temperature.

Because of the different nature of these three polarization processes the response of a dielectric solid to an applied electric field will strongly depend on the frequency of the field. The resonance of the electronic excitation in insulators (dielectrics) takes place in the ultraviolet part of the spectrum; the characteristic frequency of the lattice vibrations is located in the infrared, while the orientation of dipoles requires fields of much lower frequencies (below 10^{10} Hz). This response to electric fields of different frequencies is shown in Figure 1. Values of the electronic polarizabilities for selected atoms and ions are given in Table 1.

REFERENCES

1. Kittel, C., *Introduction to Solid State Physics*, Fourth Edition, John Wiley & Sons, New York, 1971.
2. Lerner, R.G., and Trigg, G.L., Editors, *Encyclopedia of Physics, Second Edition*, VCH Publishers, New York, 1990.
3. Ralls, K.M., Courtney, T.H., and Wulff, J., *An Introduction to Materials Science and Engineering*, John Wiley & Sons, New York, 1976.

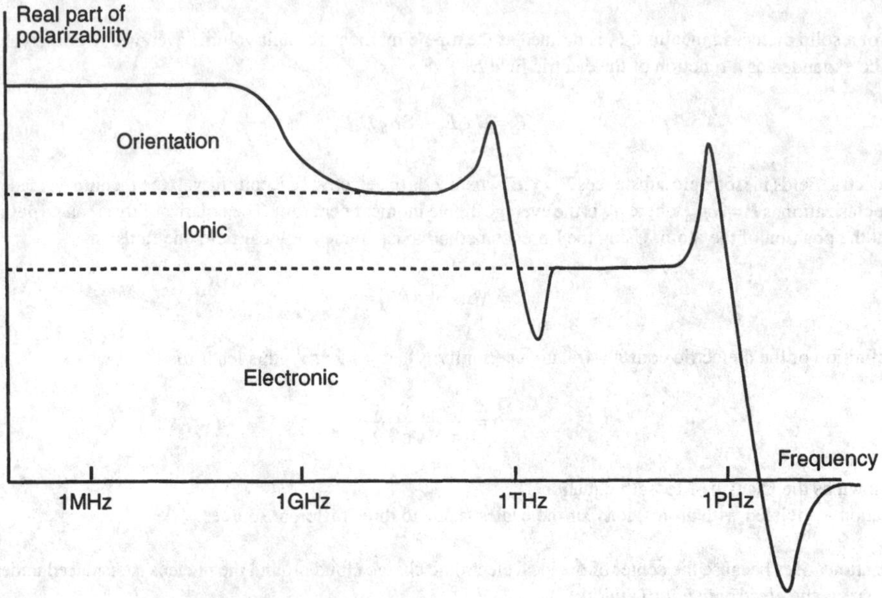

Figure 1. Schematic graph of the frequency dependence of the different contributions to polarizability.

TABLE 1
Electronic Polarizabilities in Units of 10^{-24} cm^3

						He 0.201
Li^+ 0.029	**Be^{2+}** 0.008	**B^{3+}** 0.003	**C^{4+}** 0.0013	**O^{2-}** 3.88	**F^-** 1.04	**Ne** 0.39
Na^+ 0.179	**Mg^{2+}** 0.094	**Al^{3+}** 0.052	**Si^{4+}** 0.0165	**S^{2-}** 10.2	**Cl^-** 3.66	**Ar** 1.62
K^+ 0.83	**Ca^{2+}** 0.47	**Sc^{3+}** 0.286	**Ti^{4+}** 0.185	**Se^{2-}** 10.5	**Br^-** 4.77	**Kr** 2.46
Rb^+ 1.40	**Sr^{2+}** 0.86	**Y^{3+}** 0.55	**Zr^{4+}** 0.37	**Te^{2-}** 14.0	**I^-** 7.1	**Xe** 3.99
Cs^+ 2.42	**Ba^{2+}** 1.55	**La^{3+}** 1.04	**Ce^{4+}** 0.73			

Data from Pauling, L., *Proc. R. Soc. London*, A114, 181, 1927. See also Jaswal, S.S. and Sharma, T.P., *J. Phys. Chem. Solids*, 34, 509, 1973.
Values are appropriate for cgs units. To convert to SI, use the relation

$$\alpha(\mathrm{SI})/\mathrm{C\,m^2V^{-1}} = 1.11265\cdot 10^{-16}\,\alpha(\mathrm{cgs})/\mathrm{cm}^3$$

CRYSTAL STRUCTURES AND LATTICE PARAMETERS OF ALLOTROPES OF THE ELEMENTS

H. W. King

The crystal structures of the allotropic forms of the elements are presented in terms of the Pearson symbol, the Strukturbericht designation, and the prototype of the structure. The temperatures of the phase transformations are listed in degrees Celsius and the pressures are in GPa. A consistent nomenclature is used, whereby all allotropes are labeled by Greek letters. The lattice parameters of the unit cells are given in nanometers (nm) and are considered to be accurate to ±2 in the last reported digit.

This compilation is restricted to changes in crystal structure that occur as a result of a change in temperature or pressure. Low-temperature structures are included for the diatomic and rare gases, which show many similarities with respect to the metallic elements. The elements identified with an asterisk (*) have polymorphic structures based on different molecular configurations. The crystal data given for these elements refer to the most stable structure at room temperature.

Reprinted with the permission of ASM International from T. B. Massalski, Ed., Binary Alloy Phase Diagrams, ASM International, Metals Park, Ohio, 1986; certain data on rare earth elements were provided by K. A. Gschneidner.

Element	Temperature, °C	Pressure, GPa	Pearson symbol	Space group	Strukturbericht designation	Prototype	Lattice parameters, nm a	b	c	Comment, c/a or α or β
Ac	25	atm	cF4	Fm3m	A1	Cu	0.5311	...	...	...
Ag	25	atm	cF4	Fm3m	A1	Cu	0.40857	...	...	...
αAl	25	atm	cF4	Fm3m	A1	Cu	0.40496	...	...	...
βAl	25	>20.5	hP2	$P6_3/mmc$	A3	Mg	0.2693	...	0.4398	1.6331
α'Am	25	atm	hP4	$P6_3/mmc$	A3'	αLa	0.34681	...	1.1241	2*1.621
αAm	>769	atm	cF4	Fm3m	A1	Cu	0.4894	...	...	...
βAm	>1074	atm	cI2	Im3m	A2	W	?	...	...	...
γAm	25	>15	oC4	Cmcm	A20	αU	0.3063	0.5968	0.5169	...
αAr	<−189.35	atm	cF4	Fm3m	A1	Cu	0.5316	...	...	...
(βAr)	<−189.40	atm	hP2	$P6_3/mmc$	A3	Mg	0.3760	...	0.6141	1.633
αAs	25	atm	hR2	$R\bar{3}m$	A7	αAs	0.41319	...	...	α = 54.12°
εAs	>448	atm	oC8	Cmca	...	P (black)	0.362	1.085	0.448	...
Au	25	atm	cF4	Fm3m	A1	Cu	0.40782	...	...	...
βB	25	atm	hR105	R3m	...	βB	1.017	...	...	α = 65.12°
αBa	25	atm	cI2	Im3m	A2	W	0.50227	...	...	...
βBa	25	>5.33	hP2	$P6_3/mmc$	A3	Mg	0.3901	...	0.6154	1.5775
γBa	25	>23	?	?	...	...	...	...	...	...
αBe	25	atm	hP2	$P6_3/mmc$	A3	Mg	0.22859	...	0.35845	1.5681
βBe	>1270	atm	cI2	Im3m	A2	W	0.25515	...	...	...
γBe	25	>9.3	?	...	...	...	...	...	...	...
αBi	25	atm	hR2	$R\bar{3}m$	A7	αAs	0.47460	...	...	α = 57.23°
βBi	25	>2.6	mC4	C2/m	...	βBi	0.6674	0.6117	0.3304	β = 110.33°
γBi	25	>3.0	mP3	?	...	...	0.605	0.42	0.465	β = 85.33°
σBi	25	>4.3	?	?	...	...	...	...	...	...
εBi	25	>6.5	?	?	...	...	...	...	...	...
ζBi	25	>9.0	cI2	Im3m	A2	W	0.3800	...	...	...
αBk	25	atm	hP4	$P6_3/mmc$	A3'	αLa	0.3416	...	1.1069	2*1.620
βBk	>977	atm	cF4	Fm3m	A1	Cu	0.4997	...	...	...
Br	<7.25	atm	oC8	Cmca	...	Cl	0.668	0.449	0.874	...
C (graphite)	25	atm	hP4	$P6_3/mmc$	A9	C (graphite)	0.24612	...	0.6709	2.7258
C (diamond)	25	>60	cF8	Fd3m	A4	C (diamond)	0.35669	...	...	...
C (hd)	25	HP	hP4	$P6_3/mmc$	...	C (hd)	0.2522	...	0.4119	1.633
αCa	25	atm	cF4	Fm3m	A1	Cu	0.55884	...	...	...
βCa	>443	atm	cI2	Im3m	A2	W	0.4480	...	...	...
γCa	25	>1.5	?	...	...	...	...	...	...	...
Cd	25	atm	hP2	$P6_3/mmc$	A3	Mg	0.29793	...	0.56196	1.8862
αCe	<−177	atm	cF4	Fm3m	A1	Cu	0.485	...	...	...
βCe	25	atm	hP4	$P6_3/mmc$	A3'	αLa	0.36810	...	1.1857	2*1.611
γCe	25	atm	cF4	Fm3m	A1	Cu	0.51610	...	...	...
δ-Ce	>726	atm	cI2	Im3m	A2	W	0.412	...	...	...
α'Ce	25	>5.4	oC4	Cmcm	A20	αU	0.3049	0.5998	0.5215	...
αCf	25	atm	hP4	$P6_3/mmc$	A3'	αLa	0.339	...	1.1015	2*1.625
βCf	>590	atm	cF4	Fm3m	A1	Cu	?	...	...	...
Cl	25	atm	oC8	Cmca	...	Cl	0.624	0.448	0.826	...
αCm	25	atm	hP4	$P6_3/mmc$	A3'	αLa	0.3496	...	1.1331	2*1.621
βCm	>1277	atm	cF4	Fm3m	A1	Cu	0.4382	...	...	...
εCo	25	atm	hP2	$P6_3/mmc$	A3	Mg	0.25071	...	0.40686	1.6228
αCo	>422	atm	cF4	Fm3m	A1	Cu	0.35447	...	...	...
αCr	25	atm	cI2	Im3m	A2	W	0.28848	...	...	...
α'Cr	25	HP	tI2	I4/mmm	...	α'Cr	0.2882	...	0.2887	1.002
αCs	25	atm	cI2	Im3m	A2	W	0.6141	...	...	...
βCs	25	>2.37	cF4	Fm3m	A1	Cu	0.6465	...	...	...
β'Cs	25	>4.22	cF4	Fm3m	A1	Cu	0.5800	...	...	...
γCs	25	>4.27	?	...	...	...	...	...	...	...
Cu	25	atm	cF4	Fm3m	A1	Cu	0.36146	...	...	...
α'Dy	<−187	atm	oC4	Cmcm	...	α'Dy	0.3595	0.6184	0.5678	...
αDy	25	atm	hP2	$P6_3/mmc$	A3	Mg	0.35915	...	0.56501	1.5732

Element	Temperature, °C	Pressure, GPa	Pearson symbol	Space group	Strukturbericht designation	Prototype	Lattice parameters, nm			Comment, c/a or α or β
							a	b	c	
βDy	>1381	atm	cI2	Im3m	A2	W	0.403	···	···	···
γDy	25	>7.5	hR3	$R\bar{3}m$	···	αSm	0.3436	···	2.483	4.5*1.606
Er	25	atm	hP2	$P6_3/mmc$	A3	Mg	0.35592	···	0.55850	1.5692
αEs	25	atm	hP4	$P6_3/mmc$	A3'	αLa	?	···	···	···
βEs	?	atm	cF4	Fm3m	A1	Cu	?	···	···	···
Eu	25	atm	cI2	Im3m	A2	W	0.45827	···	···	···
αF	<−227.6	atm	mC8	C2/c	···	αF	0.550	0.338	0.728	β = 102.17°
βF	<−219.67	atm	cP16	Pm3n	···	γO	0.667	···	···	···
αFe	25	atm	cI2	Im3m	A2	W	0.28665	···	···	···
γFe	>912	atm	cF4	Fm3m	A1	Cu	0.36467	···	···	···
σFe	>1394	atm	cI2	Im3m	A2	W	0.29315	···	···	···
εFe	25	>13	hP2	$P6_3/mmc$	A3	Mg	0.2468	···	0.396	1.603
αGa	25	atm	oC8	Cmca	A11	αGa	0.45186	0.76570	0.45258	···
βGa	25	>1.2	tI2	I4/mmm	A6	In	0.2808	···	0.4458	1.588
γGa	−53	>3.0	oC40	Cmcm	···	γGa	1.0593	1.3523	0.5203	···
αGd	25	atm	hP2	$P6_3/mmc$	A3	Mg	0.36336	···	0.57810	1.5910
βGd	>1235	atm	cI2	Im3m	A2	W	0.406	···	···	···
γGd	25	>3.0	hR3	R3m	···	αSm	0.361	···	2.603	4*1.60
αGe	25	atm	cF8	Fd3m	A4	C (diamond)	0.56574	···	···	···
βGe	25	>12	tI4	$I4_1/amd$	A5	βSn	0.4884	···	0.2692	0.551
γGe	25	>12 → atm	tP12	$P4_12_12$	···	σGe	0.593	···	0.698	1.18
σGe	LT	>12	cI16	Im3m	···	γSi	0.692	···	···	···
αH	<−271.9	atm	cF4	Fm3m	A1	Cu	0.5338	···	···	···
βH	<−259.34	atm	hP2	$P6_3/mmc$	A3	Mg	0.3776	···	0.6162	1.632
αHe	<−268.94	atm	hP2	$P6_3/mmc$	A3	Mg	0.3555	···	0.5798	1.631
βHe	>−258	0.125	cF4	Fm3m	A1	Cu	0.4240	···	···	···
γHe	<−271.47	0.03	cI2	Im3m	A2	W	0.4110	···	···	···
αHf	25	atm	hP2	$P6_3/mmc$	A3	Mg	0.31946	···	0.50510	1.5811
βHf	>1995	atm	cI2	Im3m	A2	W	0.3610	···	···	···
αHg	<−38.84	atm	hR1	$R\bar{3}m$	A10	αHg	0.3005	···	···	α = 70.53°
βHg	<−194	HP	tI2	I4/mmm	···	βHg	0.3995	···	0.2825	0.707
γHg	<−194	c.w.	hR1	?	···	···	···	···	···	···
αHo	25	atm	hP2	$P6_3/mmc$	A3	Mg	0.35778	···	0.56178	1.5702
βHo	25	>7.5	hR3	$R\bar{3}m$	···	αSm	0.334	···	2.45	4.5*1.63
I	25	atm	oC8	Cmca	···	Cl	0.72697	0.47903	0.97942	···
In	25	atm	tI2	I4/mmm	A6	In	0.3253	···	0.49470	1.5210
Ir	25	atm	cF4	Fm3m	A1	Cu	0.38392	···	···	···
K	25	atm	cI2	Im3m	A2	W	0.5321	···	···	···
Kr	<−157.39	atm	cF4	Fm3m	A1	Cu	0.5810	···	···	···
αLa	25	atm	hP4	$P6_3/mmc$	A3'	αLa	0.37740	···	1.2171	2*1.6125
βLa	>310	atm	cF4	Fm3m	A1	Cu	0.5303	···	···	···
γLa	>865	atm	cI2	Im3m	A2	W	0.426	···	···	···
β'La	25	>2.0	cF4	Fm3m	A1	Cu	0.517	···	···	···
αLi	<−193	atm	hP2	$P6_3/mmc$	A3	Mg	0.3111	···	0.5093	1.637
βLi	25	atm	cI2	Im3m	A2	W	0.35093	···	···	···
γLi	<−201	c.w.	cF4	Fm3m	A1	Cu	0.4388	···	···	···
Lu	25	atm	hP2	$P6_3/mmc$	A3	Mg	0.35052	···	0.55494	1.5832
Mg	25	atm	hP2	$P6_3/mmc$	A3	Mg	0.32094	···	0.52107	1.6236
αMn	25	atm	cI58	$I\bar{4}3m$	A12	αMn	0.89126	···	···	···
βMn	>710	atm	cP20	$P4_132$	A13	βMn	0.63152	···	···	···
γMn	>1079	atm	cF4	Fm3m	A1	Cu	0.3860	···	···	···
σMn	>1143	atm	cI2	Im3m	A2	W	0.3080	···	···	···
Mo	25	atm	cI2	Im3m	A2	W	0.31470	···	···	···
αN	<−237.6	atm	cP8	Pa3	···	αN	0.5661	···	···	···
βN	<−210.00	atm	hP4	$P6_3/mmc$	···	βN	0.4050	···	0.6604	1.631
γN	<−253	>3.3	tP4	$P4_2/mnm$	···	γN	0.3957	···	0.5109	1.291
αNa	<−233	atm	hP2	$P6_3/mmc$	A3	Mg	0.3767	···	0.6154	1.634
βNa	25	atm	cI2	Im3m	A2	W	0.42906	···	···	···
Nb	25	atm	cI2	Im3m	A2	W	0.33004	···	···	···
αNd	25	atm	hP4	$P6_3/mmc$	A3'	αLa	0.36582	···	1.17966	2*1.6124
βNd	>863	atm	cI2	Im3m	A2	W	0.413	···	···	···
γNd	25	>5.0	cF4	Fm3m	A1	Cu	0.480	···	···	···
Ne	<−243.59	atm	cF4	Fm3m	A1	Cu	0.4462	···	···	···
Ni	25	atm	cF4	Fm3m	A1	Cu	0.35240	···	···	···
αNp	25	atm	oP8	Pnma	A_c	αNp	0.6663	0.4723	0.4887	···
βNp	>280	atm	tP4	$P42_12$	A_d	βNp	0.4883	···	0.3389	0.694
γNp	>576	atm	cI2	Im3m	A2	W	0.352	···	···	···
αO	<−243.3	atm	mC4	C2m	···	αO	0.5403	0.3429	0.5086	β = 132.53°
βO	<−229.6	atm	hR2	$R\bar{3}m$	···	βO	0.4210	···	···	α = 46.27°
γO	<−218.79	atm	cP16	Pm3n	···	γO	0.683	···	···	···
Os	25	atm	hP2	$P6_3/mmc$	A3	Mg	0.27341	···	0.43918	1.6063
P (black)	25	atm	oC8	Cmca	···	P (black)	0.33136	1.0478	0.43763	···
αPa	25	atm	tI2	I4/mmm	A_a	αPa	0.3921	···	0.3235	0.825
βPa	>1170	atm	cI2	Im3m	A2	W	0.381	···	···	···
αPb	25	atm	cF4	Fm3m	A1	Cu	0.49502	···	···	···
βPb	25	>10.3	hP2	$P6_3/mmc$	A3	Mg	0.3265	···	0.5387	1.650

CRYSTAL STRUCTURES AND LATTICE PARAMETERS OF ALLOTROPES OF THE ELEMENTS (continued)

Element	Temperature, °C	Pressure, GPa	Pearson symbol	Space group	Strukturbericht designation	Prototype	Lattice parameters, nm *a*	*b*	*c*	Comment, *c/a* or α or β
Pd	25	atm	*cF*4	$Fm3m$	*A*1	Cu	0.38903	…	…	…
αPm	25	atm	*hP*4	$P6_3/mmc$	*A*3′	αLa	0.365	…	1.165	2*1.60
βPm	>890	atm	*cI*2	$Im3m$	*A*2	W	(0.410)	…	…	…
αPo	25	atm	*cP*1	$Pm3m$	A_h	αPo	0.3366	…	…	…
βPo	>54	atm	*hR*1	$R\bar{3}m$	…	βPo	0.3373	…	…	α = 98.08°
αPr	25	atm	*hP*4	$P6_3/mmc$	*A*3′	αLa	0.36721	…	1.18326	2*1.6111
βPr	>795	atm	*cI*2	$Im3m$	*A*2	W	0.413	…	…	…
γPr	25	>4.0	*cF*4	$Fm3m$	*A*1	Cu	0.488	…	…	…
Pt	25	atm	*cF*4	$Fm3m$	*A*1	Cu	0.39236	…	…	…
αPu	25	atm	*mP*16	$P2_1/m$	…	αPu	0.6183	0.4822	1.0963	β = 101.97°
βPu	>125	atm	*mI*34	$I2/m$	…	βPu	0.9284	1.0463	0.7859	β = 92.13°
γPu	>215	atm	*oF*8	$Fddd$	…	γPu	0.31587	0.57682	1.0162	…
σPu	>320	atm	*cF*4	$Fm3m$	*A*1	Cu	0.46371	…	…	…
σ′Pu	>463	atm	*tI*2	$I4/mmm$	*A*6	In	0.33261	…	0.44630	1.3418
εPu	>483	atm	*cI*2	$Im3m$	*A*2	W	0.36343	…	…	…
Ra	25	atm	*cI*2	$Im3m$	*A*2	W	0.5148	…	…	…
αRb	25	atm	*cI*2	$Im3m$	*A*2	W	0.5705	…	…	…
βRb	25	>1.08	?	…	…	…	…	…	…	…
γRb	25	>2.05	?	…	…	…	…	…	…	…
Re	25	atm	*hP*2	$P6_3/mmc$	*A*3	Mg	0.27609	…	0.4458	1.6145
Rh	25	atm	*cF*4	$Fm3m$	*A*1	Cu	0.38032	…	…	…
Ru	25	atm	*hP*2	$P6_3/mmc$	*A*3	Mg	0.27058	…	0.42816	1.5824
αS	25	atm	*oF*128	$Fddd$	*A*16	αS	1.0464	1.28660	2.44860	…
αSb	25	atm	*hR*2	$R\bar{3}m$	*A*7	αAs	0.45067	…	…	α = 57.11°
βSb	25	>5.0	*cP*1	$Pm3m$	A_h	αPo	0.2992	…	…	…
γSb	25	>7.5	*hP*2	$P6_3/mmc$	*A*3	Mg	0.3376	…	0.5341	1.582
σSb	25	>14.0	*mP*3	?	…	…	0.556	0.404	0.422	β = 86.0°
αSc	25	atm	*hP*2	$P6_3/mmc$	*A*3	Mg	0.33088	…	0.52680	1.5921
βSc	>1337	atm	*cI*2	$Im3m$	*A*2	W	(0.373)	…	…	…
γSe	25	atm	*hP*3	$P3_121$	*A*8	γSe	0.43659	…	0.49537	1.1346
αSi	25	atm	*cF*8	$Fd3m$	*A*4	C (diamond)	0.54306	…	…	…
βSi	25	>9.5	*tI*4	$I4_1/amd$	*A*5	βSn	0.4686	…	0.2585	0.552
γSi	25	>16.0	*cI*16	$Im3m$	…	γSi	0.6636	…	…	…
σSi	25	>16 → atm	*hP*4	$P6_3/mmc$	*A*3′	αLa	0.380	…	0.628	1.653
αSm	25	atm	*hR*3	$R\bar{3}m$	…	αSm	0.36290	…	2.6207	4*1.6048
βSm	>734	atm	*hP*2	$P6_3/mmc$	*A*3	Mg	0.36630	…	0.58448	1.5956
γ′Sm	>922	atm	*cI*2	$Im3m$	*A*2	W	(0.410)	…	…	…
σSm	25	>4.0	*hP*4	$P6_3/mmc$	*A*3′	αLa	0.3618	…	1.166	2*1.611
αSn	<13	atm	*cF*8	$Fd3m$	*A*4	C (diamond)	0.64892	…	…	…
βSn	25	atm	*tI*4	$I4_1/amd$	*A*5	βSn	0.58318	…	0.31818	0.5456
γSn	25	>9.0	*tI*2	?	…	γSn	0.370	…	0.337	0.91
αSr	25	atm	*cF*4	$Fm3m$	*A*1	Cu	0.6084	…	…	…
βSr	>547	atm	*cI*2	$Im3m$	*A*2	W	0.487	…	…	…
β′Sr	25	>3.5	*cI*2	$Im3m$	*A*2	W	0.4437	…	…	…
Ta	25	atm	*cI*2	$Im3m$	*A*2	W	0.33030	…	…	…
α′Tb	<−53	atm	*oC*4	$Cmcm$	…	α′Dy	0.3605	0.6244	0.5706	…
αTb	25	atm	*hP*2	$P6_3/mmc$	*A*3	Mg	0.36055	…	0.56966	1.5800
βTb	>1289	atm	*cI*2	$Im3m$	*A*2	W	(0.407)	…	…	…
γTb	25	>6.0	*hR*3	$R\bar{3}m$	…	αSm	0.341	…	2.45	4*1.60
Tc	25	atm	*hP*2	$P6_3/mmc$	*A*3	Mg	0.2738	…	0.4393	1.604
αTe	25	atm	*hP*3	$P3_121$	*A*8	γSe	0.44566	…	0.59264	1.3298
βTe	25	>2.0	*hR*2	$R\bar{3}m$	*A*7	αAs	0.469	…	…	α = 53.30°
γTe	25	>7.0	*hR*1	$R\bar{3}m$	…	βPo	0.3002	…	…	α = 103.3°
αTh	25	atm	*cF*4	$Fm3m$	*A*1	Cu	0.50842	…	…	…
βTh	>1360	atm	*cI*2	$Im3m$	*A*2	W	0.411	…	…	…
αTi	25	atm	*hP*2	$P6_3/mmc$	*A*3	Mg	0.29506	…	0.46835	1.5873
βTi	>882	atm	*cI*2	$Im3m$	*A*2	W	0.33065	…	…	…
ωTi	25	HP → atm	*hP*3	$P6/mmm$	…	ωTi	0.4625	…	0.2813	0.6082
αTl	25	atm	*hP*2	$P6_3/mmc$	*A*3	Mg	0.34566	…	0.55248	1.5983
βTl	>230	atm	*cI*2	$Im3m$	*A*2	W	0.3879	…	…	…
γTl	25	HP	*cF*4	$Fm3m$	*A*1	Cu	?	…	…	…
Tm	25	atm	*hP*2	$P6_3/mmc$	*A*3	Mg	0.35375	…	0.55540	1.5700
αU	25	atm	*oC*4	$Cmcm$	*A*20	αU	0.28537	0.58695	0.49548	…
βU	>668	atm	*tP*30	$P4_2/mnm$	A_b	βU	1.0759	…	0.5656	0.526
γU	>776	atm	*cI*2	$Im3m$	*A*2	W	0.3524	…	…	…
V	25	atm	*cI*2	$Im3m$	*A*2	W	0.30240	…	…	…
W	25	atm	*cI*2	$Im3m$	*A*2	W	0.31652	…	…	…
Xe	<−111.76	atm	*cF*4	$Fm3m$	*A*1	Cu	0.6350	…	…	…
αY	25	atm	*hP*2	$P6_3/mmc$	*A*3	Mg	0.36482	…	0.57318	1.5711
βY	>1478	atm	*cI*2	$Im3m$	*A*2	W	(0.410)	…	…	…
αYb	<−3	atm	*hP*2	$P6_3/mmc$	*A*3	Mg	0.38799	…	0.63859	1.6459
βYb	25	atm	*cF*4	$Fm3m$	*A*1	Cu	0.54848	…	…	…
γYb	>795	atm	*cI*2	$Im3m$	*A*2	W	0.444	…	…	…
Zn	25	atm	*hP*2	$P6_3/mmc$	*A*3	Mg	0.26650	…	0.49470	1.8563
αZr	25	atm	*hP*2	$P6_3/mmc$	*A*3	Mg	0.32316	…	0.51475	1.5929
βZr	>863	atm	*cI*2	$Im3m$	*A*2	W	0.36090	…	…	…
ωZr	25	HP → atm	*hP*2	$P6/mmm$	…	ωTi	0.5036	…	0.3109	0.617

LATTICE ENERGIES

H. D. B. Jenkins and H. K. Roobottom

THERMOCHEMICAL CYCLE AND CALCULATED VALUES

Table 1 contains calculated values of the lattice energies (total lattice potential energies), U_{POT}, of crystalline salts, M_aX_b. U_{POT} is expressed in units of kilojoules per mole, kJ mol^{-1}. M and X can be either simple or complex ions. Substances are arranged by chemical class.

Also listed in the table is the lattice energy, U_{POT}^{BFHC}, obtained from the application of the Born - Fajans - Haber cycle (BHFC) described below, using the "Standard Thermochemical Properties of Chemical Substances" table in Section 5 of this *Handbook*, References 1 through 4, and certain other data which are given in Table 3 below.

The lattice enthalpy, ΔH_L, is given by the cycle:

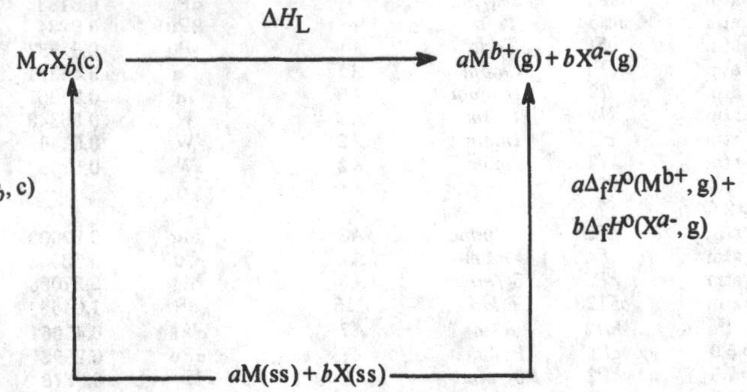

where (ss) is the standard state of the element concerned.

The lattice enthalpy, ΔH_L, is obtained using the equation:

$$\Delta H_L = a\Delta_f H^o(M^{b+}, g) + b\Delta_f H^o(X^{a-}, g) - \Delta_f H^o(M_aX_b, c)$$

and is futher related to the total lattice potential energy, U_{POT}, by the relationship:

$$\Delta H_L = U_{POT} + \left[a\left(\frac{n_M}{2}-2\right)+b\left(\frac{n_X}{2}-2\right)\right]RT$$

where n_M and n_X equal 3 for monatomic ions, 5 for linear polyatomic ions and 6 for polyatomic non-linear ions.

METHOD OF ESTIMATION OF VALUES NOT TABULATED

In cases where the lattice energy is not tabulated and we want to furnish an estimate, then the Kapustinskii equation[5] can be used to obtain a value (in kJ mol^{-1}):

$$U_{POT} = \frac{121.4 z_a z_b \nu}{(r_a + r_b)}\left(1 - \frac{0.0345}{(r_a + r_b)}\right)$$

where z_a and z_b are the moduli of the charges on the ν ions in hte lattice and r_a and r_b (in nm) are the thermochemical radii given in Table 2. The r_a for metal ions is taken to be the Goldschmidt[6] radius.

To cite an example, if we wish to estimate the lattice energy of the salt $[NH_4^+][HF_2^-]$ using the above procedure, we see that Table 2 gives the thermochemical radius (r_a) for NH_4^+ to be 0.136 nm and that for HF_2^- (r_b) to be 0.172 nm. The lattice potential energy is then estimated to be 700 kJ mol^{-1} compared with the calculated value of 705 kJ mol^{-1}and the Born - Fajans - Haber cycle value of 658 kJ mol^{-1}.

REFERENCES

1. Wagman, D. D., Evans, W. H., Parker, V. B., Schumm, R. H., Halow, I., Bailey, S. M., Churney, K. L., and Nuttall, R. L., *The NBS Tables of Chemical Thermodynamic Properties, J. Phys. Chem. Ref. Data*, Vol. 11, Suppl. 2, 1982.
2. Chase, M. W., Davies, C. A., Downey, J. R., Frurip, D. J., McDonald, R. A., and Syverud, A. N., *JANAF Thermochemical Tables, Third Edition, J. Phys. Chem. Ref. Data*, Vol. 14, Suppl. 1, 1985.
3. Lias, S. G., Bartmess, J. E., Liebman, J. F., Holmes, J. L., Levin, R. D., and Mallard, W. G., *Gas-Phase Ion and Neutral Thermochemistry, J. Phys. Chem. Ref. Data*, Vol. 17, Suppl. 1, 1988.
4. Jenkins, H. D. B., and Pratt, K. F., *Adv. Inorg. Chem. Radiochem., 22*, 1, 1978.
5. Kapustinskii, A. F., *Quart. Rev., 10*, 283-294., 1956.
6. Goldschmidt, V. M., *Skrifter Norske Videnskaps-Akad.* Oslo, I, Mat.-Naturn. Kl, 1926. See also Dasent, W. E., *Inorganic Energetics*, 2nd ed., Cambridge University Press, 1982.
7. Jenkins, H. D. B., Roobottom, H. K., Passmore, J., and Glasser, L., *J. Chem. Education*, in press.

Table 1
LATTICE ENERGIES (kJ mol^{-1})

Substance	Calc. U_{POT}	U_{POT}^{BHFC}
Acetates		
$Li(CH_3COO)$	-	843
$Na(CH_3COO)$	828	807
$K(CH_3COO)$	749	726
$Rb(CH_3COO)$	715	-
$Cs(CH_3COO)$	682	-
Acetylides		
CaC_2	2911	2902
SrC_2	2788	2782
BaC_2	2647	2652
Azides		
LiN_3	861	875
NaN_3	770	784
KN_3	697	-
RbN_3	674	691
CsN_3	665	674
AgN_3	854	910
TlN_3	689	742
$Ca(N_3)_2$	2186	2316
$Sr(N_3)_2$	2056	2187
$Ba(N_3)_2$	2021	-
$Mn(N_3)_2$	2408	2348
$Cu(N_3)_2$	2730	2738
$Zn(N_3)_2$	2840	2970
$Cd(N_3)_2$	2446	2576
$Pb(N_3)_2$	-	2300
Bihalide Salts		
$LiHF_2$	821	847
$NaHF_2$	755	748
KHF_2	657	660
$RbHF_2$	627	631
$CsHF_2$	607	-
NH_4HF_2	705	658
$CsHCl_2$	601	-
Me_4NHCl_2	427	-
Et_4NHCl_2	346	-
Bu_4NHCl_2	290	-
Bicarbonates		
$NaHCO_3$	820	656
$KHCO_3$	741	573
$RbHCO_3$	707	522
$CsHCO_3$	678	520
NH_4HCO_3	-	577
Borides		
CaB_6	5146	-
SrB_6	5104	-
BaB_6	5021	-
YB_6	7447	-
LaB_6	7406	-
CeB_6	10083	-
PrB_6	7447	—
NdB_6	7447	-
PmB_6	7406	-
SmB_6	7447	-
EuB_6	5104	-
GdB_6	7489	-
TbB_6	7489	-
DyB_6	7489	-
HoB_6	7489	-
ErB_6	7489	-
TmB_6	7489	-
YbB_6	5146	-
LuB_6	7489	-
ThB_6	10167	-
Borohydrides		
$LiBH_4$	778	-
$NaBH_4$	703	694
KBH_4	655	638
$RbBH_4$	648	-
$CsBH_4$	628	-
Borohalides		
$LiBF_4$	699	749
$NaBF_4$	657	674
KBF_4	611	616
$RbBF_4$	577	590
$CsBF_4$	556	565
NH_4BF_4	582	-
$KBCl_4$	506	497
$RbBCl_4$	489	486
$CsBCl_4$	473	-
Carbonates		
Li_2CO_3	2523	2254
Na_2CO_3	2301	2016
K_2CO_3	2084	1846
Rb_2CO_3	2000	1783
Cs_2CO_3	1920	1722
$MgCO_3$	3138	3122
$CaCO_3$	2804	2811
$SrCO_3$	2720	2688
$BaCO_3$	2615	2554
$MnCO_3$	3046	3092
$FeCO_3$	3121	3169
$CoCO_3$	3443	3235
$CuCO_3$	3494	-
$ZnCO_3$	3121	3273
$CdCO_3$	2929	3052
$SnCO_3$	2904	-
$PbCO_3$	2728	2750
Cyanates		
$LiNCO$	849	-
$NaNCO$	807	816
$KNCO$	726	734
$RbNCO$	692	-
$CsNCO$	661	-
NH_4NCO	724	-
Cyanides		
$LiCN$	874	
$NaCN$	766	759
KCN	692	686
$RbCN$	638	-
$CsCN$	601	-
$Ca(CN)_2$	2268	2240

Table 1
LATTICE ENERGIES (kJ mol^{-1}) (continued)

Substance	Calc. U_{POT}	U_{POT}^{BHFC}
$Sr(CN)_2$	2138	-
$Ba(CN)_2$	2001	2009
NH_4CN	617	691
AgCN	(741)	935
$Zn(CN)_2$	2809	2817
$Cd(CN)_2$	2583	2591
Formates		
$Li(HCO_2)$	865	-
$Na(HCO_2)$	791	804
$K(HCO_2)$	713	722
$Rb(HCO_2)$	685	-
$Cs(HCO_2)$	651	-
$NH_4(HCO_2)$	715	-
Germanates		
Mg_2GeO_4	7991	-
Ca_2GeO_4	7301	7306
Sr_2GeO_4	6987	-
Ba_2GeO_4	6653	6643
Halates		
$LiBrO_3$	883	880
$NaBrO_3$	803	791
$KBrO_3$	740	722
$RbBrO_3$	720	705
$CsBrO_3$	694	681
$NaClO_3$	770	785
$KClO_3$	711	721
$RbClO_3$	690	703
$CsClO_3$	-	679
$LiIO_3$	975	974
$NaIO_3$	883	876
KIO_3	820	780
$RbIO_3$	791	-
$CsIO_3$	761	-
Halides		
LiF	1030	1049
LiCl	834	864
LiBr	788	820
LiI	730	764
NaF	910	930
NaCl	769	790
NaBr	732	754
NaI	682	705
KF	808	829
KCl	701	720
KBr	671	691
KI	632	650
RbF	774	795
RbCl	680	695
RbBr	651	668
RbI	617	632
CsF	744	759
CsCl	657	670
CsBr	632	647
CsI	600	613
FrF	715	-
FrCl	632	-
FrBr	611	-
FrI	582	-
CuCl	992	996
CuBr	969	978
CuI	948	966
AgF	953	974
AgCl	910	918
AgBr	897	905
AgI	881	892
AuCl	1013	1066
AuBr	1029	1059
AuI	1027	1070
InCl	-	764
InBr	-	767
InI	-	733
TlF	-	850
TlCl	738	751
TlBr	720	734
TlI	692	710
Me_4NCl	566	-
Me_4NBr	553	-
Me_4NI	544	-
PH_4Br	616	-
PH_4I	590	-
BeF_2	3464	3526
$BeCl_2$	3004	3033
$BeBr_2$	2950	2914
BeI_2	2780	2813
MgF_2	2926	2978
$MgCl_2$	2477	2540
$MgBr_2$	2406	2451
MgI_2	2293	2340
CaF_2	2640	2651
$CaCl_2$	2268	2271
$CaBr_2$	2132	-
CaI_2	1971	2087
SrF_2	2476	2513
$SrCl_2$	2142	2170
SrI_2	1984	1976
BaF_2	2347	2373
$BaCl_2$	2046	2069
$BaBr_2$	1971	1995
BaI_2	1862	1890
RaF_2	2284	-
$RaCl_2$	2004	-
$RaBr_2$	1929	-
RaI_2	1803	-
$ScCl_2$	2380	-
$ScBr_2$	2291	-
ScI_2	2201	-
TiF_2	2724	-
$TiCl_2$	2439	2514
$TiBr_2$	2360	2430
TiI_2	2259	2342
VCl_2	2607	2593
VBr_2	-	2534

Table 1
LATTICE ENERGIES (kJ mol^{-1}) (continued)

Substance	Calc. U_{POT}	U_{POT}^{BHFC}
VI_2	-	2470
CrF_2	2778	2939
$CrCl_2$	2540	2601
$CrBr_2$	2377	2536
CrI_2	2269	2440
$MoCl_2$	2737	2746
$MoBr_2$	2742	2753
MoI_2	2630	-
MnF_2	2644	-
$MnCl_2$	2510	2551
$MnBr_2$	2448	2482
MnI_2	2212	-
FeF_2	2849	2967
$FeCl_2$	2569	2641
$FeBr_2$	2515	2577
FeI_2	2439	2491
CoF_2	3004	3042
$CoCl_2$	2707	2706
$CoBr_2$	2640	2643
CoI_2	2569	2561
NiF_2	3098	3089
$NiCl_2$	2753	2786
$NiBr_2$	2729	2721
NiI_2	2607	2637
$PdCl_2$	2778	2818
$PdBr_2$	2741	2751
PdI_2	2748	2760
CuF_2	3046	3102
$CuCl_2$	2774	2824
$CuBr_2$	2715	2774
CuI_2	2640	-
AgF_2	2942	2967
ZnF_2	3021	3053
$ZnCl_2$	2703	2748
$ZnBr_2$	2648	2689
ZnI_2	2581	2619
CdF_2	2809	2830
$CdCl_2$	2552	2565
$CdBr_2$	2507	2517
CdI_2	2441	2455
HgF_2	2757	-
$HgCl_2$	2657	2664
$HgBr_2$	2628	2639
HgI_2	2628	2624
SnF_2	2551	-
$SnCl_2$	2297	2310
$SnBr_2$	2251	2256
SnI_2	2193	2206
PbF_2	2535	2543
$PbCl_2$	2270	2282
$PbBr_2$	2219	2230
PbI_2	2163	2177
ScF_3	5492	5540
$ScCl_3$	4874	4901
$ScBr_3$	4729	4761
ScI_3	4640	-
YF_3	4983	-
YCl_3	4506	4524
YI_3	4240	4258
TiF_3	5644	-
$TiCl_3$	5134	5153
$TiBr_3$	5012	5023
TiI_3	4845	-
$ZrCl_3$	-	4791
$ZrBr_3$	-	4758
ZrI_3	-	4591
VF_3	5895	-
VCl_3	5322	5329
VBr_3	5214	5224
VI_3	5121	5136
$NbCl_3$	5062	-
$NbBr_3$	4980	-
NbI_3	4860	-
CrF_3	6033	6065
$CrCl_3$	5518	5529
$CrBr_3$	5355	-
CrI_3	5275	5294
MoF_3	6459	-
$MoCl_3$	5246	5253
$MoBr_3$	5156	-
MoI_3	5073	-
MnF_3	6017	-
$MnCl_3$	5544	-
$MnBr_3$	5448	-
MnI_3	5330	-
$TcCl_3$	5270	-
$TcBr_3$	5215	-
TcI_3	5188	-
FeF_3	5870	-
$FeCl_3$	5364	5436
$FeBr_3$	5333	5347
FeI_3	5117	-
$RuCl_3$	5245	5257
$RuBr_3$	5223	5232
RuI_3	5222	5235
CoF_3	5991	-
$RhCl_3$	5641	5665
IrF_3	(6112)	-
$IrBr_3$	(4794)	-
NiF_3	(6111)	-
AuF_3	(5777)	-
$AuCl_3$	(4605)	-
$ZnCl_3$	5832	-
$ZnBr_3$	5732	-
ZnI_3	5636	-
AlF_3	5924	6252
$AlCl_3$	5376	5513
$AlBr_3$	5247	5360
AlI_3	5070	5227
GaF_3	5829	6238
$GaCl_3$	5217	5665
$GaBr_3$	4966	5569

Table 1
LATTICE ENERGIES (kJ mol^{-1}) (continued)

Substance	Calc. U_{POT}	U_{POT}^{BHFC}
GaI_3	4611	5496
$InCl_3$	4736	5183
$InBr_3$	4535	5117
InI_3	4234	5001
TlF_3	5493	-
$TlCl_3$	5258	5278
$TlBr_3$	5171	-
TlI_3	5088	-
$AsBr_3$	5497	5365
AsI_3	4824	5295
SbF_3	5295	5324
$SbCl_3$	5032	4857
$SbBr_3$	4954	4776
SbI_3	4867	4692
$BiCl_3$	4689	4707
BiI_3	3774	-
LaF_3	4682	-
$LaCl_3$	4263	4242
$LaBr_3$	4209	-
LaI_3	3916	3986
$CeCl_3$	4394	4348
CeI_3	-	4061
$PrCl_3$	4322	4387
PrI_3	-	4101
$NdCl_3$	4343	4415
$SmCl_3$	4376	4450
$EuCl_3$	4393	4490
$GdCl_3$	4406	4495
$DyCl_3$	4481	4529
$HoCl_3$	4501	4572
$ErCl_3$	4527	4591
$TmCl_3$	4548	4608
TmI_3	-	4340
$YbCl_3$	-	4651
$AcCl_3$	4096	-
UCl_3	4243	-
$NpCl_3$	4268	-
$PuCl_3$	4289	-
$PuBr_3$	(3959)	-
$AmCl_3$	4293	-
TiF_4	10012	9908
$TiCl_4$	9431	-
$TiBr_4$	9288	9059
TiI_4	9108	8918
ZrF_4	8853	8971
$ZrCl_4$	8021	8144
$ZrBr_4$	7661	7984
ZrI_4	7155	7801
MoF_4	8795	-
$MoCl_4$	8556	9603
$MoBr_4$	8510	9500
MoI_4	8427	-
$SnCl_4$	8355	8930
$SnBr_4$	7970	8852
PbF_4	9519	-
CrF_2Cl	5795	-
CrF_2Br	5753	-
CrF_2I	5669	-
$CrCl_2Br$	5448	-
$CrCl_2I$	5381	5429
$CrBr_2I$	5330	5370
CuFCl	2891	-
CuFBr	2853	-
CuFI	2803	-
CuClBr	2753	-
CuClI	2694	-
CuBrI	2669	-
FeF_2Cl	5711	-
FeF_2Br	5653	-
FeF_2I	5569	-
$FeCl_2Br$	5339	-
$FeCl_2I$	5272	-
$FeBr_2I$	5209	-
$LiIO_2F_2$	845	-
$NaIO_2F_2$	766	756
KIO_2F_2	699	689
$RbIO_2F_2$	674	-
$CsIO_2F_2$	636	-
$NH_4IO_2F_2$	678	-
$AgIO_2F_2$	736	685
Hydrides		
LiH	916	918
NaH	807	807
KH	711	713
RbH	686	684
CsH	648	653
VH	1184	(1344)
NbH	1163	(1633)
PdH	979	1368
CuH	828	1254
TiH	996	1407
ZrH	916	1590
HfH	904	-
LaH	828	-
TaH	1021	-
CrH	1050	-
NiH	929	-
PtH	937	-
AgH	941	-
AuH	1033	1108
TlH	745	-
GeH	950	-
PbH	778	-
BeH_2	3205	3306
MgH_2	2791	2718
CaH_2	2410	2406
SrH_2	2250	2265
BaH_2	2121	2133
ScH_2	2711	2744
YH_2	(2598)	2733
LaH_2	2380	2522
CeH_2	2414	2509

Table 1
LATTICE ENERGIES (kJ mol^{-1}) (continued)

Substance	Calc. U_{POT}	U_{POT}^{BHFC}
PrH_2	2448	2405
NdH_2	2464	2394
PmH_2	2519	-
SmH_2	2510	2389
GdH_2	2494	2651
AcH_2	2372	-
ThH_2	2711	2738
PuH_2	2519	-
AmH_2	2544	-
TiH_2	2866	2864
ZrH_2	2711	2999
CuH_2	2941	-
ZnH_2	2870	-
HgH_2	2707	-
AlH_3	5924	5969
FeH_3	5724	-
ScH_3	5439	-
YH_3	5063	4910
LaH_3	4895	4493
FeH_3	5724	-
GaH_3	5690	-
InH_3	5092	-
TlH_3	5092	-
Hydroselenides		
NaHSe	703	732
KHSe	644	712
RbHSe	623	689
CsHse	598	669
Hydrosulphides		
LiHS	768	862
NaHS	723	771
RbHS	655	682
CsHS	628	657
NH_4HS	661	718
$Ca(HS)_2$	2184	(2171)
$Sr(HS)_2$	2063	-
$Ba(HS)_2$	1979	(1956)
Hydroxides		
LiOH	1021	1028
NaOH	887	892
KOH	789	796
RbOH	766	765
CsOH	721	732
$Be(OH)_2$	3477	3620
$Mg(OH)_2$	2870	2998
$Ca(OH)_2$	2506	2637
$Sr(OH)_2$	2330	2174
$Ba(OH)_2$	2142	2330
$Ti(OH)_2$	-	2953
$Mn(OH)_2$	2909	3008
$Fe(OH)_2$	2653	3044
$Co(OH)_2$	2786	3109
$Ni(OH)_2$	2832	3186
$Pd(OH)_2$	-	3189
$Cu(OH)_2$	2870	3229
CuOH	1006	-
AgOH	918	845
AuOH	1033	-
TlOH	705	874
$Zn(OH)_2$	2795	3151
$Cd(OH)_2$	2607	2909
$Hg(OH)_2$	2669	-
$Sn(OH)_2$	2489	2721
$Pb(OH)_2$	2376	-
$Sc(OH)_3$	5063	5602
$Y(OH)_3$	4707	-
$La(OH)_3$	4443	-
$Cr(OH)_3$	5556	6299
$Mn(OH)_3$	6213	-
$Al(OH)_3$	5627	-
$Ga(OH)_3$	5732	6368
$In(OH)_3$	5280	-
$Tl(OH)_3$	5314	-
$Ti(OH)_4$	9456	-
$Zr(OH)_4$	8619	-
$Mn(OH)_4$	10933	-
$Sn(OH)_4$	9188	9879
Imides		
CaNH	3293	-
SrNH	3146	-
BaNH	2975	-
Metavanadates		
Li_3VO_4	3945	-
Na_3VO_4	3766	-
K_3VO_4	3376	-
Rb_3VO_4	3243	-
Cs_3VO_4	3137	-
Nitrates		
$LiNO_3$	848	854
$NaNO_3$	755	763
KNO_3	685	694
$RbNO_3$	662	671
$CsNO_3$	648	650
$AgNO_3$	820	832
$TlNO_3$	690	707
$Mg(NO_3)_2$	2481	2521
$Ca(NO_3)_2$	2268	2247
$Sr(NO_3)_2$	2176	2151
$Ba(NO_3)_2$	2062	2035
$Mn(NO_3)_2$	2318	2478
$Fe(NO_3)_2$	-	(2580)
$Co(NO_3)_2$	2560	2647
$Ni(NO_3)_2$	-	2729
$Cu(NO_3)_2$	-	2739
$Zn(NO_3)_2$	2376	2649
$Cd(NO_3)_2$	2238	2462
$Sn(NO_3)_2$	2155	2254
$Pb(NO_3)_2$	2067	2208
Nitrides		
ScN	7547	7506
LaN	6876	6793
TiN	8130	8033

Table 1
LATTICE ENERGIES (kJ mol⁻¹) (continued)

Substance	Calc. U_{POT}	U_{POT}^{BHFC}
ZrN	7633	7723
VN	8283	8233
NbN	7939	8022
CrN	8269	8358
Nitrites		
$NaNO_2$	774	772
KNO_2	699	687
$RbNO_2$	724	765
$CsNO_2$	690	-
Oxides		
Li_2O	2799	2814
Na_2O	2481	2478
K_2O	2238	2232
Rb_2O	2163	2161
Cs_2O	2131	2063
Cu_2O	3273	3189
Ag_2O	3002	2910
Tl_2O	2659	2575
LiO_2	(878)	(872)
NaO_2	799	821
KO_2	741	751
RbO_2	706	721
CsO_2	679	696
Li_2O_2	2592	2557
Na_2O_2	2309	22717
K_2O_2	2114	2064
Rb_2O_2	2025	1994
Cs_2O_2	1948	1512
MgO_2	3356	3526
CaO_2	3144	3132
SrO_2	3037	2977
KO_3	697	707
BeO	4514	4443
MgO	3795	3791
CaO	3414	3401
SrO	3217	3223
BaO	3029	3054
TiO	3832	3811
VO	3932	3863
MnO	3724	3745
FeO	3795	3865
CoO	3837	3910
NiO	3908	4010
PdO	3736	-
CuO	4135	4050
ZnO	4142	3971
CdO	3806	-
HgO	3907	-
GeO	3919	-
SnO	3652	-
PbO	3520	-
Sc_2O_3	13557	13708
Y_2O_3	12705	-
La_2O_3	12452	-
Ce_2O_3	12661	-
Pr_2O_3	12703	-
Nd_2O_3	12736	-
Pm_2O_3	12811	-
Sm_2O_3	12878	-
Eu_2O_3	12945	-
Gd_2O_3	12996	-
Tb_2O_3	13071	-
Dy_2O_3	13138	-
Ho_2O_3	13180	-
Er_2O_3	13263	-
Tm_2O_3	13322	-
Yb_2O_3	13380	-
Lu_2O_3	13665	-
Ac_2O_3	12573	-
Ti_2O_3	-	14149
V_2O_3	15096	14520
Cr_2O_3	15276	14957
Mn_2O_3	15146	15035
Fe_2O_3	14309	14774
Al_2O_3	15916	-
Ga_2O_3	15590	15220
In_2O_3	13928	-
Pb_2O_3	(14841)	-
CeO_2	9627	-
ThO_2	10397	-
PaO_2	10573	-
VO_2(g)	10644	-
NpO_2	10707	-
PuO_2	10786	-
AmO_2	10799	-
CmO_2	10832	-
TiO_2	12150	-
ZrO_2	11188	-
MoO_2	11648	-
MnO_2	12970	-
SiO_2	13125	-
GeO_2	12828	-
SnO_2	11807	-
PbO_2	11217	-
Perchlorates		
$LiClO_4$	709	715
$NaClO_4$	643	641
$KClO_4$	599	595
$RbClO_4$	564	576
$CsClO_4$	636	550
NH_4ClO_4	583	580
$Ca(ClO_4)_2$	1958	1971
$Sr(ClO_4)_2$	1862	1862
$Ba(ClO_4)_2$	1795	1769
Permanganates		
$NaMnO_4$	661	-
$KMnO_4$	607	-
$RbMnO_4$	586	-
$CsMnO_4$	565	-
$Ca(MnO_4)_2$	1937	-
$Sr(MnO_4)_2$	1845	-
$Ba(MnO_4)_2$	1778	-

Table 1
LATTICE ENERGIES (kJ mol⁻¹) (continued)

Substance	Calc. U_{POT}	U_{POT}^{BHFC}
Phosphates		
$Mg_3(PO_4)_2$	11632	11407
$Ca_3(PO_4)_2$	10602	10479
$Sr_3(PO_4)_2$	10125	10075
$Ba_3(PO_4)_2$	9652	9654
$MnPO_4$	7397	-
$FePO_4$	7251	7300
BPO_4	8201	-
$AlPO_4$	7427	7507
$GaPO_4$	7381	-
Selenides		
Li_2Se	2364	-
Na_2Se	2130	-
K_2Se	1933	-
Rb_2Se	1837	-
Cs_2Se	1745	-
Ag_2Se	2686	-
Tl_2Se	2209	-
$BeSe$	3431	-
$MgSe$	3071	-
$CaSe$	2858	2862
$SrSe$	2736	-
$BaSe$	2611	-
$MnSe$	3176	-
Selenites		
Li_2SeO_3	2171	-
Na_2SeO_3	1950	1916
K_2SeO_3	1774	1749
Rb_2SeO_3	1715	1675
Cs_2SeO_3	1640	-
Tl_2SeO_3	1879	-
Ag_2SeO_3	2113	2148
$BeSeO_3$	3322	-
$MgSeO_3$	3012	2998
$CaSeO_3$	2732	-
$SrSeO_3$	2586	2588
$BaSeO_3$	2460	2451
Selenates		
Li_2SeO_4	2054	-
Na_2SeO_4	1879	-
K_2SeO_4	1732	-
Rb_2SeO_4	1686	-
Cs_2SeO_4	1615	-
Cu_2SeO_4	2201	-
Ag_2SeO_4	2033	-
Tl_2SeO_4	1766	-
Hg_2SeO_4	2163	-
$BeSeO_4$	3448	-
$MgSeO_4$	2895	-
$CaSeO_4$	2632	-
$SrSeO_4$	2489	-
Sulphides		
Li_2S	2464	2472
Na_2S	2192	2203
K_2S	1979	(2052)
Rb_2S	1929	1949
Cs_2S	1892	1850
$(NH_4)_2S$	2008	(2026)
Cu_2S	2786	2865
Ag_2S	2606	2677
Au_2S	2908	-
Tl_2S	2298	2258
Sulphates		
Li_2SO_4	2229	2142
Na_2SO_4	1827	1938
K_2SO_4	1700	1796
Rb_2SO_4	1636	1748
Cs_2SO_4	1596	1658
$(NH_4)_2SO_4$	1766	1777
Cu_2SO_4	2276	2166
Ag_2SO_4	2104	1989
Tl_2SO_4	1828	1722
Hg_2SO_4	-	2127
$CaSO_4$	2489	2480
$SrSO_4$	2577	2484
$BaSO_4$	2469	2374
$MnSO_4$	2920	2825
Ternary Salts		
Cs_2CuCl_4	1393	-
Rb_2ZnCl_4	1529	-
Cs_2ZnCl_4	1492	-
Rb_2ZnBr_4	1498	-
Cs_2ZnBr_4	1454	-
Cs_2ZnI_4	1386	-
$CsGaCl_4$	494	-
$NaAlCl_4$	556	-
$CsAlCl_4$	486	-
$NaFeCl_4$	492	-
Rb_2CoCl_4	1447	-
Cs_2CoCl_4	1391	-
K_2PtCl_4	1574	1550
Cs_2GeF_6	1573	-
$(NH_4)_2GeF_6$	1657	-
Cs_2GeCl_6	1404	1419
K_2HfCl_6	1345	1461
K_2IrCl_6	1442	1440
Na_2MoCl_6	1526	1504
K_2MoCl_6	1418	1412
Rb_2MoCl_6	1399	1399
Cs_2MoCl_6	1347	1347
K_2NbCl_6	1375	1398
Rb_2NbCl_6	1371	1385
Cs_2NbCl_6	1381	1344
K_2OsCl_6	1447	1447
Cs_2OsCl_6	1409	-
K_2OsBr_6	1396	-
K_2PdCl_6	1481	1493
Rb_2PdCl_6	1449	-
Cs_2PdCl_6	1426	-
Rb_2PbCl_6	1343	1343

Table 1
LATTICE ENERGIES (kJ mol^{-1}) (continued)

Substance	Calc. U_{POT}	U_{POT}^{BHFC}
Cs_2PbCl_6	1344	-
$(NH_4)_2PbCl_6$	1355	-
K_2PtCl_6	1468	1471
Rb_2PtCl_6	1464	-
Cs_2PtCl_6	1444	-
$(NH_4)_2PtCl_6$	1468	-
Tl_2PtCl_6	1546	-
Ag_2PtCl_6	1773	1881
$BaPtCl_6$	2047	2070
K_2PtBr_6	1423	1392
Ag_2PtBr_6	1791	2276
K_2PtI_6	1421	-
K_2ReCl_6	1416	1442
Rb_2ReCl_6	1414	-
Cs_2ReCl_6	1398	-
K_2ReBr_6	1375	1375
K_2SiF_6	1670	1765
Rb_2SiF_6	1639	1673
Cs_2SiF_6	1604	1498
Tl_2SiF_6	1675	-
K_2SnCl_6	1363	1390
Rb_2SnCl_6	1361	1363
Cs_2SnCl_6	1358	-
Tl_2SnCl_6	1437	-
$(NH_4)_2SnCl_6$	1370	1344
Rb_2SnBr_6	1309	-
Cs_2SnBr_6	1306	-
Rb_2SnI_6	1226	-
Cs_2SnBr_6	1243	-
K_2TeCl_6	1318	1320
Rb_2TeCl_6	1321	-
Cs_2TeCl_6	1323	-
Tl_2TeCl_6	1392	-
$(NH_4)_2TeCl_6$	1318	-
K_2RuCl_6	1451	-
Rb_2CoF_6	1688	-
Cs_2CoF_6	1632	-
K_2NiF_6	1721	-
Rb_2NiF_6	1688	-
Rb_2SbCl_6	1357	-
Rb_2SeCl_6	1409	-
Cs_2SeCl_6	1397	-
$(NH_4)_2SeCl_6$	1420	-
$(NH_4)_2PoCl_6$	1338	-
Cs_2PoBr_6	1286	-
Cs_2CrF_6	1603	-
Rb_2MnF_6	1688	-
Cs_2MnF_6	1620	-
K_2MnCl_6	1462	-
Rb_2MnCl_6	1451	-
$(NH_4)_2MnCl_6$	1464	-
Cs_2TeBr_6	1306	-
Cs_2TeI_6	1246	-
K_2TiCl_6	1412	1447
Rb_2TiCl_6	1415	1416
Cs_2TiCl_6	1402	1384
Tl_2TiCl_6	1560	1553
K_2TiBr_6	1379	1379
Rb_2TiBr_6	1341	1331
Cs_2TiBr_6	1339	1306
Na_2UBr_6	1504	-
K_2UBr_6	1484	-
Rb_2UBr_6	1473	-
Cs_2UBr_6	1459	-
K_2WCl_6	1398	1423
Rb_2WCl_6	1397	1434
Cs_2WCl_6	1392	1366
K_2WBr_6	1408	1408
Rb_2WBr_6	1361	1391
Cs_2WBr_6	1362	1332
K_2ZrCl_6	1339	1371
Rb_2ZrCl_6	1341	-
Cs_2ZrCl_6	1339	1307
Tellurides		
Li_2Te	2212	-
Na_2Te	1997	2095
K_2Te	1830	-
Rb_2Te	1837	-
Cs_2Te	1745	-
Cu_2Te	2706	2683
Ag_2Te	2607	2600
Tl_2Te	2084	2172
BeTe	3319	-
MgTe	2878	3081
CaTe	2721	-
Thiocyanates		
LiCNS	764	(765)
NaCNS	682	682
KCNS	623	616
RbCNS	623	619
CsCNS	623	568
NH_4CNS	605	611
$Ca(CNS)_2$	2184	2118
$Sr(CNS)_2$	2063	1957
$Ba(CNS)_2$	1979	1852
$Mn(CNS)_2$	2280	2351
$Zn(CNS)_2$	2335	2560
$Cd(CNS)_2$	2201	2374
$Hg(CNS)_2$	2146	2492
$Sn(CNS)_2$	2117	2142
$Pb(CNS)_2$	2058	-
Vanadates		
$LiVO_3$	810	-
$NaVO_3$	761	-
KVO_3	686	-
$RbVO_3$	657	-
$CsVO_3$	628	-

Table 2
THERMOCHEMICAL RADII (nm)

Ion	Radius	
Singly Charged Anions		
AgF_4^-	0.231	± 0.019
$AlBr_4^-$	0.321	± 0.023
$AlCl_4^-$	0.317	± 0.019
AlF_4^-	0.214	± 0.023
AlH_4^-	0.226	± 0.019
AlI_4^-	0.374	± 0.019
AsF_6^-	0.243	± 0.019
AsO_2^-	0.211	± 0.019
$Au(CN)_2^-$	0.266	± 0.019
$AuCl_4^-$	0.288	± 0.019
AuF_4^-	0.240	± 0.019
AuF_6^-	0.235	± 0.038
$B(OH)_4^-$	0.229	± 0.019
BF_4^-	0.205	± 0.019
BH_4^-	0.205	± 0.019
Br^-	0.190	± 0.019
BrF_4^-	0.231	± 0.019
BrO_3^-	0.214	± 0.019
$CF_3SO_3^-$	0.230	± 0.049
$CH_3CO_2^-$	0.194	± 0.019
Cl^-	0.168	± 0.019
ClO_2^-	0.195	± 0.019
ClO_3^-	0.208	± 0.019
ClO_4^-	0.225	± 0.019
$ClS_2O_6^-$	0.260	± 0.049
CN^-	0.187	± 0.023
$Cr_3O_8^-$	0.276	± 0.019
$CuBr_4^-$	0.315	± 0.019
F^-	0.126	± 0.019
$FeCl_4^-$	0.317	± 0.019
$GaCl_4^-$	0.328	± 0.019
H^-	0.148	± 0.019
$H_2AsO_4^-$	0.227	± 0.019
$H_2PO_4^-$	0.213	± 0.019
HCO_2^-	0.200	± 0.019
HCO_3^-	0.207	± 0.019
HF_2^-	0.172	± 0.019
HSO_4^-	0.221	± 0.019
I^-	0.211	± 0.019
I_2Br^-	0.261	± 0.019
I_3^-	0.272	± 0.019
I_4^-	0.300	± 0.019
IBr_2^-	0.251	± 0.019
ICl_2^-	0.235	± 0.019
ICl_4^-	0.307	± 0.019
IO_2F_2	0.233	± 0.019
IO_3^-	0.218	± 0.019
IO_4^-	0.231	± 0.019
IrF_6^-	0.242	± 0.019
MnO_4^-	0.220	± 0.019
MoF_6^-	0.241	± 0.019
$MoOF_5^-$	0.241	± 0.019
N_3^-	0.180	± 0.019
NCO^-	0.193	± 0.019
$NbCl_6^-$	0.338	± 0.049
NbF_6^-	0.254	± 0.019
$Nb_2F_{11}^-$	0.311	± 0.038
NbO_3^-	0.194	± 0.019
NH_2^-	0.168	± 0.019
$NH_2CH_2COO^-$	0.252	± 0.019
NO_2^-	0.187	± 0.019
NO_3^-	0.200	± 0.019
O_2^-	0.165	± 0.019
O_3^-	0.199	± 0.034
OH^-	0.152	± 0.019
OsF_6^-	0.252	± 0.020
PaF_6^-	0.249	± 0.019
PdF_6^-	0.252	± 0.019
PF_6^-	0.242	± 0.019
PO_3^-	0.204	± 0.019
PtF_6^-	0.247	± 0.019
PuF_5^-	0.239	± 0.019
ReF_6^-	0.240	± 0.019
ReO_4^-	0.227	± 0.019
RuF_6^-	0.242	± 0.019
S_6^-	0.305	± 0.019
SCN^-	0.209	± 0.019
$SbCl_6^-$	0.320	± 0.019
SbF_6^-	0.252	± 0.019
$Sb_2F_{11}^-$	0.312	± 0.038
$Sb_3F_{14}^-$	0.374	± 0.038
$SeCl_5^-$	0.258	± 0.038
$SeCN^-$	0.230	± 0.019
SeH^-	0.195	± 0.019
SH^-	0.191	± 0.019
SO_3F^-	0.214	± 0.019
$S_3N_3^-$	0.231	± 0.038
$S_3N_3O_4^-$	0.252	± 0.038
$TaCl_6^-$	0.352	± 0.019
TaF_6^-	0.250	± 0.019
TaO_3^-	0.192	± 0.019
UF_6^-	0.301	± 0.019
VF_6^-	0.235	± 0.019
VO_3^-	0.201	± 0.019
WCl_6^-	0.337	± 0.019
WF_6^-	0.246	± 0.019
WOF_5^-	0.241	± 0.019
Doubly Charged Anions		
AmF_6^{2-}	0.255	± 0.019
$Bi_2Br_8^{2-}$	0.392	± 0.055
$Bi_6Cl_{20}^{2-}$	0.501	± 0.073
$CdCl_4^{2-}$	0.307	± 0.019
$CeCl_6^{2-}$	0.352	± 0.019
CeF_6^{2-}	0.249	± 0.019
CO_3^{2-}	0.189	± 0.019
$CoCl_4^{2-}$	0.306	± 0.019
CoF_4^{2-}	0.209	± 0.019
CoF_6^{2-}	0.256	± 0.019
$Cr_2O_7^{2-}$	0.292	± 0.019
CrF_6^{2-}	0.253	± 0.019
CrO_4^{2-}	0.229	± 0.019

Table 2
THERMOCHEMICAL RADII (nm) (continued)

Ion	Radius	
$CuCl_4^{2-}$	0.304	± 0.019
CuF_4^{2-}	0.213	± 0.019
$GeCl_6^{2-}$	0.335	± 0.019
GeF_6^{2-}	0.244	± 0.019
HfF_6^{2-}	0.248	± 0.019
HgI_4^{2-}	0.377	± 0.019
$IrCl_6^{2-}$	0.332	± 0.019
$MnCl_6^{2-}$	0.314	± 0.031
MnF_4^{2-}	0.219	± 0.019
MnF_6^{2-}	0.241	± 0.019
$MoBr_6^{2-}$	0.364	± 0.019
$MoCl_6^{2-}$	0.338	± 0.019
MoF_6^{2-}	0.274	± 0.019
MoO_4^{2-}	0.231	± 0.019
$NbCl_6^{2-}$	0.343	± 0.019
NH^{2-}	0.128	± 0.019
$Ni(CN)_4^{2-}$	0.322	± 0.019
NiF_4^{2-}	0.211	± 0.019
NiF_6^{2-}	0.249	± 0.019
O^{2-}	0.141	± 0.019
O_2^{2-}	0.167	± 0.019
$OsBr_6^{2-}$	0.365	± 0.019
$OsCl_6^{2-}$	0.336	± 0.019
OsF_6^{2-}	0.276	± 0.019
$PbCl_4^{2-}$	0.279	± 0.019
$PbCl_6^{2-}$	0.347	± 0.019
PbF_6^{2-}	0.268	± 0.019
$PdBr_6^{2-}$	0.354	± 0.019
$PdCl_4^{2-}$	0.313	± 0.019
$PdCl_6^{2-}$	0.333	± 0.019
PdF_6^{2-}	0.252	± 0.019
$PoBr_6^{2-}$	0.380	± 0.019
PoI_6^{2-}	0.428	± 0.019
$Pt(NO_2)_3Cl_3^{2-}$	0.364	± 0.019
$Pt(NO_2)_4Cl_2^{2-}$	0.383	± 0.019
$Pt(OH)_2^{2-}$	0.333	± 0.019
$Pt(SCN)_6^{2-}$	0.451	± 0.019
$PtBr_4^{2-}$	0.324	± 0.019
$PtBr_6^{2-}$	0.363	± 0.019
$PtCl_4^{2-}$	0.307	± 0.019
$PtCl_6^{2-}$	0.333	± 0.019
PtF_6^{2-}	0.245	± 0.019
$PuCl_6^{2-}$	0.349	± 0.019
$ReBr_6^{2-}$	0.371	± 0.019
$ReCl_6^{2-}$	0.337	± 0.019
ReF_6^{2-}	0.256	± 0.019
ReF_8^{2-}	0.276	± 0.019
ReH_9^{2-}	0.257	± 0.019
ReI_6^{2-}	0.421	± 0.026
RhF_6^{2-}	0.240	± 0.019
$RuCl_6^{2-}$	0.336	± 0.019
RuF_6^{2-}	0.248	± 0.019
S^{2-}	0.189	± 0.019
$S_2O_3^{2-}$	0.251	± 0.019
$S_2O_4^{2-}$	0.262	± 0.019
$S_2O_5^{2-}$	0.270	± 0.019
$S_2O_6^{2-}$	0.283	± 0.019
$S_2O_7^{2-}$	0.275	± 0.019
$S_2O_8^{2-}$	0.291	± 0.019
$S_3O_6^{2-}$	0.302	± 0.019
$S_4O_6^{2-}$	0.325	± 0.019
$S_6O_6^{2-}$	0.382	± 0.019
ScF_6^{2-}	0.276	± 0.019
Se^{2-}	0.181	± 0.019
$SeBr_6^{2-}$	0.363	± 0.019
$SeCl_6^{2-}$	0.336	± 0.019
SeO_4^{2-}	0.229	± 0.019
SiF_6^{2-}	0.248	± 0.019
SiO_3^{2-}	0.195	± 0.019
SmF_4^{2-}	0.218	± 0.019
$Sn(OH)_6^{2-}$	0.279	± 0.020
$SnBr_6^{2-}$	0.374	± 0.019
$SnCl_6^{2-}$	0.345	± 0.019
SnF_6^{2-}	0.265	± 0.019
SnI_6^{2-}	0.427	± 0.019
SO_3^{2-}	0.204	± 0.019
SO_4^{2-}	0.218	± 0.019
$TcBr_6^{2-}$	0.363	± 0.019
$TcCl_6^{2-}$	0.337	± 0.019
TcF_6^{2-}	0.244	± 0.019
TcH_9^{2-}	0.260	± 0.019
TcI_6^{2-}	0.419	± 0.019
Te^{2-}	0.220	± 0.019
$TeBr_6^{2-}$	0.383	± 0.019
$TeCl_6^{2-}$	0.353	± 0.019
TeI_6^{2-}	0.430	± 0.019
TeO_4^{2-}	0.238	± 0.019
$Th(NO_3)_6^{2-}$	0.424	± 0.019
$ThCl_6^{2-}$	0.360	± 0.019
ThF_6^{2-}	0.263	± 0.019
$TiBr_6^{2-}$	0.356	± 0.019
$TiCl_6^{2-}$	0.335	± 0.019
TiF_6^{2-}	0.252	± 0.019
UCl_6^{2-}	0.354	± 0.019
UF_6^{2-}	0.256	± 0.019
VO_3^{2-}	0.204	± 0.019
WBr_6^{2-}	0.363	± 0.019
WCl_6^{2-}	0.339	± 0.019
WO_4^{2-}	0.237	± 0.019
$WOCl_5^{2-}$	0.334	± 0.019
$ZnBr_4^{2-}$	0.335	± 0.019
$ZnCl_4^{2-}$	0.306	± 0.019
ZnF_4^{2-}	0.219	± 0.019
ZnI_4^{2-}	0.384	± 0.019
$ZrBr_4^{2-}$	0.334	± 0.019
$ZrCl_4^{2-}$	0.306	± 0.019
$ZrCl_6^{2-}$	0.348	± 0.019
ZrF_6^{2-}	0.258	± 0.019
Multi-Charged Anions		
AlH_6^{3-}	0.256	± 0.042
AsO_4^{3-}	0.237	± 0.042
$CdBr_6^{4-}$	0.374	± 0.038

Table 2
THERMOCHEMICAL RADII (nm) (continued)

Ion	Radius	
$CdCl_6^{4-}$	0.352	± 0.038
CeF_6^{3-}	0.278	± 0.038
CeF_7^{3-}	0.282	± 0.038
$Co(CN)_6^{3-}$	0.349	± 0.038
$Co(NO_2)_6^{3-}$	0.343	± 0.038
$CoCl_5^{3-}$	0.320	± 0.038
CoF_6^{3-}	0.258	± 0.042
$Cr(CN)_6^{3-}$	0.351	± 0.038
CrF_6^{3-}	0.254	± 0.042
$Cu(CN)_4^{3-}$	0.312	± 0.038
$Fe(CN)_6^{3-}$	0.347	± 0.038
FeF_6^{3-}	0.298	± 0.042
HfF_7^{3-}	0.277	± 0.042
InF_6^{3-}	0.268	± 0.038
$Ir(CN)_6^{3-}$	0.347	± 0.038
$Ir(NO_2)_6^{3-}$	0.338	± 0.038
$Mn(CN)_6^{3-}$	0.350	± 0.038
$Mn(CN)_6^{5-}$	0.401	± 0.042
$MnCl_6^{4-}$	0.349	± 0.038
N^{3-}	0.180	± 0.042
$Ni(NO_2)_6^{3-}$	0.342	± 0.038
$Ni(NO_2)_6^{4-}$	0.383	± 0.038
NiF_6^{3-}	0.250	± 0.042
O^{3-}	0.288	± 0.038
P^{3-}	0.224	± 0.042
PaF_8^{3-}	0.299	± 0.042
PO_4^{3-}	0.230	± 0.042
PrF_6^{3-}	0.281	± 0.038
$Rh(NO_2)_6^{3-}$	0.345	± 0.038
$Rh(SCN)_6^{3-}$	0.428	± 0.042
TaF_8^{3-}	0.284	± 0.042
TbF_7^{3-}	0.290	± 0.038
$Tc(CN)_6^{5-}$	0.410	± 0.042
ThF_7^{3-}	0.282	± 0.042
$TiBr_6^{3-}$	0.315	± 0.038
TlF_6^{3-}	0.271	± 0.038
UF_7^{3-}	0.285	± 0.042
YF_6^{3-}	0.275	± 0.038
ZrF_7^{3-}	0.273	± 0.038
Singly Charged Cations		
$N(CH_3)_4^+$	0.234	± 0.019
$N_2H_5^+$	0.158	± 0.019
$N_2H_6^{2+}$	0.158	± 0.029
$NH(C_2H_5)_3^+$	0.274	± 0.019
$NH_3C_2H_5^+$	0.193	± 0.019
$NH_3C_3H_7^+$	0.225	± 0.019
$NH_3CH_3^+$	0.177	± 0.019
NH_3OH^+	0.147	± 0.019
NH_4^+	0.136	± 0.019
$NH_3C_2H_4OH^+$	0.203	± 0.019
$As_3S_4^+$	0.244	± 0.027
$As_3Se_4^+$	0.253	± 0.027
$AsCl_4^+$	0.221	± 0.027
Br_2^+	0.155	± 0.027
Br_3^+	0.204	± 0.027
Br_3^-	0.238	± 0.027

Ion	Radius	
Br_5^+	0.229	± 0.027
$BrClCNH_2^+$	0.175	± 0.027
BrF_2^+	0.183	± 0.027
BrF_4^+	0.172	± 0.027
$C_{10}F_8^+$	0.265	± 0.027
$C_6F_6^+$	0.228	± 0.027
$Cl(SNSCN)_2^+$	0.347	± 0.027
$Cl_2C{=}NH_2^+$	0.173	± 0.027
Cl_2F^+	0.165	± 0.027
Cl_3^+	0.182	± 0.027
ClF_2^+	0.147	± 0.027
ClO_2^+	0.118	± 0.027
$GaBr_4^-$	0.317	± 0.038
I_2^+	0.185	± 0.027
I_3^+	0.225	± 0.027
I_5^+	0.263	± 0.027
IBr_2^+	0.196	± 0.027
ICl_2^+	0.175	± 0.036
IF_6^+	0.209	± 0.027
$N(S_3N_2)_2^+$	0.258	± 0.027
$N(SCl)_2^+$	0.186	± 0.027
$N(SeCl)_2^+$	0.246	± 0.027
$N(SF_2)_2^+$	0.214	± 0.027
N_2F^+	0.156	± 0.027
NO^+	0.145	± 0.027
NO_2^+	0.153	± 0.027
O_2^+	0.140	± 0.027
$O_2(SCCF_3Cl)_2^+$	0.275	± 0.027
$ONCH_3CF_3^+$	0.200	± 0.027
$OsOF_5^-$	0.246	± 0.038
$P(CH_3)_3Cl^+$	0.197	± 0.027
$P(CH_3)_3D^+$	0.196	± 0.027
PCl_4^+	0.235	± 0.027
$ReOF_5^-$	0.245	± 0.038
$S(CH_3)_2Cl^+$	0.207	± 0.027
$S(N(C_2H_5)_3)_3^+$	0.439	± 0.027
$S_2(CH_3)_2Cl^+$	0.265	± 0.027
$S_2(CH_3)_2CN^+$	0.223	± 0.027
$S_2(CH_3)_3^+$	0.233	± 0.027
$S_2Br_5^+$	0.267	± 0.027
S_2N^+	0.159	± 0.034
$S_2N_2C_2H_3^+$	0.211	± 0.027
$S_2NC_2(PhCH_3)_2^+$	0.310	± 0.027
$S_2NC_3H_4^+$	0.218	± 0.027
$S_2NC_4H_8^+$	0.225	± 0.027
$S_3(CH_3)_3^+$	0.239	± 0.027
$S_3Br_3^+$	0.245	± 0.027
$S_3C_3H_7^+$	0.199	± 0.027
$S_3C_4F_6^+$	0.261	± 0.027
$S_3CF_3CN^+$	0.263	± 0.027
$S_3Cl_3^+$	0.233	± 0.027
$S_3N_2^+$	0.201	± 0.027
$S_3N_2Cl^+$	0.232	± 0.027
$S_4N_3^+$	0.231	± 0.027
$S_4N_3(Ph)_2^+$	0.316	± 0.027
$S_4N_4H^+$	0.178	± 0.027

Table 2
THERMOCHEMICAL RADII (nm) (continued)

Ion	Radius	
$S_5N_5^+$	0.257	± 0.027
S_7I^+	0.262	± 0.027
$Sb(NPPh_3)_4^+$	0.518	± 0.027
SBr_3^+	0.220	± 0.027
$SCH_3O_2^+$	0.183	± 0.027
$SCH_3P(CH_3)_3^+$	0.248	± 0.027
$SCH_3PCH_3Cl_2^+$	0.205	± 0.027
$SCl(C_2H_5)_2^+$	0.207	± 0.027
$SCl_2CF_3^+$	0.207	± 0.027
$SCl_2CH_3^+$	0.204	± 0.027
SCl_3^+	0.185	± 0.027
$Se_3Br_3^+$	0.253	± 0.027
$Se_3Cl_3^+$	0.245	± 0.027
$Se_3N_2^+$	0.288	± 0.042
$Se_3NC_{l2}^+$	0.163	± 0.027
Se_6I^+	0.260	± 0.027
$SeBr_3^+$	0.182	± 0.027
$SeCl_3^+$	0.192	± 0.027
SeF_3^+	0.179	± 0.027
SeI_3^+	0.238	± 0.027
SeN_2Cl^+	0.196	± 0.027
$SeNCl_2^+$	0.157	± 0.027
$(SeNMe_3)_3^+$	0.406	± 0.027
$SeS_2N_2^+$	0.282	± 0.042
$SF(C_6F_5)_2^+$	0.294	± 0.027
$SF_2CF_3^+$	0.198	± 0.027
$SF_2N(CH_3)_2^+$	0.210	± 0.027
SF_3^+	0.172	± 0.027
$SFS(C(CF_3)_2)_2^+$	0.275	± 0.027
$SH_2C_3H_7^+$	0.210	± 0.027
SN^+	0.158	± 0.027
$SNCl_5(CH_3CN)^-$	0.290	± 0.038
$(SNPMe_3)_3^+$	0.308	± 0.027
$SNSC(CH_3)N^+$	0.225	± 0.027
$SNSC(CN)CH^+$	0.209	± 0.027
$SNSC(Ph)N^+$	0.251	± 0.027
$SNSC(Ph)NS_3N_2^+$	0.327	± 0.027
$SNSC(PhCH_3)N^+$	0.264	± 0.027
$(Te(N(SiMe_3)_2)_2^+$	0.371	± 0.027
$Te(N_3)_3^+$	0.226	± 0.027
$Te_4Nb_3OTe_2I_6^+$	0.407	± 0.027
$TeBr_3^+$	0.235	± 0.027
$TeCl_3^+$	0.216	± 0.027
$TeCl_3(15\text{-crown-}5)^+$	0.282	± 0.027
TeI_3^+	0.243	± 0.027
$Xe_2F_{11}^+$	0.266	± 0.027
$Xe_2F_3^+$	0.221	± 0.027
XeF^+	0.174	± 0.027
XeF_3^+	0.183	± 0.027
XeF_5^+	0.186	± 0.027
$XeOF_3^+$	0.186	± 0.027
Doubly Charged Cations		
$Co_2S_2(CO)_6^{2+}$	0.263	± 0.035
$FeW(Se)_2(CO)^{2+}$	0.260	± 0.035
I_4^{2+}	0.207	± 0.035
$Mo(Te_3)(CO)_4^{2+}$	0.234	± 0.035
S_{19}^{2+}	0.292	± 0.035
$S_2(S(CH_3)_2)_2^{2+}$	0.230	± 0.035
$S_2I_4^{2+}$	0.231	± 0.035
$S_3N_2^{2+}$	0.184	± 0.035
$S_3NCCNS_3^{2+}$	0.220	± 0.035
S_3Se^{2+}	0.326	± 0.035
$S_4N_4^{2+}$	0.186	± 0.035
$S_6N_4^{2+}$	0.232	± 0.035
S_8^{2+}	0.182	± 0.035
Se_{10}^{2+}	0.253	± 0.035
Se_{17}^{2+}	0.236	± 0.035
Se_{19}^{2+}	0.296	± 0.035
$Se_2I_4^{2+}$	0.218	± 0.035
$Se_3N_2^{2+}$	0.182	± 0.035
Se_4^{2+}	0.152	± 0.035
$Se_4S_2N_4^{2+}$	0.224	± 0.035
Se_8^{2+}	0.186	± 0.035
$SeN_2S_2^{2+}$	0.182	± 0.035
$(SNP(C_2H_5)_3)_2^{2+}$	0.312	± 0.035
$TaBr_6^-$	0.351	± 0.049
$Te(trtu)_4^{2+}$	0.328	± 0.035
$Te(tu)_4^{2+}$	0.296	± 0.035
$Te_2(esu)_4Br_2^{2+}$	0.356	± 0.035
$Te_2(esu)_4Cl_2^{2+}$	0.361	± 0.035
$Te_2(esu)_4I_2^{2+}$	0.342	± 0.035
$Te_2Se_2^{2+}$	0.192	± 0.035
$Te_2Se_4^{2+}$	0.222	± 0.035
$Te_2Se_8^{2+}$	0.252	± 0.035
$Te_3S_3^{2+}$	0.217	± 0.035
Te_3Se^{2+}	0.193	± 0.035
Te_4^{2+}	0.169	± 0.035
Te_8^{2+}	0.187	± 0.035
$W(CO)_4(h3\text{-}Te)^{2+}$	0.234	± 0.035
$W_2(CO)_{10}Se_4^{2+}$	0.290	± 0.035
Multi-Charged Cations		
I_{15}^{3+}	0.442	± 0.051
$Te_2(su)_6^{4+}$	0.453	± 0.034

Ligand abbreviations: su = selenourea; esu = ethyleneselenourea; tu = thiourea; ph = phenyl.

Table 3
ANCILLARY THERMOCHEMICAL DATA (kJ mol⁻¹)

Species	State	$\Delta_f H^o$
AsO_4^{3-}	g	(289)
BrO_3^-	g	-145
ClO_4^-	g	-344
CN^-	g	66
CO_3^{2-}	g	-321
$Fe(NO_3)_2$	c	(-448)
HF_2^-	g	-774
$HfCl_6^{2-}$	g	-1640
$IO_2F_2^-$	g	-693
IO_3^-	g	-208
$IrCl_6^{2-}$	g	-785
$LiCH_3O_2$	c	(-745)
$NbCl_6^{2-}$	g	-1224
$NH_2CH_2CO_2^-$	g	-564
O_2^{2-}	g	553
$PdCl_6^{2-}$	g	-749
PO_4^-	g	291
$PtCl_6^{2-}$	g	-774
$ReBr_6^{2-}$	g	-689
$ReCl_6^{2-}$	g	-919
$Ti(OH)_2$	c	-778

THE MADELUNG CONSTANT AND CRYSTAL LATTICE ENERGY

If U is the crystal lattice energy and M is the Madelung constant, then[a]

$$U = \frac{N M z_i z_j e^2}{r}(1 - 1/n)$$

Substance	Ion type	Crystal form[b]	M
Sodium chloride, NaCl	M^+, X^-	FCC	1.74756
Cesium chloride, CsCl	M^+, X^-	BCC	1.76267
Calcium chloride, $CaCl_2$	M^{++}, $2X^-$	Cubic	2.365
Calcium fluoride (fluorite), CaF_2	M^{++}, $2X^-$	Cubic	2.51939
Cadmium chloride, $CdCl_2$	M^{++}, $2X^-$	Hexagonal	2.244[c]
Cadmium iodide (α), CdI_2	M^{++}, $2X^-$	Hexagonal	2.355[c]
Magnesium fluoride, MgF_2	M^{++}, $2X^-$	Tetragonal	2.381[c]
Cuprous oxide (cuprite), Cu_2O	$2M^+$, X^{--}	Cubic	2.22124
Zinc oxide, ZnO	M^{++}, X^{--}	Hexagonal	1.4985[c]
Sphalerite (zinc blende), ZnS	M^{++}, X^{--}	FCC	1.63806
Wurtzite, ZnS	M^{++}, X^{--}	Hexagonal	1.64132[c]
Titanium dioxide (anatase), TiO_2	M^{4+}, $2X^{--}$	Tetragonal	2.400[c]
Titanium dioxide (rutile), TiO_2	M^{4+}, $2X^{--}$	Tetragonal	2.408[c]
β-Quartz, SiO_2	M^{4+}, $2X^{--}$	Hexagonal	2.2197[c]
Corundum, Al_2O_3	$2M^{3+}$, $3X^{--}$	Rhombohedral	4.1719

[a] N is Avogadro's number, z_i and z_j are the integral charges on the ions(in units of e), and e is the charge on the electron in electrostatic units ($e = 4.803 \times 10^{-10}$ esu). r is the shortest distance between cation-anion pairs in centimeters. Then U is in ergs (1 erg = 10^{-7} J).

[b] FCC = face centered cubic; BCC = body centered cubic.

[c] For tetragonal and hexagonal crystals the value of M depends on the details of the lattice parameters.

The Born Exponent, n is:

Ion type	n
He, Li^+	5
Ne, Na^+, F^-	7
Ar, K^+, Cu^+, Cl^-	9
Kr, Rb^+, Ag^+, Br^-	10
Xe, Cs^+, Au^+, I^-	12

For a crystal with a mixed-ion type, an average of the values of n in this table is to be used (6 for LiF, for example).

ELASTIC CONSTANTS OF SINGLE CRYSTALS

H. P. R. Frederikse

This table gives selected values of elastic constants for single crystals. The values believed most reliable were selected from the original literature. The substances are arranged by crystal system and, within each system, alphabetically by name. A reference to the original literature is given for each value; a useful compilation of published values from many sources may be found in Reference 1 below.

Data are given for the single-crystal density and for the elastic constants c_{ij}, in units of 10^{11} N/m², which is equivalent to 10^{12} dyn/cm².

GENERAL REFERENCES

1. Simmons, G., and Wang, H., *Single Crystal Elastic Constants and Calculated Aggregate Properties: A Handbook, Second Edition,* The MIT Press, Cambridge, MA, 1971.
2. Gray, D.E., Ed., *American Institute of Physics Handbook, Third Edition*, McGraw-Hill, New York, 1972.

CUBIC CRYSTALS

Name	Formula	ρ/g cm^{-3}	*T*/K	Ref.	C_{11}	C_{12}	C_{44}
Aluminum	Al	2.6970	298	1	1.0675	0.6041	0.2834
Aluminum antimonide	AlSb	4.3600	300	2	0.8939	0.4427	0.4155
Ammonium bromide	NH_4Br	2.4314	300	3	0.3414	0.0782	0.0722
Ammonium chloride	NH_4Cl	1.5279	290	4	0.3814	0.0866	0.0903
Argon	Ar	1.7710	4.2	5	0.0529	0.0135	0.0159
Barium fluoride	BaF_2	4.8860	298	6	0.9199	0.4157	0.2568
Barium nitrate	$Ba(NO_3)_2$	3.2560	293	7	0.2925	0.2065	0.1277
Calcium fluoride	CaF_2	3.810	298	8	1.6420	0.4398	0.8406
Calcium telluride	CaTe	5.8544	298	9	0.5351	0.3681	0.1994
Cesium	Cs	1.9800	78	10	0.0247	0.0206	0.0148
Cesium bromide	CsBr	4.4560	298	11	0.3063	0.0807	0.0750
Cesium chloride	CsCl	3.9880	298	11	0.3644	0.0882	0.0804
Cesium iodide	CsI	4.5250	298	11	0.2446	0.0661	0.0629
Chromite	$FeCr_2O_4$	4.4500	RT	12	3.2250	1.4370	1.1670
Chromium	Cr	7.20	298	13	3.398	0.586	0.990
Cobalt oxide	CoO	6.44	298	14	2.6123	1.4699	0.8300
Cobalt zinc ferrite	$CoZnFeO_2$	5.43	303	12	2.660	1.530	0.780
Copper	Cu	8.932	298	15	1.683	1.221	0.757
Gallium antimonide	GaSb	5.6137	298	16	0.8839	0.4033	0.4316
Gallium arsenide	GaAs	5.3169	298	17	1.1877	0.5372	0.5944
Gallium phosphide	GaP	4.1297	300	18	1.4120	0.6253	0.7047
Garnet (yttrium-iron)	$Y_3Fe_2(FeO_4)_3$	5.17	298	19	2.680	1.106	0.766
Germanium	Ge	5.313	298	20	1.2835	0.4823	0.6666
Gold	Au	19.283	296.5	21	1.9244	1.6298	0.4200
Indium antimonide	InSb	5.7890	298	22	0.6720	0.3670	0.3020
Indium arsenide	InAs	5.6720	293	23	0.8329	0.4526	0.3959
Indium phosphide	InP	4.78	RT	24	1.0220	0.5760	0.4600
Iridium	Ir	22.52	300	25	5.80	2.42	2.56
Iron	Fe	7.8672	298	26	2.26	1.40	1.16
Lead	Pb	11.34	296	27	0.4966	0.4231	0.1498
Lead fluoride	PbF_2	7.79	300	28	0.8880	0.4720	0.2454
Lead nitrate	$Pb(NO_3)_2$	4.547	293	29	0.3729	0.2765	0.1347
Lead telluride	PbTe	8.2379	303.2	30	1.0795	0.0764	0.1343
Lithium	Li	0.5326	298	31	0.1350	0.1144	0.0878
Lithium bromide	LiBr	3.47	RT	32	0.3940	0.1880	0.1910
Lithium chloride	LiCl	2.068	295	33	0.4927	0.2310	0.2495
Lithium fluoride	LiF	2.638	RT	34	1.1397	0.4767	0.6364
Lithium iodide	LiI	4.061	RT	32	0.2850	0.1400	0.1350
Magnesium oxide	MgO	3.579	298	20	2.9708	0.9536	1.5613
Magnetite	Fe_3O_4	5.18	RT	32	2.730	1.060	0.971
Manganese oxide	MnO	5.39	298	35	2.23	1.20	0.79
Mercury telluride	HgTe	8.079	290	36	0.548	0.381	0.204
Molybdenum	Mo	10.2284	273	37	4.637	1.578	1.092

CUBIC CRYSTALS (continued)

Name	Formula	ρ/g cm⁻³	T/K	Ref.	C_{11}	C_{12}	C_{44}
Nickel	Ni	8.91	298	15	2.481	1.549	1.242
Niobium	Nb	8.578	300	38	2.4650	1.3450	0.2873
Palladium	Pd	12.038	300	39	2.2710	1.7604	0.7173
Platinum	Pt	21.50	300	40	3.4670	2.5070	0.7650
Potassium	K	0.851	295	41	0.0370	0.0314	0.0188
Potassium bromide	KBr	2.740	298	11	0.3468	0.0580	0.0507
Potassium chloride	KCl	1.984	298	11	0.4069	0.0711	0.0631
Potassium cyanide	KCN	1.553	RT	32	0.1940	0.1180	0.0150
Potassium fluoride	KF	2.480	295	33	0.6490	0.1520	0.1232
Potassium iodide	KI	3.128	300	42	0.2710	0.0450	0.0364
Pyrite	FeS_2	5.016	RT	43	3.818	0.310	1.094
Rubidium	Rb	1.58	170	44	0.0296	0.0250	0.0171
Rubidium bromide	RbBr	3.350	300	45	0.3152	0.0500	0.0380
Rubidium chloride	RbCl	2.797	300	45	0.3624	0.0612	0.0468
Rubidium iodide	RbI	3.551	300	45	0.2556	0.0382	0.0278
Silicon	Si	2.331	298	46	1.6578	0.6394	0.7962
Silver	Ag	10.50	300	47	1.2399	0.9367	0.4612
Silver bromide	AgBr	5.585	300	48	0.5920	0.3640	0.0616
Sodium	Na	0.971	299	49	0.0739	0.0622	0.0419
Sodium bromate	$NaBrO_3$	3.339	RT	32	0.5450	0.1910	0.1500
Sodium bromide	NaBr	3.202	300	33	0.3970	0.1001	0.0998
Sodium chlorate	$NaClO_3$	2.485	RT	50	0.4920	0.1420	0.1160
Sodium chloride	NaCl	2.163	298	11	0.4947	0.1288	0.1287
Sodium fluoride	NaF	2.804	300	51	0.9700	0.2380	0.2822
Sodium iodide	NaI	3.6689	300	52	0.3007	0.0912	0.0733
Spinel	$MgAl_2O_4$	3.6193	298	53	2.9857	1.5372	1.5758
Strontium fluoride	SrF_2	4.277	300	54	1.2350	0.4305	0.3128
Strontium nitrate	$Sr(NO_3)_2$	2.989	293	29	0.4255	0.2921	0.1590
Strontium oxide	SrO	4.99	300	55	1.601	0.435	0.590
Strontium titanate	$SrTiO_3$	5.123	RT	56	3.4817	1.0064	4.5455
Tantalum	Ta	16.626	298	57	2.6023	1.5446	0.8255
Tantalum carbide	TaC	14.65	RT	58	5.05	0.73	0.79
Thallium bromide	TlBr	7.4529	298	59	0.3760	0.1458	0.0757
Thorium	Th	11.694	300	60	0.7530	0.4890	0.4780
Thorium oxide	ThO_2	9.991	298	61	3.670	1.060	0.797
Tin telluride	SnTe	6.445	300	62	1.1250	0.0750	0.1172
Titanium carbide	TiC	4.940	RT	107	5.00	1.13	1.75
Tungsten	W	19.257	297	64	5.2239	2.0437	1.6083
Uranium carbide	UC	13.63	300	65	3.200	0.850	0.647
Uranium dioxide	UO_2	10.97	298	66	3.960	1.210	0.641
Vanadium	V	6.022	300	67	2.287	1.190	0.432
Zinc selenide	ZnSe	5.262	298	68	0.8096	0.4881	0.4405
Zinc sulfide	ZnS	4.088	298	68	1.0462	0.6534	0.4613
Zinc telluride	ZnTe	5.636	298	68	0.7134	0.4078	0.3115
Zirconium carbide	ZrC	6.606	298	63	4.720	0.987	1.593

TETRAGONAL CRYSTALS

Name	Formula	ρ/g cm⁻³	T/K	Ref.	C_{11}	C_{12}	C_{13}	C_{16}	C_{33}	C_{44}	C_{66}
Ammonium dihydrogen arsenate (ADA)	$NH_4H_2AsO_4$	2.3110	298	69	0.6747	-0.106	0.1652		0.3022	0.0685	0.0639
Ammonium dihydrogen phosphate (ADP)	$NH_4H_2PO_4$	1.8030	293	69	0.6200	-0.050	0.1400		0.3000	0.0910	0.0610
Barium titanate	$BaTiO_3$	5.9988	298	70	2.7512	1.7897	1.5156		1.6486	0.5435	1.1312
Calcium molybdate	$CaMoO_4$	4.255	298	79	1.447	0.664	0.466	0.134	1.265	0.369	0.451
Indium	In	7.300	RT	71	0.4450	0.3950	0.4050		0.4440	0.0655	0.1220
Magnesium fluoride	MgF_2	3.177	RT	72	1.237	0.732	0.536		1.770	0.552	0.978
Nickel sulfate hexahydrate	$NiSO_4 \cdot 6H_2O$	2.070	RT	73	0.3209	0.2315	0.0209		0.2931	0.1156	0.1779
Potassium dihydrogen arsenate (KDA)	KH_2AsO_4	2.867	RT	12	0.530	-0.060	-0.020		0.370	0.120	0.070
Potassium dihydrogen phosphate (KDP)	KH_2PO_4	2.388	RT	71	0.7140	-0.049	0.1290		0.5620	0.1270	0.0628
Rubidium dihydrogen phosphate (RDP)	RbH_2PO_4	2.800	298	74	0.5562	-0.064	0.0279		0.4398	0.1142	0.0350
Rutile	TiO_2	4.260	298	75	2.7143	1.7796	1.4957		4.8395	1.2443	1.9477
Tellurium oxide	TeO_2	5.99	RT	76	0.5320	0.4860	0.2120		1.0850	0.2440	0.5520
Tin (white)	Sn	7.29	288	77	0.7529	0.6156	0.4400		0.9552	0.2193	0.2336
Zircon	$ZrSiO_4$	4.70	RT	78	2.585	1.791	1.542		3.805	0.733	1.113

ORTHORHOMBIC CRYSTALS

Name	Formula	ρ/g cm^{-3}	*T*/K	Ref.	C_{11}	C_{12}	C_{13}	C_{22}	C_{23}	C_{33}	C_{44}	C_{55}	C_{66}
Acenaphthene	$C_{12}H_{10}$	1.220	293	80	0.1380	0.0210	0.0410	0.1262	0.0460	0.1117	0.0265	0.0290	0.0185
Ammonium sulfate	$(NH_4)_2SO_4$	1.774	293	81	0.3607	0.1651	0.1580	0.2981	0.1456	0.3534	0.1025	0.0717	0.0974
Aragonite	$CaCO_3$	2.93	RT	82	1.5958	0.3663	0.0197	0.8697	0.1597	0.8503	0.4132	0.2564	0.4274
Barite	$BaSO_4$	4.40	RT	82	0.8941	0.4614	0.2691	0.7842	0.2676	1.0548	0.1190	0.2874	0.2778
Benzene	C_6H_6	1.061	250	83	0.0614	0.0352	0.0401	0.0656	0.0390	0.0583	0.0197	0.0378	0.0153
Benzophenone	$(C_6H_5)_2CO$	1.219	RT	32	0.1070	0.0550	0.0169	0.1000	0.0321	0.0710	0.0203	0.0155	0.0353
Bronzite	$(MgFe)SiO_3$	3.38	RT	78	1.876	0.686	0.605	1.578	0.561	2.085	0.700	0.592	0.544
Calcium sulfate	$CaSO_4$	2.962	RT	84	0.9382	0.1650	0.1520	1.845	0.3173	1.1180	0.3247	0.2653	0.0926
Celestite	$SrSO_3$	3.96	RT	12	1.044	0.773	0.605	1.061	0.619	1.286	0.135	0.279	0.266
Cesium sulfate	Cs_2SO_4	4.243	293	81	0.4490	0.1958	0.1815	0.4283	0.1800	0.3785	0.1326	0.1319	0.1323
Fosterite	Mg_2SiO_4	3.224	298	85	3.2848	0.6390	0.6880	1.9980	0.7380	2.3530	0.6515	0.8120	0.8088
Iodic acid	HIO_3	4.630	RT	73	0.3030	0.1194	0.1169	0.5448	0.0548	0.4359	0.1835	0.2193	0.1736
Lithium ammonium													
tartrate	$LiNH_4C_4H_4O_6{\cdot}4H_2O$	1.71	RT	12	0.3864	0.1655	0.0875	0.5393	0.2007	0.3624	0.1190	0.0667	0.2326
Magnesium sulfate													
heptahydrate	$MgSO_4{\cdot}7H_2O$	1.68	RT	86	0.325	0.174	0.182	0.288	0.182	0.315	0.078	0.156	0.090
Natrolite	$(Na,Al)SiO_3$	2.25	RT	78	0.716	0.261	0.297	0.632	0.297	1.378	0.196	0.248	0.423
Nickel sulfate													
heptahydrate	$NiSO_4{\cdot}7H_2O$	1.948	RT	86	0.353	0.198	0.201	0.311	0.201	0.335	0.091	0.172	0.099
Olivine	$(MgFe)SiO_4$	3.324	RT	87	3.240	0.590	0.790	1.980	0.780	2.490	0.667	0.810	0.793
Potassium													
pentaborate	$KB_5O_8{\cdot}4H_2O$	1.74	RT	71	0.582	0.229	0.174	0.359	0.231	0.255	0.164	0.046	0.057
Potassium sulfate	K_2SO_4	2.665	293	81	0.5357	0.1999	0.2095	0.5653	0.1990	0.5523	0.195	0.1879	0.1424
Rochelle salt	$NaK(C_4H_4O_6){\cdot}4H_2O$	1.79	RT	71	0.255	0.141	0.116	0.381	0.146	0.371	0.134	0.032	0.098
Rubidium sulfate	Rb_2SO_4	3.621	293	81	0.5029	0.1965	0.1999	0.5098	0.1925	0.4761	0.1626	0.1589	0.1407
Sodium ammonium													
tartrate	$NaNH_4C_4H_4O_6{\cdot}4H_2O$	1.587	RT	12	0.3685	0.2725	0.3083	0.5092	0.3472	0.5541	0.1058	0.0303	0.0870
Sodium tartrate	$Na_2C_4H_4O_6{\cdot}2H_2O$	1.794	RT	12	0.461	0.286	0.320	0.547	0.352	0.665	0.124	0.031	0.098
Strontium formate													
dihydrate	$Sr(CHO_2)_2{\cdot}2H_2O$	2.25	RT	12	0.4391	0.1037	-0.149	0.3484	-0.014	0.3746	0.1538	0.1075	0.1724
Sulfur	S	2.07	RT	12	0.240	0.133	0.171	0.205	0.159	0.483	0.043	0.087	0.076
Thallium sulfate	$TlSO_4$	6.776	293	81	0.4106	0.2573	0.2288	0.3885	0.2174	0.4268	0.1125	0.1068	0.0751
Topaz	$Al_2SiO_3(OH,F)_2$	3.52	RT	82	2.8136	1.2582	0.8464	3.8495	0.8815	2.9452	1.0811	1.3298	1.3089
Uranium (alpha)	U	19.0453	293	88	2.1486	0.4622	0.2176	1.9983	1.0764	2.6763	1.2479	0.7379	0.7454
Zinc sulfate													
heptahydrate	$ZnSO_4{\cdot}7H_2O$	1.970	RT	86	0.3320	0.1720	0.2000	0.2930	0.1980	0.3200	0.0780	0.1530	0.0830

ELASTIC CONSTANTS OF SINGLE CRYSTALS (continued)

MONOCLINIC CRYSTALS

Name	Formula	ρ/g cm^{-3}	*T*/K	Ref.	C_{11}	C_{12}	C_{13}	C_{15}	C_{22}
Aegirine	$(NaFe)Si_2O_6$	3.50	RT	89	1.858	0.685	0.707	0.098	1.813
Anthracene	$C_{14}H_{10}$	1.258	RT	90	0.0852	0.0672	0.0590	-0.0192	0.1170
Cobalt sulfate heptahydrate	$CoSO_4 \cdot 7H_2O$	1.948	RT	86	0.335	0.205	0.158	0.016	0.378
Diopside	$(CaMg)Si_2O_6$	3.31	RT	91	2.040	0.884	0.0883	-0.193	1.750
Dipotassium tartrate	$KHC_4H_4O_6$	1.97	RT	12	0.4294	0.1399	0.3129	-0.0105	0.3460
Feldspar (microceine)	$KAlSi_3O_8$	2.56	RT	92	0.664	0.438	0.259	-0.033	1.710
Ferrous sulfate heptahydrate	$FeSO_4 \cdot 7H_2O$	1.898	RT	86	0.349	0.208	0.174	-0.020	0.376
Lithium sulfate monohydrate	$Li_2SO_4 \cdot H_2O$	2.221	RT	32	0.5250	0.1715	0.1730	-0.0196	0.5060
Naphthalene	$C_{10}H_8$	1.127	RT	93	0.0780	0.0445	0.0340	-0.006	0.0990
Potassium tartrate	$K_2C_4H_4O_6$	1.987	RT	32	0.3110	0.1720	0.1690	0.0287	0.3900
Sodium thiosulfate	$Na_2S_2O_3$	1.7499	RT	12	0.3323	0.1814	0.1875	0.0225	0.2953
Stilbene	$(C_6H_5CH)_2$	1.60	RT	94	0.0930	0.0570	0.0670	-0.003	0.0920
Triglycine sulfate (TGS)	$(NH_2CH_2COOH)_3 \cdot H_2SO_4$	1.68	RT	32	0.4550	0.1720	0.1980	-0.030	0.3210

Name	C_{23}	C_{25}	C_{33}	C_{35}	C_{44}	C_{46}	C_{55}	C_{66}
Aegirine	0.626	0.094	2.344	0.214	0.692	0.077	0.510	0.474
Anthracene	0.0375	-0.0170	0.1522	-0.0187	0.0272	0.0138	0.0242	0.0399
Cobalt sulfate heptahydrate	0.158	-0.018	0.371	-0.047	0.060	0.016	0.058	0.101
Diopside	0.482	-0.196	2.380	-0.336	0.675	-0.113	0.588	0.705
Dipotassium tartrate	0.1173	0.0176	0.6816	0.0294	0.0961	-0.0044	0.1270	0.0841
Feldspar (microceine)	0.192	-0.148	1.215	-0.131	0.143	-0.015	0.238	0.361
Ferrous sulfate heptahydrate	0.172	-0.019	0.360	-0.014	0.064	0.001	0.056	0.096
Lithium sulfate monohydrate	0.0368	0.0571	0.5400	-0.0254	0.1400	-0.0054	0.1565	0.2770
Naphthalene	0.0230	-0.0270	0.1190	0.0290	0.0330	-0.0050	0.0210	0.0415
Potassium tartrate	0.1330	0.0182	0.5540	0.0710	0.0870	0.0072	0.1040	0.0826
Sodium thiosulfate	0.1713	0.0983	0.4590	-0.0678	0.0569	-0.0268	0.1070	0.0598
Stilbene	0.0485	-0.005	0.0790	-0.005	0.0325	0.0050	0.0640	0.0245
Triglycine sulfate (TGS)	0.2080	-0.0036	0.2630	-0.0500	0.0950	-0.0026	0.1110	0.0620

HEXAGONAL CRYSTALS

Name	Formula	ρ/g cm⁻³	T/K	Ref.	C_{11}	C_{12}	C_{13}	C_{33}	C_{55}
Apatite	$Ca_5(PO_4)_3(OH,F,Cl)$	3.218	RT	12	1.667	0.131	0.655	1.396	0.663
Beryl	$Be_3Al_2Si_6O_{18}$	2.68	RT	12	2.800	0.990	0.670	2.480	0.658
Beryllium	Be	1.8477	300	95	2.923	0.267	0.140	3.364	1.625
Beryllium oxide	BeO	3.01	RT	96	4.70	1.68	1.19	4.94	1.53
Cadmium	Cd	8.652	300	97	1.1450	0.3950	0.3990	0.5085	0.1985
Cadmium selenide	CdSe	5.655	298	68	0.7046	0.4516	0.3930	0.8355	0.1317
Cadmium sulfide	CdS	4.824	298	98	0.8431	0.5208	0.4567	0.9183	0.1458
Cobalt	Co	8.836	298	99	3.071	1.650	1.027	3.581	0.755
Dysprosium	Dy	8.560	298	100	0.7466	0.2616	0.2233	0.7871	0.2427
Erbium	Er	9.064	298	100	0.8634	0.3050	0.2270	0.8554	0.2809
Gadolinium	Gd	7.888	298	101	0.6667	0.2499	0.2132	0.7191	0.2089
Hafnium	Hf	12.727	298	102	1.881	0.772	0.661	1.969	0.557
Ice	H_2O(solid)	0.920	250	103	0.1410	0.0660	0.0624	0.1515	0.0288
Indium	In	7.2788	300	104	0.4535	0.4006	0.4151	0.4515	0.0651
Magnesium	Mg	1.7364	298	105	0.5950	0.2612	0.2180	0.6155	0.1635
Rhenium	Re	21.024	298	100	6.1820	2.7530	2.0780	6.8350	1.6060
Ruthenium	Ru	12.3615	298	100	5.6260	1.8780	1.6820	6.2420	1.8060
Thallium	Tl	11.560	300	106	0.4080	0.3540	0.2900	0.5280	0.0726
Titanium	Ti	4.5063	298	102	1.6240	0.9200	0.6900	1.8070	0.4670
Titanium diboride	TiB_2	4.95	RT	107	6.90	4.10	3.20	4.40	2.50
Yttrium	Y	4.472	300	108	0.7790	0.2850	0.2100	0.7690	0.2431
Zinc	Zn	7.134	295	109	1.6368	0.3640	0.5300	0.6347	0.3879
Zinc oxide	ZnO	5.6760	298	110	2.0970	1.2110	1.0510	2.1090	0.4247
Zinc sulfide	ZnS	4.089	298	96	1.2420	0.6015	0.4554	1.4000	0.2864
Zirconium	Zr	6.505	298	102	1.434	0.728	0.653	1.648	0.320

TRIGONAL CRYSTALS

Name	Formula	ρ/g cm⁻³	T/K	Ref.	C_{11}	C_{12}	C_{13}	C_{14}	C_{33}	C_{44}
Aluminum oxide	Al_2O_3	3.986	300	111	4.9735	1.6397	1.1220	-0.2358	4.9911	1.4739
Aluminum phosphate	$AlPO_4$	2.556	RT	73	1.0503	0.2934	0.6927	-0.1271	1.3353	0.2314
Antimony	Sb	6.70	295	112	1.0130	0.3450	0.2920	0.2090	0.4500	0.3930
Bismuth	Bi	9.80	295	112	0.6370	0.2490	0.2470	0.0717	0.3820	0.1123
Calcite	$CaCO_3$	2.712	300	113	1.4806	0.5578	0.5464	-0.2058	0.8557	0.3269
Hematite	Fe_2O_3	5.240	RT	82	2.4243	0.5464	0.1542	-0.1247	2.2734	0.8569
Lithium niobate	$LiNbO_3$	4.70	RT	114	2.030	0.530	0.750	0.090	2.450	0.600
Lithium tantalate	$LiTaO_3$	7.45	RT	114	2.330	0.470	0.800	-0.110	2.750	0.940
Quartz	SiO_2	2.6485	298	115	0.8680	0.0704	0.1191	-0.1804	1.0575	0.5820
Selenium	Se	4.838	300	116	0.1870	0.0710	0.2620	0.0620	0.7410	0.1490
Sodium nitrate	$NaNO_3$	2.27	RT	12	0.8670	0.1630	0.1600	0.0820	0.3740	0.2130
Tourmaline		3.05	RT	82	2.7066	0.6927	0.0872	-0.0774	1.6070	0.6682

REFERENCES

1. Thomas, J. F., *Phys. Rev.*, 175, 955-962, 1968.
2. Bolef, D. I. and M. Menes, *J. Appl. Phys.*, 31, 1426-1427, 1960.
3. Garland, C. W. and C. F. Yarnell, *J. Chem. Phys.*, 44, 1112-1120, 1966.
4. Garland, C. W. and R. Renard, *J. Chem. Phys.*, 44, 1130-1139, 1966.

5. Gsänger, M., H. Egger and E. Lüscher, *Phys. Letters*, 27A, 695-696, 1968.
6. Wong, C. and D. E. Schuele, *J. Phys. Chem. Solids*, 29, 1309-1330, 1968.
7. Haussühl, S., *Phys. Stat. Sol.*, 3, 1072-1076, 1963.
8. Wong, C. and D. E. Schuele, *J. Phys. Chem. Solids*, 28, 1225-1231, 1967.
9. McSkimin, H. J. and D. G. Thomas, *J. Appl. Phys.*, 33, 56-59, 1962.
10. Kollarits, F. J. and J. Trivisonno, *J. Phys. Chem. Solids*, 29, 2133-2139, 1968.
11. Slagle, D. D. and H. A. McKinstry, *J. Appl. Phys.*, 38, 446-458, 1967.
12. Hearmon, R. F. S., *Adv. Phys.*, 5, 323-382, 1956.
13. Sumer, A. and J. F. Smith, *J. Appl. Phys.*, 34, 2691-2694, 1963.
14. Alexandrov, K. S. et. al., *Sov. Phys. Sol. State*, 10, 1316-1321, 1968.
15. Epstein, S. G. and O. N. Carlson, *Acta Metal.*, 13, 487-491, 1965.
16. McSkimin, H. J., et. al., *J. Appl. Phys.*, 39, 4127-4128, 1968.
17. McSkimin, H. J., et. al., *J. Appl. Phys.*, 38, 2362-2364, 1967.
18. Weil, R. and W. O. Groves, *J. Appl. Phys.*, 39, 4049-4051, 1968.
19. Bateman, T. B., *J. Appl. Phys.*, 37, 2194-2195, 1966.
20. Bogardus, E. H., *J. Appl. Phys.*, 36, 2504-2513, 1965.
21. Golding, B., S. C. Moss and B. L. Averbach, *Phys. Rev.*, 158, 637-645, 1967.
22. Bateman, T. B., H. J. McSkimin and J. M. Whelan, *J. Appl. Phys.*, 30, 544-545, 1959.
23. Gerlich, D., *J. Appl. Phys.*, 35, 3062, 1964.
24. Hickernell, F. S. and W. R. Gayton, *J. Appl. Phys.*, 37, 462, 1966.
25. MacFarlane, R. E., et. al., *Phys. Letters*, 20, 234-235, 1966.
26. Leese, J. and A. E. Lord Jr., *J. Appl. Phys.*, 39, 3986-3988, 1968.
27. Miller, R. A. and D. E. Schuele, *J. Phys. Chem. Solids*, 30, 589-600, 1969.
28. Wasilik, J. H. and M. L. Wheat, *J. Appl. Phys.*, 36, 791-793, 1965.
29. Haussühl, S., *Phys. Stat. Sol.*, 3, 1072-1076, 1963.
30. Houston, B., et. al., *J. Appl. Phys.*, 39, 3913-3916, 1968.
31. Trivisonno, J. and C. S. Smith, *Acta Metal.*, 9, 1064-1071, 1961.
32. Alexandrov, K. S. and T. V. Ryzhova, *Sov. Phys. Cryst.*, 6, 228-252, 1961.
33. Lewis, J. T., A. Lehoczky and C. V. Briscoe, *Phys. Rev.*, 161, 877-887, 1967.
34. Drabble, J. R. and R. E. B. Strathen, *Proc. Phys. Soc.*, 92, 1090-1995, 1967.
35. Oliver, D. W., *J. Appl. Phys.*, 40, 893, 1969.
36. Alper, T., and G. A. Saunders, *J. Phys. Chem. Solids*, 28, 1637-1642, 1967.
37. Dickinson, J. M. and P. E. Armstrong, *J. Appl. Phys.*, 38, 602-606, 1967.
38. Bolef, D. I., *J. Appl. Phys.*, 32, 100-105, 1961.
39. Rayne, J. A., *Phys. Rev.*, 112, 1125-1130, 1958.
40. MacFarlane, R. E., et. al., *Phys. Letters*, 18, 91-92, 1965.
41. Smith, P. A. and C. S. Smith, *J. Phys. Chem. Solids*, 26, 279-289, 1965.
42. Norwood, M. H. and C. V. Briscoe, *Phys. Rev.*, 112, 45-48, 1958.
43. Simmons, G. and F. Birch, *J. Appl. Phys.*, 34, 2736-2738, 1963.
44. Gutman, E. J. and J. Trivisonno, *J. Phys. Chem. Sol.*, 28, 805-809, 1967.
45. Ghafelehbashi, M., et. al., *J. Appl. Phys.*, 41, 652-666, 1970.
46. McSkimin, H. J. and P. Andreatch Jr., *J. Appl. Phys.*, 35, 2161-2165, 1964.
47. Neighbours, J. R. and G. A. Alers, *Phys. Rev.*, 111, 707-712, 1958.
48. Hidshaw, W., J. T. Lewis, and C. V. Briscoe, *Phys. Rev.*, 163, 876-881, 1967.
49. Daniels, W. B., *Phys. Rev.*, 119, 1246-1252, 1960.
50. Viswanathan, R., *J. Appl. Phys.*, 37, 884-886, 1966.
51. Miller, R. A. and C. S. Smith, *J. Phys. Chem. Sol.*, 25, 1279-1292, 1964.
52. Claytor, R. N. and B. J. Marshall, *Phys. Rev.*, 120, 332-334, 1960.
53. Schreiber, E., *J. Appl. Phys.*, 38, 2508-2511, 1967.
54. Gerlich, D., *Phys. Rev.*, 136, A1366-A1368, 1964.
55. Johnston, D. L., P. H. Thrasher and R. J. Kearney, *J. Appl. Phys.*, 41, 427-428, 1970.
56. Poindexter, E. and A. A. Giardini, *Phys. Rev.*, 110, 1069, 1958.
57. Soga, N., *J. Appl. Phys.*, 37, 3416-3420, 1966.
58. Bartlett, R. W. and C. W. Smith, *J. Appl. Phys.*, 38, 5428-5429, 1967.
59. Morse, G. E. and A. W. Lawson, *J. Phys. Chem. Sol.*, 28, 939-950, 1967.
60. Armstrong, P. E., O. N. Carlson and J. F. Smith, *J. Appl. Phys.*, 30, 36-41, 1959.
61. Macedo, P. M., W. Capps and J. B. Wachtman, *J. Am. Cer. Soc.*, 47, 651, 1964.
62. Beattie, A. G., *J. Appl. Phys.*, 40, 4818-4821, 1969.
63. Chang, R. and L. J. Graham, *J. Appl Phys.*, 37, 3778-3783, 1966.
64. Lowrie, R. and A. M. Gonas, *J. Appl. Phys.*, 38, 4505-4509. 1967.
65. Graham, L. J., H. Nadler and R. Chang, *J. Appl. Phys.*, 34, 1572-1573, 1963.

66. Wachtman, J. B. Jr., et. al., *J. Nucl. Mat.*, 16, 39-41, 1965.
67. Bolef, D. I., *J. Appl. Phys.*, 32, 100-105, 1961.
68. Berlincourt, D., H. Jaffe and L. R. Shiozawa, *Phys. Rev.*, 129, 1009-1017, 1963.
69. Adhav. R. S. J. *Acoust. Soc. Am.*, 43, 835-838, 1968.
70. Berlincourt, D. and H. Jaffe, *Phys. Rev.*, 111, 143-148, 1958.
71. Huntington, H. B., in *Solid State Pysics, Vol. 7*, Seitz, F., and Turnbull, D., Ed., pp. 213-285; Academic Press, New York 1958.
72. Cutler, H. R., J. J. Gibson and K. A. McCarthy, *Sol. State Comm.*, 6, 431-433, 1968.
73. Mason, W. P., *Piezoelectric Crystals and Their Application to Ultrasonics*, D. Van Nostrand Co., Inc., New York, 1950.
74. Adhav, R. S., *J. Appl. Phys.*, 40, 2725-2727, 1969.
75. Manghnani, M. H., *J. Geophys. Res.*, 74, 4317-4328, 1969.
76. Uchida, N. and Y. Ohmachi, *J. Appl Phys.*, 40, 4692-4695, 1969.
77. House, D. G. and E. Y. Vernon, *Br. J. Appl. Phys.*, 11, 254-259, 1960.
78. Ryzhova, T. V., et. al., *Bull. Acad. Sci. USSR, Earth Phys. Ser.*, English Transl., no. 2, 111-113, 1966.
79. Alton, W. J. and A. J. Barlow, *J. Appl. Phys.*, 38, 3817-3820, 1967.
80. Michard, F., et. al., *C. R. Acad. Sci., Paris*, 265, 565-567, 1967.
81. Haussühl, S., *Acta Cryst.*, 18, 839-842, 1965.
82. Hearmon, R. F. S., *Rev. Mod. Phys.*, 18, 409-440, 1946.
83. Heseltine, J. C. W., D. W. Elliott and O. B. Wilson, *J. Chem. Phys.*, 40, 2584-2587, 1964.
84. Schwerdtner, W. M., et. al., *Canad. J. Earth Sci.*, 2, 673-683, 1965.
85. Kumazawa, M. and O. L. Anderson, *J. Geophys. Res.*, 74, 5961-5972, 1969 .
86. Alexandrov, K. S., et. al., *Sov. Phys. Cryst.*, 7, 753-755, 1963.
87. Verma, R. K., *J. Geophys. Soc.*, 65, 757-766, 1960.
88. McSkimin, H. J. and E. S. Fisher, *J. Appl. Phys.*, 31, 1627-1639, 1960.
89. Alexandrov, K. S. and T.V. Ryzhova, *Bull. Acad. Sci. USSR, Geophys. Ser.*, English Transl., no.8, 871-875, 1961.
90. Afanaseva, G. K., et. al, *Phys. Stat. Sol.*, 24, K61-K63, 1967.
91. Alexandrov, K. S., et. al., *Sov. Phys. Cryst.*, 8, 589-591, 1964.
92. Alexandrov, K. S. and T. V Ryzhova, *Bull Acad. Sci. USSR, Geophys. Ser.*, English Transl., no.2, 129-131, 1962.
93. Alexandrov, K. S., et. al., *Sov. Phys. Cryst.*, 8, 164-166, 1963.
94. Teslenko, V. F., et. al., *Sov. Phys. Cryst.*, 10, 744-747, 1966.
95. Smith, J. F. and C. L. Arbogast, *J. Appl. Phys.*, 31, 99-102, 1960.
96. Cline, C. F., H. L. Dunegan and G. M. Henderson, *J. Appl. Phys.*, 38, 1944-1948, 1967.
97. Chang, Y. A. and L. Himmel, *J. Appl. Phys.*, 37, 3787-3790, 1966.
98. Gerlich, D., *J. Phys. Chem. Solids*, 28, 2575-2579, 1967.
99. McSkimin, H. J., *J. Appl. Phys.*, 26, 406-409, 1955.
100. Fisher, E. S. and D. Dever, *Trans. Met. Soc. AIME*, 239, 48-57, 1967.
101. Fisher, E. S. and D. Dever, *Proc. Conf. Rare Earth Res.*, 6th, Gatlinburg, Tenn., 522-533, 1967.
102. Fisher, E. S. and C. J. Renken, Phys. Rev., 135, A482-A494, 1964.
103. Proctor, T. M. Jr., *J. Acoust. Soc. Am.*, 39, 972-977, 1966.
104. Chandrasekhar, B. S. and J. A. Rayne, *Phys. Rev.*, 124, 1011-1041, 1961.
105. Wazzan, A. R. and L. B. Robinson, *Phys. Rev.*, 155, 586-594, 1967.
106. Ferris, R. W., et. al., *J. Appl. Phys.*, 34, 768-770, 1963.
107. Gilman, J. J. and B. W. Roberts, *J. Appl. Phys.*, 32, 1405, 1961.
108. Smith, J. F. and J. A. Gjevre, *J. Appl. Phys.*, 31, 645-647, 1960.
109. Alers, G. A. and J. R. Neighbours, *J. Phys. Chem. Solids*, 7, 58-64, 1908.
110. Bateman, T. B., *J. Appl. Phys.*, 33, 3309-3312, 1962.
111. Tefft, W. E., *J. Res. Natl. Bur. Stand.*, 70A, 277-280, 1966.
112. DeBretteville, Jr., A. et. al., *Phys. Rev.*, 148, 575-579, 1966.
113. Dandekar, D. P. and A. L. Ruoff, *J. Appl. Phys.*, 39, 6004-6009, 1968.
114. Warner, A. W., M. Onoe and G. A. Coquin, *J. Acoust. Soc. Am.*, 42, 1223-1231, 1967.
115. McSkimin, H. J., P. Andreatch and R. N. Thurston, *J. Appl. Phys.*, 36, 1624-1632, 1965.
116. Mort, J., *J. Appl. Phys.*, 38, 3414-3415, 1967.

ELECTRICAL RESISTIVITY OF PURE METALS

The first part of this table gives the electrical resistivity, in units of $10^{-8}\ \Omega$ m, for 28 common metallic elements as a function of temperature. The data refer to polycrystalline samples. The number of significant figures indicates the accuracy of the values. However, at low temperatures (especially below 50 K) the electrical resistivity is extremely sensitive to sample purity. Thus the low-temperature values refer to samples of specified purity and treatment. The references should be consulted for further information on this point, as well as for values at additional temperatures.

The second part of the table gives resistivity values in the neighborhood of room temperature for other metallic elements that have not been studied over an extended temperature range.

REFERENCES

1. C. Y. Ho, et al., *J. Phys. Chem. Ref. Data*, 12, 183—322, 1983; 13, 1069—1096, 1984; 13, 1097—1130, 1984, 13, 1131—1172, 1984.
2. R. A. Matula, *J. Phys Chem. Ref. Data*, 8, 1147—1298, 1979.
3. T. C. Chi, *J. Phys. Chem. Ref. Data*, 8, 339—438, 1979; 8, 439—498, 1979.
4. K. H. Hellwege, Ed., *Landolt-Börnstein Numerical Data and Functional Relationships in Science and Technology*, Group III, Vol. 15, Subvolume a, Springer-Verlag, Heidelberg, 1982.
5. L. A. Hall, *Survey of Electrical Resistivity Measurements on 16 Pure Metals in the Temperature Range 0 to 273 K*, NBS Technical Note 365, U.S. Superintendent of Documents, 1968.

ELECTRICAL RESISTIVITY IN $10^{-8}\ \Omega$ m

T/K	Aluminum	Barium	Beryllium	Calcium	Cesium	Chromium	Copper
1	0.000100	0.081	0.0332	0.045	0.0026		0.00200
10	0.000193	0.189	0.0332	0.047	0.243		0.00202
20	0.000755	0.94	0.0336	0.060	0.86		0.00280
40	0.0181	2.91	0.0367	0.175	1.99		0.0239
60	0.0959	4.86	0.067	0.40	3.07		0.0971
80	0.245	6.83	0.075	0.65	4.16		0.215
100	0.442	8.85	0.133	0.91	5.28	1.6	0.348
150	1.006	14.3	0.510	1.56	8.43	4.5	0.699
200	1.587	20.2	1.29	2.19	12.2	7.7	1.046
273	2.417	30.2	3.02	3.11	18.7	11.8	1.543
293	2.650	33.2	3.56	3.36	20.5	12.5	1.678
298	2.709	34.0	3.70	3.42	20.8	12.6	1.712
300	2.733	34.3	3.76	3.45	21.0	12.7	1.725
400	3.87	51.4	6.76	4.7		15.8	2.402
500	4.99	72.4	9.9	6.0		20.1	3.090
600	6.13	98.2	13.2	7.3		24.7	3.792
700	7.35	130	16.5	8.7		29.5	4.514
800	8.70	168	20.0	10.0		34.6	5.262
900	10.18	216	23.7	11.4		39.9	6.041

T/K	Gold	Hafnium	Iron	Lead	Lithium	Magnesium	Manganese
1	0.0220	1.00	0.0225		0.007	0.0062	7.02
10	0.0226	1.00	0.0238		0.008	0.0069	18.9
20	0.035	1.11	0.0287		0.012	0.0123	54
40	0.141	2.52	0.0758		0.074	0.074	116
60	0.308	4.53	0.271		0.345	0.261	131
80	0.481	6.75	0.693	4.9	1.00	0.557	132
100	0.650	9.12	1.28	6.4	1.73	0.91	132
150	1.061	15.0	3.15	9.9	3.72	1.84	136
200	1.462	21.0	5.20	13.6	5.71	2.75	139
273	2.051	30.4	8.57	19.2	8.53	4.05	143
293	2.214	33.1	9.61	20.8	9.28	4.39	144
298	2.255	33.7	9.87	21.1	9.47	4.48	144
300	2.271	34.0	9.98	21.3	9.55	4.51	144
400	3.107	48.1	16.1	29.6	13.4	6.19	147
500	3.97	63.1	23.7	38.3		7.86	149

ELECTRICAL RESISTIVITY OF PURE METALS (continued)

T/K	Gold	Hafnium	Iron	Lead	Lithium	Magnesium	Manganese
600	4.87	78.5	32.9			9.52	151
700	5.82		44.0			11.2	152
800	6.81		57.1			12.8	
900	7.86					14.4	

T/K	Molybdenum	Nickel	Palladium	Platinum	Potassium	Rubidium	Silver
1	0.00070	0.0032	0.0200	0.002	0.0008	0.0131	0.00100
10	0.00089	0.0057	0.0242	0.0154	0.0160	0.109	0.00115
20	0.00261	0.0140	0.0563	0.0484	0.117	0.444	0.0042
40	0.0457	0.068	0.334	0.409	0.480	1.21	0.0539
60	0.206	0.242	0.938	1.107	0.90	1.94	0.162
80	0.482	0.545	1.75	1.922	1.34	2.65	0.289
100	0.858	0.96	2.62	2.755	1.79	3.36	0.418
150	1.99	2.21	4.80	4.76	2.99	5.27	0.726
200	3.13	3.67	6.88	6.77	4.26	7.49	1.029
273	4.85	6.16	9.78	9.6	6.49	11.5	1.467
293	5.34	6.93	10.54	10.5	7.20	12.8	1.587
298	5.47	7.12	10.73	10.7	7.39	13.1	1.617
300	5.52	7.20	10.80	10.8	7.47	13.3	1.629
400	8.02	11.8	14.48	14.6			2.241
500	10.6	17.7	17.94	18.3			2.87
600	13.1	25.5	21.2	21.9			3.53
700	15.8	32.1	24.2	25.4			4.21
800	18.4	35.5	27.1	28.7			4.91
900	21.2	38.6	29.4	32.0			5.64

T/K	Sodium	Strontium	Tantalum	Tungsten	Vanadium	Zinc	Zirconium
1	0.0009	0.80	0.10	0.000016		0.0100	0.250
10	0.0015	0.80	0.102	0.000137	0.0145	0.0112	0.253
20	0.016	0.92	0.146	0.00196	0.039	0.0387	0.357
40	0.172	1.70	0.751	0.0544	0.304	0.306	1.44
60	0.447	2.68	1.65	0.266	1.11	0.715	3.75
80	0.80	3.64	2.62	0.606	2.41	1.15	6.64
100	1.16	4.58	3.64	1.02	4.01	1.60	9.79
150	2.03	6.84	6.19	2.09	8.2	2.71	17.8
200	2.89	9.04	8.66	3.18	12.4	3.83	26.3
273	4.33	12.3	12.2	4.82	18.1	5.46	38.8
293	4.77	13.2	13.1	5.28	19.7	5.90	42.1
298	4.88	13.4	13.4	5.39	20.1	6.01	42.9
300	4.93	13.5	13.5	5.44	20.2	6.06	43.3
400		17.8	18.2	7.83	28.0	8.37	60.3
500		22.2	22.9	10.3	34.8	10.82	76.5
600		26.7	27.4	13.0	41.1	13.49	91.5
700		31.2	31.8	15.7	47.2		104.2
800		35.6	35.9	18.6	53.1		114.9
900			40.1	21.5	58.7		123.1

Element	*T*/K	Electrical resistivity 10^{-8} Ω m
Antimony	273	39
Bismuth	273	107
Cadmium	273	6.8
Cerium (β, hex)	290—300	82.8
Cerium (γ, cub)	298	74.4
Cobalt	273	5.6
Dysprosium	290—300	92.6
Erbium	290—300	86.0
Europium	290—300	90.0
Gadolinium	290—300	131
Gallium	273	13.6
Holmium	290—300	81.4
Indium	273	8.0
Iridium	273	4.7
Lanthanum	290—300	61.5
Lutetium	290—300	58.2
Mercury	298	96.1
Neodymium	290—300	64.3
Niobium	273	15.2
Osmium	273	8.1
Polonium	273	40
Praseodymium	290—300	70.0
Promethium	290—300	75 est.
Protactinium	273	17.7
Rhenium	273	17.2
Rhodium	273	4.3
Ruthenium	273	7.1
Samarium	290—300	94.0
Scandium	290—300	56.2
Terbium	290—300	115
Thallium	273	15
Thorium	273	14.7
Thulium	290—300	67.6
Tin	273	11.5
Titanium	273	39
Uranium	273	28
Ytterbium	290—300	25.0
Yttrium	290—300	59.6

ELECTRICAL RESISTIVITY OF SELECTED ALLOYS

These values were obtained by fitting all available measurements to a theoretical formulation describing the temperature and composition dependence of the electrical resistivity of metals. Some of the values listed here fall in regions of temperature and composition where no actual measurements exist. Details of the procedure may be found in the reference.

Values of the resistivity are given in units of 10^{-8} Ω m. General comments in the preceding table for pure metals also apply here.

REFERENCE

C.Y. Ho, et al., *J. Phys. Chem. Ref. Data*, 12, 183-322, 1983.

Aluminum-Copper

Wt % Al	100 K	273 K	293 K	300 K	350 K	400 K
99[a]	0.531	2.51	2.74	2.82	3.38	3.95
95[a]	0.895	2.88	3.10	3.18	3.75	4.33
90[b]	1.38	3.36	3.59	3.67	4.25	4.86
85[b]	1.88	3.87	4.10	4.19	4.79	5.42
80[b]	2.34	4.33	4.58	4.67	5.31	5.99
70[b]	3.02	5.03	5.31	5.41	6.16	6.94
60[b]	3.49	5.56	5.88	5.99	6.77	7.63
50[b]	4.00	6.22	6.55	6.67	7.55	8.52
40[c]		7.57	7.96	8.10	9.12	10.2
30[c]		11.2	11.8	12.0	13.5	15.2
25[f]		16.3	17.2	17.6	19.8	22.2
15[h]			12.3			
10[g]	8.71	10.8	11.0	11.1	11.7	12.3
5[c]	7.92	9.43	9.61	9.68	10.2	10.7
1[b]	3.22	4.46	4.60	4.65	5.00	5.37

Aluminum-Magnesium

Wt % Al	100 K	273 K	293 K	300 K	350 K	400 K
99[c]	0.958	2.96	3.18	3.26	3.82	4.39
95[c]	3.01	5.05	5.28	5.36	5.93	6.51
90[c]	5.42	7.52	7.76	7.85	8.43	9.02
10[b]	14.0	17.1	17.4	17.6	18.4	19.2
5[b]	9.93	13.1	13.4	13.5	14.3	15.2
1[a]	2.78	5.92	6.25	6.37	7.20	8.03

Copper-Gold

Wt % Cu	100 K	273 K	293 K	300 K	350 K	400 K
99[c]	0.520	1.73	1.86	1.91	2.24	2.58
95[c]	1.21	2.41	2.54	2.59	2.92	3.26
90[c]	2.11	3.29	4.42	3.46	3.79	4.12
85[c]	3.01	4.20	4.33	4.38	4.71	5.05
80[c]	3.95	5.15	5.28	5.32	5.65	5.99
70[c]	5.91	7.12	7.25	7.30	7.64	7.99
60[c]	8.04	9.18	9.13	9.36	9.70	10.05
50[c]	9.88	11.07	11.20	11.25	11.60	11.94
40[c]	11.44	12.70	12.85	12.90	13.27	13.65
30[c]	12.43	13.77	13.93	13.99	14.38	14.78
25[c]	12.59	13.93	14.09	14.14	14.54	14.94
15[c]	11.38	12.75	12.91	12.96	13.36	13.77
10[c]	9.33	10.70	10.86	10.91	11.31	11.72
5[c]	5.91	7.25	7.41	7.46	7.87	8.28
1[c]	2.00	3.40	3.57	3.62	4.03	4.45

Copper-Nickel

Wt % Cu	100 K	273 K	293 K	300 K	350 K	400 K
99[c]	1.45	2.71	2.85	2.91	3.27	3.62
95[c]	6.19	7.60	7.71	7.82	8.22	8.62
90[c]	12.08	13.69	13.89	13.96	14.40	14.81
85[c]	18.01	19.63	19.83	19.90	20.32	20.70
80[c]	23.89	25.46	25.66	25.72	26.12	26.44
70[i]	35.73	36.67	36.72	36.76	36.85	36.89
60[i]	45.76	45.43	45.38	43.35	45.20	45.01
50[i]	50.22	50.19	50.05	50.01	49.73	49.50
40[c]	36.77	47.42	47.73	47.82	48.28	48.49
30[i]	26.73	40.19	41.79	42.34	44.51	45.40
25[c]	22.22	33.46	35.11	35.69	39.67	42.81
15[c]	13.49	22.00	23.35	23.85	27.60	31.38
10[c]	9.28	16.65	17.82	18.26	21.51	25.19
5[c]	5.20	11.49	12.50	12.90	15.69	18.78
1[c]	1.81	7.23	8.08	8.37	10.63	13.18

Copper-Palladium

Wt % Cu	100 K	273 K	293 K	300 K	350 K	400 K
99[c]	0.91	2.10	2.23	2.27	2.59	2.92
95[c]	2.99	4.21	4.35	4.40	4.74	5.08
90[c]	5.69	6.89	7.03	7.08	7.41	7.74
85[c]	8.30	9.48	9.61	9.66	10.01	10.36
80[c]	10.74	11.99	12.12	12.16	12.51	12.87
70[c]	15.67	16.87	17.01	17.06	17.41	17.78
60[c]	20.45	21.73	21.87	21.92	22.30	22.69
50[c]	26.07	27.62	27.79	27.86	28.25	28.64
40[c]	33.53	35.31	35.51	35.57	36.03	36.47
30[c]	45.03	46.50	46.66	46.71	47.11	47.47
25[c]	44.12	46.25	46.45	46.52	46.99	47.43
15[c]	31.79	36.52	36.99	37.16	38.28	39.35
10[c]	23.00	28.90	29.51	29.73	31.19	32.56
5[c]	13.09	20.00	20.75	21.02	22.84	24.54
1[c]	8.97	11.90	12.67	12.93	14.82	16.68

Copper-Zinc

Wt % Cu	100 K	273 K	293 K	300 K	350 K	400 K
99[b]	0.671	1.84	1.97	2.02	2.36	2.71
95[b]	1.54	2.78	2.92	2.97	3.33	3.69
90[b]	2.33	3.66	3.81	3.86	4.25	4.63
85[b]	2.93	4.37	4.54	4.60	5.02	5.44
80[b]	3.44	5.01	5.19	5.26	5.71	6.17
70[b]	4.08	5.87	6.08	6.15	6.67	7.19

Gold-Palladium

Wt % Au	100 K	273 K	293 K	300 K	350 K	400 K
99[c]	1.31	2.69	2.86	2.91	3.32	3.73
95[c]	3.88	5.21	5.35	5.41	5.79	6.17
90[i]	6.70	8.01	8.17	8.22	8.56	8.93
85[b]	9.14	10.50	10.66	10.72	11.10	11.48
80[b]	11.23	12.75	12.93	12.99	13.45	13.93
70[c]	16.44	18.23	18.46	18.54	19.10	19.67
60[b]	24.64	26.70	26.94	27.02	27.63	28.23

Gold-Palladium (continued)

Wt % Au	100 K	273 K	293 K	300 K	350 K	400 K
50[a]	23.09	27.23	27.63	27.76	28.64	29.42
40[a]	19.40	24.65	25.23	25.42	26.74	27.95
30[b]	14.94	20.82	21.49	21.72	23.35	24.92
25[b]	12.72	18.86	19.53	19.77	21.51	23.19
15[a]	8.54	15.08	15.77	16.01	17.80	19.61
10[a]	6.54	13.25	13.95	14.20	16.00	17.81
5[a]	4.58	11.49	12.21	12.46	14.26	16.07
1[a]	3.01	10.07	10.85	11.12	12.99	14.80

Gold-Silver

Wt % Au	100 K	273 K	293 K	300 K	350 K	400 K
99[b]	1.20	2.58	2.75	2.80	3.22	3.63
95[a]	3.16	4.58	4.74	4.79	5.19	5.59
90[j]	5.16	6.57	6.73	6.78	7.19	7.58
85[j]	6.75	8.14	8.30	8.36	8.75	9.15
80[j]	7.96	9.34	9.50	9.55	9.94	10.33
70[j]	9.36	10.70	10.86	10.91	11.29	11.68
60[j]	9.61	10.92	11.07	11.12	11.50	11.87
50[j]	8.96	10.23	10.37	10.42	10.78	11.14
40[j]	7.69	8.92	9.06	9.11	9.46	9.81
30[a]	6.15	7.34	7.47	7.52	7.85	8.19
25[a]	5.29	6.46	6.59	6.63	6.96	7.30
15[a]	3.42	4.55	4.67	4.72	5.03	5.34
10[a]	2.44	3.54	3.66	3.71	4.00	4.31
5[i]	1.44	2.52	2.64	2.68	2.96	3.25
1[b]	0.627	1.69	1.80	1.84	2.12	2.42

Iron-Nickel

Wt % Fe	100 K	273 K	293 K	300 K	400 K
99[a]	3.32	10.9	12.0	12.4	18.7
95[c]	10.0	18.7	19.9	20.2	26.8
90[c]	14.5	24.2	25.5	25.9	33.2
85[c]	17.5	27.8	29.2	29.7	37.3
80[c]	19.3	30.1	31.6	32.2	40.0
70[b]	20.9	32.3	33.9	34.4	42.4
60[c]	28.6	53.8	57.1	58.2	73.9
50[d]	12.3	28.4	30.6	31.4	43.7
40[d]	7.73	19.6	21.6	22.5	34.0
30[c]	5.97	15.3	17.1	17.7	27.4
25[b]	5.62	14.3	15.9	16.4	25.1
15[c]	4.97	12.6	13.8	14.2	21.1
10[c]	4.20	11.4	12.5	12.9	18.9
5[c]	3.34	9.66	10.6	10.9	16.1
1[b]	1.66	7.17	7.94	8.12	12.8

Silver-Palladium

Wt % Ag	100 K	273 K	293 K	300 K	350 K	400 K
99[b]	0.839	1.891	2.007	2.049	2.35	2.66
95[b]	2.528	3.58	3.70	3.74	4.04	4.34
90[b]	4.72	5.82	5.94	5.98	6.28	6.59
85[k]	6.82	7.92	8.04	8.08	8.38	8.68
80[k]	8.91	10.01	10.13	10.17	10.47	10.78

Silver-Palladium (continued)

Wt % Ag	100 K	273 K	293 K	300 K	350 K	400 K
70[k]	13.43	14.53	14.65	14.69	14.99	15.30
60[i]	19.4	20.9	21.1	21.2	21.6	22.0
50[k]	29.3	31.2	31.4	31.5	32.0	32.4
40[m]	40.8	42.2	42.2	42.2	42.3	42.3
30[b]	37.1	40.4	40.6	40.7	41.3	41.7
25[k]	32.4	36.67	37.06	37.19	38.1	38.8
15[i]	21.0	27.08	26.68	27.89	29.3	30.6
10[i]	14.95	21.69	22.39	22.63	24.3	25.9
5[b]	8.91	15.98	16.72	16.98	18.8	20.5
1[a]	3.97	11.06	11.82	12.08	13.92	15.70

[a] Uncertainty in resistivity is ± 2%.
[b] Uncertainty in resistivity is ± 3%.
[c] Uncertainty in resistivity is ± 5%.
[d] Uncertainty in resistivity is ± 7% below 300 K and ± 5% at 300 and 400 K.
[e] Uncertainty in resistivity is ± 7%.
[f] Uncertainty in resistivity is ± 8%.
[g] Uncertainty in resistivity is ± 10%.
[h] Uncertainty in resistivity is ± 12%.
[i] Uncertainty in resistivity is ± 4%.
[j] Uncertainty in resistivity is ± 1%.
[k] Uncertainty in resistivity is ± 3% up to 300 K and ± 4% above 300 K.
[m] Uncertainty in resistivity is ± 2% up to 300 K and ± 4% above 300 K.

PERMITTIVITY (DIELECTRIC CONSTANT) OF INORGANIC SOLIDS

H. P. R. Frederikse

This table lists the permittivity ϵ, frequently called the dielectric constant, of a number of inorganic solids. When the material is not isotropic, the individual components of the permittivity are given. A superscript S indicates a measurement made under constant strain ("clamped" dielectric constant). If the constraint is removed, the measurement yields ϵ^T, the "unclamped" or free dielectric constant.

The temperature of the measurement is given when available; the symbol r.t. indicates a value at nominal room temperature. The frequency of the measurement is given in the last column (i.r. indicates a measurement in the infrared).

Substances are listed in alphabetical order by chemical formula.

REFERENCE

Young, K. F. and Frederikse, H. P. R., *J. Phys. Chem. Ref. Data*, 2, 313, 1973.

Formula	Name	ϵ_{ijk}	T/K	ν/Hz
Ag_3AsS_3	Silver thioarsenate (Proustite)	$\epsilon^T_{11} = 16.5$, $\epsilon^S_{11} = 14.5$	r.t.	2×10^7
		$\epsilon^T_{33} = 20.0$, $\epsilon^S_{33} = 18.0$	r.t.	2×10^7
AgBr	Silver bromide	12.50	r.t.	
AgCN	Silver cyanide	5.6	r.t.	10^6
AgCl	Silver chloride	11.15	r.t.	
$AgNO_3$	Silver nitrate	9.0	293	5×10^5
$AgNa(NO_2)_2$	Silver sodium nitrite	4.5 ± 0.5	r.t.	9.4×10^9
Ag_2O	Silver oxide	8.8	r.t.	
$(AlF)_2SiO_4$	Aluminum fluosilicate (topaz)	$\epsilon_{11} = 6.62$	297	7×10^3
		$\epsilon_{22} = 6.58$	297	7×10^3
		$\epsilon_{33} = 6.95$	297	7×10^3
Al_2O_3	Aluminum oxide (alumina)	$\epsilon_{11} = \epsilon_{22} = 9.34$	298	10^2—8×10^9
		$\epsilon_{33} = 11.54$	298	10^2—8×10^9
$AlPO_4$	Aluminum phosphate	$\epsilon^T_{11} = 6.05$	r.t.	10^3
AlSb	Aluminum antimonide	11.21	300	i.r.
AsF_3	Arsenic trifluoride	5.7	r.t.	
BN	Boron nitride	7.1	r.t.	i.r.
$BaCO_3$	Barium carbonate	8.53	291	2×10^5
$Ba(COOH)_2$	Barium formate	$\epsilon_{11} = 7.9$	r.t.	10^3
		$\epsilon_{22} = 5.9$	r.t.	10^3
		$\epsilon_{33} = 7.5$	r.t.	10^3
$BaCl_2$	Barium chloride	9.81	r.t.	
$BaCl_2 \cdot 2H_2O$	Barium chloride dihydrate	9.00	r.t.	10^3
BaF_2	Barium fluoride	7.32	292	5×10^2—10^{11}
$Ba(NO_3)_2$	Barium nitrate	4.95	292	2×10^5
$Ba_2NaNb_5O_{15}$	Barium sodium niobate ("Bananas")	$\epsilon^S_{11} = 222$, $\epsilon^T_{11} = 235$	296	10^4
		$\epsilon^S_{22} = 227$, $\epsilon^T_{22} = 247$	296	
		$\epsilon^S_{33} = 32$, $\epsilon^T_{33} = 51$	296	
BaO	Barium oxide (baria)	34 ± 1	248, 333	60×10^7
BaO_2	Barium peroxide	10.7	r.t.	2×10^6
BaS	Barium sulfide	19.23	r.t.	7.25×10^6
$BaSO_4$	Barium sulfate	11.4	288	10^8
$BaSnO_3$	Barium stannate	18	298	25×10^5
$BaTiO_3$	Barium titanate	$\epsilon^T_{11} = 3600$	298	10^5
		$\epsilon^S_{11} = 2300$	298	2.5×10^8
		$\epsilon^T_{33} = 150$	298	10^5
		$\epsilon^S_{33} = 80$	298	2.5×10^8
$Ba_6Ti_2Nb_8O_{30}$	Barium titanium niobate	$\epsilon_{11} = \epsilon_{22} \approx 190$	298	
		$\epsilon_{33} \approx 220$	298	
$BaWO_4$	Barium tungstate	$\epsilon_{11} = \epsilon_{22} = 35.5 \pm 0.2$	297.5	1.6×10^3
		$\epsilon_{33} = 37.2 \pm 0.2$	297.5	1.6×10^3
$BaZrO_3$	Barium zirconate	43	r.t.	

Formula	Name	ϵ_{ijk}	T/K	ν/Hz
$Be_3Al_2Si_6O_{18}$	Beryllium aluminum silicate (Beryl)	$\epsilon_{33} = 5.95$	297	7×10^3
		$\epsilon_{11} = \epsilon_{22} = 6.86$	297	7×10^3
$BeCO_3$	Beryllium carbonate	9.7	291	2×10^5
BeO	Beryllium oxide (beryllia)	7.35 ± 0.2	293	2×10^6
$BiFeO_3$	Bismuth iron oxide	40 ± 3	300	9.4×10^9
$Bi_{12}GeO_{20}$	Bismuth germanite	$\epsilon^S_{11} = 38$	r.t.	
$Bi(GeO_4)_3$	Bismuth germanate	16	293	
Bi_2O_3	Bismuth sesquioxide	18.2	r.t.	2×10^6
$Bi_4Ti_3O_{12}$	Bismuth titanate	112	r.t.	10^3
C	Diamond			
	Type I	5.87 ± 0.19	300	10^3
	Type IIa	5.66 ± 0.04	300	10^3
$C_4H_4O_6$	Tartaric acid	$\epsilon_{11} = \epsilon_{22} = 4.3$	298	
		$\epsilon_{33} = 4.5$	298	
		$\epsilon_{13} = 0.55$	298	
$C_6H_{14}N_2O_6$	Ethylene diamine tartrate (EDT)	$\epsilon^T_{11} = 5.0$	293	
		$\epsilon^T_{22} = 8.3$	293	
		$\epsilon^T_{33} = 6.0$	293	
		$\epsilon^T_{13} = 0.7$	293	
$C_6H_{12}O_6NaBr$	Dextrose sodium bromide	$\epsilon^T_{11} = 4.0$	r.t.	10^3
$(CH_3NH_3)Al(SO_4)_2 \cdot 2H_2O$	Methyl ammonium alum (MASD)	19	197	
$Ca_2B_6O_{11} \cdot 5H_2O$	Colemanite	$\epsilon_{11} = 20$	293	10^3
		$\epsilon_{33} = 25$	293	10^3
$CaCO_3$	Calcium carbonate	$\epsilon_{11} = 8.67$	r.t.	9.4×10^{10}
		$\epsilon_{22} = 8.69$	r.t.	9.4×10^{10}
		$\epsilon_{33} = 8.31$	r.t.	9.4×10^{10}
$CaCeO_3$	Calcium cerate	21	r.t.	
CaF_2	Calcium fluoride	6.81	300	5×10^2—10^{11}
$CaMoO_4$	Calcium molybdate	$\epsilon_{11} = \epsilon_{22} = 24.0 \pm 0.2$	297.5	<10
		$\epsilon_{33} = 20.0 \pm 0.2$	297.5	<10
$Ca(NO_3)_2$	Calcium nitrate	6.54	292	2×10^5
$CaNb_2O_6$	Calcium niobate	$\epsilon_{11} = 22.8 \pm 1.9$	r.t.	$(5$—$500) \times 10^3$
$Ca_2Nb_2O_7$	Calcium pyroniobate	~45	r.t.	5×10^7
CaO	Calcium oxide	11.8 ± 0.3	283	2×10^6
CaS	Calcium sulfide	6.699	r.t.	7.25×10^6
$CaSO_4 \cdot 2H_2O$	Calcium sulfate dihydrate	$\epsilon_{11} = 5.10$	r.t.	
		$\epsilon_{22} = 5.24$	r.t.	
		$\epsilon_{33} = 10.30$	r.t.	
$CaTiO_3$	Calcium titanate	165	r.t.	
$CaWO_4$	Calcium tungstate	$\epsilon_{11} = \epsilon_{22} = 11.7 \pm 0.1$	297.5	1.59×10^3
		$\epsilon_{33} = 9.5 \pm 0.2$	297.5	1.59×10^3
Cd_3As_2	Cadmium arsenide	$\epsilon_{33} = 18.5$	4	
$CdBr_2$	Cadmium bromide	8.6	293	5×10^5
CdF_2	Cadmium fluoride	8.33 ± 0.08	300	10^5—10^7
CdS	Cadmium sulfide	$\epsilon_{11} = \epsilon_{22} = 8.7$	300	i.r.
		$\epsilon_{33} = 9.25$	300	i.r.
		$\epsilon_{11} = \epsilon_{22} = 8.37$	8	i.r.
		$\epsilon_{33} = 9.00$	8	i.r.
		$\epsilon^T_{11} = 8.48$	77	10^4
		$\epsilon^T_{33} = 9.48$	77	10^4
		$\epsilon^S_{11} = 9.02, \epsilon^T_{11} = 9.35$	298	10^4
		$\epsilon^S_{33} = 9.53, \epsilon^T_{33} = 10.33$	298	10^4
CdSe	Cadmium selenide	$\epsilon^S_{11} = 9.53, \epsilon^T_{11} = 9.70$	298	10^4
		$\epsilon^S_{33} = 10.2, \epsilon^T_{33} = 10.65$	298	10^4
CdTe	Cadmium telluride	$\epsilon_{11} = \epsilon_{22} = 10.60 \pm 0.15$	297	i.r.
		$\epsilon_{33} = 7.05 \pm 0.05$	297	i.r.
$Cd_2Nb_2O_7$	Cadmium pyroniobate	500—580	293	10^3
CeO_2	Cerium oxide	7.0	r.t.	2×10^6

Formula	Name	ϵ_{ijk}	T/K	ν/Hz
$CoNb_2O_6$	Cobalt niobate	$\epsilon_{11} = 18.4 \pm 0.6$	r.t.	$(5—500) \times 10^3$
		$\epsilon_{22} = 21.4 \pm 1.1$	r.t.	$(5—500) \times 10^3$
		$\epsilon_{33} = 33.0 \pm 0.7$	r.t.	$(5—500) \times 10^3$
CoO	Cobalt oxide	12.9	298	$10^2—10^{10}$
Cr_2O_3	Chromic sesquioxide	$\epsilon_{11} = \epsilon_{22} = 13.3$	298.5	10^3
		$\epsilon_{33} = 11.9$	298.5	10^3
		8	315 (T_N)	6×10^{10}
$CsAl(SO_4)_2 \cdot 12H_2O$	Cesium alum	5.0	r.t.	$20—20 \times 10^3$
CsBr	Cesium bromide	6.38	298	1.6×10^3
Cs_2CO_3	Cesium carbonate	6.53	291	2×10^5
CsCl	Cesium chloride	7.2	298	
$Cs_2H_2AsO_4$	Cesium dihydrogen arsenate (CDA)	4.8	273	9.5×10^9
$Cs_2H_2PO_4$	Cesium dihydrogen phosphate (CDP)	6.15	285	9.5×10^9
$CsH_3(SeO_3)_2$	Cesium trihydrogen selenite	$\epsilon_{11} = 80$	273	10^5
		$\epsilon_{22} = 63$	273	10^5
		$\epsilon_{33} = 12$	273	10^5
CsI	Cesium iodide	6.31	298	1.6×10^3
$CsNO_3$	Cesium nitrate	$\epsilon_{11} = \epsilon_{22} = 9.4$	r.t.	5×10^5
		$\epsilon_{33} = 8.3$	r.t.	5×10^5
$CsPbCl_3$	Cesium lead chloride	14.37	300	$10^5—10^6$
CuBr	Cuprous bromide	8.0	293	5×10^5
CuCl	Cuprous chloride	9.8 ± 0.5	r.t.	10^3
CuO	Cupric oxide	18.1	r.t.	2×10^6
Cu_2O	Cuprous oxide (Cuprite)	7.60 ± 0.06	r.t.	10^5
$CuSO_4 \cdot 5H_2O$	Cupric sulfate pentahydrate	6.60	r.t.	
EuF_2	Europium fluoride	7.7 ± 0.2	298	$(1—300) \times 10^3$
$Eu_2(MoO_4)_3$	Europium molybdate	9.5	298	
EuS	Europium sulfide	13.10 ± 0.04	80	$5 \times 10^2—10^5$
FeO	Ferrous oxide	14.2	r.t.	2×10^6
Fe_2O_3	Ferric sesquioxide	4.5	r.t.	$10^5—10^7$
Fe_2O_3-α	Ferric sesquioxide (hematite)	12		6×10^{10}
Fe_3O_4	Ferrosoferric oxide (magnetite)	20	r.t.	$10^5—10^7$
GaAs	Gallium arsenide	13.13	300	
		12.90	4	i.r.
GaP	Gallium phosphide	11.1	r.t.	
		10.75 ± 0.1	1.6	i.r.
GaSb	Gallium antimonide	15.69	r.t.	
		15.7	4	i.r.
$Gd_2(MoO_4)_3$	Gadolinium molybdate	$\epsilon^T = 10$	298	
		$\epsilon^S = 9.5$	298	10^3
Ge	Germanium	16.0 ± 0.3	4	9.2×10^9
		15.8 ± 0.2	r.t.	$500—3 \times 10^{10}$
GeO_2	Germanium dioxide	$\epsilon_{11} = \epsilon_{22} = 7.44$	r.t.	i.r.
HIO_3	Iodic acid	$\epsilon_{11} = 7.5$	r.t.	10^3
		$\epsilon_{22} = 12.4$	r.t.	10^3
		$\epsilon_{33} = 8.1$	r.t.	10^3
$HNH_4(ClCH_2COO)_2$	Hydrogen ammonium dichloroacetate	$\epsilon_{[102]} = 5.9$	r.t.	10^5
H_2O	Ice I (P = 0 kbar)	99	243	
	Ice III (P = 3 kbar)	117	243	
	Ice V(P = 5 kbar)	114	243	
	Ice VI (P = 8 kbar)	193	243	
HgCl	Mercurous chloride (Calumel)	$\epsilon_{11} = \epsilon_{22} = 14.0$	r.t.	10^{12}
$HgCl_2$	Mercuric chloride	6.5	r.t.	10^{12}
HgS	Mercurous sulfide (Cinnabar)	$\epsilon_{11} = \epsilon_{22} = 18.0$	r.t.	i.r.
		$\epsilon_{33} = 32.5$	r.t.	i.r.
HgSe	Mercurous selenide	25.6	r.t.	$10^4—10^6$
I_2	Iodine	$\epsilon_{11} = 6$	r.t.	$5 \times 10^4—10^7$
		$\epsilon_{22} = 3$	r.t.	$5 \times 10^4—10^7$
		$\epsilon_{33} = 40$	r.t.	$5 \times 10^4—10^7$

PERMITTIVITY (DIELECTRIC CONSTANT) OF INORGANIC SOLIDS (continued)

Formula	Name	ϵ_{ijk}	T/K	ν/Hz
InAs	Indium arsenide	14.55 ± 0.3	r.t.	i.r.
		15.15	4	i.r.
InP	Indium phosphide	12.61	r.t.	i.r.
InSb	Indium antimonide	17.88	4	i.r.
$KAl(SO_4)_2 \cdot 12H_2O$	Potassium alum	6.5	r.t.	20—20 × 10^3
KBr	Potassium bromide	4.88	300	
		4.53	4.2	
$KBrO_3$	Potassium bromate	7.3	r.t.	2 × 10^6
KCN	Potassium cyanide	6.15	r.t.	2 × 10^6
K_2CO_3	Potassium carbonate	4.96	291	2 × 10^5
$K_2C_4H_4O_6 \cdot {}^1/_2\ H_2O$	Dipotassium tartrate (DKT)	ϵ_{11} = 6.44	r.t.	
		ϵ_{22} = 5.80	r.t.	
		ϵ_{33} = 6.49	r.t.	
		ϵ_{13} = 0.005	r.t.	
KCl	Potassium chloride	4.86 ± 0.02	r.t.	5 × 10^3
		4.50	4.2	
$KClO_3$	Potassium chlorate	5.1	r.t.	2 × 10^6
$KClO_4$	Potassium perchlorate	5.9	r.t.	2 × 10^6
K_2CrO_4	Potassium chromate	7.3	r.t.	6 × 10^7
$KCr(SO_4)_2 \cdot 12H_2O$	Potassium chrome alum	6.5	100—240	175 × 10^3
KD_2AsO_4	Potassium dideuterium arsenate (KDDA)	ϵ_{11} = 70	298	
		ϵ_{33} = 31	298	
KD_2PO_4	Potassium dideuterium phosphate (KDDP)	50 ± 2	297	10^3
KF	Potassium fluoride	6.05		2 × 10^6
KH_2AsO_4	Potassium dihydrogen arsenate (KDA)	ϵ_{11} = 60	298	
		ϵ_{33} = 24	298	
KH_2PO_4	Potassium dihydrogen phosphate (KDP)	46	298	10^3
		ϵ_{11} = 42	r.t.	
		ϵ_{33} = 21	r.t.	
K_2HPO_4	Dipotassium monohydrogen orthophosphate	9.05	r.t.	2 × 10^6
KI	Potassium iodide	5.00	r.t.	9.4 × 10^{10}
KIO_3	Potassium iodate	170	255	10^5
		10	293	10^5
		$\epsilon_{[101]}$ ≈ 40,70	r.t.	10^5
		16.85	r.t.	2 × 10^6
$(K,H)Al_3(SiO_4)_3$	Mica (muscovite)	5.4	299	10^2—3 × 10^9
$(K,H)Mg_3Al(SiO_4)_3$	Mica (Canadian)	ϵ_{11} = ϵ_{22} = 6.9	298	10^2—10^4
		ϵ_{33} = 7.3	298	10^4
KNO_2	Potassium nitrite	25	305	
KNO_3	Potassium nitrate	4.37	293	2 × 10^5
$KNbO_3$	Potassium niobate	700	r.t.	
K_3PO_4	Potassium orthophosphate	7.75	r.t.	2 × 10^6
KSCN	Potassium thiocyanate	7.9	r.t.	2 × 10^6
K_2SO_4	Potassium sulfate	6.4	r.t.	2 × 10^6
$K_2S_3O_6$	Potassium trithionate	5.7	293	1.8 × 10^6
$K_2S_4O_6$	Potassium tetrathionate	5.5	293	1.8 × 10^6
$K_2S_5O_6 \cdot H_2O$	Potassium pentathionate	7.8	293	1.8 × 10^6
$K_2S_6O_6$	Potassium hexathionate	7.8	293	1.8 × 10^6
K_2SeO_4	Potassium selenate	ϵ_{11} = 5.9	r.t.	10^3
		ϵ_{22} = 7.7	r.t.	10^3
$KSr_2Nb_5O_{15}$	Potassium strontium niobate	ϵ_{11} = ϵ_{11} ≈ 1200	298	
		ϵ_{33} ≈ 800	298	
$KTaNbO_3$	Potassium tantalate niobate (KTN)	34,000	273	10^4
		6,000	293	10^4
$KTaO_3$	Potassium tantalate	242	298	2 × 10^5
$LaScO_3$	Lanthanum scandate	30	r.t.	
LiBr	Lithium bromide	12.1	r.t.	2 × 10^6
Li_2CO_3	Lithium carbonate	4.9	291	2 × 10^5

PERMITTIVITY (DIELECTRIC CONSTANT) OF INORGANIC SOLIDS (continued)

Formula	Name	ϵ_{ijk}	*T*/K	ν/Hz
LiCl	Lithium chloride	11.05	r.t.	2×10^6
LiD	Lithium deuteride	14.0 ± 0.5	r.t.	i.r.
LiF	Lithium fluoride	9.00	298	10^2—10^7
		9.11	353	10^2—10^7
$LiGaO_2$	Lithium metagallate	$\epsilon^T_{11} = 7.0$, $\epsilon^T_{22} = 6.0$	r.t.	
		$\epsilon^T_{33} = 9.5$	r.t.	
		$\epsilon^S_{11} = 6.8$, $\epsilon^S_{22} = 5.8$	r.t.	
Li^6H	Lithium-6 hydride	13.2 ± 0.5	r.t.	
Li^7H	Lithium-7 hydride	12.9 ± 0.5	r.t.	
$LiH_3(SeO_3)_2$	Lithium trihydrogen selenite	29	298	10^4
		$\epsilon_{11} = 13.0$	r.t.	
		$\epsilon_{22} = 12.9$	r.t.	
		$\epsilon_{33} = 46$	r.t.	
LiI	Lithium iodide	11.03	r.t.	2×10^6
$LiIO_3$	Lithium iodate	$\epsilon_{11} = \epsilon_{22} = 65$	294.5	10^3
		$\epsilon_{33} = 554$	298	
$LiNH_4C_4H_4O_6 \cdot H_2O$	Lithium ammonium tartrate (LAT)	$\epsilon^T_{11} = 7.2$	298	
		$\epsilon^T_{22} = 8.0$	298	
		$\epsilon^T_{33} = 6.9$	298	
$LiNa_3CrO_4 \cdot 6H_2O$	Lithium trisodium chromate	8.0	r.t.	10^3
$LiNa_3MoO_4 \cdot 6H_2O$	Lithium trisodium molybdate	$\epsilon_{11} = 6.7$	r.t.	10^3
		$\epsilon_{33} = 5.3$	r.t.	10^3
$LiNbO_3$	Lithium niobate	$\epsilon_{11} = \epsilon_{22} = 82$	298	10^5
		$\epsilon_{33} = 30$	298	10^5
$Li_2SO_4 \cdot H_2O$	Lithium sulfate monohydrate	$\epsilon_{11} = 5.6$	298	
		$\epsilon_{22} = 10.3$	298	
		$\epsilon_{33} = 6.5$	298	
		$\epsilon_{13} = 0.07$	298	
$LiTaO_3$	Lithium tantalate	$\epsilon_{11} = \epsilon_{22} = 53$	r.t.	10^5
		$\epsilon_{33} = 46$	r.t.	10^5
		$\epsilon^S_{11} = \epsilon^S_{22} = 41$	r.t.	
		$\epsilon^S_{33} = 43$	r.t.	
		$\epsilon^T_{11} = \epsilon^T_{22} = 51$	r.t.	
		$\epsilon^T_{33} = 45$	r.t.	
$LiTlC_4H_4O_6 \cdot H_2O$	Lithium thallium tartrate (LTT)	$\epsilon_{11} \approx 20$	80	
$Mg_3B_7O_{13}Cl$	Magnesium borate monochloride (boracite)	$\epsilon_{11} = 14.1$	r.t.	5×10^5
$MgCO_3$	Magnesium carbonate	8.1	291	2×10^5
$MgNb_2O_6$	Magnesium niobate	$\epsilon_{11} = 16.4 \pm 0.5$	r.t.	$(5—500) \times 10^3$
		$\epsilon_{22} = 20.9 \pm 0.5$	r.t.	$(5—500) \times 10^3$
		$\epsilon_{33} = 32.4 \pm 0.5$	r.t.	$(5—500) \times 10^3$
MgO	Magnesium oxide (Periclase)	9.65	298	10^2—10^8
$(MgO)_xAl_2O_3$	Spinel	8.6	r.t.	
$MgSO_4$	Magnesium sulfate	8.2	r.t.	
$MgSO_4 \cdot 7H_2O$	Magnesium sulfate septa hydrate	5.46	r.t.	
$MgTiO_3$	Magnesium titanate	13.5	r.t.	
$MgWO_4$	Magnesium tungstate	$\epsilon_{11} = 18.0 \pm 1$	r.t.	$(5—500) \times 10^3$
		$\epsilon_{22} = 18.0 \pm 1$	r.t.	$(5—500) \times 10^3$
$MnNb_2O_6$	Manganese niobate	$\epsilon_{11} = 17.4 \pm 2$	r.t.	$(5—500) \times 10^3$
		$\epsilon_{22} = 16.1 \pm 0.5$	r.t.	$(5—500) \times 10^3$
		$\epsilon_{33} = 30.7 \pm 1$	r.t.	$(5—500) \times 10^3$
MnO	Manganese oxide (Pyrolusite)	12.8	r.t.	6×10^{10}
MnO_2	Manganese dioxide	$\sim 10^4$	298	10^4
Mn_2O_3	Manganese sesquioxide	8	r.t.	6×10^{10}
$MnWO_4$	Manganese tungstate	$\epsilon_{11} = 19.3 \pm 1.3$	r.t.	$(5—500) \times 10^3$
		$\epsilon_{22} = 14.3 \pm 0.5$	r.t.	$(5—500) \times 10^3$
		$\epsilon_{33} = 16.5 \pm 1.1$	r.t.	$(5—500) \times 10^3$
$N(CH_3)_4HgBr_3$	Tetramethylammonium tribromo mercurate (TTM)	~ 10	233—373	

Formula	Name	ϵ_{ijk}	T/K	ν/Hz
$N(CH_3)_4HgI_3$	Tetramethylammonium triiodo mercurate (TTM)	~10	233—373	
$N_4(CH_2)_6$	Hexamethylene tetramine (HMTA)	2.6 ± 0.2	r.t.	10^9—10^{10}
$(ND_4)_2BeF_4$	Deuteroammonium fluoberyllate	ϵ_{11} = 10	r.t.	
		ϵ_{22} = 9	r.t.	
		ϵ_{33} = 9	r.t.	
$(ND_4)_2SO_4$	Deuteroammonium sulfate	ϵ_{11} = 9	r.t.	
		ϵ_{22} = 10	r.t.	
		ϵ_{33} = 9	r.t.	
$(NH_2 \cdot CH_2COOH)_3 \cdot H_2SO_4$	Triglycine sulfate (TGS)	ϵ_{11} = 9	273	10^4
		ϵ_{22} = 30	273	10^4
		ϵ_{33} = 6.5	273	10^4
$(NH_2 \cdot CH_2COOH)_3 \cdot H_2SeO_4$	Triglycine selenate (TGSe)	200	293	1.6×10^3
$(NH_2 \cdot CH_2\ COOH)_3 \cdot H_2BeF_4$	Triglycine fluorberyllate (TGFB)	ϵ_{22} = 12	273	10^4
$NH_4Al(SO_4)_2 \cdot 12H_2O$	Ammonium alum	6	r.t.	10^{12}
$(NH_4)_2BeF_4$	Ammonium fluorberyllate	ϵ_{11} = ϵ_{22} = 7.8	123	10^5
		ϵ_{33} = 7.1	123	10^5
		ϵ_{11} = ϵ_{22} = 8.8	293	10^5
		ϵ_{33} = 9.2	293	10^5
NH_4Br	Ammonium bromide	7.1	r.t.	7×10^5
NH_4I	Ammonium iodide	9.8	r.t.	
$(NH_4)_2C_2H_6O_6$	Ammonium tartrate	ϵ_{11} = 6.45	r.t.	10^3
		ϵ_{22} = 6.8	r.t.	10^3
		ϵ_{33} = 6.0	r.t.	10^3
$(NH_4)_2Cd_2(SO_4)_3$	Ammonium cadmium sulfate	10.0	r.t.	10^4
NH_4Cl	Ammonium chloride	6.9	r.t.	7×10^5
$NH_4(ClCH_2COO)$	Ammonium monochloroacetate	5	r.t.	2×10^6
$NH_4Cr(SO_4)_2 \cdot 12H_2O$	Ammonium chrome alum	6.5	r.t.	175×10^3
NH_4HSO_4	Ammonium bisulfate	165	273	5×10^4
$NH_4H_2AsO_4$	Ammonium dihydrogen arsenate (ADA)	5.1	265	9.5×10^9
		ϵ_{11} = ϵ_{22} = 85	298	10^3
		ϵ_{33} = 22	298	
$NH_4H_2PO_4$	Ammonium dihydrogen phosphate (ADP)	ϵ_{11} = ϵ_{22} = 57.1 ± 0.6	294.5	10^5—35×10^9
		ϵ_{33} = 14.0 ± 0.3	294	10^5—36×10^9
$ND_4D_2PO_4$	Ammonium dideuterium phosphate (ADDP)	ϵ_{11} = ϵ_{22} = 74, ϵ_{33} = 24	300	
NH_4NO_3	Ammonium nitrate	10.7	322	(5—50) × 10^3
$(NH_4)_2SO_4$	Ammonium sulfate	ϵ_{11} = ϵ_{22} = 8.0	123	10^5
		ϵ_{33} = 6.3	123	10^5
		ϵ_{11} = ϵ_{22} = 10.0	293	10^5
		ϵ_{33} = 9.3	293	10^5
$(NH_4)_2UO_2(C_2O_4)_2$	Ammonium uranyl oxalate	8.03	r.t.	10^4—3.3×10^9
$(NH_4)_2UO_2(C_2O_4)_2 \cdot 3H_2O$	Ammonium uranyl oxalate trihydrate	6.06	r.t.	10^4—3.3×10^9
NaBr	Sodium bromide	6.44	298	1.6×10^3
$NaBrO_3$	Sodium bromate	ϵ^T_{11} = 5.70	298	10^3
NaCN	Sodium cyanide	7.55	293	10^5
$NaCO_3$	Sodium carbonate	8.75	291	2×10^5
$NaCO_3 \cdot 10H_2O$	Sodium carbonate decahydrate	5.3	r.t.	6×10^7
NaCl	Sodium chloride	5.9	298	10^2—10^7
		5.45	4.2	
$NaClO_3$	Sodium chlorate	ϵ^T_{11} = 5.76	301	10^3
		5.28	r.t.	10^3
$NaClO_4$	Sodium perchlorate	5.76	r.t.	10^3
NaF	Sodium fluoride	5.08 ± 0.02	r.t.	5×10^3
$NaH_3(SeO_3)_2$	Sodium trihydrogen selenite	ϵ_{11} ≈ 75	273	2×10^5
$NaD_3(SeO_3)_2$	Sodium trideuterium selenite	ϵ_{11} ≈ 220	273	2×10^5
NaI	Sodium iodide	7.28 ± 0.03	r.t.	
$NaK(C_4H_2D_2O_6) \cdot 4D_2O$	Sodium potassium tartrate tetradeutrate (double deuterated Rochelle salt)	ϵ_{11} = 70	273	10^3
		ϵ_{22} = 8.9	273	10^3

Formula	Name	ϵ_{ijk}	*T*/K	*ν*/Hz
$NaK(C_4H_4O_6) \cdot 4H_2O$	Sodium potassium tartrate tetrahydrate	$\epsilon_{11} = 170$	273	10^3
	(Rochelle salt)	$\epsilon_{22} = 9.1$	273	10^3
$NaNH_4(C_4H_4O_6) \cdot 4H_2O$	Sodium ammonium tartrate	$\epsilon_{11} = 8.4$	298	
	(Ammonium Rochelle salt)	$\epsilon_{22} = 9.2$	298	
		$\epsilon_{33} = 9.5$	298	
$NaNbO_3$	Sodium niobate	$\epsilon_{33} = 670 \pm 13$	r.t.	
		$\epsilon_{11} = \epsilon_{22} = 76 \pm 2$	r.t.	
$NaNO_2$	Sodium nitrite	$\epsilon_{11} = 7.4$	r.t.	5×10^5
		$\epsilon_{22} = 5.5$	r.t.	5×10^5
		$\epsilon_{33} = 5.0$	r.t.	5×10^5
$NaNO_3$	Sodium nitrate	6.85	292	2×10^5
$NaSO_4$	Sodium sulfate	7.90	r.t.	
$NaSO_4 \cdot 10H_2O$	Sodium sulfate decahydrate	5.0	r.t.	
$Na_2S_2O_3 \cdot 5H_2O$	Sodium sulfate pentahydrate	7	250—290	$300—10^4$
$Na_2UO_2(C_2O_4)_2$	Sodium uranyl oxalate	5.18	r.t.	
$NdAlO_3$	Neodymium aluminate	17.5	r.t.	
$NdScO_3$	Neodymium scandate	27	r.t.	
$Ni_3B_7O_{13}I$	Nickel iodine boracite	$\epsilon_{11} = 14$	260	
$NiNb_2O_6$	Nickel niobate	$\epsilon_{11} = 16.0 \pm 0.5$	r.t.	$(5—500) \times 10^3$
		$\epsilon_{22} = 23.8 \pm 1.8$	r.t.	$(5—500) \times 10^3$
		$\epsilon_{33} = 31.3 \pm 2.5$	r.t.	$(5—500) \times 10^3$
NiO	Nickel oxide	11.9	298	10^5
$NiSO_4 \cdot 6H_2O$	Nickel sulfate hexahydrate	$\epsilon_{11} = 6.2$	r.t.	
		$\epsilon_{33} = 6.8$	r.t.	
$NiWO_4$	Nickel tungstate	$\epsilon_{11} = 17.4 \pm 2.4$	r.t.	$(5—500) \times 10^3$
		$\epsilon_{22} = 13.6 \pm 1.0$	r.t.	$(5—500) \times 10^3$
		$\epsilon_{33} = 19.7 \pm 0.6$	r.t.	$(5—500) \times 10^3$
P	Phosphorous (red)	4.1	r.t.	10^8
	Phosphorous (yellow)	3.6	r.t.	10^8
$[P(CH_3)_4]HgBr_3$	Tetramethylphosphonium tribromo mercurate (TTM)	~10	233—373	
$PbBr_2$	Lead bromide	>30	293	$(0.5—3) \times 10^6$
$PbCO_3$	Lead carbonate	18.6	288	10.8
$Pb(C_2H_3O_2)_2$	Lead acetate	2.6	290—295	10^6
$PbCl_2$	Lead chloride	33.5	273	$(0.5—3) \times 10^6$
Pb_2CoWO_6	Lead cobalt tungstate	~250	r.t.	
PbF_2	Lead fluoride	26.3	r.t.	
$PbHfO_3$	Lead hafnate	390	300	10^5
		185	400	
PbI_2	Lead iodide	20.8	293	$(0.5—3) \times 10^6$
$Pb_3MgNb_2O_9$	Lead magnesium niobate	10,000	297	
$PbMoO_4$	Lead molybdate	$\epsilon_{11} = 34.0 \pm 0.4$	297.5	1.6×10^3
		$\epsilon_{33} = 40.6 \pm 0.2$	297.5	1.6×10^3
$Pb(NO_3)_2$	Lead nitrate	16.8	r.t.	$(0.5—3) \times 10^6$
$PbNb_2O_6$	Lead niobate	$\epsilon_{33}^T = 180$	298	
PbO	Lead oxide	25.9	r.t.	2×10^6
PbS	Lead sulfide (Galena)	190	77	i.r.
		200 ± 35	r.t.	i.r.
$PbSO_4$	Lead sulfate	14.3	290—295	10^6
$PbSe$	Lead selenide	280	r.t.	i.r.
$PbTa_2O_6$	Lead metatantalate	$\epsilon_{11} = \epsilon_{22} \approx 300$	r.t.	10^4
		$\epsilon_{33} = 150$	r.t.	10^4
$PbTe$	Lead telluride	450	r.t.	i.r.
		40	77	$10^4—15 \times 10^4$
		430	4.2	$10^4—15 \times 10^4$
$PbTiO_3$	Lead titanate	~200	r.t.	10^3
$PbWO_4$	Lead tungstate	$\epsilon_{11} = \epsilon_{22} = 23.6 \pm 0.3$	297.5	1.59×10^3
		$\epsilon_{33} = 31.0 \pm 0.4$	297.5	1.59×10^3
$Pb(Zn_{1/3}Nb_{2/3})O_3$	Lead zinc niobate	7	300	10^3, 300×10^3
$PbZrO_3$	Lead zirconate	200	400	

PERMITTIVITY (DIELECTRIC CONSTANT) OF INORGANIC SOLIDS (continued)

Formula	Name	ϵ_{ijk}	*T*/K	*ν*/Hz
$RbAl(SO_4)_2 \cdot 12H_2O$	Rubidium alum	5.1	r.t.	10^{12}
$RbBr$	Rubidium bromide	4.83	300	
Rb_2CO_3	Rubidium carbonate	4.87 ± 0.02	r.t.	5×10^3
$RbCl$	Rubidium chloride	4.91 ± 0.02	r.t.	5×10^3
$RbCr(SO_4)_2 \cdot 12H_2O$	Rubidium chrome alum	5.0	r.t.	10^{12}
RbF	Rubidium fluoride	5.91	r.t.	2×10^6
$RbHSO_4$	Rubidium bisulfate	$\epsilon_{11} = 7$	r.t.	10^5
		$\epsilon_{22} = 8$	r.t.	10^5
		$\epsilon_{33} = 10$	r.t.	10^5
RbH_2AsO_4	Rubidium dihydrogen arsenate (RDA)	3.90	273	9.5×10^9
RbH_2PO_4	Rubidium dihydrogen phosphate (RDP)	6.15	285	9.5×10^9
RbI	Rubidium iodide	4.94 ± 0.02	r.t.	5×10^3
$RbInSO_4$	Rubidium indium sulfate	6.85	r.t.	
$RbNO_3$	Rubidium nitrate	20—380	433—488	10^6
		30	488—538	10^6
S	Sulfur	$\epsilon_{11} = 3.75$	298	10^2—10^3
		$\epsilon_{22} = 3.95$	298	10^2—10^3
		$\epsilon_{33} = 4.44$	298	10^2—10^3
	sublimed	3.69	298	10^2—10^3
$SC(NH_2)_2$	Thiourea	$\epsilon_{11} = \epsilon_{33} \approx 3$	77—300	10^3
		$\epsilon_{22} = 35$	300	10^3
Sb_2O_3	Antimonous sesquioxide	12.8	r.t.	$(1.5—2) \times 10^3$
Sb_2S_3	Antimonous sulfide (stibnite)	$\epsilon_{11} = \epsilon_{22} = 15$	r.t.	10^3
		$\epsilon_{33} = 180$	r.t.	10^3
Sb_2Se_3	Antimonous selenide	~110	r.t.	$(10—16.5) \times 10^9$
$SbSI$	Antimonous sulfide iodide	2000	273	10^5
		$\epsilon_{11} = \epsilon_{22} \approx 25$	r.t.	10^3—10^5
		$\epsilon_{33} \approx 5 \times 10^4$	295	10^3—10^5
Se	Selenium (monocrystal)	$\epsilon_{11} = \epsilon_{22} = 11$	300	24×10^9
		$\epsilon_{33} = 21$	300	24×10^9
	(amorphous)	6.0	298	10^2—10^{10}
Si	Silicon	12.1	4.2	10^7—10^9
SiC	Silicon carbide			
	cubic	9.72	r.t.	i.r.
	6H	$\epsilon_{11} = \epsilon_{22} = 9.66$	r.t.	i.r.
		$\epsilon_{33} = 10.03$	r.t.	i.r.
		9.7 ± 0.1	1.8	i.r.
Si_3N_4	Silicon nitride	4.2 (film)	r.t.	10^3
SiO	Silicon monoxide	5.8	r.t.	10^3
SiO_2	Silicon dioxide	$\epsilon_{11} = 4.42$	r.t.	9.4×10^{10}
		$\epsilon_{22} = 4.41$	r.t.	9.4×10^{10}
		$\epsilon_{33} = 4.60$	r.t.	9.4×10^{10}
$Sm_2(MoO_4)_3$	Samarium molybdate	12	298	
SnO_2	Stannic dioxide	$\epsilon_{11} = \epsilon_{22} = 14 \pm 2$	r.t.	10^4—10^{10}
		$\epsilon_{33} = 9.0 \pm 0.5$	r.t.	10^4—10^{10}
$SnSb$	Tin antimonide	147	r.t.	10^4—10^6
$SnTe$	Tin telluride	1770 ± 300	r.t.	i.r.
$Sr(COOH)_2 \cdot 2H_2O$	Strontium formate dihydrate	6.1	r.t.	10^3
$SrCO_3$	Strontium carbonate	8.85	298	2×10^5
$SrCl_2$	Strontium chloride	9.19	r.t.	
$Sr_4Cl_2 \cdot 6H_2O$	Strontium chloride hexahydrate	8.52	r.t.	
SrF_2	Strontium fluoride	6.50	300	5×10^2—10^{11}
$SrMoO_4$	Strontium molybdate	$\epsilon_{11} = \epsilon_{22} = 31.7 \pm 0.2$	297.5	1.59×10^3
		$\epsilon_{33} = 41.7 \pm 0.2$	297.5	1.59×10^3
$Sr(NO_3)_2$	Strontium nitrate	5.33	292	2×10^5
$Sr_2Nb_2O_7$	Strontium niobate	$\epsilon_{11} = 75$	r.t.	10^3
		$\epsilon_{22} = 46$	r.t.	10^3
		$\epsilon_{33} = 43$	r.t.	10^3

Formula	Name	ϵ_{ijk}	T/K	ν/Hz
SrO	Strontium oxide	13.3 ± 0.3	273	2×10^6
SrS	Strontium sulfide	11.3	r.t.	7.25×10^6
$SrSO_4$	Strontium sulfate	11.5	r.t.	
$SrTiO_3$	Strontium titanate	332	298	10^3
		2080	78	10^3
$SrWO_4$	Strontium tungstate	$\epsilon_{11} = \epsilon_{22} = 25.7 \pm 0.2$	297.5	1.6×10^3
		$\epsilon_{33} = 34.1 \pm 0.2$	297.5	1.6×10^3
Ta_2O_5	Tantalum pentoxide (tantala)			
	α phase	$\epsilon_{11} = \epsilon_{22} = 30$	77	10^3
		$\epsilon_{33} = 65$	77	10^3
	β phase	24	292	10^3
$Tb(MoO_4)_3$	Terbium molybdate	11	298	
		$\epsilon_{11} = \epsilon_{22} = 33$	100—200	9.4×10^9
		$\epsilon_{33} = 53$	100—200	9.4×10^9
Te	Tellurium	$\epsilon_{11} = \epsilon_{22} = 33$	r.t.	
		$\epsilon_{33} = 54$	r.t.	
	polycrystalline	27.5	r.t.	i.r.
	monocrystalline	28.0	r.t.	i.r.
ThO_2	Thorium dioxide	18.9 ± 0.4	r.t.	3×10^5
TiO_2	Titanium dioxide (rutile)	$\epsilon_{11} = \epsilon_{22} = 86$	300	10^4—10^6
		$\epsilon_{33} = 170$	300	10^4—10^6
Ti_2O_3	Titanium sesquioxide	30	77	6×10^{10}
$TlBr$	Thallium bromide	30	293	10^3—10^7
$TlCl$	Thallous chloride	32.2 ± 0.2	293	10^3—10^5
TlI	Thallous iodide (orthorhombic)	20.7 ± 0.2	293	10^4
		37.3	193	10^7
$TlNO_3$	Thallous nitrate	16.5	293	5×10^5
$TlSO_4$	Thallous sulfate	25.5	293	5×10^5
UO_2	Uranium dioxide	24	r.t.	3×10^5
WO_3	Tungsten trioxide	300		
$YMnO_3$	Yttrium manganate	20	r.t.	2×10^7
Y_2O_3	Yttrium sesquioxide	10	r.t.	10^6
$YbMnO_3$	Ytterbium manganate	20	r.t.	2×10^7
Yb_2O_3	Ytterbium sesquioxide	5.0 (film)	r.t.	10^3
ZnO	Zinc monoxide	$\epsilon^S_{11} = 8.33$	r.t.	
		$\epsilon^S_{33} = 8.84$	r.t.	
		$\epsilon^T_{11} = 9.26$	r.t.	
		$\epsilon^T_{33} = 11.0$	r.t.	
		$\epsilon_{11} = 9.26$	r.t.	
		$\epsilon_{33} = 8.2$	r.t.	
		8.15	r.t.	i.r.
ZnS	Zinc sulfide	$\epsilon^S_{11} = 8.08 \pm 2\%$	77	10^4
		$\epsilon^S_{11} = 8.32 \pm 2\%$	298	10^4
		$\epsilon^T_{11} = 8.14 \pm 2\%$	77	10^4
		$\epsilon^T_{11} = 8.37 \pm 2\%$	298	10^4
$ZnSe$	Zinc selenide	$\epsilon^T_{11} = \epsilon^S_{11} = 9.12 \pm 2\%$	298	10^4
$ZnTe$	Zinc telluride	$\epsilon^T_{11} = \epsilon^S_{11} = 10.10 \pm 2\%$	r.t.	
$ZnWO_4$	Zinc tungstate	$\epsilon_{22} = 16.1 \pm 0.5$	r.t.	$(5—500) \times 10^3$
ZrO_2	Zirconium dioxide (zirconia)	12.5	r.t.	2×10^6

CURIE TEMPERATURE OF SELECTED FERROELECTRIC CRYSTALS

H. P. R. Frederikse

The following table lists the major ferroelectric crystals and their Curie temperatures, T_C.

REFERENCE

Young, K. F. and Frederikse, H. P. R., *J. Phys. Chem. Ref. Data*, 2, 313, 1973.

Name or acronym	Formula	T_C/K
Potassium dihydrogen phosphate group		
KDP	KH_2PO_4	123
KDA	KH_2AsO_4	97
KDDP	KD_2PO_4	213
KDDA	KD_2AsO_4	162
RDP	RbH_2PO_4	146
RDA	RbH_2AsO_4	111
RDDP	RbD_2PO_4	218
RDDA	RbD_2AsO_4	178
CDP	CsH_2PO_4	159
CDA	CsH_2AsO_4	143
CDDA	CsD_2AsO_4	212
Rochelle salt group		
Rochelle salt	$NaKC_4H_4O_6 \cdot 4H_2O$	255-297
Deuterated Rochelle salt	$NaKC_4H_2D_2O_6 \cdot 4H_2O$	251-308
Ammonium Rochelle salt	$NaNH_4C_4H_4O_6 \cdot 4H_2O$	109
LAT	$LiNH_4C_4H_4O_6 \cdot H_2O$	106
Triglycine sulfate group		
TGS	$(NH_2CH_2COOH)_3 \cdot H_2SO_4$	322
TGSe	$(NH_2CH_2COOH)_3 \cdot H_2SeO_4$	295
TGFB	$(NH_2CH_2COOH)_3 \cdot H_2BeF_4$	346
AFB	$(NH_4)_2BeF_4$	176
HADA	$HNH_4(ClCH_2COO)_2$	128
Perovskites and related compounds		
Barium titanate	$BaTiO_3$	406, 278, 193
Lead titanate	$PbTiO_3$	765
Potassium niobate	$KNbO_3$	712
Potassium tantalate niobate	$KTa_{2/3}Nb_{1/3}O_3$	241, 220, 170
Lithium niobate	$LiNBO_3$	1483
Lithium tantalate	$LiTaO_3$	891
Barium titanium niobate	$Ba_6Ti_2Nb_8O_{30}$	521
Ba Na niobate ("Bananas")	$Ba_2NaNb_5O_{15}$	833
Potassium iodate	KIO_3	485, 343, 257-263, 83
Lithium iodate	$LiIO_3$	529
Potassium nitrate	KNO_3	397
Sodium nitrate	$NaNO_3$	548
Rubidium nitrate	$RbNO_3$	437-487

Name or acronym	Formula	T_C/K
Miscellaneous compounds		
Cesium trihydrogen selenite	$CsH_3(SeO_3)_2$	143
Lithium trihydrogen selenite	$LiH_3(SeO_3)_2$	$T_C > T_{mp}$
Potassium selenate	K_2SeO_4	93
Methyl ammonium alum (MASD)	$CH_3NH_3Al(SO_4)_2◊12H_2O$	177
Ammonium cadmium sulfate	$(NH_4)_2Cd_2(SO_4)_3$	95
Ammonium bisulfate	$(NH_4)HSO_4$	271
Ammonium sulfate	$(NH_4)_2SO_4$	224
Ammonium nitrate	NH_4NO_3	398, 357, 305, 255
Colemanite	$CaB_3O_4(OH)_3◊H_2O$	266
Cadmium pyroniobite	$Cd_2Nb_2O_7$	185
Gadolinium molybdate	$Gd_2(MoO_4)_3$	432

PROPERTIES OF ANTIFERROELECTRIC CRYSTALS

H. P. R. Frederikse

Some important antiferroelectric crystals are listed here with their Curie Temperatures T_C. The last column gives the constant T_0 which appears in the Curie-Weiss law describing the dielectric constant of these materials above the Curie Temperature:

$$\varepsilon = \text{const.}/(T - T_0)$$

Name or acronym	Formula	T_C/K	T_0/K
ADP	$NH_4H_2PO_4$	148	
ADA	$NH_4H_2AsO_4$	216	
ADDP	$NH_4D_2PO_4$	242, 245	
ADDA	$NH_4D_2AsO_4$	299	
A_dDDP	$ND_4D_2PO_4$	243	
A_dDDA	$ND_4D_2AsO_4$	304	
Sodium niobate	$NaNbO_3$	911, 793	
Lead hafnate	$PbHfO_3$	476	378
Lead zirconate	$PbZrO_3$	503	475
Lead metaniobate	$PbNb_2O_6$	843	530
Lead metatantalate	$PbTa_2O_6$	543	533
Tungsten trioxide	WO_3	1010	
Potassium strontium niobate	$KSr_2Nb_5O_{15}$	427	413
Sodium nitrite	$NaNO_2$	437	437
Sodium trihydrogen selenite	$NaH_3(SeO_3)_2$	193	192
Sodium trideuterium selenite	$NaD_3(SeO_3)_2$	271	245
Ammonium trihydrogen periodate	$(NH_4)_2H_3IO_6$	245	

DIELECTRIC CONSTANTS OF GLASSES

Type	Dielectric constant at 100 MHz (20°C)	Volume resistivity (350°C megohm-cm)	Loss factor[a]
Corning 0010	6.32	10	0.015
Corning 0080	6.75	0.13	0.058
Corning 0120	6.65	100	0.012
Pyrex 1710	6.00	2,500	0.025
Pyrex 3320	4.71	—	0.019
Pyrex 7040	4.65	80	0.013
Pyrex 7050	4.77	16	0.017
Pyrex 7052	5.07	25	0.019
Pyrex 7060	4.70	13	0.018
Pyrex 7070	4.00	1,300	0.0048
Vycor 7230	3.83	—	0.0061
Pyrex 7720	4.50	16	0.014
Pyrex 7740	5.00	4	0.040
Pyrex 7750	4.28	50	0.011
Pyrex 7760	4.50	50	0.0081
Vycor 7900	3.9	130	0.0023
Vycor 7910	3.8	1,600	0.00091
Vycor 7911	3.8	4,000	0.00072
Corning 8870	9.5	5,000	0.0085
G. E. Clear (silica glass)	3.81	4,000—30,000	0.00038
Quartz (fused)	3.75 4.1 (1 MHz)	—	0.0002 (1 MHz)

[a] Power factor × dielectric constant equals loss factor.

PROPERTIES OF SUPERCONDUCTORS

L. I. Berger and B. W. Roberts

The following tables include superconductive properties of selected elements, compounds, and alloys. Individual tables are given for thin films, elements at high pressures, superconductors with high critical magnetic fields, and high critical temperature superconductors.

The historically first observed and most distinctive property of a superconductive body is the near total loss of resistance at a critical temperature (T_c) that is characteristic of each material. Figure 1(a) below illustrates schematically two types of possible transitions. The sharp vertical discontinuity in resistance is indicative of that found for a single crystal of a very pure element or one of a few well annealed alloy compositions. The broad transition, illustrated by broken lines, suggests the transition shape seen for materials that are not homogeneous and contain unusual strain distributions. Careful testing of the resistivity limit for superconductors shows that it is less than 4×10^{-23} ohm cm, while the lowest resistivity observed in metals is of the order of 10^{-13} ohm cm. If one compares the resistivity of a superconductive body to that of copper at room temperature, the superconductive body is at least 10^{17} times less resistive.

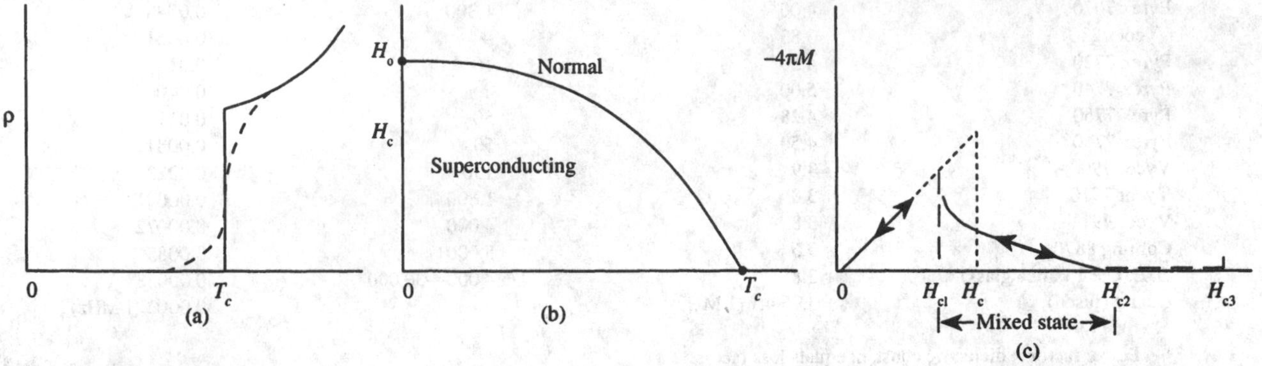

FIGURE 1. Physical properties of superconductors. (a) Resistivity vs. temperature for a pure and perfect lattice (solid line); impure and/or imperfect lattice (broken line). (b) Magnetic-field temperature dependence for Type-I or "soft" superconductors. (c) Schematic magnetization curve for "hard" or Type-II superconductors.

The temperature interval ΔT_c, over which the transition between the normal and superconductive states takes place, may be of the order of as little as 2×10^{-5} K or several K in width, depending on the material state. The narrow transition width was attained in 99.9999% pure gallium single crystals.

A Type-I superconductor below T_c, as exemplified by a pure metal, exhibits perfect diamagnetism and excludes a magnetic field up to some critical field H_c, whereupon it reverts to the normal state as shown in the *H-T* diagram of Figure 1(b).

The magnetization of a typical high-field superconductor is shown in Figure 1(c). The discovery of the large current-carrying capability of Nb_3Sn and other similar alloys has led to an extensive study of the physical properties of these alloys. In brief, a high-field superconductor, or Type-II superconductor, passes from the perfect diamagnetic state at low magnetic fields to a mixed state and finally to a sheathed state before attaining the normal resistive state of the metal. The magnetic field values separating the four stages are given as H_{c1}, H_{c2}, and H_{c3}. The superconductive state below H_{c1} is perfectly diamagnetic, identical to the state of most pure metals of the "soft" or Type-I superconductor. Between H_{c1} and H_{c2} a "mixed superconductive state" is found in which fluxons (a minimal unit of magnetic flux) create lines of normal flux in a superconductive matrix. The volume of the normal state is proportional to $-4\pi M$ in the "mixed state" region. Thus at H_{c2} the fluxon density has become so great as to drive the interior volume of the superconductive body completely normal. Between H_{c2} and H_{c3} the superconductor has a sheath of current-carrying superconductive material at the body surface, and above H_{c3} the normal state exists. With several types of careful measurement, it is possible to determine H_{c1}, H_{c2}, and H_{c3}. Table 6 contains some of the available data on high-field superconductive materials.

High-field superconductive phenomena are also related to specimen dimension and configuration. For example, the Type-I superconductor, Hg, has entirely different magnetization behavior in high magnetic fields when contained in the very fine sets of filamentary tunnels found in an unprocessed Vycor glass. The great majority of superconductive materials are Type-II. The elements in very pure form and a very few precisely stoichiometric and well annealed compounds are Type I with the possible exceptions of vanadium and niobium.

Metallurgical Aspects. The sensitivity of superconductive properties to the material state is most pronounced and has been used in a reverse sense to study and specify the detailed state of alloys. The mechanical state, the homogeneity, and the presence of impurity atoms and other electron-scattering centers are all capable of controlling the critical temperature and the current-carrying capabilities in high-magnetic fields. Well annealed specimens tend to show sharper transitions than those that are strained or inhomogeneous. This sensitivity to mechanical state underlines a general problem in the tabulation of properties for superconductive materials. The occasional divergent values of the critical temperature and of the critical fields quoted for a Type-II superconductor may lie in the variation in sample preparation. Critical temperatures of materials studied early in the history of superconductivity must be evaluated in light of the probable metallurgical state of the material, as well as the availability of less pure starting elements. It has been noted that recent work has given extended consideration to the metallurgical aspects of sample preparation.

Symbols in tables: T_c: Critical temperature; H_o: Critical magnetic field in the $T = 0$ limit; θ_D: Debye temperature; and γ: Electronic specific heat.

TABLE 1
Selective Properties of Superconductive Elements

Element	T_c(K)	H_o(oersted)	θ_D(K)	γ(mJ mol^{-1}K^{-1})
Al	1.175 ± 0.002	104.9 ± 0.3	420	1.35
Am* (α,?)	0.6			
Am* (β,?)	1.0			
Be	0.026			0.21
Cd	0.517 ± 0.002	28 ± 1	209	0.69
Ga	1.083 ± 0.001	58.3 ± 0.2	325	0.60
Ga (β)	5.9, 6.2	560		
Ga (γ)	7	950, HF[a]		
Ga (Δ)	7.85	815, HF		
Hf	0.128	12.7		2.21
Hg (α)	4.154 ± 0.001	411 ± 2	87, 71.9	1.81
Hg (β)	3.949	339	93	1.37
In	3.408 ± 0.001	281.5 ± 2	109	1.672
Ir	0.1125 ± 0.001	16 ± 0.05	425	3.19
La (α)	4.88 ± 0.02	800 ± 10	151	9.8
La (β)	6.00 ± 0.1	1096, 1600	139	11.3
Lu	0.1 ± 0.03	350 ± 50		
Mo	0.915 ± 0.005	96 ± 3	460	1.83
Nb	9.25 ± 0.02	2060 ± 50, HF	276	7.80
Os	0.66 ± 0.03	70	500	2.35
Pa	1.4			
Pb	7.196 ± 0.006	803 ± 1	96	3.1
Re	1.697 ± 0.006	200 ± 5	4.5	2.35
Ru	0.49 ± 0.015	69 ± 2	580	2.8
Sn	3.722 ± 0.001	305 ± 2	195	1.78
Ta	4.47 ± 0.04	829 ± 6	258	6.15
Tc	7.8 ± 0.1	1410, HF	411	6.28
Th	1.38 ± 0.02	1.60 ± 3	165	4.32
Ti	0.40 ± 0.04	56	415	3.3
Tl	2.38 ± 0.02	178 ± 2	78.5	1.47
U	0.2			
V	5.40 ± 0.05	1408	383	9.82
W	0.0154 ± 0.0005	1.15 ± 0.03	383	0.90
Zn	0.85 ± 0.01	54 ± 0.3	310	0.66
Zr	0.61 ± 0.15	47	290	2.77
Zr (ω)	0.65, 0.95			

TABLE 2
Range of Critical Temperatures Observed for Superconductive Elements in Thin Films Condensed Usually at Low Temperatures

Element	T_c Range (K)	Comments	Element	T_c Range (K)	Comments
Al	1.15—5.7	HF[a]	Nb	2.0—10.1	
Be	5—9.75	HF	Pb	1.8—7.5	
Bi	6.17—6.6		Re	1.7—7	
Cd			Sn	3.5—6	
(Disordered)	0.79—0.91		Ta	<1.7—4.51	HF[a]
(Ordered)	0.53—0.59		Tc	4.6—7.7	
Ga	2.5—8.5	HF	Ti	1.3 Max	
Hg	3.87—4.5		Tl	2.33—2.96	
In	2.2—5.6	HF	V	1.8—6.02	
La	3.55—6.74		W	<1.0—4.1	
Mo	3.3—8.0		Zn	0.77—1.9	

[a] HF denotes high magnetic field superconductive properties.

TABLE 3
Elements Exhibiting Superconductivity Under or After Application of High Pressure

Element	T_c Range (K)	Pressure (kbar)
Al	1.98—0.075	0—62
As	0.31—0.5	220—140
	0.2—0.25	140—100
Ba II	1—1.8	55—85
III	1.8—5	85—144
IV	4.5—5.4	144—190
Bi II	3.9	25—27
III	6.55—7.25	28—38
IV	7.0, 8.7—6.0	43, 43—62
V	6.7, 8.3	48—80
VI	8.55	90, 92—101
VII(?)	8.2	30
Ce (α)	0.020—0.045	20—35
Ce (α′)	1.9—1.3	45—125
Cs V	1.5	>125
Ga II	6.38	≥35
II′	7.5	≥35 then P removed
Ge	5.35	115
La	5.5—12.9	0—210
Lu	0.022—1.0	45—190
P	5.8	170
Pb II	3.55	160
Re II	2.3 Max.	"Plastic" compression
Sb (prepared 120 kbar, held below 77K)	2.6—2.7	
Sb II	3.55—3.40	85—150
Se II	6.75, 6.95	130
Si	6.7—7.1	120—130
Sn II	5.2—4.85	125—160
III	5.30	113
Te II	2.4—5.1	38—55
	4.1—4.2	53—62
IV	4.72—4	63—80
()	3.3—2.8	100—260
Tl (cubic form)	1.45	35
(hexagonal form)	1.95	35
U	2.4—0.4	10—85
Y	1.7—2.5	110—160
Zr (omega form, metastable)	1—1.7	60—130

TABLE 4
Superconductive Compounds and Alloys

All compositions are denoted on an atomic basis, i.e., AB, AB_2, or AB_3 for compounds, unless noted. Solid solutions or odd compositions may be denoted as A_zB_{1-z} or A_zB. A series of three or more alloys is indicated as A_xB_{1-x} or by actual indication of the atomic fraction range, such as $A_{0-0.6}B_{1-0.4}$. The critical temperature of such a series of alloys is denoted by a range of values or possibly the maximum value.

The selection of the critical temperature from a transition in the effective permeability, or the change in resistance, or possibly the incremental changes in frequency observed by certain techniques is not often obvious from the literature. Most authors choose the mid-point of such curves as the probable critical temperature of the idealized material, while others will choose the highest temperature at which a deviation from the normal state property is observed. In view of the previous discussion concerning the variability of the superconductive properties as a function of purity and other metallurgical aspects, it is recommended that appropriate literature be checked to determine the most probable critical temperature or critical field of a given alloy.

A very limited amount of data on critical fields, H_o, is available for these compounds and alloys; these values are given at the end of the table.

A. SUPERCONDUCTORS WITH T_c< 10 K

Substance	T_c, K	Crystal structure type
$Ag_{3.3}Al$	0.34	A12-cI58 (Mn)
$Ag_xAl_yZn_{1-x-y}$	0.15	Cubic
$AgBi_2$	2.87—3.0	
$Ag_7F_{0.25}N_{0.75}O_{10.25}$	0.85—0.90	
Ag_2F	0.0.066	
Ag_7FO_8	0.3	Cubic
$Ag_{0.8-0.3}Ga_{0.2-0.7}$	6.5—8	
Ag_4Ge	0.85	Hex., c.p.
$Ag_{0.438}Hg_{0.562}$	0.64	$D8_2$
$AgIn_2$	~2.4	C16

TABLE 4
Superconductive Compounds and Alloys (continued)

Substance	T_c, K	Crystal structure type
$Ag_{0.1}In_{0.9}Te$ ($n = 1.4 \times 10^{22}$)*	1.2—1.89	B1
$Ag_{0.2}In_{0.8}Te$ ($n = 1.07 \times 10^{22}$)	0.77—1.00	B1
AgLa	0.94	B2-cP2 (CsCl)
AgLa (9.5 kbar)	1.2	B2
AgLu	0.33	B2-cP2
$AgMo_4S_5$	9.1	hR15 (Mo_6PbS_8)
$Ag_{1.2}Mo_6Se_8$	5.9	Same
Ag_7NO_{11}	1.04	Cubic
Ag_xPb_{1-x}	7.2 max.	
Ag_4Sn	0.1	h**
Ag_xSn_{1-x}	1.5—3.7	
Ag_xSn_{1-x} (film)	2.0—3.8	
$AgTe_3$	2.6	Cubic
AgTh	2.2	C16-tI12 (Al_2Cu)
$AgTh_2$	2.26	C16
$Ag_{0.03}Tl_{0.97}$	2.67	
$Ag_{0.94}Tl_{0.06}$	2.32	
AgY	0.33	B2-cP2 (CsCl)
Ag_xZn_{1-x}	0.5—0.845	
$AlAu_4$	0.4—0.7	Like A13
Al_2Au	0.1	C1-cF12 (CaF_2)
Al_2CMo_3	9.8—10.2	A13+trace 2nd. phase)
Al_2CaSi	5.8	
$Al_{0.131}Cr_{0.088}V_{0.781}$	1.46	Cubic
$AlGe_2$	1.75	
Al_2Ge_2U	1.6	LI_2-cP4 (Cu_3Au)
$AlLa_3$	5.57	DO_{19}
Al_2La	3.23	C15
Al_2Lu	1.02	C15-cF24 (Cu_2Mg)
Al_3Mg_2	0.84	F.C.C.
$AlMo_3$	0.58	A15
$AlMo_6Pd$	2.1	
AlN	1.55	B4
Al_2NNb_3	1.3	A13
Al_3Nb	0.64	tI8 (Al_3Ti)
AlOs	0.39	B2
Al_3Os	5.90	
AlPb (film)	1.2—7	
Al_2Pt	0.48—0.55	C1
Al_5Re_{24}	3.35	A12
AlSb	2.8	B4-tI4 (Sn)
Al_2Sc	1.02	C15-cF24 (Cu_2Mg)
Al_2Si_2U	1.34	LI_2-cP4 (Cu_3Au)
$AlTh_2$	0.1	C16-tI12 (Al_2Cu)
Al_3Th	0.75	DO_{19}
$Al_xTi_yV_{1-x-y}$	2.05—3.62	Cubic
$Al_{0.108}V_{0.892}$	1.82	Cubic
Al_2Y	0.35	C15-cF24 (Cu_2Mg)
Al_3Yb	0.94	LI_2-cP4 (Cu_3Au)
Al_xZn_{1-x}	0.5—0.845	
$AlZr_3$	0.73	LI_2
AsBiPb	9.0	
AsBiPbSb	9.0	
AsHfOs	3.2	C22-hP9 (Fe_2P)
AsHfRu	4.9	same

TABLE 4
Superconductive Compounds and Alloys (continued)

Substance	T_c, K	Crystal structure type
$As_{0.33}InTe_{0.67}$ (n = 1.24 × 10^{22})	0.85—1.15	B1
$As_{0.5}InTe_{0.5}$ (n = 0.97 × 10^{22})	0.44—0.62	B1
As_4La_3	0.6	cI28 (Th_3P_4)
$AsNb_3$	0.3	$L1_2$-tP32
$As_{0.50}Ni_{0.06}Pd_{0.44}$	1.39	C2
$AsNi_{0.25}Pd_{0.75}$	1.6	$B8_1$-hP4 (NiAs)
AsOsZr	8.0	C22-hP9 (Fe_2P)
AsPb	8.4	
$AsPd_2$ (low-temp. phase)	0.60	Hexagonal
$AsPd_2$ (high-temp. phase)	1.70	C22
$AsPd_5$	0.46	Complex
As_3Pd_5	1.9	
AsRh	0.58	B31
$AsRh_{1.4-1.6}$	< 0.03—0.56	Hexagonal
AsSn	4.10	
AsSn (n = 2.14 × 10^{22})	3.41—3.65	B1
$As_{\sim 2}Sn_{\sim 3}$	3.5—3.6; 1.21—1.17	
As_3Sn_4 (n = 0.56 × 10^{22})	1.16—1.19	Rhombohedral
AsV_3	0.20	A15-cP8 (Cr_3Si)
Au_5Ba	0.4—0.7	$D2_d$
AuBe	2.64	B20
Au_2Bi	1.80	C15
Au_5Ca	0.34—0.38	$C15_b$
$AuGa_2$	1.6	C1-cF12 (CaF_2)
AuGa	1.2	B31
$Au_{0.40-0.92}Ge_{0.60-0.08}$	<0.32—1.63	Complex
$AuIn_2$	0.2	C1-cF12
AuIn	0.4—0.6	Complex
AuLu	<0.35	B2
$AuNb_3$	1.2	A2
$AuPb_2$	3.15	
$AuPb_2$ (film)	4.3	
$AuPb_3$	4.40	
$AuPb_3$ (film)	4.25	
Au_2Pb	1.18; 6—7	C15
$AuSb_2$	0.58	C2
AuSn	1.25	$B8_1$
Au_xSn_{1-x} (film)	2.0—3.8	
Au_5Sn	0.7—1.1	A3
$AuTa_{4.3}$	0.55	A15-cP8 (Cr_3Si)
Au_3Te_5	1.62	Cubic
$AuTh_2$	3.08	C16
AuTl	1.92	
AuV_3	0.74	A15
Au_xZn_{1-x}	0.50—0.845	
$AuZn_3$	1.21	Cubic
Au_xZr_y	1.7—2.8	A3
$AuZr_3$	0.92	A15
$B_2Ba_{0.67}Pt_3$	5.60	hP12 (B_2BaPt_3)
$BCMo_2$	5.4	Orthorhombic
$BCMo_2$	5.3—7.0	Same
$B_2Ca_{0.67}Pt_3$	1.57	hP12
B_4ErIr_4	2.1	tP18 (B_4CeCo_4)
B_4ErRh_4	4.3	oC108 (B_4LuRh_4)
B_4ErRh_4	8.7	tP18 (B_4CeCo_4)

TABLE 4
Superconductive Compounds and Alloys (continued)

Substance	T_c, K	Crystal structure type
BHf	3.1	Cubic
B_4HoIr_4	2.0	tP18
B_4HoRh_4	1.4	oC108
B_2Ir_3La	1.65	hP6 ($CaCu_5$)
B_2Ir_3Th	2.09	Same
B_4Ir_4Tm	1.6	tP18
B_6La	5.7	
B_2LaRh_3	2.82	hP6
$B_{12}Lu$	0.48	
B_2LuOs	2.66	oP16 (B_2LuRu)
B_2LuOs_3	4.62	hP6
B_4LuRh_4	6.2	oC108
B_2LuRu	9.86	oP16
B_4LuRu_4	2.0	tI72 (B_4LuRu_4)
BMo	0.5 (extrapol.)	
BMo_2	4.74	C16
BNb	8.25	B_f
B_4NdRh_4	5.3	tP18
B_2OsSc	1.34	oP16
B_2OsY	2.22	oP16
$B_2Pt_3Sr_{0.67}$	2.78	hP12 (B_2BaPt_3)
BRe_2	2.80; 4.6	
$B_4Rh_{3.4}Ru_{0.6}$	8.38	tI72
B_4Rh_4Sm	2.7	tP18
B_4Rh_4Th	4.3	Same
B_4Rh_4Tm	9.8	Same
B_4Rh_4Tm	5.4	oC108
$B_{0.3}Ru_{0.7}$	2.58	$D10_2$
B_4Ru_4Sc	7.2	tI72
B_2Ru_3Th	1.79	hP6
B_2Ru_3Y	2.85	Same
B_2Ru Y	7.80	oP16
B_4Ru_4Y	1.4	tI72
$B_{12}Sc$	0.39	
BTa	4.0	B_f
BTa_2	3.12	C16-tI12 (Al_2Cu)
B_6Th	0.74	
BW_2	3.1	C16
B_6Y	6.5—7.1	
$B_{12}Y$	4.7	
BZr	3.4	Cubic
$B_{12}Zr$	5.82	
$BaBi_3$	5.69	Tetragonal
$Ba_2Mo_{15}Se_{19}$	2.75	hP15 (Mo_6PbS_8)
$Ba_xO_3Sr_{1-x}Ti$ ($n = 4.2 \times 10^{19}$)	<0.1—0.55	
$Ba_{0.13}O_3W$	1.9	Tetragonal
$Ba_{0.14}O_3W$	<1.25—2.2	Hexagonal
$BaRh_2$	6.0	C15
$Be_{22}Mo$	2.51	Cubic ($Be_{22}Re$)
$Be_8Nb_5Zr_2$	5.2	
$Be_{0.98—0.92}Re_{0.02—0.08}$ (quenched)	9.5—9.75	Cubic
$Be_{0.957}Re_{0.043}$	9.62	Cubic ($Be_{22}Re$)
BeTc	5.21	Cubic
$Be_{22}W$	4.12	Cubic ($Be_{22}Re$)

TABLE 4
Superconductive Compounds and Alloys (continued)

Substance	T_c, K	Crystal structure type
$Be_{13}W$	4.1	Tetragonal
Bi_3Ca	2.0	
$Bi_{0.5}Cd_{0.13}Pb_{0.25}Sn_{0.12}$ (weight fractions)	8.2	
BiCo	0.42—0.49	
Bi_2Cs	4.75	C15
Bi_xCu_{1-x} (electrodeposited)	2.2	
BiCu	1.33—1.40	
Bi_3Fe	1.0	m**
$Bi_{0.019}In_{0.981}$	3.86	
$Bi_{0.05}In_{0.95}$	4.65	α-phase
$Bi_{0.10}In_{0.90}$	5.05	Same
$Bi_{0.15—0.30}In_{0.85—0.70}$	5.3—5.4	α- and β-phases
$Bi_{0.34—0.48}In_{0.66—0.52}$	4.0—4.1	
Bi_3In_5	4.1	
$BiIn_2$	5.65	β-phase
Bi_2Ir	1.7—2.3	
Bi_2Ir (quenched)	3.0—3.96	
BiK	3.6	
Bi_2K	3.58	C15
BiLi	2.47	$L1_o$, α-phase
$Bi_{4—9}Mg$	0.7—~1.0	
Bi_3Mo	3—3.7	
BiNa	2.25	$L1_o$
$BiNb_3$	4.5	A15-cP8 (Cr_3Si)
$BiNb_3$ (high pressure and temperature)	3.05	A15
BiNi	4.25	$B8_1$
Bi_3Ni	4.06	Orthorhombic
$BiNi_{0.5}Rh_{0.5}$	3.0	$B8_1$-hP4 (AsNi)
$Bi_{0.5}NiSb_{0.5}$	2.0	Same
$Bi_{1-0}Pb_{0-1}$	7.26—9.14	
$Bi_{1-0}Pb_{0-1}$ (film)	7.25—8.67	
$Bi_{0.05—0.40}Pb_{0.95—0.60}$	7.35—8.4	H.C.P. to ε-phase
Bi_2Pb	4.25	t**
BiPbSb	8.9	
$Bi_{0.5}Pb_{0.31}Sn_{0.19}$ (weight fractions)	8.5	
$Bi_{0.5}Pb_{0.25}Sn_{0.25}$	8.5	
$BiPd_2$	4.0	
$Bi_{0.4}Pd_{0.6}$	3.7—4	Hexagonal, ordered
BiPd	3.7	Orthorhombic
Bi_2Pd	1.70	Monoclinic, α-phase
Bi_2Pd	4.25	Tetragonal, β-phase
$BiPd_{0.45}Pt_{0.55}$	3.7	$B8_1$-hP4 (NiAs)
BiPdSe	1.0	C2
BiPdTe	1.2	C2
BiPt	1.21	$B8_1$
$Bi_{0.1}PtSb_{0.9}$	2.05; 1.5	$B8_1$-hP4 (NiAs)
BiPtSe	1.45	C2
BiPtTe	1.15	C2
Bi_2Pt	0.155	Hexagonal
Bi_2Rb	4.25	C15
$BiRe_2$	1.9—2.2	
BiRh	2.06	$B8_1$

TABLE 4
Superconductive Compounds and Alloys (continued)

Substance	T_c, K	Crystal structure type
Bi_3Rh	3.2	Orthorhombic (NiB_3)
Bi_4Rh	2.7	Hexagonal
BiRu	5.7	m**
Bi_3Sn	3.6—3.8	
BiSn	3.8	
Bi_xSn_y	3.85—4.18	
Bi_3Sr	5.62	$L1_2$
Bi_3Te	0.75—1.0	
Bi_5Tl_3	6.4	
$Bi_{0.26}Tl_{0.74}$	4.4	Cubic, disordered
$Bi_{0.26}Tl_{0.74}$	4.15	$L1_2$, ordered (?)
Bi_2Y_3	2.25	
Bi_3Zn	0.8—0.9	
$Bi_{0.3}Zr_{0.7}$	1.51	
$BiZr_3$	2.4—2.8	
$BrMo_6Se_7$	7.1	hP15 (Mo_6PbS_8)
$Br_3Mo_6Se_5$	7.1	Same
CCs_x	0.020—0.135	Hexagonal
CFe_3	1.30	DO_{11}-oP16 (Fe_3C)
$CGaMo_2$	3.7—4.1	Hexagonal
$CHf_{0.5}Mo_{0.5}$	3.4	B1
$CHf_{0.3}Mo_{0.7}$	5.5	B1
$CHf_{0.25}Mo_{0.75}$	6.6	B1
$CHf_{0.7}Nb_{0.3}$	6.1	B1
$CHf_{0.6}Nb_{0.4}$	4.5	B1
$CHf_{0.5}Nb_{0.5}$	4.8	B1
$CHf_{0.4}Nb_{0.6}$	5.6	B1
$CHf_{0.25}Nb_{0.75}$	7.0	B1
$CHf_{0.2}Nb_{0.8}$	7.8	B1
$CHf_{0.9-0.1}Ta_{0.1-0.9}$	5.0—9.0	B1
CK (excess K)	0.55	Hexagonal
C_8K	0.39	Hexagonal
C_2La	1.66	tI6 (CaC_2)
C_2Lu	3.33	Same
$C_{0.40-0.44}Mo_{0.60-0.56}$	9—13	
C_3MoRe	3.8	B1-cF8
$C_{0.6}Mo_{4.8}Si_3$	7.6	$D8_8$
$CMo_{0.2}Ta_{0.8}$	7.5	B1
$CMo_{0.5}Ta_{0.5}$	7.7	B1
$CMo_{0.75}Ta_{0.25}$	8.5	B1
$CMo_{0.8}Ta_{0.2}$	8.7	B1
$CMo_{0.85}Ta_{0.15}$	8.9	B1
CMo_xV_{1-x}	2.9—9.3	B1
CMo_xZr_{1-x}	9.8	B1
$C_{0.984}Nb$	9.8	B1
CNb_2	9.1	
CNb_xTi_{1-x}	<4.2—8.8	B1
$CNb_{0.1-0.9}Zr_{0.9-0.1}$	4.2—8.4	B1
CRb_x (Au)	0.023—0.151	Hexagonal
$CRe_{0.06}W$	5.0	
CRu	2.00	hP2 (CW)
$C_{0.98}7Ta$	9.7	
$C_{0.848-0.987}$	2.04—9.7	
CTa (film)	5.09	B1
CTa_2	3.26	L'_3

TABLE 4
Superconductive Compounds and Alloys (continued)

Substance	T_c, K	Crystal structure type
$CTa_{0.4}Ti_{0.6}$	4.8	B1
$Cta_{1—0.4}W_{0—0.6}$	8.5—10.5	B1
$CTa_{0.2—0.9}Zr_{0.8—0.1}$	4.6—8.3	B1
CTc (excess C)	3.85	Cubic
$CTi_{0.5—0.7}W_{0.5—0.3}$	6.7—2.1	B1
CW	1.0	
CW_2	2.74	L'_3
CW_2	5.2	F.C.C.
C_2Y	3.88	tI6 (CaC_2)
$Ca_3Co_4Sn_{13}$	5.9	cP40 ($Pr_3Rh_2Sn_{13}$)
$Ca_3Ge_{13}Rh_4$	2.1	Same
CaHg	1.6	B2-cP2 (CsCl)
$CaHg_3$	1.6	hP8 (Ni_3Sn)
$CaIr_2$	6.15	C15
$Ca_3Ir_4Sn_{13}$	7.1	cP40
$Ca_xO_3Sr_{1-x}Ti$ ($n = 3.7—11 \times 10^{19}$)	<0.1—0.55	
$Ca_{0.1}O_3W$	1.4—3.4	Hexagonal
CaPb	7.0	
$CaRh_2$	6.40	C15
$CaRh_{1.2}Sn_{4.5}$	8.7	cP40
$CaTl_3$	2.0	B2-cP2
$Cd_{0.3—0.5}Hg_{0.7—0.5}$	1.70—1.92	
CdHg	1.77; 2.15	Tetragonal
$Cd_{0.0075—0.05}In_{0.9925—0.95}$	3.24—3.36	Tetragonal
$Cd_{0.97}Pb_{0.03}$	4.2	
CdSn	3.65	
$Cd_{0.17}Tl_{0.83}$	2.3	
$Cd_{0.18}Tl_{0.82}$	2.54	
$CeCo_2$	0.84	C15
$CeCo_{1.67}Ni_{0.33}$	0.46	C15
$CeCo_{1.67}Rh_{0.33}$	0.47	C15
$Ce_xGd_{1-x}Ru_2$	3.2—5.2	C15
$CeIr_3$	3.34	
$CeIr_5$	1.82	
$Ce_{0.005}La_{0.995}$	4.6	
Ce_xLa_{1-x}	1.3—6.3	
$Ce_xPr_{1-x}Ru_2$	1.4—5.3	C15
Ce_xPt_{1-x}	0.7—1.55	
$CeRu_2$	6.0	C15
$Ce_3Mo_6Se_5$	5.7	hR15 (Mo_6PbS_8)
$Ce_2Mo_6Te_6$	1.7	Same
$Co_xFe_{1-x}Si_2$	1.4 (max.)	C1
$CoHf_2$	0.56	$E9_3$
$CoLa_3$	4.28	
$Co_4La_3Sn_{13}$	2.8	cP40
$CoLu_3$	~0.35	
Co_xLuSn_y	1.5	cP40
$Co_{0—0.01}Mo_{0.8}Re_{0.2}$	2—10	
$Co_{0.02—0.10}Nb_3Rh_{0.98—0.90}$	2.28—1.90	A15
$Co_xNi_{1-x}Si_2$	1.4 (max.)	C1
$Co_{0.5}Rh_{0.5}Si_2$	2.5	
$Co_xRh_{1-x}Si_2$	3.65 (max.)	
$Co_{\sim0.3}So_{\sim0.7}$	~0.35	
$Co_4Sc_5Si_{10}$	5.0	tP38 ($Co_4Sc_5Si_{10}$)
$CoSi_2$	1.40; 1.22	C1

TABLE 4
Superconductive Compounds and Alloys (continued)

Substance	T_c, K	Crystal structure type
Co_xSn_yYb	2.5	cP40
Co_3Th_7	1.83	$D10_2$
Co_xTi_{1-x}	2.8 (max.)	Co in α-Ti
Co_xTi_{1-x}	3.8 (max.)	Co in β-Ti
$CoTi_2$	3.44	$E9_3$
CoTi	0.71	A2
CoU	1.7	B2, distorted
CoU_6	2.29	$D2_c$
$Co_{0.28}Y_{0.72}$	0.34	
CoY_3	<0.34	
$CoZr_2$	6.3	C16
$Co_{0.1}Zr_{0.9}$	3.9	A3
$Cr_{0.6}Ir_{0.4}$	0.4	H.C.P.
$Cr_{0.65}Ir_{0.35}$	0.59	H.C.P.
$Cr_{0.7}Ir_{0.3}$	0.76	H.C.P.
$Cr_{0.72}Ir_{0.28}$	0.83	
Cr_3Ir	0.45	A15
$Cr_{0-0.1}Nb_{1-0.9}$	4.6—9.2	A2
$Cr_{0.80}Os_{0.20}$	2.5	Cubic
Cr_3Os	4.68	A15-cP8 (Cr_3Si)
Cr_xRe_{1-x}	1.2—5.2	
$Cr_{0.4}Re_{0.6}$	2.15	$D8_b$
$Cr_{0.8-0.6}Rh_{0.2-0.4}$	0.5—1.10	A3
Cr_3Rh	0.3	A15-cP8
Cr_3Ru (annealed)	3.3	A15
Cr_2Ru	2.02	$D8_b$
Cr_3Ru_2	2.10	$D8_b$-tP30 (CrFe)
$Cr_{0.1-0.5}Ru_{0.9-0.5}$	0.34—1.65	A3
Cr_xTi_{1-x}	3.6 (max.)	Cr in α-Ti
Cr_xTi_{1-x}	4.2 (max.)	Cr in β-Ti
$Cr_{0.1}Ti_{0.3}V_{0.6}$	5.6	
$Cr_{0.0175}U_{0.9825}$	0.75	β-phase
$Cs_{0.32}O_3W$	1.12	Hexagonal
$Cu_{0.15}In_{0.85}$ (film)	3.75	
$Cu_{0.04-0.08}In_{0.94-0.92}$	4.4	
CuLa	5.85	
$Cu_2Mo_6O_2S_6$	9	hR15 (Mo_6PbS_8)
$Cu_2Mo_6Se_8$	5.9	Same
Cu_xPb_{1-x}	5.7—7.7	
CuS	1.62	B18
CuS_2	1.48—1.53	C18
CuSSe	1.5—2.0	C18
$CuSe_2$	2.3—2.43	C18
CuSeTe	1.6—2.0	C18
Cu_xSn_{1-x}	3.2—3.7	
Cu_xSn_{1-x} (film, made at 10K)	3.6—7	
Cu_xSn_{1-x} (film, made at 300K)	2.8—3.7	
$CuTe_2$	<1.25—1.3	C18
$CuTh_2$	3.49	C16
$Cu_{0-0.027}V$	3.9—5.3	A2
CuY	0.33	B2-cP2 (CsCl)
Cu_xZn_{1-x}	0.5—0.845	
$DyMo_6S_8$	2.1	hR15
Er_xLa_{1-x}	1.4—6.3	
$ErMo_6S_8$	2.2	hR15

TABLE 4
Superconductive Compounds and Alloys (continued)

Substance	T_c, K	Crystal structure type
$ErMo_6Se_8$	6.2	hR15
$Fe_3Lu_2Si_5$	6.1	tP40 ($Fe_3Sc_2Si_5$)
$Fe_{0-0.04}Mo_{0.8}Re_{0.2}$	1—10	
$Fe_{0.05}Ni_{0.05}Zr_{0.90}$	~3.9	
Fe_3Re_2	6.55	$D8_b$-tP30 (FeCr)
$Fe_3Sc_2Si_5$	4.52	tP40
Fe_3Si_5Tm	1.3	Same
$Fe_3Si_5Y_2$	2.4	Same
Fe_3Th_7	1.86	D10
Fe_xTi_{1-x}	3.2 (max.)	Fe in α-Ti
Fe_xTi_{1-x}	3.7 (max.)	Fe in β-Ti
$Fe_xTi_{0.6}V_{1-x}$	6.8 (max.)	
FeU_6	3.86	$D2_c$
$Fe_{0.1}Zr_{0.9}$	1.0	A3
$Ga_{0.5}Ge_{0.5}Nb_3$	7.3	A15
Ga_2Ge_2U	0.87	B2-cP2
$GaHf_2$	0.21	C16-tI12 (Al_2Cu)
$GaLa_3$	5.84	
Ga_3Lu	2.3	B2-cP2
Ga_2Mo	9.5	
$GaMo_3$	0.76	A15
GaN (black)	5.85	B4
$Ga_{0.7}Pt_{0.3}$	2.9	C1
GaPt	1.74	B20
GaSb (120kbar, 77K, annealed)	4.24	A5
GaSb (unannealed)	~5.9	
$Ga_{0-1}Sn_{1-0}$ (quenched)	3.47—4.18	
$Ga_{0-1}Sn_{1-0}$ (annealed)	2.6—3.85	
GaTe	0.17	mC24 (GaTe)
Ga_5V_2	3.55	Tetragonal (Mn_2Hg_5)
$GaV_{4.5}$	9.15	
Ga_3Zr	1.38	
Ga_3Zr_5	3.8	$D8_8$-hP16 (Mn_5Si_3)
Gd_xLa_{1-x}	< 1.0—5.5	
$GdMo_6S_8$	3.5	hR15
$GdMo_6Se_8$	5.6	hR15
$Gd_xOs_2Y_{1-x}$	1.4—4.7	
$Gd_xRu_2Th_{1-x}$	3.6 (max.)	C15
$Ge_{10}As_4Y_5$	9.06	tP38 ($Co_4Sc_5Si_{10}$)
GeIr	4.7	B31
GeIrLa	1.64	tI12 (LaPtSi)
$Ge_{10}Ir_4Lu_5$	2.60	tP38
$Ge_{10}Ir_4Y_5$	2.62	tP38
Ge_2La	1.49; 2.2	Orthorhombic, distorted (Mn_2Hg_5)
GeLaPt	3.53	tI12
$Ge_{13}Lu_3Os_4$	3.6	cP40 ($Pr_3Rh_2Sn_{13}$)
$Ge_{10}Lu_5Rh_4$	2.79	tP38
$Ge_{13}Lu_3Ru_4$	2.3	cP40
$GeMo_3$	1.43	A15
$GeNb_2$	1.9	
$Ge_{0.29}Nb_{0.71}$	6	A15
GePt	0.40	B31
Ge_3Rh_5	2.12	Orthorhombic, related to $InNi_2$

TABLE 4
Superconductive Compounds and Alloys (continued)

Substance	T_c, K	Crystal structure type
GeRh	0.96	B31-oP8 (MnP)
$Ge_{13}Rh_4Sc_3$	1.9	c P40
$Ge_{10}Rh_4Y_5$	1.35	tP38
$Ge_{13}Ru_4Y_3$	1.7	cP40
Ge_2So	1.3	
$GeTa_3$	8.0	A15-cP8 (Cr_3Si)
Ge_3Te_4 (n = 1.06 × 10^{22})	1.55—1.80	Rhombohedral
Ge_xTe_{1-x} (n = 8.5—64 × 10^{20})	0.07—0.41	R1
GeV_3	6.01	A15
Ge_2Y	3.80	C_c
$Ge_{1.62}Y$	2.4	
Ge_2Zr	0.30	oC12 ($ZrSi_2$)
$GeZr_3$	0.4	$L1_2$-tP32 (Ti_3P)
$H_{0.33}Nb_{0.67}$	7.28	B.C.C.
$H_{0.1}Nb_{0.9}$	7.38	Same
$H_{0.05}Nb_{0.95}$	7.83	Same
$H_{0.12}Ta_{0.88}$	2.81	B.C.C.
$H_{0.08}Ta_{0.92}$	3.26	Same
$H_{0.04}Ta_{0.96}$	3.62	Same
HfIrSi	3.50	C37-cP12 (Co_2Si)
$HfMo_2$	0.05	hP24 (Ni_2Mn)
$HfN_{0.989}$	6.6	B1
$Hf_{0-0.5}Nb_{1-0.5}$	8.3—9.5	A2
$Hf_{0.75}Nb_{0.25}$	>4.2	
$HfOs_2$	2.69	C14
HfOsP	6.1	C22-hP9 (Fe_2P)
HfPRu	9.9	Same
$HfRe_2$	4.80	C14
$Hf_{0.14}Re_{0.86}$	5.86	A12
$Hf_{0.99-0.96}Rh_{0.01-0.04}$	0.85—1.51	
$Hf_{0-0.55}Ta_{1-0.45}$	4.4—6.5	A2
HfV_2	8.9—9.6	C15
Hg_xIn_{1-x}	3.14—4.55	
HgIn	3.81	
Hg_2K	1.20	Orthorhombic
Hg_3K	3.18	
Hg_4K	3.27	
Hg_8K	3.42	
Hg_3Li	1.7	Hexagonal
$HgMg_3$	0.17	hP8 (Na_3As)
Hg_2Mg	4.0	tI6 ($MoSi_2$)
Hg_3Mg_5	0.48	$D8_b$-hP16 (Mn_5Si_3)
Hg_2Na	1.62	Hexagonal
Hg_4Na	3.05	
Hg_xPb_{1-x}	4.14—7.26	
HgSn	4.2	
Hg_xTl_{1-x}	2.30—4.19	
Hg_5Tl_2	3.86	
Ho_xLa_{1-x}	1.3—6.3	
$Ho_{1.2}Mo_6Se_8$	6.1	$D10_2$-hR12 (Be_3Nb)
$In_{1-0.86}Mg_{0-0.14}$	3.395—3.363	
$In_2Mo_6Te_6$	2.6	hR15 (Mo_6PbS_8)
$InNb_3$ (high pressure and temp.)	4—8; 9.2	A15
$In_{0.5}Nb_3Zr_{0.5}$	6.4	
$In_{0.11}O_3W$	< 1.25—2.8	Hexagonal

TABLE 4
Superconductive Compounds and Alloys (continued)

Substance	T_c, K	Crystal structure type
$In_{0.95-0.85}Pb_{0.05-0.15}$	3.6—5.05	
$In_{0.98-0.91}Pb_{0.02-0.09}$	3.45—4.2	
InPb	6.65	
InPd	0.7	B2
InSb (quenched from 170 kbar into liquid N_2)	4.8	Like A5
InSb	2.1	B4
$(InSb)_{0.95-0.10}Sn_{0.05-0.90}$ (various heat treatments)	3.8—5.1	
$(InSb)_{0-0.07}Sn_{1-0.93}$	3.67—3.74	
In_3Sn	~5.5	
In_xSn_{1-x}	3.4—7.3	
$In_{0.82-1}Te$ ($n = 0.83—1.71 \times 10^{22}$)	1.02—3.45	B1
$In_{1.000}Te_{1.002}$	3.5—3.7	B1
In_3Te_4 ($n = 4.7 \times 10^{21}$)	1.15—1.25	Rhombohedral
In_xTl_{1-x}	2.7—3.374	
$In_{0.8}Tl_{0.2}$	3.223	
$In_{0.62}Tl_{0.38}$	2.760	
$In_{0.78-0.69}Tl_{0.22-0.31}$	3.18—3.32	Tetragonal
$In_{0.69-0.62}Tl_{0.31-0.38}$	2.98—3.3	F.C.C.
Ir_2La	0.48	C15
Ir_3La	2.32	$D10_2$
Ir_3La_7	2.24	$D10_2$
Ir_5La	2.13	
$IrLaSi_2$	2.03	oC16 ($CeNiSi_2$)
$IrLaSi_3$	2.7	tI10 ($BaNiSn_3$)
Ir_2Lu	2.47	C15
Ir_3Lu	2.89	C15
$Ir_4Lu_5Si_{10}$	3.9	tP38 ($Co_4Sc_5Si_{10}$)
IrMo	< 1.0	A3
$IrMo_3$	9.6	A15
$IrMo_3$	6.8	$D8_b$
$IrNb_3$	1.9	A15
$Ir_{0.4}Nb_{0.6}$	9.8	$D8_b$
$Ir_{0.37}Nb_{0.63}$	2.32	$D8_b$
IrNb	7.9	$D8_b$
$Ir_{1.15}Nb_{0.85}$	4.6	oP12 (IrTa)
$Ir_{0.02}Nb_3Rh_{0.98}$	2.43	A15
$Ir_{0.05}Nb_3Rh_{0.95}$	2.38	A15
$Ir_{0.287}O_{0.14}Ti_{0.573}$	5.5	$E9_3$
$Ir_{0.265}O_{0.035}Ti_{0.65}$	2.30	$E9_3$
Ir_xOs_{1-x}	0.3—0.98	
$Ir_{1.5}Os_{0.5}$	2.4	C14
IrOsY	2.6	C15
IrSiY	2.70	C37-oP12 (Co_2Si)
IrSiZr	2.04	Same
Ir_2Sc	2.07	C15
$Ir_{2.5}Sc$	2.46	C15
$Ir_4Sc_5Si_{10}$	8.46	tP38
Ir_2Si_2Th	2.14	tI10
$IrSi_3Th$	1.75	tI10
IrSiTh	6.50	tI12 (LaPtSi)
Ir_2Si_2Y	2.60	tI10 (Al4Ba)
$Ir_4Si_{10}Y_5$	3.10	tP38
$Ir_3Si_5Y_2$	2.83	oI40

TABLE 4
Superconductive Compounds and Alloys (continued)

Substance	T_c, K	Crystal structure type
$IrSn_2$	0.65—0.78	C1
Ir_2Sr	5.70	C15
Ir_7Ta_{13}	1.2	$D8_b$-tP30 (FeCr)
$Ir_{0.5}Te_{0.5}$	~3	
$IrTe_3$	1.18	C2
IrTh	< 0.37	B_f
Ir_2Th	6.50	C15
Ir_3Th	4.71	
Ir_3Th_7	1.52	$D10_2$
Ir_5Th	3.93	$D2_d$
$IrTi_3$	5.40	A15
IrV_2	1.39	A15
IrW_3	3.82	
$Ir_{0.28}W_{0.72}$	4.49	
Ir_2Y	2.18; 1.38	C15
$Ir_{0.69}Y_{0.31}$	1.98; 1.44	C15
$Ir_{0.70}Y_{0.30}$	2.16	C15
Ir_2Y_3	1.61	
Ir_3Y	3.50	$D10_2$-hR13 (Be_3Nb)
Ir_xY_{1-x}	0.3—3.7	
Ir_2Zr	4.10	C15
$Ir_{0.1}Zr_{0.9}$	5.5	A3
$K_2Mo_{15}S_{19}$	3.32	hR15
$K_{0.27—0.31}O_3W$	0.50	Hexagonal
$K_{0.40—0.57}O_3W$	1.5	Tetragonal
$La_{0.55}Lu_{0.45}$	2.2	Hexagonal, La type
$La_{0.8}Lu_{0.2}$	3.4	Same
$LaMg_2$	1.05	C15
$LaMo_6S_8$	7.1	hR15
LaN	1.35	
$LaOs_2$	6.5	C15
$LaPt_2$	0.46	C15
$La_{0.28}Pt_{0.72}$	0.54	C15
LaPtSi	3.48	tI12
$LaRh_3$	2.60	
$LaRh_5$	1.62	
La_7Rh_3	2.58	$D10_2$
$LaRhSi_2$	3.42	oC16 ($CeNiSi_2$)
$La_2Rh_3Si_5$	4.45	oI40 ($Co_3Si_5U_2$)
$LaRhSi_3$	2.7	tI10 ($BaNiSn_3$)
$LaRh_2Si_2$	3.90	tI10 (Al_4Ba)
$LaRu_2$	1.63	C15
La_3S_4	6.5	$D7_3$
La_3Se_4	8.6	$D7_3$
$LaSi_2$	2.3	C_c
La_xY_{1-x}	1.7—5.4	
LaZn	1.04	B2
$Li_xMo_6S_8$	4.2	hR15
LiPb	7.2	
$LuOs_2$	3.49	C14
$Lu_{0.275}Rh_{0.725}$	1.27	C15
$LuRh_5$	0.49	
$Lu_5Rh_4Si_{10}$	3.95	tP38 ($Co_4So_5Si_{10}$)
$LuRu_2$	0.86	C14
$Mg_{1.14}Mo_{6.6}S_8$	3.5	hR15

TABLE 4
Superconductive Compounds and Alloys (continued)

Substance	T_c, K	Crystal structure type
Mg2Nb	5.6	
$Mg_{\sim0.47}Tl_{\sim0.53}$	2.75	B2
MgZn	0.9	A3-oP4 (AuCd)
Mn_xTi_{1-x}	2.3 (max.)	Mn in -Ti
Mn_xTi_{1-x}	1.1—3.0	Mn in -Ti
MnU_6	2.32	$D2_c$
Mo_2N	5.0	F.C.C.
$Mo_6Na_2S_8$	8.6	hR15
Mo_xNb_{1-x}	0.016—9.2	
$Mo_{5.25}Nb_{0.75}Se_8$	6.2	hR15
Mo_6NdSa_8	8.2	hR15
Mo_3Os	7.2	A15
$Mo_{0.62}Cs_{0.38}$	5.65	$D8_b$
Mo_3P	5.31	DO_e
$Mo_6Pb_{1.2}Se_8$	6.75	hR15
$Mo_{0.5}Pd_{0.5}$	3.52	A3
Mo_6PrSe_8	9.2	hR15
MoRe	7.8	$D8_b$-tP30
$MoRe_3$	9.25; 9.89	A12
Mo_xRe_{1-x}	1.2—12.2	
$Mo_{0.42}Re_{0.58}$	6.35	$D8_b$
MoRh	1.97	A3
Mo_xRh_{1-x}	1.5—8.2	B.C.C.
MoRu	9.5—10.5	A3
$Mo_{0.61}Ru_{0.39}$	7.18	$D8_b$
$Mo_{0.2}Ru_{0.8}$	1.66	A3
Mo_3Ru_2	7.0	$D8_b$-tP30
$Mo_4Ru_2Te_8$	1.7	hR15
Mo_6S_8	1.85	hR15
Mo_6S_8Sc	3.6	hR15
$Mo_6S_8Sm_{1.2}$	2.9	hR15
Mo_6S_8Tb	2.0	hR15
Mo_6S_8Tl	8.7	hR15
$Mo_6S_8Tm_{1.2}$	2.1	hR15
$Mo_6S_8Y_{1.2}$	3.0	hR15
Mo_6S_8Yb	9.2	hR15
$Mo_{6.6}S_8Zn_{11}$	3.6	hR15
Mo_3Sb_4	2.1	
Mo_6Se_8	6.3	hR15
$Mo_6Se_8Sm_{1.2}$	6.8	hR15
$Mo_6Se_8Sn_{1.2}$	6.8	hR15
Mo_6Se_8Tb	5.7	hR15
Mo_3Se_3Tl	4.0	hP14
$Mo_6Se_8Tm_{1.2}$	6.3	hR15
Mo_6Se_8Yb	6.2	hR15
Mo_3Si	1.30	A15
$MoSi_{0.7}$	1.34	
Mo_xSiV_{3-x}	4.54—16.0	A15
$Mo_{5.25}Ta_{0.75}Te_8$	1.7	hR15
Mo_6Te_8	1.7	hR15
$Mo_{0.16}Ti_{0.84}$	4.18; 4.25	
$Mo_{0.913}Ti_{0.087}$	2.95	
$Mo_{0.04}Ti_{0.96}$	2.0	Cubic
$Mo_{0.025}Ti_{0.975}$	1.8	
Mo_xU_{1-x}	0.7—2.1	

TABLE 4
Superconductive Compounds and Alloys (continued)

Substance	T_c, K	Crystal structure type
Mo_xV_{1-x}	0—~5.3	
Mo_2Zr	4.25—4.75	C15
NNb (film)	6—9	B1
$N_xO_yTi_z$	2.9—5.6	Cubic
$N_xO_yV_z$	5.8—8.2	Cubic
$N_{0.34}Re$	4—5	F.C.C.
NTa (film)	4.84	B1
$N_{0.6—0.987}Ti$	<1.17—5.8	B1
$N_{0.82—0.99}V$	2.9—7.9	B1
NZr	9.8	B1
$N_{0.906—0.984}Zr$	3.0—9.5	B1
$Na_{0.28—0.35}O_3W$	0.56	Tetragonal
$Na_{0.28}Pb_{0.72}$	7.2	
NbO	1.25	
$NbOs_2$	2.52	A12
Nb_3Os	1.05	A15
$Nb_{0.6}Os_{0.4}$	1.89; 1.78	$D8_b$
$Nb_3Os_{0.02—0.10}Rh_{0.98—0.90}$	2.42—2.30	A15
Nb_3P	1.8	$L1_2tP32$ (Ti_3P)
NbPRh	4.08	C37-oP12 (Co_2Si)
$Nb_{0.6}Pd_{0.4}$	1.60	$D8_f$ plus cubic
$Nb_3Pd_{0.02—0.10}Rh_{0.92—0.90}$	2.49—2.55	A15
$Nb_{0.62}Pt_{0.38}$	4.21	$D8_b$
Nb_5Pt_3	3.73	$D8_b$
$Nb_3Pt_{0.02—0.98}Rh_{0.98—0.02}$	2.52—9.6	A15
$NbRe_3$	5.27	$D8_b$-tP30 (FeCr)
$Nb_{0.38—0.18}Re_{0.62—0.82}$	2.43—9.70	A15
NbRe	3.8	$D8_b$-tP30
NbReSi	5.1	oI36 (FeTiSi)
Nb_3Rh	2.64	A15
$Nb_{0.6}Rh_{0.40}$	4.21	$D8_b$ plus other
$Nb_{0.9}Rh_{1.1}$	3.07	A3-oP4 (AuCd)
$Nb_3Rh_{0.98—0.90}Ru_{0.02—0.10}$	2.42—2.44	A15
Nb_xRu_{1-x}	1.2—4.8	
NbRuSi	2.65	oI36
NbS_2	6.1—6.3	Hexagonal, $NbSe_2$ type
NbS_2	5.0—5.5	Hexagonal, three-layer type
Nb_3Sb	0.2	$L1_2$-tP32 (Ti_3P)
$Nb_3Sb_{0—0.7}Sn_{1—0.3}$	6.8—18	A15
$NbSe_2$	5.15—5.62	Hexagonal
$Nb_{1—1.05}Se_2$	2.2—7.0	Same
Nb_3Se_4	2.0	hP14
Nb_3Si	1.5	$L1_2$
Nb_3SiSnV_3	4.0	
$NbSn_2$	2.60	Orthorhombic
Nb_6Sn_5	2.8	oI44 (Sn_5Ti_6)
NbSnTaV	6.2	A15
$NbSnV_2$	5.5	A15
Nb_2SnV	9.8	A15
Nb_xTa_{1-x}	4.4—9.2	A2
Nb_3Te_4	1.8	hP14
Nb_xTi_{1-x}	0.6—9.8	
$Nb_{0.6}Ti_{0.4}$	9.8	
Nb_xU_{1-x}	1.95 (max.)	
$Nb_{0.88}V_{0.12}$	5.7	A2

TABLE 4
Superconductive Compounds and Alloys (continued)

Substance	T_c, K	Crystal structure type
$Nb_{0.5}V_{1.5}Zr$	4.3	C15-hP12 ($MgZn_2$)
$Ni_{0.3}Th_{0.7}$	1.98	$D10_2$
$NiZr_2$	1.52	
$Ni_{0.1}Zr_{0.9}$	1.5	A3
$O_3Rb_{0.27—0.29}W$	1.98	Hexagonal
OSn	3.81	tP4 (PbO)
O_3SrTi (n = 1.7—12.0 × 10^{19})	0.12—0.37	
O_3SrTi (n = 10^{18}—10^{21})	0.05—0.47	
O_3SrTi (n = 10^{20})	0.47	
$O_3Sr_{0.08}W$	2—4	Hexagonal
OTi	0.58	
$O_3Tl_{0.30}W$	2.0—2.14	Hexagonal
OV_3Zr_3	7.5	$E9_3$
OW_3 (film)	3.35; 1.1	A15
OsPti	1.2	C22-hP9 (Fe_2P)
OsPZr	7.4	Same
OsReY	2.0	C14
Os_2Sc	4.6	C14
OsTa	1.95	A12
Os_3Th_7	1.51	$D10_2$
Os_xW_{1-x}	0.9—4.1	
OsW_3	~3	
Os_2Y	4.7	C14
Os_2Zr	3.0	C14
Os_xZr_{1-x}	1.5—5.6	
PPb	7.8	
OsW_2	3.81	$D8_b$-tP30 (FeCr)
$PPd_{3.0—3.2}$	<0.35—0.7	DO_{11}
P_3Pd_7 (high temperature)	1.0	Rhombohedral
P_3Pd_7 (low temperature)	0.70	Complex
PRh	1.22	
PRh_2	1.3	C1
P_4Rh_5	1.22	oP28 ($CaFe_2O_4$)
PRhTa	4.41	C37-oP12 (Co_2Si)
PRhZr	1.55	Same
PRuTi	1.3	C22-hP9 (Fe_2P)
PRuZr	3.46	C37-oP12
PW_3	2.26	DO_e
Pb_2Pd	2.95	C16
Pb_4Pt	2.80	Related to C16
Pb_2Rh	2.66	C16
PbSb	6.6	
PbTe (plus 0.1 w/o Pb)	5.19	
PbTe (plus 0.1 w/o Te)	5.24—5.27	
$PbTl_{0.27}$	6.43	
$PbTl_{0.17}$	6.73	
$PbTl_{0.12}$	6.88	
$PbTl_{0.075}$	6.98	
$PbTl_{0.04}$	7.06	
$Pb_{1—0.26}Tl_{0—0.74}$	7.20—3.68	
$PbTl_2$	3.75—4.1	
Pb_3Zr_5	4.60	$D8_8$
$PbZr_3$	0.76	A15
$Pd_{0.9}Pt_{0.1}Te_2$	1.65	C6
$Pd_{0.05}Ru_{0.05}Zr_{0.9}$	~9	

TABLE 4
Superconductive Compounds and Alloys (continued)

Substance	T_c, K	Crystal structure type
$Pd_{2.2}S$ (quenched)	1.63	Cubic
$PdSb_2$	1.25	C2
PdSb	1.5	$B8_1$
PdSbSe	1.0	C2
PdSbTe	1.2	C2
Pd_4Se	0.42	Tetragonal
$Pd_{6-7}Se$	0.66	Like Pd_4Te
$Pd_{2.8}Se$	2.3	
Pd_xSe_{1-x}	2.5 (max.)	
PdSi	0.93	B31
PdSn	0.41	B31
$PdSn_2$	3.34	
Pd_2Sn	0.41	C37
Pd_3Sn	0.47—0.64	$B8_2$
Pd_2SnTm	1.77	DO_3-cF16 (BiF_3)
Pd_2SnY	4.92	Same
Pd_2SnYb	1.79	Same
PdTe	2.3; 3.85	$B8_1$
$PdTe_{1.02-1.08}$	2.56—1.88	$B8_1$
$PdTe_2$	1.69	C6
$PdTe_{2.1}$	1.89	C6
$PdTe_{2.3}$	1.85	C6
$Pd_{1.1}Te$	4.07	$B8_1$
Pd_3Te	0.76	cI2 (W)
$PdTh_2$	0.85	C16
$Pd_{0.1}Zr_{0.9}$	7.5	A3
PtSb	2.1	$B8_1$
PtSi	0.88	B31
PtSn	0.37	$B8_1$
$PtSn_4$	2.38	C16-oC20 ($PdSn_4$)
Pt_3Ta_7	1.5	$D8_b$-tP30
$PtTa_3$	0.4	A15-cP8 (Cr_3Si)
PtTe	0.59	Orthorhombic
PtTh	0.44	B_f
Pt_3Th_7	0.98	$D10_2$
Pt_5Th	3.13	
$PtTi_3$	0.58	A15
$Pt_{0.02}U_{0.98}$	0.87	β-phase
$PtV_{2.5}$	1.36	A15
PtV_3	2.87—3.20	A15
$PtV_{3.5}$	1.26	A15
$Pt_{0.5}W_{0.5}$	1.45	A1
Pt_xW_{1-x}	0.4—2.7	
Pt_2Y_3	0.90	
Pt_2Y	1.57; 1.70	C15
Pt_3Y_7	0.82	$D10_2$
PtZr	3.0	A3
Re_2Sc	4.2	C15-hP12 ($MgZn_2$)
$Re_{24}Sc_5$	2.2	A12-cI58 (Mg)
ReSiTa	4.4	oI36 (FeTiSi)
$Re_3Si_5Y_2$	1.76	tP40 ($Fe_3Sc_2Si_5$)
Re_3Ta_2	1.4	$D8_b$-tP30 (FeCr)
$Re_{0.64}Ta_{0.36}$	1.46	A12
Re_3Ta	6.78	A12-cI58 (Mn)
$Re_{24}Ti_5$	6.60	A12

TABLE 4
Superconductive Compounds and Alloys (continued)

Substance	T_c, K	Crystal structue type
Re_xTi_{1-x}	6.6 (max.)	
$Re_{0.76}V_{0.24}$	4.52	$D8_b$
Re_3V	6.26	$D8_b$-tP30
$Re_{0.92}V_{0.08}$	6.8	A3
$Re_{0.6}W_{0.4}$	6.0	
$Re_{0.5}W_{0.5}$	5.12	$D8_b$
$Re_{13}W_{12}$	5.2	$D8_b$-tP30
Re_3W	9.0	A12-cI58
Re_2Y	1.83	C14
Re_2Zr	5.9	C14
Re_3Zr	7.40	A12-cI58
Re_6Zr	7.40	Same
$Rh_{17}S_{15}$	5.8	Cubic
$Rh_{\sim0.24}Sc_{\sim0.76}$	0.88; 0.92	
$Rh_4Sc_5Si_{10}$	8.54	tP38
$Rh_4Sc_3Sn_{13}$	4.5	cP40
Rh_xSe_{1-x}	6.0 (max.)	
$RhSi_3Th$	1.76	tI10
$Rh_{0.86}Sc_{1.04}Th$	6.45	tI12
Rh_2Si_2Y	3.11	tI10
$Rh_3Si_5Y_2$	2.70	oI40
$Rh_4Sn_{13}Sr_3$	4.3	cP40
Rh_xSn_yTh	1.9	cI2 (W)
Rh_xSn_yTm	2.3	cP40
$Rh_4Sn_{13}Y_3$	3.2	cP40
Rh_2Sr	6.2	C15
$Rh_{0.4}Ta_{0.6}$	2.35	$D8_b$
$RhTe_2$	1.51	C2
$Rh_{0.67}Te_{0.33}$	0.49	
Rh_xTe_{1-x}	1.51 (max.)	
RhTh	0.36	B_f
Rh_3Th_7	2.15	$D10_2$
Rh_5Th	1.07	
Rh_xTi_{1-x}	2.25—3.95	
$Rh_{0.02}U_{0.98}$	0.96	
RhV_3	0.38	A15
RhW	~3.4	A3
RhY_3	0.65	
Rh_2Y_3	1.48	
Rh_3Y	1.07	C15
Rh_5Y	0.56	
Rh_3Y_7	0.32	hP20 (Fe_3Th_7)
$Rh_{0.005}Zr$ (annealed)	5.8	
$Rh_{0-0.45}Zr_{1-0.55}$	2.1—10.8	
$Rh_{0.1}Zr_{0.9}$	9.0	H.C.P.
Ru_2Sc	1.67	C14
RuSiTa	3.15	oI36
Ru_3Si_2Th	3.98	hP12
Ru_3Si_2Y	3.51	hP12
$Ru_{1.1}Sn_{3.1}Y$	1.3	cP40
Ru_2Th	3.56	C15
RuTi	1.07	B2
$Ru_{0.05}Ti_{0.95}$	2.5	
$Ru_{0.1}Ti_{0.9}$	3.5	
$Ru_xTi_{0.6}V_y$	6.6 (max.)	

TABLE 4
Superconductive Compounds and Alloys (continued)

Substance	T_c, K	Crystal structue type
Ru_3U	0.15	$L1_2$-cP4
$Ru_{0.45}V_{0.55}$	4.0	B2
RuW	7.5	A3
Ru_2Y	1.52	C14
Ru_2Zr	1.84	C14
$Ru_{0.1}Zr_{0.9}$	5.7	A3
STh	0.5	B1-cF8 (NaCl)
SbSn	1.30—1.42	B1 or distorted
$SbTa_3$	0.72	A15-cP8 (Cr_3Si)
$SbTi_3$	5.8	Same
Sb_2Ti_7	5.2	
$Sb_{0.01-0.03}V_{0.99-0.97}$	3.76—2.63	A2
SbV_3	0.80	A15
SeTh	1.7	B1-cF8
$SiMo_3$	1.4	A15-cP8
Si_2Th	3.2	C_c, α-phase
Si_2Th	2.4	C32, β-phase
$SiV_{2.7}Ru_{0.3}$	2.9	A15
Si_2W_3	2.8; 2.84	
$SiZr_3$	0.5	$L1_2$-tP32 (Ti_3P)
$Sn_{0.174-0.104}Ta_{0.826-0.896}$	6.5—< 4.2	A15
$SnTa_3$	8.35	A15, highly ordered
$SnTa_3$	6.2	A15, partially ordered
$SnTaV_2$	2.8	A15
$SnTa_2V$	3.7	A15
Sn_xTe_{1-x} ($n = 10.5$—20×10^{20})	0.07—0.22	B1
Sn_3Th	3.33	$L1_2$-cP4
$SnTi_3$	5.80	A15-cP8
Sn_xTl_{1-x}	2.37—5.2	
SnV_3	3.8	A15
$Sn_{0.02-0.057}V_{0.98-0.943}$	2.87—~1.6	A2
$SnZr_3$	0.92	A15-cP8
$Ta_{0.025}Ti_{0.975}$	1.3	Hexagonal
$Ta_{0.05}Ti_{0.95}$	2.9	Hexagonal
$Ta_{0.05-0.75}V_{0.95-0.25}$	4.30—2.65	A2
$Ta_{0.8-1}W_{0.2-0}$	1.2—4.4	A2
$Tc_{0.1-0.4}W_{0.9-0.6}$	1.25—7.18	Cubic
$Tc_{0.50}W_{0.50}$	7.52	α plus
$Tc_{0.60}W_{0.40}$	7.88	plus α
Tc_6Zr	9.7	A12
TeY	1.02	B1-cF8
$ThTl_3$	0.87	$L1_2$-cP4
$Th_{0-0.55}Y_{1-0.45}$	1.2—1.8	
$Ti_{0.70}V_{0.30}$	6.14	Cubic
Ti_xV_{1-x}	0.2—7.5	
$Ti_{0.5}Zr_{0.5}$ (annealed)	1.23	
$Ti_{0.5}Zr_{0.5}$ (quenched)	2.0	
Tl_3Y	1.52	$L1_2$-cP4
V_2Zr	8.80	C15
$V_{0.26}Zr_{0.74}$	5.9	
W_2Zr	2.16	C15
YZn	0.33	B2-cP2 (CsCl)

* n denotes current carriers concentration in cm^{-3}.

TABLE 4
Superconductive Compounds and Alloys (continued)

B. SUPERCONDUCTORS WITH T_c >10K

Substance	T_c,K	Crystal structure type
Al_2CMo_3	10.0	A13
$Al_{0.5}Ge_{0.5}Nb$	12.6	A15
$Al_{\sim0.8}Ge_{\sim0.2}Nb_3$	20.7	A15
$AlNb_3$	18.0	A15 (Cr_3Si)
$AlNb_3$	12.0	($FeCr$)
Al_xNb_{1-x}	<4.2—13.5	$D8_b$
Al_xNb_{1-x}	12—17.5	A15
$Al_{0.27}Nb_{0.73-0.48}V_{0-0.25}$	14.5—17.5	A15
$Al\,Nb_xV_{1-x}$	4.4—13.5	
$Al_{0.1}Si_{0.9}V_3$	14.05	
AlV_3	11.8	A15 (Cr_3Si)
$AuNb_3$	11.5	A15
$Au_{0-0.3}Nb_{1-0.7}$	1.1—11.0	
$Au_{0.02-0.98}Nb_3Rh_{0.98-0.02}$	2.53—10.9	A15
$AuNb_{3(1-x)}V_{3x}$	1.5—11.0	A15
$B_{0.03}C_{0.51}Mo_{0.47}$	12.5	
B_4LuRh_4	11.7	(B_4CeCo_4)
B_2LuRu	10	
B_4Rh_4Y	11.3	(B_4CeCo_4)
$B_{0.1}Si_{0.9}V_3$	15.8	A15
$BaBi_{0.2}O_3Pb_{0.8}$	13.2	
$Ba_2CaCu_2O_8Tl_2$	120	
$Ba_2Cu_3LaO_6$	80	
$Ba_2Cu_3O_7Tm$	101	
$Ba_2Cu_3O_7Y$	90	
$(Ba,La)_2CuO_4$	36	A15 (K_2NiF_4)
$Bi_2CaCu_2O_8Sr_2$	110	
$Br_2Mo_6S_6$	13.8	(Mo_6PbS_8)
C_3La	11.0	(C_3Pu_2)
CMo	14.3	B1 ($NaCl$)
CMo_2	12.2	o**
$C_{0.5}Mo_xNb_{1-x}$	10.8—12.5	B1
CMo_xTi_{1-x}	10.2(max)	B1
$CMo_{0.83}Ti_{0.17}$	10.2	B1
$C_{0-0.38}N_{1-0.62}Ta$	10.0—11.3	
CNb (whiskers)	7.5—10.5	
CNb	11.5	B1
$C_{0.7-1.0}Nb_{0.3-0}$	6—11	B1
CNb_xTa_{1-x}	8.2—13.9	
$CNb_{0.6-0.9}W_{0.4-0.1}$	12.5—11.6	B1
$C_{0.1}Si_{0.9}V_3$	16.4	A15
CTa	10.3	B1
$CTa_{1-0.4}W_{0-0.6}$	8.5—10.5	B1
$C_{0.66}Th_{0.13}Y_{0.21}$	17	(C_3Pu_2)
C_3Y_2	11.5	(C_3Pu_2)
CW	10	B1
$(Ca,La)_2CuO_4$	18	(K_2NiF_4)
$Cu(La,Sr)_2O_4$	39	
$Cu_{1.8}Mo_6S_8$	10.8	(Mo_6PbS_8)
$Cr_{0.3}SiV_{2.7}$	11.3	A15
$GaNb_3$	14.5	A15 (Cr_3Si)
$Ga_xNb_3Sn_{1-x}$	14—18.37	A15

TABLE 4
Superconductive Compounds and Alloys (continued)

Substance	T_c,K	Crystal structure type
GaV_3	16.8	A15
$GaV_{2.1—3.5}$	6.3—14.45	A15
$GeNb_3$	23.2	A15
$GeNb_3$ (quenched)	6—17	A15
$Ge_xNb_3Sn_{1-x}$	17.6—18.0	A15
$Ge_{0.5}Nb_3Sn_{0.5}$	11.3	
$Ge_{0.1}Si_{0.9}V_3$	14.0	A15
GeV_3	11	A15
$InLa_3$	9.83; 10.4	LI_2 ($AuCu_3$)
$InLa_3$ (0—35 kbar)	9.75—10.55	
$In_{0—0.3}Nb_3Sn_{1—0.7}$	18.0—18.19	A15
InV_3	13.9	A15
$Ir_{0.4}Nb_{0.6}$	10	(FeCr)
$LaMo_6Se_8$	11.4	(Mo_6PbS_8)
LiO_4Ti_2	13.7	($A1_2MgO_4$)
MoN	12; 14.8	h*
Mo_3Os	12.7	A15
$Mo_6Pb_{0.9}S_{7.5}$	15.2	(Mo_6PbS_8)
Mo_3Re	10.0; 15	A15
Mo_xRe_{1-x}	1.2—12.2	
$Mo_{0.52}Re_{0.48}$	11.1	
$Mo_{0.57}Re_{0.43}$	14.0	
$Mo_{\sim0.60}Re_{0.395}$	10.6	
MoRu	9.5—10.5	A3
Mo_3Ru	10.6	A15
Mo_6Se_8T1	12.2	(Mo_6PbS_8)
$Mo_{0.3}SiV_{2.7}$	11.7	A15
Mn_3Si	12.5	A15
Mo_3Tc	15	A15
$Mo_{0.3}Tc_{0.7}$	12.0	A15
Mo_xTc_{1-x}	10.8—15.8	
$MoTc_3$	15.8	
NNb (whiskers)	10—14.5	
NNb (diffusion wires)	16.10	
$N_{0.988}Nb$	14.9; 17.3	B1
$N_{0.824—0.988}Nb$	14.4—15.3	B1
$N_{0.7—0.795}Nb$	11.3—12.9	
NNb_xO_y	13.5—17.0	B1
NNb_xO_y	6.0—11	
$N_{100—42w/o}Nb_{0—58w/o}Ti$	15—16.8	
$N_{100—75w/o}Nb_{0—25w/o}Zr$	12.5—16.35	
NNb_xZr_{1-x}	9.8—13.8	B1
$N_{0.93}Nb_{0.85}Zr_{0.15}$	13.8	B1
NTa	12—14	B1
NZr	10.7	B1
Nb_3Pt	10.9	A15
$Nb_{0.18}Re_{0.82}$	10	(Mn)
Nb_3Si	19	A15
$Nb_{0.3}SiV_{2.7}$	12.8	A15
Nb_3Sn	18.05	A15
$Nb_{0.8}Sn_{0.2}$	18.18; 18.5	A15
Nb_xSn_{1-x} (film)	2.6—18.5	o*
Nb_3Sn_2	16.6	t*
$NbSnTa_2$	10.8	A15
Nb_2SnTa	16.4	A15

TABLE 4
Superconductive Compounds and Alloys (continued)

Substance	T_c,K	Crystal structure type
$Nb_{2.5}SnTa_{0.5}$	17.6	A15
$Nb_{2.75}SnTa_{0.25}$	17.8	A15
$Nb_{3x}SnTa_{3(1-x)}$	6.0—18.0	
$Nb_2SnTa_{0.5}V_{0.5}$	12.2	A15
$NbTc_3$	10.5	A12
$Nb_{0.75}Zr_{0.25}$	10.8	
$Nb_{0.66}Zr_{0.33}$	10.8	
$PbTa_3$	17	A15
$RhTa_3$	10	A15
$RhZr_2$	10.8; 11.3	C16 (Al_2Cu)
$Rh_{0-0.45}Zr_{1-0.55}$	2.1—10.8	
$SiTi_{0.3}V_{2.7}$	10.9	A15
SiV_3	17.1	A15
$SiV_{2.7}Zr_{0.3}$	13.2	A15

TABLE 5
Critical Field Data

Substance	H_o oersteds	Substance	H_o oersteds
Ag_2F	2.5	InSb	1100
Ag_7NO_{11}	57	In_xTl_{1-x}	252—284
Al_2CMo_3	1700	$In_{0.8}Tl_{0.2}$	252
$BaBi_3$	740	$Mg_{0.47}Tl_{0.53}$	220
Bi_2Pt	10	$Mo_{0.16}Ti_{0.84}$	<985
Bi_3Sr	530	$NbSn_2$	620
Bi_5Tl_3	>400	$PbTl_{0.27}$	756
CdSn	>266	$PbTl_{0.17}$	796
$CoSi_2$	105	$PbTl_{0.12}$	849
$Cr_{0.1}Ti_{0.3}V_{0.6}$	1360	$PbTl_{0.075}$	880
$In_{1-0.86}Mg_{0-0.14}$	272.4—259.2	$PbTl_{0.04}$	864

TABLE 6
High Critical Magnetic-Field Superconductive Compounds and Alloys

Substance	T_c, K	H_{c1}, kOe	H_{c2}, kOe	H_{c3}, kOe	T_{obs}, K[a]
Al_2CMo_3	9.8—10.2	0.091	156		1.2
$AlNb_3$		0.375			
$Ba_xO_3Sr_{1-x}Ti$	<0.1—0.55	0.0039 max.			
$Bi_{0.5}Cd_{0.1}Pb_{0.27}Sn_{0.13}$			>24		3.06
Bi_xPb_{1-x}	7.35—8.4	0.122 max.	30 max.		4.2
$Bi_{0.56}Pb_{0.44}$	8.8		15		4.2
$Bi_{7.5w/o}Pb_{92.5w/o}$[b]			2.32		
$Bi_{0.099}Pb_{0.901}$		0.29	2.8		
$Bi_{0.02}Pb_{0.98}$		0.46	0.73		
$Bi_{0.53}Pb_{0.32}Sn_{0.16}$			>25		3.06
$Bi_{1-0.93}Sn_{0-0.07}$			0—0.032		3.7
Bi_5Tl_3	6.4		>5.6		3.35
C_8K (excess K)	0.55		0.160 (H⊥c)		0.32
			0.730 (H∥c)		0.32

TABLE 6
High Critical Magnetic-Field Superconductive Compounds and Alloys (continued)

Substance	T_c, K	H_{c1}, kOe	H_{c2}, kOe	H_{c3}, kOe	T_{obs}, K[a]
C_8K	0.39		0.025 (H⊥c)		0.32
			0.250 (H∥c)		0.32
$C_{0.44}Mo_{0.56}$	12.5—13.5	0.087	98.5		1.2
CNb	8—10	0.12	16.9		4.2
$CNb_{0.4}Ta_{0.6}$	10—13.6	0.19	14.1		1.2
CTa	9—11.4	0.22	4.6		1.2
$Ca_xO_3Sr_{1-x}Ti$	<0.1—0.55	0.002—0.004			
$Cd_{0.1}Hg_{0.9}$ (by weight)		0.23	0.34		2.04
$Cd_{0.05}Hg_{0.95}$		0.28	0.31		2.16
$Cr_{0.10}Ti_{0.30}V_{0.60}$	5.6	0.071	84.4		0
GaN	5.85	0.725			4.2
Ga_xNb_{1-x}			>28		4.2
GaSb (annealed)	4.24		2.64		3.5
$GaV_{1.95}$	5.3		73[e]		
$GaV_{2.1-3.5}$	6.3—14.45		230—300[d]		0
GaV_3		0.4	350[e]		0
			500[d]		
$GaV_{4.5}$	9.15		121[c]		0
Hf_xNb_y			>52—>102		1.2
Hf_xTa_y			>28—>86		1.2
$Hg_{0.05}Pb_{0.95}$		0.235	2.3		
$Hg_{0.101}Pb_{0.899}$		0.23	4.3		4.2
$Hg_{0.15}Pb_{0.85}$	6.75		>13		2.93
$In_{0.98}Pb_{0.02}$	3.45	0.1		0.12	2.76
$In_{0.96}Pb_{0.04}$	3.68	0.1	0.12	0.25	2.94
$In_{0.94}Pb_{0.06}$	3.90	0.095	0.18	0.35	3.12
$In_{0.913}Pb_{0.087}$	4.2	~10.17	0.55	2.65	
$In_{0.316}Pb_{0.684}$		0.155	3.7		4.2
$In_{0.17}Pb_{0.83}$			2.8	5.5	4.2
$In_{1.000}Te_{1.002}$	3.5—3.7		1.2[c]		0
$In_{0.95}Tl_{0.05}$		0.263	0.263		3.3
$In_{0.90}Tl_{0.10}$		0.257	0.257		3.25
$In_{0.83}Tl_{0.17}$		0.242	0.39		3.21
$In_{0.75}Tl_{0.25}$		0.216	0.50		3.16
LaN	1.35	0.45			0.76
La_3S_4	6.5	≈0.15	>25		1.3
La_3Se_4	8.6	≈0.2	>25		1.25
$Mo_{0.52}Re_{0.48}$	11.1		14—21	22—33	4.2
			18—28	37—43	1.3
$Mo_{0.6}Re_{0.395}$	10.6		14—20	20—37	4.2
			19—26	26—37	1.3
$Mo_{0.5}Ti_{0.5}$			75[c]		0
$Mo_{0.16}Ti_{0.84}$	4.18	0.028	98.7[c]		0
			36—38		3.0
$Mo_{0.913}Ti_{0.087}$	2.95	0.060	15		4.2
$Mo_{0.1-0.3}U_{0.9-0.7}$	1.85—2.06		>25		
$Mo_{0.17}Zr_{0.83}$			30		
$N_{(12.8\ w/o)}Nb$	15.2		>9.5		13.2
NNb (wires)	16.1		153[c]		0
			132		4.2
			95		8
			53		12
NNb_xO_{1-x}	13.5—17.0		38		

TABLE 6
High Critical Magnetic-Field Superconductive Compounds and Alloys (continued)

Substance	T_c, K	H_{c1}, kOe	H_{c2}, kOe	H_{c3}, kOe	T_{obs}, K[a]
NNb_xZr_{1-x}	9.8—13.8		4- >130		4.2
$N_{0.93}Nb_{0.85}Zr_{0.15}$	13.8		>130		4.2
$Na_{0.086}Pb_{0.914}$		0.19	6.0		
$Na_{0.016}Pb_{0.984}$		0.28	2.05		
Nb	9.15		2.020		1.4
			1.710		4.2
Nb		0.4—1.1	3—5.5		4.2
Nb (unstrained)		1.1—1.8	3.40	6—9.1	4.2
Nb (strained)		1.25—1.92	3.44	6.0—8.7	4.2
Nb (cold-drawn wire)		2.48	4.10	≈10	4.2
Nb (film)			>25		4.2
NbSc			>30		
Nb_3Sn		0.170	221		4.2
			70		14.15
			54		15
			34		16
			17		17
$Nb_{0.1}Ta_{0.9}$		0.084	0.154		4.195
$Nb_{0.2}Ta_{0.8}$			10		4.2
$Nb_{0.65\text{-}0.73}Ta_{0.02\text{-}0.10}Zr_{0.25}$			>70—>90		4.2
Nb_xTi_{1-x}			148 max.		1.2
			120 max.		4.2
$Nb_{0.222}U_{0.778}$		1.98	23		1.2
Nb_xZr_{1-x}			127 max.		1.2
			94 max.		4.2
O_3SrTi	0.43	0.0049[c]	0.504[c]		0
O_3SrTi	0.33	0.00195[c]	0.420[c]		0
$PbSb_{1\ w/o}$(quenched)			>1.5		4.2
$PbSb_{1\ w/o}$(annealed)			>0.7		4.2
$PbSb_{2.8\ w/o}$(quenched)			>2.3		4.2
$PbSb_{2.8\ w/o}$(annealed)			>0.7		4.2
$Pb_{0.871}Sn_{0.129}$		0.45	1.1		
$Pb_{0.965}Sn_{0.035}$		0.53	0.56		
$Pb_{1\text{-}0.26}Tl_{0\text{-}0.74}$	7.20—3.68		2—6.9[c]		0
$PbTl_{0.17}$	6.73		4.5[c]		0
$Re_{0.26}W_{0.74}$			>30		
$Sb_{0.93}Sn_{0.07}$			0.12		3.7
SiV_3	17.0	0.55	156[e]		
Sn_xTe_{1-x}		0.00043—0.00236	0.005—0.0775		0.012—0.079
Ta (99.95%)		0.425	1.850		1.3
		0.325	1.425		2.27
		0.275	1.175		2.66
		0.090	0.375		3.72
$Ta_{0.5}Nb_{0.5}$			3.55		4.2
$Ta_{0.65\text{-}0}Ti_{0.35\text{-}1}$	4.4—7.8		>14—138		1.2
$Ta_{0.5}Ti_{0.5}$			138		1.2
Te	3.3	0.25[c]			0
Tc_xW_{1-x}	5.75—7.88		8—44		4.2
Ti				2.7	4.2
$Ti_{0.75}V_{0.25}$	5.3	0.029[c]	199[c]		0
$Ti_{0.775}V_{0.225}$	4.7	0.024[c]	172[c]		0
$Ti_{0.615}V_{0.385}$	7.07	0.050	34		4.2
$Ti_{0.516}V_{0.484}$	7.20	0.062	28		4.2
$Ti_{0.415}V_{0.585}$	7.49	0.078	25		4.2

TABLE 6
High Critical Magnetic-Field Superconductive Compounds and Alloys (continued)

Substance	T_c, K	H_{c1}, kOe	H_{c2}, kOe	H_{c3}, kOe	T_{obs}, K[a]
$Ti_{0.12}V_{0.88}$			17.3	28.1	4.2
$Ti_{0.09}V_{0.91}$			14.3	16.4	4.2
$Ti_{0.06}V_{0.94}$			8.2	12.7	4.2
$Ti_{0.03}V_{0.97}$			3.8	6.8	4.2
Ti_xV_{1-x}			108 max.		1.2
V	5.31	0.8	3.4		1.79
		0.75	3.15		2
		0.45	2.2		3
		0.30	1.2		4
$V_{0.26}Zr_{0.74}$	≈5.9	0.238			1.05
		0.227			1.78
		0.185			3.04
		0.165			3.5
W (film)	1.7—4.1		>34		1

[a] Temperature of critical field measurement.
[b] w/o denotes weight percent.
[c] Extrapolated.
[d] Linear extrapolation.
[e] Parabolic extrapolation.

REFERENCES

1. B. W. Roberts, in *Superconductive Materials and Some of Their Properties. Progress in Cryogenics*, Vol. IV, 1964, pp. 160—231.
2. B. W. Roberts, Superconductive Materials and Some of Their Properties, NBS Technical Notes 408 and 482, U.S. Government Printing Office, 1966 and 1969; B. W. Roberts, *J. Phys. Chem. Ref. Data*, 5, 581, 1976.
3. B. W. Roberts, Properties of Selected Superconductive Materials, 1978 Supplement, NBS Technical Note 983, 1978.
4. T. Claeson, *Phys. Rev.*, 147, 340, 1966.
5. C. J. Raub, W. H. Zachariasen, T. H. Geballe, and B. T. Matthias, *J. Phys. Chem. Solids*, 24, 1093, 1963.
6. T. H. Geballe, B. T. Matthias, V. B. Compton, E. Corenzwit, G. W. Hull, Jr., and L. D. Longinotti, *Phys. Rev.*, 1A, 119, 1965.
7. C. J. Raub, V. B. Compton, T. H. Geballe, B. T. Matthias, J. P. Maita, and G. W. Hull, Jr., *J. Phys. Chem. Solids*, 26, 2051, 1965.
8. R. D. Blaugher, J. K. Hulm, and P. N. Yocom, *J. Phys. Chem. Solids*, 26, 2037, 1965.
9. T. Claeson and H. L. Luo, *J. Phys. Chem. Solids*, 27, 1081, 1966.
10. S. C. Ng and B. N. Brockhouse, *Solid State Comm.*, 5, 79, 1967.
11. O. I. Shulishova and I. A. Shcherbak, *Izv. AN SSSR, Neorg. Materials*, 3, 1495, 1967.
12. T. F. Smith and H. L. Luo, *J. Phys. Chem. Solids*, 28, 569, 1967.
13. A. C. Lawson, *J. Less-Common Metals*, 23, 103, 1971.
14. R. Chevrel, M. Sergent, and J. Prigent, *J. Solid State Chem.*, 3, 515, 1971.
15. M. Marezio, P. D. Dernier, J. P. Remeika, and B. T. Matthias, *Mat. Res. Bull.*, 8, 657, 1973.
16. J. K. Hulm and R. D., *Blaugher in Superconductivity in d– and f–Band Metals*, D. H. Douglass,Ed., American Institute of Physics, 4, 1, 1972.
17. R. N. Shelton, A. C. Lawson, and D. C. Johnston, *Mat. Res. Bull.*, 10, 297, 1975.
18. H. D. Wiesinger, *Phys. Status Sol.*, 41A, 465, 1977.
19. O. Fisher, *Applied Phys.*, 16, 1, 1978.
20. D. C. Johnston, *Solid State Comm.*, 24, 699, 1977.
21. H. C. Ku and R. H. Shelton, *Mat. Res. Bull.*, 15, 1441, 1980.
22. H. Barz, *Mat. Res. Bull.*, 15, 1489, 1980.
23. G. P. Espinosa, A. S. Cooper, H. Barz, and J. P. Remeika, *Mat. Res. Bull.*, 15, 1635, 1980.
24. E. M. Savitskii, V. V. Baron, Yu. V. Efimov, M. I. Bychkova, and L. F. Myzenkova, in *Superconducting Materials*, Plenum Press, 1981, p. 107.
25. R. Fluckiger and R. Baillif, in Topics in *Current Physics*, O. Fischer and M. B. Maple, Eds., Springer Verlag, 34, 113, 1982.
26. R. N. Shelton, in *Superconductivity in d– and f–Band Metals*, W. Buckel and W. Weber, Eds., Kernforschungszentrum, Karlsruhe, 1982, p. 123.
27. D. C. Johnston and H. F. Braun, *Topics in Current Phys.*, 32, 11, 1982.
28. R. Chevrel and M. Sergent, *Topics in Current Phys.*, 32, 25, 1982.

29. G. P. Espinosa, A. S. Cooper, and H. Barz, *Mat. Res. Bull.*, 17, 963, 1982.
30. R. Muller, R. N. Shelton, J. W. Richardson, Jr., and R. A. Jacobson, *J. Less–Comm. Met.*, 92, 177, 1983.
31. You–Xian Zhao and Shou–An He, in *High Pressure in Science and Technology*, North Holland, 22, 51, 1983.
32. You–Xian Zhao and Shou–An He, *Solid State Comm.*, 24, 699, 1983.
33. G. P. Meisner and H. C. Ku, *Appl. Phys.*, A31, 201, 1983.
34. R. J. Cava, D. W. Murphy, and S. M. Zahurak, *J. Electrochem. Soc.*, 130, 2345, 1983.
35. R. N. Shelton, *J. Less–Comm. Met.*, 94, 69, 1983.
36. B. Chevalier, P. Lejay, B. Lloret, Wang Xian–Zhong, J. Etourneau, and P. Hagenmuller, *Annales de Chemie*, 9, 191, 1984.
37. G. Venturini, M. Meot–Meyer, E. McRae, J. F. Mareche, and B. Rogues, *Mat. Res. Bull.*, 19, 1647, 1984.
38. J. M. Tarascon, F. G. DiSalvo, D. W. Murphy, G. Hull, and J. V. Waszczak, *Phys. Rev.*, 29B, 172, 1984.
39. G. V. Subba and G. Balakrishnan, *Bull. Mat. Sci.*, 6, 283, 1984.
40. B. Batlog, *Physica*, 126B, 275, 1984.
41. M. J. Johnson, Ames Lab (USA) Report IS-T-1140, 1984.
42. I. M. Chapnik, *J. Mat. Sci. Lett.*, 4, 370, 1985.
43. W. Rong–Yao, L. Qi–Guang, and Z. Xiao, *Phys. Status Sol.*, 90A, 763, 1985.
44. W. Xian–Zhong, B. Chevalier, J. Etourneau, and P. Hagenmuller, *Mat. Res. Bull.*, 20, 517, 1985.
45. H. R. Ott, F. Hulliger, H. Rudigier, and Z. Fisk, *Phys. Rev.*, 31B, 1329, 1985.
46. P. Villars and L. D. Calver, *Pearson's Handbook of Crystallographic Data for Intermetallic Phases*, Vol. 1—3, ASM, 1985.
47. G. V. Subba Rao, K. Wagner, G. Balakhrishnan, J. Jakani, W. Paulus, and R. Scollhorn, *Bull. Mat. Sci.*, 7, 215, 1985.
48. J. G. Bednorz and K. A. Muller, *Zs. Physik*, B64, 189, 1986.
49. W. Rong–Yao, *Phys. Status Sol.*, 94A, 445, 1986.
50. H. D. Yang, R. N. Shelton, and H. F. Braun, *Phys. Rev.*, 33B, 5062, 1986.
51. G. Venturini, M. Kanta, E. McRae, J. F. Mareche, B. Malaman, and B. Roques, *Mat. Res. Bull.*, 21, 1203, 1986.
52. W. Rong–Yao, *J. Mat. Sci. Lett.*, 5, 87, 1986.
53. M. K. Wu, J. R. Ashburn, C. J. Torng, P. H. Hor, R. L. Meng, L. Gao, Z. J. Huang, Y. Q. Wang, and C. W. Chu, *Phys. Rev. Lett.*, 58, 908, 1987.
54. R. J. Cava, R. B. Van Dover, B. Batlog, and E. A. Rietman, *Phys. Rev. Lett.*, 58, 408, 1987.
55. L. C. Porter, T. J. Thorn, U. Geiser, A. Umezawa, H. H. Wang, W. K. Kwok, H–C. I. Kao, M. R. Monaghan, G. W. Crabtree, K. D. Carlson, and J. M. Williams, *Inorg. Chem.*, 26, 1645, 1987.
56. A. M. Kini, U. Geiser, H–C. I. Kao, K. D. Carlson, H. H. Wang, M. R. Monaghan, and K. M. Williams, *Inorg. Chem.*, 26, 1834, 1987.
57. T. Penney, S. von Molnar, D. Kaiser, F. Holtzberg, and A. W. Kleinsasser, *Phys. Rev.*, B38, 2918, 1988.
58. Y. K. Tao, J. S. Swinnea, A. Manthiram, J. S. Kim, J. B. Goodenoug, and H. Steinfink, *J. Mat. Res.*, 3, 248, 1988.
59. G. G. Peterson, B. R. Weinberger, L. Lynds, and H. A. Krasinski, *J. Mat. Res.*, 3, 605, 1988.
60. J. B. Torrance, Y. Tokura, A. Nazzai, and S. S. P. Parkin, *Phys. Rev. Lett.*, 60, 542, 1988.
61. K. Kourtakis, M. Robbins, P. K. Gallagher, and T. Teifel, *J. Mat. Res.*, 4, 1289, 1989.
62. J. C. Phillips, *Physics of High-T_c Superconductors*, Academic Press, 1989, p. 336.
63. Shui Wai Lin and L. I. Berger, *Rev. Sci. Instrum.*, 60, 507, 1989.
64. M. Tinkham, *Introduction to Superconductivity,* McGraw–Hill, New York, 1975.
65. O. Fischer and M.B. Maple, Eds., *Topics in Current Physics,* Volume 32: Superconductivity in Ternary Compounds I; Volume 34: Superconductivity in Ternary Compounds II, Springer–Verlag, Berlin, 1982.
66. K. J. Dunn and F. P. Bundy, *Phys. Rev.,* B25, 194, 1982.
67. A. Barone and G. Paterno, *Physics and Applications of the Josephson Effect,* Wiley, New York, 1982.
68. D. H. Douglass, Ed., *Superconductivity in d- and f-band Metals,* Plenum Press, New York, 1976.
69. D. M. Ginsberg, Ed., *Physical Properties of High Temperature Superconductors,* (Volume II, 1990; Volume III, 1992; Volume V, 1996), World Scientific, Singapore.
70. T. Ishiguro and K. Yamji, *Organic Superconductors,* Springer Verlag, Berlin, 1990.
71. Sh. Okada, K. Shimizu, T. C. Kobayashi, K. Amaya, and Sh. Endo., *J. Phys. Soc. Jpn.*, 65, 1924, 1996.
72. A. Bourdillon and N. X. Tan Bourdillon, *High Temperature Superconductors: Processing and Science,* Academic Press, 1994.
73. J. M. Williams, J. R. Ferraro, R. J. Thorn, K. Carlson, U. Geiser, H. H. Wang, A. M. Kini, and M.-H. Whangbo, *Organic Superconductors (Including Fullerenes): Synthesis, structure, Properties, and Theory,* Prentice–Hall, 1992.

HIGH TEMPERATURE SUPERCONDUCTORS

C. N. R. Rao and A. K. Raychaudhuri

The following tables give properties of a number of high temperature superconductors. Table 1 lists the crystal structure (space group and lattice constants) and the critical transition temperature T_c for the more important high temperature superconductors so far studied. Table 2 gives energy gap, critical current density, and penetration depth in the superconducting state. Table 3 gives electrical and thermal properties of some of these materials in the normal state. The tables were prepared in November 1992 and updated in November 1994.

REFERENCES

1. Ginsburg, D.M., Ed., *Physical Properties of High-Temperature Superconductors*, Vols. I—III, World Scientific, Singapore, 1989—1992.
2. Rao, C.N.R., Ed., *Chemistry of High-Temperature Superconductors*, World Scientific, Singapore, 1991.
3. Shackelford, J.F., *The CRC Materials Science and Engineering Handbook*, CRC Press, Boca Raton, 1992, 98—99 and 122—123.
4. Kaldis, E., Ed., *Materials and Crystallographic Aspects of HT$_c$-Superconductivity*, Kluwer Academic Publ., Dordrecht, The Netherlands, 1992.
5. Malik, S.K. and Shah, S.S., Ed., *Physical and Material Properties of High Temperature Superconductors*, Nova Science Publ., Commack, N.Y., 1994.
6. Chmaissem, O. et. al., *Physica*, C230, 231—238, 1994.
7. Antipov, E.V. et. al., *Physica*, C215, 1—10, 1993.

Table 1
Structural Parameters and Approximate T_c Values of High-Temperature Superconductors

Material	Structure	T_c/K (maximum value)
$La_2CuO_{4+\delta}$	Bmab; a = 5.355, b = 5.401, c = 13.15 Å	39
$La_{2-x}Sr_x(Ba_x)CuO_4$	I4/mmm; a = 3.779, c = 13.23 Å	35
$La_2Ca_{1-x}Sr_xCu_2O_6$	I4/mmm; a = 3.825, c = 19.42 Å	60
$YBa_2Cu_3O_7$	Pmmm; a = 3.821, b = 3.885, c = 11.676 Å	93
$YBa_2Cu_4O_8$	Ammm; a = 3.84, b = 3.87, c = 27.24 Å	80
$Y_2Ba_4Cu_7O_{15}$	Ammm; a = 3.851, b = 3.869, c = 50.29 Å	93
$Bi_2Sr_2CuO_6$	Amaa; a = 5.362, b = 5.374, c = 24.622 Å	10
$Bi_2CaSr_2Cu_2O_8$	A_2aa; a = 5.409, b = 5.420, c = 30.93 Å	92
$Bi_2Ca_2Sr_2Cu_3O_{10}$	A_2aa; a = 5.39, b = 5.40, c = 37 Å	110
$Bi_2Sr_2(Ln_{1-x}Ce_x)_2Cu_2O_{10}$	P4/mmm; a = 3.888, c = 17.28 Å	25
$Tl_2Ba_2CuO_6$	A_2aa; a = 5.468, b = 5.472, c = 23.238 Å; I4/mmm; a = 3.866, c = 23.239 Å	92
$Tl_2CaBa_2Cu_2O_8$	I4/mmm; a = 3.855, c = 29.318 Å	119
$Tl_2Ca_2Ba_2Cu_3O_{10}$	I4/mmm; a = 3.85, c = 35.9 Å	128
$Tl(BaLa)CuO_5$	P4/mmm; a = 3.83, c = 9.55 Å	40
$Tl(SrLa)CuO_5$	P4/mmm; a = 3.7, c = 9 Å	40
$(Tl_{0.5}Pb_{0.5})Sr_2CuO_5$	P4/mmm; a = 3.738, c = 9.01 Å	40
$TlCaBa_2Cu_2O_7$	P4/mmm; a = 3.856, c = 12.754 Å	103
$(Tl_{0.5}Pb_{0.5})CaSr_2Cu_2O_7$	P4/mmm; a = 3.80, c = 12.05 Å	90
$TlSr_2Y_{0.5}Ca_{0.5}Cu_2O_7$	P4/mmm; a = 3.80, c = 12.10 Å	90
$TlCa_2Ba_2Cu_3O_8$	P4/mmm; a = 3.853, c = 15.913 Å	110
$(Tl_{0.5}Pb_{0.5})Sr_2Ca_2Cu_3O_9$	P4/mmm; a = 3.81, c = 15.23 Å	120
$TlBa_2(La_{1-x}Ce_x)_2Cu_2O_9$	I4/mmm; a = 3.8, c = 29.5 Å	40
$Pb_2Sr_2La_{0.5}Ca_{0.5}Cu_3O_8$	Cmmm; a = 5.435, b = 5.463, c = 15.817 Å	70
$Pb_2(Sr,La)_2Cu_2O_6$	$P22_12$; a = 5.333, b = 5.421, c = 12.609 Å	32
$(Pb,Cu)Sr_2(La,Ca)Cu_2O_7$	P4/mmm; a = 3.820, c = 11.826 Å	50
$(Pb,Cu)(Sr,Eu)(Eu,Ce)Cu_2O_x$	I4/mmm; a = 3.837, c = 29.01 Å	25
$Nd_{2-x}Ce_xCuO_4$	I4/mmm; a = 3.95, c = 12.07 Å	30
$Ca_{1-x}Sr_xCuO_2$	P4/mmm; a = 3.902, c = 3.35 Å	110
$Sr_{1-x}Nd_xCuO_2$	P4/mmm; a = 3.942, c = 3.393 Å	40
$Ba_{0.6}K_{0.4}BiO_3$	Pm3m; a = 4.287 Å	31
Rb_2CsC_{60}	a = 14.493 Å	31
$NdBa_2Cu_3O_7$	Pmmm; a = 3.878, b = 3.913, c = 11.753	58

HIGH TEMPERATURE SUPERCONDUCTORS (continued)

Table 1
Structural Parameters and Approximate T_c Values of High-Temperature Superconductors (continued)

Material	Structure	T_c/K (maximum value)
$SmBaSrCu_3O_7$	I4/mmm; $a = 3.854$, $c = 11.62$	84
$EuBaSrCu_3O_7$	I4/mmm; $a = 3.845$, $c = 11.59$	88
$GdBaSrCu_3O_7$	I4/mmm; $a = 3.849$, $c = 11.53$	86
$DyBaSrCu_3O_7$	Pmmm; $a = 3.802$, $b = 3.850$, $c = 11.56$	90
$HoBaSrCu_3O_7$	Pmmm; $a = 3.794$, $b = 3.849$, $c = 11.55$	87
$ErBaSrCu_3O_7$ (multiphase)	Pmmm; $a = 3.787$, $b = 3.846$, $c = 11.54$	82
$TmBaSrCu_3O_7$ (multiphase)	Pmmm; $a = 3.784$, $b = 3.849$, $c = 11.55$	88
$YBaSrCu_3O_7$	Pmmm; $a = 3.803$, $b = 3.842$, $c = 11.54$	84
$HgBa_2CuO_4$	I4/mmm; $a = 3.878$, $c = 9.507$	94
$HgBa_2CaCu_2O_6$ (annealed in O_2)	I4/mmm; $a = 3.862$, $c = 12.705$	127
$HgBa_2Ca_2Cu_3O_8$	Pmmm; $a = 3.85$, $c = 15.85$	133
$HgBa_2Ca_3Cu_4O_{10}$	Pmmm; $a = 3.854$, $c = 19.008$	126

Table 2
Superconducting Properties

J_c (0): Critical current density extrapolated to 0 K
λ_{ab}: Penetration depth in *a-b* plane
k_B: Boltzmann constant

Material	Form	Energy gap (Δ) $2\Delta_{pp}/k_BT_c$*	Energy gap (Δ) $2\Delta_{fit}/k_BT_c$†	$10^{-6} \times J_c$ (0)/A cm^{-2}	λ_{ab}/Å
$Y Ba_2Cu_3O_7$	Single Crystal	5–6	4–5	30 (film)	1400
$Bi_2Sr_2CaCu_2O_8$	Single Crystal	8–9	5.5–6.5	2	2700
$Tl_2Ba_3CaCu_2O_8$	Ceramic	6–7	4–6	10 (film, 80 K)	2000
$La_{2-x}Sr_xCuO_4$, $x = 0.15$	Ceramic	7–9	4–6		
$Nd_{2-x}Ce_xCuO_4$	Ceramic	8	4–5	0.2 (film)	

* Obtained from peak to peak value.
† Obtained from fit to BCS-type relation.

Table 3
Normal State Properties

ρ_{ab}: Resistivity in the *a-b* plane
ρ_c: Resistivity along the *c* axis
+ve: ρ_c has positive temperature coefficient of resistivity
−ve: ρ_c has negative temperature coefficient of resistivity
n_H: Hall density
k: Thermal conductivity
in plane: Along *a-b* plane
out of plane: Perpendicular to *a-b* plane

Material	Form	$\rho_{ab}/\mu\Omega$ cm		ρ_c/mΩ cm	$d\rho_c/dT$	$10^{-21} \times n_H$/cm^{-3}		*k*/(mW/cm K) at 300 K	
		300 K	100 K	300 K		300 K	100 K	in plane	out of plane
$YBa_2Cu_3O_7$	Single Crystal	110	35	5	+ve	11–16	4–6	120	3
	film	200–300	60–100			5–9	2–3		
$YBa_2Cu_4O_8$	Single Crystal	75	20	10	−ve	14			
	film	100–200	20–50			22	17		
$Bi_2Sr_2CuO_6$	Single Crystal	300	150	5000	−ve	6	5		
$Bi_2Sr_2CaCu_2O_8$	Single Crystal	150	50	>1000	−ve	4	3	60	8
$Tl_2Ba_2CuO_6$	Single Crystal	300–400	50–75	200–300	+ve	3.1	2.5		
$Tl_2Ba_2Ca_2Cu_3O_{10}$	Ceramic	***	**				≈ 2*		
$La_{2-x}Sr_xCuO_4$, $x = 0.12$	Single Crystal	900	350	200	+ve for T >225 K	2.5			
$La_{2-x}Sr_xCuO_4$, $x = 0.20$	Single Crystal	400	200	80	+ve for T >150 K	10		50 (for $x = 0.04$)	20
	film	400	160			8.4	6.3		
$Nd_{2-x}Ce_xCuO_4$, $x = 0.17$	Single Crystal	500	275			53	17		
$x = 0.15$	film	140–180	35			32	11	250 (for $x = 0.15$)	

* At 200 K
** ρ ~0.4 mΩ cm at 120 K
*** ρ ~1.5 mΩ cm at 300 K

ORGANIC SUPERCONDUCTORS

H.P.R. Frederikse

Although the vast majority of organic compounds are insulators, a small number of organic solids show considerable electrical conductivity. Some of these materials appear to be superconductors. The superconducting organics fall primarily into two groups: those containing fulvalenes (pentagonal rings containing sulfur or selenium) and those based on fullerenes, involving the nearly spherical cluster C_{60}.

The transition temperatures T_c of the fulvalene derivatives are shown in Table 1. The abbreviations of the various molecular groups are listed in Table 2 and their chemical structures are depicted in Figure 1. Most of the T_c's are between 1 and 12 K. Several of the compounds only show superconductivity under pressure.

The fullerenes are A_3C_{60} compounds, where A represents a single or a combination of alkali atoms. The C_{60} cluster is shown in Figure 2a, while Figure 2b illustrates how the alkali atoms fit into the A_3C_{60} molecule to form the A15 crystallographic structure. Their superconducting transition temperatures range from 8 to 31.3 K (see Table 3).

REFERENCES

1. Ishigura, T. and Yamaji, K., *Organic Superconductors*, Springer-Verlag, Berlin, 1990.
2. Williams, Jack M. et al., *Organic Superconductors (Including Fullerenes)*, Prentice Hall, Englewood Cliffs, N.J., 1992.
3. *The Fullerenes*, Ed.: Krato, H.W., Fisher, J.E., and Cox, D.E., Pergamon Press, Oxford, 1993.
4. Schluter, M. et al., in *The Fullerenes* (Ref. 3), p. 303.

Table 1
Critical Pressure and Maximum Critical Temperature of Organic Superconductors

Material	P_c/kbar	T_c/K	Material	P_c/kbar	T_c/K
$(TMTSF)_2PF_6$	6.5	1.2	β-$(ET)_2IBr_2$	0	2.8
$(TMTSF)_2AsF_6$	9	1.3	β-$(ET)_2AuI_2$	0	4.8
$(TMTSF)_2SbF_6$	11	0.4	$(ET)_4Hg_{2.89}Cl_8$	0	4.2
$(TMTSF)_2TaF_6$	12	1.4	$(ET)_4Hg_{2.89}Br_8$	12	1.8
$(TMTSF)_2ClO_4$	0	1.4	$(ET)_3Cl_2(H_2O)_2$	16	2
$(TMTSF)_2ReO_4$	9.5	1.3	κ-$(ET)_2Cu(NCS)_2$	0	10.4
$(TMTSF)_2FSO_3$	5	3	κ-$(d\text{-}ET)_2Cu(NCS)_2$	0	11.4
$(ET)_4(ReO_4)_2$	4.5	2	$(DMET)_2Au(CN)_2$	1.5	0.9
β_L-$(ET)_2I_3$	0	1.4	$(DMET)_2AuI_2$	5	0.6
β_H-$(ET)_2I_3$	0	8.1	$(DMET)_2AuBr_2$	0	1.9
γ-$(ET)_3I_{2.5}$	0	2.5	$(DMET)_2AuCl_2$	0	0.9
ε-$(ET)_2I_3(I_8)_{0.5}$	0	2.5	$(DMET)_2I_3$	0	0.6
α-$(ET)_2I_3I_2$-doped	0	3.3	$(DMET)_2IBr_2$	0	0.7
α_t-$(ET)_2I_3$	0	8	$(MDT\text{-}TTF)_2AuI_2$	0	3.5
$\varepsilon\rightarrow\beta$-$(ET)_2I_3$[a]	0	6	$TTF[Ni(dmit)_2]_2$	2	1.6[b]
θ-$(ET)_2I_3$	0	3.6	$TTF[Pd(dmit)_2]_2$	20	6.5
κ-$(ET)_2I_3$	0	3.6	$(CH_3)_4N[Ni(dmit)_2]_2$	7	5

[a] Converted form ε-type to β-type by thermal treatment.
[b] For 7 kbar.

From Ishigura, T. and Yamaji, K., *Organic Superconductors*, Springer-Verlag, Berlin, 1990. With permission.

Table 2
List of Symbols and Abbreviations

TTF	tetrathiafulvalene
TMTSF	tetramethyltetraselenafulvalene
BEDT-TTF or "ET"	bis(ethylenedithio)tetrathiafulvalene
MDT-TTF	methylenedithiotetrathiafulvalene
DMET	[dimethyl(ethylenedithio)diselenadithiafulvalene]
dmit	4,5-dimercapto-1,3-dithiole-2-thione
T_c	transition temperature to superconducting state
P_c	minimum pressure required for superconducting transition

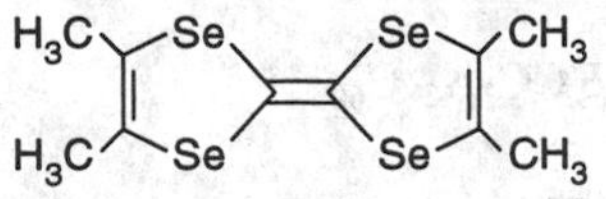

TMTSF

Tetramethyltetraselenafulvalene

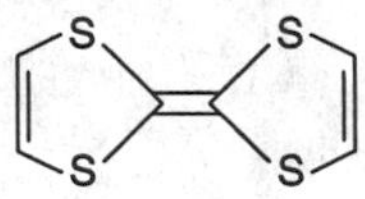

TTF

Tetrathiafulvalene

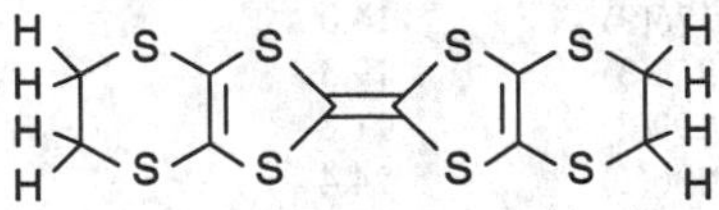

BEDT – TTF or ET

Bis(ethylenedithio)tetrathiafulvalene

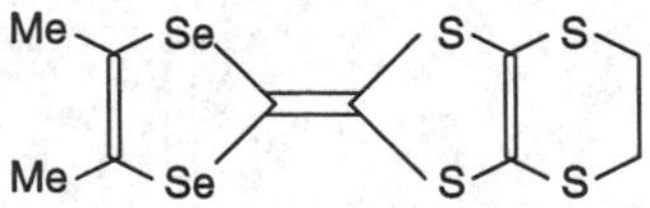

DMET

Dimethyl(ethylenedithio)diselenadithiafulvalene

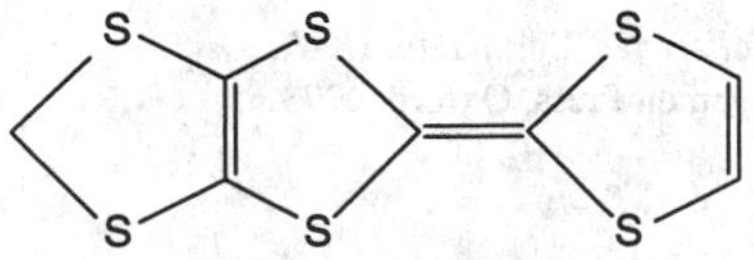

MDT – TTF

Methylenedithiotetrathiafulvalene

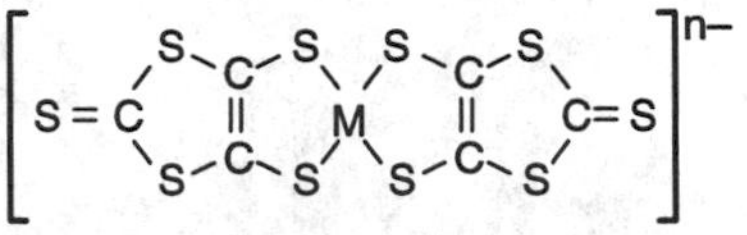

M=Ni, Pd, Pt

$M(dmit)_2^{2-}$

Ligand is 4,5-dimercapto-1.3-dithiole-2-thione

FIGURE 1. Structures of various donor molecules and acceptor species.

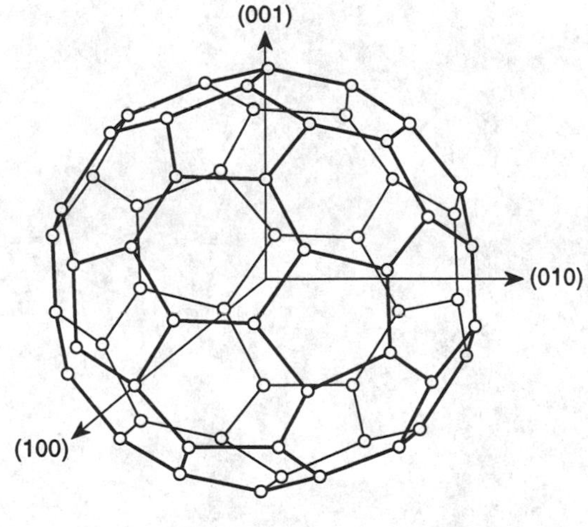

a

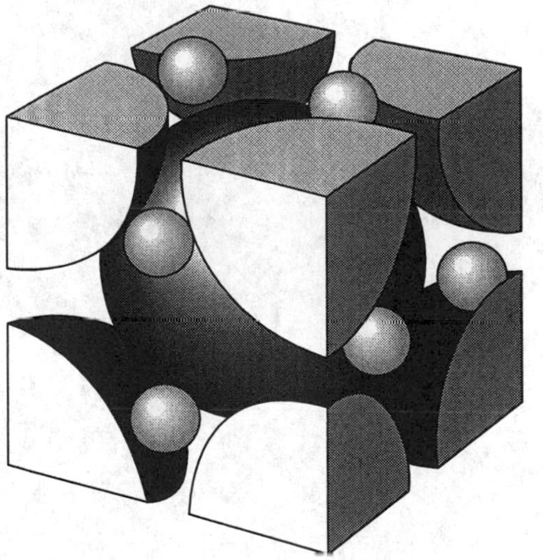

b

FIGURE 2. (a) C_{60} cluster placed in a fcc lattice. Each crystal axis crosses a double bond shared by two hexagons. (b) A hypothetical A_3C_{60} with the A15 structure. The structure can be seen to be an ordered defect structure of A_6C_{60}.

Table 3
Unit Cell and T_c for FCC-A_3 C_{60}

	Lattice parameter(s) (Å)	T_c/K
$Na_2Rb_{0.5}Cs_{0.5}C_{60}$	14.148(3)	8.0
Na_2CsC_{60} No. 1[a]	14.132(2)	10.5
Na_2CsC_{60} No. 2[a]	14.176(9)	14.0
K_3C_{60}	14.253(3)	19.3
K_2RbC_{60}	14.299(2)	21.8
Rb_2KC_{60} No. 1[a]	14.336(1)	24.4
Rb_2KC_{60} No. 2[a]	14.364(5)	26.4
Rb_3C_{60}	14.436(2)	29.4
Rb_2CsC_{60}	14.493(2)	31.3

[a] Samples labeled No. 1 and No. 2 have the same nominal composition.

From Schluter, M et. al., *The Fullerenes*, Ed.: Krato, H.W., Fisher, J.E., and Cox, D.E., Pergamon Press, Oxford, 1993. With permission.

PROPERTIES OF SEMICONDUCTORS

L. I. Berger and B. R. Pamplin

The term "semiconductor" is applied to a material in which electric current is carried by electrons or holes and whose electrical conductivity, when extremely pure, rises exponentially with temperature and may be increased from its low "intrinsic" value by many orders of magnitude by "doping" with electrically active impurities.

Semiconductors are characterized by an energy gap in the allowed energies of electrons in the material which separates the normally filled energy levels of the *valence band* (where "missing" electrons behave like positively charged current carriers "holes") and the *conduction band* (where electrons behave rather like a gas of free negatively charged carriers with an effective mass dependent on the material and the direction of the electrons' motion). This energy gap depends on the nature of the material and varies with direction in anisotropic crystals. It is slightly dependent on temperature and pressure, and this dependence is usually almost linear at normal temperatures and pressures.

Data are presented in three tables. Table I "General Properties of Semiconductors" lists the main crystallographic and semiconducting properties of a large number of semiconducting materials in three main categories: "Tetrahedral Semiconductors" in which every atom is tetrahedrally co-ordinated to four nearest neighbor atoms (or atomic sites) as for example in the diamond structure; "Octahedral Semiconductors" in which every atom is octahedrally co-ordinated to six nearest neighbor atoms — as for examples the halite structure; and "Other Semiconductors."

Table II gives more detailed information about some better known semiconductors, while Table III gives some information about the electronic energy band structure parameters of the best known materials.

Table I
PHYSICO-CHEMICAL PROPERTIES OF SEMICONDUCTORS (LISTED BY CRYSTAL STRUCTURE)

Substance	Molecular mass	Average atomic mass	Lattice parameters (Å, room temp.)	Density (g/cm³)	Melting point (K)	Microhardness, N/mm² (M-Mohs Scale)	Specific heat, J/kg·K (300 K)	Debye temp. (K)	Coefficient of thermal linear expansion [10^{-6} K^{-1} (300K)]	Thermal conductivity [mW/cm·K (300K)]
PART A. ADAMANTINE SEMICONDUCTORS										
§A1. Diamond Structure Elements (Strukturbericht symbol A4, Space Group Fd3m-O_h^7)										
C		12.01	3.56683	3.51	≈3850 Transition to graphite > 980	10(M)	471.5	2340	1.18	9900(I) 23200(IIA) 13600(IIB)
Si		28.09	5.43072	2.3283	1685 ± 2	11270	702	645	2.49	1240
Ge		72.59	5.65754	5.3234	1231	7644	321.9	374	6.1	640
α-Sn		118.69	6.4912	5.765 (281 K)	505.2 (Tr. 286.4)		213	230	5.4 (220 K)	
§A2. Sphalerite (Zinc Blende) Structure Compounds (Strukturbericht symbol B3 Space Group F $\bar{4}$ 3m-T_d^2)										
I VII Compounds										
CuF	82.54	41.27	4.255		1181					
CuCl	98.99	49.49	5.4057	3.53	695	2.3 (M)	490	240	12.1	8.4
CuBr	143.36	71.73	5.6905	4.98	770	2.5(M)	381	207	15.4	12.5
CuI	190.46	95.23	6.60427	5.63	878	192	276	181	19.2	16.8
AgBr	187.78	93.89		6.473	>1570 (Tr. 410)	2.5(M)	270			
AgI	234.77	117.39	6.502	5.67	831	2.5(M)	232	134	−2.5	4.2
II VI Compounds										
BeS	41.08	20.54	4.865	2.36						
BeSe	87.97	43.99	5.139	4.315						
BeTe	136.61	68.31	5.626	5.090						
BePo	(218)	(109)	5.838	7.3						
ZnO	81.37	40.69	4.63	5.675	2248	5.0 (M)	494	416	2.9	234
ZnS	97.43	48.72	5.4093	4.079	2100 (Tr. 1295)	1780	472	530	6.36	251
ZnSe	144.34	72.17	5.6676	5.42	1790	1350	339	400	7.2	140
ZnTe	192.99	96.5	6.101	6.34	1568	900	264	223	8.19	108
ZnPo	(274)	(137)	6.309							
CdS	144.46	72.23	5.832	4.826	1750	1250	330	219	4.7	200
CdSe	191.36	95.68	6.05	5.674	1512	1300	255	181	3.8	90
CdTe	240.00	120.00	6.477	5.86	1365	600	205	200	4.9	58.5
CdPo	(321)	(161)	6.665							
HgS	232.65	116.33	5.8517	7.73	1820	3(M)	210			
HgSe	279.55	139.78	6.084	8.25	1070	2.5(M)	178	151	5.46	10
HgTe	328.19	164.10	6.4623	8.17	943	300	164	242	4.6	20
III V Compounds										
BN	24.82	12.41	3.615	3.49	≈3300	10(M)	793	≈1900		200
BP(L.T.)	41.78	20.87	4.538	2.9	≈2800	37000		≈980		
BAs	85.73	42.87	4.777		≈2300	19000		≈625		

Table I
PHYSICO-CHEMICAL PROPERTIES OF SEMICONDUCTORS (LISTED BY CRYSTAL STRUCTURE) (continued)

Substance	Molecular mass	Average atomic mass	Lattice parameters (Å, room temp.)	Density (g/cm^3)	Melting point (K)	Microhardness, N/mm^2 (M-Mohs Scale)	Specific heat, J/kg·K (300 K)	Debye temp. (K)	Coefficient of thermal linear expansion [10^{-6} K^{-1} (300K)]	Thermal conductivity [mW/cm·K (300K)]
AlP	57.95	28.98	5.451	2.42	≈2100	5.5(M)		588		920
AlAs	101.90	50.95	5.6622	3.81	2013	5000		417	3.5	840
AlSb	148.73	74.37	6.1355	4.218	1330	4000		292	4.2	600
GaP	100.69	50.35	5.4505	4.13	1750	9450		446	5.3	752
GaAs	144.64	72.32	5.65315	5.316	1510	7500		344	5.4	560
GaSb	191.47	95.74	6.0954	5.619	980	4480	320	265	6.1	270
InP	145.79	72.90	5.86875	4.787	1330	4100		321	4.6	800
InAs	189.74	94.87	6.05838	5.66	1215	3300	268	249	4.7	290
InSb	236.57	118.29	6.47877	5.775	798	2200	144	202	4.7	160
Other sphalerite structure compounds										
MnS	87.0	43.5	5.011							
MnSe	133.9	66.95	5.82							
β-SiC	40.1	20.1	4.348	3.21	3070					
Ga_2Se_3	376.32	75.26	5.429	4.92	1020	3160			8.9	50
Ga_2Te_3	522.24	104.45	5.899	5.75	1063	2370				47
In_2Te_3(H.T.)	608.44	121.7	6.150	5.8	940	1660				69
$MgGeP_2$	158.84	39.71	5.652							
$ZnSnP_2$	246.00	61.5	5.65		1200					
$ZnSnAs_2$(H.T.)	333.90	82.38	5.851	5.53	1050					76
$ZnSnSb_2$	427.56	106.89	6.281	5.67	870	2500				76

§A3. Wurtzite (Zincite) Structure Compounds (Strukturbericht symbol B4, Space Group P 6_3mc-C^4_{6v})

Substance	Molecular mass	Average atomic mass	Lattice parameters (Å, room temp.)	Density (g/cm^3)	Melting point (K)	Microhardness, N/mm^2 (M-Mohs Scale)	Specific heat, J/kg·K (300 K)	Debye temp. (K)	Coefficient of thermal linear expansion [10^{-6} K^{-1} (300K)]	Thermal conductivity [mW/cm·K (300K)]
I VII Compounds										
CuCl	99.0	49.5	3.91 6.42		T_c680K					
CuBr	143.46	71.73	4.06 6.66		T_c658K					
CuI	190.46	95.23	4.31 7.09							
AgI	234.80	117.40	4.580 7.494							
II VI Compounds										
BeO	25.01	12.51	2.698 4.380		2800					
MgTe	151.9	76.0	4.54 7.39	3.85	≈2800					
ZnO	81.37	40.69	3.24950 5.2069	5.66	2250					600
ZnS	97.43	48.72	3.8140 6.2576	4.1	2100					460
ZnTe	192.99	46.50	4.27 6.99		1568					
CdS	144.46	72.23	4.1348 6.7490	4.82	1748					401
CdSe	191.36	95.68	4.299 7.010	5.66	1512					316
CdTe	240.00	120.00	4.57 7.47							
III V Compounds										
BP(H.T.)	41.79	20.90	3.562 5.900							
AlN	40.99	20.50	3.111 4.978	3.26	≈2500					823
GaN	83.73	41.87	3.190 5.189	6.10	1500					656
InN	128.83	64.42	3.533 5.693	6.88	1200					556
Other wurtzite structure compounds										
MnS	87.0	43.5	3.985 6.45	3.248						
MnSe	133.9	66.95	4.12 6.72							
SiC	40.1	20.1	3.076 5.048							
MnTe	182.54	91.27	4.078 6.701							
Al_2S_3	150.14	30.03	3.579 5.829	2.55	1400					
Al_2Se_3	290.84	58.17	3.890 6.30	3.91	1250					

§A4. Chalcopyrite Structure Compounds (Strukturbericht symbol $E1_1$, Space Group I $\bar{4}$ 2d-D^{12}_{2d})

Substance	Molecular mass	Average atomic mass	Lattice parameters (Å, room temp.)	Density (g/cm^3)	Melting point (K)	Microhardness, N/mm^2 (M-Mohs Scale)	Specific heat, J/kg·K (300 K)	Debye temp. (K)	Coefficient of thermal linear expansion [10^{-6} K^{-1} (300K)]	Thermal conductivity [mW/cm·K (300K)]
I III VI_2 Compounds										
$CuAlS_2$	154.65	38.66	5.323 10.44	3.47	2500					
$CuAlSe_2$	248.45	62.11	5.617 10.92	4.70	2260					
$CuAlTe_2$	345.73	86.43	5.976 11.80	5.50	2550					
$CuGaS_2$	197.39	49.35	5.360 10.49	4.35	2300					
$CuGaSe_2$	291.19	72.80	5.618 11.01	5.56	1970	4200		275	5.4	42
$CuGaTe_2$	388.47	97.12	6.013 11.93	5.99	2400	3500			6.9	27
$CuInS_2$	242.49	60.62	5.528 11.08	4.75	1400	2550				
$CuInSe_2$	336.29	84.07	5.785 11.56	5.77	1600	2050			6.6	37
$CuInTe_2$	433.57	108.39	6.179 12.365	6.10	1660	400		195	7.1	49
$CuTlS_2$	332.05	83.01	5.580 11.17	6.32						
$CuTlSe_2$(L.T.)	425.85	106.46	5.844 11.65	7.11	900					
$CuFeS_2$	183.51	45.88	5.25 10.32	4.088						
$CuFeSe_2$	277.31	69.33			850					

Table I
PHYSICO-CHEMICAL PROPERTIES OF SEMICONDUCTORS (LISTED BY CRYSTAL STRUCTURE) (continued)

Substance	Molecular mass	Average atomic mass	Lattice parameters (Å, room temp.)		Density (g/cm³)	Melting point (K)	Microhardness, N/mm² (M-Mohs Scale)	Specific heat, J/kg·K (300 K)	Debye temp. (K)	Coefficient of thermal linear expansion [10^{-6} K^{-1} (300K)]	Thermal conductivity [mW/cm·K (300K)]
$CuLaS_2$	266.58	66.65	5.65	10.86							
$AgAlS_2$	198.97	49.74	5.707	10.28	3.94						
$AgAlSe_2$	292.77	73.19	5.968	10.77	5.07	1220					
$AgAlTe_2$	390.05	97.51	6.309	11.85	6.18	1000					
$AgGaS_2$	241.71	60.43	5.755	10.28	4.72						
$AgGaSe_2$	335.51	83.88	5.985	10.90	5.84	1120	4400				
$AgGaTe_2$	432.79	108.2	6.301	11.96	6.05	990	1800		212		10
$AgInS_2$(L.T.)	286.87	71.70	5.828	11.19	5.00		2250				
$AgInSe_2$	380.61	95.15	6.102	11.69	5.81	1053	1850				30
$AgInTe_2$	477.89	119.47	6.42	12.59	6.12	965				9.49, 0.69	
$AgFeS_2$	227.83	56.96	5.66	10.30	4.53						
II IV V_2 Compounds											
$ZnSiP_2$	155.40	38.85	5.400	10.441	3.39	1640	11000				
$ZnGeP_2$	199.90	49.98	5.465	10.771	4.17	1295	8100				180
$ZnSnP_2$	246.00	61.5					6500				
$CdSiP_2$	202.43	50.61	5.678	10.431	4.00	≈1470	10500		282		
$CdGeP_2$	246.94	61.74	5.741	10.775	4.48	1049	5650				110
$CdSnP_2$	243.03	73.26	5.900	11.518			5000		195		140
$ZnSiAs_2$	242.20	60.55	5.61	10.88	4.70	1311	9200				
$ZnGeAs_2$	287.80	71.95	5.672	11.153	5.32	1150	6800		263		110
$ZnSnAs_2$	333.90	83.48	5.8515	11.704	5.53	1048	4550		271		150
$CdSiAs_2$	290.34	72.58	5.884	10.882			6850				
$CdGeAs_2$	334.83	83.71	5.9427	11.2172	5.60	938	4700				48
$CdSnAs_2$	380.93	95.23	6.0944	11.9182	5.72	880	3450				40

§A5. Other Ternary Semiconductors with Tetrahedral Coordination

Substance	Molecular mass	Average atomic mass	Lattice parameters (Å, room temp.)		Density (g/cm³)	Melting point (K)	Microhardness, N/mm² (M-Mohs Scale)	Specific heat, J/kg·K (300 K)	Debye temp. (K)	Coefficient of thermal linear expansion [10^{-6} K^{-1} (300K)]	Thermal conductivity [mW/cm·K (300K)]
I_2 IV VI_3 Compounds											
Cu_2SiS_3(H.T.)	251.36	41.89	3.684	6.004	3.81	1200					23
Cu_2SiS_3(L.T.)			5.290	10.156	3.63						
Cu_2SiTe_3	537.98	89.66	5.93		5.47						
Cu_2GeS_3(H.T.)	295.88	49.31	5.317		4.45	1210	4550	510	254	7.2	12
Cu_2GeS_3(L.T.)			5.327	5.215	4.46						
Cu_2GeSe_3	436.56	72.76	5.589	5.485	5.57	1030	3840	340	168	8.4	24
Cu_2GeTe_3	582.51	97.09	5.958	5.935	5.92		2890				130
Cu_2SnS_3	341.98	57.00	5.436		5.02	1110	2770	440	214	7.8	28
$CuSnSe_3$	482.66	80.44	5.687		5.94	960	2510	310	148	8.9	35
Cu_2SnTe_3	628.61	104.77	6.048		6.51	680	1970				144
Ag_2GeSe_3	525.21	87.54									
Ag_2SnSe_3	571.31	95.22									
Ag_2GeTe_3	671.13	111.86									
Ag_2SnTe_3	717.23	119.54									
I_3 V VI_4 Compounds											
Cu_3PS_4	349.85	40.73	7.44	6.19							
Cu_3AsS_4	393.79	49.22	6.43	6.14	4.37					3.2	30.2
Cu_3AsSe_4	581.37	72.67	5.570	10.957	5.61				169	9.5	19
Cu_3SbS_4	440.64	55.08	5.38	16.76	4.90						
Cu_3SbSe_4	628.22	78.53	5.654	11.256	6.0				131	12.4	14.6
I IV_2 V_3 Compounds											
$CuSi_2P_3$	212.64	35.44	5.25								
$CuGe_2P_3$	301.65	50.28	5.375		4.318	1113	8500	429	8.21	37.6	
$AgGe_2P_3$	345.97	57.66				1015	6150				

§A6. "Defect Chalcopyrite" Structure Compounds (Strukturbericht symbol E3, Space Group I $\bar{4}$-S_4^2)

Substance	Molecular mass	Average atomic mass	Lattice parameters (Å, room temp.)		Density (g/cm³)	Melting point (K)
$ZnAl_2Se_4$	435.18	62.17	5.503	10.90	4.37	
$ZnAl_2Te_4$(?)	629.74	84.96	5.904	12.05	4.95	
$ZnGa_2S_4$(?)	333.06	47.58	5.274	10.44	3.80	
$ZnGa_2Se_4$(?)	520.66	74.38	5.496	10.99	5.21	
$ZnGa_2Te_4$(?)	715.22	102.17	5.937	11.87	5.67	
$ZnIn_2Se_4$	610.86	87.27	5.711	11.42	5.44	1250
$ZnIn_2Te_4$	805.42	115.06	6.122	12.24	5.83	1075
$CdAl_2S_4$	294.61	42.09	5.564	10.32	3.06	
$CdAl_2Se_4$	482.21	68.89	5.747	10.68	4.54	
$CdAl_2Te_4$(?)	676.77	97.68	6.011	12.21	5.10	

Table I
PHYSICO-CHEMICAL PROPERTIES OF SEMICONDUCTORS (LISTED BY CRYSTAL STRUCTURE) (continued)

Substance	Molecular mass	Average atomic mass	Lattice parameters (Å, room temp.)		Density (g/cm³)	Melting point (K)	Microhardness, N/mm² (M-Mohs Scale)	Specific heat, J/kg·K (300 K)	Debye temp. (K)	Coefficient of thermal linear expansion [10^{-6} K^{-1} (300K)]	Thermal conductivity [mW/cm·K (300K)]
$CdGa_2S_4$	380.09	54.30	5.577	10.08	4.03						
$CdGa_2Se_4$	567.69	81.10	5.743	10.73	5.32						
$CdGa_2Te_4$	762.25	108.89	6.093	11.81	5.77						
$CdIn_2Te_4$	852.45	121.78	6.205	12.41	5.9	1060					
$HgAl_2S_4$	382.79	54.68	5.488	10.26	4.11						
$HgAl_2Se_4$	570.39	82.48	5.708	10.74	5.05						
$HgAl_2Te_4$(?)	764.48	109.28	6.004	12.11	5.81						
$HgGa_2S_4$	468.27	66.90	5.507	10.23	5.00						
$HgGa_2Se_4$	655.87	93.70	5.715	10.78	6.18						
$HgIn_2Se_4$	746.07	106.58	5.764	11.80	6.3	1100					
$HgIn_2Te_4$(?)	940.63	134.38	6.186	12.37	6.3	980					
§A7. Other Adamantine Compounds											
αSiC	40.1	20.1	3.0817 15.1183		3.21	3070					
$Hg_5Ga_2Te_8$	2163.19	144.21	6.235								
$Hg_5In_2Te_8$	2253.39	150.23	6.328								
$CdIn_2Se_4$	657.89	93.98	a = c = 5.823								
PART B. OCTAHEDRAL SEMICONDUCTORS											
§B1. HALITE STRUCTURE SEMICONDUCTORS (Strukturbericht symbol B1, Space Group Fm3m-O_h^5)											
GeTe	200.19	100.1	5.98		6.14						
SnSe	197.65	98.83	6.020			1133					
SnTe	246.29	123.15	6.313		6.45	1080 (max)					91
PbS	239.26	119.63	5.9362		7.61	1390					23
PbSe	286.16	143.08	6.1243		8.15	1340					17
PbTe	334.8	167.4	6.454		8.16	1180					23
Selected other binary halites											
BiSe	287.94	143.97	5.99		7.98	880					
BiTe	336.58	168.29	6.47								
EuSe	230.92	115.46	6.191			2300					2.4
GdSe	236.21	118.11	5.771			2400					
NiD	60.71	30.35	4.1684		6.6	2260					
CdO	128.41	64.21	4.6953			1700					7
SrS	119.68	59.84	6.0199		3.643	3000					
PART C. OTHER SEMICONDUCTORS											
§C1. Antifluorite Structure Compounds (Fm3m - O_h^5)											
Mg_2Si	76.70	25.57	6.338		1.88	1375				11.5	
Mg_2Ge	121.20	40.4	6.380		3.08	1388				15.0	
Mg_2Sn	167.3	55.77	6.765		3.53	1051				9.9	92
Mg_2Pb	225.81	85.27	6.836		5.1	823				10.0	
§C2. Tetradymite Structure Compounds ($\bar{R}3m - D_{3d}^5$)											
Sb_2Te_3	626.3	125.26	4.25	30.3	6.44	895					
Bi_2Se_3	654.84	130.97	4.14	28.7	7.51	979	167				24
Bi_2Te_3	800.76	160.15	4.38	30.45	7.73	858	155	16			30
§C3. Skutterudite Structure Compounds (Im3 - T_h^5)											
CoP_3	151.85	37.96	7.7073			>1270					
$CoAs_3$	286.70	71.65	8.2060		6.73	1230					
$CoSb_3$	424.18	106.05	9.0385			1123			307		50
$NiAs_3$	283.45	70.86	8.330		6.43						
RhP_3	195.83	48.96	7.9951			>1470					
$RhAs_3$	327.67	81.92	8.4427			>1270					
$RhSb_3$	468.16	117.04	9.2322			1170					100
IrP_3	285.14	71.29	8.0151		7.36	>1470					
$IrAs_3$	416.98	104.25	8.4673		9.12	>1470					
$IrSb_3$	557.47	139.37	9.2533		9.35	1170			303		90
§C4. Selected Multinary Compounds											
$AgSbSe_2$	387.54	96.88	5.786		6.60	910					10.5
$AgSbTe_2$ (or $Ag_{19}Sb_{29}Te_{52}$)	484.82	121.2	6.078		7.12	830					86, 0.3
$AgBiS_2$(H.T.)	380.97	95.24	5.648								
$AgBiSe_2$(H.T.)	474.77	118.69	5.82								
$AgBiTe_2$(H.T.)	572.05	143.01	6.155								
Cu_2CdSnS_4	486.43	60.80	5.586		10.83						

Table I
PHYSICO-CHEMICAL PROPERTIES OF SEMICONDUCTORS (LISTED BY CRYSTAL STRUCTURE) (continued)

Substance	Molecular mass	Average atomic mass	Lattice parameters (Å, room temp.)		Density (g/cm^3)	Melting point (K)	Microhardness, N/mm^2 (M-Mohs Scale)	Specific heat, J/kg·K (300 K)	Debye temp. (K)	Coefficient of thermal linear expansion $[10^{-6}\ K^{-1}$ (300K)]	Thermal conductivity [mW/cm·K (300K)]
§C5. Some Elemental Semiconductors											
B		10.81	4.91	12.6	2.34	2348	9.5(M)	1277	1370	8.3	600
Se(gray)		78.96	4.36	4.95	4.81	493	350	292.6		(‖C) 17.89 (⊥C) 74.09	(‖C) 45.2 (⊥C) 13.1
Te		127.6	4.45	5.91	6.23	723		196.5		16.8	(‖C) 33.8 (⊥C) 19.7

Table II
BASIC THERMODYNAMIC, ELECTRICAL, AND MAGNETIC PROPERTIES OF SEMICONDUCTORS (LISTED BY CRYSTAL STRUCTURE)

Substance	Heat of formation [kJ/mole (300K)]	Volume compressibility ($10^{-10}\ m^2/N$)	Static dielectric constant	Atomic magnetic susceptibility (10^{-6} CGS)	Index of refraction	Minimum room temperature energy gap (eV)	Mobility (Room temp.) ($cm^2/V{\cdot}s$) Electrons	Holes	Optical transition	Remarks
PART A. ADAMANTINE SEMICONDUCTORS										
§A1. Diamond Structure Elements (Strukturbericht symbol A4, Space Group Fd3m-O_h^7)										
C	714.4	18	5.7	−5.88	2.419 (589 nm)	5.4	1800	1400	i*	
Si	324	0.306	11.8	−3.9	3.49 (589 nm)	1.107	1900	500	i	
Ge	291	0.768	16	−0.12	3.99 (589 nm)	0.67	3800	1820	i	
α-Sn	267.5		24		2.75 (589 nm)	0.0; 0.8	2500	2400		
§A2. Sphalerite (Zinc Blende) Structure Compounds (Strukturbericht symbol B3 Space Group F $\overline{4}$ 3m-T_d^2)										
I VII Compounds										
CuF										
CuCl	481	0.26	7.9		1.93	3.17			d	Nantokite
CuBr	481	0.26	7.9		2.12	2.91			d	
CuI	439	0.27	6.5		2.346	2.95			d	Marshite
AgBr	486		12.4		2.253	2.50	4000		i	Bromirite
AgI	389	0.41	10		2.22	2.22	30		d	Miersite
II VI Compounds										
BeS					4.17				i	
BeSe					3.61				i	
BeTe					1.45		20		d	
BePo										
ZnO										See A3
ZnS	477		8.9	−9.9	2.356	3.54	180	5(400°C)	d	See also A3
ZnSe	422		9.2		2.89	2.58	540	28	d	
ZnTe	376		10.4		3.56	2.26	340	100	d	
ZnPo										
CdS										See A3
CdSe										See A3
CdTe	339		7.2		2.50	1.44	1200	50	d	
CdPo										
HgS					2.85		250		d	Metacinnabarite
HgSe	247					2.10 (α)	20000	≈1.5	s	Tiemannite
HgTe	242					−0.06	25000	350	s	Coloradoite
III V Compounds										
BN	815					4.6				Borazone
BP(L.T.)						≈2.1	500	70		Ignites 470K
BAs						≈1.5				
AlP						2.45	80		i	
AlAs	627		10.9			2.16	1200	420	i	
AlSb	585	0.571	11		3.2	1.60	200—400	550	i	
GaP	635	0.110	11.1	−13.8	3.2	2.24	300	150	i	
GaAs	535	0.771	13.2	−16.2	3.30	1.35	8800	400	d	
GaSb	493	0.457	15.7	−14.2	3.8	0.67	4000	1400	d	
InP	560	0.735	12.4	−22.8	3.1	1.27	4600	150	d	
InAs	477	0.549	14.6	−27.7	3.5	0.36	33000	460	d	
InSb	447	0.442	17.7	−32.9	3.96	0.163	78000	750	d	

* i = indirect, d = direct, s = semimetal.

Table II
BASIC THERMODYNAMIC, ELECTRICAL, AND MAGNETIC PROPERTIES OF SEMICONDUCTORS (LISTED BY CRYSTAL STRUCTURE) (continued)

Substance	Heat of formation [kJ/mole (300K)]	Volume compressibility (10^{-10} m²/N)	Static dielectric constant	Atomic magnetic susceptibility (10^{-6} CGS)	Index of refraction	Minimum room temperature energy gap (eV)	Mobility (Room temp.) (cm²/V·s) Electrons	Holes	Optical transition	Remarks
Other sphalerite structure compounds										
MnS										See also §A3
MnSe										See also §A3
β-SiC					2.697	2.3	4000			
Ga_2Te_3	271			−13.5		1.35	50			
In_2Te_3(H.T.)	198			−13.6		1.04	50			
$MgGeP_2$										E1—T_d^{12}
$ZnSnP_2$						2.1				Same
$ZnSnAs_2$(H.T.)						≈0.7				Same
$ZnSnSb_2$						0.4				Same

§A3. Wurtzite (Zincite) Structure Compounds (Strukturbericht symbol B4, Space Group P 6_3mc-C^4_{6v})

Substance	Heat of formation [kJ/mole (300K)]	Volume compressibility (10^{-10} m²/N)	Static dielectric constant	Atomic magnetic susceptibility (10^{-6} CGS)	Index of refraction	Minimum room temperature energy gap (eV)	Mobility (Room temp.) (cm²/V·s) Electrons	Holes	Optical transition	Remarks
I VII Compounds										
CuCl										
CuBr										
CuI										
AgI						2.63				Iodargirite
II VI Compounds										
BeO										
MgTe										
ZnO	−350					3.2	180			
ZnS	−206					3.67				
ZnTe	−163									
CdS			8.45; 9.12		2.32	2.42	350	40	d	Greenockide
CdSe						1.74	900	50	d	Cadmoselite
CdTe						1.50	650			
III V Compounds										
BP(H.T.)										
AlN						6.02				
GaN						3.34				
InN						2.0				
Other wurtzite structure compounds										
MnS										
MnSe										
SiC					2.654					
MnTe						≈1.0				
Al_2S_3	426					4.1				
Al_2Se_3	367					3.1				

§A4. Chalcopyrite Structure Compounds (Strukturbericht symbol $E1_1$, Space Group I $\bar{4}$ 2d-D^{12}_{2d})

Substance	Heat of formation [kJ/mole (300K)]	Volume compressibility (10^{-10} m²/N)	Static dielectric constant	Atomic magnetic susceptibility (10^{-6} CGS)	Index of refraction	Minimum room temperature energy gap (eV)	Mobility (Room temp.) (cm²/V·s) Electrons	Holes	Optical transition	Remarks
I III VI_2 Compounds										
$CuAlS_2$		0.106				2.5				
$CuAlSe_2$						1.1				
$CuAlTe_2$						0.88				
$CuGaS_2$		0.106				2.38				
$CuGaSe_2$		0.141				0.96, 1.63				
$CuGaTe_2$		0.227				0.82, 1.0				
$CuInS_2$		0.141				1.2				
$CuInSe_2$		0.187				0.86, 0.92				
$CuInTe_2$		0.278				0.95				
$CuTlS_2$										
$CuTlSe_2$(L.T.)						1.07				
$CuFeS_2$						0.53				Chalcopyrite
$CuFeSe_2$						0.16				
$CuLaS_2$										
$AgAlS_2$										
$AgAlSe_2$						0.7				
$AgAlTe_2$						0.56				
$AgGaS_2$		0.150				1.66				
$AgGaSe_2$		0.182				1.1				
$AgGaTe_2$		0.280				1.9				
$AgInS_2$(L.T.)		0.185				1.18				
$AgInSe_2$		0.238				0.96, 0.52				
$AgInTe_2$		0.338								
$AgFeS_2$										

Table II
BASIC THERMODYNAMIC, ELECTRICAL, AND MAGNETIC PROPERTIES OF SEMICONDUCTORS (LISTED BY CRYSTAL STRUCTURE) (continued)

Substance	Heat of formation [kJ/mole (300K)]	Volume compressibility (10^{-10} m²/N)	Static dielectric constant	Atomic magnetic susceptibility (10^{-6} CGS)	Index of refraction	Minimum room temperature energy gap (eV)	Mobility (Room temp.) (cm²/V·s) Electrons	Holes	Optical transition	Remarks
II IV V_2 Compounds										
$ZnSiP_2$	312					2.3	1000			
$ZnGeP_2$	293					2.2				
$ZnSnP_2$	275					1.45				
$CdSiP_2$		0.103				2.2	1000			
$CdGeP_2$	289					1.8				
$CdSnP_2$	270					1.5				
$ZnSiAs_2$	290					1.7		50		
$ZnGeAs_2$	271			−14.4		0.85				
$ZnSnAs_2$	252			−18.4		0.65		300		Disorders at 910K
$CdSiAs_2$		0.143				1.6				
$CdGeAs_2$	266			−23.4		0.53	70	25		Disorders at 903
$CdSnAs_2$	247		13.7	−21.5		0.26	22000	250		

§A5. Other Ternary Semiconductors with Tetrahedral Coordination

I_2 IV VI_3 Compounds										
Cu_2SiS_3(H.T.)										Wurtzite
Cu_2SiS_3(L.T.)										Tetragonal
Cu_2SiTe_3										Cubic
Cu_2GeS_3(H.T.)				−18.7						Cubic
Cu_2GeS_3(L.T.)							360			Tetragonal
Cu_2GeSe_3	211.5			−21.3		0.94	238			Same
Cu_2GeTe_3	190.2			−23.4						Same
Cu_2SnS_3				−18.2		0.91	405			Cubic
$CuSnSe_3$				−21.0		0.66	870			Cubic
Cu_2SnTe_3				−28.4						Cubic
Ag_2GeSe_3				−29.6		0.91 (77K)				
Ag_2SnSe_3				−29.5		0.81				
Ag_2GeTe_3				−31.4		0.25				
Ag_2SnTe_3				−31.0		0.08				
I_3 V VI_4 Compounds										
Cu_3PS_4										Enargite
Cu_3AsS_4	269.6			−15.8		1.24				
Cu_3AsSe_4	161.3			−13.1		0.88				Famatinite
Cu_3SbS_4				−8.3		0.74				Famatinite
Cu_3SbSe_4	127.1			−20.5		0.31				
I IV_2 V_3 Compounds										
$CuSi_2P_3$										El
$CuGe_2P_3$		0.12				0.9				El
$AgGe_2P_3$										

§A6. "Defect Chalcopyrite" Structure Compounds (Strukturbericht symbol E3, Space Group I $\bar{4}$-S_4^2)

$ZnAl_2Se_4$										
$ZnAl_2Te_4$(?)										
$ZnGa_2S_4$(?)						≈3.4				
$ZnGa_2Se_4$(?)						≈2.2				
$ZnGa_2Te_4$(?)						1.35				
$ZnIn_2Se_4$	206					1.82	35			
$ZnIn_2Te_4$	198					1.2				
$CdAl_2S_4$										
$CdAl_2Se_4$										
$CdAl_2Te_4$(?)										
$CdGa_2S_4$	256					3.44	60			
$CdGa_2Se_4$	216					2.43	33			
$CdGa_2Te_4$										
$CdIn_2Te_4$	195					(1.26 or 0.9)	4000			
$HgAl_2S_4$										
$HgAl_2Se_4$										
$HgAl_2Te_4$(?)										
$HgGa_2S_4$	249					2.84				
$HgGa_2Se_4$	204					1.95	400			
$HgIn_2Se_4$	196					0.6	290			
$HgIn_2Te_4$(?)	188					0.86	200			

Table II
BASIC THERMODYNAMIC, ELECTRICAL, AND MAGNETIC PROPERTIES OF SEMICONDUCTORS (LISTED BY CRYSTAL STRUCTURE) (continued)

Substance	Heat of formation [kJ/mole (300K)]	Volume compressibility (10^{-10} m²/N)	Static dielectric constant	Atomic magnetic susceptibility (10^{-6} CGS)	Index of refraction	Minimum room temperature energy gap (eV)	Mobility (Room temp.) (cm²/V·s) Electrons	Holes	Optical transition	Remarks
§A7. Other Adamantine Compounds										
αSiC			10.2	−6.4	2.67	2.86	400			6H structure
$Hg_5Ga_2Te_8$										B3 with superlattice
$Hg_5In_2Te_8$						0.7	2000			B3 with superlattice
$CdIn_2Se_4$						1.55				
PART B. OCTAHEDRAL SEMICONDUCTORS										
§B1. HALITE STRUCTURE SEMICONDUCTORS (Strukturbericht symbol B1, Space Group Fm3m-O_h^5)										
GeTe										
SnSe										
SnTe										
PbS	435					0.5	600	600		
PbSe	393		161			0.37	1000	900		
PbTe	393		280			0.26	1600	600		Altaite
			360			0.25				
Selected other binary halites										
BiSe										
BiTe						0.4				
EuSe										
GdSe						1.8	4			
NiD						2.0 or 3.7	100			
CdO	531					2.5				
SrSW						4.1				
PART C. OTHER SEMICONDUCTORS										
§C1. Antifluorite Structure Compounds (Fm3m - O_h^5)										
Mg_2Si	79.08					0.77	405	70		
Mg_2Ge						0.74	520	110		
Mg_2Sn	76.57					0.36	320	260		
Mg_2Pb	52.72					0.1				
§C2. Tetradymite Structure Compounds ($R\bar{3}m$ − D_{3d}^5)										
Sb_2Te_3						0.3		360		
Bi_2Se_3						0.35	600			
Bi_2Te_3						0.21	1140	680		R3m (166)
§C3. Skutterudite Structure Compounds (Im3-T_h^5)										
CoP_3						0.43				
$CoAs_3$						0.69		~4000		
$CoSb_3$						0.63	70	~3000		
RhP_3								700		
$RhAs_3$						0.85		~3000		
$RhSb_3$						0.80		~7000		
$IrSb_3$						1.18		1500		
§C4. Selected Multinary Compounds										
$AgSbSe_2$						0.58				
$AgSbTe_2$ (or $Ag_{19}Sb_{29}Te_{52}$)						0.7, 0.27				
$AgBiS_2$(H.T.)										
$AgBiSe_2$(H.T.)										
$AgBiTe_2$(H.T.)										
Cu_2CdSnS_4						1.16	<2			
§C5. Some Elemental Semiconductors										
B	397.1			−6.7	3.4	1.55	10			
Se(gray)			6.6 (0.1 GHz)	− 22.1	2.5	1.5		5		$P3_121(152)$
Te				−39.5	3.3	0.33	1700	1200		Same

Table III
SEMICONDUCTING PROPERTIES OF SELECTED MATERIALS

Substance	Minimum Energy Gap (eV) R.T.	Minimum Energy Gap (eV) 0 K	$\frac{dE_g}{dT}$ x 10^4 eV/°C	$\frac{dE_g}{dP}$ x 10^6 eV·cm²/kg	Density of States Electron Effective Mass m_{d_n} (m_o)	Electron Mobility and Temperature Dependence μ_n cm²/V·s	Electron Mobility and Temperature Dependence −x	Density of States Hole Effective Mass m_{d_p} (m_o)	Hole Mobility and Temperature Dependence μ_p cm²/V·s	Hole Mobility and Temperature Dependence −x
Si	1.107	1.153	−2.3	−2.0	1.1	1,900	2.6	0.56	500	2.3
Ge	0.67	0.744	−3.7	+7.3	0.55	3,800	1.66	0.3	1,820	2.33
αSn	0.08	0.094	−0.5		0.02	2,500	1.65	0.3	2,400	2.0
Te	0.33				0.68	1,100		0.19	560	
III–V Compounds										
AlAs	2.2	2.3				1,200			420	
AlSb	1.6	1.7	−3.5	−1.6	0.09	200	1.5	0.4	500	1.8
GaP	2.24	2.40	−5.4	−1.7	0.35	300	1.5	0.5	150	1.5
GaAs	1.35	1.53	−5.0	+9.4	0.068	9,000	1.0	0.5	500	2.1
GaSb	0.67	0.78	−3.5	+12	0.050	5,000	2.0	0.23	1,400	0.9
InP	1.27	1.41	−4.6	+4.6	0.067	5,000	2.0		200	2.4
InAs	0.36	0.43	−2.8	+8	0.022	33,000	1.2	0.41	460	2.3
InSb	0.165	0.23	−2.8	+15	0.014	78,000	1.6	0.4	750	2.1
II–VI Compounds										
ZnO	3.2		−9.5	+0.6	0.38	180	1.5			
ZnS	3.54		−5.3	+5.7		180			5(400°C)	
ZnSe	2.58	2.80	−7.2	+6		540			28	
ZnTe	2.26			+6		340			100	
CdO	2.5 ± .1		−6		0.1	120				
CdS	2.42		−5	+3.3	0.165	400		0.8		
CdSe	1.74	1.85	−4.6		0.13	650	1.0	0.6		
CdTe	1.44	1.56	−4.1	+8	0.14	1,200		0.35	50	
HgSe	0.30				0.030	20,000	2.0			
HgTe	0.15		−1		0.017	25,000		0.5	350	
Halite Structure Compounds										
PbS	0.37	0.28	+4		0.16	800		0.1	1,000	2.2
PbSe	0.26	0.16	+4		0.3	1,500		0.34	1,500	2.2
PbTe	0.25	0.19	+4	−7	0.21	1,600		0.14	750	2.2
Others										
ZnSb	0.50	0.56			0.15	10				1.5
CdSb	0.45	0.57	−5.4		0.15	300			2,000	1.5
Bi_2S_3	1.3					200			1,100	
Bi_2Se_3	0.27					600			675	
Bi_2Te_3	0.13		−0.95		0.58	1,200	1.68	1.07	510	1.95
Mg_2Si		0.77	−6.4		0.46	400	2.5		70	
Mg_2Ge		0.74	−9			280	2		110	
Mg_2Sn	0.21	0.33	−3.5		0.37	320			260	
Mg_3Sb_2		0.32				20			82	
Zn_3As_2	0.93					10	1.1		10	
Cd_3As_2	0.55				0.046	100,000	0.88			
GaSe	2.05		3.8						20	
GaTe	1.66	1.80	−3.6			14	−5			
InSe	1.8					900				
TlSe	0.57		−3.9		0.3	30		0.6	20	1.5
$CdSnAs_2$	0.23				0.05	25,000	1.7			
Ga_2Te_3	1.1	1.55	−4.8							
α-In_2Te_3	1.1	1.2			0.7				50	1.1
β-In_2Te_3	1.0								5	
$Hg_5In_2Te_8$	0.5								11,000	
SnO_2									78	

Table IV
BAND PROPERTIES OF SEMICONDUCTORS

PART A. DATA ON VALENCE BANDS OF SEMICONDUCTORS (ROOM TEMPERATURES)

Substance	Band Curvature Effective Mass: Heavy Holes (Expressed as fraction of free electron mass)	Light Holes	"Split-off" Band Holes	Energy Separation of "Split-off" Band (eV)	Measured (Light) Hole Mobility cm²/V·s
Semiconductors with Valence Band Maximum at the Center of the Brillouin Zone ('Γ')					
Si	0.52	0.16	0.25	0.044	500
Ge	0.34	0.043	0.08	0.3	1,820
Sn	0.3				2,400
AlAs					
AlSb	0.4			0.7	550
GaP				0.13	100
GaAs	0.8	0.12	0.20	0.34	400
GaSb	0.23	0.06		0.7	1,400
InP				0.21	150
InAs	0.41	0.025	0.083	0.43	460
InSb	0.4	0.015		0.85	750
CdTe	0.35				50
HgTe	0.5				350

Table IV
BAND PROPERTIES OF SEMICONDUCTORS (continued)

Semiconductors with Multiple Valence Band Maxima

Substance	Number of Equivalent Valleys and Direction	Band Curvature Effective Masses Longitudinal m_L	Transverse m_T	Anisotropy $K = m_L/m_T$	Measured (Light) Hole Mobility $cm^2/V{\cdot}s$
PbSe	4 "L" [111]	0.095	0.047	2.0	1,500
PbTe	4 "L" [111]	0.27	0.02	10	750
Bi_2Te_3	6	0.207	~0.045	4.5	515

PART B. DATA ON CONDUCTION BANDS OF SEMICONDUCTORS (Room Temperature Data)

Single Valley Semiconductors

Substance	Energy Gap (eV)	Effective Mass (m_o)	Mobility ($cm^2/V.s$)	Comments
GaAs	1.35	0.067	8,500	3(or 6?) equivalent [100] valleys 0.36 eV above this maximum with a mobility of ~50
InP	1.27	0.067	5,000	3(or 6?) equivalent [100] valleys 0.4 eV above this minimum.
InAs	0.36	0.022	33,000	equivalent valleys ~1.0 eV above this minimum.
InSb	0.165	0.014	78,000	
CdTe	1.44	0.11	1,000	4(or 8?) equivalent [111] valleys 0.51 eV above this minimum.

Multivalley Semiconductors

Substance	Energy Gap	Number of Equivalent Valleys and Direction	Band Curvature Effective Mass Longitudinal m_L	Transverse m_T	Anisotropy $K = m_L/m_T$	Comments
Si	1.107	6 in [100] "Δ"	0.90	0.192	4.7	
Ge	0.67	4 in [111] at "L"	1.588	0.0815	19.5	
GaSb	0.67	as Ge (?)	~1.0	~0.2	~5	
PbSe	0.26	4 in [111] at "L"	0.085	0.05	1.7	
PbTe	0.25	4 in [111] at "L"	0.21	0.029	5.5	
Bi_2Te_3	0.13	6			~0.05	

Table V
RESISTIVITY OF SEMICONDUCTING MINERALS

Mineral	ρ (ohm · m)
Diamond (C)	2.7
Sulfides	
Argentite, Ag_2S	1.5 to 2.0×10^{-3}
Bismuthinite, Bi_2S_3	3 to 570
Bornite, $Fe_2S_3 \cdot nCu_2S$	1.6 to 6000×10^{-6}
Chalcocite, Cu_2S	80 to 100×10^{-6}
Chalcopyrite, $Fe_2S_3 \cdot Cu_2S$	150 to 9000×10^{-6}
Covellite, CuS	0.30 to 83×10^{-6}
Galena, PbS	6.8×10^{-6} to 9.0×10^{-2}
Haverite, MnS_2	10 to 20
Marcasite, FeS_2	1 to 150×10^{-3}
Metacinnabarite, 4HgS	2×10^{-6} to 1×10^{-3}
Millerite, NiS	2 to 4×10^{-7}
Molybdenite, MoS_2	0.12 to 7.5
Pentlandite, $(Fe, Ni)_9S_8$	1 to 11×10^{-6}
Pyrrhotite, Fe_7S_8	2 to 160×10^{-6}
Pyrite, FeS_2	1.2 to 600×10^{-1}
Sphalerite, ZnS	2.7×10^{-3} to 1.2×10^{4}
Antimony-sulfur compounds	
Berthierite, $FeSb_2S_4$	0.0083 to 2.0
Boulangerite, $Pb_5Sb_4S_{11}$	2×10^{3} to 4×10^{4}
Cylindrite, $Pb_3Sn_4Sb_2S_{14}$	2.5 to 60
Franckeite, $Pb_5Sn_3Sb_2S_{14}$	1.2 to 4
Hauchecornite, $Ni_9(Bi, Sb)_2S_8$	1 to 83×10^{-6}
Jamesonite, $Pb_4FeSb_6S_{14}$	0.020 to 0.15
Tetrahedrite, Cu_3SbS_3	0.30 to 30,000
Arsenic-sulfur compounds	
Arsenopyrite, FeAsS	20 to 300×10^{-6}
Cobaltite, CoAsS	6.5 to 130×10^{-3}
Enargite, Cu_3AsS_4	0.2 to 40×10^{-3}
Gersdorffite, NiAsS	1 to 160×10^{-6}
Glaucodote, (Co, Fe)AsS	5 to 100×10^{-6}
Antimonide	
Dyscrasite, Ag_3Sb	0.12 to 1.2×10^{-6}
Arsenides	
Allemonite, $SbAs_3$	70 to 60,000
Lollingite, $FeAs_2$	2 to 270×10^{-6}
Nicollite, NiAs	0.1 to 2×10^{-6}
Skutterudite, $CoAs_3$	1 to 400×10^{-6}
Smaltite, $CoAs_2$	1 to 12×10^{-6}
Tellurides	
Altaite, PbTe	20 to 200×10^{-6}
Calavarite, $AuTe_2$	6 to 12×10^{-6}
Coloradoite, HgTe	4 to 100×10^{-6}
Hessite, Ag_2Te	4 to 100×10^{-6}
Nagyagite, $Pb_6Au(S, Te)_{14}$	20 to 80×10^{-6}
Sylvanite, $AgAuTe_4$	4 to 20×10^{-6}
Oxides	
Braunite, Mn_2O_3	0.16 to 1.0
Cassiterite, SnO_2	4.5×10^{-4} to 10,000
Cuprite, Cu_2O	10 to 50
Hollandite, $(Ba, Na, K)Mn_8O_{16}$	2 to 100×10^{-3}
Ilmenite, $FeTiO_3$	0.001 to 4
Magnetite, Fe_3O_4	52×10^{-6}
Manganite, MnO·OH	0.018 to 0.5
Melaconite, CuO	6000
Psilomelane, $KMnO \cdot MnO_2 \cdot nH_2O$	0.04 to 6000
Pyrolusite, MnO_2	0.007 to 30
Rutile, TiO_2	29 to 910
Uraninite, UO	1.5 to 200

From Carmichael, R. S., ed., *Handbook of Physical Properties of Rocks*, Vol. I, CRC Press, 1982.

REFERENCES

1. Beer, A. C., *Galvanomagnetic Effects in Semiconductors*, Academic Press, 1963.
2. Goryunova, N. A., *The Chemistry of Diamond-Like Semiconductors*, The MIT Press, 1965.
3. Abrikosov, N. Kh., Bankina, V. F., Poretskaya, L. E., Shelimova, L. E., and Skudnova, E. V., *Semiconducting II-VI, IV-VI, and V-VI Compounds*, Plenum Press, 1969.
4. Berger, L. I. and Prochukhan, V. D., *Ternary Diamond-Like Semiconductors*, Cons. Bureau/Plenum Press, 1969.
5. Shay, J. L. and Wernick, J. H., *Ternary Chalcopyrite Semiconductors: Growth, Electronic Properties, and Applications*, Pergamon Press, 1975.
6. Bergman, R., *Thermal Conductivity in Solids*, Clarendon, Oxford, 1976.
7. Handbook of Semiconductors, Vol. 1, Moss, T. S. and Paul, W., Eds., *Band Theory and Transport Properties*; Vol. 2, Moss, T. S. and Balkanski, M., Eds., *Optical Properties of Solids*; Vol. 3, Moss, T. S. and Keller, S. P., Eds., *Materials Properties and Preparation*, North Holland Publ. Co., 1980.
8. Böer, K. W., *Survey of Semiconductor Physics*, Van Nostrand Reinhold, 1990.
9. Rowe, D. M., Ed., *CRC Handbook of Thermoelectrics*, CRC Press, Boca Raton, FL, 1995.
10. Berger, L. I., *Semiconductor Materials*, CRC Press, Boca Raton, FL, 1997.
11. Glazov, V. M., Chizhevskaya, S. N., and Glagoleva, N. N., *Liquid Semiconductors*, Plenum Press, New York, 1969.
12. Phillips, J. C., *Bonds and Bands in Semiconductors*, Academic Press, New York, 1973.
13. Harrison, W. A., *Electronic Structure and the Properties of Solids*, Freeman Publ. House, San Francisco, 1980.
14. Balkanski, M., Ed., *Optical Properties of Solids*, North-Holland, Amsterdam, 1980.
15. *Landolt-Börnstein. Numerical Data and Functional Relationships in Science and Technology, New Series, Group III: Crystal and Solid State Physics*, Hellwege, K.-H., and Madelung, O., Eds., Volumes 17 and 22, Springer Verlag, Berlin, 1984 (and further).
16. Shklovskii, B. L., and Efros, A. L., *Electronic Processes in Doped Semiconductors*, Springer Verlag, Berlin, 1984.
17. Cohen, M. L., and Chelikowsky, J. R., *Electronic Structure and Optical Properties of Semiconductors*, Springer Verlag, New York, 1988.
18. Glass, J. T., Messier, R. F., and Fujimori, N., Eds., *Diamond, Silicon Carbide, and Related Wide Bandgap Semiconductors*, MRS Symposia Proc. 1652, Mater. Res. Soc., Pittsburgh, 1990.
19. Palik, E., Ed., *Handbook of Optical Constants of Solids II*, Academic Press, New York, 1991.
20. Reed, M., Ed., *Semiconductors and Semimetals*, Volume 35, Academic Press, Boston, 1992.
21. Haug, H., and Koch, S. W., *Quantum Theory of the Optical and Electronic Properties of Semiconductors*, 2nd Edition, World Scientific, Singapore, 1993.
22. Lockwood, D. J., Ed., *Proc. 22nd Intl. Conf. on the Physics of Semiconductors, Vancouver, 1994*, World Scientific, Singapore, 1994.
23. Morelli, D. T., Caillat, T., Fleurial, J.-P., Borschchevsky, A., Vandersande, J., Chen, B., and Uher, C., *Phys. Rev.*, B51, 9622, 1995.
24. Caillat, T., Borshchevsky, A., and Fleurial, J.-P., *J. Appl. Phys.*, 80, 4442, 1996.
25. Fleurial, J.-P., Caillat, T., and Borshchevsky, A., *Proc. XVI Intl. Conf. Thermoelectrics*, Dresden, Germany, August 26–29, 1997 (in print).
26. Borshchevsky A. et al., U.S. Patents 5,610,366 (March 1997) and 5,831,286 (March 1998).

DIFFUSION DATA FOR SEMICONDUCTORS

B. L. Sharma

The diffusion coefficient D in many semiconductors may be expressed by an Arrhenius-type relation

$$D = D_o \exp(-Q/kT)$$

where D_o is a frequency factor, Q is the activation energy for diffusion, k is the Boltzmann constant, and T is the absolute temperature. This table lists D_o and Q for various diffusants in common semiconductors.

Abbreviations used in the table are

AES — Auger Electron Spectroscopy
DLTS — Deep Level Transient Spectroscopy
SEM — Scanning Electron Microscopy
SIMS — Secondary Ion Mass Spectrometry
D(c) — Concentration Dependent Diffusion Coefficient
D_{max} — Maximum Diffusion Coefficient

(f) — Fast Diffusion Component
(i) — Interstitial Diffusion Component
(s) — Slow Diffusion Component
(||) — Parallel to c Direction
(⊥) — Perpendicular to c Direction

Semiconductor	Diffusant	Frequency factor, D_o (cm²/s)	Activation energy, Q (eV)	Temperature range (°C)	Method of measurement	Ref.
Si	H	6×10^{-1}	1.03	120—1207	Electrical and SIMS	1
	Li	2.5×10^{-3}	0.65	25—1350	Electrical	2
	Na	1.65×10^{-3}	0.72	530—800	Electrical and flame photometry	3
	K	1.1×10^{-3}	0.76	740—800	Electrical and flame photometry	3
	Cu	4×10^{-2}	1.0	800—1100	Radioactive	4
		4.7×10^{-3}	0.43 (i)	300—700	Radioactive	5
	Ag	2×10^{-3}	1.6	1100—1350	Radioactive	6
	Au	2.4×10^{-4}	0.39 (i)	700—1300	Radioactive	7
		2.75×10^{-3}	2.05 (s)			
	Be	($D \sim 10^{-7}$)	—	1050	Electrical	8
	Ca	($D \sim 6 \times 10^{-14}$)	—	1100	Electrical and SIMS	1
	Zn	1×10^{-1}	1.4	980—1270	Electrical	9
	B	2.46	3.59	1100—1250	Electrical	10
		2.4×10^{1}	3.87	840—1250	Electrical	11
	Al	1.38	3.41	1119—1390	Electrical	12
		1.8	3.2	1025—1175	Electrical	13
	Ga	3.74×10^{-1}	3.39	1143—1393	Electrical	12
		6×10^{1}	3.89	900—1050	Radioactive	14
	In	7.85×10^{-1}	3.63	1180—1389	Electrical	12
		1.94×10^{1}	3.86	1150—1242	Radioactive	15
	Tl	1.37	3.7	1244—1338	Electrical	12
		1.65×10^{1}	3.9	1105—1360	Electrical	16
	Sc	8×10^{-2}	3.2	1100—1250	Radioactive	1
	Ce	($D \sim 3.9 \times 10^{-13}$)	—	1050	SIMS	1
	Pr	2.5×10^{-7}	1.74	1100—1280	Electrical	1
	Pm	7.5×10^{-9}	1.2 (s)	730—1270	Radioactive	1
		4.2×10^{-12}	0.13 (f)			
	Er	2×10^{-3}	2.9	1100—1250	Radioactive	1
	Tm	8×10^{-3}	3.0	1100—1280	Radioactive	1
	Yb	2.8×10^{-5}	0.95	947—1097	Neutron activation	1
	Ti	1.45×10^{-2}	1.79	950—1200	DLTS	17
	C	3.3×10^{-1}	2.92	1070—1400	Radioactive	18
	Si (self)	1.54×10^{2}	4.65	855—1175	SIMS	19
		1.6×10^{3}	4.77	1200—1400	Radioactive	20
	Ge	3.5×10^{-1}	3.92	855—1000	Radioactive	21
		2.5×10^{3}	4.97	1030—1302	Radioactive	21
		7.55×10^{3}	5.08	1100—1300	SIMS	22
	Sn	3.2×10^{1}	4.25	1050—1294	Neutron activation	23

DIFFUSION DATA FOR SEMICONDUCTORS (continued)

Semiconductor	Diffusant	Frequency factor, D_o (cm²/s)	Activation energy, Q (eV)	Temperature range (°C)	Method of measurement	Ref.
	N	2.7×10^{-3}	2.8	800—1200	Out Diffusion; SIMS	1
	P	2.02×10^{1}	3.87	1100—1250	Electrical	10
		1.1	3.4	900—1200	Radioactive	24
		7.4×10^{-2}	3.3	1130—1405	Electrical	25
	As	6.0×10^{1}	4.2	950—1350	Radioactive	26
		6.55×10^{-2}	3.44	1167—1394	Electrical	27
		2.29×10^{1}	4.1	900—1250	Electrical	28
	Sb	1.29×10^{1}	3.98	1190—1398	Radioactive	29
		2.14×10^{-1}	3.65	1190—1405	Electrical	27
	Bi	1.03×10^{3}	4.64	1220—1380	Electrical	16
		1.08	3.85	1190—1394	Electrical	27
	Cr	1×10^{-2}	1	1100—1250	Radioactive	30
	Mo	$(D \sim 2 \times 10^{-10})$	—	1000	DLTS	1
	W	$(D \sim 10^{-12})$	—	1100	DLTS	1
	O	7×10^{-2}	2.44	700—1250	SIMS	31
		1.4×10^{-1}	2.53	700—1160	SIMS	32
	S	5.95×10^{-3}	1.83	975—1200	Radioactive	33
	Se	9.5×10^{-1}	2.6	1050—1250	Electrical	34
	Te	5×10^{-1}	3.34	900—1250	SIMS	1
	Mn	6.9×10^{-4}	0.63	900—1200	Radioactive	35
	Fe	1.3×10^{-3}	0.68	30—1250	Radioactive	36
	Co	2×10^{-3}	0.69	700—1300	Radioactive	37
	Ni	2×10^{-3}	0.47	800—1300	Radioactive	38
	Ru	$(D \sim 5 \times 10^{-7}$ — $5 \times 10^{-6})$	—	1000—1280	Electrical	1
	Rh	$(D \sim 10^{-6}$—$10^{-4})$	—	1000—1200	Electrical	39
	Pd	2.95×10^{-4}	0.22 (i)	702—1320	Nuclear Activation	1
	Pt	1.5×10^{2}	2.22	800—1000	Electrical	1
	Os	$(D \sim 2 \times 10^{-6})$	—	1280	Electrical	40
	Ir	4.2×10^{-2}	1.3	950—1250	Electrical	41
Ge	Li	1.3×10^{-3}	0.46	350—800	Electrical	42
		9.1×10^{-3}	0.57	800—500	Electrical	43
	Na	3.95×10^{-1}	2.03	700—850	Radioactive	44
	Cu	1.9×10^{-4}	0.18 (i)	750—900	Radioactive	45
		4×10^{-2}	0.99 (s)	600—700		
		4×10^{-3}	0.33 (i)	350—750	Radioactive	5
	Ag	4.4×10^{-2}	1.0 (i)	700—900	Radioactive	46, 47
		4×10^{-2}	2.23 (s)	800—900	Radioactive	48
	Au	2.25×10^{2}	2.5	600—900	Radioactive	49
	Be	5×10^{-1}	2.5	720—900	Electrical	50
	Mg	$(D \sim 8 \times 10^{-9})$	—	900	Electrical	1
	Zn	5	2.7	600—900	Radioactive and electrical	51
	Cd	1.75×10^{9}	4.4	760—915	Radioactive	52
	B	1.8×10^{9}	4.55	600—900	Electrical	51
	Al	1.0×10^{3}	3.45	554—905	SIMS	53
		$\sim 1.6 \times 10^{2}$	~3.24	750—850	Electrical	54
	Ga	1.4×10^{2}	3.35	554—916	SIMS	55
		3.4×10^{1}	3.1	600—900	Electrical	51
	In	1.8×10^{4}	3.67	554—919	SIMS	56
		3.3×10^{1}	3.02	700—855	Radioactive	57
	Tl	1.7×10^{3}	3.4	800—930	Radioactive	58
	Si	2.4×10^{-1}	2.9	650—900	(γ) resonance	59
	Ge (self)	2.48×10^{1}	3.14	549—891	Radioactive	60
		7.8	2.95	766—928	Radioactive	61
	Sn	1.7×10^{-2}	1.9	—	Radioactive	45
	P	3.3	2.5	600—900	Electrical	51

DIFFUSION DATA FOR SEMICONDUCTORS (continued)

Semiconductor	Diffusant	Frequency factor, D_o (cm^2/s)	Activation energy, Q (eV)	Temperature range (°C)	Method of measurement	Ref.
	As	2.1	2.39	700—900	Electrical	62
	Sb	3.2	2.41	700—855	Radioactive	57
		1.0×10^{1}	2.5	600—900	Radioactive and electrical	51
	Bi	3.3	2.57	650—850	—	63
	O	4×10^{-1}	2.08	—	Optical	64
	S	($D \sim 10^{-9}$)	—	920	—	65
	Se	($D \sim 10^{-10}$)	—	920	—	65
	Te	5.6	2.43	750—900	Radioactive	66
	Fe	1.3×10^{-1}	1.08	750—900	Radioactive	67
	Co	1.6×10^{-1}	1.12	750—850	Radioactive	47
	Ni	8×10^{-1}	0.9	670—900	Electrical	68
GaAs	Li	5.3×10^{-1}	1.0	250—500	Electrical and chemical	69
	Cu	3×10^{-2}	0.53	100—500	Radioactive	69
		6×10^{-2}	0.98	450—750	Ultrasonic	69
		1.5×10^{-3}	0.6	800—1000	Radioactive	69
	Ag	4×10^{-4}	0.8	500—1150	Radioactive	69
	Au	1×10^{-3}	1.0	740—1025	Radioactive	69
	Be	7.3×10^{-6}	1.2	800—990	Electrical	69
	Mg	4×10^{-5}	1.22	800—1200	Electrical	69
	Zn	1.5×10^{1}	2.49	600—980	Radioactive	69
		2.5×10^{-1}	3.0	750—1000	Radioactive	69
	Cd	1.3×10^{-3}	2.2	800—1100	Radioactive	69
		5×10^{-2}	2.43	868—1149	Radioactive	69
	Hg	($D \sim 5 \times 10^{-14}$)	—	1100	Radioactive	69
	Al	($D \sim 4 \times 10^{-18}$—10^{-14})	4.3	850—1100	AES	70
	Ga (self)	4×10^{-5}	2.6	1025—1100	Radioactive	69
		1×10^{7}	5.6	1125—1230	Radioactive	69
	In	($D \sim 7 \times 10^{-11}$)	—	1000	Radioactive	69
	C	($D \sim 1.04 \times 10^{-16}$)	—	825	SIMS	69
	Si	1.1×10^{-1}	2.5	850—1050	SIMS	69
	Ge	1.6×10^{-5}	2.06	650—850	SIMS	69
	Sn	6×10^{-4}	2.5	1060—1200	Radioactive	69
		1×10^{-5}	2	800—1000	Radioactive	69
	P	($D \sim 10^{-12}$—10^{-10})	2.9	800—1150	Reflectance measurements	69
	As (self)	7×10^{-1}	3.2	—	Radioactive	69
	Cr	2.04×10^{-6}	0.83 (f)	750—1000	SIMS	69
			1.7 (s)	700—900		
		7.9×10^{-3}	2.2	800—1100	Chemical analysis	69
	O	2×10^{-3}	1.1	700—900	Mass spectroscopy	69
	S	1.85×10^{-2}	2.6	1000—1300	Radioactive	69
		1.1×10^{1}	2.95	750—900	Electrical	69
	Se	3×10^{3}	4.16	1025—1200	Radioactive	69
	Te	1.5×10^{-1}	3.5	1000—1150	Radioactive	69
	Mn	6.5×10^{-1}	2.49	850—1100	Radioactive	69
	Fe	4.2×10^{-2}	1.8	850—1150	Radioactive	69
		2.2×10^{-3}	2.32	750—1050	Radioactive	69
	Co	5×10^{2}	2.5	800—1000	Radioactive	69
		1.2×10^{-1}	2.64	750—1050	Radioactive	69
	Tm	2.3×10^{-16}	1.0	800—1000	Radioactive	69
GaSb	Li	2.3×10^{-4}	1.9 (s)	527—657	Electrical and flame photometry	69
		1.2×10^{-1}	0.7 (f)	277—657		
	Cu	4.7×10^{-3}	0.9	470—650	Radioactive	69

DIFFUSION DATA FOR SEMICONDUCTORS (continued)

Semiconductor	Diffusant	Frequency factor, D_o (cm²/s)	Activation energy, Q (eV)	Temperature range (°C)	Method of measurement	Ref.
	Zn	($D \sim 2 \times 10^{-13}$— 1×10^{-11})	2	510—600	Radioactive	69
	Cd	1.5×10^{-6}	0.72	640—800	Electrical	69
	Ga (self)	3.2×10^{3}	3.15	658—700	Radioactive	69
	In	1.2×10^{-7}	0.53	320—650	Radioactive	69
	Sn	2.4×10^{-5}	0.8	320—650	Radioactive	69
		1.3×10^{-5}	1.1	500—650	Radioactive	69
	Sb (self)	3.4×10^{4}	3.45	658—700	Radioactive	69
	Se	($D \sim 2.4 \times 10^{-13}$— 1.37×10^{-11})	—	400—500	Radioactive	69
	Te	3.8×10^{-4}	1.20	320—650	Radioactive	69
	Fe	5×10^{-2}	1.9 (I)	500—650	Radioactive	69
		5×10^{2}	2.3 (II)	500—650		
GaP	Ag	—	—	1000—1300	Radioactive	69
	Au	8	2.5 (I)	1050—1250	Radioactive	69
		20	2.4 (II)	1100—1250	Diffusion (I) A face and (II) B face	
	Be	($D_{max} \sim 2.4 \times 10^{-9}$— 8.5×10^{-8})	—	900—1000	Atomic absorption analysis	69
	Mg	5×10^{-5}	1.4	700—1050	Electrical	69
	Zn	1.0	2.1	700—1300	Radioactive	69
	Ge	—	—	900—1000	Radioactive	69
	Cr	6.2×10^{-4}	1.2	900—1130	Radioactive; ESR	69
	S	3.2×10^{3}	4.7	1120—1305	Radioactive	69
	Mn	2.1×10^{9}	4.7	$T < 950$	Radioactive; ESR	69
		1.1×10^{-6}	0.9	950—1130		
	Fe	1.6×10^{-1}	2.3	980—1180	Radioactive	69
	Co	2.8×10^{-3}	2.9	850—1100	Radioactive	69
InP	Cu	3.8×10^{-3}	0.69	600—900	Radioactive	69
	Ag	3.6×10^{-4}	0.59	500—900	Radioactive	69
	Au	1.32×10^{-5}	0.48	600—820	Radioactive	69
		1.37×10^{-4}	0.73	600—900	Radioactive	69
	Zn	1.6×10^{-8}	0.3	750—900	Electrical	69
		($D \sim 2 \times 10^{-9}$— 4×10^{-8})	—	700—900	Radioactive	69
	Cd	1.8	1.9	700—900	Radioactive	69
		1.1×10^{-7}	0.72	700—900	Electrical	69
		($D \sim 7 \times 10^{-13}$— 2×10^{-10})	—	450—650	Electrical	69
	In (self)	1×10^{5}	3.85	830—990	Radioactive	69
	Sn	($D \sim 3 \times 10^{-8}$)	—	550	Etching and cathodo-luminescence	69
	P (self)	7×10^{10}	5.65	900—1000	Radioactive	69
	Cr	—	—	600—900	Radioactive	69
	S	3.6×10^{-4}	1.94	585—708	Electrical	69
	Se	($D \sim 2 \times 10^{-8}$)	—	550	Cathodoluminescence	69
	Mn	—	2.9	650—750	SIMS	69
	Fe	3	2	600—950	Radioactive	69
		6.8×10^{5}	3.4	600—700	SIMS	69
	Co	9×10^{-1}	1.8	600—950	Radioactive	69
InAs	Cu	3.6×10^{-3}	0.52	342—875	Radioactive	69
		2.2×10^{-2}	0.54	525—890	Radioactive	69
	Ag	7.3×10^{-4}	0.26	450—900	Radioactive	69
	Au	5.8×10^{-3}	0.65	600—900	Radioactive	69
	Mg	1.98×10^{-6}	1.17	600—900	Electrical	69
	Zn	4.2×10^{-3}	0.96	600—900	Radioactive	69
		3.11×10^{-3}	1.17	600—900	Electrical	69

Semiconductor	Diffusant	Frequency factor, D_o (cm^2/s)	Activation energy, Q (eV)	Temperature range (°C)	Method of measurement	Ref.
	Cd	7.4×10^{-4}	1.15	650—900	Radioactive	69
	Hg	1.45×10^{-5}	1.32	650—850	Radioactive	69
	In (self)	6×10^{5}	4.0	740—900	Radioactive	69
	Ge	3.74×10^{-6}	1.17	600—900	Electrical	69
	Sn	1.49×10^{-6}	1.17	600—900	Electrical	69
	As (self)	3×10^{7}	4.45	740—900	Radioactive	69
	S	6.78	2.2	600—900	Electrical	69
	Se	12.6	2.2	600—900	Electrical	69
	Te	3.43×10^{-5}	1.28	600—900	Electrical	69
InSb	Li	7×10^{-4}	0.28	0—210	Electrical	69
	Cu	9×10^{-4}	1.08	200—500	Radioactive	69
		3×10^{-5}	0.37	230—490	Radioactive	69
	Ag	1×10^{-7}	0.25	440—510	Radioactive	69
	Au	7×10^{-4}	0.32	140—510	Radioactive	69
	Zn	5×10^{-1}	1.35	362—508	Radioactive	69
		—	1.5	355—455	SIMS	69
	Cd	1×10^{-5}	1.1	250—500	Radioactive	69
		1.3×10^{-4}	1.2	360—500	Electrical	69
	Hg	4×10^{-6}	1.17	425—500	Radioactive	69
	In (self)	6×10^{-7}	1.45	400—500	Radioactive	69
		1.8×10^{13}	4.3	475—517	Radioactive	69
	Sn	5.5×10^{-8}	0.75	390—512	Radioactive	69
	Pb	($D \sim 2.7 \times 10^{-15}$)	—	500	Radioactive	71
	Sb (self)	5.35×10^{-4}	1.91	400—500	Radioactive	69
		3.1×10^{13}	4.3	475—517	Radioactive	69
	S	9×10^{-2}	1.4	360—500	Electrical	69
	Se	1.6	1.87	380—500	Electrical	69
	Te	1.7×10^{-7}	0.57	300—500	Radioactive	69
	Fe	1×10^{-7}	0.25	440—510	Radioactive	69
	Co	2.7×10^{-11}	0.39	420—500	Radioactive	69
AlAs	Ga	($D \sim 2 \times 10^{-18}$—10^{-15})	3.6	850—1100	AES	70
	Zn	($D \sim 9 \times 10^{-11}$)	—	557	SEM	69
AlSb	Cu	3.5×10^{-3}	0.36	150—500	Radioactive	69
	Zn	3.3×10^{-1}	1.93	660—860	Radioactive	69
	Cd	D(c) ~ 4×10^{-12}—3×10^{-10}	—	900	Radioactive	69
	Al (self)	2	1.88	570—620	X-ray	69
	Sb (self)	1	1.7	570—620	X-ray	69
ZnS	Cu	2.6×10^{-3}	0.79	470—750	Radioactive	69
		4.3×10^{-4}	0.64	250—1200	Electroluminescence	69
		9.75×10^{-3}	1.04	400—800	Luminescence	69
	Au	1.75×10^{-4}	1.16	500—800	Radioactive	69
	Zn (self)	3×10^{-4}	1.5	925<T<940	Radioactive	69
		1.5×10^{4}	3.26	940<T<1030		
		1×10^{16}	6.5	1030<T<1075		
	Cd	($D \sim 10^{-10}$)	—	1100	Luminescence	72
	Al	5.69×10^{-4}	1.28	800—1000	Luminescence	69
	In	3×10^{1}	2.2	750—1000	Radioactive	69
	S (self)	2.16×10^{4}	3.15	600—800	Radioactive	69
		8×10^{-5}	2.2	740—1100	Radioactive	69
	Se	($D \sim 5 \times 10^{-13}$)	—	1070	X-ray microprobe	69
	Mn	2.3×10^{3}	2.46	500—800	Radioactive	69
ZnSe	Li	2.66×10^{-6}	0.49	950—980	Electrical	69
	Cu	1×10^{-4}	0.66	400—800	Luminescence	69
		1.7×10^{-5}	0.56	200—570	Radioactive	69
	Ag	2.2×10^{-2}	1.18	400—800	Luminescence	69

Semiconductor	Diffusant	Frequency factor, D_o (cm^2/s)	Activation energy, Q (eV)	Temperature range (°C)	Method of measurement	Ref.
	Zn (self)	9.8	3.0	760—1150	Radioactive	69
	Cd	6.39×10^{-4}	1.87	700—950	Photoluminescence	69
	Al	2.3×10^{-2}	1.8	800—1100	Luminescence	69
	Ga	1.81×10^{2}	3.0	900—1100	Luminescence	69
		—	1.3	700—850	Electron probe	69
	In	$(D \sim 2 \times 10^{-12})$	—	940	—	69
	S	$(D \sim 8 \times 10^{-12})$	—	1060	X-ray microprobe	69
	Se (self)	1.3×10^{1}	2.5	860—1020	Radioactive	69
		2.3×10^{-1}	2.7	1000—1050	Radioactive	69
	Ni	$(D \sim 1.5 \times 10^{-8}$— $1.7 \times 10^{-7})$	—	740—910	Luminescence	69
ZnTe	Li	2.9×10^{-2}	1.22 (s)	400—700	Nuclear and chemical analysis	69
		1.7×10^{-4}	0.78 (f)			
	Zn (self)	2.34	2.56	760—860	Radioactive	69
		1.4×10^{1}	2.69	667—1077	Radioactive	69
	Al	—	2.0	700—1000	Electrical and optical	69
	In	4	1.96	1100—1300	Radioactive	69
	Te (self)	2×10^{4}	3.8	727—977	Radioactive	69
CdS	Li	3×10^{-6}	0.68	610—960	Microhardness	69
	Na	$(D \sim 3 \times 10^{-7})$	—	800	Radioactive	69
	Cu	1.5×10^{-3}	0.76	400—700	Radioactive	69
		1.2×10^{-2}	1.05	300—700	Ultrasonic	69
		8×10^{-5}	0.72	20—200	Electrical	69
	Ag	2.5×10^{1}	1.2 (s)	300—500	Radioactive	69
		2.4×10^{-1}	0.8 (f)			
	Au	2×10^{2}	1.8	500—800	Radioactive	69
	Zn	1.27×10^{-9}	0.86 (s)	720—1000	Radioactive	69
		1.22×10^{-8}	0.66 (f)			
	Cd (self)	3.4	2.0	700—1100	Radioactive	69
	Ga	—	—	667—967	Optical and microprobe	69
	In	6×10^{1}	2.3 (‖)	650—930	Radioactive, optical and microprobe	69
		1×10^{1}	2.03 (⊥)			
	P	6.5×10^{-4}	1.6	800—1100	Radioactive	69
	S (self)	1.6×10^{-2}	2.05	800—900	Radioactive	69
		—	2.4	750—1050	Radioactive	69
	Se	$(D \sim 1.2 \times 10^{-9})$	—	900	Radioactive	69
	Te	1.3×10^{-7}	10.4	700—1000	Radioactive	69
	Cl	$(D \sim 3 \times 10^{-10})$	—	800	Electrical	69
	I	$(D \sim 5 \times 10^{-12})$	—	1000	Radioactive	69
	Ni	6.75×10^{-3}	10.9	570—900	Luminescence	69
	Yb	$(D \sim 1.3 \times 10^{-9})$	—	960	Photoluminescence	69
CdSe	Ag	2×10^{-4}	0.53	22—400	Ultrasonic	69
	Cd (self)	1.6×10^{-3}	1.5	700—1000	Radioactive	69
		6.3×10^{-2}	1.25 (I)	600—900	Radioactive;	69
		4.12×10^{-2}	2.18 (II)	600—900	(I) saturated Cd and (II) saturated Se pressure	
	P	$(D \sim 5.3 \times 10^{-12}$— $6 \times 10^{-11})$	—	900—1000	Radioactive	69
	Se (self)	2.6×10^{3}	1.55	700—1000	Radioactive; saturated Se pressure	69
CdTe	Li	$(D \sim 1.5 \times 10^{-10})$	—	300	Ion microprobe	69
	Cu	3.7×10^{-4}	0.67	97—300	Radioactive	69
		8.2×10^{-8}	0.64	290—350	Ion backscattering	69
	Ag	—	—	700—800	Electrical and photo-	

DIFFUSION DATA FOR SEMICONDUCTORS (continued)

Semiconductor	Diffusant	Frequency factor, D_o (cm²/s)	Activation energy, Q (eV)	Temperature range (°C)	Method of measurement	Ref.
					luminescence	69
	Au	6.7×10^{1}	2.0	600—1000	Radioactive	69
	Cd (self)	1.26	2.07	700—1000	Radioactive	69
		3.26×10^{2}	2.67 (I)	650—900	Radioactive;	69
		1.58×10^{1}	2.44 (II)		(I) saturated Cd and (II) saturated Te pressure	
	In	8×10^{-2}	1.61	650—1000	Radioactive	69
		$1.17 \times {}^{2}$	2.21 (I)	500—850	Radioactive; (I) saturated	
		6.48×10^{-4}	1.15 (II)		Cd and (II) saturated Te pressure	69
	Sn	8.3×10^{-2}	2.2	700—925	Radioactive	69
	P	($D \sim 1.2 \times 10^{-10}$)	—	900	Radioactive	69
	As	—	—	850	—	69
	O	5.6×10^{-9}	1.22	200—650	Mass spectrometry	69
		6.0×10^{-10}	0.29	650—900		
	Se	1.7×10^{-4}	1.35	700—1000	Radioactive	69
	Te (self)	8.54×10^{-7}	1.42 (I)	600—900	Radioactive; (I) saturated Cd and (II) saturated Te pressure	69
		1.66×10^{-4}	1.38 (II)	500—800		
	Cl	7.1×10^{-2}	1.6	520—800	Radioactive	69
	Fe	($D \sim 4 \times 10^{-8}$)	0.77	900	Radioactive	69
HgSe	Sb	6.3×10^{-5}	0.85	540—630	Radioactive	69
	Se (self)	—	—	200—400	Radioactive	69
HgTe	Ag	6×10^{-4}	0.8	250—350	Radioactive	69
	Zn	5×10^{-8}	0.6	250—350	Radioactive	69
	Cd	3.1×10^{-4}	0.66	250—350	Radioactive	69
	Hg (self)	2×10^{-8}	0.6	200—350	Radioactive	69
	In	6×10^{-6}	0.9	200—300	Radioactive	69
	Sn	1.72×10^{-6}	0.66 (s)	200—300	Radioactive	69
		1.8×10^{-3}	0.80 (f)			
	Te (self)	10^{-6}	1.4	200—400	Radioactive	69
	Mn	1.5×10^{-4}	1.3	250—350	Radioactive	69
PbS	Cu	4.6×10^{-4}	0.36	150—450	Electrical	69
		5×10^{-3}	0.31	100—400	Electrical	69
	Pb (self)	8.6×10^{-5}	1.52	500—800	Radioactive	69
	S (self)	6.8×10^{-5}	1.38	500—750	Radioactive	69
	Ni	1.78×10^{1}	0.95	200—500	Electrical	69
PbSe	Na	1.5×10^{1}	1.74 (s)	400—850	Radioactive	69
		5.6×10^{-6}	0.4 (f)			
	Cu	2×10^{-5}	0.31	93—520	Radioactive	69
	Ag	7.4×10^{-4}	0.35	400—850	Radioactive	69
	Pb (self)	4.98×10^{-6}	0.83	400—800	Radioactive	69
	Sb	3.4×10^{-1}	2.0	650—850	Radioactive	69
	Se (self)	2.1×10^{-5}	1.2	650—850	Radioactive	69
	Cl	1.6×10^{-8}	0.45	400—850	Radioactive	69
	Ni	($D \sim 1 \times 10^{-10}$)	—	700	Radioactive	69
PbTe	Na	1.7×10^{-1}	1.91	600—850	Radioactive	69
	Sn	3.1×10^{-2}	1.56	500—800	Radioactive	69
	Pb (self)	2.9×10^{-5}	0.6	250—500	Radioactive	69
	Sb	4.9×10^{-2}	1.54	500—800	Radioactive	69
	Te	2.7×10^{-6}	0.75	500—800	Radioactive	69
	Cl	($D > 2.3 \times 10^{-10}$)	—	700	Radioactive	69
	Ni	($D > 1 \times 10^{-6}$)	—	700	Radioactive	69

REFERENCES

1. N. A. Stolwijk and H. Bracht, in *Diffusion in Semiconductors and Non-Metallic Solids,* D. L. Beke, Ed., Springer-Verlag, Berlin, 1998, 2-1.
2. E. M. Pell, *Phys. Rev.*, 119, 1960; 119, 1014, 1960.
3. L. Svob, *Solid State Electron,* 10, 991, 1967.
4. B. I. Boltaks and I. I. Sosinov, *Zh. Tekh. Fiz.*, 28, 3, 1958.
5. R. N. Hall and J. N. Racette, *J. Appl. Phys.*, 35, 379, 1964.
6. B. I. Boltaks and Hsueh Shih-Yin, *Sov. Phys. Solid State,* 2, 2383, 1961.
7. W. R. Wilcox and T. J. LaChapelle, *J. Appl. Phys.*, 35, 240, 1964.
8. E. A. Taft and R. O. Carlson, *J. Electrochem. Soc.*, 117, 711, 1970.
9. R. Sh. Malkovich and N. A. Alimbarashvili, *Sov. Phys. Solid State,* 4, 1725, 1963.
10. R. N. Ghoshtagore, *Solid State Electron,* 15, 1113, 1972.
11. C. Hill, *Semiconductor Silicon 1981,* H. R. Huff, R. J. Kreiger, and Y. Takeishi, Eds., p. 988, *Electrochem. Soc.*, 1981.
12. R. N. Ghoshtagore, *Phys. Rev. B,* 3, 2507, 1971.
13. W. Rosnowski, *J. Electrochem. Soc.*, 125, 957, 1978.
14. J. S. Makris and B. J. Masters, *J. Appl. Phys.*, 42, 3750, 1971.
15. M. F. Millea, *J. Phys. Chem. Solids,* 27, 315, 1965 (refer Reference 2).
16. C. S. Fuller and J. A. Ditzenberger, *J. Appl. Phys.*, 27, 544, 1956.
17. S. Hocine and D. Mathiot, *Appl. Phys. Lett.*, 53, 1269, 1988.
18. R. C. Newman and J. Wakefield, *J. Phys. Chem. Solids,* 19, 230, 1961.
19. L. Kalinowski and R. Seguin, *Appl. Phys. Lett.*, 35, 211, 1979; *Appl. Phys. Lett.*, 36, 171, 1980.
20. R. F. Peart, *Phys. Stat. Sol.*, 15, K 119, 1966.
21. G. Hettich, H. Mehrer and K. Maler, *Inst. Phys. Conf. Ser.*, 46, 500, 1979.
22. M. Ogina, Y. Oana and M. Watanabe, *Phys. Stat. Sol. (a),* 72, 535, 1982.
23. T. H. Yeh, S. M. Hu, and R. H. Kastl, *Appl. Phys.*, 39, 4266, 1968.
24. I. Franz and W. Langheinrich, *Solid State Electron,* 14, 835, 1971.
25. R. N. Ghoshtagore, *Hys. Rev. B,* 3, 389, 1971.
26. B. J. Masters and J. M. Fairfield, *J. Appl. Phys.*, 40, 2390, 1969.
27. R. N. Goshtagore, *Phys. Rev. B,* 3, 397, 1971.
28. R. S. Fair and J. C. C. Tsai, *J. Electrochem. Soc.*, 122, 1689, 1975.
29. J. J. Rohan, N. E. Pickering and J. Kennedy, *J. Electrochem. Soc.*, 106, 705, 1969.
30. W. Wuerker, K. Roy, and J. Hesse, *Matsr. Res. Bull.*, 9, 971, 1974.
31. J. C. Mikkelsen, Jr., *Appl. Phys. Lett.*, 40, 336, 1982.
32. S. Tang Lee and D. Nicols, *Appl. Phys. Lett.*, 47, 1001, 1985.
33. P. L. Gruzin, S. V. Zemskii, A. D. Bullkin, and N. M. Makarov, *Sov. Phys. Sem.*, 7, 1241, 1974.
34. N. S. Zhdanovich and Yu. I. Kozlov, *Svoistva Legir, Poluprovodn.*, V. S. Zemskov, Ed., Nauka, Moscow, 1977, 115-120; *Fiz Tekh. Poluprovod.*, 9, 1594, 1975.
35. D. Gilles, W. Bergholze, and W. Schroeter, *J. Appl. Phys.*, 59, 3590, 1986.
36. E. R. Weber, *Appl. Phys. A,* 30, 1, 1983.
37. E. R. Weber, Properties of Silicon, EMIS Datareviews Ser. No. 4, INSPEC Publications, 1988, 409-451.
38. M. K. Bakhadyrkhanov, S. Zainabidinov, and A. Khamidov, *Sov. Phys. Sem.*, 14, 243, 1980.
39. S. A. Azimov, M. S. Yunosov, F. K. Khatamkulov, and G. Nasyrov, *Poluprovod.*, N. Kh. Abrikosov and V. S. Zemskov, Eds., Nauka, Moscow, 1975, 21-23.
40. S. A. Azimov, M. S. Yunosov, G. Nurkuziev, and F. R. Karimov, *Sov. Phys. Sem.*, 12, 981, 1978.
41. S. A. Azimov, B. V. Umarov, and M. S. Yunusov, *Sov. Phys. Sem.*, 10, 842, 1976.
42. C. S. Fuller and J. A. Ditzenberger, *Phys. Rev.*, 91, 193, 1953.
43. B. Pratt and F. Friedman, *J. Appl. Phys.*, 37, 1893, 1966.
44. M. Stojic, V. Spiric, and D. Kostoski, *Inst. Phys. Conf. Ser.*, 31, 304, 1976.
45. B. I. Boltaks, *Diffusion in Semiconductors,* Inforsearch, London, 1963, 162.
46. A. A. Bugai, V. E. Kosenko, and E. G. Miselyuk, *Zh. Tekh. Fiz.*, 27, 67, 1957.
47. L. Y. Wei, *J. Phys. Chem. Solids,* 18, 162, 1961.
48. V. E. Kosenko, *Sov. Phys. Solid State,* 4, 42, 1962.
49. W. C. Dunlap, Jr., *Phys. Rev.*, 97, 614, 1955
50. Yu. I. Belyaev and V. A. Zhidkov, *Sov. Phys. Solid State,* 3, 133, 1961.
51. W. C. Dunlap, Jr. *Phys. Rev.*, 94, 1531, 1954.
52. V. E. Kosenko, *Sov. Phys. Solid State,* 1, 1481, 1960.
53. P. Dorner, W. Gust, A. Lodding, H. Odelius, B. Predel, and U. Roll, *Acta Metall.*, 30, 941, 1982.
54. W. Meer and D. Pommerrening, *Z. Agnew. Phys.*, 23, 369, 1967.
55. U. Sodervall, H. Odelius, A. Lodding, U. Roll, B. Predel, W. Gust, and P. Dorner, *Phil. Mag. A,* 54, 539, 1986.
56. P. Dorner, W. Gust, A. Lodding, H. Odelius, B. Predel, and U. Roll, *Z. Metalkd.*, 73, 325, 1982.

57. P. V. Pavlov, *Sov. Phys. Solid State,* 8, 2377, 1967.
58. V. I. Tagirov and A. A. Kuliev, *Sov. Phys. Solid State,* 4, 196, 1962.
59. J. Raisanen, J. Hirvonen, and A. Anttila, *Solid State Electron.,* 24, 333, 1981.
60. C. Vogel, G. Hettich, and H. Mehrer, *J. Phys. C.,* 16, 6197, 1983.
61. H. Letaw, Jr., W. M. Portnoy, and L. Slifkin, *Phys. Rev.,* 102, 363, 1956.
62. W. Bosenberg, *Z. Naturforsch.,* 10a, 285, 1955.
63. V. M. Glazov and V. S. Zemskov, Physicochemical Principles of Semiconductor Doping, Israel Program for Scientific Translation, Jerusalem, 1968.
64. J. W. Corbett, R. S. McDonald, and G. D. Watkins, *J. Phys. Chem. Solids,* 25, 873, 1964.
65. W. W. Tyler, *J. Phys. Chem. Solids,* 8, 59, 1959.
66. V. D. Ignatkov and V. E. Kosenko, *Sov. Phys. Solid State,* 4, 1193, 1962.
67. A. A. Bugal, V. E. Kosenko, and E. G. Miseluk, *Zh. Tekh. Fiz.,* 27, 210, 1957.
68. F. van der Maesen and J. A. Brenkman, *Phillips Res. Rep.,* 9, 255, 1954.
69. M. B. Dutt and B. L. Sharma, in *Diffusion in Semiconductors and Non-Metallic Solids,* D. L. Beke, Ed., Springer-Verlag, Berlin, 1998, 3-1.
70. L. L. Chang and A. Koma, *Appl. Physics Lett.,* 29, 138, 1976.
71. D. L. Kendall, *Semiconductors and Semimetals,* Vol. 4, R. K. Willardson and A. C. Beer, Eds., Academic, 1968, 255.
72. H. J. Biter and F. Williams, *J. Luminescence,* 3, 395, 1971.

PROPERTIES OF MAGNETIC MATERIALS

H. P. R. Frederikse

Glossary of Symbols

Quantity	Symbol	Units SI	Units emu
Magnetic field	H	A m^{-1}	Oe (oersted)
Magnetic induction	B	T (tesla)	G (gauss)
Magnetization	M	A m^{-1}	emu cm^{-3}
Spontaneous magnetization	M_s	A m^{-1}	emu cm^{-3}
Saturation magnetization	M_0	A m^{-1}	emu cm^{-3}
Magnetic flux	Φ	Wb (weber)	maxwell
Magnetic moment	m, μ	A m^2	erg/G
Coercive field	H_c	A m^{-1}	Oe
Remanence	B_r	T	G
Saturation magnetic polarization	J_s	T	G
Magnetic susceptibility	χ		
Magnetic permeability	μ	H m^{-1} (henry/meter)	
Magnetic permeability of free space	μ_0	H m^{-1}	
Saturation magnetostriction	λ ($\Delta l/l$)		
Curie temperature	T_C	K	K
Néel temperature	T_N	K	K

Magnetic moment $\mu = \gamma\hbar J = g\, \mu_B\, J$
where
γ = gyromagnetic ratio; J = angular momentum; g = spectroscopic splitting factor (~2)
μ_B = bohr magneton = $9.2741 \cdot 10^{-24}$ J/T = $9.2741 \cdot 10^{-21}$ erg/G

Earth's magnetic field $H = 56$ A m^{-1} = 0.7 Oe
For iron: $M_0 = 1.7 \cdot 10^6$ A m^{-1}; $B_r = 0.8 \cdot 10^6$ A m^{-1}
1 Oe = $(1000/4\pi)$ A m^{-1}; 1 G = 10^{-4} T; 1 emu cm^{-3} = 10^3 A m^{-1}
1 maxwell = 10^{-8} Wb
$\mu_0 = 4\pi \cdot 10^{-7}$ H m^{-1}

Relation Between Magnetic Induction and Magnetic Field

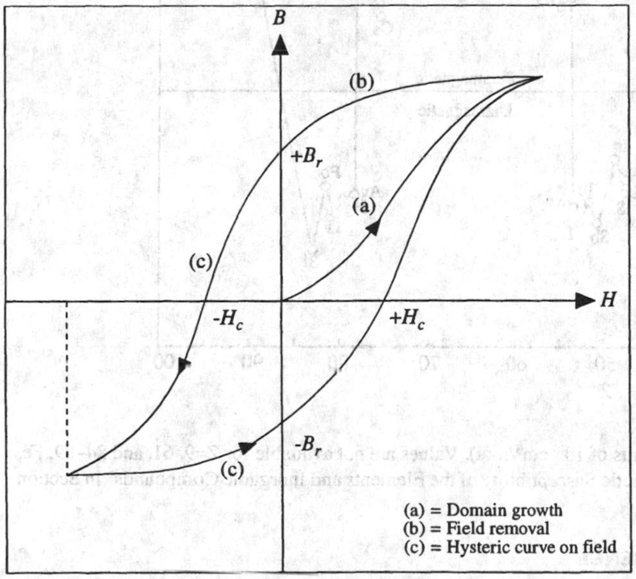

Figure 1. Typical curve representing the dependence of magnetic induction B on magnetic field H for a ferromagnetic material. When H is first applied, B follows curve a as the favorably oriented magnetic domains grow. This curve flattens as saturation is approached. When H is then reduced, B follows curve **b**, but retains a finite value (the remanence B_r) at $H = 0$. In order to demagnetize the material, a negative field $-H_c$ (where H_c is called the coercive field or coercivity) must be applied. As H is further decreased and then increased to complete the cycle (curve c), a hysteresis loop is obtained. The area within this loop is a measure of the energy loss per cycle for a unit volume of the material.

Relation Between Magnetic Induction and Magnetic Field (continued)

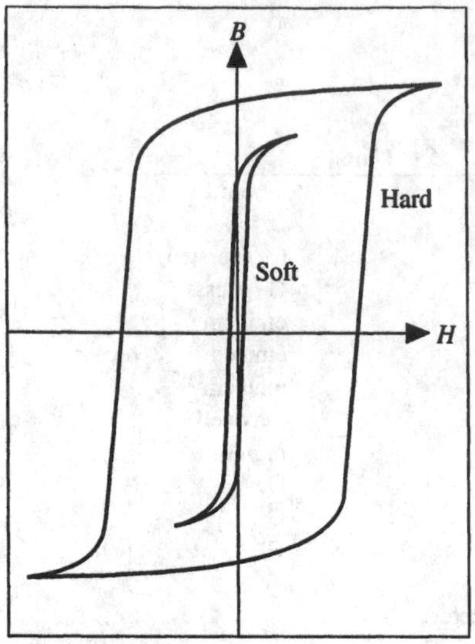

Figure 2. Schematic curve illustrating the *B* vs. *H* dependence for hard and soft magnetic materials. Hard materials have a larger remanence and coercive field, and a correspondingly large hysteresis loss.

REFERENCE

Ralls, K.M., Courtney, T.H., and Wulff, J., *Introduction to Materials Science and Engineering*, J. Wiley & Sons, New York, 1976, p. 577, 582. With permission.

Magnetic Susceptibility of the Elements

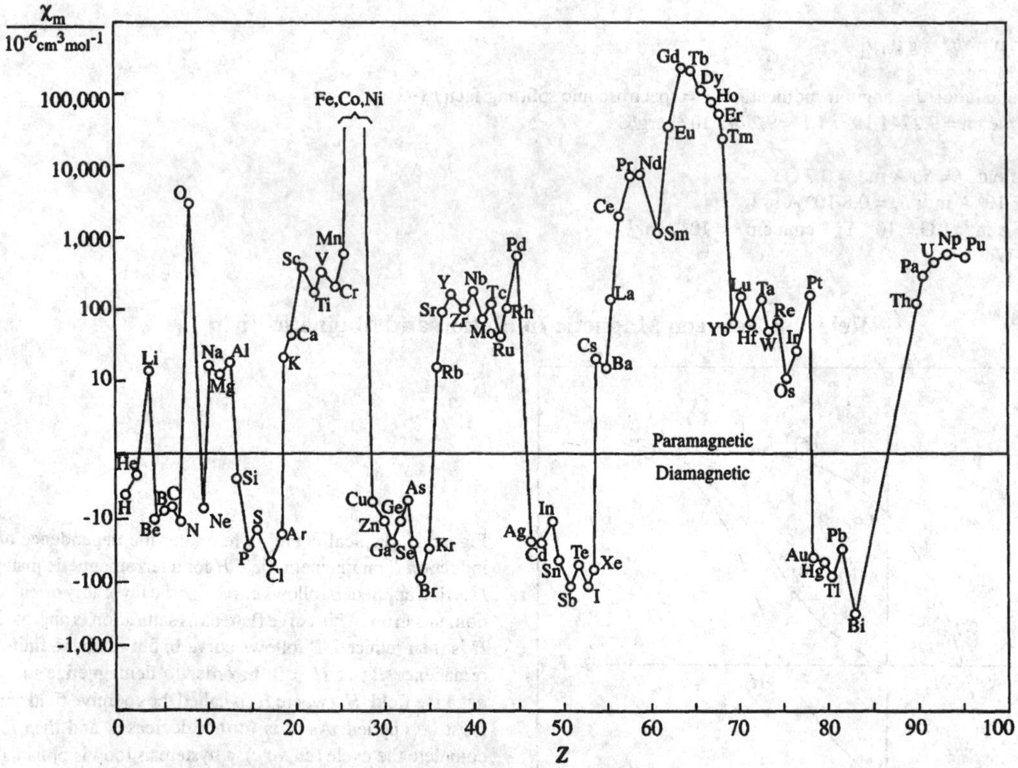

Figure 3. Molar susceptibility of the elements at room temperature (cgs units of 10^{-6} cm^3/mol). Values are not available for Z=9, 61, and 84–89; Fe, Co, and Ni (Z=26–28) are ferromagnetic. Data taken from the table "Magnetic Susceptibility of the Elements and Inorganic Compounds" in Section 4.

REFERENCE

Gray, D.E., Ed., *American Institute of Physics Handbook, Third Edition*, McGraw Hill, New York, 1972, p. 5-224. With permission.

Ground State of Ions with Partly Filled *d* or *f* Shells

Z	Element	*n*	*S*	*L*	*J*	Gr. state	p_{calc}[a]	p_{calc}[b]	p_{meas}
22	Ti^{3+}	1	1/2	2	3/2	$^2D_{3/2}$	1.73	1.55	1.8
23	V^{4+}	1	1/2	2	3/2	$^2D_{3/2}$	1.73	1.55	1.8
23	V^{3+}	2	1	3	2	$^3F_2$	2.83	1.63	2.8
23	V^{2+}	3	3/2	3	3/2	$^4F_{3/2}$	3.87	0.77	3.8
24	Cr^{3+}	3	3/2	3	3/2	$^4F_{3/2}$	3.87	0.77	3.7
25	Mn^{4+}	3	3/2	3	3/2	$^4F_{3/2}$	3.87	0.77	4.0
24	Cr^{2+}	4	2	2	0	$^5D_0$	4.90	0	4.9
25	Mn^{3+}	4	2	2	0	$^5D_0$	4.90	0	5.0
25	Mn^{2+}	5	5/2	0	5/2	$^6S_{5/2}$	5.92	5.92	5.9
26	Fe^{3+}	5	5/2	0	5/2	$^6S_{5/2}$	5.92	5.92	5.9
26	Fe^{2+}	6	2	2	4	$^5D_4$	4.90	6.70	5.4
27	Co^{2+}	7	3/2	3	9/2	$^4F_{9/2}$	3.87	6.54	4.8
28	Ni^{2+}	8	1	3	4	$^3F_4$	2.83	5.59	3.2
29	Cu^{2+}	9	1/2	2	5/2	$^2D_{5/2}$	1.73	3.55	1.9

Z	Element	*n*	*S*	*L*	*J*	Gr. state	p_{calc}[c]	p_{meas}
58	Ce^{3+}	1	1/2	3	5/2	$^2F_{5/2}$	2.54	2.4
59	Pr^{3+}	2	1	5	4	$^3H_4$	3.58	3.5
60	Nd^{3+}	3	3/2	6	9/2	$^4I_{9/2}$	3.62	3.5
61	Pm^{3+}	4	2	6	4	$^5I_4$	2.68	
62	Sm^{3+}	5	5/2	5	5/2	$^6H_{5/2}$	0.84	1.5
63	Eu^{3+}	6	3	3	0	$^7F_0$	0.0	3.4
64	Gd^{3+}	7	7/2	0	7/2	$^8S_{7/2}$	7.94	8.0
65	Tb^{3+}	8	3	3	6	$^7F_6$	9.72	9.5
66	Dy^{3+}	9	5/2	5	15/2	$^6H_{15/2}$	10.63	10.6
67	Ho^{3+}	10	2	6	8	$^5I_8$	10.60	10.4
68	Er^{3+}	11	3/2	6	15/2	$^4I_{15/2}$	9.59	9.5
69	Tm^{3+}	12	1	5	6	$^3H_6$	7.57	7.3
70	Yb^{3+}	13	1/2	3	7/2	$^2F_{7/2}$	4.54	4.5

[a] $p_{calc} = 2[S(S+1)]^{1/2}$
[b] $p_{calc} = 2[J(J+1)]^{1/2}$
[c] $p_{calc} = g[J(J+1)]^{1/2}$

REFERENCES

1. Jiles, D., *Magnetism and Magnetic Materials*, Chapman & Hall, London, 1991, p. 243.
2. Kittel, C., *Introduction to Solid State Physics, 6th Edition*, J. Wiley & Sons, New York, 1986, pp. 405—406.
3. Ashcroft, N.W. and Mermin, N.D., *Solid State Physics,* Holt, Rinehart, and Winston, New York, 1976, p. 652.

Ferro- and Antiferromagnetic Elements

M_0 is the saturation magnetization at $T = 0$ K
n_B is the number of Bohr magnetons per atom
T_C is the Curie temperature
T_N is the Néel temperature

	M_0/gauss	n_B	T_C/K	T_N/K	Comments
Fe	1752	2.22	1043		
Co	1446	1.72	1388		
Ni	510	0.62	627		
Cr				311	
Mn				100	
Ce				12.5	c-Axis antiferromagnetic
Nd				19.2	Basal plane modulation on hexagonal sites

Ferro- and Antiferromagnetic Elements (continued)

	M_0/gauss	n_B	T_C/K	T_N/K	Comments
				7.8	Cubic sites order (periodicity different from high-T phase)
Sm				106	Ordering on hexagonal sites
				13.8	Cubic site order
Eu				90.5	Spiral along cube axis
Gd	1980	7	293		
Tb		9	220		Basal plane ferromagnet
				230.2	Basal plane spiral
Dy		10	87		Basal plane ferromagnet
				176	Basal plane spiral
Ho		10	20		Bunched cone structure
				133	Basal plane spiral
Er		9	32		c-Axis ferrimagnetic cone structure
				80	c-Axis modulated structure
Tm		7	32		c-Axis ferrimagnetic cone structure
				56	c-Axis modulated structure

REFERENCES

1. Ashcroft, N.W., and Mermin, N.D., *Solid State Physics,* Holt, Rinehart, and Winston, New York, 1976, p.652.
2. Gschneidner, K.A., and Eyring, L., *Handbook on the Physics and Chemistry of Rare Earths*, North Holland Publishing Co., Amsterdam, 1978.

Selected Ferromagnetic Compounds

M_0 is the saturation magnetization at $T = 293$ K
T_C is the Curie temperature

Compound	M_0/gauss	T_C/K	Crystal system
MnB	152	578	orthorh(FeB)
MnAs	670	318	hex(FeB)
MnBi	620	630	hex(FeB)
MnSb	710	587	hex(FeB)
M_4N	183	743	
MnSi		34	cub(FeSi)
CrTe	247	339	hex(NiAs)
$CrBr_3$	270	37	hex(BiI_3)
CrI_3		68	hex(BiI_3)
CrO_2	515	386	tetr(TiO_2)
EuO	1910*	77	cub
EuS	1184*	16.5	cub
$GdCl_3$	550*	2.2	orthorh
FeB		598	orthorh
Fe_2B		1043	tetr ($CuAl_2$)
$FeBe_5$		75	cub($MgCu_2$)
Fe_3C		483	orthorh
FeP		215	orthorh (MnP)

* At $T = 0$ K

REFERENCES

1. Kittel, C., *Introduction to Solid State Physics, 6th Edition*, J. Wiley & Sons, New York, 1986.
2. Ashcroft, N.W., and Mermin, N.D., *Solid State Physics*, Holt, Rinehart, and Winston, New York, 1976.

Magnetic Properties of High-Permeability Metals and Alloys (Soft)

μ_i is the initial permeability
μ_m is the maximum permeability
H_c is the coercive force
J_s is the saturation polarization
W_H is the hysteresis loss per cycle
T_C is the Curie temperature

Material	Composition (mass %)	μ_i/μ_0	μ_m/μ_0	H_c/A m^{-1}	J_s/T	W_H/J m^{-3}	T_C/K
Iron	Commercial 99Fe	200	6000	70	2.16	500	1043
Iron	Pure 99.9Fe	25000	350000	0.8	2.16	60	1043
Silicon-iron	96Fe-4Si	500	7000	40	1.95	50-150	1008
Silicon-iron (110) [001]	97Fe-3Si	9000	40000	12	2.01	35-140	1015
Silicon-iron {100} <100>	97Fe-3Si		100000	6	2.01		1015
Mild steel	Fe-0.1C-0.1Si-0.4Mn	800	1100	200			
Hypernik	50Fe-50Ni	4000	70000	4	1.60	22	753
Deltamax {100} <100>	50Fe-50Ni	500	200000	16	1.55		773
Isoperm {100} <100>	50Fe-50Ni	90	100	480	1.60		
78 Permalloy	78Ni-22Fe	4000	100000	4	1.05	50	651
Supermalloy	79Ni-16Fe-5Mo	100000	1000000	0.15	0.79	2	673
Mumetal	77Ni-16Fe-5Cu-2Cr	20000	100000	4	0.75	20	673
Hyperco	64Fe-35Co-0.5Cr	650	10000	80	2.42	300	1243
Permendur	50Fe-50Co	500	6000	160	2.46	1200	1253
2V-Permendur	49Fe-49Co-2V	800	4000	160	2.45	600	1253
Supermendur	49Fe-49Co-2V		60000	16	2.40	1150	1253
25Perminvar	45Ni-30Fe-25Co	400	2000	100	1.55		
7Perminvar	70Ni-23Fe-7Co	850	4000	50	1.25		
Perminvar (magnet. annealed)	43Ni-34Fe-23Co		400000	2.4	1.50		
Alfenol (or Alperm)	84Fe-16Al	3000	55000	3.2	0.8		723
Alfer	87Fe-13Al	700	3700	53	1.20		673
Aluminum-Iron	96.5Fe-3.5Al	500	19000	24	1.90		
Sendust	85Fe-10Si-5Al	36000	120000	1.6	0.89		753

REFERENCES

1. McCurrie, R.A., *Structure and Properties of Ferromagnetic Materials*, Academic Press, London, 1994, p. 42.
2. Gray, D.E., Ed., *American Institute of Physics Handbook, Third Edition*, McGraw Hill, New York, 1972, p. 5-224.

Applications of High-Permeability Materials

Applications	Requirements
	POWER APPLICATIONS
Distribution and power transformers	Low core losses, high permeability, high saturation magnetic polarization
High-quality motors and generators, stators and armatures, switched-mode power supplies	
	INSTRUMENT TRANSFORMERS
Audiofrequency transformers	Low core losses, high permeability, high magnetic polarization
Pulse transformers	High permeability

Applications of High-Permeability Materials (continued)

Applications	Requirements
CORES FOR INDUCTOR COILS	
Audiofrequency	Low hysteresis, high permeability
Carrier frequency	Very low hysteresis and eddy current loss
Radiofrequency	High permeability at low fields
MISCELLANEOUS	
Relays, switches; Earth leakage circuit	High permeability, low remanence, low coercivity
Magnetic shielding	Low core loss for AC applications
Magnetic recording heads	High initial permeability, low or zero remanence
Magnetic amplifiers; Saturable reactors; Saturable transformers; Transformer cores	Rectangular hysteresis loops, low hysteresis loss
Magnetic shunts for temperature compensation in magnetic circuits	Low Curie temperature, appropriate decrease in permeability with increase in temperature
Electromagnets in indicating instruments, fire detection, quartz watches, electromechanical devices	High permeability, high saturation magnetic polarization
Magnetic yokes in permanent magnet devices, such as lifting and holding magnets, loudspeakers	High permeability, high saturation magnetic polarization

REFERENCE

McCurrie, R.A., *Structure and Properties of Ferromagnetic Materials*, Academic Press, London, 1994. With permission.

Saturation Magnetostriction of Selected Materials

The tabulated parameter λ_s is related to the fractional change in length $\Delta l/l$ by $\Delta l/l = (3/2)\lambda_s(\cos^2\theta - 1/3)$, where θ is the angle of rotation.

Material	$\lambda_s \times 10^6$
Iron	-7
Fe - 3.2% Si	+9
Nickel	-33
Cobalt	-62
45 Permalloy, 45% Ni - 55% Fe	+27
Permalloy, 82% Ni - 18% Fe	0
Permendur, 49% Co - 49% Fe - 2% V	+70
Alfer, 87% Fe - 13% Al	+30
Magnetite, Fe_3O_4	+40
Cobalt ferrite, $CoFe_2O_4$	-110
$SmFe_2$	-1560
$TbFe_2$	+1753
$Tb_{0.3}Dy_{0.7}Fe_{1.93}$ (Terfenol D)	+2000
$Fe_{66}Co_{18}B_{15}Si$ (amorphous)	+35
$Co_{72}Fe_3B_6A_{13}$ (amorphous)	0

REFERENCE

McCurrie, R.A., *Structure and Properties of Ferromagnetic Materials*, Academic Press, London, 1994, p. 91; additional data provided by A.E. Clark, Adelphi, MD.

Properties of Various Permanent Magnetic Materials (Hard)

B_r is the remanence
$_BH_c$ is the flux coercivity
$_iH_c$ is the intrinsic coercivity
$(BH)_{max}$ is the maximum energy product
T_C is the Curie temperature
T_{max} is the maximum operating temperature

Composition	B_r/T	$_BH_c/10^3$ A m^{-1}	$_iH_c/10^3$ A m^{-1}	$(BH)_{max}$/ kJ m^{-3}	T_C/°C	T_{max}/°C
Alnico1 20Ni;12Al;5Co	0.72		35	25		
Alnico2 17Ni;10Al;12.5Co;6Cu	0.72		40-50	13-14		
Alnico3 24-30Ni;12-14Al;0-3Cu	0.5-0.6		40-54	10		
Alnico4 21-28Ni;11-13Al;3-5Co;2-4Cu	0.55-0.75		36-56	11-12		
Alnico5 14Ni;8Al;24Co;3Cu	1.25	53	54	40	850	520
Alnico6 16Ni;8Al;24Co;3Cu;2Ti	1.05		75	52		
Alnico8 15Ni;7Al;35Co;4Cu;5Ti	0.83	1.6	160	45		
Alnico9 15Ni;7Al;35Co;4Cu;5Ti	1.10	1.45	1.45	75	850	520
Alnico12 13.5Ni;8Al;24.5Co;2Nb	1.20		64	76.8		
$BaFe_{12}O_{19}$ (Ferroxdur)	0.4	1.6	192	29	450	400
$SrFe_{12}O_{19}$	0.4	2.95	3.3	30	450	400
$LaCo_5$	0.91			164	567	
$CeCo_5$	0.77			117	380	
$PrCo_5$	1.20			286	620	
$NdCo_5$	1.22			295	637	
$SmCo_5$	1.00	7.9	696	196	700	250
$Sm(Co_{0.76}Fe_{0.10}Cu_{0.14})_{6.8}$	1.04	4.8	5	212	800	300
$Sm(Co_{0.65}Fe_{0.28}Cu_{0.05}Zr_{0.02})_{7.7}$	1.2	10	16	264	800	300
$Nd_2Fe_{14}B$ sintered	1.22	8.4	1120	280	300	100
Fe;52Co;14V (Vicalloy II)	1.0	42		28	700	500
Fe;24Cr;15Co;3Mo (anisotropic)	1.54	67		76	630	500
Fe;28Cr;10.5Co (Chromindur II)	0.98	32		16	630	500
Fe;23Cr;15Co;3V;2Ti	1.35	4		44	630	500
Cu;20Ni;20Fe (Cunife)	0.55	4		12	410	350
Cu;21Ni;29Fe (Cunico)	0.34	0.5		8		
Pt;23Co	0.64	4		76	480	350
Mn;29.5Al;0.5C (anisotropic)	0.61	2.16	2.4	56	300	120

REFERENCES

1. McCurrie, R.A., *Structure and Properties of Ferromagnetic Materials*, Academic Press, London, 1994, p. 204.
2. Gray, D.E., Ed., *American Institute of Physics Handbook, Third Edition*, McGraw Hill, New York, 1972, p. 5-165.
3. Jiles, D., *Magnetism and Magnetic Materials,* Chapman & Hall, London, 1991.

Selected Ferrites

J_s is the saturation magnetic polarization
T_C is the Curie temperature
ΔH is the line width

Material	J_s/T	T_C/°C	ΔH/ kA m^{-1}	Applications
Spinels				
γ-Fe_2O_3	0.52	575		
Fe_3O_4	0.60	585		
$NiFe_2O_4$	0.34	575	350	Microwave devices
$MgFe_2O_4$	0.14	440	70	
$NiZnFe_2O_4$	0.50	375	120	Transformer cores
$MnFe_2O_4$	0.50	300	50	Microwave devices

Selected Ferrites (continued)

Material	J_s/T	T_C/°C	ΔH/ kA m^{-1}	Applications
$NiCoFe_2O_4$	0.31	590	140	Microwave devices
$NiCoAlFe_2O_4$	0.15	450	330	Microwave devices
$NiAl_{0.35}Fe_{1.65}O_4$	0.12	430	67	Microwave devices
$NiAlFe_2O_4$	0.05	1860	32	Microwave devices
$Mg_{0.9}Mn_{0.1}Fe_2O_4$	0.25	290	56	Microwave devices
$Ni_{0.5}Zn_{0.5}Al_{0.8}Fe_{1.2}O_4$	0.14		17	Microwave devices
$CuFe_2O_4$	0.17	455		Electromechanical transducers
$CoFe_2O_4$	0.53	520		
$LiFe_5O_8$	0.39	670		Microwave devices
Garnets				
$Y_3Fe_5O_{12}$	0.178	280	55	Microwave devices
$Y_3Fe_5O_{12}$ (single crys.)	0.178	292	0.5	Microwave devices
$(Y,Al)_3Fe_5O_{12}$	0.12	250	80	Microwave devices
$(Y,Gd)_3Fe_5O_{12}$	0.06	250	150	Microwave devices
$Sm_3Fe_5O_{12}$	0.170	305		Microwave devices
$Eu_3Fe_5O_{12}$	0.116	293		Microwave devices
$GdFe_5O_{12}$	0.017	291		Microwave devices
Hexagonal crystals				
$BaFe_{12}O_{19}$	0.45	430	1.5	Permanent magnets
$Ba_3Co_2Fe_{24}O_{41}$	0.34	470	12	Microwave devices
$Ba_2Zn_2Fe_{12}O_{22}$	0.28	130	25	Microwave devices
$Ba_3Co_{1.35}Zn_{0.65}Fe_{24}O_{41}$		390	16	Microwave devices
$Ba_2Ni_2Fe_{12}O_{22}$	0.16	500	8	Microwave devices
$SrFe_{12}O_{19}$	0.4	450		Permanent magnets

REFERENCE

McCurrie, R.A., *Structure and Properties of Ferromagnetic Materials*, Academic Press, London, 1994.

Spinel Structure (AB_2O_4)

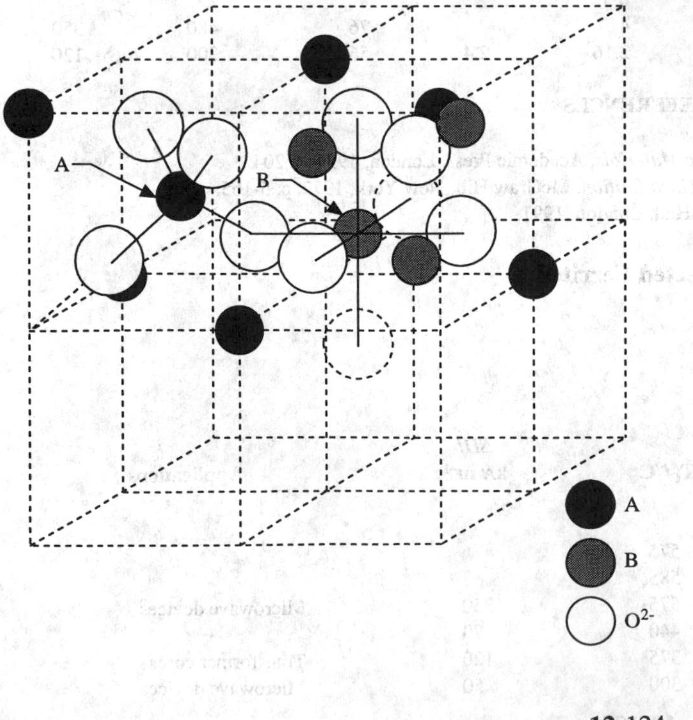

Figure 4. Arrangement of metal ions in the two octants A and B, showing tetrahedrally (A) and octahedrally (B) coordinated sites. (Reprinted from McCurrie, R.A., *Ferromagnetic Materials*, Academic Press, London, 1994. With permission.)

PROPERTIES OF MAGNETIC MATERIALS (continued)

Selected Antiferromagnetic Solids

T_N is the Néel temperature

Material	Structure	T_N/K	Material	Structure	T_N/K
Binary oxides			NiAs and related structures		
MnO	cub(fcc)	122	CrAs	orth	300
FeO	cub(fcc)	198	CrSb	hex	705-723
CoO	cub(fcc)	291	CrSe	hex	300
NiO	cub(fcc)	525	MnTe	hex	320-323
α-Mn_2O_3	cub	90	NiS	hex	263
CuO	monocl	230	CrS	monocl	460
UO_2	cub	30.8	Rutile and related structures		
Er_2O_3	cub	3.4	CoF_2	tetr	38
Gd_2O_3	cub	1.6	CrF_2	monocl	53
Perovskites			FeF_2	tetr	79
$LaCrO_3$	orth	282	MnF_2	tetr	67
$LaMnO_3$	orth	100	NiF_2	tetr	83
$LaFeO_3$	orth	750	$CrCl_2$	orth	20
$NdCrO_3$	orth	224	MnO_2	tetr	84
$NdFeO_3$	orth	760	FeOF	tetr	315
$YbCrO_3$	orth	118	Corundum and related structures		
$CaMnO_3$	cub	110	Cr_2O_3	rhomb	318
$EuTiO_3$	cub	5.3	α-Fe_2O_3	rhomb	948
$YCrO_3$	orth	141	$FeTiO_3$	rhomb	68
$BiFeO_3$	cub*	673	$MnTiO_3$	rhomb	41
$KCoF_3$	cub	125	$CoTiO_3$	rhomb	38
$KMnF_3$	cub*	88.3	VF_3 and related structures		
$KFeF_3$	cub	115	CoF_3	rhomb	460
$KNiF_3$	cub	275	CrF_3	rhomb	80
$NaMnF_3$	cub*	60	FeF_3	rhomb	394
$NaNiF_3$	orth	149	MnF_3	monocl	43
$RbMnF_3$	cub	82	MoF_3	rhomb	185
Spinels			Miscellaneous		
Co_3O_4	cub	40	K_2NiF_4	tetr	97
$NiCr_2O_4$	tetr	65	MnI_2	hex	3.4
$ZnCr_2O_4$	cub	15	$CoUO_4$	orth	12
$ZnFe_2O_4$	cub	9	$CaMn_2O_4$	orth	225
$GeFe_2O_4$	cub	10	CrN	cub*	273
MgV_2O_4	cub	45	CeC_2	tetr	33
$MnGa_2O_4$	cub	33	FeSn	hex	373
			Mn_2P	hex	103

* Distorted.

REFERENCES

1. Gray, D.E., Ed., *American Institute of Physics Handbook, Third Edition*, McGraw Hill, New York, 1972, p. 5-168 to 183.
2. Kittel, C., *Introduction to Solid State Physics, 6th Edition*, J. Wiley & Sons, New York, 1986.
3. Ashcroft, N.W., and Mermin, N.D., *Solid State Physics*, Holt, Rinehart, and Winston, New York, 1976, p. 697.

ORGANIC MAGNETS

J.S. Miller

Magnetic ordering, e.g., ferromagnetism, like superconductivity, is a property of a solid, not of an individual molecule or ion, and very rarely occurs for organic compounds. In contrast to superconductivity, where all electron spins pair to form a perfect diamagnetic material, magnetic ordering requires unpaired electron spins; hence, superconductivity and ferromagnetism are mutually exclusive.

The vast majority of organic compounds are diamagnetic (i.e., all electron spins are paired), and a relative few possess unpaired electrons (designated by an arrow, ↑) and are paramagnetic (PM), i.e., they are oriented in random directions. A few organic solids, however, exhibit strong magnetic behavior and magnetically order as ferromagnets (FO) with all spins aligned in the same direction. In some cases the spins align in the opposite direction and compensate to form an antiferromagnet (AF). In some cases these spins are not opposed to each other and do not compensate and lead to a canted antiferromagnet or weak ferromagnet (WF). If the number of spins that align in one direction differs from the number of spins that align in the opposite direction, the spins cannot compensate and a ferrimagnet (FI) results. Metamagnets (MM) are antiferromagnets in which all the spins become aligned like a ferromagnet in an applied magnetic field. Above the ordering or critical temperature, T_c, all magnets are paramagnets (PM). Organic magnets all possess electron spins in *p*-orbitals, but these may be in conjunction with metal ion-based spins.

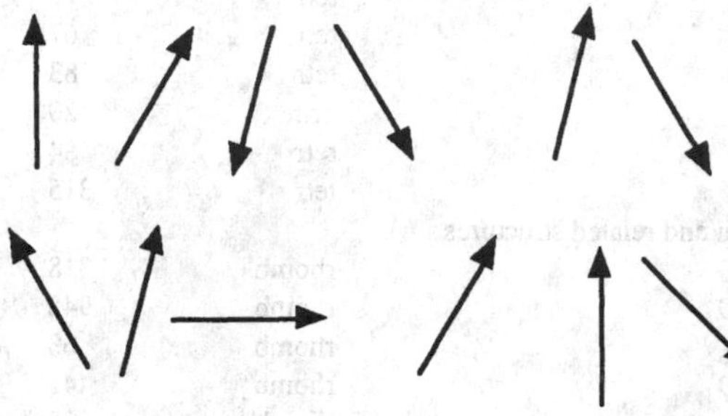

Paramagnet (PM) (random) arrangement of spins

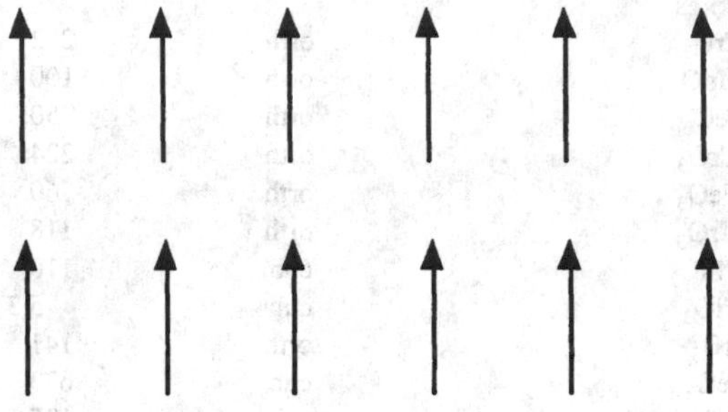

Ferromagnetic (FO) ordering of spins

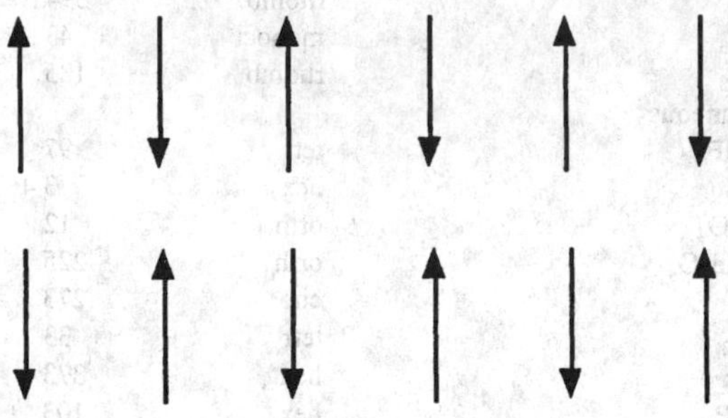

Antiferromagnetic (AF) ordering of spins

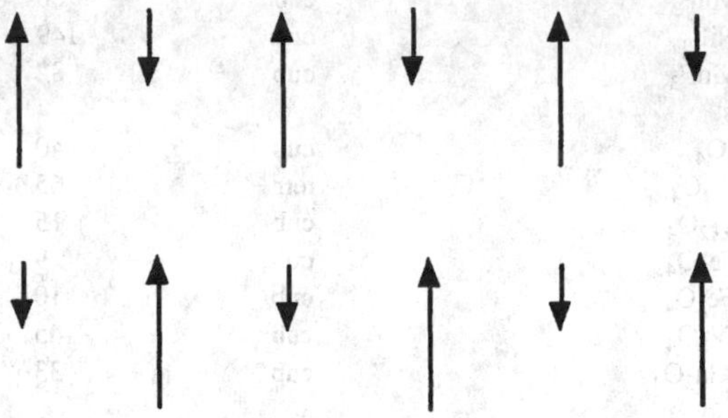

Ferrimagnetic (FI) ordering of spins

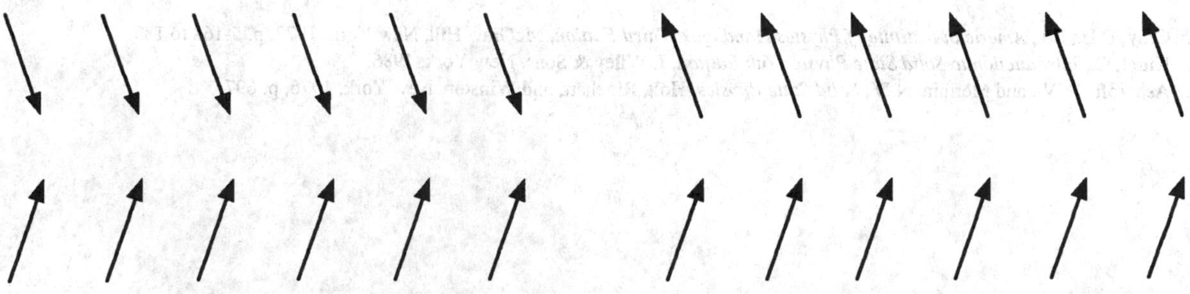

Canted antiferromagnet or weak ferromagnet (WF) ordering of spins

Figure 1. Schematic illustration of the different types of magnetic behavior.

Summary of the Critical Temperature, T_c, Saturation Magnetization, M_s, Coercive Field, H_{cr}, and Remanent Magnetization, M_r, for Selected Organic-Based Magnets

Magnet	Type	T_c/K	M_s/A m^{-1}	H_{cr}/T	M_r/A m^{-1}
α-1,3,5,7-Tetramethyl-2,6-diazaadamantane-*N,N'*-doxyl	FO	1.48	48,300	<0.00001	—
β-2-(4'-Nitrophenyl)-4,4,5,5-tetramethyl-4,5-dihydro-1H-imidazol-1-oxyl-3-*N*-oxide	FO	0.6	22,300	0.00008	<200
$\{Fe^{III}[C_5(CH_3)_5]_2\}[TCNE]$	FO	4.8	37,600	0.10	2,300
$\{Mn^{III}[C_5(CH_3)_5]_2\}[TCNE]$	FO	8.8	58,200	0.12	3,700
$\{Cr^{III}[C_5(CH_3)_5]_2\}[TCNE]$	FO	3.65	46,300	—	—
α-$\{Fe^{III}[C_5(CH_3)_5]_2\}[TCNQ]$	MM	2.55	34,200	—	—
β-$\{Fe^{III}[C_5(CH_3)_5]_2\}[TCNQ]$	FO	3.0	21,600	—	—
Tanol subarate	MM	0.38	20,700	—	—
$NCC_6F_4CN_2S_2$	WF	35.5	45	0.00009	—
$Mn^{II}(hfac)_2NITC_2H_5$	FI	7.8	39,400	0.03	27,600
$Mn^{II}(hfac)_2NIT(i\text{-}C_3H_8)$	FI	7.6	42,400	<0.0005	<420
$[Mn(hfac)_2]_3[\{ON[C_6H_3(t\text{-}C(CH_3)_3]_2NO]_2\}$	FI	46	24,400	—	—
$[MnTPP][TCNE]\cdot 2C_6H_5CH_3$	FI	13	18,400	2.4	10,300
$V[TCNE]_x.yCH_2Cl_2$ ($x \sim 2$; $y \sim 0.5$)	FI	~400	28,200	0.0015 - 0.006	1,650
$Mn[TCNE]_x.yCH_2Cl_2$ ($x \sim 2$; $y \sim 0.5$)	FI	75	52,000	0.002	270
$Fe[TCNE]_x.yCH_2Cl_2$ ($x \sim 2$; $y \sim 0.5$)	FI	97	46,300	0.23	3
$Co[TCNE]_x.yCH_2Cl_2$ ($x \sim 2$; $y \sim 0.5$)	FI	44	22,000	0.65	—

List of Symbols and Abbreviations

M_s	Saturation magnetization at 2 K
H_{cr}	Coercive Field
T_c	Critical Temperature
M_r	Remanent magnetization at 2 K
TCNE	Tetracyanoethylene
TCNQ	7,7,8,8-Tetracyano-*p*-quinodimethane
hfac	Hexafluoroacetonate
NIT	Nitronyl nitroxide
FO	Ferromagnet
FI	Ferrimagnet
MM	Metamagnet
WF	Weak ferromagnet

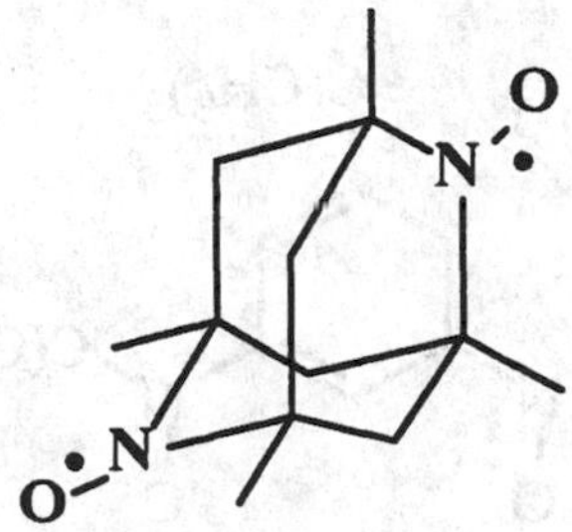

1,3,5,7-Tetramethyl-2,6-diazaadamantane-N,N'-doxyl

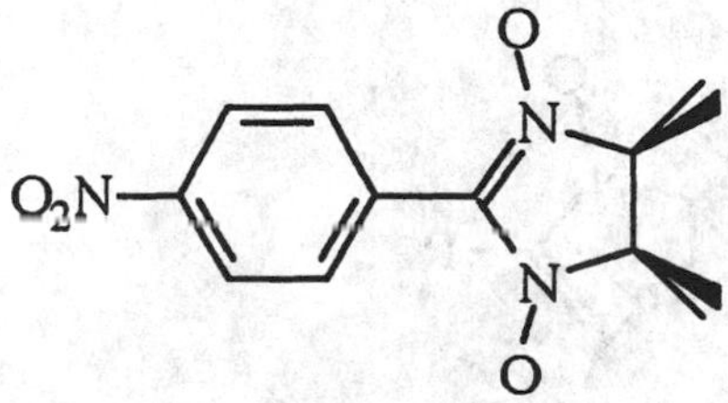

2-(4'-Nitrophyenyl)-4,4,5,5-tetramethyl-4,5-dihydro-1H-imidazol-1-oxyl-3-*N*-oxide

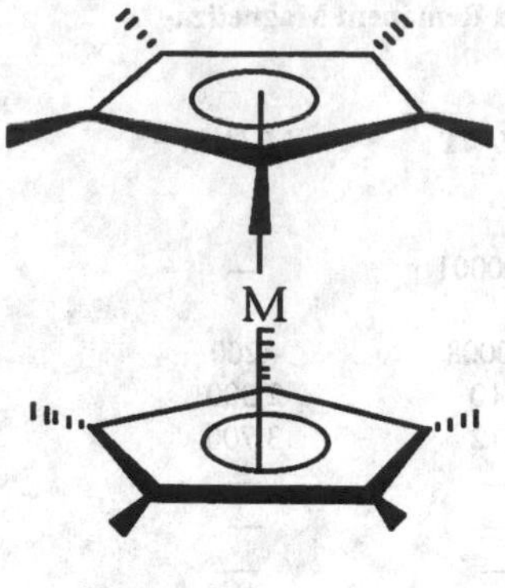

$M[C_5(CH_3)_5]_2(M = Cr, Mn, Fe)$

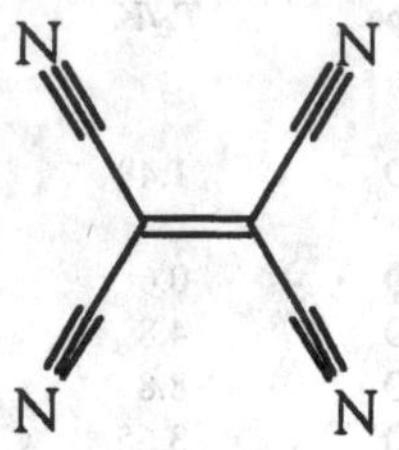

TCNE

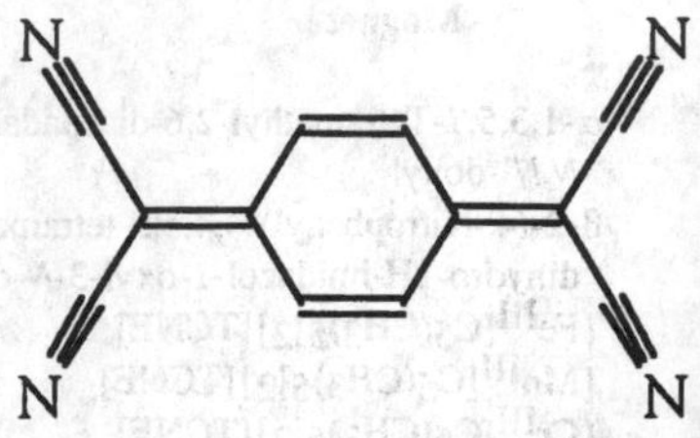

TCNQ

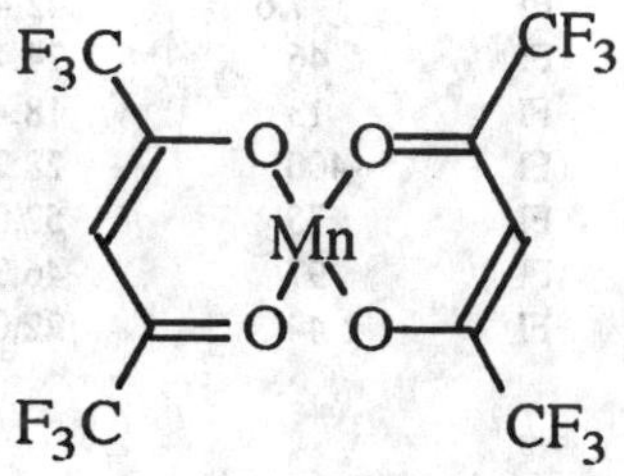

$Mn(hfac)_2$

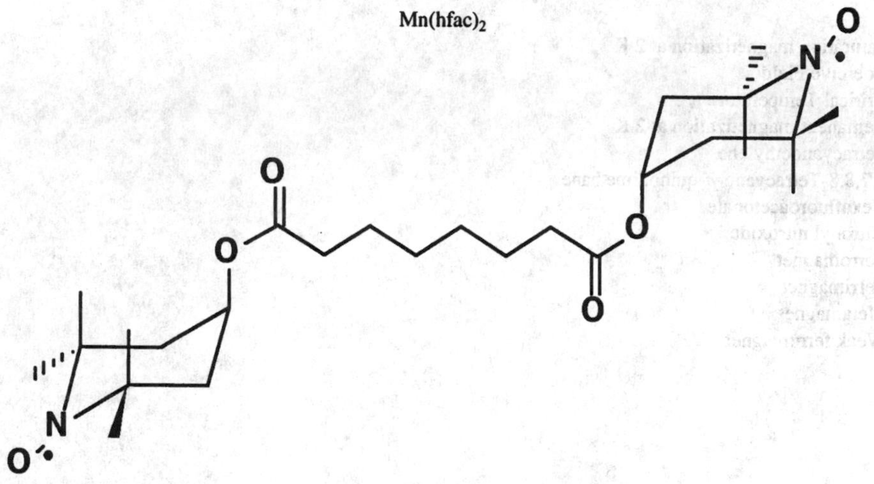

Tanol subarate

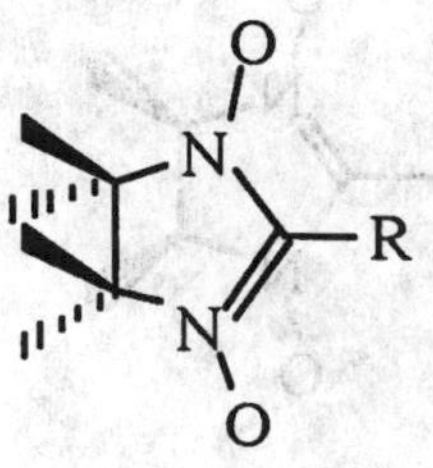

NITR ($R = C_2H_5$, $i\text{-}C_3H_8$, $n\text{-}C_3H_8$)

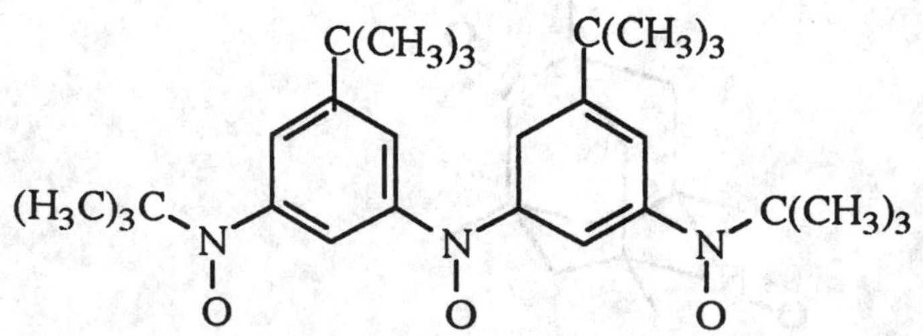

$\{ON[C_6H_3(t\text{-}C(CH_3)_3]_2NO]_2\}$

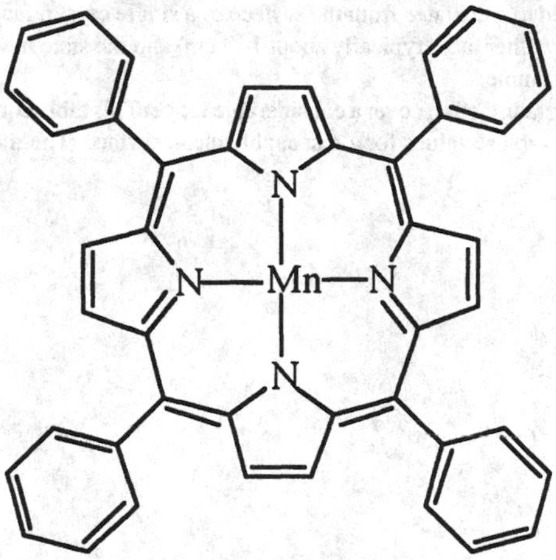

MnTPP

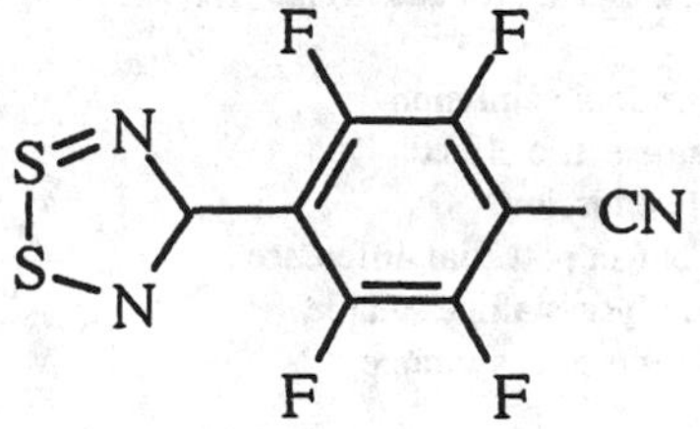

$NCC_6F_4CN_2S_2$

REFERENCES

1. Miller, J. S. and Epstein, A. J., *Angew. Chem. Internat. Ed.*, 33, 385, 1994.
2. Chiarelli, R., Rassat, A., Dromzee, Y., Jeannin, Y., Novak, M. A., and Tholence, J. L., *Phys. Scrip.*, T49, 706, 1993.
3. Kinoshita, M., *Jap. J. Appl. Phys.*, 33, 5718, 1994.
4. Gatteschi, D., *Adv. Mat.*, *6*, 635, 1994.
5. Miller, J. S. and Epstein, A. J., *J. Chem. Soc., Chem. Commun.*, 1319, 1998.
6. Broderick, W. E., Eichorn, D. M., Lu, X., Toscano, P. J., Owens, S. M. and Hoffman, B. M., *J. Am. Chem. Soc.*, 117, 3641, 1995.
7. Banister, A. J., Bricklebank, N., Lavander, I., Rawson, J., Gregory, C. I., Tanner, B. K., Clegg, W. J., Elsegood, M. R., and Palacio, F., *Angew. Chem. Internat. Ed.*, 35, 2533, 1996.

ELECTRON WORK FUNCTION OF THE ELEMENTS

The electron work function Φ is a measure of the minimum energy required to extract an electron from the surface of a solid. It is defined more precisely as the energy difference between the state in which an electron has been removed to a distance from the surface of a single crystal face that is large enough that the image force is negligible but small compared to the distance to any other face (typically about 10^{-4} cm) and the state in which the electron is in the bulk solid. In general, Φ differs for each face of a monocrystalline sample.

Since Φ is dependent on the cleanliness of the surface, measurements reported in the literature often cover a considerable range. This table contains selected values for the electron work function of the elements which may be regarded as typical values for a reasonably clean surface. The method of measurement is indicated for each value. The following abbreviations appear:

TE — Thermionic emission
PE — Photoelectric effect
FE — Field emission
CPD — Contact potential difference
polycr — Polycrystalline sample
amorp — Amorphous sample

Values in parentheses are only approximate.

REFERENCES

1. Hölzl, J., and Schulte, F. K., Work Functions of Metals, in *Solid Surface Physics*, Höhler, G., Editor, Springer-Verlag, Berlin, 1979.
2. Riviere, J. C., Work Function: Measurements and Results, in *Solid State Surface Science, Vol.1*, Green, M., Editor, Decker, New York, 1969.
3. Michaelson, H. B., *J. Appl. Phys.*, 48, 4729, 1977.

Element	Plane	Φ/eV	Method	Element	Plane	Φ/eV	Method	Element	Plane	Φ/eV	Method
Ag	100	4.64	PE		210	5.00	PE	Ru	polycr	4.71	PE
	110	4.52	PE	K	polycr	2.29	PE	Sb	amorp	4.55	
	111	4.74	PE	La	polycr	3.5	PE		100	4.7	
Al	100	4.20	PE	Li	polycr	2.93	FE	Sc	polycr	3.5	PE
	110	4.06	PE	Lu	polycr	(3.3)	CPD	Se	polycr	5.9	PE
	111	4.26	PE	Mg	polycr	3.66	PE	Si	n	4.85	CPD
As	polycr	(3.75)	PE	Mn	polycr	4.1	PE		p 100	(4.91)	CPD
Au	100	5.47	PE	Mo	100	4.53	PE		p 111	4.60	PE
	110	5.37	PE		110	4.95	PE	Sm	polycr	2.7	PE
	111	5.31	PE		111	4.55	PE	Sn	polycr	4.42	CPD
B	polycr	(4.45)	TH		112	4.36	PE	Sr	polycr	(2.59)	TH
Ba	polycr	2.52	TH		114	4.50	PE	Ta	polycr	4.25	TH
Be	polycr	4.98	PE		332	4.55	PE		100	4.15	TH
Bi	polycr	4.34	PE	Na	polycr	2.36	PE		110	4.80	TH
C	polycr	(5.0)	CPD	Nb	001	4.02	TH		111	4.00	TH
Ca	polycr	2.87	PE		110	4.87	TH	Tb	polycr	3.0	PE
Cd	polycr	4.08	CPD		111	4.36	TH	Te	polycr	4.95	PE
Ce	polycr	2.9	PE		112	4.63	TH	Th	polycr	3.4	TH
Co	polycr	5.0	PE		113	4.29	TH	Ti	polycr	4.33	PE
Cr	polycr	4.5	PE		116	3.95	TH	Tl	polycr	(3.84)	CPD
Cs	polycr	1.95	PE		310	4.18	TH	U	polycr	3.63	PE
Cu	100	5.10	FE	Nd	polycr	3.2	PE		100	3.73	PE
	110	4.48	PE	Ni	100	5.22	PE		110	3.90	PE
	111	4.94	PE		110	5.04	PE		113	3.67	PE
	112	4.53	PE		111	5.35	PE	V	polycr	4.3	PE
Eu	polycr	2.5	PE	Os	polycr	5.93	PE	W	polycr	4.55	CPD
Fe	100	4.67	PE	Pb	polycr	4.25	PE		100	4.63	FE
	111	4.81	PE	Pd	polycr	5.22	PE		110	5.22	FE
Ga	polycr	4.32	PE		111	5.6	PE		111	4.45	FE
Gd	polycr	2.90	CPD	Pt	polycr	5.64	PE		113	4.46	FE
Ge	polycr	5.0	CPD		110	5.84	FE		116	4.32	TH
Hf	polycr	3.9	PE		111	5.93	FE	Y	polycr	3.1	PE
Hg	liquid	4.475	PE		320	5.22	FE	Zn	polycr	3.63	PE
In	polycr	4.09	PE		331	5.12	FE		polycr	(4.9)	CPD
Ir	100	5.67	PE	Rb	polycr	2.261	PE	Zr	polycr	4.05	PE
	110	5.42	PE	Re	polycr	4.72	TE				
	111	5.76	PE	Rh	polycr	4.98	PE				

SECONDARY ELECTRON EMISSION

The secondary emission yield, or secondary emission ratio, δ, is the average number of secondary electrons emitted from a bombarded material for every incident primary electron. It is a function of the primary electron energy E_p. The maximum yield δ_{max} corresponds to a primary electron energy E_{pmax} (see figure). The two primary electron energies corresponding to a yield of unity are denoted the first and second crossovers (E_I and E_{II}). An insulating target, or a conducting target that is electrically floating, will charge positively or negatively depending on the primary electron energy. For $E_I < E_p < E_{II}$, $\delta > 1$ and the surface charges positively provided there is a collector present that is positive with respect to the target. For $E_p < E_I$ or $E_p > E_{II}$, $\delta < 1$, and the surface charges negatively with respect to the potential of the source of primary electrons.

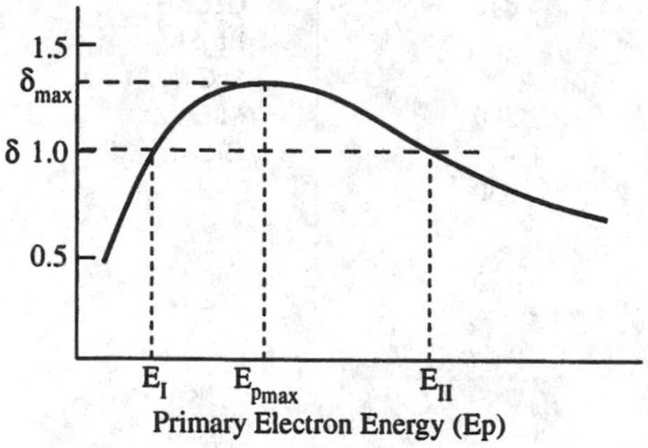

Element	δ_{max}	E_{pmax} (eV)	E_I (eV)	E_{II} (eV)
Ag	1.5	800	200	>2000
Al	1.0	300	300	300
Au	1.4	800	150	>2000
B	1.2	150	50	600
Ba	0.8	400	None	None
Bi	1.2	550	None	None
Be	0.5	200	None	None
C (diamond)	2.8	750	None	>5000
C (graphite)	1.0	300	300	300
C (soot)	0.45	500	None	None
Cd	1.1	450	300	700
Co	1.2	600	200	None
Cs	0.7	400	None	None
Cu	1.3	600	200	1500
Fe	1.3	400	120	1400
Ga	1.55	500	75	None
Ge	1.15	500	150	900
Hg	1.3	600	350	>1200
K	0.7	200	None	None

Element	δ_{max}	E_{pmax} (eV)	E_I (eV)	E_{II} (eV)
Li	0.5	85	None	None
Mg	0.95	300	None	None
Mo	1.25	375	150	1200
Na	0.82	300	None	None
Nb	1.2	375	150	1050
Ni	1.3	550	150	>1500
Pb	1.1	500	250	1000
Pd	>1.3	>250	120	None
Pt	1.8	700	350	3000
Rb	0.9	350	None	None
Sb	1.3	600	250	2000
Si	1.1	250	125	500
Sn	1.35	500	None	None
Ta	1.3	600	250	>2000
Th	1.1	800	None	None
Ti	0.9	280	None	None
Tl	1.7	650	70	>1500
W	1.4	650	250	>1500
Zr	1.1	350	None	None

Compound	δ_{max}	E_{pmax} (eV)
Alkali halides		
CsCl	6.5	
KBr (crystal)	14	1800
KCl (crystal)	12	1600
KCl (layer)	7.5	1200
KI (crystal)	10	1600
KI (layer)	5.6	
LiF (crystal)	8.5	
LiF (layer)	5.6	700
NaBr (crystal)	24	1800
NaBr (layer)	6.3	
NaCl (crystal)	14	1200

Compound	δ_{max}	E_{pmax} (eV)
NaCl (layer)	6.8	600
NaF (crystal)	14	1200
NaF (layer)	5.7	
NaI (crystal)	19	1300
NaI (layer)	5.5	
RbCl (layer)	5.8	
Oxides		
Ag_2O	1.0	
Al_2O_3 (layer)	2—9	
BaO (layer)	2.3—4.8	400
BeO	3.4	2000
CaO	2.2	500

Compound	δ_{max}	E_{pmax} (eV)
Cu_2O	1.2	400
MgO (crystal)	20—25	1500
MgO (layer)	3—15	400—1500
MoO_2	1.2	
SiO_2 (quartz)	2.1—4	400
SnO_2	3.2	640
Sulfides		
MoS_2	1.1	
PbS	1.2	500
WS_2	1.0	
ZnS	1.8	350

Compound	δ_{max}	E_{pmax} (eV)
Others		
BaF_2 (layer)	4.5	
CaF_2 (layer)	3.2	
$BiCs_3$	6	1000
BiCs	1.9	1000
GeCs	7	700
Rb_3Sb	7.1	450
$SbCs_3$	6	700
Mica	2.4	350
Glasses	2—3	300—450

OPTICAL PROPERTIES OF SELECTED ELEMENTS

J. H. Weaver and H. P. R. Frederikse

These tables list the index of refraction *n*, the extinction coefficient *k*, and the normal incidence reflection R ($\phi = 0$) as a function of photon energy *E*, which is expressed in electron volts (eV). To convert the energy in eV to wavelength in μm, use $\lambda = 1.2398/E$. To compute the dielectric function $\tilde{\varepsilon} = \varepsilon_1 + i\varepsilon_2$ from the complex index of refraction $\tilde{N} = n + ik$, use $\varepsilon_1 = n^2 - k^2$ and $\varepsilon_2 = 2nk$.

The optical constants in these tables are abridged from three more extensive tabulations:

- *Optical Properties of Metals* (OPM), Volumes I and II, *Physics Data, Nr.* 18-1 and 18-2, J. H. Weaver, C. Krafka, D. W. Lynch, and E. E. Koch, Fachinformationzentrum, Karlsruhe, Germany.
- *Handbook of Optical Constants* (HOC), Vol. I, 1985, and Vol. II, 1991. Edited by E. D. Palik, published by Academic Press, Inc.
- *American Institute of Physics Handbook* (AIPH), 3rd Edition, Coord. Editor D. E. Gray, published by McGraw-Hill Book Co., New York, 1972.

The first two of these major sources provide detailed comparisons of all optical data available in the literature at the time of the compilation. For critical applications the reader should refer to the original work. References for individual metals and semiconductors are listed at the end of the tables. Generally, tabulated values for the optical properties are accurate to better than 10%. Data in parentheses are extrapolated or interpolated values. For most elements the spectral range covered is from the far infrared (0.010 or 0.10 eV) to the far ultraviolet (10, 30 or 300 eV). The intervals between successive energies in the tables are chosen in such a way that the major spectral features are preserved.

Very small values of *k* are expressed in exponential notation, e.g., 1.23E-5 means 1.23×10^{-5}.

The following table is convenient for identifying the energy entries in these tables with the corresponding wavelengths:

λ	*E*/eV	λ	*E*/eV
1 mm	0.00124	6000 Å	2.066
500 μm	0.00248	5000 Å	2.480
100 μm	0.01240	4000 Å	3.100
50 μm	0.02480	3000 Å	4.133
10 μm	0.12398	2000 Å	6.199
5 μm	0.24797	1000 Å	12.398
1 μm	1.240	400 Å	30.996

Energy (eV)	*n*	*k*	$R(\phi = 0)$
Aluminium[1]			
0.040	98.595	203.701	0.9923
0.050	74.997	172.199	0.9915
0.060	62.852	150.799	0.9906
0.070	53.790	135.500	0.9899
0.080	45.784	123.734	0.9895
0.090	39.651	114.102	0.9892
0.100	34.464	105.600	0.9889
0.125	24.965	89.250	0.9884
0.150	18.572	76.960	0.9882
0.175	14.274	66.930	0.9879
0.200	11.733	59.370	0.9873
0.250	8.586	48.235	0.9858
0.300	6.759	40.960	0.9844
0.350	5.438	35.599	0.9834
0.400	4.454	31.485	0.9826
0.500	3.072	25.581	0.9817
0.600	2.273	21.403	0.9806
0.700	1.770	18.328	0.9794
0.800	1.444	15.955	0.9778
0.900	1.264	14.021	0.9749
1.000	1.212	12.464	0.9697
1.100	1.201	11.181	0.9630
1.200	1.260	10.010	0.9521
1.300	1.468	8.949	0.9318
1.400	2.237	8.212	0.8852
1.500	2.745	8.309	0.8678
1.600	2.625	8.597	0.8794
1.700	2.143	8.573	0.8972
1.800	1.741	8.205	0.9069
1.900	1.488	7.821	0.9116
2.000	1.304	7.479	0.9148
2.200	1.018	6.846	0.9200
2.400	0.826	6.283	0.9228
2.600	0.695	5.800	0.9238
2.800	0.598	5.385	0.9242
3.000	0.523	5.024	0.9241
3.200	0.460	4.708	0.9243
3.400	0.407	4.426	0.9245
3.600	0.363	4.174	0.9246
3.800	0.326	3.946	0.9247
4.000	0.294	3.740	0.9248
4.200	0.267	3.552	0.9248
4.400	0.244	3.380	0.9249
4.600	0.223	3.222	0.9249
4.800	0.205	3.076	0.9249
5.000	0.190	2.942	0.9244
6.000	0.130	2.391	0.9257
6.500	0.110	2.173	0.9260
7.000	0.095	1.983	0.9262
7.500	0.082	1.814	0.9265
8.000	0.072	1.663	0.9269
8.500	0.063	1.527	0.9272
9.000	0.056	1.402	0.9277
9.500	0.049	1.286	0.9282
10.000	0.044	1.178	0.9286
10.500	0.040	1.076	0.9293
11.000	0.036	0.979	0.9298
11.500	0.033	0.883	0.9283
12.000	0.033	0.791	0.9224
12.500	0.034	0.700	0.9118
13.000	0.038	0.609	0.8960
13.500	0.041	0.517	0.8789
14.000	0.048	0.417	0.8486
14.200	0.053	0.373	0.8312
14.400	0.058	0.327	0.8102
14.600	0.067	0.273	0.7802
14.800	0.086	0.211	0.7202
15.000	0.125	0.153	0.6119
15.200	0.178	0.108	0.4903
15.400	0.234	0.184	0.3881
15.600	0.280	0.073	0.3182
15.800	0.318	0.065	0.2694
16.000	0.351	0.060	0.2326
16.200	0.380	0.055	0.2031
16.400	0.407	0.050	0.1789
16.750	0.448	0.045	0.1460
17.000	0.474	0.042	0.1278
17.250	0.498	0.040	0.1129
17.500	0.520	0.038	0.1005
17.750	0.540	0.036	0.0899
18.000	0.558	0.035	0.0809
18.500	0.591	0.032	0.0664
19.000	0.620	0.030	0.0554
19.500	0.646	0.028	0.0467
20.000	0.668	0.027	0.0398
20.500	0.689	0.025	0.0342
21.000	0.707	0.024	0.0296
21.500	0.724	0.023	0.0258
22.000	0.739	0.022	0.0226
22.500	0.753	0.021	0.0199
23.000	0.766	0.021	0.0177
23.500	0.778	0.020	0.0157
24.000	0.789	0.019	0.0140
24.500	0.799	0.018	0.0126
25.000	0.809	0.018	0.0113

OPTICAL PROPERTIES OF SELECTED ELEMENTS (continued)

Energy (eV)	n	k	R(ϕ = 0)
25.500	0.817	0.017	0.0102
26.000	0.826	0.016	0.0092
27.000	0.840	0.015	0.0076
28.000	0.854	0.014	0.0063
29.000	0.865	0.014	0.0053
30.000	0.876	0.013	0.0044
35.000	0.915	0.010	0.0020
40.000	0.940	0.008	0.0010
45.000	0.957	0.007	0.0005
50.000	0.969	0.006	0.0003
55.000	0.979	0.005	0.0001
60.000	0.987	0.004	0.0000
65.000	0.995	0.004	0.0000
70.000	1.006	0.004	0.0000
72.500	1.025	0.004	0.0002
75.000	1.011	0.024	0.0002
77.500	1.008	0.025	0.0002
80.000	1.007	0.024	0.0002
85.000	1.007	0.028	0.0002
90.000	1.005	0.031	0.0002
95.000	0.999	0.036	0.0003
100.000	0.991	0.030	0.0002
110.000	0.994	0.025	0.0002
120.000	0.991	0.024	0.0002
130.000	0.987	0.021	0.0001
140.000	0.989	0.016	0.0001
150.000	0.990	0.015	0.0001
160.000	0.989	0.014	0.0001
170.000	0.989	0.011	0.0001
180.000	0.990	0.010	0.0000
190.000	0.990	0.009	0.0000
200.000	0.991	0.007	0.0000
220.000	0.992	0.006	0.0000
240.000	0.993	0.005	0.0000
260.000	0.993	0.004	0.0000
280.000	0.994	0.003	0.0000
300.000	0.995	0.002	0.0000
Carbon (diamond)[2]			
0.06199	2.3741		0.166
0.06888	2.3741		0.166
0.07749	2.3745		0.166
0.08856	2.3750		0.166
0.1033	2.3757		0.166
0.1240	2.3765		0.166
0.1550	2.3772		0.166
0.1907		3.1 E-05	
0.2066	2.3779	5.7 E-05	0.166
0.22		1.21E-04	
0.23		2.36E-04	
0.24		3.82E-04	
0.25		5.21E-04	
0.26		2.96E-04	
0.27		4.39E-04	
0.28		2.75E-04	
0.29		7.82E-05	
0.30		1.32E-04	
0.31	2.3787	1.30E-04	0.167
0.32		1.11E-04	
0.33		2.99E-05	
0.34		1.89E-05	
0.35		2.11E-05	
0.36		2.47E-05	
0.37		2.80E-05	
0.38		3.11E-05	

Energy (eV)	n	k	R(ϕ = 0)
0.39		3.67E-05	
0.40		3.58E-05	
0.41		3.25E-05	
0.4133	2.3795		0.167
0.42		2.94E-05	
0.43		2.87E-05	
0.44		3.14E-05	
0.45		3.62E-05	
0.46		3.22E-05	
0.47		1.57E-05	
0.48		6.17E-06	
0.4959	2.3801		0.167
0.6199	2.3813		0.167
0.8266	2.3837		0.167
1.240	2.3905		0.168
1.378	2.3934		0.169
1.459	2.3953		0.169
1.550	2.3975		0.169
1.653	2.4003		0.170
1.771	2.4036		0.170
1.889	2.4073		0.171
1.926	2.4084		0.171
2.066	2.4133		0.171
2.105	2.4147		0.172
2.271	2.4210		0.173
2.480	2.4299		0.174
2.650	2.4380		0.175
2.845		3.82E-07	
3.100	2.4627		0.178
3.434	2.4849		0.182
3.576	2.4955		0.183
3.961		8.97E-07	
4.160	2.5465		0.190
4.511		1.29E-06	
4.8187	2.6205	1.47E-06	0.200
5.00	2.6383		0.203
5.30		2.98E-06	
5.35		6.45E-06	
5.40		1.04E-05	
5.50		3.41E-05	
5.55		5.48E-04	
5.60	2.740	1.48E-03	0.216
5.80	2.780	5.02E-03	0.222
6.00	2.826	7.99E-03	0.228
6.10	2.852	8.62E-03	0.231
6.20	2.879	9.30E-03	0.235
6.30	2.910	9.74E-03	0.239
6.40	2.944	9.87E-03	0.243
6.50	2.985	1.10E-02	0.248
6.60	3.031	1.47E-02	0.254
6.70	3.085	2.20E-02	0.261
6.80	3.146	3.44E-02	0.268
6.90	3.220	5.24E-02	0.277
7.00	3.322	9.35E-02	0.289
7.10	3.444	0.210	0.304
7.15	3.464	0.307	0.308
7.20	3.437	0.388	0.307
7.30	3.376	0.473	0.303
7.40	3.335	0.515	0.300
7.50	3.321	0.533	0.299
7.60	3.306	0.592	0.300
7.80	3.276	0.659	0.300
8.00	3.251	0.712	0.300
8.25	3.232	0.765	0.301
8.50	3.228	0.806	0.303

Energy (eV)	n	k	R(ϕ = 0)
8.75	3.247	0.855	0.308
9.00	3.272	0.910	0.314
9.25	3.308	0.978	0.322
9.50	3.348	1.055	0.331
9.75	3.398	1.147	0.342
10.00	3.453	1.258	0.355
10.25	3.514	1.403	0.371
10.50	3.565	1.581	0.389
10.75	3.600	1.813	0.411
11.00	3.582	2.078	0.434
11.25	3.507	2.380	0.460
11.50	3.346	2.693	0.488
11.75	3.090	2.986	0.518
12.00	2.736	3.228	0.551
12.20	2.383	3.354	0.580
12.40	1.983	3.382	0.610
12.60	1.532	3.265	0.641
12.80	1.312	2.953	0.627
13.00	1.223	2.722	0.604
13.50	1.129	2.379	0.557
14.00	1.070	2.178	0.526
14.50	1.018	2.034	0.504
15.00	0.972	1.929	0.489
15.50	0.917	1.845	0.482
16.00	0.861	1.767	0.477
16.50	0.805	1.692	0.474
17.00	0.753	1.619	0.471
17.50	0.707	1.546	0.467
18.00	0.665	1.476	0.463
18.50	0.626	1.408	0.459
19.00	0.589	1.341	0.455
19.50	0.557	1.273	0.449
20.00	0.527	1.203	0.442
21.00	0.487	1.052	0.413
22.00	0.518	0.888	0.330
23.00	0.597	0.850	0.270
24.00	0.586	0.829	0.268
25.00	0.562	0.787	0.265
26.00	0.538	0.736	0.260
27.00	0.516	0.679	0.252
28.00	0.501	0.616	0.239
29.00	0.494	0.552	0.221
30.00	0.493	0.490	0.201
Cesium (evaporated)[3]			
2.145	0.264	1.123	0.631
2.271	0.278	0.950	0.561
2.845	0.425	0.438	0.235
3.064	0.540	0.320	0.127
3.397	0.671	0.233	0.057
3.966	0.827	0.174	0.018
4.889	0.916	0.143	0.007
Chromium[4]			
0.06	21.19	42.00	0.962
0.10	11.81	29.76	0.955
0.14	15.31	26.36	0.936
0.18	8.73	25.37	0.53
0.22	5.30	20.62	0.954
0.26	3.91	17.12	0.951
0.30	3.15	14.28	0.943
0.42	3.47	8.97	0.862
0.54	3.92	7.06	0.788
0.66	3.96	5.95	0.736
0.78	4.13	5.03	0.680

Energy (eV)	n	k	R(ϕ = 0)
0.90	4.43	4.60	0.650
1.00	4.47	4.43	0.639
1.12	4.53	4.31	0.631
1.24	4.50	4.28	0.629
1.36	4.42	4.30	0.631
1.46	4.31	4.32	0.632
1.77	3.84	4.37	0.639
2.00	3.48	4.36	0.644
2.20	3.18	4.41	0.656
2.40	2.75	4.46	0.677
2.60	2.22	4.36	0.698
2.80	1.80	4.06	0.703
3.00	1.54	3.71	0.695
3.20	1.44	3.40	0.670
3.40	1.39	3.24	0.657
3.60	1.26	3.12	0.661
3.80	1.12	2.95	0.660
4.00	1.02	2.76	0.651
4.20	0.94	2.58	0.639
4.40	0.90	2.42	0.620
4.50	0.89	2.35	0.607
4.60	0.88	2.28	0.598
4.70	0.86	2.21	0.586
4.80	0.86	2.13	0.572
4.90	0.86	2.07	0.557
5.00	0.85	2.01	0.542
5.10	0.86	1.94	0.523
5.20	0.87	1.87	0.503
5.40	0.93	1.80	0.466
5.60	0.95	1.74	0.443
5.80	0.97	1.74	0.437
6.00	0.94	1.73	0.444
6.20	0.89	1.69	0.446
6.40	0.85	1.66	0.447
6.60	0.80	1.59	0.444
6.80	0.75	1.51	0.439
7.00	0.74	1.45	0.425
7.20	0.71	1.39	0.414
7.40	0.69	1.33	0.404
7.60	0.66	1.23	0.378
7.80	0.67	1.15	0.347
8.00	0.68	1.07	0.315
8.20	0.71	1.00	0.278
8.50	0.74	0.92	0.235
9.0	0.83	0.81	0.170
9.50	0.92	0.74	0.132
10.00	0.98	0.73	0.120
10.50	1.01	0.72	0.112
11.00	1.05	0.69	0.103
11.50	1.09	0.69	0.100
12.00	1.13	0.70	0.101
12.50	1.15	0.73	0.108
13.00	1.15	0.77	0.119
13.50	1.12	0.80	0.128
14.00	1.09	0.82	0.135
14.50	1.03	0.82	0.142
15.00	1.00	0.82	0.143
15.50	0.96	0.80	0.141
16.00	0.92	0.77	0.139
16.50	0.31	0.75	0.134
17.00	0.90	0.73	0.132
17.50	0.88	0.72	0.130
18.00	0.87	0.70	0.129
18.50	0.84	0.69	0.130
19.00	0.82	0.68	0.131
20.00	0.77	0.64	0.130
20.5	0.76	0.63	0.129
21.0	0.74	0.58	0.121
21.5	0.72	0.55	0.116
22.0	0.71	0.52	0.112
22.5	0.70	0.50	0.109
23.0	0.69	0.48	0.105
23.5	0.68	0.45	0.101
24.0	0.68	0.43	0.096
24.5	0.67	0.39	0.089
25.0	0.68	0.36	0.080
25.5	0.68	0.33	0.072
26.0	0.70	0.31	0.063
26.5	0.71	0.28	0.055
27.0	0.72	0.26	0.048
27.5	0.73	0.25	0.043
28.0	0.75	0.23	0.037
29.0	0.77	0.22	0.032
30.0	0.78	0.21	0.030

Cobalt, single crystal, $\vec{E} \parallel \hat{c}$[5]

Energy (eV)	n	k	R(ϕ = 0)
0.10	6.71	37.87	0.982
0.15	4.66	25.47	0.973
0.20	3.55	18.78	0.962
0.25	3.98	14.59	0.933
0.30	4.04	12.16	0.907
0.40	4.24	9.13	0.847
0.50	4.41	7.19	0.782
0.60	4.91	6.13	0.729
0.70	5.24	5.85	0.713
0.80	5.17	5.89	0.716
0.90	4.94	5.95	0.720
1.00	4.46	5.86	0.722
1.10	4.07	5.61	0.715
1.20	3.81	5.36	0.706
1.30	3.60	5.20	0.701
1.40	3.37	5.09	0.701
1.50	3.10	4.96	0.701
1.60	2.84	4.77	0.697
1.70	2.66	4.57	0.690
1.80	2.45	4.41	0.687
1.90	2.31	4.18	0.675
2.00	2.21	4.00	0.664
2.10	2.13	3.85	0.654
2.20	2.07	3.70	0.642
2.30	2.01	3.59	0.634
2.40	1.95	3.49	0.627
2.50	1.88	3.40	0.622
2.60	1.81	3.32	0.618
2.70	1.73	3.24	0.615
2.80	1.66	3.13	0.607
2.90	1.61	3.05	0.600
3.00	1.55	2.96	0.594
3.20	1.46	2.80	0.579
3.40	1.38	2.64	0.563
3.60	1.31	2.48	0.544
3.80	1.28	2.33	0.519
4.00	1.26	2.20	0.495
4.20	1.25	2.10	0.471
4.40	1.24	2.01	0.452
4.60	1.24	1.94	0.435
4.80	1.23	1.88	0.423
5.00	1.22	1.83	0.411
5.20	1.21	1.79	0.403
5.40	1.19	1.77	0.399
5.60	1.16	1.75	0.400
5.80	1.10	1.73	0.406
6.00	1.03	1.68	0.407
6.20	0.97	1.62	0.401
6.40	0.94	1.53	0.386
6.60	0.91	1.46	0.368
6.80	0.91	1.38	0.345
7.00	0.91	1.32	0.326
7.00	0.91	1.26	0.305
7.40	0.92	1.21	0.286
7.60	0.93	1.17	0.269
7.80	0.94	1.13	0.253
8.00	0.95	1.09	0.239

Cobalt, single crystal, $\vec{E} \perp \hat{c}$[5]

Energy (eV)	n	k	R(ϕ = 0)
0.10	5.83	32.36	0.979
0.15	4.24	21.37	0.965
0.20	3.87	15.53	0.042
0.30	4.34	10.01	0.865
0.40	4.66	7.39	0.785
0.50	5.17	5.75	0.709
0.60	5.77	5.17	0.682
0.70	6.15	5.20	0.685
0.80	6.08	5.61	0.702
0.90	5.57	5.93	0.715
1.00	4.83	5.94	0.721
1.10	4.31	5.60	0.711
1.20	4.02	5.34	0.701
1.30	3.78	5.16	0.694
1.40	3.55	5.05	0.692
1.50	3.26	4.93	0.692
1.60	3.03	4.74	0.687
1.70	2.83	4.60	0.684
1.80	2.61	4.45	0.683
1.90	2.41	4.27	0.677
2.00	2.25	4.09	0.670
2.10	2.13	3.89	0.659
2.20	2.04	3.72	0.646
2.30	1.99	3.56	0.632
2.40	1.95	3.44	0.620
2.50	1.90	3.34	0.611
2.60	1.86	3.26	0.605
2.70	1.79	3.19	0.602
2.80	1.72	3.11	0.596
2.90	1.66	3.03	0.591
3.00	1.60	2.94	0.586
3.20	1.50	2.78	0.571
3.40	1.42	2.62	0.553
3.60	1.36	2.47	0.533
3.80	1.33	2.33	0.511
4.00	1.31	2.21	0.488
4.20	1.28	2.12	0.471
4.40	1.27	2.03	0.452
4.60	1.26	1.95	0.435
4.80	1.25	1.90	0.423
5.00	1.24	1.84	0.411
5.20	1.22	1.80	0.403
5.40	1.21	1.78	0.399
5.60	1.17	1.76	0.400
5.80	1.11	1.74	0.406
6.00	1.04	1.69	0.407
6.20	0.98	1.62	0.401
6.40	0.94	1.54	0.386
6.60	0.92	1.46	0.368
6.80	0.91	1.38	0.345

Energy (eV)	*n*	*k*	*R*(ϕ = 0)
7.00	0.91	1.32	0.326
7.20	0.91	1.26	0.305
7.40	0.92	1.21	0.285
7.60	0.93	1.17	0.269
7.80	0.94	1.13	0.253

Copper[6]

Energy (eV)	*n*	*k*	*R*(ϕ = 0)
0.10	29.69	71.57	0.980
0.50	1.71	17.63	0.979
1.00	0.44	8.48	0.976
1.50	0.26	5.26	0.965
1.70	0.22	4.43	0.958
1.75	0.21	4.25	0.956
1.80	0.21	4.04	0.952
1.85	0.22	3.85	0.947
1.90	0.21	3.67	0.943
2.00	0.27	3.24	0.910
2.10	0.47	2.81	0.814
2.20	0.83	2.60	0.673
2.30	1.04	2.59	0.618
2.40	1.12	2.60	0.602
2.60	1.15	2.50	0.577
2.80	1.17	2.36	0.545
3.00	1.18	2.21	0.509
3.20	1.23	2.07	0.468
3.40	1.27	1.95	0.434
3.60	1.31	1.87	0.407
3.80	1.34	1.81	0.387
4.00	1.34	1.72	0.364
4.20	1.42	1.64	0.336
4.40	1.49	1.64	0.329
4.60	1.52	1.67	0.334
4.80	1.53	1.71	0.345
5.00	1.47	1.78	0.366
5.20	1.38	1.80	0.380
5.40	1.28	1.78	0.389
5.60	1.18	1.74	0.391
5.80	1.10	1.67	0.389
6.00	1.04	1.59	0.380
6.50	0.96	1.37	0.329
7.00	0.97	1.20	0.271
7.50	1.00	1.09	0.230
8.00	1.03	1.03	0.206
8.50	1.03	0.98	0.189
9.00	1.03	0.92	0.171
9.50	1.03	0.87	0.154
10.00	1.04	0.82	0.139
11.00	1.07	0.75	0.118
12.00	1.09	0.73	0.111
13.00	1.08	0.72	0.109
14.00	1.06	0.72	0.111
14.50	1.03	0.72	0.111
15.00	1.01	0.71	0.111
15.50	0.98	0.69	0.109
16.00	0.95	0.67	0.106
17.00	0.91	0.62	0.097
18.00	0.89	0.56	0.084
19.00	0.88	0.51	0.071
20.00	0.88	0.45	0.059
21.00	0.90	0.41	0.048
22.00	0.92	0.38	0.040
23.00	0.94	0.37	0.035
24.00	0.96	0.37	0.035
25.00	0.96	0.40	0.040
26.00	0.92	0.40	0.044
27.00	0.88	0.38	0.043
28.00	0.86	0.35	0.039
29.00	0.85	0.30	0.032
30.00	0.86	0.26	0.025
31.00	0.88	0.24	0.020
32.00	0.89	0.22	0.017
33.00	0.90	0.21	0.015
34.00	0.91	0.20	0.014
35.00	0.92	0.20	0.013
36.00	0.92	0.19	0.012
37.00	0.92	0.19	0.011
38.00	0.93	0.18	0.010
39.00	0.93	0.17	0.009
40.00	0.93	0.17	0.009
41.00	0.94	0.16	0.008
42.00	0.94	0.16	0.007
43.00	0.94	0.15	0.007
44.00	0.95	0.15	0.007
45.00	0.95	0.15	0.006
46.00	0.95	0.15	0.006
47.00	0.95	0.14	0.006
48.00	0.95	0.14	0.006
49.00	0.95	0.14	0.005
50.00	0.95	0.13	0.005
51.00	0.95	0.13	0.005
52.00	0.95	0.13	0.005
53.00	0.96	0.12	0.004
54.00	0.96	0.12	0.004
55.00	0.96	0.12	0.004
56.00	0.96	0.11	0.004
57.00	0.96	0.11	0.004
58.00	0.96	0.11	0.004
59.00	0.97	0.11	0.003
60.00	0.97	0.11	0.003
61.00	0.97	0.11	0.003
62.00	0.97	0.11	0.003
63.00	0.96	0.10	0.003
64.00	0.96	0.10	0.003
65.00	0.97	0.10	0.003
66.00	0.97	0.10	0.003
67.00	0.97	0.09	0.003
68.00	0.97	0.09	0.002
69.00	0.97	0.09	0.002
70.00	0.97	0.09	0.002
75.00	0.98	0.09	0.002
80.00	0.98	0.09	0.002
85.00	0.97	0.09	0.002
90.00	0.96	0.08	0.002

Gallium (liquid)[7]

Energy (eV)	*n*	*k*	*R*(ϕ = 0)
1.425	2.40	9.20	0.900
1.550	2.09	8.50	0.898
1.771	1.65	7.60	0.898
2.066	1.25	6.60	0.897
2.480	0.89	5.60	0.898
3.100	0.59	4.50	0.896

Germanium, single crystal[8]

Energy (eV)	*n*	*k*	*R*(ϕ = 0)
0.01240	(4.0065)	3.00E-03	0.361
0.01364	4.0063	2.40E-03	0.361
0.01488	(4.0060)	1.70E-03	0.361
0.01612	(4.0060)	1.55E-03	0.361
0.01736	(4.0060)	1.50E-03	0.361
0.01860		1.50E-03	
0.01984		1.60E-03	
0.02108		1.60E-03	
0.02232		1.55E-03	
0.02356		1.53E-03	
0.02480		1.50E-03	
0.02604		1.25E-03	
0.02728		8.50E-04	
0.02852		6.50E-04	
0.02976		7.00E-04	
0.03100	3.9827	8.50E-04	0.358
0.03224		1.55E-03	
0.03348		2.75E-03	
0.03472		3.55E-03	
0.03596	(3.9900)	3.05E-03	0.359
0.03720		2.75E-03	
0.03844		2.70E-03	
0.03968	(3.9930)	2.90E-03	0.359
0.04092		2.95E-03	
0.04215		3.20E-03	
0.04339		6.30E-03	
0.04463		3.40E-03	
0.04587	(3.9955)	2.50E-03	0.360
0.04711		2.10E-03	
0.04835		2.00E-03	
0.04959		8.00E-04	
0.05083		1.40E-03	
0.05207		1.35E-03	
0.05331		1.10E-03	
0.05455		8.00E-04	
0.05579		6.00E-04	
0.05703		9.0 E-04	
0.05827		6.5 E-04	
0.05951		4.6 E-04	
0.06075		4.0 E-04	
0.06199	3.9992	3.98E-04	0.360
0.06323		4.0 E-04	
0.06447		4.3 E-04	
0.06571		4.4 E-04	
0.06695	(4.0000)	4.3 E-04	0.360
0.06819		3.1 E-04	
0.06943		3.3 E-04	
0.07067		3.8 E-04	
0.07191		3.3 E-04	
0.07315		2.5 E-04	
0.07439		1.9 E-04	
0.07514		1.58E-04	
0.07749	4.0009	9.55E-05	0.360
0.07999	4.0011	1.71E-04	0.360
0.08266	4.0013	9.78E-05	0.360
0.08551	4.0015	5.77E-05	0.360
0.08920		3.98E-05	
0.09460		4.59E-05	
0.09840		3.51E-05	
0.1	4.0063	3.70E-05	0.361
0.2	4.0108		0.361
0.3	4.0246		0.362
0.4	4.0429		0.364
0.5	(4.074)		0.367
0.6	(4.104)	6.58E-07	0.370
0.7	4.180	1.27E-04	0.377
0.8	4.275	5.67E-03	0.385
0.9	4.285	7.45E-02	0.386
1.0	4.325	8.09E-02	0.390
1.1	4.385	0.103	0.395
1.2	4.420	0.123	0.398
1.3	4.495	0.167	0.405

Energy (eV)	*n*	*k*	$R(\phi = 0)$
1.4	4.560	0.190	0.411
1.5	4.635	0.298	0.418
1.6	4.763	0.345	0.428
1.7	4.897	0.401	0.439
1.8	5.067	0.500	0.453
1.9	5.380	0.540	0.475
2.0	5.588	0.933	0.495
2.1	5.748	1.634	0.523
2.2	5.283	2.049	0.516
2.3	5.062	2.318	0.519
2.4	4.610	2.455	0.508
2.5	4.340	2.384	0.492
2.6	4.180	2.309	0.480
2.7	4.082	2.240	0.471
2.8	4.035	2.181	0.464
2.9	4.037	2.140	0.461
3.0	4.082	2.145	0.463
3.1	4.141	2.215	0.471
3.2	4.157	2.340	0.482
3.3	4.128	2.469	0.490
3.4	4.070	2.579	0.497
3.5	4.020	2.667	0.502
3.6	3.985	2.759	0.509
3.7	3.958	2.863	0.517
3.8	3.936	2.986	0.527
3.9	3.920	3.137	0.539
4.0	3.905	3.336	0.556
4.1	3.869	3.614	0.579
4.2	3.745	4.009	0.612
4.3	3.338	4.507	0.659
4.4	2.516	4.669	0.705
4.5	1.953	4.297	0.713
4.6	1.720	3.960	0.702
4.7	1.586	3.709	0.690
4.8	1.498	3.509	0.677
4.9	1.435	3.342	0.664
5.0	1.394	3.197	0.650
5.1	1.370	3.073	0.636
5.2	1.364	2.973	0.622
5.3	1.371	2.897	0.609
5.4	1.383	2.854	0.600
5.5	1.380	2.842	0.598
5.6	1.360	2.846	0.602
5.7	1.293	2.163	0.479
5.8	1.209	2.873	0.632
5.9	1.108	2.813	0.641
6.0	1.30	2.34	0.517
6.5	1.10	2.05	0.489
7.0	1.00	1.80	0.448
7.5		1.60	
8.0	0.92	1.40	0.348
8.5	0.92	1.20	0.282
9.0	0.92	1.14	0.262
9.5		1.00	
10.0	0.93	0.86	0.167
20.0		0.237	
22.0		0.179	
24.0		0.144	
26.0		0.110	
28.0		0.0747	
30.0		0.1020	
32.0		0.0999	
34.0		0.0856	
36.0		0.0740	
38.0		0.0651	
40.0		0.0604	

Gold, electropolished, Au (110)[9]

Energy (eV)	*n*	*k*	$R(\phi = 0)$
0.10	8.17	82.83	0.995
0.20	2.13	41.73	0.995
0.30	0.99	27.82	0.995
0.40	0.59	20.83	0.995
0.50	0.39	16.61	0.994
0.60	0.28	13.78	0.994
0.70	0.22	11.75	0.994
0.80	0.18	10.21	0.993
0.90	0.15	9.01	0.993
1.00	0.13	8.03	0.992
1.20	0.10	6.54	0.991
1.40	0.08	5.44	0.989
1.60	0.08	4.56	0.986
1.80	0.09	3.82	0.979
2.00	0.13	3.16	0.953
2.10	0.18	2.84	0.925
2.20	0.24	2.54	0.880
2.40	0.50	1.86	0.647
2.50	0.82	1.59	0.438
2.60	1.24	1.54	0.331
2.70	1.43	1.72	0.356
2.80	1.46	1.77	0.368
2.90	1.50	1.79	0.368
3.00	1.54	1.80	0.369
3.10	1.54	1.81	0.371
3.20	1.54	1.80	0.368
3.30	1.55	1.78	0.362
3.40	1.56	1.76	0.356
3.50	1.58	1.73	0.349
3.60	1.62	1.73	0.346
3.70	1.64	1.75	0.351
3.80	1.63	1.79	0.360
3.90	1.59	1.81	0.366
4.00	1.55	1.81	0.369
4.10	1.51	1.79	0.368
4.20	1.48	1.78	0.367
4.30	1.45	1.77	0.368
4.40	1.41	1.76	0.370
4.50	1.35	1.74	0.370
4.60	1.30	1.69	0.364
4.70	1.27	1.64	0.354
4.80	1.25	1.59	0.344
4.90	1.23	1.54	0.332
5.00	1.22	1.49	0.319
5.20	1.21	1.40	0.295
5.40	1.21	1.33	0.275
5.60	1.21	1.27	0.256
5.80	1.21	1.20	0.236
6.00	1.22	1.14	0.218
6.20	1.24	1.09	0.203
6.40	1.25	1.05	0.190
6.60	1.27	1.01	0.177
6.80	1.30	0.97	0.167
7.00	1.34	0.95	0.162
7.20	1.36	0.95	0.161
7.40	1.38	0.96	0.164
7.60	1.38	0.98	0.169
7.80	1.35	0.99	0.171
8.00	1.31	0.96	0.165
8.20	1.30	0.92	0.155
8.40	1.30	0.89	0.147
8.60	1.31	0.88	0.144
8.80	1.31	0.86	0.140
9.00	1.30	0.83	0.133
9.20	1.31	0.81	0.126
9.40	1.33	0.78	0.122
9.60	1.36	0.78	0.121
9.80	1.37	0.79	0.124
10.00	1.37	0.80	0.126
10.20	1.36	0.80	0.127
10.40	1.35	0.80	0.125
10.60	1.34	0.79	0.123
10.80	1.34	0.77	0.120
11.00	1.34	0.76	0.116
11.20	1.34	0.74	0.113
11.40	1.35	0.73	0.111
11.60	1.36	0.72	0.109
11.80	1.38	0.71	0.108
12.00	1.39	0.71	0.109
12.40	1.44	0.73	0.115
12.80	1.45	0.79	0.127
13.20	1.42	0.84	0.137
13.60	1.37	0.86	0.140
14.00	1.33	0.86	0.140
14.40	1.29	0.86	0.139
14.80	1.26	0.84	0.135
15.20	1.24	0.83	0.132
15.60	1.22	0.81	0.127
16.00	1.21	0.79	0.123
16.40	1.20	0.78	0.119
16.80	1.19	0.76	0.116
17.20	1.19	0.75	0.114
17.60	1.19	0.74	0.111
18.00	1.19	0.74	0.109
18.40	1.19	0.73	0.109
18.80	1.20	0.74	0.110
19.20	1.21	0.76	0.116
19.60	1.21	0.80	0.125
20.00	1.18	0.83	0.133
20.40	1.14	0.85	0.141
20.80	1.10	0.87	0.149
21.20	1.05	0.88	0.156
21.60	1.00	0.88	0.162
22.00	0.94	0.86	0.164
22.40	0.89	0.83	0.163
22.80	0.85	0.79	0.157
23.20	0.82	0.75	0.149
23.60	0.80	0.70	0.138
24.00	0.80	0.66	0.125
24.40	0.80	0.62	0.113
24.80	0.80	0.58	0.101
25.20	0.82	0.56	0.090
25.60	0.83	0.54	0.084
26.00	0.84	0.52	0.079
26.40	0.85	0.51	0.074
26.80	0.85	0.50	0.071
27.20	0.86	0.49	0.068
27.60	0.86	0.49	0.065
28.00	0.87	0.48	0.063
28.40	0.88	0.48	0.062
28.80	0.88	0.48	0.062
29.20	0.88	0.48	0.062
29.60	0.87	0.48	0.064
30.00	0.86	0.48	0.064

Hafnium, single crystal, $\vec{E} \parallel \hat{c}$[10]

Energy (eV)	*n*	*k*	$R(\phi = 0)$
0.52	1.48	4.11	0.747

Energy (eV)	*n*	*k*	*R*(φ = 0)
0.56	1.84	3.29	0.615
0.60	2.34	2.62	0.486
0.66	3.21	2.13	0.428
0.70	3.70	2.03	0.441
0.76	4.31	2.10	0.476
0.80	4.61	2.31	0.504
0.86	4.71	2.70	0.533
0.90	4.64	2.85	0.541
0.95	4.54	2.96	0.545
1.00	4.45	3.00	0.545
1.10	4.28	3.08	0.547
1.20	4.08	3.10	0.544
1.30	3.87	3.04	0.536
1.40	3.72	2.95	0.525
1.50	3.60	2.85	0.514
1.60	3.52	2.73	0.500
1.70	3.52	2.61	0.488
1.80	3.57	2.56	0.485
1.90	3.63	2.59	0.489
2.00	3.65	2.67	0.498
2.10	3.64	2.81	0.511
2.20	3.53	2.99	0.526
2.30	3.34	3.09	0.534
2.40	3.15	3.11	0.537
2.50	2.99	3.13	0.540
2.60	2.83	3.12	0.542
2.70	2.68	3.10	0.542
2.80	2.54	3.08	0.543
2.90	2.40	3.04	0.544
3.00	2.27	3.00	0.544
3.10	2.14	2.95	0.544
3.20	2.00	2.89	0.544
3.30	1.87	2.79	0.538
3.40	1.78	2.68	0.528
3.50	1.71	2.58	0.517
3.60	1.66	2.48	0.503
3.70	1.63	2.40	0.491
3.80	1.60	2.33	0.481
3.90	1.56	2.27	0.473
4.00	1.52	2.21	0.466
4.10	1.48	2.14	0.455
4.20	1.45	2.07	0.442
4.30	1.43	2.01	0.431
4.40	1.41	1.95	0.420
4.50	1.39	1.89	0.407
4.60	1.39	1.83	0.394
4.70	1.39	1.79	0.382
4.80	1.38	1.75	0.373
4.90	1.38	1.71	0.364
5.00	1.37	1.68	0.356
5.20	1.36	1.61	0.341
5.40	1.35	1.55	0.324
5.60	1.35	1.51	0.314
5.80	1.32	1.48	0.308
6.00	1.28	1.41	0.295
6.20	1.26	1.35	0.278
6.40	1.26	1.28	0.258
6.60	1.27	1.22	0.240
6.80	1.28	1.16	0.224
7.00	1.31	1.13	0.212
7.20	1.33	1.10	0.204
7.40	1.34	1.07	0.197
7.60	1.36	1.05	0.191
7.80	1.37	1.02	0.183
8.00	1.40	1.01	0.179
8.20	1.43	1.01	0.178
8.40	1.45	1.01	0.180
8.60	1.47	1.02	0.183
8.80	1.48	1.04	0.186
9.00	1.49	1.07	0.193
9.20	1.50	1.10	0.201
9.40	1.48	1.14	0.211
9.60	1.46	1.18	0.222
9.80	1.41	1.21	0.230
10.00	1.36	1.22	0.235
10.20	1.32	1.22	0.238
10.40	1.28	1.22	0.240
10.60	1.24	1.21	0.241
10.80	1.20	1.20	0.242
11.00	1.16	1.19	0.242
11.20	1.13	1.17	0.241
11.40	1.10	1.16	0.241
11.60	1.07	1.14	0.239
11.80	1.04	1.12	0.238
12.00	1.02	1.10	0.236
12.40	0.96	1.06	0.232
12.80	0.92	1.01	0.225
13.20	0.88	0.96	0.218
13.60	0.84	0.90	0.205
14.00	0.83	0.83	0.186
14.40	0.83	0.80	0.172
14.80	0.81	0.76	0.167
15.20	0.79	0.70	0.153
15.60	0.79	0.64	0.132
16.00	0.83	0.60	0.111
16.40	0.81	0.60	0.114
16.80	0.79	0.55	0.105
17.20	0.79	0.50	0.089
17.60	0.80	0.46	0.077
18.00	0.81	0.42	0.064
18.40	0.84	0.38	0.051
18.80	0.87	0.34	0.040
19.00	0.89	0.33	0.036
19.60	0.93	0.32	0.030
20.00	0.94	0.31	0.027
20.60	0.97	0.30	0.023
21.00	0.99	0.29	0.022
21.60	1.01	0.28	0.020
22.00	1.03	0.28	0.020
22.60	1.06	0.28	0.020
23.00	1.07	0.28	0.021
23.60	1.09	0.29	0.022
24.00	1.09	0.30	0.023
24.60	1.10	0.31	0.024

Hafnium, single crystal, $\vec{E} \perp \hat{c}$ [10]

Energy (eV)	*n*	*k*	*R*(φ = 0)
0.52	2.25	4.65	0.723
0.56	2.34	3.66	0.623
0.60	2.84	2.89	0.512
0.66	3.71	2.35	0.469
0.70	4.26	2.21	0.482
0.76	4.97	2.33	0.521
0.80	5.41	2.62	0.554
0.86	5.46	3.36	0.593
0.90	5.22	3.62	0.601
0.95	4.95	3.72	0.602
1.00	4.76	3.76	0.602
1.10	4.43	3.80	0.601
1.20	4.07	3.74	0.594
1.30	3.79	3.55	0.578
1.40	3.61	3.36	0.561
1.50	3.55	3.13	0.540
1.60	3.58	3.01	0.529
1.70	3.63	2.98	0.526
1.80	3.66	3.02	0.530
1.90	3.63	3.14	0.541
2.00	3.51	3.26	0.551
2.10	3.35	3.33	0.558
2.20	3.18	3.36	0.563
2.30	2.99	3.39	0.568
2.40	2.78	3.35	0.569
2.50	2.65	3.26	0.562
2.60	2.54	3.22	0.560
2.70	2.42	3.17	0.559
2.80	2.31	3.13	0.558
2.90	2.20	3.08	0.558
3.00	2.08	3.05	0.561
3.10	1.94	2.98	0.560
3.20	1.83	2.88	0.555
3.30	1.74	2.78	0.547
3.40	1.68	2.69	0.538
3.50	1.62	2.61	0.529
3.60	1.57	2.52	0.519
3.70	1.53	2.45	0.510
3.80	1.49	2.38	0.501
3.90	1.45	2.32	0.493
4.00	1.41	2.25	0.484
4.10	1.38	2.18	0.474
4.20	1.35	2.11	0.462
4.30	1.33	2.05	0.451
4.40	1.31	1.99	0.438
4.50	1.30	1.93	0.427
4.60	1.29	1.88	0.415
4.70	1.28	1.82	0.402
4.80	1.28	1.77	0.389
4.90	1.27	1.73	0.379
5.00	1.27	1.69	0.367
5.20	1.27	1.62	0.349
5.40	1.27	1.57	0.335
5.60	1.26	1.52	0.322
5.80	1.24	1.48	0.313
6.00	1.21	1.42	0.302
6.20	1.19	1.36	0.285
6.40	1.18	1.29	0.265
6.60	1.19	1.22	0.244
6.80	1.21	1.18	0.230
7.00	1.22	1.14	0.217
7.20	1.23	1.10	0.206
7.40	1.26	1.06	0.194
7.60	1.28	1.04	0.187
7.80	1.30	1.02	0.180
8.00	1.33	1.00	0.174
8.20	1.35	0.99	0.173
8.40	1.38	0.99	0.173
8.60	1.40	1.00	0.174
8.80	1.42	1.02	0.178
9.00	1.43	1.04	0.184
9.20	1.45	1.08	0.193
9.40	1.43	1.12	0.204
9.60	1.40	1.16	0.214
9.80	1.37	1.19	0.223
10.00	1.32	1.21	0.230
10.20	1.27	1.21	0.234
10.40	1.23	1.20	0.235
10.60	1.19	1.20	0.237

Energy (eV)	*n*	*k*	$R(\phi = 0)$
10.80	1.15	1.19	0.237
11.00	1.12	1.17	0.237
11.20	1.08	1.16	0.237
11.40	1.05	1.14	0.236
11.60	1.03	1.12	0.235
11.80	1.00	1.10	0.233
12.00	0.97	1.08	0.231
12.40	0.92	1.04	0.226
12.80	0.88	0.99	0.219
13.20	0.83	0.94	0.211
13.60	0.80	0.88	0.196
14.00	0.79	0.81	0.177
14.40	0.80	0.77	0.160
14.80	0.77	0.73	0.154
15.20	0.76	0.68	0.140
15.60	0.76	0.61	0.119
16.00	0.81	0.58	0.099
16.40	0.78	0.57	0.102
16.80	0.77	0.53	0.092
17.20	0.77	0.48	0.077
17.60	0.79	0.44	0.065
18.00	0.80	0.39	0.053
18.40	0.82	0.36	0.041
18.80	0.86	0.33	0.032
19.00	0.88	0.32	0.030
19.60	0.91	0.31	0.025
20.00	0.93	0.30	0.023
20.60	0.96	0.29	0.021
21.00	0.97	0.29	0.020
21.60	1.00	0.28	0.019
22.00	1.01	0.28	0.019
22.60	1.03	0.27	0.018
23.00	1.05	0.28	0.019
23.60	1.06	0.28	0.020
24.00	1.07	0.29	0.021
24.60	1.09	0.30	0.022

Iridium[11]

Energy (eV)	*n*	*k*	$R(\phi = 0)$
0.10	28.49	60.62	0.975
0.15	15.32	45.15	0.973
0.20	9.69	35.34	0.972
0.25	6.86	28.84	0.969
0.30	5.16	24.25	0.967
0.35	4.11	20.79	0.964
0.40	3.42	18.06	0.960
0.45	3.05	15.82	0.954
0.50	2.98	14.06	0.944
0.60	2.79	11.58	0.925
0.70	2.93	9.78	0.895
0.80	3.14	8.61	0.862
0.90	3.19	7.88	0.840
1.00	3.15	7.31	0.822
1.10	3.04	6.84	0.808
1.20	2.96	6.41	0.791
1.30	2.85	6.07	0.779
1.40	2.72	5.74	0.767
1.50	2.65	5.39	0.750
1.60	2.68	5.08	0.728
1.70	2.69	4.92	0.716
1.80	2.64	4.81	0.710
1.90	2.57	4.68	0.704
2.00	2.50	4.57	0.699
2.10	2.40	4.48	0.697
2.20	2.29	4.38	0.695
2.30	2.18	4.26	0.692
2.40	2.07	4.14	0.689
2.50	1.98	4.00	0.682
2.60	1.91	3.86	0.673
2.70	1.85	3.73	0.665
2.80	1.81	3.61	0.655
2.90	1.77	3.51	0.646
3.00	1.73	3.43	0.640
3.20	1.62	3.26	0.629
3.40	1.53	3.05	0.610
3.60	1.52	2.81	0.573
3.80	1.61	2.69	0.541
4.00	1.64	2.68	0.535
4.20	1.58	2.71	0.549
4.40	1.45	2.68	0.561
4.60	1.31	2.60	0.567
4.80	1.18	2.49	0.570
5.00	1.10	2.35	0.559
5.20	1.04	2.22	0.543
5.40	1.00	2.09	0.522
5.60	0.98	1.98	0.499
5.80	0.96	1.86	0.474
6.00	0.95	1.78	0.454
6.20	0.94	1.68	0.427
6.40	0.94	1.59	0.401
6.60	0.94	1.50	0.375
6.80	0.95	1.42	0.345
7.00	0.97	1.34	0.318
7.20	0.99	1.27	0.290
7.40	1.02	1.20	0.262
7.60	1.03	1.14	0.241
7.80	1.08	1.06	0.208
8.00	1.13	1.03	0.191
8.20	1.18	1.00	0.179
8.40	1.22	0.98	0.171
8.60	1.26	0.96	0.164
8.80	1.29	0.95	0.160
9.00	1.33	0.94	0.157
9.20	1.36	0.95	0.159
9.40	1.39	0.95	0.161
9.60	1.42	0.97	0.163
9.80	1.44	0.99	0.169
10.00	1.45	1.01	0.175
10.20	1.45	1.04	0.182
10.40	1.44	1.07	0.187
10.60	1.43	1.09	0.193
10.80	1.41	1.12	0.200
11.00	1.38	1.13	0.206
11.20	1.34	1.14	0.208
11.40	1.31	1.13	0.208
11.60	1.28	1.12	0.206
11.80	1.25	1.10	0.203
12.00	1.24	1.08	0.199
12.40	1.21	1.05	0.191
12.80	1.19	1.01	0.181
13.20	1.18	0.98	0.173
13.60	1.17	0.95	0.165
14.00	1.16	0.91	0.155
14.40	1.17	0.88	0.147
14.80	1.18	0.87	0.142
15.20	1.19	0.84	0.136
15.60	1.20	0.83	0.133
16.00	1.21	0.83	0.131
16.40	1.23	0.82	0.129
16.80	1.25	0.82	0.127
17.20	1.28	0.83	0.131
17.60	1.30	0.87	0.140
18.00	1.30	0.93	0.154
18.40	1.27	0.97	0.166
18.80	1.24	1.00	0.176
19.20	1.20	1.03	0.187
19.60	1.15	1.05	0.197
20.00	1.10	1.06	0.205
20.50	1.04	1.05	0.210
21.00	0.99	1.04	0.215
21.50	0.94	1.02	0.220
22.00	0.89	1.00	0.222
22.50	0.84	0.99	0.228
23.00	0.79	0.96	0.232
23.50	0.76	0.92	0.228
24.00	0.73	0.87	0.223
24.50	0.70	0.83	0.218
25.00	0.69	0.79	0.209
25.50	0.68	0.76	0.200
26.00	0.67	0.72	0.192
26.50	0.67	0.69	0.181
27.00	0.66	0.66	0.174
27.50	0.66	0.63	0.166
28.00	0.66	0.61	0.158
28.50	0.66	0.59	0.151
29.00	0.65	0.57	0.148
29.50	0.64	0.55	0.145
30.00	0.64	0.53	0.140
32.00	0.62	0.44	0.119
34.00	0.64	0.35	0.091
36.00	0.69	0.27	0.059
38.00	0.73	0.24	0.044
40.00	0.76	0.22	0.034

Iron[5]

Energy (eV)	*n*	*k*	$R(\phi = 0)$
0.10	6.41	33.07	0.978
0.15	6.26	22.82	0.956
0.20	3.68	18.23	0.958
0.26	4.98	13.68	0.911
0.30	4.87	12.05	0.892
0.36	4.68	10.44	0.867
0.40	4.42	9.75	0.858
0.50	4.14	8.02	0.817
0.60	3.93	6.95	0.783
0.70	3.78	6.17	0.752
0.80	3.65	5.60	0.725
0.90	3.52	5.16	0.700
1.00	3.43	4.79	0.678
1.10	3.33	4.52	0.660
1.20	3.24	4.26	0.641
1.30	3.16	4.07	0.626
1.40	3.12	3.87	0.609
1.50	3.05	3.77	0.601
1.60	3.00	3.60	0.585
1.70	2.98	3.52	0.577
1.80	2.92	3.46	0.573
1.90	2.89	3.37	0.563
2.00	2.85	3.36	0.563
2.10	2.80	3.34	0.562
2.20	2.74	3.33	0.563
2.30	2.65	3.34	0.567
2.40	2.56	3.31	0.567
2.50	2.46	3.31	0.570
2.60	2.34	3.30	0.576
2.70	2.23	3.25	0.575
2.80	2.12	3.23	0.580

Energy (eV)	*n*	*k*	$R(\phi = 0)$
2.90	2.01	3.17	0.580
3.00	1.88	3.12	0.583
3.10	1.78	3.04	0.580
3.20	1.70	2.96	0.576
3.30	1.62	2.87	0.572
3.40	1.55	2.79	0.565
3.50	1.50	2.70	0.556
3.60	1.47	2.63	0.548
3.70	1.43	2.56	0.542
3.83	1.38	2.49	0.534
4.00	1.30	2.39	0.527
4.17	1.26	2.27	0.510
4.33	1.23	2.18	0.494
4.50	1.20	2.10	0.482
4.67	1.16	2.02	0.470
4.83	1.14	1.93	0.451
5.00	1.14	1.87	0.435
5.17	1.12	1.81	0.425
5.33	1.11	1.75	0.408
5.50	1.09	1.17	0.401
5.67	1.09	1.65	0.383
5.83	1.10	1.61	0.373
6.00	1.09	1.59	0.366
6.17	1.08	1.57	0.365
6.33	1.04	1.55	0.365
6.50	1.02	1.51	0.358
6.67	1.00	1.47	0.351
6.83	0.97	1.43	0.346
7.00	0.96	1.39	0.333
7.17	0.94	1.35	0.327
7.33	0.94	1.30	0.311
7.50	0.94	1.26	0.298
7.67	0.94	1.23	0.288
7.83	0.94	1.21	0.279
8.00	0.94	1.18	0.272
8.17	0.94	1.16	0.265
8.33	0.94	1.14	0.258
8.50	0.94	1.12	0.251
8.67	0.94	1.10	0.246
8.83	0.92	1.08	0.240
9.00	0.93	1.07	0.236
9.17	0.92	1.06	0.233
9.33	0.91	1.04	0.231
9.50	0.90	1.02	0.226
9.67	0.90	1.00	0.221
9.83	0.89	0.99	0.218
10.00	0.88	0.97	0.213
10.17	0.87	0.94	0.203
10.33	0.87	0.91	0.196
10.50	0.87	0.89	0.189
10.67	0.88	0.87	0.179
10.83	0.89	0.85	0.170
11.00	0.91	0.83	0.162
11.17	0.92	0.83	0.159
11.33	0.93	0.84	0.159
11.50	0.93	0.84	0.160
11.67	0.93	0.84	0.162
11.83	0.92	0.84	0.163
12.00	0.91	0.84	0.163
12.17	0.90	0.84	0.165
12.33	0.89	0.83	0.164
12.50	0.98	0.83	0.165
12.67	0.87	0.82	0.166
12.83	0.86	0.81	0.166
13.00	0.85	0.80	0.162

Energy (eV)	*n*	*k*	$R(\phi = 0)$
13.17	0.84	0.79	0.161
13.33	0.84	0.78	0.160
13.50	0.83	0.77	0.159
13.67	0.82	0.76	0.157
13.83	0.81	0.75	0.154
14.00	0.81	0.73	0.151
14.17	0.80	0.72	0.149
14.33	0.80	0.71	0.146
14.50	0.79	0.79	0.144
14.67	0.79	0.69	0.141
14.83	0.78	0.67	0.138
15.00	0.78	0.66	0.135
15.17	0.78	0.65	0.131
15.33	0.78	0.64	0.238
15.50	0.77	0.63	0.126
15.67	0.77	0.62	0.123
15.83	0.77	0.61	0.119
16.00	0.77	0.60	0.116
16.17	0.78	0.58	0.112
16.33	0.78	0.58	0.110
16.50	0.78	0.57	0.107
16.67	0.77	0.56	0.106
16.83	0.78	0.55	0.103
17.00	0.78	0.55	0.102
17.17	0.78	0.54	0.100
17.33	0.78	0.54	0.098
17.50	0.77	0.53	0.097
17.67	0.77	0.52	0.095
17.83	0.78	0.51	0.092
18.00	0.78	0.51	0.091
18.17	0.78	0.51	0.090
18.33	0.78	0.50	0.089
18.50	0.77	0.50	0.089
18.67	0.77	0.50	0.088
18.83	0.77	0.49	0.087
19.00	0.77	0.49	0.087
19.17	0.76	0.49	0.088
19.33	0.76	0.48	0.087
19.50	0.75	0.47	0.086
19.67	0.75	0.47	0.085
19.83	0.75	0.46	0.084
20.00	0.74	0.45	0.083
20.17	0.74	0.44	0.081
20.33	0.74	0.44	0.081
20.50	0.74	0.42	0.080
20.67	0.73	0.43	0.079
20.83	0.73	0.42	0.078
21.00	0.73	0.41	0.077
21.17	0.72	0.40	0.076
21.33	0.72	0.39	0.074
21.50	0.72	0.38	0.073
21.67	0.72	0.38	0.071
21.83	0.72	0.37	0.070
22.00	0.72	0.36	0.068
22.17	0.71	0.35	0.067
22.33	0.72	0.34	0.064
22.50	0.72	0.34	0.063
22.67	0.72	0.33	0.062
22.83	0.72	0.32	0.059
23.00	0.72	0.31	0.058
23.17	0.72	0.30	0.056
23.33	0.72	0.29	0.054
23.50	0.73	0.28	0.050
23.67	0.73	0.28	0.049
23.83	0.74	0.27	0.047

Energy (eV)	*n*	*k*	$R(\phi = 0)$
24.00	0.74	0.27	0.045
24.17	0.74	0.26	0.044
24.33	0.74	0.26	0.043
24.50	0.74	0.25	0.042
24.67	0.75	0.25	0.040
24.83	0.75	0.24	0.039
25.00	0.75	0.24	0.038
26.00	0.76	0.21	0.031
27.00	0.78	0.18	0.026
28.00	0.79	0.16	0.021
29.00	0.81	0.14	0.017
30.00	0.82	0.13	0.014
Lithium[12]			
0.14	0.659	38.0	0.998
0.54	0.661	12.6	0.984
0.75	0.561	7.68	0.963
1.05	0.448	5.58	0.946
1.35	0.338	4.36	0.935
1.65	0.265	3.55	0.925
1.95	0.221	2.94	0.913
2.25	0.206	2.48	0.892
2.55	0.217	2.11	0.854
2.85	0.247	1.82	0.797
3.15	0.304	1.60	0.715
3.45	0.334	1.45	0.656
3.75	0.345	1.32	0.611
4.05	0.346	1.21	0.578
4.35	0.333	1.11	0.557
4.65	0.317	1.01	0.540
4.95	0.302	0.906	0.520
5.25	0.299	0.795	0.484
5.55	0.310	0.688	0.434
5.85	0.342	0.594	0.365
6.15	0.376	0.522	0.306
6.45	0.408	0.460	0.256
6.75	0.440	0.407	0.214
7.05	0.466	0.364	0.183
7.35	0.492	0.320	0.155
7.65	0.517	0.282	0.131
7.95	0.545	0.246	0.109
8.25	0.572	0.214	0.091
8.55	0.601	0.189	0.075
8.85	0.624	0.163	0.063
9.15	0.657	0.144	0.050
9.45	0.680	0.130	0.042
9.75	0.708	0.119	0.034
10.1	0.726	0.108	0.029
10.4	0.743	0.102	0.025
10.6	0.753	0.080	0.022
Magnesium (evaporated)[13]			
2.145	0.48	3.71	0.880
2.270	0.57	3.47	0.843
2.522	0.53	2.92	0.805
2.845	0.52	2.65	0.777
3.064	0.52	2.05	0.681
5.167	0.10	1.60	0.894
5.636	0.15	1.50	0.832
6.200	0.20	1.40	0.765
6.889	0.25	1.30	0.693
7.750	0.20	1.20	0.722
8.857	0.15	0.95	0.730
10.335	0.25	0.40	0.419

Energy (eV)	*n*	*k*	$R(\phi = 0)$
Manganese[14]			
0.64	3.89	5.95	0.738
0.77	3.78	5.41	0.710
0.89	3.65	5.02	0.688
1.02	3.48	4.74	0.673
1.14	3.30	4.53	0.662
1.26	3.10	4.35	0.653
1.39	2.97	4.18	0.643
1.51	2.83	4.03	0.634
1.64	2.70	3.91	0.627
1.76	2.62	3.78	0.617
1.88	2.56	3.65	0.606
2.01	2.51	3.54	0.596
2.13	2.47	3.43	0.585
2.26	2.39	3.33	0.577
2.38	2.32	3.23	0.567
2.50	2.25	3.14	0.559
2.63	2.19	3.06	0.552
2.75	2.11	2.98	0.545
2.88	2.06	2.90	0.536
3.00	2.00	2.82	0.528
3.12	1.96	2.74	0.518
3.25	1.92	2.67	0.509
3.37	1.89	2.59	0.498
3.50	1.89	2.51	0.484
3.62	1.87	2.45	0.475
3.74	1.86	2.38	0.463
3.87	1.86	2.32	0.451
3.99	1.86	2.25	0.438
4.12	1.86	2.19	0.427
4.24	1.85	2.14	0.417
4.36	1.85	2.08	0.406
4.49	1.86	2.03	0.395
4.61	1.85	1.99	0.388
4.74	1.84	1.94	0.378
4.86	1.83	1.91	0.372
4.98	1.82	1.86	0.362
5.11	1.82	1.82	0.354
5.23	1.81	1.79	0.348
5.36	1.78	1.76	0.342
5.48	1.74	1.73	0.337
5.60	1.73	1.70	0.331
5.73	1.72	1.67	0.325
5.85	1.70	1.64	0.319
5.98	1.67	1.61	0.313
6.10	1.63	1.58	0.307
6.22	1.62	1.55	0.301
6.35	1.59	1.52	0.295
6.47	1.55	1.50	0.292
6.60	1.48	1.47	0.288
Mercury (liquid)[15]			
0.2	13.99	14.27	0.869
0.3	11.37	11.95	0.846
0.4	9.741	10.65	0.830
0.5	8.528	9.805	0.818
0.6	7.574	9.195	0.808
0.8	6.086	8.312	0.796
1.0	4.962	7.643	0.789
1.2	4.050	7.082	0.786
1.4	3.324	6.558	0.785
1.6	2.746	6.054	0.783
1.8	2.284	5.582	0.782
2.0	1.910	5.150	0.782
2.2	1.620	4.751	0.780
2.4	1.384	4.407	0.779
2.6	1.186	4.090	0.779
2.8	1.027	3.802	0.779
3.0	0.898	3.538	0.777
3.2	0.798	3.294	0.773
3.4	0.713	3.074	0.770
3.6	0.644	2.860	0.763
3.8	0.589	2.665	0.755
4.0	0.542	2.502	0.749
4.2	0.507	2.341	0.738
4.4	0.477	2.195	0.727
4.6	0.452	2.058	0.715
4.8	0.431	1.929	0.701
5.0	0.414	1.806	0.685
5.2	0.401	1.687	0.666
5.4	0.394	1.569	0.642
5.6	0.386	1.454	0.617
5.7	0.386	1.396	0.601
5.8	0.386	1.341	0.585
5.9	0.385	1.287	0.569
6.0	0.386	1.232	0.551
6.1	0.388	1.176	0.531
6.2	0.390	1.118	0.510
6.3	0.399	1.058	0.481
6.4	0.412	1.002	0.450
6.5	0.428	0.949	0.418
6.6	0.436	0.898	0.392
6.7	0.438	0.836	0.367
6.8	0.459	0.756	0.320
6.9	0.510	0.676	0.255
7.0	0.585	0.617	0.191
7.1	0.663	0.589	0.148
7.2	0.717	0.584	0.128
7.3	0.769	0.575	0.111
7.4	0.817	0.574	0.100
7.5	0.860	0.580	0.094
7.6	0.893	0.597	0.093
7.8	0.929	0.623	0.096
8.0	0.946	0.639	0.098
8.2	0.952	0.645	0.099
8.4	0.953	0.638	0.097
8.6	0.956	0.624	0.093
8.8	0.965	0.607	0.087
9.0	0.975	0.588	0.082
9.2	0.988	0.568	0.076
9.4	1.009	0.548	0.069
9.6	1.044	0.541	0.066
9.8	1.061	0.557	0.069
10.0	1.062	0.567	0.071
10.2	1.054	0.569	0.072
10.4	1.045	0.561	0.070
10.6	1.041	0.550	0.068
10.8	1.039	0.537	0.065
11.0	1.039	0.523	0.062
11.5	1.050	0.491	0.055
12.0	1.064	0.467	0.050
12.5	1.078	0.445	0.045
13.0	1.092	0.430	0.042
13.5	1.104	0.416	0.040
14.0	1.115	0.404	0.038
14.5	1.125	0.394	0.037
15.0	1.135	0.383	0.035
15.5	1.146	0.374	0.034
16.0	1.159	0.368	0.034
16.5	1.170	0.367	0.034
17.0	1.177	0.367	0.034
17.5	1.184	0.366	0.034
18.0	1.191	0.367	0.035
18.5	1.195	0.367	0.035
19.0	1.200	0.366	0.035
19.5	1.208	0.364	0.035
Molybdenum[16]			
0.10	18.53	68.51	0.985
0.15	8.78	47.54	0.985
0.20	5.10	35.99	0.985
0.25	3.36	28.75	0.984
0.30	2.44	23.80	0.983
0.34	2.00	20.84	0.982
0.38	1.70	18.44	0.980
0.42	1.57	16.50	0.978
0.46	1.46	14.91	0.975
0.50	1.37	13.55	0.971
0.54	1.35	12.36	0.966
0.58	1.34	11.34	0.960
0.62	1.38	10.44	0.952
0.66	1.43	9.67	0.942
0.70	1.48	8.99	0.932
0.74	1.51	8.38	0.921
0.78	1.60	7.83	0.906
0.82	1.64	7.35	0.892
0.86	1.70	6.89	0.876
9.90	1.74	6.48	0.859
1.00	1.94	5.58	0.805
1.10	2.15	4.85	0.743
1.20	2.44	4.22	0.671
1.30	2.77	3.74	0.608
1.40	3.15	3.40	0.562
1.50	3.53	3.30	0.550
1.60	3.77	3.41	0.562
1.70	3.84	3.51	0.570
1.80	3.81	3.58	0.576
1.90	3.74	3.58	0.576
2.00	3.68	3.52	0.571
2.10	3.68	3.45	0.565
2.20	3.76	3.41	0.562
2.30	3.79	3.61	0.578
2.40	3.59	3.78	0.594
2.50	3.36	3.73	0.591
2.60	3.22	3.61	0.582
2.70	3.13	3.51	0.573
2.80	3.08	3.42	0.565
2.90	3.05	3.33	0.566
3.00	3.04	3.27	0.550
3.10	3.03	3.21	0.544
3.20	3.05	3.18	0.540
3.30	3.06	3.18	0.540
3.40	3.06	3.19	0.541
3.50	3.06	3.21	0.543
3.60	3.05	3.23	0.546
3.70	3.04	3.27	0.550
3.80	3.04	3.31	0.554
3.90	3.04	3.40	0.564
4.00	3.01	3.51	0.576
4.20	2.77	3.77	0.610
4.40	2.39	3.88	0.640
4.60	2.06	3.84	0.658
4.80	1.75	3.76	0.678
5.00	1.46	3.62	0.695
5.20	1.22	3.42	0.706

Energy (eV)	n	k	R(φ = 0)
5.40	1.07	3.20	0.706
5.60	0.96	2.99	0.700
5.80	0.89	2.80	0.688
6.00	0.85	2.64	0.674
6.20	0.81	2.50	0.660
6.40	0.79	2.36	0.641
6.60	0.78	2.24	0.619
6.80	0.78	2.13	0.592
7.00	0.80	2.04	0.568
7.20	0.81	1.98	0.548
7.40	0.81	1.95	0.542
7.60	0.75	1.90	0.552
7.80	0.71	1.81	0.542
8.00	0.69	1.73	0.530
8.20	0.67	1.65	0.512
8.40	0.66	1.57	0.495
8.60	0.65	1.49	0.475
8.80	0.65	1.41	0.450
9.00	0.65	1.33	0.420
9.20	0.67	1.25	0.385
9.40	0.69	1.19	0.355
9.60	0.71	1.12	0.320
9.80	0.74	1.05	0.285
10.00	0.77	0.99	0.250
10.20	0.81	0.93	0.217
10.40	0.86	0.88	0.188
10.60	0.91	0.83	0.162
10.80	0.98	0.79	0.138
11.00	1.05	0.77	0.125
11.20	1.12	0.78	0.123
11.40	1.18	0.80	0.125
11.60	1.23	0.85	0.135
11.80	1.25	0.89	0.145
12.00	1.26	0.92	0.154
12.40	1.25	0.98	0.168
12.80	1.23	1.00	0.178
13.20	1.20	1.02	0.185
13.60	1.17	1.02	0.187
14.00	1.15	1.01	0.185
14.40	1.13	1.00	0.182
14.80	1.13	0.99	0.179
15.00	1.14	0.99	0.179
15.60	1.15	1.01	0.184
16.00	1.14	1.04	0.194
16.60	1.10	1.10	0.216
17.00	1.04	1.12	0.233
17.60	0.94	1.14	0.257
18.00	0.87	1.12	0.270
18.60	0.77	1.08	0.283
19.00	0.71	1.02	0.284
19.60	0.66	0.94	0.275
20.00	0.64	0.89	0.264
20.60	0.62	0.81	0.245
21.00	0.61	0.77	0.234
21.60	0.61	0.71	0.215
22.00	0.60	0.69	0.207
22.60	0.59	0.63	0.195
23.00	0.58	0.60	0.185
23.60	0.58	0.53	0.166
24.00	0.58	0.49	0.151
24.60	0.60	0.43	0.124
25.00	0.62	0.39	0.106
25.60	0.66	0.35	0.085
26.00	0.68	0.33	0.072
26.50	0.71	0.31	0.060

Energy (eV)	n	k	R(φ = 0)
27.00	0.73	0.29	0.050
27.50	0.76	0.28	0.041
28.00	0.79	0.27	0.036
28.50	0.81	0.26	0.031
29.00	0.83	0.26	0.028
29.50	0.86	0.26	0.025
30.00	0.88	0.26	0.023
31.00	0.92	0.29	0.024
32.00	0.92	0.32	0.030
33.00	0.90	0.33	0.032
34.00	0.91	0.34	0.034
35.00	0.87	0.37	0.043
36.00	0.82	0.34	0.043
37.00	0.81	0.30	0.038
38.00	0.81	0.27	0.033
39.00	0.82	0.25	0.029
40.00	0.83	0.23	0.025
Nickel[17]			
0.10	9.54	45.82	0.983
0.15	5.45	30.56	0.978
0.20	4.12	22.48	0.969
0.25	4.25	17.68	0.950
0.30	4.19	15.05	0.934
0.35	4.03	13.05	0.918
0.40	3.84	11.43	0.900
0.50	4.03	9.64	0.864
0.60	3.84	8.35	0.835
0.70	3.59	7.48	0.813
0.80	3.38	6.82	0.794
0.90	3.18	6.23	0.774
1.00	3.06	5.74	0.753
1.10	2.97	5.38	0.734
1.20	2.85	5.10	0.721
1.30	2.74	4.85	0.708
1.40	2.65	4.63	0.695
1.50	2.53	4.47	0.688
1.60	2.43	4.31	0.679
1.70	2.28	4.18	0.677
1.80	2.14	4.01	0.670
1.90	2.02	3.82	0.659
2.00	1.92	3.65	0.649
2.10	1.85	3.48	0.634
2.20	1.80	3.33	0.620
2.30	1.75	3.19	0.605
2.40	1.71	3.06	0.590
2.50	1.67	2.93	0.575
2.60	1.65	2.81	0.557
2.70	1.64	2.71	0.542
2.80	1.63	2.61	0.525
2.90	1.62	2.52	0.509
3.00	1.61	2.44	0.495
3.10	1.61	2.36	0.480
3.20	1.61	2.30	0.467
3.30	1.61	2.23	0.454
3.40	1.62	2.17	0.441
3.50	1.63	2.11	0.428
3.60	1.64	2.07	0.416
3.70	1.66	2.02	0.405
3.80	1.69	1.99	0.397
3.90	1.72	1.98	0.393
4.00	1.73	1.98	0.392
4.20	1.74	2.01	0.396
4.40	1.71	2.06	0.409
4.60	1.63	2.09	0.421

Energy (eV)	n	k	R(φ = 0)
4.80	1.53	2.11	0.435
5.00	1.40	2.10	0.449
5.20	1.27	2.04	0.454
5.40	1.16	1.94	0.449
5.60	1.09	1.83	0.435
5.80	1.04	1.73	0.417
6.20	1.00	1.54	0.371
6.40	1.01	1.46	0.345
6.60	1.01	1.40	0.325
6.80	1.02	1.35	0.308
7.00	1.03	1.30	0.291
7.20	1.03	1.27	0.282
7.40	1.03	1.24	0.273
7.60	1.02	1.22	0.265
7.80	1.01	1.18	0.256
8.00	1.01	1.15	0.248
8.20	1.00	1.13	0.242
8.40	0.99	1.11	0.235
8.60	0.98	1.08	0.228
8.80	0.97	1.05	0.220
9.00	0.97	1.01	0.211
9.20	0.96	0.99	0.203
9.40	0.95	0.96	0.194
9.60	0.95	0.93	0.185
9.80	0.95	0.89	0.175
10.00	0.95	0.87	0.166
10.20	0.95	0.83	0.155
10.40	0.95	0.80	0.145
10.60	0.97	0.76	0.129
10.80	0.99	0.75	0.123
11.00	1.01	0.73	0.115
11.25	1.04	0.72	0.111
11.50	1.05	0.71	0.109
11.75	1.07	0.71	0.108
12.00	1.07	0.71	0.108
12.25	1.07	0.71	0.107
12.50	1.08	0.71	0.106
12.75	1.08	0.71	0.106
13.00	1.08	0.71	0.105
13.25	1.08	0.71	0.105
13.50	1.07	0.70	0.105
13.75	1.07	0.70	0.105
14.00	1.07	0.71	0.106
14.25	1.06	0.70	0.106
14.50	1.05	0.70	0.106
14.75	1.04	0.70	0.107
15.00	1.03	0.70	0.107
15.25	1.02	0.69	0.106
15.50	1.01	0.69	0.105
15.75	1.00	0.68	0.104
16.00	0.99	0.67	0.103
16.50	0.98	0.66	0.101
17.00	0.96	0.64	0.098
17.50	0.94	0.63	0.096
18.00	0.92	0.61	0.092
18.50	0.91	0.58	0.087
19.00	0.90	0.56	0.082
19.50	0.90	0.54	0.077
20.00	0.89	0.51	0.071
20.50	0.89	0.49	0.066
21.00	0.90	0.47	0.061
21.50	0.91	0.46	0.057
22.00	0.91	0.45	0.055
22.50	0.91	0.44	0.053
23.00	0.92	0.44	0.051

Energy (eV)	n	k	R(ϕ = 0)
23.50	0.91	0.44	0.052
24.00	0.90	0.43	0.051
24.50	0.90	0.43	0.051
25.00	0.89	0.42	0.050
26.00	0.88	0.39	0.046
27.00	0.87	0.37	0.042
28.00	0.87	0.35	0.040
29.00	0.86	0.34	0.037
30.00	0.86	0.32	0.034
35.00	0.86	0.24	0.022
40.00	0.87	0.18	0.014
45.00	0.88	0.13	0.008
50.00	0.92	0.10	0.004
60.00	0.96	0.08	0.002
65.00	0.98	0.09	0.002
68.00	0.96	0.12	0.004
70.00	0.94	0.11	0.004
75.00	0.94	0.09	0.003
80.00	0.94	0.07	0.002
90.00	0.94	0.06	0.002

Niobium[18]

Energy (eV)	n	k	R(ϕ = 0)
0.12	15.99	53.20	0.979
0.20	7.25	34.14	0.976
0.24	5.47	28.88	0.975
0.28	4.26	24.95	0.974
0.35	3.11	20.03	0.970
0.45	2.28	15.58	0.964
0.55	1.83	12.67	0.956
0.65	1.57	10.59	0.947
0.75	1.41	9.00	0.935
0.85	1.35	7.74	0.918
0.95	1.35	6.70	0.893
1.05	1.44	5.86	0.857
1.15	1.55	5.18	0.814
1.25	1.65	4.63	0.768
1.35	1.76	4.13	0.715
1.45	1.95	3.68	0.650
1.55	2.15	3.37	0.595
1.65	2.36	3.13	0.552
1.75	2.54	2.99	0.527
1.85	2.69	2.89	0.510
1.95	2.82	2.86	0.505
2.05	2.89	2.87	0.505
2.15	2.92	2.87	0.505
2.25	2.93	2.87	0.505
2.35	2.92	2.88	0.506
2.45	2.89	2.90	0.509
2.55	2.83	2.92	0.512
2.65	2.74	2.90	0.511
2.75	2.66	2.86	0.507
2.85	2.58	2.80	0.500
3.00	2.51	2.68	0.485
3.10	2.48	2.60	0.475
3.20	2.45	2.53	0.465
3.30	2.44	2.45	0.453
3.40	2.46	2.38	0.442
3.50	2.48	2.33	0.435
3.60	2.52	2.29	0.428
3.70	2.56	2.27	0.426
3.80	2.59	2.28	0.427
3.90	2.62	2.29	0.429
4.00	2.64	2.33	0.434
4.20	2.64	2.42	0.447
4.40	2.53	2.56	0.467
4.60	2.39	2.56	0.470
4.80	2.32	2.52	0.465
5.00	2.26	2.57	0.475
5.20	2.16	2.62	0.487
5.40	2.00	2.68	0.505
5.60	1.81	2.67	0.518
5.80	1.63	2.60	0.522
6.00	1.49	2.49	0.520
6.20	1.38	2.38	0.512
6.40	1.31	2.25	0.496
6.60	1.26	2.14	0.480
6.80	1.24	2.04	0.460
7.00	1.23	1.96	0.441
7.20	1.22	1.91	0.430
7.40	1.20	1.88	0.427
7.60	1.14	1.85	0.430
7.80	1.07	1.78	0.428
8.00	1.02	1.69	0.412
8.20	1.00	1.60	0.390
8.40	0.99	1.51	0.365
8.60	0.99	1.43	0.340
8.70	0.99	1.39	0.328
8.80	1.00	1.36	0.315
9.00	1.01	1.29	0.290
9.20	1.04	1.22	0.265
9.40	1.07	1.18	0.245
9.60	1.10	1.13	0.227
9.80	1.13	1.09	0.209
10.00	1.18	1.05	0.194
10.20	1.23	1.04	0.187
10.40	1.27	1.04	0.185
10.60	1.30	1.06	0.190
10.80	1.32	1.08	0.195
11.00	1.32	1.10	0.200
11.20	1.31	1.12	0.204
11.40	1.30	1.13	0.207
11.60	1.28	1.13	0.209
11.80	1.27	1.13	0.210
12.00	1.25	1.12	0.209
12.40	1.24	1.10	0.204
12.80	1.24	1.09	0.200
13.20	1.24	1.09	0.201
13.60	1.23	1.12	0.208
14.00	1.20	1.13	0.216
14.40	1.16	1.15	0.225
14.80	1.11	1.16	0.234
15.00	1.08	1.16	0.238
15.60	0.99	1.14	0.247
16.00	0.92	1.11	0.250
16.60	0.85	1.04	0.245
17.00	0.80	0.99	0.240
17.20	0.79	0.96	0.236
17.40	0.77	0.93	0.230
17.80	0.75	0.87	0.217
18.00	0.74	0.85	0.209
18.60	0.73	0.77	0.185
19.00	0.72	0.72	0.170
19.60	0.72	0.66	0.150
20.00	0.72	0.62	0.137
20.60	0.71	0.55	0.119
21.00	0.72	0.50	0.100
21.60	0.75	0.43	0.075
22.00	0.78	0.40	0.063
22.60	0.82	0.35	0.045
23.00	0.85	0.33	0.038
23.60	0.88	0.30	0.029
24.00	0.91	0.29	0.025
24.60	0.94	0.28	0.022
25.00	0.96	0.27	0.020
25.60	0.99	0.26	0.018
26.00	1.00	0.26	0.017
26.60	1.03	0.25	0.016
27.00	1.04	0.25	0.015
27.60	1.06	0.25	0.015
28.00	1.08	0.24	0.015
28.60	1.11	0.24	0.016
29.00	1.13	0.25	0.017
29.60	1.16	0.26	0.020
30.00	1.18	0.28	0.023
31.00	1.18	0.31	0.026
32.00	1.20	0.34	0.031
33.00	1.21	0.38	0.038
34.00	1.20	0.42	0.044
35.20	1.17	0.47	0.051
36.00	1.15	0.50	0.056
37.50	1.07	0.53	0.064
39.50	0.95	0.50	0.063
40.50	0.92	0.47	0.059

Osmium (Polycrystalline)[9]

Energy (eV)	n	k	R(ϕ = 0)
0.10	4.08	50.23	0.994
0.15	2.90	33.60	0.990
0.20	2.44	25.11	0.985
0.25	2.35	19.99	0.977
0.30	2.23	16.54	0.969
0.35	2.33	14.06	0.955
0.40	2.45	12.32	0.940
0.45	2.43	11.02	0.927
0.50	2.41	9.97	0.913
0.55	2.33	9.12	0.901
0.60	2.21	8.37	0.890
0.65	2.11	7.68	0.877
0.70	2.02	7.04	0.862
0.75	2.00	6.46	0.842
0.80	2.00	5.95	0.820
0.85	2.01	5.51	0.796
0.90	2.03	5.10	0.769
0.95	2.05	4.74	0.742
1.00	2.09	4.41	0.712
1.10	2.15	3.84	0.651
1.20	2.16	3.35	0.592
1.30	2.25	2.77	0.506
1.40	2.49	2.23	0.419
1.50	2.84	1.80	0.369
1.60	3.36	1.62	0.379
1.70	3.70	1.75	0.411
1.80	3.78	1.83	0.423
1.90	3.81	1.75	0.418
2.00	3.98	1.60	0.418
2.10	4.26	1.54	0.432
2.20	4.58	1.62	0.457
2.30	4.84	1.76	0.479
2.40	5.10	2.01	0.506
2.50	5.28	2.38	0.532
2.60	5.36	2.82	0.557
2.70	5.30	3.29	0.580
2.80	5.07	3.78	0.603
2.90	4.65	4.18	0.624
3.00	4.05	4.40	0.639
3.20	3.29	3.96	0.614

Energy (eV)	n	k	$R(\phi = 0)$
3.40	2.93	3.79	0.607
3.60	2.75	3.45	0.577
3.80	2.73	3.32	0.562
4.00	2.71	3.34	0.565
4.20	2.53	3.44	0.584
4.40	2.24	3.44	0.599
4.60	2.01	3.31	0.598
4.80	1.88	3.19	0.592
5.00	1.74	3.12	0.596
5.20	1.58	3.00	0.597
5.40	1.46	2.88	0.593
5.60	1.36	2.77	0.589
5.80	1.27	2.65	0.582
6.00	1.20	2.54	0.575
6.20	1.13	2.44	0.571
6.40	1.06	2.33	0.562
6.60	1.01	2.21	0.548
6.80	0.97	2.11	0.532
7.00	0.95	2.00	0.514
7.20	0.92	1.91	0.497
7.40	0.91	1.81	0.476
7.60	0.90	1.72	0.451
7.80	0.90	1.63	0.426
8.00	0.91	1.55	0.400
8.20	0.91	1.48	0.375
8.40	0.94	1.40	0.344
8.60	0.96	1.34	0.319
8.80	0.98	1.29	0.296
9.00	1.01	1.24	0.274
9.20	1.04	1.19	0.255
9.40	1.08	1.16	0.238
9.60	1.10	1.14	0.229
9.80	1.13	1.11	0.217
10.00	1.16	1.10	0.209
10.20	1.19	1.08	0.203
10.30	1.20	1.08	0.201
10.40	1.22	1.08	0.200
10.50	1.23	1.09	0.201
10.60	1.24	1.10	0.203
10.80	1.25	1.11	0.206
11.00	1.24	1.13	0.213
11.20	1.23	1.14	0.217
11.40	1.19	1.15	0.223
11.60	1.17	1.12	0.216
11.80	1.16	1.10	0.211
12.00	1.15	1.08	0.205
12.40	1.14	1.03	0.191
12.80	1.15	1.01	0.183
13.20	1.16	0.98	0.174
13.60	1.17	0.97	0.170
14.00	1.17	0.96	0.169
14.40	1.16	0.94	0.165
14.80	1.16	0.91	0.156
15.20	1.17	0.89	0.148
15.60	1.20	0.86	0.140
16.00	1.25	0.87	0.140
16.40	1.28	0.90	0.147
16.80	1.28	0.94	0.157
17.20	1.27	0.97	0.167
17.60	1.26	1.01	0.178
18.00	1.23	1.04	0.189
18.40	1.19	1.08	0.200
18.80	1.14	1.10	0.210
19.20	1.10	1.10	0.219
19.60	1.05	1.11	0.227
20.00	0.96	1.10	0.239
20.40	0.93	1.09	0.240
20.80	0.89	1.05	0.240
21.20	0.86	1.02	0.237
21.60	0.83	0.99	0.235
22.00	0.80	0.96	0.230
22.40	0.78	0.93	0.226
22.80	0.77	0.90	0.220
23.20	0.75	0.88	0.217
23.60	0.75	0.86	0.211
24.00	0.73	0.84	0.209
24.40	0.72	0.82	0.207
24.80	0.70	0.80	0.205
25.20	0.69	0.77	0.202
25.60	0.67	0.75	0.199
26.00	0.66	0.72	0.195
26.40	0.65	0.69	0.189
26.80	0.63	0.66	0.183
27.20	0.65	0.62	0.165
28.00	0.64	0.59	0.156
28.40	0.64	0.57	0.148
28.80	0.65	0.55	0.140
29.20	0.65	0.53	0.134
29.60	0.65	0.51	0.128
30.00	0.65	0.49	0.121
31.00	0.65	0.45	0.111
32.00	0.66	0.41	0.095
33.00	0.68	0.37	0.079
34.00	0.70	0.34	0.068
35.00	0.72	0.31	0.057
36.00	0.74	0.29	0.048
37.00	0.77	0.27	0.040
38.00	0.79	0.26	0.035
39.00	0.81	0.26	0.031
40.00	0.84	0.26	0.026

Palladium[19]

Energy (eV)	n	k	$R(\phi = 0)$
0.10	4.13	54.15	0.994
0.15	3.13	35.82	0.990
0.20	3.07	26.59	0.983
0.26	3.11	20.15	0.971
0.30	3.56	17.27	0.955
0.36	3.98	14.41	0.932
0.40	4.27	13.27	0.916
0.46	4.27	12.11	0.902
0.50	4.10	11.44	0.896
0.56	3.92	10.49	0.883
0.60	3.80	9.96	0.876
0.72	3.51	8.70	0.854
0.80	3.35	8.06	0.840
1.00	2.99	6.89	0.811
1.10	2.81	6.46	0.800
1.20	2.65	6.10	0.790
1.30	2.50	5.78	0.781
1.40	2.34	5.50	0.774
1.50	2.17	5.22	0.767
1.60	2.08	4.95	0.755
1.70	2.00	4.72	0.745
1.80	1.92	4.54	0.737
1.90	1.82	4.35	0.729
2.00	1.75	4.18	0.721
2.10	1.67	4.03	0.714
2.20	1.60	3.88	0.707
2.30	1.53	3.75	0.700
2.40	1.47	3.61	0.693
2.50	1.41	3.48	0.685
2.60	1.37	3.36	0.676
2.70	1.32	3.25	0.668
2.80	1.29	3.13	0.658
2.90	1.26	3.03	0.648
3.00	1.23	2.94	0.639
3.10	1.20	2.85	0.630
3.20	1.17	2.77	0.622
3.30	1.14	2.68	0.613
3.40	1.12	2.60	0.602
3.50	1.10	2.52	0.591
3.60	1.08	2.45	0.581
3.70	1.07	2.38	0.570
3.80	1.06	2.31	0.558
3.90	1.05	2.25	0.547
4.00	1.03	2.19	0.537
4.20	1.04	2.09	0.510
4.40	1.03	2.01	0.493
4.60	1.03	1.94	0.476
4.80	1.01	1.90	0.470
5.00	0.96	1.86	0.472
5.20	0.90	1.79	0.474
5.40	0.85	1.70	0.463
5.60	0.81	1.62	0.449
5.80	0.78	1.54	0.437
6.00	0.76	1.45	0.418
6.20	0.74	1.37	0.397
6.40	0.73	1.29	0.375
6.60	0.72	1.21	0.350
6.80	0.73	1.13	0.316
7.00	0.73	1.05	0.287
7.20	0.75	0.98	0.255
7.40	0.77	0.91	0.223
7.60	0.79	0.85	0.195
7.80	0.83	0.78	0.163
8.00	0.88	0.73	0.133
8.20	0.94	0.70	0.117
8.40	0.96	0.70	0.114
8.60	1.00	0.65	0.097
8.80	1.04	0.65	0.094
9.00	1.07	0.64	0.090
9.50	1.12	0.65	0.089
10.00	1.14	0.65	0.088
10.50	1.16	0.65	0.087
11.00	1.18	0.64	0.086
11.50	1.19	0.65	0.087
12.00	1.20	0.66	0.089
12.50	1.19	0.67	0.091
13.00	1.18	0.67	0.091
13.50	1.18	0.67	0.092
14.00	1.17	0.67	0.093
14.50	1.15	0.68	0.095
15.00	1.13	0.69	0.098
15.50	1.10	0.68	0.096
16.00	1.08	0.66	0.092
16.50	1.06	0.63	0.086
17.00	1.07	0.61	0.081
17.50	1.06	0.61	0.080
18.00	1.07	0.59	0.077
18.50	1.07	0.59	0.077
19.00	1.08	0.59	0.077
19.50	1.08	0.61	0.080
20.00	1.07	0.65	0.090
20.50	1.03	0.67	0.098
21.00	0.99	0.67	0.103

OPTICAL PROPERTIES OF SELECTED ELEMENTS (continued)

Energy (eV)	n	k	R(φ = 0)
21.50	0.95	0.66	0.103
22.00	0.91	0.64	0.103
22.50	0.88	0.62	0.101
23.00	0.86	0.59	0.097
23.50	0.85	0.56	0.091
24.00	0.84	0.54	0.086
25.00	0.81	0.51	0.084
26.40	0.80	0.43	0.066
27.80	0.81	0.38	0.052
29.20	0.82	0.35	0.046
Platinum[20]			
0.10	13.21	44.72	0.976
0.15	8.18	31.16	0.969
0.20	5.90	23.95	0.962
0.25	4.70	19.40	0.954
0.30	3.92	16.16	0.945
0.35	3.28	13.66	0.936
0.40	2.81	11.38	0.922
0.45	3.03	9.31	0.882
0.50	3.91	7.71	0.813
0.55	4.58	7.14	0.777
0.60	5.13	6.75	0.753
0.65	5.52	6.66	0.746
0.70	5.71	6.83	0.751
0.75	5.57	7.02	0.759
0.80	5.31	7.04	0.762
0.85	5.05	6.98	0.763
0.90	4.77	6.91	0.765
0.95	4.50	6.77	0.763
1.00	4.25	6.62	0.762
1.10	3.86	6.24	0.753
1.20	3.55	5.92	0.746
1.30	3.29	5.61	0.736
1.40	3.10	5.32	0.725
1.50	2.92	5.07	0.716
1.60	2.76	4.84	0.706
1.70	2.63	4.64	0.697
1.80	2.51	4.43	0.686
1.90	2.38	4.26	0.678
2.00	2.30	4.07	0.664
2.10	2.23	3.92	0.654
2.20	2.17	3.77	0.642
2.30	2.10	3.67	0.636
2.40	2.03	3.54	0.626
2.50	1.96	3.42	0.616
2.60	1.91	3.30	0.605
2.70	1.87	3.20	0.595
2.80	1.83	3.10	0.585
2.90	1.79	3.01	0.575
3.00	1.75	2.92	0.565
3.20	1.68	2.76	0.546
3.40	1.63	2.62	0.527
3.60	1.58	2.48	0.507
3.80	1.53	2.37	0.491
4.00	1.49	2.25	0.472
4.20	1.45	2.14	0.452
4.40	1.43	2.04	0.432
4.60	1.39	1.95	0.415
4.80	1.38	1.85	0.392
5.00	1.36	1.76	0.372
5.20	1.36	1.67	0.350
5.40	1.36	1.61	0.332
5.60	1.36	1.54	0.315
5.80	1.36	1.47	0.295

Energy (eV)	n	k	R(φ = 0)
6.00	1.38	1.40	0.276
6.20	1.39	1.35	0.261
6.40	1.42	1.29	0.246
6.60	1.45	1.26	0.236
6.80	1.48	1.24	0.231
7.00	1.50	1.24	0.230
7.20	1.50	1.25	0.231
7.40	1.49	1.23	0.228
7.60	1.48	1.22	0.225
7.80	1.48	1.20	0.221
8.00	1.47	1.18	0.216
8.20	1.47	1.17	0.212
8.40	1.47	1.15	0.209
8.60	1.47	1.14	0.205
8.80	1.47	1.13	0.202
9.00	1.48	1.12	0.200
9.20	1.49	1.11	0.198
9.40	1.49	1.12	0.200
9.60	1.49	1.13	0.203
9.80	1.48	1.15	0.207
10.00	1.46	1.15	0.209
10.20	1.43	1.16	0.211
10.40	1.40	1.15	0.210
10.60	1.37	1.14	0.207
10.80	1.35	1.12	0.203
11.00	1.33	1.10	0.199
11.20	1.31	1.08	0.194
11.40	1.30	1.06	0.188
11.60	1.29	1.04	0.183
11.80	1.29	1.01	0.177
12.00	1.29	1.00	0.173
12.40	1.29	0.97	0.165
12.80	1.29	0.94	0.158
13.20	1.31	0.93	0.155
13.60	1.31	0.93	0.155
14.00	1.31	0.93	0.155
14.40	1.30	0.93	0.156
14.80	1.27	0.93	0.157
15.20	1.27	0.93	0.155
15.60	1.25	0.92	0.151
16.00	1.24	0.89	0.146
16.50	1.24	0.87	0.142
17.00	1.25	0.86	0.138
17.50	1.27	0.85	0.135
18.00	1.31	0.88	0.142
18.50	1.30	0.94	0.157
19.00	1.28	0.99	0.171
19.50	1.23	1.03	0.184
20.00	1.18	1.06	0.197
20.50	1.11	1.09	0.212
21.00	1.03	1.10	0.226
21.50	0.94	1.08	0.238
22.00	0.87	1.04	0.240
22.50	0.81	0.98	0.235
23.00	0.77	0.92	0.226
23.50	0.75	0.87	0.213
24.00	0.74	0.82	0.201
24.50	0.73	0.77	0.187
25.00	0.73	0.73	0.174
25.50	0.73	0.70	0.162
26.00	0.74	0.67	0.150
26.50	0.74	0.65	0.142
27.00	0.74	0.63	0.136
27.50	0.74	0.62	0.130
28.00	0.75	0.60	0.125

Energy (eV)	n	k	R(φ = 0)
28.50	0.75	0.59	0.121
29.00	0.75	0.58	0.118
29.50	0.74	0.58	0.120
30.00	0.73	0.58	0.124
Potassium[21]			
0.55	0.139	7.10	0.989
0.58	0.119	6.72	0.990
0.63	0.106	6.32	0.990
0.67	0.091	5.79	0.990
0.73	0.079	5.30	0.989
0.81	0.066	4.75	0.989
0.92	0.056	4.19	0.988
1.05	0.044	3.58	0.987
1.23	0.040	3.04	0.985
1.44	0.040	2.56	0.979
1.65	0.044	2.19	0.970
1.87	0.050	1.84	0.955
2.07	0.053	1.62	0.943
2.27	0.049	1.43	0.938
2.45	0.046	1.28	0.933
2.64	0.043	1.14	0.928
2.82	0.043	1.02	0.919
2.95	0.041	0.898	0.913
3.06	0.041	0.799	0.905
3.40	0.052	0.549	0.852
3.71	0.089	0.288	0.719
3.97	0.287	0.091	0.310
4.00	0.34	0.08	0.245
4.065	0.38	0.07	0.204
4.133	0.41	0.07	0.177
4.203	0.45	0.06	0.145
4.275	0.48	0.06	0.125
4.350	0.52	0.05	0.101
4.428	0.55	0.05	0.085
4.509	0.58	0.05	0.072
4.592	0.61	0.05	0.060
4.679	0.64	0.04	0.049
4.769	0.66	0.04	0.043
4.862	0.68	0.04	0.037
4.959	0.70	0.04	0.032
5.061	0.72	0.04	0.027
5.166	0.74	0.04	0.023
5.276	0.76	0.04	0.019
5.391	0.78	0.04	0.016
5.510	0.79	0.05	0.015
5.637	0.81	0.05	0.012
5.767	0.83	0.05	0.009
6.048	0.85	0.05	0.007
6.199	0.87	0.05	0.006
6.358	0.88	0.05	0.005
6.526	0.90	0.06	0.004
6.702	0.91	0.06	0.003
6.888	0.92	0.06	0.003
7.085	0.92	0.06	0.003
7.293	0.93	0.06	0.002
7.514	0.93	0.06	0.002
7.749	0.94	0.06	0.002
7.999	0.94	0.06	0.002
8.260	0.94	0.06	0.002
8.551	0.94	0.06	0.002
8.856	0.94	0.05	0.002
9.184	0.94	0.05	0.002
9.537	0.94	0.04	0.001
9.919	0.94	0.04	0.001

Energy (eV)	*n*	*k*	$R(\phi = 0)$
10.33	0.94	0.03	0.001
11.0		0.03	
12.0		0.028	

Rhenium, single crystal, $\vec{E} \parallel \hat{c}$[9]

Energy (eV)	*n*	*k*	$R(\phi = 0)$
0.10	6.06	51.03	0.991
0.15	4.66	33.96	0.984
0.20	4.16	25.36	0.975
0.25	4.03	20.10	0.962
0.30	4.37	16.69	0.943
0.35	4.50	14.53	0.925
0.40	4.53	12.96	0.909
0.45	4.53	11.78	0.893
0.50	4.53	10.88	0.878
0.55	4.50	10.26	0.867
0.60	4.29	9.75	0.861
0.65	4.07	9.35	0.856
0.70	3.80	8.94	0.853
0.75	3.48	8.55	0.850
0.80	3.21	8.10	0.846
0.85	2.96	7.68	0.841
0.90	2.73	7.24	0.835
0.95	2.56	6.79	0.826
1.00	2.45	6.36	0.813
1.10	2.38	5.61	0.778
1.20	2.35	5.02	0.742
1.30	2.39	4.54	0.702
1.40	2.44	4.13	0.662
1.50	2.50	3.79	0.624
1.60	2.59	3.49	0.587
1.70	2.70	3.27	0.557
1.80	2.82	3.10	0.535
1.90	2.90	3.00	0.520
2.00	2.97	2.91	0.510
2.10	3.03	2.86	0.504
2.20	3.06	2.84	0.501
2.30	3.07	2.82	0.499
2.40	3.06	2.81	0.498
2.50	3.02	2.80	0.497
2.60	2.96	2.77	0.493
2.70	2.89	2.68	0.482
2.80	2.89	2.57	0.468
2.90	2.99	2.47	0.457
3.00	3.11	2.57	0.470
3.20	2.90	2.68	0.482
3.40	2.83	2.50	0.459
3.60	2.93	2.48	0.457
3.80	2.86	2.56	0.467
4.00	2.81	2.51	0.460
4.20	2.86	2.55	0.466
4.40	2.81	2.74	0.489
4.60	2.56	2.83	0.504
4.80	2.41	2.71	0.493
5.00	2.39	2.68	0.488
5.20	2.34	2.75	0.500
5.40	2.20	2.81	0.515
5.60	2.02	2.84	0.530
5.80	1.83	2.80	0.538
6.00	1.65	2.71	0.541
6.20	1.54	2.59	0.532
6.40	1.45	2.50	0.526
6.80	1.32	2.31	0.508
7.00	1.26	2.23	0.500
7.20	1.20	2.15	0.493
7.40	1.16	2.06	0.480
7.60	1.12	1.99	0.470
7.80	1.08	1.89	0.454
8.00	1.05	1.80	0.435
8.20	1.05	1.71	0.411
8.40	1.05	1.62	0.386
8.60	1.06	1.55	0.360
8.80	1.09	1.48	0.336
9.00	1.11	1.43	0.317
9.20	1.13	1.39	0.301
9.40	1.16	1.34	0.281
9.60	1.18	1.32	0.274
9.80	1.20	1.29	0.264
10.00	1.23	1.26	0.252
10.20	1.25	1.25	0.246
10.40	1.28	1.25	0.242
10.60	1.29	1.25	0.242
10.80	1.30	1.26	0.244
11.00	1.30	1.27	0.247
11.20	1.29	1.28	0.249
11.40	1.28	1.28	0.252
11.60	1.26	1.28	0.252
11.80	1.24	1.26	0.249
12.00	1.23	1.24	0.244
12.40	1.22	1.21	0.237
12.80	1.21	1.18	0.230
13.20	1.22	1.16	0.222
13.60	1.22	1.13	0.215
14.00	1.24	1.12	0.209
14.40	1.27	1.11	0.204
14.80	1.29	1.15	0.213
15.20	1.29	1.19	0.225
15.60	1.26	1.22	0.236
16.00	1.23	1.25	0.248
16.40	1.19	1.27	0.259
16.80	1.14	1.29	0.269
17.00	1.12	1.30	0.275
17.40	1.07	1.30	0.286
18.00	0.99	1.30	0.300
18.40	0.93	1.29	0.311
18.80	0.87	1.28	0.321
19.20	0.81	1.25	0.330
19.60	0.77	1.21	0.332
20.00	0.73	1.18	0.333
20.40	0.70	1.14	0.332
20.80	0.67	1.11	0.332
21.20	0.64	1.08	0.334
21.60	0.61	1.04	0.335
22.00	0.58	1.01	0.340
22.40	0.55	0.97	0.341
22.80	0.53	0.93	0.338
23.20	0.51	0.89	0.334
23.60	0.50	0.85	0.329
24.00	0.48	0.80	0.319
24.40	0.48	0.76	0.207
24.80	0.47	0.72	0.296
25.20	0.47	0.68	0.282
25.60	0.47	0.65	0.270
26.00	0.47	0.61	0.255
26.40	0.48	0.57	0.240
26.80	0.48	0.54	0.225
27.20	0.49	0.51	0.208
27.60	0.50	0.48	0.193
28.00	0.51	0.45	0.176
29.00	0.54	0.39	0.145
30.00	0.57	0.33	0.114
31.00	0.62	0.29	0.086
32.00	0.66	0.26	0.065
33.00	0.68	0.24	0.054
34.00	0.72	0.21	0.041
35.00	0.76	0.20	0.031
36.00	0.79	0.20	0.025
37.00	0.82	0.19	0.021
38.00	0.85	0.20	0.018
39.00	0.89	0.21	0.016
40.00	0.88	0.26	0.022
42.00	0.88	0.26	0.022
44.00	0.89	0.29	0.026
46.00	0.85	0.32	0.035
48.00	0.82	0.30	0.036
50.00	0.80	0.30	0.038
52.00	0.78	0.30	0.044
54.00	0.72	0.30	0.055
56.00	0.66	0.24	0.061
58.00	0.65	0.16	0.055

Rhenium, single crystal, $\vec{E} \perp \hat{c}$[9]

Energy (eV)	*n*	*k*	$R(\phi = 0)$
0.10	4.25	42.83	0.991
0.15	3.28	28.08	0.984
0.20	3.28	20.66	0.971
0.25	3.47	16.27	0.951
0.30	3.73	13.44	0.926
0.35	3.93	11.54	0.900
0.40	3.99	10.15	0.875
0.45	4.17	9.03	0.846
0.50	4.34	8.26	0.821
0.55	4.45	7.73	0.801
0.60	4.53	7.40	0.788
0.65	4.44	7.26	0.784
0.70	4.13	7.09	0.784
0.75	3.77	6.75	0.779
0.80	3.55	6.32	0.766
0.85	3.39	5.95	0.752
0.90	3.26	5.61	0.737
0.95	3.17	5.27	0.719
1.00	3.09	4.96	0.701
1.10	3.05	4.39	0.658
1.20	3.08	3.89	0.613
1.30	3.20	3.56	0.578
1.40	3.23	3.38	0.559
1.50	3.23	3.12	0.532
1.60	3.29	2.88	0.507
1.70	3.38	2.72	0.491
1.80	3.47	2.59	0.480
1.90	3.54	2.50	0.473
2.00	3.63	2.43	0.469
2.10	3.74	2.40	0.470
2.20	3.83	2.38	0.472
2.30	3.93	2.44	0.481
2.40	4.00	2.55	0.492
2.50	4.01	2.70	0.505
2.60	3.90	2.84	0.514
2.70	3.74	2.92	0.517
2.80	3.57	2.88	0.511
2.90	3.49	2.75	0.497
3.00	3.53	2.71	0.493
3.20	3.55	2.84	0.506
3.40	3.34	2.88	0.508
3.60	3.25	2.83	0.501
3.80	3.24	2.84	0.502
4.00	3.19	2.94	0.513

Energy (eV)	n	k	R(φ = 0)
4.20	3.05	3.06	0.526
4.40	2.88	3.15	0.539
4.60	2.67	3.18	0.548
4.80	2.44	3.17	0.554
5.00	2.25	3.12	0.556
5.20	2.10	3.04	0.555
5.40	1.96	2.96	0.553
5.60	1.84	2.88	0.551
5.80	1.73	2.81	0.549
6.00	1.61	2.74	0.549
6.20	1.51	2.64	0.545
6.40	1.42	2.56	0.541
6.80	1.28	2.37	0.526
7.00	1.22	2.28	0.517
7.20	1.16	2.19	0.508
7.40	1.12	2.08	0.493
7.60	1.12	1.98	0.468
7.80	1.08	1.93	0.463
8.00	1.05	1.83	0.443
8.20	1.05	1.74	0.418
8.40	1.05	1.66	0.397
8.60	1.06	1.58	0.372
8.80	1.07	1.52	0.351
9.00	1.09	1.46	0.327
9.20	1.11	1.41	0.309
9.40	1.14	1.36	0.290
9.60	1.17	1.31	0.273
9.80	1.20	1.27	0.258
10.00	1.24	1.24	0.244
10.20	1.29	1.22	0.234
10.40	1.33	1.23	0.233
10.60	1.36	1.25	0.238
10.80	1.38	1.28	0.245
11.00	1.37	1.31	0.253
11.20	1.36	1.33	0.259
11.40	1.33	1.34	0.264
11.60	1.31	1.34	0.266
11.80	1.28	1.33	0.266
12.00	1.26	1.32	0.264
12.40	1.23	1.29	0.257
12.80	1.22	1.26	0.251
13.20	1.20	1.23	0.245
13.60	1.19	1.20	0.236
14.00	1.20	1.16	0.225
14.40	1.22	1.13	0.214
14.80	1.27	1.12	0.207
15.20	1.31	1.17	0.218
15.60	1.31	1.23	0.234
16.00	1.28	1.28	0.251
16.40	1.24	1.33	0.270
16.80	1.17	1.37	0.288
17.00	1.14	1.38	0.297
17.40	1.06	1.39	0.314
18.00	0.95	1.38	0.334
18.40	0.88	1.36	0.346
18.80	0.82	1.33	0.355
19.20	0.76	1.29	0.360
19.60	0.72	1.25	0.363
20.00	0.67	1.21	0.369
20.40	0.64	1.15	0.364
20.80	0.61	1.10	0.357
21.20	0.60	1.06	0.349
21.60	0.58	1.02	0.342
22.00	0.57	0.98	0.336
22.40	0.56	0.95	0.328

Energy (eV)	n	k	R(φ = 0)
22.80	0.55	0.92	0.325
23.20	0.53	0.89	0.322
23.60	0.52	0.85	0.317
24.00	0.50	0.82	0.314
24.40	0.49	0.79	0.309
24.80	0.48	0.75	0.303
25.20	0.47	0.72	0.295
25.60	0.47	0.68	0.286
26.00	0.46	0.64	0.276
26.40	0.46	0.61	0.263
26.80	0.46	0.57	0.249
27.20	0.47	0.53	0.231
27.60	0.48	0.50	0.216
28.00	0.49	0.47	0.198
29.00	0.51	0.41	0.164
30.00	0.55	0.34	0.129
31.00	0.59	0.29	0.097
32.00	0.64	0.26	0.072
33.00	0.67	0.24	0.060
34.00	0.70	0.22	0.047
35.00	0.74	0.20	0.036
36.00	0.77	0.19	0.029
37.00	0.80	0.19	0.023
38.00	0.84	0.19	0.018
39.00	0.88	0.21	0.016
40.00	0.87	0.25	0.023
42.00	0.87	0.25	0.023
44.00	0.88	0.28	0.026
46.00	0.84	0.31	0.035
48.00	0.82	0.30	0.036
50.00	0.80	0.30	0.039
52.00	0.77	0.30	0.044
54.00	0.71	0.29	0.055
56.00	0.66	0.23	0.061
58.00	0.64	0.16	0.055
Rhodium[11]			
0.10	18.48	69.43	0.986
0.20	8.66	37.46	0.977
0.30	5.85	25.94	0.967
0.40	4.74	19.80	0.955
0.50	4.20	16.07	0.941
0.60	3.87	13.51	0.925
0.70	3.67	11.72	0.908
0.80	3.63	10.34	0.887
0.90	3.62	9.36	0.867
1.00	3.71	8.67	0.848
1.10	3.67	8.26	0.837
1.20	3.51	7.94	0.832
1.30	3.26	7.63	0.829
1.40	3.01	7.31	0.827
1.50	2.78	6.97	0.823
1.60	2.60	6.64	0.818
1.70	2.42	6.33	0.813
1.80	2.30	6.02	0.805
1.90	2.20	5.76	0.798
2.00	2.12	5.51	0.789
2.10	2.05	5.30	0.780
2.20	2.00	5.11	0.772
2.30	1.94	4.94	0.765
2.40	1.90	4.78	0.756
2.50	1.88	4.65	0.748
2.60	1.85	4.55	0.743
2.70	1.80	4.49	0.742
2.90	1.63	4.36	0.748

Energy (eV)	n	k	R(φ = 0)
3.00	1.53	4.29	0.753
3.10	1.41	4.20	0.760
3.20	1.30	4.09	0.764
3.30	1.20	3.97	0.767
3.40	1.11	3.84	0.769
3.50	1.04	3.71	0.768
3.60	0.99	3.58	0.764
3.70	0.95	3.45	0.759
3.80	0.91	3.34	0.753
3.90	0.88	3.23	0.747
4.00	0.86	3.12	0.739
4.20	0.83	2.94	0.722
4.40	0.80	2.76	0.706
4.60	0.78	2.60	0.684
4.80	0.79	2.46	0.659
5.00	0.79	2.34	0.635
5.20	0.79	2.23	0.613
5.40	0.80	2.14	0.591
5.60	0.80	2.06	0.573
5.80	0.79	2.00	0.561
6.00	0.76	1.93	0.556
6.20	0.73	1.85	0.544
6.40	0.70	1.77	0.534
6.60	0.68	1.69	0.518
6.80	0.67	1.60	0.498
7.00	0.66	1.52	0.476
7.20	0.66	1.43	0.452
7.40	0.66	1.35	0.423
7.60	0.67	1.27	0.394
7.80	0.68	1.20	0.363
8.00	0.69	1.12	0.329
8.20	0.71	1.04	0.288
8.40	0.74	0.97	0.252
8.60	0.78	0.89	0.212
8.80	0.83	0.83	0.179
9.00	0.88	0.77	0.148
9.20	0.95	0.73	0.125
9.40	1.01	0.71	0.110
9.60	1.07	0.69	0.102
9.80	1.12	0.69	0.098
10.00	1.17	0.69	0.098
10.60	1.26	0.73	0.106
11.00	1.29	0.76	0.113
11.60	1.32	0.80	0.124
12.00	1.32	0.82	0.127
12.60	1.32	0.82	0.129
13.00	1.32	0.83	0.131
13.60	1.32	0.85	0.134
14.00	1.32	0.86	0.138
14.60	1.30	0.89	0.144
15.00	1.28	0.90	0.147
15.60	1.25	0.90	0.147
16.00	1.24	0.89	0.147
16.50	1.23	0.88	0.145
17.00	1.22	0.88	0.144
17.50	1.22	0.87	0.143
18.00	1.23	0.88	0.145
18.50	1.25	0.92	0.155
19.00	1.24	0.98	0.172
19.50	1.18	1.05	0.193
20.00	1.10	1.09	0.213
20.50	1.00	1.09	0.230
21.00	0.91	1.05	0.234
21.50	0.86	1.00	0.228
22.00	0.83	0.95	0.219

Energy (eV)	*n*	*k*	$R(\phi = 0)$
22.50	0.81	0.92	0.214
23.00	0.79	0.90	0.213
23.50	0.75	0.87	0.214
24.00	0.73	0.84	0.210
24.50	0.70	0.81	0.208
25.00	0.69	0.77	0.202
25.50	0.67	0.74	0.195
26.00	0.66	0.70	0.188
26.50	0.65	0.66	0.176
27.00	0.65	0.64	0.168
27.50	0.65	0.61	0.159
28.00	0.65	0.59	0.152
29.00	0.65	0.54	0.137
30.00	0.66	0.51	0.127
31.00	0.64	0.49	0.127
32.00	0.61	0.44	0.126
33.00	0.60	0.37	0.110
34.00	0.65	0.30	0.074
35.00	0.69	0.28	0.058
36.00	0.73	0.27	0.049
37.00	0.74	0.28	0.047
38.00	0.74	0.27	0.045
39.00	0.75	0.25	0.041

Ruthenium, single crystal, $\vec{E} \parallel \hat{c}$[9]

Energy (eV)	*n*	*k*	$R(\phi = 0)$
0.10	11.50	51.38	0.984
0.20	5.93	27.14	0.970
0.30	4.33	18.50	0.953
0.40	3.60	13.97	0.933
0.50	3.18	11.04	0.909
0.60	3.28	8.89	0.865
0.70	3.62	7.73	0.822
0.80	3.42	7.02	0.801
0.90	3.25	6.12	0.766
1.00	3.39	5.33	0.715
1.10	3.66	4.83	0.675
1.20	3.84	4.57	0.654
1.30	3.94	4.38	0.638
1.40	4.02	4.19	0.624
1.50	4.16	4.07	0.614
1.60	4.33	4.08	0.615
1.70	4.42	4.21	0.624
1.80	4.40	4.38	0.636
1.90	4.29	4.61	0.651
2.00	4.04	4.81	0.667
2.10	3.69	4.90	0.679
2.20	3.35	4.82	0.683
2.30	3.09	4.70	0.681
2.40	2.89	4.55	0.677
2.50	2.74	4.40	0.671
2.60	2.64	4.25	0.663
2.70	2.58	4.14	0.656
2.80	2.54	4.05	0.650
2.90	2.48	4.03	0.650
3.00	2.38	4.03	0.656
3.10	2.26	4.00	0.661
3.20	2.13	3.96	0.666
3.30	2.00	3.91	0.671
3.40	1.87	3.83	0.673
3.50	1.76	3.74	0.674
3.60	1.66	3.65	0.675
3.70	1.57	3.55	0.673
3.80	1.49	3.45	0.672
3.90	1.42	3.35	0.668
4.00	1.37	3.24	0.661
4.20	1.29	3.08	0.649
4.40	1.22	2.93	0.639
4.60	1.16	2.79	0.628
4.80	1.11	2.67	0.617
5.00	1.06	2.56	0.607
5.20	1.01	2.46	0.600
5.40	0.95	2.35	0.593
5.60	0.92	2.23	0.576
5.80	0.90	2.14	0.559
6.00	0.88	2.05	0.545
6.20	0.87	1.98	0.531
6.40	0.84	1.91	0.521
6.60	0.82	1.84	0.510
6.80	0.79	1.77	0.500
7.00	0.76	1.69	0.489
7.20	0.75	1.61	0.472
7.40	0.73	1.54	0.455
7.60	0.73	1.46	0.433
7.80	0.73	1.39	0.411
8.00	0.72	1.33	0.391
8.20	0.72	1.26	0.366
8.40	0.73	1.20	0.342
8.60	0.74	1.14	0.318
8.80	0.74	1.08	0.295
9.00	0.75	1.02	0.267
9.20	0.77	0.97	0.243
9.40	0.79	0.91	0.217
9.60	0.82	0.86	0.190
9.80	0.85	0.81	0.167
10.00	0.88	0.76	0.144
10.20	0.92	0.72	0.125
10.40	0.96	0.69	0.110
10.60	1.01	0.67	0.100
10.80	1.05	0.66	0.094
11.00	1.09	0.65	0.090
11.20	1.12	0.65	0.088
11.40	1.15	0.65	0.087
11.60	1.18	0.65	0.088
11.80	1.21	0.66	0.090
12.00	1.23	0.67	0.092
12.40	1.26	0.69	0.098
12.80	1.27	0.72	0.104
13.20	1.28	0.74	0.108
13.60	1.28	0.75	0.111
14.00	1.28	0.76	0.114
14.40	1.27	0.76	0.114
14.80	1.27	0.76	0.114
15.00	1.27	0.76	0.114
15.60	1.28	0.77	0.115
16.00	1.30	0.78	0.118
16.50	1.32	0.80	0.123
17.00	1.34	0.85	0.136
17.50	1.32	0.93	0.155
18.00	1.26	0.99	0.173
18.50	1.18	1.02	0.185
19.00	1.11	1.02	0.192
19.50	1.05	1.02	0.199
20.00	0.99	1.02	0.208
20.50	0.92	0.99	0.212
21.00	0.86	0.94	0.209
21.50	0.83	0.90	0.203
22.00	0.81	0.86	0.193
23.00	0.77	0.79	0.182
24.00	0.74	0.74	0.171
25.00	0.71	0.69	0.163
26.00	0.68	0.63	0.154
27.00	0.67	0.57	0.140
28.00	0.66	0.51	0.124
29.00	0.67	0.46	0.107
30.00	0.67	0.43	0.097
31.00	0.67	0.37	0.084
32.00	0.69	0.33	0.070
33.00	0.71	0.30	0.058
34.00	0.73	0.27	0.048
35.00	0.75	0.25	0.039
36.00	0.77	0.24	0.035
37.00	0.79	0.23	0.039
38.00	0.80	0.22	0.027
39.00	0.82	0.22	0.024
40.00	0.83	0.22	0.022

Ruthenium, single crystal, $\vec{E} \perp \hat{c}$[5]

Energy (eV)	*n*	*k*	$R(\phi = 0)$
0.10	11.85	50.81	0.983
0.20	6.68	27.18	0.966
0.30	4.94	18.92	0.950
0.40	3.90	14.51	0.933
0.50	3.27	11.63	0.915
0.60	2.98	9.54	0.888
0.70	2.82	7.99	0.856
0.80	2.73	6.71	0.815
0.90	2.82	5.54	0.751
1.00	3.17	4.59	0.670
1.10	3.69	3.91	0.604
1.20	4.28	3.66	0.585
1.30	4.66	3.72	0.593
1.40	4.86	3.79	0.601
1.50	4.99	3.89	0.609
1.60	5.08	4.03	0.618
1.70	5.12	4.22	0.629
1.80	5.10	4.45	0.642
1.90	4.96	4.78	0.660
2.00	4.61	5.06	0.677
2.10	4.21	5.09	0.682
2.20	3.94	5.00	0.681
2.30	3.69	4.97	0.684
2.40	3.44	4.88	0.684
2.50	3.27	4.77	0.681
2.60	3.14	4.66	0.677
2.70	3.06	4.59	0.674
2.80	2.99	4.59	0.676
2.90	2.87	4.64	0.686
3.00	2.64	4.69	0.701
3.10	2.40	4.64	0.710
3.20	2.18	4.55	0.717
3.30	2.00	4.43	0.721
3.40	1.84	4.30	0.723
3.50	1.71	4.16	0.723
3.60	1.60	4.03	0.722
3.70	1.50	3.90	0.721
3.80	1.41	3.77	0.718
3.90	1.35	3.64	0.713
4.00	1.29	3.53	0.707
4.20	1.21	3.31	0.694
4.40	1.16	3.13	0.679
4.60	1.13	2.97	0.662
4.80	1.09	2.86	0.652
5.00	1.03	2.75	0.648
5.20	0.97	2.64	0.643
5.40	0.91	2.52	0.635
5.60	0.88	2.40	0.622

Energy (eV)	*n*	*k*	$R(\phi = 0)$
5.80	0.86	2.29	0.605
6.00	0.84	2.20	0.591
6.20	0.82	2.11	0.576
6.40	0.81	2.04	0.564
6.60	0.78	1.97	0.556
6.80	0.76	1.89	0.545
7.00	0.73	1.82	0.538
7.20	0.70	1.75	0.527
7.40	0.68	1.67	0.513
7.60	0.67	1.59	0.496
7.80	0.66	1.51	0.476
8.00	0.66	1.44	0.454
8.20	0.65	1.36	0.430
8.40	0.66	1.29	0.403
8.60	0.66	1.22	0.378
8.80	0.68	1.15	0.346
9.00	0.69	1.09	0.317
9.20	0.70	1.02	0.286
9.40	0.73	0.95	0.251
9.60	0.77	0.89	0.216
9.80	0.82	0.84	0.185
10.00	0.86	0.81	0.163
10.20	0.90	0.77	0.143
10.40	0.94	0.74	0.127
10.60	0.99	0.72	0.115
10.80	1.04	0.71	0.108
11.00	1.08	0.70	0.104
11.20	1.11	0.70	0.102
11.40	1.14	0.70	0.101
11.60	1.17	0.71	0.102
11.80	1.20	0.72	0.104
12.00	1.22	0.73	0.107
12.40	1.25	0.76	0.113
12.80	1.26	0.78	0.118
13.20	1.27	0.81	0.124
13.60	1.27	0.83	0.129
14.00	1.26	0.84	0.132
14.40	1.25	0.84	0.132
14.80	1.25	0.84	0.133
15.00	1.25	0.84	0.133
15.60	1.25	0.85	0.134
16.00	1.27	0.85	0.134
16.50	1.28	0.89	0.145
17.00	1.28	0.94	0.158
17.50	1.25	1.00	0.175
18.00	1.19	1.04	0.190
18.50	1.12	1.05	0.200
19.00	1.07	1.05	0.205
19.50	1.02	1.04	0.212
20.00	0.97	1.04	0.219
20.50	0.91	1.03	0.228
21.00	0.85	1.01	0.234
21.50	0.80	0.97	0.234
22.00	0.77	0.94	0.233
23.00	0.71	0.87	0.229
24.00	0.67	0.79	0.218
25.00	0.64	0.73	0.205
26.00	0.61	0.66	0.194
27.00	0.60	0.59	0.177
28.00	0.60	0.53	0.155
29.00	0.61	0.48	0.134
30.00	0.62	0.45	0.123
31.00	0.61	0.40	0.114
32.00	0.63	0.34	0.093
33.00	0.65	0.31	0.077
34.00	0.67	0.28	0.065
35.00	0.70	0.26	0.054
36.00	0.72	0.25	0.047
37.00	0.73	0.23	0.041
38.00	0.75	0.22	0.035
39.00	0.77	0.22	0.031
40.00	0.79	0.22	0.028

Selenium, single crystal, $E \parallel \vec{\hat{c}}$ [22]

Energy (eV)	*n*	*k*	$R(\phi = 0)$
0.01364	2.914	0.248	0.242
0.01488	3.175	9.95E-02	0.272
0.01612	3.263	2.13E-03	0.282
0.01736	3.306	3.81E-02	0.287
0.01860	3.330	7.04E-03	0.290
0.01984	3.346	4.23E-02	0.291
0.02108	3.358	3.40E-03	0.293
0.02232	3.366	5.31E-02	0.294
0.02356	3.372	1.96E-03	0.294
0.02480	3.377	2.39E-02	0.295
0.02604	3.380		0.295
0.02728		1.16E-02	
0.02976		7.96E-03	
0.03224		8.57E-03	
0.03472		2.70E-02	
0.03720	3.397	1.72E-02	0.297
0.04463		1.13E-02	
0.04959	3.403	2.79E-03	0.298
0.05703		1.56E-03	
0.06199	3.405	1.35E-03	0.298
0.06819		5.79E-04	
0.07439	3.407	4.44E-04	0.298
0.08059		4.41E-04	
0.08679	3.408	4.32E-04	0.298
0.09299		2.44E-04	
0.09919	3.409	3.23E-04	0.299
0.1116	3.409	2.87E-04	0.299
0.1240	3.410	2.71E-04	0.299
0.2480	3.417	2.67E-04	0.299
0.3720	3.427	1.90E-04	0.301
0.4959	3.442	1.41E-04	0.302
0.6199	3.462	1.12E-04	0.304
0.7439	3.486	9.42E-05	0.307
0.8679	3.516	8.07E-05	0.310
0.9919	3.551	7.11E-05	0.314
1.116	3.592	6.37E-05	0.319
1.240	3.640	5.81E-05	0.324
1.50		1.33E-04	
1.60		1.59E-04	
1.70		6.27E-04	
1.80	4.46	2.20E-02	0.402
2.0	4.79	0.76	0.438
2.2	4.49	1.19	0.431
2.4	4.28	1.21	0.417
2.6	4.40	1.32	0.430
2.8	4.59	1.70	0.462
3.0	4.44	2.29	0.490
3.2	3.92	2.59	0.493
3.4	3.69	2.76	0.502
3.6	3.39	3.01	0.521
3.8	(3.00)		
4.0	(2.65)		
4.2	(2.30)		
4.5	1.92	2.78	0.528
5.0	1.50	2.31	0.482
6.0	1.57	1.49	0.288
7.0	1.84	1.45	0.276
8.0	1.35	1.68	0.353
9.0	1.35	1.64	0.342
10.0	0.92	1.07	0.238
12.0	1.00	1.10	0.232
14.0	0.81	0.91	0.211
16.0	0.65	0.61	0.160
18.0	0.65	0.48	0.120
20.0	0.69	0.36	0.076
22.0	0.81	0.25	0.030
24.0	0.91	0.18	0.011
26.0	0.86	0.15	0.012
28.0	0.85	0.13	0.011
30.0	0.87	0.11	0.008

Selenium, single crystal, $E \perp \vec{\hat{c}}$ [22]

Energy (eV)	*n*	*k*	$R(\phi = 0)$
0.01364	2.854	0.0239	0.231
0.01488	2.932	0.0325	0.241
0.01612	3.140	0.1750	0.269
0.01736	2.959	1.3300	0.321
0.01860	2.111	0.2550	0.133
0.01984	2.356	0.0746	0.164
0.02108	2.462	0.0276	0.178
0.02232	2.502	0.0442	0.184
0.02356	2.543	0.0097	0.190
0.02480	2.550	0.0239	0.191
0.02604	2.582		0.195
0.02728	2.600	0.0101	0.198
0.02976	2.576	9.95E-03	0.194
0.03224	2.598	1.16E-02	0.197
0.03472	2.607	1.68E-02	0.199
0.03720	2.613	1.54E-02	0.199
0.04463		1.17E-02	
0.04959	2.627	3.58E-03	0.201
0.05703		8.65E-04	
0.06199	2.632	2.07E-03	0.202
0.06819		2.89E-04	
0.07439	2.635	1.59E-04	0.202
0.08059		1.35E-04	
0.08679	2.636	1.42E-04	0.202
0.09299		1.04E-04	
0.09919	2.637	8.95E-05	0.203
0.1116	2.638	8.84E-05	0.203
0.1240	2.639	8.51E-05	0.203
0.2480	2.645	5.97E-05	0.204
0.3720	2.652	5.44E-05	0.205
0.4959	2.654	4.58E-05	0.205
0.6199	2.675	3.82E-05	0.208
0.7439	2.692	3.32E-05	0.210
0.8679	2.713	2.96E-05	0.213
0.9919	2.739	2.69E-05	0.216
1.116	2.772	2.48E-05	0.221
1.240	2.816	2.31E-05	0.226
1.50		7.37E-05	
1.60		8.63E-05	
1.70		3.60E-04	
1.80	3.32	0.11	0.289
2.00	3.38	0.65	0.310
2.20	3.07	0.73	0.282
2.40	2.93	0.61	0.259
2.60	3.00	0.53	0.263
2.80	3.12	0.58	0.279
3.00	3.30	0.70	0.305
3.20	3.35	1.01	0.328
3.40	3.22	1.24	0.334

Energy (eV)	n	k	$R(\phi = 0)$
3.60	3.06	1.47	0.344
3.80	2.84	1.66	0.351
4.00	2.51	1.81	0.356
4.20	2.18	1.83	0.352
4.50	1.75	1.94	0.382
5.00	1.25	1.50	0.316
6.00	1.32	0.73	0.107
7.00	1.62	0.61	0.105
8.00	1.81	0.69	0.135
9.00	1.66	1.02	0.182
10.00	1.72	0.95	0.171
12.00	1.25	1.02	0.181
14.00	0.98	0.92	0.178
16.00	0.68	0.96	0.274
18.00	0.61	0.65	0.191
20.00	0.73	0.48	0.094
22.00	0.78	0.39	0.060
24.00	0.78	0.32	0.046
26.00	0.78	0.26	0.036
28.00	0.80	0.19	0.023
30.00	0.79	0.14	0.020

Silicon, single crystal[23]

Energy (eV)	n	k	$R(\phi = 0)$
0.01240	3.4185	2.90E-04	0.300
0.01488	3.4190	2.30E-04	0.300
0.01736	3.4192	1.90E-04	0.300
0.01984	3.4195	1.70E-04	0.300
0.02480	3.4197		0.300
0.03100	3.4199		0.300
0.04092	3.4200		0.300
0.04463		1.08E-04	
0.04959	3.4201	9.15E-05	0.300
0.05703		1.56E-04	
0.06199	3.4204	2.86E-04	0.300
0.06943		3.84E-04	
0.07439		7.16E-04	
0.08059	(3.4207)	1.52E-04	0.300
0.08679		1.02E-04	
0.09299		2.59E-04	
0.09919		1.77E-04	
0.1054		1.53E-04	
0.1116		2.02E-04	
0.1178		1.22E-04	
0.1240	3.4215	6.76E-05	0.300
0.1364		5.49E-05	
0.1488		2.41E-05	
0.1612		2.49E-05	
0.1736	(3.4230)	1.68E-05	0.300
0.1798		2.45E-05	
0.1860		2.66E-06	
0.1922		1.74E-06	
0.1984		8.46E-07	
0.2046		5.64E-07	
0.2108	(3.4244)	4.17E-07	0.300
0.2170		4.05E-07	
0.2232		3.94E-07	
0.2294		3.26E-07	
0.2356		2.97E-07	
0.2418		2.82E-07	
0.2480	3.4261	1.99E-07	0.300
0.3100	3.4294		0.301
0.3626	3.4327		0.301
0.4568	3.4393	2.50E-09	0.302
0.6199	3.4490		0.303
0.8093	3.4784		0.306
1.033	3.5193		0.311
1.1	(3.5341)	1.30E-05	0.312
1.2		1.80E-04	
1.3		2.26E-03	
1.4		7.75E-03	
1.5	3.673	5.00E-03	0.327
1.6	3.714	8.00E-03	0.331
1.7	3.752	1.00E-02	0.335
1.8	3.796	0.013	0.340
1.9	3.847	0.016	0.345
2.0	3.906	0.022	0.351
2.1	3.969	0.030	0.357
2.2	4.042	0.032	0.364
2.3	4.123	0.048	0.372
2.4	4.215	0.060	0.380
2.5	4.320	0.073	0.390
2.6	4.442	0.090	0.400
2.7	4.583	0.130	0.412
2.8	4.753	0.163	0.426
2.9	4.961	0.203	0.442
3.0	5.222	0.269	0.461
3.1	5.570	0.387	0.486
3.2	6.062	0.630	0.518
3.3	6.709	1.321	0.561
3.4	6.522	2.705	0.592
3.5	5.610	3.014	0.575
3.6	5.296	2.987	0.564
3.7	5.156	3.058	0.563
3.8	5.065	3.182	0.568
3.9	5.016	3.346	0.577
4.0	5.010	3.587	0.591
4.1	5.020	3.979	0.614
4.2	4.888	4.639	0.652
4.3	4.086	5.395	0.703
4.4	3.120	5.344	0.726
4.5	2.451	5.082	0.740
4.6	1.988	4.678	0.742
4.7	1.764	4.278	0.728
4.8	1.658	3.979	0.710
4.9	1.597	3.749	0.693
5.0	1.570	3.565	0.675
5.1	1.571	3.429	0.658
5.2	1.589	3.354	0.646
5.3	1.579	3.353	0.647
5.4	1.471	3.366	0.663
5.5	1.340	3.302	0.673
5.6	1.247	3.206	0.675
5.7	1.180	3.112	0.673
5.8	1.133	3.045	0.672
5.9	1.083	2.982	0.673
6.0	1.010	2.909	0.677
6.5	0.847	2.73	0.688
7.0	0.682	2.45	0.691
7.5	0.563	2.21	0.693
8.0	0.478	2.00	0.691
8.5	0.414	1.82	0.688
9.0	0.367	1.66	0.683
9.5	0.332	1.51	0.672
10.0	0.306	1.38	0.661
12.0	0.257	0.963	0.590
14.0	0.275	0.641	0.460
16.0	0.345	0.394	0.297
18.0	0.455	0.219	0.159
20.0	0.567	0.0835	0.079
22.14	0.675	0.0405	0.038
24.31	0.752	0.0243	0.020
26.38	0.803	0.0178	0.012
28.18	0.834	0.0152	0.008
30.24	0.860	0.0138	0.006
31.79	0.877	0.0132	0.004
34.44	0.899	0.0121	0.003
36.47	0.913	0.0113	0.002
38.75	0.925	0.0104	0.002
40.00	0.930	0.0100	0.001

Silver[6]

Energy (eV)	n	k	$R(\phi = 0)$
0.10	9.91	90.27	0.995
0.20	2.84	45.70	0.995
0.30	1.41	30.51	0.994
0.40	0.91	22.89	0.993
0.50	0.67	18.32	0.992
1.00	0.28	9.03	0.987
1.50	0.27	5.79	0.969
2.00	0.27	4.18	0.944
2.50	0.24	3.09	0.914
3.00	0.23	2.27	0.864
3.25	0.23	1.86	0.816
3.50	0.21	1.42	0.756
3.60	0.23	1.13	0.671
3.70	0.30	0.77	0.475
3.77	0.53	0.40	0.154
3.80	0.73	0.30	0.053
3.90	1.30	0.36	0.040
4.00	1.61	0.60	0.103
4.10	1.73	0.85	0.153
4.20	1.75	1.06	0.194
4.30	1.73	1.13	0.208
4.50	1.69	1.28	0.238
4.75	1.61	1.34	0.252
5.00	1.55	1.36	0.257
5.50	1.45	1.34	0.257
6.00	1.34	1.28	0.246
6.50	1.25	1.18	0.225
7.00	1.18	1.06	0.196
7.50	1.14	0.91	0.157
8.00	1.16	0.75	0.114
9.00	1.33	0.56	0.074
10.00	1.46	0.56	0.082
11.00	1.52	0.56	0.088
12.00	1.61	0.59	0.100
13.00	1.66	0.64	0.112
14.00	1.72	0.78	0.141
14.50	1.64	0.88	0.152
15.00	1.56	0.92	0.156
16.00	1.42	0.91	0.151
17.00	1.33	0.86	0.139
18.00	1.28	0.80	0.124
19.00	1.27	0.75	0.111
20.00	1.29	0.71	0.103
21.00	1.35	0.75	0.112
21.50	1.37	0.80	0.124
22.00	1.34	0.87	0.141
22.50	1.26	0.93	0.157
23.00	1.17	0.94	0.163
23.50	1.10	0.93	0.165
24.00	1.04	0.90	0.165
24.50	0.99	0.87	0.160
25.00	0.95	0.83	0.154
25.50	0.91	0.78	0.144
26.00	0.90	0.74	0.133

Energy (eV)	n	k	R(ϕ = 0)
26.50	0.89	0.69	0.121
27.00	0.89	0.65	0.109
27.50	0.89	0.62	0.099
28.00	0.90	0.59	0.090
28.50	0.91	0.57	0.084
29.00	0.92	0.56	0.079
30.00	0.93	0.54	0.074
31.00	0.93	0.53	0.072
32.00	0.92	0.53	0.072
33.00	0.90	0.51	0.071
34.00	0.88	0.49	0.067
35.00	0.86	0.45	0.061
36.00	0.89	0.44	0.055
38.00	0.89	0.39	0.043
40.00	0.90	0.37	0.039
42.00	0.90	0.35	0.036
44.00	0.90	0.33	0.033
46.00	0.90	0.32	0.031
48.00	0.89	0.31	0.030
50.00	0.88	0.29	0.027
52.00	0.89	0.28	0.024
54.00	0.88	0.17	0.024
56.00	0.87	0.26	0.024
58.00	0.87	0.24	0.021
60.00	0.87	0.22	0.018
62.00	0.88	0.21	0.016
64.00	0.88	0.21	0.016
66.00	0.88	0.21	0.016
68.00	0.87	0.21	0.017
70.00	0.83	0.20	0.021
72.00	0.85	0.18	0.016
74.00	0.85	0.17	0.014
76.00	0.85	0.16	0.013
78.00	0.85	0.15	0.013
80.00	0.85	0.14	0.012
85.00	0.85	0.11	0.011
90.00	0.85	0.08	0.009
95.00	0.86	0.06	0.007
100.00	0.87	0.04	0.005
Sodium[24]			
0.55	0.262	9.97	0.990
0.58	0.241	9.45	0.989
0.63	0.207	8.80	0.990
0.67	0.175	8.09	0.990
0.73	0.147	7.42	0.990
0.81	0.123	6.67	0.989
0.92	0.099	5.82	0.989
1.05	0.078	5.11	0.989
1.23	0.064	4.35	0.987
1.44	0.053	3.72	0.986
1.65	0.050	3.22	0.983
1.87	0.049	2.76	0.978
2.07	0.053	2.48	0.971
2.27	0.059	2.23	0.961
2.45	0.063	2.07	0.953
2.64	0.066	1.88	0.943
2.82	0.068	1.76	0.936
2.95	0.068	1.63	0.928
3.06	0.069	1.54	0.921
3.20	0.065	1.47	0.921
3.40	0.061	1.33	0.916
3.71	0.055	1.13	0.908
3.97	0.049	1.01	0.908
6.199	0.390		0.193
6.358	0.454		0.141
6.526	0.485		0.120
6.702	0.533		0.093
6.888	0.574		0.073
7.130	0.616		0.056
7.328	0.641		0.048
7.583	0.674		0.038
7.847	0.700		0.031
8.015	0.710		0.029
8.634	0.762		0.018
9.143	0.800		0.012
9.709	0.819		0.010
10.20	0.843		0.007
11.08	0.870		0.005
11.83	0.887		0.004
12.73	0.907		0.002
13.05	0.913		0.002
13.42	0.914		0.002
13.73	0.917		0.002
14.07	0.922		0.002
14.83	0.934		0.001
15.05	0.936		0.001
15.46	0.942		0.001
16.21	0.948		0.001
18.10	0.964		0.000
21.12	0.979		0.000
25.51	0.993		0.000
26.95	1.00		0.000
27.68	1.01		0.000
28.37	1.01		0.000
29.52	1.02		0.000
Tantalum[16]			
0.10	10.14	66.39	0.984
0.15	9.45	46.41	0.9834
0.20	5.77	35.46	0.982
0.26	3.67	27.53	0.981
0.30	2.87	23.90	0.980
0.38	2.03	18.87	0.978
0.50	1.37	14.26	0.974
0.58	1.15	12.19	0.970
0.70	0.96	9.92	0.962
0.78	0.89	8.77	0.956
0.90	0.84	7.38	0.942
1.00	0.89	6.47	0.992
1.10	0.93	5.75	0.899
1.20	0.98	5.14	0.872
1.30	1.00	4.62	0.842
1.40	1.04	4.15	0.805
1.50	1.09	3.73	0.762
1.60	1.15	3.33	0.707
1.70	1.24	2.95	0.640
1.80	1.35	2.60	0.560
1.90	1.57	2.24	0.460
2.00	1.83	1.99	0.388
2.10	2.10	1.84	0.354
2.20	2.36	1.81	0.351
2.30	2.56	1.86	0.365
2.40	2.68	1.92	0.378
2.50	2.75	1.98	0.388
2.60	2.80	2.02	0.395
2.70	2.84	2.08	0.405
2.80	2.85	2.14	0.412
2.90	2.84	2.20	0.420
3.00	2.81	2.24	0.425
3.20	2.73	2.31	0.432
3.40	2.61	2.33	0.435
3.60	2.49	2.30	0.430
3.80	2.40	2.22	0.418
4.00	2.36	2.14	0.406
4.20	2.35	2.06	0.392
4.40	2.39	2.01	0.384
4.60	2.45	2.00	0.384
4.80	2.53	2.06	0.394
5.00	2.58	2.20	0.416
5.20	2.52	2.44	0.450
5.40	2.31	2.61	0.480
5.60	2.06	2.67	0.501
5.80	1.83	2.63	0.510
6.00	1.63	2.56	0.515
6.20	1.48	2.45	0.512
6.40	1.37	2.33	0.504
6.60	1.29	2.22	0.492
6.80	1.23	2.11	0.478
7.00	1.18	2.01	0.462
7.20	1.15	1.91	0.445
7.40	1.13	1.82	0.425
7.60	1.12	1.75	0.406
7.80	1.11	1.68	0.390
8.00	1.11	1.61	0.370
8.20	1.12	1.55	0.350
8.40	1.13	1.50	0.332
8.60	1.14	1.45	0.317
8.80	1.17	1.41	0.301
9.00	1.19	1.40	0.294
9.20	1.21	1.38	0.289
9.40	1.21	1.38	0.287
9.60	1.21	1.38	0.285
9.80	1.21	1.37	0.285
10.00	1.20	1.37	0.286
10.20	1.19	1.37	0.286
10.40	1.18	1.37	0.287
10.60	1.16	1.36	0.288
10.80	1.15	1.36	0.289
11.00	1.13	1.35	0.290
11.20	1.11	1.35	0.292
11.40	1.09	1.34	0.293
11.60	1.07	1.33	0.294
11.80	1.05	1.32	0.295
12.00	1.02	1.31	0.296
12.20	1.00	1.29	0.295
12.40	0.98	1.28	0.294
12.60	0.96	1.26	0.292
12.80	0.94	1.24	0.289
13.00	0.93	1.22	0.286
13.60	0.91	1.16	0.272
14.00	0.90	1.15	0.272
14.60	0.85	1.15	0.285
15.00	0.80	1.13	0.293
15.60	0.72	1.08	0.301
16.00	0.68	1.04	0.304
16.60	0.63	0.97	0.301
17.00	0.60	0.92	0.296
17.60	0.60	0.92	0.296
18.00	0.55	0.79	0.274
18.60	0.53	0.71	0.254
19.00	0.53	0.65	0.236
19.60	0.53	0.57	0.207
20.00	0.54	0.52	0.185
20.60	0.55	0.44	0.153

Energy (eV)	n	k	$R(\phi = 0)$
21.00	0.57	0.39	0.127
21.60	0.64	0.34	0.089
22.00	0.64	0.32	0.081
22.60	0.69	0.27	0.058
23.00	0.73	0.24	0.043
23.60	0.80	0.26	0.033
24.00	0.80	0.26	0.034
24.60	0.82	0.25	0.029
25.00	0.83	0.25	0.026
25.60	0.86	0.24	0.022
26.00	0.88	0.25	0.022
26.60	0.87	0.26	0.023
27.00	0.87	0.25	0.022
27.60	0.89	0.23	0.019
28.00	0.90	0.23	0.017
28.60	0.91	0.22	0.015
29.00	0.92	0.22	0.014
29.60	0.94	0.22	0.014
30.00	0.95	0.22	0.014
31.00	0.97	0.23	0.014
32.00	0.98	0.24	0.015
33.00	0.98	0.25	0.015
34.00	0.99	0.25	0.016
35.00	0.99	0.26	0.017
36.00	0.99	0.27	0.018
37.00	0.99	0.28	0.019
38.00	0.98	0.28	0.021
39.00	0.97	0.29	0.022
40.00	0.95	0.29	0.023

Tellurium, $\vec{E} \parallel \hat{c}$[25]

Energy (eV)	n	k	$R(\phi = 0)$
0.01364	4.82	0.118	0.431
0.01488	5.26	0.0505	0.463
0.01612	5.47	0.0278	0.477
0.01736	5.59	0.0174	0.485
0.01860		0.0796	
0.01984		0.0696	
0.02108		0.0749	
0.02232		0.1900	
0.02356		0.2220	
0.02480		0.0716	
0.02604		0.0682	
0.02728		0.0832	
0.02976		0.0149	
0.03224		2.14E-03	
0.03472		1.71E-02	
0.03720	5.94	3.71E-03	0.507
0.03968		2.44E-03	
0.04339	5.96	1.59E-03	0.508
0.04711		7.85E-04	
0.05083		7.38E-04	
0.05579		3.89E-04	
0.06199	5.98	3.09E-04	0.509
0.07439		2.52E-04	
0.08679		2.96E-04	
0.09919		3.68E-04	
0.12400	6.246	3.34E-04	0.524
0.15500	6.253		0.525
0.20660	6.286		0.526
0.24800	6.316	7.48E-05	0.528
0.31	6.372	1.18E-05	0.531
0.35		4.93E-04	
0.41		6.74E-03	
0.5	6.53	2.30E-02	0.539
0.6	6.71	7.50E-02	0.549
0.7	7.00	0.24	0.563
0.8	7.23	0.48	0.574
0.9	7.48	0.94	0.589
1.0	7.70	1.56	0.606
1.2	6.99	2.22	0.593
1.4	7.11	2.46	0.604
1.6	6.75	2.91	0.606
1.8	6.89	3.70	0.637
2.0	4.67	4.67	0.654
2.2	4.94	5.16	0.681
2.4	3.94	5.08	0.686
2.6	3.25	4.77	0.681
2.8	2.73	4.42	0.674
3.0	2.30	4.16	0.674
3.5	1.69	3.44	0.646
4.0	1.33	2.64	0.571
4.5	1.32	1.96	0.428
5.0	1.63	1.60	0.312
5.5	1.72	1.57	0.302
6.0	1.73	1.45	0.276
6.5	1.78	1.36	0.257
7.0	1.83	1.36	0.257
7.5	1.72	1.51	0.289
8.0	1.54	1.37	0.260
8.5	1.55	1.23	0.226
9.0	0.99	0.93	0.179
9.5	1.47	1.25	0.233
10.0	0.86	0.86	0.181
11.0	0.80	0.77	0.165
12.0	0.79	0.76	0.164
14.0	0.67	0.59	0.146
16.0	0.59	0.49	0.147
18.0	0.48	0.31	0.160
20.0	0.74	0.20	0.035
22.0	0.83	0.18	0.018
24.0	0.85	0.15	0.013
26.0	0.87	0.12	0.009
28.0	0.89	0.090	0.006
30.0	0.90	0.045	0.003

Tellurium, $\vec{E} \perp \hat{c}$[25]

Energy (eV)	n	k	$R(\phi = 0)$
0.01364	2.61	0.2980	0.204
0.01488	3.65	0.0894	0.325
0.01612	4.10	0.0535	0.370
0.01736	4.63	0.4990	0.420
0.01860		0.1170	
0.01984		0.0343	
0.02108	(4.42)	0.0421	0.398
0.02232		0.1060	
0.02356		0.0880	
0.02480		0.0458	
0.02604		0.0928	
0.02728		0.0886	
0.02976		0.0232	
0.03224		3.06E-03	
0.03472		1.25E-02	
0.03720	4.71	2.65E-03	0.422
0.03968		1.89E-03	
0.04339	4.74	1.41E-03	0.425
0.04711		8.38E-04	
0.05083		6.79E-04	
0.05579		1.59E-04	
0.06199	4.77	1.16E-04	0.427
0.07439		7.23E-05	
0.08679		5.34E-05	
0.09919		4.28E-05	
0.1240	4.796	3.18E-05	0.429
0.1550	4.809		0.430
0.2066	4.838		0.432
0.2480	4.864	2.19E-05	0.434
0.31	4.929	3.18E-05	0.439
0.35		7.89E-02	
0.41		0.149	
0.5	4.90		0.437
0.6	4.93		0.439
0.7	4.95	0.11	0.441
0.8	5.10	0.13	0.452
0.9	5.22	0.22	0.461
1.0	5.35	0.45	0.472
1.2	5.17	0.63	0.462
1.4	5.56	0.63	0.488
1.6	5.88	1.15	0.517
1.8	6.10	1.80	0.545
2.0	5.94	2.69	0.571
2.2	5.10	3.61	0.594
2.4	4.24	3.77	0.593
2.6	3.57	3.75	0.591
2.8	3.03	3.63	0.588
3.0	2.51	3.39	0.578
3.5	1.72	2.70	0.532
4.0	1.32	2.01	0.440
4.5	1.28	1.28	0.251
5.0	1.47	0.82	0.132
5.5	1.74	0.51	0.104
6.0	1.94	0.39	0.118
6.5	2.19	0.32	0.148
7.0	2.48	0.40	0.192
7.5	2.60	0.69	0.226
8.0	2.59	0.91	0.245
8.5	2.39	1.00	0.235
9.0	1.11	1.24	0.259
9.5	2.08	1.11	0.224
10.0	0.99	1.04	0.215
11.0	0.84	1.01	0.237
12.0	0.87	0.87	0.182
14.0	0.59	0.87	0.282
16.0	0.64	0.55	0.144
18.0	0.52	0.41	0.161
20.0	0.50	0.38	0.165
22.0	0.56	0.29	0.110
24.0	0.54	0.25	0.113
26.0	0.50	0.20	0.127
28.0	0.48	0.17	0.135
30.0	0.46	0.088	0.140

Titanium (Polycrystalline)[14]

Energy (eV)	n	k	$R(\phi = 0)$
0.10	5.03	23.38	0.965
0.15	3.00	15.72	0.954
0.20	2.12	11.34	0.939
0.25	2.05	8.10	0.890
0.30	6.39	9.94	0.833
0.35	2.74	6.21	0.792
0.40	2.49	4.68	0.708
0.45	3.35	3.25	0.545
0.50	4.43	3.22	0.555
0.60	4.71	3.77	0.597
0.70	4.38	3.89	0.603
0.80	4.04	3.82	0.596
0.90	3.80	3.65	0.582
1.00	3.62	3.52	0.570

Energy (eV)	*n*	*k*	*R*(φ = 0)
1.10	3.47	3.40	0.560
1.20	3.35	3.30	0.550
1.30	3.28	3.25	0.546
1.40	3.17	3.28	0.549
1.50	2.98	3.32	0.557
1.60	2.74	3.30	0.559
1.70	2.54	3.23	0.557
1.80	2.36	3.11	0.550
1.90	2.22	2.99	0.540
2.00	2.11	2.88	0.530
2.10	2.01	2.77	0.520
2.20	1.92	2.67	0.509
2.30	1.86	2.56	0.495
2.40	1.81	2.47	0.483
2.50	1.78	2.39	0.471
2.60	1.75	2.34	0.462
2.70	1.71	2.29	0.456
2.80	1.68	2.25	0.451
2.90	1.63	2.21	0.447
3.00	1.59	2.17	0.444
3.10	1.55	2.15	0.442
3.20	1.50	2.12	0.442
3.30	1.44	2.09	0.442
3.40	1.37	2.06	0.443
3.50	1.30	2.01	0.443
3.60	1.24	1.96	0.441
3.70	1.17	1.90	0.436
3.80	1.11	1.83	0.430
3.85	1.08	1.78	0.423
3.90	1.06	1.73	0.413
4.00	1.04	1.62	0.389
4.20	1.05	1.45	0.333
4.40	1.13	1.33	0.284
4.60	1.17	1.29	0.265
4.80	1.21	1.23	0.244
5.00	1.24	1.21	0.236
5.20	1.27	1.20	0.228
5.40	1.17	1.16	0.228
5.60	1.24	1.21	0.234
5.80	1.21	1.22	0.241
6.00	1.15	1.21	0.244
6.20	1.11	1.18	0.240
6.40	1.08	1.14	0.232
6.60	1.04	1.06	0.212
6.80	1.05	1.02	0.198
7.00	1.06	0.97	0.182
7.20	1.07	0.95	0.175
7.40	1.11	0.94	0.167
7.60	1.09	0.92	0.165
7.80	1.11	0.93	0.165
8.00	1.10	0.94	0.169
8.20	1.10	0.95	0.171
8.40	1.08	0.95	0.175
8.60	1.04	0.96	0.181
8.80	1.02	0.95	0.181
9.00	1.00	0.94	0.182
9.20	0.97	0.93	0.182
9.40	0.95	0.91	0.181
9.60	0.94	0.90	0.179
9.80	0.91	0.88	0.179
10.00	0.89	0.88	0.180
10.20	0.86	0.85	0.178
10.40	0.85	0.83	0.175
10.60	0.81	0.79	0.167
10.80	0.80	0.76	0.162

Energy (eV)	*n*	*k*	*R*(φ = 0)
11.00	0.79	0.72	0.152
11.20	0.81	0.69	0.139
11.40	0.81	0.69	0.139
11.60	0.79	0.68	0.139
11.80	0.78	0.67	0.137
12.00	0.77	0.65	0.132
12.80	0.76	0.55	0.106
13.20	0.76	0.52	0.097
13.60	0.76	0.48	0.087
14.00	0.77	0.45	0.077
14.40	0.77	0.42	0.069
14.80	0.79	0.38	0.058
15.20	0.79	0.36	0.052
15.60	0.79	0.32	0.045
16.00	0.83	0.31	0.037
16.40	0.84	0.28	0.030
16.80	0.87	0.27	0.025
17.20	0.90	0.25	0.020
17.60	0.93	0.25	0.017
18.00	0.94	0.24	0.165
18.40	0.94	0.23	0.017
18.80	0.95	0.24	0.016
19.20	0.96	0.25	0.016
19.60	0.97	0.25	0.017
20.00	0.98	0.27	0.018
20.40	0.98	0.27	0.019
20.60	1.00	0.29	0.020
21.20	0.99	0.31	0.023
21.60	0.99	0.31	0.024
22.00	0.98	0.32	0.025
22.40	0.98	0.33	0.027
22.80	0.97	0.33	0.028
23.20	0.96	0.34	0.030
23.60	0.95	0.35	0.031
24.00	0.92	0.35	0.033
24.5	0.91	0.34	0.032
25.0	0.91	0.33	0.032
25.5	0.89	0.33	0.032
26.0	0.89	0.33	0.032
26.5	0.88	0.32	0.032
27.0	0.86	0.31	0.032
27.5	0.85	0.30	0.033
28.0	0.84	0.29	0.033
28.5	0.82	0.26	0.029
29.0	0.83	0.25	0.027
30.0	0.84	0.22	0.022

Tungsten[27]

Energy (eV)	*n*	*k*	*R*(φ = 0)
0.10	14.06	54.71	0.983
0.20	3.87	28.30	0.981
0.25	2.56	22.44	0.980
0.30	1.83	18.32	0.979
0.34	1.71	15.71	0.973
0.38	1.86	13.88	0.963
0.42	1.92	12.63	0.954
0.46	1.69	11.59	0.952
0.50	1.40	10.52	0.952
0.54	1.23	9.45	0.948
0.58	1.17	8.44	0.938
0.62	1.28	7.52	0.917
0.66	1.45	6.78	0.888
0.70	1.59	6.13	0.856
0.74	1.83	5.52	0.810
0.78	2.12	5.00	0.759
0.82	2.36	4.61	0.710
0.86	2.92	4.37	0.661
0.90	3.11	4.44	0.660
0.94	3.15	4.43	0.658
0.98	3.15	4.36	0.653
1.00	3.14	4.32	0.649
1.10	3.05	4.04	0.627
1.20	3.00	3.64	0.590
1.30	3.12	3.24	0.545
1.40	3.29	2.96	0.515
1.50	3.48	2.79	0.500
1.60	3.67	2.68	0.494
1.70	3.84	2.79	0.507
1.80	3.82	2.91	0.518
1.90	3.70	2.94	0.518
2.00	3.60	2.89	0.512
2.10	3.54	2.84	0.506
2.20	3.49	2.76	0.497
2.30	3.49	2.72	0.494
2.40	3.45	2.72	0.493
2.50	3.38	2.68	0.487
2.60	3.34	2.62	0.480
2.70	3.31	2.55	0.472
2.80	3.31	2.49	0.466
2.90	3.32	2.45	0.461
3.00	3.35	2.42	0.459
3.10	3.39	2.41	0.460
3.20	3.43	2.45	0.465
3.30	3.45	2.55	0.476
3.40	3.39	2.66	0.485
3.50	3.24	2.70	0.488
3.60	3.13	2.67	0.482
3.70	3.05	2.62	0.476
3.80	2.99	2.56	0.468
3.90	2.96	2.50	0.460
4.00	2.95	2.43	0.451
4.20	3.02	2.33	0.440
4.40	3.13	2.32	0.442
4.60	3.24	2.41	0.455
4.80	3.33	2.57	0.475
5.00	3.40	2.85	0.505
5.20	3.27	3.27	0.548
5.40	2.92	3.58	0.586
5.60	2.43	3.70	0.618
5.80	2.00	3.61	0.637
6.00	1.70	3.42	0.643
6.20	1.47	3.24	0.646
6.40	1.32	3.04	0.640
6.60	1.21	2.87	0.631
6.80	1.12	2.70	0.619
7.00	1.06	2.56	0.607
7.20	1.01	2.43	0.593
7.40	0.98	2.30	0.573
7.60	0.95	2.18	0.556
7.80	0.93	2.06	0.533
8.00	0.94	1.95	0.505
8.20	0.94	1.86	0.481
8.40	0.96	1.76	0.449
8.60	0.99	1.70	0.422
8.80	1.01	1.65	0.401
9.00	1.01	1.60	0.388
9.20	1.02	1.55	0.369
9.40	1.03	1.50	0.352
9.60	1.05	1.44	0.329
9.80	1.09	1.38	0.307
10.00	1.13	1.34	0.287

Energy (eV)	*n*	*k*	*R*(ϕ = 0)
10.20	1.19	1.33	0.274
10.40	1.24	1.34	0.270
10.60	1.27	1.36	0.274
10.80	1.29	1.39	0.282
11.00	1.28	1.42	0.290
11.20	1.27	1.44	0.297
11.40	1.25	1.46	0.305
11.60	1.22	1.48	0.313
11.80	1.20	1.48	0.318
12.00	1.16	1.48	0.323
12.40	1.10	1.47	0.329
12.80	1.04	1.44	0.333
13.20	0.98	1.40	0.332
13.60	0.94	1.35	0.325
14.00	0.91	1.28	0.312
14.40	0.90	1.23	0.296
14.80	0.90	1.17	0.276
15.20	0.93	1.13	0.255
15.60	0.97	1.12	0.246
16.00	0.98	1.14	0.249
16.40	0.97	1.17	0.260
16.80	0.94	1.19	0.273
17.20	0.90	1.21	0.289
17.60	0.85	1.21	0.304
18.00	0.80	1.20	0.317
18.40	0.74	1.18	0.330
18.80	0.69	1.15	0.340
19.20	0.64	1.11	0.347
19.60	0.60	1.07	0.353
20.00	0.56	1.02	0.354
20.40	0.54	0.97	0.350
20.80	0.52	0.92	0.342
21.20	0.50	0.87	0.331
21.60	0.50	0.82	0.318
22.00	0.49	0.77	0.303
22.40	0.49	0.73	0.287
22.80	0.49	0.69	0.272
23.20	0.49	0.66	0.263
23.60	0.48	0.62	0.252
24.00	0.49	0.57	0.234
24.40	0.50	0.53	0.213
24.80	0.51	0.49	0.191
25.20	0.53	0.46	0.171
25.60	0.55	0.43	0.150
26.00	0.57	0.40	0.132
26.40	0.59	0.38	0.117
26.80	0.61	0.37	0.105
27.00	0.62	0.36	0.099
27.50	0.64	0.34	0.085
28.00	0.67	0.32	0.073
28.50	0.69	0.31	0.065
29.00	0.71	0.30	0.057
29.50	0.73	0.30	0.052
30.00	0.75	0.29	0.047
31.00	0.78	0.29	0.042
32.00	0.79	0.29	0.040
33.00	0.82	0.28	0.033
34.00	0.84	0.29	0.032
35.00	0.85	0.31	0.033
36.00	0.85	0.32	0.036
37.00	0.84	0.33	0.039
38.00	0.83	0.33	0.040
39.00	0.81	0.33	0.042
40.00	0.80	0.33	0.045

Energy (eV)	*n*	*k*	*R*(ϕ = 0)
Vanadium[9]			
0.10	12.83	45.89	0.978
0.20	3.90	24.30	0.975
0.28	2.13	17.35	0.973
0.36	1.54	13.32	0.966
0.44	1.28	10.74	0.957
0.52	1.16	8.93	0.945
0.60	1.10	7.59	0.929
0.68	1.07	6.54	0.909
0.76	1.08	5.67	0.882
0.80	1.10	5.30	0.864
0.90	1.18	4.50	0.811
1.00	1.34	3.80	0.730
1.10	1.60	3.26	0.632
1.20	1.93	2.88	0.543
1.30	2.25	2.71	0.498
1.40	2.48	2.72	0.491
1.50	2.57	2.79	0.499
1.60	2.57	2.84	0.507
1.70	2.52	2.88	0.512
1.80	2.45	2.88	0.515
1.90	2.36	2.85	0.514
2.00	2.34	2.81	0.509
2.10	2.31	2.78	0.506
2.20	2.28	2.80	0.510
2.30	2.23	2.83	0.516
2.40	2.15	2.88	0.528
2.50	2.02	2.91	0.540
2.60	1.89	2.92	0.552
2.70	1.74	2.89	0.561
2.80	1.61	2.85	0.569
2.90	1.48	2.80	0.577
3.00	1.36	2.73	0.582
3.20	1.16	2.55	0.585
3.40	0.99	2.37	0.586
3.60	0.87	2.17	0.575
3.80	0.80	1.96	0.547
4.00	0.78	1.76	0.503
4.20	0.80	1.60	0.449
4.40	0.83	1.47	0.400
4.60	0.87	1.38	0.355
4.80	0.90	1.31	0.326
5.00	0.91	1.26	0.304
5.25	0.93	1.18	0.271
5.50	0.94	1.14	0.258
5.75	0.96	1.09	0.235
6.00	0.98	1.06	0.223
6.25	0.97	1.02	0.212
6.50	0.97	0.98	0.199
6.75	0.97	0.94	0.185
7.00	0.98	0.91	0.175
7.33	0.97	0.89	0.170
7.66	0.98	0.87	0.162
8.00	0.98	0.85	0.155
8.33	0.98	0.81	0.146
8.66	0.98	0.81	0.145
9.00	0.96	0.79	0.142
9.50	0.94	0.77	0.136
10.00	0.91	0.74	0.133
10.50	0.89	0.71	0.126
11.00	0.87	0.65	0.112
11.50	0.88	0.58	0.091
12.00	0.90	0.58	0.089
12.50	0.89	0.57	0.086
13.00	0.88	0.55	0.082

Energy (eV)	*n*	*k*	*R*(ϕ = 0)
13.50	0.87	0.53	0.079
14.00	0.86	0.51	0.075
14.50	0.86	0.49	0.070
15.00	0.86	0.47	0.065
15.50	0.86	0.46	0.062
16.00	0.85	0.45	0.061
16.50	0.84	0.43	0.059
17.00	0.84	0.41	0.056
17.50	0.83	0.40	0.054
18.00	0.82	0.38	0.051
18.50	0.82	0.37	0.048
19.00	0.82	0.35	0.045
19.50	0.82	0.34	0.043
20.00	0.81	0.32	0.041
20.50	0.81	0.31	0.038
21.00	0.81	0.29	0.036
21.50	0.81	0.28	0.033
22.00	0.81	0.27	0.032
22.50	0.81	0.25	0.029
23.00	0.82	0.24	0.027
23.50	0.82	0.23	0.025
24.00	0.82	0.22	0.024
24.50	0.83	0.21	0.022
25.00	0.83	0.20	0.020
25.50	0.83	0.19	0.019
26.00	0.83	0.18	0.018
26.50	0.84	0.17	0.016
27.00	0.84	0.16	0.015
27.50	0.85	0.16	0.014
28.00	0.85	0.15	0.013
28.50	0.86	0.14	0.012
29.00	0.86	0.14	0.011
29.50	0.86	0.13	0.010
30.00	0.87	0.13	0.009
31.00	0.88	0.12	0.008
32.00	0.90	0.11	0.007
33.00	0.90	0.10	0.005
34.00	0.91	0.10	0.005
35.00	0.92	0.09	0.004
36.00	0.94	0.10	0.004
37.00	0.94	0.10	0.004
38.00	0.95	0.11	0.004
39.00	0.95	0.12	0.004
40.00	0.95	0.13	0.005
Zinc, $\vec{E} \parallel \hat{c}$[28]			
0.7514	1.9241	7.5619	0.883
0.827	1.7921	6.9973	0.874
0.866	1.5571	6.7753	0.881
0.952	1.4824	6.2296	0.868
0.992	1.5762	5.8843	0.847
1.033	1.5407	5.3192	0.823
1.078	1.5853	4.9013	0.793
1.127	1.7768	4.5307	0.748
1.181	1.9808	4.2004	0.701
1.240	2.8821	3.4766	0.575
1.305	3.2039	3.0042	0.520
1.377	2.9459	3.5761	0.584
1.459	3.2523	4.2447	0.640
1.550	3.8086	4.6212	0.657
1.653	3.7577	4.6239	0.659
1.722	3.5908	4.4614	0.650
1.823	3.4234	4.3232	0.642
1.937	3.0132	3.9974	0.624
1.984	1.8562	3.9706	0.690

OPTICAL PROPERTIES OF SELECTED ELEMENTS (continued)

Energy (eV)	*n*	*k*	$R(\phi = 0)$
2.066	1.4856	4.0555	0.737
2.094	1.2525	3.9961	0.762
2.119	1.0017	3.8683	0.789
2.275	0.7737	3.9129	0.832
2.445	0.6395	3.4013	0.821
2.666	0.4430	3.1379	0.851
2.917	0.3589	2.8140	0.853
3.220	0.3069	2.5088	0.847
3.594	0.2737	2.1737	0.828
4.065	0.2510	1.8528	0.799
4.678	0.2354	1.6357	0.776
Zinc, $\vec{E} \perp \hat{c}$[28]			
0.751	1.4469	7.4158	0.905
0.827	1.4744	6.9688	0.892
0.866	1.3628	6.6886	0.892
0.952	1.3165	6.2212	0.881
0.992	1.3835	5.8910	0.863
1.033	1.2889	5.4001	0.850
1.078	1.3095	4.9025	0.822
1.127	1.6897	4.4062	0.746
1.181	1.9701	4.0176	0.684
1.240	2.8717	3.2873	0.555
1.305	3.3991	2.7684	0.497
1.377	3.1807	3.4709	0.569
1.459	3.5064	4.1994	0.630
1.550	4.1241	4.7768	0.664
1.653	4.0269	4.8027	0.667
1.722	3.9369	4.6356	0.657
1.823	3.7549	4.3042	0.635
1.937	3.4512	4.1942	0.631
1.984	3.2515	4.2980	0.644
2.066	2.0802	4.7231	0.738
2.094	1.7084	4.7923	0.774
2.119	1.3329	4.4751	0.791
2.275	0.9725	4.2879	0.825
2.455	0.7568	3.7627	0.824
2.666	0.5470	3.4277	0.845
2.917	0.4774	3.0476	0.834
3.220	0.3911	2.7463	0.835
3.594	0.3147	2.3041	0.821
4.065	0.3013	2.0077	0.789
4.678	0.2806	1.7997	0.770
Zirconium (Polycrystalline)[28]			
0.10	6.18	1.76	0.300
0.15	3.37	1.30	0.123
0.20	2.34	1.08	0.058
0.26	2.24	1.06	0.052
0.30	2.59	1.14	0.073
0.36	3.17	1.26	0.110
0.40	3.09	1.24	0.105
0.46	3.36	1.30	0.123
0.50	4.13	1.44	0.175
0.56	5.01	1.58	0.231
0.60	5.18	1.61	0.242
0.70	4.54	1.51	0.202
0.80	4.03	1.42	0.168
0.90	3.74	1.37	0.149
0.96	3.69	1.36	0.145
1.00	3.66	1.35	0.143
1.10	3.65	1.35	0.142
1.20	3.53	1.33	0.134
1.30	3.25	1.27	0.116
1.40	3.10	1.25	0.106
1.50	3.02	1.23	0.100
1.60	2.88	1.20	0.091
1.70	2.68	1.16	0.078
1.80	2.49	1.12	0.067
2.00	2.14	1.03	0.047
2.10	1.99	1.00	0.040
2.20	1.87	0.97	0.034
2.30	1.78	0.94	0.030
2.40	1.71	0.92	0.027
2.50	1.62	0.90	0.024
2.60	1.54	0.88	0.022
2.70	1.46	0.86	0.019
2.80	1.40	0.84	0.018
2.90	1.34	0.82	0.016
3.00	0.30	0.81	0.016
3.10	1.26	0.80	0.015
3.30	1.19	0.77	0.014
3.40	1.16	0.76	0.013
3.50	1.13	0.75	0.013
3.60	1.10	0.74	0.013
3.70	1.07	0.73	0.013
3.80	1.04	0.72	0.012
3.90	1.01	0.71	0.012
4.00	0.98	0.70	0.012
4.20	0.94	0.68	0.013
4.40	0.89	0.67	0.013
4.60	0.85	0.65	0.014
4.80	0.81	0.64	0.014
5.00	0.78	0.63	0.015
5.20	0.77	0.62	0.016
5.40	0.77	0.62	0.016
5.60	0.80	0.63	0.014
5.80	0.87	0.66	0.013
6.00	1.00	0.71	0.012
6.20	1.11	0.75	0.013
6.40	1.23	0.78	0.014
6.60	1.33	0.81	0.016
6.80	1.42	0.84	0.018
7.00	1.49	0.86	0.020
7.20	1.54	0.88	0.022
7.40	1.58	0.89	0.023
7.60	1.61	0.90	0.024
7.80	1.63	0.90	0.025
8.00	1.66	0.91	0.026
8.20	0.67	0.91	0.026
8.40	1.68	0.92	0.026
8.60	1.68	0.92	0.026
8.80	1.66	0.91	0.026
9.00	1.65	0.91	0.025
9.20	1.63	0.90	0.025
9.40	1.60	0.89	0.024
9.60	1.57	0.89	0.023
9.80	1.52	0.87	0.021
10.00	1.47	0.86	0.020
10.20	1.42	0.84	0.018
10.40	1.35	0.82	0.016
10.50	1.32	0.81	0.016
10.60	1.28	0.80	0.015
10.80	1.23	0.78	0.014
11.00	1.19	0.77	0.014
11.20	1.16	0.76	0.013
11.40	1.13	0.75	0.013
11.60	1.11	0.74	0.013
11.80	1.09	0.74	0.013
12.00	1.08	0.73	0.013
12.40	1.05	0.72	0.012
12.80	1.01	0.71	0.012
13.20	0.98	0.70	0.012
13.60	0.95	0.69	0.013
14.00	0.92	0.68	0.013
14.40	0.89	0.67	0.013
14.80	0.90	0.67	0.013
15.20	0.92	0.68	0.013
15.60	0.95	0.69	0.013
16.00	0.98	0.70	0.012
16.40	1.01	0.71	0.012
16.80	1.04	0.72	0.012
17.20	1.09	0.74	0.013
17.60	1.13	0.75	0.013
18.00	1.17	0.76	0.014
18.40	1.21	0.78	0.014
18.80	1.24	0.79	0.014
19.20	1.27	0.80	0.015
19.60	1.29	0.80	0.015
20.00	1.30	0.81	0.015
20.60	1.29	0.80	0.015
21.00	1.27	0.80	0.015
21.60	1.23	0.78	0.014
22.00	1.20	0.77	0.014
22.60	1.15	0.76	0.013
23.00	1.12	0.75	0.013
23.60	1.08	0.73	0.013
24.00	1.05	0.73	0.013
24.60	1.02	0.71	0.012
25.00	1.00	0.71	0.012
25.60	0.97	0.69	0.012
26.00	0.95	0.69	0.013
26.60	0.91	0.67	0.013
27.00	0.88	0.66	0.013
27.60	0.84	0.65	0.014
28.00	0.83	0.64	0.014
28.60	0.82	0.64	0.014
29.00	0.81	0.64	0.014
29.60	0.82	0.64	0.014
30.00	0.82	0.64	0.014

REFERENCES

1. Shiles, E., Sasaki, T., Inokuti, M., and Smith, D. Y., *Phys. Rev. Sect. B*, 22, 1612, 1980.
2. Edwards, D. F., and Philipp, H. R., in *HOC-I*, p.665.
3. Ives, H. E., and Briggs, N. B., *J. Opt. Soc. Am.*, 27, 395, 1937.
4. Bos, L. W., and Lynch, D. W., *Phys. Rev. Sect. B*, 2, 4567, 1970.
5. Weaver, J. H., Colavita, E., Lynch, D. W., and Rosei, R., *Phys. Rev. Sect. B*, 19, 3850, 1979.
6. Hagemann, H. J., Gudat, W., and Kunz, C., *J. Opt. Soc. Am.*, 65, 742, 1975.
7. Schulz, L. G., *J. Opt. Soc. Am.*, 47, 64, 1957.
8. Potter, R. F., in *HOC-I*, p.465.
9. Olson, C. G., Lynch, D. W., and Weaver, J. H., unpublished.
10. Lynch, D. W., Olson, C. G., and Weaver, J. H., unpublished.
11. Weaver, J. H., Olson, C. G., and Lynch, D. W., *Phys. Rev. Sect. B*, 15, 4115, 1977.
12. Lynch, D. W., and Hunter, W. R., in *HOC-II*, p.345.
13. Priol, M. A., Daudé, A., and Robin, S., *Compt. Rend.*, 264, 935, 1967.
14. Johnson, P. B., and Christy, R. W., *Phys. Rev. Sect. B*, 9, 5056, 1974.
15. Arakawn, E. T., and Inagaki, T., in *HOC-II*, p.461.
16. Weaver, J. H., Lynch, D. W., and Olson, D. G., *Phys. Rev. Sect. B*, 10, 501, 1973.
17. Lynch, D. W., Rosei, R., and Weaver, J. H., *Solid State Commun.*, 9, 2195, 1971.
18. Weaver, J. H., Lynch, D. W., and Olson, C. G., *Phys. Rev. Sect. B*, 7, 4311, 1973.
19. Weaver, J. H., and Benbow, R. L., *Phys. Rev. Sect. B*, 12, 3509, 1975.
20. Weaver, J. H., *Phys. Rev.,Sect. B*, 11, 1416, 1975.
21. Lynch, D. W., and Hunter, W. R., in *HOC-II*, p.364.
22. Palik, E. D., in *HOC-II*, p. 691.
23. Edwards, D. F., in *HOC-I*, p. 547.
24. Lynch, D. W., and Hunter, W. R., in *HOC-II*, p.354.
25. Palik, E. D., in *HOC-II*, p. 709.
26. Lynch, D. W., Olson, C. G., and Weaver, J. H., *Phys. Rev. Sect. B*, 11, 3671, 1975.
27. Weaver, J. H., Lynch, D. W., and Olson, C. G., *Phys. Rev. Sect. B*, 12, 1293, 1975.
28. Lanham, A. P., and Terherne, D. M., *Proc. Phys. Soc.*, 83, 1059, 1964.

OPTICAL PROPERTIES OF SELECTED INORGANIC AND ORGANIC SOLIDS

L. I. Berger

Optical properties of materials are closely related to their dielectric properties. The complex dielectric function (relative permittivity) of a material is equal to

$$\varepsilon(\omega) = \varepsilon'(\omega) - j\varepsilon''(\omega),$$

where $\varepsilon'(\omega)$ and $\varepsilon''(\omega)$ are its real and imaginary parts, respectively, and ω is the angular frequency of the applied electric field. For a non-absorbing medium, the index of refraction is $n = (\varepsilon\mu)^{1/2}$, where μ is the relative magnetic permeability of the medium (material); in the majority of dielectrics, $\mu \cong 1$.

For many applications, the most important optical properties of materials are the index of refraction, the extinction coefficient, k, and the reflectivity, R. The common index of refraction of a material is equal to the ratio of the phase velocity of propagation of an electromagnetic wave of a given frequency in vacuum to that in the material. Hence, $n \geqq 1$. The optical properties of highly conductive materials like metals and semiconductors (at photon energy range above the energy gap) differ from those of optically transparent media. Free electrons absorb the incident electromagnetic wave in a thin surface layer (a few hundred nanometers thick) and then release the absorbed energy in the form of secondary waves reflected from the surface. Thus, the light reflection becomes very strong; for example, highly conductive sodium reflects 99.8% of the incident wave (at 589 nm). Introduction of the effective index of refraction, $n_{eff} = (\varepsilon')^{1/2} = n - jk$, where $\varepsilon' = \varepsilon - j\delta/\omega\,\varepsilon_o$, δ is the electrical conductivity of the material in S/m, and $\varepsilon_o = 8.8542 \cdot 10^{-12}$ F/m is the permittivity of vacuum, allows one to apply the expressions of the optics of transparent media to the conductive materials. It is clear that the effective index of refraction may be smaller than 1. For example, $n = 0.05$ for pure sodium and $n = 0.18$ for pure silver (at 589.3 nm). At very high photon energies, the quantum effects, such as the internal photoeffect, start playing a greater role, and the optical properties of these materials become similar to those of insulators (low reflectance, existence of Brewster's angle, etc.).

The extinction coefficient characterizes absorption of the electromagnetic wave energy in the process of propagation of a wave through a material. The wave intensity, I, after it passes a distance x in an isotropic medium is equal to

$$I = I_0 \exp(-\alpha x),$$

where I_0 is the intensity at $x = 0$ and α is called the absorption coefficient. For many applications, the extinction coefficient, k, which is equal to

$$k = \alpha\frac{\lambda}{4\pi},$$

where λ is the wavelength of the wave in the medium, is more commonly used for characterization of the electromagnetic losses in materials.

Reflection of an electromagnetic wave from the interface between two media depends on the media indices of refraction and on the angle of incidence. It is characterized by the reflectivity, which is equal to the ratio of the intensity of the wave reflected back into the first medium to the intensity of the wave approaching the interface. For polarized light and two non-absorbing media,

$$R = \frac{(N_1 - N_2)^2}{(N_1 + N_2)^2},$$

where $N_1 = n_1/\cos\theta_1$ and $N_2 = n_2/\cos\theta_2$ for the wave polarized in the plane of incidence, and $N_1 = n_1\cos\theta_1$ and $N_2 = n_2\cos\theta_2$ for the wave polarized normal to the plane of incidence; θ_1 and θ_2 are the angles between the normal to the interface in the point of incidence and the directions of the beams in the first and second medium, respectively. The reflectivity at normal incidence in this case is

$$R = [(n_1 - n_2)/(n_1 + n_2)]^2$$

For any two opaque (absorbing) media, the normal incidence reflectivity is

$$R = \frac{(n_1 - n_2)^2 + k_2^2}{(n_1 + n_2)^2 + k_2^2}.$$

In the majority of experiments, the first medium is air ($n \approx 1$), and hence,

$$R = \frac{(1 - n)^2 + k^2}{(1 + n)^2 + k^2}.$$

The data on n and k in the following table are abridged from the sources listed in the references. The reflectivity at normal incidence, R, has been calculated from the last equation. For convenience, the energy E, wavenumber $\bar{\nu}$, and wavelength λ are given for the incidence radiation.

E/eV	$\tilde{\nu}$/cm^{-1}	λ/μm	n	n_a	n_c	k	k_a	k_c	R	R_a	R_c
Crystalline Arsenic Selenide (As_2Se_3) [Ref. 1]*											
2.194	17700	0.565					0.30				
2.168	17480	0.572					0.25				
2.141	17270	0.579					0.20				
2.123	17120	0.584					0.17				
2.098	16920	0.591					0.13				
2.094	16890	0.592						0.26			
2.091	16860	0.593						0.26			
2.073	16720	0.598					0.10	0.23			
2.060	16610	0.602						0.20			
2.049	16530	0.605					0.079	0.17			
2.036	16420	0.609						0.15			
2.023	16310	0.613						0.12			
2.013	16230	0.616					0.050				
2.009	16210	0.617						0.097			
2.000	16130	0.620						0.082			
1.987	16030	0.624						0.063			
1.977	15940	0.627					0.031				
1.974	15920	0.628						0.051			
1.962	15820	0.632						0.038			
1.953	15750	0.635						0.030			
1.949	15720	0.636					0.020				
1.937	15630	0.640						0.022			
1.925	15530	0.644						0.017			
1.922	15500	0.645					0.012				
1.905	15360	0.651					$8.6 \cdot 10^{-3}$				
1.893	15270	0.655					6.4				
1.881	15170	0.659					5.2				
1.859	14990	0.667					3.1				
1.848	14900	0.671						$1.7 \cdot 10^{-3}$			
1.845	14880	0.672					2.0				
1.842	14860	0.673						$1.2 \cdot 10^{-3}$			
1.831	14770	0.677					$1.3 \cdot 10^{-3}$	$9.0 \cdot 10^{-4}$			
1.826	14730	0.679						6.4			
1.821	14680	0.681						4.7			
1.818	14660	0.682					$8.6 \cdot 10^{-4}$				
1.815	14640	0.683						3.4			
1.807	14580	0.686					5.5				
1.802	14530	0.688					4.1				
0.06199	500.0	20.0		3.2	2.9		$1.7 \cdot 10^{-3}$	$1.8 \cdot 10^{-3}$		0.27	0.24
0.05904	476.2	21.0		3.1	2.9		$2.1 \cdot 10^{-3}$	$2.2 \cdot 10^{-3}$		0.26	0.24
0.05636	454.5	22.0		3.1	2.9		$2.5 \cdot 10^{-3}$	$2.6 \cdot 10^{-3}$		0.26	0.24
0.05391	434.8	23.0		3.1	2.9		$3.0 \cdot 10^{-3}$	$3.1 \cdot 10^{-3}$			
0.04592	370.4	27.0		3.0	2.8		$6.3 \cdot 10^{-3}$	$6.4 \cdot 10^{-3}$		0.25	0.22
0.04428	357.1	28.0		3.0	2.8		$7.6 \cdot 10^{-3}$	$7.7 \cdot 10^{-3}$		0.25	0.22
0.04275	344.8	29.0		3.0	2.8		0.0092	0.0093		0.25	0.22
0.04133	333.3	30.0		3.0	2.7		0.011	0.011		0.25	0.21
0.03542	285.7	35.0		2.7	2.5			0.037	0.034	0.21	0.18
0.03100	250.0	40.0		1.9	1.7			0.38	1.0	0.19	0.18
0.03061	247.0	40.5		2.0	2.6			0.33	0.95	0.12	0.25
0.03024	244.0	41.0		1.7	2.4			0.41	0.46	0.088	0.18
0.02883	232.6	43.0		1.2	1.3			2.2	0.94	0.50	0.16
0.02850	229.9	43.5		1.6	1.2			2.8	1.4	0.56	0.29
0.02818	227.3	44.0		2.3	1.2		3.3	2.0		0.58	0.48
0.02755	222.2	45.0		4.2	2.0		2.5	3.3		0.50	0.60
0.02480	200.0	50.0		6.5	4.0		3.6	0.26		0.62	0.36
0.02254	181.8	55.0		4.5	3.5		0.17	0.10		0.40	0.31

E/eV	$\bar{\nu}$ /cm^{-1}	λ/μm	*n*	n_a	n_c	*k*	k_a	k_c	*R*	R_a	R_c
0.02066	166.7	60.0		4.0	3.2		0.089	0.10		0.36	0.27
0.01907	153.8	65.0		3.8	3.1		0.097	0.16		0.34	0.26
0.01771	142.9	70.0		3.6	3.0		0.19	0.30		0.32	0.25
0.01653	133.3	75.0		3.7	3.0		0.41	0.44		0.34	0.26
0.01550	125.0	80.0		3.8	3.1		0.29	0.40		0.34	0.27
0.01459	117.6	85.0		3.6	2.9		0.20	0.34		0.32	0.24
0.01378	111.1	90.0		3.2	2.6		0.43	0.49		0.28	0.21
0.01305	105.3	95.0		4.7	3.0		1.5	1.5		0.46	0.34
0.01240	100.0	100.0		4.4	2.7		0.22	0.81		0.40	0.25
0.01181	95.24	105.0		4.2	3.0		0.094	3.9		0.38	0.62
0.01127	90.91	110.0		4.1	5.3		0.059	0.70		0.37	0.47
0.01033	83.33	120.0		3.9	4.2		0.034	0.13		0.35	0.38
0.009537	76.92	130.0		3.9	4.0		0.024	0.069		0.35	0.36
0.008856	71.43	140.0		3.9	3.8		0.019	0.048		0.35	0.34
0.007749	63.50	160.0		3.8	3.7		0.014	0.032		0.34	0.33
0.006888	55.55	180.0		3.8	3.7		0.011	0.024		0.34	0.33
0.006199	50.0	200.0		3.8	3.6		0.0091	0.019		0.34	0.32

*Indices a and c relate to the radiation electric field parallel to the a and c axes of the crystal, respectively.

Vitreous Arsenic Selenide (As_2Se_3) [Ref. 1]

E/eV	$\bar{\nu}$ /cm^{-1}	λ/μm	*n*	n_a	n_c	*k*	k_a	k_c	*R*	R_a	R_c
2.056	16580	0.603				0.12					
2.026	16340	0.612				0.11					
2.006	16180	0.618				0.099					
1.990	16050	0.623				9.0					
1.925	15530	0.644				5.6					
1.826	14730	0.679				1.4					
1.810	14600	0.685				0.012					
1.794	14470	0.691				0.0089					
1.771	14290	0.700				6.2					
1.715	13830	0.723				2.6					
1.701	13720	0.729				0.0022					
1.647	13280	0.753				0.00046					
1.629	13140	0.761	3.07			4.0			0.62		
1.596	12870	0.777	3.06			2.7			0.49		
1.579	12740	0.785	3.05			1.9			0.39		
1.562	12590	0.794	3.05			0.00013			0.26		
1.544	12450	0.803	3.04			0.000094			0.25		
1.529	12330	0.811	3.03			6.3			0.78		
1.512	12200	0.820	3.03			4.2			0.64		
1.494	12050	0.830	3.02			2.8			0.50		
1.476	11910	0.840	3.01			1.8			0.38		
1.378	11110	0.90	2.98								
1.240	10000	1.00	2.93								
1.127	9091	1.10	2.90								
1.051	8475	1.18	2.89								
1.033	8333	1.20	2.88								
0.2555	1900	5.05				$1.6 \cdot 10^{-7}$					
0.2380	1919	5.21				$9.9 \cdot 10^{-8}$					
0.2344	1890	5.29				$1.1 \cdot 10^{-7}$					
0.1345	1085	9.22				4.4					
0.1339	1080	9.26				3.7					
0.1333	1075	9.30				4.4					
0.1308	1055	9.48				4.5					
0.1215	980	10.20				8.9					
0.1203	970	10.31				$9.9 \cdot 10^{-7}$					
0.1196	965	10.36				$1.0 \cdot 10^{-6}$					

E/eV	$\bar{\nu}$/cm^{-1}	λ/μm	n	n_a	n_c	k	k_a	k_c	R	R_a	R_c
0.1178	950	10.53				1.1					
0.1116	900	11.11				1.8					
0.1004	810	12.35				4.9					
0.09919	800	12.50				$7.0 \cdot 10^{-6}$					
0.09795	790	12.66				$1.0 \cdot 10^{-5}$					
0.09671	780	12.82				1.5					
0.09299	750	13.33				3.7					
0.08555	690	14.49				6.9					
0.08431	680	14.71				5.9					
0.08059	650	15.38				6.1					
0.07811	630	15.87				6.3					
0.07687	620	16.13				7.7					
0.07563	610	16.39				7.8					
0.07439	600	16.67				$9.3 \cdot 10^{-5}$					
0.07315	590	16.95	2.8			$1.2 \cdot 10^{-4}$			0.22		
0.07191	580	17.24	2.8			1.4			0.32		
0.07067	570	17.54	2.8			1.8			0.37		
0.06943	560	17.86	2.8			2.8			0.50		
0.06633	535	18.69	2.8			5.2			0.73		
0.06571	530	18.87	2.8			$7.2 \cdot 10^{-4}$			0.22		
0.06509	525	19.05	2.8			$1.2 \cdot 10^{-3}$			0.22		
0.06447	520	19.23	2.8			1.7			0.35		
0.06075	490	20.41	2.7			4.9			0.71		
0.06024	485.9	20.58	2.7			5.2			0.73		
0.05331	430	23.26	2.7			1.4			0.31		
0.05269	425	23.53	2.7			$1.1 \cdot 10^{-3}$			0.21		
0.05207	420	23.81	2.7			$8.5 \cdot 10^{-4}$			0.21		
0.05145	415	24.10	2.7			7.3			0.84		
0.05083	410	24.39	2.7			8.3			0.87		
0.05021	405	24.69	2.7			$9.4 \cdot 10^{-4}$			0.21		
0.04959	400	25.0	2.7			$1.2 \cdot 10^{-3}$			0.21		
0.04862	392.2	25.5	2.6			1.6			0.33		
0.04679	377.4	26.5	2.6			5.0			0.73		
0.04592	370.4	27.0	2.6			$8.0 \cdot 10^{-3}$			0.20		
0.04509	363.6	27.5	2.6			$1.2 \cdot 10^{-2}$			0.20		
0.04428	357.1	28.0	2.6			1.7			0.34		
0.03875	312.5	32.0	2.5			8.2			0.87		
0.03815	307.7	32.5	2.5			$9.3 \cdot 10^{-3}$			0.18		
0.03757	303.0	33.0	2.4			0.11			0.17		
0.02988	241.0	41.5	2.2			0.89			0.20		
0.02952	238.1	42.0	2.2			1.0			0.22		
0.02725	219.8	45.5	3.2			1.8			0.39		
0.02362	190.5	52.5	3.6			0.30			0.32		
0.01937	156.2	64.0	3.2			0.10			0.27		
0.01922	155.0	64.5	3.2			$9.6 \cdot 10^{-2}$			0.27		
0.01907	153.8	65.0	3.2			9.4			0.88		
0.01734	139.9	71.5	3.1			8.7			0.87		
0.01653	133.3	75.0	3.1			9.4			0.88		
0.01642	132.5	75.5	3.1			0.096			0.26		
0.01494	120.5	83.0	3.0			0.15			0.25		
0.01246	100.5	99.5	3.2			0.60			0.26		
0.007606	61.35	163.0	3.3			0.12			0.29		
0.006199	50.00	200.0	3.2								
0.004592	37.04	270.0	3.1			0.072			0.26		
0.002799	22.57	443.0	3.0			4.5			0.67		
0.001826	14.73	679.0	3.0			2.8			0.50		
0.001273	10.27	974.0	3.0			2.1			0.41		
0.0006491	5.236	1910.0	3.0			$1.1 \cdot 10^{-2}$			0.25		

E/eV	$\bar{\nu}$ /cm^{-1}	λ/μm	n	n_a	n_c	k	k_a	k_c	R	R_a	R_c
0.0004376	3.530	2833.0	3.0			$7.5 \cdot 10^{-3}$			0.25		
0.0002903	2.341	4271.0	3.0			5.0			0.71		
0.0001716	1.384	7224.0	3.0			3.1			0.53		
0.00009047	0.7297	13704	3.0			$1.6 \cdot 10^{-3}$			0.25		
0.00005621	0.4534	22056	3.0			$9.9 \cdot 10^{-4}$			0.25		
0.00002774	0.2237	44699	3.0			5.2			0.72		
0.00001439	0.1161	86153	3.0			2.6			0.47		

Vitreous Arsenic Sulfide (As_2S_3) - [Ref. 2]

E/eV	$\bar{\nu}$ /cm^{-1}	λ/μm	n	n_a	n_c	k	k_a	k_c	R	R_a	R_c
4.959	40000	0.2500	2.48			1.21			0.27		
3.100	25000	0.40	3.09			0.34			0.27		
2.48	20000	0.4999	2.83			0.013			0.23		
1.879	15150	0.66	2.59			$1.7 \cdot 10^{-6}$			0.20		
1.240	10000	1.0	2.48			$2.4 \cdot 10^{-7}$			0.18		
0.6199	5000	2.0	2.43						0.17		
0.3100	2500	4.0	2.41						0.17		
0.2480	2000	5.0	2.41						0.17		
0.1736	1400	7.143	2.40			$7.4 \cdot 10^{-7}$			0.17		
0.1240	1000	10.00	2.38			$1.3 \cdot 10^{-4}$			0.17		
0.09299	750	13.33	2.35			$3.0 \cdot 10^{-3}$			0.16		
0.07439	600	16.67	2.31			$4.6 \cdot 10^{-4}$			0.16		
0.04959	400.0	25.0	1.79			0.2			0.085		
0.03757	303.0	33.0	3.59			1.4			0.38		
0.03100	250.0	40.0	2.98			0.15			0.25		
0.02480	200.0	50	2.66			0.11			0.21		
0.02066	166.7	60	2.64			0.57			0.22		
0.01771	142.9	70	2.99			0.17			0.25		
0.01550	125.0	80	2.89			0.14			0.24		
0.01378	111.1	90	2.84			0.12			0.23		
0.01240	100	100	2.81			0.10			0.23		
0.008183	66	152	2.76			0.072			0.22		
0.004029	32.5	308	2.74			0.044			0.22		
0.002418	19.5	513	2.74			0.031			0.22		
0.001984	16	625	2.74			0.025			0.22		
0.001048	8.45	1180	2.73			$8.8 \cdot 10^{-3}$			0.22		
0.0001033	0.833	12000	2.73			$1.3 \cdot 10^{-3}$			0.22		
$4.129 \cdot 10^{-12}$	$3.33 \cdot 10^{-8}$	$3 \cdot 10^{-11}$	2.73						0.22		

Cadmium Telluride (CdTe) - [Ref. 3]

E/eV	$\bar{\nu}$ /cm^{-1}	λ/μm	n	n_a	n_c	k	k_a	k_c	R	R_a	R_c
4.9	39520	0.2530	2.48			2.04			0.39		
4.1	33070	0.3024	2.33			1.59			0.32		
3.9	31460	0.3179	2.57			1.90			0.37		
3.5	28230	0.3542	2.89			1.52			0.34		
3.1	25000	0.4000	3.43			1.02			0.34		
3.0	24200	0.4133	3.37			0.861			0.32		
2.755	22220	0.45	3.080			0.485			0.27		
2.75	22180	0.4509	3.23			0.636			0.29		
2.610	21050	0.475	3.045								
2.5	20160	0.4959	3.14			0.525			0.28		
2.25	18150	0.5510	3.05			0.411			0.26		
1.771	14290	0.70	2.861			0.210			0.23		
1.512	12200	0.82	2.880			0.040			0.23		
1.50	12100	0.8266	2.98			0.319			0.25		
1.475	11900	0.840	2.905			0.00134			0.24		
1.47	11860	0.8434				0.000671					
1.465	11820	0.8463				3.37					

E/eV	$\bar{\nu}$ /cm^{-1}	λ/μm	*n*	n_a	n_c	*k*	k_a	k_c	*R*	R_a	R_c
1.46	11780	0.8492				1.89					
1.459	11760	0.850	2.948						0.24		
1.455	11740	0.8521				$1.08 \cdot 10^{-4}$					
1.45	11690	0.8551	2.9565			$5.10 \cdot 10^{-5}$			0.24		
1.445	11650	0.8580				2.73					
1.442	11630	0.860	2.952						0.24		
1.44	11610	0.8610	2.9479			1.37			0.32		
1.43	11530	0.8670	2.9402						0.24		
1.30	10490	0.9537	2.8720						0.23		
1.24	10000	1.0	2.840						0.23		
1.20	9679	1.033	2.8353						0.23		
1.10	8872	1.127	2.8050						0.23		
1.00	8065	1.240	2.7793						0.22		
0.90	7259	1.378	2.7537						0.22		
0.80	6452	1.550	2.7384						0.22		
0.70	5646	1.771	2.7223						0.21		
0.60	4839	2.066	2.7086						0.21		
0.50	4033	2.480	2.6972						0.21		
0.40	3226	3.100	2.6878						0.21		
0.30	2420	4.133	2.6800						0.21		
0.20	1613	6.199	2.6722						0.21		
0.10	806.5	12.40	2.6535						0.20		
0.09	725.9	13.78	2.6482						0.20		
0.06819	550	18.18	2.623						0.20		
0.0573	462	21.6				$3.8 \cdot 10^{-6}$					
0.05	403.3	24.80	2.5801						0.19		
0.0469	378	26.5				$8.0 \cdot 10^{-5}$					
0.04592	370.3	27				$9.88 \cdot 10^{-5}$					
0.04133	333.3	30	2.55916			$2.86 \cdot 10^{-4}$			0.19		
0.04092	330	30.30	2.531			3.34			0.57		
0.03720	300	33.33	2.494			4.97			0.73		
0.03647	294.1	34.00				8.93					
0.03596	290	34.48	2.478			$5.77 \cdot 10^{-3}$			0.18		
0.03493	281.7	35.5				7.91					
0.03472	280	35.71	2.459			6.76			0.83		
0.03100	250	40	2.378			$1.18 \cdot 10^{-2}$			0.17		
0.02917	235.3	42.5				6.93					
0.02852	230	43.48	2.289			1.87			0.36		
0.02728	220	45.45	2.224			$2.47 \cdot 10^{-2}$			0.14		
0.02604	210	47.62	2.137			$3.4 \cdot 10^{-2}$			0.13		
0.02480	200	50.00	2.013			$4.97 \cdot 10^{-2}$			0.11		
0.02384	192.3	52.0				6.21					
0.01798	145	68.97	1.8			5.2			0.79		
0.01736	140	71.43	6.778			4.50			0.66		
0.01550	125	80.0	4.598			0.294			0.41		
0.01364	110	90.91	3.868			$9.47 \cdot 10^{-2}$			0.35		
0.01240	100	100	3.649			$5.68 \cdot 10^{-2}$			0.32		
0.009919	80	125	3.415			0.0262			0.30		
0.008679	70	142.9	3.348			0.0189			0.29		
0.007439	60	166.7	3.299			1.39			0.35		
0.006199	50	200	3.263			1.03			0.32		
0.004959	40	250	3.236			$7.52 \cdot 10^{-3}$			0.28		
0.003720	30	333.3	3.217						0.28		
0.023015	18.563		538.71			3.2096			0.28		
0.001550	12.50	800				6.18					

E/eV	$\bar{\nu}$ /cm^{-1}	λ/μm	n	n_a	n_c	k	k_a	k_c	R	R_a	R_c
						Gallium Arsenide (GaAs) - [Ref. 4]					
155		0.007999				0.0181					
145		0.008551				0.0203					
130		0.009537				0.0224					
110		0.01127				0.0278					
90		0.01378				0.0323					
70		0.01771				0.0376					
40		0.03100				0.0426					
23		0.05391	1.037			0.228					
7.0		0.1771	1.063			1.838					
6.0	48390	0.2066	1.264			2.472			0.61		
5.00	40330	0.2480	2.273			4.084			0.67		
4.00	32260	0.3100	3.601			1.920			0.42		
3.00	24200	0.4133	4.509			1.948			0.47		
2.50	20160	0.4959	4.333			0.441			0.39		
2.00	16130	0.6199	3.878			0.211			0.35		
1.80	14520	0.8888	3.785			0.151			0.34		
1.60	12900	0.7749	3.700			0.091			0.33		
1.50	12100	0.8266	3.666			0.080			0.33		
1.40	11290	0.8856	3.6140			$1.69 \cdot 10^{-3}$			0.32		
1.20	9679	1.033	3.4920						0.31		
1.00	8065	1.240	3.4232						0.30		
0.80	6452	1.550	3.3737						0.29		
0.50	4033	2.480	3.3240						0.29		
0.25	2016	4.959	3.2978						0.29		
0.15	1210	8.266	3.2831						0.28		
0.100	806.5	12.40	3.2597			$4.93 \cdot 10^{-6}$			0.28		
0.090	725.9	13.78	3.2493			$1.64 \cdot 10^{-5}$			0.28		
0.070	564.6	17.71	3.2081			$2.32 \cdot 10^{-4}$			0.28		
0.060	483.9	20.66	3.1609			$3.45 \cdot 10^{-3}$			0.27		
0.0495	399.2	25.05	3.058			$2.07 \cdot 10^{-3}$			0.26		
0.03968	320	31.25	2.495			$2.43 \cdot 10^{-2}$			0.18		
0.03496	282	35.46	0.307			$294 \cdot 10^{-2}$					
0.02976	240	41.67	4.57			$4.26 \cdot 10^{-2}$			0.41		
0.02066	166.7	60	3.77			$3.89 \cdot 10^{-3}$			0.34		
0.01550	125	80	3.681			$1.84 \cdot 10^{-3}$			0.33		
0.008266	66.67	150	3.62			$2.14 \cdot 10^{-3}$			0.32		
0.002480	20	500	3.607			$1.3 \cdot 10^{-3}$			0.32		
0.001240	10	1000	3.606						0.32		
						Gallium Phosphide (GaP) - [Ref. 5]					
154.0		0.00805				$1.7 \cdot 10^{-2}$					
110.0		0.0113				$2.15 \cdot 10^{-2}$					
100.0		0.0124				$215 \cdot 10^{-2}$					
80.0		0.0155				$3.0 \cdot 10^{-2}$					
50.0		0.0248				$4.7 \cdot 10^{-2}$					
27.0		0.0460				$9.0 \cdot 10^{-2}$					
25.0		0.0496				0.122					
20.0		0.0620				0.180					
15.0		0.0826	0.748			0.628					
5.5	44360	0.2254	1.543			3.556			0.68		
4.68	37750	0.2649	4.181			2.634			0.50		
3.50	28230	0.3542	5.050			0.819			0.46		
3.00	24200	0.4133	4.081			0.224			0.37		
2.78	22420	0.4460	3.904			0.103			0.35		

E/eV	$\bar{\nu}$ /cm^{-1}	λ/µm	n	n_a	n_c	k	k_a	k_c	R	R_a	R_c
2.621	21140	0.473	3.73			$6.37 \cdot 10^{-3}$			0.33		
2.480	20000	0.500	3.590			$2.47 \cdot 10^{-3}$			0.32		
2.18	17580	0.5687	3.411			$2.8 \cdot 10^{-7}$			0.30		
2.000	16130	0.62	3.3254						0.29		
1.6	12900	0.7749	3.209						0.28		
1.240	10000	1.0	3.1192						0.26		
0.6888	5556	1.8	3.0439						0.26		
0.4769	3846	2.6	3.0271						0.25		
0.1907	1538	6.5	2.995			$4.29 \cdot 10^{-4}$			0.25		
0.1550	1250	8.0	2.984						0.25		
0.1240	1000	10	2.964						0.25		
0.06199	500	20	2.615			$7.16 \cdot 10^{-3}$			0.20		
0.03100	250	40	3.594			$1.81 \cdot 10^{-2}$			0.32		
0.02480	200	50	3.461			$5.77 \cdot 10^{-3}$			0.30		
0.01727	139.27	71.80	3.3922			$4.34 \cdot 10^{-3}$			0.30		
0.01168	94.21	106.1	3.3621			$4.26 \cdot 10^{-3}$			0.29		
0.006199	50.00	200	3.3447			$1.3 \cdot 10^{-4}$			0.29		
0.004133	33.33	300	3.3413						0.29		
0.001240	10.00	1000	3.3319						0.29		

Indium Antimonide (InSb) - [Ref. 6]

E/eV	$\bar{\nu}$ /cm^{-1}	λ/µm	n	n_a	n_c	k	k_a	k_c	R	R_a	R_c
155		0.007999				$4.77 \cdot 10^{-3}$					
60		0.02066				$7.30 \cdot 10^{-2}$					
25		0.04959	1.15			.015					
24		0.05166	1.15			0.18					
15		0.08266	0.97			0.230					
10		0.1240	0.74			0.88					
5.00	40330	0.2480	1.307			2.441			0.53		
4.50	36290	0.2755	1.443			2.894			0.60		
4.00	32260	0.3100	2.632			3.694			0.61		
3.34	26940	0.3712	3.528			2.280			0.45		
2.84	22910	0.4366	3.340			2.021			0.45		
1.80	14520	0.6888	4.909			1.396			0.47		
1.50	12100	0.8266	4.418			0.643			0.41		
0.6	4839	2.066	4.03						0.36		
0.2480	2000	5.0	4.14			$9.1 \cdot 10^{-2}$			0.37		
0.1907	1538	6.5	4.30			$6.3 \cdot 10^{-2}$			0.39		
0.1653	1333	7.5	4.18			$2.7 \cdot 10^{-2}$			0.38		
0.06199	500	20.00	3.869			$2.0 \cdot 10^{-3}$			0.35		
0.03100	250	40.00	2.98			$2.6 \cdot 10^{-3}$			0.25		
0.02480	200	50.00	2.22			0.165			0.14		
0.02244	181	55.25	3.05			7.59			0.84		
0.02207	178	56.18	9.61			4.20			0.70		
0.02033	164	60.98	4.94			0.140			0.44		
0.01054	85	117.6	2.12			0.423			0.14		
0.005579	45	222.2	1.02			5.59			0.88		
0.001860	15	666.7	6.03			17.9			0.93		
0.001240	10	1000	10.7			24.0			0.94		

Indium Arsenide (InAs) - [Ref. 7]

E/eV	$\bar{\nu}$ /cm^{-1}	λ/µm	n	n_a	n_c	k	k_a	k_c	R	R_a	R_c
25		0.04959				1.139			0.168		
20		0.06199				1.125			0.225		
15		0.08266				0.894			0.336		
10		0.1240				0.835			1.071		
6	48390	0.2066	1.434			2.112			0.45		
5.0	40330	0.2480	1.524			2.871			0.58		

E/eV	$\bar{\nu}$ /cm^{-1}	λ/μm	n	n_a	n_c	k	k_a	k_c	R	R_a	R_c
4.0	32260	0.3100	3.313			1.799			0.39		
3.5	28230	0.3542	3.008			1.754			0.37		
3.0	24200	0.4133	3.197			2.034			0.41		
2.5	20160	0.4959	4.364			1.786			0.45		
2.44	19680	0.5081	4.489			1.446			0.44		
1.86	15000	0.6666	3.889			0.554			0.36		
1.8	14520	0.6888	3.851			0.530			0.35		
1.7	13710	0.7293	3.798			0.493			0.35		
1.6	12900	0.7749	3.755			0.463			0.34		
1.5	12100	0.8266	3.714			0.432			0.34		
1.2	9679	1.033	3.613						0.32		
1.0	8065	1.240	3.548						0.31		
0.6	4839	2.066				0.161					
0.35	2823	3.542	3.608			$9.58 \cdot 10^{-3}$			0.32		
0.32	2581	3.875	3.512			$1.23 \cdot 10^{-4}$			0.31		
0.20	1613	6.199	3.427						0.30		
0.1240	1000	10.00	3.402						0.30		
0.06199	500	20.00	3.334						0.29		
0.04959	400	25.00	3.264						0.28		
0.04339	350	28.57	3.182			$5.46 \cdot 10^{-3}$			0.27		
0.03720	300	33.33	2.988						0.25		
0.03100	250	40.00	1.970			$6.37 \cdot 10^{-2}$			0.11		
0.02765	222	44.84	5.90			6.53			0.74		
0.02480	200	50.00	6.91			0.30			0.56		
0.01984	160	62.50	5.27			0.41			0.47		
0.01860	150	66.67	5.27			0.51			0.47		
0.01736	140	71.43	3.99			$1.1 \cdot 10^{-2}$			0.36		
0.01488	120	83.33	3.91			$6.6 \cdot 10^{-3}$			0.35		
0.01240	100	100.0	3.85			$4.3 \cdot 10^{-3}$			0.35		
0.009919	80	125.0	3.817						0.34		
0.007439	60	166.7	3.793						0.34		
0.004959	40	250.0	3.778						0.34		
0.002480	20	500	3.769						0.37		
0.001240	10	1000	3.766						0.34		

Indium Phosphide (InP) - [Ref. 8]

E/eV	$\bar{\nu}$ /cm^{-1}	λ/μm	n	n_a	n_c	k	k_a	k_c	R	R_a	R_c
20		0.06199	0.793			0.494					
15		0.08266	0.695			0.574					
10		0.1240	0.806			1.154					
5.5	44360	0.2254	1.426			2.562			0.79		
5.0	40330	0.2480	2.131			3.495			0.61		
4.0	32260	0.3100	3.141			1.730			0.38		
3.0	24200	0.4133	4.395			1.247			0.43		
2.0	16130	0.6199	3.549			0.317			0.32		
1.5	12100	0.8266	3.456			0.203			0.31		
1.25	10085	0.9915	3.324						0.29		
1.00	8068	1.239	3.220						0.28		
0.50	4034	2.479	3.114						0.26		
0.30	2420	4.131	3.089						0.26		
0.10	806.8	12.39	3.012						0.25		
0.075	605.1	16.53	2.932						0.24		
0.060	484.1	20.66	2.780			$1.46 \cdot 10^{-2}$			0.22		
0.050	403.4	24.79	2.429			$3.35 \cdot 10^{-2}$			0.17		
0.03992	322	31.06	0.307			3.57					
0.03496	282	35.46	3.89			0.282			0.35		
0.03100	250	40.00	4.27			$3.0 \cdot 10^{-2}$			0.39		
0.02728	220	45.45	3.93			$1.3 \cdot 10^{-2}$			0.35		

E/eV	$\bar{\nu}$/cm^{-1}	λ/μm	n	n_a	n_c	k	k_a	k_c	R	R_a	R_c
0.02480	200	50.0	3.81			$8.7 \cdot 10^{-3}$			0.34		
0.02418	195	51.28	3.19						0.27		
0.02232	180	55.56	3.19						0.27		
0.01860	150	66.67	3.65						0.32		
0.01240	100	100	3.57						0.32		
0.009919	80	125.0	3.551						0.31		
0.007439	60	166.7	3.538						0.31		
0.004959	40	250.0	3.529						0.31		
0.002480	20	500	3.523						0.31		
0.001240	10	1000.0	3.522						0.31		
Lead Selenide (PbSe) - [Ref. 9]											
14.5		0.08551	0.72			0.20					
10		0.1240	0.68			0.50					
5	40330	0.2480	0.54			1.2					
2.0	16130	0.6199	3.65			2.9			0.51		
1.65	13310	0.7514	4.51			1.73			0.46		
1.5	12100	0.8266	4.64			2.64			0.52		
1.0	8065	1.240	4.65			1.1			0.44		
0.75	6049	1.653				0.269					
0.62	5001	2.000	4.59			0.770			0.42		
0.48	3871	2.583	4.90						0.44		
0.40	3226	3.100	4.91						0.44		
0.32	2581	3.875	4.98			0.173			0.44		
0.20	1613	6.199	4.82						0.43		
0.1190	960	10.42	4.74			$1.20 \cdot 10^{-3}$			0.42		
0.09919	800	12.50	4.72			$2.09 \cdot 10^{-3}$			0.42		
0.07935	640	15.63	4.68			$4.12 \cdot 10^{-3}$			0.42		
0.05951	480	20.83	4.59			$1.00 \cdot 10^{-2}$			0.41		
0.04959	400	25.00	4.49			$1.77 \cdot 10^{-2}$			0.40		
0.03968	320	31.25	4.31			$3.62 \cdot 10^{-2}$			0.39		
0.02976	240	41.67	3.89			$9.61 \cdot 10^{-2}$			0.24		
0.01984	160	62.50	2.34			0.56			0.18		
0.009919	80	125.0	1.73			7.38			0.88		
0.007935	64	156.3	2.91			10.1			0.90		
0.004959	40	250.0	11.2			14.6			0.88		
0.002480	20	500.0	12.6			12.2					
0.001736	14	714.3	14.1			16.6					
0.001240	10	1000	17.4			21.1					
Lead Sulfide (PbS) - [Ref. 10]											
150		0.008266				$3.86 \cdot 10^{-3}$					
125		0.009919				$5.59 \cdot 10^{-3}$					
100		0.01240				$1.54 \cdot 10^{-2}$					
80		0.01550				$2.88 \cdot 10^{-2}$					
60		0.02066				$6.17 \cdot 10^{-2}$					
25		0.04959	0.845			0.171					
18.0		0.06888	0.846			0.294					
14.0		0.08856	0.651			0.665					
10.0		0.1240	0.879			1.050					
4.95	39920	0.2505	1.52			2.10			0.43		
4.0	32260	0.3100	1.73			2.83			0.55		
3.00	24200	0.4133	3.88			3.00			0.53		
2.90	23390	0.4275	4.12			2.70			0.51		
2.75	22180	0.4509	4.25			2.33			0.48		
2.55	20570	0.4862	4.35			2.00			0.47		

E/eV	$\tilde{\nu}$ /cm^{-1}	λ/μm	*n*	n_a	n_c	*k*	k_a	k_c	*R*	R_a	R_c
2.00	16130	0.6199	4.29			1.48			0.43		
1.60	12910	0.7749	4.62			0.94			0.43		
1.24	10000	1.00	4.43			0.597			0.41		
1.03	8333	1.2	4.30			0.458			0.39		
0.650	5263	1.9	4.24			0.318			0.39		
0.496	4000	2.5	4.30			0.235			0.39		
0.400	3226	3.1	4.30			$2.27{\cdot}10^{-2}$			0.39		
0.3100	2500	4.0	4.16			$6.38{\cdot}10^{-4}$			0.38		
0.2480	2000	5	4.115			$9.25{\cdot}10^{-4}$			0.37		
0.1240	1000	10	4.01			$6.32{\cdot}10^{-3}$			0.36		
0.1033	833.3	12	3.90			$1.14{\cdot}10^{-2}$			0.35		
0.08059	650	15.38	3.90						0.35		
0.06819	550	18.18	3.81						0.34		
0.04959	400	25.00	3.53						0.31		
0.03720	300	33.33	2.99						0.25		
0.02480	200.0	50	0.514			1.59					
0.01378	111.1	90	1.175			8.48			0.94		
0.01240	100.0	100	1.79			10.51			0.94		
0.008856	71.43	140	17.41			17.94			0.89		
0.006199	50.0	200	16.27			2.20			0.79		
0.003100	25.00	400	12.96			0.495			0.73		
0.001653	13.33	750	12.44			0.228			0.72		
0.001240	10.00	1000	12.35			0.167			0.72		
0.0006199	5.000	2000	12.27			0.0815			0.72		

Lead Telluride (PbTe) - [Ref. 11]

E/eV	$\tilde{\nu}$ /cm^{-1}	λ/μm	*n*	n_a	n_c	*k*	k_a	k_c	*R*	R_a	R_c
150		0.008266				$2.37{\cdot}10^{-3}$					
125		0.009919				$9.71{\cdot}10^{-3}$					
100		0.01240				$4.39{\cdot}10^{-2}$					
75		0.01653				$6.43{\cdot}10^{-2}$					
50		0.02480				$6.87{\cdot}10^{-2}$					
30		0.04133				$7.77{\cdot}10^{-2}$					
15		0.08266	0.72			0.17					
10		0.1240	0.66			0.60					
7.5		0.1653	0.8			0.92					
5.0	40330	0.2480	0.72			1.0					
3.0	24200	0.4133	1.0			2.2					
2.5	20160	0.4959	1.35			2.86			0.61		
1.5	12100	0.8266	3.8			3.1			0.53		
1.0	8065	1.240	4.55			2.2			0.49		
0.80	6452	1.550	6.25			0.71			0.53		
0.60	4839	2.066	6.10			0.521			0.52		
0.40	3226	3.100	6.075			0.331			0.52		
0.30	2420	4.133	5.95			$3.55{\cdot}10^{-2}$			0.51		
0.20	1613	6.199	5.77						0.50		
0.15	1210	8.266	5.76						0.50		
0.1017	820	12.20	5.47			$9.16{\cdot}10^{-3}$			0.48		
0.08927	720	13.89	5.38			$1.37\ 10^{-2}$			0.47		
0.06943	560	17.86	5.13			$3.06{\cdot}10^{-2}$			0.45		
0.04959	400	25.00	4.50			$9.6{\cdot}10^{-2}$			0.40		
0.03968	320	31.25	3.58			0.23			0.32		
0.02976	240	41.67	1.01			1.9					
0.009919	80	125.0	2.95			16.6			0.96		
0.007439	60	166.7	4.9			22.5			0.96		
0.006199	50	200.0	6.9			27.2			0.97		
0.004959	40	250.0	11.6			34.8			0.97		
0.003720	30	333.3	27.7			35.7			0.95		

E/eV	$\bar{\nu}$ /cm^{-1}	λ/μm	*n*	n_a	n_c	*k*	k_a	k_c	*R*	R_a	R_c
0.002480	20	500.0	27.6			39.1			0.95		
0.001240	10	1000	45.1			57.8			0.97		

Lithium Fluoride (LiF) - [Ref. 12]

E/eV	$\bar{\nu}$ /cm^{-1}	λ/μm	*n*	n_a	n_c	*k*	k_a	k_c	*R*	R_a	R_c
2000		$6.199 \cdot 10^{-4}$	0.9999347			$4.33 \cdot 10^{-6}$					
1496		$8.287 \cdot 10^{-4}$	0.999883			$1.28 \cdot 10^{-5}$					
1016		$1.220 \cdot 10^{-3}$	0.999757			$5.18 \cdot 10^{-5}$					
725		$1.710 \cdot 10^{-3}$	0.999643			$1.62 \cdot 10^{-4}$					
504		$2.460 \cdot 10^{-3}$	0.999162			$4.96 \cdot 10^{-5}$					
303		$4.092 \cdot 10^{-3}$	0.99752			$3.12 \cdot 10^{-4}$					
250		$4.959 \cdot 10^{-3}$	0.99632			$6.17 \cdot 10^{-5}$					
200		$6.199 \cdot 10^{-3}$				$2.12 \cdot 10^{-3}$					
150		$8.265 \cdot 10^{-3}$	0.9899			$3.54 \cdot 10^{-3}$					
100		$1.240 \cdot 10^{-2}$	0.9801			$1.32 \cdot 10^{-2}$					
75		$1.653 \cdot 10^{-2}$				$2.63 \cdot 10^{-2}$					
50		$2.480 \cdot 10^{-2}$				$7.89 \cdot 10^{-2}$					
25		$4.959 \cdot 10^{-2}$	0.558			0.521					
20		$6.199 \cdot 10^{-2}$	1.20			0.58			0.10		
15.1		$8.211 \cdot 10^{-2}$	1.08			0.68			0.10		
13		$9.537 \cdot 10^{-2}$	1.04			1.64					
12.0		0.1033	2.28			0.11			0.15		
11.0		0.1127	1.77			$8.07 \cdot 10^{-7}$			0.08		
10.00		0.12398	1.606			$7.70 \cdot 10^{-7}$			0.05		
9		0.1375	1.53						0.04		
7		0.1771	1.46								
4.959	40000	0.250	1.4189						0.03		
4.000	32260	0.31	1.4073						0.03		
2.952	23810	0.42	1.3978						0.03		
2.000	16130	0.62	1.3915						0.03		
0.9919	8000	1.25	1.3851								
0.7999	6452	1.55	1.3858						0.03		
0.4959	4000	2.5	1.3731						0.02		
0.4000	3226	3.1	1.3650								
0.3100	2500	4.0	1.3493								
0.2480	2000	5.0	1.3266			$1.8 \cdot 10^{-6}$			0.02		
0.2000	1613	6.2	1.2912								
0.1698	1370	7.3	1.2499								
0.1494	1205	8.3	1.2036								
0.1240	1000	10.0	1.1005			$2.6 \cdot 10^{-3}$					
0.1127	909.1	11.0	1.0208			$8.0 \cdot 10^{-3}$					
0.1033	833.3	12.0				$1.9 \cdot 10^{-2}$					
0.09537	769.2	13.0				$3.7 \cdot 10^{-2}$					
0.08679	700	14.29	0.508			$7.74 \cdot 10^{-2}$					
0.07439	600	16.67	0.124			0.804					
0.06199	500	20.00	0.306			1.47			0.68		
0.05579	450	22.22	0.191			1.88			0.85		
0.04959	400	25.00	0.208			2.71			0.91		
0.03720	300	33.33	8.76			3.91			0.68		
0.03100	250	40.00	4.64			0.287			0.42		
0.02480	200	50.00	3.69			0.102			0.33		
0.01240	100.0	100	3.067			0.106			0.26		
0.06199	50.0	200	3.067			$4.0 \cdot 10^{-2}$			0.26		
0.04959	40.00	250	3.067			$2.2 \cdot 10^{-2}$			0.26		
0.02480	20.00	500	3.067			$6.3 \cdot 10^{-3}$					
0.01378	11.11	900				$3.1 \cdot 10^{-3}$					
$4.798 \cdot 10^{-4}$	3.870	2584	3.023			$1.19 \cdot 10^{-3}$			0.25		
$1.464 \cdot 10^{-4}$	1.181	8469	3.023			$6.20 \cdot 10^{-4}$			0.25		

E/eV	$\bar{\nu}$ /cm^{-1}	λ/μm	n	n_a	n_c	k	k_a	k_c	R	R_a	R_c
$4.053 \cdot 10^{-5}$	0.3269	30590	3.023			$2.63 \cdot 10^{-4}$			0.25		
$1.861 \cdot 10^{-7}$	$1.501 \cdot 10^{-3}$	$6.662 \cdot 10^{6}$	3.018			$1.6 \cdot 10^{-5}$					
$3.718 \cdot 10^{-8}$	$2.999 \cdot 10^{-4}$	$3.335 \cdot 10^{7}$	3.018			$1.6 \cdot 10^{-5}$					
						Potassium Chloride (KCl) - [Ref. 13]					
2860.3		$4.3347 \cdot 10^{-4}$				$3.93 \cdot 10^{-6}$					
2855.3		$4.3423 \cdot 10^{-4}$				$3.39 \cdot 10^{-6}$					
2849.3		$4.3514 \cdot 10^{-4}$				$4.61 \cdot 10^{-6}$					
2835.8		$4.3721 \cdot 10^{-4}$				$5.85 \cdot 10^{-6}$					
2832.3		$4.3775 \cdot 10^{-4}$				$5.85 \cdot 10^{-6}$					
2829.8		$4.3814 \cdot 10^{-4}$				$1.57 \cdot 10^{-6}$					
2828.3		$4.3837 \cdot 10^{-4}$				$4.19 \cdot 10^{-7}$					
219		$5.661 \cdot 10^{-3}$				$1.82 \cdot 10^{-3}$					
215		$5.767 \cdot 10^{-3}$				$1.84 \cdot 10^{-3}$					
212.5		$5.834 \cdot 10^{-3}$				$2.19 \cdot 10^{-3}$					
211		$5.876 \cdot 10^{-3}$				$1.82 \cdot 10^{-3}$					
185.1		$6.7 \cdot 10^{-3}$	0.99874						$1.01 \cdot 10^{-3}$		
109.7		$1.13 \cdot 10^{-2}$	0.99578						$4.22 \cdot 10^{-3}$		
43		0.02883	0.96			$3.0 \cdot 10^{-2}$					
40		0.03179	0.925			$1.8 \cdot 10^{-2}$					
29.9		0.04147	0.756			0.145					
20.1		0.06168	0.910			0.495					
15.1		0.08211	0.965			0.344					
10.0		0.1240	1.16			0.38			0.035		
9.0		0.1378	1.99			0.50			0.13		
8.0		0.1550	1.15			0.46			0.048		
7.0		0.1771	2.0			$8.46 \cdot 10^{-7}$			0.11		
6.199	50000	0.20	1.71739						0.070		
4.959	40000	0.25	1.58972								
3.999	32260	0.31	1.54005								
2.952	23810	0.42	1.50701								
2.695	21740	0.46	1.50115						0.040		
2.616	21100	0.474				$7.6 \cdot 10^{-11}$					
2.384	19230	0.52	1.49501								
2.066	16670	0.60	1.48969						0.039		
1.550	12500	0.80	1.48291						0.038		
1.033	8333	1.2	1.47813						0.037		
0.5166	4167	2.4	1.47464						0.037		
0.2480	2000	5.0	1.47048						0.036		
0.2000	1.613	6.2	1.46796						0.036		
0.1512	1220	8.2	1.46260						0.035		
0.09999	806.5	12.4	1.44611						0.033		
0.07560	609.8	16.4	1.42295						0.030		
0.04959	400.0	25.0	1.34059			$6.57 \cdot 10^{-4}$			0.021		
0.03999	322.6	31.0	1.2431						0.012		
0.02976	240	41.67	0.85			0.16					
0.02728	220	45.45	0.53			0.35					
0.02232	180	55.56	0.31			1.05					
0.01860	150	66.67	0.44			4.0					
0.01612	130	76.92	4.1			0.32			0.37		
0.01240	100	100.0	2.7			0.11			0.21		
0.008679	70	142.9	2.4			$9.2 \cdot 10^{-2}$			0.17		
0.006199	50	200.0	2.2						0.14		
0.001240	10.00	1000				$9.0 \cdot 10^{-3}$					
0.0006199	5.000	2000				$3.7 \cdot 10^{-3}$					
0.0004133	3.333	3000				$2.0 \cdot 10^{-3}$					

E/eV	ν̄ /cm⁻¹	λ/μm	n	n_a	n_c	k	k_a	k_c	R	R_a	R_c
Silicon Dioxide (Glass) - [Ref. 14]											
2000		$6.199 \cdot 10^{-4}$	0.99993			$1.503 \cdot 10^{-5}$					
1860		$6.665 \cdot 10^{-4}$	0.99991			$1.936 \cdot 10^{-5}$					
1609		$7.705 \cdot 10^{-4}$	0.99989			$9.941 \cdot 10^{-6}$					
1496		$8.287 \cdot 10^{-4}$	0.99987			$1.308 \cdot 10^{-5}$					
1204		$1.030 \cdot 10^{-3}$	0.99980			$2.916 \cdot 10^{-5}$					
1093		$1.134 \cdot 10^{-3}$	0.99975			$4.155 \cdot 10^{-5}$					
1016		$1.220 \cdot 10^{-3}$	0.99971			$5.423 \cdot 10^{-5}$					
798		$1.554 \cdot 10^{-3}$	0.99954			$1.289 \cdot 10^{-4}$					
597		$2.077 \cdot 10^{-3}$	0.99917			$3.560 \cdot 10^{-4}$					
396		$3.131 \cdot 10^{-3}$	0.99812			$4.04 \cdot 10^{-4}$					
303		$4.092 \cdot 10^{-3}$	0.99678			$9.91 \cdot 10^{-4}$					
201		$6.168 \cdot 10^{-3}$	0.99269			$3.63 \cdot 10^{-3}$					
151.2		$8.2 \cdot 10^{-3}$	0.9871			$7.3 \cdot 10^{-3}$					
99.99		$1.24 \cdot 10^{-2}$	0.9813			$7.0 \cdot 10^{-3}$					
49.59		$2.50 \cdot 10^{-2}$	0.9164			$6.5 \cdot 10^{-2}$					
40.00		$3.10 \cdot 10^{-2}$	0.907			$9.2 \cdot 10^{-2}$					
31.00		$4.00 \cdot 10^{-2}$	0.851			0.156					
25.00		0.04959	0.733			0.325					
20.00		0.06199	0.859			0.585					
15.00		0.08266	1.168			0.711			0.10		
13.00		0.09537	1.368			0.747			0.11		
11.00		0.1127	1.739			0.569			0.11		
10.00		0.1240	2.330			0.323			0.17		
9.00		0.1378	1.904			$1.89 \cdot 10^{-2}$			0.097		
7.00		0.1771	1.600						0.053		
6.00	48390	0.2066	1.543						0.046		
4.9939	40278.4	0.248272	1.50841						0.041		
4.1034	33096.1	0.302150	1.48719						0.038		
3.0640	24712.3	0.404656	1.46961						0.036		
2.5504	20570.5	0.486133	1.46313						0.035		
2.4379	19662.5	0.508582	1.46187						0.035		
2.2705	18312.5	0.546074	1.46008						0.035		
2.1489	17332.3	0.576959	1.45885						0.035		
2.1411	17269.2	0.579065	1.45877						0.035		
2.1102	17019.5	0.587561	1.45847						0.035		
2.1041	16970.4	0.589262	1.45841						0.035		
1.9257	15531.6	0.643847	1.45671						0.035		
1.8892	15237.6	0.656272	1.45637						0.035		
1.8566	14974.2	0.667815	1.45608						0.034		
1.7549	14153.9	0.706519	1.45515						0.034		
1.4550	11735.6	0.852111	1.45248						0.034		
1.0985	8860.06	1.12866	1.44888						0.034		
0.60243	4858.9	2.0581	1.43722						0.032		
0.35354	2851.4	3.5070	1.40568						0.028		
0.2976	2400	4.176	1.383			$1.07 \cdot 10^{-4}$			0.026		
0.2728	2200	4.545	1.365			$2.56 \cdot 10^{-4}$			0.024		
0.2480	2000	5.000	1.342			$3.98 \cdot 10^{-3}$			0.021		
0.2232	1800	5.556	1.306			$5.63 \cdot 10^{-3}$					
0.1984	1600	6.250	1.239			$6.52 \cdot 10^{-3}$					
0.1736	1400	7.143	1.053			$1.06 \cdot 10^{-2}$					
0.1674	1350	7.407	0.9488			$1.48 \cdot 10^{-2}$					
0.1612	1300	7.692	0.7719			$3.72 \cdot 10^{-2}$					
0.1500	1210	8.265	0.4530			0.704			0.30		
0.1401	1130	8.850	0.3563			1.53			0.66		
0.1302	1050	9.524	2.760			1.65			0.35		
0.1209	975	10.26	2.448			0.231			0.18		

E/eV	$\bar{\nu}$ /cm^{-1}	λ/μm	n	n_a	n_c	k	k_a	k_c	R	R_a	R_c
0.1091	880	11.36	1.784			$7.75 \cdot 10^{-2}$			0.079		
0.09919	800	12.50	1.753			0.343			0.089		
0.08989	725	13.79	1.698			0.175			0.071		
0.06943	560	17.86	1.337			0.298			0.036		
0.06199	500	20.00	0.6616						0.882		
0.04959	400	25.0	2.739			0.397			0.23		
0.03720	300	33.33	2.210			$6.7 \cdot 10^{-2}$			0.14		
0.01240	100	100.0	1.967			$1.59 \cdot 10^{-2}$			0.11		
0.007439	60	166.7	1.959			$8.62 \cdot 10^{-3}$			0.11		
0.002480	20	500.0	1.955			$7.96 \cdot 10^{-3}$			0.10		

Silicon Monoxide (Noncrystalline) - [Ref. 15]

E/eV	$\bar{\nu}$ /cm^{-1}	λ/μm	n	n_a	n_c	k	k_a	k_c	R	R_a	R_c
25		0.04959	0.8690			0.2717					
20		0.06199	0.8853			0.4919					
17.5		0.07085	0.9825			0.5961					
15		0.08266	1.132			0.6651			0.092		
12.5		0.09919	1.283			0.6523			0.090		
10		0.1240	1.378			0.6843			0.10		
7.5		0.1653	1.593			0.7473			0.12		
5	40330	0.2480	2.001			0.6052			0.15		
4	32260	0.3100	2.141			0.4006			0.15		
3	24200	0.4133	2.116			0.1211			0.13		
2.8	22580	0.4428	2.085			0.08374			0.12		
2.6	20970	0.4769	2.053			0.05544			0.12		
2.4	19360	0.5166	2.021			0.03533			0.11		
2.2	17740	0.5636	1.994			0.02153			0.11		
2	16130	0.6199	1.969			0.01175			0.11		
1.8	14520	0.6888	1.948			0.00523			0.10		
1.6	12900	0.7749	1.929			0.00151			0.10		
1.240	10000	1.000	1.87						0.092		
0.6199	5000	2.000	1.84						0.087		
0.3100	2500	4.000	1.80						0.082		
0.2480	2000	5.000	1.75						0.074		
0.2066	1667	6.000	1.70						0.067		
0.1771	1492	7.000	1.60						0.053		
0.1653	1333	7.500	1.42								
0.1459	1176	8.500	0.90			0.18					
0.1305	1053	9.500	1.20			1.20			0.024		
0.1240	1000	10.00	2.00			1.38			0.27		
0.1181	952.4	10.50	2.85			0.90			0.27		
0.1153	930.2	10.75	2.86			0.58			0.25		
0.1127	909.1	11.00	2.82			0.40			0.24		
0.1078	869.6	11.50	2.50			0.20			0.19		
0.1033	833.3	12.00	2.13			0.14			0.13		
0.09537	769.2	13.00	2.04			0.20			0.12		
0.08856	714.3	14.00	2.01			0.30			0.12		

Noncrystalline Silicon Nitride (Si_3N_4) - [Ref. 16]

E/eV	$\bar{\nu}$ /cm^{-1}	λ/μm	n	n_a	n_c	k	k_a	k_c	R	R_a	R_c
24		0.05166	0.655			0.420			0.28		
23		0.05391	0.625			0.481			0.22		
22		0.05636	0.611			0.560			0.16		
21		0.05904	0.617			0.647			0.19		
20		0.06199	0.635			0.743			0.21		
19		0.06526	0.676			0.841			0.23		
18		0.06888	0.735			0.936			0.26		
17		0.07293	0.810			1.03			0.25		

E/eV	$\bar{\nu}$ /cm^{-1}	λ/μm	*n*	n_a	n_c	*k*	k_a	k_c	*R*	R_a	R_c
16		0.07749	0.902			1.11			0.26		
15		0.08266	1.001			1.18			0.26		
14		0.08856	1.111			1.26			0.26		
13		0.09537	1.247			1.35			0.27		
12	96790	0.1033	1.417			1.43			0.28		
11	88720	0.1127	1.657			1.52			0.29		
10.5	84690	0.1181	1.827			1.53			0.29		
10	80650	0.1240	2.000			1.49			0.29		
9.5	76620	0.1305	2.162			1.44			0.28		
9	72590	0.1378	2.326			1.32			0.27		
8	64520	0.1550	2.651			0.962			0.26		
7	56460	0.1771	2.752			0.493			0.23		
6	48390	0.2066	2.541			0.102			0.19		
5	40330	0.2480	2.278			$4.9 \cdot 10^{-3}$			0.15		
4.75	38310	0.2610	2.234			$1.2 \cdot 10^{-3}$			0.15		
4.5	36290	0.2755	2.198			$2.2 \cdot 10^{-4}$			0.14		
4	32260	0.3100	2.141						0.13		
3.5	28230	0.3542	2.099						0.13		
3	24200	0.4133	2.066						0.12		
2.5	20160	0.4959	2.041						0.12		
2	16130	0.6199	2.022						0.11		
1.5	12100	0.8266	2.008						0.11		
1	8065	1.240	1.998						0.11		
						Sodium Chloride (NaCl) - [Ref. 17]					
209.5		$5.918 \cdot 10^{-3}$				$2.54 \cdot 10^{-3}$					
206		$6.019 \cdot 10^{-3}$				$2.62 \cdot 10^{-3}$					
203		$6.107 \cdot 10^{-3}$				$2.08 \cdot 10^{-3}$					
200		$6.199 \cdot 10^{-3}$				$1.92 \cdot 10^{-3}$					
26.0		0.04769	0.83			0.15			0.015		
25.0		0.04959	0.83			0.18			0.018		
22.0		0.05636	0.83			0.31			0.057		
20.0		0.06199	0.88			0.34			0.036		
18.0		0.06888	0.89			0.33			0.033		
16.1		0.07700	0.74			0.45			0.084		
14.0		0.08856	0.98			0.89			0.17		
12.0		0.1033	1.22			0.79			0.12		
10.0		0.1240	1.55			0.71			0.12		
8.00		0.1550	1.38			1.10			0.20		
6.00	48390	0.2066	1.75						0.074		
5.00	40330	0.2480	1.65						0.060		
2.952	23810	0.42	1.56324						0.048		
2.480	20000	0.50	1.55157						0.047		
2.214	17860	0.56	1.54613						0.046		
2.000	16130	0.62	1.54228						0.045		
1.771	14290	0.70	1.53865						0.045		
1.675	13510	0.74	1.53728						0.045		
1.550	12500	0.80	1.53560						0.045		
1.240	10000	1.00	1.53200						0.044		
1.033	8333	1.2	1.53000						0.044		
0.6888	5556	1.8	1.52712						0.043		
0.4959	4000	2.5	1.52531						0.043		
0.4000	3226	3.1	1.52395						0.043		
0.3263	2632	3.8	1.52226			$(1.8 \pm 0.2) \cdot 10^{-9}$			0.043		
0.2952	2381	4.2	1.52121						0.043		
0.2755	2222	4.5	1.52036						0.043		
0.2480	2000	5.0	1.51883						0.042		

E/eV	$\bar{\nu}$ /cm^{-1}	λ/μm	*n*	n_a	n_c	*k*	k_a	k_c	*R*	R_a	R_c
0.1240	1000	10.0	1.49473						0.039		
0.1033	833.3	12.0	1.48000						0.037		
0.08856	714.3	14.0	1.46188						0.035		
0.07749	625.0	16.0	1.4399						0.033		
0.06888	555.5	18.0	1.41364						0.029		
0.06199	500.0	20.0	1.3822						0.026		
0.04959	400	25.0	1.27			$3.5 \cdot 10^{-3}$			0.014		
0.04215	340	29.41	1.12			$1.7 \cdot 10^{-2}$			0.0032		
0.03720	300	33.33	0.85			0.85			0.18		
0.03410	275	36.36	0.59			0.22			0.084		
0.03286	265	37.74	0.42			0.50			0.26		
0.03224	260	38.46	0.45			0.45			0.22		
0.02480	200	50.00	0.14			1.99			0.89		
0.02108	170	58.82	1.35			6.03			0.87		
0.01984	160	62.50	6.92			2.14			0.59		
0.01922	155	64.52	5.50			0.87			0.49		
0.01860	150	66.67	4.52			0.380			0.41		
0.01736	140	71.43	3.72			0.219			0.33		
0.01612	130	76.92	3.31			0.135			0.29		
0.01488	120	83.33	3.02			0.110			0.25		
0.01240	100	100.0	2.74			0.087			0.22		
0.009919	80	125.0	2.57			0.077			0.19		
0.07439	60	166.7	2.48			0.055			0.18		
0.04959	40	250.00	2.44			0.041			0.18		
0.002480	20	500.0	2.43			0.024			0.17		
0.001240	10	1000	2.43			0.006			0.17		
0.001033	8.333	1200				$8.8 \cdot 10^{-3}$					
0.0006888	5.556	1800				$5.4 \cdot 10^{-3}$					
0.0006199	5.000	2000	2.43						0.17		
0.0004959	4.000	2500				$4.4 \cdot 10^{-3}$					
0.0004797	3.869	2584	2.43			$2.1 \cdot 10^{-3}$			0.17		
0.0003875	3.125	3200				$3.3 \cdot 10^{-3}$					
0.0001464	1.181	8469	2.43			$5.8 \cdot 10^{-4}$			0.17		
0.00004053	0.3269	30590	2.43			$2.5 \cdot 10^{-4}$					

Cubic Zinc Sulfide (ZnS) - [Ref. 18]

E/eV	$\bar{\nu}$ /cm^{-1}	λ/μm	*n*	n_a	n_c	*k*	k_a	k_c	*R*	R_a	R_c
2000		$6.199 \cdot 10^{-4}$	0.999904			$1.76 \cdot 10^{-5}$					
1204		$1.030 \cdot 10^{-3}$	0.999777			$1.00 \cdot 10^{-4}$					
1016		$1.220 \cdot 10^{-3}$	0.999838			$3.61 \cdot 10^{-5}$					
901		$1.376 \cdot 10^{-3}$	0.999647			$5.42 \cdot 10^{-5}$					
798		$1.554 \cdot 10^{-3}$	0.999520			$8.28 \cdot 10^{-5}$					
707		$1.754 \cdot 10^{-3}$	0.999372			$1.25 \cdot 10^{-4}$					
597		$2.077 \cdot 10^{-3}$	0.999160			$2.19 \cdot 10^{-4}$					
377		$9.50 \cdot 10^{-3}$	0.99789			$9.50 \cdot 10^{-4}$					
201		$6.168 \cdot 10^{-3}$	0.99553			$4.82 \cdot 10^{-3}$					
100		$1.240 \cdot 10^{-2}$	0.99061			$1.17 \cdot 10^{-2}$					
61.99		$2.000 \cdot 10^{-2}$	0.964			$3.32 \cdot 10^{-2}$			$6.2 \cdot 10^{-4}$		
41.33		$3.000 \cdot 10^{-2}$	0.941			$5.10 \cdot 10^{-2}$					
31.00		$4.000 \cdot 10^{-2}$	0.847			$9.95 \cdot 10^{-2}$					
24.80		$5.000 \cdot 10^{-2}$	0.796			0.171			$2.2 \cdot 10^{-2}$		
17.71		$7.000 \cdot 10^{-2}$	0.747			0.431			$7.7 \cdot 10^{-2}$		
13.78		$9.000 \cdot 10^{-2}$	0.758			0.824			0.20		
12.40		0.1000	0.862			0.876			0.19		
9.919		0.125	1.02			1.36			0.31		
8.266		0.150	1.41			1.47			0.29		
6.199		0.200	2.32			1.62			0,32		
6.00	48390	0.2066	2.24			1.65			0.59		

E/eV	$\bar{\nu}$ /cm^{-1}	λ/μm	*n*	n_a	n_c	*k*	k_a	k_c	*R*	R_a	R_c
4.00	32260	0.3100	2.70			0.44			0.22		
3.00	24200	0.4133	2.54			$4{\cdot}10^{-2}$			0.19		
2.50	20160	0.4959	2.42			$3{\cdot}10^{-2}$			0.17		
2.30	18550	0.5391	2.3950						0.17		
2.00	16130	0.6199	2.3576						0.16		
1.75	14110	0.7085	2.3319						0.16		
1.55	12500	0.7999	2.3146			$3.50{\cdot}10^{-6}$			0.16		
1.40	11290	0.8856	2.3033						0.16		
1.240	10000	1.000	2.2907			$3.02{\cdot}10^{-6}$			0.15		
1.00	8065	1.240	2.2795						0.15		
0.80	6452	1.550	2.2706						0.15		
0.6199	5000	2.000	2.2631			$6.2{\cdot}10^{-6}$			0.15		
0.45	3629	2.755	2.2587						0.15		
0.30	2420	4.133	2.2529						0.15		
0.20	1613	6.199	2.2443						0.15		
0.1550	1250	8.0	2.2213			$4.5{\cdot}10^{-6}$			0.14		
0.1240	1000	10.00	2.1986			$8.8{\cdot}10^{-6}$			0.14		
0.100	806.5	12.4	2.1969						0.14		
0.09	725.9	13.78	2.1793						0.14		
0.07999	645.2	15.5	2.1518			$3.82{\cdot}10^{-3}$			0.14		
0.07	564.6	17.71	2.1040						0.13		
0.06075	490	20.41	2.03			$8.0{\cdot}10^{-3}$			0.12		
0.05	403.3	24.80	1.6866						0.065		
0.03546	286	34.97	3.29			$8.3{\cdot}10^{-2}$			0.28		
0.03472	280	35.71	9.54			$5.2{\cdot}10^{-2}$			0.66		
0.02480	200	50.00	3.48			$3.1{\cdot}10^{-2}$			0.31		
0.01240	100	100.0	3.06			$5.8{\cdot}10^{-3}$			0.26		
0.004955	40	250.0	2.903			$6.2{\cdot}10^{-3}$			0.24		
0.004339	35	285.7	2.899			$7.0{\cdot}10^{-3}$			0.24		
0.003720	30	333.3	2.896						0.24		
0.003100	25	400.0	2.894						0.24		
0.002480	20	500.0	2.892						0.24		
0.001860	15	666.7	2.890						0.24		

Polytetrafluoroethylene (Teflon) - [Ref. 19]

E/eV	$\bar{\nu}$ /cm^{-1}	λ/μm	*n*	n_a	n_c	*k*	k_a	k_c	*R*	R_a	R_c
4.960	40000	0.250							0.970		
4.769	38462	0.260							0.972		
4.593	37037	0.270							0.975		
4.426	35714	0.280							0.978		
4.276	34483	0.290							0.980		
4.133	33333	0.300							0.983		
4.000	32258	0.310							0.986		
3.875	31250	0.320							0.988		
3.758	30303	0.330							0.990		
3.647	29412	0.340							0.991		
3.543	28571	0.350							0.992		
3.444	27778	0.360							0.992		
3.351	27027	0.370							0.993		
2.255	18182	0.550							0.993		
2.067	16667	0.600							0.992		
1.378	11111	0.900							0.992		
1.305	10526	0.950							0.991		
1.078	8696	1.150							0.991		
1.033	8333	1.200							0.990		
0.9920	8000	1.250							0.990		
0.9538	7692	1.300							0.989		
0.9185	7407	1.350							0.988		

E/eV	$\bar{\nu}$ /cm^{-1}	λ/μm	n	n_a	n_c	k	k_a	k_c	R	R_a	R_c
0.8857	7143	1.400							0.988		
0.8552	6897	1.450							0.989		
0.8267	6667	1.500							0.989		
0.8000	6452	1.550							0.988		
0.7750	6250	1.600							0.988		
0.7515	6061	1.650							0.987		
0.7294	5882	1.700							0.986		
0.7086	5714	1.750							0.986		
0.6889	5556	1.800							0.985		
0.6703	5405	1.850							0.980		
0.6526	5263	1.900							0.978		
0.6359	51282	1.950							0.978		
0.6200	5000	2.000							0.970		
0.6049	4878	2.050							0.959		
0.5905	4762	2.100							0.951		
0.5767	4651	2.150							0.946		
0.5636	4545	2.200							0.966		
0.5511	44444	2.250							0.965		
0.5487	44247	2.260							0.964		
0.5439	4386	2.280							0.963		
0.5415	4367	2.290							0.961		
0.5368	4329	2.310							0.959		
0.5345	4310	2.320							0.957		
0.5322	4292	2.330							0.956		
0.5299	4274	2.340							0.954		
0.5277	4255	2.350							0.951		
0.5232	4219	2.370							0.950		
0.5188	4184	2.390							0.949		
0.5167	4167	2.400							0.947		
0.5061	4082	2.450							0.946		
0.4960	4000	2.500							0.945		

REFERENCES

1. Arsenic Selenide
   D. J. Treacy in *Handbook of Optical Constants of Solids,* E. D. Palik, Editor, Academic Press, 1985, p. 623. (Hereafter abbreviated as *HOCS.*)
   R. Zallen, R. E. Drews, R. L. Emerald, and M. L. Slade, *Phys. Rev. Lett.* 26, 1564 (1971)
   R. Zallen, M. L. Slade, and A. T. Ward, *Phys. Rev.* B 3, 4257 (1971).
   U. Strom and P. C. Taylor, *Phys. Rev.* B 16, 5512 (1977).
   G. Lucovsky, *Phys. Rev.* B 6, 1480 (1972).
   C. T. Moynihan, P. B. Macedo, M. S. Maklad, R. K. Mohr, and R. E. Howard, *J. Non-Cryst. Solids*, 17, 369 (1975).
   Y. Ohmachi, *J. Opt. Soc. Am.* 63, 630 (1973).
2. Arsenic Sulfide
   D. J. Treacy in *HOCS,* 1985, p. 641.
   P. A. Young, *J. Phys.* C 4, 93 (1971).
   W. S. Rodny, I. H. Malitson, and T. A. King, *J. Opt. Soc. Am.* 48, 633 (1958).
   R. Zallen, R.E. Drew, R. L. Emerald, and M.L. Slade, *Phys. Rev. Lett.* 26, 1564 (1971).
   M. S. Maklad, R. K. Mohr, R. E. Howard, P. B. Macedo, and C. T. Moynihan, *Solid State Commun.* 15, 855 (1974).
   P. B. Klein, P. C. Taylor, and D. J. Treacy, *Phys. Rev.* B16, 4511 (1977).
   G. Lucovsky, *Phys. Rev.* B 6, 1480 (1972).
3. Cadmium Telluride
   E. D. Palik in *HOCS,* 1985, p. 409.
   D. T. F. Marple and H. Ehrenreich, *Phys. Lett.* 8, 87 (1962).
   T. H. Myers, S. W. Edwards, and J. F. Schetzina, *J. Appl. Phys.* 52, 4231 (1981).
   D. T. F. Marple, *Phys. Rev.* 150, 728 (1966).
   A. N. Pikhtin and A. D. Yas'kov, *Sov. Phys. Semicond.* 12, 622 (1978).

L. S. Ladd, *Infrared Phys.* 6, 145 (1966).
J. E. Harvey and W. L. Wolfe, *J. Opt. Soc. Am.* 65, 1267 (1975).
A. Manabe, A. Mitsuishi, and H. Yoshinaga, *Jpn. J. Appl. Phys.* 6, 593 (1967).
A. Manabe, A. Mitsuishi, H. Oshinaga, Y. Ueda, and H. Sei, *Technol. Rep. Osaka Univ. Jpn.* 17, 263 (1967).
J. R. Birch and D. K. Murrey, *Infrared Phys.* 18, 283 (1978).

4. Gallium Arsenide
E. D. Palik in *HOCS,* 1985, p. 429.
M. Cardona, W. Gudat, B. Sonntag, and P. Y. Yu, in *Proc. Intl. Conf. Phys. Semicond.,* 10th. Cambridge, 1970, p. 208. US Atom. Energy Commission, Oak Ridge, TN, 1970.
H. R. Philipp and H. Ehrenreich, *Phys. Rev.* 129, 1550 (1963).
J. B. Theeten, D. E. Aspnes, and R. P. H. Chang, *J. Appl. Phys.* 49, 6097 (1978).
H. C. Casey, D. D. Sell, and K. W. Wecht, *J. Appl. Phys.* 46, 250 (1975).
A. H. Kachare, W. G. Spitzer, F. K. Euler, and A. Kahan, *J. Appl. Phys.* 45, 2938 (1974).
R. T. Holm, J. W. Gibson, and E. D. Palik, *J. Appl. Phys.* 48, 212 (1977).
W. Cochran, S. J. Fray, F. A. Johnson, J. E. Quarrington, and N. Williams, *J. Appl. Phys. Suppl.* 32, 2102 (1961).
C. P. Christensen, R. Joiner, S. K. T. Nieh, and W. H. Steier, *J. Appl. Phys.* 45, 4957 (1974).
R. H. Stolen, Phys. Rev. B 11, 767 (1975); *Appl. Phys. Lett.* 15, 74 (1969).

5. Gallium Phosphide
A. Borghesi and G. Guizzetti in *HOCS,* 1985, p. 445.
M. Cardona, W. Gudat, B. Sonntag, and P. Y. Yu, *Proc. Intl. Conf. Phys. Semicond.* Cambridge, 1970, p. 208. US Atom. Energy Commission, Oak Ridge, TN, 1970.
M. Cardona, W. Gudat, E. E. Koch, M. Skibowski, B. Sonntag, and P. Yu. *Phys. Rev. Lett.* 25, 659 (1970).
S. E. Stokowski and D. D. Sell, *Phys. Rev.* B 5, 1636 (1972).
S. A. Abagyan, G. A. Ivanov, Y. E. Shanurin, and V. I. Amosov, *Sov. Phys. Semicond.* 5, 889 (1971).
P. G. Dean, G. Kaminsky, and R. B. Zetterstorm, *J. Appl. Phys.* 38, 3551 (1967).
D. E. Aspnes and A. A. Studna, *Phys. Rev.* B 27, 985 (1983).

6. Indium Antimonide
R. T. Holm in *HOCS,* 1985, p. 491.
M. Cardona, W. Gudat, B. Sonntag, and P. Y. Yu, Proc. *Int. Conf. Phys. Semicond.,* 10th. Cambridge, 1970, p. 208. US Atom. Comm., Oak Ridge, TN, 1970.
H. R. Philipp and H. Ehrenreich, *Phys. Rev.* 129, 1550 (1963).
D. E. Aspnes and A. A. Studna, *Phys. Rev.* B 27, 985 (1983).
T. S. Moss, S. D. Smith, and T. D. F. Hawkins, *Proc. Phys. Soc. London* 70B, 776 (1957).
H. Yoshinaga and R. A. Oetjen, *Phys. Rev.* 101, 526 (1956).
R. B. Sanderson, *J. Phys. Chem. Solids* 26, 803 (1965).

7. Indium Arsenide
E. D. Palick and R. T. Holm in *HOCS,* 1985, p. 479.
H. R. Philipp and H. Ehrenreich, *Phys. Rev.* 129, 1550 (1963).
B. O. Seraphin and H. E. Bennett in *Semiconductors and Semimetals* (R. K. Willardson and A. C. Beer, Eds.), vol. 3, Academic, 1967, p. 499.
D. E. Aspnes and A. A. Studna, *Phys. Rev.* B 27, 985 (1983).
J. R. Dixon and J. M. Ellis, *Phys. Rev.* 123, 1560 (1961).
A. Memon, T. J. Parker, and J. R. Birch, *Proc. SPIE,* 289, 20 (1981).

8. Indium Phosphide
O. J. Glembocki and H. Piller in *HOCS,* 1985, p. 503.
M. Cardona, *J. Appl. Phys.* 32, 958 (1961); 36, 2181 (1965).
D. E. Aspnes and A. A. Studna, *Phys. Rev.* B 27, 985 (1983).
G. D. Pettit and W. J. Turner, *J. Appl. Phys.* 36, 2081 (1965).
R. Newman, *Phys. Rev.* 111, 1518 (1958).
W. N. Reynolds, M. T. Lilburne, and R. M. Dell, *Proc. Phys. Soc. London* 71, 416 (1958).
H. Jamshidi and T. J. Parker, Int. Meet. Infrared Mm. Waves, 7th., Marseilles, 1983.

9. Lead Selenide
G. Bauer and H. Krenn in *HOCS,* 1985, p. 517.
M. Cardona and D. L. Greenaway, *Phys. Rev.* A 133, 1685 (1964).
T. S. Moss, *Optical Properties of Semiconductors,* Butterworth, 1959, p. 189.
J. N. Zemel, J. D. Jensen, and R. B. Schoolar, *Phys. Rev.* A 140, 330 (1965).
W. W. Scanlon, *J. Phys. Chem. Solids,* 8, 423 (1959).
K. V. Vyatkin and A. P. Shotov, *Sov. Phys. Semicond.* 14, 785 (1980); Fiz. Tekh. Poluprovodn. 14, 1331 (1980).

10. Lead Sulfide
G. Guizzetti and A. Borghesi in *HOCS,* 1985, p. 525.
M. Cardona and R. Haensel, *Phys. Rev.* B 1, 2605 (1970).

M. Cardona and D. L. Greenaway, *Phys. Rev.* A 133, 1685 (1964).
M. Cardona, C. M. Penchina, E. E. Koch, and P. Y. Yu, *Phys. Status Solidi* B 53, 327 (1972).
P. R. Wessel, *Phys. Rev.* 153, 836 (1967).
C. E. Rossi and W. Paul, *J. Appl. Phys.* 38, 1803 (1967).
J. N. Zemel, J. D. Jensen, and R. B. Schoolar, *Phys. Rev.* A 140, 330 (1965).

11. Lead Telluride
G. Bauer and H. Krenn in *HOCS,* 1985, p. 535.
M. Cardona and R. Haensel, *Phys. Rev.* B 1, 2605 (1970).
M. Cardona and D. L. Greenaway, *Phys. Rev.* 133, A1685 (1964).
D. M. Korn and R. Braunstein, *Phys. Rev.* B 5, 4837 (1972).
W. W. Scanlon, *J. Phys. Chem. Solids* 8, 423 (1959).
J. N. Zemel, J. D. Jensen, and R. B. Schoolar, *Phys. Rev.* 140, A330 (1965).

12. Lithium Fluoride
E. D. Palik and W. R. Hunter in *HOCS,* 1985, p. 675.
B. L. Henke, P. Lee, T. J. Tanaka, R. L. Shimabukuro, and B. K. Fujikawa, *Low Energy X-ray Diagnostics-1981* (D. T. Attwood and B. L. Henke, Eds.), AIP Conf. Proc. No. 75, 1981.
A. P. Lukirskii, E. P. Savinov, O. A. Ershov, and Y. F. Shepelev, *Opt. Spektrosk.* 16, 168 (1964); 16, 310 (1964).
F. C. Brown, C. Gahwiller, A. B. Kunz, and N. O. Lipari, *Phys. Rev. Lett.* 25, 927 (1970).
A. Milgram and M. P. Givens, *Phys. Rev.* 125, 1506 (1962).
T. Tomiki and T. Miyata, *J. Phys. Soc. Jpn.* 27, 658 (1969).
A. Kachare, G. Andermann, and L. R. Brantley, *J. Phys. Chem. Solids* 33, 467 (1972).

13. Potassium Chloride
E. D. Palik in *HOCS,* 1985, p. 703.
O. Aita, I. Nagakura, and T. Sagawa, *J. Phys. Soc. Jpn.* 30, 1414 (1971).
A. P. Lukirskii, E. P. Savinov, O. A. Ershov, and Y. F. Shepelev, *Opt. Spectrosc.* 16, 168 (1964); Opt. Spektrosk. 16, 310 (1964).
T. Tomika, *J. Phys. Soc. Jpn.* 22, 463 (1967).
M. Antinori, A. Balzarotti, and M. Piacentini, *Phys. Rev.* B 7, 1541 (1973).
H. H. Li, *J. Phys. Chem. Ref. Data* 5, 329 (1976).
S. D. Allen and J. A. Harrington, *Appl. Opt.* 17, 1679 (1978).
K. W. Johnson and E. E. Bell, *Phys. Rev.* 139A, 1295 (1965).

14. Silicon Dioxide
H. R. Philipp in *HOCS,* 1985, p. 749.
J. Rife and J. Osantowski, *J. Opt. Soc. Am.* 70, 1513 (1980).
B. L. Henke, P. Lee, T. J. Tanaka, R. L. Shimabukuro, and B. K. Fujikawa, *Low Energy X-ray Diagnostics-1981* (D. T. Attwood and B. L. Henke, Eds.), AIP Conf. Proc. No. 75, 1981.
H. R. Philipp, *Solid State Commun.* 4, 73 (1966); *J. Phys. Chem. Solids,* 32, 1935 (1971).
P. L. Lamy, *Appl. Opt.* 16, 2212 (1977).
H. R. Philipp, *J. Appl. Phys.* 50 1053 (1979).
D. G. Drummond, *Proc. Roy. Soc. London,* 153, 328 (1935).

15. Silicon Monoxide
H. R. Philipp in *HOCS,* 1985, p. 765.
H. R. Philipp, *J. Phys. Chem. Solids,* 32, 1935 (1971).
G. Hass and C. D. Salzberg, *J. Opt. Soc. Am.* 44, 181 (1954).
E. Cremer, T. Kraus, and E. Ritter, *Zs. Electrochem.* 62, 939 (1958).
A. P. Bradford, G. Hass, M. McFarland, and E. Ritter, *Appl. Opt.* 4, 971 (1965).

16. Silicon Nitride
H. R. Philipp in *HOCS,* 1985, p. 771.
H. R. Philipp, *J. Electrochem. Soc.* 120, 295 (1973).
J. B. Theeten, D. E. Aspnes, F. Simondet, M. Errman, and P. C. Mürau, *J. Appl. Phys.* 52, 6788 (1981).
J. Bauer, *Phys. Status Solidi,* A 39, 411 (1977).

17. Sodium Chloride
J. E. Eldridge and E. D. Palik in *HOCS,* p. 775.
J. A. Harrington, C. J. Duthler, F. W. Patten, and M. Hass, *Solid State Commun.* 18, 1043 (1976).
T. Miyata and T. Tomiki, *J. Phys. Soc. Jpn.* 24, 1286 (1968); ibid., 22, 209 (1967).
D. M. Roessler and W. C. Walker, *J. Opt. Soc. Am.* 58, 279 (1968).
D. M. Roessler and W. C. Walker, *Phys. Rev.* 166, 599 (1968).
S. Allen and J. A. Harrington, *Appl. Opt.* 17, 1679 (1978).
O. Aita, I. Nagakura, and T. Sagawa, *J. Phys. Soc. Jpn.* 30, 1414 (1971).

18. Zinc Sulfide
E. D. Palik and A. Addamiano in *HOCS,* 1985, p. 597.

B. L. Henke, P. L. Lee, T. J. Tanaka, R. L. Shimabukuro, and B. F. Fujikawa, *Low Energy X-ray Diagnostics-1981* (D. T. Attwood and B. L. Henke, Eds.), *AIP Conf. Proc.* No. 75, 1981.
M. Cardona and G. Harbeke, *Phys. Rev.* 137, A1467 (1965).
Eastman Kodak, Publ. No. U-72, Rochester, New York (1981).
C. A. Klein and R. N. Donadio, *J. Appl. Phys.* 51, 797 (1980).
T. Deutsch, *Proc. Int. Conf. Phys. Semicond.*, 6th Exeter 1962, p. 505. The Inst. of Physics and the Physical Soc., London, 1962.
A. Manabe, A. Mitsuishi, and H. Yoshinaga, *Jpn. J. Appl. Phys.* 6, 593 (1967).
W. W. Piper, D. T. F. Marple, and P. D. Johnson, *Phys. Rev.* 110, 323 (1958).

19. Polytetrafluoroethylene
J. W. L. Thomas (NIST), Private communication.
NIST Certificate, STM 2044.
P. Y. Barnes, E. A. Early, and A. C. Parr, *NIST Special Publ. 250-48,* NIST Measurement Services: Spectral Reflectance.
Diffuse Reflectance Coatings and Materials Sections, Labsphere Catalog, 1996.
A. Arecchi and C. Ryder (Labsphere, North Sutten, NJ), private communication.

ELASTO-OPTIC, ELECTRO-OPTIC, AND MAGNETO-OPTIC CONSTANTS

When a crystal is subjected to a stress field, an electric field, or a magnetic field, the resulting optical effects are in general dependent on the orientation of these fields with respect to the crystal axes. It is useful, therefore, to express the optical properties in terms of the refractive index ellipsoid (or indicatrix):

$$\frac{x^2}{n_x^2} + \frac{y^2}{n_y^2} + \frac{z^2}{n_z^2} = 1$$

or

$$\sum_{ij} B_{ij} x_i y_j = 1 \quad (i, j = 1, 2, 3)$$

where

$$B_{ij} = \left[\frac{1}{\varepsilon}\right]_{ij} = \left[\frac{1}{n^2}\right]_{ij}$$

ε is the dielectric constant or permeability; the quantity B_{ij} has the name impermeability.

A crystal exposed to a *stress* **S** will show a change of its impermeability. The photo-elastic (or elasto-optic) constants, P_{ijkl}, are defined by

$$\Delta\left[\frac{1}{\varepsilon}\right]_{ij} = \Delta\left[\frac{1}{n^2}\right]_{ij} = \sum_{kl} P_{ijkl} S_{kl}$$

where n is the refractive index and S_{kl} are the strain tensor elements; the P_{ijkl} are the elements of a 4th rank tensor.

When a crystal is subjected to an *electric field* ***E*** two possible changes of the refractive index may occur depending on the symmetry of the crystal.

1. All materials, including isotropic solids and polar liquids, show an electro-optic birefringence (Kerr effect) which is proportional to the square of the electric field, ***E***:

$$\left[\frac{1}{n^2}\right]_{ij} = \sum_{k,l=1,2,3} K_{ijkl} E_k E_l = \sum_{k,l=1,2,3} g_{ijkl} P_k P_l$$

   where E_k and E_l are the components of the electric field and P_k and P_l the electric polarizations. The coefficients, K_{ijkl}, are the quadratic electro-optic coefficients, while the constants g_{ijkl} are known as the Kerr constants.

2. The other electro-optic effect only occurs in the 20 piezo-electric crystal classes (no center of symmetry). This effect is known as the Pockels effect. The optical impermeability changes linearly with the static field

$$\Delta\left[\frac{1}{n^2}\right]_{ij} = \sum_{k} r_{ij,k} E_k$$

The coefficients $r_{ij,k}$ have the name (linear) electro-optic coefficients.

The values of the electro-optic coefficients depend on the boundary conditions. If the superscripts T and S denote respectively the conditions of zero stress (free) and zero strain (clamped) one finds:

$$r_{ij}^{\mathrm{T}} = r_{ij}^{\mathrm{S}} + q_{ik}^{\mathrm{E}} e_{jk} = r_{ij}^{\mathrm{S}} + P_{ik}^{\mathrm{E}} d_{jk}$$

where $e_{jk} = (\partial T_k/\partial E_j)_S$ and $d_{jk} = (\partial S_k/\partial E_j)_T$ are the appropriate piezo-electric coefficients.

The interaction between a *magnetic field* and a light wave propagating in a solid or in a liquid gives rise to a rotation of the plane of polarization. This effect is known as *Faraday rotation*. It results from a difference in propagation velocity for left and right circular polarized light.

The Faraday rotation, θ_F, is linearly proportional to the magnetic field H:

$$\theta_F = VlH$$

where l is the light path length and V is the *Verdet* constant (minutes/oersted·cm).

For ferromagnetic, ferrimagnetic, and antiferromagnetic materials the magnetic field in the above expression is replaced by the magnetization M and the magneto-optic coefficient in this case is known as the Kund constant K:

Specific Faraday rotation $F = KM$

In the tables below the *Faraday rotation* is listed at the saturation magnetization per unit length, together with the absorption coefficient α, the temperature T, the critical temperature T_C (or T_N), and the wavelength of the measurement.

In the tables which follow, the properties are presented in groups:

- Elasto-optic coefficients (photoelastic constants)
- Linear electro-optic coefficients (Pockels constants)
- Quadratic electro-optic coefficients (Kerr constants)
- Magneto-optic coefficients:
  - Verdet constants
  - Faraday rotation parameters

Within each group, materials are classified by crystal system or physical state. References are given at the end of each group of tables.

ELASTO-OPTIC COEFFICIENTS (PHOTOELASTIC CONSTANTS)

Name Cubic (43m, 432, m3m)	Formula	λ/μm	p_{11}	p_{12}	p_{44}	p_{11}-p_{12}	Ref.
Sodium fluoride	NaF	0.633	0.08	0.20	-0.03	-0.12	1
Sodium chloride	NaCl	0.589	0.115	0.159	-0.011	-0.042	2
Sodium bromide	NaBr	0.589	0.148	0.184	-0.0036	-0.035	1
Sodium iodide	NaI	0.589	—	—	0.0048	-0.0141	3
Potassium fluoride	KF	0.546	0.26	0.20	-0.029	0.06	1
Potassium chloride	KCl	0.633	0.22	0.16	-0.025	0.06	4
Potassium bromide	KBr	0.589	0.212	0.165	-0.022	0.047	5
Potassium iodide	KI	0.590	0.212	0.171	—	0.041	6
Rubidium chloride	RbCl	0.589	0.288	0.172	-0.041	0.116	7,8
Rubidium bromide	RbBr	0.589	0.293	0.185	-0.034	0.108	7,8
Rubidium iodide	RbI	0.589	0.262	0.167	-0.023	0.095	7,8
Lithium fluoride	LiF	0.589	0.02	0.13	-0.045	-0.11	5
Lithium chloride	LiCl	0.589	—	—	-0.0177	-0.0407	3
Ammonium chloride	NH_4Cl	0.589	0.142	0.245	0.042	-0.103	9
Cadmium telluride	CdTe	1.06	-0.152	-0.017	-0.057	-0.135	10
Calcium fluoride	CaF_2	0.55-0.65	0.038	0.226	0.0254	-0.183	11
Copper chloride	CuCl	0.633	0.120	0.250	-0.082	-0.130	12
Copper bromide	CuBr	0.633	0.072	0.195	-0.083	-0.123	12
Copper iodide	CuI	0.633	0.032	0.151	-0.068	-0.119	12
Diamond	C	0.540-0.589	-0.278	0.123	-0.161	-0.385	13
Germanium	Ge	3.39	-0.151	-0.128	-0.072	-0.023	14
Gallium arsenide	GaAs	1.15	-0.165	-0.140	-0.072	-0.025	15
Gallium phosphide	GaP	0.633	-0.151	-0.082	-0.074	-0.069	15
Strontium fluoride	SrF_2	0.633	0.080	0.269	0.0185	-0.189	16
Strontium titanate	$SrTiO_3$	0.633	0.15	0.095	0.072	—	17
KRS-5	Tl(Br,I)	0.633	-0.140	0.149	-0.0725	-0.289	18,20
KRS-6	Tl(Br,Cl)	0.633	-0.451	-0.337	-0.164	-0.114	19,20
Zinc sulfide	Zn	0.633	0.091	-0.01	0.075	0.101	15

Rare Gases	Formula	λ/μm	p_{11}	p_{12}	p_{44}	p_{11}-p_{12}	Ref.
Neon (T = 24.3 K)	Ne	0.488	0.157	0.168	0.004	-0.011	21
Argon (T = 82.3 K)	Ar	0.488	0.256	0.302	0.015	-0.046	22
Krypton (T = 115.6 K)	Kr	0.488	0.34	0.34	0.037	0	21
Xenon (T = 160.5 K)	Xe	0.488	0.284	0.370	0.029	-0.086	22

Garnets	Formula	λ/μm	p_{11}	p_{12}	p_{44}	p_{11}-p_{12}	Ref.
GGG	$Gd_3Ga_5O_{12}$	0.514	-0.086	-0.027	-0.078	-0.059	23
YIG	$Y_3Fe_5O_{12}$	1.15	0.025	0.073	0.041	—	15
YGG	$Y_3Ga_5O_{12}$	0.633	0.091	0.019	0.079	—	17
YAG	$Y_3Al_5O_{12}$	0.633	-0.029	0.0091	-0.0615	-0.038	15

ELASTO-OPTIC COEFFICIENTS (PHOTOELASTIC CONSTANTS) continued)

Name Cubic (23, m3)	Formula	λ/μm	p_{11}	p_{12}	p_{44}	p_{13}	Ref.
Barium nitrate	Ba(NO3)2	0.589	—	p_{11}-p_{22} = 0.992	-0.0205	p_{11}-p_{13} = 0.713	13
Lead nitrate	$Pb(NO_3)_2$	0.589	0.162	0.24	-0.0198	0.20	24,25
Sodium bromate	$NaBrO_3$	0.589	0.185	0.218	-0.0139	0.213	26
Sodium chlorate	$NaClO_3$	0.589	0.162	0.24	-0.0198	0.20	26
Strontium nitrate	$Sr(NO_3)_2$	0.41	0.178	0.362	-0.014	0.316	27

Hexagonal (mmc,6mm)	Formula	λ/μm	p_{11}	p_{12}	p_{13}	p_{31}	p_{33}	p_{44}	Ref.
Beryl	$Be_3Al_2Si_6O_{18}$	0.589	0.0099	0.175	0.191	0.313	0.023	-0.152	28
Cadmium sulfide	CdS	0.633	-0.142	-0.066	-0.057	-0.041	-0.20	-0.099	15,2
Zinc oxide	ZnO	0.633	±0.222	±0.099	-0.111	±0.088	-0.235	0.0585	30
Zinc sulfide	ZnS	0.633	-0.115	0.017	0.025	0.0271	-0.13	-0.0627	31

Trigonal (3m,32,$\bar{3}$m)	Formula	λ/μm	p_{11}	p_{12}	p_{13}	p_{14}	p_{31}
Sapphire	Al_2O_3	0.644	-0.23	-0.03	0.02	0.00	-0.04
Calcite	$CaCO_3$	0.514	0.062	0.147	0.186	-0.011	0.241
Lithium niobate	$LiNbO_3$	0.633	±0.034	±0.072	±0.139	±0.066	±0.178
Lithium tantalate	$LiTaO_3$	0.633	-0.081	0.081	0.093	-0.026	0.089
Cinnabar	HgS	0.633			±0.445		
Quartz	SiO_2	0.589	0.16	0.27	0.27	-0.030	0.29
Proustite	Ag_3AsS_3	0.633	±0.10	±0.19	±0.22		±0.24
Sodium nitrite	$NaNO_3$	0.633		±0.21	±0.215	±0.027	±0.25
Tellurium	Te	10.6	0.155	0.130	—	—	—

Trigonal (3m,32,$\bar{3}$m) (continued)	p_{33}	p_{41}	p_{44}	Ref.
Sapphire	-0.20	0.01	-0.10	15,32
Calcite	0.139	-0.036	-0.058	33
Lithium niobate	+-0.060	±0.154	±0.300	15,34
Lithium tantalate	-0.044	-0.085	0.028	15,35
Cinnabar	+-0.115	—	—	36
Quartz	0.10	-0.047	-0.079	37
Proustite	+-0.20	—	—	38
Sodium nitrite		0.055	-0.06	39
Tellurium		—	—	15

Tetragonal (4/mmm,$\bar{4}$2m,422)	Formula	λ/μm	p_{11}	p_{12}	p_{13}	p_{31}
Ammonium dihydrogen phosphate	ADP	0.589	0.319	0.277	0.169	0.197
Barium titanate	$BaTiO_3$	0.633	0.425	—	—	—
Cesium dihydrogen arsenate	CDA	0.633	0.267	0.225	0.200	0.195
Magnesium fluoride	MgF_2	0.546	—	—	—	—
Calomel	Hg_2Cl_2	0.633	±0.551	±0.440	±0.256	±0.137
Potassium dihydrogen phosphate	KDP	0.589	0.287	0.282	0.174	0.241
Rubidium dihydrogen arsenate	RDA	0.633	0.227	0.239	0.200	0.205
Rubidium dihydrogen phosphate	RDP	0.633	0.273	0.240	0.218	0.210
Strontium barium niobate	$Sr_{0.75}Ba_{0.25}Nb_2O_6$	0.633	0.16	0.10	0.08	0.11
Strontium barium niobate	$Sr_{0.5}Ba_{0.5}Nb_2O_6$	0.633	0.06	0.08	0.17	0.09
Tellurium oxide	TeO_2	0.633	0.0074	0.187	0.340	0.090
Rutile	TiO_2	0.633	0.017	0.143	-0.139	-0.080

ELASTO-OPTIC COEFFICIENTS (PHOTOELASTIC CONSTANTS) (continued)

Name Tetragonal (4/mmm,$\bar{4}$2m,422) (continued)	p_{33}	p_{44}	p_{66}	Ref.
Ammonium dihydrogen phosphate	0.167	-0.058	-0.091	40
Barium titanate	—	—	—	41
Cesium dihydrogen arsenate	0.227	—	—	42
Magnesium fluoride	—	±0.0776	±0.0488	43
Calomel	±0.010	—	±0.047	44
Potassium dihydrogen phosphate	0.122	-0.019	-0.064	45
Rubidium dihydrogen arsenate	0.182	—	—	41
Rubidium dihydrogen phosphate	0.208	—	—	41
Strontium barium niobate	0.47	—	—	46
Strontium barium niobate	0.23	—	—	46
Tellurium oxide	0.240	-0.17	-0.046	47
Rutile	-0.057	-0.009	-0.060	48

Tetragonal (4,$\bar{4}$,4/m)	Formula	λ/μm	p_{11}	p_{12}	p_{13}	p_{16}	p_{31}
Cadmium molybdate	$CdMoO_4$	0.633	0.12	0.10	0.13	—	0.11
Lead molybdate	$PbMoO_4$	0.633	0.24	0.24	0.255	0.017	0.175
Sodium bismuth molybdate	$NaBi(MoO_4)_2$	0.633	0.243	0.205	0.25	—	0.21

Tetragonal (4,$\bar{4}$,4/m) (continued)	p_{33}	p_{44}	p_{45}	p_{61}	p_{66}	Ref.
Cadmium molybdate	0.18	—	—	—	—	49
Lead molybdate	0.300	0.067	-0.01	0.013	0.05	52
Sodium bismuth molybdate	0.29	—	—	—	—	—

Orthorhombic (222,m22,mmm)	Formula	λ/μm	p_{11}	p_{12}	p_{13}	p_{21}	p_{22}	p_{23}
Ammonium chlorate	NH_4ClO_4	0.633	—	0.24	0.18	0.23	—	0.20
Ammonium sulfate	$(NH_4)_2SO_4$	0.633	0.26	0.19	±0.260	±0.230	±0.27	±0.254
Rochelle salt	$NaKC_4H_4O_6$	0.589	0.35	0.41	0.42	0.37	0.28	0.34
Iodic acid (α)	HIO_3	0.633	0.302	0.496	0.339	0.263	0.412	0.304
Sulfur (α)	S	0.633	0.324	0.307	0.268	0.272	0.301	0.310
Barite	$BaSO_4$	0.589	0.21	0.25	0.16	0.34	0.24	0.19
Topaz	$Al_2SiO_4(OH,F)_2$	—	-0.085	0.069	0.052	0.095	-0.120	0.065

Orthorhombic (222,m22,mmm) (continued)	p_{31}	p_{32}	p_{33}	p_{44}	p_{55}	p_{66}	Ref.
Ammonium chlorate	0.19	0.18	±0.02	<±0.02	—	±0.04	51
Ammonium sulfate	0.20	±0.26	0.26	0.015	±0.0015	0.012	52
Rochelle salt	0.36	0.35	0.36	-0.030	0.0046	-0.025	53
Iodic Acid (α)	0.251	0.345	0.336	0.084	-0.030	0.098	54
Sulfur (α)	0.203	0.232	0.270	0.143	0.019	0.118	54
Barite	0.28	0.22	0.31	0.002	-0.012	0.037	55
Topaz	0.095	0.085	-0.083	-0.095	-0.031	0.098	28

Monoclinic (2,m,2/m)	Formula	λ/μm			
Taurine	$C_2H_7NO_3S$	0.589	$p_{11} = 0.313$ $p_{12} = 0.251$ $p_{13} = 0.270$ $p_{15} = -0.10$	$p_{25} = -0.0025$ $p_{31} = 0.362$ $p_{32} = 0.275$ $p_{33} = 0.308$	$p_{51} = -0.014$ $p_{52} = 0.006$ $p_{53} = 0.0048$ $p_{55} = 0.047$

ELASTO-OPTIC COEFFICIENTS (PHOTOELASTIC CONSTANTS) (continued)

Name					
Monoclinic (2,m,2/m)	**Formula**	**λ/μm**			
Taurine (continued)			p_{21} = 0.281	p_{35} = -0.003	p_{64} = 0.0024
			p_{22} = 0.252	p_{44} = 0.0025	p_{66} = 0.0028
			p_{23} = 0.272	p_{46} = -0.0056	

Isotropic	Formula	λ/μm	p_{11}	p_{12}	p_{44}	Ref.
Fused silica	SiO_2	0.633	0.121	0.270	-0.075	15
Water	H_2O	0.633	±0.31	±0.31		15
Polystyrene		0.633	±0.30	±0.31		25
Lucite		0.633	±0.30	0.28		25
Orpiment	As_2S_3-glass	1.15	0.308	0.299	0.0045	15
Tellurium oxide	TeO_2-glass	0.633	0.257	0.241	0.0079	56
Laser glasses	LGS-247-2	0.488	±0.168	±0.230		57
	LGS-250-3		±0.135	±0.198		
	LGS-1		±0.214	±0.250		
	KGSS-1621		±0.205	±0.239		
Dense flint glasses (examples)	LaSF	0.633	0.088	0.147	-0.030	58
	SF_4		0.215	0.243	-0.014	
	U10502		0.172	0.179	-0.004	
	$TaFd_7$		0.099	0.138	-0.020	

REFERENCES

A. Narasimhamurty, T. S., *Photoelastic and Electro-Optic Properties of Crystals*, Plenum Press, New York, 1981; pp. 290-293.
B. Weber, M. J., Ed., *CRC Handbook of Laser Science and Technology*, Volume IV, Part 2, CRC Press, Boca Raton, FL, 1986; pp. 324-331.
1. Petterson, H. E., *J. Opt. Soc. Am.*, 63, 1243, 1973.
2. Burstein, E. and Smith, P. L., *Phys. Rev.*, 74, 229, 1948.
3. Pakhnev, A. V., et al., *Sov. Phys. J. (transl.)*, 18, 1662, 1975.
4. Feldman, A., Horovitz, D., and Waxler, R. M., *Appl. Opt.*, 16, 2925, 1977.
5. Iyengar, K. S., *Nature (London)*, 176, 1119, 1955.
6. Bansigir, K. G. and Iyengar, K. S., *Acta Crystallogr.*, 14, 727, 1961.
7. Pakhev, A. V., et al., *Sov. Phys. J. (transl.)*, 20, 648, 1975.
8. Bansigir, K. G., *Acta Crystallogr.*, 23, 505, 1967.
9. Krishna Rao, K. V. and Krishna Murty, V. G., *Ind. J. Phys.*, 41, 150, 1967.
10. Weil, R. and Sun, M. J., *Proc. Int. Symp. CdTe (Detectors)*, Strasbourg Centre de Rech. Nucl., 1971, XIX-1 to 6, 1972.
11. Schmidt, E. D. D. and Vedam, K., *J. Phys. Chem. Solids*, 27, 1563, 1966.
12. Biegelsen, D. K., et al., *Phys. Rev. B*, 14, 3578, 1976.
13. Helwege, K. H., *Landolt-Börnstein, New Series Group III*, Vol. II, Springer-Verlag Berlin, 1979.
14. Feldman, A., Waxler, R. M., and Horovitz, D., *J. Appl. Phys.*, 49, 2589, 1978.
15. Dixon, R. W., *J. Appl. Phys.*, 38, 5149, 1967.
16. Shabin, O. V., et al., *Sov. Phys. Sol. State (transl.)*, 13, 3141, 1972.
17. Reintjes, J. and Schultz, M. B., *J. Appl. Phys.*, 39, 5254, 1968.
18. Rivoallan, L. and Favre, F., *Opt. Commun.*, 8, 404, 1973.
19. Rivoallan, L. and Favre, F., *Opt. Commun.*, 11, 296, 1974.
20. Afanasev, I. I., et al., *Sov. J. Opt. Technol.*, 46, 663, 1979.
21. Rand, S. C., et al., *Phys. Rev. B*, 19, 4205, 1979.
22. Sipe, J. E., *Can J. Phys.*, 56, 199, 1978.
23. Christyi, I. L., et al., *Sov. Phys. Sol. State (transl.)*, 17, 922, 1975.
24. Narasimhamurty, T. S., *Curr. Sci. (India)*, 23, 149, 1954.
25. Smith, T. M. and Korpel, A., *IEEE J. Quant. Electron.*, QE-1, 283, 1965.
26. Narasimhamurty, T. S., *Proc. Ind. Acad. Sci.*, A40, 164, 1954.
27. Rabman, A., *Bhagarantam Commem. Vol.*, Bangalore Print. and Publ., 173, 1969.
28. Eppendahl, R., *Ann. Phys. (IV)*, 61, 591, 1920.
29. Laurenti, J. P. and Rouzeyre, M., *J. Appl. Phys.*, 52, 6484, 1981.
30. Sasaki, H., et al., *J. Appl. Phys.*, 47, 2046, 1976.
31. Uchida, N. and Saito, S., *J. Appl. Phys.*, 43, 971, 1972.

ELASTO-OPTIC COEFFICIENTS (PHOTOELASTIC CONSTANTS) (continued)

32. Waxler, R. M. and Farabaugh, E. M., *J. Res. Natl. Bur. Stand.*, A74, 215, 1970.
33. Nelson, D. F., Lazay, P. D., and Lax, M., *Phys. Rev.*, B6, 3109, 1972.
34. O'Brien, R. J., Rosasco, G. J., and Weber, A., *J. Opt. Soc. Am.*, 60, 716, 1970.
35. Avakyants, L. P., et al., *Sov. Phys.*, 18, 1242, 1976.
36. Sapriel, J., *Appl. Phys. Litt.*, 19, 533, 1971.
37. Narasimhamurty, T. S., *J. Opt. Soc. Am.*, 59, 682, 1969.
38. Zubrinov, I. I., et al., *Sov. Phys. Sol. State (transl.)*, 15, 1921, 1974.
39. Kachalov, O. V. and Shpilko, I. O., *Sov. Phys. JETP (transl.)*, 35, 957, 1972.
40. Narasimhamurty, T. S., et al., *J. Mater. Sci.*, 8, 577, 1973.
41. Tada, K. and Kikuchi, K., *Jpn. J. Appl. Phys.*, 19, 1311, 1980.
42. Aleksandrov, K. S., et al., *Sov. Phys. Sol. State (transl.)*, 19, 1090, 1977.
43. Afanasev, I. I., et al., *Sov. Phys. Sol. State (transl.)*, 17, 2006, 1975.
44. Silvestrova, I. M., et al., *Sov. Phys. Cryst. (transl.)*, 20, 649, 1975.
45. Veerabhadra Rao, K. and Narasimhamurty, T. S., *J. Mater. Sci.*, 10, 1019, 1975.
46. Venturini, E. L., et al., *J. Appl. Phys.*, 40, 1622, 1969.
47. Vehida, N. and Ohmachi, Y., *J. Appl. Phys.*, 40, 4692, 1969.
48. Grimsditch, M. H. and Ramdus, A. K., *Phys. Rev. B*, 22, 4094, 1980.
49. Schinke, D. P. and Viehman,W., unpublished Data.
50. Coquin, G. A., et al., *J. Appl. Phys.*, 42, 2162, 1971.
51. Vasquez, F., et al., *J. Phys. Chem. Solids*, 37, 451, 1976.
52. Luspin, Y. and Hauret, G., *C.R.Ac. Sci. Paris*, B274, 995 1972.
53. Narasimhamurty, T. S., *Phys. Rev.*, 186, 945, 1969.
54. Haussühl, S. and Weber, H. J., *Z. Kristall.*, 132, 266, 1970.
55. Vedam, K., *Proc. Ind. Ac. Sci.*, A34, 161, 1951.
56. Yano, T., Fukumoto, A., and Watanabe, A., *J. Appl. Phys.*, 42, 3674, 1971.
57. Manenkov, A. A. and Ritus, A. I., *Sov. J. Quant. Electr.*, 8, 78, 1978.
58. Eschler, H. and Weidinger, F., *J. Appl. Phys.*, 46, 65, 1975.

LINEAR ELECTRO-OPTIC COEFFICIENTS

Name

Cubic ($\bar{4}3m$)	Formula	λ/μm	r_{41} pm/V
Cuprous bromide	CuBr	0.525	0.85
Cuprous chloride	CuCl	0.633	3.6
Cuprous iodide	CuI	0.55	-5.0
Eulytite (BSO)	$Bi_4Si_3O_{12}$	0.63	0.54
Germanium eulytite (BGO)	$Bi_4Ge_3O_{12}$	0.63	1.0
Gallium arsenide	GaAs	10.6	1.6
Gallium phosphide	GaP	0.56	-1.07
Hexamethylenetetramine	$C_6H_{12}N_4$	0.633	0.78
Sphalerite	ZnS	0.65	2.1
Zinc selenide	ZnSe	0.546	2.0
Zinc telluride	ZnTe	3.41	4.2
Cadmium telluride	CdTe	3.39	6.8

Cubic (23)	Formula	λ/μm	r_{41} pm/V
Ammonium chloride (77 K)	NH_4Cl	—	1.5
Ammonium cadmium langbeinite	$(NH_4)_2Cd_2(SO_4)_3$	0.546	0.70
Ammonium manganese langbeinite	$(NH_4)_2Mn_2(SO_4)_3$	0.546	0.53
Thallium cadmium langbeinite	$Tl_2Cd_2(SO_4)_3$	0.546	0.37
Potassium magnesium langbeinite	$K_2Mg_2(SO_4)_3$	0.546	0.40
Bismuth monogermanate	$Bi_{12}GeO_{20}$	—	3.3
Bismuth monosilicate	$Bi_{12}SiO_{20}$	—	3.3
Sodium chlorate	$NaClO_3$	0.589	0.4
Sodium uranyl acetate	$NaUO_2(CH_3COO)_3$	0.546	0.87

LINEAR ELECTRO-OPTIC COEFFICIENTS (continued)

Name

Cubic (23)	Formula	λ/μm	r_{41} pm/V
Trenhydrobromide	$N(CH_2CH_2NH_2)_3$ 3HBr	—	1.5
Trenhydrochloride	$N(CH_2CH_2NH_2)_3$ 3HCl	—	1.7

Tetragonal ($\bar{4}$2m)	Formula	T_{tran} K	r_{41} pm/V	r_{63} pm/V
Ammonium dihydrogen phosphate (ADP)	$NH_4H_2PO_4$	148	24.5	-8.5
Ammonium dideuterium phosphate (AD*P)	$NH_4D_2PO_4$	242	—	11.9
Ammonium dihydrogen arsenate (ADA)	$NH_4H_2AsO_4$	—	—	9.2
Cesium dihydrogen arsenate (CsDA)	CsH_2AsO_4	143	—	18.6
Cesium dideuterium arsenate (CsD*A)	CsD_2AsO_4	212	—	36.6
Potassium dihydrogen phosphate (KDP)	KH_2PO_4	123	8.6	-10.5
Potassium dideuterium phosphate (KD*P)	KD_2PO_4	222	8.8	23.8
Potassium dihydrogen arsenate (KDA)	KH_2AsO_4	97	12.5	10.9
Potassium dideuterium arsenate (KD*A)	KD_2AsO_4	162	—	18.2
Rubidium dihydrogen phosphate (RDP)	RbH_2PO_4	147	—	15.5
Rubidium dihydrogen arsenate (RDA)	RbH_2AsO_4	110	—	13.0
Rubidium dideuterium arsenate (RD*A)	RbD_2AsO_4	178	—	21.4

Tetragonal (4mm)	Formula	T_{tran} K	r_{13} pm/V	r_{33} pm/V	r_{51} pm/V
Barium titanate	$BaTiO_3$	406	8	28	—
Potassium lithium niobate	$K_3Li_2Nb_5O_{15}$	693	8.9	5.9	—
Lead titanate	$PbTiO_3$	765	13.8	5.9	—
Strontium barium niobate (SBN75)	$Sr_{0.75}Ba_{0.25}Nb_2O_6$	330	6.7	1340	42
Strontium barium niobate (SBN46)	$Sr_{0.46}Ba_{0.54}Nb_2O_6$	602	~180	35	—

Hexagonal (6mm)	Formula	r_{13} pm/V	r_{33} pm/V	r_{42} pm/V	r_{51} pm/V
Greenockite	CdS	3.1	2.9	2.0	3.7
Greenockite (const. strain)	CdS	1.1	2.4	—	—
Wurzite	ZnS	0.9	1.8	—	—
Zincite	ZnO	-1.4	+2.6	—	—

Hexagonal (6)	Formula	r_{13} pm/V	r_{33} pm/V	r_{42} pm/V	r_{51} pm/V
Lithium iodate	$LiIO_3$	4.1	6.4	1.4	3.3
Lithium potassium sulfate	$LiKSO_4$	r_{13}-r_{33} = 1.6	—	—	—

Trigonal (3m)	Formula	T_{tran} K	r_{13} pm/V	r_{22} pm/V	r_{33} pm/V	r_{42} pm/V
Cesium nitrate	$CsNO_3$	425	—	0.43	—	—
Lithium niobate	$LiNbO_3$	1483	8.6	7.0	30.8	28
Lithium tantalate	$LiTaO_3$	890	8.4	—	30.5	—
Lithium sodium sulfate	$LiNaSO_4$	—	—	<0.02	—	—
Tourmaline	—	—	—	0.3	—	—

Trigonal (32)	Formula	T_{tran} K	r_{11} pm/V	r_{41} pm/V
Cesium tartrate	$Cs_2C_4H_4O_6$	—	1.0	—
Cinnabar	HgS	659	3.1	1.5

LINEAR ELECTRO-OPTIC COEFFICIENTS (continued)

Name Trigonal (32)	Formula	T_{tran} K	r_{11} pm/V	r_{41} pm/V
Potassium dithionate	$K_2S_2O_6$	—	0.26	—
Strontium dithionate	$SrS_2O_6·4H_2O$	—	0.1	—
Quartz	SiO_2	1140	-0.47	0.2
Selenium	Se	398	2.5	

Orthorhombic (222)	Formula	T_{tran} K	r_{41} pm/V	r_{52} pm/V	r_{63} pm/V
Ammonium oxalate	$(NH_4)_2C_2O_4·4H_2O$	—	230	330	250
Rochelle salt	$KNaC_4H_4O_6·4H_2O$	$T_u = 297$ $T_l = 255$	-2.0	-1.7	+0.32

Orthorhombic (mm2)	Formula	Ttrans K	r_{13} pm/V	r_{23} pm/V	r_{33} pm/V	r_{42} pm/V	r_{51} pm/V
Barium sodium niobate (BSN)	Ba_2NaNbO_{15}	833	15	13	48	92	90
Potassium niobate	$KNbO_3$	476	28	1.3	64	380	105

Monoclinic (2)	Formula	T_{trans} K	r_{22} pm/V	r_{32} pm/V
Calcium pyroniobate	$Ca_2Nb_2O_7$	—	0.33	13.7
Triglycine sulfate (TGS)	$(NH_2CH_2COOH)_3·H_2SO_4$	322	7.2	13.6

REFERENCES

1. Narasimhamurty, T. S., *Photoelastic and Electro-Optic Properties of Crystals*, Plenum Press, New York, 1981, pp. 405-407.
2. Weber, M. J., Ed., *CRC Handbook of Laser Science and Technology*, Vol. IV, CRC Press, Boca Raton, FL, 1986, pp. 258-278.

QUADRATIC ELECTRO-OPTIC COEFFICIENTS

Kerr Constants of Ferroelectric Crystals[1,2]

Name	Formula	T_{tran} K	λ μm	g_{11} 10^{10} esu	g_{12} 10^{10} esu	g_{11}-g_{12} 10^{10} esu	g_{44} 10^{10} esu
Barium titanate	$BaTiO_3$	406	0.633	1.33	-0.11	1.44	
Strontium titanate	$SrTiO_3$	—	0.633	—	—	1.56	—
Potassium tantalate niobate	$KTa_{0.65}Nb_{0.35}O_3$	330	0.633	1.50	-0.42	1.92	1.63
Potassium tantalate	$KTaO_3$	13	0.633	—	—	1.77	1.33
Lithium niobate	$LiNbO_3$	1483	—	0.94	0.25	0.7	0.6
Lithium tantalate	$LiTaO_3$	938	—	1.0	0.17	0.8	0.7
Barium sodium niobate (BSN)	$Ba_{0.8}Na_{0.4}Nb_{2}O_{6}$	833	—	1.55	0.44	1.11	

Kerr Constants of Selected Liquids[2]

K is the Kerr constant at a wavelength of 589 nm and at room temperature; ε is the static dielectric constant; T_m is the melting point; and T_b is the normal boiling point

Name	Molecular formula	*K* 10^{-7} esu	ε	T_m °C	T_b °C
Carbon disulfide	CS_2	+3.23	2.63	-111.5	+46.3
Acetone	C_3H_6O	+16.3	21.0	-94.8	+56.1
Methyl ethyl ketone	C_4H_8O	+13.6	18.56	-86.67	+79.6

QUADRATIC ELECTRO-OPTIC COEFFICIENTS (continued)

Kerr Constants of Selected Liquids (continued)[2]

Name	Molecular formula	K 10^{-7} esu	ε	T_m °C	T_b °C
Pyridine	C_5H_5N	+20.4	13.26	-42	+115.23
Ethyl cyanoacetate	$C_5H_7NO_2$	+38.8	31.6	-22.5	205
o-Dichlorobenzene	$C_6H_4Cl_2$	+42.6	10.12	-16.7	180
Benzenesulfonyl chloride	$C_6H_5ClO_2S$	+89.9	28.90	+14.5	247
Nitrobenzene	$C_6H_5NO_2$	+326	35.6	+5.7	210.8
Ethyl 3-aminocrotonate	$C_6H_{11}NO_2$	+31.0	—	+33.9	210
Paraldehyde	$C_6H_{12}O_3$	-23.0	14.7	+12.6	124
			12.0[a]		
Benzaldehyde	C_7H_6O	+80.8	17.85	-26	179.05
			14.1[a]		
p-Chlorotoluene	C_7H_7Cl	+23.0	6.25	+7.5	162.4
o-Nitrotoluene	$C_7H_7NO_2$	+174	26.26	-10	222.3
m-Nitrotoluene	$C_7H_7NO_2$	+177	24.95	+15.5	232
p-Nitrotoluene	$C_7H_7NO_2$	+222	22.2	+51.6	238.3
Benzyl alcohol	C_7H_8O	-15.4	11.92	-15.3	205.8
			10.8[a]		
m-Cresol	C_7H_8O	+21.2	12.44	+11.8	202.27
			5.0[a]		
m-Chloroacetophenone	C_8H_7ClO	+69.1			
Acetophenone	C_8H_8O	+66.6	17.44	+19.7	202.3
			15.8[a]		
Quinoline	C_9H_7N	+15.0	9.16	-14.78	237.16
Ethyl salicylate	$C_9H_{10}O_3$	+19.6	8.48	+1.3	231.5
Carvone	$C_{10}H_{14}O$	+23.6	11.2	<0	230
Ethyl benzoylacetate	$C_{11}H_{12}O_3$	+16.0	13.50	<0	270
Water	H_2O	+4.0	80.10	0.00	100.0

[a] Dielectric constant at radiofrequencies (10^8-10^9 Hz).

REFERENCES

1. Narasimhamurty, T. S., *Photoelastic and Electro-Optic Properties of Crystals*, Plenum Press, New York, 1981, p. 408.
2. Gray, D. E., Ed., *AIP Handbook of Physics*, McGraw Hill, New York, 1972, p. 6-241.

MAGNETO-OPTIC CONSTANTS

Verdet Constants of Non-Magnetic Crystals[1]

V is the Verdet constant; n is the refractive index; and λ is the wavelength

Material	T K	λ nm	n	V min/Oe cm
Al_2O_3	300	546.1	1.771	0.0240
	300	589.3	1.768	0.0210
$BaTaO_3$	403	427		0.95
	403	496		0.38
	403	620		0.18
	403	826		0.072
$Bi_4Ge_3O_{12}$	300	442	2.077	0.289
	300	632.8	2.048	0.099
	300	1064	2.031	0.026
C (diamond)	300	589.3	2.417	0.0233
$CaCO_3$	300	589.3	1.658	0.019
CaF_2	300	589.3	1.434	0.0088
$Cd_{0.55}Mn_{0.45}Te$	300	632.8		6.87

MAGNETO-OPTIC CONSTANTS (continued)

Verdet Constants of Non-Magnetic Crystals (continued)[1]

Material	*T* K	λ nm	*n*	*V* min/Oe cm
CuCl	300	546.1	1.93	0.20
GaSe	298	632.8		0.80
$KAl(SO_4)_2 \cdot 12H_2O$	300	589.3	1.456	0.0124
KBr	300	546.1	1.564	0.0500
	300	589.3	1.560	0.0425
KCl	300	589.3	1.490	0.0275
KI	300	546.1	1.673	0.083
	300	589.3	1.666	0.070
$KTaO_3$	296	352		0.44
	296	413		0.19
	296	496		0.096
	296	620		0.051
	296	826		0.022
LaF_3	300	325	1.639	0.054
(H‖c)	300	442	1.615	0.028
	300	632.8	1.601	0.012
	300	1064	1.592	0.006
$MgAl_2O_4$	300	589.3	1.718	0.021
$NH_4AlSO_4 \cdot 12H_2O$	300	589.3	1.459	0.0128
NH_4Br	300	589.3	1.711	0.0504
NH_4Cl	300	546.1		0.0410
	300	589.3	1.643	0.0362
NaBr	300	546.1		0.0621
NaCl	300	546.1		0.0410
	300	589.3	1.544	0.0345
$NaClO_3$	300	546.1		0.0105
	300	589.3	1.515	0.0081
$NiSO_4 \cdot 6H_2O$	297	546.1		0.0256
	297	589.3	1.511	0.0221
SiO_2	300	546.1	1.546	0.0195
	300	589.3	1.544	0.0166
$SrTiO_3$	298	413	2.627	0.78
	298	496		0.31
	298	620		0.14
	298	826		0.066
ZnS	300	546.1		0.287
	300	589.3	2.368	0.226
ZnSe	300	476	2.826	1.50
	300	496	2.759	1.04
	300	514	2.721	0.839
	300	587	2.627	0.529
	300	632.8	2.592	0.406

Verdet Constants of Rare-Earth Aluminum Garnets at Various Wavelengths[1]

The absorption coefficient α for these materials ranges from 0.2 to 0.6 cm^{-1} at 300 K

Material	T/K	V in min/Oe cm							
		λ = 405 nm	450 nm	480 nm	520 nm	546 nm	578 nm	635 nm	670 nm
$Tb_2Al_5O_{12}$	300	-2.266	-1.565	-1.290	-1.039	-0.912	-0.787	-0.620	-0.542
	77		-102.16	-83.45	-3.425	-3.051	-2.603	-2.008	-1.815
	4.2				-64.80	-58.35	-53.77	48.39	-45.15
	1.45		-200.95	-172.52	-139.28	-125.07	-111.27	97.47	-93.42
$Dy_3Al_5O_{12}$	300	-1.241	-0.942	-0.803	-0.667	-0.592	-0.518	-0.411	-0.359
$Ho_3Al_5O_{12}$	300	-0.709	-0.320	-0.260	-0.335	-0.304	-0.299		-0.206

MAGNETO-OPTIC CONSTANTS (continued)

Verdet Constants of Rare-Earth Aluminum Garnets at Various Wavelengths (continued)[1]

Material	T/K	V in min/Oe cm							
		λ = 405 nm	450 nm	480 nm	520 nm	546 nm	578 nm	635 nm	670 nm
$Er_3Al_5O_{12}$	300	-0.189	-0.240	-0.154	-0.162	-0.157	-0.145	-0.105	-0.089
$Tm_3Al_5O_{12}$	300	+0.151	+0.103	+0.093	0.076	0.069	+0.059	+0.048	
$Yb_3Al_5O_{12}$	298	0.287	0.215	0.186	0.140	0.133	0.116	0.094	
	77	0.718	0.540	0.481	0.393	0.342	0.302	0.239	

Verdet Constants for KDP-Type Crystals[1]

Measurements refer to T = 298 K and λ = 632.8 nm, with **k** ∥ [001]

Material	V min/Oe cm
KH_2PO_4 (KDP)	0.0124
$KH_{0.3}D_{1.7}PO_4$ (KD*P)	0.145
$NH_4H_2PO_4$ (ADP)	0.138
KH_2AsO_4 (KDA)	0.238
$KH_{0.1}D_{1.9}AsO_4$ (KD*A)	0.245
$NH_4H_2AsO_4$ (ADH)	0.244

Verdet Constants of Gases[2]

Values refer to T = 0°C and P = 101.325 kPa (760 mmHg); n_D is the refractive index at a wavelength of 589 nm

Gas	$(n_D - 1) \times 10^3$	$10^6 \times V$ min/Oe cm
He	0.036	+0.40
Ar	2.81	+9.36
H_2		+6.29
N_2	0.297	+6.46
O_2	0.272	+5.69
Air	0.293	+6.27
Cl_2	0.773	+31.9
HCl	0.447	+21.5
H_2S	0.63	+41.5
NH_3	0.376	+19.0
CO	0.34	+11.0
CO_2	0.45	+9.39
NO	0.297	-58
CH4	0.444	+17.4
n-C4H10		+44.0

Verdet Constants of Liquids[2]

n_D is the refractive index at a wavelength of 589 nm and a temperature of 20°C, unless otherwise indicated. V is the Verdet constant

Liquid	λ/nm	T/°C	$10^2 \times V$ min/Oe cm	n_D
P	589	33	+13.3	
S	589	114	+8.1	1.929 (110°C)
H_2O	589	20	+1.309	1.3328
D_2O	589	19.7	+1.257	1.3384
H_3PO_4	578	97.4	+1.35	
CS_2	589	20	+4.255	1.6255
CCl_4	578-589	25.1	+1.60	1.463 (15°C)
$SbCl_5$	578	18	+7.45	1.601 (14°C)
$TiCl_4$	578	17	-1.65	1.61
$TiBr_4$	578	46	-5.3	
Methanol	589	18.7	+0.958	1.3289
Acetone	578-589	20.0	+1.116	1.3585
Toluene	578-589	15.0	+2.71	1.4950
Benzene	578-589	15.0	+3.00	1.5005
Chlorobenzene	589	15	+2.92	1.5246
Nitrobenzene	589	15	+2.17	1.5523
Bromoform	589	17.9	+3.13	1.5960

MAGNETO-OPTIC CONSTANTS (continued)

Verdet Constants of Rare Earth Paramagnetic Crystals[1]

n is the refractive index, and *V* is the Verdet constant at the wavelength and temperature indicated

Rare Earth	Host	*T*/K	λ/nm	*n*	*V* min/Oe cm
Ce^{3+}(30%)	CaF_2	300	325	1.516	-0.956
		300	442	1.502	-0.297
		300	633	1.494	-0.111
		300	1064	1.489	-0.035
Ce^{3+}	CeF_3	300	442	1.613	-1.05
		300	633	1.598	-0.406
		77	633		-1.418
		300	1064	1.589	-0.113
Pr^{3+}(5%)	CaF_2	300	266	1.471	-0.172
		300	325	1.461	-0.0818
		300	442	1.451	-0.0089
		300	633	1.445	-0.0168
		300	1064	1.441	-0.0045
Nd^{3+}(2.9%)	CaF_2	4.2	426		-0.19
Nd^{3+}	NdF_3	300	442	1.60	-0.553
		290	633	1.59	-0.209
		77	633		-0.755
		300	1064	1.58	-0.097
Eu^{3+}(3%)	CaF_2	4.2	430		29
		4.2	440		22
Eu^{2+}	EuF_2	300	450		-4.5
		300	500		-2.6
		300	550		-1.6
		300	600		-1.1
		300	650		-0.8
		300	1064		-0.19
Tb^{3+}	KTb_3F_{10}	300	325	1.531	-2.174
		300	442	1.518	-0.933
		300	633	1.510	-0.386
		77	633		-1.94
		300	1064	1.505	-0.114
Tb^{3+}	$LiTbF_4$	300	325	1.493	-1.9
		300	442	1.481	-0.98
		300	633	1.473	-0.44
		300	1064	1.469	-0.13
Tb^{3+}	$Tb_3Ga_5O_{12}$	300	500	1.989	-0.749
		300	570	1.981	-0.581
		300	633	1.976	-0.461
		300	830	1.967	-0.21
		300	1060	1.954	-0.12

MAGNETO-OPTIC CONSTANTS (continued)

Verdet Constants of Paramagnetic Glasses[1]

The Verdet constant *V* is given at room temperature for the wavelengths indicated

Rare earth phosphate glasses of composition $R_2O_3{\cdot}xP_2O_5$, where *x* is given in the second column

		Verdet constant *V* in min/Oe cm									
R	*x*	λ = 405 nm	λ = 436 nm	λ = 480 nm	λ = 500 nm	λ = 520 nm	λ = 546 nm	λ = 578 nm	λ = 600 nm	λ = 635 nm	λ = 670 nm
La		0.037	0.030	0.024	0.022	0.020	0.018	0.015	-0.014	0.013	—
Ce	2.67	-0.672	0.510	-0.366	-0.326	-0.287	-0.253	-0.217	-0.197	-0.173	-0.150
Pr	3.09	-0.447	-0.332	-0.283	-0.261	-0.236	-0.208	-0.182	-0.170	-0.150	-0.132
Nd	2.92	-0.250	-0.209	-0.167	-0.155	-0.136	-0.134	-0.094	-0.080	-0.080	-0.071
Sm	2.87	0.026	0.024	0.020	0.020	0.017	0.015	0.014	0.012	0.011	0.010
Eu	2.93	-0.025	-0.017	-0.010	-0.006	-0.006	-0.005	-0.004	-0.003	-0.002	-0.002
Gd	3.01	0.018	0.015	0.014	0.012	0.012	0.011	0.011	0.010	0.009	0.009
Tb	2.94	-0.560	-0.458	-0.357	-0.323	-0.295	-0.261	-0.226	-0.206	-0.190	-0.164
Dy	2.51	-0.540	-0.453	-0.359	-0.331	-0.301	0.268	-0.237	-0.217	-0.197	-0.173
Ho	2.94	-0.299	-0.313	-0.156	-0.153	-0.138	-0.138	-0.119	-0.110	-0.098	-0.084
Er	3.01	-0.139	-0.121	-0.100	-0.111	-0.095	-0.062	-0.060	-0.057	-0.051	-0.044
Tm	2.79	0.019	0.013	0.012	0.009	0.008	0.006	0.005	0.004	0.004	0.007
Yb	3.01	0.087	0.072	0.056	0.050	0.045	0.041	0.036	0.032	0.029	0.024

The following are rare earth borate glasses with composition:
for La and Pr: $R_2O_3{\cdot}xP_2O_5$; for Tb-Pr and Dy-Pr: $R_2O_3{\cdot}xB_2O_3$; and for other elements: $R_2O_3{\cdot}0.85La_3O_3{\cdot}xB_2O_3$.

R	*x*	λ = 405 nm	λ = 436 nm	λ = 480 nm	λ = 500 nm	λ = 520 nm	λ = 546 nm	λ = 578 nm	λ = 600 nm	λ = 635 nm	λ = 670 nm
La	3.04	0.043	0.036	0.029	0.026	0.023	0.022	0.019	0.018	0.016	0.014
Pr-La	5.44	-0.380	-0.307	-0.230	-0.220	-0.201	-0.178	-0.153	-0.146	-0.128	-0.110
Nd-La	5.41	-0.180	-0.147	-0.120	-0.111	-0.096	-0.094	-0.100	-0.059	-0.056	-0.046
Sm-La	4.97	0.032	0.030	0.025	0.024	0.022	0.019	0.017	0.016	0.014	0.012
Eu-La	4.69	-0.081	-0.060	-0.038	-0.033	-0.029	-0.024	0.019	-0.016	0.014	-0.012
Gd-La	4.71	0.032	0.026	0.024	0.022	0.021	0.020	0.018	0.017	0.015	0.013
Tb-La	4.73	-0.512	-0.419	-0.319	-0.288	-0.262	-0.234	-0.205	-0.186	-0.167	-0.142
Dy-La	4.88	-0.436	-0.361	-0.299	-0.273	-0.246	-0.220	-0.193	-0.177	-0.159	-0.138
Ho-La	4.36	-0.269	-0.252	-0.123	-0.131	-0.112	-0.128	-0.104	-0.096	—	-0.074
Er-La	4.50	-0.093	-0.078	-0.068	-0.082	—	-0.045	-0.042	-0.040	-0.035	-0.034
Tm-La	4.75	0.060	0.046	0.039	0.034	0.031	0.026	0.023	0.021	0.018	0.016
Yb-La	8.58	0.115	0.094	0.073	0.066	0.060	0.054	0.046	0.043	0.037	0.033
Tb-Pr	4.99	-0.940	-0.786	-0.560	-0.536	-0.489	-0.436	-0.380	-0.348	-0.306	-0.265
Dy-Pr	4.63	-0.850	—	—	-0.497	-0.465	-0.413	-0.358	-0.332	-0.290	-0.252
Pr	2.56	-0.843	-0.646	-0.471	-0.480	-0.432	-0.390	-0.334	-0.317	-0.271	-0.243

MAGNETO-OPTIC CONSTANTS (continued)

Verdet Constants of Diamagnetic Glasses[1]

The Verdet constant *V* is given at room temperature for the wavelengths indicated

Glass type	Composition (wt. %)	Verdet constant V in min/Oe cm			
		λ = 325 nm	λ = 442 nm	λ = 633 nm	λ = 1064 nm
SiO_2	100% SiO_2			0.013	
B_2O_3	100% B_2O_3			0.010	
CdO	47.5% CdO, 52.5% P_2O_5	0.079	0.033	0.022	
ZnO	36.4% ZnO, 63.6% P_2O_5	0.072	0.044	0.020	
TeO_2	88.9% TeO_2, 11.1% P_2O_5		0.196	0.076	0.022
ZrF_4	63.1% ZrF_4, 14.9% BaF_2, 7.2% LaF_3, 1.9% AlF_3, 9.1% PbF_2, 3.8% LiF			0.011	

Glass type	Composition (wt. %)	λ = 700 nm	λ = 853 nm	λ = 1060 nm
Bi_2O_3	95% Bi_2O_3, 5% B_2O_3	0.086	0.051	0.033
PbO	95% PbO, 5% B_2O_3	0.093	0.061	0.031
	82% PbO, 18% SiO_2	0.077	0.045	0.027
	50% PbO, 15% K_2O, 35% SiO_2	0.032	0.020	0.011
Tl_2O	95% Tl_2O, 5% B_2O_3	0.092	0.061	0.032
	82% Tl_2O, 18% SiO_2	0.100	0.067	0.043
	50% Tl_2O, 15% K_2O, 35% SiO_2	0.036	0.022	0.012
SnO	76% SnO, 13% B_2O_3, 11% SiO_2	0.071	0.046	0.026
TeO_3	75% TeO_2, 25% Sb_2O_3	0.076	0.052	0.032
	80% TeO_2, 20% $ZnCl_2$	0.073	0.046	0.025
	84% TeO_2, 16% BaO	0.056	0.041	0.029
	70% TeO_2, 30% WO_3	0.052	0.035	0.022
	20% TeO_2, 80% PbO	0.128	0.075	0.048
Sb_2O_3	25% Sb_2O_3, 75% TeO_2	0.076	0.050	0.032
	75% Sb_2O_3, 75% Cs_2O, 5% Al_2O_3	0.074	0.044	0.025
	75% Sb_2O_3, 10% Cs_2O, 10% Rb_2O, 5% Al_2O_3	0.078	0.052	0.030

MAGNETO-OPTIC CONSTANTS (continued)

Verdet Constants of Commercial Glasses[1]

This table gives the density, ρ, refractive index at 589 nm, n_D, and Verdet constant, V, for the wavelengths indicated; the data refer to room temperature

Glass type	ρ g/cm³	n_D	V in min/Oe cm				
			λ = 365.0 nm	λ = 404.7 nm	λ = 435.8 nm	λ = 546.1 nm	λ = 578.0 nm
BSC	2.49	1.5096	0.0499	0.0392	0.0333	0.02034	0.01798
HC	2.53	1.5189	0.0561	0.0440	0.0372	0.0225	0.01995
LBC	2.87	1.5406	0.0609	0.0477	0.0403	0.0245	0.0216
LF	3.23	1.5785	0.1143	0.0850	0.0693	0.0394	0.0344
BLF	3.48	1.6047	0.1112	0.0832	0.0685	0.0393	0.0344
DBC	3.56	1.6122	0.0662	0.0517	0.0435	0.0261	0.0231
DF	3.63	1.6203	0.1473	0.1076	0.0872	0.0485	0.0423
EDF	3.9	1.6533	0.1725	0.1248	0.1007	0.0556	0.0483

The composition of the glasses in weight percent is:

Glass type	SiO_2	B_2O_3	K_2O	CaO	Al_2O_3	As_2O_3	Na_2O	BaO	ZnO	PbO
BSC	69.6	6.7	20.5	2.9	0.3	0.1	—	—	—	—
HC	72.0	—	10.1	11.4	0.3	0.2	6.1	—	—	—
LBC	57.1	1.8	13.7	0.3	0.2	0.1	—	26.9	—	—
LF	52.5	—	9.5	0.3	0.2	0.1	—	—	—	37.6
BLF	45.2	—	7.8	—	—	0.4	—	16.0	8.3	22.2
DBC	36.2	7.7	0.2	0.2	3.5	0.7	—	44.6	6.7	—
DF	46.3	—	1.1	0.3	0.2	0.1	5.0	—	—	47.0
EDF	40.6	—	7.5	0.2	0.2	0.2	0.1	—	—	51.5

REFERENCES

1. Weber, M. J., *CRC Handbook of Laser Science and Technology*, Vol. IV, Part 2, CRC Press, Boca Raton, FL, 1988, p. 299-310.
2. Gray, D. E., Ed., *American Institute of Physics Handbook*, Third Edition, McGraw Hill, New York, 1972, p. 6-230.

FARADAY ROTATION

Ferro-, Ferri-, and Antiferromagnetic Solids

Material	T_c K	$4\pi M_s$ gauss	F deg/cm	α cm⁻¹	$2F/\alpha$	T K	λ nm
Fe	1043	21,800	4.4×10^5	6.5×10^5	1.4	300	500
			6.5×10^5	5.0×10^5	2.6	300	1000
			7×10^5	4.2×10^5	3.3	300	1500
			7×10^5	3.5×10^5	4.0	300	2000
Co	1390	18,200	2.9×10^5	—	—	300	500
			5.5×10^5	6.1×10^5	1.8	300	1000
			5.5×10^5	4.5×10^5	2.4	300	1500
			5.5×10^5	3.6×10^5	2.7	300	2000
Ni	633	6,400	0.8×10^5	—	—	300	500
			2.6×10^5	5.8×10^5	0.9	300	1000
			1.5×10^5	4.8×10^5	0.6	300	1500
			1×10^5	4.1×10^5	0.25	300	2000
Permalloy (Ni/Fe = 82/18)	803	10,700	1.2×10^5	6×10^5	0.4	300	500

FARADAY ROTATION (continued)

Ferro-, Ferri-, and Antiferromagnetic Solids (continued)

Material	T_c K	$4\pi M_s$ gauss	F deg/cm	α cm^{-1}	$2F/\alpha$	T K	λ nm
Ni/Fe = 100/0		6,000	1.2×10^5	7.05×10^5	0.34	300	632.8
Ni/Fe = 80/20		10,800	2.2×10^5	7.10×10^5	0.62	300	632.8
Ni/Fe = 60/40		14,900	2.9×10^5	7.54×10^5	0.77	300	632.8
Ni/Fe = 40/60		14,400	2.2×10^5	8.17×10^5	0.54	300	632.8
Ni/Fe = 20/80		19,400	3.3×10^5	8.10×10^5	0.81	300	632.8
Ni/Fe = 0/100	639	21,600	3.5×10^5	8.13×10^5	0.86	300	632.8
MnBi		7,700	4.2×10^5	6.1×10^5	1.4	300	450
			7.5×10^5	4.2×10^5	3.6	300	900
MnAs	313	—	0.44×10^5	5.0×10^5	0.174	300	500
			0.62×10^5	4.4×10^5	0.28	300	900
CrTe	334	1015	0.5×10^5	2.0×10^5	0.5	300	550
			0.4×10^5	1.2×10^5	0.7	300	900
FeRh	333	—	0.9×10^5	3.3×10^5	0.56	348	700
$Y_3Fe_5O_{12}$ (YIG)	560	2500	2400	1500	3.2	300	555
			1250	1400	1.8	300	625
			750	450	3.3	300	770
			175	<0.06	$>3 \times 10^3$	300	5000 to 1500
$Gd_3Fe_5O_{12}$ (GdIG)	Tn = 564 T = 286	7300	-2000	6000	0.6	300	500
			-1050	900	2.3	300	600
			-300	100	6.0	300	800
			-80	70	2.3	300	1000
$NiFe_2O_4$	858	3350	2.0×10^4	5.9×10^4	0.7	300	286
			-1.0×10^4	10×10^4	0.2	300	500
			-120	38	6	300	1500
			+75	15	10	300	3000
			+110	32	7	300	5000
$CoFe_2O_4$	793	4930	2.75×10^4	12×10^4	0.5	300	286
			3.6×10^4	17×10^4	0.4	300	400
			-2.5×10^4	6×10^4	0.8	300	660
$MgFe_2O_4$	593-713[e]	1450[e]	-60	100	1	300	2500
			0	12	0	300	4000
			+35	6	11	300	6000
$Li_{0.5}Fe_{2.5}O_4$	863-953[e]	3240[e] to 3900	-440	150	6	300	1500
			+10	85	0.2	300	3000
			+110	44	5	300	5000
			+135	80	3	300	7000
$BaFe_{12}O_{19}$	723	—	-50	-38	3	300	2000
			+75	20	7.5	300	3000
			+150	20	15	300	5000
			+165	22	15	300	7000
$Ba_2Zn_2Fe_{12}O_{19}$	—	—	90	120	1.5	300	5000
			75	65	2.0	300	7000
$RbNiF_3$	220	1250	360	35	20	77	450[a]
			70	10	14	77	600[a]
			310	70	9	77	800[a]
			75	25	6	77	1000[a]
$RbNi_{0.75}Co_{0.25}F_3$	109	—	180	9	40	77	600[b]
$RbFeF_3$	102	—	3400	7	900	82	300[c]
			1600	3	1100	82	400[c]
			620	1.5	830	82	600[c]
			300	2.5	240	82	800[c]
FeF_3	365	40 at 300 K	670	14	95	300	349[d]
			180	4.4	82	300	522.5[d]
$CrCl_3$	16.8	3880	2000	200	20	1.5	410

FARADAY ROTATION (continued)

Ferro-, Ferri-, and Antiferromagnetic Solids (continued)

Material	T_c K	$4\pi M_s$ gauss	F deg/cm	α cm⁻¹	$2F/\alpha$	T K	λ nm
			-500	300	3	1.5	450
			-1000	70	30	1.5	590
$CrBr_3$	32.5	3390	3×10^5	3×10^3	200	1.5	478
			1.6×10^5	1.4×10^4	23	1.5	500
CrI_3	68	2690	1.1×10^5	6.3×10^3	35	1.5	970
			0.8×10^5	3×10^3	53	1.5	1000
$FeBO_3$	348	115	3200	140	45	300	500
		at 300 K	450	38	24	300	700
EuO	69	23700	-1.0×10^5	0.5×10^4	40	5	1100
			5×10^5	9.7×10^4	10	5	700
			0.5×10^5	7.8×10^4	1.3	5	500
			3×10^4	>0.5	~105	20	2500
			660	>1.0	1300	20	10600
EuS	16.3	—	-1.6×10^5	0	—	6	825
			-9.6×10^5	3.3×10^4	58	6	690
			$+5.5 \times 10^5$	1.2×10^5	9.2	6	563
EuSe	7.0	13,200	1.45×10^5	80	3600	4.2	750
			0.95×10^5	60	3170	4.2	800

[a] Measured along the C-axis (magnetic hard axis).
[b] Measured along the C-axis (magnetic easy axis).
[c] Measured along the C-axis ([100]-direction at room temperature).
[d] Strong natural birefringence interferes with the Faraday effect.
[e] Depends on heat treatment.

REFERENCE

1. Weber, M. J., Ed., *CRC Handbook of Laser Science and Technology,* Vol. IV, Part 2, CRC Press, Boca Raton, FL, 1988, pp. 288-296.

NONLINEAR OPTICAL CONSTANTS

H. P. R. Frederikse

The relation between the polarization density P of a dielectric medium and the electric field E is linear when E is small, but becomes nonlinear as E acquires values comparable with interatomic electric fields (10^5 to 10^8 V/cm). Under these conditions the relation between P and E can be expanded in a Taylor's series

$$P = \varepsilon_0 \chi^{(1)} E + 2\chi^{(2)} E^2 + 4\chi^{(3)} E^3 + \cdots \quad (1)$$

where ε_o is the permittivity of free space, while $\chi^{(1)}$ is the linear and $\chi^{(2)}$, $\chi^{(3)}$ etc. the nonlinear optical susceptibilities.

If we consider two optical fields, the first $E_j^{\omega_1}$ (along the j-direction at frequency ω_1) and the second $E_k^{\omega_2}$ (along the k-direction at frequency ω_2) one can write the second term of the Taylor's series as follows

$$P_i(\omega_1\omega_2) = 2\chi_{ijk}^{\omega_3=\omega_1\pm\omega_2} E_j^{\omega_1} E_k^{\omega_2}$$

When $\omega_1 \neq \omega_2$ the (parametric) mixing of the two fields gives rise to two new polarizations at the frequencies $\omega_3 = \omega_1 + \omega_2$ and $\omega_3' = \omega_1 - \omega_2$. When the two frequencies are equal, $\omega_1 = \omega_2 = \omega$, the result is Second Harmonic Generation (SHG) $\chi_{ijk}(2\omega, \omega, \omega)$, while equal and opposite frequencies, $\omega_1 = \omega$ and $\omega_2 = -\omega$ leads to Optical Rectification (OR): $\chi_{ijk}(0,\omega,-\omega)$. In the SHG case the following convention is adopted: the second order nonlinear coefficient d is equal to one half of the second order nonlinear susceptibility

$$d_{ijk} = 1/2\chi^{(2)}$$

Because of the symmetry of the indices j and k one can replace these two by a single index (subscript) m. Consequently the notation for the SHG nonlinear coefficient in reduced form is d_{im} where m takes the values 1 to 6. Only noncentrosymmetric crystals can possess a nonvanishing d_{ijk} tensor (third rank). The unit of the SHG coefficients is m/V (in the MKSQ/SI system).

In centrosymmetric media the dominant nonlinearity is of the third order. This effect is represented by the third term in the Taylor's series (Equation 1); it is the result of the interaction of a number of optical fields (one to three) producing a new frequency $\omega_4 = \omega_1 + \omega_2 + \omega_3$. The third order polarization is given by

$$P_j(\omega_1\omega_2\omega_3) = g_4 \chi_{jklm} E_k^{\omega_1} E_l^{\omega_2} E_m^{\omega_3}$$

Third Harmonic Generation (THG) is achieved when $\omega_1 = \omega_2 = \omega_3 = \omega$. In this case the constant $g_4 = 1/4$. The third order nonlinear coefficient C is related to the third order susceptibility as follows

$$C_{jklm} = 1/4\chi_{jklm}$$

This coefficient is a fourth rank tensor. In the THG case the matrices must be invariant under permutation of the indices k, l, and m; as a result the notation for the third order nonlinear coefficient can be simplified to C_{jn}. The unit of C_{jn} is $m^2 \cdot V^{-2}$ (in the MKSQ/SI system).

Applications of second order nonlinear optical materials include the generation of higher (up to sixth) optical harmonics, the mixing of monochromatic waves to generate sum or difference frequencies (frequency conversion), the use of two monochromatic waves to amplify a third wave (parametric amplification) and the addition of feedback to such an amplifier to create an oscillation (parametric oscillation).

Third order nonlinear optical materials are used for THG, self-focusing, four wave mixing, optical amplification, and optical conjugation. Many of these effects — as well as the variation and modulation of optical propagation caused by mechanical, electric, and magnetic fields (see the preceeding table on "Elasto-Optic, Electro-Optic, and Magneto-Optic Constants") are used in the areas of optical communication, optical computing, and optical imaging.

REFERENCES (NONLINEAR OPTICS)

1. *Handbook of Laser Science and Technology*, Vol. 111, Part 1; Ed.: Marvin J. Weber, Publ.: CRC Press, Inc., Boca Raton, FL, 1986.
2. Dmitriev, V.G., Gurzadyan, G.G., and Nikogosyan, D., *Handbook of Nonlinear Optical Crystals*, Springer-Verlag, Berlin, 1991.
3. Shen, Y.R., *The Principles of Nonlinear Optics*, John Wiley, New York, 1984.
4. Yariv, A., *Quantum Electronics*, 3rd edition, John Wiley, New York, 1988.
5. Bloembergen, N., *Nonlinear Optics,* W.A. Benjamin, New York, 1965.
6. Zernike F. and Midwinter, J.E., *Applied Nonlinear Optics*, John Wiley, New York, 1973.
7. Hopf, F.A. and Stegeman, G.I., *Applied Classical Electrodynamics*, Volume 2: Nonlinear Optics, John Wiley, New York, 1986.
8. *Nonlinear Optical Properties of Organic Molecules and Crystals,* Eds.: D.S. Chemla and J. Zyss, Publ.: Academic Press, Orlando, FL, 1987.
9. *Optical Phase Conjugation*, Ed.: R.A. Fisher, Publ.: Academic Press, New York, 1983.
10. Zyss, J., *Molecular Nonlinear Optics: Materials, Devices and Physics*, Academic Press, Boston, 1994.
11. Nonlinear Optics, 5 articles in *Physics Today, (Am. Inst. of Phys.)*, Vol. 47, No. 5, May, 1994.

Selected SHG Coefficients of NLO Crystals*

Material	Symmetry class	$d_{im} \times 10^{12}$ m/V	λ μm
GaAs	$\bar{4}3$ m	$d_{14} = 134.1 \pm 42$	10.6
GaP	$\bar{4}3$ m	$d_{14} = 71.8 \pm 12.3$	1.058
InAs	$\bar{4}3$ m	$d_{14} = 364 \pm 47$	1.058
		$d_{14} = 210$	10.6
ZnSe	$\bar{4}3$ m	$d_{14} = 78.4 \pm 29.3$	10.6
		$d_{36} = 26.6 \pm 1.7$	1.058
β-ZnS	$\bar{4}3$ m	$d_{14} = 30.6 \pm 8.4$	10.6
		$d_{36} = 20.7 \pm 1.3$	1.058
ZnTe	$\bar{4}3$ m	$d_{14} = 92.2 \pm 33.5$	10.6
		$d_{14} = 83.2 \pm 8.4$	1.058
		$d_{36} = 89.6 \pm 5.7$	1.058
CdTe	$\bar{4}3$ m	$d_{14} = 167.6 \pm 63$	10.6
Bi_4GeO_{12}	$\bar{4}3$ m	$d_{14} = 1.28$	1.064
$N_4(CH_2)_6$ (hexamine)	$\bar{4}3$ m	$d_{14} = 4.1$	1.06
$LiIO_3$	6	$d_{33} = -7.02$	1.06
		$d_{31} = -5.53 \pm 0.3$	1.064
ZnO	6 mm	$d_{33} = -5.86 \pm 0.16$	1.058
		$d_{31} = 1.76 \pm 0.16$	1.058
		$d_{15} = 1.93 \pm 0.16$	1.058
α-ZnS	6 mm	$d_{33} = 11.37 \pm 0.07$	1.058
		$d_{33} = 37.3 \pm 12.6$	10.6
		$d_{31} = -18.9 \pm 6.3$	10.6
		$d_{15} = 21.37 \pm 8.4$	10.6
CdS	6 mm	$d_{33} = 25.8 \pm 1.6$	1.058
		$d_{31} = -13.1 \pm 0.8$	1.058
		$d_{15} = 14.4 \pm 0.8$	1.058
CdSe	6 mm	$d_{33} = 54.5 \pm 12.6$	10.6
		$d_{31} = -26.8 \pm 2.7$	10.6
$BaTiO_3$	4 mm	$d_{33} = 6.8 \pm 1.0$	1.064
		$d_{31} = 15.7 \pm 1.8$	1.064
		$d_{15} = 17.0 \pm 1.8$	1.064
$PbTiO_3$	4 mm	$d_{33} = 7.5 \pm 1.2$	1.064
		$d_{31} = 37.6 \pm 5.6$	1.064
		$d_{15} = 33.3 \pm 5$	1.064
$K_3Li_2Nb_5O_{15}$	4 mm	$d_{33} = 11.2 \pm 1.6$	1.064
		$d_{31} = 6.18 \pm 1.28$	1.064
		$d_{15} = 5.45 \pm 0.54$	1.064
$K_{0.8}Na_{0.2}Ba_2Nb_5O_{15}$	4 mm	$d_{31} = 13.6 \pm 1.6$	1.064
$SrBaNb_5O_{15}$	4 mm	$d_{33} = 11.3 \pm 3.3$	1.064
		$d_{31} = 4.31 \pm 1.32$	1.064
		$d_{15} = 5.98 \pm 2$	1.064
$NH_4H_2PO_4$ (ADP)	$\bar{4}2$ m	$d_{36} = 0.53$	1.064
		$d_{36} = 0.85$	0.694
KH_2PO_4 (KDP)	$\bar{4}2$ m	$d_{36} = 0.44$	1.064
		$d_{36} = 0.47 \pm 0.07$	0.694
KD_2PO_4 (KD*P)	$\bar{4}2$ m	$d_{36} = 0.38 \pm 0.016$	1.058
		$d_{36} = 0.34 \pm 0.06$	0.694
		$d_{14} = 0.37$	1.058
KH_2AsO_4 (KDA)	$\bar{4}2$ m	$d_{36} = 0.43 \pm 0.025$	1.06
		$d_{36} = 0.39 \pm 0.4$	0.694
$CdGeAs_2$	$\bar{4}2$ m	$d_{36} = 351 \pm 105$	10.6
$AgGaS_2$	$\bar{4}2$ m	$d_{36} = 18 \pm 2.7$	10.6
$AgGaSe_2$	$\bar{4}2$ m	$d_{36} = 37.4 \pm 6.0$	10.6
$(NH_2)_2CO$ (urea)	$\bar{4}2$ m	$d_{36} = 1.3$	1.06
$AlPO_4$	32	$d_{11} = 0.35 \pm 0.03$	1.058
Sc	32	$d_{11} = 97 \pm 25$	10.6

Selected SHG Coefficients of NLO Crystals (continued)*

Material	Symmetry class	$d_{im} \times 10^{12}$ m/V	λ µm
Te	32	$d_{11} = 650 \pm 30$	10.6
SiO_2 (quartz)	32	$d_{11} = 0.335$	1.064
HgS	32	$d_{11} = 50.3 \pm 17$	10.6
$(C_6H_5CO)_2$ [benzil]	32	$d_{11} = 3.6 \pm 0.5$	1.064
β-BaB_2O_4 [BBO]	3 m	$d_{22} = 2.22 \pm 0.09$	1.06
		$d_{31} = 0.16 \pm 0.08$	1.06
$LiNbO_3$	3 m	$d_{33} = 34.4$	1.06
		$d_{31} = -5.95$	1.06
		$d_{22} = 2.76$	1.06
$LiTaO_3$	3 m	$d_{33} = -16.4 \pm 2$	1.058
		$d_{31} = -1.07 \pm 0.2$	1.058
		$d_{22} = +1.76 \pm 0.2$	1.058
Ag_3AsS_3 [proustite]	3 m	$d_{31} = 11.3 \pm 2.5$	10.6
		$d_{22} = 18.0 \pm 2.5$	10.6
Ag_3SbS_3 [pyrargerite]	3m	$d_{31} = 12.6 \pm 4$	10.6
		$d_{22} = 13.4 \pm 4$	10.6
α-HIO_3	222	$d_{36} = 5.15 \pm 0.16$	1.064
$NO_2 \cdot CH_3NOC_5H_4 \cdot$ (POM)	222	$d_{36} = 6.4 \pm 1.0$	1.064
$Ba_2NaNb_5O_{15}$ [Banana]	mm 2	$d_{33} = -17.6 \pm 1.28$	1.064
		$d_{31} = -12.8 \pm 1.28$	1.064
$C_6H_4(NO_2)_2$ [MDB]	mm 2	$d_{33} = 0.74$	1.064
		$d_{32} = 2.7$	1.064
		$d_{31} = 1.78$	1.064
$Gd_2(MoO_4)_3$	mm 2	$d_{33} = -0.044 \pm 0.008$	1.064
		$d_{32} = +2.42 \pm 0.36$	1.064
		$d_{31} = -2.49 \pm 0.37$	1.064
$KNbO_3$	mm 2	$d_{33} = -19.58 \pm 1.03$	1.064
		$d_{32} = +11.34 \pm 1.03$	1.064
		$d_{31} = -12.88 \pm 1.03$	1.064
$KTiOPO_4$ [KTP]	mm 2	$d_{33} = 13.7$	1.06
		$d_{32} = \pm 5.0$	1.06
		$d_{31} = \pm 6.5$	1.06
$NO_2C_6H_4 \cdot NH_2$ [mNA]	mm 2	$d_{33} = 13.12 \pm 1.28$	1.064
		$d_{32} = 1.02 \pm 0.22$	1.064
		$d_{31} = 12.48 \pm 1.28$	1.064
$C_{10}H_{12}N_3O_6$ [MAP]	2	$d_{23} = 10.67 \pm 1.3$	1.064
		$d_{22} = 11.7 \pm 1.3$	1.064
		$d_{21} = 2.35 \pm 0.5$	1.064
		$d_{25} = -0.35 \pm 0.3$	1.064
$(NH_2CH_2COOH)_3H_2SO_4$ [TGS]	2	$d_{23} = 0.32$	0.694

* These data are taken from References 1 and 2.

Selected THG Coefficients of Some NLO Materials*

Material	NLO process	$C_{jn} \times 10^{20}$ m²/V⁻²	λ µm
$NH_4H_2PO_4$ [ADP]	(–3ω,ω,ω,ω)	$C_{11} = 0.0104$	1.06
		$C_{18} = 0.0098$	1.06
C_6H_6 [benzene]	(–3ω,ω,ω,ω)	$C_{11} = 0.0184 \pm 0.0042$	1.89
$CdGeAs_2$	(–3ω,ω,ω,ω)	$C_{11} = 182 \pm 84$	10.6
p-type: 5×10^{16} cm⁻³		$C_{16} = 175$	10.6
		$C_{18} = -35$	10.6
$C_{40}H_{56}$ [β-carotene]	(–3ω,ω,ω,ω)	C_{11} 0.263 ± 0.08	1.89
GaAs	(–3ω,ω,ω,–ω)	$C_{11} = 62 \pm 31$	1.06
high-resistivity			
Ge	(–3ω,ω,ω,–ω)	$C_{11} = 23.5 \pm 12$	1.06
$LiIO_3$	(–3ω,ω,ω,–ω)	$C_{12} = 0.2285$	1.06
		$C_{35} = 6.66 \pm 1$	1.06
KBr	(–3ω,ω,ω,–ω)	$C_{11} = 0.0392$	1.06
		$C_{18}/C_{11} = 0.3667$	1.06
KCl	(–3ω,ω,ω,–ω)	$C_{11} = 0.0168$	1.06
		$C_{18}/C_{11} = 0.28$	1.06
KH_2PO_4 [KDP]	(–3ω,ω,ω,–ω)	$C_{11}–3C_{18} = 0.04$	1.06
Si	(–3ω,ω,ω,–ω)	$C_{11} = 82.8 \pm 25$	1.06
p-type: 10^{14} cm⁻³			
NaCl	(–3,ω,ω,ω,–ω)	$C_{11} = 0.0168$	1.06
		$C_{18}/C_{11} = 0.4133$	1.06
NaF	(–3ω,ω,ω,–ω)	$C_{11} = 0.0035$	1.06

* These data are taken from Reference 1.

PHASE DIAGRAMS
H. P. R. Frederikse

A phase is a structurally homogeneous portion of matter. Regardless of the number of chemical constituents of a gas, there is only one vapor phase. This is true also for the liquid form of a pure substance, although a mixture of several liquid substances may exist as one or several phases, depending on the interactions among the substances. On the other hand a pure solid may exist in several phases at different temperatures and pressures because of differences in crystal structure (Reference 1). At the phase transition temperature, T_{tr}, the chemical composition of the solid remains the same, but often a change in the physical properties will take place. Such changes are found in ferroelectric crystals (example $BaTiO_3$) which develop a spontaneous polarization below T_{tr}, in superconductors (example Pb) which loose all electrical resistance below the transition point, and in many other classes of solids.

In quite a few cases it is difficult to bring about the phase transition, and the high- (or low-) temperature phase persists in its metastable form. Many liquids remain in the liquid state for shorter or longer periods of time when cooled below the melting point (supercooling). However, often the slightest disturbance will cause solidification. Persistence of the high temperature phase in solid-solid transitions is usually of much longer duration. An example of this behavior is found in white tin; although gray tin is the thermodynamically stable form below T_{tr} (286.4 K), the metal remains in its undercooled, white tin state all the way to $T = 0$ K, and crystals of gray tin are very difficult to produce.

A *phase diagram* is a map which indicates the areas of stability of the various phases as a function of external conditions (temperature and pressure). Pure materials, such as mercury, helium, water, and methyl alcohol are considered one-component systems and they have *unary* phase diagrams. The equilibrium phases in two-component systems are presented in *binary* phase diagrams. Because many important materials consist of three, four, and more components, many attempts have been made to deduce their multicomponent phase diagrams. However, the vast majority of systems with three or more components are very complex, and no overall maps of the phase relationships have been worked out.

It has been shown during the last 20 to 25 years that very useful partial phase diagrams of complex systems can be obtained by means of thermodynamic modeling (References 2, 3). Especially for complicated, multicomponent alloy systems the CALPHAD method has proved to be a successful approach for producing valuable portions of very intricate phase diagrams (Reference 4). With this method thermodynamic descriptions of the free energy functions of various phases are obtained which are consistent with existing (binary) phase diagram information and other thermodynamic data. Extrapolation methods are then used to extend the thermodynamic functions into a ternary system. Comparison of the results of this procedure with available experimental data is then used to fine-tune the phase diagram and add ternary interaction functions if necessary. In principle this approximation strategy can be extended to four, five, and more component systems.

The nearly two dozen phase diagrams shown below present the reader with examples of some important types of single and multicomponent systems, especially for ceramics and metal alloys. This makes it possible to draw attention to certain features like the kinetic aspects of phase transitions (see Figure 22, which presents a time-temperature-transformation, or TTT, diagram for the precipitation of α-phase particles from the β-phase in a Ti-Mo alloy; Reference 1, pp.358-360). The general references listed below and the references to individual figures contain phase diagrams for many additional systems.

GENERAL REFERENCES

1. Ralls, K.M., Courtney, T.H., and Wulff, J., *Introduction to Materials Science and Engineering*, Chapters 16 and 17, John Wiley & Sons, New York, 1976.
2. Kaufman, L., and Bernstein, H., *Computer Calculation of Phase Diagrams*, Academic Press, New York, 1970.
3. Kattner, U.R., Boettinger, W.J.B., and Coriell, S.R., *Z. Metallkd.*, 87, 9, 1996.
4. Dinsdale, A.T., Editor, *CALPHAD*, Vol. 1–20, Pergamon Press, Oxford, 1977–1996 and continuing.
5. Baker, H., Editor, *ASM Handbook, Volume 3: Alloy Phase Diagrams*, ASM International, Materials Park, OH, 1992.
6. Massalski, T.B., Editor, *Binary Alloy Phase Diagrams, Second Edition*, ASM International, Materials Park, OH, 1990.
7. Roth. R.S., Editor, *Phase Diagrams for Ceramists*, Vol. I (1964) to Volume XI (1995), American Ceramic Society, Waterville, OH.

REFERENCES TO INDIVIDUAL PHASE DIAGRAMS

Figure 1. Carbon: Reference 7, Vol. X (1994), Figure 8930. Reprinted with permission.

Figure 2. Si-Ge : Ref.5, p. 2.231. Reprinted with permission.

Figure 3. H_2O (ice): See figure.

Figure 4. SiO_2: Reference 7, Vol. XI (1995), Figure 9174. Reprinted with permission.

Figure 5. Fe-O: Darken, L.S., and Gurry, R.W., *J. Am. Chem. Soc.*, 68, 798, 1946. Reprinted with permission.

Figure 6. Ti-O: Reference 5, p. 2.324. Reprinted with permission.

Figure 7. BaO-TiO_2: Reference 7, Vol. III (1975), Figure 4302. Reprinted with permission.

Figure 8. MgO-Al_2O_3: Reference 7, Vol. XI (1995), Figure 9239. Reprinted with permission.

Figure 9. Y_2O_3-ZrO_2: Reference 7, Vol. XI (1995), Figure 9348. Reprinted with permission.

Figure 10. Si-N-Al-O (Sialon): Reference 7, Vol. X (1994), Figure 8759. Reprinted with permission.

Figure 11. PbO-ZrO_2-TiO_2 (PZT): Reference 7, Vol. III (1975), Figure 4587. Reprinted with permission.

Figure 12. Al-Si-Ca-O: Reference 7 (1964), Vol. I, Figure 630. Reprinted with permission.

Figure 13. Y-Ba-Cu-O: Whitler, J.D., and Roth, R.S., *Phase Diagrams for High T_c Superconductors*, Figure S-082, American Ceramic Society, Waterville, OH, 1990. Reprinted with permission.

Figure 14. Al-Cu: Reference 5, p. 2.44. Reprinted with permission.

Figure 15. Fe-C: Ralls, K.M., Courtney, T.H., and Wulff, J., *Introduction to Materials Science and Engineering*, Figure 16.13 , John Wiley & Sons, New York, 1976. Reprinted with permission.

Figure 16. Fe-Cr: Reference 5, p. 2.152. Reprinted with permission.

Figure 17. Cu-Sn: Reference 5, p. 2.178. Reprinted with permission.
Figure 18. Cu-Ni: Reference 5, p. 2.173. Reprinted with permission.
Figure 19. Pb-Sn (solder): Reference 5, p. 2.335. Reprinted with permission.
Figure 20. Cu-Zn (brass): Subramanian, P.R., Chakrabarti, D.J., and Laughlin, D.E., Editors, *Phase Diagrams of Binary Copper Alloys*, p. 487, ASM International, Materials Park, OH, 1994. Reprinted with permission.
Figure 21. Co-Sm: Reference 5, p. 2.148. Reprinted with permission.
Figure 22. Ti-Mo: Reference 5, p. 2.296; Reference 1, p. 359. Reprinted with permission.
Figure 23: Fe-Cr-Ni: Reference 5, Figure 48. Reprinted with permission.

Figure 1

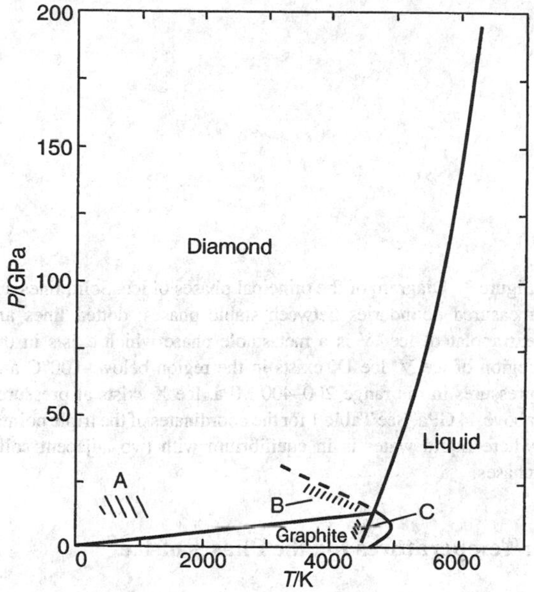

Figure 1. Phase diagram of carbon. (A) Martensitic transition: hex graphite → hex diamond. (B) Fast graphite-to-diamond transition. (C) Fast diamond-to-graphite transition.

Figure 2

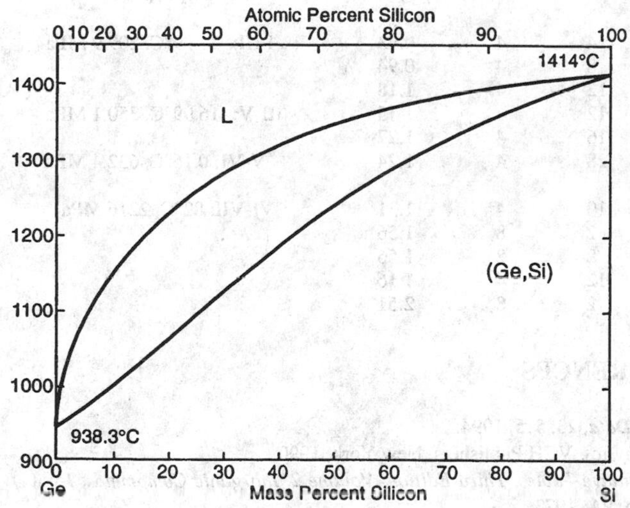

Figure 2. Si-Ge system.

Phase	Composition, mass % Si	Pearson symbol	Space group
(Ge,Si)	0 to 100	*cF*8	$Fd\bar{3}m$
High-pressure phases			
GeII	—	*tI*4	$I4_1/amd$
SiII	—	*tI*4	$I4_1/amd$

FIGURE 3

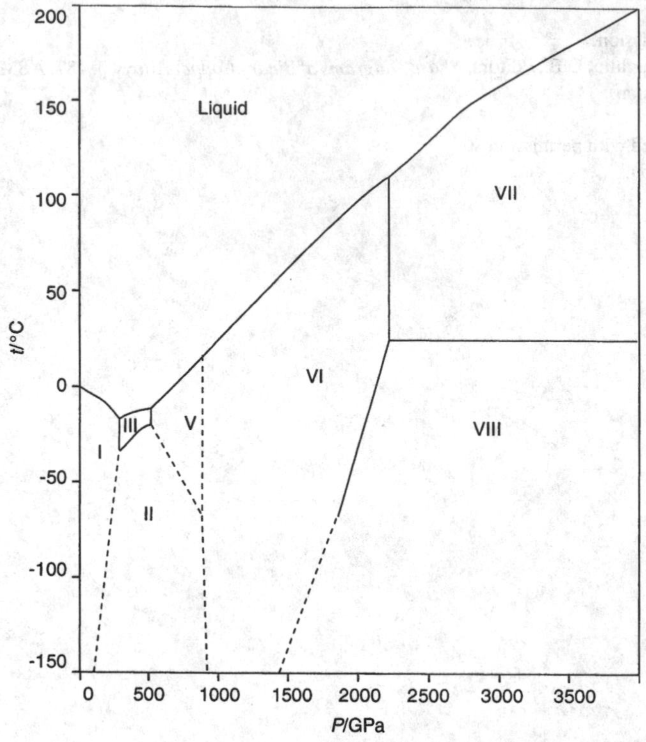

Figure 3. Diagram of the principal phases of ice. Solid lines are measured boundaries between stable phases; dotted lines are extrapolated. Ice IV is a metastable phase which exists in the region of ice V. Ice IX exists in the region below -100°C and pressures in the range 200–400 MPa. Ice X exists at pressures above 44 GPa. See Table 1 for the coordinates of the triple points, where liquid water is in equilibrium with two adjacent solid phases.

Table 1. Crystal Structure, Density, and Transition Temperatures for the Phases of Ice

Phase	Crystal system	Cell parameters	Z	n	ρ/g cm^{-3}	Triple ponts
Ih	Hexagonal	$a = 4.513$; $c = 7352$	4	4	0.93	I-III: -21.99°C, 209.9 MPa
Ic	Cubic	$a = 6.35$	8	4	0.94	
II	Rhombohedral	$a = 7.78$; $\alpha = 113.1°$	12	4	1.18	
III	Tetragonal	$a = 6.73$; $c = 6.83$	12	4	1.15	III-V: -16.99°C, 350.1 MPa
IV	Rhombohedral	$a = 7.60$; $\alpha = 70.1°$	16	4	1.27	
V	Monoclinic	$a = 9.22$; $b = 7.54$, $c = 10.35$; $\beta = 109.2°$	28	4	1.24	V-VI: 0.16°C, 632.4 MPa
VI	Tetragonal	$a = 6.27$; $c = 5.79$	10	4	1.31	VI-VII: 82°C, 2216 MPa
VII	Cubic	$a = 3.41$	2	8	1.56	
VIII	Tetragonal	$a = 4.80$; $c = 6.99$	8	8	1.56	
IX	Tetragonal	$a = 6.73$; $c = 6.83$	12	4	1.16	
X	Cubic	$a = 2.83$	2	8	2.51	

REFERENCES

1. Wagner, W., Saul, A., and Pruss, A., *J. Phys. Chem. Ref. Data*, 23, 515, 1994.
2. Lerner, R.G. and Trigg, G.L., Editors, *Encyclopedia of Physics,* VCH Publishers, New York, 1990.
3. Donnay, J.D.H. and Ondik, H.M, *Crystal Data Determinative Tables, Third Edition, Volume 2, Inorganic Compounds,* Joint Committee on Powder Diffraction Standards, Swarthmore, PA, 1973.
4. Hobbs, P.V., *Ice Physics,* Oxford University Press, Oxford, 1974.

Figure 4

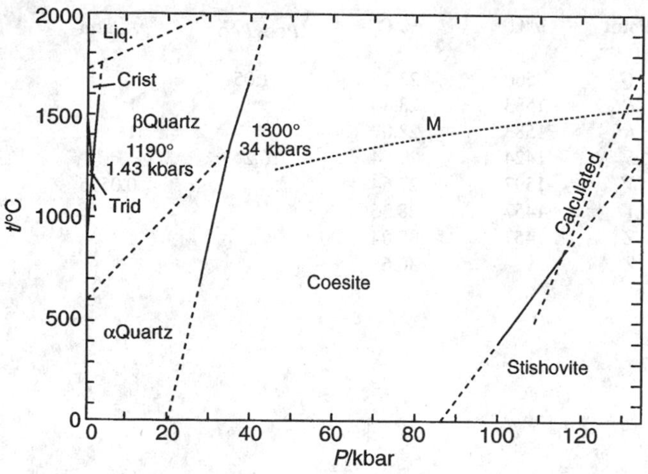

Figure 4. SiO_2 system. Crist = cristobalite; Trid = tridymite.

Figure 5

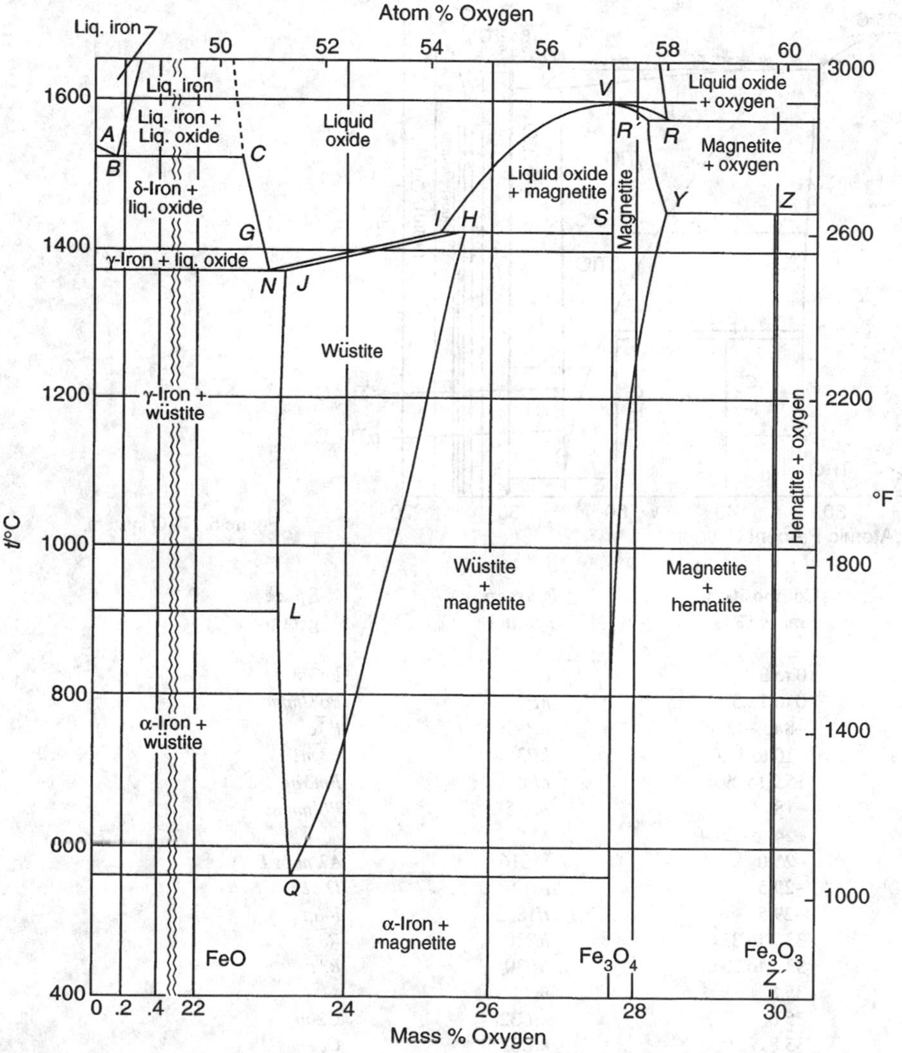

Figure 5. Fe-O system.

Figure 5 (continued)

Point	t/°C	% O	p_{CO_2}/p_{CO}
A	1539		
B	1528	0.16	0.209
C	1528	22.60	0.209
G	1400[a]	22.84	0.263
H	1424	25.60	16.2
I	1424	25.31	16.2
J	1371	23.16	0.282
L	911[a]	23.10	0.447
N	1371	22.91	0.282

Point	t/°C	% O	p_{CO_2}/p_{CO}	p_{O_2}/atm
Q	560	23.26	1.05	
R	1583	28.30		1
R′	1583	28.07		1
S	1424	27.64	16.2	
V	1597	27.64		0.0575
Y	1457	28.36		1
Z	1457	30.04		1
Z′		30.6		

[a] Values for pure iron.

Figure 6

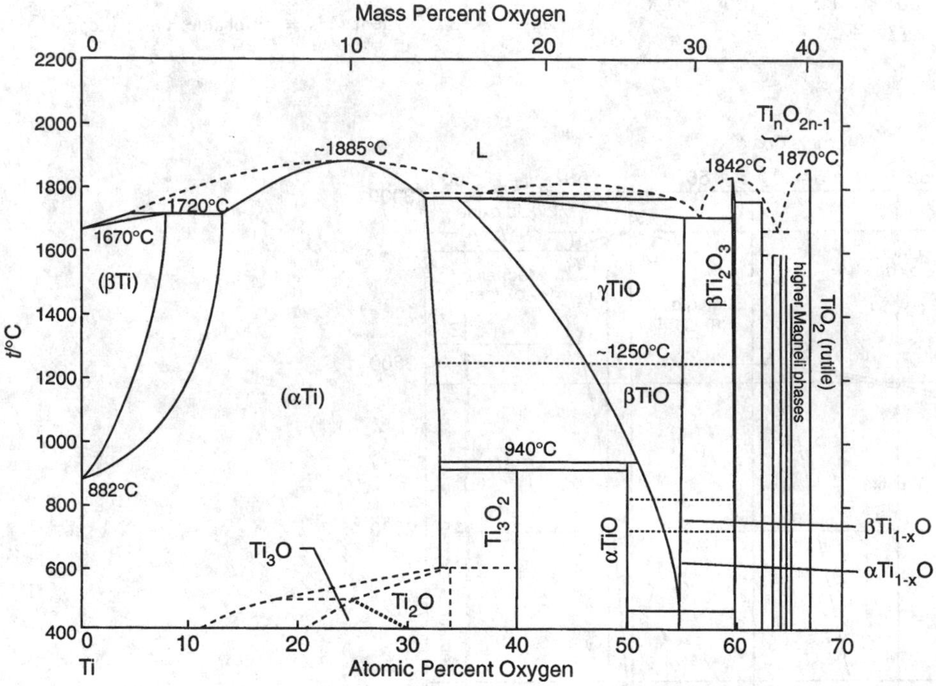

Figure 6. Ti-O system.

Phase	Composition, mass % O	Pearson symbol	Space group
(βTi)	0 to 3	*cI*2	$Im\bar{3}m$
(αTi)	0 to 13.5	*hP*2	*P63/mmc*
Ti_3O	~8 to ~13	*hP*~16	$P\bar{3}c$
Ti_2O	~10 to 14.4	*hP*3	$P\bar{3}m1$
γTiO	15.2 to 29.4	*cF*8	$Fm\bar{3}m$
Ti_3O_2	~18	*hP*~5	*P6/mmm*
βTiO	~24 to ~29.4	*c***	—
αTiO	~25.0	*mC*16	*A2/m* or *B*/**
$\beta Ti_{1-x}O$	~29.5	*oI*12	*I*222
$\alpha Ti_{1-x}O$	~29.5	*tI*18	*I4/m*
βTi_2O_3	33.2 to 33.6	*hR*30	$R\bar{3}c$
αTi_2O_3	33.2 to 33.6	*hR*30	$R\bar{3}c$
βTi_3O_5	35.8	*m***	—
αTi_3O_5	35.8	*mC*32	*C2/m*
$\alpha' Ti_3O_5$	35.8	*mC*32	*Cc*

Figure 6 (continued)

Phase	Composition, mass % O	Pearson symbol	Space group
γTi_4O_7	36.9	*aP*44	$P\bar{1}$
βTi_4O_7	36.9	*aP*44	$P\bar{1}$
$\alpha Ti4O_7$	36.9	*aP*44	$P\bar{1}$
γTi_5O_9	37.6	*aP*28	$P\bar{1}$
βTi_6O_{11}	38.0	*aC*68	$A\bar{1}$
Ti_7O_{13}	38.3	*aP*40	$P\bar{1}$
Ti_8O_{15}	38.5	*aC*92	$A\bar{1}$
Ti_9O_{17}	38.7	*aP*52	$P\bar{1}$
Rutile TiO_2	40.1	*tP*6	$P4_2/mnm$
Metastable phases			
Anatase	—	*tI*12	$I4_1/amd$
Brookite	—	*oP*24	*Pbca*
High-pressure phases			
TiO_2-II	—	*oP*12	*Pbcn*
TiO_2-III	—	*hP*~48	—

FIGURE 7

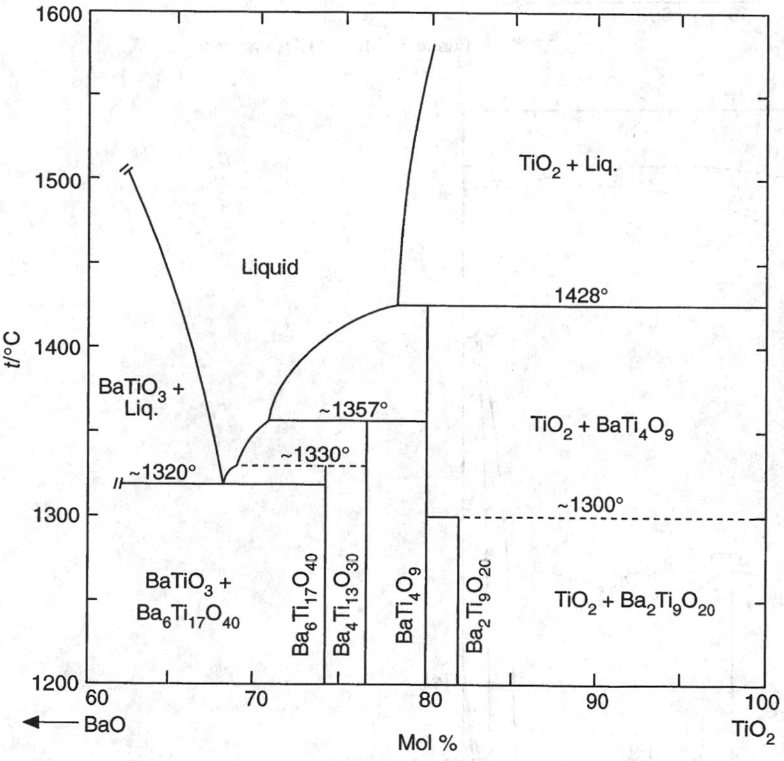

Figure 7. BaO-TiO_2 system.

Figure 8

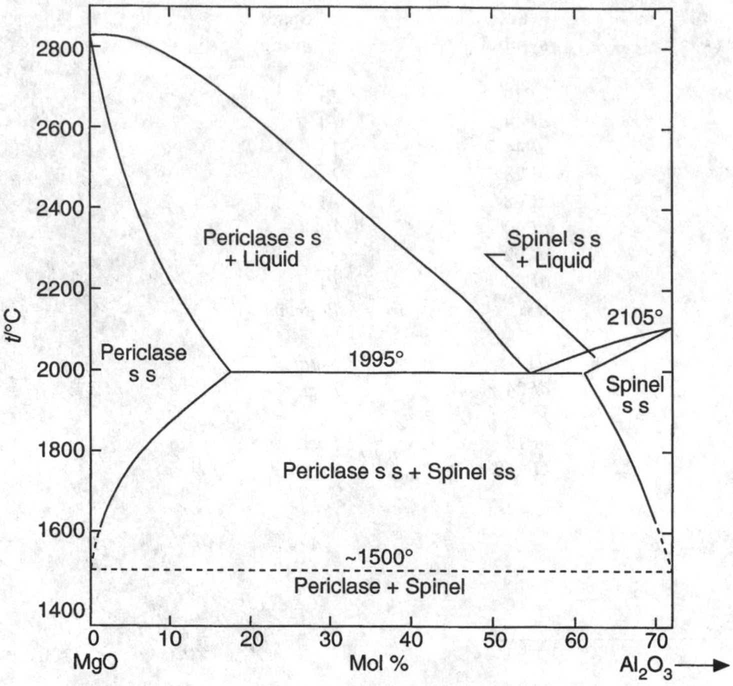

Figure 8. $MgO-Al_2O_3$ system.

Figure 9

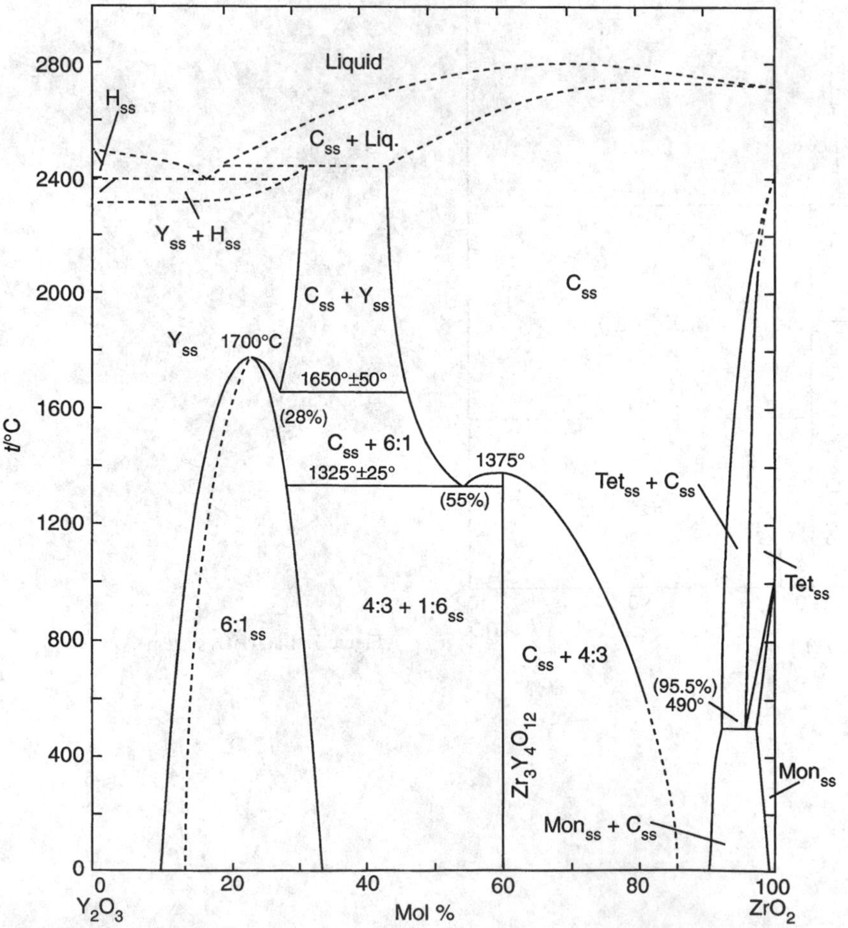

Figure 9. Y_2O_3-ZrO_2 system. C_{ss} = cubic ZrO_2 ss (fluorite-type ss); Y_{ss} = cubic Y_2O_3 ss; Tet_{ss} = tetragonal ZrO_2 ss; Mon_{ss} = monoclinic ZrO_2 ss; H_{ss} = hexagonal Y_2O_3 ss; 3:4 = $Zr_3Y_4O_{12}$; 1:6 = ZrY_6O_{11} ss.

Figure 10

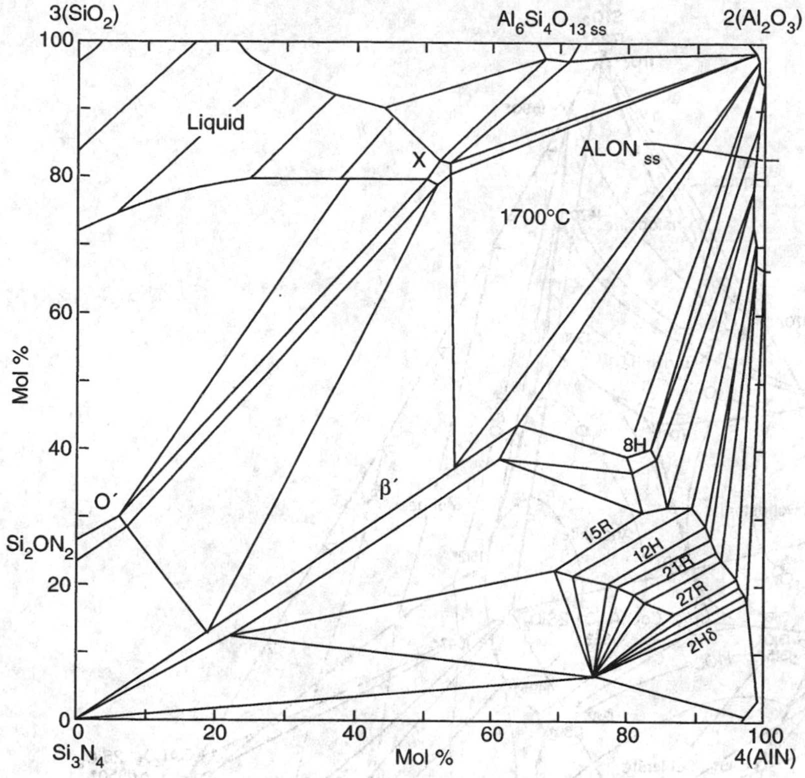

Figure 10. $3(SiO_2)$-Si_3N_4-$4(AlN)$-$2(Al_2O_3)$ system. "Behavior" diagram at 1700°C. The labels 8H, 15R, 12H, 21R, 27R, $2H^{\delta}$ indicate defect AlN polytypes. β′ = 3-sialon ($Si_{6-x}Al_xO_xN_{8-x}$); O′ = sialon of Si_2ON_2 type; X = $SiAlO_2N$ ("nitrogen mullite"). ALON ss = aluminum oxynitride ss extending from approximately Al_7O_9N to Al_3O_3N.

Figure 11

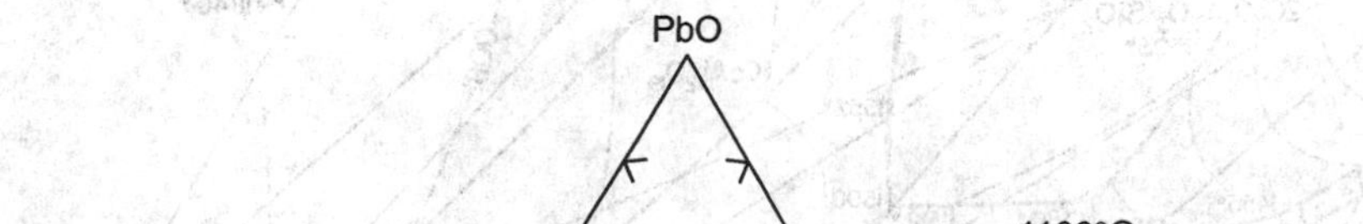

Figure 11. PbO-ZrO_2-TiO_2 (PZT) system, subsolidus at 1100°C. $P = PbO$; $T = TiO_2$; $Z = ZrO_2$.

Figure 12

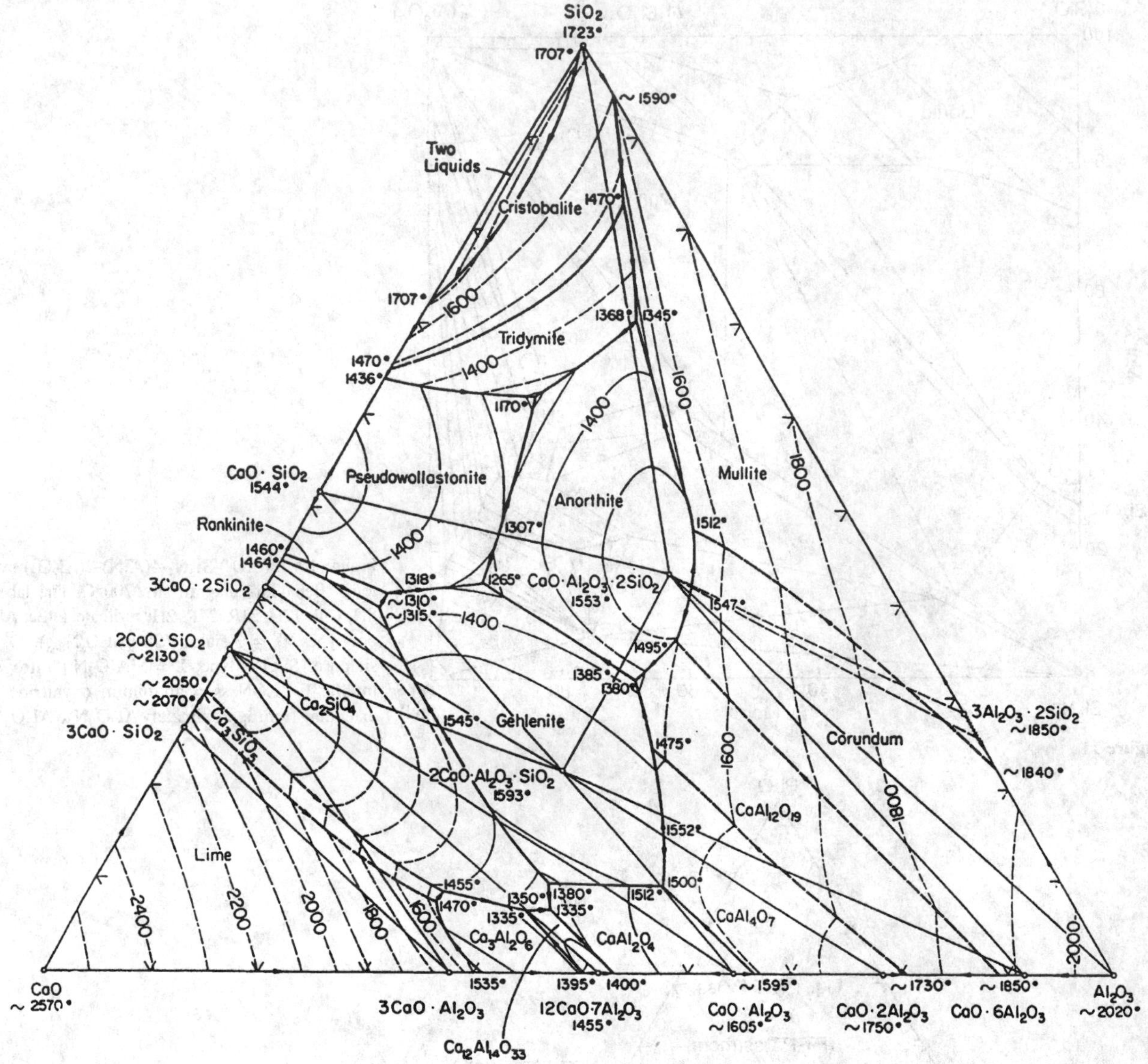

Figure 12. CaO-Al_2O_3-SiO_2 system (temperatures in °C).

Crystalline Phases

Notation	Oxide formula
Cristobalite } Tridymite }	SiO_2
Pseudowollastonite	$CaO{\cdot}SiO_2$
Rankinite	$3CaO{\cdot}2SiO_2$
Lime	CaO
Corundum	Al_2O_3
Mullite	$3Al_2O_3{\cdot}2SiO_2$
Anorthite	$CaO{\cdot}Al_2O_3{\cdot}2SiO_2$
Gehlenite	$2CaO{\cdot}Al_2O_3{\cdot}SiO_2$

Temperatures up to approximately 1550°C are on the Geophysical Laboratory Scale; those above 1550°C are on the 1948 International Scale.

Figure 13

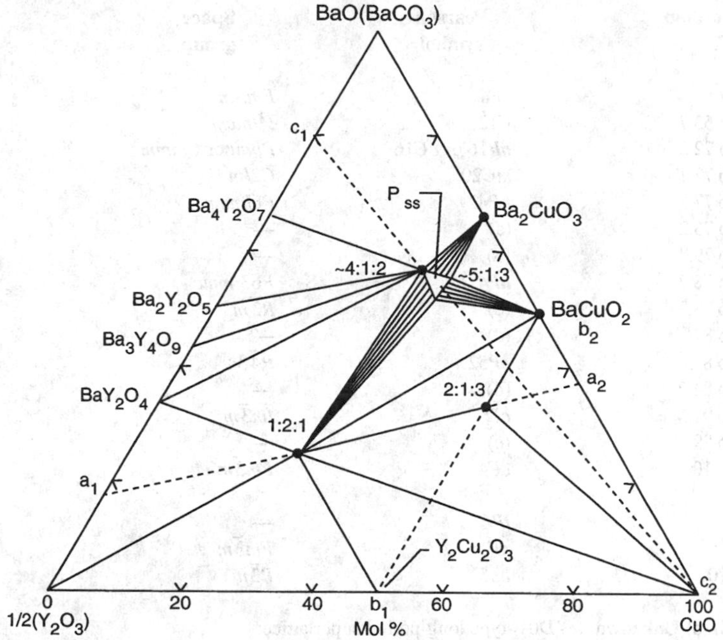

Figure 13. BaO-Y_2O_3-CuO system. 2:1:3 = $Ba_2YCu_3O_{7-x}$; 1:2:1 = BaY_2CuO_5; 4:1:2 = $Ba_4YCu_2O_{7.5+x}$; and 5:1:3 = $Ba_5YCu_3O_{9.5+x}$. The superconducting 2:1:3 phase was prepared using barium peroxide.

Figure 14

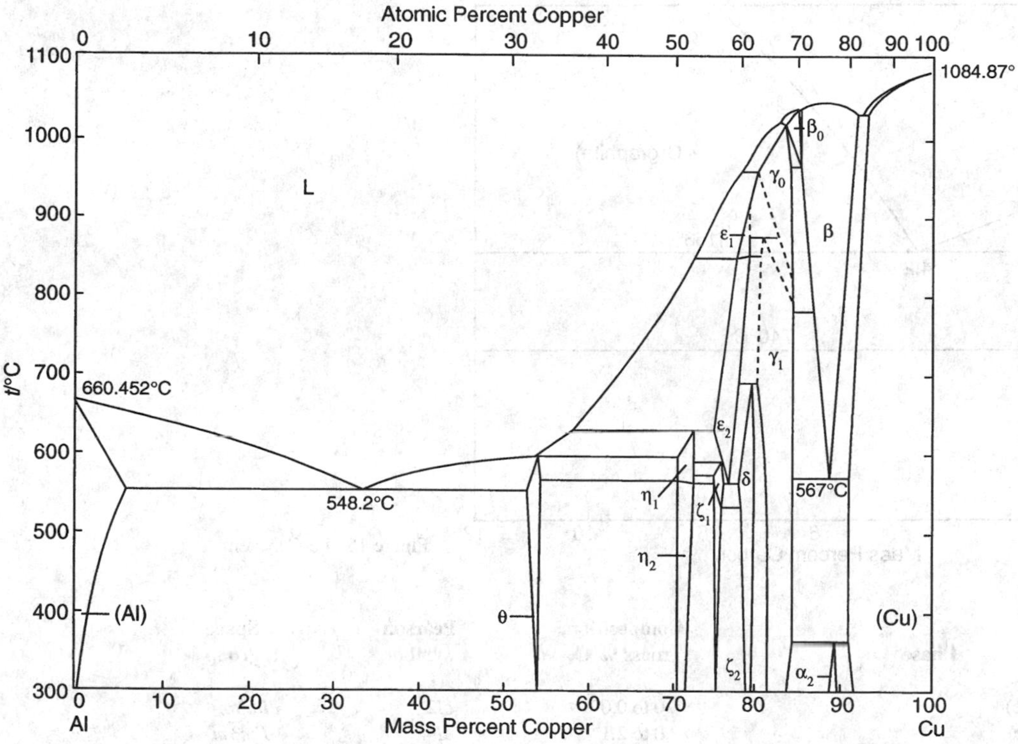

Figure 14. Al-Cu system.

Figure 14 (continued)

Phase	Composition, wt % Cu	Pearson symbol	Space group
(Al)	0 to 5.65	*cF*4	$Fm\bar{3}m$
θ	52.5 to 53.7	*tI*12	*I*4/*mcm*
η_1	70.0 to 72.2	*oP*16 or *oC*16	*Pban* or *Cmmm*
η_2	70.0 to 72.1	*mC*20	*C*2/*m*
ζ_1	74.4 to 77.8	*hP*42	*P*6/*mmm*
ζ_2	74.4 to 75.2	(a)	—
ε_1	77.5 to 79.4	(b)	—
ε_2	72.2 to 78.7	*hP*4	*P*63/*mmc*
δ	77.4 to 78.3	(c)	$R\bar{3}m$
γ_0	77.8 to 84	(d)	—
γ_1	79.7 to 84	*cP*52	$P\bar{4}3m$
β_0	83.1 to 84.7	(d)	—
β	85.0 to 91.5	*cI*2	$Im\bar{3}m$
α_2	88.5 to 89	(e)	—
(Cu)	90.6 to 100	*cF*4	$Fm\bar{3}m$
Metastable phases			
θ′	—	*tP*6	—
β′	—	*cF*16	$Fm\bar{3}m$
Al_3Cu_2	61 to 70	*hp*5	$P\bar{3}m1$

(a) Monoclinic? (b) Cubic? (c) Rhombohedral. (d) Unknown. (e) $D0_{22}$-type long-period superlattice.

Figure 15

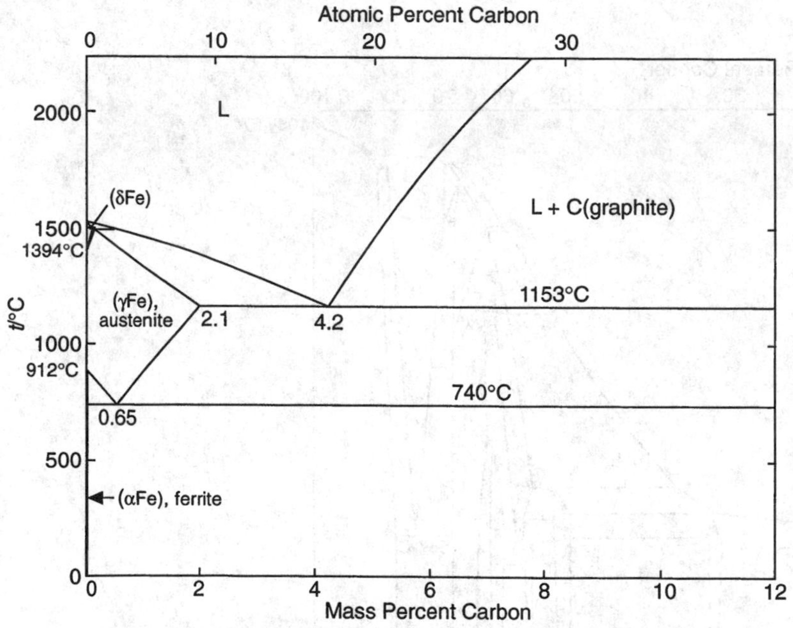

Figure 15. Fe-C system.

Phase	Composition, mass % C	Pearson symbol	Space group
(δFe)	0 to 0.09	*cI*2	$Im\bar{3}m$
(γFe)	0 to 2.1	*cF*4	$Fm\bar{3}m$
(αFe)	0 to 0.021	*cI*2	$Im\bar{3}m$
(C)	100	*hP*4	$P6_3/mmc$
Metastable/high-pressure phases			
(εFe)	0	*hP*2	$P6_3/mmc$
Martensite	< 2.1	*tI*4	*I*4/*mmm*

Figure 15 (continued)

Phase	Composition, mass % C	Pearson symbol	Space group
Fe_4C	5.1	*cP*5	$P\bar{4}3m$
Fe_3C (θ)	6.7	*oP*16	*Pnma*
Fe_5C_2 (χ)	7.9	*mC*28	*C*2/*c*
Fe_7C_3	8.4	*hP*20	$P6_3mc$
Fe_7C_3	8.4	*oP*40	*Pnma*
Fe_2C (η)	9.7	*oP*6	*Pnnm*
Fe_2C (ε)	9.7	*hP**	$P6_322$
Fe_2C	9.7	*hP**	$P\bar{3}m1$
(C)	100	*cF*8	$Fd\bar{3}m$

Figure 16

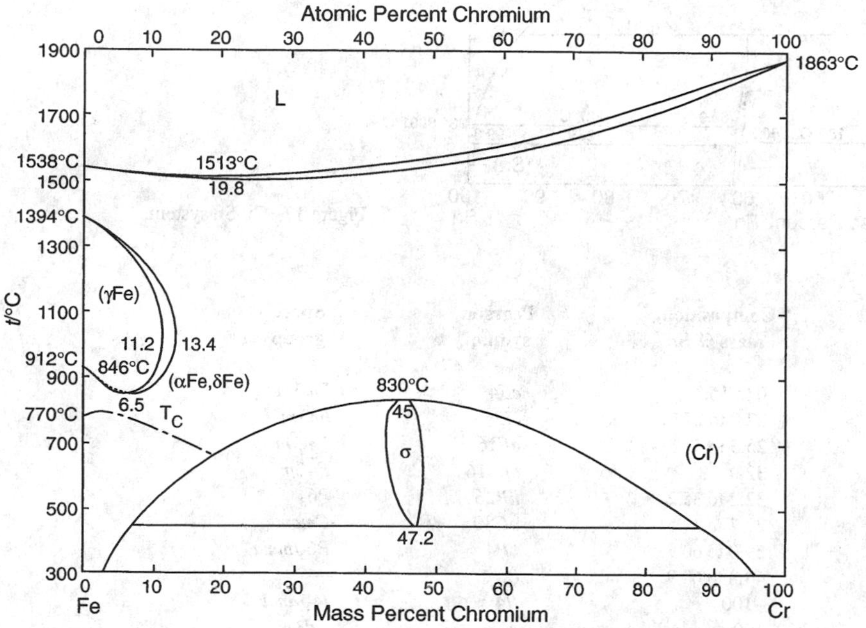

Figure 16. Fe-Cr system.

Phase	Composition, mass % Cr	Pearson symbol	Space group
(aFe, Cr)	0 to 100	*cI*2	$Im\bar{3}m$
(γFe)	0 to 11.2	*cF*4	$Fm\bar{3}m$
σ	42.7 to 48.2	*tP*30	$P4_2/mnm$

Figure 17

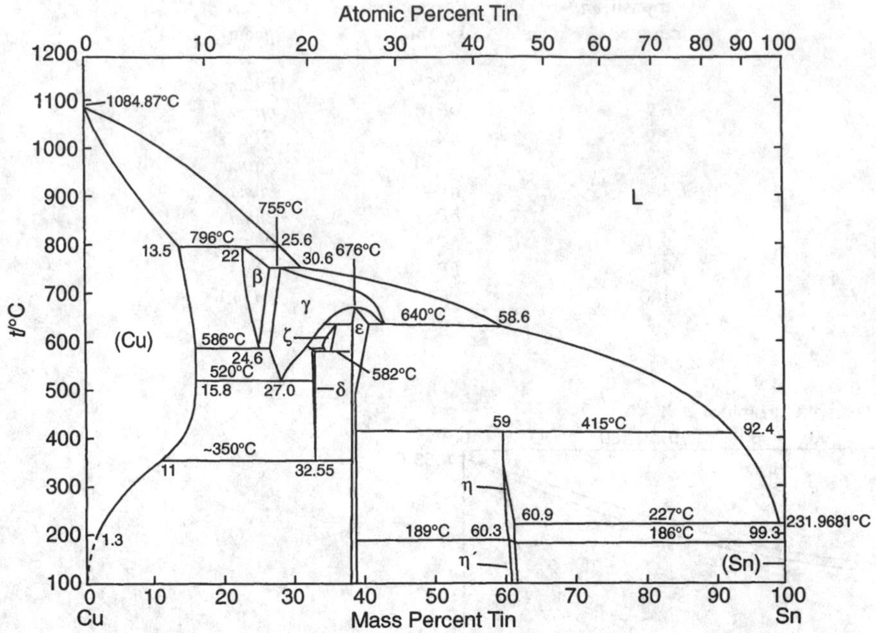

Figure 17. Cu-Sn system.

Phase	Composition, mass % Sn	Pearson symbol	Space group
α	0 to 15.8	*cF*4	$Fm\bar{3}m$
β	22.0 to 27.0	*cI*2	$Im\bar{3}m$
γ	25.5 to 41.5	*cF*16	$Fm\bar{3}m$
δ	32 to 33	*cF*416	$F\bar{4}3m$
ζ	32.2 to 35.2	*hP*26	$P6_3$
ε	27.7 to 39.5	*oC*80	*Cmcm*
η	59.0 to 60.9	*hP*4	$P6_3/mmc$
η′	44.8 to 60.9	(a)	—
(βSn)	~100	*tI*4	$I4_1/amd$
(αSn)	100	*cF*8	$Fd\bar{3}m$

(a) Hexagonal; superlattice based on NiAs-type structure.

Figure 18

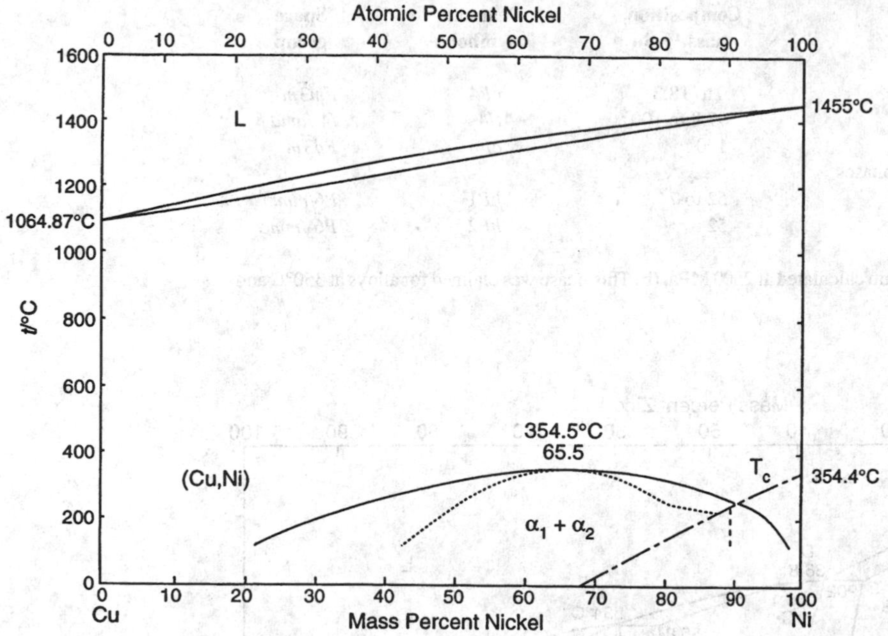

Figure 18. Cu-Ni system.

Phase	Composition, mass % Ni	Pearson symbol	Space group
(Cu, Ni) (above 354.5°C)	0 to 100	*cF*4	$Fm\bar{3}m$

Figure 19

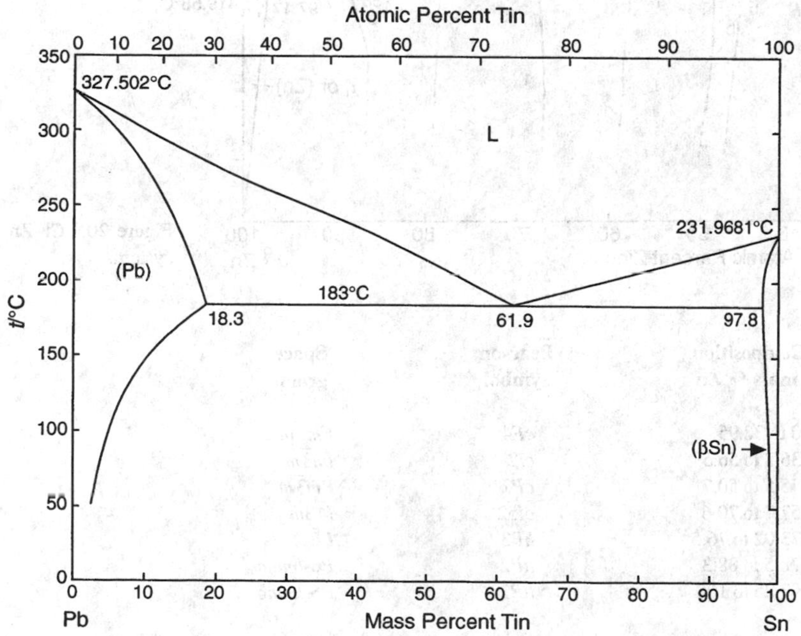

Figure 19. Pb-Sn system.

Figure 19 (continued)

Phase	Composition, mass % Sn	Pearson symbol	Space group
(Pb)	0 to 18.3	*cF*4	$Fm\bar{3}m$
(βSn)	97.8 to 100	*tI*4	$I4_1/amd$
(αSn)	100	*cF*8	$Fd\bar{3}m$
High-pressure phases			
ε(a)	52 to 74	*hP*1	$P6/mmm$
ε′(b)	52	*hP*2	$P6_3/mmc$

(a) From phase diagram calculated at 2500 MPa. (b) This phase was claimed for alloys at 350°C and 5500 MPa.

Figure 20

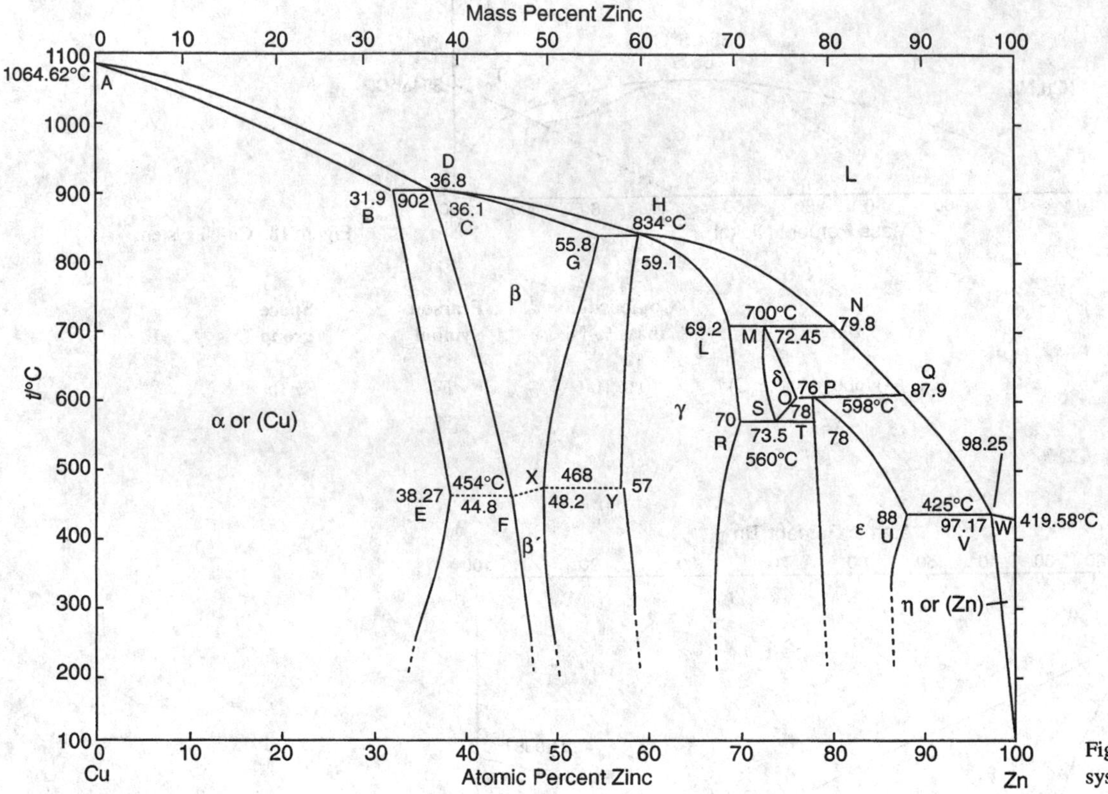

Figure 20. Cu-Zn system.

Phase	Composition, mass % Zn	Pearson symbol	Space group
α or (Cu)	0 to 38.95	*cF*4	$Fm\bar{3}m$
β	36.8 to 56.5	*cI*2	$Im\bar{3}m$
β′	45.5 to 50.7	*cP*2	$Pm\bar{3}m$
γ	57.7 to 70.6	*cI*52	$I\bar{4}3m$
δ	73.02 to 76.5	*hP*3	$P\bar{6}$
ε	78.5 to 88.3	*hP*2	$P6_3/mmc$
η or (Zn)	97.25 to 100	*hP*2	$P6_3/mmc$

Figure 21

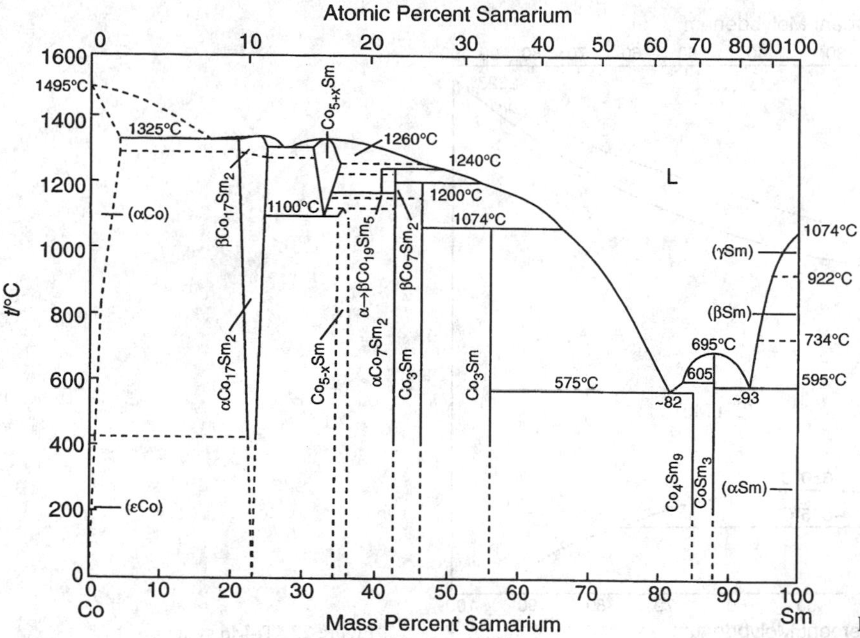

Figure 21. Co-Sm system.

Phase	Composition, mass % Sm	Pearson symbol	Space group
(αCo)	0 to ~3.7	*cF*4	$Fm\bar{3}m$
(εCo)	~0	*hP*2	$P6_3/mmc$
$\beta Co_{17}Sm_2$	~23.0	*hP*38	$P6_3/mmc$
$\alpha Co_{17}Sm_2$	~23.0	*hR*19	$R\bar{3}m$
		*hP*8	$P6/mmm$
$Co_{5+x}Sm$	~33 to 34	—	—
$Co_{5-x}Sm$	~34 to 35	—	—
$Co_{19}Sm_5$	~40.1	*hR*24	$R\bar{3}m$
		*hP*48	$P6_3/mmc$
αCo_7Sm_2	~42.1	*hR*18	$R\bar{3}m$
βCo_7Sm_2	~42.1	*hP*36	$P6_3/mmc$
Co_3Sm	46	*hR*12	$R\bar{3}m$
Co_2Sm	56.0	*hR*4	$R\bar{3}m$
		*cF*24	$Fd\bar{3}m$
Co_4Sm_9	~85.1	*o***	—
$CoSm_3$	88	*oP*16	$Pnma$
(γSm)	~100	*cI*2	$Im\bar{3}m$
(βSm)	~100	*hP*2	$P6_3/mmc$
(αSm)	~100	*hR*3	$R\bar{3}m$
Other reported phases			
Co_5Sm	~33.8	*hP*6	$P6/mmm$
Co_2Sm_5	~86.4	*mC*28	$C2/c$

Figure 22

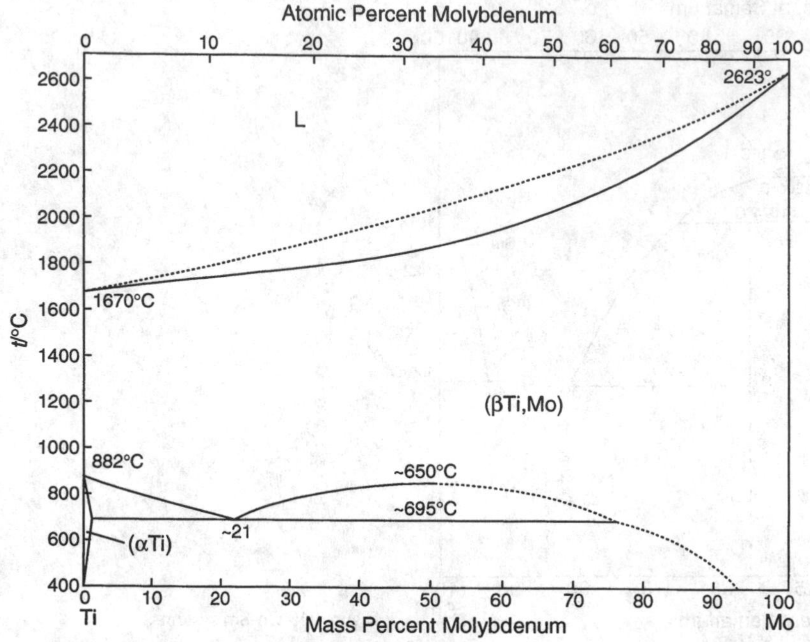

Figure 22. Ti-Mo system.

Phase	Composition, mass % Mo	Pearson symbol	Space group
(βTi, Mo)	0 to 100	*cI*2	$Im\bar{3}m$
(αTi)	0 to 0.8	*hP*2	$P6_3/mmc$
α′	(a)	*hP*2	$P6_3/mmc$
α″	(a)	*oC*4	*Cmcm*
ω	(a)	*hP*3	*P*6/*mmm*

(a) Metastable.

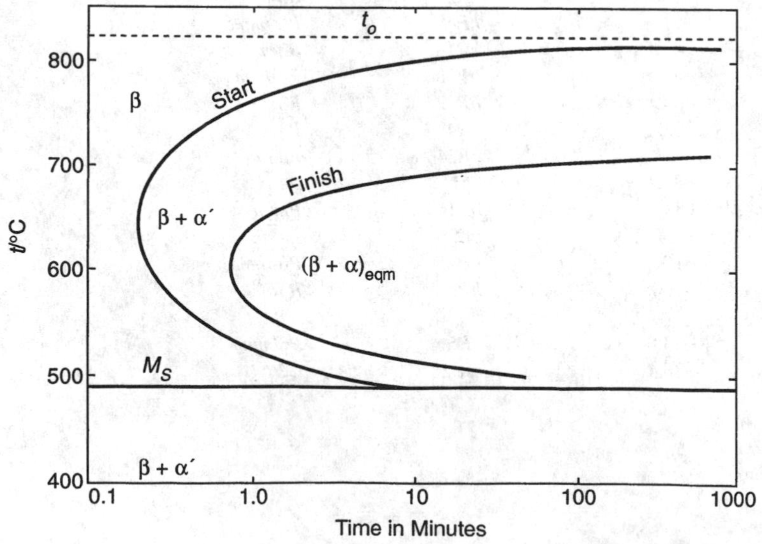

Experimental time-temperature-transformation (TTT) diagram for Ti-Mo. The start and finish times of the isothermal precipitation reaction vary with temperature as a result of the temperature dependence of the nucleation and growth processes. Precipitation is complete, at any temperature, when the equilibrium fraction of α is established in accordance with the lever rule. The solid horizontal line represents the athermal (or nonthermally activated) martensitic transformation that occurs when the β phase is quenched.

Figure 23

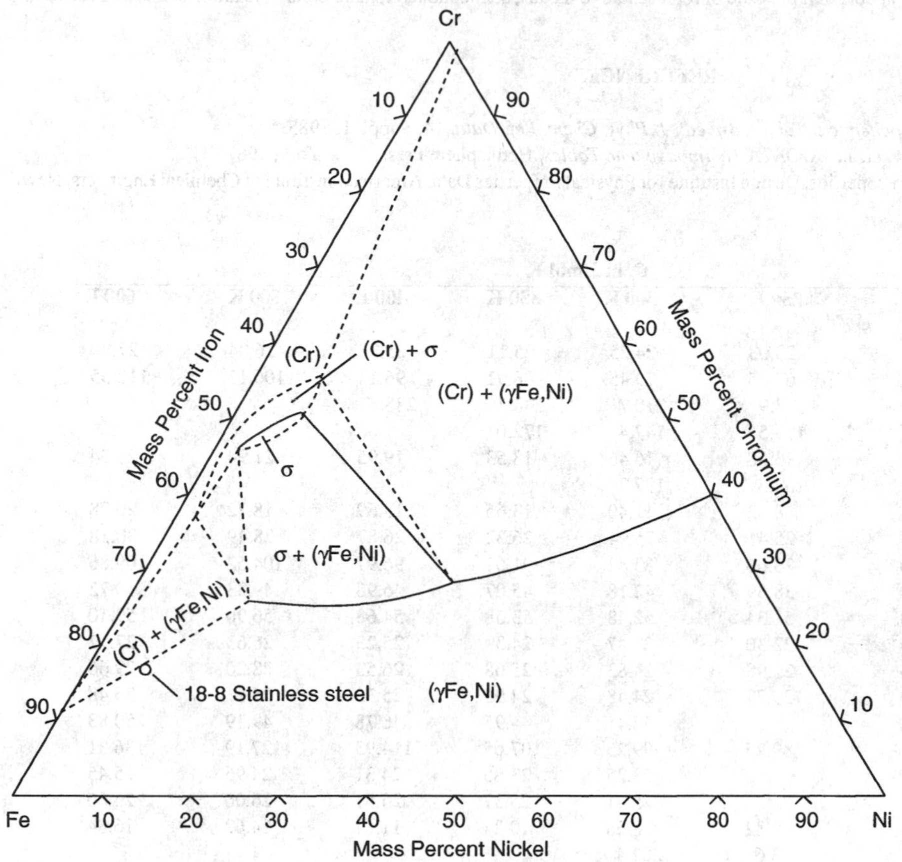

Figure 23. The isothermal section at 900°C (1652°F) of the iron-chromium-nickel ternary phase diagram, showing the nominal composition of 18-8 stainless steel.

HEAT CAPACITY OF SELECTED SOLIDS

This table gives the molar heat capacity at constant pressure of representative metals, semiconductors, and other crystalline solids as a function of temperature in the range 200 to 600 K.

REFERENCES

1. Chase, M. W., et al., *JANAF Thermochemical Tables, 3rd ed., J. Phys. Chem. Ref. Data,* 14, Suppl. 1, 1985.
2. Garvin, D., Parker, V. B., and White, H. J., *CODATA Thermodynamic Tables,* Hemisphere Press, New York, 1987.
3. DIPPR Database of Pure Compound Properties, Design Institute for Physical Properties Data, American Institute of Chemical Engineers, New York, 1987.

	C_p in J/mol K						
Name	**200 K**	**250 K**	**300 K**	**350 K**	**400 K**	**500 K**	**600 K**
Aluminum	21.33	23.08	24.25	25.11	25.78	26.84	27.89
Aluminum oxide	51.12	67.05	79.45	88.91	96.14	106.17	112.55
Anthracene	138.6	173.9	210.7	248.8	288.4		
Benzoic acid	102.7	123.5	147.4	172.0			
Beryllium	9.98	13.58	16.46	18.53	19.95	21.94	23.34
Biphenyl	131.0	162.5	197.2				
Boron	5.99	8.82	11.40	13.65	15.69	18.72	20.78
Calcium	24.54	25.41	25.94	26.32	26.87	28.49	30.38
Calcium carbonate	66.50	75.66	83.82	91.51	96.97	104.52	109.86
Calcium oxide	33.64	38.59	42.18	45.07	46.98	49.33	50.72
Cesium chloride	50.13	51.34	52.48	53.58	54.68	56.90	59.10
Chromium	19.86	22.30	23.47	24.39	25.23	26.63	27.72
Cobalt	22.23	23.98	24.83	25.68	26.53	28.20	29.66
Copper	22.63	23.77	24.48	24.95	25.33	25.91	26.48
Copper oxide	34.80		42.41	44.95	46.78	49.19	50.83
Copper sulfate	77.01	89.25	99.25	107.65	114.93	127.19	136.31
Germanium			23.25	23.85	24.31	24.96	25.45
Gold			25.41	25.37	25.51	26.06	26.65
Graphite	5.01	6.82	8.58	10.24	11.81	14.62	16.84
Hexachlorobenzene	162.7	183.6	202.4				
Iodine	51.57	53.24	54.51	58.60			
Iron	21.59	23.74	25.15	26.28	27.39	29.70	32.05
Lead	25.87	26.36	26.85	27.30	27.72	28.55	29.40
Lithium	21.57	23.42	24.64	25.96	27.60	29.28	
Lithium chloride	43.35	46.08	48.10	49.66	50.97	53.34	55.59
Magnesium	22.72	24.02	24.90	25.57	26.14	27.17	28.18
Magnesium oxide			37.38	40.59	42.77	45.56	47.30
Manganese	23.05	24.95	26.35	27.52	28.53	30.29	31.90
Naphthalene	105.8	134.1	167.8	204.1			
Potassium	27.00	28.01	29.60				
Potassium chloride	48.44	50.10	51.37	52.31	53.08	54.71	56.35
Silicon	15.64	18.22	20.04	21.28	22.14	23.33	24.15
Silicon dioxide	32.64	39.21	44.77	49.47	53.43	59.64	64.42
Silver			25.36	25.55	25.79	26.36	26.99
Sodium	22.45	27.01	28.20	30.14			
Sodium chloride	46.89	48.85	50.21	51.25	52.14	53.96	55.81
Tantalum	24.08	24.86	25.31	25.60	25.84	26.35	26.84
Titanium	22.37	24.07	25.28	26.17	26.86	27.88	28.60
Tungsten	22.49	23.69	24.30	24.65	24.92	25.36	25.79
Vanadium	21.88	23.70	24.93	25.68	26.23	26.94	27.49
Zinc	24.05	25.02	25.45	25.88	26.35	27.39	28.59
Zirconium	23.87	24.69	25.22	25.61	25.93	26.56	27.28

THERMAL AND PHYSICAL PROPERTIES OF PURE METALS

This table gives the following properties for the metallic elements:

t_m: Melting point in °C
t_b: Normal boiling point in °C, at a pressure of 101.325 kPa (760 Torr)
$\Delta_{fus} H$: Enthalpy of fusion at the melting point in J/g
ρ_{25}: Density at 25°C in g/cm^3
α: Coefficient of linear expansion at 25°C in K^{-1} (the quantity listed is $10^6 \times \alpha$)
c_p: Specific heat capacity at constant pressure at 25°C in J/g K
λ: Thermal conductivity at 27°C in W/cm K

REFERENCES

1. Dinsdale, A. T., *CALPHAD*, 15, 317, 1991 (melting points, enthalpy of fusion).
2. Touloukian, Y. S., *Thermophysical Properties of Matter*, Vol. 12, Thermal Expansion, IFI/Plenum, New York, 1975 (coefficient of expansion, density).
3. Ho, C. Y., Powell, R. W., and Liley, P. E., *J. Phys. Chem. Ref. Data*, 3, Suppl. 1, 1974 (thermal conductivity).
4. Cox, J. D., Wagman, D. D., and Medvedev, V. A., *CODATA Key Values for Thermodynamics*, Hemisphere Publishing Corp., New York, 1989 (heat capacity).
5. Glushko, V. P., Ed., *Thermal Constants of Substances*, VINITI, Moscow, (enthalpy of fusion, heat capacity).
6. Wagman, D. D., et. al., *The NBS Tables of Chemical Thermodynamic Properties, J. Phys. Chem. Ref. Data*, 11, Suppl. 2, 1982 (heat capacity).
7. Chase, M. W., et. al., *JANAF Thermochemical Tables*, 3rd ed., J. Phys. Chem. Ref. Data, 14, Suppl. 1, 1985 (heat capacity, enthalpy of fusion).
8. Gschneidner, K. A., *Bull. Alloy Phase Diagrams*, 11, 216—224, 1990 (various properties of the rare earth metals).
9. Hellwege, K. H., Ed., *Landolt Börnstein, Numerical Values and Functions in Physics, Chemistry, Astronomy, Geophysics, and Technology*, Vol. 2, Part 1, Mechanical-Thermal Properties of State, 1971 (density).
10. *Physical Encyclopedic Dictionary*, Vol. 1–5, Encyclopedy Publishing House, Moscow, 1960–66.

Metal (symbol)	Atomic weight	t_m °C	t_b °C	$\Delta_{fus} H$ J/g	ρ_{25} g/cm^3	$\alpha \times 10^6$ K^{-1}	c_p J/g K	λ W/cm K
Actinium (Ac)		1051	3198		10			
Aluminum (Al)	26.98	660.32	2519	399.9	2.70	23.1	0.897	2.37
Antimony (Sb)	121.76	630.63	1587	162.5	6.68	11.0	0.207	0.243
Barium (Ba)	137.33	727	1897	51.8	3.62	20.6	0.204	0.184
Beryllium (Be)	9.01	1287	2471	876.0	1.85	11.3	1.825	2.00
Bismuth (Bi)	208.98	271.40	1564	53.3	9.79	13.4	0.122	0.0787
Cadmium (Cd)	112.41	321.07	767	55.2	8.69	30.8	0.232	0.968
Calcium (Ca)	40.08	842	1484	213.1	1.54	22.3	0.647	2.00
Cerium (Ce)	140.11	798	3443	39.0	6.77	6.3	0.192	0.113
Cesium (Cs)	132.91	28.44	671	15.7	1.93	97	0.242	0.359
Chromium (Cr)	52.00	1907	2671	404	7.15	4.9	0.449	0.937
Cobalt (Co)	58.93	1495	2927	272.5	8.86	13.0	0.421	1.00
Copper (Cu)	63.55	1084.62	2562	203.5	8.96	16.5	0.385	4.01
Dysprosium (Dy)	162.50	1412	2567	68.1	8.55	9.9	0.170	0.107
Erbium (Er)	167.26	1529	2868	119	9.07	12.2	0.168	0.145
Europium (Eu)	151.96	822	1529	60.6	5.24	35.0	0.182	0.139[a]
Gadolinium (Gd)	157.25	1313	3273	63.6	7.90	9.4[b]	0.236	0.105
Gallium (Ga)	69.72	29.76	2204	80.0	5.91	18	0.371	0.406
Gold (Au)	196.97	1064.18	2856	64.6	19.3	14.2	0.129	3.17
Hafnium (Hf)	178.49	2233	4603	152.4	13.3	5.9	0.144	0.230
Holmium (Ho)	164.93	1474	2700	103[a]	8.80	11.2	0.165	0.162
Indium (In)	114.82	156.60	2072	28.6	7.31	32.1	0.233	0.816
Iridium (Ir)	192.22	2446	4428	213.9	22.5	6.4	0.131	1.47
Iron (Fe)	55.85	1538	2861	247.3	7.87	11.8	0.449	0.802
Lanthanum (La)	138.91	918	3464	44.6	6.15	12.1	0.195	0.134
Lead (Pb)	207.20	327.46	1749	23.1	11.3	28.9	0.129	0.353
Lithium (Li)	6.94	180.5	1342	432	0.534	46	3.582	0.847
Lutetium (Lu)	174.97	1663	3402	126[a]	9.84	9.9	0.154	0.164
Magnesium (Mg)	24.30	650	1090	348.9	1.74	24.8	1.023	1.56

THERMAL AND PHYSICAL PROPERTIES OF PURE METALS (continued)

Metal (symbol)	Atomic weight	t_m °C	t_b °C	$\Delta_{fus}H$ J/g	ρ_{25} g/cm³	$\alpha \times 10^6$ K⁻¹	c_p J/g K	λ W/cm K
Manganese (Mn)	54.94	1246	2061	235.0	7.3	21.7	0.479	0.0782
Mercury (Hg)	200.59	-38.83	356.73	11.4	13.5336	60.4	0.140	0.0834
Molybdenum (Mo)	95.94	2623	4639	390.7	10.2	4.8	0.251	1.38
Neodymium (Nd)	144.24	1021	3074	49.5	7.01	9.6	0.190	0.165
Neptunium (Np)		644		13.5	20.2			0.063
Nickel (Ni)	58.69	1455	2913	290.3	8.90	13.4	0.444	0.907
Niobium (Nb)	92.91	2477	4744	323	8.57	7.3	0.265	0.537
Osmium (Os)	190.23	3033	5012	304.1	22.59	5.1	0.130	0.876
Palladium (Pd)	106.42	1554.9	2963	157.3	12.0	11.8	0.246	0.718
Platinum (Pt)	195.08	1768.4	3825	113.6	21.5	8.8	0.133	0.716
Plutonium (Pu)		640	3228	11.6	19.7	46.7		0.0674
Polonium (Po)		254	962		9.20	23.5		0.20
Potassium (K)	39.10	63.38	759	59.6	0.89	83.3	0.757	1.024
Praseodymium (Pr)	140.91	931	3520	48.9	6.77	6.7	0.193	0.125
Promethium (Pm)		1042	3000[a]		7.26	11[a]	0.19[a]	0.15[a]
Protactinium (Pa)	231.04	1572		53.4	15.4			
Radium (Ra)		700			5			
Rhenium (Re)	186.21	3186	5596	324.5	20.8	6.2	0.137	0.479
Rhodium (Rh)	102.91	1964	3695	258.4	12.4	8.2	0.243	1.50
Rubidium (Rb)	85.47	39.30	688	25.6	1.53		0.363	0.582
Ruthenium (Ru)	101.07	2334	4150	381.8	12.1	6.4	0.238	1.17
Samarium (Sm)	150.36	1074	1794	57.3	7.52	12.7	0.197	0.133
Scandium (Sc)	44.96	1541	2836	314	2.99	10.2	0.568	0.158
Silver (Ag)	107.87	961.78	2162	104.6	10.5	18.9	0.235	4.29
Sodium (Na)	22.99	97.72	883	113.1	0.97	71	1.228	1.41
Strontium (Sr)	87.62	777	1382	84.8	2.64	22.5	0.301	0.353
Tantalum (Ta)	180.95	3017	5458	202.1	16.4	6.3	0.140	0.575
Technetium (Tc)		2157	4265	339.7	11			0.506
Terbium (Tb)	158.93	1356	3230	67.9	8.23	10.3	0.182	0.111
Thallium (Tl)	204.38	304	1473	20.3	11.8	29.9	0.129	0.461
Thorium (Th)	232.04	1750	4788	59.5	11.7	11.0	0.113	0.540
Thulium (Tm)	168.93	1545	1950	99.7	9.32	13.3	0.160	0.169
Tin (Sn)	118.71	231.93	2602	60.4	7.26	22.0	0.228	0.666
Titanium (Ti)	47.88	1668	3287	295.6	4.51	8.6	0.523	0.219
Tungsten (W)	183.84	3422	5555	284.5	19.3	4.5	0.132	1.74
Uranium (U)	238.03	1135	4131	38.4	19.1	13.9	0.116	0.276
Vanadium (V)	50.94	1910	3407	422	6.0	8.4	0.489	0.307
Ytterbium (Yb)	173.04	819	1196	44.3	6.90	26.3	0.155	0.385
Yttrium (Y)	88.91	1522	3345	128	4.47	10.6	0.298	0.172
Zinc (Zn)	65.39	419.53	907	108.1	7.14	30.2	0.388	1.16
Zirconium (Zr)	91.22	1855	4409	230.2	6.52	5.7	0.278	0.227

[a] Estimated.

[b] At 100°C.

THERMAL CONDUCTIVITY OF METALS AND SEMICONDUCTORS AS A FUNCTION OF TEMPERATURE

This table gives the temperature dependence of the thermal conductivity of several metals and of carbon, germanium, and silicon. For graphite, separate entries are given for the thermal conductivity parallel (∥) and perpendicular (⊥) to the layer planes. The thermal conductivity of all these materials is very sensitive to impurities at low temperatures, especially below 100 K. Therefore, the values given here should be regarded as typical values for a highly purified specimen; the thermal conductivity of different specimens can vary by more than an order of magnitude in the low-temperature range. See Reference 2 for details.

REFERENCES

1. Ho, C. Y., Powell, R. W., and Liley, P. E., *J. Phys. Chem. Ref. Data*, 1, 279, 1972.
2. White, G. K., and Minges, M. L., *Thermophysical Properties of Some Key Solids*, CODATA Bulletin No. 59, 1985.

Thermal Conductivity in W/cm K

				Carbon (C)						
				Diamond (type)			Pyrolytic graphite			
T/K	Ag	Al	Au	I	IIa	IIb	∥	⊥	Cr	Cu
1	39.4	41.1	5.46						0.402*	42.2
2	78.3	81.8	10.9	0.0138*	0.033*	0.0200*			0.803	84.0
3	115	121	16.1	0.0461	0.111	0.0676			1.20	125
4	147	157	20.9	0.108	0.261	0.160			1.60	162
5	172	188	25.2	0.206	0.494	0.307			2.00	195
6	187	213	28.5	0.344	0.820	0.510			2.39	222
7	193	229	30.9	0.523	1.24	0.778			2.27	239
8	190	237	32.3	0.762	1.77	1.12			3.14	248
9	181	239	32.7	1.05	2.41	1.53			3.50	249
10	168	235	32.4	1.40	3.17	2.03	0.811	0.0116	3.85	243
15	96.0	176	24.6	3.96	8.65	5.66			5.24	171
20	51.0	117	15.8	7.87	16.8	11.2	4.20	0.0397	5.93	108
30	19.3	49.5	7.55	18.8	38.9	26.5	9.86	0.0786	5.49	44.5
40	10.5	24.0	5.15	29.4	65.9	44.0	16.4	0.120	4.25	21.7
50	7.0	13.5	4.21	35.3	92.1	59.1	23.1	0.152	3.17	12.5
60	5.5	8.5	3.74	37.4	112	67.5	29.8	0.173	2.48	8.29
70	4.97	5.85	3.48	36.9	119	69.1	36.6	0.181	2.07	6.47
80	4.71	4.32	3.32	35.1	117	65.7	42.8	0.181	1.84	5.57
90	4.60	3.42	3.28	32.7	109	60.0	47.5	0.176	1.69	5.08
100	4.50	3.02	3.27	30.0	100	54.2	49.7	0.168	1.59	4.82
150	4.32	2.48	3.25	19.5	60.2	32.5	45.1	0.125	1.29	4.29
200	4.30	2.37	3.23	14.1	40.3	22.6	32.3	0.0923	1.11	4.13
250	4.29	2.35	3.21	11.0	29.7	17.0	24.4	0.0711	1.00	4.06
300	4.29	2.37	3.17	8.95	23.0	13.5	19.5	0.0570	0.937	4.01
350	4.27	2.40	3.14	7.55*	18.5*	11.1*	16.2	0.0477	0.929	3.96
400	4.25	2.40	3.11	6.5*	15.4*	9.32*	13.9	0.0409	0.909	3.93
500	4.19	2.36	3.04				10.8	0.0322	0.860	3.86
600	4.12	2.31	2.98				8.92	0.0268	0.807	3.79
800	3.96	2.18	2.84				6.67	0.0201	0.713	3.66
1000	3.79		2.70				5.34	0.0160	0.654	3.52
1200	3.61*		2.55				4.48	0.0134	0.619	3.39
1400							3.84	0.0116	0.588	
1600							3.33	0.0100	0.556	
1800							2.93	0.00895	0.526*	
2000							2.62	0.00807	0.494*	

T/K	Fe	Ge[a]	Mg	Ni	Pb	Pt	Si[a]	Sn	Ti	W
1	1.71	0.274	9.86	2.17	27.9	2.31	0.0693*	183	0.0144*	14.4
2	3.42	2.06	19.6	4.34	44.6	4.60	0.454	323	0.0288*	28.7
3	5.11	5.35	29.0	6.49	35.8	6.79	1.38	297	0.0432	42.8
4	6.77	8.77	37.6	8.59	22.2	8.8	2.97	181	0.0575	56.3
5	8.39	11.6	45.0	10.6	13.8	10.5	5.27	117	0.0719	68.7
6	9.93	13.9	50.8	12.5	8.10	11.8	8.23	76	0.0863	79.5
7	11.4	15.5	54.7	14.2	4.86	12.6	11.7	52	0.101	88.0
8	12.7	16.6	56.7	15.8	3.20	12.9	15.5	36	0.115	93.8
9	13.9	17.3	57.0	17.1	2.30	12.8	19.5	26	0.129	96.8
10	14.8	17.7	55.8	18.1	1.78	12.3	23.3	19.3	0.143	97.1
15	17.0	17.3	41.1	19.5	0.845	8.41	41.6	6.3	0.212	72.0
20	15.4	14.9	27.2	16.5	0.591	4.95	49.8	3.2	0.275	40.5
30	10.0	10.8	12.9	9.56	0.477	2.15	48.1	1.79	0.365	14.4
40	6.23	7.98	7.19	5.82	0.451	1.39	35.3	1.33	0.390	6.92
50	4.05	6.15	4.65	4.00	0.436	1.09	26.8	1.15	0.374	4.27
60	2.85	4.87	3.27	3.08	0.425	0.947	21.1	1.04	0.355	3.14
70	2.16	3.93	2.49	2.50	0.416	0.862	16.8	0.96	0.340	2.58
80	1.75	3.25	2.02	2.10	0.409	0.815	13.4	0.915	0.326	2.29
90	1.50	2.70	1.78	1.83	0.403	0.789	10.8	0.880	0.315	2.17
100	1.34	2.32	1.69	1.64	0.397	0.775	8.84	0.853	0.305	2.08
150	1.04	1.32	1.61	1.22	0.379	0.740	4.09	0.779	0.270	1.92
200	0.94	0.968	1.59	1.07	0.367	0.726	2.64	0.733	0.245	1.85
250	0.865	0.749	1.57	0.975	0.360	0.718	1.91	0.696	0.229	1.80
300	0.802	0.599	1.56	0.907	0.353	0.716	1.48	0.666	0.219	1.74
350	0.744	0.495	1.55	0.850	0.347	0.717	1.19	0.642	0.210	1.67
400	0.695	0.432	1.53	0.802	0.340	0.718	0.989	0.622	0.204	1.59
500	0.613	0.338	1.51	0.722	0.328	0.723	0.762	0.596	0.197	1.46
600	0.547	0.273	1.49	0.656	0.314	0.732	0.619		0.194	1.37
800	0.433	0.198	1.46*	0.676		0.756	0.422		0.197	1.25
1000	0.323	0.174		0.718		0.787	0.312		0.207	1.18
1200	0.283	0.174		0.762		0.826	0.257		0.220	1.12
1400	0.312			0.804		0.871	0.235		0.236	1.08
1600	0.330					0.919	0.221		0.253	1.04
1800	0.345*					0.961			0.270*	1.01
2000						0.994*				0.98

[a] Values below 300 K are typical values.

* Extrapolated.

THERMAL CONDUCTIVITY OF ALLOYS AS A FUNCTION OF TEMPERATURE

This table lists the thermal conductivity of selected alloys at various temperatures. The indicated compositions refer to weight percent. Since the thermal conductivity is sensitive to exact composition and processing history, especially at low temperatures, these values should be considered approximate.

REFERENCES

1. Powell, R. L., and Childs, G. E., in *American Institute of Physics Handbook, 3rd Edition*, Gray, D. E., Ed., McGraw-Hill, New York, 1972.
2. Ho, C. Y., et al., *J. Phys. Chem. Ref. Data*, 7, 959, 1978.

Thermal conductivity in W/m K

	Alloy	4 K	20 K	77 K	194 K	273 K	373 K	573 K	973 K
Aluminum:	1100	50	240	270	220	220			
	2024	3.2	17	56	95	130			
	3003	11	58	140	150	160			
	5052	4.8	25	77	120	140			
	5083, 5086	3	17	55	95	120			
	Duralumin	5.5	30	91	140	160	180		
Bismuth:	Rose metal		5.5	8.3	14	16			
	Wood's metal	4	17	23					
Copper:	electrolytic tough pitch	330	1300	550	400	390	380	370	350
	free cutting, leaded	200	800	460	380	380			
	phosphorus, deoxidized	7.5	42	120	190	220			
	brass, leaded	2.3	12	39	70	120			
	bronze, 68% Cu; 32% Zn	2.3	16	48	92	110			
	beryllium	2	17	36	70	90	113	172	
	german silver	0.75	7.5	17	20	23	25	30	40
	silicon bronze A		3.4	11	23	30			
	manganin	0.48	3.2	14	17	22			
	constantan	0.9	8.6	17	19	22			
Ferrous:	commercial pure iron	15	72	106	82	76	66	54	34
	plain carbon steel(AISI 1020)	13	20	58	65	65			
	plain carbon steel(AISI 1095)		8.5	31	41	45			
	3% Ni; 0.7% Cr; 0.6% Mo		6	22		33	35	36	30
	4% Si					20	24	28	26
	stainless steel	0.3	2	8	13	14	16	19	25
	27% Ni; 15% Cr		1.7	55		11	12	16	21
Gold:	colbalt thermocouple	1.2	8.6	20					
	65% Au; 35% Ag		12	24		61	89		
Indium:	85.5% In; 14.5% Pb	1.9	7.8	24	41				
Lead:	60% Pb; 40% Sn (soft solder)		28	44					
	64.35% Pb; 35.65% In	0.8	3.26	9.1		20.2			
Nickel:	80% Ni; 20% Cr					12	14	17	23
	contracid	0.2	2	7.3	9.5	13			
	inconel	0.5	4.2	12.5	13	15	16	19	26
	monel	0.9	7.1	15	20	21	24	30	43
Platinum:	90% Pt; 10% Ir					31	31.4		
	90% Pt; 10% Rh					30.1	30.5		
Silver:	silver solder		12	34	58				
	normal Ag thermocouple	48	230	310					
Tin:	60% Sn; 40% Pb	16	55	51					
Titanium:	5.5% Al; 2.5% Sn;0.2% Fe		1.8	4.3	6.4	7.8	8.4	10.8	
	4.7% Mn; 3.99% Al; 0.14% C		1.7	4.5	6.5	8.5			

THERMAL CONDUCTIVITY OF CRYSTALLINE DIELECTRICS

This table lists the thermal conductivity of a number of crystalline dielectrics, including some which find use as optical materials. Values are given at temperatures for which data are available.

REFERENCE

Powell, R. L., and Childs, G. E., in *American Institute of Physics Handbook, 3rd Edition*, Gray, D. E., Ed., McGraw-Hill, New York, 1972.

Material	*T*/K	Ther. cond. W/m K
AgCl	223	1.3
	273	1.2
	323	1.1
	373	1.1
Al,B silicate (tourmaline)	398	2.9
‖ to *c* axis	540	3.2
	723	3.5
Al,Be silicate (beryl)	315	6.4
Al,F silicate (topaz)	315	17.7
‖ to *c* axis	358	15.6
	417	13.3
Al,Fe silicate (garnet)	315	35.8
	358	35.4
	377	35.6
Al_2O_3 (sapphire):		
36° to *c* axis	4.2	110
	20	3500
	35	6000
	77	1100
⊥ to *c* axis	373	2.6
	523	3.9
	773	5.8
Al_2O_3 (sintered)	4.2	0.5
	20	23
	77	150
	194	48
	273	35
	373	26
	973	8
Ar	8	6.0
	10	3.7
	20	1.4
	77	0.31
As_2S_3 (glass)	283	0.16
	323	0.21
	373	0.27
BN	1047	36.2
	1475	22.7
	1928	21.9
	2111	18.5
BaF_2	225	20
	260	13.4
	305	10.9
	370	10.5
$BaTiO_3$	5	4.2
	30	24.0
	40	25.0
	100	12.0
	250	4.8
	300	6.2

Material	*T*/K	Ther. cond. W/m K
BeO	4.2	0.3
	20	16
	77	270
	373	210
	573	120
	1273	29
Bi_2Te_3	80	6.4
	204	2.8
	303	3.6
	370	4.6
C (diamond)	4.2	13
type I	20	800
	77	3550
	194	1450
	273	1000
$CaCO_3$		
‖ to *c* axis	83	25
	273	5.5
⊥ to *c* axis	83	17
	194	6.5
	273	4.6
	373	3.6
CaF_2	83	39
	223	18
	273	10
	323	9.2
	373	9
$CaWO_4$ (scheelite)	422	11.3
CdTe	160	7.0
	297	3.6
	422	2.9
CsBr	223	1.2
	273	0.94
	323	0.81
	373	0.77
CsI	223	1.4
	273	1.2
	323	1
	373	0.95
Cu_2O (cuprite)	102	3.74
	163	7.76
	299	5.58
	360	4.86
Fe_3O_4 (magnetite)	4.5	27.4
	20.5	293.0
	126.5	7.4
	304	7.0
Glass:		
phoenix	4.2	0.095
	20	0.13
	77	0.37

Material	T/K	Ther. cond. W/m K
plastic perspex	4.2	0.058
	20	0.074
pyrex	77	0.44
	194	0.88
	273	1
H_2 (para + 0.5% ortho)	2.5	100
	3	150
	4	200
	6	30
	10	3
H_2O (ice)	173	3.5
	223	2.8
	273	2.2
He^3 (high pressure)	0.6	25
	1	2
	1.5	0.57
	2	0.21
He^4 (high pressure)	0.5	42
	0.8	120
	1	24
	2	0.18
I_2	300	0.45
	325	0.42
	350	0.4
KBr	2	150
	4.2	360
	100	12
	273	5
	323	4.8
	373	4.8
KCl	4.2	500
	25	140
	80	35
	194	10
	273	7.0
	323	6.5
	373	6.3
KI	4.2	700
	80	13
	194	4.6
	273	3.1
Kr	4.2	0.48
	10	1.7
	20	1.2
	77	0.36
LaF_3	78	7.8
	197	5.0
	274	5.4
LiF	4.2	620
	20	1800
	77	150
$MgO{\cdot}Al_2O_3$ (spinel)	373	13
	773	8.5
MnO	4.2	0.25
	40	55
	120	8
	573	3.5
NaCl	4.2	440
	20	300
	77	30
	273	6.4
	323	5.6
	373	5.4
NaF	5	1100
	50	250
	100	90
Ne	2	3.0
	3	4.6
	4.2	4.2
	10	0.8
	20	0.3
NH_4Cl	77	17
	194	23
	230	38
	273	27
$NH_4H_2PO_4$		
‖ to optic axis	315	0.71
	339	0.71
⊥ to optic axis	313	1.26
	342	1.34
NiO	4.2	5.9
	40	400
	194	82
SiO_2 (quartz)		
‖ to *c* axis	20	720
	194	20
	273	12
⊥ to *c* axis	20	370
	194	10
	273	6.8
SiO_2 (fused silica)	4.2	0.25
	20	0.7
	77	0.8
	194	1.2
	273	1.4
	373	1.6
	673	1.8
$SrTiO_3$	5	2.4
	30	21.0
	40	19.2
	100	18.5
	250	12.5
	300	11.2
TlBr	316	0.59
TlCl	311	0.75
TiO_2 (rutile)		
‖ to optic axis	4.2	200
	20	1000
	273	13
⊥ to optic axis	4.2	160
	20	690
	273	9

THERMAL CONDUCTIVITY OF CERAMICS AND OTHER INSULATING MATERIALS

Thermal conductivity values for ceramics, refractory oxides, and miscellaneous insulating materials are given here. The thermal conductivity refers to samples with density indicated in the second column. Since most of these materials are highly variable, the values should only be considered as a rough guide.

REFERENCES

1. Powell, R. L., and Childs, G. E., in *American Institute of Physics Handbook, 3rd Edition*, Gray, D. E., Ed., McGraw-Hill, New York, 1972.
2. Perry, R. H., and Green, D., *Perry's Chemical Engineers' Handbook, Sixth Edition*, McGraw-Hill, New York, 1984.

Material	Dens. g/cm³	*t* °C	Ther. cond. W/m K
Alumina (Al_2O_3)	3.8	100	30
		400	13
		1300	6
		1800	7.4
	3.5	100	17
		800	7.6
Al_2O_3 + MgO		100	15
		400	10
		1000	5.6
Asbestos	0.4	-100	0.07
		100	1
		0	0.09
Asbestos + 85% MgO	0.3	30	0.08
Asphalt	2.1	20	0.06
Beryllia (BeO)	2.8	100	210
		400	90
		1000	20
		1800	15
	1.85	50	64
		200	40
		600	23
Brick, dry	1.54	0	0.04
Brick, refractory:			
alosite		1000	1.3
aluminous	1.99	400	1.2
		1000	1.3
diatomaceous	0.77	100	0.2
		500	0.24
	0.4	100	0.08
		500	0.1
fireclay	2	400	1
		1000	1.2
silicon carbide	2	200	2
		600	2.4
vermiculite	0.77	200	0.26
		600	0.31
Calcium oxide		100	16
		400	9
		1000	7.5
Cement mortar	2	90	0.55
Charcoal	0.2	20	0.055
Coal	1.35	20	0.26
Concrete	1.6	0	0.8
Cork	0.05	0	0.03
		100	0.04
	0.35	0	0.06
		100	0.08
Cotton wool	0.08	30	0.04

Material	Dens. g/cm³	*t* °C	Ther. cond. W/m K
Diatomite	0.2	0	0.05
		400	0.09
	0.5	0	0.09
		400	0.16
Ebonite	1.2	0	0.16
Felt, flax	0.2	30	0.05
	0.3	30	0.04
Fuller's earth	0.53	30	0.1
Glass wool	0.2	-200 to 20	0.005
		50	0.04
		100	0.05
		300	0.08
Graphite			
100 mesh	0.48	40	0.18
20-40 mesh	0.7	40	1.29
Linoleum cork	0.54	20	0.08
Magnesia (MgO)		100	36
		400	18
		1200	5.8
		1700	9.2
MgO + SiO_2		100	5.3
		400	3.5
		1500	2.3
Mica:			
muscovite		100	0.72
		300	0.65
		600	0.69
phlogopite		100	0.66
Canadian		300	0.19
		600	0.2
Micanite		30	0.3
Mineral wool	0.15	30	0.04
Perlite, expanded	0.1	-200 to 20	0.002
Plastics:			
bakelite	1.3	20	1.4
celluloid	1.4	30	0.02
polystyrene foam	0.05	-200 to 20	0.033
mylar foil	0.05	-200 to 20	0.0001
nylon		-253	0.10
		-193	0.23
		25	0.30
polytetrafluoroethylene		-253	0.13
		-193	0.16
		25	0.26
		230	2.5
urethane foam	0.07	20	0.06
Porcelain		90	1

Material	Dens. g/cm³	t °C	Ther. cond. W/m K
Rock:			
basalt		20	2
chalk		20	0.92
granite	2.8	20	2.2
limestone	2	20	1
sandstone	2.2	20	1.3
slate, ⊥		95	1.4
slate, ‖		95	2.5
Rubber:			
sponge	0.2	20	0.05
92 percent		25	0.16
Sand, dry	1.5	20	0.33
Sawdust	0.2	30	0.06
Shellac		20	0.23
Silica aerogel	0.1	-200 to 20	0.003
Snow	0.25	0	0.16
Steel wool	0.1	55	0.09
Thoria (ThO_2)		100	10
		400	5.8
		1500	2.4
Titanium dioxide		100	6.5
		400	3.8
		1200	3.3
Uranium dioxide		100	9.8
		400	5.5
		1000	3.4
Wood:			
balsa, ⊥	0.11	30	0.04
fir, ⊥	0.54	20	0.14
fir, ‖	0.54	20	0.35
oak		20	0.16
plywood		20	0.11
pine, ⊥	0.45	60	0.11
pine, ‖	0.45	60	0.26
walnut, ⊥	0.65	20	0.14
Wool	0.09	30	0.04
Zinc oxide		200	17
		800	5.3
Zirconia (ZrO_2)		100	2
		400	2
		1500	2.5
Zirconia + silica		200	5.6
		600	4.6
		1500	3.7

THERMAL CONDUCTIVITY OF GLASSES

This table gives the composition of various types of glasses and the thermal conductivity *k* as a function of temperature. Because of the variability of glasses, the data should be regarded as only approximate.

Type of glass	Composition: SiO_2 (wt%)	Composition: Other oxides (wt%)		*t* °C	*k* W/m K
Vitreous silica	100			–150	0.85
				–100	1.05
				–50	1.20
				0	1.30
				50	1.40
				100	1.50
Vycor glass	96	B_2O_3	3	–100	1.00
				0	1.25
				100	1.40
Pyrex type chemically-resistant borosilicate glasses	80–81	B_2O_3	12–13	–100	0.90
		Na_2O	4	0	1.10
		Al	2	100	1.25
Borosilicate crown glasses	60–65	B_2O_3	15–20	–100	0.65–0.75
				0	0.90–0.95
				100	1.00–1.05
	65–70	B_2O_3	10–15	–100	0.75–0.80
				0	0.95–1.00
				100	1.05–1.15
	70–75	B_3O_3	5–10	–100	0.80–0.85
				0	1.05–1.10
				100	1.15–1.20
Zinc crown glasses (i)	55–65	ZnO	5–15	–100	0.88–0.92
		Remainder:		0	1.10–1.15
		B_2O_3, Al_2O_3		100	1.15–1.25
		ZnO	5–15	–100	0.60–0.70
		Remainder:		0	0.70–0.90
		Na_2O, K_2O		100	0.85–0.95
		ZnO	15–25	–100	0.88–0.92
		Remainder:		0	1.10–1.15
		B_2O_3, Al_2O_3		100	1.15–1.20
		ZnO	15–25	–100	0.65–0.80
		Remainder:		0	0.85–0.95
		Na_2O, K_2O		100	0.90–1.05
Zinc crown glasses (ii)	65–75	ZnO	5–15	–100	0.88–0.92
		Remainder:		0	1.15–1.15
		B_2O_3, Al_2O_3		100	1.20–1.30
		ZnO	5–15	–100	0.70–0.85
		Remainder:		0	0.90–1.05
		Na_2O, K_2O		100	1.00–1.15

Type of glass	Composition: SiO_2 (wt%)	Other oxides (wt%)		t °C	k W/m K
		ZnO	15–25	–100	0.90–0.95
		Remainder:		0	1.15–1.15
		B_2O_3, Al_2O_3		100	1.20–1.25
		ZnO	15–25	–100	0.65–0.85
		Remainder:		0	0.85–1.00
		Na_2O, K_2O		100	1.05–1.20
Barium crown glasses	31	B_2O_3	12	–100	0.55
		Al_2O_3	8	0	0.70
		BaO	48	100	0.80
	41	B_2O_3	6	–100	0.60
		Al_2O_3	2	0	0.75
		ZnO	8	100	0.85
		BaO	43		
	47	B_2O_3	4	–100	0.65
		Na_2O	1	0	0.75
		K_2O	7	100	0.90
		ZnO	8		
		BaO	32		
	65	B_2O_3	2	–100	0.70
		Na_2O	5	0	0.90
		K_2O	15	100	1.00
		ZnO	2		
		BaO	10		
Borate glasses					
Borate flint glass	9	B_2O_3	36	–100	0.55
		Na_2O	1	0	0.65
		K_2O	2	100	0.80
		PbO	36		
		Al_2O_3	10		
		ZnO	6		
Borate flint glass	0	B_2O_3	56	–100	0.50
		Al_2O_3	12	0	0.65
		PbO	32	100	0.85
Borate flint glass	0	B_2O_3	43	–100	0.40
		Al_2O_3	5	0	0.55
		PbO	52	100	0.70
Borate glass	4	B_2O_3	55	–100	0.65
		Al_2O_3	14	0	0.80
		PbO	11	100	0.90
		K_2O	4		
		ZnO	12		
Borate crown glass	0	B_2O_3	64	–100	0.50
		Na_2O	8	0	0.65
		K_2O	3	100	0.85
		BaO	4		
		PbO	3		
		Al_2O_3	18		

Type of glass	Composition: SiO$_2$ (wt%)	Other oxides (wt%)		t °C	k W/m K
Light borate crown glass	0	B_2O_3	69	–100	0.55
		Na_2O	8	0	0.70
		BaO	5	100	0.90
		Al_2O_3	18		
Zinc borate glass	0	B_2O_3	40	–100	0.65
		ZnO	60	0	0.75
				100	0.85
Phosphate crown glasses					
Potash phosphate glass	0	P_2O_5	70	0	0.75
		B_2O_3	3	100	0.85
		K_2O	12		
		Al_2O_3	10		
		MgO	4		
Baryta phosphate glass	0	P_2O_5	60	45	0.75
		B_2O_3	3		
		Al_2O_3	8		
		BaO	28		
Soda-lime glasses	75	Na_2O	17	–100	0.75
		CaO	8	0	0.95
				100	1.10
	75	Na_2O	12	–100	0.90
		CaO	13	0	1.10
				100	1.15
	72	Na_2O	15	–100	0.80
		CaO	11	0	1.00
		Al_2O_3	2	100	1.15
	65	Na_2O	25	–100	0.65
		CaO	10	0	0.85
				100	0.95
	65	Na_2O	15	–100	0.85
		CaO	20	0	1.00
				100	1.10
	60	Na_2O	20	–100	0.75
		CaO	20	0	0.90
				100	1.00
Other crown glasses					
Crown glass	75	Na_2O	9	–100	0.80
		K_2O	11	0	1.00
		CaO	5	100	1.10
High dispersion crown glass	68	Na_2O	16	–100	0.65
		ZnO	3	0	0.85
		PbO	13	100	1.00

Type of glass	Composition: SiO_2 (wt%)	Other oxides (wt%)		t °C	k W/m K
Miscellaneous flint glasses					
(i) Silicate flint glasses					
Light flint glasses	65	PbO	25	−100	0.65–0.70
		Others	10	0	0.88–0.92
				100	1.00–1.05
	55	PbO	35	−100	0.60–0.65
		Others	10	0	0.75–0.85
				100	0.88–0.92
Ordinary flint glass	45	PbO	45	−100	0.50–0.60
		Others	10	0	0.65–0.75
				100	0.80–0.85
Heavy flint glass	35	PbO	60	−100	0.45–0.50
		Others	5	0	0.60–0.65
				100	0.70–0.75
Very heavy flint glasses	25	PbO	73	−100	0.40–0.45
		Others	2	0	0.55–0.60
				100	0.63–0.67
	20	PbO	80	−100	0.40
				0	0.50
				100	0.60
(ii) Borosilicate flint glass	33	B_2O_3	31	−100	0.65
		PbO	25	0	0.85
		Al_2O_3	7	100	0.95
		K_2O	3		
		Na_2O	1		
(iii) Barium flint glass	50	BaO	24	−100	0.60
		PbO	6	0	0.70
		K_2O	8	100	0.85
		Na_2O	3		
		ZnO	8		
		Sb_2O_3	1		
Other glasses					
Potassium glass	59	K_2O	33	50	0.88–0.92
		CaO	8		
Iron glasses	63	Fe_2O_3	10	−100	0.80
		Na_2O	17	0	0.95
		MgO	4	100	1.05
		CaO	3		
		Al_2O_3	2		
	67	Fe_2O_3	15	0	0.88—0.92
		Na_2O_3	18	100	1.00—1.05
	62	Fe_2O_3	20	0	0.85—0.90
		Na_2O	18	100	0.95—1.00
Rock glasses					
Obsidian				0	1.35
Artificial diabase				100	1.25

COMMERCIAL METALS AND ALLOYS

This table gives typical values of mechanical, thermal, and electrical properties of several common commercial metals and alloys. Values refer to ambient temperature (0 to 25°C). All values should be regarded as typical, since these properties are dependent on the particular type of alloy, heat treatment, and other factors. Values for individual specimens can vary widely.

REFERENCES

1. *ASM Metals Reference Book, Second Edition*, American Society for Metals, Metals Park, OH, 1983.
2. Lynch, C. T., *CRC Practical Handbook of Materials Science*, CRC Press, Boca Raton, FL, 1989.
3. Shackelford, J. F., and Alexander, W., *CRC Materials Science and Engineering Handbook*, CRC Press, Boca Raton, FL, 1991.

Common name	Thermal conductivity W/cm K	Density g/cm^3	Coeff. of linear expansion 10^{-6}/°C	Electrical resistivity μΩ cm	Modulus of elasticity GPa	Tensile strength MPa	Approx. melting point °C
Ingot iron	0.7	7.86	11.7	9.7	205	-	1540
Plain carbon steel AISI-SAE 1020	0.52	7.86	11.7	18	205	450	1515
Stainless steel type 304	0.15	7.9	17.3	72	195	550	1425
Cast gray iron	0.47	7.2	10.5	67	90	180	1175
Malleable iron		7.3	12	30	170	345	1230
Hastelloy C	0.12	8.94	11.3	125	200	780	1350
Inconel	0.15	8.25	11.5	103	200	800	1370
Aluminum alloy 3003, rolled	1.9	2.73	23.2	3.7	70	110	650
Aluminum alloy 2014, annealed	1.9	2.8	23.0	3.4	70	185	650
Aluminum alloy 360	1.5	2.64	21.0	7.5	70	325	565
Copper, electrolytic (ETP)	3.9	8.94	16.5	1.7	120	300	1080
Yellow brass (high brass)	1.2	8.47	20.3	6.4	100	300-800	930
Aluminum bronze	0.7	7.8	16.4	12	120	400-600	1050
Beryllium copper 25	0.8	8.23	17.8	7	130	500-1400	925
Cupronickel 30%	0.3	8.94	16.2		150	400-600	1200
Red brass, 85%	1.6	8.75	18.7	11	90	300-700	1000
Chemical lead	0.35	11.34	29.3	21	13	17	327
Antimonial lead (hard lead)	0.3	10.9	26.5	23	20	47	290
Solder 50-50	0.5	8.89	23.4	15	-	42	215
Magnesium alloy AZ31B	1.0	1.77	26	9	45	260	620
Monel	0.3	8.84	14.0	58	180	545	1330
Nickel (commercial)	0.9	8.89	13.3	10	200	460	1440
Cupronickel 55-45 (constantan)	0.2	8.9	18.8	49	160	-	1260
Titanium (commercial)	1.8	4.5	8.5	43	110	330-500	1670
Zinc (commercial)	1.1	7.14	32.5	6	-	130	419
Zirconium (commercial)	0.2	6.5	5.85	41	95	450	1855

HARDNESS OF MINERALS AND CERAMICS

There are several hardness scales for describing the resistance of a material to indentation or scratching. This table lists a number of common materials in order of increasing hardness. Values are given, when available, on three different hardness scales: the original Mohs Scale (range 1 to 10); the modified Mohs Scale (range 1 to 15), and the Knoop Hardness Scale. In the last case, a load of 100 g is assumed.

REFERENCE

Shackelford, J. F. and Alexander, W., *CRC Materials Science and Engineering Handbook*, CRC Press, Boca Raton, FL, 1991.

Material	Formula	Mohs	Modified mohs	Knoop
Graphite	C	0.5		
Talc	$3MgO{\cdot}4SiO_2{\cdot}H_2O$	1	1	
Alabaster	$CaSO_4{\cdot}2H_2O$	1.7		
Gypsum	$CaSO_4{\cdot}2H_2O$	2	2	32
Halite (rock salt)	$NaCl$	2		
Stibnite (antimonite)	Sb_2S_3	2.0		
Galena	PbS	2.5		
Mica		2.8		
Calcite	$CaCO_3$	3	3	135
Barite	$BaSO_4$	3.3		
Marble		3.5		
Aragonite	$CaCO_3$	3.5		
Dolomite	$CaMg(CO_3)_2$	3.5		
Fluorite	CaF_2	4	4	163
Magnesia	MgO	5		370
Apatite	$CaF_2{\cdot}3Ca_3(PO_4)_2$	5	5	430
Opal		5		
Feldspar (orthoclase)	$K_2O{\cdot}Al_2O{\cdot}6SiO_2$	6	6	560
Augite		6		
Hematite	Fe_2O_3	6		750
Magnetite	Fe_3O_4	6		
Rutile	TiO_2	6.2		
Pyrite	FeS_2	6.3		
Agate	SiO_2	6.5		
Uranium dioxide	UO_2	6.7		600
Silica (fused)	SiO_2		7	
Quartz	SiO_2	7	8	820
Flint		7		
Silicon	Si	7		
Andalusite	Al_2OSiO_4	7.5		
Zircon	$ZrSiO_4$	7.5		
Zirconia	ZrO_2			1200
Aluminum nitride	AlN			1225
Beryl	$Be_3Al_2Si_6O_{18}$	7.8		
Beryllia	BeO			1300
Topaz	$Al_2SiO_4(OH,F)_2$	8	9	1340
Garnet	$Al_2O_3{\cdot}3FeO{\cdot}3SiO_2$		10	1360
Emery	Al_2O_3 (impure)	8		
Zirconium nitride	ZrN	8+		1510
Zirconium boride	ZrB_2			1560
Titanium nitride	TiN	9		1770
Zirconia (fused)	ZrO_2		11	
Tantalum carbide	TaC			1800
Tungsten carbide	WC			1880
Corundum (alumina)	Al_2O_3	9		2025
Zirconium carbide	ZrC			2150
Alumina (fused)	Al_2O_3		12	
Beryllium carbide	Be_2C			2400

HARDNESS OF MINERALS AND CERAMICS (continued)

Material	Formula	Mohs	Modified mohs	Knoop
Titanium carbide	TiC			2470
Carborundum (silicon carbide)	SiC	9.3	13	2500
Aluminum boride	AlB			2500
Tantalum boride	TaB_2			2600
Boron carbide	B_4C		14	2800
Boron	B	9.5		
Titanium boride	TiB_2			2850
Diamond	C	10	15	7000

Section 13
Polymer Properties

Nomenclature for Organic Polymers 13-1
Solvents for Common Polymers 13-6
Glass Transition Temperature for Selected Polymers 13-7
Dielectric Constant of Selected Polymers 13-15
Pressure-Volume-Temperature Relationship for Polymer Melts 13-16

NOMENCLATURE FOR ORGANIC POLYMERS

Robert B. Fox and Edward S. Wilks

Organic polymers have traditionally been named on the basis of the monomer used, a hypothetical monomer, or a semi-systematic structure. Alternatively, they may be named in the same way as organic compounds, i.e., on the basis of a structure as drawn. The former method, often called "source-based nomenclature" or "monomer-based nomenclature", sometimes results in ambiguity and multiple names for a single material. The latter method, termed "structure-based nomenclature", generates a sometimes cumbersome unique name for a given polymer, independent of its source. Within their limitations, both types of names are acceptable and well-documented.[1] The use of stereochemical descriptors with both types of polymer nomenclature has been published.[2]

Traditional Polymer Names

Monomer-Based Names

"Polystyrene" is the name of a homopolymer made from the single monomer styrene. When the name of a monomer comprises two or more words, the name should be enclosed in parentheses, as in "poly(methyl methacrylate)" or "poly(4-bromostyrene)" to identify the monomer more clearly. This method can result in several names for a given polymer: thus, "poly(ethylene glycol)", "poly(ethylene oxide)", and "poly(oxirane)" describe the same polymer. Sometimes, the name of a hypothetical monomer is used, as in "poly(vinyl alcohol)". Even though a name like "polyethylene" covers a multitude of materials, the system does provide understandable names when a single monomer is involved in the synthesis of a single polymer. When one monomer can yield more than one polymer, e.g. 1,3-butadiene or acrolein, some sort of structural notation must be used to identify the product, and one is not far from a formal structure-based name.

Copolymers, Block Polymers, and Graft Polymers. When more than one monomer is involved, monomer-based names are more complex. Some common polymers have been given names based on an apparent structure, as with "poly(ethylene terephthalate)". A better system has been approved by the IUPAC.[1] With this method, the arrangement of the monomeric units is introduced through use of an italicized connective placed between the names of the monomers. For monomer names represented by A, B, and C, the various types of arrangements are shown in Table 1.

Table 1. IUPAC Source-Based Copolymer Classification

No.	Copolymer Type	Connective	Example
1	Unspecified or unknown	*-co-*	poly(A-*co*-B)
2	Random (obeys Bernoullian distribution)	*-ran-*	poly(A-*ran*-B)
3	Statistical (obeys known statistical laws)	*-stat-*	poly(A-*stat*-B)
4	Alternating (for two monomeric units)	*-alt-*	poly(A-*alt*-B)
5	Periodic (ordered sequence for 2 or more monomeric units)	*-per-*	poly(A-*per*-B-*per*-C)
6	Block (linear block arrangement)	*-block-*	polyA-*block*-polyB
7	Graft (side chains connected to main chains)	*-graft-*	polyA-*graft*-polyB

Table 2 contains examples of common or semi-systematic names of copolymers. The systematic names of comonomers may also be used; thus, the polyacrylonitrile-*block*-polybutadiene-*block*-polystyrene polymer in Table 2 may also be named poly(prop-2-enenitrile)-*block*-polybuta-1,3-diene-*block*-poly(ethenylbenzene). IUPAC does not require alphabetized names of comonomers within a polymer name; many names are thus possible for some copolymers.

These connectives may be used in combination and with small, non-repeating (i.e. non-polymeric) junction units; see, for example, Table 2, line 8. A long dash may be used in place of the connective *-block-*; thus, in Table 2, the polymers of lines 7 and 8 may also be written as shown on lines 9 and 10.

Table 2. Examples of Source-Based Copolymer Nomenclature

No.	Copolymer name
1	poly(propene-*co*-methacrylonitrile)
2	poly[(acrylic acid)-*ran*-(ethyl acrylate)]
3	poly(butene-*stat*-ethylene-*stat*-styrene)
4	poly[(sebacic acid)-*alt*-butanediol]
5	poly[(ethylene oxide)-*per*-(ethylene oxide)-*per*-tetrahydrofuran]
6	polyisoprene-*graft*-poly(methacrylic acid)
7	polyacrylonitrile-*block*-polybutadiene-*block*-polystyrene
8	polystyrene-*block*-dimethylsilylene-*block*-polybutadiene
9	polyacrylonitrile—polybutadiene—polystyrene
10	polystyrene—dimethylsilylene—polybutadiene

IUPAC also recommends an alternative scheme for naming copolymers that comprises use of "copoly" as a prefix followed by the names of the comonomers, a solidus (an oblique stroke) to separate comonomer names, and addition before "copoly" of any applicable connectives listed in Table 2 except *-co-*.

Table 3 gives the same examples shown in Table 2 but with the alternative format. Comonomer names need not be parenthesized.

NOMENCLATURE FOR ORGANIC POLYMERS (continued)

Table 3. Examples of Source-Based Copolymer Nomenclature (Alternative Format)

No.	Polymer name
1	copoly(propene/methacrylonitrile)
2	*ran*-copoly(acrylic acid/ethyl acrylate)
3	*stat*-copoly(butene/ethylene/styrene)
4	*alt*-copoly(sebacic acid/butanediol)
5	*block*-copoly(acrylonitrile/butadiene/styrene)
6	*per*-copoly(ethylene oxide/ethylene oxide/tetrahydrofuran)
7	*graft*-copoly(isoprene/methacrylic acid)

Source-based nomenclature for non-linear macromolecules and macromolecular assemblies is covered by a 1997 IUPAC document.[11] The types of polymers in these classes, together with their connectives, are given in Table 4; the terms shown may be used as connectives, prefixes, or both to designate the features present.

Table 4. Connectives for Non-Linear Macromolecules and Macromolecular Assemblies

No.	Type	Connective
1	Branched (type unspecified)	branch
2	Branched with branch point of functionality f	f-branch
3	Comb	comb
4	Cross-link	ι (Greek iota)
5	Cyclic	cyclo
6	Interpenetrating polymer network	ipn
7	Long-chain branched	l-branch
8	Network	net
9	Polymer blend	blend
10	Polymer-polymer complex	compl
11	Semi-interpenetrating polymer network	sipn
12	Short-chain branched	sh-branch
13	Star	star
14	Star with f arms	f-star

Non-linear polymers are named by using the italicized connective as a *prefix* to the source-based name of the polymer component or components to which the prefix applies; some examples are listed in Table 5.

Table 5. Non-Linear Macromolecules

No.	Polymer Name	Polymer Structural Features
1	poly(methacrylic acid)-*comb*-polyacrylonitrile	Comb polymer with a poly(methacrylic acid) backbone and polyacrylonitrile side chains
2	*comb*-poly[ethylene-*stat*-(vinyl chloride)]	Comb polymer with unspecified backbone composition and statistical ethylene/vinyl chloride copolymer side chains
3	polybutadiene-*comb*-(polyethylene; polypropene)	Comb polymer with butadiene backbone and side chains of polyethylene and polypropene
4	*star*-(polyA; polyB; polyC; polyD; polyE)	Star polymer with arms derived from monomers A, B, C, D, and E, respectively
5	*star*-(polyA-*block*-polyB-*block*-polyC)	Star polymer with every arm comprising a tri-block segment derived from comonomers A, B, and C
6	*star*-poly(propylene oxide)	A star polymer prepared from propylene oxide
7	5-*star*-poly(propylene oxide)	A 5-arm star polymer prepared from propylene oxide
8	*star*-(polyacrylonitrile; polypropylene) (M_r 10000: 25000)	A star polymer containing polyacrylonitrile arms of MW 10000 and polypropylene arms of MW 25000

Macromolecular assemblies held together by forces other than covalent bonds are named by inserting the appropriate italicized connective between names of individual components; Table 6 gives examples.

Table 6. Examples of Polymer Blends and Nets

No.	Polymer Name
1	polyethylene-*blend*-polypropene
2	poly(methacrylic acid)-*blend*-poly(ethyl acrylate)
3	*net*-poly(4-methylstyrene-*ι*-divinylbenzene)
4	*net*-poly[styrene-*alt*-(maleic anhydride)]-*ι*-(polyethylene glycol; polypropylene glycol)
5	*net*-poly(ethyl methacrylate)-*sipn*-polyethylene
6	[*net*-poly(butadiene-*stat*-styrene)]-*ipn*-[*net*-poly(4-methylstyrene-*ι*-divinylbenzene)]

Structure-Based Polymer Nomenclature

Regular Single-Strand Polymers

Structure-based nomenclature has been approved by the IUPAC[4] and is currently being updated; it is used by *Chemical Abstracts*.[5] Monomer names are not used. To the extent that a polymer chain can be described by a repeating unit in the chain, it can be named "poly(repeating unit)". For regular single-strand polymers, "repeating unit" is a bivalent group; for regular double-strand (ladder and spiro) polymers, "repeating unit" is usually a tetravalent group.[9]

Since there are usually many possible repeating units in a given chain, it is necessary to select one, called the "constitutional repeating unit" (CRU) to provide a unique and unambiguous name, "poly(CRU)", where "CRU" is a recitation of the names of successive units as one proceeds through the CRU from left to right. For this purpose, a portion of the main chain structure that includes at least two repeating sequences is written out. These sequences will typically be composed of bivalent subunits such as -CH_2-, -O-, and groups from ring systems, each of which can be named by the usual nomenclature rules.[6,7]

Where a chain is simply one long sequence comprising repetition of a single subunit, that subunit is itself the CRU, as in "poly(methylene)" or "poly(1,4-phenylene)". In chains having more than one kind of subunit, a seniority system is used to determine the beginning of the CRU and the direction in which to move along the main chain atoms (following the shortest path in rings) to complete the CRU. Determination of the first, most senior, subunit, is based on a descending order of seniority: (1) heterocyclic rings, (2) hetero atoms, (3) carbocyclic rings, and, lowest, (4) acyclic carbon chains.

Within each of these classes, there is a further order of seniority that follows the usual rules of nomenclature.

Heterocycles: A nitrogen-containing ring system is senior to a ring system not containing nitrogen.[4,9] Further descending order of seniority is determined by:

(i) the highest number of rings in the ring system
(ii) the largest individual ring in the ring system
(iii) the largest number of hetero atoms
(iv) the greatest variety of hetero atoms

Hetero atoms: The senior bivalent subunit is the one nearest the top, right-hand corner of the Periodic Table; the order of seniority is: O, S, Se, Te, N, P, As, Sb, Bi, Si, Ge, Sn, Pb, B, Hg.

Carbocycles: Seniority[4] is determined by:

(i) the highest number of rings in the ring system
(ii) the largest individual ring in the ring system
(iii) degree of ring saturation; an unsaturated ring is senior to a saturated ring of the same size

Carbon chains: Descending order of seniority is determined by:

(i) chain length (longer is senior to shorter)
(ii) highest degree of unsaturation
(iii) number of substituents (higher number is senior to lower number)
(iv) ascending order of locants
(v) alphabetical order of names of substituent groups

Among equivalent ring systems, preference is given to the one having lowest locants for the free valences in the subunit, and among otherwise identical ring systems, the one having least hydrogenation is senior. Lowest locants in unsaturated chains are also given preference. Lowest locants for substituents are the final determinant of seniority.

Direction within the repeating unit depends upon the shortest path, which is determined by counting main chain atoms, both cyclic and acyclic, from the most senior subunit to another subunit of the same kind or to a subunit next lower in seniority. When identification and orientation of the CRU have been accomplished, the CRU is named by writing, in sequence, the names of the largest possible subunits within the CRU from left to right. For example, the main chain of the polymer traditionally named "poly(ethylene terephthalate)" has the structure shown in Figure 1.

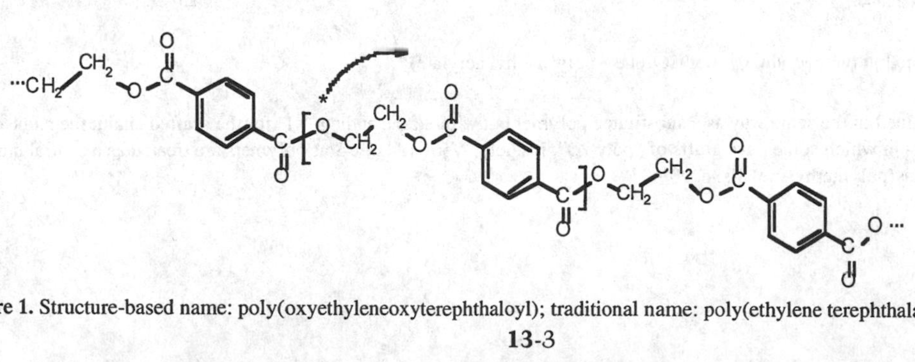

Figure 1. Structure-based name: poly(oxyethyleneoxyterephthaloyl); traditional name: poly(ethylene terephthalate)

NOMENCLATURE FOR ORGANIC POLYMERS (continued)

The CRU in Figure 1 is enclosed in brackets and read from left to right. It is selected because (1) either backbone oxygen atom qualifies as the "most senior subunit", (2) the shortest path length from either -O- to the other -O- is via the ethylene subunit. Orientation of the CRU is thus defined by (1) beginning at the -O- marked with an asterisk, and (2) reading in the direction of the arrow. The structure-based name of this polymer is therefore "poly(oxyethyleneoxyterephthaloyl)", not much longer than the traditional name and much more adaptable to the complexities of substitution. As organic nomenclature evolves, more systematic names may be used for subunits, e.g. "ethane-1,2-diyl" instead of "ethylene". IUPAC still prefers "ethylene" for the -CH_2-CH_2- unit, however, but also accepts "ethane-1,2-diyl".

Structure-based nomenclature can also be used when the CRU backbone has no carbon atoms. An example is the polymer traditionally named "poly(dimethylsiloxane)", which on the basis of structure would be named "poly(oxydimethylsilylene)" or "poly(oxydimethylsilanediyl)". This nomenclature method has also been applied to inorganic and coordination polymers[8] and to double-strand (ladder and spiro) organic polymers.[9]

Irregular Single-Strand Polymers

Polymers that cannot be described by the repetition of a single CRU or comprise units not all connected identically in a directional sense can also be named on a structure basis.[10] These include copolymers, block and graft polymers, and star polymers. They are given names of the type "poly(A/B/C...)", where A, B, C, etc. are the names of the component constitutional units, the number of which are minimized. The constitutional units may include regular or irregular blocks as well as atoms or atomic groupings, and each is named by the method described above or by the rules of organic nomenclature.

The solidus denotes an unspecified arrangement of the units within the main chain.[10] For example, a statistical copolymer derived from styrene and vinyl chloride with the monomeric units joined head-to-tail is named "poly(l-chloroethylene/l-phenylethylene)". A polymer obtained by 1,4- polymerization and both head-to-head and head-to-tail 1,2- polymerization of 1,3-butadiene would be named "poly(but-1-ene-1,4-diyl/l-vinylethylene/2-vinylethylene)".[12] In graphic representations of these polymers, shown in Figure 2, the hyphens or dashes at each end of each CRU depiction are shown *completely within* the enclosing parentheses; this indicates that they are not necessarily the terminal bonds of the macromolecule.

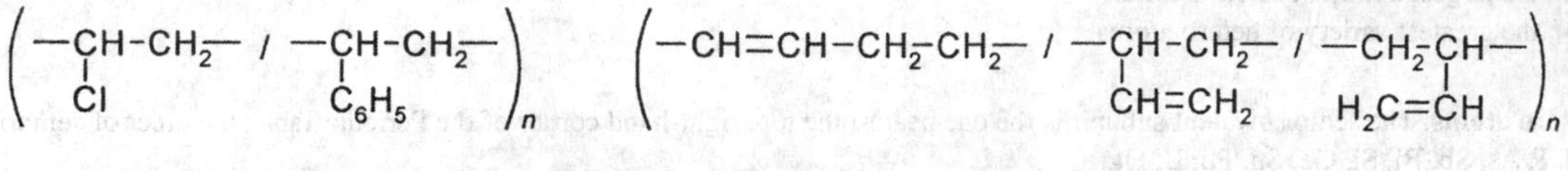

Figure 2. Graphic Representations of Copolymers

A long hyphen is used to separate components in names of block polymers, as in "poly(A)—poly(B)—poly(C)", or "poly(A)—X—poly(B)" in which X is a non-polymeric junction unit, e.g. dimethylsilylene.

In graphic representations of these polymers, the blocks are shown connected when the bonding is known (Figure 3, for example); when the bonding between the blocks is unknown, the blocks are separated by solidi and are shown *completely within* the outer set of enclosing parentheses (Figure 4, for example).[10,13]

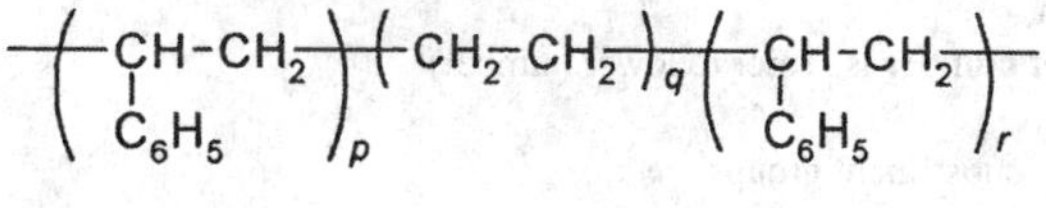

Figure 3. polystyrene—polyethylene—polystyrene

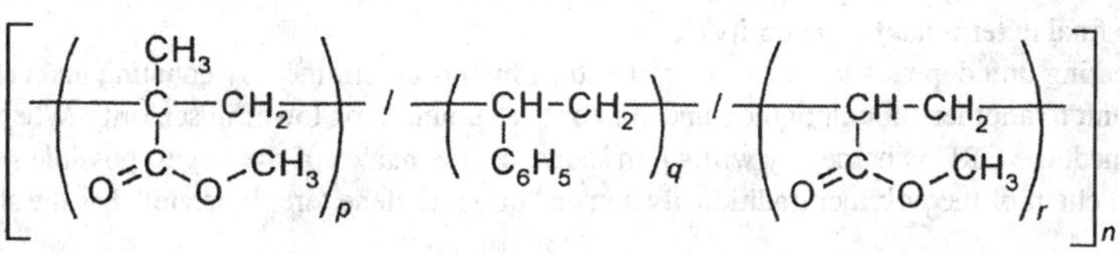

Figure 4. poly[poly(methyl methacrylate)—polystyrene—poly(methyl acrylate)]

Graft polymers are named in the same way as a substituted polymer but without the ending "yl" for the grafted chain; the name of a regular polymer, comprising Z units in which some have grafts of "poly(A)", is "poly[Z/poly(A)Z]". Star polymers are treated as a central unit with substituent blocks, as in "tetrakis(polymethylene)silane".[10,13]

NOMENCLATURE FOR ORGANIC POLYMERS (continued)

Other Nomenclature Articles and Publications

In addition to the *Chemical Abstracts* and IUPAC documents cited above and listed below, other articles on polymer nomenclature are available. A 1999 article lists significant documents on polymer nomenclature published during the last 50 years in books, encyclopedias, and journals by *Chemical Abstracts*, IUPAC, and individual authors.[14] A comprehensive review of source-based and structure-based nomenclature for all of the major classes of polymers,[15] and a short tutorial on the correct identification, orientation, and naming of most commonly encountered constitutional repeating units were both published in 2000.[16]

References and Notes

1. International Union of Pure and Applied Chemistry, *Compendium of Macromolecular Nomenclature,* Blackwell Scientific Publications, Oxford, 1991.
2. International Union of Pure and Applied Chemistry, Stereochemical Definitions and Notations Relating to Polymers (Recommendations 1980), *Pure Appl. Chem.*, **53**, 733-752 (1981).
3. International Union of Pure and Applied Chemistry, Source-Based Nomenclature for Copolymers (Recommendations 1985), *Pure Appl. Chem.*, **57**, 1427-1440 (1985).
4. International Union of Pure and Applied Chemistry, Nomenclature of Regular Single-Strand Organic Polymers (Recommendations 1975), *Pure Appl. Chem.*, **48**, 373-385 (1976).
5. Chemical Abstracts Service, Naming and Indexing of Chemical Substances for Chemical Abstracts, Appendix IV, *Chemical Abstracts 1999 Index Guide.*
6. International Union of Pure and Applied Chemistry, *A Guide to IUPAC Nomenclature of Organic Compounds* (1993), Blackwell Scientific Publications, Oxford, 1993.
7. International Union of Pure and Applied Chemistry, *Nomenclature of Organic Chemistry, Sections A, B, C, D, E, F, and H,* Pergamon Press, Oxford, 1979.
8. International Union of Pure and Applied Chemistry, Nomenclature of Regular Double-Strand and Quasi-Single-Strand Inorganic and Coordination Polymers (Recommendations 1984), *Pure Appl. Chem.*, **57**, 149-168 (1985).
9. International Union of Pure and Applied Chemistry, Nomenclature of Regular Double-Strand (Ladder and Spiro) Organic Polymers (Recommendations 1993), *Pure Appl. Chem.*, **65**, 1561-1580 (1993).
10. International Union of Pure and Applied Chemistry, Structure-Based Nomenclature for Irregular Single-Strand Organic Polymers (Recommendations 1994), *Pure Appl. Chem.*, **66**, 873-889 (1994).
11. International Union of Pure and Applied Chemistry, "Source-Based Nomenclature for Non-Linear Macromolecules and Macromolecular Assemblies (Recommendations 1997)." *Pure Appl. Chem.*, **69**, 2511-2521 (1997).
12. Poly(1,3-butadiene) obtained by polymerization of 1,3-butadiene in the so-called 1,4- mode is frequently drawn incorrectly in publications as $-(CH_2-CH=CH-CH_2)_n-$; the double bond should be assigned the lowest locant possible, i.e. the structure should be drawn as $-(CH=CH-CH_2-CH_2)_n-$.
13. International Union of Pure and Applied Chemistry, "Graphic Representations (Chemical Formulae) of Macromolecules (Recommendations 1994)." *Pure Appl. Chem.*, **66**, 2469-2482 (1994).
14. Wilks, E. S. Macromolecular Nomenclature Note No. 17: "Whither Nomenclature?" *Polym. Prepr.* **40**(2), 6-11 (1999); also available at www.chem.umr.edu/~poly/nomenclature.html.
15. Wilks, E. S. "Polymer Nomenclature: The Controversy Between Source-Based and Structure-Based Representations (A Personal Perspective)." *Prog. Polym. Sci.* **25**, 9-100 (2000).
16. Wilks, E. S. Macromolecular Nomenclature Note No. 18: "SRUs: Using the Rules." *Polym. Prepr.* **41**(1), 6a-11a (2000); also available at www.chem.umr.edu/~poly/nomenclature.html; a .pdf format version is also available.

SOLVENTS FOR COMMON POLYMERS

Abbreviations: HC: hydrocarbons; MEK: methyl ethyl ketone; THF: tetrahydrofuran; DMF: dimethylformamide; DMSO: dimethylsulfoxide

Polyethylene (HDPE)	HC and halogenated HC
Polypropylene (atactic)	HC and halogenated HC
Polybutadiene	HC, THF, ketones
Polystyrene	ethylbenzene, $CHCl_3$, CCl_4, THF, MEK
Polyacrylates	aromatic HC, chlorinated HC, THF, esters, ketones
Polymethacrylates	aromatic HC, chlorinated HC, THF, esters, MEK
Polyacrylamide	water
Poly(vinyl ethers)	halogenated HC, MEK, butanol
Poly(vinyl alcohol)	glycols (hot), DMF
Poly(vinyl acetate)	aromatic HC, chlorinated HC, THF, esters, DMF
Poly(vinyl chloride)	THF, DMF, DMSO
Poly(vinylidene chloride)	THF (hot), dioxane, DMF
Poly(vinyl fluoride)	DMF, DMSO (hot)
Polyacrylonitrile	DMF, DMSO
Poly(oxyethylene)	aromatic HC, $CHCl_3$, alcohols, esters, DMF
Poly(2,6-dimethylphenylene oxide)	aromatic HC, halogenated HC
Poly(ethylene terephthalate)	phenol, DMSO (hot)
Polyurethanes (linear)	aromatic HC, THF, DMF
Polyureas	phenol, formic acid
Polysiloxanes	HC, THF, DMF
Poly[bis(2,2,2-trifluoroethoxy)-phosphazene]	THF, ketones, ethyl acetate

GLASS TRANSITION TEMPERATURE FOR SELECTED POLYMERS

Robert B. Fox

Polymer names are based on the IUPAC structure-based nomenclature system described in the table "Naming Organic Polymers". Within each category, names are listed in alphabetical order. Source-based and trivial names are also given (in italics) for the most common polymers. The table does not include polymers for which T_g is not clearly defined because of variability of structure or because of reactions taking place near the glass transition.

All values of T_g cited in this table have been determined by differential scanning calorimetry (DSC) except those values indicated by:

(D) dynamic method
(Dil) dilatometry
(M) mechanical method

Polymer name	Glass transition temperature (T_g/K)
ACYCLIC CARBON CHAINS	
Polyalkadienes	
Poly(alkenylene) *Polyalkadiene* –[$CH=CHCH_2CH_2$]–	
Poly(*cis*-1-butenylene) *cis-1,3-polybutadiene* [PBD]	171
Poly(*trans*-1-butenylene) *trans-1,3-polybutadiene* [PBD]	215
Poly(1-chloro-*cis*-1-butenylene) *cis-1,3-polychloroprene*	253
Poly(1-chloro-*trans*-1-butenylene) *trans-1,3-polychloroprene*	233
Poly(1-methyl-*cis*-1-butenylene) *cis-1,3-polyisoprene*	200
Poly(1-methyl-*trans*-1-butenylene) *trans-1,3-polyisoprene*	207
Poly(1,4,4-trifluoro-1-butenylene)	238
Polyalkenes	
Poly(alkylethylene) *Poly(alkylethylene)* -[$RCHCH_2$]-	
Poly(1-benzylethylene)	333
Poly(1-butylethylene)	223
Poly(1-cyclohexylethylene) (atactic)	393
Poly(1-cyclohexylethylene) (isotactic)	406 (D)
Poly(1,1-dimethylethylene) *Polyisobutylene* [PIB]	200
Poly(ethylene)	148
Poly(methylene)	155
Poly(1-phenethylethylene)	283
Poly(propylene) (isotactic)	272
Poly(propylene) (syndiotactic)	ca. 265
Poly[1-(2-pyridyl)ethylene]	377
Poly[1-(4-pyridyl)ethylene]	415
Poly(1-vinylethylene)	273
Polyacrylics	
Poly[1-(alkoxycarbonyl)ethylene] *Poly(alkyl acrylate)* –[$(ROCO)CHCH_2$]–	
Poly[1-(benzyloxycarbonyl)ethylene]	279
Poly[1-(butoxycarbonyl)ethylene] *Poly(butyl acrylate)* [PBA]	219 (M)

Polymer name	Glass transition temperature (T_g/K)
Poly[1-(*sec*-butoxycarbonyl)ethylene]	251
Poly[1-(butoxycarbonyl)-1-cyanoethylene]	358
Poly[1-(butylcarbamoyl)ethylene]	319 (M)
Poly(1-carbamoylethylene)	438
Polyacrylamide [PAM]	
Poly(1-carboxyethylene)	379
Poly(acrylic acid) [PAA]	
Poly[1-(2-chlorophenoxycarbonyl)ethylene]	326
Poly[1-(4-chlorophenoxycarbonyl)ethylene]	331
Poly[1-(4-cyanobenzyloxycarbonyl)ethylene]	317
Poly[1-(2-cyanoethoxycarbonyl)ethylene]	277
Poly[1-(cyanomethoxycarbonyl)ethylene)]	433 Dil
Poly[1-(4-cyanophenoxycarbonyl)ethylene]	363
Poly[1-(cyclohexyloxycarbonyl)ethylene]	292
Poly[1-(2,4-dichlorophenoxycarbonyl)ethylene]	333
Poly[1-(dimethylcarbamoyl)ethylene]	362
Poly[1-(ethoxycarbonyl)ethylene]	249
Poly(ethyl acrylate) [PEA]	
Poly[1-(ethoxycarbonyl)-1-fluoroethylene]	316
Poly[1-(2-ethoxycarbonylphenoxycarbonyl)ethylene]	303
Poly[1-(3-ethoxycarbonylphenoxycarbonyl)ethylene]	297
Poly[1-(4-ethoxycarbonylphenoxycarbonyl)ethylene]	310
Poly[1-(2-ethoxyethoxycarbonyl)ethylene]	223
Poly[1-(3-ethoxypropoxycarbonyl)ethylene]	218
Poly[1-(isopropoxycarbonyl)ethylene]	267-270
Poly[1-(methoxycarbonyl)ethylene]	283
Poly(methyl acrylate) [PMA]	
Poly[1-(2-methoxycarbonylphenoxycarbonyl)ethylene]	319
Poly[1-(3-methoxycarbonylphenoxycarbonyl)ethylene]	311
Poly[1-(4-methoxycarbonylphenoxycarbonyl)ethylene]	340
Poly[1-(2-methoxyethoxycarbonyl)ethylene]	223
Poly[1-(4-methoxyphenoxycarbonyl)ethylene]	324
Poly[1-(3-methoxypropoxycarbonyl)ethylene]	198
Poly[1-(2-naphthyloxycarbonyl)ethylene]	358
Poly[1-(pentachlorophenoxycarbonyl)ethylene]	420
Poly[1-(phenethoxycarbonyl)ethylene]	270
Poly[1-(phenoxycarbonyl)ethylene]	330
Poly[1-(*m*-tolyloxycarbonyl)ethylene]	298
Poly[1-(*o*-tolyloxycarbonyl)ethylene]	325
Poly[1-(*p*-tolyloxycarbonyl)ethylene]	316
Poly[1-(2,2,2-trifluorethoxycarbonyl)ethylene]	263
Polymethacrylics	
Poly[1-(alkoxycarbonyl)-1-methylethylene] *Poly(alkyl methacrylate)* –[(ROCO)(Me)CCH_2]–	
Poly[1-(benzyloxycarbonyl)-1-methylethylene]	327
Poly[1-(2-bromoethoxycarbonyl)-1-methylethylene]	325
Poly[(1-(butoxycarbonyl)-1-methylethylene]	293
Poly(butyl methacrylate) [PBMA]	
Poly[1-(*sec*-butoxycarbonyl)-1-methylethylene]	333
Poly[1-(*tert*-butoxycarbonyl)-1-methylethylene)]	391
Poly[1-(2-chloroethoxycarbonyl)-1-methylethylene]	ca 315
Poly[1-(2-cyanoethoxycarbonyl)-1-methylethylene]	364
Poly[1-(4-cyanophenoxycarbonyl)-1-methylethylene]	428
Poly[1-(cyclohexyloxycarbonyl)-1-methylethylene] (atactic)	356
Poly[1-(cyclohexyloxycarbonyl)-1-methylethylene)] (isotactic)	324

Polymer name	Glass transition temperature (T_g/K)
Poly[1-(dimethylaminoethoxycarbonyl)-1-methylethylene]	292
Poly[1-(ethoxycarbonyl)-1-ethylethylene]	300
Poly[1-(ethoxycarbonyl)-1-methylethylene] (atactic) *Poly(ethyl methacrylate)* [PEMA]	338
Poly[1-(ethoxycarbonyl)-1-methylethylene] (isotactic)	285
Poly[1-(ethoxycarbonyl)-1-methylethylene)] (syndiotactic)	339
Poly[1-(hexyloxycarbonyl)-1-methylethylene]	268
Poly[1-(isobutoxycarbonyl)-1-methylethylene]	326
Poly[1-(isopropoxycarbonyl)-1-methylethylene]	354
Poly[1-(methoxycarbonyl)-1-methylethylene] (atactic) *Poly(methyl methacrylate)* [PMMA]	378
Poly[1-(methoxycarbonyl)-1-methylethylene)] (isotactic)	311
Poly[1-(methoxycarbonyl)-1-methylethylene)] (syndiotactic)	378
Poly[1-(4-methoxycarbonylphenoxy)-1-methylethylene]	379
Poly[1-(methoxycarbonyl)-1-phenylethylene)] (atactic)	391
Poly[1-(methoxycarbonyl)-1-phenylethylene)] (isotactic)	397
Poly[1-methyl-1-(phenethoxycarbonyl)ethylene]	299
Poly[1-methyl-1-(phenoxycarbonyl)ethylene]	383
Polyvinyl ethers, alcohols, and ketones	
Poly(1-alkoxyethylene) *Poly(alkyl vinyl ether)* –[$ROCHCH_2$]–	
Poly(1-hydroxyethylene) *Poly(vinyl alcohol)* –[$HOCHCH_2$]–	
Poly(1-alkanoylethylene) *Poly(alkyl vinyl ketone)* –[$RCOCHCH_2$]–	
Poly(1-butoxyethylene)	218
Poly(1-*sec*-butoxyethylene)	253
Poly(1-*tert*-butoxyethylene)	361
Poly[1-(butylthio)ethylene]	253
Poly(1-ethoxyethylene)	230
Poly[1-(4-ethylbenzoyl)ethylene]	325
Poly(1-hydroxyethylene) *Poly(vinyl alcohol)* [PVA]	358 (D)
Poly(hydroxymethylene)	407
Poly(1-isopropoxyethylene)	270
Poly[1-(4-methoxybenzoyl)ethylene]	319 (M)
Poly(1-methoxyethylene) *Poly(methyl vinyl ether)* [PMVE]	242
Poly[1-(methylthio)ethylene]	272
Poly(1-propoxyethylene)	224
Poly[1-(trifluoromethoxy)trifluoroethylene]	268
Polyvinyl halides and nitriles	
Poly(1-haloethylene) *Poly(vinyl halide)* –[$XCHCH_2$]–	
Poly(1-cyanoethylene) *Poly(acrylonitrile)* –[$NCCHCH_2$]–	
Poly(1-chloroethylene) *Poly(vinyl chloride)* [PVC]	354
Poly(chlorotrifluoroethylene)	373
Poly(1-cyanoethylene) *Polyacrylonitrile* [PAN]	370
Poly(1-cyano-1-methylethylene) *Polymethacrylonitrile*	393
Poly(1,1-dichloroethylene) *Poly(vinylidene chloride)*	255
Poly(1,1-difluoroethylene) *Poly(vinylidene fluoride)*	ca 233
Poly(1-fluoroethylene) *Poly(vinyl fluoride)*	314 (M)

Polymer name	Glass transition temperature (T_g/K)
Poly(1-hexafluoropropylene)	425
Poly[1-(2-iodoethyl)ethylene]	343
Poly(tetrafluoroethylene)	(160)
Poly[1-(trifluoromethyl)ethylene]	300
Polyvinyl esters	
Poly[1-(alkanoyloxy)ethylene] *Poly(vinyl alkanoate)* –[RCOOCHCH$_2$]–	
Poly(1-acetoxyethylene) *Poly(vinyl acetate)* [PVAc]	305
Poly[1-(benzoyloxy)ethylene]	344
Poly[1-(4-bromobenzoyloxy)ethylene]	365
Poly[1-(2-chlorobenzoyloxy)ethylene]	335
Poly[1-(3-chlorobenzoyloxy)ethylene]	338
Poly[1-(4-chlorobenzoyloxy)ethylene]	357
Poly[1-(cyclohexanoyloxy)ethylene]	349 (M)
Poly[1-(4-ethoxybenzoyloxy)ethylene]	343
Poly[1-(4-ethylbenzoyloxy)ethylene]	326
Poly[1-(4-isopropylbenzoyloxy)ethylene]	342
Poly[1-(2-methoxybenzoyloxy)ethylene]	338
Poly[1-(3-methoxybenzoyloxy)ethylene]	ca 317
Poly[1-(4-methoxybenzoyloxy)ethylene]	360
Poly[1-(4-methylbenzoyloxy)ethylene]	343
Poly[1-(4-nitrobenzoyloxy)ethylene]	395
Poly[1-(propionoyloxy)ethylene]	283 (M)
Polystyrenes	
Poly(1-phenylethylene) *Polystyrene* –[C$_6$H$_5$CHCH$_2$]–	
Poly[1-(4-acetylphenyl)ethylene]	389 (M)
Poly[1-(4-benzoylphenyl)ethylene]	371 (M)
Poly[1-(4-bromophenyl)ethylene]	391
Poly[1-(4-butoxyphenyl)ethylene]	ca 320 (M)
Poly[1-(4-butoxycarbonylphenyl)ethylene]	349 (M)
Pol[(1-(4-butylphenyl)ethylene]	279
Poly[1-(4-carboxyphenyl)ethylene]	386 (M)
Poly[1-(2-chlorophenyl)ethylene]	392
Poly[1-(3-chlorophenyl)ethylene]	363
Poly[1-(4-chlorophenyl)ethylene]	383
Poly[1-(2,4-dichlorophenyl)ethylene]	406
Poly[1-(2,5-dichlorophenyl)ethylene]	379
Poly[1-(2,6-dichlorophenyl)ethylene]	440
Poly[1-(3,4-dichlorophenyl)ethylene]	401
Poly[1-(2,4-dimethylphenyl)ethylene]	385
Poly[1-(4-(dimethylamino)phenyl)ethylene]	398 (M)
Poly[1-(4-ethoxyphenyl)ethylene]	ca 359 (M)
Poly[1-(4-ethoxycarbonylphenyl)ethylene]	367 (M)
Poly[1-(4-fluorophenyl)ethylene]	368
Poly[1-(4-iodophenyl)ethylene]	429
Poly[1-(4-methoxyphenyl)ethylene]	386
Poly[1-(4-methoxycarbonylphenyl)ethylene]	386 (M)
Poly(1-methyl-1-phenylethylene) *Poly(α-methylstyrene)*	373
Poly[1-(2-(methylamino)phenyl)ethylene]	462 (M)
Poly(1-phenylethylene) *Polystyrene* [PS]	373

Polymer name	Glass transition temperature (T_g/K)
Poly[1-(4-propoxyphenyl)ethylene]	343 (M)
Poly[1-(4-propoxycarbonylphenyl)ethylene]	365 (M)
Poly(1-*o*-tolylethylene)	409
CHAINS WITH CARBOCYCLIC UNITS	
Poly(arylenealkylene) $-[-Ar-(CH_2)_n]-$	
Poly[1-(2-bromo-1,4-phenylene)ethylene]	353 (M)
Poly[1-(2-chloro-1,4-phenylene)ethylene]	343 (M)
Poly[1-(2-cyano-1,4-phenylene)ethylene]	363 (M)
Poly[1-(2,5-dimethyl-1,4-phenylene)ethylene]	373 (M)
Poly[1-(2-ethyl-1,4-phenylene)ethylene]	298 (M)
Poly[1-(1,4-naphthylene)ethylene]	433 (M)
Poly[1-(1,4-phenylene)ethylene]	ca 353 (M)
CHAINS WITH HETEROATOM UNITS	
Main chain oxide units	
Poly(oxyalkylene) *Poly(alkylene oxide)* $-[O(CH_2)_n]-$	
Poly[oxy(1,1-bis(chloromethyl)trimethylene)]	265
Poly[oxy(1-(bromomethyl)ethylene)]	259
Poly[oxy(1-(butoxymethyl)ethylene)]	194
Poly[oxy(1-butylethylene)]	203
Poly[oxy(1-*tert*-butylethylene)]	308
Poly[oxy(1-(chloromethyl)ethylene)] *Poly(epichlorohydrin)*	251
Poly[oxy(2,6-dimethoxy-1,4-phenylene)]	440
Poly[oxy(1,1-dimethylethylene)]	264
Poly[oxy(2,6-dimethyl-1,4-phenylene)]	482
Poly[oxy(2,6-diphenyl-1,4-phenylene)]	493
Poly[oxy(1-ethylethylene)]	203
Poly(oxyethylidene) *Polyacetaldehyde*	243
Poly[oxy(1-(methoxymethyl)ethylene)]	211
Poly[oxy(2-methyl-6-phenyl-1,4-phenylene)]	428
Poly[oxy(1-methyltrimethylene)]	223 (D)
Poly[oxy(2-methyltrimethylene)]	218
Poly(oxy-1,4-phenylene) *Poly(phenylene oxide)* [PPO]	358
Poly[oxy(1-phenylethylene)]	313
Poly(oxytetramethylene) *Poly(tetrahydrofuran)* [PTMO]	189
Poly(oxytrimethylene)	195
Main-chain ester or anhydride units	
Poly(oxyalkyleneoxyalkanedioyl) *Poly(alkylene alkanedioate)* $[O(CH_2)_mOCO(CH_2)_nCO]-$	
Poly(oxyadipoyloxydecamethylene)	217
Poly(oxyadipoyloxy-1,4-phenyleneisopropylidene-1,4-phenylene)	341
Poly(oxycarbonyloxy-1,4-phenylene-isopropylidene-1,4-phenylene) *Bisphenol A polycarbonate*	422
Poly(oxycarbonylpentamethylene)	213
Poly(oxycarbonyl-1,4-phenylenemethylene-1,4-phenylene)	395
Poly(oxycarbonyl-1,4-phenyleneisopropylidene-1,4-phenylene)	333

Polymer name	Glass transition temperature (T_g/K)
Poly[oxy(2,6-dimethyl-1,4-phenyleneisopropylidene-3,5-dimethyl-1,4-phenylene)oxysebacoyl]	318
Poly(oxyethylenecarbonyl-1,4-cyclohexylenecarbonyl) (trans)	291
Poly(oxyethyleneoxycarbonyl-1,4-naphthylenecarbonyl)	337
Poly(oxyethyleneoxycarbonyl-1,5-naphthylenecarbonyl)	344
Poly(oxyethyleneoxycarbonyl-2,6-naphthylenecarbonyl)	386
Poly(oxyethyleneoxycarbonyl-2,7-naphthylenecarbonyl)	392
Poly(oxyethyleneoxyterephthaloyl) *Poly(ethylene terephthalate)* [PET]	342
Poly(oxyisophthaloyl)	403 (D)
Poly(oxy(1-oxo-2,2-dimethyltrimethylene)) *Poly(pivalolactone)*	263
Poly(oxy-1,4-phenyleneisopropylidene-1,4-phenyleneoxysebacoyl)	280
Poly(oxy-1,4-phenyleneoxy-1,4-phenyleneoxy-carbonyl-1,4-phenylene) [PEEK]	416
Poly(oxypropyleneoxyterephthaloyl)	341
Poly[oxyterephthaloyloxy(2,6-dimethyl-1,4-phenyleneisopropylidene-3,5-dimethyl-1,4-(D)phenylene)]	498
Poly(oxyterephthaloyloxyoctamethylene)	318 (D)
Poly(oxyterephthaloyloxy-1,4-phenyleneisopropylidene-1,4-phenylene) *Poly(bisphenol A terephthalate)*	478
Poly(oxytetramethyleneoxyterephthaloyl) *Poly(butylene terephthalate)* [PBT]	323
Main-chain amide units	
Poly(iminoalkyleneiminoalkanedioyl) *Poly(alkylene alkanediamide)*–[$NH(CH_2)_mNHCO(CH_2)_nCO$]–	
Poly(iminoadipoyliminodecamethylene) *Nylon 10,6*	313
Poly(iminoadipoyliminohexamethylene) *Nylon 6,6*	ca 323
Poly(iminoadipoyliminooctamethylene) *Nylon 8,6*	318
Poly[iminoadipoyliminotrimethylene(methylimino)trimethylene]	278
Poly(iminocarbonyl-1,4-cyclohexylenemethylene)	466
Poly[iminocarbonyl-1,4-phenylene(2-oxoethylene)iminohexamethylene]	377
Poly(iminoethylene-1,4-phenyleneethyleneiminosebacoyl)	378 (D)
Poly(iminohexamethyleneiminoazelaoyl) *Nylon 6,9*	331
Poly(iminohexamethyleneiminododecanedioyl) *Nylon 6, 12*	319
Poly(iminohexamethyleneiminopimeloyl) *Nylon 6,7*	331
Poly(iminohexamethyleneiminosebacoyl) *Nylon 6,10*	323
Poly(iminohexamethyleneiminosuberoyl) *Nylon 6,8*	330
Poly(iminoisophthaloylimino-4,4′-biphenylylene)	558
Poly(iminoisophthaloyliminohexamethylene)	390
Poly(iminoisophthaloyliminomethylene-1,4-cyclohexylenemethylene)	481
Poly(iminoisophthaloyliminomethylene-1,3-phenylenemethylene)	438 (M)
Poly[iminomethylene(2,5-dimethyl-1,4-phenylene)methyleneiminosuberoyl]	351
Poly(imino-1,5-naphthyleneiminoisophthaloyl)	598
Poly(imino-1,5-naphthyleneiminoterephthaloyl)	578
Poly(iminooctamethyleneiminodecanedioyl) *Nylon 8,10*	333
Poly(iminooxalyliminohexamethylene) *Nylon 6,2*	430
Poly[imino(1-oxohexamethylene)] *Nylon 6*	326

Polymer name	Glass transition temperature (T_g/K)
Poly[imino(1-oxodecamethylene)]	315
Nylon 10	
Poly[imino(1-oxoheptamethylene)]	325
Nylon 7	
Poly[imino(1-oxo-3-methyltrimethylene]	369
Poly[imino(1-oxononamethylene)]	319
Nylon 9	
Poly[imino(1-oxooctamethylene)]	323
Nylon 8	
Poly[imino(1-oxotrimethylene)]	384
Nylon 3	
Poly(iminopentamethyleneiminoadipoyl)	318
Nylon 5,6	
Poly[iminopentamethyleneiminocarbonyl-1,4-phenylene(2-oxoethylene)]	376
Poly(imino-1,3-phenyleneiminoisophthaloyl)	553 (M)
Poly(imino-1,4-phenyleneiminoterephthaloyl)	618
Poly(iminopimeloyliminoheptamethylene)	328
Nylon 7,7	
Poly(iminoterephthaloylimino-4,4′-biphenylylene)	613
Poly(iminotetramethyleneiminoadipoyl)	316
Nylon 4,6	
Poly[iminotetramethyleneiminocarbonyl-1,4-phenylene(2-oxoethylene)]	357
Poly(iminotrimethyleneiminoadipoyliminotrimethylene)	307
Poly[iminotrimethyleneiminocarbonyl-1,4-phenylene(2-oxoethylene)]	382
Poly(oxy-1,4-phenyleneiminoterephthaloyl-imino-1,4-phenylene)	613
Poly(sulfonylimino-1,4-phenyleneiminoadipoylimino-1,4-phenylene)	467
Main-chain urethane units	
Poly(oxyalkyleneoxycarbonyliminoalkyleneiminocarbonyl)–[$O(CH_2)_mOCONH(CH_2)_nNHCO$]–	
Poly(oxyethyleneoxycarbonyliminohexamethyleneiminocarbonyl)	329
Poly[oxyethyleneoxycarbonylimino(6-methyl-1,3-phenylene)iminocarbonyl]	325
Poly(oxyethyleneoxycarbonylimino-1,4-phenylenemethylene-1,4-phenyleneiminocarbonyl)	412
Poly(oxyhexamethyleneoxycarbonyliminohexamethyleneiminocarbonyl)	332
Poly[oxyhexamethyleneoxycarbonylimino(6-methyl-1,3-phenylene)iminocarbonyl]	305
Poly(oxyhexamethyleneoxycarbonylimino-1,4-phenylenemethylene-1,4-phenyleneiminocarbonyl)	364
Poly(oxyoctamethyleneoxycarbonyliminohexamethyleneiminocarbonyl)	331
Poly[oxyoctamethyleneoxycarbonylimino(6-methyl-1,3-phenylene)iminocarbonyl]	337
Poly(oxyoctamethyleneoxycarbonylimino-1,4-phenylenemethylene-1,4-phenyleneiminocarbonyl)	352
Poly(oxytetramethyleneoxycarbonyliminohexamethyleneiminocarbonyl)	332
Poly[oxytetramethyleneoxycarbonylimino(6-methyl-1,3-phenylene)iminocarbonyl]	315
Poly(oxytetramethyleneoxycarbonylimino-1,4-phenylenemethylene-1,4-phenyleneiminocarbonyl)	382
Main-chain siloxanes	
Poly[oxy(dialkylsilylene)] *Poly(dialkylsiloxane)* –[$O(R_2Si)$]–	
Poly[oxy(dimethylsilylene)]	148
Poly(dimethylsiloxane) [PDMS]	
Poly[oxy(dimethylsilylene)oxy-1,4-phenylene]	363 (M)
Poly[oxy(dimethylsilylene)oxy-1,4-phenyleneisopropylidene-1,4-phenylene]	318 (M)
Poly[oxy(diphenylsilylene)]	238
Poly(diphenylsiloxane)	
Poly[oxy(diphenylsilylene)-1,3-phenylene]	ca 331
Poly[oxy((methyl)phenylsilylene)]	187
Poly[oxy((methyl)-3,3,3-trifluoropropylsilylene]	<193

Polymer name	Glass transition temperature (T_g/K)
Main-chain sulfur-containing units	
Poly(dithioethylene)	223
Poly(dithiomethylene-1,4-phenylenemethylene)	296
Poly(oxy-4,4′-biphenylylene-1,4-phenylenesulfonyl-1,4-phenylene)	503 (M)
Poly(oxycarbonyloxy-1,4-phenylenethio-1,4-phenylene)	ca 383
Poly(oxyethylenedithioethylene)	220 (M)
Poly[oxy(2-hydroxytrimethylene)oxy-1,4-phenylenesulfonyl-1,4-phenylene]	428
Poly(oxymethyleneoxyethylenedithioethylene)	214
Poly(oxy-1,4-phenylenesulfinyl-1,4-phenyleneoxy-1,4-phenylenecarbonyl-1,4-phenylene)	478 (M)
Poly(oxy-1,4-phenylenesulfinyl-1,4-phenyleneoxy-1,4-phenyleneisopropylidene-1,4-phenylene)	438 (M)
Poly(oxy-1,4-phenylenesulfonyl-1,4-phenylene)	487
Poly(oxy-1,4-phenylenesulfonyl-4,4′-biphenylylenesulfonyl-1,4-phenylene)	533
Poly[oxy-1,4-phenylenesulfonyl-1,4-phenyleneoxy(2,6-dimethyl-1,4-phenylene)isopropylidene (3,5-dimethyl-1,4-phenylene)]	508 (M)
Poly(oxy-1,4-phenylenesulfonyl-1,4-phenyleneoxy-1,4-phenylenecarbonyl-1,4-phenylene)	478 (M)
Poly[oxy-1,4-phenylenesulfonyl-1,4-phenyleneoxy-1,4-phenylene(hexafluoroisopropylidene)1,4-phenylene]	478 (M)
Poly(oxy-1,4-phenylenesulfonyl-1,4-phenyleneoxy-1,4-phenyleneisopropylidene-1,4-phenylene)	449
Poly(oxy-1,4-phenylenesulfonyl-1,4-phenyleneoxy-1.4-phenylenemethylene-1,4-phenylene)	453 (M)
Poly(oxy-1,4-phenylenesulfonyl-1,4-phenyleneoxy-1.4-phenylenethio-1,4-phenylene)	448 (M)
Poly(oxy-1,4-phenylenesulfonyl-1,4-phenyleneoxyterephthaloyl)	522
Poly(oxytetramethylenedithiotetramethylene)	197
Poly(sulfonyl-1,2-cyclohexylene)	401
Poly(sulfonyl-1,3-cyclohexylene)	381
Poly(sulfonyl-1,4-phenylenemethylene-1,4-phenylene)	497
Poly(thio-1,3-cyclohexylene)	221
Poly[thio(difluoromethylene)]	155
Poly(thioethylene)	223
Poly[thio(1-ethylethylene]	218
Poly[thio(1-methyl-3-oxotrimethylene)]	285
Poly[thio(1-methyltrimethylene)]	214
Pol[(thio(1-oxohexamethylene)]	292
Poly(thio-1,4-phenylene)	370
Poly(thiopropylene)	226
Main-chain heterocyclic units	
Poly(1,3-dioxa-4,6-cyclohexylenemethylene)	378
Poly(vinyl formal)	
Poly[(2,6-dioxopiperidine-1,4-diyl)trimethylene]	363
Poly[(2-methyl-1,3-dioxa-4,6-cyclohexylene)methylene]	355
Poly(vinyl acetal)	
Poly(1,4-piperazinediylcarbonyloxyethyleneoxycarbonyl)	333
Poly(1,4-piperazinediylisophthaloyl)	465 (M)
Poly[(2-propyl-1,3-dioxa-4,6-cyclohexylene)methylene]	322
Poly(vinyl butyral)	
Poly(3,6-pyridazinediyloxy-1,4-phenyleneisopropylidene-1,4-phenyleneoxy)	453 (M)
Poly(2,5-pyridinediylcarbonyliminohexamethyleneiminocarbonyl)	322

DIELECTRIC CONSTANT OF SELECTED POLYMERS

This table lists typical values of the dielectric constant (more properly called relative permittivity) of some important polymers. Values are given for frequencies of 1 kHz, 1 MHz, and 1 GHz; in most cases the dielectric constant at frequencies below 1 kHz does not differ significantly from the value at 1 kHz. Since the dielectric constant of a polymeric material can vary with density, degree of crystallinity, and other details of a particular sample, the values given here should be regarded only as typical or average values.

REFERENCES

1. Gray, D.E., Ed., *American Institute of Physics Handbook, Third Edition*, p. 5-132, McGraw Hill, New York, 1972.
2. Anderson, H.L., Editor, *A Physicist's Desk Reference*, American Institute of Physics, New York, 1989.
3. Brandrup, J., and Immergut, E.H., *Polymer Handbook, Third Edition*, John Wiley & Sons, New York, 1989.

Name	*t*/°C	1 kHz	1 MHz	1 GHz
Polyacrylonitrile	25	5.5	4.2	
Polyamides (nylons)	25	3.50	3.14	2.8
	84	11	4.4	2.8
Polybutadiene	25	2.5		
Polycarbonate	23	2.92	2.8	
Polychloroprene (neoprene)	25	6.6	6.3	4.2
Polychlorotrifluoroethylene	23	2.65	2.46	2.39
Polyethylene	23	2.3		
Poly(ethylene terephthalate) (Mylar)	23	3.25	3.0	2.8
Polyisoprene (natural rubber)	27	2.6	2.5	2.4
Poly(methyl methacrylate)	27	3.12	2.76	2.6
	80	3.80	2.7	2.6
Polyoxymethylene (polyformaldehyde)	25	3.8		
Poly(phenylene oxide)	23	2.59	2.59	
Polypropylene	25	2.3	2.3	2.3
Polystyrene	25	2.6	2.6	2.6
Polysulfones	25	3.13	2.10	
Polytetrafluoroethylene (teflon)	25	2.1	2.1	2.1
Poly(vinyl acetate)	50		3.5	
	150		8.3	
Poly(vinyl chloride)	25	3.39	2.9	2.8
	100	5.3	3.3	2.7
Poly(vinylidene chloride)	23	4.6	3.2	2.7
Poly(vinylidene fluoride)	23	12.2	8.9	4.7

PRESSURE-VOLUME-TEMPERATURE RELATIONSHIP FOR POLYMER MELTS

Christian Wohlfarth

Numerous theoretical equations of state for polymer liquids have been developed. These, at the minimum, have to provide accurate fitting functions to experimental data. However, for the purpose of this table, the empirical Tait equation along with a polynomial expression for the zero pressure isobar is used. This equation is able to represent the experimental data for the melt state within the limits of experimental errors, i.e., the maximum deviations between measured and calculated specific volumes are about 0.001-0.002 cm^3/g.

The general form of the Tait equation is:

$$V(P,T) = V(0,T)\{1 - C \ln[1 + P/B(T)]\} \quad (1)$$

where the coefficient C is usually taken to be a universal constant equal to 0.0894. T is the absolute temperature in K and P the pressure in MPa. The volume V is the specific volume in cm^3/g. The Tait parameter $B(T)$ has the very simple meaning that it is inversely proportional to the compressibility κ at constant temperature and zero pressure:

$$\kappa(0,T) = -[1/V(0,T)](dV/dP) = C/B(T) \quad (2)$$

The $B(T)$ function is usually given by:

$$B(T) = B_0 \exp[-B_1(T\text{-}273.15)] \quad (3)$$

but, sometimes a polynomial expression is used:

$$B(T) = b_0 + b_1(T\text{-}273.15) + b_2(T\text{-}273.15)^2 \quad (4)$$

The zero-pressure isobar $V(0,T)$ is usually given by:

$$V(0,T) = A_0 + A_1(T\text{-}273.15) + A_2(T\text{-}273.15)^2 \quad (5)$$

where A_0, A_1, A_2 are specific constants for a given polymer (the expression T-273.15 is used because fitting to the zero-pressure isobar is usually done in terms of Celsius temperature). Other forms for $V(0,T)$ are also found in the literature, such as

$$V(0,T) = A_3 \exp[A_4(T\text{-}273.15)] \quad (6)$$

or

$$V(0,T) = A_5 \exp(A_6 T^{1.5}) \quad (7)$$

where A_3 and A_4 or A_5 and A_6 are again specific constants for a given polymer.

The Tait equation is particularly useful to calculate derivative quantities, such as the isothermal compressibility and the thermal expansivity coefficients. The isothermal compressibility $\kappa(P,T)$ is derived from equation (1) as:

$$\kappa(P,T) = -(1/V)(dV/dP) = 1/\{[P + B(T)][1/C - \ln(1 + P/B(T))]\} \quad (8)$$

and the thermal expansivity $\alpha(P,T)$ as:

$$\alpha(P,T) = (1/V)(dV/dT) = \alpha(0,T) - PB_1\kappa(P,T) \quad (9)$$

where $\alpha(0,T)$ represents the thermal expansivity at zero (atmospheric) pressure and is calculated from any suitable fit for the zero-pressure volume, such as equations (5) through (7) above.

Because polymer melt PVT-behavior depends only slightly on polymer molar mass above the oligomeric region, usually no information is given in the original literature for the average molar mass of the polymers.

Table 1 summarizes the polymers or copolymers considered here and the experimental ranges of pressure and temperature over which data are available. In Table 2 the Tait-equation functions, with parameters obtained from the fit, are given for 90 polymer or copolymer melts.

REFERENCES

1. Zoller, P., *J. Appl. Polym. Sci.* 23, 1051-1056, 1979.
2. Starkweather, H. W., Jones, G. A., and Zoller, P., *J. Polym. Sci., Pt. B Polym. Phys.* 26, 257-266,1988.
3. Fakhreddine, Y. A., and Zoller, P., *J. Polym. Sci., Pt. B Polym. Phys.* 29, 1141-1146, 1991.
4. Rodgers, P. A., *J. Appl. Polym. Sci.* 48, 1061-1080, 1993.
5. Rodgers, P. A., *J. Appl. Polym. Sci.* 48, 2075-2083, 1993.
6. Yi, Y. X., and Zoller, P., *J. Polym. Sci., Pt. B Polym. Phys.* 31, 779-788, 1993.
7. Callaghan, T. A., and Paul, D. R., *Macromolecules* 26, 2439–2450, 1993.
8. Wang, Y. Z., Hsieh, K. H., Chen, L. W.,and Tseng, H. C., *J. Appl. Polym. Sci.* 53, 1191-1201, 1994.
9. Privalko, V. P., Arbuzova, A. P., Korskanov, V. V., and Zagdanskaya, N. E., *Polym. Intern.* 35, 161-169, 1994.
10. Sachdev, V. K., Yashi, U., and Jain, R. K., *J. Polym. Sci., Pt. B Polym. Phys.* 36, 841-850, 1998.

Table 1
Names of the Polymers, Abbreviation Used, and Range of Experimental Data Applied in the Determination of the Equation Constants

Polymer	Symbol	*T*/K	*P*/MPa	Ref.
Ethylene/propylene copolymer (50 wt%)	EP50	413-523	0.1-63	4
Ethylene/vinyl acetate copolymer				
18 wt% vinyl acetate	EVA18	385-491	0.1-177	4
25 wt% vinyl acetate	EVA25	367-506	0.1-177	4
28 wt% vinyl acetate	EVA28	367-508	0.1-177	4
40 wt% vinyl acetate	EVA40	348-508	0.1-177	4
Polyamide-6	PA6	509-569	0.1-196	4
Polyamide-11	PA11	478-542	0.1-200	5
Polyamide-66	PA66	519-571	0.1-196	4
cis-1,4-Polybutadiene	cPBD	277-328	0.1-284	4
Polybutadiene, 8% 1,2-content	PBD-8	298-473	0.1-200	6
Polybutadiene, 24% 1,2-content	PBD-24	298-473	0.1-200	6
Polybutadiene, 40% 1,2-content	PBD-40	298-473	0.1-200	6
Polybutadiene, 50% 1,2-content	PBD-50	298-473	0.1-200	6
Polybutadiene, 87% 1,2-content	PBD-87	298-473	0.1-200	6
Poly(1-butene), isotactic	iPB	406-519	0.1-196	4
Poly(butyl methacrylate)	PnBMA	307-473	0.1-200	4
Poly(butylene terephthalate)	PBT	508-576	0.1-200	3
Poly(ε-caprolactone)	PCL	373-421	0.1-200	4
Polycarbonate-bisphenol-A	PC	424-613	0.1-177	4
Polycarbonate-bisphenol-chloral	BCPC	428-557	0.1-200	4
Polycarbonate-hexafluorobisphenol-A	HFPC	432-553	0.1-200	4
Polycarbonate-tetramethylbisphenol-A	TMPC	491-563	0.1-160	4
Poly(cyclohexyl methacrylate)	PcHMA	396-471	0.1-200	4
Poly(2,5-dimethylphenylene oxide)	PPO	473-593	0.1-177	4
Poly(dimethyl siloxane)	PDMS	298-343	0.1-100	4
Poly(dimethyl siloxane) $M_n = 1000$	PDMS-10	304-420	0.1-250	10
Poly(dimethyl siloxane) $M_n = 4000$	PDMS-40	298-418	0.1-250	10
Poly(dimethyl siloxane) $M_n = 6000$	PDMS-60	291-423	0.1-250	10
Poly(epichlorohydrin)	PECH	333-413	0.1-200	4
Poly(ether ether ketone)	PEEK	619-671	0.1-200	4
Poly(ethyl acrylate)	PEA	310-490	0.1-196	4
Poly(ethyl methacrylate)	PEMA	386-434	0.1-196	4
Polyethylene, high density	HDPE	413-476	0.1-196	4
Polyethylene, linear	LPE	415-473	0.1-200	4
Polyethylene, linear, high MW	HMLPE	410-473	0.1-200	4
Polyethylene, branched	BPE	398-471	0.1-200	4
Polyethylene, low density	LDPE	394-448	0.1-196	4
Polyethylene, low density, type A	LDPE-A	385-498	0.1-196	1
Polyethylene, low density, type B	LDPE-B	385-498	0.1-196	1
Polyethylene, low density, type C	LDPE-C	385-498	0.1-196	1
Poly(ethylene oxide)	PEO	361-497	0.1- 68	4
Poly(ethylene terephthalate)	PET	547-615	0.1-196	4
Poly(4-hexylstyrene)	P4HS	303-403	30-100	4
Polyisobutylene	PIB	326-383	0.1-100	4
Polyisoprene, 8% 3,4-content	PI-8	298-473	0.1-200	6
Polyisoprene, 14% 3,4-content	PI-14	298-473	0.1-200	6
Polyisoprene, 41% 3,4-content	PI-41	298-473	0.1-200	6
Polyisoprene, 56% 3,4-content	PI-56	298-473	0.1-200	6
Poly(methyl acrylate)	PMA	310-493	0.1-196	4

Table 1
Names of the Polymers, Abbreviation Used, and Range of Experimental Data Applied in the Determination of the Equation Constants

Polymer	Symbol	*T*/K	*P*/MPa	Ref.
Poly(methyl methacrylate)	PMMA	387-432	0.1-200	4
Poly(4-methyl-1-pentene)	P4MP	514-592	0.1-196	4
Poly(α-methylstyrene)	PαMS	473-533	0.1-170	7
Poly(*o*-methylstyrene)	PoMS	412-471	0.1-180	4
Polyoxymethylene	POM	463-493	0.1-196	2
Phenoxy[a]	PH	341-573	0.1-177	4
Polysulfone[b]	PSF	475-644	0.1-196	4
Polyarylate[c]	PAr	450-583	0.1-177	4
Polypropylene, atactic	aPP	353-393	0.1-100	4
Polypropylene, isotactic	iPP	443-570	0.1-196	4
Polystyrene	PS	388-469	0.1-200	4
Poly(tetrafluoroethylene)	PTFE	603-645	0.1- 39	4
Poly(tetrahydrofuran)	PTHF	335-439	0.1- 78	4
Poly(vinyl acetate)	PVAc	308-373	0.1- 80	4
Poly(vinyl chloride)	PVC	373-423	0.1-200	4
Poly(vinyl methyl ether)	PVME	303-471	0.1-200	4
Poly(vinylidene fluoride)	PVdF	451-521	0.1-200	5
Styrene/acrylonitrile copolymer				
2.7 wt% acrylonitrile	SAN3	378-539	0.1-200	4
5.7 wt% acrylonitrile	SAN6	370-540	0.1-200	4
15.3 wt% acrylonitrile	SAN15	405-531	0.1-200	4
18.0 wt% acrylonitrile	SAN18	377-528	0.1-200	4
40 wt% acrylonitrile	SAN40	373-543	0.1-200	4
70 wt% acrylonitrile	SAN70	373-544	0.1-200	4
Styrene/butadiene copolymer				
10 wt% styrene	SBR10	393-533	0.1-196	8
23.5 wt% styrene	SBR23	393-533	0.1-196	8
60 wt% styrene	SBR60	393-533	0.1-196	8
85 wt% styrene	SBR85	393-533	0.1-196	8
Styrene/methyl methacrylate copolymer				
20 wt% methyl methacrylate	SMMA20	383-543	0.1-200	4
60 wt% methyl methacrylate	SMMA60	383-543	0.1-200	4
N-Vinylcarbazole/4-ethylstyrene copolymer				
50 mol% ethylstyrene	VCES50	393-443	30-100	9
N-Vinylcarbazole/4-hexylstyrene copolymer				
80 mol% hexylstyrene	VCHS80	313-423	30-100	9
67 mol% hexylstyrene	VCHS67	333-423	30-100	9
60 mol% hexylstyrene	VCHS60	383-453	30-100	9
50 mol% hexylstyrene	VCHS50	373-443	30-100	9
40 mol% hexylstyrene	VCHS40	423-493	30-100	9
33 mol% hexylstyrene	VCHS33	463-523	30-100	9
20 mol% hexylstyrene	VCHS20	473-523	30-100	9
N-Vinylcarbazole/4-octylstyrene copolymer				
50 mol% octylstyrene	VCOS50	403-453	30-100	9
N-Vinylcarbazole/4-pentylstyrene copolymer				
50 mol% pentylstyrene	VCPS50	383-443	30-100	9

[a]Phenoxy = Poly(oxy-2-hydroxytrimethyleneoxy-1,4-phenyleneisopropylidene-1,4-phenylene)
[b]Polysulfone = Poly(oxy-1,4-phenylenesulfonyl-1,4-phenyleneoxy-1,4-phenyleneisopropylidene-1,4-phenylene)
[c]Polyarylate = Poly(oxyterephthaloyl/isophthaloyl T/I=50/50)oxy-1,4-phenyleneisopropylidene-1,4-phenylene

Table 2
Tait Equation Parameter Functions for Polymer Melts

Polymer	$V(0,T)$/cm³g⁻¹	$B(T)$/MPa
EP50	$1.2291 + 5.799 \cdot 10^{-5}(T-273.15) + 1.964 \cdot 10^{-6}(T-273.15)^2$	$487.0 \exp[-8.103 \cdot 10^{-3}(T-273.15)]$
EVA18	$1.02391 \exp(2.173 \cdot 10^{-5}T^{1.5})$	$188.2 \exp[-4.537 \cdot 10^{-3}(T-273.15)]$
EVA25	$1.00416 \exp(2.244 \cdot 10^{-5}T^{1.5})$	$184.4 \exp[-4.734 \cdot 10^{-3}(T-273.15)]$
EVA28	$1.00832 \exp(2.241 \cdot 10^{-5}T^{1.5})$	$183.5 \exp[-4.457 \cdot 10^{-3}(T-273.15)]$
EVA40	$1.06332 \exp(2.288 \cdot 10^{-5}T^{1.5})$	$205.1 \exp[-4.989 \cdot 10^{-3}(T-273.15)]$
PA6	$0.7597 \exp[4.701 \cdot 10^{-4}(T-273.15)]$	$376.7 \exp[-4.660 \cdot 10^{-3}(T-273.15)]$
PA11	$0.9581 \exp[6.664 \cdot 10^{-4}(T-273.15)]$	$254.7 \exp[-4.178 \cdot 10^{-3}(T-273.15)]$
PA66	$0.7657 \exp[6.600 \cdot 10^{-4}(T-273.15)]$	$316.4 \exp[-5.040 \cdot 10^{-3}(T-273.15)]$
cPBD	$1.0970 \exp[6.600 \cdot 10^{-4}(T-273.15)]$	$177.7 \exp[-3.593 \cdot 10^{-3}(T-273.15)]$
PBD-8	$1.1004 + 6.718 \cdot 10^{-4}(T-273.15) + 6.584 \cdot 10^{-7}(T-273.15)^2$	$200.0 \exp[-4.606 \cdot 10^{-3}(T-273.15)]$
PBD-24	$1.1049 + 6.489 \cdot 10^{-4}(T-273.15) + 7.099 \cdot 10^{-7}(T-273.15)^2$	$193.0 \exp[-4.519 \cdot 10^{-3}(T-273.15)]$
PBD-40	$1.1013 + 6.593 \cdot 10^{-4}(T-273.15) + 5.776 \cdot 10^{-7}(T-273.15)^2$	$188.0 \exp[-4.437 \cdot 10^{-3}(T-273.15)]$
PBD-50	$1.1037 + 5.955 \cdot 10^{-4}(T-273.15) + 7.789 \cdot 10^{-7}(T-273.15)^2$	$183.0 \exp[-4.425 \cdot 10^{-3}(T-273.15)]$
PBD-87	$1.1094 + 6.729 \cdot 10^{-4}(T-273.15) + 4.470 \cdot 10^{-7}(T-273.15)^2$	$175.0 \exp[-4.538 \cdot 10^{-3}(T-273.15)]$
iPB	$1.1417 \exp[6.751 \cdot 10^{-4}(T-273.15)]$	$167.5 \exp[-4.533 \cdot 10^{-3}(T-273.15)]$
PnBMA	$0.9341 + 5.5254 \cdot 10^{-4}(T-273.15) + 6.5803 \cdot 10^{-6}(T-273.15)^2 + 1.5691 \cdot 10^{-10}(T-273.15)^3$	$226.7 \exp[-5.344 \cdot 10^{-3}(T-273.15)]$
PBT	$0.9640 - 1.017 \cdot 10^{-3}(T-273.15) + 3.065 \cdot 10^{-6}(T-273.15)^2$	$263.0 \exp[-3.444 \cdot 10^{-3}(T-273.15)]$
PCL	$0.9049 \exp[6.392 \cdot 10^{-4}(T-273.15)]$	$189.0 \exp[-3.931 \cdot 10^{-3}(T-273.15)]$
PC	$0.73565 \exp(1.859 \cdot 10^{-5}T^{1.5})$	$310.0 \exp[-4.078 \cdot 10^{-3}(T-273.15)]$
BCPC	$0.6737 + 3.634 \cdot 10^{-4}(T-273.15) + 2.370 \cdot 10^{-7}(T-273.15)^2$	$363.4 \exp[-4.921 \cdot 10^{-3}(T-273.15)]$
HFPC	$0.6111 + 4.898 \cdot 10^{-4}(T-273.15) + 1.730 \cdot 10^{-7}(T-273.15)^2$	$236.6 \exp[-5.156 \cdot 10^{-3}(T-273.15)]$
TMPC	$0.8497 + 5.073 \cdot 10^{-4}(T-273.15) + 3.832 \cdot 10^{-7}(T-273.15)^2$	$231.4 \exp[-4.242 \cdot 10^{-3}(T-273.15)]$
PcHMA	$0.8793 + 4.0504 \cdot 10^{-4}(T-273.15) + 7.774 \cdot 10^{-7}(T-273.15)^2 - 7.7534 \cdot 10^{-10}(T-273.15)^3$	$295.2 \exp[-5.220 \cdot 10^{-3}(T-273.15)]$
PPO	$0.78075 \exp(2.151 \cdot 10^{-5}T^{1.5})$	$227.8 \exp[-4.290 \cdot 10^{-3}(T-273.15)]$
PDMS	$1.0079 \exp[9.121 \cdot 10^{-4}(T-273.15)]$	$89.4 \exp[-5.701 \cdot 10^{-3}(T-273.15)]$
PDMS-10	$0.8343 + 5.991 \cdot 10^{-4}(T-273.15) + 5.734 \cdot 10^{-7}(T-273.15)^2$	$542.63 \exp[-6.69 \cdot 10^{-3}(T-273.15)]$
PDMS-40	$0.8018 + 7.072 \cdot 10^{-4}(T-273.15) + 3.635 \cdot 10^{-7}(T-273.15)^2$	$482.73 \exp[-6.09 \cdot 10^{-3}(T-273.15)]$
PDMS-60	$0.8146 + 5.578 \cdot 10^{-4}(T-273.15) + 5.774 \cdot 10^{-7}(T-273.15)^2$	$482.73 \exp[-6.09 \cdot 10^{-3}(T-273.15)]$
PECH	$0.7216 \exp[5.825 \cdot 10^{-4}(T-273.15)]$	$238.3 \exp[-4.171 \cdot 10^{-3}(T-273.15)]$
PEEK	$0.7158 \exp[6.690 \cdot 10^{-4}(T-273.15)]$	$388.0 \exp[-4.124 \cdot 10^{-3}(T-273.15)]$
PEA	$0.8756 \exp[7.241 \cdot 10^{-4}(T-273.15)]$	$193.2 \exp[-4.839 \cdot 10^{-3}(T-273.15)]$
PEMA	$0.8614 \exp[7.468 \cdot 10^{-4}(T-273.15)]$	$260.9 \exp[-5.356 \cdot 10^{-3}(T-273.15)]$
HDPE	$1.1595 + 8.0394 \cdot 10^{-4}(T-273.15)$	$179.9 \exp[-4.739 \cdot 10^{-3}(T-273.15)]$
LPE	$0.9172 \exp[7.806 \cdot 10^{-4}(T-273.15)]$	$176.7 \exp[-4.661 \cdot 10^{-3}(T-273.15)]$
HMLPE	$0.8992 \exp[8.502 \cdot 10^{-4}(T-273.15)]$	$168.3 \exp[-4.292 \cdot 10^{-3}(T-273.15)]$
BPE	$0.9399 \exp[7.341 \cdot 10^{-4}(T-273.15)]$	$177.1 \exp[-4.699 \cdot 10^{-3}(T-273.15)]$
LDPE	$1.1944 + 2.841 \cdot 10^{-4}(T-273.15) + 1.872 \cdot 10^{-6}(T-273.15)^2$	$202.2 \exp[-5.243 \cdot 10^{-3}(T-273.15)]$
LDPE-A	$1.1484 \exp[6.950 \cdot 10^{-4}(T-273.15)]$	$192.9 \exp[-4.701 \cdot 10^{-3}(T-273.15)]$
LDPE-B	$1.1524 \exp[6.700 \cdot 10^{-4}(T-273.15)]$	$196.6 \exp[-4.601 \cdot 10^{-3}(T-273.15)]$
LDPE-C	$1.1516 \exp[6.730 \cdot 10^{-4}(T-273.15)]$	$186.7 \exp[-4.391 \cdot 10^{-3}(T-273.15)]$
PEO	$0.8766 \exp[7.087 \cdot 10^{-4}(T-273.15)]$	$207.7 \exp[-3.947 \cdot 10^{-3}(T-273.15)]$
PET	$0.6883 + 5.90 \cdot 10^{-4}(T-273.15)$	$369.7 \exp[-4.150 \cdot 10^{-3}(T-273.15)]$
P4HS	$0.8251 + 6.77 \cdot 10^{-4}T$	$103.1 \exp[-2.417 \cdot 10^{-3}(T-273.15)]$
PIB	$1.0750 \exp[5.651 \cdot 10^{-4}(T-273.15)]$	$200.3 \exp[-4.329 \cdot 10^{-3}(T-273.15)]$
PI-8	$1.1030 + 6.488 \cdot 10^{-4}(T-273.15) + 5.125 \cdot 10^{-7}(T-273.15)^2$	$188.0 \exp[-4.541 \cdot 10^{-3}(T-273.15)]$
PI-14	$1.0943 + 6.293 \cdot 10^{-4}(T-273.15) + 6.231 \cdot 10^{-7}(T-273.15)^2$	$202.0 \exp[-4.653 \cdot 10^{-3}(T-273.15)]$
PI-41	$1.0951 + 6.188 \cdot 10^{-4}(T-273.15) + 6.629 \cdot 10^{-7}(T-273.15)^2$	$199.0 \exp[-4.622 \cdot 10^{-3}(T-273.15)]$
PI-56	$1.0957 + 6.655 \cdot 10^{-4}(T-273.15) + 5.661 \cdot 10^{-7}(T-273.15)^2$	$200.0 \exp[-4.644 \cdot 10^{-3}(T-273.15)]$
PMA	$0.8365 \exp[6.795 \cdot 10^{-4}(T-273.15)]$	$235.8 \exp[-4.493 \cdot 10^{-3}(T-273.15)]$
PMMA	$0.8254 + 2.8383 \cdot 10^{-4}(T-273.15) + 7.792 \cdot 10^{-7}(T-273.15)^2$	$287.5 \exp[-4.146 \cdot 10^{-3}(T-273.15)]$
P4MP	$1.4075 - 9.095 \cdot 10^{-4}(T-273.15) + 3.497 \cdot 10^{-6}(T-273.15)^2$	$37.67 + 0.2134(T-273.15)] - 7.0445 \cdot 10^{-4}(T-273.15)^2$
PαMS	$0.89365 + 3.4864 \cdot 10^{-4}(T-273.15) + 5.0184 \cdot 10^{-7}(T-273.15)^2$	$297.7 \exp[-4.074 \cdot 10^{-3}(T-273.15)]$
PoMS	$0.9396 \exp[5.306 \cdot 10^{-4}(T-273.15)]$	$261.9 \exp[-4.114 \cdot 10^{-3}(T-273.15)]$
POM	$0.7484 \exp[6.770 \cdot 10^{-4}(T-273.15)]$	$305.6 \exp[-4.326 \cdot 10^{-3}(T-273.15)]$
PH	$0.76644 \exp(1.921 \cdot 10^{-5}T^{1.5})$	$359.9 \exp[-4.378 \cdot 10^{-3}(T-273.15)]$

Table 2
Tait Equation Parameter Functions for Polymer Melts

Polymer	$V(0,T)$/cm^{3}g^{-1}	$B(T)$/MPa
PSF	$0.7644 + 3.419\cdot10^{-4}(T-273.15) + 3.126\cdot10^{-7}(T-273.15)^2$	$365.9 \exp[-3.757\cdot10^{-3}(T-273.15)]$
PAr	$0.73381 \exp(1.626\cdot10^{-5}T^{1.5})$	$296.9 \exp[-3.375\cdot10^{-3}(T-273.15)]$
aPP	$1.1841 - 1.091\cdot10^{-4}(T-273.15) + 5.286\cdot10^{-6}(T-273.15)^2$	$162.1 \exp[-6.604\cdot10^{-3}(T-273.15)]$
iPP	$1.1606 \exp[6.700\cdot10^{-4}(T-273.15)]$	$149.1 \exp[-4.177\cdot10^{-3}(T-273.15)]$
PS	$0.9287 \exp[5.131\cdot10^{-4}(T-273.15)]$	$216.9 \exp[-3.319\cdot10^{-3}(T-273.15)]$
PTFE	$0.3200 + 9.5862\cdot10^{-4}(T-273.15)$	$425.2 \exp[-9.380\cdot10^{-3}(T-273.15)]$
PTHF	$1.0043 \exp[6.691\cdot10^{-4}(T-273.15)]$	$178.6 \exp[-4.223\cdot10^{-3}(T-273.15)]$
PVAc	$0.82496 + 5.820\cdot10^{-4}(T-273.15) + 2.940\cdot10^{-7}(T-273.15)^2$	$204.9 \exp[-4.346\cdot10^{-3}(T-273.15)]$
PVC	$0.7196 + 5.581\cdot10^{-5}(T-273.15) + 1.468\cdot10^{-6}(T-273.15)^2$	$294.2 \exp[-5.321\cdot10^{-3}(\mathrm{T}-273.15)]$
PVME	$0.9585 \exp[6.653\cdot10^{-4}(T-273.15)]$	$215.8 \exp[-4.588\cdot10^{-3}(T-273.15)]$
PVdF	$0.5790 \exp[8.051\cdot10^{-4}(T-273.15)]$	$244.0 \exp[-5.210\cdot10^{-3}(T-273.15)]$
SAN3	$0.9233 + 3.936\cdot10^{-4}(T-273.15) + 5.685\cdot10^{-7}(T-273.15)^2$	$239.8 \exp[-4.376\cdot10^{-3}(T-273.15)]$
SAN6	$0.9211 + 4.370\cdot10^{-4}(T-273.15) + 5.846\cdot10^{-7}(T-273.15)^2$	$226.9 \exp[-4.286\cdot10^{-3}(T-273.15)]$
SAN15	$0.9044 + 4.207\cdot10^{-4}(T-273.15) + 4.077\cdot10^{-7}(T-273.15)^2$	$238.4 \exp[-3.943\cdot10^{-3}(T-273.15)]$
SAN18	$0.9016 + 4.036\cdot10^{-4}(T-273.15) + 4.206\cdot10^{-7}(T-273.15)^2$	$240.4 \exp[-3.858\cdot10^{-3}(T-273.15)]$
SAN40	$0.8871 + 3.406\cdot10^{-4}(T-273.15) + 4.938\cdot10^{-7}(\mathrm{T}-273.15)^2$	$289.3 \exp[-4.431\cdot10^{-3}(T-273.15)]$
SAN70	$0.8528 + 3.616\cdot10^{-4}(T-273.15) + 2.634\cdot10^{-7}(\mathrm{T}-273.15)^2$	$335.4 \exp[-3.923\cdot10^{-3}(T-273.15)]$
SBR10	$0.9053 \exp(2.437\cdot10^{-5}T^{1.5})$	$530.3 \exp[-3.99\cdot10^{-3}(T-273.15)]$
SBR23	$0.8986 \exp(2.317\cdot10^{-5}T^{1.5})$	$551.6 \exp[-4.17\cdot10^{-3}(T-273.15)]$
SBR60	$0.8812 \exp(2.031\cdot10^{-5}T^{1.5})$	$486.0 \exp[-4.34\cdot10^{-3}(T-273.15)]$
SBR85	$0.8704 \exp(1.846\cdot10^{-5}T^{1.5})$	$356.7 \exp[-4.24\cdot10^{-3}(T-273.15)]$
SMMA20	$0.9063 + 3.570\cdot10^{-4}(T-273.15) + 6.532\cdot10^{-7}(T-273.15)^2$	$232.0 \exp[-4.143\cdot10^{-3}(T-273.15)]$
SMMA60	$0.8610 + 3.350\cdot10^{-4}(T-273.15) + 6.980\cdot10^{-7}(T-273.15)^2$	$261.0 \exp[-4.611\cdot10^{-3}(T-273.15)]$
VCES50	$0.6676 + 6.63\cdot10^{-4}T$	$5281.7 \exp[-9.264\cdot10^{-3}(T-273.15)]$
VCHS80	$0.7753 + 6.17\cdot10^{-4}T$	$247.6 \exp[-2.604\cdot10^{-3}(T-273.15)]$
VCHS67	$0.8028 + 6.50\cdot10^{-4}T$	$581.7 \exp[-4.553\cdot10^{-3}(T-273.15)]$
VCHS60	$0.8213 + 6.23\cdot10^{-4}T$	$229.1 \exp[-2.133\cdot10^{-3}(T-273.15)]$
VCHS50	$0.7827 + 5.05\cdot10^{-4}T$	$136.0 \exp[-1.083\cdot10^{-3}(T-273.15)]$
VCHS40	$0.7805 + 4.92\cdot10^{-4}T$	$155.0 \exp[-1.605\cdot10^{-3}(T-273.15)]$
VCHS33	$0.7710 + 4.86\cdot10^{-4}T$	$460.4 \exp[-3.453\cdot10^{-3}(T-273.15)]$
VCHS20	$0.6416 + 5.42\cdot10^{-4}T$	$489.8 \exp[-3.193\cdot10^{-3}(T-273.15)]$
VCOS50	$0.7081 + 7.40\cdot10^{-4}T$	$666.5 \exp[-4.503\cdot10^{-3}(T-273.15)]$
VCPS50	$0.7814 + 4.36\cdot10^{-4}T$	$880.1 \exp[-4.393\cdot10^{-3}(T-273.15)]$

Section 14
Geophysics, Astronomy, and Acoustics

Astronomical Constants 14-1
Properties of the Solar System 14-2
Satellites of the Planets 14-4
Interstellar Molecules 14-6
Mass, Dimensions, and other Parameters of the Earth 14-9
Geological Time Scale 14-11
Acceleration Due to Gravity 14-12
Density, Pressure, and Gravity as a Function of Depth within the Earth 14-13
Ocean Pressure as a Function of Depth and Latitude 14-14
Properties of Seawater 14-15
Abundance of Elements in the Earth's Crust and in the Sea 14-17
Solar Spectral Irradiance 14-18
U.S. Standard Atmosphere (1976) 14-19
Geographical and Seasonal Variation in Solar Radiation 14-26
Infrared Absorption by the Earth's Atmosphere 14-27
Atmospheric Concentration of Carbon Dioxide, 1958-1996 14-28
Mean Temperatures in the United States, 1900-1992 14-30
Global Temperature Trend, 1866-1996 14-32
Atmospheric Electricity 14-33
Speed of Sound in Various Media 14-41
Attenuation and Speed of Sound in Air as a Function of Humidity and Frequency 14-43
Speed of Sound in Dry Air 14-44
Musical Scales 14-45
Characteristics of Human Hearing 14-47

ASTRONOMICAL CONSTANTS

Victor Abalakin

The constants in this table are based primarilarly on the set of constants adopted by the International Astronomical Union (IAU) in 1976. Updates have been made when new data were available. All values are given in SI Units; thus masses are expressed in kilograms and distances in meters.

The astronomical unit of time is a time interval of one day (1 d) equal to 86400 s. An interval of 36525 d is one Julian century (1 cy).

REFERENCES

1. Seidelmann, P. K., *Explanatory Supplement to the Astronomical Almanac*, University Science Books, Mill Valley, CA, 1990.
2. Lang, K. R., *Astrophysical Data: Planets and Stars*, Springer-Verlag, New York, 1992.

Defining constants	
Gaussian gravitational constant	$k = 0.01720209895\ m^3\ kg^{-1}\ s^{-2}$
Speed of light	$c = 299792458\ m\ s^{-1}$
Primary constants	
Light-time for unit distance (1 AU)	$\tau_A = 499.004782$ s
Equatorial radius of earth	$a_e = 6378140$ m
Equatorial radius of earth (IUGG value)	$a_e = 6378136$ m
Dynamical form-factor for earth	$J_2 = 0.001082626$
Geocentric gravitational constant	$GE = 3.986005 \times 10^{14}\ m^3 s^{-2}$
Constant of gravitation	$G = 6.672 \times 10^{-11}\ m^3 kg^{-1} s^{-2}$
Ratio of mass of moon to that of earth	$\mu = 0.01230002$ $1/\mu = 81.300587$
General precession in longitude, per Julian century, at standard epoch J2000	$\rho = 5029''.0966$
Obliquity of the ecliptic at standard epoch J2000	$\varepsilon = 23°26'21''.448$
Derived constants	
Constant of nutation at standard epoch J2000	$N = 9''.2025$
Unit distance (AU = $c\tau_A$)	$AU = 1.49597870 \times 10^{11}$ m
Solar parallax ($\pi_0 = \arcsin(a_e/AU)$)	$\pi_0 = 8''.794148$
Constant of aberration for standard epoch J2000	$\kappa = 20''.49552$
Flattening factor for the earth	$f = 1/298.257 = 0.00335281$
Heliocentric gravitational constant ($GS = A^3k^2/D^2$)	$GS = 1.32712438 \times 10^{20}\ m^3\ s^{-2}$
Ratio of mass of sun to that of the earth (S/E) = (GS)/(GE))	$S/E = 332946.0$
Ratio of mass of sun to that of earth + moon	$(S/E)/(1 + \mu) = 328900.5$
Mass of the sun ($S = (GS)/G$)	$S = 1.9891 \times 10^{30}$ kg
Ratios of mass of sun to masses of the planets	
Mercury	6023600
Venus	408523.5
Earth + moon	328900.5
Mars	3098710
Jupiter	1047.355
Saturn	3498.5
Uranus	22869
Neptune	19314
Pluto	3000000

PROPERTIES OF THE SOLAR SYSTEM

The following tables give various properties of the planets and characteristics of their orbits in the solar system. Certain properties of the sun and of the earth's moon are also included.

Explanations of the column headings:

- *Den.*: mean density in g/cm^3
- *Radius*: radius at the equator in km
- *Flattening*: degree of oblateness, defined as $(r_e - r_p)/r_e$, where r_e and r_p are the equatorial and polar radii, respectively
- *Potential coefficients*: coefficients in the spherical harmonic representation of the gravitational potential U by the equation

$$U(r,\phi) = (GM/r)\,[1 - \Sigma\, J_n(a/r)^n\, P_n(\sin\phi)]$$

where G is the gravitational constant, r the distance from the center of the planet, a the radius of the planet, M the mass, ϕ the latitude, and P_n the Legendre polynomial of degree n.

- *Gravity*: acceleration due to gravity at the surface
- *Escape velocity*: velocity needed at the surface of the planet to escape the gravitational pull
- *Dist. to sun*: semi-major axis of the elliptical orbit (1 AU = 1.496×10^8 km)
- ε: eccentricity of the orbit
- *Ecliptic angle*: angle between the planetary orbit and the plane of the earth's orbit around the sun
- *Inclin.*: angle between the equatorial plane and the plane of the planetary orbit
- *Rot. period*: period of rotation of the planet measured in earth days
- *Albedo*: ratio of the light reflected from the planet to the light incident on it
- T_{sur}: mean temperature at the surface
- P_{sur}: pressure of the atmosphere at the surface

The following general information on the solar system is of interest:

Mass of the earth = $M_e = 5.9742 \times 10^{24}$ kg
Total mass of planetary system = 2.669×10^{27} kg = 447 M_e
Total angular momentum of planetary system = 3.148×10^{43} kg m^2/s
Total kinetic energy of the planets = 1.99×10^{35} J
Total rotational energy of planets = 0.7×10^{35} J

Properties of the sun:

Mass = 1.9891×10^{30} kg = 332946.0 M_e
Radius = 6.9599×10^8 m
Surface area = 6.087×10^{18} m^2
Volume = 1.412×10^{27} m^3
Mean density = 1.409 g/cm^3
Gravity at surface = 27398 cm/s^2
Escape velocity at surface = 6.177×10^5 m/s
Effective temperature = 5780 K
Total radiant power emitted (luminosity) = 3.86×10^{26} W
Surface flux of radiant energy = 6.340×10^7 W/m^2
Flux of radiant energy at the earth (Solar Constant) = 1373 W/m^2

REFERENCES

1. Seidelmann, P. K., Editor, *Explanatory Supplement to the Astronomical Almanac*, University Science Books, Mill Valley, CA, 1992.
2. Lang, K. R., *Astrophysical Data: Planets and Stars*, Springer-Verlag, New York, 1992.
3. Allen, C. W., *Astrophysical Quantities, Third Edition*, Athlone Press, London, 1977.

PROPERTIES OF THE SOLAR SYSTEM (continued)

Planet	Mass 10^{24} kg	Den. g/cm^3	Radius km	Flattening	Potential coeffients $10^3 J_2$	$10^6 J_3$	$10^6 J_4$	Gravity cm/s^2	Escape vel. km/s
Mercury	0.33022	5.43	2439.7	0				370	4.25
Venus	4.8690	5.24	6051.9	0	0.027			887	10.4
Earth	5.9742	5.515	6378.140	0.00335364	1.08263	-2.54	-1.61	980	11.2
(Moon)	0.073483	3.34	1738	0	0.2027			162	2.37
Mars	0.64191	3.94	3397	0.00647630	1.964	36		371	5.02
Jupiter	1898.8	1.33	71492	0.0648744	14.75	-580		2312	59.6
Saturn	568.50	0.70	60268	0.0979624	16.45	-1000		896	35.5
Uranus	86.625	1.30	25559	0.0229273	12			777	21.3
Neptune	102.78	1.76	24764	0.0171	4			1100	23.3
Pluto	0.015	1.1	1151	0				72	1.1

Planet	Dist. to sun AU	ε	Ecliptic angle	Inclin.	Rot. period d	Albedo	No. of satellites
Mercury	0.38710	0.2056	7.00°	0°	58.6462	0.106	0
Venus	0.72333	0.0068	3.39°	177.3°	-243.01	0.65	0
Earth	1.00000	0.0167		23.45°	0.99726968	0.367	1
(Moon)				6.68°	27.321661	0.12	
Mars	1.52369	0.0933	1.85°	25.19°	1.02595675	0.150	2
Jupiter	5.20283	0.048	1.31°	3.12°	0.41354	0.52	16
Saturn	9.53876	0.056	2.49°	26.73°	0.4375	0.47	18
Uranus	19.19139	0.046	0.77°	97.86°	-0.65	0.51	15
Neptune	30.06107	0.010	1.77°	29.56°	0.768	0.41	8
Pluto	39.52940	0.248	17.15°	118°	-6.3867	0.3	1

Planet	T_{sur} K	P_{sur} bar	Atmospheric composition CO_2	N_2	O_2	H_2O	H_2	He	Ar	Ne	CO
Mercury	440	2×10^{-15}					2%	98%			
Venus	730	90	96.4%	3.4%	69 ppm	0.1%			4 ppm		20 ppm
Earth	288	1	0.03%	78.08%	20.95%	0 to 3%			0.93%	18 ppm	1 ppm
Mars	218	0.007	95.32%	2.7%	0.13%	0.03%			1.6%	3 ppm	0.07%
Jupiter	129						86.1%	13.8%			
Saturn	97						92.4%	7.4%			
Uranus	58						89%	11%			
Neptune	56						89%	11%			
Pluto	50	1×10^{-5}									

SATELLITES OF THE PLANETS

This table gives characteristics of the known satellites of the planets. The parameters covered are:

- Orbital period in units of earth days. An R following the value indicates a retrograde motion.
- Distance from the planet, as measured by the semi-major axis of the orbit.
- Eccentricity of the orbit.
- Inclination of the satellite orbit with respect to the equator of the planet.
- Mass of the satellite relative to the planet.
- Radius of the satellite in km.
- Mean density of the satellite.
- Geometric albedo, which is a measure of the fraction of incident sunlight reflected by the satellite.

REFERENCES

1. Seidelmann, P. K., Editor, *Explanatory Supplement to the Astronomical Almanac,* University Science Books, Mill Valley, CA, 1992.
2. Lang, K. R., *Astrophysical Data: Planets and Stars*, Springer-Verlag, New York, 1992.
3. Burns, J. A., and Matthews, M. S., Eds., *Satellites*, University of Arizona Press, Tucson, 1986.

Planet	Satellite		Orb. Period d	Distance 10^3 km	Eccentricity	Inclination	Rel. mass	Radius km	Den. g/cm^3	Albedo
Earth		Moon	27.321661	384.400	0.054900489	18.28-28.58°	0.01230002	1738	3.34	0.12
Mars	I	Phobos	0.31891023	9.378	0.015	1.0°	1.5×10^{-8}	13.5 × 10.8 × 9.4	<2	0.06
	II	Deimos	1.2624407	23.459	0.0005	0.9-2.7°	3×10^{-9}	7.5 × 6.1 × 5.5	<2	0.07
Jupiter	I	Io	1.769137786	422	0.004	0.04°	4.68×10^{-5}	1815	3.55	0.61
	II	Europa	3.551181041	671	0.009	0.47°	2.52×10^{-5}	1569	3.04	0.64
	III	Ganymede	7.15455296	1070	0.002	0.21°	7.80×10^{-5}	2631	1.93	0.42
	IV	Callisto	16.6890184	1883	0.007	0.51°	5.66×10^{-5}	2400	1.83	0.20
	V	Amalthea	0.49817905	181	0.003	0.40°	3.8×10^{-9}	135 × 83 × 75		0.05
	VI	Himalia	250.5662	11480	0.15798	27.63°	5.0×10^{-9}	93		0.03
	VII	Elara	259.6528	11737	0.20719	24.77°	4×10^{-10}	38		0.03
	VIII	Pasiphae	735 R	23500	0.378	145°	1×10^{-10}	25		
	IX	Sinope	758 R	23700	0.275	153°	0.4×10^{-10}	18		
	X	Lysithea	259.22	11720	0.107	29.02°	0.4×10^{-10}	18		
	XI	Carme	692 R	22600	0.20678	164°	0.5×10^{-10}	20		
	XII	Ananke	631 R	21200	0.16870	147°	0.2×10^{-10}	15		
	XIII	Leda	238.72	11094	0.14762	26.07°	0.03×10^{-10}	8		
	XIV	Thebe	0.6745	222	0.015	0.8°	4×10^{-10}	55 × 45		0.05
	XV	Adrastea	0.29826	129			0.1×10^{-10}	12.5 × 10 × 7.5		0.05
	XVI	Metis	0.294780	128			0.5×10^{-10}	20		0.05
Saturn	I	Mimas	0.942421813	185.52	0.0202	1.53°	8.0×10^{-8}	196	1.44	0.5
	II	Enceladus	1.370217855	238.02	0.00452	1.86°	1.3×10^{-7}	250	1.13	1.0
	III	Tethys	1.887802160	294.66	0.00000	1.86°	1.3×10^{-6}	530	1.20	0.9
	IV	Dione	2.736914742	377.40	0.002230	0.02°	1.85×10^{-6}	560	1.41	0.7
	V	Rhea	4.517500436	527.04	0.00100	0.35°	4.4×10^{-6}	765	1.33	0.7
	VI	Titan	15.94542068	1221.83	0.029192	0.33°	2.38×10^{-4}	2575	1.88	0.21
	VII	Hyperion	21.2766088	1481.1	0.104	0.43°	3×10^{-8}	205 × 130 × 110		0.3
	VIII	Iapetus	79.3301825	3561.3	0.02828	14.72°	3.3×10^{-6}	730	1.15	0.2
	IX	Phoebe	550.48 R	12952	0.16326	177°	7×10^{-10}	110		0.06

SATELLITES OF THE PLANETS (continued)

Planet	Satellite		Orb. Period d	Distance 10^3 km	Eccentricity	Inclination	Rel. mass	Radius km	Den. g/cm^3	Albedo
	X	Janus	0.6945	151.472	0.007	0.14°		110 × 100 × 80		0.8
	XI	Epimetheus	0.6942	151.422	0.009	0.34°		70 × 60 × 50		0.8
	XII	Helene	2.7369	377.40	0.005	0.0°		18 × 16 × 15		0.7
	XIII	Telesto	1.8878	294.66				17 × 14 × 13		0.5
	XIV	Calypso	1.8878	294.66				17 × 11 × 11		0.6
	XV	Atlas	0.6019	137.670	0.000	0.3°		20 × 10		0.9
	XVI	Prometheus	0.6130	139.353	0.003	0.0°		70 × 50 × 40		0.6
	XVII	Pandora	0.6285	141.700	0.004	0.0°		55 × 45 × 35		0.9
	XVIII	Pan	0.5750	133.583				10		0.5
Uranus	I	Ariel	2.52037935	191.02	0.0034	0.3°	1.56×10^{-5}	579	1.55	0.34
	II	Umbriel	4.1441772	266.30	0.0050	0.36°	1.35×10^{-5}	586	1.58	0.18
	III	Titania	8.7058717	435.91	0.0022	0.14°	4.06×10^{-5}	790	1.69	0.27
	IV	Oberon	13.4632389	583.52	0.0008	0.10°	3.47×10^{-5}	762	1.64	0.24
	V	Miranda	1.41347925	129.39	0.0027	4.2°	0.08×10^{-5}	240	1.25	0.27
	VI	Cordelia	0.335033	49.77	<0.001	0.1°		13		0.07
	VII	Ophelia	0.376409	53.79	0.010	0.1°		15		0.07
	VIII	Bianca	0.434577	59.17	<0.001	0.2°		21		0.07
	IX	Cressida	0.463570	61.78	<0.001	0.0°		31		0.07
	X	Desdemona	0.473651	62.68	<0.001	0.2°		27		0.07
	XI	Juliet	0.493066	64.35	<0.001	0.1°		42		0.07
	XII	Portia	0.513196	66.09	<0.001	0.1°		54		0.07
	XIII	Rosalind	0.558459	69.94	<0.001	0.3°		27		0.07
	XIV	Belinda	0.623525	75.26	<0.001	0.0°		33		0.07
	XV	Puck	0.761832	86.01	<0.001	0.31°		77		0.07
Neptune	I	Triton	5.8768541 R	354.76	0.000016	157.345°	2.09×10^{-4}	1353	2.05	0.7
	II	Nereid	360.13619	5513.4	0.7512	27.6°	2×10^{-7}	170		0.4
	III	Naiad	0.294396	117.6	<0.001	4.74°		29		0.06
	IV	Thalassa	0.311485	73.6	<0.001	0.21°		40		0.06
	V	Despina	0.334655	52.6	<0.001	0.07°		74		0.06
	VI	Galatea	0.428745	62.0	<0.001	0.05°		79		0.06
	VII	Larissa	0.554654	50.0	<0.0014	0.20°		104 × 89		0.06
	VIII	Proteus	1.122315	48.2	<0.001	0.55°		218 × 208 × 201		0.06
Pluto	I	Charon	6.38725	19.6	<0.001	99°	0.22	593		0.5

INTERSTELLAR MOLECULES

Frank J. Lovas and Lewis E. Snyder

A number of molecules have been detected in the interstellar medium, in circumstellar envelopes around evolved stars, and comae and tails of comets through observation of their microwave, infrared, or optical spectra. The following list gives the molecules and the particular isotopic species that have been reported so far. Molecules are listed by molecular formula in the Hill order.

REFERENCES

1. Lovas, F. J., "Recommended Rest Frequencies for Observed Interstellar Molecule Microwave Transitions - 1991 Revision", *J. Phys. Chem. Ref. Data* 21, 181-272, 1992.

2. Snyder, L. E., "Cometary Molecules", Internat. Astron. Union Symposium No. 150, *Astrochemistry of Cosmic Phenomena*, ed. P.D. Singh, Kluwer Academic Publishers, Dordrecht, The Netherlands, pp. 427-434 (1992).

Molecular formula	Name	Isotopic species
AlCl	Aluminum monochloride	AlCl[a]
		$Al^{37}Cl$[a]
AlF	Aluminum monofluoride	AlF[a]
CH	Methylidyne	CH
CH^+	Methyliumylidene	CH^+
CHN	Hydrogen cyanide	HCN
		$H^{13}CN$
		$HC^{15}N$
		DCN
CHN	Hydrogen isocyanide	HNC
		$H^{15}NC$
		$HN^{13}C$
		DNC
		$D^{15}NC$
CHNO	Isocyanic acid	HNCO
		DNCO
CHNS	Isothiocyanic acid	HNCS
CHO	Oxomethyl	HCO
CHO^+	Oxomethylium	HCO^+
		$H^{13}CO^+$
		$HC^{17}O^+$
		$HC^{18}O^+$
		DCO^+
		$D^{13}CO^+$
CHO^+	Hydroxymethylidyne	HOC^+
$CHOS^+$	Protonated carbonyl sulfide	$HOCS^+$
CHO_2^+	Protonated carbon dioxide	$HOCO^+$
CHS^+	Thiooxomethylium	HCS^+
CH_2	Methylene	CH_2
CH_2N^+	Iminomethylium	$HCNH^+$
CH_2N	Methylene amidogen	CH_2N
CH_2N_2	Cyanamide	NH_2CN
CH_2O	Formaldehyde	H_2CO
		$H_2{}^{13}CO$
		$H_2C^{18}O$
		HDCO
		D_2CO
CH_2O_2	Formic acid	HCOOH
		$H^{13}COOH$
		HCOOD
		DCOOH
CH_2S	Thioformaldehyde	H_2CS
		$H_2{}^{13}CS$
		$H_2C^{34}S$
		HDCS
CH_3	Methyl	CH_3 [a]
CH_3N	Methanimine	CH_2NH
		$^{13}CH_2NH$
CH_3NO	Formamide	NH_2CHO
		$NH_2{}^{13}CHO$
CH_3O^+	Hydroxy methylium ion	H_2COH^+
CH_4	Methane	CH_4
CH_4O	Methanol	CH_3OH
		$^{13}CH_3OH$
		$CH_3{}^{18}OH$
		CH_2DOH
		CH_3OD
CH_4S	Methanethiol	CH_3SH
CH_5N	Methylamine	CH_3NH_2
CMgN	Magnesium cyanide	MgCN[a]
CMgN	Magnesium isocyanide	MgNC[a]
CN	Cyanide radical	CN
		^{13}CN
		$C^{15}N$
CN^+	Cyanide radical ion	CN^+ [b]
CNNa	Sodium cyanide	NaCN[a]
CO	Carbon monoxide	CO
		^{13}CO
		$C^{17}O$
		$C^{18}O$
		$^{13}C^{18}O$
CO^+	Carbon monoxide ion	CO^+
COS	Carbon oxysulfide	OCS
		$OC^{34}S$
		$O^{13}CS$
		^{18}OCS
CO_2	Carbon dioxide	CO_2
CO_2^+	Carbon dioxide ion	CO_2^+ [b]
CP	Carbon phosphide	CP[a]
CS	Carbon monosulfide	CS
		$C^{33}S$
		$C^{34}S$
		$C^{36}S$
		^{13}CS
		$^{13}C^{34}S$
CSi	Silicon carbide	SiC[a]

INTERSTELLAR MOLECULES (continued)

Molecular formula	Name	Isotopic species
C_2	Dicarbon	C_2
C_2H	Ethynyl	C_2H
		^{13}CCH
		$C^{13}CH$
		C_2D
C_2HN	Cyanomethylene	HCCN
C_2H_2	Acetylene	HCCH
C_2H_2N	Cyanomethyl	CH_2CN
C_2H_2O	Ketene	H_2CCO
C_2H_3N	Acetonitrile	CH_3CN
		$^{13}CH_3CN$
		$CH_3{}^{13}CN$
		$CH_3C^{15}N$
		CH_2DCN
C_2H_3N	Isocyanomethane	CH_3NC
C_2H_4O	Acetaldehyde	CH_3CHO
C_2H_4O	Ethylene oxide	*c*-C_2H_4O [c]
$C_2H_4O_2$	Methyl formate	CH_3OCHO
$C_2H_4O_2$	Acetic acid	CH_3COOH
$C_2H_4O_2$	Glycolaldehyde	CH_2OHCHO
C_2H_6	Ethane	CH_3CH_3
C_2H_6O	*trans*-Ethanol	*t*-CH_3CH_2OH
C_2H_6O	*gauche*-Ethanol	*g*-CH_3CH_2OH
C_2H_6O	Dimethyl ether	CH_3OCH_3
C_2O	Oxoethenylidene	CCO
C_2S	Thioxoethenylidene	CCS
		$CC^{34}S$
C_2Si	Silicon dicarbide	SiC_2
		$^{29}SiC_2$
		$^{30}SiC_2$
		$Si^{13}CC$
C_3	Tricarbon	C_3
C_3H	Cyclopropenylidyne	*c*-C_3H [c]
C_3H	Propenylidyne	*l*-C_3H [d]
C_3HN	Cyanoacetylene	HCCCN
		$H^{13}CCCN$
		$HC^{13}CCN$
		$HCC^{13}CN$
		$HCCC^{15}N$
		DCCCN
C_3HN	Isocyanoacetylene	HCCNC
C_3HN	1,2-Propadienylidene, 3-imino	HNCCC
C_3H_2	Cyclopropenylidene	*c*-C_3H_2 [c]
		$H^{13}CCCH$
		$HC^{13}CCH$
		C_3HD
C_3H_2	Propadienylidene	H_2CCC
$C_3H_2N^+$	Protonated cyanoacetylene	$HCCCNH^+$
C_3H_2O	2-Propynal	HCCCHO
C_3H_3N	Acrylonitrile	CH_2CHCN
C_3H_4	Propyne	CH_3CCH
		$CH_3C^{13}CH$
		$^{13}CH_3CCH$
		CH_2DCCH
C_3H_5N	Propanenitrile	CH_3CH_2CN
C_3H_6O	Acetone	$(CH_3)_2CO$
C_3N	Cyanoethynyl	CCCN
C_3O	1,2-Propadienylidene, 3-oxo	CCCO
C_3S	1,2-Propadienylidene, 3-thioxo	CCCS
C_3Si	Silicon tricarbon	SiC_3
C_4H	1,3-Butadiynyl radical	HCCCC
C_4H_2	Butatrienylidene	H_2CCCC
C_4H_2	1,3-Butadiyne	HCCCCH [a]
C_4H_3N	2-Butynenitrile	CH_3CCCN
C_4Si	Silicon tetracarbide	SiC_4 [a]
C_5	Pentacarbon	C_5 [a]
C_5H	2,4-Pentadiynylidyne	HCCCCC
C_5HN	2,4-Pentadiynenitrile	HCCCCCN
		$H^{13}CCCCCN$
		$HC^{13}CCCCN$
		$HCC^{13}CCCN$
		$HCCC^{13}CCN$
		$HCCCC^{13}CN$
		DCCCCCN
C_5H_4	1,3-Pentadiyne	CH_3C_4H
C_5N	1,3-Butadiynylium, 4-cyano	C_5N
C_6H	1,3,5-Hexatriynyl	HCCCCCC
C_6H_2	1,3,5-Hexatriyne	HCCCCCCH
C_6H_2	Hexatrienyl	$H_2CCCCCC$
C_7H	2,4,6-Heptatriynylidyne	HCCCCCCC
C_7HN	2,4,6-Heptatriynenitrile	HC_7N
C_8H	1,3,5,7-Octatetraynyl	HC_8
C_9HN	2,4,6,8-Nonatetraynenitrile	HC_9N
$C_{11}HN$	Cyanodecapentyne	$HC_{11}N$
ClH	Hydrogen chloride	$H^{35}Cl$
		$H^{37}Cl$
ClK	Potassium chloride	$K^{35}Cl$ [a]
		$K^{37}Cl$
ClNa	Sodium chloride	$Na^{35}Cl$ [a]
		$Na^{37}Cl$ [a]
FH	Hydrogen fluoride	HF
HLi	Lithium Hydride	^{7}LiH
HN	Imidogen	HN
HNO	Nitrosyl hydride	HNO
HN_2^+	Hydrodinotrogen(1+)	N_2H^+
		$^{15}NNH^+$
		$N^{15}NH^+$
		N_2D^+
HO	Hydroxyl	OH
		^{17}OH
		^{18}OH
HO^+	Hydroxyl ion	HO^+ [b]
HS	Mercapto	SH
H_2	Hydrogen	H_2
H_2N	Amidogen	NH_2
H_2O	Water	H_2O
		$H_2{}^{18}O$
		HDO
H_2O^+	Oxoniumyl	H_2O^+
H_2S	Hydrogen sulfide	H_2S
		$H_2{}^{34}S$
		HDS
H_3^+	Trihydrogen ion	H_3^+
		H_2D^+
H_3N	Ammonia	NH_3
		$^{15}NH_3$

INTERSTELLAR MOLECULES (continued)

Molecular formula	Name	Isotopic species
		NH_2D
		NHD_2
H_3O^+	Oxonium hydride	H_3O^+
H_4Si	Silane	SiH_4[a]
NO	Nitric oxide	NO
NP	Phosphorous nitride	NP
NS	Nitrogen sulfide	NS
		$N^{34}S$
NSi	Silicon nitride	SiN[a]
N_2^+	Nitrogen ion	N_2^+[b]
N_2O	Nitrous oxide	N_2O
OS	Sulfur monoxide	SO
		^{34}SO
		^{33}SO
		$S^{18}O$
OS^+	Sulfur monoxide ion	SO^+
OSi	Silicon monoxide	SiO
		^{29}SiO
		^{30}SIO
O_2S	Sulfur dioxide	SO_2
		$^{33}SO_2$
		$^{34}SO_2$
		$OS^{18}O$
SSi	Silicon monosulfide	SiS
		$Si^{33}S$
		$Si^{34}S$
		^{29}SiS
		^{30}SiS
S_2	Disulfur	S_2[b]

l- before the isotopic species indicates a linear configuration, while *c*- indicates a cyclic molecule.

[a] Reported only in circumstellar clouds.

[b] Reported only in comets.

MASS, DIMENSIONS, AND OTHER PARAMETERS OF THE EARTH

This table is a collection of data on various properties of the Earth. Most of the values are given in SI units. Note that 1 AU (astronomical unit) = 149,597,870 km.

REFERENCES

1. Seidelmann, P. K., Editor, *Explanatory Supplement to the Astronomical Almanac*, University Science Books, Mill Valley, CA, 1992.
2. Lang, K. R., *Astrophysical Data: Planets and Stars*, Springer-Verlag, New York, 1992.

Quantity	Symbol	Value	Unit
Mass	M	$5.9742 \cdot 10^{27}$	g
Major orbital semi-axis	a_{orb}	1.000000	AU
		$1.4959787 \cdot 10^{8}$	km
Distance from sun at perihelion	r_{π}	0.9833	AU
Distance from sun at aphelion	r_{α}	1.0167	AU
Moment of perihelion passage	T_{π}	Jan. 2, 4 h 52 min	
Moment of aphelion passage	T_{α}	July 4, 5 h 05 min	
Siderial rotation period around sun	P_{orb}	$31.5581 \cdot 10^{6}$	s
		365.25636	d
Mean rotational velocity	U_{orb}	29.78	km/s
Mean equatorial radius	$\overline{a}$	6378.140	km
Mean polar compression (flattening factor)	α	1/298.257	
Difference in equatorial and polar semi-axes	$a - c$	21.385	km
Compression of meridian of major equatorial axis	α_a	1/295.2	
Compression of meridian of minor equatorial axis	α_b	1/298.0	
Equatorial compression	ε	1/30 000	
Difference in equatorial semi-axes	$a - b$	213	m
Difference in polar semi-axes	$c_N - c_S$	~70	m
Polar asymmetry	η	$\sim 1.10^{-5}$	
Mean acceleration of gravity at equator	g_e	9.78036	m/s^2
Mean acceleration of gravity at poles	g_p	9.83208	m/s^2
Difference in acceleration of gravity at pole and at equator	$g_p - g_e$	5.172	cm/s^2
Mean acceleration of gravity for entire surface of terrestrial ellipsoid	g	9.7978	m/s^2
Mean radius	R	6371.0	km
Area of surface	S	$5.10 \cdot 10^{8}$	km^2
Volume	V	$1.0832 \cdot 10^{12}$	km^3
Mean density	ρ	5.515	g/cm^3
Siderial rotational period	P	86,164.09	s
Rotational angular velocity	ω	$7.292116 \cdot 10^{-5}$	rad/s
Mean equatorial rotational velocity	v	0.46512	km/s
Rotational angular momentum	L	$5.861 \cdot 10^{33}$	J s
Rotational energy	E	$2.137 \cdot 10^{29}$	J
Ratio of centrifugal force to force of gravity at equator	q_c	0.0034677 = 1/288	
Moment of inertia	I	$8.070 \cdot 10^{37}$	kg m^2
Relative braking of earth's rotation due to tidal friction	$\Delta\omega_e/\omega$	$-4.2 \cdot 10^{-8}$	century^{-1}
Relative secular acceleration of earth's rotation	$\Delta\omega_i/\omega$	$+1.4 \cdot 10^{-8}$	century^{-1}
Not secular braking of earth's rotation	$\Delta\omega/\omega$	$-2.8 \cdot 10^{-8}$	century^{-1}
Probable value of total energy of tectonic deformation of earth	E_t	$\sim 1 \cdot 10^{23}$	J/century
Secular loss of heat of earth through radiation into space	$\Delta' E_k$	$1 \cdot 10^{23}$	J/century
Portion of earth's kinetic energy transformed into heat as a result of lunar and solar tides in the hydrosphere	$\Delta'' E_k$	$1.3 \cdot 10^{23}$	J/century

MASS, DIMENSIONS, AND OTHER PARAMETERS OF THE EARTH (continued)

Quantity	Symbol	Value	Unit
Differences in duration of days in March and August	ΔP	0.0025 (March-August)	s
Corresponding relative annual variation in earth's rotational velocity	$\Delta^*\omega/\omega$	$2.9 \cdot 10^{-8}$ (Aug.-March)	
Presumed variation in earth's radius between August and March	$\Delta^* R$	–9.2 (Aug.-March)	cm
Annual variation in level of world ocean	Δh_o	~10 (Sept.-March)	cm
Area of continents	S_C	$1.49 \cdot 10^8$	km^2
		29.2	% of surface
Area of world ocean	S_o	$3.61 \cdot 10^8$	km^2
		70.8	% of surface
Mean height of continents above sea level	h_C	875	m
Mean depth of world ocean	h_o	3794	m
Mean thickness of lithosphere within the limits of the continents	$h_{c.l.}$	35	km
Mean thickness of lithosphere within the limits of the ocean	$h_{o.l.}$	4.7	km
Mean rate of thickening of continental lithosphere	$\Delta h/\Delta t$	10 - 40	m/10^6 y
Mean rate of horizontal extension of continental lithosphere	$\Delta l/\Delta t$	0.75 - 20	km/10^6 y
Mass of crust	m_l	$2.36 \cdot 10^{22}$	kg
Mass of mantle		$4.05 \cdot 10^{24}$	kg
Amount of water released from the mantle and core in the course of geological time		$3.40 \cdot 10^{21}$	kg
Total reserve of water in the mantle		$2 \cdot 10^{23}$	kg
Present content of free and bound water in the earth's lithosphere		$2.4 \cdot 10^{21}$	kg
Mass of hydrosphere	m_h	$1.664 \cdot 10^{21}$	kg
Amount of oxygen bound in the earth's crust		$1.300 \cdot 10^{21}$	kg
Amount of free oxygen		$1.5 \cdot 10^{18}$	kg
Mass of atmosphere	m_a	$5.136 \cdot 10^{18}$	kg
Mass of biosphere	m_b	$1.148 \cdot 10^{16}$	kg
Mass of living matter in the biosphere		$3.6 \cdot 10^{14}$	kg
Density of living matter on dry land		0.1	g/cm^2
Density of living matter in ocean		$15 \cdot 10^{-8}$	g/cm^3
Age of the earth		$4.55 \cdot 10^9$	y
Age of oldest rocks		$4.0 \cdot 10^9$	y
Age of most ancient fossils		$3.4 \cdot 10^9$	y

GEOLOGICAL TIME SCALE

Period or Epoch	Beginning and end, in 10^6 years	Key events
	Cenozoic era	
Quaternian		
Contemporary	0–10,000 y ± 2,000 y	
Pleistocene	10,000–1,000,000 y ± 50,000 y	Homo Erectus breakout
Tertiary		
Pliocene	1.8–5.3	Ape man fossils
Miocene	5–25	Origin of grass
Oligocene	25–37	Rise of cats, dogs, pigs
Eocene	37–55	Debut of hoofed mammals
Paleocene	55–67	Earliest primates
	Mesozoic era	
Cretaceous	67–138	Demise of dinosaurs
Jurassic	138–208	First birds
Triassic	208–245	Appearance of dinosaurs
	Paleozoic era	
Permian	245–290	Flowers, insect pollination
Carboniferous	290–360	First conifers
Devonian	360–410	First vertebrates ashore
Silurian	410–435	Spore-bearing plants
Ordovician	435–520	First animals ashore
Cambrian	520–570	Vertebrates appear
	Pre-Cambrian	
Pre-Cambrian III (Proterozoic)	570–2500	First plants, jellyfish
Pre-Cambrian II (Archean)	2500–3800	Photosynthetic bacteria
Pre-Cambrian I (Hadean)	3800–4450	Earth formed 4600 million years ago

Reference: Calder, N., *Timescale - An Atlas of the Fourth Dimension,* Viking Press, New York, 1983.

ACCELERATION DUE TO GRAVITY

The acceleration due to gravity is tabulated here as a function of latitude and height above the earth's surface. Values were calculated from the expression

$$g/(\mathrm{m/s^2}) = 9.780356\ (1 + 0.0052885 \sin^2 \phi - 0.0000059 \sin^2 2\phi) - 0.003086\ H$$

where ϕ is the latitude and H is the height in kilometers.

REFERENCE

Jursa, A. S., Ed., *Handbook of Geophysics and the Space Environment,* 4th ed., Air Force Geophysics Laboratory, 1985, p. 14-17.

ϕ	$H = 0$	$H = 1$ km	$H = 5$ km	$H = 10$ km
0	9.78036	9.77727	9.76493	9.74950
5	9.78075	9.77766	9.76532	9.74989
10	9.78191	9.77882	9.76648	9.75105
15	9.78381	9.78072	9.76838	9.75295
20	9.78638	9.78330	9.77095	9.75552
25	9.78956	9.78647	9.77413	9.75870
30	9.79324	9.79016	9.77781	9.76238
35	9.79732	9.79424	9.78189	9.76646
40	9.80167	9.79858	9.78624	9.77081
45	9.80616	9.80307	9.79073	9.77530
50	9.81065	9.80757	9.79522	9.77979
55	9.81501	9.81193	9.79958	9.78415
60	9.81911	9.81602	9.80368	9.78825
65	9.82281	9.81972	9.80738	9.79195
70	9.82601	9.82292	9.81058	9.79515
75	9.82860	9.82551	9.81317	9.79774
80	9.83051	9.82743	9.81508	9.79965
85	9.83168	9.82860	9.81625	9.80082
90	9.83208	9.82899	9.81665	9.80122

DENSITY, PRESSURE, AND GRAVITY AS A FUNCTION OF DEPTH WITHIN THE EARTH

This table gives the density ρ, pressure *p*, and acceleration due to gravity *g* as a function of depth below the earth's surface, as calculated from the model of the structure of the earth in Reference 1. The model assumes a radius of 6371 km for the earth. The boundary between the crust and mantle (the Mohorovicic discontinuity) is taken as 21 km, while in reality it varies considerable with location.

REFERENCES

1. Anderson, D. L., and Hart, R. S., *J. Geophys. Res.*, 81, 1461, 1976.
2. Carmichael, R. S., *CRC Practical Handbook of Physical Properties of Rocks and Minerals*, p.467, CRC Press, Boca Raton, FL, 1989.

Depth km	ρ g/cm³	*p* kbar	*g* cm/s²
Crust			
0	1.02	0	981
3	1.02	3	982
3	2.80	3	982
21	2.80	5	983
Mantle (solid)			
21	3.49	5	983
41	3.51	12	983
61	3.52	19	984
81	3.48	26	984
101	3.44	33	984
121	3.40	39	985
171	3.37	56	987
221	3.34	73	989
271	3.37	89	991
321	3.47	106	993
371	3.59	124	994
571	3.95	199	999
871	4.54	328	997
1171	4.67	466	992
1471	4.81	607	991
1771	4.96	752	994
2071	5.12	903	1002
2371	5.31	1061	1017
2671	5.45	1227	1042
2886	5.53	1352	1069
Outer core (liquid)			
2886	9.96	1352	1069
2971	10.09	1442	1050
3371	10.63	1858	953
3671	11.00	2154	874
4071	11.36	2520	760
4471	11.69	2844	641
4871	11.99	3116	517
5156	12.12	3281	427
Inner core (solid)			
5156	12.30	3281	427
5371	12.48	3385	355
5771	12.52	3529	218
6071	12.53	3592	122
6371	12.58	3617	0

OCEAN PRESSURE AS A FUNCTION OF DEPTH AND LATITUDE

The following table is based upon an ocean model which takes into account the equation of state of standard seawater and the dependence on latitude of the acceleration of gravity. The tabulated pressure value is the excess pressure over the ambient atmospheric pressure at the surface.

REFERENCES

1. *International Oceanographic Tables, Volume 4*, Unesco Technical Papers in Marine Science No. 40, Unesco, Paris, 1987.
2. Saunders, P.M., and Fofonoff, N.P., *Deep-Sea Res.* 23, 109-111, 1976.

Pressure in MPa at the Specified Latitude

Depth (meters)	0°	15°	30°	45°	60°	75°	90°
0	0.0000	0.0000	0.0000	0.0000	0.0000	0.0000	0.0000
500	5.0338	5.0355	5.0404	5.0471	5.0537	5.0586	5.0605
1000	10.0796	10.0832	10.0930	10.1064	10.1198	10.1296	10.1333
1500	15.1376	15.1431	15.1577	15.1778	15.1980	15.2127	15.2182
2000	20.2076	20.2148	20.2344	20.2613	20.2882	20.3080	20.3153
2500	25.2895	25.2985	25.3231	25.3568	25.3905	25.4153	25.4244
3000	30.3831	30.3940	30.4236	30.4641	30.5047	30.5345	30.5453
3500	35.4886	35.5012	35.5358	35.5832	35.6307	35.6654	35.6782
4000	40.6056	40.6201	40.6598	40.7140	40.7683	40.8082	40.8229
4500	45.7342	45.7505	45.7952	45.8564	45.9176	45.9626	45.9791
5000	50.8742	50.8924	50.9421	51.0102	51.0785	51.1285	51.1469
5500	56.0255	56.0456	56.1004	56.1755	56.2508	56.3059	56.3262
6000	61.1882	61.2100	61.2700	61.3521	61.4344	61.4947	61.5168
6500	66.3619	66.3857	66.4508	66.5399	66.6292	66.6947	66.7187
7000	71.5467	71.5724	71.6427	71.7388	71.8352	71.9059	71.9318
7500	76.7426	76.7701	76.8456	76.9488	77.0523	77.1282	77.1560
8000	81.9493	81.9788	82.0594	82.1697	82.2804	82.3614	82.3911
8500	87.1669	87.1983	87.2841	87.4016	87.5193	87.6057	87.6373
9000	92.3950	92.4284	92.5194	92.6440	92.7689	92.8606	92.8941
9500	97.6346	97.6698	97.7661	97.8978	98.0300	98.1269	98.1624
10000	102.8800	102.9170	103.0185	103.1572	103.2961	103.3981	103.4355

PROPERTIES OF SEAWATER

In addition to the dependence on temperature and pressure, the physical properties of seawater vary with the concentration of the dissolved constituents. A convenient parameter for describing the composition is the salinity, S, which is defined in terms of the electrical conductivity of the seawater sample. The defining equation for the practical salinity is:

$$S = a_0 + a_1K^{1/2} + a_2K + a_3K^{3/2} + a_4K^2 + a_5K^{5/2}\ ,$$

where K is the ratio of the conductivity of the seawater sample at 15°C and atmospheric pressure to the conductivity of a potassium chloride solution in which the mass fraction of KCl is 0.0324356, at the same temperature and pressure. The values of the coefficients are:

$$a_0 = 0.080 \qquad a_3 = 14.0941$$
$$a_1 = -0.1692 \qquad a_4 = -7.0261$$
$$a_2 = 25.3851 \qquad a_5 = 2.7081$$
$$\Sigma\, a_i = 35.0000$$

Thus when $K = 1$, $S = 35$ exactly (S is normally quoted in units of ‰, i.e., parts per thousand). The value of S can be roughly equated with the mass of dissolved material in grams per kilogram of seawater. Salinity values in the oceans at mid latitudes typically fall between 34 and 36.

The freezing point of seawater at normal atmospheric pressure varies with salinity as follows:

S	0	5	10	15	20	25	30	35	40
t_f/°C	0.000	-0.274	-0.542	-0.812	-1.083	-1.358	-1.636	-1.922	-2.212

The first table below gives several properties of seawater as a function of temperature for a salinity of 35. The second gives electrical conductivity as a function of salinity at several temperatures, and the third lists typical concentrations of the main constituents of seawater as a function of salinity.

REFERENCES

1. *The Practical Salinity Scale 1978 and the International Equation of State of Seawater 1980*, Unesco Technical Papers in Marine Science No. 36, Unesco, Paris, 1981; sections No. 37, 38, 39, and 40 in this series give background papers and detailed tables.
2. Kennish, M.J., *CRC Practical Handbook of Marine Science*, CRC Press, Boca Raton, FL, 1989.
3. Poisson, A. *IEEE J. Ocean. Eng.* OE-5, 50, 1981.

Properties of Seawater as a Function of Temperature at Salinity S=35 and Normal Atmospheric Pressure (100 kPa)

ρ = density in g/cm³
$\beta = (1/\rho)\,(d\rho/dS)$ = fractional change in density per unit change in salinity
$\alpha = (1/\rho)\,(d\rho/dt)$ = fractional change in density per unit change in temperature (°C⁻¹)
κ = electrical conductivity in S/cm
η = viscosity in mPa s (equal to cP)
c_p = specific heat in J/kg
v = speed of sound in m/s

t/°C	ρ/g cm^{-3}	$10^7\cdot\beta$	$10^7\cdot\alpha$/°C^{-1}	κ/S cm^{-1}	η/mPa s	c_p/J kg^{-1} °C^{-1}	v/m s^{-1}
0	1.028106	7854	526	0.029048	1.892	3986.5	1449.1
5	1.027675	7717	1136	0.033468	1.610		
10	1.026952	7606	1668	0.038103	1.388	3986.3	1489.8
15	1.025973	7516	2141	0.042933	1.221		
20	1.024763	7444	2572	0.047934	1.085	3993.9	1521.5
25	1.023343	7385	2970	0.053088	0.966		
30	1.021729	7338	3341	0.058373	0.871	4000.7	1545.6
35	1.019934	7300	3687				
40		7270	4004			4003.5	1563.2

PROPERTIES OF SEAWATER (continued)

Electrical Conductivity of Seawater in S/cm as a Function of Temperature and Salinity

t/°C	$S = 5$	$S = 10$	$S = 15$	$S = 20$	$S = 25$	$S = 30$	$S = 35$	$S = 40$
0	0.004808	0.009171	0.013357	0.017421	0.021385	0.025257	0.029048	0.032775
5	0.005570	0.010616	0.015441	0.020118	0.024674	0.029120	0.033468	0.037734
10	0.006370	0.012131	0.017627	0.022947	0.028123	0.033171	0.038103	0.042935
15	0.007204	0.013709	0.019905	0.025894	0.031716	0.037391	0.042933	0.048355
20	0.008068	0.015346	0.022267	0.028948	0.035438	0.041762	0.047934	0.053968
25	0.008960	0.017035	0.024703	0.032097	0.039276	0.046267	0.053088	0.059751
30	0.009877	0.018771	0.027204	0.035330	0.043213	0.050888	0.058373	0.065683

Composition of Seawater and Ionic Strength at Various Salinities

Concentration of major constituents expressed as molality (moles per kilogram of H_2O)

Constituent	$S = 30$	$S = 35$	$S = 40$
Cl^-	0.482	0.562	0.650
Br^-	0.00074	0.00087	0.00100
F^-		0.00007	
SO_4^{2-}	0.0104	0.0114	0.0122
HCO_3^-	0.00131	0.00143	0.00100
$NaSO_4^-$	0.0085	0.0108	0.0139
KSO_4^-	0.00010	0.00012	0.00015
Na^+	0.405	0.472	0.544
K^+	0.00892	0.01039	0.01200
Mg^{2+}	0.0413	0.0483	0.0561
Ca^{2+}	0.00131	0.00143	0.00154
Sr^+	0.00008	0.00009	0.00011
$MgHCO_3^+$	0.00028	0.00036	0.00045
$MgSO_4$	0.00498	0.00561	0.00614
$CaSO_4$	0.00102	0.00115	0.00126
$NaHCO_3$	0.00015	0.00020	0.00024
H_3BO_3	0.00032	0.00037	0.00042
Ionic strength (mol/kg)	0.5736	0.6675	0.7701

ABUNDANCE OF ELEMENTS IN THE EARTH'S CRUST AND IN THE SEA

This table gives the estimated abundance of the elements in the continental crust (in mg/kg, equivalent to parts per million by mass) and in seawater near the surface (in mg/L). Values represent the median of reported measurements. The concentrations of the less abundant elements may vary with location by several orders of magnitude.

REFERENCES

1. Carmichael, R. S., Ed., *CRC Practical Handbook of Physical Properties of Rocks and Minerals,* CRC Press, Boca Raton, FL, 1989.
2. Bodek,I., et al., *Environmental Inorganic Chemistry*, Pergamon Press, New York, 1988.
3. Ronov, A. B., and Yaroshevsky, A. A.,"Earth's Crust Geochemistry", in *Encyclopedia of Geochemistry and Environmental Sciences,* Fairbridge, R. W.,Ed., Van Nostrand, New York, 1969.

	Abundance			Abundance	
Element	Crust mg/kg	Sea mg/L	Element	Crust mg/kg	Sea mg/L
Ac	5.5×10^{-10}		N	1.9×10^{1}	5×10^{-1}
Ag	7.5×10^{-2}	4×10^{-5}	Na	2.36×10^{4}	1.08×10^{4}
Al	8.23×10^{4}	2×10^{-3}	Nb	2.0×10^{1}	1×10^{-5}
Ar	3.5	4.5×10^{-1}	Nd	4.15×10^{1}	2.8×10^{-6}
As	1.8	3.7×10^{-3}	Ne	5×10^{-3}	1.2×10^{-4}
Au	4×10^{-3}	4×10^{-6}	Ni	8.4×10^{1}	5.6×10^{-4}
B	1.0×10^{1}	4.44	O	4.61×10^{5}	8.57×10^{5}
Ba	4.25×10^{2}	1.3×10^{-2}	Os	1.5×10^{-3}	
Be	2.8	5.6×10^{-6}	P	1.05×10^{3}	6×10^{-2}
Bi	8.5×10^{-3}	2×10^{-5}	Pa	1.4×10^{-6}	5×10^{-11}
Br	2.4	6.73×10^{1}	Pb	1.4×10^{1}	3×10^{-5}
C	2.00×10^{2}	2.8×10^{1}	Pd	1.5×10^{-2}	
Ca	4.15×10^{4}	4.12×10^{2}	Po	2×10^{-10}	1.5×10^{-14}
Cd	1.5×10^{-1}	1.1×10^{-4}	Pr	9.2	6.4×10^{-7}
Ce	6.65×10^{1}	1.2×10^{-6}	Pt	5×10^{-3}	
Cl	1.45×10^{2}	1.94×10^{4}	Ra	9×10^{-7}	8.9×10^{-11}
Co	2.5×10^{1}	2×10^{-5}	Rb	9.0×10^{1}	1.2×10^{-1}
Cr	1.02×10^{2}	3×10^{-4}	Re	7×10^{-4}	4×10^{-6}
Cs	3	3×10^{-4}	Rh	1×10^{-3}	
Cu	6.0×10^{1}	2.5×10^{-4}	Rn	4×10^{-13}	6×10^{-16}
Dy	5.2	9.1×10^{-7}	Ru	1×10^{-3}	7×10^{-7}
Er	3.5	8.7×10^{-7}	S	3.50×10^{2}	9.05×10^{2}
Eu	2.0	1.3×10^{-7}	Sb	2×10^{-1}	2.4×10^{-4}
F	5.85×10^{2}	1.3	Sc	2.2×10^{1}	6×10^{-7}
Fe	5.63×10^{4}	2×10^{-3}	Se	5×10^{-2}	2×10^{-4}
Ga	1.9×10^{1}	3×10^{-5}	Si	2.82×10^{5}	2.2
Gd	6.2	7×10^{-7}	Sm	7.05	4.5×10^{-7}
Ge	1.5	5×10^{-5}	Sn	2.3	4×10^{-6}
H	1.40×10^{3}	1.08×10^{5}	Sr	3.70×10^{2}	7.9
He	8×10^{-3}	7×10^{-6}	Ta	2.0	2×10^{-6}
Hf	3.0	7×10^{-6}	Tb	1.2	1.4×10^{-7}
Hg	8.5×10^{-2}	3×10^{-5}	Te	1×10^{-3}	
Ho	1.3	2.2×10^{-7}	Th	9.6	1×10^{-6}
I	4.5×10^{-1}	6×10^{-2}	Ti	5.65×10^{3}	1×10^{-3}
In	2.5×10^{-1}	2×10^{-2}	Tl	8.5×10^{-1}	1.9×10^{-5}
Ir	1×10^{-3}		Tm	5.2×10^{-1}	1.7×10^{-7}
K	2.09×10^{4}	3.99×10^{2}	U	2.7	3.2×10^{-3}
Kr	1×10^{-4}	2.1×10^{-4}	V	1.20×10^{2}	2.5×10^{-3}
La	3.9×10^{1}	3.4×10^{-6}	W	1.25	1×10^{-4}
Li	2.0×10^{1}	1.8×10^{-1}	Xe	3×10^{-5}	5×10^{-5}
Lu	8×10^{-1}	1.5×10^{-7}	Y	3.3×10^{1}	1.3×10^{-5}
Mg	2.33×10^{4}	1.29×10^{3}	Yb	3.2	8.2×10^{-7}
Mn	9.50×10^{2}	2×10^{-4}	Zn	7.0×10^{1}	4.9×10^{-3}
Mo	1.2	1×10^{-2}	Zr	1.65×10^{2}	3×10^{-5}

SOLAR SPECTRAL IRRADIANCE

The solar luminosity (total radiant power emitted) is $3.86 \cdot 10^{26}$ W, of which 1373 W/m^2 reaches the top of the earth's atmosphere. To a zeroth approximation the sun can be considered a black body with an effective temperature of 5780 K, which implies a peak in the radiation at around 0.520 μm (5200 Å). The actual solar spectral emission is more complex, especially at ultraviolet and shorter wavelengths. The graph below, which was taken from Reference 1, summarizes the solar irradiance at the top of the atmosphere in the range 0.3 to 10 μm.

REFERENCES

1. Jursa, A.S., ed., *Handbook of Geophysics and the Space Environment*, Air Force Geophysics Laboratory, 1985.
2. Pierce, A.K., and Allen, R.G., "The Solar Spectrum between 0.3 and 10 μm", in *The Solar Output and its Variation*, White, O.R., Ed., Colorado Associated University Press, Boulder, CO, 1977.
3. Lang, K.R., *Astrophysical Data. Planets and Stars*, Springer-Verlag, New York, 1992.

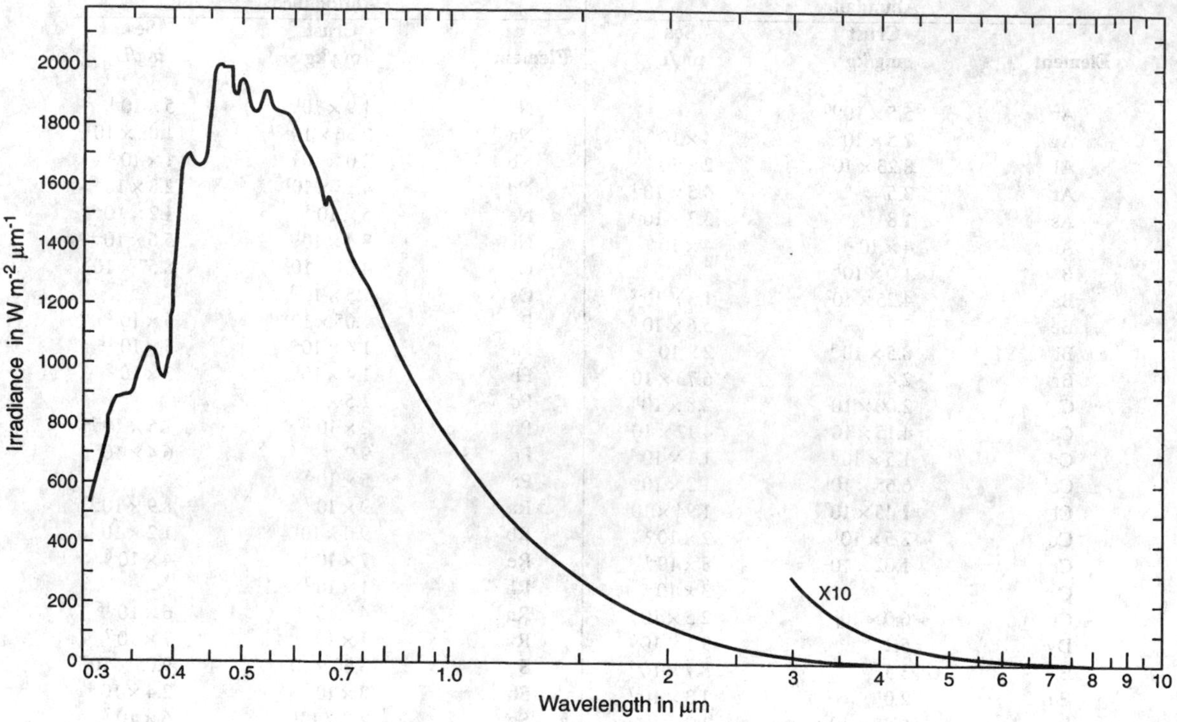

U.S. STANDARD ATMOSPHERE (1976)

A Standard Atmosphere is a hypothetical vertical distribution of atmospheric temperature, pressure, and density which is roughly representative of year-round, midlatitude conditions. Typical uses are to serve as a basis for pressure altimeter calibrations, aircraft performance calculations, aircraft and rocket design, ballistic tables, meteorological diagrams, and various types of atmospheric modeling. The air is assumed to be dry and to obey the perfect gas law and the hydrostatic equation which, taken together, relate temperature, pressure, and density with vertical position. The atmosphere is considered to rotate with the earth and to be an average over the diurnal cycle, the semiannual variation, and the range from active to quiet geomagnetic and sunspot conditions.

The U.S. Standard Atmosphere, 1976 is an idealized, steady-state representation of mean annual conditions of the earth's atmosphere from the surface to 1000 km at latitude 45°N, as it is assumed to exist during a period with moderate solar activity. The defining meteorological elements are sea-level temperature and pressure and a temperature-height profile to 1000 km. The 1976 Standard Atmosphere uses the following sea-level values which have been standard for many decades:

Temperature — 288.15 K (15°C)
Pressure — 101325 Pa (1013.25 mbar, 760 mm of Hg, or 29.92 in. of Hg)
Density — 1225 g/m^3 (1.225 g/L)
Mean molar mass — 28.964 g/mol

The parameters included in this condensed version of the U.S. Standard Atmosphere are:

Z — Height (geometric) above mean sea level in meters
T — Temperature in kelvins
P — Pressure in pascals (1 Pa = 0.01 millibars)
ρ — Density in kilograms per cubic meter (1 kg/m^3 = 1 g/L)
n — Number density in molecules per cubic meter
ν — Mean collision frequency in collisions per second
l — Mean free path in meters
η — Absolute viscosity in pascal seconds (1 Pa s = 1000 cP)
k — Thermal conductivity in joules per meter second kilogram (W/m K)
v_s — Speed of sound in meters per second
g — Acceleration of gravity in meters per second square

The sea-level composition (percent by volume) is taken to be:

N_2 — 78.084%	He — 0.000524
O_2 — 20.9476	Kr — 0.000114
Ar — 0.934	Xe — 0.0000087
CO_2 — 0.0314	CH_4 — 0.0002
Ne — 0.001818	H_2 — 0.00005

The T and P columns for the troposphere and lower stratosphere were generated from the following formulas:

	***T*/K**	***P*/Pa**
$H \leq 1100$ m	$288.15 - 0.0065\,H$	$101325(288.15/T)^{-5.25577}$
$11000 \text{ m} < H \leq 20000$ m	216.65	$22632\,e^{-0.00015768832(H-11000)}$
$20000 \text{ m} < H \leq 32000$ m	$216.65 + 0.0010(H-20000)$	$5474.87(216.65/T)^{34.16319}$

where $H = rZ/(r + Z)$ is the geopotential height in meters and r is the mean earth radius at 45° N latitude, taken as 6356766 m. For altitudes up to 32 km, $\rho = 0.003483677(P/T)$ in the units used here. Formulas for the other quantities may be found in the references.

REFERENCES

1. COESA, U.S. Standard Atmosphere, 1976, U.S. Government Printing Office, Washington, D.C., 1976.
2. Jursa, A.S., ed., *Handbook of Geophysics and the Space Environment*, Air Force Geophysics Laboratory, 1985.

Z/m	T/K	P/Pa	ρ/kg m^{-3}	n/m^{-3}	ν/s^{-1}	l/m	η/Pa s	k/J m^{-1}s^{-1}K^{-1}	v_s/m s^{-1}	g/m s^{-2}
-5000	320.68	1.778E+05	1.931	4.015E+25	1.151E+10	4.208E-08	1.942E-05	0.02788	359.0	9.822
-4500	317.42	1.685E+05	1.849	3.845E+25	1.096E+10	4.395E-08	1.927E-05	0.02763	357.2	9.830
-4000	314.17	1.596E+05	1.770	3.680E+25	1.044E+10	4.592E-08	1.912E-05	0.02738	355.3	9.819
-3500	310.91	1.511E+05	1.693	3.520E+25	9.933E+09	4.800E-08	1.897E-05	0.02713	353.5	9.818
-3000	307.66	1.430E+05	1.619	3.366E+25	9.448E+09	5.019E-08	1.882E-05	0.02688	351.6	9.816
-2500	304.41	1.352E+05	1.547	3.217E+25	8.982E+09	5.252E-08	1.867E-05	0.02663	349.8	9.814
-2000	301.15	1.278E+05	1.478	3.102E+25	8.623E+09	5.447E-08	1.852E-05	0.02638	347.9	9.813
-1500	297.90	1.207E+05	1.411	2.935E+25	8.106E+09	5.757E-08	1.836E-05	0.02613	346.0	9.811
-1000	294.65	1.139E+05	1.347	2.801E+25	7.693E+09	6.032E-08	1.821E-05	0.02587	344.1	9.810
-500	291.40	1.075E+05	1.285	2.672E+25	7.298E+09	6.324E-08	1.805E-05	0.02562	342.2	9.808
0	288.15	1.013E+05	1.225	2.547E+25	6.919E+09	6.633E-08	1.789E-05	0.02533	340.3	9.807
500	284.90	9.546E+04	1.167	2.427E+25	6.556E+09	6.961E-08	1.774E-05	0.02511	338.4	9.805
1000	281.65	8.988E+04	1.112	2.311E+25	6.208E+09	7.310E-08	1.758E-05	0.02485	336.4	9.804
1500	278.40	8.456E+04	1.058	2.200E+25	5.874E+09	7.680E-08	1.742E-05	0.02459	334.5	9.802
2000	275.15	7.950E+04	1.007	2.093E+25	5.555E+09	8.073E-08	1.726E-05	0.02433	332.5	9.801
2500	271.91	7.469E+04	0.957	1.990E+25	5.250E+09	8.491E-08	1.710E-05	0.02407	330.6	9.799
3000	268.66	7.012E+04	0.909	1.891E+25	4.959E+09	8.937E-08	1.694E-05	0.02381	328.6	9.797
3500	265.41	6.579E+04	0.863	1.795E+25	4.680E+09	9.411E-08	1.678E-05	0.02355	326.6	9.796
4000	262.17	6.166E+04	0.819	1.704E+25	4.414E+09	9.917E-08	1.661E-05	0.02329	324.6	9.794
4500	258.92	5.775E+04	0.777	1.616E+25	4.160E+09	1.046E-07	1.645E-05	0.02303	322.6	9.793
5000	255.68	5.405E+04	0.736	1.531E+25	3.918E+09	1.103E-07	1.628E-05	0.02277	320.6	9.791
5500	252.43	5.054E+04	0.697	1.450E+25	3.687E+09	1.165E-07	1.612E-05	0.02250	318.5	9.790
6000	249.19	4.722E+04	0.660	1.373E+25	3.467E+09	1.231E-07	1.595E-05	0.02224	316.5	9.788
6500	245.94	4.408E+04	0.664	1.299E+25	3.258E+09	1.302E-07	1.578E-05	0.02197	314.4	9.787
7000	242.70	4.111E+04	0.590	1.227E+25	3.058E+09	1.377E-07	1.561E-05	0.02170	312.3	9.785
7500	239.46	3.830E+04	0.557	1.159E+25	2.869E+09	1.458E-07	1.544E-05	0.02144	310.2	9.784
8000	236.22	3.565E+04	0.526	1.093E+25	2.689E+09	1.545E-07	1.527E-05	0.02117	308.1	9.782
8500	232.97	3.315E+04	0.496	1.031E+25	2.518E+09	1.639E-07	1.510E-05	0.02090	306.0	9.781
9000	229.73	3.080E+04	0.467	9.711E+24	2.356E+09	1.740E-07	1.493E-05	0.02063	303.9	9.779
9500	226.49	2.858E+04	0.440	9.141E+24	2.202E+09	1.848E-07	1.475E-05	0.02036	301.7	9.777
10000	223.25	2.650E+04	0.414	8.598E+24	2.056E+09	1.965E-07	1.458E-05	0.02009	299.5	9.776
10500	220.01	2.454E+04	0.389	8.079E+24	1.918E+09	2.091E-07	1.440E-05	0.01982	297.4	9.774
11000	216.77	2.270E+04	0.365	7.585E+24	1.787E+09	2.227E-07	1.422E-05	0.01954	295.2	9.773
11500	216.65	2.098E+04	0.337	7.016E+24	1.653E+09	2.408E-07	1.422E-05	0.01953	295.1	9.771
12000	216.65	1.940E+04	0.312	6.486E+24	1.528E+09	2.605E-07	1.422E-05	0.01953	295.1	9.770
12500	216.65	1.793E+04	0.288	5.996E+24	1.412E+09	2.818E-07	1.422E-05	0.01953	295.1	9.768
13000	216.65	1.658E+04	0.267	5.543E+24	1.306E+09	3.048E-07	1.422E-05	0.01953	295.1	9.767
13500	216.65	1.533E+04	0.246	5.124E+24	1.207E+09	3.297E-07	1.422E-05	0.01953	295.1	9.765
14000	216.65	1.417E+04	0.228	4.738E+24	1.116E+09	3.566E-07	1.422E-05	0.01953	295.1	9.764
14500	216.65	1.310E+04	0.211	4.380E+24	1.032E+09	3.857E-07	1.422E-05	0.01953	295.1	9.762
15000	216.65	1.211E+04	0.195	4.049E+24	9.538E+08	4.172E-07	1.422E-05	0.01953	295.1	9.761
16000	216.65	1.035E+04	0.166	3.461E+24	8.153E+08	4.881E-07	1.422E-05	0.01953	295.1	9.758
17000	216.65	8.850E+03	0.142	2.959E+24	6.969E+08	5.710E-07	1.422E-05	0.01953	295.1	9.754
18000	216.65	7.565E+03	0.122	2.529E+24	5.958E+08	6.680E-07	1.422E-05	0.01953	295.1	9.751
19000	216.65	6.467E+03	0.104	2.162E+24	5.093E+08	7.814E-07	1.422E-05	0.01953	295.1	9.748

U.S. STANDARD ATMOSPHERE (1976) (continued)

Z/m	T/K	P/Pa	ρ/kg m^{-3}	n/m^{-3}	ν/s^{-1}	l/m	η/Pa s	k/J m^{-1}s^{-1}K^{-1}	v_s/m s^{-1}	g/m s^{-2}
20000	216.65	5.529E+03	8.891E-02	1.849E+24	4.354E+08	9.139E-07	1.422E-05	0.01953	295.1	9.745
21000	217.58	4.729E+03	7.572E-02	1.574E+24	3.716E+08	1.073E-06	1.427E-05	0.01961	295.1	9.742
22000	218.57	4.048E+03	6.451E-02	1.341E+24	3.173E+08	1.260E-06	1.432E-05	0.01970	296.4	9.739
23000	219.57	3.467E+03	5.501E-02	1.144E+24	2.712E+08	1.477E-06	1.438E-05	0.01978	297.1	9.736
24000	220.56	2.972E+03	4.694E-02	9.759E+23	2.319E+08	1.731E-06	1.443E-05	0.01986	297.7	9.733
25000	221.55	2.549E+03	4.008E-02	8.334E+23	1.985E+08	2.027E-06	1.448E-05	0.01995	298.4	9.730
26000	222.54	2.188E+03	3.426E-02	7.123E+23	1.700E+08	2.372E-06	1.454E-05	0.02003	299.1	9.727
27000	223.54	1.880E+03	2.930E-02	6.092E+23	1.458E+08	2.773E-06	1.459E-05	0.02011	299.7	9.724
28000	224.53	1.610E+03	2.508E-02	5.214E+23	1.250E+08	3.240E-06	1.465E-05	0.02020	300.4	9.721
29000	225.52	1.390E+03	2.148E-02	4.466E+23	1.073E+08	3.783E-06	1.470E-05	0.02028	301.1	9.718
30000	226.51	1.197E+03	1.841E-02	3.828E+23	9.219E+07	4.414E-06	1.475E-05	0.02036	301.7	9.715
31000	227.50	1.031E+03	1.579E-02	3.283E+23	7.925E+07	5.146E-06	1.481E-05	0.02044	302.4	9.712
32000	228.49	8.891E+02	1.356E-02	2.813E+23	6.818E+07	5.995E-06	1.486E-05	0.02053	303.0	9.709
33000	230.97	7.673E+02	1.157E-02	2.406E+23	5.852E+07	7.021E-06	1.499E-05	0.02073	304.7	9.706
34000	233.74	6.634E+02	9.887E-03	2.056E+23	5.030E+07	8.218E-06	1.514E-05	0.02096	306.5	9.703
35000	236.51	5.746E+02	8.463E-03	1.760E+23	4.331E+07	9.601E-06	1.529E-05	0.02119	308.3	9.700
36000	239.28	4.985E+02	7.258E-03	1.509E+23	3.736E+07	1.120E-05	1.543E-05	0.02142	310.1	9.697
38000	244.82	3.771E+02	5.367E-03	1.116E+23	2.794E+07	1.514E-05	1.572E-05	0.02188	313.7	9.690
40000	250.35	2.871E+02	3.996E-03	8.308E+22	2.104E+07	2.034E-05	1.601E-05	0.02233	317.2	9.684
42000	255.88	2.200E+02	2.995E-03	6.227E+22	1.594E+07	2.713E-05	1.629E-05	0.02278	320.7	9.678
44000	261.40	1.695E+02	2.259E-03	4.697E+22	1.215E+07	3.597E-05	1.657E-05	0.02323	324.1	9.672
46000	266.93	1.313E+02	1.714E-03	3.564E+22	9.318E+06	4.740E-05	1.685E-05	0.02376	327.5	9.666
48000	270.65	1.023E+02	1.317E-03	2.738E+22	7.208E+06	6.171E-05	1.704E-05	0.02397	329.8	9.660
50000	270.65	7.978E+01	1.027E-03	2.135E+22	5.620E+06	7.913E-05	1.703E-05	0.02397	329.8	9.654
52000	269.03	6.221E+01	8.056E-04	1.675E+22	4.397E+06	1.009E-04	1.696E-05	0.02384	328.8	9.648
54000	263.52	4.834E+01	6.390E-04	1.329E+22	3.452E+06	1.272E-04	1.660E-05	0.02340	325.4	9.642
56000	258.02	3.736E+01	5.045E-04	1.049E+22	2.696E+06	1.611E-04	1.640E-05	0.02296	322.0	9.636
58000	252.52	2.872E+01	3.963E-04	8.239E+21	2.095E+06	2.051E-04	1.612E-05	0.02251	318.6	9.632
60000	247.02	2.196E+01	3.097E-04	6.439E+21	1.620E+06	2.624E-04	1.584E-05	0.02206	315.1	9.624
65000	233.29	1.093E+01	1.632E-04	3.393E+21	8.294E+05	4.979E-04	1.512E-05	0.02093	306.2	9.609
70000	219.59	5.221	8.283E-05	1.722E+21	4.084E+05	9.810E-04	1.438E-05	0.01978	297.1	9.594
75000	208.40	2.388	3.992E-05	8.300E+20	1.918E+05	2.035E-03	1.376E-05	0.01883	289.4	9.579
80000	198.64	1.052	1.846E-05	3.838E+20	8.656E+04	4.402E-03	1.321E-05	0.01800	282.5	9.564
85000	188.89	4.457E-01	8.220E-06	1.709E+20	3.766E+04	9.886E-03	1.265E-05	0.01716	275.5	9.550
90000	186.87	1.836E-01	3.416E-06	7.116E+19	1.560E+04	2.370E-02				9.535
95000	188.42	7.597E-02	1.393E-06	2.920E+19	6.440E+03	5.790E-02				9.520
100000	195.08	3.201E-02	5.604E-07	1.189E+19	2.680E+03	1.420E-01				9.505
110000	240.00	7.104E-03	9.708E-08	2.144E+18	5.480E+02	7.880E-01				9.476
120000	360.00	2.538E-03	2.222E-08	5.107E+17	1.630E+02	3.310				9.447
130000	469.27	1.251E-03	8.152E-09	1.930E+17	7.100E+01	8.800				9.418
140000	559.63	7.203E-04	3.831E-09	9.322E+16	3.800E+01	1.800E+01				9.389
150000	634.39	4.542E-04	2.076E-09	5.186E+16	2.300E+01	3.300E+01				9.360
160000	696.29	3.040E-04	1.233E-09	3.162E+16	1.500E+01	5.300E+01				9.331
170000	747.57	2.121E-04	7.815E-10	2.055E+16	1.000E+01	8.200E+01				9.302
180000	790.07	1.527E-04	5.194E-10	1.400E+16	7.200	1.200E+02				9.274

U.S. STANDARD ATMOSPHERE (1976) (continued)

Z/m	T/K	P/Pa	ρ/kg m^{-3}	n/m^{-3}	ν/s^{-1}	l/m	η/Pa s	k/J m^{-1}s^{-1}K^{-1}	v_s/m s^{-1}	g/m s^{-2}
190000	825.16	1.127E-04	3.581E-10	9.887E+15	5.200	1.700E+02				9.246
200000	854.56	8.474E-05	2.541E-10	7.182E+15	3.900	2.400E+02				9.218
220000	899.01	5.015E-05	1.367E-10	4.040E+15	2.300	4.200E+02				9.162
240000	929.73	3.106E-05	7.858E-11	2.420E+15	1.400	7.000E+02				9.106
260000	950.99	1.989E-05	4.742E-11	1.515E+15	9.300E-01	1.100E+03				9.051
280000	965.75	1.308E-05	2.971E-11	9.807E+14	6.100E-01	1.700E+03				8.997
300000	976.01	8.770E-06	1.916E-11	6.509E+14	4.200E-01	2.600E+03				8.943
320000	983.16	5.980E-06	1.264E-11	4.405E+14	2.900E-01	3.800E+03				8.889
340000	988.15	4.132E-06	8.503E-12	3.029E+14	2.000E-01	5.600E+03				8.836
360000	991.65	2.888E-06	5.805E-12	2.109E+14	1.400E-01	8.000E+03				8.784
380000	994.10	2.038E-06	4.013E-12	1.485E+14	1.000E-01	1.100E+04				8.732
400000	995.83	1.452E-06	2.803E-12	1.056E+14	7.200E-02	1.600E+04				8.680
450000	998.22	6.447E-07	1.184E-12	4.678E+13	3.300E-02	3.600E+04				8.553
500000	999.24	3.024E-07	5.215E-13	2.192E+13	1.600E-02	7.700E+04				8.429
550000	999.67	1.514E-07	2.384E-13	1.097E+13	8.400E-03	1.500E+05				8.307
600000	999.85	8.213E-08	1.137E-13	5.950E+12	4.800E-03	2.800E+05				8.188
650000	999.93	4.887E-08	5.712E-14	3.540E+12	3.100E-03	4.800E+05				8.072
700000	999.97	3.191E-08	3.070E-14	2.311E+12	2.200E-03	7.300E+05				7.958
750000	999.98	2.260E-08	1.788E-14	1.637E+12	1.700E-03	1.000E+06				7.846
800000	999.99	1.704E-08	1.136E-14	1.234E+12	1.400E-03	1.400E+06				7.737
850000	1000.00	1.342E-08	7.824E-15	9.717E+11	1.200E-03	1.700E+06				7.630
900000	1000.00	1.087E-08	5.759E-15	7.876E+11	1.000E-03	2.100E+06				7.525
950000	1000.00	8.982E-09	4.453E-15	6.505E+11	8.700E-04	2.600E+06				7.422
1000000	1000.00	7.514E-09	3.561E-15	5.442E+11	7.500E-04	3.100E+06				7.322

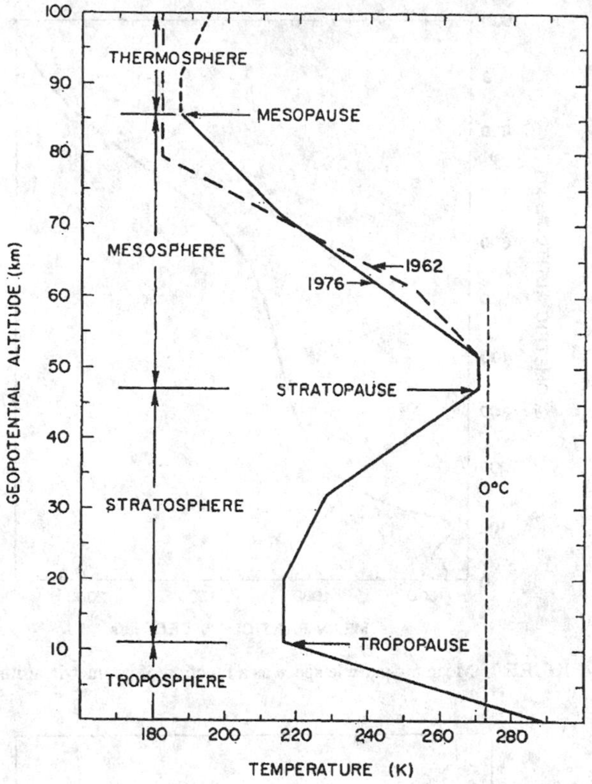

FIGURE 1. Temperature-height profile for U.S. Standard Atmosphere.

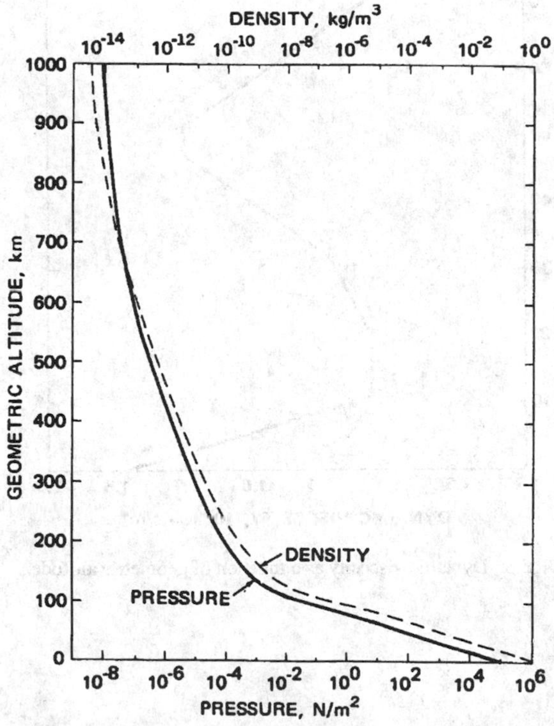

FIGURE 2. Total pressure and mass density as a function of geometric altitude.

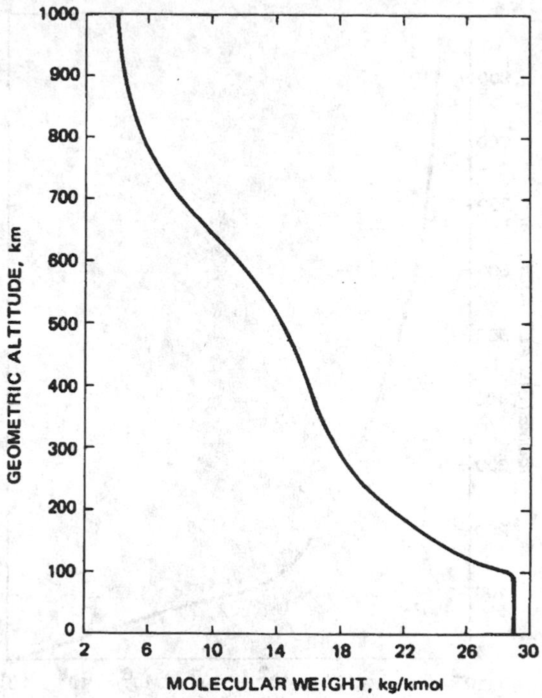

FIGURE 3. Mean molecular weight as a function of geometric altitude.

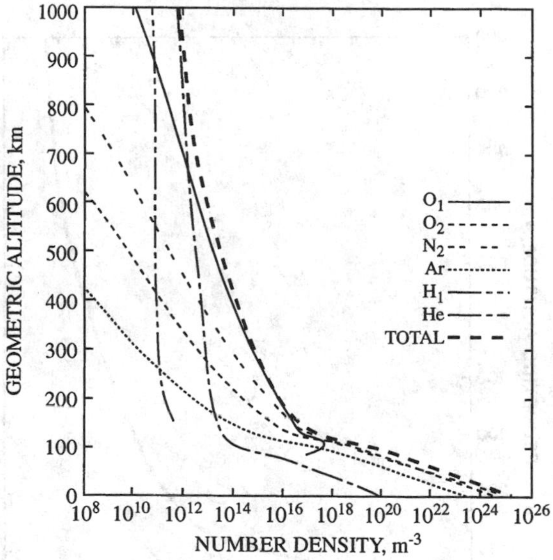

FIGURE 4. Number density of individual species and total number density as a function of geometric altitude.

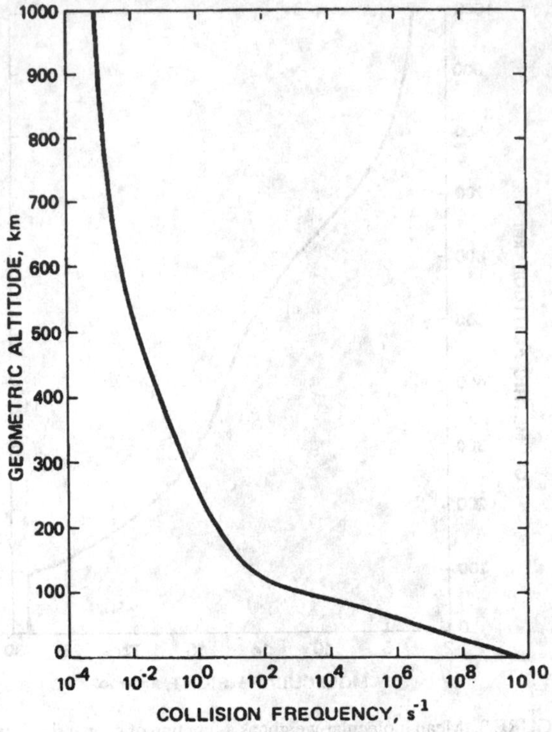

FIGURE 5. Collision frequency as a function of geometric altitude.

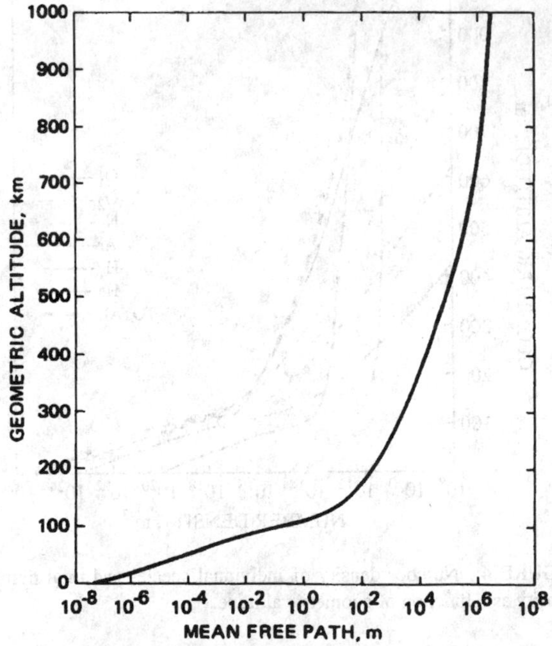

FIGURE 6. Mean free path as a function of geometric altitude.

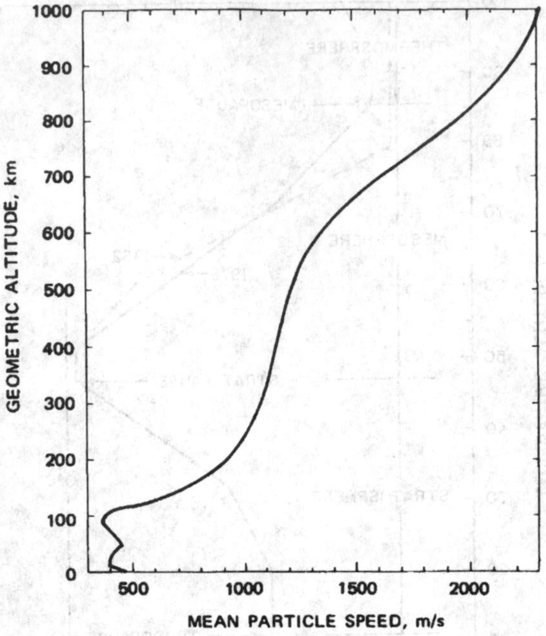

FIGURE 7. Mean air-particle speed as a function of geometric altitude

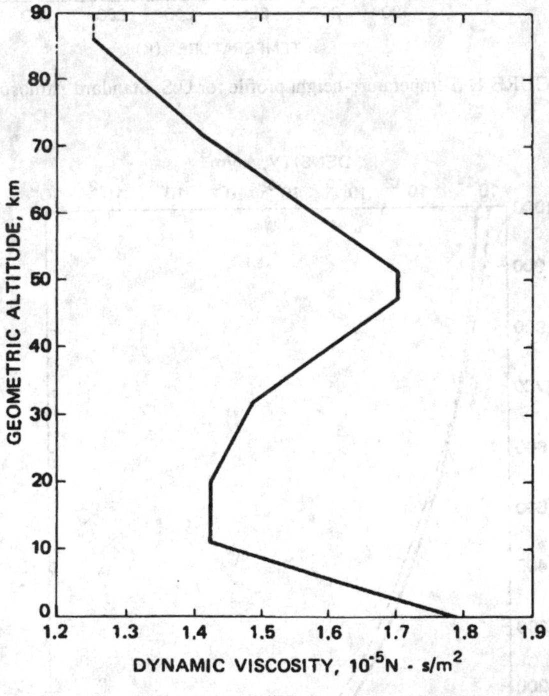

FIGURE 8. Dynamic viscosity as a function of geometric altitude.

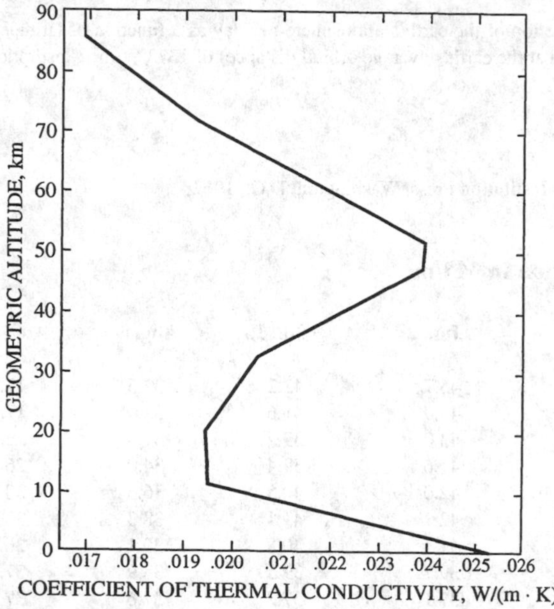

FIGURE 9. Coefficient of thermal conductivity as a function of geometric altitude.

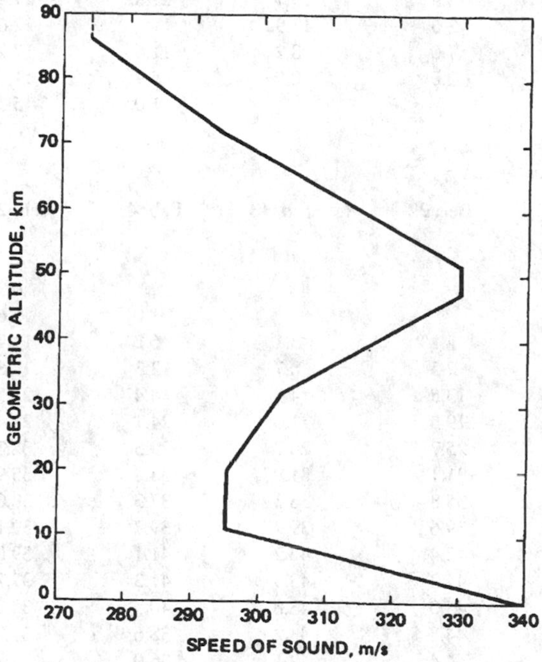

FIGURE 10. Speed of sound as a function of geometric altitude.

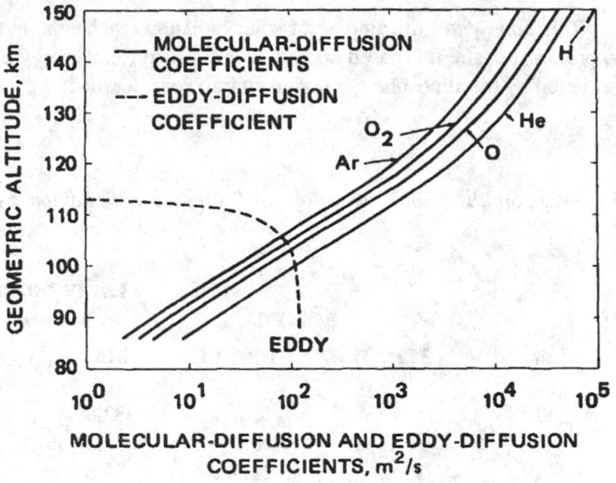

FIGURE 11. Molecular-diffusion and eddy-diffusion coefficients as a function of geometric altitude.

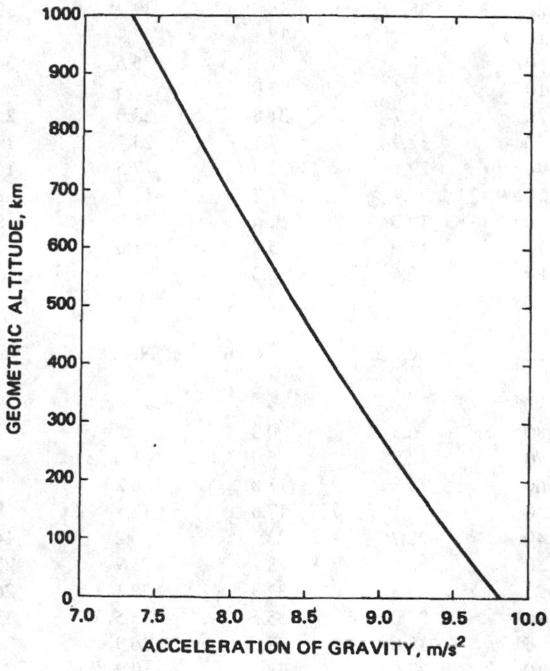

FIGURE 12. Acceleration of gravity as a function of geometric altitude.

GEOGRAPHICAL AND SEASONAL VARIATION IN SOLAR RADIATION

This table gives the amount of solar radiation reaching a unit area at the top of the earth's atmosphere per day as a function of latitude and approximate date. It is based upon a solar constant (total energy per unit area at the earth's average orbital distance) of 1373 W/m^2. Absorption of radiation by the atmosphere is not taken into consideration.

REFERENCE

List, R.J., *Smithsonian Meteorological Tables, Seventh Edition*, Smithsonian Institution Press, Washington, D.C., 1962.

Daily Solar Radiation in MJ/m^2

Lat.	Mar. 21	Apr. 13	May 6	May 29	Jun. 2	Jul. 15	Aug. 8	Aug. 31
90°		18.0	32.8	42.4	45.7	42.2	32.5	17.7
80	6.6	18.0	32.3	41.8	45.0	41.6	32.0	17.7
70	13.0	22.3	31.8	39.9	43.0	39.7	31.5	22.0
60	19.0	27.0	34.4	39.7	41.6	39.4	34.0	26.7
50	24.4	31.1	36.8	40.7	42.0	40.5	36.5	30.8
40	29.1	34.3	38.6	41.3	42.1	41.1	38.3	33.9
30	32.9	36.7	39.4	41.1	41.4	40.8	39.1	36.3
20	35.7	38.0	39.2	39.7	39.7	39.5	38.9	37.5
10	37.4	38.1	37.9	37.4	37.1	37.2	37.6	37.7
0	38.0	37.1	35.5	34.1	33.5	34.0	35.2	36.6
–10	37.4	35.0	32.3	30.0	29.2	29.9	32.0	34.6
–20	35.7	31.8	28.0	25.2	24.1	25.1	27.8	31.5
–30	32.9	27.8	23.1	19.7	18.5	19.7	22.8	27.4
–40	29.1	22.8	17.5	14.0	12.6	13.9	17.4	22.6
–50	24.4	17.3	11.7	8.2	7.0	8.2	11.6	17.2
–60	19.0	11.4	5.9	2.9	2.0	2.9	5.9	11.3
–70	13.0	5.4	1.0				1.0	5.3
–80	6.6	0.3						0.3
–90								

Lat.	Sep. 23	Oct. 16	Nov. 8	Nov. 30	Dec. 22	Jan. 13	Feb. 4	Feb. 26
90°								
80	6.5	0.3						0.3
70	12.9	5.5	1.0				1.0	5.6
60	18.8	11.6	6.2	3.1	2.1	3.1	6.2	11.7
50	24.1	17.6	12.1	8.7	7.5	8.7	12.3	17.8
40	28.7	23.1	18.2	14.8	13.5	14.9	18.4	23.5
30	32.5	28.2	23.9	20.9	19.8	21.0	24.1	28.4
20	35.3	32.3	29.1	26.6	25.7	26.7	29.3	32.7
10	37.0	35.5	33.5	31.8	31.1	31.9	33.8	35.9
0	37.6	37.6	36.9	36.1	35.8	36.3	37.3	38.0
–10	37.0	38.6	39.4	39.5	39.6	39.7	39.7	39.1
–20	35.3	38.5	40.7	42.0	42.4	42.2	41.1	39.0
–30	32.5	37.2	40.9	43.3	44.2	43.5	41.3	37.7
–40	28.7	34.8	40.1	43.6	45.0	43.8	40.5	35.2
–50	24.1	31.5	38.3	43.1	44.8	43.2	38.6	31.9
–60	18.8	27.3	35.7	41.9	44.4	42.1	36.0	27.7
–70	12.9	22.6	33.0	42.2	45.9	42.4	33.3	22.9
–80	6.5	18.2	33.5	44.2	48.1	44.4	33.8	18.4
–90		18.2	34.0	44.8	48.8	45.1	34.4	18.4

INFRARED ABSORPTION BY THE EARTH'S ATMOSPHERE

This graph summarizes the absorption by various atmospheric constituents in the wavelength range from 1 to 13 μm (wavenumber range 10,000 to 770 cm^{-1}). The vertical scale is in arbitrary units and does not take into account the wavelength variation of either the solar background radiation or the infrared detector response. Thus the intensities of the absorption bands have only qualitative significance.

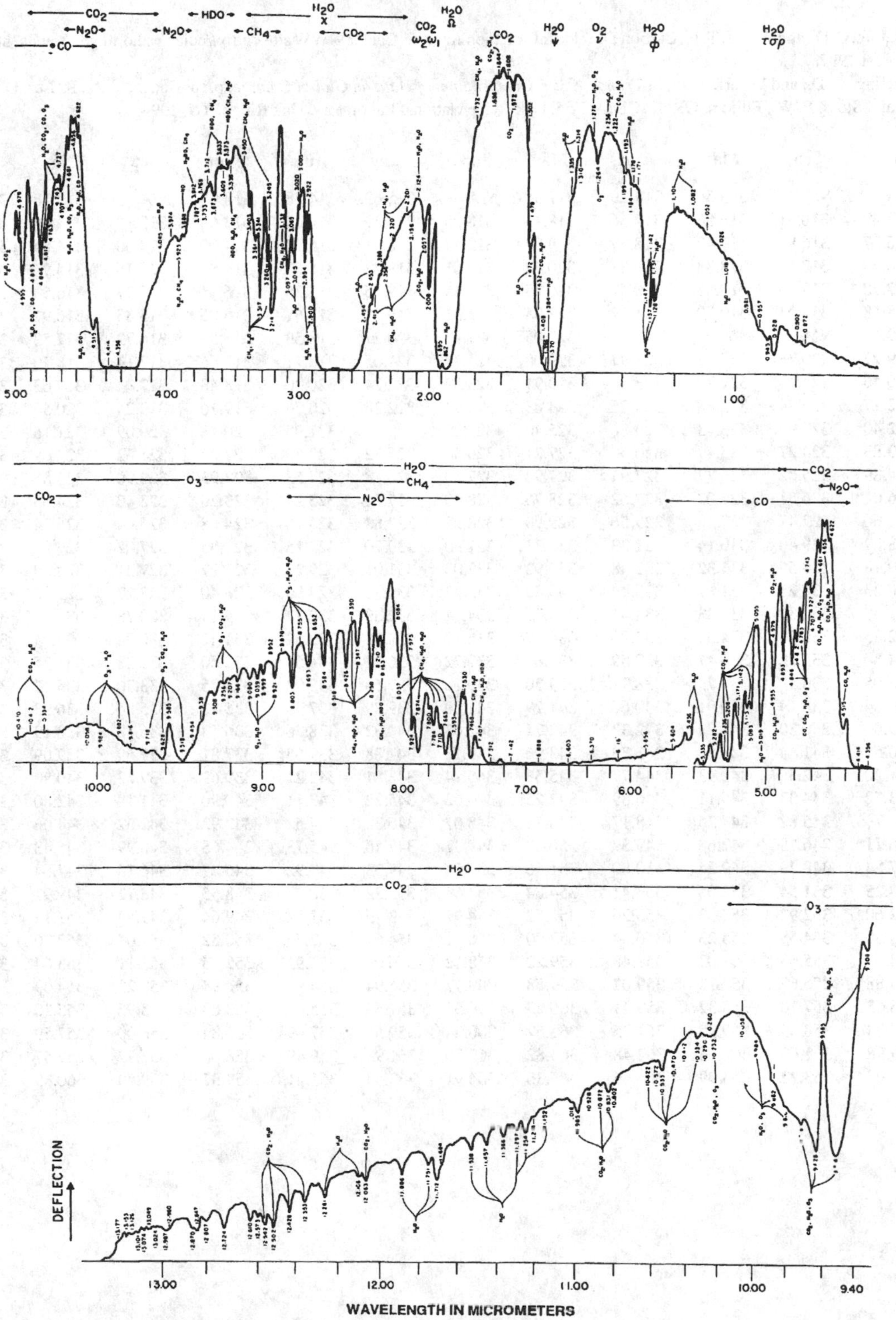

ATMOSPHERIC CONCENTRATION OF CARBON DIOXIDE, 1958-1996

The data in this table were taken at Mauna Loa Observatory in Hawaii and represent averages adjusted to the 15th of each month. The concentration of CO_2 is given in parts per million by volume. Data from other measurement sites may be found in the references.

REFERENCES

1. Keeling, C. D., and Whorf, T. P., Carbon Dioxide Information Analysis Center WWW site <http://cdiac.esd.ornl.gov/trends_html/trends/co2>, August 1997.
2. Keeling, C. D., and Whorf, T. P., in *Trends '93: A Compendium of Data on Global Change*, p.16, Boden, T. A., Kaiser, D. P., Sepanski, R. J., and Stoss, F. W., Editors, ORNL/CDIAC-65, Oak Ridge National Laboratory, Oak Ridge, TN, 1994.

Year	Jan.	Feb.	Mar.	Apr.	May	June	July	Aug.	Sept.	Oct.	Nov.	Dec.	Annual
1958	N/A	N/A	315.56	317.29	317.34	N/A	315.69	314.78	313.05	N/A	313.18	314.50	N/A
1959	315.42	316.31	316.50	317.56	318.13	318.00	316.39	314.65	313.68	313.18	314.66	315.43	315.83
1960	316.27	316.81	317.42	318.87	319.87	319.43	318.01	315.74	314.00	313.68	314.84	316.03	316.75
1961	316.73	317.54	318.38	319.31	320.42	319.61	318.42	316.63	314.84	315.16	315.94	316.85	317.49
1962	317.78	318.40	319.53	320.42	320.85	320.45	319.45	317.25	316.11	315.27	316.53	317.53	318.30
1963	318.58	318.93	319.70	321.22	322.08	321.31	319.58	317.61	316.05	315.83	316.91	318.20	318.83
1964	319.41	N/A	N/A	N/A	322.05	321.73	320.27	318.53	316.54	316.72	317.53	318.55	N/A
1965	319.27	320.28	320.73	321.97	322.00	321.71	321.05	318.71	317.66	317.14	318.70	319.25	319.87
1966	320.46	321.43	322.23	323.54	323.91	323.59	322.24	320.20	318.48	317.94	319.63	320.87	321.21
1967	322.17	322.34	322.88	324.25	324.83	323.93	322.38	320.76	319.10	319.24	320.56	321.80	322.02
1968	322.40	322.99	323.73	324.86	325.40	325.20	323.98	321.95	320.18	320.09	321.16	322.74	322.89
1969	323.83	324.27	325.47	326.50	327.21	326.54	325.72	323.50	322.22	321.62	322.69	323.95	324.46
1970	324.89	325.82	326.77	327.97	327.90	327.50	326.18	324.53	322.93	322.90	323.85	324.96	325.52
1971	326.01	326.51	327.02	327.62	328.76	328.40	327.20	325.27	323.20	323.40	324.63	325.85	326.16
1972	326.60	327.47	327.58	329.56	329.90	328.92	327.88	326.16	324.68	325.04	326.34	327.39	327.29
1973	328.37	329.40	330.14	331.33	332.31	331.90	330.70	329.15	327.35	327.02	327.99	328.48	329.51
1974	329.18	330.55	331.32	332.48	332.92	332.08	331.01	329.23	327.27	327.21	328.29	329.41	330.08
1975	330.23	331.25	331.87	333.14	333.80	333.43	331.73	329.90	328.40	328.17	329.32	330.59	330.99
1976	331.58	332.39	333.33	334.41	334.71	334.17	332.89	330.77	329.14	328.78	330.14	331.52	331.98
1977	332.75	333.24	334.53	335.90	336.57	336.10	334.76	332.59	331.42	330.98	332.24	333.68	333.73
1978	334.80	335.22	336.47	337.59	337.84	337.72	336.37	334.51	332.60	332.38	333.75	334.78	335.34
1979	336.05	336.59	337.79	338.71	339.30	339.12	337.56	335.92	333.75	333.70	335.12	336.56	336.68
1980	337.84	338.19	339.91	340.60	341.29	341.00	339.39	337.43	335.72	335.84	336.93	338.04	338.52
1981	339.06	340.30	341.21	342.33	342.74	342.08	340.32	338.26	336.52	336.68	338.19	339.44	339.76
1982	340.57	341.44	342.53	343.39	343.96	343.18	341.88	339.65	337.81	337.69	339.09	340.32	340.96
1983	341.20	342.35	342.93	344.77	345.58	345.14	343.81	342.21	339.69	339.82	340.98	342.82	342.61
1984	343.52	344.33	345.11	346.89	347.25	346.62	345.22	343.11	340.90	341.18	342.80	344.04	344.25
1985	344.79	345.82	347.25	348.17	348.74	348.07	346.38	344.52	342.92	342.62	344.06	345.38	345.73
1986	346.11	346.78	347.68	349.37	350.03	349.37	347.76	345.73	344.68	343.99	345.48	346.72	346.97
1987	347.84	348.29	349.24	350.80	351.66	351.08	349.33	347.92	346.27	346.18	347.64	348.78	348.75
1988	350.25	351.54	352.05	353.41	354.04	353.63	352.22	350.27	348.55	348.72	349.91	351.18	351.31
1989	352.60	352.92	353.53	355.26	355.52	354.97	353.75	351.52	349.64	349.83	351.14	352.37	352.75
1990	353.50	354.55	355.23	356.04	357.00	356.07	354.67	352.76	350.82	351.04	352.70	354.07	354.04
1991	354.59	355.63	357.03	358.48	359.22	358.12	356.06	353.92	352.05	352.11	353.64	354.89	355.48
1992	355.88	356.63	357.72	359.07	359.58	359.17	356.94	354.92	352.94	353.23	354.09	355.33	356.29
1993	356.63	357.10	358.32	359.41	360.23	359.55	357.53	355.48	353.67	353.95	355.30	356.78	356.99
1994	358.34	358.89	359.95	361.25	361.67	360.94	359.55	357.49	355.84	356.00	357.59	359.05	358.88
1995	359.98	361.03	361.66	363.48	363.82	363.30	361.93	359.49	358.08	357.77	359.57	360.69	360.90
1996	362.03	363.22	363.99	364.66	365.36	364.91	363.58	361.41	359.37	359.50	360.65	362.21	362.57

CO_2 Concentration in Parts per Million (Annual Average)

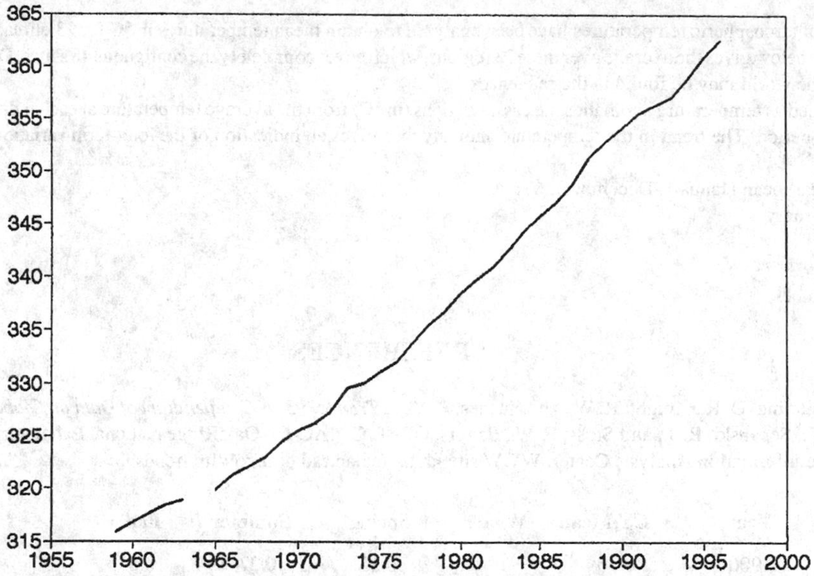

MEAN TEMPERATURES IN THE UNITED STATES, 1900-1992

Historical records of atmospheric temperatures have been analyzed to obtain mean temperatures in °C for 23 climatically distinct regions of the United States. The table below gives the average over these 23 regions, which cover completely the contiguous 48 states. Data for the individual regions and for other parts of the world may be found in the references.

The data are presented as temperature anomalies, i.e., as deviations (in °C) from the average temperature at each individual recording station over a 1961-1990 reference period. The trend in the temperature anomaly thus gives an indication of the long-term variation in average temperatures.

CY Mean: Calendar year mean (January-December)
Winter: December-February
Spring: March-May
Summer: June-August
Fall: September-November

REFERENCES

1. Karl, T. R., Easterling, D. R., Knight, R. W., and Hughes, P. Y., in *Trends '93: A Compendium of Data on Global Change*, p. 686, Boden, T. A., Kaiser, D. P., Sepanski, R. J., and Stoss, F. W., Editors, ORNL/CDIAC-65, Oak Ridge National Laboratory, Oak Ridge, TN, 1994.
2. Carbon Dioxide Information Analysis Center, WWW site <http://cdiac.esd.ornl.gov/ftp/trends93>.

Year	CY Mean	Winter	Spring	Summer	Fall
1900	0.46		0.21	0.27	0.85
1901	-0.21	0.07	-0.46	0.48	-0.28
1902	-0.14	-0.69	0.48	-0.58	0.12
1903	-0.77	-0.84	0.10	-0.83	-1.02
1904	-0.72	-1.86	-0.39	-0.92	-0.21
1905	-0.45	-1.86	0.67	-0.32	-0.25
1906	-0.04	0.23	-0.69	-0.42	-0.10
1907	-0.23	1.09	-0.61	-0.85	-0.39
1908	-0.11	0.73	0.36	-0.64	-0.59
1909	-0.36	0.82	-1.06	0.06	0.01
1910	-0.14	-2.08	0.95	-0.55	0.13
1911	0.02	0.52	0.20	-0.19	-0.62
1912	-0.88	-1.50	-0.75	-0.76	-0.51
1913	-0.23	-0.74	-0.65	-0.07	0.22
1914	-0.05	0.48	0.04	0.14	0.32
1915	-0.11	-0.37	-0.18	-1.16	0.31
1916	-0.77	-0.29	-0.36	-0.30	-1.23
1917	-1.34	-1.93	-1.75	-0.73	-1.04
1918	-0.14	-2.02	0.30	0.03	-0.09
1919	-0.16	0.69	0.00	0.12	-0.13
1920	-0.37	-0.83	-0.96	-0.67	0.02
1921	0.87	1.56	0.83	0.52	0.32
1922	0.01	-0.44	0.15	0.16	0.41
1923	-0.10	0.23	-1.02	-0.07	-0.09
1924	-1.01	0.13	-1.27	-0.64	-0.53
1925	0.20	-0.44	0.57	0.05	-0.41
1926	-0.01	0.97	-0.58	-0.27	0.02
1927	0.20	1.11	0.41	-0.83	0.83
1928	-0.08	-0.40	-0.28	-0.43	-0.08
1929	-0.68	-1.94	0.30	-0.39	-0.78
1930	-0.12	0.07	0.09	-0.07	-0.25
1931	0.81	1.16	-0.71	0.51	1.41
1932	-0.14	1.75	-0.58	0.26	-0.78
1933	0.35	-0.60	-0.08	0.45	0.54
1934	0.86	1.45	0.77	0.86	0.82
1935	0.12	0.84	0.12	0.31	-0.47
1936	-0.10	-2.23	0.48	1.00	-0.32
1937	-0.13	-0.65	-0.24	0.66	-0.03
1938	0.71	1.31	0.98	0.39	0.08
1939	0.38	0.36	0.34	0.18	0.15
1940	0.06	0.03	-0.10	0.18	0.12

MEAN TEMPERATURES IN THE UNITED STATES, 1901-1992 (continued)

Year	CY Mean	Winter	Spring	Summer	Fall
1941	0.79	1.56	0.44	0.28	0.85
1942	0.01	0.35	0.22	0.08	0.06
1943	-0.17	0.20	-0.46	0.54	-0.79
1944	0.04	0.61	-0.38	-0.20	0.43
1945	-0.02	0.30	0.25	-0.47	0.08
1946	0.53	-0.26	1.21	-0.21	0.18
1947	0.10	0.47	-0.42	0.03	0.85
1948	-0.22	-0.67	-0.01	0.02	-0.15
1949	0.05	-0.77	0.39	0.33	0.23
1950	-0.37	0.39	-1.02	-0.86	0.13
1951	-0.45	0.05	-0.69	-0.30	-0.70
1952	0.03	0.68	-0.40	0.49	-1.22
1953	0.61	1.91	0.03	0.31	0.31
1954	0.57	1.47	-0.22	0.36	0.70
1955	-0.25	-0.33	0.12	0.19	-0.60
1956	-0.05	-0.16	-0.42	0.01	-0.37
1957	0.45	1.02	0.44	0.25	-0.13
1958	0.14	0.93	0.06	0.11	0.45
1959	0.12	-0.60	0.16	0.50	-0.52
1960	-0.27	0.43	-0.98	0.01	0.56
1961	-0.01	-0.02	-0.17	0.18	-0.29
1962	-0.04	-0.52	-0.09	-0.41	0.65
1963	-0.06	-1.35	0.61	0.12	1.38
1964	-0.27	-1.30	-0.24	-0.08	-0.40
1965	-0.03	-0.07	-0.56	-0.42	0.28
1966	-0.34	-0.31	-0.25	-0.07	-0.13
1967	-0.13	0.23	-0.05	-0.37	-0.32
1968	-0.28	-0.31	-0.16	-0.12	0.01
1969	-0.06	-0.36	-0.54	0.12	-0.16
1970	-0.12	-0.17	-0.44	0.32	-0.07
1971	-0.04	-0.08	-0.99	0.01	0.56
1972	-0.13	0.20	0.26	-0.19	0.05
1973	0.48	-0.23	0.47	0.22	0.90
1974	0.16	0.52	0.74	-0.32	-0.39
1975	-0.09	0.63	-0.79	0.03	-0.20
1976	-0.62	0.88	-0.09	-0.54	-1.72
1977	0.32	-1.95	1.07	0.60	0.68
1978	-0.37	-1.31	0.23	0.09	0.21
1979	-0.53	-2.92	0.09	-0.21	-0.07
1980	0.18	0.72	-0.25	0.43	-0.04
1981	0.64	0.90	0.80	0.57	0.36
1982	-0.08	-0.86	0.03	-0.11	0.06
1983	0.40	2.33	-0.36	0.58	0.94
1984	0.21	-0.78	-0.30	0.31	0.07
1985	-0.26	-0.78	1.24	-0.23	0.05
1986	0.93	0.22	1.22	0.45	0.49
1987	0.67	1.52	0.97	0.33	-0.17
1988	-0.07	-0.26	-0.06	0.57	0.04
1989	-0.30	-0.28	0.36	0.12	-0.27
1990	0.72	0.41	0.72	0.41	0.66
1991	0.77	0.32	1.36	0.56	-0.31
1992		2.48	0.82	-0.70	

GLOBAL TEMPERATURE TREND, 1866-1996

This table and graph summarize the trend in annual mean global temperature from 1886 to 1996. The temperatures were calculated from mean temperature anomalies by assuming an absolute global mean of 14.00 °C (which is the best estimate for the 1951-1980 period). The estimated uncertainty in the year-to-year change in the recent years is about 0.07 °C.

REFERENCES

1. NASA Goddard Institute for Space Studies, World Wide Web site, www.giss.nasa.gov/data/gistemp/
2. Hansen, J., and Wilson, H., *Climatic Change*, 25, 285, 1993.

Year	*t*/°C	Year	*t*/°C	Year	*t*/°C	Year	*t*/°C	Year	*t*/°C
1866	13.52	1897	13.85	1928	14.03	1959	14.05	1990	14.47
1867	13.42	1898	13.63	1929	13.82	1960	14.00	1991	14.42
1868	13.52	1899	13.78	1930	13.93	1961	14.08	1992	14.13
1869	13.64	1900	14.00	1931	14.03	1962	14.03	1993	14.21
1870	13.57	1901	13.92	1932	14.02	1963	14.02	1994	14.31
1871	13.60	1902	13.76	1933	13.86	1964	13.74	1995	14.47
1872	13.76	1903	13.64	1934	14.02	1965	13.85	1996	14.36
1873	13.81	1904	13.51	1935	13.93	1966	13.91		
1874	13.51	1905	13.73	1936	14.02	1967	13.98		
1875	13.41	1906	13.81	1937	14.12	1968	13.88		
1876	13.58	1907	13.57	1938	14.13	1969	14.03		
1877	13.89	1908	13.70	1939	13.97	1970	14.04		
1878	13.98	1909	13.70	1940	14.14	1971	13.89		
1879	13.50	1910	13.71	1941	14.12	1972	13.93		
1880	13.68	1911	13.71	1942	14.09	1973	14.19		
1881	13.67	1912	13.72	1943	14.03	1974	13.93		
1882	13.61	1913	13.77	1944	14.11	1975	13.95		
1883	13.61	1914	13.96	1945	13.96	1976	13.78		
1884	13.35	1915	13.98	1946	14.01	1977	14.16		
1885	13.58	1916	13.76	1947	14.12	1978	14.09		
1886	13.60	1917	13.58	1948	13.99	1979	14.14		
1887	13.48	1918	13.65	1949	13.93	1980	14.28		
1888	13.70	1919	13.83	1950	13.85	1981	14.38		
1889	13.88	1920	13.83	1951	13.98	1982	14.07		
1890	13.60	1921	13.90	1952	14.04	1983	14.29		
1891	13.58	1922	13.87	1953	14.13	1984	14.11		
1892	13.61	1923	13.84	1954	13.93	1985	14.11		
1893	13.55	1924	13.85	1955	13.94	1986	14.16		
1894	13.65	1925	13.85	1956	13.83	1987	14.32		
1895	13.64	1926	14.10	1957	14.10	1988	14.35		
1896	13.81	1927	13.96	1958	14.10	1989	14.25		

Annual Mean Global Temperature in °C

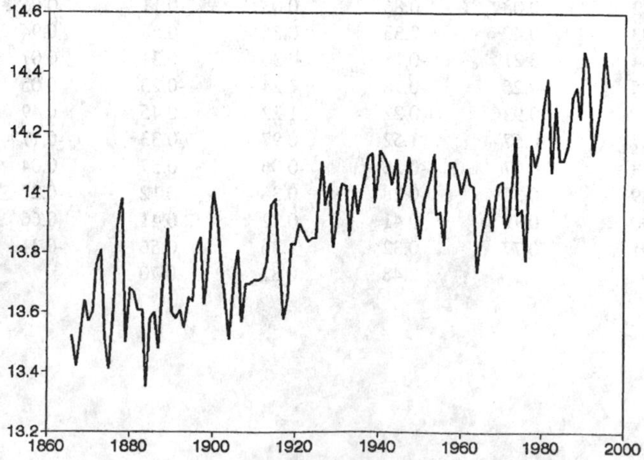

ATMOSPHERIC ELECTRICITY

Hans Dolezalek, Hannes Tammet, John Latham, and Martin A. Uman

I. SURVEY AND GLOBAL CIRCUIT

Hans Dolezalek

The science of atmospheric electricity originated in 1752 by an experimental proof of a related earlier hypothesis (that lightning is an electrical event). In spite of a large effort, in part by such eminent physicists as Coulomb, Lord Kelvin, and many others, an overall, proven theory able to generate models with sufficient resolution is not yet available. Generally accepted and encompassing text books are now more than 20 years old. The voluminous proceedings of the, so far, nine international atmospheric electricity conferences (1954 to 1992) give much valuable detail and demonstrate impressive progress, as do a number of less comprehensive textbooks published in the last 20 years, but a general theory as indicated above is not yet created. Only now, certain related measuring techniques and mathematical possibilities are emerging.

Applications to practical purposes do exist in the field of lightning research (including the electromagnetic radiation emanating from lightning) by the establishment of lightning-location networks and by the now developing possibility to detect electrified clouds which pose hazards to aircraft. Application of atmospheric electricity to other parts of meteorology seems to be promising but so far has seldom been instituted. Because some atmospheric electric signals propagate around the earth and because of the existence of a global circuit, applications for the monitoring of global change processes and conditions are now being proposed. Significant secular changes in the global circuit would indicate a change in the global climate; the availability of many old data (about a span of 100 years) could help detect a long term trend.

The concept of the "global circuit" is based on the theory of the global spherical capacitor: both, the solid (and liquid) earth as one electrode, and the high atmospheric layers (about the ionosphere) as the other, are by orders of magnitude more electrically conductive than the atmosphere between them. According to the "classical picture of atmospheric electricity", this capacitor is continuously charged by the common action of all thunderstorms to a d.c. voltage difference of several hundred kilovolts, the earth being negative. The much smaller but still existing conductivity of the atmosphere allows a current flowing from the ionosphere to the ground, integrated for all sink areas of the whole earth, of the order of 1.5 kA. In this way, a global circuit is created with many generators and sink-areas both interspaced and distributed over the whole globe, all connected to two nodes: ionosphere and ground. Within the scope of the global circuit, for each location, the current density (order of several pA/m^2) is determined by the voltage difference between ionosphere and ground (which is the same for all locations but varying in time) and the columnar resistance reaching from the ground up to the ionosphere (in the order of $10^{17}\ \Omega\ m^2$).

Natural processes, especially meteorological processes and some human activity, which produce or move electric charges ("space charges") or affect the ion distribution, constitute local generators and thereby "local circuits", horziontally and/or in parallel or antiparallel to the local part of the global circuit. In many cases, the local currents are much stronger than the global ones, making the measurement of the global current at a given location and/or during a period of time very difficult or, often, impossible. The strongest local circuits usually occur with certain weather conditions (precipitation, fog, high wind, blown-up dust or snow, heavy cloudiness) which make measurement of the global circuit impossible everywhere; but even in their absence local generators exist in varying magnitudes and of different characters. The separation of the local and global shares in the measured values of current density is a central problem of the science of atmospheric electricity. Aerological measurements are of high value in this regard.

The above description is within the "classical picture" of atmospheric electricity, a group of hypotheses to explain the electricifation of the atmosphere. It is probably fundamentally correct but certainly not complete; it has not yet been confirmed by systems of measurements resulting in no inner contradictions. In particular, extraterrestrial influences must be permitted; their general significance is still under debate.

Within this "classical picture" a kind of electric standard atmosphere may be constructed as shown in Table 1.

Values with a star, *, are rough average values from measurement. A star in parentheses, (*), points to a typical value from one or a few measurements. All other values have been calculated from starred values, under the assumption that at 2 km 50% and at 12 km 90% of the columnar resistance is reached. Voltage drop along one of the partial columns can be calculated by subtracting the value for the lower column from that of the upper one. Columnar resistances, conductances, and capacitances are valid for that particular part of the column which is indicated at the left. Capacitances are calculated with the formula for plate capacitors, and this fact must be considered also for the time constants for columns.

According to measurements, U, the potential difference between 0 m and 65 km may vary by a factor of approximately 2. The total columnar resistance, R_c, is estimated to vary up to a factor of 3, the variation being due to either reduction of conductivity in the exchange layer (about lowest 2 km of this table) or to the presence of high mountains; in both cases the variation is caused in the troposphere. Smaller variations in the stratosphere and mesosphere are being discussed because of aerosols there. The air-earth current density in fair weather varies by a factor of 3 to 6 accordingly. Conductivity near the ground varies by a factor of about 3 but only decreasing; increase of conductivity due to extraordinary radioactivity is a singular event. The field strength near the ground varies as a consequence of variations of air-earth current density and conductivity from about 1/3 to about 10 times of the value quoted in the table. Conductivity near the ground shows a diurnal and an annual variation which depends strongly on the locality: air-earth current density shows a diurnal and annual variation because the earth-ionosphere potential difference undergoes such variations, and also because the columnar resistance is supposed to have a diurnal and probably an annual variation.

Conductivities and air-earth current densities on high mountains are greater than at sea level by factors of up to 10. Conductivity decreases when atmospheric humidity increases. Values for space charges are not quoted because measurements are too few to allow calculation of average values. Values of parameters over the oceans are still rather uncertain.

Theoretically, in fair-weather conditions, Ohm's law must be fulfilled for the electric field, the conduction current density, and the electrical conductivity of the atmosphere. Deviations point to shortcomings in the applied measuring techniques. Data which are representative for a large area (in the extreme, "globally representative data", i.e. data on the global circuit), can on the ground be obtained only by stations on an open plane and only if local generators are either small or constant or are independently measured. Certain measurements with instrumented aircraft provide globally representative information valid for the period of the actual measurement.

TABLE 1
Electrical Parameters of the Clear (Fair-Weather) Atmosphere, Pertinent to the Classical Picture of Atm. Electricity (Electric Standard Atmosphere)

Part of atmosphere for which the values are calculated (elements are in free, cloudless atmosphere)	Currents, I, in A; and current densities, i, in A/m²	Potential differences, U, in V; field strength E in V/m; U = 0 at sea level	Resistances, R, in Ω; columnar resistances, R_c, in Ω m² and resistivities, ρ, in Ω m	Conductances, G, in Ω^{-1}; columnar conductances G_c, in Ω^{-1} m^{-2}; total conductivities, γ, in Ω^{-1} m^{-1}	Capacitances, C, in F; columnar capacitances, C_c, in F m^{-2} and capacitivities, ε, in F m^{-1}	Time constants τ, in seconds
Volume element at about sea level, 1 m³	$i = 3 \times 10^{-12}$*	$E_0 = 1.2 \times 10^{2}$*	$\rho_0 = 4 \times 10^{13}$	$\gamma_0 = 2.5 \times 10^{-14}$	$\varepsilon_0 = 8.9 \times 10^{-12}$*	$\tau_0 = 3.6 \times 10^{2}$
Lower column of 1 m² cross section from sea level to 2 km height	$i = 3 \times 10^{-12}$	At upper end: $U_1 = 1.8 \times 10^{5}$	$R_{c1} = 6 \times 10^{16}$	$G_{c1} = 1.7 \times 10^{-17}$	$C_{c1} = 4.4 \times 10^{-15}$	$\tau_{c1} = 2.6 \times 10^{2}$
Volume element at about 2 km height, 1 m³	$i = 3 \times 10^{-12}$	$E_2 = 6.6 \times 10^{1}$	$\rho_2 = 2.2 \times 10^{13(*)}$	$\gamma_2 = 4.5 \times 10^{-14}$	$\varepsilon_2 = 8.9 \times 10^{-12}$	$\tau_2 = = 2 \times 10^{2}$
Center column of 1 m² cross section from 2 to 12 km	$i = 3 \times 10^{-12}$	At upper end: $U_m = 3.15 \times 10^{5}$	$R_{cm} = 4.5 \times 10^{16}$	$G_{cm} = 5 \times 10^{-17}$	$C_{cm} = 8.8 \times 10^{-16}$	$\tau_{cm} = 1.8 \times 10^{1}$
Volume element at about 12 km height, 1 m³	$i = 3 \times 10^{-12}$	$E_{12} = 3.9 \times 10^{0}$	$\rho_{12} = 1.3 \times 10^{12(*)}$	$\gamma_{12} = 7.7 \times 10^{-13}$	$\varepsilon_{12} = 8.9 \times 10^{-12}$	$\tau_{12} = 1.2 \times 10^{1}$
Upper column of 1 m² cross section from 12 to 65 km height	$i = 3 \times 10^{-12}$	At upper end: $U_u = 3.5 \times 10^{5}$	$R_{cu} = 1.5 \times 10^{16}$	$G_{cu} = 2.5 \times 10^{-17}$	$C_{cu} = 1.67 \times 10^{-16}$	$\tau_{cm} = 6.7 \times 10^{0}$
Whole column of 1 m² cross section from 0 to 65 km height	$i = 3 \times 10^{-12}$	At upper end: $U = 3.5 \times 10^{5}$	$R_c = 1.2 \times 10^{17}$	$G_c = 8.3 \times 10^{-18}$	$C_c = 1.36 \times 10^{-16}$	$\tau_c = 1.64 \times 10^{1}$
Total spherical capacitor area: 5 × 10¹⁴ m²	$1 = 1.5 \times 10^{3}$	$U = 3.5 \times 10^{5}$*	$R = 2.4 \times 10^{2}$	$G = 4.2 \times 10^{-3}$	$C = 6.8 \times 10^{-2}$	$\tau = 1.64 \times 10^{1}$

Note: All currents and fields listed are part of the global circuit, i.e., circuits of local generators are not included. Values are subject to variations due to latitude and altitude of the point of observation above sea level, locality with respect to sources of disturbances, meteorological and climatological factors, and man-made changes. For more explanations, see text.

ATMOSPHERIC ELECTRICITY (continued)

II. AIR IONS

Hannes Tammet

The term " air ions" signifies all airborne particles which are the carriers of the electrical current in the air and have drift velocities determined by the electric field.

The probability of electrical dissociation of molecules in the atmospheric air under thermodynamic equilibrium is near to zero. The average ionization at the ground level over the ocean is $2 \cdot 10^6$ ion pairs $m^{-3}s^{-1}$. This ionization is produced mainly by cosmic rays. Over the continents the ionizing radiation from soil and from radioactive substances in the air each add about $4 \cdot 10^6$ $m^{-3}s^{-1}$. The total average ionization rate of 10^7 $m^{-3}s^{-1}$ is equivalent to 17 μR/h which is a customary expression of the background level of the ionizing radiations. The ionization rate over the ground varies in space due to the radioactivity of soil, and in time depending on the exchange of air between the atmosphere and radon-containing soil. Radioactive pollution increases the ionization rate. A temporary increase of about 10 times was registered in Sweden after the Chernobyl accident in 1986. The emission of Kr^{85} from nuclear power plants can noticeably increase the global ionization rate in the next century. The ionization rate decreases with altitude near the ground and increases at higher altitudes up to 15 km, where it has a maximum about $5 \cdot 10^7$ $m^{-3}s^{-1}$. Solar X-ray and extreme UV radiation cause a new increase at altitudes over 60 km.

Local sources of air ions are point discharges in strong electric fields, fluidization of charged drops from waves, etc.

The enhanced chemical activity of an ion results in a chain of ion-molecule reactions with the colliding neutrals, and, in the first microsecond of the life of an air ion, a charged molecular cluster called the *cluster ion* is formed. According to theoretical calculations in the air free from exotic trace gases the following cluster ions should be dominant:

$$NO_3^- \cdot (HNO_3) \cdot H_2O,\ NO_2^- \cdot (H_2O)_2,\ NO_3^- \cdot H_2O,\ O_2^- \cdot (H_2O)_4,\ O_2^- \cdot (H_2O)_5,$$
$$H_3O^+ \cdot (H_2O)_6,\ NH_4^+ \cdot (H_2O)_2,\ NH_4^+ \cdot (H_2O),\ H_3O^+ \cdot (H_2O)_5,\ NH_4^+ \cdot NH_3$$

A measurable parameter of air ions is the electrical mobility k, characterizing the drift velocity in the unit electric field. The mobility is inversely proportional to the density of air, and the results of measurements are as a rule reduced to normal conditions. According to mobility the air ions are called: fast or small or light ions with mobility $k > 5 \cdot 10^{-5}$ $m^2V^{-1}s^{-1}$, intermediate ions, and slow or large or heavy ions with mobility $k < 10^{-6}$ $m^2V^{-1}s^{-1}$. The boundary between intermediate and slow ions is conventional.

Cluster ions are fast ions. The masses of cluster ions may be measured with mass spectrometers, but the possible ion-molecule reactions during the passage of the air through nozzles to the vacuum chamber complicate the measurement. Mass and mobility of cluster ions are highly correlated. The experimental results[5] can be expressed by the empirical formula

$$m \approx \frac{850\ \mathrm{u}}{\left[0.3 + k/\left(10^{-4}\,\mathrm{m^2V^{-1}s^{-1}}\right)\right]^3}$$

where u is the unified atomic mass unit.

The value of the transport cross-section of a cluster ion is needed to calculate its mobility according to the kinetic theory of Chapman and Enskog. The theoretical estimation of transport cross-sections is rough and cannot be used to identify the chemical structures of cluster ions. Mass spectrometry is the main technique of identification of cluster ions.[2]

Märk and Castleman (1986) presented an overview of over 1000 publications on the experimental studies of cluster ions. Most of them present information about ions of millisecond age range. The low concentration makes it difficult to get detailed information about masses and mobilities of the natural atmospheric ions at ground level. The results of a 1-year continuous measurement[6] are as follows:

	+ ions	- ions	unit
Average mobility	1.36	1.56	10^{-4} $m^2V^{-1}s^{-1}$
The corresponding mass	190	130	u
The corresponding diameter	0.69	0.61	nm
The average concentration	400	360	10^6 m^{-3}
The corresponding conductivity	8.7	9.0	fS

The distribution of tropospheric cluster ions according to the mobility and estimated mass is depicted in Figure 1.

The problems and results of direct mass spectrometry of natural cluster ions are analysed by Eisele (1986) for ground level and by Meyerott, Reagan and Joiner (1980) for stratospheric measurements. Air ions in the high atmosphere are a subject of ionospheric physics.

During its lifetime (about 1 min), a cluster ion at ground level collides with nearly 10^{12} molecules. Thus the cluster ions are able to concentrate trace gases of very low concentration if they have an extra high electron or proton affinity. For example, Eisele (1986) demonstrated that a considerable fraction of positive atmospheric cluster ions in the unpolluted atmosphere at ground level probably consist of a molecule derived from pyridine. The concentration of these constituents is estimated to be about 10^{-12}. Therefore, air-ion mass and mobility spectrometry is considered as a promising technique for trace analysis in the air. Mass and mobility spectrometry of millisecond-age air ions has been developed as a technique of chemical analysis known as "plasma chromatography" (Carr, 1984). The sensitivity of the detection grows with the age of the cluster ions measured.

The mechanisms of annihilation of cluster ions are ion-ion recombination (on the average 3%) and sedimentation on aerosol particles (on the average 97% of cluster ions at ground level). The result of the combination of a cluster ion and neutral particle is a charged particle called an *aerosol ion*. In conditions of detailed thermodynamic equilibrium the probability that a spherical particle of diameter d carries q elementary charges is calculated from the Boltzmann distribution:

$$p_q(d) = (2\,\pi d/d_0)^{1/2} \exp(-q^2\,d_0/2d)$$

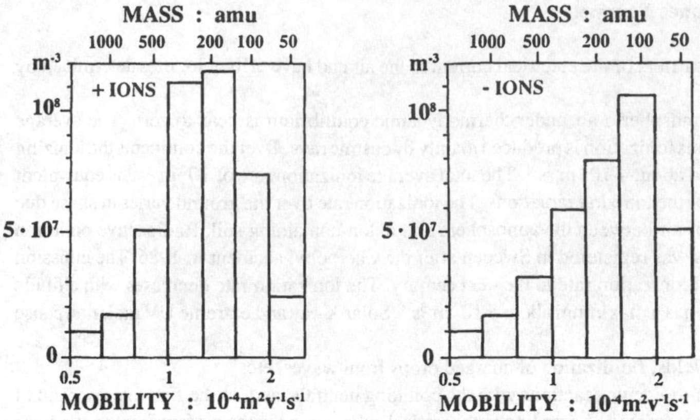

Figure 1. Average mobility and mass spectra of natural tropospheric cluster ions. Concentrations of the mobility fractions were measured in a rural site every 5 min over 1 year.[6] Ion mass is estimated according to the above empirical formula.

where d_0 = 115 nm (at 18°C). The supposition about the detailed equilibrium is an approximation and the formula is not valid for particles less than d_0. On the basis of numerical calculations by Hoppel and Frick (1990) the following charge probabilities can be derived:

d	3	10	30	100	300	1000	3000	nm
p_0	98	90	70	42	24	14	8	%
$p_{-1}+p_1$	2	10	30	48	41	25	15	%
$p_{-2}+p_2$	0	0	0	10	23	21	14	%
$p_{q>2}$	0	0	0	0	12	40	63	%
k_1	15000	1900	250	28	5.1	1.11	0.33	$10^{-9}m^2V^{-1}s^{-1}$

The last line of the table presents the mobility of a particle carrying one elementary charge. The distribution of the atmospheric aerosol ions over mobility is demonstrated in Figure 2.

Although the concentration of aerosol in continental air at ground level is an order of magnitude higher than the concentration of cluster ions, the mobilities of aerosol ions are so small that their percentage in air conductivity is less than 1%.

A specific class of aerosol ions are condensed aerosol ions produced as a result of the condensation of gaseous matter on the cluster ions. In aerosol physics the process is called ion-induced nucleation; it is considered as one among the processes of gas-to-particle conversion. The condensed aerosol ions have an inherent charge. Their sizes and mobilities are between the sizes and mobilities of cluster ions and of ordinary aerosol ions. Water and standard constituents of atmospheric air are not able to condense on the cluster ions in the real atmosphere. Thus the concentration of condensed aerosol ions depends on the trace constituents in the air and is very low in unpolluted air. Knowledge about condensed aerosol ions is poor because of measurement difficulties.

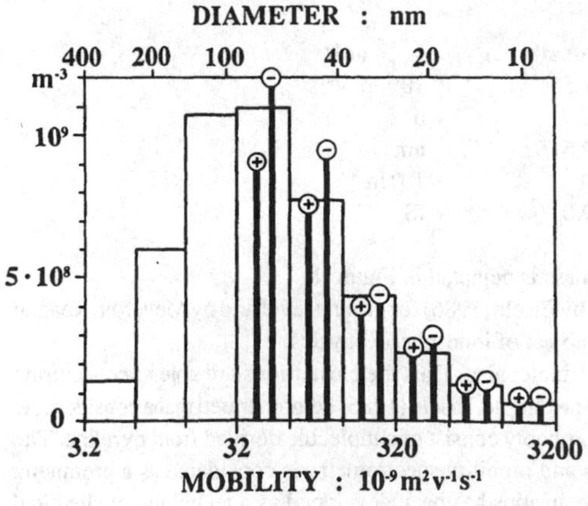

Figure 2. Mobility and size spectra of tropospheric aerosol ions.[6] The wide bars mark the fraction concentrations theoretically estimated on the basis of the standard size distribution of tropospheric aerosol. The pin bars with head + and - mark average values of positive and negative aerosol ion fraction concentrations measured in a rural site every 5 min during 4 months.

REFERENCES

1. Carr, T. W., Ed., *Plasma Chromatography*, Plenum Press, New York and London, XII + 259 pp., 1984.
2. Eisele, F. L., Identification of Tropospheric Ions, *J. Geophys. Res.*, vol. 91, no. D7, pp. 7897-7906, 1986.

3. Hoppel, W. A., and Frick, G. M., The Nonequilibrium Character of the Aerosol Charge Distributions Produced by Neutralizers, *Aerosol Sci. Technol.*, vol. 12, no. 3, pp. 471-496, 1990.
4. Mark, T. D., and Castleman, A. W., Experimental Studies on Cluster Ions, in *Advances in Atomic and Molecular Physics*, vol. 20, pp. 65.-172, Academic Press, 1985.
5. Meyerott, R. E., Reagan, J. B., and Joiner, R. G., The Mobility and Concentration of Ions and the Ionic Conductivity in the Lower Stratosphere, *J. Geophys. Res.*, vol. 85, no. A3, pp. 1273-1278, 1980.
6. Salm, J., Tammet, H., Iher, H., and Hörrak, U., Atmospheric Electrical Measurements in Tahkuse, Estonia (in Russian), in *Voprosy Atmosfernogo Elektrichestva*, pp. 168-175, Gidrometeoizdat, Leningrad, 1990.

III. THUNDERSTORM ELECTRICITY
John Latham

The development of improved radar techniques and instruments for in-cloud electrical and physical measurements, coupled with a much clearer recognition by the research community that establishment of the mechanism or mechanisms responsible for electric field development in thunderclouds, culminating in lightning, is inextricably linked to the concomitant dynamical and microphysical evolution of the clouds, has led to significant progress over the past decade.

Field studies indicate that in most thunderclouds the electrical development is associated with the process of glaciation, which can occur in a variety of incompletely understood ways. In the absence of ice, field growth is slow, individual hydrometeor charges are low, and lightning is produced only rarely. Precipitation — in the solid form, as graupel — also appears to be a necessary ingredient for significant electrification, as does significant convective activity and mixing between the clouds and their environments, via entrainment.

Increasingly, the view is being accepted that charge transfer leading to field-growth is largely a consequence of rebounding collisions between graupel pellets and smaller vapor-grown ice crystals, followed by the separation under gravity of these two types of hydrometeor. These collisions occur predominantly within the temperature range -15 to -30°C, and for significant charge transfer need to occur in the presence of supercooled cloud droplets.

The field evidence is inconsistent with an inductive mechanism, and extensive laboratory studies indicate that the principal charging mechanism is non-inductive and associated — in ways yet to be identified — with differences in surface characteristics of the interacting hydrometeors.

Laboratory studies indicate that the two most favored sites for corona emission leading to the lightning discharge are the tips of ephemeral liquid filaments, produced during the glancing collisions of supercooled raindrops, and protuberances on large ice crystals or graupel pellets. The relative importance of these alternatives will depend on the hydrometeor characteristics and the temperature in the regions of strongest fields; these features are themselves dependent on air-mass characteristics and climatological considerations.

A recently identified but unresolved question is why, in continental Northern Hemisphere thunderclouds at least, the sign of the charge brought to ground by lightning is predominantly negative in summer but more evenly balanced in winter.

IV. LIGHTNING
Martin A. Uman

From both ground-based weather-station data and satellite measurements, it has been estimated that there are about 100 lightning discharges, both cloud and ground flashes, over the whole earth each second; representing an average global lightning flash density of about 6 $km^{-2}yr^{-1}$. Most of this lightning occurs over the earth's land masses. For example, in central Florida, where thunderstorms occur about 90 days/yr, the flash density for discharges to earth is about 15 $km^{-2}yr^{-1}$. Some tropical areas of the earth have thunderstorms up to 300 days/yr.

Lightning can be defined as a transient, high-current electric discharge whose path length is measured in kilometers and whose most common source is the electric charge separated in the ordinary thunderstorm or cumulonimbus cloud. Well over half of all lightning discharges occur totally within individual thunderstorm clouds and are referred to as intracloud discharges. Cloud-to-ground lightning, however, has been studied more extensively than any other lightning form because of its visibility and its more practical interest. Cloud-to-cloud and cloud-to-air discharges are less common than intracloud or cloud-to-ground lightning.

Lightning between the cloud and earth can be categorized in terms of the direction of motion, upward or downward, and the sign of the charge, positive or negative, of the developing discharge (called a *leader*) which initiates the overall event. Over 90% of the worldwide cloud-to-ground discharges is initiated in the thundercloud by downward-moving negatively-charged leaders and subsequently results in the lowering of negative charge to earth. Cloud-to-ground lightning can also be initiated by downward-moving positive leaders, less than 10% of the worldwide cloud-to-ground lightning being of this type although the exact percentage is a function of season and latitude. Lightning between cloud and ground can also be initiated by leaders which develop upward from the earth. These upward-initiated discharges are relatively rare, may be of either polarity, and generally occur from mountaintops and tall man-made structures.

We discuss next the most common type of cloud-to-ground lightning. A negative cloud-to-ground discharge or *flash* has an overall duration of some tenths of a second and is made up of various components, among which are typically three or four high-current pulses called *strokes*. Each stroke lasts about a millisecond, the separation time between strokes being typically several tens of milliseconds. Such lightning often appears to "flicker" because the human eye can just resolve the individual light pulse associated with each stroke. A drawing of the components of a negative cloud-to-ground flash is found in Figure 3. Some values for salient parameters are found in Table 1. The negatively-charged *stepped leader* initiates the first stroke in a flash by propagating from cloud to ground through virgin air in a series of discrete steps. Photographically observed leader steps in clear

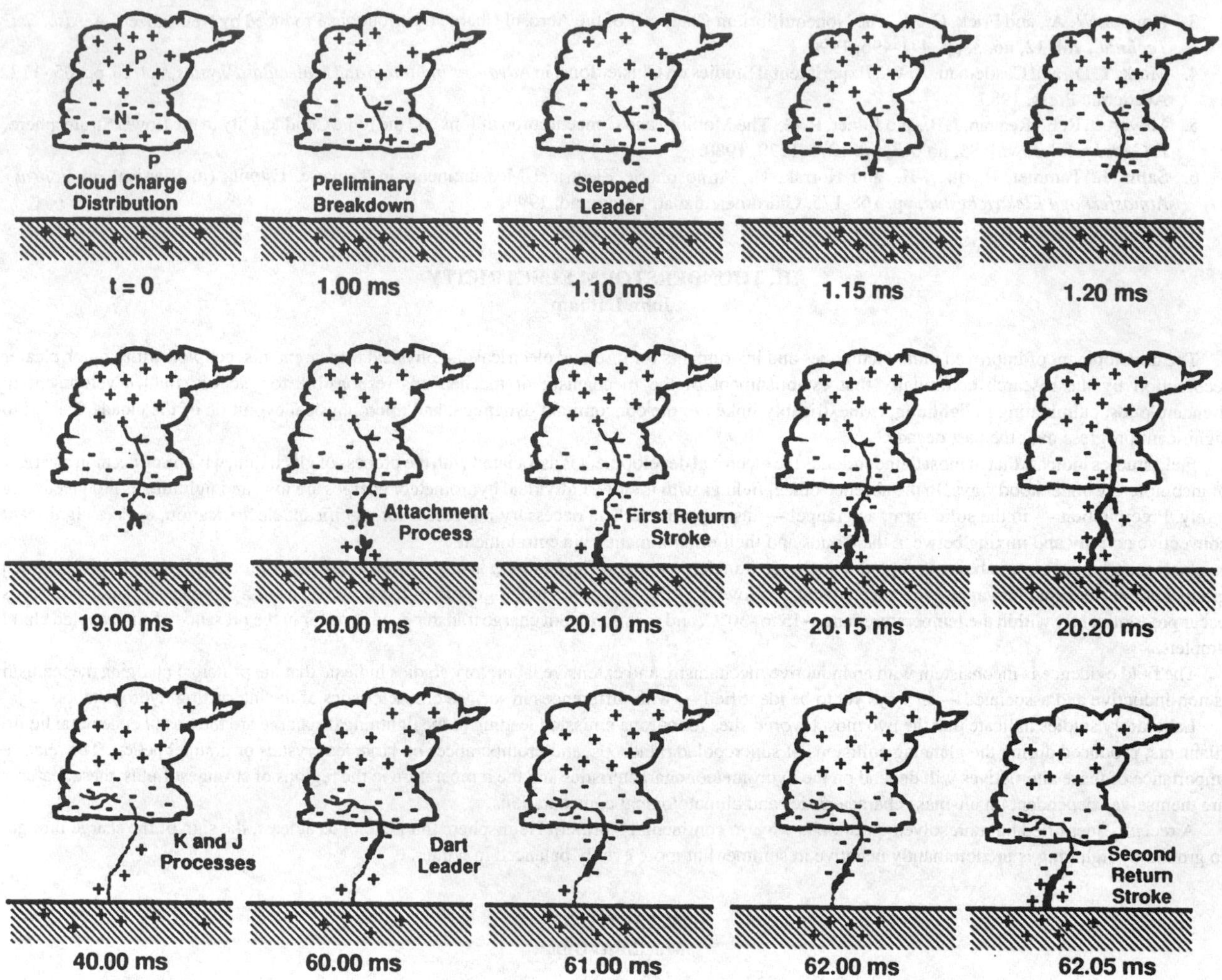

Figure 3. Sequence of steps in cloud-to-ground lightning.

air are typically 1 μs in duration and tens of meters in length, with a pause time between steps of about 50 μs. A fully developed stepped leader lowers up to 10 or more coulombs of negative cloud charge toward ground in tens of milliseconds with an average downward speed of about 2×10^5 m/s. The average leader current is in the 100 to 1000 A range. The steps have pulse currents of at least 1 kA. Associated with these currents are electric- and magnetic-field pulses with widths of about 1 μs or less and risetimes of about 0.1 μs or less. The stepped leader, during its trip toward ground, branches in a downward direction, resulting in the characteristic downward-branched geometrical structure commonly observed. The electric potential of the bottom of the negatively-charged leader channel with respect to ground has a magnitude in excess of 10^7 V. As the leader tip nears ground, the electric field at sharp objects on the ground or at irregularities of the ground itself exceeds the breakdown value of air, and one or more upward-moving discharges (often called upward leaders) are initiated from those points, thus beginning the *attachment process*. An understanding of the physics of the attachment process is central to an understanding of the operation of lightning protection of ground-based objects and the effects of lightning on humans and animals, since it is the attachment process that determines where the lightning connects to objects on the ground and the value of the early currents which flow. When one of the upward-moving discharges from the ground (or from a lightning rod or an individual) contacts the tip of the downward-moving stepped leader, typically some tens of meters above the ground, the leader tip is effectively connected to ground potential. The negatively-charged leader channel is then discharged to earth when a ground potential wave, referred to as the first *return stroke,* propagates continuously up the leader path. The upward speed of a return stroke near the ground is typically near one third the speed of light, and the speed decreases with height. The first return stroke produces a peak current near ground of typically 30 kA, with a time from zero to peak of a few microseconds. Currents measured at the ground fall to half of the peak value in about 50 μs, and currents of the order of hundreds of amperes may flow for times of a few milliseconds up to several hundred milliseconds. The longer-lasting currents are known as *continuing currents*. The rapid release of return stroke energy heats the leader channel to a temperature near 30,000 K and creates a high-pressure channel which expands and generates the shock waves that eventually become thunder, as further discussed later. The return stroke effectively lowers to ground the charge originally deposited onto the stepped-leader channel and additionally initiates the lowering of other charge which may be available to the top of its channel. First return-stroke electric fields exhibit a microsecond scale rise to peak with a typical peak value of 5 V/m, normalized to a distance of 100 km by an inverse distance relationship. Roughly half of the field rise to peak, the so-called "fast transition", takes place in tenths of a microsecond, an observation that can only be made if the field propagation is over a highly conducting surface such as salt water.

After the first return-stroke current has ceased to flow, the flash, including charge motion in the cloud, may end. The lightning is then called a single-stroke flash. On the other hand, if additional charge is made available to the top of the channel, a continuous or *dart leader* may propagate down the residual first-stroke channel at a typical speed of about 1×10^7 m/s. The dart leader lowers a charge of the order of 1 C by virtue of a current of about 1 kA. The dart leader then initiates the second (or any subsequent) return stroke. Subsequent return-stroke currents generally have faster zero-to-peak rise times than do first-stroke currents, but similar maximum rates of change, about 100 kA/μs. Some leaders begin as dart leaders, but toward the end of their trip toward ground become stepped leaders. These leaders are known as *dart-stepped leaders* and may have different ground termination points (and separate upward leaders) from the first stroke. Most often the dart-stepped leaders are associated with the second stroke of the flash. Nearly half of all flashes exhibit more than one termination point on ground with the distance between separate terminations being up to several kilometers. Subsequent return-stroke radiated electric and magnetic fields are similar to, but usually a factor of two or so smaller, than first return-stroke fields. About one third of all multiple-stroke flashes has at least one subsequent stroke which is larger than the first stroke.

Cloud-to-ground flashes that lower positive charge, though not common, are of considerable practical interest because their peak currents and total charge transfer can be much larger than for the more common negative ground flash. The largest recorded peak currents, those in the 200- to 300-kA range, are due to the return strokes of positive lightning. Such positive flashes to ground are initiated by downward-moving leaders which do not exhibit the distinct steps of their negative counterparts. Rather, they show a luminosity which is more or less continuous but modulated in intensity. Positive flashes are generally composed of a single stroke followed by a period of continuing current. Positive flashes are probably initiated from the upper positive charge in the thundercloud charge dipole when that cloud charge is horizontally separated from the negative charge beneath it, the source of the usual negative cloud-to-ground lightning. Positive flashes are relatively common in winter thunderstorms (snow storms), which produce few flashes overall, and are relatively uncommon in summer thunderstorms. The fraction of positive lightning in summer thunderstorms apparently increases with increasing latitude and with increasing height of the ground above sea level.

Distant lightning return stroke fields are often referred to as sferics (called "atmospherics" in the older literature). The peak in the sferics frequency spectrum is near 5 kHz due to the bipolar or ringing nature of the distant return-stroke electromagnetic signal and to the effects of propagation.

Thunder, the acoustic radiation associated with lightning, is sometimes divided into the categories "audible", sounds that one can hear, and "infrasonic", below a few tens of hertz, a frequency range that is inaudible. This division is made because it is thought that the mechanisms that produce audible and infrasonic thunder are different. Audible thunder is thought to be due to the expansion of a rapidly heated return stroke channel, as noted earlier, whereas infrasonic thunder is thought to be associated with the conversion to sound of the energy stored in the electrostatic field of the thundercloud when lightning rapidly reduces that cloud field.

The technology of artificially initiating lightning by firing upward small rocket trailing grounded wire of a few hundred meters length has been well-developed during the past decade. Such "triggered" flashes are similar to natural upward-initiated discharges from tall structure. They often contain subsequent strokes which, when they occur, are similar to the subsequent strokes in natural lightning. These triggered subsequent strokes have been the subject of considerable recent research.

Also in the past 10 years or so sophisticated lightning locating equipment has been installed throughout the world. For example, all ground flashes in the U.S. are now centrally monitored for research, for better overall weather prediction, and for hazard warning for aviation, electric utilities and other lightning-sensitive facilities.

Information on lightning physics can be found in M. A. Uman, *The Lightning Discharge*, Academic Press, San Diego, 1987; on lightning death and injury in *Medical Aspects of Lightning Injury*, editors C. Andrews, M. A. Cooper, M. Darveniza, and D. Mackerras, CRC Press, 1992. Ground flash location information for the U.S., in real time or archived, is available from Geomet Data Service of Tucson, AZ, which is also a source of the names of providers of those data in other countries.

Table 2 has data for cloud-to-ground lightning discharges bringing negative charge to earth. The values listed are intended to convey a rough feeling for the various physical parameters of lightning. No great accuracy is claimed since the results of different investigators are often not in good agreement. These values may, in fact, depend on the particular environment in which the lightning discharge is generated. The choice of some of the entries in the table is arbitrary.

TABLE 2
Data for Cloud-to-Ground Lightning Discharges

	Minimum[a]	Representative values	Maximum[a]
Stepped leader			
Length of step, m	3	50	200
Time interval between steps, μs	30	50	125
Average speed of propagation of stepped leader, m/s[b]	1.0×10^5	2.0×10^5	3.0×10^6
Charged deposited on stepped-leader channel, coulombs	3	5	20
Dart leader			
Speed of propagation, m/s[b]	1.0×10^6	1.0×10^7	2.4×10^7
Charged deposited on dart-leader channel, coulombs	0.2	1	6
Return stroke[c]			
Speed of propagation, m/s[b]	2.0×10^7	1.0×10^8	2.0×10^8
Maximum current rate of increase, kA/μs	<1	100	400
Time to peak current, μs	<1	2	30
Peak current, kA	2	30	200
Time to half of peak current, μs	10	50	250
Charge transferred excluding continuing current, coulombs	0.02	3	20
Channel length, km	2	5	15
Lightning flash			
Number of strokes per flash	1	4	26
Time interval between strokes in absence of continuing current, ms	3	60	100
Time duration of flash, s	10^{-2}	0.5	2
Charge transferred including continuing current, coulombs	3	30	200

[a] The words maximum and minimum are used in the sense that most measured values fall between these limits.
[b] Speeds of propagation are generally determined from photographic data and are "two-dimensional". Since many lightning flashes are not vertical, values stated are probably slight underestimates of actual values.
[c] First return strokes have longer times to current peak and generally larger charge transfer than do subsequent return strokes.

Adapted from Uman, M. A., *Lightning*, Dover Paperbook, New York, 1986, and Uman, M. A., *The Lightning Discharge*, Academic Press, San Diego, 1987.

SPEED OF SOUND IN VARIOUS MEDIA

The speed of sound in various solids, liquids, and gases is given in these tables. While only a single parameter v is needed for liquids and gases, sound propagation in isotropic solids is characterized by three velocity parameters. For a solid of infinite extent (or of finite extent if all dimensions are much larger than a wavelength, there are two relevant quantities,

v_l: velocity of longitudinal waves
v_s: velocity of shear waves.

For a cylindrical rod with diameter much smaller than a wavelength,

v_{ext}: velocity of extensional waves along the rod. (Torsional waves in the rod are propagated at the same speed as sheer waves in an infinite solid.)

Table 1 lists values for a variety of solid materials. Table 2 covers gases liquids and gases; values for cryogenic liquids are given at the normal boiling point. Table 3 gives the speed of sound in pure water and in seawater of salinity $S = 3.5\%$ as a function of temperature. All values are in meters per second and are given for normal atmospheric pressure.

REFERENCES

1. Gray, D.E., Ed., *American Institute of Physics Handbook, Third Edition*, McGraw Hill, New York, 1972.
2. Anderson, H.L., Ed., *A Physicist's Desk Reference*, American Institute of Physics, New York, 1989.
3. Younglove, B.A., *Thermophysical Proeprties of Fluids. Part I, J. Phys. Chem. Ref. Data,* 11, Suppl. 1, 1982.
4. Younglove, B.A., and Ely, J.F., *Thermophysical Properties of Fluids. Part II, J. Phys. Chem. Ref. Data*, 16, 577, 1987.
5. Haar, L., Gallagher, J.S., and Kell, G.S., *NBS/NRC Steam Tables,* Hemisphere Publishing Corp., New York, 1984.
6. Mason, W.P., *Physical Acoustics and the Properties of Solids*, D. Van Nostrand Co., Princeton, N.J., 1958.
7. *Landolt-Börnstein, Numerical Data and Functional Relationships in Science and Technology, New Series, II/5, Molecular Acoustics*, Springer-Verlag, Heidelberg, 1967.

TABLE 1
Speed of Sound in Solids at Room Temperature

Name	v_l/m s^{-1}	v_s/m s^{-1}	v_{ext}/m s^{-1}
Metals			
Aluminum, rolled	6420	3040	5000
Berylium	12890	8880	12870
Brass (70 Cu, 30 Zn)	4700	2110	3480
Constantan	5177	2625	4270
Copper, annealed	4760	2325	3810
Copper, rolled	5010	2270	3750
Duralumin 17S	6320	3130	5150
Gold, hard-drawn	3240	1200	2030
Iron, cast	4994	2809	4480
Iron, electrolytic	5950	3240	5120
Iron, Armco	5960	3240	5200
Lead, annealed	2160	700	1190
Lead, rolled	1960	690	1210
Magnesium, annealed	5770	3050	4940
Molybdenum	6250	3350	5400
Monel metal	5350	2720	4400
Nickel	6040	3000	4900
Platinum	3260	1730	2800
Silver	3650	1610	2680
Steel (1% C)	5940	3220	5180
Steel, 347 Stainless	5790	3100	5000
Steel, K9	5940	3250	5250
Tin, rolled	3320	1670	2730
Titanium	6070	3125	5090
Tungsten, annealed	5220	2890	4620
Tungsten, drawn	5410	2640	4320
Zinc, rolled	4210	2440	3850
Other materials			
Fused silica	5968	3764	5760
Glass, heavy silicate flint	3980	2380	3720
Glass, light borate crown	5100	2840	4540
Glass, pyrex	5640	3280	5170
Lucite	2680	1100	1840
Nylon 6-6	2620	1070	1800
Polyethylene	1950	540	920
Polystyrene	2350	1120	1840
Rubber, butyl	1830		
Rubber, gum	1550		
Rubber, neoprene	1600		
Tungsten carbide	6655	3980	6220

TABLE 2
Speed of Sound in Liquids and Gases

Name	t/°C	v/m s⁻¹
Liquids		
Acetone	20	1203
Argon	-185.9	813
Benzene	25	1310
Bromobenzene	20	1169
Butane	-0.5	1034
1-Butanol	20	1258
Carbon disulphide	25	1140
Chlorobenzene	20	1311
Cyclohexane	19	1280
1-Decene	20	1250
Diethyl ether	25	976
Ethane	-88.6	1326
Ethanol	20	1162
Ethylene	-103.8	1309
Ethylene glycol	25	1658
Fluorobenzene	20	1183
Glycerol	25	1904
Helium	-268.9	180
Heptane	20	1162
1-Heptene	20	1128
Hexane	20	1083
Hydrogen	-252.9	1101
Iodobenzene	20	1114
Mercury	25	1450
Methane	-161.5	1337
Methanol	20	1121
Nitrobenzene	25	1463
Nitrogen	-195.8	939
1-Nonene	20	1218
Octane	20	1197
1-Octene	20	1184
Oxygen	-183.0	906
1-Pentadecene	20	1351
Pentane	20	1008
Propane	-42.1	1158
1-Propanol	20	1223
Tetrachloromethane	25	930
Trichloromethane	25	987
1-Undecene	20	1275
Water	25	1497
Water (sea, $S = 3.5\%$)	25	1535
Gases at 1 atm		
Air, dry	25	346
Ammonia	0	415
Argon	27	323
Carbon monoxide	0	338
Carbon dioxide	0	259
Chlorine	0	206
Deuterium	0	890
Ethane	27	312
Ethylene	27	331
Helium	0	965
Hydrogen	27	1310
Hydrogen bromide	0	200
Hydrogen chloride	0	296
Hydrogen iodide	0	157
Hydrogen sulfide	0	289
Methane	27	450
Neon	0	435
Nitric oxide	10	325
Nitrogen	27	353
Nitrous oxide	0	263
Oxygen	27	330
Sulfur dioxide	0	213
Water (steam)	100	473

TABLE 3
Speed of Sound in Water and Seawater ($S = 3.5\%$) at Different Temperatures

	v/m s⁻¹	
t/°C	Water	Seawater
0	1401.0	1449.4
10	1447.8	1490.4
20	1483.2	1522.2
25	1497.4	1535.1
30	1509.5	1546.2
40	1528.4	
50	1541.4	
60	1549.5	
70	1553.2	
80	1552.8	

ATTENUATION AND SPEED OF SOUND IN AIR AS A FUNCTION OF HUMIDITY AND FREQUENCY

This table gives the attenuation and speed of sound as a function of frequency at various values of relative humidity. All values refer to still air at 20°C.

REFERENCES

1. Tables of Absorption and Velocity of Sound in Still Air at 68°F (20°C), AD-738576, National Technical Information Service, Springfield, VA.
2. Evans, L. B., Bass, H. E., and Sutherland, L. C., *J. Acoust. Soc. Am.,* 51, 1565, 1972.

Frequency (Hz)	Attenuation (dB/km)	Speed (m/s)
Relative humidity 0%		
20	0.51	343.477
40	1.07	343.514
50	1.26	343.525
63	1.43	343.536
100	1.67	343.550
200	1.84	343.559
400	1.96	343.561
630	2.11	343.562
800	2.27	343.562
1250	2.82	343.562
2000	4.14	343.562
4000	8.84	343.564
6300	14.89	343.565
10000	26.28	343.566
12500	35.81	343.566
16000	52.15	343.567
20000	75.37	343.567
40000	267.01	343.567
63000	644.66	343.567
80000	1032.14	343.567
Relative humidity 30%		
20	0.03	343.807
40	0.11	343.808
50	0.17	343.810
63	0.25	343.810
100	0.50	343.814
200	1.01	343.821
400	1.59	343.826
630	2.24	343.827
800	2.85	343.828
1250	5.09	343.828
2000	10.93	343.829
4000	38.89	343.831
6300	90.61	343.836
10000	204.98	343.846
12500	294.08	343.854
16000	422.51	343.865
20000	563.66	343.877
40000	1110.97	343.911
63000	1639.47	343.924
80000	2083.08	343.929

Frequency (Hz)	Attenuation (dB/km)	Speed (m/s)
Relative humidity 60%		
20	0.02	344.182
40	0.06	344.183
50	0.09	344.183
63	0.15	344.184
100	0.34	344.185
200	0.99	344.190
400	1.94	344.197
630	2.57	344.200
800	2.94	344.201
1250	4.01	344.202
2000	6.55	344.203
4000	18.73	344.204
6300	42.51	344.204
10000	101.84	344.206
12500	155.67	344.208
16000	247.78	344.211
20000	373.78	344.215
40000	1195.37	344.238
63000	2220.64	344.262
80000	2951.71	344.274
Relative humidity 100%		
20	0.01	344.685
40	0.04	344.685
50	0.06	344.685
63	0.09	344.685
100	0.22	344.686
200	0.77	344.689
400	2.02	344.695
630	3.05	344.699
800	3.57	344.701
1250	4.59	344.704
2000	6.29	344.705
4000	13.58	344.706
6300	27.72	344.706
10000	63.49	344.706
12500	96.63	344.707
16000	154.90	344.708
20000	237.93	344.709
40000	884.28	344.718
63000	1973.62	344.731
80000	2913.01	344.742

SPEED OF SOUND IN DRY AIR

The values in this table were calculated from the equation of state for dry air (average molecular weight 28.96) treated as a real gas. Values refer to standard atmospheric pressure. The speed of sound varies only slightly with pressure; at two atmospheres and -100°C the value decreases by 0.13%, while at two atmospheres and 80°C the speed increases by 0.04%.

REFERENCE

Sytchev, V.V., Vasserman, A.A., Kozlov, A.D., Spiridonov, G.A., and Tsymarny, V.A., *Thermodynamic Properties of Air*, Hemisphere Publishing Corp., New York, 1987.

t/°C	v_s/m s^{-1}	t/°C	v_s/m s^{-1}	t/°C	v_s/m s^{-1}
-100	263.5	-35	309.5	30	349.1
-95	267.3	-30	312.7	35	352.0
-90	271.1	-25	315.9	40	354.8
-85	274.8	-20	319.1	45	357.6
-80	278.5	-15	322.3	50	360.4
-75	282.1	-10	325.4	55	363.2
-70	285.7	-5	328.4	60	365.9
-65	289.2	0	331.5	65	368.6
-60	292.7	5	334.5	70	371.3
-55	296.1	10	337.5	75	374.0
-50	299.5	15	340.4	80	376.7
-45	302.9	20	343.4		
-40	306.2	25	346.3		

MUSICAL SCALES

EQUAL TEMPERED CHROMATIC SCALE
A_4 = 440 Hz

American Standard pitch. Adopted by the American Standards Association in 1936

Note	Frequency	Note	Frequency	Note	Frequency	Note	Frequency
C_0	16.35	C_2	65.41	C_4	261.63	C_6	1046.50
$C\#_0$	17.32	$C\#_2$	69.30	$C\#_4$	277.18	$C\#_6$	1108.73
D_0	18.35	D_2	73.42	D_4	293.66	D_6	1174.66
$D\#_0$	19.45	$D\#_2$	77.78	$D\#_4$	311.13	$D\#_6$	1244.51
E_0	20.60	E_2	82.41	E_4	329.63	E_6	1318.51
F_0	21.83	F_2	87.31	F_4	349.23	F_6	1396.91
$F\#_0$	23.12	$F\#_2$	92.50	$F\#_4$	369.99	$F\#_6$	1479.98
G_0	24.50	G_2	98.00	G_4	392.00	G_6	1567.98
$G\#_0$	25.96	$G\#_2$	103.83	$G\#_4$	415.30	$G\#_6$	1661.22
A_0	27.50	A_2	110.00	A_4	440.00	A_6	1760.00
$A\#_0$	29.14	$A\#_2$	116.54	$A\#_4$	466.16	$A\#_6$	1864.66
B_0	30.87	B_2	123.47	B_4	493.88	B_6	1975.53
C_1	32.70	C_3	130.81	C_5	523.25	C_7	2093.00
$C\#_1$	34.65	$C\#_3$	138.59	$C\#_5$	554.37	$C\#_7$	2217.46
D_1	36.71	D_3	146.83	D_5	587.33	D_7	2349.32
$D\#_1$	38.89	$D\#_3$	155.56	$D\#_5$	622.25	$D\#_7$	2489.02
E_1	41.20	E_3	164.81	E_5	659.26	E_7	2637.02
F_1	43.65	F_3	174.61	F_5	698.46	F_7	2793.83
$F\#_1$	46.25	$F\#_3$	185.00	$F\#_5$	739.99	$F\#_7$	2959.96
G_1	49.00	G_3	196.00	G_5	783.99	G_7	3135.96
$G\#_1$	51.91	$G\#_3$	207.65	$G\#_5$	830.61	$G\#_7$	3322.44
A_1	55.00	A_3	220.00	A_5	880.00	A_7	3520.00
$A\#_1$	58.27	$A\#_3$	233.08	$A\#_5$	932.33	$A\#_7$	3729.31
B_1	61.74	B_3	246.94	B_5	987.77	B_7	3951.07
						C_8	4186.01

EQUAL TEMPERED CHROMATIC SCALE
A_4 = 435 Hz

International Pitch, adopted 1891

Note	Frequency	Note	Frequency	Note	Frequency	Note	Frequency
C_0	16.17	C_2	64.66	C_4	258.65	C_6	1034.61
$C\#_0$	17.13	$C\#_2$	68.51	$C\#_4$	274.03	$C\#_6$	1096.13
D_0	18.15	D_2	72.58	D_4	290.33	D_6	1161.31
$D\#_0$	19.22	$D\#_2$	76.90	$D\#_4$	307.59	$D\#_6$	1230.37
E_0	20.37	E_2	81.47	E_4	325.88	E_6	1303.53
F_0	21.58	F_2	86.31	F_4	345.26	F_6	1381.04
$F\#_0$	22.86	$F\#_2$	91.45	$F\#_4$	365.79	$F\#_6$	1463.16
G_0	24.22	G_2	96.89	G_4	387.54	G_6	1550.16
$G\#_0$	25.66	$G\#_2$	102.65	$G\#_4$	410.59	$G\#_6$	1642.34
A_0	27.19	A_2	108.75	A_4	435.00	A_6	1740.00
$A\#_0$	28.80	$A\#_7$	115.22	$A\#_4$	460.87	$A\#_6$	1843.47
B_0	30.52	B_2	122.07	B_4	488.27	B_6	1953.08
C_1	32.33	C_3	129.33	C_5	517.31	C_7	2069.22
$C\#_1$	34.25	$C\#_3$	137.02	$C\#_5$	548.07	$C\#_7$	2192.26
D_1	36.29	D_3	145.16	D_5	580.66	D_7	2322.62
$D\#_1$	38.45	$D\#_3$	153.80	$D\#_5$	615.18	$D\#_7$	2460.73
E_1	40.74	E_3	162.94	E_5	651.76	E_7	2607.05
F_1	43.16	F_3	172.63	F_5	690.52	F_7	2762.08
$F\#_1$	45.72	$F\#_3$	182.89	$F\#_5$	731.58	$F\#_7$	2926.32
G_1	48.44	G_3	193.77	G_5	775.08	G_7	3100.33

EQUAL TEMPERED CHROMATIC SCALE
A_4 = 435 Hz

International Pitch, adopted 1891

Note	Frequency	Note	Frequency	Note	Frequency	Note	Frequency
$G\#_1$	51.32	$G\#_3$	205.29	$G\#_5$	821.17	$G\#_7$	3284.68
A_1	54.38	A_3	217.50	A_5	870.00	A_7	3480.00
$A\#_1$	57.61	$A\#_3$	230.43	$A\#_5$	921.73	$A\#_7$	3686.93
B_1	61.03	B_3	244.14	B_5	976.54	B_7	3906.17
						C_8	4138.44

SCIENTIFIC OR JUST SCALE
C_4 = 256 Hz

Note	Frequency	Note	Frequency	Note	Frequency	Note	Frequency
C_0	16	C_2	64	C_4	256	C_6	1024
D_0	18	D_2	72	D_4	288	D_6	1152
E_0	20	E_2	80	E_4	320	E_6	1280
F_0	21.33	F_2	85.33	F_4	341.33	F_6	1365.33
G_0	24	G_2	96	G_4	384	G_6	1536
A_0	26.67	A_2	106.67	A_4	426.67	A_6	1706.67
B_0	30	B_2	120	B_4	480	B_6	1920
C_1	32	C_3	128	C_5	512	C_7	2048
D_1	36	D_3	144	D_5	576	D_7	2304
E_1	40	E_3	160	E_5	640	E_7	2560
F_1	42.67	F_3	170.67	F_5	682.67	F_7	2730.67
G_1	48	G_3	192	G_5	768	G_7	3072
A_1	53.33	A_3	213.33	A_5	853.33	A_7	3413.33
B_1	60	B_3	240	B_5	960	B_7	3840
						C_8	4096

CHARACTERISTICS OF HUMAN HEARING

The human ear is sensitive to sound waves with frequencies in the range from a few hertz to almost 20 kHz. Auditory response is usually expressed in terms of the *loudness level* of a sound, which is a measure of the sound pressure. The reference level, which is given in the unit *phon,* is a pure tone of frequency 1000 Hz with sound pressure of 20 mPa (in cgs units, $2 \cdot 10^{-4}$ dyn/cm^2); loudness level is usually expressed in decibels (dB) relative to this reference level. If a normal observer perceives an arbitrary sound to be equally loud as this reference sound, the sound is said to have the loudness level of the reference. The sensitivity of the typical human ear ranges from about 0 dB, the threshold loudness level, to about 140 dB, the level at which pain sets in. The minimum detectable level thus represents a sound wave of pressure 20 mPa and intensity (power density) 10^{-16} W/cm^2.

The following figure illustrates the frequency dependence of the threshold for an average young adult.

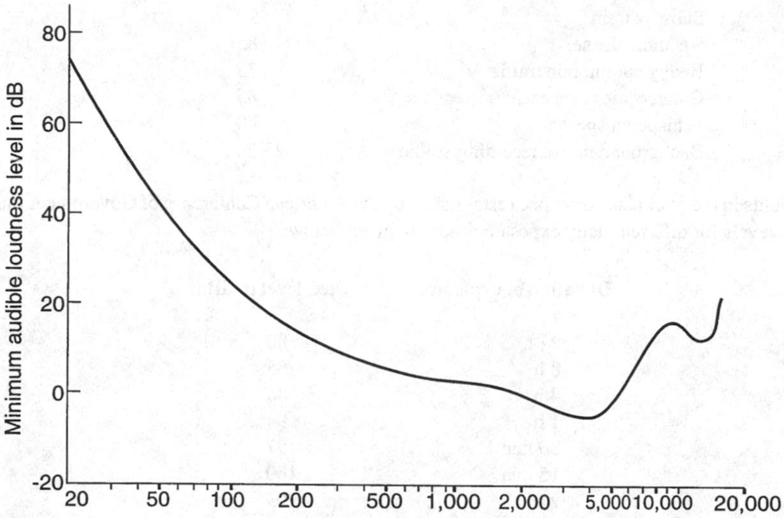

The relation between loudness level and frequency for a typical person is expressed by the following table:

Sound pressure level in dB relative to 20 mPa	Frequency in Hz 125	500	1000	4000	8000	10000
10			10	18		
20		16	20	28	11	
30	4	27	30	37	21	17
40	17	39	40	45	30	26
50	34	52	50	54	38	35
60	52	65	60	64	47	44
70	70	76	70	73	56	54
80	86	86	80	83	66	64
90	98	96	90	94	77	74
100	108	105	100	106	88	86

Thus, a 10,000 Hz tone at a pressure level of 50 dB seems equally loud as a 1000 Hz tone at a pressure of 35 dB.

The term *noise* refers to any unwanted sound, either a pure tone or a mixture of frequencies. Since the sensitivity of the ear is frequency dependent, as illustrated by the above table, noise level is expressed in a frequency-weighted scale, known as A-weighting. Decibel readings on this scale are designated as dBa. Typical noise levels from various sources are illustrated in this table:

Source	Noise level in dBa
Rocket engine	200
Jet aircraft engine	160
Light aircraft, cruising	140
Tractor, 150 hp	115
Electric motor, 100 hp at 2600 rpm	105
Pneumatic drill	100
Subway train	90
Vacuum cleaner	85
Heavy automobile traffic	75
Conversational speech	65
Whispered speech	40
Background noise, recording studio	25-30

Recommended noise thresholds in the workplace have been established by the American Conference of Government Industrial Hygenists. Some examples of the maximum safe levels for different daily exposure times are given below.

Duration of exposure	Max. level in dBa
24 h	80
8 h	85
4 h	88
1 h	94
30 min	97
15 min	100
2 min	109
28 s	115
0.11 s	139

No exposure greater than 140 dBa is permitted. Further details may be found in Reference 3.

REFERENCES

1. Anderson, H. L., Editor, *A Physicist's Desk Reference*, American Institute of Physics, New York, 1989, chap. 2.
2. Gray, D. E., Ed., *American Institute of Physics Handbook, Third Edition*, McGraw Hill, New York, 1972, chap. 3.
3. *Threshold Limit Values for Chemical Substances and Physical Agents; Biological Exposure Indices,* 1999 Edition, American Conference of Governmental Industrial Hygienists, 1330 Kemper Meadow Drive, Cincinnati, OH 45240-1634.

Section 15
Practical Laboratory Data

Standard ITS-90 Thermocouple Tables 15-1
Properties of Common Laboratory Solvents 15-14
Dependence of Boiling Point on Pressure 15-19
Ebullioscopic Constants for Calculation of Boiling Point Elevation 15-20
Cryoscopic Constants for Calculation of Freezing Point Depression 15-21
Freezing Point Lowering by Electrolytes in Aqueous Solution 15-22
Determination of Relative Humidity from Dew Point 15-23
Determination of Relative Humidity from Wet and Dry Bulb Temperatures 15-24
Constant Humidity Solutions 15-25
Standard Salt Solutions for Humidity Calibration 15-26
Low Temperature Baths for Maintaining Constant Temperature 15-26
Wire Tables 15-27
Characteristics of Particles and Particle Dispersoids 15-28
Density of Various Solids 15-29
Dielectric Strength of Insulating Materials 15-30
Allocation of Frequencies in the Radio Spectrum 15-36

STANDARD ITS-90 THERMOCOUPLE TABLES

The Instrument Society of America (ISA) has assigned standard letter designations to a number of thermocouple types having specified emf-temperature relations. These designations and the approximate metal compositions which meet the required relations, as well as the useful temperature ranges, are given below:

Type B	(Pt + 30% Rh) vs. (Pt + 6% Rh)	0 to 1820°C
Type E	(Ni + 10% Cr) vs. (Cu + 43% Ni)	-270 to 1000°C
Type J	Fe vs. (Cu + 43%Ni)	-210 to 1200°C
Type K	(Ni + 10% Cr) vs. (Ni + 2% Al + 2% Mn + 1% Si)	-270 to 1372°C
Type N	(Ni + 14% Cr + 1.5% Si) vs. (Ni + 4.5% Si + 0.1% Mg)	-270 to 1300°C
Type R	(Pt + 13% Rh) vs. Pt	-50 to 1768°C
Type S	(Pt + 10% Rh) vs. Pt	-50 to 1768°C
Type T	Cu vs. (Cu + 43% Ni)	-270 to 400°C

The compositions are given in weight percent, and the positive leg is listed first. It should be emphasized that the standard letter designations do not imply a precise composition but rather that the specified emf-temperature relation is satisfied.

The first set of tables below lists, for each thermocouple type, the emf as a function of temperature on the International Temperature Scale of 1990 (ITS-90). The coefficients in the equation used to generate the table are also given. The second set of tables gives the inverse relationships, i.e., the coefficients in the polynomial equation which expresses the temperature as a function of thermocouple emf. The accuracy of these equations is also stated.

Further details and tables at closer intervals may be found in Reference 1.

REFERENCES

1. Burns, G. W., Seroger, M. G., Strouse, G. F., Croarkin, M. C., and Guthrie, W. F., *Temperature-Electromotive Force Reference Functions and Tables for the Letter-Designated Thermocouple Types Based on the ITS-90*, Natl. Inst. Stand. Tech. (U.S.) Monogr. 175, 1993.
2. Schooley, J. F., *Thermometry*, CRC Press, Boca Raton, FL, 1986.

Type B thermocouples: emf-temperature (°C) reference table and equations.

Thermocouple emf as a Function of Temperature in Degrees Celsius (ITS-90)

emf in Millivolts — **Reference Junctions at 0 °C**

°C	0	10	20	30	40	50	60	70	80	90	100
0	0.000	−0.002	−0.003	−0.002	−0.000	0.002	0.006	0.011	0.017	0.025	0.033
100	0.033	0.043	0.053	0.065	0.078	0.092	0.107	0.123	0.141	0.159	0.178
200	0.178	0.199	0.220	0.243	0.267	0.291	0.317	0.344	0.372	0.401	0.431
300	0.431	0.462	0.494	0.527	0.561	0.596	0.632	0.669	0.707	0.746	0.787
400	0.787	0.828	0.870	0.913	0.957	1.002	1.048	1.095	1.143	1.192	1.242
500	1.242	1.293	1.344	1.397	1.451	1.505	1.561	1.617	1.675	1.733	1.792
600	1.792	1.852	1.913	1.975	2.037	2.101	2.165	2.230	2.296	2.363	2.431
700	2.431	2.499	2.569	2.639	2.710	2.782	2.854	2.928	3.002	3.078	3.154
800	3.154	3.230	3.308	3.386	3.466	3.546	3.626	3.708	3.790	3.873	3.957
900	3.957	4.041	4.127	4.213	4.299	4.387	4.475	4.564	4.653	4.743	4.834
1000	4.834	4.926	5.018	5.111	5.205	5.299	5.394	5.489	5.585	5.682	5.780
1100	5.780	5.878	5.976	6.075	6.175	6.276	6.377	6.478	6.580	6.683	6.786
1200	6.786	6.890	6.995	7.100	7.205	7.311	7.417	7.524	7.632	7.740	7.848
1300	7.848	7.957	8.066	8.176	8.286	8.397	8.508	8.620	8.731	8.844	8.956
1400	8.956	9.069	9.182	9.296	9.410	9.524	9.639	9.753	9.868	9.984	10.099
1500	10.099	10.215	10.331	10.447	10.563	10.679	10.796	10.913	11.029	11.146	11.263
1600	11.263	11.380	11.497	11.614	11.731	11.848	11.965	12.082	12.199	12.316	12.433
1700	12.433	12.549	12.666	12.782	12.898	13.014	13.130	13.246	13.361	13.476	13.591
1800	13.591	13.706	13.820								

Temperature Ranges and Coefficients of Equations Used to Compute the Above Table

The equations are of the form: $E = c_0 + c_1 t + c_2 t^2 + c_3 t^3 + \ldots c_n t^n$, where E is the emf in millivolts, t is the temperature in degrees Celsius (ITS-90), and c_0, c_1, c_2, c_3, *etc.* are the coefficients. These coefficients are extracted from NIST Monograph 175.

		0 °C to 630.615 °C	630.615 °C to 1820 °C
c_0	=	0.000 000 000 0 . . .	−3.893 816 862 1 . . .
c_1	=	$-2.465\,081\,834\,6 \times 10^{-4}$	$2.857\,174\,747\,0 \times 10^{-2}$
c_2	=	$5.904\,042\,117\,1 \times 10^{-6}$	$-8.488\,510\,478\,5 \times 10^{-5}$
c_3	=	$-1.325\,793\,163\,6 \times 10^{-9}$	$1.578\,528\,016\,4 \times 10^{-7}$
c_4	=	$1.566\,829\,190\,1 \times 10^{-12}$	$-1.683\,534\,486\,4 \times 10^{-10}$
c_5	=	$-1.694\,452\,924\,0 \times 10^{-15}$	$1.110\,979\,401\,3 \times 10^{-13}$
c_6	=	$6.299\,034\,709\,4 \times 10^{-19}$	$-4.451\,543\,103\,3 \times 10^{-17}$
c_7	=	. . . .	$9.897\,564\,082\,1 \times 10^{-21}$
c_8	=	. . . .	$-9.379\,133\,028\,9 \times 10^{-25}$

Type E thermocouples: emf-temperature (°C) reference table and equations.

Thermocouple emf as a Function of Temperature in Degrees Celsius (ITS-90)

emf in Millivolts — Reference Junctions at 0 °C

°C	0	−10	−20	−30	−40	−50	−60	−70	−80	−90	−100
−200	−8.825	−9.063	−9.274	−9.455	−9.604	−9.718	−9.797	−9.835			
−100	−5.237	−5.681	−6.107	−6.516	−6.907	−7.279	−7.632	−7.963	−8.273	−8.561	−8.825
0	0.000	−0.582	−1.152	−1.709	−2.255	−2.787	−3.306	−3.811	−4.302	−4.777	−5.237

°C	0	10	20	30	40	50	60	70	80	90	100
0	0.000	0.591	1.192	1.801	2.420	3.048	3.685	4.330	4.985	5.648	6.319
100	6.319	6.998	7.685	8.379	9.081	9.789	10.503	11.224	11.951	12.684	13.421
200	13.421	14.164	14.912	15.664	16.420	17.181	17.945	18.713	19.484	20.259	21.036
300	21.036	21.817	22.600	23.386	24.174	24.964	25.757	26.552	27.348	28.146	28.946
400	28.946	29.747	30.550	31.354	32.159	32.965	33.772	34.579	35.387	36.196	37.005
500	37.005	37.815	38.624	39.434	40.243	41.053	41.862	42.671	43.479	44.286	45.093
600	45.093	45.900	46.705	47.509	48.313	49.116	49.917	50.718	51.517	52.315	53.112
700	53.112	53.908	54.703	55.497	56.289	57.080	57.870	58.659	59.446	60.232	61.017
800	61.017	61.801	62.583	63.364	64.144	64.922	65.698	66.473	67.246	68.017	68.787
900	68.787	69.554	70.319	71.082	71.844	72.603	73.360	74.115	74.869	75.621	76.373
1000	76.373										

Temperature Ranges and Coefficients of Equations Used to Compute the Above Table

The equations are of the form: $E = c_0 + c_1t + c_2t^2 + c_3t^3 + \ldots c_nt^n$, where E is the emf in millivolts, t is the temperature in degrees Celsius (ITS-90), and c_0, c_1, c_2, c_3, *etc.* are the coefficients. These coefficients are extracted from NIST Monograph 175.

		−270 °C to 0 °C	0 °C to 1000 °C
c_0	=	0.000 000 000 0 . . .	0.000 000 000 0 . . .
c_1	=	$5.866\,550\,870\,8 \times 10^{-2}$	$5.866\,550\,871\,0 \times 10^{-2}$
c_2	=	$4.541\,097\,712\,4 \times 10^{-5}$	$4.503\,227\,558\,2 \times 10^{-5}$
c_3	=	$-7.799\,804\,868\,6 \times 10^{-7}$	$2.890\,840\,721\,2 \times 10^{-8}$
c_4	=	$-2.580\,016\,084\,3 \times 10^{-8}$	$-3.305\,689\,665\,2 \times 10^{-10}$
c_5	=	$-5.945\,258\,305\,7 \times 10^{-10}$	$6.502\,440\,327\,0 \times 10^{-13}$
c_6	=	$-9.321\,405\,866\,7 \times 10^{-12}$	$-1.919\,749\,550\,4 \times 10^{-16}$
c_7	=	$-1.028\,760\,553\,4 \times 10^{-13}$	$-1.253\,660\,049\,7 \times 10^{-18}$
c_8	=	$-8.037\,012\,362\,1 \times 10^{-16}$	$2.148\,921\,756\,9 \times 10^{-21}$
c_9	=	$-4.397\,949\,739\,1 \times 10^{-18}$	$-1.438\,804\,178\,2 \times 10^{-24}$
c_{10}	=	$-1.641\,477\,635\,5 \times 10^{-20}$	$3.596\,089\,948\,1 \times 10^{-28}$
c_{11}	=	$-3.967\,361\,951\,6 \times 10^{-23}$	. . . .
c_{12}	=	$-5.582\,732\,872\,1 \times 10^{-26}$	. . . .
c_{13}	=	$-3.465\,784\,201\,3 \times 10^{-29}$	. . . .

Type J thermocouples: emf-temperature (°C) reference table and equations.

Thermocouple emf as a Function of Temperature in Degrees Celsius (ITS-90)

emf in Millivolts — **Reference Junctions at 0 °C**

°C	0	−10	−20	−30	−40	−50	−60	−70	−80	−90	−100
−200	−7.890	−8.095									
−100	−4.633	−5.037	−5.426	−5.801	−6.159	−6.500	−6.821	−7.123	−7.403	−7.659	−7.890
0	0.000	−0.501	−0.995	−1.482	−1.961	−2.431	−2.893	−3.344	−3.786	−4.215	−4.633

°C	0	10	20	30	40	50	60	70	80	90	100
0	0.000	0.507	1.019	1.537	2.059	2.585	3.116	3.650	4.187	4.726	5.269
100	5.269	5.814	6.360	6.909	7.459	8.010	8.562	9.115	9.669	10.224	10.779
200	10.779	11.334	11.889	12.445	13.000	13.555	14.110	14.665	15.219	15.773	16.327
300	16.327	16.881	17.434	17.986	18.538	19.090	19.642	20.194	20.745	21.297	21.848
400	21.848	22.400	22.952	23.504	24.057	24.610	25.164	25.720	26.276	26.834	27.393
500	27.393	27.953	28.516	29.080	29.647	30.216	30.788	31.362	31.939	32.519	33.102
600	33.102	33.689	34.279	34.873	35.470	36.071	36.675	37.284	37.896	38.512	39.132
700	39.132	39.755	40.382	41.012	41.645	42.281	42.919	43.559	44.203	44.848	45.494
800	45.494	46.141	46.786	47.431	48.074	48.715	49.353	49.989	50.622	51.251	51.877
900	51.877	52.500	53.119	53.735	54.347	54.956	55.561	56.164	56.763	57.360	57.953
1000	57.953	58.545	59.134	59.721	60.307	60.890	61.473	62.054	62.634	63.214	63.792
1100	63.792	64.370	64.948	65.525	66.102	66.679	67.255	67.831	68.406	68.980	69.553
1200	69.553										

Temperature Ranges and Coefficients of Equations Used to Compute the Above Table

The equations are of the form: $E = c_0 + c_1t + c_2t^2 + c_3t^3 + \ldots c_nt^n$, where E is the emf in millivolts, t is the temperature in degrees Celsius (ITS-90), and c_0, c_1, c_2, c_3, *etc.* are the coefficients. These coefficients are extracted from NIST Monograph 175.

		−210 °C to 760 °C	760 °C to 1200 °C
c_0	=	0.000 000 000 0 . . .	2.964 562 568 1 X 10^{2}
c_1	=	5.038 118 781 5 X 10^{-2}	−1.497 612 778 6 . . .
c_2	=	3.047 583 693 0 X 10^{-5}	3.178 710 392 4 X 10^{-3}
c_3	=	−8.568 106 572 0 X 10^{-8}	−3.184 768 670 1 X 10^{-6}
c_4	=	1.322 819 529 5 X 10^{-10}	1.572 081 900 4 X 10^{-9}
c_5	=	−1.705 295 833 7 X 10^{-13}	−3.069 136 905 6 X 10^{-13}
c_6	=	2.094 809 069 7 X 10^{-16}	....
c_7	=	−1.253 839 533 6 X 10^{-19}	....
c_8	=	1.563 172 569 7 X 10^{-23}	....

Type K thermocouples: emf-temperature (°C) reference table and equations.

Thermocouple emf as a Function of Temperature in Degrees Celsius (ITS-90)

emf in Millivolts — Reference Junctions at 0 °C

°C	0	−10	−20	−30	−40	−50	−60	−70	−80	−90	−100
−200	−5.891	−6.035	−6.158	−6.262	−6.344	−6.404	−6.441	−6.458			
−100	−3.554	−3.852	−4.138	−4.411	−4.669	−4.913	−5.141	−5.354	−5.550	−5.730	−5.891
0	0.000	−0.392	−0.778	−1.156	−1.527	−1.889	−2.243	−2.587	−2.920	−3.243	−3.554

°C	0	10	20	30	40	50	60	70	80	90	100
0	0.000	0.397	0.798	1.203	1.612	2.023	2.436	2.851	3.267	3.682	4.096
100	4.096	4.509	4.920	5.328	5.735	6.138	6.540	6.941	7.340	7.739	8.138
200	8.138	8.539	8.940	9.343	9.747	10.153	10.561	10.971	11.382	11.795	12.209
300	12.209	12.624	13.040	13.457	13.874	14.293	14.713	15.133	15.554	15.975	16.397
400	16.397	16.820	17.243	17.667	18.091	18.516	18.941	19.366	19.792	20.218	20.644
500	20.644	21.071	21.497	21.924	22.350	22.776	23.203	23.629	24.055	24.480	24.905
600	24.905	25.330	25.755	26.179	26.602	27.025	27.447	27.869	28.289	28.710	29.129
700	29.129	29.548	29.965	30.382	30.798	31.213	31.628	32.041	32.453	32.865	33.275
800	33.275	33.685	34.093	34.501	34.908	35.313	35.718	36.121	36.524	36.925	37.326
900	37.326	37.725	38.124	38.522	38.918	39.314	39.708	40.101	40.494	40.885	41.276
1000	41.276	41.665	42.053	42.440	42.826	43.211	43.595	43.978	44.359	44.740	45.119
1100	45.119	45.497	45.873	46.249	46.623	46.995	47.367	47.737	48.105	48.473	48.838
1200	48.838	49.202	49.565	49.926	50.286	50.644	51.000	51.355	51.708	52.060	52.410
1300	52.410	52.759	53.106	53.451	53.795	54.138	54.479	54.819			

Temperature Ranges and Coefficients of Equations Used to Compute the Above Table

The equations are of the form: $E = c_0 + c_1t + c_2t^2 + c_3t^3 + \ldots c_nt^n$, where E is the emf in millivolts, t is the temperature in degrees Celsius (ITS-90), and c_0, c_1, c_2, c_3, *etc.* are the coefficients. In the 0 °C to 1372 °C range there is also an exponential term that must be evaluated and added to the equation. The exponential term is of the form: $c_0 e^{c_1 (t - 126.9686)^2}$, where t is the temperature in °C, e is the natural logarithm base, and c_0 and c_1 are the coefficients. These coefficients are extracted from NIST Monograph 175.

	−270 °C to 0 °C	0 °C to 1372 °C	0 °C to 1372 °C (exponential term)
c_0 =	0.000 000 000 0 . . .	$-1.760\,041\,368\,6 \times 10^{-2}$	$1.185\,976 \times 10^{-1}$
c_1 =	$3.945\,012\,802\,5 \times 10^{-2}$	$3.892\,120\,497\,5 \times 10^{-2}$	$-1.183\,432 \times 10^{-4}$
c_2 =	$2.362\,237\,359\,8 \times 10^{-5}$	$1.855\,877\,003\,2 \times 10^{-5}$	....
c_3 =	$-3.285\,890\,678\,4 \times 10^{-7}$	$-9.945\,759\,287\,4 \times 10^{-8}$	....
c_4 =	$-4.990\,482\,877\,7 \times 10^{-9}$	$3.184\,094\,571\,9 \times 10^{-10}$	....
c_5 =	$-6.750\,905\,917\,3 \times 10^{-11}$	$-5.607\,284\,488\,9 \times 10^{-13}$	....
c_6 =	$-5.741\,032\,742\,8 \times 10^{-13}$	$5.607\,505\,905\,9 \times 10^{-16}$	....
c_7 =	$-3.108\,887\,289\,4 \times 10^{-15}$	$-3.202\,072\,000\,3 \times 10^{-19}$	....
c_8 =	$-1.045\,160\,936\,5 \times 10^{-17}$	$9.715\,114\,715\,2 \times 10^{-23}$	....
c_9 =	$-1.988\,926\,687\,8 \times 10^{-20}$	$-1.210\,472\,127\,5 \times 10^{-26}$	....
c_{10} =	$-1.632\,269\,748\,6 \times 10^{-23}$	....	....

Type N thermocouples: emf-temperature (°C) reference table and equations.

Thermocouple emf as a Function of Temperature in Degrees Celsius (ITS-90)

emf in Millivolts **Reference Junctions at 0 °C**

°C	0	−10	−20	−30	−40	−50	−60	−70	−80	−90	−100
−200	−3.990	−4.083	−4.162	−4.226	−4.277	−4.313	−4.336	−4.345			
−100	−2.407	−2.612	−2.808	−2.994	−3.171	−3.336	−3.491	−3.634	−3.766	−3.884	−3.990
0	0.000	−0.260	−0.518	−0.772	−1.023	−1.269	−1.509	−1.744	−1.972	−2.193	−2.407

°C	0	10	20	30	40	50	60	70	80	90	100
0	0.000	0.261	0.525	0.793	1.065	1.340	1.619	1.902	2.189	2.480	2.774
100	2.774	3.072	3.374	3.680	3.989	4.302	4.618	4.937	5.259	5.585	5.913
200	5.913	6.245	6.579	6.916	7.255	7.597	7.941	8.288	8.637	8.988	9.341
300	9.341	9.696	10.054	10.413	10.774	11.136	11.501	11.867	12.234	12.603	12.974
400	12.974	13.346	13.719	14.094	14.469	14.846	15.225	15.604	15.984	16.366	16.748
500	16.748	17.131	17.515	17.900	18.286	18.672	19.059	19.447	19.835	20.224	20.613
600	20.613	21.003	21.393	21.784	22.175	22.566	22.958	23.350	23.742	24.134	24.527
700	24.527	24.919	25.312	25.705	26.098	26.491	26.883	27.276	27.669	28.062	28.455
800	28.455	28.847	29.239	29.632	30.024	30.416	30.807	31.199	31.590	31.981	32.371
900	32.371	32.761	33.151	33.541	33.930	34.319	34.707	35.095	35.482	35.869	36.256
1000	36.256	36.641	37.027	37.411	37.795	38.179	38.562	38.944	39.326	39.706	40.087
1100	40.087	40.466	40.845	41.223	41.600	41.976	42.352	42.727	43.101	43.474	43.846
1200	43.846	44.218	44.588	44.958	45.326	45.694	46.060	46.425	46.789	47.152	47.513
1300	47.513										

Temperature Ranges and Coefficients of Equations Used to Compute the Above Table

The equations are of the form: $E = c_0 + c_1t + c_2t^2 + c_3t^3 + \ldots c_nt^n$, where E is the emf in millivolts, t is the temperature in degrees Celsius (ITS-90), and c_0, c_1, c_2, c_3, *etc.* are the coefficients. These coefficients are extracted from NIST Monograph 175.

		−270 °C to 0 °C	0 °C to 1300 °C
c_0	=	0.000 000 000 0 . . .	0.000 000 000 0 . . .
c_1	=	$2.615\,910\,596\,2 \times 10^{-2}$	$2.592\,939\,460\,1 \times 10^{-2}$
c_2	=	$1.095\,748\,422\,8 \times 10^{-5}$	$1.571\,014\,188\,0 \times 10^{-5}$
c_3	=	$-9.384\,111\,155\,4 \times 10^{-8}$	$4.382\,562\,723\,7 \times 10^{-8}$
c_4	=	$-4.641\,203\,975\,9 \times 10^{-11}$	$-2.526\,116\,979\,4 \times 10^{-10}$
c_5	=	$-2.630\,335\,771\,6 \times 10^{-12}$	$6.431\,181\,933\,9 \times 10^{-13}$
c_6	=	$-2.265\,343\,800\,3 \times 10^{-14}$	$-1.006\,347\,151\,9 \times 10^{-15}$
c_7	=	$-7.608\,930\,079\,1 \times 10^{-17}$	$9.974\,533\,899\,2 \times 10^{-19}$
c_8	=	$-9.341\,966\,783\,5 \times 10^{-20}$	$-6.086\,324\,560\,7 \times 10^{-22}$
c_9	=	. . . .	$2.084\,922\,933\,9 \times 10^{-25}$
c_{10}	=	. . . .	$-3.068\,219\,615\,1 \times 10^{-29}$

Type R thermocouples: emf-temperature (°C) reference table and equations.

Thermocouple emf as a Function of Temperature in Degrees Celsius (ITS-90)

emf in Millivolts **Reference Junctions at 0 °C**

°C	0	−10	−20	−30	−40	−50	−60	−70	−80	−90	−100
0	0.000	−0.051	−0.100	−0.145	−0.188	−0.226					

°C	0	10	20	30	40	50	60	70	80	90	100
0	0.000	0.054	0.111	0.171	0.232	0.296	0.363	0.431	0.501	0.573	0.647
100	0.647	0.723	0.800	0.879	0.959	1.041	1.124	1.208	1.294	1.381	1.469
200	1.469	1.558	1.648	1.739	1.831	1.923	2.017	2.112	2.207	2.304	2.401
300	2.401	2.498	2.597	2.696	2.796	2.896	2.997	3.099	3.201	3.304	3.408
400	3.408	3.512	3.616	3.721	3.827	3.933	4.040	4.147	4.255	4.363	4.471
500	4.471	4.580	4.690	4.800	4.910	5.021	5.133	5.245	5.357	5.470	5.583
600	5.583	5.697	5.812	5.926	6.041	6.157	6.273	6.390	6.507	6.625	6.743
700	6.743	6.861	6.980	7.100	7.220	7.340	7.461	7.583	7.705	7.827	7.950
800	7.950	8.073	8.197	8.321	8.446	8.571	8.697	8.823	8.950	9.077	9.205
900	9.205	9.333	9.461	9.590	9.720	9.850	9.980	10.111	10.242	10.374	10.506
1000	10.506	10.638	10.771	10.905	11.039	11.173	11.307	11.442	11.578	11.714	11.850
1100	11.850	11.986	12.123	12.260	12.397	12.535	12.673	12.812	12.950	13.089	13.228
1200	13.228	13.367	13.507	13.646	13.786	13.926	14.066	14.207	14.347	14.488	14.629
1300	14.629	14.770	14.911	15.052	15.193	15.334	15.475	15.616	15.758	15.899	16.040
1400	16.040	16.181	16.323	16.464	16.605	16.746	16.887	17.028	17.169	17.310	17.451
1500	17.451	17.591	17.732	17.872	18.012	18.152	18.292	18.431	18.571	18.710	18.849
1600	18.849	18.988	19.126	19.264	19.402	19.540	19.677	19.814	19.951	20.087	20.222
1700	20.222	20.356	20.488	20.620	20.749	20.877	21.003				

Temperature Ranges and Coefficients of Equations Used to Compute the Above Table

The equations are of the form: $E = c_0 + c_1t + c_2t^2 + c_3t^3 + \ldots c_nt^n$, where E is the emf in millivolts, t is the temperature in degrees Celsius (ITS-90), and c_0, c_1, c_2, c_3, *etc.* are the coefficients. These coefficients are extracted from NIST Monograph 175.

	−50 °C to 1064.18 °C	1064.18 °C to 1664.5 °C	1664.5 °C to 1768.1 °C
c_0 =	0.000 000 000 00 . . .	2.951 579 253 16 . . .	1.522 321 182 09 X 10^{2}
c_1 =	5.289 617 297 65 X 10^{-3}	−2.520 612 513 32 X 10^{-3}	−2.688 198 885 45 X 10^{-1}
c_2 =	1.391 665 897 82 X 10^{-5}	1.595 645 018 65 X 10^{-5}	1.712 802 804 71 X 10^{-4}
c_3 =	−2.388 556 930 17 X 10^{-8}	−7.640 859 475 76 X 10^{-9}	−3.458 957 064 53 X 10^{-8}
c_4 =	3.569 160 010 63 X 10^{-11}	2.053 052 910 24 X 10^{-12}	−9.346 339 710 46 X 10^{-15}
c_5 =	−4.623 476 662 98 X 10^{-14}	−2.933 596 681 73 X 10^{-16}	. . . .
c_6 =	5.007 774 410 34 X 10^{-17}	. . . .	. . . .
c_7 =	−3.731 058 861 91 X 10^{-20}	. . . .	. . . .
c_8 =	1.577 164 823 67 X 10^{-23}	. . . .	. . . .
c_9 =	−2.810 386 252 51 X 10^{-27}	. . . .	. . . .

Type S thermocouples: emf-temperature (°C) reference table and equations.

Thermocouple emf as a Function of Temperature in Degrees Celsius (ITS-90)

emf in Millivolts — Reference Junctions at 0 °C

°C	0	−10	−20	−30	−40	−50	−60	−70	−80	−90	−100
0	0.000	−0.053	−0.103	−0.150	−0.194	−0.236					

°C	0	10	20	30	40	50	60	70	80	90	100
0	0.000	0.055	0.113	0.173	0.235	0.299	0.365	0.433	0.502	0.573	0.646
100	0.646	0.720	0.795	0.872	0.950	1.029	1.110	1.191	1.273	1.357	1.441
200	1.441	1.526	1.612	1.698	1.786	1.874	1.962	2.052	2.141	2.232	2.323
300	2.323	2.415	2.507	2.599	2.692	2.786	2.880	2.974	3.069	3.164	3.259
400	3.259	3.355	3.451	3.548	3.645	3.742	3.840	3.938	4.036	4.134	4.233
500	4.233	4.332	4.432	4.532	4.632	4.732	4.833	4.934	5.035	5.137	5.239
600	5.239	5.341	5.443	5.546	5.649	5.753	5.857	5.961	6.065	6.170	6.275
700	6.275	6.381	6.486	6.593	6.699	6.806	6.913	7.020	7.128	7.236	7.345
800	7.345	7.454	7.563	7.673	7.783	7.893	8.003	8.114	8.226	8.337	8.449
900	8.449	8.562	8.674	8.787	8.900	9.014	9.128	9.242	9.357	9.472	9.587
1000	9.587	9.703	9.819	9.935	10.051	10.168	10.285	10.403	10.520	10.638	10.757
1100	10.757	10.875	10.994	11.113	11.232	11.351	11.471	11.590	11.710	11.830	11.951
1200	11.951	12.071	12.191	12.312	12.433	12.554	12.675	12.796	12.917	13.038	13.159
1300	13.159	13.280	13.402	13.523	13.644	13.766	13.887	14.009	14.130	14.251	14.373
1400	14.373	14.494	14.615	14.736	14.857	14.978	15.099	15.220	15.341	15.461	15.582
1500	15.582	15.702	15.822	15.942	16.062	16.182	16.301	16.420	16.539	16.658	16.777
1600	16.777	16.895	17.013	17.131	17.249	17.366	17.483	17.600	17.717	17.832	17.947
1700	17.947	18.061	18.174	18.285	18.395	18.503	18.609				

Temperature Ranges and Coefficients of Equations Used to Compute the Above Table

The equations are of the form: $E = c_0 + c_1t + c_2t^2 + c_3t^3 + \ldots c_nt^n$, where E is the emf in millivolts, t is the temperature in degrees Celsius (ITS-90), and c_0, c_1, c_2, c_3, *etc.* are the coefficients. These coefficients are extracted from NIST Monograph 175.

		−50 °C to 1064.18 °C	1064.18 °C to 1664.5 °C	1664.5 °C to 1768.1 °C
c_0	=	0.000 000 000 00 . . .	1.329 004 440 85 . . .	$1.466\,282\,326\,36 \times 10^{2}$
c_1	=	$5.403\,133\,086\,31 \times 10^{-3}$	$3.345\,093\,113\,44 \times 10^{-3}$	$-2.584\,305\,167\,52 \times 10^{-1}$
c_2	=	$1.259\,342\,897\,40 \times 10^{-5}$	$6.548\,051\,928\,18 \times 10^{-6}$	$1.636\,935\,746\,41 \times 10^{-4}$
c_3	=	$-2.324\,779\,686\,89 \times 10^{-8}$	$-1.648\,562\,592\,09 \times 10^{-9}$	$-3.304\,390\,469\,87 \times 10^{-8}$
c_4	=	$3.220\,288\,230\,36 \times 10^{-11}$	$1.299\,896\,051\,74 \times 10^{-14}$	$-9.432\,236\,906\,12 \times 10^{-15}$
c_5	=	$-3.314\,651\,963\,89 \times 10^{-14}$	....	....
c_6	=	$2.557\,442\,517\,86 \times 10^{-17}$	....	....
c_7	=	$-1.250\,688\,713\,93 \times 10^{-20}$	....	....
c_8	=	$2.714\,431\,761\,45 \times 10^{-24}$	....	....

Type T thermocouples: emf-temperature (°C) reference table and equations.

Thermocouple emf as a Function of Temperature in Degrees Celsius (ITS-90)

emf in Millivolts — Reference Junctions at 0 °C

°C	0	−10	−20	−30	−40	−50	−60	−70	−80	−90	−100
−200	−5.603	−5.753	−5.888	−6.007	−6.105	−6.180	−6.232	−6.258			
−100	−3.379	−3.657	−3.923	−4.177	−4.419	−4.648	−4.865	−5.070	−5.261	−5.439	−5.603
0	0.000	−0.383	−0.757	−1.121	−1.475	−1.819	−2.153	−2.476	−2.788	−3.089	−3.379

°C	0	10	20	30	40	50	60	70	80	90	100
0	0.000	0.391	0.790	1.196	1.612	2.036	2.468	2.909	3.358	3.814	4.279
100	4.279	4.750	5.228	5.714	6.206	6.704	7.209	7.720	8.237	8.759	9.288
200	9.288	9.822	10.362	10.907	11.458	12.013	12.574	13.139	13.709	14.283	14.862
300	14.862	15.445	16.032	16.624	17.219	17.819	18.422	19.030	19.641	20.255	20.872
400	20.872										

Temperature Ranges and Coefficients of Equations Used to Compute the Above Table

The equations are of the form: $E = c_0 + c_1t + c_2t^2 + c_3t^3 + \dots c_nt^n$, where E is the emf in millivolts, t is the temperature in degrees Celsius (ITS-90), and c_0, c_1, c_2, c_3, *etc.* are the coefficients. These coefficients are extracted from NIST Monograph 175.

		−270 °C to 0 °C	0 °C to 400 °C
c_0	=	0.000 000 000 0 . . .	0.000 000 000 0 . . .
c_1	=	$3.874\,810\,636\,4 \times 10^{-2}$	$3.874\,810\,636\,4 \times 10^{-2}$
c_2	=	$4.419\,443\,434\,7 \times 10^{-5}$	$3.329\,222\,788\,0 \times 10^{-5}$
c_3	=	$1.184\,432\,310\,5 \times 10^{-7}$	$2.061\,824\,340\,4 \times 10^{-7}$
c_4	=	$2.003\,297\,355\,4 \times 10^{-8}$	$-2.188\,225\,684\,6 \times 10^{-9}$
c_5	=	$9.013\,801\,955\,9 \times 10^{-10}$	$1.099\,688\,092\,8 \times 10^{-11}$
c_6	=	$2.265\,115\,659\,3 \times 10^{-11}$	$-3.081\,575\,877\,2 \times 10^{-14}$
c_7	=	$3.607\,115\,420\,5 \times 10^{-13}$	$4.547\,913\,529\,0 \times 10^{-17}$
c_8	=	$3.849\,393\,988\,3 \times 10^{-15}$	$-2.751\,290\,167\,3 \times 10^{-20}$
c_9	=	$2.821\,352\,192\,5 \times 10^{-17}$	. . . .
c_{10}	=	$1.425\,159\,477\,9 \times 10^{-19}$	. . . .
c_{11}	=	$4.876\,866\,228\,6 \times 10^{-22}$	. . . .
c_{12}	=	$1.079\,553\,927\,0 \times 10^{-24}$	. . . .
c_{13}	=	$1.394\,502\,706\,2 \times 10^{-27}$	. . . .
c_{14}	=	$7.979\,515\,392\,7 \times 10^{-31}$	. . . .

Type B thermocouples: coefficients (c_i) of polynomials for the computation of temperatures in °C as a function of the thermocouple emf in various temperature and emf ranges.

Temperature Range:	250 °C to 700 °C	700 °C to 1820 °C
emf Range:	0.291 mV to 2.431 mV	2.431 mV to 13.820 mV
c_0 =	$9.842\,332\,1 \times 10^{1}$	$2.131\,507\,1 \times 10^{2}$
c_1 =	$6.997\,150\,0 \times 10^{2}$	$2.851\,050\,4 \times 10^{2}$
c_2 =	$-8.476\,530\,4 \times 10^{2}$	$-5.274\,288\,7 \times 10^{1}$
c_3 =	$1.005\,264\,4 \times 10^{3}$	9.916 080 4 . . .
c_4 =	$-8.334\,595\,2 \times 10^{2}$	−1.296 530 3 . . .
c_5 =	$4.550\,854\,2 \times 10^{2}$	$1.119\,587\,0 \times 10^{-1}$
c_6 =	$-1.552\,303\,7 \times 10^{2}$	$-6.062\,519\,9 \times 10^{-3}$
c_7 =	$2.988\,675\,0 \times 10^{1}$	$1.866\,169\,6 \times 10^{-4}$
c_8 =	−2.474 286 0 . . .	$-2.487\,858\,5 \times 10^{-6}$

NOTE—The above coefficients are extracted from NIST Monograph 175 and are for an expression of the form shown in Section 10.3.2. They yield approximate values of temperature that agree within ± 0.03 °C with the values given in Table 10.2.

Type E thermocouples: coefficients (c_i) of polynomials for the computation of temperatures in °C as a function of the thermocouple emf in various temperature and emf ranges.

Temperature Range:	−200 °C to 0 °C	0 °C to 1000 °C
emf Range:	−8.825 mV to 0.0 mV	0.0 mV to 76.373 mV
c_0 =	0.000 000 0 . . .	0.000 000 0 . . .
c_1 =	$1.697\,728\,8 \times 10^{1}$	$1.705\,703\,5 \times 10^{1}$
c_2 =	$-4.351\,497\,0 \times 10^{-1}$	$-2.330\,175\,9 \times 10^{-1}$
c_3 =	$-1.585\,969\,7 \times 10^{-1}$	$6.543\,558\,5 \times 10^{-3}$
c_4 =	$-9.250\,287\,1 \times 10^{-2}$	$-7.356\,274\,9 \times 10^{-5}$
c_5 =	$-2.608\,431\,4 \times 10^{-2}$	$-1.789\,600\,1 \times 10^{-6}$
c_6 =	$-4.136\,019\,9 \times 10^{-3}$	$8.403\,616\,5 \times 10^{-8}$
c_7 =	$-3.403\,403\,0 \times 10^{-4}$	$-1.373\,587\,9 \times 10^{-9}$
c_8 =	$-1.156\,489\,0 \times 10^{-5}$	$1.062\,982\,3 \times 10^{-11}$
c_9 =	. . . .	$-3.244\,708\,7 \times 10^{-14}$

NOTE—The above coefficients are extracted from NIST Monograph 175 and are for an expression of the form shown in Section 10.3.2. They yield approximate values of temperature that agree within ± 0.02 °C with the values given in Table 10.4.

Type J thermocouples: coefficients (c_i) of polynomials for the computation of temperatures in °C as a function of the thermocouple emf in various temperature and emf ranges.

Temperature Range:	−210 °C to 0 °C	0 °C to 760 °C	760 °C to 1200 °C
emf Range:	−8.095 mV to 0.0 mV	0.0 mV to 42.919 mV	42.919 mV to 69.553 mV
c_0 =	0.000 000 0 . . .	0.000 000 . . .	$-3.113\ 581\ 87 \times 10^{3}$
c_1 =	$1.952\ 826\ 8 \times 10^{1}$	$1.978\ 425 \times 10^{1}$	$3.005\ 436\ 84 \times 10^{2}$
c_2 =	−1.228 618 5 . . .	$-2.001\ 204 \times 10^{-1}$	−9.947 732 30 . . .
c_3 =	−1.075 217 8 . . .	$1.036\ 969 \times 10^{-2}$	$1.702\ 766\ 30 \times 10^{-1}$
c_4 =	$-5.908\ 693\ 3 \times 10^{-1}$	$-2.549\ 687 \times 10^{-4}$	$-1.430\ 334\ 68 \times 10^{-3}$
c_5 =	$-1.725\ 671\ 3 \times 10^{-1}$	$3.585\ 153 \times 10^{-6}$	$4.738\ 860\ 84 \times 10^{-6}$
c_6 =	$-2.813\ 151\ 3 \times 10^{-2}$	$-5.344\ 285 \times 10^{-8}$	. . . .
c_7 =	$-2.396\ 337\ 0 \times 10^{-3}$	$5.099\ 890 \times 10^{-10}$	. . . .
c_8 =	$-8.382\ 332\ 1 \times 10^{-5}$	. . . .	. . . .

NOTE—The above coefficients are extracted from NIST Monograph 175 and are for an expression of the form shown in Section 10.3.2. They yield approximate values of temperature that agree within ± 0.05 °C with the values given in Table 10.6.

Type K thermocouples: coefficients (c_i) of polynomials for the computation of temperatures in °C as a function of the thermocouple emf in various temperature and emf ranges.

Temperature Range:	−200 °C to 0 °C	0 °C to 500 °C	500 °C to 1372 °C
emf Range:	−5.891 mV to 0.0 mV	0.0 mV to 20.644 mV	20.644 mV to 54.886 mV
c_0 =	0.000 000 0 . . .	0.000 000 0 . . .	$-1.318\ 058 \times 10^{2}$
c_1 =	$2.517\ 346\ 2 \times 10^{1}$	$2.508\ 355 \times 10^{1}$	$4.830\ 222 \times 10^{1}$
c_2 =	−1.166 287 8 . . .	$7.860\ 106 \times 10^{-2}$	−1.646 031 . . .
c_3 =	−1.083 363 8 . . .	$-2.503\ 131 \times 10^{-1}$	$5.464\ 731 \times 10^{-2}$
c_4 =	$-8.977\ 354\ 0 \times 10^{-1}$	$8.315\ 270 \times 10^{-2}$	$-9.650\ 715 \times 10^{-4}$
c_5 =	$-3.734\ 237\ 7 \times 10^{-1}$	$-1.228\ 034 \times 10^{-2}$	$8.802\ 193 \times 10^{-6}$
c_6 =	$-8.663\ 264\ 3 \times 10^{-2}$	$9.804\ 036 \times 10^{-4}$	$-3.110\ 810 \times 10^{-8}$
c_7 =	$-1.045\ 059\ 8 \times 10^{-2}$	$-4.413\ 030 \times 10^{-5}$	. . . . . .
c_8 =	$-5.192\ 057\ 7 \times 10^{-4}$	$1.057\ 734 \times 10^{-6}$	. . . . . .
c_9 =	. . . .	$-1.052\ 755 \times 10^{-8}$	. . . . . .

NOTE—The above coefficients are extracted from NIST Monograph 175 and are for an expression of the form shown in Section 10.3.2. They yield approximate values of temperature that agree within ± 0.05 °C with the values given in Table 10.8.

Type N thermocouples: coefficients (c_i) of polynomials for the computation of temperatures in °C as a function of the thermocouple emf in various temperature and emf ranges.

Temperature Range:	−200 °C to 0 °C	0 °C to 600 °C	600 °C to 1300 °C
emf Range:	−3.990 mV to 0.0 mV	0.0 mV to 20.613 mV	20.613 mV to 47.513 mV
c_0 =	0.000 000 0 . . .	0.000 00 . . .	1.972 485 X 10^1
c_1 =	3.843 684 7 X 10^1	3.868 96 X 10^1	3.300 943 X 10^1
c_2 =	1.101 048 5 . . .	−1.082 67 . . .	−3.915 159 X 10^{-1}
c_3 =	5.222 931 2 . . .	4.702 05 X 10^{-2}	9.855 391 X 10^{-3}
c_4 =	7.206 052 5 . . .	−2.121 69 X 10^{-6}	−1.274 371 X 10^{-4}
c_5 =	5.848 858 6 . . .	−1.172 72 X 10^{-4}	7.767 022 X 10^{-7}
c_6 =	2.775 491 6 . . .	5.392 80 X 10^{-6}	. . . .
c_7 =	7.707 516 6 X 10^{-1}	−7.981 56 X 10^{-8}	. . . .
c_8 =	1.158 266 5 X 10^{-1}	. . . .	. . . .
c_9 =	7.313 886 8 X 10^{-3}	. . . .	. . . .

NOTE—The above coefficients are extracted from NIST Monograph 175 and are for an expression of the form shown in Section 10.3.2. They yield approximate values of temperature that agree within ± 0.04 °C with the values given in Table 10.10.

Type R thermocouples: coefficients (c_i) of polynomials for the computation of temperatures in °C as a function of the thermocouple emf in various temperature and emf ranges.

Temperature Range:	−50 °C to 250 °C	250 °C to 1200 °C	1064 °C to 1664.5 °C	1664.5 °C to 1768.1 °C
emf Range:	−0.226 mV to 1.923 mV	1.923 mV to 13.228 mV	11.361 mV to 19.739 mV	19.739 mV to 21.103 mV
c_0 =	0.000 000 0 . . .	1.334 584 505 X 10^1	−8.199 599 416 X 10^1	3.406 177 836 X 10^4
c_1 =	1.889 138 0 X 10^2	1.472 644 573 X 10^2	1.553 962 042 X 10^2	−7.023 729 171 X 10^3
c_2 =	−9.383 529 0 X 10^1	−1.844 024 844 X 10^1	−8.342 197 663 . . .	5.582 903 813 X 10^2
c_3 =	1.306 861 9 X 10^2	4.031 129 726 . . .	4.279 433 549 X 10^{-1}	−1.952 394 635 X 10^1
c_4 =	−2.270 358 0 X 10^2	−6.249 428 360 X 10^{-1}	−1.191 577 910 X 10^{-2}	2.560 740 231 X 10^{-1}
c_5 =	3.514 565 9 X 10^2	6.468 412 046 X 10^{-2}	1.492 290 091 X 10^{-4}	. . . .
c_6 =	−3.895 390 0 X 10^2	−4.458 750 426 X 10^{-3}	. . . .	. . . .
c_7 =	2.823 947 1 X 10^2	1.994 710 149 X 10^{-4}	. . . .	. . . .
c_8 =	−1.260 728 1 X 10^2	−5.313 401 790 X 10^{-6}	. . . .	. . . .
c_9 =	3.135 361 1 X 10^1	6.481 976 217 X 10^{-8}	. . . .	. . . .
c_{10} =	−3.318 776 9 . . .	. . . .	. . . .	. . . .

NOTE—The above coefficients are extracted from NIST Monograph 175 and are for an expression of the form shown in Section 10.3.2. They yield approximate values of temperature that agree within ± 0.02 °C with the values given in Table 10.12.

Type S thermocouples: coefficients (c_i) of polynomials for the computation of temperatures in °C as a function of the thermocouple emf in various temperature and emf ranges.

Temperature Range:	−50 °C to 250 °C	250 °C to 1200 °C	1064 °C to 1664.5 °C	1664.5 °C to 1768.1 °C
emf Range:	−0.235 mV to 1.874 mV	1.874 mV to 11.950 mV	10.332 mV to 17.536 mV	17.536 mV to 18.693 mV
c_0 =	0.000 000 00 . . .	1.291 507 177 X 10^1	−8.087 801 117 X 10^1	5.333 875 126 X 10^4
c_1 =	1.849 494 60 X 10^2	1.466 298 863 X 10^2	1.621 573 104 X 10^2	−1.235 892 298 X 10^4
c_2 =	−8.005 040 62 X 10^1	−1.534 713 402 X 10^1	−8.536 869 453 . . .	1.092 657 613 X 10^3
c_3 =	1.022 374 30 X 10^2	3.145 945 973 . . .	4.719 686 976 X 10^{-1}	−4.265 693 686 X 10^1
c_4 =	−1.522 485 92 X 10^2	−4.163 257 839 X 10^{-1}	−1.441 693 666 X 10^{-2}	6.247 205 420 X 10^{-1}
c_5 =	1.888 213 43 X 10^2	3.187 963 771 X 10^{-2}	2.081 618 890 X 10^{-4}	. . . .
c_6 =	−1.590 859 41 X 10^2	−1.291 637 500 X 10^{-3}	. . . .	. . . .
c_7 =	8.230 278 80 X 10^1	2.183 475 087 X 10^{-5}	. . . .	. . . .
c_8 =	−2.341 819 44 X 10^1	−1.447 379 511 X 10^{-7}	. . . .	. . . .
c_9 =	2.797 862 60 . . .	8.211 272 125 X 10^{-9}	. . . .	. . . .

NOTE—The above coefficients are extracted from NIST Monograph 175 and are for an expression of the form shown in Section 10.3.2. They yield approximate values of temperature that agree within ± 0.02 °C with the values given in Table 10.14.

Type T thermocouples: coefficients (c_i) of polynomials for the computation of temperatures in °C as a function of the thermocouple emf in various temperature and emf ranges.

Temperature Range:	−200 °C to 0 °C	0 °C to 400 °C
emf Range:	−5.603 mV to 0.0 mV	0.0 mV to 20.872 mV
c_0 =	0.000 000 0 . . .	0.000 000 . . .
c_1 =	2.594 919 2 X 10^1	2.592 800 X 10^1
c_2 =	−2.131 696 7 X 10^{-1}	−7.602 961 X 10^{-1}
c_3 =	7.901 869 2 X 10^{-1}	4.637 791 X 10^{-2}
c_4 =	4.252 777 7 X 10^{-1}	−2.165 394 X 10^{-3}
c_5 =	1.330 447 3 X 10^{-1}	6.048 144 X 10^{-5}
c_6 =	2.024 144 6 X 10^{-2}	−7.293 422 X 10^{-7}
c_7 =	1.266 817 1 X 10^{-3}	. . . .

NOTE—The above coefficients are extracted from NIST Monograph 175 and are for an expression of the form shown in Section 10.3.2. They yield approximate values of temperature that agree within ± 0.04 °C with the values given in Table 10.16.

PROPERTIES OF COMMON LABORATORY SOLVENTS

This table give properties of 200 organic solvents which are frequently used in laboratory and industrial applications. Compounds are listed in alphabetical order by the most common name; synonyms are given in some cases. The properties tabulated are:

MF: Molecular formula
CAS RN: Chemical Abstracts Service Registry Number
M_r: Molecular weight
t_m: Melting point in °C
t_b: Normal boiling point in °C
ρ: Density in g/cm^3 at the temperature in °C indicated by the superscript
c_p: Specific heat capacity of the liquid at constant pressure at 25°C in J/g K
vp: Vapor pressure at 25°C in kPa (1 kPa = 7.50 mmHg)
μ: Electric dipole moment in debye units. Values in parentheses are measurements on the pure liquid or in solution; these are less reliable than the other values, which were obtained in the gas phase.
FP: Flash point temperature in °C. The fact that no flash point is listed does not necessarily mean that the substance is nonflammable, because some liquids will burn if the quantity is large or impurities are present.
Fl. Lim.: Flammable (explosive) range in air in percent by volume
Ign. Temp.: Autoignition temperature in °C
TLV: Threshold limit for allowable airborne concentration, given in parts per million by volume at 25°C and atmospheric pressure (see table "Threshold Limit Values for Airborne Contaminants" in Section 16)

REFERENCES

1. Lide, D.R., *Handbook of Organic Solvents*, CRC Press, Boca Raton, FL, 1994.
2. Lide, D.R., and Kehiaian, H.V., *CRC Handbook of Thermophysical and Thermochemical Data*, CRC Press, Boca Raton, FL, 1994.
3. Riddick, J.A., Bunger, W.B., and Sakano, T.K., *Organic Solvents, Fourth Edition*, John Wiley & Sons, New York, 1986.
4. *Fire Protection Guide to Hazardous Materials*, 10th Edition , National Fire Protection Association, Quincy, MA, 1991.
5. Urben, P.G., Ed., *Bretherick's Handbook of Reactive Chemical Hazards, 5th Edition,* Butterworth-Heinemann, Oxford, 1995.

Name	MF	CAS RN	M_r	t_m/°C	t_b/°C	ρ/g cm^{-3}	c_p/J g^{-1}K^{-1}	vp/kPa	μ/D	FP/°C	Fl. Lim.	Ign. Temp./°C	TLV
Acetal (1,1-Diethoxyethane)	$C_6H_{14}O_2$	105-57-7	118.18	-100	102.2	0.8254^{20}	2.01	3.68	(1.4)	-21	2-10%	230	
Acetic acid	$C_2H_4O_2$	64-19-7	60.05	17	118	1.0492^{20}	2.06	2.07	1.70	39	4-20%	463	10
Acetone	C_3H_6O	67-64-1	58.08	-95	56	0.7899^{20}	2.18	30.8	2.88	-20	3-13%	465	750
Acetonitrile	C_2H_3N	75-05-8	41.05	-44	82	0.7857^{20}	2.23	11.8	3.92	6	3-16%	524	40
Acetylacetone	$C_5H_8O_2$	123-54-6	100.12	-23	138	0.9721^{25}	2.08	1.02	(2.8)	34		340	
Acrylonitrile	C_3H_3N	107-13-1	53.06	-83.5	77.3	0.8060^{20}	2.05	14.1	3.87	0	3-17%	481	2
Adiponitrile	$C_6H_8N_2$	111-69-3	108.14	1	295	0.9676^{20}	1.19	<0.01		93	2-5%	550	2
Allyl alcohol	C_3H_6O	107-18-6	58.08	-129	97.0	0.8540^{20}	2.39	3.14	1.60	21	3-18%	378	2
Allylamine	C_3H_7N	107-11-9	57.10	-88.2	53.3	0.758^{20}		33.1	1.2	-29	2-22%	374	
2-Aminoisobutanol	$C_4H_{11}NO$	124-68-5	89.14	25.5	165.5	0.934^{20}				67			
Benzal chloride	$C_7H_6Cl_2$	98-87-3	161.03	-17	205	1.26^{25}		0.06	(2.1)				
Benzaldehyde	C_7H_6O	100-52-7	106.12	-26	179.0	1.0415^{10}	1.62	0.17	(3.0)	63		192	
Benzene	C_6H_6	71-43-2	78.11	6	80	0.8765^{20}	1.74	12.7	0	-11	1-8%	498	10
Benzonitrile	C_7H_5N	100-47-0	103.12	-12.7	191.1	1.0093^{15}	1.60	0.11	4.18				
Benzyl chloride	C_7H_7Cl	100-44-7	126.59	-45	179	1.1004^{20}	1.44	0.16	(1.8)	67	1%-	585	1
Bromochloromethane	CH_2BrCl	74-97-5	129.38	-87.9	68.0	1.9344^{20}	0.41	19.5	(1.7)				200
Bromoform (Tribromomethane)	$CHBr_3$	75-25-2	252.73	8.0	149	2.899^{15}	0.52	0.73	0.99	83			0.5

PROPERTIES OF COMMON LABORATORY SOLVENTS (continued)

Name	MF	CAS RN	M_r	t_m/°C	t_b/°C	ρ/g cm^{-3}	c_p/J g^{-1}K^{-1}	vp/kPa	μ/D	FP/°C	Fl. Lim.	Ign. Temp./°C	TLV
Butyl acetate	$C_6H_{12}O_2$	123-86-4	116.16	-78	126	0.8825^{20}	1.96	1.66	(1.9)	22	2-8%	425	150
Butyl alcohol	$C_4H_{10}O$	71-36-3	74.12	-90	118	0.8098^{20}	2.39	0.86	1.66	37	1-11%	343	50
sec-Butyl alcohol	$C_4H_{10}O$	78-92-2	74.12	-114.7	99.5	0.8063^{20}	2.66	2.32	(1.8)	24	2-10%	405	100
tert-Butyl alcohol	$C_4H_{10}O$	75-65-0	74.12	26	82	0.7887^{20}	2.97	5.52	(1.7)	11	2-8%	478	100
Butylamine	$C_4H_{11}N$	109-73-9	73.14	-49	77	0.7414^{20}	2.45	12.2	1.0	-12	2-10%	312	5
tert-Butylamine	$C_4H_{11}N$	75-64-9	73.14	-67	44	0.6958^{20}	2.63	48.4	(1.3)	-9	2-9%	380	
Butyl methyl ketone	$C_6H_{12}O$	591-78-6	100.16	-56	128	0.8113^{20}	2.13	1.54	(2.7)	25	1-8%	423	5
p-tert-Butyltoluene	$C_{11}H_{16}$	98-51-1	148.25	-52	190	0.8612^{20}		0.09	≈0	68			10
γ-Butyrolactone	$C_4H_6O_2$	96-48-0	86.09	-43.3	204	1.1284^{16}	1.64	0.43	4.27	98			
Caprolactam	$C_6H_{11}NO$	105-60-2	113.16	69	270		1.38	<0.01	(3.9)	125			5
Carbon disulfide	CS_2	75-15-0	76.14	-112	46	1.2632^{20}	1.00	48.2	0	-30	1-50%	90	10
Carbon tetrachloride	CCl_4	56-23-5	153.82	-23	77	1.5940^{20}	0.85	15.2	0				5
1-Chloro-1,1-difluoroethane	$C_2H_3ClF_2$	75-68-3	100.50	-131	-10	1.107^{25}	1.30	351	2.14				
Chlorobenzene	C_6H_5Cl	108-90-7	112.56	-45	132	1.1058^{20}	1.33	1.6	1.69	28	1-10%	593	10
Chloroform	$CHCl_3$	67-66-3	119.38	-64	61	1.4832^{20}	0.96	26.2	1.04				10
Chloropentafluoroethane	C_2ClF_5	76-15-3	154.47	-99	-38	1.5678^{-42}	1.19	912	0.52				1000
Cumene (Isopropylbenzene)	C_9H_{12}	98-82-8	120.19	-96.0	152	0.8618^{20}	1.75	0.61	0.79	36	1-7%	424	50
Cyclohexane	C_6H_{12}	110-82-7	84.16	7	81	0.7785^{20}	1.84	13.0	≈0	-20	1-8%	245	300
Cyclohexanol	$C_6H_{12}O$	108-93-0	100.16	25	161	0.9624^{20}	2.08	0.10		68	1-9%	300	50
Cyclohexanone	$C_6H_{10}O$	108-94-1	98.14	-31	155	0.9478^{20}	1.86	0.53	2.87	44	1-9%	420	25
Cyclohexylamine	$C_6H_{13}N$	108-91-8	99.18	-18	134	0.8191^{20}		1.20	(1.3)	31	1-9%	293	10
Cyclopentane	C_5H_{10}	287-92-3	70.13	-93.8	49.3	0.7457^{20}	1.84	42.3	≈0	<-7	2%-	361	600
Cyclopentanone	C_5H_8O	120-92-3	84.12	-51.3	130.5	0.9487^{20}	1.84	1.55	3.3	26			
p-Cymene	$C_{10}H_{14}$	99-87-6	134.22	-69	177	0.8573^{20}	1.76	0.19	≈0	47	1-6%	436	
cis-Decalin	$C_{10}H_{18}$	493-01-6	138.25	-42.9	195.8	0.8965^{20}	1.68	0.10	≈0				
trans-Decalin	$C_{10}H_{18}$	493-02-7	138.25	-30.3	187.3	0.8699^{20}	1.65	0.16	≈0	54	1-5%	255	
Diacetone alcohol	$C_6H_{12}O_2$	123-42-2	116.16	-44	168	0.9387^{20}	1.91	0.22	(3.2)	58	2-7%	643	50
1,2-Dibromoethane	$C_2H_4Br_2$	106-93-4	187.86	9.9	131.6	2.1791^{20}	0.72	1.55	(1.2)				
Dibromofluoromethane	$CHBr_2F$	1868-53-7	191.83	-78	64.9	2.421^{20}							
Dibromomethane	CH_2Br_2	74-95-3	173.83	-52.5	97	2.4969^{20}	0.61	6.12	1.43				
1,2-Dibromotetrafluoroethane	$C_2Br_2F_4$	124-73-2	259.82	-110.4	47.3	2.149^{25}	0.69	43.4					
Dibutylamine	$C_8H_{19}N$	111-92-2	129.25	-62	160	0.7670^{20}	2.27	0.34	(1.0)	47	1-6%		
o-Dichlorobenzene	$C_6H_4Cl_2$	95-50-1	147.00	-17	180	1.3059^{20}	1.10	0.18	2.50	66	2-9%	648	25
1,1-Dichloroethane	$C_2H_4Cl_2$	75-34-3	98.96	-97	57	1.1757^{20}	1.28	30.5	2.06	-17	5-11%	458	100
1,2-Dichloroethane	$C_2H_4Cl_2$	107-06-2	98.96	-36	84	1.2351^{20}	1.30	10.6	(1.8)	13	6-16%	413	10
1,1-Dichloroethylene	$C_2H_2Cl_2$	75-35-4	96.94	-122.5	31.6	1.213^{20}	1.15	80.0	1.34	-15	7-16%	570	5
cis-1,2-Dichloroethylene	$C_2H_2Cl_2$	156-59-2	96.94	-80	60	1.2837^{20}	1.20	26.8	1.90	6	3-15%	460	200
trans-1,2-Dichloroethylene	$C_2H_2Cl_2$	156-60-5	96.94	-50	49	1.2565^{20}	1.20	44.2	0	2	6-13%	460	200
Dichloroethyl ether	$C_4H_8Cl_2O$	111-44-4	143.01	-52	179	1.22^{20}	1.54	0.14	(2.6)	55	3%-	369	5
Dichloromethane	CH_2Cl_2	75-09-2	84.93	-95	40	1.3266^{20}	1.19	58.2	1.60		13-23%	556	50
1,2-Dichloropropane	$C_3H_6Cl_2$	78-87-5	112.99	-100	96	1.1560^{20}	1.32	6.62	(1.8)	16	3-15%	557	75
1,2-Dichlorotetrafluoroethane	$C_2Cl_2F_4$	76-14-2	170.92	-94	4	1.518^{4}	0.96	215	0.5				1000
Diethanolamine	$C_4H_{11}NO_2$	111-42-2	105.14	28	269	1.0966^{20}	2.22	<0.01	(2.8)	172	2-13%	662	0.46
Diethylamine	$C_4H_{11}N$	109-89-7	73.14	-50	55	0.7056^{20}	2.31	30.1	0.92	-23	2-10%	312	5
Diethyl carbonate	$C_5H_{10}O_3$	105-58-8	118.13	-43	126	0.9752^{20}	1.80	1.63	1.10	25			
Diethylene glycol	$C_4H_{10}O_3$	111-46-6	106.12	-10	246	1.1197^{15}	2.31	<0.01	(2.3)	124	2-17%	224	
Diethylene glycol dimethyl ether	$C_6H_{14}O_3$	111-96-6	134.18	-68	162	0.9434^{20}	2.04	0.31	(2.0)	67			
Diethylene glycol monoethyl ether	$C_6H_{14}O_3$	111-90-0	134.18		196	0.9885^{20}	2.24	0.02	(1.6)	96			

PROPERTIES OF COMMON LABORATORY SOLVENTS (continued)

Name	MF	CAS RN	M_r	t_m/°C	t_b/°C	ρ/g cm^{-3}	c_p/J g^{-1}K^{-1}	vp/kPa	μ/D	FP/°C	Fl. Lim.	Ign. Temp./°C	TLV
Diethylene glycol monoethyl ether acetate	$C_8H_{16}O_4$	112-15-2	176.21	-25	218.5	1.0096^{20}		0.03	(1.8)	110		425	
Diethylene glycol monomethyl ether	$C_5H_{12}O_3$	111-77-3	120.15		193	1.035^{20}	2.26	0.02	(1.6)	96	1-23%	240	
Diethylenetriamine	$C_4H_{13}N_3$	111-40-0	103.17	-39	207	0.9569^{20}	2.46	0.03	(1.9)	98	2-7%	358	1
Diethyl ether	$C_4H_{10}O$	60-29-7	74.12	-116	34	0.7138^{20}	2.33	71.7	1.15	-45	2-36%	180	400
Diisobutyl ketone	$C_9H_{18}O$	108-83-8	142.24	-42	169	0.8062^{20}	2.09	0.23	(2.7)	49	1-7%	396	25
Diisopropyl ether	$C_6H_{14}O$	108-20-3	102.18	-87	69	0.7241^{20}	2.12	19.9	1.13	-28	1-8%	443	250
N,N-Dimethylacetamide	C_4H_9NO	127-19-5	87.12	-20	165	0.9366^{25}	2.02	0.07	(3.7)	70	2-12%	490	10
Dimethylamine	C_2H_7N	124-40-3	45.08	-92	7	0.6804^{0}	3.05	203	1.01	20	3-14%	400	5
Dimethyl disulfide	$C_2H_6S_2$	624-92-0	94.20	-85	109.8	1.0625^{20}	1.55	3.82	(1.8)	24			
N,N-Dimethylformamide	C_3H_7NO	68-12-2	73.09	-60	153	0.944^{25}	2.06	0.44	3.82	58	2-15%	445	10
Dimethyl sulfoxide	C_2H_6OS	67-68-5	78.14	19	189	1.1014^{20}	1.96	0.08	3.96	95	3-42%	215	
1,4-Dioxane	$C_4H_8O_2$	123-91-1	88.11	12	101	1.0337^{20}	1.74	4.95	0	12	2-22%	180	25
1,3-Dioxolane	$C_3H_6O_2$	646-06-0	74.08	-95	78	1.060^{20}	1.59	14.6	1.19	2			
Dipentene	$C_{10}H_{16}$	7705-14-8	136.24	-95.5	178	0.8402^{21}	1.83	0.26		45		237	
Epichlorohydrin	C_3H_5ClO	106-89-8	92.52	-26	116	1.1812^{20}	1.42	2.2	(1.8)	31	4-21%	411	2
Ethanolamine (Glycinol)	C_2H_7NO	141-43-5	61.08	11	171	1.0180^{20}	3.20	0.05	(2.3)	86	3-24%	410	3
Ethyl acetate	$C_4H_8O_2$	141-78-6	88.11	-84	77	0.9003^{20}	1.94	12.6	1.78	-4	2-12%	426	400
Ethyl acetoacetate	$C_6H_{10}O_3$	141-97-9	130.14	-45	180.8	1.0368^{10}	1.91	0.09		57	1-10%	295	
Ethyl alcohol	C_2H_6O	64-17-5	46.07	-114	78	0.7893^{20}	2.44	7.87	1.69	13	3-19%	363	1000
Ethylamine	C_2H_7N	75-04-7	45.08	-81	17	0.686^{17}	2.88	142	1.22	<-18	4-14%	385	5
Ethylbenzene	C_8H_{10}	100-41-4	106.17	-95	136	0.8670^{20}	1.73	1.28	0.59	21	1-7%	432	100
Ethyl bromide	C_2H_5Br	74-96-4	108.97	-118.6	38.5	1.4604^{20}	0.93	62.5	2.03		7-8%	511	5
Ethyl chloride	C_2H_5Cl	75-00-3	64.51	-139	12	0.909^{12}	1.62	160	2.05	-50	4-15%	519	1000
Ethylene carbonate	$C_3H_4O_3$	96-49-1	88.06	36.4	248	1.3214^{39}	1.52	<0.01	(4.9)	143			
Ethylenediamine	$C_2H_8N_2$	107-15-3	60.10	11	117	0.8979^{20}	2.87	1.62	1.99	40	3-12%	385	10
Ethylene glycol	$C_2H_6O_2$	107-21-1	62.07	-13	197	1.1088^{20}	2.41	0.01	2.28	111	3-22%	398	50
Ethylene glycol diethyl ether	$C_6H_{14}O_2$	629-14-1	118.18	-74	119.4	0.8484^{20}	2.19	4.33		35			
Ethylene glycol dimethyl ether	$C_4H_{10}O_2$	110-71-4	90.12	-58	85	0.8691^{20}	2.14	9.93		-2		202	
Ethylene glycol monobutyl ether	$C_6H_{14}O_2$	111-76-2	118.18	-75	168	0.9015^{20}	2.38	0.15	(2.1)	69	4-13%	238	25
Ethylene glycol monoethyl ether	$C_4H_{10}O_2$	110-80-5	90.12	-70	135	0.9297^{20}	2.34	0.71	(2.1)	43	3-18%	235	5
Ethylene glycol ethyl ether acetate	$C_6H_{12}O_3$	111-15-9	132.16	-62	156	0.9740^{20}	2.85	0.24	(2.2)	56	2-8%	379	5
Ethylene glycol monomethyl ether	$C_3H_8O_2$	109-86-4	76.10	-85	124	0.9647^{20}	2.25	1.31	2.36	39	2-14%	285	5
Ethylene glycol momomethyl ether acetate	$C_5H_{10}O_3$	110-49-6	118.13	-70	143	1.0074^{19}	2.62	0.67	(2.1)	49	2-12%	392	5
Ethyl formate	$C_3H_6O_2$	109-94-4	74.08	-80	54	0.9168^{20}	2.02	32.3	1.9	-20	3-16%	455	100
Furan	C_4H_4O	110-00-9	68.08	-86	31	0.9514^{20}	1.69	80.0	0.66	<0	2-14%		
Furfural	$C_5H_4O_2$	98-01-1	96.09	-37	162	1.1594^{20}	1.70	0.29	(3.5)	60	2-19%	316	2
Furfuryl alcohol	$C_5H_6O_2$	98-00-0	98.10	-31	171	1.1296^{20}	2.08	0.10	(1.9)	75	2-16%	491	10
Glycerol	$C_3H_8O_3$	56-81-5	92.09	18	290	1.2613^{20}	2.38	<0.01	(2.6)	199	3-19%	370	
Heptane	C_7H_{16}	142-82-5	100.20	-91	98	0.6837^{20}	2.24	6.09	≈0	-4	1-7%	204	400
1-Heptanol	$C_7H_{16}O$	111-70-6	116.20	-34	176.4	0.8219^{20}	2.34						
Hexane	C_6H_{14}	110-54-3	86.18	-95	69	0.6548^{25}	2.27	20.2	≈0	-22	1-8%	225	50
1-Hexanol (Caproyl alcohol)	$C_6H_{14}O$	111-27-3	102.18	-44.6	157.6	0.8136^{20}	2.35	0.11		63			
Hexylene glycol	$C_6H_{14}O_2$	107-41-5	118.18	-50	197	0.923^{15}	2.84	<0.01	(2.9)	102	1-9%	306	25
Hexyl methyl ketone	$C_8H_{16}O$	111-13-7	128.21	-16	172.5	0.820^{20}	2.13		(2.7)	52			
Isobutyl acetate	$C_6H_{12}O_2$	110-19-0	116.16	-99	117	0.8712^{20}	2.01	2.39	(1.9)	18	1-11%	421	150
Isobutyl alcohol	$C_4H_{10}O$	78-83-1	74.12	-108	108	0.8018^{20}	2.44	1.39	1.64	28	2-11%	415	50

PROPERTIES OF COMMON LABORATORY SOLVENTS (continued)

Name	MF	CAS RN	M_r	t_m/°C	t_b/°C	ρ/g cm^{-3}	c_p/J g^{-1}K^{-1}	vp/kPa	μ/D	FP/°C	Fl. Lim.	Ign. Temp./°C	TLV
Isobutylamine	$C_4H_{11}N$	78-81-9	73.14	-87	68	0.724^{25}	2.50	19.0	(1.3)	-9	2-12%	378	
Isopentyl acetate	$C_7H_{14}O_2$	123-92-2	130.19	-79	143	0.876^{15}	1.91	0.73	(1.9)	25	1-8%	360	100
Isophorone	$C_9H_{14}O$	78-59-1	138.21	-8	215	0.9255^{20}	1.83	0.06		84	1-4%	460	5
Isopropyl acetate	$C_5H_{10}O_2$	108-21-4	102.13	-73	89	0.8718^{20}	1.95	8.1		2	2-8%	460	250
Isopropyl alcohol	C_3H_8O	67-63-0	60.10	-90	82	0.7855^{20}	2.58	6.02	1.56	12	2-13%	399	400
Isoquinoline	C_9H_7N	119-65-3	129.16	26.47	243.2	1.0910^{30}	1.52		2.73				
d-Limonene (Citrene)	$C_{10}H_{16}$	5989-27-5	136.24	-97	178	0.8411^{20}	1.83	0.28		49			
2,6-Lutidine	C_7H_9N	108-48-5	107.16	-6.1	144.1	0.9226^{20}	1.73	0.75	(1.7)				
Mesitylene	C_9H_{12}	108-67-8	120.19	-45	165	0.8652^{20}	1.74	0.33	0	50	1-5%	559	25
Mesityl oxide	$C_6H_{10}O$	141-79-7	98.14	-59	130	0.8653^{20}	2.17	1.47	(2.8)	31	1-7%	344	15
Methyl acetate	$C_3H_6O_2$	79-20-9	74.08	-98	57	0.9342^{20}	1.92	28.8	1.72	-10	3-16%	454	200
Methylal	$C_3H_8O_2$	109-87-5	76.10	-105	42	0.8593^{20}	2.12	53.1	(0.7)	-32	2-14%	237	1000
Methyl alcohol	CH_4O	67-56-1	32.04	-98	65	0.7914^{20}	2.53	16.9	1.70	11	6-36%	464	200
Methylamine	CH_5N	74-89-5	31.06	-93	-6	0.656^{25}	3.29	353	1.31	0	5-21%	430	5
Methyl benzoate	$C_8H_8O_2$	93-58-3	136.15	-15	199	1.0933^{15}	1.63	0.05	(1.9)	83			
Methylcyclohexane	C_7H_{14}	108-87-2	98.19	-127	101	0.7694^{20}	1.88	6.18	≈0	-4	1-7%	250	400
Methyl ethyl ketone	C_4H_8O	78-93-3	72.11	-87	80	0.8054^{20}	2.20	12.6	2.78	-9	1-11%	404	200
N-Methylformamide	C_2H_5NO	123-39-7	59.07	-3.8	199.5	1.011^{19}	2.10		3.83				
Methyl formate	$C_2H_4O_2$	107-31-3	60.05	-99	32	0.9742^{20}	1.98	78.1	1.77	-19	5-23%	449	100
Methyl iodide	CH_3I	74-88-4	141.94	-66.4	42.5	2.279^{20}	0.89	53.9	1.62				2
Methyl isobutyl ketone	$C_6H_{12}O$	108-10-1	100.16	-84	116	0.7978^{20}	2.13	2.64		18	1-8%	448	50
Methyl isopentyl ketone	$C_7H_{14}O$	110-12-3	114.19		144	0.888^{20}		0.69		36	1-8%	191	50
2-Methylpentane	C_6H_{14}	107-83-5	86.18	-153.7	60.2	0.650^{25}	2.25	28.2	≈0	<-29	1-7%	264	
4-Methyl-2-pentanol	$C_6H_{14}O$	108-11-2	102.18	-90	132	0.8075^{20}	2.67	0.70		41	1-6%		25
Methyl pentyl ketone	$C_7H_{14}O$	110-43-0	114.19	-35	151	0.8111^{20}	2.04	0.49	(2.6)	39	1-8%	393	50
Methyl propyl ketone	$C_5H_{10}O$	107-87-9	86.13	-77	102	0.809^{20}	2.14	4.97	(2.7)	7	2-8%	452	200
N-Methyl-2-pyrrolidone	C_5H_9NO	872-50-4	99.13	-24	202	1.0230^{25}	3.11	0.04	(4.1)	96	1-10%	346	
Morpholine	C_4H_9NO	110-91-8	87.12	-5	128	1.0005^{20}	1.89	1.34	1.55	37	1-11%	290	20
Nitrobenzene	$C_6H_5NO_2$	98-95-3	123.11	6	211	1.2037^{20}	1.51	0.03	4.22	88	2-9%	482	1
Nitroethane	$C_2H_5NO_2$	79-24-3	75.07	-90	114	1.0448^{25}	1.79	2.79	3.23	28	3-17%	414	100
Nitromethane	CH_3NO_2	75-52-5	61.04	-29	101	1.1371^{20}	1.75	4.79	3.46	35	7-22%	418	20
1-Nitropropane	$C_3H_7NO_2$	108-03-2	89.09	-108	131.1	0.9961^{25}	1.97	1.36	3.66	36	2%-	421	25
2-Nitropropane	$C_3H_7NO_2$	79-46-9	89.09	-91	120	0.9821^{25}	1.91	2.3	3.73	24	3-11%	428	10
Octane	C_8H_{18}	111-65-9	114.23	-57	126	0.6986^{25}	2.23	1.86	≈0	13	1-7%	206	300
1-Octanol	$C_8H_{18}O$	111-87-5	130.23	-15.5	195.1	0.8262^{25}	2.34	0.01	(1.8)	81			
Pentachloroethane	C_2HCl_5	76-01-7	202.29	-29	160	1.6796^{20}	0.86	0.48	0.92				
Pentamethylene glycol	$C_5H_{12}O_2$	111-29-5	104.15	-18	239	0.9914^{20}	3.08		(2.5)	129		335	
Pentane	C_5H_{12}	109-66-0	72.15	-130	36	0.6262^{20}	2.32	68.3	≈0	<-40	2-8%	260	600
1-Pentanol	$C_5H_{12}O$	71-41-0	88.15	-79	138	0.8144^{20}	2.36	0.26	(1.7)	33	1-10%	300	
Pentyl acetate	$C_7H_{14}O_2$	628-63-7	130.19	-71	149	0.8756^{20}	2.00	0.60	1.75	16	1-8%	360	100
2-Picoline	C_6H_7N	109-06-8	93.13	-67	129	0.9443^{20}	1.70	1.5	1.85	39		538	
α-Pinene	$C_{10}H_{16}$	80-56-8	136.24	-64	156	0.8539^{25}		0.64		35		275	
β-Pinene	$C_{10}H_{16}$	127-91-3	136.24	-61.5	166	0.860^{25}		0.61		38		275	
Piperidine	$C_5H_{11}N$	110-89-4	85.15	-11	106	0.8606^{20}	2.11	4.28	(1.2)	16	1-10%		
Propanenitrile	C_3H_5N	107-12-0	55.08	-93	97	0.7818^{20}	2.17	6.14	4.05	2	3-14%	512	
Propyl acetate	$C_5H_{10}O_2$	109-60-4	102.13	-93	102	0.8878^{20}	1.92	4.49	(1.8)	13	2-8%	450	200
Propyl alcohol	C_3H_8O	71-23-8	60.10	-126	97	0.8035^{20}	2.39	2.76	1.55	23	2-14%	412	200
Propylamine	C_3H_9N	107-10-8	59.11	-83	47	0.7173^{20}	2.75	42.1	1.17	-37	2-10%	318	
Propylbenzene	C_9H_{12}	103-65-1	120.19	-99.5	159.2	0.8620^{20}	1.79		≈0	30	1-6%	450	

PROPERTIES OF COMMON LABORATORY SOLVENTS (continued)

Name	MF	CAS RN	M_r	t_m/°C	t_b/°C	ρ/g cm^{-3}	c_p/J g^{-1}K^{-1}	vp/kPa	μ/D	FP/°C	Fl. Lim.	Ign. Temp./°C	TLV
Propylene glycol	$C_3H_8O_2$	57-55-6	76.10	-60	188	1.0361^{20}	2.51	0.02	(2.2)	99	3-13%	371	
Pseudocumene	C_9H_{12}	95-63-6	120.19	-44	169	0.8758^{20}	1.79	0.30	≈0	44	1-6%	500	25
Pyridine	C_5H_5N	110-86-1	79.10	-42	115	0.9819^{20}	1.68	2.76	2.21	20	2-12%	482	5
Pyrrole	C_4H_5N	109-97-7	67.09	-23.4	129.7	0.9698^{20}	1.90	1.10	1.74	39			
Pyrrolidine	C_4H_9N	123-75-1	71.12	-57.8	86.5	0.8586^{20}	2.20	8.40	(1.6)	3			
2-Pyrrolidone	C_4H_7NO	616-45-5	85.11	25	251	1.120^{20}	1.99		(3.5)	129			
Quinoline	C_9H_7N	91-22-5	129.16	-14.78	237.1	1.0977^{15}	1.51		2.29			480	
Styrene	C_8H_8	100-42-5	104.15	-31	145	0.9060^{20}	1.75	0.81		31	1-7%	490	50
Sulfolane	$C_4H_8O_2S$	126-33-0	120.17	28	287	1.2723^{18}	1.50	<0.01	(4.8)	177			
α-Terpinene	$C_{10}H_{16}$	99-86-5	136.24		174	0.8375^{19}							
1,1,1,2-Tetrachloro-2,2-difluoroethane	$C_2Cl_4F_2$	76-11-9	203.83	40.6	91.5	1.649^{25}		7.36					500
1,1,2,2-Tetrachloro-1,2-difluoroethane	$C_2Cl_4F_2$	76-12-0	203.83	26	93	1.6447^{25}	0.85	7.51					500
1,1,1,2-Tetrachloroethane	$C_2H_2Cl_4$	630-20-6	167.85	-70	131	1.5406^{20}	0.92	1.6			5-12%		
1,1,2,2-Tetrachloroethane	$C_2H_2Cl_4$	79-34-5	167.85	-44	146	1.5953^{20}	0.97	0.62	1.32		20-54%		1
Tetrachloroethylene	C_2Cl_4	127-18-4	165.83	-22	121	1.6227^{20}	0.86	2.42	0				50
Tetraethylene glycol	$C_8H_{18}O_5$	112-60-7	194.23	-6.2	328	1.1285^{15}	2.21			182			
Tetrahydrofuran	C_4H_8O	109-99-9	72.11	-108	65	0.8892^{20}	1.72	21.6	1.75	-14	2-12%	321	200
1,2,3,4-Tetrahydronaphthalene	$C_{10}H_{12}$	119-64-2	132.21	-36	208	0.9660^{25}	1.65	0.05	≈0	71	1-5%	385	
Tetrahydropyran	$C_5H_{10}O$	142-68-7	86.13	-45	88	0.8814^{20}	1.82	9.54	1.74	-20			
Tetramethylsilane	$C_4H_{12}Si$	75-76-3	88.22	-99.0	26.6	0.648^{19}	2.31	94.2	0				
Toluene	C_7H_8	108-88-3	92.14	-95	111	0.8669^{20}	1.70	3.79	0.37	4	1-7%	480	50
o-Toluidine	C_7H_9N	95-53-4	107.16	-16.3	200.3	0.9984^{20}	1.96	0.04	(1.6)	85		482	2
Triacetin	$C_9H_{14}O_6$	102-76-1	218.21	-78	259	1.1583^{20}	1.76	<0.01		138	1%-	433	
Tributylamine	$C_{12}H_{27}N$	102-82-9	185.35	-70	217	0.7770^{20}		0.01	(0.8)	86	1-5%		
1,1,1-Trichloroethane	$C_2H_3Cl_3$	71-55-6	133.40	-30	74	1.3390^{20}	1.08	16.5	1.76		8-13%	537	350
1,1,2-Trichloroethane	$C_2H_3Cl_3$	79-00-5	133.40	-37	114	1.4397^{20}	1.13	3.1	(1.4)	32	6-28%	460	10
Trichloroethylene	C_2HCl_3	79-01-6	131.39	-85	87	1.4642^{20}	0.95	9.91	(0.8)	32	8-11%	420	50
Trichlorofluoromethane	CCl_3F	75-69-4	137.37	-111	24	1.478^{24}	0.89	106	0.46				1000
1,1,2-Trichlorotrifluoroethane	$C_2Cl_3F_3$	76-13-1	187.38	-35	48	1.5635^{25}	0.91	44.8					1000
Triethanolamine	$C_6H_{15}NO_3$	102-71-6	149.19	21	335	1.1242^{20}	2.61	<0.01	(3.6)	179	1-10%		0.5
Triethylamine	$C_{10}H_{22}O_2$	121-44-8	101.19	-115	89	0.7275^{20}	2.17	7.70	0.66	-7	1-8%	249	1
Triethylene glycol	$C_6H_{14}O_4$	112-27-6	150.17	-7	285	1.1274^{15}	2.18			177	1-9%	371	
Triethyl phosphate	$C_6H_{15}O_4P$	78-40-0	182.16	-56.4	215.5	1.0695^{20}			(3.1)	115		454	
Trimethylamine	C_3H_9N	75-50-3	59.11	-117	3	0.627^{25}	2.33	215	0.61	-7	2-12%	190	5
Trimethylene glycol	$C_3H_8O_2$	504-63-2	76.10	-26.7	214.4	1.0538^{20}			(2.5)			400	
Trimethyl phosphate	$C_3H_9O_4P$	512-56-1	140.08	-46	197.2	1.2144^{20}		0.11	(3.2)	107			
Veratrole	$C_8H_{10}O_2$	91-16-7	138.17	22.5	206	1.0810^{25}			(1.3)				
o-Xylene	C_8H_{10}	95-47-6	106.17	-25	144	0.8802^{10}	1.75	0.88	0.64	32	1-7%	463	100
m-Xylene	C_8H_{10}	108-38-3	106.17	-48	139	0.8642^{20}	1.72	1.13	≈0	27	1-7%	527	100
p-Xylene	C_8H_{10}	106-42-3	106.17	13	138	0.8611^{20}	1.71	1.19	0	27	1-7%	528	100

DEPENDENCE OF BOILING POINT ON PRESSURE

The normal boiling point of a liquid is defined as the temperature at which the vapor pressure reaches standard atmospheric pressure, 101.325 kPa. The change in boiling point with pressure may be calculated from the representation of the vapor pressure by the Antoine Equation,

$$\ln p = A_1 - A_2/(T + A_3)$$

where p is the vapor pressure, T the absolute temperature, and A_1, A_2, and A_3 are constants. This table, which has been calculated using the Antoine constants in Reference 1, gives values of $\Delta t/\Delta p$ for a number of liquids, in units of both °C/kPa and °C/mmHg. The correction to the boiling point is generally accurate to 0.1 to 0.2 °C as long as the pressure is within 10% of standard atmospheric pressure.

A slightly less accurate estimate of $\Delta t/\Delta p$ may be obtained from the Claussius-Clapeyron equation, with the assumption that the change in volume upon vaporization equals the ideal-gas volume of the vapor. This leads to the equation

$$\Delta t/\Delta p = RT_b^2/p_0\Delta_{vap}H(T_b)$$

where R is the molar gas constant, p_0 is 101.325 kPa, T_b is the normal boiling point temperature (absolute), and $\Delta_{vap}H(T_b)$ is the molar enthalpy of vaporization at the normal boiling point. Values of the last quantity may be obtained from the table "Enthalpy of Vaporization" in Section 6.

REFERENCE

1. Lide, D.R., and Kehiaian, H.V., *CRC Handbook of Thermophysical and Thermochemical Data*, CRC Press, Boca Raton, FL, 1994, pp. 49-59.

Compound	t_b °C	$\Delta t/\Delta p$ °C/kPa	°C/mmHg
Acetaldehyde	20.1	0.261	0.0348
Acetic acid	117.9	0.324	0.0432
Acetone	56.0	0.289	0.0385
Acetonitrile	81.6	0.316	0.0421
Ammonia	-33.33	0.198	0.0264
Aniline	184.1	0.378	0.0504
Anisole	153.7	0.367	0.0489
Benzaldehyde	179.0	0.392	0.0523
Benzene	80.0	0.321	0.0428
Bromine	58.8	0.300	0.0400
Butane	-0.5	0.267	0.0356
1-Butanol	117.7	0.278	0.0371
Carbon disulfide	46.2	0.304	0.0405
Chlorine	-34.04	0.224	0.0299
Chlorobenzene	131.7	0.365	0.0487
1-Chlorobutane	78.6	0.321	0.0428
Chloroethane	12.3	0.262	0.0349
Chloroethylene	-13.3	0.241	0.0321
Cyclohexane	80.7	0.328	0.0437
Cyclohexanol	160.8	0.344	0.0459
Cyclohexanone	155.4	0.382	0.0509
Decane	174.1	0.388	0.0517
Dibutyl ether	140.2	0.363	0.0484
Dichloromethane	39.6	0.276	0.0368
Diethyl ether	34.5	0.278	0.0371
Dimethyl sulfoxide	189.0	0.379	0.0505
1,4-Dioxane	101.5	0.321	0.0428
Dipropyl ether	90.0	0.326	0.0435
Ethanol	78.2	0.249	0.0332
Ethyl acetate	77.1	0.300	0.0400
Ethylene glycol	197.3	0.331	0.0441
Heptane	98.5	0.336	0.0448
Hexafluorobenzene	80.2	0.305	0.0407
Hexane	68.7	0.314	0.0419
1-Hexanol	157.6	0.318	0.0424
Hydrogen fluoride	20.1	0.276	0.0368
Iodomethane	42.5	0.291	0.0388
Isobutane	-11.7	0.254	0.0339
Methanol	64.6	0.251	0.0335
Methyl acetate	56.8	0.282	0.0376
Methyl formate	31.7	0.582	0.0776
N-Methylaniline	196.2	0.396	0.0528
N-Methylformamide	199.5	0.371	0.0495
Nitrobenzene	210.8	0.418	0.0557
Nitromethane	101.1	0.320	0.0427
1-Octanol	195.1	0.360	0.0480
Pentane	36.0	0.289	0.0385
1-Pentanol	137.9	0.296	0.0395
Phenol	181.8	0.349	0.0465
Propane	-42.1	0.224	0.0299
1-Propanol	97.2	0.261	0.0348
2-Propanol	82.3	0.247	0.0329
Pyridine	115.2	0.340	0.0453
Pyrrole	129.7	0.330	0.0440
Pyrrolidine	86.5	0.309	0.0412
Styrene	145.1	0.369	0.0492
Sulfur dioxide	-10.05	0.221	0.0295
Tetrachloroethylene	121.3	0.354	0.0472
Tetrachloromethane	76.8	0.325	0.0433
Toluene	110.6	0.353	0.0471
Trichloroethylene	87.2	0.330	0.0440
Trichloromethane	61.1	0.302	0.0403
Trimethylamine	2.8	0.248	0.0331
Water	100.0	0.276	0.0368
o-Xylene	144.5	0.373	0.0497
m-Xylene	139.1	0.368	0.0491
p-Xylene	138.3	0.369	0.0492

EBULLIOSCOPIC CONSTANTS FOR CALCULATION OF BOILING POINT ELEVATION

The boiling point T_b of a dilute solution of a non-volatile, non-dissociating solute is elevated relative to that of the pure solvent. If the solution is ideal (i.e., follows Raoult's Law), the amount of elevation depends only on the number of particles of solute present. Hence the change in boiling point ΔT_b can be expressed as

$$\Delta T_b = E_b\, m_2$$

where m_2 is the molality (moles of solute per kilogram of solvent) and E_b is the Ebullioscopic Constant, a characteristic property of the solvent. The Ebullioscopic Constant may be calculated from the relation

$$E_b = R\, T_b^2\, M/\Delta_{vap}H$$

where R is the molar gas constant, T_b is the normal boiling point temperature (absolute) of the solvent, M the molar mass of the solvent, and $\Delta_{vap}H$ the molar enthalpy (heat) of vaporization of the solvent at its normal boiling point.

This table lists E_b values for some common solvents, as calculated from data in the table "Enthalpy of Vaporization" in Section 6.

Compound	E_b/K kg mol^{-1}
Acetic acid	3.22
Acetone	1.80
Acetonitrile	1.44
Aniline	3.82
Anisole	4.20
Benzaldehyde	4.24
Benzene	2.64
1-Butanol	2.17
Carbon disulfide	2.42
Chlorobenzene	4.36
1-Chlorobutane	3.13
Cyclohexane	2.92
Cyclohexanol	3.5
Decane	6.10
Dichloromethane	2.42
Diethyl ether	2.20
Dimethyl sulfoxide	3.22
1,4-Dioxane	3.01
Ethanol	1.23
Ethyl acetate	2.82
Ethylene glycol	2.26
Heptane	3.62
Hexane	2.90
Iodomethane	4.31
Methanol	0.86
Methyl acetate	2.21
N-Methylaniline	4.3
N-Methylformamide	2.2
Nitrobenzene	5.2
Nitromethane	2.09
1-Octanol	5.06
Phenol	3.54
1-Propanol	1.66
2-Propanol	1.58
Pyridine	2.83
Pyrrole	2.33
Pyrrolidine	2.32
Tetrachloroethylene	6.18
Tetrachloromethane	5.26
Toluene	3.40
Trichloroethylene	4.52
Trichloromethane	3.80
Water	0.513
o-Xylene	4.25

CRYOSCOPIC CONSTANTS FOR CALCULATION OF FREEZING POINT DEPRESSION

The freezing point T_f of a dilute solution of a non-volatile, non-dissociating solute is depressed relative to that of the pure solvent. If the solution is ideal (i.e., follows Raoult's Law), this lowering is a function only of the number of particles of solute present. Thus the absolute value of the lowering of freezing point ΔT_f can be expressed as

$$\Delta T_f = E_f m_2$$

where m_2 is the molality (moles of solute per kilogram of solvent) and E_f is the Cryoscopic Constant, a characteristic property of the solvent. The Cryoscopic Constant may be calculated from the relation

$$E_f = R\, T_f^2\, M/\Delta_{fus}H$$

where R is the molar gas constant, T_b is the freezing point temperature (absolute) of the solvent, M the molar mass of the solvent, and $\Delta_{fus}H$ the molar enthalpy (heat) of fusion of the solvent.

This table lists cryscopic constants for selected substances, as calculated from data in the table "Enthalpy of Fusion" in Section 6.

Compound	E_f/K kg mol^{-1}
Acetamide	3.92
Acetic acid	3.63
Acetophenone	5.16
Aniline	5.23
Benzene	5.07
Benzonitrile	5.35
Benzophenone	8.58
(+)-Camphor	37.8
1-Chloronaphthalene	7.68
o-Cresol	5.92
m-Cresol	7.76
p-Cresol	7.20
Cyclohexane	20.8
Cyclohexanol	42.2
cis-Decahydronaphthalene	6.42
trans-Decahydronaphthalene	4.70
Dibenzyl ether	6.17
p-Dichlorobenzene	7.57
Diethanolamine	3.16
Dimethyl sulfoxide	3.85
1,4-Dioxane	4.63
Diphenylamine	8.38
Ethylene glycol	3.11
Formamide	4.25
Formic acid	2.38
Glycerol	3.56
Methylcyclohexane	2.60
Naphthalene	7.45
Nitrobenzene	6.87
Phenol	6.84
Pyridine	4.26
Quinoline	6.73
Succinonitrile	19.3
1,1,2,2-Tetrabromoethane	21.4
1,1,2,2-Tetrachloro-1,2-difluoroethane	41.0
Toluene	3.55
p-Toluidine	4.91
Tribromomethane	15.0
Water	1.86
p-Xylene	4.31

FREEZING POINT LOWERING BY ELECTROLYTES IN AQUEOUS SOLUTION

REFERENCE

Forsythe, W. E., *Smithsonian Physical Tables, Ninth Edition*, Smithsonian Institution, Washington, 1956.

Compound	Lowering of freezing point of water (in °C) as function of molality (mol/kg)									
	0.05	**0.10**	**0.25**	**0.50**	**0.75**	**1.00**	**1.50**	**2.00**	**2.50**	**3.00**
$CaCl_2$	0.25	0.49	1.27	2.66	4.28	6.35	10.78	15.27	20.42	28.08
$CuSO_4$	0.13	0.23	0.47	0.96						
HCl	0.18	0.36	0.90	1.86	2.90	4.02	6.63	9.94		
HNO_3	0.18	0.35	0.88	1.80	2.78	3.80	5.98	8.34	10.95	13.92
H_2SO_4	0.20	0.39	0.96	1.95	3.04	4.28	7.35	11.35	16.32	
KBr	0.18	0.36	0.92	1.78						
KCl	0.17	0.35	0.86	1.68	2.49	3.29	4.88	6.50	8.14	9.77
KNO_3	0.17	0.33	0.78	1.47	2.11	2.66				
K_2SO_4	0.23	0.43	1.01	1.87						
LiCl	0.18	0.35	0.88	1.80	2.78					
$MgSO_4$	0.13	0.24	0.55	1.01	1.50	2.08	3.41			
NH_4Cl	0.17	0.34	0.85	1.70	2.55					
NaCl	0.18	0.35	0.85	1.68	2.60					
$NaNO_3$	0.18	0.36	0.80	1.62	2.63	3.10				

DETERMINATION OF RELATIVE HUMIDITY FROM DEW POINT

The relative humidity of a water vapor-air mixture is defined as 100 times the partial pressure of water divided by the saturation vapor pressure of water at the same temperature. The relative humidity may be determined from the dew point t_{dew}, which is the temperature at which liquid water first condenses when the mixture is cooled from an initial temperature t. This table gives relative humidity as a function of the dew point depression $t - t_{dew}$ for several values of the dew point. Values are calculated from the vapor pressure table in Section 6.

$t - t_{dew}$	t_{dew}/°C -10	0	10	20	30
0.0	100	100	100	100	100
0.2	99	99	99	99	99
0.4	97	97	97	98	98
0.6	95	96	96	96	97
0.9	94	94	95	95	96
1.0	92	93	94	94	94
1.2	91	92	92	93	93
1.4	90	90	91	92	92
1.6	88	89	90	91	91
1.8	87	88	89	90	90
2.0	86	87	88	88	89
2.2	84	85	86	87	89
2.4	83	84	85	86	87
2.6	82	83	84	85	86
2.8	80	82	83	84	85
3.0	79	81	82	83	84
3.2	78	80	81	82	83
3.4	77	79	80	81	82
3.6	76	77	79	80	82
3.8	75	76	78	79	81
4.0	73	75	77	78	80
4.2	72	74	76	77	79
4.4	71	73	75	77	78
4.6	70	72	74	76	77
4.8	69	71	73	75	76
5.0	69	70	72	74	75
5.2	67	69	71	73	75
5.4	66	68	70	72	74
5.6	65	67	69	71	73
5.9	64	66	69	70	72
6.0	63	66	68	70	71
6.2	62	65	67	69	71
6.4	61	64	66	68	70
6.6	60	63	65	67	69
6.8	60	62	64	66	68
7.0	59	61	63	66	68
7.2	58	60	63	65	67
7.4	57	60	62	64	66
7.6	56	59	61	63	65
7.8	55	58	60	63	65
8.0	54	57	60	62	64
8.2	54	56	59	61	63
8.4	53	56	58	60	63
8.6	52	55	57	60	62
8.8	51	54	57	59	61
9.0	51	53	56	58	61
9.2	50	53	55	58	60
9.4	49	52	55	57	59
9.6	48	51	54	56	59
9.8	48	51	53	56	58
10.0	47	50	53	55	57
10.5	45	48	51	54	56
11.0	44	47	49	52	55
11.5	42	45	48	51	53
12.0	41	44	47	49	52
12.5	39	42	45	48	50
13.0	38	41	44	46	49
13.5	37	40	43	45	48
14.0	35	38	41	44	47
14.5	34	37	40	43	45
15.0	33	36	39	42	44
15.5	32	35	38	40	
16.0	31	34	37	39	
16.5	30	33	36	38	
17.0	29	32	35	37	
17.5	28	31	34	36	
18.0	27	30	33	35	
18.5	26	29	32	34	
19.0	25	28	31	33	
19.5	24	27	30	33	
20.0	24	26	29	32	
21.0	22	25	27	30	
22.0	21	23	26	29	
23.0	19	22	24	27	
24.0	18	21	23	26	
25.0	17	19	22	24	
26.0	16	18	21	23	
27.0	15	17	20	22	
28.0	14	16	19	21	
29.0	13	15	18	20	
30.0	12	14	17	19	

DETERMINATION OF RELATIVE HUMIDITY FROM WET AND DRY BULB TEMPERATURES

Relative humidity may be determined by comparing temperature readings of wet and dry bulb thermometers. The following table, extracted from more extensive U.S. National Weather Service tables, gives the relative humidity as a function of air temperature t_d (dry bulb) and the difference $t_d - t_w$ between dry and wet bulb temperatures. The data assume a pressure near normal atmospheric pressure and an instrumental configuration with forced ventilation.

	$(t_d - t_w)$/°C											
t_d/°C	**0.5**	**1.0**	**1.5**	**2.0**	**2.5**	**3.0**	**3.5**	**4.0**	**4.5**	**5.0**	**5.5**	**6.0**
-10	83	67	51	35	19							
-8	86	71	57	43	29	15						
-6	88	74	61	49	37	25	8					
-4	89	77	66	55	44	33	23	12				
-2	90	79	69	60	50	40	31	22	12			
0	91	81	72	64	55	46	38	29	21	13	5	
2	91	84	76	68	60	52	44	37	29	22	14	7
4	92	85	78	71	63	57	49	43	36	29	22	16
6	93	86	79	73	66	60	54	48	41	35	29	24
8	93	87	81	75	69	63	57	51	46	40	35	29
10	94	88	82	77	71	66	60	55	50	44	39	34
12	94	89	83	78	73	68	63	58	53	48	43	39
14	95	90	85	79	75	70	65	60	56	51	47	42
16	95	90	85	81	76	71	67	63	58	54	50	46
18	95	91	86	82	77	73	69	65	61	57	53	49
20	96	91	87	83	78	74	70	66	63	59	55	51
22	96	92	87	83	80	76	72	68	64	61	57	54
24	96	92	88	84	80	77	73	69	66	62	59	56
26	96	92	88	85	81	78	74	71	67	64	61	58
28	96	93	89	85	82	78	75	72	69	65	62	59
30	96	93	89	86	83	79	76	73	70	67	64	61
35	97	94	90	87	84	81	78	75	72	69	67	64
40	97	94	91	88	85	82	80	77	74	72	69	67

	$(t_d - t_w)$/°C											
t_d/°C	**6.5**	**7.0**	**7.5**	**8.0**	**8.5**	**9.0**	**10.0**	**11.0**	**12.0**	**13.0**	**14.0**	**15.0**
4	9											
6	17	11	5									
8	24	19	14	8								
10	29	24	20	15	10	6						
12	34	29	25	21	16	12	5					
14	38	34	30	26	22	18	10					
16	42	38	34	30	26	23	15	8				
18	45	41	38	34	30	27	20	14	7			
20	48	44	41	37	34	31	24	18	12	6		
22	50	47	44	40	37	34	28	22	17	11	6	
24	53	49	46	43	40	37	31	26	20	15	10	5
26	54	51	49	46	43	40	34	29	24	19	14	10
28	56	53	51	48	45	42	37	32	27	22	18	13
30	58	55	52	50	47	44	39	35	30	25	21	17
32	60	57	54	51	49	46	41	37	32	28	24	20
34	61	58	56	53	51	48	43	39	35	30	26	23
36	62	59	57	54	52	50	45	41	37	33	29	25
38	63	61	58	56	54	51	47	43	39	35	31	27
40	64	62	59	57	54	53	48	44	40	36	33	29

CONSTANT HUMIDITY SOLUTIONS

Anthony Wexler

An excess of a water soluble salt in contact with its saturated solution and contained within an enclosed space produces a constant relative humidity and water vapor pressure according to

$$RH = A\exp(B/T)$$

where *RH* is the percent relative humidity (generally accurate to ±2 %), *T* is the temperature in kelvin, and the constants *A* and *B* and the range of valid temperatures are given in the table below. The vapor pressure, *p*, can be calculated from

$$p = (RH/100) \times p_0$$

where p_0 is the vapor pressure of pure water at temperature *T* as given in the table in Section 6 titled "Vapor Pressure of Water from 0 to 370°C".

REFERENCES

1. Wexler, A. S. and Seinfeld, J. H., *Atmospheric Environment*, 25A, 2731, 1991.
2. Greenspan, L., *J. Res. National Bureau of Standards*, 81A, 89, 1977.
3. Broul, et al., *Solubility of Inorganic Two-Component Systems*, Elsevier, New York, 1981.
4. Wagman, D. D. et al., *J. Phys. Chem. Ref. Data*, Vol. 11, Suppl. 2, 1982.

Compound	Temperature range (°C)	*RH* 25°C	*A*	*B*
$NaOH \cdot H_2O$	15—60	6	5.48	27
$LiBr \cdot 2H_2O$	10—30	6	0.23	996
$ZnBr_2 \cdot 2H_2O$	5—30	8	1.69	455
$KOH \cdot 2H_2O$	5—30	9	0.014	1924
$LiCl \cdot H_2O$	20—65	11	14.53	–75
$CaBr_2 \cdot 6H_2O$	11—22	16	0.17	1360
$LiI \cdot 3H_2O$	15—65	18	0.15	1424
$CaCl_2 \cdot 6H_2O$	15—25	29	0.11	1653
$MgCl_2 \cdot 6H_2O$	5—45	33	29.26	34
$NaI \cdot 2H_2O$	5—45	38	3.62	702
$Ca(NO_3)_2 \cdot 4H_2O$	10—30	51	1.89	981
$Mg(NO_3)_2 \cdot 6H_2O$	5—35	53	25.28	220
$NaBr \cdot 2H_2O$	0—35	58	20.49	308
NH_4NO_3	10—40	62	3.54	853
KI	5—30	69	29.35	254
$SrCl_2 \cdot 6H_2O$	5—30	71	31.58	241
$NaNO_3$	10—40	74	26.94	302
NaCl	10—40	75	69.20	25
NH_4Cl	10—40	79	35.67	235
KBr	5—25	81	40.98	203
$(NH_4)_2SO_4$	10—40	81	62.06	79
KCl	5—25	84	49.38	159
$Sr(NO_3)_2 \cdot 4H_2O$	5—25	85	28.34	328
$BaCl_2 \cdot 2H_2O$	5—25	90	69.99	75
CsI	5—25	91	70.77	75
KNO_3	0—50	92	43.22	225
K_2SO_4	10—50	97	86.75	34

STANDARD SALT SOLUTIONS FOR HUMIDITY CALIBRATION

Saturated aqueous solutions of inorganic salts are convenient secondary standards for calibration of instruments for measurement of relative humidity. The International Union of Pure and Applied Chemistry has recommended salt solutions for calibrations in the range of 10% to 90% relative humidity, and the American Society for Testing and Materials has published similar standards. The data in this table are taken from the IUPAC recommendations, except for K_2CO_3 and K_2SO_4, which are ASTM recommendations.

Details on the preparation and use of these standards may be found in References 1 and 2. Data for other salts are given in Reference 3.

REFERENCES

1. Marsh, K. N., Editor, *Recommended Reference Materials for the Realization of Physicochemical Properties*, Blackwell Scientific Publications, Oxford, 1987, pp.157-162.
2. *Standard Practice for Maintaining Constant Relative Humidity by Means of Aqueous Solutions*, ASTM Standard E 104-85, Reapproved 1991.
3. Greenspan, L., *J. Res. Nat. Bur. Stand.*, 81A, 89, 1977.

Relative Humidity in %

t/°C	LiCl	$MgCl_2$	K_2CO_3	$Mg(NO_3)_2$	NaCl	KCl	K_2SO_4
0		33.66±0.33	43.1±0.7	60.35±0.55	75.51±0.34	88.61±0.53	98.8±2.1
5		33.60±0.28	43.1±0.5	58.86±0.43	75.65±0.27	87.67±0.45	98.5±0.9
10		33.47±0.24	43.1±0.4	57.36±0.33	75.67±0.22	86.77±0.39	98.2±0.8
15		33.30±0.21	43.2±0.3	55.87±0.27	75.61±0.18	85.92±0.33	97.9±0.6
20	11.31±0.31	33.07±0.18	43.2±0.3	54.38±0.23	75.47±0.14	85.11±0.29	97.6±0.5
25	11.30±0.27	32.78±0.16	43.2±0.4	52.89±0.22	75.29±0.12	84.34±0.26	97.3±0.5
30	11.28±0.24	32.44±0.14	43.2±0.5	51.40±0.24	75.09±0.11	83.62±0.25	97.0±0.4
35	11.25±0.22	32.05±0.13		49.91±0.29	74.87±0.12	82.95±0.25	96.7±0.4
40	11.21±0.21	31.60±0.13		48.42±0.37		82.32±0.25	96.4±0.4
45	11.16±0.21	31.10±0.13		46.93±0.47		81.74±0.28	96.1±0.4
50	11.10±0.22	30.54±0.14		45.44±0.60		81.20±0.31	95.8±0.5
55	11.03±0.23	29.93±0.16				80.70±0.35	
60	10.95±0.26	29.26±0.18				80.25±0.41	
65	10.86±0.29	28.54±0.21				79.85±0.48	
70	10.75±0.33	27.77±0.25				79.49±0.57	
75	10.64±0.38	26.94±0.29				79.17±0.66	
80	10.51±0.44	26.05±0.34				78.90±0.77	

LOW TEMPERATURE BATHS FOR MAINTAINING CONSTANT TEMPERATURE

A liquid-solid slurry is a convenient means of maintaining a constant temperature environment below room temperature. The following is a list of readily available organic liquids suitable for this purpose, arranged in order of their melting (freezing) points t_m. The normal boiling points t_b are also given.

Compound	t_m/°C	t_b/°C
Isopentane (2-Methylbutane)	-159.9	27.8
Methylcyclopentane	-142.5	71.8
3-Chloropropene (Allyl chloride)	-134.5	45.1
Pentane	-129.7	36.0
Allyl alcohol	-129	97.0
Ethanol	-114.1	78.2
Carbon disulfide	-111.5	46
Isobutyl alcohol	-108	107.8
Toluene	-94.9	110.6
Acetone	-94.8	56.0
Ethyl acetate	-83.6	77.1
Dry ice + acetone	-78	
p-Cymene	-68.9	177.1
Trichloromethane (Chloroform)	-63.6	61.1
N-Methylaniline	-57	196.2
Chlorobenzene	-45.2	131.7
Anisole	-37.5	153.7
Bromobenzene	-30.6	156.0
Tetrachloromethane (Carbon tetrachloride)	-23	76.8
Benzonitrile	-12.7	191.1

WIRE TABLES

The resistance per unit length of wires of various metals is tabulated here. Values were calculated from resistivity values in the tables "Electrical Resistivity of Pure Metals" and "Electrical Resistivity of Selected Alloys", which appear in Section 12. In practice, resistance may vary because of differing heat treatments and metal composition. The values in the table refer to 20°C, but values at other temperatures may be calculated from the following resistivity data:

Metal	Resistivity in 10^{-8} Ω m at temperature			
	0°C	20°C	25°C	100°C
Aluminum	2.417	2.650	2.709	3.56
Brass (70% Cu, 30% Zn)	5.87	6.08	6.13	6.91
Constantan (60% Cu, 40% Ni)	45.43	45.38	45.35	45.11
Copper	1.543	1.678	1.712	2.22
Nichrome (79% Ni, 21% Cr)	107.3	107.5	107.6	108.3
Platinum	9.6	10.5	10.7	13.6
Silver	1.467	1.587	1.617	2.07
Tungsten	4.82	5.28	5.39	7.18

Resistance per unit length at 20°C in Ω/m

B & S Gauge	Diameter (mm)	Aluminum	Brass	Constantan	Copper	Nichrome	Platinum	Silver	Tungsten
0	8.252	0.000495	0.00114	0.00848	0.000314	0.0201	0.00196	0.000297	0.00099
2	6.543	0.000788	0.00181	0.0135	0.000499	0.0320	0.00312	0.000472	0.00157
4	5.189	0.00125	0.00287	0.0214	0.000793	0.0508	0.00496	0.000750	0.00250
6	4.115	0.00199	0.00457	0.0341	0.00126	0.0808	0.00789	0.00119	0.00397
8	3.264	0.00317	0.00727	0.0542	0.00200	0.128	0.0125	0.00190	0.00631
10	2.588	0.00504	0.0115	0.0863	0.00319	0.204	0.0200	0.00302	0.0100
12	2.053	0.00800	0.0184	0.137	0.00507	0.325	0.0317	0.00479	0.0159
14	1.628	0.0127	0.0292	0.218	0.00806	0.516	0.0504	0.00762	0.0254
16	1.291	0.0202	0.0464	0.347	0.0128	0.821	0.0802	0.0121	0.0403
18	1.024	0.0322	0.0738	0.551	0.0204	1.30	0.127	0.0193	0.0641
20	0.8118	0.0512	0.117	0.877	0.0324	2.08	0.203	0.0307	0.102
22	0.6439	0.0814	0.187	1.39	0.0515	3.30	0.322	0.0487	0.162
24	0.5105	0.129	0.297	2.22	0.0820	5.25	0.513	0.0775	0.258
26	0.4049	0.206	0.472	3.52	0.130	8.35	0.815	0.123	0.410
28	0.3211	0.327	0.751	5.60	0.207	13.3	1.30	0.196	0.652
30	0.2548	0.520	1.19	8.90	0.329	21.1	2.06	0.311	1.03
32	0.2019	0.828	1.90	14.2	0.524	33.6	3.28	0.496	1.65
34	0.1601	1.32	3.02	22.5	0.833	53.4	5.22	0.788	2.62
36	0.1270	2.09	4.80	35.8	1.32	84.9	8.29	1.25	4.17
38	0.1007	3.33	7.63	57.0	2.11	135	13.2	1.99	6.63
40	0.07988	5.29	12.1	90.5	3.35	214	20.9	3.17	10.5

CHARACTERISTICS OF PARTICLES AND PARTICLE DISPERSOIDS

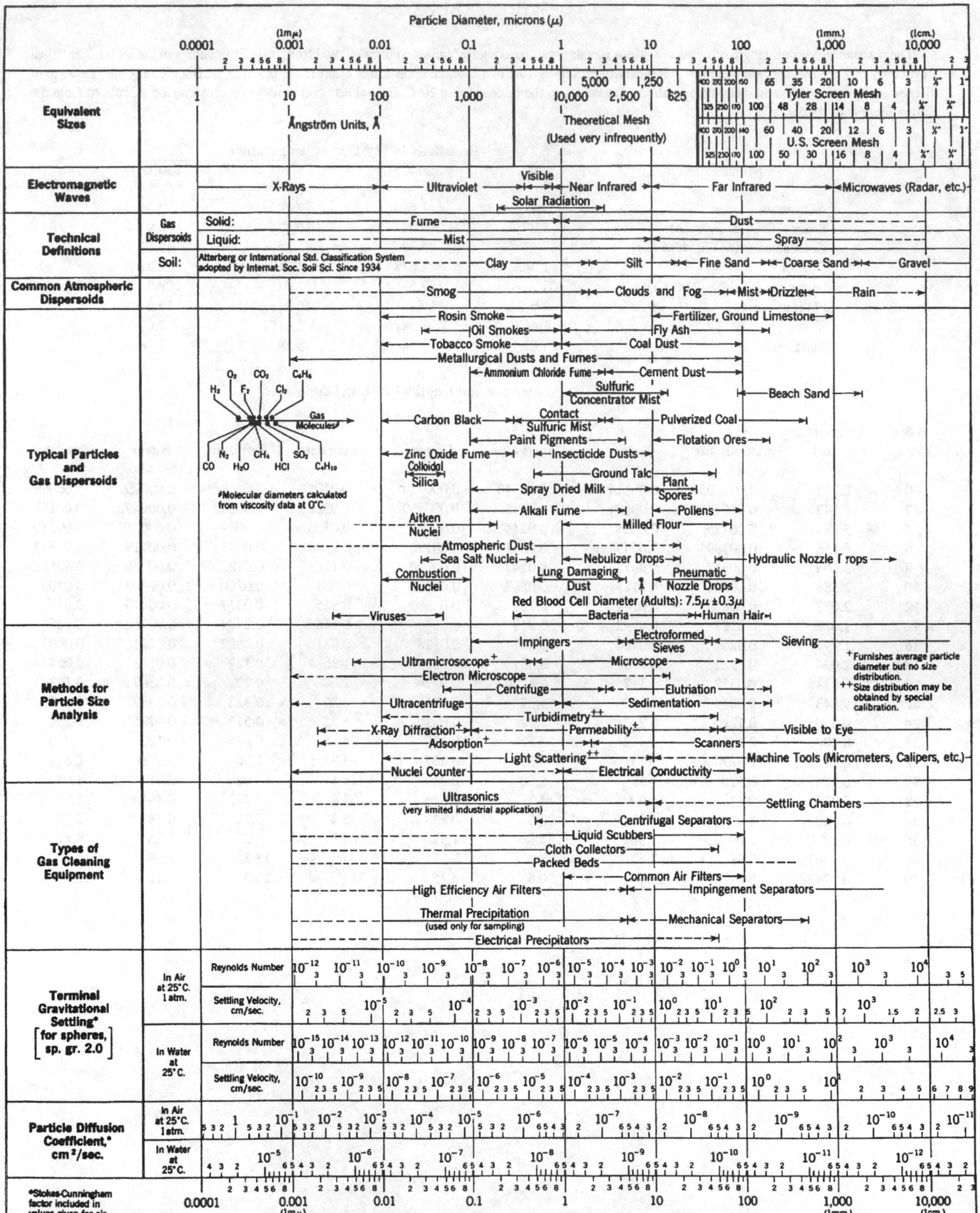

C. E. Lapple, Stanford Research Institute Journal, Vol. 5, p.95 (Third Quarter, 1961)

DENSITY OF VARIOUS SOLIDS

This table gives the range of density for miscellaneous solid materials whose characteristics depend on the source or method of preparation.

REFERENCES

1. Forsythe, W. E., *Smithsonian Physical Tables, Ninth Edition*, Smithsonian Institution, Washington, 1956.
2. Kaye, G. W. C., and Laby, T. H., *Tables of Physical and Chemical Constants, 16th Edition,* Longman, London, 1995.
3. Brandrup, J., and Immergut, E. H., *Polymer Handbook, Third Edition*, John Wiley & Sons, New York, 1989.

Material	ρ/ g cm^{-3}
Agate	2.5-2.7
Alabaster,	
carbonate	2.69-2.78
sulfate	2.26-2.32
Albite	2.62-2.65
Amber	1.06-1.11
Amphiboles	2.9-3.2
Anorthite	2.74-2.76
Asbestos	2.0-2.8
Asbestos slate	1.8
Asphalt	1.1-1.5
Basalt	2.4-3.1
Beeswax	0.96-0.97
Beryl	2.69-2.70
Biotite	2.7-3.1
Bone	1.7-2.0
Brasses	8.44-8.75
Brick	1.4-2.2
Bronzes	8.74-8.89
Butter	0.86-0.87
Calamine	4.1-4.5
Calcspar	2.6-2.8
Camphor	0.99
Cardboard	0.69
Celluloid	1.4
Cement, set	2.7-3.0
Chalk	1.9-2.8
Charcoal,	
oak	0.57
pine	0.28-0.44
Cinnabar	8.12
Clay	1.8-2.6
Coal,	
anthracite	1.4-1.8
bituminous	1.2-1.5
Coke	1.0-1.7
Copal	1.04-1.14
Cork	0.22-0.26
Corundum	3.9-4.0
Diamond	3.51
Dolomite	2.84
Ebonite	1.15
Emery	4.0
Epidote	3.25-3.50
Feldspar	2.55-2.75
Flint	2.63
Fluorite	3.18
Galena	7.3-7.6
Garnet	3.15-4.3
Gelatin	1.27
Glass,	
common	2.4-2.8
lead	3-4
Pyrex	2.23
Granite	2.64-2.76
Graphite	2.30-2.72
Gum arabic	1.3-1.4
Gypsum	2.31-2.33
Hematite	4.9-5.3
Hornblende	3.0
Ice	0.917
Iron, cast	7.0-7.4
Ivory	1.83-1.92
Kaolin	2.6
Leather, dry	0.86
Lime, slaked	1.3-1.4
Limestone	2.68-2.76
Linoleum	1.18
Magnetite	4.9-5.2
Malachite	3.7-4.1
Marble	2.6-2.84
Meerschaum	0.99-1.28
Mica	2.6-3.2
Muscovite	2.76-3.00
Ochre	3.5
Opal	2.2
Paper	0.7-1.15
Paraffin	0.87-0.91
Peat blocks	0.84
Pitch	1.07
Polyamides	1.15-1.25
Polyethylene	0.92-0.97
Poly(methyl methacrylate)	1.19
Polypropylene	0.91-0.94
Polystyrene	1.06-1.12
Polytetrafluoroethylene	2.28-2.30
Poly(vinyl acetate)	1.19
Poly(vinyl chloride)	1.39-1.42
Porcelain	2.3-2.5
Porphyry	2.6-2.9
Pyrite	4.95-5.10
Quartz	2.65
Resin	1.07
Rock salt	2.18
Rubber,	
hard	1.19
soft	1.1
pure gum	0.91-0.93
Neoprene	1.23-1.25
Sandstone	2.14-2.36
Serpentine	2.50-2.65
Silica, fused,	2.21
Silicon carbide	3.16
Slag	2.0-3.9
Slate	2.6-3.3
Soapstone	2.6-2.8
Solder	8.7-9.4
Starch	1.53
Steel, stainless	7.8
Sugar	1.59
Talc	2.7-2.8
Tallow, beef	0.94
Tar	1.02
Topaz	3.5-3.6
Tourmaline	3.0-3.2
Tungsten carbide	14.0-15.0
Wax, sealing	1.8
Wood (seasoned)	
alder	0.42-0.68
apple	0.66-0.84
ash	0.65-0.85
balsa	0.11-0.14
bamboo	0.31-0.40
basswood	0.32-0.59
beech	0.70-0.90
birch	0.51-0.77
blue gum	1.00
box	0.95-1.16
butternut	0.38
cedar	0.49-0.57
cherry	0.70-0.90
dogwood	0.76
ebony	1.11-1.33
elm	0.54-0.60
hickory	0.60-0.93
holly	0.76
juniper	0.56
larch	0.50-0.56
locust	0.67-0.71
logwood	0.91
mahogany	0.66-0.85
maple	0.62-0.75
oak	0.60-0.90
pear	0.61-0.73
pine, pitch	0.83-0.85
white	0.35-0.50
yellow	0.37-0.60
plum	0.66-0.78
poplar	0.35-0.50
satinwood	0.95
spruce	0.48-0.70
sycamore	0.40-0.60
teak, Indian	0.66-0.98
walnut	0.64-0.70
water gum	1.00
willow	0.40-0.60
Wood's metal	9.70

DIELECTRIC STRENGTH OF INSULATING MATERIALS

L. I. Berger

The loss of the dielectric properties by a sample of a gaseous, liquid, or solid insulator as a result of application to the sample of an electric field* greater than a certain critical magnitude is called *dielectric breakdown.* The critical magnitude of electric field at which the breakdown of a material takes place is called the *dielectric strength* of the material (or *breakdown voltage*). The dielectric strength of a material depends on the specimen thickness (as a rule, thin films have greater dielectric strength than that of thicker samples of a material), the electrode shape**, the rate of the applied voltage increase, the shape of the voltage vs. time curve, and the medium surrounding the sample, e.g., air or other gas (or a liquid — for solid materials only).

Breakdown in Gases

The current carriers in gases are free electrons and ions generated by external radiation. The equilibrium concentration of these particles at normal pressure is about 10^3 cm^{-3}, and hence the electrical conductivity is very small, of the order of 10^{-16} - 10^{-15} S/cm. But in a strong electric field, these particles acquire kinetic energy along their free pass, large enough to ionize the gas molecules. The new charged particles ionize more molecules; this avalanche-like process leads to formation between the electrodes of channels of conducting plasma (streamers), and the electrical resistance of the space between the electrodes decreases virtually to zero.

Because the dielectric strength (breakdown voltage) of gases strongly depends on the electrode geometry and surface condition and the gas pressure, it is generally accepted to present the data for a particular gas as a fraction of the dielectric strength of either nitrogen or sulfur hexafluoride measured at the same conditions. In Table 1, the data are presented in comparison with the dielectric strength of nitrogen, which is considered equal to 1.00. For convenience to the reader, a few average magnitudes of the dielectric strength of some gases are expressed in kilovolts per millimeter. The data in the table relate to the standard conditions, unless indicated otherwise.

Breakdown in Liquids

If a liquid is pure, the breakdown mechanism in it is similar to that in gases. If a liquid contains liquid impurities in the form of small drops with greater dielectric constant than that of the main liquid, the breakdown is the result of formation of ellipsoids from these drops by the electric field. In a strong enough electric field, these ellipsoids merge and form a high-conductivity channel between the electrodes. The current increases the temperature in the channel, liquid boils, and the current along the steam canal leads to breakdown. Formation of a conductive channel (bridge) between the electrodes is observed also in liquids with solid impurities. If a liquid contains gas impurities in the form of small bubbles, breakdown is the result of heating of the liquid in strong electric fields. In the locations with the highest current density, the liquid boils, the size of the gas bubbles increases, they merge and form gaseous channels between the electrodes, and the breakdown medium is again the gas plasma.

Breakdown in Solids

It is known that the current in solid insulators does not obey Ohm's law in strong electric fields. The current density increases almost exponentially with the electric field, and at a certain field magnitude it jumps to very high magnitudes at which a specimen of a material is destroyed. The two known kinds of electric breakdown are thermal and electrical breakdowns. The former is the result of material heating by the electric current. Destruction of a sample of a material happens when, at a certain voltage, the amount of heat produced by the current exceeds the heat release through the sample surface; the breakdown voltage in this case is proportional to the square root of the ratio of the thermal conductivity and electrical conductivity of the material. The electrical breakdown results from the tunneling of the charge carriers from electrodes or from the valence band or from the impurity levels into the conduction band, or by the impact ionization. The tunnel effect breakdown happens mainly in thin layers, e.g., in thin p-n junctions. Otherwise, the impact ionization mechanism dominates. For this mechanism, the dielectric strength of an insulator can be estimated using Boltzmann's kinetic equation for electrons in a crystal.

In the following tables, the dielectric strength values are for room temperature and normal atmospheric pressure, unless indicated otherwise.

* The unit of electric field in the SI system is newton per coulomb or volt per meter.

** For example, the U.S. standard ASTM D149 is based on use of symmetrical electrodes, while per U.K. standard BS2918 one electrode is a plane and the other is a rod with the axis normal to the plane.

Table 1
Dielectric Strength of Gases

Material	Dielectric* Strength	Ref.
Nitrogen, N_2	1.00	
Hydrogen, H_2	0.50	1,2
Helium, He	0.15	1
Oxygen, O_2	0.92	2
Air	0.97	6
Air (flat electrodes), kV/mm	3.0	3
Air, kV/mm	0.4-0.7	4
Air, kV/mm	1.40	5
Neon, Ne	0.25	1
	0.16	2
Argon, Ar	0.18	2
Chlorine, Cl_2	1.55	1
Carbon monoxide, CO	1.02	1
	1.05	2
Carbon dioxide, CO_2	0.88	1
	0.82	2
	0.84	6
Nitrous oxide, N_2O	1.24	2
Sulfur dioxide, SO_2	2.63	2
	2.68	6
Sulfur monochloride, S_2Cl_2 (at 12.5 Torr)	1.02	1
Thionyl fluoride, SOF_2	2.50	1
Sulfur hexafluoride, SF_6	2.50	1
	2.63	2
Sulfur hexafluoride, SF_6, kV/mm	8.50	7
	9.8	8
Perchloryl fluoride, ClO_3F	2.73	1
Tetrachloromethane, CCl_4	6.33	1
	6.21	2
Tetrafluoromethane, CF_4	1.01	1
Methane, CH_4	1.00	1
	1.13	2
Bromotrifluoromethane, CF_3Br	1.35	1
	1.97	2
Bromomethane, CH_3Br	0.71	2
Chloromethane, CH_3Cl	1.29	2
Iodomethane, CH_3I	3.02	2
Iodomethane, CH_3I, at 370 Torr	2.20	7
Dichloromethane, CH_2Cl_2	1.92	2
Dichlorodifluoromethane, CCl_2F_2	2.42	1
	2.63	2,6
Chlorotrifluoromethane, $CClF_3$	1.43	1
	1.53	2

Material	Dielectric* Strength	Ref.
Trichlorofluoromethane, CCl_3F	3.50	1
	4.53	2
Trichloromethane, $CHCl_3$	4.2	1
	4.39	2
Methylamine, CH_3NH_2	0.81	1
Difluoromethane, CH_2F_2	0.79	2
Trifluoromethane, CHF_3	0.71	2
Bromochlorodifluoromethane, CF_2ClBr	3.84	2
Chlorodifluoromethane, $CHClF_2$	1.40	1
	1.11	2
Dichlorofluoromethane, $CHCl_2F$	1.33	1
	2.61	2
Chlorofluoromethane, CH_2ClF	1.03	1
Hexafluoroethane, C_2F_6	1.82	1
	2.55	2
Ethyne (Acetylene), C_2H_2	1.10	1
	1.11	2
Chloropentafluoroethane, C_2ClF_5	2.3	1
	3.0	6
Dichlorotetrafluoroethane, $C_2Cl_2F_4$	2.52	1
Chlorotrifluoroethylene, C_2ClF_3	1.82	2
1,1,1-Trichloro-2,2,2-trifluoroethane	6.55	2
1,1,2-Trichloro-1,2,2-trifluoroethane	6.05	2
Chloroethane, C_2H_5Cl	1.00	1
1,1-Dichloroethane	2.66	2
Trifluoroacetonitrile, CF_3CN	3.5	1
Acetonitrile, CH_3CN	2.11	2
Dimethylamine, $(CH_3)_2NH$	1.04	1
Ethylamine, $C_2H_5NH_2$	1.01	1
Ethylene oxide (oxirane), CH_3CHO	1.01	1
Perfluoropropene, C_3F_6	2.55	2
Octafluoropropane, C_3F_8	2.19	1
	2.47	2
3,3,3-Trifluoro-1-propene, CH_2CHCF_3	2.11	2
Pentafluoroisocyanoethane, C_2F_5NC	4.5	1
1,1,1,4,4,4-Hexafluoro-2-butyne, CF_3CCCF_3	5.84	2
Octafluorocyclobutane, C_4F_8	3.34	2
1,1,1,2,3,4,4,4-Octafluoro-2-butene	2.8	1
Decafluorobutane, C_4F_{10}	3.08	1
Perfluorobutanenitrile, C_3F_7CN	5.5	1
Perfluoro-2-methyl-1,3-butadiene, C_5F_8	5.5	1
Hexafluorobenzene, C_6F_6	2.11	2
Perfluorocyclohexane, C_6F_{12}, (saturated vapor)	6.18	2

*Relative to nitrogen, unless units of kV/mm are indicated.

Table 2
Dielectric Strength of Liquids

Material	Dielectric strength kV/mm	Ref.
Helium, He, liquid, 4.2 K	10	9
Static	10	11
Dynamic	5	11
	23	12
Nitrogen, N_2, liquid, 77K		
Coaxial cylinder electrodes	20	10
Sphere to plane electrodes	60	10
Water, H_2O, distilled	65-70	13
Carbon tetrachloride, CCl_4	5.5	14
	16.0	15
Hexane, C_6H_{14}	42.0	16
Two 2.54 cm diameter spherical electrodes, 50.8 μm space	156	17,18
Cyclohexane, C_6H_{12}	42-48	16
2-Methylpentane, C_6H_{14}	149	17,18
2,2-Dimethylbutane, C_6H_{14}	133	17,18
2,3-Dimethylbutane, C_6H_{14}	138	17,18
Benzene, C_6H_6	163	17,18
Chlorobenzene, C_6H_5Cl	7.1	14
	18.8	15
2,2,4-Trimethylpentane, C_8H_{18}	140	17,18
Phenylxylylethane	23.6	19
Heptane, C_7H_{16}	166	17,18
2,4-Dimethylpentane, C_7H_{16}	133	17,18
Toluene, $C_6H_5CH_3$	199	17,18
	46	16
	12.0	14
	20.4	15
Octane, C_8H_{18}	16.6	14
	20.4	15
	179	17,18
Ethylbenzene, C_8H_{10}	226	17,18
Propylbenzene, C_9H_{12}	250	17,18
Isopropylbenzene, C_9H_{12}	238	17,18
Decane, $C_{10}H_{22}$	192	17,18
Synthetic Paraffin Mixture		
Synfluid 2cSt PAO	29.5	37
Butylbenzene, $C_{10}H_{14}$	275	17,18
Isobutylbenzene, $C_{10}H_{14}$	222	17,18
Silicone oils—polydimethylsiloxanes, $(CH_3)_3Si\text{-}O\text{-}[Si(CH_3)_2]_x\text{-}O\text{-}Si(CH_3)_3$		
Polydimethylsiloxane silicone fluid	15.4	20
Dimethyl silicone	24.0	21,22
Phenylmethyl silicone	23.2	22
Silicone oil, Basilone M50	10-15	23
Mineral insulating oils	11.8	6
Polybutene oil for capacitors	13.8	6
Transformer dielectric liquid	28-30	6
Isopropylbiphenyl capacitor oil	23.6	6
Transformer oil	110.7	24
Transformer oil Agip ITE 360	9-12.6	23
Perfluorinated hydrocarbons		
Fluorinert FC 6001	8.0	23
Fluorinert FC 77	10.7	23
Perfluorinated polyethers		
Galden XAD (Mol. wt. 800)	10.5	23
Galden D40 (Mol. wt. 2000)	10.2	23
Castor oil	65	25

Table 3
Dielectric Strength of Solids

Material	Dielectric strength kV/mm	Ref
Sodium chloride, NaCl, crystalline	150	26
Potassium bromide, KBr, crystalline	80	26
Ceramics		
Alumina (99.9% Al_2O_3)	13.4	6,27a
Aluminum silicate, Al_2SiO_5	5.9	6
Berillia (99% BeO)	13.8	6,27b
Boron nitride, BN	37.4	6
Cordierite, $Mg_2Al_4Si_5O_{18}$	7.9	6,27c
Forsterite, Mg_2SiO_4	9.8	28
Porcelain	35-160	26
Steatite, $Mg_3Si_4O_{11}\cdot H_2O$	9.1-15.4	6
Titanates of Mg, Ca, Sr, Ba, and Pb	20-120	3
Barium titanate, glass bonded	>30	36
Zirconia, ZrO_2	11.4	29
Glasses		
Fused silica, SiO_2	470-670	26
Alkali-silicate glass	200	26
Standard window glass	9.8-13.8	28
Micas		
Muscovite, ruby, natural	118	6
Phlogopite, amber, natural	118	6
Fluorophlogopite, synthetic	118	6
Glass-bonded mica	14.0-15.7	6
Thermoplastic Polymers		
Polypropylene	23.6	6
Amide polymer nylon 6/6, dry	23.6	6
Polyamide-imide copolymer	22.8	6
Modified polyphenylene oxide	21.7	6
Polystyrene	19.7	6
Polymethyl methacrylate	19.7	6
Polyetherimide	18.9	6
Amide polymer nylon 11(dry)	16.7	6
Polysulfone	16.7	6
Styrene-acrylonitrile copolymer	16.7	6
Acrylonitrile-butadiene-styrene	16.7	6
Polyethersulfone	15.7	6
Polybutylene terephthalate	15.7	6
Polystyrene-butadiene copolymer	15.7	6
Acetal homopolymer	15.0	6
Acetal copolymer	15.0	6

Table 3
Dielectric Strength of Solids (continued)

Material	Dielectric strength kV/mm	Ref.
Polyphenylene sulfide	15.0	6
Polycarbonate	15.0	6
Acetal homopolymer resin (molding resin)	15.0	6
Acetal copolymer resin	15.0	6
Thermosetting Molding Compounds		
Glass-filled allyl (Type GDI-30 per MIL-M-14G)	15.7	6
Glass-filled epoxy, electrical grade	15.4	6
Glass-filled phenolic (Type GPI-100 per MIL-M-14G)	15.0	6
Glass-filled alkyd/polyester (Type MAI-60 per MIL-M-14G)	14.8	6
Glass-filled melamine (Type MMI-30 per MIL-M-14G)	13.4	6
Extrusion Compounds for High-Temperature Insulation		
Polytetrafluoroethylene	19.7	6
Perfluoroalkoxy polymer	21.7	6
Fluorinated ethylene-propylene copolymer	19.7	6
Ethylene-tetrafluoroethylene copolymer	15.7	6
Polyvinylidene fluoride	10.2	6
Ethylene-chlorotrifluoroethylene copolymer	19.3	6
Polychlorotrifluoroethylene	19.7	6
Extrusion Compounds for Low-Temperature Insulation		
Polyvinyl chloride		
Flexible	11.8-15.7	30
Rigid	13.8-19.7	30
Polyethylene	18.9	28
Polyethylene, low-density	21.7	6
	300	31
Polyethylene, high-density	19.7	6
Polypropylene/polyethylene copolymer	23.6	6
Embedding Compounds		
Basic epoxy resin: bisphenol-A/epichlorohydrin polycondensate	19.7	6
Cycloaliphatic epoxy: alicyclic diepoxy carboxylate	19.7	6
Polyetherketone	18.9	30
Polyurethanes		
Two-component, polyol-cured	25.4	6
Two-part solventless, polybutylene-based	24.0	6
Silicones		
Clear two-part heat curing eletrical grade silicone embedding resin	21.7	6
Red insulating enamel (MIL-E-22118)		
Dry	47.2	6
Wet	11.8	6
Enamels		
Red enamel, fast cure		
Standard conditions	78.7	6
Immersion conditions	47.2	6
Black enamel		
Standard conditions	70.9	6
Immersion conditions	47.2	6

Material	Dielectric strength kV/mm	Ref.
Varnishes		
Vacuum-pressure impregnated baking type solventless polyester varnish		
Rigid, two-part	70.9	6
Semiflexible high-bond thixotropic	78.7	6
Rigid high-bond high-flash freon-resistant	68.9	6
Baking type epoxy varnish		
Solventless, rigid, low viscosity, one-part	90.6	6
Solventless, semiflexible, one-part	82.7	6
Solventless, semirigid, chemical resistant, low dielectric constant	106.3	6
Solvable, for hermetic electric motors	181.1	6
Polyurethane coating		
Clear conformal, fast cure		
Standard conditions	78.7	6
Immersion conditions	47.2	6
Insulating Films and Tapes		
Low-density polyethylene film (40 μm thick)	300	31
Poly-*p*-xylylene film	410-590	32
Aromatic polymer films		
Kapton H (Du Pont)	389-430	33
Ultem (GE Plastic and Roem AG)	437-565	33
Hostaphan (Hoechst AG)	338-447	33
Amorphous Stabar K2000 (ICI film)	404-422	33
Stabar S100 (ICI film)	353-452	33
Polyetherimide film (26 μm)	486	34
Parylene N/D (poly-*p*-xylylene/poly-dichloro-*p*-xylylene) 25 μm film	275	6
Cellulose acetate film	157	6
Cellulose triacetate film	157	6
Polytetrafluoroethylene film	87-173	6
Perfluoroalkoxy film	157-197	6
Fluorinated ethylene-propylene copolymer film	197	6
Ethylene-tetrafluoroethylene film	197	6
Ethylene-chlorotrifluoroethylene copolymer film	197	6
Polychlorotrifluoroethylene film	118-153.5	6
High-voltage rubber insulating tape	28	6
Composites		
Isophthalic polyester (vinyl toluene monomer) filled with		
Calcium carbonate, $CaCO_3$	15.0	38
Gypsum, $CaSO_4$	14.4	38
Alumina trihydrate	15.4	38
Clay	14.4	38
BPA fumarate polyester (vinyl toluene monomer) filled with		
Calcium carbonate	6.1	38
Gypsum	5.9	38
Alumina trihydrate	11.8	38
Clay	12.6	38

Table 3
Dielectric Strength of Solids (continued)

Material	Dielectric strength kV/mm	Ref.
Polysulfone resin—30% glass fiber	16.5-18.7	38
Polyamid resin (Nylon 66)—30% carbon fiber	13.0	38
Polyimide thermoset resin, glass reinforced	12.0	39
Polyester resin (thermoplastic)—40% glass fiber	20.0	38
Epoxy resin (diglycidyl ether of bisphenol A), glass reinforced	16.0	40
Various Insulators		
Rubber, natural	100-215	26
Butyl rubber	23.6	6
Neoprene	15.7-27.6	6
Silicone rubber	26-36	6
Room-temperature vulcanized silicone rubber	9.2-10.9	35
Ureas (from carbamide to tetraphenylurea)	11.8-15.7	28
Dielectric papers		
Aramid paper, calendered	28.7	6
Aramid paper, uncalendered	12.2	6
Aramid with Mica	39.4	6

REFERENCES

1. Vijh, A. K. *IEEE Trans.*, EI-12, 313, 1997.
2. Brand, K. P., *IEEE Trans.*, EI-17, 451, 1982.
3. *Encyclopedic Dictionary in Physics*, Vedensky, B. A. and Vul, B. M., Eds., Vol. 4, Soviet Encyclopedia Publishing House, Moscow, 1965.
4. Kubuki, M., Yoshimoto, R., Yoshizumi, K., Tsuru, S., and Hara, M., *IEEE Trans.*, DEI-4, 92, 1997.
5. Al-Arainy, A. A. Malik, N. H., and Cureshi, M. I., *IEEE Trans.*, DEI-1, 305, 1994.
6. Shugg, W. T., *Handbook of Electrical and Electronic Insulating Materials*, Van Nostrand Reinhold, New York, 1986.
7. Devins, J. C., *IEEE Trans.*, EI-15, 81, 1980.
8. Xu, X., Jayaram, S., and Boggs, S. A., *IEEE Trans.*, DEI-3, 836, 1996.
9. Okubo, H., Wakita, M., Chigusa, S., Nayakawa, N., and Hikita, M., *IEEE Trans.*, DEI-4, 120, 1997.
10. Hayakawa, H., Sakakibara, H., Goshima, H., Hikita, M., and Okubo, H., *IEEE Trans.*, DEI-4, 127, 1997.
11. Okubo, H., Wakita, M., Chigusa, S., Hayakawa, N., and Hikita, M., *IEEE Trans.*, DEI-4, 220, 1997.
12. Von Hippel, A. R., *Dielectric Materials and Applications*, MIT Press, Cambridge, MA, 1954.
13. Jones, H. M. and Kunhards, E. E., *IEEE Trans.*, DEI-1, 1016, 1994.
14. Nitta, Y. and Ayhara, Y., *IEEE Trans.*, EI-11, 91, 1976.
15. Gallagher, T. J., *IEEE Trans.*, EI-12, 249, 1977.
16. Wong, P. P. and Forster, E. O., in *Dielectric Materials. Measurements and Applications*, IEE Conf. Publ. 177, 1, 1979.
17. Kao, K. C. *IEEE Trans.*, EI-11, 121, 1976.
18. Sharbaugh, A. H., Crowe, R. W., and Cox, E. B., *J. Appl. Phys.*, 27, 806, 1956.
19. Miller, R. L., Mandelcorn, L., and Mercier, G. E., in *Proc. Intl. Conf. on Properties and Applications of Dielectric Materials*, Xian, China, June 24-28, 1985; cited in Ref. 6, p. 492.
20. Hakim, R. M., Oliver, R. G., and St-Onge, H., *IEEE Trans.*, EI-12, 360, 1977.
21. Hosticka, C., *IEEE Trans.*, 389, 1977.
22. Yasufuku, S., Umemura, T., and Ishioka, Y., *IEEE Trans.*, EI-12, 402, 1977.
23. Forster, E. O., Yamashita, H., Mazzetti, C., Pompini, M., Caroli, L., and Patrissi, S., *IEEE Trans.*, DEI-1, 440, 1994.
24. Bell, W. R., *IEEE Trans.*, 281, 1977.
25. Ramu, T. C. and Narayana Rao, Y., in *Dielectric Materials. Measurements and Applications*, IEE Conf. Publ. 177, 37.
26. Skanavi, G. I., *Fizika Dielektrikov; Oblast Silnykh Polei* (Physics of Dielectrics; Strong Fields). Gos. Izd. Fiz. Mat. Nauk (State Publ. House for Phys. and Math. Scis.), Moscow, 1958.
27. Kleiner, R. N., in *Practical Handbook of Materials Science*, Lynch, C. T., Ed., CRC Press, 1989; 27a: p. 304; 27b: p.300; 27c: p. 316.
28. *Materials Selector Guide. Materials and Methods*, Reinhold Publ., New York, 1973.
29. Flinn, R. A. and Trojan, P. K., *Engineering Materials and Their Applications*, 2nd ed., Houghton Mifflin, 1981, p. 614.
30. Lynch, C. T., Ed., *Practical Handbook of Materials Science*, CRC Press, Boca Raton, FL, 1989.
31. Suzuki, H., Mukai, S., Ohki, Y., Nakamichi, Y., and Ajiki, K., *IEEE Trans.*, DEI-4, 238, 1997.
32. Mori, T., Matsuoka, T., and Muzitani, T., *IEEE Trans.*, DEI-1, 71, 1994.
33. Bjellheim, P. and Helgee, B., *IEEE Trans.*, DEI-1, 89, 1994.
34. Zheng, J. P., Cygan, P. J., and Jow, T. R., *IEEE Trans.*, DEI-3, 144, 1996.
35. Danukas, M. G., *IEEE Trans.*, DEI-1, 1196, 1994.
36. Burn, I. and Smithe, D. H., *J. Mater. Sci.*, 7, 339, 1972.
37. Hope, K.D., Chevron Chemical, Private Communication.

38. *Engineering Materials Handbook*, vol. 1, Composites, C.A. Dostal, Ed., ASM Intl., 1987.
39. 1985 Materials Selector, *Mater. Eng.*, (12) 1984.
40. *Modern Plastics Encyclopedia*, McGraw-Hill, v. 62 (No. 10A) 1985–1986.

Review Literature on the Subject

R1. Kuffel, E. and Zaengl, W. S., *HV Engineering Fundamentals*, Pergamon, 1989.
R2. Kok, J. A., *Electrical Breakdown of Insulating Liquids*, Phillips Tech. Library, Cleaver-Hum, Longon, 1961.
R3. Gallagher, T. J., *Simple Dielectric Liquids*, Clarendon, Oxford, 1975.
R4. Meek, J. M. and Craggs, J. D., Eds., *Electric Breakdown in Gases*, John Wiley & Sons, 1976.
R5. Von Hippel, A. R., *Dielectric Materials and Applications*, MIT Press, Cambridge, MA, 1954.

ALLOCATION OF FREQUENCIES IN THE RADIO SPECTRUM

In the United States the National Telecommunications and Information Administration (NTIA) has responsibility for assigning each portion of the radio spectrum (9 kHz to 300 GHz) for different uses. These assignments must be compatible with the rules of the International Telecommunications Union (ITU), to which the United States is bound by treaty. The current assignments are given in a wall chart (Reference 1) and may also be found on the NTIA web site (Reference 2). The list below summarizes the broad features of the spectrum allocation, with particular attention to those sections of scientific interest. The references should be consulted for details of the allocations in the frequency bands listed here, which in some cases are quite complex.

REFERENCES

1. *United States Frequency Allocations*, 1996 Spectrum Wall Chart, Stock No. 003-000-00652-2, U. S. Government Printing Office, P. O. Box 371954, Pittsburgh, PA 15250-7954.
2. http://www.ntia.doc.gov/osmhome/allochrt.html

Frequency range	Allocation
9 - 19.95 kHz	Maritime communication, navigation
19.95 - 20.05 kHz	Standard frequency and time signal (also at 60 kHz and 2.5, 5, 10, 15, 20, 25 MHz)
20.05 - 535 kHz	Maritime and aeronautical communication, navigation
535 - 1605 kHz	AM radio broadcasting
1605 - 3500 kHz	Mobile communication and navigation, amateur radio (1800-1900 kHz)
3.5 - 4.0 MHz	Amateur radio
4.0 - 5.95 MHz	Mobile communication
5.95 - 13.36 MHz	Mobile communication, amateur, short-wave broadcasting
13.36 - 13.41 MHz	Radioastronomy
13.41 - 25.55 MHz	Mobile communication, amateur, short-wave broadcasting
25.55 - 25.67 MHz	Radioastronomy
25.67 - 37.5 MHz	Mobile communication, amateur, short-wave broadcasting
37.5 -38.25 MHz	Radioastronomy
38.25 - 50.0 MHz	Mobile communication
50.0 - 54.0 MHz	Amateur
54.0 - 72.0 MHz	TV channels 2-4
72.0 - 73.0 MHz	Mobile communication
73.0 - 74.6 MHz	Radioastronomy
74.6 - 76.0 MHz	Mobile communication
76.0 - 88.0 MHz	TV channels 5-6
88.0 - 108.0 MHz	FM radio broadcasting
108.0 - 118.0 MHz	Aeronautical navigation
118.0 - 174.0 MHz	Mobile communication, space research, meteorological satellites
174.0 - 216.0 MHz	TV channels 7-13
216.0 - 400.05 MHz	Mobile communication
400.05 - 400.15 MHz	Standard frequency and time satellite (also 20 and 25 GHz)
400.15 - 406.1 MHz	Meteorological aids (radiosonde)
406.1 - 410.0 MHz	Radioastronomy
410.0 - 470.0 MHz	Mobile communication, amateur
470.0 - 512.0 MHz	TV channels 14-20
512.0 - 608.0 MHz	TV channels 21-36
608.0 - 614.0 MHz	Radioastronomy
614.0 - 806.0 MHz	TV channels 38-69
806 -1400 MHz	Mobile communication, navigation
1400 - 1427 MHz	Radioastronomy, space research
1427 - 1660 MHz	Various navigation and satellite applications
1660 - 1710 MHz	Radioastronomy, space research, meteorology
1710 - 2655 MHz	Various navigation and satellite applications
2655 - 2700 MHz	Radioastronomy, space research
2.7 - 4.99 GHz	Various navigation and satellite applications
4.99 - 5.0 GHz	Radioastronomy, space research
5.0 - 10.6 GHz	Various navigation and satellite applications
10.6 - 10.7 GHz	Radioastronomy, space research
10.7 - 15.35 GHz	Various navigation and satellite applications
15.35 - 15.4 GHz	Radioastronomy, space research
15.4 - 22.21 GHz	Various navigation and satellite applications

ALLOCATION OF FREQUENCIES IN THE RADIO SPECTRUM (continued)

Frequency range	Allocation
22.21 - 22.5 GHz	Radioastronomy, space research
22.25 - 23.6 GHz	Various navigation and satellite applications
23.6 - 24.0 GHz	Radioastronomy, space research
24.0 - 31.3 GHz	Various navigation and satellite applications
31.3 - 31.8 GHz	Radioastronomy, space research
31.8 - 42.5 GHz	Various navigation and satellite applications
42.5 - 43.5 GHz	Radioastronomy
43.5 - 51.4 GHz	Various navigation and satellite applications
51.4 - 54.25 GHz	Radioastronomy, space research
54.25 - 58.2 GHz	Space research
58.2 - 59.0 GHz	Radioastronomy, space research
59.0 - 64.0 GHz	Satellite applications
64.0 - 65.0 GHz	Radioastronomy, space research
65.0 - 72.77 GHz	Various navigation and satellite applications
72.77 - 72.91 GHz	Radioastronomy, space research
72.91 - 86.0 GHz	Various navigation and satellite applications
86.0 - 92.0 GHz	Radioastronomy, space research
92.0 - 105.0 GHz	Various navigation and satellite applications
105.0 - 116.0 GHz	Radioastronomy, space research
116.0 - 164.0 GHz	Various navigation and satellite applications
164.0 - 168.0 GHz	Radioastronomy, space research
168.0 - 182.0 GHz	Various navigation and satellite applications
182.0 - 185.0 GHz	Radioastronomy, space research
185.0 - 217.0 GHz	Various navigation and satellite applications
217.0 - 231.0 GHz	Radioastronomy, space research
231.0 - 265.0 GHz	Various navigation and satellite applications
265.0 - 275.0 GHz	Radioastronomy
275.0 - 300.0 GHz	Mobile communications

Section 16
Health and Safety Information

Handling and Disposal of Chemicals in Laboratories **16**-1
Flammability of Chemical Substances **16**-16
Threshold Limits for Airborne Contaminants **16**-32
Octanol-Water Partition Coefficients **16**-43
Protection Against Ionizing Radiation **16**-48
Annual Limits on Intakes of Radionuclides **16**-49
Chemical Carcinogens **16**-53

HANDLING AND DISPOSAL OF CHEMICALS IN LABORATORIES

Robert Joyce and Blaine C. McKusick

The following material has been extracted from two books prepared under the auspices of the Committee on Hazardous Substances in the Laboratory of the National Academy of Sciences — National Research Council. Readers are referred to these books for full details:

Prudent Practices for Handling Hazardous Chemicals in Laboratories, National Academy Press, Washington, 1981.
Prudent Practices for Disposal of Chemicals from Laboratories, National Academy Press, Washington, 1983.

The permission of the National Academy Press to use these extracts is gratefully acknowledged.

INCOMPATIBLE CHEMICALS

The term "incompatible chemicals" refers to chemicals that can react with each other

- Violently
- With evolution of substantial heat
- To produce flammable products
- To produce toxic products

Good laboratory safety practice requires that incompatible chemicals be stored, transported, and disposed of in ways that will prevent their coming together in the event of an accident. Tables 1 and 2 give some basic guidelines for the safe handling of acids, bases, reactive metals, and other chemicals. Neither of these tables is exhaustive, and additional information on incompatible chemicals can be found in the following references.

1. L. Bretherick, *Handbook of Reactive Chemical Hazards*, 3rd ed., Butterworths, London–Boston, 1985.
2. L. Bretherick, Ed., *Hazards in the Chemical Laboratory*, 3rd ed., Royal Society of Chemistry, London, 1981.
3. *Manual of Hazardous Chemical Reactions, A Compilation of Chemical Reactions Reported to be Potentially Hazardous,* National Fire Protection Association, NFPA 491M, 1975, NFPA, 470 Atlantic Avenue, Boston, MA 02210.

TABLE 1
General Classes of Incompatible Chemicals

A	B
Acids	Bases, reactive metals
Oxidizing agents[a]	Reducing agents[a]
Chlorates	Ammonia, anhydrous and aqueous
Chromates	Carbon
Chromium trioxide	Metals
Dichromates	Metal hydrides
Halogens	Nitrites
Halogenating agents	Organic compounds
Hydrogen peroxide	Phosphorus
Nitric acid	Silicon
Nitrates	Sulfur
Perchlorates	
Peroxides	
Permanganates	
Persulfates	

[a] The examples of oxidizing and reducing agents are illustrative of common laboratory chemicals; they are not intended to be exhaustive.

TABLE 2
Examples of Incompatible Chemicals

Chemical	Is incompatible with
Acetic acid	Chromic acid, nitric acid, hydroxyl compounds, ethylene glycol, perchloric acid, peroxides, permanaganates
Acetylene	Chlorine, bromine, copper, fluorine, silver, mercury
Acetone	Concentrated nitric and sulfuric acid mixtures
Alkali and alkaline earth metals (such as powdered aluminum or magnesium, calcium, lithium, sodium, potassium)	Water, carbon tetrachloride or other chlorinated hydrocarbons, carbon dioxide, halogens
Ammonia (anhydrous)	Mercury (in manometers, for example), chlorine, calcium hypochlorite, iodine, bromine, hydrofluoric acid (anhydrous)
Ammonium nitrate	Acids, powdered metals, flammable liquids, chlorates, nitrites, sulfur, finely divided organic or combustible materials
Aniline	Nitric acid, hydrogen peroxide
Arsenical materials	Any reducing agent
Azides	Acids
Bromine	See Chlorine
Calcium oxide	Water
Carbon (activated)	Calcium hypochlorite, all oxidizing agents
Carbon tetrachloride	Sodium
Chlorates	Ammonium salts, acids, powdered metals, sulfur, finely divided organic or combustible materials
Chromic acid and chromium troixide	Acetic acid, naphthalene, camphor, glycerol, alcohol, flammable liquids in general
Chlorine	Ammonia, acetylene, butadiene, butane, methane, propane (or other petroleum gases), hydrogen, sodium carbide, benzene, finely divided metals, turpentine
Chlorine dioxide	Ammonia, methane, phosphine, hydrogen sulfide
Copper	Acetylene, hydrogen peroxide
Cumene hydroperoxide	Acids (organic or inorganic)
Cyanides	Acids
Flammable liquids	Ammonium nitrate, chromic acid, hydrogen peroxide, nitric acid, sodium peroxide, halogens
Fluorine	Everything
Hydrocarbons (such as butane, propane, benzene)	Fluorine, chlorine, bromine, chromic acid, sodium peroxide
Hydrocyanic acid	Nitric acid, alkali
Hydrofluoric acid (anhydrous)	Ammonia (aqueous or anhydrous)
Hydrogen peroxide	Copper, chromium, iron, most metals or their salts, alcohols, acetone, organic materials, aniline, nitromethane, combustible matierals
Hydrogen sulfide	Fuming nitric acid, oxidizing gases
Hypochlorites	Acids, activated carbon
Iodine	Acetylene, ammonia (aqueous or anhydrous), hydrogen
Mercury	Acetylene, fulminic acid, ammonia
Nitrates	Sulfuric acid
Nitric acid (concentrated)	Acetic acid, aniline, chromic acid, hydrocyanic acid, hydrogen sulfide, flammable liquids, flammable gases, copper, brass, any heavy metals
Nitrites	Acids
Nitroparaffins	Inorganic bases, amines
Oxalic acid	Silver, mercury

TABLE 2
Examples of Incompatible Chemicals (continued)

Chemical	Is incompatible with
Oxygen	Oils, grease, hydrogen, flammable liquids, solids, or gases
Perchloric acid	Acetic anhydride, bismuth and its alloys, alcohol, paper, wood, grease, oils
Peroxides, organic	Acids (organic or mineral), avoid friction, store cold
Phosphorus (white)	Air, oxygen, alkalis, reducing agents
Potassium	Carbon tetrachloride, carbon dioxide, water
Potassium chlorate	Sulfuric and other acids
Potassium perchlorate (see also chlorates)	Sulfuric and other acids
Potassium permanganate	Glycerol, ethylene glycol, benzaldehyde, surfuric acid
Selenides	Reducing agents
Silver	Acetylene, oxalic acid, tartartic acid, ammonium compounds, fulminic acid
Sodium	Carbon tetrachloride, carbon dioxide, water
Sodium nitrite	Ammonium nitrate and other ammonium salts
Sodium peroxide	Ethyl or methyl alcohol, glacial acetic acid, acetic anhydride, benzaldehyde, carbon disulfide, glycerin, ethylene glycol, ethyl acetate, methyl acetate, furfural
Sulfides	Acids
Sulfuric acid	Potassium chlorate, potassium perchlorate, potassium permanganate (similar compounds of light metals, such as sodium, lithium)
Tellurides	Reducing agents

EXPLOSION HAZARDS

Table 3 lists some common classes of laboratory chemicals that have potential for producing a violent explosion when subjected to shock or friction. These chemicals should never be disposed of as such, but should be handled by procedures given in *Prudent Practices for Disposal of Chemicals from Laboratories*, National Academy Press, 1983, chapters 6 and 7. Additional information on these, as well as on some less common classes of explosives, can be found in L. Bretherick, *Handbook of Reactive Chemical Hazards*, 3rd ed., Butterworths, London–Boston, 1985.

Table 4 lists some illustrative combinations of common laboratory reagents that can produce explosions when they are brought together or that form reaction products that can explode without any apparent external initiating action. This list is not exhaustive, and additional information on potentially explosive reagent combinations can be found in Manual of Hazardous Chemical Reactions, A Compilation of Chemical Reactions Reported to be Potentially Hazardous, National Fire Protection Association, NFPA 491M, 1975, NFPA, 470 Atlantic Avenue, Boston, MA 02210.

WATER-REACTIVE CHEMICALS

Table 5 lists some common laboratory chemicals that react violently with water and that should always be stored and handled so that they do not come into contact with liquid water or water vapor. Procedures for decomposing laboratory quantities are given in Prudent Practices for Disposal of Chemicals from Laboratories, chapter 6; the pertinent section of that chapter is given in parentheses.

PYROPHORIC CHEMICALS

Many members of the classes of readily oxidized, common laboratory chemicals listed in Table 6 ignite spontaneously in air. A more extensive list can be found in L. Bretherick, *Handbook of Reactive Chemical Hazards*, 3rd ed., Butterworths, London-Boston, 1985. Pyrophoric chemicals should be stored in tightly closed containers under an inert atmosphere (or, for some, an inert liquid), and all transfers and manipulations of them must be carried out under an inert atmosphere or liquid. Suggested procedures for decomposing them are given in Prudent Practices for Disposal of Chemicals from Laboratories, chapter 6; the pertinent section of that chapter is given in parentheses.

TABLE 3
Shock–Sensitive Compounds

Acetylenic compounds, especially polyacetylenes, haloacetylenes, and heavy metal salts of acetylenes (copper, silver, and mercury salts are particularly sensitive)
Acyl nitrates
Alkyl nitrates, particularly polyol nitrates such as nitrocellulose and nitroglycerine
Alkyl and acyl nitrites
Alkyl perchlorates
Amminemetal oxosalts: metal compounds with coordinated ammonia, hydrazine, or similar nitrogenous donors and ionic perchlorate, nitrate, permanganate, or other oxidizing group
Azides, including metal, nonmetal, and organic azides
Chlorite salts of metals, such as $AgClO_2$ and $Hg(ClO_2)_2$
Diazo compounds such as CH_2N_2
Diazonium slats, when dry
Fulminates (silver fulminate, AgCNO, can form in the reaction mixture from the Tollens' test for aldehydes if it is allowed to stand for some time; this can be prevented by adding dilute nitric acid to the test mixture as soon as the test has been completed)
Hydrogen peroxide becomes increasingly treacherous as the concentration rises above 30%, forming explosive mixtures with organic materials and decomposing violently in the presence of traces of transition metals
N–Halogen compounds such as difluoroamino compounds and halogen azides
N–Nitro compounds such as *N*–nitromethylamine, nitrourea, nitroguanidine, and nitric amide
Oxo salts of nitrogenous bases: perchlorates, dichromates, nitrates, iodates, chlorites, chlorates, and permanganates of ammonia, amines, hydroxylamine, guanidine, etc.
Perchlorate salts. Most metal, nonmetal, and amine perchlorates can be detonated and may undergo violent reaction in contact with combustible materials
Peroxides and hydroperoxides, organic (see Chapter 6, Section II.P)
Peroxides (solid) that crystallize from or are left from evaporation of peroxidizable solvents (see Chapter 6 and Appendix I)
Peroxides, transition–metal salts
Picrates, especially salts of transition and heavy metals, such as Ni, Pb, Hg, Cu, and Zn; picric acid is explosive but is less sensitive to shock or friction than its metal salts and is relatively safe as a water–wet paste (see Chapter 7)
Polynitroalkyl compounds such as tetranitromethane and dinitroacetonitrile
Polynitroaromatic compounds, especially polynitro hydrocarbons, phenols, and amines

TABLE 4
Potentially Explosive Combinations of Some Common Reagents

Acetone + chloroform in the presence of base
Acetylene + copper, silver, mercury, or their salts
Ammonia (including aqueous solutions) + Cl_2, Br_2, or I_2
Carbon disulfide + sodium azide
Chlorine + an alcohol
Chloroform or carbon tetrachloride + powdered Al or Mg
Decolorizing carbon + an oxidizing agent
Diethyl ether + chlorine (including a chlorine atmosphere)
Dimethyl sulfoxide + an acyl halide, $SOCl_2$, or $POCl_3$
Dimethyl sulfoxide + CrO_3
Ethanol + calcium hypochlorite
Ethanol + silver nitrate
Nitric acid + acetic anhydride or acetic acid
Picric acid + a heavy–metal salt, such as of Pb, Hg, or Ag
Silver oxide + ammonia + ethanol
Sodium + a cholrinated hydrocarbon
Sodium hypochlorite + an amine

TABLE 5
Water–Reactive Chemicals

Alkali metals (III.D)
Alkali metal hydrides (III.C.2)
Alkali metal amides (III.C.7)
Metal alkyls, such as lithium alkyls and aluminum alkyls (IV.A)
Grignard reagents (IV.A)
Halides of nonmetals, such as BCl_3, BF_3, PCl_3, PCl_5, $SiCl_4$, S_2Cl_2 (III.F)
Inorganic acid halides, such as $POCl_3$, $SOCl_2$, SO_2Cl_2 (III.F)
Anhydrous metal halides, such as $AlCl_3$, $TiCl_4$, $ZrCl_4$, $SnCl_4$ (III.E)
Phosphorus pentoxide (III.I)
Calcium carbide (IV.E)
Organic acid halides and anhydrides of low molecular weight (II.J)

TABLE 6
Classes of Pyrophoric Chemicals

Grignard reagents, RMgX (IV.A)
Metal alkyls and aryls, such as RLi, RNa, R_3Al, R_2Zn (IV.A)
Metal carbonyls, such as $Ni(CO)_4$, $Fe(CO)_5$, $Co_2(CO)_8$ (IV.B)
Alkali metals such as Na, K (III.D.1)
Metal powders, such as Al, Co, Fe, Mg, Mn, Pd, Pt, Ti, Sn, Zn, Zr (III.D.2)
Metal hydrides, such as NaH, $LiAlH_4$ (IV.C.2)
Nonmetal hydrides, such as B_2H_6 and other boranes, PH_3, AsH_3 (III.G)
Nonmetal alkyls, such as R_3B, R_3P, R_3As (IV.C)
Phosphorus (white) (III.H)

HAZARDS FROM PEROXIDE FORMATION

Many common laboratory chemicals can form peroxides when allowed access to air over a period of time. A single opening of a container to remove some of the contents can introduce enough air for peroxide formation to occur. Some types of compounds form peroxides that are treacherously and violently explosive in concentrated solution or as solids. Accordingly, peroxide–containing liquids should never be evaporated near to or to dryness. Peroxide formation can also occur in many polymerizable unsaturated compounds, and these peroxides can initiate a runaway, sometimes explosive, polymerization reaction. Procedures for testing for peroxides and for removing small amounts from laboratory chemicals are given in Prudent Practices for Disposal of Chemicals from Laboratories, chapter 6, Section II.P.

Table 7 provides a list of structural characteristics in organic compounds that can peroxidize. These structures are listed in approximate order of decreasing hazard. Reports of serious incidents involving the last five structural types are extremely rare, but these structures are listed because laboratory workers should be aware that they can form peroxides that can influence the course of experiments in which they are used.

Table 8 gives examples of common laboratory chemicals that are prone to form peroxides on exposure to air. The lists are not exhaustive, and analogous organic compounds that have any of the structural features given in Table 7 should be tested for peroxides before being used as solvents or reagents, or before being distilled. The recommended retention times begin with the date of synthesis or of opening the original container.

DISPOSAL OF TOXIC CHEMICALS

It is often desirable to precipitate toxic cations or hazardous anions from solution to facilitate recovery or disposal. Table 9 lists precipitants for many common cations, and Table 10 gives precipitants for some hazardous anions. Many cations can be precipitated as sulfides by adding sodium sulfide solution (preferable to the highly toxic hydrogen sulfide) to a neutral solution of the cation (Table 11). Control of pH is important because some sulfides will redissolve in excess sulfide ion. After precipitation, excess sulfide can be destroyed by addition of hypochlorite.

Most metal cations are precipitated as hydroxides or oxides at high pH. Since many of these precipitates will redissolve in excess base, it is often necessary to control pH. Table 12 shows the recommended pH range for precipitating many cations in their most common oxidation state. The notation "1 N" in the right–hand column indicates that the precipitate will not dissolve in 1 N sodium hydroxide (pH 14).

The distinctions between high and low toxicity or hazard are based on toxicological and other data, and are relative. There is no implication of a sharp distinction between high and low, or that any cations or anions are totally without hazard.

TABLE 7
Types of Chemicals That Are Prone to Form Peroxides

A. Organic structures (in approximate order of decreasing hazard)

1. 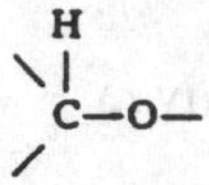 Ethers and acetals with a hydrogen atoms

2. 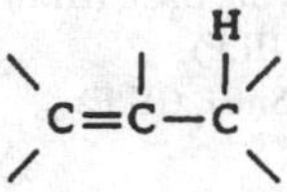 Olefins with allylic hydrogen atoms

3. 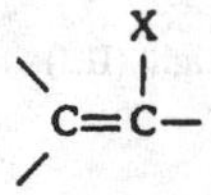 Chloroolefins and fluoroolefins

4. 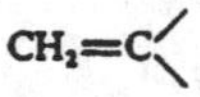 Vinyl halides, esters, and ethers

5. 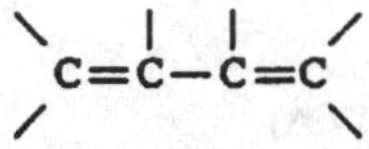 Dienes

6. 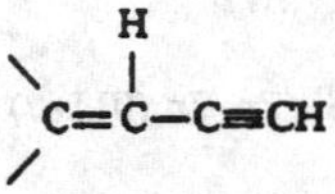 Vinylacetylenes with α hydrogen atoms

7. 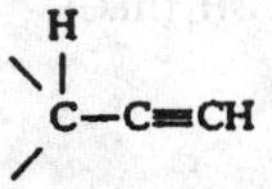 Alkylacetylenes with α hydrogen atoms

8. 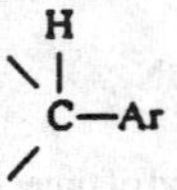 Alkylarenes that contain tertiary hydrogen atoms

9. 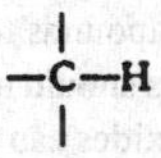 Alkanes and cycloalkanes that contain tertiary hydrogen atoms

10. 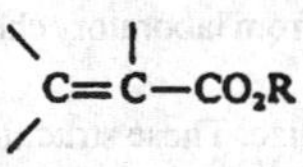 Acrylates and methacrylates

11. 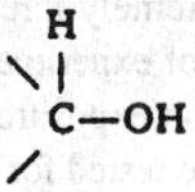 Secondary alcohols

12. 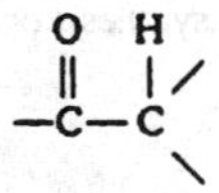 Ketones that contain a hydrogen atoms

13. 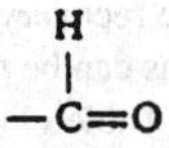 Aldehydes

14. 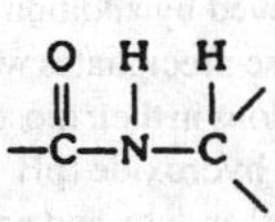 Ureas, amides, and lactams that have a hydrogen atom on a carbon atom attached to nitrogen

TABLE 7
Types of Chemicals That Are Prone to Form Peroxides (continued)

B. Inorganic substances

1. Alkali metals, especially potassium, rubidium, and cesium (see Chapter 6, Section III.D)
2. Metal amides (see Chapter 6, Section III.C.7)
3. Organometallic compounds with a metal atom bonded to carbon (see Chapter 6, Section IV)
4. Metal alkoxides

TABLE 8
Common Peroxide-Forming Chemicals

LIST A

Severe Peroxide Hazard on Storage with Exposure to Air

Discard within 3 months

- Diisopropyl ether (isopropyl ether)
- Divinylacetylene (DVA)[a]
- Potassium metal
- Potassium amide
- Sodium amide (sodamide)
- Vinylidene chloride (1,1–dichloroethylene)[a]

LIST B

Peroxide Hazard on Concentration; Do Not Distill or Evaporate Without First Testing for the Presence of Peroxides

Discard or test for peroxides after 6 months

- Acetaldehyde diethyl acetal (acetal)
- Cumene (isopropylbenzene)
- Cyclohexene
- Cyclopentene
- Decalin (decahydronaphthalene)
- Diacetylene
- Dicyclopentadiene
- Diethyl ether (ether)
- Diethylene glycol dimethyl ether (diglyme)
- Dioxane
- Ethylene glycol dimethyl ether (glyme)
- Ethylene glycol ether acetates
- Ethylene glycol monoethers (cellosolves)
- Furan
- Methylacetylene
- Methylcyclopentane
- Methyl isobutyl ketone
- Tetrahydrofuran (THF)
- Tetralin (tetrahydronaphthalene)
- Vinyl ethers[a]

LIST C

Hazard of Rapid Polymerization Initiated by Internally Formed Peroxides[a]

a. Normal Liquids; Discard or test for peroxides after 6 months[b]

- Chloroprene (2–chloro–1,3–butadiene)[c]
- Styrene
- Vinyl acetate
- Vinylpyridine

b. Normal Gases; Discard after 12 months[d]

- Butadiene[c]
- Tetrafluoroethylene (TFE)[c]
- Vinylacetylene (MVA)[c]
- Vinyl chloride

[a] Polymerizable monomers should be stored with a polymerization inhibitor from which the monomer can be separated by distillation just before use.

[b] Although common acrylic monomers such as acrylonitrile, acrylic acid, ethyl acrylate, and methyl methacrylate can form peroxides, they have not been reported to develop hazardous levels in normal use and storage.

[c] The hazard from peroxides in these compounds is substantially greater when they are stored in the liquid phase, and if so stored without an inhibitor they should be considered as in LIST A.

TABLE 8
Common Peroxide-Forming Chemicals (continued)

[d] Although air will not enter a gas cylinder in which gases are stored under pressure, these gases are sometimes transferred from the original cylinder to another in the laboratory, and it is difficult to be sure that there is no residual air in the receiving cylinder. An inhibitor should be put into any such secondary cylinder before one of these gases is transferred into it; the supplier can suggest inhibitors to be used. The hazard posed by these gases is much greater if there is a liquid phase in such a secondary container, and even inhibited gases that have been put into a secondary container under conditions that create a liquid phase should be discarded within 12 months.

Note: Laboratory workers should label all containers of peroxidizable solvents or reagents with one of the following:

[LIST A]

Peroxidizable compound

	Received	Opened
Date	______	______

Discard 3 months after opening

[LISTS B AND C]

Peroxidizable compound

	Received	Opened
Date	______	______

Discard or test for peroxides
6 months after opening

TABLE 9
Relative Toxicity of Cations

High toxic hazard	Precipitant[a]	Low toxic hazard	Precipitant[a]
Antimony	OH^-, S^{2-}	Aluminum	OH^-
Arsenic	S^{2-}	Bismuth	OH^-, S^{2-}
Barium	SO_4^{2-}, CO_3^{2-}	Calcium	SO_4^{2-}, CO_3^{2-}
Beryllium	OH^-	Cerium	OH^-
Cadmium	OH^-, S^{2-}	Cesium	
Chromium (III)[b]	OH^-	Copper[c]	OH^-, S^{2-}
Cobalt (II)[b]	OH^-, S^{2-}	Gold	OH^-, S^{2-}
Gallium	OH^-	Iron[c]	OH^-, S^{2-}
Germanium	OH^-, S^{2-}	Lanthanides	OH^-
Hafnium	OH^-	Lithium	
Indium	OH^-, S^{2-}	Magnesium	OH^-
Iridium	OH^-, S^{2-}	Molybdenum (VI)[b,d]	
Lead	OH^-, S^{2-}	Niobium (V)	OH^-
Manganese (II)[b]	OH^-, S^{2-}	Palladium	OH^-, S^{2-}
Mercury	OH^-, S^{2-}	Potassium	
Nickel	OH^-, S^{2-}	Rubidium	
Osmium (IV)[b,e]	OH^-, S^{2-}	Scandium	OH^-
Platinum (II)[b]	OH^-, S^{2-}	Sodium	
Rhenium (VII)[b]	S^{2-}	Strontium	SO_4^{2-}, CO_3^{2-}
Rhodium (III)[b]	OH^-, S^{2-}	Tantalum	OH^-
Ruthenium (III)[b]	OH^-, S^{2-}	Tin	OH^-, S^{2-}
Selenium	S^{2-}	Titanium	OH^-
Silver	Cl^-, OH^-, S^{2-}	Yttrium	OH^-
Tellurium	S^{2-}	Zinc[c]	OH^-, S^2
Thallium	OH^-, S^{2-}	Zirconium	OH^-
Tungsten (VI)[b,d]			
Vanadium	OH^-, S^{2-}		

TABLE 9
Relative Toxicity of Cations (continued)

[a] Precipitants are listed in order of preference:
- OH^- = base (sodium hydroxide or sodium carbonate)
- S^{2-} = sulfide
- Cl^- = chloride
- SO_4^{2-} = sulfate
- CO_3^{2-} = carbonate

[b] The precipitant is for the indicated valence state.

[c] Maximum tolerance levels have been set for these low–toxicity ions by the U.S. Public Health Service, and large amounts should not be put into public sewer systems. The small amounts typically used in laboratories will not normally affect water supplies.

[d] These ions are best precipitated as calcium molybdate or calcium tungstate.

[e] CAUTION: OsO_4, a volatile, extremely poisonous substance, is formed from almost any osmium compound under acid conditions in the presence of air.

TABLE 10
Relative Hazard of Anions

High–hazard anions			
Ion	**Hazard type**[a]	**Precipitant**	**Low-hazard anions**
Aluminum hydride, AlH_4	F	—	Bisulfite, HSO_3^-
Amide, NH_2^-	F, E[b]	—	Borate, BO_3^{3-}, $B_4O_7^{2-}$
Arsenate, AsO_3^-, AsO_4^{3-}	T	Cu^{2+}, Fe^{2+}	Bromide, Br^-
Arsenite, AsO_2^-, AsO_3^{3-}	T	Pb^{2+}	Carbonate, CO_3^{2-}
Azide, N_3^-	E, T	—	Chloride, Cl^-
Borohydride, BH_4^-	F	—	Cyanate, OCN^-
Bromate, BrO_3^-	O, E	—	Hydroxide, OH^-
Chlorate, ClO_3^-	O, E	—	Iodide, I^-
Chromate, CrO_4^{2-}, $Cr_2O_7^{2-}$	T, O	[c]	Oxide, O^{2-}
Cyanide, CN^-	T	—	Phosphate, PO_4^{3-}
Ferricyanide, $Fe(CN)_6^{3-}$	T	Fe^{2+}	Sulfate, SO_4^{2}
Ferrocyanide, $Fe(CN)_6^{4-}$	T	Fe^{3+}	Sulfite, SO_3^{2-}
Fluoride, F^-	T	Ca^{2+}	Thiocyanate, SCN^-
Hydride, H^-	F	—	
Hydroperoxide, O_2H^-	O, E	—	
Hydrosulfide, SH^-	T	—	
Hypochlorite, OCl^-	O	—	
Iodate, IO_3^-	O, E	—	
Nitrate, NO_3^-	O	—	
Nitrite, NO_2^-	T, O	—	
Perchlorate, ClO_4^-	O, E	—	
Permanganate, MnO_4^-	T, O	[d]	
Peroxide, O_2^{2-}	O, E	—	
Persulfate, $S_2O_8^{2-}$	O	—	
Selenate, SeO_4^{2-}	T	Pb^{2+}	
Selenide, Se^{2-}	T	Cu^{2+}	
Sulfide, S^{2-}	T	[e]	

[a] Toxic, T: oxidant, O; flammable, F; explosive, E.

[b] Metal amides readily form explosive peroxides on exposure to air.

[c] Reduce and precipitate as Cr(III); see Table 9.

[d] Reduce and precipitate as Mn(II); see Table 9.

[e] See Table 11.

TABLE 11
Precipitation of Sulfides

Precipitated at pH 7	Not precipitated at low pH	Forms a soluble complex at high pH
Ag^{+}		
As^{3+}[a]		X
Au^{+}[a]		X
Bi^{3+}		
Cd^{2+}		
Co^{2+}	X	
Cr^{3+}[a]		
Cu^{2+}		
Fe^{2+}[a]	X	
Ge^{2+}		X
Hg^{2+}		X
In^{3+}	X	
Ir^{4+}		X
Mn^{2+}[a]	X	
Mo^{3+}		X
Ni^{2+}	X	
Os^{4+}		
Pb^{2+}		
Pd^{2+}[a]		
Pt^{2+}[a]		X
Re^{4+}		
Rh^{2+}[a]		
Ru^{4+}		
Sb^{3+}[a]		X
Se^{2+}		X
Sn^{2+}		X
Te^{4+}		X
Tl^{+}[a]	X	
V^{4+}[a]		
Zn^{2+}	X	

[a] Higher oxidation states of this ion are reduced by sulfide ion and precipitated as this sulfide.

TABLE 12
pH Range for Precipitation of Metal Hydroxides and Oxides

Ion	pH range (scale 1–10)
Ag^{1+}	9 → 1 N
Al^{3+}	7–8
As^{3+}	Not precipitated (precipitate as sulfide)
As^{5+}	Not precipitated (precipitate as sulfide)
Au^{3+}	7–8
Be^{2+}	7–8
Bi^{3+}	7 → 1 N
Cd^{2+}	7 → 1 N
Co^{2+}	8 → 1 N
Cr^{3+}	7 → 1 N
Cu^{1+}	9 → 1 N

TABLE 12
pH Range for Precipitation of Metal Hydroxides and Oxides (continued)

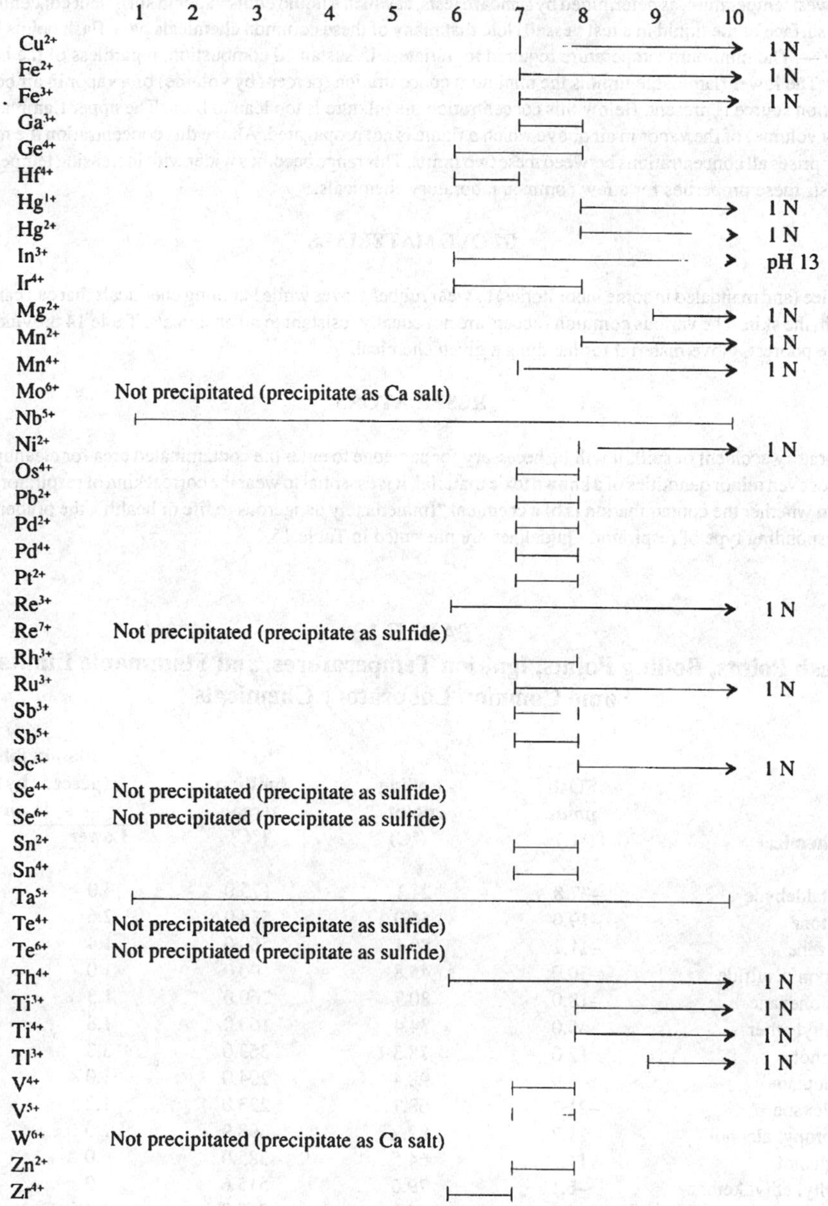

REFERENCES

L. Erdey, *Gravimetric Analysis*, Part II, Pergamon Press, New York, 1965.

D. T. Burns, A. Towsend, and A. H. Carter, *Inorganic Reaction Chemistry*, Vol. 2, Ellis Horwood, New York, 1981.

FIRE HAZARDS

Flammable solvents are a common source of laboratory fires. The relative ease with which some common laboratory solvents can be ignited is indicated by the following properties.

Flash Point — The lowest temperature, as determined by standard tests, at which a liquid emits vapor in sufficient concentration to form an ignitable mixture with air near the surface of the liquid in a test vessel. Note that many of these common chemicals have flash points below room temperature.

Ignition Temperature — The minimum temperature required to initiate self–sustained combustion, regardless of the heat source.

Flammable Limits — The lower flammable limit is the minimum concentration (percent by volume) of a vapor in air below which a flame is not propagated when an ignition source is present. Below this concentration the mixture is too lean to burn. The upper flammable limit is the maximum concentration (percent by volume) of the vapor in air above which a flame is not propagated. Above this concentration the mixture is too rich to burn. The flammable range comprises all concentrations between these two limits. This range becomes wider with increasing temperature and in oxygen–rich atmospheres. Table 13 lists these properties for a few common laboratory chemicals.

GLOVE MATERIALS

It is good safety practice (and mandated in some laboratories) to wear rubber gloves while handling chemicals that can cause injury when in contact with, or absorbed through, the skin. The various common rubbers are not equally resistant to all chemicals. Table 14 provides guidelines for selecting the best, and avoiding the poorest, glove material for handling a given chemical.

RESPIRATORS

In the event of a laboratory accident or spill, it will be necessary for someone to enter the contaminated area for cleanup. If significant quantities of a chemical are spilled, or even minor quantities of a known toxic material, it is essential to wear the correct kind of respirator equipment when entering the area. If it is not known whether the contamination is of a chemical "Immediately dangerous to life or health", the prudent course is to assume that it is, and to use the corresponding type of respirator. Guidelines are presented in Table 15.

TABLE 13
Flash Points, Boiling Points, Ignition Temperatures, and Flammable Limits of Some Common Laboratory Chemicals

Chemical	Flash point (°C)	Boiling point (°C)	Ignition temp. (°C)	Flammable limit (percent by volume in air)	
				Lower	Upper
Acetaldehyde	–37.8	21.1	175.0	4.0	60.0
Acetone	–19.0	56.0	538.0	2.6	12.8
Benzene	–11.1	80.1	560.0	1.4	8.0
Carbon disulfide	–30.0	45.8	90.0	1.0	44.0
Cyclohexane	–18.0	80.7	260.0	1.3	8.0
Diethyl ether	–45.0	34.4	160.0	1.8	48.0
Ethanol	12.0	78.3	363.0	3.3	19.0
n–Heptane	–3.9	98.4	204.0	1.0	6.7
n–Hexane	–21.7	68.7	223.0	1.2	7.5
Isopropyl alcohol	11.7	82.2	398.9	2.0	12.0
Methanol	11.1	64.5	385.0	6.0	36.5
Methyl ethyl ketone	–6.1	79.6	515.6	1.9	11.0
Pentane	–40.0	36.1	260.0	1.4	7.8
Styrene	31.0	145.0	490.0	1.1	6.1
Toluene	4.4	110.6	530.0	1.3	7.0
p–Xylene	25.0	132.4	529.0	1.1	7.0

Note: For a more extensive listing, see the table "Properties of Common Solvents" in Section 15.

TABLE 14
Resistance to Chemicals of Common Glove Materials
(E = Excellent, G = Good, F = Fair, P = Poor)

Chemical	Natural rubber	Neoprene	Nitrile	Vinyl
Acetaldehyde	G	G	E	G
Acetic acid	E	E	E	E
Acetone	G	G	G	F
Acrylonitrile	P	G	—	F
Ammonium hydroxide (sat)	G	E	E	E
Aniline	F	G	E	G
Benzaldehyde	F	F	E	G
Benzene[a]	P	F	G	F
Benzyl chloride[a]	F	P	G	P
Bromine	G	G	—	G
Butane	P	E	—	P
Butyraldehyde	P	G	—	G
Calcium hypochlorite	P	G	G	G
Carbon disulfide	P	P	G	F
Carbon tetrachloride[a]	P	F	G	F
Chlorine	G	G	—	G
Chloroacetone	F	E	—	P
Chloroform[a]	P	F	G	P
Chromic acid	P	F	F	E
Cyclohexane	F	E	—	P
Dibenzyl ether	F	G	—	P
Dibutyl phthalate	F	G	—	P
Diethanolamine	F	E	—	E
Diethyl ether	F	G	E	P
Dimethyl sulfoxide[b]	—	—	—	—
Ethyl acetate	F	G	G	F
Ethylene dichloride[a]	P	F	G	
Ethylene glycol	G	G	E	E
Ethylene trichloride[a]	P	P	—	P
Fluorine	G	G	—	G
Formaldehyde	G	E	E	E
Formic acid	G	E	E	E
Glycerol	G	G	E	E
Hexane	P	E	—	P
Hydrobromic acid (40%)	G	E	—	E
Hydrochloric acid (conc)	G	G	G	E
Hydrofluoric acid (30%)	G	G	G	E
Hydrogen peroxide	G	G	G	E
Iodine	G	G	—	G
Methylamine	G	G	E	E
Methyl cellosolve	F	E	—	P
Methyl chloride[a]	P	E	—	P
Methyl ethyl ketone	F	G	G	P
Methylene chloride[a]	F	F	G	F
Monoethanolamine	F	E	—	E
Morpholine	F	E	—	E
Naphthalene[a]	G	G	E	G
Nitric acid (conc)	P	P	P	G
Perchloric acid	F	G	F	E
Phenol	G	E	—	E
Phosphoric acid	G	E	—	E
Potassium hydroxide (sat)	G	G	G	E

TABLE 14
Resistance to Chemicals of Common Glove Materials
(E = Excellent, G = Good, F = Fair, P = Poor) (continued)

Chemical	Natural rubber	Neoprene	Nitrile	Vinyl
Propylene dichloride[a]	P	F	—	P
Sodium hydroxide	G	G	G	E
Sodium hypochlorite	G	P	F	G
Sulfuric acid (conc)	G	G	F	G
Toluene[a]	P	F	G	F
Trichloroethylene[a]	P	F	G	F
Tricresyl phosphate	P	F	—	F
Triethanolamine	F	E	E	E
Trinitrotoluene	P	E	—	P

[a] Aromatic and halogenated hydrocarbons will attack all types of natural and synthetic glove materials. Should swelling occur, the user should change to fresh gloves and allow the swollen gloves to dry and return to normal.

[b] No data on the resistance to dimethyl sulfoxide of natural rubber, neoprene, nitrile rubber, or vinyl materials are available; the manufacturer of the substance recommends the use of butyl rubber gloves.

TABLE 15
Guide for Selection of Respirators

Type of hazard	Type of respirator
Oxygen deficiency	Self–contained breathing apparatus Hose mask with blower Combination of air–line respirator and auxiliary self–contained air supply or air–storage receiver with alarm
Gas and vapor contaminants Immediately dangerous to life or health	Self–contained breathing apparatus Hose mask with blower Air–purifying full–facepiece respirator with chemical canister (gas mask) Self–rescue mouthpiece respirator (for escape only) Combination of air–line respirator and auxiliary self–contained air supply or air–storage receiver with alarm
Not immediately dangerous to life or health	Air–line respirator Hose mask with blower Air–purifying half–mask or mouthpiece respirator with chemical cartridge
Particulate Contaminants Immediately dangerous to life or health	Self–contained breathing apparatus Hose mask with blower Air–purifying full–facepiece respirator with appropriate filter Self–rescue mouthpiece respirator (for escape only) Combination of air–line respirator and auxiliary self–contained air supply or air–storage receiver with alarm

TABLE 15
Guide for Selection of Respirators (continued)

Type of hazard	Type of respirator
Not immediately dangerous to life or health	Air–purifying half–mask or mouthpiece respirator with filter pad or cartridge Air–line respirator Air–line abrasive–blasting respirator Hose mask with blower
Combination of gas, vapor, and particulate contaminants Immediately dangerous to life or health	Self–contained breathing apparatus Hose mask with blower Air–purifying full–facepiece respirator with chemical canister and appropriate filter (gas mask with filter) Self–rescue mouthpiece respirator (for escape only) Combination of air–line respirator and auxiliary self–contained air supply or air–storage receiver with alarm
Not immediately dangerous to life or health	Air–line respirator Hose mask without blower Air–purifying half–mask or mouthpiece respirator with chemical cartridge and appropriate filter

Source: ANSI Standard Z88.2 (1969).

FLAMMABILITY OF CHEMICAL SUBSTANCES

This table gives properties related to the flammability of about 900 chemical substances. The properties listed are:

t_B : Normal boiling point in °C (at 101.325 kPa pressure).

FP: Flash point, which is the minimum temperature at which the vapor pressure of a liquid is sufficient to form an ignitable mixture with air near the surface of the liquid. Flash point is not an intrinsic physical property but depends on the conditions of measurement (see Reference 1).

Fl. Limits: Flammable limits (often called explosive limits), which specify the range of concentration of the vapor in air (in percent by volume) for which a flame can propagate. Below the lower flammable limit, the gas mixture is too lean to burn; above the upper flammable limit, the mixture is too rich. Values refer to ambient temperature and pressure and are dependent on the precise test conditions. A ? indicates that one of the limits is not known.

IT: Ignition temperature (sometimes called autoignition temperature), which is the minimum temperature required for self-sustained combustion in the absence of an external ignition source. As in the case of flash point, the value depends on specified test conditions.

Even in cases where very careful measurements of flash point have been replicated in several laboratories, observed values can differ by 3 to 6°C (Reference 4). For more typical measurements, larger uncertainties should be assumed in both flash points and autoignition temperatures. The absence of a flash point entry in this table does not mean that the substance is nonflammable, but only that no reliable value is available.

Compounds are listed by molecular formula following the Hill convention. Substances not containing carbon are listed first, followed by those that contain carbon. To locate an organic compound by name or CAS Registry Number when the molecular formula is not known, use the table "Physical Constants of Organic Compounds" in Section 3 and its indexes to determine the molecular formula.

REFERENCES

1. *Fire Protection Guide to Hazardous Materials, 11th Edition*, National Fire Protection Association, Quincy, MA, 1994.
2. Urben, P.G., Editor, *Bretherick's Handbook of Reactive Chemical Hazards, 5th Edition*, Butterworth-Heinemann, Oxford, 1995.
3. Daubert, T.E., Danner, R.P., Sibul, H.M., and Stebbins, C.C., *Physical and Thermodynamic Properties of Pure Compounds: Data Compilation*, extant 1994 (core with 4 supplements), Taylor & Francis, Bristol, PA.
4. *Report of Investigation: Flash Point Reference Materials*, National Institute of Standards and Technology, Standard Reference Materials Program, Gaithersburg, MD, 1995.

Mol. Form.	Name	t_B/°C	FP/°C	Fl. Limits	IT/°C
	Compounds not containing carbon				
B_2H_6	Diborane	-92.4	-90	1-98%	≈40
B_5H_9	Pentaborane(9)	60	30	0.4-?	35
BrH_3Si	Bromosilane	1.9	<0		≈20
Br_3HSi	Tribromosilane	109			≈20
Cl_2H_2Si	Dichlorosilane	8.3		4.1-99%	36
Cl_3HSi	Trichlorosilane	33	-50		104
GeH_4	Germane	-88.1			≈20
Ge_2H_6	Digermane	30.8			≈50
H_2	Hydrogen	-252.8		4-74%	
H_2S	Hydrogen sulfide	-59.55		4-44%	260
H_2S_2	Hydrogen disulfide	70.7	<22		
H_2Te	Hydrogen telluride	-2			-50
H_3N	Ammonia	-33.33		16-25%	
H_3P	Phosphine	-87.75		1.8-?	
H_4N_2	Hydrazine	113.55	38	5-100%	
H_4P_2	Diphosphine	63.5			≈20
H_4Si	Silane	-111.9	-112	1.4-?	≈20
H_6Si_2	Disilane	-14.3	-14		≈20
H_8Si_3	Trisilane	52.9	<0		≈20
P	Phosphorus (white)	280.5			38
	Compounds containing carbon				
CHN	Hydrogen cyanide	26	-18	6-40%	538
CH_2Cl_2	Dichloromethane	40		13-23%	556
CH_2N_2	Cyanamide		141		

FLAMMABILITY OF CHEMICAL SUBSTANCES (continued)

Mol. Form	Name	t_B/°C	FP/°C	Fl. Limits	IT/°C
CH_2O	Formaldehyde	-19.1	85	7.0-73%	424
$(CH_2O)_x$	Paraformaldehyde		70	7.0-73%	300
CH_2O_2	Formic acid	101	50	18-57%	434
CH_3Br	Bromomethane	3.5		10-16%	537
CH_3Cl	Chloromethane	-24.0		8.1-17.4%	632
CH_3Cl_3Si	Methyltrichlorosilane	65.6	-9	7.6->20%	>404
CH_3NO	Formamide	220	154		
CH_3NO_2	Nitromethane	101.1	35	7.3-?	418
CH_4	Methane	-161.5		5.0-15.0%	537
CH_4Cl_2Si	Dichloromethylsilane	41	-9	6.0-55%	316
CH_4O	Methanol	64.6	11	6.0-36%	464
CH_4S	Methanethiol	5.9	-18	3.9-21.8%	
CH_5N	Methylamine	-6.3	0	4.9-20.7%	430
CH_6N_2	Methylhydrazine	87.5	-8	2.5-92%	194
CO	Carbon monoxide	-191.5		12.5-74%	609
COS	Carbon oxysulfide	-50		12-29%	
CS_2	Carbon disulfide	46	-30	1.3-50.0%	90
C_2ClF_3	Chlorotrifluoroethylene	-27.8		8.4-16.0%	
C_2F_4	Tetrafluoroethylene	-75.9		10.0-50.0%	200
C_2HCl_3	Trichloroethylene	87.2		8-10.5%	420
C_2HCl_3O	Dichloroacetyl chloride	108	66		
C_2H_2	Acetylene	-84.7		2.5-100%	305
$C_2H_2Cl_2$	1,1-Dichloroethylene	31.6	-28	6.5-15.5%	570
$C_2H_2Cl_2$	*cis*-1,2-Dichloroethylene	60.1	6	3-15%	460
$C_2H_2Cl_2$	*trans*-1,2-Dichloroethylene	48.7	2	6-13%	460
$C_2H_2F_2$	1,1-Difluoroethylene	-85.7		5.5-21.3%	
C_2H_3Br	Bromoethylene	15.8		9-15%	530
C_2H_3Cl	Chloroethylene	-13.3	-78	3.6-33.0%	472
$C_2H_3ClF_2$	1-Chloro-1,1-difluoroethane	-9.7		6-18%	632
C_2H_3ClO	Acetyl chloride	50.7	4		390
$C_2H_3Cl_2NO_2$	1,1-Dichloro-1-nitroethane	123.5	76		
$C_2H_3Cl_3$	1,1,1-Trichloroethane	74.0		8-10.5%	500
$C_2H_3Cl_3$	1,1,2-Trichloroethane	113.8	32	6-28%	460
$C_2H_3Cl_3Si$	Trichlorovinylsilane	91.5	21		
C_2H_3F	Fluoroethylene	-72		2.6-21.7%	
C_2H_3N	Acetonitrile	81.6	6	3.0-16.0%	524
C_2H_3NO	Methyl isocyanate	39.5	-7	5.3-26%	534
C_2H_4	Ethylene	-103.7		2.7-36%	450
$C_2H_4ClNO_2$	1-Chloro-1-nitroethane	124.5	56		
$C_2H_4Cl_2$	1,1-Dichloroethane	57.4	-17	5.4-11.4%	458
$C_2H_4Cl_2$	1,2-Dichloroethane	83.5	13	6.2-16%	413
C_2H_4O	Acetaldehyde	20.1	-39	4.0-60%	175
C_2H_4O	Ethylene oxide	10.6	-20	3.0-100%	429
$C_2H_4O_2$	Acetic acid	117.9	39	4.0-19.9%	463
$C_2H_4O_2$	Methyl formate	31.7	-19	4.5-23%	449
$C_2H_4O_3$	Ethaneperoxoic acid	110	41		
C_2H_5Br	Bromoethane	38.5		6.8-8.0%	511
C_2H_5Cl	Chloroethane	12.3	-50	3.8-15.4%	519
C_2H_5ClO	Ethylene chlorohydrin	128.6	60	4.9-15.9%	425
$C_2H_5Cl_3Si$	Trichloroethylsilane	100.5	22		
C_2H_5N	Ethyleneimine	56	-11	3.3-54.8%	320
$C_2H_5NO_2$	Nitroethane	114.0	28	3.4-17%	414
$C_2H_5NO_2$	Ethyl nitrite	18	-35	4.0-50%	90
$C_2H_5NO_3$	Ethyl nitrate	87.2	10	4-?	
C_2H_6	Ethane	-88.6		3.0-12.5%	472
$C_2H_6Cl_2Si$	Dichlorodimethylsilane	70.3	<21	3.4-9.5%	
C_2H_6O	Ethanol	78.2	13	3.3-19%	363
C_2H_6O	Dimethyl ether	-24.8	-41	3.4-27.0%	350
C_2H_6OS	2-Mercaptoethanol	158	74		
C_2H_6OS	Dimethyl sulfoxide	189	95	2.6-42%	215
$C_2H_6O_2$	Ethylene glycol	197.3	111	3.2-22%	398

Mol. Form.	Name	t_B/°C	FP/°C	Fl. Limits	IT/°C
$C_2H_6O_4S$	Dimethyl sulfate		83		188
C_2H_6S	Ethanethiol	35.1	-17	2.8-18.0%	300
C_2H_6S	Dimethyl sulfide	37.3	-37	2.2-19.7%	206
$C_2H_6S_2$	Dimethyl disulfide	109.8	24		
C_2H_7N	Ethylamine	16.5	-16	3.5-14%	385
C_2H_7N	Dimethylamine	6.8	20	2.8-14.4%	400
C_2H_7NO	Ethanolamine	171	86	3.0-23.5%	410
$C_2H_8N_2$	1,2-Ethanediamine	117	40	2.5-12.0%	385
$C_2H_8N_2$	1,1-Dimethylhydrazine	63.9	-15	2-95%	249
C_2N_2	Cyanogen	-21.1		6.6-32%	
C_3H_3Br	3-Bromo-1-propyne	89	10	3.0-?	324
C_3H_3N	2-Propenenitrile	77.3	0	3.0-17.0%	481
C_3H_4	Propyne	-23.2		2.1-12.5%	
C_3H_4ClN	3-Chloropropanenitrile	175.5	76		
$C_3H_4Cl_2$	2,3-Dichloropropene	94	15	2.6-7.8%	
C_3H_4O	Propargyl alcohol	113.6	36		
C_3H_4O	Acrolein	52.6	-26	2.8-31%	220
$C_3H_4O_2$	Propenoic acid	141	50	2.4-8.0%	438
$C_3H_4O_2$	2-Oxetanone	162	74	2.9-?	
$C_3H_4O_3$	Ethylene carbonate	248	143		
C_3H_5Br	3-Bromopropene	70.1	-1	4.4-7.3%	295
C_3H_5Cl	2-Chloropropene	22.6	-37	4.5-16%	
C_3H_5Cl	3-Chloropropene	45.1	-32	2.9-11.1%	485
C_3H_5ClO	Epichlorohydrin	118	31	3.8-21.0%	411
C_3H_5ClO	Propanoyl chloride	80	12		
$C_3H_5ClO_2$	2-Chloropropanoic acid	185	107		500
$C_3H_5ClO_2$	Ethyl chloroformate	95	16		500
$C_3H_5ClO_2$	Methyl chloroacetate	129.5	57	7.5-18.5%	
$C_3H_5Cl_2NO_2$	1,1-Dichloro-1-nitropropane	145	66		
$C_3H_5Cl_3$	1,2,3-Trichloropropane	157	71	3.2-12.6%	
$C_3H_5Cl_3Si$	Trichloro-2-propenylsilane	117.5	35		
C_3H_5N	Propanenitrile	97.1	2	3.1-14%	512
C_3H_5NO	3-Hydroxypropanenitrile	221	129		
$C_3H_5N_3O_9$	Trinitroglycerol				270
C_3H_6	Propene	-47.6		2.0-11.1%	455
C_3H_6	Cyclopropane	-32.8		2.4-10.4%	498
$C_3H_6ClNO_2$	1-Chloro-1-nitropropane	142	62		
$C_3H_6ClNO_2$	2-Chloro-2-nitropropane		57		
$C_3H_6Cl_2$	1,2-Dichloropropane	96.4	21	3.4-14.5%	557
$C_3H_6Cl_2O$	1,3-Dichloro-2-propanol	176	74		
$C_3H_6N_2$	Dimethylcyanamide	163.5	71		
C_3H_6O	Allyl alcohol	97.0	21	2.5-18.0%	378
C_3H_6O	Methyl vinyl ether	5.5			287
C_3H_6O	Propanal	48	-30	2.6-17%	207
C_3H_6O	Acetone	56.0	-20	2.5-12.8%	465
C_3H_6O	Methyloxirane	35	-37	3.1-27.5%	449
$C_3H_6O_2$	Propanoic acid	141.1	52	2.9-12.1%	465
$C_3H_6O_2$	Ethyl formate	54.4	-20	2.8-16.0%	455
$C_3H_6O_2$	Methyl acetate	56.8	-10	3.1-16%	454
$C_3H_6O_2$	1,3-Dioxolane	78	2		
$C_3H_6O_3$	Dimethyl carbonate	90.5	19		
$C_3H_6O_3$	1,3,5-Trioxane	114.5	45	3.6-29%	414
C_3H_7Br	1-Bromopropane	71.1			490
C_3H_7Cl	1-Chloropropane	46.5	<-18	2.6-11.1%	520
C_3H_7Cl	2-Chloropropane	35.7	-32	2.8-10.7%	593
C_3H_7ClO	2-Chloro-1-propanol	133.5	52		
C_3H_7ClO	1-Chloro-2-propanol	127	52		
$C_3H_7Cl_3Si$	Trichloropropylsilane	123.5	37		
C_3H_7N	Allylamine	53.3	-29	2.2-22%	374
C_3H_7NO	*N,N*-Dimethylformamide	153	58	2.2-15.2%	445
$C_3H_7NO_2$	1-Nitropropane	131.1	36	2.2-?	421

Mol. Form.	Name	t_B/°C	FP/°C	Fl. Limits	IT/°C
$C_3H_7NO_2$	2-Nitropropane	120.2	24	2.6-11.0%	428
$C_3H_7NO_3$	Propyl nitrate	110	20	2-100%	175
C_3H_8	Propane	-42.1	-104	2.1-9.5%	450
C_3H_8O	1-Propanol	97.2	23	2.2-13.7%	412
C_3H_8O	2-Propanol	82.3	12	2.0-12.7%	399
C_3H_8O	Ethyl methyl ether	7.4	-37	2.0-10.1%	190
$C_3H_8O_2$	1,2-Propylene glycol	187.6	99	2.6-12.5%	371
$C_3H_8O_2$	1,3-Propylene glycol	214.4			400
$C_3H_8O_2$	Ethylene glycol monomethyl ether	124.1	39	1.8-14%	285
$C_3H_8O_2$	Dimethoxymethane	42	-32	2.2-13.8%	237
$C_3H_8O_3$	Glycerol	290	199	3-19%	370
$C_3H_9BO_3$	Trimethyl borate	67.5	-8		
C_3H_9ClSi	Trimethylchlorosilane	60	-28		395
C_3H_9N	Propylamine	47.2	-37	2.0-10.4%	318
C_3H_9N	Isopropylamine	31.7	-37		402
C_3H_9N	Trimethylamine	2.8	-5	2.0-11.6%	190
C_3H_9NO	3-Amino-1-propanol	187.5	80		
C_3H_9NO	1-Amino-2-propanol	159.4	77		374
C_3H_9NO	*N*-Methyl-2-ethanolamine	158	74		
$C_3H_9O_3P$	Trimethyl phosphite	111.5	54		
$C_3H_9O_4P$	Trimethyl phosphate	197.2	107		
$C_3H_{10}N_2$	1,3-Propanediamine	139.8	24		
C_4Cl_6	Hexachloro-1,3-butadiene	215			610
$C_4H_2O_3$	Maleic anhydride	202	102	1.4-7.1%	477
C_4H_4	1-Buten-3-yne	5.1		21-100%	
$C_4H_4N_2$	Succinonitrile	266	132		
C_4H_4O	Furan	31.5	-36	2.3-14.3%	
$C_4H_4O_2$	Diketene	126.1	34		
C_4H_4S	Thiophene	84.0	-1		
C_4H_5Cl	2-Chloro-1,3-butadiene	59.4	-20	4.0-20.0%	
C_4H_5N	2-Butenenitrile	120.5	16		
C_4H_5N	Methylacrylonitrile	90.3	1	2-6.8%	
C_4H_5N	Pyrrole	129.7	39		
C_4H_6	1,3-Butadiene	-4.4		2.0-12.0%	420
C_4H_6	2-Butyne	26.9	-31	1.4-?	
C_4H_6O	Divinyl ether	28.3	<-30	1.7-27%	360
C_4H_6O	Ethoxyacetylene	50	<-7		
C_4H_6O	*trans*-2-Butenal	102.2	13	2.1-15.5%	232
C_4H_6O	3-Buten-2-one	81.4	-7	2.1-15.6%	491
C_4H_6O	Vinyloxirane	68	<-50		
$C_4H_6O_2$	Methacrylic acid	162.5	77	1.6-8.8%	68
$C_4H_6O_2$	Vinyl acetate	72.5	-8	2.6-13.4%	402
$C_4H_6O_2$	Methyl acrylate	80.7	-3	2.8-25%	468
$C_4H_6O_2$	2,3-Butanedione	88	27		
$C_4H_6O_2$	gamma-Butyrolactone	204	98		
$C_4H_6O_3$	Acetic anhydride	139.5	49	2.7-10.3%	316
$C_4H_6O_3$	Propylene carbonate	242	135		
$C_4H_6O_6$	*L*-Tartaric acid		210		425
C_4H_7Br	1-Bromo-2-butene	104.5		4.6-12.0%	
$C_4H_7BrO_2$	Ethyl bromoacetate	168.5	48		
C_4H_7Cl	2-Chloro-1-butene	58.5	-19	2.3-9.3%	
C_4H_7Cl	3-Chloro-2-methylpropene	71.5	-12	3.2-8.1%	
C_4H_7ClO	2-Chloroethyl vinyl ether	108	27		
$C_4H_7ClO_2$	Ethyl chloroacetate	144.3	64		
C_4H_7N	Butanenitrile	117.6	24	1.6-?	501
C_4H_7N	2-Methylpropanenitrile	103.9	8		482
C_4H_7NO	Acetone cyanohydrin		74	2.2-12.0%	688
C_4H_7NO	2-Pyrrolidone	251	129		
C_4H_8	1-Butene	-6.2		1.6-10.0%	385
C_4H_8	*cis*-2-Butene	3.7		1.7-9.0%	325
C_4H_8	*trans*-2-Butene	0.8		1.8-9.7%	324

Mol. Form.	Name	t_B/°C	FP/°C	Fl. Limits	IT/°C
C_4H_8	Isobutene	-6.9		1.8-9.6%	465
C_4H_8	Cyclobutane	12.6	<10	1.8-?	
$C_4H_8Cl_2$	1,2-Dichlorobutane	124.1			275
$C_4H_8Cl_2$	1,4-Dichlorobutane	161	52		
$C_4H_8Cl_2O$	Bis(2-chloroethyl) ether	178.5	55	2.7-?	369
C_4H_8O	2-Buten-1-ol	121.5	27	4.2-35.3%	349
C_4H_8O	2-Methyl-2-propenol	114.5	33		
C_4H_8O	Ethyl vinyl ether	35.5	<-46	1.7-28%	202
C_4H_8O	1,2-Epoxybutane	63.4	-22	1.7-19%	439
C_4H_8O	Butanal	74.8	-22	1.9-12.5%	218
C_4H_8O	Isobutanal	64.5	-18	1.6-10.6%	196
C_4H_8O	2-Butanone	79.5	-9	1.4-11.4%	404
C_4H_8O	Tetrahydrofuran	65	-14	2-11.8%	321
C_4H_8OS	1,4-Oxathiane	147	42		
$C_4H_8O_2$	Butanoic acid	163.7	72	2.0-10.0%	443
$C_4H_8O_2$	2-Methylpropanoic acid	154.4	56	2.0-9.2%	481
$C_4H_8O_2$	Propyl formate	80.9	-3		455
$C_4H_8O_2$	Isopropyl formate	68.2	-6		485
$C_4H_8O_2$	Ethyl acetate	77.1	-4	2.0-11.5%	426
$C_4H_8O_2$	Methyl propanoate	79.8	-2	2.5-13%	469
$C_4H_8O_2$	3-Hydroxybutanal		66		250
$C_4H_8O_2$	1,4-Dioxane	101.5	12	2.0-22%	180
$C_4H_8O_2S$	Sulfolane	287.3	177		
$C_4H_8O_3$	Methyl lactate	144.8	49	2.2-?	385
$C_4H_8O_3$	Ethylene glycol monoacetate	188	102		
C_4H_9Br	1-Bromobutane	101.6	18	2.6-6.6%	265
C_4H_9Br	2-Bromobutane	91.2	21		
C_4H_9Cl	1-Chlorobutane	78.6	-12	1.9-10.1%	240
C_4H_9Cl	2-Chlorobutane	68.2	-10		
C_4H_9Cl	1-Chloro-2-methylpropane	68.5	-6	2.0-8.7%	
C_4H_9Cl	2-Chloro-2-methylpropane	50.9	0		
$C_4H_9Cl_3Si$	Butyltrichlorosilane	148.5	54		
C_4H_9N	Pyrrolidine	86.5	3		
C_4H_9NO	*N*-Ethylacetamide	205	110		
C_4H_9NO	*N,N*-Dimethylacetamide	165	70	1.8-11.5%	490
C_4H_9NO	Butanal oxime	154	58		
C_4H_9NO	2-Butanone oxime	152.5	≈70		
C_4H_9NO	Morpholine	128	37	1.4-11.2%	290
$C_4H_9NO_2$	*N*-Acetylethanolamine		179		460
$C_4H_9NO_3$	Butyl nitrate	133	36		
C_4H_{10}	Butane	-0.5	-60	1.9-8.5%	287
C_4H_{10}	Isobutane	-11.7	-87	1.8-8.4%	460
$C_4H_{10}N_2$	Piperazine	146	81		
$C_4H_{10}O$	1-Butanol	117.7	37	1.4-11.2%	343
$C_4H_{10}O$	2-Butanol	99.5	24	1.7-9.8%	405
$C_4H_{10}O$	2-Methyl-1-propanol	107.8	28	1.7-10.6%	415
$C_4H_{10}O$	2-Methyl-2-propanol	82.4	11	2.4-8.0%	478
$C_4H_{10}O$	Diethyl ether	34.5	-45	1.9-36.0%	180
$C_4H_{10}O$	Methyl propyl ether	39.1	-20	2.0-14.8%	
$C_4H_{10}O_2$	1,2-Butanediol	190.5	40		
$C_4H_{10}O_2$	1,3-Butanediol	207.5	121		395
$C_4H_{10}O_2$	1,4-Butanediol	235	121		
$C_4H_{10}O_2$	2,3-Butanediol	182.5			402
$C_4H_{10}O_2$	Ethylene glycol monoethyl ether	135	43	3-18%	235
$C_4H_{10}O_2$	Ethylene glycol dimethyl ether	85	-2		202
$C_4H_{10}O_2$	*tert*-Butyl hydroperoxide		27		
$C_4H_{10}O_2S$	2,2′-Thiodiethanol	282	160		298
$C_4H_{10}O_3$	Diethylene glycol	245.8	124	2-17%	224
$C_4H_{10}O_4S$	Diethyl sulfate	208	104		436
$C_4H_{10}S$	1-Butanethiol	98.5	2		
$C_4H_{10}S$	2-Butanethiol	85	-23		

Mol. Form.	Name	t_B/°C	FP/°C	Fl. Limits	IT/°C
$C_4H_{10}S$	2-Methyl-1-propanethiol	88.5	2		
$C_4H_{10}S$	2-Methyl-2-propanethiol	64.3	<-29		
$C_4H_{10}Se$	Diethyl selenide	108		2.5-?	
$C_4H_{11}N$	Butylamine	77.0	-12	1.7-9.8%	312
$C_4H_{11}N$	*sec*-Butylamine	63.5	-9		
$C_4H_{11}N$	*tert*-Butylamine	44.0	-9	1.7-8.9%	380
$C_4H_{11}N$	Isobutylamine	67.7	-9	2-12%	378
$C_4H_{11}N$	Diethylamine	55.5	-23	1.8-10.1%	312
$C_4H_{11}NO$	2-Amino-1-butanol	178	74		
$C_4H_{11}NO$	2-Amino-2-methyl-1-propanol	165.5	67		
$C_4H_{11}NO_2$	Diethanolamine	268.8	172	2-13%	662
$C_4H_{12}Sn$	Tetramethylstannane	78	-12	1.9-?	
$C_4H_{13}N_3$	Diethylenetriamine	207	98	2-6.7%	358
$C_5H_4O_2$	Furfural	161.7	60	2.1-19.3%	316
C_5H_5N	Pyridine	115.2	20	1.8-12.4%	482
C_5H_6	2-Methyl-1-buten-3-yne	32	<-7		
$C_5H_6N_2$	2-Methylpyrazine	137	50		
C_5H_6O	3-Methylfuran	66	-30		
$C_5H_6O_2$	Furfuryl alcohol	171	75	1.8-16.3%	491
C_5H_7N	1-Methylpyrrole	115	16		
C_5H_7NO	2-Furanmethanamine	145.5	37		
$C_5H_7NO_2$	Ethyl cyanoacetate	205	110		
C_5H_8	2-Methyl-1,3-butadiene	34.0	-54	1.5-8.9%	395
C_5H_8	1-Pentyne	40.1	<-20		
C_5H_8	Cyclopentene	44.2	-29		395
C_5H_8O	3-Methyl-3-buten-2-one	98		1.8-9.0%	
C_5H_8O	Cyclopentanone	130.5	26		
C_5H_8O	3,4-Dihydro-2H-pyran	86	-18		
$C_5H_8O_2$	Allyl acetate	103.5	22		374
$C_5H_8O_2$	Isopropenyl acetate	94	26		432
$C_5H_8O_2$	Vinyl propanoate	91.2	1		
$C_5H_8O_2$	Ethyl acrylate	99.4	10	1.4-14%	372
$C_5H_8O_2$	Methyl methacrylate	100.5	10	1.7-8.2%	
$C_5H_8O_2$	2,4-Pentanedione	138	34		340
$C_5H_8O_3$	Methyl acetoacetate	171.7	77		280
C_5H_9NO	*N*-Methyl-2-pyrrolidone	202	96	1-10%	346
C_5H_{10}	1-Pentene	29.9	-18	1.5-8.7%	275
C_5H_{10}	*cis*-2-Pentene	36.9	<-20		
C_5H_{10}	*trans*-2-Pentene	36.3	<-20		
C_5H_{10}	2-Methyl-1-butene	31.2	-20		
C_5H_{10}	3-Methyl-1-butene	20.1	-7	1.5-9.1%	365
C_5H_{10}	2-Methyl-2-butene	38.5	-20		
C_5H_{10}	Cyclopentane	49.3	-25	1.5-?	361
$C_5H_{10}Cl_2$	1,5-Dichloropentane	179	>27		
$C_5H_{10}N_2$	3-(Dimethylamino)propanenitrile	173	65		
$C_5H_{10}O$	Cyclopentanol	140.4	51		
$C_5H_{10}O$	Pentanal	103	12		222
$C_5H_{10}O$	2-Pentanone	102.2	7	1.5-8.2%	452
$C_5H_{10}O$	3-Pentanone	101.9	13	1.6-?	450
$C_5H_{10}O$	Tetrahydropyran	88	-20		
$C_5H_{10}O$	2-Methyltetrahydrofuran	78	-11		
$C_5H_{10}O_2$	Pentanoic acid	186.1	96		400
$C_5H_{10}O_2$	3-Methylbutanoic acid	176.5			416
$C_5H_{10}O_2$	Butyl formate	106.1	18	1.7-8.2%	322
$C_5H_{10}O_2$	Isobutyl formate	98.2	5	2-9%	320
$C_5H_{10}O_2$	Propyl acetate	101.5	13	1.7-8%	450
$C_5H_{10}O_2$	Isopropyl acetate	88.6	2	1.8-8%	460
$C_5H_{10}O_2$	Ethyl propanoate	99.1	12	1.9-11%	440
$C_5H_{10}O_2$	Methyl butanoate	102.8	14		
$C_5H_{10}O_2$	3-Ethoxypropanal	135.2	38		
$C_5H_{10}O_2$	Tetrahydrofurfuryl alcohol	178	75	1.5-9.7%	282

Mol. Form.	Name	t_B/°C	FP/°C	Fl. Limits	IT/°C
$C_5H_{10}O_3$	Diethyl carbonate	126	25		
$C_5H_{10}O_3$	Ethylene glycol monomethyl ether acetate	143	49	1.5-12.3%	392
$C_5H_{10}O_3$	Ethyl lactate	154.5	46	1.5-?	400
$C_5H_{11}Br$	1-Bromopentane	129.8	32		
$C_5H_{11}Cl$	1-Chloropentane	107.8	13	1.6-8.6%	260
$C_5H_{11}Cl$	2-Chloro-2-methylbutane	85.6		1.5-7.4%	345
$C_5H_{11}Cl$	1-Chloro-3-methylbutane	98.9	<21	1.5-7.4%	
$C_5H_{11}Cl_3Si$	Trichloropentylsilane	172	63		
$C_5H_{11}N$	Piperidine	106.2	16		
$C_5H_{11}N$	*N*-Methylpyrrolidine	81	-14		
$C_5H_{11}NO$	4-Methylmorpholine	116	24		
$C_5H_{11}NO_2$	Isopentyl nitrite	99.2			210
C_5H_{12}	Pentane	36.0	-40	1.4-8.0%	260
C_5H_{12}	Isopentane	27.8	-51	1.4-7.6%	420
C_5H_{12}	Neopentane	9.4	-65	1.4-7.5%	450
$C_5H_{12}N_2$	1-Methylpiperazine	138	42		
$C_5H_{12}N_2O$	Tetramethylurea	176.5	77		
$C_5H_{12}O$	1-Pentanol	137.9	33	1.2-10.0%	300
$C_5H_{12}O$	2-Pentanol	119.3	34	1.2-9.0%	343
$C_5H_{12}O$	3-Pentanol	116.2	41	1.2-9.0%	435
$C_5H_{12}O$	2-Methyl-1-butanol	128	50		385
$C_5H_{12}O$	3-Methyl-1-butanol	131.1	43	1.2-9.0%	350
$C_5H_{12}O$	2-Methyl-2-butanol	102.4	19	1.2-9.0%	437
$C_5H_{12}O$	3-Methyl-2-butanol	112.9	38		
$C_5H_{12}O$	2,2-Dimethyl-1-propanol	113.5	37		
$C_5H_{12}O$	Ethyl propyl ether	63.2	<-20	1.7-9.0%	
$C_5H_{12}O_2$	1,5-Pentanediol	239	129		335
$C_5H_{12}O_2$	2-Isopropoxyethanol	145	33		
$C_5H_{12}O_2$	2,2-Dimethyl-1,3-propanediol	208	129		399
$C_5H_{12}O_3$	Diethylene glycol monomethyl ether	193	96	1.38-22.7%	240
$C_5H_{12}S$	1-Pentanethiol	126.6	18		
$C_5H_{12}S$	3-Methyl-2-butanethiol		3		
$C_5H_{13}N$	Pentylamine	104.3	-1	2.2-22%	
$C_5H_{13}N$	Butylmethylamine	91	13		
$C_6H_2Cl_4$	1,2,4,5-Tetrachlorobenzene	244.5	155		
$C_6H_3ClN_2O_4$	1-Chloro-2,4-dinitrobenzene	315	194	2.0-22%	
$C_6H_3Cl_3$	1,2,4-Trichlorobenzene	213.5	105	2.5-6.6%	571
$C_6H_4ClNO_2$	1-Chloro-4-nitrobenzene	242	127		
$C_6H_4Cl_2$	*o*-Dichlorobenzene	180	66	2.2-9.2%	648
$C_6H_4Cl_2$	*m*-Dichlorobenzene	173	72		
$C_6H_4Cl_2$	*p*-Dichlorobenzene	174	66		
$C_6H_4Cl_2O$	2,4-Dichlorophenol	210	114		
C_6H_5Br	Bromobenzene	156.0	51		565
C_6H_5Cl	Chlorobenzene	131.7	28	1.3-9.6%	593
C_6H_5ClO	*o*-Chlorophenol	174.9	64		
C_6H_5ClO	*p*-Chlorophenol	220	121		
$C_6H_5Cl_2N$	3,4-Dichloroaniline	272	166		
$C_6H_5Cl_3Si$	Trichlorophenylsilane	201	91		
C_6H_5F	Fluorobenzene	84.7	-15		
$C_6H_5NO_2$	Nitrobenzene	210.8	88	1.8-?	482
$C_6H_5N_3O_4$	2,4-Dinitroaniline		224		
C_6H_6	1,5-Hexadien-3-yne	85	<-20	1.5-?	
C_6H_6	Benzene	80.0	-11	1.2-7.8%	498
$C_6H_6N_2O_2$	*p*-Nitroaniline	332	199		
C_6H_6O	Phenol	181.8	79	1.8-8.6%	715
$C_6H_6O_2$	1,2-Benzenediol	245	127		
$C_6H_6O_2$	Resorcinol		127	1.4-?	608
$C_6H_6O_2$	*p*-Hydroquinone	287	165		516
C_6H_7N	Aniline	184.1	70	1.3-11%	615
C_6H_7N	2-Methylpyridine	129.3	39		538
C_6H_7N	4-Methylpyridine	145.3	57		

FLAMMABILITY OF CHEMICAL SUBSTANCES (continued)

Mol. Form.	Name	t_B/°C	FP/°C	Fl. Limits	IT/°C
C_6H_8ClN	Aniline, hydrochloride		193		
$C_6H_8Cl_2O_2$	Hexanedioyl dichloride		72		
$C_6H_8N_2$	Adiponitrile	295	93	1.0-?	550
$C_6H_8N_2$	*o*-Phenylenediamine	257	156	1.5-?	
$C_6H_8N_2$	Phenylhydrazine	243.5	88		
$C_6H_8N_2$	2,5-Dimethylpyrazine	155	64		
C_6H_8O	2,5-Dimethylfuran	93.5	7		
$C_6H_8O_4$	Dimethyl maleate	202	113		
C_6H_{10}	1,4-Hexadiene	65	-21	2.0-6.1%	
C_6H_{10}	2-Methyl-1,3-pentadiene	75.8	-12		
C_6H_{10}	4-Methyl-1,3-pentadiene	76.5	-34		
C_6H_{10}	2-Hexyne	84.5	-10		
C_6H_{10}	Cyclohexene	82.9	-12	1.2-?	310
$C_6H_{10}O$	Diallyl ether	94	-7		
$C_6H_{10}O$	Cyclohexanone	155.4	44	1.1-9.4%	420
$C_6H_{10}O$	Mesityl oxide	130	31	1.4-7.2%	344
$C_6H_{10}O_2$	Vinyl butanoate	116.7	20	1.4-8.8%	
$C_6H_{10}O_2$	Ethyl 2-butenoate	136.5	2		
$C_6H_{10}O_2$	Ethyl methacrylate	117	20		
$C_6H_{10}O_2$	2,5-Hexanedione	194	79		499
$C_6H_{10}O_3$	Ethyl acetoacetate	180.8	57	1.4-9.5%	295
$C_6H_{10}O_3$	Propanoic anhydride	170	63	1.3-9.5%	285
$C_6H_{10}O_4$	Adipic acid	337.5	196		420
$C_6H_{10}O_4$	Diethyl oxalate	185.7	76		
$C_6H_{10}O_4$	Ethylene glycol diacetate	190	88	1.6-8.4%	482
$C_6H_{11}Cl$	Chlorocyclohexane	142	32		
$C_6H_{11}NO$	Caprolactam	270	125		
$C_6H_{11}NO_2$	Nitrocyclohexane	205	88		
$C_6H_{11}NO_2$	4-Acetylmorpholine		113		
C_6H_{12}	1-Hexene	63.4	-26	1.2-6.9%	253
C_6H_{12}	*cis*-2-Hexene	68.8	-21		
C_6H_{12}	2-Methyl-1-pentene	62.1	-28		300
C_6H_{12}	4-Methyl-1-pentene	53.9	-7		300
C_6H_{12}	4-Methyl-*cis*-2-pentene	56.3	-32		
C_6H_{12}	4-Methyl-*trans*-2-pentene	58.6	-29		
C_6H_{12}	2-Ethyl-1-butene	64.7	<-20		315
C_6H_{12}	2,3-Dimethyl-1-butene	55.6	<-20		360
C_6H_{12}	2,3-Dimethyl-2-butene	73.3	<-20		401
C_6H_{12}	Cyclohexane	80.7	-20	1.3-8%	245
C_6H_{12}	Methylcyclopentane	71.8	-29	1.0-8.35%	258
C_6H_{12}	Ethylcyclobutane	70.8	-15	1.2-7.7%	210
C_6H_{12}	2-Methyl-2-pentene	67.3	<-7		
$C_6H_{12}Cl_2O_2$	1,2-Bis(2-chloroethoxy)ethane	232	121		
$C_6H_{12}O$	*cis*-3-Hexen-1-ol	156.5	54		
$C_6H_{12}O$	Butyl vinyl ether	94	-9		255
$C_6H_{12}O$	Isobutyl vinyl ether	83	-9		
$C_6H_{12}O$	Hexanal	131	32		
$C_6H_{12}O$	2-Ethylbutanal		21	1.2-7.7%	
$C_6H_{12}O$	2-Methylpentanal	117	17		199
$C_6H_{12}O$	2-Hexanone	127.6	25	1-8%	423
$C_6H_{12}O$	3-Hexanone	123.5	35	1-8%	
$C_6H_{12}O$	4-Methyl-2-pentanone	116.5	18	1.2-8.0%	448
$C_6H_{12}O$	Cyclohexanol	160.8	68	1 9%	300
$C_6H_{12}O_2$	Hexanoic acid	205.2	102		380
$C_6H_{12}O_2$	2-Methylpentanoic acid	195.6	107		378
$C_6H_{12}O_2$	Diethylacetic acid	194	99		400
$C_6H_{12}O_2$	Pentyl formate	130.4	26		
$C_6H_{12}O_2$	Butyl acetate	126.1	22	1.7-7.6%	425
$C_6H_{12}O_2$	*sec*-Butyl acetate	112	31	1.7-9.8%	
$C_6H_{12}O_2$	Isobutyl acetate	116.5	18	1.3-10.5%	421
$C_6H_{12}O_2$	Propyl propanoate	122.5	79		

Mol. Form.	Name	t_B/°C	FP/°C	Fl. Limits	IT/°C
$C_6H_{12}O_2$	Ethyl butanoate	121.5	24		463
$C_6H_{12}O_2$	Ethyl 2-methylpropanoate	110.1	13		
$C_6H_{12}O_2$	Diacetone alcohol	167.9	58	1.8-6.9%	643
$C_6H_{12}O_3$	Ethylene glycol monoethyl ether acetate	156.4	56	2-8%	379
$C_6H_{12}O_3$	Paraldehyde	124.3	36	1.3-?	238
$C_6H_{12}S$	Cyclohexanethiol	158.9	43		
$C_6H_{13}Cl$	1-Chlorohexane	135	35		
$C_6H_{13}N$	Cyclohexylamine	134	31	1.9-9.4%	293
$C_6H_{13}NO$	*N*-Butylacetamide	229	116		
$C_6H_{13}NO$	2,6-Dimethylmorpholine	146.6	44		
$C_6H_{13}NO$	*N*-Ethylmorpholine	138.5	32		
$C_6H_{13}NO_2$	4-Morpholineethanol	227	99		
C_6H_{14}	Hexane	68.7	-22	1.1-7.5%	225
C_6H_{14}	2-Methylpentane	60.2	<-29	1.0-7.0%	264
C_6H_{14}	3-Methylpentane	63.2	-7	1.2-7.0%	278
C_6H_{14}	2,2-Dimethylbutane	49.7	-48	1.2-7.0%	405
C_6H_{14}	2,3-Dimethylbutane	57.9	-29	1.2-7.0%	405
$C_6H_{14}N_2O$	1-Piperazineethanol	246	124		
$C_6H_{14}O$	1-Hexanol	157.6	63		
$C_6H_{14}O$	2-Methyl-1-pentanol	149	54	1.1-9.65%	310
$C_6H_{14}O$	4-Methyl-2-pentanol	131.6	41	1.0-5.5%	
$C_6H_{14}O$	2-Ethyl-1-butanol	147	57		
$C_6H_{14}O$	Dipropyl ether	90.0	21	1.3-7.0%	188
$C_6H_{14}O$	Diisopropyl ether	68.5	-28	1.4-7.9%	443
$C_6H_{14}O$	Butyl ethyl ether	92.3	4		
$C_6H_{14}O_2$	2,5-Hexanediol	218	110		
$C_6H_{14}O_2$	2-Methyl-2,4-pentanediol	197.1	102	1-9%	306
$C_6H_{14}O_2$	Ethylene glycol monobutyl ether	168.4	69	4-13%	238
$C_6H_{14}O_2$	1,1-Diethoxyethane	102.2	-21	1.6-10.4%	230
$C_6H_{14}O_2$	Ethylene glycol diethyl ether	119.4	27		205
$C_6H_{14}O_3$	1,2,6-Hexanetriol		191		
$C_6H_{14}O_3$	Diethylene glycol monoethyl ether	196	96		
$C_6H_{14}O_3$	Diethylene glycol dimethyl ether	162	67		
$C_6H_{14}O_3$	Trimethylolpropane		149		
$C_6H_{14}O_4$	Triethylene glycol	285	177	0.9-9.2%	371
$C_6H_{15}N$	Hexylamine	132.8	29		
$C_6H_{15}N$	Butylethylamine	107.5	18		
$C_6H_{15}N$	Dipropylamine	109.3	17		299
$C_6H_{15}N$	Diisopropylamine	83.9	-1	1.1-7.1%	316
$C_6H_{15}N$	Triethylamine	89	-7	1.2-8.0%	249
$C_6H_{15}NO_2$	Diisopropanolamine	250	127		374
$C_6H_{15}NO_3$	Triethanolamine	335.4	179	1-10%	
$C_6H_{15}N_3$	1-Piperazineethanamine	220	93		
$C_6H_{15}O_4P$	Triethyl phosphate	215.5	115		454
$C_6H_{16}N_2$	*N,N*-Diethylethylenediamine	144	46		
$C_7H_3ClF_3NO_2$	1-Chloro-4-nitro-2-(trifluoromethyl)benzene	232	135		
$C_7H_4ClF_3$	1-Chloro-2-(trifluoromethyl)benzene	152.2	59		
$C_7H_4F_3NO_2$	1-Nitro-3-(trifluoromethyl)benzene	202.8	103		
C_7H_5ClO	Benzoyl chloride	197.2	72		
C_7H_5ClO	4-Chlorobenzaldehyde	213.5	88		
$C_7H_5Cl_3$	(Trichloromethyl)benzene	221	127		211
$C_7H_5F_3$	(Trifluoromethyl)benzene	102.1	12		
$C_7H_6N_2O_4$	1-Methyl-2,4-dinitrobenzene		207		
C_7H_6O	Benzaldehyde	179.0	63		192
$C_7H_6O_2$	Benzoic acid	249.2	121		570
$C_7H_6O_2$	Salicylaldehyde	197	78		
$C_7H_6O_3$	Salicylic acid		157	1.1-?	540
C_7H_7Br	*o*-Bromotoluene	181.7	79		
C_7H_7Br	*p*-Bromotoluene	184.3	85		
C_7H_7Cl	(Chloromethyl)benzene	179	67	1.1-?	585
$C_7H_7NO_2$	*o*-Nitrotoluene	222	106		

FLAMMABILITY OF CHEMICAL SUBSTANCES (continued)

Mol. Form.	Name	t_B/°C	FP/°C	Fl. Limits	IT/°C
$C_7H_7NO_2$	*m*-Nitrotoluene	232	106		
$C_7H_7NO_2$	*p*-Nitrotoluene	238.3	106		
C_7H_8	Toluene	110.6	4	1.1-7.1%	480
C_7H_8	Bicyclo[2.2.1]hepta-2,5-diene	89.5	-21		
C_7H_8O	*o*-Cresol	191.0	81	1.4-?	599
C_7H_8O	*m*-Cresol	202.2	86	1.1-?	558
C_7H_8O	*p*-Cresol	201.9	86	1.1-?	558
C_7H_8O	Benzyl alcohol	205.3	93		436
C_7H_8O	Anisole	153.7	52		475
$C_7H_8O_2$	4-Methoxyphenol	243	132		421
$C_7H_8O_3S$	*p*-Toluenesulfonic acid		184		
C_7H_9N	*o*-Methylaniline	200.3	85		482
C_7H_9N	*p*-Methylaniline	200.4	87		482
C_7H_9NO	*o*-Anisidine	224	118		
$C_7H_{10}O$	3-Cyclohexene-1-carboxaldehyde	105	57		
$C_7H_{10}O_4$	3,3-Diacetoxy-1-propene	180	82		
C_7H_{12}	4-Methylcyclohexene	102.7	-1		
$C_7H_{12}O_2$	Butyl acrylate	145	29	1.7-9.9%	292
$C_7H_{12}O_2$	Isobutyl acrylate	132	30		427
$C_7H_{12}O_2$	Cyclohexyl formate	162	51		
$C_7H_{12}O_4$	Diethyl malonate	200	93		
C_7H_{14}	1-Heptene	93.6	-1		260
C_7H_{14}	*trans*-2-Heptene	98	<0		
C_7H_{14}	Cycloheptane	118.4	<21	1.1-6.7%	
C_7H_{14}	Methylcyclohexane	100.9	-4	1.2-6.7%	250
C_7H_{14}	Ethylcyclopentane	103.5	<21	1.1-6.7%	260
$C_7H_{14}O$	2-Heptanone	151.0	39	1.1-7.9%	393
$C_7H_{14}O$	3-Heptanone	147	46		
$C_7H_{14}O$	4-Heptanone	144	49		
$C_7H_{14}O$	5-Methyl-2-hexanone	144	36	1.0-8.2%	191
$C_7H_{14}O$	*cis*-2-Methylcyclohexanol	165	65		296
$C_7H_{14}O$	*trans*-2-Methylcyclohexanol	167.5	65		296
$C_7H_{14}O$	*cis*-3-Methylcyclohexanol	174.5	70		295
$C_7H_{14}O$	*trans*-3-Methylcyclohexanol	174.5	70		295
$C_7H_{14}O$	*cis*-4-Methylcyclohexanol	173	70		295
$C_7H_{14}O$	*trans*-4-Methylcyclohexanol	174	70		295
$C_7H_{14}O_2$	Pentyl acetate	149.2	16	1.1-7.5%	360
$C_7H_{14}O_2$	Isopentyl acetate	142.5	25	1.0-7.5%	360
$C_7H_{14}O_2$	*sec*-Pentyl acetate	130.5	32		
$C_7H_{14}O_2$	Butyl propanoate	146.8	32		426
$C_7H_{14}O_2$	Propyl butanoate	143.0	37		
$C_7H_{15}NO_2$	Ethyl *N*-butylcarbamate	202	92		
C_7H_{16}	Heptane	98.5	-4	1.05-6.7%	204
C_7H_{16}	2-Methylhexane	90.0	-1	1.0-6.0%	280
C_7H_{16}	3-Methylhexane	92	-4		280
C_7H_{16}	2,3-Dimethylpentane	89.7	-56	1.1-6.7%	335
C_7H_{16}	2,4-Dimethylpentane	80.4	-12		
C_7H_{16}	2,2,3-Trimethylbutane	80.8	<0		412
$C_7H_{16}N_2O$	4-Morpholinepropanamine	220	104		
$C_7H_{16}O$	2-Heptanol	159	71		
$C_7H_{16}O$	3-Heptanol	157	60		
$C_7H_{16}O$	2,4-Dimethyl-3-pentanol	138.7	49		
$C_7H_{16}O$	2,3,3-Trimethyl-2-butanol	131	<0		375
$C_7H_{17}N$	Heptylamine	156	54		
$C_7H_{18}N_2$	*N,N*-Diethyl-1,3-propanediamine	168.5	59		
$C_8H_4O_3$	Phthalic anhydride	295	152	1.7-10.5%	570
$C_8H_6O_4$	Phthalic acid		168		
$C_8H_6O_4$	Terephthalic acid		260		496
C_8H_7ClO	α-Chloroacetophenone	247	118		
C_8H_7N	Benzeneacetonitrile	233.5	113		
C_8H_8	Styrene	145	31	0.9-6.8%	490

FLAMMABILITY OF CHEMICAL SUBSTANCES (continued)

Mol. Form.	Name	t_B/°C	FP/°C	Fl. Limits	IT/°C
C_8H_8O	Phenyloxirane	194.1	74		498
C_8H_8O	Benzeneacetaldehyde	195	71		
C_8H_8O	Acetophenone	202	77		570
$C_8H_8O_2$	Benzeneacetic acid	265.5	>100		
$C_8H_8O_2$	Phenyl acetate	196	80		
$C_8H_8O_2$	Methyl benzoate	199	83		
$C_8H_8O_2$	2-Methoxybenzaldehyde	243.5	118		
$C_8H_8O_3$	Methyl salicylate	222.9	96		454
C_8H_9Cl	1-Chloro-4-ethylbenzene	184.4	64		
C_8H_9NO	Acetanilide	304	169		530
$C_8H_9NO_2$	Methyl 2-aminobenzoate	256	>100		
C_8H_{10}	Ethylbenzene	136.1	21	0.8-6.7%	432
C_8H_{10}	*o*-Xylene	144.5	32	0.9-6.7%	463
C_8H_{10}	*m*-Xylene	139.1	27	1.1-7.0%	527
C_8H_{10}	*p*-Xylene	138.3	27	1.1-7.0%	528
$C_8H_{10}O$	*p*-Ethylphenol	217.9	104		
$C_8H_{10}O$	Benzeneethanol	218.2	96		
$C_8H_{10}O$	α-Methylbenzyl alcohol	205	93		
$C_8H_{10}O$	Phenetole	169.8	63		
$C_8H_{10}O$	Benzyl methyl ether	170	135		
$C_8H_{10}O$	4-Methylanisole	175.5	60		
$C_8H_{10}O_2$	2-Phenoxyethanol	245	121		
$C_8H_{11}N$	*N*-Ethylaniline	203.0	85		
$C_8H_{11}N$	*N,N*-Dimethylaniline	194.1	63		371
$C_8H_{11}N$	2,3-Xylidine	221.5	97	1.0-?	
$C_8H_{11}N$	2,6-Xylidine	215	96		
$C_8H_{11}N$	α-Methylbenzylamine	187	79		
$C_8H_{11}N$	5-Ethyl-2-picoline	178.3	68	1.1-6.6%	
$C_8H_{11}NO$	*N*-Phenylethanolamine	279.5	152		
$C_8H_{11}NO$	*o*-Phenetidine	232.5	115		
$C_8H_{11}NO$	*p*-Phenetidine	254	116		
C_8H_{12}	1,5-Cyclooctadiene	150.8	35		
C_8H_{12}	4-Vinylcyclohexene	128	16		269
$C_8H_{12}O_4$	Diethyl maleate	223	121		350
$C_8H_{12}O_4$	Diethyl fumarate	214	104		
$C_8H_{14}O_2$	Cyclohexyl acetate	173	58		335
$C_8H_{14}O_2$	Butyl methacrylate	160	52		
$C_8H_{14}O_3$	Butanoic anhydride	200	54	0.9-5.8%	279
$C_8H_{14}O_3$	2-Methylpropanoic anhydride	183	59	1.0-6.2%	329
$C_8H_{14}O_3$	Butyl acetoacetate		85		
$C_8H_{14}O_4$	Ethyl succinate	217.7	90		
$C_8H_{14}O_5$	Diethylene glycol diacetate	200	135		
$C_8H_{14}O_6$	Diethyl tartrate	281	93		
$C_8H_{15}ClO$	Octanoyl chloride	195.6	82		
C_8H_{16}	1-Octene	121.2	21		230
C_8H_{16}	2,4,4-Trimethyl-1-pentene	101.4	-5	0.8-4.8%	391
C_8H_{16}	2,4,4-Trimethyl-2-pentene	104.9	2		305
C_8H_{16}	Ethylcyclohexane	131.9	35	0.9-6.6%	238
C_8H_{16}	*cis*-1,2-Dimethylcyclohexane	129.8	16		304
C_8H_{16}	*trans*-1,2-Dimethylcyclohexane	123.5	11		304
C_8H_{16}	*cis*-1,4-Dimethylcyclohexane	124.4	16		
C_8H_{16}	Propylcyclopentane	131			269
$C_8H_{16}O$	Octanal	171	52		
$C_8H_{16}O$	2-Ethylhexanal	163	44	0.85-7.2%	190
$C_8H_{16}O$	2-Octanone	172.5	52		
$C_8H_{16}O_2$	Hexyl acetate	171.5	45		
$C_8H_{16}O_2$	sec-Hexyl acetate	147.5	45		
$C_8H_{16}O_2$	2-Ethylbutyl acetate	162.5	54		
$C_8H_{16}O_2$	Pentyl propanoate	168.6	41		378
$C_8H_{16}O_2$	Butyl butanoate	166	53		

FLAMMABILITY OF CHEMICAL SUBSTANCES (continued)

Mol. Form.	Name	t_B/°C	FP/°C	Fl. Limits	IT/°C
$C_8H_{16}O_2$	Isobutyl butanoate	156.9	50		
$C_8H_{16}O_2$	Isobutyl isobutanoate	148.6	38	0.96-7.59%	432
$C_8H_{16}O_2$	Ethyl hexanoate	167	49		
$C_8H_{16}O_2$	1,4-Cyclohexanedimethanol	283	167		316
$C_8H_{16}O_3$	Pentyl lactate		79		
$C_8H_{16}O_4$	Diethylene glycol monoethyl ether acetate	218.5	110		425
$C_8H_{17}Cl$	1-Chlorooctane	181.5	70		
$C_8H_{17}Cl$	3-(Chloromethyl)heptane	172	60		
C_8H_{18}	Octane	125.6	13	1.0-6.5%	206
C_8H_{18}	2,3-Dimethylhexane	115.6	7		438
C_8H_{18}	2,4-Dimethylhexane	109.5	10		
C_8H_{18}	3-Ethyl-2-methylpentane	115.6	<21		460
C_8H_{18}	2,2,3-Trimethylpentane	110	<21		346
C_8H_{18}	2,2,4-Trimethylpentane	99.2	-12		418
C_8H_{18}	2,3,3-Trimethylpentane	114.8	<21		425
$C_8H_{18}O$	1-Octanol	195.1	81		
$C_8H_{18}O$	2-Octanol	180	88		
$C_8H_{18}O$	2-Ethyl-1-hexanol	184.6	73	0.88-9.7%	231
$C_8H_{18}O$	Dibutyl ether	140.2	25	1.5-7.6%	194
$C_8H_{18}O_2$	2-Ethyl-1,3-hexanediol	244	127		360
$C_8H_{18}O_2$	2,2,4-Trimethyl-1,3-pentanediol	235	113		346
$C_8H_{18}O_2$	Di-*tert*-butyl peroxide	111	18		
$C_8H_{18}O_3$	Diethylene glycol diethyl ether	188	82		
$C_8H_{18}O_4$	2,5,8,11-Tetraoxadodecane	216	111		
$C_8H_{18}O_5$	Tetraethylene glycol	328	182		
$C_8H_{18}S$	1-Octanethiol	199.1	69		
$C_8H_{18}S$	Dibutyl sulfide	185	76		
$C_8H_{19}N$	Octylamine	179.6	60		
$C_8H_{19}N$	Dibutylamine	159.6	47	1.1-6%	
$C_8H_{19}N$	Diisobutylamine	139.6	29		
$C_8H_{19}N$	2-Ethylhexylamine	169.2	60		
$C_8H_{20}O_4Si$	Ethyl silicate	168.8	52		
$C_8H_{23}N_5$	Tetraethylenepentamine	341.5	163		321
$C_9H_6N_2O_2$	Toluene-2,4-diisocyanate	251	127	0.9-9.5%	
C_9H_7N	Quinoline	237.1			480
C_9H_{10}	*o*-Methylstyrene	169.8	53	0.8-11.0%	538
C_9H_{10}	*m*-Methylstyrene	164	53	0.8-11.0%	538
C_9H_{10}	*p*-Methylstyrene	172.8	53	0.8-11.0%	538
C_9H_{10}	Isopropenylbenzene	165.4	54	1.9-6.1%	574
$C_9H_{10}O$	1-Phenyl-1-propanone	217.5	99		
$C_9H_{10}O$	4-Methylacetophenone	226	96		
$C_9H_{10}O_2$	Ethyl benzoate	212	88		490
$C_9H_{10}O_2$	Benzyl acetate	213	90		460
$C_9H_{10}O_2$	Methyl 2-phenylacetate	216.5	91		
$C_9H_{11}NO$	4-Methylacetanilide	307	168		
C_9H_{12}	Propylbenzene	159.2	30	0.8-6.0%	450
C_9H_{12}	Isopropylbenzene	152.4	36	0.9-6.5%	424
C_9H_{12}	*o*-Ethyltoluene	165.2			440
C_9H_{12}	*m*-Ethyltoluene	161.3			480
C_9H_{12}	*p*-Ethyltoluene	162			475
C_9H_{12}	1,2,3-Trimethylbenzene	176.1	44	0.8-6.6%	470
C_9H_{12}	1,2,4-Trimethylbenzene	169.3	44	0.9-6.4%	500
C_9H_{12}	1,3,5-Trimethylbenzene	164.7	50	1-5%	559
$C_9H_{12}O$	α–Ethylbenzyl alcohol	219	100		
$C_9H_{12}O_2$	Ethylene glycol monobenzyl ether	256	129		352
$C_9H_{12}O_3S$	Ethyl *p*-toluenesulfonate		158		
$C_9H_{13}N$	Amphetamine	203	<100		
$C_9H_{14}O$	Phorone	197.5	85		
$C_9H_{14}O$	Isophorone	215.2	84	0.8-3.8%	460
$C_9H_{14}O_6$	Triacetin	259	138	1.0-?	433
C_9H_{16}	Octahydroindene	167			296

FLAMMABILITY OF CHEMICAL SUBSTANCES (continued)

Mol. Form.	Name	t_B/°C	FP/°C	Fl. Limits	IT/°C
$C_9H_{16}O_2$	Allyl hexanoate	186	66		
C_9H_{18}	1-Nonene	146.9	26		
C_9H_{18}	Propylcyclohexane	156.7			248
C_9H_{18}	Isopropylcyclohexane	154.8			283
C_9H_{18}	Butylcyclopentane	156.6			250
$C_9H_{18}O$	2-Nonanone	195.3	60	0.9-5.9%	360
$C_9H_{18}O$	Diisobutyl ketone	169.4	49	0.8-7.1%	396
$C_9H_{18}O_2$	Pentyl butanoate	186.4	57		
$C_9H_{18}O_2$	Isopentyl butanoate	179	59		
$C_9H_{18}O_2$	Butyl 3-methylbutanoate		53		
C_9H_{20}	Nonane	150.8	31	0.8-2.9%	205
C_9H_{20}	3-Ethyl-4-methylhexane	140	24		
C_9H_{20}	4-Ethyl-2-methylhexane	133.8	<21	0.7-?	280
C_9H_{20}	2,2,5-Trimethylhexane	124.0	13		
C_9H_{20}	3,3-Diethylpentane	146.3		0.7-5.7%	290
C_9H_{20}	3-Ethyl-2,4-dimethylpentane	136.7	390		
C_9H_{20}	2,2,3,3-Tetramethylpentane	140.2	<21	0.8-4.9%	430
C_9H_{20}	2,2,3,4-Tetramethylpentane	133.0	<21		
$C_9H_{21}BO_3$	Triisopropyl borate	140	28		
$C_9H_{21}N$	Tripropylamine	156	41		
$C_9H_{21}NO_3$	Triisopropanolamine		160		320
$C_{10}H_7Cl$	1-Chloronaphthalene	259	121		>558
$C_{10}H_8$	Naphthalene	217.9	79	0.9-5.9%	526
$C_{10}H_8O$	2-Naphthol	285	153		
$C_{10}H_9N$	1-Naphthalenamine	300.8	157		
$C_{10}H_{10}O_2$	Safrole	234.5	100		
$C_{10}H_{10}O_4$	Dimethyl phthalate	283.7	146	0.9-?	490
$C_{10}H_{10}O_4$	Dimethyl isophthalate	282	138		
$C_{10}H_{10}O_4$	Dimethyl terephthalate	288	153		518
$C_{10}H_{11}NO_2$	Acetoacetanilide		185		
$C_{10}H_{12}$	1,2,3,4-Tetrahydronaphthalene	207.6	71	0.8-5.0%	385
$C_{10}H_{12}O_2$	Isopropyl benzoate	216	99		
$C_{10}H_{12}O_2$	Ethyl phenylacetate	227	99		
$C_{10}H_{14}$	Butylbenzene	183.3	71	0.8-5.8%	410
$C_{10}H_{14}$	*sec*-Butylbenzene	173.3	52	0.8-6.9%	418
$C_{10}H_{14}$	*tert*-Butylbenzene	169.1	60	0.7-5.7%	450
$C_{10}H_{14}$	Isobutylbenzene	172.7	55	0.8-6.0%	427
$C_{10}H_{14}$	*p*-Cymene	177.1	47	0.7-5.6%	436
$C_{10}H_{14}$	1,2,3,4-Tetramethylbenzene	205	74		427
$C_{10}H_{14}$	1,2,3,5-Tetramethylbenzene	198	71		427
$C_{10}H_{14}$	1,2,4,5-Tetramethylbenzene	196.8	54		
$C_{10}H_{14}$	*o*-Diethylbenzene	184	57		395
$C_{10}H_{14}$	*m*-Diethylbenzene	181.1	56		450
$C_{10}H_{14}$	*p*-Diethylbenzene	183.7	55	0.7-6.0%	430
$C_{10}H_{14}O$	Butyl phenyl ether	210	82		
$C_{10}H_{14}O_2$	4-*tert*-Butyl-1,2-benzenediol	285	130		
$C_{10}H_{15}N$	*N*-Butylaniline	243.5	107		
$C_{10}H_{15}N$	*N,N*-Diethylaniline	216.3	85		630
$C_{10}H_{15}NO_2$	*N*-Phenyl-*N,N*-diethanolamine		196	0.7-?	387
$C_{10}H_{16}$	Dipentene	178	45		237
$C_{10}H_{16}$	*d*-Limonene	178	45	0.7-6.1%	237
$C_{10}H_{16}$	α-Pinene	156.2	33		255
$C_{10}H_{16}$	β-Pinene	166	38		275
$C_{10}H_{16}$	β-Phellandrene	171.5	49		
$C_{10}H_{16}O$	Camphor	207.4	66	0.6-3.5%	466
$C_{10}H_{18}$	*trans*-Decahydronaphthalene	187.3	54	0.7-5.4%	255
$C_{10}H_{18}O$	Borneol		66		
$C_{10}H_{18}O$	Linalol	198	71		
$C_{10}H_{18}O$	α-Terpineol	220	90		
$C_{10}H_{18}O$	Cineole	176.4	48		

Mol. Form.	Name	t_B/°C	FP/°C	Fl. Limits	IT/°C
$C_{10}H_{18}O$	*trans*-Geraniol	230	>100		
$C_{10}H_{18}O_4$	Dibutyl oxalate	241	104		
$C_{10}H_{19}NO_2$	*N-tert*-Butylaminoethyl methacrylate		96		
$C_{10}H_{20}$	1-Decene	170.5	<55		235
$C_{10}H_{20}$	Butylcyclohexane	180.9			246
$C_{10}H_{20}$	Isobutylcyclohexane	171.3			274
$C_{10}H_{20}$	*tert*-Butylcyclohexane	171.5			342
$C_{10}H_{20}O$	Citronellol	224	96		
$C_{10}H_{20}O_2$	2-Ethylhexyl acetate	199	71	0.76-8.14%	268
$C_{10}H_{20}O_2$	Ethyl octanoate	208.5	79		
$C_{10}H_{21}N$	*N*-Butylcyclohexanamine		93		
$C_{10}H_{22}$	Decane	174.1	51	0.8-5.4%	210
$C_{10}H_{22}$	2-Methylnonane	167.1			210
$C_{10}H_{22}$	3-Ethyloctane	166.5			230
$C_{10}H_{22}$	4-Ethyloctane	163.7			229
$C_{10}H_{22}O$	1-Decanol	231.1	82		288
$C_{10}H_{22}O$	Dipentyl ether	190	57		170
$C_{10}H_{22}O_2$	Ethylene glycol dibutyl ether	203.3	85		
$C_{10}H_{22}O_5$	Tetraethylene glycol dimethyl ether	275.3	141		
$C_{10}H_{22}S$	Dipentyl sulfide		85		
$C_{10}H_{23}N$	Decylamine	220.5	99		
$C_{10}H_{23}N$	Dipentylamine	202.5	51		
$C_{11}H_{10}$	1-Methylnaphthalene	244.7			529
$C_{11}H_{12}O_3$	Ethyl benzoylacetate		141		
$C_{11}H_{14}O_2$	Butyl benzoate	250.3	107		
$C_{11}H_{16}$	*p-tert*-Butyltoluene	190	68		
$C_{11}H_{16}$	Pentylbenzene	205.4	66		
$C_{11}H_{16}$	1,3-Diethyl-5-methylbenzene	205			455
$C_{11}H_{16}$	Pentamethylbenzene	232	93		427
$C_{11}H_{16}O$	4-*tert*-Butyl-2-methylphenol	237	118		
$C_{11}H_{17}N$	*p-tert*-Pentylaniline	260.5	102		
$C_{11}H_{20}O_2$	2-Ethylhexyl acrylate		82		252
$C_{11}H_{22}$	Pentylcyclohexane	203.7			239
$C_{11}H_{22}O$	2-Undecanone	231.5	89		
$C_{11}H_{22}O_2$	Nonyl acetate	210	68		
$C_{11}H_{24}$	Undecane	195.9	69		
$C_{11}H_{24}$	2-Methyldecane	189.3			225
$C_{11}H_{24}O$	2-Undecanol	228	113		
$C_{12}H_9Br$	4-Bromo-1,1'-Biphenyl	310	144		
$C_{12}H_{10}$	Biphenyl	256.1	113	0.6-5.8%	540
$C_{12}H_{10}Cl_2Si$	Dichlorodiphenylsilane	305	142		
$C_{12}H_{10}O$	*o*-Phenylphenol	286	124		530
$C_{12}H_{10}O$	Diphenyl ether	258.0	112	0.8-1.5%	618
$C_{12}H_{11}N$	2-Aminobiphenyl	299			450
$C_{12}H_{11}N$	Diphenylamine	302	153		634
$C_{12}H_{12}$	1-Ethylnaphthalene	258.6			480
$C_{12}H_{14}O_4$	Diethyl phthalate	295	161	0.7-?	457
$C_{12}H_{14}O_4$	Diethyl terephthalate	302	117		
$C_{12}H_{16}$	Cyclohexylbenzene	240.1	99		
$C_{12}H_{16}O_3$	Pentyl salicylate	270	132		
$C_{12}H_{17}NO$	*N*-Butyl-*N*-phenylacetamide	281	141		
$C_{12}H_{18}$	1,5,9-Cyclododecatriene	240	71		
$C_{12}H_{20}O_4$	Dibutyl maleate	280	141		
$C_{12}H_{22}O_4$	Dimethyl sebacate		145		
$C_{12}H_{22}O_6$	Dibutyl tartrate	320	91		284
$C_{12}H_{23}N$	Dicyclohexylamine		>99		
$C_{12}H_{24}$	1-Dodecene	213.8	79		
$C_{12}H_{24}O_2$	Ethyl decanoate	241.5	>100		
$C_{12}H_{25}Br$	1-Bromododecane	276	144		
$C_{12}H_{26}$	Dodecane	216.3	74	0.6-?	203
$C_{12}H_{26}O$	1-Dodecanol	259	127		275

Mol. Form.	Name	t_B/°C	FP/°C	Fl. Limits	IT/°C
$C_{12}H_{26}O$	2-Butyl-1-octanol	246.5	110		
$C_{12}H_{26}O_3$	Diethylene glycol dibutyl ether	256	118		310
$C_{12}H_{26}S$	1-Dodecanethiol	277	128		
$C_{12}H_{27}BO_3$	Tributyl borate	234	93		
$C_{12}H_{27}N$	Tributylamine	216.5	63		
$C_{12}H_{27}O_4P$	Tributyl phosphate	289	146		
$C_{13}H_{12}$	2-Methylbiphenyl	255.5	137		502
$C_{13}H_{12}$	Diphenylmethane	265.0	130		485
$C_{13}H_{14}N_2$	*p,p′*-Diaminodiphenylmethane	398	220		
$C_{13}H_{26}$	1-Tridecene	232.8	79		
$C_{13}H_{26}O$	2-Tridecanone	263	107		
$C_{13}H_{28}$	Tridecane	235.4	79		
$C_{13}H_{28}O$	1-Tridecanol		121		
$C_{14}H_8O_2$	9,10-Anthracenedione	377	185		
$C_{14}H_{10}$	Anthracene	339.9	121	0.6-?	540
$C_{14}H_{10}$	Phenanthrene	340	171		
$C_{14}H_{12}O_2$	Benzyl benzoate	323.5	148		480
$C_{14}H_{12}O_3$	Benzyl salicylate	320	>100		
$C_{14}H_{14}$	1,1-Diphenylethane	272.6	>100		440
$C_{14}H_{14}O$	Dibenzyl ether	298	135		
$C_{14}H_{16}$	1-Butylnaphthalene	289.3	360		
$C_{14}H_{16}N_2O_2$	*o*-Dianisidine		206		
$C_{14}H_{23}N$	*N,N*-Dibutylaniline	274.8	110		
$C_{14}H_{28}$	1-Tetradecene	233	110		235
$C_{14}H_{30}$	Tetradecane	253.5	112	0.5-?	200
$C_{14}H_{30}O$	1-Tetradecanol	289	141		
$C_{15}H_{18}$	1-Pentylnaphthalene	307	124		
$C_{15}H_{24}$	Nonylbenzene	280.5	99		
$C_{15}H_{24}O$	2,6-Di-*tert*-butyl-4-methylphenol	265	127		
$C_{15}H_{26}O_6$	Tributyrin	307.5	180	0.5-?	407
$C_{15}H_{33}N$	Tripentylamine	242.5	102		
$C_{16}H_{14}O$	1,3-Diphenyl-2-buten-1-one	342.5	177		
$C_{16}H_{18}$	2-Butyl-1,1'-biphenyl		>100		430
$C_{16}H_{22}O_4$	Dibutyl phthalate	340	157	0.5-?	402
$C_{16}H_{26}$	Decylbenzene	298	107		
$C_{16}H_{34}$	Hexadecane	286.8	136		202
$C_{16}H_{34}O$	Dioctyl ether	283	>100		205
$C_{16}H_{35}N$	Bis(2-ethylhexyl)amine		132		
$C_{17}H_{20}N_2O$	*N,N′*-Diethylcarbanilide		150		
$C_{17}H_{34}O$	2-Heptadecanone	320	120		
$C_{17}H_{36}O$	1-Heptadecanol	333	154		
$C_{18}H_{14}$	*o*-Terphenyl	332	163		
$C_{18}H_{14}$	*m*-Terphenyl	363	191		
$C_{18}H_{15}O_3P$	Triphenyl phosphite	360	218		
$C_{18}H_{15}O_4P$	Triphenyl phosphate		220		
$C_{18}H_{15}P$	Triphenylphosphine		180		
$C_{18}H_{30}$	Dodecylbenzene	328	140		
$C_{18}H_{32}O_7$	Butyl citrate		157		368
$C_{18}H_{34}O_2$	Oleic acid	360	189		363
$C_{18}H_{34}O_4$	Dibutyl sebacate	344.5	178	0.4-?	365
$C_{18}H_{36}O_2$	Stearic acid		196		395
$C_{18}H_{37}Cl_3Si$	Trichlorooctadecylsilane		89		
$C_{18}H_{38}$	Octadecane	316.3	>100		227
$C_{18}H_{38}O$	1-Octadecanol				450
$C_{19}H_{16}$	Triphenylmethane	359	>100		
$C_{19}H_{38}O$	2-Nonadecanone		124		
$C_{19}H_{38}O_2$	Methyl stearate	443	153		
$C_{19}H_{40}$	Nonadecane	329.9	>100		230
$C_{20}H_{14}O_4$	Diphenyl phthalate		224		
$C_{20}H_{28}$	1-Decylnaphthalene	379	177		
$C_{20}H_{42}$	Eicosane	343	>100		232

Mol. Form.	Name	t_B/°C	FP/°C	Fl. Limits	IT/°C
$C_{21}H_{21}O_4P$	Tri-*o*-cresyl phosphate	410	225		385
$C_{21}H_{26}O_3$	4-Octylphenyl salicylate		216		416
$C_{21}H_{32}O_2$	Methyl abietate		180		
$C_{22}H_{42}O_2$	Butyl oleate		180		
$C_{22}H_{42}O_4$	Bis(2-ethylhexyl) adipate		206	0.4-?	377
$C_{22}H_{44}O_2$	Butyl stearate	343	160		355
$C_{23}H_{46}O_2$	Pentyl stearate		185		
$C_{24}H_{20}Sn$	Tetraphenylstannane	420	232		
$C_{24}H_{38}O_4$	Bis(2-ethylhexyl) phthalate	384	218		
$C_{25}H_{48}O_4$	Bis(2-ethylhexyl) azelate	227		0.3-?	374

THRESHOLD LIMITS FOR AIRBORNE CONTAMINANTS

Several organizations recommend limits of exposure to airborne contaminants in the workplace. These include the Occupational Safety and Health Administration (OSHA), the National Institute for Occupational Safety and Health (NIOSH), and the non-governmental organization, American Conference of Governmental Industrial Hygienists (ACGIH). The threshold limit value (TLV) for a substance is defined as the concentration level under which the majority of workers may be repeatedly exposed, day after day, without adverse effects. The TLV recommendations are given in two forms:

- Time-weighted average (TWA) concentration for a normal 8-h workday and 40-h workweek.
- Short-term exposure limit (STEL), which should not be exceeded for more than 15 min.

Both kinds of limits are specified for some substances.

The following table gives threshold limit values for a number of substances that may be encountered in the atmosphere of a chemical laboratory or industrial facility. All values refer to the concentration in air at 25°C and normal atmospheric pressure. Data for gases are given both in parts per million by volume (ppm) and in mass concentration (mg/m^3). Values for liquids refer to mists or aerosols, and those for solids to dusts or fumes; both are stated in mg/m^3. A "C" following a value indicates a ceiling limit which should not be exceeded even for very brief periods because of acute toxic effects of the substance.

Substances are listed by systematic name, which is followed by molecular formula in the Hill format and Chemical Abstracts Service Registry Number. Common synonyms are given in brackets [ ] for some compounds.

REFERENCES

1. *2000 TLV's and BEI's,* American Conference of Governmental Industrial Hygienists, 1330 Kemper Meadow Drive, Cincinnati, OH 45240-1634, 2000.
2. *NIOSH Pocket Guide to Chemical Hazards*, U.S. Department of Health and Human Services, National Institute for Occupational Health and Safety, U.S. Government Printing Office, Washington, DC, 1994.
3. *Chemical Information Manual*, U.S. Department of Labor, Occupational Safety and Health Administration, Washington, DC, 1991.

Substance	Molecular Formula	CAS Reg. No.	Time-Weighted Average		Short-Term Exposure Limit	
			ppm	mg/m^3	ppm	mg/m^3
Abate [Temephos]	$C_{16}H_{20}O_6P_2S_3$	3383-96-8		10		
Acetaldehyde	C_2H_4O	75-07-0			25 C	45 C
Acetic acid	$C_2H_4O_2$	64-19-7	10	25	15	37
Acetic anhydride	$C_4H_6O_3$	108-24-7	5	21		
Acetone	C_3H_6O	67-64-1	500	1188	750	1780
Acetone cyanohydrin	C_4H_7NO	75-86-5			4.7 C	5 C
Acetonitrile	C_2H_3N	75-05-8	40	67	60	101
Acetophenone	C_8H_8O	98-86-2	10	49		
2-(Acetyloxy)benzoic acid [Aspirin]	$C_9H_8O_4$	50-78-2		5		
Acrolein [2-Propenal]	C_3H_4O	107-02-8			0.1 C	0.23 C
Acrylamide	C_3H_5NO	79-06-1		0.03		
Acrylic acid [2-Propenoic acid]	$C_3H_4O_2$	79-10-7	2	5.9		
Acrylonitrile [Propenenitrile]	C_3H_3N	107-13-1	2	4.3		
Adipic acid	$C_6H_{10}O_4$	124-04-9		5		
Adiponitrile	$C_6H_8N_2$	111-69-3	2	9		
Aldrin	$C_{12}H_8Cl_6$	309-00-2		0.25		
Allyl alcohol [2-Propen-1-ol]	C_3H_6O	107-18-6	0.5	1.2		
Allyl glycidyl ether	$C_6H_{10}O_2$	106-92-3	1	5		
Allyl propyl disulfide	$C_6H_{12}S_2$	2179-59-1	2	12	3	18
Aluminum (metal dust)	Al	7429-90-5		10		
Aluminum oxide	Al_2O_3	1344-28-1		10		
4-Amino-3,5,6-trichloropyridinecarboxlic acid [Picloram]	$C_6H_3Cl_3N_2O_2$	1918-02-1		10		
Ammonia	H_3N	7664-41-7	25	17	35	24
Ammonium chloride	ClH_4N	12125-02-9		10		20
Ammonium perfluorooctanoate	$C_8H_4F_{15}NO_2$	3825-26-1		0.01		
Ammonium sulfamate	$H_6N_2O_3S$	7773-06-0		10		
Aniline	C_6H_7N	62-53-3	2	7.6		
Antimony	Sb	7440-36-0		0.5		
Arsenic	As	7440-38-2		0.01		
Arsine	AsH_3	7784-42-1	0.05	0.16		

THRESHOLD LIMITS FOR AIRBORNE CONTAMINANTS (continued)

Substance	Molecular Formula	CAS Reg. No.	Time-Weighted Average ppm	Time-Weighted Average mg/m³	Short-Term Exposure Limit ppm	Short-Term Exposure Limit mg/m³
Atrazine	$C_8H_{14}ClN_5$	1912-24-9		5		
Azinphos-methyl	$C_{10}H_{12}N_3O_3PS_2$	86-50-0		0.2		
Barium	Ba	7440-39-3		0.5		
Barium sulfate	BaO_4S	7727-43-7		10		
Benomyl	$C_{14}H_{18}N_4O_3$	17804-35-2	0.84	10		
Benzene	C_6H_6	71-43-2	0.5	1.6	2.5	8
1,3-Benzenedimethanamine [*m*-Xylene diamine]	$C_8H_{12}N_2$	1477-55-0				0.1 C
Benzenethiol [Phenyl mercaptan]	C_6H_6S	108-98-5	0.5	2.3		
p-Benzoquinone [Quinone]	$C_6H_4O_2$	106-51-4	0.1	0.44		
Benzoyl chloride	C_7H_5ClO	98-88-4			0.5 C	2.8 C
Benzoyl peroxide	$C_{14}H_{10}O_4$	94-36-0		5		
Benzyl acetate	$C_9H_{10}O_2$	140-11-4	10	61		
Beryllium	Be	7440-41-7		0.002		0.01
Biphenyl	$C_{12}H_{10}$	92-52-4	0.2	1.3		
Bis(4-amino-3-chlorophenyl)methane [4,4-Methylene bis(2-chloroaniline)]	$C_{13}H_{12}Cl_2N_2$	101-14-4	0.01	0.11		
Bis(2-chloroethyl) ether [2,2'-Dichlorethyl ether]	$C_4H_8Cl_2O$	111-44-4	5	29	10	58
Bis(chloromethyl) ether	$C_2H_4Cl_2O$	542-88-1	0.001	0.0047		
Bis(2-dimethylaminoethyl) ether [DMAEE]	$C_8H_{20}N_20$	3033-62-3	0.05	0.33	0.15	1.0
Bis(2-ethylhexyl) phthalate [Di-*sec*-octyl phthalate]	$C_{24}H_{38}O_4$	117-81-7		5		10
Bismuth telluride	Bi_2Te_3	1304-82-1		10		
Boron oxide	B_2O_3	1303-86-2		10		
Boron tribromide	BBr_3	10294-33-4			1 C	10 C
Boron trifluoride	BF_3	7637-07-2			1 C	2.8 C
Bromacil	$C_9H_{13}BrN_2O_2$	314-40-9		10		
Bromine	Br_2	7726-95-6	0.1	0.66	0.2	1.3
Bromine pentafluoride	BrF_5	7789-30-2	0.1	0.72		
Bromochloromethane [Halon 1011]	CH_2BrCl	74-97-5	200	1060		
2-Bromo-2-chloro-1,1,1-trifluoroethane [Halothane]	$C_2HBrClF_3$	151-67-7	50	404		
Bromoethane [Ethyl bromide]	C_2H_5Br	74-96-4	5	22		
Bromoethene [Vinyl bromide]	C_2H_3Br	593-60-2	0.5	2.2		
Bromomethane [Methyl bromide]	CH_3Br	74-83-9	1	3.9		
Bromotrifluoromethane	$CBrF_3$	75-63-8	1000	6090		
1,3-Butadiene	C_4H_6	106-99-0	2	4.4		
Butane	C_4H_{10}	106-97-8	800	1900		
1-Butanethiol [Butyl mercaptan]	$C_4H_{10}S$	109-79-5	0.5	1.8		
1-Butanol	$C_4H_{10}O$	71-36-3			50 C	152 C
2-Butanol [*sec*-Butyl alcohol]	$C_4H_{10}O$	78-92-2	100	303		
2-Butanone [Methyl ethyl ketone]	C_4H_8O	78-93-3	200	590	300	885
trans-2-Butenal [Crotonaldehyde]	C_4H_6O	4170-30-3			0.3 C	0.9 C
3-Buten-2-one	C_4H_6O	78-94-4			0.2 C	0.6 C
Butyl acetate	$C_6H_{12}O_2$	123-86-4	150	713	200	950
sec-Butyl acetate	$C_6H_{12}O_2$	105-46-4	200	950		
tert-Butyl acetate	$C_6H_{12}O_2$	540-88-5	200	950		
Butyl acrylate	$C_7H_{12}O_2$	141-32-2	2	10		
Butylamine	$C_4H_{11}N$	109-73-9			5 C	15 C
tert-Butyl chromate	$C_8H_{18}CrO_4$	1189-85-1				0.1 C
Butyl glycidyl ether	$C_7H_{14}O_2$	2426-08-6	25	133		
Butyl lactate	$C_7H_{14}O_3$	138-22-7	5	30		
o-*sec*-Butylphenol	$C_{10}H_{14}O$	89-72-5	5	31		
p-*tert*-Butyltoluene	$C_{11}H_{16}$	98-51-1	1	6.1		
Cadmium	Cd	7440-43-9		0.01		
Calcium carbonate	$CCaO_3$	1317-65-3		10		

Substance	Molecular Formula	CAS Reg. No.	Time-Weighted Average ppm	Time-Weighted Average mg/m³	Short-Term Exposure Limit ppm	Short-Term Exposure Limit mg/m³
Calcium chromate	$CaCrO_4$	13765-19-0		0.003		
Calcium cyanamide	$CCaN_2$	156-62-7		0.5		
Calcium hydroxide	CaH_2O_2	1305-62-0		5		
Calcium metasilicate	CaO_3Si	1344-95-2		10		
Calcium oxide	CaO	1305-78-8		2		
Calcium sulfate	CaO_4S	7778-18-9		10		
Camphor	$C_{10}H_{16}O$	76-22-2	2	12	4	24
Caprolactam	$C_6H_{11}NO$	105-60-2	5 (gas)	1 (solid)	10 (gas)	3 (solid)
Captafol	$C_{10}H_9Cl_4NO_2S$	2425-06-1		0.1		
Captan	$C_9H_8Cl_3NO_2S$	133-06-2		5		
Carbaryl	$C_{12}H_{11}NO_2$	63-25-2		5		
Carbofuran	$C_{12}H_{15}NO_3$	1563-66-2		0.1		
Carbon black	C	1333-86-4		3.5		
Carbon dioxide	CO_2	124-38-9	5000	9000	30,000	54,000
Carbon disulfide	CS_2	75-15-0	10	31		
Carbon monoxide	CO	630-08-0	25	29		
Carbonyl chloride [Phosgene]	CCl_2O	75-44-5	0.1	0.40		
Carbonyl fluoride	CF_2O	353-50-4	2	5.4	5	13
Cesium hydroxide	CsHO	21351-79-1		2		
Chlordane	$C_{10}H_6Cl_8$	57-74-9		0.5		
Chlorine	Cl_2	7782-50-5	0.5	1.5	1	2.9
Chlorine dioxide	ClO_2	10049-04-4	0.1	0.28	0.3	0.83
Chlorine trifluoride	ClF_3	7790-91-2			0.1 C	0.38 C
Chloroacetaldehyde	C_2H_3ClO	107-20-0			1 C	3.2 C
Chloroacetone	C_3H_5ClO	78-95-5			1 C	3.8 C
α-Chloroacetophenone	C_8H_7ClO	532-27-4	0.05	0.32		
Chloroacetyl chloride	$C_2H_2Cl_2O$	79-04-9	0.05	0.23	0.15	0.69
Chlorobenzene	C_6H_5Cl	108-90-7	10	46		
o-Chlorobenzylidene malononitrile	$C_{10}H_5ClN_2$	2698-41-1			0.05 C	0.39 C
2-Chloro-1,3-butadiene [Chloroprene]	C_4H_5Cl	126-99-8	10	36		
Chlorodifluoromethane	$CHClF_2$	75-45-6	1000	3540		
Chloroethane [Ethyl chloride]	C_2H_5Cl	75-00-3	100	264		
2-Chloroethanol [Ethylene chlorohydrin]	C_2H_5ClO	107-07-3			1 C	3.3 C
Chloroethene [Vinyl chloride]	C_2H_3Cl	75-01-4	1	2.5		
Chloromethane [Methyl chloride]	CH_3Cl	74-87-3	50	103	100	207
(Chloromethyl)benzene [Benzyl chloride]	C_7H_7Cl	100-44-7	1	5.2		
1-Chloro-4-nitrobenzene	$C_6H_4ClNO_2$	100-00-5	0.1	0.64		
1-Chloro-1-nitropropane	$C_3H_6ClNO_2$	600-25-9	2	10		
Chloropentafluoroethane	C_2ClF_5	76-15-3	1000	6320		
2-Chloropropanoic acid	$C_3H_5ClO_2$	598-78-7	0.1	0.44		
3-Chloropropene [Allyl chloride]	C_3H_5Cl	107-05-1	1	3	2	6
2-Chlorostyrene	C_8H_7Cl	2039-87-4	50	283	75	425
o-Chlorotoluene	C_7H_7Cl	95-49-8	50	259		
Chlorpyrifos	$C_9H_{11}Cl_3NO_3PS$	2921-88-2		0.2		
Chromium	Cr	7440-47-3		0.5		
Chromyl chloride	Cl_2CrO_2	14977-61-8	0.025	0.16		
Clopidol	$C_7H_7Cl_2NO$	2971-90-6		10		
Cobalt	Co	7440-48-4		0.02		
Cobalt carbonyl	$C_8Co_2O_8$	10210-68-1		0.1		
Cobalt hydrocarbonyl	C_4HCoO_4	16842-03-8		0.1		
Copper	Cu	7440-50-8		0.2		
Cresol (all isomers)	C_7H_8O	1319-77-3	5	22		
Crufomate	$C_{12}H_{19}ClNO_3P$	299-86-5		5		
Cyanamide	CH_2N_2	420-04-2		2		
Cyanogen	C_2N_2	460-19-5	10	21		
Cyanogen chloride	CClN	506-77-4			0.3 C	0.75 C
Cyclohexane	C_6H_{12}	110-82-7	300	1030		

Substance	Molecular Formula	CAS Reg. No.	Time-Weighted Average ppm	Time-Weighted Average mg/m^3	Short-Term Exposure Limit ppm	Short-Term Exposure Limit mg/m^3
Cyclohexanol	$C_6H_{12}O$	108-93-0	50	206		
Cyclohexanone	$C_6H_{10}O$	108-94-1	25	100		
Cyclohexene	C_6H_{10}	110-83-8	300	1010		
Cyclohexylamine	$C_6H_{13}N$	108-91-8	10	41		
Cyclonite [Hexahydro-1,3,5-trinitro-1,3,5-triazine]	$C_3H_6N_6O_6$	121-82-4		0.5		
1,3-Cyclopentadiene	C_5H_6	542-92-7	75	203		
Cyclopentane	C_5H_{10}	287-92-3	600	1720		
Cyhexatin	$C_{18}H_{34}OSn$	13121-70-5		5		
Decaborane(14)	$B_{10}H_{14}$	17702-41-9	0.05	0.25	0.15	0.75
Diacetone alcohol	$C_6H_{12}O_2$	123-42-2	50	238		
4,4'-Diaminodiphenylmethane [4,4-Methylene dianiline]	$C_{13}H_{14}N_2$	101-77-9	0.1	0.81		
Diazinon	$C_{12}H_{21}N_2O_3PS$	333-41-5		0.1		
Diazomethane	CH_2N_2	334-88-3	0.2	0.34		
Diborane	B_2H_6	19287-45-7	0.1	0.11		
Dibromodifluoromethane	CBr_2F_2	75-61-6	100	858		
2-Dibutylaminoethanol	$C_{10}H_{23}NO$	102-81-8	0.5	3.5		
2,6-Di-*tert*-butyl-4-methylphenol	$C_{15}H_{24}O$	128-37-0		10		
Dibutylphenyl phosphate	$C_{14}H_{23}O_4P$	2528-36-1	0.3	3.5		
Dibutyl phosphate	$C_8H_{19}O_4P$	107-66-4	1	8.6	2	17
Dibutyl phthalate	$C_{16}H_{22}O_4$	84-74-2		5		
Dichloroacetylene	C_2Cl_2	7572-29-4			0.1 C	0.39 C
o-Dichlorobenzene	$C_6H_4Cl_2$	95-50-1	25	150	50	301
p-Dichlorobenzene	$C_6H_4Cl_2$	106-46-7	10	60		
1,4-Dichloro-2-butene (unspecified isomer)	$C_4H_6Cl_2$	764-41-0	0.005	0.026		
Dichlorodifluoromethane	CCl_2F_2	75-71-8	1000	4950		
1,3-Dichloro-5,5-dimethyl hydantoin	$C_5H_6Cl_2N_2O_2$	118-52-5		0.2		0.4
Dichlorodiphenyltrichloroethane [DDT]	$C_{14}H_9Cl_5$	50-29-3		1		
1,1-Dichloroethane [Ethylidene dichloride]	$C_2H_4Cl_2$	75-34-3	100	405		
1,2-Dichloroethane [Ethylene dichloride]	$C_2H_4Cl_2$	107-06-2	10	40		
1,1-Dichloroethene [Vinylidene chloride]	$C_2H_2Cl_2$	75-35-4	5	20		
1,2-Dichloroethylene (both isomers)	$C_2H_2Cl_2$	540-59-0	200	793		
Dichlorofluoromethane	$CHCl_2F$	75-43-4	10	42		
Dichloromethane [Methylene chloride]	CH_2Cl_2	75-09-2	50	174		
1,1-Dichloro-1-nitroethane	$C_2H_3Cl_2NO_2$	594-72-9	2	12		
(2,4-Dichlorophenoxy)acetic acid	$C_8H_6Cl_2O_3$	94-75-7		10		
1,2-Dichloropropane	$C_3H_6Cl_2$	78-87-5	75	347	110	508
2,2-Dichloropropanoic acid	$C_3H_4Cl_2O_2$	75-99-0		5		
1,3-Dichloropropene (both isomers)	$C_3H_4Cl_2$	542-75-6	1	4.5		
1,2-Dichloro-1,1,2,2-tetrafluoroethane	$C_2Cl_2F_4$	76-14-2	1000	7000		
Dichlorvos	$C_4H_7Cl_2O_4P$	62-73-7	0.1	0.90		
Dicrotophos	$C_8H_{16}NO_5P$	141-66-2		0.25		
m-Dicyanobenzene [*m*-Phthalodinitrile]	$C_8H_4N_2$	626-17-5		5		
Dicyclopentadiene	$C_{10}H_{12}$	77-73-6	5	27		
Dieldrin	$C_{12}H_8Cl_6O$	60-57-1		0.25		
Diethanolamine	$C_4H_{11}NO_2$	111-42-2	0.46	2		
Diethylamine	$C_4H_{11}N$	109-89-7	5	15	15	45
2-Diethylaminoethanol	$C_6H_{15}NO$	100-37-8	2	9.6		
Diethylenetriamine [Bis(2-amimoethyl)amine]	$C_4H_{13}N_3$	111-40-0	1	4.2		
Diethyl ether	$C_4H_{10}O$	60-29-7	400	1210	500	1520
Diethyl phthalate	$C_{12}H_{14}O_4$	84-66-2		5		
1,1-Difluoroethene	$C_2H_2F_2$	75-38-7	500	1310		
Diglycidyl ether	$C_6H_{10}O_3$	2238-07-5	0.1	0.53		
Diisopropylamine	$C_6H_{15}N$	108-18-9	5	21		
Diisopropyl ether	$C_6H_{14}O$	108-20-3	250	1040	310	1300
Dimethoxymethane [Methylal]	$C_3H_8O_2$	109-87-5	1000	3110		

THRESHOLD LIMITS FOR AIRBORNE CONTAMINANTS (continued)

Substance	Molecular Formula	CAS Reg. No.	Time-Weighted Average ppm	Time-Weighted Average mg/m³	Short-Term Exposure Limit ppm	Short-Term Exposure Limit mg/m³
Dimethyl mercury	C_2H_6Hg	593-74-8		0.01		0.03
N,N-Dimethylacetamide	C_4H_9NO	127-19-5	10	36		
Dimethylamine	C_2H_7N	124-40-3	5	9.2	15	27.6
N,N-Dimethylaniline	$C_8H_{11}N$	121-69-7	5	25	10	50
2,2-Dimethylbutane	C_6H_{14}	75-83-2	500	1760	1000	3500
2,3-Dimethylbutane	C_6H_{14}	79-29-8	500	1760	1000	3500
N,N-Dimethylformamide	C_3H_7NO	68-12-2	10	30		
2,6-Dimethyl-4-heptanone [Diisobutyl ketone]	$C_9H_{18}O$	108-83-8	25	145		
1,1-Dimethylhydrazine	$C_2H_8N_2$	57-14-7	0.01	0.025		
Dimethyl phthalate	$C_{10}H_{10}O_4$	131-11-3		5		
Dimethyl sulfate	$C_2H_6O_4S$	77-78-1	0.1	0.52		
Dinitrobenzene (all isomers)	$C_6H_4N_2O_4$	25154-54-5	0.15	1.0		
Dinitrotoluene (all isomers)	$C_7H_6N_2O_4$	25321-14-6		0.2		
1,4-Dioxane	$C_4H_8O_2$	123-91-1	20	72		
Dioxathion	$C_{12}H_{26}O_6P_2S_4$	78-34-2		0.2		
Diphenylamine	$C_{12}H_{11}N$	122-39-4		10		
Diphenyl ether	$C_{12}H_{10}O$	101-84-8	1	7	2	14
4,4'-Diphenylmethane diisocyanate	$C_{15}H_{10}N_2O_2$	101-68-8	0.005	0.051		
Dipropylene glycol monomethyl ether	$C_7H_{16}O_3$	34590-94-8	100	600	150	900
Diquat	$C_{12}H_{12}N_2$	231-36-7		0.5		
Disulfiram	$C_{10}H_{20}N_2S_4$	97-77-8		2		
Disulfoton	$C_8H_{19}O_2PS_3$	298-04-4		0.1		
Diuron	$C_9H_{10}Cl_2N_2O$	330-54-1		10		
Divinyl benzene (all isomers)	$C_{10}H_{10}$	1321-74-0	10	53		
Endosulfan	$C_9H_6Cl_6O_3S$	115-29-7		0.1		
Endrin	$C_{12}H_8Cl_6O$	72-20-8		0.1		
Enflurane	$C_3H_2ClF_5O$	13838-16-9	75	566		
Epichlorohydrin [(Chloromethyl)oxirane]	C_3H_5ClO	106-89-8	0.5	1.9		
1,2-Epoxy-4-(epoxyethyl)cyclohexane [Vinylcyclohexene dioxide]	$C_8H_{12}O_2$	106-87-6	0.1	0.57		
1,2-Ethanediamine [Ethylenediamine]	$C_2H_8N_2$	107-15-3	10	25		
Ethanethiol [Ethyl mercaptan]	C_2H_6S	75-08-1	0.5	1.3		
Ethanol	C_2H_6O	64-17-5	1000	1880		
Ethanolamine	C_2H_7NO	141-43-5	3	7.5	6	15
Ethion	$C_9H_{22}O_4P_2S_4$	563-12-2		0.4		
Ethoxydimethylsilane	$C_4H_{12}OSi$	14857-34-2	0.5	2.1	1.5	6.4
Ethyl acetate	$C_4H_8O_2$	141-78-6	400	1440		
Ethyl acrylate	$C_5H_8O_2$	140-88-5	5	20	15	61
Ethylamine	C_2H_7N	75-04-7	5	9.2	15	27.6
Ethylbenzene	C_8H_{10}	100-41-4	100	434	125	543
Ethyl *tert*-butyl ether [ETBE]	$C_6H_{14}O$	637-92-3	5	20		
Ethylene glycol	$C_2H_6O_2$	107-21-1				100 C
Ethylene glycol dinitrate	$C_2H_4N_2O_6$	628-96-6	0.05	0.31		
Ethylene glycol monobutyl ether [2-Butoxyethanol]	$C_6H_{14}O_2$	111-76-2	20	97		
Ethylene glycol monoethyl ether [2-Ethoxyethanol]	$C_4H_{10}O_2$	110-80-5	5	18		
Ethylene glycol monoethyl ether acetate [2-Ethoxyethyl acetate]	$C_6H_{12}O_3$	111-15-9	5	27		
Ethylene glycol monomethyl ether [2-Methoxyethanol]	$C_3H_8O_2$	109-86-4	5	16		
Ethylene glycol monomethyl ether acetate [2-Methoxyethyl acetate]	$C_5H_{10}O_3$	110-49-6	5	24		
Ethyleneimine	C_2H_5N	151-56-4	0.5	0.88		
Ethylene oxide [Oxirane]	C_2H_4O	75-21-8	1	1.8		
Ethyl formate	$C_3H_6O_2$	109-94-4	100	303		
Ethylidene norbornene	C_9H_{12}	16219-75-3			5 C	25 C

THRESHOLD LIMITS FOR AIRBORNE CONTAMINANTS (continued)

Substance	Molecular Formula	CAS Reg. No.	Time-Weighted Average		Short-Term Exposure Limit	
			ppm	mg/m^3	ppm	mg/m^3
N-Ethylmorpholine	$C_6H_{13}NO$	100-74-3	5	24		
Ethyl *p*-nitrophenyl benzenethiophosphate [EPN]	$C_{14}H_{14}NO_4PS$	2104-64-5		0.1		
Ethyl silicate	$C_8H_{20}O_4Si$	78-10-4	10	85		
Fenamiphos	$C_{13}H_{22}NO_3PS$	22224-92-6		0.1		
Fensulfothion	$C_{11}H_{17}O_4PS_2$	115-90-2		0.1		
Fenthion	$C_{10}H_{15}O_3PS_2$	55-38-9		0.2		
Ferbam	$C_9H_{18}FeN_3S_6$	14484-64-1		10		
Ferrocene [Dicyclopentadienyl iron]	$C_{10}H_{10}Fe$	102-54-5		10		
Fluorine	F_2	7782-41-4	1	1.6	2	3.1
Fluorine monoxide [Oxygen difluoride]	F_2O	7783-41-7			0.05 C	0.11 C
Fonofos	$C_{10}H_{15}OPS_2$	944-22-9		0.1		
Formaldehyde	CH_2O	50-00-0			0.3 C	0.37 C
Formamide	CH_3NO	75-12-7	10	18		
Formic acid	CH_2O_2	64-18-6	5	9.4	10	19
Furfural [2-Furaldehyde]	$C_5H_4O_2$	98-01-1	2	7.9		
Furfuryl alcohol [2-Furanmethanol]	$C_5H_6O_2$	98-00-0	10	40	15	60
Germane [Germanium tetrahydride]	GeH_4	7782-65-2	0.2	0.63		
Glycerol	$C_3H_8O_3$	56-81-5		10		
Graphite	C	7440-44-0		2		
Hafnium	Hf	7440-58-6		0.5		
Heptachlor	$C_{10}H_5Cl_7$	76-44-8		0.05		
Heptane	C_7H_{16}	142-82-5	400	1640	500	2050
2-Heptanone [Methyl pentyl ketone]	$C_7H_{14}O$	110-43-0	50	233		
3-Heptanone [Ethyl butyl ketone]	$C_7H_{14}O$	106-35-4	50	233	75	350
4-Heptanone [Dipropyl ketone]	$C_7H_{14}O$	123-19-3	50	233		
Hexachlorobenzene	C_6Cl_6	118-74-1		0.002		
Hexachloro-1,3-butadiene	C_4Cl_6	87-68-3	0.02	0.21		
1,2,3,4,5,6-Hexachlorocyclohexane [Lindane]	$C_6H_6Cl_6$	58-89-9		0.5		
Hexachloro-1,3-cyclopentadiene	C_5Cl_6	77-47-4	0.01	0.11		
Hexachloroethane [Perchloroethane]	C_2Cl_6	67-72-1	1	9.7		
Hexachloronaphthalene (all isomers)	$C_{10}H_2Cl_6$	1335-87-1		0.2		
Hexamethylene diisocyanate	$C_8H_{12}N_2O_2$	822-06-0	0.005	0.034		
Hexane	C_6H_{14}	110-54-3	50	176		
1,6-Hexanediamine [Hexamethylenediamine]	$C_6H_{16}N_2$	124-09-4	0.5	2.3		
2-Hexanone [Butyl methyl ketone]	$C_6H_{12}O$	591-78-6	5	20	10	40
1-Hexene	C_6H_{12}	592-41-6	30	103		
sec-Hexyl acetate	$C_8H_{16}O_2$	108-84-9	50	295		
Hydrazine	H_4N_2	302-01-2	0.01	0.013		
Hydrazoic acid	HN_3	7782-79-8			0.11 C	0.19 C
Hydrogen bromide	BrH	10035-10-6			3 C	9.9 C
Hydrogen chloride	ClH	7647-01-0			5 C	7.5 C
Hydrogen cyanide	CHN	74-90-8			4.7 C	5 C
Hydrogen fluoride	FH	7664-39-3			3 C	2.3 C
Hydrogen peroxide	H_2O_2	7722-84-1	1	1.4		
Hydrogen selenide	H_2Se	7783-07-5	0.05	0.16		
Hydrogen sulfide	H_2S	7783-06-4	10	14	15	21
p-Hydroquinone [1,4-Benzenediol]	$C_6H_6O_2$	123-31-9		2		
2-Hydroxypropyl acrylate	$C_6H_{10}O_3$	999-61-1	0.5	2.8		
Indene	C_9H_8	95-13-6	10	48		
Indium	In	7440-74-6		0.1		
Iodine	I_2	7553-56-2			0.1 C	1.0 C
Iodomethane [Methyl iodide]	CH_3I	74-88-4	2	12		
Iron(III) oxide	Fe_2O_3	1309-37-1		5		
Iron pentacarbonyl	C_5FeO_5	13463-40-6	0.1	0.23	0.2	0.45
Isobutyl acetate	$C_6H_{12}O_2$	110-19-0	150	713		
Isopentane	C_5H_{12}	78-78-4	600	1770		

THRESHOLD LIMITS FOR AIRBORNE CONTAMINANTS (continued)

Substance	Molecular Formula	CAS Reg. No.	Time-Weighted Average ppm	Time-Weighted Average mg/m³	Short-Term Exposure Limit ppm	Short-Term Exposure Limit mg/m³
Isopentyl acetate [Isoamyl acetate]	$C_7H_{14}O_2$	123-92-2	100	532		
Isophorone	$C_9H_{14}O$	78-59-1			5 C	28 C
Isophorone diisocyanate	$C_{12}H_{18}N_2O_2$	4098-71-9	0.005	0.045		
Isopropenylbenzene [α-Methyl styrene]	C_9H_{10}	98-83-9	50	242	100	483
2-Isopropoxyethanol	$C_5H_{12}O_2$	109-59-1	25	106		
Isopropyl acetate	$C_5H_{10}O_2$	108-21-4	250	1040	310	1290
Isopropylamine	C_3H_9N	75-31-0	5	12	10	24
N-Isopropylaniline	$C_9H_{13}N$	768-52-5	2	11		
Isopropylbenzene [Cumene]	C_9H_{12}	98-82-8	50	246		
Isopropyl glycidyl ether	$C_6H_{12}O_2$	4016-14-2	50	238	75	356
Kaolin		1332-58-7		2		
Ketene	C_2H_2O	463-51-4	0.5	0.86	1.5	2.6
Lead	Pb	7439-92-1		0.05		
Lead(II) arsenate	$As_2O_8Pb_3$	7784-40-9		0.15		
Lead(II) chromate	CrO_4Pb	7758-97-6		0.075		
Lithium hydride	HLi	7580-67-8		0.025		
Magnesium carbonate [Magnesite]	$CMgO_3$	546-93-0		10		
Magnesium oxide	MgO	1309-48-4		10		
Malathion	$C_{10}H_{19}O_6PS_2$	121-75-5		10		
Maleic anhydride	$C_4H_2O_3$	108-31-6	0.1	4		
Manganese	Mn	7439-96-5		0.2		
Manganese cyclopentadienyl tricarbonyl	$C_8H_5MnO_3$	12079-65-1		0.4		
Mercury	Hg	7439-97-6		0.025		
Mesityl oxide	$C_6H_{10}O$	141-79-7	15	60	25	100
Methacrylic acid [2-Methylpropenoic acid]	$C_4H_6O_2$	79-41-4	20	70		
Methanethiol [Methyl mercaptan]	CH_4S	74-93-1	0.5	0.98		
Methanol	CH_4O	67-56-1	200	262	250	328
Methomyl	$C_5H_{10}N_2O_2S$	16752-77-5		2.5		
o-Methoxyaniline [*o*-Anisidine]	C_7H_9NO	90-04-0	0.1	0.5		
p-Methoxyaniline [*p*-Anisidine]	C_7H_9NO	104-94-9	0.1	0.5		
Methoxychlor	$C_{16}H_{15}Cl_3O_2$	72-43-5		10		
4-Methoxyphenol	$C_7H_8O_2$	150-76-5		5		
Methyl acetate	$C_3H_6O_2$	79-20-9	200	606	250	757
Methyl acrylate	$C_4H_6O_2$	96-33-3	2	7		
2-Methylacrylonitrile	C_4H_5N	126-98-7	1	2.7		
Methylamine	CH_5N	74-89-5	5	6.4	15	19
o-Methylaniline [*o*-Toluidine]	C_7H_9N	95-53-4	2	8.8		
m-Methylaniline [*m*-Toluidine]	C_7H_9N	108-44-1	2	8.8		
p-Methylaniline [*p*-Toluidine]	C_7H_9N	106-49-0	2	8.8		
N-Methylaniline	C_7H_9N	100-61-8	0.5	2.2		
3-Methyl-1-butanol [Isoamyl alcohol]	$C_5H_{12}O$	123-51-3	100	361	125	452
3-Methyl-2-butanone [Methyl isopropyl ketone]	$C_5H_{10}O$	563-80-4	200	705		
Methyl *tert*-butyl ether [MTBE]	$C_5H_{12}O$	1634-04-4	40	144		
Methyl 2-cyanoacrylate	$C_5H_5NO_2$	137-05-3	0.2	0.9		
Methylcyclohexane	C_7H_{14}	108-87-2	400	1610		
Methylcyclohexanol (all isomers)	$C_7H_{14}O$	25639-42-3	50	234		
2-Methylcyclohexanone	$C_7H_{12}O$	583-60-8	50	229	75	344
2-Methylcyclopentadienyl manganese tricarbonyl	$C_9H_7MnO_3$	12108-13-3		0.8		
Methyl demeton	$C_6H_{15}O_3PS_2$	8022-00-2		0.5		
2-Methyl-3,5-dinitrobenzamide [Dinitolmide]	$C_8H_7N_3O_5$	148-01-6		5		
2-Methyl-4,6-dinitrophenol [Dinitro-*o*-cresol]	$C_7H_6N_2O_5$	534-52-1		0.2		
Methylene bis(4-cyclohexylisocyanate)	$C_{15}H_{22}N_2O_2$	5124-30-1	0.005	0.054		
Methyl ethyl ketone peroxide	$C_8H_{18}O_2$	1338-23-4			0.2 C	1.5 C
Methyl formate	$C_2H_4O_2$	107-31-3	100	246	150	368
6-Methyl-1-heptanol [Isooctyl alcohol]	$C_8H_{18}O$	26952-21-6	50	266		
5-Methyl-3-heptanone	$C_8H_{16}O$	541-85-5	25	131		

THRESHOLD LIMITS FOR AIRBORNE CONTAMINANTS (continued)

Substance	Molecular Formula	CAS Reg. No.	Time-Weighted Average ppm	Time-Weighted Average mg/m³	Short-Term Exposure Limit ppm	Short-Term Exposure Limit mg/m³
5-Methyl-2-hexanone [Methyl isopentyl ketone]	$C_7H_{14}O$	110-12-3	50	234		
Methylhydrazine	CH_6N_2	60-34-4	0.01	0.019		
Methyl isocyanate	C_2H_3NO	624-83-9	0.02	0.047		
Methyl methacrylate	$C_5H_8O_2$	80-62-6	50	205	100	410
Methyloxirane [1,2-Propylene oxide]	C_3H_6O	75-56-9	20	48		
Methyl parathion	$C_8H_{10}NO_5PS$	298-00-0		0.2		
2-Methylpentane	C_6H_{14}	107-83-5	500	1760	1000	3500
3-Methylpentane	C_6H_{14}	96-14-0	500	1760	1000	3500
2-Methyl-2,4-pentanediol [Hexylene glycol]	$C_6H_{14}O_2$	107-41-5			25 C	121 C
4-Methyl-2-pentanol [Methyl isobutyl carbinol]	$C_6H_{14}O$	108-11-2	25	104	40	167
4-Methyl-2-pentanone [Isobutyl methyl ketone]	$C_6H_{12}O$	108-10-1	50	205	75	307
2-Methyl-1-propanol [Isobutyl alcohol]	$C_4H_{10}O$	78-83-1	50	152		
2-Methyl-2-propanol [*tert*-Butyl alcohol]	$C_4H_{10}O$	75-65-0	100	303		
Methylstyrene (all isomers)	C_9H_{10}	25013-15-4	50	242	100	483
N-Methyl-*N*,2,4,6-tetranitroaniline [Tetryl]	$C_7H_5N_5O_8$	479-45-8		1.5		
Metribuzin	$C_8H_{14}N_4OS$	21087-64-9		5		
Mevinphos	$C_7H_{13}O_6P$	7786-34-7	0.01	0.092	0.03	0.27
Mica		12001-26-2		3		
Molybdenum	Mo	7439-98-7		10		
Monocrotophos	$C_7H_{14}NO_5P$	6923-22-4		0.25		
Morpholine	C_4H_9NO	110-91-8	20	71		
Naled	$C_4H_7Br_2Cl_2O_4P$	300-76-5		3		
Naphthalene	$C_{10}H_8$	91-20-3	10	52	15	79
1-Naphthalenylthiourea [ANTU]	$C_{11}H_{10}N_2S$	86-88-4		0.3		
Neopentane	C_5H_{12}	463-82-1	600	1770		
Nickel	Ni	7440-02-0		1.5		
Nickel carbonyl	C_4NiO_4	13463-39-3	0.05	0.12		
Nickel(III) sulfide	Ni_3S_2	12035-72-2		0.14		
Nicotine	$C_{10}H_{14}N_2$	54-11-5		0.5		
Nitrapyrin	$C_6H_3Cl_4N$	1929-82-4		10		20
Nitric acid	HNO_3	7697-37-2	2	5.2	4	10
Nitric oxide	NO	10102-43-9	25	31		
p-Nitroaniline	$C_6H_6N_2O_2$	100-01-6		3		
Nitrobenzene	$C_6H_5NO_2$	98-95-3	1	5		
Nitroethane	$C_2H_5NO_2$	79-24-3	100	307		
Nitrogen dioxide	NO_2	10102-44-0	3	5.6	5	9.4
Nitrogen trifluoride	F_3N	7783-54-2	10	29		
Nitromethane	CH_3NO_2	75-52-5	20	50		
1-Nitropropane	$C_3H_7NO_2$	108-03-2	25	91		
2-Nitropropane	$C_3H_7NO_2$	79-46-9	10	36		
Nitrotoluene (all isomers)	$C_7H_7NO_2$	1321-12-6	2	11		
Nitrous oxide	N_2O	10024-97-2	50	90		
Nonane (all isomers)	C_9H_{20}	111-84-2	200	1050		
Octachloronaphthalene	$C_{10}Cl_8$	2234-13-1		0.1		0.3
Octane (all isomers)	C_8H_{18}	111-65-9	300	1400	375	1750
Osmium(VIII) oxide [Osmium tetroxide]	O_4Os	20816-12-0	0.0002	0.0016	0.0006	0.0047
Oxalic acid	$C_2H_2O_4$	144-62-7		1		2
2-Oxetanone [β-Propiolactone]	$C_3H_4O_2$	57-57-8	0.5	1.5		
Oxiranemethanol [Glycidol]	$C_3H_6O_2$	556-52-5	2	6.1		
Ozone	O_3	10028-15-6	0.1	0.2		
Paraquat	$C_{12}H_{14}N_2$	4685-14-7		0.5		
Parathion	$C_{10}H_{14}NO_5PS$	56-38-2		0.1		
Pentaborane(9)	B_5H_9	19624-22-7	0.005	0.013	0.015	0.039
Pentachloronaphthalene (unspecified isomer)	$C_{10}H_3Cl_5$	1321-64-8		0.5		
Pentachloronitrobenzene	$C_6Cl_5NO_2$	82-68-8		0.5		
Pentachlorophenol	C_6HCl_5O	87-86-5		0.5		
Pentaerythritol	$C_5H_{12}O_4$	115-77-5		10		

THRESHOLD LIMITS FOR AIRBORNE CONTAMINANTS (continued)

Substance	Molecular Formula	CAS Reg. No.	Time-Weighted Average ppm	Time-Weighted Average mg/m³	Short-Term Exposure Limit ppm	Short-Term Exposure Limit mg/m³
Pentanal [Valeraldehyde]	$C_5H_{10}O$	110-62-3	50	176		
Pentane	C_5H_{12}	109-66-0	600	1770	750	2210
Pentanedial [Glutaraldehyde]	$C_5H_8O_2$	111-30-8			0.05 C	0.2 C
2-Pentanone [Methyl propyl ketone]	$C_5H_{10}O$	107-87-9	200	705	250	881
3-Pentanone [Diethyl ketone]	$C_5H_{10}O$	96-22-0	200	700	300	1050
Pentyl acetate (all isomers)	$C_7H_{14}O_2$	628-63-7	50	265	100	530
Perchloromethyl mercaptan	CCl_4S	594-42-3	0.1	0.76		
Perchloryl fluoride	$ClFO_3$	7616-94-6	3	13	6	25
Perfluoroacetone [Hexafluoroacetone]	C_3F_6O	684-16-2	0.1	0.68		
Perfluoroisobutene	C_4F_8	382-21-8			0.01 C	0.082 C
Phenol	C_6H_6O	108-95-2	5	19		
10*H*-Phenothiazine	$C_{12}H_9NS$	92-84-2		5		
Phenylenediamine (all isomers)	$C_6H_8N_2$	25265-76-3		0.1		
Phenyl glycidyl ether	$C_9H_{10}O_2$	122-60-1	0.1	0.6		
Phenylhydrazine	$C_6H_8N_2$	100-63-0	0.1	0.44		
Phenylphosphine	C_6H_7P	638-21-1			0.05 C	0.23 C
Phorate	$C_7H_{17}O_2PS_3$	298-02-2		0.05		0.2
Phosphine	H_3P	7803-51-2	0.3	0.42	1	1.4
Phosphoric acid	H_3O_4P	7664-38-2		1		3
Phosphorus (white)	P	7723-14-0	0.02	0.1		
Phosphorus(III) chloride [Phosphorus trichloride]	Cl_3P	7719-12-2	0.2	1.1	0.5	2.8
Phosphorus(V) chloride [Phosphorus pentachloride]	Cl_5P	10026-13-8	0.1	0.85		
Phosphorus(V) oxychloride [Phosphoryl chloride]	Cl_3OP	10025-87-3	0.1	0.63		
Phosphorus(V) sulfide	P_2S_5	1314-80-3		1		3
Phthalic anhydride	$C_8H_4O_3$	85-44-9	1	6.1		
Piperazine dihydrochloride	$C_4H_{12}Cl_2N_2$	142-64-3		5		
2-Pivaloyl-1,3-indandione [Pindone]	$C_{14}H_{14}O_3$	83-26-1		0.1		
Platinum	Pt	7440-06-4		1		
Potassium hydroxide	HKO	1310-58-3		2 C		
Propane	C_3H_8	74-98-6	2500	4500		
Propanoic acid	$C_3H_6O_2$	79-09-4	10	30		
1-Propanol	C_3H_8O	71-23-8	200	492	250	614
2-Propanol [Isopropyl alcohol]	C_3H_8O	67-63-0	400	983	500	1230
Propargyl alcohol [2-Propyn-1-ol]	C_3H_4O	107-19-7	1	2.3		
Propoxur	$C_{11}H_{15}NO_3$	114-26-1		0.5		
Propyl acetate	$C_5H_{10}O_2$	109-60-4	200	835	250	1040
1,2-Propylene glycol dinitrate	$C_3H_6N_2O_6$	6423-43-4	0.05	0.34		
Propylene glycol monomethyl ether	$C_4H_{10}O_2$	107-98-2	100	369	150	553
Propyleneimine	C_3H_7N	75-55-8	2	4.7		
Propyl nitrate	$C_3H_7NO_3$	627-13-4	25	107	40	172
Propyne [Methylacetylene]	C_3H_4	74-99-7	1000	1640		
2-Pyridinamine [2-Aminopyridine]	$C_5H_6N_2$	504-29-0	0.5	1.9		
Pyridine	C_5H_5N	110-86-1	5	16		
Pyrocatechol [Catechol]	$C_6H_6O_2$	120-80-9	5	23		
Resorcinol	$C_6H_6O_2$	108-46-3	10	45	20	90
Rhodium	Rh	7440-16-6		1		
Ronnel	$C_8H_8Cl_3O_3PS$	299-84-3		10		
Rotenone	$C_{23}H_{22}O_6$	83-79-4		5		
Selenium	Se	7782-49-2		0.2		
Selenium hexafluoride	F_6Se	7783-79-1	0.12	0.95		
Sesone	$C_8H_7Cl_2NaO_5S$	136-78-7		10		
Silane	H_4Si	7803-62-5	5	6.6		
Silicon	Si	7440-21-3		10		
Silicon carbide	CSi	409-21-2		10		

THRESHOLD LIMITS FOR AIRBORNE CONTAMINANTS (continued)

Substance	Molecular Formula	CAS Reg. No.	Time-Weighted Average ppm	Time-Weighted Average mg/m^3	Short-Term Exposure Limit ppm	Short-Term Exposure Limit mg/m^3
Silicon dioxide (α-quartz)	O_2Si	14808-60-7		0.05		
Silicon dioxide (tridymite)	O_2Si	15468-32-3		0.05		
Silicon dioxide (cristobalite)	O_2Si	14464-46-1		0.05		
Silicon dioxide (vitreous)	O_2Si	60676-86-0		0.1		
Silver	Ag	7440-22-4		0.1		
Sodium azide	N_3Na	26628-22-8				0.29 C
Sodium fluoroacetate	$C_2H_2FNaO_2$	62-74-8		0.05		
Sodium hydrogen sulfite	$HNaO_3S$	7631-90-5		5		
Sodium hydroxide	HNaO	1310-73-2				2 C
Sodium metabisulfite	$Na_2O_5S_2$	7681-57-4		5		
Sodium pyrophosphate	$Na_4O_7P_2$	7722-88-5		5		
Sodium tetraborate decahydrate	$B_4H_{20}Na_2O_{17}$	1303-96-4		5		
Stibine	H_3Sb	7803-52-3	0.1	0.51		
Strontium chromate	CrO_4Sr	7789-06-2		0.002		
Strychnine	$C_{21}H_{22}N_2O_2$	57-24-9		0.15		
Styrene	C_8H_8	100-42-5	20	85	40	170
Sucrose	$C_{12}H_{22}O_{11}$	57-50-1		10		
Sulfotep	$C_8H_{20}O_5P_2S_2$	3689-24-5		0.2		
Sulfur chloride	Cl_2S_2	10025-67-9			1 C	5.5 C
Sulfur decafluoride	$F_{10}S_2$	5714-22-7			0.01 C	0.10 C
Sulfur dioxide	O_2S	7446-09-5	2	5.2	5	13
Sulfur hexafluoride	F_6S	2551-62-4	1000	6000		
Sulfur tetrafluoride	F_4S	7783-60-0			0.1 C	0.44 C
Sulfuric acid	H_2O_4S	7664-93-9		1		3
Sulfuryl fluoride	F_2O_2S	2699-79-8	5	21	10	42
Sulprofos	$C_{12}H_{19}O_2PS_3$	35400-43-2		1		
Talc		14807-96-6		2		
Tantalum	Ta	7440-25-7		5		
Tantalum(V) oxide	O_5Ta_2	1314-61-0		5		
Tellurium	Te	13494-80-9		0.1		
Tellurium hexafluoride	F_6Te	7783-80-4	0.02	0.10		
Terephthalic acid	$C_8H_6O_4$	100-21-0		10		
Terphenyl (all isomers)	$C_{18}H_{14}$	26140-60-3			0.53 C	5 C
1,1,2,2-Tetrabromoethane [Acetylene tetrabromide]	$C_2H_2Br_4$	79-27-6	1	14		
Tetrabromomethane [Carbon tetrabromide]	CBr_4	558-13-4	0.1	1.4	0.3	4.1
1,1,1,2-Tetrachloro-2,2-difluoroethane	$C_2Cl_4F_2$	76-11-9	500	4170		
1,1,2,2-Tetrachloro-1,2-difluoroethane	$C_2Cl_4F_2$	76-12-0	500	4170		
1,1,2,2-Tetrachloroethane	$C_2H_2Cl_4$	79-34-5	1	6.9		
Tetrachloroethene [Perchloroethylene]	C_2Cl_4	127-18-4	25	170	100	685
Tetrachloromethane [Carbon tetrachloride]	CCl_4	56-23-5	5	31	10	63
Tetrachloronaphthalene (all isomers)	$C_{10}H_4Cl_4$	1335-88-2		2		
Tetraethyl lead	$C_8H_{20}Pb$	78-00-2		0.1		
Tetraethyl pyrophosphate [TEPP]	$C_8H_{20}O_7P_2$	107-49-3		0.05		
Tetrahydrofuran [Oxolane]	C_4H_8O	109-99-9	200	590	250	737
Tetramethyl lead	$C_4H_{12}Pb$	75-74-1		0.15		
Tetramethyl silicate	$C_4H_{12}O_4Si$	681-84-5	1	6		
Tetramethyl succinonitrile	$C_8H_{12}N_2$	3333-52-6	0.5	2.8		
Tetranitromethane	CN_4O_8	509-14-8	0.005	0.04		
Thallium	Tl	7440-28-0		0.1		
4,4'-Thiobis(6-*tert*-butyl-*m*-cresol)	$C_{22}H_{30}O_2S$	96-69-5		10		
Thioglycolic acid	$C_2H_4O_2S$	68-11-1	1	3.8		
Thionyl chloride	Cl_2OS	7719-09-7			1 C	4.9 C
Thiram	$C_6H_{12}N_2S_4$	137-26-8		1		
Tin	Sn	7440-31-5		2		
Titanium(IV) oxide [Titanium dioxide]	O_2Ti	13463-67-7		10		
Toluene	C_7H_8	108-88-3	50	188		

Substance	Molecular Formula	CAS Reg. No.	Time-Weighted Average ppm	Time-Weighted Average mg/m^3	Short-Term Exposure Limit ppm	Short-Term Exposure Limit mg/m^3
Toluene-2,4-diisocyanate	$C_9H_6N_2O_2$	584-84-9	0.005	0.036	0.02	0.14
1*H*-1,2,4-Triazol-3-amine	$C_2H_4N_4$	61-82-5		0.2		
Tribromomethane [Bromoform]	$CHBr_3$	75-25-2	0.5	5.2		
Tributyl phosphate	$C_{12}H_{27}O_4P$	126-73-8	0.2	2.2		
Trichloroacetic acid	$C_2HCl_3O_2$	76-03-9	1	6.7		
1,2,4-Trichlorobenzene	$C_6H_3Cl_3$	120-82-1			5 C	37 C
1,1,1-Trichloroethane [Methyl chloroform]	$C_2H_3Cl_3$	71-55-6	350	1910	450	2460
1,1,2-Trichloroethane	$C_2H_3Cl_3$	79-00-5	10	55		
Trichloroethene	C_2HCl_3	79-01-6	50	269	100	537
Trichlorofluoromethane	CCl_3F	75-69-4			1000 C	5620 C
Trichloromethane [Chloroform]	$CHCl_3$	67-66-3	10	49		
(Trichloromethyl)benzene [Benzotrichloride]	$C_7H_5Cl_3$	98-07-7			0.01 C	0.08 C
Trichloronaphthalene (all isomers)	$C_{10}H_5Cl_3$	1321-65-9		5		
Trichloronitromethane [Chloropicrin]	CCl_3NO_2	76-06-2	0.1	0.67		
2,4,5-Trichlorophenoxyacetic acid	$C_8H_5Cl_3O_3$	93-76-5		10		
1,2,3-Trichloropropane	$C_3H_5Cl_3$	96-18-4	10	60		
1,1,2-Trichloro-1,2,2-trifluoroethane	$C_2Cl_3F_3$	76-13-1	1000	7670	1250	9590
Tri-*o*-cresyl phosphate	$C_{21}H_{21}O_4P$	78-30-8		0.1		
Triethanolamine	$C_6H_{15}NO_3$	102-71-6		5		
Triethylamine	$C_6H_{15}N$	121-44-8	1	4.1	3	12
Triiodomethane [Iodoform]	CHI_3	75-47-8	0.6	10		
Trimellitic anhydride [1,2,4-Benzenetricarboxylic anhydride]	$C_9H_4O_5$	552-30-7				0.04 C
Trimethylamine	C_3H_9N	75-50-3	5	12	15	36
Trimethylbenzene (all isomers)	C_9H_{12}	25551-13-7	25	123		
Trimethyl phosphite	$C_3H_9O_3P$	121-45-9	2	10		
Trinitroglycerol [Nitroglycerin]	$C_3H_5N_3O_9$	55-63-0	0.05	0.46		
2,4,6-Trinitrophenol [Picric acid]	$C_6H_3N_3O_7$	88-89-1		0.1		
2,4,6-Trinitrotoluene [TNT]	$C_7H_5N_3O_6$	118-96-7		0.1		
Triphenylamine	$C_{18}H_{15}N$	603-34-9		5		
Triphenyl phosphate	$C_{18}H_{15}O_4P$	115-86-6		3		
Tungsten	W	7440-33-7		5		10
Uranium	U	7440-61-1		0.2		0.6
Vanadium(V) oxide	O_5V_2	1314-62-1		0.05		
Vinyl acetate	$C_4H_6O_2$	108-05-4	10	35	15	53
4-Vinylcyclohexene	C_8H_{12}	100-40-3	0.1	0.44		
Warfarin	$C_{19}H_{16}O_4$	81-81-2		0.1		
Xylene (all isomers)	C_8H_{10}	1330-20-7	100	434	150	651
Xylidine (all isomers)	$C_8H_{11}N$	1300-73-8	0.5	2.5		
Yttrium	Y	7440-65-5		1		
Zinc chloride	Cl_2Zn	7646-85-7		1		2
Zinc chromate, basic	CrH_2O_4Zn	13530-65-9		0.045		
Zinc oxide	OZn	1314-13-2		5		10
Zirconium	Zr	7440-67-7		5		10

OCTANOL-WATER PARTITION COEFFICIENTS

The octanol-water partition coefficient, *P*, is a widely used parameter for correlating biological effects of organic substances. It is a property of the two-phase system in which water and 1-octanol are in equilibrium at a fixed temperature and the substance is distributed between the water-rich and octanol-rich phases. *P* is defined as the ratio of the equilibrium concentration of the substance in the octanol-rich phase to that in the water-rich phase, in the limit of zero concentration. In general, *P* tends to be large for compounds with extended non-polar structures (such as long chain or multi-ring hydrocarbons) and small for compounds with highly polar groups. Thus *P* (or, in its more common form of expression, log *P*) provides a measure of the lipophilic vs. hydrophilic nature of a compound, which is an important consideration in assessing the potential toxicity. A discussion of methods of measurement and accuracy considerations for log *P* may be found in Reference 1.

This table gives selected values of log *P* for about 450 organic compounds, including many of environmental importance. All values refer to a nominal temperature of 25°C. The source of each value is indicated in the last column. These references contain data on many more compounds than are included here.

Compounds are listed by molecular formula following the Hill convention. To locate a compound by name or CAS Registry Number when the molecular formula is not known, use the table "Physical Constants of Organic Compounds" in Section 3 and its indexes to determine the molecular formula.

REFERENCES

1. Sangster, J., *J. Phys. Chem. Ref. Data*, 18, 1111, 1989.
2. Mackay, D., Shiu, W.Y., and Ma, K.C., *Illustrated Handbook of Physical-Chemical Properties and Environmental Fate for Organic Chemicals*, Lewis Publishers/CRC Press, Boca Raton, FL, 1992.
3. Shiu, W.Y., and Mackay, D., *J. Phys. Chem. Ref. Data*, 15, 911, 1986.
4. Pinsuwan, S., Li, L., and Yalkowsky, S.H., *J. Chem. Eng. Data*, 40, 623, 1995.
5. *Solubility Data Series, International Union of Pure and Applied Chemistry, Vol. 20*, Pergamon Press, Oxford, 1985.
6. *Solubility Data Series, International Union of Pure and Applied Chemistry, Vol. 38*, Pergamon Press, Oxford, 1985.
7. Miller, M.M., Ghodbane, S., Wasik, S.P., Tewari, Y.B., and Martire, D.E., *J. Chem. Eng. Data*, 29, 184, 1984.

Mol. Form.	Name	log *P*	Ref.
CCl_2F_2	Dichlorodifluoromethane	2.16	2
CCl_3F	Trichlorofluoromethane	2.53	2
CCl_4	Tetrachloromethane	2.64	2
$CHBr_3$	Tribromomethane	2.38	2
$CHCl_3$	Trichloromethane	1.97	2
CH_2BrCl	Bromochloromethane	1.41	2
CH_2Br_2	Dibromomethane	2.3	2
CH_2Cl_2	Dichloromethane	1.25	2
CH_2F_2	Difluoromethane	0.20	1
CH_2I_2	Diiodomethane	2.5	2
CH_2O	Formaldehyde	0.35	1
CH_2O_2	Formic acid	-0.54	1
CH_3Br	Bromomethane	1.19	2
CH_3Cl	Chloromethane	0.91	2
CH_3F	Fluoromethane	0.51	1
CH_3I	Iodomethane	1.5	2
CH_3NO	Formamide	-1.51	1
CH_3NO_2	Nitromethane	-0.33	1
CH_4O	Methanol	-0.74	1
CH_5N	Methylamine	-0.57	1
$C_2Cl_3F_3$	1,1,2-Trichlorotrifluoroethane	3.16	2
C_2Cl_4	Tetrachloroethylene	2.88	2
C_2Cl_6	Hexachloroethane	4.00	4
C_2HCl_3	Trichloroethylene	2.53	2
C_2HCl_5	Pentachloroethane	2.89	2
$C_2H_2Cl_2$	1,1-Dichloroethylene	2.13	2
$C_2H_2Cl_2$	*cis*-1,2-Dichloroethylene	1.86	2
$C_2H_2Cl_2$	*trans*-1,2-Dichloroethylene	1.93	2
$C_2H_2Cl_4$	1,1,2,2-Tetrachloroethane	2.39	2
C_2H_3Cl	Chloroethylene	1.38	2
$C_2H_3Cl_3$	1,1,1-Trichloroethane	2.49	2
$C_2H_3Cl_3$	1,1,2-Trichloroethane	2.38	2
C_2H_3N	Acetonitrile	-0.34	1
$C_2H_4Cl_2$	1,1-Dichloroethane	1.79	2
$C_2H_4Cl_2$	1,2-Dichloroethane	1.48	2
C_2H_4O	Acetaldehyde	0.45	1
C_2H_4O	Ethylene oxide	-0.30	1
$C_2H_4O_2$	Acetic acid	-0.17	1
C_2H_5Br	Bromoethane	1.6	2
C_2H_5Cl	Chloroethane	1.43	2
C_2H_5I	Iodoethane	2	2
C_2H_5NO	Acetamide	-1.26	1
$C_2H_5NO_2$	Nitroethane	0.18	1
C_2H_6O	Ethanol	-0.30	1
C_2H_6O	Dimethyl ether	0.10	1
C_2H_6OS	Dimethyl sulfoxide	-1.35	1
$C_2H_6O_2S$	Dimethyl sulfone	-1.41	1
C_2H_7N	Ethylamine	-0.13	1
C_2H_7N	Dimethylamine	-0.38	1
C_3H_3N	2-Propenenitrile	0.25	1
$C_3H_4Cl_2$	*cis*-1,3-Dichloropropene	2.03	2
C_3H_4O	Propargyl alcohol	-0.38	1
C_3H_4O	Acrolein	-0.01	1
C_3H_5Br	3-Bromopropene	1.79	1
C_3H_5ClO	Epichlorohydrin	0.30	2
$C_3H_5Cl_3$	1,2,3-Trichloropropane	2.63	2
C_3H_5N	Propanenitrile	0.16	1
C_3H_5NO	Acrylamide	-0.78	1
$C_3H_6Cl_2$	1,2-Dichloropropane	2.0	2
C_3H_6O	Allyl alcohol	0.17	1
C_3H_6O	Propanal	0.59	1
C_3H_6O	Acetone	-0.24	1
C_3H_6O	Methyloxirane	0.03	1
$C_3H_6O_2$	Propanoic acid	0.33	1
$C_3H_6O_2$	Methyl acetate	0.18	1
C_3H_7Br	1-Bromopropane	2.1	2
C_3H_7Br	2-Bromopropane	1.9	2
C_3H_7Cl	1-Chloropropane	2.04	1
C_3H_7Cl	2-Chloropropane	1.90	1
C_3H_7I	1-Iodopropane	2.5	2

OCTANOL-WATER PARTITION COEFFICIENTS (continued)

Mol. Form.	Name	log *P*	Ref.
C_3H_7N	Allylamine	0.03	1
C_3H_7NO	*N,N*-Dimethylformamide	-1.01	1
C_3H_7NO	*N*-Methylacetamide	-1.05	1
$C_3H_7NO_2$	1-Nitropropane	0.87	1
C_3H_8O	1-Propanol	0.25	1
C_3H_8O	2-Propanol	0.05	1
C_3H_8S	1-Propanethiol	1.81	1
C_3H_9N	Propylamine	0.48	1
C_3H_9N	Isopropylamine	0.26	1
C_3H_9N	Ethylmethylamine	0.15	1
C_3H_9N	Trimethylamine	0.16	1
C_4H_4O	Furan	1.34	1
C_4H_4S	Thiophene	1.81	1
C_4H_5N	Pyrrole	0.75	1
C_4H_6	1,3-Butadiene	1.99	1
C_4H_6	2-Butyne	1.46	1
C_4H_6O	2,5-Dihydrofuran	0.46	1
$C_4H_6O_2$	Methacrylic acid	0.93	1
$C_4H_6O_2$	Vinyl acetate	0.73	1
$C_4H_6O_2$	Methyl acrylate	0.80	1
C_4H_7N	Butanenitrile	0.60	1
C_4H_8	*cis*-2-Butene	2.33	1
C_4H_8	*trans*-2-Butene	2.31	1
C_4H_8	Isobutene	2.35	1
$C_4H_8Cl_2O$	Bis(2-chloroethyl) ether	1.12	2
C_4H_8O	Ethyl vinyl ether	1.04	1
C_4H_8O	Butanal	0.88	1
C_4H_8O	2-Butanone	0.29	1
C_4H_8O	Tetrahydrofuran	0.46	1
$C_4H_8O_2$	Butanoic acid	0.79	1
$C_4H_8O_2$	Propyl formate	0.83	1
$C_4H_8O_2$	Ethyl acetate	0.73	1
C_4H_9Br	1-Bromobutane	2.75	1
C_4H_9Cl	1-Chlorobutane	2.64	2
C_4H_9F	1-Fluorobutane	2.58	1
C_4H_9I	1-Iodobutane	3	2
C_4H_9N	Pyrrolidine	0.46	1
C_4H_9NO	Butanamide	-0.21	1
C_4H_9NO	*N,N*-Dimethylacetamide	-0.77	1
$C_4H_9NO_2$	1-Nitrobutane	1.47	1
C_4H_{10}	Isobutane	2.8	2
$C_4H_{10}O$	1-Butanol	0.84	1
$C_4H_{10}O$	2-Butanol	0.65	1
$C_4H_{10}O$	2-Methyl-1-propanol	0.76	1
$C_4H_{10}O$	2-Methyl-2-propanol	0.35	1
$C_4H_{10}O$	Diethyl ether	0.89	1
$C_4H_{10}S$	1-Butanethiol	2.28	1
$C_4H_{10}S$	Diethyl sulfide	1.95	1
$C_4H_{11}N$	Butylamine	0.86	1
$C_4H_{11}N$	*tert*-Butylamine	0.40	1
$C_4H_{11}N$	Diethylamine	0.58	1
C_5H_5N	Pyridine	0.65	1
C_5H_6O	2-Methylfuran	1.85	1
C_5H_7N	1-Methylpyrrole	1.21	1
C_5H_8	1,4-Pentadiene	2.48	1
C_5H_8	1-Pentyne	1.98	1
$C_5H_8O_2$	Methyl methacrylate	1.38	1
$C_5H_8O_2$	Ethyl acrylate	1.32	1
C_5H_9N	Pentanenitrile	0.94	1
C_5H_{10}	1-Pentene	2.2	2
C_5H_{10}	Cyclopentane	3.00	1
$C_5H_{10}O$	2-Pentanone	0.84	1
$C_5H_{10}O$	3-Pentanone	0.82	1
$C_5H_{10}O$	3-Methyl-2-butanone	0.56	1
$C_5H_{10}O$	Tetrahydropyran	0.82	1
$C_5H_{10}O$	2-Methyltetrahydrofuran	1.85	2
$C_5H_{10}O_2$	Pentanoic acid	1.39	1
$C_5H_{10}O_2$	Propyl acetate	1.24	1
$C_5H_{10}O_2$	Ethyl propanoate	1.21	1
$C_5H_{10}O_3$	Diethyl carbonate	1.21	1
$C_5H_{11}Br$	1-Bromopentane	3.37	1
$C_5H_{11}F$	1-Fluoropentane	2.33	1
$C_5H_{11}N$	Piperidine	0.84	1
$C_5H_{11}NO_2$	1-Nitropentane	2.01	1
C_5H_{12}	Pentane	3.45	1
C_5H_{12}	Neopentane	3.11	1
$C_5H_{12}O$	1-Pentanol	1.51	1
$C_5H_{12}O$	2-Pentanol	1.25	1
$C_5H_{12}O$	3-Pentanol	1.21	1
$C_5H_{12}O$	3-Methyl-1-butanol	1.28	1
$C_5H_{12}O$	2-Methyl-2-butanol	0.89	1
$C_5H_{12}O$	3-Methyl-2-butanol	1.28	1
$C_5H_{12}O$	2,2-Dimethyl-1-propanol	1.31	1
$C_5H_{12}O$	Methyl *tert*-butyl ether	0.94	1
$C_5H_{13}N$	Pentylamine	1.49	1
C_6Cl_6	Hexachlorobenzene	5.47	5
C_6HCl_5	Pentachlorobenzene	5.03	5
C_6HCl_5O	Pentachlorophenol	5.07	4
$C_6H_2Cl_4$	1,2,3,4-Tetrachlorobenzene	4.55	5
$C_6H_2Cl_4$	1,2,3,5-Tetrachlorobenzene	4.65	5
$C_6H_2Cl_4$	1,2,4,5-Tetrachlorobenzene	4.51	5
$C_6H_3Cl_3$	1,2,3-Trichlorobenzene	4.04	5
$C_6H_3Cl_3$	1,2,4-Trichlorobenzene	3.98	5
$C_6H_3Cl_3$	1,3,5-Trichlorobenzene	4.02	5
$C_6H_4Cl_2$	*o*-Dichlorobenzene	3.38	5
$C_6H_4Cl_2$	*m*-Dichlorobenzene	3.48	5
$C_6H_4Cl_2$	*p*-Dichlorobenzene	3.38	5
$C_6H_4Cl_2O$	2,4-Dichlorophenol	3.23	4
C_6H_5Br	Bromobenzene	2.99	2
C_6H_5Cl	Chlorobenzene	2.84	1
C_6H_5F	Fluorobenzene	2.27	2
C_6H_5I	Iodobenzene	3.28	2
$C_6H_5NO_2$	Nitrobenzene	1.85	1
C_6H_6	Benzene	2.13	1
C_6H_6O	Phenol	1.48	4
C_6H_6S	Benzenethiol	2.52	1
C_6H_7N	Aniline	0.90	1
C_6H_7N	2-Methylpyridine	1.11	1
C_6H_7N	3-Methylpyridine	1.20	1
C_6H_7N	4-Methylpyridine	1.22	1
C_6H_8	1,4-Cyclohexadiene	2.3	2
C_6H_8O	5-Hexyn-2-one	0.58	1
C_6H_8O	2-Cyclohexen-1-one	0.61	1
C_6H_8O	2-Ethylfuran	2.40	1
C_6H_{10}	1,5-Hexadiene	2.8	2
C_6H_{10}	1-Hexyne	2.73	2
C_6H_{10}	Cyclohexene	2.86	1
$C_6H_{10}O$	5-Hexen-2-one	1.02	1
$C_6H_{10}O$	Cyclohexanone	0.81	1
$C_6H_{10}O_2$	Ethyl methacrylate	1.94	1

OCTANOL-WATER PARTITION COEFFICIENTS (continued)

Mol. Form.	Name	log *P*	Ref.
$C_6H_{11}Br$	Bromocyclohexane	3.20	1
$C_6H_{11}N$	Hexanenitrile	1.66	1
C_6H_{12}	1-Hexene	3.40	1
C_6H_{12}	4-Methyl-1-pentene	2.5	2
C_6H_{12}	Cyclohexane	3.44	1
C_6H_{12}	Methylcyclopentane	3.37	2
$C_6H_{12}O$	Cyclohexanol	1.23	1
$C_6H_{12}O$	Hexanal	1.78	1
$C_6H_{12}O$	2-Hexanone	1.38	1
$C_6H_{12}O$	4-Methyl-2-pentanone	1.31	1
$C_6H_{12}O_2$	Hexanoic acid	1.92	1
$C_6H_{12}O_2$	Butyl acetate	1.82	1
$C_6H_{13}Br$	1-Bromohexane	3.80	1
$C_6H_{13}N$	Cyclohexylamine	1.49	1
C_6H_{14}	Hexane	4.00	1
C_6H_{14}	3-Methylpentane	3.60	2
C_6H_{14}	2,2-Dimethylbutane	3.82	1
C_6H_{14}	2,3-Dimethylbutane	3.85	2
$C_6H_{14}O$	1-Hexanol	2.03	1
$C_6H_{14}O$	2-Hexanol	1.76	1
$C_6H_{14}O$	3-Hexanol	1.65	1
$C_6H_{14}O$	3,3-Dimethyl-2-butanol	1.48	1
$C_6H_{14}O$	Dipropyl ether	2.03	1
$C_6H_{14}O$	Diisopropyl ether	1.52	1
$C_6H_{15}N$	Hexylamine	2.06	1
$C_6H_{15}N$	Dipropylamine	1.67	1
$C_6H_{15}N$	Triethylamine	1.45	1
$C_7H_5BrO_2$	2-Bromobenzoic acid	2.20	4
$C_7H_5BrO_2$	3-Bromobenzoic acid	2.87	4
$C_7H_5BrO_2$	4-Bromobenzoic acid	2.86	4
C_7H_5N	Benzonitrile	1.56	1
C_7H_6O	Benzaldehyde	1.48	1
$C_7H_6O_2$	Benzoic acid	1.88	4
$C_7H_6O_2$	Phenyl formate	1.26	1
$C_7H_6O_3$	Salicylic acid	2.20	4
C_7H_7Br	(Bromomethyl)benzene	2.92	1
C_7H_7Cl	o-Chlorotoluene	3.42	1
C_7H_7Cl	*m*-Chlorotoluene	3.28	1
C_7H_7Cl	*p*-Chlorotoluene	3.33	1
C_7H_7Cl	(Chloromethyl)benzene	2.30	1
$C_7H_7NO_2$	*p*-Nitrotoluene	2.42	1
C_7H_8	Toluene	2.73	1
C_7H_8	1,3,5-Cycloheptatriene	2.63	2
C_7H_8O	*o*-Cresol	1.98	1
C_7H_8O	*m*-Cresol	1.98	1
C_7H_8O	*p*-Cresol	1.97	1
C_7H_8O	Benzyl alcohol	1.05	1
C_7H_8O	Anisole	2.11	1
C_7H_9N	Benzylamine	1.09	1
C_7H_9N	*o*-Methylaniline	1.32	1
C_7H_9N	*m*-Methylaniline	1.40	1
C_7H_9N	*p*-Methylaniline	1.39	1
C_7H_9N	*N*-Methylaniline	1.66	1
C_7H_{14}	1-Heptene	3.99	1
C_7H_{14}	Methylcyclohexane	3.88	1
$C_7H_{14}O$	2-Heptanone	1.98	1
$C_7H_{14}O$	5-Methyl-2-hexanone	1.88	1
$C_7H_{15}Br$	1-Bromoheptane	4.36	1
$C_7H_{15}Cl$	1-Chloroheptane	4.15	1
$C_7H_{15}I$	1-Iodoheptane	4.70	1
C_7H_{16}	Heptane	4.50	1
$C_7H_{16}O$	1-Heptanol	2.62	1
$C_7H_{16}O$	2-Heptanol	2.31	1
$C_7H_{16}O$	3-Heptanol	2.24	1
$C_7H_{16}O$	4-Heptanol	2.22	1
$C_7H_{17}N$	Heptylamine	2.57	1
C_8H_6	Phenylacetylene	2.40	1
C_8H_6O	Benzofuran	2.67	1
C_8H_6S	Benzo[b]thiophene	3.12	1
C_8H_7N	Benzeneacetonitrile	1.56	1
C_8H_7N	Indole	2.14	1
C_8H_8	Styrene	3.05	1
C_8H_8O	Acetophenone	1.63	1
C_8H_8O	2-Methylbenzaldehyde	2.26	1
C_8H_8O	Benzeneacetaldehyde	1.78	1
C_8H_8O	2,3-Dihydrobenzofuran	2.14	1
C_8H_8O	Phenyloxirane	1.61	1
$C_8H_8O_2$	*o*-Toluic acid	2.32	4
$C_8H_8O_2$	*m*-Toluic acid	2.37	1
$C_8H_8O_2$	*p*-Toluic acid	2.34	1
$C_8H_8O_2$	Benzeneacetic acid	1.41	1
$C_8H_8O_2$	Phenyl acetate	1.49	1
$C_8H_8O_2$	Methyl benzoate	2.20	1
C_8H_{10}	Ethylbenzene	3.15	1
C_8H_{10}	*o*-Xylene	3.12	1
C_8H_{10}	*m*-Xylene	3.20	1
C_8H_{10}	*p*-Xylene	3.15	1
$C_8H_{10}O$	*o*-Ethylphenol	2.47	1
$C_8H_{10}O$	*m*-Ethylphenol	2.50	1
$C_8H_{10}O$	*p*-Ethylphenol	2.50	1
$C_8H_{10}O$	2,4-Xylenol	2.35	1
$C_8H_{10}O$	2,5-Xylenol	2.34	1
$C_8H_{10}O$	2,6-Xylenol	2.36	1
$C_8H_{10}O$	3,4-Xylenol	3.23	1
$C_8H_{10}O$	3,5-Xylenol	2.35	1
$C_8H_{10}O$	Benzeneethanol	1.36	1
$C_8H_{10}O$	α-Methylbenzyl alcohol	1.42	1
$C_8H_{10}O$	3-Methylbenzenemethanol	1.60	1
$C_8H_{10}O$	4-Methylbenzenemethanol	1.58	1
$C_8H_{10}O$	Phenetole	2.51	1
$C_8H_{10}O$	Benzyl methyl ether	1.35	1
$C_8H_{10}O$	2-Methylanisole	2.74	1
$C_8H_{10}O$	3-Methylanisole	2.66	1
$C_8H_{10}O$	4-Methylanisole	2.81	1
$C_8H_{11}N$	*p*-Ethylaniline	1.96	1
$C_8H_{11}N$	*N,N*-Dimethylaniline	2.31	1
$C_8H_{11}N$	Benzeneethanamine	1.41	1
$C_8H_{14}O_2$	Butyl methacrylate	2.88	1
$C_8H_{15}N$	Octanenitrile	2.75	1
C_8H_{16}	1-Octene	4.57	1
C_8H_{16}	Cyclooctane	4.45	2
$C_8H_{16}O$	2-Octanone	2.37	1
$C_8H_{16}O_2$	Octanoic acid	3.05	1
$C_8H_{17}Br$	1-Bromooctane	4.89	1
C_8H_{18}	Octane	5.15	1
$C_8H_{18}O$	1-Octanol	3.07	1
$C_8H_{18}O$	2-Octanol	2.90	1
$C_8H_{18}O$	4-Octanol	2.68	1
$C_8H_{18}O$	Dibutyl ether	3.21	1
C_9H_7N	Quinoline	2.03	1

OCTANOL-WATER PARTITION COEFFICIENTS (continued)

Mol. Form.	Name	log P	Ref.
C_9H_7N	Isoquinoline	2.08	1
C_9H_8	Indene	2.92	1
$C_9H_8O_2$	*trans*-Cinnamic acid	2.13	1
C_9H_9N	Benzenepropanenitrile	1.72	1
C_9H_{10}	Indan	3.33	1
$C_9H_{10}O$	1-Phenyl-1-propanone	2.19	1
$C_9H_{10}O$	1-Phenyl-2-propanone	1.44	1
$C_9H_{10}O$	4-Methylacetophenone	2.19	1
$C_9H_{10}O_2$	2-Phenylpropanoic acid	1.80	1
$C_9H_{10}O_2$	Benzyl acetate	1.96	1
$C_9H_{10}O_2$	4-Methylphenyl acetate	2.11	1
$C_9H_{10}O_2$	Ethyl benzoate	2.64	1
C_9H_{12}	Propylbenzene	3.69	1
C_9H_{12}	Isopropylbenzene	3.66	1
C_9H_{12}	*o*-Ethyltoluene	3.53	1
C_9H_{12}	*p*-Ethyltoluene	3.63	2
C_9H_{12}	1,2,3-Trimethylbenzene	3.60	1
C_9H_{12}	1,2,4-Trimethylbenzene	3.63	1
C_9H_{12}	1,3,5-Trimethylbenzene	3.42	1
$C_9H_{12}O$	2-Propylphenol	2.93	1
$C_9H_{12}O$	4-Propylphenol	3.20	1
$C_9H_{12}O$	2,3,6-Trimethylphenol	2.67	1
$C_9H_{12}O$	2,4,6-Trimethylphenol	2.46	1
$C_9H_{12}O$	Benzenepropanol	1.88	1
$C_9H_{13}N$	*N,N*-Dimethylbenzylamine	1.98	1
$C_9H_{13}N$	Amphetamine	1.76	1
C_9H_{18}	1-Nonene	5.15	1
$C_9H_{18}O$	2-Nonanone	3.16	1
$C_9H_{18}O$	5-Methyl-2-octanone	2.92	1
C_9H_{20}	Nonane	5.65	1
$C_9H_{20}O$	1-Nonanol	4.02	1
$C_9H_{21}N$	Tripropylamine	2.79	1
$C_{10}H_7Cl$	1-Chloronaphthalene	3.90	1
$C_{10}H_7Cl$	2-Chloronaphthalene	3.98	1
$C_{10}H_8$	Naphthalene	3.34	4
$C_{10}H_8$	Azulene	3.22	1
$C_{10}H_8O$	1-Naphthol	2.84	1
$C_{10}H_8O$	2-Naphthol	2.70	1
$C_{10}H_{12}O_2$	Isopropyl benzoate	3.18	1
$C_{10}H_{14}$	Butylbenzene	4.26	1
$C_{10}H_{14}$	*tert*-Butylbenzene	4.11	1
$C_{10}H_{14}$	Isobutylbenzene	4.01	2
$C_{10}H_{14}$	*p*-Cymene	4.10	1
$C_{10}H_{14}$	1,2,4,5-Tetramethylbenzene	4.10	2
$C_{10}H_{14}$	1,2,3,4-Tetramethylbenzene	4.00	1
$C_{10}H_{14}$	1,2,3,5-Tetramethylbenzene	4.10	1
$C_{10}H_{14}O$	4-Butylphenol	3.65	1
$C_{10}H_{20}O$	2-Decanone	3.77	1
$C_{10}H_{20}O_2$	Decanoic acid	4.09	1
$C_{10}H_{22}$	Decane	6.25	1
$C_{10}H_{22}O$	1-Decanol	4.57	1
$C_{11}H_9N$	4-Phenylpyridine	2.59	1
$C_{11}H_{10}$	1-Methylnaphthalene	3.87	1
$C_{11}H_{10}$	2-Methylnaphthalene	4.00	1
$C_{11}H_{16}$	Pentylbenzene	4.90	1
$C_{11}H_{16}$	Pentamethylbenzene	4.56	1
$C_{11}H_{22}O$	2-Undecanone	4.09	1
$C_{11}H_{22}O_2$	Methyl decanoate	4.41	1
$C_{12}Cl_{10}$	Decachlorobiphenyl	8.26	3
$C_{12}HCl_9$	2,2′,3,3′,4,5,5′,6,6′-Nonachlorobiphenyl	8.16	3
$C_{12}H_2Cl_8$	2,2′,3,3′,5,5′,6,6′-Octachlorobiphenyl	7.10	3
$C_{12}H_3Cl_7$	2,2′,3,3′,4,4′,6-Heptachlorobiphenyl	6.70	3
$C_{12}H_4Cl_6$	2,2′,3,3′,4,4′-Hexachlorobiphenyl	7.00	3
$C_{12}H_4Cl_6$	2,2′,4,4′,6,6′-Hexachlorobiphenyl	7.00	3
$C_{12}H_4Cl_6$	2,2′,3,3′,6,6′-Hexachlorobiphenyl	6.70	3
$C_{12}H_5Cl_5$	2,3,4,5,6-Pentachlorobiphenyl	6.30	3
$C_{12}H_5Cl_5$	2,2′,4,5,5′-Pentachlorobiphenyl	6.40	3
$C_{12}H_6Cl_4$	2,3,4,5-Tetrachlorobiphenyl	5.72	3
$C_{12}H_6Cl_4$	2,2′,4′,5-Tetrachlorobiphenyl	5.73	7
$C_{12}H_7Cl_3$	2,4,5-Trichlorobiphenyl	5.60	3
$C_{12}H_7Cl_3$	2,4,6-Trichlorobiphenyl	5.47	3
$C_{12}H_8Cl_2$	2,5-Dichlorobiphenyl	5.10	3
$C_{12}H_8Cl_2$	2,6-Dichlorobiphenyl	5.00	3
$C_{12}H_8O$	Dibenzofuran	4.12	1
$C_{12}H_9Cl$	2-Chlorobiphenyl	4.52	1
$C_{12}H_9Cl$	3-Chlorobiphenyl	4.58	1
$C_{12}H_9Cl$	4-Chlorobiphenyl	4.61	1
$C_{12}H_9N$	Carbazole	3.72	1
$C_{12}H_{10}$	Acenaphthene	3.96	4
$C_{12}H_{10}$	Biphenyl	3.76	6
$C_{12}H_{10}N_2$	Azobenzene	3.82	1
$C_{12}H_{10}O$	Diphenyl ether	4.21	1
$C_{12}H_{10}S$	Diphenyl sulfide	4.45	1
$C_{12}H_{11}N$	Diphenylamine	3.44	4
$C_{12}H_{12}$	1-Ethylnaphthalene	4.40	1
$C_{12}H_{12}$	1,2-Dimethylnaphthalene	4.31	1
$C_{12}H_{12}$	1,4-Dimethylnaphthalene	4.37	1
$C_{12}H_{14}O$	4-Phenylcyclohexanone	2.45	1
$C_{12}H_{18}$	Hexylbenzene	5.52	1
$C_{12}H_{18}$	Hexamethylbenzene	4.69	4
$C_{12}H_{22}O$	Cyclododecanone	4.10	1
$C_{12}H_{24}O_2$	Dodecanoic acid	4.6	1
$C_{12}H_{26}O$	1-Dodecanol	5.13	1
$C_{13}H_8O$	9H-Fluoren-9-one	3.58	1
$C_{13}H_9N$	Acridine	3.40	1
$C_{13}H_{10}$	9H-Fluorene	4.20	4
$C_{13}H_{10}O$	Benzophenone	3.18	1
$C_{13}H_{10}O_2$	Phenyl benzoate	3.59	1
$C_{13}H_{11}NO$	*N*-Phenylbenzamide	2.62	1
$C_{13}H_{12}$	Diphenylmethane	4.14	1
$C_{13}H_{12}$	4-Methylbiphenyl	4.63	1
$C_{13}H_{12}O$	Diphenylmethanol	2.67	1
$C_{13}H_{12}O$	Benzyl phenyl ether	3.79	1
$C_{14}H_{10}$	Anthracene	4.56	4
$C_{14}H_{10}$	Phenanthrene	4.52	4
$C_{14}H_{12}$	*trans*-Stilbene	4.81	1
$C_{14}H_{12}$	1-Methylfluorene	4.97	1
$C_{14}H_{12}O$	2-Phenylacetophenone	3.18	1
$C_{14}H_{12}O_2$	Benzyl benzoate	3.97	1
$C_{14}H_{14}$	1,2-Diphenylethane	4.70	1
$C_{14}H_{14}$	4,4′-Dimethylbiphenyl	5.09	1
$C_{14}H_{22}$	Octylbenzene	6.30	1
$C_{14}H_{28}O_2$	Tetradecanoic acid	6.1	1
$C_{15}H_{12}$	2-Methylanthracene	5.15	2
$C_{15}H_{12}$	9-Methylanthracene	5.07	1
$C_{15}H_{12}$	1-Methylphenanthrene	5.14	2
$C_{16}H_{10}$	Fluoranthene	5.07	4
$C_{16}H_{10}$	Pyrene	5.08	4
$C_{16}H_{14}$	9,10-Dimethylanthracene	5.69	1
$C_{16}H_{32}O_2$	Hexadecanoic acid	7.17	1

Mol. Form.	Name	log *P*	Ref.
$C_{17}H_{12}$	11H-Benzo[a]fluorene	5.40	1
$C_{17}H_{12}$	11H-Benzo[b]fluorene	5.75	1
$C_{18}H_{12}$	Benz[a]anthracene	5.91	1
$C_{18}H_{12}$	Chrysene	5.73	4
$C_{18}H_{12}$	Naphthacene	5.76	1
$C_{18}H_{12}$	Triphenylene	5.49	4
$C_{18}H_{15}N$	Triphenylamine	5.74	1
$C_{18}H_{30}O_2$	Linolenic acid	6.46	1
$C_{18}H_{32}O_2$	Linoleic acid	7.05	1
$C_{18}H_{34}O_2$	Oleic acid	7.64	1
$C_{18}H_{36}O_2$	Stearic acid	8.23	1
$C_{19}H_{16}O$	Triphenylmethanol	3.68	1
$C_{20}H_{12}$	Perylene	6.25	1
$C_{20}H_{12}$	Benzo[a]pyrene	6.20	4
$C_{20}H_{32}O_2$	Arachidonic acid	6.98	1
$C_{20}H_{40}O_2$	Arachidic acid	9.29	1
$C_{21}H_{16}$	1,2-Dihydro-3-methylbenz[j]aceanthrylene	6.75	1
$C_{22}H_{12}$	Benzo[ghi]perylene	6.90	1
$C_{24}H_{12}$	Coronene	6.05	4

PROTECTION AGAINST IONIZING RADIATION

The following data and rules of thumb are helpful in estimating the penetrating capability of and danger of exposure to various types of ionizing radiation. More precise data should be used for critical applications.

Alpha Particles

Alpha particles of at least 7.5 MeV are required to penetrate the epidermis, the protective layer of skin, 0.07 mm thick.

Electrons

Electrons of at least 70 keV are required to penetrate the epidermis, the protective layer of skin, 0.07 mm thick.

The range of electrons in g/cm^2 is approximately equal to the maximum energy (E) in MeV divided by 2.

The range of electrons in air is about 3.65 m per MeV; for example, a 3 MeV electron has a range of about 11 m in air.

A chamber wall thickness of 30 mg/cm^2 will transmit 70% of the initial fluence of 1 MeV electrons and 20% of that of 0.4 MeV electrons.

When electrons of 1 to 2 MeV pass through light materials such as water, aluminum, or glass, less than 1% of their energy is dissipated as bremsstrahlung.

The bremsstrahlung from 1 Ci of ^{32}P aqueous solution in a glass bottle is about 1 mR/h at 1 meter distance.

When electrons from a 1 Ci source of ^{90}Sr - ^{90}Y are absorbed, the bremsstrahlung hazard is approximately equal to that presented by the gamma radiation from 12 mg of radium. The average energy of the bremsstrahlung is about 300 keV.

Gamma Rays

The air-scattered radiation (sky-shine) from a 100 Ci ^{60}Co source placed 1 ft behind a 4 ft high shield is about 100 mrad/h at 6 ft from the outside of the shield.

Within ±20% for point source gamma emitters with energies between 0.07 and 4 MeV, the exposure rate (R/h) at 1 ft is $6C{\cdot}E{\cdot}n$ where C is the activity in curies, E is the energy in MeV, and n is the number of gammas per disintegration.

Neutrons

An approximate HVL (thickness of absorber for which the neutron flux falls to half its initial value) for 1 MeV neutrons is 3.2 cm of paraffin; that for 5 MeV neutrons is 6.9 cm of paraffin).

Miscellaneous

The activity of any radionuclide is reduced to less than 1% after 7 half-lives (i.e., $2^{-7} = 0.8\%$).

For nuclides with a half-life greater than 6 days, the change in activity in 24 hours will be less than 10%.

10 HVL (half-value layers) attenuates approximately by 10^{-3}.

There is 0.64 mm^3 of radon gas at STP in transient equilibrium with 1 Ci of radium.

The natural background from all sources in most parts of the world leads to an equivalent dose rate of about 0.04 to 4 mSv per year for the average person. About 84% of this comes from terrestrial sources, the remainder from cosmic rays. The U. S. average is about 3.6 mSv/yr but can range up to 50 mSv/yr in some areas. A passenger in a plane flying at 12,000 meters receives 5 μSv/hr from cosmic rays (as compared to about 0.03 μSv/hr at sea level).

The ICRP recommended exposure limit to man-made sources of ionizing radiation (Reference 2) is 20 mSv/yr averaged over 5 years, with the dose in any one year not to exceed 50 mSv.

A whole-body dose of about 3 Gy over a short time interval will typically lead to 50% mortality in 30 days assuming no medical treatment.

Units

The gray (Gy) is the SI unit of absorbed dose; it is a measure of the mean energy imparted to a sample of irradiated matter, divided by the mass of the sample. Gy is a special name for the SI unit J/kg.

The sievert (Sv) is the SI unit of equivalent dose, which is defined as the absorbed dose multiplied by a weighting factor that expresses the long-term biological risk from low-level chronic exposure to a specified type of radiation. The Sv is another special name for J/kg.

1 curie (Ci) = $3.7{\cdot}10^{10}$ becquerel (Bq); i.e., $3.7{\cdot}10^{10}$ disintegrations per second.

1 roentgen (R) = $2.58{\cdot}10^{-4}$ coulomb per kilogram (C/kg); a measure of the charge (positive or negative) liberated by x-ray or gamma radiation in air, divided by the mass of air.

1 rad = 0.01 Gy

1 rem = 0.01 Sv

REFERENCES

1. Padikal, T.N., and Fivozinsky, S.P., *Medical Physics Data Book, National Bureau of Standards Handbook 138*, U. S. Government Printing Office, Washington, D.C., 1981.
2. *1990 Recommendations of the International Commission on Radiological Protection*, ICRP Publication 60, *Annals of the ICRP*, Pergamon Press, Oxford, 1991.
3. *Radiation: Doses, Effects, Risks*, United Nations Sales No. E.86.III.D.4, 1985.
4. *Review of Particle Properties, Phys. Rev. D*, 50, 1173, 1994 (p. 1268).

ANNUAL LIMITS ON INTAKES OF RADIONUCLIDES

K. F. Eckerman

The following table lists, for workers, the annual limits on oral and inhalation intakes (ALI) for selected radionuclides based on the occupational radiation protection guidance of the International Commission on Radiological Protection (References 1 and 2). An intake of one ALI corresponds to an annual whole body dose of 0.02 Sv (2 rem).

The ALI is expressed in the SI unit of activity, the becquerel (Bq), and in the conventional unit, the microcurie (μCi); 1 μCi = $3.7 \cdot 10^4$ Bq. The chemical form of inhaled radionuclides is, in most instances, stated in terms of the rate of absorption to blood from the lungs and the fractional absorption from the small intestine. Type F, M, and S denote chemical forms which are absorbed from the lungs at rates characterized as fast, moderate, and slow, respectively. The time to absorb 90% of the deposited radionuclide, in the absence of radioactive decay, corresponds to about 10 minutes, 150 days, and 7000 days for Type F, M, and S compounds, respectively. Type F compounds can be considered to be more soluble than M or S, S being the most insoluble. Chemical form consideration for ingestion is specified by the fractional absorption from the small intestine, denoted as f_1. The f_1 values range from 10^{-5} to 1. Higher fractional absorption is associated with greater solubility of the compound.

REFERENCES

1. *1990 Recommendations of the International Commission on Radiological Protection, ICRP Publication 60, Annals of the ICRP 21, (1—3),* Pergamon Press, Oxford, 1991.
2. *Dose Coefficients for Intakes of Radionuclides by Workers, ICRP Publication 68, Annals of the ICRP*, 24(4), Pergamon Press, Oxford, 1995.

		Inhalation intakes			Oral intakes		
	Physical half-life	Chemical form Type/f_1	ALI Bq	ALI μCi	Chemical form f_1	ALI Bq	ALI μCi
^{3}H	12.3 y	HT gas	1.1E+13	3.0E+08	1.000	1.1E+13	3.0E+08
		HTO vapor	1.1E+09	3.0E+04			
^{11}C	0.340 h	CO	1.7E+10	4.5E+05	1.000	8.3E+08	2.3E+04
		CO_2	9.1E+09	2.5E+05			
		Organic compounds	6.2E+09	1.7E+05			
^{14}C	5730 y	CO	2.5E+10	6.8E+05	1.000	3.4E+07	9.3E+02
		CO_2	3.1E+09	8.3E+04			
		Organic compounds	3.4E+07	9.3E+02			
^{18}F	1.83 h	F 1.000	3.7E+08	1.0E+04	1.000	4.1E+08	1.1E+04
		M 1.000	2.2E+08	6.1E+03			
		S 1.000	2.2E+08	5.8E+03			
^{22}Na	2.60 y	F 1.000	1.0E+07	2.7E+02	1.000	6.3E+06	1.7E+02
^{24}Na	15.0 h	F 1.000	3.8E+07	1.0E+03	1.000	4.7E+07	1.3E+03
^{32}P	14.3 d	F 0.800	1.8E+07	4.9E+02	0.800	8.3E+06	2.3E+02
		M 0.800	6.9E+06	1.9E+02			
^{35}S	87.4 d	Inorganic compounds					
		F 0.800	2.5E+08	6.8E+03	0.800	1.4E+08	3.9E+03
		M 0.800	1.8E+07	4.9E+02	0.100	1.1E+08	2.8E+03
		Vapor	1.7E+08	4.5E+03			
		Organic compounds			1.000	2.6E+07	7.0E+02
^{42}K	12.4 h	F 1.000	1.0E+08	2.7E+03	1.000	4.7E+07	1.3E+03
^{43}K	22.6 h	F 1.000	7.7E+07	2.1E+03	1.000	8.0E+07	2.2E+03
^{45}Ca	163 d	M 0.300	8.7E+06	2.4E+02	0.300	2.6E+07	7.1E+02
^{47}Ca	4.53 d	M 0.300	9.5E+06	2.6E+02	0.300	1.3E+07	3.4E+02
^{51}Cr	27.7 d	F 0.100	6.7E+08	1.8E+04	0.100	5.3E+08	1.4E+04
		M 0.100	5.9E+08	1.6E+04	0.010	5.4E+08	1.5E+04
		S 0.100	5.6E+08	1.5E+04			
^{54}Mn	312 d	F 0.100	1.8E+07	4.9E+02	0.100	2.8E+07	7.6E+02
		M 0.100	1.7E+07	4.5E+02			
^{52}Fe	8.28 h	F 0.100	2.9E+07	7.8E+02	0.100	1.4E+07	3.9E+02
		M 0.100	2.1E+07	5.7E+02			
^{55}Fe	2.70 y	F 0.100	2.2E+07	5.9E+02	0.100	6.1E+07	1.6E+03
		M 0.100	6.1E+07	1.6E+03			

	Physical half-life	Inhalation intakes: Chemical form Type/f_1	Inhalation ALI Bq	Inhalation ALI μCi	Oral intakes: Chemical form f_1	Oral ALI Bq	Oral ALI μCi
^{59}Fe	44.5 d	F 0.100	6.7E+06	1.8E+02	0.100	1.1E+07	3.0E+02
		M 0.100	6.3E+06	1.7E+02			
^{57}Co	271 d	M 0.100	5.1E+07	1.4E+03	0.100	9.5E+07	2.6E+03
		S 0.050	3.3E+07	9.0E+02	0.050	1.1E+08	2.8E+03
^{58}Co	70.8 d	M 0.100	1.4E+07	3.9E+02	0.100	2.7E+07	7.3E+02
		S 0.050	1.2E+07	3.2E+02	0.050	2.9E+07	7.7E+02
^{60}Co	5.27 y	M 0.100	2.8E+06	7.6E+01	0.100	5.9E+06	1.6E+02
		S 0.050	1.2E+06	3.2E+01	0.050	8.0E+06	2.2E+02
^{64}Cu	12.7 h	F 0.500	2.9E+08	7.9E+03	0.500	1.7E+08	4.5E+03
		M 0.500	1.3E+08	3.6E+03			
		S 0.500	1.3E+08	3.6E+03			
^{59}Ni	75000 y	F 0.050	9.1E+07	2.5E+03	0.050	3.2E+08	8.6E+03
		M 0.050	2.1E+08	5.8E+03			
		Vapor	2.4E+07	6.5E+02			
^{63}Ni	96.0 y	F 0.050	3.8E+07	1.0E+03	0.050	1.3E+08	3.6E+03
		M 0.050	6.5E+07	1.7E+03			
		Vapor	1.0E+07	2.7E+02			
^{65}Zn	244 d	S 0.500	7.1E+06	1.9E+02	0.500	5.1E+06	1.4E+02
^{67}Ga	3.26 d	F 0.001	1.8E+08	4.9E+03	0.001	1.1E+08	2.8E+03
		M 0.001	7.1E+07	1.9E+03			
^{68}Ga	1.13 h	F 0.001	4.1E+08	1.1E+04	0.001	2.0E+08	5.4E+03
		M 0.001	2.5E+08	6.7E+03			
^{68}Ge	288 d	F 1.000	2.4E+07	6.5E+02	1.000	1.5E+07	4.2E+02
		M 1.000	2.5E+06	6.8E+01			
^{75}Se	120 d	F 0.800	1.4E+07	3.9E+02	0.800	7.7E+06	2.1E+02
		M 0.800	1.2E+07	3.2E+02	0.050	4.9E+07	1.3E+03
^{79}Se	65000 y	F 0.800	1.3E+07	3.4E+02	0.800	6.9E+06	1.9E+02
		M 0.800	6.5E+06	1.7E+02	0.050	5.1E+07	1.4E+03
^{86}Rb	18.6 d	F 1.000	1.5E+07	4.2E+02	1.000	7.1E+06	1.9E+02
^{85}Sr	64.8 d	F 0.300	3.6E+07	9.7E+02	0.300	3.6E+07	9.7E+02
		S 0.010	3.1E+07	8.4E+02	0.010	6.1E+07	1.6E+03
^{87m}Sr	2.80 h	F 0.300	9.1E+08	2.5E+04	0.300	6.7E+08	1.8E+04
		S 0.010	5.7E+08	1.5E+04	0.010	6.1E+08	1.6E+04
^{89}Sr	50.5 d	F 0.300	1.4E+07	3.9E+02	0.300	7.7E+06	2.1E+02
		S 0.010	3.6E+06	9.7E+01	0.010	8.7E+06	2.4E+02
^{90}Sr	29.1 y	F 0.300	6.7E+05	1.8E+01	0.300	7.1E+05	1.9E+01
		S 0.010	2.6E+05	7.0E+00	0.010	7.4E+06	2.0E+02
^{99}Mo	2.75 d	F 0.800	5.6E+07	1.5E+03	0.800	2.7E+07	7.3E+02
		S 0.050	1.8E+07	4.9E+02	0.050	1.7E+07	4.5E+02
^{99m}Tc	6.02 h	F 0.800	1.0E+09	2.7E+04	0.800	9.1E+08	2.5E+04
		M 0.800	6.9E+08	1.9E+04			
^{99}Tc	213000 y	F 0.800	5.0E+07	1.4E+03	0.800	2.6E+07	6.9E+02
		M 0.800	6.3E+06	1.7E+02			
^{106}Ru	1.01 y	F 0.050	2.0E+06	5.5E+01	0.050	2.9E+06	7.7E+01
		M 0.050	1.2E+06	3.2E+01			
		S 0.050	5.7E+05	1.5E+01			
^{111}In	2.83 d	F 0.020	9.1E+07	2.5E+03	0.020	6.9E+07	1.9E+03
		M 0.020	6.5E+07	1.7E+03			
^{113m}In	1.66 h	F 0.020	1.1E+09	2.8E+04	0.020	7.1E+08	1.9E+04
		M 0.020	6.3E+08	1.7E+04			
^{113}Sn	115 d	F 0.020	2.5E+07	6.8E+02	0.020	2.7E+07	7.4E+02
		M 0.020	1.1E+07	2.8E+02			
^{123}I	13.2 h	F 1.000	1.8E+08	4.9E+03	1.000	9.5E+07	2.6E+03
		Vapor	9.5E+07	2.6E+03			
^{125}I	60.1 d	F 1.000	2.7E+06	7.4E+01	1.000	1.3E+06	3.6E+01
		Vapor	1.4E+06	3.9E+01			
^{129}I	$1.57 \cdot 10^7$ y	F 1.000	3.9E+05	1.1E+01	1.000	1.8E+05	4.9E+00

		Inhalation intakes			Oral intakes		
	Physical half-life	Chemical form Type/f_1	ALI Bq	ALI μCi	Chemical form f_1	ALI Bq	ALI μCi
		Vapor	2.1E+05	5.6E+00			
^{131}I	8.04 d	F 1.000	1.8E+06	4.9E+01	1.000	9.1E+05	2.5E+01
		Vapor	1.0E+06	2.7E+01			
^{129}Cs	1.34 d	F 1.000	2.5E+08	6.7E+03	1.000	3.3E+08	9.0E+03
^{134}Cs	2.06 y	F 1.000	2.1E+06	5.6E+01	1.000	1.1E+06	2.8E+01
^{136}Cs	13.1 d	F 1.000	1.1E+07	2.8E+02	1.000	6.7E+06	1.8E+02
^{137}Cs	30.0 y	F 1.000	3.0E+06	8.1E+01	1.000	1.5E+06	4.2E+01
^{141}Ce	32.5 d	M 5.0E-04	7.4E+06	2.0E+02	5.0E-04	2.8E+07	7.6E+02
		S 5.0E-04	6.5E+06	1.7E+02			
^{144}Ce	284 d	M 5.0E-04	8.7E+05	2.4E+01	5.0E-04	3.8E+06	1.0E+02
		S 5.0E-04	6.9E+05	1.9E+01			
^{133}Ba	10.7 y	F 0.100	1.1E+07	3.0E+02	0.100	2.0E+07	5.4E+02
^{140}Ba	12.7 d	F 0.100	1.3E+07	3.4E+02	0.100	8.0E+06	2.2E+02
^{169}Yb	32.0 d	M 5.0E-04	9.5E+06	2.6E+02	5.0E-04	2.8E+07	7.6E+02
		S 5.0E-04	8.3E+06	2.3E+02			
^{198}Au	2.69 d	F 0.100	5.1E+07	1.4E+03	0.100	2.0E+07	5.4E+02
		M 0.100	2.0E+07	5.5E+02			
		S 0.100	1.8E+07	4.9E+02			
^{198m}Au	2.30 d	F 0.100	3.4E+07	9.2E+02	0.100	1.5E+07	4.2E+02
		M 0.100	1.0E+07	2.7E+02			
		S 0.100	1.1E+07	2.8E+02			
^{197}Hg	2.67 d	Inorganic compounds					
		F 0.400	2.4E+08	6.4E+03	1.000	2.0E+08	5.5E+03
					0.400	1.2E+08	3.2E+03
		Vapor	4.5E+06	1.2E+02			
		Organic compounds					
		F 0.020	2.0E+08	5.4E+03	0.020	8.7E+07	2.4E+03
		M 0.020	7.1E+07	1.9E+03			
^{203}Hg	46.6 d	Inorganic compounds					
		F 0.400	2.7E+07	7.2E+02	1.000	1.1E+07	2.8E+02
					0.400	1.8E+07	4.9E+02
		Vapor	2.9E+06	7.7E+01			
		Organic compounds					
		F 0.020	3.4E+07	9.2E+02	0.020	3.7E+07	1.0E+03
		M 0.020	1.1E+07	2.8E+02			
^{201}Tl	3.04 d	F 1.000	2.6E+08	7.1E+03	1.000	2.1E+08	5.7E+03
^{210}Pb	22.3 y	F 0.200	1.8E+04	4.9E-01	0.200	2.9E+04	7.9E-01
^{207}Bi	38.0 y	F 0.050	2.4E+07	6.4E+02	0.050	1.5E+07	4.2E+02
		M 0.050	6.3E+06	1.7E+02			
^{210}Po	138 d	F 0.100	2.8E+04	7.6E-01	0.100	8.3E+04	2.3E+00
		M 0.100	9.1E+03	2.5E-01			
^{224}Ra	3.66 d	M 0.200	8.3E+03	2.3E-01	0.200	3.1E+05	8.3E+00
^{226}Ra	1600 y	M 0.200	1.7E+03	4.5E-02	0.200	7.1E+04	1.9E+00
^{228}Ra	5.75 y	M 0.200	1.2E+04	3.2E-01	0.200	3.0E+04	8.1E-01
^{228}Th	1.91 y	M 5.0E-04	8.7E+02	2.4E-02	5.0E-04	2.9E+05	7.7E+00
		S 2.0E-04	6.3E+02	1.7E-02	2.0E-04	5.7E+05	1.5E+01
^{230}Th	77000 y	M 5.0E-04	7.1E+02	1.9E-02	5.0E-04	9.5E+04	2.6E+00
		S 2.0E-04	2.8E+03	7.5E-02	2.0E-04	2.3E+05	6.2E+00
^{232}Th	$1.40 \cdot 10^{10}$ y	M 5.0E-04	6.9E+02	1.9E-02	5.0E-04	9.1E+04	2.5E+00
		S 2.0E-04	1.7E+03	4.5E-02	2.0E-04	2.2E+05	5.9E+00
^{234}U	$2.44 \cdot 10^{5}$ y	F 0.020	3.1E+04	8.4E-01	0.020	4.1E+05	1.1E+01
		M 0.020	9.5E+03	2.6E-01	0.002	2.4E+06	6.5E+01
		S 0.002	2.9E+03	7.9E-02			

ANNUAL LIMITS ON INTAKES OF RADIONUCLIDES (continued)

	Physical half-life	Inhalation intakes: Chemical form Type/f_1	Inhalation intakes: ALI Bq	Inhalation intakes: ALI μCi	Oral intakes: Chemical form f_1	Oral intakes: ALI Bq	Oral intakes: ALI μCi
^{235}U	7.04·10^8 y	F 0.020	3.3E+04	9.0E-01	0.020	4.3E+05	1.2E+01
		M 0.020	1.1E+04	3.0E-01	0.002	2.4E+06	6.5E+01
		S 0.002	3.3E+03	8.9E-02			
^{238}U	4.47·10^9 y	F 0.020	3.4E+04	9.3E-01	0.020	4.5E+05	1.2E+01
		M 0.020	1.3E+04	3.4E-01	0.002	2.6E+06	7.1E+01
		S 0.002	3.5E+03	9.5E-02			
^{237}Np	2.14·10^6 y	M 5.0E-04	1.3E+03	3.6E-02	5.0E-04	1.8E+05	4.9E+00
^{239}Np	2.36 d	M 5.0E-04	1.8E+07	4.9E+02	5.0E-04	2.5E+07	6.8E+02
^{238}Pu	87.7 y	M 5.0E-04	6.7E+02	1.8E-02	5.0E-04	8.7E+04	2.4E+00
		S 1.0E-05	1.8E+03	4.9E-02	1.0E-05	2.3E+06	6.1E+01
					1.0E-04	4.1E+05	1.1E+01
^{239}Pu	24100 y	M 5.0E-04	6.3E+02	1.7E-02	5.0E-04	8.0E+04	2.2E+00
		S 1.0E-05	2.4E+03	6.5E-02	1.0E-05	2.2E+06	6.0E+01
					1.0E-04	3.8E+05	1.0E+01
^{241}Pu	14.4 y	M 5.0E-04	3.4E+04	9.3E-01	5.0E-04	4.3E+06	1.2E+02
		S 1.0E-05	2.4E+05	6.4E+00	1.0E-05	1.8E+08	4.9E+03
					1.0E-04	2.1E+07	5.6E+02
^{241}Am	432 y	M 5.0E-04	7.4E+02	2.0E-02	5.0E-04	1.0E+05	2.7E+00
^{244}Cm	18.1 y	M 5.0E-04	1.2E+03	3.2E-02	5.0E-04	1.7E+05	4.5E+00
^{252}Cf	2.64 y	M 5.0E-04	1.5E+03	4.2E-02	5.0E-04	2.2E+05	6.0E+00

CHEMICAL CARCINOGENS

The following substances are listed in the *9th Report on Carcinogens 2000*, released by the National Institute of Environmental Health Sciences (NIEHS) under the National Toxicology Program (NTP). Substances are grouped in two classes:

- Known to be human carcinogens: There is sufficient evidence of carcinogenicity from studies in humans which indicates a causal relationship between exposure to the substance and human cancer.

- Reasonably anticipated to be human carcinogens: There is limited evidence of carcinogenicity from studies in humans which indicates that causal interpretation is credible, but that alternative explanations, such as chance, bias, or confounding factors, could not be adequately excluded; or there is sufficient evidence of carcinogenicity from studies in experimental animals.

The NTP report also lists many poorly defined materials such as soots, tars, mineral oils, coke oven emissions, etc. These materials are not included here.

The table lists the name normally used in the *Handbook of Chemistry and Physics*, followed by additional names by which the substance is known. In many cases the primary name given here is different from that used in the NTP report; however, names used in the NTP report appear in the *Other names* column. The Chemical Abstracts Service Registry Number (CAS RN), given in the last column, is taken from the NTP report. Extensive details on each substance are given in the reference.

REFERENCE

Public Health Service, National Toxicology Program, *9th Report on Carcinogens*, OCR Services, Inc., P.O. Box 12510, Research Triangle Park, NC 27709-2510. Also available on the Internet at http://ehis.niehs.nih.gov/roc/

Substance	Other names	CAS RN
Known to be Human Carcinogens		
Aflatoxins		1402-68-2
4-Aminobiphenyl	*p*-Biphenylamine	92-67-1
Arsenic compounds, inorganic		
Asbestos		1332-21-4
Azathioprine	1H-Purine, 6-[(1-methyl-4-nitro-1H-imidazol-5-yl)thio]-	446-86-6
Benzene		71-43-2
p-Benzidine	[1,1'-Biphenyl]-4,4'-diamine	92-87-5
Bis(2-chloroethyl) sulfide	Mustard gas	505-60-2
Bis(chloromethyl) ether		542-88-1
1,3-Butadiene		106-99-0
1,4-Butanediol dimethylsulfonate	Myleran; Busulfan	55-98-1
Cadmium and cadmium compounds		7440-43-9
Cadmium chloride		10108-64-2
Cadmium oxide		1306-19-0
Cadmium sulfate		10124-36-4
Cadmium sulfide		1306-23-6
Chlorambucil		305-03-3
Chloroethene	Vinyl chloride; Chloroethylene	75-01-4
1-(2-Chloroethyl)-3-(4-methylcyclohexyl)-1-nitrosourea	MeCCNU; Urea,	13909-09-6
Chloromethyl methyl ether		107-30-2
Chromium hexavalent compounds		
Cyclophosphamide	2H-1,3,2-Oxazaphosphorin-2-amine, *N,N*-bis(2-chloroethyl)tetrahydro-, 2-oxide	50-18-0
Cyclosporin A	Cyclosporine	59865-13-3
Diethylstilbestrol		56-53-1
Direct Black 38, disodium salt		1937-37-7
Direct Blue 6, tetrasodium salt		2602-46-2
Erionite		66733-21-9
Ethylene oxide	Oxirane	75-21-8
Lead(II) chromate		7758-97-6
Melphalan	*L*-Phenylalanine, 4-[bis(2-chloroethyl)amino]-	148-82-3
Methoxsalen (with UV therapy)	PUVA; 9-Methoxy-7H-furo[3,2-g][1]benzopyran-7-one	298-81-7
2-Naphthylamine	2-Aminonaphthalene; β-Naphthylamine	91-59-8
Piperazine estrone sulfate	Estrone, hydrogen sulfate, compd. with piperazine	7280-37-7
Radon		10043-92-2
Silicon dioxide (respirable size)	Quartz; Silica	14808-60-7
Silicon dioxide (respirable size)	Cristobalite; Silica	14464-46-1
Silicon dioxide (respirable size)	Tridymite; Silica	15468-32-3

CHEMICAL CARCINOGENS (continued)

Substance	Other names	CAS RN
Sodium equilin sulfate	Estra-1,3,5(10),7-tetraen-17-one, 3-hydroxy, hydrogen sulfate, sodium salt	16680-47-0
Sodium estrone sulfate	Estrone, hydrogen sulfate, sodium salt	438-67-5
Strontium chromate		7789-06-2
Tamoxifen		10540-29-1
Thorium(IV) oxide	Thorium dioxide	1314-20-1
Triethylenethiophosphoramide	Thiotepa; Tris(1-aziridinyl)phosphine, sulfide	52-24-4
Zinc chromate, basic		13530-65-9
Reasonably Anticipated to be Human Carcinogens		
Acetaldehyde	Ethanal	75-07-0
2-(Acetylamino)fluorene		53-96-3
Acrylamide	2-Propenamide	79-06-1
Acrylonitrile	Propenenitrile	107-13-1
Adriamycin, hydrochloride	Doxorubicin hydrochloride	25316-40-9
2-Amino-9,10-anthracenedione	2-Aminoanthraquinone	117-79-3
1-Amino-2-methyl-9,10-anthracenedione	1-Amino-2-methylanthraquinone	82-28-0
Arochlor 1254	Polychlorinated biphenyls (54% Cl)	11097-69-1
Arochlor 1260	Polychlorinated biphenyls (60% Cl)	11096-82-5
Azacitidine	5-Azacytidine; 1,3,5-Triazine-2(1H)-one, 4-amino-1-beta-*D*-ribofuranosyl-	320-67-2
Benz[a]anthracene		56-55-3
Benzo[b]fluoranthene	Benz[e]acephenanthrylene	205-99-2
Benzo[j]fluoranthene		205-82-3
Benzo[k]fluoranthene	2,3,1',8'-Binaphthylene	207-08-9
Benzo[a]pyrene		50-32-8
Beryllium and certain beryllium compounds		7440-41-7
Beryllium aluminum alloy		12770-50-2
Beryllium aluminum metasilicate	Beryl	1302-52-9
Beryllium chloride		7787-47-5
Beryllium fluoride		7787-49-7
Beryllium hydroxide		13327-32-7
Beryllium oxide	Beryllia	1304-56-9
Beryllium phosphate		13598-15-7
Beryllium sulfate		13510-49-1
Beryllium sulfate tetrahydrate		7787-56-6
Beryllium zinc silicate		39413-47-3
2,2'-Bioxirane	Diepoxybutane	1464-53-5
Bis(4-amino-3-chlorophenyl)methane	4,4-Methylene-bis(2-chloraniline); MBOCA	101-14-4
Bis(2-chloroethyl)methylamine	Nitrogen mustard hydrochloride	55-86-7
N,N'-Bis(2-chloroethyl)-N-nitrosourea	BCNU; Carmustine	154-93-8
Bis[4-(dimethylamino)phenyl]methane	Michler's Base; 4,4-Methylenebis(*N,N*-dimethylbenzenamine)	101-61-1
1,3-Bis(2,3-epoxypropoxy)benzene	Diglycidyl resorcinol ether	101-90-6
Bis(2-ethylhexyl) phthalate	DEHP; Di(2-ethylhexyl) phthalate	117-81-7
Bromodichloromethane		75-27-4
tert-Butyl-4-hydroxyanisole	BHA; Butylated hydroxyanisole	25013-16-5
N-Butyl-*N*-(4-hydroxybutyl)nitrosamine	*N*-Nitrosobutyl-*N*-(4-hydroxybutyl)amine	3817-11-6
Chlorendic acid	5-Norbornene-2,3-dicarboxylic acid, 1,4,5,6,7,7-hexachloro-	115-28-6
Chlorinated paraffins (C_{12}, 60% Cl)		108171-26-2
4-Chloro-1,2-benzenediamine	4-Chloro-*o*-phenylenediamine	95-83-0
2-Chloro-1,3-butadiene	Chloroprene	126-99-8
1-(2-Chloroethyl)-3-cyclohexyl-1-nitrosourea	CCNU; Lomustine; Belustine	13010-47-4
4-Chloro-2-methylaniline	*p*-Chloro-*o*-toluidine	95-69-2
4-Chloro-2-methylaniline hydrochloride	*p*-Chloro-*o*-toluidine hydrochloride	3165-93-3
1-Chloro-2-methylpropene	Dimethylvinyl chloride	513-37-1
3-Chloro-2-methylpropene		563-47-3
Chlorozotocin	*D*-Glucose, 2-[[[(2-chloroethyl)nitrosoamino]carbonyl]amino]-2-deoxy-	54749-90-5
C.I. Basic Red 9, monohydrochloride		569-61-9
Cupferron		135-20-6

Substance	Other names	CAS RN
Dacarbazine	1H-Imidazole-4-carboxamide, 5-(3,3-dimethyl-1-triazenyl)-	4342-03-4
Decabromobiphenyl		13654-09-6
cis-Diaminedichloroplatinum	Cisplatin	15663-27-1
2,4-Diaminoanisole sulfate	1,3-Benzenediamine, 4-methoxy, sulfate	39156-41-7
4,4'-Diaminodiphenyl ether	4,4-Oxydianiline	101-80-4
4,4'-Diaminodiphenylmethane	4,4'-Methylenedianiline	101-77-9
Dibenz[a,h]acridine		226-36-8
Dibenz[a,j]acridine		224-42-0
Dibenz[a,h]anthracene		53-70-3
7H-Dibenzo[c,g]carbazole		194-59-2
Dibenzo[a,e]pyrene	Naphtho[1,2,3,4-def]chrysene	192-65-4
Dibenzo[a,h]pyrene	Dibenzo[b,def]chrysene	189-64-0
Dibenzo[a,i]pyrene	Benzo[rst]pentaphene	189-55-9
Dibenzo[a,l]pyrene	Dibenzo[def,p]chrysene	191-30-0
1,2-Dibromo-3-chloropropane		96-12-8
1,2-Dibromoethane	Ethylene dibromide; EDB	106-93-4
2,3-Dibromo-1-propanol, phosphate (3:1)	Tris(2,3-dibromopropyl) phosphate	126-72-7
p-Dichlorobenzene	1,4-Dichlorobenzene	106-46-7
3,3'-Dichloro-*p*-benzidine	[1,1'-Biphenyl]-4,4'-diamine, 3,3'-dichloro-	91-94-1
3,3'-Dichloro-*p*-benzidine dihydrochloride	3,3'-Dichloro-[1,1'-biphenyl]-4,4'-diamine dihydrochloride	612-83-9
1,2-Dichloroethane	Ethylene dichloride	107-06-2
Dichloromethane	Methylene chloride	75-09-2
1,3-Dichloropropene (unspecified isomer)		542-75-6
Diethyl sulfate		64-67-5
2,3-Dihydro-6-propyl-2-thioxo-4(1H)-pyrimidinone	Propylthiouracil	51-52-5
1,8-Dihydroxy-9,10-anthracenedione	Danthron; 1,8-Dihydroxyanthraquinone	117-10-2
3,3'-Dimethoxybenzidine	Dianisidine	119-90-4
p-(Dimethylamino)azobenzene		60-11-7
2',3-Dimethyl-4-aminoazobenzene	*o*-Aminoazotoluene; 4-*o*-Tolylazo-*o*-toluidine	97-56-3
Dimethylcarbamic chloride	Dimethylcarbamoyl chloride	79-44-7
1,1-Dimethylhydrazine	UDMH	57-14-7
Dimethyl sulfate		77-78-1
1,6-Dinitropyrene		42397-64-8
1,8-Dinitropyrene		42397-65-9
1,4-Dioxane		123-91-1
1,2-Diphenylhydrazine	Hydrazobenzene	122-66-7
Disperse Blue No. 1	9,10-Anthracenedione, 1,4,5,8-tetraamino-	2475-45-8
Epichlorohydrin	(Chloromethyl)oxirane	106-89-8
1,2-Epoxy-4-(epoxyethyl)cyclohexane	4-Vinyl-1-cyclohexene dioxide	106-87-6
Estra-1,3,5(10)-triene-3,17-diol, (17β)-	Estradiol-17β	50-28-2
Estrone		53-16-7
N-(4-Ethoxyphenyl)acetamide	Phenacetin	62-44-2
Ethyl carbamate	Urethane	51-79-6
Ethyl methanesulfonate		62-50-0
N-Ethyl-*N*-nitrosourea	ENU; *N*-Nitroso-*N*-ethylurea	759-73-9
Formaldehyde		50-00-0
Furan		110-00-9
Hexabromobiphenyl (unspecified isomer)	Firemaster FF-1	67774-32-7
Hexachlorobenzene	Perchlorobenzene	118-74-1
1,2,3,4,5,6-Hexachlorocyclohexane, (1α,2α,3β,4α,5α,6β)	Lindane; γ-Hexachlorocyclohexane	58-89-9
1,2,3,4,5,6-Hexachlorocyclohexane, (1α,2α,3β,4α,5β,6β)	α-Hexachlorocyclohexane	319-84-6
1,2,3,4,5,6-Hexachlorocyclohexane, (1α,2β,3α,4β,5α,6β)	β-Hexachlorocyclohexane	319-85-7
Hexachlorocyclohexane (unspecified isomer)		608-73-1
Hexachloroethane	Perchloroethane	67-72-1
Hexamethylphosphoric triamide	Hexamethylphosphoramide; Tris(dimethylamino)phosphine oxide	680-31-9

Substance	Other names	CAS RN
Hydrazine		302-01-2
Hydrazine sulfate		10034-93-2
2-Imidazolidinethione	Ethylene thiourea	96-45-7
Indeno[1,2,3-cd]pyrene	1,10-(1,2-Phenylene)pyrene	193-39-5
Kanechlor 500	Polychlorinated biphenyls	37317-41-2
Kepone	Chlordecone	143-50-0
Lead(II) acetate		301-04-2
Lead(II) phosphate		7446-27-7
Mestranol	19-Norpregna-1,3,5(10)-trien-20-yn-17-ol, 3-methoxy-, (17 α)-	72-33-3
o-Methoxyaniline hydrochloride	*o*-Anisidine hydrochloride	134-29-2
2-Methoxy-5-methylaniline	*p*-Cresidine; 5-Methyl-*o*-anisidine	120-71-8
o-Methylaniline	*o*-Toluidine	95-53-4
o-Methylaniline hydrochloride	*o*-Toluidine hydrochloride	636-21-5
2-Methyl-1,3-butadiene	Isoprene	78-79-5
5-Methylchrysene		3697-24-3
4,4-Methylenedianiline dihydrochloride	Benzenamine, 4,4'-methylenedi-, dihydrochloride	13552-44-8
Methyl methanesulfonate		66-27-3
N-Methyl-*N'*-nitro-*N*-nitrosoguanidine		70-25-7
N-Methyl-*N*-nitrosourea	*N*-Nitroso-*N*-methylurea	684-93-5
Methyloxirane	1,2-Propylene oxide	75-56-9
Metronidazole	2-Methyl-5-nitro-1H-imidazole-1-ethanol	443-48-1
Mirex	1,3,4-Metheno-1H-cyclobuta[cd]pentalene, 1,1a,2,2,3,3a,4,5,5,5a,5b,6-dodecachlorooctahydro-	2385-85-5
Nickel		7440-02-0
Nickel(II) acetate		373-02-4
Nickel(II) carbonate		3333-67-3
Nickel carbonyl		13463-39-3
Nickel(II) hydroxide		12054-48-7
Nickel hydroxide (unspecified oxidation state)		11113-74-9
Nickelocene	Bis(η5-2,4-cyclopentadien-1-yl)nickel	1271-28-9
Nickel(II) oxide		1313-99-1
Nickel(III) sulfide	Nickel subsulfide	12035-72-2
Nitrilotriacetic acid	*N*,*N*-Bis(carboxymethyl)glycine	139-13-9
2-Nitroanisole	1-Methoxy-2-nitrobenzene	91-23-6
6-Nitrochrysene		7496-02-8
Nitrofen	Benzene, 2,4-dichloro-1-(4-nitrophenoxy)-	1836-75-5
2-Nitropropane		79-46-9
1-Nitropyrene		5522-43-0
4-Nitropyrene		57835-92-4
N-Nitrosobutyl-*N*-(3-carboxypropyl)amine	Butyl(3-carboxypropyl)nitrosoamine	38252-74-3
N-Nitrosodibutylamine		924-16-3
N-Nitrosodiethanolamine	Ethanol, 2,2'-(nitrosoimino)-	1116-54-7
N-Nitrosodiethylamine	DEN; Diethylnitrosamine	55-18-5
N-Nitrosodimethylamine	DMN; Dimethylnitrosamine	62-75-9
4-(*N*-Nitrosomethylamino)-1-(3-pyridyl)-1-butanone	NNK; Ketone, 3-pyridyl-3-(*N*-methyl-*N*-nitrosamino)propyl	64091-91-4
N-Nitroso-*N*-methylvinylamine	Ethenamine, *N*-methyl-*N*-nitroso-	4549-40-0
4-Nitrosomorpholine	*N*-Nitrosomorpholine	59-89-2
N-Nitrosonornicotine		16543-55-8
N-Nitrosopiperidine	1-Nitrosopiperidine	100-75-4
N-Nitroso-*N*-propyl-1-propanamine	*N*-Nitrosodipropylamine	621-64-7
N-Nitrosopyrrolidine		930-55-2
N-Nitrososarcosine	Glycine, *N*-methyl-*N*-nitroso-	13256-22-9
Norethisterone	19-Norpregn-4-en-20-yn-3-one, 17-hydroxy-, (17 α)-	68-22-4
19-Norpregna-1,3,5(10)-trien-20-yne-3, 17-diol, (17α)-	Ethinylestradiol	57-63-6
Ochratoxin A		303-47-9
Octabromobiphenyl (unspecified isomer)		61288-13-9
2-Oxetanone	β-Propiolactone	57-57-8

Substance	Other names	CAS RN
Oxiranemethanol	Glycidol	556-52-5
Oxymetholone	Androstan-3-one, 17-hydroxy-2-(hydroxymethylene)-17-methyl-	434-07-1
Phenazopyridine hydrochloride	2,6-Pyridinediamine, 3-(phenylazo)-, monohydrochloride	136-40-3
Phenolphthalein	3,3-Bis(4-hydroxyphenyl)-1(3H)-isobenzofuranone	77-09-8
Phenoxybenzamine hydrochloride	Benzenemethanamine, *N*-(2-chloroethyl)-*N*-(1-methyl-2-phenoxyethyl)-, hydrochloride	63-92-3
Phenytoin	5,5-Diphenyl-2,4-imidazolidinedione	57-41-0
Polychlorinated biphenyls	PCBs	1336-36-3
Procarbazine hydrochloride		366-70-1
Progesterone	Pregn-4-ene-3,20-dione	57-83-0
1,3-Propane sultone	1,2-Oxathiolane, 2,2-dioxide	1120-71-4
Propyleneimine	2-Methylaziridine	75-55-8
Reserpine		50-55-5
Safrole	5-(2-Propenyl)-1,3-benzodioxole	94-59-7
Selenium sulfide		7446-34-6
Streptozotocin	*D*-Glucopyranose, 2-deoxy-2-[[(methylnitrosoamino)carbonyl]amino]-	18883-66-4
Sulfallate	*N,N*-Diethyldithiocarbamic acid, 2-chloroallyl ester	95-06-7
2,3,7,8-Tetrachlorodibenzo-*p*-dioxin	TCDD; Dioxin	1746-01-6
Tetrachloroethene	Perchloroethylene	127-18-4
Tetrachloromethane	Carbon tetrachloride	56-23-5
Tetrafluoroethene	Tetrafluoroethylene	116-14-3
N,N,N',N'-Tetramethyl-4,4'-diaminobenzophenone	Bis(dimethylamino)benzophenone; Michler's Ketone	90-94-8
Tetranitromethane		509-14-8
Thioacetamide		62-55-5
Thiourea		62-56-6
o-Tolidine	3,3-Dimethylbenzidine	119-93-7
Toluene-2,4-diamine	2,4-Diaminotoluene	95-80-7
Toluene diisocyanate (unspecified isomer)		26471-62-5
Toxaphene	Polychlorocamphene	8001-35-2
1H-1,2,4-Triazol-3-amine	Amitrole	61-82-5
1,1,1-Trichloro-2,2-bis(4-chlorophenyl)ethane	DDT; Dichlorodiphenyltrichloroethane	50-29-3
Trichloroethene	Trichloroethylene	79-01-6
Trichloromethane	Chloroform	67-66-3
(Trichloromethyl)benzene	Benzotrichloride	98-07-7
2,4,6-Trichlorophenol		88-06-2
1,2,3-Trichloropropane		96-18-4

Appendix A
Mathematical Tables

Miscellaneous Mathematical Constants A-1
Exponential and Hyperbolic Functions and their Common Logarithms A-2
Natural Trigonometric Functions to Four Places A-7
Relation of Angular Functions in Terms of One Another A-11
Derivatives A-12
Integration A-15
Integrals A-20
Differential Equations A-65
Fourier Series A-75
The Fourier Transforms A-80
Series Expansion A-84
Vector Analysis A-87
Moment of Inertial for Various Bodies of Mass A-96
Bessel Functions A-96
The Gamma Function A-99
The Beta Function A-101
The Error Function A-101
Orthogonal Polynomials A-102
Normal Probability Function A-104
Percentage Points, Student's *t*-Distribution A-105
Percentage Points, Chi-Square Distribution A-105
Percentage Points, F-Distribution A-106

MISCELLANEOUS MATHEMATICAL CONSTANTS

π CONSTANTS

$\pi =$	3.14159 26535 89793 23846 26433 83279 50288 41971 69399 37511
$1/\pi =$	0.31830 98861 83790 67153 77675 26745 02872 40689 19291 48091
$\pi^2 =$	9.86960 44010 89358 61883 44909 99876 15113 53136 99407 24079
$\log_e \pi =$	1.14472 98858 49400 17414 34273 51353 05871 16472 94812 91531
$\log_{10} \pi =$	0.49714 98726 94133 85435 12682 88290 89887 36516 78324 38044
$\log_{10}\sqrt{2\pi} =$	0.39908 99341 79057 52478 25035 91507 69595 02099 34102 92128

CONSTANTS INVOLVING e

$e =$	2.71828 18284 59045 23536 02874 71352 66249 77572 47093 69996
$1/e =$	0.36787 94411 71442 32159 55237 70161 46086 74458 11131 03177
$e^2 =$	7.38905 60989 30650 22723 04274 60575 00781 31803 15570 55185
$M = \log_{10} e =$	0.43429 44819 03251 82765 11289 18916 60508 22943 97005 80367
$1/M = \log_e 10 =$	2.30258 50929 94045 68401 79914 54684 36420 76011 01488 62877
$\log_{10} M =$	9.63778 43113 00536 78912 29674 98645 −10

π^e AND e^π CONSTANTS

$\pi^e =$	22.45915 77183 61045 47342 71522
$e^\pi =$	23.14069 26327 79269 00572 90864
$e^{-\pi} =$	0.04321 39182 63772 24977 44177
$e^{1/2\pi} =$	4.81047 73809 65351 65547 30357
$i^i = e^{-1/2\pi} =$	0.20787 95763 50761 90854 69556

NUMERICAL CONSTANTS

$\sqrt{2} =$	1.41421 35623 73095 04880 16887 24209 69807 85696 71875 37695
$\sqrt[3]{2} =$	1.25992 10498 94873 16476 72106 07278 22835 05702 51464 70151
$\log_e 2 =$	0.69314 71805 59945 30941 72321 21458 17656 80755 00134 36026
$\log_{10} 2 =$	0.30102 99956 63981 19521 37388 94724 49302 67881 89881 46211
$\sqrt{3} =$	1.73205 08075 68877 29352 74463 41505 87236 69428 05253 81039
$\sqrt[3]{3} =$	1.44224 95703 07408 38232 16383 10780 10958 83918 69253 49935
$\log_e 3 =$	1.09861 22886 68109 69139 52452 36922 52570 46474 90557 82275
$\log_{10} 3 =$	0.47712 12547 19662 43729 50279 03255 11530 92001 28864 19070

OTHER CONSTANTS

Euler's Constant $\gamma =$	0.57721 56649 01532 86061
$\log_e \gamma =$	−0.54953 93129 81644 82234
Golden Ratio $\phi =$	1.61803 39887 49894 84820 45868 34365 63811 77203 09180

EXPONENTIAL AND HYPERBOLIC FUNCTIONS AND THEIR COMMON LOGARITHMS

x	e^x Value	e^x $\log_{10}$	e^{-x} (value)	sinh x Value	sinh x $\log_{10}$	cosh x Value	cosh x $\log_{10}$	tanh x (value)
0.00	1.0000	0.00000	1.00000	0.0000	$-\infty$	1.0000	0.00000	0.00000
0.01	1.0101	.00434	0.99005	.0100	$\bar{2}.00001$	1.0001	.00002	.01000
0.02	1.0202	.00869	.98020	.0200	$\bar{2}.30106$	1.0002	.00009	.02000
0.03	1.0305	.01303	.97045	.0300	$\bar{2}.47719$	1.0005	.00020	.02999
0.04	1.0408	.01737	.96079	.0400	$\bar{2}.60218$	1.0008	.00035	.03998
0.05	1.0513	.02171	.95123	.0500	$\bar{2}.69915$	1.0013	.00054	.04996
0.06	1.0618	.02606	.94176	.0600	$\bar{2}.77841$	1.0018	.00078	.05993
0.07	1.0725	.03040	.93239	.0701	$\bar{2}.84545$	1.0025	.00106	.06989
0.08	1.0833	.03474	.92312	.0801	$\bar{2}.90355$	1.0032	.00139	.07983
0.09	1.0942	.03909	.91393	.0901	$\bar{2}.95483$	1.0041	.00176	.08976
0.10	1.1052	.04343	.90484	.1002	$\bar{1}.00072$	1.0050	.00217	.09967
0.11	1.1163	.04777	.89583	.1102	$\bar{1}.04227$	1.0061	.00262	.10956
0.12	1.1275	.05212	.88692	.1203	$\bar{1}.08022$	1.0072	.00312	.11943
0.13	1.1388	.05646	.87809	.1304	$\bar{1}.11517$	1.0085	.00366	.12927
0.14	1.1503	.06080	.86936	.1405	$\bar{1}.14755$	1.0098	.00424	.13909
0.15	1.1618	.06514	.86071	.1506	$\bar{1}.17772$	1.0113	.00487	.14889
0.16	1.1735	.06949	.85214	.1607	$\bar{1}.20597$	1.0128	.00554	.15865
0.17	1.1853	.07383	.84366	.1708	$\bar{1}.23254$	1.0145	.00625	.16838
0.18	1.1972	.07817	.83527	.1810	$\bar{1}.25762$	1.0162	.00700	.17808
0.19	1.2092	.08252	.82696	.1911	$\bar{1}.28136$	1.0181	.00779	.18775
0.20	1.2214	.08686	.81873	.2013	$\bar{1}.30392$	1.0201	.00863	.19738
0.21	1.2337	.09120	.81058	.2115	$\bar{1}.32541$	1.0221	.00951	.20697
0.22	1.2461	.09554	.80252	.2218	$\bar{1}.34592$	1.0243	.01043	.21652
0.23	1.2586	.09989	.79453	.2320	$\bar{1}.36555$	1.0266	.01139	.22603
0.24	1.2712	.10423	.78663	.2423	$\bar{1}.38437$	1.0289	.01239	.23550
0.25	1.2840	.10857	.77880	.2526	$\bar{1}.40245$	1.0314	.01343	.24492
0.26	1.2969	.11292	.77105	.2629	$\bar{1}.41986$	1.0340	.01452	.25430
0.27	1.3100	.11726	.76338	.2733	$\bar{1}.43663$	1.0367	.01564	.26362
0.28	1.3231	12160	.75578	.2837	$\bar{1}.45282$	1.0395	.01681	.27291
0.29	1.3364	.12595	.74826	.2941	$\bar{1}.46847$	1.0423	.01801	.28213
0.30	1.3499	.13029	.74082	.3045	$\bar{1}.48362$	1.0453	.01926	.29131
0.31	1.3634	.13463	.73345	.3150	$\bar{1}.49830$	1.0484	.02054	.30044
0.32	1.3771	.13897	.72615	.3255	$\bar{1}.51254$	1.0516	.02187	.30951
0.33	1.3910	.14332	.71892	.3360	$\bar{1}.52637$	1.0549	.02323	.31852
0.34	1.4049	.14766	.71177	.3466	$\bar{1}.53981$	1.0584	.02463	.32748
0.35	1.4191	.15200	.70469	.3572	$\bar{1}.55290$	1.0619	.02607	.33638
0.36	1.4333	.15635	.69768	.3678	$\bar{1}.56564$	1.0655	.02755	.34521
0.37	1.4477	.16069	.69073	.3785	$\bar{1}.57807$	1.0692	.02907	.35399
0.38	1.4623	.16503	.68386	.3892	$\bar{1}.59019$	1.0731	.03063	.36271
0.39	1.4770	.16937	.67706	.4000	$\bar{1}.60202$	1.0770	.03222	.37136
0.40	1.4918	.17372	.67032	.4108	$\bar{1}.61358$	1.0811	.03385	.37995
0.41	1.5063	.17806	.66365	.4216	$\bar{1}.62488$	1.0852	.03552	.33847
0.42	1.5220	.18240	.65705	.4325	$\bar{1}.63594$	1.0895	.03723	.39693
0.43	1.5373	.18675	.65051	.4434	$\bar{1}.64677$	1.0939	.03897	.40532
0.44	1.5527	.19109	.64404	.4543	$\bar{1}.65738$	1.0984	.04075	.41364
0.45	1.5683	.19543	.63763	.4653	$\bar{1}.66777$	1.1030	.04256	.42190
0.46	1.5841	.19978	.63128	.4764	$\bar{1}.67797$	1.1077	.04441	.43008
0.47	1.6000	.20412	.62500	.4875	$\bar{1}.68797$	1.1125	.04630	.43820
0.48	1.6161	.20846	.61878	.4986	$\bar{1}.69779$	1.1174	.04822	.44624
0.49	1.6323	.21280	.61263	.5098	$\bar{1}.70744$	1.1225	.05018	.45422
0.50	1.6487	.21715	.60653	.5211	$\bar{1}.71692$	1.1276	.05217	.46212
0.51	1.6653	.22149	.60050	.5324	$\bar{1}.72624$	1.1329	.05419	.46995
0.52	1.6820	.22583	.59452	.5438	$\bar{1}.73540$	1.1383	.05625	.47770
0.53	1.6989	.23018	.58860	.5552	$\bar{1}.74442$	1.1438	.05834	.48538
0.54	1.7160	.23452	.58275	.5666	$\bar{1}.75330$	1.1494	.06046	.49299
0.55	1.7333	.23886	.57695	.5782	$\bar{1}.76204$	1.1551	.06262	.50052
0.56	1.7507	.24320	.57121	.5897	$\bar{1}.77065$	1.1609	.06481	.50798
0.57	1.7683	.24755	.56553	.6014	$\bar{1}.77914$	1.1669	.06703	.51536
0.58	1.7860	.25189	.55990	.6131	$\bar{1}.78751$	1.1730	.06929	.52267
0.59	1.8040	.25623	.55433	.6248	$\bar{1}.79576$	1.1792	.07157	.52990
0.60	1.8221	.26058	.54881	.6367	$\bar{1}.80390$	1.1855	.07389	.53705
0.61	1.8404	.26492	.54335	.6485	$\bar{1}.81194$	1.1919	.07624	.54413
0.62	1.8589	.26926	.53794	.6605	$\bar{1}.81987$	1.1984	.07861	.55113
0.63	1.8776	.27361	.53259	.6725	$\bar{1}.82770$	1.2051	.08102	.55805
0.64	1.8965	.27795	.52729	.6846	$\bar{1}.83543$	1.2119	.08346	.56490
0.65	1.9155	.28229	.52205	.6967	$\bar{1}.84308$	1.2188	.08593	.57167
0.66	1.9348	.28664	.51685	.7090	$\bar{1}.85063$	1.2258	.08843	.57836
0.67	1.9542	.29098	.51171	.7213	$\bar{1}.85809$	1.2330	.09095	.58498
0.68	1.9739	.29532	.50662	.7336	$\bar{1}.86548$	1.2402	.09351	.59152
0.69	1.9937	.29966	.50158	.7461	$\bar{1}.87278$	1.2476	.09609	.59798
0.70	2.0138	.30401	.49659	.7586	$\bar{1}.88000$	1.2552	.09870	.60437
0.71	2.0340	.30835	.49164	.7712	$\bar{1}.88715$	1.2628	.10134	.61068
0.72	2.0544	.31269	.48675	.7838	1.89423	1.2706	.10401	.61691

EXPONENTIAL AND HYPERBOLIC FUNCTIONS AND THEIR COMMON LOGARITHMS (continued)

	e^x		e^{-x}	sinh x		cosh x		tanh x
x	Value	$\log_{10}$	(value)	Value	$\log_{10}$	Value	$\log_{10}$	(value)
0.73	2.0751	.31703	.48191	.7966	$\bar{1}.90123$	1.2785	.10670	.62307
0.74	2.0959	.32138	.47711	.8094	$\bar{1}.90817$	1.2865	.10942	.62915
0.75	2.1170	.32572	.47237	.8223	$\bar{1}.91504$	1.2947	.11216	.63515
0.76	2.1383	.33006	.46767	.8353	$\bar{1}.92185$	1.3030	.11493	.64108
0.77	2.1598	.33441	.46301	.8484	$\bar{1}.92859$	1.3114	.11773	.64693
0.78	2.1815	.33875	.45841	.8615	$\bar{1}.93527$	1.3199	.12055	.65721
0.79	2.2034	.34309	.45384	.8748	$\bar{1}.94190$	1.3286	.12340	.65841
0.80	2.2255	.34744	.44933	.8881	$\bar{1}.94846$	1.3374	.12627	.66404
0.81	2.2479	.35178	.44486	.9015	$\bar{1}.95498$	1.3464	.12917	.66959
0.82	2.2705	.35612	.44043	.9150	$\bar{1}.96144$	1.3555	.13209	.67507
0.83	2.2933	.36046	.43605	.9286	$\bar{1}.96784$	1.3647	.13503	.68048
0.84	2.3164	.36481	.43171	.9423	$\bar{1}.97420$	1.3740	.13800	.68581
0.85	2.3396	.36915	.42741	.9561	$\bar{1}.98051$	1.3835	.14099	.69107
0.86	2.3632	.37349	.42316	.9700	$\bar{1}.98677$	1.3932	.14400	.69626
0.87	2.3869	.37784	.41895	.9840	$\bar{1}.99299$	1.4029	.14704	.70137
0.88	2.4100	.38218	.41478	.9981	1.99916	1.4128	.15009	.70642
0.89	2.4351	.38652	.41066	1.0122	0.00528	1.4229	.15317	.71139
0.90	2.4596	.39087	.40657	1.0265	.01137	1.4331	.15627	.21630
0.91	2.4843	39521	.40242	1.0409	.01741	1.4434	.15939	.72113
0.92	2.5093	.39955	.39852	1.0554	.02341	1.4539	.16254	.72590
0.93	2.5345	.40389	.39455	1.0700	.02937	1.4645	.16570	.73059
0.94	2.5600	.40824	.39063	1.0847	.03530	1.4753	.16888	.73522
0.95	2.5857	.41258	.38674	1.0995	.04119	1.4862	.17208	.73978
0.96	2.6117	.41692	.38289	1.1144	.04704	1.4973	.17531	.74428
0.97	2.6379	.42127	.37908	1.1294	.05286	1.5085	.17855	.74870
0.98	2.6645	.42561	.37531	1.1446	.05864	1.5199	.18181	.75307
0.99	2.6912	.42995	.37158	1.1598	.06439	1.5314	.18509	.75736
1.00	2.7183	.43429	.36788	1.1752	.07011	1.5431	.18839	.76159
1.01	2.7456	.43864	.36422	1.1907	.07580	1.5549	.19171	.76576
1.02	2.7732	.44298	.36060	1.2063	.06146	1.5669	.19504	.76987
1.03	2.8011	.44732	.35701	1.2220	.08708	1.5790	.19839	.77391
1.04	2.8292	.45167	.35345	1.2379	.09268	1.5913	.20176	.77789
1.05	2.8577	.45601	.34994	1.2539	.09825	1.6038	.20515	.78181
1.06	2.8864	.46035	.34646	1.2700	.10379	1.6164	.20855	.78566
1.07	2.9154	.46470	.34301	1.2862	.10930	1.6292	.21197	.78946
1.08	2.9447	.46904	.33960	1.3025	.11479	1.6421	.21541	.79320
1.09	2.9743	.47338	.33622	1.3190	.12025	1.6552	.21886	.79688
1.10	3.0042	.47772	.33287	1.3356	.12569	1.6685	.22233	.80050
1.11	3.0344	.48207	.32956	1.3524	.13111	1.6820	.22582	.80406
1.12	3.0659	.48641	.32628	1.3693	.13649	1.6956	.22931	.80757
1.13	3.0957	.49075	.32303	1.3863	.14186	1.7083	.23283	.81102
1.14	3.1268	.49510	.31982	1.4035	.14720	1.7233	.23636	.81441
1.15	3.1582	.49944	.31644	1.4208	.15253	1.7374	.23990	.81775
1.16	3.1899	.50378	.31349	1.4382	.15783	1.7517	.24346	.82104
1.17	3.2220	.50812	.31037	1.4558	.16311	1.7662	.24703	.82427
1.18	3.2544	.51247	.30728	1.4735	.16836	1.7808	.25062	.82745
1.19	3.2871	.51681	.30422	1.4914	.17360	1.7957	.25422	.83058
1.20	3.3201	.52115	.30119	1.5095	.17882	1.8107	.25784	.83365
1.21	3.3535	.52550	.29820	1.5276	.18402	1.8258	.26146	.83668
1.22	3.3872	.52984	.29523	1.5460	.18920	1.8412	.26510	.83965
1.23	3.4212	.53418	.29229	1.5645	.19437	1.8568	.26876	.84258
1.24	3.4556	.53853	.28938	1.5831	.19951	1.8725	.27242	.83546
1.25	3.4903	.54287	.28650	1.6019	.20464	1.8884	.27610	.84828
1.26	3.5254	.54721	.28365	1.6209	.20975	1.9045	.27979	.85106
1.27	3.5609	.55155	.28083	1.6400	.21485	1.9208	.28349	.85380
1.28	3.5996	.55590	.27804	1.6593	.21993	1.9373	.28721	.85648
1.29	3.6328	.56024	.27527	1.6788	.22499	1.9540	.29093	.85913
1.30	3.6693	.56458	.27253	1.6984	.23004	1.9709	.29467	.86172
1.31	3.7062	.56893	.26982	1.7182	.23507	1.9880	.29842	.86428
1.32	3.7434	.57327	.26714	1.7381	.24009	2.0053	.30217	.86678
1.33	3.7810	.57761	.26448	1.7583	.24509	2.0228	.30594	.86925
1.34	3.8190	.58195	.26185	1.7786	.25008	2.0404	.30972	.87167
1.35	3.8574	.58630	.25924	1.7991	.25505	2.0583	.31352	.87405
1.36	3.8962	.59064	.25666	1.8198	.26002	2.0764	.31732	.87639
1.37	3.9354	.59498	.25411	1.8406	.26496	2.0947	.32113	.87869
1.38	3.9749	.59933	.25158	1.8617	.26990	2.1132	.32495	.88095
1.39	4.0149	.60367	.24908	1.8829	.27482	2.1320	.32878	.88317
1.40	4.0552	.60801	.24660	1.9043	.27974	2.1509	.33262	.88535
1.41	4.0960	.61236	.24414	1.9259	.28464	2.1700	.33647	.88749
1.42	4.1371	.61670	.24171	1.9477	.28952	2.1894	.34033	.88960
1.43	4.1787	.62104	.23931	1.9697	.29440	2.2090	.34420	.89167

EXPONENTIAL AND HYPERBOLIC FUNCTIONS AND THEIR COMMON LOGARITHMS (continued)

x	e^x Value	e^x $\log^{10}$	e^{-x} (value)	sinh x Value	sinh x $\log^{10}$	cosh x Value	cosh x $\log^{10}$	tanh x (value)
1.44	4.2207	.62538	.23693	1.9919	.29926	2.2288	.34807	.89370
1.45	4.2631	.62973	.23457	2.0143	.30412	2.2488	.35196	.89569
1.46	4.3060	.63407	.23224	2.0369	.30896	2.2691	.35585	.89765
1.47	4.3492	.63841	.22993	2.0597	.31379	2.2896	.35976	.89958
1.48	4.3929	.64276	.22764	2.0827	.31862	2.3103	.36367	.90147
1.49	4.4371	.64710	.22537	2.1059	.32343	2.3312	.36759	.90332
1.50	4.4817	.65144	.22313	2.1293	.32823	2.3524	.37151	.90515
1.51	4.5267	.65578	.22091	2.1529	.33303	2.3738	.37545	.90694
1.52	4.5722	.66013	.21871	2.1768	.33781	2.3955	.37939	.90870
1.53	4.6182	.66447	.21654	2.2008	.34258	2.4174	.38334	.91042
1.54	4.6646	.66881	.21438	2.2251	.34735	2.4395	.38730	.91212
1.55	4.7115	.67316	.21225	2.2496	.35211	2.4619	.39126	.91379
1.56	4.7588	.67750	.21014	2.2743	.35686	2.4845	.39524	.91542
1.57	4.8066	.68184	.20805	2.2993	.36160	2.5073	.39921	.91703
1.58	4.8550	.68619	.20598	2.3245	.36633	2.5305	.40320	.91860
1.59	4.9037	.69053	.20393	2.3499	.37105	2.5538	.40719	.92015
1.60	4.9530	.69487	.20190	2.3756	.37577	2.5775	.41119	.92167
1.61	5.0028	.69921	.19989	2.4015	.38048	2.6013	.41520	.92316
1.62	5.0531	.70356	.19790	2.4276	.38518	2.6255	.41921	.92462
1.63	5.1039	.70790	.19593	2.4540	.38987	2.6499	.42323	.92606
1.64	5.1552	.71224	.19398	2.4806	.39456	2.6746	.42725	.92747
1.65	5.2070	.71659	.19205	2.5075	.39923	2.6995	.43129	.92886
1.66	5.2593	.72093	.19014	2.5346	.40391	2.7247	.43532	.93022
1.67	5.3122	.72527	.18825	2.5620	.40857	2.7502	.43937	.93155
1.68	5.3656	.72961	.18637	2.5896	.41323	2.7760	.44341	.93286
1.69	5.4195	.73396	.18452	2.6175	.41788	2.8020	.44747	.93415
1.70	5.4739	.73830	.18268	2.6456	.42253	2.8283	.45153	.93541
1.71	5.5290	.74264	.18087	2.6740	.42717	2.8549	.45559	.93665
1.72	5.5845	.74699	.17907	2.7027	.43180	2.8818	.45966	.93786
1.73	5.6407	.75133	.17728	2.7317	.43643	2.9090	.46374	.93906
1.74	5.6973	.75567	.17552	2.7609	.44105	2.9364	.46782	.94023
1.75	5.7546	.76002	.17377	2.7904	.44567	2.9642	.47191	.94138
1.76	5.8124	.76436	.17204	2.8202	.45028	2.9922	.47600	.94250
1.77	5.8709	.76870	.17033	2.8503	.45488	3.0206	.48009	.94361
1.78	5.9299	.77304	.16864	2.8806	.45948	3.0492	.48419	.94470
1.79	5.9895	.77739	.16696	2.9112	.46408	3.0782	.48830	.94576
1.80	6.0496	.78173	.16530	2.9422	.46867	3.1075	.49241	.94681
1.81	6.1104	.78607	.16365	2.9734	.47325	3.1371	.49652	.94783
1.82	6.1719	.79042	.16203	3.0049	.47783	3.1669	.50064	.94884
1.83	6.2339	.79476	.16041	3.0367	.48241	3.1972	.50476	.94983
1.84	6.2965	.79910	.15882	3.0689	.48698	3.2277	.50889	.95080
1.85	6.3598	.80344	.15724	3.1013	.49154	3.2585	.51302	.95175
1.86	6.4237	.80779	.15567	3.1340	.49610	3.2897	.51716	.95268
1.87	6.4383	.81213	.15412	3.1671	.50066	3.3212	.52130	.95359
1.88	6.5535	.81647	.15259	3.2005	.50521	3.3530	.52544	.95449
1.89	6.6194	.82082	.15107	3.2341	.50976	3.3852	.52959	.95537
1.90	6.6859	.82516	.14957	3.2682	.51430	3.4177	.53374	.95624
1.91	6.7531	.82950	.14808	3.3025	.51884	3.4506	.53789	.95709
1.92	6.8210	.83385	.14661	3.3372	.52338	3.4838	.54205	.95792
1.93	6.8895	.83819	.14515	3.3722	.52791	3.5173	.54621	.95873
1.94	6.9588	.84253	.14370	3.4075	.53244	3.5512	.55038	.95953
1.95	7.0287	.84687	.14227	3.4432	.53696	3.5855	.55455	.96032
1.96	7.0993	.85122	.14086	3.4792	.54148	3.6201	.55872	.96109
1.97	7.1707	.85556	.13946	3.5156	.54600	3.6551	.56290	96185
1.98	7.2427	.85990	.13807	3.5923	.55051	3.6904	.56707	.96259
1.99	7.3155	.86425	.13670	3.5894	.55502	3.7261	.57126	.96331
2.00	7.3891	.86859	.13534	3.6269	.55953	3.7622	.57544	.96403
2.01	7.4633	.87293	.13399	3.6647	.56403	3.7987	.57963	.96473
2.02	7.5383	.87727	.13266	3.7028	.56853	3.8335	.58382	.96541
2.03	7.6141	.88162	.13134	3.7414	.57303	3.8727	.58802	.96609
2.04	7.6906	.88596	.13003	3.7803	.57753	3.9103	.59221	.96675
2.05	7.7679	.89030	.12873	3.8196	.58202	3.9483	.59641	.96740
2.06	7.8460	.89465	.12745	3.8593	.58650	3.9867	.60061	.96803
2.07	7.9248	.89899	.12619	3.8993	.59099	4.0255	.60482	.96865
2.08	8.0045	.90333	.12493	3.9398	.59547	4.0647	.60903	.96926
2.09	8.0849	.90768	.12369	3.9806	.59995	4.1043	.61324	.96986
2.10	8.1662	.91202	.12246	4.0219	.60443	4.1443	.61745	.97045
2.11	8.2482	.91636	.12124	4.0635	.60890	4.1847	.62167	.97103
2.12	8.3311	.92070	.12003	4.1056	.61337	4.2256	.62589	.97159
2.13	8.4149	.92505	.11884	4.1480	.61784	4.2669	.63011	.97215
2.14	8.4994	.92939	.11765	4.1909	.62231	4.3085	.63433	.97269
2.15	8.5849	.93373	.11648	4.2342	62677	4.3507	.63856	.97323

EXPONENTIAL AND HYPERBOLIC FUNCTIONS AND THEIR COMMON LOGARITHMS (continued)

	e^x			sinh x		cosh x		tanh x
x	Value	$\log_{10}$	(value)	Value	$\log_{10}$	Value	$\log_{10}$	(value)
2.16	8.6711	.93808	.11533	4.2779	.63123	4.3932	.64278	.97375
2.17	8.7583	.94242	.11418	4.3221	.63569	4.4362	.64701	.97426
2.18	8.8463	.94676	.11304	4.3666	.64015	4.4797	.65125	.97477
2.19	8.9352	.95110	.11192	4.4116	.64460	4.5236	.65548	.97526
2.20	9.0250	.95545	.11080	4.4571	.64905	4.5679	.65972	.97574
2.21	9.1157	.95979	.10970	4.5030	.65350	4.6127	.66396	.97622
2.22	9.2073	.96413	.10861	4.5494	.65795	4.6580	.66820	.97668
2.23	9.2999	.96848	.10753	4.5962	.66240	4.7037	.67244	.97714
2.24	9.3933	.97282	.10646	4.6434	.66684	4.7499	.67668	.97759
2.25	9.4877	.97716	.10540	4.6912	.67128	4.7966	.68093	.97803
2.26	9.5831	.98151	.10435	4.7394	.67572	4.8437	.68518	.97846
2.27	9.6794	.98585	.10331	4.7880	.68016	4.8914	.68943	.97888
2.28	9.7767	.99019	.10228	4.8372	.68459	4.9395	.69368	.97929
7.29	9.8749	.99453	.10127	4.8868	.68903	4.9881	.69794	97970
2.30	9.9742	.99888	.10026	4.9370	.69346	5.0372	.70219	.98010
2.31	10.074	1.00322	.09926	4.9876	.69789	5.0868	.70645	.98049
2.32	10.176	1.00756	.09827	5.0387	.70232	5.1370	.71071	.98087
2.33	10.278	1.01191	.09730	5.0903	.70675	5.1876	.71497	.98124
2.34	10.381	1.01625	.09633	5.1425	.71117	5.2388	.71923	.98161
2.35	10.486	1.02059	.09537	5.1951	.71559	5.2905	.72349	.98197
2.36	10.591	1.02493	.09442	5.2483	.72002	5.3427	.72776	.98233
2.37	10.697	1.02928	.09348	5.3020	.72444	5.3954	.73203	.98267
2.38	10.805	1.03362	.09255	5.3562	.72885	5.4487	.73630	.98301
2.39	10.913	1.03796	.09163	5.4109	.73327	5.5026	.74056	.98335
2.40	11.023	1.04231	.09072	5.4662	.73769	5.5569	.74484	.98367
2.41	11.134	1.04665	.08982	5.5221	.74210	5.6119	.74911	.98400
2.42	11.246	1.05099	.08892	5.5785	.74652	5.6674	.75338	.98431
2.43	11.359	1.05534	08804	5.6354	.75093	5.7235	.75766	.98462
2.44	11.473	1.05968	.08716	5.6929	.75534	5.7801	.76194	.98492
2.45	11.588	1.06402	.08629	5.7510	.75975	5.8373	.76621	.98522
2.46	11.705	1.06836	.08543	5.8097	.76415	5.8951	.77049	.98551
2.47	11.822	1.07271	.08458	5.8689	.76856	5.9535	.77477	.98579
2.48	11.941	1.07705	.08374	5.9288	.77296	6.0125	.77906	.98607
2.49	12.061	1.08139	.08291	5.9892	.77737	6.0721	.78334	.98635
2.50	12.182	1.08574	.08208	6.0502	.78177	6.1323	.78762	.98661
2.51	12.305	1.09008	.08127	6.1118	.78617	6.1931	.79191	.98688
2.52	12.429	1.09442	.08046	6.1741	.79057	6.2545	.79619	.98714
2.53	12.554	1.09877	.07966	6.2369	.79497	6.3166	.80048	.98739
2.54	12.680	1.10311	.07887	6.3004	.79937	6.3793	.80477	.98764
2.55	12.807	1.10745	.07808	6.3645	.80377	6.4426	.80906	.98788
2.56	12.936	1.11179	.07730	6.4293	.80816	6.5066	.81335	.98812
2.57	13.066	1.11614	.07654	6.4946	.81256	6.5712	.81764	.98835
2.58	13.197	1.12048	.07577	6.5607	.81695	6.6365	.82194	.98858
2.59	13.330	1.12482	.07502	6.6274	.82134	6.7024	.82623	.98881
2.60	13.464	1.12917	.07427	6.6947	.82573	6.7690	.83052	.98903
2.61	13.599	1.13351	.07353	6.7628	.83012	6.8363	.83482	.98924
2.62	13.736	1.13785	.07280	6.8315	.83451	6.9043	.83912	.98946
2.63	13.874	1.14219	.07208	6.9008	.83890	6.9729	.84341	.98966
2.64	14.013	1.14654	.07136	6.9709	.84329	7.0423	.84771	98987
2.65	14.154	1.15008	.07065	7.0417	.84768	7.1123	.85201	.99007
2.66	14.296	1.15522	.06995	7.1132	.85206	7.1831	.85631	.99026
2.67	14.440	1.15957	.06925	7.1854	.85645	7.2546	.86061	.99045
2.68	14.585	1.16391	.06856	7.2583	.86083	7.3268	.86492	.99064
2.69	14.732	1.16825	.06788	7.3319	.86522	7.3998	.86922	.99083
2.70	14.880	1.17260	.06721	7.4063	.86960	7.4735	.87352	.99101
2.71	15.029	1.17694	.06654	7.4814	.87398	7.5479	.87783	.99118
2.72	15.180	1.18128	.06587	7.5572	.87836	7.6231	.88213	.99136
2.73	15.333	1.18562	.06522	7.6338	.88274	7.6991	.89644	.99153
2.74	15.487	1.18997	.06457	7.7112	.88712	7.7758	.89074	.99170
2.75	15.643	1.19431	06393	7.7894	.89150	7.8533	.89505	.99186
2.76	15.800	1.19865	.06329	7.8683	.89588	7.9316	.89936	.99202
2.77	15.959	1.20300	.06266	7.9480	.90026	8.0106	.90367	.99218
2.78	16.119	1.20734	.06201	0.0205	.90463	8.0905	.90798	.99233
2.79	16.281	1.21168	.06142	8.1098	.90901	8.1712	.91229	.99248
2.80	16.445	1.21602	.06081	8.1919	.91339	8.2527	.91660	.99263
2.81	16.610	1.22037	.06020	8.2749	.91776	8.3351	92091	.99278
2.82	16.777	1.22471	.05961	8.3586	.92213	8.4182	.92522	.99292
2.83	16.945	1.22905	.05901	8.4432	.92651	8.5022	.92953	.99306
2.84	17.116	1.23340	.05843	8.5287	.93088	8.5871	.93385	.99320
2.85	17.288	1.23774	.05784	8.6150	.93525	8.6728	.93816	.99333
2.86	17.462	1.24208	.05727	8.7021	.93963	8.7594	.94247	.99346

EXPONENTIAL AND HYPERBOLIC FUNCTIONS AND THEIR COMMON LOGARITHMS (continued)

	e^x		e^{-x}	sinh x		cosh x		tanh x
x	Value	$\log^{10}$	(value)	Value	$\log^{10}$	Value	$\log^{10}$	(value)
2.87	17.637	1.24643	.05670	8.7902	.94400	8.8469	.94679	.99359
2.88	17.814	1.25077	.05613	8.8791	.94837	8.9352	95110	.99372
2.89	17.993	1.25511	.05558	8.9689	.95274	9.0244	.95542	.99384
2.90	18.174	1.25945	.05502	9.0596	.95711	9.1146	.95974	.99396
2.91	18.357	1.26380	.05448	9.1512	.96148	9.2056	.96405	.99408
2.92	18.541	1.26814	.05393	9.2437	.96584	9.2976	.96837	.99420
2.93	18.728	1.27248	.05340	9.3371	.97021	9.3905	.97269	.99431
2.94	18.916	1.27683	.05287	9.4315	.97458	9.4844	.97701	.99443
2.95	19.106	1.28117	.05234	9.5268	.97895	9.5791	.98133	.99454
2.96	19.298	1.28551	.05182	9.6231	.98331	9.6749	.98565	.99464
2.97	19.492	1.28985	.05130	9.7203	.98768	9.7716	.98997	.99475
2.98	19.688	1.29420	.05079	9.8185	.99205	9.8693	.99429	.99485
2.99	19.886	1.29854	.05029	9.9177	.99641	9.9680	.99861	.99496
3.00	20.086	1.30288	.04979	10.018	1.00078	10.068	1.00293	0.99505
3.05	21.115	1.32460	.04736	10.534	1.02259	10.581	1.02454	0.99552
3.10	22.198	1.34631	.04505	11.076	1.04440	11.122	1.04616	0.99595
3.15	23.336	1.36803	.04285	11.647	1.06620	11.690	1.06779	0.99633
3.20	24.533	1.38974	.04076	12.246	1.08799	12.287	1.08943	0.99668
3.25	25.790	1.41146	.03877	12.876	1.10977	12.915	1.11108	0.99700
3.30	27.113	1.43317	.03688	13.538	1.13155	13.575	1.13273	0.99728
3.35	28.503	1.45489	.03508	14.234	1.15332	14.269	1.15439	0.99754
3.40	29.964	1.47660	.03337	14.965	1.17509	14.999	1.17605	0.99777
3.45	31.500	1.49832	.03175	15.734	1.19685	15.766	1.19772	0.99799
3.50	33.115	1.52003	.03020	16.543	1.21860	16.573	1.21940	0.99818
3.55	34.813	1.54175	.02872	17.392	1.24036	17.421	1.24107	0.99835
3.60	36.598	1.56346	.02732	18.286	1.26211	18.313	1.26275	0.99851
3.65	38.475	1.58517	02599	19.224	1.28385	19.250	1.28444	9.99865
3.70	40.447	1.60689	.02472	20.211	1.30559	20.236	1.30612	0.99878
3.75	42.521	1.62860	.02352	21.249	1.32733	21.272	1.32781	0.99889
3.80	44.701	1.65032	.02237	22.339	1.34907	22.362	1.34951	0.99900
3.85	46.993	1.67203	.02128	23.486	1.37081	23.507	1.37120	0.99909
3.90	49.402	1.69375	.02024	24.691	1.39254	24.711	1.39290	0.99918
3.95	51.935	1.71546	.01925	25.958	1.41427	25.977	1.41459	0.09926
4.00	54.598	1.73718	.01832	27.290	1.43600	27.308	1.43629	0.99933
4.10	60.340	1.78061	.01657	30.162	1.47946	30.178	1.47970	0.99945
4.20	66.686	1.82404	.01500	33.336	1.52291	33.351	1.52310	0.99955
4.30	73.700	1.86747	.01357	36.843	1.56636	36.857	1.56652	0.99963
4.40	81.451	1.91090	.01227	40.719	1.60980	40.732	1.60993	0.99970
4.50	90.017	1.95433	.01111	45.003	1.65324	45.014	1.65335	0.99975
4.60	99.484	1.99775	.01005	49.737	1.69668	49.747	1.69677	0.99980
4.70	109.95	2.04118	.00910	54.969	1.74012	54.978	1.74019	0.99983
4.80	121.51	2.08461	.00823	60.751	1.78355	60.759	1.78361	0.99986
4.90	134.29	2.12804	.00745	67.141	1.82699	67.149	1.82704	0.99989
5.00	148.41	2.17147	.00674	74.203	1.87042	74.210	1.87046	0.99991
5.10	164.02	2.21490	.00610	82.008	1.91389	82.014	1.91389	0.99993
5.20	181.27	2.25833	.00552	90.633	1.95729	90.639	1.95731	0.99994
5.30	200.34	2.30176	.00499	100.17	2.00074	100.17	2.00074	0.99995
5.40	221.41	2.34519	.00452	110.70	2.04415	110.71	2.04417	0.99996
5.50	244.69	2.38862	.00409	122.34	2.08758	122.35	2.08760	0.99997
5.60	270.43	2.43205	.00370	135.21	2.13101	135.22	2.13103	0.99997
5.70	298.87	2.47548	.00335	149.43	2.17444	149.44	2.17445	0.99998
5.80	330.30	2.51891	.00303	165.15	2.21787	165.15	2.21788	0.99998
5.90	365.04	2.56234	.00274	182.52	2.26130	182.52	2.26131	0.99998
6.00	403.43	2.60577	.00248	201.71	2.30473	201.72	2.30474	0.99999
6.25	518.01	2.71434	.00193	259.01	2.41331	259.01	2.41331	0.99999
6.50	665.14	2.82291	.00150	332.57	2.52188	332.57	2.52189	1.00000
6.75	854.06	2.93149	.00117	427.03	2.63046	427.03	2.63046	1.00000
7.00	1096.6	3.04006	.00091	548.32	2.73904	548.32	2.73903	1.00000
7.50	1808.0	3.25721	.00055	904.02	2.95618	904.02	2.95618	1.00000
8.00	2981.0	3.47436	.00034	1490.5	3.17333	1490.5	3.17333	1.00000
8.50	4914.8	3.69150	.00020	2457.4	3.39047	2457.4	3.39047	1.00000
9.00	8103.1	3.90865	.00012	4051.5	3.60762	4051.5	3.60762	1.00000
9.50	13360.	4.12580	.00007	6679.9	3.82477	6679.9	3.82477	1.00000
10.00	22026.	4.34294	.00005	11013.	4.04191	11013.	4.04191	1.00000

NATURAL TRIGONOMETRIC FUNCTIONS TO FOUR PLACES

X radians	X degrees	sin x	cos x	tan x	cot x	sec x	csc x		
.0000	**0° 00′**	.0000	1.0000	.0000	—	1.000	—	**90° 00′**	1.5708
.0029	10	.0029	1.0000	.0029	343.8	1.000	343.8	50	1.5679
.0058	20	.0058	1.0000	.0058	171.9	1.000	171.9	40	1.5650
.0087	30	.0087	1.0000	.0087	114.6	1.000	114.6	30	1.5621
.0116	40	.0116	.9999	.0116	85.94	1.000	85.95	20	1.5592
.0145	50	.0145	.9999	.0145	68.75	1.000	68.76	10	1.5563
.0175	**1° 00′**	.0175	.9998	.0175	57.29	1.000	57.30	**89° 00′**	1.5533
.0204	10	.0204	.9998	.0204	49.10	1.000	49.11	50	1.5504
.0233	20	.0233	.9997	.0233	42.96	1.000	42.98	40	1.5475
.0262	30	.0262	.9997	.0262	38.19	1.000	38.20	30	1.5446
.0291	40	.0291	.9996	.0291	34.37	1.000	34.38	20	1.5417
.0320	50	.0320	.9995	.0320	31.24	1.001	31.26	10	1.5388
.0349	**2° 00′**	.0349	.9994	.0349	28.64	1.001	28.65	**88° 00′**	1.5359
.0378	10	.0378	.9993	.0378	26.43	1.001	26.45	50	1.5330
.0407	20	.0407	.9992	.0407	24.54	1.001	24.56	40	1.5301
.0436	30	.0436	.9990	.0437	22.90	1.001	22.93	30	1.5272
.0465	40	.0465	.9989	.0466	21.47	1.001	21.49	20	1.5243
.0495	50	.0494	.9988	.0495	20.21	1.001	20.23	10	1.5213
.0524	**3° 00′**	.0523	.9986	.0524	19.08	1.001	19.11	**87° 00′**	1.5184
.0553	10	.0552	.9985	.0553	18.07	1.002	18.10	50	1.5155
.0582	20	.0581	.9983	.0582	17.17	1.002	17.20	40	1.5126
.0611	30	.0610	.9981	.0612	16.35	1.002	16.38	30	1.5097
.0640	40	.0640	.9980	.0641	15.60	1.002	15.64	20	1.5068
.0669	50	.0669	.9978	.0670	14.92	1.002	14.96	10	1.5039
.0698	**4° 00′**	.0698	.9976	.0699	14.30	1.002	14.34	**86° 00′**	1.5010
.0727	10	.0727	.9974	.0729	13.73	1.003	13.76	50	1.4981
.0756	20	.0756	.9971	.0758	13.20	1.003	13.23	40	1.4952
.0785	30	.0785	.9969	.0787	12.71	1.003	12.75	30	1.4923
.0814	40	.0814	.9967	.0816	12.25	1.003	12.29	20	1.4893
.0844	50	.0843	.9964	.0846	11.83	1.004	11.87	10	1.4864
.0873	**5° 00′**	.0872	.9962	.0875	11.43	1.004	11.47	**85° 00′**	1.4835
.0902	10	.0901	.9959	.0904	11.06	1.004	11.10	50	1.4806
.0931	20	.0929	.9957	.0934	10.71	1.004	10.76	40	1.4777
.0960	30	.0958	.9954	.0963	10.39	1.005	10.43	30	1.4748
.0989	40	.0987	.9951	.0992	10.08	1.005	10.13	20	1.4719
.1018	50	.1016	.9948	.1022	9.788	1.005	9.839	10	1.4690
.1047	**6° 00′**	.1045	.9945	.1051	9.514	1.006	9.567	**84° 00′**	1.4661
.1076	10	.1074	.9942	.1080	9.255	1.006	9.309	50	1.4632
.1105	20	.1103	.9939	.1110	9.010	1.006	9.065	40	1.4603
.1134	30	.1132	.9936	.1139	8.777	1.006	8.834	30	1.4573
.1164	40	.1161	.9932	.1169	8.556	1.007	8.614	20	1.4544
.1193	.50	.1190	.9929	.1198	8.345	1.007	8.405	10	1.4515
.1222	**7° 00′**	.1219	.9925	.1228	8.144	1.008	8.206	**83° 00′**	1.4486
.1251	10	.1248	.9922	.1257	7.953	1.008	8.016	50	1.4457
.1280	20	.1276	.9918	.1287	7.770	1.008	7.834	40	1.4428
.1309	30	.1305	.9914	.1317	7.596	1.009	7.661	30	1.4399
.1338	40	.1334	.9911	.1346	7.429	1.009	7.496	20	1.4370
.1367	50	.1363	.9907	.1376	7.269	1.009	7.337	10	1.4341
.1396	**8° 00′**	.1392	.9903	.1405	7.115	1.010	7.185	**82° 00′**	1.4312
.1425	10	.1421	.9899	.1435	6.968	1.010	7.040	50	1.4283
.1454	20	.1449	.9894	.1465	6.827	1.011	6.900	40	1.4254
.1484	30	.1478	.9890	.1495	6.691	1.011	6.765	30	1.4224
.1513	40	.1507	.9886	.1524	6.561	1.012	6.636	20	1.4195
.1542	50	.1536	.9881	.1554	6.435	1.012	6.512	10	1.4166
.1571	**9° 00′**	.1564	.9877	.1584	6.314	1.012	6.392	**81° 00′**	1.4137
.1600	10	.1593	.9872	.1614	6.197	1.013	6.277	50	1.4108
.1629	20	.1622	.9868	.1644	6.084	1.013	6.166	40	1.4079
.1658	30	.1650	.9863	.1673	5.976	1.014	6.059	30	1.4050
.1687	40	.1679	.9858	.1703	5.871	1.014	5.955	20	1.4021
.1716	50	.1708	.9853	.1733	5.769	1.015	5.855	10	1.3992
.1745	**10° 00′**	.1736	.9848	.1763	5.671	1.015	5.759	**80° 00′**	1.3963
.1774	10	.1765	.9843	.1793	5.576	1.016	5.665	50	1.3934
.1804	20	.1794	.9838	.1823	5.485	1.016	5.575	40	1.3904
.1833	30	.1822	.9833	.1853	5.396	1.017	5.487	30	1.3875
.1862	40	.1851	.9827	.1883	5.309	1.018	5.403	20	1.3846
.1891	50	.1880	.9822	.1914	5.226	1.018	5.320	10	1.3817
.1920	**11° 00′**	.1908	.9816	.1944	5.145	1.019	5.241	**79° 00′**	1.3788
.1949	10	.1937	.9811	.1974	5.066	1.019	5.164	50	1.3759
.1978	20	.1965	.9805	.2004	4.989	1.020	5.089	40	1.3730
.2007	30	.1994	.9799	.2035	4.915	1.020	5.016	30	1.3701
		cos x	sin x	cot x	tan x	csc x	sec x	x degrees	x radians

NATURAL TRIGONOMETRIC FUNCTIONS TO FOUR PLACES (continued)

x radians	x degrees	sin x	cos x	tan x	cot x	sec x	csc x		
.2036	40	.2022	.9793	.2065	3.843	1.021	4.945	20	1.3672
.2065	50	.2051	.9787	.2095	4.773	1.022	4.876	10	1.3643
.2094	**12° 00′**	.2079	.9781	.2126	4.705	1.022	4.810	**78° 00′**	1.3614
.2123	10	.2108	.9775	.2156	4.638	1.023	4.745	50	1.3584
.2153	20	.2136	.9769	.2186	4.574	1.024	4.682	40	1.3555
.2182	30	.2164	.9763	.2217	4.511	1.025	4.620	30	1.3526
.2211	40	.2193	.9757	.2247	4.449	1.025	4.560	20	1.3497
.2240	50	.2221	.9750	.2278	4.390	1.026	4.502	10	1.3468
.2269	**13° 00′**	.2250	.9744	.2309	4.331	1.026	4.445	**77° 00′**	1.3439
.2298	10	.2278	.9737	.2339	4.275	1.027	4.390	50	1.3410
.2327	20	.2306	.9730	.2370	4.219	1.028	4.336	40	1.3381
.2356	30	.2334	.9724	.2401	4.165	1.028	4.284	30	1.3352
.2385	40	.2363	.9717	.2432	4.113	1.029	4.232	20	1.3323
.2414	50	.2391	.9710	.2462	4.061	1.030	4.182	10	1.3294
.2443	**14° 00′**	.2419	.9703	.2493	4.011	1.031	4.134	**76° 00′**	1.3265
.2473	10	.2447	.9696	.2524	3.962	1.031	4.086	50	1.3235
.2502	20	.2476	.9689	.2555	3.914	1.032	4.039	40	1.3206
.2531	30	.2404	.9681	.2586	3.867	1.033	3.994	30	1.3177
.2560	40	.2532	.9674	.2617	3.821	1.034	3.950	20	1.3148
.2589	50	.2560	.9667	.2648	3.776	1.034	3.906	10	1.3119
.2618	**15° 00′**	.2588	.9659	.2679	3.732	1.035	3.864	**75° 00′**	1.3090
.2647	10	.2616	.9652	.2711	3.689	1.036	3.822	50	1.3061
.2676	20	.2644	.9644	.2732	3.647	1.037	3.782	40	1.3032
.2705	30	.2672	.9636	.2773	3.606	1.038	3.742	30	1.3003
.2734	40	.2700	.9628	.2805	3.566	1.039	3.703	20	1.2974
.2763	50	.2728	.9621	2836	3.526	1.039	3.665	10	1.2945
.2793	**16° 00′**	.2756	.9613	.2867	3.487	1.040	3.628	**74° 00′**	1.2915
.2822	10	.2784	.9605	.2899	3.450	1.041	3.592	50	1.2886
.2851	20	.2812	.9596	.2931	3.412	1.042	3.556	40	1.2857
.2880	30	.2840	.9588	.2962	3.376	1.043	3.521	30	1.2828
.2909	40	.2868	.9580	.2994	3.340	1.044	3.487	20	1.2799
.2938	50	.2896	.9572	.3026	3.305	1.045	3.453	10	1.2770
.2967	**17° 00′**	.2924	.9563	.3057	3.271	1.046	3.420	**73° 00′**	1.2741
.2996	10	.2952	.9555	.3089	3.237	1.047	3.388	50	1.2712
.3025	20	.2979	.9546	.3121	4.204	1.048	3.356	40	1.2683
.3054	30	.3007	.9537	.3153	3.172	1.049	3.326	30	1.2654
.3083	40	.3035	.9528	.3185	3.140	1.049	3.295	20	1.2625
.3113	50	.3062	.9520	.3217	3.108	1.050	3.265	10	1.2595
.3142	**18° 00′**	.3090	.9511	.3249	3.078	1.051	3.236	**72° 00′**	1.2566
.3171	10	.3118	.9502	.3281	3.047	1.052	3.207	50	1.2537
.3200	20	.3145	.9492	.3314	3.018	1.053	3.179	40	1.2508
.3229	30	.3173	.9483	.3346	2.989	1.054	3.152	30	1.2479
.3258	40	.3201	.9474	.3378	2.960	1.056	3.124	20	1.2450
.3287	50	.3228	.9465	.3411	2.932	1.057	3.098	10	1.2421
.3316	**19° 00′**	.3256	.9455	.3443	2.904	1.058	3.072	**71° 00′**	1.2392
.3345	10	.3283	.9446	.3476	2.877	1.059	3.046	50	1.2363
.3374	20	.3311	.9436	.3508	2.850	1.060	3.021	40	1.2334
.3403	30	.3338	.9426	.3541	2.824	1.061	2.996	30	1.2305
.3432	40	.3365	.9417	.3574	2.798	1.062	2.971	20	1.2275
.3462	50	.3393	.9407	.3607	2.773	1.063	2.947	10	1.2246
.3491	**20° 00′**	.3420	.9397	.3640	2.747	1.064	2.924	**70° 00′**	1.2217
.3520	10	.3448	.9387	.3673	2.723	1.065	2.901	50	1.2188
.3599	20	.3475	.9377	.3706	2.699	1.066	2.878	40	1.2159
.3578	30	.3502	.9367	.3739	2.675	1.068	2.855	30	1.2130
.3607	40	.3529	.9356	.3772	2.651	1.069	2.833	20	1.2101
.3636	50	.3557	.9346	.3805	2.628	1.070	2.812	10	1.2072
.3665	**21° 00′**	.3584	.9336	.3839	2.605	1.071	2.790	**69° 00′**	1.2043
.3694	10	.3611	.9325	.3872	2.583	1.072	2.769	50	1.2014
.3723	20	.3638	.9315	.3906	2.560	1.074	2.749	40	1.1985
.3752	30	.3665	.9304	.3939	2.539	1.075	2.729	30	1.1956
.3782	40	.3692	.9293	.3973	2.517	1.076	2.709	20	1.1926
.3811	50	.3719	.9283	.4006	2.496	1.077	2.689	10	1.1897
.3840	**22° 00′**	.3746	.9272	.4040	2.475	1.079	2.669	**68° 00′**	1.1868
.3869	10	.3773	.9261	.4074	2.455	1.080	2.650	50	1.1839
.3898	20	.3800	.9250	.4108	2.434	1.081	2.632	40	1.1810
.3927	30	.3827	.9239	.4142	2.414	1.082	2.613	30	1.1781
.3956	40	.3854	.9228	.4176	2.394	1.084	2.595	20	1.1752
.3985	50	.3881	.9216	.4210	2.375	1.085	2.577	10	1.1723
.4014	**23° 00′**	.3907	.9205	.4245	2.356	1.086	2.559	**67° 00′**	1.1694
.4043	10	.3934	.9194	.4279	2.337	1.088	2.542	50	1.1665
		cos x	sin x	cot x	tan x	csc x	sec x	x degrees	x radians

NATURAL TRIGONOMETRIC FUNCTIONS TO FOUR PLACES (continued)

x radians	x degrees	sin x	cos x	tan x	cot x	sec x	csc x		
.4072	20	.3961	.9182	.4314	2.318	1.089	2.525	40	1.1636
.4102	30	.3987	.9171	.4348	2.300	1.090	2.508	30	1.1606
4131	40	.4014	.9159	.4383	2.282	1.092	2.491	20	1.1577
.4160	50	4041	.9147	.4417	2.264	1.093	2.475	10	1.1548
.4189	**24° 00′**	.4067	.9135	.4452	2.246	1.095	2.459	**66° 00′**	1.1519
.4218	10	.4094	.9124	.4487	2.229	1.096	2.443	50	1.1490
.4247	20	.4120	.9112	.4522	2.211	1.097	2.427	40	1.1461
.4276	30	.4147	.9100	.4557	2.194	1.099	2.411	30	1.1432
.4305	40	.4173	.9088	.4592	2.177	1.100	2.396	20	1.1403
.4334	50	.4200	.9075	.4628	2.161	1.102	2.381	10	1.1374
.4363	**25° 00′**	.4226	.9063	.4663	2.145	1.103	2.366	**65° 00′**	1.1345
.4392	10	.4253	.9051	.4699	2.128	1.105	2.352	50	1.1316
.4422	20	.4279	.9038	.4734	2.112	1.106	2.337	40	1.1286
.4451	30	.4305	.9026	.4770	2.097	1.108	2.323	30	1.1257
.4480	40	.4331	.9013	.4806	2.081	1.109	2.309	20	1.1228
.4509	50	.4358	.9001	.4841	2.066	1.111	2.295	10	1.1199
.4538	**26° 00′**	.4384	.8988	.4877	2.050	1.113	2.281	**64° 00′**	1.1170
.4567	10	.4410	.8975	.4913	2.035	1.114	2.268	50	1.1141
.4596	20	.4436	.8962	.4950	2.020	1.116	2.254	40	1.1112
.4625	30	.4462	.8949	.4986	2.006	1.117	2.241	30	1.1083
.4654	40	.4488	.8936	.5022	1.991	1.119	2.228	20	1.1054
.4683	50	.4514	.8923	.5059	1.977	1.121	2.215	10	1.1025
.4712	**27° 00′**	.4540	.8910	.5095	1.963	1.122	2.203	**63° 00′**	1.0996
.4741	10	.4566	.8897	.5132	1.949	1.124	2.190	50	1.0966
.4771	20	.4592	.8884	.5169	1.935	1.126	2.178	40	1.0937
.4800	30	.4617	.8870	.5206	1.921	1.127	2.166	30	1.0908
.4829	40	.4643	.8857	.5243	1.907	1.129	2.154	20	1.0879
.4858	50	.4669	.8843	.5280	1.894	1.131	2.142	10	1.0850
.4887	**28° 00′**	.4695	.8829	.5317	1.881	1.133	2.130	**62° 00′**	1.0821
.4916	10	.4720	.8816	.5354	1.868	1.134	2.118	50	1.0792
.4945	20	.4746	.8802	.5392	1.855	1.136	2.107	40	1.0763
.4974	30	.4772	.8788	.5430	1.842	1.138	2.096	30	1.0734
.5003	40	.4797	.8774	.5467	1.829	1.140	2.085	20	1.0705
.5032	50	.4823	.8760	.5505	1.816	1.142	2.074	10	1.0676
.5061	**29° 00′**	.4848	.8746	.5543	1.804	1.143	2.063	**61° 00′**	1.0647
.5091	10	.4874	.8732	.5581	1.792	1.145	2.052	50	1.0617
.5120	20	.4899	.8718	.5619	1.780	1.147	2.041	40	1.0588
.5149	30	.4924	.8704	.5658	1.767	1.149	2.031	30	1.0559
.5178	40	.4950	.8689	.5696	1.756	1.151	2.020	20	1.0530
.5207	50	.4975	.8675	.5735	1.744	1.153	2.010	10	1.0501
.5236	**30° 00′**	.5000	.8660	.5774	1.732	1.155	2.000	**60° 00′**	1.0472
.5265	10	.5025	.8646	.5812	1.720	1.157	1.990	50	1.0443
.5294	20	.5050	.8631	.5851	1.709	1.159	1.980	40	1.0414
.5323	30	.5075	.8616	.5890	1.698	1.161	1.970	30	1.0385
.5352	40	.5100	.8601	.5930	1.686	1.163	1.961	20	1.0356
.5381	50	.5125	.8587	.5969	1.675	1.165	1.951	10	1.0327
.5411	**31° 00′**	.5150	.8572	.6009	1.664	1.167	1.942	**59° 00′**	1.0297
.5440	10	.5175	.8557	.6048	1.653	1.169	1.932	50	1.0268
.5469	20	.5200	.8542	.6088	1.643	1.171	1.923	40	1.0239
.5498	30	.5225	.8526	.6128	1.632	1.173	1.914	30	1.0210
.5527	40	.5250	.8511	.6168	1.621	1.175	1.905	20	1.0181
.5556	50	.5275	.8496	.6208	1.611	1.177	1.896	10	1.0152
.5585	**32° 00′**	.5299	8480	.6249	1.600	1.179	1.887	**58° 00′**	1.0123
.5614	10	.5324	.8465	.6289	1.590	1.181	1.878	50	1.0094
.5643	20	.5348	.8450	.6330	1.580	1.184	1.870	40	1.0065
.5672	30	.5373	.8434	.6371	1.570	1.186	1.861	30	1.0036
.5701	40	.5398	.8418	.6412	1.560	1.188	1.853	20	1.0007
.5730	50	.5422	.8403	.6453	1.550	1.190	1.844	10	.9977
.5760	**33° 00′**	.5446	.8397	.6494	1.540	1.192	1.836	**57° 00′**	.9948
.5789	10	.5471	.8371	.6536	1.530	1.195	1.828	50	.9919
.5818	20	.5495	.8355	.6577	1.520	1.197	1.820	40	.9890
.5847	30	.5519	.8339	.6619	1.511	1.199	1.812	30	.9861
.5876	40	.5544	.8323	.6661	1.501	1.202	1.804	20	.9832
.5905	50	.5568	.8307	.6703	1.492	1.204	1.796	10	.9003
.5934	**34° 00′**	.5592	.8290	.6745	1.483	1.206	1.788	**56° 00′**	.9774
.5963	10	.5616	.8274	.6787	1.473	1.209	1.781	50	.9745
.5992	20	.5640	.8258	.6830	1.464	1.211	1.773	40	.9716
.6021	30	.5664	.8241	.6873	1.455	1.213	1.766	30	.9687
.6050	40	.5688	.8225	.6916	1.446	1.216	1.758	20	.9657
.6080	50	.5712	.8208	.6959	1.437	1.218	1.751	10	.9628
		cos x	**sin x**	**cot x**	**tan x**	**csc x**	**sec x**	**x degrees**	**x radians**

NATURAL TRIGONOMETRIC FUNCTIONS TO FOUR PLACES (continued)

x radians	x degrees	sin x	cos x	tan x	cot x	sec x	csc x		
.6109	**35° 00′**	.5736	.8192	.7002	1.428	1.221	1.743	**55° 00′**	.9599
.6138	10	.5760	.8175	.7046	1.419	1.223	1.736	50	.9570
.6167	20	.5783	.8158	.7089	1.411	1.226	1.729	40	.9541
.6196	30	.5807	.8141	.7133	1.402	1.228	1.722	30	.9512
.6225	40	.5831	.8124	.7177	1.393	1.231	1.715	20	.9483
.6254	50	.5854	.8107	.7221	1.385	1.233	1.708	10	.9454
.6283	**36° 00′**	.5878	.8090	.7265	1.376	1.236	1.701	**54° 00′**	.9425
.6312	10	.5901	.8073	.7310	1.368	1.239	1.695	50	.9396
.6341	20	.5925	.8056	.7355	1.360	1.241	1.688	40	.9367
.6370	30	.5948	.8039	.7400	1.351	1.244	1.681	30	.9338
.6400	40	.5972	.8021	.7445	1.343	1.247	1.675	20	.9308
.6429	50	.5995	.8004	.7490	1.335	1.249	1.668	10	.9279
.6458	**37° 00′**	.6018	.7986	.7536	1.327	1.252	1.662	**53° 00′**	.9250
.6487	10	.6041	.7969	.7581	1.319	1.255	1.655	50	.9221
.6516	20	.6065	.7951	.7627	1.311	1.258	1.649	40	.9192
.6545	30	.6088	.7934	.7673	1.303	1.260	1.643	30	.9163
.6574	40	.6111	.7916	.7720	1.295	1.263	1.636	20	.9134
.6603	50	.6134	.7898	.7766	1.288	1.266	1.630	10	.9105
.6632	**38° 00′**	.6157	.7880	.7813	1.280	1.269	1.624	**52° 00′**	.9076
.6661	10	.6180	.7862	.7860	1.272	1.272	1.618	50	.9047
.6690	20	.6202	.7844	.7907	1.265	1.275	1.612	40	.9018
.6720	30	.6225	.7826	.7954	1.257	1.278	1.606	30	.8988
.6749	40	.6248	.7808	.8002	1.250	1.281	1.601	20	.8959
.6778	50	.6271	.7790	.8050	1.242	1.284	1.595	10	.8930
.6807	**39° 00′**	.6293	.7771	.8098	1.235	1.287	1.589	**51° 00′**	.8901
.6836	10	.6316	.7753	.8146	1.228	1.290	1.583	50	.8872
.6865	20	.6338	.7735	.8195	1.220	1.293	1.578	40	.8843
.6894	30	.6361	.7716	.8243	1.213	1.296	1.572	30	.8814
.6923	40	.6383	.7698	.8292	1.206	1.299	1.567	20	.8785
.6952	50	.6406	.7679	.8342	1.199	1.302	1.561	10	.8756
.6981	**40° 00′**	.6428	.7660	.8391	1.192	1.305	1.556	**50° 00′**	.8727
.7010	10	.6450	.7642	.8441	1.185	1.309	1.550	50	.8698
.7039	20	.6472	.7623	.8491	1.178	1.312	1.545	40	.8668
.7069	30	.6494	.7604	.8541	1.171	1.315	1.540	30	.8639
.7098	40	.6517	.7585	.8591	1.164	1.318	1.535	20	.8610
.7127	50	.6539	.7566	.8642	1.157	1.322	1.529	10	.8581
.7156	**41° 00′**	.6561	.7547	.8693	1.150	1.325	1.524	**49° 00′**	.8552
.7185	10	.6583	.7528	.8744	1.144	1.328	1.519	50	.8523
.7214	20	.6604	.7509	.8796	1.137	1.332	1.514	40	.8494
.7243	30	.6626	.7490	.8847	1.130	1.335	1.509	30	.8465
.7272	40	.6648	.7470	.8899	1.124	1.339	1.504	20	.8436
.7301	50	.6670	.7451	.8952	1.117	1.342	1.499	10	.8407
.7330	**42° 00′**	.6691	.7431	.9004	1.111	1.346	1.494	**48° 00′**	.8378
.7359	10	.6713	.7412	.9057	1.104	1.349	1.490	50	.8348
7389	20	.6734	.7392	.9110	1.098	1.353	1.485	40	.8319
.7418	30	.6756	.7373	.9163	1.091	1.356	1.480	30	.8290
.7447	40	.6777	.7353	.9217	1.085	1.360	1.476	20	.8261
.7476	50	.6799	.7333	.9271	1.079	1.364	1.471	10	.8232
.7505	**43° 00′**	.6820	.7314	.9325	1.072	1.367	1.466	**47° 00′**	.8203
.7534	10	.6841	.7294	.9380	1.066	1.371	1.462	50	.8174
.7563	20	.6862	.7274	.9435	1.060	1.375	1.457	40	.8145
.7592	30	.6884	.7254	.9490	1.054	1.379	1.453	30	.8116
.7621	40	.6905	.7234	.9545	1.048	1.382	1.448	20	.8087
.7650	50	.6926	.7214	.9601	1.042	1.386	1.444	10	.8058
.7679	**44° 00′**	.6947	.7193	.9657	1.036	1.390	1.440	**46° 00′**	.8029
.7709	10	.6967	.7173	.9713	1.030	1.394	1.435	50	.7999
.7738	20	.6988	.7153	.9770	1.024	1.398	1.431	40	.7970
.7767	30	.7009	.7133	.9827	1.018	1.402	1.427	30	.7941
.7796	40	.7030	.7112	.9884	1.012	1.406	1.423	20	.7912
.7825	50	.7050	.7092	.9942	1.006	1.410	1.418	10	.7883
.7854	**45° 00′**	.7071	.7071	1.0000	1.0000	1.414	1.414	**45° 00′**	.7854
		cos x	**sin x**	**cot x**	**tan x**	**csc x**	**sec x**	**x degrees**	**x radians**

RELATION OF ANGULAR FUNCTIONS IN TERMS OF ONE ANOTHER

TRIGONOMETRIC FUNCTIONS

Function	sin α	cos α	tan α	cot α	sec α	csc α
sin α	$\sin\alpha$	$\pm\sqrt{1-\cos^2\alpha}$	$\dfrac{\tan\alpha}{\pm\sqrt{1+\tan^2\alpha}}$	$\dfrac{1}{\pm\sqrt{1+\cot^2\alpha}}$	$\dfrac{\pm\sqrt{\sec^2\alpha-1}}{\sec\alpha}$	$\dfrac{1}{\csc\alpha}$
cos α	$\pm\sqrt{1-\sin^2\alpha}$	$\cos\alpha$	$\dfrac{1}{\pm\sqrt{1+\tan^2\alpha}}$	$\dfrac{\cot\alpha}{\pm\sqrt{1+\cot^2\alpha}}$	$\dfrac{1}{\sec\alpha}$	$\dfrac{\pm\sqrt{\csc^2\alpha-1}}{\csc\alpha}$
tan α	$\dfrac{\sin\alpha}{\pm\sqrt{1-\sin^2\alpha}}$	$\dfrac{\pm\sqrt{1-\cos^2\alpha}}{\cos\alpha}$	$\tan\alpha$	$\dfrac{1}{\cot\alpha}$	$\pm\sqrt{\sec^2\alpha-1}$	$\dfrac{1}{\pm\sqrt{\csc^2\alpha-1}}$
cot α	$\dfrac{\pm\sqrt{1-\sin^2\alpha}}{\sin\alpha}$	$\dfrac{\cos\alpha}{\pm\sqrt{1-\cos^2\alpha}}$	$\dfrac{1}{\tan\alpha}$	$\cot\alpha$	$\dfrac{1}{\pm\sqrt{\sec^2\alpha-1}}$	$\pm\sqrt{\csc^2\alpha-1}$
sec α	$\dfrac{1}{\pm\sqrt{1-\sin^2\alpha}}$	$\dfrac{1}{\cos\alpha}$	$\pm\sqrt{1+\tan^2\alpha}$	$\dfrac{\pm\sqrt{1+\cot^2\alpha}}{\cot\alpha}$	$\sec\alpha$	$\dfrac{\csc\alpha}{\pm\sqrt{\csc^2\alpha-1}}$
csc α	$\dfrac{1}{\sin\alpha}$	$\dfrac{1}{\pm\sqrt{1-\cos^2\alpha}}$	$\dfrac{\pm\sqrt{1+\tan^2\alpha}}{\tan\alpha}$	$\pm\sqrt{1+\cot^2\alpha}$	$\dfrac{\sec\alpha}{\pm\sqrt{\sec^2\alpha-1}}$	$\csc\alpha$

Note: The choice of sign depends upon the quadrant in which the angle terminates.

HYPERBOLIC FUNCTIONS

Function	sinh x	cosh x	tanh x
sinh x =	$\sinh x$	$+\sqrt{\cosh^2 x-1}$	$\dfrac{\tanh x}{\sqrt{1-\tanh^2 x}}$
cosh x =	$\sqrt{1+\sinh^2 x}$	$\cosh x$	$\dfrac{1}{\sqrt{1-\tanh^2 x}}$
tanh x =	$\dfrac{\sinh x}{\sqrt{1+\sinh^2 x}}$	$\pm\dfrac{\sqrt{\cosh^2 x-1}}{\cosh x}$	$\tanh x$
cosech x =	$\dfrac{1}{\sinh x}$	$\pm\dfrac{1}{\sqrt{\cosh^2 x-1}}$	$\dfrac{\sqrt{1-\tanh^2 x}}{\tanh x}$
sech x =	$\dfrac{1}{\sqrt{1+\sinh^2 x}}$	$\dfrac{1}{\cosh x}$	$\sqrt{1-\tanh^2 x}$
coth x =	$\dfrac{\sqrt{1+\sinh^2 x}}{\sinh x}$	$\dfrac{\pm\cosh x}{\sqrt{\cosh^2 x-1}}$	$\dfrac{1}{\tanh x}$

Function	cosech x	sech x	coth x
sinh x =	$\dfrac{1}{\mathrm{cosech}\, x}$	$\pm\dfrac{\sqrt{1-\mathrm{sech}^2 x}}{\mathrm{sech}\, x}$	$\dfrac{\pm 1}{\sqrt{\coth^2 x-1}}$
cosh x =	$\pm\dfrac{\sqrt{\mathrm{cosech}^2 x+1}}{\mathrm{cosech}\, x}$	$\dfrac{1}{\mathrm{sech}\, x}$	$\pm\dfrac{\coth x}{\sqrt{\coth^2 x-1}}$
tanh x =	$\dfrac{1}{\sqrt{\mathrm{cosech}^2 x+1}}$	$\pm\sqrt{1-\mathrm{sech}^2 x}$	$\dfrac{1}{\coth x}$
cosech x =	$\mathrm{cosech}\, x$	$\pm\dfrac{\mathrm{sech}\, x}{\sqrt{1-\mathrm{sech}^2\, zsx}}$	$\pm\dfrac{\sqrt{\coth^2 x-1}}{1}$
sech x =	$\pm\dfrac{\mathrm{cosec}\, x}{\sqrt{\mathrm{cosech}^2 x+1}}$	$\mathrm{sech}\, x$	$\pm\dfrac{\sqrt{\coth^2 x-1}}{\coth x}$
coth x =	$\sqrt{\mathrm{cosech}^2 x+1}$	$\pm\dfrac{1}{\sqrt{1-\mathrm{sech}^2 x}}$	$\coth x$

Whenever two signs are shown, choose + sign if x is positive, − sign if x is negative.

Derivatives*

In the following formulas u, v, w represent functions of x, while a, c, n represent fixed real numbers. All arguments in the trigonometric functions are measured in radians, and all inverse trigonometric and hyperbolic functions represent principal values.

1. $\frac{d}{dx}(a) = 0$

2. $\frac{d}{dx}(x) = 1$

3. $\frac{d}{dx}(au) = a\frac{du}{dx}$

4. $\frac{d}{dx}(u + v - w) = \frac{du}{dx} + \frac{dv}{dx} - \frac{dw}{dx}$

5. $\frac{d}{dx}(uv) = u\frac{dv}{dx} + v\frac{du}{dx}$

6. $\frac{d}{dx}(uvw) = uv\frac{dw}{dx} + vw\frac{du}{dx} + uw\frac{dv}{dx}$

7. $\frac{d}{dx}\left(\frac{u}{v}\right) = \frac{v\frac{du}{dx} - u\frac{dv}{dx}}{v^2} = \frac{1}{v}\frac{du}{dx} - \frac{u}{v^2}\frac{dv}{dx}$

8. $\frac{d}{dx}(u^n) = nu^{n-1}\frac{du}{dx}$

9. $\frac{d}{dx}(\sqrt{u}) = \frac{1}{2\sqrt{u}}\frac{du}{dx}$

10. $\frac{d}{dx}\left(\frac{1}{u}\right) = -\frac{1}{u^2}\frac{du}{dx}$

11. $\frac{d}{dx}\left(\frac{1}{u^n}\right) = -\frac{n}{u^{n+1}}\frac{du}{dx}$

12. $\frac{d}{dx}\left(\frac{u^n}{v^m}\right) = \frac{u^{n-1}}{v^{m+1}}\left(nv\frac{du}{dx} - mu\frac{dv}{dx}\right)$

13. $\frac{d}{dx}(u^n v^m) = u^{n-1}v^{m-1}\left(nv\frac{du}{dx} + mu\frac{dv}{dx}\right)$

14. $\frac{d}{dx}[f(u)] = \frac{d}{du}[f(u)] \cdot \frac{du}{dx}$

* Let $y = f(x)$ and $\frac{dy}{dx} = \frac{d[f(x)]}{dx} = f'(x)$ define respectively a function and its derivative for any value x in their common domain. The differential for the function at such a value x is accordingly defined as

$$dy = d[f(x)] = \frac{dy}{dx}dx = \frac{d[f(x)]}{dx}dx = f'(x)\,dx$$

Each derivative formula has an associated differential formula. For example, formula 6 above has the differential formula

$$d(uvw) = uv\,dw + vw\,du + uw\,dv$$

15. $\frac{d^2}{dx^2}[f(u)] = \frac{df(u)}{du} \cdot \frac{d^2u}{dx^2} + \frac{d^2f(u)}{du^2} \cdot \left(\frac{du}{dx}\right)^2$

16. $\frac{d^n}{dx^n}[uv] = \binom{n}{0} v \frac{d^n u}{dx^n} + \binom{n}{1} \frac{dv}{dx} \frac{d^{n-1}u}{dx^{n-1}} + \binom{n}{2} \frac{d^2v}{dx^2} \frac{d^{n-2}u}{dx^{n-2}}$

$+ \cdots + \binom{n}{k} \frac{d^k v}{dx^k} \frac{d^{n-k}u}{dx^{n-k}} + \cdots + \binom{n}{n} u \frac{d^n v}{dx^n}$

where $\binom{n}{r} = \frac{n!}{r!(n-r)!}$ the binomial coefficient, n non-negative integer and $\binom{n}{0} = 1$.

17. $\frac{du}{dx} = \frac{1}{\frac{dx}{du}}$ if $\frac{dx}{du} \neq 0$

18. $\frac{d}{dx}(\log_a u) = (\log_a e)\frac{1}{u}\frac{du}{dx}$

19. $\frac{d}{dx}(\log_e u) = \frac{1}{u}\frac{du}{dx}$

20. $\frac{d}{dx}(a^u) = a^u(\log_e a)\frac{du}{dx}$

21. $\frac{d}{dx}(e^u) = e^u\frac{du}{dx}$

22. $\frac{d}{dx}(u^v) = vu^{v-1}\frac{du}{dx} + (\log_e u)u^v\frac{dv}{dx}$

23. $\frac{d}{dx}(\sin u) = \frac{du}{dx}(\cos u)$

24. $\frac{d}{dx}(\cos u) = -\frac{du}{dx}(\sin u)$

25. $\frac{d}{dx}(\tan u) = \frac{du}{dx}(\sec^2 u)$

26. $\frac{d}{dx}(\cot u) = -\frac{du}{dx}(\csc^2 u)$

27. $\frac{d}{dx}(\sec u) = \frac{du}{dx} \sec u \cdot \tan u$

28. $\frac{d}{dx}(\csc u) = -\frac{du}{dx} \csc u \cdot \cot u$

29. $\frac{d}{dx}(\text{vers } u) = \frac{du}{dx} \sin u$

30. $\frac{d}{dx}(\arcsin u) = \frac{1}{\sqrt{1-u^2}}\frac{du}{dx}, \quad \left(-\frac{\pi}{2} \leq \arcsin u \leq \frac{\pi}{2}\right)$

31. $\frac{d}{dx}(\text{arc cos } u) = -\frac{1}{\sqrt{1-u^2}}\frac{du}{dx}, \qquad (0 \le \text{arc cos } u \le \pi)$

32. $\frac{d}{dx}(\text{arc tan } u) = \frac{1}{1+u^2}\frac{du}{dx}, \qquad \left(-\frac{\pi}{2} < \text{arc tan } u < \frac{\pi}{2}\right)$

33. $\frac{d}{dx}(\text{arc cot } u) = -\frac{1}{1+u^2}\frac{du}{dx}, \qquad (0 \le \text{arc cot } u \le \pi)$

34. $\frac{d}{dx}(\text{arc sec } u) = \frac{1}{u\sqrt{u^2-1}}\frac{du}{dx}, \qquad \left(0 \le \text{arc sec } u < \frac{\pi}{2}, -\pi \le \text{arc sec } u < -\frac{\pi}{2}\right)$

35. $\frac{d}{dx}(\text{arc csc } u) = -\frac{1}{u\sqrt{u^2-1}}\frac{du}{dx}, \qquad \left(0 < \text{arc csc } u \le \frac{\pi}{2}, -\pi < \text{arc csc } u \le -\frac{\pi}{2}\right)$

36. $\frac{d}{dx}(\text{arc vers } u) = \frac{1}{\sqrt{2u-u^2}}\frac{du}{dx}, \qquad (0 \le \text{arc vers } u \le \pi)$

37. $\frac{d}{dx}(\sinh u) = \frac{du}{dx}(\cosh u)$

38. $\frac{d}{dx}(\cosh u) = \frac{du}{dx}(\sinh u)$

39. $\frac{d}{dx}(\tanh u) = \frac{du}{dx}(\text{sech}^2 u)$

40. $\frac{d}{dx}(\coth u) = -\frac{du}{dx}(\text{csch}^2 u)$

41. $\frac{d}{dx}(\text{sech } u) = -\frac{du}{dx}(\text{sech } u \cdot \tanh u)$

42. $\frac{d}{dx}(\text{csch } u) = -\frac{du}{dx}(\text{csch } u \cdot \coth u)$

43. $\frac{d}{dx}(\sinh^{-1} u) = \frac{d}{dx}[\log(u + \sqrt{u^2+1})] = \frac{1}{\sqrt{u^2+1}}\frac{du}{dx}$

44. $\frac{d}{dx}(\cosh^{-1} u) = \frac{d}{dx}[\log(u + \sqrt{u^2-1})] = \frac{1}{\sqrt{u^2-1}}\frac{du}{dx}, \qquad (u > 1, \cosh^{-1} u > 0)$

45. $\frac{d}{dx}(\tanh^{-1} u) = \frac{d}{dx}\left[\frac{1}{2}\log\frac{1+u}{1-u}\right] = \frac{1}{1-u^2}\frac{du}{dx}, \qquad (u^2 < 1)$

46. $\frac{d}{dx}(\coth^{-1} u) = \frac{d}{dx}\left[\frac{1}{2}\log\frac{u+1}{u-1}\right] = \frac{1}{1-u^2}\frac{du}{dx}, \qquad (u^2 > 1)$

47. $\frac{d}{dx}(\text{sech}^{-1} u) = \frac{d}{dx}\left[\log\frac{1+\sqrt{1-u^2}}{u}\right] = -\frac{1}{u\sqrt{1-u^2}}\frac{du}{dx}, \qquad (0 < u < 1, \text{sech}^{-1} u > 0)$

48. $\frac{d}{dx}(\text{csch}^{-1} u) = \frac{d}{dx}\left[\log\frac{1+\sqrt{1+u^2}}{u}\right] = -\frac{1}{|u|\sqrt{1+u^2}}\frac{du}{dx}$

49. $\frac{d}{dq}\int_p^q f(x)\,dx = f(q), \qquad [p \text{ constant}]$

50. $\frac{d}{dp}\int_p^q f(x)\,dx = -f(p), \qquad [q \text{ constant}]$

51. $\frac{d}{da}\int_p^q f(x,a)\,dx = \int_p^q \frac{\partial}{\partial a}[f(x,a)]\,dx + f(q,a)\frac{dq}{da} - f(p,a)\frac{dp}{da}$

INTEGRATION

The following is a brief discussion of some integration techniques. A more complete discussion can be found in a number of good text books. However, the purpose of this introduction is simply to discuss a few of the important techniques which may be used, in conjunction with the integral table which follows, to integrate particular functions.

No matter how extensive the integral table, it is a fairly uncommon occurrence to find in the table the exact integral desired. Usually some form of transformation will have to be made. The simplest type of transformation, and yet the most general, is substitution. Simple forms of substitution, such as $y = ax$, are employed almost unconsciously by experienced users of integral tables. Other substitutions may require more thought. In some sections of the tables, appropriate substitutions are suggested for integrals which are similar to, but not exactly like, integrals in the table. Finding the right substitution is largely a matter of intuition and experience.

Several precautions must be observed when using substitutions:

1. Be sure to make the substitution in the dx term, as well as everywhere else in the integral.
2. Be sure that the function substituted is one-to-one and continuous. If this is not the case, the integral must be restricted in such a way as to make it true. See the example following.
3. With definite integrals, the limits should also be expressed in terms of the new dependent variable. With indefinite integrals, it is necessary to perform the reverse substitution to obtain the answer in terms of the original independent variable. This may also be done for definite integrals, but it is usually easier to change the limits.

Example:

$$\int \frac{x^4}{\sqrt{a^2 - x^2}}\,dx$$

Here we make the substitution $x = |a| \sin\theta$. Then $dx = |a| \cos\theta\, d\theta$, and

$$\sqrt{a^2 - x^2} = \sqrt{a^2 - a^2 \sin^2\theta} = |a|\sqrt{1 - \sin^2\theta} = |a \cos\theta|$$

Notice the absolute value signs. It is very important to keep in mind that a square root radical always denotes the positive square root, and to assure the sign is always kept positive. Thus $\sqrt{x^2} = |x|$. Failure to observe this is a common cause of errors in integration.

Notice also that the indicated substitution is not a one-to-one function, that is, it does not have a unique inverse. Thus we must restrict the range of θ in such a way as to make the function one-to-one. Fortunately, this is easily done by solving for θ

$$\theta = \sin^{-1}\frac{x}{|a|}$$

and restricting the inverse sine to the principal values, $-\frac{\pi}{2} \le \theta \le \frac{\pi}{2}$.

Thus the integral becomes

$$\int \frac{a^4 \sin^4 \theta |a| \cos \theta \, d\theta}{|a| \, |\cos \theta|}$$

Now, however, in the range of values chosen for θ, $\cos \theta$ is always positive. Thus we may remove the absolute value signs from $\cos \theta$ in the denominator. (This is one of the reasons that the principal values of the inverse trigonometric functions are defined as they are.)

Then the $\cos \theta$ terms cancel, and the integral becomes

$$a^4 \int \sin^4 \theta \, d\theta$$

By application of integral formulas 299 and 296, we integrate this to

$$-a^4 \frac{\sin^3 \theta \cos \theta}{4} - \frac{3a^4}{8} \cos \theta \sin \theta + \frac{3a^4}{8} \theta + C$$

We now must perform the inverse substitution to get the result in terms of x. We have

$$\theta = \sin^{-1} \frac{x}{|a|}$$

$$\sin \theta = \frac{x}{|a|}$$

Then

$$\cos \theta = \pm\sqrt{1 - \sin^2 \theta} = \pm\sqrt{1 - \frac{x^2}{a^2}} = \pm\frac{\sqrt{a^2 - x^2}}{|a|}.$$

Because of the previously mentioned fact that $\cos \theta$ is positive, we may omit the $\pm$ sign. The reverse substitution then produces the final answer

$$\int \frac{x^4}{\sqrt{a^2 - x^2}} dx = -\tfrac{1}{4}x^3\sqrt{a^2 - x^2} - \tfrac{3}{8}a^2x\sqrt{a^2 - x^2} + \frac{3a^4}{8} \sin^{-1} \frac{x}{|a|} + C.$$

Any rational function of x may be integrated, if the denominator is factored into linear and irreducible quadratic factors. The function may then be broken into partial fractions, and the individual partial fractions integrated by use of the appropriate formula from the integral table. See the section on partial fractions for further information.

Many integrals may be reduced to rational functions by proper substitutions. For example,

$$z = \tan \frac{x}{2}$$

will reduce any rational function of the six trigonometric functions of x to a rational function of z. (Frequently there are other substitutions which are simpler to use, but this one will always work. See integral formula number 484.)

Any rational function of x and $\sqrt{ax + b}$ may be reduced to a rational function of z by making the substitution

$$z = \sqrt{ax + b}.$$

Other likely substitutions will be suggested by looking at the form of the integrand.

The other main method of transforming integrals is integration by parts. This involves applying formula number 5 or 6 in the accompanying integral table. The critical factor in this method is the choice of the functions u and v. In order for the method to be successful, $v = \int dv$ and $\int v \, du$ must be easier to integrate than the original integral. Again, this choice is largely a matter of intuition and experience.

Example:

$$\int x \sin x \, dx$$

Two obvious choices are $u = x$, $dv = \sin x \, dx$, or $u = \sin x$, $dv = x \, dx$. Since a preliminary mental calculation indicates that $\int v \, du$ in the second choice would be more, rather than less, complicated than the original integral (it would contain x^2), we use the first choice.

$$u = x \qquad du = dx$$

$$dv = \sin x \, dx \qquad v = -\cos x$$

$$\int x \sin x \, dx = \int u \, dv = uv - \int v \, du = -x \cos x + \int \cos x \, dx$$

$$= \sin x - x \cos x$$

Of course, this result could have been obtained directly from the integral table, but it provides a simple example of the method. In more complicated examples the choice of u and v may not be so obvious, and several different choices may have to be tried. Of course, there is no guarantee that any of them will work.

Integration by parts may be applied more than once, or combined with substitution. A fairly common case is illustrated by the following example.

Example:

$$\int e^x \sin x \, dx$$

Let

$$u = e^x \qquad \text{Then} \quad du = e^x \, dx$$

$$dv = \sin x \, dx \qquad v = -\cos x$$

$$\int e^x \sin x \, dx = \int u \, dv = uv - \int v \, du = -e^x \cos x + \int e^x \cos x \, dx$$

In this latter integral,

$$\text{let} \quad u = e^x \qquad \text{Then} \quad du = e^x \, dx$$

$$dv = \cos x \, dx \qquad v = \sin x$$

$$\int e^x \sin x \, dx = -e^x \cos x + \int e^x \cos x \, dx = -e^x \cos x + \int u \, dv$$

$$= -e^x \cos x + uv - \int v \, du$$

$$= -e^x \cos x + e^x \sin x - \int e^x \sin x \, dx$$

This looks as if a circular transformation has taken place, since we are back at the same integral we started from. However, the above equation can be solved algebraically for the required integral:

$$\int e^x \sin x \, dx = \tfrac{1}{2}(e^x \sin x - e^x \cos x)$$

In the second integration by parts, if the parts had been chosen as $u = \cos x$, $dv = e^x \, dx$, we would indeed have made a circular transformation, and returned to the starting place.

In general, when doing repeated integration by parts, one should never choose the function u at any stage to be the same as the function v at the previous stage, or a constant times the previous v.

The following rule is called the extended rule for integration by parts. It is the result of $n + 1$ successive applications of integration by parts.

If

$$g_1(x) = \int g(x)\,dx, \qquad g_2(x) = \int g_1(x)\,dx,$$

$$g_3(x) = \int g_2(x)\,dx, \ldots, g_m(x) = \int g_{m-1}(x)\,dx, \ldots,$$

then

$$\int f(x)\cdot g(x)\,dx = f(x)\cdot g_1(x) - f'(x)\cdot g_2(x) + f''(x)\cdot g_3(x) - + \cdots$$
$$+ (-1)^n f^{(n)}(x) g_{n+1}(x) + (-1)^{n+1}\int f^{(n+1)}(x) g_{n+1}(x)\,dx.$$

A useful special case of the above rule is when $f(x)$ is a polynomial of degree n. Then $f^{(n+1)}(x) = 0$, and

$$\int f(x)\cdot g(x)\,dx = f(x)\cdot g_1(x) - f'(x)\cdot g_2(x) + f''(x)\cdot g_3(x) - + \cdots + (-1)^n f^{(n)}(x) g_{n+1}(x) + C$$

Example:

If $f(x) = x^2$, $g(x) = \sin x$

$$\int x^2 \sin x\,dx = -x^2 \cos x + 2x \sin x + 2 \cos x + C$$

Another application of this formula occurs if

$$f''(x) = af(x) \quad \text{and} \quad g''(x) = bg(x),$$

where a and b are unequal constants. In this case, by a process similar to that used in the above example for $\int e^x \sin x\,dx$, we get the formula

$$\int f(x)g(x)\,dx = \frac{f(x)\cdot g'(x) - f'(x)\cdot g(x)}{b - a} + C$$

This formula could have been used in the example mentioned. Here is another example.

Example:

If $f(x) = e^{2x}$, $g(x) = \sin 3x$, then $a = 4$, $b = -9$, and

$$\int e^{2x} \sin 3x\,dx = \frac{3\,e^{2x}\cos 3x - 2\,e^{2x}\sin 3x}{-9 - 4} + C = \frac{e^{2x}}{13}(2 \sin 3x - 3 \cos 3x) + C$$

The following additional points should be observed when using this table.

1. A constant of integration is to be supplied with the answers for indefinite integrals.
2. Logarithmic expressions are to base $e = 2.71828\ldots$, unless otherwise specified, and are to be evaluated for the absolute value of the arguments involved therein.
3. All angles are measured in radians, and inverse trigonometric and hyperbolic functions represent principal values, unless otherwise indicated.
4. If the application of a formula produces either a zero denominator or the square root of a negative number in the result, there is usually available another form of the answer which

avoids this difficulty. In many of the results, the excluded values are specified, but when such are omitted it is presumed that one can tell what these should be, especially when difficulties of the type herein mentioned are obtained.

5. When inverse trigonometric functions occur in the integrals, be sure that any replacements made for them are strictly in accordance with the rules for such functions. This causes little difficulty when the argument of the inverse trigonometric function is positive, since then all angles involved are in the first quadrant. However, if the argument is negative, special care must be used. Thus if $u > 0$,

$$\sin^{-1} u = \cos^{-1}\sqrt{1-u^2} = \csc^{-1}\frac{1}{u}, \text{ etc.}$$

However, if $u < 0$,

$$\sin^{-1} u = -\cos^{-1}\sqrt{1-u^2} = -\pi - \csc^{-1}\frac{1}{u}, \text{ etc.}$$

See the section on inverse trigonometric functions for a full treatment of the allowable substitutions.

6. In integrals 340–345 and some others, the right side includes expressions of the form

$$A \tan^{-1} [B + C \tan f(x)].$$

In these formulas, the $\tan^{-1}$ does not necessarily represent the principal value. Instead of always employing the principal branch of the inverse tangent function, one must instead use that branch of the inverse tangent function upon which $f(x)$ lies for any particular choice of x.

Example:

$$\int_0^{4\pi} \frac{dx}{2+\sin x} = \frac{2}{\sqrt{3}}\tan^{-1}\frac{2\tan\frac{x}{2}+1}{\sqrt{3}}\Bigg]_0^{4\pi}$$

$$= \frac{2}{\sqrt{3}}\left[\tan^{-1}\frac{2\tan 2\pi+1}{\sqrt{3}} - \tan^{-1}\frac{2\tan 0+1}{\sqrt{3}}\right]$$

$$= \frac{2}{\sqrt{3}}\left[\frac{13\pi}{6} - \frac{\pi}{6}\right] = \frac{4\pi}{\sqrt{3}} = \frac{4\sqrt{3}\pi}{3}$$

Here

$$\tan^{-1}\frac{2\tan 2\pi+1}{\sqrt{3}} = \tan^{-1}\frac{1}{\sqrt{3}} = \frac{13\pi}{6},$$

since $f(x) = 2\pi$; and

$$\tan^{-1}\frac{2\tan 0+1}{\sqrt{3}} = \tan^{-1}\frac{1}{\sqrt{3}} = \frac{\pi}{6},$$

since $f(x) = 0$.

7. B_n and E_n where used in Integrals represents the Bernoulli and Euler numbers as defined in tables of Bernoulli and Euler polynomials contained in certain mathematics reference and handbooks, as for example, Beyer, W. H., *Handbook of Mathematical Sciences,* 5th ed., CRC Press, Inc., West Palm Beach 1978, 577–583.

ELEMENTARY FORMS

1. $\int a\,dx = ax$

2. $\int a\cdot f(x)\,dx = a\int f(x)\,dx$

3. $\int \phi(y)\,dx = \int \frac{\phi(y)}{y'}\,dy$, where $y' = \frac{dy}{dx}$

4. $\int (u+v)\,dx = \int u\,dx + \int v\,dx$, where u and v are any functions of x

5. $\int u\,dv = u\int dv - \int v\,du = uv - \int v\,du$

6. $\int u\frac{dv}{dx}dx = uv - \int v\frac{du}{dx}dx$

7. $\int x^n\,dx = \frac{x^{n+1}}{n+1}$, except $n = -1$

8. $\int \frac{f'(x)\,dx}{f(x)} = \log f(x)$, $(df(x) = f'(x)\,dx)$

9. $\int \frac{dx}{x} = \log x$

10. $\int \frac{f'(x)\,dx}{2\sqrt{f(x)}} = \sqrt{f(x)}$, $(df(x) = f'(x)\,dx)$

11. $\int e^x\,dx = e^x$

12. $\int e^{ax}\,dx = e^{ax}/a$

13. $\int b^{ax}\,dx = \frac{b^{ax}}{a\log b}$, $(b > 0)$

14. $\int \log x\,dx = x\log x - x$

15. $\int a^x \log a\,dx = a^x$, $(a > 0)$

16. $\int \frac{dx}{a^2+x^2} = \frac{1}{a}\tan^{-1}\frac{x}{a}$

17. $$\int \frac{dx}{a^2-x^2} = \begin{cases} \frac{1}{a}\tanh^{-1}\frac{x}{a} \\ \text{or} \\ \frac{1}{2a}\log\frac{a+x}{a-x}, \quad (a^2 > x^2) \end{cases}$$

18. $$\int \frac{dx}{x^2-a^2} = \begin{cases} -\frac{1}{a}\coth^{-1}\frac{x}{a} \\ \text{or} \\ \frac{1}{2a}\log\frac{x-a}{x+a}, \quad (x^2 > a^2) \end{cases}$$

19. $$\int \frac{dx}{\sqrt{a^2 - x^2}} = \begin{cases} \sin^{-1} \dfrac{x}{|a|} \\ \text{or} \\ -\cos^{-1} \dfrac{x}{|a|}, \quad (a^2 > x^2) \end{cases}$$

20. $$\int \frac{dx}{\sqrt{x^2 \pm a^2}} = \log(x + \sqrt{x^2 \pm a^2})$$

21. $$\int \frac{dx}{x\sqrt{x^2 - a^2}} = \frac{1}{|a|} \sec^{-1} \frac{x}{a}$$

22. $$\int \frac{dx}{x\sqrt{a^2 \pm x^2}} = -\frac{1}{a} \log\left(\frac{a + \sqrt{a^2 \pm x^2}}{x}\right)$$

FORMS CONTAINING $(a + bx)$

For forms containing $a + bx$, but not listed in the table, the substitution $u = \frac{a + bx}{x}$ may prove helpful.

23. $$\int (a + bx)^n \, dx = \frac{(a + bx)^{n+1}}{(n + 1)b}, \quad (n \neq -1)$$

24. $$\int x(a + bx)^n \, dx$$
$$= \frac{1}{b^2(n + 2)}(a + bx)^{n+2} - \frac{a}{b^2(n + 1)}(a + bx)^{n+1}, \quad (n \neq -1, -2)$$

25. $$\int x^2(a + bx)^n \, dx = \frac{1}{b^3}\left[\frac{(a + bx)^{n+3}}{n + 3} - 2a\frac{(a + bx)^{n+2}}{n + 2} + a^2\frac{(a + bx)^{n+1}}{n + 1}\right]$$

26. $$\int x^m(a + bx)^n \, dx = \begin{cases} \dfrac{x^{m+1}(a + bx)^n}{m + n + 1} + \dfrac{an}{m + n + 1}\displaystyle\int x^m(a + bx)^{n-1} \, dx \\ \text{or} \\ \dfrac{1}{a(n + 1)}\Big[-x^{m+1}(a + bx)^{n+1} \\ \qquad + (m + n + 2)\displaystyle\int x^m(a + bx)^{n+1} \, dx\Big] \\ \text{or} \\ \dfrac{1}{b(m + n + 1)}\Big[x^m(a + bx)^{n+1} - ma\displaystyle\int x^{m-1}(a + bx)^n \, dx\Big] \end{cases}$$

27. $$\int \frac{dx}{a + bx} = \frac{1}{b} \log(a + bx)$$

28. $$\int \frac{dx}{(a + bx)^2} = -\frac{1}{b(a + bx)}$$

29. $$\int \frac{dx}{(a + bx)^3} = -\frac{1}{2b(a + bx)^2}$$

30. $$\int \frac{x \, dx}{a + bx} = \begin{cases} \dfrac{1}{b^2}[a + bx - a \log(a + bx)] \\ \text{or} \\ \dfrac{x}{b} - \dfrac{a}{b^2} \log(a + bx) \end{cases}$$

31. $$\int \frac{x\,dx}{(a+bx)^2} = \frac{1}{b^2}\left[\log(a+bx) + \frac{a}{a+bx}\right]$$

32. $$\int \frac{x\,dx}{(a+bx)^n} = \frac{1}{b^2}\left[\frac{-1}{(n-2)(a+bx)^{n-2}} + \frac{a}{(n-1)(a+bx)^{n-1}}\right], \quad n \neq 1, 2$$

33. $$\int \frac{x^2\,dx}{a+bx} = \frac{1}{b^3}\left[\frac{1}{2}(a+bx)^2 - 2a(a+bx) + a^2\log(a+bx)\right]$$

34. $$\int \frac{x^2\,dx}{(a+bx)^2} = \frac{1}{b^3}\left[a+bx - 2a\log(a+bx) - \frac{a^2}{a+bx}\right]$$

35. $$\int \frac{x^2\,dx}{(a+bx)^3} = \frac{1}{b^3}\left[\log(a+bx) + \frac{2a}{a+bx} - \frac{a^2}{2(a+bx)^2}\right]$$

36. $$\int \frac{x^2\,dx}{(a+bx)^n} = \frac{1}{b^3}\left[\frac{-1}{(n-3)(a+bx)^{n-3}} + \frac{2a}{(n-2)(a+bx)^{n-2}} - \frac{a^2}{(n-1)(a+bx)^{n-1}}\right], \quad n \neq 1, 2, 3$$

37. $$\int \frac{dx}{x(a+bx)} = -\frac{1}{a}\log\frac{a+bx}{x}$$

38. $$\int \frac{dx}{x(a+bx)^2} = \frac{1}{a(a+bx)} - \frac{1}{a^2}\log\frac{a+bx}{x}$$

39. $$\int \frac{dx}{x(a+bx)^3} = \frac{1}{a^3}\left[\frac{1}{2}\left(\frac{2a+bx}{a+bx}\right)^2 + \log\frac{x}{a+bx}\right]$$

40. $$\int \frac{dx}{x^2(a+bx)} = -\frac{1}{ax} + \frac{b}{a^2}\log\frac{a+bx}{x}$$

41. $$\int \frac{dx}{x^3(a+bx)} = \frac{2bx-a}{2a^2x^2} + \frac{b^2}{a^3}\log\frac{x}{a+bx}$$

42. $$\int \frac{dx}{x^2(a+bx)^2} = -\frac{a+2bx}{a^2x(a+bx)} + \frac{2b}{a^3}\log\frac{a+bx}{x}$$

FORMS CONTAINING $c^2 \pm x^2$, $x^2 - c^2$

43. $$\int \frac{dx}{c^2+x^2} = \frac{1}{c}\tan^{-1}\frac{x}{c}$$

44. $$\int \frac{dx}{c^2-x^2} = \frac{1}{2c}\log\frac{c+x}{c-x}, \quad (c^2 > x^2)$$

45. $$\int \frac{dx}{x^2-c^2} = \frac{1}{2c}\log\frac{x-c}{x+c}, \quad (x^2 > c^2)$$

46. $$\int \frac{x\,dx}{c^2 \pm x^2} = \pm\frac{1}{2}\log(c^2 \pm x^2)$$

47. $$\int \frac{x\,dx}{(c^2 \pm x^2)^{n+1}} = \mp\frac{1}{2n(c^2 \pm x^2)^n}$$

48. $$\int \frac{dx}{(c^2 \pm x^2)^n} = \frac{1}{2c^2(n-1)}\left[\frac{x}{(c^2 \pm x^2)^{n-1}} + (2n-3)\int \frac{dx}{(c^2 \pm x^2)^{n-1}}\right]$$

49. $$\int \frac{dx}{(x^2 - c^2)^n} = \frac{1}{2c^2(n-1)}\left[-\frac{x}{(x^2-c^2)^{n-1}} - (2n-3)\int \frac{dx}{(x^2-c^2)^{n-1}}\right]$$

50. $$\int \frac{x\,dx}{x^2-c^2} = \frac{1}{2}\log(x^2-c^2)$$

51. $$\int \frac{x\,dx}{(x^2 - c^2)^{n+1}} = -\frac{1}{2n(x^2 - c^2)^n}$$

FORMS CONTAINING $a + bx$ and $c + dx$

$$u = a + bx, \qquad v = c + dx, \qquad k = ad - bc$$

If $k = 0$, then $v = \frac{c}{a}u$

52. $$\int \frac{dx}{u \cdot v} = \frac{1}{k} \cdot \log\left(\frac{v}{u}\right)$$

53. $$\int \frac{x\,dx}{u \cdot v} = \frac{1}{k}\left[\frac{a}{b}\log(u) - \frac{c}{d}\log(v)\right]$$

54. $$\int \frac{dx}{u^2 \cdot v} = \frac{1}{k}\left(\frac{1}{u} + \frac{d}{k}\log\frac{v}{u}\right)$$

55. $$\int \frac{x\,dx}{u^2 \cdot v} = \frac{-a}{bku} - \frac{c}{k^2}\log\frac{v}{u}$$

56. $$\int \frac{x^2\,dx}{u^2 \cdot v} = \frac{a^2}{b^2ku} + \frac{1}{k^2}\left[\frac{c^2}{d}\log(v) + \frac{a(k - bc)}{b^2}\log(u)\right]$$

57. $$\int \frac{dx}{u^n \cdot v^m} = \frac{1}{k(m-1)}\left[\frac{-1}{u^{n-1} \cdot v^{m-1}} - (m + n - 2)b\int \frac{dx}{u^n \cdot v^{m-1}}\right]$$

58. $$\int \frac{u}{v}\,dx = \frac{bx}{d} + \frac{k}{d^2}\log(v)$$

59. $$\int \frac{u^m\,dx}{v^n} = \begin{cases} \dfrac{-1}{k(n-1)}\left[\dfrac{u^{m+1}}{v^{n-1}} + b(n - m - 2)\displaystyle\int \frac{u^m}{v^{n-1}}\,dx\right] \\ \text{or} \\ \dfrac{-1}{d(n - m - 1)}\left[\dfrac{u^m}{v^{n-1}} + mk\displaystyle\int \frac{u^{m-1}}{v^n}\,dx\right] \\ \text{or} \\ \dfrac{-1}{d(n-1)}\left[\dfrac{u^m}{v^{n-1}} - mb\displaystyle\int \frac{u^{m-1}}{v^{n-1}}\,dx\right] \end{cases}$$

FORMS CONTAINING $(a + bx^n)$

60. $$\int \frac{dx}{a + bx^2} = \frac{1}{\sqrt{ab}}\tan^{-1}\frac{x\sqrt{ab}}{a}, \qquad (ab > 0)$$

61. $$\int \frac{dx}{a + bx^2} = \begin{cases} \dfrac{1}{2\sqrt{-ab}}\log\dfrac{a + x\sqrt{-ab}}{a - x\sqrt{-ab}}, & (ab < 0) \\ \text{or} & \\ \dfrac{1}{\sqrt{-ab}}\tanh^{-1}\dfrac{x\sqrt{-ab}}{a}, & (ab < 0) \end{cases}$$

62. $$\int \frac{dx}{a^2 + b^2x^2} = \frac{1}{ab}\tan^{-1}\frac{bx}{a}$$

63. $$\int \frac{x\,dx}{a + bx^2} = \frac{1}{2b}\log(a + bx^2)$$

64. $$\int \frac{x^2\,dx}{a + bx^2} = \frac{x}{b} - \frac{a}{b}\int \frac{dx}{a + bx^2}$$

65. $$\int \frac{dx}{(a + bx^2)^2} = \frac{x}{2a(a + bx^2)} + \frac{1}{2a}\int \frac{dx}{a + bx^2}$$

66. $$\int \frac{dx}{a^2 - b^2x^2} = \frac{1}{2ab} \log \frac{a + bx}{a - bx}$$

67. $$\int \frac{dx}{(a + bx^2)^{m+1}} = \begin{cases} \dfrac{1}{2ma} \dfrac{x}{(a + bx^2)^m} + \dfrac{2m - 1}{2ma} \displaystyle\int \dfrac{dx}{(a + bx^2)^m} \\ \text{or} \\ \dfrac{(2m)!}{(m!)^2} \left[\dfrac{x}{2a} \displaystyle\sum_{r=1}^{m} \dfrac{r!(r - 1)!}{(4a)^{m-r}(2r)!(a + bx^2)^r} + \dfrac{1}{(4a)^m} \displaystyle\int \dfrac{dx}{a + bx^2} \right] \end{cases}$$

68. $$\int \frac{x\,dx}{(a + bx^2)^{m+1}} = -\frac{1}{2bm(a + bx^2)^m}$$

69. $$\int \frac{x^2\,dx}{(a + bx^2)^{m+1}} = \frac{-x}{2mb(a + bx^2)^m} + \frac{1}{2mb} \int \frac{dx}{(a + bx^2)^m}$$

70. $$\int \frac{dx}{x(a + bx^2)} = \frac{1}{2a} \log \frac{x^2}{a + bx^2}$$

71. $$\int \frac{dx}{x^2(a + bx^2)} = -\frac{1}{ax} - \frac{b}{a} \int \frac{dx}{a + bx^2}$$

72. $$\int \frac{dx}{x(a + bx^2)^{m+1}} = \begin{cases} \dfrac{1}{2am(a + bx^2)^m} + \dfrac{1}{a} \displaystyle\int \dfrac{dx}{x(a + bx^2)^m} \\ \text{or} \\ \dfrac{1}{2a^{m+1}} \left[\displaystyle\sum_{r=1}^{m} \dfrac{a^r}{r(a + bx^2)^r} + \log \dfrac{x^2}{a + bx^2} \right] \end{cases}$$

73. $$\int \frac{dx}{x^2(a + bx^2)^{m+1}} = \frac{1}{a} \int \frac{dx}{x^2(a + bx^2)^m} - \frac{b}{a} \int \frac{dx}{(a + bx^2)^{m+1}}$$

74. $$\int \frac{dx}{a + bx^3} = \frac{k}{3a} \left[\frac{1}{2} \log \frac{(k + x)^3}{a + bx^3} + \sqrt{3} \tan^{-1} \frac{2x - k}{k\sqrt{3}} \right] \qquad \left(k = \sqrt[3]{\frac{a}{b}} \right)$$

75. $$\int \frac{x\,dx}{a + bx^3} = \frac{1}{3bk} \left[\frac{1}{2} \log \frac{a + bx^3}{(k + x)^3} + \sqrt{3} \tan^{-1} \frac{2x - k}{k\sqrt{3}} \right] \qquad \left(k = \sqrt[3]{\frac{a}{b}} \right)$$

76. $$\int \frac{x^2\,dx}{a + bx^3} = \frac{1}{3b} \log (a + bx^3)$$

77. $$\int \frac{dx}{a + bx^4} = \frac{k}{2a} \left[\frac{1}{2} \log \frac{x^2 + 2kx + 2k^2}{x^2 - 2kx + 2k^2} + \tan^{-1} \frac{2kx}{2k^2 - x^2} \right]$$

$$\left(ab > 0, k = \sqrt[4]{\frac{a}{4b}} \right)$$

78. $$\int \frac{dx}{a + bx^4} = \frac{k}{2a} \left[\frac{1}{2} \log \frac{x + k}{x - k} + \tan^{-1} \frac{x}{k} \right], \qquad \left(ab < 0, k = \sqrt[4]{-\frac{a}{b}} \right)$$

79. $$\int \frac{x\,dx}{a + bx^4} = \frac{1}{2bk} \tan^{-1} \frac{x^2}{k}, \qquad \left(ab > 0, k = \sqrt{\frac{a}{b}} \right)$$

80. $$\int \frac{x\,dx}{a + bx^4} = \frac{1}{4bk} \log \frac{x^2 - k}{x^2 + k}, \qquad \left(ab < 0, k = \sqrt{-\frac{a}{b}} \right)$$

81. $$\int \frac{x^2\,dx}{a + bx^4} = \frac{1}{4bk} \left[\frac{1}{2} \log \frac{x^2 - 2kx + 2k^2}{x^2 + 2kx + 2k^2} + \tan^{-1} \frac{2kx}{2k^2 - x^2} \right]$$

$$\left(ab > 0, k = \sqrt[4]{\frac{a}{4b}} \right)$$

82. $\int \frac{x^2\,dx}{a+bx^4} = \frac{1}{4bk}\left[\log\frac{x-k}{x+k} + 2\tan^{-1}\frac{x}{k}\right], \quad \left(ab<0, k=\sqrt[4]{-\frac{a}{b}}\right)$

83. $\int \frac{x^3\,dx}{a+bx^4} = \frac{1}{4b}\log(a+bx^4)$

84. $\int \frac{dx}{x(a+bx^n)} = \frac{1}{an}\log\frac{x^n}{a+bx^n}$

85. $\int \frac{dx}{(a+bx^n)^{m+1}} = \frac{1}{a}\int\frac{dx}{(a+bx^n)^m} - \frac{b}{a}\int\frac{x^n\,dx}{(a+bx^n)^{m+1}}$

86. $\int \frac{x^m\,dx}{(a+bx^n)^{p+1}} = \frac{1}{b}\int\frac{x^{m-n}\,dx}{(a+bx^n)^p} - \frac{a}{b}\int\frac{x^{m-n}\,dx}{(a+bx^n)^{p+1}}$

87. $\int \frac{dx}{x^m(a+bx^n)^{p+1}} = \frac{1}{a}\int\frac{dx}{x^m(a+bx^n)^p} - \frac{b}{a}\int\frac{dx}{x^{m-n}(a+bx^n)^{p+1}}$

88.
$$\int x^m(a+bx^n)^p\,dx = \begin{cases} \frac{1}{b(np+m+1)}\left[x^{m-n+1}(a+bx^n)^{p+1} - a(m-n+1)\int x^{m-n}(a+bx^n)^p\,dx\right] \\ \text{or} \\ \frac{1}{np+m+1}\left[x^{m+1}(a+bx^n)^p + anp\int x^m(a+bx^n)^{p-1}\,dx\right] \\ \text{or} \\ \frac{1}{a(m+1)}\left[x^{m+1}(a+bx^n)^{p+1} - (m+1+np+n)b\int x^{m+n}(a+bx^n)^p\,dx\right] \\ \text{or} \\ \frac{1}{an(p+1)}\left[-x^{m+1}(a+bx^n)^{p+1} + (m+1+np+n)\int x^m(a+bx^n)^{p+1}\,dx\right] \end{cases}$$

FORMS CONTAINING $c^3 \pm x^3$

89. $\int \frac{dx}{c^3 \pm x^3} = \pm\frac{1}{6c^2}\log\frac{(c\pm x)^3}{c^3\pm x^3} + \frac{1}{c^2\sqrt{3}}\tan^{-1}\frac{2x\mp c}{c\sqrt{3}}$

90. $\int \frac{dx}{(c^3\pm x^3)^2} = \frac{x}{3c^3(c^3\pm x^3)} + \frac{2}{3c^3}\int\frac{dx}{c^3\pm x^3}$

91. $\int \frac{dx}{(c^3\pm x^3)^{n+1}} = \frac{1}{3nc^3}\left[\frac{x}{(c^3\pm x^3)^n} + (3n-1)\int\frac{dx}{(c^3\pm x^3)^n}\right]$

92. $\int \frac{x\,dx}{c^3\pm x^3} = \frac{1}{6c}\log\frac{c^3\pm x^3}{(c\pm x)^3} \pm \frac{1}{c\sqrt{3}}\tan^{-1}\frac{2x\mp c}{c\sqrt{3}}$

93. $\int \frac{x\,dx}{(c^3\pm x^3)^2} = \frac{x^2}{3c^3(c^3\pm x^3)} + \frac{1}{3c^3}\int\frac{x\,dx}{c^3\pm x^3}$

94. $$\int \frac{x\,dx}{(c^3 \pm x^3)^{n+1}} = \frac{1}{3nc^3}\left[\frac{x^2}{(c^3 \pm x^3)^n} + (3n-2)\int \frac{x\,dx}{(c^3 \pm x^3)^n}\right]$$

95. $$\int \frac{x^2\,dx}{c^3 \pm x^3} = \pm\frac{1}{3}\log(c^3 \pm x^3)$$

96. $$\int \frac{x^2\,dx}{(c^3 \pm x^3)^{n+1}} = \mp\frac{1}{3n(c^3 \pm x^3)^n}$$

97. $$\int \frac{dx}{x(c^3 \pm x^3)} = \frac{1}{3c^3}\log\frac{x^3}{c^3 \pm x^3}$$

98. $$\int \frac{dx}{x(c^3 \pm x^3)^2} = \frac{1}{3c^3(c^3 \pm x^3)} + \frac{1}{3c^6}\log\frac{x^3}{c^3 \pm x^3}$$

99. $$\int \frac{dx}{x(c^3 \pm x^3)^{n+1}} = \frac{1}{3nc^3(c^3 \pm x^3)^n} + \frac{1}{c^3}\int \frac{dx}{x(c^3 \pm x^3)^n}$$

100. $$\int \frac{dx}{x^2(c^3 \pm x^3)} = -\frac{1}{c^3x} \mp \frac{1}{c^3}\int \frac{x\,dx}{c^3 \pm x^3}$$

101. $$\int \frac{dx}{x^2(c^3 \pm x^3)^{n+1}} = \frac{1}{c^3}\int \frac{dx}{x^2(c^3 \pm x^3)^n} \mp \frac{1}{c^3}\int \frac{x\,dx}{(c^3 \pm x^3)^{n+1}}$$

FORMS CONTAINING $c^4 \pm x^4$

102. $$\int \frac{dx}{c^4 + x^4} = \frac{1}{2c^3\sqrt{2}}\left[\frac{1}{2}\log\frac{x^2 + cx\sqrt{2} + c^2}{x^2 - cx\sqrt{2} + c^2} + \tan^{-1}\frac{cx\sqrt{2}}{c^2 - x^2}\right]$$

103. $$\int \frac{dx}{c^4 - x^4} = \frac{1}{2c^3}\left[\frac{1}{2}\log\frac{c+x}{c-x} + \tan^{-1}\frac{x}{c}\right]$$

104. $$\int \frac{x\,dx}{c^4 + x^4} = \frac{1}{2c^2}\tan^{-1}\frac{x^2}{c^2}$$

105. $$\int \frac{x\,dx}{c^4 - x^4} = \frac{1}{4c^2}\log\frac{c^2 + x^2}{c^2 - x^2}$$

106. $$\int \frac{x^2\,dx}{c^4 + x^4} = \frac{1}{2c\sqrt{2}}\left[\frac{1}{2}\log\frac{x^2 - cx\sqrt{2} + c^2}{x^2 + cx\sqrt{2} + c^2} + \tan^{-1}\frac{cx\sqrt{2}}{c^2 - x^2}\right]$$

107. $$\int \frac{x^2\,dx}{c^4 - x^4} = \frac{1}{2c}\left[\frac{1}{2}\log\frac{c+x}{c-x} - \tan^{-1}\frac{x}{c}\right]$$

108. $$\int \frac{x^3\,dx}{c^4 \pm x^4} = \pm\frac{1}{4}\log(c^4 \pm x^4)$$

FORMS CONTAINING $(a + bx + cx^2)$

$$X = a + bx + cx^2 \text{ and } q = 4ac - b^2$$

If $q = 0$, then $X = c\left(x + \frac{b}{2c}\right)^2$, and formulas starting with 23 should be used in place of these.

109. $$\int \frac{dx}{X} = \frac{2}{\sqrt{q}}\tan^{-1}\frac{2cx + b}{\sqrt{q}}, \qquad (q > 0)$$

110. $$\int \frac{dx}{X} = \begin{cases} \dfrac{-2}{\sqrt{-q}}\tanh^{-1}\dfrac{2cx + b}{\sqrt{-q}} \\ \text{or} \\ \dfrac{1}{\sqrt{-q}}\log\dfrac{2cx + b - \sqrt{-q}}{2cx + b + \sqrt{-q}}, \qquad (q < 0) \end{cases}$$

111. $$\int \frac{dx}{X^2} = \frac{2cx + b}{qX} + \frac{2c}{q}\int \frac{dx}{X}$$

112. $\displaystyle\int \frac{dx}{X^3} = \frac{2cx+b}{q}\left(\frac{1}{2X^2}+\frac{3c}{qX}\right)+\frac{6c^2}{q^2}\int\frac{dx}{X}$

113. $$\int \frac{dx}{X^{n+1}} = \begin{cases} \dfrac{2cx+b}{nqX^n}+\dfrac{2(2n-1)c}{qn}\displaystyle\int\frac{dx}{X^n} \\ \text{or} \\ \dfrac{(2n)!}{(n!)^2}\left(\dfrac{c}{q}\right)^n\left[\dfrac{2cx+b}{q}\displaystyle\sum_{r=1}^{n}\left(\dfrac{q}{cX}\right)^r\left(\dfrac{(r-1)!r!}{(2r)!}\right)+\displaystyle\int\frac{dx}{X}\right] \end{cases}$$

114. $\displaystyle\int \frac{x\,dx}{X} = \frac{1}{2c}\log X - \frac{b}{2c}\int\frac{dx}{X}$

115. $\displaystyle\int \frac{x\,dx}{X^2} = -\frac{bx+2a}{qX} - \frac{b}{q}\int\frac{dx}{X}$

116. $\displaystyle\int \frac{x\,dx}{X^{n+1}} = -\frac{2a+bx}{nqX^n} - \frac{b(2n-1)}{nq}\int\frac{dx}{X^n}$

117. $\displaystyle\int \frac{x^2}{X}\,dx = \frac{x}{c} - \frac{b}{2c^2}\log X + \frac{b^2-2ac}{2c^2}\int\frac{dx}{X}$

118. $\displaystyle\int \frac{x^2}{X^2}\,dx = \frac{(b^2-2ac)x+ab}{cqX} + \frac{2a}{q}\int\frac{dx}{X}$

119. $$\int \frac{x^m\,dx}{X^{n+1}} = -\frac{x^{m-1}}{(2n-m+1)cX^n} - \frac{n-m+1}{2n-m+1}\cdot\frac{b}{c}\int\frac{x^{m-1}\,dx}{X^{n+1}} + \frac{m-1}{2n-m+1}\cdot\frac{a}{c}\int\frac{x^{m-2}\,dx}{X^{n+1}}$$

120. $\displaystyle\int \frac{dx}{xX} = \frac{1}{2a}\log\frac{x^2}{X} - \frac{b}{2a}\int\frac{dx}{X}$

121. $\displaystyle\int \frac{dx}{x^2X} = \frac{b}{2a^2}\log\frac{X}{x^2} - \frac{1}{ax} + \left(\frac{b^2}{2a^2}-\frac{c}{a}\right)\int\frac{dx}{X}$

122. $\displaystyle\int \frac{dx}{xX^n} = \frac{1}{2a(n-1)X^{n-1}} - \frac{b}{2a}\int\frac{dx}{X^n} + \frac{1}{a}\int\frac{dx}{xX^{n-1}}$

123. $$\int \frac{dx}{x^mX^{n+1}} = -\frac{1}{(m-1)ax^{m-1}X^n} - \frac{n+m-1}{m-1}\cdot\frac{b}{a}\int\frac{dx}{x^{m-1}X^{n+1}} - \frac{2n+m-1}{m-1}\cdot\frac{c}{a}\int\frac{dx}{x^{m-2}X^{n+1}}$$

FORMS CONTAINING $\sqrt{a+bx}$

124. $\displaystyle\int \sqrt{a+bx}\,dx = \frac{2}{3b}\sqrt{(a+bx)^3}$

125. $\displaystyle\int x\sqrt{a+bx}\,dx = -\frac{2(2a-3bx)\sqrt{(a+bx)^3}}{15b^2}$

126. $\displaystyle\int x^2\sqrt{a+bx}\,dx = \frac{2(8a^2-12abx+15b^2x^2)\sqrt{(a+bx)^3}}{105b^3}$

127. $$\int x^m\sqrt{a+bx}\,dx = \begin{cases} \dfrac{2}{b(2m+3)}\left[x^m\sqrt{(a+bx)^3} - ma\displaystyle\int x^{m-1}\sqrt{a+bx}\,dx\right] \\ \text{or} \\ \dfrac{2}{b^{m+1}}\sqrt{a+bx}\displaystyle\sum_{r=0}^{m}\dfrac{m!(-a)^{m-r}}{r!(m-r)!(2r+3)}(a+bx)^{r+1} \end{cases}$$

128. $$\int \frac{\sqrt{a+bx}}{x}dx = 2\sqrt{a+bx} + a\int \frac{dx}{x\sqrt{a+bx}}$$

129. $$\int \frac{\sqrt{a+bx}}{x^2}dx = -\frac{\sqrt{a+bx}}{x} + \frac{b}{2}\int \frac{dx}{x\sqrt{a+bx}}$$

130. $$\int \frac{\sqrt{a+bx}}{x^m}dx = -\frac{1}{(m-1)a}\left[\frac{\sqrt{(a+bx)^3}}{x^{m-1}} + \frac{(2m-5)b}{2}\int \frac{\sqrt{a+bx}}{x^{m-1}}dx\right]$$

131. $$\int \frac{dx}{\sqrt{a+bx}} = \frac{2\sqrt{a+bx}}{b}$$

132. $$\int \frac{x\,dx}{\sqrt{a+bx}} = -\frac{2(2a-bx)}{3b^2}\sqrt{a+bx}$$

133. $$\int \frac{x^2\,dx}{\sqrt{a+bx}} = \frac{2(8a^2-4abx+3b^2x^2)}{15b^3}\sqrt{a+bx}$$

134. $$\int \frac{x^m\,dx}{\sqrt{a+bx}} = \begin{cases} \dfrac{2}{(2m+1)b}\left[x^m\sqrt{a+bx} - ma\displaystyle\int \frac{x^{m-1}\,dx}{\sqrt{a+bx}}\right] \\ \text{or} \\ \dfrac{2(-a)^m\sqrt{a+bx}}{b^{m+1}}\displaystyle\sum_{r=0}^{m} \frac{(-1)^r m!(a+bx)^r}{(2r+1)r!(m-r)!a^r} \end{cases}$$

135. $$\int \frac{dx}{x\sqrt{a+bx}} = \frac{1}{\sqrt{a}}\log\left(\frac{\sqrt{a+bx}-\sqrt{a}}{\sqrt{a+bx}+\sqrt{a}}\right), \quad (a>0)$$

136. $$\int \frac{dx}{x\sqrt{a+bx}} = \frac{2}{\sqrt{-a}}\tan^{-1}\sqrt{\frac{a+bx}{-a}}, \quad (a<0)$$

137. $$\int \frac{dx}{x^2\sqrt{a+bx}} = -\frac{\sqrt{a+bx}}{ax} - \frac{b}{2a}\int \frac{dx}{x\sqrt{a+bx}}$$

138. $$\int \frac{dx}{x^n\sqrt{a+bx}} = \begin{cases} -\dfrac{\sqrt{a+bx}}{(n-1)ax^{n-1}} - \dfrac{(2n-3)b}{(2n-2)a}\displaystyle\int \frac{dx}{x^{n-1}\sqrt{a+bx}} \\ \text{or} \\ \dfrac{(2n-2)!}{[(n-1)!]^2}\left[-\dfrac{\sqrt{a+bx}}{a}\displaystyle\sum_{r=1}^{n-1} \frac{r!(r-1)!}{x^r(2r)!}\left(-\frac{b}{4a}\right)^{n-r-1}\right. \\ \qquad\left. + \left(-\dfrac{b}{4a}\right)^{n-1}\displaystyle\int \frac{dx}{x\sqrt{a+bx}}\right] \end{cases}$$

139. $$\int (a+bx)^{\pm\frac{n}{2}}\,dx = \frac{2(a+bx)^{\frac{2\pm n}{2}}}{b(2\pm n)}$$

140. $$\int x(a+bx)^{\pm\frac{n}{2}}\,dx = \frac{2}{b^2}\left[\frac{(a+bx)^{\frac{4\pm n}{2}}}{4\pm n} - \frac{a(a+bx)^{\frac{2\pm n}{2}}}{2\pm n}\right]$$

141. $$\int \frac{dx}{x(a+bx)^{\frac{m}{2}}} = \frac{1}{a}\int \frac{dx}{x(a+bx)^{\frac{m-2}{2}}} - \frac{b}{a}\int \frac{dx}{(a+bx)^{\frac{m}{2}}}$$

142. $\displaystyle\int \frac{(a+bx)^{\frac{n}{2}}\,dx}{x} = b\int (a+bx)^{\frac{n-2}{2}}\,dx + a\int \frac{(a+bx)^{\frac{n-2}{2}}}{x}\,dx$

143. $\displaystyle\int f(x, \sqrt{a+bx})\,dx = \frac{2}{b}\int f\left(\frac{z^2-a}{b}, z\right) z\,dz, \qquad (z = \sqrt{a+bx})$

FORMS CONTAINING $\sqrt{a+bx}$ and $\sqrt{c+dx}$

$$u = a + bx \qquad v = c + dx \qquad k = ad - bc$$

If $k = 0$, then $v = \frac{c}{a}u$, and formulas starting with 124 should be used in place of these.

144. $\displaystyle\int \frac{dx}{\sqrt{uv}} = \begin{cases} \dfrac{2}{\sqrt{bd}} \tanh^{-1} \dfrac{\sqrt{bduv}}{bv}, \; bd > o,\; k < o \\ \text{or} \\ \dfrac{2}{\sqrt{bd}} \tanh^{-1} \dfrac{\sqrt{bduv}}{du}, \; bd > o,\; k > o. \\ \text{or} \\ \dfrac{1}{\sqrt{bd}} \log \dfrac{(bv + \sqrt{bduv})^2}{v}, \qquad (bd > 0) \end{cases}$

145. $\displaystyle\int \frac{dx}{\sqrt{uv}} = \begin{cases} \dfrac{2}{\sqrt{-bd}} \tan^{-1} \dfrac{\sqrt{-bduv}}{bv} \\ \text{or} \\ -\dfrac{1}{\sqrt{-bd}} \sin^{-1}\left(\dfrac{2bdx + ad + bc}{|k|}\right), \qquad (bd < 0) \end{cases}$

146. $\displaystyle\int \sqrt{uv}\,dx = \frac{k + 2bv}{4bd}\sqrt{uv} - \frac{k^2}{8bd}\int \frac{dx}{\sqrt{uv}}$

147. $\displaystyle\int \frac{dx}{v\sqrt{u}} = \begin{cases} \dfrac{1}{\sqrt{kd}} \log \dfrac{d\sqrt{u} - \sqrt{kd}}{d\sqrt{u} + \sqrt{kd}} \\ \text{or} \\ \dfrac{1}{\sqrt{kd}} \log \dfrac{(d\sqrt{u} - \sqrt{kd})^2}{v}, \qquad (kd > 0) \end{cases}$

148. $\displaystyle\int \frac{dx}{v\sqrt{u}} = \frac{2}{\sqrt{-kd}} \tan^{-1} \frac{d\sqrt{u}}{\sqrt{-kd}} \qquad (kd < 0)$

149. $\displaystyle\int \frac{x\,dx}{\sqrt{uv}} = \frac{\sqrt{uv}}{bd} - \frac{ad + bc}{2bd}\int \frac{dx}{\sqrt{uv}}$

150. $\displaystyle\int \frac{dx}{v\sqrt{uv}} = \frac{-2\sqrt{uv}}{kv}$

151. $\displaystyle\int \frac{v\,dx}{\sqrt{uv}} = \frac{\sqrt{uv}}{b} - \frac{k}{2b}\int \frac{dx}{\sqrt{uv}}$

152. $\displaystyle\int \sqrt{\frac{v}{u}}\,dx = \frac{v}{|v|}\int \frac{v\,dx}{\sqrt{uv}}$

153. $\displaystyle\int v^m \sqrt{u}\,dx = \frac{1}{(2m+3)d}\left(2v^{m+1}\sqrt{u} + k\int \frac{v^m\,dx}{\sqrt{u}}\right)$

154. $\int \frac{dx}{v^m\sqrt{u}} = -\frac{1}{(m-1)k}\left(\frac{\sqrt{u}}{v^{m-1}} + \left(m - \frac{3}{2}\right)b \int \frac{dx}{v^{m-1}\sqrt{u}}\right)$

155. $$\int \frac{v^m\,dx}{\sqrt{u}} = \begin{cases} \dfrac{2}{b(2m+1)}\left[v^m\sqrt{u} - mk\displaystyle\int \frac{v^{m-1}}{\sqrt{u}}\,dx\right] \\ \text{or} \\ \dfrac{2(m!)^2\sqrt{u}}{b(2m+1)!}\displaystyle\sum_{r=0}^{m}\left(-\frac{4k}{b}\right)^{m-r}\frac{(2r)!}{(r!)^2}v^r \end{cases}$$

FORMS CONTAINING $\sqrt{x^2 \pm a^2}$

156. $\int \sqrt{x^2 \pm a^2}\,dx = \frac{1}{2}[x\sqrt{x^2 \pm a^2} \pm a^2 \log(x + \sqrt{x^2 \pm a^2})]$

157. $\int \frac{dx}{\sqrt{x^2 \pm a^2}} = \log(x + \sqrt{x^2 \pm a^2})$

158. $\int \frac{dx}{x\sqrt{x^2 - a^2}} = \frac{1}{|a|}\sec^{-1}\frac{x}{a}$

159. $\int \frac{dx}{x\sqrt{x^2 + a^2}} = -\frac{1}{a}\log\left(\frac{a + \sqrt{x^2 + a^2}}{x}\right)$

160. $\int \frac{\sqrt{x^2 + a^2}}{x}\,dx = \sqrt{x^2 + a^2} - a\log\left(\frac{a + \sqrt{x^2 + a^2}}{x}\right)$

161. $\int \frac{\sqrt{x^2 - a^2}}{x}\,dx = \sqrt{x^2 - a^2} - |a|\sec^{-1}\frac{x}{a}$

162. $\int \frac{x\,dx}{\sqrt{x^2 \pm a^2}} = \sqrt{x^2 \pm a^2}$

163. $\int x\sqrt{x^2 \pm a^2}\,dx = \frac{1}{3}\sqrt{(x^2 \pm a^2)^3}$

164. $$\int \sqrt{(x^2 \pm a^2)^3}\,dx = \frac{1}{4}\left[x\sqrt{(x^2 \pm a^2)^3} \pm \frac{3a^2x}{2}\sqrt{x^2 \pm a^2} + \frac{3a^4}{2}\log(x + \sqrt{x^2 \pm a^2})\right]$$

165. $\int \frac{dx}{\sqrt{(x^2 \pm a^2)^3}} = \frac{\pm x}{a^2\sqrt{x^2 \pm a^2}}$

166. $\int \frac{x\,dx}{\sqrt{(x^2 \pm a^2)^3}} = \frac{-1}{\sqrt{x^2 \pm a^2}}$

167. $\int x\sqrt{(x^2 \pm a^2)^3}\,dx = \frac{1}{5}\sqrt{(x^2 \pm a^2)^5}$

168. $\int x^2\sqrt{x^2 \pm a^2}\,dx = \frac{x}{4}\sqrt{(x^2 \pm a^2)^3} \mp \frac{a^2}{8}x\sqrt{x^2 \pm a^2} - \frac{a^4}{8}\log(x + \sqrt{x^2 \pm a^2})$

169. $\int x^3\sqrt{x^2 + a^2}\,dx = (\frac{1}{5}x^2 - \frac{2}{15}a^2)\sqrt{(a^2 + x^2)^3}$

170. $\int x^3\sqrt{x^2 - a^2}\,dx = \frac{1}{5}\sqrt{(x^2 - a^2)^5} + \frac{a^2}{3}\sqrt{(x^2 - a^2)^3}$

171. $\int \frac{x^2\,dx}{\sqrt{x^2 \pm a^2}} = \frac{x}{2}\sqrt{x^2 \pm a^2} \mp \frac{a^2}{2}\log(x + \sqrt{x^2 \pm a^2})$

172. $\int \frac{x^3\,dx}{\sqrt{x^2 \pm a^2}} = \frac{1}{3}\sqrt{(x^2 \pm a^2)^3} \mp a^2\sqrt{x^2 \pm a^2}$

173. $\int \frac{dx}{x^2\sqrt{x^2 \pm a^2}} = \mp\frac{\sqrt{x^2 \pm a^2}}{a^2x}$

174. $\int \frac{dx}{x^3\sqrt{x^2 + a^2}} = -\frac{\sqrt{x^2 + a^2}}{2a^2x^2} + \frac{1}{2a^3}\log\frac{a + \sqrt{x^2 + a^2}}{x}$

175. $\int \frac{dx}{x^3\sqrt{x^2 - a^2}} = \frac{\sqrt{x^2 - a^2}}{2a^2x^2} + \frac{1}{2|a^3|}\sec^{-1}\frac{x}{a}$

176. $\int x^2\sqrt{(x^2 \pm a^2)^3}\,dx = \frac{x}{6}\sqrt{(x^2 \pm a^2)^5} \mp \frac{a^2x}{24}\sqrt{(x^2 \pm a^2)^3} - \frac{a^4x}{16}\sqrt{x^2 \pm a^2}$

$$\mp \frac{a^6}{16}\log(x + \sqrt{x^2 \pm a^2})$$

177. $\int x^3\sqrt{(x^2 \pm a^2)^3}\,dx = \frac{1}{7}\sqrt{(x^2 \pm a^2)^7} \mp \frac{a^2}{5}\sqrt{(x^2 \pm a^2)^5}$

178. $\int \frac{\sqrt{x^2 \pm a^2}\,dx}{x^2} = -\frac{\sqrt{x^2 \pm a^2}}{x} + \log(x + \sqrt{x^2 \pm a^2})$

179. $\int \frac{\sqrt{x^2 + a^2}}{x^3}\,dx = -\frac{\sqrt{x^2 + a^2}}{2x^2} - \frac{1}{2a}\log\frac{a + \sqrt{x^2 + a^2}}{x}$

180. $\int \frac{\sqrt{x^2 - a^2}}{x^3}\,dx = -\frac{\sqrt{x^2 - a^2}}{2x^2} + \frac{1}{2|a|}\sec^{-1}\frac{x}{a}$

181. $\int \frac{\sqrt{x^2 \pm a^2}}{x^4}\,dx = \mp\frac{\sqrt{(x^2 \pm a^2)^3}}{3a^2x^3}$

182. $\int \frac{x^2\,dx}{\sqrt{(x^2 \pm a^2)^3}} = \frac{-x}{\sqrt{x^2 \pm a^2}} + \log(x + \sqrt{x^2 \pm a^2})$

183. $\int \frac{x^3\,dx}{\sqrt{(x^2 \pm a^2)^3}} = \sqrt{x^2 \pm a^2} \pm \frac{a^2}{\sqrt{x^2 \pm a^2}}$

184. $\int \frac{dx}{x\sqrt{(x^2 + a^2)^3}} = \frac{1}{a^2\sqrt{x^2 + a^2}} - \frac{1}{a^3}\log\frac{a + \sqrt{x^2 + a^2}}{x}$

185. $\int \frac{dx}{x\sqrt{(x^2 - a^2)^3}} = -\frac{1}{a^2\sqrt{x^2 - a^2}} - \frac{1}{|a^3|}\sec^{-1}\frac{x}{a}$

186. $\int \frac{dx}{x^2\sqrt{(x^2 \pm a^2)^3}} = -\frac{1}{a^4}\left[\frac{\sqrt{x^2 \pm a^2}}{x} + \frac{x}{\sqrt{x^2 \pm a^2}}\right]$

187. $\int \frac{dx}{x^3\sqrt{(x^2 + a^2)^3}} = -\frac{1}{2a^2x^2\sqrt{x^2 + a^2}} - \frac{3}{2a^4\sqrt{x^2 + a^2}}$

$$+ \frac{3}{2a^5}\log\frac{a + \sqrt{x^2 + a^2}}{x}$$

188. $\int \frac{dx}{x^3\sqrt{(x^2 - a^2)^3}} = \frac{1}{2a^2x^2\sqrt{x^2 - a^2}} - \frac{3}{2a^4\sqrt{x^2 - a^2}} - \frac{3}{2|a^5|}\sec^{-1}\frac{x}{a}$

189. $\int \frac{x^m}{\sqrt{x^2 \pm a^2}}\,dx = \frac{1}{m}x^{m-1}\sqrt{x^2 \pm a^2} \mp \frac{m-1}{m}a^2\int \frac{x^{m-2}}{\sqrt{x^2 \pm a^2}}\,dx$

190. $$\int \frac{x^{2m}}{\sqrt{x^2 \pm a^2}}\,dx = \frac{(2m)!}{2^{2m}(m!)^2}\left[\sqrt{x^2 \pm a^2}\sum_{r=1}^{m}\frac{r!(r-1)!}{(2r)!}(\mp a^2)^{m-r}(2x)^{2r-1} + (\mp a^2)^m \log\left(x + \sqrt{x^2 \pm a^2}\right)\right]$$

191. $$\int \frac{x^{2m+1}}{\sqrt{x^2 \pm a^2}}\,dx = \sqrt{x^2 \pm a^2}\sum_{r=0}^{m}\frac{(2r)!(m!)^2}{(2m+1)!(r!)^2}(\mp 4a^2)^{m-r}x^{2r}$$

192. $$\int \frac{dx}{x^m\sqrt{x^2 \pm a^2}} = \mp\frac{\sqrt{x^2 \pm a^2}}{(m-1)a^2x^{m-1}} \mp \frac{(m-2)}{(m-1)a^2}\int \frac{dx}{x^{m-2}\sqrt{x^2 \pm a^2}}$$

193. $$\int \frac{dx}{x^{2m}\sqrt{x^2 \pm a^2}} = \sqrt{x^2 \pm a^2}\sum_{r=0}^{m-1}\frac{(m-1)!m!(2r)!2^{2m-2r-1}}{(r!)^2(2m)!(\mp a^2)^{m-r}x^{2r+1}}$$

194. $$\int \frac{dx}{x^{2m+1}\sqrt{x^2 + a^2}} = \frac{(2m)!}{(m!)^2}\left[\frac{\sqrt{x^2 + a^2}}{a^2}\sum_{r=1}^{m}(-1)^{m-r+1}\frac{r!(r-1)!}{2(2r)!(4a^2)^{m-r}x^{2r}} + \frac{(-1)^{m+1}}{2^{2m}a^{2m+1}}\log\frac{\sqrt{x^2 + a^2} + a}{x}\right]$$

195. $$\int \frac{dx}{x^{2m+1}\sqrt{x^2 - a^2}} = \frac{(2m)!}{(m!)^2}\left[\frac{\sqrt{x^2 - a^2}}{a^2}\sum_{r=1}^{m}\frac{r!(r-1)!}{2(2r)!(4a^2)^{m-r}x^{2r}} + \frac{1}{2^{2m}|a|^{2m+1}}\sec^{-1}\frac{x}{a}\right]$$

196. $$\int \frac{dx}{(x-a)\sqrt{x^2 - a^2}} = -\frac{\sqrt{x^2 - a^2}}{a(x-a)}$$

197. $$\int \frac{dx}{(x+a)\sqrt{x^2 - a^2}} = \frac{\sqrt{x^2 - a^2}}{a(x+a)}$$

198. $$\int f\left(x, \sqrt{x^2 + a^2}\right)dx = a\int f(a\tan u, a\sec u)\sec^2 u\,du, \qquad \left(u = \tan^{-1}\frac{x}{a}, a > 0\right)$$

199. $$\int f\left(x, \sqrt{x^2 - a^2}\right)dx = a\int f(a\sec u, a\tan u)\sec u\tan u\,du, \qquad \left(u = \sec^{-1}\frac{x}{a}, a > 0\right)$$

FORMS CONTAINING $\sqrt{a^2 - x^2}$

200. $$\int \sqrt{a^2 - x^2}\,dx = \frac{1}{2}\left[x\sqrt{a^2 - x^2} + a^2\sin^{-1}\frac{x}{|a|}\right]$$

201. $$\int \frac{dx}{\sqrt{a^2 - x^2}} = \begin{cases} \sin^{-1}\dfrac{x}{|a|} \\ \text{or} \\ -\cos^{-1}\dfrac{x}{|a|} \end{cases}$$

202. $$\int \frac{dx}{x\sqrt{a^2 - x^2}} = -\frac{1}{a}\log\left(\frac{a + \sqrt{a^2 - x^2}}{x}\right)$$

203. $$\int \frac{\sqrt{a^2 - x^2}}{x}\,dx = \sqrt{a^2 - x^2} - a\log\left(\frac{a + \sqrt{a^2 - x^2}}{x}\right)$$

204. $$\int \frac{x\,dx}{\sqrt{a^2 - x^2}} = -\sqrt{a^2 - x^2}$$

205. $$\int x\sqrt{a^2 - x^2}\,dx = -\tfrac{1}{3}\sqrt{(a^2 - x^2)^3}$$

206. $\int \sqrt{(a^2 - x^2)^3}\, dx = \frac{1}{4}\left[x\sqrt{(a^2 - x^2)^3} + \frac{3a^2x}{2}\sqrt{a^2 - x^2} + \frac{3a^4}{2}\sin^{-1}\frac{x}{|a|}\right]$

207. $\int \frac{dx}{\sqrt{(a^2 - x^2)^3}} = \frac{x}{a^2\sqrt{a^2 - x^2}}$

208. $\int \frac{x\, dx}{\sqrt{(a^2 - x^2)^3}} = \frac{1}{\sqrt{a^2 - x^2}}$

209. $\int x\sqrt{(a^2 - x^2)^3}\, dx = -\frac{1}{5}\sqrt{(a^2 - x^2)^5}$

210. $\int x^2\sqrt{a^2 - x^2}\, dx = -\frac{x}{4}\sqrt{(a^2 - x^2)^3} + \frac{a^2}{8}\left(x\sqrt{a^2 - x^2} + a^2\sin^{-1}\frac{x}{|a|}\right)$

211. $\int x^3\sqrt{a^2 - x^2}\, dx = (-\frac{1}{5}x^2 - \frac{2}{15}a^2)\sqrt{(a^2 - x^2)^3}$

212. $\int x^2\sqrt{(a^2 - x^2)^3}\, dx = -\frac{1}{6}x\sqrt{(a^2 - x^2)^5} + \frac{a^2x}{24}\sqrt{(a^2 - x^2)^3}$

$$+ \frac{a^4x}{16}\sqrt{a^2 - x^2} + \frac{a^6}{16}\sin^{-1}\frac{x}{|a|}$$

213. $\int x^3\sqrt{(a^2 - x^2)^3}\, dx = \frac{1}{7}\sqrt{(a^2 - x^2)^7} - \frac{a^2}{5}\sqrt{(a^2 - x^2)^5}$

214. $\int \frac{x^2\, dx}{\sqrt{a^2 - x^2}} = -\frac{x}{2}\sqrt{a^2 - x^2} + \frac{a^2}{2}\sin^{-1}\frac{x}{|a|}$

215. $\int \frac{dx}{x^2\sqrt{a^2 - x^2}} = -\frac{\sqrt{a^2 - x^2}}{a^2x}$

216. $\int \frac{\sqrt{a^2 - x^2}}{x^2}\, dx = -\frac{\sqrt{a^2 - x^2}}{x} - \sin^{-1}\frac{x}{|a|}$

217. $\int \frac{\sqrt{a^2 - x^2}}{x^3}\, dx = -\frac{\sqrt{a^2 - x^2}}{2x^2} + \frac{1}{2a}\log\frac{a + \sqrt{a^2 - x^2}}{x}$

218. $\int \frac{\sqrt{a^2 - x^2}}{x^4}\, dx = -\frac{\sqrt{(a^2 - x^2)^3}}{3a^2x^3}$

219. $\int \frac{x^2\, dx}{\sqrt{(a^2 - x^2)^3}} = \frac{x}{\sqrt{a^2 - x^2}} - \sin^{-1}\frac{x}{|a|}$

220. $\int \frac{x^3\, dx}{\sqrt{a^2 - x^2}} = -\frac{2}{3}(a^2 - x^2)^{\frac{3}{2}} - x^2(a^2 - x^2)^{\frac{1}{2}} = -\frac{1}{3}\sqrt{a^2 - x^2}(x^2 + 2a^2)$

221. $\int \frac{x^3\, dx}{\sqrt{(a^2 - x^2)^3}} = 2(a^2 - x^2)^{\frac{1}{2}} + \frac{x^2}{(a^2 - x^2)^{\frac{1}{2}}} = -\frac{a^2}{\sqrt{a^2 - x^2}} + \sqrt{a^2 - x^2}$

222. $\int \frac{dx}{x^3\sqrt{a^2 - x^2}} = -\frac{\sqrt{a^2 - x^2}}{2a^2x^2} - \frac{1}{2a^3}\log\frac{a + \sqrt{a^2 - x^2}}{x}$

223. $\int \frac{dx}{x\sqrt{(a^2 - x^2)^3}} = \frac{1}{a^2\sqrt{a^2 - x^2}} - \frac{1}{a^3}\log\frac{a + \sqrt{a^2 - x^2}}{x}$

224. $\int \frac{dx}{x^2\sqrt{(a^2 - x^2)^3}} = \frac{1}{a^4}\left[-\frac{\sqrt{a^2 - x^2}}{x} + \frac{x}{\sqrt{a^2 - x^2}}\right]$

225. $\int \frac{dx}{x^3\sqrt{(a^2 - x^2)^3}} = -\frac{1}{2a^2x^2\sqrt{a^2 - x^2}} + \frac{3}{2a^4\sqrt{a^2 - x^2}}$

$$-\frac{3}{2a^5}\log\frac{a + \sqrt{a^2 - x^2}}{x}$$

226. $$\int \frac{x^m}{\sqrt{a^2-x^2}}\,dx = -\frac{x^{m-1}\sqrt{a^2-x^2}}{m} + \frac{(m-1)a^2}{m}\int \frac{x^{m-2}}{\sqrt{a^2-x^2}}\,dx$$

227. $$\int \frac{x^{2m}}{\sqrt{a^2-x^2}}\,dx = \frac{(2m)!}{(m!)^2}\left[-\sqrt{a^2-x^2}\sum_{r=1}^{m}\frac{r!(r-1)!}{2^{2m-2r+1}(2r)!}a^{2m-2r}x^{2r-1} + \frac{a^{2m}}{2^{2m}}\sin^{-1}\frac{x}{|a|}\right]$$

228. $$\int \frac{x^{2m+1}}{\sqrt{a^2-x^2}}\,dx = -\sqrt{a^2-x^2}\sum_{r=0}^{m}\frac{(2r)!(m!)^2}{(2m+1)!(r!)^2}(4a^2)^{m-r}x^{2r}$$

229. $$\int \frac{dx}{x^m\sqrt{a^2-x^2}} = -\frac{\sqrt{a^2-x^2}}{(m-1)a^2x^{m-1}} + \frac{m-2}{(m-1)a^2}\int \frac{dx}{x^{m-2}\sqrt{a^2-x^2}}$$

230. $$\int \frac{dx}{x^{2m}\sqrt{a^2-x^2}} = -\sqrt{a^2-x^2}\sum_{r=0}^{m-1}\frac{(m-1)!m!(2r)!2^{2m-2r-1}}{(r!)^2(2m)!a^{2m-2r}x^{2r+1}}$$

231. $$\int \frac{dx}{x^{2m+1}\sqrt{a^2-x^2}} = \frac{(2m)!}{(m!)^2}\left[-\frac{\sqrt{a^2-x^2}}{a^2}\sum_{r=1}^{m}\frac{r!(r-1)!}{2(2r)!(4a^2)^{m-r}x^{2r}} + \frac{1}{2^{2m}a^{2m+1}}\log\frac{a-\sqrt{a^2-x^2}}{x}\right]$$

232. $$\int \frac{dx}{(b^2-x^2)\sqrt{a^2-x^2}} = \frac{1}{2b\sqrt{a^2-b^2}}\log\frac{(b\sqrt{a^2-x^2}+x\sqrt{a^2-b^2})^2}{b^2-x^2}, \quad (a^2 > b^2)$$

233. $$\int \frac{dx}{(b^2-x^2)\sqrt{a^2-x^2}} = \frac{1}{b\sqrt{b^2-a^2}}\tan^{-1}\frac{x\sqrt{b^2-a^2}}{b\sqrt{a^2-x^2}}, \quad (b^2 > a^2)$$

234. $$\int \frac{dx}{(b^2+x^2)\sqrt{a^2-x^2}} = \frac{1}{b\sqrt{a^2+b^2}}\tan^{-1}\frac{x\sqrt{a^2+b^2}}{b\sqrt{a^2-x^2}}$$

235. $$\int \frac{\sqrt{a^2-x^2}}{b^2+x^2}\,dx = \frac{\sqrt{a^2+b^2}}{|b|}\sin^{-1}\frac{x\sqrt{a^2+b^2}}{|a|\sqrt{x^2+b^2}} - \sin^{-1}\frac{x}{|a|}$$

236. $$\int f(x, \sqrt{a^2-x^2})\,dx = a\int f(a\sin u, a\cos u)\cos u\,du, \quad \left(u = \sin^{-1}\frac{x}{a},\ a > 0\right)$$

FORMS CONTAINING $\sqrt{a + bx + cx^2}$

$$X = a + bx + cx^2,\ q = 4ac - b^2, \text{ and } k = \frac{4c}{q}$$

If $q = 0$, then $\sqrt{X} = \sqrt{c}\left|x + \frac{b}{2c}\right|$

237. $$\int \frac{dx}{\sqrt{X}} = \begin{cases} \frac{1}{\sqrt{c}}\log(2\sqrt{cX} + 2cx + b) \\ \text{or} \\ \frac{1}{\sqrt{c}}\sinh^{-1}\frac{2cx+b}{\sqrt{q}}, \quad (c > 0) \end{cases}$$

238. $$\int \frac{dx}{\sqrt{X}} = -\frac{1}{\sqrt{-c}}\sin^{-1}\frac{2cx+b}{\sqrt{-q}}, \quad (c < 0)$$

239. $$\int \frac{dx}{X\sqrt{X}} = \frac{2(2cx+b)}{q\sqrt{X}}$$

240. $$\int \frac{dx}{X^2\sqrt{X}} = \frac{2(2cx+b)}{3q\sqrt{X}}\left(\frac{1}{X} + 2k\right)$$

241. $$\int \frac{dx}{X^n\sqrt{X}} = \begin{cases} \dfrac{2(2cx+b)\sqrt{X}}{(2n-1)qX^n} + \dfrac{2k(n-1)}{2n-1}\displaystyle\int \frac{dx}{X^{n-1}\sqrt{X}} \\ \text{or} \\ \dfrac{(2cx+b)(n!)(n-1)!4^n k^{n-1}}{q[(2n)!]\sqrt{X}} \displaystyle\sum_{r=0}^{n-1} \frac{(2r)!}{(4kX)^r(r!)^2} \end{cases}$$

242. $$\int \sqrt{X}\,dx = \frac{(2cx+b)\sqrt{X}}{4c} + \frac{1}{2k}\int \frac{dx}{\sqrt{X}}$$

243. $$\int X\sqrt{X}\,dx = \frac{(2cx+b)\sqrt{X}}{8c}\left(X + \frac{3}{2k}\right) + \frac{3}{8k^2}\int \frac{dx}{\sqrt{X}}$$

244. $$\int X^2\sqrt{X}\,dx = \frac{(2cx+b)\sqrt{X}}{12c}\left(X^2 + \frac{5X}{4k} + \frac{15}{8k^2}\right) + \frac{5}{16k^3}\int \frac{dx}{\sqrt{X}}$$

245. $$\int X^n\sqrt{X}\,dx = \begin{cases} \dfrac{(2cx+b)X^n\sqrt{X}}{4(n+1)c} + \dfrac{2n+1}{2(n+1)k}\displaystyle\int X^{n-1}\sqrt{X}\,dx \\ \text{or} \\ \dfrac{(2n+2)!}{[(n+1)!]^2(4k)^{n+1}}\left[\dfrac{k(2cx+b)\sqrt{X}}{c}\displaystyle\sum_{r=0}^{n}\frac{r!(r+1)!(4kX)^r}{(2r+2)!} + \int \frac{dx}{\sqrt{X}}\right] \end{cases}$$

246. $$\int \frac{x\,dx}{\sqrt{X}} = \frac{\sqrt{X}}{c} - \frac{b}{2c}\int \frac{dx}{\sqrt{X}}$$

247. $$\int \frac{x\,dx}{X\sqrt{X}} = -\frac{2(bx+2a)}{q\sqrt{X}}$$

248. $$\int \frac{x\,dx}{X^n\sqrt{X}} = -\frac{\sqrt{X}}{(2n-1)cX^n} - \frac{b}{2c}\int \frac{dx}{X^n\sqrt{X}}$$

249. $$\int \frac{x^2\,dx}{\sqrt{X}} = \left(\frac{x}{2c} - \frac{3b}{4c^2}\right)\sqrt{X} + \frac{3b^2-4ac}{8c^2}\int \frac{dx}{\sqrt{X}}$$

250. $$\int \frac{x^2\,dx}{X\sqrt{X}} = \frac{(2b^2-4ac)x + 2ab}{cq\sqrt{X}} + \frac{1}{c}\int \frac{dx}{\sqrt{X}}$$

251. $$\int \frac{x^2\,dx}{X^n\sqrt{X}} = \frac{(2b^2-4ac)x + 2ab}{(2n-1)cqX^{n-1}\sqrt{X}} + \frac{4ac+(2n-3)b^2}{(2n-1)cq}\int \frac{dx}{X^{n-1}\sqrt{X}}$$

252. $$\int \frac{x^3\,dx}{\sqrt{X}} = \left(\frac{x^2}{3c} - \frac{5bx}{12c^2} + \frac{5b^2}{8c^3} - \frac{2a}{3c^2}\right)\sqrt{X} + \left(\frac{3ab}{4c^2} - \frac{5b^3}{16c^3}\right)\int \frac{dx}{\sqrt{X}}$$

253. $$\int \frac{x^n\,dx}{\sqrt{X}} = \frac{1}{nc}x^{n-1}\sqrt{X} - \frac{(2n-1)b}{2nc}\int \frac{x^{n-1}\,dx}{\sqrt{X}} - \frac{(n-1)a}{nc}\int \frac{x^{n-2}\,dx}{\sqrt{X}}$$

254. $$\int x\sqrt{X}\,dx = \frac{X\sqrt{X}}{3c} - \frac{b(2cx+b)}{8c^2}\sqrt{X} - \frac{b}{4ck}\int \frac{dx}{\sqrt{X}}$$

255. $$\int xX\sqrt{X}\,dx = \frac{X^2\sqrt{X}}{5c} - \frac{b}{2c}\int X\sqrt{X}\,dx$$

256. $\int xX^n\sqrt{X}\,dx = \frac{X^{n+1}\sqrt{X}}{(2n+3)c} - \frac{b}{2c}\int X^n\sqrt{X}\,dx$

257. $\int x^2\sqrt{X}\,dx = \left(x - \frac{5b}{6c}\right)\frac{X\sqrt{X}}{4c} + \frac{5b^2-4ac}{16c^2}\int\sqrt{X}\,dx$

258. $\int\frac{dx}{x\sqrt{X}} = -\frac{1}{\sqrt{a}}\log\frac{2\sqrt{aX}+bx+2a}{x}, \qquad (a>0)$

259. $\int\frac{dx}{x\sqrt{X}} = \frac{1}{\sqrt{-a}}\sin^{-1}\left(\frac{bx+2a}{|x|\sqrt{-q}}\right), \qquad (a<0)$

260. $\int\frac{dx}{x\sqrt{X}} = -\frac{2\sqrt{X}}{bx}, \qquad (a=0)$

261. $\int\frac{dx}{x^2\sqrt{X}} = -\frac{\sqrt{X}}{ax} - \frac{b}{2a}\int\frac{dx}{x\sqrt{X}}$

262. $\int\frac{\sqrt{X}\,dx}{x} = \sqrt{X} + \frac{b}{2}\int\frac{dx}{\sqrt{X}} + a\int\frac{dx}{x\sqrt{X}}$

263. $\int\frac{\sqrt{X}\,dx}{x^2} = -\frac{\sqrt{X}}{x} + \frac{b}{2}\int\frac{dx}{x\sqrt{X}} + c\int\frac{dx}{\sqrt{X}}$

FORMS INVOLVING $\sqrt{2ax - x^2}$

264. $\int\sqrt{2ax-x^2}\,dx = \frac{1}{2}\left[(x-a)\sqrt{2ax-x^2} + a^2\sin^{-1}\frac{x-a}{|a|}\right]$

265. $\int\frac{dx}{\sqrt{2ax-x^2}} = \begin{cases}\cos^{-1}\frac{a-x}{|a|} \\ \text{or} \\ \sin^{-1}\frac{x-a}{|a|}\end{cases}$

266. $\int x^n\sqrt{2ax-x^2}\,dx = \begin{cases} -\frac{x^{n-1}(2ax-x^2)^{3/2}}{n+2} + \frac{(2n+1)a}{n+2}\int x^{n-1}\sqrt{2ax-x^2}\,dx \\ \text{or} \\ \sqrt{2ax-x^2}\left[\frac{x^{n+1}}{n+2} - \sum_{r=0}^{n}\frac{(2n+1)!(r!)^2a^{n-r+1}}{2^{n-r}(2r+1)!(n+2)!n!}x^r\right] \\ \qquad + \frac{(2n+1)!a^{n+2}}{2^n n!(n+2)!}\sin^{-1}\frac{x-a}{|a|}\end{cases}$

267. $\int\frac{\sqrt{2ax-x^2}}{x^n}\,dx = \frac{(2ax-x^2)^{3/2}}{(3-2n)ax^n} + \frac{n-3}{(2n-3)a}\int\frac{\sqrt{2ax-x^2}}{x^{n-1}}\,dx$

268. $\int\frac{x^n\,dx}{\sqrt{2ax-x^2}} = \begin{cases}\frac{-x^{n-1}\sqrt{2ax-x^2}}{n} + \frac{a(2n-1)}{n}\int\frac{x^{n-1}}{\sqrt{2ax-x^2}}\,dx \\ \text{or} \\ -\sqrt{2ax-x^2}\sum_{r=1}^{n}\frac{(2n)!r!(r-1)!a^{n-r}}{2^{n-r}(2r)!(n!)^2}x^{r-1} \\ \qquad + \frac{(2n)!a^n}{2^n(n!)^2}\sin^{-1}\frac{x-a}{|a|}\end{cases}$

269. $$\int \frac{dx}{x^n\sqrt{2ax - x^2}} = \begin{cases} \dfrac{\sqrt{2ax - x^2}}{a(1 - 2n)x^n} + \dfrac{n - 1}{(2n - 1)a}\displaystyle\int \frac{dx}{x^{n-1}\sqrt{2ax - x^2}} \\ \text{or} \\ -\sqrt{2ax - x^2}\displaystyle\sum_{r=0}^{n-1} \frac{2^{n-r}(n - 1)!n!(2r)!}{(2n)!(r!)^2a^{n-r}x^{r+1}} \end{cases}$$

270. $$\int \frac{dx}{(2ax - x^2)^{\frac{3}{2}}} = \frac{x - a}{a^2\sqrt{2ax - x^2}}$$

271. $$\int \frac{x\,dx}{(2ax - x^2)^{\frac{3}{2}}} = \frac{x}{a\sqrt{2ax - x^2}}$$

MISCELLANEOUS ALGEBRAIC FORMS

272. $$\int \frac{dx}{\sqrt{2ax + x^2}} = \log\left(x + a + \sqrt{2ax + x^2}\right)$$

273. $$\int \sqrt{ax^2 + c}\,dx = \frac{x}{2}\sqrt{ax^2 + c} + \frac{c}{2\sqrt{a}}\log\left(x\sqrt{a} + \sqrt{ax^2 + c}\right), \qquad (a > 0)$$

274. $$\int \sqrt{ax^2 + c}\,dx = \frac{x}{2}\sqrt{ax^2 + c} + \frac{c}{2\sqrt{-a}}\sin^{-1}\left(x\sqrt{-\frac{a}{c}}\right), \qquad (a < 0)$$

275. $$\int \sqrt{\frac{1 + x}{1 - x}}\,dx = \sin^{-1}x - \sqrt{1 - x^2}$$

276. $$\int \frac{dx}{x\sqrt{ax^n + c}} = \begin{cases} \dfrac{1}{n\sqrt{c}}\log\dfrac{\sqrt{ax^n + c} - \sqrt{c}}{\sqrt{ax^n + c} + \sqrt{c}} \\ \text{or} \\ \dfrac{2}{n\sqrt{c}}\log\dfrac{\sqrt{ax^n + c} - \sqrt{c}}{\sqrt{x^n}}, \qquad (c > 0) \end{cases}$$

277. $$\int \frac{dx}{x\sqrt{ax^n + c}} = \frac{2}{n\sqrt{-c}}\sec^{-1}\sqrt{-\frac{ax^n}{c}}, \qquad (c < 0)$$

278. $$\int \frac{dx}{\sqrt{ax^2 + c}} = \frac{1}{\sqrt{a}}\log\left(x\sqrt{a} + \sqrt{ax^2 + c}\right), \qquad (a > 0)$$

279. $$\int \frac{dx}{\sqrt{ax^2 + c}} = \frac{1}{\sqrt{-a}}\sin^{-1}\left(x\sqrt{-\frac{a}{c}}\right), \qquad (a < 0)$$

280. $$\int (ax^2 + c)^{m+\frac{1}{2}}\,dx = \begin{cases} \dfrac{x(ax^2 + c)^{m+\frac{1}{2}}}{2(m + 1)} + \dfrac{(2m + 1)c}{2(m + 1)}\displaystyle\int (ax^2 + c)^{m-\frac{1}{2}}\,dx \\ \text{or} \\ x\sqrt{ax^2 + c}\displaystyle\sum_{r=0}^{m} \frac{(2m + 1)!(r!)^2c^{m-r}}{2^{2m-2r+1}m!(m + 1)!(2r + 1)!}(ax^2 + c)^r \\ \quad + \dfrac{(2m + 1)!c^{m+1}}{2^{2m+1}m!(m + 1)!}\displaystyle\int \frac{dx}{\sqrt{ax^2 + c}} \end{cases}$$

281. $$\int x(ax^2 + c)^{m+\frac{1}{2}}\,dx = \frac{(ax^2 + c)^{m+\frac{3}{2}}}{(2m + 3)a}$$

282. $$\int \frac{(ax^2+c)^{m+\frac{1}{2}}}{x}\,dx = \begin{cases} \dfrac{(ax^2+c)^{m+\frac{1}{2}}}{2m+1} + c\displaystyle\int \frac{(ax^2+c)^{m-\frac{1}{2}}}{x}\,dx \\ \text{or} \\ \sqrt{ax^2+c}\displaystyle\sum_{r=0}^{m} \frac{c^{m-r}(ax^2+c)^r}{2r+1} + c^{m+1}\displaystyle\int \frac{dx}{x\sqrt{ax^2+c}} \end{cases}$$

283. $$\int \frac{dx}{(ax^2+c)^{m+\frac{1}{2}}} = \begin{cases} \dfrac{x}{(2m-1)c(ax^2+c)^{m-\frac{1}{2}}} + \dfrac{2m-2}{(2m-1)c}\displaystyle\int \frac{dx}{(ax^2+c)^{m-\frac{1}{2}}} \\ \text{or} \\ \dfrac{x}{\sqrt{ax^2+c}}\displaystyle\sum_{r=0}^{m-1} \frac{2^{2m-2r-1}(m-1)!\,m!\,(2r)!}{(2m)!\,(r!)^2 c^{m-r}(ax^2+c)^r} \end{cases}$$

284. $$\int \frac{dx}{x^m\sqrt{ax^2+c}} = -\frac{\sqrt{ax^2+c}}{(m-1)cx^{m-1}} - \frac{(m-2)a}{(m-1)c}\int \frac{dx}{x^{m-2}\sqrt{ax^2+c}}$$

285. $$\int \frac{1+x^2}{(1-x^2)\sqrt{1+x^4}}\,dx = \frac{1}{\sqrt{2}}\log\frac{x\sqrt{2}+\sqrt{1+x^4}}{1-x^2}$$

286. $$\int \frac{1-x^2}{(1+x^2)\sqrt{1+x^4}}\,dx = \frac{1}{\sqrt{2}}\tan^{-1}\frac{x\sqrt{2}}{\sqrt{1+x^4}}$$

287. $$\int \frac{dx}{x\sqrt{x^n+a^2}} = -\frac{2}{na}\log\frac{a+\sqrt{x^n+a^2}}{\sqrt{x^n}}$$

288. $$\int \frac{dx}{x\sqrt{x^n-a^2}} = -\frac{2}{na}\sin^{-1}\frac{a}{\sqrt{x^n}}$$

289. $$\int \sqrt{\frac{x}{a^3-x^3}}\,dx = \frac{2}{3}\sin^{-1}\left(\frac{x}{a}\right)^{\frac{3}{2}}$$

FORMS INVOLVING TRIGONOMETRIC FUNCTIONS

290. $$\int (\sin ax)\,dx = -\frac{1}{a}\cos ax$$

291. $$\int (\cos ax)\,dx = \frac{1}{a}\sin ax$$

292. $$\int (\tan ax)\,dx = -\frac{1}{a}\log\cos ax = \frac{1}{a}\log\sec ax$$

293. $$\int (\cot ax)\,dx = \frac{1}{a}\log\sin ax = -\frac{1}{a}\log\csc ax$$

294. $$\int (\sec ax)\,dx = \frac{1}{a}\log(\sec ax + \tan ax) = \frac{1}{a}\log\tan\left(\frac{\pi}{4}+\frac{ax}{2}\right)$$

295. $$\int (\csc ax)\,dx = \frac{1}{a}\log(\csc ax - \cot ax) = \frac{1}{a}\log\tan\frac{ax}{2}$$

296. $$\int (\sin^2 ax)\,dx = -\frac{1}{2a}\cos ax\sin ax + \frac{1}{2}x = \frac{1}{2}x - \frac{1}{4a}\sin 2ax$$

297. $$\int (\sin^3 ax)\,dx = -\frac{1}{3a}(\cos ax)(\sin^2 ax + 2)$$

298. $$\int (\sin^4 ax)\,dx = \frac{3x}{8} - \frac{\sin 2ax}{4a} + \frac{\sin 4ax}{32a}$$

299. $$\int (\sin^n ax)\,dx = -\frac{\sin^{n-1} ax\cos ax}{na} + \frac{n-1}{n}\int (\sin^{n-2} ax)\,dx$$

300. $\int (\sin^{2m} ax)\,dx = -\frac{\cos ax}{a}\sum_{r=0}^{m-1}\frac{(2m)!(r!)^2}{2^{2m-2r}(2r+1)!(m!)^2}\sin^{2r+1} ax + \frac{(2m)!}{2^{2m}(m!)^2}x$

301. $\int (\sin^{2m+1} ax)\,dx = -\frac{\cos ax}{a}\sum_{r=0}^{m}\frac{2^{2m-2r}(m!)^2(2r)!}{(2m+1)!(r!)^2}\sin^{2r} ax$

302. $\int (\cos^2 ax)\,dx = \frac{1}{2a}\sin ax\cos ax + \frac{1}{2}x = \frac{1}{2}x + \frac{1}{4a}\sin 2ax$

303. $\int (\cos^3 ax)\,dx = \frac{1}{3a}(\sin ax)(\cos^2 ax + 2)$

304. $\int (\cos^4 ax)\,dx = \frac{3x}{8} + \frac{\sin 2ax}{4a} + \frac{\sin 4ax}{32a}$

305. $\int (\cos^n ax)\,dx = \frac{1}{na}\cos^{n-1} ax\sin ax + \frac{n-1}{n}\int (\cos^{n-2} ax)\,dx$

306. $\int (\cos^{2m} ax)\,dx = \frac{\sin ax}{a}\sum_{r=0}^{m-1}\frac{(2m)!(r!)^2}{2^{2m-2r}(2r+1)!(m!)^2}\cos^{2r+1} ax + \frac{(2m)!}{2^{2m}(m!)^2}x$

307. $\int (\cos^{2m+1} ax)\,dx = \frac{\sin ax}{a}\sum_{r=0}^{m}\frac{2^{2m-2r}(m!)^2(2r)!}{(2m+1)!(r!)^2}\cos^{2r} ax$

308. $\int \frac{dx}{\sin^2 ax} = \int (\csc^2 ax)\,dx = -\frac{1}{a}\cot ax$

309. $\int \frac{dx}{\sin^m ax} = \int (\csc^m ax)\,dx = -\frac{1}{(m-1)a}\cdot\frac{\cos ax}{\sin^{m-1} ax} + \frac{m-2}{m-1}\int \frac{dx}{\sin^{m-2} ax}$

310. $\int \frac{dx}{\sin^{2m} ax} = \int (\csc^{2m} ax)\,dx = -\frac{1}{a}\cos ax\sum_{r=0}^{m-1}\frac{2^{2m-2r-1}(m-1)!m!(2r)!}{(2m)!(r!)^2\sin^{2r+1} ax}$

311. $\int \frac{dx}{\sin^{2m+1} ax} = \int (\csc^{2m+1} ax)\,dx =$

$$-\frac{1}{a}\cos ax\sum_{r=0}^{m-1}\frac{(2m)!(r!)^2}{2^{2m-2r}(m!)^2(2r+1)!\sin^{2r+2} ax} + \frac{1}{a}\cdot\frac{(2m)!}{2^{2m}(m!)^2}\log\tan\frac{ax}{2}$$

312. $\int \frac{dx}{\cos^2 ax} = \int (\sec^2 ax)\,dx = \frac{1}{a}\tan ax$

313. $\int \frac{dx}{\cos^n ax} = \int (\sec^n ax)\,dx = \frac{1}{(n-1)a}\cdot\frac{\sin ax}{\cos^{n-1} ax} + \frac{n-2}{n-1}\int \frac{dx}{\cos^{n-2} ax}$

314. $\int \frac{dx}{\cos^{2m} ax} = \int (\sec^{2m} ax)\,dx = \frac{1}{a}\sin ax\sum_{r=0}^{m-1}\frac{2^{2m-2r-1}(m-1)!m!(2r)!}{(2m)!(r!)^2\cos^{2r+1} ax}$

315. $\int \frac{dx}{\cos^{2m+1} ax} = \int (\sec^{2m+1} ax)\,dx =$

$$\frac{1}{a}\sin ax\sum_{r=0}^{m-1}\frac{(2m)!(r!)^2}{2^{2m-2r}(m!)^2(2r+1)!\cos^{2r+2} ax} + \frac{1}{a}\cdot\frac{(2m)!}{2^{2m}(m!)^2}\log(\sec ax + \tan ax)$$

316. $\int (\sin mx)(\sin nx)\,dx = \frac{\sin(m-n)x}{2(m-n)} - \frac{\sin(m+n)x}{2(m+n)}, \quad (m^2 \neq n^2)$

317. $\int (\cos mx)(\cos nx)\,dx = \frac{\sin(m-n)x}{2(m-n)} + \frac{\sin(m+n)x}{2(m+n)}, \quad (m^2 \neq n^2)$

318. $\int (\sin ax)(\cos ax)\,dx = \frac{1}{2a}\sin^2 ax$

319. $\int (\sin mx)(\cos nx)\,dx = -\frac{\cos(m-n)x}{2(m-n)} - \frac{\cos(m+n)x}{2(m+n)}, \qquad (m^2 \neq n^2)$

320. $\int (\sin^2 ax)(\cos^2 ax)\,dx = -\frac{1}{32a}\sin 4ax + \frac{x}{8}$

321. $\int (\sin ax)(\cos^m ax)\,dx = -\frac{\cos^{m+1} ax}{(m+1)a}$

322. $\int (\sin^m ax)(\cos ax)\,dx = \frac{\sin^{m+1} ax}{(m+1)a}$

323. $$\int (\cos^m ax)(\sin^n ax)\,dx = \begin{cases} \frac{\cos^{m-1} ax \sin^{n+1} ax}{(m+n)a} + \frac{m-1}{m+n}\int (\cos^{m-2} ax)(\sin^n ax)\,dx \\ \text{or} \\ -\frac{\sin^{n-1} ax \cos^{m+1} ax}{(m+n)a} + \frac{n-1}{m+n}\int (\cos^m ax)(\sin^{n-2} ax)\,dx \end{cases}$$

324. $$\int \frac{\cos^m ax}{\sin^n ax}\,dx = \begin{cases} -\frac{\cos^{m+1} ax}{(n-1)a\sin^{n-1} ax} - \frac{m-n+2}{n-1}\int \frac{\cos^m ax}{\sin^{n-2} ax}\,dx \\ \text{or} \\ \frac{\cos^{m-1} ax}{a(m-n)\sin^{n-1} ax} + \frac{m-1}{m-n}\int \frac{\cos^{m-2} ax}{\sin^n ax}\,dx \end{cases}$$

325. $$\int \frac{\sin^m ax}{\cos^n ax}\,dx = \begin{cases} \frac{\sin^{m+1} ax}{a(n-1)\cos^{n-1} ax} - \frac{m-n+2}{n-1}\int \frac{\sin^m ax}{\cos^{n-2} ax}\,dx \\ \text{or} \\ -\frac{\sin^{m-1} ax}{a(m-n)\cos^{n-1} ax} + \frac{m-1}{m-n}\int \frac{\sin^{m-2} ax}{\cos^n ax}\,dx \end{cases}$$

326. $\int \frac{\sin ax}{\cos^2 ax}\,dx = \frac{1}{a\cos ax} = \frac{\sec ax}{a}$

327. $\int \frac{\sin^2 ax}{\cos ax}\,dx = -\frac{1}{a}\sin ax + \frac{1}{a}\log\tan\left(\frac{\pi}{4} + \frac{ax}{2}\right)$

328. $\int \frac{\cos ax}{\sin^2 ax}\,dx = -\frac{1}{a\sin ax} = -\frac{\csc ax}{a}$

329. $\int \frac{dx}{(\sin ax)(\cos ax)} = \frac{1}{a}\log\tan ax$

330. $\int \frac{dx}{(\sin ax)(\cos^2 ax)} = \frac{1}{a}\left(\sec ax + \log\tan\frac{ax}{2}\right)$

331. $\int \frac{dx}{(\sin ax)(\cos^n ax)} = \frac{1}{a(n-1)\cos^{n-1} ax} + \int \frac{dx}{(\sin ax)(\cos^{n-2} ax)}$

332. $\int \frac{dx}{(\sin^2 ax)(\cos ax)} = -\frac{1}{a}\csc ax + \frac{1}{a}\log\tan\left(\frac{\pi}{4} + \frac{ax}{2}\right)$

333. $\int \frac{dx}{(\sin^2 ax)(\cos^2 ax)} = -\frac{2}{a}\cot 2ax$

$$334.\quad \int \frac{dx}{\sin^m ax \cos^n ax} = \begin{cases} -\dfrac{1}{a(m-1)(\sin^{m-1} ax)(\cos^{n-1} ax)} \\ \qquad + \dfrac{m+n-2}{m-1}\displaystyle\int \frac{dx}{(\sin^{m-2} ax)(\cos^n ax)} \\ \text{or} \\ \dfrac{1}{a(n-1)\sin^{m-1} ax \cos^{n-1} ax} \\ \qquad - \dfrac{m+n-2}{n-1}\displaystyle\int \frac{dx}{\sin^m ax \cos^{n-2} ax} \end{cases}$$

$$335.\quad \int \sin(a+bx)\,dx = -\frac{1}{b}\cos(a+bx)$$

$$336.\quad \int \cos(a+bx)\,dx = \frac{1}{b}\sin(a+bx)$$

$$337.\quad \int \frac{dx}{1 \pm \sin ax} = \mp \frac{1}{a}\tan\left(\frac{\pi}{4} \mp \frac{ax}{2}\right)$$

$$338.\quad \int \frac{dx}{1+\cos ax} = \frac{1}{a}\tan\frac{ax}{2}$$

$$339.\quad \int \frac{dx}{1-\cos ax} = -\frac{1}{a}\cot\frac{ax}{2}$$

$${}^*340.\quad \int \frac{dx}{a+b\sin x} = \begin{cases} \dfrac{2}{\sqrt{a^2-b^2}}\tan^{-1}\dfrac{a\tan\frac{x}{2}+b}{\sqrt{a^2-b^2}} \\ \text{or} \\ \dfrac{1}{\sqrt{b^2-a^2}}\log\dfrac{a\tan\frac{x}{2}+b-\sqrt{b^2-a^2}}{a\tan\frac{x}{2}+b+\sqrt{b^2-a^2}} \end{cases}$$

$${}^*341.\quad \int \frac{dx}{a+b\cos x} = \begin{cases} \dfrac{2}{\sqrt{a^2-b^2}}\tan^{-1}\dfrac{\sqrt{a^2-b^2}\tan\frac{x}{2}}{a+b} \\ \text{or} \\ \dfrac{1}{\sqrt{b^2-a^2}}\log\left(\dfrac{\sqrt{b^2-a^2}\tan\frac{x}{2}+a+b}{\sqrt{b^2-a^2}\tan\frac{x}{2}-a-b}\right) \end{cases}$$

$${}^*342.\quad \int \frac{dx}{a+b\sin x+c\cos x}$$

$$= \begin{cases} \dfrac{1}{\sqrt{b^2+c^2-a^2}}\log\dfrac{b-\sqrt{b^2+c^2-a^2}+(a-c)\tan\frac{x}{2}}{b+\sqrt{b^2+c^2-a^2}+(a-c)\tan\frac{x}{2}}, & \text{if } a^2 < b^2+c^2,\ a \neq c \\ \text{or} \\ \dfrac{2}{\sqrt{a^2-b^2-c^2}}\tan^{-1}\dfrac{b+(a-c)\tan\frac{x}{2}}{\sqrt{a^2-b^2-c^2}}, & \text{if } a^2 > b^2+c^2 \\ \text{or} \\ \dfrac{1}{a}\left[\dfrac{a-(b+c)\cos x-(b-c)\sin x}{a-(b-c)\cos x+(b+c)\sin x}\right], & \text{if } a^2 = b^2+c^2,\ a \neq c. \end{cases}$$

*See note 6 on page A-19.

*343. $\int \frac{\sin^2 x\,dx}{a+b\cos^2 x} = \frac{1}{b}\sqrt{\frac{a+b}{a}}\tan^{-1}\left(\sqrt{\frac{a}{a+b}}\tan x\right) - \frac{x}{b},\qquad (ab > 0, \text{ or } |a| > |b|)$

*344. $\int \frac{dx}{a^2\cos^2 x + b^2\sin^2 x} = \frac{1}{ab}\tan^{-1}\left(\frac{b\tan x}{a}\right)$

*345. $\int \frac{\cos^2 cx}{a^2 + b^2\sin^2 cx}\,dx = \frac{\sqrt{a^2+b^2}}{ab^2c}\tan^{-1}\frac{\sqrt{a^2+b^2}\tan cx}{a} - \frac{x}{b^2}$

346. $\int \frac{\sin cx\cos cx}{a\cos^2 cx + b\sin^2 cx}\,dx = \frac{1}{2c(b-a)}\log(a\cos^2 cx + b\sin^2 cx)$

347. $\int \frac{\cos cx}{a\cos cx + b\sin cx}\,dx = \int \frac{dx}{a + b\tan cx} =$

$$\frac{1}{c(a^2+b^2)}[acx + b\log(a\cos cx + b\text{ si}$$

348. $\int \frac{\sin cx}{a\sin cx + b\cos cx}\,dx = \int \frac{dx}{a + b\cot cx} =$

$$\frac{1}{c(a^2+b^2)}[acx - b\log(a\sin cx + b\cos x)]$$

*349. $$\int \frac{dx}{a\cos^2 x + 2b\cos x\sin x + c\sin^2 x} = \begin{cases} \frac{1}{2\sqrt{b^2-ac}}\log\frac{c\tan x + b - \sqrt{b^2-ac}}{c\tan x + b + \sqrt{b^2-ac}} & (b^2 > ac) \\ \text{or} & \\ \frac{1}{\sqrt{ac-b^2}}\tan^{-1}\frac{c\tan x + b}{\sqrt{ac-b^2}}, & (b_2 < ac) \\ \text{or} & \\ -\frac{1}{c\tan x + b}, & (b^2 = ac) \end{cases}$$

350. $\int \frac{\sin ax}{1 \pm \sin ax}\,dx = \pm x + \frac{1}{a}\tan\left(\frac{\pi}{4} \mp \frac{ax}{2}\right)$

351. $\int \frac{dx}{(\sin ax)(1 \pm \sin ax)} = \frac{1}{a}\tan\left(\frac{\pi}{4} \mp \frac{ax}{2}\right) + \frac{1}{a}\log\tan\frac{ax}{2}$

352. $\int \frac{dx}{(1+\sin ax)^2} = -\frac{1}{2a}\tan\left(\frac{\pi}{4} - \frac{ax}{2}\right) - \frac{1}{6a}\tan^3\left(\frac{\pi}{4} - \frac{ax}{2}\right)$

353. $\int \frac{dx}{(1-\sin ax)^2} = \frac{1}{2a}\cot\left(\frac{\pi}{4} - \frac{ax}{2}\right) + \frac{1}{6a}\cot^3\left(\frac{\pi}{4} - \frac{ax}{2}\right)$

354. $\int \frac{\sin ax}{(1+\sin ax)^2}\,dx = -\frac{1}{2a}\tan\left(\frac{\pi}{4} - \frac{ax}{2}\right) + \frac{1}{6a}\tan^3\left(\frac{\pi}{4} - \frac{ax}{2}\right)$

355. $\int \frac{\sin ax}{(1-\sin ax)^2}\,dx = -\frac{1}{2a}\cot\left(\frac{\pi}{4} - \frac{ax}{2}\right) + \frac{1}{6a}\cot^3\left(\frac{\pi}{4} - \frac{ax}{2}\right)$

356. $\int \frac{\sin x\,dx}{a + b\sin x} = \frac{x}{b} - \frac{a}{b}\int \frac{dx}{a + b\sin x}$

357. $\int \frac{dx}{(\sin x)(a + b\sin x)} = \frac{1}{a}\log\tan\frac{x}{2} - \frac{b}{a}\int \frac{dx}{a + b\sin x}$

358. $\int \frac{dx}{(a + b\sin x)^2} = \frac{b\cos x}{(a^2-b^2)(a + b\sin x)} + \frac{a}{a^2-b^2}\int \frac{dx}{a + b\sin x}$

*See note 6 on page A-19.

359. $$\int \frac{\sin x\,dx}{(a + b\sin x)^2} = \frac{a\cos x}{(b^2 - a^2)(a + b\sin x)} + \frac{b}{b^2 - a^2}\int \frac{dx}{a + b\sin x}$$

*360. $$\int \frac{dx}{a^2 + b^2\sin^2 cx} = \frac{1}{ac\sqrt{a^2 + b^2}}\tan^{-1}\frac{\sqrt{a^2 + b^2}\tan cx}{a}$$

*361. $$\int \frac{dx}{a^2 - b^2\sin^2 cx} = \begin{cases} \dfrac{1}{ac\sqrt{a^2 - b^2}}\tan^{-1}\dfrac{\sqrt{a^2 - b^2}\tan cx}{a}, & (a^2 > b^2) \\ \text{or} & \\ \dfrac{1}{2ac\sqrt{b^2 - a^2}}\log\dfrac{\sqrt{b^2 - a^2}\tan cx + a}{\sqrt{b^2 - a^2}\tan cx - a}, & (a^2 < b^2) \end{cases}$$

362. $$\int \frac{\cos ax}{1 + \cos ax}\,dx = x - \frac{1}{a}\tan\frac{ax}{2}$$

363. $$\int \frac{\cos ax}{1 - \cos ax}\,dx = -x - \frac{1}{a}\cot\frac{ax}{2}$$

364. $$\int \frac{dx}{(\cos ax)(1 + \cos ax)} = \frac{1}{a}\log\tan\left(\frac{\pi}{4} + \frac{ax}{2}\right) - \frac{1}{a}\tan\frac{ax}{2}$$

365. $$\int \frac{dx}{(\cos ax)(1 - \cos ax)} = \frac{1}{a}\log\tan\left(\frac{\pi}{4} + \frac{ax}{2}\right) - \frac{1}{a}\cot\frac{ax}{2}$$

366. $$\int \frac{dx}{(1 + \cos ax)^2} = \frac{1}{2a}\tan\frac{ax}{2} + \frac{1}{6a}\tan^3\frac{ax}{2}$$

367. $$\int \frac{dx}{(1 - \cos ax)^2} = -\frac{1}{2a}\cot\frac{ax}{2} - \frac{1}{6a}\cot^3\frac{ax}{2}$$

368. $$\int \frac{\cos ax}{(1 + \cos ax)^2}\,dx = \frac{1}{2a}\tan\frac{ax}{2} - \frac{1}{6a}\tan^3\frac{ax}{2}$$

369. $$\int \frac{\cos ax}{(1 - \cos ax)^2}\,dx = \frac{1}{2a}\cot\frac{ax}{2} - \frac{1}{6a}\cot^3\frac{ax}{2}$$

370. $$\int \frac{\cos x\,dx}{a + b\cos x} = \frac{x}{b} - \frac{a}{b}\int \frac{dx}{a + b\cos x}$$

371. $$\int \frac{dx}{(\cos x)(a + b\cos x)} = \frac{1}{a}\log\tan\left(\frac{x}{2} + \frac{\pi}{4}\right) - \frac{b}{a}\int \frac{dx}{a + b\cos x}$$

372. $$\int \frac{dx}{(a + b\cos x)^2} = \frac{b\sin x}{(b^2 - a^2)(a + b\cos x)} - \frac{a}{b^2 - a^2}\int \frac{dx}{a + b\cos x}$$

373. $$\int \frac{\cos x}{(a + b\cos x)^2}\,dx = \frac{a\sin x}{(a^2 - b^2)(a + b\cos x)} - \frac{b}{a^2 - b^2}\int \frac{dx}{a + b\cos x}$$

*374. $$\int \frac{dx}{a^2 + b^2 - 2ab\cos cx} = \frac{2}{c(a^2 - b^2)}\tan^{-1}\left(\frac{a + b}{a - b}\tan\frac{cx}{2}\right)$$

*375. $$\int \frac{dx}{a^2 + b^2\cos^2 cx} = \frac{1}{ac\sqrt{a^2 + b^2}}\tan^{-1}\frac{a\tan cx}{\sqrt{a^2 + b^2}}$$

*376. $$\int \frac{dx}{a^2 - b^2\cos^2 cx} = \begin{cases} \dfrac{1}{ac\sqrt{a^2 - b^2}}\tan^{-1}\dfrac{a\tan cx}{\sqrt{a^2 - b^2}}, & (a^2 > b^2) \\ \text{or} & \\ \dfrac{1}{2ac\sqrt{b^2 - a^2}}\log\dfrac{a\tan cx - \sqrt{b^2 - a^2}}{a\tan cx + \sqrt{b^2 - a^2}}, & (b^2 > a^2) \end{cases}$$

377. $$\int \frac{\sin ax}{1 \pm \cos ax}\,dx = \mp\frac{1}{a}\log(1 \pm \cos ax)$$

*See note 6 on page A-19.

378. $\int \frac{\cos ax}{1 \pm \sin ax}\, dx = \pm \frac{1}{a} \log (1 \pm \sin ax)$

379. $\int \frac{dx}{(\sin ax)(1 \pm \cos ax)} = \pm \frac{1}{2a(1 \pm \cos ax)} + \frac{1}{2a} \log \tan \frac{ax}{2}$

380. $\int \frac{dx}{(\cos ax)(1 \pm \sin ax)} = \mp \frac{1}{2a(1 \pm \sin ax)} + \frac{1}{2a} \log \tan \left(\frac{\pi}{4} + \frac{ax}{2}\right)$

381. $\int \frac{\sin ax}{(\cos ax)(1 \pm \cos ax)}\, dx = \frac{1}{a} \log (\sec ax \pm 1)$

382. $\int \frac{\cos ax}{(\sin ax)(1 \pm \sin ax)}\, dx = -\frac{1}{a} \log (\csc ax \pm 1)$

383. $\int \frac{\sin ax}{(\cos ax)(1 \pm \sin ax)}\, dx = \frac{1}{2a(1 \pm \sin ax)} \pm \frac{1}{2a} \log \tan \left(\frac{\pi}{4} + \frac{ax}{2}\right)$

384. $\int \frac{\cos ax}{(\sin ax)(1 \pm \cos ax)}\, dx = -\frac{1}{2a(1 \pm \cos ax)} \pm \frac{1}{2a} \log \tan \frac{ax}{2}$

385. $\int \frac{dx}{\sin ax \pm \cos ax} = \frac{1}{a\sqrt{2}} \log \tan \left(\frac{ax}{2} \pm \frac{\pi}{8}\right)$

386. $\int \frac{dx}{(\sin ax \pm \cos ax)^2} = \frac{1}{2a} \tan \left(ax \mp \frac{\pi}{4}\right)$

387. $\int \frac{dx}{1 + \cos ax \pm \sin ax} = \pm \frac{1}{a} \log \left(1 \pm \tan \frac{ax}{2}\right)$

388. $\int \frac{dx}{a^2 \cos^2 cx - b^2 \sin^2 cx} = \frac{1}{2abc} \log \frac{b \tan cx + a}{b \tan cx - a}$

389. $\int x(\sin ax)\, dx = \frac{1}{a^2} \sin ax - \frac{x}{a} \cos ax$

390. $\int x^2(\sin ax)\, dx = \frac{2x}{a^2} \sin ax - \frac{a^2x^2 - 2}{a^3} \cos ax$

391. $\int x^3(\sin ax)\, dx = \frac{3a^2x^2 - 6}{a^4} \sin ax - \frac{a^2x^3 - 6x}{a^3} \cos ax$

392. $$\int x^m \sin ax\, dx = \begin{cases} -\frac{1}{a} x^m \cos ax + \frac{m}{a} \int x^{m-1} \cos ax\, dx \\ \text{or} \\ \cos ax \sum_{r=0}^{\left[\frac{m}{2}\right]} (-1)^{r+1} \frac{m!}{(m-2r)!} \cdot \frac{x^{m-2r}}{a^{2r+1}} \\ \quad + \sin ax \sum_{r=0}^{\left[\frac{m-1}{2}\right]} (-1)^r \frac{m!}{(m-2r-1)!} \cdot \frac{x^{m-2r-1}}{a^{2r+2}} \end{cases}$$

Note: $[s]$ means greatest integer $\leq s$; $[3\frac{1}{2}] = 3$, $[\frac{1}{2}] = 0$, etc.

393. $\int x(\cos ax)\, dx = \frac{1}{a^2} \cos ax + \frac{x}{a} \sin ax$

394. $\int x^2(\cos ax)\, dx = \frac{2x \cos ax}{a^2} + \frac{a^2x^2 - 2}{a^3} \sin ax$

395. $\int x^3(\cos ax)\, dx = \frac{3a^2x^2 - 6}{a^4} \cos ax + \frac{a^2x^3 - 6x}{a^3} \sin ax$

$$396.\quad \int x^m(\cos ax)\,dx = \begin{cases} \dfrac{x^m \sin ax}{a} - \dfrac{m}{a}\displaystyle\int x^{m-1}\sin ax\,dx \\ \text{or} \\ \sin ax \displaystyle\sum_{r=0}^{\left[\frac{m}{2}\right]} (-1)^r \frac{m!}{(m-2r)!}\cdot\frac{x^{m-2r}}{a^{2r+1}} \\ \quad + \cos ax \displaystyle\sum_{r=0}^{\left[\frac{m-1}{2}\right]} (-1)^r \frac{m!}{(m-2r-1)!}\cdot\frac{x^{m-2r-1}}{a^{2r+2}} \end{cases}$$

See note integral 392.

$$397.\quad \int \frac{\sin ax}{x}\,dx = \sum_{n=0}^{\infty} (-1)^n \frac{(ax)^{2n+1}}{(2n+1)(2n+1)!}$$

$$398.\quad \int \frac{\cos ax}{x}\,dx = \log x + \sum_{n=1}^{\infty} (-1)^n \frac{(ax)^{2n}}{2n(2n)!}$$

$$399.\quad \int x(\sin^2 ax)\,dx = \frac{x^2}{4} - \frac{x\sin 2ax}{4a} - \frac{\cos 2ax}{8a^2}$$

$$400.\quad \int x^2(\sin^2 ax)\,dx = \frac{x^3}{6} - \left(\frac{x^2}{4a} - \frac{1}{8a^3}\right)\sin 2ax - \frac{x\cos 2ax}{4a^2}$$

$$401.\quad \int x(\sin^3 ax)\,dx = \frac{x\cos 3ax}{12a} - \frac{\sin 3ax}{36a^2} - \frac{3x\cos ax}{4a} + \frac{3\sin ax}{4a^2}$$

$$402.\quad \int x(\cos^2 ax)\,dx = \frac{x^2}{4} + \frac{x\sin 2ax}{4a} + \frac{\cos 2ax}{8a^2}$$

$$403.\quad \int x^2(\cos^2 ax)\,dx = \frac{x^3}{6} + \left(\frac{x^2}{4a} - \frac{1}{8a^3}\right)\sin 2ax + \frac{x\cos 2ax}{4a^2}$$

$$404.\quad \int x(\cos^3 ax)\,dx = \frac{x\sin 3ax}{12a} + \frac{\cos 3ax}{36a^2} + \frac{3x\sin ax}{4a} + \frac{3\cos ax}{4a^2}$$

$$405.\quad \int \frac{\sin ax}{x^m}\,dx = -\frac{\sin ax}{(m-1)x^{m-1}} + \frac{a}{m-1}\int\frac{\cos ax}{x^{m-1}}\,dx$$

$$406.\quad \int \frac{\cos ax}{x^m}\,dx = -\frac{\cos ax}{(m-1)x^{m-1}} - \frac{a}{m-1}\int\frac{\sin ax}{x^{m-1}}\,dx$$

$$407.\quad \int \frac{x}{1\pm\sin ax}\,dx = \mp\frac{x\cos ax}{a(1\pm\sin ax)} + \frac{1}{a^2}\log(1\pm\sin ax)$$

$$408.\quad \int \frac{x}{1+\cos ax}\,dx = \frac{x}{a}\tan\frac{ax}{2} + \frac{2}{a^2}\log\cos\frac{ax}{2}$$

$$409.\quad \int \frac{x}{1-\cos ax}\,dx = -\frac{x}{a}\cot\frac{ax}{2} + \frac{2}{a^2}\log\sin\frac{ax}{2}$$

$$410.\quad \int \frac{x+\sin x}{1+\cos x}\,dx = x\tan\frac{x}{2}$$

$$411.\quad \int \frac{x-\sin x}{1-\cos x}\,dx = -x\cot\frac{x}{2}$$

$$412.\quad \int \sqrt{1-\cos ax}\,dx = -\frac{2\sin ax}{a\sqrt{1-\cos ax}} = -\frac{2\sqrt{2}\cos\left(\frac{ax}{2}\right)}{a}$$

$$413.\quad \int \sqrt{1+\cos ax}\,dx = \frac{2\sin ax}{a\sqrt{1+\cos ax}} = \frac{2\sqrt{2}\sin\left(\frac{ax}{2}\right)}{a}$$

$$414.\quad \int \sqrt{1+\sin x}\,dx = \pm 2\left(\sin\frac{x}{2} - \cos\frac{x}{2}\right),$$

[use + if $(8k-1)\frac{\pi}{2} < x \leq (8k+3)\frac{\pi}{2}$, otherwise −; k an integer]

415. $\int \sqrt{1 - \sin x}\, dx = \pm 2\left(\sin\frac{x}{2} + \cos\frac{x}{2}\right),$

[use + if $(8k-3)\frac{\pi}{2} < x \leq (8k+1)\frac{\pi}{2}$, otherwise −; k an integer]

416. $\int \frac{dx}{\sqrt{1 - \cos x}} = \pm \sqrt{2} \log \tan \frac{x}{4},$

[use + if $4k\pi < x < (4k+2)\pi$, otherwise −; k an integer]

417. $\int \frac{dx}{\sqrt{1 + \cos x}} = \pm \sqrt{2} \log \tan \left(\frac{x+\pi}{4}\right),$

[use + if $(4k-1)\pi < x < (4k+1)\pi$, otherwise −; k an integer]

418. $\int \frac{dx}{\sqrt{1 - \sin x}} = \pm \sqrt{2} \log \tan \left(\frac{x}{4} - \frac{\pi}{8}\right),$

[use + if $(8k+1)\frac{\pi}{2} < x < (8k+5)\frac{\pi}{2}$, otherwise −; k an integer]

419. $\int \frac{dx}{\sqrt{1 + \sin x}} = \pm \sqrt{2} \log \tan \left(\frac{x}{4} + \frac{\pi}{8}\right),$

[use + if $(8k-1)\frac{\pi}{2} < x < (8k+3)\frac{\pi}{2}$, otherwise −; k an integer]

420. $\int (\tan^2 ax)\, dx = \frac{1}{a} \tan ax - x$

421. $\int (\tan^3 ax)\, dx = \frac{1}{2a} \tan^2 ax + \frac{1}{a} \log \cos ax$

422. $\int (\tan^4 ax)\, dx = \frac{\tan^3 ax}{3a} - \frac{1}{a} \tan x + x$

423. $\int (\tan^n ax)\, dx = \frac{\tan^{n-1} ax}{a(n-1)} - \int (\tan^{n-2} ax)\, dx$

424. $\int (\cot^2 ax)\, dx = -\frac{1}{a} \cot ax - x$

425. $\int (\cot^3 ax)\, dx = -\frac{1}{2a} \cot^2 ax - \frac{1}{a} \log \sin ax$

426. $\int (\cot^4 ax)\, dx = -\frac{1}{3a} \cot^3 ax + \frac{1}{a} \cot ax + x$

427. $\int (\cot^n ax)\, dx = -\frac{\cot^{n-1} ax}{a(n-1)} - \int (\cot^{n-2} ax)\, dx$

428. $\int \frac{x}{\sin^2 ax}\, dx = \int x(\csc^2 ax)\, dx = -\frac{x \cot ax}{a} + \frac{1}{a^2} \log \sin ax$

429. $\int \frac{x}{\sin^n ax}\, dx = \int x(\csc^n ax)\, dx = -\frac{x \cos ax}{a(n-1)\sin^{n-1} ax}$

$$- \frac{1}{a^2(n-1)(n-2)\sin^{n-2} ax} + \frac{(n-2)}{(n-1)} \int \frac{x}{\sin^{n-2} ax}\, dx$$

430. $\int \frac{x}{\cos^2 ax}\, dx = \int x(\sec^2 ax)\, dx = \frac{1}{a} x \tan ax + \frac{1}{a^2} \log \cos ax$

431. $\int \frac{x}{\cos^n ax}\, dx = \int x(\sec^n ax)\, dx = \frac{x \sin ax}{a(n-1)\cos^{n-1} ax}$

$$- \frac{1}{a^2(n-1)(n-2)\cos^{n-2} ax} + \frac{n-2}{n-1} \int \frac{x}{\cos^{n-2} ax}\, dx$$

432. $\int \frac{\sin ax}{\sqrt{1 + b^2 \sin^2 ax}} dx = -\frac{1}{ab} \sin^{-1} \frac{b \cos ax}{\sqrt{1 + b^2}}$

433. $\int \frac{\sin ax}{\sqrt{1 - b^2 \sin^2 ax}} dx = -\frac{1}{ab} \log (b \cos ax + \sqrt{1 - b^2 \sin^2 ax})$

434. $\int (\sin ax)\sqrt{1 + b^2 \sin^2 ax}\, dx = -\frac{\cos ax}{2a} \sqrt{1 + b^2 \sin^2 ax} - \frac{1 + b^2}{2ab} \sin^{-1} \frac{b \cos ax}{\sqrt{1 + b^2}}$

435. $\int (\sin ax)\sqrt{1 - b^2 \sin^2 ax}\, dx = -\frac{\cos ax}{2a} \sqrt{1 - b^2 \sin^2 ax} - \frac{1 - b^2}{2ab} \log (b \cos ax + \sqrt{1 - b^2 \sin^2 ax})$

436. $\int \frac{\cos ax}{\sqrt{1 + b^2 \sin^2 ax}} dx = \frac{1}{ab} \log (b \sin ax + \sqrt{1 + b^2 \sin^2 ax})$

437. $\int \frac{\cos ax}{\sqrt{1 - b^2 \sin^2 ax}} dx = \frac{1}{ab} \sin^{-1} (b \sin ax)$

438. $\int (\cos ax) \sqrt{1 + b^2 \sin^2 ax}\, dx = \frac{\sin ax}{2a} \sqrt{1 + b^2 \sin^2 ax} + \frac{1}{2ab} \log (b \sin ax + \sqrt{1 + b^2 \sin^2 ax})$

439. $\int (\cos ax) \sqrt{1 - b^2 \sin^2 ax}\, dx = \frac{\sin ax}{2a} \sqrt{1 - b^2 \sin^2 ax} + \frac{1}{2ab} \sin^{-1} (b \sin ax)$

440. $\int \frac{dx}{\sqrt{a + b \tan^2 cx}} = \frac{\pm 1}{c\sqrt{a - b}} \sin^{-1} \left(\sqrt{\frac{a - b}{a}} \sin cx \right), \qquad (a > |b|)$

[use + if $(2k - 1)\frac{\pi}{2} < x \leq (2k + 1)\frac{\pi}{2}$, otherwise − ; k an integer]

FORMS INVOLVING INVERSE TRIGONOMETRIC FUNCTIONS

441. $\int (\sin^{-1} ax)\, dx = x \sin^{-1} ax + \frac{\sqrt{1 - a^2x^2}}{a}$

442. $\int (\cos^{-1} ax)\, dx = x \cos^{-1} ax - \frac{\sqrt{1 - a^2x^2}}{a}$

443. $\int (\tan^{-1} ax)\, dx = x \tan^{-1} ax - \frac{1}{2a} \log (1 + a^2x^2)$

444. $\int (\cot^{-1} ax)\, dx = x \cot^{-1} ax + \frac{1}{2a} \log (1 + a^2x^2)$

445. $\int (\sec^{-1} ax)\, dx = x \sec^{-1} ax - \frac{1}{a} \log (ax + \sqrt{a^2x^2 - 1})$

446. $\int (\csc^{-1} ax)\, dx = x \csc^{-1} ax + \frac{1}{a} \log (ax + \sqrt{a^2x^2 - 1})$

447. $\int \left(\sin^{-1} \frac{x}{a} \right) dx = x \sin^{-1} \frac{x}{a} + \sqrt{a^2 - x^2}, \qquad (a > 0)$

448. $\int \left(\cos^{-1} \frac{x}{a} \right) dx = x \cos^{-1} \frac{x}{a} - \sqrt{a^2 - x^2}, \qquad (a > 0)$

449. $\int \left(\tan^{-1}\frac{x}{a}\right) dx = x\tan^{-1}\frac{x}{a} - \frac{a}{2}\log(a^2 + x^2)$

450. $\int \left(\cot^{-1}\frac{x}{a}\right) dx = x\cot^{-1}\frac{x}{a} + \frac{a}{2}\log(a^2 + x^2)$

451. $\int x[\sin^{-1}(ax)]\, dx = \frac{1}{4a^2}[(2a^2x^2 - 1)\sin^{-1}(ax) + ax\sqrt{1 - a^2x^2}]$

452. $\int x[\cos^{-1}(ax)]\, dx = \frac{1}{4a^2}[(2a^2x^2 - 1)\cos^{-1}(ax) - ax\sqrt{1 - a^2x^2}]$

453. $\int x^n[\sin^{-1}(ax)]\, dx = \frac{x^{n+1}}{n+1}\sin^{-1}(ax) - \frac{a}{n+1}\int \frac{x^{n+1}\, dx}{\sqrt{1 - a^2x^2}}, \quad (n \neq -1)$

454. $\int x^n[\cos^{-1}(ax)]\, dx = \frac{x^{n+1}}{n+1}\cos^{-1}(ax) + \frac{a}{n+1}\int \frac{x^{n+1}\, dx}{\sqrt{1 - a^2x^2}}, \quad (n \neq -1)$

455. $\int x(\tan^{-1} ax)\, dx = \frac{1 + a^2x^2}{2a^2}\tan^{-1} ax - \frac{x}{2a}$

456. $\int x^n(\tan^{-1} ax)\, dx = \frac{x^{n+1}}{n+1}\tan^{-1} ax - \frac{a}{n+1}\int \frac{x^{n+1}}{1 + a^2x^2}\, dx$

457. $\int x(\cot^{-1} ax)\, dx = \frac{1 + a^2x^2}{2a^2}\cot^{-1} ax + \frac{x}{2a}$

458. $\int x^n(\cot^{-1} ax)\, dx = \frac{x^{n+1}}{n+1}\cot^{-1} ax + \frac{a}{n+1}\int \frac{x^{n+1}}{1 + a^2x^2}\, dx$

459. $\int \frac{\sin^{-1}(ax)}{x^2}\, dx = a\log\left(\frac{1 - \sqrt{1 - a^2x^2}}{x}\right) - \frac{\sin^{-1}(ax)}{x}$

460. $\int \frac{\cos^{-1}(ax)\, dx}{x^2} = -\frac{1}{x}\cos^{-1}(ax) + a\log\frac{1 + \sqrt{1 - a^2x^2}}{x}$

461. $\int \frac{\tan^{-1}(ax)\, dx}{x^2} = -\frac{1}{x}\tan^{-1}(ax) - \frac{a}{2}\log\frac{1 + a^2x^2}{x^2}$

462. $\int \frac{\cot^{-1} ax}{x^2}\, dx = -\frac{1}{x}\cot^{-1} ax - \frac{a}{2}\log\frac{x^2}{a^2x^2 + 1}$

463. $\int (\sin^{-1} ax)^2\, dx = x(\sin^{-1} ax)^2 - 2x + \frac{2\sqrt{1 - a^2x^2}}{a}\sin^{-1} ax$

464. $\int (\cos^{-1} ax)^2\, dx = x(\cos^{-1} ax)^2 - 2x - \frac{2\sqrt{1 - a^2x^2}}{a}\cos^{-1} ax$

465. $$\int (\sin^{-1} ax)^n\, dx = \begin{cases} x(\sin^{-1} ax)^n + \dfrac{n\sqrt{1 - a^2x^2}}{a}(\sin^{-1} ax)^{n-1} - n(n-1)\displaystyle\int (\sin^{-1} ax)^{n-2}\, dx \\ \text{or} \\ \displaystyle\sum_{r=0}^{\left[\frac{n}{2}\right]} (-1)^r \frac{n!}{(n-2r)!} x(\sin^{-1} ax)^{n-2r} + \displaystyle\sum_{r=0}^{\left[\frac{n-1}{2}\right]} (-1)^r \frac{n!\sqrt{1 - a^2x^2}}{(n-2r-1)!\,a}(\sin^{-1} ax)^{n-2r-1} \end{cases}$$

Note: $[s]$ means greatest integer $\leq s$. Thus $[3.5]$ means 3; $[5] = 5$, $[\frac{1}{2}] = 0$.

466. $$\int (\cos^{-1} ax)^n\,dx = \begin{cases} x(\cos^{-1} ax)^n - \dfrac{n\sqrt{1-a^2x^2}}{a}(\cos^{-1} ax)^{n-1} \\ \qquad -n(n-1)\displaystyle\int (\cos^{-1} ax)^{n-2}\,dx \\ \text{or} \\ \displaystyle\sum_{r=0}^{\left[\frac{n}{2}\right]} (-1)^r \frac{n!}{(n-2r)!} x(\cos^{-1} ax)^{n-2r} \\ \quad \displaystyle\sum_{r=0}^{\left[\frac{n-1}{2}\right]} (-1)^r \frac{n!\sqrt{1-a^2x^2}}{(n-2r-1)!\,a}(\cos^{-1} ax)^{n-2r-1} \end{cases}$$

467. $$\int \frac{1}{\sqrt{1-a^2x^2}}(\sin^{-1} ax)\,dx = \frac{1}{2a}(\sin^{-1} ax)^2$$

468. $$\int \frac{x^n}{\sqrt{1-a^2x^2}}(\sin^{-1} ax)\,dx = -\frac{x^{n-1}}{na^2}\sqrt{1-a^2x^2}\sin^{-1} ax + \frac{x^n}{n^2a} + \frac{n-1}{na^2}\int \frac{x^{n-2}}{\sqrt{1-a^2x^2}}\sin^{-1} ax\,dx$$

469. $$\int \frac{1}{\sqrt{1-a^2x^2}}(\cos^{-1} ax)\,dx = -\frac{1}{2a}(\cos^{-1} ax)^2$$

470. $$\int \frac{x^n}{\sqrt{1-a^2x^2}}(\cos^{-1} ax)\,dx = -\frac{x^{n-1}}{na^2}\sqrt{1-a^2x^2}\cos^{-1} ax - \frac{x^n}{n^2a} + \frac{n-1}{na^2}\int \frac{x^{n-2}}{\sqrt{1-a^2x^2}}\cos^{-1} ax\,dx$$

471. $$\int \frac{\tan^{-1} ax}{a^2x^2+1}\,dx = \frac{1}{2a}(\tan^{-1} ax)^2$$

472. $$\int \frac{\cot^{-1} ax}{a^2x^2+1}\,dx = -\frac{1}{2a}(\cot^{-1} ax)^2$$

473. $$\int x\sec^{-1} ax\,dx = \frac{x^2}{2}\sec^{-1} ax - \frac{1}{2a^2}\sqrt{a^2x^2-1}$$

474. $$\int x^n\sec^{-1} ax\,dx = \frac{x^{n+1}}{n+1}\sec^{-1} ax - \frac{1}{n+1}\int \frac{x^n\,dx}{\sqrt{a^2x^2-1}}$$

475. $$\int \frac{\sec^{-1} ax}{x^2}\,dx = -\frac{\sec^{-1} ax}{x} + \frac{\sqrt{a^2x^2-1}}{x}$$

476. $$\int x\csc^{-1} ax\,dx = \frac{x^2}{2}\csc^{-1} ax + \frac{1}{2a^2}\sqrt{a^2x^2-1}$$

477. $$\int x^n\csc^{-1} ax\,dx = \frac{x^{n+1}}{n+1}\csc^{-1} ax + \frac{1}{n+1}\int \frac{x^n\,dx}{\sqrt{a^2x^2-1}}$$

478. $$\int \frac{\csc^{-1} ax}{x^2}\,dx = -\frac{\csc^{-1} ax}{x} - \frac{\sqrt{a^2x^2-1}}{x}$$

FORMS INVOLVING TRIGONOMETRIC SUBSTITUTIONS

479. $$\int f(\sin x)\,dx = 2\int f\left(\frac{2z}{1+z^2}\right)\frac{dz}{1+z^2}, \qquad \left(z = \tan\frac{x}{2}\right)$$

480. $$\int f(\cos x)\,dx = 2\int f\left(\frac{1-z^2}{1+z^2}\right)\frac{dz}{1+z^2}, \qquad \left(z = \tan\frac{x}{2}\right)$$

*481. $\int f(\sin x)\,dx = \int f(u)\frac{du}{\sqrt{1-u^2}}, \qquad (u = \sin x)$

*482. $\int f(\cos x)\,dx = -\int f(u)\frac{du}{\sqrt{1-u^2}}, \qquad (u = \cos x)$

*483. $\int f(\sin x, \cos x)\,dx = \int f(u, \sqrt{1-u^2})\frac{du}{\sqrt{1-u^2}}, \qquad (u = \sin x)$

484. $\int f(\sin x, \cos x)\,dx = 2\int f\left(\frac{2z}{1+z^2}, \frac{1-z^2}{1+z^2}\right)\frac{dz}{1+z^2}, \qquad \left(z = \tan\frac{x}{2}\right)$

LOGARITHMIC FORMS

485. $\int (\log x)\,dx = x\log x - x$

486. $\int x(\log x)\,dx = \frac{x^2}{2}\log x - \frac{x^2}{4}$

487. $\int x^2(\log x)\,dx = \frac{x^3}{3}\log x - \frac{x^3}{9}$

488. $\int x^n(\log ax)\,dx = \frac{x^{n+1}}{n+1}\log ax - \frac{x^{n+1}}{(n+1)^2}$

489. $\int (\log x)^2\,dx = x(\log x)^2 - 2x\log x + 2x$

490. $\int (\log x)^n\,dx = \begin{cases} x(\log x)^n - n\int(\log x)^{n-1}\,dx, & (n \neq -1) \\ \text{or} \\ (-1)^n n!\,x\sum_{r=0}^{n}\frac{(-\log x)^r}{r!} \end{cases}$

491. $\int \frac{(\log x)^n}{x}\,dx = \frac{1}{n+1}(\log x)^{n+1}$

492. $\int \frac{dx}{\log x} = \log(\log x) + \log x + \frac{(\log x)^2}{2\cdot 2!} + \frac{(\log x)^3}{3\cdot 3!} + \cdots$

493. $\int \frac{dx}{x\log x} = \log(\log x)$

494. $\int \frac{dx}{x(\log x)^n} = -\frac{1}{(n-1)(\log x)^{n-1}}$

495. $\int \frac{x^m\,dx}{(\log x)^n} = -\frac{x^{m+1}}{(n-1)(\log x)^{n-1}} + \frac{m+1}{n-1}\int\frac{x^m\,dx}{(\log x)^{n-1}}$

496. $\int x^m(\log x)^n\,dx = \begin{cases} \frac{x^{m+1}(\log x)^n}{m+1} - \frac{n}{m+1}\int x^m(\log x)^{n-1}\,dx \\ \text{or} \\ (-1)^n\frac{n!}{m+1}x^{m+1}\sum_{r=0}^{n}\frac{(-\log x)^r}{r!(m+1)^{n-r}} \end{cases}$

497. $\int x^p\cos(b\ln x)\,dx = \frac{x^{p+1}}{(p+1)^2+b^2}\cdot[b\sin(b\ln x) + (p+1)\cos(b\ln x)] + c$

498. $\int x^p\sin(b\ln x)\,dx = \frac{x^{p+1}}{(p+1)^2+b^2}\cdot[(p+1)\sin(b\ln x) - b\cos(b\ln x)] + c$

499. $\int [\log(ax+b)]\,dx = \frac{ax+b}{a}\log(ax+b) - x$

*The square roots appearing in these formulas may be plus or minus, depending on the quadrant of x. Care must be used to give them the proper sign.

500. $\int \frac{\log(ax+b)}{x^2}dx = \frac{a}{b}\log x - \frac{ax+b}{bx}\log(ax+b)$

501. $\int x^m[\log(ax+b)]\,dx = \frac{1}{m+1}\left[x^{m+1} - \left(-\frac{b}{a}\right)^{m+1}\right]\log(ax+b)$

$$- \frac{1}{m+1}\left(-\frac{b}{a}\right)^{m+1}\sum_{r=1}^{m+1}\frac{1}{r}\left(-\frac{ax}{b}\right)^r$$

502. $\int \frac{\log(ax+b)}{x^m}dx = -\frac{1}{m-1}\frac{\log(ax+b)}{x^{m-1}} + \frac{1}{m-1}\left(-\frac{a}{b}\right)^{m-1}\log\frac{ax+b}{x}$

$$+ \frac{1}{m-1}\left(-\frac{a}{b}\right)^{m-1}\sum_{r=1}^{m-2}\frac{1}{r}\left(-\frac{b}{ax}\right)^r, (m > 2)$$

503. $\int \left[\log\frac{x+a}{x-a}\right]dx = (x+a)\log(x+a) - (x-a)\log(x-a)$

504. $\int x^m\left[\log\frac{x+a}{x-a}\right]dx = \frac{x^{m+1}-(-a)^{m+1}}{m+1}\log(x+a) - \frac{x^{m+1}-a^{m+1}}{m+1}\log(x-a)$

$$+ \frac{2a^{m+1}}{m+1}\sum_{r=1}^{\left[\frac{m+1}{2}\right]}\frac{1}{m-2r+2}\left(\frac{x}{a}\right)^{m-2r+2}$$

See note integral 392.

505. $\int \frac{1}{x^2}\left[\log\frac{x+a}{x-a}\right]dx = \frac{1}{x}\log\frac{x-a}{x+a} - \frac{1}{a}\log\frac{x^2-a^2}{x^2}$

506. $\int (\log X)\,dx =$

$$\left(x + \frac{b}{2c}\right)\log X - 2x + \frac{\sqrt{4ac-b^2}}{c}\tan^{-1}\frac{2cx+b}{\sqrt{4ac-b^2}}, \quad (b^2 - 4ac < 0)$$

or

$$\left(x + \frac{b}{2c}\right)\log X - 2x + \frac{\sqrt{b^2-4ac}}{c}\tanh^{-1}\frac{2cx+b}{\sqrt{b^2-4ac}}, \quad (b^2 - 4ac > 0)$$

where

$X = a + bx + cx^2$

507. $\int x^n(\log X)\,dx = \frac{x^{n+1}}{n+1}\log X - \frac{2c}{n+1}\int\frac{x^{n+2}}{X}dx - \frac{b}{n+1}\int\frac{x^{n+1}}{X}dx$

where $X = a + bx + cx^2$

508. $\int [\log(x^2+a^2)]\,dx = x\log(x^2+a^2) - 2x + 2a\tan^{-1}\frac{x}{a}$

509. $\int [\log(x^2-a^2)]\,dx = x\log(x^2-a^2) - 2x + a\log\frac{x+a}{x-a}$

510. $\int x[\log(x^2 \pm a^2)]\,dx = \frac{1}{2}(x^2 \pm a^2)\log(x^2 \pm a^2) - \frac{1}{2}x^2$

511. $\int [\log(x + \sqrt{x^2 \pm a^2})]\,dx = x\log(x + \sqrt{x^2 \pm a^2}) - \sqrt{x^2 \pm a^2}$

512. $\int x[\log(x + \sqrt{x^2 \pm a^2})]\,dx = \left(\frac{x^2}{2} \pm \frac{a^2}{4}\right)\log(x + \sqrt{x^2 \pm a^2}) - \frac{x\sqrt{x^2 \pm a^2}}{4}$

513. $\int x^m[\log(x + \sqrt{x^2 \pm a^2})]\,dx = \frac{x^{m+1}}{m+1}\log(x + \sqrt{x^2 \pm a^2})$

$$- \frac{1}{m+1}\int\frac{x^{m+1}}{\sqrt{x^2 \pm a^2}}dx$$

514. $$\int \frac{\log(x + \sqrt{x^2 + a^2})}{x^2}\,dx = -\frac{\log(x + \sqrt{x^2 + a^2})}{x} - \frac{1}{a}\log\frac{a + \sqrt{x^2 + a^2}}{x}$$

515. $$\int \frac{\log(x + \sqrt{x^2 - a^2})}{x^2}\,dx = -\frac{\log(x + \sqrt{x^2 - a^2})}{x} + \frac{1}{|a|}\sec^{-1}\frac{x}{a}$$

516. $$\int x^n \log(x^2 - a^2)\,dx = \frac{1}{n+1}\left[x^{n+1}\log(x^2 - a^2) - a^{n+1}\log(x - a) - (-a)^{n+1}\log(x + a) - 2\sum_{r=0}^{\left[\frac{n}{2}\right]}\frac{a^{2r}x^{n-2r+1}}{n - 2r + 1}\right]$$

See note integral 392.

EXPONENTIAL FORMS

517. $$\int e^x\,dx = e^x$$

518. $$\int e^{-x}\,dx = -e^{-x}$$

519. $$\int e^{ax}\,dx = \frac{e^{ax}}{a}$$

520. $$\int x\,e^{ax}\,dx = \frac{e^{ax}}{a^2}(ax - 1)$$

521. $$\int x^m e^{ax}\,dx = \begin{cases} \dfrac{x^m e^{ax}}{a} - \dfrac{m}{a}\displaystyle\int x^{m-1} e^{ax}\,dx \\ \text{or} \\ e^{ax}\displaystyle\sum_{r=0}^{m}(-1)^r\frac{m!x^{m-r}}{(m-r)!a^{r+1}} \end{cases}$$

522. $$\int \frac{e^{ax}\,dx}{x} = \log x + \frac{ax}{1!} + \frac{a^2x^2}{2\cdot 2!} + \frac{a^3x^3}{3\cdot 3!} + \cdots$$

523. $$\int \frac{e^{ax}}{x^m}\,dx = -\frac{1}{m-1}\frac{e^{ax}}{x^{m-1}} + \frac{a}{m-1}\int \frac{e^{ax}}{x^{m-1}}\,dx$$

524. $$\int e^{ax}\log x\,dx = \frac{e^{ax}\log x}{a} - \frac{1}{a}\int \frac{e^{ax}}{x}\,dx$$

525. $$\int \frac{dx}{1 + e^x} = x - \log(1 + e^x) = \log\frac{e^x}{1 + e^x}$$

526. $$\int \frac{dx}{a + be^{px}} = \frac{x}{a} - \frac{1}{ap}\log(a + be^{px})$$

527. $$\int \frac{dx}{ae^{mx} + be^{-mx}} = \frac{1}{m\sqrt{ab}}\tan^{-1}\left(e^{mx}\sqrt{\frac{a}{b}}\right), \qquad (a > 0, b > 0)$$

528. $$\int \frac{dx}{ae^{mx} - be^{-mx}} = \begin{cases} \dfrac{1}{2m\sqrt{ab}}\log\dfrac{\sqrt{a}\,e^{mx} - \sqrt{b}}{\sqrt{a}\,e^{mx} + \sqrt{b}} \\ \text{or} \\ \dfrac{-1}{m\sqrt{ab}}\tanh^{-1}\left(\sqrt{\dfrac{a}{b}}\,e^{mx}\right), \qquad (a > 0, b > 0) \end{cases}$$

529. $$\int (a^x - a^{-x})\,dx = \frac{a^x + a^{-x}}{\log a}$$

530. $$\int \frac{e^{ax}}{b + ce^{ax}}\,dx = \frac{1}{ac}\log(b + ce^{ax})$$

531. $$\int \frac{x\,e^{ax}}{(1 + ax)^2}\,dx = \frac{e^{ax}}{a^2(1 + ax)}$$

532. $\int x e^{-x^2}\, dx = -\frac{1}{2} e^{-x^2}$

533. $\int e^{ax}[\sin(bx)]\, dx = \frac{e^{ax}[a \sin(bx) - b \cos(bx)]}{a^2 + b^2}$

534. $\int e^{ax}[\sin(bx)][\sin(cx)]\, dx = \frac{e^{ax}[(b-c)\sin(b-c)x + a\cos(b-c)x]}{2[a^2 + (b-c)^2]} - \frac{e^{ax}[(b+c)\sin(b+c)x + a\cos(b+c)x]}{2[a^2 + (b+c)^2]}$

535. $\int e^{ax}[\sin(bx)][\cos(cx)]\, dx = \begin{cases} \frac{e^{ax}[a\sin(b-c)x - (b-c)\cos(b-c)x]}{2[a^2 + (b-c)^2]} + \frac{e^{ax}[a\sin(b+c)x - (b+c)\cos(b+c)x]}{2[a^2 + (b+c)^2]} \\ \text{or} \\ \frac{e^{ax}}{\rho}[(a\sin bx - b\cos bx)[\cos(cx-\alpha)] - c(\sin bx)\sin(cx-\alpha)] \\ \text{where} \\ \rho = \sqrt{(a^2+b^2-c^2)^2 + 4a^2c^2}, \\ \rho\cos\alpha = a^2 + b^2 - c^2, \qquad \rho\sin\alpha = 2ac \end{cases}$

536. $\int e^{ax}[\sin(bx)][\sin(bx+c)]\, dx$

$$= \frac{e^{ax}\cos c}{2a} - \frac{e^{ax}[a\cos(2bx+c) + 2b\sin(2bx+c)]}{2(a^2+4b^2)}$$

537. $\int e^{ax}[\sin(bx)][\cos(bx+c)]\, dx$

$$= \frac{-e^{ax}\sin c}{2a} + \frac{e^{ax}[a\sin(2bx+c) - 2b\cos(2bx+c)]}{2(a^2+4b^2)}$$

538. $\int e^{ax}[\cos(bx)]\, dx = \frac{e^{ax}}{a^2+b^2}[a\cos(bx) + b\sin(bx)]$

539. $\int e^{ax}[\cos(bx)][\cos(cx)]\, dx = \frac{e^{ax}[(b-c)\sin(b-c)x + a\cos(b-c)x]}{2[a^2+(b-c)^2]} + \frac{e^{ax}[(b+c)\sin(b+c)x + a\cos(b+c)x]}{2[a^2+(b+c)^2]}$

540. $\int e^{ax}[\cos(bx)][\cos(bx+c)]\, dx$

$$= \frac{e^{ax}\cos c}{2a} + \frac{e^{ax}[a\cos(2bx+c) + 2b\sin(2bx+c)]}{2(a^2+4b^2)}$$

541. $\int e^{ax}[\cos(bx)][\sin(bx+c)]\, dx$

$$= \frac{e^{ax}\sin c}{2a} + \frac{e^{ax}[a\sin(2bx+c) - 2b\cos(2bx+c)]}{2(a^2+4b^2)}$$

542. $\int e^{ax}[\sin^n bx]\, dx = \frac{1}{a^2+n^2b^2}\left[(a\sin bx - nb\cos bx)e^{ax}\sin^{n-1} bx + n(n-1)b^2 \int e^{ax}[\sin^{n-2} bx]\, dx\right]$

543. $\int e^{ax}[\cos^n bx]\, dx = \frac{1}{a^2+n^2b^2}\left[(a\cos bx + nb\sin bx)e^{ax}\cos^{n-1} bx + n(n-1)b^2 \int e^{ax}[\cos^{n-2} bx]\, dx\right]$

544. $$\int x^m e^x \sin x\,dx = \frac{1}{2}x^m e^x(\sin x - \cos x) - \frac{m}{2}\int x^{m-1} e^x \sin x\,dx + \frac{m}{2}\int x^{m-1} e^x \cos x\,dx$$

545. $$\int x^m e^{ax}[\sin bx]\,dx = \begin{cases} x^m e^{ax}\dfrac{a\sin bx - b\cos bx}{a^2+b^2} - \dfrac{m}{a^2+b^2}\displaystyle\int x^{m-1} e^{ax}(a\sin bx - b\cos bx)\,dx \\ \text{or} \\ e^{ax}\displaystyle\sum_{r=0}^{m}\dfrac{(-1)^r m!\,x^{m-r}}{\rho^{r+1}(m-r)!}\sin[bx-(r+1)\alpha] \\ \text{where} \\ \rho = \sqrt{a^2+b^2}, \quad \rho\cos\alpha = a, \quad \rho\sin\alpha = b \end{cases}$$

546. $$\int x^m e^x \cos x\,dx = \frac{1}{2}x^m e^x(\sin x + \cos x) - \frac{m}{2}\int x^{m-1} e^x \sin x\,dx - \frac{m}{2}\int x^{m-1} e^x \cos x\,dx$$

547. $$\int x^m e^{ax}\cos bx\,dx = \begin{cases} x^m e^{ax}\dfrac{a\cos bx + b\sin bx}{a^2+b^2} - \dfrac{m}{a^2+b^2}\displaystyle\int x^{m-1} e^{ax}(a\cos bx + b\sin bx)\,dx \\ \text{or} \\ e^{ax}\displaystyle\sum_{r=0}^{m}\dfrac{(-1)^r m!\,x^{m-r}}{\rho^{r+1}(m-r)!}\cos[bx-(r+1)\alpha] \\ \text{where} \\ \rho = \sqrt{a^2+b^2}, \quad \rho\cos\alpha = a, \quad \rho\sin\alpha = b \end{cases}$$

548. $$\int e^{ax}(\cos^m x)(\sin^n x)\,dx = \begin{cases} \dfrac{e^{ax}\cos^{m-1}x\sin^n x[a\cos x + (m+n)\sin x]}{(m+n)^2+a^2} \\ \quad -\dfrac{na}{(m+n)^2+a^2}\displaystyle\int e^{ax}(\cos^{m-1}x)(\sin^{n-1}x)\,dx \\ \quad +\dfrac{(m-1)(m+n)}{(m+n)^2+a^2}\displaystyle\int e^{ax}(\cos^{m-2}x)(\sin^n x)\,dx \\ \text{or} \\ \dfrac{e^{ax}\cos^m x\sin^{n-1}x[a\sin x - (m+n)\cos x]}{(m+n)^2+a^2} \\ \quad +\dfrac{ma}{(m+n)^2+a^2}\displaystyle\int e^{ax}(\cos^{m-1}x)(\sin^{n-1}x)\,dx \\ \quad +\dfrac{(n-1)(m+n)}{(m+n)^2+a^2}\displaystyle\int e^{ax}(\cos^m x)(\sin^{n-2}x)\,dx \\ \text{or} \\ \dfrac{e^{ax}(\cos^{m-1}x)(\sin^{n-1}x)(a\sin x\cos x + m\sin^2 x - n\cos^2 x)}{(m+n)^2+a^2} \\ \quad +\dfrac{m(m-1)}{(m+n)^2+a^2}\displaystyle\int e^{ax}(\cos^{m-2}x)(\sin^n x)\,dx \\ \quad +\dfrac{n(n-1)}{(m+n)^2+a^2}\displaystyle\int e^{ax}(\cos^m x)(\sin^{n-2}x)\,dx \\ \text{or} \\ \dfrac{e^{ax}(\cos^{m-1}x)(\sin^{n-1}x)(a\cos x\sin x + m\sin^2 x - n\cos^2 x)}{(m+n)^2+a^2} \\ \quad +\dfrac{m(m-1)}{(m+n)^2+a^2}\displaystyle\int e^{ax}(\cos^{m-2}x)(\sin^{n-2}x)\,dx \\ \quad +\dfrac{(n-m)(n+m-1)}{(m+n)^2+a^2}\displaystyle\int e^{ax}(\cos^m x)(\sin^{n-2}x)\,dx \end{cases}$$

549. $$\int x\,e^{ax}(\sin bx)\,dx = \frac{x\,e^{ax}}{a^2+b^2}(a\sin bx - b\cos bx) - \frac{e^{ax}}{(a^2+b^2)^2}[(a^2-b^2)\sin bx - 2ab\cos bx]$$

550. $$\int x\,e^{ax}(\cos bx)\,dx = \frac{x\,e^{ax}}{a^2+b^2}(a\cos bx + b\sin bx) - \frac{e^{ax}}{(a^2+b^2)^2}[(a^2-b^2)\cos bx + 2ab\sin bx]$$

551. $$\int \frac{e^{ax}}{\sin^n x}dx = -\frac{e^{ax}[a\sin x + (n-2)\cos x]}{(n-1)(n-2)\sin^{n-1}x} + \frac{a^2+(n-2)^2}{(n-1)(n-2)}\int \frac{e^{ax}}{\sin^{n-2}x}dx$$

552. $$\int \frac{e^{ax}}{\cos^n x}dx = -\frac{e^{ax}[a\cos x - (n-2)\sin x]}{(n-1)(n-2)\cos^{n-1}x} + \frac{a^2+(n-2)^2}{(n-1)(n-2)}\int \frac{e^{ax}}{\cos^{n-2}x}dx$$

553. $$\int e^{ax}\tan^n x\,dx = e^{ax}\frac{\tan^{n-1}x}{n-1} - \frac{a}{n-1}\int e^{ax}\tan^{n-1}x\,dx - \int e^{ax}\tan^{n-2}x\,dx$$

HYPERBOLIC FORMS

554. $$\int (\sinh x)\,dx = \cosh x$$

555. $$\int (\cosh x)\,dx = \sinh x$$

556. $\int (\tanh x)\,dx = \log \cosh x$

557. $\int (\coth x)\,dx = \log \sinh x$

558. $\int (\operatorname{sech} x)\,dx = \tan^{-1}(\sinh x)$

559. $\int \operatorname{csch} x\,dx = \log \tanh\left(\frac{x}{2}\right)$

560. $\int x(\sinh x)\,dx = x\cosh x - \sinh x$

561. $\int x^n(\sinh x)\,dx = x^n \cosh x - n\int x^{n-1}(\cosh x)\,dx$

562. $\int x(\cosh x)\,dx = x\sinh x - \cosh x$

563. $\int x^n(\cosh x)\,dx = x^n \sinh x - n\int x^{n-1}(\sinh x)\,dx$

564. $\int (\operatorname{sech} x)(\tanh x)\,dx = -\operatorname{sech} x$

565. $\int (\operatorname{csch} x)(\coth x)\,dx = -\operatorname{csch} x$

566. $\int (\sinh^2 x)\,dx = \frac{\sinh 2x}{4} - \frac{x}{2}$

567. $$\int (\sinh^m x)(\cosh^n x)\,dx = \begin{cases} \frac{1}{m+n}(\sinh^{m+1} x)(\cosh^{n-1} x) \\ \quad + \frac{n-1}{m+n}\int (\sinh^m x)(\cosh^{n-2} x)\,dx \\ \text{or} \\ \frac{1}{m+n}\sinh^{m-1} x \cosh^{n+1} x \\ \quad - \frac{m-1}{m+n}\int (\sinh^{m-2} x)(\cosh^n x)\,dx, \quad (m+n \neq 0) \end{cases}$$

568. $$\int \frac{dx}{(\sinh^m x)(\cosh^n x)} = \begin{cases} -\frac{1}{(m-1)(\sinh^{m-1} x)(\cosh^{n-1} x)} \\ \quad - \frac{m+n-2}{m-1}\int \frac{dx}{(\sinh^{m-2} x)(\cosh^n x)}, \quad (m \neq 1) \\ \text{or} \\ \frac{1}{(n-1)\sinh^{m-1} x \cosh^{n-1} x} \\ \quad + \frac{m+n-2}{n-1}\int \frac{dx}{(\sinh^m x)(\cosh^{n-2} x)}, \quad (n \neq 1) \end{cases}$$

569. $\int (\tanh^2 x)\,dx = x - \tanh x$

570. $\int (\tanh^n x)\,dx = -\frac{\tanh^{n-1} x}{n-1} + \int (\tanh^{n-2} x)\,dx, \quad (n \neq 1)$

571. $\int (\operatorname{sech}^2 x)\,dx = \tanh x$

572. $\int (\cosh^2 x)\,dx = \frac{\sinh 2x}{4} + \frac{x}{2}$

573. $\int (\coth^2 x)\,dx = x - \coth x$

574. $\int (\coth^n x)\,dx = -\dfrac{\coth^{n-1} x}{n-1} + \int \coth^{n-2} x\,dx, \qquad (n \neq 1)$

575. $\int (\operatorname{csch}^2 x)\,dx = -\operatorname{ctnh} x$

576. $\int (\sinh mx)(\sinh nx)\,dx = \dfrac{\sinh (m+n)x}{2(m+n)} - \dfrac{\sinh (m-n)x}{2(m-n)}, \qquad (m^2 \neq n^2)$

577. $\int (\cosh mx)(\cosh nx)\,dx = \dfrac{\sinh (m+n)x}{2(m+n)} + \dfrac{\sinh (m-n)x}{2(m-n)}, \qquad (m^2 \neq n^2)$

578. $\int (\sinh mx)(\cosh nx)\,dx = \dfrac{\cosh (m+n)x}{2(m+n)} + \dfrac{\cosh (m-n)x}{2(m-n)}, \qquad (m^2 \neq n^2)$

579. $\int \left(\sinh^{-1} \dfrac{x}{a}\right) dx = x \sinh^{-1} \dfrac{x}{a} - \sqrt{x^2 + a^2}, \qquad (a > 0)$

580. $\int x\left(\sinh^{-1} \dfrac{x}{a}\right) dx = \left(\dfrac{x^2}{2} + \dfrac{a^2}{4}\right) \sinh^{-1} \dfrac{x}{a} - \dfrac{x}{4}\sqrt{x^2 + a^2}, \qquad (a > 0)$

581. $\int x^n(\sinh^{-1} x)\,dx = \dfrac{x^{n+1}}{n+1} \sinh^{-1} x - \dfrac{1}{n+1}\int \dfrac{x^{n+1}}{(1+x^2)^{\frac{1}{2}}}\,dx, \qquad (n \neq -1)$

582. $\int \left(\cosh^{-1} \dfrac{x}{a}\right) dx = \begin{cases} x \cosh^{-1} \dfrac{x}{a} - \sqrt{x^2 - a^2}, & \left(\cosh^{-1} \dfrac{x}{a} > 0\right) \\ \text{or} & \\ x \cosh^{-1} \dfrac{x}{a} + \sqrt{x^2 - a^2}, & \left(\cosh^{-1} \dfrac{x}{a} < 0\right), \quad (a > 0) \end{cases}$

583. $\int x\left(\cosh^{-1} \dfrac{x}{a}\right) dx = \dfrac{2x^2 - a^2}{4} \cosh^{-1} \dfrac{x}{a} - \dfrac{x}{4}(x^2 - a^2)^{\frac{1}{2}}$

584. $\int x^n(\cosh^{-1} x)\,dx = \dfrac{x^{n+1}}{n+1} \cosh^{-1} x - \dfrac{1}{n+1}\int \dfrac{x^{n+1}}{(x^2-1)^{\frac{1}{2}}}\,dx, \qquad (n \neq -1)$

585. $\int \left(\tanh^{-1} \dfrac{x}{a}\right) dx = x \tanh^{-1} \dfrac{x}{a} + \dfrac{a}{2} \log (a^2 - x^2), \qquad \left(\left|\dfrac{x}{a}\right| < 1\right)$

586. $\int \left(\coth^{-1} \dfrac{x}{a}\right) dx = x \coth^{-1} \dfrac{x}{a} + \dfrac{a}{2} \log (x^2 - a^2), \qquad \left(\left|\dfrac{x}{a}\right| > 1\right)$

587. $\int x\left(\tanh^{-1} \dfrac{x}{a}\right) dx = \dfrac{x^2 - a^2}{2} \tanh^{-1} \dfrac{x}{a} + \dfrac{ax}{2}, \qquad \left(\left|\dfrac{x}{a}\right| < 1\right)$

588. $\int x^n\left(\tanh^{-1} x\right) dx = \dfrac{x^{n+1}}{n+1} \tanh^{-1} x - \dfrac{1}{n+1}\int \dfrac{x^{n+1}}{1 - x^2}\,dx, \qquad (n \neq -1)$

589. $\int x\left(\coth^{-1} \dfrac{x}{a}\right) dx = \dfrac{x^2 - a^2}{2} \coth^{-1} \dfrac{x}{a} + \dfrac{ax}{2}, \qquad \left(\left|\dfrac{x}{a}\right| > 1\right)$

590. $\int x^n(\coth^{-1} x)\,dx = \dfrac{x^{n+1}}{n+1} \coth^{-1} x + \dfrac{1}{n+1}\int \dfrac{x^{n+1}}{x^2 - 1}\,dx, \qquad (n \neq -1)$

591. $\int (\operatorname{sech}^{-1} x)\,dx = x \operatorname{sech}^{-1} x + \sin^{-1} x$

592. $\int x \operatorname{sech}^{-1} x\,dx = \dfrac{x^2}{2} \operatorname{sech}^{-1} x - \dfrac{1}{2}\sqrt{1 - x^2}$

593. $\int x^n \operatorname{sech}^{-1} x\,dx = \dfrac{x^{n+1}}{n+1} \operatorname{sech}^{-1} x + \dfrac{1}{n+1}\int \dfrac{x^n}{(1 - x^2)^{\frac{1}{2}}}\,dx, \qquad (n \neq -1)$

594. $\int \operatorname{csch}^{-1} x\,dx = x \operatorname{csch}^{-1} x + \dfrac{x}{|x|} \sinh^{-1} x$

595. $\int x \operatorname{csch}^{-1} x\, dx = \frac{x^2}{2} \operatorname{csch}^{-1} x + \frac{1}{2}\frac{x}{|x|}\sqrt{1+x^2}$

596. $\int x^n \operatorname{csch}^{-1} x\, dx = \frac{x^{n+1}}{n+1} \operatorname{csch}^{-1} x + \frac{1}{n+1}\frac{x}{|x|}\int \frac{x^n}{(x^2+1)^{\frac{1}{2}}} dx, \quad (n \neq -1)$

DEFINITE INTEGRALS

597. $\int_0^\infty x^{n-1} e^{-x}\, dx = \int_0^1 \left(\log \frac{1}{x}\right)^{n-1} dx = \frac{1}{n}\prod_{m=1}^{\infty} \frac{\left(1+\frac{1}{m}\right)^n}{1+\frac{n}{m}}$

$= \Gamma(n), n \neq 0, -1, -2, -3, \ldots$ (Gamma Function)

598. $\int_0^\infty t^n p^{-t}\, dt = \frac{n!}{(\log p)^{n+1}}, \quad (n = 0, 1, 2, 3, \ldots \text{ and } p > 0)$

599. $\int_0^\infty t^{n-1} e^{-(a+1)t}\, dt = \frac{\Gamma(n)}{(a+1)^n}, \quad (n > 0, a > -1)$

600. $\int_0^1 x^m \left(\log \frac{1}{x}\right)^n dx = \frac{\Gamma(n+1)}{(m+1)^{n+1}}, \quad (m > -1, n > -1)$

601. $\Gamma(n)$ is finite if $n > 0$, $\Gamma(n+1) = n\Gamma(n)$

602. $\Gamma(n) \cdot \Gamma(1-n) = \frac{\pi}{\sin n\pi}$

603. $\Gamma(n) = (n-1)!$ if $n =$ integer > 0

604. $\Gamma(\frac{1}{2}) = 2\int_0^\infty e^{-t^2} dt = \sqrt{\pi} = 1.7724538509 \cdots = (-\frac{1}{2})!$

605. $\Gamma(n + \frac{1}{2}) = \frac{1 \cdot 3 \cdot 5 \ldots (2n-1)}{2^n}\sqrt{\pi} \quad n = 1, 2, 3, \ldots$

606. $\Gamma(-n + \frac{1}{2}) = \frac{(-1)^n 2^n \sqrt{\pi}}{1 \cdot 3 \cdot 5 \ldots (2n-1)} \quad n = 1, 2, 3, \ldots$

607. $\int_0^1 x^{m-1}(1-x)^{n-1}\, dx = \int_0^\infty \frac{x^{m-1}}{(1+x)^{m+n}} dx = \frac{\Gamma(m)\Gamma(n)}{\Gamma(m+n)} = B(m, n)$

(Beta function)

608. $B(m, n) = B(n, m) = \frac{\Gamma(m)\Gamma(n)}{\Gamma(m+n)}$, where m and n are any positive real numbers.

609. $\int_a^b (x-a)^m (b-x)^n\, dx = (b-a)^{m+n+1} \frac{\Gamma(m+1) \cdot \Gamma(n+1)}{\Gamma(m+n+2)},$

$(m > -1, n > -1, b > a)$

610. $\int_1^\infty \frac{dx}{x^m} = \frac{1}{m-1}, \quad [m > 1]$

611. $\int_0^\infty \frac{dx}{(1+x)x^p} = \pi \csc p\pi, \quad [p < 1]$

612. $\int_0^\infty \frac{dx}{(1-x)x^p} = -\pi \cot p\pi, \quad [p < 1]$

613. $\int_0^\infty \frac{x^{p-1}\, dx}{1+x} = \frac{\pi}{\sin p\pi}$

$= B(p, 1-p) = \Gamma(p)\Gamma(1-p), \quad [0 < p < 1]$

614. $\int_0^\infty \frac{x^{m-1}\, dx}{1+x^n} = \frac{\pi}{n \sin \frac{m\pi}{n}}, \quad [0 < m < n]$

615. $\displaystyle\int_0^\infty \frac{x^a\,dx}{(m+x^b)^c} = \frac{m^{\frac{a+1-bc}{b}}}{b}\left[\frac{\Gamma\left(\frac{a+1}{b}\right)\Gamma\left(c-\frac{a+1}{b}\right)}{\Gamma(c)}\right]$

$\left(a>-1,\ b>0,\ m>0,\ c>\frac{a+1}{b}\right)$

616. $\displaystyle\int_0^\infty \frac{dx}{(1+x)\sqrt{x}} = \pi$

617. $\displaystyle\int_0^\infty \frac{a\,dx}{a^2+x^2} = \frac{\pi}{2}$, if $a>0$; 0, if $a=0$; $-\frac{\pi}{2}$, if $a<0$

618. $\displaystyle\int_0^a (a^2-x^2)^{\frac{n}{2}}\,dx = \frac{1}{2}\int_{-a}^a (a^2-x^2)^{\frac{n}{2}}\,dx = \frac{1\cdot 3\cdot 5\ldots n}{2\cdot 4\cdot 6\ldots(n+1)}\cdot\frac{\pi}{2}\cdot a^{n+1}$ (n odd)

619. $\displaystyle\int_0^a x^m(a^2-x^2)^{\frac{n}{2}}\,dx = \begin{cases} \frac{1}{2}a^{m+n+1}B\left(\frac{m+1}{2},\frac{n+2}{2}\right) \\ \text{or} \\ \frac{1}{2}a^{m+n+1}\dfrac{\Gamma\left(\frac{m+1}{2}\right)\Gamma\left(\frac{n+2}{2}\right)}{\Gamma\left(\frac{m+n+3}{2}\right)} \end{cases}$

620. $\displaystyle\int_0^{\pi/2} (\sin^n x)\,dx = \begin{cases} \displaystyle\int_0^{\pi/2} (\cos^n x)\,dx \\ \text{or} \\ \dfrac{1\cdot 3\cdot 5\cdot 7\ldots(n-1)}{2\cdot 4\cdot 6\cdot 8\ldots(n)}\dfrac{\pi}{2}, & (n \text{ an even integer, } n\neq 0) \\ \text{or} \\ \dfrac{2\cdot 4\cdot 6\cdot 8\ldots(n-1)}{1\cdot 3\cdot 5\cdot 7\ldots(n)}, & (n \text{ an odd integer, } n\neq 1) \\ \text{or} \\ \dfrac{\sqrt{\pi}}{2}\dfrac{\Gamma\left(\frac{n+1}{2}\right)}{\Gamma\left(\frac{n}{2}+1\right)}, & (n>-1) \end{cases}$

621. $\displaystyle\int_0^\infty \frac{\sin mx\,dx}{x} = \frac{\pi}{2}$, if $m>0$; 0, if $m=0$; $-\frac{\pi}{2}$, if $m<0$

622. $\displaystyle\int_0^\infty \frac{\cos x\,dx}{x} = \infty$

623. $\displaystyle\int_0^\infty \frac{\tan x\,dx}{x} = \frac{\pi}{2}$

624. $\displaystyle\int_0^\pi \sin ax\cdot\sin bx\,dx = \int_0^\pi \cos ax\cdot\cos bx\,dx = 0,$ ($a\neq b$; a, b integers)

625. $\displaystyle\int_0^{\pi/a} [\sin(ax)][\cos(ax)]\,dx = \int_0^\pi [\sin(ax)][\cos(ax)]\,dx = 0$

626. $\displaystyle\int_0^\pi [\sin(ax)][\cos(bx)]\,dx = \frac{2a}{a^2-b^2}$, if $a-b$ is odd, or 0 if $a-b$ is even

627. $\displaystyle\int_0^\infty \frac{\sin x\cos mx\,dx}{x}$

$= 0$, if $m<-1$ or $m>1$; $\frac{\pi}{4}$, if $m=\pm 1$; $\frac{\pi}{2}$, if $m^2<1$

628. $\int_0^\infty \frac{\sin ax \sin bx}{x^2} dx = \frac{\pi a}{2}, \qquad (a \le b)$

629. $\int_0^\pi \sin^2 mx\, dx = \int_0^\pi \cos^2 mx\, dx = \frac{\pi}{2}$

630. $\int_0^\infty \frac{\sin^2 (px)}{x^2} dx = \frac{\pi p}{2}$

631. $\int_0^\infty \frac{\sin x}{x^p} dx = \frac{\pi}{2\Gamma(p) \sin (p\pi/2)}, \qquad 0 < p < 1$

632. $\int_0^\infty \frac{\cos x}{x^p} dx = \frac{\pi}{2\Gamma(p) \cos (p\pi/2)}, \qquad 0 < p < 1$

633. $\int_0^\infty \frac{1 - \cos px}{x^2} dx = \frac{\pi p}{2}$

634. $\int_0^\infty \frac{\sin px \cos qx}{x} dx = \left\{0, \quad q > p > 0; \quad \frac{\pi}{2}, \quad p > q > 0; \quad \frac{\pi}{4}, \quad p = q > 0\right\}$

635. $\int_0^\infty \frac{\cos (mx)}{x^2 + a^2} dx = \frac{\pi}{2|a|} e^{-|ma|}$

636. $\int_0^\infty \cos (x^2)\, dx = \int_0^\infty \sin (x^2)\, dx = \frac{1}{2}\sqrt{\frac{\pi}{2}}$

637. $\int_0^\infty \sin ax^n\, dx = \frac{1}{na^{1/n}} \Gamma(1/n) \sin \frac{\pi}{2n}, \qquad n > 1$

638. $\int_0^\infty \cos ax^n\, dx = \frac{1}{na^{1/n}} \Gamma(1/n) \cos \frac{\pi}{2n}, \qquad n > 1$

639. $\int_0^\infty \frac{\sin x}{\sqrt{x}} dx = \int_0^\infty \frac{\cos x}{\sqrt{x}} dx = \sqrt{\frac{\pi}{2}}$

640. (a) $\int_0^\infty \frac{\sin^3 x}{x} dx = \frac{\pi}{4}$ (b) $\int_0^\infty \frac{\sin^3 x}{x^2} dx \frac{3}{4} \log 3$

641. $\int_0^\infty \frac{\sin^3 x}{x^3} dx = \frac{3\pi}{8}$

642. $\int_0^\infty \frac{\sin^4 x}{x^4} dx = \frac{\pi}{3}$

643. $\int_0^{\pi/2} \frac{dx}{1 + a \cos x} = \frac{\cos^{-1} a}{\sqrt{1 - a^2}}, \qquad (a < 1)$

644. $\int_0^\pi \frac{dx}{a + b \cos x} = \frac{\pi}{\sqrt{a^2 - b^2}}, \qquad (a > b \ge 0)$

645. $\int_0^{2\pi} \frac{dx}{1 + a \cos x} = \frac{2\pi}{\sqrt{1 - a^2}}, \qquad (a^2 < 1)$

646. $\int_0^\infty \frac{\cos ax - \cos bx}{x} dx = \log \frac{b}{a}$

647. $\int_0^{\pi/2} \frac{dx}{a^2 \sin^2 x + b^2 \cos^2 x} = \frac{\pi}{2ab}$

648. $\int_0^{\pi/2} \frac{dx}{(a^2 \sin^2 x + b^2 \cos^2 x)^2} = \frac{\pi(a^2 + b^2)}{4a^3 b^3}, \qquad (a, b > 0)$

649. $\int_0^{\pi/2} \sin^{n-1} x \cos^{m-1} x\, dx = \frac{1}{2} \mathrm{B}\left(\frac{n}{2}, \frac{m}{2}\right), \qquad$ m and n positive integers

650. $\int_0^{\pi/2} (\sin^{2n+1}\theta)\, d\theta = \dfrac{2\cdot 4\cdot 6\ldots(2n)}{1\cdot 3\cdot 5\ldots(2n+1)}$, $\quad (n = 1, 2, 3 \ldots)$

651. $\int_0^{\pi/2} (\sin^{2n}\theta)\, d\theta = \dfrac{1\cdot 3\cdot 5\ldots(2n-1)}{2\cdot 4\ldots(2n)}\left(\dfrac{\pi}{2}\right)$, $\quad (n = 1, 2, 3 \ldots)$

652. $\int_0^{\pi/2} \dfrac{x}{\sin x}\, dx = 2\left\{\dfrac{1}{1^2} - \dfrac{1}{3^2} + \dfrac{1}{5^2} - \dfrac{1}{7^2} + \ldots\right\}$

653. $\int_0^{\pi/2} \dfrac{dx}{1 + \tan^m x} = \dfrac{\pi}{4}$

654. $\int_0^{\pi/2} \sqrt{\cos\theta}\, d\theta = \dfrac{(2\pi)^{3/2}}{[\Gamma(\frac{1}{4})]^2}$

655. $\int_0^{\pi/2} (\tan^h\theta)\, d\theta = \dfrac{\pi}{2\cos\left(\dfrac{h\pi}{2}\right)}$, $\quad (0 < h < 1)$

656. $\int_0^{\infty} \dfrac{\tan^{-1}(ax) - \tan^{-1}(bx)}{x}\, dx = \dfrac{\pi}{2}\log\dfrac{a}{b}$, $\quad (a, b > 0)$

657. The area enclosed by a curve defined through the equation $x^{\frac{b}{c}} + y^{\frac{b}{c}} = a^{\frac{b}{c}}$ where $a > 0$, c a positive odd integer and b a positive even integer is given by

$$\frac{\left[\Gamma\left(\frac{c}{b}\right)\right]^2}{\Gamma\left(\frac{2c}{b}\right)}\left(\frac{2ca^2}{b}\right)$$

658. $I = \iiint\limits_R x^{h-1} y^{m-1} z^{n-1}\, dv$, where R denotes the region of space bounded by the co-ordinate planes and that portion of the surface $\left(\dfrac{x}{a}\right)^p + \left(\dfrac{y}{b}\right)^q + \left(\dfrac{z}{c}\right)^k = 1$, which lies in the first octant and where $h, m, n, p, q, k, a, b, c$, denote positive real numbers is given by

$$\int_0^a x^{h-1}\, dx \int_0^{b\left[1-\left(\frac{x}{a}\right)^p\right]^{\frac{1}{q}}} y^m\, dy \int_0^{c\left[1-\left(\frac{x}{a}\right)^p-\left(\frac{y}{b}\right)^q\right]^{\frac{1}{k}}} z^{n-1}\, dz$$

$$= \frac{a^h b^m c^n}{pqk} \frac{\Gamma\left(\frac{h}{p}\right)\Gamma\left(\frac{m}{q}\right)\Gamma\left(\frac{n}{k}\right)}{\Gamma\left(\frac{h}{p} + \frac{m}{q} + \frac{n}{k} + 1\right)}$$

659. $\int_0^{\infty} e^{-ax}\, dx = \dfrac{1}{a}$, $\quad (a > 0)$

660. $\int_0^{\infty} \dfrac{e^{-ax} - e^{-bx}}{x}\, dx = \log\dfrac{b}{a}$, $\quad (a, b > 0)$

661. $\int_0^{\infty} x^n e^{-ax}\, dx = \begin{cases} \dfrac{\Gamma(n+1)}{a^{n+1}}, & (n > -1, a > 0) \\ \text{or} & \\ \dfrac{n!}{a^{n+1}}, & (a > 0, n \text{ positive integer}) \end{cases}$

662. $\int_0^{\infty} x^n \exp(-ax^p)\, dx = \dfrac{\Gamma(k)}{pa^k}$, $\quad \left(n > -1, p > 0, a > 0, k = \dfrac{n+1}{p}\right)$

663. $\int_0^{\infty} e^{-a^2x^2}\, dx = \dfrac{1}{2a}\sqrt{\pi} = \dfrac{1}{2a}\Gamma\left(\dfrac{1}{2}\right)$, $\quad (a > 0)$

664. $\int_0^{\infty} x e^{-x^2}\, dx = \frac{1}{2}$

665. $\int_0^\infty x^2 e^{-x^2}\,dx = \frac{\sqrt{\pi}}{4}$

666. $\int_0^\infty x^{2n} e^{-ax^2}\,dx = \frac{1\cdot 3\cdot 5\ldots(2n-1)}{2^{n+1}a^n}\sqrt{\frac{\pi}{a}}$

667. $\int_0^\infty x^{2n+1} e^{-ax^2}\,dx = \frac{n!}{2a^{n+1}}, \qquad (a > 0)$

668. $\int_0^1 x^m e^{-ax}\,dx = \frac{m!}{a^{m+1}}\left[1 - e^{-a}\sum_{r=0}^{m}\frac{a^r}{r!}\right]$

669. $\int_0^\infty e^{\left(-x^2-\frac{a^2}{x^2}\right)}\,dx = \frac{e^{-2a}\sqrt{\pi}}{2}, \qquad (a \geq 0)$

670. $\int_0^\infty e^{-nx}\sqrt{x}\,dx = \frac{1}{2n}\sqrt{\frac{\pi}{n}}$

671. $\int_0^\infty \frac{e^{-nx}}{\sqrt{x}}\,dx = \sqrt{\frac{\pi}{n}}$

672. $\int_0^\infty e^{-ax}(\cos mx)\,dx = \frac{a}{a^2+m^2}, \qquad (a > 0)$

673. $\int_0^\infty e^{-ax}(\sin mx)\,dx = \frac{m}{a^2+m^2}, \qquad (a > 0)$

674. $\int_0^\infty x\,e^{-ax}[\sin(bx)]\,dx = \frac{2ab}{(a^2+b^2)^2}, \qquad (a > 0)$

675. $\int_0^\infty x\,e^{-ax}[\cos(bx)]\,dx = \frac{a^2-b^2}{(a^2+b^2)^2}, \qquad (a > 0)$

676. $\int_0^\infty x^n e^{-ax}[\sin(bx)]\,dx = \frac{n![(a+ib)^{n+1}-(a-ib)^{n+1}]}{2i(a^2+b^2)^{n+1}}, \qquad (i^2 = -1, a > 0)$

677. $\int_0^\infty x^n e^{-ax}[\cos(bx)]\,dx = \frac{n![(a-ib)^{n+1}+(a+ib)^{n+1}]}{2(a^2+b^2)^{n+1}}, \qquad (i^2 = -1, a > 0)$

678. $\int_0^\infty \frac{e^{-ax}\sin x}{x}\,dx = \cot^{-1} a, \qquad (a > 0)$

679. $\int_0^\infty e^{-a^2x^2}\cos bx\,dx = \frac{\sqrt{\pi}}{2a}\exp\left(-\frac{b^2}{4a^2}\right), \qquad (ab \neq 0)$

680. $\int_0^\infty e^{-t\cos\phi}\,t^{b-1}\sin(t\sin\phi)\,dt = [\Gamma(b)]\sin(b\phi), \qquad \left(b > 0, -\frac{\pi}{2} < \phi < \frac{\pi}{2}\right)$

681. $\int_0^\infty e^{-t\cos\phi}\,t^{b-1}[\cos(t\sin\phi)]\,dt = [\Gamma(b)]\cos(b\phi), \qquad \left(b > 0, -\frac{\pi}{2} < \phi < \frac{\pi}{2}\right)$

682. $\int_0^\infty t^{b-1}\cos t\,dt = [\Gamma(b)]\cos\left(\frac{b\pi}{2}\right), \qquad (0 < b < 1)$

683. $\int_0^\infty t^{b-1}(\sin t)\,dt = [\Gamma(b)]\sin\left(\frac{b\pi}{2}\right), \qquad (0 < b < 1)$

684. $\int_0^1 (\log x)^n\,dx = (-1)^n\cdot n!$

685. $\int_0^1 \left(\log\frac{1}{x}\right)^{\frac{1}{2}}\,dx = \frac{\sqrt{\pi}}{2}$

686. $\int_0^1 \left(\log\frac{1}{x}\right)^{-\frac{1}{2}}\,dx = \sqrt{\pi}$

687. $\int_0^1 \left(\log \frac{1}{x}\right)^n dx = n!$

688. $\int_0^1 x \log(1 - x)\, dx = -\frac{3}{4}$

689. $\int_0^1 x \log(1 + x)\, dx = \frac{1}{4}$

690. $\int_0^1 x^m (\log x)^n\, dx = \frac{(-1)^n n!}{(m + 1)^{n+1}}, \qquad m > -1, n = 0, 1, 2, \ldots$

If $n \neq 0, 1, 2, \ldots$ replace $n!$ by $\Gamma(n + 1)$.

691. $\int_0^1 \frac{\log x}{1 + x}\, dx = -\frac{\pi^2}{12}$

692. $\int_0^1 \frac{\log x}{1 - x}\, dx = -\frac{\pi^2}{6}$

693. $\int_0^1 \frac{\log(1 + x)}{x}\, dx = \frac{\pi^2}{12}$

694. $\int_0^1 \frac{\log(1 - x)}{x}\, dx = -\frac{\pi^2}{6}$

695. $\int_0^1 (\log x)[\log(1 + x)]\, dx = 2 - 2\log 2 - \frac{\pi^2}{12}$

696. $\int_0^1 (\log x)[\log(1 - x)]\, dx = 2 - \frac{\pi^2}{6}$

697. $\int_0^1 \frac{\log x}{1 - x^2}\, dx = -\frac{\pi^2}{8}$

698. $\int_0^1 \log\left(\frac{1 + x}{1 - x}\right) \cdot \frac{dx}{x} = \frac{\pi^2}{4}$

699. $\int_0^1 \frac{\log x\, dx}{\sqrt{1 - x^2}} = -\frac{\pi}{2} \log 2$

700. $\int_0^1 x^m \left[\log\left(\frac{1}{x}\right)\right]^n dx = \frac{\Gamma(n + 1)}{(m + 1)^{n+1}}, \qquad \text{if } m + 1 > 0, n + 1 > 0$

701. $\int_0^1 \frac{(x^p - x^q)\, dx}{\log x} = \log\left(\frac{p + 1}{q + 1}\right), \qquad (p + 1 > 0, q + 1 > 0)$

702. $\int_0^1 \frac{dx}{\sqrt{\log\left(\frac{1}{x}\right)}} = \sqrt{\pi}$, (same as integral 686)

703. $\int_0^\infty \log\left(\frac{e^x + 1}{e^x - 1}\right) dx = \frac{\pi^2}{4}$

704. $\int_0^{\pi/2} (\log \sin x)\, dx = \int_0^{\pi/2} \log \cos x\, dx = -\frac{\pi}{2} \log 2$

705. $\int_0^{\pi/2} (\log \sec x)\, dx = \int_0^{\pi/2} \log \csc x\, dx = \frac{\pi}{2} \log 2$

706. $\int_0^{\pi} x(\log \sin x)\, dx = -\frac{\pi^2}{2} \log 2$

707. $\int_0^{\pi/2} (\sin x)(\log \sin x)\, dx = \log 2 - 1$

708. $\int_0^{\pi/2} (\log \tan x)\, dx = 0$

709. $\int_0^{\pi} \log (a \pm b \cos x)\, dx = \pi \log \left(\frac{a + \sqrt{a^2 - b^2}}{2}\right), \qquad (a \geq b)$

710. $\int_0^{\pi} \log (a^2 - 2ab \cos x + b^2)\, dx = \begin{cases} 2\pi \log a, & a \geq b > 0 \\ 2\pi \log b, & b \geq a > 0 \end{cases}$

711. $\int_0^{\infty} \frac{\sin ax}{\sinh bx} dx = \frac{\pi}{2b} \tanh \frac{a\pi}{2b}$

712. $\int_0^{\infty} \frac{\cos ax}{\cosh bx} dx = \frac{\pi}{2b} \operatorname{sech} \frac{a\pi}{2b}$

713. $\int_0^{\infty} \frac{dx}{\cosh ax} = \frac{\pi}{2a}$

714. $\int_0^{\infty} \frac{x\, dx}{\sinh ax} = \frac{\pi^2}{4a^2}$

715. $\int_0^{\infty} e^{-ax}(\cosh bx)\, dx = \frac{a}{a^2 - b^2}, \qquad (0 \leq |b| < a)$

716. $\int_0^{\infty} e^{-ax}(\sinh bx)\, dx = \frac{b}{a^2 - b^2}, \qquad (0 \leq |b| < a)$

717. $\int_0^{\infty} \frac{\sinh ax}{e^{bx} + 1} dx = \frac{\pi}{2b} \csc \frac{a\pi}{b} - \frac{1}{2a}$

718. $\int_0^{\infty} \frac{\sinh ax}{e^{bx} - 1} dx = \frac{1}{2a} - \frac{\pi}{2b} \cot \frac{a\pi}{b}$

719. $\int_0^{\pi/2} \frac{dx}{\sqrt{1 - k^2 \sin^2 x}} = \frac{\pi}{2}\left[1 + \left(\frac{1}{2}\right)^2 k^2 + \left(\frac{1 \cdot 3}{2 \cdot 4}\right)^2 k^4 + \left(\frac{1 \cdot 3 \cdot 5}{2 \cdot 4 \cdot 6}\right)^2 k^6 + \cdots\right], \text{ if } k^2 < 1$

720. $\int_0^{\pi/2} \sqrt{1 - k^2 \sin^2 x}\, dx = \frac{\pi}{2}\left[1 - \left(\frac{1}{2}\right)^2 k^2 - \left(\frac{1 \cdot 3}{2 \cdot 4}\right)^2 \frac{k^4}{3} - \left(\frac{1 \cdot 3 \cdot 5}{2 \cdot 4 \cdot 6}\right)^2 \frac{k^6}{5} - \cdots\right], \text{ if } k^2 < 1$

721. $\int_0^{\infty} e^{-x} \log x\, dx = -\gamma = -0.5772157\ldots$

722. $\int_0^{\infty} e^{-x^2} \log x\, dx = -\frac{\sqrt{\pi}}{4}(\gamma + 2 \log 2)$

723. $\int_0^{\infty} \left(\frac{1}{1 - e^{-x}} - \frac{1}{x}\right) e^{-x}\, dx = \gamma = 0.5772157\ldots$ [Euler's Constant]

724. $\int_0^{\infty} \frac{1}{x}\left(\frac{1}{1 + x} - e^{-x}\right) dx = \gamma = 0.5772157\ldots$

For n even:

725. $\int \cos^n x\, dx = \frac{1}{2^{n-1}} \sum_{k=0}^{\frac{n}{2}-1} \binom{n}{k} \frac{\sin (n - 2k)\, x}{(n - 2k)} + \frac{1}{2^n} \binom{n}{\frac{n}{2}} x$

$$726.\ \int \sin^n x\,dx = \frac{1}{2^{n-1}} \sum_{k=0}^{\frac{n}{2}-1} \binom{n}{k} \frac{\sin\left[(n-2k)\left(\frac{\pi}{2}-x\right)\right]}{2k-n} + \frac{1}{2^n}\binom{n}{\frac{n}{2}} x$$

For n odd:

$$727.\ \int \cos^n x\,dx = \frac{1}{2^{n-1}} \sum_{k=0}^{\frac{n-1}{2}} \binom{n}{k} \frac{\sin(n-2k)x}{(n-2k)}$$

$$728.\ \int \sin^n x\,dx = \frac{1}{2^{n-1}} \sum_{k=0}^{\frac{n-1}{2}} \binom{n}{k} \frac{\sin\left[(n-2k)\left(\frac{\pi}{2}-x\right)\right]}{2k-n}$$

DIFFERENTIAL EQUATIONS

SPECIAL FORMULAS

Certain types of differential equations occur sufficiently often to justify the use of formulas for the corresponding particular solutions. The following set of tables I to XIV covers all first, second, and nth order ordinary linear differential equations with constant coefficients for which the right members are of the form $P(x)e^{rx}\sin sx$ or $P(x)e^{rx}\cos sx$, where r and s are constants and $P(x)$, is a polynomial of degree n.

When the right member of a reducible linear partial differential equation with constant coefficients is not zero, particular solutions for certain types of right members are contained in tables XV to XXI. In these tables both F and P are used to denote polynomials, and it is assumed that no denominator is zero. In any formula the roles of x and y may be reversed throughout, changing a formula in which x dominates to one in which y dominates. Tables XIX, XX, XXI are applicable whether the equations are reducible or not. The symbol $\binom{m}{n}$ stands for $\frac{m!}{(m-n)!n!}$ and is the $n+1$ st coefficient in the expansion of $(a+b)^m$. Also $0! = 1$ by definition.

The tables as herewith given are those contained in the text *Differential Equations* by Ginn and Company (1955) and are published with their kind permission and that of the author, Professor Frederick H. Steen.

Solution of Linear Differential Equations with Constant Coefficients

Any linear differential equation with constant coefficients may be written in the form

$$p(D)y = R(x)$$

where D is the differential operation

$$Dy = \frac{dy}{dx}$$

$p(D)$ is a polynomial in D,
y is the dependent variable,
x is the independent variable,
$R(x)$ is an arbitrary function of x.

A power of D represents repeated differentiation, that is

$$D^n y = \frac{d^n y}{dx^n}$$

For such an equation, the general solution may be written in the form

$$y = y_c + y_p$$

where y_p is any particular solution, and y_c is called the *complementary function*. This complementary function is defined as the general solution of the *homogeneous equation*, which is the original differential equation with the right side replaced by zero, i.e.

$$p(D)y = 0$$

The complementary function y_c may be determined as follows:

1. Factor the polynomial $p(D)$ into real and complex linear factors, just as if D were a variable instead of an operator.
2. For each nonrepeated linear factor of the form $(D - a)$, where a is real, write down a term of the form

$$ce^{ax}$$

where c is an arbitrary constant.

3. For each repeated real linear factor of the form $(D - a)^n$, write down n terms of the form

$$c_1e^{ax} + c_2xe^{ax} + c_3x^2e^{ax} + \cdots + c_nx^{n-1}e^{ax}$$

where the c_i's are arbitrary constants.

4. For each non-repeated conjugate complex pair of factors of the form $(D - a + ib)(D - a - ib)$, write down 2 terms of the form

$$c_1e^{ax}\cos bx + c_2e^{ax}\sin bx$$

5. For each repeated conjugate complex pair of factors of the form $(D - a + ib)^n(D - a - ib)^n$, write down $2n$ terms of the form

$$c_1e^{ax}\cos bx + c_2e^{ax}\sin bx + c_3xe^{ax}\cos bx + c_4xe^{ax}\sin bx + \cdots$$
$$+ c_{2n-1}x^{n-1}e^{ax}\cos bx + c_{2n}x^{n-1}e^{ax}\sin bx$$

6. The sum of all the terms thus written down is the complementary function y_c.

To find the particular solution y_p, use the following tables, as shown in the examples. For cases not shown in the tables, there are various methods of finding y_p. The most general method is called *variation of parameters*. The following example illustrates the method:

Find y_p for $(D^2 - 4)y = e^x$.

This example can be solved most easily by use of equation 63 in the tables following. However it is given here as an example of the method of variation of parameters.

The complementary function is

$$y_c = c_1e^{2x} + c_2e^{-2x}$$

To find y_p, replace the constants in the complementary function with unknown functions,

$$y_p = ue^{2x} + ve^{-2x}$$

We now prepare to substitute this assumed solution into the original equation. We begin by taking all the necessary dervatives:

$$y_p = ue^{2x} + ve^{-2x}$$
$$y_p' = 2ue^{2x} - 2ve^{-2x} + u'e^{2x} + v'e^{-2x}$$

For each derivative of y_p except the highest, we set the sum of all the terms containing u' and v' to 0. Thus the above equation becomes

$$u'e^{2x} + v'e^{-2x} = 0 \quad \text{and} \quad y_p' = 2ue^{2x} - 2ve^{-2x}$$

Continuing to differentiate, we have

$$y_p'' = 4ue^{2x} + 4ve^{-2x} + 2u'e^{2x} - 2v'e^{-2x}$$

When we substitute into the original equation, all the terms not containing u' or v' cancel out. This is a consequence of the method by which y_p was set up.

Thus all that is necessary is to write down the terms containing u' or v' in the highest order derivative of y_p, multiply by the constant coefficient of the highest power of D in $p(D)$, and set it equal to $R(x)$. Together with the previous terms in u' and v' which were set equal to 0, this gives us as many linear equations in the first derivatives of the unknown functions as there are unknown functions. The first derivatives may then be solved for by algebra, and the unknown functions found by integration. In the present example, this becomes

$$u'e^{2x} + v'e^{-2x} = 0$$
$$2u'e^{2x} - 2v'e^{-2x} = e^x$$

We eliminate v' and u' separately, getting

$$4u'e^{2x} = e^x$$
$$4v'e^{-2x} = -e^x$$

Thus

$$u' = \tfrac{1}{4}e^{-x}$$
$$v' = -\tfrac{1}{4}e^{3x}$$

Therefore, by integrating

$$u = -\tfrac{1}{4}e^{-x}$$
$$v = -\tfrac{1}{12}e^{3x}$$

A constant of integration is not needed, since we need only one particular solution. Thus

$$y_p = ue^{2x} + ve^{-2x} = -\tfrac{1}{4}e^{-x}e^{2x} - \tfrac{1}{12}e^{3x}e^{-2x}$$
$$= -\tfrac{1}{4}e^x - \tfrac{1}{12}e^x = -\tfrac{1}{3}e^x$$

and the general solution is

$$y = y_c + y_p = c_1e^{2x} + c_2e^{-2x} - \tfrac{1}{3}e^x$$

The following examples illustrate the use of the tables.

Example 1. Solve $(D^2 - 4)y = \sin 3x$.

Substitution of $q = -4$, $s = 3$ in formula 24 gives

$$y_p = \frac{\sin 3x}{-9 - 4}$$

wherefore the general solution is

$$y = c_1 e^{2x} + c_2 e^{-2x} - \frac{\sin 3x}{13}$$

Example 2. Obtain a particular solution of $(D^2 - 4D + 5)y = x^2 e^{3x} \sin x$.

Applying formula 40 with $a = 2, b = 1, r = 3, s = 1$, $P(x) = x^2$, $s + b = 2$, $s - b = 0$, $a - r = -1$, $(a - r)^2 + (s + b)^2 = 5$, $(a - r)^2 + (s - b)^2 = 1$, we have

$$y_p = \frac{e^{3x}\sin x}{2}\left[\left(\frac{2}{5} - \frac{0}{1}\right)x^2 + \left(\frac{2(-1)2}{25} - \frac{2(-1)0}{1}\right)2x + \left(\frac{3\cdot 1\cdot 2 - 2^3}{125} - \frac{3\cdot 1\cdot 0 - 0}{1}\right)2\right]$$

$$- \frac{e^{3x}\cos x}{2}\left[\left(\frac{-1}{5} - \frac{-1}{1}\right)x^2 + \left(\frac{1-4}{25} - \frac{1-0}{1}\right)2x + \left(\frac{-1 - 3(-1)4}{125} - \frac{-1 - 3(-1)0}{1}\right)2\right]$$

$$= \left(\tfrac{1}{5}x^2 - \tfrac{4}{25}x - \tfrac{2}{125}\right)e^{3x}\sin x + \left(-\tfrac{2}{5}x^2 + \tfrac{28}{25}x - \tfrac{136}{125}\right)e^{3x}\cos x$$

The special formulas effect a very considerable saving of time in problems of this type.

Example 3. Obtain a particular solution of $(D^2 - 4D + 5)y = x^2 e^{2x} \cos x$. (Compare with Example 2.)

Formula 40 is not applicable here since for this equation $r = a$, $s = b$, wherefore the denominator $(a - r)^2 + (s - b)^2 = 0$. We turn instead to formula 44. Substituting $a = 2$, $b = 1$, $P(x) = x^2$ and replacing sin by cos, cos by $-\sin$, we obtain

$$y_p = \frac{e^{2x}\cos x}{4}\left(x^2 - \tfrac{2}{4}\right) + \frac{e^{2x}\sin x}{2}\int \left(x^2 - \tfrac{1}{2}\right)dx$$

$$= \left(\frac{x^2}{4} - \frac{1}{8}\right)e^{2x}\cos x + \left(\frac{x^3}{6} - \frac{x}{4}\right)e^{2x}\sin x$$

which is the required solution.

Example 4. Find z_p for $(D_x - 3D_y)z = \ln(y + 3x)$.

Referring to Table XV we note that formula 69 (not 68) is applicable. This gives

$$z_p = x \ln(y + 3x)$$

It is easily seen that $-y/3 \ln(y + 3x)$ would serve equally well.

Example 5. Solve $(D_x + 2D_y - 4)z = y\cos(y - 2x)$.

Since R in formula 76 contains a polynomial in x, not y, we rewrite the given equation in the form

$$\left(D_y + \tfrac{1}{2}D_x - 2\right)z = \tfrac{1}{2}y\cos(y - 2x)$$

Then

$$z_c = e^{2y}F\left(x - \tfrac{1}{2}y\right) = e^{2y}f(2x - y)$$

and by the formula

$$z_p = -\tfrac{1}{2}\cos(y - 2x)\cdot\left(\frac{y}{2} + \frac{\frac{1}{2}}{2}\right)$$

$$= -\tfrac{1}{8}(2y + 1)\cos(y - 2x)$$

Example 6. Find z_p for $(D_x + 4D_y)^3 z = (2x - y)^2$.

Using formula 79, we obtain

$$z_p = \frac{\int\int\int u^2 du^3}{[2 + 4(-1)]^3} = \frac{u^5}{5\cdot 4\cdot 3\cdot(-8)} = -\frac{(2x - y)^5}{480}$$

Example 7. Find z_p for $(D_x^3 + 5D_x^2 D_y - 7D_x + 4)z = e^{2x+3y}$.

By formula 87

$$z_p = \frac{e^{2x+3y}}{2^3 + 5\cdot 2^2\cdot 3 - 7\cdot 2 + 4} = \frac{e^{2x+3y}}{58}$$

Example 8. Find z_p for

$$(D_x^4 + 6D_x^3 D_y + D_x D_y + D_y^2 + 9)z = \sin(3x + 4y)$$

Since every term in the left member is of even degree in the two operators D_x and D_y, formula 90 is applicable. It gives

$$z_p = \frac{\sin(3x + 4y)}{(-9)^2 + 6(-9)(-12) + (-12) + (-16) + 9}$$

$$= \frac{\sin(3x + 4y)}{710}$$

TABLE I: $(D - a)y = R$

R	y_p
1. e^{rx}	$\dfrac{e^{rx}}{r - a}$
2. $\sin sx$*	$-\dfrac{a \sin sx + s \cos sx}{a^2 + s^2} = -\dfrac{1}{\sqrt{a^2 + s^2}} \sin\left(sx + \tan^{-1}\dfrac{s}{a}\right)$
3. $P(x)$	$-\dfrac{1}{a}\left[P(x) + \dfrac{P'(x)}{a} + \dfrac{P''(x)}{a^2} + \cdots + \dfrac{P^{(n)}(x)}{a^n}\right]$
4. $e^{rx} \sin sx$*	Replace a by $a - r$ in formula 2 and multiply by e^{rx}.
5. $P(x)e^{rx}$	Replace a by $a - r$ in formula 3 and multiply by e^{rx}.
6. $P(x) \sin sx$*	$-\sin sx\left[\dfrac{a}{a^2 + s^2}P(x) + \dfrac{a^2 - s^2}{(a^2 + s^2)^2}P'(x) + \dfrac{a^3 - 3as^2}{(a^2 + s^2)^3}P''(x) + \cdots + \dfrac{a^k - \binom{k}{2}a^{k-2}s^2 + \binom{k}{4}a^{k-4}s^4 - \cdots}{(a^2 + s^2)^k}P^{(k-1)}(x) + \cdots\right]$ $-\cos sx\left[\dfrac{s}{a^2 + s^2}P(x) + \dfrac{2as}{(a^2 + s^2)^2}P'(x) + \dfrac{3a^2s - s^3}{(a^2 + s^2)^3}P''(x) + \cdots + \dfrac{\binom{k}{1}a^{k-1}s - \binom{k}{3}a^{k-3}s^3 + \cdots}{(a^2 + s^2)^k}P^{(k-1)}(x) + \cdots\right]$
7. $P(x)e^{rx} \sin sx$*	Replace a by $a - r$ in formula 6 and multiply by e^{rx}.
8. e^{ax}	xe^{ax}
9. $e^{ax} \sin sx$*	$-\dfrac{e^{ax} \cos sx}{s}$
10. $P(x)e^{ax}$	$e^{ax}\displaystyle\int P(x)\,dx$
11. $P(x)e^{ax} \sin sx$*	$\dfrac{e^{ax} \sin sx}{s}\left[\dfrac{P'(x)}{s^3} - \dfrac{P'''(x)}{s^3} + \dfrac{P^{v}(x)}{s^5} - \cdots\right] - \dfrac{e^{ax} \cos sx}{s}\left[P(x) - \dfrac{P''(x)}{s^2} + \dfrac{P^{iv}(x)}{s^4} - \cdots\right]$

* For cos sx in R replace "sin" by "cos" and "cos" by "− sin" in y_p.

$$D^n = \frac{d^n}{dx^n} \qquad \binom{m}{n} = \frac{m!}{(m - n)!n!} \qquad 0! = 1$$

TABLE II: $(D - a)^2y = R$

R	y_p
12. e^{rx}	$\dfrac{e^{rx}}{(r - a)^2}$
13. $\sin sx$*	$\dfrac{1}{(a^2 + s^2)^2}[(a^2 - s^2)\sin sx + 2as \cos sx] = \dfrac{1}{a^2 + s^2}\sin\left(sx + \tan^{-1}\dfrac{2as}{a^2 - s^2}\right)$
14. $P(x)$	$\dfrac{1}{a^2}\left[P(x) + \dfrac{2P'(x)}{a} + \dfrac{3P''(x)}{a^2} + \cdots + \dfrac{(n + 1)P^{(n)}(x)}{a^n}\right]$
15. $e^{rx} \sin sx$*	Replace a by $a - r$ in formula 13 and multiply by e^{rx}.
16. $P(x)e^{rx}$	Replace a by $a - r$ in formula 14 and multiply by e^{rx}.
17. $P(x) \sin sx$*	$\sin sx\left[\dfrac{a^2 - s^2}{(a^2 + s^2)^2}P(x) + 2\dfrac{a^3 - 3as^2}{(a^2 + s^2)^3}P'(x) + 3\dfrac{a^4 - 6a^2s^2 + s^4}{(a^2 + s^2)^4}P''(x) + \cdots + (k - 1)\dfrac{a^k - \binom{k}{2}a^{k-2}s^2 + \binom{k}{4}a^{k-4}s^4 - \cdots}{(a^2 + s^2)^k}P^{(k-2)}(x) + \cdots\right]$ $+\cos sx\left[\dfrac{2as}{(a^2 + s^2)^2}P(x) + 2\dfrac{3a^2s - s^3}{(a^2 + s^2)^3}P'(x) + 3\dfrac{4a^3s - 4as^3}{(a^2 + s^2)^4}P''(x) + \cdots + (k - 1)\dfrac{\binom{k}{1}a^{k-1}s - \binom{k}{3}a^{k-3}s^3 + \cdots}{(a^2 + s^2)^k}P^{(k-2)}(x) + \cdots\right]$
18. $P(x)e^{rx} \sin sx$*	Replace a by $a - r$ in formula 17 and multiply by e^{rx}.
19. e^{ax}	$\frac{1}{2}x^2e^{ax}$
20. $e^{ax} \sin sx$*	$-\dfrac{e^{ax} \sin sx}{s^2}$
21. $P(x)e^{ax}$	$e^{ax}\displaystyle\iint P(x)dx\,dx$
22. $P(x)e^{ax} \sin sx$*	$-\dfrac{e^{ax} \sin sx}{s^2}\left[P(x) - \dfrac{3P''(x)}{s^2} + \dfrac{5P^{iv}(x)}{s^4} - \dfrac{7P^{vi}(x)}{s^6} + \cdots\right] - \dfrac{e^{ax} \cos sx}{s^2}\left[\dfrac{2P'(x)}{s} - \dfrac{4P'''(x)}{s^3} + \dfrac{6P^{v}(x)}{s^5} - \cdots\right]$

* For cos sx in R replace "sin" by "cos" and "cos" by "− sin" in y_p.

TABLE III: $(D^2 + q)y = R$

R	y_p
23. e^{rx}	$\dfrac{e^{rx}}{r^2 + q}$
24. $\sin sx$*	$\dfrac{\sin sx}{-s^2 + q}$
25. $P(x)$	$\dfrac{1}{q}\left[P(x) - \dfrac{P''(x)}{q} + \dfrac{P^{iv}(x)}{q^2} - \cdots + (-1)^k \dfrac{P^{(2k)}(x)}{q^k} \cdots\right]$
26. $e^{rx}\sin sx$*	$\dfrac{(r^2 - s^2 + q)e^{rx}\sin sx - 2rse^{rx}\cos sx}{(r^2 - s^2 + q)^2 + (2rs)^2} = \dfrac{e^{rx}}{\sqrt{(r^2 - s^2 + q)^2 + (2rs)^2}}\sin\left[sx - \tan^{-1}\dfrac{2rs}{r^2 - s^2 + q}\right]$
27. $P(x)e^{rx}$	$\dfrac{e^{rx}}{r^2 + q}\left[P(x) - \dfrac{2r}{r^2 + q}P'(x) + \dfrac{3r^2 - q}{(r^2 + q)^2}P''(x) - \dfrac{4r^3 - 4qr}{(r^2 + q)^3}P'''(x) + \cdots + (-1)^{k-1}\dfrac{\binom{k}{1}r^{k-1} - \binom{k}{3}r^{k-3}q + \binom{k}{5}r^{k-5}q^2 - \cdots}{(r^2 + q)^{k-1}}P^{(k-1)}(x) + \cdots\right]$
28. $P(x)\sin sx$*	$\dfrac{\sin sx}{(-s^2 + q)}\left[P(x) - \dfrac{3s^2 + q}{(-s^2 + q)^2}P''(x) + \dfrac{5s^4 + 10s^2q + q^2}{(-s^2 + q)^4}P^{iv}(x) + \cdots + (-1)^k\dfrac{\binom{2k+1}{1}s^{2k} + \binom{2k+1}{3}s^{2k-2}q + \binom{2k+1}{5}s^{2k-4}q^2 + \cdots}{(-s^2 + q)^{2k}}P^{(2k)}(x) + \cdots\right]$ $- \dfrac{s\cos sx}{(-s^2 + q)}\left[\dfrac{2P'(x)}{(-s^2 + q)} - \dfrac{4s^2 + 4q}{(-s^2 + q)^3}P'''(x) + \cdots + (-1)^{k+1}\dfrac{\binom{2k}{1}s^{2k-2} + \binom{2k}{3}s^{2k-4}q + \cdots}{(-s^2 + q)^{2k-1}}P^{(2k-1)}(x) + \cdots\right]$

TABLE IV: $(D^2 + b^2)y = R$

R	y_p
29. $\sin bx$*	$-\dfrac{x\cos bx}{2b}$
30. $P(x)\sin bx$*	$\dfrac{\sin bx}{(2b)^2}\left[P(x) - \dfrac{P''(x)}{(2b)^2} + \dfrac{P^{iv}(x)}{(2b)^4} - \cdots\right] - \dfrac{\cos bx}{2b}\int\left[P(x) - \dfrac{P''(x)}{(2b)^2} + \cdots\right]dx$

* For $\cos sx$ in R replace "sin" by "cos" and "cos" by "− sin" in y_p.

TABLE V: $(D^2 + pD + q)y = R$

R	y_p
31. e^{rx}	$\dfrac{e^{rx}}{r^2 + pr + q}$
32. $\sin sx$*	$\dfrac{(q - s^2)\sin sx - ps\cos sx}{(q - s^2)^2 + (ps)^2} = \dfrac{1}{\sqrt{(q - s^2)^2 + (ps)^2}}\sin\left(sx - \tan^{-1}\dfrac{ps}{q - s^2}\right)$
33. $P(x)$	$\dfrac{1}{q}\left[P(x) - \dfrac{p}{q}P'(x) + \dfrac{p^2 - q}{q^2}P''(x) - \dfrac{p^3 - 2pq}{q^3}P'''(x) + \cdots + (-1)^n\dfrac{p^n - \binom{n-1}{1}p^{n-2}q + \binom{n-2}{2}p^{n-4}q^2 - \cdots}{q^n}P^{(n)}(x)\right]$
34. $e^{rx}\sin sx$*	Replace p by $p + 2r$, q by $q + pr + r^2$ in formula 32 and multiply by e^{rx}.
35. $P(x)e^{rx}$	Replace p by $p + 2r$, q by $q + pr + r^2$ in formula 33 and multiply by e^{rx}.

TABLE VI: $(D - b)(D - a)y = R$

R	y_p
36. $P(x)\sin sx$*	$\dfrac{\sin sx}{b - a}\left[\left(\dfrac{a}{a^2 + s^2} - \dfrac{b}{b^2 + s^2}\right)P(x) + \left(\dfrac{a^2 - s^2}{(a^2 + s^2)^2} - \dfrac{b^2 - s^2}{(b^2 + s^2)^2}\right)P'(x) + \left(\dfrac{a^3 - 3as^2}{(a^2 + s^2)^3} - \dfrac{b^3 - 3bs^2}{(b^2 + s^2)^3}\right)P''(x) + \cdots\right]$ $+ \dfrac{\cos sx}{b - a}\left[\left(\dfrac{s}{a^2 + s^2} - \dfrac{s}{b^2 + s^2}\right)P(x) + \left(\dfrac{2as}{(a^2 + s^2)^2} - \dfrac{2bs}{(b^2 + s^2)^2}\right)P'(x) + \left(\dfrac{3a^2s - s^3}{(a^2 + s^2)^3} - \dfrac{3b^2s - s^3}{(b^2 + s^2)^3}\right)P''(x) + \cdots\right]^{\dagger}$
37. $P(x)e^{rx}\sin sx$*	Replace a by $a - r$, b by $b - r$ in formula 36 and multiply by e^{rx}.
38. $P(x)e^{ax}$	$\dfrac{e^{ax}}{a - b}\left[\int P(x)dx + \dfrac{P(x)}{(b - a)} + \dfrac{P'(x)}{(b - a)^2} + \dfrac{P''(x)}{(b - a)^3} + \cdots + \dfrac{P^{(n)}(x)}{(b - a)^{n+1}}\right]$

* For $\cos sx$ in R replace "sin" by "cos" and "cos" by "− sin" in y_p.

† For additional terms, compare with formula 6.

TABLE VII: $(D^2 - 2aD + a^2 + b^2)y = R$

R — y_p

39. $P(x)\sin sx$*

$$\frac{\sin sx}{2b}\left[\left(\frac{s+b}{a^2+(s+b)^2} - \frac{s-b}{a^2+(s-b)^2}\right)P(x) + \left(\frac{2a(s+b)}{[a^2+(s+b)^2]^2} - \frac{2a(s-b)}{[a^2+(s-b)^2]^2}\right)P'(x) + \left(\frac{3a^2(s+b)-(s+b)^3}{[a^2+(s+b)^2]^3} - \frac{3a^2(s-b)-(s-b)^3}{[a^2+(s-b)^2]^3}\right)P''(x) + \cdots\right]$$
$$- \frac{\cos sx}{2b}\left[\left(\frac{a}{a^2+(s+b)^2} - \frac{a}{a^2+(s-b)^2}\right)P(x) + \left(\frac{a^2-(s+b)^2}{[a^2+(s+b)^2]^2} - \frac{a^2-(s-b)^2}{[a^2+(s-b)^2]^2}\right)P'(x) + \left(\frac{a^3-3a(s+b)^2}{[a^2+(s+b)^2]^3} - \frac{a^3-3a(s-b)^2}{[a^2+(s-b)^2]^3}\right)P''(x) + \cdots\right]^{\dagger}$$

40. $P(x)e^{rx}\sin sx$* — Replace a by $a - r$ in formula 39 and multiply by e^{rx}.

41. $P(x)e^{ax}$ — $\frac{e^{ax}}{b^2}\left[P(x) - \frac{P''(x)}{b^2} + \frac{P^{iv}(x)}{b^4} - \cdots\right]$

42. $e^{ax}\sin sx$* — $\frac{e^{ax}\sin sx}{-s^2+b^2}$

43. $e^{ax}\sin bx$* — $-\frac{xe^{ax}\cos bx}{2b}$

44. $P(x)e^{ax}\sin bx$* — $\frac{e^{ax}\sin bx}{(2b)^2}\left[P(x) - \frac{P''(x)}{(2b)^2} + \frac{P^{iv}(x)}{(2b)^4} - \cdots\right] - \frac{e^{ax}\cos bx}{2b}\int\left[P(x) - \frac{P''(x)}{(2b)^2} + \frac{P^{iv}(x)}{(2b)^4} - \cdots\right]dx$

* For $\cos sx$ in R replace "sin" by "cos" and "cos" by "− sin" in y_p.
† For additional terms, compare with formula 6.

TABLE VIII: $f(D)y = [D^n + a_{n-1}D^{n-1} + \cdots + a_1D + a_0]y = R$

R — y_p

45. e^{rx} — $\frac{e^{rx}}{f(r)}$

46. $\sin sx$* — $\frac{[a_0 - a_2s^2 + a_4s^4 - \cdots]\sin sx - [a_1s - a_3s^3 + a_5s^5 + \cdots]\cos sx}{[a_0 - a_2s^2 + a_4s^4 - \cdots]^2 + [a_1s - a_3s^3 + a_5s^5 - \cdots]^2}$

TABLE IX: $f(D^2)y = R$

47. $\sin sx$* — $\frac{\sin sx}{f(-s^2)} = \frac{\sin sx}{a_0 - a_2s^2 + \cdots \pm s^{2n}}$

TABLE X: $(D - a)^n y = R$

48. e^{rx} — $\frac{e^{rx}}{(r-a)^n}$

49. $\sin sx$* — $\frac{(-1)^n}{(a^2+s^2)^n}\{[a^n - \binom{n}{2}a^{n-2}s^2 + \binom{n}{4}a^{n-4}s^4 - \cdots]\sin sx + [\binom{n}{1}a^{n-1}s - \binom{n}{3}a^{n-3}s^3 + \cdots]\cos sx\}$

50. $P(x)$ — $\frac{(-1)^n}{a^n}\left[P(x) + \binom{n}{1}\frac{P'(x)}{a} + \binom{n+1}{2}\frac{P''(x)}{a^2} + \binom{n+2}{3}\frac{P'''(x)}{a^3} + \cdots\right]$

51. $e^{rx}\sin sx$* — Replace a by $a - r$ in formula 49 and multiply by e^{rx}.

52. $e^{rx}P(x)$ — Replace a by $a - r$ in formula 50 and multiply by e^{rx}.

53. $P(x)\sin sx$*

$$(-1)^n \sin sx[A_nP(x) + \binom{n}{1}A_{n+1}P'(x) + \binom{n+1}{2}A_{n+2}P''(x) + \binom{n+2}{3}A_{n+3}P'''(x) + \cdots]$$
$$+ (-1)^n \cos sx[B_nP(x) + \binom{n}{1}B_{n+1}P'(x) + \binom{n+1}{2}B_{n+2}P''(x) + \binom{n+2}{3}B_{n+3}P'''(x) + \cdots]$$
$$A_1 = \frac{a}{a^2+s^2},\ A_2 = \frac{a^2-s^2}{(a^2+s^2)^2},\ \cdots,\ A_k = \frac{a^k - \binom{k}{2}a^{k-2}s^2 + \binom{k}{4}a^{k-4}s^4 - \cdots}{(a^2+s^2)^k}.$$
$$B_1 = \frac{s}{a^2+s^2},\ B_2 = \frac{2as}{(a^2+s^2)^2},\ \cdots,\ B_k = \frac{\binom{k}{1}a^{k-1}s - \binom{k}{3}a^{k-3}s^3 + \cdots}{(a^2+s^2)^k}.$$

54. $P(x)e^{rx}\sin sx$* — Replace a by $a - r$ in formula 53 and multiply by e^{rx}.

55. $e^{ax}P(x)$ $\qquad e^{ax}\int\int\cdots\int P(x)dx^n$

56. $P(x)e^{ax}\sin sx$* $\qquad \frac{(-1)^{\frac{n-1}{2}}e^{ax}\sin sx}{s^n}\left[\binom{n}{n-1}\frac{P'(x)}{s}-\binom{n+2}{n-1}\frac{P'''(x)}{s^3}+\binom{n+4}{n-1}\frac{P^{v}(x)}{s^5}-\cdots\right]$

$$+\frac{(-1)^{\frac{n+1}{2}}e^{ax}\cos sx}{s^n}\left[\binom{n-1}{n-1}P(x)-\binom{n+1}{n-1}\frac{P''(x)}{s^2}+\binom{n+3}{n-1}\frac{P^{iv}(x)}{s^4}-\cdots\right] \text{ (n odd)}$$

$$\frac{(-1)^{\frac{n}{2}}e^{ax}\sin sx}{s^n}\left[\binom{n-1}{n-1}P(x)-\binom{n+1}{n-1}\frac{P''(x)}{s^2}+\binom{n+3}{n-1}\frac{P^{iv}(x)}{s^4}-\cdots\right]$$

$$+\frac{(-1)^{\frac{n}{2}}e^{ax}\cos sx}{s^n}\left[\binom{n}{n-1}\frac{P'(x)}{s}-\binom{n+2}{n-1}\frac{P'''(x)}{s^3}+\binom{n+4}{n-1}\frac{P^{v}(x)}{s^5}-\cdots\right] \text{ (n even)}$$

* For cos sx in R replace "sin" by "cos" and "cos" by "− sin" in y_p.

TABLE XI: $(D-a)^n f(D)y = R$

57. e^{ax} $\qquad \frac{x^n}{n!}\cdot\frac{e^{ax}}{f(a)}$

* For cos sx in R replace "sin" by "cos" and "cos" by "− sin" in y_p.

TABLE XII: $(D^2+q)^n y = R$

R	y_p
58. e^{rx}	$e^{rx}/(r^2+q)^n$
59. $\sin sx$*	$\sin sx/(q-s^2)^n$
60. $P(x)$	$\frac{1}{q^n}\left[P(x)-\binom{n}{1}\frac{P''(x)}{q}+\binom{n+1}{2}\frac{P^{iv}(x)}{q^2}-\binom{n+2}{3}\frac{P^{vi}(x)}{q^3}+\cdots\right]$
61. $e^{rx}\sin sx$*	$\frac{e^{rx}}{(A^2+B^2)^n}\{[A^n-\binom{r}{2}A^{n-2}B^2+\binom{n}{4}A^{n-4}B^4-\cdots]\sin sx-[\binom{n}{1}A^{n-1}B-\binom{n}{3}A^{n-3}B^3+\cdots]\cos sx\}$ $A=r^2-s^2+q,\quad B=2rs$

TABLE XIII: $(D^2+b^2)^n y = R$

62. $\sin bx$* $\qquad (-1)^{\frac{n+1}{2}}\frac{x^n\cos bx}{n!(2b)^n}$ (n odd), $\qquad (-1)^{\frac{n}{2}}\frac{x^n\sin bx}{n!(2b)^n}$ (n even)

TABLE XIV: $(D^n-q)y = R$

63. e^{rx} $\qquad e^{rx}/(r^n-q)$

64. $P(x)$ $\qquad -\frac{1}{q}\left[P(x)\frac{P^{(n)}(x)}{q}+\frac{P^{(2n)}(x)}{q^2}+\cdots\right]$

65. $\sin sx$* $\qquad -\frac{q\sin sx+(-1)^{\frac{n-1}{2}}s^n\cos sx}{q^2+s^{2n}}$ (n odd), $\qquad \frac{\sin sx}{(-s^2)^{n/2}-q}$ (n even)

66. $e^{rx}\sin sx$* $\qquad \frac{Ae^{rx}\sin sr-Be^{rx}\cos sx}{A^2+B^2}=\frac{e^{rx}}{\sqrt{A^2+B^2}}\sin\left(sx-\tan^{-1}\frac{B}{A}\right)$

$$A=[r^n-\binom{n}{2}r^{n-2}s^2+\binom{n}{4}r^{n-4}s^4-\cdots]-q,\qquad B=[\binom{n}{1}r^{n-1}s-\binom{n}{3}r^{n-3}s^3+\cdots]$$

* For cos sx in R replace "sin" by "cos" and "cos" by "− sin" in y_p.

TABLE XV: $(D_x+mD_y)z = R$

R	z_p
67 e^{ax+by}	$\frac{e^{ax+by}}{a+mb}$
68. $f(ax+by)$	$\frac{\int f(u)du}{a+mb}$, $u=ax+by$
69. $f(y-mx)$	$xf(y-mx)$
70. $\phi(x,y)f(y-mx)$	$f(y-mx)\int\phi(x,a+mx)dx$ ($a=y-mx$ after integration)

TABLE XVI: $(D_x + mD_y - k)z = R$

71. e^{ax+by} — $\dfrac{e^{ax+by}}{a+mb-k}$

72. $\sin(ax+by)$* — $-\dfrac{(a+bm)\cos(ax+by)+k\sin(ax+by)}{(a+bm)^2+k^2}$

73. $e^{\alpha x+\beta y}\sin(ax+by)$* — Replace k in 72 by $k-\alpha-m\beta$ and multiply by $e^{\alpha x+\beta y}$

74. $e^{kx}f(ax+by)$ — $\dfrac{e^{kx}\int f(u)du}{a+mb}$, $u = ax+by$

75. $f(y-mx)$ — $-\dfrac{f(y-mx)}{k}$

76. $P(x)f(y-mx)$ — $-\dfrac{1}{k}f(y-mx)\left[P(x)+\dfrac{P'(x)}{k}+\dfrac{P''(x)}{k^2}+\cdots+\dfrac{P^{(n)}(x)}{k^n}\right]$

77. $e^{kx}f(y-mx)$ — $xe^{kx}f(y-mx)$

* For cos $(ax+by)$ replace "sin" by "cos," and "cos" by "− sin" in z_p.

$$D_x = \frac{\partial}{\partial x};\qquad D_y = \frac{\partial}{\partial y};\qquad D_x^k D_y^r = \frac{\partial^{k+r}}{\partial_x^k \partial_y^r}$$

TABLE XVII: $(D_x + mD_y)^n z = R$

R	z_p
78. e^{ax+by}	$\dfrac{e^{ax+by}}{(a+mb)^n}$
79. $f(ax+by)$	$\dfrac{\int\int\cdots\int f(u)du^n}{(a+mb)^n}$, $u = ax+by$
80. $f(y-mx)$	$\dfrac{x^n}{n!}f(y-mx)$
81. $\phi(x,y)f(y+mx)$	$f(y-mx)\int\int\cdots\int\phi(x, a+mx)dx^n$ ($a = y-mx$ after integration)

TABLE XVIII: $(D_x + mD_y - k)^n z = R$

82. e^{ax+by} — $\dfrac{e^{ax+by}}{(a+mb-k)^n}$

83. $f(y-mx)$ — $\dfrac{(-1)^n f(y-mx)}{k^n}$

84. $P(x)f(y-mx)$ — $\dfrac{(-1)^n}{k^n}f(y-mx)\left[P(x)+\binom{n}{1}\dfrac{P'(x)}{k}+\binom{n+1}{2}\dfrac{P''(x)}{k^2}+\binom{n+2}{3}\dfrac{P'''(x)}{k^3}+\cdots\right]$

85. $e^{kx}f(ax+by)$ — $\dfrac{e^{kx}\int\int\cdots\int f(u)du^n}{(a+mb)^n}$, $u = ax+by$

86. $e^{kx}f(y-mx)$ — $\dfrac{x^n}{n!}e^{kx}f(y-mx)$

TABLE XIX: $[D_x^n + a_1D_x^{n-1}D_y + a_2D_x^{n-2}D_y^2 + \cdots + a^nD_y^n]z = R$

87. e^{ax+by} — $\dfrac{e^{ax+by}}{a+a_1a^{n-1}b+a_2a^{n-2}b^2+\cdots+a_nb^n}$

88. $f(ax+by)$ — $\dfrac{\int\int\cdots\int f(u)du^n}{a^n+a_1a^{n-1}b+a_2a^{n-2}b^2+\cdots+a^nb^n}$, $(u = ax+by)$

TABLE XX: $F(D_x, D_y)z = R$

89. e^{ax+by} — $\dfrac{e^{ax+by}}{F(a,b)}$

TABLE XXI: $F(D_x^2, D_xD_y, D_y^2)z = R$

90. $\sin(ax+by)$* — $\dfrac{\sin(ax+by)}{F(-a^2, -ab, -b^2)}$

* For cos $(ax+by)$ replace "sin" by "cos", and "cos" by "−sin" in z_p.

DIFFERENTIAL EQUATIONS

Differential equation	Method of solution
Separation of variables $f_1(x)\,g_1(y)\,dx + f_2(x)\,g_2(y)\,dy = 0$	$\int \frac{f_1(x)}{f_2(x)}\,dx + \int \frac{g_2(y)}{g_1(y)}\,dy = c$
Exact equation $M(x,y)dx + N(x,y)dy = 0$ where $\partial M/\partial y = \partial N/\partial x$	$\int M\partial x + \int\left(N - \frac{\partial}{\partial y}\int M\partial x\right)dy = c$ where ∂x indicates that the integration is to be performed with respect to x keeping y constant.
Linear first order equation $\frac{dy}{dx} + P(x)y = Q(x)$	$ye^{\int Pdx} = \int Qe^{\int Pdx}\,dx + c$
Bernoulli's equation $\frac{dy}{dx} + P(x)y = Q(x)y^n$	$ve^{(1-n)\int Pdx} = (1-n)\int Qe^{(1-n)\int Pdx}\,dx + c$ where $v = y^{1-n}$. If $n = 1$, the solution is $\ln y = \int (Q - P)\,dx + c$
Homogeneous equation $\frac{dy}{dx} = F\left(\frac{y}{x}\right)$	$\ln x = \int \frac{dv}{F(v) - v} + c$ where $v = y/x$. If $F(v) = v$, the solution is $y = cx$
Reducible to homogeneous $(a_1x + b_1y + c_1)\,dx + (a_2x + b_2y + c_2)\,dy = 0$ $\frac{a_1}{a_2} \neq \frac{b_1}{b_2}$	Set $u = a_1x + b_1y + c_1$ $v = a_2x + b_2y + c_2$ Eliminate x and y and the equation becomes homogenous
Reducible to separable $(a_1x + b_1y + c_1)\,dx + (a_2x + b_2y + c_2)\,dy = 0$ $\frac{a_1}{a_2} = \frac{b_1}{b_2}$	Set $u = a_1x + b_1y$ Eliminate x or y and equation becomes separable

$y\,F(xy)\,dx + x\,G(xy)\,dy = 0$	$\ln x = \int \frac{G(v)\,dv}{v\{G(v) - F(v)\}} + c$ where $v = xy$. If $G(v) = F(v)$, the solution is $xy = c$.
Linear, homogeneous second order equation $\frac{d^2y}{dx^2} + b\frac{dy}{dx} + cy = 0$ b, c are real constants	Let m_1, m_2 be the roots of $m^2 + bm + c = 0$. Then there are 3 cases: Case 1. m_1, m_2 real and distinct: $y = c_1 e^{m_1 x} + c_2 e^{m_2 x}$ Case 2. m_1, m_2 real and equal: $y = c_1 e^{m_1 x} + c_2 x e^{m_1 x}$ Case 3. $m_1 = p + qi$, $m_2 = p - qi$: $y = e^{px}(c_1 \cos qx + c_2 \sin qx)$ where $p = -b/2$, $q = \sqrt{4c - b^2}/2$
Linear, nonhomogeneous second order equation $\frac{d^2y}{dx^2} + b\frac{dy}{dx} + cy = R(x)$ b, c are real constants	There are 3 cases corresponding to those immediately above: Case 1. $y = c_1 e^{m_1 x} + c_2 e^{m_2 x} + \frac{e^{m_1 x}}{m_1 - m_2}\int e^{-m_1 x} R(x)\,dx + \frac{e^{m_2 x}}{m_2 - m_1}\int e^{-m_2 x} R(x)\,dx$ Case 2. $y = c_1 e^{m_1 x} + c_2 x e^{m_1 x} + x e^{m_1 x}\int e^{-m_1 x} R(x)\,dx - e^{m_1 x}\int x e^{-m_1 x} R(x)\,dx$ Case 3. $y = e^{px}(c_1 \cos qx + c_2 \sin qx) + \frac{e^{px}\sin qx}{q}\int e^{-px} R(x)\cos qx\,dx - \frac{e^{px}\cos qx}{q}\int e^{-px} R(x)\sin qx\,dx$

DIFFERENTIAL EQUATIONS (Continued)

Euler or Cauchy equation $x^2 \frac{d^2y}{dx^2} + bx\frac{dy}{dx} + cy = S(x)$	Putting $x = e^t$, the equation becomes $\frac{d^2y}{dt^2} + (b-1)\frac{dy}{dt} + cy = S(e^t)$ and can then be solved as a linear second order equation.
Bessel's equation $x^2 \frac{d^2y}{dx^2} + x\frac{dy}{dx} + (\lambda^2 x^2 - n^2)y = 0$	$y = c_1 J_n(\lambda x) + c_2 Y_n(\lambda x)$
Transformed Bessel's equation $x^2 \frac{d^2y}{dx^2} + (2p+1)x\frac{dy}{dx} + (\alpha^2 x^{2r} + \beta^2)y = 0$	$y = x^{-p}\left\{c_1 J_{q/r}\left(\frac{\alpha}{r}x^r\right) + c_2 Y_{q/r}\left(\frac{\alpha}{r}x^r\right)\right\}$ where $q = \sqrt{p^2 - \beta^2}$.
Legendre's equation $(1-x^2)\frac{d^2y}{dx^2} - 2x\frac{dy}{dx} + n(n+1)y = 0$	$y = c_1 P_n(x) + c_2 Q_n(x)$

FOURIER SERIES

If $f(x)$ is a bounded periodic function of period 2L (i.e. $f(x + 2L) = f(x)$, and satisfies the *Dirichlet conditions:*

A. In any period $f(x)$ is continuous, except possibly for a finite number of jump discontinuities.
B. In any period $f(x)$ has only a finite number of maxima and minima.

Then $f(x)$ may be represented by the *Fourier series*

$$\frac{a_0}{2} + \sum_{n=1}^{\infty}\left(a_n \cos\frac{n\pi x}{L} + b_n \sin\frac{n\pi x}{L}\right)$$

where a_n and b_n are as determined below. This series will converge to $f(x)$ at every point where $f(x)$ is continuous, and to

$$\frac{f(x^+) + f(x^-)}{2}$$

(i.e. the average of the left-hand and right-hand limits) at every point where $f(x)$ has a jump discontinuity.

$$a_n = \frac{1}{L}\int_{-L}^{L} f(x)\cos\frac{n\pi x}{L}\,dx, \quad n = 0, 1, 2, 3, \ldots,$$

$$b_n = \frac{1}{L}\int_{-L}^{L} f(x)\sin\frac{n\pi x}{L}\,dx, \quad n = 1, 2, 3, \ldots$$

We may also write

$$a_n = \frac{1}{L}\int_{\alpha}^{\alpha+2L} f(x)\cos\frac{n\pi x}{L}\,dx \text{ and } b_n = \frac{1}{L}\int_{\alpha}^{\alpha+2L} f(x)\sin\frac{n\pi x}{L}\,dx$$

where α is any real number. Thus if $\alpha = 0$,

$$a_n = \frac{1}{L}\int_{0}^{2L} f(x)\cos\frac{n\pi x}{L}\,dx, \quad n = 0, 1, 2, 3, \ldots,$$

$$b_n = \frac{1}{L}\int_{0}^{2L} f(x)\sin\frac{n\pi x}{L}\,dx, \quad n = 1, 2, 3, \ldots$$

2. If in addition to the above restrictions, $f(x)$), is even (i.e., $f(-x) = f(x)$ the Fourier series reduces to

$$\frac{a_0}{2} + \sum_{n=1}^{\infty} a_n \cos \frac{n\pi x}{L}$$

That is, $b_n = 0$. In this case, a simpler formula for a_n is

$$a_n = \frac{2}{L} \int_0^L f(x) \cos \frac{n\pi x}{L} \, dx, \quad n = 0, 1, 2, 3, \ldots$$

3. If in addition to the restrictions in (1), $f(x)$ is an odd function (i.e., $f(-x) = -f(x)$), then the Fourier series reduces to

$$\sum_{n=1}^{\infty} b_n \sin \frac{n\pi x}{L}$$

That is, $a_n = 0$. In this case, a simpler formula for the b_n is

$$b_n = \frac{2}{L} \int_0^L f(x) \sin \frac{n\pi x}{L} \, dx, \quad n = 1, 2, 3, \ldots$$

4. If in addition to the restrictions in (2) above, $f(x) = -f(L - x)$, then a_n will be 0 for all even values of n, including $n = 0$. Thus in this case, the expansion reduces to

$$\sum_{m=1}^{\infty} a_{2m-1} \cos \frac{(2m - 1)\pi x}{L}$$

5. If in addition to the restrictions in (3) above, $f(x)$, $= f(L - x)$, then b_n will be 0 for all even values of n. Thus in this case, the expansion reduces to

$$\sum_{m=1}^{\infty} b_{2m-1} \sin \frac{(2m - 1)\pi x}{L}$$

(The series in (4) and (5) are known as *odd-harmonic series,* since only the odd harmonics appear. Similar rules may be stated for even-harmonic series, but when a series appears in the even-harmonic form, it means that $2L$ has not been taken as the smallest period of $f(x)$. Since any integral multiple of a period is also a period, series obtained in this way will also work, but in general computation is simplified if $2L$ is taken to be the smallest period.)

6. If we write the Euler definitions for cos θ and sin θ, we obtain the complex form of the Fourier Series known either as the "Complex Fourier Series" or the "Exponential Fourier Series" of $f(x)$. It is represented as

$$f(x) = \frac{1}{2} \sum_{n=-\infty}^{n=+\infty} c_n e^{i\omega_n x}$$

where

$$c_n = \frac{1}{L} \int_{-L}^{L} f(x) e^{-i\omega_n x} dx, \quad n = 0, \pm 1, \pm 2, \pm 3, \ldots$$

with $\omega_n = \dfrac{n\pi}{L}, \quad n = 0, \pm 1, \pm 2, \ldots$

The set of coefficients $\{c_n\}$ is often referred to as the Fourier spectrum.

7. If both sine and cosine terms are present and if $f(x)$ is of period $2L$ and expandable by a Fourier series, it can be represented as

$$f(x) = \frac{a_0}{2} + \sum_{n=1}^{\infty} c_n \sin\left(\frac{n\pi x}{L} + \phi_n\right), \text{ where } a_n = c_n \sin \phi_n,$$

$$b_n = c_n \cos \phi_n, \quad c_n = \sqrt{a_n^2 + b_n^2}, \quad \phi_n = \arctan\left(\frac{a_n}{b_n}\right)$$

It can also be represented as

$$f(x) = \frac{a_0}{2} + \sum_{n=1}^{\infty} c_n \cos\left(\frac{n\pi x}{L} + \phi_n\right), \text{ where } a_n = c_n \cos \phi_n,$$

$$b_n = -c_n \sin \phi_n, \quad c_n = \sqrt{a_n^2 + b_n^2}, \quad \phi_n = \arctan\left(-\frac{b_n}{a_n}\right)$$

where ϕ_n is chosen so as to make a_n, b_n, and c_n hold.

8. The following table of trigonometric identities should be helpful for developing Fourier Series.

	n	n even	n odd	$n/2$ odd	$n/2$ even
$\sin n\pi$	0	0	0	0	0
$\cos n\pi$	$(-1)^n$	$+1$	-1	$+1$	$+1$
*$\sin \frac{n\pi}{2}$		0	$(-1)^{(n-1)/2}$	0	0
*$\cos \frac{n\pi}{2}$		$(-1)^{n/2}$	0	-1	$+1$
$\sin \frac{n\pi}{4}$			$\frac{\sqrt{2}}{2}(-1)^{(n^2+4n+11)/8}$	$(-1)^{(n-2)/4}$	0

*A useful formula for $\sin \frac{n\pi}{2}$ and $\cos \frac{n\pi}{2}$ is given by

$$\sin \frac{n\pi}{2} = \frac{(i)^{n+1}}{2}[(-1)^n - 1] \text{ and } \cos \frac{n\pi}{2} = \frac{(i)^n}{2}[(-1)^n + 1], \text{ where } i^2 = -1.$$

AUXILIARY FORMULAS FOR FOURIER SERIES

$$1 = \frac{4}{\pi}\left[\sin \frac{\pi x}{k} + \frac{1}{3}\sin \frac{3\pi x}{k} + \frac{1}{5}\sin \frac{5\pi x}{k} + \cdots\right] \qquad [0 < x < k]$$

$$x = \frac{2k}{\pi}\left[\sin \frac{\pi x}{k} - \frac{1}{2}\sin \frac{2\pi x}{k} + \frac{1}{3}\sin \frac{3\pi x}{k} - \cdots\right] \qquad [-k < x < k]$$

$$x = \frac{k}{2} - \frac{4k}{\pi^2}\left[\cos \frac{\pi x}{k} + \frac{1}{3^2}\cos \frac{3\pi x}{k} + \frac{1}{5^2}\cos \frac{5\pi x}{k} + \cdots\right] \qquad [0 < x < k]$$

$$x^2 = \frac{2k^2}{\pi^3}\left[\left(\frac{\pi^2}{1} - \frac{4}{1}\right)\sin \frac{\pi x}{k} - \frac{\pi^2}{2}\sin \frac{2\pi x}{k} + \left(\frac{\pi^2}{3} - \frac{4}{3^3}\right)\sin \frac{3\pi x}{k} - \frac{\pi^2}{4}\sin \frac{4\pi x}{k} + \left(\frac{\pi^2}{5} - \frac{4}{5^3}\right)\sin \frac{5\pi x}{k} + \cdots\right] \qquad [0 < x < k]$$

$$x^2 = \frac{k^2}{3} - \frac{4k^2}{\pi^2}\left[\cos \frac{\pi x}{k} - \frac{1}{2^2}\cos \frac{2\pi x}{k} + \frac{1}{3^2}\cos \frac{3\pi x}{k} - \frac{1}{4^2}\cos \frac{4\pi x}{k} + \cdots\right] \qquad [-k < x < k]$$

$$1 - \frac{1}{3} + \frac{1}{5} - \frac{1}{7} + \cdots = \frac{\pi}{4}$$

$$1 + \frac{1}{2^2} + \frac{1}{3^2} + \frac{1}{4^2} + \cdots = \frac{\pi^2}{6}$$

$$1 - \frac{1}{2^2} + \frac{1}{3^2} - \frac{1}{4^2} + \cdots = \frac{\pi^2}{12}$$

$$1 + \frac{1}{3^2} + \frac{1}{5^2} + \frac{1}{7^2} + \cdots = \frac{\pi^2}{8}$$

$$\frac{1}{2^2} + \frac{1}{4^2} + \frac{1}{6^2} + \frac{1}{8^2} + \cdots = \frac{\pi^2}{24}$$

FOURIER EXPANSIONS FOR BASIC PERIODIC FUNCTIONS

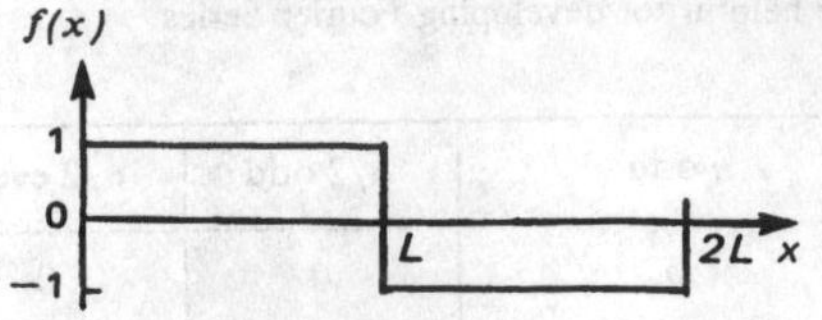

$$f(x) = \frac{4}{\pi} \sum_{n=1,3,5\ldots} \frac{1}{n} \sin \frac{n\pi x}{L}$$

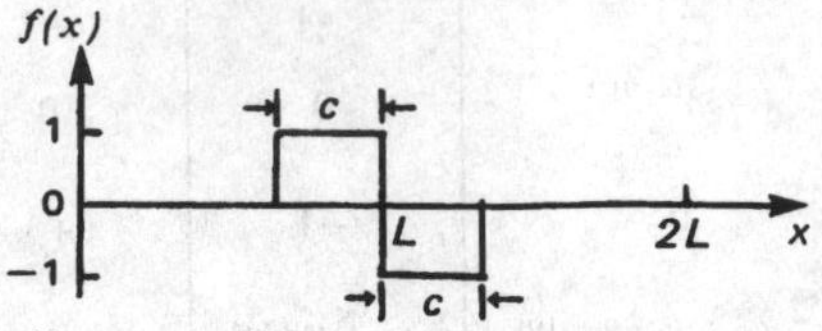

$$f(x) = \frac{2}{\pi} \sum_{n=1}^{\infty} \frac{(-1)^n}{n} \left(\cos \frac{n\pi c}{L} - 1\right) \sin \frac{n\pi x}{L}$$

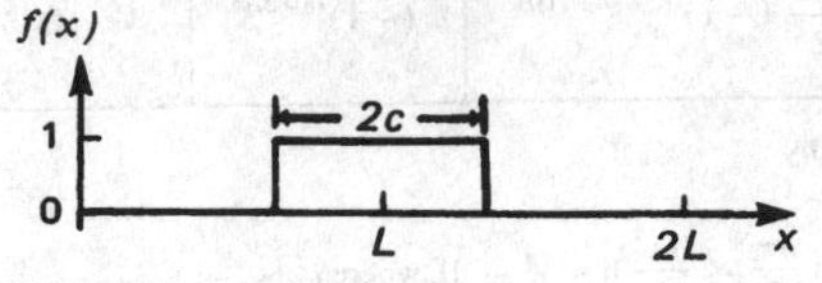

$$f(x) = \frac{c}{L} + \frac{2}{\pi} \sum_{n=1}^{\infty} \frac{(-1)^n}{n} \sin \frac{n\pi c}{L} \cos \frac{n\pi x}{L}$$

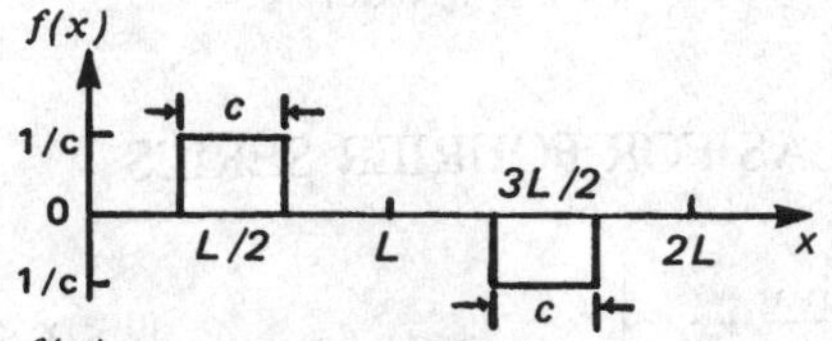

$$f(x) = \frac{2}{L} \sum_{n=1}^{\infty} \sin \frac{n\pi}{2} \, \frac{\sin\left(\frac{1}{2} n\pi c/L\right)}{\frac{1}{2} n\pi c/L} \sin \frac{n\pi x}{L}$$

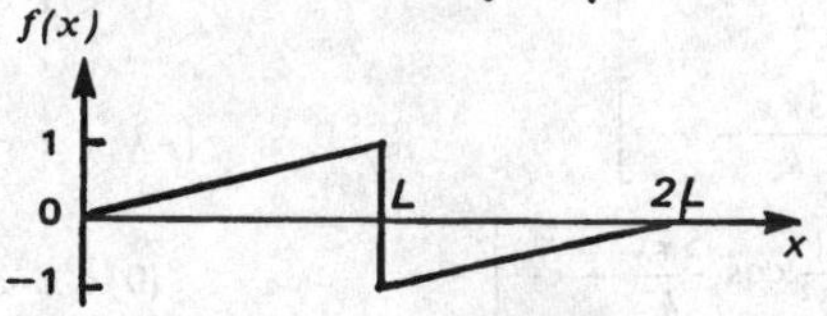

$$f(x) = \frac{2}{\pi} \sum_{n=1}^{\infty} \frac{(-1)^{n+1}}{n} \sin \frac{n\pi x}{L}$$

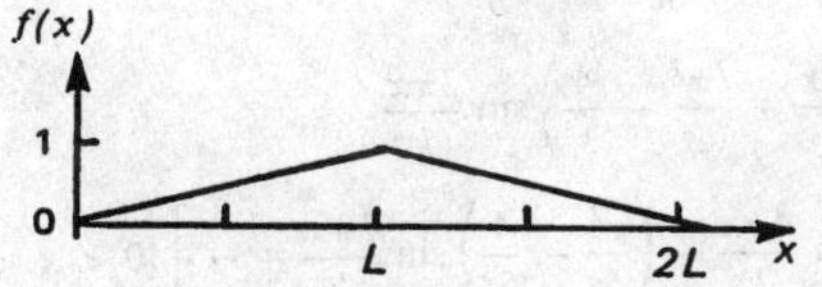

$$f(x) = \frac{1}{2} - \frac{4}{\pi^2} \sum_{n=1,3,5\ldots} \frac{1}{n^2} \cos \frac{n\pi x}{L}$$

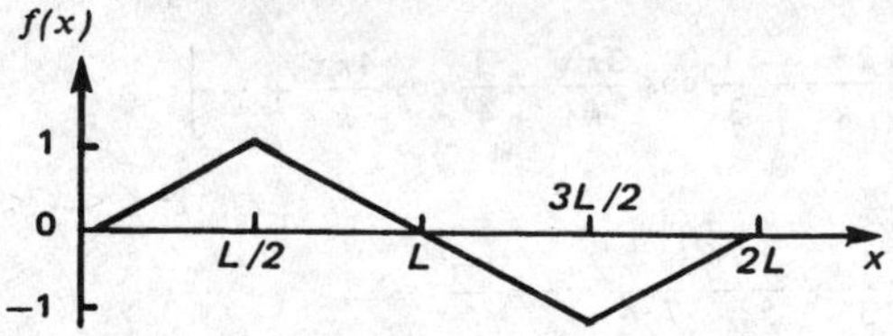

$$f(x) = \frac{8}{\pi^2} \sum_{n=1,3,5\ldots} \frac{(-1)^{(n-1)/2}}{n^2} \sin \frac{n\pi x}{L}$$

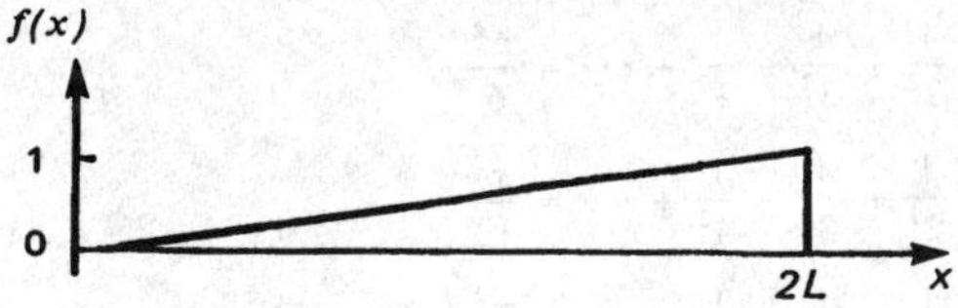

$$f(x) = \frac{1}{2} - \frac{1}{\pi} \sum_{n=1}^{\infty} \frac{1}{n} \sin \frac{n\pi x}{L}$$

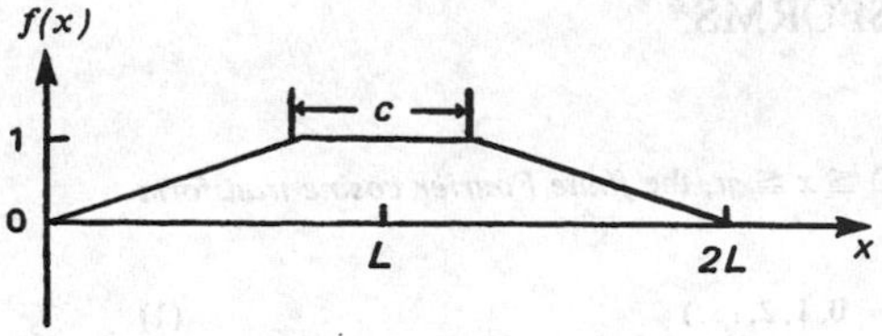

$$f(x) = \frac{1}{2}(1 + a) + \frac{2}{\pi^2(1-a)} \sum_{n=1}^{\infty} \frac{1}{n^2} [(-1)^n \cos n\pi a - 1] \cos\frac{n\pi x}{L}; \left(a = \frac{c}{2L}\right)$$

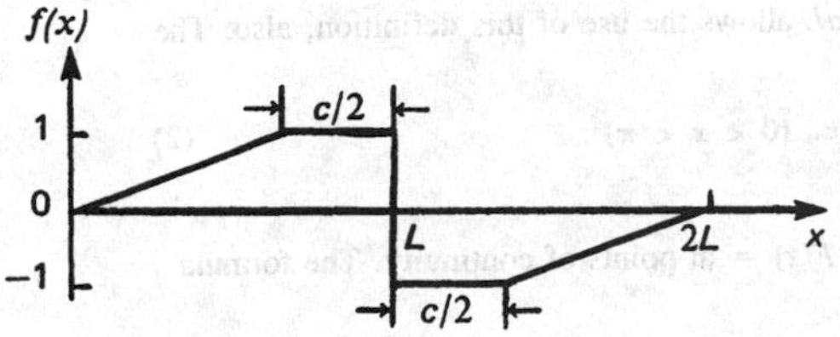

$$f(x) = \frac{2}{\pi} \sum_{n=1}^{\infty} \frac{(-1)^{n-1}}{n} \left[1 + \frac{\sin n\pi a}{n\pi(1-a)}\right] \sin\frac{n\pi x}{L}; \left(a = \frac{c}{2L}\right)$$

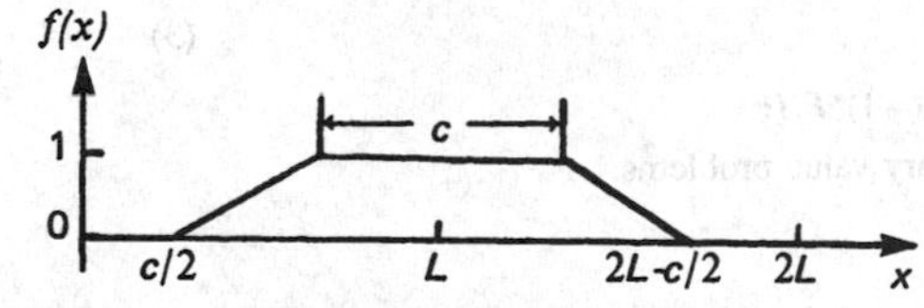

$$f(x) = \frac{1}{2} - \frac{4}{\pi^2(1-2a)} \sum_{n=1,3,5\ldots} \frac{1}{n^2} \cos n\pi a \cos\frac{n\pi x}{L}; \left(a = \frac{c}{2L}\right)$$

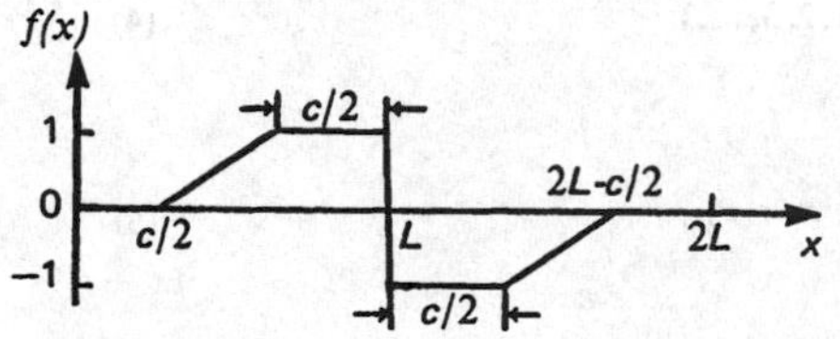

$$f(x) = \frac{2}{\pi} \sum_{n=1}^{\infty} \frac{(-1)^n}{n} \left[1 + \frac{1+(-1)^n}{n\pi(1-2a)} \sin n\pi a\right] \sin\frac{n\pi x}{L}; \left(a = \frac{c}{2L}\right)$$

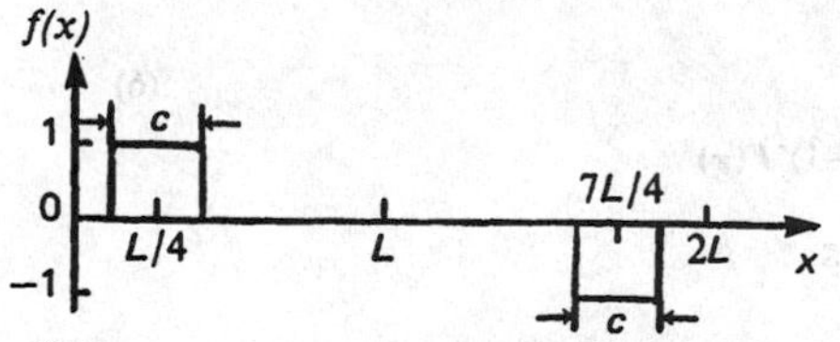

$$f(x) = \frac{4}{\pi} \sum_{n=1}^{\infty} \frac{1}{n} \sin\frac{n\pi}{4} \sin n\pi a \sin\frac{n\pi x}{L}; \left(a = \frac{c}{2L}\right)$$

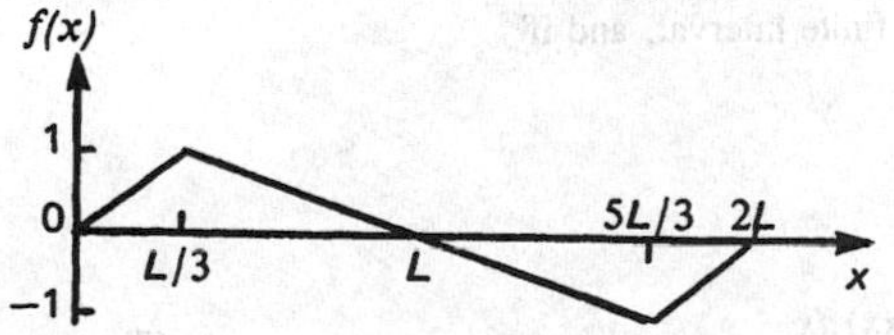

$$f(x) = \frac{9}{\pi^2} \sum_{n=1}^{\infty} \frac{1}{n^2} \sin\frac{n\pi}{3} \sin\frac{n\pi x}{L}; \left(a = \frac{c}{2L}\right)$$

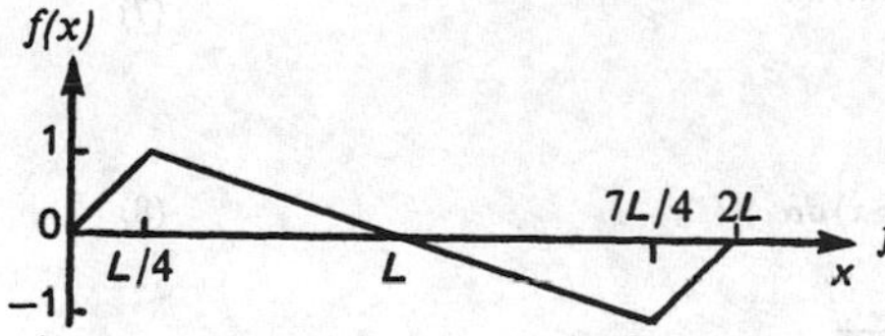

$$f(x) = \frac{32}{3\pi^2} \sum_{n=1}^{\infty} \frac{1}{n^2} \sin\frac{n\pi}{4} \sin\frac{n\pi x}{L}; \left(a = \frac{c}{2L}\right)$$

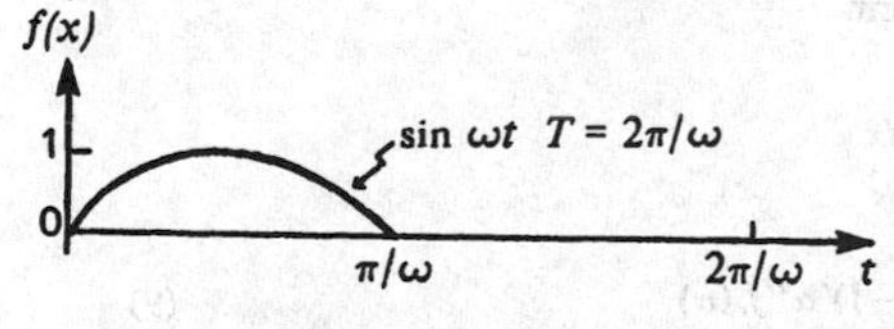

$$f(x) = \frac{1}{\pi} + \frac{1}{2} \sin\omega t - \frac{2}{\pi} \sum_{n=2,4,6\ldots} \frac{1}{n^2-1} \cos n\omega t$$

Extracted from graphs and formulas, pages 372, 373, Differential Equations in Engineering Problems, Salvadori and Schwarz, published by Prentice-Hall, Inc., 1954.

THE FOURIER TRANSFORMS*

R. E. Gaskell

For a piecewise continuous function $F(x)$ over a finite interval $0 \leqq x \leqq \pi$, the *finite Fourier cosine transform* of $F(x)$ is

$$f_c(n) = \int_0^\pi F(x) \cos nx\, dx \quad (n = 0, 1, 2, \ldots) \tag{1}$$

If x ranges over the interval $0 \leqq x \leqq L$, the substitution $x' = \pi x/L$ allows the use of this definition, also. The inverse transform is written.

$$\bar{F}(x) = \frac{1}{\pi} f_c(0) - \frac{2}{\pi} \sum_{n=1}^{\infty} f_c(n) \cos nx \quad (0 < x < \pi) \tag{2}$$

where $F(x) = \dfrac{[F(x + o) + F(x - o)]}{2}$. We observe that $F(x) = F(x) =$ at points of continuity. The formula

$$\begin{aligned} f_c^{(2)}(n) &= \int_0^\pi F''(x) \cos nx\, dx \\ &= -n^2 f_c(n) - F'(0) + (-1)^n F'(\pi) \end{aligned} \tag{3}$$

makes the finite Fourier cosine transform useful in certain boundary value problems.

Analogously, the *finite Fourier sine transform* of $F(x)$ is

$$f_s(n) = \int_0^\pi F(x) \sin nx\, dx \quad (n = 1, 2, 3, \ldots) \tag{4}$$

and

$$\bar{F}(x) = \frac{2}{\pi} \sum_{n=1}^{\infty} f_s(n) \sin nx \quad (0 < x < \pi) \tag{5}$$

Corresponding to (3) we have

$$\begin{aligned} f_s^{(2)}(n) &= \int_0^\pi F''(x) \sin nx\, dx \\ &= -n^2 f_s(n) - nF(0) - n(-1)^n F(\pi) \end{aligned} \tag{6}$$

Fourier Transforms

If $F(x)$ is defined for $x \geqq 0$ and is piecewise continuous over any finite interval, and if

$$\int_0^\infty F(x)\, dx$$

is absolutely convergent, then

$$f_c(\alpha) = \sqrt{\frac{2}{\pi}} \int_0^\infty F(x) \cos(\alpha x)\, dx \tag{7}$$

is the *Fourier cosine transform* of $F(x)$. Furthermore.

$$\bar{F}(x) = \sqrt{\frac{2}{\pi}} \int_0^\infty f_c(\alpha) \cos(\alpha x)\, d\alpha \tag{8}$$

If $\lim_{x\to\infty} \dfrac{d^n F}{dx^n} = 0$, an important property of the Fourier cosine transform

$$\begin{aligned} f_c^{(2r)}(\alpha) &= \sqrt{\frac{2}{\pi}} \int_0^\infty \left(\frac{d^{2r} F}{dx^{2r}}\right) \cos(\alpha x)\, dx \\ &= -\sqrt{\frac{2}{\pi}} \sum_{n=0}^{r-1} (-1)^n a_{2r-2n-1} \alpha^{2n} + (-1)^r \alpha^{2r} f_c(\alpha) \end{aligned} \tag{9}$$

where $\lim_{x\to\infty} \dfrac{d^r F}{dx^r} = a_r$, makes it useful in the solution of many problems.

Under the same conditions.

$$f_s(\alpha) = \sqrt{\frac{2}{\pi}} \int_0^\infty F(x) \sin(\alpha x)\, dx \tag{10}$$

* From Beyer, W. H., Ed., *CRC Handbook of Mathematical Sciences*, 5th ed., CRC Press, Boca Raton, 1978, 592—598. With permission.

defines the *Fourier sine transform* of $F(x)$, and

$$\bar{F}(x) = \sqrt{\frac{2}{\pi}} \int_0^\infty f_s(\alpha) \sin(\alpha x)\, d\alpha \tag{11}$$

Corresponding to (9) we have

$$f_s^{(2r)}(\alpha) = \sqrt{\frac{2}{\pi}} \int_0^\infty \frac{d^{2r}F}{dx^{2r}} \sin(\alpha x)\, dx$$

$$= -\sqrt{\frac{2}{\pi}} \sum_{n=1}^{r} (-1)^n \alpha^{2n-1} a_{2r-2n} + (-1)^{r-1} \alpha^{2r} f_s(\alpha) \tag{12}$$

Similarly, if $F(x)$ is defined for $-\infty < x < \infty$, and if $\int_{-\infty}^{\infty} F(x)\, dx$ is absolutely convergent, then

$$f(\alpha) = \frac{1}{\sqrt{2\pi}} \int_{-\infty}^{\infty} F(x) e^{i\alpha x}\, dx \tag{13}$$

is the *Fourier transform* of $F(x)$, and

$$\bar{F}(x) = \frac{1}{\sqrt{2\pi}} \int_{-\infty}^{\infty} f(\alpha) e^{-i\alpha x}\, d\alpha \tag{14}$$

Also, if

$$\lim_{|x| \to \infty} \left| \frac{d^n F}{dx^n} \right| = 0 \quad (n = 1, 2, \ldots, r-1)$$

then

$$f^{(r)}(\alpha) = \frac{1}{\sqrt{2\pi}} \int_{-\infty}^{\infty} F^{(r)}(x) e^{i\alpha x}\, dx = (-i\alpha)^r f(\alpha) \tag{15}$$

Finite Sine Transforms

	$f_s(n)$	$F(x)$
1	$f_s(n) = \int_0^\pi F(x) \sin nx\, dx \; (n = 1, 2, \cdots)$	$F(x)$
2	$(-1)^{n+1} f_s(n)$	$F(\pi - x)$
3	$\frac{1}{n}$	$\frac{\pi - x}{\pi}$
4	$\frac{(-1)^{n+1}}{n}$	$\frac{x}{\pi}$
5	$\frac{1 - (-1)^n}{n}$	1
6	$\frac{2}{n^2} \sin \frac{n\pi}{2}$	$\begin{cases} x & \text{when } 0 < x < \pi/2 \\ \pi - x & \text{when } \pi/2 < x < \pi \end{cases}$
7	$\frac{(-1)^{n+1}}{n^3}$	$\frac{x(\pi^2 - x^2)}{6\pi}$
8	$\frac{1 - (-1)^n}{n^3}$	$\frac{x(\pi - x)}{2}$
9	$\frac{\pi^2(-1)^{n-1}}{n} - \frac{2[1 - (-1)^n]}{n^3}$	x^2
10	$\pi(-1)^n \left(\frac{6}{n^3} - \frac{\pi^2}{n}\right)$	x^3
11	$\frac{n}{n^2 + c^2}[1 - (-1)^n e^{c\pi}]$	e^{cx}
12	$\frac{n}{n^2 + c^2}$	$\frac{\sinh c(\pi - x)}{\sinh c\pi}$
13	$\frac{n}{n^2 - k^2} \; (k \neq 0, 1, 2, \cdots)$	$\frac{\sin k(\pi - x)}{\sin k\pi}$
14	$\begin{cases} \frac{\pi}{2} & \text{when } n = m \\ 0 & \text{when } n \neq m \end{cases} \quad (m = 1, 2, \cdots)$	$\sin mx$

	$f_s(n)$	$F(x)$
15	$\frac{n}{n^2 - k^2}[1 - (-1)^n \cos k\pi]$ $(k \neq 1, 2, \cdots)$	$\cos kx$
16	$\begin{cases} \frac{n}{n^2 - m^2}[1 - (-1)^{n+m}] & \text{when } n \neq m = 1, 2, \cdots \\ 0 & \text{when } n = m \end{cases}$	$\cos mx$
17	$\frac{n}{(n^2 - k^2)^2}$ $(k \neq 0, 1, 2, \cdots)$	$\frac{\pi \sin kx}{2k \sin^2 k\pi} - \frac{x \cos k(\pi - x)}{2k \sin k\pi}$
18	$\frac{b^n}{n}$ $(\|b\| \leqq 1)$	$\frac{2}{\pi} \arctan \frac{b \sin x}{1 - b \cos x}$
19	$\frac{1 - (-1)^n}{n} b^n$ $(\|b\| \leq 1)$	$\frac{2}{\pi} \arctan \frac{2b \sin x}{1 - b^2}$

Finite Cosine Transforms

	$f_c(n)$	$F(x)$
1	$f_c(n) = \int_0^\pi F(x) \cos nx\, dx$ $(n = 0, 1, 2, \cdots)$	$F(x)$
2	$(-1)^n f_c(n)$	$F(\pi - x)$
3	0 when $n = 1, 2, \cdots$; $f_c(0) = \pi$	1
4	$\frac{2}{n} \sin \frac{n\pi}{2}$; $f_c(0) = 0$	$\begin{cases} 1 & \text{when } 0 < x < \pi/2 \\ -1 & \text{when } \pi/2 < x < \pi \end{cases}$
5	$-\frac{1 - (-1)^n}{n^2}$; $f_c(0) = \frac{\pi^2}{2}$	x
6	$\frac{(-1)^n}{n^2}$; $f_c(0) = \frac{\pi^2}{6}$	$\frac{x^2}{2\pi}$
7	$\frac{1}{n^2}$; $f_c(0) = 0$	$\frac{(\pi - x)^2}{2\pi} - \frac{\pi}{6}$
8	$3\pi^2 \frac{(-1)^n}{n^2} - 6 \frac{1 - (-1)^n}{n^4}$; $f_c(0) = \frac{\pi^4}{4}$	x^3
9	$\frac{(-1)^n e^c \pi - 1}{n^2 + c^2}$	$\frac{1}{c} e^{cx}$
10	$\frac{1}{n^2 + c^2}$	$\frac{\cosh c(\pi - x)}{c \sinh c\pi}$
11	$\frac{k}{n^2 - k^2}[(-1)^n \cos \pi k - 1]$ $(k \neq 0, 1, 2, \cdots)$	$\sin kx$
12	$\frac{(-1)^{n+m} - 1}{n^2 - m^2}$; $f_c(m) = 0$ $(m = 1, 2, \cdots)$	$\frac{1}{m} \sin mx$
13	$\frac{1}{n^2 - k^2}$ $(k \neq 0, 1, 2, \cdots)$	$-\frac{\cos k(\pi - x)}{k \sin k\pi}$
14	0 when $n = 1, 2, \cdots$; $f_c(m) = \frac{\pi}{2}$ $(m = 1, 2, \cdots)$	$\cos mx$

Fourier Sine Transforms*

	$F(x)$	$f_s(\alpha)$
1	$\begin{cases} 1 & (0 < x < a) \\ 0 & (x > a) \end{cases}$	$\sqrt{\frac{2}{\pi}}\left[\frac{1 - \cos \alpha}{\alpha}\right]$
2	x^{p-1} $(0 < p < 1)$	$\sqrt{\frac{2}{\pi}} \frac{\Gamma(p)}{\alpha^p} \sin \frac{p\pi}{2}$
3	$\begin{cases} \sin x & (0 < x < a) \\ 0 & (x > a) \end{cases}$	$\frac{1}{\sqrt{2\pi}}\left[\frac{\sin[a(1 - \alpha)]}{1 - \alpha} - \frac{\sin[a(1 + \alpha)]}{1 + \alpha}\right]$
4	e^{-x}	$\sqrt{\frac{2}{\pi}}\left[\frac{\alpha}{1 + \alpha^2}\right]$
5	$xe^{-x^2/2}$	$\alpha e^{-\alpha^2/2}$
6	$\cos \frac{x^2}{2}$	$\sqrt{2}\left[\sin \frac{\alpha^2}{2} C\left(\frac{\alpha^2}{2}\right) - \cos \frac{\alpha^2}{2} S\left(\frac{\alpha^2}{2}\right)\right]^*$

	$F(x)$	$f_s(\alpha)$
7	$\sin \frac{x^2}{2}$	$\sqrt{2}\left[\cos \frac{\alpha^2}{2}\, C\left(\frac{\alpha^2}{2}\right) + \sin \frac{\alpha^2}{2}\, S\left(\frac{\alpha^2}{2}\right)\right]^*$

* $C(y)$ and $S(y)$ are the Fresnel integrals

$$C(y) = \frac{1}{\sqrt{2\pi}} \int_0^y \frac{1}{\sqrt{t}} \cos t \, dt,$$

$$S(y) = \frac{1}{\sqrt{2\pi}} \int_0^y \frac{1}{\sqrt{t}} \sin t \, dt$$

* More extensive tables of the Fourier sine and cosine transforms can be found in Fritz Oberhettinger, *Tabellen zur-Fourier Transformation*, Springer, 1957.

Fourier Cosine Transforms

	$F(x)$	$f_c(\alpha)$
1	$\begin{cases} 1 & (0 < x < a) \\ 0 & (x > a) \end{cases}$	$\sqrt{\frac{2}{\pi}}\frac{\sin a\alpha}{\alpha}$
2	$x^{p-1} \quad (0 < p < 1)$	$\sqrt{\frac{2}{\pi}}\frac{\Gamma(p)}{\alpha^p}\cos\frac{p\pi}{2}$
3	$\begin{cases} \cos x & (0 < x < a) \\ 0 & (x > a) \end{cases}$	$\frac{1}{\sqrt{2\pi}}\left[\frac{\sin[a(1-\alpha)]}{1-\alpha} + \frac{\sin[a(1+\alpha)]}{1+\alpha}\right]$
4	e^{-x}	$\sqrt{\frac{2}{\pi}}\left(\frac{1}{1+\alpha^2}\right)$
5	$e^{-x^2/2}$	$e^{-\alpha^2/2}$
6	$\cos\frac{x^2}{2}$	$\cos\left(\frac{\alpha^2}{2} - \frac{\pi}{4}\right)$
7	$\sin\frac{x^2}{2}$	$\cos\left(\frac{\alpha^2}{2} + \frac{\pi}{4}\right)$

Fourier Transforms*

	$F(x)$	$f(\alpha)$
1	$\frac{\sin ax}{x}$	$\begin{cases} \sqrt{\frac{\pi}{2}} & \lvert\alpha\rvert < a \\ 0 & \lvert\alpha\rvert > a \end{cases}$
2	$\begin{cases} e^{iwx} & (p < x < q) \\ 0 & (x < p,\ x > q) \end{cases}$	$\frac{i}{\sqrt{2\pi}}\frac{e^{ip(w+\alpha)} - e^{iq(w+\alpha)}}{(w+\alpha)}$
3	$\begin{cases} e^{-cx+iwx} & (x > 0) \\ 0 & (x < 0) \end{cases} \quad (c > 0)$	$\frac{i}{\sqrt{2\pi}(w+\alpha+ic)}$
4	$e^{-px^2} \quad R(p) > 0$	$\frac{1}{\sqrt{2p}}e^{-\alpha^2/4p}$
5	$\cos px^2$	$\frac{1}{\sqrt{2p}}\cos\left[\frac{\alpha^2}{4p} - \frac{\pi}{4}\right]$
6	$\sin px^2$	$\frac{1}{\sqrt{2p}}\cos\left[\frac{\alpha^2}{4p} + \frac{\pi}{4}\right]$
7	$\lvert x\rvert^{-p} \quad (0 < p < 1)$	$\sqrt{\frac{2}{\pi}}\frac{\Gamma(1-p)\sin\frac{p\pi}{2}}{\lvert\alpha\rvert^{(1-p)}}$
8	$\frac{e^{-a\lvert x\rvert}}{\sqrt{\lvert x\rvert}}$	$\frac{\sqrt{\sqrt{(a^2+\alpha^2)}+a}}{\sqrt{a^2+\alpha^2}}$
9	$\frac{\cosh ax}{\cosh \pi x} \quad (-\pi < a < \pi)$	$\sqrt{\frac{2}{\pi}}\frac{\cos\frac{a}{2}\cosh\frac{\alpha}{2}}{\cosh\alpha + \cos a}$
10	$\frac{\sinh ax}{\sinh \pi x} \quad (-\pi < a < \pi)$	$\frac{1}{\sqrt{2\pi}}\frac{\sin a}{\cosh\alpha + \cos a}$
11	$\begin{cases} \frac{1}{\sqrt{a^2 - x^2}} & (\lvert x\rvert < a) \\ 0 & (\lvert x\rvert > a) \end{cases}$	$\sqrt{\frac{\pi}{2}}J_0(a\alpha)$

	$F(x)$	$f(\alpha)$
12	$\dfrac{\sin[b\sqrt{a^2+x^2}]}{\sqrt{a^2+x^2}}$	$\begin{cases} 0 & (\lvert\alpha\rvert > b) \\ \sqrt{\frac{\pi}{2}}\, J_0(a\sqrt{b^2-\alpha^2}) & (\lvert\alpha\rvert < b) \end{cases}$
13	$\begin{cases} P_n(x) & (\lvert x\rvert < 1) \\ 0 & (\lvert x\rvert > 1) \end{cases}$	$\dfrac{i^n}{\sqrt{\alpha}} J_{n+\frac{1}{2}}(\alpha)$
14	$\begin{cases} \dfrac{\cos[b\sqrt{a^2-x^2}]}{\sqrt{a^2-x^2}} & (\lvert x\rvert < a) \\ 0 & (\lvert x\rvert > a) \end{cases}$	$\sqrt{\frac{\pi}{2}}\, J_0(a\sqrt{a^2+b^2})$
15	$\begin{cases} \dfrac{\cosh[b\sqrt{a^2-x^2}]}{\sqrt{a^2-x^2}} & (\lvert x\rvert < a) \\ 0 & (\lvert x\rvert > a) \end{cases}$	$\sqrt{\frac{\pi}{2}}\, J_0(a\sqrt{\alpha^2-b^2})$

* More extensive tables of Fourier transforms can be found in W. Magnus and F. Oberhettinger, *Formulas and Theorems of the Special Functions of Mathematical Physics*. Chelsea, 1949, 116—120.

The following functions appear among the entries of the tables on transforms.

Function	Definition	Name
$Ei(x)$	$\int_{-\infty}^{x} \frac{e^v}{v}\,dv$; or sometimes defined as $-Ei(-x) = \int_x^{\infty} \frac{e^{-v}}{v}\,dv$	Sine, Cosine, and Exponential Integral tables pages 548–556
$Si(x)$	$\int_0^x \frac{\sin v}{v}\,dv$	Sine, Cosine, and Exponential Integral tables pages 548–556
$Ci(x)$	$\int_{\infty}^{x} \frac{\cos v}{v}\,dv$; or sometimes defined as negative of this integral	Sine, Cosine, and Exponential Integral tables pages 548–546
$erf(x)$	$\frac{2}{\sqrt{\pi}} \int_0^x e^{-v^2}\,dv$	Error function
$erfc(x)$	$1 - erf(x) = \frac{2}{\sqrt{\pi}} \int_x^{\infty} e^{-v^2}\,dv$	Complementary function to error function
$L_n(x)$	$\frac{e^x}{n!} \frac{d^n}{dx^n}(x^n e^{-x})$, $n = 0, 1, \cdots$	Laguerre polynomial of degree n

SERIES EXPANSION

The expression in parentheses following certain of the series indicates the region of convergence. If not otherwise indicated it is to be understood that the series converges for all finite values of x.

BINOMIAL

$$(x+y)^n = x^n + nx^{n-1}y + \frac{n(n-1)}{2!}x^{n-2}y^2 + \frac{n(n-1)(n-2)}{3!}x^{n-3}y^3 + \cdots \quad (y^2 < x^2)$$

$$(1 \pm x)^n = 1 \pm nx + \frac{n(n-1)x^2}{2!} \pm \frac{n(n-1)(n-2)x^3}{3!} + \cdots \quad (x^2 < 1)$$

$$(1 \pm x)^{-n} = 1 \mp nx + \frac{n(n+1)x^2}{2!} \mp \frac{n(n+1)(n+2)x^3}{3!} + \cdots \quad (x^2 < 1)$$

$$(1 \pm x)^{-1} = 1 \mp x + x^2 \mp x^3 + x^4 \mp x^5 + \cdots \quad (x^2 < 1)$$

$$(1 \pm x)^{-2} = 1 \mp 2x + 3x^2 \mp 4x^3 + 5x^4 \mp 6x^5 + \cdots \quad (x^2 < 1)$$

REVERSION OF SERIES

Let a series be represented by

$$y = a_1x + a_2x^2 + a_3x^3 + a_4x^4 + a_5x^5 + a_6x^6 + \cdots \quad (a_1 \neq 0)$$

to find the coefficients of the series

$$x = A_1y + A_2y^2 + A_3y^3 + A_4y^4 + \cdots$$

$$A_1 = \frac{1}{a_1} \qquad A_2 = -\frac{a_2}{a_1^3} \qquad A_3 = \frac{1}{a_1^5}(2a_2^2 - a_1a_3)$$

$$A_4 = \frac{1}{a_1^7}(5a_1a_2a_3 - a_1^2a_4 - 5a_2^3)$$

$$A_5 = \frac{1}{a_1^9}(6a_1^2a_2a_4 + 3a_1^2a_3^2 + 14a_2^4 - a_1^3a_5 - 21a_1a_2^2a_3)$$

$$A_6 = \frac{1}{a_1^{11}}(7a_1^3a_2a_5 + 7a_1^3a_3a_4 + 84a_1a_2^3a_3 - a_1^4a_6 - 28a_1^2a_2^2a_4 - 28a_1^2a_2a_3^2 - 42a_2^5)$$

$$A_7 = \frac{1}{a_1^{13}}(8a_1^4a_2a_6 + 8a_1^4a_3a_5 + 4a_1^4a_4^2 + 120a_1^2a_2^3a_4 + 180a_1^2a_2^2a_3^2 + 132a_2^6 - a_1^5a_7 - 36a_1^3a_2^2a_5 - 72a_1^3a_2a_3a_4 - 12a_1^3a_3^3 - 330a_1a_2^4a_3)$$

TAYLOR

1. $$f(x) = f(a) + (x-a)f'(a) + \frac{(x-a)^2}{2!}f''(a) + \frac{(x-a)^3}{3!}f'''(a) + \cdots + \frac{(x-a)^n}{n!}f^{(n)}(a) + \cdots \text{ (Taylor's Series)}$$

(Increment form)

2. $$f(x+h) = f(x) + hf'(x) + \frac{h^2}{2!}f''(x) + \frac{h^3}{3!}f'''(x) + \cdots$$
$$= f(h) + xf'(h) + \frac{x^2}{2!}f''(h) + \frac{x^3}{3!}f'''(h) + \cdots$$

3. If $f(x)$ is a function possessing derivatives of all orders throughout the interval $a \leqq x \leqq b$, then there is a value X, with $a < X < b$, such that

$$f(b) = f(a) + (b-a)f'(a) + \frac{(b-a)^2}{2!}f''(a) + \cdots + \frac{(b-a)^{n-1}}{(n-1)!}f^{(n-1)}(a) + \frac{(b-a)^n}{n!}f^{(n)}(X)$$

$$f(a+h) = f(a) + hf'(a) + \frac{h^2}{2!}f''(a) + \cdots + \frac{h^{n-1}}{(n-1)!}f^{(n-1)}(a) + \frac{h^n}{n!}f^{(n)}(a + \theta h), \quad b = a + h,\ 0 < \theta < 1.$$

or

$$f(x) = f(a) + (x-a)f'(a) + \frac{(x-a)^2}{2!}f''(a) + \cdots + (x-a)^{n-1}\frac{f^{(n-1)}(a)}{(n-1)!} + R_n,$$

where

$$R_n = \frac{f^{(n)}[a + \theta \cdot (x-a)]}{n!}(x-a)^n, \quad 0 < \theta < 1.$$

The above forms are known as Taylor's series with the remainder term.

4. *Taylor's series for a function of two variables*

If $\left(h\frac{\partial}{\partial x} + k\frac{\partial}{\partial y}\right)f(x,y) = h\frac{\partial f(x,y)}{\partial x} + k\frac{\partial f(x,y)}{\partial y}$;

$$\left(h\frac{\partial}{\partial x} + k\frac{\partial}{\partial y}\right)^2 f(x,y) = h^2\frac{\partial^2 f(x,y)}{\partial x^2} + 2hk\frac{\partial^2 f(x,y)}{\partial x\,\partial y} + k^2\frac{\partial^2 f(x,y)}{\partial y^2}$$

etc., and if $\left(h\frac{\partial}{\partial x} + k\frac{\partial}{\partial y}\right)^n f(x,y)\Big|_{\substack{x=a\\y=b}}$ with the bar and subscripts means that after differentiation we are to replace x by a and y by b,

$$f(a+h, b+k) = f(a,b) + \left(h\frac{\partial}{\partial x} + k\frac{\partial}{\partial y}\right)f(x,y)\Big|_{\substack{x=a\\y=b}} + \cdots + \frac{1}{n!}\left(h\frac{\partial}{\partial x} + k\frac{\partial}{\partial y}\right)^n f(x,y)\Big|_{\substack{x=a\\y=b}} + \cdots$$

MACLAURIN

$$f(x) = f(0) + xf'(0) + \frac{x^2}{2!}f''(0) + \frac{x^3}{3!}f'''(0) + \cdots + x^{n-1}\frac{f^{(n-1)}(0)}{(n-1)!} + R_n,$$

where

$$R_n = \frac{x^n f^{(n)}(\theta x)}{n!}, \quad 0 < \theta < 1.$$

EXPONENTIAL

$$e = 1 + \frac{1}{1!} + \frac{1}{2!} + \frac{1}{3!} + \frac{1}{4!} + \cdots$$

$$e^x = 1 + x + \frac{x^2}{2!} + \frac{x^3}{3!} + \frac{x^4}{4!} + \cdots \qquad \text{(all real values of } x\text{)}$$

$$a^x = 1 + x\log_e a + \frac{(x\log_e a)^2}{2!} + \frac{(x\log_e a)^3}{3!} + \cdots$$

$$e^x = e^a\left[1 + (x-a) + \frac{(x-a)^2}{2!} + \frac{(x-a)^3}{3!} + \cdots\right]$$

LOGARITHMIC

$$\log_e x = \frac{x-1}{x} + \frac{1}{2}\left(\frac{x-1}{x}\right)^2 + \frac{1}{3}\left(\frac{x-1}{x}\right)^3 + \cdots \qquad (x > \tfrac{1}{2})$$

$$\log_e x = (x-1) - \tfrac{1}{2}(x-1)^2 + \tfrac{1}{3}(x-1)^3 - \cdots \qquad (2 \ge x > 0)$$

$$\log_e x = 2\left[\frac{x-1}{x+1} + \frac{1}{3}\left(\frac{x-1}{x+1}\right)^3 + \frac{1}{5}\left(\frac{x-1}{x+1}\right)^5 + \cdots\right] \qquad (x > 0)$$

$$\log_e(1+x) = x - \tfrac{1}{2}x^2 + \tfrac{1}{3}x^3 - \tfrac{1}{4}x^4 + \cdots \qquad (-1 < x \le 1)$$

$$\log_e(n+1) - \log_e(n-1) = 2\left[\frac{1}{n} + \frac{1}{3n^3} + \frac{1}{5n^5} + \cdots\right]$$

$$\log_e(a+x) = \log_e a + 2\left[\frac{x}{2a+x} + \frac{1}{3}\left(\frac{x}{2a+x}\right)^3 + \frac{1}{5}\left(\frac{x}{2a+x}\right)^5 + \cdots\right]$$

$$(a > 0,\ -a < x < +\infty)$$

$$\log_e\frac{1+x}{1-x} = 2\left[x + \frac{x^3}{3} + \frac{x^5}{5} + \cdots + \frac{x^{2n-1}}{2n-1} + \cdots\right], \qquad -1 < x < 1$$

$$\log_e x = \log_e a + \frac{(x-a)}{a} - \frac{(x-a)^2}{2a^2} + \frac{(x-a)^3}{3a^3} - + \cdots, \qquad 0 < x \leqq 2a$$

TRIGONOMETRIC

$$\sin x = x - \frac{x^3}{3!} + \frac{x^5}{5!} - \frac{x^7}{7!} + \cdots \qquad \text{(all real values of } x\text{)}$$

$$\cos x = 1 - \frac{x^2}{2!} + \frac{x^4}{4!} - \frac{x^6}{6!} + \cdots \qquad \text{(all real values of } x\text{)}$$

$$\tan x = x + \frac{x^3}{3} + \frac{2x^5}{15} + \frac{17x^7}{315} + \frac{62x^9}{2835} + \cdots + \frac{(-1)^{n-1}2^{2n}(2^{2n}-1)B_{2n}}{(2n)!}x^{2n-1} + \cdots,$$

$$\left[x^2 < \frac{\pi^2}{4}, \text{ and } B_n \text{ represents the } n\text{th Bernoulli number.}\right]$$

$$\cot x = \frac{1}{x} - \frac{x}{3} - \frac{x^3}{45} - \frac{2x^5}{945} - \frac{x^7}{4725} - \cdots - \frac{(-1)^{n+1}2^{2n}}{(2n)!}B_{2n}x^{2n-1} - \cdots,$$

$$[x^2 < \pi^2, \text{ and } B_n \text{ represents the } n\text{th Bernoulli number.}]$$

$$\sec x = 1 + \frac{x^2}{2} + \frac{5}{24}x^4 + \frac{61}{720}x^6 + \frac{277}{8064}x^8 + \cdots + \frac{(-1)^n}{(2n)!}E_{2n}x^{2n} + \cdots,$$

$$\left[x^2 < \frac{\pi^2}{4}, \text{ and } E_n \text{ represents the } n\text{th Euler number.}\right]$$

$$\csc x = \frac{1}{x} + \frac{x}{6} + \frac{7}{360}x^3 + \frac{31}{15{,}120}x^5 + \frac{127}{604{,}800}x^7 + \cdots$$

$$+ \frac{(-1)^{n+1}2(2^{2n-1}-1)}{(2n)!}B_{2n}x^{2n-1} + \cdots,$$

$$[x^2 < \pi^2, \text{ and } B_n \text{ represents } n\text{th Bernoulli number.}]$$

$$\sin x = x\left(1 - \frac{x^2}{\pi^2}\right)\left(1 - \frac{x^2}{2^2\pi^2}\right)\left(1 - \frac{x^2}{3^2\pi^2}\right)\cdots \qquad (x^2 < \infty)$$

$$\cos x = \left(1 - \frac{4x^2}{\pi^2}\right)\left(1 - \frac{4x^2}{3^2\pi^2}\right)\left(1 - \frac{4x^2}{5^2\pi^2}\right)\cdots \qquad (x^2 < \infty)$$

$$\sin^{-1}x = x + \frac{x^3}{2\cdot 3} + \frac{1\cdot 3}{2\cdot 4\cdot 5}x^5 + \frac{1\cdot 3\cdot 5}{2\cdot 4\cdot 6\cdot 7}x^7 + \cdots \qquad \left(x^2 < 1,\ -\frac{\pi}{2} < \sin^{-1}x < \frac{\pi}{2}\right)$$

$$\cos^{-1}x = \frac{\pi}{2} - \left(x + \frac{x^3}{2\cdot 3} + \frac{1\cdot 3}{2\cdot 4\cdot 5}x^5 + \frac{1\cdot 3\cdot 5x^7}{2\cdot 4\cdot 6\cdot 7} + \cdots\right) \qquad (x^2 < 1,\ 0 < \cos^{-1}x < \pi)$$

$$\tan^{-1}x = x - \frac{x^3}{3} + \frac{x^5}{5} - \frac{x^7}{7} + \cdots \qquad (x^2 < 1)$$

$$\tan^{-1}x = \frac{\pi}{2} - \frac{1}{x} + \frac{1}{3x^3} - \frac{1}{5x^5} + \frac{1}{7x^7} - \cdots \qquad (x > 1)$$

$$\tan^{-1}x = -\frac{\pi}{2} - \frac{1}{x} + \frac{1}{3x^3} - \frac{1}{5x^5} + \frac{1}{7x^7} - \cdots \qquad (x < -1)$$

$$\cot^{-1}x = \frac{\pi}{2} - x + \frac{x^3}{3} - \frac{x^5}{5} + \frac{x^7}{7} - \cdots \qquad (x^2 < 1)$$

$$\log_e \sin x = \log_e x - \frac{x^2}{6} - \frac{x^4}{180} - \frac{x^6}{2835} - \cdots \qquad (x^2 < \pi^2)$$

$$\log_e \cos x = -\frac{x^2}{2} - \frac{x^4}{12} - \frac{x^6}{45} - \frac{17x^8}{2520} - \cdots \qquad \left(x^2 < \frac{\pi^2}{4}\right)$$

$$\log_e \tan x = \log_e x + \frac{x^2}{3} + \frac{7x^4}{90} + \frac{62x^6}{2835} + \cdots \qquad \left(x^2 < \frac{\pi^2}{4}\right)$$

$$e^{\sin x} = 1 + x + \frac{x^2}{2!} - \frac{3x^4}{4!} - \frac{8x^5}{5!} - \frac{3x^6}{6!} + \frac{56x^7}{7!} + \cdots$$

$$e^{\cos x} = e\left(1 - \frac{x^2}{2!} + \frac{4x^4}{4!} - \frac{31x^6}{6!} + \cdots\right)$$

$$e^{\tan x} = 1 + x + \frac{x^2}{2!} + \frac{3x^3}{3!} + \frac{9x^4}{4!} + \frac{37x^5}{5!} + \cdots \qquad \left(x^2 < \frac{\pi^2}{4}\right)$$

$$\sin x = \sin a + (x - a)\cos a - \frac{(x-a)^2}{2!}\sin a - \frac{(x-a)^3}{3!}\cos a + \frac{(x-a)^4}{4!}\sin a + \cdots$$

VECTOR ANALYSIS

Definitions

Any quantity which is completely determined by its magnitude is called a *scalar*. Examples of such are mass, density, temperature, etc. Any quantity which is completely determined by its magnitude and direction is called a *vector*. Examples of such are velocity, acceleration, force, etc. A vector quantity is represented by a directed line segment, the length of which represents the magnitude of the vector. A vector quantity is usually represented by a boldfaced letter such as **V**. Two vectors $\mathbf{V}_1$ and $\mathbf{V}_2$ are equal to one another if they have equal magnitudes and are acting in the same directions. A negative vector, written as $-\mathbf{V}$, is one which acts in the opposite direction to **V**, but is of equal magnitude to it. If we represent the magnitude of **V** by v, we write $|\mathbf{V}| = v$. A vector parallel to **V**, but equal to the reciprocal of its magnitude is written as $\mathbf{V}^{-1}$ or as $\frac{1}{\mathbf{V}}$.

The *unit vector* $\frac{\mathbf{V}}{v}$ $(v \neq 0)$ is that vector which has the same direction as **V**, but has a magnitude of unity (sometimes represented as $\mathbf{V}_0$ or $\hat{\mathbf{v}}$).

Vector Algebra

The vector sum of $\mathbf{V}_1$ and $\mathbf{V}_2$ is represented by $\mathbf{V}_1 + \mathbf{V}_2$. The vector sum of $\mathbf{V}_1$ and $-\mathbf{V}_2$, or the difference of the vector $\mathbf{V}_2$ from $\mathbf{V}_1$ is represented by $\mathbf{V}_1 - \mathbf{V}_2$.

If r is a scalar, then $r\mathbf{V} = \mathbf{V}r$, and represents a vector r times the magnitude of **V**, in the same direction as **V** if r is positive, and in the opposite direction if r is negative. If r and s are scalars, $\mathbf{V}_1$, $\mathbf{V}_2$, $\mathbf{V}_3$, vectors, then the following rules of scalars and vectors hold:

$$\mathbf{V}_1 + \mathbf{V}_2 = \mathbf{V}_2 + \mathbf{V}_1$$

$$(r + s)\mathbf{V}_1 = r\mathbf{V}_1 + s\mathbf{V}_1; \qquad r(\mathbf{V}_1 + \mathbf{V}_2) = r\mathbf{V}_1 + r\mathbf{V}_2$$

$$\mathbf{V}_1 + (\mathbf{V}_2 + \mathbf{V}_3) = (\mathbf{V}_1 + \mathbf{V}_2) + \mathbf{V}_3 = \mathbf{V}_1 + \mathbf{V}_2 + \mathbf{V}_3$$

Vectors in Space

A plane is described by two distinct vectors $\mathbf{V}_1$ and $\mathbf{V}_2$. Should these vectors not intersect each other, then one is displaced parallel to itself until they do (fig. 1.) Any other vector **V** lying in this plane is given by

$$\mathbf{V} = r\mathbf{V}_1 + s\mathbf{V}_2$$

A *position vector* specifies the position in space of a point relative to a fixed origin. If therefore $\mathbf{V}_1$ and $\mathbf{V}_2$ are the position vectors of the points A and B, relative to the origin O, then any point P on the line AB has a position vector **V** given by

$$\mathbf{V} = r\mathbf{V}_1 + (1 - r)\mathbf{V}_2$$

The scalar "r" can be taken as the parametric representation of P since $r = 0$ implies $P = B$ and $r = 1$ implies $P = A$. (fig. 2). If P divides the line AB in the ratio $r:s$ then

$$\mathbf{V} = \left(\frac{r}{r+s}\right)\mathbf{V}_1 + \left(\frac{s}{r+s}\right)\mathbf{V}_2$$

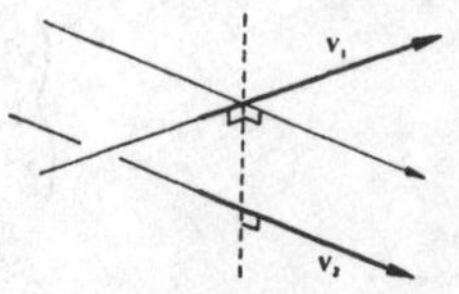

Fig. 1.

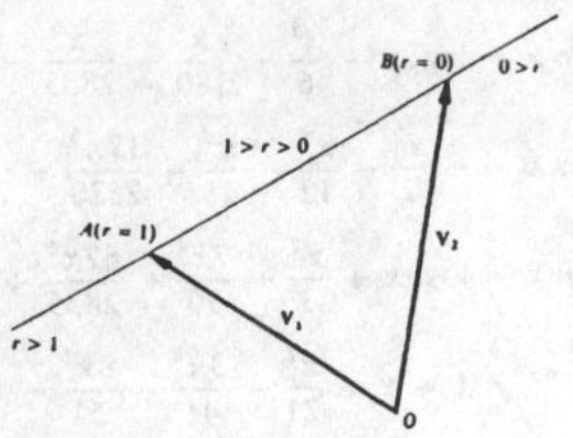

Fig. 2.

The vectors $\mathbf{V}_1, \mathbf{V}_2, \mathbf{V}_3, \ldots, \mathbf{V}_n$ are said to be *linearly dependent* if there exist scalars $r_1, r_2, r_3, \ldots, r_n$, not all zero, such that

$$r_1\mathbf{V}_1 + r_2\mathbf{V}_2 + \cdots + r_n\mathbf{V}_n = 0$$

A vector $\mathbf{V}$ is linearly dependent upon the set of vectors $\mathbf{V}_1, \mathbf{V}_2, \mathbf{V}_3, \ldots, \mathbf{V}_n$ if

$$\mathbf{V} = r_1\mathbf{V}_1 + r_2\mathbf{V}_2 + r_3\mathbf{V}_3 + \cdots + r_n\mathbf{V}_n$$

Three vectors are linearly dependent if and only if they are co-planar.

All points in space can be uniquely determined by linear dependence upon three *base vectors* i.e. three vectors any one of which is linearly independent of the other two. The simplest set of base vectors are the unit vectors along the coordinate Ox, Oy and Oz axes. These are usually designated by $\mathbf{i}$, $\mathbf{j}$ and $\mathbf{k}$ respectively.

If $\mathbf{V}$ is a vector in space, and a, b and c are the respective magnitudes of the projections of the vector along the axes then

$$\mathbf{V} = a\mathbf{i} + b\mathbf{j} + c\mathbf{k}$$

and $$v = \sqrt{a^2 + b^2 + c^2}$$

and the direction cosines of $\mathbf{V}$ are

$$\cos\alpha = a/v, \quad \cos\beta = b/v, \quad \cos\gamma = c/v.$$

The law of addition yields

$$\mathbf{V}_1 + \mathbf{V}_2 = (a_1 + a_2)\mathbf{i} + (b_1 + b_2)\mathbf{j} + (c_1 + c_2)\mathbf{k}$$

The Scalar, Dot, or Inner Product of Two Vectors $\mathbf{V}_1$ and $\mathbf{V}_2$

This product is represented as $\mathbf{V}_1 \cdot \mathbf{V}_2$ and is defined to be equal to $v_1v_2 \cos\theta$, where θ is the angle from $\mathbf{V}_1$ to $\mathbf{V}_2$, i.e.,

$$\mathbf{V}_1 \cdot \mathbf{V}_2 = v_1v_2 \cos\theta$$

The following rules apply for this product:

$$\mathbf{V}_1 \cdot \mathbf{V}_2 = a_1a_2 + b_1b_2 + c_1c_2 = \mathbf{V}_2 \cdot \mathbf{V}_1$$

It should be noted that this verifies that scalar multiplication is commutative.

$$(\mathbf{V}_1 + \mathbf{V}_2) \cdot \mathbf{V}_3 = \mathbf{V}_1 \cdot \mathbf{V}_3 + \mathbf{V}_2 \cdot \mathbf{V}_3$$
$$\mathbf{V}_1 \cdot (\mathbf{V}_2 + \mathbf{V}_3) = \mathbf{V}_1 \cdot \mathbf{V}_2 + \mathbf{V}_1 \cdot \mathbf{V}_3$$

If $\mathbf{V}_1$ is perpendicular to $\mathbf{V}_2$ then $\mathbf{V}_1 \cdot \mathbf{V}_2 = 0$, and if $\mathbf{V}_1$ is parallel to $\mathbf{V}_2$ then $\mathbf{V}_1 \cdot \mathbf{V}_2 = v_1v_2 = rw_1^2$

In particular

$$\mathbf{i} \cdot \mathbf{i} = \mathbf{j} \cdot \mathbf{j} = \mathbf{k} \cdot \mathbf{k} = 1,$$

and $$\mathbf{i} \cdot \mathbf{j} = \mathbf{j} \cdot \mathbf{k} = \mathbf{k} \cdot \mathbf{i} = 0$$

The Vector or Cross Product of Vectors $\mathbf{V}_1$ and $\mathbf{V}_2$

This product is represented as $\mathbf{V}_1 \times \mathbf{V}_2$ and is defined to be equal to $v_1v_2(\sin\theta)\mathbf{1}$, where θ is the angle from $\mathbf{V}_1$ to $\mathbf{V}_2$ and $\mathbf{1}$ is a unit vector perpendicular to the plane of $\mathbf{V}_1$ and $\mathbf{V}_2$ and so directed that a right-handed screw driven in the direction of $\mathbf{1}$ would carry $\mathbf{V}_1$ into $\mathbf{V}_2$, i.e.,

$$\mathbf{V}_1 \times \mathbf{V}_2 = v_1v_2(\sin\theta)\mathbf{1}$$

and $$\tan\theta = \frac{|\mathbf{V}_1 \times \mathbf{V}_2|}{\mathbf{V}_1 \cdot \mathbf{V}_2}$$

The following rules apply for vector products:

$$\mathbf{V}_1 \times \mathbf{V}_2 = -\mathbf{V}_2 \times \mathbf{V}_1$$
$$\mathbf{V}_1 \times (\mathbf{V}_2 + \mathbf{V}_3) = \mathbf{V}_1 \times \mathbf{V}_2 + \mathbf{V}_1 \times \mathbf{V}_3$$
$$(\mathbf{V}_1 + \mathbf{V}_2) \times \mathbf{V}_3 = \mathbf{V}_1 \times \mathbf{V}_3 + \mathbf{V}_2 \times \mathbf{V}_3$$
$$\mathbf{V}_1 \times (\mathbf{V}_2 \times \mathbf{V}_3) = \mathbf{V}_2(\mathbf{V}_3 \cdot \mathbf{V}_1) - \mathbf{V}_3(\mathbf{V}_1 \cdot \mathbf{V}_2)$$
$$\mathbf{i} \times \mathbf{i} = \mathbf{j} \times \mathbf{j} = \mathbf{k} \times \mathbf{k} = 0.\mathbf{1} \text{ (zero vector)}$$
$$= \mathbf{0}$$

$$\mathbf{i} \times \mathbf{j} = \mathbf{k}, \qquad \mathbf{j} \times \mathbf{k} = \mathbf{i}, \qquad \mathbf{k} \times \mathbf{i} = \mathbf{j}$$

If $\mathbf{V}_1 = a_1\mathbf{i} + b_1\mathbf{j} + c_1\mathbf{k}$, $\mathbf{V}_2 = a_2\mathbf{i} + b_2\mathbf{j} + c_2\mathbf{k}$, $\mathbf{V}_3 = a_3\mathbf{i} + b_3\mathbf{j} + c_3\mathbf{k}$,

then

$$\mathbf{V}_1 \times \mathbf{V}_2 = \begin{vmatrix} \mathbf{i} & \mathbf{j} & \mathbf{k} \\ a_1 & b_1 & c_1 \\ a_2 & b_2 & c_2 \end{vmatrix} = (b_1c_2 - b_2c_1)\mathbf{i} + (c_1a_2 - c_2a_1)\mathbf{j} + (a_1b_2 - a_2b_1)\mathbf{k}$$

It should be noted that, since $\mathbf{V}_1 \times \mathbf{V}_2 = -\mathbf{V}_2 \times \mathbf{V}_1$, the vector product is not commutative.

Scalar Triple Product

There is only one possible interpretation of the expression $\mathbf{V}_1 \cdot \mathbf{V}_2 \times \mathbf{V}_3$ and that is $\mathbf{V}_1 \cdot (\mathbf{V}_2 \times \mathbf{V}_3)$ which is obviously a scalar.

Further $\mathbf{V}_1 \cdot (\mathbf{V}_2 \times \mathbf{V}_3) = (\mathbf{V}_1 \times \mathbf{V}_2) \cdot \mathbf{V}_3 = \mathbf{V}_2 \cdot (\mathbf{V}_3 \times \mathbf{V}_1)$

$$= \begin{vmatrix} a_1 & b_1 & c_1 \\ a_2 & b_2 & c_2 \\ a_3 & b_3 & c_3 \end{vmatrix}$$

$$= v_1v_2v_3 \cos\phi \sin\theta,$$

Where θ is the angle between $\mathbf{V}_2$ and $\mathbf{V}_3$ and ϕ is the angle between $\mathbf{V}_1$ and the normal to the plane of $\mathbf{V}_2$ and $\mathbf{V}_3$.

This product is called the *scalar triple product* and is written as $[\mathbf{V}_1\mathbf{V}_2\mathbf{V}_3]$.

The determinant indicates that it can be considered as the volume of the parallelepiped whose three determining edges are $\mathbf{V}_1$, $\mathbf{V}_2$ and $\mathbf{V}_3$.

It also follows that cyclic permutation of the subscripts does not change the value of the scalar triple product so that

$$[\mathbf{V}_1\mathbf{V}_2\mathbf{V}_3] = [\mathbf{V}_2\mathbf{V}_3\mathbf{V}_1] = [\mathbf{V}_3\mathbf{V}_1\mathbf{V}_2]$$

$$\text{but } [\mathbf{V}_1\mathbf{V}_2\mathbf{V}_3] = -[\mathbf{V}_2\mathbf{V}_1\mathbf{V}_3] \text{ etc.} \qquad \text{and } [\mathbf{V}_1\mathbf{V}_1\mathbf{V}_2] \equiv 0 \text{ etc.}$$

Given three non-coplanar reference vectors $\mathbf{V}_1$, $\mathbf{V}_2$ and $\mathbf{V}_3$, the *reciprocal system is* given by $\mathbf{V}_1^*$, $\mathbf{V}_2^*$ and $\mathbf{V}_3^*$, where

$$1 = v_1v_1^* = v_2v_2^* = v_3v_3^*$$

$$0 = v_1v_2^* = v_1v_3^* = v_2v_1^* \text{ etc.}$$

$$\mathbf{V}_1^* = \frac{\mathbf{V}_2 \times \mathbf{V}_3}{[\mathbf{V}_1\mathbf{V}_2\mathbf{V}_3]}, \quad \mathbf{V}_2^* = \frac{\mathbf{V}_3 \times \mathbf{V}_1}{[\mathbf{V}_1\mathbf{V}_2\mathbf{V}_3]}, \quad \mathbf{V}_3^* = \frac{\mathbf{V}_1 \times \mathbf{V}_2}{[\mathbf{V}_1\mathbf{V}_2\mathbf{V}_3]}$$

The system $\mathbf{i}, \mathbf{j}, \mathbf{k}$ is its own reciprocal.

Vector Triple Product

The product $\mathbf{V}_1 \times (\mathbf{V}_2 \times \mathbf{V}_3)$ defines the *vector triple product*. Obviously, in this case, the brackets are vital to the definition.

$$\mathbf{V}_1 \times (\mathbf{V}_2 \times \mathbf{V}_3) = (\mathbf{V}_1 \cdot \mathbf{V}_3)\mathbf{V}_2 - (\mathbf{V}_1 \cdot \mathbf{V}_2)\mathbf{V}_3$$

$$= \begin{vmatrix} \mathbf{i} & \mathbf{j} & \mathbf{k} \\ a_1 & b_1 & c_1 \\ \begin{vmatrix} b_2 & c_2 \\ b_3 & c_3 \end{vmatrix} & \begin{vmatrix} c_2 & a_2 \\ c_3 & a_3 \end{vmatrix} & \begin{vmatrix} a_2 & b_2 \\ a_3 & b_3 \end{vmatrix} \end{vmatrix}$$

i.e. it is a vector, perpendicular to $\mathbf{V}_1$, lying in the plane of $\mathbf{V}_2$, $\mathbf{V}_3$.

Similarly

$$(\mathbf{V}_1 \times \mathbf{V}_2) \times \mathbf{V}_3 = \begin{vmatrix} \mathbf{i} & \mathbf{j} & \mathbf{k} \\ \begin{vmatrix} b_1 & c_1 \\ b_2 & c_2 \end{vmatrix} & \begin{vmatrix} c_1 & a_1 \\ c_2 & a_2 \end{vmatrix} & \begin{vmatrix} a_1 & b_1 \\ a_2 & b_2 \end{vmatrix} \\ a_3 & b_3 & c_3 \end{vmatrix}$$

$$\mathbf{V}_1 \times (\mathbf{V}_2 \times \mathbf{V}_3) + \mathbf{V}_2 \times (\mathbf{V}_3 \times \mathbf{V}_1) + \mathbf{V}_3 \times (\mathbf{V}_1 \times \mathbf{V}_2) \equiv 0$$

If $\mathbf{V}_1 \times (\mathbf{V}_2 \times \mathbf{V}_3) = (\mathbf{V}_1 \times \mathbf{V}_2) \times \mathbf{V}_3$ then $\mathbf{V}_1$, $\mathbf{V}_2$, $\mathbf{V}_3$ form an *orthogonal set*. Thus $\mathbf{i}, \mathbf{j}, \mathbf{k}$ form an orthogonal set.

Geometry of the Plane, Straight Line and Sphere

The position vectors of the fixed points A, B, C, D relative to O are $\mathbf{V}_1$, $\mathbf{V}_2$, $\mathbf{V}_3$, $\mathbf{V}_4$ and the position vector of the variable point P is $\mathbf{V}$.

The vector form of the equation of the straight line through A parallel to $\mathbf{V}_2$ is

$$\mathbf{V} = \mathbf{V}_1 + r\mathbf{V}_2$$

$$\text{or} \quad (\mathbf{V} - \mathbf{V}_1) = r\mathbf{V}_2$$

$$\text{or} \quad (\mathbf{V} - \mathbf{V}_1) \times \mathbf{V}_2 = 0$$

while that of the plane through A perpendicular to $\mathbf{V}_2$ is

$$(\mathbf{V} - \mathbf{V}_1) \cdot \mathbf{V}_2 = 0$$

The equation of the line AB is

$$\mathbf{V} = r\mathbf{V}_1 + (1 - r)\mathbf{V}_2$$

and those of the bisectors of the angles between $\mathbf{V}_1$ and $\mathbf{V}_2$ are

$$\mathbf{V} = r\left(\frac{\mathbf{V}_1}{v} \pm \frac{\mathbf{V}_2}{v_2}\right)$$

$$\text{or} \quad \mathbf{V} = r(\hat{\mathbf{v}}_1 \pm \hat{\mathbf{v}}_2)$$

The perpendicular from C to the line through A parallel to $\mathbf{V}_2$ has as its equation

$$\mathbf{V} = \mathbf{V}_1 - \mathbf{V}_3 - \hat{\mathbf{v}}_2 \cdot (\mathbf{V}_1 - \mathbf{V}_3)\hat{\mathbf{v}}_2.$$

The condition for the intersection of the two lines,

$$\mathbf{V} = \mathbf{V}_1 + r\mathbf{V}_3$$

$$\text{and} \quad \mathbf{V} = \mathbf{V}_2 + s\mathbf{V}_4$$

$$\text{is} \quad [(\mathbf{V}_1 - \mathbf{V}_2)\mathbf{V}_3\mathbf{V}_4] = 0.$$

The common perpendicular to the above two lines is the line of intersection of the two planes

$$[(\mathbf{V} - \mathbf{V}_1)\mathbf{V}_3(\mathbf{V}_3 \times \mathbf{V}_4)] = 0$$

$$\text{and} \quad [(\mathbf{V} - \mathbf{V}_2)\mathbf{V}_4(\mathbf{V}_3 \times \mathbf{V}_4)] = 0$$

and the length of this perpendicular is

$$\frac{[(\mathbf{V}_1 - \mathbf{V}_2)\mathbf{V}_3\mathbf{V}_4]}{|\mathbf{V}_3 \times \mathbf{V}_4|}.$$

The equation of the line perpendicular to the plane ABC is

$$\mathbf{V} = \mathbf{V}_1 \times \mathbf{V}_2 + \mathbf{V}_2 \times \mathbf{V}_3 + \mathbf{V}_3 \times \mathbf{V}_1$$

and the distance of the plane from the origin is

$$\frac{[\mathbf{V}_1\mathbf{V}_2\mathbf{V}_3]}{|(\mathbf{V}_2 - \mathbf{V}_1) \times (\mathbf{V}_3 - \mathbf{V}_1)|}.$$

In general the vector equation

$$\mathbf{V} \cdot \mathbf{V}_2 = r$$

defines the plane which is perpendicular to $\mathbf{V}_2$, and the perpendicular distance from A to this plane is

$$\frac{r - \mathbf{V}_1 \cdot \mathbf{V}_2}{v_2}.$$

The distance from A, measured along a line parallel to $\mathbf{V}_3$, is

$$\frac{r - \mathbf{V}_1 \cdot \mathbf{V}_2}{\mathbf{V}_2 \cdot \hat{\mathbf{v}}_3} \quad \text{or} \quad \frac{r - \mathbf{V}_1 \cdot \mathbf{V}_2}{v_2 \cos\theta}$$

where θ is the angle beween $\mathbf{V}_2$ and $\mathbf{V}_3$.

(If this plane contains the point C then $r = \mathbf{V}_3 \cdot \mathbf{V}_2$ and if it passes through the origin then $r = 0$.)

Given two planes

$$\mathbf{V} \cdot \mathbf{V}_1 = r$$

$$\mathbf{V} \cdot \mathbf{V}_2 = s$$

then any plane through the line of intersection of these two planes is given by

$$\mathbf{V} \cdot (\mathbf{V}_1 + \lambda\mathbf{V}_2) = r + \lambda s$$

where λ is a scalar parameter. In particular $\lambda = \pm v_1/v_2$ yields the equation of the two planes bisecting the angle between the given planes.

The plane through A parallel to the plane of $\mathbf{V}_2$, $\mathbf{V}_3$ is

$$\mathbf{V} = \mathbf{V}_1 + r\mathbf{V}_2 + s\mathbf{V}_3$$

$$\text{or} \quad (\mathbf{V} - \mathbf{V}_1) \cdot \mathbf{V}_2 \times \mathbf{V}_3 = 0$$

$$\text{or} \quad [\mathbf{V}\mathbf{V}_2\mathbf{V}_3] - [\mathbf{V}_1\mathbf{V}_2\mathbf{V}_3] = 0$$

so that the expansion in rectangular Cartesian coordinates yields

$$\begin{vmatrix} (x - a_1) & (y - b_1) & (z - c_1) \\ a_2 & b_2 & c_2 \\ a_3 & b_3 & c_3 \end{vmatrix} = 0 \qquad (\mathbf{V} \equiv x\mathbf{i} + y\mathbf{j} + z\mathbf{k})$$

which is obviously the usual linear equation in x, y and z.

The plane through AB parallel to $\mathbf{V}_3$ is given by

$$[(\mathbf{V} - \mathbf{V}_1)(\mathbf{V}_1 - \mathbf{V}_2)\mathbf{V}_3] = 0$$

$$\text{or} \quad [\mathbf{V}\mathbf{V}_2\mathbf{V}_3] - [\mathbf{V}\mathbf{V}_1\mathbf{V}_3] - [\mathbf{V}_1\mathbf{V}_2\mathbf{V}_3] = 0$$

The plane through the three points A, B and C is

$$\mathbf{V} = \mathbf{V}_1 + s(\mathbf{V}_2 - \mathbf{V}_1) + t(\mathbf{V}_3 - \mathbf{V}_1)$$

or $\mathbf{V} = r\mathbf{V}_1 + s\mathbf{V}_2 + t\mathbf{V}_3 \qquad (r + s + t \equiv 1)$

or $[(\mathbf{V} - \mathbf{V}_1)(\mathbf{V}_1 - \mathbf{V}_2)(\mathbf{V}_2 - \mathbf{V}_3)] = 0$

or $[\mathbf{V}\mathbf{V}_1\mathbf{V}_2] + [\mathbf{V}\mathbf{V}_2\mathbf{V}_3] + [\mathbf{V}\mathbf{V}_3\mathbf{V}_1] - [\mathbf{V}_1\mathbf{V}_2\mathbf{V}_3] = 0$

For four points A, B, C, D to be coplanar, then

$$r\mathbf{V}_1 + s\mathbf{V}_2 + t\mathbf{V}_3 + u\mathbf{V}_4 \equiv 0 \equiv r + s + t + u$$

The following formulae relate to a sphere when the vectors are taken to lie in three dimensional space and to a circle when the space is two dimensional. For a circle in three dimensions take the intersection of the sphere with a plane.

The equation of a sphere with center O and radius OA is

$$\mathbf{V} \cdot \mathbf{V} = v_1^2 \qquad (\text{not } \mathbf{V} = \mathbf{V}_1)$$

or $(\mathbf{V} - \mathbf{V}_1) \cdot (\mathbf{V} + \mathbf{V}_1) = 0$

while that of a sphere with center B radius v_1 is

$$(\mathbf{V} - \mathbf{V}_2) \cdot (\mathbf{V} - \mathbf{V}_2) = v_1^2$$

or $\mathbf{V} \cdot (\mathbf{V} - 2\mathbf{V}_2) = v_1^2 - v_2^2$

If the above sphere passes through the origin then

$$\mathbf{V} \cdot (\mathbf{V} - 2\mathbf{V}_2) = 0$$

(note that in two dimensional polar coordinates this is simply)

$$r = 2a \cdot \cos \theta$$

while in three dimensional Cartesian coordinates it is

$$x^2 + y^2 + z^2 - 2(a_2x + b_2y + c_2x) = 0.$$

The equation of a sphere having the points A and B as the extremities of a diameter is

$$(\mathbf{V} - \mathbf{V}_1) \cdot (\mathbf{V} - \mathbf{V}_2) = 0.$$

The square of the length of the tangent from C to the sphere with center B and radius v_1 is given by

$$(\mathbf{V}_3 - \mathbf{V}_2) \cdot (\mathbf{V}_3 - \mathbf{V}_2) = v_1^2$$

The condition that the plane $\mathbf{V} \cdot \mathbf{V}_3 = s$ is tangential to the sphere $(\mathbf{V} - \mathbf{V}_2) \cdot (\mathbf{V} - \mathbf{V}_2) = v_1^2$ is

$$(s - \mathbf{V}_3 \cdot \mathbf{V}_2) \cdot (s - \mathbf{V}_3 \cdot \mathbf{V}_2) = v_1^2 v_3^2.$$

The equation of the tangent plane at D, on the surface of sphere $(\mathbf{V} - \mathbf{V}_2) \cdot (\mathbf{V} - \mathbf{V}_2) = v_1^2$, is

$$(\mathbf{V} - \mathbf{V}_4) \cdot (\mathbf{V}_4 - \mathbf{V}_2) = 0$$

or $\mathbf{V} \cdot \mathbf{V}_4 - \mathbf{V}_2 \cdot (\mathbf{V} + \mathbf{V}_4) = v_1^2 - v_2^2$

The condition that the two circles $(\mathbf{V} - \mathbf{V}_2) \cdot (\mathbf{V} - \mathbf{V}_2) = v_1^2$ and $(\mathbf{V} - \mathbf{V}_4) \cdot (\mathbf{V} - \mathbf{V}_4) = v_3^2$ intersect orthogonally is clearly

$$(\mathbf{V}_2 - \mathbf{V}_4) \cdot (\mathbf{V}_2 - \mathbf{V}_4) = v_1^2 + v_3^2$$

The polar plane of D with respect to the circle

$$(\mathbf{V} - \mathbf{V}_2) \cdot (\mathbf{V} - \mathbf{V}_2) = v_1^2 \quad \text{is}$$

$$\mathbf{V} \cdot \mathbf{V}_4 - \mathbf{V}_2 \cdot (\mathbf{V} + \mathbf{V}_4) = v_1^2 - v_2^2$$

Any sphere through the intersection of the two spheres $(\mathbf{V} - \mathbf{V}_2) \cdot (\mathbf{V} - \mathbf{V}_2) = v_1^2$ and $(\mathbf{V} - \mathbf{V}_4) \cdot (\mathbf{V} - \mathbf{V}_1) = v_3^2$ is given by

$$(\mathbf{V} - \mathbf{V}_2) \cdot (\mathbf{V} - \mathbf{V}_2) + \lambda(\mathbf{V} - \mathbf{V}_4) \cdot (\mathbf{V} - \mathbf{V}_4) = v_1^2 + \lambda v_3^2$$

while the radical plane of two such spheres is

$$\mathbf{V} \cdot (\mathbf{V}_2 - \mathbf{V}_4) = -\tfrac{1}{2}(v_1^2 - v_2^2 - v_3^2 + v_4^2)$$

Differentiation of Vectors

If $\mathbf{V}_1 = a_1\mathbf{i} + b_1\mathbf{j} + c_1\mathbf{k}$, and $\mathbf{V}_2 = a_2\mathbf{i} + b_2\mathbf{j} + c_2\mathbf{k}$ and if $\mathbf{V}_1$ and $\mathbf{V}_2$ are functions of the scalar t, then

$$\frac{d}{dt}(\mathbf{V}_1 + \mathbf{V}_2 + \cdots) = \frac{d\mathbf{V}_1}{dt} + \frac{d\mathbf{V}_2}{dt} + \cdots,$$

where $\dfrac{d\mathbf{V}_1}{dt} = \dfrac{da_1}{dt}\mathbf{i} + \dfrac{db_1}{dt}\mathbf{j} + \dfrac{dc_1}{dt}\mathbf{k}$, etc.

$$\frac{d}{dt}(\mathbf{V}_1 \cdot \mathbf{V}_2) = \frac{d\mathbf{V}_1}{dt} \cdot \mathbf{V}_2 + \mathbf{V}_1 \cdot \frac{d\mathbf{V}_2}{dt}$$

$$\frac{d}{dt}(\mathbf{V}_1 \times \mathbf{V}_2) = \frac{d\mathbf{V}_1}{dt} \times \mathbf{V}_2 + \mathbf{V}_1 \times \frac{d\mathbf{V}_2}{dt}$$

$$\mathbf{V} \cdot \frac{d\mathbf{V}}{dt} = v \cdot \frac{dv}{dt}$$

In particular, if $\mathbf{V}$ is a vector of constant length then the right hand side of the last equation is identically zero showing that $\mathbf{V}$ is perpendicular to its derivative.

The derivatives of the triple products are

$$\frac{d}{dt}[\mathbf{V}_1\mathbf{V}_2\mathbf{V}_3] = \left[\left(\frac{d\mathbf{V}_1}{dt}\right)\mathbf{V}_2\mathbf{V}_3\right] + \left[\mathbf{V}_1\left(\frac{d\mathbf{V}_2}{dt}\right)\mathbf{V}_3\right] + \left[\mathbf{V}_1\mathbf{V}_2\left(\frac{d\mathbf{V}_3}{dt}\right)\right]$$

and

$$\frac{d}{dt}\left\{\mathbf{V}_1 \times (\mathbf{V}_2 \times \mathbf{V}_3)\right\} = \left(\frac{d\mathbf{V}_1}{dt}\right) \times (\mathbf{V}_2 \times \mathbf{V}_3) + \mathbf{V}_1$$

$$\times \left(\left(\frac{d\mathbf{V}_2}{dt}\right) \times \mathbf{V}_3\right) + \mathbf{V}_1 \times \left(\mathbf{V}_2 \times \left(\frac{d\mathbf{V}_3}{dt}\right)\right)$$

Geometry of Curves in Space

s = the *length of arc*, measured from some fixed point on the curve (fig. 3).

$\mathbf{V}_1$ = the position vector of the point A on the curve

$\mathbf{V}_1 + \delta\mathbf{V}_1$ = the position vector of the point P in the neighborhood of A

$\hat{\mathbf{t}}$ = the *unit tangent* to the curve at the point A, measured in the direction of s increasing.

The *normal plane* is that plane which is perpendicular to the unit tangent. The principal normal is defined as the intersection of the normal plane with the plane defined by $\mathbf{V}_1$ and $\mathbf{V}_1 + \delta\mathbf{V}_1$ in the limit as $\delta\mathbf{V}_1 - 0$.

$\hat{\mathbf{n}}$ = the *unit normal* (principal) at the point A. The plane defined by $\hat{\mathbf{t}}$ and $\hat{\mathbf{n}}$ is called the *osculating plane* (alternatively plane of curvature or local plane).

ρ = the radius of curvature at A

$\delta\theta$ = the angle subtended at the origin by $\delta\mathbf{V}_1$.

$$\kappa = \frac{d\theta}{ds} = \frac{1}{\rho}$$

$\hat{\mathbf{b}}$ = the *unit binormal* i.e. the unit vector which is parallel to $\hat{\mathbf{t}} \times \hat{\mathbf{n}}$ at the point A:

λ = the *torsion* of the curve at A

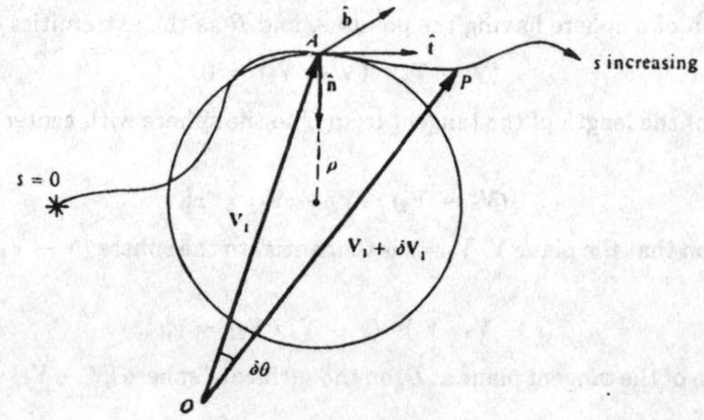

Figure 3.

Frenet's Formulae:

$$\frac{d\hat{\mathbf{t}}}{ds} = \kappa\hat{\mathbf{n}}$$

$$\frac{d\hat{\mathbf{n}}}{ds} = -\kappa\hat{\mathbf{t}} + \lambda\hat{\mathbf{b}}$$

$$\frac{d\hat{\mathbf{b}}}{ds} = -\lambda\hat{\mathbf{n}}$$

The following formulae are also applicable:

Unit tangent $\quad \hat{\mathbf{t}} = \dfrac{d\mathbf{V}_1}{ds}$

Equation of the tangent $\quad (\mathbf{V} - \mathbf{V}_1) \times \hat{\mathbf{t}} = 0$

or $\quad \mathbf{V} = \mathbf{V}_1 + q\hat{\mathbf{t}}$

Unit normal $\quad \hat{\mathbf{n}} = \dfrac{1 d^2\mathbf{V}_1}{\kappa ds^2}$

Equation of the normal plane $\quad (\mathbf{V} - \mathbf{V}_1) \cdot \hat{\mathbf{t}} = 0$

Equation of the normal $\quad (\mathbf{V} - \mathbf{V}_1) \times \hat{\mathbf{n}} = 0$

or $\quad \mathbf{V} = \mathbf{V}_1 + r\hat{\mathbf{n}}$

Unit binormal $\quad \hat{\mathbf{b}} = \hat{\mathbf{t}} \times \hat{\mathbf{n}}$

Equation of the binormal $\quad (\mathbf{V} - \mathbf{V}_1) \times \hat{\mathbf{b}} = 0$

or $\quad \mathbf{V} = \mathbf{V}_1 + u\hat{\mathbf{b}}$

or $\quad \mathbf{V} = \mathbf{V}_1 + w\dfrac{d\mathbf{V}_1}{ds} \times \dfrac{d^2\mathbf{V}_1}{ds^2}$

Equation of the osculating plane:

$$[(\mathbf{V} - \mathbf{V}_1)\hat{\mathbf{t}}\hat{\mathbf{n}}] = 0$$

or

$$\left[(\mathbf{V} - \mathbf{V}_1)\left(\frac{d\mathbf{V}_1}{ds}\right)\left(\frac{d^2\mathbf{V}_1}{ds^2}\right)\right] = 0$$

A *geodetic line* on a surface is a curve, the osculating plane of which is everywhere normal to the surface.

The differential equation of the geodetic is

$$[\hat{\mathbf{n}}d\mathbf{V}_1\, d^2\mathbf{V}_1] = 0$$

Differential Operators—Rectangular Coordinates

$$dS = \frac{\partial S}{\partial x}\cdot dx + \frac{\partial S}{\partial y}\cdot dy + \frac{\partial S}{\partial z}\cdot dz$$

By definition

$$\nabla \equiv \text{del} \equiv \mathbf{i}\frac{\partial}{\partial x} + \mathbf{j}\frac{\partial}{\partial y} + \mathbf{k}\frac{\partial}{\partial z}$$

$$\nabla^2 \equiv \text{Laplacian} \equiv \frac{\partial^2}{\partial x^2} + \frac{\partial^2}{\partial y^2} + \frac{\partial^2}{\partial z^2}$$

If S is a scalar function, then

$$\nabla S \equiv \text{grad } S \equiv \frac{\partial S}{dx}\mathbf{i} + \frac{\partial S}{dy}\mathbf{j} + \frac{\partial S}{dz}\mathbf{k}$$

Grad S defines both the direction and magnitude of the maximum rate of increase of S at any point. Hence the name *gradient* and also its vectorial nature. ∇S is independent of the choice of rectangular coordinates.

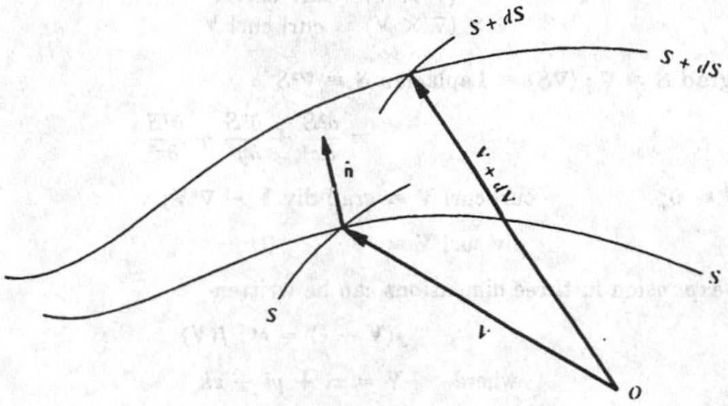

Figure 4

$$\nabla S = \frac{\partial S}{\partial n}\hat{\mathbf{n}}$$

where $\hat{\mathbf{n}}$ is the unit normal to the surface S = constant, in the direction of S increasing. The total derivative of S at a point having the position vector $\mathbf{V}$ is given by (fig. 4)

$$dS = \frac{\partial S}{\partial n}\hat{\mathbf{n}}\cdot d\mathbf{V}$$

$$= d\mathbf{V}\cdot\nabla S$$

and the directional derivative of S in the direction of $\mathbf{U}$ is

$$\mathbf{U}\cdot\nabla S = \mathbf{U}\cdot(\nabla S) = (\mathbf{U}\cdot\nabla)S$$

Similarly the directional derivative of the vector $\mathbf{V}$ in the direction of $\mathbf{U}$ is

$$(\mathbf{U}\cdot\nabla)\mathbf{V}$$

The *distributive* law holds for finding a gradient. Thus if S and T are scalar functions

$$\nabla(S + T) = \nabla S + \nabla T$$

The *associative* law becomes the rule for differentiating a product:

$$\nabla(ST) = S\nabla T + T\nabla S$$

If $\mathbf{V}$ is a vector function with the magnitudes of the components parallel to the three coordinate axes V_x, V_y, V_z, then

$$\nabla\cdot\mathbf{V} \equiv \text{div }\mathbf{V} \equiv \frac{\partial V_x}{\partial x} + \frac{\partial V_y}{\partial y} + \frac{\partial V_z}{\partial z}$$

The divergence obeys the distributive law. Thus, if $\mathbf{V}$ and $\mathbf{U}$ are vectors functions, then

$$\nabla\cdot(\mathbf{V} + \mathbf{U}) = \nabla\cdot\mathbf{V} + \nabla\cdot\mathbf{U}$$

$$\nabla\cdot(S\mathbf{V}) = (\nabla S)\cdot\mathbf{V} + S(\nabla\cdot\mathbf{V})$$

$$\nabla\cdot(\mathbf{U}\times\mathbf{V}) = \mathbf{V}\cdot(\nabla\times\mathbf{U}) - \mathbf{U}\cdot(\nabla\times\mathbf{V})$$

As with the gradient of a scalar, the divergence of a vector is invariant under a transformation from one set of rectangular coordinates to another.

$$\nabla\times\mathbf{V} \equiv \text{curl }\mathbf{V} \text{ (sometimes } \nabla\wedge\mathbf{V} \text{ or rot } \mathbf{V})$$

$$\equiv \left(\frac{\partial V_z}{\partial y} - \frac{\partial V_y}{\partial z}\right)\mathbf{i} + \left(\frac{\partial V_x}{\partial z} - \frac{\partial V_z}{\partial x}\right)\mathbf{j} + \left(\frac{\partial V_y}{\partial x} - \frac{\partial V_x}{\partial y}\right)\mathbf{k}$$

$$\equiv \begin{vmatrix} \mathbf{i} & \mathbf{j} & \mathbf{k} \\ \frac{\partial}{\partial x} & \frac{\partial}{\partial y} & \frac{\partial}{\partial z} \\ V_x & V_y & V_z \end{vmatrix}$$

The *curl* (or *rotation*) of a vector is a vector which is invariant under a transformation from one set of rectangular coordinates to another.

$$\nabla \times (\mathbf{U} + \mathbf{V}) = \nabla \times \mathbf{U} + \nabla \times \mathbf{V}$$
$$\nabla \times (S\mathbf{V}) = (\nabla S) \times \mathbf{V} + S(\nabla \times \mathbf{V})$$
$$\nabla \times (\mathbf{U} \times \mathbf{V}) = (\mathbf{V} \cdot \nabla)\mathbf{U} - (\mathbf{U} \cdot \nabla)\mathbf{V} + \mathbf{U}(\nabla \cdot \mathbf{V}) - \mathbf{V}(\nabla \cdot \mathbf{U})$$

$$\text{grad }(\mathbf{U} \cdot \mathbf{V}) = \nabla(\mathbf{U} \cdot \mathbf{V})$$
$$= (\mathbf{V} \cdot \nabla)\mathbf{U} + (\mathbf{U} \cdot \nabla)\mathbf{V} + \mathbf{V} \times (\nabla \times \mathbf{U}) + \mathbf{U} \times (\nabla \times \mathbf{V})$$

If
$$\mathbf{V} = V_x\mathbf{i} + V_y\mathbf{j} + V_z\mathbf{k}$$
$$\nabla \cdot \mathbf{V} = \nabla V_x \cdot \mathbf{i} + \nabla V_y \cdot \mathbf{j} + \nabla V_z \cdot \mathbf{k}$$
and
$$\nabla \times \mathbf{V} = \nabla V_x \times \mathbf{i} + \nabla V_y \times \mathbf{j} + \nabla V_z \times \mathbf{k}$$

The operator ∇ can be used more than once. The number of possibilities where ∇ is used twice are

$$\nabla \cdot (\nabla \theta) \equiv \text{div grad } \theta$$
$$\nabla \times (\nabla \theta) \equiv \text{curl grad } \theta$$
$$\nabla(\nabla \cdot \mathbf{V}) \equiv \text{grad div } \mathbf{V}$$
$$\nabla \cdot (\nabla \times \mathbf{V}) \equiv \text{div curl } \mathbf{V}$$
$$\nabla \times (\nabla \times \mathbf{V}) \equiv \text{curl curl } \mathbf{V}$$

Thus: div grad $S \equiv \nabla \cdot (\nabla S) \equiv$ Laplacian $S \equiv \nabla^2 S$

$$\equiv \frac{\partial^2 S}{\partial x^2} + \frac{\partial^2 S}{\partial y^2} + \frac{\partial^2 S}{\partial z^2}$$

curl grad $S \equiv 0$; curl curl $\mathbf{V} \equiv$ grad div $\mathbf{V} - \nabla^2\mathbf{V}$;

div curl $\mathbf{V} \equiv 0$

Taylor's expansion in three dimensions can be written

$$f(\mathbf{V} + \boldsymbol{\varepsilon}) = e^{\boldsymbol{\varepsilon} \cdot \nabla} f(\mathbf{V})$$

where $\mathbf{V} = x\mathbf{i} + y\mathbf{j} + z\mathbf{k}$

and $\boldsymbol{\varepsilon} = h\mathbf{i} + l\mathbf{j} + m\mathbf{k}$

(note the analogy with $f_p = e^{phD} f_0$ in finite difference methods).

Orthogonal Curvilinear Coordinates

If at a point P there exist three uniform point functions u, v and w so that the surfaces u = const., v = const., and w = const., intersect in three distinct curves through P then the surfaces are called the *coordinate surfaces* through P. The three lines of intersection are referred to as the *coordinate lines* and their tangents a, b, and c as the *coordinate axes*. When the coordinate axes form an orthogonal set the system is said to define *orthogonal curvilinear coordinates* at P.

Consider an infinitesimal volume enclosed by the surfaces u, v, w, $u + du$, $v + dv$, and $w + dw$ (fig. 5).

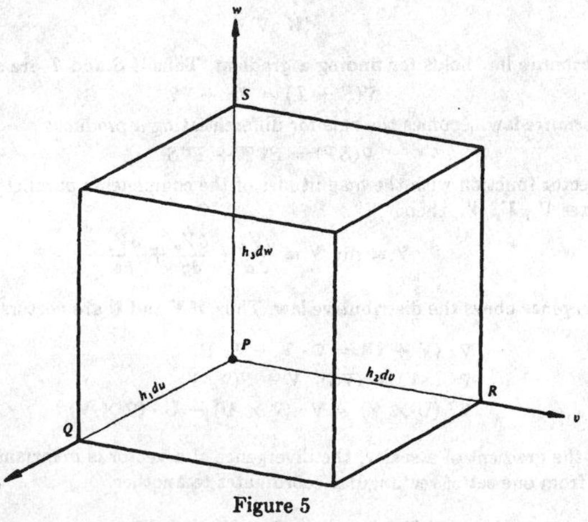

Figure 5

The surface $PRS \equiv u$ = const., and the face of the curvilinear figure immediately opposite this is $u + du$ = const. etc.

In terms of these surface constants

$$P = P(u,v,w)$$
$$Q = Q(u + du,v,w) \quad \text{and} \quad PQ = h_1 du$$

$$R = R(u,v + dv,w) \qquad PR = h_2 dv$$
$$S = S(u,v,w + dw) \qquad PS = h_3 dw$$

where h_1, h_2, and h_3 are functions of u, v, and w.

In rectangular Cartesians $\mathbf{i}$, $\mathbf{j}$, $\mathbf{k}$

$$h_1 = 1, \qquad h_2 = 1, \qquad h_3 = 1.$$

$$\frac{\hat{\mathbf{a}}}{h_1}\frac{\partial}{\partial u} = \mathbf{i}\frac{\partial}{\partial x}, \qquad \frac{\hat{\mathbf{b}}}{h_2}\frac{\partial}{\partial v} = \mathbf{j}\frac{\partial}{\partial y}, \qquad \frac{\hat{\mathbf{c}}}{h_3}\frac{\partial}{\partial w} = \mathbf{k}\frac{\partial}{\partial z}.$$

In cylindrical coordinates $\hat{\mathbf{r}}$, $\hat{\boldsymbol{\phi}}$, $\hat{\mathbf{k}}$

$$h_1 = 1, \qquad h_2 = r \qquad h_3 = 1.$$

$$\frac{\hat{\mathbf{a}}}{h_1}\frac{\partial}{\partial u} = \hat{\mathbf{r}}\frac{\partial}{\partial r}, \qquad \frac{\hat{\mathbf{b}}}{h_2}\frac{\partial}{\partial v} = \frac{\hat{\boldsymbol{\phi}}}{r}\frac{\partial}{\partial \phi} \qquad \frac{\hat{\mathbf{c}}}{h_3}\frac{\partial}{\partial w} = \hat{\mathbf{k}}\frac{\partial}{\partial z}$$

In spherical coordinates $\hat{\mathbf{r}}$, $\hat{\boldsymbol{\theta}}$, $\hat{\boldsymbol{\phi}}$

$$h_1 = 1, \qquad h_2 = r, \qquad h_3 = r\sin\theta$$

$$\frac{\hat{\mathbf{a}}}{h_1}\frac{\partial}{\partial u} = \hat{\mathbf{r}}\frac{\partial}{\partial r}, \qquad \frac{\mathbf{b}}{h_2}\frac{\partial}{\partial v} = \frac{\hat{\boldsymbol{\phi}}}{r}\frac{\partial}{\partial \theta}, \qquad \frac{\hat{\mathbf{c}}}{h_3}\frac{\partial}{\partial w} = \frac{\hat{\boldsymbol{\phi}}}{r\sin\theta}\frac{\partial}{\partial \phi}$$

The general expressions for grad, div and curl together with those for ∇^2 and the directional derivative are, in orthogonal curvilinear coordinates, given by

$$\nabla S = \frac{\hat{\mathbf{a}}}{h_1}\frac{\partial S}{\partial u} + \frac{\hat{\mathbf{b}}}{h_2}\frac{\partial S}{\partial v} + \frac{\hat{\mathbf{c}}}{h_3}\frac{\partial S}{\partial w}$$

$$(\mathbf{V}\cdot\nabla)S = \frac{V_1}{h_1}\frac{\partial S}{\partial u} + \frac{V_2}{h_2}\frac{\partial S}{\partial v} + \frac{V_3}{h_3}\frac{\partial S}{\partial w}$$

$$\nabla\cdot\mathbf{V} = \frac{1}{h_1h_2h_3}\left\{\frac{\partial}{\partial u}(h_2h_3V_1) + \frac{\partial}{\partial v}(h_3h_1V_2) + \frac{\partial}{\partial w}(h_1h_2V_3)\right\}$$

$$\nabla\times\mathbf{V} = \frac{\hat{\mathbf{a}}}{h_2h_3}\left\{\frac{\partial}{\partial v}(h_3V_3) - \frac{\partial}{\partial w}(h_2V_2)\right\} + \frac{\hat{\mathbf{b}}}{h_3h_1}\left\{\frac{\partial}{\partial w}(h_1V_1) - \frac{\partial}{\partial u}(h_3V_3)\right\} + \frac{\hat{\mathbf{c}}}{h_1h_2}\left\{\frac{\partial}{\partial u}(h_2V_2) - \frac{\partial}{\partial v}(h_1V_1)\right\}$$

$$\nabla^2 S = \frac{1}{h_1h_2h_3}\left\{\frac{\partial}{\partial u}\left(\frac{h_2h_3}{h_1}\frac{\partial S}{\partial u}\right) + \frac{\partial}{\partial v}\left(\frac{h_3h_1}{h_2}\frac{\partial S}{\partial v}\right) + \frac{\partial}{\partial w}\left(\frac{h_1h_2}{h_3}\frac{\partial S}{\partial w}\right)\right\}$$

FORMULAS OF VECTOR ANALYSIS

	Rectangular coordinates	Cylindrical coordinates	Spherical coordinates
Conversion to rectangular coordinates		$x = r\cos\varphi \quad y = r\sin\varphi \quad z = z$	$x = r\cos\varphi\sin\theta \quad y = r\sin\varphi\sin\theta \quad z = r\cos\theta$
Gradient	$\nabla\phi = \frac{\partial\phi}{\partial x}\mathbf{i} + \frac{\partial\phi}{\partial y}\mathbf{j} + \frac{\partial\phi}{\partial z}\mathbf{k}$	$\nabla\phi = \frac{\partial\phi}{\partial r}\mathbf{r} + \frac{1}{r}\frac{\partial\phi}{\partial\varphi}\boldsymbol{\phi} + \frac{\partial\phi}{\partial z}\mathbf{k}$	$\nabla\phi = \frac{\partial\phi}{\partial r}\mathbf{r} + \frac{1}{r}\frac{\partial\phi}{\partial\theta}\boldsymbol{\theta} + \frac{1}{r\sin\theta}\frac{\partial\phi}{\partial\varphi}\boldsymbol{\phi}$
Divergence	$\nabla\cdot\mathbf{A} = \frac{\partial A_x}{\partial x} + \frac{\partial A_y}{\partial y} + \frac{\partial A_z}{\partial z}$	$\nabla\cdot\mathbf{A} = \frac{1}{r}\frac{\partial(rA_r)}{\partial r} + \frac{1}{r}\frac{\partial A_\varphi}{\partial\varphi} + \frac{\partial A_z}{\partial z}$	$\nabla\cdot\mathbf{A} = \frac{1}{r^2}\frac{\partial(r^2A_r)}{\partial r} + \frac{1}{r\sin\theta}\frac{\partial(A_\theta\sin\theta)}{\partial\theta} + \frac{1}{r\sin\theta}\frac{\partial A_\varphi}{\partial\varphi}$
Curl	$\nabla\times\mathbf{A} = \begin{vmatrix}\mathbf{i} & \mathbf{j} & \mathbf{k}\\ \frac{\partial}{\partial x} & \frac{\partial}{\partial y} & \frac{\partial}{\partial z}\\ A_x & A_y & A_z\end{vmatrix}$	$\nabla\times\mathbf{A} = \begin{vmatrix}\frac{1}{r}\mathbf{r} & \boldsymbol{\phi} & \frac{1}{r}\mathbf{k}\\ \frac{\partial}{\partial r} & \frac{\partial}{\partial\varphi} & \frac{\partial}{\partial z}\\ A_r & rA_\varphi & A_z\end{vmatrix}$	$\nabla\times\mathbf{A} = \begin{vmatrix}\frac{\mathbf{r}}{r^2\sin\theta} & \frac{\boldsymbol{\theta}}{r\sin\theta} & \frac{\boldsymbol{\phi}}{r}\\ \frac{\partial}{\partial r} & \frac{\partial}{\partial\theta} & \frac{\partial}{\partial\varphi}\\ A_r & rA_\theta & rA_\varphi\sin\theta\end{vmatrix}$
Laplacian	$\nabla^2\phi = \frac{\partial^2\phi}{\partial x^2} + \frac{\partial^2\phi}{\partial y^2} + \frac{\partial^2\phi}{\partial z^2}$	$\nabla^2\phi = \frac{1}{r}\frac{\partial}{\partial r}\left(r\frac{\partial\phi}{\partial r}\right) + \frac{1}{r^2}\frac{\partial^2\phi}{\partial\varphi^2} + \frac{\partial^2\phi}{\partial z^2}$	$\nabla^2\phi = \frac{1}{r^2}\frac{\partial}{\partial r}\left(r^2\frac{\partial\phi}{\partial r}\right) + \frac{1}{r^2\sin\theta}\frac{\partial}{\partial\theta}\left(\sin\theta\frac{\partial\phi}{\partial\theta}\right) + \frac{1}{r^2\sin^2\theta}\frac{\partial^2\phi}{\partial\varphi^2}$

Transformation of Integrals

s = the distance along some curve "C" in space and is measured from some fixed point.
S = a surface area
V = a volume contained by a specified surface
$\hat{\mathbf{t}}$ = the unit tangent to C at the point P
$\hat{\mathbf{n}}$ = the unit outward pointing normal
F = some vector function
ds = the vector element of curve ($= \hat{\mathbf{t}}\, ds$)
dS = the vector element of surface ($= \hat{\mathbf{n}}\, dS$)

Then

$$\int_{(c)} \mathbf{F} \cdot \hat{\mathbf{t}}\, ds = \int_{(c)} \mathbf{F} \cdot \mathbf{ds}$$

and when

$$\mathbf{F} = \nabla\phi$$

$$\int_{(c)} (\nabla\phi) \cdot \hat{\mathbf{t}}\, ds = \int_{(c)} d\phi$$

Gauss' Theorem (Green's Theorem)

When S defines a closed region having a volume V

$$\iiint_{(v)} (\nabla \cdot \mathbf{F})\, dV = \iint_{(s)} (\mathbf{F} \cdot \hat{\mathbf{n}})\, dS = \iint_{(s)} \mathbf{F} \cdot \mathbf{dS}$$

also

$$\iiint_{(v)} (\nabla\phi)\, dV = \iint_{(s)} \phi\hat{\mathbf{n}}\, dS$$

and

$$\iiint_{(v)} (\nabla \times \mathbf{F})\, dV = \iint_{(s)} (\hat{\mathbf{n}} \times \mathbf{F})\, dS$$

Stokes' Theorem

When C is closed and bounds the open surface S.

$$\iint_{(s)} \hat{\mathbf{n}} \cdot (\nabla \times \mathbf{F})\, dS = \int_{(c)} \mathbf{F} \cdot \mathbf{ds}$$

also

$$\iint_{(s)} (\hat{\mathbf{n}} \times \nabla\phi)\, dS = \int_{(c)} \phi\, \mathbf{ds}$$

Green's Theorem

$$\iint_{(s)} (\nabla\phi \cdot \nabla\theta)\, dS = \iint_{(s)} \phi\hat{\mathbf{n}} \cdot (\nabla\theta)\, dS = \iiint_{(v)} \phi(\nabla^2\theta)\, dV$$

$$= \iint_{(s)} \theta \cdot \hat{\mathbf{n}}(\nabla\phi)\, dS = \iiint_{(v)} \theta(\nabla^2\phi)\, dV$$

MOMENT OF INERTIA FOR VARIOUS BODIES OF MASS

The mass of the body is indicated by m

Body	Axis	Moment of inertia	Body	Axis	Moment of inertia
Uniform thin rod	Normal to the length, at one end	$m\frac{l^2}{3}$	Spherical shell, very thin, mean radius, r	Any diameter	$m\frac{2}{3}r^2$
Uniform thin rod	Normal to the length, at the center	$m\frac{l^2}{12}$	Right circular cylinder of radius r, length l	The longitudinal axis of the solid	$m\frac{r^2}{2}$
Thin rectangular sheet, sides a and b	Through the center parallel to b	$m\frac{a^2}{12}$	Right circular cylinder of radius r, length l	Transverse diameter	$m\left(\frac{r^2}{4} + \frac{l^2}{12}\right)$
Thin rectangular sheet, sides a and b	Through the center perpendicular to the sheet	$m\frac{a^2 + b^2}{12}$	Hollow circular cylinder, length l, radii r_1 and r_2	The longitudinal axis of the figure	$m\frac{(r_1^2 + r_2^2)}{2}$
Thin circular sheet of radius r	Normal to the plate through the center	$m\frac{r^2}{2}$	Thin cylindrical shell, length l, mean radius, r	The longitudinal axis of the figure	mr^2
Thin circular sheet of radius r	Along any diameter	$m\frac{r^2}{4}$	Hollow circular cylinder, length l, radii r_1 and r_2	Transverse diameter	$m\left[\frac{r_1^2 + r_2^2}{4} + \frac{l^2}{12}\right]$
Thin circular ring. Radii r_1 and r_2	Through center normal to plane of ring	$m\frac{r_1^2 + r_2^2}{2}$	Hollow circular cylinder, length l, very thin, mean radius	Transverse diameter	$m\left(\frac{r^2}{2} + \frac{l^2}{12}\right)$
Thin circular ring. Radii r_1 and r_2	Any diameter	$m\frac{r_1^2 + r_2^2}{4}$	Elliptic cylinder, length l, transverse semiaxes a and b	Longitudinal axis	$m\left(\frac{a^2 + b^2}{4}\right)$
Rectangular parallelopiped, edges a, b, and c	Through center perpendicular to face ab, (parallel to edge c)	$m\frac{a^2 + b^2}{12}$	Right cone, altitude h, radius of base r	Axis of the figure	$m\frac{3}{10}r^2$
Sphere, radius r	Any diameter	$m\frac{2}{5}r^2$	Spheroid of revolution, equatorial radius r	Polar axis	$m\frac{2r^2}{5}$
Spherical shell, external radius, r_1, internal radius r_2	Any diameter	$m\frac{2}{5}\frac{(r_1^5 - r_2^5)}{(r_1^3 - r_2^3)}$	Ellipsoid, axes $2a$, $2b$, $2c$	Axis $2a$	$m\frac{(b^2 + c^2)}{5}$

Bessel Functions*

1. Bessel's differential equation for a real variable x is

$$x^2\frac{d^2y}{dx^2} + x\frac{dy}{dx} + (x^2 - n^2)y = 0$$

* From Beyer, W. H., Ed., *CRC Handbook of Mathematical Sciences*, 5th ed., CRC Press, Boca Raton, 1978, 500—503. With permission.

2. When n is not an integer, two independent solutions of the equation are $J_n(x)$, $J_{-n}(x)$, where

$$J_n(x) = \sum_{k=0}^{\infty} \frac{(-1)^k}{k!\,\Gamma(n+k+1)}\left(\frac{x}{2}\right)^{n+2k}$$

3. If n is an integer $J_n(x) = (-1)^n J_n(x)$, where

$$J_n(x) = \frac{x^n}{2^n n!}\left\{1 - \frac{x^2}{2^2 \cdot 1!(n+1)} + \frac{x^4}{2^4 \cdot 2!(n+1)(n+2)} - \frac{x^6}{2^6 \cdot 3!(n+1)(n+2)(n+3)} + \cdots\right\}$$

4. For $n = 0$ and $n = 1$, this formula becomes

$$J_0(x) = 1 - \frac{x^2}{2^2(1!)^2} + \frac{x^4}{2^4(2!)^2} - \frac{x^6}{2^6(3!)^2} + \frac{x^8}{2^8(4!)^2} - \cdots$$

$$J_1(x) = \frac{x}{2} - \frac{x^3}{2^3 \cdot 1!2!} + \frac{x^5}{2^5 \cdot 2!3!} - \frac{x^7}{2^7 \cdot 3!4!} + \frac{x^9}{2^9 \cdot 4!5!} - \cdots$$

5. When x is large and positive, the following asymptotic series may be used

$$J_0(x) = \left(\frac{2}{\pi x}\right)^{\frac{1}{2}}\left\{P_0(x)\cos\left(x - \frac{\pi}{4}\right) - Q_0(x)\sin\left(x - \frac{\pi}{4}\right)\right\}$$

$$J_1(x) = \left(\frac{2}{\pi x}\right)^{\frac{1}{2}}\left\{P_1(x)\cos\left(x - \frac{3\pi}{4}\right) - Q_1(x)\sin\left(x - \frac{3\pi}{4}\right)\right\}$$

where

$$P_0(x) \sim 1 - \frac{1^2 \cdot 3^2}{2!(8x)^2} + \frac{1^2 \cdot 3^2 \cdot 5^2 \cdot 7^2}{4!(8x)^4} - \frac{1^2 \cdot 3^2 \cdot 5^2 \cdot 7^2 \cdot 9^2 \cdot 11^2}{6!(8x)^6} + \cdots$$

$$Q_0(x) \sim -\frac{1^2}{1!8x} + \frac{1^2 \cdot 3^2 \cdot 5^2}{3!(8x)^3} - \frac{1^2 \cdot 3^2 \cdot 5^2 \cdot 7^2 \cdot 9^2}{5!(8x)^5} + \cdots$$

$$P_1(x) \sim 1 + \frac{1^2 \cdot 3 \cdot 5}{2!(8x)^2} - \frac{1^2 \cdot 3^2 \cdot 5^2 \cdot 7 \cdot 9}{4!(8x)^4} + \frac{1^2 \cdot 3^2 \cdot 5^2 \cdot 7^2 \cdot 9^2 \cdot 11 \cdot 13}{6!(8x)^6} - + \cdots$$

$$Q_1(x) \sim \frac{1 \cdot 3}{1!8x} - \frac{1^2 \cdot 3^2 \cdot 5 \cdot 7}{3!(8x)^3} + \frac{1^2 \cdot 3^2 \cdot 5^2 \cdot 7^2 \cdot 9 \cdot 11}{5!(8x)^5} - \cdots$$

[In $P_1(x)$ the signs alternate from + to − after the first term]

6. If $x > 25$, it is convenient to use the formulas

$$J_0(x) = A_0(x)\sin x + B_0(x)\cos x$$

$$J_1(x) = B_1(x)\sin x - A_1(x)\cos x$$

where

$$A_0(x) = \frac{P_0(x) - Q_0(x)}{(\pi x)^{\frac{1}{2}}} \quad \text{and} \quad A_1(x) = \frac{P_1(x) - Q_1(x)}{(\pi x)^{\frac{1}{2}}}$$

$$B_0(x) = \frac{P_0(x) + Q_0(x)}{(\pi x)^{\frac{1}{2}}} \quad \text{and} \quad B_1(x) = \frac{P_1(x) + Q_1(x)}{(\pi x)^{\frac{1}{2}}}$$

7. The zeros of $J_0(x)$ and $J_1(x)$

If $j_{0,s}$ and $j_{1,s}$ are the sth zeros of $J_0(x)$ and $J_1(x)$ respectively, and if $a = 4_s - 1$, $b = 4_s + 1$

$$j_{0,s} \sim \frac{1}{4}\pi a\left\{1 + \frac{2}{\pi^2 a^2} - \frac{62}{3\pi^4 a^4} + \frac{15{,}116}{15\pi^6 a^6} - \frac{12{,}554{,}474}{105\pi^8 a^8} + \frac{8{,}368{,}654{,}292}{315\pi^{10} a^{10}} - + \cdots\right\}$$

$$j_{1,s} \sim \frac{1}{4}\pi b\left\{1 - \frac{6}{\pi^2 b^2} + \frac{6}{\pi^4 b^4} - \frac{4716}{5\pi^6 b^6} + \frac{3{,}902{,}418}{35\pi^8 b^8} - \frac{895{,}167{,}324}{35\pi^{10} b^{10}} + \cdots\right\}$$

$$J_1(j_{0,s}) \sim \frac{(-1)^{s+1} 2^{\frac{1}{2}}}{\pi a^{\frac{1}{2}}}\left\{1 - \frac{56}{3\pi^4 a^4} + \frac{9664}{5\pi^6 a^6} - \frac{7{,}381{,}280}{21\pi^8 a^8} + \cdots\right\}$$

$$J_0(j_{1,s}) \sim \frac{(-1)^s 2^{\frac{1}{2}}}{\pi b^{\frac{1}{2}}}\left\{1 + \frac{24}{\pi^4 b^4} - \frac{19{,}584}{10\pi^6 b^6} + \frac{2{,}466{,}720}{7\pi^8 b^8} - \cdots\right\}$$

8. Table of zeros for $J_0(x)$ and $J_1(x)$

$J_1(\alpha_n) = 0 \qquad J_0(\beta_n) = 0$

Roots α_n	$J_1(\alpha_n)$	Roots β_n	$J_0(\beta_n)$
2.4048	0.5191	0.0000	1.0000
5.5201	−0.3403	3.8317	−0.4028
8.6537	0.2715	7.0156	0.3001
11.7915	−0.2325	10.1735	−0.2497
14.9309	0.2065	13.3237	0.2184
18.0711	−0.1877	16.4706	−0.1965
21.2116	0.1733	19.6159	0.1801

9. Recurrence formulas

$$J_{n-1}(x) + J_{n+1}(x) = \frac{2n}{x} J_n(x) \qquad nJ_n(x) + xJ_n'(x) = xJ_{n-1}(x)$$

$$J_{n-1}(x) - J_{n+1}(x) = 2J_n'(x) \qquad nJ_n(x) - xJ_n'(x) = xJ_{n+1}(x)$$

10. If J_n is written for $J_n(x)$ and $J_n^{(k)}$ is written for $\frac{d^k}{dx^k}\{J_n(x)\}$, then the following derivative relationships are important

$$J_0^{(r)} = -J_1^{(r-1)}$$

$$J_0^{(2)} = -J_0 + \frac{1}{x} J_1 = \frac{1}{2}(J_2 - J_0)$$

$$J_0^{(3)} = \frac{1}{x} J_0 + \left(1 - \frac{2}{x^2}\right) J_1 = \frac{1}{4}(-J_3 + 3J_1)$$

$$J_0^{(4)} = \left(1 - \frac{3}{x^2}\right) J_0 - \left(\frac{2}{x} - \frac{6}{x^3}\right) J_1 = \frac{1}{8}(J_4 - 4J_2 + 3J_0), \text{ etc.}$$

11. Half order Bessel functions

$$J_{\frac{1}{2}}(x) = \sqrt{\frac{2}{\pi x}} \sin x$$

$$J_{-\frac{1}{2}}(x) = \sqrt{\frac{2}{\pi x}} \cos x$$

$$J_{n+\frac{3}{2}}(x) = -x^{n+\frac{1}{2}} \frac{d}{dx}\{x^{-(n+\frac{1}{2})} J_{n+\frac{1}{2}}(x)\}$$

$$J_{n-\frac{1}{2}}(x) = x^{-(n+\frac{1}{2})} \frac{d}{dx}\{x^{n+\frac{1}{2}} J_{n+\frac{1}{2}}(x)\}$$

n	$\left(\frac{\pi x}{2}\right)^{\frac{1}{2}} J_{n+\frac{1}{2}}(x)$	$\left(\frac{\pi x}{2}\right)^{\frac{1}{2}} J_{-(n+\frac{1}{2})}(x)$
0	$\sin x$	$\cos x$
1	$\frac{\sin x}{x} - \cos x$	$-\frac{\cos x}{x} - \sin x$
2	$\left(\frac{3}{x^2} - 1\right)\sin x - \frac{3}{x}\cos x$	$\left(\frac{3}{x^2} - 1\right)\cos x + \frac{3}{x}\sin x$
3	$\left(\frac{15}{x^3} - \frac{6}{x}\right)\sin x - \left(\frac{15}{x^2} - 1\right)\cos x$	$-\left(\frac{15}{x^3} - \frac{6}{x}\right)\cos x - \left(\frac{15}{x^2} - 1\right)\sin x$
	etc.	

12. Additional solutions to Bessel's equation are

$Y_n(x)$ (also called Weber's function, and sometimes denoted by $N_n(x)$)

$H_n^{(1)}(x)$ and $H_n^{(2)}(x)$ (also called Hankel functions)

These solutions are defined as follows

$$Y_n(x) = \begin{cases} \dfrac{J_n(x)\cos(n\pi) - J_{-n}(x)}{\sin(n\pi)} & n \text{ not an integer} \\[2ex] \lim\limits_{\nu \to n} \dfrac{J_\nu(x)\cos(\nu\pi) - J_{-\nu}(x)}{\sin(\nu\pi)} & n \text{ an integer} \end{cases}$$

$$H_n^{(1)}(x) = J_n(x) + iY_n(x)$$

$$H_n^{(2)}(x) = J_n(x) - iY_n(x)$$

The additional properties of these functions may all be derived from the above relations and the known properties of $J_n(x)$.

13. Complete solutions to Bessel's equation may be written as

$$c_1 J_n(x) + c_2 J_{-n}(x) \qquad \text{if } n \text{ is not an integer}$$

or

$$\left.\begin{array}{l} c_1 J_n(x) + c_2 Y_n(x) \\ \\ c_1 H_n^{(1)}(x) + c_2 H_n^{(2)}(x) \end{array}\right\} \text{for any value of } n$$

or

14. The modified (or hyperbolic) Bessel's differential equation is

$$x^2 \frac{d^2y}{dx^2} + x\frac{dy}{dx} - (x^2 + n^2)y = 0$$

15. When n is not an integer, two independent solutions of the equation are $I_n(x)$ and $I_{-n}(x)$, where

$$I_n(x) = \sum_{k=0}^{\infty} \frac{1}{k!\Gamma(n+k+1)}\left(\frac{x}{2}\right)^{n+2k}$$

16. If n is an integer,

$$I_n(x) = I_{-n}(x) = \frac{x^n}{2^n n!}\left\{1 + \frac{x^2}{2^2 \cdot 1!(n+1)} + \frac{x^4}{2^4 \cdot 2!(n+1)(n+2)} + \frac{x^6}{2^6 \cdot 3!(n+1)(n+2)(n+3)} + \cdots\right\}$$

17. For $N = 0$ and $n = 1$, this formula becomes

$$I_0(x) = 1 + \frac{x^2}{2^2(1!)^2} + \frac{x^4}{2^4(2!)^2} + \frac{x^6}{2^6(3!)^2} + \frac{x^8}{2^8(4!)^2} + \cdots$$

$$I_1(x) = \frac{x}{2} + \frac{x^3}{2^3 \cdot 1!2!} + \frac{x^5}{2^5 \cdot 2!3!} + \frac{x^7}{2^7 \cdot 3!4!} + \frac{x^9}{2^9 \cdot 4!5!} + \cdots$$

18. Another solution to the modified Bessel's equation is

$$K_n(x) = \begin{cases} \dfrac{1}{2}\pi \dfrac{I_{-n}(x) - I_n(x)}{\sin(n\pi)} & n \text{ not an integer} \\ \lim\limits_{\nu \to n} \dfrac{1}{2}\pi \dfrac{I_{-\nu}(x) - I_\nu(x)}{\sin(\nu\pi)} & n \text{ an integer} \end{cases}$$

This function is linearly independent of $I_n(x)$ for all values of n. Thus the complete solution to the modified Bessel's equation may be written as

$$c_1 I_n(x) + c_2 I_{-n}(x) \qquad n \text{ not an integer}$$

or

$$c_1 I_n(x) + c_2 K_n(x) \qquad \text{any } n$$

19. The following relations hold among the various Bessel functions:

$$I_n(z) = i^{-m} J_m(iz)$$

$$Y_n(iz) = (i)^{n+1} I_n(z) - \frac{2}{\pi} i^{-n} K_n(z)$$

Most of the properties of the modified Bessel function may be deduced from the known properties of $J_n(x)$ by use of these relations and those previously given.

20. Recurrence formulas

$$I_{n-1}(x) - I_{n+1}(x) = \frac{2n}{x} I_n(x) \qquad I_{n-1}(x) + I_{n+1}(x) = 2I_n'(x)$$

$$I_{n-1}(x) - \frac{n}{x} I_n(x) = I_n'(x) \qquad I_n'(x) = I_{n+1}(x) + \frac{n}{x} I_n(z)$$

The Gamma Function*

Deflnlulon. $\Gamma (n - \int_0^\infty t^{n-1} e^{-t} dt \;\; n > 0$

Recursion Formula: $\Gamma(n + 1 = n\Gamma(n)$
$\Gamma(n) + 1) = n!$ if $n = 0,1,2, \ldots$ where $0! = 1$
For $n < 0$ the gamma function can be defined by using

* From Beyer, W. H., Ed., *CRC Handbook of Mathematical Sciences,* 5th ed., CRC Press, Boca Raton, 1978, 484—485. With permision.

$$\Gamma(n) = \frac{\Gamma(n+1)}{n}$$

Graph:

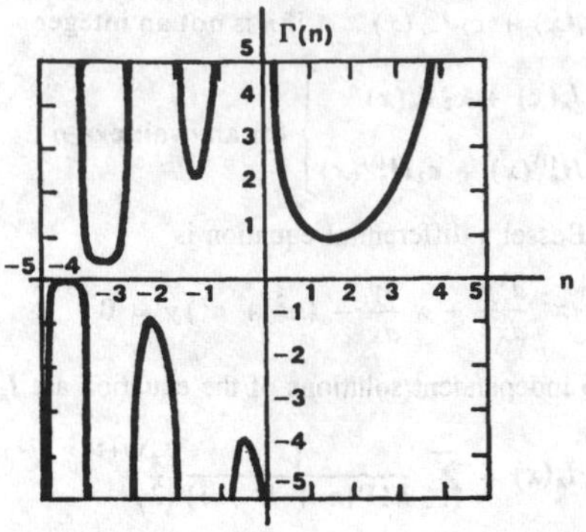

Special Values:

$$\Gamma(\tfrac{1}{2}) = \sqrt{\pi}$$

$$\Gamma(m + \tfrac{1}{2}) = \frac{1 \cdot 3 \cdot 5 \cdots (2m-1)}{2^m}\sqrt{\pi} \qquad m = 1,2,3,\ldots$$

$$\Gamma(-m + \tfrac{1}{2}) = \frac{(-1)^m 2^m \sqrt{\pi}}{1 \cdot 3 \cdot 5 \cdots (2m-1)} \qquad m = 1,2,3,\ldots$$

Definition:

$$\Gamma(x+1) = \lim_{k \to \infty} \frac{1 \cdot 2 \cdot 3 \cdots k}{(x+1)(x+2)\cdots(x+k)} k^x$$

$$\frac{1}{\Gamma(x)} = x e^{\gamma x} \prod_{m=1}^{\infty} \left\{ \left(1 + \frac{x}{m}\right) e^{-x/m} \right\}$$

This is an infinite product representation for the gamma function where γ is Euler's constant.

Properties:

$$\Gamma'(1) = \int_0^{\infty} e^{-x} \ln x \, dx = -\gamma$$

$$\frac{\Gamma'(x)}{\Gamma(x)} = -\gamma + \left(\frac{1}{1} - \frac{1}{x}\right) + \left(\frac{1}{2} - \frac{1}{x+1}\right) + \ldots + \left(\frac{1}{n} - \frac{1}{x+n-1}\right) + \ldots$$

$$\Gamma(x+1) = \sqrt{2\pi x}\, x^x e^{-x} \left\{ 1 + \frac{1}{12x} + \frac{1}{288x^2} - \frac{139}{51{,}840x^3} + \ldots \right\}$$

This is called *Stirling's asymptotic series.*

If we let $x = n$ a positive integer, then a useful approximation for $n!$ where n is large (e.g., $n > 10$) is given by *Stirling's formula*

$$n! \approx \sqrt{2\pi n}\, n^n e^{-n}$$

The Gamma Function*

Values of $\Gamma(n) = \int_0^\infty e^{-x} x^{n-1}\, dx;\ \Gamma(n+1) = n\Gamma(n)$

n	$\Gamma(n)$	n	$\Gamma(n)$	n	$\Gamma(n)$	n	$\Gamma(n)$
1.00	1.00000	1.25	.90640	1.50	.88623	1.75	.91906
1.01	.99433	1.26	.90440	1.51	.88659	1.76	.92137
1.02	.98884	1.27	.90250	1.52	.88704	1.77	.92376
1.03	.98355	1.28	.90072	1.53	.88757	1.78	.92623
1.04	.97844	1.29	.89904	1.54	.88818	1.79	.92877
1.05	.97350	1.30	.89747	1.55	.88887	1.80	.93138
1.06	.96874	1.31	.89600	1.56	.88964	1.81	.93408
1.07	.96415	1.32	.89464	1.57	.89049	1.82	.93685
1.08	.95973	1.33	.89338	1.58	.89142	1.83	.93969
1.09	.95546	1.34	.89222	1.59	.89243	1.84	.94261
1.10	.95135	1.35	.89115	1.60	.89352	1.85	.94561
1.11	.94740	1.36	.89018	1.61	.89468	1.86	.94869
1.12	.94359	1.37	.88931	1.62	.89592	1.87	.95184
1.13	.93993	1.38	.88854	1.63	.89724	1.88	.95507
1.14	.93642	1.39	.88785	1.64	.89864	1.89	.95838
1.15	.93304	1.40	.88726	1.65	.90012	1.90	.96177
1.16	.92980	1.41	.88676	1.66	.90167	1.91	.96523
1.17	.92670	1.42	.88636	1.67	.90330	1.92	.96877
1.18	.92373	1.43	.88604	1.68	.90500	1.93	.97240
1.19	.92089	1.44	.88581	1.69	.90678	1.94	.97610
1.20	.91817	1.45	.88566	1.70	.90864	1.95	.97988
1.21	.91558	1.46	.88560	1.71	.91057	1.96	.98374
1.22	.91311	1.47	.88563	1.72	.91258	1.97	.98768
1.23	.91075	1.48	.88575	1.73	.91466	1.98	.99171
1.24	.90852	1.49	.88595	1.74	.91683	1.99	.99581
						2.00	1.00000

* For large positive values of x, $\Gamma(x)$ approximates Stirling's asymptotic series

$$x^x e^{-x} \sqrt{\frac{2\pi}{x}} \left[1 + \frac{1}{12x} + \frac{1}{288x^2} - \frac{139}{51840x^3} - \frac{571}{2488320x^4} + \cdot\cdot\right]$$

The Beta Function*

Definition: $B(m,n) = \int_0^1 t^{m-1}(1-t)^{m-1}\, dt \quad m > 0, n > 0$

Relationship with Gamma Function: $B(m,n) = \dfrac{\Gamma(m)\Gamma(n)}{\Gamma(m+n)}$

Properties:

$$B(m,n) = B(n,m)$$

$$B(m,n) = 2\int_0^{\pi/2} \sin^{2m-1}\theta \cos^{2n-1}\theta\, d\theta$$

$$B(m,n) = \int_0^\infty \frac{t^{m-1}}{(1+t)^{m+n}}\, dt$$

$$B(m,n) = r^n(r+1)^m \int_0^1 \frac{t^{m-1}(1-t)^{n-1}}{(r+t)^{m+n}}\, dt$$

The Error Function

Definition: $\operatorname{erf} x = \dfrac{2}{\sqrt{\pi}} \int_0^x e^{-t^2}\, dt$

Series: $\operatorname{erf} x = \dfrac{2}{\sqrt{\pi}}\left(x - \dfrac{x^3}{3} + \dfrac{1}{2!}\dfrac{x^5}{5} - \dfrac{1}{3!}\dfrac{x^7}{7} + \ldots\right)$

Property: $\operatorname{erf} x = -\operatorname{erf}(-x)$

Relationship with Normal Probability Function f(t): $\int_0^x f(t)\, dt = \frac{1}{2} \operatorname{erf}\left(\frac{x}{\sqrt{2}}\right)$

To evaluate *erf* (2.3), one proceeds as follows: Since $\frac{x}{\sqrt{2}} = 2.3$, one finds $x = (2.3)(\sqrt{2}) = 3.25$. In the normal probability function table (page A-104), one finds the entry 0.4994 opposite the value 3.25. Thus *erf* (2.3) = 2(0.4994) = 0.9988.

* From Beyer, W. H., Ed., *CRC Handbook of Mathematical Sciences*, 5th ed., CRC Press, Boca Raton, 1978, 499. With permission.

$$erfc\, z = 1 - erf\, z = \frac{2}{\sqrt{\pi}} \int_z^\infty e^{-t^2}\, dt$$

is known as the complementary error function.

Orthogonal Polynomials*

I

Name: Legendre *Symbol:* $P_n(x)$ *Interval:* $[-1,1]$
Differential Equation: $(1 - x^2)\, y'' - 2\, xy' + n(n + 1)\, y = 0$

$$y = P_n(x)$$

Explicit Expression: $P_n(x) = \frac{1}{2^n} \sum_{m=0}^{[n/2]} (-1)^m \binom{n}{m} \binom{2n - 2m}{n} x^{n-2m}$

Recurrence Relation: $(n + 1)\, P_{n+1}(x) = (2n + 1)\, x\, P_n(x) - nP_{n-1}(x)$
Weight: 1 *Standardization:* $P_n(1) = 1$

Norm: $\int_{-1}^{+1} [P_n(x)]^2\, dx = \frac{2}{2n + 1}$

Rodrigues' Formula: $P_n(x) = \frac{(-1)^n}{2^n n!} \frac{d^n}{dx^n} \{(1 - x^2)^n\}$

Generating Function: $R^{-1} = \sum_{n=0}^{\infty} P_n(x) z^n;\ -1 < x < 1,\ |z| < 1,$

$$R = \sqrt{1 - 2xz + z^2}$$

Inequality: $|P_n(x)| \leq 1,\ -1 \leq x \leq 1.$

II

Name: Tschebysheff, First Kind *Symbol:* $T_n(x)$ *Interval:* $[-1,1]$
Differential Equation: $(1 - x^2)y - xy' + n^2 y = 0$

$$y = T_n(x)$$

Explicit Expression: $\frac{n}{2} \sum_{m=0}^{[n/2]} (-1)^m \frac{(n - m - 1)!}{m!(n - 2m)!} (2x)^{n-2m} = \cos(n \arccos x) = T_n(x)$

Recurrence Relation: $T_{n+1}(x) = 2xT_n(x) - T_{n-1}(x)$
Weight: $(1 - x^2)^{-1/2}$ *Standardization:* $T_n(1) = 1$

Norm: $\int_{-1}^{+1} (1 - x^2)^{-1/2} [T_n(x)]^2\, dx = \begin{cases} \pi/2, & n \neq 0 \\ \pi, & n = 0 \end{cases}$

Rodrigues' Formula: $\frac{(-1)^n (1 - x^2)^{1/2} \sqrt{\pi}}{2^{n+1} \Gamma(n + \frac{1}{2})} \frac{d^n}{dx^n} \{(1 - x^2)^{n-(1/2)}\} = T_n(x)$

Generating Function: $\frac{1 - xz}{1 - 2xz + z^2} = \sum_{n=0}^{\infty} T_n(x) z^n,\ -1 < x < 1,\ |z| < 1$

Inequality: $|T_n(x)| \leq 1,\ -1 \leq x \leq 1.$

III

Name: Tschebysheff, Second Kind *Symbol:* $U_n(x)$ *Interval:* $[-1,1]$
Differential Equation: $(1 - x^2)\, y'' - 3\, xy' + n(n + 2)y = 0$

$$y = U_n(x)$$

Explicit Expression: $U_n(x) = \sum_{m=0}^{[n/2]} (-1)^m \frac{(m - n)!}{m!(n - 2m)!} (2x)^{n-2m}$

$$U_n(\cos\theta) = \frac{\sin[(n + 1)\theta]}{\sin\theta}$$

* From Beyer, W. H., Ed., *CRC Handbook of Mathematical Sciences*, 5th ed., CRC Press, Boca Raton, 1978, 557—560. With permission.

Recurrence Relation: $U_{n+1}(x) = 2xU_n(x) - U_{n-1}(x)$

Weight: $(1 - x^2)^{1/2}$ Standardization: $U_n(1) = n + 1$

Norm: $\int_{-1}^{+1} (1 - x^2)^{1/2}[U_n(x)]^2\,dx = \frac{\pi}{2}$

Rodrigues' Formula: $U_n(x) = \frac{(-1)^n (n+1)\sqrt{\pi}}{(1 - x^2)^{1/2} 2^{n+1} \Gamma(n + \frac{3}{2})} \frac{d^n}{dx^n} \{(1 - x^2)^{n+(1/2)}\}$

Generating Function: $\frac{1}{1 - 2xz + z^2} = \sum_{n=0}^{\infty} U_n(x) z^n, \; -1 < x < 1, \; |z| < 1$

Inequality: $|U_n(x)| \leqslant n + 1, \; -1 \leqslant x \leqslant 1.$

IV

Name: Jacobi Symbol: $P_n^{(\alpha,\beta)}(x)$ Interval: $[-1, 1]$

Differential Equation:

$$(1 - x^2)y'' + [\beta - \alpha - (\alpha + \beta + 2)x]y' + n(n + \alpha + \beta + 1)y = 0$$

$$y = P_n^{(\alpha,\beta)}(x)$$

Explicit Expression: $P_n^{(\alpha,\beta)}(x) = \frac{1}{2^n} \sum_{m=0}^{n} \binom{n+\alpha}{m} \binom{n+\beta}{n-m} (x - 1)^{n-m}(x + 1)^m$

Recurrence Relation:
$$\begin{aligned} 2(n+1)(n+\alpha+\beta+1)(2n+\alpha+\beta) P_{n+1}^{(\alpha,\beta)}(x) &= (2n+\alpha+\beta+1)[(\alpha^2 - \beta^2) + (2n+\alpha+\beta+2) \\ &\quad \times (2n+\alpha+\beta)x] P_n^{(\alpha,\beta)}(x) \\ &\quad - 2(n+\alpha)(n+\beta)(2n+\alpha+\beta+2) P_{n-1}^{(\alpha,\beta)}(x) \end{aligned}$$

Weight: $(1 - x)^\alpha (1 + x)^\beta;\; \alpha, \beta > 1$ Standardization: $P_n^{(\alpha,\beta)}(x) = \binom{n+\alpha}{n}$

Norm: $\int_{-1}^{+1} (1 - x)^\alpha (1 + x)^\beta [P_n^{(\alpha,\beta)}(x)]^2\,dx = \frac{2^{\alpha+\beta+1}\Gamma(n+\alpha+1)\Gamma(n+\beta+1)}{(2n+\alpha+\beta+1)n!\,\Gamma(n+\alpha+\beta+1)}$

Rodrigues' Formula: $P_n^{(\alpha,\beta)}(x) = \frac{(-1)^n}{2^n n!(1-x)^\alpha(1+x)^\beta} \frac{d^n}{dx^n} \{(1-x)^{n+\alpha}(1+x)^{n+\beta}\}$

Generating Function: $R^{-1}(1 - z + R)^{-\alpha}(1 + z + R)^{-\beta} = \sum_{n=0}^{\infty} 2^{-\alpha-\beta} P_n^{(\alpha,\beta)}(x) z^n,$

$$R = \sqrt{1 - 2xz + z^2}, \; |z| < 1$$

Inequality:
$$\max_{-1 \le x \le 1} |P_n^{(\alpha,\beta)}(x)| = \begin{cases} \binom{n+q}{n} \sim n^q \text{ if } q = \max(\alpha,\beta) \ge -\frac{1}{2} \\ |P_n^{(\alpha,\beta)}(x')| \sim n^{-1/2} \text{ if } q < -\frac{1}{2} \\ x' \text{ is one of the two maximum points nearest} \\ \frac{\beta - \alpha}{\alpha + \beta + 1} \end{cases}$$

V

Name: Generalized Laguerre Symbol: $L_n^{(\alpha)}(x)$ Interval: $[0, \infty]$

Differential Equation: $xy'' + (\alpha + 1 - x)y' + ny = 0$

$$y = L_n^{(\alpha)}(x)$$

Explicit Expression: $L_n^{(\alpha)}(x) = \sum_{m=0}^{n} (-1)^m \binom{n+\alpha}{n-m} \frac{1}{m!} x^m$

Recurrence Relation: $(n + 1) L_{n+1}^{(\alpha)}(x) = [(2n + \alpha + 1) - x] L_n^{(\alpha)}(x) - (n + \alpha) L_{n-1}^{(\alpha)}(x)$

Weight: $x^\alpha e^{-x}, \; \alpha > -1$ Standardization: $L_n^{(\alpha)}(x) = \frac{(-1)^n}{n!} x^n + \cdots$

Norm: $\int_0^\infty x^\alpha e^{-x} [L_n^{(\alpha)}(x)]^2\,dx = \frac{\Gamma(n + \alpha + 1)}{n!}$

Rodrigues' Formula: $L_n^{(\alpha)}(x) = \frac{1}{n!\,x^\alpha e^{-x}} \frac{d^n}{dx^n} \{x^{n+\alpha} e^{-x}\}$

Generating Function: $(1 - z)^{-\alpha-1} \exp\left(\frac{xz}{z-1}\right) = \sum_{n=0}^{\infty} L_n^{(\alpha)}(x) z^n$

Inequality: $|L_n^{(\alpha)}(x)| \le \frac{\Gamma(n+\alpha+1)}{n!\,\Gamma(\alpha+1)} e^{x/2}; \quad \begin{matrix} x \ge 0 \\ \alpha > 0 \end{matrix}$

$$|L_n^{(\alpha)}(x)| \le \left[2 - \frac{\Gamma(\alpha+n+1)}{n!\,\Gamma(\alpha+1)}\right] e^{x/2}; \quad \begin{matrix} x \ge 0 \\ -1 < \alpha < 0 \end{matrix}$$

Orthogonal Polynomials

Name: Hermite *Symbol:* $H_n(x)$ *Interval:* $[-\infty, \infty]$
Differential Equation: $y'' - 2xy' + 2ny = 0$

Explicit Expression: $H_n(x) = \sum_{m=0}^{[n/2]} \frac{(-1)^m \, n! \, (2x)^{n-2m}}{m! \, (n-2m)!}$

Recurrence Relation: $H_{n+1}(x) = 2x\,H_n(x) - 2nH_{n-1}(x)$
Weight: e^{-x2} *Standardization:* $H_n(1) = 2^n x^n + \ldots$

Norm: $\int_{-\infty}^{\infty} e^{-x^2} [H_n(x)]^2 \, dx = 2^n \, n! \sqrt{\pi}$

Rodriques' Formula: $H_n(x) = (-1)^n e^{x^2} \frac{d^n}{dx^n} (e^{-x^2})$

Generating Function: $e^{-z^2 + 2zx} = \sum_{n=0}^{\infty} H_n(x) \frac{z^n}{n!}$

Inequality: $|H_n(x)| < e^{\frac{x^2}{2}} k \, 2^{n/2} \sqrt{n!} \quad k \approx 1.086435$

NORMAL PROBABILITY FUNCTION

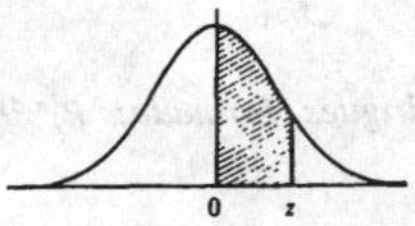

Areas under the Standard Normal Curve from 0 to z

z	0	1	2	3	4	5	6	7	8	9
0.0	.0000	.0040	.0080	.0120	.0160	.0199	.0239	.0279	.0319	.0359
0.1	.0398	.0438	.0478	.0517	.0557	.0596	.0636	.0675	.0714	.0754
0.2	.0793	.0832	.0871	.0910	.0948	.0987	.1026	.1064	.1103	.1141
0.3	.1179	.1217	.1255	.1293	.1331	.1368	.1406	.1443	.1480	.1517
0.4	.1554	.1591	.1628	.1664	.1700	.1736	.1772	.1808	.1844	.1879
0.5	.1915	.1950	.1985	.2019	.2054	.2088	.2123	.2157	.2190	.2224
0.6	.2258	.2291	.2324	.2357	.2389	.2422	.2454	.2486	.2518	.2549
0.7	.2580	.2612	.2652	.2673	.2704	.2734	.2764	.2794	.2823	.2852
0.8	.2881	.2910	.2939	.2967	.2996	.3023	.3051	.3078	.3106	.3133
0.9	.3159	.3186	.3212	.3238	.3264	.3289	.3315	.3340	.3365	.3389
1.0	.3413	.3438	.3461	.3485	.3508	.3531	.3554	.3577	.3599	.3621
1.1	.3643	.3665	.3686	.3708	.3729	.3749	.3770	.3790	.3810	.3830
1.2	.3849	.3869	.3888	.3907	.3925	.3944	.3962	.3980	.3997	.4015
1.3	.4032	.4049	.4066	.4082	.4099	.4115	.4131	.4147	.4162	.4177
1.4	.4192	.4207	.4222	.4236	.4251	.4265	.4279	.4292	.4306	.4319
1.5	.4332	.4345	.4357	.4370	.4382	.4394	.4406	.4418	.4429	.4441
1.6	.4452	.4463	.4474	.4484	.4495	.4505	.4515	.4525	.4535	.4545
1.7	.4554	.4564	.4573	.4582	.4591	.4599	.4608	.4616	.4625	.4633
1.8	.4641	.4649	.4656	.4664	.4671	.4678	.4686	.4693	.4699	.4706
1.9	.4713	.4719	.4726	.4732	.4738	.4744	.4750	.4756	.4761	.4767
2.0	.4772	.4778	.4783	.4788	.4793	.4798	.4803	.4808	.4812	.4817
2.1	.4821	.4826	.4830	.4834	.4838	.4842	.4846	.4850	.4854	.4857
2.2	.4861	.4864	.4868	.4871	.4875	.4878	.4881	.4884	.4887	.4890
2.3	.4893	.4896	.4898	.4901	.4904	.4906	.4909	.4911	.4913	.4916
2.4	.4918	.4920	.4922	.4925	.4927	.4929	.4931	.4932	.4934	.4936
2.5	.4938	.4940	.4941	.4943	.4945	.4946	.4948	.4949	.4951	.4952
2.6	.4953	.4955	.4956	.4957	.4959	.4960	.4961	.4962	.4963	.4964
2.7	.4965	.4966	.4967	.4968	.4969	.4970	.4971	.4972	.4973	.4974
2.8	.4974	.4975	.4976	.4977	.4977	.4978	.4979	.4979	.4980	.4981
2.9	.4981	.4982	.4982	.4983	.4984	.4984	.4985	.4985	.4986	.4986
3.0	.4987	.4987	.4987	.4988	.4988	.4989	.4989	.4989	.4990	.4990
3.1	.4990	.4991	.4991	.4991	.4992	.4992	.4992	.4992	.4993	.4993
3.2	.4993	.4993	.4994	.4994	.4994	.4994	.4994	.4995	.4995	.4995
3.3	.4995	.4995	.4995	.4996	.4996	.4996	.4996	.4996	.4996	.4997
3.4	.4997	.4997	.4997	.4997	.4997	.4997	.4997	.4997	.4997	.4998
3.5	.4998	.4998	.4998	.4998	.4998	.4998	.4998	.4998	.4998	.4998
3.6	.4998	.4998	.4999	.4999	.4999	.4999	.4999	.4999	.4999	.4999
3.7	.4999	.4999	.4999	.4999	.4999	.4999	.4999	.4999	.4999	.4999
3.8	.4999	.4999	.4999	.4999	.4999	.4999	.4999	.4999	.4999	.4999
3.9	.5000	.5000	.5000	.5000	.5000	.5000	.5000	.5000	.5000	.5000

F(z) below refers to area under Standard Normal Curve from $-\infty$ to z

z	1.282	1.645	1.960	2.326	2.576	3.090
F(z)	.90	.95	.975	.99	.995	.999
2[1 − F(z)]	.20	.10	.05	.02	.01	.002

PERCENTAGE POINTS, STUDENT'S *t*-DISTRIBUTION

This table gives values of t such that

$$F(t) = \int_{-\infty}^{t} \frac{\Gamma\left(\frac{n+1}{2}\right)}{\sqrt{n\pi}\,\Gamma\left(\frac{n}{2}\right)} \left(1+\frac{x^2}{n}\right)^{-\frac{n+1}{2}} dx$$

for n, the number of degrees of freedom, equal to 1, 2, . . ., 30, 40, 60, 120, ∞; and for $F(t)$ = 0.60, 0.75, 0.90, 0.95, 0.975, 0.99, 0.995, and 0.9995. The *t*-distribution is symmetrical, so that $F(-t) = 1 - F(t)$

n \ F	.60	.75	.90	.95	.975	.99	.995	.9995
1	.325	1.000	3.078	6.314	12.706	31.821	63.657	636.619
2	.289	.816	1.886	2.920	4.303	6.965	9.925	31.598
3	.277	.765	1.638	2.353	3.182	4.541	5.841	12.924
4	.271	.741	1.533	2.132	2.776	3.747	4.604	8.610
5	.267	.727	1.476	2.015	2.571	3.365	4.032	6.869
6	.265	.718	1.440	1.943	2.447	3.143	3.707	5.959
7	.263	.711	1.415	1.895	2.365	2.998	3.499	5.408
8	.262	.706	1.397	1.860	2.306	2.896	3.355	5.041
9	.261	.703	1.383	1.833	2.262	2.821	3.250	4.781
10	.260	.700	1.372	1.812	2.228	2.764	3.169	4.587
11	.260	.697	1.363	1.796	2.201	2.718	3.106	4.437
12	.259	.695	1.356	1.782	2.179	2.681	3.055	4.318
13	.259	.694	1.350	1.771	2.160	2.650	3.012	4.221
14	.258	.692	1.345	1.761	2.145	2.624	2.977	4.140
15	.258	.691	1.341	1.753	2.131	2.602	2.947	4.073
16	.258	.690	1.337	1.746	2.120	2.583	2.921	4.015
17	.257	.689	1.333	1.740	2.110	2.567	2.898	3.965
18	.257	.688	1.330	1.734	2.101	2.552	2.878	3.922
19	.257	.688	1.328	1.729	2.093	2.539	2.861	3.883
20	.257	.687	1.325	1.725	2.086	2.528	2.845	3.850
21	.257	.686	1.323	1.721	2.080	2.518	2.831	3.819
22	.256	.686	1.321	1.717	2.074	2.508	2.819	3.792
23	.256	.685	1.319	1.714	2.069	2.500	2.807	3.767
24	.256	.685	1.318	1.711	2.064	2.492	2.797	3.745
25	.256	.684	1.316	1.708	2.060	2.485	2.787	3.725
26	.256	.684	1.315	1.706	2.056	2.479	2.779	3.707
27	.256	.684	1.314	1.703	2.052	2.473	2.771	3.690
28	.256	.683	1.313	1.701	2.048	2.467	2.763	3.674
29	.256	.683	1.311	1.699	2.045	2.462	2.756	3.659
30	.256	.683	1.310	1.697	2.042	2.457	2.750	3.646
40	.255	.681	1.303	1.684	2.021	2.423	2.704	3.551
60	.254	.679	1.296	1.671	2.000	2.390	2.660	3.460
120	.254	.677	1.289	1.658	1.980	2.358	2.617	3.373
∞	.253	.674	1.282	1.645	1.960	2.326	2.576	3.291

* This table is abridged from the "Statistical Tables" of R. A. Fisher and Frank Yates published by Oliver & Boyd. Ltd., Edinburgh and London, 1938. It is here published with the kind permission of the authors and their publishers.

PERCENTAGE POINTS, CHI-SQUARE DISTRIBUTION

This table gives values of χ^2 such that

$$F(\chi^2) = \int_0^{\chi^2} \frac{1}{2^{\frac{n}{2}}\,\Gamma\left(\frac{n}{2}\right)} x^{\frac{n-2}{2}} e^{-\frac{x}{2}} dx$$

for n, the number of degrees of freedom, equal to 1, 2, . . . , 30. For $n > 30$, a normal approximation is quite accurate. The expression $\sqrt{2\chi^2} - \sqrt{2n-1}$ is approximately normally distributed as the standard normal distribution. Thus χ^2_α, the α-point of the distribution, may be computed by the formula

$$\chi^2_\alpha = \tfrac{1}{2}[x_\alpha + \sqrt{2n-1}]^2,$$

where x_α is the α-point of the cumulative normal distribution. For even values of n, $F(\chi^2)$ can be written as

$$1 - F(\chi^2) = \sum_{x=0}^{x'-1} \frac{e^{-\lambda}\lambda^x}{x!}$$

with $\lambda = \frac{1}{2}\chi^2$ and $x' = \frac{1}{2}n$. Thus the cumulative Chi-Square distribution is related to the cumulative Poisson distribution.

Another approximate formula for large n

$$\chi^2_\alpha = n\left(1 - \frac{2}{9n} + z_\alpha\sqrt{\frac{2}{9n}}\right)^3$$

n = degrees of freedom

z_α = the normal deviate, (the value of x for which $F(x)$ = the desired percentile).

x	1.282	1.645	1.960	2.326	2.576	3.090
$F(x)$	.90	.95	.975	.99	.995	.999

$\chi^2_{.99} = 60[1 - 0.00370 + 2.326(0.06086)]^3 = 88.4$ is the 99th percentile for 60 degrees of freedom.

$$F(\chi^2) = \int_0^{\chi^2} \frac{1}{2^{\frac{n}{2}}\Gamma\left(\frac{n}{2}\right)} x^{\frac{n-2}{2}} e^{-\frac{x}{2}}\, dx$$

n \ F	.005	.010	.025	.050	.100	.250	.500	.750	.900	.950	.975	.990	.995
1	.0000393	.000157	.000982	.00393	.0158	.102	.455	1.32	2.71	3.84	5.02	6.63	7.88
2	.0100	.0201	.0506	.103	.211	.575	1.39	2.77	4.61	5.99	7.38	9.21	10.6
3	.0717	.115	.216	.352	.584	1.21	2.37	4.11	6.25	7.81	9.35	11.3	12.8
4	.207	.297	.484	.711	1.06	1.92	3.36	5.39	7.78	9.49	11.1	13.3	14.9
5	.412	.554	.831	1.15	1.61	2.67	4.35	6.63	9.24	11.1	12.8	1[illegible].1	16.7
6	.676	.872	1.24	1.64	2.20	3.45	5.35	7.84	10.6	12.6	14.4	16.8	18.5
7	.989	1.24	1.69	2.17	2.83	4.25	6.35	9.04	12.0	14.1	16.0	18.5	20.3
8	1.34	1.65	2.18	2.73	3.49	5.07	7.34	10.2	13.4	15.5	17.5	20.1	22.0
9	1.73	2.09	2.70	3.33	4.17	5.90	8.34	11.4	14.7	16.9	19.0	21.7	23.6
10	2.16	2.56	3.25	3.94	4.87	6.74	9.34	12.5	16.0	18.3	20.5	23.2	25.2
11	2.60	3.05	3.82	4.57	5.58	7.58	10.3	13.7	17.3	19.7	21.9	24.7	26.8
12	3.07	3.57	4.40	5.23	6.30	8.44	11.3	14.8	18.5	21.0	23.3	26.2	28.3
13	3.57	4.11	5.01	5.89	7.04	9.30	12.3	16.0	19.8	22.4	24.7	27.7	29.8
14	4.07	4.66	5.63	6.57	7.79	10.2	13.3	17.1	21.1	23.7	26.1	29.1	31.3
15	4.60	5.23	6.26	7.26	8.55	11.0	.4.3	18.2	22.3	25.0	27.5	30.6	32.8
16	5.14	5.81	6.91	7.96	9.31	11.9	15.3	19.4	23.5	26.3	28.8	32.0	34.3
17	5.70	6.41	7.56	8.67	10.1	12.8	16.3	20.5	24.8	27.6	30.2	33.4	35.7
18	6.26	7.01	8.23	9.39	10.9	13.7	17.3	21.6	26.0	28.9	31.5	34.8	37.2
19	6.84	7.63	8.91	10.1	11.7	14.6	18.3	22.7	27.2	30.1	32.9	36.2	38.6
20	7.43	8.26	9.59	10.9	12.4	15.5	19.3	23.8	28.4	31.4	34.2	37.6	40.0
21	8.03	8.90	10.3	11.6	13.2	16.3	20.3	24.9	29.6	32.7	35.5	38.9	41.4
22	8.64	9.54	11.0	12.3	14.0	17.2	21.3	26.0	30.8	33.9	36.8	40.3	42.8
23	9.26	10.2	11.7	13.1	14.8	18.1	22.3	27.1	32.0	35.2	38.1	41.6	44.2
24	9.89	10.9	12.4	13.8	15.7	19.0	23.3	28.2	33.2	36.4	39.4	43.0	45.6
25	10.5	11.5	13.1	14.6	16.5	19.9	24.3	29.3	34.4	37.7	40.6	44.3	46.9
26	11.2	12.2	13.8	15.4	17.3	20.8	25.3	30.4	35.6	38.9	41.9	45.6	48.3
27	11.8	12.9	14.6	16.2	18.1	21.7	26.3	31.5	36.7	40.1	43.2	47.0	49.6
28	12.5	13.6	15.3	16.9	18.9	22.7	27.3	32.6	37.9	41.3	44.5	48.3	51.0
29	13.1	14.3	16.0	17.7	19.8	23.6	28.3	33.7	39.1	42.6	45.7	49.6	52.3
30	13.8	15.0	16.8	18.5	20.6	24.5	29.3	34.8	40.3	43.8	47.0	50.9	53.7

PERCENTAGE POINTS, *F*-DISTRIBUTION

This table gives values of F such that

$$F(F) = \int_0^F \frac{\Gamma\left(\frac{m+n}{2}\right)}{\Gamma\left(\frac{m}{2}\right)\Gamma\left(\frac{n}{2}\right)} m^{\frac{m}{2}} n^{\frac{n}{2}} x^{\frac{m-2}{2}} (n + mx)^{-\frac{m+n}{2}}\, dx$$

for selected values of m, the number of degrees of freedom of the numerator of F; and for selected values of n, the number of degrees of freedom of the denominator of F. The table also provides values corresponding to $F(F)$ = .10, .05, .025, .01, .005, .001 since $F_{1-\alpha}$ for m and n degrees of freedom is the reciprocal of F_α for n and m degrees of freedom. Thus

$$F_{.05}(4, 7) = \frac{1}{F_{.95}(7, 4)} = \frac{1}{6.09} = .164\ .$$

$$F(F) = \int_0^F \frac{\Gamma\left(\frac{m+n}{2}\right)}{\Gamma\left(\frac{m}{2}\right)\Gamma\left(\frac{n}{2}\right)} m^{\frac{m}{2}} n^{\frac{n}{2}} x^{\frac{m}{2}-1} (n+mx)^{-\frac{m+n}{2}} dx = .90$$

n \ m	1	2	3	4	5	6	7	8	9	10	12	15	20	24	30	40	60	120	∞
1	39.86	49.50	53.59	55.83	57.24	58.20	58.91	59.44	59.86	60.19	60.71	61.22	61.74	62.00	62.26	62.53	62.79	63.06	63.33
2	8.53	9.00	9.16	9.24	9.29	9.33	9.35	9.37	9.38	9.39	9.41	9.42	9.44	9.45	9.46	9.47	9.47	9.48	9.49
3	5.54	5.46	5.39	5.34	5.31	5.28	5.27	5.25	5.24	5.23	5.22	5.20	5.18	5.18	5.17	5.16	5.15	5.14	5.13
4	4.54	4.32	4.19	4.11	4.05	4.01	3.98	3.95	3.94	3.92	3.90	3.87	3.84	3.83	3.82	3.80	3.79	3.78	3.76
5	4.06	3.78	3.62	3.52	3.45	3.40	3.37	3.34	3.32	3.30	3.27	3.24	3.21	3.19	3.17	3.16	3.14	3.12	3.10
6	3.78	3.46	3.29	3.18	3.11	3.05	3.01	2.98	2.96	2.94	2.90	2.87	2.84	2.82	2.80	2.78	2.76	2.74	2.72
7	3.59	3.26	3.07	2.96	2.88	2.83	2.78	2.75	2.72	2.70	2.67	2.63	2.59	2.58	2.56	2.54	2.51	2.49	2.47
8	3.46	3.11	2.92	2.81	2.73	2.67	2.62	2.59	2.56	2.54	2.50	2.46	2.42	2.40	2.38	2.36	2.34	2.32	2.29
9	3.36	3.01	2.81	2.69	2.61	2.55	2.51	2.47	2.44	2.42	2.38	2.34	2.30	2.28	2.25	2.23	2.21	2.18	2.16
10	3.29	2.92	2.73	2.61	2.52	2.46	2.41	2.38	2.35	2.32	2.28	2.24	2.20	2.18	2.16	2.13	2.11	2.08	2.06
11	3.23	2.86	2.66	2.54	2.45	2.39	2.34	2.30	2.27	2.25	2.21	2.17	2.12	2.10	2.08	2.05	2.03	2.00	1.97
12	3.18	2.81	2.61	2.48	2.39	2.33	2.28	2.24	2.21	2.19	2.15	2.10	2.06	2.04	2.01	1.99	1.96	1.93	1.90
13	3.14	2.76	2.56	2.43	2.35	2.28	2.23	2.20	2.16	2.14	2.10	2.05	2.01	1.98	1.96	1.93	1.90	1.88	1.85
14	3.10	2.73	2.52	2.39	2.31	2.24	2.19	2.15	2.12	2.10	2.05	2.01	1.96	1.94	1.91	1.89	1.86	1.83	1.80
15	3.07	2.70	2.49	2.36	2.27	2.21	2.16	2.12	2.09	2.06	2.02	1.97	1.92	1.90	1.87	1.85	1.82	1.79	1.76
16	3.05	2.67	2.46	2.33	2.24	2.18	2.13	2.09	2.06	2.03	1.99	1.94	1.89	1.87	1.84	1.81	1.78	1.75	1.72
17	3.03	2.64	2.44	2.31	2.22	2.15	2.10	2.06	2.03	2.00	1.96	1.91	1.86	1.84	1.81	1.78	1.75	1.72	1.69
18	3.01	2.62	2.42	2.29	2.20	2.13	2.08	2.04	2.00	1.98	1.93	1.89	1.84	1.81	1.78	1.75	1.72	1.69	1.66
19	2.99	2.61	2.40	2.27	2.18	2.11	2.06	2.02	1.98	1.96	1.91	1.86	1.81	1.79	1.76	1.73	1.70	1.67	1.63
20	2.97	2.59	2.38	2.25	2.16	2.09	2.04	2.00	1.96	1.94	1.89	1.84	1.79	1.77	1.74	1.71	1.68	1.64	1.61
21	2.96	2.57	2.36	2.23	2.14	2.08	2.02	1.98	1.95	1.92	1.87	1.83	1.78	1.75	1.72	1.69	1.66	1.62	1.59
22	2.95	2.56	2.35	2.22	2.13	2.06	2.01	1.97	1.93	1.90	1.86	1.81	1.76	1.73	1.70	1.67	1.64	1.60	1.57
23	2.94	2.55	2.34	2.21	2.11	2.05	1.99	1.95	1.92	1.89	1.84	1.80	1.74	1.72	1.69	1.66	1.62	1.59	1.55
24	2.93	2.54	2.33	2.19	2.10	2.04	1.98	1.94	1.91	1.88	1.83	1.78	1.73	1.70	1.67	1.64	1.61	1.57	1.53
25	2.92	2.53	2.32	2.18	2.09	2.02	1.97	1.93	1.89	1.87	1.82	1.77	1.72	1.69	1.66	1.63	1.59	1.56	1.52
26	2.91	2.52	2.31	2.17	2.08	2.01	1.96	1.92	1.88	1.86	1.81	1.76	1.71	1.68	1.65	1.61	1.58	1.54	1.50
27	2.90	2.51	2.30	2.17	2.07	2.00	1.95	1.91	1.87	1.85	1.80	1.75	1.70	1.67	1.64	1.60	1.57	1.53	1.49
28	2.89	2.50	2.29	2.16	2.06	2.00	1.94	1.90	1.87	1.84	1.79	1.74	1.69	1.66	1.63	1.59	1.56	1.52	1.48
29	2.89	2.50	2.28	2.15	2.06	1.99	1.93	1.89	1.86	1.83	1.78	1.73	1.68	1.65	1.62	1.58	1.55	1.51	1.47
30	2.88	2.49	2.28	2.14	2.05	1.98	1.93	1.88	1.85	1.82	1.77	1.72	1.67	1.64	1.61	1.57	1.54	1.50	1.46
40	2.84	2.44	2.23	2.09	2.00	1.93	1.87	1.83	1.79	1.76	1.71	1.66	1.61	1.57	1.54	1.51	1.47	1.42	1.38
60	2.79	2.39	2.18	2.04	1.95	1.87	1.82	1.77	1.74	1.71	1.66	1.60	1.54	1.51	1.48	1.44	1.40	1.35	1.29
120	2.75	2.35	2.13	1.99	1.90	1.82	1.77	1.72	1.68	1.65	1.60	1.55	1.48	1.45	1.41	1.37	1.32	1.26	1.19
∞	2.71	2.30	2.08	1.94	1.85	1.77	1.72	1.67	1.63	1.60	1.55	1.49	1.42	1.38	1.34	1.30	1.24	1.17	1.00

$F = \frac{s_1^2}{s_2^2} = \frac{S_1}{m} \Big/ \frac{S_2}{n}$, where $s_1^2 = S_1/m$ and $s_2^2 = S_2/n$ are independent mean squares estimating a common variance σ^2 and based on m and n degrees of freedom, respectively.

$$F(F) = \int_0^F \frac{\Gamma\left(\frac{m+n}{2}\right)}{\Gamma\left(\frac{m}{2}\right)\Gamma\left(\frac{n}{2}\right)} m^{\frac{m}{2}} n^{\frac{n}{2}} x^{\frac{m}{2}-1} (n+mx)^{-\frac{m+n}{2}} dx = .95$$

n \ m	1	2	3	4	5	6	7	8	9	10	12	15	20	24	30	40	60	120	∞
1	161.4	199.5	215.7	224.6	230.2	234.0	236.8	238.9	240.5	241.9	243.9	245.9	248.0	249.1	250.1	251.1	252.2	253.3	254.3
2	18.51	19.00	19.16	19.25	19.30	19.33	19.35	19.37	19.38	19.40	19.41	19.43	19.45	19.45	19.46	19.47	19.48	19.49	19.50
3	10.13	9.55	9.28	9.12	9.01	8.94	8.89	8.85	8.81	8.79	8.74	8.70	8.66	8.64	8.62	8.59	8.57	8.55	8.53
4	7.71	6.94	6.59	6.39	6.26	6.16	6.09	6.04	6.00	5.96	5.91	5.86	5.80	5.77	5.75	5.72	5.69	5.66	5.63
5	6.61	5.79	5.41	5.19	5.05	4.95	4.88	4.82	4.77	4.74	4.68	4.62	4.56	4.53	4.50	4.46	4.43	4.40	4.36
6	5.99	5.14	4.76	4.53	4.39	4.28	4.21	4.15	4.10	4.06	4.00	3.94	3.87	3.84	3.81	3.77	3.74	3.70	3.67
7	5.59	4.74	4.35	4.12	3.97	3.87	3.79	3.73	3.68	3.64	3.57	3.51	3.44	3.41	3.38	3.34	3.30	3.27	3.23
8	5.32	4.46	4.07	3.84	3.69	3.58	3.50	3.44	3.39	3.35	3.28	3.22	3.15	3.12	3.08	3.04	3.01	2.97	2.93
9	5.12	4.26	3.86	3.63	3.48	3.37	3.29	3.23	3.18	3.14	3.07	3.01	2.94	2.90	2.86	2.83	2.79	2.75	2.71
10	4.96	4.10	3.71	3.48	3.33	3.22	3.14	3.07	3.02	2.98	2.91	2.85	2.77	2.74	2.70	2.66	2.62	2.58	2.54
11	4.84	3.98	3.59	3.36	3.20	3.09	3.01	2.95	2.90	2.85	2.79	2.72	2.65	2.61	2.57	2.53	2.49	2.45	2.40
12	4.75	3.89	3.49	3.26	3.11	3.00	2.91	2.85	2.80	2.75	2.69	2.62	2.54	2.51	2.47	2.43	2.38	2.34	2.30
13	4.67	3.81	3.41	3.18	3.03	2.92	2.83	2.77	2.71	2.67	2.60	2.53	2.46	2.42	2.38	2.34	2.30	2.25	2.21
14	4.60	3.74	3.34	3.11	2.96	2.85	2.76	2.70	2.65	2.60	2.53	2.46	2.39	2.35	2.31	2.27	2.22	2.18	2.13
15	4.54	3.68	3.29	3.06	2.90	2.79	2.71	2.64	2.59	2.54	2.48	2.40	2.33	2.29	2.25	2.20	2.16	2.11	2.07
16	4.49	3.63	3.24	3.01	2.85	2.74	2.66	2.59	2.54	2.49	2.42	2.35	2.28	2.24	2.19	2.15	2.11	2.06	2.01
17	4.45	3.59	3.20	2.96	2.81	2.70	2.61	2.55	2.49	2.45	2.38	2.31	2.23	2.19	2.15	2.10	2.06	2.01	1.96
18	4.41	3.55	3.16	2.93	2.77	2.66	2.58	2.51	2.46	2.41	2.34	2.27	2.19	2.15	2.11	2.06	2.02	1.97	1.92
19	4.38	3.52	3.13	2.90	2.74	2.63	2.54	2.48	2.42	2.38	2.31	2.23	2.16	2.11	2.07	2.03	1.98	1.93	1.88
20	4.35	3.49	3.10	2.87	2.71	2.60	2.51	2.45	2.39	2.35	2.28	2.20	2.12	2.08	2.04	1.99	1.95	1.90	1.84
21	4.32	3.47	3.07	2.84	2.68	2.57	2.49	2.42	2.37	2.32	2.25	2.18	2.10	2.05	2.01	1.96	1.92	1.87	1.81
22	4.30	3.44	3.05	2.82	2.66	2.55	2.46	2.40	2.34	2.30	2.23	2.15	2.07	2.03	1.98	1.94	1.89	1.84	1.78
23	4.28	3.42	3.03	2.80	2.64	2.53	2.44	2.37	2.32	2.27	2.20	2.13	2.05	2.01	1.96	1.91	1.86	1.81	1.76
24	4.26	3.40	3.01	2.78	2.62	2.51	2.42	2.36	2.30	2.25	2.18	2.11	2.03	1.98	1.94	1.89	1.84	1.79	1.73
25	4.24	3.39	2.99	2.76	2.60	2.49	2.40	2.34	2.28	2.24	2.16	2.09	2.01	1.96	1.92	1.87	1.82	1.77	1.71
26	4.23	3.37	2.98	2.74	2.59	2.47	2.39	2.32	2.27	2.22	2.15	2.07	1.99	1.95	1.90	1.85	1.80	1.75	1.69
27	4.21	3.35	2.96	2.73	2.57	2.46	2.37	2.31	2.25	2.20	2.13	2.06	1.97	1.93	1.88	1.84	1.79	1.73	1.67
28	4.20	3.34	2.95	2.71	2.56	2.45	2.36	2.29	2.24	2.19	2.12	2.04	1.96	1.91	1.87	1.82	1.77	1.71	1.65
29	4.18	3.33	2.93	2.70	2.55	2.43	2.35	2.28	2.22	2.18	2.10	2.03	1.94	1.90	1.85	1.81	1.75	1.70	1.64
30	4.17	3.32	2.92	2.69	2.53	2.42	2.33	2.27	2.21	2.16	2.09	2.01	1.93	1.89	1.84	1.79	1.74	1.68	1.62
40	4.08	3.23	2.84	2.61	2.45	2.34	2.25	2.18	2.12	2.08	2.00	1.92	1.84	1.79	1.74	1.69	1.64	1.58	1.51
60	4.00	3.15	2.76	2.53	2.37	2.25	2.17	2.10	2.04	1.99	1.92	1.84	1.75	1.70	[illegible]	[illegible]	[illegible]	1.47	1.39
120	[illegible]	3.07	[illegible]	[illegible]	[illegible]	[illegible]	2.09	2.02	1.96	1.91	1.83	1.75	1.66	1.61	1.55	1.50	1.43	1.35	1.25
∞	3.84	3.00	2.60	2.37	2.21	2.10	2.01	1.94	1.88	1.83	1.75	1.67	1.57	1.52	1.46	1.39	1.32	1.22	1.00

$F = \frac{s_1^2}{s_2^2} = \frac{S_1}{m} \Big/ \frac{S_2}{n}$, where $s_1^2 = S_1/m$ and $s_2^2 = S_2/n$ are independent mean squares estimating a common variance σ^2 and based on m and n degrees of freedom, respectively.

$$F(F) = \int_0^F \frac{\Gamma\left(\frac{m+n}{2}\right)}{\Gamma\left(\frac{m}{2}\right)\Gamma\left(\frac{n}{2}\right)} m^{\frac{m}{2}} n^{\frac{n}{2}} x^{\frac{m}{2}-1} (n + mx)^{-\frac{m+n}{2}} dx = .975$$

n \ m	1	2	3	4	5	6	7	8	9	10	12	15	20	24	30	40	60	120	∞
1	647.8	799.5	864.2	899.6	921.8	937.1	948.2	956.7	963.3	968.6	976.7	984.9	993.1	997.2	1001	1006	1010	1014	1018
2	38.51	39.00	39.17	39.25	39.30	39.33	39.36	39.37	39.39	39.40	39.41	39.43	39.45	39.46	39.46	39.47	39.48	39.49	39.50
3	17.44	16.04	15.44	15.10	14.88	14.73	14.62	14.54	14.47	14.42	14.34	14.25	14.17	14.12	14.08	14.04	13.99	13.95	13.90
4	12.22	10.65	9.98	9.60	9.36	9.20	9.07	8.98	8.90	8.84	8.75	8.66	8.56	8.51	8.46	8.41	8.36	8.31	8.26
5	10.01	8.43	7.76	7.39	7.15	6.98	6.85	6.76	6.68	6.62	6.52	6.43	6.33	6.28	6.23	6.18	6.12	6.07	6.02
6	8.81	7.26	6.60	6.23	5.99	5.82	5.70	5.60	5.52	5.46	5.37	5.27	5.17	5.12	5.07	5.01	4.96	4.90	4.85
7	8.07	6.54	5.89	5.52	5.29	5.12	4.99	4.90	4.82	4.76	4.67	4.57	4.47	4.42	4.36	4.31	4.25	4.20	4.14
8	7.57	6.06	5.42	5.05	4.82	4.65	4.53	4.43	4.36	4.30	4.20	4.10	4.00	3.95	3.89	3.84	3.78	3.73	3.67
9	7.21	5.71	5.08	4.72	4.48	4.32	4.20	4.10	4.03	3.96	3.87	3.77	3.67	3.61	3.56	3.51	3.45	3.39	3.33
10	6.94	5.46	4.83	4.47	4.24	4.07	3.95	3.85	3.78	3.72	3.62	3.52	3.42	3.37	3.31	3.26	3.20	3.14	3.08
11	6.72	5.26	4.63	4.28	4.04	3.88	3.76	3.66	3.59	3.53	3.43	3.33	3.23	3.17	3.12	3.06	3.00	2.94	2.88
12	6.55	5.10	4.47	4.12	3.89	3.73	3.61	3.51	3.44	3.37	3.28	3.18	3.07	3.02	2.96	2.91	2.85	2.79	2.72
13	6.41	4.97	4.35	4.00	3.77	3.60	3.48	3.39	3.31	3.25	3.15	3.05	2.95	2.89	2.84	2.78	2.72	2.66	2.60
14	6.30	4.86	4.24	3.89	3.66	3.50	3.38	3.29	3.21	3.15	3.05	2.95	2.84	2.79	2.73	2.67	2.61	2.55	2.49
15	6.20	4.77	4.15	3.80	3.58	3.41	3.29	3.20	3.12	3.06	2.96	2.86	2.76	2.70	2.64	2.59	2.52	2.46	2.40
16	6.12	4.69	4.08	3.73	3.50	3.34	3.22	3.12	3.05	2.99	2.89	2.79	2.68	2.63	2.57	2.51	2.45	2.38	2.32
17	6.04	4.62	4.01	3.66	3.44	3.28	3.16	3.06	2.98	2.92	2.82	2.72	2.62	2.56	2.50	2.44	2.38	2.32	2.25
18	5.98	4.56	3.95	3.61	3.38	3.22	3.10	3.01	2.93	2.87	2.77	2.67	2.56	2.50	2.44	2.38	2.32	2.26	2.19
19	5.92	4.51	3.90	3.56	3.33	3.17	3.05	2.96	2.88	2.82	2.72	2.62	2.51	2.45	2.39	2.33	2.27	2.20	2.13
20	5.87	4.46	3.86	3.51	3.29	3.13	3.01	2.91	2.84	2.77	2.68	2.57	2.46	2.41	2.35	2.29	2.22	2.16	2.09
21	5.83	4.42	3.82	3.48	3.25	3.09	2.97	2.87	2.80	2.73	2.64	2.53	2.42	2.37	2.31	2.25	2.18	2.11	2.04
22	5.79	4.38	3.78	3.44	3.22	3.05	2.93	2.84	2.76	2.70	2.60	2.50	2.39	2.33	2.27	2.21	2.14	2.08	2.00
23	5.75	4.35	3.75	3.41	3.18	3.02	2.90	2.81	2.73	2.67	2.57	2.47	2.36	2.30	2.24	2.18	2.11	2.04	1.97
24	5.72	4.32	3.72	3.38	3.15	2.99	2.87	2.78	2.70	2.64	2.54	2.44	2.33	2.27	2.21	2.15	2.08	2.01	1.94
25	5.69	4.29	3.69	3.35	3.13	2.97	2.85	2.75	2.68	2.61	2.51	2.41	2.30	2.24	2.18	2.12	2.05	1.98	1.91
26	5.66	4.27	3.67	3.33	3.10	2.94	2.82	2.73	2.65	2.59	2.49	2.39	2.28	2.22	2.16	2.09	2.03	1.95	1.88
27	5.63	4.24	3.65	3.31	3.08	2.92	2.80	2.71	2.63	2.57	2.47	2.36	2.25	2.19	2.13	2.07	2.00	1.93	1.85
28	5.61	4.22	3.63	3.29	3.06	2.90	2.78	2.69	2.61	2.55	2.45	2.34	2.23	2.17	2.11	2.05	1.98	1.91	1.83
29	5.59	4.20	3.61	3.27	3.04	2.88	2.76	2.67	2.59	2.53	2.43	2.32	2.21	2.15	2.09	2.03	1.96	1.89	1.81
30	5.57	4.18	3.59	3.25	3.03	2.87	2.75	2.65	2.57	2.51	2.41	2.31	2.20	2.14	2.07	2.01	1.94	1.87	1.79
40	5.42	4.05	3.46	3.13	2.90	2.74	2.62	2.53	2.45	2.39	2.29	2.18	2.07	2.01	1.94	1.88	1.80	1.72	1.64
60	5.29	3.93	3.31	3.01	2.79	2.63	2.51	2.41	2.33	2.27	2.17	2.06	1.94	1.88	1.82	1.74	1.67	1.58	1.48
120	5.15	3.80	3.23	2.89	2.67	2.52	2.39	2.30	2.22	2.16	2.05	1.94	1.82	1.76	1.69	1.61	1.53	1.43	1.31
∞	5.02	3.69	3.12	2.79	2.57	2.41	2.29	2.19	2.11	2.05	1.94	1.83	1.71	1.64	1.57	1.48	1.39	1.27	1.00

$F = \frac{s_1^2}{s_2^2} = \frac{S_1}{m} \Big/ \frac{S_2}{n}$, where $s_1^2 = S_1/m$ and $s_2^2 = S_2/n$ are independent mean squares estimating a common variance σ^2 and based on m and n degrees of freedom, respectively.

$$F(F) = \int_0^F \frac{\Gamma\left(\frac{m+n}{2}\right)}{\Gamma\left(\frac{m}{2}\right)\Gamma\left(\frac{n}{2}\right)} m^{\frac{m}{2}} n^{\frac{n}{2}} x^{\frac{m}{2}-1} (n + mx)^{-\frac{m+n}{2}} dx = .99$$

n \ m	1	2	3	4	5	6	7	8	9	10	12	15	20	24	30	40	60	120	∞
1	4052	4999.5	5403	5625	5764	5859	5928	5982	6022	6056	6106	6157	6209	6235	6261	6287	6313	6339	6366
2	98.50	99.00	99.17	99.25	99.30	99.33	99.36	99.37	99.39	99.40	99.42	99.43	99.45	99.46	99.47	99.47	99.48	99.49	99.50
3	34.12	30.82	29.46	28.71	28.24	27.91	27.67	27.49	27.35	27.23	27.05	26.87	26.69	26.60	26.50	26.41	26.32	26.22	26.13
4	21.20	18.00	16.69	15.98	15.52	15.21	14.98	14.80	14.66	14.55	14.37	14.20	14.02	13.93	13.84	13.75	13.65	13.56	13.46
5	16.26	13.27	12.06	11.39	10.97	10.67	10.46	10.29	10.16	10.05	9.89	9.72	9.55	9.47	9.38	9.29	9.20	9.11	9.02
6	13.75	10.92	9.78	9.15	8.75	8.47	8.26	8.10	7.98	7.87	7.72	7.56	7.40	7.31	7.23	7.14	7.06	6.97	6.88
7	12.25	9.55	8.45	7.85	7.46	7.19	6.99	6.84	6.72	6.62	6.47	6.31	6.16	6.07	5.99	5.91	5.82	5.74	5.65
8	11.26	8.65	7.59	7.01	6.63	6.37	6.18	6.03	5.91	5.81	5.67	5.52	5.36	5.28	5.20	5.12	5.03	4.95	4.86
9	10.56	8.02	6.99	6.42	6.06	5.80	5.61	5.47	5.35	5.26	5.11	4.96	4.81	4.73	4.65	4.57	4.48	4.40	4.31
10	10.04	7.56	6.55	5.99	5.64	5.39	5.20	5.06	4.94	4.85	4.71	4.56	4.41	4.33	4.25	4.17	4.08	4.00	3.91
11	9.65	7.21	6.22	5.67	5.32	5.07	4.89	4.74	4.63	4.54	4.40	4.25	4.10	4.02	3.94	3.86	3.78	3.69	3.60
12	9.33	6.93	5.95	5.41	5.06	4.82	4.64	4.50	4.39	4.30	4.16	4.01	3.86	3.78	3.70	3.62	3.54	3.45	3.36
13	9.07	6.70	5.74	5.21	4.86	4.62	4.44	4.30	4.19	4.10	3.96	3.82	3.66	3.59	3.51	3.43	3.34	3.25	3.17
14	8.86	6.51	5.56	5.04	4.69	4.46	4.28	4.14	4.03	3.94	3.80	3.66	3.51	3.43	3.35	3.27	3.18	3.09	3.00
15	8.68	6.36	5.42	4.89	4.56	4.32	4.14	4.00	3.89	3.80	3.67	3.52	3.37	3.29	3.21	3.13	3.05	2.96	2.87
16	8.53	6.23	5.29	4.77	4.44	4.20	4.03	3.89	3.78	3.69	3.55	3.41	3.26	3.18	3.10	3.02	2.93	2.84	2.75
17	8.40	6.11	5.18	4.67	4.34	4.10	3.93	3.79	3.68	3.59	3.46	3.31	3.16	3.08	3.00	2.92	2.83	2.75	2.65
18	8.29	6.01	5.09	4.58	4.25	4.01	3.84	3.71	3.60	3.51	3.37	3.23	3.08	3.00	2.92	2.84	2.75	2.66	2.57
19	8.18	5.93	5.01	4.50	4.17	3.94	3.77	3.63	3.52	3.43	3.30	3.15	3.00	2.92	2.84	2.76	2.67	2.58	2.49
20	8.10	5.85	4.94	4.43	4.10	3.87	3.70	3.56	3.46	3.37	3.23	3.09	2.94	2.86	2.78	2.69	2.61	2.52	2.42
21	8.02	5.78	4.87	4.37	4.04	3.81	3.64	3.51	3.40	3.31	3.17	3.03	2.88	2.80	2.72	2.64	2.55	2.46	2.36
22	7.95	5.72	4.82	4.31	3.99	3.76	3.59	3.45	3.35	3.26	3.12	2.98	2.83	2.75	2.67	2.58	2.50	2.40	2.31
23	7.88	5.66	4.76	4.26	3.94	3.71	3.54	3.41	3.30	3.21	3.07	2.93	2.78	2.70	2.62	2.54	2.45	2.35	2.26
24	7.82	5.61	4.72	4.22	3.90	3.67	3.50	3.36	3.26	3.17	3.03	2.89	2.74	2.66	2.58	2.49	2.40	2.31	2.21
25	7.77	5.57	4.68	4.18	3.85	3.63	3.46	3.32	3.22	3.13	2.99	2.85	2.70	2.62	2.54	2.45	2.36	2.27	2.17
26	7.72	5.53	4.64	4.14	3.82	3.59	3.42	3.29	3.18	3.09	2.96	2.81	2.66	2.58	2.50	2.42	2.33	2.23	2.13
27	7.68	5.49	4.60	4.11	3.78	3.56	3.39	3.26	3.15	3.06	2.93	2.78	2.63	2.55	2.47	2.38	2.29	2.20	2.10
28	7.64	5.45	4.57	4.07	3.75	3.53	3.36	3.23	3.12	3.03	2.90	2.75	2.60	2.52	2.44	2.35	2.26	2.17	2.06
29	7.60	5.42	4.54	4.04	3.73	3.50	3.33	3.20	3.09	3.00	2.87	2.73	2.57	2.49	2.41	2.33	2.23	2.14	2.03
30	7.56	5.39	4.51	4.02	3.70	3.47	3.30	3.17	3.07	2.98	2.84	2.70	2.55	2.47	2.39	2.30	2.21	2.11	2.01
40	7.31	5.18	4.31	3.83	3.51	3.29	3.12	2.99	2.89	2.80	2.66	2.52	2.37	2.29	2.20	2.11	2.02	1.92	1.80
60	7.08	4.98	4.13	3.65	3.34	3.12	2.95	2.82	2.72	2.63	2.50	2.35	2.20	2.12	2.03	1.94	1.84	1.73	1.60
120	6.85	4.79	3.95	3.48	3.17	2.96	2.79	2.66	2.56	2.47	2.34	2.19	2.03	1.95	1.86	1.76	1.66	1.53	1.38
∞	6.63	4.61	3.78	3.32	3.02	2.80	2.64	2.51	2.41	2.32	2.18	2.04	1.88	1.79	1.70	1.59	1.47	1.32	1.00

$F = \frac{s_1^2}{s_2^2} = \frac{S_1}{m} \Big/ \frac{S_2}{n}$, where $s_1^2 = S_1/m$ and $s_2^2 = S_2/n$ are independent mean squares estimating a common variance σ^2 and based on m and n degrees of freedom, respectively.

$$F(F) = \int_0^F \frac{\Gamma\left(\frac{m+n}{2}\right)}{\Gamma\left(\frac{m}{2}\right)\Gamma\left(\frac{n}{2}\right)} m^{\frac{m}{2}} n^{\frac{n}{2}} x^{\frac{m}{2}-1} (n+mx)^{-\frac{m+n}{2}} dx = .995$$

n \ m	1	2	3	4	5	6	7	8	9	10	12	15	20	24	30	40	60	120	∞
1	16211	20000	21615	22500	23056	23437	23715	23925	24091	24224	24426	24630	24836	24940	25044	25148	25253	25359	25465
2	198.5	199.0	199.2	199.2	199.3	199.3	199.4	199.4	199.4	199.4	199.4	199.4	199.4	199.5	199.5	199.5	199.5	199.5	199.5
3	55.55	49.80	47.47	46.19	45.39	44.84	44.43	44.13	43.88	43.69	43.39	43.08	42.78	42.62	42.47	42.31	42.15	41.99	41.83
4	31.33	26.28	24.26	23.15	22.46	21.97	21.62	21.35	21.14	20.97	20.70	20.44	20.17	20.03	19.89	19.75	19.61	19.47	19.32
5	22.78	18.31	16.53	15.56	14.94	14.51	14.20	13.96	13.77	13.62	13.38	13.15	12.90	12.78	12.66	12.53	12.40	12.27	12.14
6	18.63	14.54	12.92	12.03	11.46	11.07	10.79	10.57	10.39	10.25	10.03	9.81	9.59	9.47	9.36	9.24	9.12	9.00	8.88
7	16.24	12.40	10.88	10.05	9.52	9.16	8.89	8.68	8.51	8.38	8.18	7.97	7.75	7.65	7.53	7.42	7.31	7.19	7.08
8	14.69	11.04	9.60	8.81	8.30	7.95	7.69	7.50	7.34	7.21	7.01	6.81	6.61	6.50	6.40	6.29	6.18	6.06	5.95
9	13.61	10.11	8.72	7.96	7.47	7.13	6.88	6.69	6.54	6.42	6.23	6.03	5.83	5.73	5.62	5.52	5.41	5.30	5.19
10	12.83	9.43	8.08	7.34	6.87	6.54	6.30	6.12	5.97	5.85	5.66	5.47	5.27	5.17	5.07	4.97	4.86	4.75	4.64
11	12.23	8.91	7.60	6.88	6.42	6.10	5.86	5.68	5.54	5.42	5.24	5.05	4.86	4.76	4.65	4.55	4.44	4.34	4.23
12	11.75	8.51	7.23	6.52	6.07	5.76	5.52	5.35	5.20	5.09	4.91	4.72	4.53	4.43	4.33	4.23	4.12	4.01	3.90
13	11.37	8.19	6.93	6.23	5.79	5.48	5.25	5.08	4.94	4.82	4.64	4.46	4.27	4.17	4.07	3.97	3.87	3.76	3.65
14	11.06	7.92	6.68	6.00	5.56	5.26	5.03	4.86	4.72	4.60	4.43	4.25	4.06	3.96	3.86	3.76	3.66	3.55	3.44
15	10.80	7.70	6.48	5.80	5.37	5.07	4.85	4.67	4.54	4.42	4.25	4.07	3.88	3.79	3.69	3.58	3.48	3.37	3.26
16	10.58	7.51	6.30	5.64	5.21	4.91	4.69	4.52	4.38	4.27	4.10	3.92	3.73	3.64	3.54	3.44	3.33	3.22	3.11
17	10.38	7.35	6.16	5.50	5.07	4.78	4.56	4.39	4.25	4.14	3.97	3.79	3.61	3.51	3.41	3.31	3.21	3.10	2.98
18	10.22	7.21	6.03	5.37	4.96	4.66	4.44	4.28	4.14	4.03	3.86	3.68	3.50	3.40	3.30	3.20	3.10	2.99	2.87
19	10.07	7.09	5.92	5.27	4.85	4.56	4.34	4.18	4.04	3.93	3.76	3.59	3.40	3.31	3.21	3.11	3.00	2.89	2.78
20	9.94	6.99	5.82	5.17	4.76	4.47	4.26	4.09	3.96	3.85	3.68	3.50	3.32	3.22	3.12	3.02	2.92	2.81	2.69
21	9.83	6.89	5.73	5.09	4.68	4.39	4.18	4.01	3.88	3.77	3.60	3.43	3.24	3.15	3.05	2.95	2.84	2.73	2.61
22	9.73	6.81	5.65	5.02	4.61	4.32	4.11	3.94	3.81	3.70	3.54	3.36	3.18	3.08	2.98	2.88	2.77	2.66	2.55
23	9.63	6.73	5.58	4.95	4.54	4.26	4.05	3.88	3.75	3.64	3.47	3.30	3.12	3.02	2.92	2.82	2.71	2.60	2.48
24	9.55	6.66	5.52	4.89	4.49	4.20	3.99	3.83	3.69	3.59	3.42	3.25	3.06	2.97	2.87	2.77	2.66	2.55	2.43
25	9.48	6.60	5.46	4.84	4.43	4.15	3.94	3.78	3.64	3.54	3.37	3.20	3.01	2.92	2.82	2.72	2.61	2.50	2.38
26	9.41	6.54	5.41	4.79	4.38	4.10	3.89	3.73	3.60	3.49	3.33	3.15	2.97	2.87	2.77	2.67	2.56	2.45	2.33
27	9.34	6.49	5.36	4.74	4.34	4.06	3.85	3.69	3.56	3.45	3.28	3.11	2.93	2.83	2.73	2.63	2.52	2.41	2.25
28	9.28	6.44	5.32	4.70	4.30	4.02	3.81	3.65	3.52	3.41	3.25	3.07	2.89	2.79	2.69	2.59	2.48	2.37	2.29
29	9.23	6.40	5.28	4.66	4.26	3.98	3.77	3.61	3.48	3.38	3.21	3.04	2.86	2.76	2.66	2.56	2.45	2.33	2.24
30	9.18	6.35	5.24	4.62	4.23	3.95	3.74	3.58	3.45	3.34	3.18	3.01	2.82	2.73	2.63	2.52	2.42	2.30	2.18
40	8.83	6.07	4.98	4.37	3.99	3.71	3.51	3.35	3.22	3.12	2.95	2.78	2.60	2.50	2.40	2.30	2.18	2.06	1.93
60	8.49	5.79	4.73	4.14	3.76	3.49	3.29	3.13	3.01	2.90	2.74	2.57	2.39	2.29	2.19	2.08	1.96	1.83	1.69
120	8.18	5.54	4.50	3.92	3.55	3.28	3.09	2.93	2.81	2.71	2.54	2.37	2.19	2.09	1.98	1.87	1.75	1.61	1.43
∞	7.88	5.30	4.28	3.72	3.35	3.09	2.90	2.74	2.62	2.52	2.36	2.19	2.00	1.90	1.79	1.67	1.53	1.36	1.00

$F = \frac{s_1^2}{s_2^2} = \frac{S_1}{m} \Big/ \frac{S_2}{n}$, where $s_1^2 = S_1/m$ and $s_2^2 = S_2/n$ are independent mean squares estimating a common variance σ^2 and based on m and n degrees of freedom, respectively.

$$F(F) = \int_0^F \frac{\Gamma\left(\frac{m+n}{2}\right)}{\Gamma\left(\frac{m}{2}\right)\Gamma\left(\frac{n}{2}\right)} m^{\frac{m}{2}} n^{\frac{n}{2}} x^{\frac{m}{2}-1} (n+mx)^{-\frac{m+n}{2}} dx = .999$$

n \ m	1	2	3	4	5	6	7	8	9	10	12	15	20	24	30	40	60	120	∞
1	4053*	5000*	5404*	5625*	5764*	5859*	5929*	5981*	6023*	6056*	6107*	6158*	6209*	6235*	6261*	6287*	6313*	6340*	6366*
2	998.5	999.0	999.2	999.2	999.3	999.3	999.4	999.4	999.4	999.4	999.4	999.4	999.4	999.5	999.5	999.5	999.5	999.5	999.5
3	167.0	148.5	141.1	137.1	134.6	132.8	131.6	130.6	129.9	129.2	128.3	127.4	126.4	125.9	125.4	125.0	124.5	124.0	123.5
4	74.14	61.25	56.18	53.44	51.71	50.53	49.66	49.00	48.47	48.05	47.41	46.76	46.10	45.77	45.43	45.09	44.75	44.40	44.05
5	47.18	37.12	33.20	31.09	29.75	28.84	28.16	27.64	27.24	26.92	26.42	25.91	25.39	25.14	24.87	24.60	24.33	24.06	23.79
6	35.51	27.00	23.70	21.92	20.81	20.03	19.46	19.03	18.69	18.41	17.99	17.56	17.12	16.89	16.67	16.44	16.21	15.99	15.75
7	29.25	21.69	18.77	17.19	16.21	15.52	15.02	14.63	14.33	14.08	13.71	13.32	12.93	12.73	12.53	12.33	12.12	11.91	11.70
8	25.42	18.49	15.83	14.39	13.49	12.86	12.40	12.04	11.77	11.54	11.19	10.84	10.48	10.30	10.11	9.92	9.73	9.53	9.33
9	22.86	16.39	13.90	12.56	11.71	11.13	10.70	10.37	10.11	9.89	9.57	9.24	8.90	8.72	8.55	8.37	8.19	8.00	7.81
10	21.04	14.91	12.55	11.28	10.48	9.92	9.52	9.20	8.96	8.75	8.45	8.13	7.80	7.64	7.47	7.30	7.12	6.94	6.76
11	19.69	13.81	11.56	10.35	9.58	9.05	8.66	8.35	8.12	7.92	7.63	7.32	7.01	6.85	6.68	6.52	6.35	6.17	6.00
12	18.64	12.97	10.80	9.63	8.89	8.38	8.00	7.71	7.48	7.29	7.00	6.71	6.40	6.25	6.09	5.93	5.76	5.59	5.42
13	17.81	12.31	10.21	9.07	8.35	7.86	7.49	7.21	6.98	6.80	6.52	6.23	5.93	5.78	5.63	5.47	5.30	5.14	4.97
14	17.14	11.78	9.73	8.62	7.92	7.43	7.08	6.80	6.58	6.40	6.13	5.85	5.56	5.41	5.25	5.10	4.94	4.77	4.60
15	16.59	11.34	9.34	8.25	7.57	7.09	6.74	6.47	6.26	6.08	5.81	5.54	5.25	5.10	4.95	4.80	4.64	4.47	4.31
16	16.12	10.97	9.00	7.94	7.27	6.81	6.46	6.19	5.98	5.81	5.55	5.27	4.99	4.85	4.70	4.54	4.39	4.23	4.06
17	15.72	10.66	8.73	7.68	7.02	6.56	6.22	5.96	5.75	5.58	5.32	5.05	4.78	4.63	4.48	4.33	4.18	4.02	3.85
18	15.38	10.39	8.49	7.46	6.81	6.35	6.02	5.76	5.56	5.39	5.13	4.87	4.59	4.45	4.30	4.15	4.00	3.84	3.67
19	15.08	10.16	8.28	7.26	6.62	6.18	5.85	5.59	5.39	5.22	4.97	4.70	4.43	4.29	4.14	3.00	3.84	3.68	3.51
20	14.82	9.95	8.10	7.10	6.46	6.02	5.69	5.44	5.24	5.08	4.82	4.56	4.29	4.15	4.00	3.86	3.70	3.54	3.38
21	14.59	9.77	7.94	6.95	6.32	5.88	5.56	5.31	5.11	4.95	4.70	4.44	4.17	4.03	3.88	3.74	3.58	3.42	3.26
22	14.38	9.61	7.80	6.81	6.19	5.76	5.44	5.19	4.99	4.83	4.58	4.33	4.06	3.92	3.78	3.63	3.48	3.32	3.15
23	14.19	9.47	7.67	6.69	6.08	5.65	5.33	5.09	4.89	4.73	4.48	4.23	3.96	3.82	3.68	3.53	3.38	3.22	3.05
24	14.03	9.34	7.55	6.59	5.98	5.55	5.23	4.99	4.80	4.64	4.39	4.14	3.87	3.74	3.59	3.45	3.29	3.14	2.97
25	13.88	9.22	7.45	6.49	5.88	5.46	5.15	4.91	4.71	4.56	4.31	4.06	3.79	3.66	3.52	3.37	3.22	3.06	2.89
26	13.74	9.12	7.36	6.41	5.80	5.38	5.07	4.83	4.64	4.48	4.24	3.99	3.72	3.59	3.44	3.30	3.15	2.99	2.82
27	13.61	9.02	7.27	6.33	5.73	5.31	5.00	4.76	4.57	4.41	4.17	3.92	3.66	3.52	3.38	3.23	3.08	2.92	2.75
28	13.50	8.93	7.19	6.25	5.66	5.24	4.93	4.69	4.50	4.35	4.11	3.86	3.60	3.46	3.32	3.18	3.02	2.86	2.69
29	13.39	8.85	7.12	6.19	5.59	5.18	4.87	4.64	4.45	4.29	4.05	3.80	3.54	3.41	3.27	3.12	2.97	2.81	2.64
30	13.29	8.77	7.05	6.12	5.53	5.12	4.82	4.58	4.39	4.24	4.00	3.75	3.49	3.36	3.22	3.07	2.92	2.76	2.59
40	12.61	8.25	6.60	5.70	5.13	4.73	4.44	4.21	4.02	3.87	3.64	3.40	3.15	3.01	2.87	2.73	2.57	2.41	2.23
60	11.97	7.76	6.17	5.31	4.76	4.37	4.09	3.87	3.69	3.54	3.31	3.08	2.83	2.69	2.55	2.41	2.25	2.08	1.80
120	11.38	7.32	5.79	4.95	4.42	4.04	3.77	3.55	3.38	3.24	3.02	2.78	2.53	2.40	2.26	2.11	1.95	1.76	1.54
∞	10.83	6.91	5.42	4.62	4.10	3.74	3.47	3.27	3.10	2.96	2.74	2.51	2.27	2.13	1.99	1.84	1.66	1.45	1.00

* Multiply these entries by 100.

Index

The most efficient way to use this index is to look for the pertinent *property* (e.g., vapor pressure, entropy), *process* (e.g., disposal of chemicals, calibration), or *general concept* (e.g., units, radiation). Most primary entries are subdivided into several secondary entries, e.g., under heat capacity there are 17 secondary entries such as air, metals, water, etc. Primary entries will be found for certain *classes of substances*, such as alloys, elements, organic compounds, refrigerants, semiconductors, etc. Primary entries are also given for the individual chemical elements and for a few compounds such as water and carbon dioxide. However, only the most important tables are listed under these substances. Therefore, the user will find in most cases that it is best to look first for the property of interest, then examine the table or tables that are referenced.

The reference given for each index term is the inclusive pages of the pertinent table (e.g., **8**-45 to 55). The introduction to each table describes the method of ordering the substances within that table.

The Editor would be grateful for comments and suggestions on this index.

INDEX

A

AAS, definition, **12**-1 to 6
Abbreviations
  amino acids, **7**-3
  physical quantities, **2**-1 to 21
  scientific terms, **2**-29 to 36
  units, **1**-25 to 32
Absorption
  infrared, by earth's atmosphere, **14**-27
  light, by elemental solids, **12**-133 to 156
  light, by other solids, **12**-157 to 178
  microwave power, by water, **6**-15
  sound, by air, **14**-43
Abundance, isotopic, **1**-15 to 18, **11**-50 to 198
Acceleration due to gravity
  at poles and equator, **14**-9 to 10
  at various latitudes, **14**-12
  on the sun, moon, and planets, **14**-2 to 3
  standard value, **1**-1 to 11
Acid dissociation constant
  amino acids, **7**-1 to 2
  biological buffers, **7**-12
  indicators, **8**-16 to 18
  inorganic acids and bases, **8**-44 to 45
  organic acids and bases, **8**-46 to 56
  purine and pyrimidine bases, **7**-4
Acid rain, pH measurement, **8**-41 to 42
Acids
  activity coefficients, **5**-97 to 98, **5**-99 to 103
  decinormal solutions, **8**-5 to 6
  enthalpy of dilution, **5**-104
  fatty, **7**-6
  indicators, **8**-16 to 18
  inorganic, dissociation constant, **8**-44 to 45
  organic, dissociation constant, **8**-46 to 56
Acoustics
  human hearing, **14**-47 to 48
  musical scales, **14**-45 to 46
  noise levels, **14**-47 to 48
  sound velocity, **14**-41 to 42, **14**-43
Acronyms, definitions, **2**-29 to 36
Actinium: see also Elements
  electron configuration, **1**-19 to 20
  heat capacity, **4**-133
  history, occurrence, uses, **4**-1 to 36
  ionization energy, **10**-175 to 177
  isotopes and their properties, **11**-50 to 198
  physical properties, **4**-132
  thermal properties, **12**-219 to 220
Activation energy
  atmospheric reactions, **5**-106 to 120
  diffusion in semiconductors, **12**-108 to 116
Activity coefficients of acids, bases, and salts, **5**-97 to 98, **5**-99 to 103
AES, definition, **12**-1 to 6
AFM, definition, **12**-1 to 6
Air
  absorption of sound, **14**-43
  cryogenic properties, **6**-139
  density, **6**-1 to 2
  diffusion of gases in, **6**-192 to 193
  enthalpy, **6**-1 to 2
  entropy, **6**-1 to 2
  heat capacity, **6**-1 to 2
  index of refraction, **10**-220
  mean free path, **6**-46
  permittivity (dielectric constant), **6**-174 to 175
  thermal conductivity, **6**-188 to 189
  thermodynamic properties, **6**-1 to 2
  U.S. Standard Atmosphere, **14**-19 to 25
  vapor pressure, **6**-1 to 2
  velocity of sound, **14**-43
  velocity of sound, as function of temperature, **14**-44
  viscosity, **6**-180 to 181
Airborne contaminants, threshold limits, **16**-32 to 42
Albedo
  planets, **14**-2 to 3
  satellites of the planets, **14**-4 to 5
ALI for radionuclides, **16**-49 to 52
Alkali halides, secondary electron emission, **12**-131 to 132
Alkali metals: see entries for Lithium, Sodium, etc.
Alloys
  composition, **12**-117 to 125
  electrical resistivity, **12**-48 to 51
  magnetic properties, **12**-117 to 125
  phase diagrams, **12**-200 to 217
  superconducting properties, **12**-64 to 90
  thermal conductivity, **12**-223
Alloys, commercial
  electrical resistivity, **12**-232
  mechanical and thermal properties, **12**-232
  thermal conductivity, **12**-223
Alphabets: Greek, Russian, Hebrew, **2**-37
Aluminum: see also Elements
  electrical resistivity, **12**-45 to 47
  electron configuration, **1**-19 to 20
  heat capacity, **4**-133, **12**-218
  history, occurrence, uses, **4**-1 to 36
  ionization energy, **10**-175 to 177
  isotopes and their properties, **11**-50 to 198
  magnetic susceptibility, **4**-141 to 146
  molten, density, **4**-137 to 140
  physical properties, **4**-132
  resistance of wires, **15**-27
  thermal conductivity, **12**-221 to 222
  thermal properties, **12**-219 to 220
  thermodynamic properties, **5**-1 to 3
  vapor pressure, **6**-65 to 92
  wire tables, **15**-27
Americium: see also Elements
  electron configuration, **1**-19 to 20
  history, occurrence, uses, **4**-1 to 36
  ionization energy, **10**-175 to 177
  isotopes and their properties, **11**-50 to 198
Amino acids
  abbreviations and symbols, **7**-3
  dissociation constants, **7**-1 to 2
  physical properties, **7**-1 to 2
  solubility, **7**-1 to 2
  structure, **7**-3
Ampere
  definition, **1**-25 to 32
  maintained value, **1**-1 to 11
Analytical procedures
  organic reagents, **8**-8 to 15
  preparation of reagents, **8**-1 to 4
  reduction of weighings, **8**-125
  solids and surfaces, **12**-1 to 6
  volumetric calibrations, **8**-125
Ångström, definition, **1**-25 to 32
Antiferroelectric crystals, Curie temperature, **12**-62
Antiferromagnetic elements and compounds, **12**-117 to 125
Antiferromagnetic materials, Faraday rotation, **12**-179 to 195
Antimony: see also Elements
  electrical resistivity, **12**-45 to 47
  electron configuration, **1**-19 to 20
  heat capacity, **4**-133
  history, occurrence, uses, **4**-1 to 36
  ionization energy, **10**-175 to 177
  isotopes and their properties, **11**-50 to 198
  magnetic susceptibility, **4**-141 to 146
  molten, density, **4**-137 to 140
  physical properties, **4**-132
  thermal properties, **12**-219 to 220
  vapor pressure, **6**-65 to 92
Appearance potential, molecules, **10**-178 to 195
APS, definition, **12**-1 to 6
Aqueous solutions
  activity coefficients, **5**-97 to 98, **5**-99 to 103
  concentrative properties, **8**-57 to 83
  conductivity, **5**-94, **5**-95 to 97
  density, **8**-57 to 83
  diffusion of ions, **5**-95 to 97
  diffusion of non-electrolytes, **6**-195
  enthalpy, **5**-105
  freezing point depression, **8**-57 to 83
  heat capacity, **5**-85 to 88
  hydrohalogen acids, conductivity, **5**-92 to 93
  index of refraction, **8**-57 to 83
  sodium chloride, volumetric properties, **6**-8
  solubility product constant, **8**-117 to 120
  surface tension, **6**-150
  thermodynamic properties, **5**-85 to 88

vapor pressure, **6**-104
viscosity, **8**-57 to 83
Argon: see also Elements
critical constants, **6**-48 to 61
cryogenic properties, **6**-139
electron configuration, **1**-19 to 20
entropy, **5**-1 to 3
history, occurrence, uses, **4**-1 to 36
ionization energy, **10**-175 to 177
isotopes and their properties, **11**-50 to 198
magnetic susceptibility, **4**-141 to 146
mean free path, **6**-46
permittivity (dielectric constant), **6**-151 to 173, **6**-174 to 175
physical properties, **4**-132
solubility in water, **8**-86 to 89
thermal conductivity, **6**-16 to 24, **6**-188 to 189
thermodynamic properties, **6**-16 to 24
thermodynamic properties at high temperature, **5**-61 to 84
van der Waals constants, **6**-45
vapor pressure, **6**-65 to 92, **6**-93 to 103
viscosity, **6**-16 to 24, **6**-180 to 181
Arsenic: see also Elements
critical constants, **6**-48 to 61
electron configuration, **1**-19 to 20
heat capacity, **4**-133
history, occurrence, uses, **4**-1 to 36
ionization energy, **10**-175 to 177
isotopes and their properties, **11**-50 to 198
magnetic susceptibility, **4**-141 to 146
molten, density, **4**-137 to 140
physical properties, **4**-132
vapor pressure, **6**-65 to 92
Astatine: see also Elements
electron configuration, **1**-19 to 20
history, occurrence, uses, **4**-1 to 36
ionization energy, **10**-175 to 177
isotopes and their properties, **11**-50 to 198
vapor pressure, **6**-65 to 92
Astronomical constants, **14**-1
Astronomical unit (AU), **14**-1
Atmosphere
carbon dioxide concentration, **14**-28 to 29
cosmic ray background, **11**-216 to 220
electrical conductance, **14**-33 to 40
infrared absorption, **14**-27
ion mobility, **14**-33 to 40
kinetic and photochemical data, **5**-106 to 120
mass, **14**-9 to 10
planetary, composition, **14**-2 to 3
potential, electrical, **14**-33 to 40
standard (unit), **1**-25 to 32
U.S. Standard, various properties, **14**-19 to 25
Atmospheric electricity, **14**-33 to 40
Atomic mass unit (amu), **1**-1 to 11, **1**-25 to 32
Atomic masses, **1**-15 to 18, **11**-50 to 198
Atomic radius, rare earth elements, **4**-122 to 131
Atomic spectra
elements, line spectra, **10**-1 to 87
ionization energies, **10**-175 to 177
transition probabilities, **10**-88 to 146
wavelengths, **10**-1 to 87
Atomic transition probability, **10**-88 to 146
Atomic weights, **1**-12 to 14
Atoms
electron affinity, **10**-147 to 159
electron binding energy, **10**-200 to 205
electron configuration, **1**-19 to 20
ionization energies, **10**-175 to 177
masses, **1**-15 to 18
photon cross sections, **10**-207 to 211
polarizability, **10**-160 to 174
spectra, **10**-1 to 87
x-ray energy levels, **10**-196 to 199
ATR, definition, **12**-1 to 6
Autoignition temperature, **16**-16 to 31
Avogadro constant, **1**-1 to 11
Azeotropes, **6**-176 to 179

B

Bands, electromagnetic (classification), **10**-212 to 213
Barium: see also Elements
electrical resistivity, **12**-45 to 47
electron configuration, **1**-19 to 20
heat capacity, **4**-133
history, occurrence, uses, **4**-1 to 36
ionization energy, **10**-175 to 177
isotopes and their properties, **11**-50 to 198
magnetic susceptibility, **4**-141 to 146
molten, density, **4**-137 to 140
physical properties, **4**-132
thermal properties, **12**-219 to 220
vapor pressure, **6**-65 to 92
Barn, definition, **1**-25 to 32
Baryons, summary of properties, **11**-1 to 49
Bases
activity coefficients, **5**-97 to 98, **5**-99 to 103
decinormal solutions, **8**-5 to 6
indicators, **8**-16 to 18
inorganic, dissociation constant, **8**-44 to 45
organic, dissociation constant, **8**-46 to 56
purine and pyrimidine, **7**-4
Becquerel, definition, **1**-25 to 32
Beilstein references for organic compounds, **3**-1 to 330
Berkelium: see also Elements
electron configuration, **1**-19 to 20
history, occurrence, uses, **4**-1 to 36
isotopes and their properties, **11**-50 to 198
Beryllium: see also Elements
electrical resistivity, **12**-45 to 47
electron configuration, **1**-19 to 20
heat capacity, **4**-133
history, occurrence, uses, **4**-1 to 36
ionization energy, **10**-175 to 177
isotopes and their properties, **11**-50 to 198
magnetic susceptibility, **4**-141 to 146
molten, density, **4**-137 to 140
physical properties, **4**-132
thermal properties, **12**-219 to 220
thermodynamic properties, **5**-1 to 3
vapor pressure, **6**-65 to 92
Bessel functions, **A**-96 to 99
BET, definition, **12**-1 to 6
Beta function, **A**-101
Billion, billiard (definition), **1**-46
Binding energy
electrons in atoms, **10**-200 to 205
in molecules, **9**-51 to 74
Binomial series, **A**-84 to 87
Biochemical nomenclature, references, **2**-22
Biochemical reactions, redox potentials, **7**-9 to 11
Biological buffers, **7**-12
Biological materials and tissues
effect of cosmic rays, **11**-216 to 220
pH, **7**-13
Biosphere, mass of, **14**-9 to 10
Bismuth: see also Elements
electrical resistivity, **12**-45 to 47
electron configuration, **1**-19 to 20
heat capacity, **4**-133
history, occurrence, uses, **4**-1 to 36
ionization energy, **10**-175 to 177
isotopes and their properties, **11**-50 to 198
magnetic susceptibility, **4**-141 to 146
molten, density, **4**-137 to 140
physical properties, **4**-132
thermal properties, **12**-219 to 220
vapor pressure, **6**-65 to 92
Black body radiation, **10**-214 to 215
Blood, pH of, **7**-13
Bohr magneton, **1**-1 to 11
Bohr radius, **1**-1 to 11
Bohrium (element 107), **4**-1 to 36, **11**-50 to 198
Boiling point
correction to standard pressure, **15**-19
cryogenic fluids, **6**-139
D_2O, **6**-4
elements, **4**-132
elevation of, **15**-20
halocarbons, **6**-141 to 143
index for organic compounds, **3**-751 to 763
inorganic compounds, **4**-37 to 96, **6**-48 to 61, **6**-106 to 121
metals, **12**-219 to 220
organic compounds, **3**-1 to 330, **6**-48 to 61, **6**-106 to 121
organic compounds, index, **3**-751 to 763
pressure dependence, **15**-19
rare earth elements, **4**-122 to 131
solvents, **15**-14 to 18
water, as function of pressure, **6**-12
Boltzmann constant, **1**-1 to 11
Bond lengths
diatomic molecules, **9**-81 to 86
gas-phase molecules, **9**-17 to 43
organic crystals, **9**-1 to 14
organometallic compounds, **9**-15 to 16
Bonds, chemical
lengths and angles, **9**-1 to 14, **9**-15 to 16, **9**-17 to 43

strengths, **9**-51 to 74
stretching force constants, **9**-76
Born-Haber cycle, **12**-22 to 35
Boron: see also Elements
electron configuration, **1**-19 to 20
heat capacity, **4**-133
history, occurrence, uses, **4**-1 to 36
ionization energy, **10**-175 to 177
isotopes and their properties, **11**-50 to 198
magnetic susceptibility, **4**-141 to 146
molten, density, **4**-137 to 140
physical properties, **4**-132
thermodynamic properties, **5**-1 to 3
vapor pressure, **6**-65 to 92
Bosons, summary of properties, **11**-1 to 49
Brass
phase diagram, **12**-200 to 217
resistance of wires, **15**-27
thermal conductivity, **12**-223
wire tables, **15**-27
Bravais lattices, **12**-7 to 13
Breakdown voltage, **15**-30 to 35
Bromine: see also Elements
critical constants, **6**-48 to 61
electron configuration, **1**-19 to 20
heat capacity, **4**-133
history, occurrence, uses, **4**-1 to 36
ionization energy, **10**-175 to 177
isotopes and their properties, **11**-50 to 198
physical properties, **4**-132
thermodynamic properties, **5**-1 to 3
thermodynamic properties at high temperature, **5**-61 to 84
vapor pressure, **6**-65 to 92, **6**-93 to 103
Buffers
biological, **7**-12
for round values of pH, **8**-43
for seawater measurements, **8**-41 to 42
standard solutions, **8**-35 to 40, **8**-41 to 42

C

Cadmium: see also Elements
electrical resistivity, **12**-45 to 47
electron configuration, **1**-19 to 20
heat capacity, **4**-133
history, occurrence, uses, **4**-1 to 36
ionization energy, **10**-175 to 177
isotopes and their properties, **11**-50 to 198
magnetic susceptibility, **4**-141 to 146
molten, density, **4**-137 to 140
physical properties, **4**-132
thermal properties, **12**-219 to 220
thermodynamic properties, **5**-1 to 3
vapor pressure, **6**-65 to 92
Calcium: see also Elements
electrical resistivity, **12**-45 to 47
electron configuration, **1**-19 to 20
heat capacity, **4**-133
history, occurrence, uses, **4**-1 to 36
ionization energy, **10**-175 to 177
isotopes and their properties, **11**-50 to 198
magnetic susceptibility, **4**-141 to 146
molten, density, **4**-137 to 140
physical properties, **4**-132
thermal properties, **12**-219 to 220
thermodynamic properties, **5**-1 to 3
vapor pressure, **6**-65 to 92
Calibration
boiling points, to standard pressure, **15**-19
conductivity cells, **5**-91
index of refraction, **10**-219
infrared and far infrared frequencies, **10**-236 to 240
infrared frequencies, **10**-228 to 235
relative humidity, **15**-26
temperature scale, **1**-21 to 22
thermocouples, **15**-1 to 13
vapor pressure, **6**-105
volumetric, **8**-125
weighings in air, **8**-125
Californium: see also Elements
electron configuration, **1**-19 to 20
history, occurrence, uses, **4**-1 to 36
isotopes and their properties, **11**-50 to 198
Calorie, value of, **1**-34 to 45
Candela, definition, **1**-25 to 32
Carbohydrate names and symbols, **7**-7 to 8
Carbon: see also Elements
dielectric constant, **12**-52 to 60
electron configuration, **1**-19 to 20
heat capacity, **4**-133
history, occurrence, uses, **4**-1 to 36
ionization energy, **10**-175 to 177
isotopes and their properties, **11**-50 to 198
magnetic susceptibility, **4**-141 to 146
physical properties, **4**-132
thermodynamic properties, **5**-1 to 3
vapor pressure, **6**-65 to 92
Carbon dioxide
atmospheric concentration (historical), **14**-28 to 29
critical constants, **6**-48 to 61
infrared laser frequencies, **10**-228 to 235
mean free path, **6**-46
permittivity (dielectric constant), **6**-151 to 173, **6**-174 to 175
solubility in water, **8**-86 to 89
solubility in water at various pressures, **8**-90
speed of sound in, **14**-41 to 42
standard thermodynamic properties, **5**-4 to 60
sublimation pressure, **6**-62 to 64
thermal conductivity, **6**-188 to 189
thermodynamic properties at high temperature, **5**-61 to 84
van der Waals constants, **6**-45
vapor pressure, **6**-93 to 103, **6**-105
viscosity, **6**-180 to 181, **6**-187
Carcinogenic chemicals, list, **16**-53 to 57
Carrier gas properties, for chromatography, **8**-126
CARS, definition, **12**-1 to 6
CAS Registry Numbers
inorganic compounds, **4**-37 to 96
inorganic compounds, index, **4**-112 to 121
organic compounds, **3**-1 to 330
organic compounds, index, **3**-709 to 736
Celsius temperature
conversion to other scales, **1**-46
definition, **1**-25 to 32
Ceramics
breakdown voltage, **15**-30 to 35
permittivity (dielectric constant), **12**-52 to 60
phase diagrams, **12**-200 to 217
thermal conductivity, **12**-226 to 227
Cerenkov light, in cosmic ray showers, **11**-216 to 220
Cerium: see also Elements
electrical resistivity, **12**-45 to 47
electron configuration, **1**-19 to 20
heat capacity, **4**-133
history, occurrence, uses, **4**-1 to 36
ionization energy, **10**-175 to 177
isotopes and their properties, **11**-50 to 198
magnetic susceptibility, **4**-141 to 146
molten, density, **4**-137 to 140
physical properties, **4**-132
thermal properties, **12**-219 to 220
vapor pressure, **6**-65 to 92
Cesium: see also Elements
electrical resistivity, **12**-45 to 47
electron configuration, **1**-19 to 20
heat capacity, **4**-133
history, occurrence, uses, **4**-1 to 36
ionization energy, **10**-175 to 177
isotopes and their properties, **11**-50 to 198
magnetic susceptibility, **4**-141 to 146
molten, density, **4**-137 to 140
physical properties, **4**-132
thermal properties, **12**-219 to 220
thermodynamic properties, **5**-1 to 3
vapor pressure, **6**-65 to 92
CFC's, various properties, **6**-141 to 143
Characterization of materials, **12**-1 to 6
Charge
electron, **1**-1 to 11
fundamental particles, **11**-1 to 49
Chemical Abstracts Service nomenclature, **2**-22
Chemical Abstracts Service Registry Numbers: see CAS Registry Numbers
Chemical carcinogens, **16**-53 to 57
Chemical composition of human body, **7**-14
Chemical kinetics
atmospheric reactions, **5**-106 to 120
conversion factors, **1**-51
Chemical nomenclature, **2**-22
Chemical shifts, NMR
for ^{13}C, **9**-96
for protons, **9**-95
of solvents for NMR, **8**-128
Chemicals, safe handling and disposal, **16**-1 to 15
Chi-square distribution, **A**-105 to 106
Chlorine: see also Elements
critical constants, **6**-48 to 61
electron configuration, **1**-19 to 20
heat capacity, **4**-133
history, occurrence, uses, **4**-1 to 36

ionization energy, **10**-175 to 177
isotopes and their properties, **11**-50 to 198
physical properties, **4**-132
thermodynamic properties, **5**-1 to 3
thermodynamic properties at high temperature, **5**-61 to 84
vapor pressure, **6**-65 to 92, **6**-93 to 103
Chlorofluorocarbon refrigerants, **6**-141 to 143
Chromatography, carrier gas properties, **8**-126
Chromium: see also Elements
electrical resistivity, **12**-45 to 47
electron configuration, **1**-19 to 20
heat capacity, **4**-133
history, occurrence, uses, **4**-1 to 36
ionization energy, **10**-175 to 177
isotopes and their properties, **11**-50 to 198
magnetic susceptibility, **4**-141 to 146
molten, density, **4**-137 to 140
physical properties, **4**-132
thermal properties, **12**-219 to 220
vapor pressure, **6**-65 to 92
Clausius-Mosotti equation, **12**-17 to 18
Cobalt: see also Elements
electrical resistivity, **12**-45 to 47
electron configuration, **1**-19 to 20
heat capacity, **4**-133
history, occurrence, uses, **4**-1 to 36
ionization energy, **10**-175 to 177
isotopes and their properties, **11**-50 to 198
magnetic susceptibility, **4**-141 to 146
molten, density, **4**-137 to 140
physical properties, **4**-132
thermal properties, **12**-219 to 220
vapor pressure, **6**-65 to 92
CODATA fundamental constants, **1**-1 to 11
CODATA Key Values for Thermodynamics, **5**-1 to 3
Coercivity, magnetic materials, **12**-117 to 125
Collision diameter of gases, **6**-46
Collision frequency, common gases, **6**-46
Combustion, heat of, **5**-89
Composition
atmosphere, **14**-2 to 3
earth's crust, **14**-17
glasses, **12**-228 to 231
human body, **7**-14
magnetic materials, **12**-117 to 125
planetary atmospheres, **14**-2 to 3
sea water, **14**-17
U.S. Standard Atmosphere, **14**-19 to 25
Compressibility
ice, **6**-6 to 7
liquids, **6**-137 to 138
semiconductors, **12**-97 to 107
sodium chloride solutions, **6**-8
Compton wavelength (electron, proton, neutron), **1**-1 to 11
Concentrative properties of aqueous solutions, **8**-57 to 83
Conductance: see Conductivity, electrical
Conductivity, electrical
calibration standards, **5**-91
earth's atmosphere, **14**-33 to 40
electrolyte solutions, **5**-94
hydrohalogen acids, **5**-92 to 93
ions, at infinite dilution, **5**-95 to 97
potassium chloride solutions, **5**-91
seawater, **14**-15 to 16
standard solutions, **5**-91
water, **5**-90
Conductivity, thermal: see Thermal conductivity
Constantan
thermal conductivity, **12**-223
wire tables, **15**-27
Construction materials
density, **15**-29
thermal conductivity, **12**-226 to 227
Conversion factors
chemical kinetics, **1**-51
electrical resistivity, **1**-50
energy, **1**-47
gas constant, **1**-54
general, **1**-34 to 45
ionizing radiation, **1**-52 to 53
pressure, **1**-48
SI units, **1**-34 to 45
temperature scales, **1**-46
temperature, to ITS-90, **1**-23 to 24
thermal conductivity, **1**-49
Conversion formulas for concentration of solutions, **8**-20
Copper: see also Elements
electrical resistivity, **12**-45 to 47
electron configuration, **1**-19 to 20
heat capacity, **4**-133, **12**-218
history, occurrence, uses, **4**-1 to 36
ionization energy, **10**-175 to 177
isotopes and their properties, **11**-50 to 198
magnetic susceptibility, **4**-141 to 146
molten, density, **4**-137 to 140
physical properties, **4**-132
resistance of wires, **15**-27
thermal conductivity, **12**-221 to 222
thermal properties, **12**-219 to 220
thermodynamic properties, **5**-1 to 3
vapor pressure, **6**-65 to 92
wire tables, **15**-27
Copper-constantan thermocouple tables, **15**-1 to 13
Correlation charts for infrared spectra, **9**-87 to 91
Cosecant function, **A**-7 to 10
Cosine function, **A**-7 to 10
Cosmic rays, **11**-216 to 220
Cotangent function, **A**-7 to 10
Coulomb, definition, **1**-25 to 32
Critical constants
cryogenic fluids, **6**-139
elements, **4**-132, **6**-48 to 61
H_2O and D_2O, **6**-4
halocarbons, **6**-141 to 143
inorganic compounds, **6**-48 to 61
organic compounds, **6**-48 to 61
Cross section, x-ray and gamma-ray, **10**-207 to 211
Crust
composition, **14**-17
density, pressure, gravity, **14**-13
Cryogenic fluids
density, **6**-16 to 24, **6**-139
liquid helium properties, **6**-140
vapor pressure, **6**-93 to 103
various properties, **6**-139
Cryoscopic constant, **15**-21
Crystal elastic constants, **12**-37 to 44
Crystal ionic radii, **12**-14 to 16
Crystal lattice energy, **12**-22 to 35, **12**-36
Crystal structure
elements, **4**-156 to 166, **12**-19 to 21
inorganic compounds, **4**-37 to 96, **4**-156 to 166
magnetic materials, **12**-117 to 125
minerals, **4**-149 to 155, **4**-156 to 166
rare earth elements, **4**-122 to 131
semiconductors, **12**-97 to 107
superconductors, **12**-64 to 90, **12**-91 to 93
Crystal symmetry, **12**-7 to 13
Curie temperature
antiferroelectric crystals, **12**-62
ferroelectric crystals, **12**-61 to 62
magnetic materials, **12**-117 to 125
rare earth elements, **4**-122 to 131
Curie, definition, **16**-48
Curium: see also Elements
electron configuration, **1**-19 to 20
history, occurrence, uses, **4**-1 to 36
isotopes and their properties, **11**-50 to 198

D

D_2O
boiling point, **6**-4
critical constants, **6**-4
density, **6**-9
dissociation constant, **8**-85
fixed point properties, **6**-4
ion product, **8**-85
thermal conductivity, **6**-4, **6**-188 to 189
triple point constants, **6**-4
vapor pressure, **6**-4
viscosity, **6**-180 to 181
Debye equation, **6**-15
Debye temperature
rare earth elements, **4**-122 to 131
semiconductors, **12**-97 to 107
Decay mode
fundamental particles, **11**-1 to 49
nuclides, **11**-50 to 198
Decinormal solutions
oxidation and reduction reagents, **8**-7
salts, acids, and bases, **8**-5 to 6
Definitions
abbreviations and acronyms, **2**-29 to 36
scientific terms, **2**-38 to 59
Density
air, **6**-1 to 2
aqueous solutions, **8**-57 to 83
atmosphere, as function of altitude, **14**-19 to 25
carrier gases for chromatography, **8**-126

commercial metals and alloys, **12**-232
common fluids, as function of temperature and pressure, **6**-16 to 24
construction materials, **15**-29
cryogenic fluids, **6**-16 to 24, **6**-139
D_2O, **6**-4, **6**-9
earth, **14**-9 to 10
earth, as function of depth, **14**-13
elements, **4**-137 to 140
ice, **6**-6 to 7
inorganic compounds, **4**-37 to 96
liquid elements and salts, **4**-137 to 140
liquids, pressure and temperature dependence, **6**-137 to 138
mercury, **6**-144
metallic elements, **12**-219 to 220
minerals, **4**-149 to 155
miscellaneous materials, **15**-29
molten salts, **4**-137 to 140
organic compounds, **3**-1 to 330
planets, **14**-2 to 3
polymer melts, **13**-16 to 19
rare earth elements, **4**-122 to 131
rocks, **15**-29
satellites, **14**-4 to 5
seawater, **14**-15 to 16
semiconductors, **12**-97 to 107
sodium chloride solutions, **6**-8
solvents, **15**-14 to 18
steam, **6**-13 to 14
water, **6**-3, **6**-4, **6**-5, **6**-13 to 14
water, supercooled, **6**-6 to 7
wood, **15**-29
Depression of the freezing point, **8**-57 to 83, **15**-21, **15**-22
Derivatives, **A**-12 to 15
Detectors, infrared, **10**-216
Deuterium
solubility in water, **8**-86 to 89
viscosity, **6**-180 to 181
Dew point and relative humidity, **15**-23
Diamagnetic susceptibility
elements, **4**-141 to 146
inorganic compounds, **4**-141 to 146
organic compounds, **3**-764 to 767
Diamond
dielectric constant, **12**-52 to 60
optical properties, **12**-133 to 156
phase diagram, **12**-200 to 217
thermal conductivity, **12**-221 to 222
Diatomic molecules
bond lengths, **9**-81 to 86
bond strengths, **9**-51 to 74
electron affinity, **10**-147 to 159
force constants, **9**-76
polarizability, **10**-160 to 174
spectroscopic constants, **9**-81 to 86
vibrational frequencies, **9**-81 to 86
Dielectric constant
common fluids, as function of temperature and pressure, **6**-16 to 24
cryogenic fluids, temperature and pressure dependence, **6**-16 to 24
crystals, **12**-52 to 60
gases, **6**-174 to 175
glass, **12**-63
ice, **6**-6 to 7
liquids, **6**-151 to 173
plastics, **13**-15
quartz, **12**-63
rubbers, **13**-15
semiconductors, **12**-97 to 107
solids, **12**-52 to 60
solvents, **8**-127
vacuum, **1**-1 to 11
water, **6**-3
water, frequency dependence, **6**-15
Dielectric strength, insulating materials, **15**-30 to 35
Differential equations, **A**-65 to 75
Diffusion
gases, **6**-192 to 193
gases in water, **6**-194
in air, **6**-192 to 193
ions in solution, **5**-95 to 97
liquids, **6**-195
semiconductors, **12**-108 to 116
Dipole moment
electric, of molecules, **9**-44 to 50
magnetic, of nuclides, **11**-50 to 198
solvents, **15**-14 to 18
Discharges, electrical, in the atmosphere, **14**-33 to 40
Disposal of laboratory chemicals, **16**-1 to 15
Dissociation constant
acid-base indicators, **8**-16 to 18
amino acids, **7**-1 to 2
biological buffers, **7**-12
D_2O, **8**-85
inorganic acids and bases, **8**-44 to 45
inorganic salts in water, **8**-117 to 120
organic acids and bases, **8**-46 to 56
purine and pyrimidine bases, **7**-4
water, **8**-84, **8**-85
Dissociation energy of chemical bonds, **9**-51 to 74
DSC, definition, **12**-1 to 6
DTA, definition, **12**-1 to 6
Dubnium (element 105), **4**-1 to 36, **11**-50 to 198
Dysprosium: see also Elements
electrical resistivity, **12**-45 to 47
electron configuration, **1**-19 to 20
heat capacity, **4**-133
history, occurrence, uses, **4**-1 to 36
ionization energy, **10**-175 to 177
isotopes and their properties, **11**-50 to 198
magnetic susceptibility, **4**-141 to 146
molten, density, **4**-137 to 140
physical properties, **4**-132
thermal properties, **12**-219 to 220
vapor pressure, **6**-65 to 92

E

Earth
age, **14**-9 to 10
area of land and oceans, **14**-9 to 10
atmospheric composition, **14**-2 to 3
composition of crust, **14**-17
density as function of depth, **14**-13
dimensions, **14**-1, **14**-9 to 10
gravity in interior, **14**-13
mass and density, **14**-9 to 10
orbital and rotational parameters, **14**-2 to 3, **14**-9 to 10
pressure in interior, **14**-13
Ebullioscopic constant, **15**-20
ECR, definition, **12**-1 to 6
EELS, definition, **12**-1 to 6
Einsteinium: see also Elements
electron configuration, **1**-19 to 20
history, occurrence, uses, **4**-1 to 36
isotopes and their properties, **11**-50 to 198
Elastic constants, single crystal, **12**-37 to 44
Elastic modulus
commercial metals and alloys, **12**-232
rare earth elements, **4**-122 to 131
Elasto-optic constants, **12**-179 to 195
Electrical conductance: see Conductivity, electrical
Electrical resistivity: see Resistivity
Electro-optic constants, **12**-179 to 195
Electrochemical series, **8**-21 to 31
Electrode potential
general table, **8**-21 to 31
ion radicals, **8**-32 to 34
Electrolytes
activity coefficients, **5**-97 to 98, **5**-99 to 103
diffusion of ions, **5**-95 to 97
electrical conductivity, **5**-94, **5**-95 to 97
enthalpy of solution, **5**-105
freezing point lowering, **8**-57 to 83, **15**-22
Electromagnetic radiation, classification of bands, **10**-212 to 213
Electron
charge, **1**-1 to 11
in cosmic ray showers, **11**-216 to 220
magnetic moment, **1**-1 to 11, **11**-1 to 49
range in various materials, **16**-48
Electron affinity, atoms and molecules, **10**-147 to 159
Electron binding energy of the elements, **10**-200 to 205
Electron configuration
neutral atoms, **1**-19 to 20
rare earth elements, **4**-122 to 131
Electron volt, **1**-1 to 11, **1**-25 to 32
Electron work function of the elements, **12**-130
Electronegativity values, **9**-75
Electrons, secondary, emission by metals, **12**-131 to 132
Elementary charge, **1**-1 to 11
Elements
abundance of isotopes, **1**-15 to 18, **11**-50 to 198
atomic mass, **1**-15 to 18, **11**-50 to 198
atomic spectrum, **10**-1 to 87
atomic transition probability, **10**-88 to 146
atomic weight, **1**-12 to 14

boiling point, **4**-132
critical temperature, **4**-132
crystal ionic radii, **12**-14 to 16
crystal structure, **4**-156 to 166, **12**-19 to 21
density, **4**-137 to 140
electrical resistivity, **12**-45 to 47
electron affinity, **10**-147 to 159
electron binding energy, **10**-200 to 205
electron configuration, **1**-19 to 20
electronegativity, **9**-75
enthalpy of fusion, **6**-122 to 136
enthalpy of vaporization, **6**-106 to 121
gamma-ray cross sections, **10**-207 to 211
gamma-ray emission, **11**-50 to 198
general information, **4**-1 to 36
heat capacity, **4**-133
historical information, **4**-1 to 36
in seawater, **14**-17
in the earth's crust, **14**-17
in the human body, **7**-14
line spectrum, **10**-1 to 87
magnetic susceptibility, **4**-141 to 146
melting point, **4**-132
periodic table, inside front cover
photon attenuation coefficients, **10**-207 to 211
polarizability, **10**-160 to 174
radii of ions, **12**-14 to 16
reference states, **5**-4 to 60
semiconducting properties, **12**-97 to 107
superconducting properties, **12**-64 to 90
thermal conductivity, **6**-188 to 189, **12**-219 to 220, **12**-221 to 222
thermal properties, **12**-219 to 220
thermodynamic properties, **5**-4 to 60
transition probabilities, **10**-88 to 146
vapor pressure, **6**-65 to 92
vapor pressure at high temperature, **4**-134 to 136
work function, **12**-130
x-ray atomic energy levels, **10**-196 to 199
x-ray cross sections, **10**-207 to 211
Elevation of the boiling point, **15**-20
Emission, secondary electrons, **12**-131 to 132
Emissivity, rare earth metals, **4**-122 to 131
ENDOR, definition, **12**-1 to 6
Energy
conversion factors, **1**-47
crystal lattice, **12**-22 to 35, **12**-36
spectrum of cosmic rays, **11**-216 to 220
Energy, activation
chemical reactions, **5**-106 to 120
diffusion in semiconductors, **12**-108 to 116
Energy gap
semiconductors, **12**-97 to 107
superconductors, **12**-91 to 93
Energy levels, x-ray, **10**-196 to 199
Energy states of solids, **12**-1 to 6
Enthalpy
air, **6**-1 to 2
common fluids, as function of temperature and pressure, **6**-16 to 24
crystal lattices, **12**-22 to 35
steam, **6**-13 to 14
water, **6**-13 to 14
Enthalpy of combustion, **5**-89
Enthalpy of dilution, acids, **5**-104
Enthalpy of formation
aqueous systems, **5**-85 to 88
CODATA Key Values, **5**-1 to 3
free radicals, **9**-51 to 74
gaseous atoms, **9**-51 to 74
high temperature, **5**-61 to 84
inorganic compounds, **5**-4 to 60
ions, **10**-178 to 195
organic compounds, **5**-4 to 60
semiconductors, **12**-97 to 107
standard state values, **5**-4 to 60
Enthalpy of fusion
cryogenic fluids, **6**-139
elements, **6**-122 to 136
ice, **6**-6 to 7
inorganic compounds, **6**-122 to 136
metals, **12**-219 to 220
organic compounds, **6**-122 to 136
rare earth elements, **4**-122 to 131
Enthalpy of solution, electrolytes, **5**-105
Enthalpy of vaporization
cryogenic fluids, **6**-139
elements, **6**-106 to 121
ice, **6**-6 to 7
inorganic compounds, **6**-106 to 121
organic compounds, **6**-106 to 121
rare earth elements, **4**-122 to 131
water, **6**-3
Entropy
air, **6**-1 to 2
aqueous systems, **5**-85 to 88
CODATA Key Values, **5**-1 to 3
common fluids, as function of temperature and pressure, **6**-16 to 24
high temperature, **5**-61 to 84
rare earth elements, **4**-122 to 131
standard state values, **5**-4 to 60
steam, **6**-13 to 14
water, **6**-13 to 14
EPMA, definition, **12**-1 to 6
EPR, definition, **12**-1 to 6
Equation of state
Tait (for polymer melts), **13**-16 to 19
van der Waals, **6**-45
virial, **6**-25 to 44
Equilibrium constant of formation, **5**-61 to 84
Equivalent conductance: see Conductivity, electrical
Erbium: see also Elements
electrical resistivity, **12**-45 to 47
electron configuration, **1**-19 to 20
heat capacity, **4**-133
history, occurrence, uses, **4**-1 to 36
ionization energy, **10**-175 to 177
isotopes and their properties, **11**-50 to 198
magnetic susceptibility, **4**-141 to 146
molten, density, **4**-137 to 140
physical properties, **4**-132
thermal properties, **12**-219 to 220
vapor pressure, **6**-65 to 92
Error function, **A**-101 to 102
ESCA, definition, **12**-1 to 6
ESD, definition, **12**-1 to 6
Euler's constant, value of, **A**-1
Europium: see also Elements
electrical resistivity, **12**-45 to 47
electron configuration, **1**-19 to 20
heat capacity, **4**-133
history, occurrence, uses, **4**-1 to 36
ionization energy, **10**-175 to 177
isotopes and their properties, **11**-50 to 198
magnetic susceptibility, **4**-141 to 146
molten, density, **4**-137 to 140
physical properties, **4**-132
thermal properties, **12**-219 to 220
vapor pressure, **6**-65 to 92
EXAFS, definition, **12**-1 to 6
EXELFS, definition, **12**-1 to 6
Expansion coefficient
commercial metals and alloys, **12**-232
liquid helium, **6**-140
metals, **12**-219 to 220
rare earth elements, **4**-122 to 131
semiconductors, **12**-97 to 107
sodium chloride solutions, **6**-8
Explosion hazards of laboratory chemicals, **16**-1 to 15
Explosive limits, **16**-16 to 31
Exponential function, **A**-2 to 6
Exponential series, **A**-84 to 87
Exposure limits
airborne contaminants, **16**-32 to 42
ionizing radiation, **16**-48
radionuclides, **16**-49 to 52
Extinction coefficient, in solids, **12**-133 to 156, **12**-157 to 178

F

Fahrenheit temperature, conversion to other scales, **1**-46
Farad, definition, **1**-25 to 32
Faraday constant, **1**-1 to 11
Faraday effect, **12**-179 to 195
Fatty acids, physical properties, **7**-6
F-distribution, **A**-106 to 109
Fehling's solution, preparation, **8**-1 to 4
FEM, definition, **12**-1 to 6
Fermium: see also Elements
electron configuration, **1**-19 to 20
history, occurrence, uses, **4**-1 to 36
isotopes and their properties, **11**-50 to 198
Ferrimagnetic materials
Faraday rotation, **12**-179 to 195
organic, **12**-126 to 129
Ferrites, magnetic properties, **12**-117 to 125
Ferroelectric crystals
Curie temperature, **12**-61 to 62
Kerr constants, **12**-179 to 195
Ferromagnetic materials
Faraday rotation, **12**-179 to 195
organic, **12**-126 to 129
various properties, **12**-117 to 125

Ferromagnetic moment, rare earth elements, **4**-122 to 131
FIM, definition, **12**-1 to 6
Fine structure constant, **1**-1 to 11
First radiation constant, **1**-1 to 11
Fixed point properties
  cryogenic fluids, **6**-139
  water and heavy water, **6**-4
Flammability
  chemical substances, general, **16**-16 to 31
  laboratory chemicals, **16**-1 to 15
  organic solvents, **15**-14 to 18
Flash point: see Flammability
Flattening factor for the earth, **14**-1
Fluorine: see also Elements
  critical constants, **6**-48 to 61
  electron configuration, **1**-19 to 20
  heat capacity, **4**-133
  history, occurrence, uses, **4**-1 to 36
  ionization energy, **10**-175 to 177
  isotopes and their properties, **11**-50 to 198
  physical properties, **4**-132
  thermodynamic properties, **5**-1 to 3
  thermodynamic properties at high temperature, **5**-61 to 84
  vapor pressure, **6**-65 to 92, **6**-93 to 103
Fluorocarbon refrigerants, **6**-141 to 143
Foods, pH, **7**-13
Force constants for bond stretching, **9**-76
Formation, heat of: see Enthalpy of formation
  inorganic compounds, **4**-97 to 111
  organic compounds, **3**-631 to 708
Fossils, age of, **14**-9 to 10
Fourier series, **A**-75 to 79
Fourier transforms, **A**-80 to 84
Francium: see also Elements
  electron configuration, **1**-19 to 20
  history, occurrence, uses, **4**-1 to 36
  ionization energy, **10**-175 to 177
  isotopes and their properties, **11**-50 to 198
  vapor pressure, **6**-65 to 92
Free energy: see Thermodynamic properties
Free radicals
  dipole moment, **9**-44 to 50
  enthalpy of formation, **9**-51 to 74
  ionization energy, **10**-178 to 195
  reaction rates, **5**-106 to 120
  vibrational frequencies, **9**-77 to 80
Freezing point: see also Melting point
  depression of, **8**-57 to 83, **15**-21
  pressure dependence, **6**-47
  seawater, **14**-15 to 16
Freezing point depression
  aqueous solutions, **8**-57 to 83
  cryoscopic constants for various liquids, **15**-21
  electrolytes, **15**-22
Frequency
  electromagnetic radiation bands, **10**-212 to 213
  human hearing range, **14**-47 to 48
  musical scales, **14**-45 to 46
  NMR resonances, **9**-92 to 94
  radio spectrum allocations, **15**-36 to 37
  standards, infrared, **10**-228 to 235, **10**-236 to 240
FTIR, definition, **12**-1 to 6
Fundamental constants, **1**-1 to 11
Fundamental particles, **11**-1 to 49
Fundamental vibrations of molecules, **9**-77 to 80
Fusion: see Enthalpy of fusion

G

g-Factor of the electron, **1**-1 to 11
Gadolinium: see also Elements
  electrical resistivity, **12**-45 to 47
  electron configuration, **1**-19 to 20
  heat capacity, **4**-133
  history, occurrence, uses, **4**-1 to 36
  ionization energy, **10**-175 to 177
  isotopes and their properties, **11**-50 to 198
  magnetic susceptibility, **4**-141 to 146
  molten, density, **4**-137 to 140
  physical properties, **4**-132
  thermal properties, **12**-219 to 220
  vapor pressure, **6**-65 to 92
Gallium: see also Elements
  electrical resistivity, **12**-45 to 47
  electron configuration, **1**-19 to 20
  heat capacity, **4**-133
  history, occurrence, uses, **4**-1 to 36
  ionization energy, **10**-175 to 177
  isotopes and their properties, **11**-50 to 198
  magnetic susceptibility, **4**-141 to 146
  molten, density, **4**-137 to 140
  physical properties, **4**-132
  thermal properties, **12**-219 to 220
  vapor pressure, **6**-65 to 92
Gamma function, **A**-99 to 101
Gamma rays
  cross sections, for the elements, **10**-207 to 211
  energy, of nuclides, **11**-50 to 198
  in cosmic ray showers, **11**-216 to 220
  photon attenuation coefficients, **10**-207 to 211
  protection against, **16**-48
Gas chromatography, carrier gas properties, **8**-126
Gas constant, **1**-1 to 11, **1**-54
Gases
  average velocity, **6**-46
  breakdown voltage, **15**-30 to 35
  collision diameter, **6**-46
  dielectric constant, **6**-174 to 175
  diffusion, **6**-192 to 193
  diffusion in water, **6**-194
  dipole moment, **9**-44 to 50
  mean free path, **6**-46
  permittivity, **6**-174 to 175
  solubility in water, **8**-86 to 89
  speed of sound in, **14**-41 to 42
  thermal conductivity, **6**-188 to 189
  threshold limits, **16**-32 to 42
  van der Waals constants, **6**-45
  Verdet constants, **12**-179 to 195
  virial coefficients, **6**-25 to 44
  viscosity, **6**-180 to 181
Gauges, of wires, **15**-27
Gaussian gravitational constant, **14**-1
GDMS, definition, **12**-1 to 6
Genetic code, **7**-5
Geological time scale, **14**-11
Geophysical constants, **14**-9 to 10
Germanium: see also Elements
  dielectric constant, **12**-52 to 60
  electron configuration, **1**-19 to 20
  heat capacity, **4**-133, **12**-218
  history, occurrence, uses, **4**-1 to 36
  ionization energy, **10**-175 to 177
  isotopes and their properties, **11**-50 to 198
  magnetic susceptibility, **4**-141 to 146
  molten, density, **4**-137 to 140
  optical properties, **12**-133 to 156
  physical properties, **4**-132
  semiconducting properties, **12**-97 to 107
  thermal conductivity, **12**-221 to 222
  thermodynamic properties, **5**-1 to 3
  vapor pressure, **6**-65 to 92
Gibbs energy of formation
  aqueous systems, **5**-85 to 88
  biochemical species, **7**-9 to 11
  high temperature, **5**-61 to 84
  standard state values, **5**-4 to 60
Glass transition temperature of polymers, **13**-7 to 14
Glasses
  composition, **12**-228 to 231
  density, **15**-29
  dielectric constant, **12**-63
  index of refraction, **10**-217
  loss factor, **12**-63
  resistivity, **12**-63
  speed of sound in, **14**-41 to 42
  thermal conductivity, **12**-224 to 225, **12**-228 to 231
  transmittance, **10**-217
  Verdet constants, **12**-179 to 195
Global warming
  atmospheric carbon dioxide concentration, **14**-28 to 29
  mean temperatures, global, **14**-32
  mean temperatures, U. S., **14**-30 to 31
Gloves, resistance to chemicals, **16**-1 to 15
Glucose, aqueous solution properties, **8**-57 to 83
Gold: see also Elements
  electrical resistivity, **12**-45 to 47
  electron configuration, **1**-19 to 20
  heat capacity, **4**-133, **12**-218
  history, occurrence, uses, **4**-1 to 36
  ionization energy, **10**-175 to 177
  isotopes and their properties, **11**-50 to 198
  magnetic susceptibility, **4**-141 to 146
  molten, density, **4**-137 to 140
  physical properties, **4**-132
  thermal conductivity, **12**-221 to 222
  thermal properties, **12**-219 to 220
  vapor pressure, **6**-65 to 92
Graphite
  heat capacity, **12**-218

heat of combustion, **5**-89
phase diagram, **12**-200 to 217
sublimation pressure, **6**-65 to 92
thermal conductivity, **12**-221 to 222
Gravitational constant, **1**-1 to 11, **14**-1
Gravitational potential, **14**-2 to 3
Gravity, acceleration of
at poles and equator, **14**-9 to 10
at various latitudes, **14**-12
in interior of earth, **14**-13
standard value, **1**-1 to 11
Gray, definition, **1**-25 to 32, **16**-48
Greek alphabet, **2**-37

H

Hafnium: see also Elements
electrical resistivity, **12**-45 to 47
electron configuration, **1**-19 to 20
heat capacity, **4**-133
history, occurrence, uses, **4**-1 to 36
ionization energy, **10**-175 to 177
isotopes and their properties, **11**-50 to 198
magnetic susceptibility, **4**-141 to 146
molten, density, **4**-137 to 140
physical properties, **4**-132
thermal properties, **12**-219 to 220
vapor pressure, **6**-65 to 92
Hall coefficient, rare earth elements, **4**-122 to 131
Hall density, superconductors, **12**-91 to 93
Hall resistance, quantized, **1**-1 to 11
Halocarbon refrigerants, **6**-141 to 143
Hardness
ceramics, **12**-233 to 234
minerals, **4**-149 to 155, **12**-233 to 234
scales, comparison, **12**-233 to 234
semiconductors, **12**-97 to 107
various materials, **12**-233 to 234
Hartree energy, **1**-1 to 11
Hassium (element 108), **4**-1 to 36, **11**-50 to 198
Hazardous chemicals, handling and disposal, **16**-1 to 15
Hearing, characteristics, **14**-47 to 48
Heat capacity
air, **6**-1 to 2
aqueous systems, **5**-85 to 88
carrier gases for chromatography, **8**-126
common fluids, as function of temperature and pressure, **6**-16 to 24
cryogenic fluids, **6**-139
elements, **4**-133
high temperature, **5**-61 to 84
ice, **6**-6 to 7
liquid helium, **6**-140
mercury, **6**-145
metals, **12**-218, **12**-219 to 220
rare earth elements, **4**-122 to 131
seawater, **14**-15 to 16
semiconductors, **12**-97 to 107
solids, **12**-218
solvents, **15**-14 to 18
standard state values, **5**-4 to 60
steam, **6**-13 to 14
water, **6**-3
Heat conductivity: see Thermal conductivity
Heat of combustion, **5**-89
Heat of dilution: see Enthalpy of dilution
Heat of formation: see Enthalpy of formation
Heat of fusion: see Enthalpy of fusion
Heat of solution: see Enthalpy of solution
Heat of vaporization: see Enthalpy of vaporization
Hebrew alphabet, **2**-37
Helium: see also Elements
critical constants, **6**-48 to 61
cryogenic properties, **6**-139, **6**-140
density, **6**-140
electron configuration, **1**-19 to 20
enthalpy of vaporization, **6**-140
entropy, **5**-1 to 3
expansion coefficient, **6**-140
heat capacity, **6**-140
history, occurrence, uses, **4**-1 to 36
ionization energy, **10**-175 to 177
isotopes and their properties, **11**-50 to 198
liquid properties, **6**-140
magnetic susceptibility, **4**-141 to 146
mean free path, **6**-46
permittivity (dielectric constant), **6**-140, **6**-151 to 173, **6**-174 to 175
physical properties, **4**-132
solubility in water, **8**-86 to 89
surface tension, **6**-140
thermal conductivity (gas), **6**-188 to 189
thermal conductivity (liquid), **6**-140
thermal conductivity (two-phase), **6**-16 to 24
thermodynamic properties, **6**-16 to 24
van der Waals constants, **6**-45
vapor pressure, **6**-65 to 92, **6**-93 to 103, **6**-140
viscosity (gas), **6**-180 to 181
viscosity (liquid), **6**-140
viscosity (two-phase), **6**-16 to 24
Henry, definition, **1**-25 to 32
Henry's Law constant, **8**-91 to 107
Hermite polynomials, **A**-102 to 104
Hertz, definition, **1**-25 to 32
Holmium: see also Elements
electrical resistivity, **12**-45 to 47
electron configuration, **1**-19 to 20
heat capacity, **4**-133
history, occurrence, uses, **4**-1 to 36
ionization energy, **10**-175 to 177
isotopes and their properties, **11**-50 to 198
magnetic susceptibility, **4**-141 to 146
molten, density, **4**-137 to 140
physical properties, **4**-132
thermal properties, **12**-219 to 220
vapor pressure, **6**-65 to 92
Human body
chemical composition, **7**-14
pH of fluids, **7**-13
Humidity, relative
from wet and dry bulb temperatures, **15**-24
relation to dew point, **15**-23
solutions for calibration, **15**-26
solutions for constant humidity, **15**-25
Hydrocarbons
heat of combustion, **5**-89
thermophysical properties, **6**-16 to 24
Hydrogen: see also Elements
critical constants, **6**-48 to 61
cryogenic properties, **6**-139
electron configuration, **1**-19 to 20
enthalpy of fusion, **6**-122 to 136
enthalpy of vaporization, **6**-106 to 121
heat of combustion, **5**-89
history, occurrence, uses, **4**-1 to 36
ionization energy, **10**-175 to 177
isotopes and their properties, **11**-50 to 198
magnetic susceptibility, **4**-141 to 146
mean free path, **6**-46
permittivity (dielectric constant), **6**-151 to 173, **6**-174 to 175
physical properties, **4**-132
solubility in water, **8**-86 to 89
thermal conductivity, **6**-188 to 189
thermodynamic properties, **5**-1 to 3, **6**-16 to 24
thermodynamic properties at high temperature, **5**-61 to 84
van der Waals constants, **6**-45
vapor pressure, **6**-65 to 92, **6**-93 to 103
viscosity, **6**-180 to 181
Hydrosphere, mass of, **14**-9 to 10
Hyperbolic functions, **A**-2 to 6, **A**-11
Hysteresis, in magnetic materials, **12**-117 to 125

I

Ice
compressibility, **6**-6 to 7
crystal structure, **4**-156 to 166
density, **6**-6 to 7
dielectric constant, **6**-6 to 7, **12**-52 to 60
heat capacity, **6**-6 to 7
melting point, pressure dependence, **6**-12, **6**-47
phase diagram, **12**-200 to 217
phase transitions, **6**-6 to 7, **6**-12
thermal conductivity, **6**-6 to 7, **12**-224 to 225
thermal expansion coefficient, **6**-6 to 7
vapor pressure, **6**-9
ICPMS, definition, **12**-1 to 6
Ignition temperature
chemical substances, general, **16**-16 to 31
laboratory chemicals, **16**-1 to 15
solvents, **15**-14 to 18
Index of refraction
air, **10**-220
aqueous solutions, **8**-57 to 83
glass, **10**-217
inorganic liquids, **4**-147 to 148
liquids, for calibration, **10**-219
metals, **12**-133 to 156
minerals, **4**-149 to 155

organic compounds, **3**-1 to 330
semiconductors, **12**-97 to 107, **12**-133 to 156, **12**-157 to 178
solids, as function of wavelength, **12**-133 to 156, **12**-157 to 178
water, **10**-218
Indicators
acid-base, **8**-16 to 18
fluorescent, **8**-19 to 20
pH, **8**-16 to 18, **8**-19 to 20
p*K*, **8**-16 to 18
preparation, **8**-1 to 4, **8**-16 to 18
Indium: see also Elements
electrical resistivity, **12**-45 to 47
electron configuration, **1**-19 to 20
heat capacity, **4**-133
history, occurrence, uses, **4**-1 to 36
ionization energy, **10**-175 to 177
isotopes and their properties, **11**-50 to 198
magnetic susceptibility, **4**-141 to 146
molten, density, **4**-137 to 140
physical properties, **4**-132
thermal properties, **12**-219 to 220
vapor pressure, **6**-65 to 92
Infrared detectors, **10**-216
Infrared spectrum
carbon dioxide laser, **10**-228 to 235
characteristic group frequencies, **9**-87 to 91
correlation charts, **9**-87 to 91
earth's atmosphere, **14**-27
frequency standards, **10**-236 to 240
vibrational frequencies of molecules, **9**-77 to 80
Inorganic compounds
activity coefficients, **5**-97 to 98, **5**-99 to 103
boiling point, **6**-106 to 121
bond lengths and angles, **9**-17 to 43
characteristic infrared frequencies, **9**-87 to 91
crystal lattice energy, **12**-22 to 35
crystal structure, **4**-37 to 96, **4**-156 to 166
dielectric constant, **12**-52 to 60
dipole moment, **9**-44 to 50
dissociation constant in water, **8**-117 to 120
electrical conductivity, **5**-94
enthalpy of formation, **5**-4 to 60
enthalpy of fusion, **6**-122 to 136
enthalpy of solution, **5**-105
enthalpy of vaporization, **6**-106 to 121
entropy, **5**-4 to 60
Gibbs energy of formation, **5**-4 to 60
heat capacity, **5**-4 to 60
index of refraction, **4**-147 to 148
magnetic susceptibility, **4**-141 to 146
nomenclature, **2**-23 to 24
permittivity, **12**-52 to 60
physical properties, **4**-37 to 96
polarizability, **10**-160 to 174
reagents for determination, **8**-8 to 15
solubility as a function of temperature, **8**-108 to 116
solubility product constant, **8**-117 to 120
solubility, qualitative rules, **8**-122 to 124
standard thermodynamic properties, **5**-4 to 60
surface tension, **6**-146 to 149
INS, definition, **12**-1 to 6
Insulation, thermal conductivity of, **12**-226 to 227
Insulators, breakdown voltage, **15**-30 to 35
Integral tables, **A**-20 to 65
Integration, methods and techniques, **A**-15 to 19
Interatomic distances
diatomic molecules, **9**-81 to 86
gas-phase molecules, **9**-17 to 43
organic crystals, **9**-1 to 14
organometallic compounds, **9**-15 to 16
International System of Units (SI), **1**-25 to 32
International Temperature Scale (ITS-90)
conversion from IPTS-68 and IPTS-48, **1**-23 to 24
definition and fixed points, **1**-21 to 22
International Union of Pure and Applied Chemistry: see IUPAC
Interstellar molecules, **14**-6 to 8
Iodine: see also Elements
critical constants, **6**-48 to 61
dielectric constant, **12**-52 to 60
electron configuration, **1**-19 to 20
heat capacity, **4**-133
history, occurrence, uses, **4**-1 to 36
ionization energy, **10**-175 to 177
isotopes and their properties, **11**-50 to 198
magnetic susceptibility, **4**-141 to 146
physical properties, **4**-132
thermodynamic properties, **5**-1 to 3
thermodynamic properties at high temperature, **5**-61 to 84
vapor pressure, **6**-65 to 92
Ion product
D_2O, **8**-85
water, **8**-84, **8**-85
Ionic radii
in crystals, **12**-14 to 16
rare earth elements, **4**-122 to 131
Ionization constant
D_2O, **8**-85
inorganic acids and bases, **8**-44 to 45
inorganic compounds in water, **8**-117 to 120
organic acids and bases, **8**-46 to 56
water, **8**-84, **8**-85
Ionization energy
atoms and ions, **10**-175 to 177
molecules, **10**-178 to 195
rare earth elements, **4**-122 to 131
Ionization potential: See Ionization energy
Ions
aerosol, **14**-33 to 40
diffusion in aqueous solutions, **5**-95 to 97
electrical conductivity in aqueous solutions, **5**-95 to 97
enthalpy of formation, **10**-178 to 195
heat capacity, aqueous solutions, **5**-85 to 88
in the atmosphere, **14**-33 to 40
magnetic properties, **12**-117 to 125
nomenclature, **2**-23 to 24
polarizability, **12**-17 to 18
radii, in crystals, **12**-14 to 16
thermodynamic properties, aqueous solutions, **5**-85 to 88
Iridium: see also Elements
electrical resistivity, **12**-45 to 47
electron configuration, **1**-19 to 20
heat capacity, **4**-133
history, occurrence, uses, **4**-1 to 36
ionization energy, **10**-175 to 177
isotopes and their properties, **11**-50 to 198
magnetic susceptibility, **4**-141 to 146
molten, density, **4**-137 to 140
physical properties, **4**-132
thermal properties, **12**-219 to 220
vapor pressure, **6**-65 to 92
Iron: see also Elements
electrical resistivity, **12**-45 to 47
electron configuration, **1**-19 to 20
heat capacity, **4**-133, **12**-218
history, occurrence, uses, **4**-1 to 36
ionization energy, **10**-175 to 177
isotopes and their properties, **11**-50 to 198
magnetic susceptibility, **4**-141 to 146
molten, density, **4**-137 to 140
physical properties, **4**-132
thermal conductivity, **12**-221 to 222
thermal properties, **12**-219 to 220
vapor pressure, **6**-65 to 92
Iron-constantan thermocouple tables, **15**-1 to 13
Irradiance of the sun, **14**-18
Isoelectric point, amino acids, **7**-1 to 2
Isotopes, summary of properties, **11**-50 to 198
Isotopic abundance, **1**-15 to 18, **11**-50 to 198
ITS-90
conversion from IPTS-68 and IPTS-48, **1**-23 to 24
definition and fixed points, **1**-21 to 22
IUPAC
atomic weights, **1**-12 to 14
nomenclature for carbohydrates, **7**-7 to 8
nomenclature for organic substituent groups, **2**-25 to 28
nomenclature for polymers, **13**-1 to 5
symbols for physical quantities, **2**-1 to 21
vapor pressure calibration data, **6**-105

J

Jacobi polynomials, **A**-102 to 104
Josephson ratio, **1**-1 to 11
Joule, definition, **1**-25 to 32
Jupiter, orbital data and dimensions, **14**-2 to 3

K

Kelvin, definition, **1**-25 to 32
Kerr constants, **12**-179 to 195
Kilogram, definition, **1**-25 to 32
Kinetics
atmospheric reactions, **5**-106 to 120

conversion factors, **1**-51
Krypton: see also Elements
critical constants, **6**-48 to 61
cryogenic properties, **6**-139
electron configuration, **1**-19 to 20
entropy, **5**-1 to 3
history, occurrence, uses, **4**-1 to 36
ionization energy, **10**-175 to 177
isotopes and their properties, **11**-50 to 198
magnetic susceptibility, **4**-141 to 146
mean free path, **6**-46
permittivity (dielectric constant), **6**-151 to 173, **6**-174 to 175
physical properties, **4**-132
solubility in water, **8**-86 to 89
thermal conductivity, **6**-188 to 189
van der Waals constants, **6**-45
vapor pressure, **6**-65 to 92, **6**-93 to 103
viscosity, **6**-180 to 181

L

Laboratory reagents, preparation of, **8**-1 to 4
Laguerre polynomials, **A**-102 to 104
Lanthanum: see also Elements
electrical resistivity, **12**-45 to 47
electron configuration, **1**-19 to 20
heat capacity, **4**-133
history, occurrence, uses, **4**-1 to 36
ionization energy, **10**-175 to 177
isotopes and their properties, **11**-50 to 198
magnetic susceptibility, **4**-141 to 146
molten, density, **4**-137 to 140
physical properties, **4**-132
thermal properties, **12**-219 to 220
vapor pressure, **6**-65 to 92
Lasers
characteristics of various types, **10**-221 to 227
infrared, frequencies, **10**-228 to 235
Lattice constants
elements, **4**-156 to 166, **12**-19 to 21
inorganic compounds, **4**-156 to 166
minerals, **4**-156 to 166
rare earth elements, **4**-122 to 131
semiconductors, **12**-97 to 107
Lattice energy, **12**-22 to 35, **12**-36
Lawrencium: see also Elements
electron configuration, **1**-19 to 20
history, occurrence, uses, **4**-1 to 36
isotopes and their properties, **11**-50 to 198
Lead: see also Elements
electrical resistivity, **12**-45 to 47
electron configuration, **1**-19 to 20
heat capacity, **4**-133, **12**-218
history, occurrence, uses, **4**-1 to 36
ionization energy, **10**-175 to 177
isotopes and their properties, **11**-50 to 198
magnetic susceptibility, **4**-141 to 146
molten, density, **4**-137 to 140
physical properties, **4**-132
thermal conductivity, **12**-221 to 222
thermal properties, **12**-219 to 220
thermodynamic properties, **5**-1 to 3
vapor pressure, **6**-65 to 92
LEED, definition, **12**-1 to 6
Legendre polynomials, **A**-102 to 104
Leptons, summary of properties, **11**-1 to 49
Lifetime
fundamental particles, **11**-1 to 49
in laser systems, **10**-221 to 227
nuclides, **11**-50 to 198
Light, speed of, **1**-1 to 11
Lightning, **14**-33 to 40
LIMS, definition, **12**-1 to 6
Line spectra of the elements, **10**-1 to 87
Line strengths in atomic spectra, **10**-88 to 146
Line width, x-ray lines, **10**-206
Liquid air, thermodynamic properties, **6**-1 to 2
Liquid helium properties, **6**-140
Liquids
breakdown voltage, **15**-30 to 35
dielectric constant, **6**-151 to 173
diffusion, **6**-195
flammability, **15**-14 to 18
index of refraction, **4**-147 to 148, **10**-219
Kerr constants, **12**-179 to 195
permittivity, **6**-151 to 173
speed of sound in, **14**-41 to 42
surface tension, **6**-146 to 149
thermal conductivity, **6**-190 to 191
Verdet constants, **12**-179 to 195
viscosity, **6**-182 to 186
Liter, definition, **1**-25 to 32
Lithium: see also Elements
electrical resistivity, **12**-45 to 47
electron configuration, **1**-19 to 20
heat capacity, **4**-133
history, occurrence, uses, **4**-1 to 36
ionization energy, **10**-175 to 177
isotopes and their properties, **11**-50 to 198
magnetic susceptibility, **4**-141 to 146
molten, density, **4**-137 to 140
physical properties, **4**-132
thermal properties, **12**-219 to 220
thermodynamic properties, **5**-1 to 3
vapor pressure, **6**-65 to 92
Lithosphere, mass of, **14**-9 to 10
Logarithmic series, **A**-84 to 87
Log *P*, **16**-43 to 47
Loss factor, glasses, **12**-63
Loss tangent, **6**-15
Loudness level, definition, **14**-47 to 48
Low temperature liquid baths, **15**-26
Lumen, definition, **1**-25 to 32
Lutetium: see also Elements
electrical resistivity, **12**-45 to 47
electron configuration, **1**-19 to 20
heat capacity, **4**-133
history, occurrence, uses, **4**-1 to 36
ionization energy, **10**-175 to 177
isotopes and their properties, **11**-50 to 198
magnetic susceptibility, **4**-141 to 146
molten, density, **4**-137 to 140
physical properties, **4**-132
thermal properties, **12**-219 to 220
vapor pressure, **6**-65 to 92
Lux, definition, **1**-25 to 32

M

Maclaurin series, **A**-84 to 87
Madelung constant, **12**-36
Magnesium: see also Elements
electrical resistivity, **12**-45 to 47
electron configuration, **1**-19 to 20
heat capacity, **4**-133
history, occurrence, uses, **4**-1 to 36
ionization energy, **10**-175 to 177
isotopes and their properties, **11**-50 to 198
magnetic susceptibility, **4**-141 to 146
molten, density, **4**-137 to 140
thermal conductivity, **12**-221 to 222
thermal properties, **12**-219 to 220
thermodynamic properties, **5**-1 to 3
vapor pressure, **6**-65 to 92
Magnetic materials, composition and properties, **12**-117 to 125
Magnetic moment
electron, **1**-1 to 11, **11**-1 to 49
fundamental particles, **11**-1 to 49
NMR, for important nuclei, **9**-92 to 94
nuclides, **11**-50 to 198
proton, **1**-1 to 11
rare earth elements, **4**-122 to 131
Magnetic properties
alloys, **12**-117 to 125
organic magnets, **12**-126 to 129
rare earth elements, **4**-122 to 131
superconductors, **12**-64 to 90, **12**-91 to 93
Magnetic susceptibility
elements, **4**-141 to 146
inorganic compounds, **4**-141 to 146
organic compounds, **3**-764 to 767
rare earth elements, **4**-122 to 131
semiconductors, **12**-97 to 107
various materials, **12**-117 to 125
Magnetism, symbols and units, **1**-33, **12**-117 to 125
Magneton (nuclear, Bohr), **1**-1 to 11
Magneto-optic constants, **12**-179 to 195
Magnetostriction, **12**-117 to 125
Manganese: see also Elements
electrical resistivity, **12**-45 to 47
electron configuration, **1**-19 to 20
heat capacity, **4**-133
history, occurrence, uses, **4**-1 to 36
ionization energy, **10**-175 to 177
isotopes and their properties, **11**-50 to 198
magnetic susceptibility, **4**-141 to 146
molten, density, **4**-137 to 140
physical properties, **4**-132
thermal properties, **12**-219 to 220
vapor pressure, **6**-65 to 92
Mars, orbital data and dimensions, **14**-2 to 3
Mass
atmosphere, oceans, and crust, **14**-9 to 10
atomic mass unit, **1**-1 to 11
atomic, of nuclides, **11**-50 to 198
earth, moon, and sun, **14**-1
electron, proton, neutron, **1**-1 to 11

fundamental particles, **11**-1 to 49
planets, **14**-2 to 3
planets, relative to sun, **14**-1
satellites, **14**-4 to 5
sun, **14**-2 to 3
Mass spectral peaks of common solvents, **8**-129 to 134
Materials characterization, techniques, **12**-1 to 6
Mathematical constants, **A**-1
Mean free path
as function of altitude, **14**-19 to 25
common gases, **6**-46
Mechanical properties
commercial metals and alloys, **12**-232
rare earth elements, **4**-122 to 131
Meitnerium (element 109), **4**-1 to 36, **11**-50 to 198
Melting point
amino acids, **7**-1 to 2
commercial metals and alloys, **12**-232
cryogenic fluids, **6**-139
D_2O, **6**-4
depression of, **15**-21
elements, **4**-132
fatty acids, **7**-6
halocarbons, **6**-141 to 143
ice, as function of pressure, **6**-12
index for organic compounds, **3**-737 to 750
inorganic compounds, **4**-37 to 96, **6**-122 to 136
metals, **12**-219 to 220
organic compounds, **3**-1 to 330, **6**-122 to 136
organic compounds, index, **3**-737 to 750
pressure dependence, **6**-47
rare earth elements, **4**-122 to 131
semiconductors, **12**-97 to 107
solvents, **15**-14 to 18
Mendelevium: see also Elements
electron configuration, **1**-19 to 20
history, occurrence, uses, **4**-1 to 36
isotopes and their properties, **11**-50 to 198
Merck Index references for organic compounds, **3**-1 to 330
Mercury: see also Elements
compressibility, **6**-145
critical constants, **6**-48 to 61
density, **6**-144
electrical resistivity, **12**-45 to 47
electron configuration, **1**-19 to 20
heat capacity, **4**-133, **6**-145
history, occurrence, uses, **4**-1 to 36
ionization energy, **10**-175 to 177
isotopes and their properties, **11**-50 to 198
magnetic susceptibility, **4**-141 to 146
physical properties, **4**-132
sound velocity, **6**-145
specific volume, **6**-144
thermal conductivity, **6**-190 to 191
thermal expansion, **6**-145
thermal properties, **12**-219 to 220
thermodynamic properties, **5**-1 to 3
vapor pressure, **6**-65 to 92, **6**-105
Mercury (planet), orbital data and dimensions, **14**-2 to 3
Mesons, summary of properties, **11**-1 to 49
Metal oxides, secondary electron emission, **12**-131 to 132
Metals
commercial, mechanical and thermal properties, **12**-232
crystal structure, **12**-19 to 21
elastic constants, **12**-37 to 44
electrical resistivity, **12**-45 to 47
extinction coefficient, **12**-133 to 156
heat capacity, **12**-218
index of refraction, **12**-133 to 156
optical properties, **12**-133 to 156
reflection coefficient, **12**-133 to 156
secondary electron emission, **12**-131 to 132
speed of sound in, **14**-41 to 42
sublimation pressure, **6**-65 to 92
superconducting properties, **12**-64 to 90
thermal conductivity, **12**-221 to 222
thermal properties, **12**-219 to 220
Meter, definition, **1**-25 to 32
Microwave bands, classification, **10**-212 to 213
Million, milliard (definition), **1**-46
Minerals
chemical formulas, **4**-149 to 155, **4**-156 to 166
crystal structure, **4**-156 to 166
elastic constants, **12**-37 to 44
hardness, **12**-233 to 234
index of refraction, **4**-149 to 155
physical constants, **4**-149 to 155
semiconducting properties, **12**-97 to 107
solubility as a function of temperature, **8**-108 to 116
thermal conductivity, **12**-226 to 227
Mobility
in semiconductors, **12**-97 to 107
of atmospheric ions, **14**-33 to 40
Mole, definition, **1**-25 to 32
Molecular weight
amino acids, **7**-1 to 2
inorganic compounds, **4**-37 to 96
organic compounds, **3**-1 to 330
Molecules
appearance potential, **10**-178 to 195
bond lengths, **9**-1 to 14, **9**-15 to 16
bond lengths and angles, **9**-17 to 43
bond strengths, **9**-51 to 74
electron affinity, **10**-147 to 159
force constants, **9**-76
fundamental vibrational frequencies, **9**-77 to 80
ionization energy, **10**-178 to 195
polarizability, **10**-160 to 174
Molybdenum: see also Elements
electrical resistivity, **12**-45 to 47
electron configuration, **1**-19 to 20
heat capacity, **4**-133
history, occurrence, uses, **4**-1 to 36
ionization energy, **10**-175 to 177
isotopes and their properties, **11**-50 to 198
magnetic susceptibility, **4**-141 to 146
molten, density, **4**-137 to 140
physical properties, **4**-132
thermal properties, **12**-219 to 220
vapor pressure, **6**-65 to 92
Moment of inertia, formulas for, **A**-96
Moon
orbital constants and other parameters, **14**-2 to 3, **14**-4 to 5
ratio of mass to earth's mass, **14**-1
Muon
in cosmic ray showers, **11**-216 to 220
summary of properties, **11**-1 to 49
Musical scales, **14**-45 to 46

N

NAA, definition, **12**-1 to 6
Néel temperature
magnetic materials, **12**-117 to 125
rare earth elements, **4**-122 to 131
Neodymium: see also Elements
electrical resistivity, **12**-45 to 47
electron configuration, **1**-19 to 20
heat capacity, **4**-133
history, occurrence, uses, **4**-1 to 36
ionization energy, **10**-175 to 177
isotopes and their properties, **11**-50 to 198
magnetic susceptibility, **4**-141 to 146
molten, density, **4**-137 to 140
physical properties, **4**-132
thermal properties, **12**-219 to 220
vapor pressure, **6**-65 to 92
Neon: see also Elements
critical constants, **6**-48 to 61
cryogenic properties, **6**-139
electron configuration, **1**-19 to 20
entropy, **5**-1 to 3
history, occurrence, uses, **4**-1 to 36
ionization energy, **10**-175 to 177
isotopes and their properties, **11**-50 to 198
magnetic susceptibility, **4**-141 to 146
mean free path, **6**-46
permittivity (dielectric constant), **6**-151 to 173, **6**-174 to 175
physical properties, **4**-132
solubility in water, **8**-86 to 89
thermal conductivity, **6**-188 to 189
van der Waals constants, **6**-45
vapor pressure, **6**-65 to 92, **6**-93 to 103
viscosity, **6**-180 to 181
Neptune, orbital data and dimensions, **14**-2 to 3
Neptunium: see also Elements
electron configuration, **1**-19 to 20
history, occurrence, uses, **4**-1 to 36
ionization energy, **10**-175 to 177
isotopes and their properties, **11**-50 to 198
magnetic susceptibility, **4**-141 to 146
physical properties, **4**-132
Neutrino, summary of properties, **11**-1 to 49
Neutron
mass, **1**-1 to 11

range in paraffin, **16**-48
scattering and absorption, **11**-199 to 215
summary of properties, **11**-1 to 49
Neutron cross sections, **11**-199 to 215
Neutron resonance integrals, **11**-199 to 215
Newton, definition, **1**-25 to 32
Nichrome, wire tables, **15**-27
Nickel: see also Elements
electrical resistivity, **12**-45 to 47
electron configuration, **1**-19 to 20
heat capacity, **4**-133
history, occurrence, uses, **4**-1 to 36
ionization energy, **10**-175 to 177
isotopes and their properties, **11**-50 to 198
magnetic susceptibility, **4**-141 to 146
molten, density, **4**-137 to 140
physical properties, **4**-132
thermal properties, **12**-219 to 220
vapor pressure, **6**-65 to 92
Niobium: see also Elements
electrical resistivity, **12**-45 to 47
electron configuration, **1**-19 to 20
heat capacity, **4**-133
history, occurrence, uses, **4**-1 to 36
ionization energy, **10**-175 to 177
isotopes and their properties, **11**-50 to 198
magnetic susceptibility, **4**-141 to 146
physical properties, **4**-132
thermal properties, **12**-219 to 220
vapor pressure, **6**-65 to 92
Nitrogen: see also Elements
critical constants, **6**-48 to 61
cryogenic properties, **6**-139
electron configuration, **1**-19 to 20
enthalpy of vaporization, **6**-106 to 121
history, occurrence, uses, **4**-1 to 36
ionization energy, **10**-175 to 177
isotopes and their properties, **11**-50 to 198
magnetic susceptibility, **4**-141 to 146
mean free path, **6**-46
permittivity (dielectric constant), **6**-151 to 173, **6**-174 to 175
physical properties, **4**-132
solubility in water, **8**-86 to 89
thermal conductivity, **6**-16 to 24, **6**-188 to 189
thermodynamic properties, **5**-1 to 3, **6**-16 to 24
thermodynamic properties at high temperature, **5**-61 to 84
van der Waals constants, **6**-45
vapor pressure, **6**-65 to 92, **6**-93 to 103
viscosity, **6**-16 to 24, **6**-180 to 181
NMR spectrum
characteristic ^{13}C chemical shifts, **9**-96
characteristic shifts for protons, **9**-95
chemical shifts of solvents, **8**-128
nuclear moments and resonance frequencies, **9**-92 to 94
Nobelium: see also Elements
electron configuration, **1**-19 to 20
history, occurrence, uses, **4**-1 to 36
isotopes and their properties, **11**-50 to 198
Noise thresholds, **14**-47 to 48
Nomenclature
carbohydrates, **7**-7 to 8
chemical, references, **2**-22
inorganic ions and ligands, **2**-23 to 24
minerals, **4**-149 to 155
organic substituent groups and ring systems, **2**-25 to 28
physical quantities, **2**-1 to 21
polymers, **13**-1 to 5
NQR, definition, **12**-1 to 6
NRA, definition, **12**-1 to 6
Nuclear magnetic resonance: see NMR
Nuclear magneton, **1**-1 to 11
Nuclear spins and moments
for all nuclides, **11**-50 to 198
for important nuclei in NMR, **9**-92 to 94
Nucleic acids
genetic code, **7**-5
purine and pyrimidine bases, **7**-4
Nuclides, summary of properties, **11**-50 to 198

O

Oceans
abundance of chemical elements, **14**-17
pressure as a function of depth and latitude, **14**-14
Octanol-water partition coefficients, **16**-43 to 47
Ohm
definition, **1**-25 to 32
maintained value, **1**-1 to 11
Optical materials
elasto-, electro-, and magneto-optic constants, **12**-179 to 195
harmonic generation, **12**-196 to 199
index of refraction, **10**-217
nonlinear constants, **12**-196 to 199
Optical properties
glass, **10**-217
metals, **12**-133 to 156
polytetrafluoroethylene, **12**-157 to 178
semiconductors, **12**-133 to 156, **12**-157 to 178, **12**-196 to 199
solids, as function of wavelength, **12**-157 to 178
various materials, **12**-179 to 195, **12**-196 to 199
Organic compounds
boiling point index, **3**-751 to 763
bond lengths (in crystals), **9**-1 to 14
bond lengths and angles (in gas phase), **9**-17 to 43
bond strengths, **9**-51 to 74
characteristic ^{13}C chemical shifts, **9**-96
characteristic proton chemical shifts, **9**-95
classes, definitions, **2**-38 to 59
dipole moment, **9**-44 to 50
enthalpy of fusion, **6**-122 to 136
enthalpy of vaporization, **6**-106 to 121
heat of combustion, **5**-89
infrared correlation charts, **9**-87 to 91
magnetic susceptibility, **3**-764 to 767
mass spectral peaks, **8**-129 to 134
melting point index, **3**-737 to 750
nomenclature, **2**-25 to 28
physical properties, **3**-1 to 330
polarizability, **10**-160 to 174
solubility, aqueous, **8**-91 to 107
sublimation pressure, **6**-62 to 64
superconducting properties, **12**-94 to 96
surface tension, **6**-146 to 149
thermal conductivity, **6**-190 to 191
thermodynamic properties, **5**-4 to 60
Organic magnets, **12**-126 to 129
Organometallic compounds, bond lengths, **9**-15 to 16
Orthogonal polynomials, **A**-102 to 104
Oscillator strengths in atomic spectra, **10**-88 to 146
Osmium: see also Elements
electrical resistivity, **12**-45 to 47
electron configuration, **1**-19 to 20
heat capacity, **4**-133
history, occurrence, uses, **4**-1 to 36
ionization energy, **10**-175 to 177
isotopes and their properties, **11**-50 to 198
magnetic susceptibility, **4**-141 to 146
molten, density, **4**-137 to 140
physical properties, **4**-132
thermal properties, **12**-219 to 220
vapor pressure, **6**-65 to 92
Oxidation and reduction reagents, decinormal solutions, **8**-7
Oxidation-reduction potentials
biochemical species, **7**-9 to 11
general table, **8**-21 to 31
ion radicals, **8**-32 to 34
Oxygen: see also Elements
critical constants, **6**-48 to 61
cryogenic properties, **6**-139
electron configuration, **1**-19 to 20
enthalpy of fusion, **6**-122 to 136
enthalpy of vaporization, **6**-106 to 121
history, occurrence, uses, **4**-1 to 36
ionization energy, **10**-175 to 177
isotopes and their properties, **11**-50 to 198
magnetic susceptibility, **4**-141 to 146
mean free path, **6**-46
permittivity (dielectric constant), **6**-151 to 173, **6**-174 to 175
physical properties, **4**-132
solubility in water, **8**-86 to 89
thermal conductivity, **6**-16 to 24, **6**-188 to 189
thermodynamic properties, **5**-1 to 3, **6**-16 to 24
van der Waals constants, **6**-45
vapor pressure, **6**-65 to 92, **6**-93 to 103
viscosity, **6**-16 to 24, **6**-180 to 181

P

Palladium: see also Elements
electrical resistivity, **12**-45 to 47
electron configuration, **1**-19 to 20
heat capacity, **4**-133

- history, occurrence, uses, **4**-1 to 36
- ionization energy, **10**-175 to 177
- isotopes and their properties, **11**-50 to 198
- magnetic susceptibility, **4**-141 to 146
- molten, density, **4**-137 to 140
- physical properties, **4**-132
- thermal properties, **12**-219 to 220
- vapor pressure, **6**-65 to 92

Paramagnetic moment, rare earth elements, **4**-122 to 131

Paramagnetic susceptibility, elements and inorganic compounds, **4**-141 to 146

Particle size, **15**-28

Particles, fundamental, summary of properties, **11**-1 to 49

Pascal, definition, **1**-25 to 32

Pauling electronegativity scale, **9**-75

Pearson symbols, **12**-7 to 13, **12**-19 to 21

Periodic table of the elements, inside front cover

Permeability, magnetic alloys, **12**-117 to 125

Permeability of vacuum, **1**-1 to 11

Permittivity (dielectric constant)
- cryogenic fluids, temperature and pressure dependence, **6**-16 to 24
- crystals, **12**-52 to 60
- gases, **6**-174 to 175
- glass, **12**-63
- ice, **6**-6 to 7
- liquid helium, **6**-140
- liquids, **6**-151 to 173
- plastics, **13**-15
- quartz, **12**-63
- rubbers, **13**-15
- semiconductors, **12**-97 to 107
- solids, **12**-52 to 60
- solvents, **8**-127
- vacuum, **1**-1 to 11
- water, **6**-3
- water, frequency dependence, **6**-15

Peroxide formation by laboratory chemicals, **16**-1 to 15

pH
- acid-base indicators, **8**-16 to 18
- biological buffers, **7**-12
- biological materials and tissues, **7**-13
- blood, **7**-13
- definition of pH scale, **8**-35 to 40
- fluorescent indicators, **8**-19 to 20
- foods, **7**-13
- measurement in natural waters, **8**-41 to 42
- seawater, **8**-41 to 42
- solutions giving round values, **8**-43
- standards, **8**-35 to 40

Phase diagrams, **12**-200 to 217

Phase transitions
- enthalpy of fusion, **6**-122 to 136
- enthalpy of vaporization, **6**-106 to 121
- ice, **6**-6 to 7, **6**-12
- polymers, glass to crystal, **13**-7 to 14
- rare earth elements, **4**-122 to 131

Phon, definition, **14**-47 to 48

Phonon-electron coupling, rare earth elements, **4**-122 to 131

Phosphorus: see also Elements
- critical constants, **6**-48 to 61
- dielectric constant, **12**-52 to 60
- electron configuration, **1**-19 to 20
- heat capacity, **4**-133
- history, occurrence, uses, **4**-1 to 36
- ionization energy, **10**-175 to 177
- isotopes and their properties, **11**-50 to 198
- magnetic susceptibility, **4**-141 to 146
- physical properties, **4**-132
- thermodynamic properties, **5**-1 to 3
- vapor pressure, **6**-65 to 92

Photochemical data, **5**-106 to 120

Photoelastic constants, **12**-179 to 195

Photon attenuation coefficients, **10**-207 to 211

Physical constants, fundamental, **1**-1 to 11

Physical properties
- amino acids, **7**-1 to 2
- inorganic compounds, **4**-37 to 96
- minerals, **4**-149 to 155
- organic compounds, **3**-1 to 330
- semiconductors, **12**-97 to 107
- solvents, **15**-14 to 18

Physical quantities
- definitions, **2**-38 to 59
- terminology and symbols, **2**-1 to 21

Pi, value of, **A**-1

Pion, summary of properties, **11**-1 to 49

Pitch, in musical scales, **14**-45 to 46

PIXE, definition, **12**-1 to 6

p*K*
- acid-base indicators, **8**-16 to 18
- amino acids, **7**-1 to 2
- biological buffers, **7**-12
- inorganic acids and bases, **8**-44 to 45
- organic acids and bases, **8**-46 to 56
- purine and pyrimidine bases, **7**-4

Planck constant, **1**-1 to 11

Planets
- atmospheric composition, **14**-2 to 3
- general properties, **14**-2 to 3
- orbital parameters, **14**-2 to 3
- satellites, **14**-4 to 5

Plastics
- breakdown voltage, **15**-30 to 35
- density, **15**-29
- dielectric constant, **13**-15
- speed of sound in, **14**-41 to 42
- thermal conductivity, **12**-226 to 227

Platinum: see also Elements
- electrical resistivity, **12**-45 to 47
- electron configuration, **1**-19 to 20
- heat capacity, **4**-133
- history, occurrence, uses, **4**-1 to 36
- ionization energy, **10**-175 to 177
- isotopes and their properties, **11**-50 to 198
- magnetic susceptibility, **4**-141 to 146
- molten, density, **4**-137 to 140
- physical properties, **4**-132
- resistance of wires, **15**-27
- thermal conductivity, **12**-221 to 222
- thermal properties, **12**-219 to 220
- vapor pressure, **6**-65 to 92
- wire tables, **15**-27

Platinum-rhodium thermocouple tables, **15**-1 to 13

Pluto, orbital data and dimensions, **14**-2 to 3

Plutonium: see also Elements
- electron configuration, **1**-19 to 20
- heat capacity, **4**-133
- history, occurrence, uses, **4**-1 to 36
- ionization energy, **10**-175 to 177
- isotopes and their properties, **11**-50 to 198
- magnetic susceptibility, **4**-141 to 146
- molten, density, **4**-137 to 140
- physical properties, **4**-132
- thermal properties, **12**-219 to 220
- vapor pressure, **6**-65 to 92

Point groups of small molecules, **9**-77 to 80

Polarizability
- atoms and ions in solids, **12**-17 to 18
- free atoms and molecules, **10**-160 to 174

Pollutants
- airborne, limits in the workplace, **16**-32 to 42
- Henry's Law constants, **8**-91 to 107
- octanol-water partition coefficients, **16**-43 to 47
- solubility, **8**-91 to 107

Polonium: see also Elements
- electrical resistivity, **12**-45 to 47
- electron configuration, **1**-19 to 20
- heat capacity, **4**-133
- history, occurrence, uses, **4**-1 to 36
- ionization energy, **10**-175 to 177
- isotopes and their properties, **11**-50 to 198
- physical properties, **4**-132
- thermal properties, **12**-219 to 220
- vapor pressure, **6**-65 to 92

Polymers
- breakdown voltage, **15**-30 to 35
- density of melts, **13**-16 to 19
- dielectric constant, **13**-15
- glass transition temperature, **13**-7 to 14
- molar volume, **13**-16 to 19
- nomenclature, **13**-1 to 5
- solvents for, **13**-6

Polynomials, orthogonal, **A**-102 to 104

Potassium: see also Elements
- electrical resistivity, **12**-45 to 47
- electron configuration, **1**-19 to 20
- heat capacity, **4**-133
- history, occurrence, uses, **4**-1 to 36
- ionization energy, **10**-175 to 177
- isotopes and their properties, **11**-50 to 198
- magnetic susceptibility, **4**-141 to 146
- molten, density, **4**-137 to 140
- physical properties, **4**-132
- thermal properties, **12**-219 to 220
- thermodynamic properties, **5**-1 to 3
- thermodynamic properties at high temperature, **5**-61 to 84
- vapor pressure, **6**-65 to 92

Potential
- electrical, in the atmosphere, **14**-33 to 40
- oxidation-reduction, **8**-21 to 31
- oxidation-reduction, of ion radicals, **8**-32 to 34

Praseodymium: see also Elements
electrical resistivity, **12**-45 to 47
electron configuration, **1**-19 to 20
heat capacity, **4**-133
history, occurrence, uses, **4**-1 to 36
ionization energy, **10**-175 to 177
isotopes and their properties, **11**-50 to 198
magnetic susceptibility, **4**-141 to 146
molten, density, **4**-137 to 140
physical properties, **4**-132
thermal properties, **12**-219 to 220
vapor pressure, **6**-65 to 92
Precipitation of metal ions, **16**-1 to 15
Pressure
atmosphere, as function of altitude, **14**-19 to 25
conversion factors, **1**-48
earth, as function of depth, **14**-13
effect on boiling point, **15**-19
effect on freezing point, **6**-47
ocean, as function of depth, **14**-14
planetary atmospheres, **14**-2 to 3
Probability function, **A**-104
Promethium: see also Elements
electrical resistivity, **12**-45 to 47
electron configuration, **1**-19 to 20
heat capacity, **4**-133
history, occurrence, uses, **4**-1 to 36
ionization energy, **10**-175 to 177
isotopes and their properties, **11**-50 to 198
physical properties, **4**-132
thermal properties, **12**-219 to 220
Protactinium: see also Elements
electrical resistivity, **12**-45 to 47
electron configuration, **1**-19 to 20
heat capacity, **4**-133
history, occurrence, uses, **4**-1 to 36
ionization energy, **10**-175 to 177
isotopes and their properties, **11**-50 to 198
magnetic susceptibility, **4**-141 to 146
physical properties, **4**-132
thermal properties, **12**-219 to 220
Proton
in cosmic ray showers, **11**-216 to 220
magnetic moment, **1**-1 to 11
mass, **1**-1 to 11
summary of properties, **11**-1 to 49
PSD, definition, **12**-1 to 6
psia and psig, definition, **1**-48
Purine bases, properties of, **7**-4
Pyrimidine bases, properties of, **7**-4
Pyrophoric chemicals, safe handling, **16**-1 to 15

Q

Quadrupole moments
all nuclides, **11**-50 to 198
important nuclei for NMR, **9**-92 to 94
Quartz
crystallographic data, **4**-156 to 166
dielectric constant, **12**-63
loss factor, **12**-63
optical properties, **4**-149 to 155
phase diagram, **12**-200 to 217
thermal conductivity, **12**-224 to 225

R

Rad, definition, **16**-48
Radiation
black body, **10**-214 to 215
electromagnetic, classification, **10**-212 to 213
microwave, classification of bands, **10**-212 to 213
Radiation, ionizing
conversion factors, **1**-52 to 53
from nuclear decay, **11**-50 to 198
permissible intake of radionuclides, **16**-49 to 52
protection against, **16**-48
Radiation, solar
by month and latitude, **14**-26
by wavelength, **14**-18
flux, solar constant, **14**-2 to 3
Radiative transition probability, **10**-88 to 146
Radicals, free: see Free radicals
Radicals, nomenclature, **2**-23 to 24, **2**-25 to 28
Radio spectrum, **15**-36 to 37
Radioastronomy, **14**-6 to 8
Radionuclides, permissible intake, **16**-49 to 52
Radium: see also Elements
electron configuration, **1**-19 to 20
heat capacity, **4**-133
history, occurrence, uses, **4**-1 to 36
ionization energy, **10**-175 to 177
isotopes and their properties, **11**-50 to 198
physical properties, **4**-132
thermal properties, **12**-219 to 220
vapor pressure, **6**-65 to 92
Radius of ions in crystals, **12**-14 to 16
Radon: see also Elements
critical constants, **6**-48 to 61
electron configuration, **1**-19 to 20
history, occurrence, uses, **4**-1 to 36
ionization energy, **10**-175 to 177
isotopes and their properties, **11**-50 to 198
physical properties, **4**-132
solubility in water, **8**-86 to 89
vapor pressure, **6**-65 to 92, **6**-93 to 103
Rankine temperature, conversion to other scales, **1**-46
Rare earth crystals, Verdet constants, **12**-179 to 195
Rare earth metals, general properties, **4**-122 to 131
Rate constants
atmospheric reactions, **5**-106 to 120
conversion factors, **1**-51
RBS, definition, **12**-1 to 6
Reagents
decinormal solutions, **8**-5 to 6, **8**-7
disposal of, **16**-1 to 15
formulas for concentration, **8**-20
organic, for analysis, **8**-8 to 15
preparation, **8**-1 to 4
Reduction potentials
biochemical species, **7**-9 to 11
general table, **8**-21 to 31
ion radicals, **8**-32 to 34
Reference states of elements, **5**-4 to 60
Reflection coefficient of solids, **12**-133 to 156, **12**-157 to 178
Refractive index: see Index of refraction
Refractory materials
hardness, **12**-233 to 234
thermal conductivity, **12**-226 to 227
Refrigerants, various properties, **6**-141 to 143
Relative humidity
from wet and dry bulb temperatures, **15**-24
relation to dew point, **15**-23
solutions for calibration, **15**-26
solutions for constant humidity, **15**-25
Relaxation time, in water, **6**-15
Rem, definition, **16**-48
Remanence, magnetic materials, **12**-117 to 125
Resistance of wires, **15**-27
Resistivity, electrical
alloys, **12**-48 to 51
commercial metals and alloys, **12**-232
conversion factors, **1**-50
elements, **12**-45 to 47
glasses, **12**-63
pure metals, **12**-45 to 47
quartz, **12**-63
rare earth elements, **4**-122 to 131
semiconducting minerals, **12**-97 to 107
semiconductors, **12**-97 to 107
superconductors, **12**-91 to 93
Respirators, for laboratory use, **16**-1 to 15
RHEED, definition, **12**-1 to 6
Rhenium: see also Elements
electrical resistivity, **12**-45 to 47
electron configuration, **1**-19 to 20
heat capacity, **4**-133
history, occurrence, uses, **4**-1 to 36
ionization energy, **10**-175 to 177
isotopes and their properties, **11**-50 to 198
magnetic susceptibility, **4**-141 to 146
molten, density, **4**-137 to 140
physical properties, **4**-132
thermal properties, **12**-219 to 220
vapor pressure, **6**-65 to 92
Rhodium: see also Elements
electrical resistivity, **12**-45 to 47
electron configuration, **1**-19 to 20
heat capacity, **4**-133
history, occurrence, uses, **4**-1 to 36
ionization energy, **10**-175 to 177
isotopes and their properties, **11**-50 to 198
magnetic susceptibility, **4**-141 to 146
molten, density, **4**-137 to 140
physical properties, **4**-132
thermal properties, **12**-219 to 220
vapor pressure, **6**-65 to 92
Ring systems, nomenclature, **2**-25 to 28
Rochelle salts, **12**-61 to 62
Rocks
age, **14**-9 to 10

density, 15-29
thermal conductivity, 12-226 to 227
Roentgen, definition, 16-48
Rotational constants, diatomic molecules, 9-81 to 86
Rounding of numbers, 1-34 to 45
Rubbers
breakdown voltage, 15-30 to 35
density, 15-29
dielectric constant, 13-15
speed of sound in, 14-41 to 42
thermal conductivity, 12-226 to 227
Rubidium: see also Elements
electrical resistivity, 12-45 to 47
electron configuration, 1-19 to 20
heat capacity, 4-133
history, occurrence, uses, 4-1 to 36
ionization energy, 10-175 to 177
isotopes and their properties, 11-50 to 198
magnetic susceptibility, 4-141 to 146
molten, density, 4-137 to 140
physical properties, 4-132
thermal properties, 12-219 to 220
thermodynamic properties, 5-1 to 3
vapor pressure, 6-65 to 92
Russian alphabet, 2-37
Ruthenium: see also Elements
electrical resistivity, 12-45 to 47
electron configuration, 1-19 to 20
heat capacity, 4-133
history, occurrence, uses, 4-1 to 36
ionization energy, 10-175 to 177
isotopes and their properties, 11-50 to 198
magnetic susceptibility, 4-141 to 146
molten, density, 4-137 to 140
physical properties, 4-132
thermal properties, 12-219 to 220
vapor pressure, 6-65 to 92
Rutherfordium (element 104)
electron configuration, 1-19 to 20
history, occurrence, uses, 4-1 to 36
isotopes and their properties, 11-50 to 198
Rydberg constant, 1-1 to 11

S

Sackur-Tetrode constant, 1-1 to 11
Safety
chemical carcinogens, 16-53 to 57
flammability of chemicals, 16-16 to 31
laboratory practices, 16-1 to 15
radiation, 16-48, 16-49 to 52
SALI, definition, 12-1 to 6
Salinity scale for seawater, 14-15 to 16
Salts
activity coefficients, 5-97 to 98, 5-99 to 103
decinormal solutions, 8-5 to 6
electrical conductivity, 5-94
enthalpy of solution, 5-105
molten, density of, 4-137 to 140
solubility as a function of temperature, 8-108 to 116, 8-121
vapor pressure of aqueous solutions, 6-104
Samarium: see also Elements
electrical resistivity, 12-45 to 47
electron configuration, 1-19 to 20
heat capacity, 4-133
history, occurrence, uses, 4-1 to 36
ionization energy, 10-175 to 177
isotopes and their properties, 11-50 to 198
magnetic susceptibility, 4-141 to 146
molten, density, 4-137 to 140
physical properties, 4-132
thermal properties, 12-219 to 220
vapor pressure, 6-65 to 92
SANS, definition, 12-1 to 6
Saturn, orbital data and dimensions, 14-2 to 3
Scandium: see also Elements
electrical resistivity, 12-45 to 47
electron configuration, 1-19 to 20
heat capacity, 4-133
history, occurrence, uses, 4-1 to 36
ionization energy, 10-175 to 177
isotopes and their properties, 11-50 to 198
magnetic susceptibility, 4-141 to 146
molten, density, 4-137 to 140
physical properties, 4-132
thermal properties, 12-219 to 220
vapor pressure, 6-65 to 92
Seaborgium (element 106), 4-1 to 36, 11-50 to 198
Seawater
composition (elemental), 14-17
composition (ions), 14-15 to 16
density, 14-15 to 16
electrical conductivity, 14-15 to 16
freezing point, 14-15 to 16
pH measurement, 8-41 to 42
pressure as a function of depth, 14-14
salinity scale, 14-15 to 16
specific heat, 14-15 to 16
speed of sound in, 14-41 to 42
viscosity, 14-15 to 16
Secant function, A-7 to 10
Second, definition, 1-25 to 32
Second radiation constant, 1-1 to 11
Secondary electron emission, 12-131 to 132
Selenium: see also Elements
critical constants, 6-48 to 61
electron configuration, 1-19 to 20
heat capacity, 4-133
history, occurrence, uses, 4-1 to 36
ionization energy, 10-175 to 177
isotopes and their properties, 11-50 to 198
magnetic susceptibility, 4-141 to 146
molten, density, 4-137 to 140
physical properties, 4-132
vapor pressure, 6-65 to 92
SEM, definition, 12-1 to 6
Semiconductors
crystal structure, 12-19 to 21, 12-97 to 107
diffusion in, 12-108 to 116
effective mass, 12-97 to 107
elastic constants, 12-37 to 44
electrical properties, 12-97 to 107
extinction coefficient, 12-133 to 156, 12-157 to 178
index of refraction, 12-133 to 156, 12-157 to 178
minerals, resistivity of, 12-97 to 107
optical properties, 12-133 to 156, 12-157 to 178, 12-179 to 195
physical properties, 12-97 to 107
reflection coefficient, 12-133 to 156, 12-157 to 178
thermal conductivity, 12-97 to 107, 12-221 to 222
Series expansions, A-84 to 87
Shielding, from radiation, 10-207 to 211, 16-48
SI units
conversion factors to, 1-34 to 45
definitions and symbols, 1-25 to 32
prefixes, 1-25 to 32
Siemens, definition, 1-25 to 32
Sievert, definition, 1-25 to 32, 16-48
Silicon: see also Elements
dielectric constant, 12-52 to 60
electron configuration, 1-19 to 20
heat capacity, 4-133, 12-218
history, occurrence, uses, 4-1 to 36
ionization energy, 10-175 to 177
isotopes and their properties, 11-50 to 198
magnetic susceptibility, 4-141 to 146
molten, density, 4-137 to 140
optical properties, 12-133 to 156
physical properties, 4-132
semiconducting properties, 12-97 to 107
thermal conductivity, 12-221 to 222
thermodynamic properties, 5-1 to 3
thermodynamic properties at high temperature, 5-61 to 84
vapor pressure, 6-65 to 92
Silver: see also Elements
electrical resistivity, 12-45 to 47
electron configuration, 1-19 to 20
heat capacity, 4-133, 12-218
history, occurrence, uses, 4-1 to 36
ionization energy, 10-175 to 177
isotopes and their properties, 11-50 to 198
magnetic susceptibility, 4-141 to 146
molten, density, 4-137 to 140
physical properties, 4-132
resistance of wires, 15-27
thermal conductivity, 12-221 to 222
thermal properties, 12-219 to 220
thermodynamic properties, 5-1 to 3
vapor pressure, 6-65 to 92
wire tables, 15-27
SIMS, definition, 12-1 to 6
Sine function, A-7 to 10
SLAM, definition, 12-1 to 6
SMOW (standard mean ocean water), density, 6-5
SNMS, definition, 12-1 to 6
Sodium: see also Elements
electrical resistivity, 12-45 to 47
electron configuration, 1-19 to 20
heat capacity, 4-133, 12-218
history, occurrence, uses, 4-1 to 36
ionization energy, 10-175 to 177
isotopes and their properties, 11-50 to 198

magnetic susceptibility, **4**-141 to 146
molten, density, **4**-137 to 140
physical properties, **4**-132
thermal properties, **12**-219 to 220
thermodynamic properties, **5**-1 to 3
vapor pressure, **6**-65 to 92
Sodium chloride
activity coefficients, **5**-97 to 98, **5**-99 to 103
aqueous solutions, concentrative properties, **8**-57 to 83
aqueous solutions, relative humidity, **15**-26
aqueous solutions, volumetric properties, **6**-8
density of aqueous solutions, **6**-8
enthalpy of solution, **5**-105
standard thermodynamic properties, **5**-4 to 60
Solar constant, **14**-2 to 3
Solar radiation
by month and latitude, **14**-26
by wavelength, **14**-18
Solar spectrum, **14**-18
Solar system, **14**-2 to 3
Solder
phase diagram, **12**-200 to 217
thermal conductivity, **12**-223
Solids, characterization and analysis, **12**-1 to 6
Solubility
amino acids, **7**-1 to 2
carbon dioxide in water, **8**-90
fatty acids in water, **7**-6
gases in water, **8**-86 to 89
inorganic compounds, **4**-37 to 96
inorganic compounds, as function of temperature, **8**-108 to 116, **8**-121
inorganic compounds, qualitative rules, **8**-122 to 124
inorganic compounds, sparingly soluble, **8**-117 to 120
octanol-water partition coefficients, **16**-43 to 47
organic compounds, **3**-1 to 330
organic compounds in water, **8**-91 to 107
purine and pyrimidine bases, **7**-4
salts in water, **8**-121
Solubility product constant, **8**-117 to 120
Solutions
aqueous, concentrative properties, **8**-57 to 83
decinormal, oxidation reagents, **8**-7
decinormal, salts, **8**-5 to 6
density, **8**-57 to 83
diffusion of ions, **5**-95 to 97
enthalpy, for common electrolytes, **5**-105
for round values of pH, **8**-43
formulas for concentration, **8**-20
freezing point depression, **8**-57 to 83
index of refraction, **8**-57 to 83
ionic conductivity, **5**-95 to 97
viscosity, **8**-57 to 83
Solvents
dielectric constant, **8**-127
dipole moment, **15**-14 to 18
flammability, **15**-14 to 18
for NMR, chemical shifts, **8**-128
for polymers, **13**-6
for ultraviolet spectrophotometry, **8**-127
heat capacity, **15**-14 to 18
mass spectral peaks, **8**-129 to 134
physical properties, **15**-14 to 18
threshold limit in air, **15**-14 to 18
vapor pressure, **15**-14 to 18
wavelength cutoff (UV), **8**-127
Sound level, in human hearing, **14**-47 to 48
Sound velocity
air, as function of frequency, **14**-43
air, as function of humidity, **14**-43
air, as function of temperature, **14**-44
atmosphere, as function of altitude, **14**-19 to 25
fluids, **6**-16 to 24
mercury, **6**-145
various solids, liquids, and gases, **14**-41 to 42
water and seawater, **14**-41 to 42
Space group
elements, **12**-19 to 21
notation, **12**-7 to 13
Specific gravity: see Density
Specific heat: see Heat capacity
Specific volume: see also Density
mercury, **6**-144
sodium chloride solutions, **6**-8
water, **8**-125
Spectroscopic constants, diatomic molecules, **9**-81 to 86
Spectrum, infrared
calibration frequencies, **10**-228 to 235, **10**-236 to 240
correlation charts, **9**-87 to 91
fundamental vibrational frequencies, **9**-77 to 80
Spectrum, line, of the elements, **10**-1 to 87
Speed of light, **1**-1 to 11
Speed of sound
air, as function of frequency, **14**-43
air, as function of humidity, **14**-43
air, as function of temperature, **14**-44
atmosphere, as function of altitude, **14**-19 to 25
fluids, **6**-16 to 24
various solids, liquids, and gases, **14**-41 to 42
water and seawater, **14**-41 to 42
Spin
fundamental particles, **11**-1 to 49
nuclides, **11**-50 to 198
nuclides of NMR interest, **9**-92 to 94
ordering in magnetic materials, **12**-126 to 129
SPM, definition, **12**-1 to 6
SSMS, definition, **12**-1 to 6
Standard Atmosphere (U.S.), **14**-19 to 25
Standard atomic weights, **1**-12 to 14
Standard density of water, **6**-5
Standard solutions, for pH measurement, **8**-35 to 40, **8**-41 to 42
Standards
CODATA thermodynamic values, **5**-1 to 3
index of refraction, **10**-219
infrared absorption frequencies, **10**-236 to 240
infrared laser frequencies, **10**-228 to 235
temperature, **1**-21 to 22
vapor pressure, **6**-105
Steam, thermodynamic properties (see also Water), **6**-13 to 14
Steel
mechanical properties, **12**-232
thermal conductivity, **12**-223
Stefan-Boltzmann constant, **1**-1 to 11
STEM, definition, **12**-1 to 6
STM, definition, **12**-1 to 6
Stratosphere
chemical reactions, **5**-106 to 120
properties, **14**-19 to 25
Strontium: see also Elements
electrical resistivity, **12**-45 to 47
electron configuration, **1**-19 to 20
heat capacity, **4**-133
history, occurrence, uses, **4**-1 to 36
ionization energy, **10**-175 to 177
isotopes and their properties, **11**-50 to 198
magnetic susceptibility, **4**-141 to 146
molten, density, **4**-137 to 140
thermal properties, **12**-219 to 220
vapor pressure, **6**-65 to 92
Structure
amino acids, **7**-3
bond lengths in organic crystals, **9**-1 to 14
bond lengths in organometallic compounds, **9**-15 to 16
characteristic ^{13}C chemical shifts, **9**-96
characteristic infrared frequencies, **9**-87 to 91
characteristic proton chemical shifts, **9**-95
crystal, of elements, **12**-19 to 21
crystal, of superconductors, **12**-64 to 90, **12**-91 to 93
force constants, **9**-76
formulas for organic compounds, **3**-331 to 585
fundamental vibrational frequencies, **9**-77 to 80
gas-phase molecules, **9**-17 to 43
geometry of small molecules, **9**-77 to 80
solids, characterization techniques, **12**-1 to 6
Student's *t*-distribution, **A**-105
Sublimation pressure of solids, **6**-62 to 64
Sugars
aqueous solution properties, **8**-57 to 83
nomenclature, **7**-7 to 8
Sulfur: see also Elements
critical constants, **6**-48 to 61
dielectric constant, **12**-52 to 60
electron configuration, **1**-19 to 20
heat capacity, **4**-133
history, occurrence, uses, **4**-1 to 36
ionization energy, **10**-175 to 177
isotopes and their properties, **11**-50 to 198

magnetic susceptibility, **4**-141 to 146
molten, density, **4**-137 to 140
physical properties, **4**-132
thermodynamic properties, **5**-1 to 3
thermodynamic properties at high temperature, **5**-61 to 84
vapor pressure, **6**-65 to 92
Sulfuric acid
activity coefficients, **5**-97 to 98, **5**-99 to 103
concentrative properties, **8**-57 to 83
constant humidity solutions, **15**-25
vapor pressure, **6**-65 to 92
Sun
mass, dimensions, and other properties, **14**-2 to 3
radiative properties, **14**-2 to 3
spectral irradiance, **14**-18
Superconductors
electrical and magnetic properties, **12**-64 to 90, **12**-91 to 93
organic, **12**-94 to 96
rare earth elements, **4**-122 to 131
transition temperature, **12**-64 to 90, **12**-91 to 93, **12**-94 to 96
Superconductors, high temperature
general properties, **12**-91 to 93
phase diagram, **12**-200 to 217
Supercooled water, **6**-6 to 7
Surface characterization and analysis, **12**-1 to 6
Surface tension
aqueous mixtures, **6**-150
liquid helium, **6**-140
liquid rare earth metals, **4**-122 to 131
various liquids, **6**-146 to 149
water, **6**-3
Susceptibility: see Magnetic susceptibility
Symbols
amino acids, **7**-3
carbohydrates, **7**-7 to 8
magnetism, **12**-117 to 125
physical quantities, **2**-1 to 21, **2**-29 to 36
SI units, **1**-25 to 32
units, **2**-29 to 36
Symmetry of crystals, **12**-7 to 13
Synonym index, organic compounds, **3**-586 to 630

T

Tangent function, **A**-7 to 10
Tantalum: see also Elements
electrical resistivity, **12**-45 to 47
electron configuration, **1**-19 to 20
heat capacity, **4**-133
history, occurrence, uses, **4**-1 to 36
ionization energy, **10**-175 to 177
isotopes and their properties, **11**-50 to 198
magnetic susceptibility, **4**-141 to 146
molten, density, **4**-137 to 140
physical properties, **4**-132
thermal properties, **12**-219 to 220
vapor pressure, **6**-65 to 92
Taylor series, **A**-84 to 87
Technetium: see also Elements
electron configuration, **1**-19 to 20
heat capacity, **4**-133
history, occurrence, uses, **4**-1 to 36
ionization energy, **10**-175 to 177
isotopes and their properties, **11**-50 to 198
magnetic susceptibility, **4**-141 to 146
physical properties, **4**-132
thermal properties, **12**-219 to 220
vapor pressure, **6**-65 to 92
Tellurium: see also Elements
dielectric constant, **12**-52 to 60
electron configuration, **1**-19 to 20
history, occurrence, uses, **4**-1 to 36
ionization energy, **10**-175 to 177
isotopes and their properties, **11**-50 to 198
magnetic susceptibility, **4**-141 to 146
molten, density, **4**-137 to 140
physical properties, **4**-132
vapor pressure, **6**-65 to 92
Temperature
atmosphere, as function of altitude, **14**-19 to 25
baths for low temperature, **15**-26
calibration, ITS-90, **1**-21 to 22
celsius and absolute, definitions, **1**-25 to 32
conversion between scales, **1**-46
conversion to ITS-90, **1**-23 to 24
glass transition, in polymers, **13**-7 to 14
International Temperature Scale (ITS-90), **1**-21 to 22
mean global, **14**-32
mean United States, **14**-30 to 31
planetary atmospheres, **14**-2 to 3
superconducting transition, **12**-64 to 90
thermocouple tables, **15**-1 to 13
Tensile strength
commercial metals and alloys, **12**-232
rare earth elements, **4**-122 to 131
Terbium: see also Elements
electrical resistivity, **12**-45 to 47
electron configuration, **1**-19 to 20
heat capacity, **4**-133
history, occurrence, uses, **4**-1 to 36
ionization energy, **10**-175 to 177
isotopes and their properties, **11**-50 to 198
magnetic susceptibility, **4**-141 to 146
molten, density, **4**-137 to 140
physical properties, 4-132
thermal properties, **12**-219 to 220
vapor pressure, **6**-65 to 92
Terminology
inorganic ions and ligands, **2**-23 to 24
organic substituent groups and ring systems, **2**-25 to 28
physical quantities, **2**-1 to 21
polymers, **13**-1 to 5
scientific terms, definitions, **2**-38 to 59
Tesla, definition, **1**-25 to 32
TGA, definition, **12**-1 to 6
Thallium: see also Elements
electrical resistivity, **12**-45 to 47
electron configuration, **1**-19 to 20
heat capacity, **4**-133
history, occurrence, uses, **4**-1 to 36
ionization energy, **10**-175 to 177
isotopes and their properties, **11**-50 to 198
magnetic susceptibility, **4**-141 to 146
molten, density, **4**-137 to 140
physical properties, **4**-132
thermal properties, **12**-219 to 220
vapor pressure, **6**-65 to 92
Thermal conductivity
alloys, **12**-223
argon, liquid and gas, **6**-16 to 24
atmosphere, as function of altitude, **14**-19 to 25
carrier gases for chromatography, **8**-126
ceramics, **12**-226 to 227
commercial metals, **12**-223
commercial metals and alloys, **12**-232
common fluids, as function of temperature and pressure, **6**-16 to 24
construction materials, **12**-226 to 227
conversion factors, **1**-49
cryogenic solids, **12**-224 to 225
crystalline solids, **12**-224 to 225
D_2O, **6**-4
dielectric crystals, **12**-224 to 225
gases, at atmospheric pressure, **6**-188 to 189
glasses, **12**-228 to 231
helium, liquid, **6**-140
helium, liquid and gas, **6**-16 to 24
ice, **6**-6 to 7, **12**-224 to 225
insulation, **12**-226 to 227
liquids, **6**-190 to 191
mercury, **6**-190 to 191
metals, **12**-219 to 220, **12**-221 to 222
methane, liquid and gas, **6**-16 to 24
minerals, **12**-226 to 227
nitrogen, liquid and gas, **6**-16 to 24
oxygen, liquid and gas, **6**-16 to 24
plastics, **12**-226 to 227
quartz, **12**-224 to 225
rare earth elements, **4**-122 to 131
refractory materials, **12**-226 to 227
rocks, **12**-226 to 227
rubber, **12**-226 to 227
semiconductors, **12**-97 to 107, **12**-221 to 222
superconductors, **12**-91 to 93
water, **6**-3, **6**-4, **6**-190 to 191
wood, **12**-226 to 227
Thermal expansion coefficient
commercial metals and alloys, **12**-232
ice, **6**-6 to 7
liquids, **6**-137 to 138
metals, **12**-219 to 220
rare earth elements, **4**-122 to 131
semiconductors, **12**-97 to 107
sodium chloride solutions, **6**-8
Thermal neutron cross sections, **11**-199 to 215
Thermocouple calibration tables, **15**-1 to 13
Thermodynamic properties (see also Enthalpy, Heat capacity, etc.)
air, **6**-1 to 2

aqueous systems, **5**-85 to 88
argon, **6**-16 to 24
biochemical species, **7**-9 to 11
CODATA Key Values, **5**-1 to 3
common fluids, as function of temperature and pressure, **6**-16 to 24
ethane, **6**-16 to 24
helium, **6**-16 to 24
high temperature, **5**-61 to 84
hydrogen, **6**-16 to 24
inorganic compounds, **5**-4 to 60
nitrogen, **6**-16 to 24
organic compounds, **5**-4 to 60
oxygen, **6**-16 to 24
propane, **6**-16 to 24
rare earth elements, **4**-122 to 131
standard state values, **5**-4 to 60
steam, **6**-13 to 14
temperature dependence, **5**-61 to 84
water, **6**-13 to 14
Thermometers, wet and dry bulb, **15**-24
Thorium: see also Elements
electrical resistivity, **12**-45 to 47
electron configuration, **1**-19 to 20
heat capacity, **4**-133
history, occurrence, uses, **4**-1 to 36
ionization energy, **10**-175 to 177
isotopes and their properties, **11**-50 to 198
magnetic susceptibility, **4**-141 to 146
physical properties, **4**-132
thermal properties, **12**-219 to 220
thermodynamic properties, **5**-1 to 3
vapor pressure, **6**-65 to 92
Threshold limits
airborne contaminants, **16**-32 to 42
halocarbon refrigerants, **6**-141 to 143
solvents, **15**-14 to 18
Thulium: see also Elements
electrical resistivity, **12**-45 to 47
electron configuration, **1**-19 to 20
heat capacity, **4**-133
history, occurrence, uses, **4**-1 to 36
ionization energy, **10**-175 to 177
isotopes and their properties, **11**-50 to 198
magnetic susceptibility, **4**-141 to 146
molten, density, **4**-137 to 140
physical properties, **4**-132
thermal properties, **12**-219 to 220
vapor pressure, **6**-65 to 92
Thunderstorm electricity, **14**-33 to 40
Time
astronomical units, **14**-1
geological scale, **14**-11
Tin: see also Elements
electrical resistivity, **12**-45 to 47
electron configuration, **1**-19 to 20
heat capacity, **4**-133
history, occurrence, uses, **4**-1 to 36
ionization energy, **10**-175 to 177
isotopes and their properties, **11**-50 to 198
magnetic susceptibility, **4**-141 to 146
molten, density, **4**-137 to 140
physical properties, **4**-132
thermal conductivity, **12**-221 to 222
thermal properties, **12**-219 to 220
thermodynamic properties, **5**-1 to 3
vapor pressure, **6**-65 to 92
Titanium: see also Elements
electrical resistivity, **12**-45 to 47
electron configuration, **1**-19 to 20
heat capacity, **4**-133, **12**-218
history, occurrence, uses, **4**-1 to 36
ionization energy, **10**-175 to 177
isotopes and their properties, **11**-50 to 198
magnetic susceptibility, **4**-141 to 146
molten, density, **4**-137 to 140
physical properties, **4**-132
thermal properties, **12**-219 to 220
thermodynamic properties, **5**-1 to 3
vapor pressure, **6**-65 to 92
Toxicity of anions and cations, **16**-1 to 15
Transition probability, atomic, **10**-88 to 146
Transition temperature
glass, in polymers, **13**-7 to 14
superconductors, **12**-64 to 90, **12**-91 to 93, **12**-94 to 96
Transport properties: see Thermal conductivity, Viscosity, Diffusion
Trigonometric functions, **A**-7 to 10, **A**-11
Trigonometric series, **A**-84 to 87
Trillion, definition, **1**-46
Triple point constants
carbon dioxide, **6**-62 to 64
cryogenic fluids, **6**-139
D_2O, **6**-4
various compounds, **6**-62 to 64
water, **6**-4
Tschebysheff polynomials, **A**-102 to 104
Tungsten: see also Elements
electrical resistivity, **12**-45 to 47
electron configuration, **1**-19 to 20
heat capacity, **4**-133, **12**-218
history, occurrence, uses, **4**-1 to 36
ionization energy, **10**-175 to 177
isotopes and their properties, **11**-50 to 198
magnetic susceptibility, **4**-141 to 146
molten, density, **4**-137 to 140
physical properties, **4**-132
resistance of wires, **15**-27
thermal conductivity, **12**-221 to 222
thermal properties, **12**-219 to 220
vapor pressure, **6**-65 to 92
wire tables, **15**-27

U

Ultraviolet spectrophotometry, solvents for, **8**-127
Units
conversion factors, **1**-34 to 45
definitions, **2**-38 to 59
ionizing radiation, **1**-52 to 53, **16**-48
magnetic quantities, **1**-33, **12**-117 to 125
SI, definitions and symbols, **1**-25 to 32
UPS, definition, **12**-1 to 6
Uranium: see also Elements
electrical resistivity, **12**-45 to 47
electron configuration, **1**-19 to 20
heat capacity, **4**-133
history, occurrence, uses, **4**-1 to 36
ionization energy, **10**-175 to 177
isotopes and their properties, **11**-50 to 198
magnetic susceptibility, **4**-141 to 146
molten, density, **4**-137 to 140
physical properties, **4**-132
thermal properties, **12**-219 to 220
thermodynamic properties, **5**-1 to 3
vapor pressure, **6**-65 to 92
Uranus, orbital data and dimensions, **14**-2 to 3

V

Van der Waals constants for gases, **6**-45
Vanadium: see also Elements
electrical resistivity, **12**-45 to 47
electron configuration, **1**-19 to 20
heat capacity, **4**-133
history, occurrence, uses, **4**-1 to 36
ionization energy, **10**-175 to 177
isotopes and their properties, **11**-50 to 198
magnetic susceptibility, **4**-141 to 146
molten, density, **4**-137 to 140
physical properties, **4**-132
thermal properties, **12**-219 to 220
vapor pressure, **6**-65 to 92
Vapor pressure
air, **6**-1 to 2
aqueous salt solutions, **15**-26
at temperatures below 300 K, **6**-93 to 103
calibration data, **6**-105
carbon dioxide, **6**-62 to 64, **6**-93 to 103, **6**-105
cryogenic fluids, **6**-93 to 103
elements, **6**-65 to 92
elements, high temperature, **4**-134 to 136
general table, **6**-65 to 92
helium, **1**-21 to 22, **6**-93 to 103, **6**-140
ice, **6**-9, **6**-105
inorganic compounds, **6**-65 to 92
IUPAC recommended data, **6**-105
metals, at high temperatures, **4**-134 to 136
organic compounds, **6**-65 to 92
rare earth elements, **4**-122 to 131
rare gases, **6**-93 to 103
salt solutions, **6**-104
solids, **6**-62 to 64
solvents, **15**-14 to 18
water, **6**-3, **6**-10 to 11, **6**-105
water, over salt solutions, **15**-25, **15**-26
Vaporization: see Enthalpy of vaporization
Vector analysis, **A**-87 to 96
Velocity of light, **1**-1 to 11
Velocity of sound
air, as function of frequency, **14**-43
air, as function of humidity, **14**-43
air, as function of temperature, **14**-44
atmosphere, as function of altitude, **14**-19 to 25
fluids, **6**-16 to 24
various solids, liquids, and gases, **14**-41 to 42

Velocity, mean, in gases, **6**-46
Venus, orbital data and dimensions, **14**-2 to 3
Verdet constants, **12**-179 to 195
Vibrational force constants, **9**-76
Vibrational frequencies of molecules, **9**-77 to 80, **9**-81 to 86
Vibrational-rotational spectra, for frequency calibration, **10**-236 to 240
Virial coefficients of gases, **6**-25 to 44
Viscosity
- aqueous solutions, **8**-57 to 83
- argon, liquid and gas, **6**-16 to 24
- atmosphere, as function of altitude, **14**-19 to 25
- carbon dioxide, on saturation line, **6**-187
- carrier gases for chromatography, **8**-126
- common fluids, as function of temperature and pressure, **6**-16 to 24
- gases, at atmospheric pressure, **6**-180 to 181
- helium, liquid, **6**-140
- helium, liquid and gas, **6**-16 to 24
- liquids, **6**-182 to 186
- methane, liquid and gas, **6**-16 to 24
- nitrogen, liquid and gas, **6**-16 to 24
- oxygen, liquid and gas, **6**-16 to 24
- seawater, **14**-15 to 16
- water, as function of temperature, **6**-3

Volt
- definition, **1**-25 to 32
- maintained value, **1**-1 to 11

W

Water
- azeotropic mixtures, **6**-176 to 179
- boiling point, as function of pressure, **6**-12
- compressibility, **6**-137 to 138
- critical constants, **6**-4
- density, **6**-3, **6**-4, **6**-5
- density, as function of pressure, **6**-137 to 138
- density (supercooled), **6**-6 to 7
- dielectric constant, **6**-3, **6**-151 to 173, **6**-174 to 175
- dielectric constant, as function of frequency, **6**-15
- diffusion of gases, **6**-194
- dissociation constant, **8**-84, **8**-85
- electrical conductivity, **5**-90
- enthalpy of fusion, **6**-122 to 136
- enthalpy of vaporization, **6**-3
- fixed point properties, **6**-4
- freezing point, pressure dependence, **6**-47
- heat capacity, **6**-3
- index of refraction, **10**-218
- ion product, **8**-84, **8**-85
- octanol-water partition coefficients, **16**-43 to 47
- permittivity (dielectric constant), **6**-3, **6**-151 to 173, **6**-174 to 175
- permittivity, as function of frequency, **6**-15
- pH measurement, **8**-41 to 42
- speed of sound in, **14**-41 to 42
- surface tension, **6**-3
- thermal conductivity, **6**-4, **6**-188 to 189
- thermal expansion coefficient, **6**-137 to 138
- thermodynamic properties, **6**-13 to 14
- thermodynamic properties at high temperature, **5**-61 to 84
- triple point constants, **6**-4
- van der Waals constants, **6**-45
- vapor pressure, **6**-10 to 11
- vapor pressure over salt solutions, **15**-25
- viscosity, **6**-3, **6**-180 to 181
- volume of one gram, **8**-125

Water(D_2O)
- boiling point, **6**-4
- critical constants, **6**-4
- density, **6**-9
- dissociation constant, **8**-85
- fixed point properties, **6**-4
- ion product, **8**-85
- thermal conductivity, **6**-4, **6**-188 to 189
- triple point constants, **6**-4
- vapor pressure, **6**-4
- viscosity, **6**-180 to 181

Watt, definition, **1**-25 to 32
Wavelengths
- atomic spectra, **10**-1 to 87
- correction to vacuum, **10**-220
- electromagnetic radiation bands, **10**-212 to 213
- laser sources, **10**-221 to 227

Weber, definition, **1**-25 to 32
Weighings, reduction from air to vacuum, **8**-125
Width, x-ray lines, **10**-206
Wien displacement law constant, **1**-1 to 11
Wire tables, **15**-27
Wood
- density, **15**-29
- speed of sound in, **14**-41 to 42
- thermal conductivity, **12**-226 to 227

Work function, of the elements, **12**-130

X

Xenon: see also Elements
- critical constants, **6**-48 to 61
- cryogenic properties, **6**-139
- electron configuration, **1**-19 to 20
- entropy, **5**-1 to 3
- history, occurrence, uses, **4**-1 to 36
- ionization energy, **10**-175 to 177
- isotopes and their properties, **11**-50 to 198
- magnetic susceptibility, **4**-141 to 146
- mean free path, **6**-46
- permittivity (dielectric constant), **6**-151 to 173, **6**-174 to 175
- physical properties, **4**-132
- solubility in water, **8**-86 to 89
- thermal conductivity, **6**-188 to 189
- van der Waals constants, **6**-45
- vapor pressure, **6**-65 to 92, **6**-93 to 103
- viscosity, **6**-180 to 181

X-rays
- atomic energy levels, **10**-196 to 199
- attenuation coefficients, **10**-207 to 211
- cross sections, for the elements, **10**-207 to 211
- natural line width, **10**-206

Y

Young's modulus, rare earth elements, **4**-122 to 131
Ytterbium: see also Elements
- electrical resistivity, **12**-45 to 47
- electron configuration, **1**-19 to 20
- heat capacity, **4**-133
- history, occurrence, uses, **4**-1 to 36
- ionization energy, **10**-175 to 177
- isotopes and their properties, **11**-50 to 198
- magnetic susceptibility, **4**-141 to 146
- molten, density, **4**-137 to 140
- physical properties, **4**-132
- thermal properties, **12**-219 to 220
- vapor pressure, **6**-65 to 92

Yttrium: see also Elements
- electrical resistivity, **12**-45 to 47
- electron configuration, **1**-19 to 20
- heat capacity, **4**-133
- history, occurrence, uses, **4**-1 to 36
- ionization energy, **10**-175 to 177
- isotopes and their properties, **11**-50 to 198
- magnetic susceptibility, **4**-141 to 146
- molten, density, **4**-137 to 140
- physical properties, **4**-132
- thermal properties, **12**-219 to 220
- vapor pressure, **6**-65 to 92

Z

Zeotropes, **6**-176 to 179
Zinc: see also Elements
- electrical resistivity, **12**-45 to 47
- electron configuration, **1**-19 to 20
- heat capacity, **4**-133
- history, occurrence, uses, **4**-1 to 36
- ionization energy, **10**-175 to 177
- isotopes and their properties, **11**-50 to 198
- magnetic susceptibility, **4**-141 to 146
- molten, density, **4**-137 to 140
- physical properties, **4**-132
- thermal properties, **12**-219 to 220
- thermodynamic properties, **5**-1 to 3
- vapor pressure, **6**-65 to 92

Zirconium: see also Elements
- electrical resistivity, **12**-45 to 47
- electron configuration, **1**-19 to 20
- heat capacity, **4**-133
- history, occurrence, uses, **4**-1 to 36
- ionization energy, **10**-175 to 177
- isotopes and their properties, **11**-50 to 198
- magnetic susceptibility, **4**-141 to 146
- molten, density, **4**-137 to 140
- physical properties, **4**-132
- thermal properties, **12**-219 to 220
- vapor pressure, **6**-65 to 92

Virial expansion in gases, 6-46
Visual, ultraviolet, and infrared [illegible], 14-1[illegible]
Viscosity constants, 12-129 to [illegible]
Vibrational force constants, 9-76
Vibrational frequencies of molecules, 9-77 to 9-[illegible]
Vibrational/rotational spectra, for frequency calibration, 10-226 to 229
[illegible] molecules of gases, 6-23[illegible], 14
Viscosity
aqueous solutions, 8-57 to 8-8[illegible]
steam, liquid and gas, 6-[illegible] to 6-14
atmosphere, as function of altitude, 14-19 to 14-22
[illegible] solutions, 6-[illegible]
common gases, function of temperature, 6-1[illegible]
common fluids, as function of temperature and pressure, 6-[illegible] to 6-24
gases at atmospheric pressure, 6-180 to 6-182
hydrocarbon liquid, 6-102
helium, liquid and gas, 6-10 to 6-[illegible]
liquid metals, 4-120 to 4-126
methane, liquid and gas, 6-16 to 6-24
nitrogen, liquid and gas, 6-16 to 6-24
oxygen, liquid and gas, 6-16 to 6-22
seawater, 14-13 to 14-[illegible]
water, as function of temperature, 6-3 to 6-[illegible]

W

definition, 1-[illegible] to 1-[illegible]
[illegible] fundamental value, 1-[illegible]

Water
electrochemical series, 8-[illegible]
boiling point as function of pressure, 6-[illegible]
conductivity, 6-[illegible] to 6-[illegible]
critical constants, 6-4
density, 6-3 and 6-5
density, as function of pressure, 6-[illegible] to 6-[illegible]
density (supercooled), 6-6 to 6-7
dielectric constant, 6-[illegible] to 6-[illegible]
dielectric constant, as function of frequency, 6-[illegible]
diffusion of gases, 6-19[illegible]
dissociation constant, 8-[illegible], 8-83
electrical conductivity, 6-[illegible]
enthalpy of fusion, 6-[illegible]
enthalpy of vaporization, 6-[illegible]
fixed point properties, 6-[illegible]
freezing point, as function of pressure, 6-[illegible]
heat capacity, 6-[illegible]
index of refraction, 10-[illegible]
ion product, 8-[illegible] to 8-[illegible]
octanol/water partition coefficients, 16-[illegible] to 16-[illegible]
permittivity (dielectric constant), 6-[illegible] to 6-[illegible], 6-[illegible] to 6-176
permittivity, as function of frequency, 6-[illegible]
pH measurement, 8-[illegible] to 8-[illegible]
speed of sound in, 14-[illegible] to 14-[illegible]
surface tension, 6-[illegible]
thermal conductivity, 6-[illegible], 6-[illegible] to 6-[illegible]
thermal expansion coefficient, 6-[illegible] to 6-[illegible]
thermodynamic properties, 6-[illegible] to 6-[illegible]
thermodynamic properties at high temperature, 6-[illegible] to 6-[illegible]
triple point constants, 6-[illegible]
van der Waals constants, 6-[illegible]
vapor pressure, 6-10 to 6-[illegible]
vapor pressure over salt solutions, 15-[illegible]
viscosity, 6-180 to 6-181
volume of one gram, 6-[illegible]
Water (D2O)
boiling point, 6-[illegible]
critical constants, 6-4
density, 6-[illegible]
dissociation constant, 8-83
fixed point properties, 6-[illegible]
ion product, 8-[illegible]
thermal conductivity, 6-[illegible], 6-[illegible] to 6-[illegible]
triple point constants, 6-[illegible]
vapor pressure, 6-[illegible]
viscosity, 6-180 to 6-181
Wave equation, 1-[illegible] to 1-[illegible]
Wavelength
atomic spectra, 10-1 to 10-[illegible]
correction to vacuum, 10-[illegible]
for calibration of infrared spectra, 10-[illegible] to 10-[illegible]
laser sources, 10-[illegible] to 10-[illegible]
Weak acids, pH of, [illegible] to [illegible]
Weights and measures, [illegible], 1-[illegible]
Width of [illegible] lines, 10-[illegible]
[illegible] constant, 1-[illegible] to 1-[illegible]
Wien's [illegible], 15-[illegible]
Wood
density, 15-[illegible]
speed of sound in, 14-[illegible]
thermal conductivity, 12-[illegible] to 12-[illegible]
Work function of the elements, 12-[illegible]

X

Xenon, see also Elements
critical constants, 6-[illegible] to 6-[illegible]
cryogenic properties, 6-[illegible]
electron configuration, 1-[illegible] to 1-[illegible]
electron affinity, 10-[illegible] to 10-[illegible]
history, occurrence, uses, 4-1 to 4-[illegible]
ionization energy, 10-[illegible] to 10-[illegible]
isotopes and their properties, 11-50 to 11-[illegible]
magnetic susceptibility, 4-[illegible] to 4-[illegible]
mean free path, 6-[illegible]
permittivity (dielectric constant), 6-[illegible] to 6-[illegible]
physical properties, 4-[illegible]
solubility in water, 8-[illegible] to 8-[illegible]
thermal conductivity, 6-[illegible] to 6-[illegible]
van der Waals constants, 6-[illegible]
vapor pressure, 6-[illegible] to 6-[illegible]
viscosity, 6-[illegible] to 6-[illegible]

Y

Yield energy levels, 10-[illegible] to 10-[illegible]
[illegible] constants, 11-[illegible] to 11-[illegible]
[illegible], 11-[illegible] to 11-[illegible]
[illegible], 10-[illegible]
Young's modulus, for various elements, 4-[illegible]
Ytterbium, see also Elements
electrical resistivity, 12-[illegible] to 12-[illegible]
electron configuration, 1-[illegible] to 1-[illegible]
heat capacity, 4-[illegible]
history, occurrence, uses, 4-1 to 4-[illegible]
ionization energy, 10-[illegible] to 10-[illegible]
isotopes and their properties, 11-[illegible] to 11-[illegible]
magnetic susceptibility, 4-[illegible] to 4-[illegible]
molar density, 4-[illegible] to 4-[illegible]
physical properties, 4-[illegible]
thermal properties, 12-[illegible] to 12-[illegible]
vapor pressure, 6-[illegible] to 6-[illegible]
Yttrium, see also Elements
electrical resistivity, 12-[illegible] to 12-[illegible]
electron configuration, 1-[illegible] to 1-[illegible]
heat capacity, 4-[illegible]
history, occurrence, uses, 4-1 to 4-[illegible]
ionization energy, 10-[illegible] to 10-[illegible]
isotopes and their properties, 11-[illegible] to 11-[illegible]
magnetic susceptibility, 4-[illegible] to 4-[illegible]
molar density, 4-[illegible] to 4-[illegible]
physical properties, 4-[illegible]
thermal properties, 12-[illegible] to 12-[illegible]
vapor pressure, 6-[illegible] to 6-[illegible]

Z

Zeolites, 4-[illegible] to 4-[illegible]
Zinc, see also Elements
electrical resistivity, 12-[illegible] to 12-[illegible]
electron configuration, 1-[illegible] to 1-[illegible]
heat capacity, 4-[illegible]
history, occurrence, uses, 4-1 to 4-[illegible]
ionization energy, 10-[illegible] to 10-[illegible]
isotopes and their properties, 11-[illegible] to 11-[illegible]
magnetic susceptibility, 4-[illegible] to 4-[illegible]
molar density, 4-[illegible] to 4-[illegible]
physical properties, 4-[illegible]
thermal properties, 12-[illegible] to 12-[illegible]
thermodynamic properties, 5-[illegible] to 5-[illegible]
vapor pressure, 6-[illegible] to 6-[illegible]
Zirconium, see also Elements
electrical resistivity, 12-[illegible] to 12-[illegible]
electron configuration, 1-[illegible] to 1-[illegible]
heat capacity, 4-[illegible]
history, occurrence, uses, 4-1 to 4-[illegible]
ionization energy, 10-[illegible] to 10-[illegible]
isotopes and their properties, 11-[illegible] to 11-[illegible]
magnetic susceptibility, 4-[illegible] to 4-[illegible]
molar density, 4-[illegible] to 4-[illegible]
physical properties, 4-[illegible]
thermal properties, 12-[illegible] to 12-[illegible]
vapor pressure, 6-[illegible] to 6-[illegible]

STANDARD ATOMIC WEIGHTS (1995)

At. no.	Name	Symbol	Atomic weight	At. no.	Name	Symbol	Atomic weight
1	Hydrogen	H	1.00794(7)	55	Cesium	Cs	132.90545(2)
2	Helium	He	4.002602(2)	56	Barium	Ba	137.327(7)
3	Lithium	Li	6.941(2)	57	Lanthanum	La	138.9055(2)
4	Beryllium	Be	9.012182(3)	58	Cerium	Ce	140.116(1)
5	Boron	B	10.811(7)	59	Praseodymium	Pr	140.90765(2)
6	Carbon	C	12.0107(8)	60	Neodymium	Nd	144.24(3)
7	Nitrogen	N	14.00674(7)	61	Promethium	Pm	[145]
8	Oxygen	O	15.9994(3)	62	Samarium	Sm	150.36(3)
9	Fluorine	F	18.9984032(5)	63	Europium	Eu	151.964(1)
10	Neon	Ne	20.1797(6)	64	Gadolinium	Gd	157.25(3)
11	Sodium	Na	22.989770(2)	65	Terbium	Tb	158.92534(2)
12	Magnesium	Mg	24.3050(6)	66	Dysprosium	Dy	162.50(3)
13	Aluminum	Al	26.981538(2)	67	Holmium	Ho	164.93032(2)
14	Silicon	Si	28.0855(3)	68	Erbium	Er	167.26(3)
15	Phosphorus	P	30.973761(2)	69	Thulium	Tm	168.93421(2)
16	Sulfur	S	32.066(6)	70	Ytterbium	Yb	173.04(3)
17	Chlorine	Cl	35.4527(9)	71	Lutetium	Lu	174.967(1)
18	Argon	Ar	39.948(1)	72	Hafnium	Hf	178.49(2)
19	Potassium	K	39.0983(1)	73	Tantalum	Ta	180.9479(1)
20	Calcium	Ca	40.078(4)	74	Tungsten	W	183.84(1)
21	Scandium	Sc	44.955910(8)	75	Rhenium	Re	186.207(1)
22	Titanium	Ti	47.867(1)	76	Osmium	Os	190.23(3)
23	Vanadium	V	50.9415(1)	77	Iridium	Ir	192.217(3)
24	Chromium	Cr	51.9961(6)	78	Platinum	Pt	195.078(2)
25	Manganese	Mn	54.938049(9)	79	Gold	Au	196.96655(2)
26	Iron	Fe	55.845(2)	80	Mercury	Hg	200.59(2)
27	Cobalt	Co	58.933200(9)	81	Thallium	Tl	204.3833(2)
28	Nickel	Ni	58.6934(2)	82	Lead	Pb	207.2(1)
29	Copper	Cu	63.546(3)	83	Bismuth	Bi	208.98038(2)
30	Zinc	Zn	65.39(2)	84	Polonium	Po	[209]
31	Gallium	Ga	69.723(1)	85	Astatine	At	[210]
32	Germanium	Ge	72.61(2)	86	Radon	Rn	[222]
33	Arsenic	As	74.92160(2)	87	Francium	Fr	[223]
34	Selenium	Se	78.96(3)	88	Radium	Ra	[226]
35	Bromine	Br	79.904(1)	89	Actinium	Ac	[227]
36	Krypton	Kr	83.80(1)	90	Thorium	Th	232.0381(1)
37	Rubidium	Rb	85.4678(3)	91	Protactinium	Pa	231.03588(2)
38	Strontium	Sr	87.62(1)	92	Uranium	U	238.0289(1)
39	Yttrium	Y	88.90585(2)	93	Neptunium	Np	[237]
40	Zirconium	Zr	91.224(2)	94	Plutonium	Pu	[244]
41	Niobium	Nb	92.90638(2)	95	Americium	Am	[243]
42	Molybdenum	Mo	95.94(1)	96	Curium	Cm	[247]
43	Technetium	Tc	[98]	97	Berkelium	Bk	[247]
44	Ruthenium	Ru	101.07(2)	98	Californium	Cf	[251]
45	Rhodium	Rh	102.90550(2)	99	Einsteinium	Es	[252]
46	Palladium	Pd	106.42(1)	100	Fermium	Fm	[257]
47	Silver	Ag	107.8682(2)	101	Mendelevium	Md	[258]
48	Cadmium	Cd	112.411(8)	102	Nobelium	No	[259]
49	Indium	In	114.818(3)	103	Lawrencium	Lr	[262]
50	Tin	Sn	118.710(7)	104	Rutherfordium	Rf	[261]
51	Antimony	Sb	121.760(1)	105	Dubnium	Db	[262]
52	Tellurium	Te	127.60(3)	106	Seaborgium	Sg	[266]
53	Iodine	I	126.90447(3)	107	Bohrium	Bh	[264]
54	Xenon	Xe	131.29(2)	108	Hassium	Hs	[269]
				109	Meitnerium	Mt	[268]

Numbers in brackets are the mass numbers of the longest-lived isotope of elements for which a standard atomic weight cannot be defined.